# FLORA OF MISSOURI

# FLORA OF MISSOURI

JULIAN A. STEYERMARK

THE IOWA STATE UNIVERSITY PRESS
*Ames*, IOWA, U.S.A.

Iowa State University Press, Ames, Iowa, U.S.A.
Library of Congress Catalog Card Number: 62-12193
Printed in the Netherlands by G. J. Thieme, Nijmegen

**Published November, 1963**
**Second printing, 1968**
**Third printing, 1972**
**Fourth printing, 1975**
**International Standard Book Number: 0-8138-0655-0**

## THE AUTHOR

JULIAN A. STEYERMARK...
a leading plant explorer and taxonomist, has devoted thirty years to tireless exploration and classification of the vegetation of Missouri, resulting in several hundred published works, including *Spring Flora of Missouri* and co-authorship of the *Annotated Catalogue of the Flowering Plants of Missouri*. His intimate field knowledge of unique botanical areas in Missouri has led to the preservation of several natural areas maintained by the Missouri Chapter of Nature Conservancy, which he helped to organize, and his efforts in conservation of unique botanical areas in Illinois helped lead to the establishment of the Volo and Wauconda Bog nature preserves. He is the recipient of the Order of Quetzal, a meritorious award from the Guatemalan government in recognition of his explorations and contributions to the study of the flora of that country. He is botanist on the staff of the Instituto Botanico of the Ministry of Agriculture in Caracas, Venezuela, where he is actively engaged in the exploration and classification of the Venezuelan flora. Before coming to Venezuela, he was Curator of the Herbarium at the Chicago Natural History Museum, and was associated with that institution from 1937 to 1958. For his special achievements in the field of botany, he was one of the first recipients of a special plaque awarded by Washington University, St. Louis. He has led and participated in numerous botanical expeditions in Central and South America, particularly in Guatemala, Panama, Ecuador, and Venezuela. He is co-author of a many-volume work on the flora of Guatemala, and has published four volumes on new species of plants found in various parts of Venezuela.

*to Mr. Ernest Jesse Palmer,*
*life-long student of the Missouri Flora,*
*this book is humbly dedicated*

## *Table of Contents*

# *Preface*

The study and exploration of the Missouri flora has engaged the author's attention during the past thirty years. Grants from the National Science Foundation have enabled him to complete the task of amassing all the data, rechecking the herbarium material, preparing the maps showing the geographical distribution of species, and of writing the present manuscript.

Despite the intensive exploration of the state undertaken by the writer since the publication in 1935 of *An Annotated Catalogue of the Flowering Plants of Missouri* by Palmer and Steyermark (Ann. Mo. Bot. Gard. 22: 375–758), large sectors of each county in the state remain botanically unexplored. Another century of careful and thorough botanizing in Missouri should undoubtedly reveal scores of additional species previously unknown to exist in the state. At the time of publication of their *Annotated Catalogue* (*ibid.*, p. 686) Palmer and Steyermark recorded a total of 2,281 species, 371 varieties, and 78 forms, for a grand total of 2,790 taxa, referred to 707 genera. In the present enumeration, this number has been considerably increased to a total of 2,438 species, 517 subspecies and varieties, and 297 forms, altogether totalling 3351 taxa which are referred to 799 genera.

*Acknowledgments*

The records included in the present flora have involved a re-examination of all Missouri material contained in the herbaria of the Missouri Botanical Garden, Chicago Natural History Museum, and the University of Missouri, which together possess the largest assemblage of Missouri collections. Collections of certain Missouri material were also checked at the Gray Herbarium of Harvard University, the New York Botanical Garden, the United States National Museum, Southeast Missouri State College, Southwest Missouri State College, Northwest Missouri State College, Park College, William Jewell College, School of the Ozarks, Central College, and the private herbaria of Mr. Donald Rhodes, Mr. Bruce Dowling, Mr. Bill Bauer, Mr. Ernest J. Palmer, and Drs. Victor Muehlenbach, Alfred Etter, and John Dwyer. To the curators and administrators of these institutions I am deeply appreciative of their kind consideration and co-operation in making this mass of specimens available for study.

I wish especially to thank Mr. Ernest J. Palmer of Webb City, Missouri, for his counsel and guidance during the past twenty-eight years. Many other individuals have assisted in furnishing me with data and suggestions. Among them I am particularly grateful to F. J. Hermann, H. N. Moldenke, I. M. Johnston, C. E. Kobuski, Agnes Chase, Mildred Mathias, Lincoln Constance, Jason R. Swallen, Carl Epling, U. T. Waterfall, David D. Keck, Charles Heiser, Robert Long, T. R. Fisher, Hugh Iltis, William J. Dress, Edgar T. Wherry, H. W. Rickett, Lloyd Shinners, Lyman Benson, Paul Redfearn, Jr., Henry W. Radloff, Jr., Clair Kucera, Joan H. Wilce, James H. Zimmerman, Duane Isely, Walter H. Lewis, John Voigt, Robert Mohlenbrock, Jr., Bassett Maguire, Dwight M. Moore, Alan A. Beetle, Bill Bauer, Victor Muehlenbach, George B. Van Schaack, Albert Chandler, William E. Liggett, Oscar Peterson, Leslie Hubricht, George Moore, Francis Drouet, Kendall Laughlin, Harold E. Moore, Jr., and Mrs. Fredia Kreamelmeyer.

To Drs. Edward L. Clark, formerly state geologist, and Tom Beveridge, State Geologist of the Missouri Geological Survey, I am greatly indebted for the use of the Missouri topographical maps furnished me. These have been of inestimable value in connection with detailed field exploration throughout the state.

To the University of Missouri I wish to acknowledge thanks for the completion of the distribution maps of Missouri used in this work.

I wish to make grateful acknowledgment and thanks to the following publishers for permission to use certain information from their books: *Edible Wild Plants of Eastern North America* by Merritt Lyndon Fernald and Alfred Charles Kinsey (Harper & Brothers); *A Natural History of Trees of Eastern and Central United States* by Donald Culross Peattie (Houghton Mifflin Company); *Economic Botany* by A. F. Hill (McGraw-Hill Book Company, Inc.).

Finally, I wish to express particular appreciation to the Iowa State University Press and to the National Science Foundation through whose joint support this publication has been made possible.

*Illustrations*

For the privilege of reproducing various illustrations used in the present flora, I wish to extend my most grateful thanks to the following: Dr. Jason R. Swallen, Head Curator of the Department of Botany, United States National Museum, for the use of illustrations of grasses in A. S. Hitchcock's *Manual of the Grasses of the United States*; Dr. Reed C. Rollins, Director of the Gray Herbarium of Harvard University, and the editors of American Book Company for the privilege of reproducing the *Cyperaceae* and *Juncaceae* from *Gray's Manual of Botany*, eighth edition, by Merritt Lyndon Fernald; Dr. William C. Steere, Director of the New York Botanical Garden, for the privilege of reproducing various illustrations from *Manual of the Southeastern Flora* by John Kunkel Small, and from both the second edition of *An Illustrated Flora of the United States and Canada* by Nathaniel Lord Britton and Addison Brown and from the *New Britton and Brown Illustrated Flora* by Henry A. Gleason; Regents of the University of Wisconsin, copyright owners of Norman C. Fassett's *Manual of Aquatic Plants*, copyright 1940, the University of Wisconsin Press for the privilege of reproducing various illustrations; Charles Scribner's Sons for the privilege of reproducing certain drawings from *Just Weeds* by Edwin Rollins Spencer; and Dr. Duane Isely and Mrs. Elsie Froeschner for the privilege of reproducing Mrs. Froeschner's drawings of *Leguminosae* in Dr. Isely's publications, *The Leguminosae of the North-Central United States* I. *Loteae and Trifolieae*, and II. *Hedysareae*. I wish to express especial thanks to Dr. Albert E. Vatter, Jr., for assistance in photographic work.

Many of the illustrations have been drawn from living and herbarium material and were executed by the following artists: Miss Dorothy Anderson, Mr. George Burian, Miss Lila Campbell, Miss Sharon Harkless, Mr. John Hendee, Mr. Tom Kendzie, Mr. Donald Kreamelmeyer, Miss Patricia McNabb, Mr. Charles McNabb, Miss Kay Oman, Mr. Marvin Rabe, and Miss Sandra Vorbeck. The original drawings from the author's *Spring Flora of Missouri* have provided the remainder of the drawings included in the present volume.

Finally, I wish to thank Mrs. Jean Kreamelmeyer for the care used in typing the manuscript, and my wife, Cora, who has helped in many tasks connected with the completion of the present flora.

Julian A. Steyermark

# *Introduction*

The purpose of this work is to present an up-to-date account of the seed-plants, ferns, and fern allies growing wild in Missouri or introduced, naturalized, or escaped from cultivation through various sources. In the event that a given plant has seeded itself or is growing away from cultivation or habitation, therefore appearing as associated with other wild-growing species, it has been included on the assumption that (1) it manifests the ability to survive in the wild state and may persist, and (2) it may become collected and preserved by some individual curious to know the name of the particular plant. Many a present-day pest has had its inauspicious beginning as a railroad or ballast-dump plant, only to spread later to become an ubiquitous weed. As it is often difficult to determine whether or not plants casually appearing on dump heaps or along railroads and roadsides may persist and become established, in the present flora they have been given the benefit of the doubt in most cases if vouched for by a herbarium record. Some of these introduced plants have not been re-collected for several decades, but until a thorough and intensive search has been made along railroads and disturbed areas inhabited by them, the apparent present scarcity of the species may be due to lack of adequate collecting from such habitats.

## Historical Sketch

The following account is a brief historical sketch of the principal collectors in Missouri up to the present time.

The records of botanical exploration which have been preserved in the form of herbarium specimens collected in Missouri date back principally to 1810–11 when John Bradbury botanized portions of counties along the Missouri and Mississippi rivers. Edwin James on the Long Expedition collected in the counties bordering the Missouri River in 1821 and 1822, and was followed by L. C. Beck between 1826 and 1828 and by Charles Geyer in 1845 on the Nicollet Expedition. Nicolas Riehl botanized around St. Louis and Jefferson counties between 1835 and 1838. Prince Maximilian collected between 1841 and 1843 along the Missouri River, and Dr. A. Wislizenus botanized parts of the eastern Ozarks in 1847 and 1848. Dr. George Engelmann botanized the vicinity of St. Louis between 1840 and 1870. The geologist, Garland C. Broadhead, collected between 1850 and 1880 in a number of central and western Missouri counties. At about the same time G. C. Swallow was engaged in botanical collecting, especially in central Missouri, followed somewhat later by B. F. Shumard, another geologist. B. F. Galloway collected around Boone County between 1855 and 1885. S. M. Tracy was active botanically between 1880 and 1886, publishing his *Catalogue of the phaenogamous and vascular cryptogamous plants of Missouri* in 1885. Benjamin Franklin Bush, one of the most active collectors in Missouri, began in 1882 what was to become a long and fruitful botanical career that ended in 1937. About the same time between 1895 and 1915 K. K. Mackenzie was active in Clay and Jackson counties, while William Trelease botanized sections of the Ozarks and the swamp lands of the southeastern portion of the state between 1885 and 1905. George Letterman collected between 1875 and 1900 around Allenton in St. Louis County, and Henry Eggert botánized St. Louis County and various eastern Ozark and southeastern lowland counties between 1875 and 1895. Noteworthy collections were made by Colton Russell in southeastern Missouri between 1886 and 1900, by J. W. Blankinship in Oregon and Greene counties between 1885 and 1900, by Stuart Weller in Greene County in 1889 and 1890, and by T. E. Savage and Wilfred Stull from Butler County between 1897 and 1900. N. M. Glatfelter collected willows especially around St. Louis in the late 1890's. Professor E. M. Shepard collected in Greene County between 1885 and 1905. At about the same time (1884), John Kellogg, a disciple of Letterman, began collecting around Allenton, St. Louis County, and was active collecting in the state until 1940. Francis Daniels did active collecting in the vicinity of Columbia, Boone County between 1897 and 1905, publishing his *Flora of Columbia* in 1907. In 1901 Ernest Jesse Palmer collected in southwestern Missouri in Jasper and Newton counties, eventually botanizing the entire state, and he is active to the present day.

Paul C. Standley collected in Greene County between 1903 and 1905. Huron H. Smith collected in southern Missouri, especially in Iron and Wright counties between 1905 and 1908. Reverend John Davis made extensive collections in Lewis, Marion, Ralls, Pike, and St. Charles counties between 1912 and 1921. Jesse More Greenman, formerly curator of the herbarium at the Missouri Botanical Garden, collected in a few of the eastern Ozark counties between 1910 and 1925. W. H. Emig collected in Osage County in 1915–18. Harold W. Rickett botanized in the Ozarks and in Boone and Callaway counties between 1927 and 1937, in 1931 revising Daniels' *Catalogue*. F. P. Metcalf collected aquatic plants in various parts of northern and central Missouri between 1918 and 1920. C. J. Elmore collected in Clay County between 1920 and 1930. A number of Dr. Greenman's students made botanical collections in the state, beginning with Earl Sherff, who botanized the St. Louis and eastern Ozark section between 1910 and 1913, followed by Carl Epling (1920–24), Mildred Mathias (1925–29), and Robert E. Woodson, Jr. (1925–32), who made collections from various parts of the Ozarks. The present author has collected throughout the 114 counties of the state since 1926. Francis Drouet and Lisle Jeffrey were active collectors between 1932 and 1935 in Boone and Callaway counties especially, but also botanized in other counties. J. R. Singleton collected in Nodaway County in 1938 and D. R. Crookshanks in Livingston County between 1936 and 1938. William Drew collected in Boone and Callaway counties around 1940–42. Clair Kucera, from 1950 to the present, has botanized in Boone and Callaway counties and some of the Ozark region. Other modern collectors have been George Moore, collecting in Laclede County between 1935 and 1945; Bill Bauer, collecting throughout Missouri since 1930; Albert Chandler, collecting mainly in the eastern Ozarks from 1930 to 1955. L. J. Gier and his students have been busy collecting the flora of Clay County since 1940. Victor Muehlenbach has added many species of foreign origin new to the state flora as a result of his botanizing along railroad rights-of-way from 1954 to the present time.

While many other collectors have also contributed to the records, the ones cited above are responsible for the majority of the specimens encountered. Early explorers, such as Meriwether Lewis, Henri Brackenridge, and Henry R. Schoolcraft made observations but did not preserve specimens of plants seen in Missouri. The botanist Frederick Pursh (1812) was the first botanical author to describe plants collected in Missouri, based on the collections of John Bradbury, while Thomas Nuttall described a number of species from Arkansas which extend into Missouri. The work of André Michaux, published in 1803, was based on his collections east of Missouri, but included many species found in the vicinity of St. Louis.

Altogether this botanical activity has resulted in the accumulation of over 200,000 numbered collections from Missouri. Of these the largest have been made by the writer, totaling approximately 60,000 numbers, by Bush, with about 55,000 numbers, and by E. J. Palmer, totaling about 35,000. Kellogg collected about 28,000 numbered Missouri collections, followed in decreasing numbered collections by Eggert, Davis, Broadhead, Letterman, Trelease, Drouet, Woodson, Gier, and others. Despite this apparent activity, most of the 114 counties still possess large areas in need of detailed and careful field work. Each year of intensive field activity in sections previously unbotanized yields a number of species and varieties new to the flora of the state. The taxa which become added to the state flora include not only recently introduced plants brought in by railroads, highway construction, garden escapes, and by various other human agencies, but also native plants which have not been found in previous exploration, partly due to the remoteness of their locality from roads, but mostly due to the lack of sufficient collectors to cover the many yet botanically unexplored parts of the state. Because of this inadequate coverage of each county, even genera which are native to the state have not been found until the past ten years, including such distinct and phytogeographically significant ones as *Lyonia*, *Gaylussacia*, *Bartonia*, and *Obolaria*.

Some of the least worked counties include Andrew,

Atchison, Audrain, Buchanan, Caldwell, Cass, De Kalb, Holt, Johnson, Lafayette, Mercer, Pemiscot, Platte, Randolph, Ray, Saline, Schuyler, Scotland, Shelby, Sullivan, Webster, Worth, and Wright, but the counties in southeastern Missouri situated in the swampy lowlands and traversed by Crowley Ridge, as well as those of southern Missouri containing sink-hole ponds and swampy calcareous, spring-fed meadows continue to produce numerous additions to the native flora. Remote rocky bluffs, previously unbotanized ravines, glades, and rocky exposures bring in their share of botanical records. The careful application of topographical maps to areas of field work in future exploration should add numerous taxa to the flora of the state.

## General Treatment of the Flora

At the beginning of this work a general key has been provided to enable the reader to reach a given family, or, in many cases, genus. This general key has been grouped into fourteen sections in order to run down readily a given plant into its obvious or main category. After the family is reached, keys within a given family lead to genera, those under genera lead to species, and within the species keys are provided to reach subdivisions of a species. In general the keys have been designed with two objectives: (1) to make available, whenever possible, the most obvious and easily recognizable clues or aids which will furnish a ready means of identification of the plant, and (2) to enlist the use of a number of characters, combining, if necessary, essential details of foliage, flower, and fruit in order that an imperfect specimen or one possessing only certain parts will fit into at least one of the multiple characters presented. In order to facilitate as much as possible the ready identification of the plant, data have been inserted throughout the keys, wherever justifiable, concerning the relative rarity and geographical occurrence, habitat, habit, color of flowers and leaves, and characters of the foliage, items often neglected but frequently useful in the identification of a plant. Another item found in the keys is the insertion of words or explanations of terms placed within parentheses. This has been done in order to avoid the frequent, time-consuming, and distracting use of turning to a glossary at the time of identification of a given plant, and in order to provide immediate explanation of the word in question. By furnishing a glossary within the confines of the key, it is believed that the user of the book gains a more rapid knowledge of those technical or unfamiliar terms encountered and learns to associate them more quickly and more easily.

The relative amount of detail in the keys has included the chief diagnostic characters by which a given taxon may be recognized, and has eliminated the need, in most cases, for a description. For this reason and because of the limitation of space, descriptions have been omitted. Measurements have been taken from Missouri material, both dried and living. Where discrepancies exist between measurements given in the chief manuals and those found in Missouri material, such differences are noted accordingly in the text.

Throughout the work each taxon has been given the choice of botanical name, which, in my judgment, is valid according to the present International Rules of Botanical Nomenclature. If the name selected differs from that used in the eighth edition of *Gray's Manual of Botany*, the abbreviation [G] follows; if it differs from Gleason's *New Britton and Brown Illustrated Flora*, [BB] is used; if it differs from Palmer and Steyermark's *Annotated Catalogue of the Flowering Plants of Missouri*, [P & S] is inserted; and if it differs from Steyermark's *Spring Flora of Missouri*, [Steyerm.] is the abbreviation. If the name used is different from that published by a recent worker of that group, the name of the particular individual is placed in brackets, thus [Doe]. These references have been given only for the sake of ready comparison with familiar names likely to be encountered in standard regional and local Missouri works and no complete synonymy is intended.

Following the scientific name there follows, whenever applicable, one or more common names. If a common name occurs, the one most prevalent and widely used in Missouri is given, sometimes followed by other names likely to be used for the same plant. It will be noted that for many plants no common name is known. Wherever possible, the use of coined names or of those translated from the Latin has been avoided. However, where a common name has had some acceptance in general and local works, it is used in this flora. As sometimes happens, certain common names have been based upon a wrongly identified plant or upon one of misleading geographical origin, with the result that the common name used is either unreliable, nonsensical, or illogical. Since many of the established common names which are used are often rooted in ancient tradition, folk-lore, and medical tales, the immediate reason for their usage is not apparent or makes little sense with the result that they are committed to memory parrot-fashion.

The earliest and latest dates of flowering, and in some cases fruiting, are next given, based in all cases on Missouri phenological records. These will be found to differ sometimes from the earliest and latest dates found in the standard regional manuals. Data concerning the habitats occupied by the species in Missouri are presented, followed by the range known in Missouri. A map showing the geographical range within the state by counties usually accompanies each species. In all cases the variations within a given species are represented on the maps by different symbols. In the case of species with a restricted distribution within the state, herbarium specimens exist for each county indicated. However, in the case of certain species, actually twenty-two in all, which have been noted in the field as occurring in every county or in all but five counties at the most, herbarium specimens were not obtained, due to limitations of collecting time as well as herbarium space occupied. In such cases the indication of their occurrence has been based upon field rather than herbarium records. This is likewise true of the records indicated for such weeds and ubiquitous plants as *Verbascum Thapsus*, *Stellaria media*, *Abutilon Theophrasti*, and similar easily recognized taxa. In the case of species showing variations, only those records are mapped which are represented by voucher herbarium specimens. For the sake of graphic expression, the symbols are placed in the middle of each county. In most instances, the actual occurrence of a widespread species is throughout a given county so that it would not be practical to represent its area of total occupation. For all practical purposes, the fact that the species has been found in a given county is shown by its representation from that county.

The citation of specimens with full data taken from the label accompanies a statement of range when the species, variety, or form is a rare one known from only one, two, or few counties. This has also been done in the case of newly revised taxa, where the inclusion of such citations should make clear the specimens now identified as pertaining to a given variation as contrasted to previous work. The specimens have been cited in such cases with two objectives in mind: (1) to establish a scientific record of factual data which should serve as a reference for study and (2) to furnish records not only to taxonomists and plant geographers, but also to those in the fields of ecology, genetics, cytology, morphology, physiology, or horticulture, who may be interested in carrying on additional experiments with the particular taxa treated.

The general range of the species or variety as it exists outside of Missouri follows. In general, the distribution is stated to occur from east to west, except in the case of species of the western United States ranging eastward. A north to south or south to north statement follows the east to west part of the range given, depending upon whether the particular taxon is more abundant in the northern or southern portion of its range. General ranges of geographical distribution have been obtained from the latest published state records, monographs and revisions, and regional floras.

The variations and uses possessed by a given species, as well as its known poisonous properties, have been included wherever this knowledge is available.

### Discussion of Variation

It will be noted that in the present flora the names are sometimes different from those found in such standard regional works as the eighth edition of *Gray's Manual of Botany* and Gleason's *New Britton and Brown Illustrated Flora*. The reasons for this are that (1) perfect unanimity of opinion on all species does not exist among botanists, (2) more detailed studies and recent publications on particular plants have provided other criteria which have necessitated the adoption of different names from those previously used, (3) changes in rules of nomenclature have required a modification of expression in the subdivision of a species, and (4) the writer's own observations of herbarium and living material of a particular taxon both wild or under cultivation in his garden have influenced him and modified his judgment accordingly in accepting or opposing the views of other botanists.

Whenever the present writer has been unable to produce any evidence contrary to the results of another worker, he has accepted that worker's judgment. In most instances, the results of monographic work have been accepted, wherever possible, out of deference and respect to the efforts of the particular worker, who has certainly devoted much more time and labor to his special subject than I have. In other cases, however, where the results published seem to indicate insufficient data, inconclusive evidence, or omission of certain Missouri material deemed of particular importance, I have followed my own judgment. This is certainly no expression of conceit, but merely reflects one's adaptation to a situation in which it seems preferable to express one's own views rather than to accept that with which one cannot agree. This is probably a part of the principle of democratic expression manifested by botanists, and may serve to answer the complaints expressed by those who decry any disagreement among botanists in their concepts of species or who dislike any bickering or modification of previously accepted names. Such examples of differences of opinion are found mentioned throughout this flora.

In other matters involving disagreement where varying points of view exist for the same taxon, I have tried not to take any dogmatic or final stand on the subject, but have indicated the lack of finality of present judgment by suggesting that future field work and experimentation may reveal further evidence concerning the degree of distinctness of a given variation. In many instances I have admitted the manifestation of a great amount of intergradation between variations, yet have maintained the variations pending future more intensive field and experimental investigations. In other cases, there is such an absence of correlation in the variations noted that I believe that further maintenance of them cannot be justified.

### Subdivision of a Species

Since the publication in 1956 of the International Code of Botanical Nomenclature adopted by the Eighth International Botanical Congress in Paris during July, 1954, it has become necessary to change the manner of expression when indicating certain rank of subdivisions of a species. According to Article 25 of these Rules: 'For nomenclatural purposes, a species or any taxon below the rank of species is regarded as the sum of its subordinate taxa, if any. Valid publication of a subordinate taxon which does not include the nomenclatural type of the higher taxon automatically circumscribes a second taxon of the same rank which has its nomenclatural type, the type of the higher taxon and bears the same epithet.

*Example*:

The publication in 1843 of *Lycopodium inundatum* L. var. *bigelowii* Tuckerm. automatically circumscribes another variety, *Lycopodium inundatum* var. *inundatum*, the type of which is that of *Lycopodium inundatum* L.'

Also, according to Article 26 of the Rules: 'In the name of an infraspecific taxon which includes the nomenclatural type of the epithet of the next higher taxon, the epithet of this higher taxon must be repeated unaltered but ... without citation of an author's name. This epithet can no longer be used when that of the next higher taxon is changed.

*Examples*:

The combination *Lobelia spicata* Lam. var. *originalis* McVaugh, which includes the type of *Lobelia spicata* Lam., must be replaced by *Lobelia spicata* Lam. var. *spicata*.

Since under *Lobelia siphilitica* L. there is described var. *ludoviciana* A. DC., one must write *Lobelia siphilitica* L. var. *siphilitica* if only that part of *L. siphilitica* L. which includes the type is meant.'

Adherence to the above requirements will be found throughout the present flora. At first the repetition of the specific name for the typical variation may seem superfluous and cumbersome, but eventually the logic and preciseness of this form of expression becomes obvious. If a species has no subspecies, varieties, or forms, then no additional infraspecific name is required. If a form (forma) only has been described within a species, then the typical form (the one containing the nomenclatural type on which the species was based) of the species becomes 'forma' with repetition of the specific name, contrasting it with the other form or forms. If a variety has been described within a species, then the typical variety is the one containing the nomenclatural type on which the species was based and becomes 'variety' with repetition of the specific name, as suggested by the various examples given above. With a little practice, the ease of application of this mode of expression of the infraspecific categories becomes clear and precise.

### Sequence of Genera, Families, and Orders

The presentation and arrangement of genera have followed, with some exceptions, the eighth edition of Gray's *Manual of Botany* and Gleason's *New Britton and Brown Illustrated Flora*, in many respects following the latter more closely than the former, except for the presentation of the Compositae, which follows more closely the arrangement of *Gray's Manual*. The genus *Geocarpon* in the present flora, based upon floral morphological studies by Mr. E. J. Palmer and myself, is placed in the Caryophyllaceae instead of the Aizoaceae. Some of the respects in which arrangement of genera differs in the present flora from that of *Gray's Manual* are noted in the maintenance of *Najas* in the same family with *Ruppia*, *Zannichellia*, and *Potamogeton*, the merging of *Vulpia* with *Festuca*, merging of *Nyssa* with *Cornaceae*, separation of *Dracopis* from *Rudbeckia*, and merging of *Actinomeris* with *Verbesina*. The proposed systems of Charles E. Bessey in 1897 and 1915, Alfred B. Rendle in 1925, John Hutchinson and Charles Mez in 1926, Richard Wettstein in 1935, Carl Scottsberg in 1940, Alfred Gundersen in 1950, and of other botanists, have all contributed important refinements and modifications to the system of classification of Adolf Engler and Karl Prantl in their *Die Natürlichen Pflanzenfamilien*. Since the latter system is followed at present by most botanists throughout the world, it is the one adopted in the present work, although it is expected that future research in the fields of paleobotany, floral morphology, and comparative anatomy may eventually modify our present concepts of sequence and relationships of families and orders.

# *Flora and Vegetation*

## Description of the Area

Missouri lies in the middle Mississippi Valley somewhat east and a little north of the geographical center of the United States. The northern boundary of the state is along 40°30′ north latitude; most of the southern boundary is along 36°30′ north latitude, while a small section in the southeastern corner extends south to 36° north latitude. In longitude it lies between 89° on the east to 95°46′ on the west with about two-thirds of the western boundary in 94°38′ west longitude. The irregular course of the Mississippi River bounds the eastern border, except for a short distance near the northeastern corner, where the Des Moines River separates it from Iowa. The Missouri River bounds the northwestern border of the state. Missouri is bounded altogether by eight states—on the east by Illinois, Kentucky, and Tennessee, on the north by Iowa, on the south by Arkansas, and on the west by Nebraska, Kansas, and Oklahoma.

The land area occupies 69,674 square miles, of which 448 are water surface. It is about 390 miles from the most easterly to the most westerly point in the state, but nowhere is the distance in a direct line so great, the average width being 255 miles. Over most of the state the direct distance from north to south is about 285 miles, but the southeastern corner extends 35 miles farther south. In size it ranks eighteenth in the United States.

Physiographically the state may be broadly described as an undulating plain which rises gradually, or abruptly in places, from the Mississippi River towards the west and northwest, where it merges with the Great Plains. Three principal physiographic regions, (1) the Southeastern Lowlands, (2) the Ozark Plateau, and (3) the Prairie region, may be recognized in the state. (1) The unglaciated Southeastern Lowlands is a small area of low relief between 230 and 350 feet, lying in the southeastern corner of the state as a northern part of the Mississippi Embayment extension of the Coastal Plain Province. It is a generally level surface of alluvial plain, the relief of which is diversified only by Crowley Ridge, running mainly through Dunklin, Stoddard, New Madrid, and Scott counties and reaching a maximum elevation of about 450 feet. Sikeston and Malden Ridge are narrow ridges lying east of the main mass of Crowley Ridge. (2) The unglaciated Ozark Plateau, which occupies the greater part of southern Missouri and a strip just north of the Missouri River, is the most conspicuous and extensive relief feature between the Appalachians to the east and the Cordilleras to the west. Tom Sauk Mountain, near the boundary of Iron and Reynolds counties in the southeastern Ozarks, is 1,772 feet above sea level, the highest point in the state. Most of the area of the Ozark Plateau farther west has an elevation varying between 1,000 and 1,300, reaching a maximum of slightly over 1,700 feet in Wright County. On the northern, northwestern, and eastern sides there is a gradual reduction in elevation towards the valleys of the Missouri, Osage, and Mississippi rivers, diminishing to 400 feet on the east. Near the southwestern corner of the state, in Barry County, an elevation of over 1,500 feet is attained. (3) The Prairie region, occupying the remainder of the state is part of the Central Lowland Province and embraces (a) a glaciated dissected till plain lying mainly north of the Missouri River, except for small areas in St. Louis, Saline, Cooper, Lafayette, and Jackson counties, varying in altitude from 600 feet in St. Charles County to slightly over 1,200 feet at the northwestern corner of the state, and (b) an unglaciated Osage Plains section, a wedge-shaped area of nearly flat to gently rolling relief in west-central and southwestern Missouri south of the Missouri River, including various escarpments of the Bethany Falls Escarpment, the Henrietta Escarpment, Nevada Lowland, and the Cherokee Lowland.

The drainage of the state is into the Mississippi-Missouri River system. The direction of flow of the streams varies according to the elevation and topography, hardness and strike of the underlying rocks, and past geological history. In the northern part of the state the streams generally flow southward into the Missouri River, except along the eastern quarter where they flow mainly southeast into the Mississippi. South of the Missouri River nearly all the streams have their sources in the Ozark Plateau and flow in various directions. Many of them originate from large springs, and the majority are fed by springs along their courses. Most of them in the southeastern part have a generally southerly course, but along the northern side of the plateau the Meramec, Gasconade, and Osage rivers flow in mainly northerly and easterly directions. White River, with its sources in the Boston Mountains of Arkansas, is one of the larger tributaries of the Mississippi River. It flows through southwestern Missouri, eventually joining the Mississippi far to the south in Arkansas. In the southwestern corner of the state the drainage is westward through Spring River, Shoal Creek, and Elk River, which join the Neosho River farther west, the latter stream emptying into the Arkansas and eventually the Mississippi River.

Like that of all the central Mississippi Valley, the climate of Missouri is marked by great extremes and subject to sudden changes in temperature. The average annual precipitation varies from 40–45 inches over the southern part of the state, diminishing to 30–35 inches in the northwestern portion, with most of it occurring during the spring and early summer months. The southeastern section of the state receives the heaviest amount, with an annual average of about 48 inches. The maximum amount of precipitation recorded is 55.06 inches in 1927, the minimum amount recorded is 25.28 inches in 1901. Snowfall varies from an average of 16 inches in the southeastern section to about 21 inches in the northern section, but the snow rarely remains on the ground for more than a few days. Some winters pass with only a few light snows totalling not more than two or three inches. The average annual temperature in the northwestern portion of the state is 50 degrees, that in the southeastern section is 60 degrees. The hottest month is July, with an average temperature of 77 degrees Fahrenheit; the coldest month is January, with an average of 30 degrees Fahrenheit. The highest temperature ever recorded was 118 degrees Fahrenheit at Lamar, on July 18, 1936, the lowest recorded was 40 degrees below zero at Warsaw, on February 13, 1905. The average length of the growing season is 188 days, with the earliest killing frost of autumn recorded on September 18, and the latest killing frost of spring recorded on May 9. Northwest winds prevail through the autumn and winter, while south and southwest winds prevail during spring and summer. The west and northwest winds in summer are often hot and dry or at other times are storm-laden, while the south winds in summer, originating in the Gulf-Caribbean cyclone center, are usually cool. The east winds of spring and early summer are often damp and chilly.

All of the state of Missouri, with the exception of the Southeastern Lowlands of the Mississippi Embayment, has been above sea level since late Paleozoic time. Parts of the Ozark region are among the oldest lands on the continent, the pre-Cambrian igneous rocks of the St. Francois Mountain area having been part of the original land axis of North America. Sedimentary beds of sandstones, limestones, dolomites, and shales were deposited around the igneous core of the St. Francois Mountains from Cambrian to late Pennsylvanian time. Since the close of the Paleozoic Era the Ozark region has been a continuous land area. On at least two occasions (once at the end of the Cretaceous Period and another at the close of the Tertiary Period) this area was uplifted. Following the Cretaceous uplift and during the succeeding Tertiary Period, the Ozark region was worn down to a low and comparatively level plain with sluggish streams and probably large areas of swamps, covered with a forest in many respects similar to that which now occupies the Southeastern Lowlands. Subsequently a second uplift in late Tertiary time began. The probably slow and long-continued movement that elevated the region resulted in a rejuvenation of the streams, which, because of the steep grades, initiated the cutting of new valleys and produced the rough topography now found especially along the steep slopes of the Ozark dome. The extent of the peneplaned surface is reflected at the present time in the mostly accordant levels of the summits of the hills occupying the landscape.

During the late Cretaceous Period the waters of the Gulf of Mexico reached the southeastern border of the present Ozark region and deposits made along the shores appear at the surface in Scott and Stoddard counties in parts of Crowley Ridge. The Mississippi Embayment remained until a much later time, until towards the close of the Tertiary Period. The Southeastern Lowlands now occupy the area of this old embayment.

The Ozark region has never been glaciated and thus has remained a land surface continuously available for occupation by plant life since the end of the Paleozoic Era. During the Pleistocene Period of the Cenozoic Era several southward movements of the polar ice sheets moved toward Missouri, but only the early advances (Kansan and Nebraskan) penetrated the state, covering a large area north of the Missouri River in what is now designated as the Glaciated Prairie or Dissected Till Plains region. It is estimated that the Kansas ice sheet (the last to withdraw from Missouri) left the state at the earliest date about 600,000 years ago.

### Plant Regions

The principal plant regions of the state correspond in general to the location of the physiographic regions previously described, but their boundaries are not quite identical. It will be noted that wedgelike extensions of the prairie flora penetrate the Ozark region, and colonies of prairie plants occupy glades and openings in the forest of the Ozark region, while the flora of the Southeastern Lowlands pushes for short distances up the river valleys into the Ozark Plateau, and the flora of the latter extends westward into the Unglaciated Prairie region to occupy the river bluffs and rocky uplands.

#### *Southeastern Lowlands Region*

The flora over most of this region is rather uniform with the dominant species consisting of characteristic plants of the Gulf Coastal Plain. The region as a whole is Lower Austral with the southern phase of the Carolinian flora. Most of the region was originally forested with a dense and luxuriant growth of deciduous trees. In the swampy or more frequently inundated areas dominant trees and shrubs are *Taxodium distichum, Populus heterophylla, Leitneria floridana, Carya aquatica, Planera aquatica, Quercus lyrata, Q. Phellos,* and *Q. nigra, Itea virginica, Gleditsia aquatica, Acer rubrum* var. *Drummondii, Nyssa aquatica, Styrax americana, Fraxinus tomentosa,* and *Cephalanthus occidentalis* var. *pubescens,* while *Wisteria macrostachya, Vitis palmata, Trachelospermum difforme,* and *Mikania scandens* are dominant vines.

Some characteristic herbaceous and aquatic plants of this region are *Arundinaria gigantea, Rhynchospora macrostachya, Carex louisianica, Wolfiella floridana, Hymenocallis occidentalis, Iris fulva, Thalia dealbata, Cabomba caroliniana, Viola lanceolata, Ludwigia glandulosa, Eryngium prostratum, Cynosciadium digitatum, Lysimachia radicans, Asclepias perennis, Hydrolea uniflora, Justicia ovata, Diodia virginiana, Oldenlandia uniflora, Cayaponia grandifolia, Spilanthes americana* var. *repens,* and *Pluchea camphorata.*

The elevated strip of territory, known as Crowley Ridge, traverses sections of the Southeastern Lowlands. It consists mainly of sandy and gravelly terrain with some clay deposits, the soil generally being of an acid type. Geological evidence indicates that Crowley Ridge lay to the east of the Mississippi River until comparatively recently, when the channel of the river changed, and the flora would appear to lend support to this. Some of the species restricted to Crowley Ridge are of distinctly eastern or northeastern distribution and near or at their western limits, at least in this latitude. Among these, including others restricted to the Ridge, are *Aristida lanosa, Trisetum pensylvanicum, Fagus grandifolia, Polygonella americana, Liriodendron Tulipifera, Pyrus melanocarpa, Ilex opaca, Trichostema setaceum, Bartonia paniculata, Obolaria virginica, Epifagus virginiana,* and *Helianthus microcephalus.*

Some of the species found in the Southeastern Lowlands region reappear along the White River and its tributaries in the Ozarks, indicating a former connection geologically and floristically between the two regions. The flora of the Southeastern Lowlands is probably a very ancient one which may have changed only slightly since early or middle Tertiary time, when it extended far beyond its present limits. In Missouri it probably occupied the lowlands to the north and west before the last elevation of the Ozark Plateau, where traces of it are still found, and it may have covered most of the state before the advance of the Pleistocene ice sheets.

#### *The Ozark Region*

This region possesses the most diversified flora, including the greatest number of species, of any part of the state. Most of the endemic species occur in the Ozark region. The wealth and diversity of the flora can be attributed to the present topography, drainage, and surface geology, and to the geologic history of the region. A great variety of habitats has been provided by the varied edaphic environments and is further elaborated in the following brief account. Rock outcrops are frequent over the entire area and comprise chert or flint, sandstone, limestone, and such igneous rocks as granite, syenite, and porphyritic trachyte. These rocks range from the pre-Cambrian igneous rocks to the sedimentary ones of Cambrian through Pennsylvanian age. The magnesian limestones

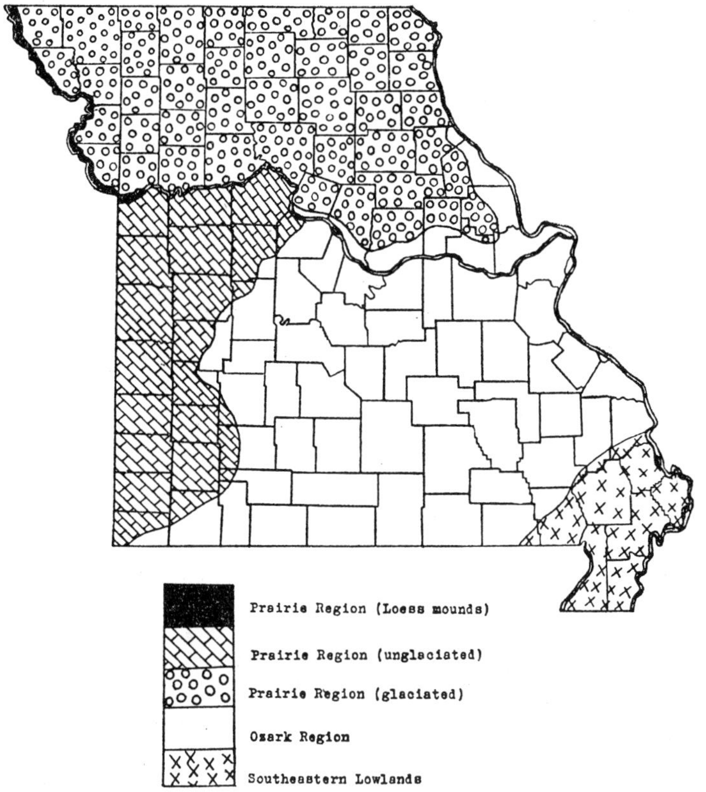


or dolomites produce an alkaline soil, characterized by the occurrence of various species, while the cherts, sandstones, and granites generally produce an acid soil frequented by particular species. Such types of rocks often come to the surface and form areas of small or large extent of open glades with little or no soil, devoid of trees and shrubs. These glades are common throughout the Ozarks and may occur on open slopes, tops of bluff escarpments, and eroded upland ridges and hilltop. They possess a characteristic herbaceous flora, with various species limited to particular types of rocks.

The limestone and dolomite glades harbor such species as *Ophioglossum Engelmanni*, *Juniperus Ashei*, *Carex Crawei*, *Eriogonum longifolium*, *Clematis Fremontii* var. *Riehlii*, *Delphinium Treleasei*, *Sedum pulchellum*, *Leavenworthia uniflora*, *Lesquerella filiformis*, *Acacia angustissima* var. *hirta*, *Baptisia australis* var. *minor*, *Psoralea esculenta*, *Euphorbia missurica*, *Phyllanthus polygonoides*, *Andrachne phyllanthoides*, *Callirhoe digitata* var. *digitata*, *Mentzelia oligosperma*, *Stenosiphon linifolius*, *Oenothera missouriensis*, *Amsonia ciliata* var. *filifolia*, *Centaurium texense*, *Evolvulus Nuttallianus*, *Heliotropium tenellum*, *Onosmodium subsetosum*, *Scutellaria Bushii*, *Satureja arkansana*, *Penstemon Cobaea* var. *purpureus*, *Galium virgatum*, *Houstonia nigricans*, *Valerianella ozarkana*, *Solidago Gattingeri*, *Aster sericeus*, *Aster oblongifolius*, *Aster ptarmicoides*, *Parthenium hispidum*, *Marshallia caespitosa* var. *signata*, *Echinacea paradoxa*, *Palafoxia callosa*, and *Rudbeckia missouriensis*. Some of these species are widespread over the greater part of the Ozark region whereas others such as *Juniperus Ashei*, *Delphinium Treleasei*, *Lesquerella filiformis*, *Acacia angustissima* var. *hirta*, *Phyllanthus polygonoides*, *Stenosiphon linifolius*, *Amsonia ciliata* var. *filifolia*, *Penstemon Cobaea* var. *purpureus*, *Marshallia caespitosa* var. *signata*, and *Palafoxia callosa* are confined to the bald knobs and limestone glades of the White River section and tributaries of the southwestern Ozarks.

Most of the characteristic species of sandstone glades are likewise found on chert and granite glades in the Ozarks, but not on limestone. Those restricted to such types of substrata include *Cheilanthes lanosa*, *Selaginella rupestris*, *Scirpus koilolepis*, *Polygonum tenue*, *Talinum parviflorum*, *Crotonopsis elliptica*, *Hypericum gentianoides*, *Oenothera linifolia*, and *Trichostema dichotomum*.

Viewed broadly, the flora of the Ozark region is a forest flora, with the oak-hickory and pine-oak type of woodland with a *Vaccinium vacillans* understory predominating on acid soils, while sugar maple, blue ash, chinquapin oak, walnut, papaw, linden, deciduous holly, southern buckthorn, and other forest species prevail on soils derived from limestone. The forest, with its associated herbaceous plants, belongs to the Carolinian flora, and floristically it is intermediate between the austral and boreal phases, with a slight predominance of southern species. The herbaceous plants of general distribution on the uplands are usually species that range from the Appalachian plateau to the grassy plains.

Aside from the correlated occurrence and restriction manifested by numerous species in the Ozarks to rocky glades, other types of habitats in the region provide equally special types of ecological niches occupied by characteristic species. Thus, the gravel bars of the Ozark streams are inhabited by some species of limited occurrence, such as *Hamamelis vernalis*, *Amsonia illustris*, and *Vernonia crinita*. In the cool waters (54–60 degrees Fahrenheit) of the Ozark springs and spring branches certain aquatic species are restricted to spring water and usually not found in Missouri outside the Ozark region, such as *Sparganium americanum*, *Potamogeton amplifolius*, *Potamogeton illinoensis*, *Anacharis Nuttallii*, *Lemna trisulca*, *Ranunculus longirostris*, *Nasturtium officinale* var. *officinale*, *Myriophyllum heterophyllum*, and *Veronica comosa*. Aquatic species generally found in the warmer waters of Ozark streams, but not in the cool water of the spring branches include *Potamogeton nodosus*, *Vallisneria americana*, *Heteranthera dubia*, and *Nuphar luteum* subsp. *ozarkanum*.

Some species present in the Ozark region are restricted in their occurrence to the protection of moist bluffs and shaded rocky ledges of ravines and canyons. Thus, it will be noted that *Zigadenus elegans*, *Trautvetteria caroliniensis*, *Galium boreale* var. *hyssopifolium*, and *Campanula rotundifolia* are known only from a few shaded, north-facing limestone bluffs, while *Lycopodium tristachyum*, *L. obscurum* var. *dendroideum*, *L. porophilum*, *L. lucidulum*, *Dennstaedtia punctilobula*, and *Viola pallens* are equally rare, but encountered only on certain sandstone ledges. Similarly, in the wet meadows developed by calcareous seepage flowing from spring branches in valleys of the Ozarks, especially along seepage of limestone outcrops along streams and their spring fed branches are rarely found *Carex suberecta*, *C. Woodii*, *Pogonia ophioglossoides*, *Liparis Loeselii*, *Filipendula rubra*, *Viola cucullata*, *Menyanthes trifoliata* var. *minor*, *Campanula aparinoides*, and *Aster puniceus* var. *firmus* f. *lucidulus*. The major portion of the distribution of the species noted above lies east and north of Missouri. All of them appear to represent relict species which have migrated southward or westward into the warmer Ozarks at the time of one of the advances of Pleistocene glaciation, subsequent to which they have persisted in these peculiarly favorable special habitats as remnants of a population able to adapt to the micro-climate or ecological niche of sufficient similarity to their more usual cooler or moister environment farther north or east to enable them to survive in their present location.

Of similar rarity and again representing a significant survival from former geological ages are the species found in and around the unique sink-hole ponds. These upland ponds, resulting from the collapse of the roof of subterranean caverns, formed in late Tertiary time of the last Ozark uplift, occupy natural depressions of the dry and more level upland portions of the Ozark Plateau. They apparently represent the last remnants of parts of the ancient swampy peneplain that formerly existed in the Ozarks prior to its last Tertiary uplift. Primarily Coastal Plain species, such as *Zizaniopsis miliacea*, *Eleocharis equisetoides*, *Scirpus Hallii*, *Echinodorus tenellus*, *Nyssa aquatica*, *Eryngium prostratum*, *Hottonia inflata*, *Hydrolea uniflora*, *Gratiola viscidula*, *Oldenlandia Boscii*, *Diodia virginiana*, *Cephalanthus occidentalis* var. *pubescens*, *Pluchea foetida*, and *Eupatorium hyssopifolium* var. *calcaratum* are isolated in and around some of the ponds. Some of the herbaceous species are known elsewhere in the state only from the Mississippi Embayment section of the Southeastern Lowlands, whereas others have their nearest stations along the Atlantic and Gulf Coastal Plains. Other species of great rarity, which range farther northward, such as *Najas gracillima*, *Potamogeton epihydrus* var. *Nuttallii*, *P. pulcher*, *Glyceria acutiflora*, *Scirpus Torreyi*, *Dulichium arundinaceum*, *Carex decomposita*, *C. alata*, *C. straminea*, *Viola lanceolata*, *Decodon verticillatus*, and *Galium tinctorium* are likewise associated with the more southern Coastal Plain species, and some of them are restricted in Missouri to sink-hole ponds. Since Pleistocene times the Xerothermic period may have further reduced the occurrence of swampy habitats in the dry upland plateau and exterminated all but a small remnant of the Coastal Plain species and other species of formerly Tertiary dispersal which today still survive in these natural ponds.

Other native plants, not previously mentioned as relicts, but whose rarity and limited occurrence in the Ozarks represent remote or isolated stations at or near the limits of the geographical range of the species, are *Cheilanthes tomentosa*, *Dryopteris Clintoniana* var. *australis*, *Juniperus Ashei*, *Calamagrostis insperata*, *Oryzopsis race-*

*mosa*, *Panicum calliphyllum*, *P. nitidum*, *P. annulum*, *Scirpus verecundus*, *S. subterminalis*, *S. Olneyi*, *Rhynchospora capillacea*, *Scleria nitida*, *S. verticillata*, *S. Muhlenbergii*, *Carex interior*, *C. leptalea*, *C. communis*, *C. aquatilis* var. *altior*, *C. stricta* var. *strictior*, *C. torta*, *C. trichocarpa*, *Juncus canadensis* var. *canadensis*, *J. subcaudatus* var. *subcaudatus*, *J. debilis*, *Zigadenus Nuttallii*, *Yucca glauca* var. *mollis*, *Smilacina stellata*, *Trillium nivale*, *Iris cristata*, *Sisyrinchium atlanticum*, *Cypripedium Reginae*, *Habenaria ciliaris*, *Calopogon pulchellus*, *Goodyera pubescens*, *Isotria medeoloides*, *I. verticillata*, *Eriogonum longifolium*, *Silene caroliniana* var. *Wherryi*, *Delphinium exaltatum*, *Berberis canadensis*, *Dicentra canadensis*, *Draba aprica*, *Saxifraga pensylvanica* var. *Forbesii*, *Sullivantia renifolia*, *Mitella diphylla*, *Philadelpus pubescens* var. *verrucosus*, *Ribes odoratum*, *Parnassia grandifolia*, *Hamamelis virginiana*, *Fragaria vesca* var. *americana*, *Waldsteinia fragarioides*, *Neviusia alabamensis*, *Desmodium humifusum*, *D. ochroleucum*, *Linum striatum*, *Andrachne phyllanthoides*, *Phyllanthus polygonoides*, *Ilex verticillata* var. *padifolia*, *Euonymus obovatus*, *Ludwigia microcarpa*, *Oenothera fruticosa* var. *linearis*, *O. perennis* var. *perennis*, *O. tetragona* var. *hybrida*, *Hydrocotyle verticillata* var. *verticillata*, *Lyonia mariana*, *Rhododendron roseum*, *Gaylussacia baccata*, *Dodecatheon amethystinum*, *Spigelia marilandica*, *Centaurium texense*, *Amsonia ciliata* var. *filifolia*, *Phlox bifida* var. *stellaria*, *P. carolina* var. *carolina*, *P. amplifolia*, *P. maculata* var. *pyramidalis*, *Phacelia ranunculacea*, *Callicarpa americana*, *Monarda clinopodia*, *Mimulus glabratus* var. *Fremontii*, *Chelone obliqua* var. *speciosa*, *Mitchella repens*, *Lonicera dioica* var. *dioica*, *Viburnum recognitum*, *Liatris spicata*, *Solidago Riddellii*, *Aster furcatus*, *Berlandiera texana* var. *texana*, *Marshallia caespitosa* var. *signata*, and *Lactuca hirsuta* var. *hirsuta* and var. *sanguinea*. The above species occupy particular habitats to which they are peculiarly adapted. Their local ranges likewise appear to be correlated with certain edaphic factors of soil, exposure, and moisture, as well as with past geological events.

A significant floral element of the Ozark flora is that found in the rugged section of the White River and its tributaries in southwestern Missouri. Although the uplands in this section are covered with oak-hickory and various kinds of forest found in other parts of the Ozarks, many rare and unusual plants are found along the river bluffs and in the deep canyons of the tributary streams, and on the limestone bald knobs and glades. Here on protected north-facing bluffs occur such predominant Alleghenian species as *Magnolia acuminata* and *Cladrastis lutea*, the latter not occurring elsewhere in the state, while on south- and west-facing exposures such trees and shrubs as *Juniperus Ashei*, *Ribes odoratum*, *Acacia angustissima* var. *hirta*, *Cotinus obovatus*, *Sapindus Drummondii*, and *Chionanthus virginica*, and the vining plant, *Cissus incisa*, are present. The majority of the species on these south- and west-facing exposures are mainly of southwestern origin, but a few of these are known from isolated eastern stations in southeastern Missouri. Here also along the bluff ledges and outcrops are found *Notholaena dealbata*, *Cheilanthes alabamensis*, and *C. tomentosa*. On ridge tops and summits of hills where chert or sandstone outcroppings produce acid soil conditions, the native chestnut, *Castanea ozarkensis*, occurs. Many of the species known from limestone glades frequent the exposed rocky areas found throughout this section of the Ozarks, and some, such as *Delphinium Treleasei*, *Phyllanthus polygonoides*, *Stenosiphon linifolius*, *Amsonia ciliata* var. *filifolia*, *Penstemon Cobaea* var. *purpureus*, *Marshallia caespitosa* var. *signata*, and *Palafoxia callosa* are known only from this portion of Missouri and often represent a western and southwestern element of the flora. The flora of this area shows a relationship with a Coastal Plain and southern element, exemplified by the occurrence in both the White River and the Southeastern Lowlands region of *Arundinaria gigantea*, *Quercus falcata*, *Phoradendron flavescens*, *Magnolia acuminata*, *Chionanthus virginica*, *Ampelopsis arborea*, *Jussiaea decurrens*, *Diodia virginiana*, and *Hymenopappus scabiosaeus*. Likewise, the flora of the White River section manifests a strong relationship with the flora of parts of the western and southwestern United States, particularly with that of Oklahoma and Texas.

### Endemic Species of the Ozarks

Some of the species occurring in the Ozark region of Missouri are known solely or chiefly from Missouri, Arkansas, and Oklahoma, and apparently have originated within this circumscribed area. Their present distribution is either confined to the area or the mass of their distribution occupies the Ozark region with outlying stations in unglaciated portions of adjacent states. Over the lapse of time since the Ozarks have been available for plant occupation since the end of the Paleozoic Era, endemic species, subspecies, and varieties may have developed from the beginning in the area itself, or, after separation from geographically related species, become gradually localized to the section in the Ozarks where now found. Among such endemics are *Juniperus Ashei*, *Tradescantia ozarkana*, *T. Ernestiana*, *T. longipes*, *Trillium pusillum* var. *ozarkanum*, *T. viride* var. *viride*, *Nemastylis Nuttallii*, *Carya texana*, *Castanea ozarkensis*, *Nuphar luteum* subsp. *ozarkanum*, *Delphinium Treleasei*, *D. carolinianum* var. *Nortonianum*, *Clematis Fremontii* var. *Riehlii*, *C. versicolor* (also in Kentucky and Tennessee), *Ranunculus Harveyi*,

*Heuchera puberula, H. missouriensis, Ribes missouriense* var. *ozarkanum, Hamamelis vernalis, Crataegus padifolia, C. Lettermani, C. sicca, C. noelensis, C. dispessa, C. latebrosa, C. nuda, Amorpha brachycarpa, Andrachne phyllanthoides, Callirhoe Papaver* var. *Bushii, C. digitata* var. *digitata, Vaccinium stamineum* var. *interius*; *Amsonia illustris, Phlox pilosa* var. *ozarkana, Onosmodium subsetosum, Scutellaria Bushii, Penstemon Cobaea* var. *purpureus, P. arkansanus, Gerardia flava* var. *calycosa, Ruellia pedunculata, Galium arkansanum, Valerianella ozarkana, V. Bushii, Vernonia crinita, Grindelia lanceolata, Solidago arguta* var. *neurolepis, S. Drummondii, S. Gattingeri, Aster anomalus, Parthenium hispidum, Rudbeckia fulgida* var. *palustris, R. missouriensis*, and *Echinacea paradoxa*.

Other species not mentioned above may also have originated in the Ozark region, but, since they occupy a much wider geographical range, are omitted.

### Geographical Distribution in General Within the Ozark Region

While many of the species confined in Missouri to the Ozark region are of widespread occurrence throughout the Ozarks, some, however, are known only from the eastern section, while others are concentrated in the western sector, and still others known only from the southernmost portion of the state. This frequently indicates the direction from which migration has originated or entered the region. For example, the dominant and characteristic element of the flora found over the southeastern part of the Ozark region is of Alleghenian relationship, and it is in this section where a number of eastern plants reach their western limit of distribution. Species of more northern range are frequently found in the Ozarks along north-facing bluffs along streams, or in deep V-shaped valleys of the more rugged parts of the region, or in cool moist springs, wet meadows, swamps, or wet ledges where constant seepage prevails. On the dry, exposed, and sunnier south-and west-facing wooded slopes, rocky open glades, and prairie openings, the plants whose main distribution is southwestern, southern, or western, comprise an important or dominant element of the vegetation. The Ozark region in this discussion has been interpreted as extending from the southernmost border of the state from McDonald County on the west to Butler, Bollinger, and Cape Girardeau counties on the east, thence north across the Missouri River to a portion of Marion County on the northeast, south and westward paralleling the Missouri River to Boone County. The counties between Marion and Boone in the region indicated have preserved for the most part many characteristic Ozark herbaceous and woody plants absent north of them. Moreover, they have preserved topographical and old weathered rock surfaces similar to that of other Ozark landscape to the south.

### *Prairie Region*

#### (a) *Glaciated Prairie or Dissected Till Plains*

This area of Missouri lying mainly north of the Missouri River, except for the inclusion of small areas in St. Louis, Saline, Cooper, Lafayette, and Jackson counties, is covered with glacial drift pertaining to the Nebraskan and Kansan Glacial age. Most of the stratified rocks occurring in this area are of Mississippian and Pennsylvanian age. Rocks of Ordovician, Silurian, and Devonian age also occur north of the Missouri River, but they occur mainly in the section assigned to the Ozark region. Much of the Glaciated Prairie area is forested, with forests of a similar type to those of the Ozarks, but with less diversification of species, a majority of the more southern species disappearing entirely from the flora. Some northern species of plants which are wholly or mostly absent from the Ozarks to the south are found in this region and include *Matteuccia Struthiopteris* var. *pensylvanica, Phalaris arundinacea, Phragmites communis* var. *Berlandieri, Calamagrostis canadensis* var. *Macouniana, Eleocharis Wolfii, Scirpus paludosus* var. *paludosus, Allium tricoccum, Smilacina stellata, Medeola virginiana, Salix gracilis* var. *textoris, Populus tremuloides, Quercus ellipsoidalis, Ulmus Thomasi, Urtica dioica* var. *procera, Arenaria lateriflora, Anemone canadensis, Cardamine Douglassii, Spiraea alba, Rosa blanda, R. virginiana, Prunus virginiana, Acer nigrum, A. Negundo* var. *violaceum, Euphorbia Geyeri, Vitis riparia, Fraxinus pennsylvanica, Asclepias Sullivantii, Collomia linearis, Chelone obliqua* var. *speciosa, Sambucus pubens, Viburnum Lentago, V. dentatum* var. *Deamii, Senecio pauperculus* var. *Balsamitae, S. aureus* var. *semicordatus*, and *Prenanthes alba*.

Forested hills with deeply dissected ravines abound in the vicinity of streams. In the valleys of the streams are often expanses of alluvial wet meadows or 'bottom prairies' characterized by species of grasses, sedges, rushes, and other herbaceous plants. Swamps and oxbow lakes are frequent in the valley bottoms. In the more level sections of the upland and on ridges, prairie is frequently encountered with grasses, sedges, legumes, composites, and other perennial herbs dominant. The prairie flora is often preserved in strips along railroads and has its center of distribution mainly farther west, whereas the flora of the woodlands appear to represent evident extensions of an eastern, northern, and southern floral element.

*Saline Springs and Salt Licks*

In portions of this region occur natural saline and mineral springs which sometimes form salt 'licks.' They are present in areas in the valleys of streams usually in territory of low relief and reach their greatest development in Saline, Randolph, Howard, and Cooper counties. The soil immediately around the springs is heavily charged with concentrated alkaline salts, and usually supports an unusual assemblage of plants, some of which are known only from such places. *Distichlis spicata*, a grass usually found in brackish soil along the coastal section of the United States, occurs at its only recorded station in the interior of North America around a salt 'lick' in the valley of Heath Creek, in Saline County. *Typha angustifolia* and *Eleocharis parvula*, occurring in marshy ground around certain saline springs, are known in Missouri chiefly from these stations. Floating in the water of the springs or their branches are clumps of *Zannichellia palustris* var. *major*, *Ruppia maritima* var. *rostrata*, *Potamogeton pectinatus*, and *Callitriche heterophylla*, as well as species of algae, such as *Chara* and *Nitella*. Other species present around the saline springs include *Typha latifolia*, *Diplachne fascicularis* and *D. acuminata*, *Echinochloa pungens*, *Scirpus americanus*, *S. paludosus* var. *paludosus*, *S. fluviatilis*, *Atriplex rosea* and *A. patula* var. *hastata*, and *Polygonum aviculare*.

*Loess Hills*

Perhaps the most unusual part of the dissected till plains are the loess hills or mounds of northwestern Missouri along the Missouri River, best developed in Atchison and Holt counties, but extending southeastward into Jackson County below Kansas City. The thick accumulation of loess and its characteristic erosion has developed a peculiar type of topography marked by terraced exposed hills or mounds separated by deep, steep-sided ravines. Many forest species extend up the deeper ravines, but rapid erosion and wind action have kept the slopes bare in most places. A flora characteristic of the region west and northwest of Missouri dominates the exposed parts of the mounds with grasses and other herbaceous plants constituting the bulk of the vegetation. Some of the species of the Great Plains reach their eastern limit of geographical range here. The characteristic species of this region include *Muhlenbergia cuspidata*, *Sporobolus airoides*, *Bouteloua hirsuta*, *B. gracilis*, *Yucca glauca* var. *glauca*, *Anemone cylindrica*, *Cleome serrulata*, *Oxytropis Lambertii*, *Astragalus lotiflorus*, *Psoralea argophylla*, *Dalea enneandra*, *Glycyrrhiza lepidota*, *Euphorbia marginata*, *E. glyptosperma*, *Oenothera serrulata*, *Gaura coccinea*, *Penstemon grandiflorus*, *Symphoricarpos occidentalis*, *Liatris punctata* var. *nebraskana*, *Heterotheca subaxillaris*, *Chrysopsis villosa* var. *canescens*, *Lygodesmia juncea*, and *Lactuca pulchella*.

The incursion of this plains flora into the state may have taken place in relatively recent times during a period some time subsequent to the final retreat of the glaciers and perhaps within the past few thousand years.

(b) *Osage Plains Section or Unglaciated Prairie*

This portion of the Interior Highland has remained unglaciated and occupies a wedge-shaped division south of the Missouri River in the western part of the state, merging irregularly and indefinitely into the western border of the Ozark Region. The Ozark flora occurring in the area is dominant on broken rocky ground along the streams, while the prairie and plains flora occupies the more level and open sections of the uplands in a wide transition zone. There is sharp contrast between the floras growing on different geological formations, which include mainly Mississippian and Pennsylvanian limestones, sandstones, chert, and shales. Broad shallow valleys containing numerous small lakes and sloughs abound, particularly in areas possessing soft and easily weathered formations of Pennsylvanian shales and sandstones. Wet meadows and bottom prairies occur in the valleys of the streams, with particularly well developed flood plains along the Grand and Marais des Cygnes rivers. On the drier or rocky level upland a true prairie flora is found, similar to that of the glaciated prairie or dissected till plains of northern Missouri, except for an increasing proportion of southwestern species appearing towards the south. In the unglaciated Osage section most of the characteristic northern species of plants found in the glaciated prairie section are absent, and there is a greater variety of trees and shrubs along the streams.

More than in any other part of the state, the prairie flora has been better preserved in this section due to the maintenance of large tracts used for grazing and hay production. Plants characteristic of this region include *Sagittaria ambigua*, *Panicum scoparium*, *P. malacophyllum*, *Schedonnardus paniculatus*, *Andropogon saccharoides*, *Eleocharis lanceolata*, *Rhynchospora Harveyi*, *Fimbristylis dichotoma*, *Carex arkansana*, *C. oklahomensis*, *Tradescantia Tharpii*, *Juncus kansanus*, *Sisyrinchium campestre* var. *kansanum*, *Anemone caroliniana*, *Lotus Purshianus*, *Ammannia auriculata*, *Rhexia interior*, *Sabatia campestris*, *Salvia azurea* var. *grandiflora*, *Gerardia fasciculata*, *Collinsia violacea*, *Gutierrezia dracunculoides*, *Aster paludosus* subsp. *hemisphericus*, *Helianthus salicifolius*, *Marshallia caespitosa* var. *caespitosa*, and *Krigia occidentalis*.

Sandstone and chert glades and bluffs in the unglaciated prairie Osage section often harbor a very specialized flora not found elsewhere in the state. Some of the unusual plants which occur on the sandstone formations are *Geocarpon minimum*, *Corydalis crystallina*, *Selenia aurea*, *Sedum Nuttallianum*, and *Saxifraga texana*. An unusual formation of extensive chert glades, known as the Grand Falls Chert of the Mississippi rocks, is developed in the valleys of Turkey Creek and Shoal Creek in Jasper and Newton counties. Rains which fill the depressions between the broken portions of the impermeable chert rocks often remain until evaporated by the sun. Under the peculiar edaphic conditions developed here, a number of plants occur which are restricted to this section. Among those restricted to this area are *Portulaca retusa* and *Lathyrus pusillus*.

### Excluded Species

Throughout this work a species has been placed in an excluded section following the listing of the last species within a given genus. The species placed in the excluded list are those which have been misidentified or attributed to the state by error or which are not supported by an herbarium record or are based upon cultivated material. A number of species previously reported by Palmer and Steyermark in their *Annotated Catalogue of Flowering Plants* are shown in the present work to have been either misidentified or represented by specimens since reinterpreted otherwise. For some species reported by Fernald in the eighth edition of Gray's *Manual of Botany* and Gleason's *New Britton and Brown Illustrated Flora* there do not appear to be any voucher herbarium specimens to support the existence of the report, and such species have been excluded.

For a number of species which Benjamin Franklin Bush reported to the writer in various personal communications, no specimens have ever been found. Bush informed me that many of his earlier collections had been sent to Dr. N. L. Britton and to George Vasey for determination. He stated that from 1888 to 1910 he sent many specimens to Britton from Dunklin, Pemiscot, Oregon, Taney, Barry, and McDonald counties, and believed that all these should be in the herbarium of the New York Botanical Garden. He further stated that all the species collected in Jackson County up to about 1907 were sent to K. K. Mackenzie to insert in the manual upon which he and Mackenzie were working, and that these should be in the Mackenzie Herbarium at The New York Botanical Garden. The following species reported to me by Bush as based upon his collections were supposed by him to be in the New York Botanical Garden Herbarium: *Jeffersonia diphylla*, *Dalea aurea*, *Ononis arvensis*, *Lupinus perennis*, *Petalostemum villosum*, *Opuntia polyacantha*, *Bifora americana* [= *B. radians*], *Hydrocotyle umbellata*, *Ammoselinum Popei*, *A. Butleri*, *Conioselinum chinense*, *Ludwigia alata*, and *Breweria aquatica*. Careful search through that herbarium by Dr. David D. Keck has failed to turn up any of the above species, a few of which are noted in the present flora as misidentifications of other species. Correspondence with Drs. Mathias and Constance, specialists in the Umbelliferae, concerning the species of that family referred to in the above list, has produced similar negative evidence.

As some of the species recorded for Missouri are known only from specimens collected for the World's Fair Collection held in Chicago in 1893, and have not been found since in the state, I have suspected that there may have been some confusion as to the origin or source of the material mounted. In this connection it is of interest to note Bush's reply in answer to the writer's query about these collections. In his letter of September 15, 1933, to the writer, Bush states, 'In 1893 I had charge of the work preparing the herbarium for the Chicago World's Fair. I had Mr. Kellogg write the labels and had all the specimens mounted at the warehouse in St. Louis. I collected the greater part of the specimens, some 2,100 species, and got several hundred from Eggert and a few from Letterman. You can depend on these specimens being from the state, unless it can be shown that they are certainly outside of our flora.' Again, in his letter of October 21, 1934, Bush wrote that 'all of the specimens [in World's Fair collection] were collected in Missouri, unless they were collected by Eggert in Illinois. Eggert had a habit of marking all specimens St. Louis, even if he got them in St. Clair Co., Illinois, but I think there are very few of these amongst the plants he gave me.' It does seem to this author that there may be doubt as to the source of some of these World's Fair collections, especially those for which no record has ever turned up in the state since.

It should also be pointed out here that a number of reports for the occurrence of Missouri species were based upon cultivated material collected in St. Louis by Dr. Earl E. Sherff. In Rhodora 54: 256. 1952, I have noted that 'The great majority of Sherff's specimens labeled "St. Louis" were (according to personal communication with Dr. Sherff) collected either from cultivated plants found on the grounds of the Missouri Botanical Garden or in other parts of St. Louis.... These specimens represented by duplicates in several herbaria have been misconstrued by other workers to represent collections of wild or spontaneous plants, which is certainly far from the real situation.' Such collections as Sherff's collection (no. 194) from St. Louis 'cited by Munz from a Gray

Herbarium specimen, was obtained, according to an examination of Dr. Sherff's notebook, on July 2, 1910, and undoubtedly was collected from cultivated plants at the same time with such other cultigens as *Hydrangea quercifolia, Sedum acre, Heuchera sanguinea,* and *Tolmiaea Menziesii.*' As was noted in the present flora, the report of *Achillea ptarmica* in Gleason's *New Britton and Brown Illustrated Flora,* was based upon a cultivated specimen collected by Sherff in St. Louis.

Directing the attention of botanists and other field workers to the above notes and data may help to eradicate some of the reports which tend to persist and accumulate in a local flora, and should caution them against accepting reports based on cultivated or nonauthenticated specimens.

### Species Whose Geographic Range Approaches Missouri

Intensive and careful future exploration of Missouri should add scores of native or introduced species which at present are unknown in the state. The following lists taken from substantiated records of adjacent states include species whose known distribution approaches reasonably close to the borders of Missouri, and thereby suggest a possible clue to their further extension into Missouri. It is hoped that such information may be a source which may serve future workers to track down additions to the Missouri flora. For each of the following species a letter separated by a dash indicates the probable part of Missouri (northeast, east, south, southeast, west, southwest, and northwest) from which the particular species may be expected. These lists are not complete, but include some of the species thought most likely as extending into the state. These lists, of course, do not provide for the numerous possibilities of introduced and adventive species, nor for the native ones which afford surprising outlying stations removed from their usual range.

*From Illinois* (records taken from Jones *et al., Vascular Plants of Illinois.* 1955, and from records by John W. Voigt and Robert H. Mohlenbrock, Jr.).

Equisetum fluviatile – NE.
Dryopteris noveboracensis – SE.
Trichomanes Boschianum – SE.
Potamogeton Friesii – SE.
Potamogeton natans – SE.
Aristida tuberculosa – NE.
Panicum yadkinense – SE.
Carex cephaloidea – NE.
Carex retrorsa – SE.
Carex Sprengelii – NE.
Carex Tuckermani – NE.
Eleocharis intermedia – E.
Eleocharis ovata – NE.
Eleocharis palustris – NE.
Scirpus Smithii – NE.
Symplocarpus foetidus – NE.
Luzula echinata – SE.
Luzula multiflora – SE.
Lilium umbellatum – NE.
Tipularia discolor – SE.
Castanea dentata – SE.
Quercus Prinus – SE.
Ulmus serotina – SE.
Aristolochia Nashii – SE.
Polygonum tomentosum – E. (intro.)
Silene armeria – E. (cult.)
Caltha palustris – NE.
Cimicifuga cordifolia – SE.
Ranunculus ambigens – SE.
Jeffersonia diphylla – SE.
Fumaria officinalis – NE. (adventive)
Alliaria officinalis – E.
Alyssum alyssoides – NE.
Cleome speciosissima – SE. (cult.)
Sedum telephioides – SE.
Heuchera parviflora var. Rugelii – SE.
Geum virginianum – SE.
Prunus nigra – NE.
Amorpha nitens – SE.
Apios Priceana – SE.
Astragalus tennesseensis – E.
Cassia hebecarpa – E.
Linum intercursum – SE.
Linum virginianum – SE.
Polygala polygama – NE.
Euphorbia helioscopia – NE.
Celastrus articulatus – SE. (intro.)
Hibiscus esculentus – SE. (intro.)
Hypericum adpressum – SE.
Hypericum denticulatum – SE.
Hypericum ellipticum – E.
Lechea minor – SE.
Lechea stricta – NE.
Viola affinis – SE.
Viola pubescens – NE.
Myriophyllum verticillatum – SE.
Callitriche palustris – SE.
Cornus stolonifera – NE.

Sanicula marilandica – NE.
Halesia carolina – SE.
Lysimachia terrestris – NE.
Lysimachia vulgaris – SE. (intro.)
Asclepias exaltata – E., NE., SE.
Breweria Pickeringii var. Pattersoni – NE.
Allocarya figurata – E.
Heliotropium curassavicum – E.
Pycnanthemum pycnanthemoides – SE.
Stachys Clingmanii – SE.
Synandra hispidula – SE.
Penstemon calycosus – E., SE.
Penstemon Deamii – SE.
Synthyris Bullii – NE.
Viburnum cassinoides var. cassinoides – SE.
Valeriana pauciflora – SE.
Artemisia Absinthium – SE., E.
Aster Shortii – NE., SE.
Aster umbellatus – NE.
Bidens coronata – NE., E., SE.
Lactuca biennis – SE.
Picris echioides – NE.
Solidago bicolor – SE.
Xanthium oviforme – E.

*From Arkansas* (records furnished by Dr. Dwight M. Moore)

Pilularia americana – SW.
Festuca dertonensis – SW.
Festuca versuta – SW.
Poa arachnifera – SE.
Anthoxanthum aristatum – SW.
Limnodea arkansana – SW.
Cynosurus echinatus – S. from Oregon and Howell counties
Erianthus contortus – SW.
Andropogon glomeratus – SE.
Sorghastrum Elliottii – SW.
Sphenopholis filiformis – S.
Axonopus affinis – SE.
Axonopus furcatus – SW.
Panicum brachyanthum – SW.
Panicum hirticaule – SW.
Panicum lucidum – SE.
Panicum Webberianum – SW.
Paspalum Boscianum – SW.
Digitaria violascens – S. of Oregon Co.
Eleocharis tortilis – S.
Juncus coriaceus – S. of Howell and Ozark counties
Nothoscordum striatum – SW.
Aletris farinosa – SW.
Sisyrinchium hastile – SW.
Tipularia discolor – S. of Howell Co.
Spiranthes montana – SW.
Carya leiodermis – S.
Ulmus crassifolia – SE.
Aristolochia reticulata – S. of Howell Co.
Stellaria uniflora – SW.
Stellaria Nuttallii – SW.
Silene ovata – SW.
Magnolia tripetala – SW.
Magnolia macrophylla – SE. (15 mi. W. of Dunklin Co. line)
Schizandra coccinea – SE.
Asimina parviflora – SE.
Delphinium newtonianum – S.
Anemone quinquefolia – S.
Drosera brevifolia – SW.
Podostemum ceratophyllum – S.
Saxifraga Palmeri – SW.
Heuchera arkansana – SW.
Philadelphus Sharpianus – S.
Philadelphus intectus var. pubigerus – S.
Alchemilla microcarpa – SW.
Agrimonia microcarpa – SW.
Rubus cuneifolius – S.
Amorpha nitens – SE.
Petalostemon Stanfieldii – SW.
Geranium columbinum – SW.
Oxalis filipes – S.
Oxalis interior – SW.
Polygala cruciata – SW.
Euphorbia Peplus – SW.
Abutilon incanum – S.
Hypericum adpressum – S.
Viola blanda – SW.
Viola canadensis – S.
Viola hastata – SW.
Viola palmata – SW.
Oenothera arenicola – S.
Trepocarpus aethusae – S.
Halesia carolina – S.
Halesia monticola var. vestita – S.
Styrax grandifolia – S.
Rhododendron canescens – S.
Rhododendron oblongifolium – S.
Rhododendron viscosum – S.
Lyonia ligustrina – S.
Vaccinium Elliottii – S.
Gentiana saponaria – SW.
Amsonia Hubrichtii – SW.

Ipomoea barbigera – SW.
Ipomoea setosa – SW.
Ipomoea trichocarpa – S.
Nemophila microcalyx – SW.
Nemophila phacelioides – SW.
Verbena Lambertii – SW.
Scutellaria cordifolia – S.
Scutellaria integrifolia – S.
Scutellaria resinosa – SE.
Bouchetia anomala – SW.
Penstemon multicaulis – SW.
Penstemon Wherryi – SW.
Penstemon tenuis – SE.
Utricularia geminiscapa – S.
Utricularia subulata – S.
Galium parisiense – SW.
Valerianella longiflora – SW.
Valerianella Nuttallii – SW.
Sphenoclea zeylanica – SE.
Aster texanus var. parviceps – S.
Carduus leptocephalus – S.
Coreopsis cardaminefolia – SE.
Coreopsis heterolepis – S.
Elephantopus tomentosus – SE.
Eupatorium album var. glandulosum – S. of Oregon Co.
Rudbeckia amplectens – S.
Senecio tomentosus – SW., SE.
Solidago bicolor – SW.
Solidago microphylla – SE.

*From Kansas* (records based on F. C. Gates, *Flora of Kansas*. 1940, and Woodson's Monograph of Asclepias, *Ann. Mo. Bot. Gard.* 41. 1954).

Sagittaria cuneata – W.
Sagittaria esculenta – W.
Tridens albescens – W.
Eragrostis secundiflora – W., NW.
Melica Porteri – W.
Alopecurus myosuroides – W., NW.
Redfieldia flexuosa – W., NW.
Agropyron cristatum – SW.
Panicum Wilcoxianum – W., NW.
Eleocharis elliptica – W., NW.
Juncus neomexicanus – W.
Allium arenicola – SW.
Cooperia Drummondii – SW.
Populus Sargentii – W.
Corylus cornuta – SW. (W. of Vernon Co.)
Comandra pallida – NW. (W. of Holt and Cass counties)
Polygonum omissum – W.
Eriogonum annuum – NW.
Alliaria officinalis – W.
Potentilla sulfurea – NW. (W. of Andrew and Platte counties)
Rosa pimpinellifolia – W.
Astragalus plattensis – W. (W. of Jackson Co.)
Dalea aurea – NW. (W. of Andrew and Holt counties)
Psoralea digitata – NW. (near Platte Co.)
Psoralea lanceolata – SW. (near McDonald Co.)
Sophora sericea – SW. (W. of Jasper Co.)
Petalostemon villosum – NW. (near Platte Co.)
Petalostemon occidentale – W.
Galactia regularis – SW.
Vicia oregana – NW. (W. of Holt and Atchison counties)
Vicia sparsifolia – W. (W. of Jackson Co.)
Zygophyllum fabago – W.
Polygala alba – NW.
Stillingia salicifolia – SW.
Euphorbia lucida – NW.
Sphaeralcea coccinea – W.
Hybanthus verticillatus – NW.
Opuntia polyacantha – W. (W. of Jackson Co.).
Opuntia tortispina – SW. (W. of Jasper Co.)
Neomamillaria missouriensis – W. (W. of Cass Co.)
Neomamillaria similis – SW. (W. of Newton Co.)
Cornus instolonea – W. (W. of Jackson and Cass counties)
Berula pusilla – NW. (W. of Andrew Co.)
Eryngium Leavenworthii – SW. (near Bates, Vernon, and Jasper counties)
Lomatium orientale – W. (W. of Bates Co.)
Lomatium villosum – NW.
Sanicula marilandica – NW.
Periploca graeca – SW.
Asclepias arenaria – NW. (near Buchanan and Jackson counties)
Asclepias speciosa – NW.
Asclepias pumila – NW.
Asclepias Engelmanniana – SW.
Asclepias Nuttalliana – W. (W. of Jackson Co.)
Ipomoea leptophylla – W.
Lippia cuneifolia – SW. (near McDonald Co.)
Lycopus uniflorus – NW. (W. of Buchanan Co.)
Stachys annua – W. (W. of Jackson Co.)
Chamaesaracha conoides – SW.
Physalis hederaefolia var. comata – NW.
Gerardia densiflora – SW.

Penstemon albidus – W., NW.
Valerianella amarella – SW. (W. of Bates and Vernon counties)
Lobelia appendiculata – SW.
Helianthus Besseyi – W. (W. of Jackson Co.)
Hymenoxys linearifolia – NW. (W. of Holt Co.)
Gnaphalium uliginosum – W. (W. of Jackson Co.)
Senecio Riddellii – W. (cult.)
Cirsium ochrocentrum – NW. (by Holt Co.)
Pyrrhopappus grandiflorus – NW. (W. of Platte Co.)
Hieracium paniculatum – W. (W. of Jackson Co.)

*From Oklahoma* (records based on publications of U. T. Waterfall)

Sphenopholis filiformis – SW.
Rivina humilis – SW.
Stellaria muscorum – SW. (by McDonald Co.)
Stellaria Nuttallii – SW.
Neptunia lutea – SW.
Dalea purpurea – SW.
Hybanthus verticillatus – SW.
Asclepias asperula var. capricornu – SW.
Nemophila phacelioides – SW.
Dichondra repens var. carolinensis – SW.
Castilleja indivisa – SW.

In addition to the above are numerous species known only from McCurtain or LeFlore counties, southeastern Oklahoma, where occur floral elements of the Coastal Plain and southeastern United States. Some of these already have been found in southern Missouri, but additional ones known from this section of Oklahoma may be expected eventually to occur in Missouri.

# *Finding County Map of Missouri*

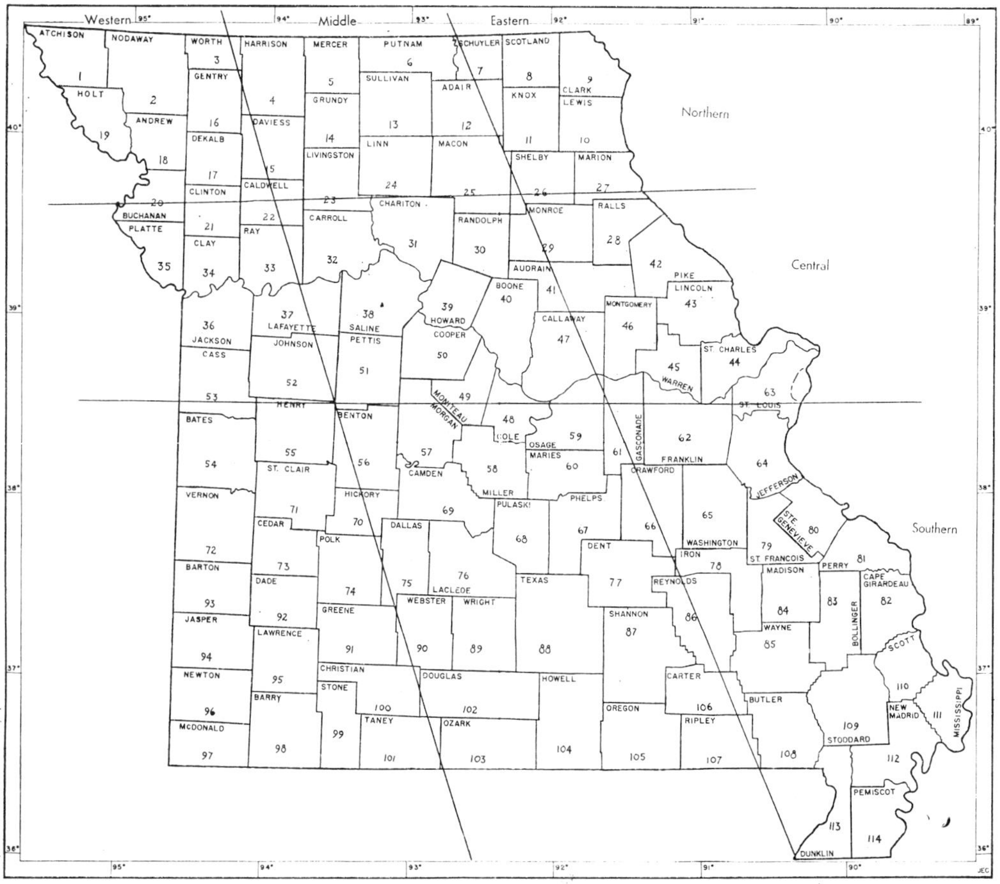


| *County* | *Number* | *County* | *Number* |
|---|---|---|---|
| Adair | 12 | Barton | 93 |
| Andrew | 18 | Bates | 54 |
| Atchison | 1 | Benton | 56 |
| Audrain | 41 | Bollinger | 83 |
| Barry | 98 | Boone | 40 |

| *County* | *Number* |
|---|---|
| Buchanan | 20 |
| Butler | 108 |
| Caldwell | 22 |
| Callaway | 47 |
| Camden | 69 |
| Cape Girardeau | 82 |
| Carroll | 32 |
| Carter | 106 |
| Cass | 53 |
| Cedar | 73 |
| Chariton | 31 |
| Christian | 100 |
| Clark | 9 |
| Clay | 34 |
| Clinton | 21 |
| Cole | 48 |
| Cooper | 50 |
| Crawford | 66 |
| Dade | 92 |
| Dallas | 75 |
| Daviess | 15 |
| DeKalb | 17 |
| Dent | 77 |
| Douglas | 102 |
| Dunklin | 113 |
| Franklin | 62 |
| Gasconade | 61 |
| Gentry | 16 |
| Greene | 91 |
| Grundy | 14 |
| Harrison | 4 |
| Henry | 55 |
| Hickory | 70 |
| Holt | 19 |
| Howard | 39 |
| Howell | 104 |
| Iron | 78 |
| Jackson | 36 |
| Jasper | 94 |
| Jefferson | 64 |
| Johnson | 52 |
| Knox | 11 |
| Laclede | 76 |
| Lafayette | 37 |
| Lawrence | 95 |
| Lewis | 10 |
| Lincoln | 43 |
| Linn | 24 |
| Livingston | 23 |
| Macon | 25 |
| Madison | 84 |
| Maries | 60 |
| Marion | 27 |
| McDonald | 97 |
| Mercer | 5 |
| Miller | 58 |
| Mississippi | 111 |
| Moniteau | 49 |
| Monroe | 29 |
| Montgomery | 46 |
| Morgan | 57 |
| New Madrid | 112 |
| Newton | 96 |
| Nodaway | 2 |
| Oregon | 105 |
| Osage | 59 |
| Ozark | 103 |
| Pemiscot | 114 |
| Perry | 81 |
| Pettis | 51 |
| Phelps | 67 |
| Pike | 42 |
| Platte | 35 |
| Polk | 74 |
| Pulaski | 68 |
| Putnam | 6 |
| Ralls | 28 |
| Randolph | 30 |
| Ray | 33 |
| Reynolds | 86 |
| Ripley | 107 |
| St. Charles | 44 |
| St. Clair | 71 |
| St. Francois | 79 |
| Ste. Genevieve | 80 |
| St. Louis | 63 |
| Saline | 38 |
| Schuyler | 7 |
| Scotland | 18 |
| Scott | 110 |
| Shannon | 87 |
| Shelby | 26 |
| Stoddard | 109 |
| Stone | 99 |
| Sullivan | 13 |
| Taney | 101 |
| Texas | 88 |
| Vernon | 72 |
| Warren | 45 |
| Washington | 65 |
| Wayne | 85 |
| Webster | 90 |
| Worth | 3 |
| Wright | 89 |

# *General Key*

## Key to Main Sections

A. Trees, shrubs, or woody vines, woody at least in the lower portions . . . . . . . . . . . *B*
  B. Leaves absent at flowering time . . . . . . . . . . . . . . . . . . Section 1, p. xxxii
  B. Leaves present at flowering time (includes those having leaves during fruiting stage) . . . . *C*
    C. Leaves scale-like, needle-like, linear, or subulate (awl-shaped) . . . . . . Section 2, p. xxxiii
    C. Leaves flat or broader than 5 mm., not needle-like, scale-like, linear, nor subulate . . . . *D*
      D. Leaves opposite (in pairs) or whorled (arranged in circles of 3 or 4) . . . Section 3, p. xxxiii
      D. Leaves alternate (the leaflets themselves alternate or opposite) . . . . . . . . . *E*
        E. Each leaf compound, divided into 2 or more separate divisions or leaflets (beginning of leaf usually indicated by the presence of a bud for next year or next season in the axil of the leaf at junction of the base of leaf or petiole [leaf-stalk] and the twig; sometimes the bud is visible only after loosening the leaf-stalk away from the twig) Section 4, p. xxxvi
        E. Each leaf simple, not divided into separate divisions or leaflets, although often deeply lobed or toothed . . . . . . . . . . . . . . . . . Section 5, p. xxxviii
A. Non-woody plants, herbs or soft-stemmed plants . . . . . . . . . . . . . . . . . *F*
  F. Plants lacking green color from all parts of plant (does not include plants which develop leaves in early spring or fall with other colors) . . . . . . . . . . . . . Section 6, p. xlvi
  F. Plants with green color on some of the parts (includes all plants not previously accounted for) . *G*
    G. Water plants floating on or submerged in the water or with only the flowers raised above the water (does not include plants with stems growing above the water and sending the leaves into the air, nor plants temporarily covered by water, but does include water plants which have become stranded on dry soil as the result of drouth or the drying up of bodies of water) . . . . . . . . . . . . . . . . . . . . . . Section 7, p. xlvi
    G. Land plants growing on dry soil or with leaves and stems normally growing above the water surface . . . . . . . . . . . . . . . . . . . . . . . . . . . *H*
      H. Green part of plant consisting of large, thick, flattened, usually spiny joints . . . . CACTACEAE, Section 8, p. l
      H. Plants not as above . . . . . . . . . . . . . . . . . . . . . . . . *I*
        I. Some or all of the leaves compound, one leaf divided to its base or center into separate, distinct, smaller subdivisions (leaflets) which are not connected with one another by green leaf tissue . . . . . . . . . . . . . . . . . Section 9, p. l
        I. All the leaves simple, either with entire (without teeth and smooth-edged), toothed, or shallowly to deeply lobed margins, but not divided into separate subdivisions (leaflets), all the divisions connected with each other by some green leaf tissue . . . *J*
          J. Leaves opposite or whorled (in circles), 2 or more at a single level of the stem . . *K*
            K. Leaf-blades entire (without teeth and smooth-edged) . . . . Section 10, p. lv
            K. Leaf-blades toothed or lobed . . . . . . . . . . . . Section 11, p. lx

J. Leaves alternate on the stem, only 1 at each level, or all arising from the very base of the plant . . . *L*
L. All leaves arising at base of plant . . . . . . . . . . . . . . . . . . Section 12, p. lxii
L. Leaves alternate on the stem (some may occur at base also) . . . . . . . . . . . . *M*
M. All the leaf-blades entire (without teeth and smooth-edged) . . . . . . Section 13, p. lxvi
M. Some or all of the leaf-blades toothed or lobed or with other irregularities . . Section 14, p. lxxv

## Section 1. **Trees, shrubs or woody vines; leaves absent at flowering time**

a. Leaf-scars and leaf-buds opposite . . . . . . . . . . . . . . . . . . . . . . . . . . *b*
b. Flower-clusters surrounded by 4 large white, pink, or rose-colored, petal-like bracts . . . . Cornus florida, in CORNACEAE, p. 1150
b. Flower-clusters not as above. . . . . . . . . . . . . . . . . . . . . . . . . . . *c*
c. Individual flowers large, violet or purple, about 5 cm. long, in panicles at the tips of the branches; fruit a capsule consisting of many, small, winged seeds . . . . . . . . Paulownia, in SCROPHULARIACEAE, p. 1330
c. Individual flowers small, greenish, yellow, yellowish, red, or purplish in dense clusters often along the sides of the branches or in the axils of the leaf-buds of the preceding year; fruit a 1–2-seeded winged samara or a 1-seeded fleshy drupe . . . . . . . . . . . . . . *d*
d. Calyx, when present, 4-parted; stamens usually 2; fruit either a single-winged samara or a fleshy 1-seeded drupe . . . . . . . . . . . . . . . . . . . . OLEACEAE, p. 1178
d. Calyx usually 5-lobed or 5 parted; stamens 3–12; fruit a 2-winged samara . . ACERACEAE, p. 1011
a. Leaf-scars and leaf-buds alternate . . . . . . . . . . . . . . . . . . . . . . . . *e*
e. At least the staminate flowers in catkins (cylindrical scaly spikes or spike-like racemes falling as one piece) or cone-like spikes . . . . . . . . . . . . . . . . . . . . . . *f*
f. Milky juice present in stems or parts of the flowers (seen after breaking or injuring these parts); calyx 4-parted; stamens 4 . . . . . . . . . . . . . . . MORACEAE, p. 560
f. Milky juice absent; calyx either absent or 2–8-parted or lobed; stamens 2–30 . . . . . . *g*
g. Styles or stigmas 3; female (pistillate) flowers solitary or clustered, not in a short, elongated, or cylindrical catkin; ovary with 3 cells and 6 ovules . . . . FAGACEAE, p. 528
g. Styles or stigmas 1 or 2; female (pistillate) flowers in a short, elongated, or cylindrical catkin; ovary with 1 or 2 cells and 1 to many ovules . . . . . . . . . . . . . *h*
h. Male (staminate) and female (pistillate) flowers separated but present on the same plant (monoecious); ovary partly or completely inferior; bract of female (pistillate) catkins usually 3- or 5-lobed at summit . . . . . . . . . . CORYLACEAE, p. 523
h. Male (staminate) and female (pistillate) flowers separated and on different plants (dioecious); ovary superior; bract of female (pistillate) catkins either not lobed or deeply cut . . . . . . . . . . . . . . . . . . . . . . . . . . *i*
i. Stipules usually present, scale-like and deciduous, or leaf-like and persistent; styles or stigmas 2; staminate flowers with 1 or more glands at the base or subtended by a cup- or saucer-like disk; fruit many-seeded; seeds with tufts of long hairs; plants found in all parts of Missouri; wood not lighter than cork . . SALICACEAE, p. 489
i. Stipules absent; style or stigma 1; staminate flowers without a gland or disk at the base; fruit 1-seeded; seeds without tufts of hairs; swamp plants known only from extreme southeastern Missouri; wood lighter than cork . . LEITNERIACEAE, p. 510
e. None of the flowers in catkins . . . . . . . . . . . . . . . . . . . . . . . *j*
j. Prickles or spines present on branches . . . . . . . . . . . . . . RUTACEAE, p. 966
j. No prickles or spines present on branches . . . . . . . . . . . . . . . . . . *k*
k. Flowers with only a calyx or corolla-like calyx present, the corolla or petals absent . . . . *l*
l. Leaves and bark spicy-fragrant; calyx 6-parted, in 2 series; stamens arranged in 3 or 4 lengths, of the male (staminate) flowers 9, of the female (pistillate) very reduced to 6 or 12–18 rudiments . . . . . . . . . . . . . . . . LAURACEAE, p. 716
l. Leaves and bark not spicy-fragrant; calyx 3–9-lobed or parted, in 1 series only; stamens arranged in 1 or 2 series, either 3–9, or 8 or 10 in 2 series . . . . . . . . . . . *m*
m. Flowers greenish, greenish-yellow, greenish tinged red, or brown-purplish; flowers arising from the sides of the branches in the axils of leaf-buds or leaf-scars; styles 2; stamens 3–9; calyx short or bell-shaped, 3–7-lobed or parted . . . . ULMACEAE, p. 553
m. Flowers yellow or yellowish; flowers arising from the tips of the branches; style 1; stamens 8 or 10; corolla tubular-funnelform, 4-toothed. . . . THYMELAEACEAE, p. 1086

k. Flowers with both calyx and corolla present . . . . . . . . . . . . . . . . . . . . . *n*
n. Petals (corolla) of flowers joined at least at their base so that the attempt to remove one portion disturbs the neighboring section of the corolla. . . . . . . . . . . . ERICACEAE, p. 1158
n. Petals separated all the way down to their base, not connected, so that one petal can be removed without tearing or disturbing the other petals . . . . . . . . . . . . . . . . . . *o*
o. Corolla irregular and pea-shaped, the petals of a flower of different size and shape; petals rose-red, rose-lavender, purple, lilac, or, rarely, white. . . . . . . . . LEGUMINOSAE, p. 865
o. Corolla regular, not pea-shaped, the petals similar in sizes and shapes; petals white, yellow, orange, or reddish, but not rose-purple or rose-red . . . . . . . . . . . . . . . *p*
p. Petals white; stamens 15–20 or more, the filaments conspicuous, elongated . . ROSACEAE, p. 792
p. Petals yellow, orange, or reddish; anther-bearing stamens 4 or 5, the filaments short and inconspicuous . . . . . . . . . . . . . . . . . . . . . . . . *q*
q. Petals 4, long and narrow, ribbon-like, 10–20 mm. long; anther-bearing stamens 4 . HAMAMELIDACEAE, p. 786
q. Petals 5, much less than 10 mm. long; anther-bearing stamens 5 . . . ANACARDIACEAE, p. 999

### Section 2. **Trees or shrubs or woody below; leaves scale-like, needle-like, linear, or subulate**

a. Leaves covered with silvery-silky hairs, finely divided into numerous segments . . Artemisia frigida in COMPOSITAE, p. 1442
a. Leaves without a hairy covering, simple, not divided into segments . . . . . . . . . . . *b*
b. Leaves compound, the main leaf divided into numerous very small leaflets (beginning of leaf usually indicated by the presence of a bud for the next year or season at the junction of the main leaf-stalk and the twig; sometimes the bud can be seen only after taking the leaf-stalk off the twig) . . . . . . . . . . . . . . . . . . . Acacia, in LEGUMINOSAE, p. 869
b. Each leaf simple, not divided into smaller separate divisions (see explanation in above alternate section) . . . . . . . . . . . . . . . . . . . . . . . . . *c*
c. A thin tissue-like stipule in the form of a sheath surrounding the base of each leaf; leaves jointed at the base; low plant up to 1.2 meters tall . . . Polygonella, in POLYGONACEAE, p. 598
c. No stipule or sheath surrounding the base of a single leaf, sheaths when present surrounding a cluster of 2–3 leaves; leaves not jointed at the base; large trees or shrubs not less than 2 meters tall . . . . . . . . . . . . . . . . . . . . . . . . *d*
d. Leaves pale or silvery-green, scale-like, without resin-ducts; flowers pink with sepals and petals; seed with a tuft of hairs at one end; fruit not a cone nor fleshy, drupe- or berry-like TAMARICACEAE, p. 1066
d. Leaves dark green if scale-like, with resin-ducts; 'flowers' cone-like or in cones, cream-colored or brownish; seed without any tuft of hairs; fruit a cone or fleshy and drupe- or berry-like . . . . . . . . . . . . . . . . . . . . PINACEAE, p. 42

### Section 3. **Trees, shrubs, or woody vines; leaves opposite or whorled**

a. Tip of twig, shoots, or branchlets tapering into a spine . . Rhamnus cathartica, in RHAMNACEAE, p. 1027
a. Tip of twig, shoot, or branchlet not tapering into a spine . . . . . . . . . . . . . . *b*
b. Each leaf compound, divided into 2 or more separate divisions or leaflets (beginning of leaf usually indicated by the presence of a bud for the next year or season at the junction of the main leaf-stalk and the twig; sometimes the bud can be seen only after taking the leaf-stalk off the twig) . . *c*
c. Leaflets 2 or 3 . . . . . . . . . . . . . . . . . . . . . . . . *d*
d. Leaflets 2; stems climbing or trailing . . . . . . . . Bignonia, in BIGNONIACEAE, p. 1370
d. Leaflets 3; trees or shrubs, not climbing or trailing . . . . . . . . . . *e*
e. Leaflets finely toothed; flowers white, showy; fruit 3-lobed, inflated . STAPHYLEACEAE, p. 1010
e. Leaflets coarsely toothed; flowers greenish, not showy; fruit with 2 spreading wings Acer Negundo, in ACERACEAE, p. 1018
c. Leaflets 5–11 . . . . . . . . . . . . . . . . . . . . . . . . *f*
f. Leaves palmately compound, all the leaflets attached to and spreading from the same point at the tip of the main leaf-stalk . . . . . . . . . . HIPPOCASTANACEAE, p. 1019
f. Leaves pinnately compound, with some leaflets attached at different levels on the leaf axis or with the leaflet at the tip on a longer stalk than the other leaflets . . . . . . . . *g*

g. Vines or stems trailing; individual flowers large, 6–8 cm. long, scarlet or orange-red; fruit pencil-like, narrowly cylindrical, 10–20 cm. long . . . . . . Campsis, in BIGNONIACEAE, p. 1368

g. Trees or shrubs, not vining or trailing; individual flowers smaller, white, greenish, greenish-yellow, or purplish; fruit not as above, but either with 1 or 2 wings or purple-black or red, fleshy and berry-like . . . . . . . . . . . . . . . . . . . . . . . . . . *h*

h. Leaves with close and crowded teeth 4–12 per cm.; flowers white; fruit fleshy and berry-like, or bladdery and 3-lobed, but not winged . . . . . . . . . . . . . . . . . . . . *i*

i. Leaflets usually 5–11; corolla 2–3 mm. long, 4.5–7 mm. broad; ovary inferior; fruit fleshy, berry-like, purple-black or red . . . . . . . . Sambucus, in CAPRIFOLIACEAE, p. 1418

i. Leaflets rarely 5 (usually 3); corolla 8–10 mm. long, 6–10 mm. broad; ovary superior; fruit bladdery, 3-lobed, green turning brown . . . . . . . . STAPHYLEACEAE, p. 1010

h. Leaves with mainly 1–4 teeth per cm.; flowers greenish, greenish-yellow, or purplish; fruit with 1 or 2 wings . . . . . . . . . . . . . . . . . . . . . . . . . *j*

j. Leaflets usually 5–11, either smooth-edged (entire) or with shallow teeth of 1–4 per cm.; stamens 2; fruit with 1 wing . . . . . . . . . . . Fraxinus, in OLEACEAE, p. 1178

j. Leaflets usually 3–5 (rarely 7) with coarse teeth of usually 1–2 per cm.; stamens 3–12; fruit with 2 wings . . . . . . . . . . . . . . . Acer Negundo, in ACERACEAE, p. 1018

b. Each leaf simple, not divided into smaller separate divisions (beginning of leaf is usually indicated by the presence of a bud for the next year or season at the junction of the main leaf-stalk and the twig; sometimes the bud can be seen only after taking the leaf-stalk off the twig) . . . . . . . *k*

k. Margins of leaves toothed or lobed . . . . . . . . . . . . . . . . . . . . . . *l*

l. Leaves very large, 15–40 cm. long, 15–40 cm. broad, broadly ovate and heart-shaped, only slightly lobed or angled . . . . . . . . . . . . . . . . . . . . . . . *m*

m. Corolla white marked with yellow and purple; anther-bearing stamens 2; leaves usually in whorls (circles) of 3; fruit a long cylindrical, narrow, pencil-like capsule . . . . . Catalpa, in BIGNONIACEAE, p. 1370

m. Corolla purple; anther-bearing stamens 4; leaves opposite, in 2's; fruit an ovoid capsule Paulownia, in SCROPHULARIACEAE, p. 1330

l. Leaves smaller, or if large, furnished with conspicuous lobes, teeth, or angles . . . . . . . *n*

n. Margins of leaves lobed, the lobes smooth-edged (entire) or toothed. . . . . . . . . . *o*

o. Lobes of leaves tapering to short- or long-pointed tips . . . . . . . .ACERACEAE, p. 1011

o. Lobes of leaves blunt or rounded and smooth-edged . . . . . CAPRIFOLIACEAE, p. 1405

n. Margins of leaves toothed but not lobed . . . . . . . . . . . . . . . . . . *p*

p. Leaves silvery green on lower surface . . . . . . . . . . . . . . . . . . *q*

q. Leaves 1–2 cm. broad, oblanceolate or tongue-shaped, more or less purple-tinged in spring and summer; buds covered with only 1 scale; fruit hairy, without wings; seeds furnished with long silky hairs; flowers in slender ascending catkins; flowers without petals. . . . . . . . . . . . . . . . . Salix purpurea, in SALICACEAE, p. 501

q. Fully grown leaves at least 6 cm. broad, triangular-ovate to somewhat round, not purple-tinged in spring and summer; buds covered with several scales; fruit glabrous with 2 spreading wings; seeds smooth without hairs; flowers reddish, in rounded or broad head-like clusters or umbels, not in catkins; flowers with petals . . . . Acer rubrum, in ACERACEAE, p. 1016

p. Leaves not silvery green on lower surface . . . . . . . . . . . . . . . . . *r*

r. Flowers white; ovary inferior, the calyx united with the base of the ovary or fruit . . . . . . . . . . . . . . . . . . . . . . . . . . . . . *s*

s. Leaves with 1–5 main nerves on each side of midrib; corolla of 4 separate petals, not united at their base; stamens 8, 10, or 20 or more; fruit a dry, many-seeded capsule . . . . . . . . . . . . . . . . . . . SAXIFRAGACEAE, p. 773

s. Leaves with 5–10 main nerves on each side of midrib; corolla of united petals joined at their base so that an attempt to remove one portion disturbs the neighboring section of the corolla; stamens 5; fruit a fleshy 1-seeded drupe . . . Viburnum, in CAPRIFOLIACEAE, p. 1412

r. Flowers bluish or pink, yellow, green, brown-purple, or combinations of yellow, green, and purple; ovary superior, the calyx not united to any part of the ovary or fruit . . . . . . . . . . . . . . . . . . . . . . . . . . . *t*

t. Only 3–4 nerves on each side of midrib of leaf; bark bitter . . . . . . Rhamnus cathartica, in RHAMNACEAE, p. 1027

t. Usually 5–8 nerves on each side of midrib; bark not bitter . . . . . . . . . . *u*

u. Rarely encountered, known only near Mo.-Ark. line in Taney County; star-shaped (stellate) hairs present on lower surface of leaf or on some part of the stem; corolla bluish or pink, the petals united into a corolla-tube and lobes; fruit violet, blue-purple, or pink. . . . . . Callicarpa, in VERBENACEAE, p. 1262

u. Without the above combination of characters; frequently encountered in various parts of Missouri; no star-shaped (stellate) hairs present; flowers yellow, green, yellow-green with purple, or brown-purple, the petals absent or separate and distinct to their base; fruit of other colors . . . . *v*

v. Flowers yellow; petals absent; leaves conspicuously narrowed at base; young branches never 4-angled; stamens 2; fruit 1-seeded . . . . . . . . . . . Forestiera, in OLEACEAE, p. 1182

v. Flowers green, purple, yellow-green with purple, or brown-purple; leaves not conspicuously narrowed at base; young branches sometimes 4-angled; stamens 4–5; fruit 3–20-seeded . . Euonymus, in CELASTRACEAE, p. 1008

k. Margins of leaves without teeth (entire) . . . . . . . . . . . . . . . . . . . . *w*

w. Plants growing in green, tough-stemmed rounded clumps as parasites on branches of trees . . LORANTHACEAE, p. 571

w. Plants not growing as parasites on tree branches . . . . . . . . . . . . . . . . . *x*

x. Leaves broadly ovate and heart-shaped, very large, 15–40 cm. long, 15–40 cm. broad. . . . . *y*

y. Corolla white marked with yellow and purple; anther-bearing stamens 2; leaves usually in whorls (circles) of 3; fruit a long cylindrical, narrow, pencil-like capsule . . . . Catalpa, in BIGNONIACEAE, p. 1370

y. Corolla purple; anther-bearing stamens 4; leaves opposite, in 2's; fruit an ovoid capsule Paulownia, in SCROPHULARIACEAE, p. 1330

x. Without the above combination of characters, the leaves much smaller and narrowed or rounded at base, or if large not heart-shaped at base . . . . . . . . . . . . . . *z*

z. Buds covered with only 1 scale; leaves silvery green and more or less purple-tinged, oblanceolate or tongue-shaped, 1–2 cm. broad; flowers without petals, separated into staminate (male) and pistillate (female) on different plants; staminate flowers with 1 or more glands at the base or subtended by a cup- or saucer-like disk; flowers in catkins Salix purpurea, in SALICACEAE, p. 501

z. Without the above combination of clusters . . . . . . . . . . . . . . . . . *1*

1. Bases of the leaves connected by a thin stipule; flowers in large dense ball-like heads with white corollas; fruits crowded into a dense, ball-like mass . . . . . . . Cephalanthus, in RUBIACEAE, p. 1398

1. Bases of the leaves not connected by a stipule; flowers of other colors, or, if white not in dense, ball-like heads; fruits not in a dense, ball-like mass . . . . . . . . . . *2*

2. Leaves with pale dots (hold leaf to light to note translucent dots); stamens numerous, 20–100, in a pompon-like mass . . . . . . . . . . . . HYPERICACEAE, p. 1054

2. Leaves without pale dots; stamens 2–10, not arranged in a pompon-like mass . . . . *3*

3. Calyx scarcely evident, of 4 minute sepals; corolla absent; flowers yellow in clusters along the sides of the twigs; fruit fleshy, 1-seeded; stamens 2–4 . . . . . Forestiera, in OLEACEAE, p. 1182

3. Without the above combination of characters; calyx conspicuous or at least evident; corolla present; flowers yellow, orange, white, cream-colored, rose-purple, or pink with green; fruit dry or fleshy, 1–several-seeded . . . . . . . . . *4*

4. Twining or trailing shrubs or vines. . . . . . . . . . . . . . . . *5*

5. Leaves rounded or short-pointed at tip; flowers 1.5–5.5 cm. long; ovary inferior, i.e., placed below base of the corolla and united with the calyx . Lonicera, in CAPRIFOLIACEAE, p. 1405

5. Leaves usually abruptly long-pointed at tip; flowers 1–1.3 cm. long; ovary superior, i.e., placed above base of the corolla and not united with the calyx Trachelospermum, in APOCYNACEAE, p. 1196

4. Upright shrubs or trees . . . . . . . . . . . . . . . . . . . *6*

6. Stamens 2 . . . . . . . . . . . . . . . . OLEACEAE, p. 1178

6. Stamens 4–5, or 10 . . . . . . . . . . . . . . . . . . *7*

7. Petals (corolla) of flowers joined into a tube so that the attempt to remove one portion disturbs the neighboring section of the corolla . . . . CAPRIFOLIACEAE, p. 1405

7. Petals separated all the way down to their base, not connected, so that one petal can be removed without tearing or disturbing the other petals . . *8*

8. Common shrubs throughout Missouri, the stems not corky; leaf-stalks (petioles) usually 1 cm. or more long; petals 4; calyx 4-toothed; stamens 4; fruit fleshy, 2-seeded . . . . . CORNACEAE, p. 1150
8. Rare shrub, known only from sink-hole ponds in Reynolds County, the stems often corky at base; leaf-stalks (petioles) practically absent; petals 5; calyx 5–7-toothed; stamens 10; fruit dry, 2–several-seeded . . . . . . . . . . . . . . . . . . . . . . . Decodon, in LYTHRACEAE, p. 1090

## Section 4. **Trees, shrubs, or woody vines; leaves alternate and compound**

a. All of the leaflets 3 . . . . . . . . . . . . . . . . . . . . . . . . . . . *b*
b. Prickles present on at least the stems or branches of the plant . . . . . . . . ROSACEAE, p. 792
b. Prickles absent on all parts of plant . . . . . . . . . . . . . . . . . . . *c*
c. Stems sprawling, trailing, leaning, climbing, or vining . . . . . . . . . . . . . . *d*
d. Leaves with both stipules (small bracts or leaf-like scales at base of main leaf or leaf-stalk) and stipels (tiny bracts or appendages at base of leaflets) . . . Pueraria, in LEGUMINOSAE, p. 956
d. Leaves without stipels, but stipules may be present . . . . . . . . . . . . *e*
e. Leaflets with many small, closely and regularly set teeth on margins (averaging 4–10 per cm.); stipules persisting and remaining attached to base of leaf-stalk (petiole); flowers large, with 5 showy petals and numerous (more than 20) stamens; fruit consisting of the fleshy calyx-tube with numerous hard, seed-like carpels within . Rosa, in ROSACEAE, p. 847
e. Leaflets with relatively few, coarse, or irregular-sized teeth along margins (averaging 1–2 per cm.) or teeth absent; stipules absent or not persisting; flowers small and inconspicuous with 4 or 5 petals and 4 or 5 stamens; fruit a 1–2-seeded fleshy dry berry or drupe . . . . . . . . . . . . . . . . . . . . . . . . . . *f*
f. Known only from southwestern Missouri; non-poisonous; leaflets thick-fleshy and succulent; flowers with 4 petals and 4 stamens; inflorescence broader than long, flat-topped in an umbel-like cyme; stems glabrous (without hairs) throughout; stems sometimes with tendrils. . . . . . . . . . . . Cissus, in VITACEAE, p. 1032
f. Throughout Missouri; poisonous to many persons; leaflets not thick-fleshy nor succulent; flowers with 5 petals and 5 stamens; inflorescence a panicle longer than broad arising along sides of stem; upper part of stem or branch hairy; plant when climbing showing aerial roots. . . . . . . . . . . . Rhus, in ANACARDIACEAE, p. 999
c. Upright shrubs . . . . . . . . . . . . . . . . . . . . . . . . . . *g*
g. Leaflets with transparent dots between the veins (hold leaf to light), smooth-edged (entire) or nearly so, the teeth if present mostly inconspicuous. . . . . . Ptelea, in RUTACEAE, p. 968
g. Leaflets lacking transparent dots, usually coarsely and conspicuously toothed, or rarely smooth-edged (entire) . . . . . . . . . . . . . . Rhus, in ANACARDIACEAE, p. 999
a. Some or all of the leaflets 5 or more to a leaf. . . . . . . . . . . . . . . . . . *h*
h. Prickles present on at least the stems or branches of the plant . . . . . . . . . . . *i*
i. Leaves palmately compound, all the leaflets attached to and spreading from the same point at the tip of the main leaf-stalk (petiole) . . . . . . . . . . . Rubus, in ROSACEAE, p. 834
i. Leaves pinnately compound, with some leaflets attached at different levels on the leaf axis. . . *j*
j. Leaves more than simply once-compound, with the lateral axes or leaf-divisions further divided into additional leaflets . . . . . . . . . . . . . . . . . . . *k*
k. Leaflets large, 5–15 cm. long, 1.5–8.5 cm. broad, with short-pointed, outwardly projecting teeth; lower surface of leaflets silvery or gray-green . Aralia spinosa, in ARALIACEAE, p. 1112
k. Leaflets small, 1–2.5 cm. long, 0.5–3 cm. broad, the margins minutely wavy or the teeth rounded; lower surface of leaflets dull green . . . . Gleditsia, in LEGUMINOSAE, p. 872
j. Leaves simply once-compound, with the lateral leaflets attached directly to the main axis and not further divided into additional leaflets . . . . . . . . . . . . . . . *l*
l. Leaflets with transparent dots between the veins (hold leaf to light); leaf lacking stipules (small bracts or leaf-like scales at base of main leaf or leaf-stalk) or stipels (tiny bracts or appendages at base of leaflets) . . . . . . . . . Zanthoxylum, in RUTACEAE, p. 966
l. Leaflets lacking transparent dots; stipules present at base of main leaf-stalk, either free or united with the stalk or minute stipels present at base of the leaf-stalks (petiolules) of the leaflets . . . . . . . . . . . . . . . . . . . . . . . . *m*
m. Leaflets without teeth (entire); stipels (minute bracts or appendages) present at base of the leaf-stalks of the leaflets, but stipules absent at base of main leaf-stalk; corolla irregular and pea-shaped, the petals of a flower of different sizes and shapes . . . Robinia, in LEGUMINOSAE, p. 906

m. Leaflets with either sharp, rounded, or wavy teeth; stipels absent at base of the leaflets, but stipules present at base of main leaf-stalk, either united to it or free at the base; corolla regular or nearly so, not pea-shaped, the petals similar in sizes and shapes . . . . . . . . . . . . *n*

n. Compound leaves with an even number of leaflets, with a pair of leaflets at the tip; leaflets with wavy margins or the teeth rounded or blunt . . . . . Gleditsia, in LEGUMINOSAE, p. 872

n. Compound leaves with an odd number of leaflets, with 1 leaflet at the tip; leaflets with sharp, acutely pointed or projecting teeth . . . . . . . . . . . . . . . . ROSACEAE, p. 792

h. Prickles absent on all parts of plant . . . . . . . . . . . . . . . . . . . . . . *o*

o. Leaves palmately compound, with all the leaflets attached to the same point at the tip of the main leaf-stalk (petiole) . . . . . . . . . . . . . Parthenocissus, in VITACEAE, p. 1032

o. Leaves pinnately compound, with some leaflets attached at different levels on the leaf axis or subdivisions of the leaf axis . . . . . . . . . . . . . . . . . . . . . . . *p*

p. Leaves more than simply once-compound, with the lateral axes or leaf-divisions further divided into additional leaflets . . . . . . . . . . . . . . . . . . . . . *q*

q. Leaflets coarsely or conspicuously toothed . . . . . . . . . . . . . . . . . *r*

r. Native vine with sprawling or climbing stems; tendrils (hair-like, curling or twisting organs) usually present . . . . . . . . . . Ampelopsis arborea, in VITACEAE, p. 1032

r. Cultivated tree; no tendrils present . . . . . . . . . . Melia, in MELIACEAE, p. 969

q. Leaflets with wavy margins or the teeth rounded or blunt or absent. . . LEGUMINOSAE, p. 865

p. Leaves simply once-compound, with the lateral leaflets attached directly to the main axis and not further divided into additional leaflets . . . . . . . . . . . . . . . . . . *s*

s. Stems trailing, sprawling, leaning, climbing or vining; tendrils (hair-like, curling or twisting organs) usually present; leaflets with coarse or conspicuous teeth . . . . . Ampelopsis arborea, in VITACEAE, p. 1032

s. Without the above combination of characters; trees, shrubs, or small plants woody at base, if vining or trailing, then no tendrils present; leaflets with or without teeth. . . . . . . *t*

t. Stipules (small bracts or leaf-like scales at base of main leaf or leaf-stalk) or stipels (tiny bracts or appendages at base of leaflets) present; style and stigma 1; calyx usually with a well-developed calyx-tube . . . . . . . . . . . . . . . LEGUMINOSAE, p. 865

t. Stipules and stipels both absent; styles or stigmas 2 or more; calyx composed of either separate sepals or these united into only a short, inconspicuous tube . . . . . . . . *u*

u. Leaves and twigs, when bruised, with a spicy odor; lower surface of leaflets dotted with glands; petals absent; male flowers arranged in drooping slender, cylindrical catkins; styles or stigmas 2 . . . . . . . . . . . . . . . JUGLANDACEAE, p. 510

u. No spicy odor or fragrance to bruised twigs or leaves; lower surface of leaflets not dotted with glands; petals present; none of the inflorescences arranged in cylindrical catkins; styles or stigmas usually 3–5, rarely 2 . . . . . . . . . . . . . . *v*

v. Compound leaves usually with an even number of leaflets or with a pair of leaflets at the tip . . . . . . . . . . . . . . . . . . . . . . . *w*

w. Known wild only in southwestern Missouri; flowers white; leaflets sessile (without a stalk) or nearly so, their leaf-stalk (petiolule) only 1–3 mm. long; leaflets usually scythe-shaped, with one side strongly curved at base, the other side narrowed to the base or straight . . . . . . . . . SAPINDACEAE, p. 1023

w. Throughout Missouri; flowers green or yellow-green; leaflets on stalks (petiolules) 4–8 mm. long; leaflets truncate (as if cut straight across), broadened, or rounded at base on both sides . . . . . . . . SIMAROUBACEAE, p. 969

v. Compound leaves usually with an odd number of leaflets or with 1 leaflet at the tip . . *x*

x. Known wild only in southwestern Missouri; flowers white . . SAPINDACEAE, p. 1023

x. Throughout Missouri; flowers greenish, greenish-yellow, yellow, or rarely pink . . . . . . . . . . . . . . . . . . . . . . . . *y*

y. Leaflets either noticeably toothed, or if smooth-edged (entire) the rachis (central axis) of leaf bordered by a green winged tissue; broken stems or leaf-stalks (petioles) with a milky or brown sticky juice; stamens 5; fruit more or less spherical, reddish, fleshy, covered with sour-tasting hairs, not winged; seed bony and hard . . . . . . . . . . . . ANACARDIACEAE, p. 999

y. Leaflets mostly smooth-edged (entire) or with a few blunt teeth toward the base, never winged along the rachis; broken stems lacking a milky or brown sticky juice; stamens 10; fruit oblong, flattened, thin, winged; seed not bony or hard. . . . . . . . . . . . . . . SIMAROUBACEAE, p. 969

## Section 5. **Trees, shrubs, or woody vines; leaves alternate and simple**

a. Leaves smooth-edged (entire) without teeth or lobes . . . *b*

b. At least part of the stem or branches bearing prickles or spines . . . *c*

c. A pair of tendrils (slender curling or twisting appendages) present at the base of some of the leaf-stalks (petioles) . . . Smilax, in LILIACEAE, p. 448

c. No tendrils present . . . *d*

d. Low plant with long arching or drooping branches, sometimes vine-like; corolla greenish-purple with a conspicuous tube 6–7 mm. long and corolla-lobes 4–5 mm. long; mature fruit orange-red or scarlet; broken twigs or leaf-stalks (petioles) not producing a milky sap . Lycium, in SOLANACEAE, p. 1322

d. Tree or tall shrub; corolla white and creamy-white, with a short inconspicuous tube 1 mm. long and corolla-lobes 1.5–2 mm. long, or flowers greenish with a corolla; mature fruit green or black; broken twigs or leaf-stalks (petioles) producing a milky sap . . . *e*

e. Leaves long taper-pointed at tip, ovate or ovate-lanceolate; flowers without a corolla, the sexes separated on different trees (dioecious), the calyx 4-parted in both male (staminate) and female (pistillate) flowers; fruit green, the size of an orange or small grapefruit . . . Maclura, in MORACEAE, p. 563

e. Leaves rounded, blunt or short-pointed at tip, oblong-obovate, narrowly obovate, or elliptic-oblong; flowers with a white or cream-colored corolla, perfect (stamens and pistil in the same flower), the calyx 5-parted; fruit black or nearly so, small, only 1–1.5 cm. long . . . SAPOTACEAE, p. 1172

b. No prickles or spines present on stems or branches . . . *f*

f. Stems sprawling, twining, trailing, or climbing . . . *g*

g. Tendrils (slender coiling or twisting appendages) present at the base of some of the leaf-stalks or at the tips of the stems or branches . . . *h*

h. A pair of tendrils present at the base of some of the leaf-stalks (petioles); fruit a small fleshy, spherical berry, without wings; sexes of the flowers separated on different plants (dioecious); segments of the flower 6; stamens 6 . . . Smilax, in LILIACEAE, p. 448

h. Tendrils present at the tips of the stems or branches; fruit with a winged calyx; flowers perfect (stamens and pistil in the same flower); segments of the flower 5; stamens 8 Brunnichia, in POLYGONACEAE, p. 600

g. No tendrils present . . . *i*

i. Leaf-blades broadly ovate to round and strongly or conspicuously heart-shaped at base, 5–15 cm. long, 5–15 cm. broad; flower strongly curved, with a somewhat S-shaped tube, the complete tube 3–5 cm. long . . . ARISTOLOCHIACEAE, p. 572

i. Leaf-blades either smaller or at most slightly heart-shaped; flower not curved, regularly symmetrical, 0.5–1.3 cm. long . . . *j*

j. Corolla purple or green-purple, the petals joined into a short to long tube so that the attempt to remove one portion disturbs the neighboring section of the corolla; fruit many-seeded . . . SOLANACEAE, p. 1322

j. Corolla greenish or greenish-white, the petals separate to the base, not connected, so that one petal can be removed without tearing or disturbing the other petals; fruit 1- or 2-seeded . . . *k*

k. Lower surface of leaves more or less hairy; stem minutely hairy, easily broken or torn; petals 6; stamens 6; mature fruit red, 1-seeded . . . Cocculus, in MENISPERMACEAE, p. 716

k. Lower surface of leaves glabrous (without hairs); stem glabrous, tough, and difficult to break; petals 5; stamens 5; mature fruit blue, 2-seeded . . . Berchemia, in RHAMNACEAE, p. 1027

f. Trees or shrubs . . . *l*

l. Stems hollow between the leaf-nodes; base of leaves with a sheath surrounding the stem; scaly prolongation (ligule) present on the upper (inner) surface of the leaf-sheath . . Arundinaria, in GRAMINEAE, p. 81

l. Without the above combination of characters; stems not hollow; leaves without any sheaths; no ligule present . . . *m*

m. Leaves heart-shaped, broadly rounded, as broad as or broader than long . . . Cercis, in LEGUMINOSAE, p. 876

m. Leaves not heart-shaped or longer than broad . . . *n*

n. Young twigs and at least lower side of leaves covered with a dense silvery or whitish hairiness. . . . *o*

o. Leaves densely silvery-scaly on at least lower side and often on both sides; outside of flowers silvery-scaly; flowers not in catkins . . . . . . . . . . . . . . . ELAEAGNACEAE, p. 1087

o. Leaves silvery- or whitish-hairy but not silvery-scaly; outside of flowers not silvery-scaly; flowers in catkins . . . . . . . . . . . . . . . . . . . . . . . SALICACEAE, p. 489

n. Young twigs and leaves not covered with a dense silvery or whitish hairiness. . . . . . . . *p*

p. Leaves with 3 main nerves arising from the same place at the very base of the leaf-blade at junction with petiole (leaf-stalk) . . . . . . . . . . . . . . . . Celtis, in ULMACEAE, p. 556

p. Leaves pinnately nerved with 1 main midnerve, the side (lateral) nerves arising at different levels above the base of the leaf-blade. . . . . . . . . . . . . . . . . . . . . . . *q*

q. Most of the leaf-blades 15–40 cm. long . . . . . . . . . . . . . . . . . . . . *r*

r. Crushed leaves and bark pleasant-smelling or spicy-tasting; leaf-blades with 2 prominent side (lateral) nerves near base; some of the leaves of the branch usually with lobes; flowers several to many in a cluster, small, greenish-yellow or yellow, 4–6 mm. long, the sexes separated on different plants (dioecious); stamens 9 in male flowers, 6 and reduced in female flowers; petals absent; sepals 6; fruit black-blue, fleshy, small, 0.8–1.1 cm. long, 1-seeded. . . . . . . . . . . . . . . . . Sassafras, in LAURACEAE, p. 716

r. Crushed leaves and bark without a pleasant smell or spicy taste; leaf-blades with the side nerves appearing equally from base to tip; none of the leaves lobed; flowers solitary, not grouped in a cluster, larger, either dull purple, reddish-purple, maroon and 2–2.5 cm. long, or green or greenish-yellow and 5–8 cm. long, perfect (the stamens and pistils in the same flower); stamens numerous, more than 20; petals 6–9; sepals 3; fruit green to brown, 5–16 cm. long, whole fruit containing several to many seeds . . . . . . . . . . . . . *s*

s. Known wild only in extreme southern Missouri; leaves oblong, $1\frac{1}{2}$–2 times as long as broad, mostly broadest at the middle or at base; leaf-buds covered by thin stipules; flowers at the tip of the twig or branch, green or greenish-yellow, the petals 5–8 cm. long; fruit cone-like . . . . . . . . . . . . . . Magnolia, in MAGNOLIACEAE, p. 669

s. Throughout most of Missouri except the extreme northern counties; leaves oblanceolate, elliptical-obovate to ovate-lanceolate, $2\frac{1}{2}$–4 times as long as broad, mostly broadest above the middle; leaf-buds naked, without stipules; flowers along the sides of the branches in the axils of last year's leaves, dull purple or reddish-purple, the outer petals 2–2.5 cm. long; fruit pulpy, resembling a small short banana . . . . . . . ANNONACEAE, p. 671

q. Most of the leaf-blades less than 15 cm. long . . . . . . . . . . . . . . . . . *t*

t. All the leaves 3.5 cm. or less (averaging 1–2 cm.) long; broken leaves or stems with milky juice; ovary 3-lobed, 3-celled; fruit a dry 3-lobed capsule with 6 seeds . . . . . . Andrachne, in EUPHORBIACEAE, p. 974

t. Without the above combination of characters; some or all of the leaves 2–15 cm. long; broken leaves or stems with or without milky juice; ovary, if 3-celled, not 3-lobed; fruit a fleshy 1–many-seeded fruit or a dry capsule not 3-lobed. . . . . . . . . . *u*

u. Some or all of the flowers in crowded slender catkins (usually narrow cylindrical inflorescences) or spikes; fruit either an acorn (a scaly cup surrounding base of fruit) or a small capsule 2–9 mm. long, if the latter the capsules crowded in a dense spike or elongated cluster . . . . . . . . . . . . . . . . . . . . . . . *v*

v. Buds covered by 1 scale having 2 coats; twigs long and slender, with few if any side branches; leaves not tipped by a bristle; small or leaf-like stipules usually persisting and present at base of leaf; both male (staminate) and female (pistillate) flowers arranged in erect or ascending catkins, the staminate and pistillate flowers on separate plants (dioecious); 1–4 glands present at base of each flower; pistillate flowers without a calyx; seeds bearing tufts of hairs; fruit 2–9 mm. long, at maturity opening into a 2-valved capsule. . . . . . . . . . . . . . . Salix, in SALICACEAE, p. 489

v. Buds covered by several scales; twigs rather stout and short, with many side branches; leaves usually tipped by a bristle (except in *Q. nigra*); stipules usually falling before the leaf matures; only the male (staminate) flowers arranged in catkins, these drooping, the male and female flowers in separate inflorescences on the same plant (monoecious); no glands present at base of flowers; pistillate flowers with a calyx; seeds not bearing hairy tufts; fruit 10–17 or more mm. long, an indehiscent (not opening) acorn. . Quercus, in FAGACEAE, p. 532

u. Flowers not arranged in catkins; fruit either fleshy and 1–many-seeded or a capsule 7–20 or more mm. long, if the latter the capsules not crowded into a dense spike or elongated cluster . . *w*

w. Buds hidden and covered by base of leaf-stalk; leaf-stalk (petiole) hollow; twigs very tough, hard to tear or break, enlarged at joints; outer bark loosely separating from the wood; flowers appearing before the leaves, pale yellow . . . . . . . . . . . . . . . . THYMELAEACEAE, p. 1086

w. Buds not covered by base of leaf-stalk; leaf-stalk (petiole) not hollow; twigs either easily broken or torn or not enlarged at joints; outer bark not loosely separating from the wood . . . . . . . . *x*

x. Bruised leaves and bark with a spicy or pleasant smell or spicy taste; flowers yellow; sepals 6; stamens 9 in the male (staminate) flowers, 6 or 15–18 and greatly reduced in female (pistillate) flowers; fruit fleshy, 1-seeded . . . . . . . . . . . . . . . . . . LAURACEAE, p. 716

x. Bruised leaves and bark not producing a pleasant smell or spicy taste (leaves somewhat fragrant in *Cotinus*); flowers white, creamy, pale yellow, pink, rose, or greenish; sepals or calyx-lobes 4–5; stamens 4–5, 8, 10, or 6–12; fruit fleshy, 1–many-seeded or a many-seeded capsule . . . . . . *y*

y. Buds naked (not covered by scales), densely hairy, slender and elongated; stipules usually present at base of leaf-stalks (petioles); bark bitter to the taste; stamens opposite the petals which are separate to the base . . . . . . . . . . . Rhamnus, in RHAMNACEAE, p. 1027

y. Buds covered by scales, short and conical or ovate, not slender and elongated; stipules absent; bark not bitter-tasting; stamens alternate with the petals, the petals separate to the base or united into a short or long tube . . . . . . . . . . . . . . . . . . . . . . . . . *z*

z. Larger leaf-blades 1–4 (rarely 5) cm. broad . . . . . . . . . . . . . . . . *1*

1. Broken leaf-stalks and young stems producing milky juice; flowers 3–3.5 mm. long; stamens opposite the corolla-lobes, the fertile stamens 5, sterile stamens 5; fruit 1-seeded SAPOTACEAE, p. 1172

1. Broken leaf-stalks and young stems not producing a milky juice; flowers 4–30 or more mm. long; stamens alternate with the corolla-lobes, the fertile stamens 5, 8, or 10; fruit with 10-many seeds . . . . . . . . . . . . . . . . . . ERICACEAE, p. 1158

z. Larger leaf-blades mainly 3.5–8 cm. broad . . . . . . . . . . . . . . . . . *2*

2. True terminal bud absent; corolla with a well-developed tube, the pistillate (female) flowers 10–18 mm. long, the staminate (male) 10–13 mm. long; stamens 8 in pistillate flowers, 16 in staminate flowers; fruit large, depressed-spherical or oblong, 2–4 cm. in diameter, sweet when ripe, reddish-orange or rarely bluish-black, 4–8 seeded, the seeds 10–15 mm. long . . . . . . . . . . . . . . . . . . . EBENACEAE, p. 1174

2. True terminal bud present; corolla either without any tube and petals all separate, or with a tube, the whole corolla with calyx not over 4 mm. long; stamens 4, 5, 10, or 5–12 in staminate flowers of *Nyssa*; fruit 0.5–1.5 cm. in diameter, sour or bitter when ripe, black or blue-black, 1–2 seeded, the seeds 4–25 mm. long . . . . . . . . . . . *3*

3. Leaf-blades 2½–4 times as long as broad, elliptic-oblong, oblong-obovate, or narrowly obovate, conspicuously narrowed to the base; broken leaf-stalks and young stems producing a milky juice; petals united at base into a very short tube . SAPOTACEAE, p. 1172

3. Leaf-blades 1½–2 times as long as broad, elliptic-ovate, obovate, elliptic, or oval; broken leaf-stalks and young stems not producing a milky juice or the juice resinous and not colored; petals separate all the way to their base or absent . . . . . . . *4*

4. Leaves rounded or blunt at summit, upper surface bluish-green; leaf-stalk and lower surface of fully grown leaf-blade glabrous (without hairs); bruised leaves somewhat fragrant; styles 3; ovary superior; some of flower-stalks with a feathery hairiness, the hairs horizontally spreading from the axis. Cotinus, in ANACARDIACEAE, p. 999

4. Leaves short- to long-pointed at summit, upper surface dark green; either the leaf-stalk and/or lower surface of fully grown leaf-blade more or less somewhat hairy; style 1; ovary inferior; flower-stalks without a feathery hairiness or glabrous (without hairs) . . . . . . . . . . . . . . . . . . . . . . . *5*

5. Main lateral (side) nerves 5–6 on each side of midrib; upper pair of lateral nerves arising about ⅓ the length of the blade from the tip and running to the tip nearly parallel to the midrib; petals 4; stamens 4; calyx 4-toothed; flowers perfect (stamens and pistil in same flower); fruit spherical, 2-seeded; pith of branchlets continuous, without cross-partitions . . . . . . CORNACEAE, p. 1150

5. Main lateral (side) nerves 6–15 on each side of midrib; upper pair of lateral nerves arising less than ⅓ the length of the blade from the tip and not running to the tip nor parallel to the midrib; petals often absent or obsolete; stamens 5–12; calyx 5-parted in male (staminate) flowers; flowers with the sexes separated on different plants (dioecious); fruit ovoid or ellipsoid, 1-seeded; pith of branchlets interrupted by cross-partitions . . . . CORNACEAE, p. 1150

a. Leaves with either teeth or lobes or both . . . . . . . . . . . . . . . . . . . . . . . . *6*

6. Leaves lobed, the lobes toothed or toothless . . . . . . . . . . . . . . . . . . . . *7*

7. Vines or plants with trailing, climbing, or coiling stems . . . . . . . . . . . . . . *8*

8. Margins of lobes toothed; some part of stem with a slender coiling or twining organ (tendril) VITACEAE, p. 1035

8. Margins of lobes without teeth (entire) or sometimes toothed; no tendrils present . . . . . *9*

9. The upper or middle lobe much larger than the 2 or 4 lobes at the base; leaf-blade pinnately nerved, the lateral (side) nerves arising at various levels on either side of midrib . . . . . . . . . . . . . Solanum Dulcamara, in SOLANACEAE, p. 1311

9. The upper or middle lobe not much, if at all, larger than the other lobes; leaf-blades palmately nerved, most of main nerves arising at very base of leaf-blade . . . . . *10*

10. At least the lower part of leaf-blade and the leaf-stalk (petiole) with some star-shaped (stellate) hairs; stems climbing or attaching themselves by aerial roots. . Hedera, in ARALIACEAE, p. 1114

10. Hairs on lower surface of leaf-blade or on leaf-stalk simple, when present; stems not climbing or attaching themselves by aerial roots . . . . . . MENISPERMACEAE, p. 714

7. Trees or shrubs, not vining or climbing, but the branches sometimes drooping or arching . . *11*

11. Lobes of leaves without teeth . . . . . . . . . . . . . . . . . . . . . . . . *12*

12. Some of the flowers (the staminate) in drooping, long and slender catkins (narrow cylindrical inflorescences); fruit an acorn (a scaly cup surrounding base of nut). . Quercus, in FAGACEAE, p. 532

12. None of the flowers in catkins; fruit not an acorn . . . . . . . . . . . *13*

13. Leaf-blade more or less broadly V-shaped, flat-topped or truncate at summit (as if cut straight across) with a shallow or deep notch or depression at center; buds covered by leaf-like stipules; flowers large, solitary, 5 cm. long, yellow with orange . . . . . . . . . . . . Liriodendron, in MAGNOLIACEAE, p. 671

13. Leaf-blade pointed or rounded, but not flat-topped or truncate at summit; buds either exposed and with scales or covered by the enlarged base of the leaf-stalk; flowers small, several or many in an inflorescence, the individual flowers at most 1.3 cm. long, either yellow, green, grayish, or reddish . . . . . . . . . . *14*

14. Bruised leaves and bark with a pleasant smell or spicy taste; bark not flaking off or exposing white patches; leaf-blades usually longer than broad, the lateral lobes ascending or upwardly pointing; buds exposed and with scales; stipules absent; flowers yellow, 1–1.3 cm. long, not in dense spherical heads; fruit fleshy, blue-black . . . . . . . . . Sassafras, in LAURACEAE, p. 716

14. Bruised leaves and bark not producing a pleasant smell or spicy taste; bark of older trees flaking off and exposing large white patches; leaf-blades mostly broader than long, the lateral lobes horizontally spreading or nearly so; buds covered by the enlarged base of the leaf-stalk; stipules present at base of leaf-stalks, often broad and leaf-like; flowers green, grayish, or reddish, the individual flowers less than 5 mm. long, in dense spherical heads; fruit dry, yellow-brown . . . . . . . . . . . . . . . . PLATANACEAE, p. 789

11. Lobes of leaves with teeth . . . . . . . . . . . . . . . . . . . . . . . . *15*

15. Leaves pinnately nerved with 1 main midnerve, the side (lateral) nerves arising at different levels above the base of the leaf-blade . . . . . . . . . . . . *16*

16. Bark of older trees flaking off and exposing large white patches; buds covered by the enlarged base of the leaf-stalk (petiole); flowers in tight spherical heads; fruit in dense spherical heads, composed of dry narrowly club-shaped nutlets . . . PLATANACEAE, p. 789

16. Bark not flaking off or exposing white patches; buds exposed, not covered by base of the leaf-stalk; flowers not in tight spherical heads; fruit not in dense spherical heads . . . . . . . . . . . . . . . . . . . . . . . . . . *17*

17. Petals absent, some of the flowers (the staminate) in drooping, long, slender catkins (narrow cylindrical inflorescences); fruit an acorn (a scaly cup surrounding base of nut); leaves large, mainly 7–25 or more cm. long . . . Quercus, in FAGACEAE, p. 532

17. Petals present, showy; flowers not in catkins; fruit fleshy or dry, but not an acorn; leaves smaller, mainly 3–10 cm. long. . . . . . . ROSACEAE, p. 792

15. Leaves palmately nerved with 3 or more main nerves arising from the same place at

the very base of the leaf-blade at junction with petiole (leaf-stalk) . . . . . . . . . . *18*

18. Leaves shaped somewhat like a 5- or 7-pointed star, the lobes long-pointed; bruised leaves fragrant; some of the 2-year old branchlets with corky-ridges. . . . . . . . . . Liquidambar, in HAMAMELIDACEAE, p. 788

18. Without the above combination of characters; leaves not fragrant when bruised, nor resembling a 5–7-pointed star; branchlets always lacking corky ridges . . . . . . . . . *19*

19. Young twigs covered with a white cottony mat of hairs; lower surface of leaves densely white-hairy; pistillate (female) flowers without a calyx . . Populus, in SALICACEAE, p. 502

19. Without the above combination of characters; young twigs and lower surface of leaves not densely white-hairy; all flowers, including any pistillate ones, with a calyx. . . . *20*

20. Broken leaf-stalks and twigs producing a milky juice; petals absent in flowering plant; either stamens only or pistils only present in a flower; flowers in either cylindrical or dense spherical catkins, the staminate (male) and pistillate (female) catkins on the same (monoecious) or different (dioecious) plants; calyx 4-parted; stamens 4 . . . . . . . . . . . . . . . . . . . . . MORACEAE, p. 560

20. No milky juice produced by broken leaf-stalks and twigs; petals present in flowering plant; both stamens and pistils present in each flower, the flowers perfect; calyx 5-lobed or of 5 sepals; stamens 5 to 40 . . . . . . . . . . . . . . . . *21*

21. Calyx enveloped at base by a circle of many narrow bractlets; stamens arranged on a long column; some of the few hairs on the lower surface and lower nerves of the leaf-blade star-shaped . . . . . . . . . Hibiscus, in MALVACEAE, p. 1053

21. Calyx lacking a circle of bractlets at the base; stamens not arranged on a long column; no star-shaped hairs present on lower surface of leaf-blades . . . . *22*

22. Flowers yellow or greenish, often nodding; stamens 5; calyx with a prominent bell-shaped or elongated tube; ovary inferior; leaves mostly as broad as or broader than long; bark not shredding . . Ribes, in SAXIFRAGACEAE, p. 785

22. Flowers white, upright; stamens 30–40; calyx with only a short tube; ovary superior; leaves mostly longer than broad or some of them as broad as long; bark shredding . . . . . . . . . . . Physocarpus, in ROSACEAE, p. 793

6. Leaves toothed only, the teeth wavy, rounded, or pointed, short or long . . . . . . . . . *23*

23. Thorns, spines, or prickles on twigs or stems . . . . . . . . . . . . . . . . . . *24*

24. Inner bark and wood yellow; sepals 6; petals 6; stamens 6; ovary superior or not surrounded by a prolonged calyx-tube. . . . . . . . . . . . Berberis, in BERBERIDACEAE, p. 711

24. Inner bark and wood not yellow; calyx 5-parted; petals 5; stamens 5–20; ovary inferior or inclosed in the calyx-tube . . . . . . . . . . . . . . . . . ROSACEAE, p. 792

23. No thorns, spines, or prickles present on twigs or stems . . . . . . . . . . . . *25*

25. Some part of stem or shoot with a slender coiling or twining organ (tendril) . VITACEAE, p. 1031

25. No tendrils present on plant . . . . . . . . . . . . . . . . . . . . . . *26*

26. Trailing, climbing, or sprawling woody vines . . . . . . . . . . . . . . *27*

27. Leaves finely toothed, pale green, long-pointed; lateral (side) nerves curving; flowers dioecious, the fertile stamens and pistils appearing on separate plants; fruit orange on outside, 3-valved, when ripe exposing scarlet pulp-covered seeds. . Celastrus, in CELASTRACEAE, p. 1010

27. Leaves wavy-edged or very slightly toothed, dark green, slightly pointed at tip; lateral (side) nerves straight and paralleling one another; flowers perfect, the stamens and pistils in the same flower; fruit blue or blackish-blue . . . . . Berchemia, in RHAMNACEAE, p. 1027

26. Trees or shrubs, not trailing, climbing, or sprawling vines. . . . . . . . . . *28*

28. At least 2 of the lowest side nerves conspicuous and arising at the very base of the leaf-blade, there joining the base of the main mid-nerve . . . . . . . . . *29*

29. Leaf-blade more or less lop-sided (asymmetrical) with one half longer or broader than the other half, most noticeable at the base. . . . . . . . *30*

30. Flowers greenish or grayish, sometimes tinged reddish, without petals at flowering time, the flowers bearing either stamens (staminate) or pistils (pistillate), but not bearing both stamens and pistils in the same flower; fruit fleshy, 1-seeded; ovary 1-celled . . . . . Celtis, in ULMACEAE, p. 556

30. Flowers yellow, orange, reddish, or white or cream-colored, with petals at flowering time, the flowers perfect, the stamens and pistils in the same flower; fruit dry, woody, 1–2-seeded; ovary 2–5-celled . . . . . . . *31*

31. Leaves wavy-edged or with rounded teeth with 1–2 teeth per cm.; flowers yellow, orange, or red, blooming either from January–April or September–December, usually preceding the appearance of the new leaves or after the fall of the old ones; anther-bearing stamens 4; petals 4, ribbon-like; fruit on a short stalk or nearly stalkless . . . . . . . . Hamamelis, in HAMAMELIDACEAE, p. 786

31. Leaves sharply or finely toothed with 4–5 teeth per cm.; flowers white or cream-colored, blooming from early June to August with the leaves; anther-bearing stamens many, more than 15; petals 5; fruit small and round on a long peduncle (stalk) bearing a pale green, narrow, leaf-like bract . . . . . . . . . . . . . . . . . TILIACEAE, p. 1042

29. Leaf-blade not lop-sided, each half the same length and width as the other half . . . . . *32*

32. Lowest lateral (side) nerves nearly parallel to the mid-nerve, running lengthwise for $\frac{3}{4}$ the length of leaf-blade . . . . . . . . . . . . Ceanothus, in RHAMNACEAE, p. 1028

32. Lowest lateral (side) nerves running at an angle away from the mid-nerve, ending at $\frac{2}{3}$ or less the length of the leaf-blade . . . . . . . . . . . . . . . . . . *33*

33. Buds naked, not covered by scales; some star-shaped (stellate) hairs present on young twigs, nerves on lower surface, or lower surface of leaf-blades; flowers with 4 ribbon-like yellow, orange, or red petals; mature fruit bearing 2 shiny black seeds . . . Hamamelis, in HAMAMELIDACEAE, p. 786

33. Buds covered by scales; hairs when present on twigs or leaves not star-shaped; flowers and fruit otherwise, not as above . . . . . . . . . . . . . . . . *34*

34. Petals present at flowering time; flowers not in catkins (slender or cylindrical inflorescences); bark shreddy; stamens 30–40 in a single flower . . . . . Physocarpus, in ROSACEAE, p. 793

34. Petals absent at flowering time; flowers in catkins; bark not shreddy; stamens 4–many . . . . . . . . . . . . . . . . . . . . . *35*

35. Broken leaf-stalks and twigs producing a milky juice; buds not resinous (not sticky); calyx 4-parted, present in both male (staminate) and female (pistillate) flowers; seeds lacking hairy tufts . . . . . . . . . MORACEAE, p. 560

35. Broken leaf-stalks and twigs not producing a milky juice; buds usually resinous (sticky); calyx absent, each flower surrounded by a cup-shaped disk and in the axil of a jagged-edged bract; seeds bearing silky tufts of hair. . . . . Populus, in SALICACEAE, p. 502

28. Lowest side nerves joining the main mid-nerve above the very base of the leaf-blade . . . . . *36*

36. Leaves evergreen, thick and leathery, with spiny edges . . Ilex opaca, in AQUIFOLIACEAE, p. 1006

36. Without the above combination of characters, if thick and evergreen, then not spiny-edged. . *37*

37. Teeth of leaf-blades rounded or blunt (obtuse), or the margin of the leaf wavy . . . . *38*

38. Buds slightly to conspicuously resinous (sticky); leaf-stalks (petioles) mostly $\frac{1}{2}$ as long as to nearly as long as length of leaf-blade, 3.8–9 cm. long; all the flowers in drooping or arching catkins (slender cylindrical inflorescences), the flowers bearing stamens (staminate) in separate inflorescences from those bearing pistils (pistillate); scales of the catkins laciniate (prominently cut); each flower of the catkin surrounded by a cup-shaped disk . . . . . . . . . . . . . Populus, in SALICACEAE, p. 502

38. Buds not resinous, or if rarely so, then the buds with only 1 scale and flowers otherwise; leaf-stalks (petioles) much less than $\frac{1}{2}$ the length of the leaf-blade, mostly $\frac{1}{25}$–$\frac{1}{4}$ the length, nearly sessile (stalkless) or up to 3.5 cm. long; flowers either not in catkins, or if in catkins, the flowers not as described in alternate section above . . . . . . *39*

39. Buds covered by 1 scale; all the flowers in catkins (slender cylindrical inflorescences), the catkins erect or ascending, the flowers bearing stamens (staminate) in separate inflorescences from those bearing pistils (pistillate); scales of catkins entire (without teeth); each flower of the catkin with 1–4 glands at the base; seeds with a tuft of silky hairs. . . . . . . . . . . . . . Salix, in SALICACEAE, p. 489

39. Buds either naked and without scales or covered by 2 to many scales, or if rarely with only 1 scale, then without the additional characters of flower and fruits given in alternate section above; flowers not in catkins, or if so, then only those containing only stamens (the staminate) in catkins and drooping; flowers otherwise than described in alternate section above; seeds without any hairy tuft . . . . . . . . . . . . . . . . . . . . . . *40*

40. Young twigs, flower- or fruit-stalks, leaf-stalk, or some part of the lower surface of leaf-blade bearing at least a few star-shaped (stellate) hairs . . . *41*

41. Buds naked, not covered by a scale; petals 4, ribbon-like, yellow, orange, or red; flowering period either from January–April or September–December, the flowers appearing preceding the new leaves or during or after the fall of the old ones; mature fruit with 2 black and shining seeds . . . . . . . . . . . Hamamelis, in HAMAMELIDACEAE, p. 786

41. Buds covered by 1 or many scales; petals either absent at flowering time or the petals united into a 5-parted white corolla, when no petals present the inflorescence greenish-yellow, grayish, or reddish; flowering period April–May, the flowers appearing with the young or older leaves; mature fruit a 1-seeded acorn (not surrounded at base by a cup) or a 3-valved dry drupe, the seed brown or grayish . . . . . . . . . . . . . . . . . . . . 42

42. Known wild from only the lowland counties of extreme southeastern Missouri; buds covered by 1 scale; leaf-stalks (petioles) 2–4 mm. long; petals united into a 5-parted white corolla; none of the flowers in catkins; fruit a 3-valved dry drupe with 1–2 seeds STYRACACEAE, p. 1176

42. Found wild in various sections of Missouri; buds covered by several or many scales; leaf-stalks (petioles) 5–30 mm. long; petals absent, some of the flowers (the male) arranged in drooping catkins (slender cylindrical inflorescences); fruit an acorn (nut surrounded at base by a cup) . . . . . . . . . . Quercus, in FAGACEAE, p. 532

40. No star-shaped hairs occurring on young twigs, flower- or fruit-stalks, leaf-stalks, or leaf-blade . . . . . . . . . . . . . . . . . . . . . . . . . . . 43

43. Petals absent at flowering time, the flowers bearing stamens (staminate) arranged in drooping catkins (slender cylindrical inflorescences); fruit not fleshy, but an acorn (nut containing 1 seed surrounded at base by a cup) . . . . . . Quercus, in FAGACEAE, p. 532

43. Petals or corolla present at flowering time, the flowers not arranged in catkins; fruit fleshy, not an acorn, and with 3 to many seeds . . . . . . . . . . . . . . . 44

44. Stamens opposite the petals; mature fruit black or black-purple, 3–4-seeded . . Rhamnus, in RHAMNACEAE, p. 1027

44. Stamens alternate with the petals or corolla-lobes; mature fruit red, green, yellow, or purplish, 4–many-seeded . . . . . . . . . . . . . . . . . . . . 45

45. Leaf-blades with the base rounded, obtuse, or curved; petals united into a tube; stamens 10; flowers perfect, the stamens and pistil in the same flower; fruit greenish, yellow, or purplish, many-seeded . . . . . . . Vaccinium, in ERICACEAE, p. 1159

45. Leaf-blades tapering or narrowed to a wedge-shaped base; petals separate to their base; stamens 4–5; flowers with the stamen-bearing flowers (staminate) on a separate plant from the pistil-bearing female flower (pistillate); fruid red, 4–9-seeded Ilex decidua, in AQUIFOLIACEAE, p. 1007

37. Teeth of leaf-blades either saw-like, or acutely short- or long-pointed . . . . . . . . . 46

46. Young twigs, flower- or fruit-stalks, leaf-stalk, or some part of the lower surface of leaf-blade bearing at least a few star-shaped (stellate) hairs . . . . . . . . . . . . . . . 47

47. Known wild from only the lowland counties of extreme southeastern Missouri; buds covered by only 1 scale; leaf-stalks (petioles) 2–4 mm. long; petals united into a 5-parted white corolla; none of the flowers in catkins; fruit a 3-valved dry drupe with 1–2 seeds STYRACACEAE, p. 1176

47. Found wild in various sections of Missouri; buds covered by several or many scales; leaf-stalks (petioles) 5–30 mm. long; petals absent, some of the flowers (the male) arranged in drooping catkins (slender cylindrical inflorescences); fruit an acorn (nut surrounded at base by a cup) . . . . . . . . . . . . . . . Quercus, in FAGACEAE, p. 532

46. No star-shaped hairs occurring on young twigs, flower- or fruit-stalks, leaf-stalks, or leaf-blade . 48

48. Buds covered by 1 scale; all the flowers in catkins (slender cylindrical inflorescences), all the catkins erect or ascending, the flowers bearing stamens (staminate) in separate inflorescences and on different plants (dioecious) from those bearing pistils (pistillate); scales of catkins entire (without teeth); each flower of the catkin with 1–4 glands at the base; seeds with a tuft of silky hairs . . . . . . . . . . . . . . . Salix, in SALICACEAE, p. 489

48. Without the above combination of characters; buds with 2 or more scales; flowers either not in catkins, or arranged other than that described in alternate section above; seeds without any hairy tuft . . . . . . . . . . . . . . . . . . . . . . . 49

49. Leaf-stalks (petioles), when fully grown, mostly 3.5–6.5 cm. long; leaf-blades with 0–4 (rarely 6) teeth on each half margin; stipules or their scars absent; flowers dioecious, the stamen-bearing flowers on one plant, the pistil-bearing flowers on another plant Nyssa aquatica, in CORNACEAE, p. 1150

49. Without the above combination of characters; leaf-stalks (petioles) nearly absent, very short, or up to 3 cm. long; leaf-blades with usually many teeth closely crowded along margins; stipules or their scars present; flowers perfect (stamens and pistil on same flower), monoecious (stamens and pistils in separate flowers on the same plant), or dioecious . . . . . . . . . . . . . . *50*

50. Flowers without petals or sepals or both. . . . . . . . . . . . . . . . . . . . . *51*

51. A gap of 7–20 mm. between the tips of the closest teeth midway along the leaf-margin, with only 1–2 teeth per cm. midway along the margin; most or all of the teeth simple and more or less equal, without smaller teeth between the main teeth; pistillate flowers or fruit surrounded by a prickly involucral envelope . . . . . . . . . . . FAGACEAE, p. 528

51. A gap of only 1–5 mm. between the tips of the closest and smallest teeth midway along the leaf-margin, with 4–11 teeth per cm. midway along the margin; most or all of the teeth unequal with smaller teeth between the larger ones; none of the flowers or fruits surrounded by prickly involucral envelope . . . . . . . . . . . . . . . . . . . . . *52*

52. Rare shrub, known wild only from Butler County, southeastern Missouri; sepals 5, free and separate to their base, persisting, green, leaf-like, and spreading after flowering, 7–12 mm. long; stamens white, numerous, more than 20 in each flower; each flower with 2–4 pistils, each pistil with 1 style. . . . . . . . Neviusia, in ROSACEAE, p. 843

52. Commonly found trees and shrubs in all sections of Missouri; calyx, when present, 2–4-parted or 3–9-lobed, with a short to conspicuous tube, usually dry and brown after flowering, less than 4 mm. long; stamens 2–9 to a flower, when present; each pistil of a flower with 2 styles or stigmas . . . . . . . . . . . . . . . . . . *53*

53. Leaf-blades often more or less lopsided (asymmetrical) with one half longer or broader than the other half, most noticeable at the base; some or all of the flowers perfect and bisexual with stamens and pistils in the same flowers; all the flowers with a calyx, the sepals united into a bell-shaped (campanulate) or top-shaped (turbinate) tube with 4–9 lobes . . . . . . . . . . . . ULMACEAE, p. 553

53. Leaf-blades usually not lopsided, each half the same length and width as the other half; flowers all unisexual and monoecious, the stamen-bearing (staminate) flowers separated from the pistil-bearing (pistillate) ones in different inflorescences on the same plant; either the staminate or the pistillate flowers without a calyx. . . . CORYLACEAE, p. 523

50. Flowers with both petals and sepals . . . . . . . . . . . . . . . . . . . . . *54*

54. Petals united into a short bell-shaped to cylindric tube so that the attempt to remove one portion disturbs the neighboring section of the corolla; stamens 10. . . . . . . . Vaccinium, in ERICACEAE, p. 1159

54. Petals free from each other, separated all the way down to their base, so that one petal can be removed without tearing or disturbing the other petals; stamens 4, 5–8, or 10–50 . *55*

55. Lower surface of leaves with a close white or tawny hairy felt . . . . . . . . Spiraea tomentosa, in ROSACEAE, p. 795

55. Lower surface of leaves pale to dark green but not covered with a close white or tawny felt . . . . . . . . . . . . . . . . . . . . . . . . . *56*

56. Flowers dioecious and unisexual, the stamen-bearing (staminate) flowers on a different plant from the pistil-bearing (pistillate) ones . . . AQUIFOLIACEAE, p. 1006

56. Flowers perfect, the stamens and pistils in the same flower . . . . . . . . *57*

57. Petals 4; calyx-lobes 4; stamens 4, opposite the petals, alternate with the calyx-lobes . . . . . . . . . . . . . . . RHAMNACEAE, p. 1026

57. Petals 5; calyx-lobes 5; stamens 5, 10–50, alternate with the petals . . . . *58*

58. Pith of branches chambered, or interrupted by partitions; stipules or their scars absent; styles or stigmas 2; stamens 5; fruit consisting of a dry 2-celled capsule . . . . . . . . . . . . . . Itea, in SAXIFRAGACEAE, p. 785

58. Pith of branches solid and continuous; stipules or their scars present; style 1 or styles 2–5 in each flower; stamens 10–60; fruit consisting of 5 dry follicles or of a fleshy pome . . . . . . . . . . ROSACEAE, p. 792

## Section 6. **Nonwoody plants lacking green color**

a. Stems twining on other plants, orange, yellow, or brownish . . . Cuscuta, in CONVOLVULACEAE, p. 1221
a. Stems upright, not twining . . . . . b
b. Plant floating on water . . . . . Utricularia, in LENTIBULARIACEAE, p. 1375
b. Land plants . . . . . c
c. Only 1 flower to a stem . . . . . d
d. All parts of plant above ground white or tinged flesh-pink . . . . . Monotropa, in PYROLACEAE, p. 1156
d. Stalks above ground yellow-brown or pale brown; corolla creamy-white tinged with lilac or purple and with 2 yellow hairy folds in the throat . . . . . Orobanche, in OROBANCHACEAE, p. 1374
c. Two or more flowers to a stem . . . . . e
e. All the petals united into a tube, so that an attempt to remove one part of the corolla results in tearing or disturbing the adjacent part . . . . . OROBANCHACEAE, p. 1372
e. Most or all of the petals or petal-like parts free and separate down to the base, so that each petal or petal-like part can be removed without tearing the rest of the flower . . . . . f
f. Flowers with 5 petals, these equal or alike in size, shape, and color; ovary superior, attached above the base of the petals and calyx . Oxalis violacea, in OXALIDACEAE, p. 959
f. Flowers with 3 outer sepals and 3 inner petals, one of the petals (the lip) different in size, shape, or color from the other parts of the flower; ovary inferior, remaining below the base of the petals and sepals . . . . . ORCHIDACEAE, p. 468

## Section 7. **Water Plants**

a. Complete plant not over 1.5 cm. broad or long . . . . . b
b. Plants with conspicuous hairs on upper surface; introduced, and rarely encountered in the wild state . . . . . Salvinia, in SALVINIACEAE, p. 17
b. Plants appearing smooth and glabrous (without hairs); commonly encountered in the wild state . . c
c. Plants with numerous, imbricated (overlapping), lobed leaf-like parts . . . . . Azolla, in SALVINIACEAE, p. 17
c. Plants with 1 or few non-lobed, non-imbricated leaf-like parts, these occurring singly or connected to form groups or masses . . . . . LEMNACEAE, p. 386
a. Complete plant, including stem and leaves, always more than an inch (2.5 cm.) broad or long. . . . d
d. Leaf-like part resembling a 4-leaved clover, with 4 leaflets at the top of a stem . MARSILEACEAE, p. 17
d. Leaf-like part not as above described . . . . . e
e. Leaves or leaf-like parts all arising directly from the base of the plant . . . . . f
f. Leaves entire (smooth-edged), without toothed or cut margins . . . . . g
g. Leaves or leaf-like parts long and narrow, similar in form to grass, ribbon, hair, thread, quills, or needle. . . . . . h
h. Leaves hair-like (capillary), less than 0.5 mm. broad . . . . . Eleocharis acicularis, in CYPERACEAE, p. 276
h. Leaves more than 0.5 mm. broad . . . . . i
i. Leaves, stems, or leaf-like parts quill-like, thread-like, or needle-like, mainly 0.5–5 mm. broad (but the stems broader and up to 12 mm. broad in the rare *Scirpus subterminalis*), usually terete (rounded or with curved sides in cross-section) or cylindrical, but if flattened, then less than 5 mm. broad . . . . . j
j. Spore-producing or reproductive parts at base of plant; no horizontal stem present; leaves usually with horizontal marks or thickenings (resulting from 4 transversely partitioned air-tubes bearing cells of the 4 thickened bast-bundles), enlarged at the base with a pouch bearing the reproductive spores . . ISOETACEAE, p. 10
j. Reproductive parts (flowers and fruit) at or near tip of elongated stem; no horizontal stem present under the soil; leaves with or without thickenings, not enlarged at the base with spores, the flowering parts at the top of the stem . . . . k

k. Sheaths at base of stem (culm) closed at the summit or united into a tube, not split down one side; flowers concealed by overlapping or spirally arranged scales; ovule and seed 1 . . . . . . . . . . . . . . . . . . . CYPERACEAE, p. 255
k. Sheaths at base of stem (culm) split or open along one side, the edges of the sheath often overlapping, but not united to form a tube; flowers evident, with 3 greenish or brownish sepals and 3 similar petals; ovules and seeds numerous . . . . JUNCACEAE, p. 404
i. Leaves not quill-like, thread-like, or needle-like, but flattened and more than 5 mm. broad Vallisneria, in HYDROCHARITACEAE, p. 67
g. Leaves not grass- or ribbon-like, if long or slender, then tapering or narrowed at least at base, or with a leaf-stalk (petiole) . . . . . . . . . . . . . . . . . . . . . *l*
l. Leaf-stalk (petiole) joining leaf-blade approximately in the middle of the blade . . . Nelumbo, in NYMPHAEACEAE, p. 668
l. Leaf-stalk (petiole) joining directly to the lower end of the leaf-blade . . . . . . . . *m*
m. Whole plant floating on the surface of the water, not rooted to the soil . . . . . . . *n*
n. Base of leaf-stalk (petiole) conspicuously enlarged and spongy; leaf-blades with parallel veins only; flowers blue, showy, 4–6 cm. long and broad, many close together in a raceme . . . . . . . . . . Eichhornia, in PONTEDERIACEAE, p. 401
n. Leaf-stalk slender throughout, not conspicuously enlarged or spongy, but the lower side of the leaf-blade often spongy; leaf-blades marked with cross veins as well as longitudinal veins; flowers white, small, 1–1.5 cm. long and broad, 1–4 . . . . Limnobium, in HYDROCHARITACEAE, p. 67
m. Plant rooted to the soil in the water . . . . . . . . . . . . . . . . . . *o*
o. Leaves arrowhead- or heart-shaped or with a deep cleft or cut at the lower end . . . *p*
p. Leaves broadly rounded at summit, the main unlobed body of fully grown leaves mostly 1–4 dm. (rarely only 0.5) broad; leaf-blades mostly rounded at summit; flowers with 4, 5, or 6 sepals, and numerous petals which are either large, or, if small, stamen-like . . . . . . . . . . . . . . . NYMPHAEACEAE, p. 664
p. Leaves usually pointed, but sometimes rounded at tip, the main unlobed body of fully grown leaves mostly less than 1 dm. broad, or, if broader, then the lobes nearly equaling length of the main body of the leaf; flowers with 3 sepals and 3 petals . . . . . . . . . . . . . . . . . . . ALISMACEAE, p. 57
o. Leaves without any cleft at the lower end . . . . . . . . . . . . . . *q*
q. Lower surface of leaf-blade often spongy with a mesh-work and often purplish; filaments of staminate flowers united into a solid central column; ovary 6–9-celled, inferior . . . . . . . . Limnobium, in HYDROCHARITACEAE, p. 67
q. Lower surface of leaf-blade not spongy nor purplish; filaments of stamens free, not united into a column; ovary 1-celled, superior . . . . . . . . ALISMACEAE, p. 57
f. Leaf-margins with rounded or scalloped teeth or cut into narrow segments or divisions. . . . . . . *r*
r. Leaf-margin scalloped with rounded teeth or wavy-toothed; plants rooted to the soil in the water . . . . . . . . . . . . . . . . . . . . . . . . . . *s*
s. Leaf-stalk (petiole) joining leaf-blade approximately in the middle of the leaf. . . . Hydrocotyle, in UMBELLIFERAE, p. 1121
s. Leaf-stalk joining directly to the cleft lower end of leaf-blade. . . . . . . . . Nymphoides, in GENTIANACEAE, p. 1194
r. Leaf-blades cut into narrow segments or divisions; plants usually floating or not rooted to the soil . . . . . . . . . . . . . . . . . . . . . . . . . . *t*
t. Main flower-stem (peduncle) inflated and swollen between its joints; leaves cut into simple divisions, not bearing any bladders; flowers with a white, regular corolla of 5 equal lobes, 5 calyx lobes, and 5 stamens . . . . . . . . . . Hottonia, in PRIMULACEAE, p. 1172
t. Main flower-stem (peduncle) hair-like, not inflated; leaves finely dissected into hair-like segments, often entangled, at least some of them bearing minute bladders; flowers with a yellow irregular corolla of 2 (upper and lower) lips, 2-lobed (upper and lower) calyx, and 2 stamens . . . . . . . . . . . . . . . . . . LENTIBULARIACEAE, p. 1375
**e.** Not all the leaves arising at the base of the plant, but also along the sides of the stem . . . . . . *u*
u. Leaves of the stem alternate. . . . . . . . . . . . . . . . . . . . . *v*
v. Leaves either toothed, deeply lobed, deeply dissected, or divided into separate segments. . . . *w*
w. Main flower-stem (peduncle) inflated and swollen between its joints . . . . . . Hottonia, in PRIMULACEAE, p. 1172
w. Main flower-stem not inflated or swollen . . . . . . . . . . . . . . . *x*

x. Leaves divided into 3–11 round, oblong, or ovate leaflets; leaves with a peppery or spicy taste . Nasturtium, in CRUCIFERAE, p. 757

x. Leaves either toothed, deeply lobed or dissected, or divided into hair- or thread-like divisions. . . *y*

y. All the leaves toothed, not deeply lobed or dissected . . . . . . . . . . . . . . . *z*

z. Stipules showing at base of leaf as free thin out-growths; leaves thin and transparent or translucent; leaves rounded or obtuse at tip, the leaf-surface wavy . . . . . . . Potamogeton, in NAJADACEAE, p. 50

z. Stipules absent; leaves thick, firm, opaque; leaves acutely pointed, leaf-surface not wavy Armoracia, in CRUCIFERAE, p. 747

y. At least the submerged leaves deeply dissected into narrow lobes, segments, or finely divided . . *1*

1. Submerged leaves finely pinnatifid or pinnately once parted (feather-like) into 5–14 undivided pairs of very narrow divisions, the main or lateral (side) divisions or segments not further divided. . . . . . . . . . . . . . . . . . HALORAGIDACEAE, p. 1108

1. Submerged leaves divided into numerous divisions (30–50 or more), the main divisions further subdivided . . . . . . . . . . . . . . . . . . . . . . . *2*

2. Leaf elongated with 1 central more or less straight axis and 2–4 times as long as broad; stem erect; usually uppermost leaves and those out of water not deeply dissected . . Armoracia, in CRUCIFERAE, p. 747

2. Leaf spreading in several directions with a forking branched axis, the complete leaf circular or as broad as long or only slightly longer than broad; stem lying in the water, more or less supported by the water or prostrate; all leaves deeply dissected . . . . . *3*

3. At least some of the leaves bearing small green or black bladders the size of a pin-head or smaller; main part of leaves not separated from stem by a broadened sheath or stipule; flowers with an irregular corolla of united petals consisting of 2 (upper and lower) lips, 2-lobed (upper and lower) calyx, and 2 stamens . . . . . . LENTIBULARIACEAE, p. 1375

3. No bladders present on any leaves; main part of leaves separated from stem by an enlarged stipule or sheath attached to petiole; flowers with a regular corolla of 5 equal-sized petals separate to their base and not connected with one another, 5 sepals, and numerous stamens . . . . . . . . . . . . . RANUNCULACEAE, p. 672

v. Leaves entire or nearly so (without teeth, lobes, or divisions) . . . . . . . . . . . . . *4*

4. Leaf-stalk (petiole) joining leaf-blade approximately in the middle of the blade; stem and lower side of floating leaves covered with a jelly-like slime . . . . . Brasenia, in NYMPHAEACEAE, p. 669

4. Leaf-stalk, when present, joining directly to the lower end of the leaf-blade; no jelly-like slime present. . . . . . . . . . . . . . . . . . . . . . . . . . . . . *5*

5. All the leaves linear, grass-, ribbon-, hair-, or thread-like, of about the same width above their base from one end to the other . . . . . . . . . . . . . . . . . . . . *6*

6. A ligule present (prolonged or protruding thin or membrane-like appendage or hairy ring extending across the inner side of leaf at the junction of the green, free, upper leafy portion, the leaf-blade, and the lower clasping part, the leaf-sheath, surrounding the stem); leaves mainly blue- or silvery-green . . . . . . . . . . . . . . . . . GRAMINEAE, p. 68

6. No ligule present, but auricles (projecting lobes at sides of base of leaves or leaf-blades) may be present; leaves light green, grass- or yellow-green, or dark green . . . . . . . *7*

7. Stipules (thin or membranous, free or united outgrowths at base of leaf or leaf-sheath) absent . . . . . . . . . . . . . . . . . . . . . . . . . *8*

8. The clasping leaf-sheath surrounding the stem closed, the margins of the sheath united, not split down one side . . . . . . . . . Scirpus, in CYPERACEAE, p. 286

8. The clasping leaf-sheath surrounding the stem open, the margins of the sheath open on one side and not united. . . . . . . . . . . . . . . . . . . . . *9*

9. Leaves pale or grass-green, rather soft and spongy, flattened, 4–15 mm. broad; flowers monoecious, the stamens and pistils separated in different inflorescences on the same plant; fruits 1–2-seeded, the fruiting portion of solid spherical bur-like heads 1.5–3.5 cm. in diameter. . . . . . . . . . . SPARGANIACEAE, p. 49

9. Leaves dark green, firm, at least the lowest ones quill-like or terete (round in cross-section), less than 4 mm. broad; flowers perfect, the stamens and pistils present in the same flower; fruits many-seeded, the fruiting portion of scattered small clusters 0.2–1 cm. in diameter . . . . . . . . . . Juncus, in JUNCACEAE, p. 404

7. Stipules present, either free from the rest of the leaf or the stem or partially or wholly united with them . . . . . . . . . . . . . . . . . . . . . *10*

10. Leaves 2–6.5 mm. broad, without a midrib; flowers yellow, the perianth 6-parted; stamens 3 Heteranthera dubia, in PONTEDERIACEAE, p. 404
10. Submerged leaves 0.3–3 mm. broad, with or without a midrib; flowers green, without a true perianth; stamens 2, or when 4 with sepal-like outgrowths from the anthers . . NAJADACEAE, p. 50
5. Some of the floating leaves or all the leaves broadened in one section more than in another, the sides curved and not parallel their whole length, the leaves not linear nor thread- nor ribbon-like . *11*
11. Stipules which form a sheath around stem at base of leaf are fringed with bristles or long hairs on their summit . . . . . . . . . . . . Polygonum, in POLYGONACEAE, p. 582
11. No bristles or long hairs on summit of stipules, if stipules are present . . . . . . . . . *12*
12. Leaves expanded at base into a sheath or with stipules, or the stipules fused to form a sheath; leaves with 3 or more parallel main nerves running from base to tip of the leaf; flowers green, purple, or blue, rarely white . . . . . . . . . . . . . . *13*
13. Stipules (thin outgrowths at base of leaf) free from part or all of the base of leaf, evident as loose appendages or as projections from leaf-base; flowers green, without a tube and with no perianth but with 4 stamens having sepal-like outgrowths; stems weak and flexible, supported by the water . . . . Potamogeton, in NAJADACEAE, p. 50
13. Stipules completely fused with the leaf-stalk (petiole) to form a sheath, with no loose outgrowths or projections from leaf-base; flowers purple or blue, rarely white, with a tube, 6-parted perianth, and 3 stamens . . Heteranthera, in PONTEDERIACEAE, p. 402
12. Leaves not expanded at base into either sheath or stipules, the stipules absent; leaves nerved like a feather with 1 main midnerve and few to several lateral (side) nerves arising at several levels from the midrib; flowers with white or yellow petals. . . . . *14*
14. All the leaves on petioles (stalks); usually more than 15 leaves on a stem; petals 5, yellow; stamens 10 . . . . Jussiaea repens var. glabrescens, in ONAGRACEAE, p. 1094
14. At most only the lowest 2–5 leaves on petioles; usually only 4–8 (up to 14) leaves on a stem; petals 4, white; stamens 6 . . . . . Cardamine, in CRUCIFERAE, p. 748

u. Leaves of the stem opposite or in whorls (circles) of 3 or more . . . . . . . . . . . . *15*
15. Leaves not dissected nor subdivided . . . . . . . . . . . . . . . . . . . *16*
16. The submerged leaves 0.2–5 mm. broad, hair-like and linear to linear-lanceolate . . . . *17*
17. Leaves usually in whorls (circles) of 3 or 4, sometimes more, rarely only 2 at a level Anacharis, in HYDROCHARITACEAE, p. 66
17. All the leaves 2 at each level, opposite one another (sometimes with bunches of smaller leaves in the axils or at the base of the larger ones giving the appearance of more than 2 leaves) . . . . . . . . . . . . . . . . . . . . . . . . *18*
18. Margins of leaves or enlarged sheathing bases with very minute teeth (use lens of 10 × magnification); leaves slightly to conspicuously enlarged at the base into a sheath . . . . . . . . . . . . . . . . . Najas, in NAJADACEAE, p. 56
18. Margins of leaves without teeth (use lens of 10 × magnification); leaves with about the same width at the base as at the main part of the leaf or tapering at the base . . . . . . . . . . . . . . . . . . . . . . . *19*
19. Leaves 0.4–1 mm. broad; midrib not conspicuous or scarcely evident; base of leaves with stipules which form a sheath; fruit toothed down one side. . . Zannichellia, in NAJADACEAE, p. 56
19. Leaves 1–4.5 mm. broad; midrib evident or conspicuous; leaves without stipules or sheaths . . . . . . . . . . . . . . . . . . . . . *20*
20. Stems with ridges or wings extending down from the base of each leaf; calyx present; flowers perfect (stamens and pistil in same flower); fruit many-seeded . . . . . . . . . . Peplis, in LYTHRACEAE, p. 1088
20. Stems without ridges running from the base of the leaf; calyx absent; flowers monoecious, the male (staminate) and female (pistillate) occurring in separate flowers on the same plant; fruit 4-lobed, 4-seeded . . . . CALLITRICHACEAE, p. 996
16. The submerged leaves mainly 8–25 mm. broad (rarely 2 mm. broad in Ludwigia palustris) . *21*
21. Leaves with several main nerves starting from the very base of the leaf-blade; leaves nearly as broad as long or only slightly longer than broad, broadly rounded at summit Bacopa, in SCROPHULARIACEAE, p. 1332
21. Leaves with 1 main midnerve and few to several lateral (side) nerves on each side of main nerve; leaves clearly longer than broad, 2–7 or more times as long as broad. . *22*

22. Leaves 7 or more times as long as broad, mainly 7–20 (rarely 4) cm. long . . . . . . Justicia, in ACANTHACEAE, p. 1377
22. Leaves 2–4 (–6) times as long as broad, 1–8 (rarely 10) cm. long . . . . . . . . . . . *23*
23. Leaves without stalks (petioles), rounded at base or the upper often clasping the stem; leaves with or without teeth . . . . . . . . . Veronica, in SCROPHULARIACEAE, p. 1328
23. Leaves tapering at base into more or less of a petiole (leaf-stalk), narrowed below the middle to the base; leaves always without teeth . . . . Ludwigia, in ONAGRACEAE, p. 1095
15. Leaves dissected or subdivided into narrow segments . . . . . . . . . . . . . . . . . *24*
24. Each joint of stem or leaf-division actually one single cell . . . . . . . . . . . . . . Chara and Nitella in CHAROPHYTA, a division of higher algae not included in this work
24. Each leaf-division and joints of stem consisting of numerous cells . . . . . . . . . . *25*
25. Each leaf-division or segment provided with tiny teeth or spiny projections . . . . Ceratophyllum demersum, in CERATOPHYLLACEAE, p. 662
25. No teeth or spiny projections on margins of leaf-divisions or segments . . . . . . . . *26*
26. Main side divisions of each leaflet remaining uncut their whole length, each leaflet only once pinnately divided or parted . . . Myriophyllum, in HALORAGIDACEAE, p. 1108
26. Main side divisions of each leaflet divided more than once, forking or branching . . . *27*
27. Main leaves opposite, 2 at a level . . . . . . . . . . . . . . . . . . *28*
28. Each of main leaves with a leaf-stalk (petiole); upper floating leaves, if present, alternate, without teeth or nearly so, long-stalked with the petiole (stalk) joining the leaf-blade near its center . . . . . . . Cabomba, in NYMPHAEACEAE, p. 669
28. Each of main leaves without a leaf-stalk (petiole) or essentially so; upper leaves, if present, opposite, sharply toothed, and without a leaf-stalk or nearly so . . . . . . . . . . . . . . Bidens Beckii, in COMPOSITAE, p. 1590
27. Main leaves in whorls (circles) of 3 or more . . . . . . . . . . . . . . *29*
29. Each of main leaves with a leaf-stalk (petiole); leaf-divisions of submerged leaves flat, ribbon-like; upper floating leaves, if present, alternate, without teeth or nearly so, long-stalked with the petiole (stalk) joining the leaf-blade near its center . . . . . . . . . . . Cabomba, in NYMPHAEACEAE, p. 669
29. Each of main leaves attached to stem more or less directly without a leaf-stalk (petiole) or essentially so; leaf-divisions of submerged leaves with curved or rather rounded sides, or slightly compressed, but not flattened nor ribbon-like; no floating leaves present, or, when above water or on surface, opposite and/or without petioles . . . . . . . . . . . . . . . . . . . . *30*
30. Intervals of stem between leaf-whorls 3–20 mm. long, the whorls crowded; stem 0.5–1.5 mm. thick; main leaf-divisions usually once or twice forked; plants with green flowers, flowering below the surface of the water, the fruit nut-like, occurring along the sides of submerged stem in axils of leaves Ceratophyllum echinatum, in CERATOPHYLLACEAE, p. 664
30. Intervals of stem between leaf-whorls mostly 20–70 mm. long, the whorls widely separated; stem 2–5 mm. thick; main leaf-divisions forked and branched several or many times; plants flowering above surface of water with orange-yellow heads of flowers . Bidens Beckii, in COMPOSITAE, p. 1590

Section 8. **Cactaceae, plants with thick, flattened, usually spiny-jointed leaf-like stems, p. 1084**

Section 9. **Herbaceous, soft-stemmed, or nonwoody plants with some or all leaves compound**

a. Plants without true flowers or seeds, reproducing by green or brown spores (without embryos) contained in sporangia and located on the lower portions of the leaf (frond) or in specially modified leaves (fronds), stalks, or branches . . . . . . . . . . . . . FERNS AND FERN ALLIES, p. 2
a. Plants with true flowers bearing stamens or carpels (pistils) or both, reproducing by seeds containing embryos . . . . . . . . . . . . . . . . . . . . . . . . . . *b*
b. The compound leaves opposite or in whorls (circles), 2 or more at one level on the stem . . . . *c*
c. Stems trailing, lying on the ground, sprawling, vining, or climbing . . . . . . . . . . *d*
d. Leaflets small, 1 cm. or less long; flowers yellow; fruits bearing prickles or tubercles (small projections from the surface) . . . . . . . . . . . . . ZYGOPHYLLACEAE, p. 964

d. Leaflets large, 3 cm. or more long; flowers white, purple, or bluish-lavender; fruits with long silky or feathery tails . . . . . . . . . . . . . Clematis, in RANUNCULACEAE, p. 705

c. Stems upright or spreading, but not trailing or vining . . . . . . . . . . . . . . . . *e*

e. Compound leaves opposite, 2 at one level . . . . . . . . . . . . . . . . . . *f*

f. Bruised stems and leaf-stalks with yellow juice; petals 4; stamens numerous . . . . Stylophorum, in PAPAVERACEAE, p. 720

f. No yellow juice in bruised stems and leaf-stalks; corolla-lobes 5 or petals none; stamens 4 or 5 . . . . . . . . . . . . . . . . . . . . . . . . . . . . *g*

g. Leaves palmately divided, all the divisions arising from one point at the very base of the leaf-blade; leaves 4–15 cm. long . . . . . . . . . Cannabis, in MORACEAE, p. 564

g. Leaves pinnately parted, the divisions arising at 1 or more places above the base of the leaf-blade; leaves 1–3 cm. long . . . . . . . . . . . . . . . . . . . . *h*

h. Corolla regular with 5 lobes more or less equal; stamens 5 . . . . . . . . . Ellisia, in HYDROPHYLLACEAE, p. 1236

h. Corolla irregular, with an upper 2-lobed lip and a lower 3-parted lip; stamens 4 . Conobea, in SCROPHULARIACEAE, p. 1333

e. Compound leaves 3 or 4 in a whorl (circle) . . . . . . . . . . . . . . . . . *i*

i. Stem more or less hairy below the leafy parts on the stem; leaves or leaf-like parts on stem more or less hairy; true petals absent, the sepals petal-like; pistils several and distinct . Anemone, in RANUNCULACEAE, p. 699

i. Stem glabrous (without hairs) or essentially so below the leafy parts on the stem; leaves or leaf-like parts on the stem glabrous or nearly so; true petals present; pistil of 1–5 united carpels . . . . . . . . . . . . . . . . . . . . . . . . . . . . *j*

j. Leaf-divisions of a narrow type, linear, oblong, or oblanceolate, 0.1–2 cm. broad; petiole (main leaf-stalk) 0.5–4 cm. long; petals 4; sepals 4; stamens 6; ovary superior (above the insertion of the petals) . . . . . . . . Dentaria, in CRUCIFERAE, p. 751

j. Leaf-divisions of a broad type, oblong-ovate, obovate-oblong, or oval, 2–6 cm. broad; petiole mainly 4–10 or more (rarely 3) cm. long; petals 5; calyx of 5 sepals or obsolete lobes; stamens 5; ovary inferior (below the insertion of the petals) . . ARALIACEAE, p. 1112

b. The compound leaves alternate with only 1 at each level of the stem or arising only at the base of the plant or both alternate and at the base . . . . . . . . . . . . . . . . . *k*

k. Compound leaves arising only at the base of the plant . . . . . . . . . . . . . . *l*

l. All the leaf-divisions hair- or thread-like and less than 0.5 mm. broad . . . . . . . . Lomatium, in UMBELLIFERAE, p. 1146

l. Leaf-divisions not hair- or thread-like, the larger ones 1.5 or more mm. broad. . . . . . . *m*

m. Leaves divided into 3 leaflets which are either entire (without teeth and smooth-edged) or shallowly toothed or lobed, but not deeply cut . . . . . . . . . . . . . . . *n*

n. Leaf-divisions entire (without teeth) . . . . . . . . . . . . . . . . . . *o*

o. Leaf-divisions 8–30 cm. (rarely 6 cm. on young leaves) long, pointed at tip; flowers on a column (spadix) surrounded by a large green or brown-purple spathe. . . Arisaema, in ARACEAE, p. 383

o. Leaf-divisions 2.5 cm. or less long, rounded or notched at summit; flowers not as in alternate (o.) above, but in heads, racemes, or umbels . . . . . . . . . . . *p*

p. Stipules (small, leaf-like parts) present at base of petiole (leaf-stalk); individual flowers irregular, some of the petals different in size and shape from other petals, in a dense rounded head; plants without underground bulbs . . LEGUMINOSAE, p. 865

p. No stipules present; individual flowers regular, all the petals equal, in an umbel; plants with an underground bulb . . . . . . . . . . . OXALIDACEAE, p. 958

n. Leaf-divisions shallowly toothed or lobed, not deeply cut . . . . . . . . . . . *q*

q. No stipules (small leaf-like appendages) present at base of petiole (leaf-stalk); margins of leaf-divisions with rounded or obscure teeth; rare plant of swampy meadows in Reynolds County, southeastern Missouri . . . . Menyanthes, in GENTIANACEAE, p. 1192

q. Stipules (small leaf-like appendages) present at base of petiole; margins of leaf-divisions with pointed or evident teeth or lobes; found in various sections of Missouri . . *r*

r. Flowers regular, the petals of equal size and shape; stamens numerous, more than 10 . . . . . . . . . . . . . . . . . . . . . ROSACEAE, p. 792

r. Flowers irregular, the petals of unequal size and shape; stamens 10 . LEGUMINOSAE, p. 865

m. Leaves divided into 4–100 divisions (leaflets), or if appearing to be cut into 3 main divisions, then these deeply cut into numerous fine segments or deeply dissected . . . . . . *s*

s. Each leaf divided into 3 main divisions which are finely cut and deeply dissected into numerous (50–150) smaller divisions . . . . . . . . . . . . . . . . . . . . . . *t*

t. Flowers very small, about 3 mm. long, in small clusters at the tips of umbrella-like rays; ovary inferior, situated below the insertion of the petals; fruit round, 2–3 mm. long . . . Erigenia, in UMBELLIFERAE, p. 1129

t. Flowers showy, broadly heart- or V-shaped, 18–30 mm. long, arranged along the sides of a slender raceme; ovary superior, situated above the insertion of the petals; fruit long and slender, 10–15 mm. long . . . . . . . . . . . . Dicentra, in FUMARIACEAE, p. 724

s. Each leaf divided into 5–23 divisions (leaflets) . . . . . . . . . . . . . . . . . *u*

u. Leaf-divisions entire (without teeth and smooth-edged) . . . . . . . . . . . . . . *v*

v. Leaflets arising at different levels on the leaf and arranged in pairs one above another, with an odd one at tip . . . . . . . . . . . . . . . . . . LEGUMINOSAE, p. 865

v. Leaflets all arising from the basal center or lower half of the leaf . . . . . . . . . *w*

w. Plant 35–60 cm. tall; leaf-divisions 10–30 cm. long (5 in young or immature leaves); flowers creamy-white on a column (spadix) protruding from a greenish spathe; petals absent . . . . . . . . . . . . . . . . . Arisaema, in ARACEAE, p. 383

w. Plants 4–25 cm. tall; leaf-divisions 1–10 cm. long; flowers blue, purple, or white; petals present . . . . . . . . . . . . . . . . . . . . . VIOLACEAE, p. 1068

u. Leaf-divisions toothed. . . . . . . . . . . . . . . . . . . . . . . . . *x*

x. Petiole (main leaf-stalk) of the compound leaf 15–50 cm. tall; 2 of the leaflets arising lower down on the leaf-axis than the other 3 . . . . . . . . . Aralia, in ARALIACEAE, p. 1112

x. Petiole (main leaf-stalk) of the leaf 3–10 cm. (up to 25 cm. in fruiting or late season plants) tall; all leaf-divisions arising from the basal center or lower half of the leaf. . . . . . . *y*

y. Flowers blue, purple, or white, irregular with some of the petals of different size and shape from the others; true petals present; stamens 5 . . . . . . . VIOLACEAE, p. 1068

y. Flowers pink, rose, white, or greenish, regular with the sepal-like petals equal in size and shape; true petals absent; stamens numerous, more than 15 . . RANUNCULACEAE, p. 672

k. The compound leaves alternate with 1 leaf at 1 or more levels of the stem, or leaves both alternate and at the base of the plant. . . . . . . . . . . . . . . . . . . . . . . *z*

z. What appears to be one flower actually is a dense mass of small flowers surrounded by green or colored scales or bracts of the involucre; anthers united into a tube; petals united into a tube; ovary inferior; ovule and seed 1 . . . . . . . . . . . . . . . . COMPOSITAE, p. 1442

z. Without the above combination of characters . . . . . . . . . . . . . . . . . . . *1*

1. Stems climbing, vining, or trailing on the ground. . . . . . . . . . . . . . . . . *2*

2. Stipules (small leaf-like appendages at base of leaf-stalk) absent or not persisting. . . . . *3*

3. Tendrils (coiling or twining hair-like or thread-like organs) not present on any part of plant . . . . . . . . . . . . . . . . . . . . . SOLANACEAE, p. 1310

3. Tendrils present on some part of plant . . . . . . . . . . . . . . . . . . *4*

4. Leaves on a plan of 3 divided 1–3 times; stamens 5; fruit a fleshy berry . VITACEAE, p. 1031

4. Leaves on a plan of 3 divided 2 times; stamens 8; fruit an inflated dry capsule. . Cardiospermum, in SAPINDACEAE, p. 1023

2. Stipules present and persisting, or, if stipules rarely absent, then the flowers are irregular with some of the petals different in size and shape from some of the others . . . . . . . *5*

5. Leaflets entire (without teeth); stamens usually 10; flowers irregular, with some of the petals different in size and shape from the others . . . . . . . LEGUMINOSAE, p. 865

5. Leaflets toothed; stamens usually 15 or more; flowers regular with all the petals of equal size and shape . . . . . . . . . . . . . . . . . . . . . ROSACEAE, p. 792

1. Stems upright, spreading, or sometimes sprawling, but not trailing or vining . . . . . . . *6*

6. Margins of the ultimate leaf-divisions entire (without teeth) . . . . . . . . . . . *7*

7. Leaf-divisions 3 . . . . . . . . . . . . . . . . . . . . . . . . *8*

8. Ovary inferior, situated below the insertion of the petals; stems usually hollow . . UMBELLIFERAE, p. 1115

8. Ovary superior, situated above the insertion of the petals; stems usually solid . . . . *9*

9. Flowers irregular with some of the petals of different size and shape from the others; sepals more or less united into a calyx-tube . . . . . LEGUMINOSAE, p. 865

9. Flowers regular with all the petals of equal size and shape; sepals free from each other, not connected at the base . . . . . . . . . . . . . . . . *10*

10. Stipules (small, leaf-like appendages at base of leaf-stalk) present. . . . . *11*

11. Leaflets as broad as or broader than long, broadest in the upper half with a notch at the summit, acid-tasting; flowers with 5 petals and 5 sepals . . . . . . . OXALIDACEAE, p. 958
11.¹Leaflets much longer than broad, broadest at the middle or in the lower half, not acid-tasting; flowers with 4 petals and 4 sepals . . . . . . . . . . . CAPPARIDACEAE, p. 768
10. Stipules absent . . . . . . . . . . . . . . . . . . . . . . . . . . 12
12. Flowers occurring solitary along the sides of the stems in the leaf-axils; petals inconspicuous, shorter than the sepals, 1.5–2 mm. long; petals 3; sepals 3 . . . . LIMNANTHACEAE, p. 998
12. Flowers occurring 2 or more together in an inflorescence at the ends of stems or branches; petals conspicuous, much longer than the sepals, 3.5–30 mm. long; petals 4 or 5; sepals 4 or 5 . . . . . . . . . . . . . . . . . . . . . . . . . . 13
13. Leaflets as broad as or broader than long, broadest in upper half with a notch at the summit, acid-tasting; flowers with 5 petals and 5 sepals . . . . . OXALIDACEAE, p. 958
13. Leaflets much longer than broad, broadest at the middle or in the lower half, not acid-tasting; flowers with 4 petals and 4 sepals . . . . . . . . . . CAPPARIDACEAE, p. 768
7. Leaf-divisions 2, 4, 5 or more . . . . . . . . . . . . . . . . . . . . 14
14. Ovary inferior, situated below the insertion of the petals; stems usually hollow . . . UMBELLIFERAE, p. 1115
14. Ovary superior, situated above the insertion of the petals; stems usually solid. . . . . . . . 15
15. Stems covered with hooked prickles . . . . . . . . . Schrankia, in LEGUMINOSAE, p. 865
15. Stems without prickles . . . . . . . . . . . . . . . . . . . . . . 16
16. Petals united at least at base into a short or long corolla-tube, the whole corolla being removable as one piece or torn if one section is disturbed . . . . POLEMONIACEAE, p. 1224
16. Petals all separate and distinct to the base, so that one petal may be removed separately in one piece . . . . . . . . . . . . . . . . . . . . . . 17
17. Stamens numerous, more than 10, mounted one above the other on a column; pistils several, the ovaries united into a ring; mature fruit separating into a ring of 1-seeded carpels . . . . . . . . . . . . . . MALVACEAE, p. 1044
17. Stamens 5–10, not arranged one above the other on a column, although often united; pistils 1–5, separate or combined but not united into a ring; mature fruit not separating into a ring of 1-seeded carpels . . . . . . . . . . . . . 18
18. A saucer-shaped, stalked, or raised gland present at the base of the leaf-stalk or between the lower pairs of leaf-divisions . . . Cassia, in LEGUMINOSAE, p. 874
18. No glands as above described present at base of leaf-stalk or between the lower pairs of leaf-divisions. . . . . . . . . . . . . . . . . . . 19
19. Stipules (leaf-like or needle-like appendages at base of leaf-stalk) usually present, although sometimes not persisting; sepals more or less united into a calyx-tube; flowers more or less irregular with some of the petals of unequal size and shape from the others or the pistil 1 with a 1-celled ovary LEGUMINOSAE, p. 865
19. Stipules usually absent; if present, without the characters as described above in alternate section; sepals more or less free from each other, not connected at the base; flowers regular with all the petals of equal size and shape; pistils either 2–5 and the ovary 1–5-celled . . . . . . . . . . . 20
20. Leaves with a strong odor and bitter taste, dotted with glands; petals yellow, minutely toothed; stamens 8–10; ovary 4–5-celled . . . . Ruta, in RUTACEAE, p. 968
20. Leaves not as above, sometimes with a pleasant spicy but not bitter taste, not dotted with glands; petals white, rose, or pink, or sometimes yellow; stamens 6 or 11–16; ovary 1–2-celled . . . . . . . . 21
21. Flowers occurring solitary along the sides of the stems in the leaf-axils; petals inconspicuous, shorter than the sepals, 1.5–2 mm. long; petals 3; sepals 3 . . . . . . . LIMNANTHACEAE, p. 998
21. Flowers occur 2 or more together in an inflorescence at the ends of the stems or branches; petals conspicuous, longer than the sepals, 3–30 mm. long; petals 4; sepals 4 . . . . . . . . . . . 22
22. Leaves palmately compound, all the leaflets arising from approximately the same point at the top of the leaf-stalk; stamens 11–16, nearly equal, some or all of them longer than the petals; ovary and fruit 1-celled . . . . . . CAPPARIDACEAE, p. 768

22. Leaves pinnately compound, some of the leaflets arising at other levels on the leaf-axis above the base; stamens 6, 2 of them shorter than the other 4, shorter than or equaling the petals; ovary and fruit 2-celled . . . . . . . . . . . . . . . . . . . . CRUCIFERAE, p. 728

6. Margins of the ultimate leaf-divisions with small or large teeth or with lobes or deeply cut portions . *23*

23. Leaf-divisions palmately compound, all arising from the tip of the main leaf-stalk or at the base of the leaf-blade . . . . . . . . . . . . . . . . . . . . . . . . . . *24*

24. Stamens and petals perigynous, inserted on a cup-like hypanthium or calyx-tube . ROSACEAE, p. 792

24. Stamens and/or petals either hypogynous and inserted on the receptacle at the base of the pistil or pistils, sometimes stamens only are present, or epigynous and inserted on top of the ovary . . . . . . . . . . . . . . . . . . . . . . . . . *25*

25. Ovary inferior, situated below the insertion of the petals; stems usually hollow. . . UMBELLIFERAE, p. 1115

25. Ovary superior, situated above the insertion of the petals; stems usually solid . . . . *26*

26. Flowers green, the stamens and pistils in separate flowers on different parts of the same plant (monoecious) . . . . . . . . . . Cannabis, in MORACEAE, p. 564

26. Flowers of other colors, the flowers usually perfect with the stamens and pistils in the same flower . . . . . . . . . . . . . . . . . . . . . . *27*

27. Petals united at least at base into a corolla-tube, the whole corolla being removable as one piece or torn if one section is disturbed . . . . . . . Phacelia, in HYDROPHYLLACEAE, p. 1236

27. Petals either absent, or when present all separate and distinct to the base, so that one petal may be removed separately in one piece . . . . . . . . . . *28*

28. Stamens mounted one above the other on a column; ovaries united into a ring; mature fruit separating into a ring of 1-seeded carpels . MALVACEAE, p. 1044

28. Stamens not arranged one above the other on a column, although often united; ovaries not united into a ring; mature fruit not separating into a ring . . . . . . . . . . . . . . . . . . . . *29*

29. Flowers more or less irregular with some of the petals of unequal size and shape from the others . . . . . . . . . . . . . . . . *30*

30. Sepals free from each other, not connected at the base; one of the sepals and 2 of the petals prolonged into a slender pointed spur (hollow, tube-like) . . . . Delphinium, in RANUNCULACEAE, p. 679

30. Sepals more or less united into a calyx-tube; no spur present. . LEGUMINOSAE, p. 865

29. Flowers regular with all the petals or petal-like sepals (when true petals are absent) of equal size and shape . . . . . . . . . . *31*

31. Flowers white, rose, or pink; petals 4; sepals 4; ovary of 2 united carpels, with 2 parietal placentae. . . . . CAPPARIDACEAE, p. 768

31. Flowers yellow, yellow and red, or white; petals when present of various numbers, but usually 5; petal-like sepals often 5, but also 3–20; ovary of 1 separate carpel with 1 placenta . RANUNCULACEAE, p. 672

23. Leaf-divisions either pinnately compound with the leaflet or leaflets arising at 1 or more levels above the base of the leaf-blade or on individual leaf-stalks arising from different portions of the main leaf . . . . . . . . . . . . . . . . . . . . . . . . . *32*

32. Petals united at least at base into a corolla-tube, the whole corolla being removable as one piece or torn if one section is disturbed . . . . . . . . . . . . . . . . . . *33*

33. Corolla yellow; anthers touching one another, arranged in a ring surrounding the style Lycopersicon, in SOLANACEAE, p. 1314

33. Corolla lavender, bluish-purple, or white; anthers separated from one another, not touching nor forming a tube around the style . . . . . . HYDROPHYLLACEAE, p. 1233

32. Petals either absent, or when present all separate and distinct to the base so that one petal may be removed separately in one piece. . . . . . . . . . . . . . . *34*

34. Stamens and petals perigynous, inserted on a cup-like hypanthium or calyx-tube. . ROSACEAE, p. 792

34. Stamens and/or petals either hypogynous and inserted on the receptacle at the base of the pistil or pistils, or epigynous and inserted on top of the ovary. . . . . . . . *35*

35. Ovary inferior, situated below the insertion of the petals . . . . . . . . . . *36*

36. Styles 5; ovary 5-celled; fruit a wine red to black-purple fleshy berry with 5 seeds . . . . . . . . . . . . Aralia racemosa, in ARALIACEAE, p. 1112

36. Styles 2; ovary 2-celled; fruit with 2 seed-like dry carpels united at first by their inner faces but usually separating when ripe . . . . . . . . . . . . . . . . UMBELLIFERAE, p. 1115

35. Ovary superior, situated above the insertion of the petals . . . . . . . . . . . . . . 37

37. Flowers more or less irregular with some of the petals of unequal size and shape from the others . 38

38. Leaves divided into 3 leaflets; leaflets with short, shallow teeth; sepals more or less united into a calyx-tube; no spur present on flower . . . . . . . . . LEGUMINOSAE, p. 865

38. Leaves divided into many leaflets, more than 3; leaflets deeply cut or parted; sepals free from each other and not united into a calyx-tube; a short or long spur (hollow, tube-like projection at one end) present on flower . . . . . . . . . . . . . . . . . 39

39. Flowers yellow or orange-yellow, 6–20 mm. long; sepals 2, equal, small and scale-like; stamens 6, their filaments united at least at base; seeds black and shining . . . . Corydalis, in FUMARIACEAE, p. 725

39. Flowers blue, purple, white, or combinations of these, 15-37 mm. long; sepals 5, unequal, showy and petal-like; stamens more than 6, their filaments not connected with one another . . . . . . . . . . . . Delphinium, in RANUNCULACEAE, p. 679

37. Flowers regular with all the petals or petal-like sepals (when true petals are absent) of equal size and shape . . . . . . . . . . . . . . . . . . . . . . . . . 40

40. Each flower with 3 petals and 3 sepals; flowers all solitary, arising along the sides of the stem in the leaf-axils; rarely encountered family . . . . . . . LIMNANTHACEAE, p. 998

40. Each flower with usually 4 or more petals (rarely 3) and 4 or more sepals (rarely 3); flowers 1 or more together usually at the top of the stem or of the branches; frequently encountered . . . . . . . . . . . . . . . . . . . . . . . . 41

41. Bruised stem or leaf-stalk with yellow, orange, or reddish juice; sepals usually 2, quickly falling from the flowers; stamens usually 16–24 . . . . . . . . PAPAVERACEAE, p. 719

41. Without the above combination of characters; no colored juice present on bruised stem or leaf-stalks; sepals mainly 4, 5, 6 or more, rarely 3; stamens 5, 6, or up to 50 . 42

42. Petals absent at flowering time, the sepals sometimes petal-like, often quickly falling from the flower; stamens 10–40 . . . . . . . . RANUNCULACEAE, p. 672

42. Petals present at flowering time, falling quickly from the flower or persistent; stamens 6, 10, or up to 40 . . . . . . . . . . . . . . . . . . 43

43. Flowers greenish-yellow; petals 6; sepals 6; stamens 6 . . . . . . . Caulophyllum, in BERBERIDACEAE, p. 711

43. Without this combination of characters; flowers usually of other colors; petals usually 4 or 5, sometimes more or 1–3; sepals usually 4 or 5, sometimes more or 3; stamens 5, 6, or up to 40. . . . . . . . . . . . . . . . . 44

44. Anther-bearing stamens usually more than 6; carpels 1 to many, distinct from the beginning . . . . . . . . . . . . . RANUNCULACEAE, p. 672

44. Anther-bearing stamens 5 or 6; carpels 2 or 5, either united completely when 2 or partly united by their styles when 5 . . . . . . . . . . 45

45. Sepals 5; petals 5; anther-bearing stamens 5; carpels 5, partly united by their styles . . . . . . . . . . Erodium, in GERANIACEAE, p. 964

45. Sepals 4; petals 4; anther-bearing stamens 6; carpels 2, completely united . . . . . . . . . . . . . . . . . CRUCIFERAE, p. 728

Section 10. **Herbaceous or nonwoody plants with opposite or whorled, simple entire leaves without teeth or lobes**

a. Plants without true flowers or seeds, reproducing by green or brown spores (without embryos) contained in sporangia and located on the lower portions of the leaf (frond) or in specially modified leaves (fronds), stalks, or branches . . . . . . . . . . . . . . FERNS AND FERN ALLIES, p. 2

a. Plants with true flowers bearing stamens or carpels (pistils) or both, reproducing by seeds containing embryos . . . . . . . . . . . . . . . . . . . . . . . . . . . b

b. What appears to be one flower actually is a dense mass of small flowers surrounded by green or colored scales or bracts of the involucre; anthers united into a tube; petals united into a tube; ovary inferior; ovule and seed 1 in each of the individual flowers . . . . . COMPOSITAE, p. 1442

b. Without the above combination of characters . . . . . . . . . . . . . . . . . . c

c. Growing in green, rounded clumps as a parasite on tree branches; branches brittle; leaves thick, leathery, evergreen; fruit white, fleshy . . . . . . . . . . LORANTHACEAE, p. 571

c. Without the above combination of characters . . . . . . . . . . . . . . . . . *d*
d. Stems prickly; flower-heads bristly with long needle-pointed bracts . . . . DIPSACACEAE, p. 1422
d. Stems not prickly; no needle-pointed bracts present . . . . . . . . . . . . . . . . . *e*
e. Bruised stem, leaf, or leaf-stalk producing a milky sap . . . . . . . . . . . . . . *f*
f. Flowers inconspicuous, greenish, or with small, white spreading petal-like appendages; true petals absent, the group of flowers surrounded by an involucre in a cup-shaped cyathium; fruit deeply 3-lobed . . . . . . . . . Euphorbia, in EUPHORBIACEAE, p. 984
f. Flowers noticeable or conspicuous, white, red, pink, green, bluish, or greenish- or brownish purple, either solitary or in flat-topped, umbrella-like, or ball-like clusters; a true corolla present; fruit a pair of 2 distinct or united carpels . . . . . . . . . . . . . . . *g*
g. Corolla-lobes deeply parted almost to the base, the corolla-tube short or inconspicuous; individual flower not bell-shaped nor tubular; filaments of stamens united into a tube and column surrounding the pistil with the anthers attached to a large stigmatic body; pollen united in waxy masses; styles distinct . . . . . . . . ASCLEPIADACEAE, p. 1200
g. Corolla-lobes not parted almost to the base, the corolla-tube usually long and conspicuous; individual flower either bell-shaped or tube-shaped or with a long tube; filaments of stamens not united, distinct; pollen loose and granular; styles united . . APOCYNACEAE, p. 1194
e. No milky sap produced by bruised stem, leaf, or leaf-stalk . . . . . . . . . . . . . *h*
h. Leaves in whorls (circles) of 3 or more, never opposite nor in 2's . . . . . . . . . . *i*
i. Only 1 whorl (circle) of leaves on the stem, no other leaves present . . . . . . . . . *j*
j. Leaves narrow, linear, or linear-spatulate, only 1.5–4 mm. broad . . POLYGALACEAE, p. 969
j. Leaves much broader, obovate, ovate, or almost round, 10–150 mm. broad . . . . . *k*
k. Leaves in whorls of 3's . . . . . . . . . . . Trillium, in LILIACEAE, p. 443
k. Leaves in whorls of 5's or 6's (rarely up to 10) . . . Isotria, in ORCHIDACEAE, p. 477
i. Two or more whorls (circles) of leaves on the stem. . . . . . . . . . . . . . *l*
l. Stems bearing 2 whorls of leaves, the lower whorl with 5–9 leaves, the upper with 3–5 leaves . . . . . . . . . . . . . . . . Medeola, in LILIACEAE, p. 443
l. Stems bearing 3 or more whorls of leaves, or if only with 2 whorls, with 3–7 leaves in each whorl . . . . . . . . . . . . . . . . . . . . . . . . . *m*
m. Stems vining, twining, or climbing; leaves in whorls of 4–7, 5–15 cm. long, 2.5–10 cm. broad . . . . . . . . . . . . . . . DIOSCOREACEAE, p. 454
m. Stems erect or spreading, but, if vining or sprawling, the leaves 6 or 8 in a whorl, 2–7 cm. long and 0.1–1 cm. broad . . . . . . . . . . . . . . . . . . *n*
n. Leaves large and showy, those at the base of plant 2–3 dm. long, the ones higher on stem 1–2 dm. long . . . . . . Swertia, in GENTIANACEAE, p. 1191
n. Leaves smaller, 0.1–1.5 dm. long . . . . . . . . . . . . . . . . . *o*
o. Leaves 4–15 cm. long; petals deeply cut or fringed on the sides and summit; stem 3–12 dm. tall . . . . . Silene stellata, in CARYOPHYLLACEAE, p. 657
o. Leaves 1–3.5 cm. long; petals, if present, not fringed; stems up to 4.5 dm. tall . . . . . . . . . . . . . . . . . . . . . . . . . *p*
p. Flowers irregular, with the sepals and petals of unequal size and shape POLYGALACEAE, p. 969
p. Flowers regular, with the sepals or petals or corolla-lobes of equal size and shape . . . . . . . . . . . . . . . . . . . . . . *q*
q. Stems square or 4-sided; petals united into a 3- or 4-lobed corolla with a tube . . . . . . . . . . . Galium, in RUBIACEAE, p. 1388
q. Stems not square; calyx 4- or 5-parted; petals absent, or, if present, 4 or 5, distinct and separate, not united . . . . . . . . . . . *r*
r. Leaves very narrow and thread-like; minute stipules present at base of leaves . . . . . . . . Spergula, in CARYOPHYLLACEAE, p. 642
r. Leaves spatulate, broadly obovate to round, usually broadest above the middle, or as broad as long, not thread-like; stipules absent. . . *s*
s. Stems and leaves minutely hairy with star-shaped hairs . . Glinus, in AIZOACEAE, p. 632
s. Stems and leaves glabrous (without hairs) . . . . . . . . *t*
t. Leaves dark green, fleshy and thick, those on sterile branches in whorls of 3 . . . . . . Sedum, in CRASSULACEAE, p. 770
t. Leaves pale green, thin, in whorls of 4–8 . Mollugo, in AIZOACEAE, p. 632

h. Some or all of the leaves in 2's in pairs opposite each other, sometimes in 1 or more whorls of 3 or more . . . . . *u*

u. Stems vining, twining, or climbing . . . . . DIOSCOREACEAE, p. 454

u. Stems erect, spreading, or lying on the ground . . . . . *v*

v. Leaves thick and fleshy, some of them in whorls (circles) of 3, the others in 2's opposite one another . . . . . *w*

w. Whorled leaves of sterile branches (those bearing leaves only) abruptly narrowed with curved margins below the middle; most of leaves of sterile branches in whorls; calyx 4-parted, not joined to the ovary; pistils 4 . . . . . Sedum, in CRASSULACEAE, p. 770

w. Whorled leaves gradually narrowed with more or less straight margins to the base and below the middle; calyx 2-parted, partly joined to the ovary; pistil 1 . . . . . Portulaca, in PORTULACACEAE, p. 633

v. Without the above combination of characters . . . . . *x*

x. Only 2 leaves (a pair opposite each other) on the same stem (does not include plants which have only 2 leaves at the base of the stem) . . . . . *y*

y. Sepals 2; petals 5; stamens 5; leaves not glaucous (with a silvery-green coating which can be rubbed off) nor brown-mottled, usually 0.2–1 cm. (sometimes up to 3.3) broad Claytonia, in PORTULACACEAE, p. 636

y. Sepals 3, petal-like; petals 3; stamens 6; leaves glaucous or brown-mottled, usually 1.3–5 cm. (sometimes only 0.5) broad . . . . . Erythronium, in LILIACEAE, p. 433

x. Four or more leaves (2 pairs or more) present on the stem . . . . . *z*

z. A white woolly or cottony covering surrounds and somewhat conceals the flowers . . . . *1*

1. Leaves without petioles (leaf-stalks) and gradually narrowed to the base, linear, narrowly lanceolate, or elliptic-lanceolate; flowers perfect with stamens and pistils in the same flower; fruiting calyx with wing-like toothed crests or spiny projections Froelichia, in AMARANTHACEAE, p. 626

1. Leaves with petioles, rounded or heart-shaped at base, elliptic, oval, or lanceolate-oblong; flowers monoecious, the stamens and pistils in separate flowers on the same plant; fruiting calyx merely 5–12-parted, without crests or spiny projections . . . Croton, in EUPHORBIACEAE, p. 976

z. No white woolly or cottony covering surrounds or conceals the flowers . . . . . *2*

2. Stems and lower or upper surface of leaves covered with a dense silvery-green hairiness made up of small stellate (star-shaped) hairs . . . . . *3*

3. Leaf-blades nearly round to broadly ovate; flowers perfect with stamens and pistils in the same flower; style 1 . . . . . Tidestromia, in AMARANTHACEAE, p. 626

3. Leaf-blades oblong to linear-lanceolate; flowers with the stamens and pistils in separate flowers on the same plant (monoecious) or on different plants (dioecious) EUPHORBIACEAE, p. 973

2. Stems and leaves lacking stellate (star-shaped) hairiness . . . . . *4*

4. Both calyx and corolla absent at time of flowering; dwarf plants up to 3.5 cm. tall growing on damp soil; leaves 2–4 mm. long . . . . . CALLITRICHACEAE, p. 996

4. Without the above combination of characters; at least the calyx present at flowering time; short or tall plants with short or long leaves . . . . . *5*

5. Leaves with either black or transparent dots (hold leaf to the light to note) . . . . *6*

6. Stamens 5–20 or more, usually in clusters or groups; styles 2–5; petals distinct and separate, not united at their base, so that each one can be removed individually without disturbing or tearing the others . . . HYPERICACEAE, p. 1054

6. Stamens 5 or 6, not clustered or grouped; style 1; petals slightly united at the very base so that an attempt to remove one tears or disturbs a neighboring section of the corolla . . . . . PRIMULACEAE, p. 1164

5. Leaves lacking transparent or black dots . . . . . *7*

7. Calyx composed of 2 united sepals; stamens 7–20; petals 5–6, not united, quickly falling . . . . . Portulaca, in PORTULACACEAE, p. 633

7. Without the above combination of characters . . . . . *8*

8. The complete inflorescence of spiked or branching panicle of white or silvery-white, very small flowers, each with 5 distinct sepals and bracts; flowers unisexual, with the stamens and pistils in separate flowers on different plants . . . . . Iresine, in AMARANTHACEAE, p. 626

8. Without the above combination of characters . . . . . *9*

9. Green sepals or colored petal-like sepals distinct, not united at the base, so that one can be removed without tearing or disturbing the others . . . . . . . . . . . . . . . . . . . . *10*
10. Petals absent at flowering time, the sepals sometimes conspicuously colored purple and petal-like . *11*
11. Sepals petal-like, purple, 1.7–2.5 cm. long; leaves large, ovate, or elliptic-lanceolate, 2.5–8.5 cm. broad; stamens numerous, more than 20 . . . . . Clematis, in RANUNCULACEAE, p. 705
11. Sepals small, green, not more than 0.5 cm. long; leaves narrow, hair-like to oblanceolate or oval, at most 1 cm. broad; stamens 2–10. . . . . . . . . . . CARYOPHYLLACEAE, p. 637
10. Petals present at flowering time . . . . . . . . . . . . . . . . . . . *12*
12. Petals united into a short or long tube, so that an attempt to remove a portion of the corolla results in tearing or disturbing a neighboring section . . . . . SCROPHULARIACEAE, p. 1328
12. Petals all separate and distinct to their base, so that one petal may be removed without tearing the others . . . . . . . . . . . . . . . . . . . . . *13*
13. Thin membrane-like stipules present between the leaves; sepals 3; petals 3; stamens 3 ELATINACEAE, p. 1064
13. No stipules present at base of or between leaves; sepals 4–5; petals 4–5 or rarely 1–3; stamens 3–10 . . . . . . . . . . . . . . . . . . . . *14*
14. Petals yellow; ovary and capsule 5-celled; stamens 5 . . . . . . LINACEAE, p. 957
14. Petals white, reddish, or greenish; ovary and capsule usually 1- or 3-celled, rarely 4–6-celled; stamens 3–25 . . . . . . . . . . . . . . . . *15*
15. Flowers minute, the 3 reddish petals shorter than the 5 unequal sepals; stamens 3–25; upper leaves alternate . . . . . . . . Lechea, in CISTACEAE, p. 1067
15. Flowers showy or usually obvious, the usual 4–5 petals (sometimes 1–3) usually equaling or longer (sometimes shorter) than the usual 4–5 equal sepals; stamens usually 8–10, sometimes 3–5; all leaves opposite . CARYOPHYLLACEAE, p. 637
9. Sepals more or less united at the base into a scarcely noticeable short tube or into a long tube, either green or brightly colored like a corolla, or fused with the ovary, the calyx-teeth sometimes inconspicuous or even absent . . . . . . . . . . . . . . . . . . . . *16*
16. Flowers without a corolla at flowering time but with a small calyx 1–6 mm. long (including calyx-tube and lobes) which may be colored other than green . . . . . . . . . . *17*
17. Lower surface of leaf-blade silvery or white with minute black or brown dots . . . . Boerhaavia, in NYCTAGINACEAE, p. 630
17. Lower surface of leaf-blade not silvery or white nor with black or brown dots . . . . . *18*
18. Calyx-lobes 4; ovary wholly inferior, the calyx-tube completely fused to the wall of the ovary, the calyx-lobes arising from the summit of the ovary . . . . . . Ludwigia, in ONAGRACEAE, p. 1095
18. Calyx-lobes 5; ovary superior to partly inferior, the calyx-tube wholly or partly free from the ovary . . . . . . . . . . . . . . . . . . . . *19*
19. Leaf-blades 1–3 mm. broad, thread-like, narrowly linear or oblong; calyx greenish, not petal-like . . . . . . . . . . . . . . . CARYOPHYLLACEAE, p. 637
19. Leaf-blades 7–50 mm. broad, obovate to suborbicular; calyx pink or purplish within, petal-like . . . . . . . . . . . Trianthema, in AIZOACEAE, p. 632
16. Flowers with a corolla at flowering time, or if without a corolla, then the calyx is corolla-like, pink or purple and 10–13 mm. long . . . . . . . . . . . . . . . . . *20*
20. Flowers with a pink or purple broadly bell-shaped, short-tubed corolla-like calyx, surrounded at base by a green 5-lobed calyx-lobe involucre, 1 or more flowers in each involucre; ovary with 1 cell and 1 ovule, surrounded by the base of the corolla-like calyx (considerable caution is required in examining these flowers not to confuse the corolla-like calyx with a true corolla) . . . . . . . . . . . . . . . Mirabilis, in NYCTAGINACEAE, p. 627
20. Flowers with a true corolla and calyx, sometimes with bracts at the base of the flowers . . . *21*
21. Corolla of separate petals distinct to their base, so that one may be removed individually without tearing or disturbing the others. . . . . . . . . . . . . . . *22*
22. Style 1; ovary usually 2–6-celled; stamens inserted on the calyx, originating near or above the middle of the calyx-tube . . . . . . . . . LYTHRACEAE, p. 1087
22. Styles 2–5; ovary usually 1-celled; stamens originating on the receptacle, attached at the very base of the calyx-tube and petals . . . . . . CARYOPHYLLACEAE, p. 637
21. Corolla of united petals joined into a short or long tube, so that an attempt to remove one petal-like corolla-lobe tears or disturbs a neighboring section . . . . . . . *23*
23. Stamens 3; calyx-teeth or lobes or sepals absent or practically so at flowering time (anthesis) . . . . . . . . . . . . . . . . . VALERIANACEAE, p. 1419

23. Stamens 2, 4, or 5; calyx-teeth or lobes or sepals present or well-developed at flowering time, if nearly absent, then the leaves evergreen and the stems creeping or lying upon the ground or substratum . . . . . . . . . . . . . . . . . . . . . . . . . *24*

24. Small or prominent stipules or a stipular thin membrane present between each of the opposite leaves at least on the uppermost or youngest parts of the stem . . . . . . . . . . *25*

25. Ovary inferior, partly or wholly united with the calyx and inserted below the base of the corolla-tube . . . . . . . . . . . . . . . . . . . . . . RUBIACEAE, p. 1386

25. Ovary superior, not united with and entirely free from the calyx, the base of the ovary inserted above the base of the corolla-tube . . . . . . . . . . LOGANIACEAE, p. 1183

24. No stipules or stipular membrane present between nor at the base of the leaves. . . . . . *26*

26. Ovary inferior, wholly united with the calyx-tube and inserted below the base of the corolla-tube . . . . . . . . . . . . . . Triosteum, in CAPRIFOLIACEAE, p. 1410

26. Ovary superior, not united with and entirely free from the calyx, the base of the ovary inserted above the base of the corolla-tube . . . . . . . . . . . . . . . . . *27*

27. Corolla regular (actinomorphic), the corolla-lobes of more or less equal size and shape . *28*

28. Calyx very irregular, some of the calyx-lobes or teeth unequal in size and shape . *29*

29. Ovary 4-lobed, appearing almost like 4 separate ovaries containing 4 ovules; fruit with 4 seeds . . . . . . . . . . . . . . . . . LABIATAE, p. 1265

29. Ovary not 4-lobed, containing many ovules; fruit with many seeds . . . Gerardia auriculata, in SCROPHULARIACEAE, p. 1361

28. Calyx more or less regular, all the lobes or teeth of more or less equal size and shape . *30*

30. Corolla yellow or yellow with purple-brown at the base within . . . . . *31*

31. Corolla deeply parted nearly to its base, the corolla-tube scarcely developed and much shorter than the conspicuous corolla-lobes; stamens opposite the corolla-lobes . . . . . Lysimachia, in PRIMULACEAE, p. 1168

31. Corolla shallowly lobed, the corolla-tube or united part of the corolla much longer than the inconspicuous corolla-lobes; stamens alternate with the corolla-lobes . . . . . . . . . . Physalis, in SOLANACEAE, p. 1314

30. Corolla of colors other than yellow . . . . . . . . . . . . . . *32*

32. Corolla deeply parted nearly to its base, with scarcely any tube developed; stamens opposite the corolla-lobes and of the same number as the lobes . Anagallis, in PRIMULACEAE, p. 1171

32. Corolla-tube or united portion either evidently developed or conspicuous, either longer than or somewhat shorter than the corolla-lobes; stamens alternate with the corolla-lobes, either fewer than the lobes or of the same number as the lobes. . . . . . . . . . . . . . . . *33*

33. Stamens of the same number as the lobes of the corolla . . . . . *34*

34. Tops of the stamens occupying different heights in the corolla, the filaments unequally inserted on the inside of the corolla-tube; ovary and fruit 3-celled; ovules 2–3 in each cell of the ovary; seeds usually 1 in each of the 3 cells of the fruit . Phlox in POLEMONIACEAE, p. 1225

34. Tops of the stamens at the same height in the corolla, equally inserted on the inside of the corolla-tube; ovary and fruit 1-celled; ovules and seeds numerous . . . . . . . GENTIANACEAE, p. 1184

33. Stamens fewer than the lobes of the corolla . . . . . . . . *35*

35. Ovary 4-lobed; fruit consisting of 4 seed-like nutlets or achenes; leaves with transparent dots or glands (hold leaf-blade against light); plant with a slightly minty odor . Isanthus, in LABIATAE, p. 1267

35. Ovary 2-celled; fruit 2-celled with a total of usually 6–20 or more (rarely 4) seeds; leaves without any transparent dots or glands (hold leaf against light); plant without any minty odor . . . . *36*

36. Leaves narrowly linear, or thread- or hair-like, usually 1–6 (rarely 8) mm. broad . . Gerardia, in SCROPHULARIACEAE, p. 1357

36. Leaves lanceolate, oblong, or ovate, 8–90 mm. broad . . . *37*

37. Corolla 20–23 mm. long, the narrow part of the tube shorter than the lobes; calyx-lobes 10–30 mm. long; ovary with numerous ovules; seeds numerous in a fruit; leaves without thickened places (cystoliths) on the surface. . Gerardia auriculata, in SCROPHULARIACEAE, p. 1361

37. Corolla mainly 25–60 mm. long (rarely 12–25 mm. long in a form of *Ruellia strepens* with corolla more or less closed), narrow part of the tube much longer than the lobes; calyx-lobes 9–12 mm. long; ovary with 2–12 ovules; seeds 2–12 in a fruit; leaves with thickened places (cystoliths) on the surface . . . . . . . . . . . . . . . . Ruellia, in ACANTHACEAE, p. 1375

27. Corolla irregular (zygomorphic), the corolla-lobes of unequal size and shape . . . . . . . . *38*

38. Stamens with 2 fertile (pollen-bearing) anthers. . . . . . . . . . . . . . . . . . . . *39*

39. Ovary 4-lobed; fruit consisting of 4 seed-like nutlets or achenes . . . . . . LABIATAE, p. 1265

39. Ovary 2-celled; fruit 2-celled with a total of 4 or many seeds . . . . . . . . . . . *40*

40. Leaves 0.5–3 cm. long, rounded, blunt, or slightly pointed at tip; corolla 5–10 mm. long; ovary with numerous ovules; seeds numerous in a fruit . . . . . . . . Lindernia, in SCROPHULARIACEAE, p. 1336

40. Leaves 2–20 cm. long, short- to long-pointed at tip; corolla 12–18 mm. long; ovary with 4–20 ovules; seeds 4–20 in a fruit. . . . . . . . . . . ACANTHACEAE, p. 1375

38. Stamens with 4 fertile (pollen-bearing) anthers. . . . . . . . . . . . . . . . . . . . *41*

41. Ovary 4-lobed; ovule 1 in each of the 4 lobes; fruit consisting of 4 seed-like nutlets or achenes. . . . . . . . . . . . . . . . . . . . . . . . . LABIATAE, p. 1265

41. Ovary 1- or 2-celled, or falsely 4-celled; ovules several to many in each cell; fruit with several to many seeds in each cell . . . . . . . . . . . . . . . . . . . . . *42*

42. At least the leaves of the upper half of the stem without stalks (sessile); commonly encountered; placentae (where the ovules or seeds are attached) of the ovary or fruit axile (situated in the center) . . . . . . . . . . . SCROPHULARIACEAE, p. 1328

42. Leaves of the upper half of the stem with stalks (petioles); rarely encountered; placentae (where the ovules or seeds are attached) of the ovary or fruit parietal (situated on the wall) . . . . . . . . . . . . . . . . . . . . . . . . . . . . . . *43*

43. Flowers and fruits borne next to the main stem at the base of the leaves (axillary); leaf-blades lanceolate or oblong-ovate; corolla 2–2.5 cm. long; fruit erect, ellipsoid, without any curved beaks, 2.5–3 cm. long. . . . . . . . . PEDALIACEAE, p. 1371

43. Flowers and fruits borne at the top of the stem and branches (terminal); leaf-blades somewhat round and heart-shaped; corolla 3–5 cm. long; fruit with 2 spreading, curved horned beaks, 8–15 cm. long . . . . . . . . . MARTYNIACEAE, p. 1372

### Section 11. **Herbaceous or nonwoody plants with opposite or whorled, simple toothed or lobed leaves**

a. Plants without true flowers or seeds, reproducing by green or brown spores (without embryos) contained in sporangia and located on the lower portions of the leaf (frond) or in specially modified leaves (fronds), stalks, or branches . . . . . . . . . . . . . FERNS AND FERN ALLIES, p. 2

a. Plants with true flowers bearing stamens or carpels (pistils) or both, reproducing by seeds containing embryos . . . . . . . . . . . . . . . . . . . . . . . . . . . . . . . *b*

b. What appears to be one flower actually is a dense mass of small flowers surrounded by green or colored scales or bracts of the involucre; anthers united into a tube or rarely scarcely united; petals united into a tube; ovary inferior; ovule and seed 1 in each of the individual flowers. COMPOSITAE, p. 1442

b. Without the above combination of characters . . . . . . . . . . . . . . . . . . . . *c*

c. Stems prickly; flower-heads bristly with long needle-pointed bracts . . . . DIPSACACEAE, p. 1422

c. Stems not prickly; no needle-pointed bracts present . . . . . . . . . . . . . . . . *d*

d. Bruised stem, leaf, or leaf-stalk producing a milky sap; fruit deeply 3-lobed; flowers produced within the base of a cup-shaped cyathium, the margin or summit of which bears 4 or 5 glands with or without colored appendages . . . . . . . . Euphorbia, in EUPHORBIACEAE, p. 984

d. No milky sap produced by bruised stem, leaf, or leaf-stalk; without the above combination of characters . . . . . . . . . . . . . . . . . . . . . . . . . . . *e*

e. Leaves in whorls (circles) of 3 or more at a level on the stem, never opposite nor in 2's . . *f*

f. Plants occurring in ground recently under water; leaves finely cut into thread-like segments or finely toothed; flowers greenish, inconspicuous, monoecious, the stamens and pistils in separate flowers on the same plant; stamens 4 or 8 . . . . . . Myriophyllum, in HALORAGIDACEAE, p. 1108

f. Plants occurring in normally dry ground or in ground not recently under water; leaves coarsely toothed, coarsely lobed, or the lobes variously toothed, smooth-edged, or with rounded teeth; flowers showy, usually lavender, purple, rose, pink, or white, rarely greenish, perfect, the stamens and pistils in the same flower; stamens 6 or many . . . . . *g*

g. Stamens 6; petals 4, lavender or purple; sepals 4 . . . . . . Dentaria, in CRUCIFERAE, p. 751

g. Stamens numerous; true petals none; colored petal-like sepals white, pink, rose-colored, or rarely green, 5–10 . . . . . . . . . . . . . . . . . . . RANUNCULACEAE, p. 672

e. Leaves appearing in pairs opposite each other, 2 at a level on the stem . . . . . . . . . . *h*

h. Only 2 leaves (a pair) on the stem . . . . . . . . . . . . . . . . . . . . . . . *i*

i. Leaves not lobed as far as the middle; petals deeply cut and fringed; calyx attached to the base of the ovary . . . . . . . . . . . . . . . . Mitella, in SAXIFRAGACEAE, p. 781

i. Leaves lobed beyond the middle and sometimes nearly to their base; petals without teeth or irregularities along margins; calyx completely free from the base of the ovary . . . . . . . *j*

j. Stem and leaves glabrous (without hairs); leaves peltate, the petiole (leaf-stalk) joining the lower surface of the leaf-blade near its center; petals 6 or 9 . . . . . . . . Podophyllum, in BERBERIDACEAE, p. 710

j. Stem or leaves hairy; leaves not peltate, petiole (leaf-stalk) joining directly to the lower end or base of the leaf-blade; petals or petal-like sepals, when petals are absent, 5–10. . . . *k*

k. Petals lavender, pink, or rose-colored; stamens 10 or rarely 5; pistil of each flower consisting of 5 carpels, the styles 5; fruits not woolly . . . . . . . GERANIACEAE, p. 962

k. Petals white or cream-colored; stamens numerous, more than 10; pistils many in each flower, more than 5, 1 style to each pistil; fruiting head and fruits (achenes) densely woolly . . . . . . . . . . . . . . Anemone, in RANUNCULACEAE, p. 699

h. Four or more leaves (2 pairs or more) on the stem . . . . . . . . . . . . . . . . *l*

l. Stems vining, climbing, or long-trailing and rooting at the nodes . . . . . . . . . . *m*

m. Leaves usually with 3–7 lobes; stipules present and persisting at base of leaf-stalks . . Humulus, in MORACEAE, p. 566

m. Leaves merely toothed; stipules absent . . . . . . . . . . . . . . . . . . . . *n*

n. Leaf-blades 2–6 cm. long; corolla of separate petals . Euonymus, in CELASTRACEAE, p. 1008

n. Leaf-blades 1.5 cm. or less long; corolla of united petals with a tube. . . . . . Veronica, in SCROPHULARIACEAE, p. 1352

l. Stems upright or spreading, but not vining, climbing, nor trailing . . . . . . . . . . *o*

o. Stamens many, more than 10; carpels (pistils) many in each flower, more than 5, entirely free from one another and spirally arranged, with the tips of some higher than those of others . . . . . . . . . . . . . Clematis Fremontii, in RANUNCULACEAE, p. 708

o. Stamens, when present, 10 or less; carpels (pistils), when present, 5 or less, free or united, all on the same level . . . . . . . . . . . . . . . . . . . . . . . . *p*

p. Flowers greenish, without a corolla at flowering time, the stamens and pistils in separate flowers on the same (monoecious) or different (dioecious) plants . . . . . . . . . . *q*

q. Stipules absent at base of leaf-stalks; plants scurfy with very small whitish or grayish scales; pistillate (female) flowers without a calyx . Atriplex, in CHENOPODIACEAE, p. 614

q. Stipules (small appendages or outgrowths) present at base of leaf-stalks; plants lacking a scurfy or scaly covering; pistillate (female) flowers with a tubular or 3–4-parted calyx . . . . . . . . . . . . . . . . . . . URTICACEAE, p. 567

p. Flowers of various colors other than green, with a corolla of free or united petals, the stamens and pistils in the same flower (flowers perfect) . . . . . . . . . . . . *r*

r. Corolla of separate petals which are not connected at their base, so that if one petal is removed from the flower, none of the others are torn . . . . . . . . . . . *s*

s. Leaves fleshy, thick, succulent; carpels 4 or 5, distinct from each other . . . Sedum, in CRASSULACEAE, p. 770

s. Leaves thin and membranous, or if firm, then not thick, fleshy, nor succulent; carpels united at flowering time into 1 pistil . . . . . . . . . . . . . . *t*

t. Stipules (membranous, thin, or small appendages) present at base of leaf-stalks . . *u*

u. Leaf-blades palmately lobed, the lobes all arising at the base of the leaf-blade; main nerves of leaf-blade all arising from the same point at base of leaf-blade; flowers and fruits on elongated flower-stalks . . . GERANIACEAE, p. 962

u. Leaf-blades shortly toothed, not lobed; main side (lateral) nerves of leaf-blade arising at various levels at the side of the main midnerve, one above another; flowers and fruits nearly stalkless in the axils of the leaves. . . Bergia, in ELATINACEAE, p. 1064

t. No stipules present at base of leaf-stalks . . . . . . . . . . . . . . . *v*

v. Petals 1.1–2.8 cm. long, deep pink, rose, or rose-lavender; leaf-blades with 3–4 conspicuous nerves arising at base of blade; stems and/or calyx-tube usually with some spreading hairs present and conspicuous . . . . . . . . . . . . . . . . . . . MELASTOMACEAE, p. 1092

v. Petals 0.2–0.65 cm. long, white or pink; leaf-blades with lateral (side) nerves joining the midrib at different levels above the base of blade; stems and calyx without conspicuous hairs . ONAGRACEAE, p. 1093

r. Corolla of united petals, united into a short or long tube and lobes, so that no one of the petal-like lobes can be removed without tearing or disturbing the rest of the corolla . . . . . . . . . . *w*

w. Some part of stem or lower surface of leaves covered with star-shaped hairs (stellate pubescence) Callicarpa, in VERBENACEAE, p. 1262

w. Stem or leaves glabrous or hairy, but without star-shaped (stellate) hairs . . . . . . . . . . *x*

x. Stamens 3; calyx-teeth or lobes or sepals absent or practically so at flowering time (anthesis); ovary inferior, situated below the base of the corolla-tube. . . . . . VALERIANACEAE, p. 1419

x. Stamens 2 or 4, rarely 5; calyx-teeth or lobes or sepals present or well-developed at flowering time (anthesis); ovary superior, the base of the ovary situated above the insertion of the corolla-tube . . . . . . . . . . . . . . . . . . . . . . . . . *y*

y. Stamens 2 . . . . . . . . . . . . . . . . . . . . . . . . . *z*

z. Ovary 4-lobed; ovules 1 in each of the 4 lobes; fruit consisting of 4 seed-like nutlets or achenes; stem often 4-angled; leaves often with a minty odor and often with transparent dots or glands (hold leaf against light) . . . . . . . . . . . LABIATAE, p. 1265

z. Ovary 2-celled; ovules many in each cell; fruit many-seeded; stem sometimes 4-angled; leaves without a minty odor and without transparent dots or glands (hold leaf against light). . . . . . . . . . . . . . . . . SCROPHULARIACEAE, p. 1328

y. Stamens 4, rarely 5 . . . . . . . . . . . . . . . . . . . . . *1*

1. Ovary 4-lobed, the style arising between the lobes; leaves often with a minty odor and often with transparent dots or glands (hold leaf against light). . . . . . LABIATAE, p. 1265

1. Ovary 1- or 2-celled, or falsely 4-celled, but not 4-lobed, the style arising from the summit of the ovary; leaves without a minty odor and without transparent dots or glands (hold leaf against light) . . . . . . . . . . . . . . . . . . *2*

2. Only 1 ovule in each of the 1–4 cells of the ovary . . . . . . . . . . . . . *3*

3. Stem usually acutely 4-angled; calyx-teeth usually not hooked at tip; flowers not reflexed (turned down) nor becoming appressed to the stem (paralleling the length) in fruit; ovary 2–4-celled; fruit with 2 or 4 nutlets . . . VERBENACEAE, p. 1257

3. Stem faintly or bluntly 4-angled with the angles curved or rounded; calyx-teeth hooked at tip; flowers reflexed (turned down) and becoming appressed to the stem (paralleling the length) in fruit; ovary only 1-celled; fruit with 1 seed . . PHRYMACEAE, p. 1380

2. Several to many ovules in each cell of the ovary . . . . . . . . . . . . *4*

4. Placentae (where the ovules or seeds are attached) of the ovary or fruit axile (situated in the center); commonly encountered . . . . SCROPHULARIACEAE, p. 1328

4. Placentae (where the ovules or seeds are attached) of the ovary or fruit parietal (situated on the wall); rarely encountered . . . . . . . . . . . . . . *5*

5. Flowers and fruits borne next to the main stem at the base of the leaves (axillary); leaf-blades lanceolate or oblong-ovate; corolla 2–2.5 cm. long; fruit erect, ellipsoid, without any curved beaks, 2.5–3 cm. long . . . . PEDALIACEAE, p. 1371

5. Flowers and fruits borne at the top of the stem and branches (terminal); leaf-blades somewhat round and heart-shaped; corolla 3–5 cm. long; fruit with 2 spreading, curved, horned beaks, 8–15 cm. long . . . . MARTYNIACEAE, p. 1372

## Section 12. **Herbaceous or nonwoody plants with all the leaves simple and arising only at the base of the plant**

a. Plants without true flowers or seeds, reproducing by green or brown spores (without embryos) contained in sporangia and located on the lower portions of the leaf (frond) or in specially modified leaves (fronds), stalks, or branches . . . . . . . . . . . . . FERNS AND FERN ALLIES, p. 2

a. Plants with true flowers bearing stamens or carpels (pistils) or both, reproducing by seeds containing embryos . . . . . . . . . . . . . . . . . . . . . . . . . *b*

b. What appears to be one flower actually is a dense mass of small flowers surrounded by green or colored scales or bracts of the involucre; anthers united into a tube or rarely scarcely united;

petals united into a tube; ovary inferior; ovule and seed 1 in each of the individual flowers . . COMPOSITAE, p. 1442

b. Without the above combination of characters . . . . . . . . . . . . . . . . . c
c. Leaves without teeth or lobes projecting from the margins . . . . . . . . . . . . . d
d. Leaves grass- or ribbon-like, long and slender or narrowly linear with parallel or nearly parallel sides . . . . . . . . . . . . . . . . . . . . . . . . . . e
e. Leaves or stem or both more or less hairy . . . . . . . . . . . . . . . . . f
f. Flowers purple, blue, rose, pink, white, or yellow, the color produced in a conspicuous or developed perianth or corolla . . . . . . . . . . . . . . . . . . . . g
g. Sepals green; petals equal in size and shape, purple, blue, rose, or pink; pollen-bearing stamens 6 or 5 . . . . . . . . . . . Tradescantia, in COMMELINACEAE, p. 392
g. No green sepals present; one of the petals of a different size and shape from any of the others, the perianth white or cream-colored; pollen-bearing stamen 1 . . . Spiranthes, in ORCHIDACEAE, p. 479
f. Flowers of tones of brown, buff, greenish, or straw color, or, if of other colors, the perianth, corolla, or calyx dry and scale-like or thin and transparent, or perianth absent or reduced to bristles or scales with the flowers then occurring in the axils of imbricated (overlapping) dry scales forming spikes or spikelets . . . . . . . . . . . . . . . . . . h
h. Perianth consisting of 3 sepals and 3 similar petals together with 3 or 6 stamens or of 4 sepals, 4 corolla-lobes, and 4 stamens . . . . . . . . . . . . . . . i
i. Flowers usually arranged in a longer than broad spike on an elongated axis; sepals 4; corolla-lobes 4; stamens 4 . . . . . . . . . . PLANTAGINACEAE, p. 1380
i. Flowers in small, stalked cylindrical heads or clusters; sepals 3; petals similar, 3; stamens 6 or 3 . . . . . . . . . . . . . Luzula, in JUNCACEAE, p. 416
h. Perianth absent or reduced to bristles or scales, the flowers in the axils of imbricated (overlapping) dry scales forming spikes or spikelets . . . . . . . . . . . j
j. Stems usually rounded (terete) or flattened, usually hollow or with the pith easily removable, but the nodes (joints) solid, closed, hard, and prominent; sheath at base of leaf usually split open or overlapping on one side; anthers versatile (attached near their middle and swinging free); seed of the fruit not separable or free from the thin wall (pericarp) of the fruit; perianth absent or apparently so, part of the flowers consisting of an outer lemma, an inner palea, and sometimes 2 minute lodicules, the whole subtended at the base by 1 or 2 scale-like bracts, the glumes, and forming a spikelet; a hairy or membranous projection, the ligule, usually present at the inner junction of leaf-sheath and leaf-blade . . GRAMINEAE, p. 68
j. Stems often 3-angled, but also rounded, flattened, or 4- or more-angled, usually solid with pith, sometimes hollow, but the nodes (joints) soft and not conspicuous; sheaths at base of leaf usually closed and not split; anthers attached by their base (basifixed) to the filament; seed of the fruit free from and not attached to the wall (pericarp) of the fruit; perianth absent or consisting of bristles or scales, the flowers each in the axil of a single scale or glume, these overlapping into spirally arranged or 2-rowed spikelets or spikes; ligule present or absent . . . . CYPERACEAE, p. 255
e. Leaves and stem glabrous (without hairs) . . . . . . . . . . . . . . . . k
k. Bruised rootstocks and leaves sweet-smelling or spicy-fragrant; flowers greenish-yellow, crowded on a thick finger-like column (spadix) arising from the side of the long, green leaf-like spathe . . . . . . . . . . . . . . . Acorus, in ARACEAE, p. 386
k. Bruised rootstocks and leaves not sweet-smelling; flowers other than described above. . . l
l. Flowers greenish-yellow, each flower with numerous pistils arranged up and down the length of an elongated, tail-like receptacle, at the base of which are 10–18 stamens, 5 petals, and 5 sepals . . . . . . . . . . Myosurus, in RANUNCULACEAE, p. 699
l. Flowers of other colors or without the above combination of characters . . . . . . m
m. Flowers of tones of brown, buff, greenish, or straw color, or, if of other colors, the perianth, corolla, or calyx dry and scale-like or thin and transparent, or perianth absent or reduced to bristles or scales, with the flowers then occurring in the axils of imbricated (overlapping) dry scales forming spikes or spikelets . . . . . . . n
n. Perianth consisting of 3 sepals and 3 similar petals together with 3 or 6 stamens or of 4 sepals, 4 corolla-lobes, and 4 stamens . . . . . . . . . . . . . o
o. Flowers usually arranged in a longer than broad spike on an elongated axis; sepals 4; corolla-lobes 4; stamens 4 . . . . . . . . PLANTAGINACEAE, p. 1380

o. Flowers in stalked spherical, hemispherical, or cylindrical heads, or occurring singly or in small groups in a branched inflorescence; sepals 3; petals similar, 3; stamens 6 or 3 . . . Juncus, in JUNCACEAE, p. 404

n. Perianth absent or reduced to bristles or scales, the flowers in the axils of imbricated (overlapping) dry scales forming spikes or spikelets . . . . . . . . . . . . . . . . . . . *p*

p. Stems usually rounded (terete) or flattened, usually hollow or with the pith easily removable, but the nodes (joints) solid, closed, hard, and prominent; sheath at base of leaf usually split open or overlapping on one side; anthers versatile (attached near their middle and swinging free); seed of the fruit not separable or free from the thin wall (pericarp) of the fruit; perianth absent or apparently so, part of the flowers consisting of an outer lemma, an inner palea, and sometimes 2 minute lodicules, the whole subtended at the base by 1 or 2 scale-like bracts, the glumes, and forming a spikelet; a hairy or membranous projection, the ligule, usually present at the inner junction of leaf-sheath and leaf-blade . GRAMINEAE, p. 68

p. Stems often 3-angled, but also rounded, flattened, or 4- or more angled, usually solid with pith, sometimes hollow, but the nodes (joints) soft and not conspicuous; sheaths at base of leaf usually closed and not split; anthers attached by their base (basifixed) to the filament; seed of the fruit free from and not attached to the wall (pericarp) of the fruit; perianth absent or consisting of bristles or scales, the flowers each in the axil of a single scale or glume, these overlapping into spirally arranged or 2-rowed spikelets or spikes; ligule present or absent . . . . . . . . . . . . . . . . . . CYPERACEAE, p. 255

m. Flowers purple, blue, lavender, rose, pink, white, or yellow, but the color produced in a conspicuous or developed perianth or corolla . . . . . . . . . . . . . . . . *q*

q. Flowers irregular, one of the petals larger and of different shape and size from the others; fertile stamen only 1 . . . . . . . . . . . . . . . . . . ORCHIDACEAE, p. 468

q. Flowers regular, all the petals or parts of each series of equal size and shape; stamens 3 or more . . *r*

r. Petals 5; sepals 2, sometimes not persisting on the flower . . . . . PORTULACACEAE, p. 633

r. Petals or petal-like lobes or divisions 3 or 6 . . . . . . . . . . . . . . . . . . *s*

s. Stamens 9 or 18 or more; carpels in each flower numerous, distinct or not united with each other . . . . . . . . . . . . . . . . . . ALISMACEAE, p. 57

s. Stamens 6 or 3; carpels in each flower 3, united into a compound 3-celled ovary . . . . *t*

t. Stamens 3 . . . . . . . . . . . . . . . . . . . . . . *u*

u. Rarely encountered; perianth divided into an outer greenish calyx and an inner yellow corolla; ovary superior, the petals attached below the base of the ovary . XYRIDACEAE, p. 392

u. Commonly encountered; perianth of similarly colored purple, lavender, blue, white, or rarely yellow divisions, and lacking a distinct green calyx; ovary inferior, the petal-like divisions attached to the top of the ovary or fruit. . . IRIDACEAE, p. 458

t. Stamens 6 . . . . . . . . . . . . . . . . . . . . . . *v*

v. Petal-like parts attached below the base of the ovary or fruit, the ovary therefore superior . . . . . . . . . . . . . . . . . . LILIACEAE, p. 418

v. Petal-like parts attached to top of ovary or fruit, the ovary therefore inferior. . AMARYLLIDACEAE, p. 455

d. Leaves not grass- or ribbon-like, but obviously broader in some parts of the leaf, not linear, the sides not parallel . . . . . . . . . . . . . . . . . . . . . *w*

w. Leaves peltate, the petiole (leaf-stalk) joining the lower surface of the leaf-blade near its center Nelumbo, in NYMPHAEACEAE, p. 668

w. Leaves not peltate, the petiole (leaf-stalk) joining directly to the lower end or base of the leaf-blade . . *x*

x. Flower more or less hidden from plain view, occurring close to or by the ground in the woodland, brownish-purplish with 3 pointed lobes; stamens 12; leaves broadly lima-bean or kidney-shaped; underground rootstock with an aromatic or spicy odor. Asarum, in ARISTOLOCHIACEAE, p. 572

x. Flowers raised above the ground in plain view; without the above combination of characters . . *y*

y. Sepals 4, dry and thinnish; corolla of 4 united petals, dry and transparent, brownish-white or greenish; stamens 4; flowers arranged in a narrow cylindrical spike much longer than broad or in a dense head; fruiting capsule circumscissile, an upper half splitting away from a lower half by a circular transverse line . . . . . . . . . PLANTAGINACEAE, p. 1380

y. Without the above combination of characters . . . . . . . . . . . . . *z*

z. Carpels (pistils) numerous, distinct or not united with each other; stamens 9–20 or more; inflorescence with often 2 or more whorls of branches, sometimes with only 1 whorl. . . . . . . . . . . . . . . . . . ALISMACEAE, p. 5

z. Carpels (pistils) 2–3, rarely 4–6, usually united into 2–3-celled ovary, rarely the carpels only slightly united; stamens 1, 3, 4, 5, 6, 8, or 10; inflorescence various, sometimes composed of a single whorl . . . . . . . . . . . . . . . . . . . . . . . . *1*

1. Stamens 3 . . . . . . . . . . . . . . . . . . . . . . . . *2*
   2. Leaves sword-shaped or broadly lanceolate, conspicuously pointed at tip; individual flower large, the tube of the flower 4.5–7 cm. long; ovary inferior with the tube of the perianth attached to the top of the ovary . . . . . . . . . . . . Iris, in IRIDACEAE, p. 460
   2. Leaves round, heart-shaped, oblong, or lance-oblong, blunt or somewhat pointed at tip; individual flower small, the tube of the perianth 0.7–2 cm. long; ovary superior, the tube of the perianth attached below the base of the ovary. . . . . . . . . . . . Heteranthera, in PONTEDERIACEAE, p. 402
1. Stamens 1, 4, 5, 6, 8, or 10 . . . . . . . . . . . . . . . . . . . . . . *3*
   3. Stamens 6 . . . . . . . . . . . . . . . . . . . . . . . . *4*
      4. Leaves very thick, fleshy, and succulent, 12–30 cm. long, mainly 3–7 cm. broad; flowers greenish-yellow, with the perianth-divisions united for more than ⅔ their length into a funnelform tube; ovary inferior with the tube of the perianth attached to the top of the ovary . . . . . . . . . . . . . . . . Agave, in AMARYLLIDACEAE, p. 457
      4. Leaves not conspicuously thick, fleshy, or succulent, or if rarely fleshy, then much smaller; flowers white, lavender, yellow, or brown-purple, the sepals and petals or perianth-divisions usually separate and distinct, not united, or if so, only very slightly at the base; ovary superior with the petals or perianth attached below the base of the ovary . . . . . . . . . . . . . . . . . . . . . . . . *5*
         5. Petals 4; sepals 4; ovary 2-celled; leaves with netted or feather-venation . . . Draba, in CRUCIFERAE, p. 743
         5. Flower usually of 6 perianth-divisions, the 3 sepals and 3 petals of the same or nearly same shape and size and color; ovary 3-celled; leaves with parallel nerves . LILIACEAE, p. 418
   3. Stamens 1, 4, 5, 8, or 10 . . . . . . . . . . . . . . . . . . . . *6*
      6. Stamen 1; ovary inferior, the base of the petals attached to the top of the ovary . . Aplectrum, in ORCHIDACEAE, p. 486
      6. Stamens 4, 5, 8, or 10; ovary superior, the base of the petals or corolla attached below the base of the ovary . . . . . . . . . . . . . . . . . . . . *7*
         7. Each flower-stem (peduncle) terminated by only 1 flower or fruit. . . . . . . . *8*
            8. Stamens 5; larger leaf-blades 3–7 cm. long, oval or ovate, somewhat heart-shaped at base; petals white, 15–20 mm. long, separate, not united into a tube . . . Parnassia, in SAXIFRAGACEAE, p. 781
            8. Stamens 4; larger leaf-blades 1–2.5 cm. long, oblong or elliptic, not subcordate; corolla pink, 2.5–3 mm. long, the petals united into a tube . . . . . . Limosella, in SCROPHULARIACEAE, p. 1338
         7. Each flower-stem (peduncle) terminated by 2 to many flowers or fruits . . . . . . *9*
            9. Leaves 2; sepals 2; petals separate and distinct, not joined at their base . . . Claytonia, in PORTULACACEAE, p. 636
            9. Leaves several or many; calyx deeply 5-cleft; petals united at their base into a short tube . . . . . . . . . . . . . . . . PRIMULACEAE, p. 1164

c. Leaves with teeth or lobes projecting from the margins . . . . . . . . . . . . . . *10*

10. Leaves without any teeth but with one or two arrowhead-shaped pointed lobes or rounded projections (auricles) from the base . . . . . . . . . . . . . . . . . . . *11*
    11. Petals conspicuous, white, 3 to each flower; sepals 3; stamens 9–20 or more; carpels (pistils) numerous, distinct or not united with each other; inflorescence with often 2 or more whorls of branches, sometimes with only 1 whorl . . . . . . . . . . . . . ALISMACEAE, p. 57
    11. Petals and sepals absent, the very small crowded greenish-white or white flowers on an elongated fleshy column (spadix) closely enveloped by a wavy-edged green spathe; stamens 6–10 . . . . . . . . . . . . . . . . . . . Peltandra, in ARACEAE, p. 384
10. Leaves with either small, medium-sized, or large teeth, and with or without lobes . . . . . *12*
    12. Sepals 4, dry and thinnish; corolla of 4 united petals, dry and transparent, brownish-white or greenish; stamens 4; flowers arranged in a narrow cylindrical spike much longer than broad or in a dense head; fruiting capsule circumscissile, an upper half splitting away from a lower half by a circular transverse line . . . . . . . . . . . . . PLANTAGINACEAE, p. 1380
    12. Without the above combination of characters . . . . . . . . . . . . . . *13*
        13. Each flower-stem (peduncle) terminated by only 1 flower or fruit . . . . . . . *14*

14. Stamens about 24; petals 6–12; bruised underground rootstock producing orange-red juice . Sanguinaria, in PAPAVERACEAE, p. 719
14. Stamens 5 or 8; petals 4 or 5; no orange-red juice produced by bruised roots or rootstocks . . *15*
 15. Petals 4, very showy, equal in size and shape, 1–4 cm. long; stamens 8; ovary inferior with the tube of calyx united with the ovary. . . . . Oenothera triloba, in ONAGRACEAE, p. 1106
 15. Petals 5, somewhat unequal, 0.7–1.8 cm. long; stamens 5; ovary superior with the distinct sepals attached below the base of the ovary . . . . . . . . . . . VIOLACEAE, p. 1068

13. Each flower-stem (peduncle) terminated by 2 to many flowers or fruits . . . . . . . . . . *16*
 16. Leaves very thick, fleshy, and succulent with finely spiny-toothed margin; flowers not separated into a distinct calyx and corolla, but with the 6 perianth-divisions united for more than $\frac{2}{3}$ their length into a funnelform tube . . . . . . . . . . . . Agave, in AMARYLLIDACEAE, p. 457
 16. Leaves not conspicuously thick, fleshy, or succulent; flowers separated into distinct calyx and corolla, the calyx 5-cleft or of 4 separate sepals, the corolla 5-parted or of 4 or 5 separate petals . . *17*
  17. Petals 4; sepals 4; stamens 6 of which 2 are shorter than the other 4 . . . CRUCIFERAE, p. 728
  17. Petals 5 or corolla 5-parted; calyx 5-cleft; stamens 4, 5, or 10, rarely 6 or 8 . . . . . . *18*
   18. Corolla 5-parted, the 5 segments joined or connected with one another at their base, so that an attempt to remove one of the segments disturbs or tears a neighboring portion; flowers in an umbel on elongated pedicels (stalks) arising from the same place at the summit of the flower-stem (scape) . . . . . . . . . . PRIMULACEAE, p. 1164
   18. Corolla of 5 separate petals, not joined or connected with one another at their base, so that one petal can be removed without tearing another; flowers in a corymb, cyme, or panicle, usually from branches appearing at more than one level on the stem of the inflorescence, or, if occurring in compact or congested clusters, then the flower-stalks very short . . . . . . . . . . . . . . . . . . SAXIFRAGACEAE, p. 773

## Section 13. **Herbaceous or nonwoody plants with alternate, simple entire leaves without teeth or lobes**

a. Plants without true flowers or seeds, reproducing by green or brown spores (without embryos) contained in sporangia and located on the lower portions of the leaf (frond) or in specially modified leaves (fronds), stalks, or branches . . . . . . . . . . . . . FERNS AND FERN ALLIES, p. 2
a. Plants with true flowers bearing stamens or carpels (pistils) or both, reproducing by seeds containing embryos . . . . . . . . . . . . . . . . . . . . . . . . . *b*
 b. What appears to be one flower actually is a dense mass of small flowers surrounded by green or colored scales or bracts of the involucre; anthers united into a tube or rarely scarcely united; petals united into a tube; ovary inferior; ovule and seed 1 in each of the individual flowers. . COMPOSITAE, p. 1442
 b. Without the above combination of characters . . . . . . . . . . . . . . . . . *c*
  c. Flowers in dense masses in an upright, long, narrow, cylindrical, constricted or divided spike, 10–35 cm. long, the lower half (pistillate) chocolate or reddish-brown, the upper half (staminate) mustard- or brownish-yellow; fruiting spikes with dense masses of down; leaves very long and strap-shaped; stem 0.75–2.7 m. ($2\frac{1}{2}$–9 ft.) tall . . . . . . . . . . . TYPHACEAE, p. 46
  c. Without the above combination of characters . . . . . . . . . . . . . . . . *d*
   d. Plants with stems vining, twining, climbing, sprawling, or trailing on the ground or over other plants or supports . . . . . . . . . . . . . . . . . . . . . *e*
    e. Stipules in the form of a thin tissue-like tubular sheath (ocrea) surrounding the stem at the base of the leaf-stalk . . . . . . . . . . . . . . POLYGONACEAE, p. 574
    e. Stipules either absent or not as above . . . . . . . . . . . . . . . . . *f*
     f. Tendrils (slender coiling or twisting appendages) present at the base of some of the leaf-stalks or at the tips of the stems or branches. . . . . . . . . . . . *g*
      g. A pair of tendrils present at the base of some of the leaf-stalks (petioles); fruit a small, fleshy, spherical berry, without wings; sexes of the flowers separated on different plants (dioecious); segments of the flower 6; stamens 6 . . . . . Smilax, in LILIACEAE, p. 448
      g. Tendrils present at the tips of the stem or branches; fruit with a winged calyx; flowers perfect (stamens and pistil in the same flower); segments of the flower 5; stamens 8. . . . . . . . . . . . Brunnichia, in POLYGONACEAE, p. 600
     f. No tendrils present . . . . . . . . . . . . . . . . . . . . . *h*

h. At least the lower part of leaf-blade and the leaf-stalk (petiole) with some star-shaped (stellate) hairs; stems climbing or attaching themselves by aerial roots . . . Hedera, in ARALIACEAE, p. 1114

h. Hairs on lower surface of leaf-blade or on leaf-stalk simple, when present; stems not climbing or attaching themselves by aerial roots . . . . . . . . . . . . . . . . . . . . *i*

i. All the main nerves arising at the same point at the base of the leaf-blade . . . . . . . *j*

j. Main nerves 9–13 (sometimes 7), arching with their ends curving inward toward tip or upper half of leaf-blade . . . . . . . . . . . . . . DIOSCOREACEAE, p. 454

j. Main nerves 5–7, the lower ones downwardly or horizontally spreading or with their ends pointing to the margins in the lower ⅔ of the leaf-blade . . . . . . . . . . Menispermum, in MENISPERMACEAE, p. 714

i. Some of the nerves arising above the base of or at various levels above the base of the leaf-blade . . *k*

k. Flowers funnelform to broadly bell-shaped, 1.5–8 cm. long, white, blue, purple, pink, or rose-red; bruised stem or leaf-stalk and sometimes the leaf-blade producing milky juice . CONVOLVULACEAE, p. 1212

k. Flowers not funnelform or bell-shaped, 0.5–1.3 cm. long, or, if larger, then with a strongly curved, somewhat S-shaped tube; no milky juice produced by bruised stem or leaf-stalk . . . *l*

l. Leaf-blades 5–15 cm. long, 5–15 cm. broad, broadly ovate to round and strongly or conspicuously heart-shaped at base; flower strongly curved with a somewhat S-shaped tube, the complete tube 3–5 cm. long; fruit a 6-valved hard capsule 4–6 cm. long with numerous flat seeds. . . . . . . . . . Aristolochia, in ARISTOLOCHIACEAE, p. 572

l. Leaf-blades usually smaller; flower not curved, regularly symmetrical, 0.5–1.3 cm. long; fruit a fleshy 1–2-seeded drupe or a soft fleshy many-seeded berry 0.7–1.5 cm. long . . . *m*

m. Corolla purple, the petals joined into a short tube so that the attempt to remove one portion disturbs the neighboring section of the corolla; bruised plant with a rank fetid odor; fruit a many-seeded berry. . . Solanum Dulcamara, in SOLANACEAE, p. 1311

m. Corolla greenish, white, or greenish-white, the petals separate to the base, not connected, so that one petal can be removed without tearing or disturbing the other petals; bruised plant lacking a rank fetid odor; fruit a 1- or 2-seeded drupe . . . . . *n*

n. All the leaf-blades as broad as or broader than long; usually 7 main nerves (including midrib) arising at very base of leaf-blade; stamens 12–24 in the staminate flowers; fruit bluish-black, about 10 mm. long. Menispermum, in MENISPERMACEAE, p. 714

n. Some or all of the leaf-blades longer than broad; usually 1–5 main nerves (including midrib) arising at very base of leaf-blade; stamens 5–6; fruit blue or red, 7–8 mm. long . . . . . . . . . . . . . . . . . . . . . . *o*

o. Lower surface of leaves more or less hairy; stem minutely hairy, easily broken or torn; petals 6; stamens 6; mature fruit red, 1-seeded . . . . . . . Cocculus, in MENISPERMACEAE, p. 716

o. Lower surface of leaves glabrous (without hairs); stem glabrous, tough, and difficult to break; petals 5; stamens 5; mature fruit blue, 2-seeded . . . . Berchemia, in RHAMNACEAE, p. 1027

d. Plants with stems upright, spreading, or hanging, but not vining, twining, climbing, or trailing . . . *p*

p. Only 1 leaf present on the stem (there may be few, many, or no leaves present at very base of plant) . . *q*

q. Flowers of tones of brown, buff, greenish, or straw-color, or, if of other colors, the perianth, corolla, or calyx dry and scale-like or thin and transparent, or perianth absent or reduced to bristles or scales, with the flowers then occurring in the axils of imbricate (overlapping) dry scales forming spikes or spikelets. . . . . . . . . . . . . . . . . . . . . *r*

r. Perianth consisting of 3 sepals, 3 similar petals, and 3 or 6 stamens . Juncus, in JUNCACEAE, p. 404

r. Perianth absent or reduced to bristles or scales, the flowers in the axils of imbricated (overlapping) dry scales forming spikes or spikelets . . . . . . . . . . . . . . . . *s*

s. Stems usually rounded (terete) or flattened, usually hollow or with the pith easily removable, but the nodes (joints) solid, closed, hard, and prominent; sheath at base of leaf usually split open or overlapping on one side; anthers versatile (attached near their middle and swinging free); seed of the fruit not separable nor free from the thin wall (pericarp) of the fruit; perianth absent or apparently so, part of the flowers consisting of an outer lemma, an inner palea, and sometimes 2 minute lodicules, the whole subtended at the base by 1 or 2 scale-like bracts, the glumes, and forming a spikelet; a hairy or membranous projection, the ligule, usually present at the inner junction of leaf-sheath and leaf-blade; leaves of the stem usually 2-ranked, arising on 2 alternating sides of the stem . . . . . . . . . . . . . . . . . GRAMINEAE, p. 68

s. Stems often 3-angled, but also rounded, flattened, or 4- or more angled, usually solid with pith, sometimes hollow, but the nodes (joints) soft and not conspicuous; sheaths at base of leaf usually closed and not split; anthers attached by their base (basifixed) to the filament; seed of the fruit free from and not attached to the wall (pericarp) of the fruit; perianth absent or consisting of bristles or scales, the flowers each in the axil of a single scale or glume, these overlapping into spirally arranged or 2-rowed spikelets or spikes; ligule present or absent; leaves of the stem 3-ranked, arising in 3 different positions or directions on the stem. . . CYPERACEAE, p. 255

q. Flowers of mainly other colors, white, purple, lavender, blue, or greenish, but the color produced in either a small or large, conspicuous or developed perianth or corolla . . . . . . . . . . *t*

t. Petals 5; sepals 5; stamens 5 . . . . . . . . . . . . Parnassia, in SAXIFRAGACEAE, p. 781

t. Petals 3 and sepals 3, or when not distinguishable into petals and sepals, the petal-like perianth of 6 free or united parts; stamens 1, 3, or 6 . . . . . . . . . . . . . . . . *u*

u. Flowers irregular, one or more of the petals or perianth-divisions of a different size and shape from the others . . . . . . . . . . . . . . . . . . . . . . . . . *v*

v. Plants growing mainly in water or in recently submerged ground; flowers blue-purple; stamens 6; ovary superior, the perianth attached below the base of the ovary . . . Pontederia, in PONTEDERIACEAE, p. 401

v. Plants usually growing on dry land or, if in moist ground, not in water; flowers white, greenish, yellowish, or rose-colored; stamen 1; ovary inferior, the petals attached to the top of the ovary. . . . . . . . . . . . . . . . . ORCHIDACEAE, p. 468

u. Flowers regular, the petals or petal-like perianth-divisions of equal size and shape. . . . . *w*

w. Stamens 3; ovary inferior with the petals or petal-like parts of the perianth attached to the top of the ovary. . . . . . . . . . . . . . . . . . . IRIDACEAE, p. 458

w. Stamens 6; ovary superior with the petals or petal-like parts of the perianth attached below the base of the ovary . . . . . . . . . . . . . . . . . LILIACEAE, p. 418

p. Two or more leaves present on the stem . . . . . . . . . . . . . . . . . . . . . *x*

x. Bruised stem, leaf-stalk, or leaf-blade producing milky juice . . . . . . . . . . . . . . *y*

y. Ovary inferior, united with the calyx-tube, the base of the corolla-tube attached to the top of the ovary . . . . . . . . . . . . . . . . . . . . CAMPANULACEAE, p. 1428

y. Ovary superior, free from and not united with the calyx . . . . . . . . . . . . . . *z*

z. Only 1 ovary in each flower . . . . . . . . . . . . . . . . . . . . . . . . *1*

1. Flowers inconspicuous, greenish or greenish-yellow or with white, spreading petal-like appendages; true petals absent, the group of flowers surrounded by an involucre in a cup-shaped cyathium; fruit deeply 3-lobed . . . . . Euphorbia, in EUPHORBIACEAE, p. 984

1. Flowers with a conspicuous white, purplish, or rose-lavender corolla of united petals and a green calyx of 5 sepals; fruit not 3-lobed . . . . . . . CONVOLVULACEAE, p. 1212

z. Ovaries 2 in each flower, the styles either distinct or united . . . . . . . . . . . . *2*

2. Corolla-lobes deeply parted almost to the base, the corolla-tube short or inconspicuous; individual flowers not bell-shaped nor tubular; filaments of stamens united into a tube and column surrounding the pistil with the anthers attached to a large stigmatic body; pollen united in waxy masses; styles distinct . . . . . . . ASCLEPIADACEAE, p. 1200

2. Corolla-lobes not parted almost to the base, but with a long and conspicuous corolla-tube; filaments of stamens not united, distinct; pollen loose and granular; styles united Amsonia, in APOCYNACEAE, p. 1194

x. Bruised stem, leaf-stalk, and leaf-blade not producing milky juice . . . . . . . . . . *3*

3. Stems, leaf-stalks, and part of the leaves covered with star-shaped (stellate) or branching hairs or scurfy scales. . . . . . . . . . . . . . . . . . . . . . . . . . *4*

4. Corolla none . . . . . . . . . . . . . . . . . . . . . . . . . . . *5*

5. Stamens and pistils occurring in the same flower; style 1; ovary 2-celled . . . . non-petalous forms in CRUCIFERAE, p. 737, 743–744 (728–768)

5. Stamens and pistils occurring in separate flowers either on the same plant (monoecious) or on different plants (dioecious); styles 3, 1–3-parted; ovary 3-celled . EUPHORBIACEAE, p. 973

4. Corolla present. . . . . . . . . . . . . . . . . . . . . . . . . . . *6*

6. Corolla of united petals forming a short or long tube, so that one petal-like lobe cannot be removed without disturbing or tearing a neighboring section; stamens 5 . . *7*

7. Inflorescence a long dense cylindrical spike with crowded flowers; stem winged by leafy tissue extending from the base of the leaves; leaves and stem densely woolly. . . . . . . . . . . . . Verbascum, in SCROPHULARIACEAE, p. 1339

7. Inflorescence with solitary or loosely arranged flowers not occurring in a long dense spike; stem not winged by leafy tissue; leaves and stem often hairy, but not densely woolly . . . Physalis, in SOLANACEAE, p. 1314

6. Corolla of separate distinct petals, so that one may be removed without tearing the neighboring petals; stamens 6 or many . . . . . . . . . . . . . . . . . . . . . . . . *8*

8. Petals 4; calyx of 4 separate sepals not united at base; stamens 6, not united . CRUCIFERAE, p. 728

8. Petals 5; calyx of 5 partially united sepals; stamens numerous, their filaments united into a tube . . . . . . . . . . . . . . . . . . . . . . . MALVACEAE, p. 1044

3. Stems, leaf-stalks, and leaves either glabrous (without hairs) or, if hairy, the hairs either simple or forked, but not star-shaped (stellate) with several rays or branches . . . . . . . . . *9*

9. Leaves either hair- or thread-like or narrowly linear, narrowly lanceolate, ribbon- or strap-shaped, the sides more or less parallel, and usually many times longer than broad . . . . . *10*

10. Plant with strong odor of onion . . . . . . . . . . . . Allium, in LILIACEAE, p. 424

10. Plant lacking an onion odor . . . . . . . . . . . . . . . . . . . . . *11*

11. Both corolla and calyx absent, the flowers, such as they are, in the axils of regularly imbricated (overlapping or shingle-like) scales . . . . . . . . . . . . . . . *12*

12. Stems usually rounded (terete) or flattened, usually hollow or with the pith easily removable, but the nodes (joints) solid, closed, hard, and prominent; sheath at base of leaf usually split open or overlapping on one side; anthers versatile (attached near their middle and swinging free); seed of the fruit not separable or free from the thin wall (pericarp) of the fruit; perianth absent or apparently so, part of the flowers consisting of an outer lemma, an inner palea, and sometimes 2 minute lodicules, the whole subtended at the base by 1 or 2 scale-like bracts, the glumes, and forming a spikelet; a hairy or membranous projection, the ligule, usually present at the inner junction of leaf-sheath and leaf-blade; leaves of the stem usually 2-ranked, arising on 2 alternating sides of the stem . . . . . . . . . . . . . GRAMINEAE, p. 68

12. Stems often 3-angled, but also rounded, flattened, or 4- or more angled, usually solid with pith, sometimes hollow, but the nodes (joints) soft and not conspicuous; sheaths at base of leaf usually closed and not split; anthers attached by their base (basifixed) to the filament; seed of the fruit free from and not attached to the wall (pericarp) of the fruit; perianth absent or consisting of bristles or scales, the flowers each in the axil of a single scale or glume, these overlapping into spirally arranged or 2-rowed spikelets or spikes; ligule present or absent; leaves of the stem 3-ranked, arising in 3 different positions or directions on the stem . . . . . . . CYPERACEAE, p. 255

11. Either corolla or calyx or both present on at least the pistil-bearing flowers, the flowers, such as they are, not in the axils of regularly imbricated scales, the petals or sepals sometimes greatly reduced in size and sometimes scale-like or dry and thin, brown, buff, or green . *13*

13. Petals absent at flowering time (anthesis), only a calyx or sepal-like parts present, the latter sometimes reduced to small scales . . . . . . . . . . . . . . . *14*

14. Flowers on a thick, fleshy, finger-like axis 4–9 cm. long becoming 0.7–2 cm. thick in fruit; bruised rootstocks fragrant and sweet-tasting . Acorus, in ARACEAE, p. 386

14. Flowers not on a fleshy axis nor the bruised rootstocks fragrant . . . . . . *15*

15. Plants usually growing in water or very wet ground; flowers in more or less spherical heads, the stamen-bearing (staminate) heads above, the pistil-bearing (pistillate) heads below . . . . . . . . . SPARGANIACEAE, p. 49

15. Plants usually occurring on dry, sandy, or rocky ground, never in water; flowers not in spherical heads . . . . . . . . . . . . . . . *16*

16. Nodes (joints) of stem covered or surrounded by a thin tube-like sheath (ocrea) formed from united stipules . . . . . . . POLYGONACEAE, p. 574

16. Stipules, if present, not forming a tube-like sheath around the nodes (joints) of the stem . . . . . . . . . . . . . . . . . . *17*

17. Stipules (small appendages or scale-like outgrowths at base of leaves) present; stamen-bearing (staminate) and pistil-bearing (pistillate) flowers separated in different flowers on the same plant (monoecious); stamens of the staminate flowers 3 or 8–16; calyx 3–6-parted . . . . . . . . . . . EUPHORBIACEAE, p. 973

17. Stipules absent; flowers usually perfect, the stamens and pistils in the same flower; stamens 1–2, mostly 5–6; calyx either 5-parted (rarely 4-parted or of only 1 sepal) or of 4 distinct separate sepals . *18*

18. Stamens 1–2, mostly 5; calyx usually 5-parted (rarely 4-parted or of only 1 sepal) . CHENOPODIACEAE, p. 600

18. Stamens 6; sepals 4, free and separate to the base . . . non-petalous forms in CRUCIFERAE, p. 728

13. Petals present at flowering time (anthesis), although sometimes very small or inconspicuous or shorter than the sepals or calyx-lobes, or sometimes green, brown, dry, or scale-like . . . . . . *19*

19. Calyx with only 2 divisions, these (the calyx-lobes) coming off with the top of the fruit . . . Portulaca, in PORTULACACEAE, p. 633

19. Calyx with 3, 4, or 5 sepals or calyx-lobes, or, if not distinguishable into calyx and corolla, then with 6 petal- or sepal- like perianth-divisions . . . . . . . . . . . . . . . *20*

20. Flower of 6 perianth-divisions, comprising 3 sepals and 3 petals either alike or different in color, size, and shape . . . . . . . . . . . . . . . . . . *21*

21. All 6 parts of the perianth dry and scale-like, usually brownish, buff, greenish, or straw colored, sometimes purplish; plants with a grass- or sedge-like appearance . JUNCACEAE, p. 4[illegible]4

21. Either 3 or 6 parts of the perianth with brightly colored petals or petal-like parts, white, lavender, purple, blue, brown-purple, rose, pink, yellowish, or sometimes green . . . *22*

22. Pistils in each flower several and unconnected with each other; petals yellow; stamens 3–10 . . . . . . . . . . . . Ranunculus, in RANUNCULACEAE, p. 683

22. Pistil in each flower 1; petals or petal-like parts of the perianth usually of other colors; stamens 1, 3, or 6 . . . . . . . . . . . . . . . . . . *23*

23. Perianth of 3 green sepals but with 3 petals of colors other than green . . COMMELINACEAE, p. 392

23. All 6 perianth divisions of green or other colors, but not distinguishable into an outer green calyx and an inner differently colored corolla . . . . . . *24*

24. Flower irregular, one of the petals of a different size and shape or color from all the other parts of the flower. . . . . . . . ORCHIDACEAE, p. 468

24. Flower regular, either all the perianth-divisions of similar size and shape, or each of the outer and inner series equal in size and shape . . . . . *25*

25. Stamens 6; ovary superior with the petals or petal-like parts of the perianth attached below the base of the ovary . . . . LILIACEAE, p. 418

25. Stamens 3; ovary inferior with the petals or petal-like parts of the perianth attached to the top of the ovary . . . . . IRIDACEAE, p. 458

20. Flower of either 4 or 5 (rarely up to 7) sepals or calyx-lobes and 4 or 5 (rarely 1–3 or 6–7) petals or corolla-lobes . . . . . . . . . . . . . . . . . . . . . *26*

26. Separate ovaries or pistils in each flower 4 or more and unconnected or slightly connected with one another at the base . . . . . . . . . . . . . . . *27*

27. Leaves fleshy and succulent; base of leaf not widened into a stipule-like thin margin; petals 4 or 5; sepals 4 or 5; stamens 8–10, inserted on the calyx rather than on the receptacle; pistils all on the same level, not spirally arranged . . . . . . . Sedum, in CRASSULACEAE, p. 770

27. Leaves not fleshy or succulent; base of leaf widened into a thin tissue-like margin; produced by a stipule-like portion; petals 1–3 (up to 5); sepals 3 or 5; stamens 3–10, inserted on the receptacle beneath the ovary; pistils spirally arranged, the tips of unequal height . . . . . . . . Ranunculus, in RANUNCULACEAE, p. 683

26. Only 1 or 2 separate ovaries or pistils in each flower . . . . . . . . . . . *28*

28. Petals separate, not united at their base, so that one petal may be removed without tearing the rest of the corolla . . . . . . . . . . . . . . . . *29*

29. Petals 4; sepals or calyx-lobes 4 . . . . . . . . . . . . . . . . *30*

30. Flowers irregular with the sepals unequal in size or shape and the petals unequal in size and shape; ovary open at the summit during the flowering period; stamens mainly 20–30, sometimes 10–20 . . . RESEDACEAE, p. 770

30. Flowers regular with the sepals equal in size or shape and petals equal in size and shape; ovary closed during the flowering period; stamens 6 or 8 . *31*

31. Stamens 6; sepals distinct and separate, not connected at the base to form a tube; ovary superior, not united with the tube of the calyx, the petals arising below the base of the ovary . . . CRUCIFERAE, p. 728

31. Stamens 8; sepals united into a tube; ovary inferior, united with the tube of the calyx, the petals arising from the top of the calyx-tube and ovary . . . . . . . . . . . . . . ONAGRACEAE, p. 1093

29. Petals 3 or 5–7; sepals or calyx-lobes 5–7 . . . . . . . . . . . . *32*

32. Leaves thick and fleshy, stiff, 1–4 cm. broad, with spiny or bristly margins;

flowers in large dense rather round heads 1.5–2.5 cm. long; ovary inferior, the petals arising from the top of the ovary . . . . . . Eryngium, in UMBELLIFERAE, p. 1122

32. Leaves thin or membranous, not stiff, very slender, 0.05–1.5 (–2) cm. broad, with non-bristly or non-spiny margins; flowers either in shorter heads or not in heads; ovary superior, the petals arising below the base of the ovary . . . . . . . . . . . . . . . . *33*

33. Leaf-like stipules appearing at the base of leaves . . . Crotalaria, in LEGUMINOSAE, p. 882

33. No leaf-like stipules present at base of leaves . . . . . . . . . . . . . . *34*

34. All the sepals equal in size and shape; petals 5; stamens 5; styles 5; ovary 5-celled LINACEAE, p. 957

34. Some of the sepals larger than the others, or some of them of unequal size and shape; petals 3, 5–7; stamens 3–25; styles or stigmas 1 or 2–6; ovary 1–2-celled . . . . *35*

35. Petals 5–7, yellow; sepals 5–7, very slightly united at base; ovary open at the summit during the flowering period . . . . . . . . RESEDACEAE, p. 770

35. Petals 3; sepals 5, sometimes the 2 inner ones petal-like, distinct and separate to their base; ovary closed during the flowering period . . . . . . . . . 36

36. The 2 inner sepals petal-like; filaments of the 6 or 8 stamens united below into a sheath; ovary 2-celled . . . . . . . . . . POLYGALACEAE, p. 969

36. All the sepals small and greenish, at most 3 mm. long; filaments of the 3–25 stamens distinct and separated, not united below; ovary 1-celled . Lechea, in CISTACEAE, p. 1067

28. Petals united into a short or long tube so that one part cannot be removed without disturbing or tearing a neighboring section . . . . . . . . . . . . . . . . . . . . *37*

37. Ovaries 2 and separate in each flower, but the stigmas united (4-lobed ovaries appearing almost like 4 separate ovaries should be looked for in alternate 37); stamens and stigma of a specially modified form differing from the usual type, the anthers attached to the stigma, and the pollen borne in special waxy masses . . . . . . . . ASCLEPIADACEAE, p. 1200

37. Ovary 1 in each flower (sometimes the deeply 4-lobed ovaries appear almost like 4 separate ovaries); anthers not attached to the stigma but sometimes forming a ring or tube around the stigma, the pollen not borne in special waxy masses . . . . . . . . . . . *38*

38. Stamens usually 8, sometimes 6, their filaments united below into a sheath; petals 3, slightly united; the 2 inner sepals petal-like. . . . . . . . . . POLYGALACEAE, p. 969

38. Stamens 4 or 5, their filaments distinct or not united below into a sheath; corolla-lobes 5; calyx-lobes green, not petal-like . . . . . . . . . . . . . . . . *39*

39. Stamens 4; corolla irregular, 2-lipped into an upper and lower division, or some of the lobes unequal in size and shape. . . . . . . . SCROPHULARIACEAE, p. 1328

39. Stamens 5; corolla regular, the lobes or folds equal in size and shape . . . . . *40*

40. Bruised stem or leaves usually producing milky juice; ovary inferior, united with the calyx-tube, the calyx-lobes and base of the corolla arising from the top of the ovary . . . . . . . . . . . . . . . . CAMPANULACEAE, p. 1428

40. Bruised stem and leaves not producing milky juice; ovary superior, not united with the calyx-tube, the base of the corolla arising below the base of the ovary . *41*

41. Leaves or stem not densely and conspicuously hairy; ovary 3-celled . . Collomia, in POLEMONIACEAE, p. 1225

41. Leaves and stem densely and conspicuously hairy; ovary 2-celled or deeply 4-lobed . . . . . . . . . . . . . . . . . . . . *42*

42. Stems many and widely or loosely spreading from the base; corolla purplish; styles 2; ovary 2-celled, not lobed. Evolvulus, in CONVOLVULACEAE, p. 1212

42. Stems usually solitary or few, usually erect; corolla white, yellow, or orange-yellow; style 1; ovary deeply 4-lobed, almost appearing like 4 separate ovaries. . . . . . . . . . . . BORAGINACEAE, p. 1242

9. Leaves not as described in alternate 9, usually broadest in the lower or upper half or in the middle with curving sides or noticeably narrowed at either end, sometimes not much longer than broad or as broad as long . . . . . . . . . . . . . . . . . . . . . . . *43*

43. Stem completely surrounded by the base of the leaf (perfoliate), and appearing to pass through the leaf . . . . . . . . . . . . . . . . . . . . . . . *44*

44. Only the upper leaves perfoliate; petals 4, white or greenish; sepals 4 . . . . . . Lepidium, in CRUCIFERAE, p. 737

44. All the leaves perfoliate; petals 3 and sepals 3, yellow, similar, or petals 5 and calyx-teeth 5 and nearly absent . . . . . . . . . . . . . . . . . . . *45*

45. Petals 3 and sepals 3, similar in size and shape; ovary superior, not united with the calyx, the base of the perianth arising below the base of the ovary . . . . Uvularia, in LILIACEAE, p. 423
45. Petals 5; calyx-teeth 5, nearly absent; ovary inferior, united with the calyx-tube, the calyx-teeth and petals arising from the top of the ovary . . . . Bupleurum, in UMBELLIFERAE, p. 1132
43. Leaves not perfoliate, although the leaf-bases may sometimes clasp the stem but not completely surround it . . . . . . . . . . . . . . . . . . . . . . . . *46*
46. Both corolla and calyx absent . . . . . . . . . . . . . . . . . . . . *47*
47. Leaves broadly heart-shaped with the side (lateral) nerves spreading at an angle and with a network of veins; bruised leaves with a fragrant, spicy odor; flowers in a long, white, tail-like spike; stamens 6 or 7 to each flower; 3–4 distinct or nearly distinct pistils to each flower; flowers naked, not in the axils of regularly imbricated (overlapping or shingle-like) scales . . . . . . . . . . . . . . . . . . . . . . SAURURACEAE, p. 488
47. Leaves not broadly heart-shaped, with parallel nerves; bruised leaves without any fragrant spicy odor; stamens usually 3, sometimes 1, 2, or rarely 6 to a flower; 1 pistil to each flower; flowers in the axils of regularly imbricated (overlapping or shingle-like) scales. . . . . *48*
48. Stems usually rounded (terete) or flattened, usually hollow or with the pith easily removable, but the nodes (joints) solid, closed, hard, and prominent; sheath at base of leaf usually split open or overlapping on one side; anthers versatile (attached near their middle and swinging free); seed of the fruit not separable or free from the thin wall (pericarp) of the fruit; perianth absent or apparently so, part of the flowers consisting of an outer lemma, an inner palea, and sometimes 2 minute lodicules, the whole subtended at the base by 1 or 2 scale-like bracts, the glumes, and forming a spikelet; a hairy or membranous projection, the ligule, usually present at the inner junction of leaf-sheath and leaf-blade; leaves of the stem usually 2-ranked, arising on 2 alternating sides of the stem . . . . . . . . . . . . . . . . . . . . GRAMINEAE, p. 68
48. Stems often 3-angled, but also rounded, flattened, or 4- or more angled, usually solid with pith, sometimes hollow, but the nodes (joints) soft and not conspicuous; sheaths at base of leaf usually closed and not split; anthers attached by their base (basifixed) to the filament; seed of the fruit free from and not attached to the wall (pericarp) of the fruit; perianth absent or consisting of bristles or scales, the flowers each in the axil of a single scale or glume, these overlapping into spirally arranged or 2-rowed spikelets or spikes; ligule present or absent; leaves of the stem 3-ranked, arising in 3 different positions or directions on the stem . . . . . . . . . . . CYPERACEAE, p. 255
46. Either corolla or calyx or both present on at least the pistil-bearing flowers, the flowers, such as they are, not in the axils of regularly imbricated scales, the petals or sepals sometimes greatly reduced in size and sometimes scale-like or dry and thin, brown, buff, or green . . . . . *49*
49. Petals absent at flowering time (anthesis), only a calyx or sepal-like parts present, the latter sometimes reduced to small scales . . . . . . . . . . . . . . . . . . *50*
50. Nodes (joints) of stem covered or surrounded by a thin tube-like sheath (ocrea) formed from united stipules . . . . . . . . . . . . . . . . POLYGONACEAE, p. 574
50. Stipules, if present, not forming a tube-like sheath around the nodes (joints) of the stem . *51*
51. Stem or part of the leaf or both covered with star-shaped (stellate) hairs or scurfy scales . . . . . . . . . . . . . . . . . . . . . . . *52*
52. Style 1 . . . . . . . . . . . . . . . . . . . *53*
53. Ovary 1-celled; stamens mainly 4 or 3–10; the 5 sepals unequal with the 2 outer ones much smaller than the 3 inner ones . . . . . . . Helianthemum, in CISTACEAE, p. 1066
53. Ovary 2-celled; stamens mainly 6; sepals 4, unequal . . . . . . . non-petalous forms of CRUCIFERAE, p. 737, 743–744 (728–768)
52. Styles 2 or 3 . . . . . . . . . . . . . . . . . . . *54*
54. Stipules absent at base of leaves; stamens in the staminate (male) or perfect flowers 1–5; style branches or styles 2 in the pistillate (female) or perfect flowers . . . . . . . . . . . . CHENOPODIACEAE, p. 600
54. Hair-like or small stipules present at base of leaves; stamens in the staminate (male) flowers 5–14; styles in the pistillate (female) flowers 3 . . EUPHORBIACEAE, p. 973
51. No star-shaped hairs or scurfy scales present on stem or leaves. . . . . . . *55*
55. Leaves heart-shaped at base; bruised root with a fragrant turpentine or spicy menthol odor; flowers on solitary stalks at base of the plant, S-shaped, bluntly

3-lobed, 1–1.5 cm. long, brown-purple . . . . . . Aristolochia, in ARISTOLOCHIACEAE, p. 572

55. Without the above combination of characters; leaves not heart-shaped at base; no turpentine or spicy root odor; flowers in clusters along the sides of or at the tip of the stem, not S-shaped, the individual flowers smaller or of other colors . . . . . . . . . . . . . . *56*

56. Styles 10 to a flower; ovary forming a ring of 10 united carpels; fruit a 10-scalloped berry; flowers in an elongated raceme . . . . . . . . . . . . PHYTOLACCACEAE, p. 630

56. Styles 1–3 to a flower, sometimes 2-cleft or fringed; ovary not forming a ring; fruit not a 10-scalloped berry; flowers not as above . . . . . . . . . . . . . . . 57

57. Stipules (small appendages or scale-like outgrowths at base of leaves) present; pistillate flowers with each of the 3 styles 2-cleft or fringed. . . . . . EUPHORBIACEAE, p. 973

57. Stipules absent; pistillate or perfect flowers with 0, 1, 2, or 3 styles, but these not cleft . *58*

58. Leaves densely white-hairy on lower surface; rarely encountered plants known only from extreme southern Missouri; stamens 9. . Eriogonum, in POLYGONACEAE, p. 574

58. Without the above combination of characters; leaves usually not densely white-hairy on lower surface, but if so, then plants commonly encountered throughout Missouri; stamens 1–5 . . . . . . . . . . . . . . . . . *59*

59. Calyx with a conspicuous bell-, urn-shaped, or cylindrical tube joined to part or all of the ovary . . . . . . . . . . . . . . . . . *60*

60. Calyx-lobes whitish and petal-like . . . . . . . SANTALACEAE, p. 572

60. Calyx-lobes green, not petal-like . . . . Ludwigia, in ONAGRACEAE, p. 1095

59. Calyx deeply parted nearly to the base or consisting of separate, distinct sepals not united at the base and without a conspicuous tube, the flowers greenish or of other colors; ovary entirely free from the calyx and not joined to it. . *61*

61. Style 1; stamens 4; calyx or perianth 4-lobed; flowers in small, loose clusters along the stem at the base of the leaves; leaves with 3 noticeable nerves (including midrib) . . . . . . Parietaria, in URTICACEAE, p. 570

61. Without the above combination of characters; styles 2 or 3, or 1 and branched, or none; stamens 1, 2, 3, 4, or 5; calyx or perianth 1–5-parted or of 3 or 5 distinct sepals; flowers arranged variously; leaves with more than 3 noticeable nerves (including midrib) or only 1-nerved or nerveless . *62*

62. Individual flowers accompanied by thin dry scarious bracts; sepals thin and dry (scarious), usually acutely or sharp- or long-pointed . AMARANTHACEAE, p. 619

62. Individual flowers not accompanied by thin dry scarious bracts, but leafy bracts may surround pistillate flowers in some cases; sepals or calyx-lobes not dry and thin (scarious) but more leaf-like in color or texture, usually blunt or rounded at tip. . . . CHENOPODIACEAE, p. 600

49. Petals present at flowering time (anthesis), although sometimes quite small . . . . . . . . *63*

63. Calyx with only 2 divisions, these (the calyx-lobes) coming off with the top of the fruit . . Portulaca, in PORTULACACEAE, p. 633

63. Calyx with 3, 4, 5, 6, or 7 sepals or calyx-lobes, or, if flower not distinguishable into calyx and corolla, then with all the 6 parts of the perianth petal-like . . . . . . . . . . . *64*

64. Flower of 6 perianth-divisions, comprising 3 sepals and 3 petals either alike or different in color, size, and shape . . . . . . . . . . . . . . . . . . *65*

65. Pistils in each flower several and unconnected with each other; petals yellow; stamens 3–10 . . . . . . . . . . . . . . Ranunculus, in RANUNCULACEAE, p. 683

65. Pistil in each flower 1; petals or petal-like parts of the perianth usually of other colors, but sometimes orange or yellow; stamens 1, 3 or 6 . . . . . . . . . . *66*

66. Perianth of 3 green sepals but with petals of colors others than green. . . . . *67*

67. Leaf-blades 8–20 cm. broad; whole plant covered with whitish powder and 1–2 m. tall, the powdery part often disappearing from the dried leaves and stems; anther-bearing stamen reduced to one-half of an anther, the other stamens petal-like in appearance; ovary inferior, situated below the base of the sepals and petals . . . . . . . . . . . . . . MARANTACEAE, p. 468

67. Leaf-blades at most 5 cm. broad; plant not covered with a whitish powder, at most 1.2 m. tall, usually less than 1 m.; anther-bearing stamens 3–6; ovary superior, situated above the insertion of the sepals and petals . COMMELINACEAE, p. 392

66. All 6 perianth-divisions of green or other colors, but not distinguishable into an outer green calyx and an inner differently colored corolla . . . . . . . . . 68

68. Flower irregular, one of the petals of a different size and shape or color from all the other parts of the flower . . . . . . . . . . . . . . . . . . . . . ORCHIDACEAE, p. 468
68. Flower regular, either all the perianth-divisions of similar size and shape or each of the outer and inner series equal in size and shape. . . . . . . . . . . . . . . . . . . . *69*
69. Stamens 6; ovary superior with the petals or petal-like parts of the perianth attached below the base of the ovary . . . . . . . . . . . . . . . . . . . LILIACEAE, p. 418
69. Stamens 3; ovary inferior with the petals or petal-like parts of the perianth attached to the top of the ovary . . . . . . . . . . . . . . . . . . . IRIDACEAE, p. 458
64. Flower of either 4 or 5 (rarely up to 7) sepals or calyx-lobes and 4 or 5 (rarely 1–3 or 6–7) petals or corolla-lobes . . . . . . . . . . . . . . . . . . . . . . . . *70*
70. Separate ovaries or pistils in each flower 4 or more, unconnected or slightly connected with one another at the base . . . . . . . . . . . . . . . . . . . . . *71*
71. Leaves fleshy and succulant; petals 4 or 5; sepals 4 or 5; stamens 8–10, inserted on the calyx rather than on the receptacle; pistils all on the same level, not spirally arranged . Sedum, in CRASSULACEAE, p. 770
71. Leaves not fleshy or succulent; petals 1–3 (up to 5); sepals 3 or 5; stamens 3–10 or more, inserted on the receptacle beneath the ovary; pistils spirally arranged, the tips of unequal height . . . . . . . . . . . . . . . . . Ranunculus, in RANUNCULACEAE, p. 683
70. Separate ovaries or pistils in each flower only 1 or 2 . . . . . . . . . . . . . . *72*
72. Petals separate, not united at their base, so that one petal may be removed without tearing the rest of the corolla . . . . . . . . . . . . . . . . . . . . . *73*
73. Petals 4; sepals or sepal-lobes 4–6 . . . . . . . . . . . . . . . . . . . *74*
74. Flowers irregular with the sepals unequal in size or shape and the petals unequal in size and shape; ovary open at the summit during the flowering period; stamens 12–40 . . . . . . . . . . . . . . . . . . . RESEDACEAE, p. 770
74. Flowers regular with the sepals equal in size or shape and the petals equal in size and shape; ovary closed during the flowering period; stamens 4, 6, or 8–12 . . . *75*
75. Stamens 6; sepals distinct and separate, not connected at the base to form a tube; ovary superior, not united with the tube of the calyx, the petals arising below the base of the ovary . . . . . . . . . . CRUCIFERAE, p. 728
75. Stamens 4 or 8–12; sepals united into a tube; ovary inferior, united with the tube of the calyx, the petals, if present, arising from the top of the calyx-tube and ovary . . . . . . . . . . . . . . . . . ONAGRACEAE, p. 1093
73. Petals 3 or 5–7; sepals or calyx-lobes 5–7 . . . . . . . . . . . . . . . *76*
76. Stamens 10 or more in a flower . . . . . . . . . . . . . . . . . *77*
77. Small, narrow scale-like stipules usually appearing at base of leaves. . . LEGUMINOSAE, p. 865
77. No leaf-stipules present at base of leaves . . . . . . . . . . . . . *78*
78. Flowers close together in a many-flowered narrow elongate inflorescence (spike); ovary open at the summit during the flowering period. RESEDACEAE, p. 770
78. Flowers well-separated in a few-flowered loosely spreading inflorescence (raceme-like corymb); ovary closed during the flowering period. . . Helianthemum, in CISTACEAE, p. 1066
76. Stamens 4–8 in a flower. . . . . . . . . . . . . . . . . . . . *79*
79. Calyx with the sepals united into a prominent cylindrical tube with 5–7 short teeth at the summit. . . . . . . . . . . Lythrum, in LYTHRACEAE, p. 1090
79. Calyx of separate prominent sepals more or less distinct to the base and not united into a cylindrical tube . . . . . . . . . . . . . . . . *80*
80. Flowers greenish, situated along the sides of the stem at the base of the leaves; narrow thread-like or linear-lanceolate stipules present at base of leaves . . . . . . . . . . . Hybanthus, in VIOLACEAE, p. 1068
80. Flowers yellow, blue, pink, rose, lavender, purplish, or white, rarely greenish, situated in few- or many-flowered inflorescences at the ends of branches or in the upper half of the plant; stipules absent . . . . . . *81*
81. Stamens 5, separate from one another, the filaments not united; styles 5; flowers regular, sepals all alike, petals all alike . . . LINACEAE, p. 957
81. Stamens 6 or 8, their filaments united into a sheath at the base; style 1; flowers irregular, the 2 inner sepals petal-like and different from the 3 other green sepals, the petals also not alike . . . . POLYGALACEAE, p. 969

72. Petals united into a short or long tube so that one part cannot be removed without disturbing or tearing a neighboring section . . . . . 82
82. Ovaries 2 and separate in each flower, but the stigmas united (4-lobed ovaries of Boraginaceae, appearing almost like 4 separate ovaries should be looked for in alternate 82); stamens and stigma of a specially modified form differing from the usual type, the anthers attached to the stigma, and the pollen borne in special waxy masses . . . . . ASCLEPIADACEAE, p. 1200
82. Ovary 1 in each flower (sometimes the deeply 4-lobed ovary of Boraginaceae appears almost like 4 separate ovaries); anthers not attached to the stigma but sometimes forming a ring or tube around the stigma, the pollen not borne in special waxy masses . . . . . 83
83. Stamens 6 or 8; the 2 inner sepals petal-like . . . . . POLYGALACEAE, p. 969
83. Stamens 4 or 5; all the calyx-divisions green or at least not petal-like . . . . . 84
84. Ovary inferior, united with the calyx-tube, the top of the ovary below the base of the corolla-tube . . . . . CAMPANULACEAE, p. 1428
84. Ovary superior, free from and not united with the calyx-tube, the base of the corolla-tube situated below the base of the ovary . . . . . 85
85. Stamens inserted opposite the corolla-lobes; ovary 1-celled . . . PRIMULACEAE, p. 1164
85. Stamens inserted alternate with the corolla-lobes; ovary 2-celled or sometimes appearing (falsely) 4-celled because of the deep lobes or false partitions . . . . 86
86. Ovary deeply 4-lobed, almost appearing like 4 separate ovaries. BORAGINACEAE, p. 1242
86. Ovary not 4-lobed . . . . . 87
87. Corolla irregular, 2-lipped into upper and lower portions, or the lobes of the corolla unequal in size and shape . . . . SCROPHULARIACEAE, p. 1328
87. Corolla regular, the lobes or divisions equal or nearly so in size and shape . 88
88. Plants growing in water, or in swampy or wet ground; spines usually present at the base of the leaves . . Hydrolea, in HYDROPHYLLACEAE, p. 1240
88. Plants usually of dry ground; no spines present . . . . . 89
89. Ovules 4 in each ovary; seeds 4 in each fruit. CONVOLVULACEAE, p. 1212
89. Ovules many in each ovary; seeds many in each fruit . . . . 90
90. Inflorescence a long dense cylindrical spike with crowded flowers; stem winged by leafy tissue extending from the base of the leaves; leaves and stem densely woolly. . . . . Verbascum Thapsus, in SCROPHULARIACEAE, p. 1339
90. Inflorescence with solitary or loosely arranged flowers not occurring in a long dense spike; stem not winged by leafy tissue; leaves and stem often hairy, but not densely woolly SOLANACEAE, p. 1310

## Section 14. **Herbaceous or nonwoody plants with alternate, simple toothed or lobed leaves or margins of leaves not completely entire**

a. Plants without true flowers or seeds, reproducing by green or brown spores (without embryos) contained in sporangia and located on the lower portions of the leaf (frond) or in specially modified leaves (fronds), stalks, or branches . . . . . FERNS AND FERN ALLIES, p. 2
a. Plants with true flowers bearing stamens or carpels (pistils) or both, reproducing by seeds containing embryos . . . . . *b*
b. What appears to be one flower actually is a dense mass of small flowers surrounded by green or colored scales or bracts of the involucre; anthers united into a tube or rarely scarcely united; petals united into a tube; ovary inferior; ovule and seed 1 in each of the individual flowers. . COMPOSITAE, p. 1442
b. Without the above combination of characters . . . . . *c*
c. Plants with stems vining, twining, climbing, sprawling, or trailing on the ground or over other plants or supports . . . . . *d*
d. Margins of leaves with teeth, and with or without lobes . . . . . *e*
e. Tendrils (slender coiling or twisting appendages) present on some part of plant . . . . *f*
f. Stipules present at base of leaf-stalks in the form of small hair- or thread-like appendages or outgrowths; flower white and purple, large and showy, 5–8 cm. across, with 5 large separate petals not united at their base . . . . PASSIFLORACEAE, p. 1083

f. Stipules absent or disappearing as the leaf matures; flowers either small, inconspicuous, greenish, whitish, yellowish, or orange, or if large, the petals all united at the base . . . . . . *g*

g. Calyx with well-developed calyx-tube and calyx-lobes; corolla with the petals united at base into a short or long tube, the corolla persisting during and slightly after flowering (anthesis); stamens usually 3, the anthers usually united or cohering, sometimes free; ovary inferior, united with the calyx-tube, and situated below the base of the corolla; some of the corolla united with calyx-tube . . . . . . . . . . CUCURBITACEAE, p. 1422

g. Calyx very small and without a developed calyx-tube and calyx-lobes; corolla with the petals either all free, separate, and not united, or, if united at the tip, then falling quickly during flowering; stamens 5, the anthers free and not uniting; ovary superior, not united with the calyx and located above the insertion of the stamens, petals, and calyx; corolla and calyx not united . . . . . . . . . . . . . . . . VITACEAE, p. 1031

e. Tendrils absent from all parts of plant . . . . . . . . . . . . . . . . . . . *h*

h. At least the lower part of leaf-blade and the leaf-stalk (petiole) with some star-shaped (stellate) hairs; stems climbing or attaching themselves by aerial roots . . Hedera, in ARALIACEAE, p. 1114

h. Hairs simple when present on lower surface of blade or on leaf-stalk; stems not climbing or attaching themselves by aerial roots . . . . . . . . . . . . . . . . . . . . *i*

i. Leaf-stalks 1–4 mm. long . . . . . . . . . . . Kickxia, in SCROPHULARIACEAE, p. 1350

i. Leaf-stalks usually more than 4 mm. long . . . . . . . . . . . . . . . . . *j*

j. Leaves rounded-heart-shaped, round at summit, 1–3 (–3.5) cm. long . . . . . . Cymbalaria, in SCROPHULARIACEAE, p. 1350

j. Leaves not as above, short- to long-pointed at summit, 3–20 cm. or more long . . . . . *k*

k. Leaves thin, palmately nerved with 5 or more main nerves arising from the same point at the base of the leaf-blade, slightly hairy on the nerves of the lower surface; stem easy to tear or break . . . . . . . Calycocarpum, in MENISPERMACEAE, p. 716

k. Leaves thick, pinnately nerved with 1 main central midrib and several pairs of side (lateral) nerves arising from the midrib, glabrous (without hairs) on the lower surface; stem tough, difficult to tear or break . . . Berchemia, in RHAMNACEAE, p. 1027

d. Margins of leaves without teeth, but somewhere with lobes . . . . . . . . . . . . . *l*

l. Tendrils (slender coiling or twisting appendages) present on some part of plant . . . . . . . *m*

m. Stipules present at base of leaf-stalk in the form of short hair-like appendages or outgrowths; corolla of separate petals, not united at their base . . . . . . . . . PASSIFLORACEAE, p. 1083

m. Stipules absent; corolla of united petals, so that one portion cannot be removed without tearing or disturbing a neighboring section . . . . . . . . . . CUCURBITACEAE, p. 1422

l. Tendrils absent from all parts of the plant . . . . . . . . . . . . . . . . . . *n*

n. Stipules in the form of a thin tissue-like tubular sheath (ocrea) surrounding the stem at the base of the leaf-stalk . . . . . . . . . . . . Polygonum, in POLYGONACEAE, p. 582

n. Stipules either absent or not as above . . . . . . . . . . . . . . . . . . . *o*

o. Flowers funnelform to broadly bell-shaped, 1.5–8 cm. long, white, blue, purple, pink, or rose-red; bruised stem or leaf-stalk and sometimes the leaf-blade producing milky juice CONVOLVULACEAE, p. 1212

o. Flowers not funnelform or bell-shaped, 0.5–1.3 cm. long; no milky juice produced by bruised stem or leaf-stalks . . . . . . . . . . . . . . . . . . . . . . *p*

p. Leaf-stalks 1–4 mm. long . . . . . . . . . . Kickxia, in SCROPHULARIACEAE, p. 1350

p. Leaf-stalks longer, usually more than 4 mm. long . . . . . . . . . . . . . . *q*

q. Corolla purple, the petals joined into a short tube so that the attempt to remove one portion disturbs the neighboring section of the corolla; bruised plant with a rank fetid odor; fruit a many-seeded berry . . Solanum Dulcamara, in SOLANACEAE, p. 1311

q. Corolla greenish, white, or greenish-white, the petals, if present, separate to the base, not connected, so that one petal can be removed without tearing or disturbing the other petals; bruised plant lacking a rank fetid odor; fruit a 1- or 2-seeded drupe . . . . . . . . . . . . . . . . . . . . . . . . . . . *r*

r. All the leaf-blades as broad as or broader than long . . . . MENISPERMACEAE, p. 714

r. Some or all of the leaf-blades longer than broad . . . . . . . . . . . *s*

s. Lower surface or nerves on lower surface of leaf more or less hairy; stem easily broken or torn; stamens, when present, 6 or 12; mature fruit 1-seeded MENISPERMACEAE, p. 714

s. Lower surface of leaves glabrous (without hairs); stem tough, difficult to break; stamens 5; mature fruit 2-seeded . . . . . Berchemia, in RHAMNACEAE, p. 1027

c. Plants with stems upright, spreading, or hanging, but not vining, twining, climbing, or trailing . . . *t*

t. Leaves reduced to minute scattered traps or scales lying upon or under the damp ground; calyx with 2 lobes united at base . . . . . . . . . . . . . . LENTIBULARIACEAE, p. 1375

t. Without the above combination of characters; leaves well-developed or at least not reduced to traps or scales; calyx with 3–5 or 10 lobes or 2–6 separate, distinct sepals . . . . . . . . . *u*

u. Stipules in the form of a thin tissue-like tubular sheath (ocrea) surrounding the stem at the base of the leaf-stalk . . . . . . . . . . . . Polygonum tenue, in POLYGONACEAE, p. 587

u. Stipules either absent or not as above . . . . . . . . . . . . . . . . . . . . *v*

v. Leaves peltate, the leaf-stalk attached to the lower surface of the leaf-blade near the center of the blade. . . . . . . . . . . . . . . . . Ricinus, in EUPHORBIACEAE, p. 982

v. Leaves not peltate, the leaf-stalk joining the leaf-blade at its lower end or leaves attached directly to the stem without any leaf-stalk (sessile) . . . . . . . . . . . . . . . *w*

w. Flowers with the stamens and pistils separated in different flowers on the same plant (monoecious) or different plants (dioecious) . . . . . . . . . . . . . . *x*

x. Stipules present at base of leaf-stalks in the form of very small, narrow, scale- or hair-like appendages or outgrowths, or if stipules are absent, then the plant with either stinging hairs present or 1 or 2 cup-shaped glands occur at base of leaf-blade . . *y*

y. Style 1 in the pistil-bearing flower; ovary 1-celled; fruit 1-seeded; stamens 5 in the stamen-bearing flower . . . . . . . . . . Laportea, in URTICACEAE, p. 568

y. Styles 3 and each branched in the pistil-bearing flower; ovary 3-celled; fruit 3-seeded; stamens either 3–5 or 8–16 in the stamen-bearing flowers . . . . EUPHORBIACEAE, p. 973

x. Without any of the characters of the preceding x; stipules absent and plants without stinging hairs or cup-shaped glands at base of leaf-blade . . . . . . . . . . *z*

z. Most of leaves finely dissected into hair- or thread-like divisions or segments; plants completely glabrous (without hairs); stamens 4 in the stamen-bearing flowers; ovary inferior, united with the calyx, 4-celled; fruit 4-lobed, with 4 seeds. Myriophyllum, in HALORAGIDACEAE, p. 1108

z. Leaves not finely dissected into thread-like segments, but if deeply cut the plants then often covered with hairs, glands, or mealy scurfiness; stamens 3–5 in the stamen-bearing flowers; ovary superior, not united with nor inserted above the calyx, 1-celled; fruit with 1 seed . . . . . . . . . . CHENOPODIACEAE, p. 600

w. Flowers perfect with the stamens and pistils in the same flower (the genus *Euphorbia* should be keyed here) . . . . . . . . . . . . . . . . . . . . . . *1*

1. Bruised stem, leaf-stalk, or leaf-blade producing milky juice; flowers surrounded by an involucre in a cup-shaped cyathium; the cup-shaped involucre with 1–5 glands, with or without colored or petal-like appendages on the margin; a stalked 3-lobed ovary with 3 styles, each 2-cleft, projects from center of flower . . . . . . Euphorbia, in EUPHORBIACEAE, p. 984

1. Without the above combination of characters; plants with or without milky juice; flowers not as above . . . . . . . . . . . . . . . . . . . . . . *2*

2. Stamens 3; calyx-tube 3-sided; petals none; flowers greenish, in the leaf-axils; plants of swamps and swampy ground . . Proserpinaca, in HALORAGIDACEAE, p. 1111

2. Without the above combination of characters; stamens 2, 4, 5, 6, 8–10 or more; calyx various; petals absent or present; flowers of various colors; plants of wet or dry ground . . . . . . . . . . . . . . . . . . . . . . . . *3*

3. Separate ovaries or pistils in each flower 3 or more and unconnected or somewhat connected with one another at the base during the flowering period (anthesis) (4-lobed ovaries of Boraginaceae appearing almost like separate ovaries should be keyed out in alternate 3) . . . . . . . . . . . *4*

4. Stamens 5–10 (rarely 12) . . . . . . . . . . . . . . . . *5*

5. Leaves very thick, fleshy, and succulent; flowers rose or red-purple . . CRASSULACEAE, p. 770

5. Leaves of papery or membranous texture, not fleshy nor succulent; flowers yellow-green . . . . . . Penthorum, in SAXIFRAGACEAE, p. 774

4. Stamens many, more than 12 . . . . . . . . . . . . . . *6*

6. Filaments of stamens united into a tube; star-shaped (stellate) hairs commonly present on some part of plant; ovaries arranged in a ring or circle . . . . . . . . . . . . . . . MALVACEAE, p. 1044

6. Filaments of stamens separate, not united into a tube; star-shaped (stellate) hairs not present on plant; ovaries spirally arranged or not in a ring . . . . . . . . . . . . . . . . . 7
   7. Leaf-like stipules 1–2 cm. long at base of leaf and not united with the leaf-stalk or leaf-blade; styles abruptly hooked or S-bent above the middle; petals and stamens inserted at or near the edge of a saucer-shaped calyx-tube (hypanthium), the sepals united into a tube . . . . Geum, in ROSACEAE, p. 831
   7. Stipules absent or in the form of dilated appendages attached to the base of the leaf-blade or leaf-stalk (petiole); styles straight or curved, but not S-bent or hooked; petals, when present, and stamens inserted directly on the receptacle together with the sepals, the sepals not united into a tube but separate to their base . . . . . . . . . . . . RANUNCULACEAE, p. 672

3. Only 1 or 2 separate ovaries or pistils in each flower (4-lobed ovaries of Boraginaceae appearing almost like separate ovaries are keyed in this section) . . . . . . . . . . . . . . . . . 8
   8. Separate ovaries or pistils in each flower 2, remaining separated from the beginning of the flowering period (anthesis) . . . . . . . . . . . . . . . SAXIFRAGACEAE, p. 773
   8. Only 1 ovary or pistil in each flower from the beginning of the flowering period (anthesis), but the upper part of the pistil may separate into distinct and separate units corresponding to the number of carpels present . . . . . . . . . . . . . . . . . . . . . . 9
      9. Upper half of pistil showing 5 or more distinct and separate beaks or tips corresponding to the number of carpels present . . . . . . . . . . . . . . . . . . . . . 10
         10. Filaments of stamens united into a column or tube; star-shaped hairs commonly present on some part of plant . . . . . . . . . . . . . . . . MALVACEAE, p. 1044
         10. Filaments of stamens free, not united into a tube; no star-shaped hairs present . . . . 11
            11. Petals 5, lavender, rose-purple, lilac, or whitish; calyx divided into distinct sepals not united at their base . . . . . . . . . . . . . . . . GERANIACEAE, p. 962
            11. Petals absent or rarely present, the flowers yellowish-green, turning orange-red in fruit; calyx deeply parted, but the segments connected at the base. . . . . . . Penthorum, in SAXIFRAGACEAE, p. 774
      9. Upper half of pistil showing as 1–4 tips. . . . . . . . . . . . . . . . . . 12
         12. Stamens 16–24 or more . . . . . . . . . . . . . . . . . . . . . . 13
            13. Corolla irregular, some of the petals unequal in size and shape from the others and prolonged at one end into a pointed slender spur . . Delphinium, in RANUNCULACEAE, p. 679
            13. Corolla regular, the petals all equal in size and shape . . . . . . . . . . . 14
               14. Filaments of stamens united into a column or tube . . . . . MALVACEAE, p. 1044
               14. Filaments of stamens free, not united into a tube . . . . . . . . . . . 15
                  15. Leaves sticking easily to cloth or skin; broken parts of plant not producing milky, yellow, or orange juice; sepals united into a tube; ovary inferior, united with calyx-tube, the petals arising from the summit of the ovary . LOASACEAE, p. 1084
                  15. Leaves not sticking to cloth or skin; broken parts of plant producing a milky, yellow, or orange juice; sepals free and separate to their base; ovary superior, free from the calyx, and inserted above the base of the petals. PAPAVERACEAE, p. 719
         12. Stamens 1–10 . . . . . . . . . . . . . . . . . . . . . . . . 16
            16. Stamens 6, 8, or 10 . . . . . . . . . . . . . . . . . . . . . 17
               17. Petals absent. . . . non-petalous forms of CRUCIFERAE, p. 737, 743–744 (728–768)
               17. Petals present . . . . . . . . . . . . . . . . . . . . . . 18
                  18. Petals 3, irregular, not all equal in size and shape; sepals irregular, the 2 inner ones colored and petal-like, the 3 outer ones smaller and green; filaments of stamens united at base into a sheath or tube. . . . . . POLYGALACEAE, p. 969
                  18. Petals 4 or 5, regular, equal in size and shape; sepals or calyx-lobes regular, equal in size and shape; filaments of stamens free and not united into a tube . 19
                     19. Stamens 6 . . . . . . . . . . . . . . . . CRUCIFERAE, p. 728
                     19. Stamens 8 or 10 . . . . . . . . . . . . . . . . . . . 20
                        20. Stipules (small scale- or leaf-like appendages or outgrowths) present at base of leaf-stalks; petals 5; calyx divided into separate and distinct 5 sepals; ovary superior, the base of the ovary inserted above the base of the petals . . . . . . . . . . . . GERANIACEAE, p. 962
                        20. Stipules absent; petals 4; the 4 sepals united into a calyx-tube and calyx-lobes; ovary inferior, united with calyx-tube, the petals arising from summit of calyx-tube and ovary . . . . . . ONAGRACEAE, p. 1093
            16. Stamens 1–5 . . . . . . . . . . . . . . . . . . . . . . . 21

21. Corolla absent at flowering time, only a calyx present . . . . . . . . . . . . . . . *22*
22. Stipules (leaf-like or scale-like appendages or outgrowths) present at base of leaf-stalks. . . cleistogamous forms of Viola, in VIOLACEAE, p. 1069
22. No stipules present. . . . . . . . . . . . . . . . . . . . . . . . . *23*
23. Sepals 4; style 1; ovules numerous in the ovary; seeds numerous in the fruit . . . . Rorippa, in CRUCIFERAE, p. 758
23. Sepals or calyx-lobes 1–5; styles 2 or 3; ovule 1 in the ovary; seed 1 in the fruit . . . CHENOPODIACEAE, p. 600
21. Corolla present at flowering time . . . . . . . . . . . . . . . . . . . . . *24*
24. Corolla of separate petals, not united at the base, so that one petal may be removed without tearing or disturbing the rest of the corolla . . . . . . . . . . . . . . . . *25*
25. One end of flower prolonged into a curved comma-shaped, slender spur; petals 2 or 3, 2 of them 2-lobed; stem hollow with watery juice . . . . . . . . BALSAMINACEAE, p. 1024
25. Flower without a spur, or ending in a blunt knob-like sac which is not curved or comma-shaped; petals 5; stem, if hollow, without a watery juice . . . . . . . . . . . *26*
26. Main leaves or leaf-nodes on the stem numerous, more than 10; petals opposite the stamens . . . . . . . . . . . . . . . . . . Ceanothus, in RHAMNACEAE, p. 1028
26. Main leaves or leaf-nodes on the stem fewer, mostly 1–10; petals alternate with the stamens . . . . . . . . . . . . . . . . . . . . . . . . *27*
27. Loose leaf- or scale-like stipules present and conspicuous at base of leaf-stalk; petals showy, conspicuous, 7–15 mm. or more long; stamens closely surrounding the ovary; corolla with a knob-like spur at one end . . . . . . . VIOLACEAE, p. 1068
27. Stipules absent or very inconspicuous, if present, the base of the leaf-stalk sometimes with a dilated thin margin; petals small, inconspicuous, at most 3 mm. long; filaments of stamens spreading or remote from ovary; corolla without a knob-like spur at one end . . . . . . . . . . . . . . . . . . . . . . *28*
28. Ovary superior, free from the calyx, the petals arising below the insertion or base of the ovary; sepals 4; stamens 4. . . . . . Rorippa, in CRUCIFERAE, p. 758
28. Ovary partly to completely inferior, the petals arising either from the summit of the ovary or between the calyx-lobes; calyx either greatly reduced to 5 minute teeth or lobes, or the calyx conspicuous and elongated into a tube with 5 lobes; stamens 5. . . . . . . . . . . . . . . . . . *29*
29. Plant completely glabrous; calyx greatly reduced, the teeth or lobes minute; ovary with 2 ovules; fruit 2-seeded; ovary completely inferior, the petals arising from the summit of the ovary . . . . . . . UMBELLIFERAE, p. 1115
29. Some part of leaf-blade, leaf- or flower-stalk, or inflorescence more or less hairy; calyx conspicuous and elongated into a calyx-tube (hypanthium) and lobes; ovary with many ovules; fruit many-seeded; ovary partly inferior, partly united with the lower portion of the ovary, the petals arising between the calyx-lobes . . . . . . . Heuchera, in SAXIFRAGACEAE, p. 776
24. Corolla of united petals joined at least at the base, forming a short or long tube, so that one part of the corolla cannot be removed without tearing or disturbing a neighboring section . . *30*
30. One end of flower prolonged into a curved, comma-shaped, slender spur; petals 2–3, 2 of them 2-lobed; stem hollow with watery juice. . . . . . . . . BALSAMINACEAE, p. 1024
30. Flower without a spur or the spur, if present, straight; corolla-lobes 4 or 5; stem usually solid. . . . . . . . . . . . . . . . . . . . . . . . . . . *31*
31. Stamens with 4 fertile (pollen-bearing) anthers . . . . . . . . . . . . *32*
32. Placentae (where the ovules or seeds are attached) of the ovary or fruit axil (situated in the center); commonly encountered plants . . SCROPHULARIACEAE, p. 1328
32. Placentae (where the ovules or seeds are attached) of the ovary or fruit parietal (situated on the wall); rarely encountered . . . . . . . . . . . . *33*
33. Flowers and fruits borne next to the main stem at the base of the leaves (axillary); leaf-blades lanceolate or oblong-ovate; corolla 2–2.5 cm. long; fruit erect, ellipsoid, without any curved beaks, 2.5–3 cm. long . . PEDALIACEAE, p. 1371
33. Flowers and fruits borne at the top of the stem and branches (terminal); leaf-blades somewhat round and heart-shaped; corolla 3–5 cm. long; fruit with 2 spreading, curved horned beaks, 8–15 cm. long. . . MARTYNIACEAE, p. 1372
31. Stamens with 5 fertile (pollen-bearing) anthers . . . . . . . . . . . . *34*

34. Ovary inferior, united with the calyx-tube, the corolla-tube arising from the summit of the ovary; bruised stem or leaf-stalk usually producing milky juice . . . . . . . . Campanulaceae, p. 1428
34. Ovary superior, free from the calyx, the corolla-tube arising below the insertion or base of the ovary; bruised stem or leaf-stalk usually not producing milky juice . . . . . . . . . . . *35*
35. Ovary deeply 4-lobed, almost appearing like 4 separate ovaries . . . . . Boraginaceae, p. 1242
35. Ovary not 4-lobed . . . . . . . . . . . . . . . . . . . . . . . . . *36*
36. Corolla yellow or yellow with brown or purple-brown center . . . . . . . . . . *37*
37. Plants with prickles . . . . . . . . . . . . . . Solanum, in Solanaceae, p. 1311
37. Plants without prickles . . . . . . . . . . . . . . . . . . . . . . 38
38. Corolla-tube very short, much shorter than the corolla-lobes; filaments of the stamens conspicuously purple-hairy; calyx divided nearly to the base, practically with no tube present . . . . . Verbascum Blattaria, in Scrophulariaceae, p. 1339
38. Corolla-tube or united part of corolla conspicuous, much longer than the slightly lobed or toothed border; filaments of the stamens mainly without hairs, sometimes slightly hairy with white hairs; calyx lobed only part way, with an evident tube Physalis, in Solanaceae, p. 1314
36. Corolla white, blue, lavender, purple, or rose-red . . . . . . . . . . . . . . *39*
39. Ovary 1-celled . . . . . . . . . . . . . . . . Hydrophyllaceae, p. 1233
39. Ovary 2–5-celled . . . . . . . . . . . . . . . . . . . . . . . . *40*
40. Filaments of the stamens conspicuously purple-hairy for most of their length . Verbascum Blattaria, in Scrophulariaceae, p. 1339
40. Filaments of the stamens either without hairs or with hairs only at very base of filament, but not conspicuously purple-hairy . . . . . . . . . . . . . . *41*
41. Anthers touching, coming together and forming a tube around the style; filaments very short, usually much shorter than the anther. . . . . . . Solanum, in Solanaceae, p. 1311
41. Anthers more or less separated from one another, not touching nor forming a tube around the style; filaments usually as long as or longer than the anthers . *42*
42. Flowers or clusters of flowers at the ends of the stems or branches; leaf-margins toothed and angled; ovary containing numerous ovules; fruit containing numerous seeds . . . . . . . . . . . Solanaceae, p. 1310
42. Flowers arising along the sides of the stem (laterally) from the axils of the leaves; leaf-margins mainly without teeth or angles, the lobes extending mainly from the base of the leaf-blade; ovary containing 4–6 ovules; ovary containing 4–6 seeds . . . . . . . . . Convolvulaceae, p. 1212

# *List of Champion trees in Missouri*

*(Furnished by Kendall Laughlin, 165 Pine Ave., Chicago, Ill.)*

| Scientific Name | Owner | Location | County | Circumference* | Height |
|---|---|---|---|---|---|
| *Taxodium distichum* | State | Big Oak Tree State Park | Mississippi | 22′8″ | — |
| *Salix nigra* | State | Big Oak Tree State Park | Mississippi | 8′9″ | — |
| *Populus deltoides* | State | Pershing State Park | Linn | 19′1″ | 81′ |
| † *Populus heterophylla* | State | Big Oak Tree State Park | Mississippi | 5′8″ | 74′ |
| *Juglans nigra* | City | Hyde Park, St. Joseph | Buchanan | 9′3″ | — |
| *Carya cordiformis* f. *cordiformis* | Elmer W. Ahmann | 413 N. Crysler, Independence | Jackson | 9′7″ | — |
| † *Carya cordiformis* f. *latifolia* | Della P. McColgan | Stoddard near Catalpa, Dexter | Stoddard | 9′5″ | 90′ |
| *Carya illinoensis* | Handy and Bill Moore | Sec. 34, T 24 N, R 16 E | Mississippi | 20′10″ | — |
| † *Carya laciniosa* | State | Big Oak Tree State Park | Mississippi | 12′10″ | 101′ |
| *Carya ovata* | City | Hyde Park, St. Joseph | Buchanan | 6′ | — |
| *Carya tomentosa* | City | Forest Park | St. Louis | 7′3″ | 62′ |
| *Carya texana* var. *texana* | City | Swope Park, Kansas City | Jackson | 5′9″ | 55′ |
| *Carpinus caroliniana* | State | Big Spring State Park | Carter | 3′11″ | — |
| *Betula nigra* | State | Pershing State Park | Linn | 8′ | — |
| *Castanea ozarkensis* | State | Roaring River State Park | Barry | 6′2″ | — |
| *Quercus alba* | Same as location | Elmwood Cemetery, Kansas City | Jackson | 11′8″ | — |
| † *Quercus* × *bebbiana* | City | Forest Park | St. Louis | 7′8″ | 62′ |
| *Quercus stellata* | Mrs. Edith L. Camp | Sassafras near St. Francis, Dexter | Stoddard | 10′2″ | — |
| | City | Elm near Vine, Dexter | Stoddard | 10′2″ | — |
| *Quercus lyrata* | State | Big Oak Tree State Park | Mississippi | 11′11″ | — |
| *Quercus macrocarpa* [1] | State | Big Oak Tree State Park | Mississippi | 17′7″ | 127′ |
| *Quercus bicolor* | City | Forest Park | St. Louis | 11′7″ | 65′ |
| *Quercus Michauxii* | State | Big Oak Tree State Park | Mississippi | 19′11″ | 126′ |
| *Quercus prinoides* var. *acuminata* | Charles F. Curry | Eastwood Hills, Kansas City | Jackson | 12′4″ | — |
| *Quercus prinoides* var. *prinoides* | City | Swope Park, Kansas City | Jackson | 1′2″ | — |
| *Quercus rubra* | Same as location | Elmwood Cemetery, Kansas City | Jackson | 12′11″ | — |
| *Quercus palustris* | State | Big Oak Tree State Park | Mississippi | 13′5″ | — |
| † *Quercus* × *mutabilis* | State | Big Oak Tree State Park | Mississippi | 10′9″ | 131′ |
| *Quercus coccinea* var. *tuberculata* | Dr. A. L. May | Cynthia St., Poplar Bluff | Butler | 6′8″ | — |

| | Scientific Name | Owner | Location | County | Circumference* | Height |
|---|---|---|---|---|---|---|
| | *Quercus Shumardii* | State | Big Oak Tree State Park | Mississippi | 13′3″ | — |
| | *Quercus velutina* | City | Hyde Park, St. Joseph | Buchanan | 12′10″ | — |
| † | *Quercus velutina* f. *missouriensis* | City | Hyde Park, St. Joseph | Buchanan | 11′6″ | 62′ |
| | *Quercus falcata* var. *falcata* | Grace Danforth | 105 Danforth St., Charleston | Mississippi | 14′10″ | 79′ |
| | *Quercus falcata* var. *pagodaefolia* | Laura Dawson | Dawson Rd., New Madrid | New Madrid | 18′11″ | 109′ |
| | *Quercus marilandica* | T. R. Seifert | 815 N. Main St., Poplar Bluff | Butler | 7′11″ | — |
| † | *Quercus* × *Bushii* | City | Forest Park | St. Louis | 8′7″ | 67′ |
| † | *Quercus imbricaria* | Laura Dawson | Dawson Rd., New Madrid | New Madrid | 11′2″ | 92′ |
| | *Quercus* × *runcinata* | State | Van Meter State Park | Saline | 7′11″ | 46′ |
| † | *Quercus* × *leana* | Ada Maupin | 5607 E. 23rd St., Kansas City | Jackson | 9′3″ | 68′ |
| | *Quercus* × *tridentata* | May V. Robertson | S. of Swope Park, Kansas City | Jackson | 2′4″ | — |
| | *Quercus Phellos* | Viverette Lee | 601 E. Cypress, Charleston | Mississippi | 10′3″ | 84′ |
| | *Ulmus americana* | State | Big Oak Tree State Park | Mississippi | 21′7″ | — |
| † | *Ulmus Thomasi* | City | Swope Park, Kansas City | Jackson | 10′11″ | 99′ |
| † | *Planera aquatica* | Lee O'Reilly | Sec. 9, T 23 N, R 16 E | Mississippi | 2′8″ | 26′ |
| | *Celtis occidentalis* var. *occidentalis* | J. Abner Beck | 507 S. Main St., Charleston | Mississippi | 12′10″ | 83′ |
| | *Celtis occidentalis* var. *canina* | City | Swope Park, Kansas City | Jackson | 10′5″ | 65′ |
| | *Celtis laevigata* | Max Friedman | 903 E. Commercial St., Charleston | Mississippi | 9′8″ | 58′ |
| | *Morus rubra* | Miss A. L. Thrower | 26-28 S. Mulberry St., Dexter | Stoddard | 8′ | — |
| | *Maclura pomifera* | Laura Dawson | Dawson Rd., New Madrid | New Madrid | 8′3″ | — |
| | *Asimina triloba* | State | Big Oak Tree State Park | Mississippi | 1′8″ | — |
| | *Sassafras albidum* | — | South of Co. Highway P | New Madrid | 9′9″ | — |
| † | *Liquidambar styraciflua* [2] | Laura Dawson | Dawson Rd., New Madrid | New Madrid | 16′11″ | 112′ |
| | *Platanus occidentalis* [3] | Estate of W. F. Wilkinson *et al.* | 89th & Olive, Kansas City | Jackson | 16′2″ | 109′ |
| † | *Pyrus coronaria* var. *lancifolia* | City | Blue Valley Park, Kansas City | Jackson | 1′10″ | 25′ |
| | *Pyrus ioensis* | City | Blue Valley Park, Kansas City | Jackson | 1′10″ | 30′ |
| | *Crataegus crus-galli* [4] | City | Swope Park, Kansas City | Jackson | 1′10″ | — |
| | *Crataegus mollis* | City | Swope Park, Kansas City | Jackson | 4′5″ | 27′ |
| | *Prunus serotina* | Nathan B. Baraban | 5925 State Line, Kansas City | Jackson | 9′1″ | — |
| | *Gymnocladus dioicus* | James P. Reed | 1054 W. 55th St., Kansas City | Jackson | 9′10″ | — |
| | *Gleditsia triacanthos* | City | Loose Park, Kansas City | Jackson | 14′4″ | 71′ |
| † | *Gleditsia triacanthos* f. *inermis* | State | Big Oak Tree State Park | Mississippi | 11′2″ | 118′ |
| † | *Gleditsia aquatica* | State | Big Oak Tree State Park | Mississippi | 4′7″ | 88′ |
| | *Cercis canadensis* | City | Union Cemetery, Kansas City | Jackson | 3′3″ | — |
| † | *Ilex decidua* | State | Big Oak Tree State Park | Mississippi | 1′1″ | 20′ |
| † | *Acer rubrum* var. *Drummondii* | Vernon Cresson | St. Francis River W. of Kennett | Dunklin | 6′5″ | 52′ |
| | *Acer nigrum* | Same as location | Elms Hotel, Excelsior Springs | Clay | 7′4″ | 77′ |
| | *Acer saccharinum* | State | Big Oak Tree State Park | Mississippi | 17′11″ | — |
| | *Acer saccharum* | City | Hyde Park, St. Joseph | Buchanan | 6′11″ | — |
| | *Acer Negundo* | City | Brush Creek east of state line, Kansas City | Jackson | 8′7″ | — |
| * | *Aesculus glabra* var. *arguta* | Same as location | Elms Hotel, Excelsior Springs | Clay | 4′7″ | 48′ |
| | *Tilia americana* | City | Krug Park, St. Joseph | Buchanan | 10′3″ | — |
| | *Nyssa aquatica* [2] | Vernon Cresson | St. Francis River W. of Kennett | Dunklin | 4′10″ | 54′ |
| | *Nyssa sylvatica* | City | 11 E. Market St., Dexter | Stoddard | 9′9″ | — |
| † | *Cornus Drummondii* | Same as location | Mt. Washington Cemetery | Jackson | 1′2″ | 24′ |

| | Scientific Name | Owner | Location | County | Circumference* | Height |
|---|---|---|---|---|---|---|
| | *Bumelia lanuginosa* var. *albicans* | State | Meramec State Park | Franklin | 2′8″ | 26′ |
| | *Diospyros virginiana* | City | Carondelet Park | St. Louis | 6′10″ | — |
| | *Fraxinus americana* | State | Big Oak Tree State Park | Mississippi | 10′4″ | — |
| † | *Fraxinus tomentosa* | State | Big Oak Tree State Park | Mississippi | 13′ | 100′ |
| † | *Fraxinus pennsylvanica* var. *lanceolata* | State | Big Oak Tree State Park | Mississippi | 14′5″ | 105′ |
| † | *Forestiera acuminata* | State | Big Oak Tree State Park | Mississippi | 2′1″ | 28′ |
| | *Catalpa speciosa* | Della P. McColgan | Stoddard near Catalpa, Dexter | Stoddard | 13′10″ | — |
| | *Viburnum prunifolium* L. | City | Swope Park, Kansas City | Jackson | 11″ | 19′ |
| † | *Viburnum rufidulum* | State | Big Oak Tree State Park | Mississippi | 2′2″ | 23′ |

* Measured at 54 inches above the ground.
† This is the largest tree of its kind in the United States.
[1] Measured jointly with Robert E. McDermott, Roy H. Degler, and Richard Holecamp.
[2] Measured jointly with Robert E. McDermott.
[3] Measured jointly with Stanley R. McLane.
[4] With yellow anthers.

*Division I*

# *Pteridophyta (Vascular Cryptogams)*

Plants of soil, in or floating on the water, or growing as epiphytes on trees or rocks. Plants belonging to this group may resemble what are popularly considered fern-like, but may also be moss-like, rush-like, or quill-like. They reproduce by spores, and not by embryos, and have no true flowers. An alternation of generations is characteristic of this division: a small green or inconspicuous *prothallium* (gametophyte) represents the haploid or sexual generation bearing antheridia and archegonia; the mostly large *sporophyte* represents the diploid or asexual generation, and has usually roots, stems, and leaves of diverse shapes and forms, with reproduction asexually by means of spores produced within *sporangia* or spore-cases (*sporocarp*). The spores may be all alike, or in some families of two types (minute microspores and larger megaspores).

About 9,000 species are known.

The following key to the main orders of Pteridophyta is adapted partly from Morton's treatment in Gleason's *New Illustrated Flora*, and partly from Fernald's eighth edition of *Gray's Manual.*

a. Plants free-floating, small, moss-like; sporangia on the lower side, enveloped by stalkless or nearly stalkless sporocarps . . . . . . . . . . . . . . Fam. SALVINIACEAE in Order Filicales (p. 16)
a. Plants rooting in the soil, or growing on rocks, bluffs, trees, or submerged in water . . . . . . . *b*
  b. Stems jointed; the leaves whorled and united into a sheath; sporangia arranged in a cone-like strobilus at the tip of the stem . . . . . . . . . . . . . . Order EQUISETALES (p. 11)
  b. Stems not jointed; the leaves not united into a sheath; sporangia either in the axils of leaves, or at the base of the leaves, or borne on the back or margin of the leaves, or in spike-like or branched special fronds . . . . . . . . . . . . . . . . . . . . . . . . . *c*
    c. Leaves small and overlapping like shingles, or grass-like and capping a short thickened stem . . . *d*
      d. Leaves grass-like and elongated; stem unbranched; sporangia sunken in the leaf-base . . Order ISOETALES (p. 10)
      d. Leaves small, less than ½ inch (13 mm.) long; stems branched; sporangia not sunken in the leaf-base . . . . . . . . . . . . . . . . . . . . . . . *e*
        e. Leaves in 4–16 ranks, without a ligule; sporangia-bearing stems cylindrical (terete); sporangia all alike, minute, numerous . . . . . . . . . Order LYCOPODIALES (p. 7)
        e. Sterile leaves in 4–6 ranks, with a transverse ligule at base; sporangia-bearing stems more or less 4-sided; sporangia consisting of 4 large megaspores or many minute microspores . . . . . . . . . . . . . . . . Order SELAGINELLALES (p. 8)
    c. Sterile leaves either not overlapping, or lacking sporangia arranged in axils of or at the bases of leaves, or borne on the back or margin of the leaves, or in spike-like or on branched special fronds, or in special hard spore-cases . . . . . . . . . . . . . . . . . . . *f*
      f. Fronds (leaf-like part) with 4 leaflets arranged like a 4-leaf clover . . . Fam. MARSILEACEAE in Order Filicales (p. 17)
      f. Fronds otherwise . . . . . . . . . . . . . . . . . . . . . . . . . *g*

g. Fertile fronds or fertile portions of the fronds similar to the sterile fronds . . . . Fam. POLYPODIACEAE in Order Filicales (p. 20)

g. Fertile fronds or fertile portions of the fronds quite unlike the sterile frond or fronds . . . . . . *h*

h. Sterile (leaf-like) portion of frond simple and without any teeth or lobes; the fertile consisting of a simple 2-rowed spike of sporangia . . . . . . . . . . Order OPHIOGLOSSALES (p. 15)

h. Sterile (leaf-like) frond or portion of frond more or less divided; fertile frond or portion of frond variously divided . . . . . . . . . . . . . . . . . . . . . . . . . . *i*

i. Leaves divided into a sterile lower half and a fertile upper panicle; leaves fleshy; leaves not coiled in bud, with sheathing leaf-bases; sporangia without an annulus, nearly stalkless . . Order OPHIOGLOSSALES (p. 14)

i. Leaves not fleshy; leaves usually coiled in bud, without sheathing leaf-bases; sporangia with an annulus, short- or long-stalked . . . . . . . . . . . . . Order FILICALES (p. 18)

The most modern classification discards the division Pteridophyta, and divides the above orders into the following natural alliances:

Class Filicineae:

| Phylum | Orders |
|---|---|
| Phylum: Lycopsida | Lycopodiales<br>Selaginellales<br>Isoetales |
| Phylum: Sphenopsida | Equisetales |
| Phylum: Pteropsida | Ophioglossales<br>Filicales |

## KEY TO FERNS AND FERN ALLIES

a. Whole plant floating free on water. . . . . . . . . . Water Fern Family (Salviniaceae) (p. 16)

a. Some part of the plant either rooted in the soil, on land, or in water, or growing on rocks or trees, not free floating . . . . . . . . . . . . . . . . . . . . . . . . . . *b*

b. Stems conspicuously jointed and finely many-ridged . . . Horsetail Family (Equisetaceae) (p. 11)

b. Stems not jointed and rigid, with green leaf-blades or leaf-like fronds . . . . . . . . . . *c*

c. Leaves small, scale-like, moss-like, needle-like, less than 2 cm. long, or, if longer, quill-like . . . *d*

d. Leaves quill- or needle-like, or in a single rosette . . . Quillwort Family (Isoetaceae) (p. 10)

d. Leaves scale- or moss-like, numerous, often overlapping . . . . . . . . . . . . . *e*

e. Leaves pale or bright green and thin on delicately creeping stems, or leaves dull, dark, or gray-green, closely overlapping, thread-like, bristly tipped, the individual stems (excluding the leaves) only 0.2–1.2 mm. thick; stems and branches at most 5 cm. tall . . . . . . . . . . . . . . . Spikemoss Family (Selaginellaceae) (p. 8)

e. Leaves otherwise, or dark green, the individual stems thicker; stems or erect branches usually 7–25 cm. tall . . . . . . . . . Clubmoss Family (Lycopodiaceae) (p. 7)

c. Leaves (at least the sterile blades) larger, flattened, and expanded, more than 2 cm. long, not quill-like . . . . . . . . . . . . . . . . . . . . . . . . . . *f*

f. Leaves divided into 4 fan-shaped leaflets, like a 4-leaf clover . . . . . . Pepperwort Family (Marsileaceae) (p. 17)

f. Leaves not divided into 4 fan-shaped leaflets . . . . . . . . . . . . . . . . . *g*

g. At least the green leaf-like part of plant (frond) simple, not toothed or lobed . . . . . *h*

h. Green leaf-like part long and narrow, long-tapering at tip, the midrib distinct . . 22. Walking fern (CAMPTOSORUS in Polypodiaceae)

h. Green leaf-like part short-pointed or rounded at tip, the midrib none or obscure . . 6. Adder's Tongue (OPHIOGLOSSUM in Ophioglossaceae)

g. Green leaf-like part of plant (frond) shallowly or deeply lobed, or divided into distinct units or segments (pinnae and pinnules) . . . . . . . . . . . . . . . . . *i*

i. Leafy part lobed, but the innermost portion (base) (part toward midrib) of at least some of the lobes connected with one another by a continuous leafy connection. . . . *j*

j. Stipe (main stem) and rachis (central axis between leafy part) split into 3 main branches . . . . . . . . . . . . . . . 12. PTERIDIUM in Polypodiaceae

j. Stipe and rachis straight or one main axis . . . . . . . . . . . . . . . . . . . *k*
k. Main leafy part with a conspicuous long tail-like tip . . . . . . . 21. ASPLENIUM (A. EBENOIDES, GRAVESII, and PINNATIFIDUM) in Polypodiaceae
k. Main leafy part lacking a conspicuous long tail-like tip, or, if somewhat long-tailed at tip, either at least the greatest part of the leafy area is deeply divided, or the leafy area 10–40 cm. wide . . . . . . . . . . . . . . . . . . . . . . . . . . . . . *l*
l. Leafy part 1–7 cm. wide, each of the lobes 2–7 mm. wide; plants growing on rocks, bluffs, or trees . . . . . . . . . . . . . . . . . . . 17. POLYPODIUM in Polypodiaceae
l. Leafy part 10–40 cm. wide, each of the lobes 10–80 mm. wide; plants growing on the ground in soil . . . . . . . . . . . . . . . . . . . . . . . . . . *m*
m. Upper side of leaf segments finely hairy (use lens); veins of segments simple or forked, not joining in a net (reticulate) (use lens or hold up to light) . . . . . 26. THELYPTERIS HEXAGONOPTERA in Polypodiaceae
m. Upper side of segments mostly glabrous (without hairs); veins netted, joining . . . . . *n*
n. At least some of the segments (usually lower ones) obviously lobed or with wavy margins, but without fine teeth on margin itself; leaf segments, at least in the upper half of leaf-blades, opposite . . . . . . . . . . 19. ONOCLEA in Polypodiaceae
n. None of the segments obviously lobed or with wavy margins, but the margins actually have fine teeth; leaf segments all alternate . 20. WOODWARDIA AREOLATA in Polypodiaceae
i. Leafy part of plant variously divided into distinct segments completely cut to the main stem (rachis) . . *o*
o. Leafy part of plant white-powdery on lower surface . . . . . . . . . . . 15. NOTHOLAENA
o. Leafy part of plant of other colors on lower surface but not white-powdery . . . . . . . . *p*
(see alternate key for completely sterile plants on p. 5)

## KEY TO FERTILE (SPORE-BEARING) PLANTS

p. Fertile (spore-bearing) part of plant on a special branch completely or partly separated from the sterile leafy part, or if not on a special branch, at least separated in some manner from the leafy part . . . . . . . . . . . . . . . . . . . . . . . . . *q*
q. Roots fleshy; leafy part broadly triangular, divided into three main parts . . . 5. BOTRYCHIUM in Ophioglossaceae
q. Roots fibrous; leafy part once divided into main segments arranged stepladder-like one above the other (pinnae) or twice compoundly divided (bipinnate) . . . . . . . . . . *r*
r. Leafy part twice compoundly divided (twice pinnate), the main branches and segments well separated from one another; fertile parts, when present, at summit of leafy branch . . . . . . . . . . 10. OSMUNDA REGALIS var. SPECTABILIS in Osmundaceae
r. Leafy part once divided into main segments arranged stepladder-like and crowded one above the other; fertile parts, when present, either on branches separate from and shorter than the leafy branches, or interrupting a regular sequence of sterile segments on a leafy branch . . . . . . . . . . . . . . . . . . . . . . . . *s*
s. Tufts of woolly hair present at the base of most leaf divisions (pinnae); margins of lobes of leaf divisions loosely fringed with hairs 10. OSMUNDA CINNAMOMEA in Osmundaceae
s. Tufts of wool not present at the base of main leaf divisions (pinnae); margins of lobes of leaf divisions without hairs . . . . . . . . . . . . . . . . . . . *t*
t. Smallest segments or pinnules mostly 2–4 mm. wide; veins of leaf segments simple, free; main pinnae long-pointed (acuminate) at tip; fertile parts arising on branches separate from and shorter than the leafy branches . . 18. MATTEUCCIA in Polypodiaceae
t. Smallest leaf segments or pinnules mostly 4.5–8 mm. wide; veins of leaf segments forking; main pinnae blunt or short-pointed at tip; fertile parts interrupting a sequence of leafy sterile segments on the leafy branch . . 10. OSMUNDA CLAYTONIANA in Osmundaceae
p. Fertile (spore-bearing) part (fruit dots) of plant usually appearing on the under surface or along or near the margins of the leafy part (frond) . . . . . . . . . . . . . . *u*
u. Leaf-blade broadly triangular, compoundly divided into three large main divisions (three times divided or nearly so) . . . . . . . . . . . . . . . . . 12. PTERIDIUM
u. Leaf-blade otherwise, not alternately divided . . . . . . . . . . . . . . . . . *v*

v. Stipe (main leaf-stalk) and rachis (central axis of leaf-blade) dark brown, reddish- or purplish-brown or black mostly from end to end . . . *w*

w. Stipe and rachis mostly lacking hairs or scales . . . *x*

x. Stipe and rachis remaining simple, straight, and unbranched; leaf-blade once-pinnate, the divisions stepladder-like one above the other, the lowermost pinnae at most deeply lobed, but not completely separated at their base . . . 21. ASPLENIUM

x. Stipe and rachis either forking into two main branches above or alternately branched, or the lowermost main leaf divisions composed of more than one pinnule . . . *y*

y. Ultimate leaf divisions (pinnules) toothed or lobed; stipe and rachis either forking into two main branches above or the lateral branches of rachis alternate . . . 13. ADIANTUM

y. Ultimate leaf divisions (pinnules) lacking teeth; stipe straight, the lower main leaf divisions (pinnae) divided into 3 or more segments (pinnules) from opposite branches . . . 14. PELLAEA GLABELLA

w. Stipe and rachis mostly with hairs . . . *z*

z. Leaf-blade once-pinnate, all the divisions remaining simple and undivided, even the lowermost pinnae only deeply lobed at most and not completely separated at their base. . 21. ASPLENIUM

z. Leaf-blade twice or thrice pinnate, or at least the lowest main divisions (pinnae) of 3–9 segments (pinnules) . . . *1*

1. Ultimate leaf-divisions (pinnules) 5–10 mm. wide; leaf-blade simply pinnate above, the lowermost pinnae with 3–9 pinnules . . . 14. PELLAEA ATROPURPUREA

1. Ultimate leaf-divisions (pinnules) at most 1.5 mm. wide; leaf-blade twice to thrice pinnate . . . 16. CHEILANTHES

v. Stipe (main leaf stalk) and/or rachis (central axis of leaf-blade) at least green or only pale brown in upper half, the lower half green, pink, straw-colored or pale brown, but never dark brown or black from one end to the other . . . *2*

2. Main leaf-divisions (pinnae) simple, not lobed nor divided to their midribs, at most only toothed. . . *3*

3. Stipe with conspicuous chaffy scales; leaf-divisions firm and thickish; some of leaf-blades of last year lying on the ground. . . . 28. POLYSTICHUM

3. Stipe mostly lacking hairs or scales, presenting a smooth appearance; leaf divisions thin, delicately membranous; all leaf-blades upright . . . 23. ATHYRIUM PYCNOCARPON

2. Main leaf divisions (pinnae) deeply lobed or divided to their midribs . . . *4*

4. No conspicuous hairs or chaffy scales present on stipe and rachis . . . *5*

5. Main leaf-divisions (pinnae) obviously alternate, 2–6, obviously long-stalked (petiolate); frond 1–7 (–17) cm. long . . . 21. ASPLENIUM RUTA-MURARIA var. CRYPTOLEPIS

5. At least the lower and middle main leaf-divisions (pinnae) opposite, or nearly or chiefly opposite, the uppermost often alternate, 9–40, stalkless (sessile) or only very short-stalked (petiolate); frond mostly 7–90 cm. (rarely as little as 2 or as much as 100 cm.) long . . . *6*

6. Main leaf-divisions (pinnae) divided all the way to the midribs, or if not quite so divided, at least some of the ultimate pinnules unequal in size and shape; fruit dots (sori) on under side of leaf roundish; membrane (indusium) over fruit dots cup-shaped. . 25. CYSTOPTERIS

6. Main leaf-divisions (pinnae) deeply lobed but not divided all the way to their midribs, the lobes connected from base to tip by continuous green tissue, the ultimate lobes all of about the same size and shape; fruit dots and membrane over them elongated, longer than broad, or rounded or kidney-shaped . . . *7*

7. Ultimate leaf-divisions minutely hairy-fringed on margins; fruit dots rounded or kidney-shaped (reniform); rootstocks slender, creeping; plants of wet meadows and spring-fed wet ground . . . 26. THELYPTERIS PALUSTRIS var. PUBESCENS

7. Ultimate leaf-divisions not hairy-fringed on margins; fruit dots elongated, somewhat curled, longer than broad; rootstock thickish; plants of rich woods . . . 23. ATHYRIUM THELYPTEROIDES

4. Obvious scales or tiny hairs or glands, or both, present on stipe and rachis, at least in early stage of growth . . . *8*

8. Tiny hairs and glands, but no scales, present on stipe and/or rachis; fruits dots appearing at the margin of the smallest leaf-divisions . . . 11. DENNSTAEDTIA

8. Obvious scales present on some part of the stipe, either when young or at maturity; hairs and glands present or absent on stipe and/or rachis; fruit dots appearing on the under surface of the ultimate leaf-divisions . . . *9*

9. Tiny long-stalked glands present on rachis (axis of leafy blade): indusium attached beneath the sorus . . . 24. WOODSIA

9. No long-stalked glands on rachis, although other kinds of glands or hairs may be present; indusium otherwise . . . . . . *10*
10. Hairs present on upper and lower surfaces or veins of leaf-divisions . . . . *11*
11. Ultimate leaf-divisions minutely hairy-fringed on margins; fruit dots rounded or kidney-shaped (reniform); rootstocks slender, creeping; plants of wet meadows and spring-fed ground . . . . 26. THELYPTERIS PALUSTRIS var. PUBESCENS
11. Ultimate leaf divisions not hairy-fringed on margins; fruit dots elongated, somewhat curved, longer than broad; rootstock thickish; plants of rich woods . . . . 23. ATHYRIUM THELYPTEROIDES
10. Hairs absent on upper and lower surfaces or veins of leaf divisions . . . . *12*
12. Fronds not evergreen, rather thin and nearly membranaceous; indusium with a linear attachment, thin and inconspicuous . . . . 23. ATHYRIUM FILIX-FEMINA
12. Fronds evergreen, rather thick; indusium conspicuous . . . . 27. DRYOPTERIS

## ALTERNATE KEY TO COMPLETELY STERILE SPECIMENS

(substitute for o. on p. 3 and continue)

p. Stipe and rachis either forking into 2 main branches above or the lateral branches of rachis alternate . . . . 13. ADIANTUM in Polypodiaceae
p. Stipe and rachis either straight, or split into 3 main branches . . . . *q*
q. Blade as a whole broadly triangular and 3-parted . . . . *r*
r. Stipe round, soft and brittle, easy to break, root fleshy . . . 5. BOTRYCHIUM in Ophioglossaceae
r. Stipe not round, fibrous, hard to tear, root fibrous . . . . 12. PTERIDIUM in Polypodiaceae
q. Blade as a whole not 3-parted . . . . *s*
s. Blade once pinnate, all the divisions simple . . . . *t*
t. Stipe and rachis dark brown, reddish- or purplish-brown or black mostly from end to end . . *u*
u. Minute teeth on edges of leaf-divisions . . . . 21. ASPLENIUM
u. No teeth present, the margins entire . . . . 14. PELLAEA GLABELLA
t. Stipe and/or rachis at least green or only pale brown in upper half, the lower half green, pink, straw-colored, or pale to dark brown, but never completely dark brown or black from one end to the other . . . . *v*
v. Dwarf ferns with leafy blade only 4–25 cm. long; leaf-blade at most 4.5 cm. wide . 21. ASPLENIUM
v. Large ferns with leafy blade mostly 25–110 cm. long; leaf-blade 5–20 cm. wide. . . . . *w*
w. Stipe with conspicuous chaffy scales; leaf-divisions firm, thickish; some of leaf-blades of last year remaining next to the ground . . . . 28. POLYSTICHUM
w. Stipe mostly lacking hairs or scales, presenting a smooth appearance; leaf-divisions thin, delicately membranous; all leaf-blades upright . . . 23. ATHYRIUM PYCNOCARPON
s. Blade more than once pinnate, at least the lower or lowermost pinnae (main horizontal divisions) divided, or all the pinnae deeply lobed (pinnatifid) or divided to their midrib into vertical divisions (pinnules) . . . . *x*
x. Stipe and rachis dark brown, reddish- or purplish-brown or black mostly from end to end . . *y*
y. Ultimate leaf-divisions (pinnules) 5–10 mm. wide; leaf simply pinnate above, or the lower and/or upper pinnae divided into 3–9 divisions . . . . 14. PELLAEA
y. Ultimate leaf-divisions (pinnules) at least 1.5 mm. wide; leaf twice to thrice pinnate. . 16. CHEILANTHES
x. Stipe and/or rachis at least green or only pale brown in upper half, the lower half green, pink, straw-colored or pale brown, but never a dark brown or black from one end to the other . . . . *z*
z. Tufts of woolly hair present at the base of most leaf-divisions . . . 10. OSMUNDA CINNAMOMEA in Osmundaceae
z. No tufts of wool present at base of leaf-divisions . . . . *1*
1. Main leaf-divisions (pinnae) all alternate, 2–6, and obviously long-stalked (petiolate); frond 1–7 (–17) cm. long . . . . 21. ASPLENIUM RUTA-MURARIA var. CRYPTOLEPIS
1. Without this combination of characters, the fronds either larger, or the main pinnae opposite, or, if alternate, the pinnae not long-stalked or more numerous . . . . *2*
2. Leaf-blade twice or thrice pinnate, the main pinnae divided to their midribs into separate unconnected divisions (pinnules) . . . . *3*

3. Ultimate leaf-divisions nearly entire, obviously well separated from one another, mostly 3–7 cm. long and 0.7–2 cm. broad . . . . . . . . . . . . 10. OSMUNDA REGALIS var. SPECTABILIS
3. Ultimate leaf-divisions mostly toothed or lobed, rather crowded, mostly much shorter and narrower . . . . . . . . . . . . . . . . . . . . . . . . . . . . . . . . *4*
4. No conspicuous hairs or chaffy scales present on stipe and rachis at maturity . . 25. CYSTOPTERIS
4. Obvious scales or tiny hairs or glands, or both, present on stipe and rachis, at least in early stage of growth . . . . . . . . . . . . . . . . . . . . . . . . . . *5*
5. Tiny hairs and glands, but no scales present on stipe and/or rachis . . . . 11. DENNSTAEDTIA
5. Obvious scales present on some part of the stipe, either when young or at maturity; hairs and glands present or absent on stipe and/or rachis . . . . . . . . . . . . . . *6*
6. Tiny long-stalked glands present on rachis (axis of leaf-blade) . . . . . 24. WOODSIA
6. No long-stalked glands on rachis, although other kinds of glands or hairs may be present . . *7*
7. Fronds not evergreen, rather thin and nearly membranaceous. 23. ATHYRIUM FILIX-FEMINA
7. Fronds evergreen, rather thick . . . . . . . . . . . . 27. DRYOPTERIS
2. Leaf-blade with the main pinnae deeply lobed (pinnatifid), but not divided to their midribs into disconnected separate divisions, the ultimate divisions remaining connected at least at the very base. . . *8*
8. Ultimate leaf-divisions with entire or nearly entire margins . . . . . . . . . . . . . . *9*
9. No conspicuous hairs or chaffy scales present on stipe and rachis at maturity . . . . . . *10*
10. Ultimate leaf-divisions minutely hairy-fringed on margins; rootstock slender and creeping; fronds not growing in a clump from a center; plants of wet meadows and moist spring-fed ground. . . . . . . . . . . . . . 26. THELYPTERIS PALUSTRIS var. PUBESCENS
10. Ultimate leaf-divisions not minutely hairy-fringed on margins; rootstocks thick; fronds growing in a clump from a center; plants of rich woodland . . . . . . . . . . *11*
11. Main horizontal pinnae long-pointed at tip; veins of ultimate leaf-divisions simple, not forked . . . . . . . . . . . . . . . 23. ATHYRIUM THELYPTEROIDES
11. Main horizontal pinnae blunt or short-pointed at tip; veins of ultimate leaf-divisions forked. . . . . . . . . . . . . . . . . . . 10. OSMUNDA CLAYTONIANA
9. Obvious scales or hairs or glands or both present on stipe and rachis . . . . . . . . . *12*
12. Ultimate leaf-divisions minutely hairy-fringed on margins; rootstock slender and creeping; fronds not growing in a clump from a center . . . 26. THELYPTERIS PALUSTRIS var. PUBESCENS
12. Ultimate leaf-divisions not minutely hairy-fringed on margins; rootstock thick; fronds growing in a clump from a center . . . . . . . . . . . . . . . . . . . . *13*
13. Fronds conspicuously narrowed at base, the lowest main pinnae conspicuously shorter than those half way up; fronds numerous, growing around the center in a vase-like arrangement . . . . . . . . . . . . . . . . . 18. MATTEUCCIA
13. Fronds slightly narrower at base, but the lowest main pinnae only slightly shorter than those half way up; fronds relatively few, grouped close to one another, but not arranged around the center in a vase-like manner . . . 23. ATHYRIUM THELYPTEROIDES
8. Ultimate leaf-divisions toothed, either minutely or conspicuously . . . . . . . . . . *14*
14. No scales or hairs present on stipe or rachis . . . . . . . . . . . . 25. CYSTOPTERIS
14. Either scales or hairs or both present on some part of stipe or rachis . . . . . . . . . *15*
15. Hairs absent on upper and lower surface or veins of leaf-divisions. . . . . . . . . *16*
16. Fronds not evergreen, rather thin and nearly membranaceous . 23. ATHYRIUM FILIX-FEMINA
16. Fronds evergreen, rather thick . . . . . . . . . . . . . 27. DRYOPTERIS
15. Hairs present on upper and/or lower surfaces or veins of leaf-divisions . . . . . . . *17*
17. Fronds triangular, the lowest pinnae obviously the longest. 27. DRYOPTERIS HEXAGONOPTERA
17. Fronds broadly oblong-lanceolate or oblong, the lowest pinnae shorter than the others . . . . . . . . . . . . . . . . . 23. ATHYRIUM THELYPTEROIDES

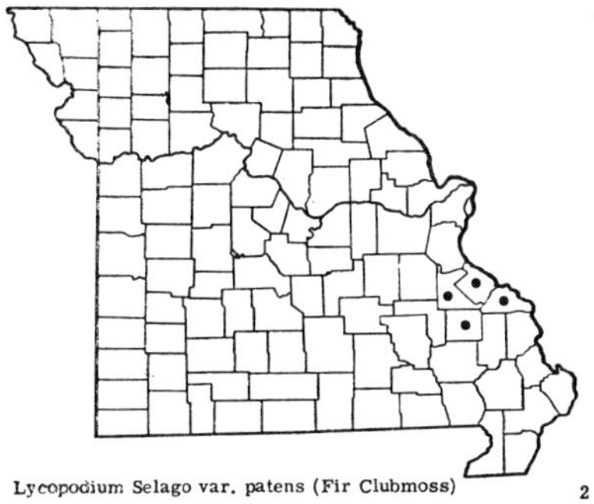

1 Lycopodium Selago var. patens (Fir Clubmoss)

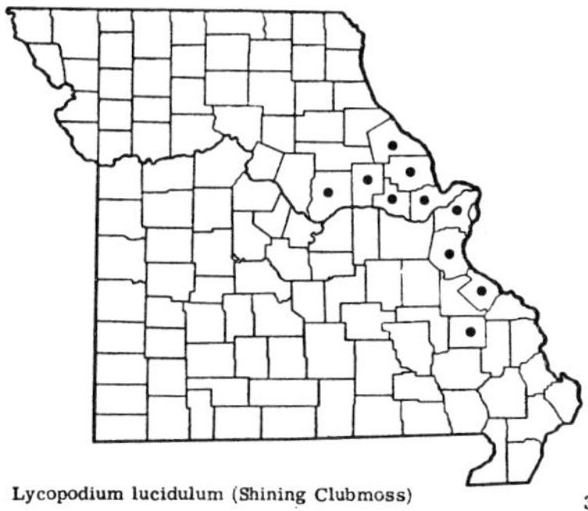

2 Lycopodium lucidulum (Shining Clubmoss)

3 Lycopodium obscurum var. dendroideum (Round-branched Ground-Pine)

## Order **LYCOPODIALES**

### Fam. **LYCOPODIACEAE** (Clubmoss Family)

#### 1. **Lycopodium** L. (Clubmoss)

a. Leafy branches flattened, the erect ones forking and fan-like; leaves scale-like, greatly reduced in size, attached to stem for most of their length . . . . . . . . . . . . . . . 4. L. TRISTACHYUM

a. Leafy branches not appearing to be flattened; leaves obviously developed, although often moss-like or needle-like, loose . . . . . . . . . . . . . . . . . . . . . . . . . . . . . *b*

b. Plant tree-like; spores in modified leaf axils crowded into a cone (strobilus) at the tip of a branch . . . . . . . . . . . . . . . . . . 3. L. OBSCURUM var. DENDROIDEUM

b. Plant erect to ascending, nearly unbranched or the stems loosely once- to few-forked; spores in axils of ordinary foliage leaves, not crowded into a cone . . . . . . . . . . . . . . *c*

c. Leaves broadest above the middle, toothed; plant with loosely creeping, rooting, leafy bases 10–40 cm. long; groups of shorter leaves alternate with groups of longer leaves on stem . 2. L. LUCIDULUM

c. Leaves broadest at base, nearly entire; plant with short rooting base 1–8 cm. long; all leaves of about same length . . . . . . . . . . . . . . . . . . . 1. L. SELAGO var. PATENS

1. **Lycopodium Selago** L. var. **patens** (Beauv.) Desv. Fir Clubmoss Map 1

*Lycopodium lucidulum* var. *porophilum* (Lloyd & Underw.) Clute [P & S]; *Lycopodium porophilum* Lloyd & Underw. [G]

Forming spores July–October.

Shaded crevices of usually La Motte sandstone bluffs, south-eastern Missouri Ozarks.

Ranges from Newfoundland to Ontario and Minnesota south to Alabama and Missouri.

2. **Lycopodium lucidulum** Michx. var. **lucidulum** Shining Clubmoss Map 2

Under 15 × magnification it will be observed that the stomata appear as small whitish dots in this species on the lower surface only, whereas in *L. Selago* var. *patens* the stomata appear on both upper and lower surfaces of leaves.

Forming spores July–October.

Shaded crevices and ledges of La Motte and St. Peter sandstone bluffs, eastern Ozark region north to Pike and west to Callaway counties, mostly near the Mississippi and Missouri rivers. North and west of St. Louis County and Jefferson County the locations are on St. Peter sandstone; the Ste. Genevieve and Madison County records are from La Motte sandstone.

Ranges from Newfoundland to Minnesota, south to South Carolina and Missouri.

3. **Lycopodium obscurum** L. var. **dendroideum** (Michx.) D. C. Eat. Map 3

Round-branched Ground Pine

Spores July–November.

Known only from Ste. Genevieve County, on shaded La Motte sandstone bluffs along upper reaches of River Aux Vases, 3½–4 mi. northwest of Avon, August 30, 1951, *Steyermark 72506*. The plants occur in a bed of sphagnum moss high on a steep bluff in a narrow ravine.

A specimen from Marion County ('Teajne's Conser.' [?], Hannibal, Dec. 15, 1909, *Rev. John Davis*) in the Mo. Bot. Gard. Herbarium, was un-

4 Lycopodium tristachyum (Ground Cedar)

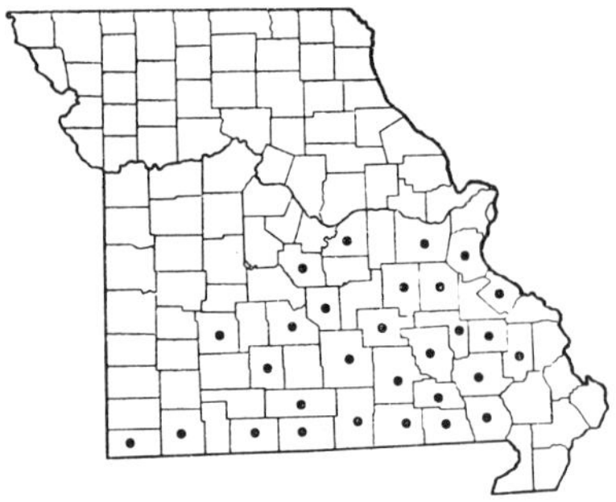

5 Selaginella apoda

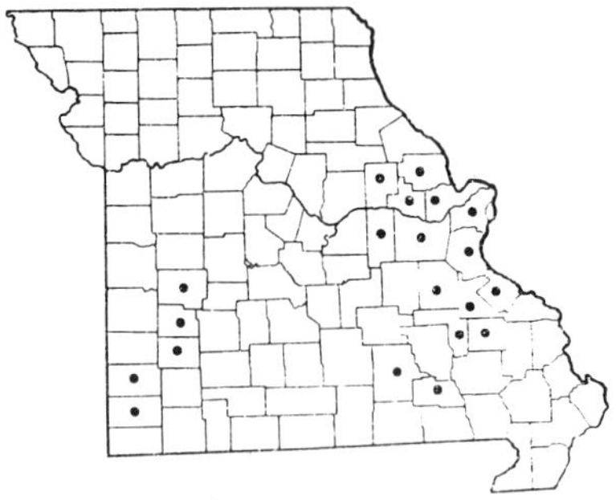

6 Selaginella rupestris

doubtedly collected in a conservatory or greenhouse from a cultivated plant, but is recorded here in order to eliminate it as a future source of error in citation of a possible native station.

Ranges from Labrador to Alaska, south to Georgia, Tennessee, Missouri, and Montana.

The isolated occurrence of this more northern species may be interpreted as a survivor or relict which has migrated southward into the Missouri Ozarks during one of the interglacial stages of the Pleistocene period.

4. **Lycopodium tristachyum** Pursh
Ground Cedar, Ground Pine — Map 4
Spores July–September.

Known only from dry wooded edge of bluffs of La Motte sandstone in pine woods in Ste. Genevieve County. Effort to relocate the plant at the Madison County station, where reported and presumably collected by *E. J. Palmer* near Mine La Motte, has failed. The only present known locality is at Pickle Springs, Ste. Genevieve County. Collections by *Kellogg 8718* and *Kellogg* s.n. are representative.

Ranges from Newfoundland to Minnesota, south to Georgia, Tennessee, and Missouri.

The collections have been identified by the latest monographer of the group, Miss Joan H. Wilce. They had been referred previously to *L. complanatum* var. *flabelliforme* Fern., differing from the latter in the well-developed ventral leaves, the bluish color of the sterile branchlets, and the many constrictions along the branchlets.

## Order **SELAGINELLALES**

### Fam. **SELAGINELLACEAE** (Spikemoss Family)

#### 2. **Selaginella** Beauv. (Spikemoss)

Stems and branches delicately creeping or flattened; leaves pale or bright green, spreading, oval or ovate; wet limestone ledges and banks . . . . . . . . . . . . . . . . . . . . 1. S. APODA

Stems with erect branches; leaves gray-green, thread-like, closely spirally arranged; dry sandstone, granitic, or chert rocks, ledges, and glades . . . . . . . . . . . . . . . . 2. S. RUPESTRIS

1. **Selaginella apoda** (L.) Fernald — Map 5
Spores May–November.

On wet limestone bluffs, along wet banks and moist ground of spring branches flowing from calcareous rocks, and on wet ground of springy meadows fed by calcareous spring branches. Throughout the Ozark region north to the Missouri River. Of expected occurrence in the Ozark border counties just north of the Missouri River.

Ranges from Maine to Quebec and Wisconsin south to Florida and Texas.

2. **Selaginella rupestris** (L.) Spring — Map 6
Spores July–October.

Predominantly found on dry sandstone (La Motte, St. Peter, Roubidoux, and those of Mississippian and Pennsylvanian age) glades and open ledges, and on dry exposed chert and igneous (granite and porphy-

---

Plate no. 1. **1.** Lycopodium obscurum var. dendroideum, × 2/5. **2.** Lycopodium lucidulum, × 2/5. **3.** Lycopodium tristachyum, × 2/5. **4.** Selaginella rupestris, × 2/5. **5.** Selaginella apoda, × 2/5. **6.** Isoetes Engelmanni, × 2/5; *a.* Megaspore, × 12. **7.** Isoetes melanopoda, × 2/5; *a.* Megaspore, × 12. **8.** Isoetes butleri, × 2/5; After Gleason, details from Small, The New York Botanical Garden.

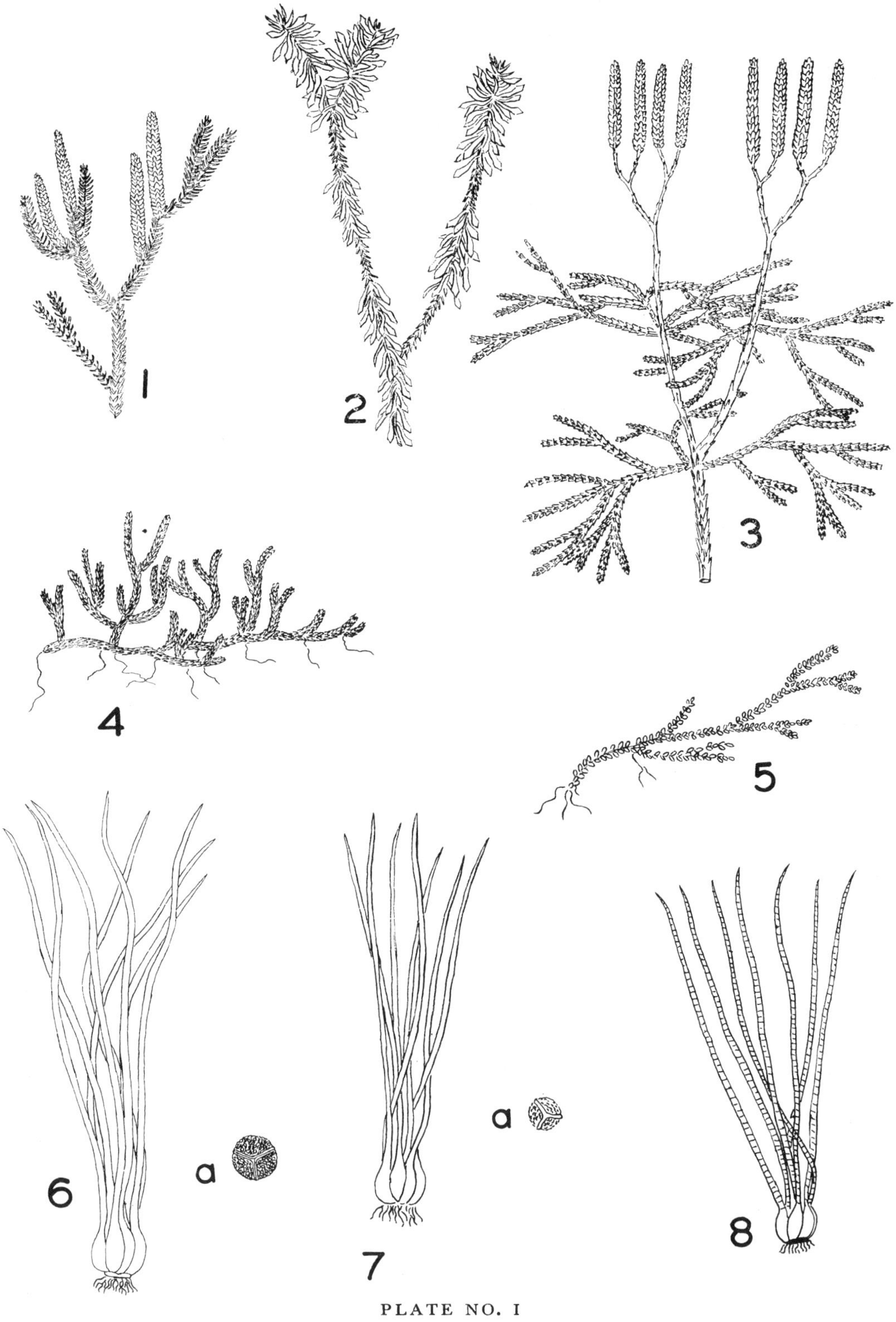


PLATE NO. I

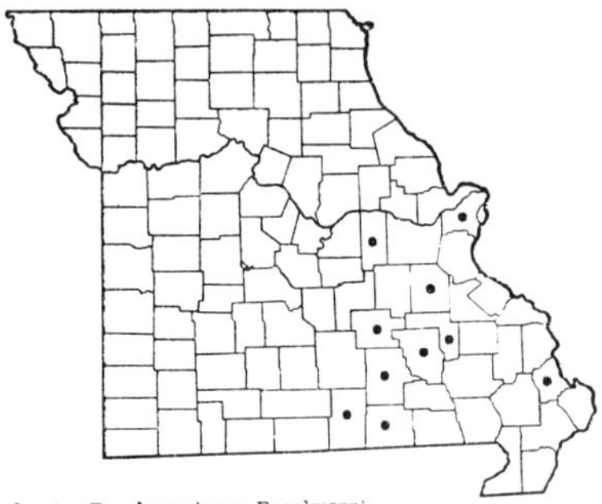
7 Isoetes Engelmanni var. Engelmanni

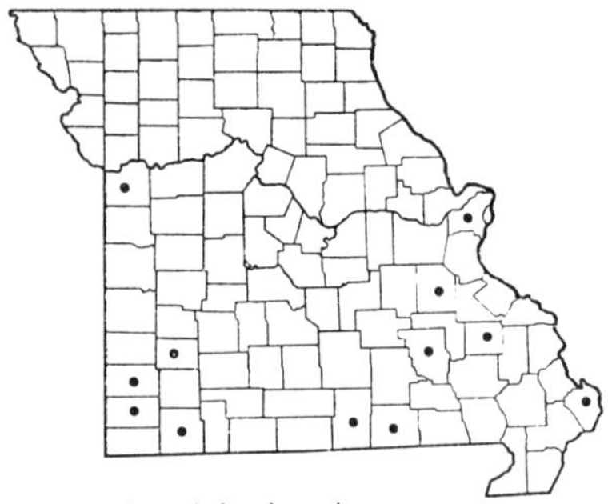
8 Isoetes melanopoda f. melanopoda

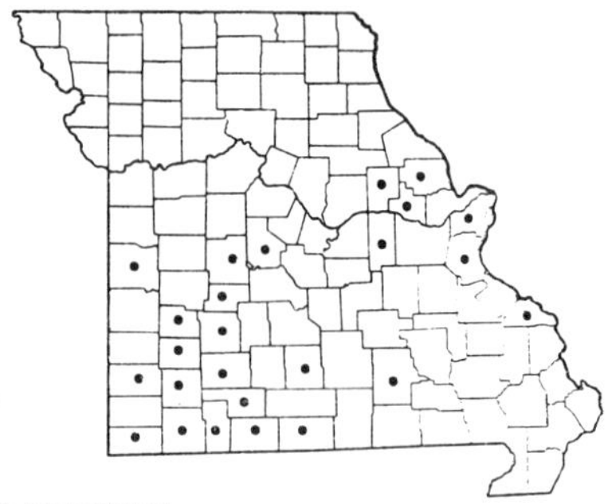
9 Isoetes Butleri

ritic trachyte) rocks. Mainly distributed in two areas: (1) eastern Ozarks north to Lincoln and Montgomery counties, and (2) unglaciated prairie on chert and sandstone of Mississippian and Pennsylvanian formations in southwestern Missouri. The Adair County reference, originally attributed to Missouri, should be eliminated. The data on the label read 'shaded damp hillside, Adair Co., Mo.,' and is in the Broadhead herbarium at the Missouri Botanical Garden Herbarium. No collector was indicated on the label. Dr. Tryon, who has recently monographed this group, has identified the specimen as one with which some *Selaginella Underwoodii* is mixed. Since *S. Underwoodii* is a species of the western states, and the 'shaded damp hillside' is a habitat unlikely for *S. rupestris,* I agree with Dr. Tryon that the locality is taken from a mistaken or confused source. Therefore, this record should be ignored.

Ranges from Nova Scotia to Ontario and Minnesota, south to Georgia, Arkansas, and Oklahoma.

The plants are subjected to long spells of summer drouth and heat on the exposed sites where they grow. At such times the plants appear dried, brittle, and lifeless. However, they respond to the slightest moisture, and become soft, dull green, and lifelike again after a rain.

## Order **ISOETALES**

### Fam. **ISOETACEAE** (Quillwort Family)

### 3. **Isoetes** L. (Quillwort)

a. Plants mostly aquatic, submerged in water or along the exsiccated margins of sink-hole ponds and ditches; leaves 1–2 mm. wide at the middle; megaspores with honeycomb network of narrow ridges; microspores smooth or nearly so . . . . . . . . . . . . . . . . . 1. I. Engelmanni

a. Plants usually growing in wet prairies or wet sandy or rocky glades; leaves 0.4–1.3 mm. wide at the middle; megaspores tuberculate; microspores spinulose or papillose . . . . . . . . . . . *b*

b. Bases of leaves dark chestnut-brown or blackish; leaves mostly 1–1.3 mm. wide at the middle; megaspores averaging 0.33 mm. in diameter; microspores 20–30 $\mu$ long, covered with spinules . 2. I. melanopoda

b. Bases of leaves usually pale or brown; leaves mostly 0.4–0.9 mm. wide at the middle; megaspores averaging 0.6 mm. in diameter; microspores 27–37 $\mu$ long, papillose . . . . . . . 3. I. Butleri

1. **Isoetes Engelmanni** A. Br. var. **Engelmanni** Map 7

This is a characteristic plant found partly or entirely submerged along the margins of sink-hole ponds in the eastern Ozark section. So far as known, the St. Louis County record is exterminated.

Ranges from Florida and Alabama, north to New Hampshire, New York, southern Indiana, southern Illinois, and eastern Missouri.

Usually much coarser in habit than the other species in Missouri, the leaves of *I. Engelmanni* average longer and broader and have a more coarsely cellular structure.

2. **Isoetes melanopoda** Gay & Durieu f. **melanopoda** Map 8

*Isoetes melanopoda* Gay & Durieu [G], [P & S]

Wet open ground and wet swales. Scattered in

southern and eastern Missouri south of the Missouri River and locally in Jackson County.

Ranges from Illinois and Iowa, south to Missouri, Oklahoma, and Texas.

*Isoetes melanopoda* may further be distinguished from *I. Butleri*, which it closely resembles, by the brown-spotted instead of brown-lineolate usually larger sporangia (0.5–3 cm. instead of 0.6–0.7 cm. long).

Another form of the species, *I. melanopoda* f. *pallida* (Engelm.) Fern., with pale leaf-bases is to be expected in Missouri.

3. **Isoetes Butleri** Engelm. Map 9

Wet prairies and wet places on rocky glades and ledges of bluffs. This is the commonest species in Missouri. It is found commonly in the western Ozarks and unglaciated prairie section of southwestern Missouri, and in the Ozark border counties northeast to Lincoln County. It occurs on both sandstone and chert as well as limestone glades which are wet in spring and become dry and hot in the summer when the plants then disappear. The species is sometimes difficult to distinguish from young, similar-looking plants with erect, aciculiform leaves of *Eleocharis compressa*, *Nothoscordum bivalve*, and other glade plants with which it is associated.

Ranges from Tennessee, Missouri, and Kansas, south to Arkansas and Oklahoma.

The leaves generally average shorter and narrower than those of *I. melanopoda* f. *melanopoda*, and it appears of firmer texture than any of the other species in Missouri.

## Order **EQUISETALES**

### Fam. **EQUISETACEAE** (Horsetail Family)

#### 4. **Equisetum** L. (Horsetail, Scouring Rush)

Plants of this genus contain a nerve poison which is known to be poisonous to grazing animals.

The stem is conspicuously jointed, with a sheath ending in a fringe of teeth at each joint, and in most species is hollow with a central cavity.

a. Teeth not jointed with the sheath; fruiting stem of a different appearance from sterile stem, fleshy, varying from whitish to pinkish-brown or yellow-brown, appearing above the ground before the slender, green, branching, sterile stems which are only 1–5 mm. thick; cones long-peduncled . 1. E. ARVENSE
a. Teeth definitely jointed; fruiting and sterile stems similar, green, stiff, mostly simple or only sparingly branched, usually 5–18 mm. thick; cones short-peduncled, the peduncle at most only slightly exceeding the subtending sheath . . . . . . . . . . . . . . . . . . . . . . . . . . *b*
  b. Stem smooth or nearly so, annual; cones rounded at tip . . . . . . . . . . . 2. E. KANSANUM
  b. Stem roughened to some degree with silica deposit, perennial; cones apiculate . . . . . . . . *c*
    c. Sheaths at least the upper ones green or pale green . . . . . . . . . . . 3. E. LAEVIGATUM
    c. All sheaths ashy-gray . . . . . . . . . . . . . . . 4. E. HYEMALE var. PSEUDOHYEMALE

1. **Equisetum arvense** L. Map 10

Field or Common Horsetail

A number of variations are known of which the following have been collected in Missouri:

a. Branches of the whorls simple . . 1a. E. ARVENSE L. var. ARVENSE f. ARVENSE
a. Branches of the whorls forking or compound *b*
  b. Sterile stem depressed or nearly flat on the ground, the branches upright . 1b. E. ARVENSE var. ARVENSE f. DIFFUSUM (A. A. Eat.) Clute
  b. Sterile stem erect or strongly ascending. . 1c. E. ARVENSE var. ARVENSE f. RAMULOSUM (Rupr.) Klinge

*Equisetum arvense* f. *nemorosum* A. Br., recognized by Palmer and Steyermark in their 1932 list, is here considered synonymous with f. *ramulosum*.

The species and its varieties are commonly found along railroad embankments, roadsides, banks of streams, and low ground, and occur throughout the state. They occur nearly throughout the United States.

There are the two following different growth stages in this horsetail: (1) the fertile, mostly unbranched, fleshy and whitish to pink- or yellow-brown stems in early spring, followed by (2) green, more or less branched, vegetative or sterile stems arising from the same rootstock.

The fertile stems bear spores from April to May. The teeth of the lower sheaths of the fertile stems are dark brown or blackish, mostly separate from each other, and have long acuminate tips.

Most cases of *Equisetum* poisoning originate from

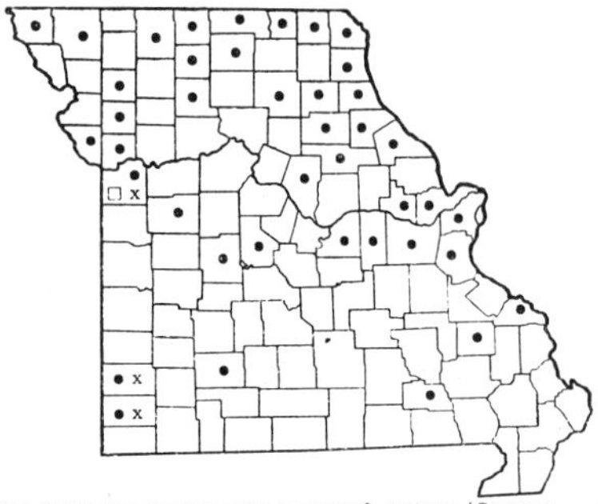

10 • Equisetum arvense var. arvense f. arvense (Common Horsetail)
10 □ Equisetum arvense var. arvense f. diffusum
10 x Equisetum arvense var. arvense f. ramulosum

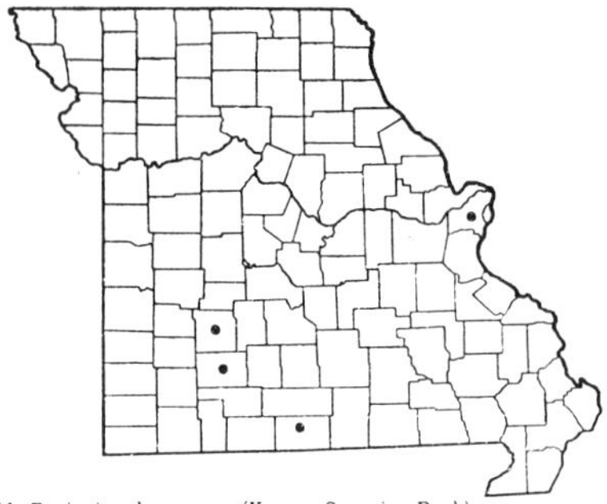

11 Equisetum kansanum (Kansas Scouring Rush)

12 Equisetum laevigatum (Smooth Scouring Rush)

*E. arvense* var. *arvense* or from one of its variations. Most trouble arises when hay including this horsetail is fed to horses. It is also known to be injurious to sheep.

2. **Equisetum kansanum** Schaffn.
Kansan Scouring Rush — Map 11
*Equisetum laevigatum* A. Br. [BB, in part]
Spores May–July.

Moist or dry banks, alluvial bottoms, and ditches. Known in Missouri only from St. Louis, Polk, and Greene counties.

Ranges from British Columbia, east to Michigan, south to Ohio, Texas, and California.

The status of the Kansan Horsetail as an acceptable species is not positive. Jones *et al.* (*Vascular Plants of Illinois*, pp. 12–13) cites evidence to show that no reliance can be placed on the characters of cone apex or habit used in distinguishing this species from *E. laevigatum*, and his opinion is shared with that of the eminent fern student, C. V. Morton (Am. Fern Jour. 43: 170. 1953). Other qualified students (Tryon *et al.*, *Ferns and Fern Allies of Wisconsin*, p. 119; Tryon, *Fern and Fern Allies of Minnesota*, p. 122; Billington, *Ferns of Michigan*, p. 83) maintain this as a distinct species. In Gleason's *New Illustrated Flora*, Morton (vol. 1, p. 14) places *E. kansanum* under *E. laevigatum* with the remarks, 'Plants with annual rather than evergreen stems have been distinguished as *E. kansanum* Schaffner.'

Only a few Missouri specimens have been identified as this species, which, at the present time, until other evidence is forthcoming, is being maintained as distinct from *E. laevigatum*. The ridges of the stem have poorly developed tubercles and are nearly smooth, as contrasted with the other species. The sheaths all have a black band at their summit but not at the base. It remains to be proved in Missouri whether the stems are definitely annual.

3. **Equisetum laevigatum** A. Br.
Smooth Scouring Rush — Map 12
*Equisetum hyemale* L. var. *intermedium* A. A. Eat.
*Equisetum hyemale* var. *intermedium* f. *proliferum* Haberer [G]
Spores March–July.

Moist ground along stream banks, alluvial bottoms, and open places. Scattered throughout the state.

Ranges from British Columbia, Michigan, and Massachusetts, south to California, Texas, Georgia, and West Virginia.

This species is intermediate between *E. hyemale* var. *pseudohyemale* (Farw.) Morton and *E. kansanum*. The lower sheaths are sometimes gray, appressed, and black-banded as in those of *E. hyemale* var. *pseudohyemale*, but the upper are green and flaring with little or no black band at base. In general, the stems are smoother than in *E. hyemale* var. *pseudohyemale*, and rougher than in *E. kansanum*.

A collection from Lawrence County (wet gravelly margins of spring brook, Big Spring, 5 mi. west northwest of Mt. Vernon, July 13, 1953, *Palmer 56131-A*)

Plate no. 2. **1.** Equisetum arvense; *a.* Sheath from fertile stem, × $1^1/_5$; *b.* Sterile stem, × $^2/_5$; *c.* Fertile stem, × $^2/_5$. **2.** Equisetum laevigatum; *a.* Sheath, × $^2/_5$; *b.* Fertile stem, × $^2/_5$. **3.** Equisetum hyemale; *a.* Fertile stem, × $^2/_5$; *b.* Equisetum hyemale var. elatum, Sheath, × $2^2/_5$; *c.* Equisetum hyemale var. pseudohyemale, Sheath, × $2^2/_5$. **4.** Botrychium dissectum var. dissectum; *a.* Sterile blade, × $^2/_5$; *b.* Fertile portion of frond, × $^2/_5$. **5.** Botrychium dissectum var. obliquum; *a.* Fertile portion of frond, × $^2/_5$; *b.* Sterile blade, × $^2/_5$. **6.** Botrychium virginianum, × $^2/_5$. **7.** Ophioglossum vulgatum var. pycnostichum; *a.* Fertile portion of frond, × $^4/_5$; *b.* Sterile blade, × $^2/_5$. **8.** Ophioglossum Engelmanni; *a.* Sterile blade, × $^2/_5$; *b.* Fertile portion of frond, × $^2/_5$. **9.** Azolla mexicana, × 6. **10.** Marsilea quadrifolia, × $^2/_5$; After Gleason, The New York Botanical Garden.

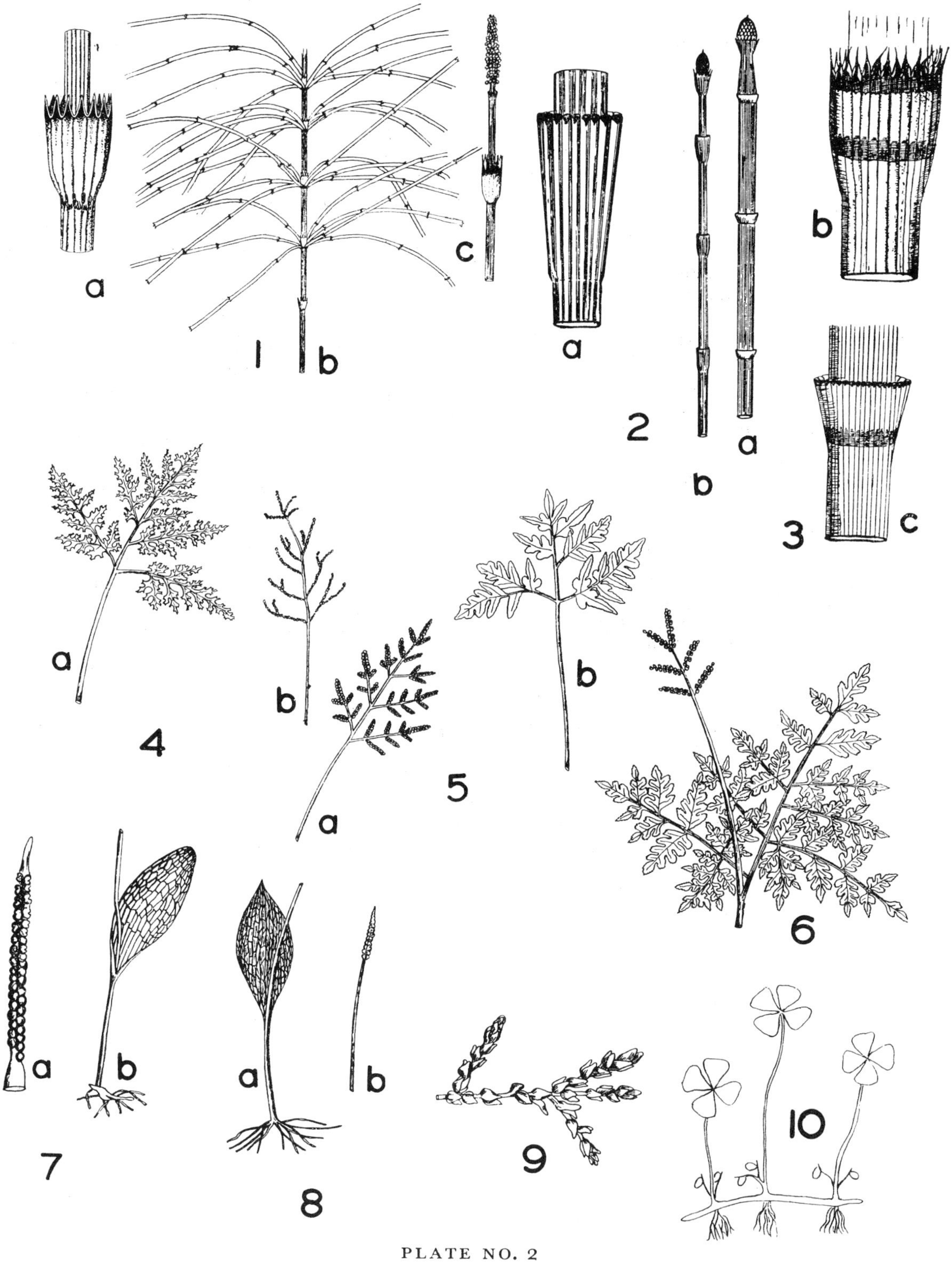


PLATE NO. 2

identified as *E. hyemale* var. *intermedium* f. *proliferum* is here placed under *E. laevigatum* and not segregated.

4. **Equisetum hyemale** L.
Winter Scouring Rush Map 13
Spores March–August.

Typical *E. hyemale* var. *hyemale* is an Eurasian plant, not represented in our flora. The Missouri varieties and forms may be distinguished as follows:

Teeth of most of sheaths quickly falling . . . . 4a. E. HYEMALE var. PSEUDOHYEMALE
Teeth of sheaths mostly persisting or falling off late in maturity. . . 4b. E. HYEMALE var. ELATUM

4a. **Equisetum hyemale** var. **pseudohyemale** (Farw.) Morton Map 13
*Equisetum hyemale* var. *affine* sensu A. A. Eaton [G], not *E. robustum* var. *affine* Engelm.
*Equisetum variegatum* [P & S], not Schleich.

The stems in this variety are generally more slender than in var. *elatum*, the diameter varying from 2–12 mm. in thickness.

Material from Jackson County, collected by Bush, misidentified previously as *E. variegatum*, is now identified as *E. hyemale* var. *affine* f. *polystachyum* Prager, which is considered in the present treatment as synonymous with *E. hyemale* var. *pseudohyemale*.

Ranges from Nova Scotia and Quebec to Ontario, south to Georgia and Texas; also Colorado to Washington and British Columbia.

4b. **Equisetum hyemale** var. **elatum** (Engelm.) Morton Map 13
*Equisetum hyemale* var. *robustum* (A. Br.) A. A. Eat. [G, P & S]
*Equisetum robustum* A. Br. [1st BB]
*Equisetum prealtum* Raf. [Pinkerton]

The stems in this variety are stouter than in var. *pseudohyemale*, the diameter averaging 6–18 mm. in thickness.

Ranges from Ontario to Idaho and Yukon, south to Florida, Texas, and California.

The above varieties of horsetail were at one time used extensively for scouring and polishing kitchen and other utensils and as a sandpaper material. In Holland a related variety is planted on the dikes to bind the soil. The young fruiting cones of this group of horsetail have been eaten, after being boiled and fried, but because of the nerve poison contained in this genus, it is recommended that the group be left untouched for gastronomic consideration.

The erect, mostly simple, dark evergreen stems occurring in dense colonies along stream banks and low alluvial ground is a familiar sight in many sections of the state where the varieties occur.

## Order **OPHIOGLOSSALES**

### Fam. **OPHIOGLOSSACEAE** (Adder's-tongue Family)

Green leaf-like portion (sterile blade) deeply split and lobed into three or more main divisions; fertile portion branched . . . . . . . . . . . . . . . . . . . . . . 5. BOTRYCHIUM
Green leaf-like portion (sterile blade) simple, entire, smooth-margined; fertile portion a simple, slender, tail-like spike . . . . . . . . . . . . . . . . . . . . . . 6. OPHIOGLOSSUM

### 5. **Botrychium** Sw. (Grape Fern)

All the Missouri species and varieties of Botrychium have a more or less triangular leaf-blade.

Leaf-like sterile blade dark green to bronze, overwintering, longstalked, the stalk arising less than half way up the main common stalk (stipe); margin of blade (under high magnification) whitish . 1. B. DISSECTUM
Leaf-like sterile blade pale or yellow-green, not overwintering, sessile or nearly so, attached about half to two-thirds way up the main common stalk (stipe); margin of blade not whitish . . . 2. B. VIRGINIANUM

1. **Botrychium dissectum** Spreng.
Grape Fern Map 14
Three varieties may be distinguished in Missouri:

a. Leaf-like sterile blade finely cut into many very narrow, notched or emarginate, somewhat rectangular teeth or little lobes, appearing lace-like. 1a. B. DISSECTUM var. DISSECTUM
a. Leaf-like sterile blade coarsely divided into nearly entire or finely toothed segments . . . *b*
b. Sterile blade firm or leathery, mostly tripinnate (the main and secondary segments again divided), the ultimate segments up to 100 or more, entire to minutely toothed, most of them rounded or broadened at the base . . . 1b. B. DISSECTUM var. OBLIQUUM
b. Sterile blade thin and membranaceous, pinnate (simply divided into the main segments) to bipinnate (main segments again divided into smaller segments), the ultimate

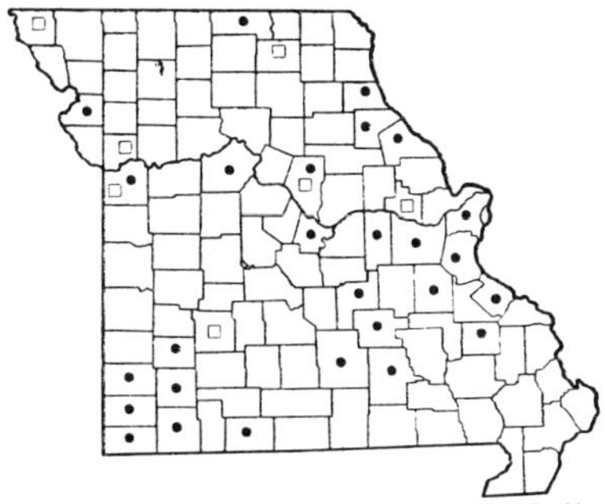

13 • Equisetum hyemale var. elatum (Winter Scouring Rush)
13 □ Equisetum hyemale var. pseudohyemale (Winter Scouring Rush)

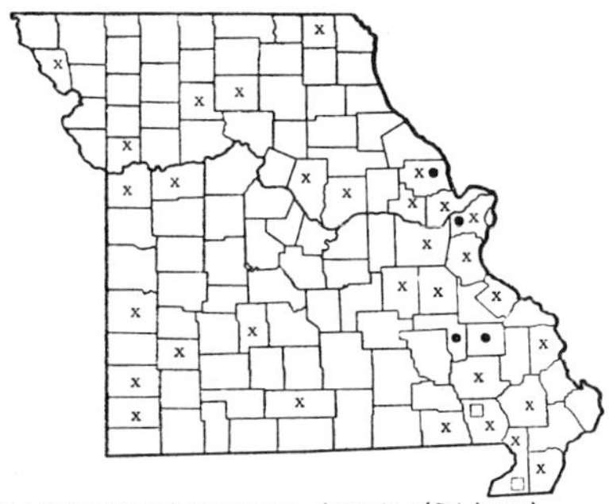

14 • Botrychium dissectum var. dissectum (Cut-leaved Grape Fern)
14 □ Botrychium dissectum var. tenuifolium
14 x Botrychium dissectum var. obliquum

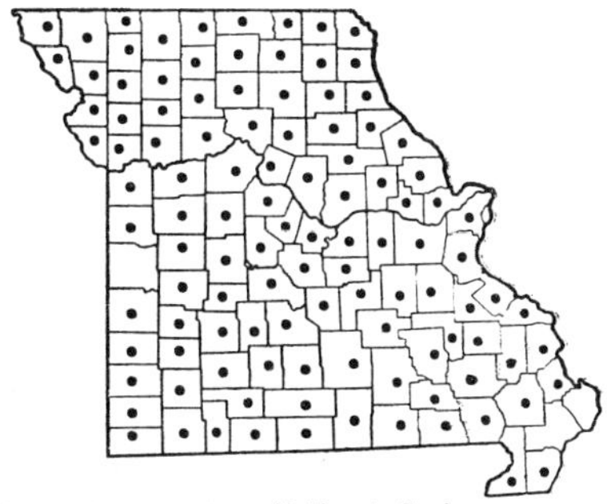

15 Botrychium virginianum (Rattlesnake Fern)

segments relatively few (5–35), mostly unlobed or only slightly lobed, mostly contracted to a somewhat narrowed base . 1c. B. DISSECTUM var. TENUIFOLIUM

1a. **Botrychium dissectum** Spreng. var. **dissectum**
Cut-leaved Grape Fern Map 14
*Botrychium dissectum* Spreng. [G, P & S]
*Botrychium obliquum* var. *dissectum* Prantl [Pinkerton]
Spores mature August–November.

This variety of the Grape Fern is rarely found in Missouri. Occurs in rich wooded ravines and narrow wooded valleys of the eastern section of the Ozarks north to Lincoln County.

Ranges from Nova Scotia to Minnesota, south to South Carolina, Tennessee, and Arkansas.

The sterile blade persists over winter, becoming bronze-russet or greenish and copper, disappearing in early summer when the new sterile and fertile blades appear.

1b. **Botrychium dissectum** var. **obliquum**
(Muhl.) Clute Map 14
*Botrychium dissectum* f. *obliquum* (Muhl.) Fern. [G]
*Botrychium obliquum* Muhl. [Pinkerton]
Spores mature August–November.

This broad-leaved variety of the Cut-leaved Grape Fern is by far the commonest in Missouri, the new sterile blades appearing in fall and persisting until the following summer.

Occurs in rich wooded ravines and creek bottoms, generally in low ground, scattered over the state, most of the present known stations occurring in eastern Missouri.

Ranges from Nova Scotia and New Brunswick, west to Minnesota, and south to Georgia, Louisiana, and Oklahoma.

1c. **Botrychium dissectum** var. **tenuifolium**
(Underw.) Farw. Map 14
*Botrychium obliquum* var. *tenuifolium* (Underw.) Gilbert [Pinkerton]
Spores mature September–November.

Known in Missouri from only two of the counties in the southeastern lowlands section, where it occurs in low wet woodland.

Ranges from Connecticut and Ohio, south to Florida, Missouri, and Texas.

2. **Botrychium virginianum** (L.) Sw. var. **virginianum** Rattlesnake Fern Map 15
*Botrychium virginianum* var. *intermedium* Butters [Pinkerton]
Spores mature May–July.

Occurs throughout Missouri in woodland in nearly every county.

Ranges from Labrador to British Columbia, south to Florida, Mexico, and California.

The thinner, paler green fronds do not persist over winter as in the preceding species.

## 6. **Ophioglossum** L. (Adder's-tongue)

Plants of wooded lowlands, valley, and creek bottoms; sterile blades blunt or rounded at tip; network of principal (towards middle of sterile blade) veins not enclosing any additional and smaller network . . . . . . . . . . . . . . . . . . . . . . . . . . 1. O. VULGATUM

Plants of limestone glades or outcroppings; sterile blades with a tiny projection at tip; network of principal veins (towards middle of sterile blade) enclosing an additional, smaller, secondary network . . . . . . . . . . . . . . . . . . . . . . . . . . 2. O. ENGELMANNI

16 • Ophioglossum vulgatum var. pycnostichum (Adder's-tongue)
16 □ Ophioglossum vulgatum var. pseudopodum (Adder's-tongue)

17 Ophioglossum Engelmanni (Adder's-tongue)

18 Salvinia natans

1. **Ophioglossum vulgatum** L. Map 16

Typical *Ophioglossum vulgatum* var. *vulgatum* is Eurasian. The two varieties found in Missouri may be keyed as follows:

Sterile blade ovate or ovate-lanceolate, mostly rounded at base, 1–4 cm. broad; sporangia transversely oblong . 1a. O. VULGATUM var. PYCNOSTICHUM

Sterile blade oblanceolate, obovate, or elliptic, mostly gradually tapering to the base, 0.7–3.2 cm. broad; sporangia suborbicular. . 1b. O. VULGATUM var. PSEUDOPODUM

1a. **Ophioglossum vulgatum** var. **pycnostichum** Fern. Map 16

*Ophioglossum vulgatum* L. in part [BB]

Spores mature April–August.

This variety is known in Missouri at present from St. Charles (*Steyermark 81994*) and Lincoln counties, north of the Missouri River.

Ranges from South Carolina, Tennessee, and Missouri, north to New Jersey, Ohio, and Indiana.

The sterile blades average generally shorter (2.5–8.5 cm. long) than in the other variety. The common stalk of the plant usually has a persistent dark tubular sheath.

1b. **Ophioglossum vulgatum** var. **pseudopodum** (Blake) Farw. Map 16

*Ophioglossum vulgatum* L. in part [BB, P & S, Pinkerton]

Spores mature April–June.

This variety is found in Missouri mostly in the valley woodlands of Crowley Ridge and swampland of counties of southeastern Missouri, locally north in the Ozarks in Crawford County (*Steyermark 83927*). Small sterile plants of this variety are often found in such habitats appearing above the surface of the woodland floor in April.

2. **Ophioglossum Engelmanni** Prantl Map 17

Spores mature March–October.

This species is found as a characteristic plant of limestone glades and outcrops throughout the Ozarks north to Monroe and Ray counties; absent from the lowlands of southeastern Missouri.

Ranges from Florida to Arizona and Mexico, north to Virginia, Ohio, Indiana, Missouri, Kansas, and Oklahoma.

Wherever found, it is usually abundant. Frequently most of the individual plants found have sterile blades only. The species appears well adapted to withstand the alternating early wet spring followed by summer desiccation conditions prevalent on the limestone openings. After summer drouths the sterile blades usually wither and die, often reappearing in late summer or fall with new growth following rainy periods.

The isolated limestone glades where the species. occurs at its northern limits in Monroe and Ray counties indicate areas which have not been subjected to Pleistocene glaciation.

## Order **FILICALES**

### Fam. **SALVINIACEAE** (Water Fern Family)

A family of floating plants in which the vascular tissue has become reduced. The small leaves, arranged on a more or less elongated axis, are roughened with papillae or tubercles.

Foliage leaves 1–1.5 cm. long, not lobed; stems simple . . . . . . . . . . . . . . 7. SALVINIA

Foliage leaves only 0.5 mm. long, minutely 2-lobed; stems pinnately branching . . . . . . 8. AZOLLA

19 Azolla mexicana (Mosquito Fern, Water Velvet)

20 Marsilea quadrifolia

21 Marsilea mucronata

## 7. **Salvinia** Adans. Salvinia

**Salvinia natans** (L.) All. Map 18

Found in sloughs in the Bois Brule bottoms of the Mississippi River, Perry County, by *C. Demetrio* in 1886, presumably as an escape from cultivation, and not known elsewhere to occur in North America. Native of Europe.

The Missouri locality has not been revisited for this plant, and it is not known whether the plants have persisted at this station.

## 8. **Azolla** Lam. Mosquito Fern, Water Velvet

**Azolla mexicana** Presl Map 19

*Azolla caroliniana* [sensu G, P & S, Pinkerton], not Willd.

Spores mature summer and fall.

Floating on the surface of quiet waters or stranded along the muddy margins of ponds and ditches, scattered in counties of eastern and western Missouri bordering the Mississippi, Missouri, St. Francis, and other large rivers.

Ranges from Wisconsin, Minnesota, Illinois, and Missouri, the western U.S., south into South America.

Plants usually appear as dense gray-green to purplish-red masses, and resemble small mosses or liverworts. Exposure to intense sunlight increases the purplish-rose coloring.

The upper leaf lobes measuring 0.7 mm. long or longer, and the greater spread of a single branched plant exceeding 1 cm. or more wide are characters used to distinguish this species from *A. caroliniana* Willd., which is not known in Missouri.

# Fam. **MARSILEACEAE** (Pepperwort Family)

## 9. **Marsilea** L.

Aquatic plants, rooted in the mud, with clover-like leaves on long slender petioles.

Leaflets glabrous; fruiting structures (sporocarps) long-stalked, glabrous . . . . . . . 1. M. QUADRIFOLIA
Leaflets sparsely hairy; fruiting structures (sporocarps) short-stalked, axillary at the base of the leaves, hairy . . . . . . . . . . . . . . . . . . . . . . . . . . . . . 2. M. MUCRONATA

1. **Marsilea quadrifolia** L. Map 20

In the general range of the species, the spores mature from June–December. At the single Missouri station known, the spores are recorded as mature in October.

Known in Missouri only from muddy margins of Basswood Lakes, T53N, R34W, northeast ¼ sect. 27, Platte County, October 11, 1953, coll. *L. J. Gier* (see Am. Fern Jour. 45: 64–65. 1955). Dr. Gier believes that the plants have been introduced here from outside sources.

A European species, sometimes cultivated, and introduced from New England to Maryland west to Iowa and Missouri.

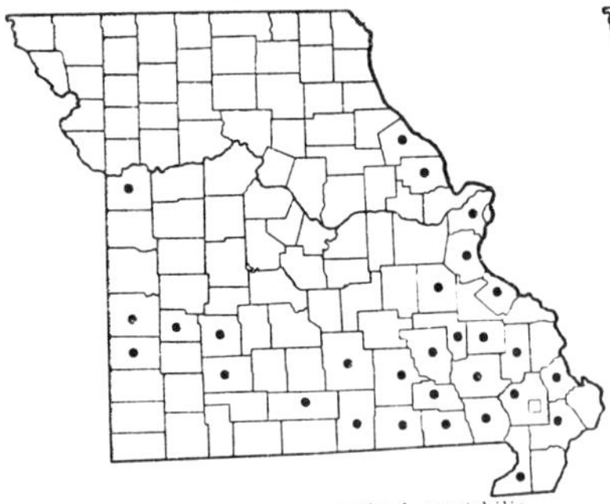

22 • Osmunda regalis var. spectabilis f. spectabilis (Royal Fern)
22 □ Osmunda regalis var. spectabilis f. anomala

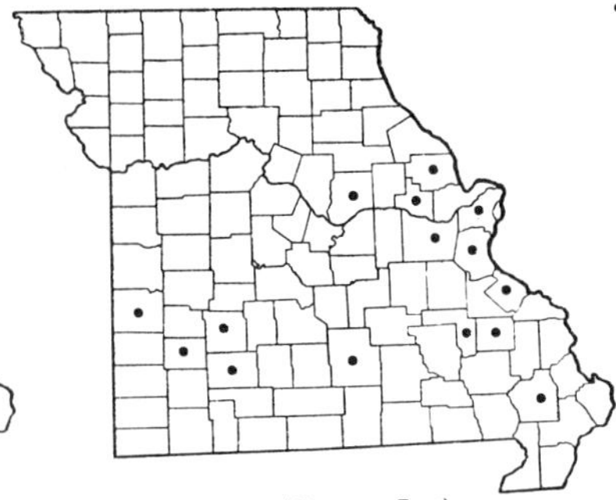

23 Osmunda cinnamomea (Cinnamon Fern)

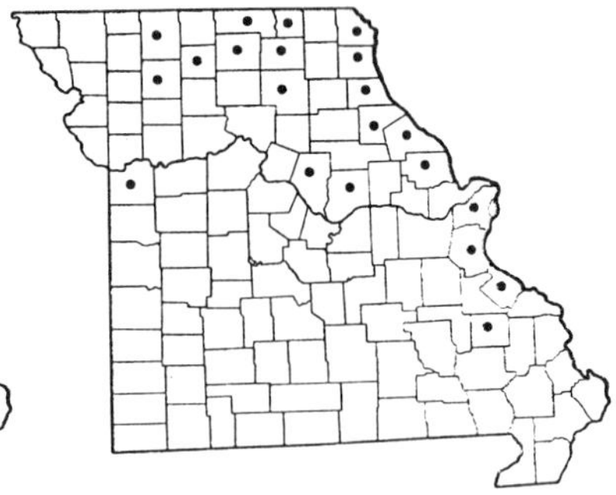

24 Osmunda Claytoniana (Interrupted Fern)

2. **Marsilea mucronata** A. Br. Map 21

*M. vestita* [of G, ed. 7, 1st BB], not Hook. & Grev.

The spores are recorded to mature in August and September. In Missouri they are known to fruit in September.

The single Missouri station is in Barton County, southwestern Missouri (shallow water at the muddy grassy margins of a small pond, about 1½ miles north of Milford, September 12, 1951, *E. J. Palmer 53101*).

Ranges from Minnesota and Saskatchewan, south to Florida, Texas, and Mexico.

## Fam. **OSMUNDACEAE** (Royal Fern Family)

### 10. **Osmunda** L. (Royal Fern)

Large ferns with the sporangia borne naked on special completely fertile fronds or parts of fronds. The spores are green and contain chlorophyll.

a. Leafy frond twice-pinnate, pinnules well separated; fertile pinnae, when present, at the upper part of the blade . . . . . . . . . . . . . . . . . . . . . . . . . . . b
  b. All the branches of the fertile portion spore-bearing . . . 1. O. REGALIS var. SPECTABILIS f. SPECTABILIS
  b. Some of the branches of the fertile portion leaf-like . . . 1. O. REGALIS var. SPECTABILIS f. ANOMALA
a. Leafy frond once-pinnate, the pinnae deeply lobed (pinnatifid), the lobes close together; fertile fronds separate, or fertile pinnae borne between sterile ones . . . . . . . . . . . . c
  c. Tufts of woolly hair present at base of each sterile pinna; sterile pinnae loosely fringed with hairs on margins; lobes of pinnae acutish; fertile fronds separate from the sterile, densely woolly, cinnamon-colored fronds. . . . . . . . . . . . . . . . . . . . . 2. O. CINNAMOMEA
  c. Tufts of woolly hair mostly absent at base of each sterile pinna; sterile pinnae smooth; lobes of pinnae more rounded, obtusish; fertile pinnae borne between (above and below) sterile ones, green to black. . . . . . . . . . . . . . . . . . . . . 3. O. CLAYTONIANA

1. **Osmunda regalis** L. var. **spectabilis** (Willd.) Gray f. **spectabilis** Royal Fern Map 22

Spores mature spring to late summer.

Wet ground along spring branches, or small creeks usually flowing through siliceous or granitic regions, the stems sometimes growing in water; also on moist sandstone bluffs and ledges. Chiefly in the eastern Ozark region northeast to Pike County and locally northwest to Jackson County.

Ranges from Newfoundland to Saskatchewan, south to Florida, Texas, and tropical America.

*Osmunda regalis* var. *spectabilis* f. *anomala* (Farw.)

Plate no. 3. **1.** Osmunda regalis var. spectabilis; *a.* Tip and middle of sterile frond, × 2/7; *b.* Portion of fruiting frond, × 2/7. **2.** Dennstaedtia punctilobula; *a.* Tip of frond, × 2/7; *b.* Middle portion of frond, × 2/7; *c.* Base of frond, × 2/7; *d.* Pinnule with fruit dots, × 6/7. **3.** Osmunda Claytoniana; *a.* Tip of frond, × 2/7; *b.* Middle portion of fruiting frond, × 2/7; *c.* Base of frond, × 2/7. **4.** Osmunda cinnamomea; *a.* Fertile frond (portion), × 2/7; *b.* Tip of sterile frond, × 2/7;, *c.* Middle portion of sterile frond, × 2/7; *d.* Base of sterile frond, × 2/7; *e.* Division of pinna showing venation, × 6/7.

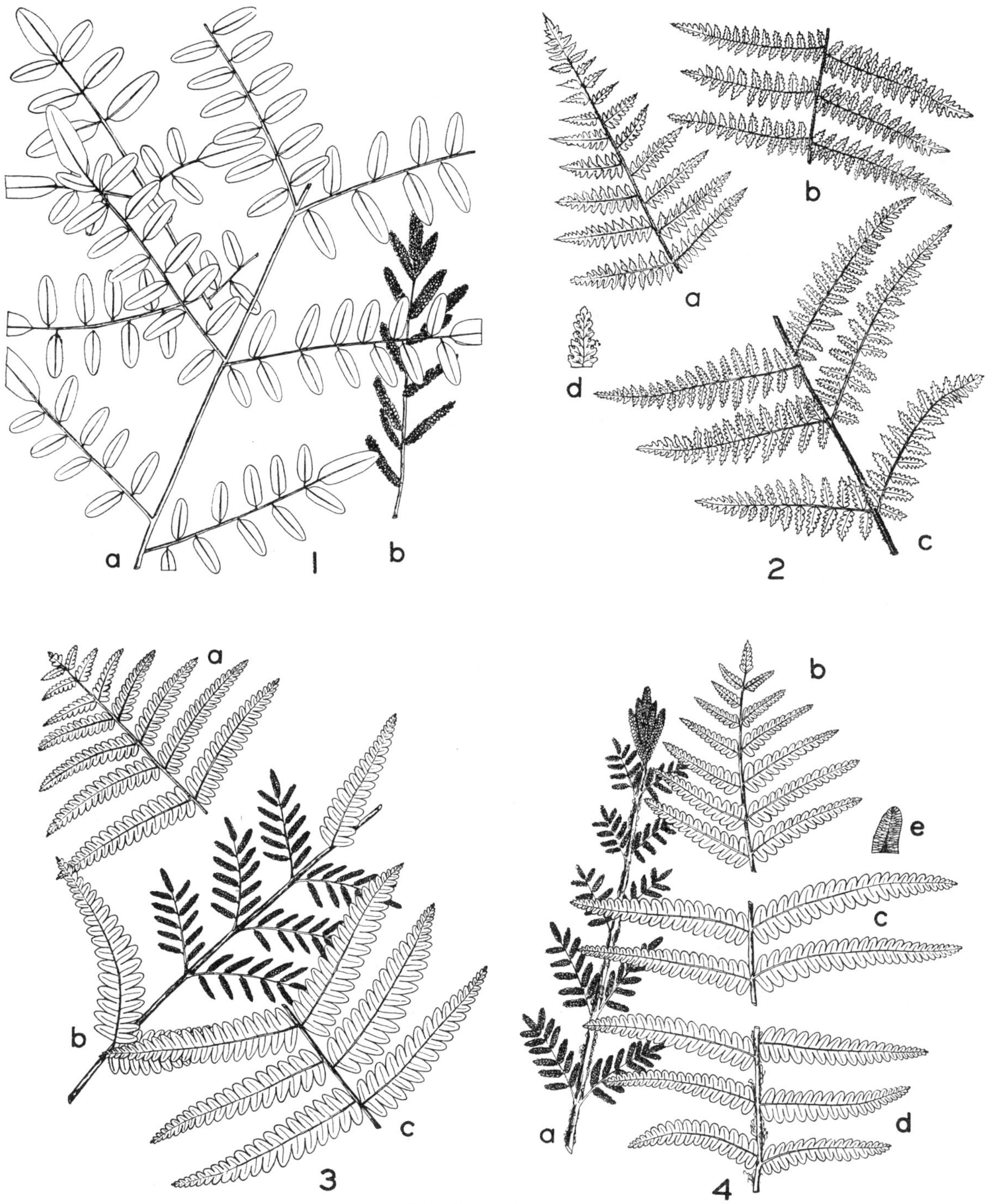


PLATE NO. 3

Harris has been found along a spring branch in the Crowley Ridge section of Stoddard County (*Steyermark 76774*).

The Royal Fern is excellent for naturalizing around shores of streams and ponds.

2. **Osmunda cinnamomea** L.
Cinnamon Fern Map 23
Spores mature May–June.

Boggy ground along small rivulets or spring branches in siliceous and granitic areas, or shaded sandstone or granitic ledges and bluffs, mostly in the eastern Ozark region north to Lincoln and Callaway counties, southeast on Crowley Ridge in Stoddard County, and locally west to Vernon County.

Ranges from Labrador to Minnesota, south to Florida, Texas, and tropical America.

In early spring the young unrolling sterile fronds ('fiddle-heads'), when not over 8 inches tall, are sometimes cooked as a vegetable, but do not compare very well in flavor with those of Ostrich Fern. The crisp, tender, central part of the crown, with its slightly nutty flavor, is sometimes relished in a raw state by those acquainted with it.

The Cinnamon Fern does well in shady wild gardens, if provided with sufficient loose, acid humus and sandy, well-drained soil.

3. **Osmunda Claytoniana** L.
Interrupted Fern Map 24
Spores mature May–July.

Rich wooded slopes of ravines, often on moist ledges of sandstone bluffs, occasionally in wet woods. Locally common north of the Missouri River northwest and west to Harrison, Daviess, and Jackson counties; absent from most of the Ozark region, where its occurrence is limited to a few of the easternmost counties.

Ranges from Newfoundland to Manitoba and Minnesota, south to Georgia and Arkansas.

This is the commonest of the larger ferns of the glaciated woodland ravines of northern Missouri. This fern does well in a shaded wild garden, if provided with loose rich humus and subneutral or somewhat acid soil.

## Fam. **POLYPODIACEAE** (Fern Family, Polypody Family)

Herbaceous plants (in Missouri) with creeping rhizomes. The fruit dots or sori are clusters of sporangia, which are stalked and open transversely, and are massed in dots, lines, or variously shaped clusters on the back or margins of the frond or its divisions. A membrane (indusium) often covers the fruit dots, or it may be absent.

### 11. **Dennstaedtia** Bernh. Hay-scented Fern

These are pale green, delicate, finely hairy ferns with slender creeping rhizomes. The indusia are cup-shaped and open terminally, while the sorus is globular, marginal, and located at the tip of a free vein or fork of the upper margin of each lobe of the pinnules.

**Dennstaedtia punctilobula** (Michx.) Moore
Hay-scented Fern Map 25
Spores July–September.

One of the rarest ferns of Missouri, occurring only on shaded ledges of La Motte sandstone bluffs and ravines in Ste. Genevieve, St. Francois, and Madison counties of the southeastern Ozarks.

Ranges from Newfoundland and Nova Scotia to Michigan, south to Georgia, Illinois, and Arkansas.

Although a common fern in the New England states, where it frequently grows in old pastures, rocky sterile slopes, often exposed to the sun, in Missouri it is confined to deep shaded canyons, and undoubtedly represents a survivor from past migration during one of the interglacial stages of the Pleistocene period.

The bruised or crushed fronds give off a sweet odor, resembling newly mown hay.

### 12. **Pteridium** Gled. Bracken

Large coarse ferns with long, creeping, hairy, underground rhizomes, stiff, erect, solitary stipes and broad, generally triangular-shaped, mostly tripinnately compound, nearly leathery fronds with mostly opposite pinnae and relatively large alternate pinnules. The sori are arranged more or less continuously along the margins of the frond, with the sporangia located under the recurved border of the leaf segment (outer false indusium) and the minute true inner indusium (membranous or of hairs).

25 Dennstaedtia punctilobula (Hay-scented Fern)

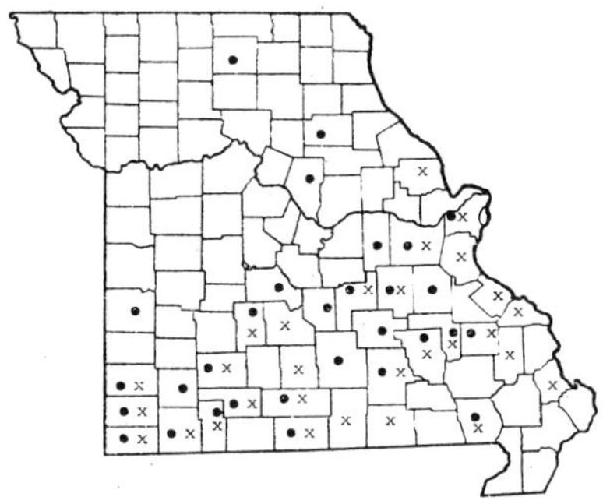

26 • Pteridium aquilinum var. latiusculum
26 × Pteridium aquilinum var. pseudocaudatum

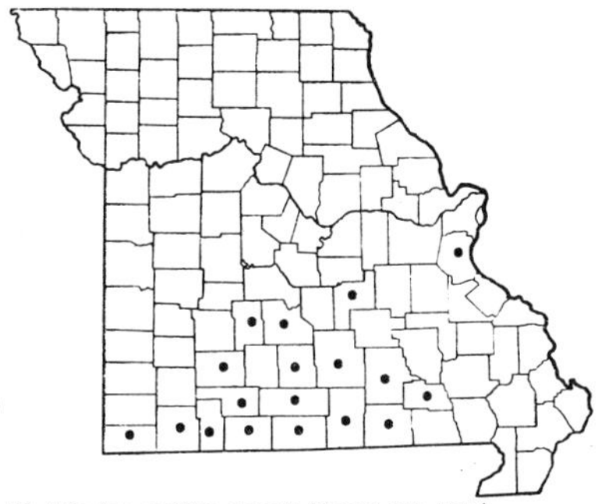

27 Adiantum capillus-veneris (Venus'-hair Fern)

**Pteridium aquilinum** (L.) Kuhn
Bracken Map 26

Only the following two varieties of the Bracken Fern are known in Missouri, but Tryon (Rh. 43: 12–15. 1941), in his revision of the genus, accounts for ten additional varieties, distributed over most of the hemispheres. Intergradation and intermediates between the following two varieties occur.

The longest entire leaf-segment or part of a leaf-segment about 4 times as long as wide; terminal leaf-segments 5–8 mm. wide; margins of the ultimate leaf-segments somewhat pubescent. . . a. P. AQUILINUM var. LATIUSCULUM

The longest entire leaf-segment or part of leaf-segment 6–9 or more times as long as wide; terminal leaf-segments mostly 2–4.5 mm. wide; margins of the ultimate leaf-segments mostly glabrous or nearly so . . . . . b. P. AQUILINUM var. PSEUDOCAUDATUM

a. **Pteridium aquilinum** (L.) Kuhn var. **latiusculum** (Desv.) Underw. Map 26
*Pteridium latiusculum* [P & S, Pinkerton]

Spores mature July–September.

Rocky or dry open woods, in acid soil, mostly confined to Ozark region in southern and east-central Missouri west to Vernon and Camden counties, locally north of the Missouri River in Boone, Monroe, and Sullivan counties.

Ranges from Newfoundland, Ontario, and South Dakota, and Wyoming, south to North Carolina, Mississippi, Missouri, Oklahoma, and locally in Mexico.

The fresh and dried fronds of this and the other variety of Bracken Fern in hay are poisonous to stock. Pigs, however, are not affected by eating the fresh rhizomes. The young leafy fronds, when thoroughly cooked, may be eaten like asparagus.

b. **Pteridium aquilinum** var. **pseudocaudatum** (Clute) Heller Map 26
*Pteridium latiusculum* var. *pseudocaudatum* [P & S, Pinkerton]

Spores mature July–September.

Rocky or dry open woods, in acid soil in the Ozark section of the state, north to Jasper, Dallas, Franklin, and Lincoln counties.

This is a more southern variety, distributed from Massachusetts, Indiana, Missouri, and Oklahoma, south to Florida and Texas.

Neither variety of Bracken in Missouri becomes the dominant and weedy plant that can form dense thickets and take over extensive patches of terrain, such as in the more northern and eastern states, especially in burnt-over sections of those areas. Some varieties of Bracken attain heights of 4 meters.

The fresh young stipes are, on occasion, tasted or chewed for their mucilaginous juice. By removing the woolly hairs from the young stipes, it is possible to use the uncurling young fronds as a cooked vegetable in the spring as an asparagus substitute. The plants in various countries have been employed in a number of ways. In some countries the fronds are used as thatching, packing material, and for glass making. The green fronds, when burned, produce potash.

## 13. **Adiantum** L. Maidenhair

Ferns with forking, delicate, finely cut fronds on a brown- or reddish-black stipe naked and scaleless above the base from a slender, creeping rhizome. The indusium is lacking, but the margin of some pinnules is folded

28 Adiantum pedatum var. pedatum (Maidenhair Fern)

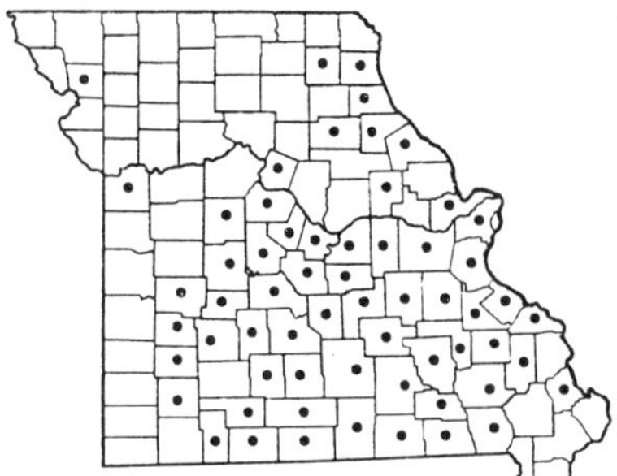

29 Pellaea glabella var. glabella (Smooth Cliff-Brake)

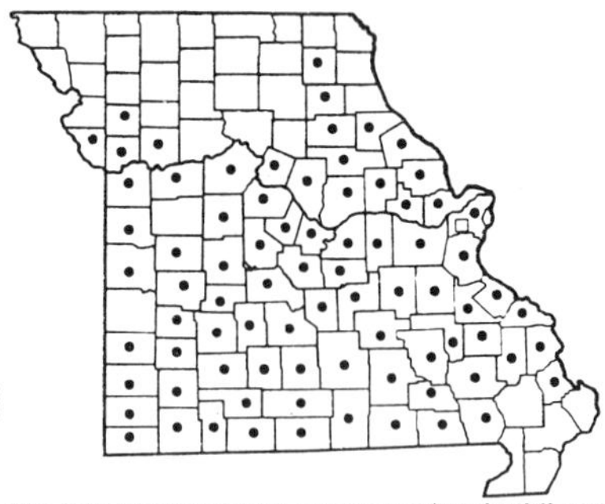

30 • Pellaea atropurpurea f. atropurpurea (Purple Cliff-Brake)
30 □ Pellaea atropurpurea f. cristata

down to form a false indusium over the marginal sporangia. (The name *adiantum* is derived from a word meaning unwetted, because the leaf-blade sheds rain.)

Leaf-blade longer than broad; main rachis with alternate branching, not forked at summit; fronds hanging from wet, usually limestone bluffs, rarely on sandstone . . . . . . . 1. A. CAPILLUS-VENERIS
Leaf-blade broader than long; main stipe palmately forking at summit, then curving into opposite directions; stipes upright with horizontal foliage; growing usually on the ground, sometimes on moist bluffs . . . . . . . . . . . . . . . . . . . . . . . . . 2. A. PEDATUM

1. **Adiantum Capillus-veneris** L.
Venus'-hair Fern Map 27
Spores mature June–August.

The usual habitat is mossy, moisture-soaked cracks of overhanging limestone bluffs. Venus'-hair Fern is confined to the Ozark section of the state; it is common in the counties drained by the Current, Eleven Points, White, North Fork, Piney, Niangua, and Gasconade rivers, north into Dallas and Phelps counties, and locally northeast to Jefferson County. The collection originally cited by Palmer & Steyermark (Am. Fern. Jour. 22: 111. 1932) from 'woods' in Saline County is based upon erroneous data and should be eliminated.

Ranges from tropical America, north in the United States to Virginia, Kentucky, Missouri, South Dakota, Colorado, Utah, and California.

The drooping habit of the fronds is characteristic.

In favorable habitats it is quite abundant, occurring in dense tiers or patches in the horizontal or vertical cracks of the strata. In dry years the fern may withstand considerable desiccation. Numerous buff-colored dried fronds may be found at the base of the clumps. Usually, however, sufficient moisture is available, and, in seasons of abundant water, the roots may be subjected to almost constantly running or dripping water.

2. **Adiantum pedatum** L. var. **pedatum**
Maidenhair Fern Map 28
Spores mature June–August.

A common fern of rich wooded slopes, ravine bottoms, and rarely occurring on moist rocky ledges and bluffs. It occurs throughout the state, and is absent only from the extreme southeastern lowland counties.

Ranges from Quebec, Ontario, and Minnesota, south to Georgia, Louisiana, and Oklahoma.

The Maidenhair Fern responds favorably to transplanting to shaded wild gardens, but needs a light humus soil.

---

Plate no. 4. **1.** Pteridium aquilinum var. latiusculum, × $^2/_7$; *a.* Lower surface of pinnae, × $^2/_7$; *b.* Margin of fertile segment rolled back to show indusium, × $^4/_7$; After Gleason, The New York Botanical Garden. **2.** Adiantum Capillus-veneris, × $^2/_7$. **3.** Adiantum pedatum, × $^2/_7$; *a.* Pinnule, showing marginal sporangia, × $^6/_7$. **4.** Pellaea atropurpurea, × $^2/_7$; *a.* Lower surface of fertile pinnule, showing continuous indusium, × 1$^3/_7$; After Gleason, The New York Botanical Garden. **5.** Notholaena dealbata, × $^2/_7$; *a.* Lower surface of pinna with confluent sporangia, × 1$^5/_7$; After Gleason, The New York Botanical Garden. **6.** Cheilanthes alabamensis, × $^2/_7$; *a.* Pinnule, lower surface, with sporangia and indusium, × 3$^3/_7$; After Gleason, The New York Botanical Garden. **7.** Cheilanthes lanosa, × $^2/_7$; *a.* Pinnule, lower surface, with sporangia, × 2$^6/_7$; After Gleason, The New York Botanical Garden. **8.** Cheilanthes tomentosa, × $^2/_7$; *a.* Pinna with pinnules, lower surface, × 2$^6/_7$; After Gleason, The New York Botanical Garden.

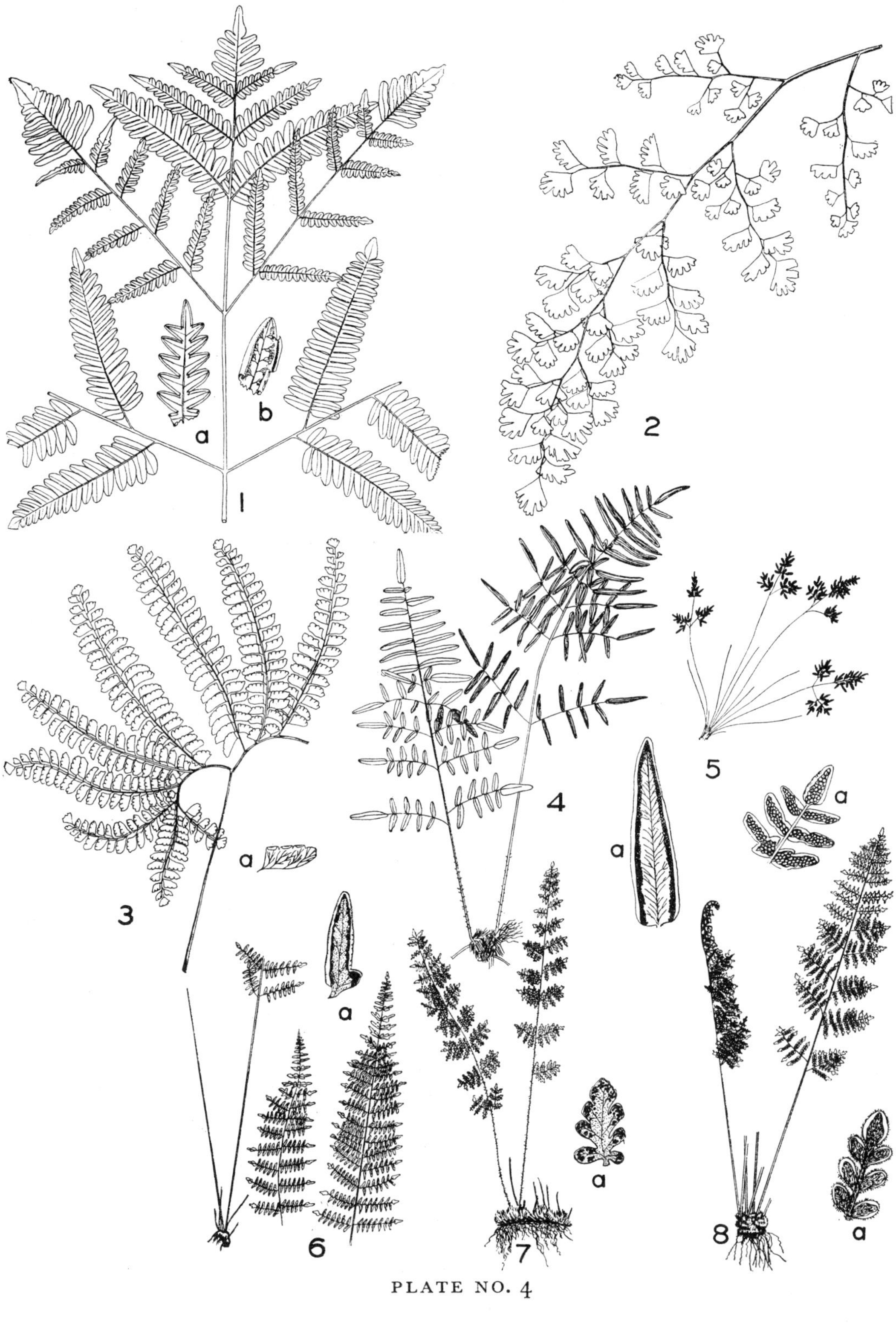


PLATE NO. 4

## 14. **Pellaea** Link Cliff-Brake

Rock ferns with evergreen grayish-blue-green, or gray-green blades.

Main stems (stipe) and rachis (slender axis to which leaf-segments are attached) without hairs or very sparsely hairy, dark reddish-brown and shining; blades bluish-green, their lowest branches at most short-stalked or nearly sessile; lowest pinnae less than twice as long as the upper pinnae . . 1. P. GLABELLA

Main stems (stipe) and rachis scurfy with hairs, dark purple-brown or blackish-purple to gray, dull; blades dull green, at least their lowest branches showing definite stalks; lowest pinnae usually 2 or more times as long as the upper pinnae . . . . . . . . . . . . . . . . . 2. P. ATROPURPUREA

1. **Pellaea glabella** Mett. ex Kuhn var. **glabella**
Smooth Cliff-Brake Map 29
*Pellaea atropurpurea* var. *Bushii* Mack. [BB]
*Pellaea atropurpurea* var. *minima* Eggert ex Farw.
Spores mature April–October.

Occurs usually on limestone cliffs or boulders throughout the Ozark region northeast to Lewis and Knox counties and locally in Andrew County, northwestern Missouri.

Ranges from Ontario to Texas, east to Vermont, Virginia, and Tennessee.

2. **Pellaea atropurpurea** (L.) Link
Purple Cliff-Brake Map 30

Main leaf-segments (pinnae) simple or pinnately divided; common type . . . 2a. P. ATROPURPUREA f. ATROPURPUREA

Main leaf-segments (pinnae) forked, crowded toward tip of frond; rare . . 2b. P. ATROPURPUREA f. CRISTATA

2a. **Pellaea atropurpurea** f. **atropurpurea**
Map 30
Spores mature June–September.

Occurs throughout the Ozark region northeast in the state to Knox County and northwest to Platte County. Ranges from British Columbia, Saskatchewan, and Vermont, south to Florida, Texas, Mexico, and Guatemala.

2b. **Pellaea atropurpurea** f. **cristata** (Trel.) Clute Map 30
*Pellaea atropurpurea* var. *cristata* Trel. [P & S, Pinkerton]

Known in the United States only from the original Missouri station, collected in 1899 on limestone rocks, Eureka, St. Louis County, by Gustav Pauls.

This form is considered synonymous with *P. atropurpurea* by Dr. Alice F. Tryon in her recent revision of the genus *Pellaea*, section Pellaea (Ann. Mo. Bot. Gard. 44: 144. 1957), but it should merit the same recognition acknowledged for similar variations of sporadic appearance among other ferns.

## 15. **Notholaena** R. Br. Cloak-Fern

Tufted rock ferns with finely divided small blades and wiry blackish stipes. The mostly solitary sporangia are borne in sori near the scarcely recurved margins of the pinnules, and an indusium is absent.

**Notholaena dealbata** (Pursh) Kunze Map 31
*Pellaea dealbata* (Pursh) Prantl [BB]
Spores mature June–September.

On dry limestone or dolomite cliffs, ledges, or boulders, chiefly in western Missouri, locally east to Howard and St. Louis counties.

Ranges from Texas and Arkansas, north to Colorado, Nebraska, and east-central Missouri.

Morton has brought evidence (Am. Fern Jour. 40: 249. 1950) to show that this fern may better be included in the genus *Pellaea* as *Pellaea dealbata* (Pursh) Kunze.

## 16. **Cheilanthes** Sw. Lip-Fern

Tufted rock ferns with dark wiry stipes and finely divided blades. An indusium-like flap (false indusium) is developed as the folded or reflexed margins of the leaf-lobes covering the separate or nearly approximate, small rounded sori.

31 Notholaena dealbata (Cloak Fern)

32 Cheilanthes alabamensis (Lip Fern)

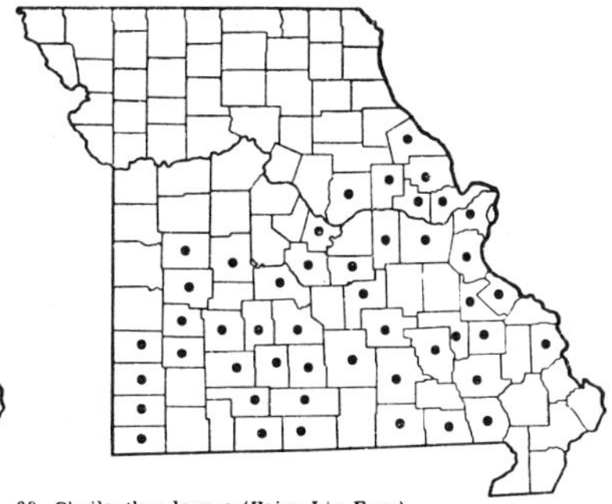

33 Cheilanthes lanosa (Hairy Lip Fern)

a. Blades hairless (glabrous) or nearly so on lower surface . . . . . . . . . . . 1. C. ALABAMENSIS
a. Blades obviously hairy (pubescent) on the lower surface . . . . . . . . . . . . . . *b*
  b. Mature stipes becoming nearly glabrous, or at most with scattered hairs; fronds mostly 4–14 cm. long, the blade mostly 2–6 cm. long; main pinnae rather dense and crowded; plants of dry limestone bluffs . . . . . . . . . . . . . . . . . . . . . . . . 4. C. FEEI
  b. Mature stipes densely woolly or with obvious spreading hairs; fronds mainly 7–50 cm. long, the blade mainly 6–25 cm. long; main pinnae well separated from each other; plants of sandstone, chert, or granite, rarely limestone . . . . . . . . . . . . . . . . . . . . . . . . . *c*
    c. Leaf-blades twice pinnate, mostly 6–15 cm. long, the pubescence on lower surface of spreading hairs; fronds averaging 12–23 cm. long; rhizome elongated, creeping; plants of sandstone, chert, or granite rocks . . . . . . . . . . . . . . . . . . . . 2. C. LANOSA
    c. Leaf-blades three times pinnate, mostly 12–25 cm. long, very densely woolly on lower surface; fronds averaging 25–40 cm. long; rhizome short, condensed; rare limestone plant. . 3. C. TOMENTOSA

1. **Cheilanthes alabamensis** (Buckl.) Kunze
Map 32

Spores mature August–October.

On dry limestone and dolomitic bluffs and mossy boulders in the southwestern Ozark section east to Ozark County.

Ranges from Georgia, north to Virginia, Tennessee, Missouri, and Oklahoma, and west to Arizona, Texas, and Mexico.

2. **Cheilanthes lanosa** (Michx.) D. C. Eaton
Hairy Lip-Fern Map 33
*Cheilanthes vestita* (Spreng.) Sw. [G]

Spores mature June–September.

On dry, often mossy, sandstone, chert, or granitic outcrops or ledges of bluffs, throughout the Ozark region north to Barton, Henry, Benton, Cole, Callaway, and Pike counties.

Ranges from Georgia and Texas, north to West Virginia, Pennsylvania, Indiana, Illinois, Missouri, and Kansas, with local stations northward in Wisconsin, New York, and Connecticut.

The leaf-blades are of a dull rusty green color. This fern can withstand long periods of drouth, during which time the fronds curl up and become brittle, but quickly revive after rains. It can be transplanted to sandy or sterile soils and rock ledges.

3. **Cheilanthes tomentosa** Link
Woolly Lip-Fern Map 34
*Cheilanthes lanosa* [G], not (Michx.) Eaton

Spores mature June–September.

The only known station for this fern was found on August 22, 1935, (*Steyermark 19530*), one-half way up exposed crevices of limestone bluffs on an escarpment at the junction of Indian Creek and White River, Stone County, southwestern Missouri. Unfortunately, the area has now been destroyed by the impounded waters of Table Rock Dam.

Ranges from Virginia and Kentucky, south and west to Georgia, Missouri, Arkansas, Texas, and Mexico.

4. **Cheilanthes Feei** Moore
Fée's Lip-Fern, Slender Lip-Fern Map 35

Spores mature June–October.

This fern has a gray-green color on the upper surface, in contrast to the other common species, *C. lanosa*, which has a dull rusty-green color with the pinnae more separated.

In crevices of west- or south-facing dry limestone and dolomitic bluffs and boulders throughout the Ozark region where it occurs west to Jasper, Cedar, Henry, Benton, and Howard counties, thence northeast to Lewis County.

34 Cheilanthes tomentosa (Woolly Lip Fern)

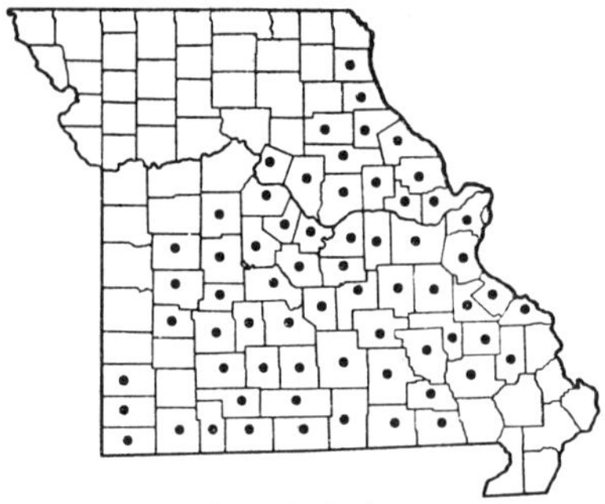

35 Cheilanthes Feei (Fee's Lip Fern)

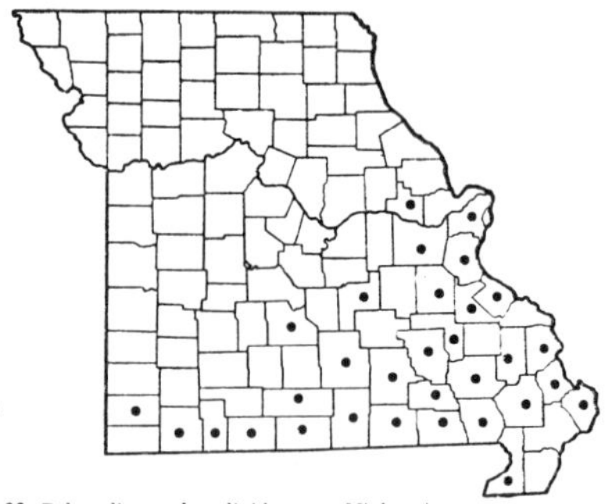

36 Polypodium polypodioides var. Michauxianum (Gray Polypody)

Ranges from Illinois and Wisconsin to British Columbia, south to Missouri, Arkansas, Texas, Arizona, and California.

During periods of extensive drouth, this fern may become as dried and shriveled-looking as some lichens, but the blades quickly become soft again after rains.

This species, so far as my experience goes, cannot be transplanted unless the original rock is taken without disturbing the fern.

### 17. **Polypodium** L. Polypody

Ferns growing upon either rocks or trees, with creeping superficial rhizomes; pinnately cut blades (not divided to the midrib), and round sori without an indusium.

Stipe and lower side of leaf-blade densely gray or gray-brown scaly; leaf-segments entire (without teeth) . . . . . . . . . . . . . . . . . . . . . . . 1. P. POLYPODIOIDES var. MICHAUXIANUM

Stipe and lower side of leaf-blade green; leaf-segments minutely toothed . . 2. P. VULGARE var. VIRGINIANUM

1. **Polypodium polypodioides** (L.) Watt var. **Michauxianum** Weath.
Resurrection Fern, Gray Polypody Map 36
Spores mature June–September.

In the lowland swamp counties of southeastern Missouri, this species is found always on the trunks and limbs of forest trees, and likewise grows epiphytically on trees along courses of streams in southern Missouri. Northward in the Ozark region it occurs usually on mossy sandstone or chert ledges or boulders of wooded bluffs. In Missouri it is confined to the Ozark section of the state, ranging north to Newton, Laclede, Phelps, Franklin, and St. Louis counties, and in Warren County in an unglaciated section of the Ozarks just north of the Missouri River.

Ranges from Guatemala, Texas, and Florida, north to Oklahoma, Iowa, Missouri, southern Illinois, Kentucky, Virginia, and Delaware.

During periods of drouth, this fern can shrivel and curl the blades, appear to be dead, but quickly revives following rains. It transplants easily, when given mossy rocks or bark, but dies out in extreme cold spells during winters of the northern United States.

2. **Polypodium vulgare** L. var. **virginianum** (L.) Eaton Common Polypody Map 37
*Polypodium virginianum* L. [G]
Spores mature July–August.

On mossy, often dry, ledges of sandstone, chert, granite, or porphyritic trachyte bluffs and boulders, usually exposed to the north or east, generally following streams or branches heading up into ravines throughout the eastern part of the Ozark section, west to Callaway, Maries, Laclede, Douglas, and Ozark counties and locally north in Marion County; absent from the swampland of southeastern Missouri.

Ranges from Newfoundland to Alaska, south to Georgia, Alabama, Tennessee, and Arkansas.

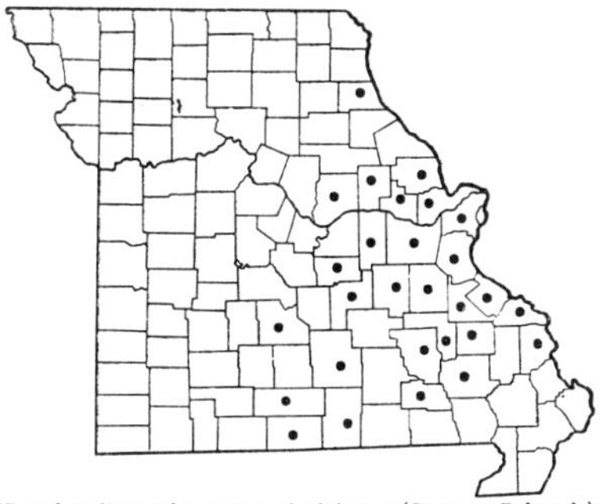

37 Polypodium vulgare var. virginianum (Common Polypody)

38 Matteuccia Struthiopteris var. pensylvanica (Ostrich Fern)

39 • Onoclea sensibilis f. sensibilis (Sensitive Fern)
39 □ Onoclea sensibilis f. obtusilobata (Sensitive Fern)

## 18. **Matteuccia** Todaro Ostrich Fern

Large terrestrial ferns, the fertile and sterile fronds unlike. The sterile fronds are once-pinnate with pinnatifid pinnae, and surround in vase-like arrangement the shorter, simply pinnate, and plume-like fertile fronds.

**Matteuccia Struthiopteris** (L.) Todaro var. **pensylvanica** (Willd.) Morton Ostrich Fern Map 38

*Pteretis pensylvanica* (Willd.) Fern [G]
*Pteretis nodulosa* (Michx.) Nieuwl. [P & S, Pinkerton]

Spores mature July–October.

In Missouri a very rare fern, found in rich alluvial woods and along banks of streams and wet ground in scattered sections of northern and west-central Missouri, where it is known from Putnam, Mercer, Linn, Daviess, and Platte counties.

Ranges from Newfoundland to Alaska, south to Virginia, Ohio, Missouri, and British Columbia.

This is one of the most beautiful and graceful native ferns, the blades of the larger sterile fronds arising in a vase-like form, abruptly narrowed at each end and widened towards the middle.

Since it is a rare fern in Missouri, it should be rigidly protected. In the northern and northeastern part of its range, where the fern is more common, the tender young leaves are gathered before they have uncoiled and cooked as a substitute for asparagus. Served with melted butter, salt and pepper, they furnish a tasty morsel.

## 19. **Onoclea** L. Sensitive Fern

Terrestrial ferns with creeping, forking rhizomes, solitary long-stalked sterile fronds with pale green deeply pinnatifid leaf-blades, and separate, stiffly erect, twice pinnate fertile fronds. The ultimate divisions (pinnules) of the fertile fronds are reduced to ball-like segments filled with sporangia.

**Onoclea sensibilis** L. Sensitive Fern Map 39

The following variations occur in Missouri:

Spore-bearing (fertile) blade completely uniform; leaf-blade completely uniform . . . a. ONOCLEA SENSIBILIS f. SENSIBILIS

Spore-bearing (fertile) blade not uniform (some of the segments leaf-like), or, if uniform, then some of the segments of the leaf-like (sterile) blade with some spore-bearing segments . b. ONOCLEA SENSIBILIS f. OBTUSILOBATA

a. **Onoclea sensibilis** L. f. **sensibilis** Map 39

Spores mature June–November.

Occurs in wet woods and thickets, and along streams and spring branches throughout the state.

Ranges from Newfoundland, Saskatchewan, Minnesota, and South Dakota, south to Florida and Texas.

With the coming of the first frosts and cold spells, the sterile leaf-blades wither and die, hence the common name for this fern. Horses fed on hay containing this fern in combination with bracken and horsetail appear to be adversely affected thereby.

b. **Onoclea sensibilis** f. **obtusilobata** (Schkuhr) Gilbert Map 39

This form, which combines characters of both fertile and sterile fronds, and in addition, has partially netted venation of the leaf-blade, has been collected from only a few counties in the state, and is scattered throughout the range of f. *sensibilis*.

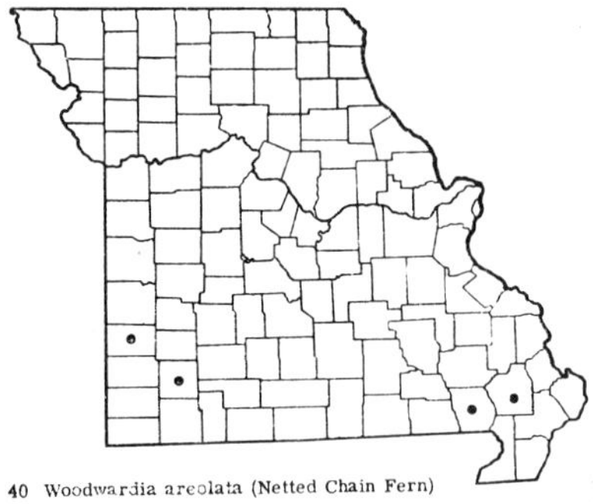

40 Woodwardia areolata (Netted Chain Fern)

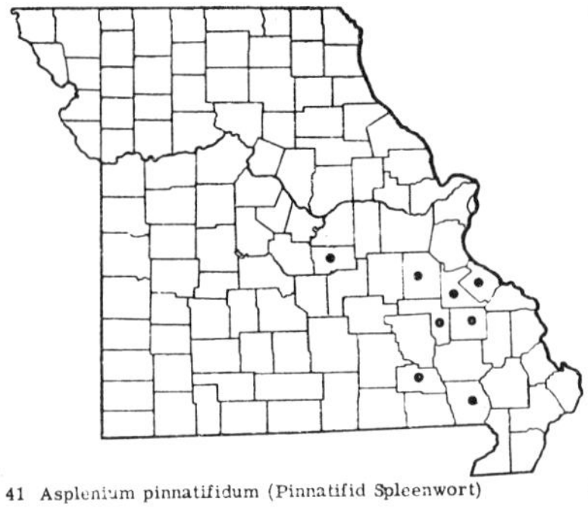

41 Asplenium pinnatifidum (Pinnatifid Spleenwort)

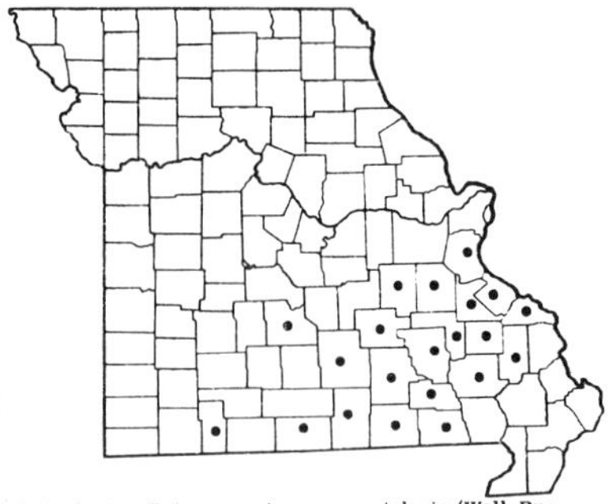

42 Asplenium Ruta-muraria var. cryptolepis (Wall-Rue Spleenwort)

### 20. **Woodwardia** Smith Chain Fern

Terrestrial ferns of wet ground with creeping rhizomes and fertile and sterile fronds on separate blades. The narrow linear fruit dots (sori) are arranged in chain-like rows parallel to the midrib of the narrow leaf-divisions of the fertile blade, the narrow indusium being attached on one side.

**Woodwardia areolata** (L.) Moore
Netted Chain Fern Map 40
Spores mature July–October.

One of the rarest ferns of Missouri. Swampy or wet ground in wooded areas. Of the four stations known, all in the southern part of the state, two were discovered in southwestern Missouri by E. J. Palmer (wet springy ground below sandstone ledge, along Horse Creek, northeast of Milford, Barton County, *Palmer 53205*; Lawrence County), and two in southeastern Missouri (*Eby*, Poplar Bluff, Butler County and *Steyermark 80393*, along sandy banks of a spring branch in sand hills of Crowley Ridge, T25N, R10E, northeast $\frac{1}{4}$ sect, 1, $3\frac{1}{2}$–$3\frac{3}{4}$ mi. [by air] southeast of Bloomfield in Stoddard County). The latter two stations are at the junction of Crowley Ridge and the swampy lowlands.

Ranges from Florida and Texas, north to Nova Scotia, Tennessee, Missouri, and Oklahoma, and locally in Michigan.

### 21. **Asplenium** L. Spleenwort

Small, evergreen, frequently rock-inhabiting ferns with mostly wiry, green to black stipes bearing several to numerous pairs of toothed or lobed segments or pinnae. The sori are usually longer than broad with a narrow indusium attached at one side.

a. Leaf-stalk (stipe) as well as main axis (rachis) of leaf-blade completely green; leaf-blade twice pinnate, the smaller divisions cut all the way to the midrib of the main pinnae, with no green connecting tissue at base; the main pinnae alternately arranged on the rachis, the lower divisions all stalked (petiolate) . . . . . . . . . . . . . . . . . . 2. A. RUTA-MURARIA var. CRYPTOLEPIS
a. Leaf-stalk (stipe) at least brown, purple-brown, or black at base or completely dark, but not completely green; main axis (rachis) of leaf-blade green, partly brown, purple-brown, black or wholly brown, purple-brown, or black from base to tip; leaf-blade simply or once-pinnate, or partly or

---

Plate no. 5. **1.** Cheilanthes Feei, × $^2/_7$; *a.* Portion of pinna, lower surface, with sporangia, × 2; After Gleason, The New York Botanical Garden. **2.** Polypodium polypodioides var. Michauxianum, × $^2/_7$; *a.* Lower portion of fertile pinna, × $^6/_7$; After Britton and Brown, The New York Botanical Garden. **3.** Polypodium vulgare var. virginianum, × $^2/_7$; *a.* Lower portion of fertile pinna, × $^6/_7$; After Britton and Brown, The New York Botanical Garden. **4.** Matteuccia Struthiopteris var. pensylvanica; *a.* Tip of sterile frond, × $^2/_7$; *b.* Middle portion of sterile frond, × $^2/_7$; *c.* Base of sterile frond, × $^2/_7$; *d.* Fertile frond, × $^2/_7$. **5.** Onoclea sensibilis, × $^2/_7$; *a.* Fertile frond, × $^2/_7$; *b.* Cluster of sporangia, × $1^5/_7$, *c.* Sporangium opening and shedding spores, × $2^6/_7$. **6.** Woodwardia areolata, × $^2/_7$; *a.* Lower portion of fertile pinna, × $^2/_7$; After Britton and Brown, The New York Botanical Garden. **7.** Asplenium pinnatifidum, × $^2/_7$; After Gleason, The New York Botanical Garden. **8.** Asplenium Ruta-muraria var. cryptolepis, × $^4/_7$; After Gleason, The New York Botanical Garden. **9.** Asplenium Trichomanes, × $^2/_7$; *a.* Fertile pinna, lower portion, × $1^5/_7$; After Gleason, The New York Botanical Garden. **10.** Asplenium resiliens, × $^2/_7$; *a.* Fertile pinna, lower portion, × $1^5/_7$; After Gleason, The New York Botanical Garden. **11.** Asplenium platyneuron, × $^2/_7$; *a.* Fertile pinna, lower portion, × 2; After Gleason, The New York Botanical Garden.

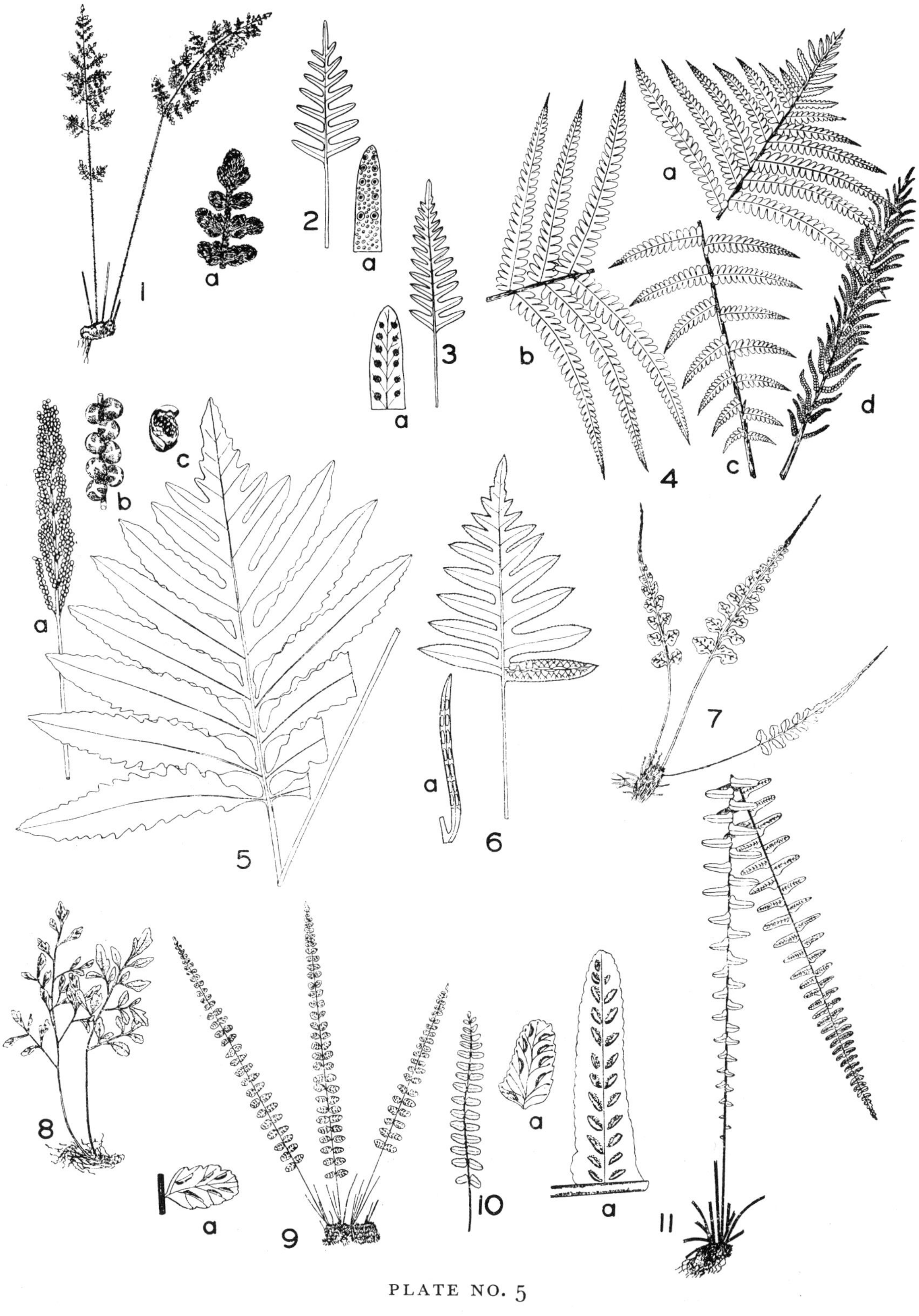


PLATE NO. 5

deeply lobed, sometimes the tip becoming prolonged and tail-like; the main pinnae opposite or alternate, without stalks (sessile) . . . . . . . . . . . . . . . . . . . . . . . . . . . *b*
b. Stipe dark brown only near base, otherwise green; rachis green from base to tip . . 1. A. PINNATIFIDUM
b. Stipe completely dark brown or black; rachis either completely dark brown or black or green in upper half or three-fourths . . . . . . . . . . . . . . . . . . . . . . . *c*
c. Stipe and rachis completely dark brown, purple-brown to black from base to tip . . . . . . *d*
d. Leaf-divisions (pinnae) alternate . . . . . . . . . . . . . . . . 5. A. PLATYNEURON
d. Leaf-divisions opposite . . . . . . . . . . . . . . . . . . . . . . . . *e*
e. Stipe and rachis purple-brown; leaf-divisions without a small ear-like lobe (auricle) at base, broadest above the base; growing on sandstone, chert, or granitic rocks. 3. A. TRICHOMANES
e. Stipe and rachis blackish; leaf-divisions with a small ear-like lobe (auricle) at base, broadest at the base; growing on limestone . . . . . . . . . . . . . 4. A. RESILIENS
c. Stipe dark brown; lower part of the rachis brown, the remainder green . . . . . . . . . *f*
f. Stipes and lower part of rachis purple-brown; main pinnae with usually somewhat wavy margins, but without fine short teeth or pointed lobes; upper part of leaf-blade conspicuously long-tapering, tail-like, and usually without prominent lobes; growing usually on limestone rocks . . . . . . . . . . . . . . . . . . . . . 6. A. × ASPLENIOIDES
f. Stipes and lower part of rachis brown; main pinnae usually with small pointed or rounded teeth on margins, not wavy-margined; upper part of leaf-blade narrowed, but not prolonged and tail-like, usually with lobes and teeth; growing usually on sandstone, chert, or granitic rocks . . . . . . . . . . . . . . . . . . . . . . . . . . *g*
g. Rachis dark brown for $\frac{1}{3}$ its length, green for remaining $\frac{2}{3}$; most of pinnae deeply lobed or cut; rachis mostly not winged . . . . . . . . . . . . . 7. A. BRADLEYI
g. Rachis barely brown at base, green most of its length; most of pinnae only slightly toothed, not deeply lobed; rachis slightly winged . . . . . . . . . 8. A. × GRAVESII

1. **Asplenium pinnatifidum** Nutt. Lobed Spleenwort, Pinnatifid Spleenwort Map 41
Spores mature July–October.

This fern seeks sheltered crevices of noncalcareous bluffs along narrow ravines and canyons at the headwaters of small creeks on shaded sandstone and granitic ledges of bluffs or boulders, in the eastern section of the Ozarks west to Maries County.

Ranges from Georgia and Oklahoma, north to Pennsylvania, Indiana, and Missouri.

This fern has sometimes been confused with the hybrid Scott's or Walking Spleenwort (*A. × ebenoides*), but may be distinguishable by the rachis, being green throughout its length, by the stipe dark only near the base, and by the habitat on noncalcareous sandstone or granitic rocks.

Dr. Wagner had produced evidence to show that this fern is a fertile tetraploid of hybrid origin between the Walking Fern (*Camptosorus rhizophyllus*) and *Asplenium montanum*. The latter species does not occur in Missouri.

2. **Asplenium Ruta-muraria** L. var. **cryptolepis** (Fern.) Wherry Wall-Rue Spleenwort Map 42
*Asplenium cryptolepis* Fern. [G, P & S, Pinkerton]
Spores mature May–September.

Shaded, mossy crevices of limestone in the upper portion of escarpments or bluffs, generally on exposures facing east or noth. Although not abundant, this tiny fern occurs fairly frequently in the eastern Ozark section west to Laclede, Texas, Ozark, and Stone counties and northeast to Jefferson County.

Ranges from Vermont, Ontario, and Michigan, south to Alabama, Illinois, Missouri, and Arkansas.

3. **Asplenium Trichomanes** L.
Maidenhair Spleenwort Map 43
Spores mature July–September.

Usually in shaded moist crevices of overhanging ledges of sandstone, chert, or granitic bluffs or boulders, rarely on calcareous rocks. Scattered throughout the southern and eastern sections of the Ozarks, north to Cedar, Polk, Phelps, Franklin, and St. Louis counties. North of the Missouri River it is known from isolated St. Peter sandstone bluffs in Lincoln County in an area which has apparently escaped glaciation.

Ranges from Nova Scotia and Manitoba to Wisconsin and North Dakota, south to Georgia, Alabama, and Texas; also found in the western states north to Alaska.

This very dainty fern may be grown if given sufficient moisture and shelter from drying winds. A porous usually noncalcareous rock seems preferred as a growing medium.

4. **Asplenium resiliens** Kunze Little Ebony Spleenwort, Blackstem Spleenwort Map 44
Spores mature May–October.

Shaded crevices of limestone and dolomitic bluffs and boulders throughout the Ozark region, following

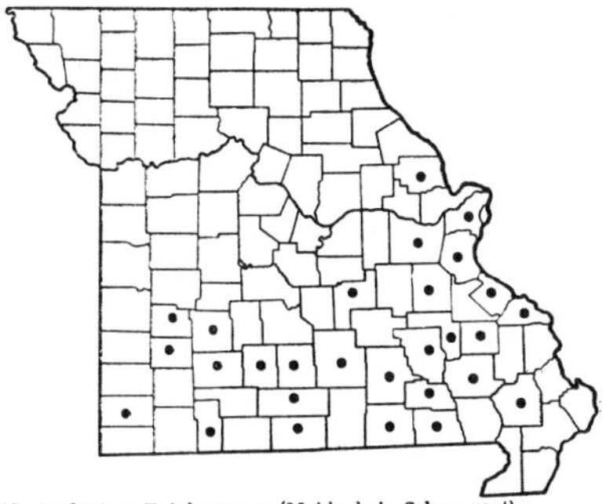

43 Asplenium Trichomanes (Maidenhair Spleenwort)

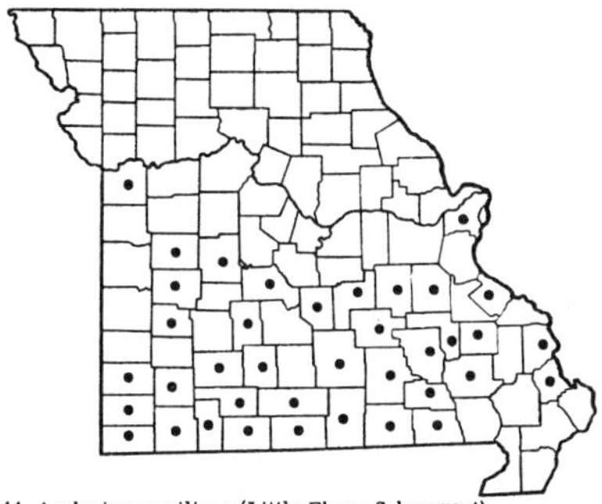

44 Asplenium resiliens (Little Ebony Spleenwort)

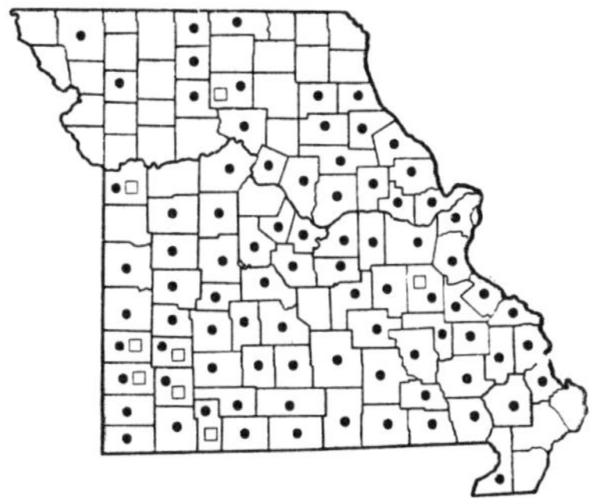

45 • Asplenium platyneuron var. platyneuron f. platyneuron (Ebony Spleenwort)

45 □ Asplenium platyneuron var. platyneuron f. serratum

courses of small and large streams, north to Jasper, Henry, Benton, Camden, Phelps, Washington, and St. Louis counties, locally northwestward in Jackson County.

Ranges from Florida, Mexico, and Arizona north to Pennsylvania, Ohio, Kentucky, southern Illinois, Missouri, Kansas, and Oklahoma; also in the West Indies and western South America.

The black stipe and rachis quickly distinguish this fern from *A. platyneuron.*

5. **Asplenium platyneuron** (L.) Oakes
Ebony Spleenwort — Map 45

The following variations occur in Missouri:

Main leaf-divisions (pinnae) narrowly ovate to oblong-lanceolate with minutely toothed margins; tip of the pinna rounded or bluntly pointed.
5a. A. PLATYNEURON var. PLATYNEURON f. PLATYNEURON

Main leaf-divisions (pinnae) linear-lanceolate, with mostly sharply toothed margins; tip of the pinna tapering to an acute tip. 5b. A. PLATYNEURON var. PLATYNEURON f. SERRATUM

5a. **Asplenium platyneuron** var. **platyneuron** f. **platyneuron** — Map 45

Spores mature May–September.

Open wooded slopes, mossy banks, and rocky wooded ledges of all kinds of rocks, but more frequently in sandstone, chert, and granitic areas. Found commonly throughout southern and central Missouri, becoming less common in the northern section of the state.

Ranges from Florida and Texas, north to Maine, Quebec, Ontario, Wisconsin, Iowa, and Colorado.

5b. **Asplenium platyneuron** var. **platyneuron** f. **serratum** (E. S. Miller) Hoffm. — Map 45

This form is known from several counties within the general range of var. *platyneuron.*

6. **Asplenium** × **ebenoides** R. R. Scott
Walking Spleenwort
× *Asplenosorus ebenoides* (R. R. Scott) Wherry [G, BB] — Map 46

Spores mature July–September.

The fern is of only rare occurrence in Missouri. The known localities in the state are from shaded limestone ledges and boulders, often at the foot of limestone bluffs along creeks, in the eastern half of the Ozarks west to Phelps and Ozark counties and northeast to St. Louis County.

Ranges from Vermont to Alabama, west to Missouri.

It has been shown that this fern is of hybrid origin between Ebony Spleenwort (*Asplenium platyneuron* var. *platyneuron*) and Walking Fern (*Camptosorus rhizophyllus*). It may also be associated with Little Ebony Spleenwort (*Asplenium resiliens*), as noted for the collection from Washington County by *Albert Chandler 401230* in the Mo. Bot. Gard. Herb.

The genus *Asplenosorus* is considered as invalid by Morton, since it was published without a Latin description. *Asplenium* × *ebenoides* and other hybrids that might be placed in *Asplenosorus* are referred by Morton to *Asplenium.*

7. **Asplenium Bradleyi** D. C. Eat.
Bradley's Spleenwort — Map 47

Spores mature June–September.

Dry sheltered crevices of sandstone, chert, or granitic cliffs, scattered throughout the Ozark region west to Vernon County, north to St. Clair, Maries, Montgomery, and Franklin counties. The Montgomery County station (*Steyermark 86127*) is the only record north of the Missouri River and occurs in the area which is Ozarkian both geologically and botanically.

Ranges from Georgia and Oklahoma, north to New York, Pennsylvania, Ohio, Kentucky, and Missouri.

46 Asplenium X ebenoides (Walking Spleenwort)

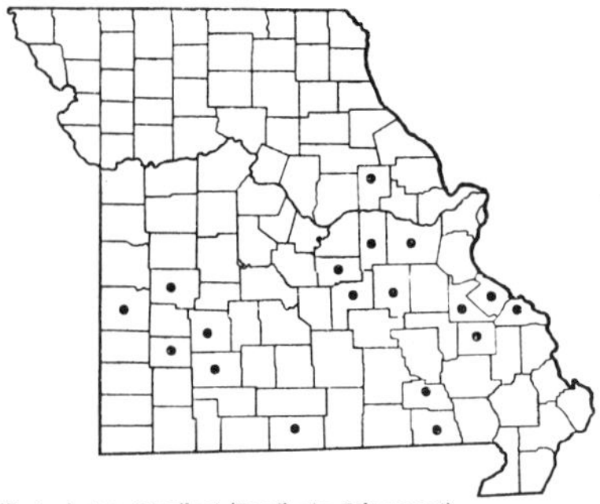

47 Asplenium Bradleyi (Bradley's Spleenwort)

48 Asplenium X Gravesii (A. Bradleyi X A. pinnatifidum)

This species is considered by Dr. Warren Wagner, Jr., to be a fertile hybrid, originating from a cross between *Asplenium montanum* (which does not occur in Missouri) and *A. platyneuron*, the Ebony Spleenwort.

8. **Asplenium** × **Gravesii** Maxon Map 48

Known in Missouri only from Madison County, in the southeastern Ozarks, where it was collected on November 10, 1898, by *Colton Russell* from cliffs of the La Motte sandstone formation near Mine La Motte. The specimen was determined by Dr. Wagner.

The fern is also known from Pennsylvania and New Jersey. It is considered a hybrid between *Asplenium pinnatifidum* and *A. Bradleyi*, combining characters of each. It differs from *A. pinnatifidum* in the dark stipe, merely acuminate instead of elongated tip of the leaf-blade, and in the sessile lowermost pinnae, whereas from *A. Bradleyi* it differs in the slightly winged rachis, and in the less cut pinnae.

It is probable that with more intensive botanical exploration of the state, other hybrids, now known to occur within the genus *Asplenium*, or between *Asplenium* and *Camptosorus*, will be found in Missouri. Those which may be expected to occur are *Asplenium kentuckiense* McCoy (*A. platyneuron* × *A. pinnatifidum*), *A. inexpectatum* E. L. Braun (*Camptosorus rhizophyllus* and *Asplenium Ruta-muraria* var. *cryptolepis*), and *A. Bradleyi* × *A. platyneuron*. The origin of these hybrids is being investigated by Dr. Warren H. Wagner, Jr., and his associates.

## 22. **Camptosorus** Link Walking Fern

Usually rock-inhabiting evergreen ferns with stipes green except for dark brown base and with simple narrow blades which taper to a very long, slender rooting tip. The veins are arranged in a rather loose network. The sori are oblong or linear with the indusium attached along one side.

**Camptosorus rhizophyllus** (L.) Link Map 49

Spores mature May–October.

Usually occurring on mossy sheltered ledges and bluffs of limestone or dolomitic bluffs, or on calcareous boulders or talus of hills, but also found on sandstone, chert, or granitic rocks. Throughout southern and central Missouri in the Ozark and unglaciated prairie region, and north in the glaciated section of the state in favored sites to Andrew, Grundy, Chariton, Randolph, Knox, and Clark counties.

Ranges from Georgia and Oklahoma, north to Quebec, Ontario, and Minnesota.

Morton (Am. Fern Jour. 45: 26. 1955; 46: 153. 1956), Wherry (Guide to Eastern Ferns. 2nd ed. p. 135. 1948, U. of Penn. Press, Philadelphia), and others have suggested that the Walking Fern be transferred to the genus *Asplenium*, and be called *A. rhizophyllum* L. There is sound justification for this

---

Plate no. 6. **1.** Asplenium × ebenoides, × $^2/_7$; *a.* Lower portion of fertile pinna, × $^6/_7$; After Gleason, The New York Botanical Garden. **2.** Asplenium Bradleyi, × $^4/_7$; *a.* Lower portion of fertile pinna, × $^6/_7$; After Gleason, The New York Botanical Garden. **3.** Asplenium × Trudelii, × 1; After Gleason, The New York Botanical Garden. **4.** Camptosorus rhizophyllus, × $^2/_7$; *a.* Portion of lower surface of fertile frond, × $^6/_7$; After Gleason, The New York Botanical Garden. **5.** Athyrium pycnocarpon, × $^2/_7$, with portions of tip, middle, and base of frond; *a.* Pinna, showing venation, × $^2/_7$. **6.** Athyrium Filix-femina var. Michauxii, × $^2/_7$; *a.* Tip of frond; *b.* Middle portion of frond; *c.* Base of frond. **7.** Woodsia obtusa, × $^2/_7$; *a.* Lower portion of fertile pinnule, × $2^6/_7$; After Gleason, The New York Botanical Garden. **8.** Cystopteris bulbifera, × $^2/_7$, with upper and lower portions of frond. **9.** Cystopteris fragilis var. fragilis, × $^2/_7$; *a.* Lower portion of fertile pinnule, × $^6/_7$. **10.** Athyrium thelypteroides, × $^2/_7$.

PLATE NO. 6

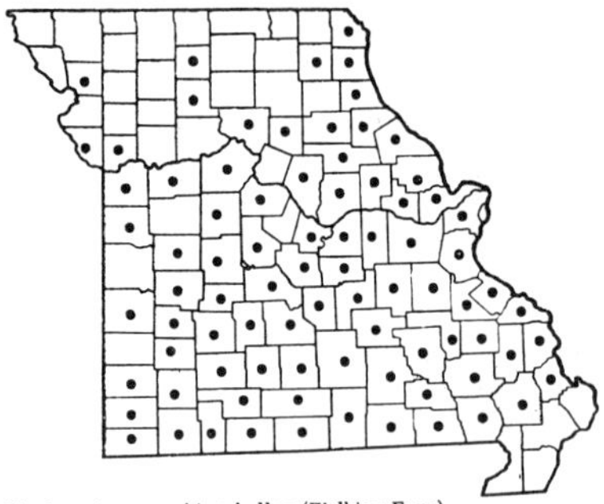

49 Camptosorus rhizophyllus (Walking Fern)

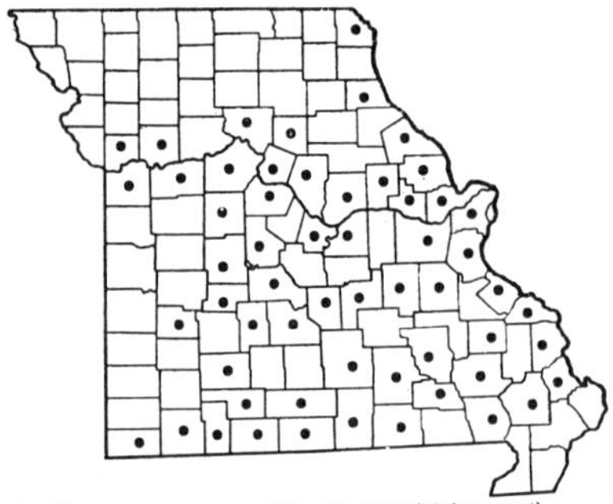

50 Athyrium pycnocarpon (Narrow-leaved Spleenwort)

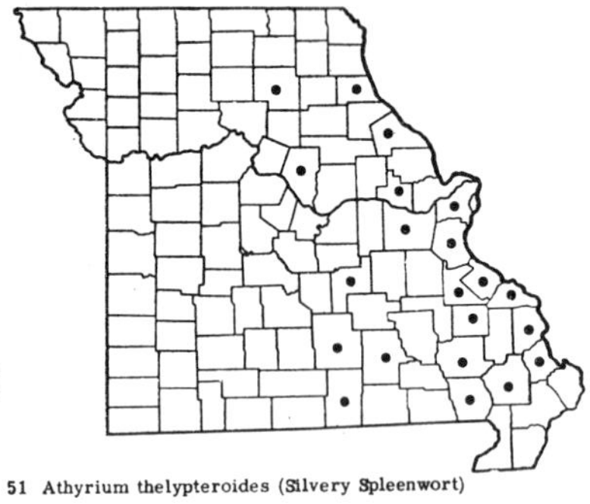

51 Athyrium thelypteroides (Silvery Spleenwort)

since, as Morton states, the elongated rooting tip is found in many species of *Asplenium*, and the veins in a network arrangement are found in several unrelated species of *Asplenium*.

This fern does well in terraria and wardian cases, and may be grown in the wild garden if given adequate shelter from drying winds and damp, mossy limestone rocks as a substrate medium.

## 23. **Athyrium** Roth.

Large ferns with soft, rather thin, deciduous fronds. Leaf-blades once to three times pinnate. Stipes greenish. Sori narrow and linear to oblong or hooked or horseshoe-shaped, covered by an indusium attached along one side.

a. Leaf-blade only simply or once pinnate, the leaf-divisions (pinnae) nearly entire (without teeth) or only slightly toothed . . . . . . . . . . . . . . . . . . . . . . . 1. A. PYCNOCARPON

a. Leaf-blade either pinnate-pinnatifid (main leaf-divisions deeply lobed but not cut all the way to their midribs) or twice or more pinnately divided . . . . . . . . . . . . . . . . . . . . *b*

b. Leaf-blade pinnate-pinnatifid (main leaf-divisions deeply lobed but not cut all the way to their midribs); ultimate segments of the pinnule more or less without teeth; rachis and midnerves of main leaf-divisions mostly with narrow scales; veins of leaf-lobes mostly not forked . 2. A. THELYPTEROIDES

b. Leaf-blade twice or more pinnate (main leaf-divisions divided to their midribs or still further divided); ultimate segments of the pinnule mostly with small teeth; rachis and midnerves of main leaf-divisions smooth; veins of leaf-lobes mostly forked . . . . . . . . . 3. A. FILIX-FEMINA

1. **Athyrium pycnocarpon** (Spreng.) Tidestr.
Narrow-leaved Spleenwort Map 50
*Athyrium angustifolium* (Michx.) Milde [P & S, Pinkerton]
Spores mature July–September.

Rich wooded slopes or wooded floors of ravines or valleys near streams or their branches. Frequent in all the Ozark section, forming luxuriant stands in the rich hill soils of the counties along the Missouri River, north along the Mississippi River to Clark County, and west to Clay, Jackson, Pettis, Benton, Cedar, and McDonald counties. Absent from much of the unglaciated and glaciated prairie region, and from the southeastern Missouri lowlands.

Ranges from Quebec to Minnesota, south to Georgia and Louisiana.

This fern grows well in a wild garden, if provided with a woodland or protected situation with light well-drained rich soil.

2. **Athyrium thelypteroides** (Michx.) Desv.
Silvery Spleenwort Map 51
*Athyrium acrostichoides* (Sw.) Diels [P & S, Pinkerton]

Rich wooded slopes or wooded floors of ravines or valleys near streams or their branches. Eastern Missouri west to Macon, Boone, Phelps, Texas, and Howell counties. This fern occupies much the same types of rich soil habitats as *A. pycnocarpon*, but is much less common.

Ranges from Nova Scotia, Minnesota, and Iowa. south to Georgia, Louisiana, and Missouri.

This species, like the narrow-leaved spleenwort, does well in the wild garden, when provided with sufficient shade, protection from wind, and light rich soil.

3. **Athyrium Filix-femina** (L.) Roth.
Lady Fern Map 52
The following three variations occur in Missouri:

52 • Athyrium Filix-femina var. Michauxii f. rubellum (Lady Fern)

52 □ Athyrium Filix-femina var. Michauxii f. elatius (Lady Fern)

52 x Athyrium Filix-femina var. asplenioides (Lady Fern)

53 Woodsia obtusa (Blunt-lobed Woodsia)

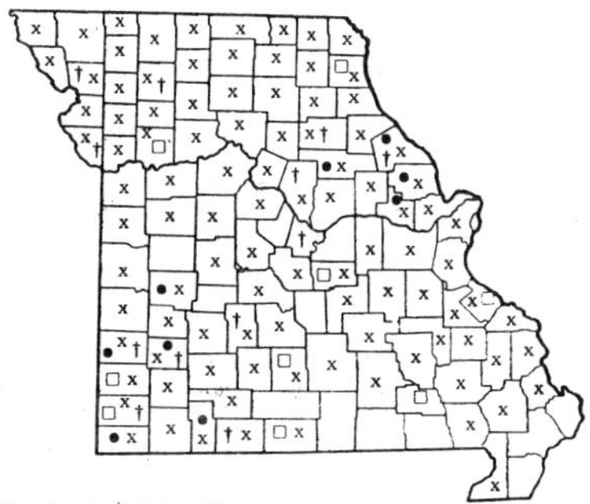

54 • Cystopteris fragilis var. fragilis f. fragilis (Common Fragile Fern)

54 † Cystopteris fragilis var. Mackayii (Common Fragile Fern)

54 x Cystopteris fragilis var. protrusa (Common Fragile Fern)

54 □ Cystopteris fragilis var. fragilis f. simulans (Common Fragile Fern)

a. Leaf-blade widest near the base, the second or third pair of pinnae from the base the longest; stalked glands usually absent from the rachis and lower side of the midrib . . 3a. A. FILIX-FEMINA var. ASPLENIOIDES

a. Leaf-blade widest near the middle, the fourth or fifth pair of pinnae from the base the longest; stalked glands usually occurring on the rachis and the lower side of the midrib. *b*

b. Sun form with fertile fronds (with spores attached to under surface) more contracted and leathery than the sterile ones; sori coming together and covering lower side of fertile leaf-divisions at maturity . 3b. A. FILIX-FEMINA var. MICHAUXII f. ELATIUS

b. Shade form with fertile and sterile fronds of more or less similar appearance; sori mostly remaining separate from one another at maturity . . 3c. A. FILIX-FEMINA var. MICHAUXII f. RUBELLUM

3a. **Athyrium Filix-femina** var. **asplenioides** (Michx.) Farw. Map 52
*Athyrium asplenioides* (Michx.) Desv. [P & S, Pinkerton]

Spores mature July–October.

Rich wooded slopes and floors of ravines, along creeks and spring branches, or in wooded valleys along larger streams, rarely on moist sandstone or granitic ledges or bluffs. Southern and central Missouri, north to Jackson, Boone, and Lincoln counties.

Ranges from Florida to Texas, north to Massachusetts, New York, Ohio, Indiana, Missouri, and Oklahoma.

This fern grows well in a wild garden when provided with a loose humusy soil and a sheltered location.

3b. **Athyrium Filix-femina** var. **Michauxii** (Spreng.) Farw. f. **elatius** (Link) Clute Map 52
*Athyrium angustum* var. *elatius* (Link) Butters [P & S]
*Athyrium angustum* in part [Pinkerton]

In similar habitats to those of var. *asplenioides*. This form is the rarest of the three variations encountered in Missouri, and occurs in only a few counties in eastern Missouri. Some authors consider it only an ecological variation, responding to a greater amount of sun.

Ranges from Quebec, Manitoba, and Minnesota, south to Maryland and Missouri.

3c. **Athyrium Filix-femina** var. **Michauxii** f. **rubellum** (Gilbert) Farwell Map 52
*Athyrium angustum* var. *rubellum* (Gilbert) Butters [G]
*Athyrium angustum* in part [Pinkerton]

Spores mature June–October.

In similar situations to those of var. *asplenioides*. The variety is comparatively rare in the Ozark region, where it is replaced by var. *asplenioides*, but is the common variety in glaciated northern Missouri, where it is found on wooded hills and in ravines bordering creeks and their tributary branches.

Ranges from Labrador to Manitoba, south to Virginia, Indiana, Illinois, Missouri, and South Dakota.

## 24. **Woodsia** R. Br. Woodsia

Low-growing, tufted ferns with pinnately divided fronds. Indusium star-shaped or lobed, arising beneath the sorus, the lobes of the indusium curving rather symmetrically around the sorus.

**Woodsia obtusa** (Spreng.) Torr.
Blunt-lobed Woodsia Map 53
Spores mature May–October.

Moist or dry, mossy, shaded, or sometimes exposed ledges or bluffs of limestone, sandstone, chert, or granitic rocks, and rocky open wooded slopes and banks. Widespread and common throughout the state, but apparently absent in the extreme northwestern part and in the southeastern lowlands.

Ranges from Nova Scotia to Minnesota, south to Florida and Texas.

Sometimes confused with *Cystopteris fragilis*, but can be distinguished by having the scales on the rachis.

## 25. **Cystopteris** Bernh. Bladder Fern

Delicate ferns with deciduous, finely pinnately cut fronds. Indusium attached at the base of and partly at one side of the somewhat round sorus, and becoming arched or hood-like over the sporangia.

Leaf-blade broadest above the base; veins of leaf segments usually ending in a tooth; tip of leaf-blade acute or acuminate; undersurface of leaf-blade never bearing rounded bulblets . . . . . . 1. C. FRAGILIS

Leaf-blade broadest at the base; veins of leaf segments running to a notch (sinus); tip of leaf-blade long-tapering; undersurface of leaf-blade often bearing rounded bulblets . . . . . . 2. C. BULBIFERA

1. **Cystopteris fragilis** (L.) Bernh.
Common Fragile Fern Map 54

This may be mistaken for *Woodsia obtusa*, but is separated from that species by the absence of hairs, scales, or glands on the leaf-blade.

The following variations occur in Missouri:

a. Plants growing in soil; rhizome long-creeping with the growing point prolonged 2–5 cm. beyond the fronds; basal pinnules of the larger main leaf-divisions (pinnae) stalked, cuneate (acutely narrowed) at base. . . 1d. C. FRAGILIS var. PROTRUSA

a. Plants usually growing on rocks or on rocky ledges; rhizome short and contracted, not long creeping, the growing point not prolonged beyond the fronds; basal pinnules of the larger main leaf-divisions (pinnae) sessile (stalkless) or nearly sessile . . . . . . . . . . *b*

b. Indusium about 0.5 mm. long, without teeth (entire) or merely shallowly cleft; basal segments of pinnule of the main leaf-divisions (pinnae) cuneate (acutely narrowed) at the base . . . . 1c. C. FRAGILIS var. MACKAYII

b. Indusium up to 1 mm. long, becoming deeply cleft at the tip; basal segments of pinnule of the main leaf-divisions (pinnae) broadened at base . . . . . . . . *c*

c. Lower and larger main-divisions (pinnae) lanceolate or narrowly lanceolate-ovate, one-third to one-half as broad as long. 1a. C. FRAGILIS var. FRAGILIS f. FRAGILIS

c. Lower and larger main-divisions (pinnae) deltoid-ovate, one-half to five-sixths as broad as long . . 1b. C. FRAGILIS var. FRAGILIS f. SIMULANS

1a. **Cystopteris fragilis** (L.) Bernh. var. **fragilis** f. **fragilis** Map 54
*Cystopteris fragilis* in part [P & S, Pinkerton]
*Cystopteris fragilis* [G]
Spores mature June–September.

Moist shaded recesses of mostly limestone and sandstone bluffs, sometimes on cherty and granitic sheltered ledges or outcrops with moist mossy pockets. This variety is found scattered over sections of the state, but is not as common as var. *protrusa*.

Ranges from Labrador to Alaska, south to North Carolina, Texas, and California; also in Eurasia.

1b. **Cystopteris fragilis** var. **fragilis** f. **simulans** Weath. Map 54

The habitat is the same as that of var. *fragilis*, with which it frequently is associated.

Known from Illinois and Tennessee to Missouri and Texas.

1c. **Cystopteris fragilis** var. **Mackayii** Lawson Map 54

The habitat is similar to that of var. *fragilis*. This variety is of frequent occurrence throughout the state, but not as common as var. *protrusa*.

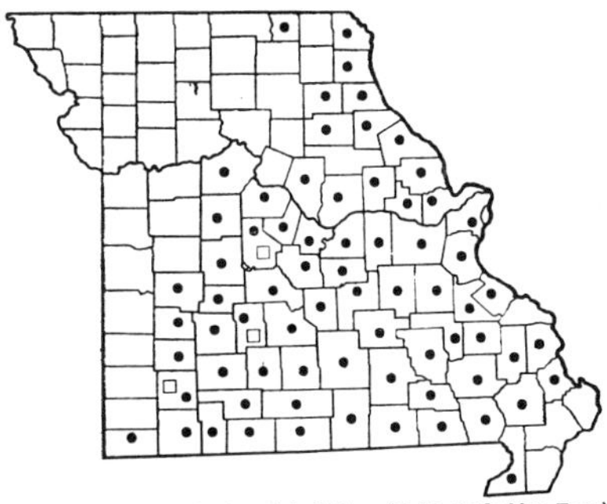

55 • Cystopteris bulbifera f. bulbifera (Bulblet Bladder Fern)

55 □ Cystopteris bulbifera f. horizontalis

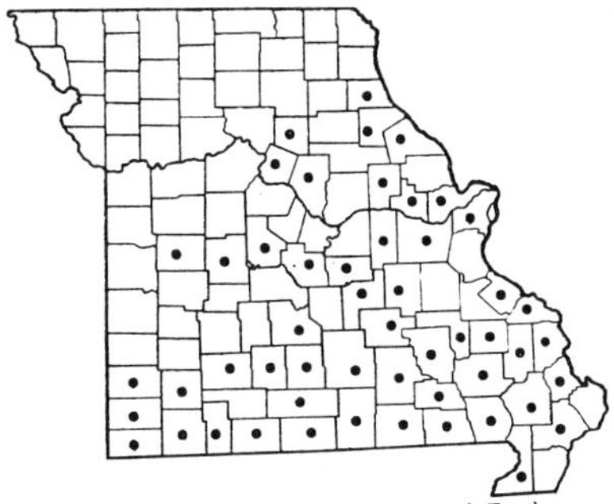

56 Thelypteris hexagonoptera (Broad Beech Fern)

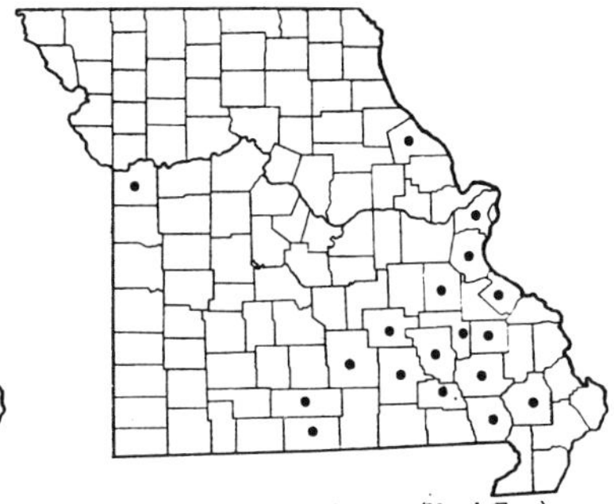

57 Thelypteris palustris var. pubescens (Marsh Fern)

Ranges from Nova Scotia to North Dakota, south to Virginia and Missouri.

1d. **Cystopteris fragilis** var. **protrusa** Weath. Map 54

Rich wooded slopes, wooded floors of ravines or of river valleys, and banks along streams. This is the commonest variety in Missouri, and is found throughout the state.

Ranges from New York to Minnesota and Nebraska, south to Georgia, Louisiana, and Oklahoma.

The habitat of this variety which is on the ground in soil, instead of on rocks, readily distinguishes it from the other varieties. This variety is easily grown in wild gardens where given sufficient shade. The fronds die down to the ground during summer.

2. **Cystopteris bulbifera** (L.) Bernh.
Bulblet Fern, Bladder Fern Map 55

The following two forms may be distinguished:

Leaf-blade long-tapering at tip; common type . . 2a. C. BULBIFERA f. BULBIFERA

Leaf-blade not long-tapering at tip; rare type . 2b. C. BULBIFERA f. HORIZONTALIS

2a. **Cystopteris bulbifera** f. **bulbifera** Map 55
*Cystopteris bulbifera* in part [P & S, Pinkerton]
Spores mature June–September.

Shaded, usually damp or mossy ledges, talus, or lower rocky portion of usually limestone bluffs, rarely on sandstone or granitic rocks. Throughout the Ozark region and counties bordering the Ozarks north to Schuyler and Clark counties, west to McDonald, Lawrence, St. Clair, Pettis, Saline, Boone, and Shelby counties; absent from the lowlands of southeastern Missouri, from most of the northwest and north-central sections, and from most of the unglaciated prairie region.

Ranges from Newfoundland, Manitoba, and Utah, south to Georgia, Arkansas, and Arizona.

Bulblets are often absent from the fronds, but, when present, offer a useful character in readily distinguishing this from the Fragile Fern (*C. fragilis* var. *fragilis* and other varieties). This fern does well in a shaded rock garden when provided with calcareous rocks in a sheltered, damp location.

2b. **Cystopteris bulbifera** f. **horizontalis**
(Lawson) Gilbert Map 55
Associated with f. *bulbifera*, but of rather infrequent occurrence.

## 26. **Thelypteris** Schmidel

Ferns with deciduous fronds and slender, creeping rhizomes. The leaf-blade has a hairy rachis. The fertile blades are similar to the sterile ones. The round sori are covered by a heart-shaped indusium, or in *T. hexagonoptera* the indusium is absent.

Ferns of dry or rich woods or wooded slopes; leaf-blades triangular, definitely broadest at the base; main leaf-divisions (pinnae) not divided all the way to the midrib (rachis), but connected at their bases by green tissue . . . . . . . . . . . . . . . . . . . . . 1. T. HEXAGONOPTERA

Ferns of swamps and wet ground along streams and in wet thickets; leaf-blades lanceolate, the lowest main leaf-divisions (pinnae) about same length or shorter than the middle ones; main leaf-divisions (pinnae) divided all the way to the midrib (rachis), not connected at their bases by any green tissue . . . . . . . . . . . . . . . . . . . . . 2. T. PALUSTRIS var. PUBESCENS

58 • Dryopteris austriaca var. spinulosa (Spinulose Shield Fern)
58 □ Dryopteris austriaca var. intermedia

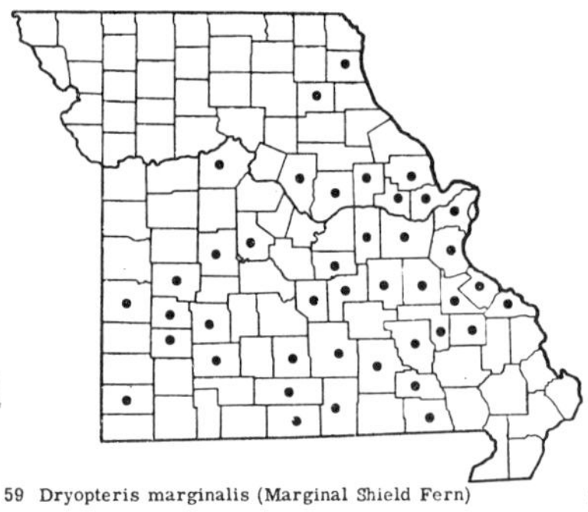

59 Dryopteris marginalis (Marginal Shield Fern)

60 Dryopteris Goldiana (Goldie's Fern)

1. **Thelypteris hexagonoptera** (Michx.) Weath. Broad Beech Fern Map 56

*Dryopteris hexagonoptera* (Michx.) Christens. [G]

Spores mature June–September.

Usually occurring in acid soils of rocky wooded, often dry slopes or floors of ravines, frequently in areas of sandstone, chert, or granitic rocks. Common throughout most of the Ozark section of the state, west to McDonald, Jasper, Laclede, Henry, and Howard counties, and north to Randolph and Marion counties.

Ranges from Florida and Texas, north to Quebec, Maine, Ontario, and Minnesota.

An indusium is absent. This fern thrives in shaded wild gardens where the soil is somewhat acid and covered with decayed leaves (oak, especially).

2. **Thelypteris palustris** Schott, var. **pubescens** (Lawson) Fern. Marsh Fern Map 57

*Dryopteris Thelypteris* (L.) Gray, var. *pubescens* (Lawson) Nakai [G]

Spores mature June–October.

Usually in swampy ground, either in woods, thickets, or wet open meadows where springs and spring branches are located. The fern is often found in wet swampy meadows kept moist by springs issuing from calcareous strata, but also occurs on more acid, moist soils. Mostly encountered in the eastern half of the Ozark region, west to Dent, Texas, and Ozark counties, and locally north and west to Pike and Jackson counties.

Ranges from Newfoundland to Manitoba, south to Georgia and Oklahoma. Other varieties occur in the southern United States, West Indies, South Africa, Eurasia, and New Zealand.

An indusium is present. Not suitable for the wild garden unless provided with a constantly moist soil.

27. **Dryopteris** Adans. Shield Fern

Ferns with mostly dark green, thickish, evergreen or nearly evergreen leaf-blades. Rhizomes stout, covered with the bases of the old leaf-stalks (stipes). Indusium horseshoe-shaped, rather conspicuous.

a. Teeth on ultimate segments of leaf-blades ending in bristle-like tips; leaf-blade mostly twice or thrice pinnate (each main division again divided to its midrib), or the smaller divisions likewise divided; at least the basal pinnules (innermost lowermost ultimate divisions) of the basal pinnae (lowermost main divisions) with stalks; lower pinnae (main leaf-divisions) conspicuously unequal-sided . . . . . . . . . . . . . . . . . 1. D. AUSTRIACA var. SPINULOSA and var. INTERMEDIA

a. Teeth on ultimate segments of leaf-blades not ending in bristle-like tips; leaf-blade with main leaf-

Plate no. 7. **1.** Thelypteris palustris var. pubescens, × 2/7; *a.* Pinna showing venation, × 2; After Gleason, The New York Botanical Garden. **2.** Dryopteris austriaca var. spinulosa; *a.* Tip of frond; *b.* Base of frond; *c.* Middle portion of frond. **3.** Thelypteris hexagonoptera, × 2/7; *a.* Lobe of portion of pinna, lower surface, × 6/7. **4.** Dryopteris Goldiana, × 2/7; *a.* Tip of frond; *b.* Middle portion of frond; *c.* Base of frond; *d.* Fertile pinnule, lower portion, × 6/7. **5.** Dryopteris Clintoniana var. australis, × 2/7; *a.* Upper part of fertile frond, lower surface; *b.* Lower part of frond, upper surface; After Gleason, The New York Botanical Garden. **6.** Dryopteris marginalis, × 2/7; *a.* Tip of frond; *b.* Middle portion of frond; *c.* Base of frond; *d.* Fertile pinnule, lower portion, × 1 1/7. **7.** Polystichum acrostichoides, × 2/7; *a.* Upper half of frond, upper surface; *b.* Lower half of frond, upper surface; *c.* Fertile pinna, lower surface, × 2/7; After Britton and Brown, The New York Botanical Garden.

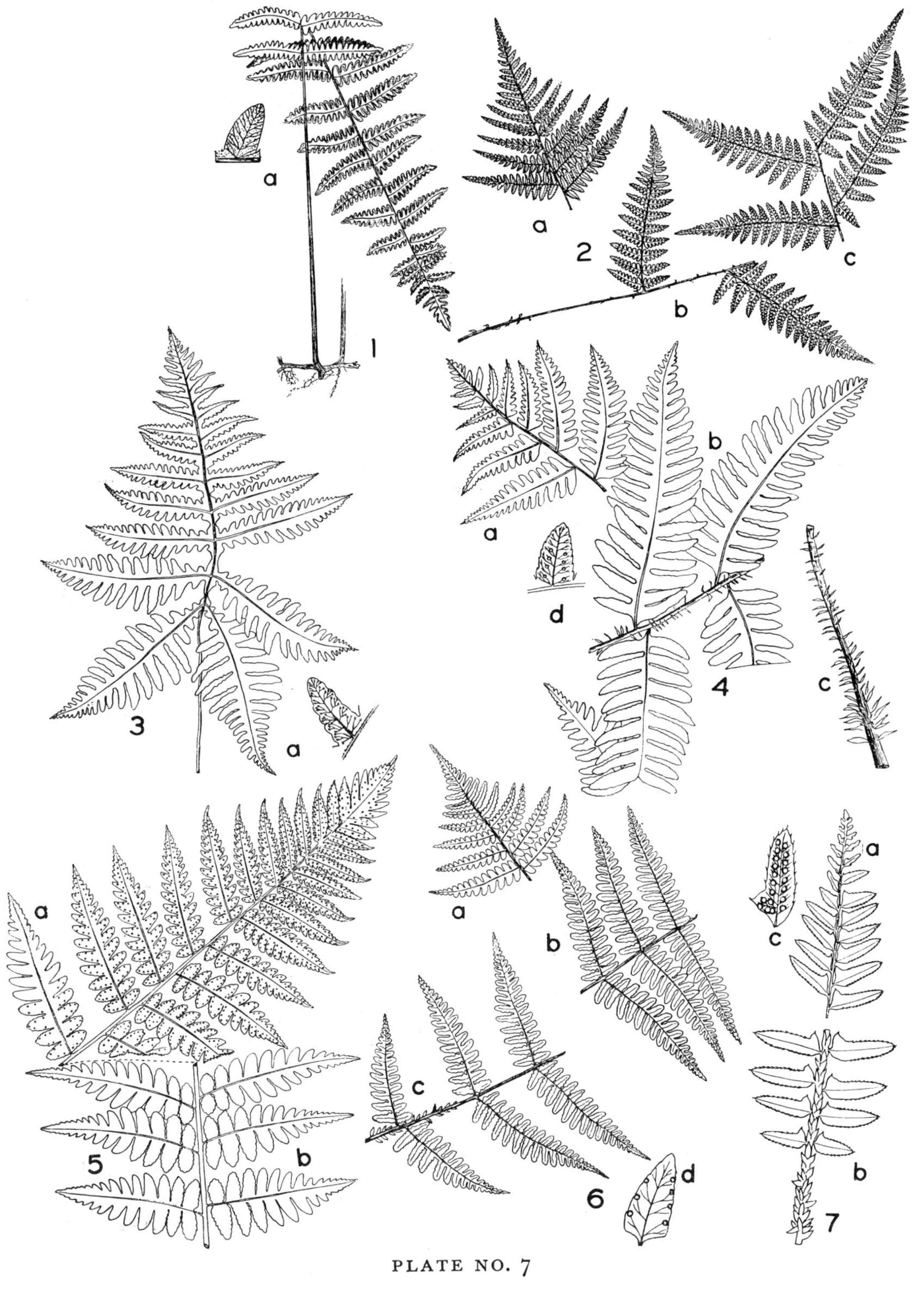


PLATE NO. 7

divisions (pinnae) deeply pinnatifid (cut nearly to their midrib) or almost twice-pinnate; basal pinnules of the basal pinnae without stalks (sessile); lower pinnae more or less equal-sided . . . . . *b*

b. Margins of ultimate segments smooth, or if toothed, the teeth blunt or rounded; sori located along the margins of the leaf segments . . . . . . . . . . . . . . . 2. D. MARGINALIS

b. Margins of ultimate segments with pointed teeth; sori nearer midribs of the leaf segments . . . . *c*

c. Lowest pinnae (main leaf-divisions) with usually 12–18 pairs of pinnules (ultimate segments), the pinnae located in the middle part of the leaf-blade with 14–22 pairs; pinnae (main leaf-divisions) broadest above the base, the inner basal (lowermost) segment of the lowest pinnae shorter than the next one . . . . . . . . . . . . . . . . . . . 3. D. GOLDIANA

c. Lowest pinnae with 20–31 pairs of pinnules (ultimate segments); pinnae (main leaf-divisions) broadest at the base, the inner basal (lowermost) segment of the lowest pinnae noticeably longer than the next one . . . . . . . . . . . . . 4. D. CLINTONIANA var. AUSTRALIS

1. **Dryópteris austriaca** (Jacq.) Woynar
Spinulose Shield Fern Map 58

The following two varieties are found in Missouri, and are semi-evergreen:

Basal (lowermost) pinnule (ultimate leaf-division) of the lower side of the lowest pinnae (main leaf-divisions) longer than the next one; indusium and leaf-blade not glandular . . . 1a. D. AUSTRIACA var. SPINULOSA

Basal pinnule of the lower side of the lowest pinnae shorter than or equaling the next one; indusium and frequently the leaf-blade glandular . . . . 1b. D. AUSTRIACA var. INTERMEDIA

1a. **Dryopteris austriaca** var. **spinulosa** (Müll.) Fiori Map 58

*Dryopteris spinulosa* (O. F. Muell.) Watt. (typical) [G]

*Thelypteris spinulosa* (Muell.) Nieuwl. [P & S, Pinkerton]

Spores mature June–October.

Shaded crevices of moist St. Peter sandstone bluffs in Jefferson County and on bald cypress knees in wooded swamp in Butler County (*Colton Russell*, October 30, 1899).

Ranges from Quebec to Minnesota and British Columbia, south to Virginia, Kentucky, and Missouri.

1b. **Dryopteris austriaca** var. **intermedia** (Muhl.) Morton Map 58

*Dryopteris spinulosa* var. *intermedia* (Muhl.) Underw. [G]

*Thelypteris spinulosa* var. *intermedia* (Muhl.) Nieuwl. [P & S, Pinkerton]

Shaded crevices of moist St. Peter sandstone bluffs in Jefferson County, moist La Motte sandstone bluffs in Ste. Genevieve and Madison counties, and wooded swamps of Stoddard and Butler counties, all the known stations occurring in eastern Missouri south of the Missouri River.

Ranges from Newfoundland, Ontario, and Minnesota, south to Georgia and Missouri.

2. **Dryopteris marginalis** (L.) Gray
Marginal Shield Fern, Evergreen Wood Fern Map 59

Spores mature June–October.

Commonly in shaded crevices of sandstone, chert, or granitic ledges and bluffs, in acid soils, more rarely on talus of rocky wooded slopes; occurring in most counties of the Ozarks on sandstone, west to Newton, Vernon, and Benton counties; in the counties bordering the Missouri River on St. Peter sandstone, locally north in Lewis and Shelby counties, and west locally in Saline County.

Ranges from Nova Scotia and Ontario to Wisconsin, south to Georgia, Arkansas, and Oklahoma.

This fern does well in wild or rock gardens, if provided with sufficient shade, protection from wind, and good drainage. It prefers an acid rocky situation.

3. **Dryopteris Goldiana** (Hook.) Gray
Goldie's Fern Map 60

Spores mature June–September.

Known from two isolated stands, one from a north-facing bluff of St. Peter sandstone in a ravine tributary to Dry Fork of Charrette Creek, 6 miles south of Warrenton in Warren County (*Steyermark 73069*), and the second from rich woods at the base of a north-facing slope of a spring-fed branch 1¼ mi. east of Marquand, in Madison County, near the Bollinger County line (*Steyermark 77555*).

Ranges from New Brunswick and Ontario to Minnesota, south to South Carolina, Tennessee, and Missouri.

This is one of the most handsome and stately ferns. The fronds are nearly evergreen, and usually last well into early winter before dying.

This fern does well in wild flower gardens or around foundation plantings if given rich soil, leaf mold, and a protected, shaded situation.

61 Dryopteris Clintoniana var. australis (Clinton's Fern)

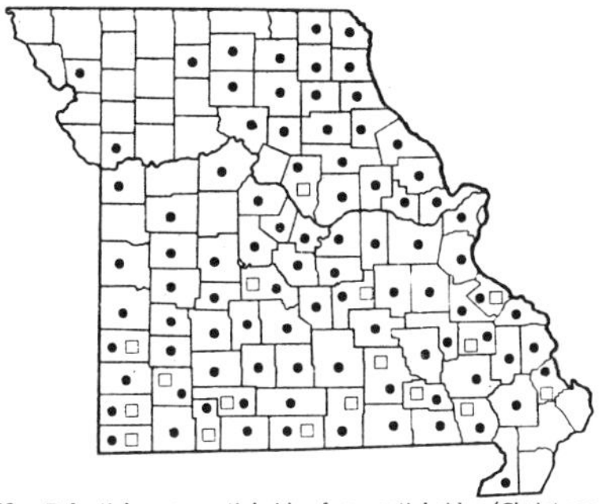

62 • Polystichum acrostichoides f. acrostichoides (Christmas Fern)

62 □ Polystichum acrostichoides f. incisum

63 Pinus nigra (Austrian Pine)

4. **Dryopteris Clintoniana** (D. C. Eaton) Dowell var. **australis** Wherry Clinton's Fern Map 61

*Dryopteris cristata* var. *Clintoniana* in part [G]

Spores mature May–July.

One of the rarest ferns of Missouri, known only from the original station discovered in 1938 in Carter County, southeastern Ozark region (*Steyermark 5333*), around outlet of Blue Spring, tributary to Big Barren Creek, 10 miles northwest of Bennett.

Ranges from the Gulf Coast to Pennsylvania, northwest to Missouri.

The status of this fern is controversial. Wherry recognized it as a variety of *D. Clintoniana*, Fernald merged it with var. *Clintoniana* under *D. cristata*, while by others it has been questioned as being a hybrid with *D. ludoviciana* (Kunze) Small. The most recent studies by S. Walker (Am. Fern Jour. 49: 108, 1959) suggest that it is a triploid hybrid, with *D. ludoviciana* as one of its parents.

## 28. **Polystichum** Roth.

Terrestrial ferns with thick evergreen pinnate (in Missouri species) blades. Indusium round, attached at the center only, the round sorus opening on all sides of the center of attachment of the indusium.

**Polystichum acrostichoides** (Michx.) Schott Christmas Fern Map 62

Two forms are recognizable in Missouri:

Main leaf-segments (pinnae) minutely toothed, with shallow, slanting teeth . 1a. P. ACROSTICHOIDES f. ACROSTICHOIDES

Main leaf-segments coarsely toothed, especially the lower ones more deeply toothed . . . . . 1b. P. ACROSTICHOIDES f. INCISUM

1a. **Polystichum acrostichoides** f. **acrostichoides** Map 62

Spores mature June–October.

On dry or moist, but well-drained wooded slopes, generally on the more acid soils, where often associated with sandstone, chert, or granitic rocks, attaining its best development on north-facing slopes, in narrow ravines, or at the base of talus slopes in acid soil; sometimes luxuriating along moist mossy banks or moist rock ledges. Occurring in all except the counties of northwestern and extreme southeastern Missouri, and absent from some northern and west-central counties.

Ranges from Nova Scotia to Wisconsin, south to Florida, Texas, and Mexico.

A satisfactory fern for the wild garden, but preferring well-drained slopes with protection from wind and sun.

1b. **Polystichum acrostichoides** f. **incisum** (Gray) Gilbert Map 62

*Polystichum acrostichoides* var. *incisum* Gray [Pinkerton]

Frequently occurring with typical f. *acrostichoides*, but less common. Scattered in sections of southern and central Missouri.

# *Division II*

# *Spermatophyta*

## Class 1. **GYMNOSPERMAE**

Ovules pollinated directly without being enclosed within an ovary, usually in an open scale.

### Order **CONIFERAE**

a. Leaves not evergreen, falling in late autumn, set singly on two sides of the stem; in swamps of southeastern Missouri . . . . . . . . . . . . . . . . . . 2. TAXODIUM in Taxodiaceae
a. Leaves evergreen, persistent, in clusters of 2–3, or scale-like . . . . . . . . . . . . . . *b*
  b. Leaves in clusters of 2 or 3, at least 2.5 cm. long . . . . . . . . . . . 1. PINUS in Pinaceae
  b. Leaves usually scale-like, but, if needle-like on some branches, less than 1 cm. long . . 3. JUNIPERUS in Cupressaceae

### **PINACEAE** (Pine Family)

#### 1. **Pinus** L. Pine

The pollen shed by species of pine is very abundant, but is mostly harmless, so far as hay fever is concerned. American Indians and Europeans have used the inner bark of various species of pine as food, in Europe mixing the ground dried bark with other flour as a type of bread. *Pinus sylvestris,* the Scotch Pine, commonly cultivated in Missouri, was especially used in northern Europe. The inner bark of various pines may be cut into strips and chewed in the spring for the spicy taste.

a. Leaves 3–7 cm. long . . . . . . . . . . . . . . . . . . . . . 3. P. VIRGINIANA
a. Leaves 7–15 cm. long. . . . . . . . . . . . . . . . . . . . . . . . *b*
  b. Native Ozark species; new leafy shoots of spring with more than 1 whorl (circle) of branchlets; young branches glaucous (with a 'bloom' that can be rubbed off) . . . . . . . . 2. P. ECHINATA
  b. Introduced tree, escaped from cultivation; new leafy shoots of spring with only 1 whorl of branchlets; young branches not glaucous . . . . . . . . . . . . . 1. P. NIGRA var. NIGRA

1. **Pinus nigra** Arnold var. **nigra**
Austrian Pine Map 63

Occasionally persisting from an original planting, as in St. Louis County, where a few trees have persisted in a wooded valley north of highway 66 west of Peerless Park.

The tree, a native of Europe, is common in cultivation. It will doubtless be found in other localities where it has adapted to a wild situation.

2. **Pinus echinata** Mill. Short-leaf Pine Map 64

On acid, non-calcareous soils, associated with sandstone, chert, or granitic rocks, in rocky wooded ravines, bluffs, steep slopes, tops of narrow

Plate no. 8. **1.** Pinus echinata, × 2/5; *a.* Male flowering branch; *b.* Branch with fruit; *c.* Female flowering branch. **2.** Pinus virginiana; *a.* Male flowering branch, × 2/5; *b.* Fruit and leafy branch, × 2/5; *c.* Female flowering branch, × 2/5. **3.** Pinus nigra; *a.* Fruit, × 2/5; *b.* Leafy branch, × 2/5; *c.* Cluster of leaves, × 2/5. **4.** Taxodium distichum, × 2/5; *a.* Male flowering branch; *b.* Leafy branch with fruit; Details from Small, The New York Botanical Garden.

PLATE NO. 8

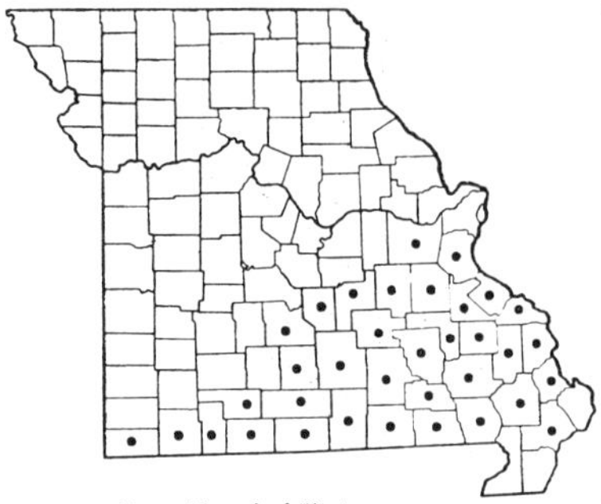

64 Pinus echinata (Short-leaf Pine)

65 Pinus virginiana (Scrub Pine)

66 • Taxodium distichum f. distichum (Bald Cypress)
66 □ Taxodium distichum f. confusum

ridges, dry uplands, or in valleys along streams where the soil has sandy, chert, or granitic components.

Confined to the Ozark region, extending west to McDonald, Christian, Wright, Laclede, and Pulaski counties, and north to Phelps, Franklin, and Jefferson counties.

Ranges from Florida and Texas, north to New York, West Virginia, Ohio, Southern Illinois, Missouri, and Oklahoma.

An important lumber tree in Missouri. Its wood pulp is employed for papermaking.

3. **Pinus virginiana** Mill. Scrub Pine Map 65

Like the Austrian Pine, this tree is planted and occasionally is found away from cultivation, having become established in a natural wild environment in Callaway County (*Steyermark 18515*) at the head of a ravine between Calwood and Fulton. Records from St. Louis County are from trees that are growing in cultivated sections of estates.

Native of the United States, ranging from Georgia to Arkansas, north to New York, New Jersey, West Virginia, Ohio, and Indiana; introduced in Missouri through plantings.

## TAXODIACEAE (Bald Cypress Family)

### 2. Taxodium Rich. Bald Cypress

**Taxodium distichum** (L.) Rich.
Bald Cypress Map 66

Two variations may be noted in Missouri:

Leaves spreading or diverging at a noticeable angle from the branches; tips of branches spreading or outwardly curving; common type . 1a. T. DISTICHUM var. DISTICHUM f. DISTICHUM

Leaves lying close to (appressed) the branches; shorter and more crowded; tips of branches upwardly curving; rare type . . 1b. T. DISTICHUM var. DISTICHUM f. CONFUSUM

1a. **Taxodium distichum** var. **distichum** f. **distichum** Map 66

Swamps, bayous, and alluvial soils along streams and lowlands in southeastern Missouri north to Cape Girardeau, Bollinger, Wayne, and Ripley counties. Reported by Beilmann as established, away from cultivation, in St. Louis County.

Ranges from Florida and Texas north to New Jersey, southern Indiana, southern Illinois, southeastern Missouri, Arkansas, and Oklahoma.

In the fall the leaves turn from a rich green to russet or cinnamon-brown.

1b. **Taxodium distichum** var. **distichum** f. **confusum** Palmer & Steyermark Map 66

Occurring within the range of and with the common type, and known from Wayne County (*Steyermark 11261*, holotype in Mo. Bot. Gard. Herbarium) and Butler County.

Known also from stations in southern Illinois and Arkansas.

This is a form which is more frequently found in sunny or exposed situations, and is probably only an ecological variant.

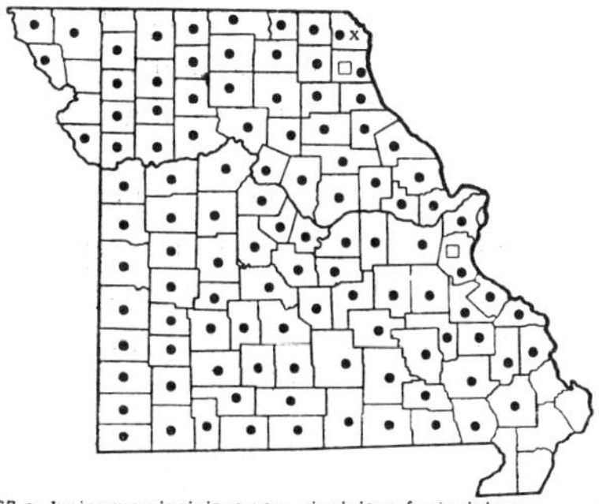

67 • Juniperus virginiana var. virginiana f. virginiana (Red Cedar)
67 x Juniperus virginiana var. virginiana f. pendula
67 □ Juniperus virginiana var. crebra

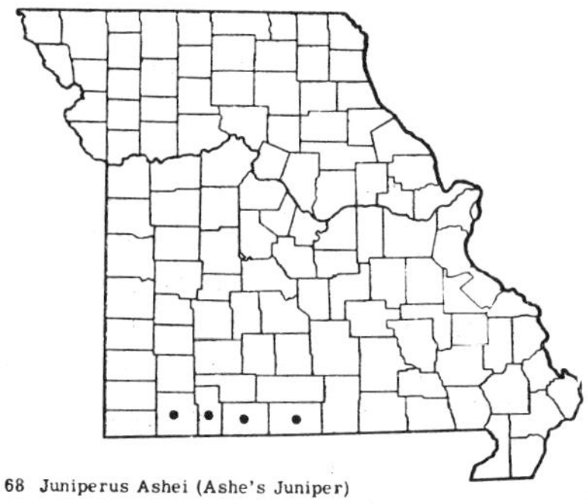

68 Juniperus Ashei (Ashe's Juniper)

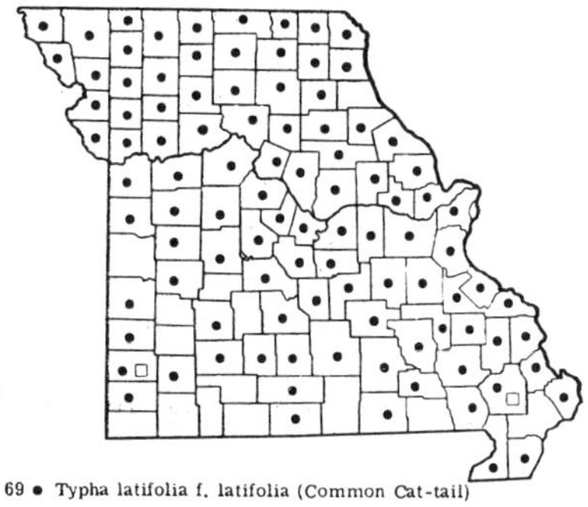

69 • Typha latifolia f. latifolia (Common Cat-tail)
69 □ Typha latifolia f. ambigua

## Fam. **CUPRESSACEAE** (Cypress Family)

### 3. **Juniperus** L. Cedar

Trees of this genus are usually either male or female (dioecious), the female bearing the fruit.

The roasted berries of some species of juniper are sometimes used as a coffee substitute. The berries of the common juniper (*Juniperus communis* var. *communis*) are used as an ingredient of gin. In parts of northern Europe an edible pulp is extracted from the berries of the juniper and often eaten with bread.

Along White River and tributaries of southwestern Missouri only; scale-like leaves minutely denticulate (toothed) when viewed with a lens; seeds 1, rarely 2 per berry, sharply pointed, 4–5 mm. in diam.; no pits on seed; conspicuous white growth (hilum) on seed; trunk of tree more or less branched near the base; fruit 6–8 mm. in diam.; glands on leaves round, raised well above the leaf . . . . . 2. J. ASHEI

Common throughout Missouri; scale-like leaves with entire (toothless) margins; seeds 1 or 2 per berry, or 3–6, blunt-tipped, 2–3 mm. in diam.; numerous pits on seed; small inconspicuous white growth (hilum) on seed; trunk of tree solitary, unbranched near base, erect; fruit 3.5–5 mm. in diam.; glands on leaves elongated or longer than broad, not raised or rarely raised above the leaf. . . . 1. J. VIRGINIANA

1. **Juniperus virginiana** L. Red Cedar Map 67

Two variations occur:

Crown of the tree ovoid (broadly egg-shaped), rather broad with widely spreading branches; seeds conspicuously pitted; common type . . 1a. J. VIRGINIANA var. VIRGINIANA

Crown of the tree narrowly pyramidal-shaped, with strongly ascending branches; seeds rather inconspicuously pitted; rarer type . 1b. J. VIRGINIANA var. CREBRA

1a. **Juniperus virginiana** var. **virginiana** Map 67

In Missouri this occurs most commonly on limestone bluffs and open glades, borders of woods, old fields, pastures, and fence rows, less commonly on sandstone, chert, or granitic rocks. Occurs in all the counties except those of the extreme southeastern lowlands.

Ranges from Alabama and Texas, north to southern New England, Kentucky, and Missouri.

The red cedar wood is employed in the manufacture of lead pencils, chests, cigar boxes, and fence posts. The wood left over from the products manufactured is one of the principal sources of cedarwood oil, which is employed in the manufacture of soaps, furniture polish, sanitary supplies, and perfumes. The refined oil, because of its high index of refraction, is used in oil immersion microscopes. An oil which comes from the leaves is used for ointments, liniments, bath salts, soap, and shoe polishes.

A variation of Red Cedar, with spreading branches and slender drooping branchlets is known as *J. virginiana* var. *pendula* Carr., and is recorded from Clark Co.

1b. **Juniperus virginiana** var. **crebra** Fern. & Grisc. Map 67

This variety with a narrowly pyramidal form is infrequently found in Missouri, and is known from limestone bluffs in Knox County (*Steyermark 10875*) along the South Fork of Fabius River, 3 miles northwest of Newark, and from Jefferson County.

Ranges farther north than typical var. *virginiana,*

extending north to Maine, Quebec, Ontario, Minnesota, and North Dakota.

Dr. Norman Fassett (Bull. Torr. Bot. Club 72: 42–46, 480–82. 1945; and 71: 410–18, 475–83. 1944) has discussed the variation of *J. virginiana* var. *virginiana* and var. *crebra*.

2. **Juniperus Ashei** Buchholz — Ashe's Juniper — Map 68

*Juniperus mexicana* Spreng. [G, BB, P & S, Steyerm.]

Limestone glades and bald knobs along bluffs of the White River and tributaries in southwestern Missouri from Barry County east to McVey Knob, Ozark County.

Also occurs in northern Arkansas, Arbuckle Mountains and bluffs of Pryor Creek in Oklahoma, and in parts of Texas, especially the Edwards Plateau section.

Dr. M. T. Hall (Ann. Mo. Bot. Gard. 39: 1–64. 1952) has discussed the variation of this species and *J. virginiana*.

## Class 2. ANGIOSPERMAE

Ovules enclosed within a closed ovary. Flowers with a perianth present or absent.

### Subclass 1. MONOCOTYLEDONEAE, Monocotyledons

Embryo with one cotyledon. Woody (vascular) bundles scattered and separate in the stem, appearing as dots scattered through the tissue. Parts of flowers often in 3's or 6's, sometimes in 2's or 4's, never in 5's. Leaves usually parallel-veined (but there are many exceptions).

### Order PANDANALES

#### Fam. TYPHACEAE (Cat-tail Family)

##### Typha L. (Cat-tail)

a. Flowering spikes continuous, the male (staminate) part not separated from the female (pistillate) part . . . . . 1a. T. LATIFOLIA f. LATIFOLIA
a. Flowering spikes with the male (staminate) part separated from the female (pistillate) part by a naked axis . . . . . b
  b. Pollen grains in 4's; surface of female (pistillate) part with minute dark or blackish markings; flowers from female (pistillate) part with flattened lance-ovate stigmas . . 1b. T. LATIFOLIA f. AMBIGUA
  b. Pollen grains single; surface of female (pistillate) part without dark or blackish markings; flowers from female (pistillate) part with thread-like stigmas . . . . . c
    c. Female (pistillate) spike 6–15 mm. in diameter; leaves 3–8 mm. wide; stigmas (from female spike) with many dark brown bracts occurring down among the hairs . . . . 2. T. ANGUSTIFOLIA
    c. Female (pistillate) spike 16–22 mm. in diameter; leaves 7–15 mm. wide; stigmas (from female spike) without any bracts or merely a few bracts occurring down among the hairs . 3. T. GLAUCA

1. **Typha latifolia** L. Common Cat-tail — Map 69

1a. **Typha latifolia** f. **latifolia** — Map 69

Flowers late May–July. This is the common type, found throughout Missouri in swamps, muddy margins of ponds and streams, and other wet places.

Ranges throughout North America, and in Eurasia and North Africa.

Although much pollen is shed from the male (staminate) spike, it does not cause hay fever.

This and other species of cat-tail are used in a number of ways. The leaves supply the rush seats for chairs which several companies today manufacture. The American Indian made floor mats from the leaves, which, together with the leaf sheaths, supplied him caulking material.

---

Plate no. 9. **1.** Juniperus virginiana; *a.* Habit greatly reduced; *b.* Normal mature leafy branch with fruit, × $^2/_7$; *c.* Young or juvenile branch, × $^2/_7$; *d.* Single leaf, × $8^4/_7$; Details from Small, The New York Botanical Garden. **2.** Juniperus Ashei; *a.* Habit, greatly reduced; *b.* Normal mature leafy branch with fruit, × $^2/_7$; *c.* Single leaf, × $8^4/_7$; Details from Small, The New York Botanical Garden. **3.** Typha latifolia, × $^2/_7$; *a.* Female flower, × $2^6/_7$; *b.* Male flower, × $2^6/_7$; Details from Small, The New York Botanical Garden. **4.** Typha angustifolia × $^2/_7$; *a.* Female flower, × $2^6/_7$; *b.* Male flower, × $2^6/_7$; Details from Small, The New York Botanical Garden. **5.** Sparganium eurycarpum; *a.* Leafy branch, × $^2/_7$; *b.* Fruiting head, × $^2/_7$; *c.* Female flower/head, × $^2/_7$; *d.* Single female flower, × $^6/_7$.

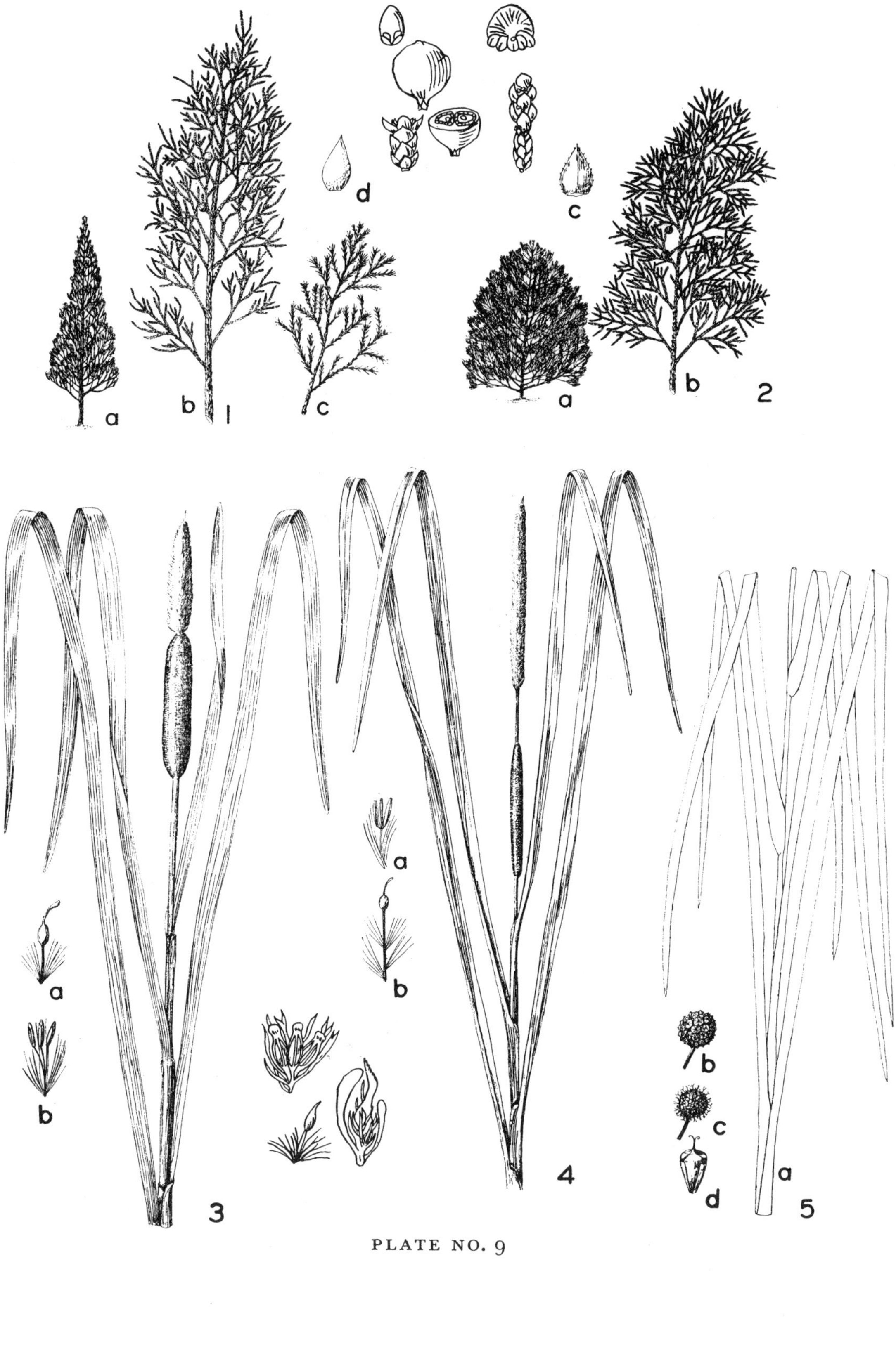


PLATE NO. 9

70 Typha angustifolia (Narrow-leaved Cat-tail)

71 Typha glauca

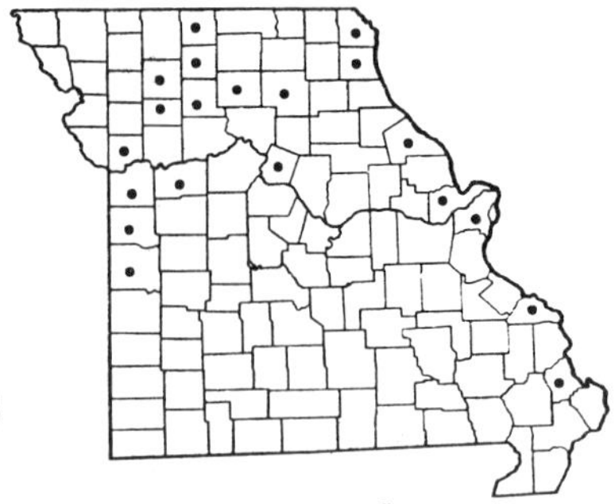

72 Sparganium eurycarpum (Bur-reed)

An attempt has been made to use the plants as a substitute for cotton, wool, and jute, for the manufacture of paper and ethyl alcohol, and the soft featherlike down from the mature fruiting spikes is used for its buoyancy in life jackets, and as a soundproofing, and heat-insulating material. As a waste material it is being considered seriously in the papermaking industry. The seeds, which contain about 18% oil, of which some 70% is linolenic acid, are being tested as a drying oil as well as an edible oil.

Various portions of the plant may be used as food. The starchy rootstocks can be eaten raw in salad or cooked as a vegetable. A highly nutritious flour, containing 57% carbohydrate, is obtained from the central part of the rootstock. A jelly is also extracted from the root. The young stems are used in early spring as a vegetable. Perhaps the most interesting part for food is prepared from the young flowering spikes. Before the pollen has developed, the young staminate spikes are steamed or boiled in water to which salt has been added, and may be pickled, or after the flowers have been separated from the hard central axis, they may be cooked with eggs, milk, buttered crumbs, or oil. The pollen from the staminate flower spikes is also used to prepare a bread or cake-stuff, about twenty-five flower spikes being needed for preparing a small cake.

1b. **Typha latifolia** f. **ambigua** (Sonder) Kronf. Map 69

Known only from Jasper (*Palmer 54921*) and Stoddard (*Steyermark 78660*) counties. In the latter locality it was found along a swampy spring branch at the base of Crowley Ridge near the junction with the flood plain, 5 miles northeast of Dexter, and occurred with the typical form of the species.

This form is sometimes confused with *T. angustifolia*, because of the separated spikes.

2. **Typha angustifolia** L.

Narrow-leaved Cat-tail Map 70

Flowers late May–July.

Borders of muddy and calcareous ponds and in shallow water of natural brine lakes. Scattered throughout the western half of the state, where very rare and local.

Ranges from Nova Scotia and Maine to Quebec and Ontario, south to South Carolina, Kentucky, Missouri, and Nebraska; also in California and Eurasia; especially common along the brackish water estuaries of the Atlantic Coast.

Narrow-leaved specimens of *T. latifolia* are often confused with this species.

Dr. Norman Fassett (Evolution 6: 367–79. 1952) has presented evidence pointing to the occurrence of hybridization between *T. latifolia* and *T. angustifolia*.

3. **Typha glauca** Godr. Map 71

*Typha angustifolia* var. *elongata* (Dudley) Wieg. [BB]

Flowers late May–July.

Thus far, known only from Platte County (marshy area near Basswood Lakes, 4 miles east of Platte City, July 12, 1956, *Radloff*), and from Boone County (border of Moore's lake of Power and Light Co., just north of highway 40, Columbia, July 31, 1958, *Steyermark 86146*, and *Dunn 12945*).

Ranges from Maine, Ontario, Iowa, and South Dakota, south to North Carolina, Alabama, and Missouri; also in California and reported from Guatemala.

Although Dr. Fassett (Fassett, N. C. and Barbara Calhoun, Introgression between *Typha latifolia* and *T. angustifolia*. Evolution 7, no. 4. December. 367–79. 1952) considers this taxon a series of individuals in hybrid swarms, other botanists, such as Drs. Hotchkiss and Dozier, who have carried on special investigations, consider *T. glauca* a distinct species (Am. Midl. Nat. 41: 237–54. 1949).

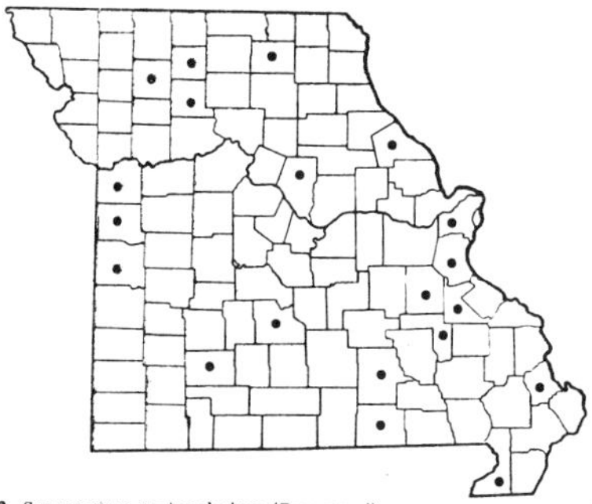

73 Sparganium androcladum (Bur-reed)

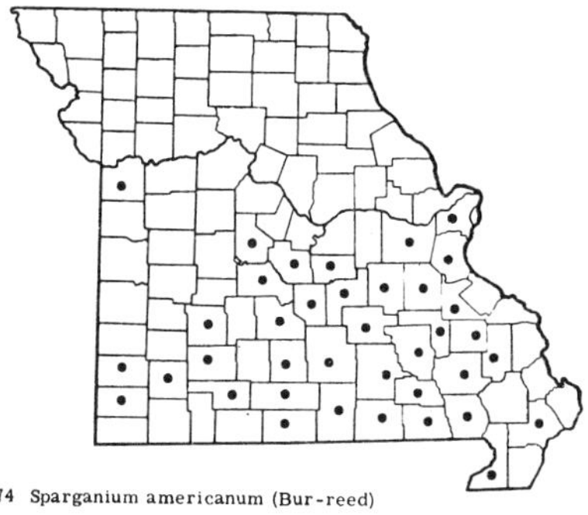

74 Sparganium americanum (Bur-reed)

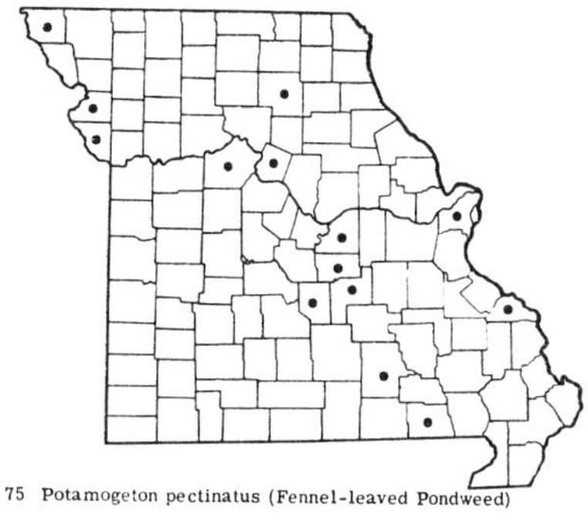

75 Potamogeton pectinatus (Fennel-leaved Pondweed)

## SPARGANIACEAE (Bur-reed Family)

### Sparganium L. Bur-reed

It is reported that the Klamath Indians dig the tubers produced in late autumn from the creeping rootstocks of some of the species of this genus, and use them as food.

The individual flowers are small and occur in separate male (staminate) or female (pistillate) globular clusters on the same plant.

a. Stigmas 2 to each female (pistillate) flower; fruits sessile, wedge-shaped and widest at summit or in the upper half. . . . . . . . . . . . . . . . . . . . . . . . 1. S. EURYCARPUM
a. Stigmas 1 to each female (pistillate) flower; fruits narrowed to a stalk-like base, and tapering above, widest around the middle, or at least not conspicuously broadened above. . . . . . . . . . . *b*
 b. Stigma 2–4 mm. long; leaves firm and stiff, at least the middle leaves keeled; fruit shining, its body 5.5–7 mm. long, its beak 4–6 mm. long . . . . . . . . . . . . 2. S. ANDROCLADUM
 b. Stigma 1–2 mm. long; leaves rather soft, flat or only inconspicuously keeled; fruit dull, its body 3–5 mm. long, its beak 3–4.5 mm. long . . . . . . . . . . . . . . . 3. S. AMERICANUM

1. **Sparganium eurycarpum** Engelm. Map 72
Flowers May–August.

Common on the margins of oxbow lakes and sloughs along the larger rivers, and in swampy ground in the valleys of the Missouri, Mississipi, South Grand and tributaries, and Grand River and tributaries in eastern, northern, and central Missouri.

Ranges from Quebec and Nova Scotia to British Columbia, south to New Jersey, Ohio, Indiana, Illinois, Missouri, Kansas, Colorado, and California.

This species has a persistent creeping rhizome and stiff erect leaves. It grows taller than the other species in Missouri, reaching a height of 1.5 meters.

2. **Sparganium androcladum** (Engelm.) Morong Map 73
Flowers May–July.

Around ponds, sloughs, and margins of oxbow lakes in river valleys, usually growing in more quiet waters than the next species. Scattered over the state, but absent from most of the Ozark section.

Ranges from Quebec to Minnesota, south to Virginia, Kentucky, Illinois, Missouri, and Oklahoma.

In this species the rhizome is lacking or inconspicuous.

3. **Sparganium americanum** Nutt. Map 74
Flowers May–early September, mostly June–July.

Common in cold-water springs and spring branches throughout the Ozark region north to St. Louis, Franklin, Maries, Miller, Morgan, Polk, and Newton counties, and locally in Jackson County.

Ranges from Newfoundland and Ontario to Minnesota and North Dakota, south to Florida, Alabama, and Missouri.

This is usually the largest aquatic plant found in and along the waters of the Ozark springs. The plants often remain submerged in a vegetative state for long periods of time in the constantly flowing spring water.

## Order **NAJADALES**

### Fam. **NAJADACEAE** (Pondweed Family)

(*Zosteraceae* in large part [G]; *Potamogetonaceae*)

a. Some or all of the leaves over 5 mm. broad, broadly oblong or ovate . . . . . . . . 1. POTAMOGETON
a. All the leaves thread-like or very narrow, less than 3 mm. broad . . . . . . . . . . *b*
b. Leaves alternate . . . . . . . . . . . . . . . . . . . . . . . *c*
c. Plants growing in salty-sulphur water; each flower and fruit on an obvious stalk; leaf widening at base into an enlarged sheath . . . . . . . . . . . . . . . 2. RUPPIA
c. Plants growing in fresh water; flowers and fruits closely clustered or arranged in a head or spike, the individual flowers without stalks (sessile); submerged leaves not noticeably widened at base into a sheath, or, if furnished with a stipule, the base not noticeably widened 1. POTAMOGETON
b. Leaves opposite . . . . . . . . . . . . . . . . . . . . . . . *d*
d. Leaves slightly or conspicuously widened at the base with a basal sheath, the leaf-margins or broadened base with tiny teeth (use lens); flowers or fruits solitary in the leaf-axils . . . 4. NAJAS
d. Leaves with about the same width at the base as at other parts of the leaf, the leaf-margins without teeth (use lens); flowers or fruits 2–4 . . . . . . . . . . . 3. ZANNICHELLIA

### 1. **Potamogeton** L. Pondweed

A few of the species in this genus possess starchy and edible fleshy rootstocks.

a. Margins of all the leaves finely and definitely toothed . . . . . . . . . . . . 2. P. CRISPUS
a. Margins of leaves either entire (smooth-edged) or teeth very minute and inconspicuous. . . . . . *b*
b. Leaves all submerged . . . . . . . . . . . . . . . . . . . . . *c*
c. Submerged leaves with sides curved, not straight and parallel . . . . . . . . . . *d*
d. Submerged leaves 27–37-nerved, arched, curved, or folded with a strongly wavy surface; margins of submerged leaves without any minute teeth (use lens); floating leaves with usually 30–51 nerves . . . . . . . . . . . . . . . . 7. P. AMPLIFOLIUS
d. Submerged leaves 7–19-nerved, flat, not arched, nor curved, nor folded, nor with a strongly wavy surface; margins of submerged leaves with tiny 1-celled teeth (use lens); floating leaves with usually 7–29 nerves . . . . . . . . . . . . . . . . 9. P. NODOSUS
c. Submerged leaves thread-like or ribbon-like with straight or parallel sides, 0.5–8 mm. wide, usually 0.5–3 mm. wide . . . . . . . . . . . . . . . . . . . . *e*
e. Submerged leaves 2–8 mm. wide, with many rows of long cells joining in a network 1–2 mm. wide or wider on each side of the midrib (hold leaves to light); stems flat . . . . . . . 6. P. EPIHYDRUS var. NUTTALLII
e. Submerged leaves 0.5–3 mm. wide, without or with 1–3 inconspicuous rows of enlarged cells forming a network less than 1 mm. wide on each side of the midrib; stems cylindrical (round when cut cross-wise) . . . . . . . . . . . . . . . . . . *f*
f. Stipules united with the base of the leaf, forming a sheath . . . . . . . . . *g*
g. Stem dichotomously forking; tree-like in aspect; flower spikes elongated, several times longer than broad, 1.5–5 cm. long, with interrupted whorls of flowers; main stalk (peduncle) supporting flower-spike elongated, 3–20 cm. long; fruits 3.5–4 mm. long. 1. P. PECTINATUS
g. Stem freely branching, but not dichotomously forking or tree-like; flower spikes oblong to almost globular, barely or somewhat longer than broad, 0.5–2 cm. long, the flowers clustered, not in interrupted whorls; main stalk (peduncle) supporting flower-spike very short or at most 1.5 cm. long; fruits 1–1.8 mm. long . . . . 5. P. DIVERSIFOLIUS
f. Stipules not united with, and free from the base of the leaf . . . . . . . . . *h*
h. Leaves with 2 minute glands at their base; main stalk (peduncle) supporting flower-spike thread-like, 1.5–9 cm. long; plants usually in ponds and quiet water of slow streams; main stems long and much-branched . . . . . . . . . . 4. P. PUSILLUS
h. Leaves without any basal glands; main stalk (peduncle) supporting flower-spike

---

Plate no. 10. **1.** Sparganium androcladum, × 2/7; *a.* Female flower-head, × 2/7. **2.** Sparganium americanum, × 2/7; *a.* Single female flower, × 1 3/7; Details from Small, The New York Botanical Garden. **3.** Potamogeton pulcher, × 2/7. **4.** Potamogeton amplifolius, × 2/7; *a.* Shallow water type; *b.* Deep water type. **5.** Potamogeton nodosus, × 2/7; *a.* Submerged type in swift-flowing water; *b.* Type with floating and submerged leaves.

PLATE NO. 10

76 Potamogeton crispus (Curly Muck-weed)

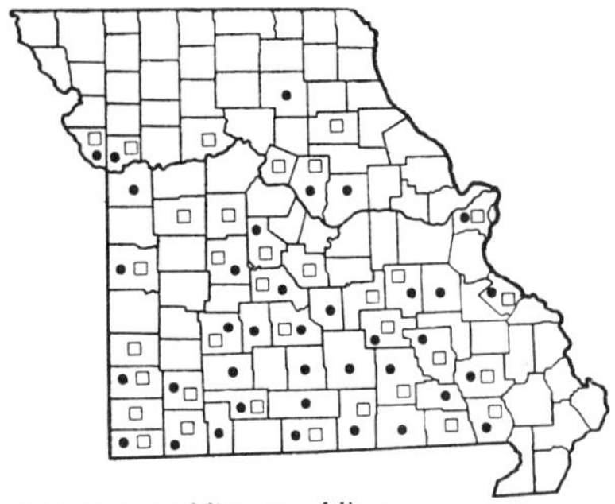

77 • Potamogeton foliosus var. foliosus
77 □ Potamogeton foliosus var. macellus

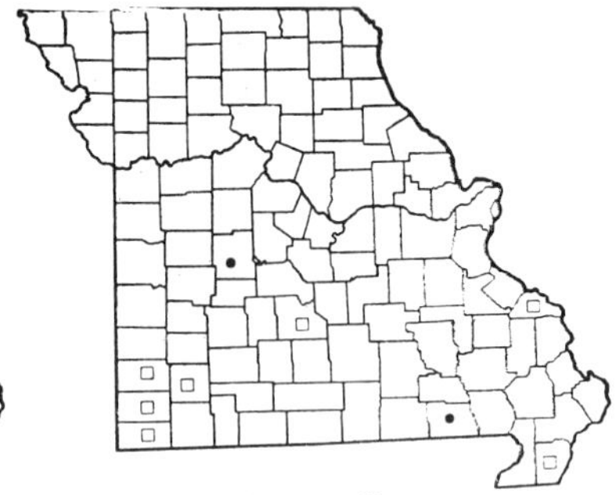

78 • Potamogeton pusillus var. pusillus
78 □ Potamogeton pusillus var. minor

slightly thickened upward, mostly 0.3–1 cm. (rarely more) long; plants usually in running water of streams and spring branches, occasionally in ponds; main stems nearly simple to loosely branched . . . . . . . . . . . . . . 3. P. FOLIOSUS

b. Some leaves floating with an oblong, elliptical, or ovate shape, much thicker than the submerged leaves, or not resembling them . . . . . . . . . . . . . . . . . . . . . *i*
  i. Submerged leaves 0.5–8 mm. wide, ribbon-like with straight or parallel sides . . . . . . . *j*
    j. Submerged leaves 0.5–1.5 mm. wide; no rows of enlarged cells forming loose network on either side of midrib; stems cylindrical . . . . . . . . . . . . 5. P. DIVERSIFOLIUS
    j. Submerged leaves 2–8 mm. wide; many rows of long cells joining in a network, 1–2 mm. wide or wider on each side of midrib (hold leaves to light); stem compressed, flat . . . 6. P. EPIHYDRUS var. NUTTALLII
  i. Submerged leaves 4–50 mm. wide, with curved, not parallel sides . . . . . . . . . . *k*
    k. Main stem and petiole (stalk of leaf) dark-spotted; at least some of the floating leaves rather cordate at base (heart-shaped); plants of sink-hole ponds of the Ozarks. . . . . 8. P. PULCHER
    k. No dark spots on stem or leaf-stalk; floating leaves narrowed or rounded, but not heart-shaped at base; plants of slow or rapid streams, spring-fed branches, ponds, and sloughs . . . . *l*
      l. Submerged leaves (25) 27–37 (–39)-nerved, arched, curved, or folded with a *strongly wavy surface*; margins of submerged leaves without any minute teeth (use lens); floating leaves with usually 30–51 (–55) nerves . . . . . . . . . . . . 7. P. AMPLIFOLIUS
      l. Submerged leaves 7–19-nerved, *flat*, not arched, curved or folded, nor with a strongly wavy surface; margins of submerged leaves with tiny 1-celled teeth (use lens); floating leaves with usually 7–29 nerves . . . . . . . . . . . . . . . . . . . . . *m*
        m. Submerged leaves 4–20 mm. wide; floating leaves usually lanceolate-oblong or narrowly elliptical; plants of larger rivers and streams . . . . . . . . 9. P. NODOSUS
        m. Submerged leaves 20–50 mm. wide; floating leaves usually ovate, broadly oblong, or oval; plants of cool spring-fed sloughs and branches. . . . . . . 10. P. ILLINOENSIS

1. **Potamogeton pectinatus** L.

Fennel-leaved Pondweed — Map 75

Flowers May–September.

In fresh- or salt-water lakes fed by springs, and in streams, scattered along the Mississippi, Missouri, Current, and Gasconade rivers and tributaries, and in artificial and natural ponds and sloughs.

Ranges from Newfoundland to Alaska, south to Florida, West Virginia, Great Lake States, Missouri, Texas, and California; also in Mexico, South America, Eurasia, and Africa.

2. **Potamogeton crispus** L.

Curly Muck-weed — Map 76

Flowers May–September.

Lakes, ponds, and sloughs fed by spring water. Known only from Crawford and Newton counties.

An introduced European species which has spread from Quebec and Ontario to Minnesota, south to Virginia and Missouri; also in California.

3. **Potamogeton foliosus** Raf. — Map 77

Flowers from June 1 to October, but mainly in July and August.

Plate no. 11. **1.** Potamogeton pectinatus, × 2/5. **2.** Potamogeton crispus, × 2/5. 3. Potamogeton illinoensis, × 2/5. **4.** Potamogeton foliosus var. foliosus, × 2/5. **5.** Potamogeton pusillus var. minor, × 2/5; Details from Small, The New York Botanical Garden.

PLATE NO. 11

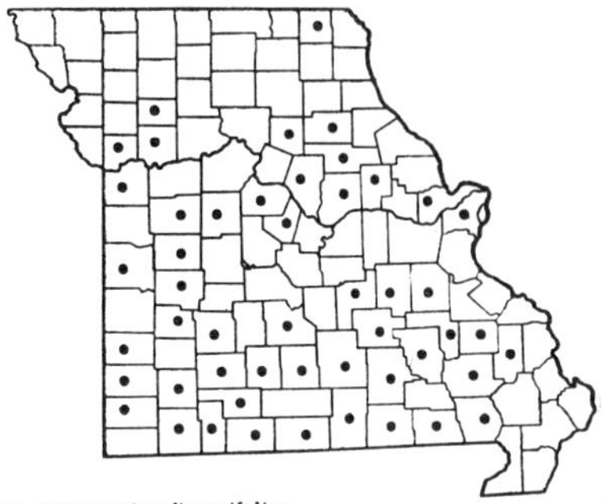

79 Potamogeton diversifolius

80 Potamogeton epihydrus var. Nuttallii

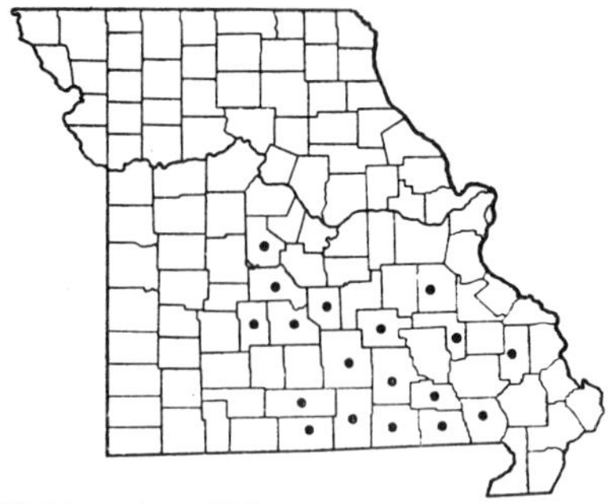

81 Potamogeton amplifolius

Two scarcely separable varieties have been recognized, but completely intergrade:

Largest leaves 3–5-nerved, 1.4–2.7 mm. wide; 1–3 rows of enlarged cells on each side of and at the base of the midnerve; stipules 7–18 mm. long; beak of fruit broad at base, 0.2–0.4 mm. long; leaves deep green to bronze . . . P. FOLIOSUS var. FOLIOSUS

Largest leaves 1–3-nerved, 0.3–1.5 mm. wide; no enlarged cells or 1 row of them on each side of midnerve below the middle; stipules 3–11 mm. long; beak of fruit slender, 0.3–0.8 mm. long; leaves bright green . . P. FOLIOSUS var. MACELLUS

Both varieties inhabit springs, or mostly spring-fed streams, although occasionally they occur in ponds and lakes. They are found usually throughout the Ozark region, extending north to Monroe, Macon, Chariton, and Platte counties.

*Potamogeton foliosus* var. *foliosus* occurs from New York to Washington south to the West Indies and Mexico; var. *macellus* Fern. extends from Nova Scotia and Quebec to British Columbia, south to Florida, Kentucky, Missouri, Kansas, New Mexico, and California.

4. **Potamogeton pusillus** L. Map 78

*Potamogeton panormitanus* Bivona-Bernardi var. *major* G. Fischer [P & S]

*Potamogeton panormitanus* var. *minor* Biv. [P & S]

Flowers about June 1 to October.

Two forms, sometimes recognized, occur in Missouri:

Main leaves 1–3 mm. wide P. PUSILLUS var. PUSILLUS

Main leaves 0.3–1 mm. wide P. PUSILLUS var. MINOR

Both varieties are quite rare and local in Missouri, where they occur usually in slow-water of streams, sloughs, and ponds, in contrast to the faster-flowing, cooler spring-fed waters associated usually with *P. foliosus*.

*Potamogeton pusillus* var. *pusillus* (*P. panormitanus* var. *major*) ranges from Quebec to Alberta and British Columbia, south to Virginia, Alabama, Louisiana, Texas, and Mexico; also in the West Indies and Eurasia; var. *minor* (Biv.) Fern. & Schub. (*P. panormitanus* var. *minor*) has nearly the same range.

5. **Potamogeton diversifolius** Raf. Map 79

Flowers late May–October.

Common in warm waters of both natural and artificial ponds, and sloughs or quiet waters of slow streams in most of central and southern Missouri, and locally north in Scotland County.

Ranges from Mexico, Texas, and Florida north to New Jersey, Pennsylvania, Wisconsin, Montana, and Oregon.

This species is very common in artificial lakes and ponds.

6. **Potamogeton epihydrus** Raf. var. **Nuttallii** (C. & S.) Fern. Map 80

Flowers usually July through September.

In Missouri known only from a sink-hole pond in the Ozarks in Reynolds County (Lily Pond, 7 miles southeast of Centerville, September 5, 1949, *Steyermark 69232*).

This variety ranges from Newfoundland to Manitoba, south to Virginia, Georgia, Tennessee, and Missouri; and from Alaska to Colorado and California.

This northern plant is one of the isolated relicts known from sink-hole ponds in Missouri, growing close to another sink-hole pond where *Decodon verticillatus*, also an isolated relict in Missouri, occurs.

7. **Potamogeton amplifolius** Tuckerm. Map 81

Flowers about June 1 to October.

Usually occurring in springs and spring-fed branches, sloughs, and lakes in the Ozark region west to

Ozark, Dallas, and Camden counties, north to Morgan, Pulaski, and Washington counties.

Ranges from Newfoundland to North Dakota, south to Virginia, Alabama, Arkansas, and Oklahoma; and from Montana to British Columbia south to Idaho and California.

8. **Potamogeton pulcher** Tuckerm.
Spotted Pondweed Map 82
Flowers about June 1 to October.

This species is characteristic of sink-hole ponds in the Ozark upland in Missouri, and is occasionally found in streams. In drouth years when the sink-ponds may become nearly dry, the stems of this species become shorter and only the upper broad leaves of the floating type are evident. The species has been collected as far north in Missouri as St. Louis and Franklin counties, but the present known stations occur farther south in the Ozarks.

Ranges from Florida to Texas north to Nova Scotia, New York, Pennsylvania, Ohio, Indiana, Illinois, Minnesota, and Oklahoma.

9. **Potamogeton nodosus** Poir. Map 83
*Potamogeton americanus* Cham. & Schl. [P & S]
*Potamogeton rotundatus* Hagstr. [P & S, Steyerm.]
Flowers late May–October.

Common in the major Ozark streams, in sloughs, and ponds north to Pike, Macon, and Platte counties.

Ranges from Alabama, Texas, and Mexico, north to New Brunswick, Ontario, Michigan, Minnesota, South Dakota, Montana, and British Columbia.

This species has much more elongated and narrower floating leaves than any of the other large-leaved species of pondweed in Missouri. It is the most common large-leaved species encountered in the Ozark streams, but avoids the colder water of the springs and spring branches.

10. **Potamogeton illinoensis** Morong.
Shining Pondweed Map 84
*Potamogeton lucens* of Am. auth., not L. [P & S, Steyerm.]
Flowers about June 1 to October.

This species often occurs with submerged leaves

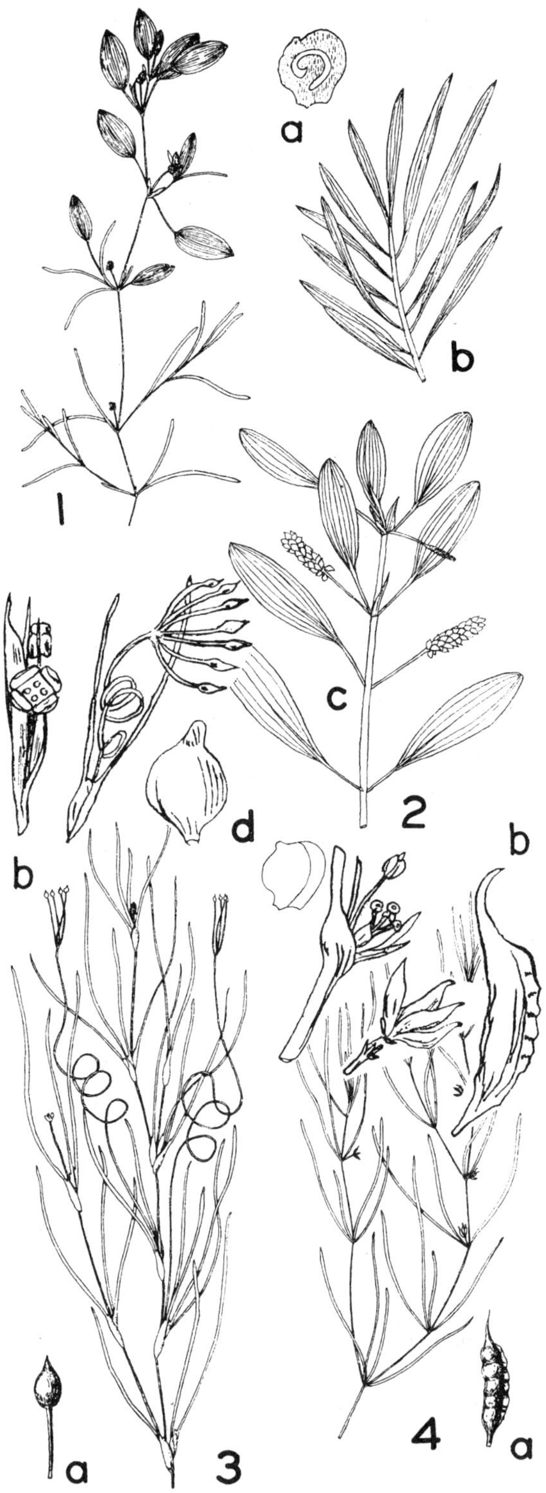


Plate no. 12. **1.** Potamogeton diversifolius, $\times {}^2/_5$. **2.** Potamogeton epihydrus var. Nuttallii, $\times {}^2/_5$; *a.* Fruit, vertical section, showing embryo, $\times 2{}^2/_5$; *b.* Submerged leaves, $\times {}^2/_5$; *c.* Floating leaves and inflorescences, $\times {}^2/_5$; *d.* Fruit, $\times 2{}^2/_5$; After Britton and Brown, The New York Botanical Garden. **3.** Ruppia maritima var. rostrata, $\times 1$; *a.* Single fruit, $\times 2{}^2/_5$; Details from Small, The New York Botanical Garden. **4.** Zannichellia palustris var. major, $\times {}^2/_5$; *a.* Single fruit, $\times 4$; Details from Small, The New York Botanical Garden.

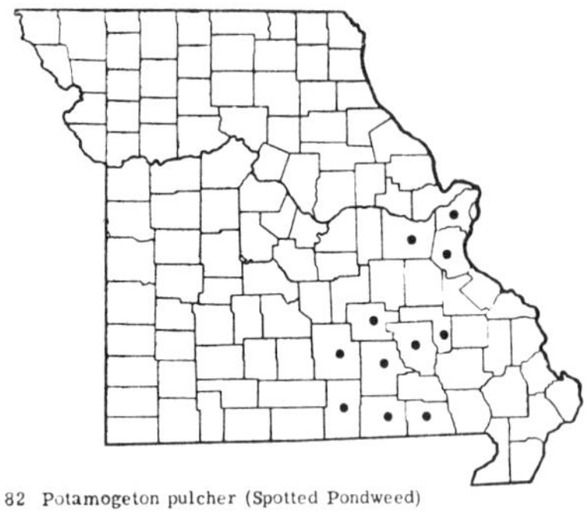

82 Potamogeton pulcher (Spotted Pondweed)

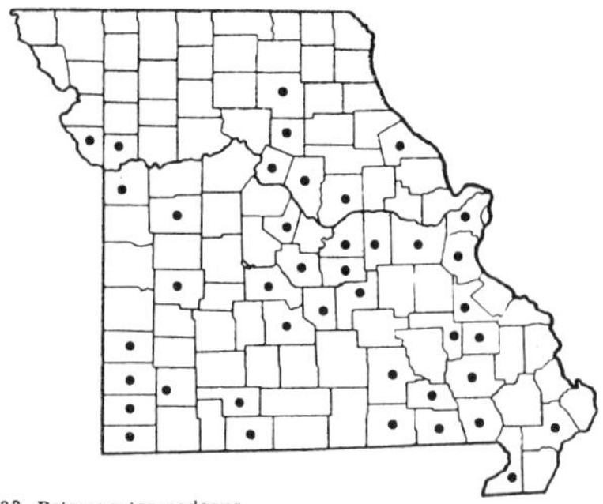

83 Potamogeton nodosus

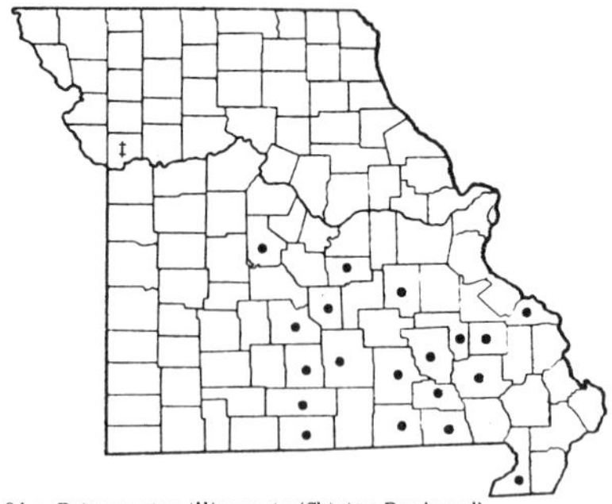

84 • Potamogeton illinoensis (Shining Pondweed)
84 ‡ Potamogeton illinoensis X Potamogeton nodosus

only. It is a characteristic large-leaved pondweed of the cool waters of springs, and spring-fed ponds and streams of the Ozark region, occurring west to Ozark, Wright, Laclede, and Morgan counties, and north to Maries, Crawford, and Perry counties.

Ranges from Quebec to British Columbia south to Florida, Ohio, Indiana, Illinois, Arkansas, Texas, Nebraska, Colorado, Utah, and California.

10a. **Potamogeton illinoensis** × **P. nodosus**
Map 84

This hybrid, cited by Ogden (Rh. 45: 185–86. 1943) has been collected from Cooley Lake, Clay County, by *Metcalf 1055* (U.S. Nat. Herb.).

## 2. **Ruppia** L. Ditch Grass

**Ruppia maritima** L. var. **rostrata** Agardh
Map 85

Flowers mostly July–October.

Known in Missouri from two stations: (1) Henry County and (2) Ralls County. The Ralls County record is from a lake formed by a salt-sulphur spring at Spalding, September 3, 1937, *Steyermark 25691*.

The var. *rostrata* ranges from Newfoundland to Quebec south to Florida, and from scattered stations inland from New York to the Pacific, south to Missouri, Oklahoma, and California; also in Mexico, the West Indies, South America, Eurasia, and Africa.

## 3. **Zannichellia** L. Horned Pondweed

**Zannichellia palustris** var. **major** (Boenn.) W. D. J. Koch
Map 86

*Zannichellia palustris* L. [BB, P & S]

Flowers late May–October.

In fresh water springs, salt and sulphur springs, marshes, ponds, and along shallow and back-water sloughs of streams. It is scattered throughout Missouri, but is commonest in the Ozark streams and spring water, and in salt-sulphur spring waters of central Missouri.

*Zannichellia palustris* var. *palustris* is widespread from Quebec to Alaska south to West Virginia, Tennessee, Texas, and Mexico, also in South America, Eurasia, and Africa; the var. *major* extends south to Florida. Since Missouri specimens have the fruits commonly dentate and pedicelled with a body 3–3.6 mm. long and a beak 1–2 mm. long they are assigned to var. *major*, as interpreted by Fernald in the eighth edition of *Gray's Manual*.

## 4. **Najas** L. Naiad

a. Base of leaf abruptly widened into coarsely toothed or jagged lobes; leaves mostly 0.1–0.3 mm. wide . . . 1. N. GRACILLIMA

a. Base of leaf gradually widened, not lobed above base and only finely toothed along margin; leaves mostly (0.4) 0.5–1 mm. wide . . . *b*

b. Plants of artificial ponds, very rare; leaves tapering to a long fine point; style 0.9 mm. or more long; seed finely and inconspicuously marked with 30–40 rows of hexagonal pits across the middle, usually shining . . . 2. N. FLEXILIS

b. Plants of streams, sloughs, and marshes, common; style 0.6 mm. or less long; tip of leaves rounded to short-pointed; seed dull, coarsely and deeply pitted, with 10–20 rows of rectangular pits across the middle . . . 3. N. GUADALUPENSIS

85 Ruppia maritima var. rostrata (Ditch Grass)

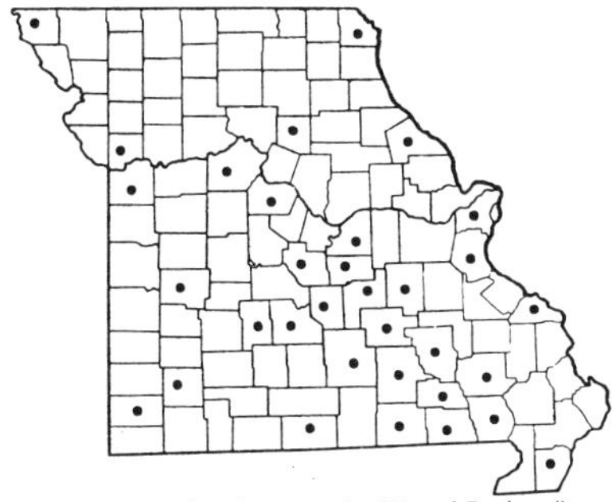
86 Zannichellia palustris var. major (Horned Pondweed)

87 Najas gracillima (Naiad)

1. **Najas gracillima** (A. Br.) Magnus Map 87
Flowers mostly July–September.

Known from only two upland sink-hole ponds in the Ozarks in Shannon and Texas counties (3½ mi. south of Licking, July 15, 1937, *Steyermark 23277*), and introduced in an artificial pond at Moberly, Audrain County (Crystal Lake, Aug. 21, 1939, *J. R. Hurt 50*).

Ranges from Maine to Minnesota, south to Virginia, Kentucky, and Missouri.

This species has more filiform leaves (0.1–0.3 mm. wide), and the areolae (honeycombed network of cells) on the surface of the seeds are slightly more elongate than in the other species of Missouri.

2. **Najas flexilis** (Willd.) Rostk. Map 88
Flowers mainly July–September.

This is included for Missouri on the basis of a fruiting collection from 'More's Lake, Columbia' in Boone County, collected by *Francis Daniels* on July 21, 1903, deposited in the herbarium of the University of Missouri. 'More's Lake' is actually the same as Moore's Lake, indicated on government topographical maps, and is the one by the power and light company plant just north of highway 40 in Columbia. The elongate style and stigma, together with the yellowish pericarp closely investing the obscurely reticulate fruit with 30–40 rows of minute subhexagonal areolae, bushy-branching habit, recurved, narrowly pointed leaves, are the chief characters relied upon in the present identification. Two other sheets in the University of Missouri Herbarium, labeled 'Najas flexilis,' were similarly collected by Daniels from 'More's Lake' on July 21, 1903, but lack mature fruit. The immature fruits of these two sheets are purplish with rather deeply marked areolae, more characteristic of *N. guadalupensis*.

Ranges from Newfoundland and Quebec to Minnesota, south to Virginia, Ohio, Indiana, Illinois, Missouri, and Iowa, and from Alberta to British Columbia south to Oregon; also in northern Europe.

3. **Najas guadalupensis** (Spreng.) Magnus Map 89
Flowers mostly July–October.

Common in shallow margins of streams, sloughs, and marshes in southern and central Missouri, commoner in the Ozarks, and north locally to Macon County.

Ranges from tropical America north to Massachusetts, New York, Pennsylvania, Ohio, Michigan, Minnesota, South Dakota, Idaho, and Oregon.

## Order **ALISMALES**

### Fam. **ALISMACEAE** (Water Plantain Family)

(*Alismataceae* [G])

a. Each flower with 6 stamens; pistils in one ring on a small flat receptacle . . . . . . . . 1. ALISMA
a. Some of the flowers with 6–20 or more stamens; pistils in several series in a head on enlarged greenish-white receptacle . . . . . . . . . . . . . . . . . . . . . . . . . *b*
  b. All flowers perfect (with both stamens and pistils in the same flower); individual mature seed-like fruits (achenes) plump with ribs or ridges, not flattened or winged; in addition to the 3 minute or leaf-like bracts at each joint where the flower-stalks originate, there are additional papery or minute outgrowths (bracteoles) . . . . . . . . . . . . . . . . . . 2. ECHINODORUS
  b. Some flowers with either stamens or pistils, but not both on the same flower; individual mature seed-like fruits (achenes) flattened or winged; only the 3 papery or leaf-like bracts present at each joint where the flower-stalks originate . . . . . . . . . . . . . . . . 3. SAGITTARIA

88 Najas flexilis (Naiad)

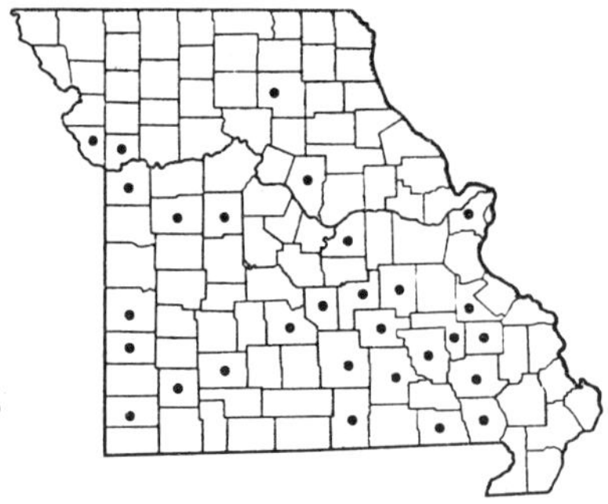

89 Najas guadalupensis (Naiad)

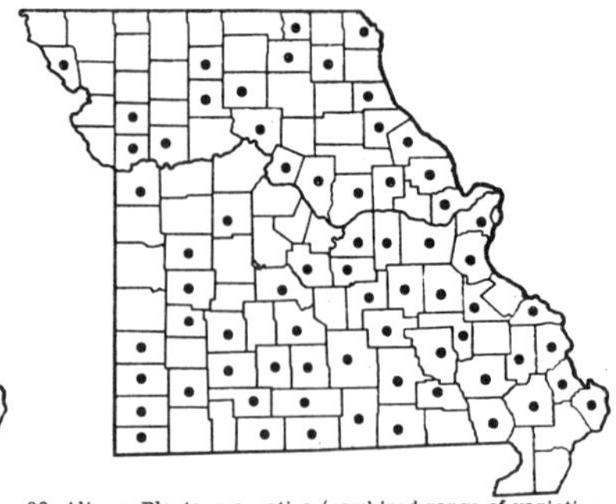

90 Alisma Plantago-aquatica (combined range of varieties shown) (Water Plantain)

## 1. **Alisma** L. Water Plantain

**Alisma Plantago-aquatica** L. Water Plantain
Map 90

Flowers late May–September.

Generally found along muddy borders of ponds, sloughs, slow streams, marshes, and temporary pools throughout Missouri.

The genus *Alisma* has been most recently studied by Dr. A. J. Hendricks, and according to him (Am. Midl. Nat. 58: 470–93. 1957), the Missouri material is divided as follows into three varieties of *A. Plantago-aquatica*:

a. Diameter of mature fruiting portion 3.5 mm. or less; fruits mostly with 1 groove . c. A. PLANTAGO-AQUATICA var. PARVIFLORUM

a. Diameter of mature fruiting portion 4–4.5 mm. or more; fruits either with 1 or 2 grooves. . *b*

b. Diameter of mature fruiting portion about 4 mm.; petals white . . . b. A. PLANTAGO-AQUATICA var. AMERICANUM

b. Diameter of mature fruiting portion 4.5 mm. or more; petals white to purple-tipped a. A. PLANTAGO-AQUATICA var. PLANTAGO-AQUATICA

a. **Alisma Plantago-aquatica** var. **Plantago-aquatica**

*Alisma Plantago-aquatica* var. *Michaletii* (Aschers. & Graebn.) Buchenau

Hendricks credits this variety to Missouri on the basis of the specimen of *Palmer 3786* from Melugen, Lawrence County.

According to Hendricks, the range extends into Europe and Asia, Africa, South America, and North America.

b. **Alisma Plantago-aquatica** var. **americanum** J. A. Schultes & Schult.

*Alisma triviale* Pursh [G]

Hendricks cites Missouri material from Butler Co. (*Stull 735*) and Maries Co. (*Steyermark 27615*).

The range is given as Asia and North America.

c. **Alisma Plantago-aquatica** var. **parviflorum** (Pursh) Torr.

*Alisma subcordatum* Raf. [P & S, G in part, BB in part]

Hendricks cites Missouri material from Livingston (*Sparling 1049B*), Laclede (*Steyermark 13908*), and Howell (Willow Springs, *Bush*) counties.

The range is indicated as the southern United States north to Vermont, Pennsylvania, Ohio, Indiana, Illinois, Missouri, Kansas, and Nebraska.

Typical *A. Plantago-aquatica* var. *Plantago-aquatica* is apparently the widespread variety which Hendricks treats under var. *Michaletii.*

The present author has not yet had an opportunity to evaluate Dr. Hendricks' treatment in the field nor to apply it to all the Missouri material which has been collected. Accordingly, the varieties presented here are being accepted, subject to future studies, and on the distribution map 90 only the combined range of the three varieties is shown.

Plants belonging to this group of varieties are sometimes dug in autumn or spring when the roots and solid base of the stem are well filled with starch, and then cooked as a starchy vegetable.

---

Plate no. 13. **1.** Najas gracillima, × $^2/_5$; *a.* Base of a single leaf, × $2^2/_5$; *b.* Cluster of flowers and portion of surrounding leaf-bases, × $1^1/_5$; Details from Small, The New York Botanical Garden. **2.** Najas guadalupensis, × $^2/_5$; *a.* Single female flower, × 12; *b.* Single leaf, × $3^3/_5$. **3.** Echinodorus cordifolius; *a.* Fruit, × $4^4/_5$; *b.* Leaf, × $^2/_5$; *c.* Inflorescence, × $^2/_5$; After Gleason, The New York Botanical Garden. **4.** Alisma Plantago-aquatica var. parviflorum, × $^2/_5$; *a.* Flower, × $3^3/_5$; Details from Small, The New York Botanical Garden.

PLATE NO. 13

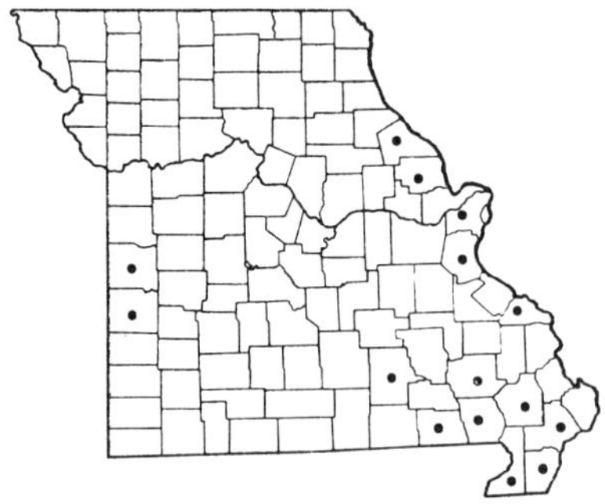

91 Echinodorus cordifolius (Burhead)

92 Echinodorus Berteroi var. lanceolatus (Burhead)

93 Echinodorus tenellus var. parvulus

## 2. **Echinodorus** Rich. Burhead

The present treatment is adapted from Dr. Fassett's recent revision of this genus (Rh. 57: 133–56; 174–88; 202–12. 1955).

a. Plants 1.5–10 cm. tall; leaves mostly narrow, linear-lanceolate to narrowly elliptic, tapering at each end, up to 12 mm. wide; stamens 6 or 9; pistils 20 or fewer in a loose head; fruits without beaks . 3. E. TENELLUS var. PARVULUS

a. Plants usually 10–60 cm. tall; leaves mostly ovate or broadly lanceolate, usually rounded or heart-shaped at base, mostly 15–150 mm. wide, but in dwarfed plants the leaves are narrowed to 5–10 mm. wide and are lanceolate; stamens 9–30; fruits with beaks . . . . . . . . . . . . . . . *b*

b. Main flower-stem (scape) upright, mostly branched; sepals with smooth veins; fruits with beaks 0.5–0.8 (–1) mm. long; transparent lines on leaves (use lens) mostly less than 1 mm. apart and often several mm. long . . . . . . . . . . . . . . . 2. E. BERTEROI var. LANCEOLATUS

b. Main flower-stem (scape) eventually lying on or touching the ground (prostrate or procumbent); sepals with ridges covered with tiny projections (papillae); fruits with beaks 0.2–0.8 mm. long; transparent lines on leaves mostly 1 mm. or more apart and rarely more than 1 mm. long. 1. E. CORDIFOLIUS

1. **Echinodorus cordifolius** (L.) Griseb. Map 91
*Echinodorus radicans* (Nutt.) Engelm. [BB, P & S]
Flowers July–September.

The main occurrence of this species is in the swampy lowland section of southeastern Missouri, along muddy wet shores of oxbow lakes of the Mississippi River north to Pike County, and of the Osage and Marmaton rivers in Vernon and Bates counties.

Ranges from Texas and Florida to the District of Columbia, north along the Mississippi Valley to Tennessee, Indiana, Illinois, Missouri, and west to Kansas and Oklahoma.

2. **Echinodorus Berteroi** (Spreng.) Fassett var. **lanceolatus** (Engelm.) Fassett Map 92
*Echinodorus rostratus* (Nutt.) Engelm. f. *lanceolatus* (Engelm.) Fern. [G]
*Echinodorus cordifolius* var. *lanceolatus* (Engelm.) Mack. & Bush [P & S]
*Echinodorus cordifolius* [BB]
Flowers July–September.

Occurring in low wet woods in the southeastern lowland section, and along sloughs and margins of oxbow lakes and backwaters of the Missouri and Mississippi rivers and their tributaries.

Ranges from Ohio, Illinois, Missouri, Iowa, and South Dakota, south to Arkansas and Texas, and also in California.

The submerged leaves of this and the preceding species often are used as decorative aquatic plants in tropical fish aquaria, sold under the name of 'Poor Man's Lace Plant.'

3. **Echinodorus tenellus** (Mart.) Buch. var. **parvulus** (Engelm.) Fassett Map 93
*Echinodorus parvulus* Engelm. [BB]
*Echinodorus tenellus* (Mart.) Buch. [G, P & S]
Flowers August–September.

Originally found in 1845 in St. Louis County along margins of ponds in the hills north, northwest, and

Plate no. 14. **1.** Sagittaria rigida f. rigida, $\times\ ^{2}/_{5}$; *a*, *b*. Two types of leaves found in Sagittaria rigida f. elliptica, $\times\ ^{2}/_{5}$. **2.** Echinodorus Berteroi var. lanceolatus, $\times\ ^{2}/_{5}$; After Gleason, details from Small, The New York Botanical Garden. **3.** Echinodorus tenellus var. parvulus, $\times\ 1^{1}/_{5}$; *a*. Flower, $\times\ 2^{4}/_{5}$; *b*. Fruit, $\times$ 4; After Britton and Brown, The New York Botanical Garden. **4.** Sagittaria montevidensis subsp. calycina, $\times\ ^{2}/_{5}$; After Gleason, details from Small, The New York Botanical Garden.

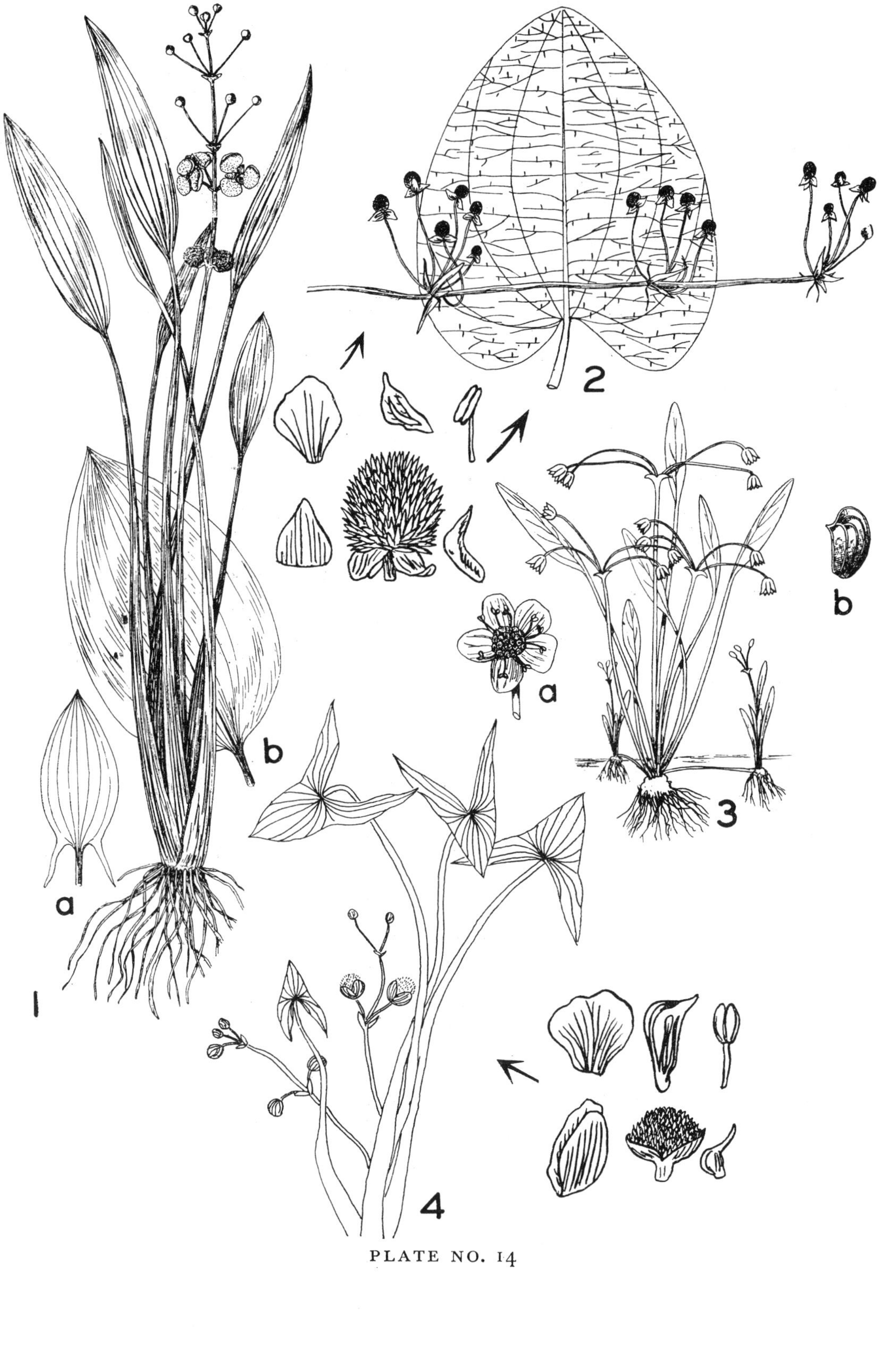


PLATE NO. 14

west of St. Louis by Dr. Engelmann. These ponds have since become destroyed. The only known Missouri station is from an upland sink-hole pond in the southern Ozarks in Howell County (Adobesee Pond, T22N, R7W, southeast sect. 36, 9 mi. southeast of West Plains, Sept. 4, 1949, *Steyermark 69124a*).

Dr. Fassett considers this Howell County station in the heart of the Ozarks as the possible ancestral area from which the North American variety of *Echinodorus tenellus* migrated to the younger coastal plain, where it is now mainly found, ranging from Mexico, Texas, and Florida, north to Massachusetts, Illinois, and Missouri. Typical *E. tenellus* var. *tenellus* is limited in its distribution to South America.

The submerged leaves of this variety and of other tropical American varieties make handsome aquatic plants for tropical fish tanks, and enter the trade under various common names, such as 'Dwarf Cellophane Plant.'

### 3. **Sagittaria** L. Arrowhead
(including *Lophotocarpus* Th. Durand [G])

In a recent revision of this genus (Mem. N.Y. Bot. Gard. 9: 179–233. 1955), Dr. Bogin has united the genus *Lophotocarpus* with *Sagittaria*. His work, in the main, is followed in the present treatment.

The tubers produced at the ends of the rhizomes in *Sagittaria* are sometimes confused with the roots of *Peltandra* and *Pontederia*. For purposes of identification, it may be pointed out here that *Peltandra* has a deep, vertical, stout root, *Sagittaria* does not have a strong deep rootstock, but sends out root-fibers from the base of the leaf-tufts, and *Pontederia* has a long creeping rhizome.

The submerged bladeless ribbon-like leaves of a number of species of the genus are commonly sold for ornamenting tropical fish tanks.

a. Sepals large and conspicuous, appressed to and surrounding the mature fruit, nearly orbicular; lower flowers perfect (with both stamens and pistils in the same flower); fruiting pedicels very thick, mostly 2–5 cm. long . . . . . . 1. S. MONTEVIDENSIS
a. Sepals not large and conspicuous, spreading or reflexed (turned down), not surrounding the fruit at maturity nor nearly orbicular; lower whorls (circles) of flowers either all pistillate (female) or all staminate (male); fruiting (pistillate) pedicels not conspicuously thick. . . . . . *b*
  b. Lowest (pistillate) heads of flowers without stalks (sessile) . . . . . 2. S. RIGIDA
  b. All heads of flowers with stalks (pedicels) . . . . . *c*
    c. Leaves not arrowhead-shaped nor with tail-like lobes at base . . . . . *d*
      d. Filaments glabrous (without hairs or scales); leaf-blades pinnately-nerved (main side nerves joining the midnerve at points one above the other, feather-like); the 3 papery or leaf-like bracts (at base of each whorl of flower-stalks) nearly separate, lanceolate, long-pointed, 9–15 mm. long . . . . . 4. S. AMBIGUA
      d. Filaments minutely hairy or roughened (use lens); leaf-blades palmately-nerved (main nerves arising from the base); the 3 papery bracts (at base of each whorl of flowers), united at their base, ovate, 3–8 mm. long . . . . . 3. S. GRAMINEA
    c. Leaves arrowhead-shaped or with tail-like lobes at base . . . . . *e*
      e. The 3 papery or thin bracts at base of each whorl of pistillate (female) flower- or fruit-stalks rounded, blunt, or only slightly pointed at tip, 1 cm. or less long; main flower-stem (scape) round in cross-section, scarcely angled; beak of mature fruit horizontal . . . . 6. S. LATIFOLIA
      e. The 3 papery or firm leaf-like bracts at base of each whorl of pistillate (female) flower- or fruit-stalks long-pointed at tip, 1.5–4 cm. long; main flower-stem (scape) angled; beak of mature fruit erect to ascending . . . . . 5. S. ENGELMANNIANA subsp. BREVIROSTRA

1. **Sagittaria montevidensis** Cham. & Schl., subsp. **calycina** (Engelm.) Bogin Map 94
*Lophotocarpus calycinus* (Engelm.) J. G. Sm. [G, P & S]
*Sagittaria montevidensis* [of BB]
Flowers late June–September.

Inhabiting shallow water and muddy margins of sloughs, ponds, and sluggish streams, where found nearly throughout Missouri; absent from many counties of northern Missouri, but of expected frequent occurrence with more intensive collecting there.

Ranges from Ohio and Michigan to Minnesota and South Dakota, south to Virginia, Alabama, Louisiana, and Texas.

Several forms, originally cited under *Lophotocarpus calycinus*, have been noted as occurring in Missouri:

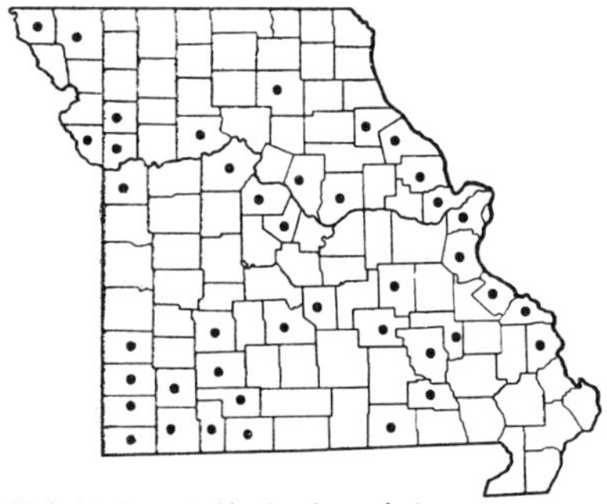

94 Sagittaria montevidensis subsp. calycina

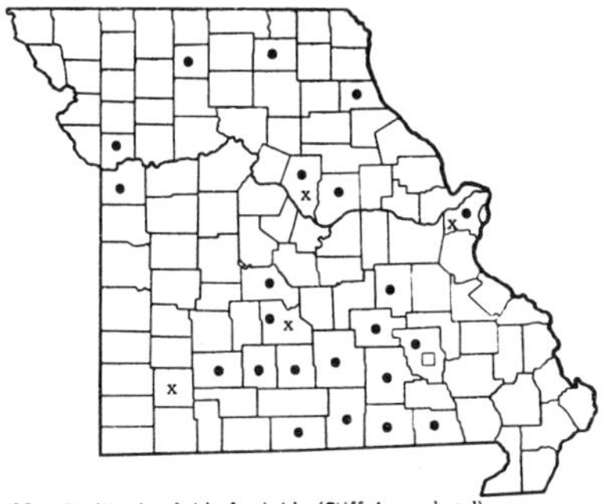

95 • Sagittaria rigida f. rigida (Stiff Arrowhead)
95 x Sagittaria rigida f. elliptica
95 □ Sagittaria rigida f. fluitans

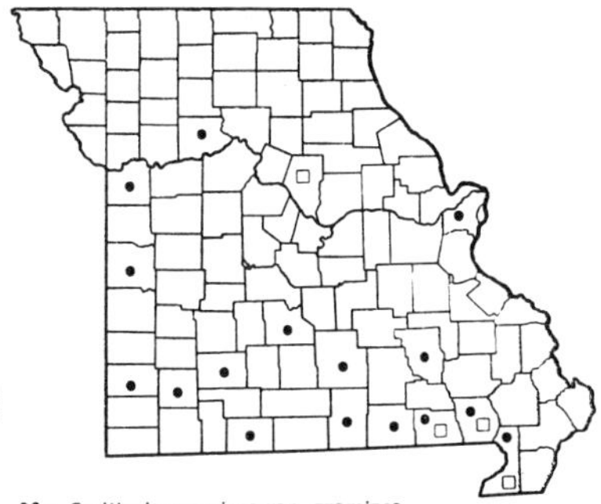

96 • Sagittaria graminea var. graminea
96 □ Sagittaria graminea var. platyphylla

*Lophotocarpus calycinus* f. *depauperatus* (Engelm.) Fern., with unlobed or barely auricled narrowly elliptic to spatulate leaves, known from Jefferson, Reynolds, Christian, and Newton counties; f. *maximus* (Engelm.) Fern., with lobed leaves 3 dm. wide and much wider than long, known from Christian County (*Palmer 64429*) and Moniteau County (*Steyermark 76596*); and f. *fluitans* (Engelm.) Steyermark from Ste. Genevieve County (*Steyermark 21097*). All of these are considered as of no taxonomic value by Dr. Bogin, and are placed as direct synonyms under *Sagittaria montevidensis* subsp. *calycinus*.

2. **Sagittaria rigida** Pursh Stiff Arrowhead Map 95
Flowers late May–October.
The following leaf forms occur in Missouri:

a. Leaves without blades, only the leaf-stalks present, 5–10 mm. wide . . . . 2c. S. RIGIDA f. FLUITANS
a. Leaves with definitely developed leaf-blades . . *b*
b. Leaf-blade scarcely wider than the leaf-stalk (petiole), linear-lanceolate to oblanceolate, unlobed, tapering at each end 2a. S. RIGIDA f. RIGIDA
b. Leaf-blades definitely wider than the leaf-stalk (petiole), broadly lanceolate, elliptical, or oval, lobed, unlobed, or lobed with 1 or 2 short basal lobes, gradually curving or rounded at the base . 2b. S. RIGIDA f. ELLIPTICA

2a. **Sagittaria rigida** Pursh f. **rigida** Map 95
*Sagittaria heterophylla* Pursh [P & S]
*Sagittaria heterophylla* var. *angustifolia* Engelm. [P & S]
*Sagittaria rigida* Pursh [G]
*Sagittaria rigida* in part [BB]

2b. **Sagittaria rigida** f. **elliptica** (Engelm.) Fern. Map 95
*Sagittaria heterophylla* var. *elliptica* Engelm. [P & S]
*Sagittaria rigida* in part [BB]

2c. **Sagittaria rigida** f. **fluitans** (Engelm.) Fern. Map 95

*Sagittaria rigida* f. *rigida* and the other forms occur throughout Missouri, where they are found mostly along the margins of natural upland sink-hole ponds, oxbow ponds of river valleys, artificial ponds, and along sloughs. Commonest in the Ozark region.

The whole series of variations range from Quebec to Maine and Minnesota, south to Virginia, Kentucky, Tennessee, Missouri, and Nebraska; also introduced in Europe.

Some authors of aquatic plants (Fassett, Fernald) have maintained the above as forms deserving recognition. Other botanists, such as Gleason, believe that the plant varies according to the habitat, while Muenscher states that all forms of leaf variation may originate from the same plant. My own experience indicates that some leaf variations are constant and characteristic of some places or occur in definite colonies. I have not found all types of leaf variations originating from the same plant, as indicated by Muenscher.

3. **Sagittaria graminea** Michx. Map 96

The following two varieties, as treated by Dr. Bogin, may be recognized in Missouri:

Fruiting heads 4–8 mm. in diameter; stalks (pedicels) of pistillate (female) flowers and of fruits slender, horizontally or upwardly spreading; petioles (leaf-stems) not noticeably 3-angled; leaves mostly with 3–5 main nerves . 3a. S. GRAMINEA var. GRAMINEA
Fruiting heads 9–14 mm. in diameter; stalks (pedicels) of pistillate (female) flowers and of fruits thickish, the fruiting pedicels slightly to noticeably downwardly curving; petioles (leaf-stems) noticeably 3-angled; leaves mostly with 5–11 main nerves . . . . 3b. S. GRAMINEA var. PLATYPHYLLA

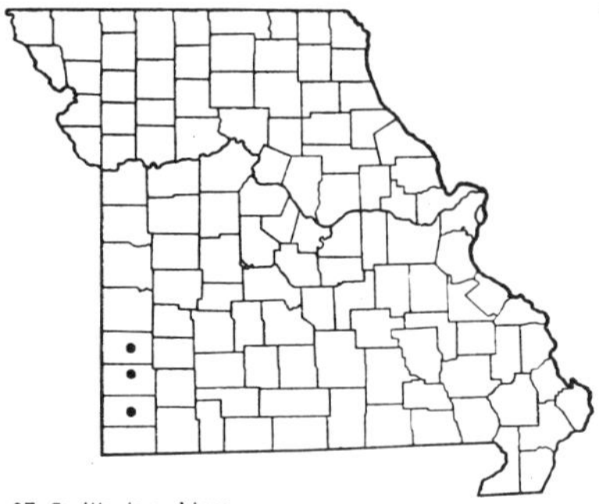

97 Sagittaria ambigua

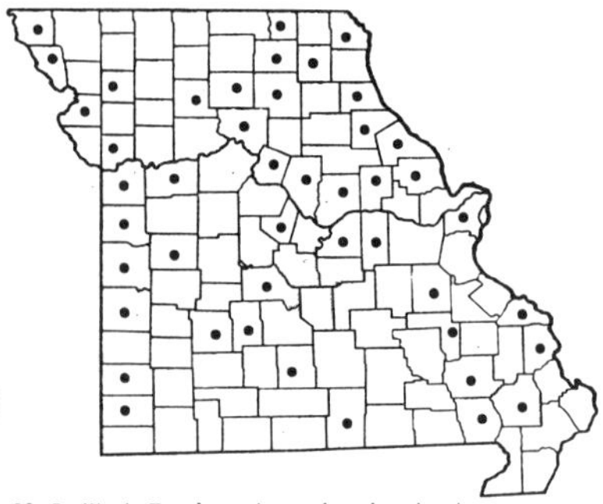

98 Sagittaria Engelmanniana subsp. brevirostra

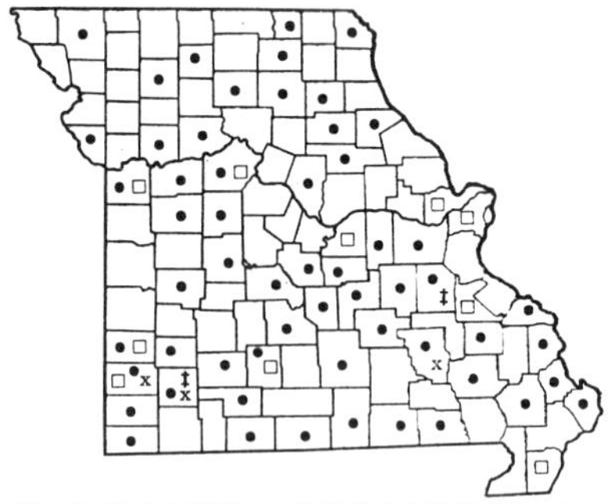

99 • Sagittaria latifolia var. latifolia f. latifolia (Common Arrowhead)
99 x Sagittaria latifolia var. latifolia f. hastata
99 ‡ Sagittaria latifolia var. latifolia f. gracilis
99 □ Sagittaria latifolia var. obtusa

3a. **Sagittaria graminea** Michx. var. **graminea** Michx. Map 96

*Sagittaria graminea* Michx. [G, P & S, BB]

Flowers late May–September.

This occurs along the margins of ponds, sloughs, and in ditches, occurring mostly in southern and central Missouri north to St. Louis, Carroll, and Jackson counties.

Ranges from Newfoundland to Ontario, south to Virginia, Florida, Ohio, Indiana, Illinois, Missouri, and Texas.

Plants are frequently encountered with slender petioles lacking leaf-blades, or with only submerged, thin, linear-lanceolate, bladeless phyllodia.

3b. **Sagittaria graminea** var. **platyphylla** Engelm. Map 96

*Sagittaria platyphylla* (Engelm.) J. G. Sm. [G, P & S, Steyerm.]

*Sagittaria graminea* [of BB]

Flowers late May–August.

Borders of ponds, sloughs, and in swampy ground, in the lowlands of southeastern Missouri, and introduced in a pond in Boone County (*Dunn 12855*).

Ranges from Missouri and Kansas to Alabama, Louisiana, and Texas.

4. **Sagittaria ambigua** J. G. Sm. Map 97

Flowers June 2–July.

Borders of ponds, sloughs, and ditches of southwestern Missouri.

Ranges from southwestern Missouri to Kansas and Oklahoma.

5. **Sagittaria Engelmanniana** subsp. **brevirostra** (Mack. & Bush) Bogin Map 98

*Sagittaria brevirostra* Mack. & Bush [G, P & S]

*Sagittaria cuneata* Sheldon [BB]

*Sagittaria longirostra* of P & S, not (Micheli) J. G. Sm.

Flowers June–September.

Common throughout Missouri in shallow water and muddy borders of sloughs, ponds, and sluggish streams.

Ranges from Ohio and Michigan to South Dakota, south to Illinois, Missouri, Arkansas, Oklahoma, and Texas.

All the Missouri material, according to Dr. Bogin, falls within *S. Engelmanniana* subsp. *brevirostra*, and includes Missouri material previously misidentified as *S. longirostra* and *S. australis*.

Bogin places *Sagittaria australis* (J. G. Sm.) Small and *S. longirostra* J. G. Sm. of *Gray's Manual*, eighth edition, under *Sagittaria Engelmanniana* subsp. *longirostra* (Micheli) Bogin.

6. **Sagittaria latifolia** Willd. Duck Potato, Common Arrowhead Map 99

Flowers early June–October.

The following forms, which occur in Missouri, have been recognized by some workers (Fernald, Fassett, Muenscher), but are placed by Dr. Bogin in synonymy under *Sagittaria latifolia* var. *latifolia*:

Plate no. 15. **1.** Sagittaria graminea var. platyphylla, $\times {}^{2}/_{7}$; *a.* Lobed type of leaf, $\times {}^{2}/_{7}$; *b.* Fruiting branch, $\times {}^{2}/_{7}$. **2.** Vallisneria americana, $\times {}^{2}/_{7}$; *a.* Male flower, $\times {}^{4}/_{7}$; *b.* Female flower, $\times {}^{4}/_{7}$; Details from Small, The New York Botanical Garden. **3.** Sagittaria graminea var. graminea, $\times {}^{2}/_{7}$. **4.** Sagittaria ambigua, $\times {}^{2}/_{7}$; *a.* Fruit, $\times 2{}^{6}/_{7}$; *b.* Stamen, $\times 2{}^{6}/_{7}$; After Britton and Brown, The New York Botanical Garden. **5.** Sagittaria Engelmanniana subsp. brevirostra, $\times {}^{2}/_{7}$; *a.* Fruit, $\times 2{}^{6}/_{7}$; After Britton and Brown, details from Small, The New York Botanical Garden. **6.** Sagittaria latifolia var. latifolia, $\times {}^{2}/_{7}$; *a.* Fruit, $\times 5{}^{5}/_{7}$; After Britton and Brown, The New York Botanical Garden. **7.** Anacharis Nuttallii, $\times {}^{2}/_{7}$; *a.* Female flower, $\times 2{}^{4}/_{7}$; Details from Small, The New York Botanical Garden. **8.** Limnobium Spongia, $\times {}^{3}/_{7}$; After Gleason, details from Small, The New York Botanical Garden.

PLATE NO. 15

a. Leaf-blades blunt (obtuse) or rounded at tip, the main part of the leaf-blade as broad as or nearly as broad as long . . . 6d. S. LATIFOLIA var. OBTUSA

a. Leaf-blades acutely pointed at tip, the main part of the leaf-blade several times longer than broad or nearly as broad as long . . . . *b*

b. Basal lobes of leaf-blade triangular; main part of leaf-blade ovate or triangular-ovate, as broad as or two-thirds as broad as long . 6a. S. LATIFOLIA var. LATIFOLIA f. LATIFOLIA

b. Basal lobes of leaf-blade narrower with parallel sides; main part of leaf-blade oblong-lanceolate to triangular-ovate, 0.2–25 cm. wide, 1/7–3/4 or more as broad as long . . . . . . . . . . . *c*

c. Main part of leaf-blade 1/16–1/8 as broad as long, linear-lanceolate to narrowly linear, 0.2–1.5 cm. wide . . 6c. S. LATIFOLIA var. LATIFOLIA f. GRACILIS

c. Main part of leaf-blade 1/7–2/5 as broad as long, narrowly triangular to oblong-lanceolate . . . . 6b. S. LATIFOLIA var. LATIFOLIA f. HASTATA

6a. **Sagittaria latifolia** var. **latifolia** f. **latifolia**
Throughout Missouri. Map 99

6b. **Sagittaria latifolia** var. **latifolia** f. **hastata** (Pursh) Robins. Map 99
Known from Reynolds, Lawrence, and Jasper counties.

6c. **Sagittaria latifolia** var. **latifolia** f. **gracilis** (Pursh) Robins. Map 99
Known from Washington and Lawrence counties.

6d. **Sagittaria latifolia** var. **obtusa** (Muhl.) Wieg. Map 99
*Sagittaria latifolia* f. *obtusa* (Muhl.) Rob. [P & S]
Occasionally found with var. *latifolia* in the same general range.

*Sagittaria latifolia* var. *latifolia* and forms, and var. *obtusa* inhabit the muddy borders of ponds, ditches, sloughs, and sluggish streams.

The species, together with its varieties and forms, ranges from Nova Scotia to British Columbia, south to Florida, California, and Mexico.

The large starchy tubers especially of this species and *S. Engelmanniana* and its subspecies are edible after baking or boiling for about half an hour. Under the name 'Wappato' the Indians of the northwestern United States ate the tubers of this and related species, while related species are cultivated for their succulence by the Chinese.

## Order **HYDROCHARITALES**

### Fam. **HYDROCHARITACEAE** (Frogbit Family)

a. Leaves heart-shaped and rounded, 30–70 mm. wide . . . . . . . . . . . . . . 3. LIMNOBIUM

a. Leaves ribbon-shaped, long and narrow, or short, small and linear to ovate, 0.3–20 mm. wide . . . . *b*

b. Leaves long and ribbon-shaped, arising from the base of the plant . . . . . . . . 2. VALLISNERIA

b. Leaves short, in numerous whorls (circles) of 3 or more (sometimes only 2 opposite each other) arising up and down the length of the stem . . . . . . . . . . . . . . 1. ANACHARIS

#### 1. **Anacharis** Rich. Waterweed, Bassweed
(*Elodea* Michx. [G])

The generic name *Elodea* was proposed for conservation, but has not been adopted.

Common species; leaves from upper half of stem 0.5–1.8 mm. wide; staminate (male) flowers sessile (without stalks); sepals of staminate flower 2 mm. long; sepals of pistillate (female) flowers 1.1–1.8 mm. long . . . . . . . . . . . . . . . . . . . . . . . . 1. A. NUTTALLII

Rare species; leaves from upper half of stem 1.2–4.5 mm. wide; staminate flowers on long thread-like peduncle (stalk); sepals of staminate flowers 3.5–5 mm. long; sepals of pistillate flowers 2.2–2.7 mm. long 2. A. CANADENSIS

1. **Anacharis Nuttallii** Planch. Map 100
*Elodea Nuttallii* (Planch.) St. John [G]
*Anacharis occidentalis* (Pursh) Victorin [P & S]
Flowers mostly July–September.

Commonly found in cool water of springs, spring branches, and margins of streams throughout the Ozark region, and also in ponds, lakes, pools, sloughs, and on the surface of stagnant water, locally north to Platte, Carroll, Grundy, Randolph, and Lincoln counties.

Ranges from Maine to Minnesota and Oregon,

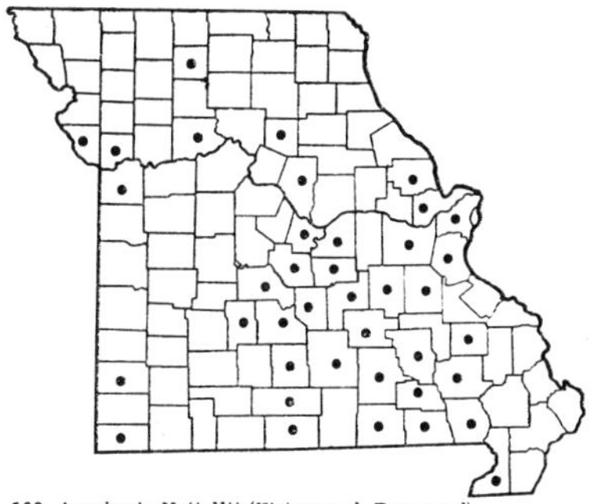

100 Anacharis Nuttallii (Waterweed, Bassweed)

101 Anacharis canadensis (Waterweed, Bassweed)

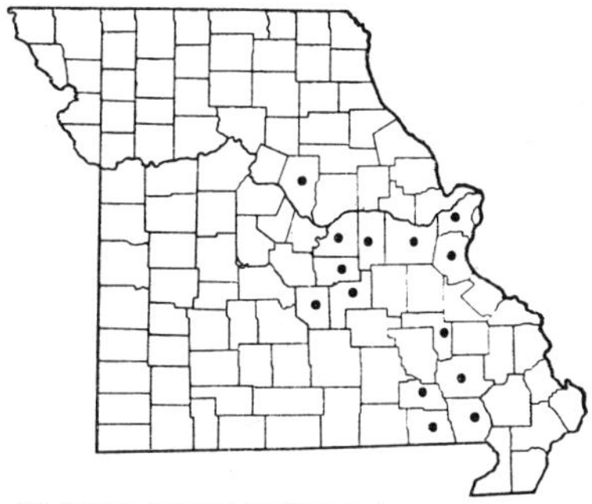

102 Vallisneria americana (Eelgrass)

south to North Carolina, Indiana, Illinois, Missouri, Kansas, and Colorado.

2. **Anacharis canadensis** (Michx.) Planch. Map 101

*Elodea canadensis* Michx. [G]

Flowers July–September.

Found in cold spring water in Wayne County (Mill Spring, Aug. 26, 1899, *Trelease*; Sept. 1, 1898, *Russell*) and in Newton County (Neosho, July 15, 1893, *Bush*).

Ranges from Quebec and Saskatchewan to Washington, south to North Carolina, Alabama, Illinois, Missouri, Oklahoma, Colorado, Utah, and California.

A third species, *Anacharis densa* (Rich.) Vict., with leaves 2–3.3 cm. long and in whorls of 4 or 6, is often planted in fish tanks and pools, and spreads vegetatively. It may be expected to occur spontaneously away from cultivation. It is introduced into the United States from South America.

## 2. **Vallisneria** L. Eelgrass, Tapegrass, Wild Celery

**Vallisneria americana** Michx. Map 102

Flowers June–September.

Found mostly in streams (Meramec, Gasconade, Black, and Current river drainage) of the Ozarks, and locally introduced north of the Missouri River in Boone County in the lake at Stephens College Country Club.

Ranges from Nova Scotia and Quebec to North Dakota, south to Florida and Texas.

This species is a favorite food of wild ducks. It and the European *V. spiralis* L. are commonly grown aquarium plants.

## 3. **Limnobium** Rich. Frogbit

**Limnobium Spongia** (Bosc) Steud.

American Frogbit Map 103

Flowers usually June–September.

In Missouri a very rare species where known only from the southeastern lowland counties (Mississippi, Pemiscot, and Dunklin), occurring there in backwater sloughs and drainage-ditches of the Mississippi and St. Francis river drainage.

Ranges from tropical America, Florida, and Texas, north in the coastal states to New Jersey, New York (Lake Ontario), and north in the Mississippi Valley to southern Illinois and southeastern Missouri.

This plant, with spongy coarsely cellular leaves, often purplish beneath, is generally found floating on the surface of quiet water; occasionally plants are found rooted in the soft mud. This species is found sometimes sold as an aquarium plant under the name of 'Giant Water Violet,' but it has not been very successfully grown as such.

## Order **GRAMINALES**

### Fam. **GRAMINEAE** (Grass Family)

This family is the most important one economically, yielding as it does the main cereal and forage crops (wheat, rice, rye, barley, oats, corn, sorghum) and sugar cane.

Flowers of this family are entirely wind-pollinated. It is fortunate that most of the grasses are relatively harmless, so far as their importance in hay fever cases. Only a small number of the species account for the occurrence of grass hay fever, the chief ones responsible, in the order of their seasonal appearance, being sweet vernal grass (*Anthoxanthum*), June grass (*Koeleria cristata*), orchard grass (*Dactylis glomerata*), which appear in May and June, timothy (*Phleum pratense*), redtop (*Agrostis alba*), and other species of *Agrostis*, appearing in June and July. Less important grasses from the standpoint of hay fever are red fescue (*Festuca rubra*), meadow fescue (*Festuca elatior*), quack grass (*Agropyron repens*), ryegrass (*Lolium perenne*), darnel (*Lolium temulentum*), and velvet grass (*Holcus lanatus*).

Simply stated, grasses have jointed stems (*culms*), which are usually solid at the joints (*nodes*) and round or compressed and usually hollow (solid in several groups) between the joints. The leaves are arranged in two vertical alternate ranks or rows, and are divided into a usually split tubular portion (the *sheath*) which surrounds the culm (exceptions to this are in *Melica, Glyceria,* and *Bromus*), a little hairy collar or membranous or firm projection (the *ligule*) at the inner junction of the sheath and blade, and the expanded or spreading flat or needle-like leafy part (the *blade*) with usually parallel nerves (when flattened).

The flowers of grasses have a specialized structure, and it is necessary to become acquainted with the elementary terminology used. A grass inflorescence or flower cluster is composed of units, known as *spikelets,* which are arranged in various ways (spikes, racemes, and panicles, and modifications of these). The stalk of each spikelet (if it has one) is the *pedicel.* The main axis of the inflorescence or the branches to which the spikelets are attached is the *rachis.*

An individual spikelet consists of 1 or 2 *florets* arranged on either side (if there are more than 1) of a central jointed axis (the *rachilla*). At the base of the spikelet are usually 2 empty bracts (*first glume* and *second glume*). Sometimes the first or second glumes are reduced or absent. Each floret has usually 2 other bracts: (1) an outer *lemma* and (2) an inner, often thinner, *palea* enclosed by the lemma. Sometimes the palea is minute or not present. Between the lemma and palea is the flower itself in which are usually three stamens (rarely 2 or 1), a 1-celled 1-ovuled ovary with 2 plume-like stigmas, and 2 (sometimes 3) tiny *lodicules,* which represent the modified sepals. The fruit is a one-seeded *caryopsis, achene,* or *utricle.*

Various modifications of the grass flower have produced the great diversity to be found within the family. Sometimes the basal end of the lemma is hardened as a *callus,* which may or may not be furnished with hairs. Sometimes a slender hair- or bristle-like appendage (awn) is attached to the glumes or lemmas. Sometimes, some of the spikelets are fertile and others are sterile or consist only of a pedicel.

The following key to the genera attempts to utilize and correlate vegetative, floral, and field characters in the identification of the grasses.

a. Plants woody or with the culms (stems) perennial, the stems not dying back to the ground but starting from an already live stem above the ground; upper leaf-blades crowded, their sheaths overlapping and bearing bristles at the top . . . . . . . . . . . . . . . . . . 1. ARUNDINARIA
a. Plants herbaceous or with the culms (stems) annual and dying back to the ground at the end of the season, with a new crop of stems arising from the ground each year either from the old root cluster or newly germinated plants . . . . . . . . . . . . . . . . . . . . . . . . . . *b*
  b. Spikelets enclosed in ball-shaped, spiny, barbed burs . . . . . . . . . . . 72. CENCHRUS
  b. Spikelets not enclosed in spiny burs . . . . . . . . . . . . . . . . . . . . *c*
    c. Many soft or silky hairs conspicuous throughout the inflorescence (flowering portion), producing a feather-duster or plume-like effect (hairs may be present on various parts of the spikelets without producing the effect as here described, so look into the contrasting category of this same letter c) . . . . . . . . . . . . . . . . . . . . . . . . . *d*
      d. Over-all color effect produced by whole inflorescence yellow-brown, due to brownish spikelets and fawn-colored hairs; ligule a prominent stiff, erect, cartilaginous stub 3–8 mm. long present at base of leaf-blade . . . . . . . . . . . . . . 78. SORGHASTRUM

d. Over-all color effect produced by whole inflorescence either white, silvery, gray, straw-color, pinkish, or silvery and purplish, but not yellow-brown; ligule not a stiff erect cartilaginous stub 3–8 mm. long, but either consisting of a band or fringe of hairs, a short membranous collar or projection at most 2 mm. long . . . . . . . . . . . . . . . . . . . . . . . *e*

e. Leaf-blades 3–7 mm. wide; culms (stems) 0.3–1.5 meters tall; culms mostly 0.2–0.6 cm. in diameter . . . . . . . . . . . . . . . . . . . . . . . . . . . *f*

f. The conspicuous hairs of the flowering portion are chiefly on the rachis and pedicels or below the glumes . . . . . . . . . . . . . . . . . . 76. ANDROPOGON

f. The conspicuous hairs of the flowering portion are on the margin of the fertile lemma, not on the rachis or pedicels . . . . . . . . . . . . . . . . . . 55. CHLORIS

e. Leaf-blades 10–70 mm. wide; culms (stems) mostly 2–4 meters tall, mostly 0.7–1.5 cm. in diameter . . . . . . . . . . . . . . . . . . . . . . . . . . . *g*

g. Basal leaves mostly absent, the conspicuous ones being the widely spreading, 2-ranked leaf-blades on the culms (stems); leaf-blades 1–7 cm. wide; no awns present on spikelets; spikelets each 3–7-flowered, at least 2 of the flowers functional in each spikelet . . . . . . . *h*

h. Conspicuous hairs present are produced from the lemmas; rachilla glabrous (without hairs); cultivated grass, rarely naturalized . . . . . . . . . . . . . 13. ARUNDO

h. Conspicuous hairs present are produced from the rachilla and overtop the lemmas; lemmas glabrous; native plants of wet ground . . . . . . . . . . . 14. PHRAGMITES

g. Basal leaves often numerous or forming dense bushy clumps, the leaf-blades on the culms not conspicuously 2-ranked or horizontally spreading; leaf-blades 0.8–2 cm. wide; lemmas with a short to long awn; spikelets each with only 1 functional flower . . . . . . . . *i*

i. Basal leaves very numerous, forming dense bushy clumps or rosettes; inflorescence (flowering portion) fan-shaped, its main axis shorter than its branches; axis and branches of inflorescence mostly glabrous, the conspicuous hairs present arising as a ring surrounding the base of each spikelet . . . . . . . . . . . 73. MISCANTHUS

i. Basal leaves few, either not conspicuous nor forming dense bushy clumps; axis of inflorescence much longer than the branches; axis and branches of inflorescence densely silky-hairy and a ring of hairs also present around the base of each spikelet. . 74. ERIANTHUS

c. Soft or silky hairs absent throughout the inflorescence, or, if present, not conspicuous nor producing a feather-duster or plume-like effect . . . . . . . . . . . . . . . . . . . *j*

j. Inflorescence consisting of 1 (or presenting the appearance of 1) dense or closely-flowered spike (shaped like a pencil, tail, broom, brush, finger, shaft, hook, curve, or match stick), the spikelets surrounding all sides of the main axis, or with only their sides or edges next to the main axis (rachis) of the inflorescence (inflorescence may be hidden or enclosed in an enlarged leaf sheath) . . *k*

k. Spreading or erect silky hairs present on joints of rachis and various pedicels (stalks) of sterile florets . . . . . . . . . . . . . . . . . . . . . . . . . . . *l*

l. A conspicuous stiff, erect, cartilaginous stub 3–8 mm. long present at base of leaf-blade; spikelets yellow- or chestnut-brown . . . . . . . . . . . . . 78. SORGHASTRUM

l. Without the above combination of characters . . . . . . . . . . 76. ANDROPOGON

k. Without the above combination of characters . . . . . . . . . . . . . . . . . *m*

m. Bristles or awns (stiff or delicate, short or long, extension or outgrowth, usually from midrib of lemma or glume) on some part of the spikelets or at the base of the spikelets (hairs attached to or covering parts of spikelets should not be judged as bristles or awns here). . . . *n*

n. Slender bristles on the outside and at the base of each spikelet, but no awns arise from the glumes, lemma, or palea of the spikelets . . . . . . . . . . . . 71. SETARIA

n. Awn (either stiff and bristly, or delicate and hair-like) attached to some part of the spikelet, but no bristles are on the outside or at the base of a spikelet (do not confuse bristle-like glumes of some species of *Elymus*, *Sitanion*, *Hystrix*, or *Hordeum*) . . . . . . . . *o*

o. Lemma ending in a 3-awned tip . . . . . . . . . . . . . 47. ARISTIDA

o. Lemma without an awn or with only 1 awn at most, or awn occurring on glumes or palea . . . . . . . . . . . . . . . . . . . . . . . . . . *p*

p. Spikelets with only one functional floret (use presence and location of glumes to judge where one spikelet ends and another begins), which may either be perfect, pistillate (pistil only), or staminate (stamens only) . . . . . . . . . . . . *q*

q. Long awns present on both glumes and lemmas . . . . . . . . . . . . *r*

r. Leaf-sheaths without any auricles (portions projecting from the edge) at the summit . . . . . . . . . . . . . . . . . . 25. HORDEUM

r. Leaf-sheaths with auricles at the summit . . . . . . . . . . . . *s*

s. Awns of lemma 6–10 cm. long . . . . . . . . . . . . . . 25. Hordeum vulgare
s. Awns of lemma at most 2.5 cm. long . . . . . . . . . . . . . . . . . 22. Elymus
q. Long awn, when present, only on glumes or lemma, not on both . . . . . . . . . . . *t*
t. Awn 11–20 cm. long; spikelets, excluding awn, 2.8–4 cm. long . . . . . . . . . 46. Stipa
t. Awn 0.05–2 cm. long; spikelets, excluding awn, 0.25–1 cm. long. . . . . . . . . . . . . *u*
u. Awn 1.5–2 cm. long; inflorescence 20–40 cm. long; plant 1–2 meters tall . . . 74. Erianthus
u. Awn 0.1–1.5 cm. long; inflorescence 2–15 cm. long; plant 0.3–1.5 meters tall . . . . . . *v*
v. Awns on glumes, not on lemmas . . . . . . . . . . . . . . . . . . . *w*
w. Awns inconspicuous, at most 1 mm. long, much shorter than remainder of spikelet 40. Phleum
w. Awns conspicuous, 2–4 mm. long, as long as or longer than remainder of spikelet 41. Muhlenbergia
v. Awns on lemmas, not on glumes . . . . . . . . . . . . . . . . . . . *x*
x. Awn arising from the back, but not the tip of the lemma . . . . . . . . . . *y*
y. A cluster of conspicuous fine hairs surrounds the lemma; plants about 1 meter tall 35. Calamagrostis
y. No cluster of prominent hairs surrounds the lemma; plants 0.1–0.8 meter tall . . . . *z*
z. Inflorescence brownish-yellow; glumes separate, not joined at the base 58. Anthoxanthum
z. Inflorescence pale green, whitish-green, or drab; glumes united at the base 39. Alopecurus
x. Awn arising from some part of the tip of the lemma . . . . . . . . . . . . *1*
1. Awn inconspicuous, at most 1 mm. long, shorter than the remainder of the spikelet 66. Eriochloa
1. Awn conspicuous, 3–15 mm. long, as long as or 2–4 times as long as remainder of spikelet . . . . . . . . . . . . . . . . . 41. Muhlenbergia

p. Spikelet consisting of 2 or more functional florets (use presence and location of glumes to judge where one spikelet ends and another begins) . . . . . . . . . . . . . . . . . . . . . *2*
2. Entire plant velvety to the touch; awn with a small hook at tip; inflorescence with a dull rose to pink color . . . . . . . . . . . . . . . . . . . . . . . . 33. Holcus
2. Entire plant not velvety to the touch; awn lacking a hook at tip; inflorescence of other colors . . . *3*
3. Midnerve of lemma projecting as a minute awn less than 1 mm. long; side nerves of lemma frequently projecting above top of lemma as minute points; palea hairy on ridges; lemma hairy on keel (narrow ridge on back) or on marginal nerves. . . . . . . . . . 16. Tridens
3. Without the above combination of characters . . . . . . . . . . . . . . . . . *4*
4. Some or all of the spikelets separated from the main axis of the inflorescence by pedicels (stalks) . . . . . . . . . . . . . . . . . . . . . . . . . . . *5*
5. Awn arising from the back of the lemma on at least one of the florets; spikelets 2-flowered 32. Arrhenatherum
5. Awn arising at the very tip of the lemma either between two teeth or lobes or simply as a simple projection from a toothless lemma; spikelets 3-many-flowered . . . . . . . *6*
6. Both glumes and lemmas with prominent long awns . . . . 25. Hordeum jubatum
6. Only the lemmas with conspicuous awns. . . . . . . . . . . . . . . . *7*
7. Glumes much longer than all the lemmas (excluding the awns), 8.5–13 mm. long; perennials with old leaves at base and with a fresh cluster of basal rosette leaves; ligule in the form of erect conspicuous hairs . . . . . . . . . . . 34. Danthonia
7. Glumes shorter than the rest of the spikelet, 2–5 mm. long; annuals with delicate roots, with few leaves, the leaves not clustered at the base . . . . . . 3. Festuca
4. Spikelets all attached directly to the rachis (sessile) . . . . . . . . . . . . . *8*
8. Only 1 spikelet at each joint (node) of the rachis . . . . . . . . . . . . . *9*
9. Both glumes and lemmas with long conspicuous awns . . . . . . . . . 22. Elymus
9. Only glumes or lemmas, but not both, with conspicuous long awns . . . . . . *10*
10. Spikelets placed with their sides to the rachis; both glumes present on each spikelet . . . . . . . . . . . . . . . . . . . . . . . . *11*
11. Spikelets cylindrical, fitting into the depressions or pockets of the rachis, appearing as part of the diameter of the rachis . . . . . . . 20. Aegilops
11. Spikelets more or less flattened, but obviously distinct from and not appearing as part of the rachis . . . . . . . . . . . . . . . . . . *12*
12. Awns conspicuous, 1.5–8 cm. long (sometimes shorter on lowest spikelet) . *13*

13. Glumes ovate, 3-nerved . . . . . . . . . . . . . . . . . 19. TRITICUM
13. Glumes subulate (very narrow), 1-nerved . . . . . . . . . . . 21. SECALE
12. Awns shorter or inconspicuous . . . . . . . . . . . . . . . . . 18. AGROPYRON
10. Spikelets placed with their narrow edge to the rachis; except for the uppermost spikelet, only one glume present (the one away from the rachis) . . . . . . . . . . . 26. LOLIUM
8. A group of 2 or 3 spikelets at each joint (node) of the rachis . . . . . . . . . . . . *14*
14. Spikelets in 3's at each joint, 1-flowered . . . . . . . . . . . . 25. HORDEUM
14. Spikelets in 2's at each joint, 2–6-flowered . . . . . . . . . . . . . . . . *15*
15. Glumes 6–8 cm. long; main axis (rachis) of inflorescence breaking up into segments upon reaching maturity; rare introduced plant . . . . . . . . . . . 23. SITANION
15. Glumes 1–4 cm. long; main axis of inflorescence persistent at maturity, not breaking up into segments; common native plants . . . . . . . . . . . . . . . . . *16*
16. Spikelets outwardly spreading, or spreading at right angles from the rachis at maturity; glumes not present or reduced to little bristles . . . . . . . 24. HYSTRIX
16. Spikelets ascending, erect, or upwardly pointing at maturity with reference to the axis of the inflorescence (spikes which are nodding or curving should be considered accordingly); glumes usually conspicuous or evident . . . . . . . . 22. ELYMUS
m. No bristles or awns present in the inflorescence (hairs attached to or covering parts of the spikelets may be looked for here, and should not be judged as bristles or awns) . . . . . . . . *17*
17. Spikelets appearing as if part of the diameter of the main axis (rachis) of the inflorescence, fitting into the depressions or sunken portions of the rachis; outermost part of spikelet (1st glume) with little pits . . . . . . . . . . . . . . . . . . . . . . . . 79. MANISURIS
17. Spikelets not as above, but distinct and separate from the rachis, and not appearing to be part of it . . . . . . . . . . . . . . . . . . . . . . . . . . . . . . *18*
18. Sexes of spikelets in separate inflorescences, the staminate (male) at the summit of the plant, the pistillate (female) lower down in the axils of the leaves . . . . . . . . 81. ZEA
18. Both stamens and pistils in the same florets . . . . . . . . . . . . . . . . . *19*
19. Lemmas with a 3-lobed (cross-like) tip . . . . . . . . . . . . . . 25. HORDEUM
19. Lemmas without a 3-lobed (cross-like) tip . . . . . . . . . . . . . . . . *20*
20. Spikelets placed with their narrow edge to the rachis; except in the uppermost spikelet, only the glume away from the rachis present (one glume present) . . . 26. LOLIUM
20. Spikelets placed with their sides to the rachis or encircling all sides of the rachis; both glumes present . . . . . . . . . . . . . . . . . . . . . . *21*
21. Each spikelet consisting of 2 or more functional florets . . . . . . . . . . *22*
22. Individual spikelets 2–5 mm. long, usually 2- or 3-flowered . . . . . . . *23*
23. Main axis of inflorescence densely short-hairy; spikelets mostly 4–5 mm. long; glumes and lemmas acutely pointed and narrowed at tip . 27. KOELERIA
23. Main axis of inflorescence glabrous or slightly scabrous; spikelets mostly 2–3.5 (rarely 4) mm. long; at least the glumes rounded or blunt at tip . . . . . . . . . . . . . . . . 28. SPHENOPHOLIS
22. Individual spikelets 7–22 mm. long, 2–5-flowered . . . . . . . . . *24*
24. Spikelets on very short or else noticeable pedicels (stalks); spikelets 5–12 mm. long . . . . . . . . . . . . . . . . 16. TRIDENS
24. Spikelets without stalks (sessile), arising directly from each joint of the rachis; spikelets 10–22 mm. long . . . . . . . . . . . . *25*
25. Two or 3 spikelets at each joint (node) of the rachis . . . 22. ELYMUS
25. Only 1 spikelet at each joint (node) of the rachis . . . . . . *26*
26. Plants producing long creeping rhizomes . . . . 18. AGROPYRON
26. Plants not producing creeping rhizomes . . . . . . . . *27*
27. Glumes lanceolate, acute; lemma with an acute or narrowly tapering tip and obscurely ridged at the middle; leaves 2–10 mm. wide; plants perennial . . . 18. AGROPYRON
27. Glumes ovate, blunt, strongly keeled toward one side; lemma rounded or blunt at tip, keeled; leaves 10–20 mm. wide; annuals . . . . . . . . . . . . . 19. TRITICUM
21. Each spikelet consisting of only one functional floret, which may be perfect, pistillate (pistil only), or staminate (stamens only) . . . . . . . . . . *28*
28. Glumes longer than the lemma and palea; ligule large or noticeable as a membranous prolonged part . . . . . . . . . . . . 59. PHALARIS

28. Glumes shorter than or at most equaling the length of the lemma and palea; ligule a zone of hairs or a collar 0.5 mm. high or none at all . . . . . . . . . . . . . . . . *29*

29. Lemma faintly 3–5-nerved; inflorescence not projecting from an inflated sheath 41. MUHLENBERGIA

29. Lemma 1-nerved; inflorescence partly or wholly enclosed in, or protruding from, an inflated sheath . . . . . . . . . . . . . . . . . . . . . . *30*

30. Ligule a zone of hairs; leaves mostly 4–10 cm. long, at most 25 times longer than broad; spike densely flowered, 6–10 mm. thick; rarely found . . . . . . . 43. HELEOCHLOA

30. Ligule minute or none; leaves usually 10–100 cm. long, averaging 40–100 times longer than broad, or, if less than 10 cm. long, then the inflorescences are 2–5 mm. thick, or loosely-flowered, or elongated; common. . . . . . . . . . . . . 42. SPOROBOLUS

j. Inflorescence not appearing as 1 dense or closely-flowered spike, but as other types of arrangements which have 1 or usually 2 or more or many separate units or branches, these separate units or branches branched or unbranched and usually several to many times longer than broad . . . . *31*

31. Inflorescence shaped like a V, pitchfork, rays of an umbrella, spokes of a wheel, fingers on a hand, toes of a bird, or all the separate units hanging at an angle like a flag, or comb-like, feather-like, stepladder-like, or flattened tree-like, all units or *all* lateral units of the inflorescence *unbranched*, and usually several to many times longer than broad, sometimes short, the spikelets usually 1-flowered (sometimes many-flowered), and usually along one side of the axis (rachis) . . . . . . . . . . . . . . . . . . . . . . . . . . . *32*

32. Inflorescence shaped like a V, pitchfork, rays of an umbrella, spokes of a wheel, fingers of a hand, or toes of a bird, these units closely crowded at the top of the stem . . . . . . *33*

33. Leaf-blades ovate to ovate-lanceolate, with curved sides, 3–4 times longer than broad 75. ARTHRAXON

33. Leaf-blades linear with more or less parallel sides, many times longer than broad . . . *34*

34. Plants 1–3 meters tall, averaging close to the height of a man . . . . . . . *35*

35. Leaf-blades 1–3.5 cm. wide; inflorescence at the top of the culm (stem), consisting of 2–4 spikes which are pistillate (female) in the lower portion; no awns present on any part of the spikelets. . . . . . . . . . . . 80. TRIPSACUM

35. Leaf-blades 0.5–1 cm. wide; inflorescence at the top of the culm (stem), consisting of 2–6 or more densely-flowered racemes with no separate pistillate (female) portion; awns present on some (fertile lemma) of the florets. 76. ANDROPOGON GERARDI

34. Low plants averaging only 0.05–0.4 meters tall, averaging about knee-high or less on a person 5′9″ tall . . . . . . . . . . . . . . . . . . . . . *36*

36. Sexes of spikelets separated, either the staminate (male) and the pistillate (female) on separate plants or in different inflorescences of the same plant . . *37*

37. Spikelets many-flowered, either all staminate (male) or all pistillate (female) on separate plants . . . . 7. ERAGROSTIS REPTANS (pistillate plant)

37. Staminate and pistillate inflorescences separated on the same plant, the staminate at the top, the pistillate lower down in the axils of the leaves; spikelets 1–2-flowered . . . . . . . . . . . . . . . . 81. ZEA

36. Sexes of spikelets not separated, spikelets perfect, with the staminate (male) and pistillate (female) parts together in the same floret . . . . . . . . *38*

38. Awns present on some part of the spikelet . . . . . . . . 55. CHLORIS

38. No awns on any part of the spikelet . . . . . . . . . . . . . *39*

39. Spikes rather thick, 5–7 mm. wide; ligule an irregularly cut collar with some long hairs at the sides . . . . . . . . . . . 49. ELEUSINE

39. Spikes slender, 0.5–2 mm. wide; ligule a band of white hairs or short membranous collar . . . . . . . . . . . . . . . . . *40*

40. All the rays (spikes) horizontally spreading or nearly so; ligule a conspicuous band of white hairs; plants forming creeping dense mats from hard scaly rhizomes or long stolons, the leaves 2–4 mm. wide . . . . . . . . . . . . . . . . 50. CYNODON

40. Most of the rays (spikes) ascending to erect, not horizontally spreading; ligule a membranous projecting collar; plants either tufted and not creeping or rooting at the lower nodes (joints) and forming loose colonies, those from the rooting type with leaves 3–10 mm. wide . . . . . . . . . . . . . . 64. DIGITARIA

32. The separate rays or branches of the inflorescence separated in such a manner as to appear pinnately branched often on two sides of a main axis, or only on one side, resembling the arrangements of a flattened feather, of a flattened tree with main branches, of a comb, of a stepladder, or of a hanging flag . . . . . . . . . . . . . . . . . . . . . . . . . . . . *41*

41. Spikes (not spikelets) *all* nodding, numerous, usually only on 1 side or on 2 sides of the main rachis of the inflorescence, 1–2 cm. long. . . . . . . . . 56. BOUTELOUA CURTIPENDULA

41. Spikes either erect, ascending, widely spreading, or the lower ones drooping, from 1, 2, or more sides of the main axis of the inflorescence . . . . . . . . . . . . . . . . . *42*

42. At least the *lowest* spikes or branches widely or horizontally spreading or drooping . . . *43*

43. Rachis of the lateral (side) racemes or spikes leaf-like, wider than and with their margins arching over the spikelets; plants growing in the water or nearly aquatic. 68. PASPALUM

43. Rachis not leaf-like, not as wide as the rows of spikelets; plants of dry or wet soils, sometimes in the water . . . . . . . . . . . . . . . . . . . . *44*

44. Some of the hairs of the spikelets with swollen or blister-like bases; ligule absent; sheaths compressed; plants often of wet soils or even growing in the water. 70. ECHINOCHLOA

44. Without the above combination of characters . . . . . . . . . . *45*

45. Plants 1–5 meters tall (or smaller in young, poor, or undernourished specimens); culms (stems) 0.5 cm. or more (usually 1.5–10 cm.) in diameter; leaf-blades 3–8 cm. wide . . . . . . . . . . . . . . . . . 81. ZEA

45. Plants 0.15–0.6 meters tall; culms (stems) 1–3 mm. in diameter; leaf-blades 1–12 mm. wide . . . . . . . . . . . . . . . . . . . . . *46*

46. Spikes 1–3, about 4.5–6 mm. wide (excluding the awn), 2–4 cm. long, rather thick and densely flowered . . . . . . . . . . 56. BOUTELOUA

46. Spikes more numerous, 0.5–2 mm. wide (excluding the awns), usually 5–15 cm. long, slender and loosely flowered . . . . . . . . . . . *47*

47. Awns absent on all part of the inflorescence . . . . . . . . . *48*

48. Spikelets 2–4-flowered . . . . . . . . . 48. LEPTOCHLOA

48. Spikelets 1-flowered . . . . . . . . . . . . . . *49*

49. Spikelets glabrous . . . . . . . . 51. SCHEDONNARDUS

49. Spikelets with short bristle-like hairs . . . . . . . 61. LEERSIA

47. Awns present on some part of the inflorescence . . . . . . . *50*

50. Plants of dry, rocky, or acid soils; inflorescence about as broad as long, dome-shaped; leaf-blades of the main stem rather stiff, mostly horizontally spreading, 4–10 cm. long, chiefly 4–6 times longer than broad; spikelets 6 mm. long; awn 4.5–9 mm. long 54. GYMNOPOGON

50. Plants of wet or alluvial soils; inflorescence mostly longer than broad; leaf-blades of main stem strongly ascending, chiefly 7–40 cm. long (except shorter in depauperate individuals), mostly 12–30 times longer than broad; spikelets up to 2.5 mm. long; awn 1 mm. or less long . . . . . . . . . 48. LEPTOCHLOA

42. All the spikes erect or ascending, or, if spreading, not horizontally so . . . . . . . *51*

51. Rachis (axis) of the lateral (side) racemes or spikes leaf-like, wider than and with their margins arching over the spikelets; plants growing in the water or nearly aquatic . 68. PASPALUM

51. Rachis (axis) not leaf-like, not as wide as the rows of spikelets; plants of dry or wet soils . *52*

52. Spreading or erect silky hairs present on joints of rachis or on various sterile pedicels (stalks) . . . . . . . . . . . . . . . . . . 76. ANDROPOGON

52. Not as above . . . . . . . . . . . . . . . . . . . . . . *53*

53. Some of the hairs of the spikelets with swollen or blister-like bases; sheaths compressed; ligule absent; plants often of wet soils, or even growing in the water. 70. ECHINOCHLOA

53. Without the above combination of characters . . . . . . . . . . . *54*

54. Awns absent on all parts of the spikelets (spikelets with minute awns only 0.5–1 mm. long should be looked for in the opposite section of this number 54). (Awns may be present only on uppermost glumes of staminate florets in no. 66, which should be looked for in opposite section of this number 54) . . . . . . . . . . . . . . . . . . . *55*

55. Spikelets 5–many-flowered . . . . . . . . . . . . . . . . . . . . . . *56*
56. Leaf-blades and culms (stems) pubescent; culms spreading; leaf-blades not stiff; spikelets 1–4 mm. wide . . . . . . . . . . . . . . . . . 7. ERAGROSTIS REPTANS
56. Leaf-blades and culms (stems) glabrous; culms upright or nearly so; leaves stiff; spikelets 4–5.5 mm. wide . . . . . . . . . . . . . . . . . . . 10. DISTICHLIS
55. Spikelets 1-flowered . . . . . . . . . . . . . . . . . . . . . . *57*
57. Some part of the spikelet either with cross-wrinkles or with veins running cross-wise or at various angles . . . . . . . . . . . . . . . . . . . . . . *58*
58. Ligule conspicuous, prolonged, membranous, 4 mm. long; both glumes cross-wrinkled, both folded; spikes numerous, overlapping, erect or strongly ascending, 1–2 cm. long 52. BECKMANNIA
58. Ligule a band of hairs 0.5–0.6 mm. high; the outside lowest glume not cross-veined; the inner glume and sterile lemma with few weak cross-veins; glumes not folded; inflorescences 2–6, ascending or spreading, not overlapping, 3–8 cm. long . . 67. BRACHIARIA PLATYPHYLLA
57. No cross wrinkles or cross veins or angled veins on the spikelet . . . . . . . . . *59*
59. Spikelets projecting conspicuously from the rachis at different angles; spikelets on short pedicels (stalks) and set at an angle to them, pointing in different directions away from the rachis, not parallel to the rachis, nor appearing flattened, nor crowded against one another . . . . . . . . . . . . . . . . . 69. PANICUM
59. Spikelets conforming to the length of the rachis, not conspicuously standing out from the rachis; spikelets either appearing flattened against or parallel to or along the rachis, or crowded so that they lie upon or overlap each other . . . . . . . . *60*
60. First glume present and rounded, truncate, or obtuse; spikelets glabrous . 69. PANICUM
60. Without the above combination of characters . . . . . . . . . . *61*
61. Spikes usually 10–50, elongated and slender, 0.5–2 mm. wide, at first ascending, at maturity widely spreading; spikelets well separated from one another . . . *62*
62. Plants of dry soils; leaf-blades 1–2 mm. wide; spikelets 4 mm. long, glabrous; spikes mostly alternate, distant and well separated from each other by gaps of as much as 8 cm. . . . . . . . . 51. SCHEDONNARDUS
62. Plants of wet or alluvial soils; leaf-blades mostly 5–10 mm. wide; spikelets 1–2 mm. long, the lemmas hairy; spikes alternate or opposite, most of them spaced rather closely together with a gap often of only 0.5–2 cm. separating them . . . . . . . . . . . . . . . 48. LEPTOCHLOA
61. Spikes or racemes usually 1–10, rather thick, 1.5–5 mm. wide, ascending to somewhat erect; spikelets closely crowded or overlapping . . . . . . . 63
63. Spikes less than 1 cm. long, 1.5 mm. or less wide. 67. BRACHIARIA ERUCAEFORMIS
63. Spikes or racemes 1–15 cm. long, 1.5–7 mm. wide . . . . . . . . *64*
64. Spikelets rounded or blunt at tip . . . . . . . . . 68. PASPALUM
64. Spikelets acutely or narrowly pointed at tip . . . . . . . . . *65*
65. Racemes usually not over five times longer than broad, mainly 1–2 cm. long; ligule absent; leaf-sheaths compressed . 70. ECHINOCHLOA
65. Racemes 10–20 or more times longer than broad, 8–12 cm. long; ligule a membranous whitish projection; leaf-sheaths not compressed . . . . . . . . . . . . . . . 68. PASPALUM
54. Awns present on some part of the spikelets or rachis (in no. 66 awns may be present only on glumes of uppermost staminate florets) . . . . . . . . . . . . . . . . . *66*
66. Spikelets 4–10-flowered . . . . . . . . . . . . . . . . . . . . *67*
67. Lemmas with two minute teeth at tip, 1–3-nerved . . . . . . . . . . 8. DIPLACHNE
67. Lemmas with a simple tapering tip, faintly 5-nerved . . . . . . . . . . 3. FESTUCA
66. Spikelets 1-flowered . . . . . . . . . . . . . . . . . . . . *68*
68. Leaf-blades 1–3 mm. wide; comb-like spikes 1–3 . . . . . . . . . . . . . *69*
69. Plants sending out horizontal branches (stolons) which develop new stems and leaves at the nodes (joints); spikes 0.5–1.5 cm. long . . . . . 57. BUCHLOE (staminate plant)
69. Plants in single clumps lacking horizontal runners; spikes 2–4 cm. long . . 56. BOUTELOUA
68. Leaf-blades mainly 4–15 (rarely 3) mm. wide; spikes 5 to many. . . . . . . . . *70*
70. Leaf-blades ovate to ovate-lanceolate with curved sides, 3–4 times longer than broad 75. ARTHRAXON
70. Leaf-blades linear to lanceolate, with parallel sides, 4–50 times or more longer than broad. *71*

71. Spikes 3–10 mm. wide, the spikelets congested, placed next to one another or conspicuously close and overlapping . . . . . 72

72. Plants with stout scaly rhizomes forming colonies; leaf-blades with sharp edges cutting the flesh, their surfaces rough; spikes conspicuously erect to ascending; spikelets conspicuously 1-sided, all slanting and crowded in the same direction parallel to one another. 53. SPARTINA

72. Without the above combination of characters . . . . . 73

73. Some part of the spikelet prominently nerved; some of the hairs on the spikelet often with a blister-like or swollen base (papillose-hispid) or rather stiff . . . 70. ECHINOCHLOA

73. No part of the spikelet nerved; all the hairs on the spikelets fine and soft . . 66. ERIOCHLOA

71. Spikes 0.5–2 mm. wide (not including the awns); spikelets loose on the rachis, not conspicuously close and overlapping . . . . . 74

74. Plants of dry, rocky, or acid soils; inflorescence about as broad as long, dome-shaped; leaf-blades of the main stem rather stiff, mostly horizontally spreading, 4–10 cm. long, chiefly 4–6 times longer than broad; spikelets 6 mm. long; awn 4.5–9 mm. long. 54. GYMNOPOGON

74. Plants of wet or alluvial soils; inflorescence mostly longer than broad; leaf-blades of main stem strongly ascending, chiefly 7–40 cm. long (except shorter in depauperate individuals), mostly 12–30 times longer than broad; spikelets up to 2.5 mm. long; awn 1 mm. or less long. 48. LEPTOCHLOA

31. Inflorescence of many other types, as stated in the contrasting sections of number 81 (pages 75 and 77), but not as defined in the other (alternative) section of this number 31 . . . . . 75

75. Spikelets either pistillate (female) or staminate (male), the two sexes not in the same floret (use fresh flowering-plants and do not confuse too mature or fruiting plants which have lost their stamens with one-sexed spikelets) . . . . . 76

76. Pistillate (female) and staminate (male) flowers on *different* plants; plants low, 3–40 cm. tall, with creeping rhizomes or stoloniferous branches (sending out horizontal runners which root at the joints) . . . . . 77

77. Inflorescence nearly hidden by and scarcely protruding from the broad bases of the upper leaves; inflorescence of 3–5 spikelets; spikelets bunched together into a dense short head about 8 mm. long; some parts of the spikelets narrowed above into a 3-lobed summit . . . . . 57. BUCHLOE (pistillate plant)

77. Inflorescence standing out and well separated from the leaves; inflorescence of separate and several or many spikelets; all parts of the spikelet with a simple unlobed summit . 78

78. Leaf-blades and culms (stems) pubescent; culms spreading; leaf-blades not stiff; spikelets 1–4 mm. wide . . . . . 7. ERAGROSTIS

78. Leaf-blades and culms (stems) glabrous; culms upright or nearly so; leaf-blades stiff; spikelets 4–5.5 mm. wide . . . . . 10. DISTICHLIS

76. Pistillate (female) and staminate (male) flowers at different locations on the same plant; plants 1–3 meters tall (or lower in poor, undernourished, or young specimens) . . . . 79

79. Common cultivated plant, sometimes growing along roadsides, woodland areas, or vacant ground; staminate (male) and pistillate (female) spikelets in separate inflorescences, the staminate at the summit of the plant, the pistillate lower down in the axils of the leaves . . . . . 81. ZEA

79. Native plants growing in the water of swamps or ponds; staminate and pistillate spikelets at different locations on the same inflorescence . . . . . 80

80. Pistillate (female) spikelets only on the uppermost upright branches of the inflorescence, the staminate (male) on the lower spreading branches; leaf-blades soft, not with sharp edges; scattered in various parts of Missouri . . . 62. ZIZANIA

80. Pistillate (female) spikelets on the same branch as the staminate (male); leaf-blades with sharp edges cutting the flesh easily; swamps and sink-hole ponds of southeastern Missouri only . . . . . 63. ZIZANIOPSIS

75. Spikelets perfect, usually both pistillate (female) and staminate (male) in the same floret (occasionally sterile or imperfect florets present) . . . . . 81

81. Inflorescence of a narrow elongated type, the side branches remaining close to the main axis of the inflorescence . . . . . 82

82. Spreading or erect brown, gray, or white hairs on joints of rachis and various sterile pedicels (stalks) or around base of each floret . . . . . 83

83. A conspicuous stiff, erect, cartilaginous stub (stipule) 3–8 mm. long present at base of leaf-blade; spikelets yellow or chestnut-brown . . . . . 78. SORGHASTRUM

83. Without the above combination of characters . . . . . 76. ANDROPOGON

82. Not as above . . . . . *84*
84. Entire plant velvety to the touch; awn with a small hook at tip; inflorescence with dull rose or pink color . . . . . 33. Holcus
84. Not as above . . . . . *85*
85. Three bristle-like awns (2 of them sometimes mere stubs or short projections) at the summit of lemma . . . . . 47. Aristida
85. Awn, if present, only 1 at summit of lemma or glumes . . . . . *86*
86. Awn 11–20 cm. long; spikelet, excluding awn, 2.8–4 cm. long . . . . . 46. Stipa
86. Awn, if present, at most 4 cm. long. . . . . . *87*
87. Prominent or minute awns (bristle- or hair-like extensions of nerves of lemmas or glumes) present on some part of the spikelets (hairs attached to or covering parts of the spikelets should not be judged as awns) . . . . . *88*
88. Each spikelet consisting of only 1 functional floret, which may be perfect, pistillate (pistil only) or staminate (stamens only) . . . . . *89*
89. Awns conspicuous, projecting noticeably beyond the length of spikelet, 3–25 mm. long . . . . . *90*
90. Spikelet, excluding the awn, 2–4.5 mm. long; spikelets relatively numerous . . . . . 41. Muhlenbergia
90. Spikelet, excluding the awn, 6–11 mm. long; spikelets relatively few . *91*
91. Awns bent or twisted at base; sterile stalked spikelets present . 77. Sorghum
91. Awns straight, not bent or twisted at base; all spikelets fertile . . *92*
92. Spikelets (excluding awn) 10–14mm. long, about 1.5–1.7mm. wide, 7–9 times longer than broad, narrowly linear-lanceolate and not noticeably wider at the middle; leaves mostly spreading, mostly 6–11 times longer than broad . 44. Brachyelytrum
92. Spikelets (excluding awn) 7–9 mm. long, 2–2.5 mm. wide, 2½–3 times longer than broad, elliptic, noticeably wider at the middle; leaves erect to ascending, mostly 15–25 times longer than broad . . . . . 45. Oryzopsis
89. Awns inconspicuous, 0.5–2 mm. long. . . . . . *93*
93. Soft hairs arise from the base of the lemma and surround the lemma. 35. Calamagrostis
93. Not as above . . . . . 38. Cinna
88. Each spikelet consisting of 2 to several florets . . . . . *94*
94. Some or all of the spikelets separated from the main rachis by their own pedicels (stalks). . . . . . *95*
95. Leaves dark green, shining, 10–18 mm. wide; plants spreading from a hard scaly rhizome (underground stem) . . . . . 9. Diarrhena
95. Not as above . . . . . *96*
96. Awn arising from the back of the lemma on at least one of the florets; spikelets 2-flowered . . . . . 32. Arrhenatherum
96. Awn arising at the very tip of the lemma, either between 2 teeth or simply as a prolongation or projection from a toothless lemma; spikelets 3-many-flowered . . . . . *97*
97. Spikelets conspicuously crowded or overlapping, and in groups of dense 1-sided cluster . . . . . 12. Dactylis
97. Spikelets not as above . . . . . *98*
98. Ligule in the form of erect conspicuous hairs; glumes much longer than all the lemmas (excluding the awns), 8.5–13 mm. long. . . . . . 34. Danthonia
98. Ligule a white or brown membranous projection; glumes shorter than the rest of the spikelet, 2–5 mm. long . 3. Festuca
94. Spikelets all attached directly to the main rachis (sessile) . . . . . *99*
99. Spikelets conspicuously crowded or overlapping, and in groups of dense, 1-sided clusters. . . . . . 12. Dactylis
99. Spikelets not as above. . . . . . *100*
100. Spikelets placed with their narrow edge to the rachis; except for the uppermost spikelet, only the glume away from the rachis present (1 glume present) . . . . . 26. Lolium

100. Spikelets placed with their sides to the main rachis or spreading away from the rachis; both glumes present . . . . . . . . . . . . . . . . . . . . . 101
101. Spikelets outwardly spreading or spreading at right angles to the rachis at maturity; spikelets 2–3 at each joint (node) of the rachis. . . . . . . . . 24. Hystrix
101. Spikelets erect, ascending, or pressed against the rachis (axis); spikelets 1 at each joint (node) of the rachis . . . . . . . . . . . . . . . . 18. Agropyron
87. No awns on any part of spikelet (lemma taper into a short stout point [cusp] in *Diarrhena* which is keyed out here and in previous part of number 87) . . . . . . . . . . . . . . 102
102. Each spikelet consisting of only 1 functional floret, which may be perfect, pistillate (pistil only) or staminate (stamens only) . . . . . . . . . . . . . . 103
103. Glumes and lemmas rounded or obtuse at summit. . . . . . . . 69. Panicum
103. Glumes and/or lemmas distinctly short- or long-pointed . . . . . . . . 104
104. Glumes longer than the rest of the spikelet (fertile and sterile lemma and palea); ligule conspicuous, 9–10 mm. long . . . . . . . . 59. Phalaris arundinacea
104. Glumes shorter than to equaling the lemma and palea; ligule minute or at most 2 mm. long . . . . . . . . . . . . . . . 41. Muhlenbergia
102. Each spikelet consisting of 2 to several florets. . . . . . . . . . . . . . 105
105. Spikelets 2–3-flowered, 2.5–4.2 mm. long . . . . . . . . . . 28. Sphenopholis
105. Spikelets 3–many-flowered, 7–22 mm. long . . . . . . . . . . . . . . 106
106. Some or all of the spikelets separated from the main rachis by their own pedicels (stalks) . . . . . . . . . . . . . . . . . . . 107
107. Spikelets conspicuously crowded or overlapping, and in groups of dense 1-sided clusters. . . . . . . . . . . . . . . . 12. Dactylis
107. Spikelets not as above. . . . . . . . . . . . . . . . 108
108. Spikelets some shade of lavender-rose; plants of sunny limestone glades and bluffs; ligule consisting of short hairs; hairs present on nerves at base of lemma . . . . . . . . . . . 16. Tridens
108. Spikelets gray-green or straw-colored; plants of water, swampy ground, or woodland; ligule a short cartilaginous collar or a thin whitish membrane; lemma glabrous . . . . . . . . . . . 109
109. Leaves dark green, shining; plants of woodland; lemmas rigid, firm, sharply pointed . . . . . . . . . . . 9. Diarrhena
109. Leaves gray- or blue-green, dull; plants of water, ponds, or swamps; lemmas soft, thin, rounded or blunt at tip . 5. Glyceria
106. Spikelets all attached directly to the main rachis (sessile) . . . . . . . 110
110. Plants growing in water of ponds, or in swampy ground; lemmas blunt or rounded at tip; glumes and lemmas thin or membranous . . 5. Glyceria
110. Plants growing in other situations, usually in drier ground; lemmas usually acute or pointed at tip; glumes and lemmas mostly firm. . . . . 111
111. Spikelets placed with their narrow edge to the rachis; except for the uppermost spikelet, only the glume away from the rachis present (1 glume present) . . . . . . . . . . . . . . 26. Lolium
111. Spikelets placed with sides to the rachis; both glumes present. 18. Agropyron
81. Inflorescence with some or all of the side branches outwardly spreading, the lower branches often longer and more spreading, the inflorescence as a whole with the shape either of a narrow to broad triangle, pyramid, spire, or Christmas tree, or as broad or broader than long (some variations of the alternative section of this number 81 are placed here, and variations normally placed at this point in the key should be looked for in the other section of this same number) . . . . . . . . 112
112. Entire plant velvety to the touch; awn with a small hook at tip; inflorescence with a dull rose or pink color . . . . . . . . . . . . . . . . . . . . . 33. Holcus
112. Without the above combination of characters. . . . . . . . . . . . . . . 113
113. Each spikelet consisting of only one functional floret, which may be perfect, pistillate (pistil only), or staminate (stamens only) . . . . . . . . . . . . . . 114
114. Spreading or ascending silky hairs present on joints of rachis or various sterile pedicels (stalks); or if not giving that appearance, then sterile pedicels present . . 115
115. One, two, or more unbranched units of an inflorescence protruding from the sides or in the axils of a leafy stem . . . . . . . . . . 76. Andropogon
115. Complete inflorescence at the tip of the stem, obviously branched into many sections . . . . . . . . . . . . . . . . . . . . . 116

116. Ligule or auricle stiff, cartilaginous, 3–8 mm. long; lower main nodes (joints) or axils of inflorescence without hairs; spikelets yellow- or chestnut-brown; some pedicels (stalks) occur without a spikelet at the tip; native grass of prairies, glades, prairie strips along railroads, and open woods . . . . . . . . . . . . 78. SORGHASTRUM

116. Ligule hairy or membranous, 1–3 mm. long; lower main nodes (joints) or axils of inflorescence hairy; spikelets often tinged with rose, or purplish; pedicellate (stalked) spikelet a staminate (male) spikelet at tip of pedicel; introduced forage grass found in fields, bottomland, along roadsides, and waste ground . . . . . . . . 77. SORGHUM

114. No spreading or ascending silky hairs present on joints of rachis nor sterile pedicels present, but a tuft of spreading hairs may arise from the base of a floret . . . . . . . . . . . . . *117*

117. Three bristle-like awns (2 of them sometimes mere stubs or short projections) at summit of lemma . . . . . . . . . . . . . . . . . . . . . . 47. ARISTIDA

117. Awn, if present, only 1 on the lemma or glume . . . . . . . . . . . . . . . . *118*

118. Awns on some part of the spikelet (hairs attached to or covering parts of the spikelet should not be judged here as awns) . . . . . . . . . . . . . . . . . . *119*

119. Spikelet 3–4 mm. wide; prominent depressions and vertical keels (ridges) on the spikelet (on lemma) with dotted roughened surface between the ridges; plant found as an escape from cultivation . . . . . . . . . . . . 60. ORYZA

119. Spikelet 0.5–2.5 mm. wide; no prominent keels or depressions or dotted roughened surface on spikelet; native plant . . . . . . . . . . . . . . . . *120*

120. Spikelet prominently nerved; hairs on spikelet with usually some blister-like or enlarged bases (pustulose); ligule not present . . . . . . 70. ECHINOCHLOA

120. Without the above combination . . . . . . . . . . . . . . . . . *121*

121. Spikelets 7–9 mm. long, relatively few and scattered in the inflorescence 45. ORYZOPSIS

121. Spikelets 2–6 mm. long, relatively many . . . . . . . . . . . . *122*

122. Spikelets quite compressed or flattened on the sides; ligule 3–5 mm. long; stamen 1 . . . . . . . . . . . . . . 38. CINNA

122. Spikelets turgid, filled out more or less equally from all sides, not noticeably flattened; ligule 0.5–3 mm. long . . . . . . . . *123*

123. Awn, if present, at the tip of the lemma; glumes mostly unequal, both glumes shorter than the lemma or usually one of them shorter than the lemma . . . . . 41. MUHLENBERGIA

123. Awn arising from some point on the back of the lemma (dorsally); lemma loosely embracing the grain; glumes mostly equal, either equaling the lemma or the first glume longer than the lemma . . . . . . . . . . . . . *124*

124. Hairs at base of lemma present or absent, but, if present, either few, short and inconspicuous, or attaining to 1/3–1/2 length of spikelet, but not subtended by a tuft of hairs from a hard projection (callus) at base of lemma; 2nd glume 1-nerved; spikelets about 2 mm. long; annuals, 2–7 dm. tall, with delicate roots . 37. AGROSTIS

124. A conspicuous tuft of spreading hairs surrounds 1/2 to nearly length of lemma and arises from a hard projection (callus) at its base; 2nd glume 3-nerved; spikelets 2.2–5.5 mm. long; perennial plants, 5–15 dm. tall, with strong roots . . . . . . 35. CALAMAGROSTIS

118. No awn present on any part of spikelet (hairs attached to or covering parts of the spikelet should not be judged as awns) . . . . . . . . . . . . . . . . *125*

125. Spikelets conspicuously flattened on the sides (laterally); main part of spikelet (lemma) with prominent keels (sharp ridges) on the sides or margins, or the keels on the margins with short bristle-like hairs . . . . . . . . . . . . . *126*

126. Surface of spikelet between the keels with dotted roughened surface; spikelets 7–10 mm. long, rather thick; plant found as an escape from cultivation . 60. ORYZA

126. Surface of spikelet between the keels not dotted roughened; keels on the margins of spikelets with conspicuous short bristle-like hairs; spikelets 3–5 mm. long, thin, and flattened; native plants growing in or near the water or in wet woodland . . . . . . . . . . . . . . . 61. LEERSIA

125. Spikelets turgid, filled out more or less equally from all sides, not noticeably flattened . . . *127*

127. Spikelets not all perfect (both stamens and pistil in a perfect floret), each group of spikelets in 2's or 3's with 1 of them sessile (without stalks) and perfect, the other or other spikelets staminate (male) or sterile and on pedicels (stalks) with a short feathery beard . . . . . . . . . . . . . . . . . . . . . . 77. SORGHUM

127. Spikelets all perfect (with stamens and pistil in the same floret), either all on short or long pedicels (stalks), but not as described above . . . . . . . . . . . . *128*

128. A hard projection (callus) present at base of lemma with a conspicuous beard of hairs surrounding the lower half of the lemma; ligule a fringe of bristles . 36. CALAMOVILFA

128. Without the above combination of characters . . . . . . . . . . . *129*

129. Outermost parts of floret (2 unequal glumes, a sterile lemma resembling a 3rd glume, and a palea) thinner and of a different texture from the thicker, more cartilaginous or leathery inner fertile lemma and palea (do not consider for this section the lemma which closely envelops the grain in *Muhlenbergia*) . . . . . . . . . . . . . . . . . . . . . *130*

130. Ligule a collar-like projection, not hairy; the whole inflorescence with a rosy-purplish color; conspicuous lines or streaks of short silky hairs flattened along the nerves and white margins of the second glume and sterile lemma; pedicels (stalks) of spikelets 3-angled at the summit . 65. LEPTOLOMA

130. Ligule a dense band of hairs or a collar fringed with hairs; the whole inflorescence usually shades of green or brownish-yellow (purplish or reddish in a few species, such as *P. capillare*, *P. virgatum*, and others); no conspicuous lines or streaks of short silky hairs on the florets; pedicels (stalks) of spikelets terete (circular when cut crosswise) . . . 69. PANICUM

129. Outermost parts of floret (glumes) about as thin as inner parts (lemma and palea) . . . . . . . . . . . . . . . . . . . . . . . . *131*

131. Ligule a close fringe or band of bristles; lemma 1-nerved . 42. SPOROBOLUS

131. Ligule a thin membranous collar; lemma 3–5-nerved . . . . . . *132*

132. Lemma usually with hairs at base (no hairs in *M. glabriflora* and *M. asperifolia*); lemma tightly embracing the grain; glumes mostly unequal, both glumes shorter than the lemma or usually one of them shorter than the lemma . . . . . . . 41. MUHLENBERGIA

132. Lemma usually without hairs at its base; lemma loosely embracing the grain; glumes mostly equal, either equaling the lemma or the first glume longer than the lemma . . . 37. AGROSTIS

113. Each spikelet consisting of 2 or more functional florets . . . . . . . . . . . . *133*

133. An awn present on some part of the spikelet (hairs attached to or covering parts of the spikelet should not be judged here as awns) . . . . . . . . . . . . . . *134*

134. Nodes (joints) of culms (stems) conspicuously bearded with spreading white hairs; leaf-sheaths inflated; upper keeled margin of palea densely bearded; the 3 nerves of lemma strongly hairy . . . . . . . . . . . . . . . . . . . . 17. TRIPLASIS

134. Without such a combination of characters . . . . . . . . . . . . . . . *135*

135. Spikelets tiny, at most 2.2 mm. long (excluding the awn), and solitary at the tips of long spreading pedicels (stalks) . . . . . . . . . . . . . . 30. AIRA

135. Spikelets much longer, 5–100 mm. long . . . . . . . . . . . . . . *136*

136. Ligule in the form of erect, stiff, noticeable hairs; glumes longer than the lemmas; spikelets 2–13 . . . . . . . . . . . . . . 34. DANTHONIA

136. Ligule not of stiff hairs, either a shallow band or rounded or triangular membranous projection or with an irregularly cut tip with or without a few short hairs; glumes either shorter than the spikelets or the lower fertile lemma, or only 2–3-flowered; spikelets more numerous . . . . . . . . *137*

137. Spikelets conspicuously crowded or overlapping, and in groups of dense 1-sided clusters, nearly sessile (stalkless); individual spikelets only slightly longer than broad or clusters of spikelets not much longer, if at all longer than broad . . . . . . . . . . . . 12. DACTYLIS

137. Without the above combination, the spikelets obviously scattered on the branches, or, if slightly clustered, on usually noticeable pedicels (stalks); spikelets either much longer than broad, or inflorsecence as a whole longer than broad . . . . . . . . . . . . . . . . *138*

138. Glumes 20 mm. or more long . . . . . . . . . . . . . . . . . . 31. AVENA
138. Glumes 0.5–15 mm. long . . . . . . . . . . . . . . . . . . . . *139*
139. Awn arising at the very tip of the lemma, either between 2 teeth, or simply as a projection or prolongation from a toothless lemma . . . . . . . . . . . *140*
140. No teeth at tip of lemma, the awn extending from the tip itself . . . . 3. FESTUCA
140. Lemma 2-toothed or lobed at tip . . . . . . . . . . . . . . . . . *141*
141. Leaf-sheaths normal, open; branches of inflorescence erect or ascending, mostly unbranched; spikelets very short-stalked, scattered or crowded, more or less parallel the length of branch; awn 1 mm. or less long . . 8. DIPLACHNE
141. Leaf-sheaths unusual, closed; branches of inflorescence mostly spreading or drooping; spikelets long-stalked, not paralleling the length of branch; awn usually 2–50 mm. long . . . . . . . . . . . . . . 2. BROMUS
139. Awn arising from the back of the lemma on at least one of the florets, either just below the tip or at or below the middle . . . . . . . . . . . . . . . . *142*
142. Lemma tapering to a tip without teeth . . . . . . . . 32. ARRHENATHERUM
142. Lemma 2-toothed at the tip . . . . . . . . . . . . . . 29. TRISETUM
133. No awn present on any part of spikelet (hairs attached to or covering parts of the spikelet should not be judged here as awns) . . . . . . . . . . . . . . . . . . . . *143*
143. Spikelets consisting of staminate (male) florets only . . 7. ERAGROSTIS REPTANS (staminate plant)
143. Spikelets with the stamens and pistil in the same floret in most or all the florets present . . *144*
144. Spikelets conspicuously crowded and overlapping, and in bunched groups of dense 1-sided clusters . . . . . . . . . . . . . . . . . . . *145*
145. Leaf-blades 7–8 mm. broad; glumes shorter than lemmas; glumes and lemmas hairy . . . . . . . . . . . . . . . . . . . . . 12. DACTYLIS
145. Leaf-blades mainly 10–20 mm. broad; glumes longer than lemmas; glumes and lemmas glabrous . . . . . . . . . . . . . 59. PHALARIS ARUNDINACEA
144. Spikelets obviously scattered on the branches, or if appearing crowded or overlapping then not in bunched groups of dense 1-sided clusters . . . . . . . . . . *146*
146. Lemmas either minutely 2-toothed or with 3 minute projections (extensions of the 3 nerves above the rounded tip) at summit (do not judge the irregularly edged and sparsely fringed rounded summit of the lemma in *Puccinellia distans* as teeth) . . . . . . . . . . . . . . . . . . . . . . . *147*
147. Three minute projections from the top of the rounded lemma present; spikelets usually reddish-purple, maroon, or dark wine-colored (yellow-green in a rare form); ligule a band of hairs 0.5–1 mm. long with an outer tuft of hairs on each side at the base of leaf-blade. . . . . . . . 16. TRIDENS
147. Two minute triangular teeth or lobes at the tip of the lemma (use magnifying lens to see either conspicuous or very minute teeth); ligule chiefly a membranous border or an irregularly cut appendage, but hairs, if present, an inconspicuous part of ligule . . . . . . . . . . . . . . *148*
148. Spikelets 5–11-flowered. . . . . . . . . . . . . . . *149*
149. At least the upper leaf-sheaths closed at summit . . . . 2. BROMUS
149. Leaf-sheaths all open . . . . . . . . . . . . . 3. FESTUCA
148. Spikelets 2–3-flowered . . . . . . . . . . . . . . . *150*
150. Glumes 20 mm. or more long; leaf-sheaths normal, open at the overlapping margins . . . . . . . . . . . . . . 31. AVENA
150. Glumes 5–8 mm. long; leaf-sheaths unusual, closed at the margins . . . . . . . . . . . . . . . . . 15. MELICA
146. Lemma with a simple pointed or rounded tip or the tip rounded and irregularly edged and fringed . . . . . . . . . . . . . . . . . . . . . *151*
151. Cobwebby hairs at base of lemma (pull off parts above the first 2 glumes to see) . . . . . . . . . . . . . . . . . . . . . . 6. POA
151. No cobwebby hairs present at base of lemma . . . . . . . . . . . *152*
152. Leaf-blades 10–25 mm. wide; spikelets very conspicuously flattened, 1–2 cm. wide . . . . . . . . . . . . . . . . . 11. UNIOLA
152. Leaf-blades 2–12 mm. wide; spikelets, if flattened, at most 8 mm. wide . *153*
153. Leaf-sheaths of at least the upper leaves closed at summit . . . *154*
154. Spikelets 3–12-flowered; all spikelets of about the same size and shape, all perfect . . . . . . . . . . 5. GLYCERIA

103 Limnobium Spongia (American Frogbit)

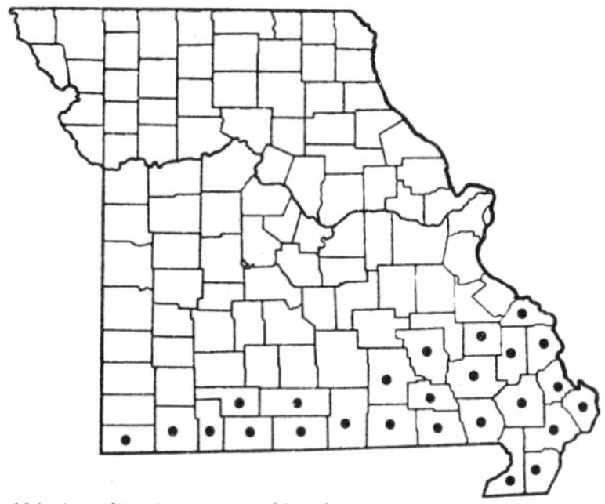

104 Arundinaria gigantea (Cane)

105 Bromus catharticus (Rescue Grass)

154. Spikelets 2–3-flowered; the uppermost floret distinctly smaller and different in size and shape (it is sterile) . . . . . . . . . . . . . . . . . . . . . 15. Melica
153. Leaf-sheaths all open . . . . . . . . . . . . . . . . . . . . . . . *155*
155. At least the second glume broadest in the upper half; spikelets 2–3-flowered. . 28. Sphenopholis
155. The glumes broadest at the base or in the lower half; spikelets usually 2–8-flowered . . . *156*
156. Lemmas faintly or not at all nerved, or with at most three prominent nerves . . . . *157*
157. Lemmas with 3 prominent or weak nerves . . . . . . . . . 7. Eragrostis
157. Lemmas with 5 very faint nerves or with only the main keel showing . . . . *158*
158. Lemmas short- to long-pointed (acute to acuminate) . . . . . . . 3. Festuca
158. Lemmas rounded or flat-topped at summit . . . . . . . . 4. Puccinellia
156. Lemmas prominently 5–9-nerved . . . . . . . . . . . . . . . . . *159*
159. Leaf-sheaths of at least the upper leaves closed; lemma rounded (convex) on the back. . . . . . . . . . . . . . . . . . . . . . 5. Glyceria
159. Leaf-sheaths all open; lemma keeled on the back (a nerved ridge in the center) . *160*
160. Spikelets V-shaped, broadest above the middle; lemmas and palea firm and rigid; stamen 1; lower lemma empty . . . . . . . . 11. Uniola laxa
160. Spikelets oblong- to ovate-lanceolate, broadest at the base or in the lower half; stamens 3; florets all perfect . . . . . . . . . . . . . 6. Poa

## Tribe I. **BAMBUSEAE**

Grasses of this tribe have the culms woody, at least at the base, usually hollow, and are always perennial. The leaf-blade is flat, frequently broad, jointed and petioled with the leaf-sheath. The spikelets are 2- to several-flowered, and the lodicules are conspicuous. The rachilla is articulated above the glumes.

### 1. **Arundinaria** Michx. Cane

**Arundinaria gigantea** (Walt.) Chapm. Map 104

Flowers April–May; in some years apparently not at all flowering.

Southern Missouri northeast to Perry, Madison, and Reynolds counties. Common in the lowlands and along the larger streams of the Ozarks of the southeastern part of the state; moist places on limestone glades, on and at the base of limestone bluffs and along the White River and tributaries in southwestern Missouri. Vast acreage once covered by the cane along White River and tributaries has now been inundated by the creation of the high Table Rock, Bull Shoals, and Norfork dams.

Ranges from Florida and Texas, north to Delaware, Ohio, Indiana, Illinois, and Missouri.

Often forming canebrakes difficult of penetration. The flowering stem has few or no leaves. The young shoots are sometimes cooked and eaten, and the mature grains (at the rare intervals when produced) are also eaten. The young parts especially of the plants furnish forage for livestock. Baskets, mats, pipestems, and fishing poles are made from the culms.

## Tribe II. **FESTUCEAE**

Grasses of this tribe have 2–many perfect flowers, the glumes are usually shorter than the lowermost floret, the awns on the lemma, when present, start from the tip or from between terminal teeth of the lemma, and the rachilla is usually articulated above the glumes.

### 2. **Bromus** L. Brome Grass

Most species of this genus shed relatively little pollen and are regarded as unimportant in hay fever.

a. At least the upper leaf-sheaths glabrous . . . . . *b*
  b. Branches of inflorescence widely spreading to drooping; lemmas more or less hairy; woodland plants . . . . . *c*
    c. Leaf-sheaths closed or nearly closed at summit, overlapping, covering 4 or more of the nodes (joints) of the culm; the sheaths densely hairy at their summit and with conspicuous horizontal spreading flanges and auricles (ear-like outgrowths); nerves of lemma rather prominent below the middle . . . . . 4. B. LATIGLUMIS f. LATIGLUMIS
    c. Leaf-sheaths V-shaped at summit, not overlapping and covering only 1–2 (rarely 3) nodes, the other nodes exposed; the sheaths not densely hairy at their summit and without any horizontal spreading flanges or auricles; nerves of lemma faint or inconspicuous below the middle . . . . . 3. B. PURGANS f. LAEVIVAGINATUS
  b. Branches of inflorescence ascending to nearly erect; lemmas glabrous or nearly so, at least not short-hairy; plants of open places . . . . . *d*
    d. First glume 1-nerved; creeping rhizomes present . . . . . 2. B. INERMIS
    d. First glume 3–5-nerved; no rhizomes present . . . . . *e*
      e. Spikelets strongly flattened with keeled lemmas, 20–38 mm. long; lemmas closely overlapping; rachilla of spikelet not exposed at maturity; lemma without any conspicuous hairy (ciliate) margin . . . . . 1. B. CATHARTICUS
      e. Spikelets rather plump, the lemma more or less rounded on the back, 10–20 mm. long; lemmas with margins folding in and exposing rachilla of spikelet at maturity; fringe of hairs (ciliate) on margin of palea conspicuous and projecting at maturity (seen on very mature spikelet) . . . . . 5. B. SECALINUS
a. Most or all of leaf-sheaths hairy . . . . . *f*
  f. Awns 12–50 mm. long, longer than their lemmas themselves . . . . . *g*
    g. Awns 30–50 mm. long; lemma 25–32 mm. long; 2nd glume 24–30 mm. long . . . 10. B. RIGIDUS
    g. Awns 10–30 mm. long; lemmas 10–22 mm. long; 2nd glume 8–18 mm. long . . . . . *h*
      h. Awns 20–30 mm. long; spikelets 4–5 cm. long (including awns); lemmas 16–22 mm. long; 2nd glume 13–18 mm. long . . . . . 11. B. STERILIS
      h. Awns 10–17 mm. long; spikelets 2–3.5 cm. long (including awns); lemmas 10–12 mm. long; 2nd glume 8–10 mm. long . . . . . 12. B. TECTORUM
  f. Awns none or up to 12 mm. long, shorter than their lemmas themselves . . . . . *i*
    i. Glumes and lemmas rather conspicuously hairy . . . . . *j*
      j. Plants of open places and waste ground; branches of inflorescence erect or strongly ascending, the pedicels (stalks of spikelets) mostly shorter than the spikelets; lemmas with prominent nerves . . . . . 6. B. MOLLIS
      j. Plants of woodland; branches of inflorescence at first erect, eventually widely spreading or drooping, the pedicels either longer than or about equaling length of spikelet; lemmas without prominent nerves . . . . . *k*
        k. Leaf-sheaths closed or nearly closed at summit, overlapping, covering 4 or more of the nodes (joints) of the culm; the sheaths hairy at their inside summit and with conspicuous horizontal spreading flanges and auricles (ear-like outgrowths); nerves of lemma rather prominent below the middle . . . . . 4. B. LATIGLUMIS f. INCANUS
        k. Leaf-sheaths V-shaped at summit, not overlapping and covering only 1–2 (rarely 3) nodes, the other nodes exposed; the sheaths not hairy at their inside summit and without

---

Plate no. 16. **1.** Arundinaria gigantea; *a.* Flowering shoot, × 3/8; *b.* Summit of culm sheath, outer and inner face, showing auricles and ligule, × 1 1/2; *c.* Two views of floret; × 1 1/2. **2.** Bromus catharticus × 3/4. **3.** Bromus inermis; *a.* Plant, × 3/8; *b.* Spikelet, × 1 7/8. **4.** Bromus latiglumis; Base of blades, × 3/4. **5.** Bromus purgans; Floret, × 3 3/4.

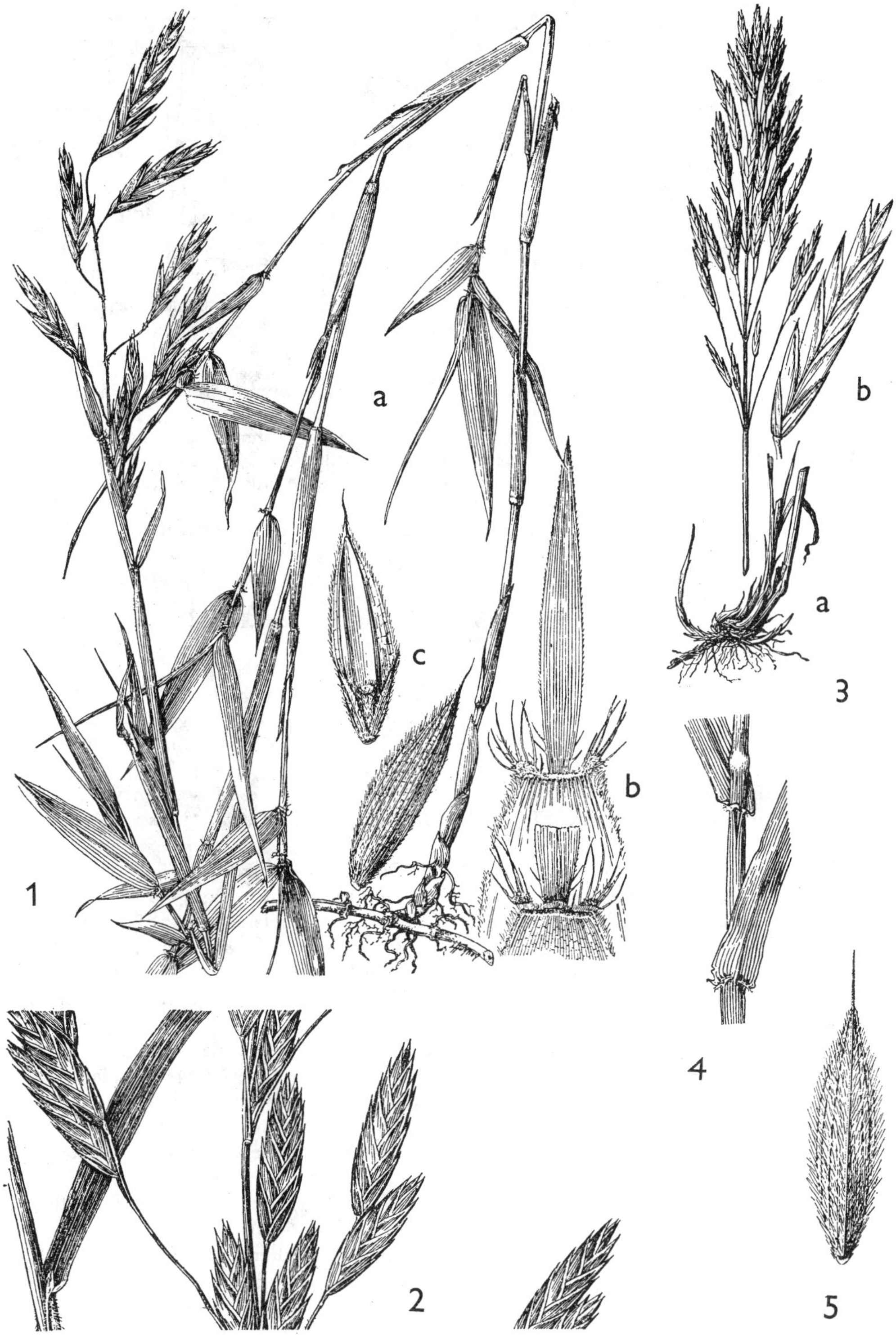


PLATE NO. 16

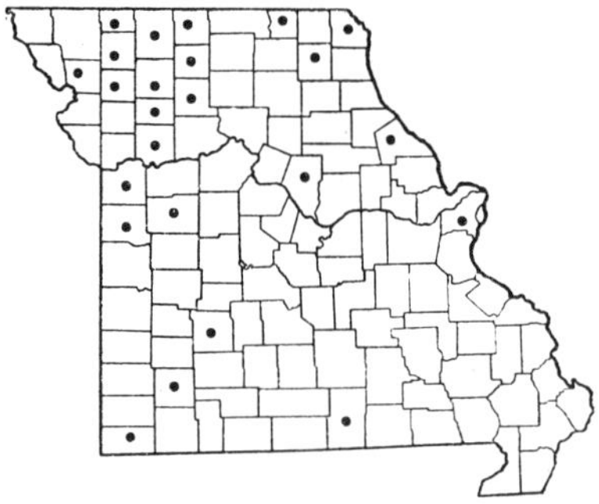

106 Bromus inermis (Smooth Brome)

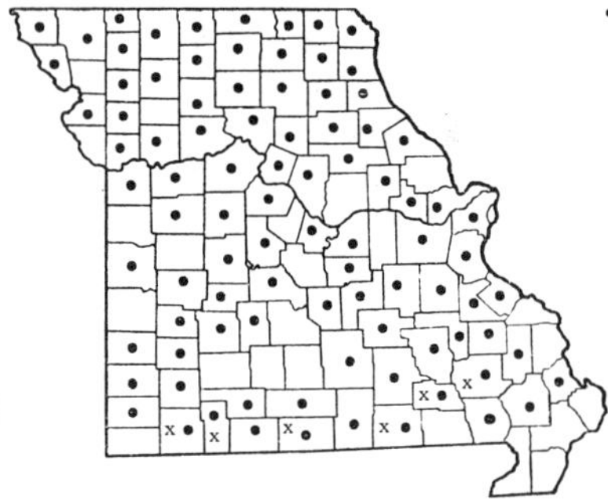

107 ● Bromus purgans f. purgans
107 x Bromus purgans f. laevivaginatus

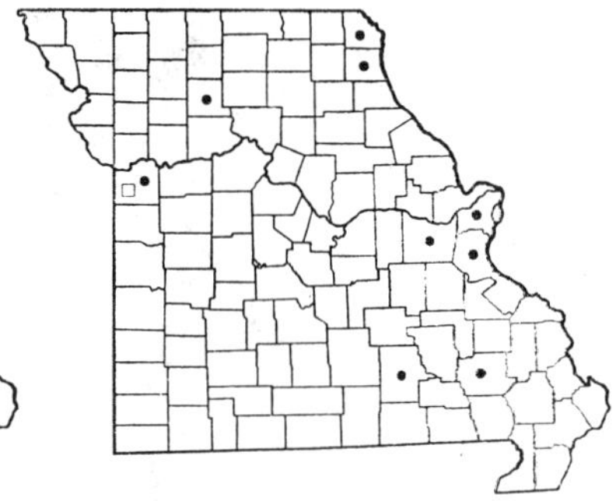

108 ● Bromus latiglumis f. latiglumis
108 □ Bromus latiglumis f. incanus

any horizontal spreading flanges or auricles; nerves of lemma faint or inconspicuous below the middle . . . . . . . . . . . . . . . . . . . 3. B. PURGANS f. PURGANS

i. Glumes or lemmas glabrous or nearly so, the hairs if present very short and inconspicuous . . . *l*

l. Branches of the inflorescence erect or ascending, more or less straight; use magnifying lens to note that hairs on leaf-sheaths are straight and rather evenly spaced and not dense enough to present a whitish appearance. . . . . . . . . . . . . 7. B. RACEMOSUS

l. Branches of the inflorescence widely spreading to drooping, rather flexuous (wavy or curved alternately in opposite directions); use magnifying lens to note that hairs on leaf-sheaths are backward-pointing, curved or tangled towards tip and dense, giving a matted whitish appearance of soft hairs . . . . . . . . . . . . . . . . . . . . . *m*

m. Awn strongly slanting or arching or spreading; anthers 1 mm. long . . . 8. B. SQUARROSUS

m. Awn straight to slightly spreading; anthers 1–4 mm. long . . . . . . . . 9. B. JAPONICUS

1. **Bromus catharticus** Vahl Rescue Grass Map 105

Flowers late April–July.

Known from Jackson County, where introduced at Sheffield (May 28, 1904, *Bush 1957*), St. Louis, and Jasper (open waste ground, Stockyard Switch, Joplin, April 30, 1957, *Palmer 64896*) counties.

Native to South America. Cultivated in the southern United States as an early winter and spring pasture, escaped and naturalized north to New York, Missouri, and Kansas.

A cathartic action is ascribed to the plant.

2. **Bromus inermis** Leyss Smooth Brome, Hungarian Brome Map 106

Flowers late May–July.

Introduced along roadsides, in fields, and waste ground, where common in northern, central, and western Missouri; mostly absent at present throughout the Ozark section.

A native of Europe, which ranges in North America from Newfoundland to British Columbia, south to Virginia, Ohio, Missouri, Kansas, and New Mexico.

This grass is successfully planted in this country for its value in pasturage and hay-production. Each year it is becoming increasingly common as a roadside waif.

By Wagnon the Missouri material is treated as *B. inermis* subsp. *inermis* (Leyss.) Wagnon.

3. **Bromus purgans** L. Map 107

Flowers late May–early July.

A native woodland species of rich or rocky ground, grassy thickets along streams, and wooded bluffs and talus, especially limestone. Typical *B. purgans* f. *purgans* is found throughout Missouri. *B. purgans* f. *laevivaginatus* Wieg., with glabrous leaf-sheaths and blades, occurs in several counties in southern Missouri.

Ranges from Quebec, New England, and Alberta, south to Florida, Louisiana, Texas, and Arizona.

This and the following species (*B. latiglumis*) are the only native species found in the state, the others all being introduced from Europe or South America.

Purgative qualities are ascribed to this species.

According to Wagnon (Rh. 52: 211–15. 1950) this species should receive the name *B. pubescens* Muhl. ex Willd.

---

Plate no. 17. **1.** Bromus japonicus, × 3/4. **2.** Bromus secalinus; *a.* Plant, × 3/8; *b.* Spikelet, × 3 3/4; *c.* Floret, × 3 3/4. **3.** Bromus racemosus; *a.* Spikelet, × 3 3/4; Lemma, × 3 3/4. **4.** Bromus mollis, × 3/4. **5.** Bromus squarrosus, × 3/4.

PLATE NO. 17

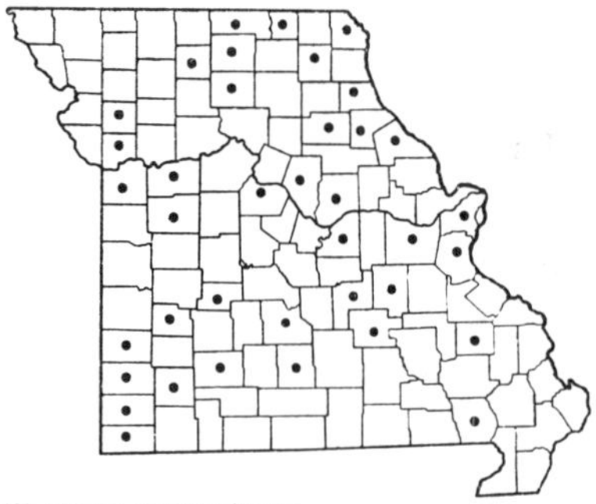

109 Bromus secalinus (Chess)

110 Bromus mollis (Soft Chess)

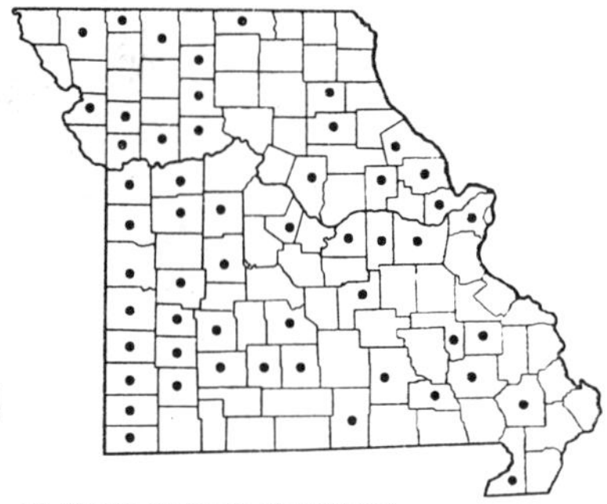

111 Bromus racemosus (Hairy Chess)

4. **Bromus latiglumis** (Shear) Hitchc. Map 108
Flowers July and August.

Rich wooded slopes and bluffs, usually in limestone areas, where rarely collected in parts of eastern, northern, and central Missouri where represented as *B. latiglumis* f. *latiglumis*. The form with pilose leaf-sheaths, *B. latiglumis* f. *incanus* (Shear) Fern., known by a single collection from Jackson County (uncommon, Little Blue, Aug. 9, 1896, *Bush 396*), has been identified by Wagnon as typical *B. latiglumis* f. *latiglumis*.

According to Wagnon, material identified as *B. latiglumis* = *B. purgans* L., whereas material previously called *B. purgans* = *B. pubescens* Muhl. ex Willd. (Rh. 52: 211–15. 1950).

5. **Bromus secalinus** L. Chess Map 109
Flowers principally May–early July.

Introduced in fields, limestone glades, prairies, roadsides, railroads, and waste places throughout Missouri.

Native to Europe and spread throughout much of the United States north into Quebec and British Columbia.

6. **Bromus mollis** L. Soft Chess Map 110
Flowers May and June principally.

Occurs along railroads and cultivated ground. Known only from St. Louis and Jackson (*Bush*, June 5, 1891) counties.

Native of Europe, and introduced in North America from Newfoundland to British Columbia, south to North Carolina, Indiana, Illinois, Missouri, Kansas, Utah, and California; most abundant in the Pacific states.

This grass is planted for its spring forage value. The perennial *B. inermis* is regarded as superior to this annual brome from the standpoint of a cover and forage crop.

7. **Bromus racemosus** L. Hairy Chess Map 111
*Bromus commutatus* Schrad. [G, P & S]
Flowers principally late May–July.

Occurs in fields, waste ground, along roadsides, and railroads throughout Missouri.

Native of Europe; introduced into the United States from the Atlantic to the Pacific Coast and north into Nova Scotia and British Columbia.

Material identified previously as *B. racemosus* L. and *B. commutatus* Schrad. is placed together here, as the two are scarcely separable as species. The present treatment agrees with that of Gleason (New Illustrated Flora, p. 105). He maintains that the form having longer lemmas (9–11 mm. long) and larger panicles with more numerous spikelets, known as *B. commutatus*, is more abundant in North America than the one (*B. racemosus*) having smaller lemmas (about 7 mm.) and smaller panicles with fewer spikelets.

8. **Bromus squarrosus** L. Chess Map 112
Flowers June–July.

In Missouri known only from St. Louis County (St. Louis, along the Terminal R. R. Assoc. track between Rutger Street and the control at Carroll St., June 17, 1956, *Muehlenbach 954*).

Native of Europe; introduced in the United States locally from Connecticut to Michigan and North Dakota to Missouri.

9. **Bromus japonicus** Thunb. Japanese Chess, Japanese Brome Map 113
*Bromus arvensis* [of G, P & S], not L.
Flowers late May–July.

Occurs in fields, waste ground, and along roadsides throughout Missouri.

---

Plate no. 18. **1.** Bromus sterilis, $\times {}^{3}/_{4}$. **2.** Bromus tectorum; *a.* Plant, $\times {}^{3}/_{8}$; *b.* Spikelet, $\times 3^{3}/_{4}$; *c.* Floret, $\times 3^{3}/_{4}$. **3.** Bromus rigidus, $\times {}^{3}/_{4}$.

PLATE NO. 18

112 Bromus squarrosus (Chess)

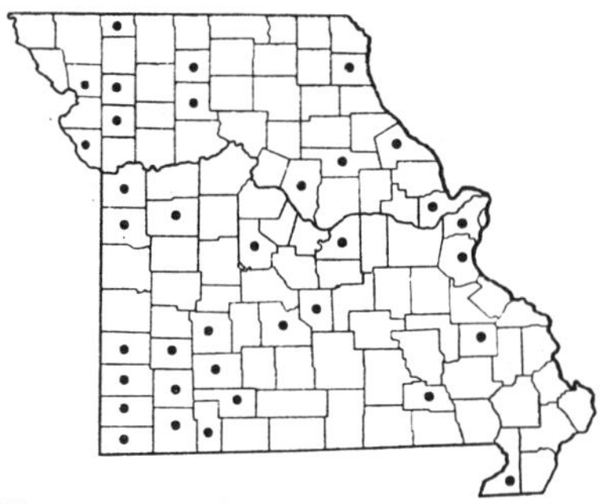

113 Bromus japonicus (Japanese Brome)

114 Bromus rigidus (Ribgut Grass)

Introduced from Europe; naturalized throughout most of the United States from Vermont to Washington, south to Florida, Texas, and California.

This is a palatable grass, which often appears in overgrazed range land, and early in the spring offers temporary pasture.

All the Missouri material, previously identified as *B. arvensis*, is here treated as *B. japonicus*. *Bromus arvensis* L. is known only from a few states in the United States, and may be recognized principally by having the lemma as long as the palea.

*Bromus japonicus*, var. *porrectus* Hack., with straighter awns, has been collected in Pike County (along road in alfalfa field, June 24, 1947, *Etter 337*). However, it is not maintained in the present treatment as distinct, and seems better regarded as synonymous with typical *B. japonicus*.

10. **Bromus rigidus** Roth. Ribgut Grass Map 114
Flowers May–July.

Found in waste ground and along railroads, where it has been found only in St. Louis (Bremen Avenue freight yard of the Terminal Railroad, St. Louis, May 30, 1955, *Muehlenbach 614*) and Jackson (Courtney, July 1, 1904, *Bush 2067*) counties.

Native of Europe; introduced into the United States, where very common in the Pacific states, but in the eastern half known locally from Massachusetts to Maryland, and Mississippi.

In the Pacific states this grass is believed to be a contributing factor in causing hay fever. The plant owes its common name from the effect produced by the long and strong awns of the lemmas, which can give bad lacerating wounds to sheep and other grazing animals.

11. **Bromus sterilis** L. Map 115
Flowers May and June.

Open or rocky ground, frequently along streams, where known only in southwestern Missouri from Newton (open ground along brook, George Washington Carver National Monument, near Diamond, May 19, 1956, *Palmer 62347*, and May 31, 1957, *Palmer 65392*), Polk (open ground between roadway and river, near Burns, May 23, 1957, *Palmer 65293*), and Wright (open rocky valley of Dove Creek, T30N, R13W, SE ¼ sect. 15, 3/4 mi. ESE of Rayborn, May 6, 1957, *Steyermark 84097*) counties.

Native of Europe; introduced into the United States from Massachusetts and Ontario south to North Carolina, Alabama, Arkansas, Colorado, and California.

Like *B. rigidus*, the awns of this species also penetrate and cause damage to the soft facial tissues of sheep and other grazing animals.

12. **Bromus tectorum** L. var. **tectorum**
Downy Chess Map 116
Flowers April–early June.

Common in fields, waste ground, along roadsides, and railroads, throughout Missouri.

Native of Europe; introduced throughout most of the United States.

This is the most common of the introduced species in Missouri and elsewhere, and comes into flower earlier than any of the other species. The inflorescences with their drooping green or reddish-purple or rose-tinted spikelets produce a striking color effect in early spring along the secondary roads and highways, where they occur usually in dense stands.

The long awns of this species easily penetrate the mouth parts or eyes of grazing sheep, cattle, and horses, causing mechanical injuries, and are especially troublesome among sheep in the Rocky Mountain states.

The form with glabrous spikelets, *B. tectorum* var. *glabratus* Spenner, should eventually be found in the state.

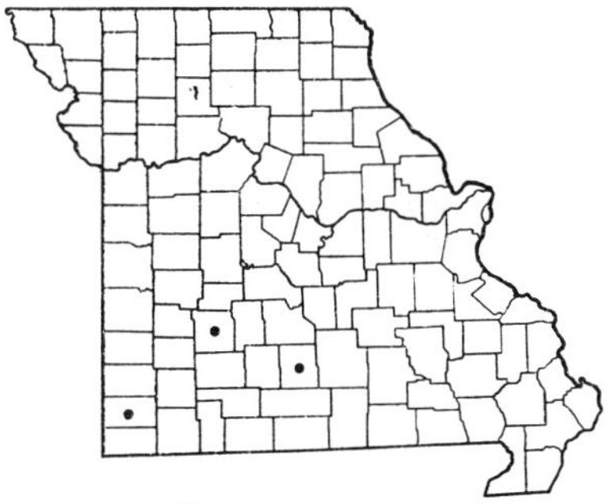

115 Bromus sterilis

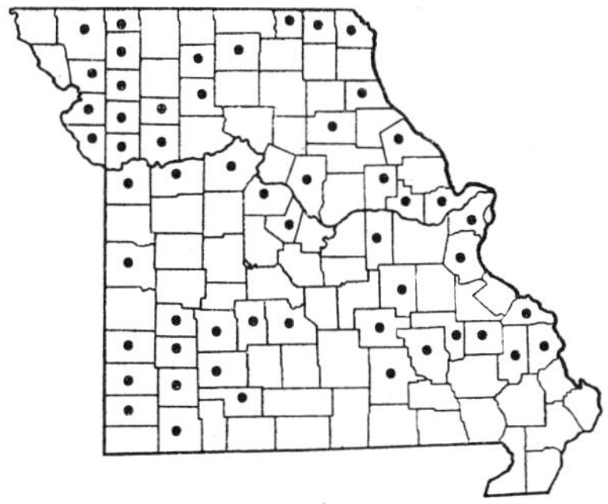

116 Bromus tectorum (Downy Chess)

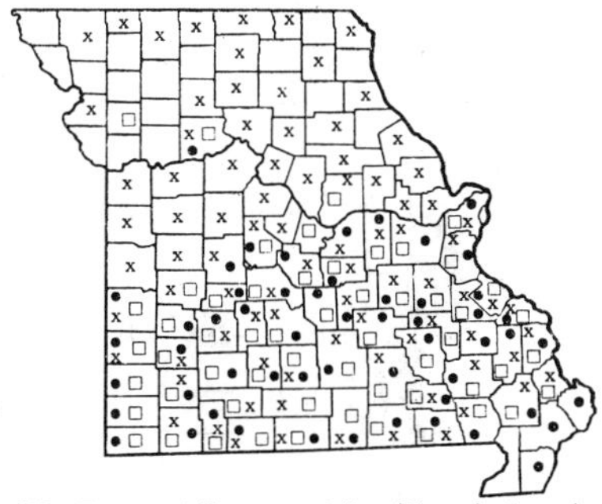

117 • Festuca octoflora var. octoflora (Six-weeks Fescue)
117 □ Festuca octoflora var. glauca
117 x Festuca octoflora var. tenella

*Excluded species*

**Bromus breviaristatus** Buckley.

This species was cited from Jackson County in Palmer & Steyermark's Annotated Catalogue. It was based upon a specimen collected by Bush (*4961* from along railroad, Sheffield), but this specimen is now identified as *B. inermis* Leyss.

## 3. **Festuca** L. Fescue

Several species of this genus are known to cause hay fever.

a. Leaf-blades flat, 4 mm. or more wide . . . . . . *b*

b. Leaf-blades with ear-like extensions (auricles) at base by junction with sheaths; spikelets 8–15 mm. long; lemmas 5.5–10 mm. long; branches of inflorescence (panicle) bearing spikelets nearly to the base . . . . . . 4. F. ELATIOR

b. Leaf-blades without ear-like extensions (auricles) at base by junction with sheaths; spikelets 4–8 mm. long; lemmas 3–5.5 mm. long; branches of inflorescence bearing spikelets above the middle . . . . . . *c*

c. Spikelets lanceolate to elliptic, 1–5 flowered, (usually 2–3) scattered and not overlapping at least along the lowest branches of inflorescence; lemmas and palea acute or nearly acute . 5. F. OBTUSA

c. Spikelets ovate, 3–6-flowered, rather densely clustered and overlapping at least along the lowest branches of inflorescence; lemma and palea obtuse (blunt at tip) . . . . 6. F. PARADOXA

a. Leaf-blades hair-like or thread-like, or very narrow, usually with the margins rolled in, 0.5–3 mm. wide, or if flattened, then at most 3 mm. wide . . . . . . *d*

d. Perennial plants with numerous crowded leaves at the base, and with the remains of the old leaf-sheaths at the base; culms (stems) usually several from a dense leafy clump; anthers 3 in each floret . . . . . . *e*

e. Leaf-sheaths at base of plant thin, reddish or purple, breaking up into shreds with the strong veins persisting as fibers; anthers 1/2–2/3 lengths of palea . . . . . . 7. F. RUBRA

e. Leaf-sheaths at base of plant firm, whitish or grayish, persistent and not breaking up into shreds; anthers 1/4–1/2 length of palea . . . . . . 8. F. OVINA

d. Annual plants with but few leaves or leaf-sheaths at the slender base; culms (stems) solitary or very few in a small tuft; anther 1 (rarely 3) in each floret . . . . . . *f*

f. Awn of lemma 9–14 mm. long; 1st glume 1–1.9 mm. long, 1/4–1/3 the length of the 2nd glume; 2nd glume 4–5 mm. long . . . . . . 3. F. MYUROS

f. Awn of lemma 1–9 mm. long; 1st glume 2 mm. long, 2/3–3/4 the length of the 2nd glume; 2nd glume 3.5 mm. long . . . . . . *g*

g. Awn of lemma 6.5–9 mm. long; lemma sparsely short-hairy . . . . . . 2. F. SCIUREA

g. Awn of lemma 1–7 mm. long (mostly 1–5 mm.); lemma glabrous to scabrous (roughened with short stiffish hairs) . . . . . . 1. F. OCTOFLORA

1. **Festuca octoflora** Walt. Six-weeks Fescue Map 117

Flowers from mid-April to mid-June.

The spikelets are green to purplish-red, often turning to the redder hues upon ripening. After the florets mature and fall from the spikelet, the glumes

118 Festuca sciurea

119 Festuca myuros

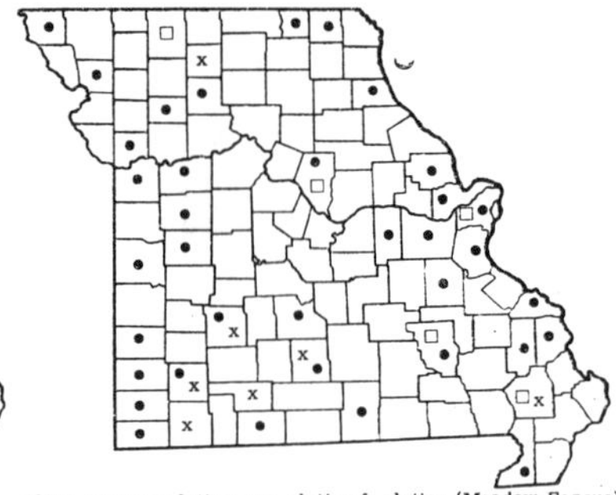

120 • Festuca elatior var. elatior f. elatior (Meadow Fescue)
120 x Festuca elatior var. elatior f. aristata
120 □ Festuca elatior var. arundinacea

and hair like culms, or simply the culms themselves, persist in the dried state during summer and fall as delicate pale to tawny-brown erect stems. The common name of this species refers to the relatively short span of life, seed formation taking place in about six weeks from time of germination of seed.

The Missouri material falls into three varieties:

a. Longer awns of lemmas 3.5–7 mm. long . . 1a. F. OCTOFLORA var. OCTOFLORA
a. Longer awns of lemmas up to 3 mm. long . *b*
 b. Awns 1–3 mm. long; 1st glume 2.3–4 mm. long . . . 1b. F. OCTOFLORA var. TENELLA
 b. Awns at most 2 mm. long or none; 1st glume 1.5–3 mm. long . 1c. F. OCTOFLORA var. GLAUCA

1a. **Festuca octoflora** var. **octoflora** Map 117

*Festuca octoflora* [of P & S]

*Festuca octoflora* var. *aristulata* Torr. ex Dewey [BB]

*Vulpia octoflora* (Walt.) Rydb. [G]

Frequently on sandstone, granite, or chert glades in the Ozark section; also on acid, sterile, and thin soils of dry open woods and slopes, prairies, fields, and leached waste ground. Mostly confined to the Ozark section, north locally to Carroll County.

Ranges from Florida and Texas, north to New Jersey, Illinois, Missouri, and Oklahoma.

1b. **Festuca octoflora** var. **tenella** (Willd.) Fern. Map 117

*Vulpia octoflora* var. *tenella* (Willd.) Fern. [G]

Same habitats as var. *octoflora*. This is the commonest variety in Missouri, occurring in most counties of the state. Of expected occurrence eventually in northwestern counties, where as yet not recorded.

Ranges from Georgia, Texas, Colorado, and California, north to Maine, Quebec, and British Columbia.

1c. **Festuca octoflora** var. **glauca** (Nutt.) Fern. Map 117

*Vulpia octoflora* var. *glauca* (Nutt.) Fern. [G]

Same habitats as the preceding varieties.

Throughout the Ozark region and northwest locally in Carroll and Clinton counties.

Ranges from Florida and New Mexico, north to Missouri, Wisconsin, South Dakota, and Wyoming.

The three varieties of *F. octoflora* intergrade in length of awns and measurements of glumes and lemmas, and are maintained in the present treatment for convenience in naming extremes of variation.

The varieties of *F. octoflora* often have inflorescences which show more branching with shorter lateral branches than shown by either *F. sciurea* or *F. myuros*. The last two species usually have a longer and more continuous inflorescence.

2. **Festuca sciurea** Nutt. Map 118

*Vulpia Elliotea* (Raf.) Fern. [G]

Flowers April–June.

Mostly in sandy soils and fields in southeastern Missouri and rocky open ground westward in Douglas and Jasper counties.

Ranges from Florida to Texas, north to New Jersey, Maryland, Missouri, and Oklahoma.

3. **Festuca myuros** L. Map 119

*Vulpia myuros* (L.) K. C. Gmel. [G]

Flowers mid-April–June, the spikelets maturing as late as July.

Waste ground and along railroads, where known from St. Louis, Boone (University of Missouri South Farm, 5 mi. south of Columbia, July 10, 1947, *R. B. Livingston*), and McDonald (open ground around artificial pond near Pineville, June 5, 1957, *Palmer 65531*,

Plate no. 19. **1.** Festuca octoflora; *a.* Plant, × 3/8; *b.* Spikelet, × 3 3/4. **2.** Festuca myuros; Soikelet, × 3 3/4. **3.** Festuca sciurea; *a.* Panicle, × 3/8; *b.* Spikelet, × 3 3/4. **4.** Festuca elatior; *a.* Plant, × 3/8; *b.* Spikelet, × 3 3/4; *c.* Floret, × 3 3/4.

PLATE NO. 19

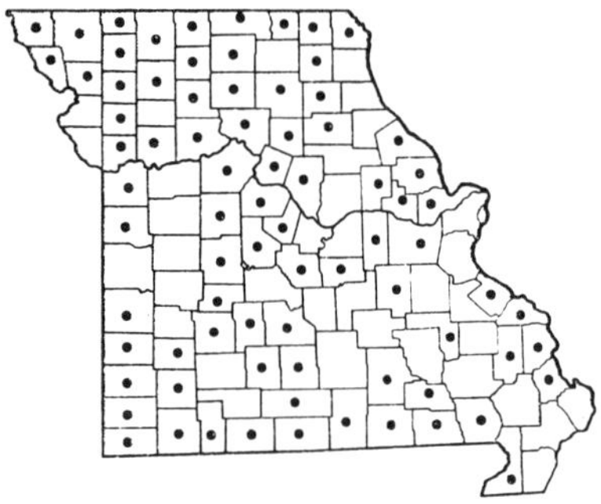

121 Festuca obtusa (Nodding Fescue)

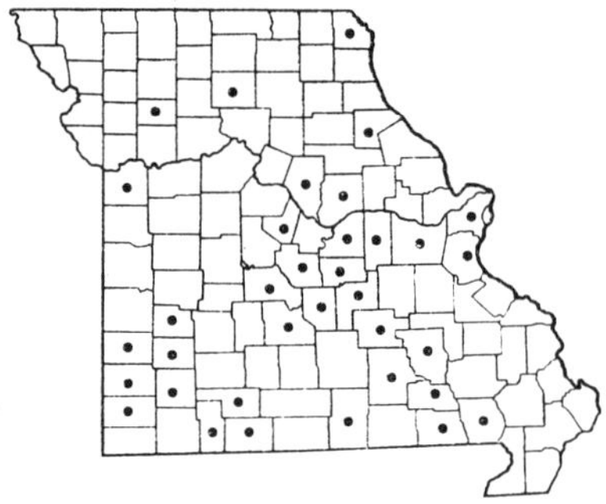

122 Festuca paradoxa

123 • Festuca rubra var. rubra (Red Fescue)
123 x Festuca rubra var. commutata (Chewings Fescue)
123 □ Festuca rubra var. arenaria

and June 29, 1957, *Palmer 65836*) counties.

This grass is a native of Europe, and has been introduced in the United States, where it ranges from Maine to Wisconsin, south to Florida, Louisiana, and Texas, also in the Pacific states.

4. **Festuca elatior** L. Meadow Fescue Map 120

Sometimes called English Bluegrass.

Flowers mainly late April through July, and continues to flower in September and October.

Missouri material may be classified into the following varieties and forms:

a. Lemmas short-awned . . . 4b. F. ELATIOR var. ELATIOR f. ARISTATA
a. Lemmas not awned. . . . . . . . . *b*
b. Spikelets 6–9-flowered; lemmas 5–7 mm. long; lowest branch of inflorescence less than 1/3 length of inflorescence; culms 2–3 mm. thick at base . . . 4a. F. ELATIOR var. ELATIOR f. ELATIOR
b. Spikelets 4–5-flowered; lemmas 7–10 mm. long; lowest branch of inflorescence 1/3–1/2 length of inflorescence; culms 3–5 mm. thick at base . . . . . 4c. F. ELATIOR var. ARUNDINACEA

4a. **Festuca elatior** var. **elatior** f. **elatior** Map 120

*Festuca elatior* [G, P & S, BB]

*Festuca pratensis* Huds.

*Festuca elatior* var. *pratensis* (Huds.) Gray [BB]

Meadows, fields, waste ground, along railroad tracks, and open woods, where found throughout Missouri.

Introduced from Europe into North America from Newfoundland to British Columbia, south to Alabama, Louisiana, Kansas, and California.

Material from St. Louis County labeled *F. pratensis* is considered in the present treatment as synonymous with *F. elatior* var. *elatior*.

This and the other varieties rate high in forage value, for which reason the grass is cultivated.

4b. **Festuca elatior** var. **elatior** f. **aristata** Holmb. Map 120

Known in Missouri from Grundy (*Steyermark 81690*), Stoddard (*Steyermark 83935*), Polk (*Palmer 65295*), Lawrence (*Palmer 51993*), Wright, Christian (*Palmer 65009*), and Barry (*Palmer 65540*) counties.

4c. **Festuca elatior** var. **arundinacea** (Schreb.) Wimmer Alta Fescue, Reed Fescue Map 120

*Festuca arundinacea* Schreb. [Hitchcock]

This variety has somewhat taller and more robust culms (3–5 mm. thick at base) than *F. elatior* var. *elatior*. It is known from Harrison, Boone, and St. Louis (near Rutger Street Warehouse, St. Louis, May 14, 1955, *Muehlenbach 576*) counties.

This is maintained as a distinct species in the United States National Herbarium.

There is considerable intergradation and overlapping of spikelet size and number of florets between *F. elatior* var. *elatior* and var. *arundinacea*.

5. **Festuca obtusa** Biehler Nodding Fescue Map 121

Flowers late April to June, maturing into July and August.

Frequents woodland, especially rich or rocky wooded slopes, bluffs, alluvial ground along streams, and waste ground. Occurs throughout Missouri.

Ranges from Nova Scotia to Manitoba, south to Florida, Mississippi, and Texas.

At first the sheaths at the base are purplish or rose-purple. The dark brown nodes of the culm are quite characteristic, although shared by a number of other grasses. The flowering branches in the beginning are erect, but later become widely spreading.

6. **Festuca paradoxa** Desv. Map 122

*Festuca Shortii* Kunth [P & S]

Flowers May–July.

Frequents dry upland or river bottom prairies, wet meadows, dry rocky slopes, margins of glades and ledges of bluffs as well as moist open ground and moist woods. Rather common in the Ozark section, north locally to Clark, Linn, Caldwell, and Jackson counties.

Ranges from Pennsylvania to Georgia, west to Minnesota, Iowa, Oklahoma, and Texas.

This species frequents drier and rockier sites on the whole than *F. obtusa*. The spikelets are usually 4–5-flowered (occasionally 3- or 6-), whereas in *F. obtusa* they are usually 3-flowered (occasionally 1–2- or 4–5-flowered).

7. **Festuca rubra** L. Red Fescue Map 123

Flowers late April–June.

This species has a number of forms and varieties used in grass mixtures for shady dry lawns. One of these forms is known as Chewings Fescue and can tolerate much shade, drouth, and poor or sterile soils.

The following variations have been found in Missouri, but others are to be expected as escapes from introduced lawn seed mixtures:

a. Lemmas hairy; spikelets 10–15 mm. long . . 7c. F. RUBRA var. ARENARIA
a. Lemma glabrous . . . . . . . . . *b*
 b. Basal offshoots mainly spreading or reclining with the summit ascending (decumbent), forming loose mats . 7a. F. RUBRA var. RUBRA
 b. Basal offshoots all upright with elongated leaves 1.5–6 dm. long and erect stems (culms) 2.5–9 dm. tall . . . 7b. F. RUBRA var. COMMUTATA

7a. **Festuca rubra** var. **rubra** Red Fescue Map 123

Known from alluvial shaded waste ground and low woods in St. Louis and Pulaski (low woods near cabin along Big Piney River, near Hooker, May 9, 1936, *Steyermark 10389*) counties.

Native to Europe, Asia, North Africa, and in North America native from Greenland and Labrador to Alaska, south to Virginia, Tennessee, Missouri, and California.

7b. **Festuca rubra** var. **commutata** Gaudin Chewings Fescue Map 123

Occasionally escaped from lawn seed mixtures. Known from Boone (escaped, forming dense sod in shaded spots, Columbia, May 28, 1956, *Kucera*) and St. Louis (Terperance Street freight yard of the Missouri Pacific Railroad, south of Rutger Street, St. Louis, June 7, 1958, *Muehlenbach 1415*; also June 28, 1958, *Muehlenbach 1430*) counties.

Ranges from Newfoundland to Michigan and North Carolina.

7c. **Festuca rubra** var. **arenaria** (Osbeck) Fries Map 123

*Festuca rubra* var. *lanuginosa* Mert. & Koch [Hitchcock]

Occasionally escaped from cultivation, and known only from St. Louis County (on grassy area on Jackson Avenue, opposite north entrance to University City Senior High School, University City, May 6, 1936, *Steyermark 10339*).

Ranges from arctic regions south to Newfoundland, Nova Scotia, Quebec, Saskatchewan, Colorado, and Oregon.

The plants form dense, bluish-green clumps, the stems spreading outwards rather than erect as in var. *commutata* and var. *rubra*.

8. **Festuca ovina** L. Sheep Fescue Map 124

Flowers May–July.

Missouri material is represented by two variations:

Spikelets 5–7 mm. long; 1st glume 2–3.8 mm. long; 2nd glume 2.5–4.6 mm. long; lemmas 3–5.6 mm. long . . . . . 8a. F. OVINA var. OVINA
Spikelets 7–10 mm. long; 1st glume 2.4–4 mm. long; 2nd glume 3.5–5.2 mm. long; lemmas 4.3–6 mm. long . . . 8b. F. OVINA var. DURIUSCULA

8a. **Festuca ovina** var. **ovina** Map 124

A specimen from Boone County (State Farm, Columbia, June 3, 1903, *B. M. Duggar*) with only detached flowering culms with spikelets and no leaves present, is the basis for the Missouri record. It was doubtfully determined at the United States National Herbarium as *F. ovina*.

Native to Europe and Asia; introduced in North America from Quebec to New Jersey west to Michigan and Missouri.

8b. **Festuca ovina** var. **duriuscula** (L.) Koch Map 124

This is rarely found in Missouri, where it has been found in St. Louis and Jackson counties.

Native to Europe; introduced in North America from Newfoundland to Minnesota, south to Pennsylvania and Missouri.

In this species the young shoots originate inside the

124 • Festuca ovina var. ovina (Sheep Fescue)
124 □ Festuca ovina var. duriuscula

125 Puccinellia distans (Alkali Grass)

126 Puccinellia airoides (Alkali Grass)

sheaths parallel to the main axis, whereas in *F. rubra* var. *rubra* and other varieties the young shoots grow out from the main axis and break through the sheath at its base.

Plants generally have a gray-green color. Since they form dense bumpy tufts, they are not a desirable lawn grass, as is Chewings Fescue (*F. rubra* var. *commutata*).

## 4. **Puccinellia** Parl. Alkali Grass

Lemmas broadly rounded or truncate at the summit, broadest near the tip, the margin (under a lens) with a minute fringe of teeth; lower branches of inflorescence usually downward spreading; ligule 0.5–1.5 mm. long . . . . . . . . . . . . . . . . . . . . . . 1. P. DISTANS

Lemmas somewhat pointed or tapering at the tip, the margin (under a lens) with minute irregular teeth; lower branches of inflorescence ascending or spreading horizontally, but not downward; ligule 1.5–2 mm. long . . . . . . . . . . . . . . . . . . . . . . . 2. P. AIROIDES

1. **Puccinellia distans** (L.) Parl. Map 125

Flowers May–June.

Known only from St. Louis County (100–200 meters north of Municipal Bridge between the right-of-way of the Terminal Railroad and the Mississippi River levee, St. Louis, May 23, 1954, *Muehlenbach 70*).

Native of Europe; naturalized in North America from New Brunswick to Delaware, inland to Ontario, Michigan, Illinois, Wisconsin, and Missouri.

2. **Puccinellia airoides** (Nutt.) Wats. & Coult. Alkali Grass Map 126

*Puccinellia Nuttalliana* (Schultes) Hitchc. [G, BB]

Flowers May–July.

Known only from St. Louis County (Carrie Avenue freight yards of Terminal Railroad Assoc., St. Louis, June 1, 1957, *Muehlenbach 1179*).

Ranges from British Columbia and the Yukon to Wisconsin, Minnesota, Texas, New Mexico, and California; introduced east in Maine, Vermont, New York, Ontario, and Missouri.

## 5. **Glyceria** R. Br. Manna Grass

The name Manna Grass refers to the use made of several European species, the seeds of which furnish a flour. It is reported that in Europe, a flour made from *Manna Seeds*, by which the seeds are known, provides the constituents of a tasty and nutritive bread. The seeds are also used as a thickening in soups and gruels. It is reported also, that geese fatten quickly when fed on grain from various species of this genus. Most species provide good forage.

Most of the species inhabit the water. The spikelets easily fall from the inflorescence when handled.

---

Plate no. 20. **1.** Festuca obtusa; *a.* Panicle, × 3/8; *b.* Floret, × 3 3/4. **2.** Festuca rubra; *a.* Plant, × 3/8; *b.* Spikelet, × 3 3/4; *c.* Floret, × 3 3/4. **3.** Festuca paradoxa; *a.* Panicle, × 3/8; *b.* Floret, × 3 3/4. **4.** Festuca ovina; *a.* Panicle, × 3/8; *b.* Floret, × 3 3/4. **5.** Puccinellia airoides; *a.* Panicle, × 3/4; *b.* Floret, × 7 1/2. **6.** Puccinellia distans; *a.* Panicle, × 3/8; *b.* Floret, × 7 1/2. **7.** Glyceria septentrionalis; *a.* Panicle, × 3/4; *b.* Floret, × 7 1/2. **8.** Glyceria septentrionalis var. arkansana; *a.* Panicle, × 3/4; *b.* Floret, × 7 1/2. **9.** Glyceria acutiflora; *a.* Panicle, × 3/4; *b.* Floret × 7 1/2.

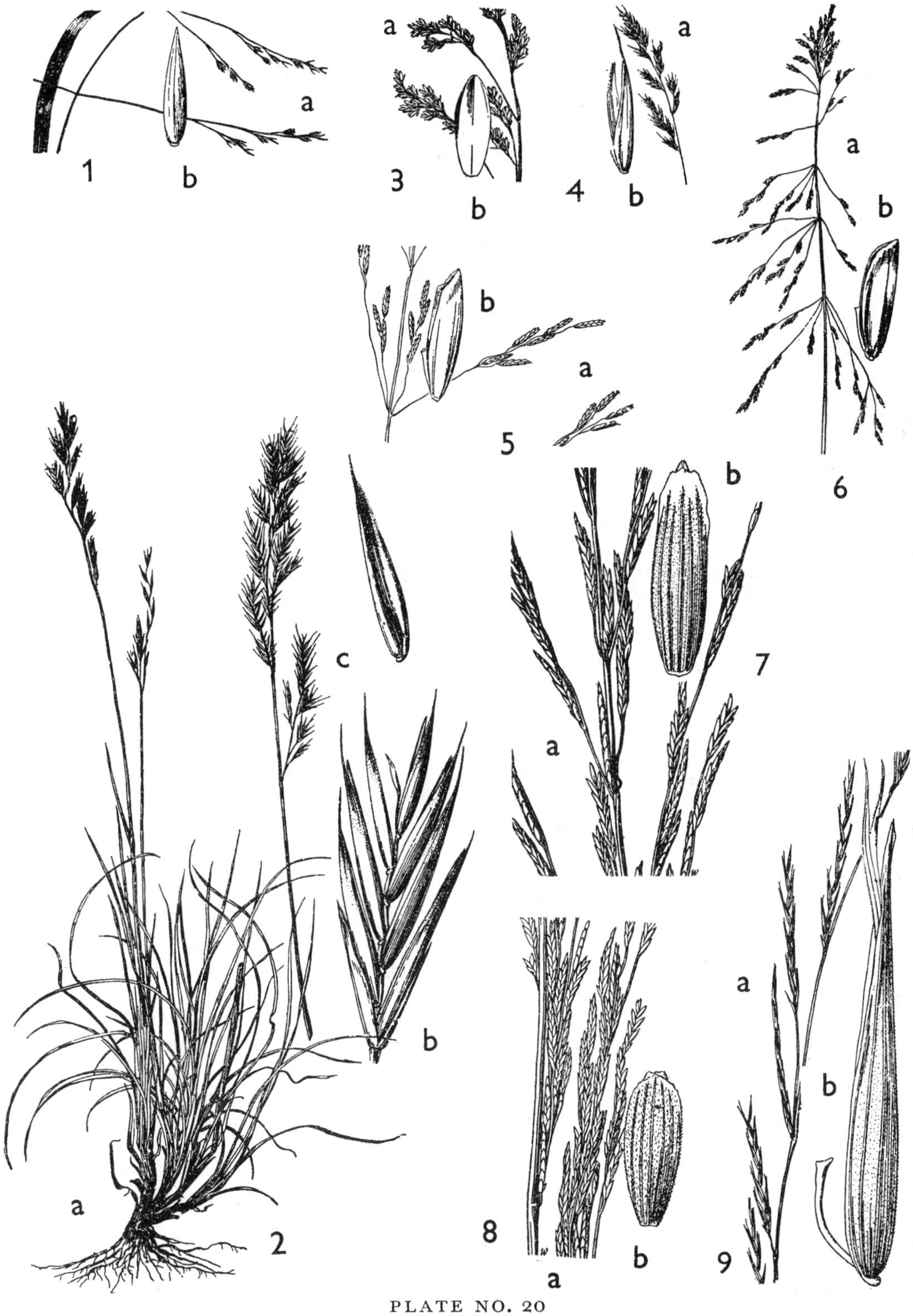


PLATE NO. 20

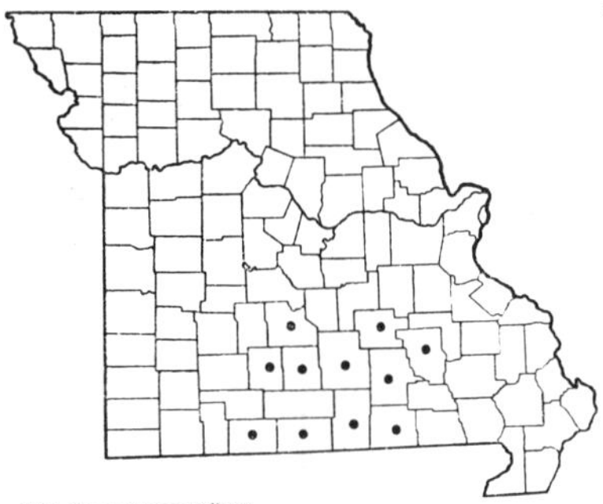

127 Glyceria acutiflora

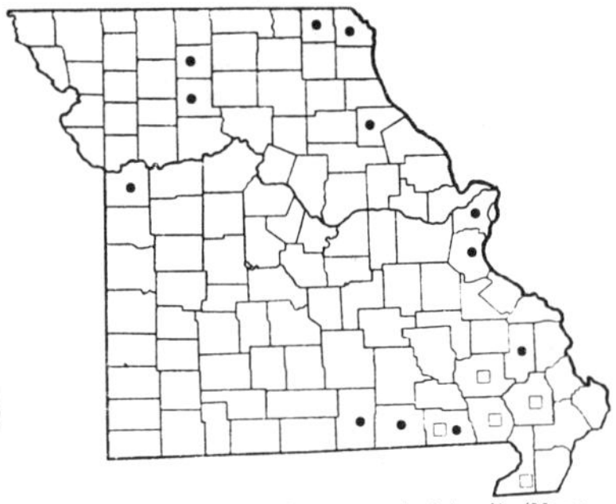

128 • Glyceria septentrionalis var. septentrionalis (Manna Grass)
128 □ Glyceria septentrionalis var. arkansana (Manna Grass)

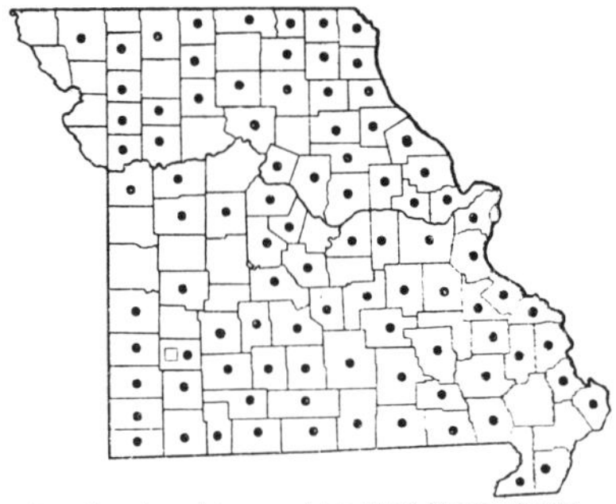

129 • Glyceria striata var. striata (Fowl Meadow Grass)
129 □ Glyceria striata var. stricta

a. Spikelets $1\frac{1}{2}$–$2\frac{1}{2}$ times longer than broad, broadly ovate or ovate-oblong, 3–6 mm. long; leaf-sheaths strongly compressed or flattened . . . . . . . . . . . . . . . . . . . . . . . *b*
  b. Leaf-sheaths closed nearly to the summit; culms erect or nearly erect; spikelets 2–4.5 mm. long 3. G. STRIATA
  b. Leaf-sheaths open; culms mostly creeping over much of their length; spikelets 5–7 mm. long 4. G. PALLIDA
a. Spikelets 4–11 times longer than broad, narrowly linear, 10–40 mm. long; leaf-sheaths terete (circular or round in cross-section) or nearly terete . . . . . . . . . . . . . . . . . . *c*
  c. Lemmas conspicuously pointed at tip, conspicuously shorter than the prominent palea; plants of sink-hole ponds of southeastern Missouri . . . . . . . . . . . . . . 1. G. ACUTIFLORA
  c. Lemmas rounded or blunt at tip, only slightly shorter than the palea; plants of sink-hole ponds and other habitats . . . . . . . . . . . . . . . . . . . . . . . . *d*
    d. Lemmas 3.6–5.5 mm. long, coriaceous (leathery) and rather firm; anthers 1–2 mm. long; leaf-blades 2–12 mm. wide . . . . . . . . . 2a. G. SEPTENTRIONALIS var. SEPTENTRIONALIS
    d. Lemmas 2.5–3 mm. long, membranaceous and rather thin; anthers 0.5–0.8 mm. long; leaf-blades 10–18 mm. wide . . . . . . . . . . . . 2b. G. SEPTENTRIONALIS var. ARKANSANA

1. **Glyceria acutiflora** Torr. Map 127

Flowers early May–July.

In Missouri this plant is exclusively confined to natural upland sink-hole ponds of the southern Ozarks north to Dent and Laclede counties, east to Reynolds and west to Webster and Taney counties. It always grows directly in the water.

Ranges from Maine to Michigan, south to Delaware, West Virginia, Indiana, Tennessee, and Missouri.

2. **Glyceria septentrionalis** Hitchc.

Floating Manna Grass Map 128

The following two varieties were distinguished above in the key to species:

2a. **Glycera septentrionalis** var. **septentrionalis** Map 128

*Glyceria septentrionalis* [of P & S, G]

Flowers late May–September.

Frequents sink-hole ponds, borders of sloughs and oxbow lakes in alluvial wooded bottoms, and swampy ground at the base of river bluffs. Rather local and scattered in northern, eastern, and central Missouri; absent from most of the Ozark section.

Ranges from Massachusetts to Ontario, south to Georgia, Kentucky, Missouri, and Texas.

2b. **Glyceria septentrionalis** var. **arkansana** (Fern.) Steyermark & Kucera Map 128

*Glyceria arkansana* Fern. [G, P & S]

Flowers May–July.

Occurs along sloughs and swampy ground principally in the lowland section of southeastern Missouri.

Ranges from New York to Illinois, south to Louisiana and Arkansas; also in Virginia.

3. **Glyceria striata** (Lam.) Hitchc.

Fowl Meadow Grass Map 129

The Missouri material may be divided into two varieties:

Commonly encountered; lemmas scarcely or inconspicuously scarious-tipped; branches of inflorescence widely spreading or even turning down in age; spikelets greenish or purplish 3a. G. STRIATA var. STRIATA

Rarely encountered; lemmas with conspicuous, broad, scarious (thin, whitish, or transparent) tips; branches of inflorescence stiffer, more ascending; spikelets purple, rarely green . . . 3b. G. STRIATA var. STRICTA

3a. **Glyceria striata** var. **striata** Map 129

Flowers May–July.

Common in wet woods, along streams, borders of ponds, drainage-ditches, sloughs, and wet depressions in prairies. Throughout Missouri.

Ranges from Newfoundland to Alberta, south to Florida, Alabama, and Texas.

3b. **Glyceria striata** var. **stricta** (Scribn.) Fern. Map 129

Known only from Dade County (wet ground in woods along Sons Creek, 4 mi. north of Greenfield, June 10, 1951, *Palmer 52200*).

This variety has a more northerly range, extending from Labrador to Alaska, south to New York, Illinois, Iowa, Missouri, South Dakota, and northern Mexico.

4. **Glyceria pallida** (Torr.) Trin.

Pale Manna Grass Map 130

Flowers May–June.

Known from swamps in Scott and Butler counties in southeastern Missouri, and in cold water of Yancy Mills Spring, Phelps County (Nov. 28, 1936, *Steyermark 20914*).

Ranges from Nova Scotia and Maine to Ontario, south to Virginia, Tennessee, and southeast Missouri.

## 6. **Poa** L. Blue Grass

Members of this genus are important for their forage value, many of them being nutritious and furnishing valuable food for livestock.

Some of the species cannot be properly determined without use of a high-power lens. This facilitates seeing the marginal and intermediate nerves and their pubescence, characters of importance used in distinguishing some of the species. The marginal nerves may be glabrous or pubescent. The intermediate nerves, which are located between the marginal nerves and midrib, may be strong or very faint, and glabrous or pubescent. Most species of *Poa* have 2 or more florets, but in those rare individuals having only 1-flowered spikelets, the tuft of wool at the base of the lemma is sufficient to identify the plant as a *Poa*.

a. Plants submerged or growing in the water; lemmas not cobwebby at base . 2a. P. ANNUA var. REPTANS, and var. AQUATICA

a. Without the above combination of characters . . . *b*

- b. Culms swollen and bulblike at base; florets usually changed into little bulblets which are dark purple at base. . . . 8. P. BULBOSA
- b. Without the above combination of characters . . . *c*
  - c. Lemmas not cobwebby at base . . . 2. P. ANNUA var. ANNUA
  - c. Lemmas cobwebby at base . . . *d*
    - d. Annual, with a soft base and a tuft of roots; plants mostly 3–20 cm. (up to 35 cm.) tall, without old leaves or dried leaf-sheaths persisting from the previous year at base 1. P. CHAPMANIANA
    - d. Tufted or creeping perennial, the base of plant either hardened or with scaly creeping rhizomes (underground stems) or offshoots or with old leaves or dried leaf-sheaths persisting at base from previous year; plants mostly 30–120 cm. tall . . . *e*
      - e. Scaly or creeping rhizomes or offshoots developed at base of plant . . . *f*
        - f. Culms flat, conspicuously flattened and 2-edged, arising from long slender rhizomes; lemmas obtuse (blunt) at tip, the intermediate nerves faint or not evident; inflorescence mainly contracted, often less than 1 cm. wide, sometimes in shade forms 2 cm. wide and with more spreading branches, the branches mainly erect to ascending or spreading, and mainly 0.5–1.8 cm. long, rarely longer . . . 3. P. COMPRESSA
        - f. Culms terete or cylindrical (circular or rounded in cross-section) or compressed, but not 2-edged, arising from scaly stolons or offshoots among crowded leaf-tufts; lemmas acute or nearly acute (pointed) at tip; the intermediate nerves evident or prominent; inflorescence more loosely branched, the branches spreading or ascending, usually 2–7 cm. wide . . . 4. P. PRATENSIS
      - e. Plants tufted, without rhizomes or offshoots developed at base . . . *g*
        - g. Leaf-sheaths with short downward-pointing hairs; ligule 4–8 mm. long; marginal nerves of lemmas glabrous . . . 5. P. TRIVIALIS

130 Glyceria pallida (Pale Manna Grass)

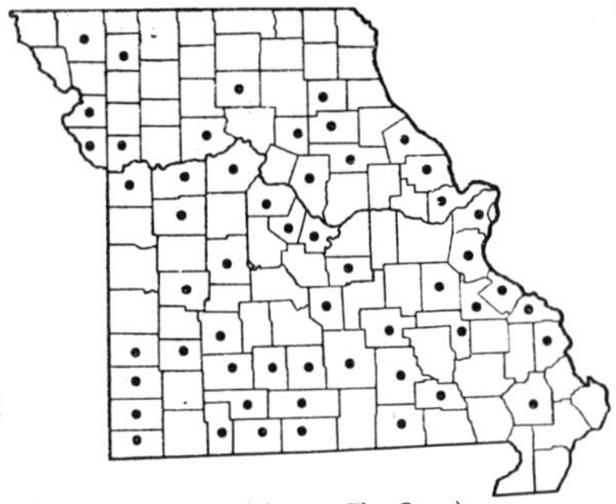

131 Poa Chapmaniana (Chapman Blue Grass)

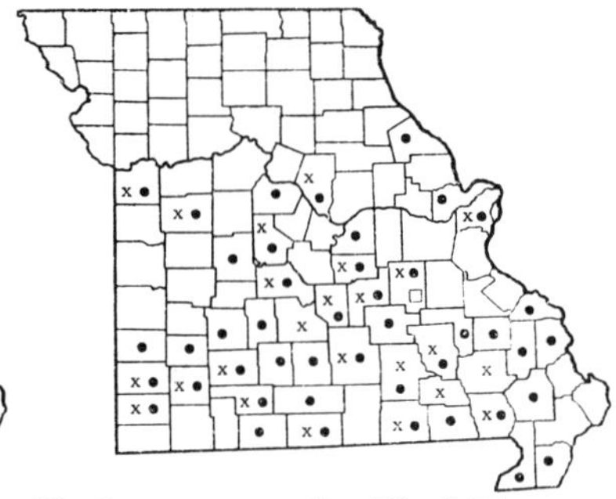

132 • Poa annua var. annua (Annual Blue Grass)
132 □ Poa annua var. aquatica
132 x Poa annua var. reptans

g. Leaf-sheaths glabrous (without hairs); ligule 0.5–5 mm. long; marginal nerves of lemmas pubescent, even if only near the base . . . . . . . . . . . . . . *h*

h. Spikelets mostly 5–6 mm. long; lemmas 3.5–4.5 mm. long; lower branches of panicle starting together from the central axis usually in 2's or 3's. . . . . . 7. P. WOLFII

h. Spikelets 2.8–4.5 mm. long; lower branches of panicle starting together from the central axis in mostly 4's to 10's (rarely 3's) . . . . . . . . . . . . . . *i*

i. Ligule conspicuous, 2–5 mm. long; inflorescence rather narrowly pyramidal or somewhat cylindric; spikelets usually with some touch of bronze or purplish; lemma of firm texture, with 3 noticeable nerves, the other (intermediate) nerves obscure . . . . . . . . . . . . . . . . . 9. P. PALUSTRIS

i. Ligule only 1–2 mm. long; inflorescence oblong-pyramidal or broadly pyramidal; spikelets silvery or pale green; lemma of soft thin texture, with 5 distinct nerves (including the ones along the very margins) . . . . . 6. P. SYLVESTRIS

1. **Poa Chapmaniana** Scribn. Chapman Blue Grass Map 131
Flowers April–May.

Fallow and sandy open fields, rocky glades of sandstone, chert, or granite, prairies, openings in dry sterile woods, along dry roads, and in waste ground, mostly in acid soils. Most common in southern and central Missouri, north in fallow fields and openings in dry acid woodlands to Nodaway, Gentry, Linn, Shelby, and Pike counties.

Ranges from Florida to Texas, north to Delaware, Indiana, Iowa, and Kansas.

This grass ordinarily grows in drier sites than *P. annua* and dries up sooner. The culms are usually more erect and the inflorescence is narrower and more contracted than in *P. annua,* and turns brownish more quickly. In fallow fields *P. Chapmaniana* is often found associated with *Cardamine parviflora* var. *arenicola, Festuca octoflora* var. *tenella, Myosurus minimus,* and *Chaerophyllum procumbens,* and on sandy glades or openings in dry upland woods it is more frequently associated with *Festuca octoflora* var. *tenella* and *Cardamine parviflora* var. *arenicola.*

2. **Poa annua** L. Annual Blue Grass Map 132
Flowers February–December.

The common variety encountered, *P. annua* var. *annua* is annual, and occurs in gardens and lawns around dwellings, in pastures, fields, waste places, along paths, and streams, and in disturbed sections of woodland. Although undoubtedly it is present in every county of the state, it has been recorded actually only from the counties of southern and central Missouri north to Pike, Boone, and Jackson counties.

Introduced from Europe; ranging throughout the United States north to Labrador and Alaska.

Two other variations of *Poa annua,* which are perennial in habit, have been found in Missouri. Such plants grow partly or completely submerged in cold running waters of the spring branches, or produce creeping stems which root at the nodes in the gravel of spring branches in the Ozark section of the state,

Plate no. 21. **1.** Glyceria striata; *a.* Plant, × $^3/_8$; *b.* Spikelet, × $3^3/_4$; *c.* Floret, × $7^1/_2$. **2.** Glyceria pallida; *a.* Plant, × $^3/_4$; *b.* Floret, × $7^1/_2$. **3.** Poa annua; *a.* Panicle, × $^3/_4$; *b.* Floret, × $7^1/_2$. **4.** Poa compressa; *a.* Panicle, × $^3/_4$; *b.* Floret, × $7^1/_2$. **5.** Poa Chapmaniana; *a.* Panicle, × $^3/_4$; *b.* Floret, × $7^1/_2$. **6.** Poa arida; *a.* Panicle, × $^3/_4$; *b.* Floret, × $7^1/_2$.

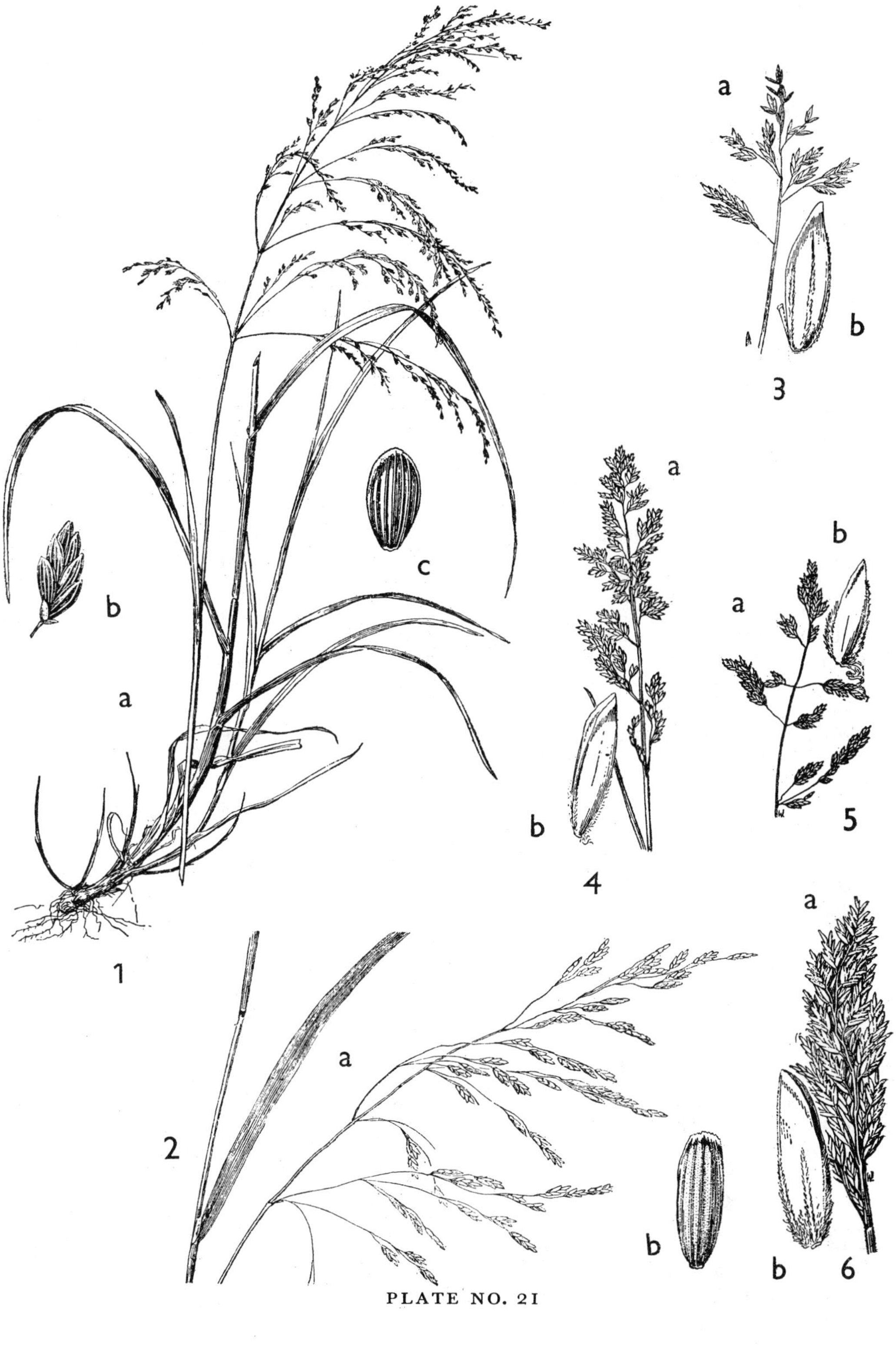


PLATE NO. 21

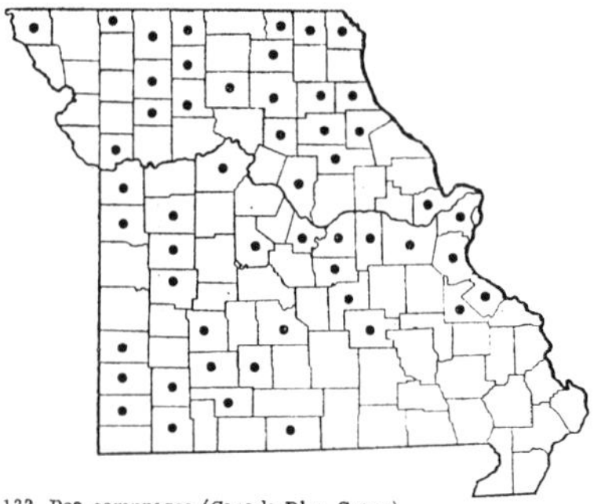

133 Poa compressa (Canada Blue Grass)

134 Poa pratensis (Kentucky Blue Grass)

* * *

135 Excluded species

and appear very different from ordinary *P. annua* var. *annua*. A collection from Steelville Spring, at Steelville, Crawford County (*Steyermark 6638*) has been identified as *P. annua* var. *aquatica* Aschers. It has a delicate and lax habit with greatly prolonged culms characteristic of this variety. The other variety encountered, *P. annua* var. *reptans* Haussknecht, with greatly elongated prostrate culms, which root at the nodes, appears to be commoner, and occurs in springs and spring branches in a number of counties of southern and central Missouri north to St. Louis, Boone, and Jackson counties. These varieties do not appear to be well marked. Both variations may be found in the partly or completely submerged state in spring water, and transitional variations occur which might be classified either as one or the other of the two perennial types. For a discussion of the aquatic varieties and their localities in Missouri, see Steyermark (Rhodora 43: 630–32. 1941).

Forty-one varieties (besides subvarieties and forms) of *P. annua* have been described. For a discussion of these the reader is referred to Ascherson, P. and P. Graebner, Synopsis der Mitteleuropäischen Flora 2, part 1: 388–90. 1921.

3. **Poa compressa** L. Canada Blue Grass

Map 133

Flowers May–August.

Occurs throughout the state in various types of soils in dry fields, pastures, waste ground, prairie openings, grazed open woodland, and along roadsides, railroads, and rocky ground. It is absent generally from wet places and dense woodland.

Native of Europe; introduced into North America, where found nearly throughout, from Newfoundland and Alaska, south to Georgia, Oklahoma, New Mexico, and California.

Canada Blue Grass is cultivated for pastures in poor soil, but is inferior to the Kentucky Blue Grass. It is being used as a grass on athletic and polo fields, where it can be cut infrequently and as high as 3–4 inches. The large amount of pollen shed by Canada Blue Grass makes it a serious cause of hay fever during the months especially from June to August.

This species sometimes causes considerable confusion in identification because of its variability with respect to the number of florets per spikelet and length of spikelet. Most manuals describe the spikelets as 3–6-flowered, and 4–6 mm. long. However, the spikelets may have as many as 9 flowers and attain 7–8 mm. in length, as in the collection of *Steyermark 85320* from Mercer County. The eighth edition of *Gray's Manual* (p. 117) correctly accounts for such extremes of variation, giving the spikelets as '3–6 (–9)-flowered, 3–8 mm. long.'

4. **Poa pratensis** L. Kentucky Blue Grass

Map 134

Flowers chiefly May–July.

Found throughout Missouri in fields, meadows, rocky glades, open woodland, along river banks, and in waste ground.

Ranges throughout the United States, north to Labrador and Alaska. Some botanists consider most of the Kentucky Blue Grass which is found in the United States as an introduction from Europe, with the possibility of its being native in Canada and the northern United States. Other botanists believe the grass to be mainly a native one, particularly in the northern half of the United States, but introduced as a cultivated species in the southern United States.

Kentucky Blue Grass is the chief lawn grass in the cooler and moister climates of the northern half of the

Plate no. 22. **1.** Poa trivialis; *a.* Panicle, × $^3/_4$; *b.* Floret, × $7^1/_2$. **2.** Poa Wolfii; *a.* Panicle, × $^3/_4$; *b.* Floret, × $7^1/_2$. **3.** Poa sylvestris; *a.* Panicle, × $^3/_4$; *b.* Floret, × $7^1/_2$. **4.** Poa pratensis; *a.* Plant, × $^3/_8$; *b.* Spikelet, × $3^3/_4$; C. Floret, × $7^1/_2$.

PLATE NO. 22

136 Poa trivialis (Rough Blue Grass)

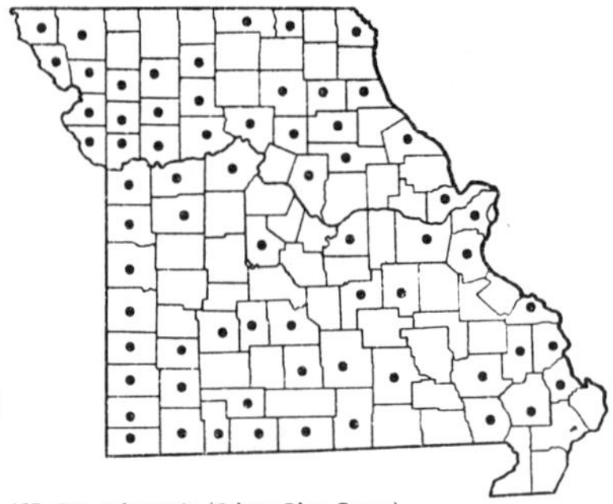
137 Poa sylvestris (Sylvan Blue Grass)

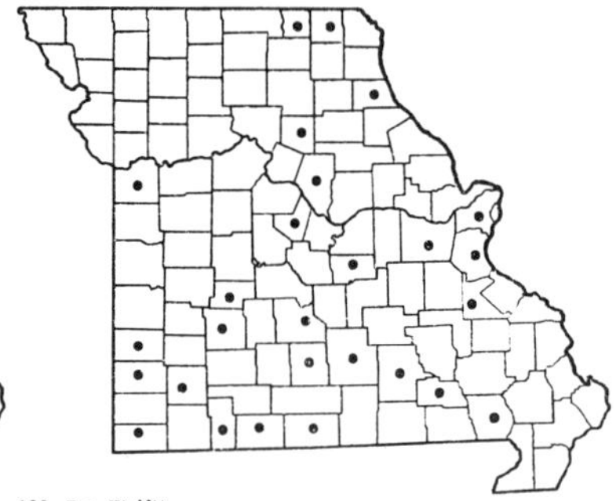
138 Poa Wolfii

United States. Some recent superior grass strains, such as Merion Blue Grass, a variety discovered on the Merion Golf Course near Philadelphia, have been developed from Kentucky Blue Grass. It is also considered one of the best pasture grasses.

Together with Orchard Grass (*Dactylis glomerata*), Kentucky Blue Grass accounts for most of the hay fever occurring in the eastern United States during May and June. Some manuals give the spikelet length as 4–5 mm. long, but the spikelets vary in Missouri material from 3–6.5 mm. long. The spikelets are often purplish or tinged with rose-purplish color.

5. **Poa trivialis** L. Rough Blue Grass Map 136
Flowers May–June.

Known only from Newton County, southwestern Missouri, where collected in wet, rocky, shaded ground, about Haddock Spring, 3 mi. southwest of Wentworth, May 14, 1955 (*Palmer 59826*). The determination has been made at the United States National Herbarium.

Native of Europe; naturalized in North America from Newfoundland and Ontario to Wisconsin, south to Georgia, Lousiana, and Kansas, and also from Alaska to California.

Some Missouri specimens, previously identified as *P. trivialis*, have proven to be *P. pratensis*. The Palmer collection, here cited, has the retrorsely scabrous sheaths characteristic of *P. trivialis*, but is atypical in having the ligule only 2–2.5 mm. instead of 4–8 mm. long, and in the marginal nerves somewhat pubescent instead of glabrous.

Rough Blue Grass is sometimes used in grass seed mixtures, especially being recommended for meadows and pastures, and in many sections is recommended as a shade grass. It thrives in damp shaded rich soils, where it produces a low-growing grassy cover.

This species is sometimes found to be an important contributing causal agent in hay fever cases.

6. **Poa sylvestris** A. Gray Sylvan Blue Grass Map 137
Flowers late April–June.

This is a common woodland species, occurring throughout the state in lowland woods along streams and river bottoms, as well as on wooded rocky slopes and in rich ravines.

Ranges from New York to Minnesota and Nebraska, south to Florida, Louisiana, and Texas.

After flowering, the lowest branches of the inflorescence become widely spreading to reflexed. The leaves of this species are soft, dark green, and flexible. The plants grow in small individual tufts scattered over the forested ground. The widely spreading panicle branches with rather few pale green spikelets clustered above the middle toward the tips are characteristic. The lemmas vary from sparsely or conspicuously pubescent to nearly glabrous between the nerves.

7. **Poa Wolfii** Scribn. Map 138
Flowers late April–June.

Like *P. sylvestris*, this species favors woodland habitats, and is found in wooded areas along streams, rocky wooded slopes and moist ledges. It is much less commonly collected than *P. sylvestris*, and has been found thus far in southern, central, and eastern Missouri.

Ranges from Virginia, Ohio, and Minnesota to Missouri.

In addition to the larger spikelets, *P. Wolfii* has the lower branches of the panicle only spreading at maturity, and not reflexed as in *P. sylvestris*. Moreover, the lemmas are acute and glabrous between the nerves in *P. Wolfii*, obtuse and often more or less pubescent or rarely glabrous between the nerves in *P. sylvestris*.

8. **Poa bulbosa** L. Bulbous Blue Grass Map 139
Flowers April–June.

Known only from Christian County, southwestern Missouri, where introduced in open ground along

139 Poa bulbosa

140 Poa palustris (Fowl Meadow Grass)

... 141 Excluded species

142 Eragrostis reptans

railway grade, Billings, April 14, 1955 (*Palmer 59430*).

Native to Europe; introduced in the United States from New York to North Carolina; Missouri, Kansas, Oklahoma, North Dakota, Utah; and in the Pacific states from Idaho to British Columbia and California.

9. **Poa palustris** L. Fowl Blue Grass Map 140

*Poa nemoralis* [of P & S], not L.

Flowers June–September.

Known only from Jackson County, western Missouri, where collected along railroad, Sheffield, June 16, 1896 (*Bush 774*).

Ranges from Newfoundland to Alaska south to Virginia, Ohio, Illinois, Missouri, Nebraska, Colorado, and California; also occurs in Europe and Asia.

The Bush specimen was previously misidentified as *P. nemoralis*.

In *P. palustris* the inflorescence bears numerous spikelets at and beyond the middle of a branch.

*Excluded species*

**Poa alsodes** Gray. A specimen collected by Bush (*Bush 13434*), and so labeled by him, from rocky woods, Blue Lick, Saline County, May 2, 1934, has lemmas distinctly 5-nerved with distinct intermediate nerves, and the anthers 1.25 mm. long. It is referred in the present flora to *P. Wolfii*.

**Poa autumnalis** Muhl. A specimen so labeled was collected by *N. M. Glatfelter* from Meramec Heights, St. Louis County, May 14, 1897. It has the lemmas 4–4.5 mm. long, webbed at base, and glabrous between nerves; the spikelets are 5 mm. long. I have identified it as *P. Wolfii*.

**Poa nemoralis** L. *Bush 774*, which formed the basis for this determination, is *P. palustris* L.

## 7. **Eragrostis** Beauv. Love Grass

This group of grasses flowers and matures fruit mainly during the summer and autumn months (*Eragrostis poaeoides* may start flowering in late May).

a. Main stems (culms) creeping and rooting at the nodes (joints) . . . . . . . . . . . . . . . . *b*

b. Rare species; inflorescence a compact and closely crowded mass of spikelets; spikelets 2–4 mm. wide; flowers with only one sex developed on a single plant, either staminate (male) or pistillate (female); lemmas more or less pubescent (with hairs), 2–4 mm. long; anthers 1.5–2 mm. long . 2. E. REPTANS

b. Common species; inflorescence a rather open and loose panicle; spikelets 1.5–2 mm. wide; flowers perfect (both stamens and pistil in the same flower); lemmas glabrous (lacking hairs), 1.5–2 mm. long; anthers 0.2–0.5 mm. long . . . . . . . . . . . . . . . . . . . . 3. E. HYPNOIDES

a. Main stems (culms) upright to spreading, but not creeping and rooting at the nodes . . . . . . . *c*

c. Tiny warts or bumps (glands lacking stalks) on margins of leaf-blades and keels (back edge) of glumes and lemmas . . . . . . . . . . . . . . . . . . . . . . . . . . *d*

d. Spikelets mostly 2.5–4 mm. wide; lemmas 2–2.6 mm. long; common species . . 8. E. CILIANENSIS

d. Spikelets 1.5–2 mm. wide; lemmas 1.7–2 mm. long; rare species . . . . . . . 9. E. POAEOIDES

c. No tiny warts or stalkless glands present on leaf margins nor on keels of glumes and lemmas . . . . *e*

e. Inflorescence narrow and contracted, the branches rather upright and crowded . 1. E. GLOMERATA

e. Inflorescence with loosely spreading branches, producing an open appearance . . . . . . . *f*

f. Inflorescence nodding; leaves mainly at the base of the plant, curved, 6–12 dm. long, long-attenuate . . . . . . . . . . . . . . . . . . . . . . 15. E. CURVULA

f. Inflorescence erect; leaves scattered on the stem as well as at the base of the plant, less than 6 dm. long . . . . . . . . . . . . . . . . . . . . . . . . *g*

g. Mature (fully grown) spikelets at most 1 mm. wide, 5–8 mm. long, 5–15 flowered; lateral nerves of lemma faint or scarcely showing; plant annual with a soft base . . . . 6. E. PILOSA

g. Without the above combination of characters; mature spikelets 1–2.5 mm. wide, but, if only 1 mm. wide, then plants are either perennial with hard or firm bases, or spikelets are usually less than 5 mm. long or less than 5-flowered . . . . . . . . . . . *h*

h. First glume 2.5–3 mm. long; at least the lowest lemma 2.5–3 mm. long . 13. E. TRICHODES

h. First glume 0.8–2 mm. long; lemmas 1.1–2.5 mm. long . . . . . . . . . . *i*

i. No hairs or essentially none on the tiny swelling (pulvinus) at junction of the lowest 2 or 3 main branches with the central axis of the inflorescence . . . . . . *j*

j. Spikelets mostly 6–15-flowered, (4–) 5–8 mm. long; lateral nerves of lemma conspicuous . . . . . . . . . . . . . . . . . . . . . . *k*

k. Spikelets spreading on pedicels spreading from the main primary branches of the inflorescence; spikelets reddish or rose-purple, 2–2.5 mm. wide; lemma 1.8–2.5 mm. long; 1st glume 1.5 mm. long; seed with a longitudinal furrow extending its whole length . . . . E. MEXICANA (see excluded species)

k. Spikelets rather appressed to and lying along or paralleling the main primary branches of the inflorescence, the pedicels (stalks of spikelets) not spreading; spikelets dull green or lead-colored, sometimes tinged with purplish, 1.3–1.7 mm. wide; lemma 1.5–1.6 mm. long; 1st glume 1 mm. long; seed not longitudinally furrowed . . . . . . . . . . . . . 7. E. PECTINACEA

j. Spikelets mainly 1–5-flowered, 1.4–4 (–5) mm. long; lateral nerves of lemma faint or scarcely showing . . . . . . . . . . . . . . . . . . *l*

l. Inflorescence more than 1/2–2/3 height of plant; culms (stems) branched only from the base; leaf-sheaths usually with hairs; most of the pedicels (stalks of spikelets) 5–10 mm. long . . . . . . . . . . 4. E. CAPILLARIS

l. Inflorescence 1/3 to less than 1/2 height of plant; culms (stems) branched at the nodes higher on the plant; leaf-sheaths glabrous (without hairs) except at the summit; pedicels (stalks of spikelets) mainly 1–3 (–5) mm. long . 5. E. FRANKII

i. Sparse to many hairs on the swelling (pulvinus) at junction of at least the lowest 2 or 3 main branches with the central axis of the inflorescence (junction at the upper branches with the main axis often glabrous) . . . . . . . . . . . *m*

m. Lemmas with conspicuous lateral nerves; spikelets 3- mainly 6–15-flowered; annuals or perennials . . . . . . . . . . . . . . . . . . . . *n*

n. Leaf-blades 4–8 mm. wide; 1st glume 1.2–1.9 mm. long; spikelets rose-purple; plants perennial with a hard base producing pointed buds; culms (stems) simple below the inflorescence; pulvinus at base of branches conspicuously long-pilose . . . . . . . . . . . . . 14. E. SPECTABILIS

n. Leaf-blades 1–3.5 mm. wide; 1st glume 1 mm. long; spikelets dull green or lead-colored, sometimes tinged with dull purplish; plants annual from a soft base; culms (stems) branched from the nodes below the inflorescence; pulvinus at base of branches sparsely or inconspicuously hairy . . . . . . *o*

o. Spikelets rather appressed to and lying along or paralleling the main primary branches of the inflorescence, the pedicels (stalks of spikelets) also appressed, not spreading; lemmas 1.5–1.6 mm. long . . 7. E. PECTINACEA

o. Spikelets spreading away from main primary branches of the inflorescence on spreading pedicels; lemmas 1.6–1.8 mm. long . . . . . 10. E. ARIDA

m. Lemmas with faint or inconspicuous lateral nerves; spikelets 7–9-flowered; perennials . . . . . . . . . . . . . . . . . . . . . *p*

---

Plate no. 23. **1.** Poa bulbosa, × 3/4. **2.** Poa palustris; *a.* Panicle, × 3/4; *b.* Floret, × 7 1/2. **3.** Eragrostis capillaris; *a.* Panicle, × 3/4; *b.* Floret, × 7 1/2. **4.** Eragrostis hypnodies; *a.* Plant, × 3/8; *b.* Floret, × 7 1/2. **5.** Eragrostis reptans; *a.* Pistillate plant (♀), × 3/8; *b.* Staminate plant (♂), × 3/8; *c.* Pistillate floret (♀), × 7 1/2; *d.* Staminate floret (♂), × 7 1/2. **6.** Eragrostis Frankii; *a.* Panicle, × 3/4; *b.* Floret, × 7 1/2. **7.** Eragrostis glomerata; *a.* Panicle, × 3/8; *b.* Spikelet, × 7 1/2; *c.* Floret, × 7 1/2.

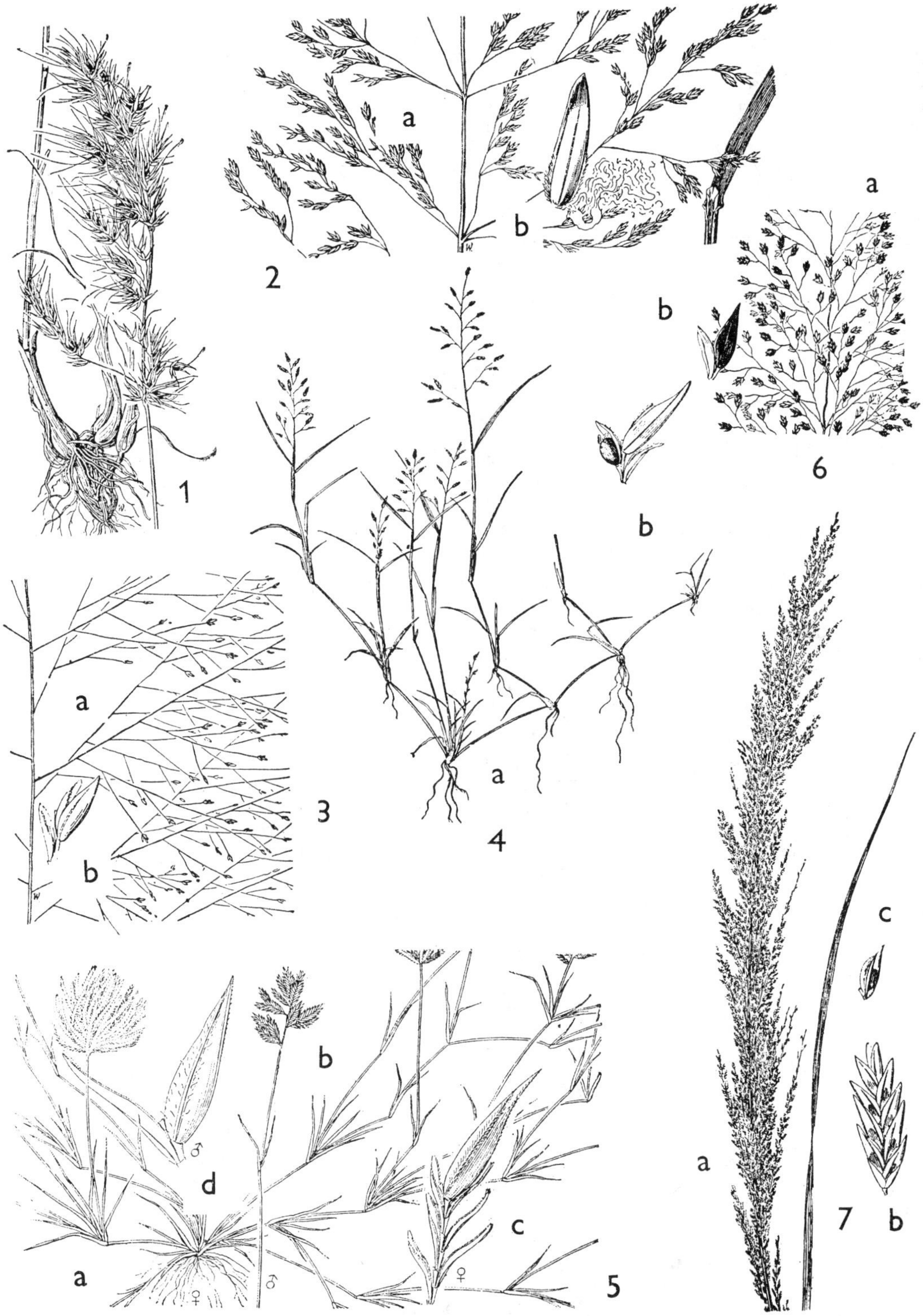


PLATE NO. 23

143 Eragrostis hypnoides (Creeping Love Grass)

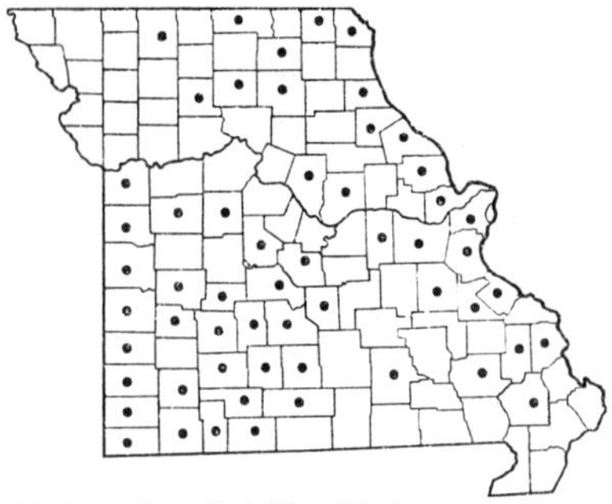

144 Eragrostis capillaris (Lace Grass)

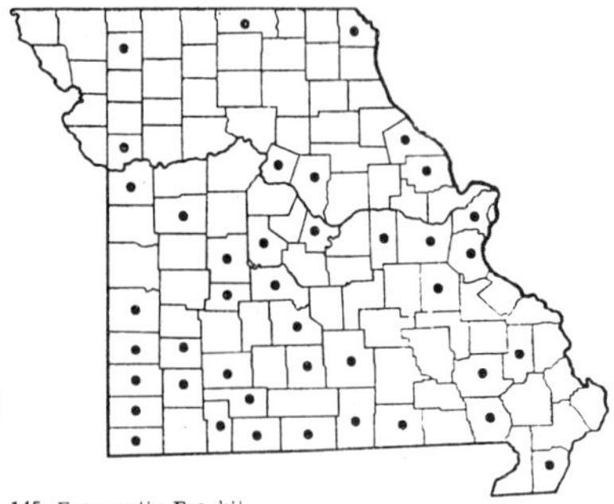

145 Eragrostis Frankii

p. Leaf-blades 4–12 mm. wide (sometimes less than 4 mm.); leaf-sheaths usually conspicuously hispid-pubescent, rarely almost glabrous in one variety; 1st glume 1.5–2 mm. long; spikelets often purplish; fully developed inflorescence 3–9 dm. long. . . . . . . . . . . . . . 11. E. HIRSUTA

p. Leaf-blades 1–4 mm. wide; leaf-sheaths glabrous except at hairy summit; 1st glume 1–1.3 mm. long; spikelets usually dull grayish- to brownish-green; fully developed inflorescence 1.5–3.5 dm. long. . . . . . . 12. E. INTERMEDIA

1. **Eragrostis glomerata** (Walt.) Dewey

This grass has been attributed to Missouri (by Gleason in *New Ill. Fl.* 1: 130. 1952) on the basis of a specimen in the New York Botanical Garden Herbarium with the following information on the label 'Unio itiner. In civitate Missouri. 1837. leg. b. Frank.' No definite locality in Missouri is mentioned by the collector, Frank, but if the specimen were actually obtained in Missouri, it is assumed that it probably originated from some part of the southeastern section of the state, as the species occurs in Arkansas. The range of *E. glomerata* is South Carolina and Arkansas south in the United States to Florida and Texas, southward to Argentina.

2. **Eragrostis reptans** (Michx.) Nees Map 142

Flowers August–October.

Sand bars of the Missouri River in Jackson (Courtney, September 21, 1930, *Bush 11979, 11978, 11973, 11971, 11975, 11976, 11972, 11969*; Sept. 5, 1935, *15141*), Cass (Pleasant Hill, 1876, *Broadhead*), and Livingston (muddy shores of lake in Grand River bottoms, 3 mi. north of Avalon, Sept. 23, 1955, *Steyermark 79879*) counties are known localities for this rare species.

Ranges from Kentucky, Illinois, Missouri, and South Dakota, south to Tennessee, Louisiana, Texas, and Mexico.

The pistillate inflorescence is dense and somewhat hemispherical in form, whereas the staminate inflorescence is more open and ovoid with fewer-flowered spikelets.

3. **Eragrostis hypnoides** (Lam.) BSP.

Creeping Love Grass Map 143

Flowers July–November.

Occurs on margins of ponds, sloughs, and moist, alluvial, usually muddy banks of river, although encountered also on moist sands. Throughout Missouri.

Ranges from southern Quebec throughout most of the United States south to Argentina.

This and the preceding species are creeping annuals. *Eragrostis hypnoides* forms characteristic small-leaved mats along flooded river banks, where it often is the dominant plant present.

4. **Eragrostis capillaris** (L.) Nees

Lace Grass Map 144

Flowers July–October.

Common throughout Missouri, except in the northwestern section, where not yet recorded, in fields, recently cleared areas, waste and cultivated ground, rocky open glades, and ledges.

Ranges from Maine to Wisconsin and Iowa, south to Georgia and Texas.

The lower leaf-sheaths and lower part of the culms (stems) are often rose-lavender or purplish. A woolly aphis frequently infests the spikelets of this species. Deam notes this type of aphis on *E. Frankii*.

This annual grass is frequently associated with and sometimes mistaken for *Panicum flexile*, *P. philadelphicum*, and other members of the *Capillare* group, but the normally 2–5-flowered spikelets at once distinguishes the species as belonging in *Eragrostis*.

146 Eragrostis pilosa (India Love Grass)

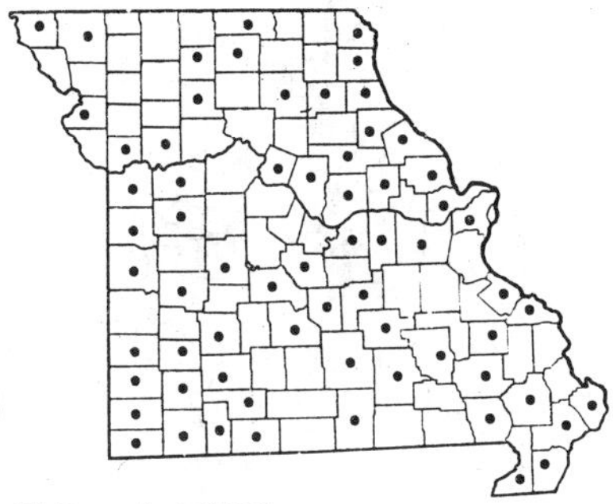

147 Eragrostis pectinacea

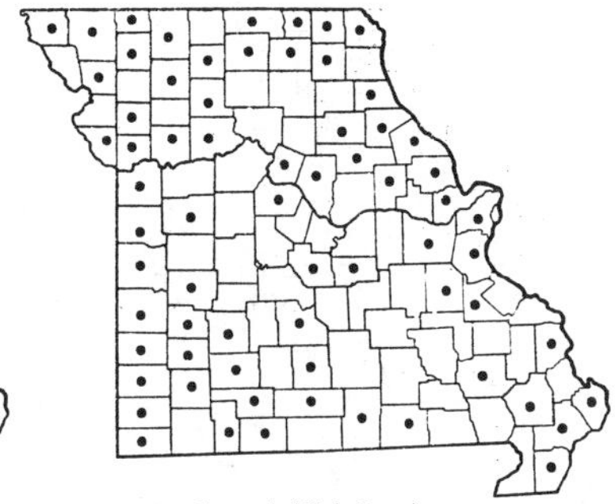

148 Eragrostis cilianensis (Stink Grass)

5. **Eragrostis Frankii** C. A. Meyer var. **Frankii** Map 145

Flowers August–October.

Throughout Missouri, but with most of the records thus far from the counties in the central and southern sections of the state, where it occurs on sandbars and mud flats of streams, and in waste and cultivated ground.

Ranges from Quebec to Ontario and Minnesota, south to Florida, Louisiana, and Kansas.

This species is sometimes confused with *E. capillaris*, but the shorter pedicels and compact elongated inflorescence serve to distinguish it from *E. capillaris*.

The var. *brevipes* Fassett, distinguished by having mainly 6–7-flowered spikelets, has not yet been recorded from Missouri.

6. **Eragrostis pilosa** (L.) Beauv.
India Love Grass Map 146
Flowers July–October.

Of infrequent occurrence in Missouri, where it is scattered over the state, occurring on damp muddy or sandy banks along streams, rocky bluffs, and waste ground.

Native of Europe; introduced into the United States from Massachusetts to Colorado, south to Florida and Texas.

The very narrow spikelets, at most 1 mm wide, with only faintly nerved lemmas, serve to distinguish this from *E. pectinacea*, with which it is most frequently confused. The following specimens of *E. pilosa* have been determined by L. H. Harvey: Cooper Co. (*Demetrio 35*); Franklin Co. (Aug. 20, 1886, *Eggert*); Jackson Co. (July 8, 1920, *Bush 9017*); Marion Co. (*Davis 3427*); McDonald Co. (*Bush 94*); St. Louis Co. (October, 1859, *Engelmann*; July 25, 1954, *Muehlenbach 274*).

7. **Eragrostis pectinacea** (Michx.) Nees Map 147
*Eragrostis diffusa* Buckl. [P & S, Hitchcock]
Flowers July–October.

Common throughout Missouri, where it occurs most frequently in open ground along streams and borders of ponds, in sandy, gravelly, or muddy soil, or in waste and cultivated ground.

As contrasted with *E. pilosa*, this is a native species ranging from Quebec to British Columbia south throughout much of the United States to Mexico; not recorded from some western states.

This species is frequently misidentified as *E. pilosa*, but the spikelets in *E. pectinacea* are wider (1.3–1.7 mm.) with conspicuous lateral nerves of the lemma, and the glumes are longer (the 1st glume 1 mm. long, the 2nd 1.5 mm. long, whereas in *E. pilosa* the 1st glume is 0.5 mm. long, the 2nd 1 mm. long).

Variable in habit (from erect to spreading to nearly prostrate culms, which may be few to numerous), in the complexity of branching of the panicle, and in the spikelets, which may vary in width, number of flowers to a spikelet, and number of spikelets.

*Eragrostis diffusa* Buckl., by some botanists considered a distinct species, by others synonymous with *E. pectinacea*, is included in this present treatment under *E. pectinacea*. On the whole, it appears to be merely a more robust-developed phase of *E. pectinacea*. The following Missouri specimens have been determined as *E. diffusa* by Dr. LeRoy H. Harvey, monographer of the genus, or are so filed at the United States National Herbarium: Barry Co. (*Palmer 52782*); Cass Co. (*Broadhead*); Franklin Co. (*Colvey*); Jackson Co. (*Bush 31, 789, 8635*); McDonald Co. (*Palmer 4115*); Pike Co. (*Davis 9034*); Pulaski Co. (*Steyermark 19342*); St. Louis Co. (Allenton, 1892, *Letterman*; St. Louis, *Eggert*).

Some of the specimens of the above group, (*Palmer 4115* and *Bush 8635*) were at one time misidentified as *E. curtipedicellata* Buckl., a species not known from Missouri. They were later referred to *E. diffusa*.

149 Eragrostis poaeoides

150 Eragrostis arida

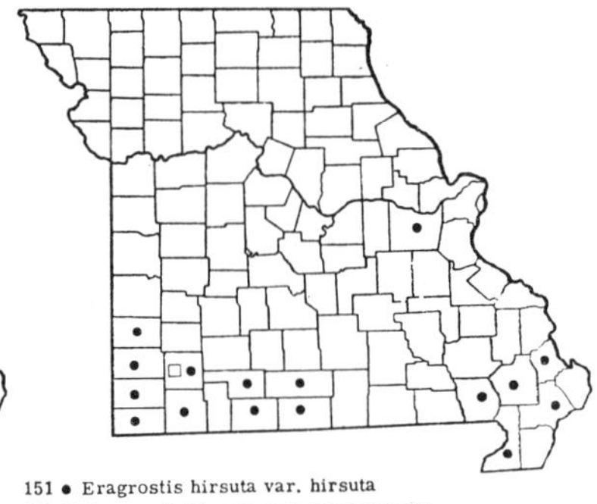

151 • Eragrostis hirsuta var. hirsuta
151 □ Eragrostis hirsuta var. laevivaginata

8. **Eragrostis cilianensis** (All.) Hitchcock
Stink Grass Map 148
*Eragrostis megastachya* (Koel.) Link [G]
Flowers June–October.

Common throughout Missouri in gardens, cultivated fields, waste ground, overgrazed dry pastures, along roads, and railroads tracks.

Native of Europe; introduced into the United States, over much of which it has spread; also introduced into Mexico and the West Indies.

This weedy annual derives its name from the somewhat disagreeable odor (to some people) of the fresh plant. The plant in either fresh or dried condition in hay is considered to be poisonous to stock, especially horses, if eaten in too large quantities. The inflorescence has a distinctive greenish-lead color.

9. **Eragrostis poaeoides** Beauv. Map 149
Flowers late May–October.

Infrequently collected in Missouri, where it is known in parts of southern and central Missouri north to St. Louis and Jackson counties. It is usually found in waste or open ground along railroad tracks.

Native of Europe; introduced into the United States, where it is scattered from Vermont to Wisconsin, south to Georgia, Texas, and California.

Similar in appearance to *E. cilianensis,* but smaller throughout, and the spikelets are more loosely flowered. The spikelets are purplish to lead-colored. The glandular-warty leaf-margins serve to distinguish this from *E. Frankii,* with which it is sometimes confused.

10. **Eragrostis arida** Hitchc. Map 150
Flowers July–October.

At present known in Missouri only from Dent, Jackson, Polk, and Stone counties, where it is found in open ground along railroads, waste places, and gravel bars of streams. The following collections are referred to this species: Dent Co. (Rhyse, *Kellogg 1359*); Jackson Co. (Courtney, *Bush 212, 803, 892,* and June 1891, and July 22, 1935); Polk Co. (Bolivar, *Palmer 64214*).

Ranges from Arkansas and Texas to Arizona and Mexico.

All the stations in Missouri of this annual species appear to be results of introductions, rather than representing indigenous (native) plants.

11. **Eragrostis hirsuta** (Michx.) Nees var. **hirsuta** Map 151
Flowers July–October.

Mostly in southern Missouri northeast to Franklin County (4 mi. northwest of Lonedell, *Steyermark 8268*), where it is found most frequently in sand washes or alluvial deposits of small valleys and creeks, and also in sandy ground along roadsides and waste places.

Ranges from Florida to Texas, north to South Carolina and inland north to Missouri; introduced northward in Virginia, Maryland, Massachusetts, and Maine.

Spikelets are sometimes purplish in this species. Plants of this perennial species and *E. trichodes* attain the tallest of any of the Love Grasses found in Missouri, culms (stems) reaching 1.3 meters in height, while the elongated large panicles become 9 dm. long and 4.5 dm. broad.

An occasional plant occurs in which the leaf-sheaths are nearly or quite glabrous throughout except for their ciliate summit and margins. Such plants have been described as *E. hirsuta* var. *laevivaginata* Fern.,

Plate no. 24. **1.** Eragrostis pilosa; *a.* Panicle, × $^{3}/_{4}$; *b.* Floret, palea, × $7^{1}/_{2}$. **2.** Eragrostis pectinacea; *a.* Panicle, × $^{3}/_{4}$; *b.* Floret, × $7^{1}/_{2}$. **3.** Eragrostis poaeoides; *a.* Panicle, × $^{3}/_{4}$; *b.* Floret, × $7^{1}/_{2}$. **4.** Eragrostis intermedia; *a.* Panicle, × $^{3}/_{4}$; *b.* Floret, × $7^{1}/_{2}$. **5.** Eragrostis cilianensis; *a.* Plant, × $^{3}/_{8}$; *b.* Spikelet, × $3^{3}/_{4}$; *c.* Floret, × $7^{1}/_{2}$. **6.** Eragrostis arida; *a* Panicle, × $^{3}/_{4}$; *b.* Floret, × $7^{1}/_{2}$. **7.** Eragrostis hirsuta; *a.* Panicle, × $^{3}/_{4}$; *b.* Floret, × $7^{1}/_{2}$.

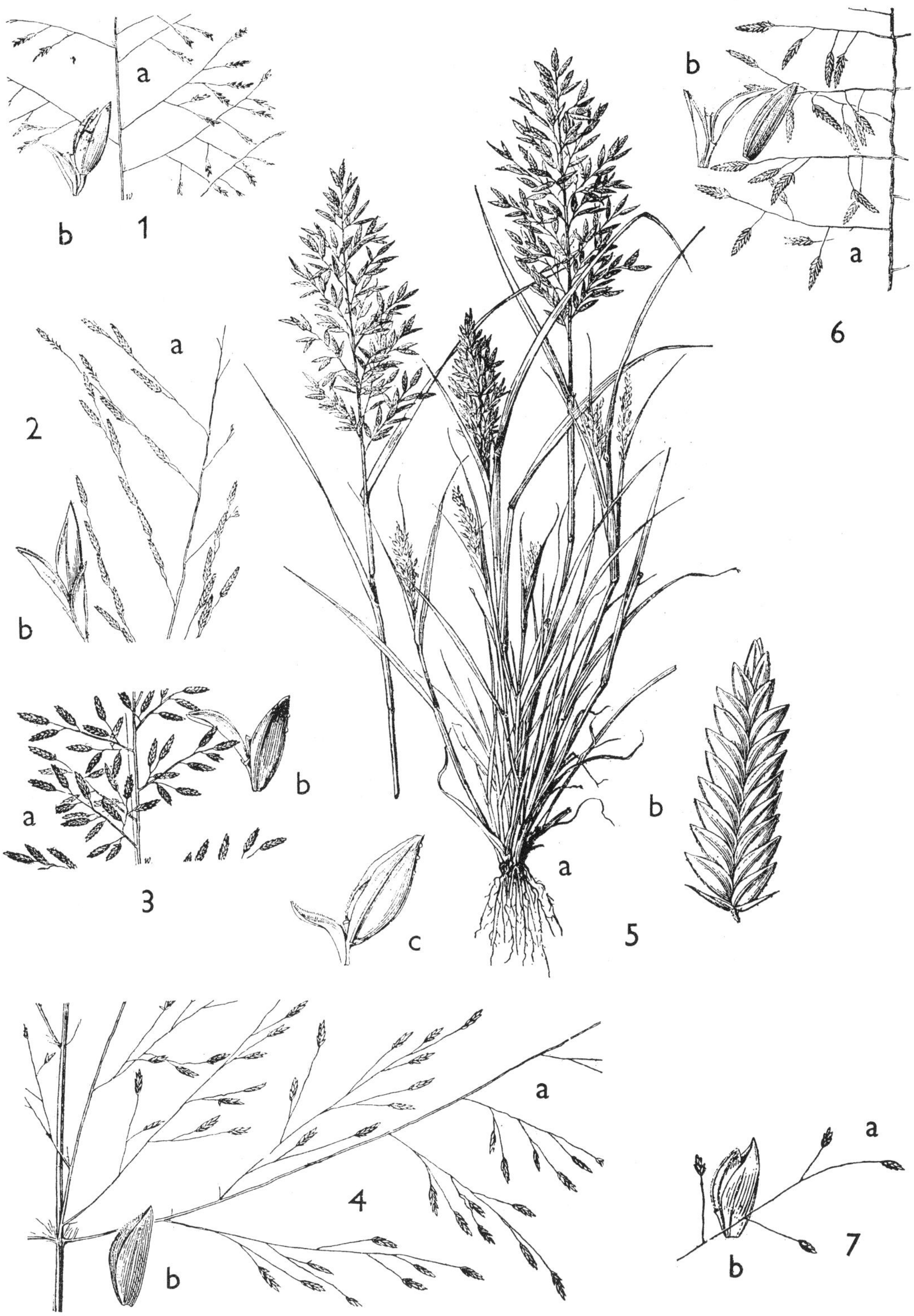


PLATE NO. 24

152 Eragrostis intermedia

153 Eragrostis trichodes (Sand Love Grass)

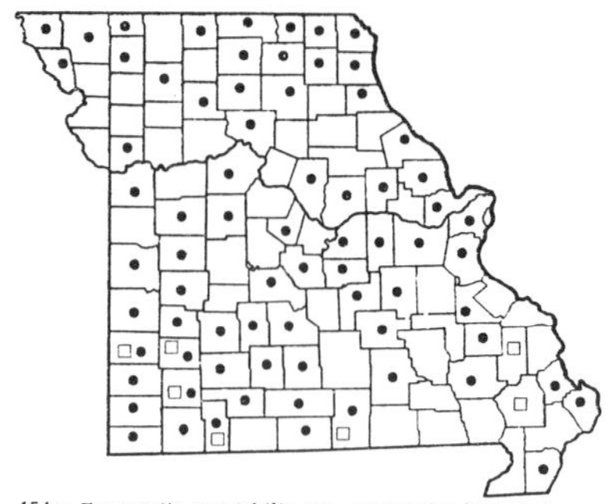

154 • Eragrostis spectabilis var. spectabilis (Purple Love Grass)

154 □ Eragrostis spectabilis var. sparsihirsuta

the range of which is from Florida to Maryland. A recent collection of *Palmer 67106* from Lawrence Co. (rocky open ground, border of woods, ½ mi. south of Aurora, Oct. 12, 1957) shows most of the leaf-sheaths nearly glabrous except for the summits and margins, but the upper part of some of the leaf-sheaths have hairs on some of the surface, more characteristic of var. *hirsuta.* Such a specimen appears to be somewhat intermediate between var. *hirsuta* and var. *laevivaginata.*

12. **Eragrostis intermedia** Hitchc. Map 152

Flowers June–October.

Rare in parts of southern and central Missouri, where known from Texas, Christian, Jackson, Johnson, Jasper, and Newton counties. It is known from rocky prairies, thin soil in chert glades, sandy or dry open ground, meadows, and waste places.

Ranges from Missouri and Kansas, south to Louisiana, Texas, Mexico, Arizona, and California; also in Georgia.

Representative collections of this species are: Jasper Co. (*Bush 9052*); Johnson Co. (*Palmer 63114*); Newton Co. (*Palmer 52452*).

Some Missouri specimens of *E. intermedia* have been misidentified as *E. lugens* Nees, a species of Mexico, Central and South America which reaches the southern borders of the United States, but has not been found in Missouri.

13. **Eragrostis trichodes** (Nutt.) Wood

Sand Love Grass Map 153

Flowers July–October.

Of scattered occurrence in southern and central Missouri north to St. Louis, Carroll, and Jackson counties, where often found in sandy soils along streams, sand washes of alluvium in valleys, or along railroads.

The following collections are representative: Jackson Co. (Courtney, *Bush 582*); St. Louis Co. (Carson Station, vicinity of St. Louis, *Eggert*; St. Louis, *Muehlenbach 1040*).

Ranges from Ohio to Nebraska, south to Louisiana and Texas.

The inflorescence of this species has a distinctly purplish color. This species is sometimes confused with the other perennial species, such as *E. hirsuta, E. spectabilis,* and *E. intermedia,* but is readily distinguished from them by the relatively larger first glume and lower lemma.

This grass attains a height of 1.5 meters, therefore, it is the tallest of the Love Grasses in Missouri. In the central Great Plains states, where known as 'ice-cream grass,' it is considered a very desirable and palatable forage grass for beef cattle.

14. **Eragrostis spectabilis** (Pursh) Steud.

Purple Love Grass Map 154

Flowers July–October.

Missouri material may be separated into the two following varietics:

Leaf-sheaths wholly glabrous (without hairs) or nearly so. . 14a. E. SPECTABILIS var. SPECTABILIS

Leaf-sheaths, at least the lowest ones, mostly hairy. 14b. E. SPECTABILIS var. SPARSIHIRSUTA

14a. **Eragrostis spectabilis** var. **spectabilis**

Occurs throughout Missouri, where it abounds along roadsides, fence-rows, sandy, rocky, or open ground, and along railroads. It is the commoner variation in the state.

Plate no. 25. **1.** Eragrostis trichodes; *a.* Panicle, × $^{3}/_{4}$; *b.* Floret, × $7^{1}/_{2}$. **2.** Eragrostis spectabilis; *a.* Plant, × $^{3}/_{8}$; *b.* Spikelet, × $3^{3}/_{4}$; *c.* Floret, × $7^{1}/_{2}$. **3.** Eragrostis curvula; *a.* Panicle, × $^{3}/_{4}$; *b.* Floret, × $7^{1}/_{2}$. **4.** Diplachne Fascicularis; *a.* Panicle, × $^{3}/_{4}$; *b, c.* Two views of floret, × $7^{1}/_{2}$. **5.** Diplachne Halei; *a.* Panicle, × $^{3}/_{4}$; *b, c.* Two views of floret, × $7^{1}/_{2}$.

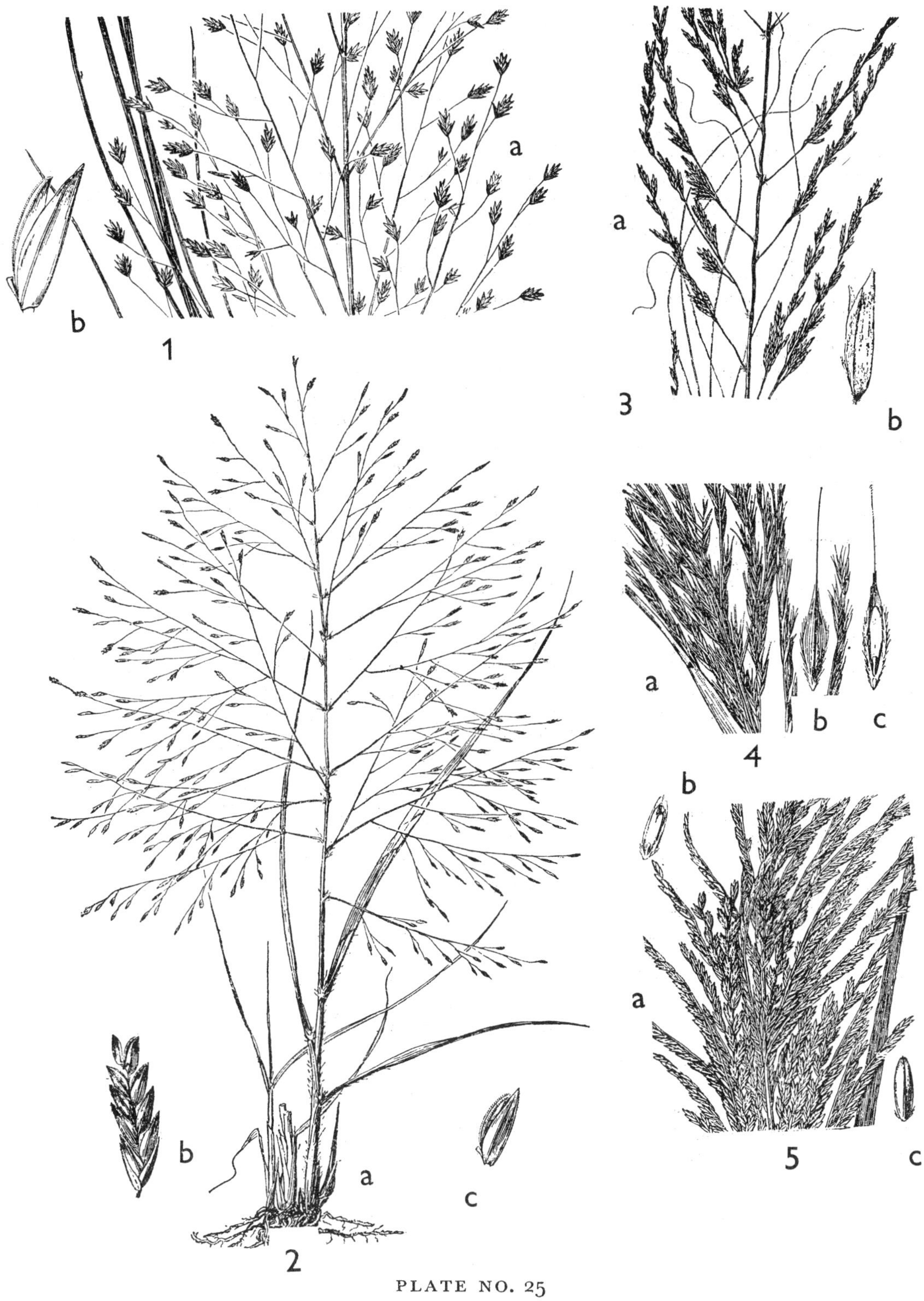


PLATE NO. 25

Ranges from Florida to Texas and Mexico, north to Massachusetts, Ohio, Michigan, and Minnesota.

14b. **Eragrostis spectabilis** var. **sparsihirsuta** Farw.

Occurs in similar situations as var. *spectabilis* and has been found in several localities in southern Missouri. It has a similar range, extending northward to Maine and South Dakota, and westward to New Mexico and Arizona.

When mature, the entire panicle breaks away from the rest of the plant and becomes blown around like a tumbleweed.

This is one of the most attractive of native grasses, with its rosy-purple color predominant in all parts of the inflorescence. The grass often occurs in dense stands which, when they occur along highways and fence-rows, produce sheets or bands of rosy-purplish hues.

The inflorescences of this species, because of their purplish color, make desirable winter bouquets for indoor adornment and for dried flower arrangements. This is also true for other species of Love Grass, whose generally delicate hair-like branches produce a graceful feathery effect.

Immature specimens of this species with small young spikelets are often mistaken for or misidentified as *E. intermedia* or *E. hirsuta*, but may be readily distinguished as *E. spectabilis* by the conspicuous nerves of the lemma, stiff scabrous pedicels (flower stalks), and scabrous keel of the lemma.

15. **Eragrostis curvula** (Schrad.) Nees
Weeping Love Grass Map 155
Flowers June–July.

Known in Missouri only from Howell County (along highway 60, dry fields and roadsides, on chert substratum, and adjacent open woods, T 27 N, R 8 W, sect. 24, July 18, 1956, *C. L. Kucera*).

A native of South Africa, this grass is occasionally cultivated for its ornamental value, and planted on poor and sandy soils for controlling erosion and improving range pasture.

It is a perennial species attaining a height of 1.5 meters. The spikelets are 5–8-flowered, about 1.5 mm. broad, dark olive green to lead gray and are crowded in a long inflorescence 10–30 cm. long, 7–16 cm. broad, with spreading or slightly ascending branches 4–7 cm. long. The leaves are usually involute and less than 2 mm. wide.

*Excluded species*

**Eragrostis mexicana** (Hornem.) Link

This species has been reported for Missouri by Gleason in *New Ill. Fl.* 1: 129. 1952. A specimen from Jackson County (Courtney, Aug. 22, 1918, *Bush 8635*) in the United States National Herbarium, at first labeled *E. curtipedicellata*, later as *E. mexicana*, and finally referred to *E. diffusa*, may be the basis for this report. This, and other Missouri specimens referred to *E. diffusa*, are treated under *E. pectinacea*.

**Eragrostis neomexicana** Vasey

This species is included under *E. mexicana* by Gleason in *New Ill. Fl.* 1: 129. 1952. In Hitchcock's *Manual of Grasses*, ed. 1, p. 157. 1935, the species is given as occurring in Missouri, the record being based upon collections from Jackson County by Bush (Courtney, Aug. 12, 1894, *789*; Oct. 2, 1892; Aug. 22, 1918, *8635*), which collections were later reidentified at United States National Herbarium as *E. diffusa*, which is considered in the present treatment as synonymous with *E. pectinacea*.

**Eragrostis lugens** Nees

Misidentified Missouri material, sometimes labeled *E. lugens*, is placed in *E. intermedia*.

## 8. **Diplachne** Beauv. Salt Meadow Grass

(*Leptochloa*, in part [BB, Hitchcock])

This genus is here considered to be a proper member of the Festuceae tribe, as maintained in Fernald's eighth edition, *Gray's Manual*, and not as one of the Chlorideae tribe, as held by Gleason in the *New Illustrated Flora* and by Hitchcock in the *Manual of Grasses*. As such, it is treated in the present work as a genus distinct from *Leptochloa*, which genus is placed in the tribe Chlorideae.

a. Lemma without any awn, the obtuse (blunt) tip at most with a short point (mucro); lower lemmas 2.2–2.9 mm. long . . . . . . . . . . 2. D. Halei
a. Lemma with a definite awn 0.5–1 mm. long from an acute to acuminate tip; lower lemmas 3–8 mm. long . . . . . . . . . . *b*

155 Eragrostis curvula (Weeping Love Grass)

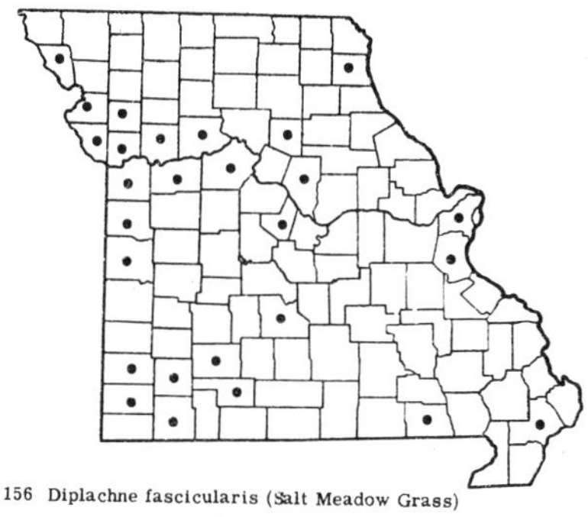

156 Diplachne fascicularis (Salt Meadow Grass)

157 Diplachne Halei

b. Culms (stems) mainly erect or ascending; spikelets densely crowded, mostly conspicuously overlapping; lemmas mostly 3–4 mm. long; 2nd glume 2–3.5 mm. long . . . . . . 1. D. FASCICULARIS

b. Culms (stems) mainly widely spreading to nearly prostrate (lying flat on ground); spikelets distant or at most slightly overlapping; lemmas 5–8 mm. long; 2nd glume 4.5–7 mm. long. . 3. D. ACUMINATA

1. **Diplachne fascicularis** (Lam.) Beauv. Map 156

*Leptochloa fascicularis* (Lam.) Gray [Hitchcock]
*Leptochloa fascicularis* var. *fascicularis* [BB]
Flowers July–October.

Inhabits soils bordering saline and brackish springs, alluvial soils of the Missouri River and other streams, depressions in rocky glades, margins of natural ponds of central Missouri, and along railroads. Mostly in central and western Missouri, and locally eastward.

Ranges from Ohio, Minnesota, South Dakota, and Washington, south to Florida and California, the West Indies, and South America.

2. **Diplachne Halei** Nash Map 157

*Leptochloa panicoides* (Presl) Hitchc. [BB, Hitchcock]
Flowers August–October.

Known only from New Madrid County, southeastern Missouri (from open sandy banks of the Mississippi River, October 3, 1955, *Palmer 61658*).

Ranges from Florida to Texas, inland in Indiana and Missouri, and south to Mexico, Guatemala, and Brazil.

3. **Diplachne acuminata** Nash Map 158

*Leptochloa fascicularis* var. *acuminata* (Nash) Gl. [BB]
*Leptochloa fascicularis* in part [Hitchcock]
Flowers July–September.

Occurs in wet soils in the vicinity of saline and brackish springs and occasionally in wet open ground along the Missouri River, mostly in central and northwestern Missouri.

Ranges in saline soils from Missouri, Nebraska, and North Dakota south to Louisiana and Texas.

By some botanists this is treated as synonymous with *D. fascicularis* or as a variety of it. It is here considered a separate species because of its larger lemmas and 2nd glume, more distantly spaced spikelets, and the characteristic spreading to prostrate culms.

Individuals of this species abound in the light gray brackish soils of various brackish-water springs in Saline, Randolph, Howard, and Cooper counties, associated with other unusual species, such as *Eleocharis parvula*, *Typha angustifolia*, and, occasionally, *Distichlis spicata* and *Scirpus paludosus*.

## 9. **Diarrhena** Beauv.

**Diarrhena americana** Beauv. Map 159

Flowers June–September.

Occurs nearly throughout Missouri, where it is apparently absent only from the southeastern counties of the state. An inhabitant of woodland, it usually occurs in rich wooded alluvial soils along rivers, at the base of limestone bluffs, and, occasionally, on moist calcareous ledges and rich upland woods and ravines.

Ranges from West Virginia, Ohio, Michigan, and South Dakota, south to Georgia, Arkansas, Oklahoma, and Texas.

The species has been divided by Gleason into two varieties, separated as follows:

Leaf-sheaths mostly pubescent; callus hairy; first glume 2.2–4 mm. long; second glume 4–5.4 mm. long; lemmas ovate, 7–10 mm. long, acuminate. a. D. AMERICANA var. AMERICANA

158 Diplachne acuminata

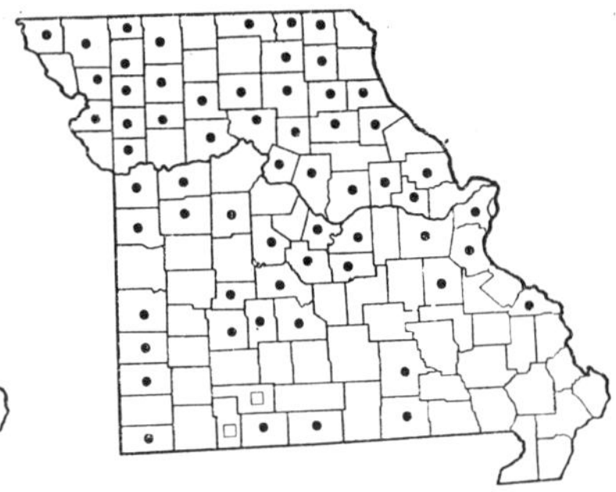

159 □ Diarrhena americana var. americana

159 ● Diarrhena americana var. obovata

160 Distichlis spicata (Seashore Salt Grass)

Leaf-sheaths glabrous or nearly so; callus glabrous; first glume 1.9–2.8 mm. long; second glume 2.4–4.3 mm. long; lemmas 5.2–6.8 mm. long, obovate, abruptly rounded into the short cusp . . b. D. AMERICANA var. OBOVATA

The species has recently been studied by Dennis Anderson, who recognizes the two varieties as meriting good varietal status, and I agree with the results of hiss tudies.

Both variations occur in Missouri, the var. *obovata* Gleason found throughout the state and var. *americana* known only in southwestern Missouri from Christian (*Steyermark 23037*) and Stone (*Steyermark 8197*) counties.

The three or four long slender leaves, nearly the length of the plant, ascend and eventually arch near the tip. By means of their creeping rootstocks the plants form dense colonies. Because of the dark green shining foliage, which remains green well into the winter, the plants are very striking and ornamental and should be more commonly used as a ground coves in shady wild gardens and for covering shady slopes.

## 10. **Distichlis** Raf. Salt Grass

Plants of this genus are dioecious (staminate and pistillate flowers on separate plants).

Growing around salt springs; margins of leaves mainly smooth, the tips rather blunt; spikelets mostly 5–12 mm. long; inflorescence condensed, the spikelets closely overlapping and hiding the pedicels (stalks of spikelets) and other spikelets; 1st glume 2–3.5 mm. long; lemmas 3.3–4.5 mm. long . . . 1. D. SPICATA

Growing along railroads and in waste ground; margins of leaves mainly minutely toothed, the tips sharply pointed; spikelets mostly 10–25 mm. long; inflorescence more open, plainly exposing the pedicels and individual spikelets; 1st glume 3.2–8 mm. long; lemmas 4.4–7.8 mm. long . . . . . 2. D. STRICTA

1. **Distichlis spicata** (L.) Greene
Seashore Salt Grass Map 160
Flowers September–October.

Known only from a salt lick in Saline County (valley of Heath's Creek, east of Elk Lick Springs, sect. 17, 3 mi. southwest of Ridge Prairie, Oct. 6, 1938, *Steyermark 21581*).

Ranges in saline soils along the coasts from Nova Scotia to Florida and Texas, British Columbia to California, Mexico, British Honduras, Cuba, and western coast of South America; inland in Missouri.

This is undoubtedly the most remarkable native grass in Missouri, representing, as it does, the only interior station for the species in the United States (Steyermark, Rhodora 42: 22–24. 1940). At the time it was discovered in 1938, it dominated several acres of a valley containing salt springs and salt licks, presenting the appearance of some salt marsh estuarier found along the coasts. It was associated with other species which inhabit saline soils, such as *Eleocharis parvula*, *Scirpus paludosus*, *Zannichellia palustris* var. *major*, and *Typha angustifolia*. The area probably represents an isolated relic habitat of a formerly prehistoric, more widespread saline region.

2. **Distichlis stricta** (Torr.) Rydb.
Inland Salt Grass Map 161
Flowers June–July.

Known only from Buchanan, Jackson, and St. Louis (*Muehlenbach 976*) counties, where it is found along railroads and in waste ground.

Ranges from Minnesota and Manitoba to British Columbia, south to Missouri, Oklahoma, Texas, California, and Mexico.

161 Distichlis stricta (Inland Salt Grass)

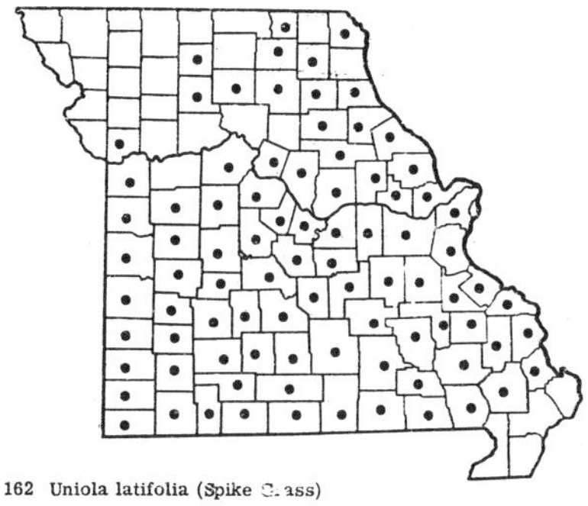

162 Uniola latifolia (Spike Grass)

163 Uniola laxa

## 11. **Uniola** L. Spike Grass

In this genus the lowermost florets of the spikelets are empty, and do not contain stamens or pistils.

Leaf-blades mainly 8–25 mm. wide; mature inflorescence with drooping branches and spikelets; spikelets 6–20-flowered, 15–40 mm. long, 10–20 mm. broad; common species . . . . . . . . 1. U. LATIFOLIA
Leaf-blades mainly 3–6 mm. wide; inflorescence with erect branches and spikelets; spikelets 1–5 (–7)-flowered, 3–9 mm. long, 2.5–5 mm. broad; known only from Ripley County. . . . . . . . 2. U. LAXA

1. **Uniola latifolia** Michx.
Broadleaf Uniola, Spike Grass Map 162
Flowers June–October.

Throughout southern and central Missouri, and in the northeastern half west to Livingston and Grundy counties. Absent from the northwestern section and the lowlands of southeastern Missouri. Frequents alluvial soils and rich woods or rocky slopes along streams, moist bluffs, especially in limestone areas.

Ranges from New Jersey, Pennsylvania, Ohio, Illinois, Missouri, and Kansas, south to Florida and Texas.

The large drooping inflorescences with their broad flattened pale spikelets provide material for decorative winter bouquets and backgrounds for flower arrangements.

2. **Uniola laxa** (L.) BSP. Map 163
Flowers July–October.

Known only from wet woods of a swale bordered by knolls in Ripley County, southeastern Missouri (T22N, R4E, $\frac{1}{4}$ southwest sect. 35, 4$\frac{3}{4}$ mi. south of Naylor, October 19, 1948, *Steyermark 66921*).

Ranges from Florida to Texas, north to New York, Pennsylvania, Kentucky, Missouri, and Oklahoma.

On account of its dark glossy green foliage, this species was confused by Bush (*Bush 6123* from Monteer, Shannon Co., Aug. 6, 1910, in N. Y. Bot. Gard. Herb.) with *Diarrhena americana*. This collection was labeled by him *Uniola laxa*.

## 12. **Dactylis** L. Orchard Grass

**Dactylis glomerata** L. Orchard Grass Map 164
Flowers May–August.

Found throughout Missouri in fields, waste ground, roadsides, fence rows, and wooded thickets bordering pastures.

Native to Europe; introduced into North America, where it is widely spread from Newfoundland to Alaska south to Florida and California.

Most of the Missouri material falls into *D. glomerata* var. *glomerata* with long-ciliate keels of the glumes and lemmas, which are glabrous on the back. A collection (fields, Grundy Co., June 27, 1937, *D. R. Crookshanks 255*) in the University of Missouri Herb. has short scabrous keels characteristic of *D. glomerata* var. *detonsa* Fries, but the short appressed hairs on the lemma indicate interbreeding with *D. glomerata* var. *ciliata* Peterm.

Spikelets vary from pale green to suffused with wine-purple. There is a striking resemblance in the field between this grass and the early flowering stage of *Phalaris arundinacea*. The broader stiffer leaves and glabrous glumes and lemmas of the latter dis-

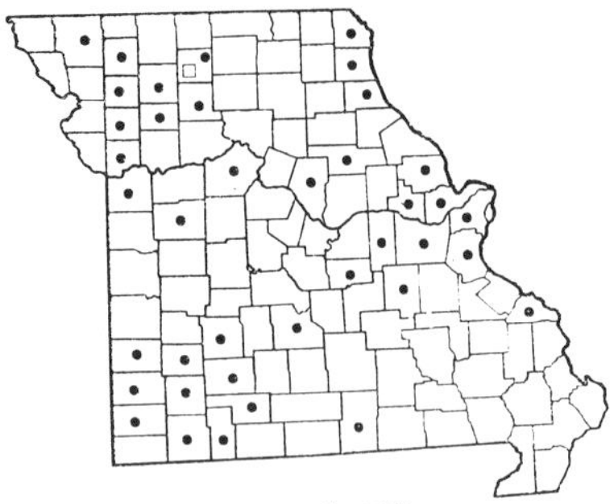

164 • Dactylis glomerata var. glomerata
164 □ Dactylis glomerata var. detonsa

165 Arundo Donax (Giant Reed)

166 Phragmites communis var. Berlandieri (Reed)

tinguish it from the relatively narrower leaves and more or less hairy glumes and lemmas of *Phragmites*, in addition to the technical differences of the number of florets in the spikelet.

This grass was formerly more commonly sown for both pasture and hay, combining drouth resistance with early pasturing properties. It is considered one of the most important hay fever-causing grasses in the United States; it sheds its pollen mainly during May and June about the same time when Blue Grass (*Poa pratensis*) is actively pollinating the air.

## 13. **Arundo** L. Giant Reed

**Arundo Donax** L. Giant Reed Map 165

Flowers September–November.

Often cultivated as an ornamental grass, but rarely escaping and persistent away from cultivation. It has escaped and persisted in Dunklin Co. (Nov. 24, 1892, *Bush*), Boone Co., and Jasper Co. (waste ground in the old mines, ½ mi. southeast of Webb City, October 8, 1953, *Palmer 57005*). At these localities it has also flowered.

Native to the Mediterranean region; introduced and escaped north in the United States to Virginia and Missouri.

In Italy and Argentina the grass is a source of pulp for paper. In Europe reeds of clarinets and organ pipes have been obtained from the culms (stems), and in some parts of the southwestern United States the culms are used as framework in the construction of adobe huts, and for screens and mats.

## 14. **Phragmites** Trin. Reed

**Phragmites communis** Trin. var. **Berlandieri** (Fourn.) Fern. Reed Map 166

*Phragmites communis* Trin. [BB, Hitchcock]

Flowers July–September.

Alluvial ground in valleys, along the larger rivers, and in low ground and swales bordering ponds and ditches in northern and central Missouri south to Callaway, Moniteau, and Bates counties.

Ranges from Quebec and Nova Scotia to British Columbia, south to Maryland, Ohio, Indiana, Illinois, Missouri, Louisiana, Texas, California, Mexico, British Honduras, and Guatemala.

The broad leaves are suggestive of corn.

In Italy and Rumania the stems are used as a source of pulp for papermaking. As with *Arundo Donax*, the stems are used in the southwestern United States as a framework in the contruction of adobe houses. The stems have also been used for thatching, nets, screens, mats, and as shafts for arrows.

Various parts of the plant may be eaten as follows: when young and beginning to unfold, the leaves can be cooked as greens. In early spring the young shoots

---

Plate no. 26. **1.** Diarrhena americana; *a.* Plant, × 3/8; *b.* Spikelet, × 3 3/4; *c.* Floret, × 3 3/4. **2.** Tridens Chapmani; *a.* Panicle, × 3/4; *b.* Floret, × 7 1/2. **3.** Tridens oklahomensis; *a.* Panicle, × 3/4; *b.* Floret, × 3 3/4. **4.** Tridens strictus; *a.* Panicle, × 3/4; *b, c.* Two views of floret, × 3 3/4. **5.** Tridens elongatus; *a.* Panicle, × 3/4; *b, c.* Two views of floret, × 3 3/4. **6.** Uniola laxa; *a.* Plant, × 3/4; *b.* Floret, × 3 3/4.

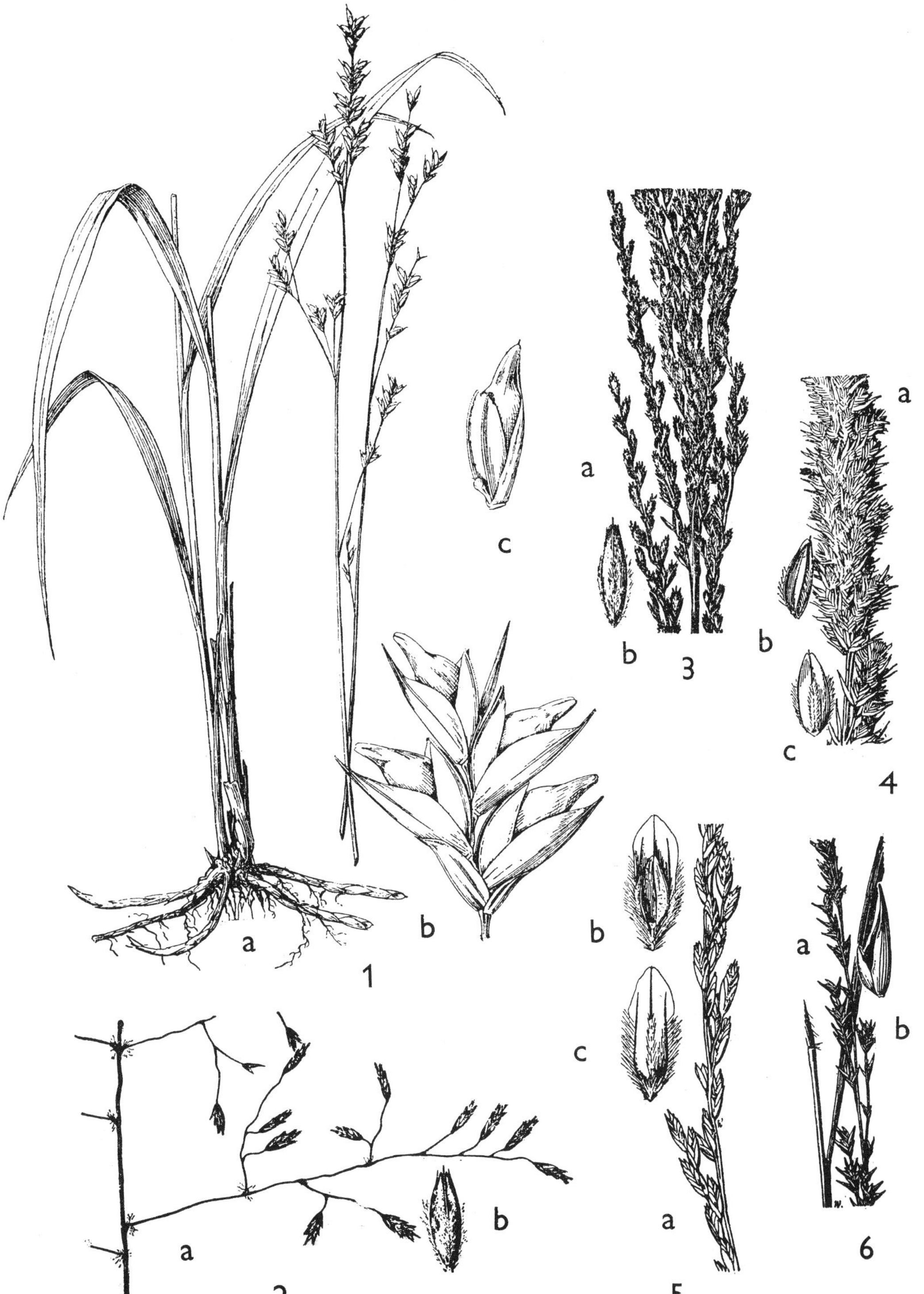


PLATE NO. 26

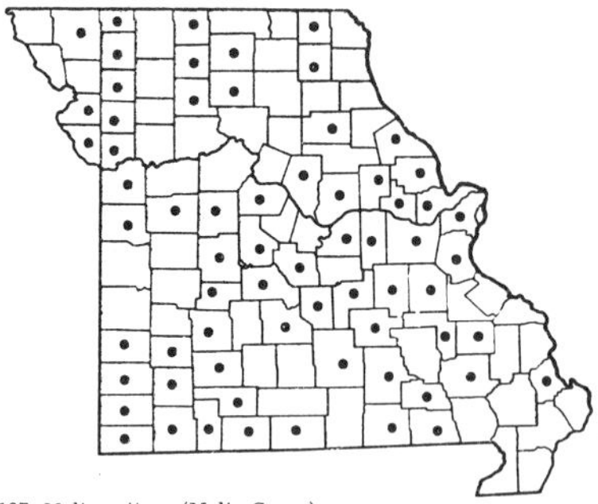

167 Melica nitens (Melic Grass)

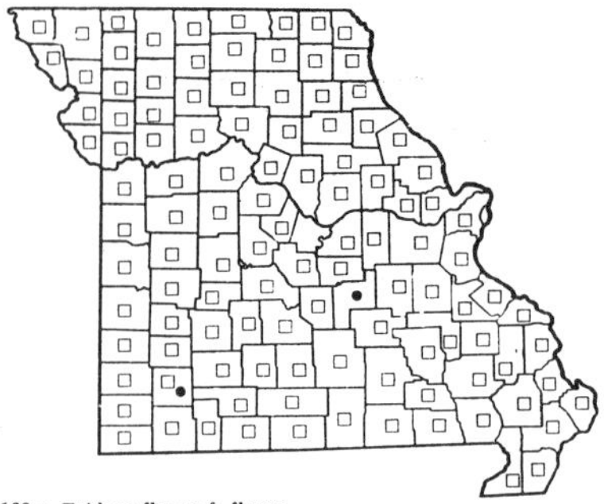

168 • Tridens flavus f. flavus
168 □ Tridens flavus f. cupreus

169 Tridens oklahomensis

can be boiled and eaten as a substitute for asparagus. At a time previous to flower formation, the stems, rich in carbohydrates and with a fleshy consistency, are collected, dried, and after being beaten into a flour, the finer part is sifted out, moistened, and roasted, and then eaten in the manner of a taffy or marshmallow. The stout rhizomes can be boiled and eaten like potatoes. The seed is nutritious. The whole grain can be cooked into a wholesome gruel.

## 15. **Melica** L. Melic Grass

**Melica nitens** Nutt. Melic Grass Map 167

*Melica Porteri* [of P & S], not Scribn.

*Melica mutica* [of P & S], not Walt.

Flowers April–July.

Common in limestone areas in open rocky woods, on bluffs, and open rocky limestone glades, and prairies. Throughout Missouri, except absent in the lowlands of southeastern Missouri. Ranges from Pennsylvania to Minnesota, south to Virginia, Kentucky, and Texas. All material in Missouri, previously identified as *M. mutica*, is here included, following Dr. Boyle's studies of the genus, under *M. nitens*.

*Excluded species*

**Melica Porteri** Scribn.

The basis for the original assignment of this species to the Missouri flora is a specimen from Jackson County (common on bluffs, at Swope Park, May 17, 1896, *Bush 856*) deposited in the United States National Herbarium. Originally labeled as *M. mutica*, it was later identified by Scribner as *M. parviflora* (Porter) Scribn., which name is synonymous with *M. Porteri*. The specimen was then taken up under the latter name and included from Missouri in Hitchcock's first edition (p. 200) of *Manual of Grasses*. It was later found to be misidentified and correctly determined as *M. nitens*. *Gray's Manual*, eighth edition and Gleason's *New Illustrated Flora*, however, have continued the error in identification, taken from Hitchcock's *Manual*. The species should be eliminated from the flora of the state.

## 16. **Tridens** Roem. & Schult

(*Triodia* of some Am. auth., not R. Br. [G, BB, Hitchcock, ed. 1])

The decision to use the generic name *Tridens* instead of *Triodia* follows the reasons given in the second edition of Hitchcock's *Manual* (p. 997), supported also by the work of Hubbard (1937), Burbridge (1946), and Shinners (1954). In the main, *Tridens*, confined to the western hemisphere, has the lemmas 3-nerved only, with the lateral nerves of the lemma marginal or nearly so, whereas *Triodia*, indigenous to Australia, consists of tussock grasses with rigidly pungent-pointed leaves, the spikelets having lemmas mostly in 3 groups of 2 or 3 nerves each, with the lateral nerves of the lemma not marginal.

---

Plate no. 27. **1.** Tridens flavus; *a.* Plant, $\times\ ^3/_8$; *b.* Spikelet, $\times\ 3^3/_4$; *c.* Floret, $\times\ 3^3/_4$. **2.** Agropyron Smithii, $\times\ ^3/_4$.

PLATE NO. 27

a. Inflorescence loose, open, the branches spreading widely . . . . . . . . . . . . . . . . . *b*
  b. Spikelets straw-colored or greenish-yellow in all stages of development. . . 1a. T. FLAVUS f. FLAVUS
  b. Spikelets dull or deep purple or rose-purple . . . . . . . . . . . . . . . . . . . . *c*
    c. Swollen base (pulvinus) of each main branch of inflorescence short-hairy on upper side only; lateral spikelets appressed along or paralleling the branches, on short stalks up to 3 mm. long at most; branches of inflorescence spreading-ascending . . . . . . . 1b. T. FLAVUS f. CUPREUS
    c. Swollen base (pulvinus) of each main branch of inflorescence surrounded by long hairs on all sides; spikelets conspicuously divergent (spreading out from the attachment), on stalks 3–20 mm. long; branches of inflorescence widely spreading and remote, presenting an open skeleton-like inflorescence . . . . . . . . . . . . . . . . . . . . . 3. T. CHAPMANI
a. Inflorescence contracted, slender, narrow, the branches ascending close to main axis . . . . . . *d*
  d. Sticky (viscid) hairs near summit of culm (stem) and on branches of inflorescence; inflorescence dark or dull purple . . . . . . . . . . . . . . . . . . . . . . 2. T. OKLAHOMENSIS
  d. No sticky (viscid) hairs either near summit of culm or on branches of inflorescence; inflorescence brownish, straw-color, gray-green, or tinged with rose-purple, but not dark or dull purple. . . . *e*
    e. Culms and leaf-sheaths smooth; spikelets 4–6 mm. long; glumes glabrous with glandular-viscid (sticky) keel; lemmas mucronate (with a tiny projection at tip) . . . . . . . 4. T. STRICTUS
    e. Culms and leaf-sheaths rough or scabrous (fine short stiff hairs); spikelets 7–12 mm. long; keel (narrow back edge) of glumes scabrous; lemmas obtuse (blunt) . . . . 5. T. ELONGATUS

1. **Tridens flavus** (L.) Hitchc.
Purpletop, Tall Redtop Map 168
*Triodia flava* (L.) Smyth [G, BB, Hitchcock, ed. 1]
Flowers July–August.

Fields, pastures, sandy open ground, rocky glades, open, often cut-over woodland, waste ground, and along roadsides. The common purple form, *T. flavus* f. *cupreus* (Jacq.) Fosb., is known from every county in Missouri. The rare form, *T. flavus* f. *flavus*, with straw-colored or greenish-yellow spikelets is known from Phelps (openings along road in valley of Kaintuck Hollow, T36N, R9W, sect. 10, $4\frac{1}{2}$–$4\frac{3}{4}$ mi. south southwest of Newburg, September 11, 1956, *Steyermark 82568*) and Lawrence (dry sandy open ground along small ravine, $1\frac{1}{2}$ mi. northeast of Nickelville, October 20, 1953, *Palmer 57099*) counties.

Occasionally plants are found which show some intergradation in color, with a slight pale purplish tinge on spikelets mostly pale greenish or straw-color. An example of this is *Palmer 65957* from Newton County.

Ranges from New Hampshire to Minnesota and Nebraska, south to Florida and Texas.

Although this grass is one of the most abundant in Missouri, it is practically absent from the northern third of nearby Illinois and Indiana.

As the seed matures, the branches of the inflorescence and upper portion of the culm secrete a sticky substance which darkens as the plant becomes exposed to dust and air particles. The hands of unknowing persons receive a greasy blackish streak upon handling the mature inflorescence.

During the early stages of growth, livestock find this grass quite palatable.

2. **Tridens oklahomensis** (Feath.) Feath.
Map 169
*Triodia oklahomensis* Feath. Rh. 40: 243. 1938.
Flowers September–October.

Known only from Boone County (moist soil, T48N, R13W, sect. 3, September, 1955, *Kucera*), where associated with *T. flavus* and *T. strictus*.

Known only from Oklahoma, where originally discovered by Mr. A. E. Wade, and from Missouri.

The spikelets, which are 6–8 mm. long, 4 mm. wide, and 7–9-flowered, are deep purple as in *T. flavus* f. *cupreus*, and resemblances to the Purpletop are also found in the viscid-sticky inflorescence, in size of plant, and habitat. The branches of the inflorescence of *T. oklahomensis* have many more flowers crowded along their branches than in *T. flavus*. Dr. Kucera (Rh. 59: 72. 1957) suggests that *T. oklahomensis* may be of hybrid origin, as it has been found in various stations where *T. flavus* and *T. strictus* were growing with it, and shows characters partaking of both these species.

3. **Tridens Chapmani** (Small) Chase Map 170
*Triodia Chapmani* (Small) Bush [G]
*Triodia flava* var. *Chapmani* (Small) Fern. & Grisc.
*Tridens flavus* var. *Chapmanii* (Small) Shinners
Flowers August–September.

Known only from Shannon County (rocky woods, September 10, 1908, *Bush 5126*).

Ranges from Florida to Texas, north to New Jersey, Virginia, Tennessee, and southeastern Missouri.

Intergradation between this and *T. flavus* has, thus far, not been noted. Dr. Fernald concluded it was a distinct species after several years of field observations.

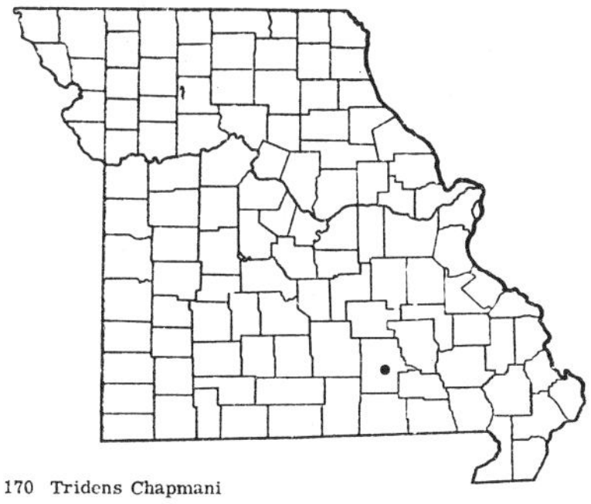

170 Tridens Chapmani

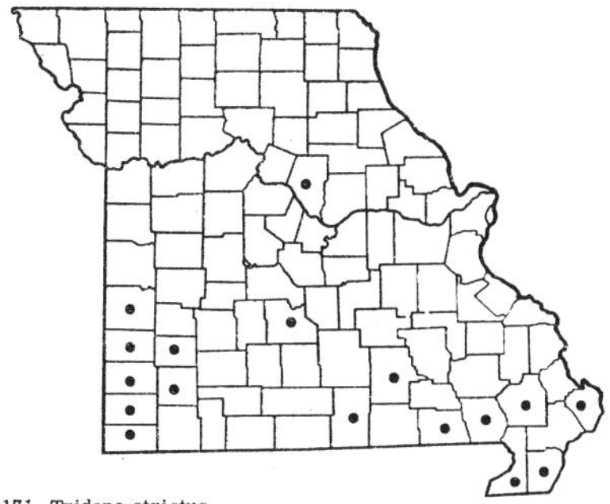

171 Tridens strictus

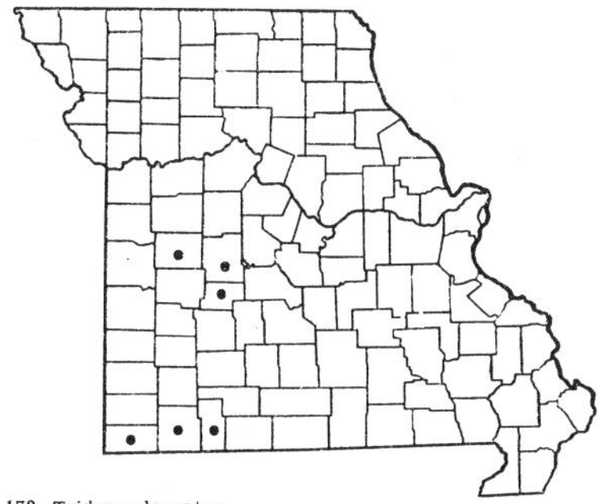

172 Tridens elongatus

Until additional Missouri collections are made and become available for study, *T. Chapmani* is, likewise, being retained as a species separate from *T. flavus*.

4. **Tridens strictus** (Nutt.) Nash Map 171
Flowers July–October.

Prairies, sandy open ground, and moist or dry open ground along ditches and roadsides in southern Missouri locally north to Boone County.

Ranges from Alabama to Texas, north to Tennessee, Missouri, and Kansas.

5. **Tridens elongatus** (Buckl.) Nash Map 172
*Triodia elongata* (Buckl.) Scribn. [G, BB, Hitchcock, ed. 1]
*Tridens muticus* (Torr.) Nash, var. *elongatus* (Buckley) Shinners
Flowers June–October.

Limestone glades and rocky ledges of limestone bluffs in southwestern Missouri north to Benton and Henry counties.

Ranges from Missouri to Colorado, south to Arkansas, Texas, and Arizona.

## 17. **Triplasis** Beauv. Sand Grass

**Triplasis purpurea** (Walt.) Chapm.
Sand Grass Map 173
Flowers August–November.

Sand dunes and flats along the Missouri and Mississippi rivers, and locally in sandy soil in Ozark and Dunklin counties.

Ranges along the coast from Maine to Florida and Texas, inland on the shores of the Great Lakes from New York and Ontario to Minnesota and Colorado, south into Illinois, Missouri, Arkansas, and Texas.

The occurrence of this grass is practically always with a habitat of pure or nearly pure sand. Along the sand dunes of the Mississippi and Missouri rivers it is commonly associated with other sand-loving species, such as *Sporobolus cryptandrus*.

Among grasses this species is unusual in having an acid taste.

## Tribe III. **HORDEAE**

Grasses of this tribe have usually sessile (stalkless) or nearly sessile spikelets arranged in rows on opposite sides of the central axis, the spikelets are usually 2- or more-flowered, or sometimes 1-flowered, and the rachilla is articulated above the glumes.

## 18. **Agropyron** Gaertn. Wheat Grass

a. No underground elongated creeping rootstocks present; anthers mainly 1–2.5 mm. long; mature spikelets breaking apart easily, the individual florets of the spikelets separating from one another; glumes persisting after lemmas fall . . . . . . . . . . . . . . . . . 3. A. TRACHYCAULUM
a. Underground elongated creeping rootstocks present; anthers mainly 3–7 mm. long (rarely shorter); mature spikelets falling as a whole, not breaking apart into the separate florets; glumes not persisting after lemmas fall . . . . . . . . . . . . . . . . . . . . . . . . . . *b*
b. Leaves silvery blue-green, stiff, deeply grooved on upper side with inrolled margins; rootstocks

173 Triplasis purpurea (Sand Grass)

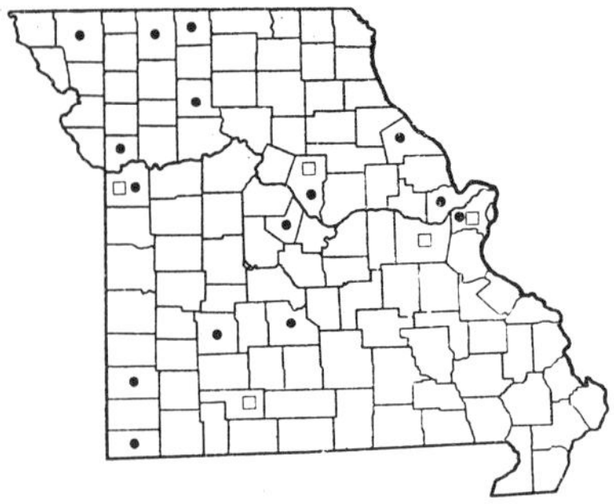

174 ● Agropyron repens var. subulatum f. subulatum (Quack Grass)
174 □ Agropyron repens var. subulatum f. Vaillantianum (Quack Grass)

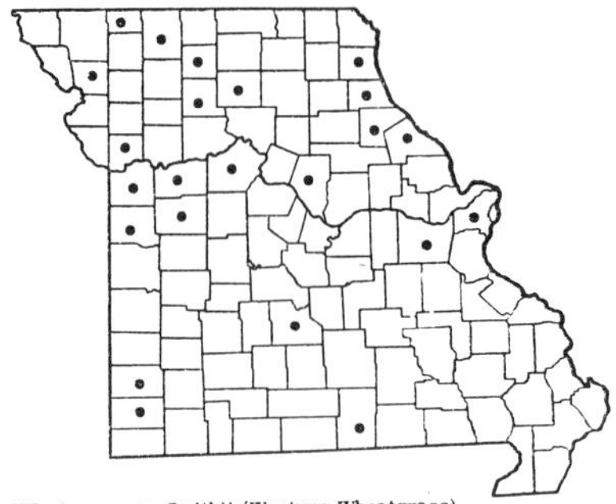

175 Agropyron Smithii (Western Wheatgrass)

dull brown or yellowish-brown; upper nodes (joints) of stems with the swollen portion shorter than broad (thick); spikelets 6–16-flowered, 15–28 mm. (excluding awns) long . . . 2. A. SMITHII

b. Leaves deep green or bluish-green, flexible and rather soft, flat and not deeply grooved; rootstocks white or yellowish-white, sometimes with lavender sheaths; upper nodes (joints) of stems with the swollen portion longer than broad (thick); spikelets mostly 3–6-flowered, mostly 10–16 mm. (excluding awns) long . . . . . . . . . . . . . . . . . . . 1. A. REPENS

1. **Agropyron repens** (L.) Beauv. Map 174
Quack Grass

Flowers late May–September.

Fields, gardens, fence rows, thickets, along roadsides, and waste ground.

Typical *A. repens* var. *repens* has not been found, thus far, in Missouri. Missouri material may be divided into the two following segregates:

No awns protruding from tip of lemma
1a. A. REPENS var. SUBULATUM f. SUBULATUM

Awns protruding from tip of lemma . 1b. A. REPENS var. SUBULATUM f. VAILLANTIANUM

1a. **A. repens** var. **subulatum** (Schreb.) Reichenb. f. **subulatum** Map 174

This is the more common variation in Missouri, and is more or less scattered over the entire state. Only a comparatively few herbarium records exist, as may be noted on the accompanying map, but undoubtedly the plants occur in nearly every county of the state.

1b. **A. repens** var. **subulatum** f. **Vaillantianum** (Wulf. & Schreb.) Fern. Map 174

This variation is occasionally found in the state. A representative specimen is *Steyermark 7888* (near Pacific, Franklin County, June 8, 1930) in the Mo. Bot. Gard. Herb.

*Agropyron repens* var. *repens* and its other varieties and forms were chiefly introduced in the early 1800's from Europe, and have since spread over most of North America. Along the north Atlantic coast there are apparently indigenous wild varieties.

During June and July when *A. repens* var. *repens* and its other varieties and forms are at the height of pollen production, they are considered to be a contributing factor in causing hay fever, but, because of the small amount of pollen produced, are of only secondary importance.

The creeping underground rootstocks have been dried and ground into a flour for use in making bread (Withering, 1830). This is a very bad weed. Once it gets established in gardens, especially vegetable gardens or strawberry beds, it is practically impossible to eradicate. Chemicals for getting rid of the grass are being developed, but as yet no positive selective chemical killer has been effective, and the only practical method thus far known is spading up each plant and picking out the roots and then burning them. Turning a couple of geese loose in plots of ground where Quack Grass is rampant is said to constitute an effective measure for eradicating it, but unfortunately the geese eat other desirable plants. The tough rootstocks may travel from 2 to 8 inches below the surface of the ground. The root system produces a heavy and packed soil, and, for this purpose, sometimes is used for stabilizing embankments and waterways. The grass is palatable to livestock and is used for pasture and hay production.

Plate no. 28. **1.** Agropyron repens; *a.* Plant, × 3/8; *b.* Spikelet, × 2 1/4; *c.* Floret, × 2 1/4. **2.** Triplasis purpurea; *a.* Plant, × 3/8; *b.* Spikelet, × 3 3/4; *c.* Floret, × 3 3/4; *d.* Cleistogamous spikelet, × 3 3/4. **3.** Agropyron trachycaulum, × 3/4.

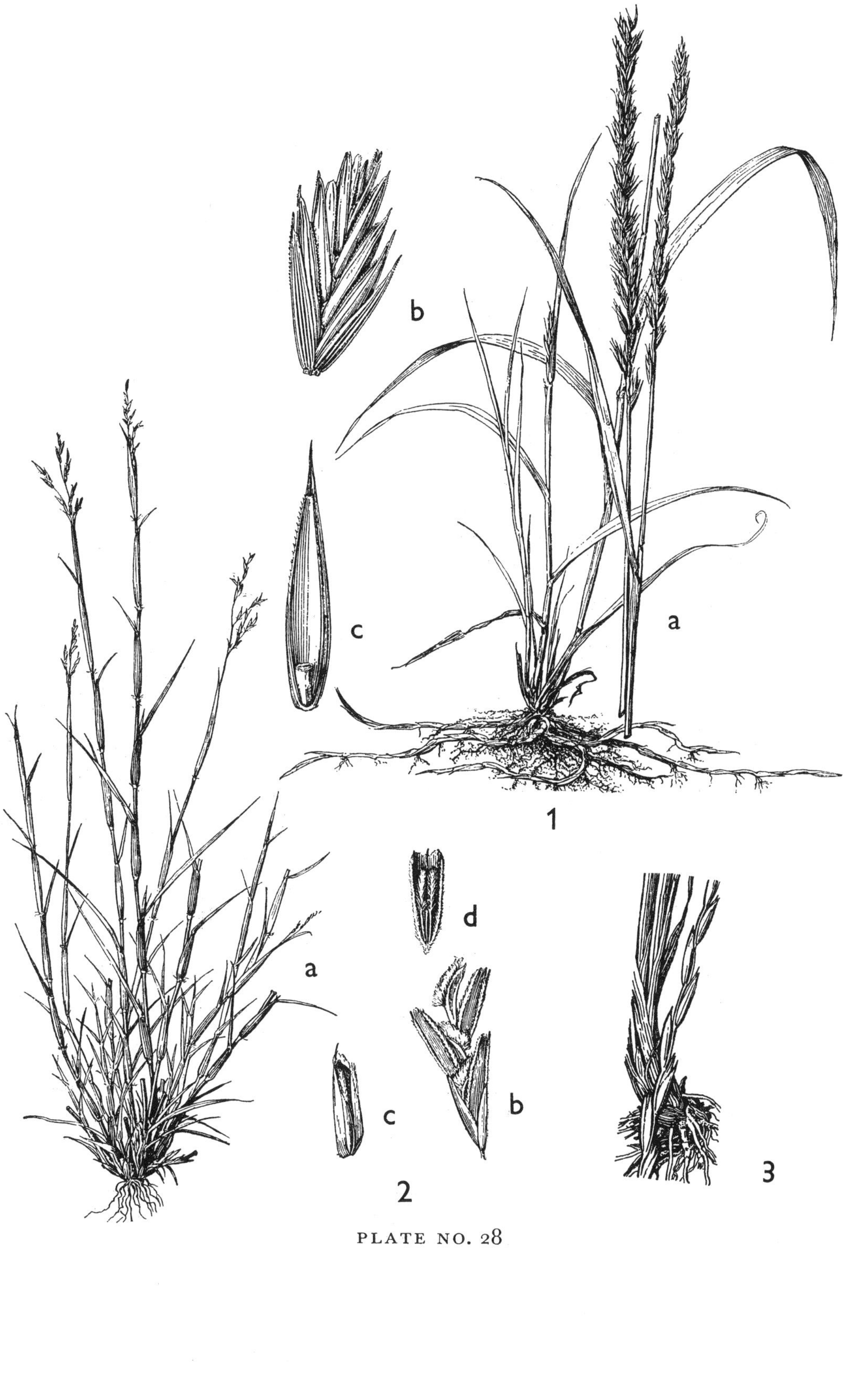


PLATE NO. 28

2. **Agropyron Smithii** Rydb.
Western Wheat Grass Map 175
Flowers May–August.

Rather common along railroad tracks in northern and central Missouri, and also in fields, prairies, along roadsides, and waste ground. Not yet recorded from southeastern Missouri.

Ranges from New York, Ontario, and Alberta to Washington, south to Tennessee, Arkansas, Texas, Arizona, and California; spreading eastward to Quebec and New Hampshire.

Missouri material previously misidentified as *A. spicatum* (Pursh) Scribn. & Smith, a western species, and as *A. pseudorepens* Scribn. & Smith (synonymous with *A. trachycaulum* var. *majus* [Vasey] Fern.), has been referred to *A. Smithii.*

This species is easily recognized by the silvery blue color of the leaves and culms, and by the stiff rigid leaves with inrolled margins (involute). The name 'Bluestem' is sometimes applied to this species.

The grass, during its period of flowering (mostly June and July) is believed to be of some importance in causing hay fever. It is considered to be nutritious, and is commonly eaten by livestock, principally in the Great Plains and Pacific Northwest states where it is native. Its hay ranks high in protein value.

3. **Agropyron trachycaulum** (Link) Steud.
Bearded Wheat Grass, Awned Wheat Grass Map 176
Flowers June–August.
Missouri material falls into two varieties:

No awns present or awns at most only half as long as the body (awnless part) of the lemma . . . 3a. A. TRACHYCAULUM var. TRACHYCAULUM
Awns present and nearly as long as or longer than the body (awnless part) of the lemma . . . 3b. A. TRACHYCAULUM var. GLAUCUM

3a. **Agropyron trachycaulum** (Link) Steud. var. **trachycaulum** Map 176
*Agropyron trachycaulum* (typical) [G]
*Agropyron pauciflorum* (Schwein.) Hitchc. [Hitchcock ed. 1]
*Agropyron tenerum* Vasey

Fields, waste ground, and cemeteries. Known thus far from a few eastern (Marion, Ralls, Pike, St. Louis) and western (Jackson, Jasper) counties.

Var. *trachycaulum* ranges from Quebec and Ontario to Alaska, south to Illinois, Iowa, Missouri, Kansas, New Mexico, and California.

3b. **Agropyron trachycaulum** var. **glaucum** (Pease & Moore) Malte Map 176
*Agropyron caninum* of Am. auth., not Beauv.

Known only from Newton County (open ground in grassy and weedy field, 2 mi. southeast of Joplin, July 26, 1957, *Palmer 66179*).

Ranges from Newfoundland to British Columbia, south to Maryland, Pennsylvania, Michigan, Wisconsin, Missouri, Nebraska, Colorado, Nevada, and California.

Missouri material has been confused with var. *unilaterale.*

*Excluded species and varieties*

**Agropyron trachycaulum** var. **unilaterale** (Cassidy) Malte

Missouri is cited in the range of this variety in Gleason's *Illustrated Flora* (p. 140), but the Missouri specimens are referred in this work to var. *trachycaulum.* Some material so identified has been placed in var. *glaucum.*

## 19. **Triticum** L. Wheat

**Triticum aestivum** L. Wheat Map 177
Flowers early May–July.

This cultivated annual has been introduced from Eurasia.

Frequently seed is scattered along roads, railroad tracks, and in waste ground, where it germinates and grows, but does not actually persist or perpetuate itself as part of the introduced flora in the manner of most other introduced species.

Because of the tendency for wheat to be self-pollinated, it can be considered only a minor cause of hay fever.

Straw from this species is used for pulp in paper making in some parts of the United States, South America, Europe, North Africa, and Asia Minor.

---

Plate no. 29. **1.** Triticum aestivum; *a.* Plant with awned spikes (bearded wheat), × $^3/_8$; *b.* Plant with nearly awnless spike (beardless wheat), × $^3/_8$; *c.* Spikelet, × $2^1/_4$; *d.* Floret, × $2^1/_4$. **2.** Aegilops cylindrica, × $^3/_8$.

PLATE NO. 29

176 • Agropyron trachycaulum var. trachycaulum (Bearded Wheat Grass)
176 □ Agropyron trachycaulum var. glaucum (Bearded Wheat Grass)

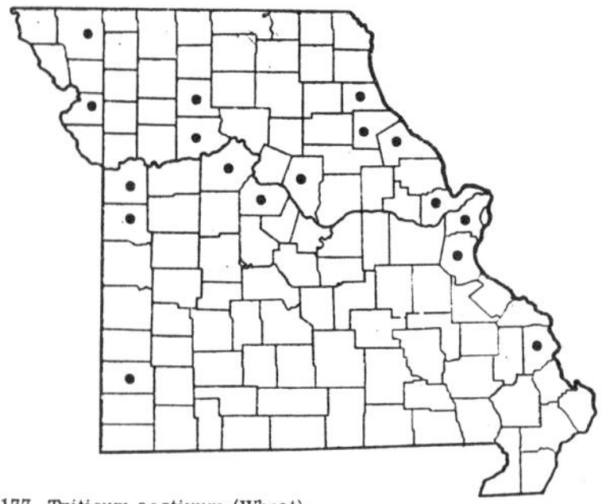

177 Triticum aestivum (Wheat)

178 Aegilops cylindrica (Goat Grass)

## 20. **Aegilops** L. Goat Grass

**Aegilops cylindrica** Host Goat Grass Map 178
Flowers May–August.

Usually found along railroad tracks, or in dry open ground along roads and waste places. Until 1935 recorded from Jackson County only. Since then it has become increasingly more common, and at present is known from eleven counties in the state.

Native of Europe; introduced into the United States, where it has extended its range from Colorado and New Mexico east to Oklahoma, Missouri, Indiana, and locally to New York.

## 21. **Secale** L. Rye

**Secale cereale** L. Rye Map 179
Flowers April–June.

Commonly cultivated annual and germinating along roadsides, railroad tracks, alluvial banks, and waste ground, but never persisting. Scattered over the state.

Originated in Eurasia.

The abundant pollen shed from Rye during its flowering season, extending mainly from May to June, is considered to be the cause of some hay fever, more especially in Europe.

The dried straw of the stems is used principally in central Europe as pulp in the making of paper.

## 22. **Elymus** L. Wild Rye

The leaf-blades have an auricle (outgrowth) at base. The spikelets are 2–6- (rarely 1-) flowered, and usually occur in 2's at each joint of the rachis, or rarely in 3's or solitary at the joint.

a. No awns or very short inconspicuous awns at most 3 mm. long present on the spikelets . . 6e. E. VIRGINICUS var. SUBMUTICUS
a. Conspicuous awns 5–40 mm. long present on the spikelets . . . . . . . . . . . . . . . . *b*
  b. Awns at maturity curved, bending, or spreading outward; palea 8.5–11 mm. long . . . . . . *c*
    c. Glumes bristle-like their entire length, not broadened above the base, not noticeably nerved, 2–15 mm. long (including the awn); spikelets 2–3-flowered . . . . . . . 3. E. INTERRUPTUS
    c. Glumes bristle-like in upper half, but broadened above base in lower half, where 3–5-nerved, 15–35 mm. long (including the awn); spikelets 2–5-flowered. . . . . . . . 4. E. CANADENSIS
  b. Awns at maturity straight or nearly straight, at least not curved or spreading outward; palea mainly 5–8 mm. long. . . . . . . . . . . . . . . . . . . . . . . . . *d*
    d. Glumes 0.8–2.5 mm. wide, flat, at least above the base, not bristle-like in the lower half; spikes mainly erect, stiff . . . . . . . . . . . . . . . . . . . . . . . *e*
      e. Glumes not thickened at base, rather thin, straight; internodes of rachis 8–10 mm. long; limestone bluffs of southwestern Missouri . . . . . . . . . . . . 1. E. GLAUCUS

---

Plate no. 30. **1.** Elymus glaucus, × 3/4. **2.** Elymus villosus, × 3/4. **3.** Elymus interruptus, × 3/4. **4.** Elymus riparius, × 3/4. **5.** Elymus virginicus, × 3/4. **6.** Secale cereale; *a.* Plant, × 3/8; *b.* Spikelet, × 2 1/4; *c.* Floret, × 3 3/4.

PLATE NO. 30

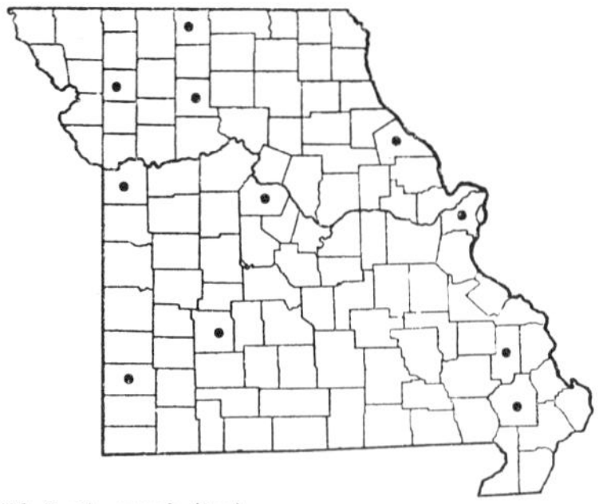

179 Secale cereale (Rye)

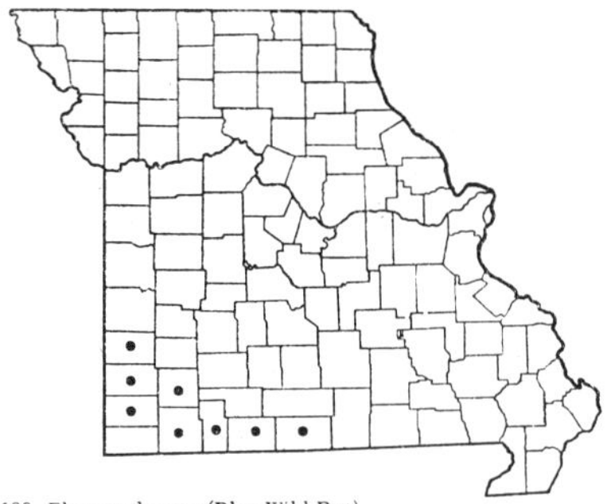

180 Elymus glaucus (Blue Wild Rye)

181 • Elymus villosus f. villosus
181 □ Elymus villosus f. arkansanus

e. Glumes thickened at base, the base strongly bowed outward, outwardly curved or arched; internodes of rachis 3–6 mm. long; throughout Missouri . . . . . . . . 6. E. VIRGINICUS

d. Glumes 0.1–0.8 mm. wide, bristle-like or hair-like throughout from base to tip; spikes mainly nodding, declining, or curving down . . . . . . . . . . . . . . . . . f

f. Spikelets 1- (rarely 2-) flowered, palea 5.2–6.7 mm. long; leaf-blades usually hairy on upper surface . . . . . . . . . . . . . . . . . 2. E. VILLOSUS

f. Spikelets 2–4-flowered; palea 7.5–8 mm. long; leaf-blades usually glabrous. . . . 5. E. RIPARIUS

1. **Elymus glaucus** Buckley
Blue Wild Rye Map 180
*Elymus Mackenzii* Bush, Am. Midl. Nat. 10: 53. 1926 (holotype, *Bush* 77 from Eagle Rock, Barry County).
Flowers late May–August.

Usually found on rocky limestone ledges of bluffs along White River and tributaries and other streams of southwest Missouri, east to Ozark County and north to Barton County.

Ranges from western New York and Ontario to Michigan, south to Missouri and Arkansas; and Alaska to California and New Mexico.

Bush originally labeled his specimens of this species as *E. Macounii* Vasey, which species does not occur in Missouri.

This is a distinctive species of the White River and adjacent southwestern Missouri area. It is mostly glabrous with 3–6-flowered spikelets, 3–5-nerved short-awned glumes which are shorter than the nearly glabrous lemma. The elongated internodes of the rachis, up to 8–10 mm. long, are much longer than are found in any other Missouri species of *Elymus*. However, internodes only 5 mm. long are commonly encountered.

2. **Elymus villosus** Muhl. Map 181
*Elymus canadensis* var. *villosus* (Muhl.) Shinners
Flowers June–September.

Rich or rocky woods, on slopes, ridges, alluvial wooded banks of streams, thickets, sometimes in prairies, and along roadsides.

Ranges from Quebec to North Dakota, south to North Carolina, Alabama, Arkansas, Oklahoma, Texas, and New Mexico.

Missouri material may be separated as follows:

Glumes and lemmas villous (bearing long soft hairs) . . . . . . 2a. E. VILLOSUS f. VILLOSUS

Glumes and lemmas glabrous or merely scabrous (very short rough hairs) . . . 2b. E. VILLOSUS f. ARKANSANUS

2a. **Elymus villosus** f. **villosus** Map 181
Common throughout Missouri.

2b. **Elymus villosus** f. **arkansanus** Map 181
(Scribn. & Ball) Fern.

Mostly in southern Missouri, locally northward to Miller, St. Charles, Pike, and Gentry counties.

The leaf-sheaths vary from pubescent to glabrous. The spikes are usually declined or nodding.

3. **Elymus interruptus** Buckl. Map 182
Flowers June–August.

Prairies and rocky wooded slopes. Known only from three counties in western Missouri, in Gentry County (natural prairie along road Z, 2.8 mi. west of Berlin, northeast ¼ sect. 33, July 30, 1954, *Steyermark*

Plate no. 31. **1.** Hordeum pusillum, × 3/4. **2.** Hordeum jubatum, × 3/4. **3.** Hystrix patula; *a.* Plant, × 3/8; *b.* Spikelet, × 2 1/4; *c.* Floret, × 2 1/4. **4.** Sphenopholis intermedia; *a.* Panicle, × 3/4; *b.* Floret, × 7 1/2; *c.* Glumes, × 7 1/2. **5.** Sphenopholis nitida; *a.* Panicle, × 3/4; *b.* Florets, × 7 1/2; *c.* Glumes, × 7 1/2.

PLATE NO. 31

182 Elymus interruptus

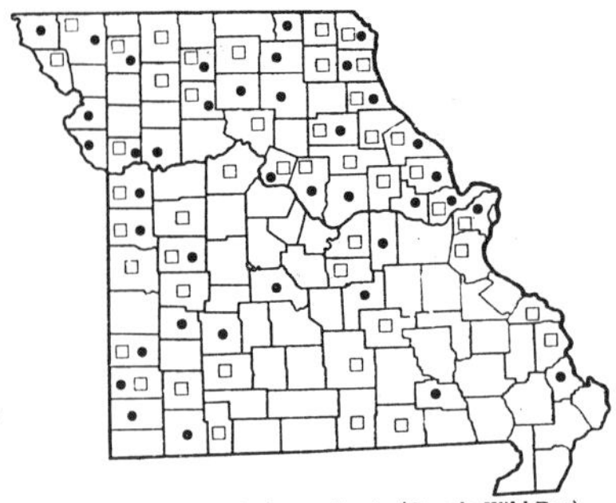

183 • Elymus canadensis f. canadensis (Canada Wild Rye)
183 □ Elymus canadensis f. glaucifolius

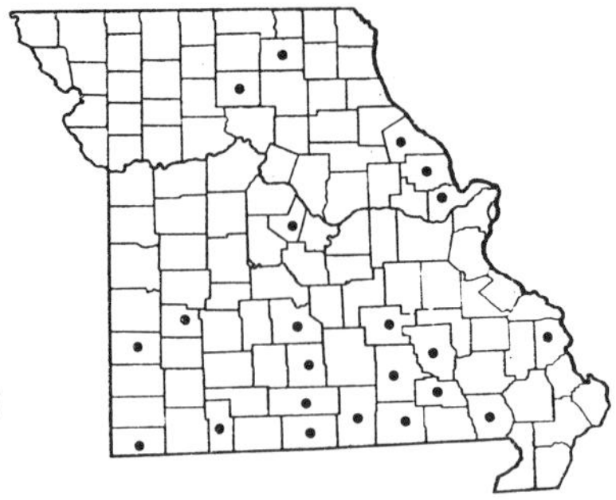

184 Elymus riparius

*76299*), Lafayette County (wooded rocky hillside along Missouri River, 1½ mi. northeast of Dover, June 12, 1953, *Palmer 55939*), and Barry County (rocky open woods, near Viola, August 26, 1957, *Palmer 66416*).

Ranges from Wisconsin to North Dakota and Wyoming, south to Tennessee, Oklahoma, Texas, and Mexico.

This resembles a hybrid between *Elymus canadensis* and *Hystrix patula* as well as suspected of being one between *Elymus canadensis* and *E. villosus*. The spike is rather loosely flowered below and the lemmas are hirsute with spreading awns 2–4 cm. long.

4. **Elymus canadensis** L. Canada Wild Rye, Nodding Wild Rye Map 183

Flowers June–October.

Prairies, rocky open woods, rocky slopes of bluffs, alluvial woods, open banks of river, and waste ground. Occurring in most parts of the state, but mostly absent, so far as available records go, from the south-eastern section.

Ranges from New Brunswick to Alaska, south to North Carolina, West Virginia, Ohio, Indiana, Illinois, Missouri, Texas, New Mexico, Arizona, and California.

Two forms are recognized, both about equally common:

Lemmas villous-hirsute (noticeably hairy) . . . .
4a. E. CANADENSIS f. CANADENSIS

Lemmas glabrous or at most scabrous . . . . .
4b. E. CANADENSIS f. GLAUCIFOLIUS (Muhl.) Fern.

Some plants are more glaucous or silvery blue-green than others. There is also considerable variation in width and pubescence of leaf-blades, and in length and pubescence of lemmas and awns.

The species is of minor importance in causing hay fever.

As the flower spikes mature, they curve or droop downward. A black fungus, ergot, sometimes infests the flowering heads of this species and of *E. virginicus*. This grass is very palatable and is eaten by all classes of livestock. It provides good forage and hay, and is nutritious.

5. **Elymus riparius** Wieg. Map 184

Flowers June–September.

Along rocky wooded slopes, bluffs, and banks of streams, where frequent in the Ozark region, and extending north locally to Pike, Adair, and Linn counties.

Ranges from Maine and Quebec to Wisconsin, south to Florida, Kentucky, and Arkansas.

The long downward-curving or nodding spikes, wide glabrous leaf-blades (1–2.2 cm. wide), glabrous leaf-sheaths, internodes of rachis 3–8 mm. long, palea 7.5–8 mm. long, and bristlelike glumes only 0.4–0.8 mm. wide are distinguishing characters of this species.

6. **Elymus virginicus** L. Wild Rye Map 185

Flowers late May–September.

A great variety of habitats are occupied by the various varieties and forms of *E. virginicus*. The most common habitats are in low woods, valley bottoms, and alluvial wooded banks of streams, but other habitats, such as prairies, open or sandy fields, rocky exposures on bluffs, wooded limestone ledges, and rocky limestone glades, are occupied.

The combined varieties and forms of *E. virginicus*, which occur throughout Missouri, range from Newfoundland to Alberta, south to Florida and Arizona.

The following variations are found in Missouri:

a. No conspicuous awns at tip of glumes or lemmas, the glumes and lemmas ending only in a short point or subulate tip, or prolonged into an awn at most 3 mm. long . 6d. E. VIRGINICUS var. SUBMUTICUS

a. Definite conspicuous awns 5–40 mm. long at tip of glumes or lemmas . . . . . . *b*

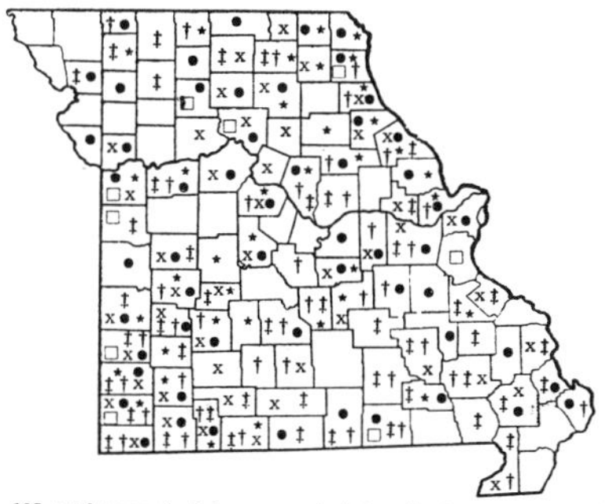

185 • Elymus virginicus var. virginicus f. virginicus (Wild Rye)
185 ⋆ Elymus virginicus var. virginicus f. hirsutiglumis.
185 x Elymus virginicus var. jejunus
185 □ Elymus virginicus var. submuticus
185 ‡ Elymus virginicus var. glabriflorus f. glabriflorus
185 † Elymus virginicus var. glabriflorus f. australis

186 Sitanion longifolium (Squirrel Tail)

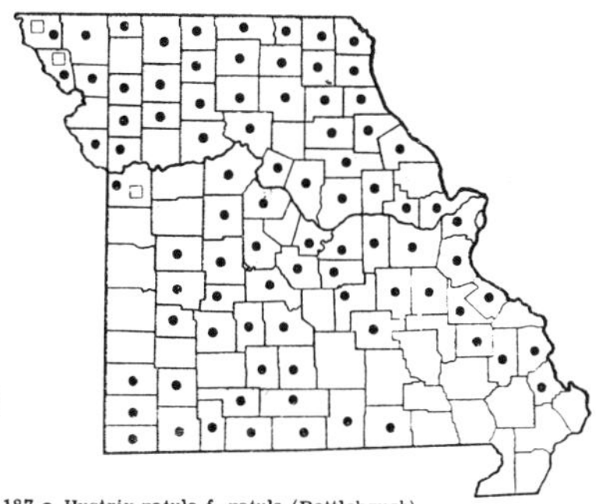

187 • Hystrix patula f. patula (Bottlebrush)
187 □ Hystrix patula f. Bigeloviana

b. Lower part of inflorescence surrounded by an enlarged or inflated uppermost leaf-sheath; glumes 1.5–2 mm. wide . . . *c*
  c. Glumes and lemmas glabrous or at most scabrous . . . . . 6a. E. VIRGINICUS var. VIRGINICUS f. VIRGINICUS
  c. Glumes and lemmas villous-hirsute. . 6b. E. VIRGINICUS var. VIRGINICUS f. HIRSUTIGLUMIS

b. The whole inflorescence long-exserted, projecting mainly 3–15 cm. beyond the little- or not-inflated uppermost leaf-sheath; glumes 0.8–1.6 mm. wide . . . . . *d*
  d. Glumes (including the awns) 1–2.7 cm. long; lemmas (including the awns) 1–3 cm. long (examine spikelets half-way up for measurements) . . 6c. E. VIRGINICUS var. JEJUNUS
  d. Glumes (including the awns) 2.7–4 cm. long; lemmas (including the awns) mainly 3.5–4.5 cm. long (examine spikelets half-way up for measurements) (spikelets near base often have shorter lengths) . . . . . . . . . . *e*
    e. Glumes and lemmas glabrous or at most scabrous . . . 6e. E. VIRGINICUS var. GLABRIFLORUS f. GLABRIFLORUS
    e. Glumes and lemmas hirsute . . . 6f. E. VIRGINICUS var. GLABRIFLORUS f. AUSTRALIS

6a. **Elymus virginicus** var. **virginicus** f. **virginicus** Map 185

*Elymus virginicus* (typical) [G, P & S]

Common throughout Missouri, especially in low and alluvial woods and banks of streams.

6b. **Elymus virginicus** var. **virginicus** f. **hirsutiglumis** (Scribn.) Fern. Map 185

*Elymus virginicus* f. *hirsutiglumis* (Scribn.) Fern. [G, P & S]

Common throughout Missouri. Often found in open sandy fields, rocky bluff exposures, prairies, open and alluvial woods.

6c. **Elymus virginicus** var. **jejunus** (Ramaley) Bush Map 185

Not recorded from northwestern Missouri, elsewhere in the state common in fields, open woods, wooded limestone ledges, prairies, glades, and waste ground.

6d. **Elymus virginicus** var. **submuticus** Hook. Map 185

The rarest of the varieties and forms of *E. virginicus* in Missouri, where scattered over the state.

6e. **Elymus virginicus** var. **glabriflorus** (Vasey) Bush f. **glabriflorus** Map 185

*Elymus virginicus* var. *glabriflorus* (Vasey) Bush [G, P & S]

Common throughout Missouri, where often found in open, rocky, rich, or alluvial woods, and on limestone glades.

6f. **Elymus virginicus** var. **glabriflorus** (Vasey) Bush f. **australis** (Scribn. & Ball) Fern. Map 185

Mostly occurring in the Ozark section of the state and in eastern Missouri. Apparently absent and not recorded from most of the glaciated prairie section of the northwestern half of the state. Occurs commonly in rocky open woods, glades, open sandy fields, and prairies.

*Elymus virginicus* is a very palatable and nutritious grass, and is eaten by all classes of livestock. It provides good hay and forage.

## 23. **Sitanion** Raf. Squirreltail

**Sitanion longifolium** J. G. Smith
Squirreltail Map 186
*Sitanion Hystrix* of many authors [BB, G, P & S], not (Nutt.) J. G. Smith
Flowers May–August.

Known only from Jackson County, western Missouri, where introduced (Sheffield, June 2, 1906, *Bush 3907*). This specimen has been determined by F. Douglas Wilson, and is in the herbarium of the New York Botanical Garden. There is also a specimen in the Missouri Botanical Garden Herbarium.

Ranges from South Dakota to British Columbia, south to Missouri, Texas, California, and Mexico.

## 24. **Hystrix** Moench Bottlebrush

The leaf-blades have tiny auricles (outgrowths) at base.

**Hystrix patula** Moench Bottlebrush Map 187
Flowers June–August.

Commonly occurring in dry rocky or open woodland, less frequently in open or waste ground. Throughout Missouri, where apparently absent from the lowland southeastern section and a few extreme western counties.

Ranges from Maine to North Dakota, south to Georgia, Tennessee, Arkansas, and Oklahoma.

Occurs in two forms in Missouri:

Lemmas glabrous . . . . H. PATULA f. PATULA
Lemmas pubescent . . H. PATULA f. BIGELOVIANA

**Hystrix patula** Moench f. **patula** Map 187
*Hystrix patula* Moench [G, P & S]
This is the common form encountered in Missouri.

**Hystrix patula** f. **Bigeloviana** (Fern.) Gl. Map 187
*Hystrix patula* var. *Bigeloviana* (Fern.) Deam [G]

Known from Atchison (*Steyermark 85716*), Holt (*Steyermark 73829*), and Jackson (*Bush*, June 22, 1890) counties of northwestern and west-central Missouri.

## 25. **Hordeum** L. Barley

The narrow glumes, with their position in front of the 3 spikelets at each node, give the appearance of 6 involucre-like bracts or bristles on a spikelet.

The central one of the 3 spikelets of a group is sessile and perfect, while the other 2 (lateral ones) are shortly stalked and sterile or reduced in size, but in some species these are sessile and mostly perfect and as large as the central spikelet.

a. No awns on lemma . . . . . . . . . . . . . . . . 3b. H. VULGARE var. TRIFURCATUM
a. Long slender awns on lemma . . . . . . . . . . . . . . . . . . . . . . . . . . . b
  b. Conspicuous auricles (loose projections or outgrowths) present at top of leaf-sheath; awns of lemma stout, flat, 1 mm. wide at the base, 0.5 mm. or less wide on most of length, much longer than the short bristle-like awns of the glumes; leaf-blades mainly 5–15 mm. wide . 3a. H. VULGARE var. VULGARE
  b. No auricles at top of leaf-sheath; awns of lemma bristle-like, 0.2–0.5 mm. wide at the base, similar to and about equaling the long awns of the glumes; leaf-blades 1–6 mm. wide . . . . . . . c
    c. Inflorescence nodding or curving down; awns curving outward or spreading at maturity; spikelets 30–70 mm. (in a rare var. *caespitosum* 15–30 mm.) long (measured from the base to the tip of awn); all glumes bristle-like throughout . . . . . . . . . . . 1. H. JUBATUM
    c. Inflorescence erect; awns erect or ascending, straight at maturity; spikelets 12–20 mm. long (measured from base to tip of awn); at least some of glumes broadened above bases . . 2. H. PUSILLUM

1. **Hordeum jubatum** L.
Squirreltail, Foxtail Barley Map 188
Flowers May–October.

Fields, pastures, roadsides, along railroads, and waste ground. Of common occurrence over the northern half of the state, but so far as records exist, the grass is scarce and in many parts absent south of the Missouri River.

Extends from Labrador to Alaska, south to Newfoundland, Virginia, Kentucky, Illinois, Missouri,

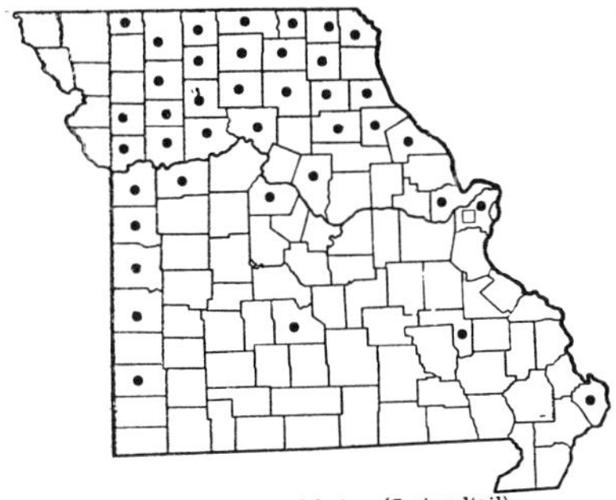

188 • Hordeum jubatum var. jubatum (Squirreltail)
188 □ Hordeum jubatum var. caespitosum

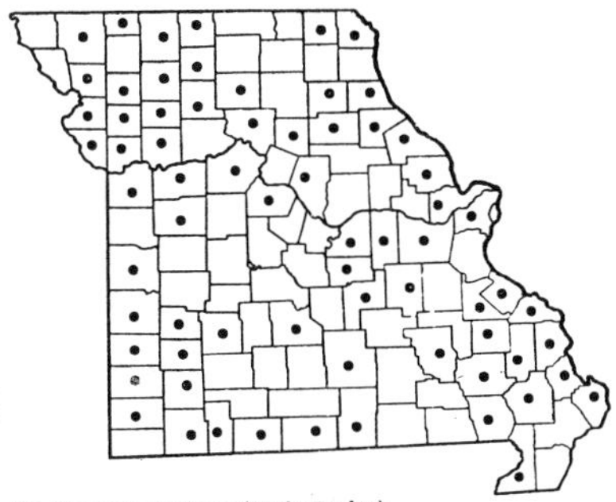

189 Hordeum pusillum (Little Barley)

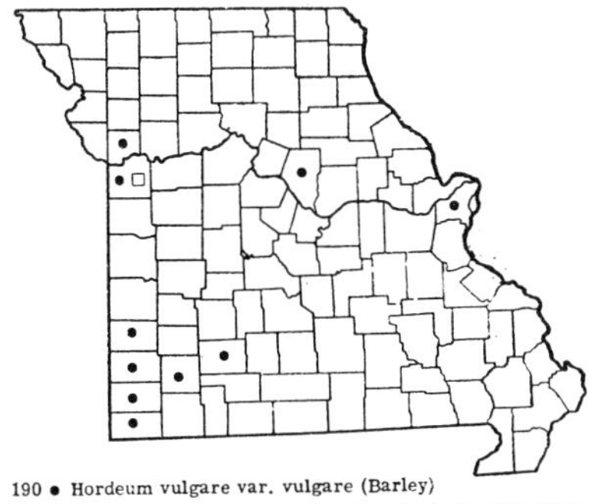

190 • Hordeum vulgare var. vulgare (Barley)
190 □ Hordeum vulgare var. trifurcatum (Beardless Barley)

Texas, California, and Mexico. Introduced into South America, Europe, and Asia.

Missouri material falls into two varieties:

Spikelets, including awns, 30–70 mm. long . . 1a. H. JUBATUM var. JUBATUM
Spikelets, including awns, 15–30 mm. long . . 1b. H. JUBATUM var. CAESPITOSUM

1a. **Hordeum jubatum** L. var. **jubatum** Map 188

*Hordeum jubatum* L. [G, BB, P & S]

Common variety encountered.

1b. **Hordeum jubatum** var. **caespitosum** (Scribn.) Hitchc. Map 188

Known only from St. Louis County (along central switching tracks, freight yard Brennan Ave., Terminal R. R. Association, St. Louis, July 4, 1956, *Muehlenbach 995*).

Sometimes called 'Bobtail Barley.' When dry, the bristly awns easily penetrate the mouth parts and eyes of grazing sheep, cattle, and horses, being especially troublesome in the Rocky Mountain states. The young inflorescence of this species sometimes takes on a beautiful rose-purple color with a glistening effect in the sunlight.

2. **Hordeum pusillum** Nutt. Little Barley Map 189

Flowers late April–late June.

Occurs in waste and cultivated ground, along roadsides, and railroad tracks throughout Missouri.

Ranges from Delaware to Florida, west to Washington and California. Also occurs in South America.

Like *H. jubatum* the bristles of this species can cause lacerations in the mouths of grazing livestock. In early spring this species affords, before the flowering heads mature, a temporary source of pasture. The grass becomes available in the southern states as pasture, also in the moist fall months when germination occurs and produces green shoots which may survive the winter.

3a. **Hordeum vulgare** L. var. **vulgare** Barley Map 190

*Hordeum sativum hexastichyon*, as listed in Daniels' Flora.

Flowers May–June.

This annual cultivated plant is occasionally found in Missouri along roadsides, railroad tracks, and in waste ground, but does not persist. Recent collections from St. Louis (right-of-way of Terminal R. R. Association under the municipal bridge, St. Louis, May 14, 1955, *Muehlenbach 579*) and Greene (open pasture, northeast end of Lake Springfield, southeast of Springfield, May 17, 1958, *Redfearn 3593*) counties are cited here.

Probably native of Asia; introduced and naturalized throughout the United States and Canada.

The dried straw from the cultivated barley is used principally in Europe for pulp in papermaking.

3b. **Hordeum vulgare** var. **trifurcatum** (Schlecht.) Alefeld Beardless Barley, Pearl Barley Map 190

*Hordeum aegiceras* (E. Meyer) Royle

This awnless variety, having the lemmas tipped with a 3-lobed appendage, has been collected as a waif only from Jackson County (fields, Courtney, June 24, 1932, *Bush 12479*).

*Excluded species*

**Hordeum montanense** Scribn. Meadow Barley

*Hordeum nodosum* [of P & S], not L.

Material previously identified as *H. nodosum* is referred in the present treatment to *H. pusillum.*

191 Lolium perenne (Perennial Rye Grass)

192 • Lolium multiflorum var. multiflorum (Italian Rye Grass)
192 □ Lolium multiflorum var. diminutum

193 • Lolium temulentum var. temulentum (Darnel)
193 □ Lolium temulentum var. macrochaeton

## 26. **Lolium** L. Rye Grass, Darnel

a. Glume (excluding awns) about as long as or longer than the spikelet . . . . . . . 3. L. TEMULENTUM
a. Glume (excluding awns) obviously shorter than the spikelet . . . . . . . . . . . . . . *b*
b. Lemmas without an awn, mainly 5–10-flowered; rachis (axis) of spike glabrous except on the angles; upper spikelets with the glume longer than the next floret just above the glume; auricles (projections of green tissue) at top of leaf-sheath short and rounded . . . . . . . 1. L. PERENNE
b. Lemmas usually with an awn, 5–20-flowered; rachis (axis) of spike roughened on the upper side away from the spikelets; upper spikelets with the glume not longer than the next floret just above the glume; auricles (projections of green tissue) at top of leaf-sheath long and pointed . 2. L. MULTIFLORUM

1. **Lolium perenne** L. Perennial Rye Grass, English Rye Grass Map 191

Flowers May–August.

Fields, meadows, and waste ground. Scattered over the state. Relatively few records of this species have been found in the state, and more intensive collecting is necessary, although it is probable that it occurs in every county.

Native of Europe; introduced into North America, where it is found from Newfoundland to Alaska, southward throughout most of the United States.

The spikelets are mainly 5–6 mm. long. This is a perennial grass, which, because of the large amount of pollen shed by it, may become a serious cause of hay fever in certain areas, such as on the Pacific coast.

This grass tolerates dry conditions. It is often planted in the spring or fall to produce a temporary new lawn, which may be spaded over as a green manure during the same or following year just before the seeding begins for a permanent lawn. This species and *L. multiflorum*, Italian Rye Grass, are important in the southern states for winter forage. In the United States they are used to some extent for producing meadows, pastures, and lawns.

2. **Lolium multiflorum** Lam. Italian Rye Grass Map 192

*Lolium perenne* var. *italicum* (A. Braun) Parnell

Flowers May–August.

Fields, meadows, and waste ground.

Native of Europe; introduced and naturalized in North America from Newfoundland to Alaska, south to Virginia and California.

Missouri material may be divided as follows:

Spikelets 10–20-flowered; glume of lowest spikelets often longer than the next floret just above the glume . . 2a. L. MULTIFLORUM var. MULTIFLORUM
Spikelets 5–9-flowered; glume of all the spikelets shorter than the next floret just above the glume. 2b. L. MULTIFLORUM var. DIMINUTUM

2a. **Lolium multiflorum** var. **multiflorum** Map 192

The commoner variety, scattered in Missouri. The spikelets (excluding the awns) are 5–7 mm. long.

2b. **Lolium multiflorum** var. **diminutum** Mutel. Map 192

Known from Buchanan (*Steyermark 70116*), Jasper (*Palmer 2408*), and Howell counties.

This variety has sometimes been confused by collectors with *Festuca elatior*.

*Lolium multiflorum* is more commonly cultivated than

Plate no. 32. **1.** Sphenopholis obtusata; *a*. Plant, × 3/8; *b*. Floret, × 7 1/2; *c*. Glumes, × 7 1/2. **2.** Lolium temulentum, × 3/8. **3.** Lolium multiflorum; *a*. Plant, × 3/8; *b*. Spikelet, × 2 1/4; *c*. Floret, × 3 3/4. **4.** Lolium perenne, × 3/8.

PLATE NO. 32

*L. perenne*. It is used chiefly to produce temporary lawns and to produce a rapid start until the more permanent grass species take over. It is often seeded along shoulders and banks bordering new highways, and sown in seed mixtures for parks or public grounds where an early new growth is desired. It is somewhat important as a cause of hay fever.

3. **Lolium temulentum** L. Darnel — Map 193
Flowers May–August.
Missouri material is represented by two varieties:

Lemmas not awned . . . 3a. L. TEMULENTUM var. TEMULENTUM
Lemmas awned . . . . . 3b. L. TEMULENTUM var. MACROCHAETON

3a. **Lolium temulentum** var. **temulentum** — Map 193

Fields and waste ground. Known mostly from western Missouri in Platte, Jackson, and McDonald counties, and from St. Louis County (*Muehlenbach 1411*).

Native of Europe; introduced into North America, where it extends from Quebec and Minnesota, south to Georgia, Louisiana, and Texas, and from Idaho and Washington south to California and Arizona.

3b. **Lolium temulentum** var. **macrochaeton** A. Br. — Map 193

Known only from St. Louis County (freight station of the Wabash Railroad, south of Carr Street, St. Louis, May 30, 1958, *Muehlenbach 1398*).

This grass is suspected of causing hay fever during its flowering season. It contains a poisonous alkaloidal substance which is toxic both to human beings and to livestock. The poisonous property is stated to become active only when the plant is infected by a parasitic fungus.

## Tribe IV. AVENEAE

In this tribe the spikelets are usually on stalks (pedicels), are usually 2- or more-flowered, the glumes are as long as or longer than the lowest lemma, the awn on the lemma, if present, is dorsal (on the back), and the spikelets are articulated usually above, sometimes below, the glumes.

### 27. Koeleria Pers.

**Koeleria cristata** (L.) Pers.
Crested Hair Grass — Map 194
Flowers late May–July.

Occurs in prairies and fields. So far as available records indicate, this species is absent from the northeastern half of the state, the southeastern lowland section, and from most of the southern and eastern Ozarks.

Ranges from Quebec to British Columbia, south to Delaware, Ohio, Indiana, Missouri, Texas, California, and Mexico..

In the western part of its range the grass is believed to be an important cause of hay fever during July to September.

The densely fine-pubescent rachis helps to distinguish this genus from *Sphenopholis* in which the rachis is glabrous or at most slightly scabrous.

### 28. Sphenopholis Scribn. Wedge Grass

The spikelets are articulated below the glumes, and ordinarily 2-flowered, rarely with a 3rd rudimentary flower.

These are all perennial, spring-flowering species, and all afford good early grazing.

a. Second (upper) lemma scabrous above the middle; anthers 1–2 mm. long . . . . . . . 3. S. NITIDA
a. Second (upper) lemma glabrous throughout; anthers 0.3–0.8 mm.. . . . . . . . . . . . *b*
  b. Inflorescences loosely flowered, the branches and pedicels visible; spikelets 3–4.2 mm. long; lemmas acute or nearly acute; second glume acute or nearly acute, narrowed at tip . . . 2. S. INTERMEDIA
  b. Inflorescence compact with crowded dense clusters of spikelets, the branches and pedicels mostly

Plate no. 33. **1.** Trisetum flavescens; *a.* Panicle, × 3/4; *b.* Floret, × 3 3/4. **2.** Trisetum pensylvanicum; *a.* Panicle, × 3/4; *b.* Glumes, × 3 3/4; *c.* Floret, × 3 3/4. **3.** Aira elegans; *a.* Panicle, × 3/4; *b.* Florets, × 7 1/2; *c.* Spikelet, × 7 1/2. **4.** Avena fatua; *a.* Plant, × 3/8; *b.* Floret, × 1 1/2; *c.* Spikelet, × 1 1/2. **5.** Avena sativa, floret, × 1 1/2.

PLATE NO. 33

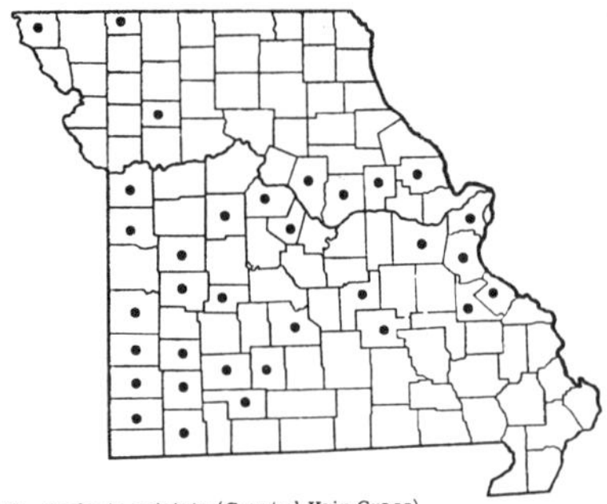

194 Koeleria cristata (Crested Hair Grass)

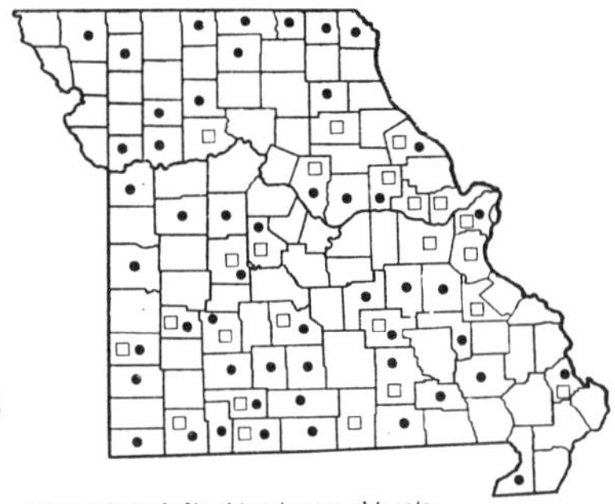

195 • Sphenopholis obtusata var. obtusata
195 □ Sphenopholis obtusata var. lobata

196 Sphenopholis intermedia

hidden by the spikelets; spikelets 2.5–3 mm. long; lemmas rounded at tip or with a hood-like arching tip; second glume rounded and hood-like at tip . . . . . . . . . . . 1. S. OBTUSATA

1. **Sphenopholis obtusata** (Michx.) Scribn. Map 195

Flowers May–July.

Prairies, rocky open ground, and bluffs.

Throughout Missouri.

The variations of this species range from Florida to Arizona, Mexico, and California, north to Maine, Ontario, Minnesota, North Dakota, Alberta, and British Columbia.

Missouri material may be distinguished as two forms:

Leaf-blades and their leaf-sheaths glabrous, or at most somewhat scabrous . . . 1a. S. OBTUSATA var. OBTUSATA

Leaf-blades or their leaf-sheaths more or less pubescent . . . 1b. S. OBTUSATA var. LOBATA

1a. **Sphenopholis obtusata** var. **obtusata** Map 195

This is the commoner variation in Missouri and occurs throughout the state.

1b. **Sphenopholis obtusata** var. **lobata** (Trin.) Scribn. Map 195

*Sphenopholis obtusata* var. *pubescens* (Scribn. & Merr.) Scribn. [G, P & S]

This is found mainly in southern and central Missouri, extending north to Pike, Monroe, and Carroll counties.

*Sphenopholis obtusata* is separated into three varieties in *Gray's Manual*, eighth edition. Too much intergradation and lack of correlation exist, however, as regards the amont of branching of the panicle, its relative length and thickness, and degree of tightness, density, and shape. Consequently, the plants with pubescent leaf-sheaths or blades have been placed together in one category, those with glabrous sheaths and blades in the other.

*Sphenopholis obtusata* or other species of the genus is occasionally confused with *Koeleria cristata*. The two genera are easily distinguished, however, *Koeleria* having the rachis of the inflorescence densely fine-pubescent and *Sphenopholis* having it glabrous or slightly scabrous. The spikelets articulate above the glumes in *Koeleria*, so that the glumes remain, whereas in *Sphenopholis* the spikelets articulate below the glumes and fall as a whole.

2. **Sphenopholis intermedia** Rydb. Map 196

Flowers May–July.

Rich or rocky woodland, moist open ground, low wet woodland and moist bluffs along streams. Throughout Missouri.

Ranges from Newfoundland to Alaska, south to Florida and west to the Pacific.

This species is predominantly a woodland inhabitant as contrasted with *S. obtusata* var. *obtusata* and var. *lobata* which are more commonly found in prairies, fields, and open ground.

3. **Sphenopholis nitida** (Biehler) Scribn. Map 197

Flowers April–July.

Rocky or sandy open woods of southern Missouri northeast to Ste. Genevieve County.

Ranges from Florida to Texas, north to Massachusetts, Vermont, New York, Ontario, Michigan, Illinois, and Missouri.

This species is found usually in acid soils in regions of sandstone, chert, or granite.

Spikelets in some Missouri specimens may attain 4 mm. in length and the lower lemma a length of 3 mm. The leaf-sheaths and leaf-blades in this species are usually more or less pubescent.

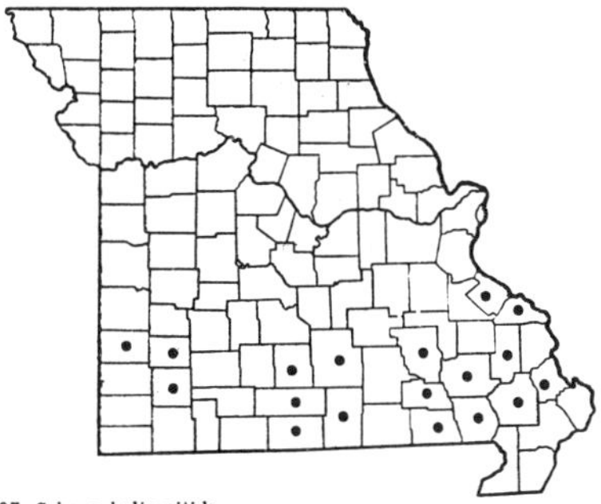

197 Sphenopholis nitida

198 Trisetum flavescens (Yellow Oats)

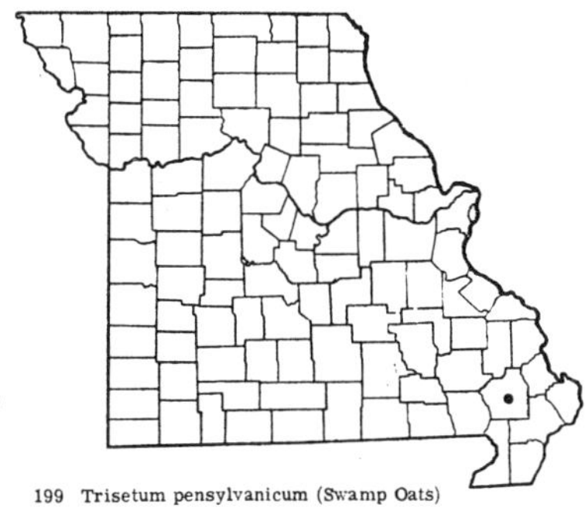

199 Trisetum pensylvanicum (Swamp Oats)

## 29. **Trisetum** L.

The spikelets are articulated both above and below the glumes in this genus.

Growing in water of small spring-fed creeks; lemmas finely papillose-scabrous; spikelets mainly 2-flowered (rarely 3-flowered); anthers about 1 mm. long . . . . . . . . . . . 2. T. PENSYLVANICUM

Dry soil of fields or roadsides; lemmas glabrous; spikelets 3–4-flowered; anthers 2.5–3 mm. long . 1. T. FLAVESCENS

1. **Trisetum flavescens** (L.) Beauv.
Yellow Oats Map 198
Flowers late May–July.

Known only from Jackson (June 6, 1891, *Bush*, in U. of Mo. herb.) and Pike (Sept. 29, 1911, *Davis 226*) counties. The *Davis* collection was originally filed under *Arrhenatherum elatius*.

Native of Europe; introduced into the United States, extending from Vermont and New York to Washington, south to Mississippi, Missouri, and Kansas.

2. **Trisetum pensylvanicum** (L.) Beauv.
Swamp Oats Map 199
Flowers April–May.

Known only from Stoddard County in southeastern Missouri (along spring-fed creek near junction of Crowley Ridge and lowland, bordered by sandy wooded ravine slopes, T25N, R11E, northwest $\frac{1}{4}$ sect. 6, $3\frac{1}{2}$ mi. southeast of Bloomfield, April 25, 1955, *Steyermark 78285*).

Ranges from Massachusetts to Ohio, south to Florida, Alabama, and Louisiana.

A very rare species, only recently found in the state, where it was growing in the water of a creek in which Alder (*Alnus serrulata*) occurred. It is a very early-flowering species, by June apparently having matured its seeds and disappearing from the location in which it occurred earlier.

## 30. **Aira** L. Hair Grass

Spikelets 2-flowered. Small annual grasses with the leaf-blades very short and narrow.

**Aira elegans** Willd. Hair Grass Map 200
Flowers May–June.

Dry sterile soil along roadsides and waste ground, where known from St. Louis County (on the embankment of the Mississippi River, north of the municipal bridge, opposite Grunden Martin Mfg. Co., St. Louis, May 13, 1956, *Muehlenbach 885*) and Oregon County (dry roadside, 4 mi. north of Thayer, June 11, 1958, *Kucera 2623*).

Native of Europe; introduced into the United States, extending from New Jersey to Florida and Texas, and in California and Oregon.

The rounded-ovoid panicle is very much branched, open, and spreading with tiny spikelets scattered at the ends of the delicate, hairlike branches.

## 31. **Avena** L. Oats

Spikelets 2–3-flowered. Annual grasses with nodding spikelets in our local species.

Lemma glabrous, without an awn or the awn straight; spikelets 2-flowered . . . . . . . 2. A. SATIVA

Lemma with brown or white hairs at the base and usually on the back, with a long awn bent near its middle; spikelets 3-flowered . . . . . . . . . . . . . . . . . . . . . . 1. A. FATUA

200 Aira elegans (Hair Grass)

201 Avena fatua (Wild Oats)

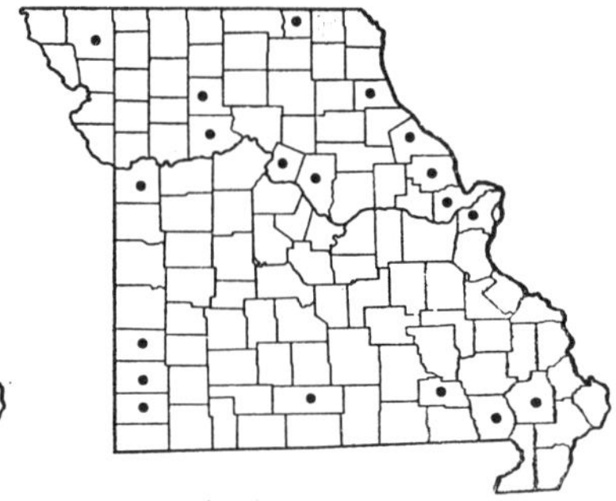

202 Avena sativa (Oats)

1. **Avena fatua** L. Wild Oats Map 201

Flowers June–October.

Occasionally found in gardens and waste ground in St. Louis (right-of-way of Mo. Pacific R. R. north of Elwood St., St. Louis, June 13, 1954, *Muehlenbach 153*), Boone (garden, Columbia, July 5, 1910, collector unknown), and Jackson (Sheffield, June 22, 1905, *Bush 3025*) counties.

Native of Europe; introduced into North America, extending from Newfoundland to British Columbia, south to Massachusetts, Pennsylvania, Tennessee, Missouri, New Mexico, and Mexico.

On the Pacific Coast, this species is regarded as an important cause of hay fever during its flowering season from July to September. The seeds of this species are used for food by Indians in the western United States.

2. **Avena sativa** L. Oats Map 202

*Avena fatua* var. *sativa* (L.) Haussknecht

*Avena nuda* L.

Flowers May–November.

Along roadsides, railroad tracks, and waste ground. This is much more common in Missouri than *A. fatua* and is scattered throughout the state.

Probably native of Europe.

It is the commonly cultivated oat, usually sown for early feed and for crop rotation, especially cultivated in the north central states of the Corn Belt. It is also sown on newly graded highways to prevent washing of the soil.

Dried straw of this species is used principally in Europe as pulp for papermaking.

*Avena nuda* L., the naked oat, with a grain which readily separates from the lemma and palea, is here included with *A. sativa* L.

## 32. **Arrhenatherum** Beauv. Oat Grass

Tall perennial grasses with loose narrow panicles. Spikelets 2-flowered, the lower one staminate (male) with a long bent awn, the upper floret perfect with only a short awn on the lemma or merely pointed.

**Arrhenatherum elatius** (L.) Mert. & Koch var. **elatius** Tall Oat Grass Map 203

Flowers late May–August.

Fields and waste ground. Known only from Jasper and Boone (open ground, field crop plots, University of Missouri farms, August 1, 1945, *W. E. Maneval*) counties.

Native of Europe; introduced into North America, extending from Newfoundland to British Columbia, south to Georgia, Louisiana, New Mexico, and California.

This grass is cultivated in the cooler and more moist regions of the northern United States as a meadow grass.

---

Plate no. 34. **1.** Distichlis spicata; *a.* Plant, $\times\ ^{2}/_{5}$; *b.* Floret, $\times$ 2. **2.** Distichlis stricta; *a.* Staminate (♂) plant, $\times\ ^{1}/_{5}$; *b.* Pistillate (♀) panicle, $\times\ ^{2}/_{5}$; *c.* Pistillate floret, $\times$ 2; *d.* Staminate floret, $\times$ 2; *e.* Staminate spikelet, $\times$ 2. **3.** Uniola latifolia; *a.* Plant, $\times\ ^{1}/_{5}$; *b.* Floret, $\times\ 1^{1}/_{5}$; *c.* Spikelet, $\times\ 1^{1}/_{5}$. **4.** Dactylis glomerata; *a.* Plant, $\times\ ^{1}/_{5}$; *b.* Floret, $\times$ 2; *c.* Spikelet, $\times$ 2. **5.** Arundo Donax; *a.* Plant, $\times\ 1^{1}/_{5}$; *b.* Floret, $\times\ 1^{1}/_{5}$; *c.* Spikelet, $\times\ 1^{1}/_{5}$. **6.** Melica nitens; *a.* Plant, $\times\ ^{2}/_{5}$; *b.* Floret, $\times$ 2. **7.** Elymus canadensis; *a.* Plant, $\times\ ^{1}/_{5}$; *b.* Spikelet, $\times$ 2; *c.* Floret, $\times$ 2. **8.** Sitanion longifolium; *a.* Plant, $\times\ ^{1}/_{5}$; *b.* Spikelet, $\times\ 1^{1}/_{5}$; *c.* Floret $\times\ 1^{1}/_{5}$. **9.** Phragmites communis; *a.* Plant, $\times\ ^{2}/_{15}$; *b.* Floret, $\times\ 1^{1}/_{5}$; *c.* Spikelet, $\times\ 1^{1}/_{5}$.

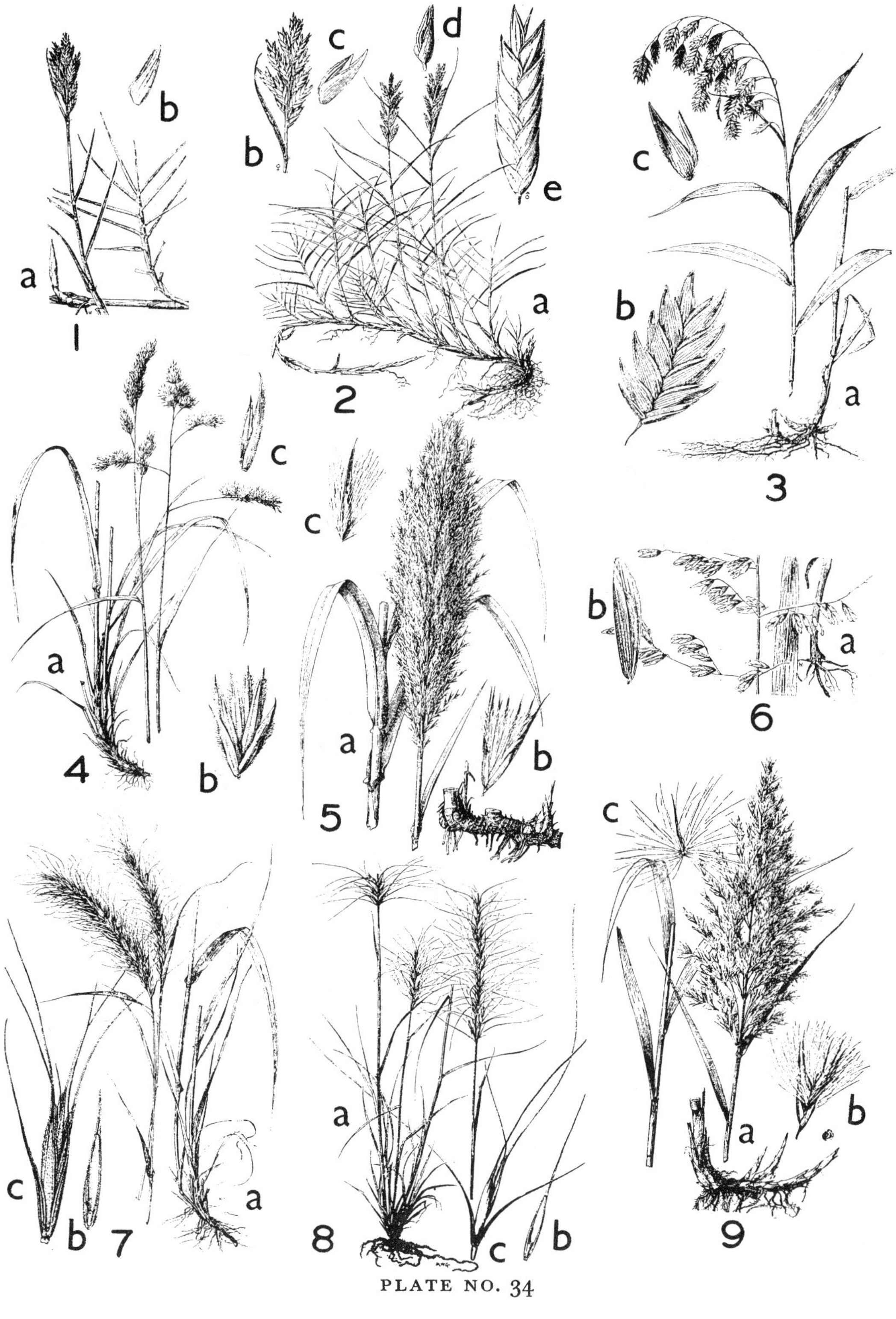


PLATE NO. 34

203 Arrhenatherum elatius var. elatius (Tall Oat Grass)

204 Holcus lanatus (Velvet Grass)

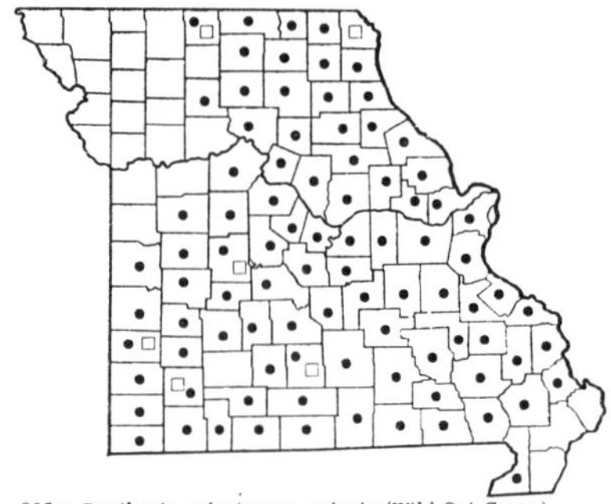

205 • Danthonia spicata var. spicata (Wild Oat Grass)
205 □ Danthonia spicata var. longipila

## 33. **Holcus** L. Velvet Grass

**Holcus lanatus** L. Velvet Grass Map 204

Flowers May–July.

Along railroads and waste ground, where known from Jackson, Vernon, and Jasper counties of western Missouri.

Native of Europe; introduced into North America, ranging from Newfoundland to Ontario, Michigan, and Iowa, south to Georgia and Louisiana; also in many Rocky Mountain and Pacific States.

The grass was introduced as a forage plant, where it has been used, principally in the southern states. It is sometimes a cause of hay fever in regions where it is more abundant, especially in California and Oregon. Sometimes it can poison livestock when eaten in a fresh or wilted stage, because of the prussic acid produced.

## 34. **Danthonia** DC. Wild Oat Grass

Spikelets usually 5–8-flowered in the local species (except for 1–12-flowered spikelets on some of the axillary spikelets), the florets crowded into a column, with the uppermost floret imperfect.

**Danthonia spicata** (L.) Beauv. Poverty Grass, Wild Oat Grass Map 205

Flowers May–August.

Most frequently found in dry acid soil of upland woods and on wooded dry ridges and slopes overlying sandstone, chert, or granitic rocks; also on rocky open glades of these types of rocks, and in rocky acid-soil prairies. Absent only from northwestern and west central and extreme southeastern Missouri.

Missouri material falls into two categories:

Column of flowers 3/4 length to equaling length of the glumes; inflorescence with mainly 2–15 spikelets; glumes firm, with 3–5 prominent nerves. 1a. D. SPICATA var. SPICATA

Column of flowers 1/2 to 3/5 length of the glumes; inflorescence relatively fewer-flowered, with 3–7 scattered spikelets; glumes thin and more or less transparent, without nerves or the nerves very faint . . . . . 1b. D. SPICATA var. LONGIPILA

1a. **Danthonia spicata** var. **spicata** Map 205

This is the common variety in the state.

Ranges from Quebec to Minnesota, south to Florida, Alabama, Tennessee, and Arkansas.

1b. **Danthonia spicata** var. **longipila** Scribn. & Merr. Map 205

Future collecting may show this variety to occur over a broader range of counties; at present known from a few scattered places.

Ranges from North Carolina and Alabama to New Mexico, north to Connecticut, Pennsylvania, Kentucky, Missouri, and Kansas.

The plants occur characteristically in scattered small clumps, the dried basal leaves persisting as curled and twisted blades. Both variations exhibit a characteristic curling or curving and inrolling of the margins of the basal leaves.

Two kinds of spikelets are found in this grass: the

---

Plate no. 35. **1.** Calamagrostis insperata; *a.* Panicle, × $^3/_4$; *b.* Floret, × $7^1/_2$; *c.* Glumes, × $7^1/_2$. **2.** Danthonia spicata; *a.* Plant, × $^3/_8$; *b.* Cleistogene, × $3^3/_4$; *c.* Floret, × $3^3/_4$; *d.* Spikelet, × $3^3/_4$. **3.** Calamagrostis canadensis var. canadensis; *a.* Plant, × $^3/_8$; *b.* Floret, × $7^1/_2$; *c.* Glumes, × $7^1/_2$. **4.** Calamagrostis canadensis var. Macouniana; *a.* Floret, × $7^1/_2$; *b.* Glumes, × $7^1/_2$. **5.** Calamagrostis inexpansa; *a.* Panicle, × $^3/_4$; *b.* Glumes, × $7^1/_2$; *c.* Floret, × $7^1/_2$.

PLATE NO. 35

206 • Calamagrostis canadensis var. canadensis (Bluejoint)
206 x Calamagrostis canadensis var. Macouniana

207 Calamagrostis inexpansa var. brevior

208 Calamagrostis insperata

ones in the visible inflorescence which terminates the culms have 5–8 florets which usually do not open when flowering (cleistogamous), while the other kind of spikelets are hidden in the leaf-sheaths and have 1–12 florets (those of the basal leaves with usually 1 floret, those from the upper axils of the leaves with as many as 12). For more details on this type of variation, see Weatherwax: Cleistogamy in two species of Danthonia. Bot. Gaz. 85: 104–9. 1928; Chase: Axillary cleistogenes in some American grasses. Am. Jour. Bot. 5: 254–58. 1918.

## Tribe V. **AGROSTIDEAE**

The grasses of this tribe are usually 1-flowered (except 2-flowered in some species of *Sporobolus, Agrostis,* and *Muhlenbergia*). The glumes are often longer than the lower lemma. Articulation of the spikelets is usually above the glumes, but in *Cinna* and *Alopecurus* (of our local genera) the articulation is below the glumes.

### 35. **Calamagrostis** Adans. Reed Bent Grass

Perennial grasses with a many-flowered, branched, open or contracted panicle.

a. Panicle loose and open, at least during time of flowering . . . . . . . . . . . 1. C. CANADENSIS
a. Panicle contracted or spike-like during time of flowering . . . . . . . . . . . . . . . . *b*
  b. Leaf-blades mainly 1–4 mm. wide; spikelets 3–4.5 mm. long; awn of lemma straight or slightly bent, not twisted, arising at about the middle of the lemma . . . . . . 2. C. INEXPANSA var. BREVIOR
  b. Leaf-blades 5–11 mm. wide; spikelets 5–5.5 mm. long (3.75–5 mm. in Mo. material); awn of lemma bent near the middle, twisted near the base (when dry), arising near the base of the lemma 3. C. INSPERATA

1. **Calamagrostis canadensis** (Michx.) Beauv. Bluejoint Map 206

Flowers late May–August.

Missouri material falls into two categories:

Spikelets 2.8–3.8 mm. long; glumes longer than the lemma, acute to acuminate . 1a. C. CANADENSIS var. CANADENSIS
Spikelets 2.2–2.8 mm. long; glumes as long as lemma or slightly longer, obtuse or somewhat acute. . . 1b. C. CANADENSIS var. MACOUNIANA

1a. **Calamagrostis canadensis** var. **canadensis** Map 206

Wet meadows and swales in prairies and prairie remnants along railroads. Scattered in parts of northern, central, and eastern Missouri, southeast to Cape Girardeau County.

Ranges from Newfoundland to Alaska and British Columbia, south to Delaware, Pennsylvania, West Virginia, Ohio, Indiana, Illinois, Missouri, Nebraska, New Mexico, and California.

1b. **Calamagrostis canadensis** var. **Macouniana** (Vasey) Stebbins Map 206

Northern and central Missouri south to Jackson, Lafayette, Randolph, and Ralls counties.

Ranges from Newfoundland to New Jersey and Pennsylvania, Michigan to Saskatchewan, west to Alberta and Washington, south to Missouri, Nebraska, and Wyoming.

2. **Calamagrostis inexpansa** Gray var. **brevior** (Vasey) Stebbins Map 207

Flowers June–August.

Swales in sandy open ground and low wet ground, sometimes in or bordering woodland where found in Jackson, Texas, and Barton (swales in sandy open ground, border of woods near Bakers, July 19, 1955, *Palmer 60766*) counties.

Ranges from Newfoundland to British Columbia, south to New Hampshire, New York, Ontario, Michigan, Indiana, Wisconsin, Iowa, Missouri, Nebraska, New Mexico, Arizona, and California.

3. **Calamagrostis insperata** Swallen Map 208

Flowers June–September.

Inhabits rocky wooded ravines and open grassy rocky slopes of bluffs in the southern part of the Ozarks, where known from Ozark County (lower chert slopes in ravine in Blue Springs Game Refuge, Gardner National Forest, Sept. 12, 1936, *Steyermark 20043*) and Douglas County (rocky, grassy, open cherty limestone slopes at east end of bluffs along Indian Creek, near Holy Cliff, 3½ mi. northeast of Topaz, July 19, 1937, *Steyermark 23350*).

Range is limited to Ohio (where first discovered) and Missouri.

Dr. Van Schaack (Rh. 56: 43. 1954) has reported on the occurrence of this species in Missouri, and on the variability shown in length of spikelets, width of leaves, and length of ligule as compared with the material originally described from Ohio.

## 36. Calamovilfa (Gray) Hack.

Tall perennial grasses with strongly creeping, scaly rhizomes, and loosely spreading panicles. The leaf-blades have a pale straw color. The ligule is a ring of short hairs.

**Calamovilfa longifolia** (Hook.) Scribn. var. **longifolia** Long-leaved Reed Grass, Sand Reed Grass Map 209

Flowers July–September.

Low and sandy open ground, scattered in four counties (St. Louis, Jefferson, Jackson, and Clay) of central Missouri.

Ranges from Ontario to British Columbia, south to Indiana, Illinois, Missouri, Kansas, and Colorado.

As the grass spreads rapidly by means of its underground stems, it has been planted in sandy blowouts and other similar areas to keep the sand from shifting or covering over areas.

## 37. Agrostis L. Bent Grass

a. Awn present, attached to lemma . . . . . *b*
- b. Inflorescence dense, contracted, the spikelets and branches remaining close to main axis. 2. A. INTERRUPTA
- b. Inflorescence open, the spikelets on branches spreading out from main axis . . . . . *c*
  - c. Awns conspicuous, longer than spikelet, 5–10 mm. long . . . . . *d*
    - d. Leaves 2–6 mm. wide; awns 5–8 mm. long; palea present, half to about as long as lemma. 1. A. SPICA-VENTI
    - d. Leaves about 1 mm. wide; awns around 10 mm. long; palea absent or very minute. 5. A. ELLIOTTIANA
  - c. Awns inconspicuous, shorter than spikelet, 3 mm. or less long . . . . . *e*
    - e. Leaf-blades mainly 1–2.5 (–3) mm. long; ligule deeply and irregularly cut; branches of inflorescence dividing near the tip or above the middle; spikelets mainly purplish, crowded near tips of branches; plants mostly of sunny open places, mostly flowering April–June 6. A. HYEMALIS
    - e. Leaf-blades mainly 3–6 mm. (rarely less) wide; ligule entire, oblong, rounded at tip; branches of inflorescence dividing near or below the middle; spikelets mainly green, appearing midway or lower on the branches; plants mostly growing in woodland, usually flowering July–October . . . . . 7. A. PERENNANS

a. No awn present on any part of spikelet . . . . . *f*
- f. Leaf blades mainly 1–2.5 (–3) mm. wide; ligule deeply, finely, and irregularly cut in upper half; branches of inflorescence forking or dividing only near or towards the tip; spikelets crowded at or near the tips of the branches . . . . . 6. A. HYEMALIS
- f. Leaf-blades mainly 3–8 mm. (rarely less) wide; ligule entire, rounded at tip (too old or damaged

209 Calamovilfa longifolia var. longifolia (Long-leaved Reed Grass)

210 Agrostis spica-venti

211 Agrostis interrupta

leaves may appear otherwise); branches of inflorescence forking or dividing near or below the middle; spikelets appearing midway or lower on the branches . . . . . . . . . . . . g

g. Palea at least half as long as lemma; plants with underground scaly rhizomes or creeping leafy stems . . . . . . . . . . . . . . . . . . . . . . h

h. Ligule of lower and middle leaves longer than broad or as long as broad; rounded at tip, 2–6 mm. long; leaves 3–8 mm. wide; inflorescence with the branches flower-bearing to below the middle . . . . . . . . . . . . . . . . . . 3. A. ALBA

h. Ligule of lower and middle leaves broader than long, truncate at tip, 0.5–1.5 mm. long; leaves 1–5 mm. long; inflorescence with the branches flower-bearing mainly above the middle . . . . . . . . . . . . . . . . . . . . . 4. A. TENUIS

g. Palea absent or very minute; plants tufted, without underground rhizomes or creeping leafy stems . . . . . . . . . . . . . . . . . . . . 7. A. PERENNAN

1. **Agrostis spica-venti** L. Map 210

Flowers May–July.

Along railroads in St. Louis (Right-of-way of Burlington R. R. south of Merchants Bridge, St. Louis, June 5, 1954, *Muehlenbach 112*) and Jackson (along railroads, becoming common, June 11, 1891, *Bush 1890*) counties.

Native of Europe; introduced into the United States, ranging from Maine to Michigan, south to Delaware, Maryland, Ohio, and Missouri.

An annual grass.

2. **Agrostis interrupta** L. Map 211

Flowers late June–July.

Known only from a collection from St. Louis County (Missouri Botanical Garden, St. Louis, June 30, 1904, *Kellogg*).

Native of Europe; introduced into the United States where known only from Missouri and on the Pacific Coast.

An annual grass.

3. **Agrostis alba** L. Redtop Map 212

*Agrostis stolonifera* var. *major* (Gaud.) Farw. [BB]

*Agrostis stolonifera* [of P & S], not L.

Flowers June–August.

Fields, along railroads, and waste ground, along spring branches, wet meadows, swampy prairies, ditches, in damp thickets, alluvial ground along streams, crevices of rocks in stream beds, and moist open places.

Ranges from Newfoundland to Yukon, south to Georgia, Louisiana, New Mexico, Arizona, and California; Eurasia.

The foliage is bluish-green. This grass and timothy (*Phleum pratense*), flowering mainly during June and July, are responsible for causing most of the hay fever cases occurring at this time in the eastern states.

Because of the dense underground root system, this grass makes a good soil binder and is used in preventing erosion on unprotected slopes. Due to its success in growing in acid soils, it has been used as a forage crop on poor soils and in worn-out fields. It is often found in lawn seed grass mixtures, being of value as a temporary quick-germinating green cover until the more permanent Colonial Bent or Kentucky Blue Grass becomes established.

A variety of this, the Creeping Bent (*A. alba* var. *palustris* [*A. stolonifera* var. *compacta*]) has leaf-blades 1–3 (rarely 5) mm. wide and a compactly-flowered

Plate no. 36. **1.** Agrostis spica-venti; *a.* Plant, × 3/8; *b.* Floret, × 3 3/4; *c.* Glumes, × 3 3/4. **2.** Agrostis interrupta; *a.*, Panicle, × 3/8; *b.* Floret, × 3 3/4; *c.* Glumes, × 3 3/4. **3.** Calamovilfa longifolia; *a.* Plant, × 3/8; *b.* Floret, × 3 3/4; *c.* Spikelet, × 3 3/4.

PLATE NO. 36

212 Agrostis alba var. alba (Redtop)

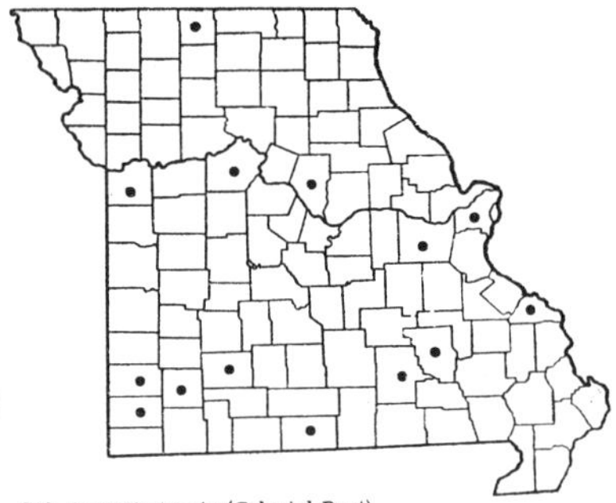

213 Agrostis tenuis (Colonial Bent)

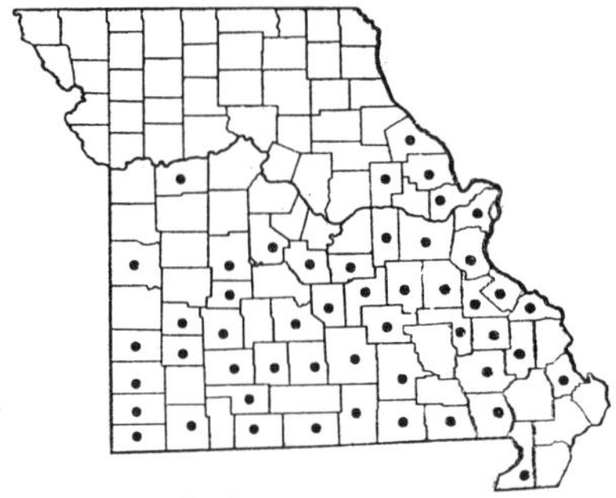

214 Agrostis Elliottiana

inflorescence with appressed, closely ascending branches. It forms dense mats by virtue of the creeping rooting stolons. Since it is resistant to salt spray, Creeping Bent is sometimes used as a lawn grass near the seashore. It is also commonly used on golf courses for putting greens, but requires cool moist conditions to keep it going.

4. **Agrostis tenuis** Sibth. Colonial Bent, Rhode Island Bent — Map 213

*Agrostis capillaris* L.

Flowers June–September.

Scattered over the state.

Fields, waste ground, roadsides, and damp thickets.

Native of Europe; introduced into the United States, ranging from Labrador to British Columbia, south to North Carolina, Ohio, Indiana, Missouri, North Dakota, and Oregon.

This grass is much used for lawns, sports turf, and pastures, and many improved strains have been developed. It does well in full sun. It is a good grass for cool, moist climates, but requires constant attention in watering, fertilizing, close mowing, and disease control to keep it thriving.

Like *A. alba*, this species is responsible for some hay fever cases.

5. **Agrostis Elliottiana** Schultes — Map 214

Flowers April–June.

Occurs in sterile acid soils exposed to the sun, often on sandstone, chert, on granite glades, sterile fields and prairies, open sandy and cherty soils, and sometimes along leached or eroded sterile slopes, in the Ozark region of southern and central Missouri north to Lafayette, Morgan, Montgomery, and Pike counties. Absent from the southeastern lowland section.

Ranges from Georgia to Texas, north to Virginia, Kentucky, Indiana, Illinois, Missouri, and Kansas.

An annual grass.

6. **Agrostis hyemalis** (Walt.) BSP. var. **hyemalis** Hair Grass, Tickle Grass, Winter Bent Grass — Map 215

*Agrostis scabra* [of P & S], not Willd.

Flowers April–June.

Occurs in prairies, fields, meadows, roadsides, and occasionally open woods throughout Missouri.

Ranges from Florida to Texas to Massachusetts, New York, Pennsylvania, Indiana, Illinois, Wisconsin, Minnesota, and Kansas.

When mature the entire inflorescence breaks off from the plant and is carried away by the wind in the manner of various tumbleweeds.

This species varies in the diameter of the inflorescence, increasing in size with maturity. Under favorable conditions of growth the culms and inflorescences develop into luxuriant specimens which are in contrast to plants growing in drier more exposed situations.

7. **Agrostis perennans** (Walt.) Tuckerm. Upland Bent — Map 216

Flowers July–October.

Found on upland open wooded slopes, low or rich woods, moist limestone bluffs, exposed sandstone rocks, prairie swales, fields, margins of streams and ponds, and moist open ground. Throughout Missouri except in the northwestern section, where apparently absent or local.

Two varieties may be recognized:

Most pedicels (stalks of spikelets) shorter than their spikelets; culms (stems) firm and erect with 3–6 leaves; spikelets more or less uniform,

Plate no. 37. **1.** Agrostis alba; *a.* Plant, × 3/8; *b,c.* Spikelets, × 3 3/4; *d.* Floret, × 3 3/4. **2.** Agrostis Elliottiana; *a.* Panicle, × 3/4; *b.* Floret, × 3 3/4; *c.* Glumes, × 3 3/4. **3.** Agrostis hyemalis; *a.* Plant, × 3/8; *b.* Floret, × 3 3/4; *c.* Glumes, × 3 3/4. **4.** Agrostis tenuis; *a.* Panicle, × 3/4; *b.* Ligule, × 3 3/4; *c.* Floret, × 3 3/4; *d.* Glumes, × 3 3/4. **5.** Agrostis perennans; *a.* Panicle, × 3/4; *b.* Floret, × 3 3/4; *c.* Glumes, × 3 3/4.

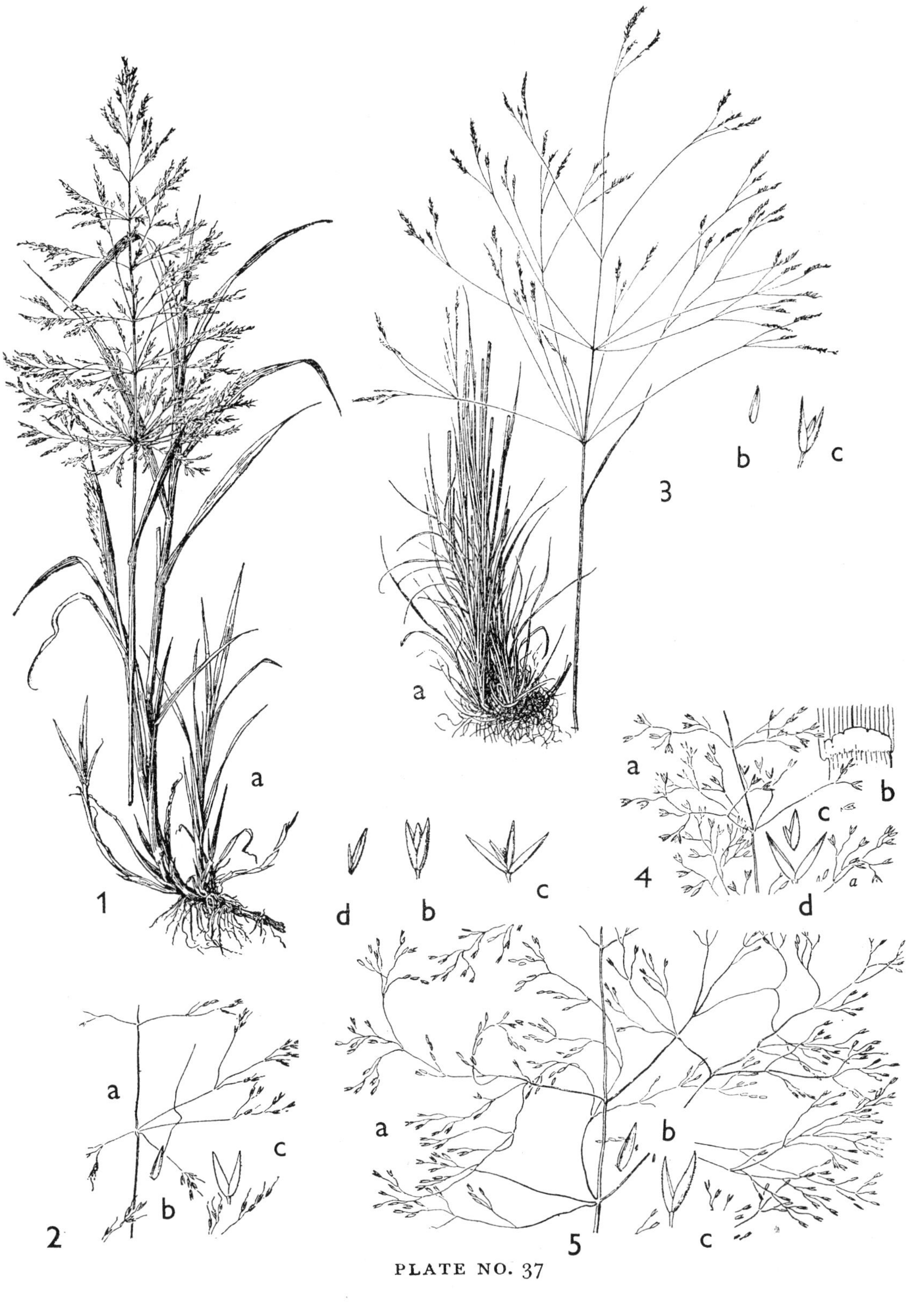


PLATE NO. 37

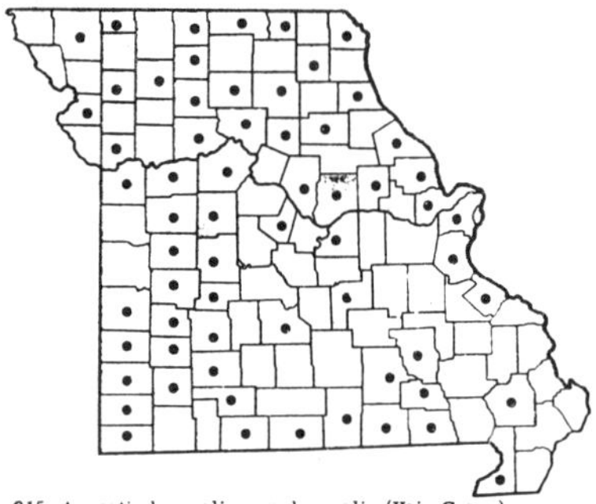

215 Agrostis hyemalis var. hyemalis (Hair Grass)

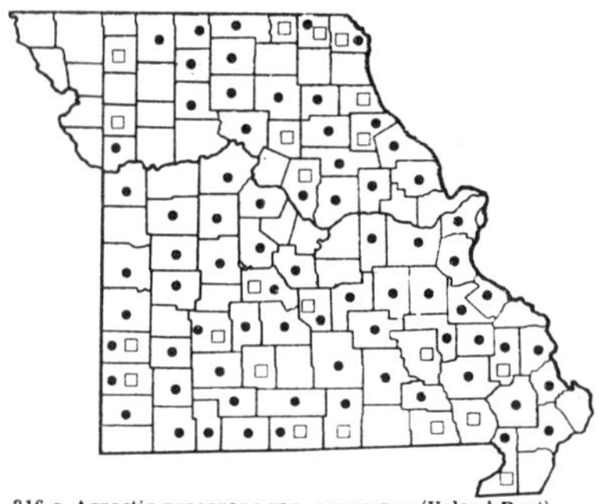

216 • Agrostis perennans var. perennans (Upland Bent)
216 □ Agrostis perennans var. aestivalis

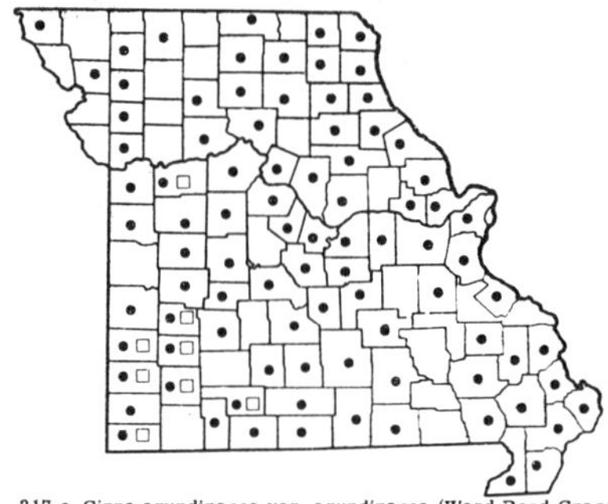

217 • Cinna arundinacea var. arundinacea (Wood Reed Grass)
217 □ Cinna arundinacea var. inexpansa

2–3 mm. long; branches of inflorescence ascending to spreading; frequent in open ground, margins of streams, and moist fields, as well as woodland. 7a. A. PERENNANS var. PERENNANS

Most pedicels 1–3 times longer than their spikelets; culms weak, erect to sprawling with 5–10 leaves; spikelets often 1.5–2 mm. long, variable in size; branches of inflorescence mainly widely spreading; inflorescence with a loose and open appearance; frequent in woodland, on shaded slopes or banks. . 7b. A. PERENNANS var. AESTIVALIS

7a. **Agrostis perennans** var. **perennans** Map 216

This has a stiffer inflorescence and is more often found in more exposed situations. However, there is great intergradation between typical var. *perennans* and var. *aestivalis* in the various characters used to separate them.

Ranges from Quebec to Minnesota, south to Florida, Louisiana, and Texas.

7b. **Agrostis perennans** var. **aestivalis** Vasey Map 216

*Agrostis Schweinitzii* Trin. [Rickett]

Scattered over the state. More frequent on wooded slopes and banks, rich shaded woods, and along streams.

Ranges from Quebec to Wisconsin, south to North Carolina, Tennessee, Mississippi, Missouri, and Kansas.

## 38. **Cinna** L. Wood Reed Grass

Tall perennial grasses with a many- and closely-flowered, gray-green to purple-tinged, often drooping inflorescence with spreading to rather ascending branches. Ligule conspicuous, membranous, transparent.

**Cinna arundinacea** L. Wood Reed Grass Map 217

Flowers July–October.

Wet woodland, meadows, damp thickets, and wooded banks of streams and ponds. Throughout Missouri.

Missouri material falls into two categories:

Branches of inflorescence ascending or spreading; spikelets 4.5–6 mm. long; both glumes scabrous-hispid . . 1a. C. ARUNDINACEA var. ARUNDINACEA

Branches of inflorescence closely contracted; spikelets 3.7–4.2 mm. long; 1st glume glabrous except for the scabrous keel 1b. C. ARUNDINACEA var. INEXPANSA

1a. **Cinna arundinacea** var. **arundinacea** Map 217

This is the common type encountered in the state.

Ranges from Maine and Quebec to Ontario and Minnesota, south to Georgia, Tennessee, Arkansas, and Texas.

1b. **Cinna arundinacea** var. **inexpansa** Fern. & Grisc. Map 217

Known from six counties in the western half of the state south of the Missouri River. Probably much more common in the southern part of the state.

Ranges from Virginia to Louisiana, Missouri, and Oklahoma. There is considerable intergradation in the varieties here treated, with sometimes little or no correlation apparent.

---

Plate no. 38. **1.** Alopecurus aequalis; *a.* Panicle, $\times\ ^{3}/_{4}$; *b.* Floret, $\times\ 7^{1}/_{2}$; *c.* Glumes, $\times\ 7^{1}/_{2}$. **2.** Cinna arundinacea; *a.* Plant, $\times\ ^{3}/_{8}$; *b.* Floret, $\times\ 7^{1}/_{2}$; *c.* Glumes, $\times\ 7^{1}/_{2}$. **3.** Alopecurus pratensis; *a.* Plant, $\times\ ^{3}/_{8}$; *b.* Floret, $\times\ 7^{1}/_{2}$; *c.* Glumes, $\times\ 7^{1}/_{2}$. **4.** Alopecurus carolinianus; *a.* Plant, $\times\ ^{3}/_{4}$; *b.* Floret, $\times\ 7^{1}/_{2}$; *c.* Glumes, $\times\ 7^{1}/_{2}$.

PLATE NO. 38

218 Alopecurus pratensis (Meadow Foxtail)

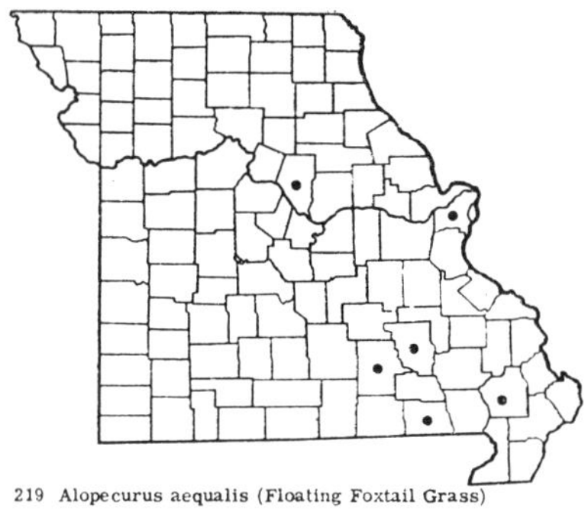

219 Alopecurus aequalis (Floating Foxtail Grass)

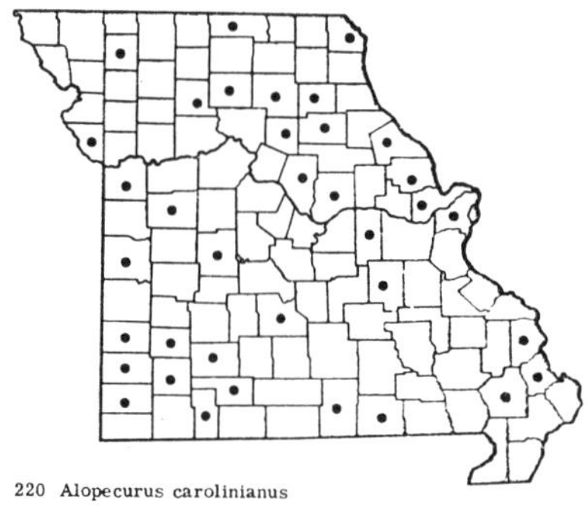

220 Alopecurus carolinianus

## 39. **Alopecurus** L. Foxtail

Low annual or perennial grasses with dense cylindric inflorescences. The ligule is membranous. The spikelets are strongly flattened laterally.

a. Spikelets about 5 mm. long; awn protruding about 5 mm.; glumes 3.5 mm. or more long, lateral nerves conspicuous . . . . . . . . . . . . . . . . . . . . . . . . 1. A. PRATENSIS
a. Spikelets 2–3 mm. long; awn not protruding from spikelet or protruding 4 mm. at most; glumes 3 mm. or less long, the lateral nerves faint or absent . . . . . . . . . . . . . . . . . . *b*
  b. Awn attached near base of lemma, twisted or bent, protruding 2–4 mm. beyond the lemma; anthers 0.4–0.6 mm. long; annuals . . . . . . . . . . . . . . . . . 3. A. CAROLINIANUS
  b. Awn attached about halfway up the lemma, straight, not protruding or protruding at most 1 mm. beyond the lemma; anthers 0.5–1 mm. long; perennials . . . . . . . . . . 2. A. AEQUALIS

1. **Alopecurus pratensis** L. Meadow Foxtail Map 218

Flowers May–August.

Occurs in fields, meadows, and waste ground.

Known only from St. Louis and Boone counties.

Native of Europe; introduced into North America, ranging from Newfoundland to Ontario, south to Georgia, Indiana, Missouri, and Kansas; and from Alaska to Idaho and Oregon.

A perennial grass, sometimes planted for its palatable and nutritious forage value in meadows. During its flowering season it can be a cause of hay fever, especially in some of the New England states, where more abundant. It flowers several weeks earlier than timothy (*Phleum pratense*), with which it can be confused.

2. **Alopecurus aequalis** Sobol.
Floating Foxtail Grass Map 219

Flowers May–September.

Occurs in swampy ground, borders of ponds and sloughs, and wet open ground, in southern and central Missouri. This is one of the grasses occasionally found around the margins of natural sink-hole ponds in the southern Ozarks.

Ranges from Newfoundland to Alaska, south to Maryland, Ohio, Indiana, Illinois, Missouri, Kansas, New Mexico, Arizona, and California.

It is a perennial species.

3. **Alopecurus carolinianus** Walt. Map 220

Flowers April–August.

Occurs in meadows, fallow fields, along moist ditches, borders of ponds, sloughs, wet ground, wet depressions in glades, and waste ground. Throughout Missouri.

Ranges from Florida to Texas to New Jersey, Pennsylvania, Ohio, Indiana, Wisconsin, Minnesota, South Dakota, Montana, and Washington.

An annual species. As it does in Indiana, this grass often grows in fallow corn fields associated with other annuals, such as *Poa Chapmaniana*, *Myosurus minimus*, *Draba brachycarpa* and *Myosotis verna*.

## 40. **Phleum** L. Timothy

Bunch grasses with a dense, cylindric, blue-green inflorescence. The ligule is conspicuous and elongated.

**Phleum pratense** L. var. **pratense** f. **pratense**
Timothy Map 221
Flowers May–August.

In meadows, along roadsides, and waste ground. Throughout Missouri.

Native of Europe; introduced and naturalized throughout North America.

A short-lived perennial grass having a thickened or bulblike base. It is the chief dry grass and hay crop of the United States. It is sown usually in the fall, and will grow on a wide variety of soils. Name originated from Timothy Hanson of Maryland who introduced the grass from Europe about 1720.

During its main flowering season in June and July, it sheds great quantities of pollen, which are responsible for the grass being one of the worst contributors to hay fever in early summer.

Plants with the spikelets changed to leafy tufts are known as var. *pratense* f. *viviparum* (S. F. Gray) Louis-Marie, and narrower-leaved specimens with the leaf-blades only 2–5 mm. wide and the inflorescence only 3–6 mm. thick are placed in var. *nodosum* (L.) Huds.

## 41. **Muhlenbergia** Schreb. Muhly

Mostly summer- and fall-flowering perennial grasses (in the local species) with erect or sprawling culms, and slender elongated or loose and open inflorescence. Ligule prominent or nearly lacking. Spikelets usually 1-flowered, occasionally 2-flowered. Fruit closely enveloped by the lemma.

a. Inflorescence loosely and openly branched, 5–20 cm. wide, broadly ellipsoid or ovoid; spikelets on pedicels (stalks) several to many times length of spikelet (excluding awn) . . . . . . . . . . *b*
 b. Awn 4–20 mm. long projecting from tip of lemma; plant growing in a clump or tuft, lacking any creeping underground stem (rhizome); leaves mainly 10–30 cm. long, 2–4 mm. wide; ligule 2–5 mm. long, triangular . . . . . . . . . . . . . . . . . . . 12. M. CAPILLARIS
 b. No awn on lemma; plant sending out 1 or more creeping elongated underground stems (rhizome); leaves short, mainly 2–7 cm. long, 1–2 mm. wide; ligule 0.5–1 mm. long, with ragged margin (erose) . . . . . . . . . . . . . . . . . . . . . . . . 2. M. ASPERIFOLIA
a. Inflorescence not loosely and openly branched as described above, but narrow, contracted, dense or many times longer than broad, usually less than 2.5 cm. thick or wide; spikelets either sessile (without stalk) or on short pedicels (stalks) usually shorter than length of spikelet (excluding awn) or on pedicels at most less than twice length of spikelet . . . . . . . . . . . . . . . . . . *c*
 c. Glumes with their awns 4.5–8 mm. long, much longer than lemma . . . . . . . 3. M. RACEMOSA
 c. Glumes 0.2–3 mm. long, either shorter than or at most only somewhat longer than lemma . . . . *d*
  d. Glumes mostly 0.1–0.5 mm. (rarely 1–2 mm.) long, 1/7 to less than 1/2 length of the body of the lemma (excluding awn of lemma or glume); culms (stems) rooting at lower nodes, lacking scaly creeping underground stem (rhizomes) . . . . . . . . . . . . . . . . . *e*
   e. Commonly found throughout Missouri; glumes about as broad as long, rounded or flat-topped at summit; 2nd glume 0.2–0.5 mm. long; lemma 1.8–2.5 mm. long (excluding awn); awns 2–5 mm. long . . . . . . . . . . . . 11a. M. SCHREBERI var. SCHREBERI
   e. Rare, known only from Barry Co.; glumes longer than broad, acute to sharp-pointed at tip; 2nd glume 1–2 mm. long; lemma 2.5–3 mm. long (excluding awn) ;awn 0.3–2 mm. long. 11b. M. SCHREBERI var. CURTISETOSA
  d. Glumes 1.4–3 mm. long, from 1/2 length to slightly longer than length of the body of the lemma (excluding awn of lemma); plants either with scaly underground creeping or elongated stems (rhizomes) or growing in a bunch or clump without rhizomes . . . . . . . . . *f*
   f. Lemma glabrous (without hair) at its base . . . . . . . . . . . . . . . . *g*
    g. Plants growing in a dense clump or tuft without any scaly underground creeping or elongated stems (rhizomes); leaves stiff, 1–2 mm. long, with inrolled margins; edges of limestone bluffs, limestone glades, and loess hills . . . . . . . . . 1. M. CUSPIDATA
    g. Plants with scaly underground creeping or elongated stems (rhizomes); leaves not stiff, 2–4 (–7) mm. wide, flat; rocky wooded hills, gravelly places, low banks, and prairies 8. M. GLABRIFLORA
   f. Lemma with a beard of hairs at the base . . . . . . . . . . . . . . . . *h*

h. Culms (stems) between the nodes, and nodes themselves glabrous, (infrequently minutely scabrous just below nodes in no. 4 [*M. sobolifera*] and no. 6 [*M. brachyphylla*]); key out such exceptions here or in other h . . . . . . . . . . . . . . . . . . . *i*

i. Ligule nearly obsolete or scarcely showing, less than 0.5 mm. long; anthers 0.8–1 mm. long; spikelets 1.5–2.7 mm. long; lemma 1.7–2.6 mm. long; culms solitary or few, unbranched or very slightly branched; plants usually of dry wooded upland or rocky or non-rocky slopes of hills . . . . . . . . . . . . . . . . 4. M. SOBOLIFERA

i. Ligule an obvious protruding membrane about 1 mm. or more long; anthers 0.3–0.8 mm. long; spikelets mostly 2.5–3.8 (rarely 2–2.5) mm. long; lemma 2–3.5 mm. long; culms usually branching from various nodes, usually with many branches; plants usually of moist low slopes along streams, river banks, bluffs, base of bluffs, and gravel bars . . . . . . . . . . . . . . . . . . . . . . . *j*

j. Glumes ovate or ovate-lanceolate, broadest at or below middle, abruptly narrowed to the slender tip, usually shorter than body of lemma (excluding awn of lemma), except in *M. brachyphylla* f. *aristata*; lemma usually awned, the awn mainly 2–8 mm. (rarely 0.5–2 mm.) long. . . . . . . . . . . . . 6. M. BRACHYPHYLLA

j. Glumes lanceolate, gradually narrowed from base (where widest) and tapering to a long slender tip, slightly shorter than or nearly equaling the lemma; lemma usually without an awn, or with an awn 4–10 mm. long . . . . 7. M. FRONDOSA

h. Culms (stems) below the nodes (turn the plant in natural growing position) more or less pubescent, the nodes themselves glabrous or pubescent . . . . . . . . . . . *k*

k. Glumes ovate or ovate-lanceolate, broadest at or below middle, abruptly narrowed to a slender tip, usually noticeably shorter than lemma (except longer in *M. brachyphylla* f. *aristata*); anthers 0.5–1.5 mm. long . . . . . . . . . . . . . . . . *l*

l. Spikelets (excluding the awn) 1.5–2.7 mm. long; lemma (excluding the awn) 1.7–2.6 mm. long, usually without an awn . . . . . . . . . 4. M. SOBOLIFERA

l. Spikelets (excluding the awn) 2.5–4 mm. long; lemma (excluding the awn) 2.5–3 mm. long, usually with an awn. . . . . . . . . . . . . . . . . . *m*

m. Nodes themselves hairy; culms and leaf-sheaths usually conspicuously or densely hairy, hairs on leaf-sheaths up to 0.5 mm. long; leaves 5–15 mm. wide; anthers 1–1.5 mm. long . . . . . . . . . . . . . . 5. M. TENUIFLORA

m. Nodes themselves glabrous; culms and leaf-sheaths mostly glabrous; leaves 3–7 mm. wide; anthers 0.5–0.8 mm. long . . . . . . . . 6. M. BRACHYPHYLLA

k. Glumes lanceolate, gradually narrowed from above base (where widest) and tapering to a long slender tip, shorter to slightly longer than lemma; anthers 0.3–0.6 mm. long . . *n*

n. Lemmas usually awned, rarely awnless; inflorescence long and slender, flexuous, the branches slender, 12–30 mm. apart; spikelets loosely overlapping, often on pedicels half as long as spikelets; glumes very gradually tapering to the tip, translucent, thin, silvery-green or whitish, mainly shorter than the lemma and unequal to nearly equal . . . . . . . . . . . . . . . . . 9. M. SYLVATICA

n. Lemmas usually awnless, more rarely with awn up to 10 mm. long; inflorescence stout and stiff, the branches often club-shaped and 10 mm. or less apart; spikelets closely overlapping, short-pedicelled to nearly sessile; glumes more abruptly tapering to the tip from a broader base, opaque, firm, green or purple, mainly as long as the lemma, and mostly equal . . . . . . . . . 10. M. MEXICANA

1. **Muhlenbergia cuspidata** (Torr.) Muhl.
Plains Muhly Map 222
Flowers July–October.

Characteristic of edges of limestone bluff escarpments bordering limestone glades in the Ozark region, and in the loess hills of northwestern Missouri; in the Ozarks occurring east to Washington and Shannon counties and north to Maries, Morgan, and Henry counties.

Ranges from Michigan to Alberta, south to Ohio, Kentucky, Illinois, Missouri, Oklahoma, and New Mexico.

This species frequents west- to south-facing, hot, dry limestone bluffs in the Ozark section, generally selecting the precipitous margins of the bluffs where it forms dense wiry clumps.

2. **Muhlenbergia asperifolia** (Nees & Meyen) Parodi Scratchgrass Map 223
Flowers late May–September.

Along railroads and in sandy open ground. Known only from St. Louis (freight yard of Burlington R. R. north of Carrie Ave., St. Louis, May 30, 1955, *Muehlenbach 619*), Jackson, and Holt counties.

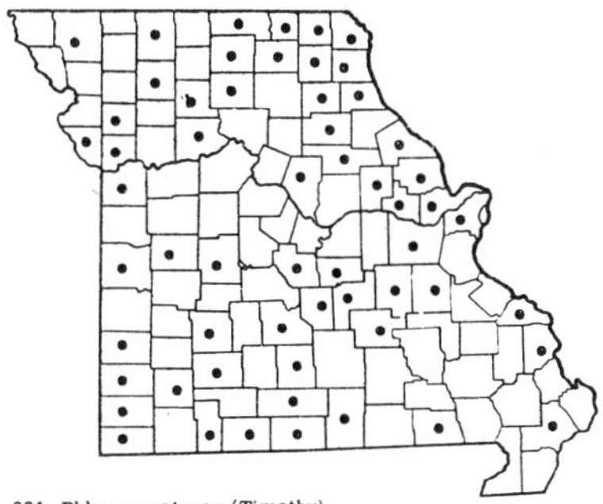

221 Phleum pratense (Timothy)

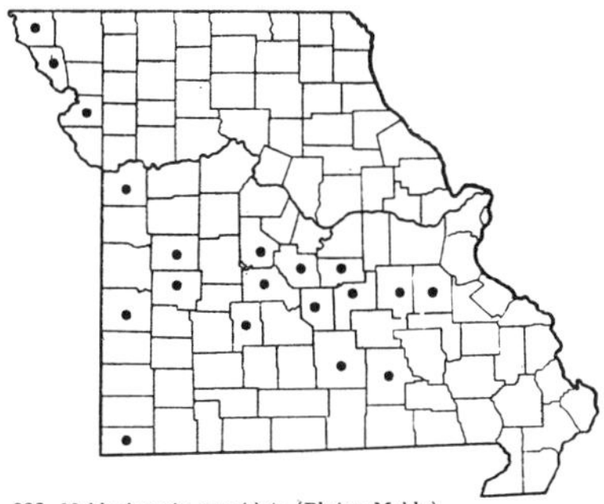

222 Muhlenbergia cuspidata (Plains Muhly)

223 Muhlenbergia asperifolia (Scratchgrass)

Ranges from Illinois, Minnesota, and Saskatchewan to British Columbia, south to Missouri, Texas, and California; also Mexico and South America.

The small spikelets, 1.5 mm. long, are purplish.

3. **Muhlenbergia racemosa** (Michx.) BSP. Map 224

Flowers August–October.

Alluvial bottomland of the Missouri and Mississippi rivers, prairies, and moist open places, in northern and central Missouri south to Jackson, Osage, Gasconade, and Jefferson counties.

Ranges from Michigan to Saskatchewan, south to Illinois, Missouri, Kansas, and New Mexico, and introduced eastward to New Hampshire.

The dense spikelike panicles vary from 4–20 mm. thick and from greenish to purple.

4. **Muhlenbergia sobolifera** (Muhl.) Trin. Map 225

Flowers July–October.

Commonly found on dry rocky or rich wooded slopes and upland wooded ridges, in limestone, but also chert, and sandy areas. It often is found in more acid soils in Missouri than the other species of the genus. It also occurs on bluffs and rarely in open situations. Throughout Missouri.

Ranges from New Hampshire to Wisconsin and Iowa, south to Virginia, Tennessee, Arkansas, and Texas.

Missouri material falls into two categories:

Lemmas without an awn . . 4a. M. SOBOLIFERA f. SOBOLIFERA

Lemmas with an awn 1–3 mm. long . . . . 4b. M. SOBOLIFERA f. SETIGERA

4a. **Muhlenbergia sobolifera** f. **sobolifera** Map 225

The common type throughout the state.

4b. **Muhlenbergia sobolifera** f. **setigera** (Scribn.) Deam Map 225

Known only from Christian County (upland woods along Pine Ridge road, 5 mi. southwest of Chadwick, July 17, 1954, *Palmer 58256*).

This form ranges from Indiana to Missouri, south to Arkansas and Texas.

This species ordinarily has solitary or separated, few, erect unbranched stems with divaricately spreading leaves arranged in two opposite directions. The rhizomes are usually remotely and loosely branched with the rhizome scales 4–6 mm. long. The spikelets are usually pale greenish-white, the glumes whitish.

5. **Muhlenbergia tenuiflora** (Willd.) BSP. Map 226

Flowers July–October.

Like *M. sobolifera*, this species usually inhabits woodland, being found generally in open rocky woods, limestone wooded slopes, and low woods along streams. It is sometimes found in exposed situations. Found in southern and central Missouri north to Jackson, Boone, and Lincoln counties.

This is a handsome grass with dark green leaves as much as 15 mm. wide, which is the greatest width reached by any of the local species. A height of 1 meter is sometimes attained by the usually unbranched erect culms. The rhizomes are loosely and remotely branched as in *M. sobolifera*, but have scales 5–10 mm. long. The hairs on the culms and leaf-sheaths are more conspicuous and longer (up to 0.5 mm. long on the leaf-sheaths) than in any of the other species in Missouri.

6. **Muhlenbergia brachyphylla** Bush Map 227

Flowers August–October.

Frequent in low moist woods, river banks, and at the base of bluffs. Throughout Missouri.

Ranges from Indiana to Nebraska, south to Missouri and Texas; reported from Virginia.

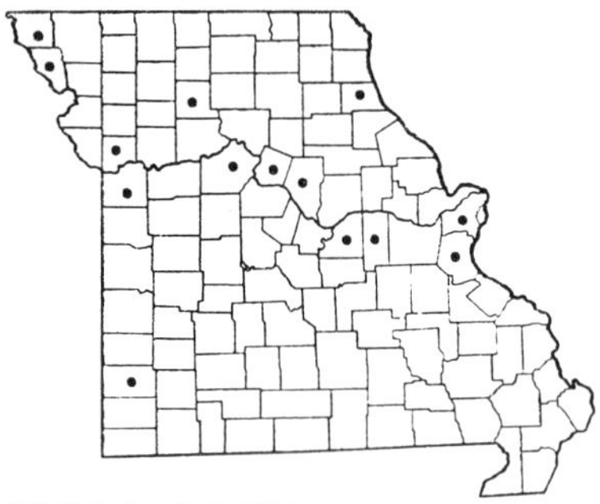

224 Muhlenbergia racemosa

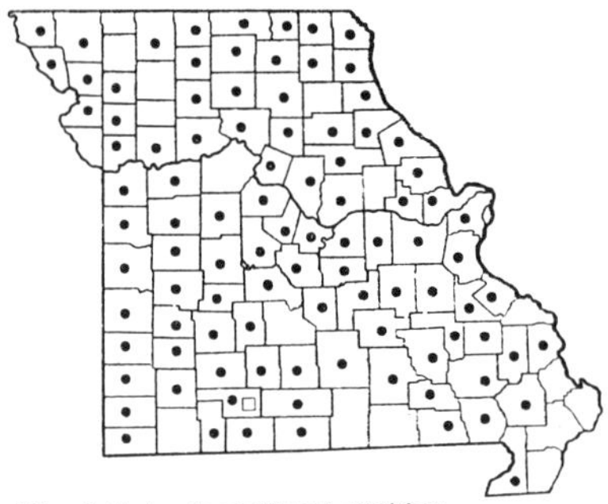

225 • Muhlenbergia sobolifera f. sobolifera
225 □ Muhlenbergia sobolifera f. setigera

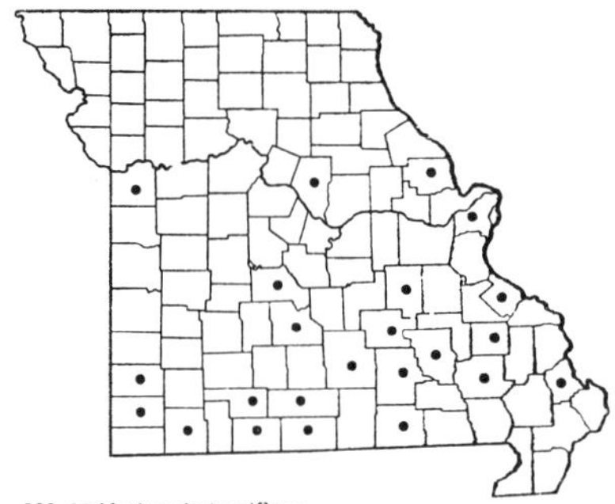

226 Muhlenbergia tenuiflora

Missouri material falls into two categories:

Glumes acuminate to shortly aristate or at most with short awns shorter than length of body of lemma (excluding awn of lemma) . . . . 6a. M. BRACHYPHYLLA f. BRACHYPHYLLA

Glumes definitely awned with awns up to 2 mm. long, longer than the length of the body of the lemma . . . 6b. M. BRACHYPHYLLA f. ARISTATA

6a. **Muhlenbergia brachyphylla** f. **brachyphylla** Map 227

The common type encountered throughout Missouri.

6b. **Muhlenbergia brachyphylla** f. **aristata** Palmer & Steyermark Map 227

Known only from the type collection from Pike County (wooded slopes along spring branch at base of high limestone bluff, vicinity of Stillhouse Cave along Clifty Fork of Spencer Creek, T53N, R4W, sect. 5, $1\frac{1}{2}$–$1\frac{3}{4}$ mi. southeast of Spenceburg, Oct. 7, 1956, *Steyermark 82921*).

There is considerable variation in *M. brachyphylla.* In addition to the variation in the apex of the glumes, the lemma may be only slightly awned, with the awn barely 0.5 mm. long, or the awn may attain a length of 8 mm. The principal terminal panicles are usually exserted, but plants frequently are found with the panicles mostly included within the leaf-sheaths.

The rhizomes are closely branched and bunched as in *M. frondosa* but the median rhizome scales are closely appressed to the surface of the rhizome and are oblong to narrowly ovate.

7. **Muhlenbergia frondosa** (Poir.) Fern. Nimble Will, Satin Grass Map 228

*M. mexicana* [of P & S], not (L.) Trin.

Flowers August–November.

Commonly occurring on low wooded slopes, ledges, base of bluffs and muddy banks along streams, gravel bars of small streams, sandy bluffs, prairie swales, moist woods, moist ground near lakes, and occasionally waste ground. In some areas becoming weedy, where difficult to eradicate (see Deam, *Grasses of Indiana*, p. 165. 1929).

Ranges from New Brunswick and Quebec to Ontario and North Dakota, south to Georgia, Tennessee, Missouri, Arkansas, and Texas.

Missouri material falls into two categories:

Lemma without an awn . . . 7a. M. FRONDOSA f. FRONDOSA

Lemma with an awn 4–10 (–20) mm. long . . 7b. M. FRONDOSA f. COMMUTATA

7a. **Muhlenbergia frondosa** f. **frondosa** Map 228

Common throughout Missouri, except for lack of records in a number of counties in the southeastern quarter.

7b. **Muhlenbergia frondosa** f. **commutata** (Scribn.) Fern. Map 228

*Muhlenbergia mexicana* f. *commutata* (Scribn.) Wieg. [P & S]

Infrequently occurring, and scattered in various parts of the state. In a specimen from Wayne County (*Steyermark 6619*) the awn of the lemma is 20 mm. long.

This species has closely branched and bunched

Plate no. 39. **1.** Muhlenbergia asperifolia; *a.* Plant, $\times\ ^3/_4$; *b.* Floret, $\times\ 7^1/_2$; *c.* Glumes, $\times\ 7^1/_2$. **2.** Muhlenbergia glabriflora; *a.* Floret, $\times\ 7^1/_2$; *b.* Glumes, $\times\ 7^1/_2$. **3.** Muhlenbergia sobolifera; *a.* Plant, $\times\ ^3/_4$; *b.* Glumes, $7^1/_2$; *c.* Floret, $\times\ 7^1/_2$. **4.** Muhlenbergia sobolifera f. setigera; *a.* Glumes, $\times\ 7^1/_2$; *b.* Floret, $\times\ 7^1/_2$. **5.** Phleum pratense; *a.* Plant, $\times\ ^3/_8$; *b.* Floret, $\times\ 7^1/_2$; *c.* Glumes, $\times\ 7^1/_2$. **6.** Muhlenbergia racemosa; *a.* Panicle, $\times\ ^3/_4$; *b.* Floret, $\times\ 7^1/_2$; *c.* Glumes, $\times\ 7^1/_2$. **7.** Muhlenbergia tenuiflora; *a.* Plant, $\times\ ^3/_4$; *b.* Floret, $\times\ 7^1/_2$; *c.* Glumes, $\times\ 7^1/_2$. **8.** Muhlenbergia cuspidata; *a.* Plant, $\times\ ^3/_4$; *b.* Floret, $\times\ 7^1/_2$; *c.* Glumes, $\times\ 7^1/_2$.

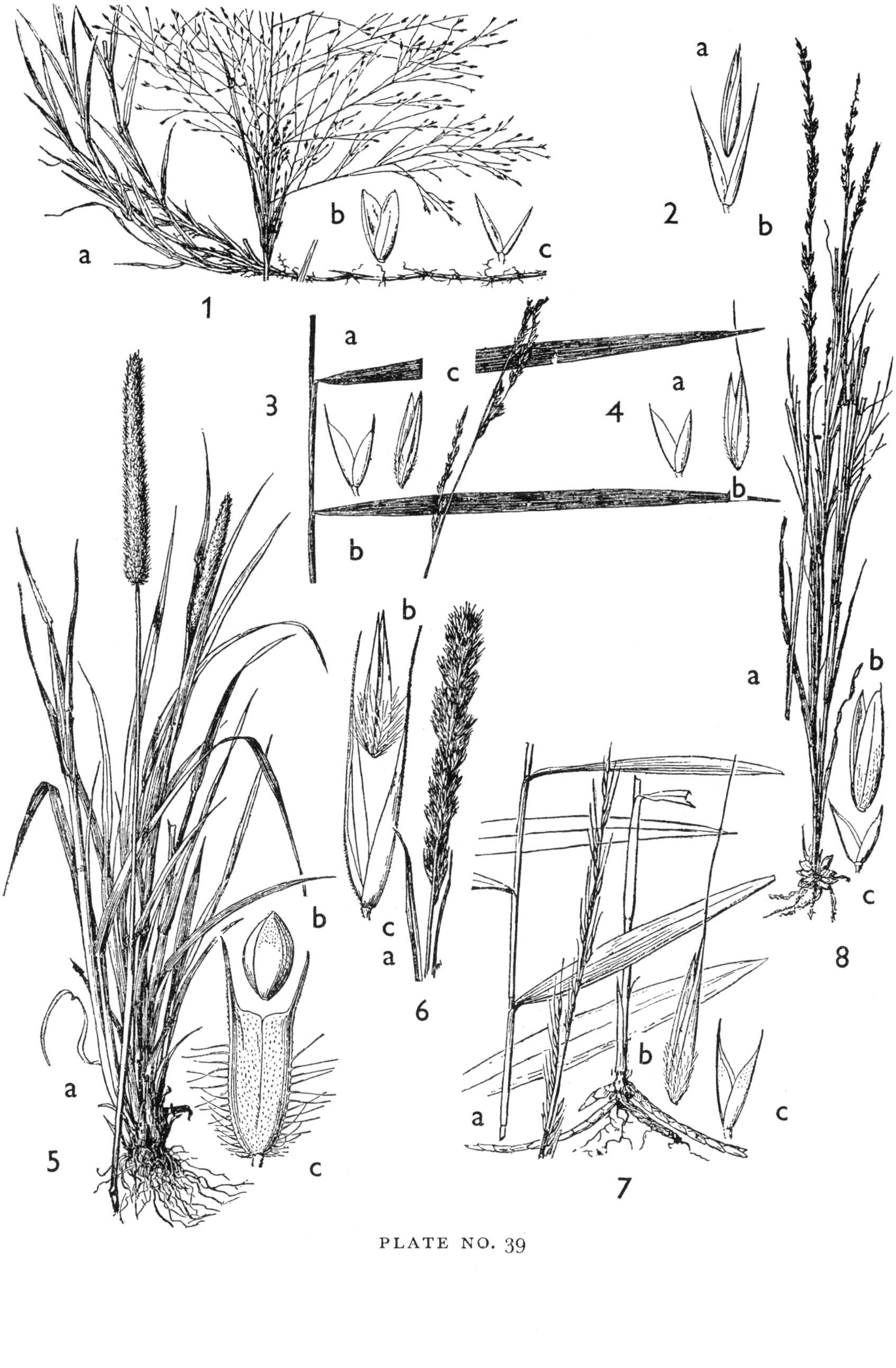


PLATE NO. 39

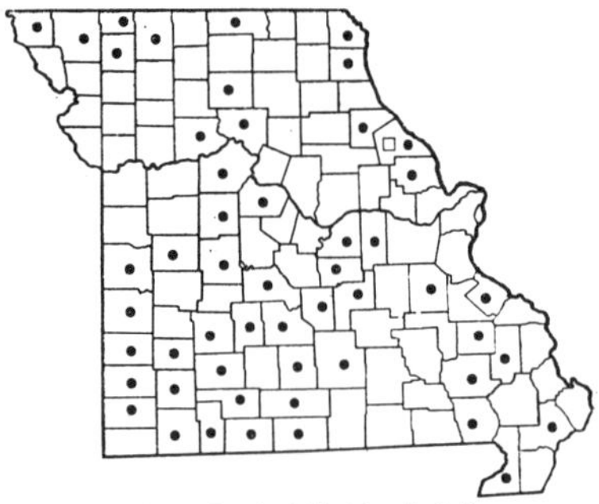

227 ● Muhlenbergia brachyphylla f. brachyphylla
227 □ Muhlenbergia brachyphylla f. aristata

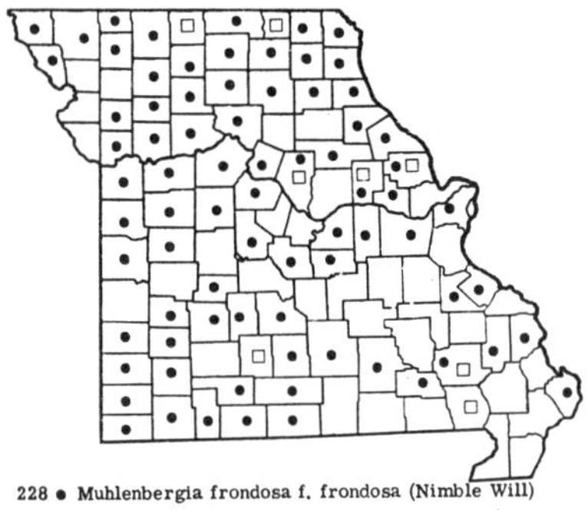

228 ● Muhlenbergia frondosa f. frondosa (Nimble Will)
228 □ Muhlenbergia frondosa f. commutata

228-A Muhlenbergia glabriflora

rhizomes. As compared with *M. brachyphylla*, the median rhizome scales in *M. frondosa* are more loosely arching and ovate instead of narrowly ovate or oblong. The stems sometimes attain a height of 1.3 meters, and vary from erect to sprawling.

8. **Muhlenbergia glabriflora** Scribn. Map 228A
Flowers August–October.

Occurs in prairies, moist open soils, and rocky slopes. Scattered in southern and central Missouri, north to St. Louis, Phelps, and Cass counties.

Ranges from Indiana, Illinois, and Missouri, south to Texas; also Maryland, Virginia, and North Carolina.

9. **Muhlenbergia sylvatica** Torr. Map 229
Flowers July–October.

Found in swampy meadows in the southeastern Ozarks, in moist thickets bordering meadows, thickets at the base of wooded bluffs, along moist crevices of sandstone or limestone bluffs, open banks bordering woods, and moist rocky woods and slopes bordering streams and occasionally upland woods. Throughout Missouri, except in some sections.

Ranges from Quebec to Ontario, Minnesota, and South Dakota, south to North Carolina, Alabama, Arkansas, and Texas.

Missouri material falls into two categories:

Lemma with an awn 3–15 mm. long . . . . 9a. MUHLENBERGIA SYLVATICA var. SYLVATICA f. SYLVATICA

Lemma without an awn . . 9b. MUHLENBERGIA SYLVATICA var. SYLVATICA f. ATTENUATA

9a. **Muhlenbergia sylvatica** var. **sylvatica** f. **sylvatica** Map 229

The common one encountered throughout the state. Apparently absent from the southeastern lowland section.

9b. **Muhlenbergia sylvatica** var. **sylvatica** f. **attenuata** (Scribn.) Palmer & Steyermark Map 229

Infrequent and scattered throughout the state.

The glumes in *M. sylvatica* are somewhat variable, but are generally shorter than the lemma (some current manuals have confused the situation by indicating in the key that the glumes are nearly or about as long as the lemma), and nearly equal (although stated to be unequal or very unequal in some manuals). The glumes are usually ovate-lanceolate or lanceolate with broad scarious margins (confusion is again created by their description in some manuals as 'linear-lanceolate to linear-attenuate'). The glumes are broader and more abruptly tapering than in *M. mexicana*, a species with which it is often misidentified.

10. **Muhlenbergia mexicana** (L.) Trin. Map 230
*M. foliosa* [of P & S], not (R & S) Trin.
Flowers August–October.

Occurs in damp thickets, moist woods, upland wooded slopes and wooded ravines, wet open swales and meadows, and waste ground.

Rather infrequent, and mostly in western Missouri, locally eastward to St. Louis, Lincoln, and Pike counties.

Ranges from Nova Scotia and Quebec to British

Plate no. 40. **1.** Muhlenbergia sylvatica; *a.* Plant, × 3/4; *b.* Floret, × 7 1/2; *c.* Glumes, × 7 1/2. **2.** Muhlenbergia Schreberi var. Schreberi; *a.* Plant, × 3/8; *b.* Floret, × 7 1/2; *c.* Glumes, × 7 1/2. **3.** Muhlenbergia Schreberi var. curtisetosa; *a.* Floret, × 7 1/2; *b.* Glumes, × 7 1/2. **4.** Muhlenbergia frondosa; *a.* Plant, × 3/4; *b.* Floret, × 7 1/2; *c.* Glumes, × 7 1/2. **5.** Muhlenbergia brachyphylla; *a.* Plant, × 3/4; *b.* Glumes, × 7 1/2; *c.* Floret, × 7 1/2. **6.** Muhlenbergia mexicana; *a.* Plant, × 3/4; *b.* Glumes, × 7 1/2; *c.* Floret, × 7 1/2.

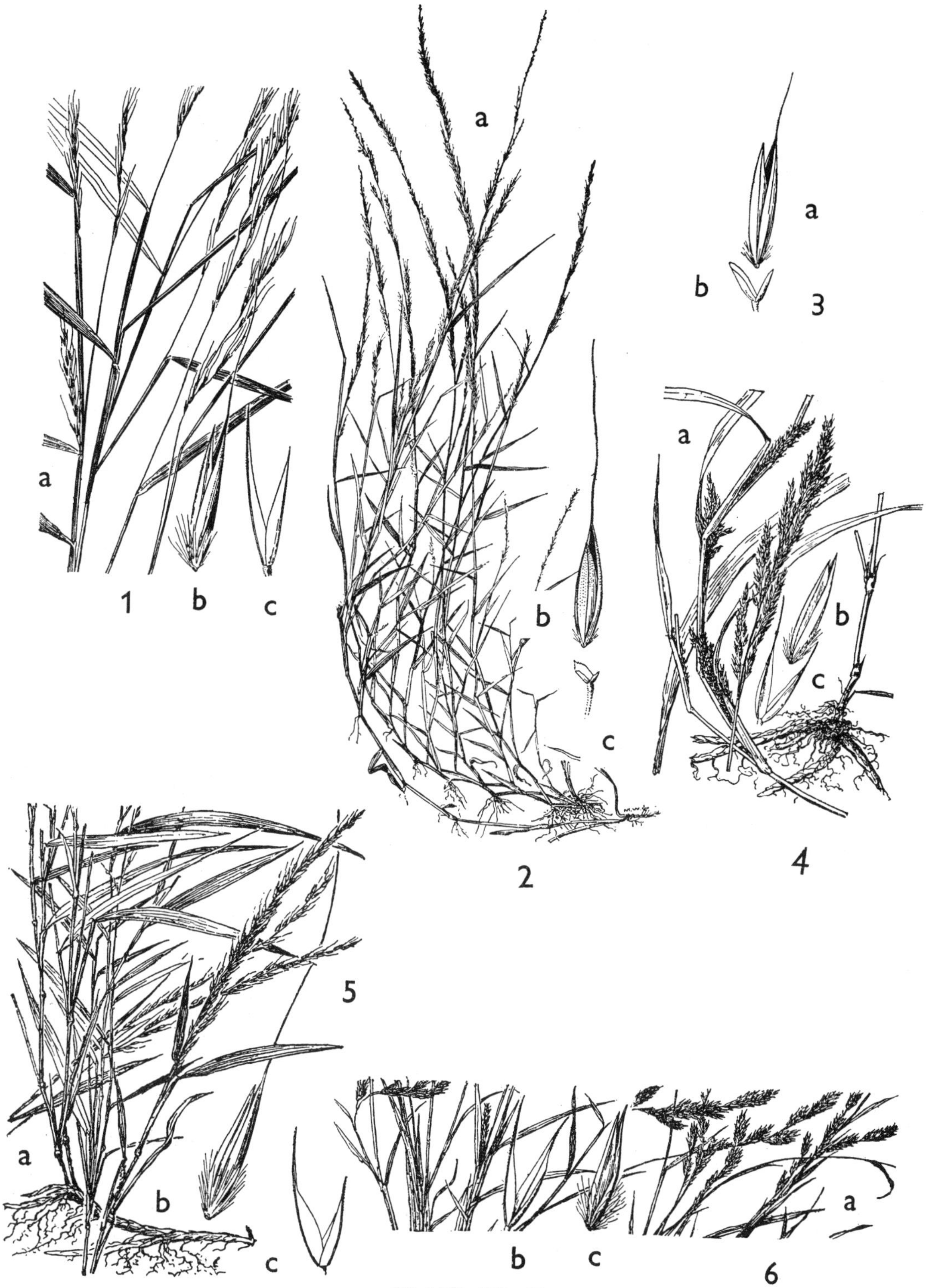


PLATE NO. 40

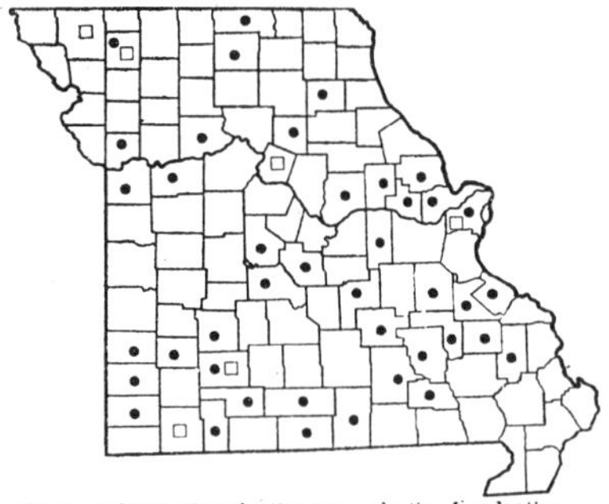

229 ○ Muhlenbergia sylvatica var. sylvatica f. sylvatica
229 □ Muhlenbergia sylvatica var. sylvatica f. attenuata

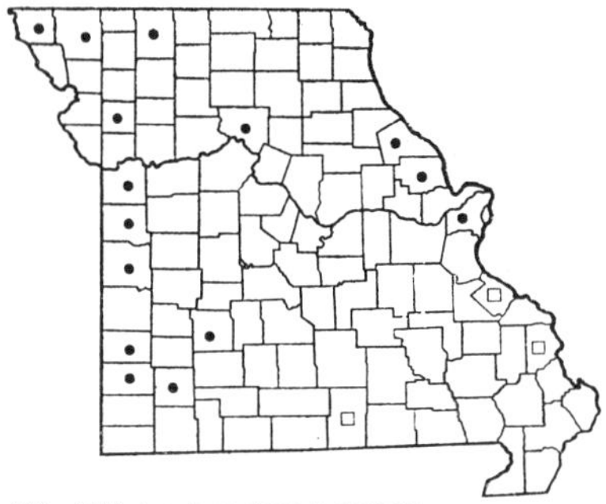

230 ● Muhlenbergia mexicana f. mexicana
230 □ Muhlenbergia mexicana f. ambigua

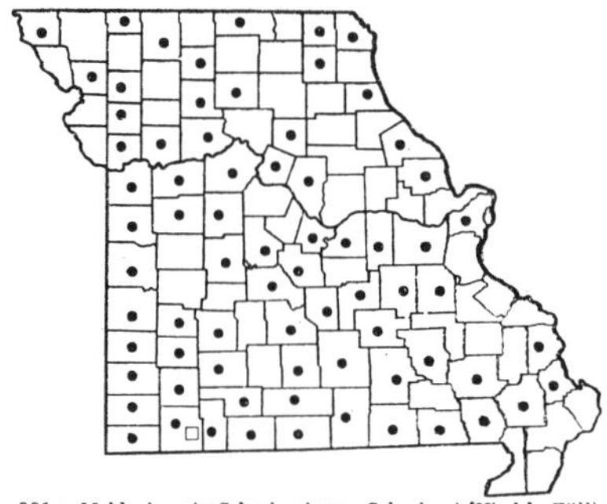

231 ● Muhlenbergia Schreberi var. Schreberi (Nimble Will)
231 □ Muhlenbergia Schreberi var. curtisetosa

Columbia, south to North Carolina, Ohio, Indiana, Illinois, Missouri, Kansas, New Mexico, Arizona, and California.

Missouri material falls into two categories:

Lemma without an awn . . 10a. M. MEXICANA f. MEXICANA

Lemma with an awn 4–10 mm. long. . . . 10b. M. MEXICANA f. AMBIGUA

10a. **Muhlenbergia mexicana** f. **mexicana** Map 230

The commoner one encountered in Missouri.

10b. **Muhlenbergia mexicana** f. **ambigua** (Torr.) Fern. Map 230

Infrequent, and known only from Ste. Genevieve, Cape Girardeau, and Howell (gravelly open places along South Fork of Spring Creek along highway A, T23N, R8W, south ½ sect. 33, 8 mi. south of West Plains, Sept. 3, 1949, *Steyermark 69076*) counties.

*Muhlenbergia mexicana* is most readily distinguished from *M. sylvatica* by the normally awnless lemmas and more densely flowered branches of the inflorescence. Both species have rather bunched and closely branching rhizomes, the scales according to observations of Shinners (Am. Midl. Nat. 26: 70. 1941) being 2–7 mm. long, whereas in *M. frondosa* they vary from 5–10 mm. long.

11. **Muhlenbergia Schreberi** Gmel. Nimble Will Map 231

Flowers July–November.

This is found in a wide variety of habitats, ranging from moist low meadows, thickets, open and waste ground, to gravelly banks, wooded sandstone slopes, and pine woods. It is weedy, often found in gardens and about dwellings, and in disturbed or cultivated areas. Throughout Missouri.

Ranges from Florida to Texas and Mexico, north to New Hampshire, New York, Ontario, Michigan, Wisconsin, Iowa, and Nebraska.

Missouri material falls into two varieties:

Commonly found; glumes as broad as long, rounded or flat-topped at summit; 2nd glume 0.2–0.5 mm. long; lemma 1.8–2.5 mm. long (excluding awn); awns 2–5 mm. long. 11a. M. SCHREBERI var. SCHREBERI

Known only from Barry Co.; glumes longer than broad, acute to sharp-pointed at tip; 2nd glume 1–2 mm. long; lemma 2.5–3 mm. long (excluding awn); awn 0.3–2 mm. long . . 11b. M. SCHREBERI var. CURTISETOSA

11a. **Muhlenbergia Schreberi** var. **Schreberi** Map 231

Throughout Missouri.

11b. **Muhlenbergia Schreberi** var. **curtisetosa** (Scribn.) Steyermark & Kucera Map 231

*Muhlenbergia Schreberi curtisetosa* Scribn. Rh. 9: 17. 1907.

*Muhlenbergia curtisetosa* (Scribn.) Bush [G, BB, P & S, Hitchcock]

Known only from Barry County, southwestern Missouri (Eagle Rock, Sept. 24, 1896, *Bush 377*). As suggested by Gleason (*New Ill. Fl.* 1: 174. 1952), *M. curtisetosa* appears to be of doubtful taxonomic status as a species and is placed in the present treatment in the varietal category with *M. Schreberi*, which it closely resembles in general appearance.

*Muhlenbergia Schreberi* forms dense colonies by virtue of the decumbent to prostrate lower parts of the culm which easily root at the lower nodes and send up slender wiry often purple smooth culms. The roots are fine and slender, and there are no creeping, elongated, scaly underground rhizomes as in most of the other local species. The long-exserted, very narrow panicles are characteristic. The glumes are so small that they might easily be overlooked.

12. **Muhlenbergia capillaris** (Lam.) Trin.
Hair Grass Map 232
Flowers September–October.

Found in acid soils usually in open woods of deciduous or pine trees, sandstone or more commonly cherty limestone glades, or openings above limestone escarpments, and openings along roads. Occurring only in the Ozark section north to Henry, Pettis, Cooper, and Franklin counties, and locally east to Cape Girardeau County. Absent from the southeastern lowland section and from many of the eastern Ozark counties.

Ranges from Florida to Texas, north to Massachusetts, Pennsylvania, Kentucky, Illinois, Missouri, and Kansas.

This is a handsome grass with its graceful rosy-purple panicle of hairlike delicate branches and pedicels, and is conspicuous when encountered in the fall. It is a rare grass, however, and not commonly seen. It is characteristic of acid soils and is difficult to transplant and establish in a wild garden.

*Excluded species*

**Muhlenbergia expansa** (DC.) Trin. Hair Grass

This species was listed by Palmer & Steyermark (1935) in their *Annotated Catalogue* as occurring in McDonald County. This report was based on a misidentified specimen confused with *M. capillaris.*

## 42. **Sporobolus** R. Br. Dropseed

Chiefly summer- and fall-flowering annual or perennial grasses with awnless spikelets. The fruit, when mature falls readily from the spikelet, and its seed separates easily from the pericarp.

a. Leaf-sheaths densely bearded at summit . . . . . 8. S. CRYPTANDRUS
a. Leaf-sheaths glabrous or if pubescent, at least not long-bearded at summit . . . . . *b*
b. Inflorescence long and narrow, contracted, spike-like, sometimes enclosed by an inflated sheath; branches of inflorescence erect to ascending close to the main axis . . . . . *c*
c. Length of inflorescence $\frac{1}{4}$–$\frac{1}{2}$ the height (length) of culm (stem); spikelets 1.5–2 mm. long; new stem-leaves 2–3 . . . . . 5. S. POIRETII
c. Length of inflorescence many times shorter than height (length) of culm (stem); spikelets 2–7 mm. long; new stem-leaves 4–12 . . . . . *d*
d. At least some part of lemma appressed-pubescent (examine carefully with lens or binocular microscope to note colorless hairs lying flat on surface of lemma) . . . . . *e*
e. Inflorescence at very top of stem usually covered over by an inflated leaf-sheath, rarely exserted; inflorescences also arising along the side (axillary) but also covered over by a leaf-sheath; annuals . . . . . 1. S. VAGINIFLORUS
e. Inflorescence at very top of stem either partly covered over at the base by an inflated leaf-sheath or entirely free and exserted from the leaf-sheath; axillary inflorescence usually absent (rarely 1 or more); perennials . . . . . 4. S. CLANDESTINUS
d. Lemma wholly glabrous . . . . .
f. Glumes nearly equal, the 2nd rarely surpassing the 1st by more than 1 mm.; glumes nearly equaling the lemma in length, 1st glume 3/4 to as long as length of lemma; inflorescence at very top of stem usually covered over by an inflated leaf-sheath, rarely exserted; inflorescences arising along the side (axillary) but also covered over by a leaf-sheath; annuals . . . . . 2. S. NEGLECTUS
f. Glumes obviously unequal, the 2nd glume surpassing the first by 1 mm. or more; at least the 1st glume 1/3–2/3 length of lemma; inflorescence at very top of stem either partly covered over at the base by an inflated leaf-sheath or entirely free and exserted from the leaf-sheath; axillary inflorescence usually absent (rarely 1 or more); perennials . 3. S. ASPER
b. Inflorescence open and loose, pyramid-shaped or ovoid, not spike-like; branches of inflorescence spreading to ascending away from main axis . . . . . *g*
g. Leaf-blades 2 mm. or less wide; spikelets 4.5–6 mm. long; 1st glume tapering to a long narrow tip . . . . . 6. S. HETEROLEPIS
g. Leaf-blades 2–5 mm. wide; spikelets 1.5–3.5 mm. long; 1st glume short-pointed (acute) to blunt (obtuse) . . . . . *h*
h. Main branches of inflorescence usually alternate; spikelets 1.5–2.5 mm. long . . 9. S. AIROIDES
h. Main branches of inflorescence opposite or in whorls; spikelets 1.5 mm. long . 7. S. PYRAMIDATUS

232 Muhlenbergia capillaris (Hair Grass)

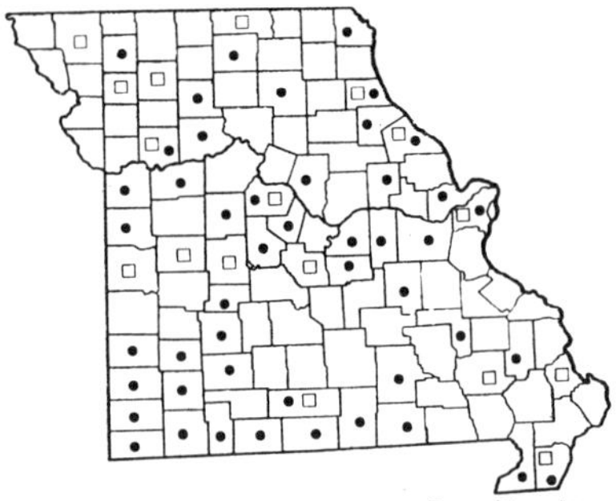

233 • Sporobolus vaginiflorus var. vaginiflorus (Poverty Grass)
233 □ Sporobolus vaginiflorus var. inaequalis

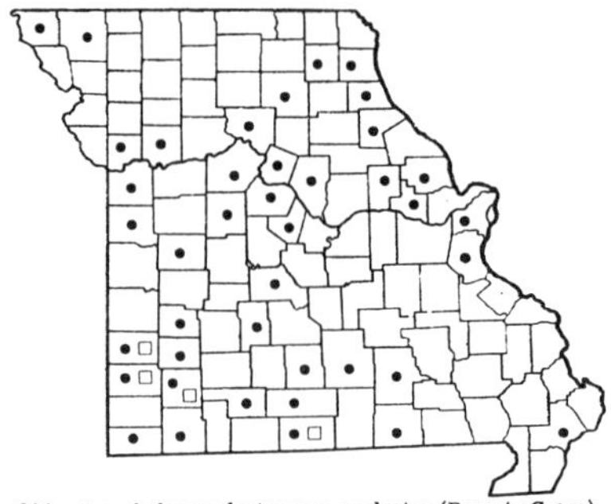

234 • Sporobolus neglectus var. neglectus (Poverty Grass)
234 □ Sporobolus neglectus var. ozarkanus

1. **Sporobolus vaginiflorus** (Torr.) Wood
Poverty Grass Map 233

Flowers August–November.

Common on limestone glades, cherty openings in woodland, overgrazed fields and prairies, along railroads, and waste ground, especially sterile soils. Throughout Missouri.

Missouri material falls into two varieties:

Palea and lemma nearly equal in length or only slightly unequal, the glumes nearly as long as palea or longer . . . . . 1a. S. VAGINIFLORUS var. VAGINIFLORUS

Palea conspicuously longer than and prolonged by its slender beak beyond the lemma, the glumes much shorter than the palea . . . . 1b. S. VAGINIFLORUS var. INAEQUALIS

1a. **Sporobolus vaginiflorus** var. **vaginiflorus**
Map 233

Common throughout Missouri.

Ranges from Nova Scotia, Maine, and Quebec, to North Dakota, south to Georgia and Texas.

1b. **Sporobolus vaginiflorus** var. **inaequalis** Fern.
Map 233

Scattered throughout Missouri.

Ranges from Maine to Ontario, south to New York, Pennsylvania, Michigan, Wisconsin, Missouri, Nebraska, and Texas.

There is some intergradation between these two variations with respect to the relative length and prolongation of the apex of the palea. The length of the spikelets is given in some manuals as 3.5–6.5 mm. However, spikelets are sometimes found measuring only 2.5 mm. long. Such spikelets in Missouri material constitute fully mature, late-season, inflated ones with the surface of the lemma prominently convex. After the fruits have fallen, however, the inflated appearance disappears, and the spikelets become longer and attain 3.5–4 mm. Shinners (Rh. 56: 29. 1954) also reports finding late-season atypical forms of *S. vaginiflorus* var. *vaginiflorus* with pubescent lemmas of spikelets only 2–2.6 mm. long. The leaf-sheaths are sometimes stated as being glabrous and often long-ciliate at the orifice. Occasionally they are pilose elsewhere on the surface of the sheath.

2. **Sporobolus neglectus** Nash Poverty Grass
Map 234

Sometimes called Dropseed or Puffsheath Dropseed

*Sporobolus vaginiflorus* var. *neglectus* (Nash) Scribner, erroneously published as a new comb. by Shinners.

Flowers August–November.

Limestone glades, cherty openings in woods, fields, along railroads, and waste ground.

Ranges from New Brunswick, Maine, and Ontario to North Dakota, south to Virginia, Tennessee, Missouri, and Texas.

Missouri material may be placed into two categories:

Spikelets 2–3.5 mm. long; lower leaf-sheaths and blades glabrous or sparsely hairy; grain (seed-like fruit) 1–1.5 mm. long . . . 2a. S. NEGLECTUS var. NEGLECTUS

Spikelets 3.5–5 mm. long; lower leaf-sheaths and blades papillose-pilose (hairs with swellings at base); grain (seed-like fruit) 1.7–2.2 mm. long . 2b. S. NEGLECTUS var. OZARKANUS

Plate no. 41. **1.** Sporobolus asper; *a.* Plant, × 3/4; *b.* Floret, × 7 1/2; *c.* Glumes, × 7 1/2. **2.** Sporobolus neglectus; *a.* Plant, × 3/4; *b.* Floret, × 7 1/2; *c.* Spikelet, × 7 1/2. **3.** Sporobolus clandestinus; *a.* Plant, × 3/4; *b.* Floret, × 7 1/2; *c.* Glumes, × 7 1/2. **4.** Sporobolus vaginiflorus; *a.* Plant, × 3/4; *b.* Floret, × 7 1/2; *c.* Glumes, × 7 1/2. **5.** Muhlenbergia capillaris; *a.* Plant, × 3/8; *b.* Floret, × 7 1/2; *c.* Glumes, × 7 1/2.

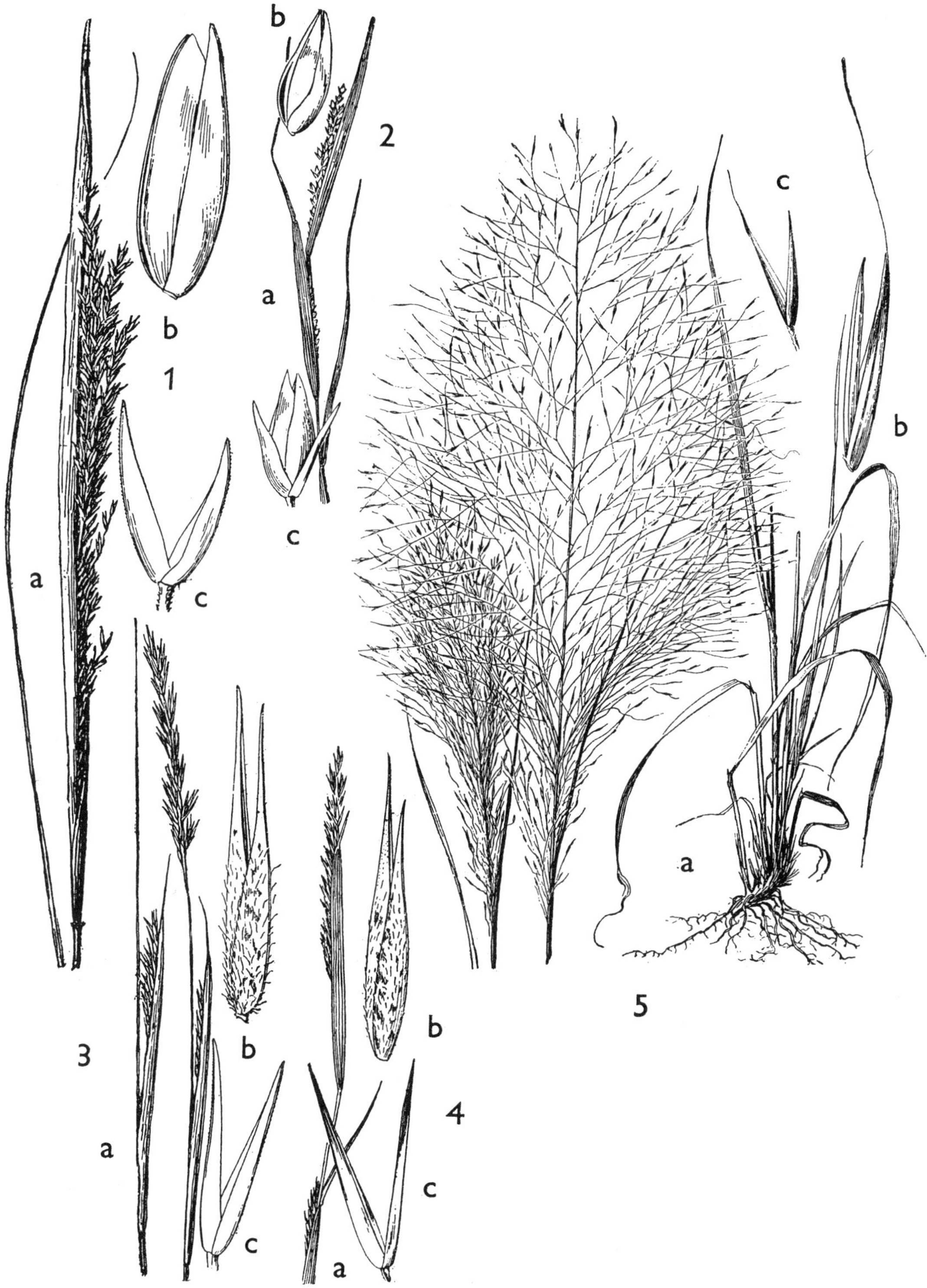


PLATE NO. 41

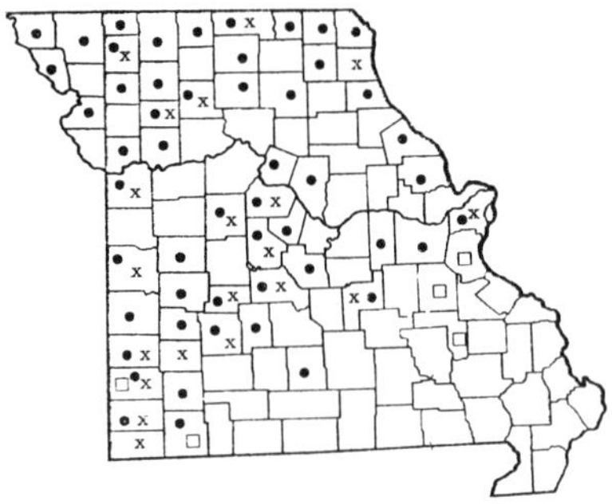

235 • Sporobolus asper var. asper
235 □ Sporobolus asper var. pilosus
235 x Sporobolus asper var. Hookeri

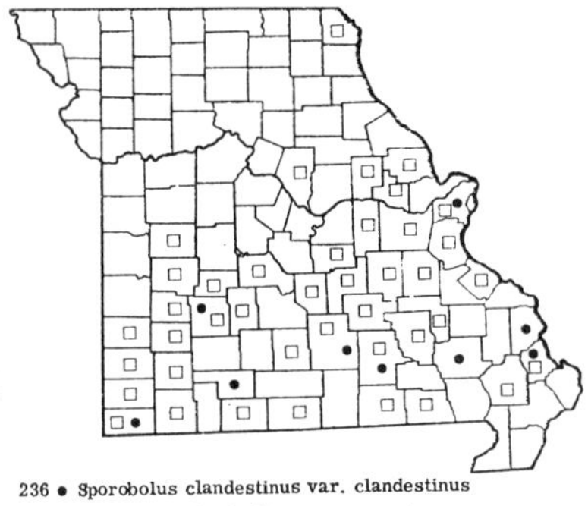

236 • Sporobolus clandestinus var. clandestinus
236 □ Sporobolus clandestinus var. canovirens

237 Sporobolus Poiretii (Smut Grass)

2a. **Sporobolus neglectus** var. **neglectus** Map 234

Common throughout Missouri. Records are absent from parts of southeastern and north central Missouri, although doubtless present.

2b. **Sporobolus neglectus** var. **ozarkanus** (Fern.). Steyermark & Kucera Map 234

*Sporobolus ozarkanus* Fern. [G, P & S]

*Sporobolus vaginiflorus* var. *ozarkanus* (Fern.) Shinners, Rh. 56: 29. 1954.

Occurs in Jasper (barrens, Webb City, September 8, 1910. *Palmer 3133, holotype, 3133 A*), Lawrence (limestone ledge, open or partly wooded hillside, near Verona, Aug. 22, 1950, *Palmer 50816*), Ozark (slopes and ledges, bald knobs, 'Bald Jesse,' near Gainesville, Oct. 10, 1927, *Palmer 33064*) and Barton counties, all in southwestern Missouri.

Plants of this group vary in their amount of pubescence shown on the leaf-sheaths and blades and in the size of their spikelets. Some spikelets of *S. neglectus* var. *neglectus* may attain 3.5 mm. in length (*Steyermark 79874*), although the manuals state the length as 2–3 mm. Specimens with conspicuous papillose leaf-sheaths and blades and larger spikelets are placed in var. *ozarkanus*, but intermediates occur, such as Palmer's collection from Barton County. The glabrity of the lemmas and their short-pointedness together with the relatively less elongated apex of the palea appear to be characters relating *S. ozarkanus* more closely with *S. neglectus* than with *S. vaginiflorus*, whereas Shinners judges *S. ozarkanus* as a variety of *S. vaginiflorus*.

The strongly ciliate orifices of the leaf-sheaths, believed by Fernald to be characteristic of *S. ozarkanus*, are also found in *S. vaginiflorus*, and in this respect *S. ozarkanus* resembles *S. vaginiflorus*. However, occasional specimens of *S. neglectus* var. *neglectus* also have ciliate orifices of the leaf-sheath. It is true, also, that the relatively longer spikelets of *S. ozarkanus* afford another point of resemblance to *S. vaginiflorus*.

On account of the above, it is admitted that *S. ozarkanus* may likely be judged and eventually proved a hybrid, intermediate between *S. vaginiflorus* var. *vaginiflorus* and *S. neglectus* var. *neglectus*.

3. **Sporobulus asper** (Michx.) Kunth Map 235

Flowers August–October.

Common along roadsides and railroads; also occurring in dry prairies and rocky open woods and glades.

Missouri material may be divided into the following varieties:

a. Inflorescence at tip of culm (terminal inflorescence) relatively thin, loose, linear-filiform, 3–6 mm. thick; spikelets mostly 3–5 mm. long; 1st glume linear-lanceolate . 3c. S. ASPER var. HOOKERI

a. Inflorescence at tip of culm (terminal inflorescence) relatively dense, stiff, thick-cylindrical, 5–18 mm. thick; spikelets 4.5–6.5 mm. long; 1st glume narrowly ovate . . . . . . *b*

b. Leaf-sheaths and blades more or less pilose with conspicuous hairs on all parts . . 3b. S. ASPER var. PILOSUS

b. Usually the upper surface of leaf-blades at the base pilose or hirsute, otherwise rough-scabrous; leaf-sheaths glabrous or the lower ones appressed-pubescent, rarely pilose at the summit . . . . 3a. S. ASPER var. ASPER

3a. **Sporobolus asper** var. **asper** Map 235

Apparently absent in most of the eastern portion of the Ozarks and the southeastern quarter section of the state, otherwise common in the remainder of Missouri.

Ranges from Quebec, Vermont, New York, and Ohio to North Dakota, south to Virginia, Tennessee, Louisiana, and Texas.

3b. **Sporobolus asper** var. **pilosus** (Vasey) Hitchc. Map 235

Scattered in a few counties of the Ozark section of the state.

Ranges from Missouri and Kansas to Texas.

This is included by Shinners (Rh. 56: 29. 1954) under *S. asper* var. *Hookeri*, but I am maintaining it as the pilose-leaved variation of plants with the dense thick-cylindrical inflorescences more characteristic of *S. asper* var. *asper*.

3c. **Sporobolus asper** var. **Hookeri** (Trin.) Vasey Map 235

*Sporobolus Drummondii* (Trin.) Vasey [G]

Usually found on glades, prairies, and rocky open woods over much the same range as *S. asper* var. *asper*, but less common north of the Missouri River. Absent from most of the eastern quarter of the state.

Ranges from Missouri to Mississippi and Texas.

Overlapping of measurements of the length of the spikelets breaks down distinctions between *S. asper* var. *asper* and the other varieties. There is also intergradation in the width of the 1st glume. For example, *Palmer 60937* from McDonald County has the slender inflorescence of *S. asper* var. *Hookeri* with spikelets 4–5 mm. long, but the 1st glume is not as narrow as in *S. asper* var. *Hookeri*.

*Sporobolus asper* var. *asper* and other varieties are drought-resistant bunch grasses. They spread into overgrazed areas where bluestem (*Andropogon scoparius*) is prevalent. The stems and long leaves of *S. asper* var. *asper* turn whitish during the winter. Often called 'flag grass' on account of the habit of the upper leaves waving in the wind which is especially noticeable on cutover prairies in the autumn.

4. **Sporobolus clandestinus** (Biehler) Hitchc. Map 236

Flowers August–October.

Missouri material falls into two categories:

Palea much prolonged beyond the lemma into a narrowed nearly awned beak; spikelets 5.5–8 mm. long . . 4a. S. CLANDESTINUS var. CLANDESTINUS

Palea and lemma nearly equal, the palea not prolonged into a nearly awned beak; spikelets 4.5–6 mm. long . 4b. S. CLANDESTINUS var. CANOVIRENS

4a. **Sporobolus clandestinus** var. **clandestinus** Map 236

*Sporobolus clandestinus* (Biehler) Hitchc. [G]

*Sporobolus asper* var. *clandestinus* (Biehler) Shinners, Rh. 56: 30. 1954.

Rocky open woods, glades, edges of bluffs, and prairies. Scattered in several counties of the Ozark section.

Ranges from Connecticut, Indiana, and Missouri to Florida and Texas.

4b. **Sporobolus clandestinus** var. **canovirens** (Nash) Steyermark & Kucera Map 236

*Sporobolus canovirens* Nash [G]

*Sporobolus asper* var. *canovirens* (Nash) Shinners, Rh 56: 30. 1954.

Rocky open woods, glades, edges of bluffs, and prairies. Much commoner than *S. clandestinus* var. *clandestinus*, throughout the Ozark region north to Henry, Boone, Montgomery, and Lincoln counties, and locally in Clark County, northeast Missouri.

Ranges from Indiana and Wisconsin to Kansas, south to Mississippi and Texas.

There is intergradation between the two variations of *S. clandestinus* in spikelet length, in the relative length between palea and lemma, and in the degree of prolongation of the palea. Extremes of the two types are found which appear very distinct, but since transitions are often difficult to place, their reduction to varietal rank appears more consistent.

As both variations possess consistently pubescent lemmas, characteristic of *S. clandestinus*, it is believed that they should be kept separate as distinct varieties of that species, rather than merged, as Shinners has done, as varieties of *S. asper*, the latter species, together with its varieties, possessing glabrous lemmas.

5. **Sporobolus Poiretii** (R & S) Hitchc.

Smut Grass Map 237

Flowers May–October.

Known only from southeastern Missouri in Dunklin County (Oct. 26, 1892, *Bush*), where introduced in cotton fields.

This species is reported from dry sandy or wet open soil, roadsides, and along ditches, from Florida to Texas, north to Virginia, Kentucky, and southeastern Missouri. Also native throughout tropical America and tropical Asia, and adventive in New Jersey.

No specimen has been found in the state since the original collection by Bush. It is probable that the species is now extinct in Missouri.

The common name for this grass is due to a black fungus, which frequently infects the inflorescence.

6. **Sporobolus heterolepis** Gray

Prairie Dropseed Map 238

Flowers August–October.

Prairies, glades, rocky cliffs, and open ground along railroads. Throughout Missouri, except absent

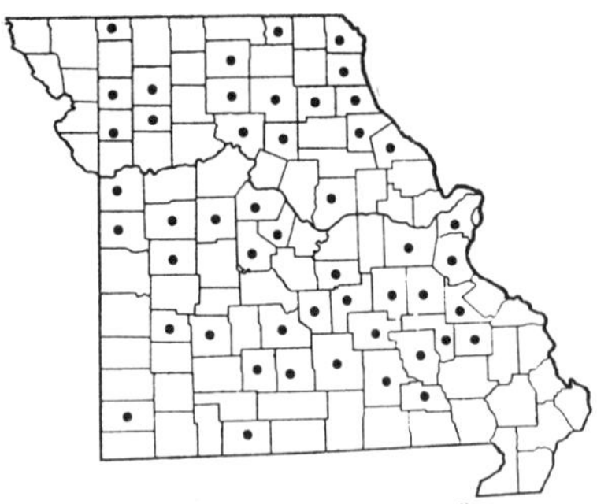

238 Sporobolus heterolepis (Prairie Dropseed)

239 Sporobolus pyramidatus

240 Sporobolus cryptandrus var. cryptandrus (Sand Dropseed)

in the southeastern lowland region, and extreme northwestern Missouri.

Ranges from Quebec to Saskatchewan, locally south to New York, Ohio, Indiana, Illinois, Arkansas, Texas and Wyoming; local eastward from Connecticut to Pennsylvania.

This grass forms conspicuous large clumps of handsome narrow, almost hairlike, long blades, which are mostly basal and curl or arch gracefully backwards. The long-exserted openly branched panicles are purple to blackish.

7. **Sporobolus pyramidatus** (Lam.) Hitchc. Map 239

*Sporobolus argutus* (Nees) Kunth [P & S]

Flowers May–October.

Known only from dry open ground where introduced in Jackson County (Westminister, July 27, 1896, *Bush 510*).

Ranges from Kansas and Colorado to Louisiana and Texas, south to the West Indies, Mexico to Nicaragua, and South America. Introduced eastward in Missouri and New York.

Not collected in the state since Bush's original collection.

8. **Sporobolus cryptandrus** (Torr.) A. Gray var. **cryptandrus** Sand Dropseed Map 240

Flowers July–October.

Usually found on dry sandy flats bordering the Missouri and Mississippi rivers, also on loess mounds of northwestern Missouri, and infrequently on exposed limestone strata; also introduced along railroads.

Absent from most parts of Missouri. Confined to those habitats indicated above, which limit the geographic range to northwestern, central, and extreme eastern Missouri south of the Missouri River. Apparently absent from most of northern Missouri, and locally south in Texas and Jasper counties.

Ranges from Quebec to Washington, south to North Carolina, Ohio, Indiana, Illinois, Louisiana, Texas, and Mexico.

This is so characteristic of sandy areas that one can almost be certain of encountering it in any area where there is pure sand. It is probable that this species will be found eventually in all the counties along the Mississippi and Missouri rivers in the state where sand bars exist.

It is reported that the Kiowa Indians employ the seeds of this species for a flour which is obtained by grinding the dried grains. It is believed that other species of this genus *Sporobolus* can be similarly employed for production of a flour.

9. **Sporobolus airoides** Torr. Alkali Sacaton, Fine top Salt Grass Map 241

Flowers May–October.

Loess hills in Atchison County, northwestern Missouri, and introduced in Jackson County (common, Sheffield, June 12, 1895, *Bush 788*).

Ranges from South Dakota to Washington, south to Missouri, Kansas, Texas, and Mexico, usually growing on alkaline flats, prairies, and sands; introduced eastward in New York.

Plate no. 42. **1.** Sporobolus cryptandrus; *a.* Plant, × $^{3}/_{8}$; *b.* Floret, × $7^{1}/_{2}$; *c.* Glumes, × $7^{1}/_{2}$. **2.** Sporobolus Poiretii; *a.* Plant, × $^{3}/_{8}$; *b.* Floret, × $7^{1}/_{2}$; *c.* Spikelet, × $7^{1}/_{2}$. **3.** Sporobolus pyramidatus; *a.* Panicle, × $^{3}/_{4}$; *b.* Floret, × $7^{1}/_{2}$; *c.* Glumes, × $7^{1}/_{2}$. **4.** Sporobolus heterolepis; *a.* Plant, × $^{3}/_{4}$; *b.* Floret with caryopsis and split palea, × $7^{1}/_{2}$; *c.* Spikelet, × $7^{1}/_{2}$.

PLATE NO. 42

241 Sporobolus airoides (Fine Top Salt Grass)

242 Heleochloa schoenoides

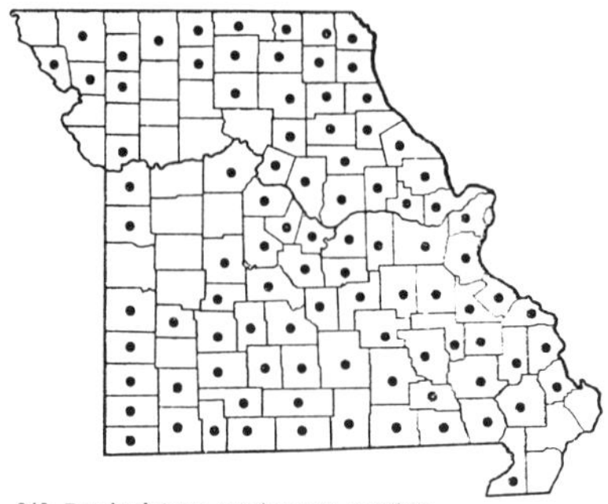

243 Brachyelytrum erectum var. erectum

## 43. **Heleochloa** Host

Low annual grasses with short leaf-blades and dense short spike-like inflorescences. Ligule a zone of hairs.

**Heleochloa schoenoides** (L.) Host Map 242

Flowers July–September.

Fields, meadows, along railroads, and waste places in St. Louis (streets of St. Louis, 1878, and 1881, *Eggert*; along railroads, *Muehlenbach*) and Franklin (meadows, Gray Summit, July 9, 1926, *Kellogg 15033*), counties.

Native of Europe; introduced into the United States, ranging from Massachusetts to Wisconsin, south to Delaware, Pennsylvania, Iowa, and Missouri.

## 44. **Brachyelytrum** Beauv.

Perennials with several slender, erect stems in a clump. Leaves flat, elongated and rather broad and spreading. Ligule truncate-erose, membranous. Inflorescence a narrow, loosely-flowered panicle.

**Brachyelytrum erectum** (Schreb.) Beauv. var. **erectum** Map 243

Flowers June–August.

Common in woodland, especially on rich or rocky steep slopes. Throughout Missouri, except in a few southeastern lowland counties.

Ranges from Massachusetts to Wisconsin and Iowa, south to Georgia, Alabama, Mississippi, and Louisiana. A more northern variety (var. *septentrionale* Babel) with glabrous or very short puberulent lemmas extends south to Minnesota and Ohio.

The spikelets fall easily and do not remain attached for long to the rachis after reaching maturity. The leaf-blades are 2-ranked, spreading out in two opposite directions.

In its vegetative appearance this grass resembles *Uniola latifolia*, but may be easily distinguished by the minutely hairy leaf-sheaths with downward-pointing (retrorse) pubescence, the narrower leaf-blades, and more slender, shorter culms (stems).

## 45. **Oryzopsis** Michx. Mountain Rice, Rice Grass

Tufted perennial grasses. Our local species have a loosely-flowered simple panicle with ascending branches. The lemma is thickened and hardened.

**Oryzopsis racemosa** (Sm.) Ricker Black-seeded Mountain Rice Map 244

Flowers late May–early September.

Wooded slopes, eastern Missouri, where known from Clark (common, Aug. 26, 1892, *Bush*) and Shannon (July 2, 1886, *Tracy*) counties.

This species is a northern relict, reaching its southwestern limit of geographical distribution in Missouri.

Ranges from Maine and Quebec to North Dakota, south to Delaware, Virginia, Kentucky, and Missouri.

The large seeds of one of the species in this genus, *O. asperifolia*, are reported to contain a starchy content desirable for making into a flour.

Plate no. 43. **1.** Stipa spartea; *a.* Plant, × $^3/_8$; *b.* Glumes, × $1^1/_2$; *c.* Floret, × $1^1/_2$. **2.** Oryzopsis racemosa; *a.* Floret, × $3^3/_4$; *b.* Panicle, × $^3/_8$. **3.** Sporobolus airoides; *a.* Plant, × $^3/_8$; *b.* Floret, × $7^1/_2$; *c.* Glumes, × $7^1/_2$.

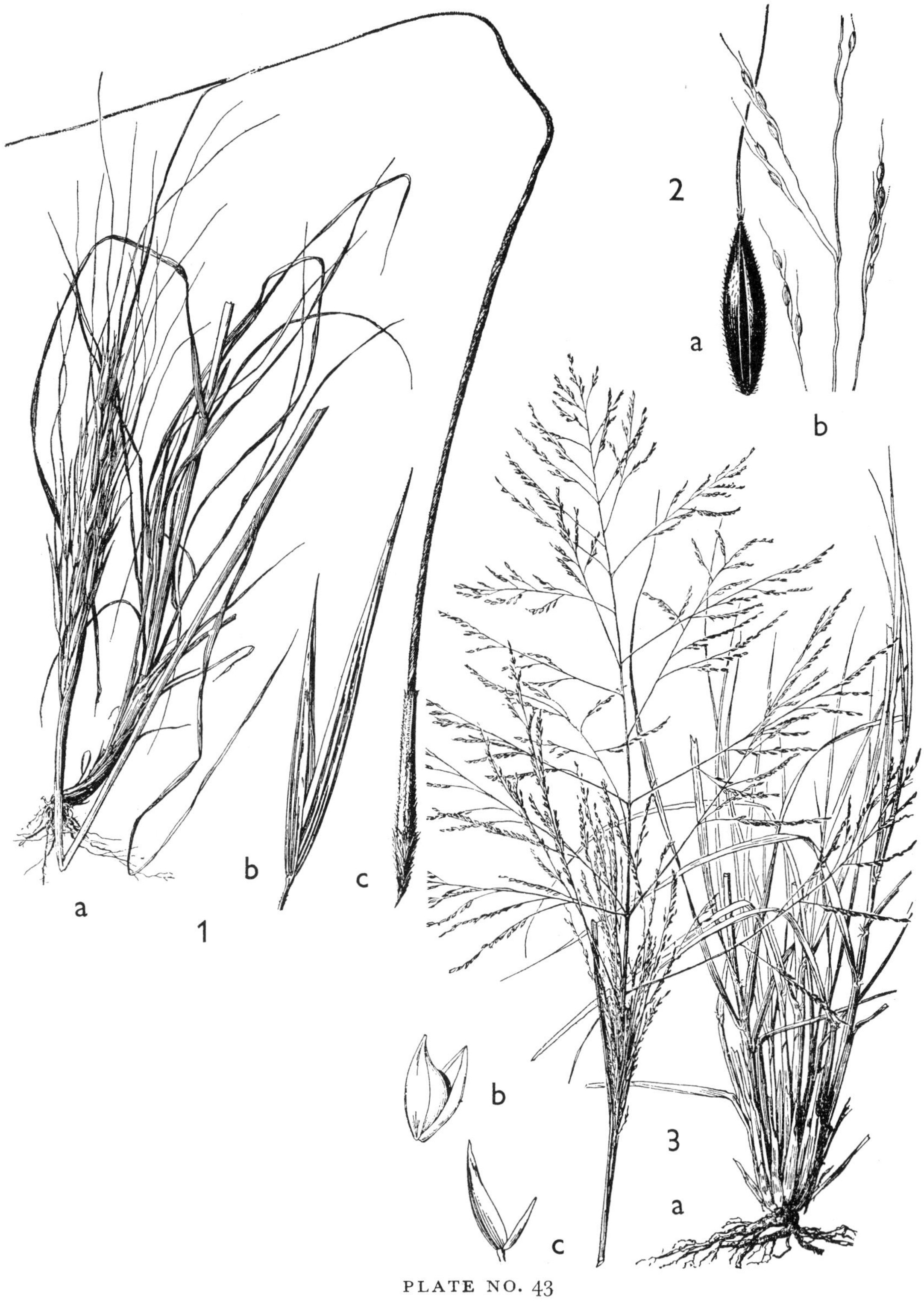


PLATE NO. 43

244 Oryzopsis racemosa (Black-seeded Mountain Rice)

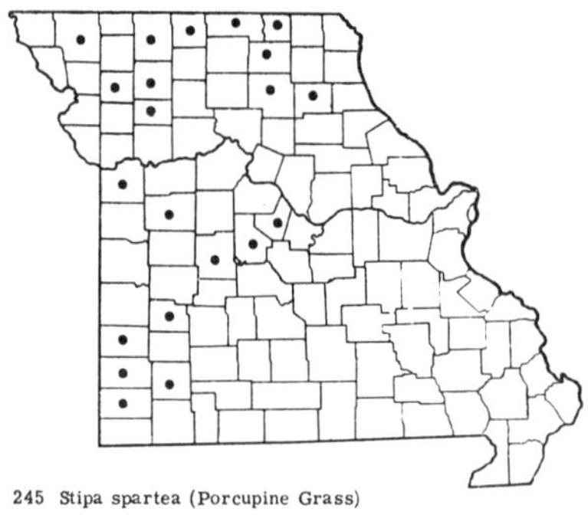
245 Stipa spartea (Porcupine Grass)

246 Aristida basiramea (Triple Awn Grass)

## 46. **Stipa** L. Porcupine Grass

Rather tall perennial grasses growing (our local species) in small tufts. Leaf-blades narrow and elongated. Panicles narrow (in the local species).

**Stipa spartea** Trin. Porcupine Grass Map 245
Flowers May–June.

Dry or rocky prairies, roadsides, and along railroads. Rather frequent in the glaciated and unglaciated prairie region of northern and western Missouri east to Schuyler, Shelby, Moniteau, Morgan, Benton, Cedar, and Lawrence counties. Absent from the Ozark and southeastern lowland region.

Ranges from Ontario to British Columbia, south to Pennsylvania, Ohio, Indiana, Illinois, Missouri, Kansas, and New Mexico.

Although the grass is highly nutritious and palatable to all livestock in its vegetative stage of growth from September to April, it can injure grazing animals during the period of seed formation (May–August) with the long-pointed awns. These get caught in the tender mouthparts, causing mechanical injury to the tissues.

The automatic twisting and untwisting of the awn as a response to change in atmospheric moisture enables the sharp-pointed seed to become buried in the soil with subsequent germination.

## 47. **Aristida** L. Triple-awned Grass, Three-awn Grass

Annual or perennial grasses, usually flowering in late summer and autumn. Leaves very narrow, often with inrolled margins and minute ciliate ligules. The glumes and lemma are narrow, the lemma ending in 3 awns, which are either separate to their base or united into a short column; sometimes the lateral ones are greatly reduced in size or lacking.

a. Lower and young leaf-sheaths with woolly or cobwebby gray-white hairs; at least lower nodes of inflorescence with tufts of wool . . . . . 8. A. LANOSA
a. Leaf-sheaths not woolly; nodes of inflorescence without tufts of wool . . . . . *b*
  b. At least 1 of the 3 awns of a floret 20–70 mm. long . . . . . *c*
    c. Glumes 6–10 mm. long; awns ascending . . . . . 7. A. INTERMEDIA
    c. The longer of the two glumes 8–30 mm. long; awns horizontally spreading or reflexed. . . . *d*
      d. First glume 12–30 mm. long, 3–5-nerved; annuals with soft bases; leaf-sheaths never hairy with long hairs . . . . . 3. A. OLIGANTHA
      d. First glume 8–14 mm. long, 1-nerved; perennials with firm hard bases; leaf-sheaths often hairy with long hairs . . . . . 9. A. PURPURASCENS
  b. Some of the 3 awns of a floret less than 18 mm. long. . . . . *e*
    e. The longer of the two glumes 8–30 mm. long . . . . . *f*
      f. None of the awns coiled at the base; perennials with firm or hard bases; leaf-sheaths often hairy with long hairs . . . . . 9. A. PURPURASCENS
      f. One of the awns of each floret coiled at base; annuals with soft bases; leaf-sheaths never hairy with long hairs . . . . . *g*
        g. Lateral awns of lemma 1–6 mm. long; the longer (upper) of the two glumes 15–28 mm. long; glumes 2–5-nerved . . . . . 4. A. RAMOSISSIMA

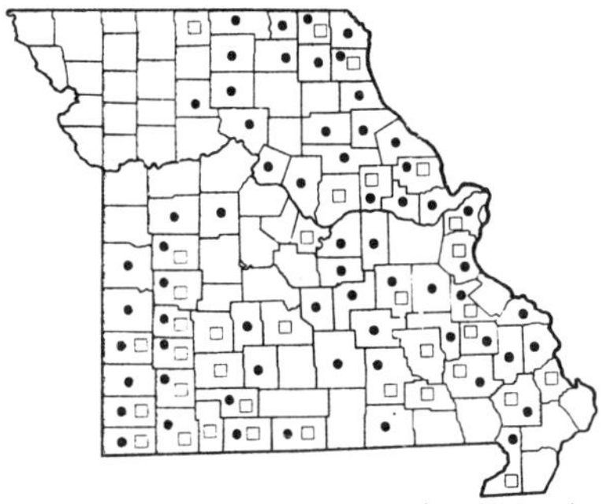

247 • Aristida dichotoma var. dichotoma (Poverty Grass)
247 □ Aristida dichotoma var. Curtissii

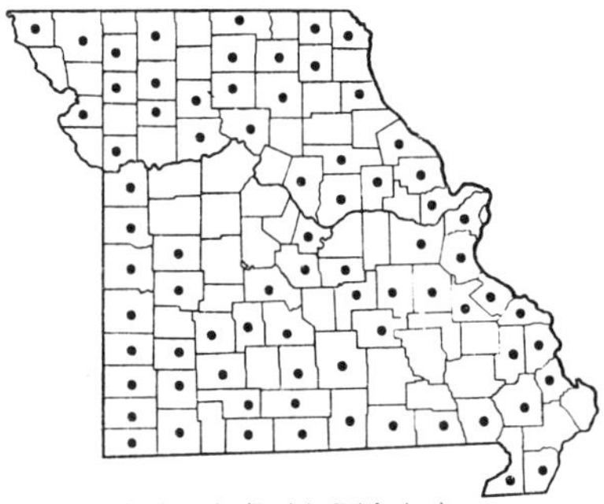

248 Aristida oligantha (Prairie Triple Awn)

249 Aristida ramosissima (Triple Awn Grass)

g. Lateral awns of lemma 5–10 mm. long; the longer (upper) of the two glumes 9–15 mm. long; glumes 1-nerved . . . . . 1. A. BASIRAMEA

e. The longer of the two glumes 5–12.5 mm. long . . . . . *h*

h. One of the awns of each floret coiled at base . . . . . 2. A. DICHOTOMA

h. None of the awns coiled at the base . . . . . *i*

i. One of the awns of each floret abruptly bent and much longer than the other 2 shorter lateral awns . . . . . 5. A. LONGESPICA

i. All 3 awns of each floret about equal in length and equally ascending . . . . . *j*

j. Awns 15–24 mm. long; awns terete (circular in cross-section); hardened part (callus) at base of lemma acute (sharp pointed), sparsely bearded; glumes nearly equal or the 1st about 4/5 length of the 2nd . . . . . 7. A. INTERMEDIA

j. Awns 8–15 mm. long; awns flat at base; hardened part (callus) at base of lemma obtuse (blunt), densely bearded; glumes unequal, the 1st 1/2–2/3 length of the 2nd. 6. A. ADSCENSIONIS

1. **Aristida basiramea** Engelm. Map 246

Flowers August–October.

Dry open ground. Scattered in southern and central Missouri.

Ranges from Michigan to North Dakota, south to Illinois, Missouri, and Kansas. Introduced from Maine to New York.

2. **Aristida dichotoma** Michx. Poverty Grass Map 247

Flowers August–October.

Occurs in rocky prairies, glades, open upland woods, eroded slopes, waste ground, and roadsides. Often found in hard, white clay soil, and hard baked ground along roads.

Missouri material falls into two variations:

Lemmas 5–7 mm. long; glumes nearly equal, 6–9 mm. long, the 1st usually longer than the lemma; longer central awn usually 3–7 mm. long . . 2a. A. DICHOTOMA var. DICHOTOMA

Lemmas 7–10 mm. long; glumes conspicuously unequal, the 2nd 8–12.5 mm. long, the 1st 1/2–3/4 as long, the 1st usually shorter than the lemma; longer central awn 6–12.5 mm. long . . . . 2b. A. DICHOTOMA var. CURTISSII

2a. **Aristida dichotoma** var. **dichotoma** Map 247

*Aristida dichotoma* Michx. [G, BB, P & S]

Southern, central, and eastern Missouri, north of the Missouri River extending west to Putnam, Sullivan, Livingston, and Chariton counties.

Ranges from Florida to Texas, north to Maine, New York, Ohio, Michigan, Illinois, Missouri, and Kansas.

Some Missouri specimens attain a height of 8 dm., although 6 dm. is given as the tallest height in *Gray's Manual*, eighth edition. The usual height is 2–4 dm.

2b. **Aristida dichotoma** var. **Curtissii** Gray Map 247

*Aristida Curtissii* (Gray) Nash [BB, P & S]

*Aristida basiramea* var. *Curtissii* (Gray) Shinners [Fassett]

Similar range to that of *A. dichotoma* var. *dichotoma*, but less common.

*Aristida dichotoma* var. *dichotoma* and var. *Curtissii*, along with other species of the genus cause mechanical injuries to the mouthparts of grazing animals when eaten in the mature state.

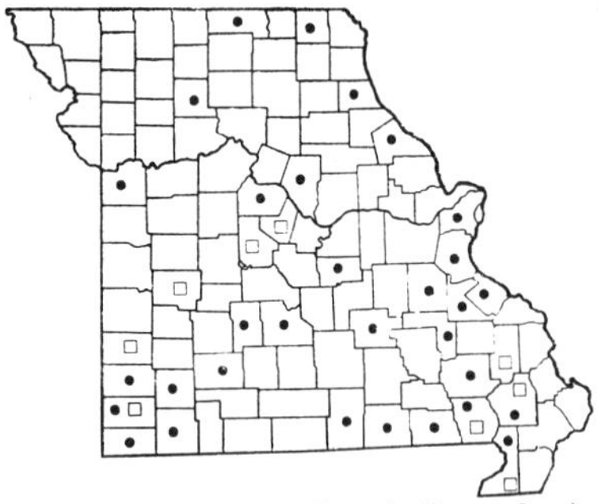

250 • Aristida longespica var. longespica (Poverty Grass)
250 □ Aristida longespica var. geniculata

251 Aristida adscensionis (Triple Awn Grass)

252 Aristida intermedia (Triple Awn Grass)

3. **Aristida oligantha** Michx.
Prairie Three Awn Grass Map 248

Rocky glades, fallow dry fields, eroded slopes, clayey roadsides, open waste ground, and along railroads. Hard, white, clay soil is a frequent habitat for this and other species of the genus.

Common throughout Missouri; this is the most frequently encountered species of *Aristida* in the state.

Ranges from Florida to Texas, north to New York, Pennsylvania, Ohio, Michigan, Wisconsin, Iowa, and South Dakota; also in California and Oregon; introduced in Massachusetts.

The lengths of the awns of this grass are stated in some of the manuals to vary from 3.5–7 cm., but some Missouri specimens have awns only 2–3 cm. long.

This grass commonly grows in overgrazed areas and places where it has no competition with other plants, such as on thin, eroded, and rocky soils. When mature this species assumes a whitish color, and becomes tough and wiry. In the mature stage its awns are responsible for mechanical injuries to grazing animals.

4. **Aristida ramosissima** Engelm. Map 249
Flowers July–October.

Dry prairies, glades, and sterile or open clayey soils. Scattered in southern and east-central Missouri north to St. Louis County. This is a rare annual species in the state, known only from a few stations.

Ranges from Indiana to Iowa, south to Tennessee, Louisiana, and Texas.

5. **Aristida longespica** Poir. Poverty Grass
Map 250

Flowers August–October.

Rocky prairies, glades, edges of bluffs, fallow sterile fields, open woods, eroded slopes, open and waste ground, especially in hard clayey soil, and along roadsides and railroads. Scattered throughout the state, except not recorded from the northwestern sector.

Missouri material falls into two variations:

Second glume 5–6 mm. long; central awn 5–13 mm. long; lateral awns 1–4 mm. long. . . .
5a. A. LONGESPICA var. LONGESPICA

Second glume 5–9 mm. long; central awn 10–21 mm. long; lateral awns 4–15 mm. long. . . .
5b. A. LONGESPICA var. GENICULATA

5a. **Aristida longespica** var. **longespica** Map 250
*Aristida longespica* Poir. [G, P & S]

The common variety in Missouri.

Ranges from Florida to Texas, north to Connecticut, Ohio, Michigan, Illinois, Missouri, and Kansas.

5b. **Aristida longespica** var. **geniculata** (Raf.) Fern. Map 250
*Aristida intermedia* according to Shinners in Rh. 56: 30. 1954, not Scribn. & Ball

Rare in southern and central Missouri, north to Morgan and Moniteau counties.

Ranges from Florida to Texas, north to New Hampshire, Vermont, New York, Ohio, Indiana, and Missouri.

6. **Aristida adscensionis** L. Map 251
Flowers July–October.

Upland prairies, wooded sandstone slopes, and sterile open ground. Rare in western Missouri, where known from a few stations south of the Missouri river north to Jackson County. Material from Maries and Madison counties in eastern Missouri, previously referred to this species, was based upon misdetermined specimens.

---

Plate no. 44. **1.** Aristida oligantha; *a.* Plant, × 3/8; *b.* Glumes, × 1 1/2; *c.* Floret, × 1 1/2. **2.** Aristida basiramea, × 3/4. **3.** Aristida longespica, × 3/4. **4.** Aristida dichotoma var. dichotoma, × 3/4. **5.** Aristida dichotoma var. Curtissii, × 3/4. **6.** Aristida intermedia, × 3/4. **7.** Aristida lanosa, × 3/4. **8.** Aristida adscensionis, × 3/4.

PLATE NO. 44

253 Aristida lanosa (Triple Awn Grass)

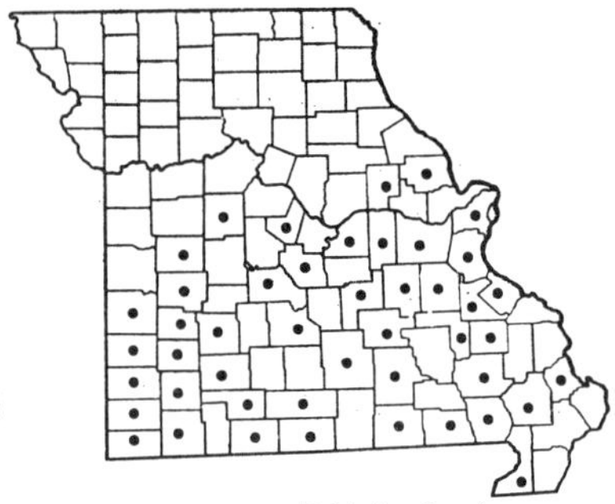

254 Aristida purpurascens (Triple Awn Grass)

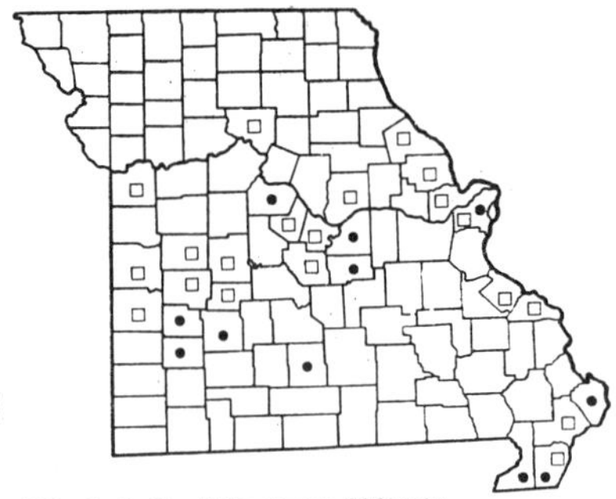

255 ● Leptochloa filiformis var. filiformis
255 □ Leptochloa filiformis var. attenuata

Ranges from Missouri, Kansas, and Colorado to Texas and California, south to Mexico, Guatemala, South America, and the West Indies; also in warmer parts of Asia and Africa.

7. **Aristida intermedia** Scribn. & Ball Map 252

*Aristida longespica* var. *geniculata* of Shinners in Rh. 56: 30. 1954, not (Raf.) Fern.

*Aristida necopina* Shinners

Flowers July–October.

Glades, rocky dry open woods, upland prairies, and sandy open ground. Scattered in a few parts of southern and central Missouri.

Ranges from Indiana, Michigan, Wisconsin, and Nebraska, south to Mississippi and Texas.

According to Shinners, much of the material previously passing as *A. intermedia* of most authors, not Scribner & Ball, is *A. necopina* Shinners, while true *A. intermedia* Scribn. & Ball, according to him, is *A. longespica* var. *geniculata* (Raf.) Fern.

8. **Aristida lanosa** Muhl. Map 253

Flowers September–November.

Known only from sandy open wooded slopes and ridges, and sandy open soil in valley openings on Crowley Ridge, southeastern Missouri, in Dunklin (east-facing slopes, Crowley Ridge, T23N, R9E, sect. 26, 6½–7 mi. northwest of Malden, October 30, 1956, *Steyermark 83363*) and in Stoddard (sandy openings in valley, Crowley Ridge, on property of Mr. Martin, T25N, R11E, northwest ¼ sect. 6, 3½–3¾ mi. southeast of Bloomfield, August 20, 1954, *Steyermark 76744*; T25N, R10E, northeast ¼ sect. 1, on property of Mr. Kidd, 3½–3¾ mi. southeast of Bloomfield, October 18, 1955, *Steyermark 80387*; sandy prairie in valley on east side of road and gully, Crowley Ridge, T26N, R10E, sect. 31, 3½ mi. southeast of Bloomfield, October 17, 1955, *Steyermark 80332, 80342, 80343, 80356*; *Bush* without locality) counties.

Ranges from Florida to Texas, north to New Jersey, West Virginia, and southeast Missouri.

This very distinct perennial woolly species of *Aristida* was originally known in Missouri by a collection of Bush labeled 'Stoddard County,' without any specific locality data. The grass was not relocated again until the author's collections in 1954–56, when it was found to be of rather frequent occurrence on open sandy dunes on the ridges and in sandy valleys of Crowley Ridge.

9. **Aristida purpurascens** Poir. Map 254

*Aristida purpurascens* var. *minor* Vasey [G]

Flowers August–October.

Occurs most frequently in dry open oak or pine-oak woodland developed over acid rock strata of chert, sandstone, or granite; sometimes found on sandstone, chert, and granite glades, and in open sterile ground. Found only in the Ozark region of southern and central Missouri, north to Lincoln, Montgomery, Osage, Moniteau, Pettis, Henry, and Vernon counties.

Ranges from Massachusetts to Michigan, Wisconsin, and Kansas, south to Florida and Texas, and British Honduras, Central America.

*Aristida purpurascens* var. *minor* is distinguished by some authors by the relatively thinner panicles, and smaller first glume 6.5–9 mm. long. It does not appear to have particular taxonomic significance, however, and is not maintained in the present treatment.

This perennial species is frequently encountered throughout the Ozark region on acid soil in oak-hickory or oak-pine woods, frequently with sour gum and red maple. The leaves are grayish-green in small

Plate no. 45. **1.** Aristida ramosissima, × 3/4. **2.** Aristida purpurascens, × 3/4. **3.** Eleusine indica; *a.* Plant, × 3/8; *b.* Seed (without pericarp), × 3 3/4; *c.* Floret, × 3 3/4; *d.* Spikelet, × 3 3/4.

PLATE NO. 45

bunches, sometimes curled at the base of the plant, as in *Danthonia spicata*. The inflorescences are somewhat purplish, accounting for the specific scientific name of the plant.

The lower leaf-sheaths vary from glabrous or sparsely hairy to rather conspicuously villous. The nearly equal awns vary from spreading to ascending, and from 1.2–3 cm. long.

## Tribe VI. **CHLORIDEAE**

The grasses of this tribe may have the spikelets 1-flowered, or 2–3-flowered. The spikelets are usually arranged in 2 rows, these usually appearing along one side of the rachis, forming simple 1-sided spikes, racemes, or are digitately branched. Articulation of the spikelets is usually above the glumes, but is below the glumes in *Spartina* and *Beckmannia*.

### 48. **Leptochloa** Beauv. Feathergrass, Sprangle Top

Annuals with elongated panicles having a feather-like or simple pinnate branching, the numerous very narrow branches with 2–4-flowered spikelets, the uppermost floret of each spikelet usually imperfect or barely developed; spikelets arranged in 2 rows along one side of the slender rachis.

The genus is maintained here in the sense used by Fernald in the eighth edition of *Gray's Manual*, consisting of the following taxa.

**Leptochloa filiformis** (Lam.) Beauv.
Red Sprangle Top Map 255
Flowers July–October.

Frequents alluvial banks, mud flats, flood plains and valleys of the larger rivers (Missouri and Mississippi) and tributaries, and moist ground of cultivated fields in valleys along streams, mainly in eastern and central Missouri.

Missouri material falls into the following categories:

Glumes merely pointed at tip, the tip not drawn out into a little awn; lemmas 1–1.3 mm. long, minutely hairy on midrib on lower margin; 2nd glume shorter than or rarely longer than the upper floret of the spikelet. . . 1a. L. FILIFORMIS var. FILIFORMIS

Glumes aristate, the tips with little awns; lemmas 0.5–1 mm. long, mainly glabrous or with only a few inconspicuous hairs; 2nd glume usually longer than upper floret of the spikelet . . 1b. L. FILIFORMIS var. ATTENUATA

1a. **Leptochloa filiformis** var. **filiformis** Map 255

Ranges from Florida to Texas and New Mexico, north to Virginia, Indiana, Illinois, Missouri, and Kansas; introduced in Massachusetts. Native also in Mexico, Central America, the West Indies, and South America.

1b. **Leptochloa filiformis** var. **attenuata** (Nutt.) Steyermark & Kucera Map 255
*Leptochloa attenuata* (Nutt.) Steud. [G, BB].

Ranges from Illinois to Louisiana and Texas.

The two variations of *Leptochloa filiformis* are quite distinct in extreme specimens. However, many intergradations are found among specimens in Missouri with both types sometimes appearing together. In general, *L. filiformis* var. *filiformis* is taller, attaining 1.2 meters in height, and the inflorescence is often larger with 20–100 stiff spikes (branches) to an inflorescence, while *L. filiformis* var. *attenuata* is usually shorter in stature, and the inflorescence is usually smaller with only 10–30 flexuous spikes to an inflorescence. Unfortunately, tall-growing plants (characteristic of *L. filiformis* var. *filiformis*) are found with aristate glumes and smaller lemmas of *L. filiformis* var. *attenuata*, while low-growing plants (characteristic of *L. filiformis* var. *attenuata*) occur with the acute glumes and larger lemmas of *L. filiformis* var. *filiformis*. The same lack of correlation is noted occasionally between the greater or lesser length of the glumes with respect to the upper floret and the height of the plant.

Deam (*Grasses of Indiana*, p. 198. 1929) also had difficulty in determining whether a specimen placed by him in *Leptochloa attenuata* should warrant specific or varietal status.

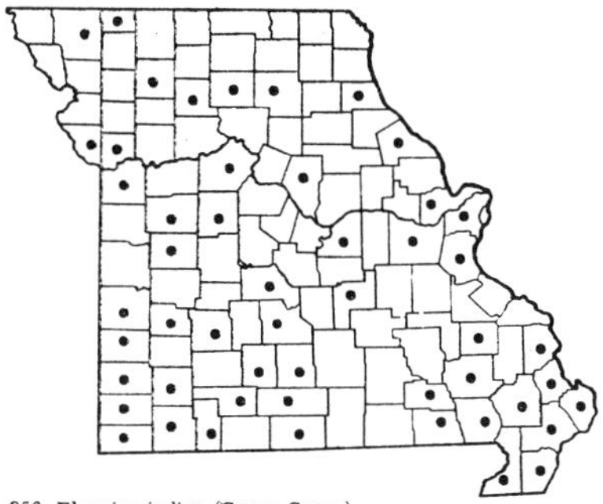
256 Eleusine indica (Goose Grass)

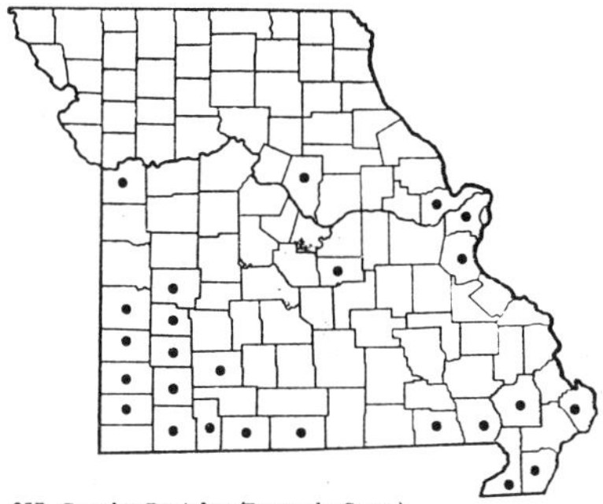
257 Cynodon Dactylon (Bermuda Grass)

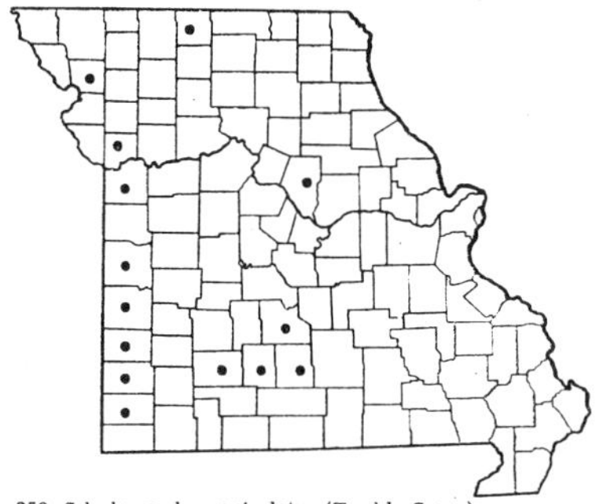
258 Schedonnardus paniculatus (Tumble Grass)

## 49. **Eleusine** Gaertn. Goose Grass

Tufted annual grasses with florets closely imbricated in two rows along 1 side of the rachis; the 3–6-flowered crowded spikelets are in finger-like spikes.

**Eleusine indica** (L.) Gaertn.
Goose Grass, Yard Grass Map 256
Flowers July–October.

Frequents cultivated and waste ground, lawns, gardens, pastures, and along roadsides and railroad tracks. Throughout Missouri.

Native of the Old World; introduced into the United States, ranging from Quebec to Minnesota and South Dakota, south to the Gulf States and west to the Pacific Coast.

Abnormal forms are sometimes found, in which some or all of the spikelets of an inflorescence have changed into leafy or vegetative shoots (*Muehlenbach 1487* from St. Louis Co.).

The seeds are said to be ground as a flour for bread or soup in times of emergency by Arabs and various nomadic peoples in Africa.

## 50. **Cynodon** Richard Bermuda Grass

Creeping perennials with 4–6 divaricately spreading finger-like spikes; the closely imbricated spikelets are 1-flowered. Ligule a ring of white hairs.

**Cynodon Dactylon** (L.) Pers.
Bermuda Grass Map 257
Flowers June–October.

Most abundant in sandy soils in southeastern and southwestern Missouri; scattered elsewhere in southern and central Missouri north to St. Charles, Boone, and Jackson counties, in open, sandy, waste, cultivated ground, along roadsides and railroad tracks.

Native of Europe; introduced into the southern United States, ranging north to New Hampshire, New York, Ohio, Michigan, northern Illinois, Iowa, Missouri, and Kansas.

Certain strains (Ormond and Tifton 57) are recommended as the best strains for lawn grasses in the southern states; other strains of Bermuda grass are used for golf greens and areas of fine turf. For best growth Bermuda grass should have full sun, slightly acid to neutral soil, regularly applied fertilizer, and close mowing.

It is not known how far north in the state Bermuda grass will succeed as a permanent lawn grass, but probably would succeed in the northern sections of the state. The grass seldom perfects seed, maintaining itself vegetatively by creeping underground rootstocks or stolons.

Because of its long flowering season, wide geographical distribution, and very light and prolific pollen, Bermuda grass is considered to be the worst hay fever grass in the United States.

## 51. **Schedonnardus** Steud.

Low-growing tufted annuals with short narrow leaves and a large open inflorescence from half to more than half the entire height of the plant with rather widely-spaced slender branches (spikes). The 1-flowered spikelets are sessile and tightly pressed in 2 rows along side of the slender axis (rachis). The ligule is membranous.

PLATE NO. 46

259 Beckmannia syzigachne

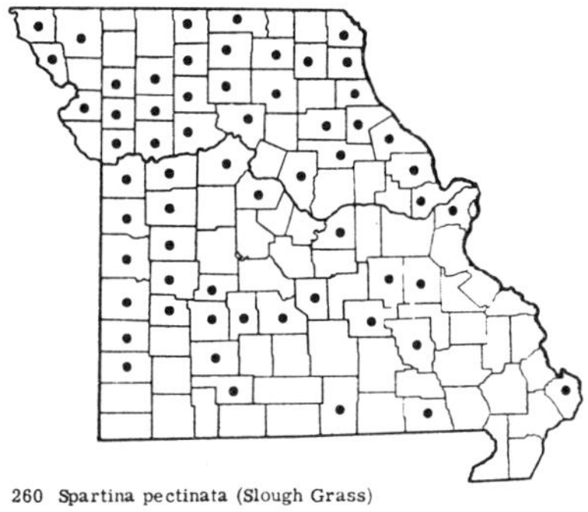

260 Spartina pectinata (Slough Grass)

261 Gymnopogon ambiguus (Beard Grass)

**Schedonnardus paniculatus** (Nutt.) Trel.
Tumble Grass Map 258
Flowers May–October.

Occurs on prairies, calcareous exposed hills and mounds, glades, open fields, waste ground, and, rarely in dry open woods. Western half of Missouri east to Mercer, Boone, Laclede, and Wright counties.

Ranges from Saskatchewan to Montana, south to Illinois, Missouri, Texas, New Mexico, and Arizona; also in Argentina.

After flowering, the entire inflorescence lengthens and curves, eventually breaking from the plant at maturity to be blown away as a tumbleweed. It invades native grass ranges in the western states often as a result of overgrazing.

## 52. **Beckmannia** Host.

Tall erect annual grasses; the flat moderately broad leaf-blades with a large ligule. Inflorescence with numerous short, crowded, upright to ascending spikes in a long and narrow panicle. The glumes are cross-wrinkled.

**Beckmannia syzigachne** (Steud.) Fern.
(scissorslike glumes) Map 259
Flowers June–August.

Wet open ground, introduced in Jackson County, western Missouri. A collection by Engelmann (swampy ground on [Massey or Murrey?] Creek, July 17, 1859) exists in the Missouri Botanical Garden Herbarium, but it is not known whether this locality is on the Illinois or Missouri side of the Mississippi River in the vicinity of St. Louis.

Ranges from Quebec to Alaska, south to Michigan, northeastern Illinois, northwestern Iowa, Kansas, New Mexico, and California; introduced eastward to New York, Pennsylvania, and Ohio, and south to Missouri. Also found in Asia.

This grass, where native, is suitable as a forage grass, and sometimes cut for hay. The seeds were gathered and parched by the Indians for food.

Sometimes called 'Slough Grass' or 'American Slough Grass,' but it should not be confused with the better-known and more widespread *Spartina pectinata,* which is often known, at least in Missouri, by the name 'Slough Grass,' here applied to *Spartina* in the present treatment.

## 53. **Spartina** Schreb. Cord Grass, Slough Grass

Perennial plants, usually forming colonies, with long underground rhizomes, long narrow leaves; and 1-flowered flattened spikelets closely overlapping in 2 rows on 1 side of the branch (rachis), the branches (spikes) usually 5–30.

**Spartina pectinata** Link Slough Grass,
Prairie Cord Grass Map 260
*Spartina pectinata* var. *Suttiei* (Farw.) Fern. [G, P & S]

Flowers late June–September.

Frequents low meadows and swampy open ground of valleys and river bottom soils, margins of sloughs and ponds, wet and dry prairies, along railroads, and

---

Plate no. 46. **1.** Schedonnardus paniculatus; *a.* Plant, × 3/8; *b.* Floret, × 3 3/4; *c.* Spikelet, × 3 3/4. **2.** Cynodon Dactylon; *a.* Plant, × 3/8; *b.* Spikelet, × 3 3/4; *c, d.* Two views of floret, × 3 3/4.

along roadsides in low ground. Common in all parts of northern and central Missouri, and along the western sectors, becoming infrequent to scarce in the southeastern and southern portions of the state. Apparently absent from a number of counties in the Ozarks and southeastern lowlands.

Ranges from Newfoundland to Alberta and Washington, south to North Carolina, West Virginia, Indiana, Illinois, Kentucky, Missouri, Arkansas, Texas, New Mexico, and Oregon. Introduced into Europe.

*Spartina pectinata* var. *Suttiei* has been separated on the basis of narrower, longer spikes (branches), 3–5 mm. wide (instead of 5–8 mm.), with appressed instead of salient awns of the spikelets. However, while selected examples and extremes exhibit these differences, there is a conspicuous absence of correlation in these characters. The most recent monographer of *Spartina*, Dr. David G. Mobberley, has concluded, likewise, that the var. *Suttiei* cannot be maintained as a variety (Iowa State College Jour. Sci. 30: 553–58. 1956), and that those who previously recognized it as distinct were 'dealing with selected extremes of a somewhat variable population rather than with discrete infraspecific taxa.'

Although Dr. Mobberley states (*loc. cit.* p. 560) that for this grass 'the primary habitat seems to be open dry prairie or high ground along railroad right-of-way and roadsides, in many sections of Missouri where I have observed it, *Spartina pectinata* is mostly found in low wet swales, bottom prairies, and low wet ground. It often forms extensive colonies in the wet bottom prairies of the Grand, Chariton, and other large rivers of northern and central Missouri. The razor-sharp rough edges of the leaf-blades can easily cut the skin of tender, unwary hands that suddenly grasp or otherwise handle them without caution.

If cut while still young and tender, this grass makes a fairly good hay. Ordinarily, the grass is cut 2 or 3 times a year to produce more yield and hay crops. The marsh hay of the Atlantic Coast, often used for purposes of packing or for bedding, is usually obtained from *S. patens*. *Spartina pectinata* was used by the early pioneers and Indians for thatching roofs and for covering haystacks and corncribs. This is not considered to be an important grass in causing hay fever.

## 54. **Gymnopogon** Beauv. Beard Grass

Perennials (in the local species) with rather stiffly spreading leaves, a minute ligule, and numerous, slender, widely spreading or reflexed branches (spikes).

**Gymnopogon ambiguus** (Michx.) BSP.

Beard Grass Map 261

Flowers July–September.

Frequents pine or oak-hickory open woodland in acid soils overlying sandstone or chert, either on dry rocky slopes or in flat sandy valley soils; also in sandy open places and glades, where limited to the Ozark sector of southern and east-central Missouri north to Barton, Polk, Phelps, Crawford, and St. Charles counties.

Ranges from Florida to Texas, north to New Jersey, Ohio, Indiana, Missouri, and Kansas.

An attractive grass with leaves resembling some species of *Panicum*.

## 55. **Chloris** Sw. Windmill Grass

Annual or perennial grasses with digitate branches (spikes) arising from the summit of the culm. The spikelets have 1 perfect flower and 1 or more empty lemmas, and are sessile in 2 rows along the triangular-sided rachis.

Some part (fertile lemma) of each spikelet prominently white-hairy on the margin at the summit, giving the whole spikes a feathery appearance; spikelets closely overlapping each other (imbricate) . . 1. C. VIRGATA

No part of the spikelet more than inconspicuously covered with hairs pressed along the surface of the fertile lemma, the spikelets not presenting a feathery or hairy appearance; spikelets not overlapping each other . . . . . . . . . . . . . . . . . . . . . . 2. C. VERTICILLATA

1. **Chloris virgata** Sw. Feather Finger Grass Map 262

Flowers July–October.

Occurs in fields and along railroads. Known only from St. Louis (Baden freight yard [Mo.-Kan.-Tex. R. R.] north of the junction with Terminal Railroad

Plate no. 47. **1.** Spartina pectinata; *a.* Plant, × $^3/_8$; *b.* Floret, × $3^3/_4$; *c.* Spikelet, × $3^3/_4$. **2.** Beckmannia syzigachne; *a.* Plant, × $^3/_8$; *b.* Floret, × $3^3/_4$; *c.* Spikelet, × $3^3/_4$.

PLATE NO. 47

262 Chloris virgata (Feather Finger Grass)

263 Chloris verticillata (Windmill Grass)

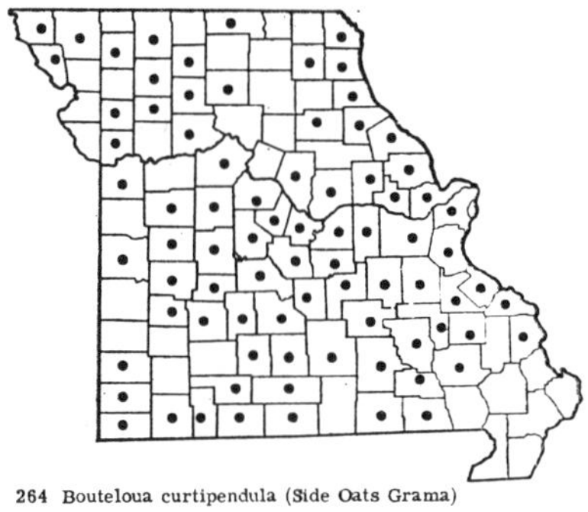

264 Bouteloua curtipendula (Side Oats Grama)

Association, St. Louis, July 8, 1956, *Muehlenbach 1002*) and Jackson (fields, Courtney, October 13, 1921, *Bush 9733*; introduced, Courtney, September 5, 1913, *Bush 7005*) counties.

Introduced in Maine, Massachusetts, New York, North and South Carolina, and Missouri; ranging from Nebraska to Nevada, south to Texas and California, Mexico, Central America (Guatemala), West Indies, and South America.

An annual grass which is weedy in the southwestern United States. It affords forage in those sections of the Southwest where abundant.

2. **Chloris verticillata** Nutt. Windmill Grass
Map 263

Flowers May–October.

Upland dry prairies, lawns, waste ground, and along railroad tracks, where scattered in parts of central and southern Missouri, but absent from the southeastern quarter; extends north to St. Louis, Boone, and Clay counties.

Ranges from Nebraska and western Missouri to Colorado, south to Louisiana and New Mexico; introduced eastward in parts of Missouri, Iowa, Illinois, Indiana, and Maryland.

When mature the entire inflorescence breaks off and is blown around like a tumbleweed. This is a perennial grass which forms clumps by rooting at the lower nodes. It invades overgrazed ranges in the southwestern states.

## 56. **Bouteloua** Lag. Grama Grass

Perennial (in local species) grasses usually in clumps. The 1 to many short spikes (branches), racemosely arranged, bear crowded, sessile spikelets in 2 rows on 1 side of a rachis.

a. Spikes 10–50 in a long erect inflorescence; plants 3–10 dm. tall . . . . . . . 1. B. CURTIPENDULA
a. Spikes 1–6; plants 1.5–5 dm. tall . . . . . . . . . . . . . . . . . . . . . . *b*
  b. Leaf-blades glabrous; axis (rachis) of the spike not extending beyond the spikelet at the tip; back of 2nd glume with few tuberculate hairs (enlarged at the base), the hairs very short (mostly less than 1 mm. long) . . . . . . . . . . . . . . . . . . . . 3. B. GRACILIS
  b. Leaf-blades with some hairs; axis (rachis) of the spike extending beyond the spikelet at the tip as a conspicuous projection; back of 2nd glume with numerous tuberculate hairs (enlarged at base), the hairs 1 mm. or more long . . . . . . . . . . . . . . . . 2. B. HIRSUTA

1. **Bouteloua curtipendula** (Michx.) Torr.
Sideoats Grama Map 264

Flowers July–September.

Frequent on limestone glades, prairies, prairie openings in rocky woodland, grassy knolls, and along railroads. Throughout Missouri except absent in the southeastern lowland sector.

Ranges from Connecticut to Ontario west to Montana, south to Georgia, Alabama, Mississippi, Louisiana, Texas, California, Mexico, Central America, and South America.

The plant receives its name from the oatlike drooping spikes which hang uniformly from the sides of the elongated axis. The ligule is a short (0.5 mm. long) band of hairs. This species is considered highly nutritious and is eaten by all classes of livestock. Since

Plate no. 48. **1.** Chloris verticillata; *a*. Panicle, × 3/4; *b*. Floret, × 3 3/4. **2.** Gymnopogon ambiguus; *a*. Plant, × 3/8; *b*. Floret, × 3 3/4; *c*. Spikelet, × 3 3/4. **3.** Chloris virgata; *a*. Plant, × 3/8; *b*. Floret, × 3 3/4; *c*. Glumes, × 3 3/4.

PLATE NO. 48

265 Bouteloua hirsuta (Hairy Grama)

266 Bouteloua gracilis (Blue Grama)

267 Buchloe dactyloides (Buffalo Grass)

it does not shed much pollen, it is not considered as of much importance in causing hay fever.

2. **Bouteloua hirsuta** Lag. Hairy Grama Map 265
Flowers July–September.

Occurs on dry exposed loess hills of Holt and Atchison counties in northwestern Missouri, and locally east in Clark County, northeastern Missouri, where it is found on eroded slopes of a natural sandy prairie near the Mississippi River (T65N, R6W, sect. 9, $1\frac{1}{2}$ mi. southeast of St. Francisville, Aug. 13, 1949, *Steyermark 68878*). At this latter unusual locality it was associated with *Bouteloua curtipendula*, *Cyperus Schweinitzii*, *Polygonum tenue*, *Astragalus distortus*, *Euphorbia glyptosperma*, and *Asclepias amplexicaulis*. On the loess mounds it is usually associated with species from the short grass plains region of the western United States, among them being *Muhlenbergia cuspidata*, *Yucca glauca*, *Anemone cylindrica*, *Oxytropis plattensis*, *Astragalus lotiflorus*, *Psoralea argophylla*, *Dalea enneandra*, *Oenothera serrulata*, *Penstemon grandiflorus*, *Castilleja sessiliflora*, *Liatris punctata*, *Lygodesmia juncea*, and *Lactuca pulchella*.

Ranges from Wisconsin to North Dakota and Colorado, south to western Illinois, Louisiana, Texas, California, and Mexico; introduced eastward in Florida.

3. **Bouteloua gracilis** (HBK.) Lag.
Blue Grama Map 266

According to Shinners (Rh. 56: 31. 1954), the proper authorship for this species is *Bouteloua gracilis* (Willd.) Lag.

Flowers July–September.

Occurs on loess hills in Atchison and Holt counties of northwestern Missouri, on prairie knolls in Clay and Jackson counties of west-central Missouri, and along railroads in St. Louis County, east-central Missouri.

Ranges from Illinois, Wisconsin, Minnesota, and Alberta to British Columbia, south to Missouri, Texas, California, and Mexico; introduced eastward.

These species, together with *B. hirsuta*, are prominent members of the 'short grass' flora of the Great Plains region, along with other species of Bouteloua. In places where abundant, *B. gracilis* may be a contributing factor of slight importance in causing hay fever. It withstands extreme drouth and is highly palatable and nutritious for all classes of livestock.

## 57. **Buchloe** Engelm. Buffalo Grass

This genus is unusual in being usually dioecious (staminate and pistillate flowers on separate plants). These grasses are perennial and spread by stolons, which root at the nodes (joints). The pistillate spikelets are 1-flowered and closely clustered in a head-like 1-sided cluster. The staminate spikelets are 2–3-flowered, and somewhat resemble those of Bouteloua, occurring in 2 rows along 1 side of the spike.

**Buchloe dactyloides** (Nutt.) Engelm.
Buffalo Grass Map 267
Flowers May–August.

Occurs on dry loess hills of Atchison and Holt counties in northwestern Missouri, and in open ground and along railroads in Clay, Jackson, and St. Louis (Baden freight yard of the Mo.-Kan.-Topeka R. R., south of the yard office, St. Louis, June 22, 1957,

Plate no. 49. **1.** Bouteloua gracilis; *a.* Plant, $\times\ ^3/_8$; *b, c.* Florets, $\times\ 3^3/_4$; *d.* Glumes, $\times\ 3^3/_4$. **2.** Bouteloua curtipendula; *a.* Plant, $\times\ ^3/_8$; *b.* Florets, $\times 3^3/_4$ *c.* Spikelet, $\times\ 3^3/_4$. **3.** Bouteloua hirsuta; *a.* Panicle, $\times\ ^3/_4$; *b.* Spikelet, $\times\ 3^3/_4$.

PLATE NO. 49

268 Anthoxanthum odoratum (Sweet Vernal Grass)

269 Phalaris canariensis (Canary Grass)

269A Phalaris brachystachys

*Muehlenbach 1219*; also July 5, 1958, *Muehlenbach 1455*) counties.

Ranges from Illinois, Iowa, Minnesota, Manitoba, Alberta, and Montana, south to Louisiana, Texas, and Mexico.

This species and *Bouteloua gracilis* exhibit the highest resistance to drouth of all the native American grasses. Buffalo grass is one of the most important, if not the most important, forage grass of the Great Plains region. In the western states the leaves and culms dry and cure on the ground, thereafter furnishing food all winter for the livestock.

Buffalo grass has been found to be unimportant so far as causing hay fever is concerned. It has been reported that the early settlers of the western United States prepared sod houses from the sod of buffalo grass.

## Tribe VII. **PHALARIDEAE**

Spikelets with 1 perfect terminal floret and, below this, 2 sterile (staminate) lemmas. Spikelets either falling in groups due to the articulation located below a group, or the spikelets individually articulated above the glumes and falling singly.

### 58. **Anthoxanthum** L. Sweet Vernal Grass

Inflorescence narrow and spike-like. Perennial grass (in local species) growing in tufts. Sterile lemmas large and awned from the back. Stamens 2. Lodicules none.

**Anthoxanthum odoratum** L.

Sweet Vernal Grass Map 268

Flowers May–August.

Along railroads and waste ground. Known only from St. Louis, Iron, Boone, and Jackson counties.

Native of Europe; introduced into North America, ranging from Newfoundland to Ontario and Michigan, south to Georgia and Louisiana, and along the Pacific Coast from Alaska southward.

This sweet-scented grass begins to flower in early May. Throughout much of its range it causes hay fever, and is, therefore, regarded as one of the most important hay fever grasses which cause trouble early in the season.

The dried plants may be boiled and used as a tea, the sweet fragrance of the plant adding to the flavor.

---

Plate no. 50. **1.** Hordeum vulgare; *a.* Plant, × 1/5; *b.* Floret, × 1 1/5; *c.* Group of spikelets, × 1 1/5; *d.* Spike of Beardless Barley, × 1/5. **2.** Koeleria cristata; *a.* Plant, × 1/5; *b.* Floret, × 4; *c.* Glumes, × 4. **3.** Arrhenatherum elatius; *a.* Plant, × 1/5; *b.* Upper floret, × 2; *c.* Spikelet, × 2. **4.** Holcus lanatus; *a.* Plant, × 1/5; *b.* Mature fertile floret, × 2; *c.* Floret, × 2; *d.* Spikelet, × 2. **5.** Heleochloa schoenoides; *a.* Plant, × 1/5; *b.* Floret, × 2; *c.* Spikelet, × 2. **6.** Brachyelytrum erectum; *a.* Plant, × 1/5; *b.* Floret, × 2; *c.* Branchlet with glumes of two spikelets, × 2. **7.** Leptochloa filiformis; *a.* Plant × 1/5; *b.* Floret × 4; *c.* Spikelet, × 4. **8.** Buchloe dactyloides; *a.* Pistillate plant, × 1/5; *b.* Pistillate floret, × 2; *c.* Pistillate spike, × 2; *d.* Staminate plant, × 1/5; *e.* Staminate spikelet, × 2. **9.** Zizania aquatica; *a.* Plant, × 1/5; *b.* Pistillate spikelet, × 4/5; *c.* Pistillate spikelet, × 2; *d.* Staminate spikelet, × 2.

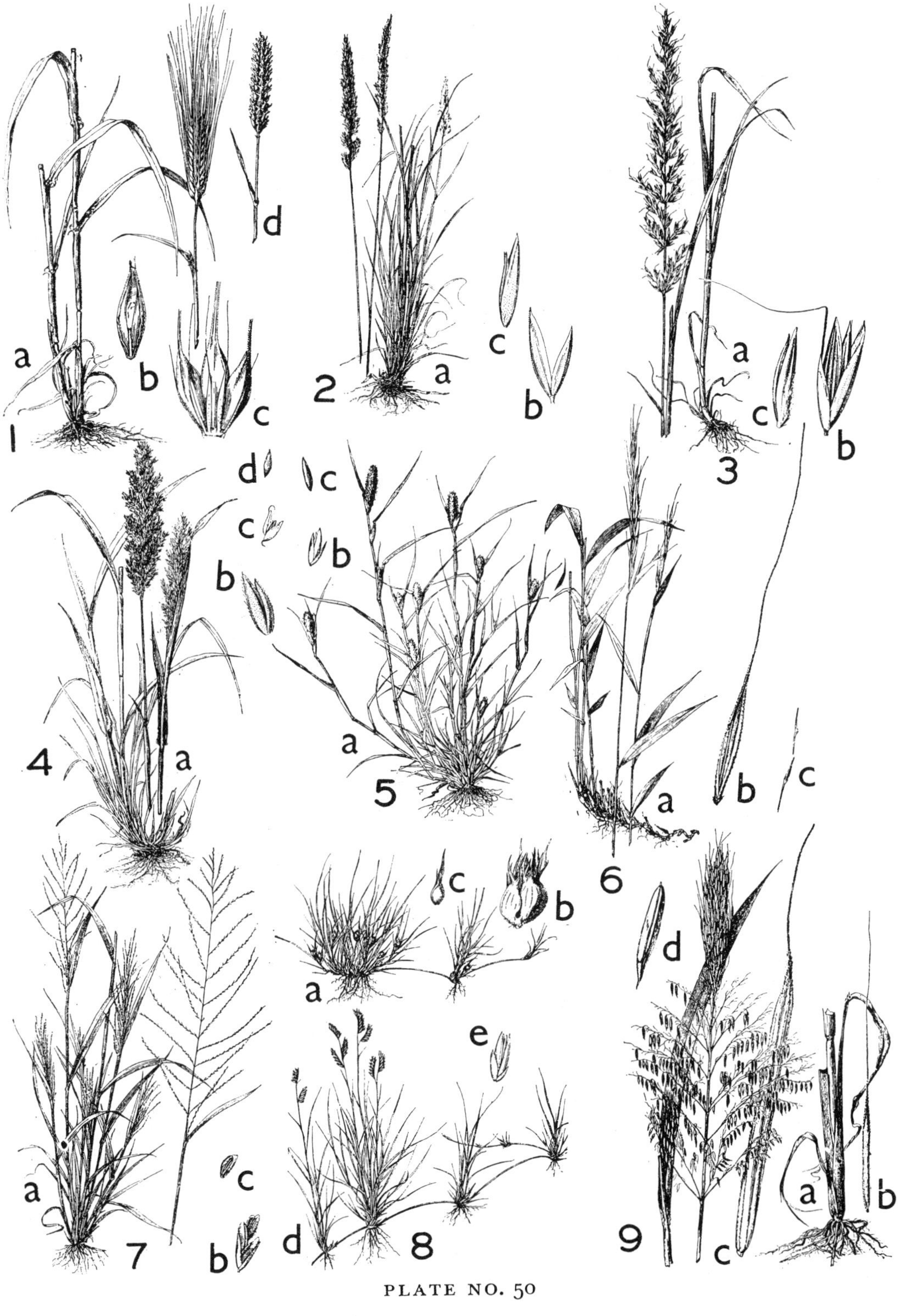


PLATE NO. 50

## 59. **Phalaris** L. Canary Grass

Annual or perennial grasses with dense, lobed, or spike-like panicles. Ligule large and mostly very thin Spikelets closely imbricated. Stamens 3. Lodicules present and noticeable.

a. Inflorescence interrupted, branching or lobed; keel of glumes without any wing; perennials with creeping underground rhizomes; leaf-blades mainly 10–20 mm. wide . . . . . . 4. P. ARUNDINACEA
a. Inflorescence continuous and dense, not interrupted; keel of glumes either with a broad wing or the wing narrow at the summit of keel; annuals, without creeping rhizomes; leaf-blades 4–10 mm. wide . . *b*
  b. Inflorescence ellipsoid, tapering at each end, broadest in the middle, 1.5–12 cm. long; keel of glumes narrowly winged at summit; fruit 1 mm. wide . . . . . . . . . . 3. P. CAROLINIANA
  b. Inflorescence ovoid or broadly oblong, broadest in the lower half, 1.5–4 cm. long; keel of glumes broadly winged; fruit 2 mm. broad . . . . . . . . . . . . . . . . . . . . . *c*
    c. Sterile lemmas 0.6 mm. long or less . . . . . . . . . . . . . . 2. P. BRACHYSTACHYS
    c. Sterile lemmas 1–2 mm. long, 1/3–1/2 as long as the fertile lemma . . . . . 1. P. CANARIENSIS

1. **Phalaris canariensis** L. Canary Grass, Birdseed Grass Map 269

Flowers June–October.

Occurs infrequently as a waif in waste ground, along streets and roadsides, and railroads. Scattered in a few places in southern and central Missouri.

Native of Europe; introduced into North America, ranging from Newfoundland to Alaska, south to Florida, Texas, and California.

The plants vary in height from 3–8 dm. Although the canary seed sold for bird food comes from this grass, the plant is only rarely encountered as an escape, and does not appear to maintain or seed itself permanently.

It is stated that in its native country a meal, ground from the seeds, is made into a bread by the inhabitants, and that in Java young tender plants are eaten raw or cooked, combined with parts of other plants, with rice.

2. **Phalaris brachystachys** Link Map 269A

Flowers May–July.

Known only from Laclede County, southern Missouri (Horseshoe Pond, June 30, 1938, *Geo. Moore*, in Chi. Nat. Hist. Mus. Herb.).

Native of the Mediterranean region; introduced in the United States, where known from Missouri, Texas, Oregon, and California.

The specimen was originally mistaken for *P. canariensis*, which it resembles, and has been identified correctly by Dennis Anderson.

3. **Phalaris caroliniana** Walt. Canary Grass Map 270

Flowers May–June.

Occurs in sandy and open fields, waste and moist ground, and along roadsides. Mainly found in western and extreme southeastern Missouri.

Ranges from Florida to Texas and Mexico, north to Maryland, Tennessee, Missouri, Kansas, Colorado, and Oregon.

The plants vary in height from 2–10 dm., but are usually 3–6 dm. tall.

4. **Phalaris arundinacea** L. Reed Canary Grass Map 271

Flowers last of April (southwest Missouri) to August.

Occurs in wet meadows, open ground along streams, open swales in alluvial ground around oxbow lakes and valleys of the large streams of northwestern Missouri and western Missouri. Common only in the northwestern and west-central sectors of the state, east to Macon and Linn counties, and locally south in Newton County (open ground along banks of Hickory Creek, Neosho, May 1, 1954, *Palmer 57441*; May 19, 1954, *Palmer 57560, 57573*; and June 29, 1954, *Palmer 58069*).

Ranges from Newfoundland to Alaska, south to Maryland, North Carolina, Kentucky, Illinois, Missouri, Oklahoma, New Mexico, Arizona, and California. Also in Eurasia.

In its early stages of flowering, this grass in the field or from a distance closely resembles the Orchard Grass (*Dactylis glomerata*). Both have crowded 1-sided groups of spikelets which vary from pale green to suffused with wine-purple. The leaves of *Dactylis* are narrower than those of *Phalaris arundinacea*, and the glumes and lemmas of *Dactylis* are more or less hairy, whereas those of *P. arundinacea* are glabrous, in addition to the technical differences.

---

Plate no. 51: **1.** Anthoxanthum odoratum; *a.* Plant, × 3/8; *b, d.* Sterile lemmas, × 3 3/4; *c.* Fertile floret, × 3 3/4; *e.* Spikelet, × 3 3/4. **2.** Phalaris canariensis; *a.* Plant, × 3/8; *b.* Floret, × 3 3/4; *c.* Spikelet, × 3 3/4. **3.** Phalaris arundinacea; *a.* Plant, × 3/4; *b.* Floret, × 3 3/4; *c.* Glumes, × 3 3/4. **4.** Phalaris brachystachys; *a.* Floret, × 3 3/4; *b.* Glumes, × 3 3/4. **5.** Phalaris caroliniana; *a.* Plant, × 3/4; *b.* Floret, × 3 3/4; *c.* Glumes, × 3 3/4.

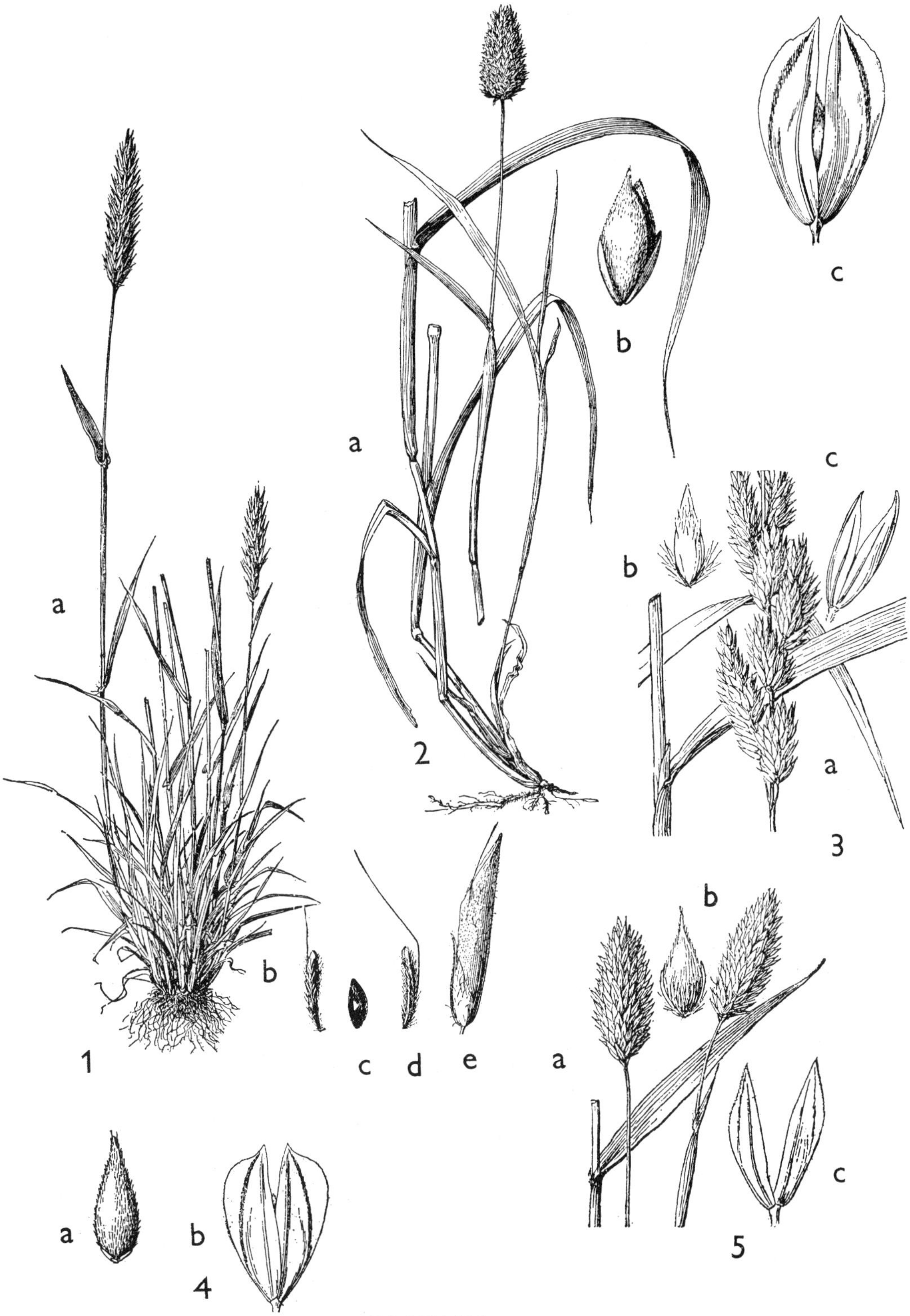


PLATE NO. 51

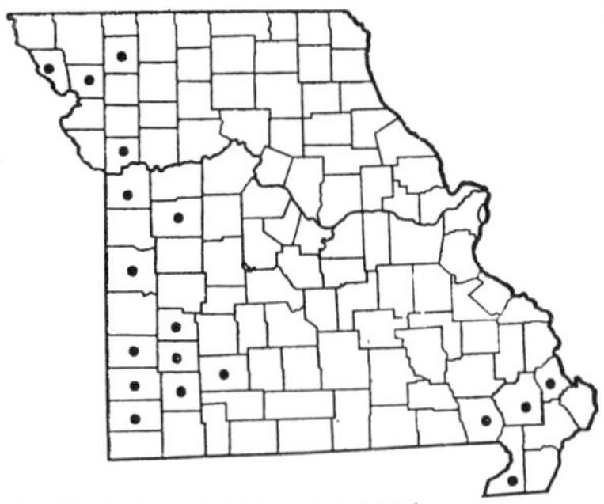

270 Phalaris caroliniana (Canary Grass)

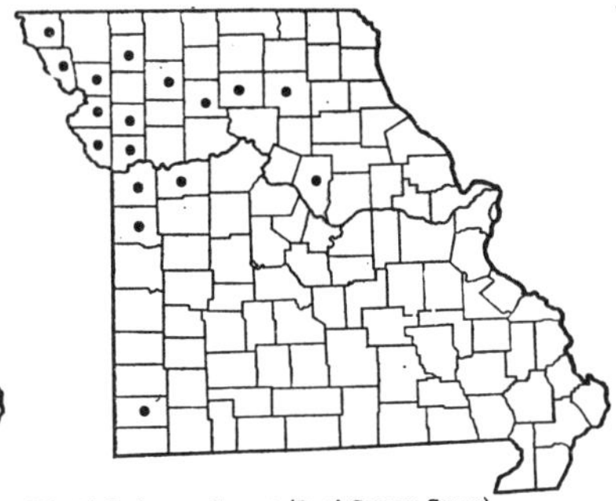

271 Phalaris arundinacea (Reed Canary Grass)

272 Oryza sativa (Rice)

This grass is an important lowland hay crop from Wisconsin to Montana. The species is sometimes planted as a forage plant, and the station in Newton County, southwestern Missouri, may represent an introduction or escape from cultivation. It is native, however, in the lowland swales and bottom meadows of northwestern Missouri. Once planted, it is considered a difficult weed to eradicate, in this respect resembling Johnson Grass.

A cultivated leaf variation, with white-striped blades, known as *P. arundinacea* var. *picta* L. (or f. *variegata* [Parnell] Druce), is sometimes escaped from gardens, but no records of such escapes are as yet known from Missouri.

## Tribe VIII. **ORYZEAE**

Grasses of this tribe have 1-flowered perfect spikelets which are strongly flattened (laterally compressed). The glumes are greatly reduced in size or are not present. Stamens 1–6. The spikelet is articulated below the glumes.

### 60. **Oryza** L. Rice.

Usually annual grasses in our region. The narrow glumes are much shorter than the 5-nerved, keeled lemma.

**Oryza sativa** L. Rice. Map 272

Flowers June–September.

Cultivated in the lowlands of southeastern Missouri, where occasionally escapes. The only known station, however, where rice has been collected as an escape in the state is from Marion County in northeastern Missouri, where a large number of plants were found in a dried-up depression (along the south side of highway 24, T59N, R6W, sect. 1, just east of the city limits of Taylor, September 22, 1956, *Steyermark 82729*).

Presumably a native of southeastern Asia; cultivated in the southern United States and throughout tropical and warm temperate regions of the world at low altitudes.

The Missouri station cited as an escape is farther north than the usual zone of cultivation. Inquiry concerning the cultivation of rice north of the Missouri River revealed the fact that hundreds of acres of rich alluvial Mississippi River flood plain north of Hannibal to the Iowa line are often cultivated for this crop.

Besides its importance as one of the principal cereals used by mankind, the dried straw is made into a pulp for paper making in China, Indonesia, Egypt, Spain, Italy, Brazil, Argentina, and the United States.

### 61. **Leersia** Sw. Cut Grass, White Grass.

Perennials, usually of wet or moist ground, the inflorescence paniculate, with the spikelets nearly sessile and arranged on 1 side of the branchlets. Glumes absent. The ligule is membranous.

a. Spikelets 3–4 mm. wide, broadly oval to nearly round, more than half as wide as long. 1. L. LENTICULARIS
a. Spikelets 1–2 mm. wide, oblong, 1/4–1/3 as wide as long . . . . . . . . . . . . . . . . . *b*

Plate no. 52. **1.** Leersia virginica, × 3/4. **2.** Leersia oryzoides; *a.* Plant, × 3/8; *b.* Spikelet, × 3 3/4. **3.** Oryza sativa; *a.* Plant, × 3/8; *b.* Spikelet, × 3 3/4. **4.** Leersia lenticularis, × 3/4.

PLATE NO. 52

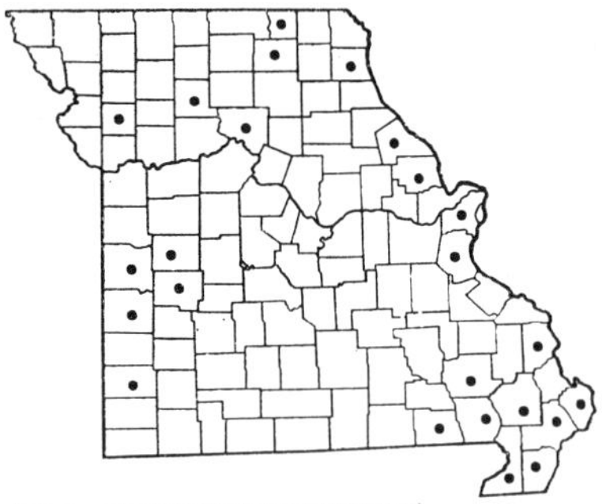

273 Leersia lenticularis (Catchfly Grass)

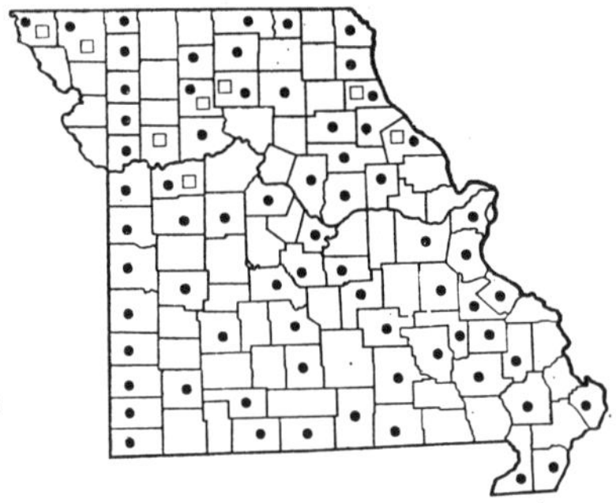

274 • Leersia oryzoides f. oryzoides (Ricecut Grass)
274 □ Leersia oryzoides f. inclusa

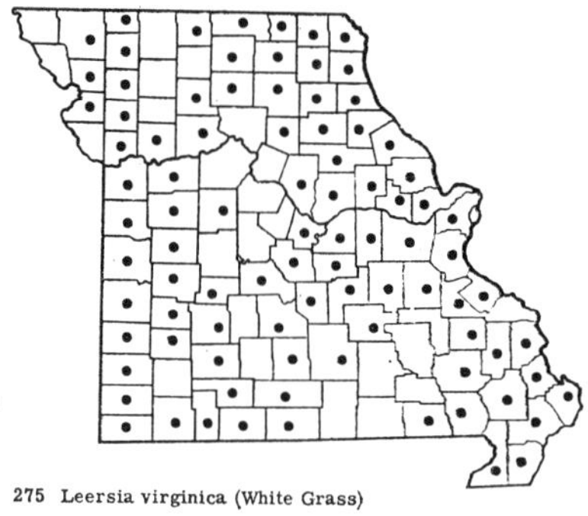

275 Leersia virginica (White Grass)

b. Spikelets 1–1.3 mm. wide, 3–4 mm. long; leaves at most finely scabrous (rough with minute hairs) on margins and this not obvious; culms compressed (flattened); lower branches of inflorescence solitary . . . . . . . . . . . . . . . . . . . . . . . . 3. L. VIRGINICA

b. Spikelets 1.5–2 mm. wide, 4.5–6 mm. long; leaves with bristly-ciliate or scabrous-hispid margins (with longer stiffer hairs) and this obvious; culms terete (circular in cross-section); lower branches of inflorescence whorled . . . . . . . . . . . . . . . . . . . . . 2. L. ORYZOIDES

1. **Leersia lenticularis** Michx.
Catchfly Grass Map 273
Flowers August–September.

Frequents border of sloughs and wooded swamps, bayous, margins of oxbow lakes and swales in river flood plain and valleys. Restricted to those sections of the state where its habitats occur: in the southeastern lowlands section, where common, sloughs and margins of wooded swamps and swales along the Mississippi River and tributaries in eastern Missouri, similar habitats especially around oxbow lakes in western and northwestern Missouri. Absent from practically the entire Ozark section.

Ranges from Florida to Texas, north to Maryland, Ohio, Indiana, Wisconsin, and Minnesota.

2. **Leersia oryzoides** (L.) Sw.
Ricecut Grass Map 274
Flowers June–October.

Borders of sloughs, lakes, swamps, spring-fed rivulets in wet meadows, open swales, and ditches, but in situations usually exposed to sunlight.

Missouri material falls into two categories:

Inflorescence completely projecting from or partly included in the upper leaf-sheath . . . . 2a. L. ORYZOIDES f. ORYZOIDES

Inflorescence wholly or mostly included in or covered up by the upper leaf-sheath . . . . 2b. L. ORYZOIDES f. INCLUSA

2a. **Leersia oryzoides** f. **oryzoides** Map 274
The common type found throughout Missouri.

2b. **Leersia oryzoides** f. **inclusa** (Wiesb.) Dörfler Map 274
Occasionally associated with the typical form, but less common.

Some authors consider this latter f. *inclusa* as a mere late-season unexpanded growth stage, not worthy of taxonomic recognition. For a discussion of this subject see Fogg (Rh. 30: 81–85. 1928).

The rough leaf-blades of *L. oryzoides* and *L. lenticularis* can easily cut or tear the flesh if grabbed quickly or without caution.

3. **Leersia virginica** Willd. White Grass
Map 275
Flowers June–October.

Frequents usually moist or somewhat damp soil in woods, in ravines, valleys, along spring branches, borders of swamps, bayous, and rivers; usually associated with low woodland. Throughout Missouri.

*Leersia virginica* var. *ovata* (Poir.) Fern. was separated by Fernald (Rh. 38: 386. 1936) from var. *virginica* on the basis of having the margin of the lemma and often its keel bristly-ciliate with cilia up to 0.6 mm. long instead of being smooth or minutely ciliolate. According to this distinction, both types of variations occur in Missouri. The variations, however, grade freely into one another and do not appear to show sufficient correlation in the material examined to be maintained.

276 Zizania aquatica var. interior (Water Rice)

277 Zizaniopsis miliacea (Water Millet)

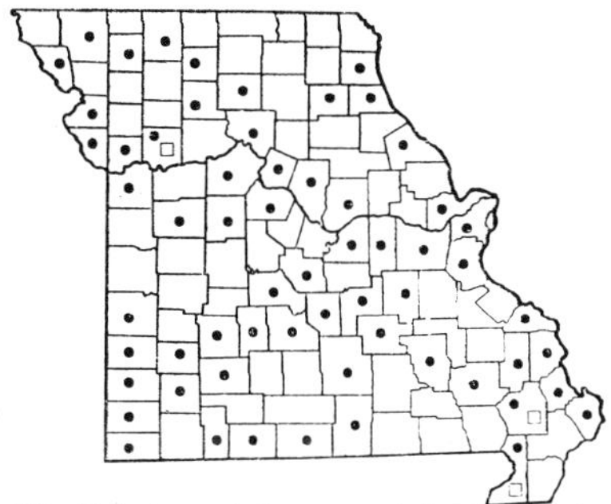

278 • Digitaria sanguinalis var. sanguinalis (Crab Grass)
278 □ Digitaria sanguinalis var. ciliaris

## Tribe IX. **ZIZANIEAE**

The grasses in this tribe are monoecious with unisexual 1-flowered spikelets (either staminate or pistillate), the pistillate ones subulate or terete with united styles; the staminate ones have 6 stamens. Glumes are absent, and the spikelets are articulated below the glumes.

### 62. **Zizania** L. Wild Rice

Tall annual grasses growing in water, with long leaves and long, narrow, terminal panicles.

**Zizania aquatica** L. var. **interior** Fassett
Water Rice, Indian Rice, Wild Rice Map 276
Flowers June–October.

Frequents swamps and borders of ponds and streams. Very rare and local in Missouri, where it occurs in favored areas along the St. Francis, Mississippi, and Missouri rivers and their tributaries, locally elsewhere in Greene and Newton counties in southwestern Missouri.

Ranges from Indiana to North Dakota, south to Missouri and Texas.

The sole collection known from northern Missouri is from Clark County (Wayland, October 12, 1920, *F. P. Metcalf 1085*) in the United States National Museum.

Although large amounts of pollen are shed by this plant, it is not an important one in the consideration of causes of hay fever.

This rice furnishes grain which is used in modern cookery as a breakfast food and as a substitute for ordinary rice usually served with fowl on special occasions. As the wild rice commands fairly high prices, it has not been used as much as it would otherwise. The grain was used by the Indians for flour for bread, for thickening soups, and to cook with wild meats.

### 63. **Zizaniopsis** Döll & Aschers. Water Millet

Tall perennial grasses, with creeping rhizomes, growing in the water.

**Zizaniopsis miliacea** (Michx.) Döll & Aschers.
Water Millet Map 277
Flowers April–August.

Frequent in sink-hole ponds of the Ozark region, and in swamps and borders of drainage ditches and bayous in the southeastern lowland section. Found only in southeastern Missouri west to Shannon and Oregon counties and north to New Madrid and Stoddard counties.

Ranges from Florida to Texas, north to Maryland, Kentucky, southeast Missouri, and Oklahoma. Also in Mexico and South America.

The leaves of this grass are razor-sharp, easily cutting the flesh when handled; they are long and erect, 3–10 dm. long and 1–3 cm. wide. The plants, which vary from 1–4 meters high, often occur in dense colonies. At a distance they somewhat resemble sugar cane.

It has been suggested that the young tips of the rhizomes of this species, when cooked, would make a desirable cooked vegetable, in much the same manner in which the Chinese use a species of *Zizania* native to China and Japan; this is either cooked with meat or cut into two or three pieces which are boiled and served with melted butter as a vegetable.

## Tribe X. **PANICEAE**

Grasses of this tribe have the articulation of the spikelets below the glumes. The spikelets consist of 1 perfect floret with 2 often thin or membranous glumes, a sterile lemma similar to the glumes in texture and resembling a 3rd glume, and a thicker rather tough fertile lemma and palea together covering the grain or seed-like fruit. Spikelets are more or less dorsally compressed.

### 64. **Digitaria** Heist. Crab Grass

Annual grasses (in the local species) with a digitate or nearly digitate inflorescence. Spikelets in 2 rows on 1 side of the rachis, 1-flowered with the 1st glume very small or lacking.

a. The axis (rachis) of each ray or spoke of the inflorescence very narrowly winged, the wings narrower than the main rib of the rachis . . . . . . . . . . . . . . . . . . 3. D. FILIFORMIS
a. The axis (rachis) of each ray or spoke of the inflorescence broadly winged, the wings wider than or as wide as the main rib of the rachis (resembling a divided highway) . . . . . . . . . . . . . . *b*
  b. At least the lower leaf-sheaths and blades more or less conspicuously hairy; spikelets 2.3–3.5 mm. long; spikelets glabrous between the smooth or finely hairy nerves . . . . . . . 1. D. SANGUINALIS
  b. Leaf-sheaths and blades mostly glabrous, a few hairs sometimes on lower sheaths; spikelets 1.7–2.2 mm. long; at least some of the spikelets with few to many short spreading hairs between the nerves 2. D. ISCHAEMUM

1. **Digitaria sanguinalis** (L.) Scop.
Crab Grass Map 278
Flowers July–November.

Waste and cultivated ground, frequently in lawns, gardens, margins of ponds, spring branches, rocky open ground, prairie openings, and along railroads and roadways. Throughout Missouri.

Native of Europe; introduced in North America, where found throughout the United States and southern Canada.

Missouri material may be divided into two varieties:

Spikelets 2.3–3 mm. long, the hairs on the marginal nerves of the sterile lemma minute . . . . 1a. D. SANGUINALIS var. SANGUINALIS
Spikelets 3–3.5 mm. long, the hairs on the marginal nerves of sterile lemma longer and more conspicuous . . . . 1b. D. SANGUINALIS var. CILIARIS

1a. **Digitaria sanguinalis** var. **sanguinalis** Map 278

The common type encountered.

1b. **Digitaria sanguinalis** var. **ciliaris** (Retz.) Parl. Map 278
*Digitaria sanguinalis* var. *marginata* (Link) Fern. [P & S]

Known from Ray, Stoddard, and Dunklin counties.

Although *D. sanguinalis* is very widespread, it does not appear to be of any importance in causing hay fever. It is considered a good forage plant for late summer, but more often is condemned because of being a weed in lawns and gardens. It often roots at the lower nodes (joints) of the stem, producing a large colony from one plant. Various chemicals (phenyl mercury acetate, potassium cyanate, and disodium methyl arsonate obtained on the market under trade names) are used to control crab grass, when it becomes a nuisance. Some chemicals, such as calcium cyanimid, applied, before seeding, kill the weed seeds before they have sprouted and also act as a fertilizer. The inflorescences are green or often turn dull purple; parts of the culms and sometimes the leaf blades are also purple or tinged with purple.

The seeds have been cultivated and used in some parts of Europe, such as Poland, as a substitute for rice.

2. **Digitaria Ischaemum** (Schreb.) Muhl.
Crab Grass Map 279
Flowers July–November.

Frequent in moist low ground of fields, cultivated valleys along river bottoms, pastures, meadows, low woodlands, roadsides, and waste ground. Found in lawns and garden soils, but less frequent, for the most part, in such places than *D. sanguinalis*.

Throughout Missouri.

Native of Europe; introduced into North America,

Plate no. 53. **1.** Leptoloma cognatum; *a.* Plant, × 3/8; *b.* Floret, × 7 1/2; *c, d.* Two views of spikelet, × 7 1/2. **2.** Digitaria filiformis; *a.* Plant, × 3/4; *b.* Floret, × 7 1/2; *c.* Spikelet, × 7 1/2. **3.** Digitaria Ischaemum; *a.* Plant, × 3/4; *b.* Floret, × 7 1/2; *c.* Spikelet, × 7 1/2. **4.** Digitaria sanguinalis; *a.* Plant, × 3/8; *b.* Floret, × 7 1/2; *c, d.* Two views of spikelet, × 7 1/2.

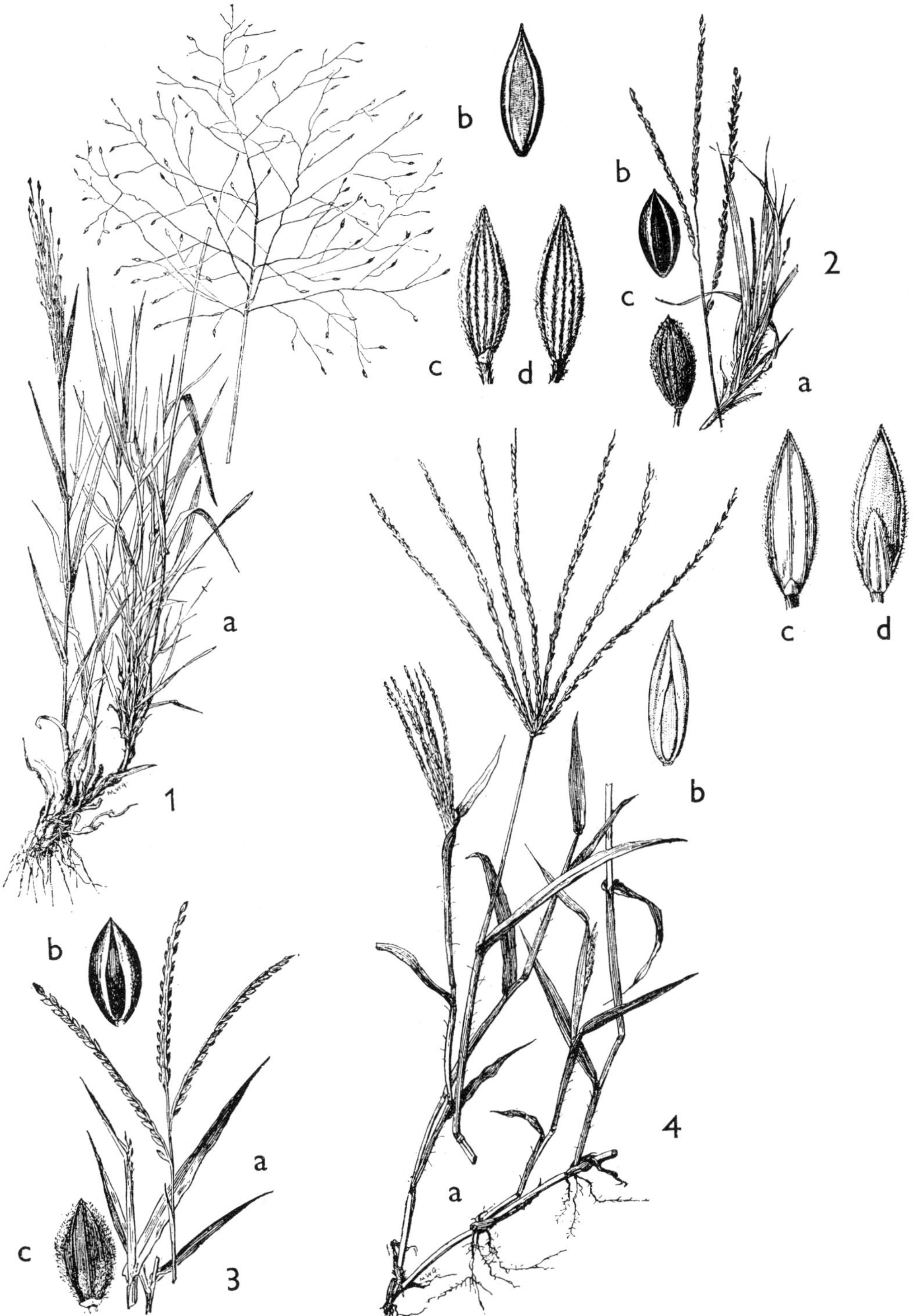


PLATE NO. 53

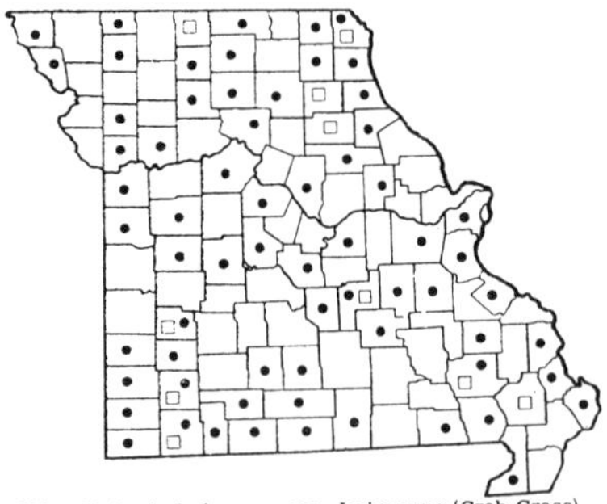

279 • Digitaria Ischaemum var. Ischaemum (Crab Grass)
279 □ Digitaria Ischaemum var. mississippiensis

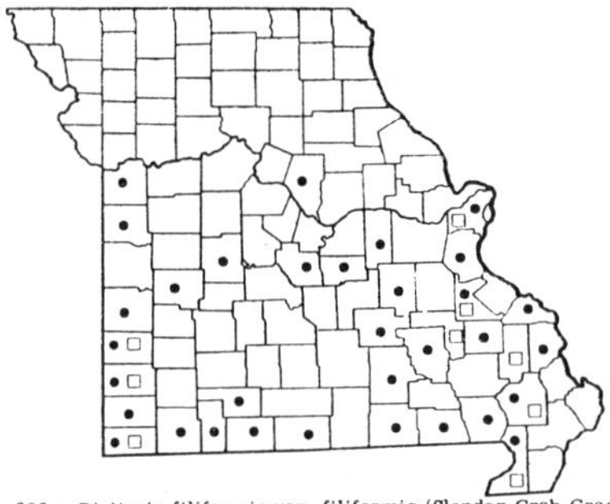

280 • Digitaria filiformis var. filiformis (Slender Crab Grass)
280 □ Digitaria filiformis var. villosa

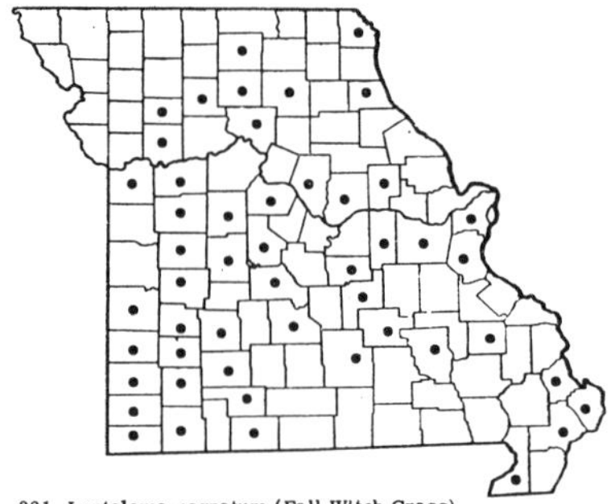

281 Leptoloma cognatum (Fall Witch Grass)

ranging from Quebec to North Dakota and Washington, south to Florida, Louisiana, Texas, and California.

Missouri material falls into two categories:

The 1–6 branches (spikelike racemes) of inflorescence 1–9 cm. long; leaf-blades 2–10 cm. long, 3–8 mm. wide. . 2a. D. ISCHAEMUM var. ISCHAEMUM
The 5–8 branches (spikelike racemes) of inflorescence 6–15 cm. long; leaf-blades up to 20 cm. long, 6–9 mm. wide . . . . 2b. D. ISCHAEMUM var. MISSISSIPPIENSIS

2a. **Digitaria Ischaemum** var. **Ischaemum** Map 279

The commoner variation found in Missouri, throughout the state.

2b. **Digitaria Ischaemum** var. **mississippiensis** (Gattinger) Fern. Map 279

Scattered throughout the state.

There is much intergradation between the two variations, which are distinct in their extremes. In general the inflorescence branches are shorter and the leaf-blades shorter and narrower in *D. Ischaemum* var. *Ischaemum*. The culms are generally more spreading than in *D. sanguinalis* var. *sanguinalis*, although they may be erect to prostrate in the extreme with rooting nodes, and the plants as a whole are often paler green, but are also tinged purple on various parts. The culms generally are more slender and shorter than those of *D. sanguinalis*.

Plants are sometimes found having some of the spikelets typically covered with hairs between the nerves, with other spikelets of the same inflorescence glabrous.

Rarely are found inflorescences which have the spikelets proliferating on branched stalks. Such specimens are *Steyermark 80682* from Barry County, southwestern Missouri.

3. **Digitaria filiformis** (L.) Koel.
Slender Crab Grass Map 280
Flowers August–October.

Occurs on sandstone and chert glades and prairies, sandy fields, valleys, and roadsides, where rather infrequent in southern and central Missouri. It is characteristic of acid sterile soils in dry open areas.

Missouri material falls into two varieties:

Spikelets 1.5–2 mm. long; lower leaf-sheaths glabrous to sparsely hairy . . . . 3a. D. FILIFORMIS var. FILIFORMIS
Spikelets 2–2.6 mm. long; lower leaf-sheaths densely and conspicuously hairy . 3b. D. FILIFORMIS var. VILLOSA

3a. **Digitaria filiformis** var. **filiformis** Map 280

The commoner variation in Missouri, extending north to Jackson, Boone, and St. Louis counties.

Ranges from Florida to Texas and Mexico, north to New Hampshire, New York, Michigan, Illinois, and Iowa.

3b. **Digitaria filiformis** var. **villosa** (Walt.) Fern. Map 280

Scattered in southern and east-central Missouri.

Ranges from Florida to Texas, north to Virginia, Tennessee, Illinois, Missouri, and Kansas.

The rachis of the inflorescence branches is much narrower than in either of the other species in Missouri. Generally, the culms are more erect and ascending, and do not possess the decumbent or rooting habit of *D. sanguinalis* or spreading to prostrate habit of *D. Ischaemum*. The culms (stems) in *D. filiformis* are the slenderest of the three species found in Missouri. Tiny gland-tipped hairs occur on the glumes and sterile lemma of this species.

282 Eriochloa contracta (Prairie Cup Grass)

283 Eriochloa gracilis var. gracilis (Cup Grass)
* * *
284 Excluded species

## 65. **Leptoloma** Chase Fall Witch Grass

Perennial grasses with spindle-shaped (fusiform) spikelets on long hair-like pedicels (stalks). The 1st glume is minute or lacking. The culms are brittle and, when mature, break up easily.

**Leptoloma cognatum** (Schultes) Chase
Fall Witch Grass Map 281
Flowers June–October.

Frequent along roadsides, waste places, railroads, prairies, glades, pastures, and sandy areas. Mostly in the southern half of the state, not recorded from northwestern Missouri, north and west to Clark, Sullivan, Livingston, and Caldwell counties.

Ranges from Florida to Mexico and Arizona, north to New Hampshire, New York, Ohio, Michigan, Wisconsin, Minnesota, and Nebraska.

This grass becomes much branched at the base, producing a sprawling many-stemmed plant. The many-branched, graceful panicle, which occupies a third to half the height of the plant, takes on a lively rose-purple color and eventually breaks away from the rest of the plant as a tumbleweed. The mass color effect of these plants along highways and pastures is quite striking.

Plants of this genus are often mistaken for species of *Panicum* from which they are separated by the ligule, which is a collarlike hairless scale, not hairy or hairy-fringed as in *Panicum*, and by the thin and flexible fertile lemma with a colorless (hyaline) margin which is not inrolled, rather than a hard firm fertile lemma inrolled on the margins as in *Panicum*.

## 66. **Eriochloa** HBK. Cup Grass

Annual or perennial grasses, with a long, narrow, dense inflorescence (panicle) of racemes appressed to the main axis. The inflorescence is often at first partly enclosed in or covered by the upper leaf-sheath.

Leaf-blade pubescent, 4–7 mm. wide; spikelets 3.5–4 mm. long . . . . . . . . . . 1. E. CONTRACTA
Leaf-blade glabrous, 5–10 mm. wide; spikelets 4–5 mm. long . . . . . . . . . . . . 2. E. GRACILIS

1. **Eriochloa contracta** Hitchc.
Prairie Cup Grass Map 282
Flowers July–October.

Cultivated and waste ground, often in alluvial ground of fields and river valleys, where rather rare in central and eastern Missouri, following the Mississippi and Missouri rivers.

Ranges from Missouri (where introduced) and Kansas to Louisiana, Texas, and New Mexico; introduced in Virginia.

The two species treated are very similar. *Eriochloa contracta*, which is the commonest of the two in Missouri, has the fruit tipped by an awn about 1 mm. long, and is an annual.

2. **Eriochloa gracilis** (Fourn.) Hitchc. var. **gracilis**
Cup Grass Map 283
Flowers July–October.

Cultivated and alluvial ground.

Known only from southeastern Missouri, in Mississippi County (open places bordering soybean field along road K, T27N, R16E, sect. 24, 7½ mi. [by air] northeast of Charleston, August 21, 1955, *Steyermark 79553*).

Ranges from Texas to California and Mexico; introduced in Missouri.

285 Brachiaria platyphylla

286 Brachiaria erucaeformis

287 Paspalum dissectum

## 67. **Brachiaria** (Trin.) Griseb.

Annual (in the local species) grasses with the inflorescence of several spreading or appressed racemes arranged racemose or pinnate along the main axis. The spikelets are placed with the back of the fruit turned away from the rachis of the racemes. First glume present.

Spikelets glabrous, 4–4.5 mm. long . . . . . . . . . . . . . . . . . . . . . 1. B. PLATYPHYLLA
Spikelets pubescent, 2–2.5 mm. long . . . . . . . . . . . . . . . . . . . . . 2. B. ERUCAEFORMIS

1. **Brachiaria platyphylla** (Griseb.) Nash Map 285

*Brachiaria extensa* Chase (Hitchcock, ed. 1)

Flowers August–November.

Cultivated low fields, sandy open knolls, and borders of wet ditches along roads in the southeastern lowland region of Missouri, where known only from Scott (edge of field in lowland and on sandy ridge top at edge of field, T27N, R13E, sect. 12, $4\frac{1}{2}$ mi. east of Vanduser, November 1, 1956, *Steyermark 83578*) and Pemiscot (openings along ditch, along road E, sect. 8, $1\frac{1}{4}$ mi. east of Cooter, October 23, 1948, *Steyermark 67054*) counties.

Ranges from Florida to Texas, north to southeastern Missouri and Oklahoma.

2. **Brachiaria erucaeformis** (J. E. Smith) Griseb. Map 286

Flowers October.

Known only from southwestern Missouri in Stone County (gravel bed of Creek, between Aunt's Creek and James River, T23N, R23–24W, sect. 31, 36, and north part of sect. 1, $4\frac{1}{2}$–5 mi. [by air] northwest of Radical, 6–7 mi. northeast of Baxter, Oct. 20, 1955, *Steyermark 80556*).

Native to the Old World (India west to Spain); escaped from cultivation in a few places in the United States.

The locality where this species was found is now exterminated, being part of the impounded waters of Table Rock Dam.

## 68. **Paspalum** L.

Perennial (in the local species) grasses with soft leaves and an inflorescence of 1–several racemes arranged digitately, pinnately, or rarely solitarily. The almost round to ovate spikelets, which articulate below the glumes, are in 2 or 4 rows on 1 side of the rachis and are flat on one side, curved on the other (plano-convex), with a somewhat flattened appearance. The first glume is usually lacking.

a. Rachis (axis) of each ray or spoke (raceme) of the inflorescence broad, winged, or leaf-like, and wider than the spikelets themselves . . . . . . . . . . . . . . . . . . . . *b*
  b. Rays (racemes) 5–50; spikelets 1.1–1.7 mm. long, with tiny hairs; leaf-blades 3–25 mm. wide (mostly 6 or more) . . . . . . . . . . . . . . . . . . . . 2. P. FLUITANS
  b. Rays (racemes) 1–4 (sometimes 5); spikelets 1.8–2.3 mm. long, glabrous (without hairs); leaf-blades 2–5 mm. wide . . . . . . . . . . . . . . . . . . 1. P. DISSECTUM

---

Plate no. 54. **1.** Paspalum dissectum; *a.* Panicle, × $^3/_4$; *b.* Floret, × $7^1/_2$; *c, d.* Two views of spikelet, × $7^1/_2$. **2.** Brachiaria platyphylla; *a.* Plant, × $^3/_8$; *b.* Floret, × $7^1/_2$; *c, d.* Two views of spikelet, × $7^1/_2$. **3.** Eriochloa gracilis; *a.* Plant, × $^3/_8$; *b.* Floret, × $7^1/_2$; *c, d.* Two views of spikelet, × $7^1/_2$. **4.** Eriochloa contracta; *a.* Panicle, × $^3/_4$; *b.* Floret, × $7^1/_2$. **5.** Brachiaria erucaeformis; *a.* Panicle, × $^3/_4$; *b.* Floret, × $7^1/_2$; *c, d.* Two views of spikelet, × $7^1/_2$.

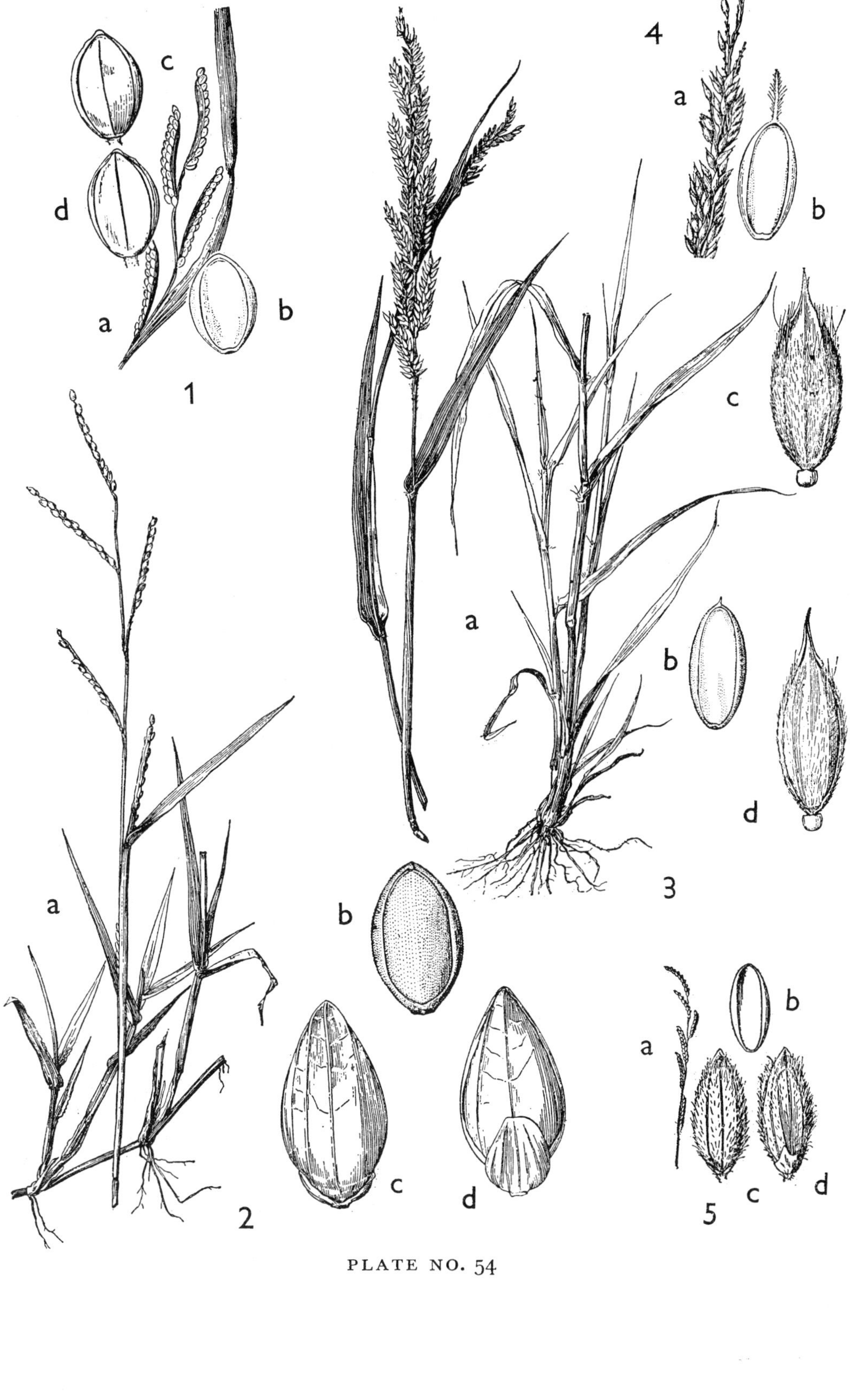


PLATE NO. 54

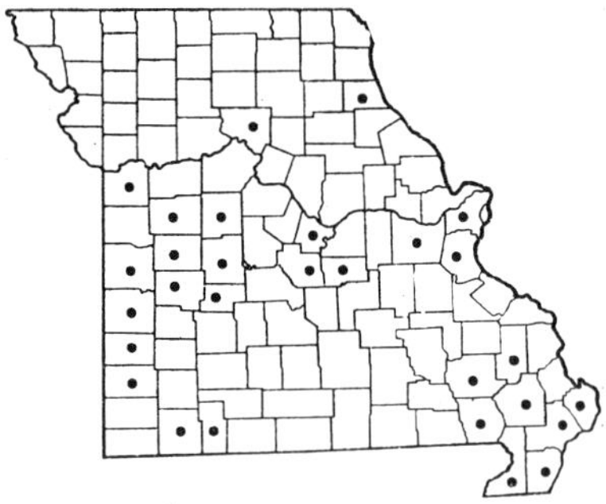

288 Paspalum fluitans

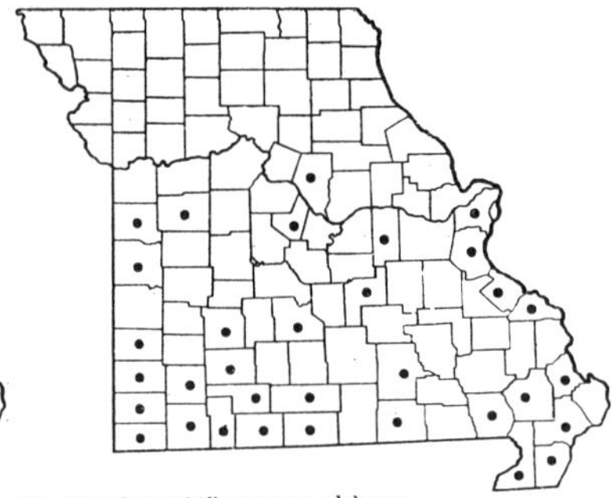

289 Paspalum pubiflorum var. glabrum

290 Paspalum setaceum var. setaceum

a. Rachis (axis) of each ray or spoke (raceme) of the inflorescence narrow or without wings, always narrower than the rows of spikelets . . . . . . . . . . . . . . . . . . . . *c*
- c. Some part of spikelet with conspicuous silky hairs; spikelet conspicuously pointed at tip (acute to acuminate); tip flexible, not ending uniformly, one part of tip (of glume and sterile lemma) distinctly longer than other part of tip (fertile lemma) . . . . . . . . . . . 7. P. DILATATUM
- c. Spikelet glabrous or at most with tiny short hairs; spikelet rounded or blunt at tip, the tip firm and ending uniformly on all sides . . . . . . . . . . . . . . . . . . . *d*
  - d. Each spikelet attached singly to 1 pedicel (stalk) along the rachis . . . . . . . . 8. P. LAEVE
  - d. Spikelets arising in pairs, the pedicel forking at the base (the 2nd spikelet sometimes poorly or not developed) . . . . . . . . . . . . . . . . . . . . *e*
    - e. Spikelets 3.6–4.3 mm. long, 2.8–3.1 mm. wide . . . . . . . . . . 9. P. FLORIDANUM
    - e. Spikelets 1.4–3.2 mm. long, 1.3–2.4 mm. wide . . . . . . . . . . . . . *f*
      - f. Spikelets 2.9–3.2 mm. long . . . . . . . . . . . . . 3. P. PUBIFLORUM
      - f. Spikelets 1.4–2.8 (–3) mm. long . . . . . . . . . . . . . . . *g*
        - g. Spikelets 1–1.6 mm. wide, ellipsoid, about 3/4 as wide as long. . . . . . 4. P. SETACEUM
        - g. Spikelets mostly 1.5–2.4 mm. wide, round or nearly round in outline. . . . . . . *h*
          - h. Spikelets densely hairy; leaves densely soft, short-hairy and mixed with long hairs; nodes (joints) of culms mostly hairy . . . . . . . . . . . . . 6. P. BUSHII
          - h. Spikelets mostly glabrous or with a few short hairs; leaves glabrous, long hairy, or minutely hairy (puberulent) and mixed with long hairs; nodes (joints) of culms glabrous . . . . . . . . . . . . . . . . . 5. P. CILIATIFOLIUM

1. **Paspalum dissectum** L. Map 287

Flowers July–October.

Muddy banks of streams, ditches, and sloughs, and margins of upland ponds. Rare, and found only in southern Missouri in the southeastern lowland section, around an upland sink-hole pond in Howell Co., and in Jasper (*Palmer 968, 969*) and Barton (*Palmer 55318*) counties of southwestern Missouri.

Ranges from Florida to Texas, north to New Jersey, Tennessee, southern Illinois and southern Missouri, and also in Cuba.

2. **Paspalum fluitans** (Ell.) Kunth Map 288

*Paspalum repens* of some authors, not Bergius [P & S, Hitchcock, ed. 1]

Wet muddy alluvial banks along streams, bayous, sloughs, and margins of lakes and ponds, especially oxbow lakes in the valleys of large rivers. Restricted to various sections of southern, central, and eastern Missouri, the southeastern lowland section, alluvial ground of the White River in southwestern Missouri (now dammed), along sections of the Missouri, Osage, Clear Creek, South Grand, Marmaton, Mariais des Cynes, and their oxbow lakes, and locally north of the Missouri River in Chariton and Marion counties.

Ranges from Florida to Texas north to Virginia, Kentucky, Indiana, Illinois, Missouri, and Kansas.

3. **Paspalum pubiflorum** Rupr. var. **glabrum** Vasey Map 289

Flowers June–September.

Occurs in moist open ground, prairie swales, meadows, borders of sandy wet ditches, sandy roadsides and open fields, alluvial woods, and along railroads, in southern and central Missouri north to Cass, Johnson, Boone, Gasconade, and St. Louis counties.

Ranges from Florida to Texas, north to North Carolina, Kentucky, Ohio, Indiana, Illinois, Missouri, and Kansas.

This species may be distinguished vegetatively by

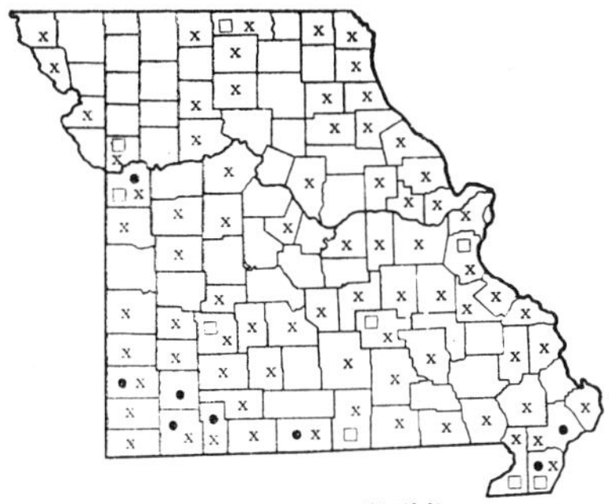

291 • Paspalum ciliatifolium var. ciliatifolium
291 x Paspalum ciliatifolium var. Muhlenbergii
291 □ Paspalum ciliatifolium var. stramineum

292 Paspalum Bushii

293 Paspalum dilatatum (Dallis Grass)

the decumbent and geniculate rooting lower nodes of the culms as contrasted with another tall-growing species, *P. floridanum*, in which the culms are erect, but do not root at the lower nodes.

Typical *P. pubiflorum* var. *pubiflorum*, with pubescent spikelets, occurs in some of the southern states, Mexico, and the West Indies.

4. **Paspalum setaceum** Michx. var. **setaceum** Map 290

Flowers June–October.

Known only from Crowley Ridge, southeastern Missouri in Stoddard County (dry sandy wooded slopes bordering ravines along and tributary to spring-fed creek near junction of Crowley Ridge and lowland, on property of Mr. Martin, T25N, R11E, NW $\frac{1}{4}$ sect. 6, $3\frac{1}{2}$ mi. southeast of Bloomfield, August 20, 1954, *Steyermark 76726*).

Ranges from Florida to Texas, north to Massachusetts, New York, Ohio, Kentucky, and Missouri.

The variety occurring in Missouri is the typical one, distinguished by the relatively small spikelets, 1.4–1.8 mm. long, 1–1.2 mm. wide, and by the densely hairy leaf-blades 2–6 mm. wide.

The culms (stems) of this species are very slender and filiform. The 1–3 slender racemes of the inflorescence have a very slender rachis, which is much narrower than any of the other species in Missouri.

5. **Paspalum ciliatifolium** Michx. Map 291

Flowers late May–October.

Occurs in prairies, roadside banks, along railroads, border of woods, grazed fields and pastures, and dry open woods. Throughout Missouri.

Missouri material falls into the following categories:

a. Upper surface of leaf-blade with 2 lengths of hair, with many minute hairs (puberulence) between longer hairs . . . 1c. P. CILIATIFOLIUM var. STRAMINEUM
a. Upper surface of leaf-blade entirely glabrous, or else long-hairy with one length of long hair. *b*
  b. Upper surface entirely glabrous . . . 1a. P. CILIATIFOLIUM var. CILIATIFOLIUM
  b. Upper surface with long hairs of about the same length . . . . 1b. P. CILIATIFOLIUM var. MUHLENBERGII

5a. **Paspalum ciliatifolium** var. **ciliatifolium** Map 291

*Paspalum ciliatifolium* Michx. [G, P & S]

Known in only a few southern counties and northwest in Jackson County.

Ranges from Florida to Texas, north to D. C., Tennessee, and Missouri.

5b. **Paspalum ciliatifolium** var. **Muhlenbergii** (Nash) Fern. Map 291

The common type found throughout the state.

Ranges from Florida to Texas, north to New Hampshire, New York, Ohio, Michigan, and Wisconsin.

5c. **Paspalum ciliatifolium** var. **stramineum** (Nash) Fern. Map 291

Scattered over the state, infrequent.

Ranges from Texas to Arizona, north to Indiana, Wisconsin, Minnesota, Nebraska, and Colorado.

Although it is possible to separate the above variations as such, intergradation is found in the amount of pubescence of the leaf-blades, height of culm, erect or spreading habit, and glabrity of the spikelet. Deam (*Grasses of Indiana*, p. 234. 1929) believes that much of the variation in leaf pubescence and habit might be due to the effects of grazing. Shinners (Rh. 56: 32. 1954) does not believe that these variations in Texas merit varietal names and places them all together as *P. ciliatifolium*.

Most of the Missouri material falls under *P. ciliatifolium* var. *Muhlenbergii*. The other variations are found infrequently.

6. **Paspalum Bushii** Nash Map 292
*Paspalum ciliatifolium* var. *stramineum* [of BB, P & S], not (Nash) Fern.

Sandy open woods, sandy open places, and in fields, rare in southern and central Missouri north to Jackson and St. Charles counties. The species is based upon a Missouri collection (Bernie, Stoddard County, *Bush 730*).

Ranges from Illinois and Missouri to Nebraska and Texas.

Some authors (Shinners, Gleason, Hitchcock) consider this species inseparable from *P. ciliatifolium* var. *stramineum*. Although variation does exist in *P. Bushii*, it is maintained in the present treatment on the basis of a combination of characters including pubescent nodes of the culms, densely soft pubescent leaf-blades on both surfaces with appressed long pubescence, and densely pubescent spikelets.

7. **Paspalum dilatatum** Poir.
Dallis Grass, Water Paspalum Map 293
Flowers May–October.

Known only from St. Louis (right-of-way of Wabash R. R., north of Palm St. along the most western track, St. Louis, September 8, 1957, *Muehlenbach 1355*; also September 1, 1958, *Muehlenbach 1478*) and Newton (along spring branch, 4 mi. south of Saginaw, August 13, 1957, *Palmer 66230*) counties.

Native of South America from Brazil to Argentina; introduced into Central America, the West Indies, and the Old World; in the United States introduced and ranging from Florida to California, north to Virginia, Tennessee, Missouri, Colorado, and Oregon.

Introduced into the southern United States about 1850 as a pasture grass. It is believed to be a contributing cause of hay fever at the time it is in flower between spring and fall.

8. **Paspalum laeve** Michx. Map 294
Flowers July–October.

Occurs in wet places in prairies, fields, marshy ground at the base of slopes, along ditches, moist open ground, borders of lakes and bayous, and along roadsides and railroads.

Missouri material may be segregated as follows:

a. Spikelets more or less circular or orbicular, 2.8–3.2 mm. wide . 8b. P. LAEVE var. CIRCULARE
a. Spikelets slightly longer than wide, 2–2.5 mm. wide . . . . . . . . . . . . *b*
  b. Leaf-sheaths and blades glabrous or nearly glabrous . . . . 8a. P. LAEVE var. LAEVE
  b. Leaf-sheaths strongly hairy; leaf-blades often quite hairy . 8c. P. LAEVE var. PILOSUM

8a. **Paspalum laeve** var. **laeve** Map 294
*Paspalum laeve* Michx. [G]
*Paspalum laeve* in part [BB]

Rare, known only from Cape Girardeau (*Steyermark 64158*), Bollinger, Stoddard (*Steyermark 66187*), Ripley (*Steyermark 64158*), and Stone counties.

Ranges from Florida to Texas, north to New Jersey, Pennsylvania, Ohio, Illinois, and Missouri.

8b. **Paspalum laeve** var. **circulare** (Nash) Fern. Map 294
*Paspalum laeve* in part [BB]

Frequent in southern and central Missouri north to Jackson, Boone, Shelby, and Lincoln counties. This is the commonest of the variations in Missouri.

Ranges from Georgia to Texas, north to Connecticut, New York, Pennsylvania, Indiana, Illinois, Missouri, and Kansas.

8c. **Paspalum laeve** var. **pilosum** Scribn. Map 294
*Paspalum laeve* in part [BB]

Scattered in southern Missouri, extending north to Barton, Polk, Phelps, and Washington counties.

Ranges from Florida to Texas north to New York, Pennsylvania, Tennessee, and Missouri.

The range of variation in *P. laeve* is considerable, with intergradation occurring as in *P. ciliatifolium* and other species. Some authors, such as Gleason in the *New Illustrated Flora*, find no correlation between the variations and do not believe the variations warrant recognition.

The branches (racemes) of the inflorescence vary from 2–8 and have a rachis about 1 mm. wide.

9. **Paspalum floridanum** Michx. Map 295
Flowers August–October.

Occurs in wet meadows, swales in prairies, margins

Plate no. 55. **1.** Paspalum dilatatum; *a.* Plant, × 3/8; *b.* Floret, × 7 1/2; *c, d.* Two views of spikelet, × 7 1/2. **2.** Paspalum fluitans; *a.* Panicle, × 3/4; *b.* Floret, × 7 1/2; *c, d.* Two views of spikelet, × 7 1/2. **3.** Paspalum setaceum; *a.* Raceme, × 3/4; *b.* Floret, × 7 1/2; *c, d.* Two views of spikelet, × 7 1/2. **4.** Paspalum Bushii; *a.* Panicle, × 3/4; *b.* Floret, × 7 1/2; *c, d.* Two views of spikelet, × 7 1/2. **5.** Paspalum pubiflorum; *a.* Panicle, × 3/4; *b.* Floret, × 7 1/2; *c, d.* Two views of spikelet, × 7 1/2. **6.** Paspalum ciliatifolium; *a.* Raceme, × 3/4; *b.* Floret, × 7 1/2; *c, d.* Two views of spikelet, × 7 1/2.

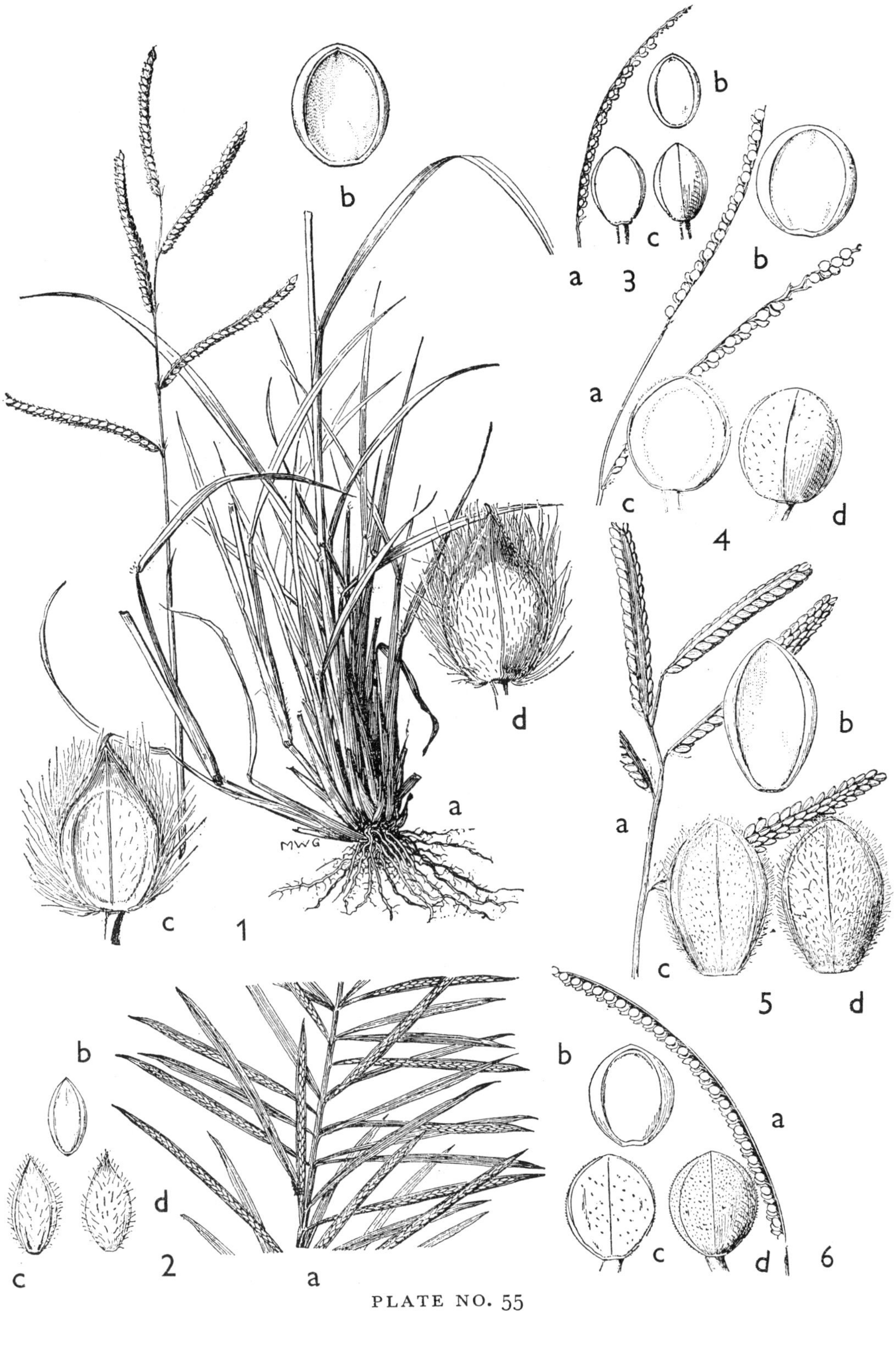


PLATE NO. 55

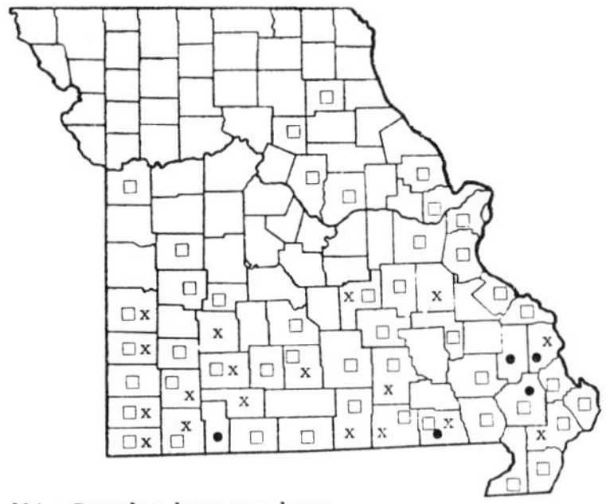

294 • Paspalum laeve var. laeve
294 □ Paspalum laeve var. circulare
294 x Paspalum laeve var. pilosum

295 • Paspalum floridanum var. floridanum
295 □ Paspalum floridanum var. glabratum

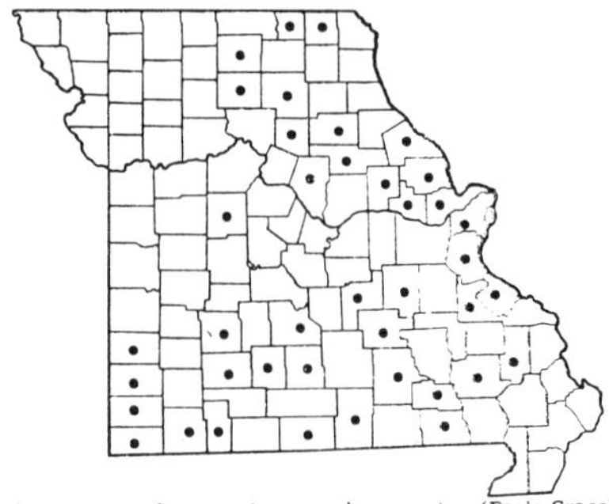

296 Panicum depauperatum var. depauperatum (Panic Grass)

of upland sink-hole ponds, sandy fields, along streams, in ravine bottoms, ditches, and roadsides.

Missouri material falls into two categories:

Leaf-sheaths hairy; leaf-blades hairy . . . .
9a. P. FLORIDANUM var. FLORIDANUM
Leaf-sheaths and blades glabrous or nearly so .
9b. P. FLORIDANUM var. GLABRATUM

9a. **Paspalum floridanum** var. **floridanum** Map 295

This is less common in Missouri than the other variety, and is known from southern Missouri only from Mississippi, Oregon, Howell, Polk, Barton, Jasper, and Newton counties.

Ranges from Florida to Texas, north to Virginia, Tennessee, Missouri, and Oklahoma.

9b. **Paspalum floridanum** var. **glabratum** Engelm. Map 295

Scattered in southern Missouri from Stoddard County to Polk and Vernon counties.

Ranges from Florida and Texas, north to New Jersey, Pennsylvania, Missouri, and Kansas.

*Paspalum floridanum* and *P. pubiflorum* are the tallest of the species of *Paspalum* in Missouri, attaining a height of about 2 meters; *P. floridanum* averages taller plants more frequently than *P. pubiflorum*. Both have robust culms.

---

Plate no. 56. **1.** Paspalum floridanum; *a.* Panicle, $\times$ $^3/_4$; *b.* Floret, $\times$ $7^1/_2$; *c,d.* Two views of spikelet, $\times$ $7^1/_2$. **2.** Panicum perlongum; *a.* Floret, $\times$ $7^1/_2$; *b, c.* Two views of spikelet, $\times$ $7^1/_2$. **3.** Panicum laxiflorum; *a.* Panicle, $\times$ $^3/_4$; *b.* Floret, $\times$ $7^1/_2$; *c, d.* Two views of spikelet, $\times$ $7^1/_2$. **4.** Panicum depauperatum; *a.* Panicle, $\times$ $^3/_4$; *b.* Floret, $\times$ $7\frac{1}{2}$. *c, d.* Two views of spikelet, $\times$ $7^1/_2$. **5.** Paspalum laeve var. circulare; *a.* Panicle, $\times$ $^3/_4$; *b.* Floret, $\times$ $7^1/_2$; *c, d.* Two views of spikelet, $\times$ $7^1/_2$. **6.** Paspalum laeve var. laeve; *a.* Plant, $\times$ $^3/_8$; *b.* Floret, $\times$ $7^1/_2$; *c, d.* Two views of spikelet, $\times$ $7^1/_2$.

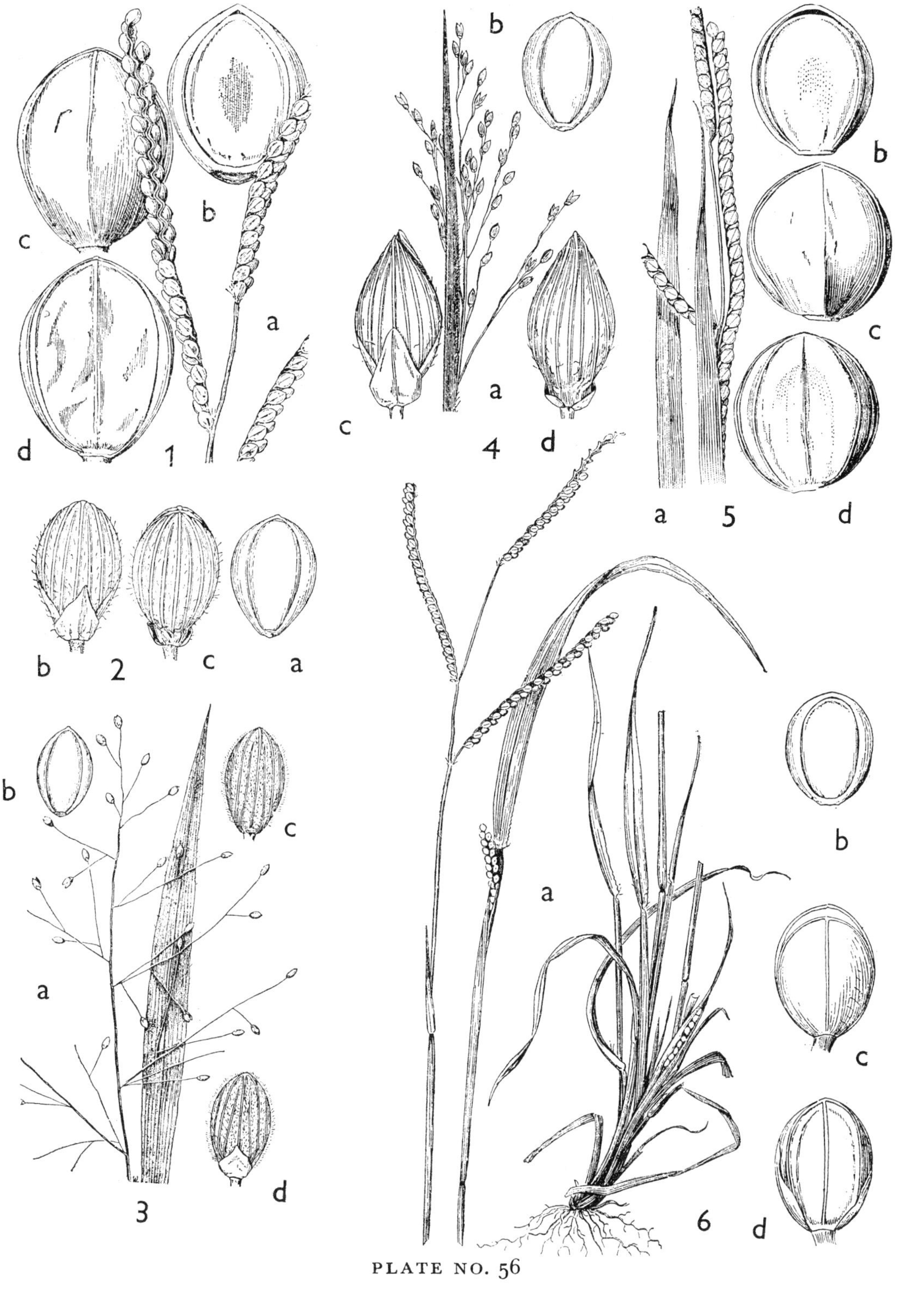


PLATE NO. 56

## 69. **Panicum** L. Panic Grass

Annual or perennial grasses, the culms (stems) usually not fracturing, the axis of the inflorescence remaining long after maturity. The 1-flowered spikelets, which articulate below the glumes, have a thin 2nd glume and sterile lemma, and a hard, firmer, thickened fertile lemma.

This is a large group, some species of which are very distinct, others quite involved and difficult to separate. Differences of 0.1–0.5 mm. in the length or width of the spikelets are important in separating a number of the species, so that great care in measuring the spikelets is especially necessary to exercise in this genus.

The genus is divided into two main groups (1) those species having the leaves at the base of the plant similar to those of the culms (stem), the plants not forming rosettes of winter leaves, and (2) those species having the leaves at the base of the plant unlike or different from those of the culm, the plants forming rosettes of winter leaves. The first group includes annuals as well as perennials, the second group perennials only. In this second group of winter-rosette forming perennials, the plants generally flower twice, at first in the spring producing the *vernal* panicles with chiefly simple unbranched culms which become much branched in summer and fall with *autumnal* panicles occurring along the sides or from the axils of the leaf-sheaths;* and, then in fall developing a rosette of leaves which are broader than those of the culm and persist over winter. Plants pertaining to the group of winter-rosette leaves are frequently collected in summer or other times of the year without these basal leaves much in evidence. This is due sometimes to the obtaining of hastily or incompletely collected specimens, and at other times due to the drying up of the lowest leaves or to collecting the plants before the rosette leaves are well developed.

a. All the spikelets densely crowded together into unbranched spike-like, erect to ascending racemes which occur alternately along the main axis of the inflorescence, the whole inflorescence giving the appearance of being pinnately arranged on 2 sides of a main axis . . . . . . . . . . . . *b*
- b. Spikelets 5–6 mm. long . . . . . . . . . . . . . . . . . . . 32. P. TEXANUM
- b. Spikelets 2.2–3.8 mm. long . . . . . . . . . . . . . . . . . . . . . *c*
  - c. Spikelets 3–3.8 mm. long, about 2 mm. wide, rounded or obtuse at tip; fruit smooth . 44. P. OBTUSUM
  - c. Spikelets 2.2–2.4 mm. long, 1.3–1.5 mm. wide, acutely pointed at tip; fruit cross-wrinkled or rugose . . . . . . . . . . . . . . . . . . . . . 31. P. GEMINATUM

a. Inflorescence (panicle) with some or all of the side branches outwardly spreading, the inflorescence as a whole with the shape either of a narrow to broad triangle, pyramid, spire, or Christmas tree, or with the lower branches often longer and more spreading, or the inflorescence as broad as or broader than long . . . . . . . . . . . . . . . . . . . . . . . . *d*
- d. Most of leaf-blades erect, crowded, and appearing to come from the base of the plant (due to the very much shortened lower internodes) . . . . . . . . . . . . . . . . . . *e*
  - e. Spikelets acutely pointed with an obvious beak at tip, mainly 3–4.5 mm. long . 1. P. DEPAUPERATUM
  - e. Spikelets mostly rounded or obtuse (blunt) at tip, not beaked, 1.7–3.5 mm. long . . . . . . *f*
    - f. Leaf-blades firm and rather thick, long, mainly 1–5 mm. wide, more than 20 times as long as wide; spikelets mainly 2–3.5 mm. long, 1.2–2 mm. wide; ligule a band of hairs 0.5–1 mm. long; culms and nodes usually pubescent . . . . . . . . . . . . . . . . *g*
      - g. Spikelets 2.7–3.5 mm. long, 1.6–2 mm. wide; branches of inflorescence erect or strongly ascending, closely appressed to the main axis of inflorescence, side branches mostly simple or little branched, the inflorescence often only 0.5–1 cm. wide; spikelets relatively few, averaging 10–20 to an inflorescence. . . . . . . . . . . . 2. P. PERLONGUM
      - g. Spikelets mainly 2–3 mm. long, 1.1–1.5 mm. wide; lower branches of inflorescence usually branched, usually spreading or ascending so that they appear separated and extending out from main axis, the inflorescence mainly 2–3 cm. wide; spikelets frequently averaging 20–35 to an inflorescence, sometimes only 15 . . . . . . . . . . . . *h*
        - h. Leaf-sheaths hairy . . . . . . . . . . 3a. P. LINEARIFOLIUM var. LINEARIFOLIUM
        - h. Leaf-sheaths entirely or mainly glabrous . . . . . 3b. P. LINEARIFOLIUM var. WERNERI
    - f. Leaf-blades thin and soft, mainly 5–12 mm. wide, mostly 10–15 times as long as wide; spikelets 1.7–2.1 (–2.6) mm. long, 0.9–1.2 mm. wide; ligule nearly obsolete, less than 0.5 mm. long; culms and lower nodes glabrous . . . . . . . . . . 4. P. LAXIFLORUM

* In the information on flowering given for each species, following the key to species, *vernal* indicates the first flowering, *autumnal* the second flowering.

d. Most of leaf-blades more or less scattered from bottom to above the middle of culm, either horizontally spreading, ascending or erect, not appearing more bunched at the base than on the culm . . *i*

i. Lemmas and glumes of spikelets themselves glabrous (but sometimes with delicate bristles at base on outside of spikelet in *P. agrostoides*); plants blooming only once during the season, all the inflorescences produced at the same period . . . . . . . . . . . . . . . . . . . . . *j*

j. Spikelets 4.5–6 mm. long . . . . . . . . . . . . . . . . . . . . . . . *k*

k. Leaf-sheaths glabrous; inflorescence as a whole erect with spreading widely spaced branches . . . . . . . . . . . . . . . . . . . . . . 39. P. VIRGATUM

k. Leaf-sheaths pubescent; inflorescence as a whole nodding or drooping with closely crowded ascending branches (not obvious in dried, pressed state where it appears erect). 38. P. MILLIACEUM

j. Spikelets less than 4.5 mm. long . . . . . . . . . . . . . . . . . . . . . *l*

l. Spikelets short-to-slender-pointed (acute to acuminate) at tip; basal leaves similar to leaves on culm (stem), all long and narrow; basal leaves often numerous, but not forming a winter rosette . . . . . . . . . . . . . . . . . . . . . . . . . . *m*

m. Most of leaf-sheath glabrous (except at very summit); nodes of culms (stems) glabrous . . *n*

n. First glume broadly triangular or rounded-deltoid, sloping to a curving or barely pointed summit, 1/5–1/3 the length of the spikelet; annuals with a tuft of roots at base; stems often conspicuously zig-zag . . . . . . . . . 33. P. DICHOTOMIFLORUM

n. First glume rather sharply pointed to a V-shaped tip, acute to acuminate, 1/3–7/8 length of spikelet; perennials; stems usually not zig-zag . . . . . . . . . . *o*

o. Culms (stems) terete (round in cross section), not compressed or flattened; leaf-sheaths rounded, not keeled; ligule densely hairy-fringed, 2–4 mm. long; spikelets appearing on different sides of the divisions of the inflorescence. . 39. P. VIRGATUM

o. Culms (stems) compressed and/or leaf-sheaths keeled; ligule minute or at most a jagged membrane 0.5–1 mm. long; spikelets either obliquely curved or turned at an angle to the stalk of spikelet or chiefly secund or nearly secund (all turned to the same side or direction) along one side of the axis of the divisions of the inflorescence . . . . . . . . . . . . . . . . . . . . . . . . . . *p*

p. Spikelets obliquely curved or turned at an angle to the pedicel (stalk) of spikelet; spikelets mainly 3–4 mm. long . . . . . . . . . . 42. P. ANCEPS

p. Spikelets straight, not curved nor turned at an angle to the pedicel; spikelets 1.7–2.8 mm. long. . . . . . . . . . . . . . . . . . . . . . *q*

q. Spikelets 2.4–2.8 mm. long; fruit with a basal stalk 0.2–0.4 mm. long. 41. P. STIPITATUM

q. Spikelets 1.7–2.4 mm. long; fruit without any basal stalk or the stalk less than 0.2 mm. long . . . . . . . . . . . . . . . . . . *r*

r. Culms wiry, slender, 1–2 mm. wide below middle; ligule minute; fruit minutely papillose-roughened; leaf-blades 1–5 (rarely 6) mm. wide; inflorescence mainly 2–12 cm. long (rarely longer) . . . . . 43. P. HIANS

r. Culms stout, 3–6 mm. wide below the middle; ligule a jagged-edged membrane 0.5–1 mm. long; fruit smooth and shining; leaf-blades mainly 5–12 (rarely 4) mm. wide; inflorescence mainly 10–40 cm. long (rarely shorter) . . . . . . . . . . . . . . . . . . 40. P. AGROSTOIDES

m. Most of leaf-sheath conspicuously hairy; nodes of culm (stem) with spreading or appressed (more or less parallel to surface) hairs . . . . . . . . . . . . . *s*

s. Spikelets chiefly secund or nearly secund (all turned to the same side or position) along one side of axis of divisions of inflorescence, obliquely turned or set at an angle to the pedicel (stalk) of spikelet; perennials with scaly creeping underground rhizomes and stolons; ligule a jagged membrane about 0.5 mm. long . . . . . 42. P. ANCEPS

s. Spikelets not turned to one side of the axis or set at an angle; annuals with a tuft of roots at base, no creeping rhizomes or stolons present; ligule a hairy fringe 1–2 mm. long . . . . . . . . . . . . . . . . . . . . . . . . . . *t*

t. Spikelets 3–4 mm. long . . . . . . . . . . . . . . . . . . . . *u*

u. Lower pulvini of inflorescence (the swellings at junction of lower branches of inflorescence with main axis) glabrous or minutely puberulent (with very short hairs); inflorescence usually less than ½ as broad as long, 2–3 times as long as thick, usually long-exserted at flowering time; leaf-blades mostly erect, mainly 2–6 (rarely 7) mm. wide; most all spikelets on pedicels (stalks) several times longer than length of spikelet; spikelets long-pointed (acuminate) at tip, lanceolate . . . . . . . . . . . . . . . . . . 34. P. FLEXILE

u. Lower pulvini of inflorescence conspicuously hispid (stiffish straight hairs); inflorescence usually as broad or broader than long, less than 2 times as long as thick, the lower branches usually enclosed in the uppermost leaf-sheath at flowering time; leaf-blades mostly spreading or drooping, mainly 5–20 mm. wide (narrower in dwarfed individuals); most of spikelets on short pedicels (stalks) shorter than or barely longer than the length of spikelet; spikelets short-acuminate at tip, lance-ovoid . . . . . . . . . . 37b. P. CAPILLARE var. OCCIDENTALE

t. Spikelets mainly 1.5–2.6 (rarely to 3) mm. long . . . . . . . . . . . . . . *v*

v. Main inflorescence usually 1/2 or more the height of the plant (except in dwarfed individuals); 1st glume definitely short-pointed (acute) or V-shaped at tip; tip of spikelet (of the lemma and 2nd glume) narrowed to a long-pointed (long-acuminate) tip; spikelet lance-ovoid . . . . . . . . . . 37a. P. CAPILLARE var. CAPILLARE

v. Main inflorescence usually not more than 1/3 the height of the plant (except in dwarfed individuals); 1st glume obtuse (blunt) to somewhat short-pointed (subacute) at tip; tip of spikelet merely short-pointed; spikelet of a broader ovoid, ellipsoid, or obovoid shape . . *w*

w. At least the lower pulvini of the inflorescence (swelling at junction of lower branches of inflorescence with main axis) short hispid, rarely glabrous; spikelets mostly occurring in 2's at the ends of the branchlets; peduncle (main supporting stalk of inflorescence) 1/3 as long as to longer than inflorescence, long-exserted . . . 36a. P. PHILADELPHICUM var. PHILADELPHICUM

w. At least the lower pulvini of the inflorescence glabrous, rarely with a few hairs; spikelets rather uniformly scattered throughout inflorescence on many of its branchlets, or racemosely arranged along the tips of most of the branchlets; peduncle about 1/4 as long as the inflorescence, the base enclosed in or shortly exserted from uppermost leaf-sheath . . . . . . . . . . . . . . . . . . . *x*

x. Spikelets 0.6–0.7 mm. wide, 1.5–2 mm. long, short-stalked and somewhat racemosely arranged along tips of most of branchlets; branches of inflorescence widely spreading, the inflorescence of a broadly ovoid to deltoid shape. 36b. P. PHILADELPHICUM var. TUCKERMANI

x. Spikelets 0.9–1.2 mm. wide, 1.8–2.5 mm. long, on conspicuous stalks and appearing regularly scattered throughout the inflorescence; branches of inflorescence spreading to ascending, the panicle of an obovoid, oval, or ellipsoid shape 35. P. GATTINGERI

l. Spikelets obtuse (blunt) or rounded at tip (very slightly pointed in *P. Bicknellii*); basal leaves of a different size and shape from those on the culm (stem), eventually becoming a winter rosette; plants blooming twice during the season, once in spring and early summer with inflorescences at the top of the culm, and later on bearing small inflorescences in the axils of the leaves, culms, or branches . . . . . . . . . . . . . . . . . . . *y*

y. Spikelets 2.9–4 mm. long, 1.5–2 mm. wide . . . . . . . . . . . 23. P. OLIGOSANTHES

y. Spikelets 1.4–2.8 mm. long, 0.7–1.2 mm. wide . . . . . . . . . . . . . . *z*

z. Spikelets 2.3–2.9 mm. long, 1.1–1.2 mm. wide; leaves 15–20 times as long as broad. 6. P. BICKNELLII

z. Spikelets 1.4–2.2 mm. long, 0.7–1 mm. wide; leaves mostly 8–12 times as long as broad . . . *1*

1. Spikelets 1.9–2.2 mm. long . . . . . . . 11a. P. DICHOTOMUM and var. BARBULATUM

1. Spikelets 1.4–1.8 mm. long . . . . . . . . . . . . . . . . *2*

2. Nodes (joints) of culm (stem) glabrous; no white spots on leaf-sheaths; ligule a brush of hairs 1.5–3 mm. long . . . . . . . . . . . . . . 12. P. SPRETUM

2. Nodes bearded; white spots on leaf-sheaths; ligule practically lacking . 8. P. MICROCARPON

i. Lemmas and glumes of spikelets themselves short- to long-hairy . . . . . . . . . . *3*

3. Spikelets 5–6 mm. long, sharp- or slender-pointed at tip; fruit cross-wrinkled or lined; inflorescence of several or more densely-flowered racemes . . . . . . . . . . . 32. P. TEXANUM

3. Spikelets 4.5 mm. or less long (rarely longer in *P. Boscii*), rounded or obtuse (blunt) at tip, never slender-pointed; fruit not as above; inflorescence not as above . . . . . . . . . . *4*

4. Ligule a conspicuous brush of stiff white hairs 2–5 mm. long protruding at summit of leaf-sheath (look for these best developed on main cauline [stem] leaves) . . . . . . . . . . *5*

5. Spikelets mostly 4–4.3 mm. long (rarely shorter), 2–2.2 mm. broad . . . . 24. P. RAVENELII

5. Spikelets 1.1–3.9 mm. long, 0.7–2 mm. broad . . . . . . . . . . . . . . *6*

6. Leaf-sheaths, culms (stems) and their nodes glabrous . . . . . . . . . . . *7*

7. Inflorescence narrowly ellipsoid or narrowly ovoid, 1/4–1/2 as wide as long; branches of inflorescence closely ascending; many or most of the lateral spikelets equaling or longer than their pedicels (stalks) . . . . . . . . . . . 12. P. SPRETUM

7. Inflorescence rhomboid (baseball-diamond-shaped) to broadly ovoid in outline, almost as wide as or as wide as long; branches of inflorescence widely spreading to ascending; most of the spikelets shorter than their pedicels . . . . . . . . . . . . . . . . . . . . *8*
8. Spikelets 1.3–2 mm. long . . . . . . . . . . . 15a. P. LANUGINOSUM var. LINDHEIMERI
8. Spikelets 1.1–1.2 mm. long . . . . . . . . . . . . . . 13. P. LONGILIGULATUM
6. Some or all of the leaf-sheaths, culms (stems) and/or their nodes pubescent . . . . . . . . *9*
9. Culms (stems), their nodes, and upper leaf-sheaths with all the hairs conspicuously long (mostly 2–6 mm. long) and horizontally spreading . . . . . . . . . . . . . *10*
10. Spikelets mainly 2.6–2.8 mm. (extremes 2.3–3 mm.) long; all leaf-blades or at least the lower ones covered with a soft (to the touch) dense velvety hairiness . . 5. P. CONSANGUINEUM
10. Spikelets 1.3–2.5 mm. long; leaf-blades, if hairy, not with a soft velvety type of hair . . . *11*
11. Spikelets 2–2.5 mm. long, 1.1–1.2 mm. wide . . . . . . . . 18a. P. VILLOSISSIMUM
11. Spikelets 1.3–2.1 mm. long, 0.8–1 mm. wide . . . . . . . . . . . . . . . *12*
12. First glume 1/6–1/3 length of spikelet, either rounded, broadly obtuse, or subtruncate at apex. . . . . . . . . . . . . . . . . . 15. P. LANUGINOSUM
12. First glume 1/3–1/2 length of spikelet, elongate-triangular, more or less acutely pointed . . . . . . . . . . . . . . . . . . . . . . . 16. P. PRAECOCIUS
9. Culms (stems), their nodes, and upper leaf-sheaths with shorter ascending or appressed (parallel to surface) hairs (mostly less than 2 mm. long), or, if some of the hairs spreading, then mostly less than 2 mm. long . . . . . . . . . . . . . . . . . . . . . . *13*
13. Spikelets mainly 2.6–2.8 mm. (extremes 2.3–3 mm.) long; all leaf-blades or at least the lower ones covered with a soft (to the touch) dense velvety hairiness . . 5. P. CONSANGUMINEUM
13. Spikelets 1.3–2.5 mm. long; leaf-blades hairy, but not soft velvety to the touch . . . . *14*
14. Spikelets 2–2.5 mm. long, 1.1–1.2 mm. wide . . 18b. P. VILLOSISSIMUM var. PSEUDOPUBESCENS
14. Spikelets 1.3–2.1 mm. long, 0.8–1 mm. wide . . . . . . . . . . . . . . *15*
15. Axis of inflorescence mostly glabrous or minutely puberulent (with short powdery-like pubescence); lower side (secondary) branches of panicle mainly simple (unbranched) . . . . . . . . . . . . . . . . . 14. P. MERIDIONALE
15. At least the lower part of the axis of the inflorescence more or less hairy with obvious hairs, more rarely without hairs; lower side branches of panicle usually forking or branching . . . . . . . . . . . . . . . . . . . . . *16*
16. First glume of spikelet 1/3–1/2 length of spikelet, as long as or longer than broad, rather acutely pointed; at least lower part of axis of inflorescence with mainly appressed pubescence (parallel to length of or pressed against axis), rarely without hairs; main leaf-sheaths of main stem usually equaling or longer than the internodes. . . . . . . . . . . 17. P. SUBVILLOSUM
16. First glume of spikelet 1/6–1/3 (usually 1/4) length of spikelet, angularly rounded, subtruncate, or broadly obtuse at tip, broader than long; at least lower part of axis of inflorescence with usually spreading hairs; main leaf-sheaths of main stem usually shorter than the internodes . . 15. P. LANUGINOSUM
4. Ligule either none or minute and nearly obsolete, or a brush of hairs less than 2 mm. long . . . . *17*
17. Culm (stem) and leaf-sheaths covered, with a close, grayish, soft velvety (to the touch) hairiness; a hairless (glabrous) sticky band present on the culm below each bearded node; culms stout, 5 mm. or more thick . . . . . . . . . . . . . . . . . . . 26. P. SCOPARIUM
17. Without the above combination of characters . . . . . . . . . . . . . . *18*
18. Nodes (joints) of culm (stem) encircled by a conspicuous beard of spreading or down-pointing hairs (densely hairy lower sheaths or sparsely hairy nodes found in *P. latifolium* sometimes confused here) . . . . . . . . . . . . . . . . . . . . *19*
19. Spikelets 3.2–5.2 mm. long, 1.8–2.2 mm. wide; leaf-blades mostly 15–40 mm. wide (rarely less) . . . . . . . . . . . . . . . . . . . . . . *20*
20. Spikelets 3.4–3.7 (rarely 3.2) mm. long; nodes (joints) of culm (stem) with short hairs less than 2 mm. long; leaf-blades mainly 2½–5 times as long as broad 29. P. LATIFOLIUM
20. Spikelets 3.7–5.2 mm. long; nodes (joints) of culm (stem) long-bearded with hairs 2 mm. or more long; leaf-blades mainly 4–7 times as long as broad . . 30. P. BOSCII
19. Spikelets 1.5–3.3 mm. (rarely up to 3.8 mm.) long, 0.7–1.8 mm. wide; leaf-blades mainly 3–13 mm. wide . . . . . . . . . . . . . . . . . . . . . *21*
21. Long hairs appearing on most of the margin of the leaf-blades; leaf-blades pale green, soft, most of them in tufts crowded toward base of plant and long and narrow (linear-lanceolate to

narrowly lanceolate); winter rosette-leaves not forming nor conspicuous; culms (stems) remaining unbranched or forking only at base . . . . . . . . . . . 4. P. LAXIFLORUM

21. Without the above combination; any long marginal hairs, if present on leaf-blades, confined near base of blade; leaf-blades dull to deep green, rather uniformly scattered or crowded on stem, not crowded toward base of plant; true winter-rosette of short leaves eventually forming or present; culms eventually forking or branching from the middle or upper nodes late in the summer and fall (spring-collected specimens may not show any branching) . . . . . . . *22*

22. Spikelets 2.9–3.8 mm. long . . . . . . . . . . . . . . . . . . . . . . . . . *23*

23. Spikelets 1.1–1.4 mm. wide; ligule inconspicuous or obsolete; leaf-blades not velvety-hairy; culms at most with appressed hairiness; axis and branches of inflorescence glabrous or sparsely hairy, without conspicuous spreading hairiness. . 7. P. CALLIPHYLLUM

23. Spikelets 1.5–1.8 mm. wide; ligule conspicuous, 1–1.5 mm. long; leaf-blades velvety-hairy; culms, axis, and branches of inflorescence with conspicuous spreading hairiness. 22. P. MALACOPHYLLUM

22. Spikelets 1.5–2.8 mm. long . . . . . . . . . . . . . . . . . . . . . . . . . *24*

24. All leaf-blades or at least the lower ones covered with a soft (to the touch) dense velvety hairiness; leaf-sheaths mainly hairy. . . . . . . . . . . . . . . . . . . . *25*

25. Main or largest leaf-blades averaging about 15 times as long as wide, 3–8 mm. wide, on the late summer and autumn plants 2–3 mm. wide; spikelets mainly 2.6–2.8 mm. long (extremes 2.3–3 mm.) . . . . . . . . . . . . . . 5. P. CONSANGUINEUM

25. Main or largest leaf-blades averaging about 10 times as long as wide, 7–13 mm. wide; spikelets 2–2.4 mm. (rarely 1.8 mm.) long . . . . . . . . 10. P. ANNULUM

24. Leaf-blades mainly glabrous, at most with some hairs on margin near base of blade; leaf-sheaths mainly glabrous. . . . . . . . . . . . . . . . . . . . . *26*

26. Spikelets 0.7 mm. wide, mostly 1.5–1.8 mm. long; plants of wet habitats 8. P. MICROCARPON

26. Spikelets 0.9–1 mm. wide, mainly 1.8–2.2 mm. (rarely 1.7) long; plants of wet or dry habitats . . . . . . . . . . . . . . . . . . . . . . . . . *27*

27. Common plants of dry or rocky woods of the Ozark region; upper leaf-sheaths not dotted with pale spots; spikelets rarely with hairs, mostly glabrous; nodes of culms lightly bearded . . . . . . 11b. P. DICHOTOMUM var. BARBULATUM

27. Rare plant of moist gravelly ground in Carter County, southeastern Mo.: upper leaf-sheaths usually dotted with pale spots; spikelets always pubescent; nodes of culm heavily bearded . . . . . . . . . . . 9. P. NITIDUM

18. Nodes (joints) of culm (stem) glabrous or with minute hairs or with hairs appressed or ascending, but not encircled by a conspicuous beard of spreading or down-pointing hairs (do not mistake sometimes densely hairy lower part of leaf-sheaths for nodes) . . . . . . . . . . . . *28*

28. Most of the leaf-sheaths (but especially the upper ones) or culms mainly hairy (older specimens of *P. clandestinum* may lose main slender part of hairs but papillate stubs at base of original hairs remain on surface of sheath); the hairs may be very minute and crisp-puberulent in *P. commutatum* var. *Ashei* . . . . . . . . . . . . . . . . . . . . . . . . *29*

29. Spikelets 1.2–2 mm. long . . . . . . . . . . . . . 19. P. COLUMBIANUM

29. Spikelets 2.5–4 mm. long . . . . . . . . . . . . . . . . . *30*

30. Principal leaf-blades mainly 17–35 mm. wide . . . . . . . . . 28. P. CLANDESTINUM

30. Principal leaf-blades mainly 5–15 mm. wide . . . . . . . . . . . . . . *31*

31. Spikelets 2.1–2.7 mm. long; pubescence of culms or leaf-sheaths minute (crisp-puberulent) . . . . . . . . . . . . . . . 27b. P. COMMUTATUM var. ASHEI

31. Spikelets 2.9–4 mm. long; pubescence of culms or leaf-sheaths of easily seen hairs . *32*

32. 1st glume ½ or more the length of the spikelet, narrowly ovate; spikelets long-hairy with hairs up to 1 mm. long; leaf-blades thin, often with papillose-hispid hairs above and below; ligule practically obsolete, less than 0.5 mm. long. 25. P. LEIBERGII

32. 1st glume less than ½ the length of the spikelet, broadly ovate; spikelets glabrous to short-hairy; leaf-blades thick and firm, glabrous or sparsely hairy above, glabrous or finely hairy with usually short appressed hairs below; ligule mostly a brush of hairs 1–1.5 mm. long . . . . . . 23. P. OLIGOSANTHES

28. Most of the leaf-sheaths, but especially the upper ones, glabrous (except for hairs which may be on the margins). . . . . . . . . . . . . . . . . . . . . . . . *33*

33. Spikelets 1.5–1.8 mm. long . . . . . . . . . . . . . . . . . . . . . . *34*

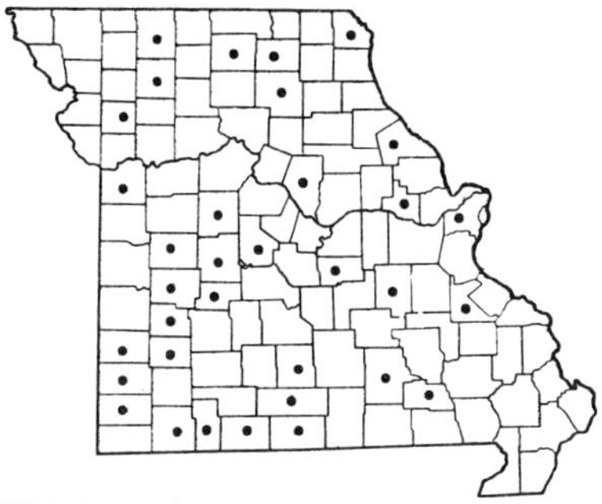

297 Panicum perlongum

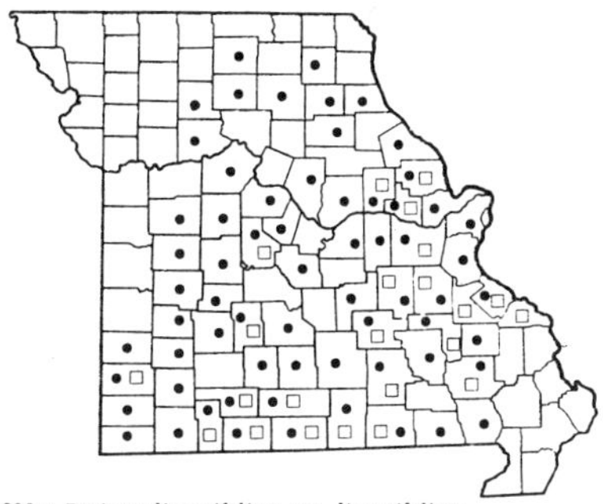

298 • Panicum linearifolium var. linearifolium
298 □ Panicum linearifolium var. Werneri

299 Panicum laxiflorum
... 300 Excluded species

34. Uppermost main leaf-blade of culm 3–15 mm. (mainly 5–10) wide, 2–11 cm. (mainly 4–9) long; main leaf-blades of the culm more or less of the same or similar size . . . . 20. P. SPHAEROCARPUM
34. Uppermost main leaf-blade of culm 10–30 mm. (mainly 15–25) wide, 8–20 cm. (mainly 10–15) long; upper 3 leaf-blades of the culm obviously longer than the ones below . . 21. P. POLYANTHES
33. Spikelets 1.9–3.7 mm. long . . . . . . . . . . . . . . . . . . . . . *35*
35. Leaves mainly 15–40 mm. (averaging 20–30) wide; spikelets mostly 3.4–3.7 mm. long (rarely shorter), 1.8–2 mm. wide; ligule about 1 mm. long . . . . . . . . . 29. P. LATIFOLIUM
35. Leaves 4–25 mm. wide; spikelets 1.9–3.2 long, 0.9–1.3 mm. wide; ligule practically obsolete or absent . . . . . . . . . . . . . . . . . . . . . . . . *36*
36. Main leaves with a base heart-shaped or somewhat clasping the stem (subamplexicaul), broadly lanceolate, much broader near the base . . . . . . . . 27a. P. COMMUTATUM
36. Main leaves long and narrow, linear or lanceolate, not much, if at all, broader at the base, the base not heart-shaped nor somewhat clasping the stem . . . . . . . . . . *37*
37. Spikelets 1.9–2.2 mm. long, 0.9–1 mm. wide; late summer and autumn stage tree-like in habit, much-branched above middle with numerous narrow leaves often only 3–4 mm. wide and no or few leaves in lower half of plant . . . . . . . . 11a. P. DICHOTOMUM
37. Spikelets 2.3–3.2 (rarely to 3.8) mm. long, 1.1–1.6 mm. wide; late summer and autumn stage with culms less or slightly branching and leaf-blades not much different from earlier spring stage, sometimes becoming bushy-crowned above the middle . . . . *38*
38. Main leaf-blades 4–8 mm. wide, about 15 times as long as wide; spikelets 2.3–2.8 mm. long . . . . . . . . . . . . . . . . . . 6. P. BICKNELLII
38. Main leaf-blades 9–15 mm. wide, about 10 times as long as wide; spikelets 2.9–3.2 mm. long . . . . . . . . . . . . . . . . . . 7. P. CALLIPHYLLUM

1. **Panicum depauperatum** Muhl. var. **depauperatum** Map 296

Flowers May–August (vernal); July–October (autumnal).

Occurs in prairies, glades, sterile open upland woods, open waste ground, and along roadsides, in the eastern half and southern third of the state west to Sullivan, Linn, Pettis, and Barton counties. Absent from the southeastern lowland counties.

Ranges from South Carolina and Georgia to Texas, north to Quebec, Ontario, and Minnesota.

The more northern var. *psilophyllum* Fern., with glabrous or sparsely pilose leaf-sheaths, has not been found in the state.

2. **Panicum perlongum** Nash Map 297

Flowers May–August (vernal); July–October (autumnal). Occurs in prairies, glades, prairie openings of upland woods, usually in acid soils associated with chert, sandstone, or granite. Throughout most of the state, except apparently absent in the southeastern lowlands and extreme northwestern counties.

Ranges from Michigan to Manitoba, south to Indiana, Illinois, Arkansas, and Texas.

The spikelets are intermediate in length between those of *P. depauperatum* and *P. linearifolium*, and, in fact, the length of spikelet measurements overlaps with that of *P. linearifolium*, which it closely resembles. Some authors have attempted to separate *P. perlongum* from *P. linearifolium* on the basis of pedicel length, with the pedicel given as shorter than the spikelets in *P. perlongum* and longer than the spikelets in *P. linearifolium*. This character does not hold at all in Missouri material, where most of the pedicels in both species are

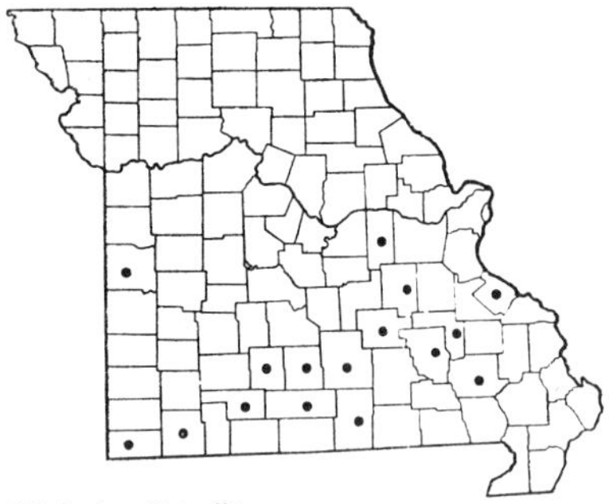

301 Panicum Bicknellii

302 Panicum calliphyllum

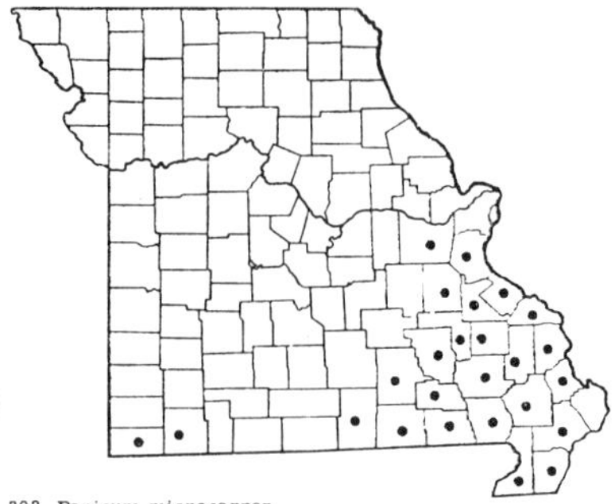

303 Panicum microcarpon

much longer than the spikelets; nor can these two species be separated on the length of the ligule.

3a. **Panicum linearifolium** Scribn. var. **linearifolium** Map 298

Flowers May–July (vernal); July–October (autumnal).

Occurs in prairies, rocky glades, open places along roads, and open woods in usually acid soils. Throughout the Ozarks and Ozark border region; absent from the southeastern lowlands, northwestern, west-central, and extreme northern counties.

Ranges from Georgia to Texas, north to Quebec, Ontario, Michigan, Wisconsin, and Minnesota.

3b. **Panicum linearifolium** var. **Werneri** (Scribn.) Fern. Map 298

In similar habitats to var. *linearifolium*, but most frequent in rocky open woodland in acid soils associated with sandstone, chert, or granite. Limited to the Ozark region, north to Lincoln, Montgomery, Morgan, and Jasper counties.

Ranges from Quebec to Minnesota, south to Virginia, Kentucky, Arkansas, and Texas.

This variety with mainly glabrous leaf-sheaths is usually readily distinguishable where encountered in Missouri.

Most of the spikelets of *P. linearifolium* range in length from 2–2.7 mm. long, but specimens up to 3 mm. long occur, thus intergrading into the minimum spikelet length of *P. perlongum*. The spikelets of *P. linearifolium* are always narrower, however. The panicle of *P. perlongum* is usually much narrower relatively than that of *P. linearifolium*, averaging usually 1/6–1/3 as broad as long, whereas in *P. linearifolium* it averages 1/2–3/4 as broad as long, but there are also intergrading forms. Usually the spikelets are more closely appressed to the main axis of the panicle in *P. perlongum* than in *P. linearifolium*, with more erect appressed simple branches.

4. **Panicum laxiflorum** Lam. Map 299

*Panicum xalapense* HBK. [P & S]

Flowers April–July (vernal); June–September (autumnal).

Occurs in low or upland woodland, and wet rocky or sandy banks, usually in acid soils associated with sandstone, chert, or granite. Eastern and southern Ozark region, north to St. Louis, Warren, and Boone counties, west to Osage, Maries, Texas, Christian, and McDonald counties.

Ranges from Maryland to Indiana, southern Illinois, Missouri, and Oklahoma, south to Florida, and Texas, Mexico, Central America, and the West Indies.

This species is easily recognized by the leaf-blades, which are pale green, soft, elongated, ciliate-margined, and mostly crowded toward the base, forming loose clumps. Sterile tufts of these long, narrow, hairy blades are frequently encountered in oak-hickory and pine-oak woodland and are readily identifiable.

5. **Panicum consanguineum** Kunth

See excluded species.

6. **Panicum Bicknellii** Nash Map 301

*Panicum Bicknellii* var. *Bicknellii* [BB]

Flowers May–July (vernal); July–September (autumnal).

Occurs in dry rocky oak-hickory or pine-oak woodland in acid soils associated with chert, sandstone, or granite, more rarely in prairie openings or prairies. Infrequent in the southern Ozark region of Missouri, and locally north in Gasconade and Bates counties.

Ranges from Quebec to Michigan, south to Georgia and Missouri.

This well-marked species has narrow, long-pointed, tapering leaf-blades which become gradually narrowed at the base. The spikelets vary from somewhat pubescent to rarely glabrous.

304 Panicum nitidum

305 Panicum annulum

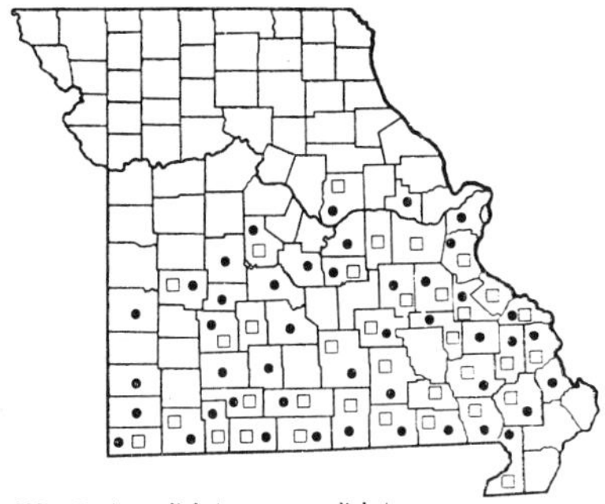

306 • Panicum dichotomum var. dichotomum
306 □ Panicum dichotomum var. barbulatum

7. **Panicum calliphyllum** Ashe Map 302
*Panicum Bicknellii* var. *calliphyllum* (Ashe) Gl. [BB]

Flowers June–July (vernal); July–September (autumnal).

Occurs in oak-hickory and pine-oak woodland, where known from two isolated localities in the Ozarks, in Callaway (*Steyermark 65453*) and Ripley (upper cherty slopes along Current River, 7 mi. northwest of Doniphan, October 19, 1936, *Steyermark 20507*) counties. The specimen cited from Callaway County has spikelets up to 3.8 mm. long. It may possibly be an atypical *P. oligosanthes* having very narrow spikelets.

Ranges from Maine to Ontario, south to Connecticut, Pennsylvania, Ohio, Michigan, and Missouri.

8. **Panicum microcarpon** Muhl. Map 303
*Panicum nitidum* Lam. var. *ramulosum* Torr. [BB]

Flowers May–September (vernal); July–October (autumnal).

Occurs in low wet woodland, margins of small streams and spring branches and low moist open ground, in the eastern quarter of the state south of the Missouri River, and locally in Barry and McDonald counties, southwestern Missouri. In the eastern sector of its Missouri range it inhabits small wooded streams and spring branches, often in sandy soil, in the Ozarks, and in wet woodlands and swampy ground of the southeastern lowland.

Ranges from Florida to Texas, north to Massachusetts, New York, Ohio, Michigan, Illinois, Missouri, and Oklahoma.

The spikelets are generally glabrous, rarely puberulent. In the autumnal stage the plants become conspicuously branched and sprawling, forming dense masses. The thin leaf-blades are usually glabrous, rarely pubescent.

9. **Panicum nitidum** Lam. Map 304
*Panicum nitidum* var. *nitidum* [BB]

Flowers May–June (vernal); July–October (autumnal).

Known only from Carter County, southeastern Missouri (gravelly ground, June 6, 1893, *Eggert 288*). This collection is in the United States National Museum, and is cited by Hitchcock and Chase (Contr. U. S. Nat. Herb. 15: 184. 1910).

Ranges from Florida to Texas, locally north to Virginia and southeast Missouri; also in the West Indies, Mexico, and Guatemala.

10. **Panicum annulum** Ashe Map 305
*Panicum annulum* var. *annulum* [BB]

Flowers late May–August (vernal); July–October (autumnal).

Known only from Carter County, southeastern Missouri (woods near Hunter, June 6 , 1893, *Eggert*).

Ranges from Georgia to Mississippi, north locally to Massachusetts, New Jersey, Pennsylvania, Tennessee, and southeastern Missouri.

11a. **Panicum dichotomum** L. var. **dichotomum** Map 306
*Panicum dichotomum* L. [G, BB in part, P & S]

Flowers May–July (vernal); June–November (autumnal).

Usually in dry, rocky open oak-hickory or pine-oak woodlands, on ridges or slopes of ravines in acid soils associated with sandstone, chert, or granite, less frequently in low rich woodlands. Common throughout the Ozark section, but confined to it.

Ranges from New Brunswick and Maine to Ontario, Michigan, Illinois, and Missouri, south to Florida and Texas.

11b. **Panicum dichotomum** var. **barbulatum** (Michx.) Wood Map 306
*Panicum barbulatum* [P & S]
*Panicum dichotomum* in part [BB]

307 Panicum longiligulatum

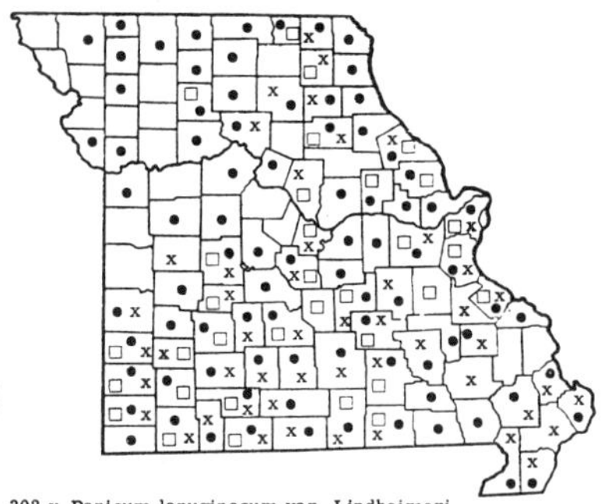

308 x Panicum lanuginosum var. Lindheimeri
308 • Panicum lanuginosum var. fasciculatum
308 □ Panicum lanuginosum var. implicatum

309 Panicum praecocius

Occurs in the same situations as var. *dichotomum* with similar range in the Ozark section.

Ranges from Georgia to Texas, north to Massachusetts, New York, Ohio, Michigan, Illinois, and Missouri.

*Panicum dichotomum* var. *dichotomum* typically has beardless nodes of the culms and, at maturity, the fruit is usually shorter than the 2nd glume and sterile lemma, as contrasted to *P. dichotomum* var. *barbulatum* in which the nodes (usually the lower) of the culm are bearded and, at maturity, the fruit is usually covered by the longer 2nd glume and sterile lemma.

Frequent intergradation is found, however, plants occurring with the lower nodes bearded but with the fruit not covered, or with the lower nodes beardless and the fruit covered. It is questionable whether the two varieties can be maintained or will have to be united under *P. dichotomum*, as some authors have done.

The treelike branched appearance with very slender culms and narrow leaves of the autumnal phase is characteristic of these two varieties.

12. **Panicum spretum** Schultes
See excluded species.

13. **Panicum longiligulatum** Nash Map 307
*Panicum lanuginosum* var. *Lindheimeri* in part [BB]

Flowers May–September (vernal); June–October (autumnal).

Known from one collection from Newton County, southwestern Missouri (open banks of an artificial pond, along highway 86, about 7 mi. east of Neosho, August 2, 1956, *Palmer 63506*).

Ranges from the West Indies and Central America (British Honduras), north to Florida and Louisiana, locally north to Missouri, Tennessee, and southeastern Pennsylvania.

It is doubtful whether this species can be maintained as distinct from *P. lanuginosum* var. *Lindheimeri* (Nash) Fern. The species is included for Missouri in the present treatment only on the basis of the identification made at the United States National Museum. Palmer's specimen is not typical, however, as most of the spikelets examined in his collection measure as much as 1.5 mm. long, thus placing it in *P. lanuginosum* var. *Lindheimeri* (as treated by Fernald and others).

14. **Panicum meridionale** Ashe
See excluded species.

15. **Panicum lanuginosum** Ell. Map 308

Flowers May–September (vernal); June–November (autumnal).

Missouri material is placed here under the following categories:

a. Axis of inflorescence mainly glabrous or at most with a few appressed hairs; leaf-blades mainly glabrous; upper leaf-sheaths mainly glabrous, sometimes sparsely hairy . . . 15a. P. LANUGINOSUM var. LINDHEIMERI
a. Axis of inflorescence, or at least the lowest part of it, pubescent with spreading hairs;

Plate no. 57. **1.** Panicum microcarpon; *a.* Plant, × 3/4; *b.* Floret, × 7 1/2; *c, d.* Two views of spikelet, × 7 1/2. **2.** Panicum consanguineum; *a.* Floret, × 7 1/2; *b, c.* Two views of spikelet, × 7 1/2. **3.** Panicum annulum; *a.* Floret, × 7 1/2; *b, c.* Two views of spikelet, × 7 1/2. **4.** Panicum nitidum; *a.* Floret, × 7 1/2; *b, c.* Two views of spikelet, × 7 1/2. **5.** Panicum longiligulatum; *a, b.* Two views of spikelet, × 7 1/2; *c.* Floret, × 7 1/2. **6.** Panicum calliphyllum; *a, b.* Two views of spikelet, × 7 1/2; *c.* Floret, × 7 1/2. **7.** Panicum dichotomum var. dichotomum; Plant, × 3/8; *b, c.* Two views of spikelet, × 7 1/2; *d.* Floret, × 7 1/2. **8.** Panicum dichotomum var. barbulatum; *a, b.* Two views of spikelet, × 7 1/2; *c.* Floret, × 7 1/2. **9.** Panicum Bicknellii; *a.* Plant, × 3/4; *b.* Floret, × 7 1/2; *c, d.* Two views of floret, × 7 1/2. **10.** Panicum linearifolium; *a, b.* Two views of spikelet, × 7 1/2; *c.* Floret, × 7 1/2.

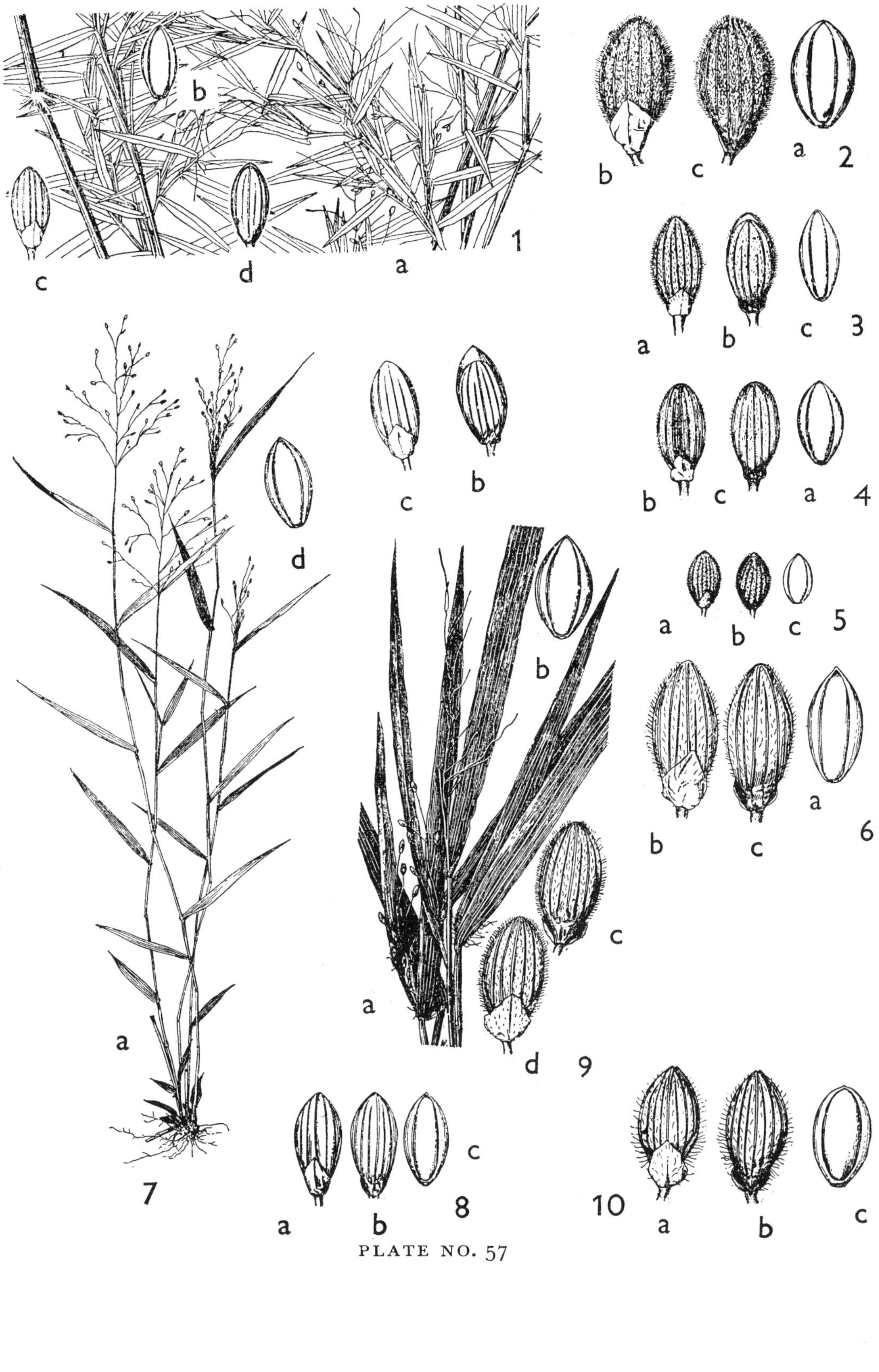


PLATE NO. 57

leaf-blades hairy or glabrous (without hairs); upper leaf-sheaths mainly hairy . . . . *b*

b. Hairs none or up to 2.9 mm. long on upper surface of leaf-blades; spikelets usually 1.6–2.1 mm. long . . 15c. P. LANUGINOSUM var. FASCICULATUM

b. Hairs mainly 3–6 mm. long on upper surface of leaf-blades; spikelets usually 1.3–1.5 mm. long. . . . 15b. P. LANUGINOSUM var. IMPLICATUM

15a. **Panicum lanuginosum** var. **Lindheimeri** (Nash) Fern. Map 308

*Panicum Lindheimeri* Nash [Hitchcock, Fassett]

*Panicum lanuginosum* var. *septentrionale* Fern. [P & S]

Usually frequents moist ground, either along margins of streams and spring branches, moist limestone and sandstone bluffs and ledges, and limestone glades, or wet meadows and prairie swales; occasionally along railroads. Mainly in the eastern and southern half of the state; not recorded from the northwestern third of the state.

Ranges from Nova Scotia and New Brunswick to Manitoba and Minnesota, south to Florida, and Texas: also Arizona and California.

This variety is most often found in wet ground of low meadows and prairie swales or moist limestone rocks along streams, where it may frequent the ledges or the limestone glades above the stream.

15b. **Panicum lanuginosum** var. **implicatum** (Scribn.) Fern. Map 308

*Panicum implicatum* Scribn. [Hitchcock, Fassett]

Occurs in dry rocky woodland, often on cherty or sandstone slopes, sandy alluvial soils in valleys, less frequently in prairie openings or along railroads. Mostly found in the Ozark section and locally northward in the state; apparently absent from the southeastern lowland and extreme northwest sectors.

Ranges from Newfoundland to Ontario and Minnesota, south to District of Columbia, Pennsylvania, Ohio, Indiana, Illinois, and Missouri.

This is the least frequently encountered of the three varieties here treated.

A number of specimens labeled var. *implicatum* possess spikelets more than 1.5 mm. long and fall in the range of var. *fasciculatum*. Such specimens combine the characters of both varieties, making it arbitrary as to which name they should take.

15c. **Panicum lanuginosum** var. **fasciculatum** (Torr.) Fern. Map 308

*Panicum lanuginosum* var. *tennesseense* (Ashe) Gl. [BB]

*Panicum implicatum* in part [Fassett], not Scribn.

*Panicum tennesseense* Ashe [Hitchcock]

*Panicum huachucae* Ashe [Hitchcock]

Occurs in many types of habitats throughout the state, where found on prairies, limestone or other rocky substrata and glades, rocky slopes along streams, ravines, small valleys, dry open woods, meadows, pastures, waste ground, and along railroads. Common throughout Missouri, and the commonest variation encountered.

Ranges from Newfoundland to Quebec, Ontario, Minnesota, South Dakota, Montana, Idaho, and British Columbia, south to Florida and California.

As may be judged by the synonymy under each of the above varieties, there is at present no agreement among botanists as to the splitting up of the *Panicum lanuginosum–P. Lindheimeri* complex. What the future consensus of botanical opinion will be with respect to this group is very uncertain. Some authors (Fassett and others) maintain as distinct species *P. Lindheimeri* for plants having leaf-sheaths glabrous or nearly so (including both *P. lanuginosum* var. *Lindheimeri* and var. *septentrionale*) and *P. implicatum* for those with hairy leaf-sheaths (including here *P. lanuginosum* var. *fasciculatum* and var. *implicatum*, *P. huachucae*, and *P. tennesseense*). Though this procedure has its merits, to follow it would indicate little or no overlapping between *P. implicatum* and *P. Lindheimeri*. This is not the case, however; although the two are quite distinct as extremes, there are too many intergradations in the range and degree of pubescence of leaf-blades and leaf-sheaths. For the same reason, it seems impossible to maintain as distinct species those plants previously named *Panicum tennesseense*, *P. huachucae*, *P. implicatum*, and *P. Lindheimeri*.

The other method of treating the above entities as varieties within one group (*P. lanuginosum*) is the procedure followed by Fernald and by Gleason, and to a lesser extent in the present flora. Gleason (*New Ill. Fl.* 1: 221. 1952) has mainly followed Fernald's treatment in the eighth edition of *Gray's Manual* (p. 215), but has divided the group to a greater extent than has Fernald, emphasizing papillose and nonpapillose types of hairs.

In the present treatment, *P. lanuginosum* var. *septentrionale*, which has been segregated from var. *Lindheimeri* on the basis of the larger spikelets (1.6–2 mm. instead of 1.3–1.6 mm. long), is placed under *P. lanuginosum* var. *Lindheimeri* as one inclusive variety with spikelets varying from 1.3–2 mm. in length, whereas variations with both types of papillose and nonpapillose hairs are included under var. *fasciculatum*.

The majority of specimens encountered in Missouri can be placed satisfactorily in one of the three

310 Panicum subvillosum

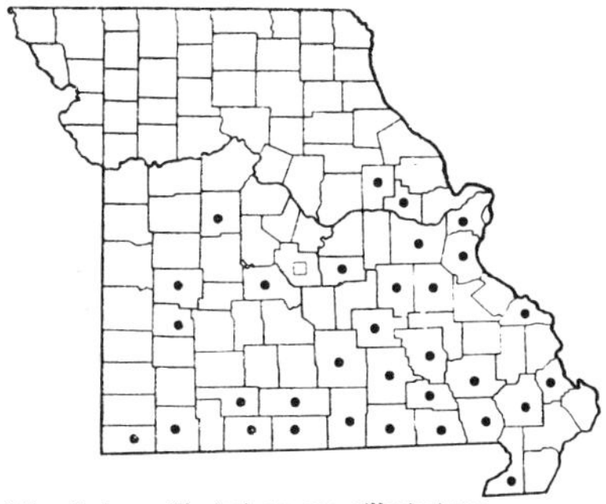

311 ● Panicum villosissimum var. villosissimum
311 □ Panicum villosissimum var. pseudopubescens

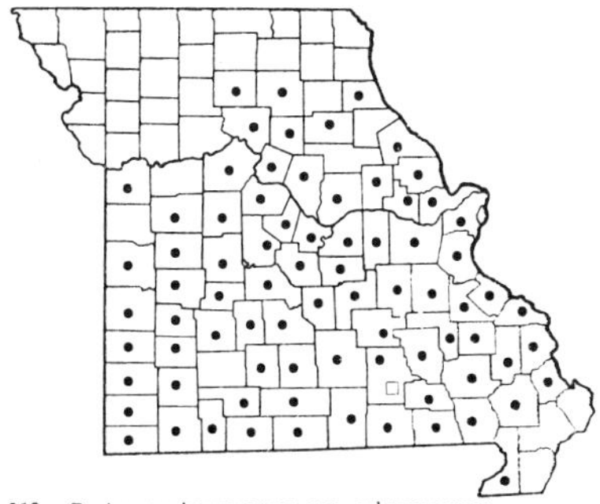

312 ● Panicum sphaerocarpon var. sphaerocarpon
312 □ Panicum sphaerocarpon var. inflatum

varieties treated in the present flora. However, an occasional one offers perplexities. An example is *Palmer 65623* (upland prairie, 4½ mi. northwest of Diamond, Newton Co., June 17, 1957) with spikelets only 1.3–1.5 mm. long, but with a dense ascending pubescence of leaf-sheaths, and both short to long hairs on upper surface of the leaf blades. While showing some characteristics of *P. lanuginosum* var. *lanuginosum*, the short spikelets and occasional long hairs (over 3 mm. long) on the upper leaf surface indicate a closer relationship with var. *implicatum*.

16. **Panicum praecocius** Hitchc. & Chase Map 309

Flowers May–July (vernal); June–September (autumnal).

Occurs in dry open woods, prairies, glades, and thickets. Widespread in Missouri.

Ranges from Ontario and Michigan to Minnesota and Nebraska, south to Illinois, Missouri, and Texas.

This species resembles *P. lanuginosum* var. *implicatum* and *P. subvillosum* in having the hairs on the upper surface of the leaf-blades mostly 3–6 mm. long. From the former it differs in the longer spikelets and longer, more acutely pointed first glume, and from the latter in the longer, more widely spreading hairs on the leaf-sheaths and on most of the culms.

The culms of this species often branch before the expansion of the first main panicle and develop a geniculate habit (bent abruptly at the nodes of the culm) as contrasted with *P. subvillosum* and *P. lanuginosum* var. *implicatum*.

17. **Panicum subvillosum** Ashe Map 310

Flowers May–August (vernal); July–November (autumnal).

Occurs mainly in dry rocky woods. Known only from southwestern Missouri, in Jasper, Newton, and McDonald (Noel, May 26, 1909, *Bush 5741, 5741A*, and *5741B*) counties.

Ranges from Nova Scotia to Saskatchewan, south to Massachusetts, New York, Pennsylvania, northern Indiana, and Missouri.

Bush's specimens were identified as *P. meridionale* Ashe by Gleason and form the basis for the inclusion of that species from Missouri in the *New Britton and Brown Illustrated Flora* (for which see under excluded species, p. 232).

18a. **Panicum villosissimum** Nash var. **villosissimum** Map 311

Flowers May–July (vernal); June–October (autumnal).

Occurs mainly in dry open or rocky thin oak-pine or oak-hickory woodland on acid soils associated with sandstone, chert, or granite, throughout the Ozark section north to St. Louis, Montgomery, Pettis, and St. Clair counties.

Ranges from Central America to eastern Mexico, and from Florida to Texas north to Massachusetts, New York, Ontario, Michigan, Wisconsin, and Minnesota.

Of the various species of panicum found in Missouri, *P. villosissimum* var. *villosissimum* is probably the one most conspicuously hairy by virtue of the abundant long spreading hairs of the culms, nodes, leaf-sheaths, and leaf-blades. In the long spreading hairs of the culms and leaf-sheaths, it resembles *P. praecocius* and *P. lanuginosum* var. *implicatum*, but has longer spikelets than those species, and generally broader leaf-blades (5–13 mm. wide) than *P. praecocius* (4–6 mm.). The hairs on the upper leaf surface of *P. villosissimum* are usually elongated, varying from usually 2–3 mm. up to 5 mm. long.

18b. **Panicum villosissimum** var. **pseudopubescens** (Nash) Fern. Map 311

Flowers May–July (vernal); June–October (autumnal).

Known only from Miller County (cherty upland

woods of ravine, tributary to Grand Glaize branch of Lake of the Ozarks, 3 mi. southwest of Brumley, Oct. 25, 1936, *Steyermark 20685*).

Ranges from Guatemala and Florida to Texas, north to Connecticut, New York, Ohio, Michigan, Wisconsin, and Minnesota.

The Missouri specimen was determined by Mrs. Chase, who judged it to be abnormal or not characteristic. Of this specimen she wrote (letter of May 27, 1937) that it 'looks as if it were a combination of late autumnal phase with a precocious winter rosette, well grown as it is in early spring. I have seen specimens of this kind in ground that has been burned over where the plant seems to produce vernal and autumnal phase at the same time.'

Possibly, a re-collection from this locality may show that the plant will have to be referred instead to var. *villosissimum*. Until such time as a verification of the species from another collection at the Miller County station or from another locality is forthcoming, the variety is being maintained tentatively for the state.

19. **Panicum columbianum** Scribn.

See excluded species.

20. **Panicum sphaerocarpon** Ell. Map 312

Flowers May–September (vernal); July–November (autumnal).

Occurs most frequently in dry or rocky open ridges or slopes in oak-hickory or pine-oak woodland in acid soils associated with sandstone, chert, or granite substrata, also on sandstone, chert, or granite glades, acid-soil prairies, and alluvial or moist ground along streams. Throughout the Ozark region and central Missouri north to Marion, Macon, Linn, Saline, and Jackson counties.

Missouri material falls into two varieties:

Common type; spikelets 1.5–1.8 mm. long; ligule obsolete or not evident; leaf-sheaths lacking viscous (sticky) spots . . 20a. P. SPHAEROCARPON var. SPHAEROCARPON

Rare, known from Shannon Co.; spikelets 1.3–1.5 mm. long; ligule evident, 0.3–1 mm. long; upper leaf-sheaths sometimes with viscous spots. 20b. P. SPHAEROCARPON var. INFLATUM

20a. **Panicum sphaerocarpon** var. **sphaerocarpon** Map 312

*Panicum sphaerocarpon* Ell. [G, P & S]

The common variety occurring in Missouri.

Ranges from Vermont and Ontario to Michigan, southern Illinois, Missouri, and Kansas, south to Florida, Texas, eastern Mexico, Central America, and northern South America.

20b. **Panicum sphaerocarpon** var. **inflatum** (Scribn. & Sm.) Hitchc. Map 312

Known only from Shannon County in southeastern Missouri (Monteer, *Bush 753*, and *Bush 747* in part).

Ranges from Florida to Texas, north to Delaware and southeastern Missouri.

21. **Panicum polyanthes** Schult. Map 313

Flowers May–August (vernal); July–November (autumnal).

Occurs along sandy and moist banks of small streams, low wet woods, moist sandstone ledges, and occasionally in wet prairie swales, generally in acid soils in the southeastern half of the state north to Lincoln and Boone counties, west to Phelps and Howell counties, and locally to Johnson County.

Ranges from Florida to Texas north to Massachusetts, Pennsylvania, Ohio, Michigan, Indiana, Illinois, Missouri, and Oklahoma.

Although very similar to *P. sphaerocarpon*, this species is usually readily separated from *P. sphaerocarpon* by the relatively longer and broader, thinner leaf-blades, which are more numerously nerved (41–115- instead of 25–67-nerved), the conspicuously smaller reduced lower leaf-blades as contrasted with the uppermost ones, and the relatively longer than broad as well as longer panicles. In general, *P. polyanthes* is found in lower wetter sites than *P. sphaerocarpon*, the latter mostly frequenting dry rocky upland open woods or rocky open situations.

---

Plate no. 58. **1.** Panicum lanuginosum var. Lindheimeri; *a.* Plant, × 3/4; *b.* Floret, × 7 1/2; *c, d.* Two views of spikelet, × 7 1/2. **2.** Panicum lanuginosum var. implicatum; *a, b.* Two views of spikelet, × 7 1/2; *c.* Floret, × 7 1/2. **3.** Panicum lanuginosum var. fasciculatum; *a, b.* Two views of spikelet, × 7 1/2; *c.* Floret, × 7 1/2. **4.** Panicum villosissimum; *a, b.* Two views of spikelet, × 7 1/2; *c.* Floret, × 7 1/2. **5.** Panicum villosissimum var. pseudopubescens; *a, b.* Two views of spikelet, × 7 1/2; *c.* Floret, × 7 1/2. **6.** Panicum subvillosum; *a, b.* Two views of spikelet, × 7 1/2; *c.* Floret, × 7 1/2. **7.** Panicum praecocius; *a, b.* Two views of spikelet, × 7 1/2; *c.* Floret, × 7 1/2. **8.** Panicum polyanthes; *a, b.* Two views of spikelet, × 7 1/2. *c.* Floret, × 10. **9.** Panicum malacophyllum; *a, b.* Two views of spikelet, × 7 1/2; *c.* Floret, × 7 1/2. **10.** Panicum sphaerocarpon; *a.* Plant, × 3/4; *b.* Floret, × 7 1/2; *c, d.* Two views of spikelet, × 7 1/2. **11.** Panicum oligosanthes var. oligosanthes; *a, b.* Two views of spikelet, × 7 1/2; *c.* Floret, × 7 1/2. **12.** Panicum oligosanthes var. Helleri; *a, b.* Two views of spikelet, × 7 1/2; *c.* Floret, × 7 1/2. **13.** Panicum oligosanthes var. Scribnerianum; *a.* Plant, × 1; *b.* Floret, × 7 1/2; *c, d.* Two views of spikelet, × 7 1/2. **14.** Panicum Ravenelii; *a, b.* Two views of spikelet, × 7 1/2; *c.* Floret, × 7 1/2.

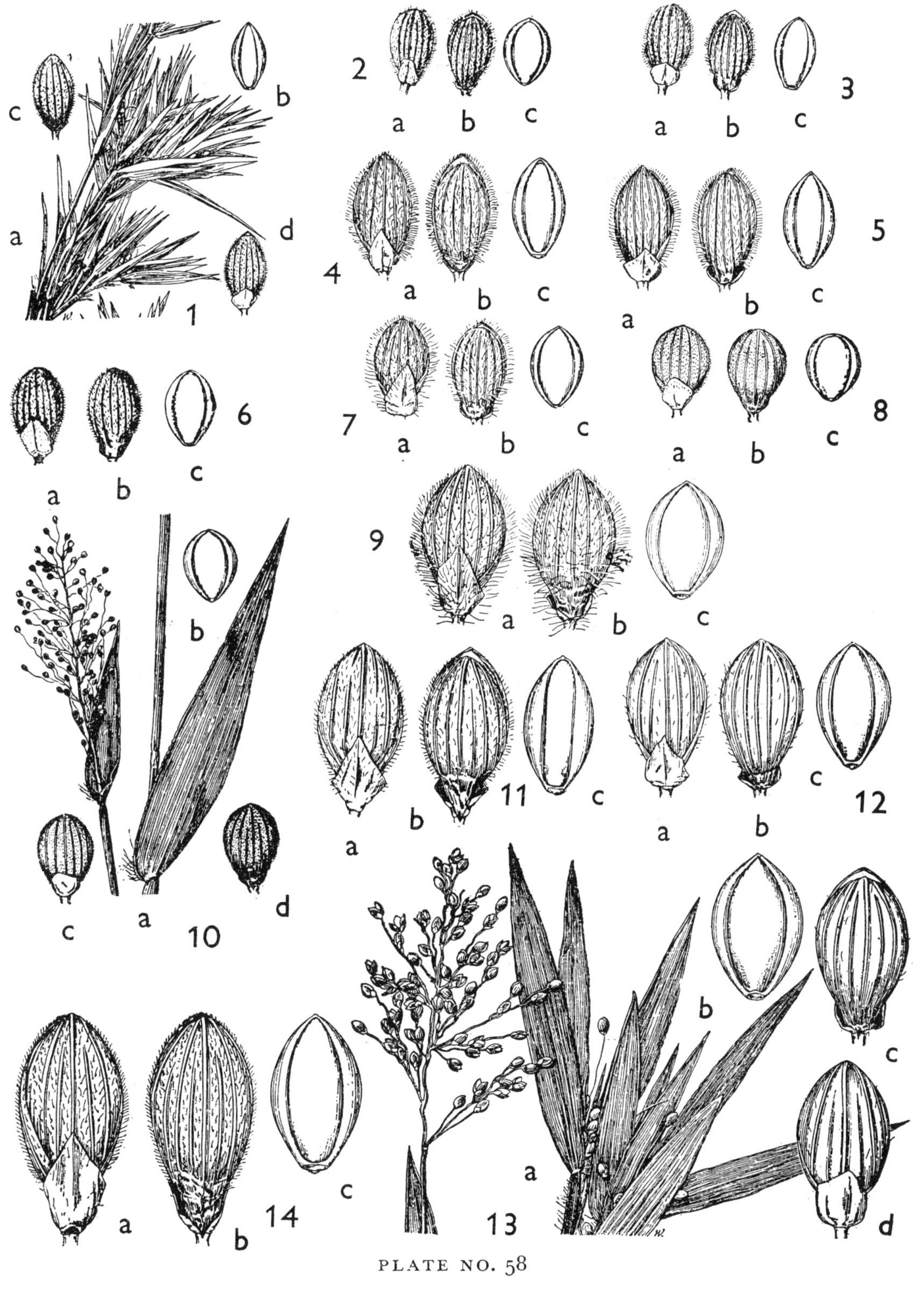
c
b
a
d
1
2
a b c
3
a b c
4
a b c
5
a b c
6
a b c
7
a b c
8
a b c
9
a b c
10
a b c d
11
a b c
12
a b c
13
a b c d
14
a b c

PLATE NO. 58

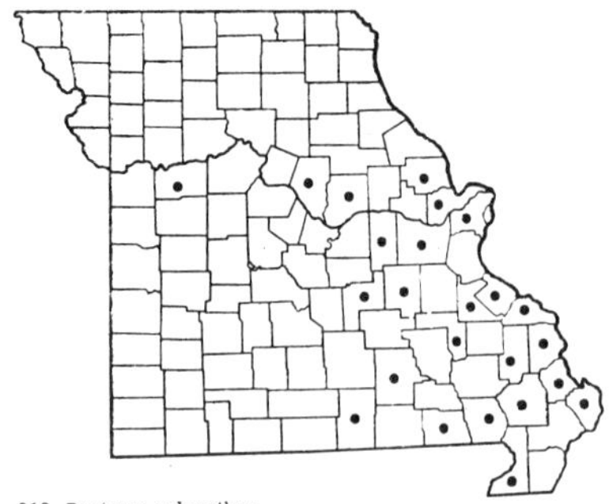

313 Panicum polyanthes

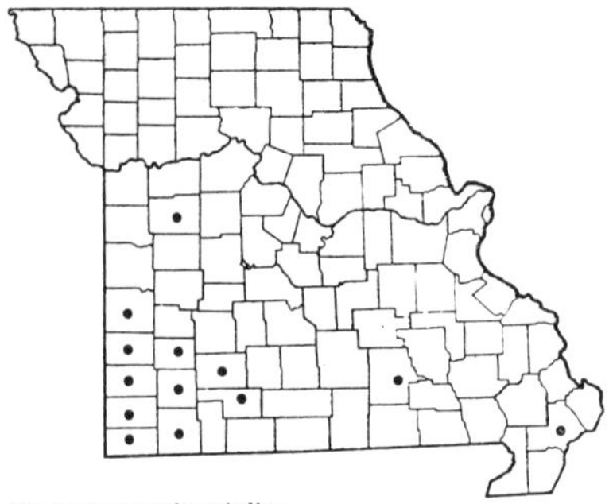

314 Panicum malacophyllum

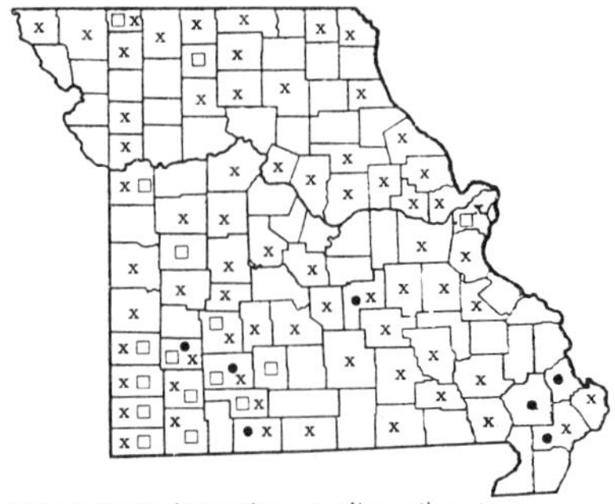

315 • Panicum oligosanthes var. oligosanthes
315 □ Panicum oligosanthes var. Helleri
315 x Panicum oligosanthes var. Scribnerianum

22. **Panicum malacophyllum** Nash Map 314

Flowers late May–July (vernal); July–September (autumnal).

Occurs usually in dry sandy or rocky open upland woods, less frequently in sandy openings or along hedge rows. Most common in southwestern Missouri, locally east to Shannon and New Madrid counties, and locally northwest to Johnson County (sandy open woods, Warrensburg, July 4, 1903, *Stigall 8*).

Ranges from Tennessee, Missouri, and Kansas, south to Arkansas, Oklahoma, and Texas.

A very distinct species, possessing a soft velvety pubescence, most conspicuous on the leaf-blades and culms, but also soft pubescent on the spikelets and to a lesser extent on the leaf-sheaths.

23. **Panicum oligosanthes** Schultes Map 315

Flowers May–July (vernal); June–November (autumnal).

Missouri material falls into the following categories:

a. Leaf-sheaths of main leaves with appressed pubescence (lying along and close against surface); spikelets 3.5–4 mm. long, on pedicels (stalks) often 5–15 mm. long . 23a. P. OLIGOSANTHES var. OLIGOSANTHES

a. Leaf-sheaths of main leaves glabrous or with hairs spreading away from surface; spikelets 2.8–3.9 mm. long, on pedicels mostly less than 5 mm. long . . . . . . . . . . . *b*
 b. Spikelets mostly 2.8–3.1 mm. long . . . 23b. P. OLIGOSANTHES var. HELLERI
 b. Spikelets mostly 3.2–3.6 mm. (rarely 3.9 mm.) long . . . 23c. P. OLIGOSANTHES var. SCRIBNERIANUM

23a. **Panicum oligosanthes** var. **oligosanthes**
*Panicum oligosanthes* Schultes [G]
Map 315

Occurs in sandy open ground, wooded sandstone or limestone slopes, and wooded rocky slopes along streams, usually in acid soils. Scattered in the southern third of the state north to Phelps (*Steyermark 20901*) and east to Scott (*Steyermark 20749*) county.

Ranges from Florida to Texas, north to New Jersey, Indiana, Illinois, Missouri, and Oklahoma.

23b. **Panicum oligosanthes** var. **Helleri** (Nash) Fern. Map 315

Occurs in rocky open woods, prairies, glades, open ground, and along railroad tracks. Frequent in the southwestern quarter of the state, rare and local east to St. Louis County and north to Grundy and Worth counties.

Ranges from Missouri to Louisiana, Oklahoma, Texas, and New Mexico.

23c. **Panicum oligosanthes** var. **Scribnerianum** (Nash) Fern. Map 315

Occupies a variety of habitats in dry open or upland rocky or sandy woods, glades, prairies, rocky banks of streams, sandy open or waste ground, and along railroads. Common throughout Missouri, and expected to be found in every county. One of the commonest and most frequently encountered *Panicum* in Missouri.

Ranges from Maine to British Columbia, south to Virginia, Georgia, Tennessee, Arkansas, Texas, Arizona, and California.

The spikelets of *P. oligosanthes* have a characteristic turgid obovate shape with 7–9 strong nerves. In the autumnal state the branches are more or less crowded and top-heavy toward the summit with rather firm elongated leaf-blades, which are not reduced in size.

Where *P. oligosanthes* var. *Scribnerianum* occurs in grazing areas, it is considered to be one of the most important of the secondary grasses for green forage palatability and nutrition.

The var. *Scribnerianum* and var. *Helleri* have leaf-

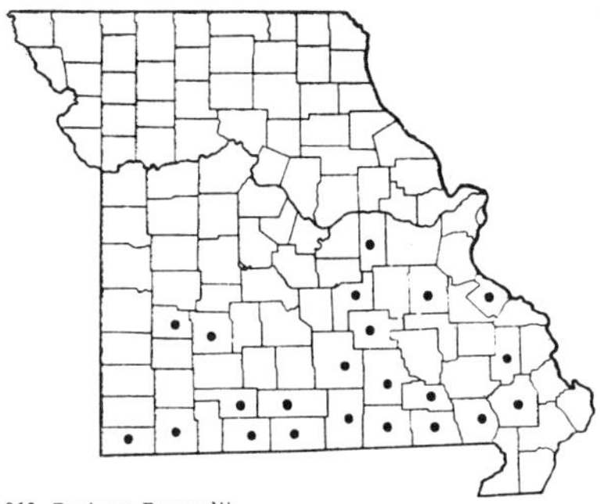

316 Panicum Ravenelii

317 Panicum Leibergii

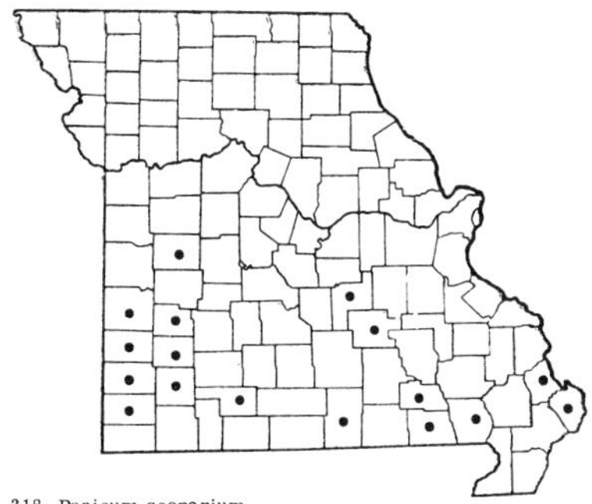

318 Panicum scoparium

sheaths, leaf-blades, and culms varying from pubescent to glabrous; the spikelets also vary from sparsely hairy to glabrous. The hairs on the leaf-sheaths of var. *Scribnerianum* and var. *Helleri* are often papillose-hirsute. The varieties intergrade into one another, and various intermediates between the named varieties are difficult to place. Some specimens, such as *Palmer 65943* (sandy open woods along Horse Creek, 1 mi. east of Milford, Barton County, July 6, 1957) have spikelets nearly 4 mm. long on pedicels 5–15 mm. long characteristic of *P. oligosanthes* var. *oligosanthes*, but with spreading pubescence on leaf-sheaths characteristic of var. *Scribnerianum*.

24. **Panicum Ravenelii** Scribn. & Merr. Map 316

Flowers May–July (vernal); June–October (autumnal).

Occurs on dry or rocky open ravine slopes and ridges in pine-oak or oak-hickory upland woodland, open rocky sterile fields, in acid soils associated with sandstone, chert, or granite substrata, rarely on limestone glades. Scattered in the Ozark region, mainly in the southern section of it, extending north to Cedar, Phelps, Gasconade, Washington, and Ste. Genevieve counties.

Ranges from Florida to Texas, north to Delaware, Maryland, Kentucky, Missouri, and Oklahoma.

This species is characterized by its generally purplish, rather stout culms covered with long appressed-ascending hairs, the thick, dull or olive green, lanceolate leaf-blades, the prominent ligule, and the relatively large pubescent spikelets.

25. **Panicum Leibergii** (Vasey) Scribn. Map 317

Flowers May–August (vernal); June–September (autumnal).

Occurs in rocky or open woods, alluvial or bottom meadows, and prairies. Rare and scattered over the state.

Ranges from New York to Saskatchewan, south to Pennsylvania, Ohio, Indiana, Illinois, and Kansas.

An infrequently collected species, but easily recognized by the combination of rather thin, narrowly lanceolate pubescent leaf-blades, conspicuously hairy leaf-sheaths with spreading papillose-hispid hairs, practically no ligule, relatively large spikelets with long spreading hairs up to 1 mm. long, and the relatively long lance-ovate, acutish 1st glume which reaches half the length of the spikelet.

26. **Panicum scoparium** Lam. Map 318

Flowers June–August (vernal); July–November (autumnal).

Occurs most commonly in sandy open prairies and prairie swales, sandstone exposures, sandy open woods, sandy roadsides, and along railroads. Scattered mainly in the southern third of the state, most frequent in the unglaciated prairie section, absent from most of the Ozark counties.

Ranges from Florida to Texas, north to Massachusetts, New York, Kentucky, Missouri, and Oklahoma.

This is one of the taller species of *Panicum* in Missouri, attaining a height of nearly 1.5 meters. In the autumnal state, the culms fork and branch with many crowded smaller branchlets bearing leaves smaller in size than those of the main stem. With its dense velvety pubescence covering the leaf-blades, leaf-sheaths, and stout culms, together with the conspicuous glabrous band below the bearded nodes, one is not likely to confuse *P. scoparium* with any other species of *Panicum*.

27a. **Panicum commutatum** Schultes var. **commutatum** Map 319

*Panicum commutatum* Schultes [G, P & S]

Flowers May–August (vernal); June–November (autumnal).

Occurs in alluvial rich woods and thickets, in the southeastern lowland section, and elsewhere on dry wooded sandstone or chert slopes and wooded banks of streams.

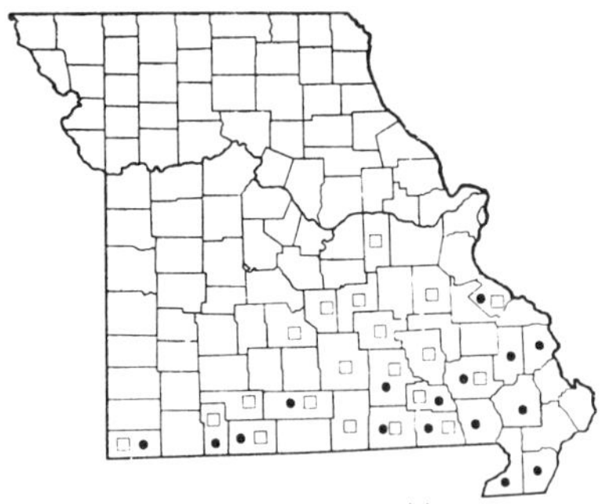

319 • Panicum commutatum var. commutatum
319 □ Panicum commutatum var. Ashei

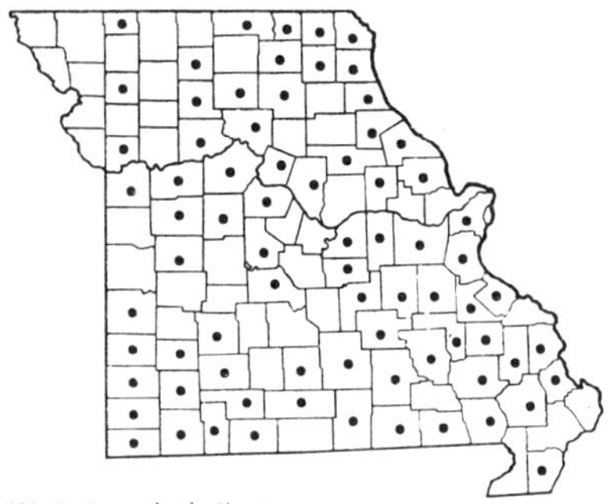

320 Panicum clandestinum

321 Panicum latifolium

Found only in southern Missouri northeast to Ste. Genevieve County, west to McDonald County, most frequent in the southeastern corner of the state, where frequent on Crowley Ridge and in the southeastern lowland section.

Ranges from Florida to Texas, north to Massachusetts, New York, Ohio, Michigan, Illinois, and Missouri.

27b. **Panicum commutatum** var. **Ashei** (Pearson) Fern. Map 319

Flowers May–July (vernal); July–November (autumnal).

Frequents slopes, crests, ridges, and ravines in dry rocky pine-oak or oak-hickory woodland in acid soils associated with sandstone, chert, or granite substrata, and also along rocky banks of streams. Restricted to the Ozark region of Missouri, extending north to Gasconade County, west to Laclede, Christian, and McDonald counties.

Ranges from Massachusetts to Michigan, Indiana, and Missouri, south to Florida, Mississippi, Arkansas, and Oklahoma.

The variety *Ashei* is distinguished from *P. commutatum* mainly by the minute puberulence covering the culms and leaf-sheaths, the narrower leaf-blades 5–11 mm. instead of 10–25 mm. wide, and the more slender wiry culms. The culms are dull purplish in both varieties, but in var. *Ashei* the leaf-blades are generally flushed with purple, whereas in var. *commutatum* they usually remain a deep green shade. The spikelets average somewhat larger in var. *commutatum* (2.2–3.2 mm. long, as compared with 2.1–2.7 mm. in var. *Ashei*).

28. **Panicum clandestinum** L. Map 320

Flowers May–September (vernal); July–November (autumnal).

Frequents moist thickets in valleys, along streams, roadsides, alluvial ground, prairie swales, open upland woods, and along railroads. Common throughout Missouri; one of the commonest species in the state.

Ranges from Quebec and Nova Scotia to Maine, Michigan, Iowa, and Kansas, south to Florida and Texas.

The papillose-hispid leaf-sheaths distinguish this species from both *P. latifolium* and *P. Boscii* with which it is sometimes confused, but especially with *P. latifolium*. Even in those specimens in which most of the hair has fallen, the papillate bases of at least the lower leaf-sheaths usually remain to indicate their presence; sometimes the upper leaf-sheaths are glabrous except for the ciliate margins.

The spikelets vary in size from 2.5–3.2 mm. long. The broadly lanceolate leaf-blades vary from usually glabrous on both sides with some scabrous hairs near the tip, to rarely pubescent above and below with scabrous and ciliate margins near the base.

29. **Panicum latifolium** L. Map 321

Flowers May–August (vernal); July–October (autumnal).

Occurs in rich or rocky woodland, on slopes of ravines and bluffs, or in wooded valleys along streams, also in upland woods and thickets. Throughout Missouri.

Ranges from Maine and Quebec to Minnesota, south to North Carolina, Tennessee, Missouri, and Kansas.

Plate no. 59. **1.** Panicum Leibergii; *a*, *b*. Two views of spikelet, $\times 7\frac{1}{2}$; *c*. Floret, $\times 7\frac{1}{2}$. **2.** Panicum commutatum var. Ashei; *a*, *b*. Two views of spikelet, $\times 7\frac{1}{2}$; *c*. Floret, $\times 7\frac{1}{2}$. **3.** Panicum clandestinum; *a*, *b*. Two views of spikelet, $\times 7\frac{1}{2}$; *c*. Floret, $\times 7\frac{1}{2}$. **4.** Panicum Boscii; *a*, *b*. Two views of spikelet, $\times 7\frac{1}{2}$; *c*. Floret, $\times 10$. **5.** Panicum scoparium; *a*. Plant, $\times \frac{3}{4}$; *b*. Floret, $\times 7\frac{1}{2}$; *c*, *d*. Two views of spikelet, $\times 7\frac{1}{2}$. **6.** Panicum commutatum var. commutatum; *a*. Plant, $\times \frac{3}{4}$; *b*. Floret, $\times 7\frac{1}{2}$; *c*, *d*. Two views of spikelet, $\times 7\frac{1}{2}$. **7.** Panicum latifolium; *a*. Plant, $\times \frac{3}{4}$; *b*. Floret, $\times 7\frac{1}{2}$; *c*. Spikelet, $\times 7\frac{1}{2}$.

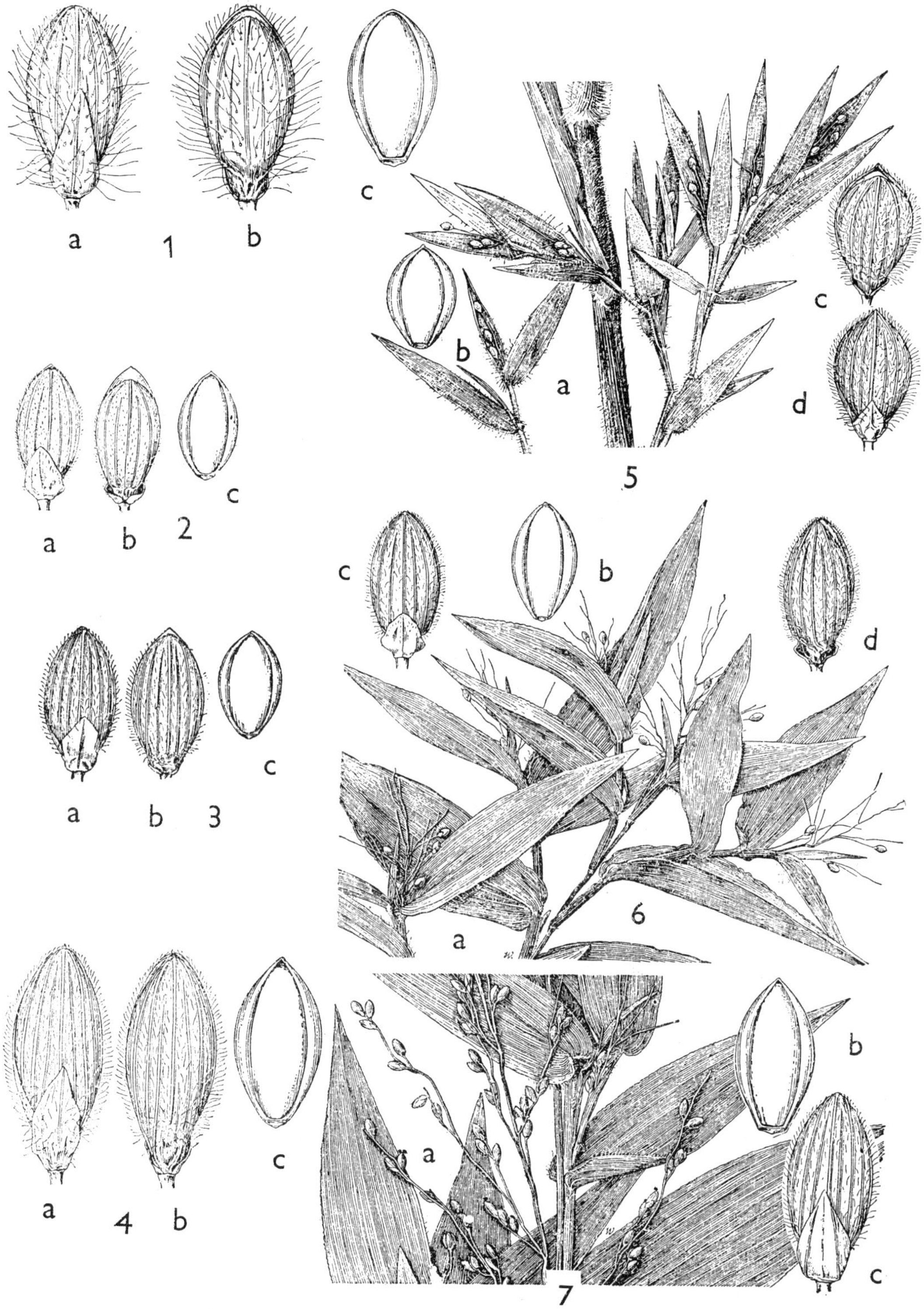


PLATE NO. 59

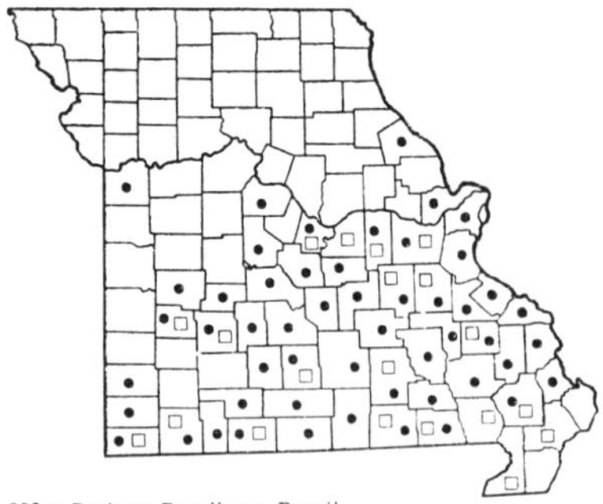

322 • Panicum Boscii var. Boscii
322 □ Panicum Boscii var. molle

323 Panicum geminatum

324 Panicum texanum (Texas Millet)

This species typically has mostly glabrous culms, leaf-sheaths and leaf-blades, the latter, ciliate at base. However, these parts may show varying amounts of hairiness on some plants. Some specimens have the lower leaf-sheaths softly hairy, but the upper ones are usually glabrous, except for ciliate margins at the base. A band or ring of gray-white hairs is found at the summit of the leaf-sheath. The nodes of the culms are usually glabrous, but the lower nodes are sometimes shortly or sparsely hairy. Spikelets vary from 3.2–3.7 mm. long and 1.8–2 mm. wide.

The species is sometimes confused with the other two broadest-leaved species of *Panicum* in Missouri, *P. clandestinum* and *P. Boscii*. Especially pubescent forms of *P. latifolium* may be distinguished from *P. Boscii* by the shorter spikelets (3.2–3.7 instead of 3.7–5.2 mm. long), and by the glabrous or short-haired nodes, the nodes in *P. Boscii* being long-bearded with hairs 2 mm. or more long. A number of specimens of *P. latifolium* from northern Missouri, misidentified as *P. Boscii*, have long hairs on the lower part of the leaf-sheaths or sparsely or short-hairy nodes. An examination of such specimens reveals either the shorter spikelets or the shorter hairiness around the nodes of *P. latifolium*. Usually glabrous *P. latifolium* may be distinguished from unusual apparently glabrous forms of *P. clandestinum* by the larger spikelets (3.2–3.7 instead of 2.5–3.2 mm.), when these are present, and, vegetatively, by the leaf-sheaths and margins of leaf-blades lacking stiff papillose hairs (hairs which arise from a swelling or bump at base), the remains of which can be observed on the surface of the leaf-sheaths or the margins of the leaf-blades in glabrate *P. clandestinum*, and by the leaf-sheaths usually being much shorter than their internodes, whereas in *P. clandestinum* the leaf-sheaths nearly equal or are longer than the internodes, the upper ones often overlapping in *P. clandestinum*.

30. **Panicum Boscii** Poir.

Flowers May–July (vernal); July–October (autumnal).

Occurs in dry open, often rocky woodland in various types of topography – upland, lowland, ridge tops, ravine slopes, or bluffs.

Missouri material may be separated as two varieties:

Leaf-blades glabrous or nearly so on lower surface; culms glabrous or sparsely or minutely hairy (puberulent) . . . 30a. P. Boscii var. Boscii
Leaf-blades velvety hairy on lower surface; culms usually soft-hairy . . 30b. P. Boscii var. molle

30a. **Panicum Boscii** var. **Boscii** Map 322

*Panicum Boscii* Poir. [G, P & S]

Common throughout the Ozark region and mostly confined to it.

Ranges from Massachusetts to Ohio, Illinois, and Missouri, south to Florida and Texas.

30b. **Panicum Boscii** var. **molle** (Vasey) Hitchc. & Chase Map 322

Known only in the Ozark, Crowley Ridge, and southeastern lowland sections of the state.

Ranges from Connecticut to southern Illinois and Missouri, south to Florida and Texas.

The two varieties of *P. Boscii* may always be distinguished from *P. clandestinum* and *P. latifolium*, the other very broad-leaved species in Missouri, by the long-bearded nodes of the culm with hairs 2 mm. or more long, and by the larger spikelets, usually 4–5.2 mm. (sometimes only 3.7). Specimens from northern Missouri, here referred to *P. latifolium*, sometimes have been mistaken for *P. Boscii*, but lack the truly long-bearded nodes and/or longer spikelets characteristic of *P. Boscii*. The leaves in *P. Boscii* usually are only 2½–5 times as long as broad, whereas in *P. latifolium* they are usually 4–7 times as long as broad. Some

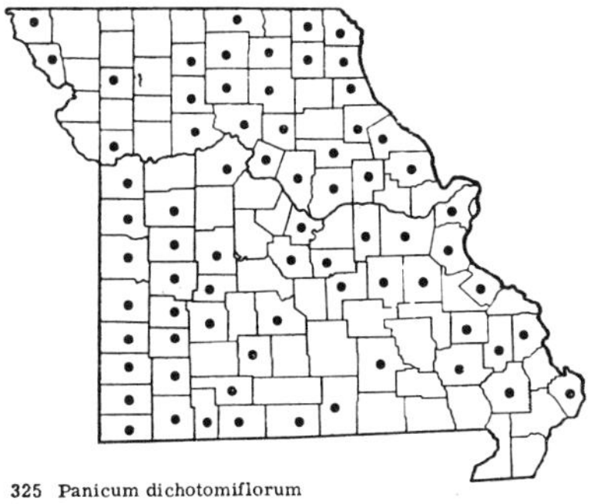
325 Panicum dichotomiflorum

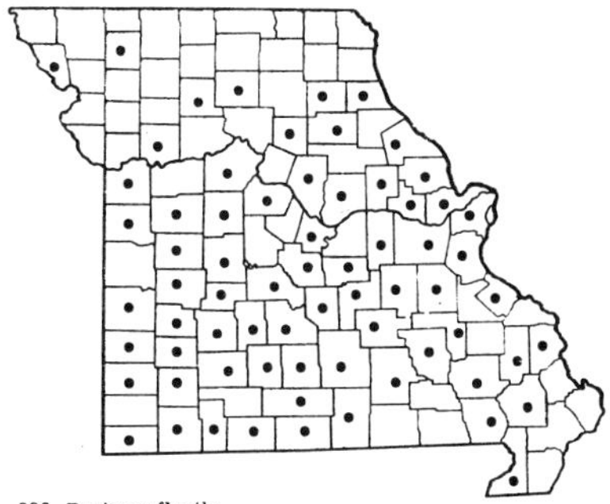
326 Panicum flexile

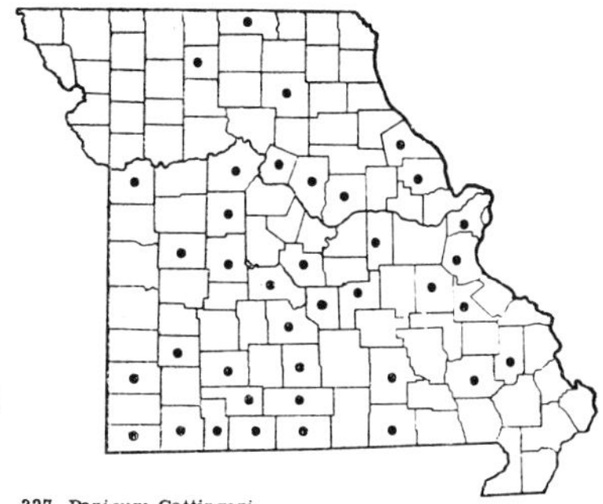
327 Panicum Gattingeri

Missouri specimens occasionally have some leaf-blades as narrow as 0.9 cm. wide, most of them, however, ranging from 1.8–3.3 cm. wide.

It is questionable whether the var. *molle* can be maintained as distinct from var. *Boscii*, as various non-correlated intergradations occur between the two (see Deam, *Grasses of Indiana*, p. 301. 1929).

31. **Panicum geminatum** Forsk. Map 323
*Paspalidium geminatum* (Forsk.) Stapf.
Flowers June–October.

Introduced along railroads, where known only from St. Louis County (right-of-way, Wabash R. R. track, north of Palm St. along the most western track, St. Louis, June 30, 1957, *Muehlenbach 1246*; same locality, September 8, 1957, *Muehlenbach 1354*).

Ranges from Florida to Texas; also Guatemala, and tropical regions of both hemispheres.

This is a mostly glabrous perennial with more or less succulent culms and rather inflated leaf-sheaths with a hairy ligule about 1 mm. long. The leaf-blades are 10–20 cm. long, 3–6 mm. wide, somewhat scabrous on the upper surface and margins. The inflorescence consists of 12–18 rather short erect or ascending racemes. No distinct winter leaf-rosettes are produced.

32. **Panicum texanum** Buckl. Texas Millet Map 324
Flowers July–November.

Introduced along railroads, where known only from St. Louis County (Baden freight yard of M.-K.-T. Railroad along the western siding near Dodridge Avenue, St. Louis, October 22, 1954, *Muehlenbach 517*).

Ranges from northern Mexico and Arizona to Texas and Florida, north to South Carolina, Alabama, Mississippi, Missouri, and Kansas.

This is a branching annual with culms which often root at the lower nodes. The culms below the nodes and below the inflorescence are generally softly hairy, as are the leaf-sheaths. The leaf-blades, from 8–20 cm. long, and 7–15 mm. wide, are also softly hairy. The axis and branches of the inflorescence are pubescent.

33. **Panicum dichotomiflorum** Michx. Map 325
Flowers June–October.

Moist open and alluvial soils and gravel bars along streams, low fields, roadside ditches, waste and cultivated ground, rocky glades, and along railroads.

Common throughout Missouri.

Ranges from Quebec to Ontario, Minnesota, and Nebraska, south to Florida, Texas, Mexico, and the West Indies; introduced in some western states.

Of the varieties described, Missouri material would fall into both var. *dichotomiflorum* and var. *geniculatum* (Wood) Fern., which are separated as follows:

Culms straight or nearly so, erect or ascending, slender, 2–3 mm. thick; leaf-sheaths scarcely or not inflated; leaf-blades 4–10 mm. wide; main inflorescence terminal, becoming long-exserted with ascending hair-like branches; spikelets loosely arranged on the branchlets . . . . P. DICHOTOMIFLORUM var. DICHOTOMIFLORUM

Culms crooked or strongly geniculate (zig-zag), depressed or low-spreading, stout, 3–6 mm. thick; leaf-sheaths usually conspicuously inflated; leaf-blades 6–20 mm. wide; inflorescences arising from most of the nodes, their bases included or only slightly exserted with widely spreading branches; spikelets crowded . . . . P. DICHOTOMIFLORUM var. GENICULATUM

Although plants are found which fit these extremes, there appears to be too much intergradation and lack of correlation in the several characters enumerated in the key to justify separating the varieties as distinct.

34. **Panicum flexile** (Gattinger) Scribn. Map 326
Flowers July–October.

Occurs in usually sterile soils of rocky open ground, glades and exposed bluff escarpments of sandstone, chert, granite, and leached limestone, eroded and open slopes, cherty and sandy open woods, low woods along creeks, gravel bars, prairies, and wet meadows in small valleys.

Common throughout southern and central Missouri, and extending north to Marion, Shelby, Linn, Gentry, and Holt (Big Lake State Park, *Steyermark 15105*, as to plant on left side of sheet).

Ranges from Florida to Texas, north to New York, Quebec, Ontario, Michigan, Iowa, and South Dakota.

This species is sometimes confused with *Eragrostis capillaris*, *Panicum capillare* var. *occidentale*, and *P. philadelphicum*, resembling each of them to some extent. The longer and long-pointed spikelets distinguish it from *P. philadelphicum*, the mostly glabrous lower pulvini of the inflorescence separate it from *P. capillare* var. *occidentale*, whereas the 1-flowered instead of 2–4-flowered spikelets distinguish it at once from *Eragrostis capillaris*.

35. **Panicum Gattingeri** Nash Map 327
*Panicum capillare* var. *campestre* Gatt. [BB]
Flowers August–October.

Occurs commonly in fallow fields, especially in alluvial soils in valleys, in dry rocky ground and glades of sandstone, chert, granite, or leached limestone, gravel bars, and wet soils bordering sloughs, streams, and thickets.

Mainly found in southern and central Missouri, but extending north locally to Macon, Putnam, and Grundy counties.

Ranges from Quebec and Massachusetts to New York, Pennsylvania, Ohio, Indiana, Illinois, and Minnesota, south to North Carolina, Georgia, Tennessee, Arkansas, and Oklahoma.

Stunted or dwarfed plants often found in moist spots on glades are sometimes misidentified as *P. philadelphicum* var. *philadelphicum*. In general appearance, this species resembles and has much in common with *P. capillare*, from which it is separated by the smaller, rather oval or ellipsoid panicle, the smaller spikelets, and the generally glabrous lower pulvini of the inflorescence. The culms, axis, and branches of the inflorescence, and spikelets of *P. Gattingeri* are usually prevailingly pale green or grass green, whereas in *P. capillare* var. *capillare* and var. *occidentale* the inflorescence and culms prevailingly become dull lavender or magenta purple.

36a. **Panicum philadelphicum** Bernh. var. **philadelphicum** Map 328
*Panicum philadelphicum* Bernh. [G, P & S]
Flowers June–October.

Occurs most frequently in rocky open ground, rocky open and upland woods, on sandstone, chert, granite, and limestone glades, eroded slopes, gravel bars, along streams, and fallow fields.

Essentially confined to the Ozark region of Missouri northeast into Ralls County in area believed to be unglaciated and Ozarkian in character of flora and topography.

Ranges from Georgia and Texas, north to New Hampshire, Quebec, New York, Ohio, Michigan, Indiana, Illinois, and Iowa.

36b. **Panicum philadelphicum** var. **Tuckermani** (Fern.) Steyermark & Schmoll Map 328
*Panicum Tuckermani* Fern. [G]
*Panicum philadelphicum* in part [BB]
Flowers July–October.

Known only from Washington County in the eastern Ozark section (alluvial gravel bar along Meramec River, vicinity of Green Cave, T40N, R1W, sect. 24 and 25, 6½–7 mi. [by air] north of Anthonies Mill, September 26, 1954, *Steyermark 77796*).

Ranges from Nova Scotia and Quebec to Minnesota, south to Virginia, Ohio, Indiana, and Missouri.

A discussion of the variation of this group and its relationships is to be found in Rhodora 41: 86–90. 1939, and in Fassett's *Grasses of Wisconsin*, p. 84. 1950. The differences between var. *philadelphicum* and var. *Tuckermani* were brought out in the key to *Panicum*. *Panicum philadelphicum* var. *Tuckermani* is maintained in the present treatment as a variety, instead of a species, because there is too much intergradation between it and var. *philadelphicum* as regards the glabrity of the pulvini, prominence and relative exsertion of the inflorescence, and number and arrangement of the spikelets in the inflorescence.

37a. **Panicum capillare** L. var. **capillare**
Witch Grass Map 329
*Panicum capillare* L. [G, P & S]
*Panicum capillare* var. *agreste* Gatt. [BB]
Flowers July–October.

Frequents various kinds of open or sunny habitats,

Plate no. 60. **1.** Panicum dichotomiflorum; *a.* Panicle, × 3/4; *b, c.* Two views of spikelet, × 7 1/2; *d.* Floret, × 7 1/2. **2.** Panicum texanum; *a.* Panicle, × 3/4; *b, c.* Two views of spikelet, × 7 1/2; *d.* Floret, × 7 1/2. **3.** Panicum Gattingeri; *a, b.* Two views of spikelet, × 7 1/2; *c.* Floret, × 7 1/2. **4.** Panicum philadelphicum; *a, b.* Two views of spikelet, × 7 1/2; *c.* Floret, × 7 1/2. **5.** Panicum philadelphicum var. Tuckermani; *a, b.* Two views of spikelet, × 10; *c.* Floret, × 10. **6.** Panicum geminatum; *a.* Plant, × 3/8; *b.* Floret, × 7 1/2; *c, d.* Two views of spikelet, × 7 1/2. **7.** Panicum miliaceum; *a, b.* Two views of spikelet, × 7 1/2; *c.* Floret, × 7 1/2.

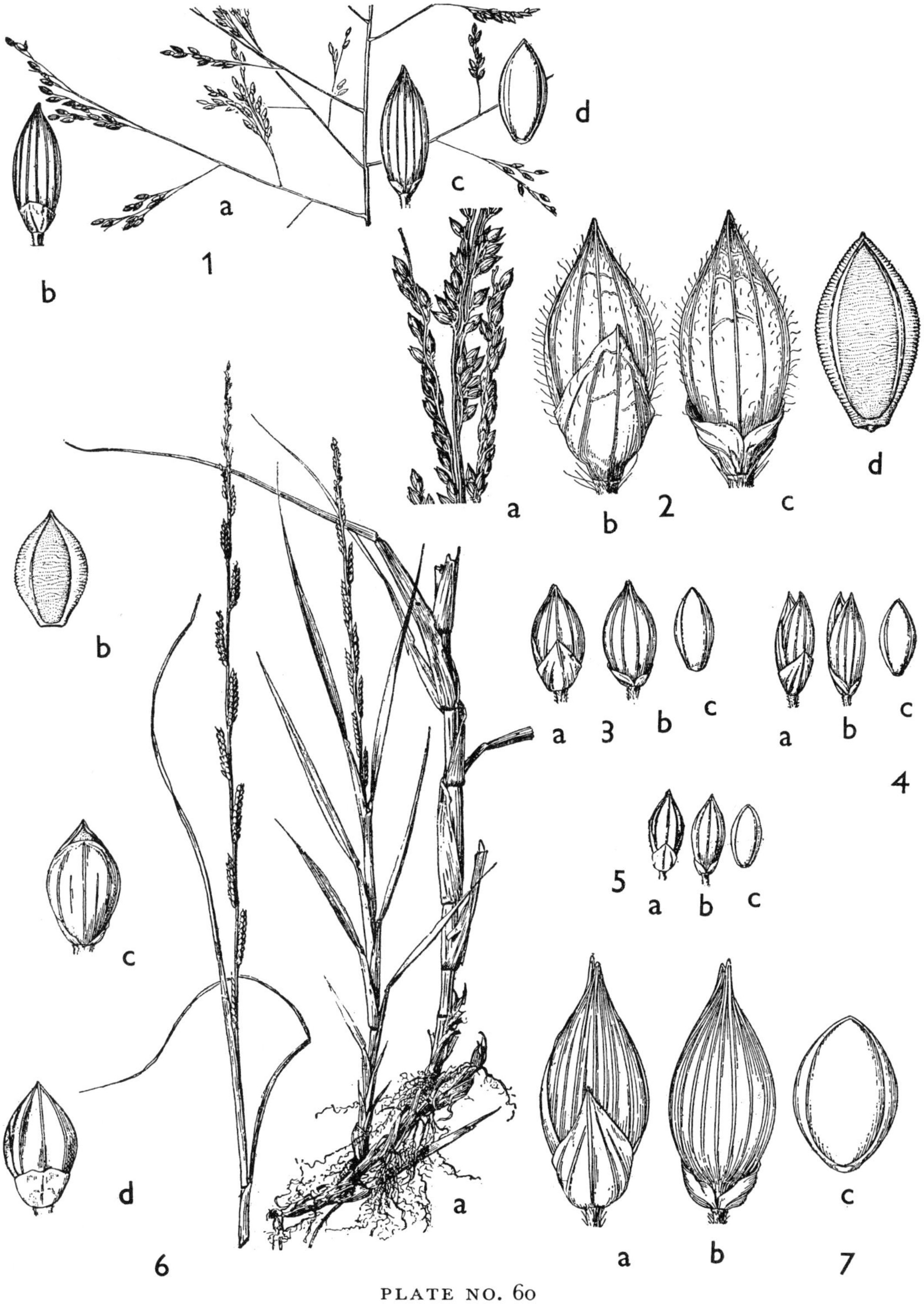


PLATE NO. 60

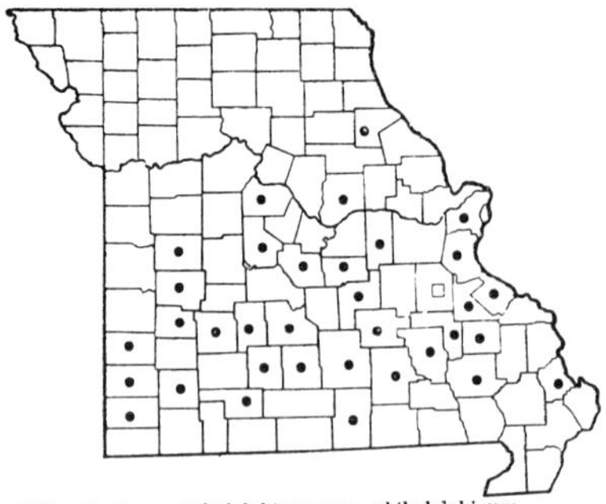

328 • Panicum philadelphicum var. philadelphicum
328 □ Panicum philadelphicum var. Tuckermani

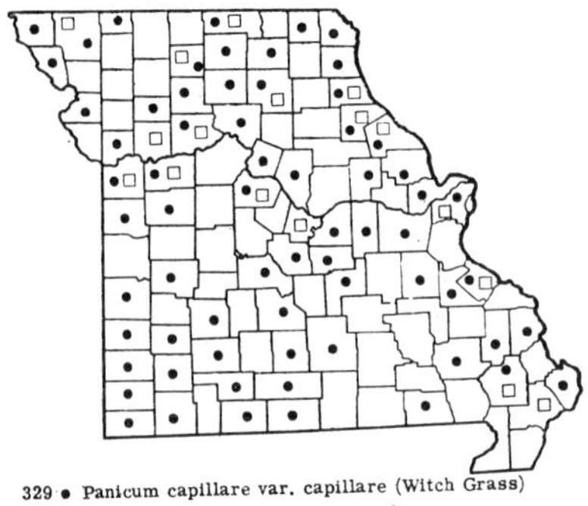

329 • Panicum capillare var. capillare (Witch Grass)
329 □ Panicum capillare var. occidentale

330 Panicum miliaceum (Broomcorn Millet)

either wet or dry, occurring in cultivated, waste, and alluvial soils, gravel bars and banks of streams, limestone, sandstone, chert, and granite glades, and along railroad tracks. Throughout Missouri.

Ranges from Florida to Texas, north to Maine, Quebec, Ontario, Manitoba, and Montana; also Bermuda.

37b. **Panicum capillare** var. **occidentale** Rydb. Map 329

Occupies similar habitats as var. *capillare*, but less common, and mostly absent from the Ozark and unglaciated prairie sections.

Ranges from Quebec to British Columbia, south to New Jersey, West Virginia, Kentucky, Missouri, Texas, Arizona, and California.

Both these varieties are distinguished in the key to *Panicum*. At maturity both varieties become magenta- or rose-purple throughout the inflorescence and also frequently on most of the culms. When occurring in quantity, the mass rose-purplish color effect is very striking. When mature, the entire sprawling inflorescence breaks away from the rest of the plant, and becomes blown about as a 'tumbleweed.'

38. **Panicum miliaceum** L. Broomcorn Millet, Hog Millet Map 330

This is the species usually referred to in Europe as Common Millet, whereas in the United States Common Millet refers usually to *Setaria italica*.

Flowers mid-June–November.

Introduced in waste and cultivated ground and along railroads. Known from central Missouri in St. Louis (*Muehlenbach 667, 1433, 1479*), Morgan (Versailles, *Parker Rodgers*), Chariton (Swan Lake Refuge, *Kucera*), and Jackson counties.

Native of the Old World; introduced into the United States where scattered over most of the eastern and central sections.

This hairy-leaved annual, which attains a height of about 1 meter, but also is often dwarfed to 1 dm. tall, usually does not persist in the wild state. Originally introduced into the northeastern United States for its forage value, the seed was also used for feeding hogs as well as in mixtures of commercial birdseed. In China the straw of this species is used as a pulp in papermaking. In parts of China and in India the seed constitutes an important cereal food, supplementing the widely-used rice.

39. **Panicum virgatum** L. var. **virgatum** Switch Grass Map 331

*Panicum virgatum* L. [G, P & S, BB]

Flowers July–September.

Frequents a wide variety of habitats, usually sunny, including dry or moist prairies, moist seepage of rocky glades and bluff escarpments, rocky banks and gravel bars of streams, wet or dry open woods, and along railroad tracks. Found throughout Missouri, except apparently absent in the extreme southeastern lowland and extreme northwestern sections.

Ranges from Quebec and New Hampshire to Saskatchewan, south to Florida, Texas, Arizona, Mexico, the West Indies, Central and South America.

All the Missouri material appears to fall into the typical var. *virgatum*, in which the rhizomes are elongate and creeping, and the spikelets are 3.5–6 mm. long, with the first glume acuminate and about 2/3 the length of the spikelet, the 2nd glume long-acuminate and definitely longer than the fertile lemma and fruit.

There is considerable variation in the habit of growth of this species in Missouri, ranging from small plants 3 dm. tall with relatively short inflorescences 5 cm. long, to tall plants 2 meters tall with large inflorescences expanding to 50 cm. long. Immature plants, with closely ascending panicle-branches and smaller young spikelets, appear quite in contrast to mature plants with spreading panicle-branches and large well-developed spikelets. Dwarfed or stunted

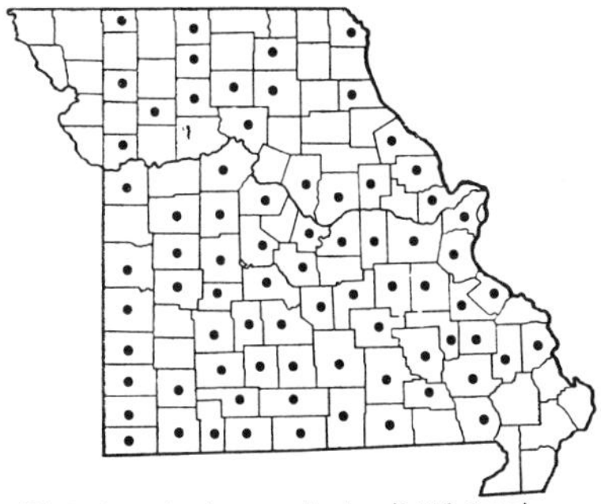

331 Panicum virgatum var. virgatum (Switch Grass)

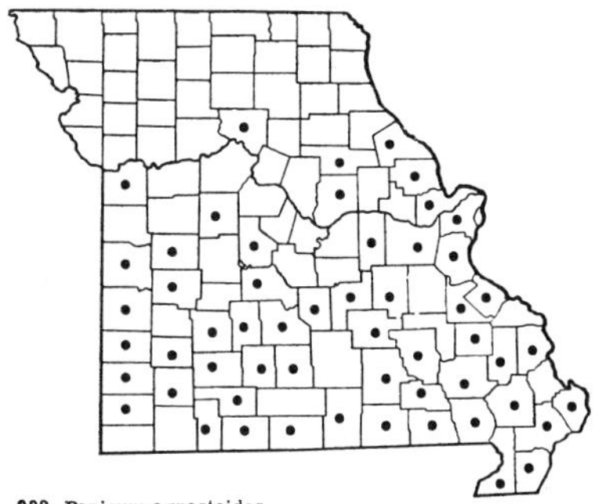

332 Panicum agrostoides

333 Panicum stipitatum

plants growing on dry limestone glades or bluff escarpments contrast markedly to those well-developed in prairie swales, moist gravel bars, and similar moist sites.

The grass is quite palatable to livestock, and is a constituent of prairie hay.

40. **Panicum agrostoides** Spreng. Map 332

*Panicum agrostoides* Spreng. (typical) [G]

*Panicum agrostoides* var. *condensum* (Nash) Fern. [G]

*Panicum agrostoides* var. *ramosius* (Mohr) Fern. [G]

Flowers July–October.

Frequents wet open places in upland prairies, swampy meadows, borders of ponds and sloughs, moist rocky open ground along creeks, occasionally occurring in moist open woods. Found throughout most of southern and central Missouri, including the southeastern lowland, and southwestern unglaciated prairie sections of the state, extending north to Pike, Audrain, Chariton, and Jackson counties.

Ranges from Maine to Indiana, Illinois, Missouri, and Kansas, south to Florida and Texas.

This species was divided by Fernald into the following three varieties:

a. Nodes of culms straw-color or pale green, not much contrasting in color with the internodes of the culm, only slightly constricted in the dried specimen; leaves firm; larger leaves with conspicuous whitish midrib 0.3–1 mm. wide at base; spikelets 0.8–1 mm. in diam., purplish or bronze (rarely green), short-pointed, ellipsoid to lance-oblong, closely imbricated (overlapping) on the branches of the inflorescence. . . . . . . . . . *b*
  - b. Panicles 0.8–2.5 dm. long, with widely spreading remote lower and middle branches; culms 2–9 dm. tall; spikelets 1.7–2.2 mm. long . P. AGROSTOIDES var. AGROSTOIDES
  - b. Panicles 0.6–4 dm. long, with ascending or closely appressed branches; spikelets 2–2.5 mm. long . P. AGROSTOIDES var. CONDENSUM

a. Nodes of culms darker than and contrasting with the internodes, conspicuously constricted in the dried specimen; leaves submembranaceous; larger leaves with slender midrib rarely 0.5 mm. wide; terminal panicles 1.5–4 dm. long, loosely open, the branchlets and branches spreading and loosely-flowered; spikelets 0.5–0.8 mm. in diam., green or lead-colored (rarely purplish), slenderly lance-elongated to a long tip . . . P. AGROSTOIDES var. RAMOSIUS

An attempt was made to apply the characters used to separate these varieties to the abundant Missouri material collected. However, so much intergradation and lack of correlation was found in numerous instances that it has not been possible to maintain these varieties in Missouri. Plants with purplish long-pointed spikelets and dark nodes were found together with those having short-pointed green or lead-colored spikelets and pale nodes. Such inconsistent combinations and other similar ones have made the retention of definite varietal characters impractical for separation of specimens.

*Panicum agrostoides* forms usually large clumps with most of the long leaves crowded toward the base.

41. **Panicum stipitatum** Nash Map 333

*Panicum agrostoides* var. *elongatum* Scribn. [BB]

Flowers August–October.

Wet banks of sloughs and streams and swampy ground. Known only from Dunklin, Wayne, and Shannon counties, southeastern Missouri.

Ranges from Georgia to Texas, north to Connecticut, New York, Pennsylvania, Ohio, Indiana, and southeastern Missouri.

No collections of this have been made since the first decade of this century.

42. **Panicum anceps** Michx. var. **anceps** Map 334

*Panicum anceps* Michx. [G, P & S]

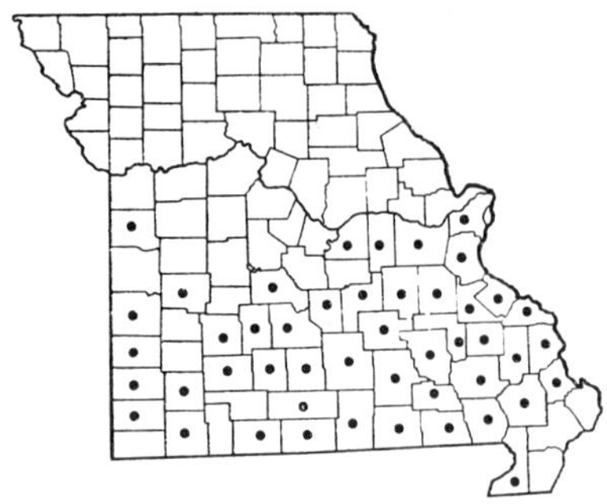

334 Panicum anceps var. anceps

335 Panicum hians

336 Panicum obtusum

Flowers June–October.

Frequents wet prairies, swampy meadows, roadside ditches, borders of streams, moist cherty or sandy ground in valleys, and occasionally low open ground in woodland. Rather common in favored wet sites throughout southern Missouri, except absent in most of the southeastern lowland section, extending north to St. Louis, Franklin, Osage, Camden, St. Clair, and Cass counties.

Ranges from Florida to Texas, north to New Jersey, Pennsylvania, Ohio, Indiana, Illinois, Missouri, and Kansas.

Another variety, var. *rhizomatum* (Hitchc. & Chase) Fern., is found from Virginia to Florida west to Texas.

A well-marked species, with spikelets much larger than *P. agrostoides* and obliquely set above the first glume, the spikelets presenting the appearance of being curved or turned to one side. Often found growing with *P. agrostoides* for which it is sometimes superficially mistaken.

43. **Panicum hians** Ell. Map 335

Flowers June–October.

Frequents swamps, streams, spring branches, swaley ground, roadside ditches, and low wet woods in the southeastern lowland section of the state.

Ranges from Florida to Texas and New Mexico, north along the Coastal Plain to Virginia, north up the Mississippi Embayment to southeast Missouri, and west to Oklahoma.

This is the only species of *Panicum* in Missouri showing a restricted Atlantic Coastal Plain-Mississippi Gulf Embayment type of geographic distribution. As a characteristic species of the Mississippi Embayment, it grows in bald cypress-swamp tupelo wet woodlands where it is associated with other species, such as *Diodia virginiana*, of similar distribution.

The very slender wiry culms of *P. hians* are much more slender than those of any other species of *Panicum* in Missouri.

44. **Panicum obtusum** HBK. Vine Mesquite Map 336

Flowers June–October.

Introduced in Jackson County (Kansas City, July 29, 1905, *Bush 3107*, and August 2, 1903, *Bush 1832*), the only known Missouri station.

Ranges from Kansas to Colorado, south to Texas, Arizona, and northern Mexico.

This species is a wiry-stemmed perennial spreading by long stolons or runners, which may reach a length of nearly 3 meters, rooting at the nodes. The nodes of the culm are conspicuously soft-hairy. The elongated leaf-blades are 2–7 mm. wide and mostly glabrous.

Because of its habit of forming dense mats of runners, this grass has been used to cover banks of streams to prevent erosion.

*Excluded Species*

5. **Panicum consanguineum** Kunth.

This species was reported by Palmer and Steyermark (Rh. 57: 311. 1955) as occurring in New Madrid County (wooded oak-hickory sandy knoll, on Crowley Ridge, T26N, R14E, sect. 34, 4½ mi. southeast of Sikeston, May 18, 1950, *Steyermark 69667*). This has been found to be a misidentified specimen of *P. malacophyllum* Nash, as determined by Dr. George B. Van Schaack. I have verified Dr. Van Schaack's determination.

---

Plate no. 61. **1.** Panicum anceps; *a.* Spikelet, × 7½; *b.* Floret, × 7½. **2.** Panicum virgatum; *a.* Plant, × ⅜; *b.* Floret, × 7½; *c, d.* Two views of spikelet, × 7½. **3.** Echinochloa colonum, × ¾. **4.** Panicum hians; *a.* Spikelet, × 7½; *b.* Floret, × 7½. **5.** Panicum stipitatum; *a, b.* Two views of spikelet, × 7½; *c.* Floret, × 7½. **6.** Panicum agrostoides; *a.* Panicle, × ¾; *b.* Floret, × 7½; *c, d.* Two views of spikelet, × 7½.

PLATE NO. 61

12. **Panicum spretum** Schultes

This species was included in the Palmer & Steyermark *Catalogue* from New Madrid and Jasper counties. The Jasper County specimen, upon re-examination, proves to be *P. lanuginosum* var. *Lindheimeri*. The specimen from New Madrid County has not been found.

14. **Panicum meridionale** Ashe

Reported from Missouri by Gleason in the *New Britton & Brown Illustrated Flora*, but no authentic material has been found to substantiate this report. Specimens so labelled, such as *Bush 5741* from McDonald Co. are *P. subvillosum*, and are so treated at the United States National Herbarium. Bush's specimens at the New York Botanical Garden (*Bush 5741, 5741A,* and *5741B*) have spikelets 1.8–1.9 mm. long with the first glume 0.8–0.9 mm. long and about ½ the length of or slightly less than ½ the length of the spikelet, longer than broad, and acute or acutish at apex; the lower portion of the axis of the panicle varies from hairs appressed-ascending to variously spreading. The hairs on the leaf-sheaths and culms vary from non-papillose to papillose. The ligule is 3–4 mm. long, the hairs on the upper surface of the erect blades are 3–4 mm. long, those on the lower surface dense, appressed, and 0.2–1 mm. long.

The larger spikelets would remove these collections immediately from consideration in *P. meridionale* in the treatments of *Panicum* by Gleason, Fernald, Hitchcock, and Deam, whereas the combination of larger spikelets, acutish longer than broad first glumes nearly ½ the length of the spikelet, and spreading-ascending to appressed hairs of the lower axis of panicle are various characters shared by *P. subvillosum*, although the Bush specimens do not especially show sheaths equaling or longer than the internodes, a character stressed by Fernald.

Although some of the same characters, as above given, are found in *P. praecocius*, the hairs of the leaf-sheaths and culms in that species are mostly 2–6 mm. long and widely spreading. Since the Bush specimens have hairs on the leaf-sheaths and culms mainly less than 2 mm. long and often ascending or even appressed (they are also spreading), this last character of pubescence of leaf-sheath and culm would place the Bush material in *P. subvillosum* rather than *P. praecocius*.

One other specimen, *Bush 15595* (woods, Baxter, Stone Co., May 27, 1936) in the Missouri Botanical Garden Herbarium, labelled by Bush as *P. meridionale*, should be referred to *P. lanuginosum* var. *fasciculatum*.

19. **Panicum columbianum** Scribn.

A specimen in the herbarium of the University of Missouri labeled 'Columbia, Mo., coll. B. T. Galloway, Aug. 1, 1884 legit Vasey.' was determined as *P. tsugetorum* Nash by either Hitchcock or Chase in Mrs. Chase's inked handwriting. Underneath this inked determination is a note by Hitchcock, 'I am uncertain about this.' The specimen is cited in Daniel's *Flora of Columbia, Missouri*, identified by Bush as '? *P. subvillosum* Ashe.'

The specimen has appressed-pubescent leaf-sheaths and lower leaf surfaces, the hairs on the lower leaf surface being short and fine, while the upper leaf surfaces are glabrous. The ligules vary from 2–4 mm. long. The spikelets are about 2 mm. long, pubescent, and with the 1st glume triangular-ovate and 1/3–3/8 the length of the spikelet. The axis of the inflorescence is glabrous or nearly so. The specimen is not typical *P. columbianum*, as the ligules attain a length of 4 mm.

Since the locality as well as the collector (both Galloway and Vasey are cited on the label) is doubtful, it seems better to exclude the species.

## 70. **Echinochloa** Beauv. Barnyard Grass

Annual grasses with long leaf-blades, compressed (flattened or folded) leaf-sheaths, and green or purple inflorescences composed of 1-flowered spikelets densely crowded along one side of thick racemes pinnately spreading or ascending from the main axis. The 1st glume is the shortest of the parts present on the spikelet. The 2nd glume and sterile lemma have a rather thin membranous texture similar to that of the 1st glume, but are much longer and about equal the length of the spikelet. The fertile lemma and palea are much thicker, coriaceous (leathery), smooth, and shining.

a. Racemes (main side branches of inflorescence) mostly 1–3 cm. long; leaf-blades mostly 2–6 (rarely to 10) mm. wide; spikelets (excluding awns) mainly 2–2.5 mm. (rarely to 2.9) long; few or no long hairs or bristles present on rachis and at nodes of inflorescence . . . . . . . . . . 1. E. COLONUM

a. Racemes mainly 3–12 cm. long (rarely shorter); leaf-blades mainly 5–30 mm. wide; spikelets (excluding awns) 2.5–4.5 mm. long; long hairs and bristles often present and common on rachis and at nodes of inflorescence . . . . . . . . . . . . . . . . . . . . . . . . *b*

b. Some part of the spikelet with a conspicuous or long awn . . . . . . . . . . . . . . *c*

c. Swollen- or blister-based stiff hairs present on all or most nerves in addition to those on marginal nerves of 2nd glume and sterile lemma; use a magnification of 15 × or more to observe that the summit of the smooth shining fertile lemma tapers into a *long* rather firm acuminate (well-pointed) or nearly acuminate tip lacking a ring of microscopic hairs; conspicuous long and bristle-like hairs few or sometimes absent at the nodes (joints) and along the rachis of the branches of the inflorescence . . . . . . . . . . . . 3a. E. MURICATA var. MURICATA

c. Hairs, when present on 2nd glume and sterile lemma, fine and slender from base to tip, the swollen- or blister-based hairs, if present at all, occurring only on marginal nerves; use a magnification of 15 × or more to observe that the summit of the smooth shining fertile lemma tapers into a *short* soft, easily bent, or wrinkled obtuse (blunt) tip with a ring of microscopic hairs present where the shining summit grades into the softer dull tip; conspicuous long and bristle-like hairs frequent at the nodes and sometimes along the rachis of the inflorescence 2b. E. CRUSGALLI var. CRUSGALLI f. LONGISETA

b. Spikelets either completely without awns, or at most with a short awn, never with a long or conspicuous awn . . . . . . . . . . . . . . . . . . . . . . . . . *d*

d. Swollen- or blister-based hairs present on all or most nerves as well as the margins of 2nd glume and sterile lemma . . . . . . . . . . . . 3d. E. MURICATA var. MICROSTACHYA

d. Swollen- or blister-based hairs either absent or present only on the marginal nerves of 2nd glume and sterile lemma . . . . . . . . . . . . . . . . . . . . . *e*

e. Use a magnification of 15 × or more to observe that the summit of the smooth shining fertile lemma tapers into a *long* firm acuminate (well-pointed) or nearly acuminate tip lacking a ring of microscopic hairs; conspicuous long and bristle-like hairs few or absent at the nodes (joints) and along rachis of the inflorescence . . . . . . . . . . . . . *f*

f. Spikelets mainly 3.3–4.5 mm. long, 1.8–2.2 mm. wide . . 3b. E. MURICATA var. LUDOVICIANA

f. Spikelets mainly 2.5–3.4 mm. long, 1.4–1.8 mm. wide . . 3c. E. MURICATA var. OCCIDENTALIS

e. Use a magnification of 15 × or more to observe that the summit of the smooth shining fertile lemma tapers into a *short* soft, easily bent, or wrinkled obtuse (blunt) tip with a ring of microscopic hairs present where the shining summit grades into the softer dull tip; conspicuous long and bristle-like hairs frequent at the nodes and sometimes along the rachis of the inflorescence . . . . . . . . . . . . . . . . . . . *g*

g. Spikelets glabrous or sterile lemma with minute sparse appressed hairs; sterile lemma soft-tipped, obtuse (blunt); inflorescence rather narrowly oblong and dense, dark brown-purple, the tips of branches turning inward; branches of inflorescence overlapping . 2c. E. CRUSGALLI var. FRUMENTACEA

g. Spikelets mostly fine and slender, setiform (of a bristly type), the swollen- or blister-based hairs, if present, occurring only on margins of 2nd glume and sterile lemma; sterile lemma firm-tipped, abruptly short-pointed (apiculate); inflorescence with a pyramidal or Christmas-tree shape (broadly ovoid, rhomboid, or lanceolate), green or purple, the branches straight; branches of inflorescence not overlapping . . . . . . . . 2a. E. CRUSGALLI var. CRUSGALLI

1. **Echinochloa colonum** (L.) Link f. **colonum**
Jungle Rice Map 337
*Echinochloa colonum* (L.) Link [G, P & S]
Flowers July–October.

Moist ground along fields, roadside ditches, and along railroads. Introduced in southeastern, east-central and west-central Missouri, in Jackson (Sheffield, September 16, 1920, *Bush 9228*), St. Louis (Burlington R. R. right-of-way beneath Merchants Bridge, St. Louis, July 5, 1954, *Muehlenbach 224*), Mississippi, and Pemiscot counties.

Native of the Old World; introduced into the United States, ranging from Florida and Texas to California, north to New England, Pennsylvania, Ohio, Illinois, and Missouri; also in Mexico, Central America, and in most tropical regions of the Old and New World.

Of all the species of *Echinochloa* in Missouri, this one is the smallest in all parts: in stature, length of leaf-blades (mainly 5–10 cm. long). Plants are cultivated in tropical Asia and Africa for their seeds, which are made into a flour for food.

2a. **Echinochloa crusgalli** (L.) Beauv. var. **crusgalli** f. **crusgalli** Barnyard Grass Map 338
*Echinochloa crusgalli* (L.) Beauv. [G, BB, P & S]
Flowers June–November.

Scattered throughout the state, where, it occurs along roadsides, moist ditches, moist ground along

337 Echinochloa colonum f. colonum (Jungle Rice)

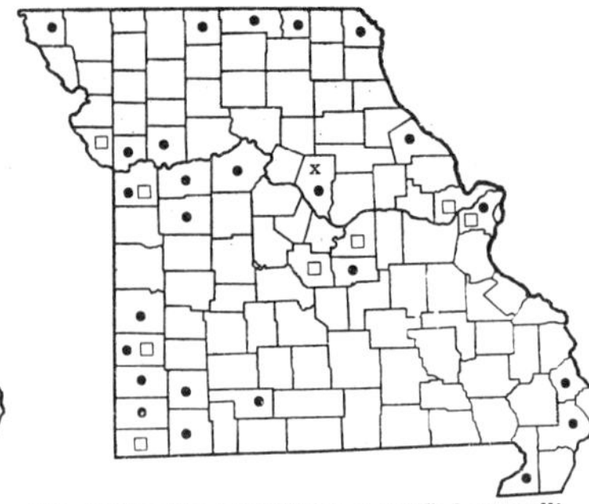

338 • Echinochloa crusgalli var. crusgalli f. crusgalli (Barnyard Grass)

338 □ Echinochloa crusgalli var. crusgalli f. longiseta

338 x Echinochloa crusgalli var. frumentacea

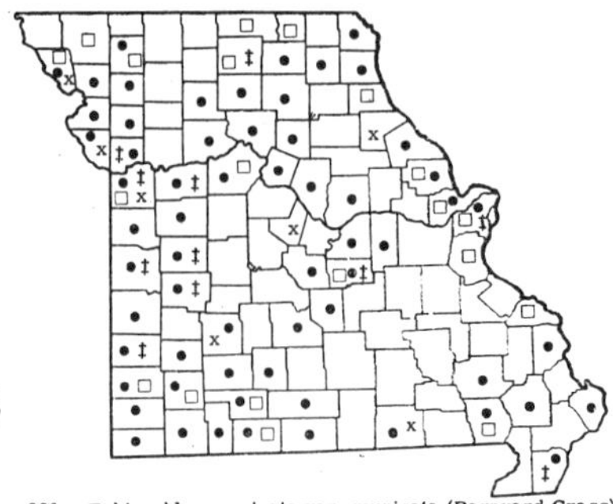

339 • Echinochloa muricata var. muricata (Barnyard Grass)

339 ‡ Echinochloa muricata var. ludoviciana (Barnyard Grass)

339 □ Echinochloa muricata var. occidentalis (Barnyard Grass)

339 x Echinochloa muricata var. microstachya (Barnyard Grass)

fields, pastures, streams, ponds, and in waste places. Apparently less common than *E. muricata.*

Native of the Old World; introduced in North America, extending from New Brunswick throughout most of the United States.

2b. **Echinochloa crusgalli** var. **crusgalli** f. **longiseta** (Trin.) Farw. Map 338

Scattered in the range of var. *crusgalli,* but less frequently collected.

2c. **Echinochloa crusgalli** var. **frumentacea** (Roxb.) Link Map 338

Japanese Barnyard Millet, Billion Dollar Grass

The name Japanese Millet, is more usually applied to *Setaria italica.*

Occasionally escaped from cultivation.

Native of eastern Asia; introduced as a forage plant in the United States.

This grass formerly had been introduced as a highly rated forage plant under the name Billion Dollar Grass. Seeds of this are said to be cultivated for food in the East Indies.

3a. **Echinochloa muricata** (Bcauv.) Fern. var. **muricata** Barnyard Grass Map 339

*Echinochloa muricata* (Michx.) Fern. as incorrect authorship, acc. to Fairbrothers, Rh. 58: 48. 1956

*Echinochloa pungens* (Poir.) Rydb. [G, BB, Fassett]

*Echinochloa crusgalli* (L.) Beauv. subsp. *muricata* (Michx.) Shinners, Rh. 56: 33. 1954.

Flowers July–October.

Frequents usually low or moist ground in the open or shade by borders of ponds, oxbow lakes in alluvial flood plains of wooded or open river bottoms, muddy banks and gravel bars of streams, dried up mud flats, waste, and cultivated ground; often a weed in gardens. Throughout the state, and apparently the commonest of the variations found.

Ranges from Maine to Minnesota, south to North Carolina, Tennessee, Arkansas, and Oklahoma.

3b. **Echinochloa muricata** var. **ludoviciana** Wieg. Map 339

*Echinochloa pungens* var. *ludoviciana* (Wieg.) Fern. & Grisc. [G, P & S]

*Echinochloa crusgalli* ssp. *muricata* var. *muricata* (Michx.) Farwell [of Shinners, in part Rh. 56: 34. 1954]

Ranges from Mississippi to New Mexico, north to Virginia, Pennsylvania, Tennessee, and Missouri.

3c. **Echinochloa muricata** var. **occidentalis** Wieg. Map 339

*Echinochloa pungens* var. *Wiegandii* Fassett [G, Fassett]

*Echinochloa pungens* var. *occidentalis* (Wieg.) Fern. & Grisc.

*Echinochloa occidentalis* (Wieg.) Rydb. [BB]

In similar situations as var. *muricata.* Scattered over the state.

Ranges from Maine to Rhode Island, Wisconsin

Plate no. 62. **1.** Zizaniopsis miliacea; *a.* Plant, × 1/5; *b.* Ripe caryopsis, × 2; *c.* Pistillate spikelet, × 2; *d.* Staminate spikelet, × 2. **2.** Panicum obtusum; *a.* Plant, × 1/5; *b.* Floret, × 10; *c.* Spikelet, × 10. **3.** Panicum capillare; *a.* Plant, × 1/5; *b.* Floret, × 4; *c, d.* Two views of spikelet, × 4. **4.** Echinochloa crusgalli; *a.* Plant, × 1/5; *b.* Floret, × 4; *c, d.* Two views of spikelet, × 4. **5.** Miscanthus sinensis; *a.* Plant, much reduced; *b.* Raceme, × 1/5; *c.* Spikelet, × 2. **6.** Andropogon virginicus; *a.* Plant, × 1/5; *b.* Spikelet with rachis joint and pedicel, × 5. **7.** Andropogon Gerardi; *a.* Plant, × 1/5; *b.* Pair of spikelets, × 2. **8.** Manisuris cylindrica; *a.* Plant, × 1/5; *b, c.* Two views of rachis joint with fertile and sterile spikelets attached.

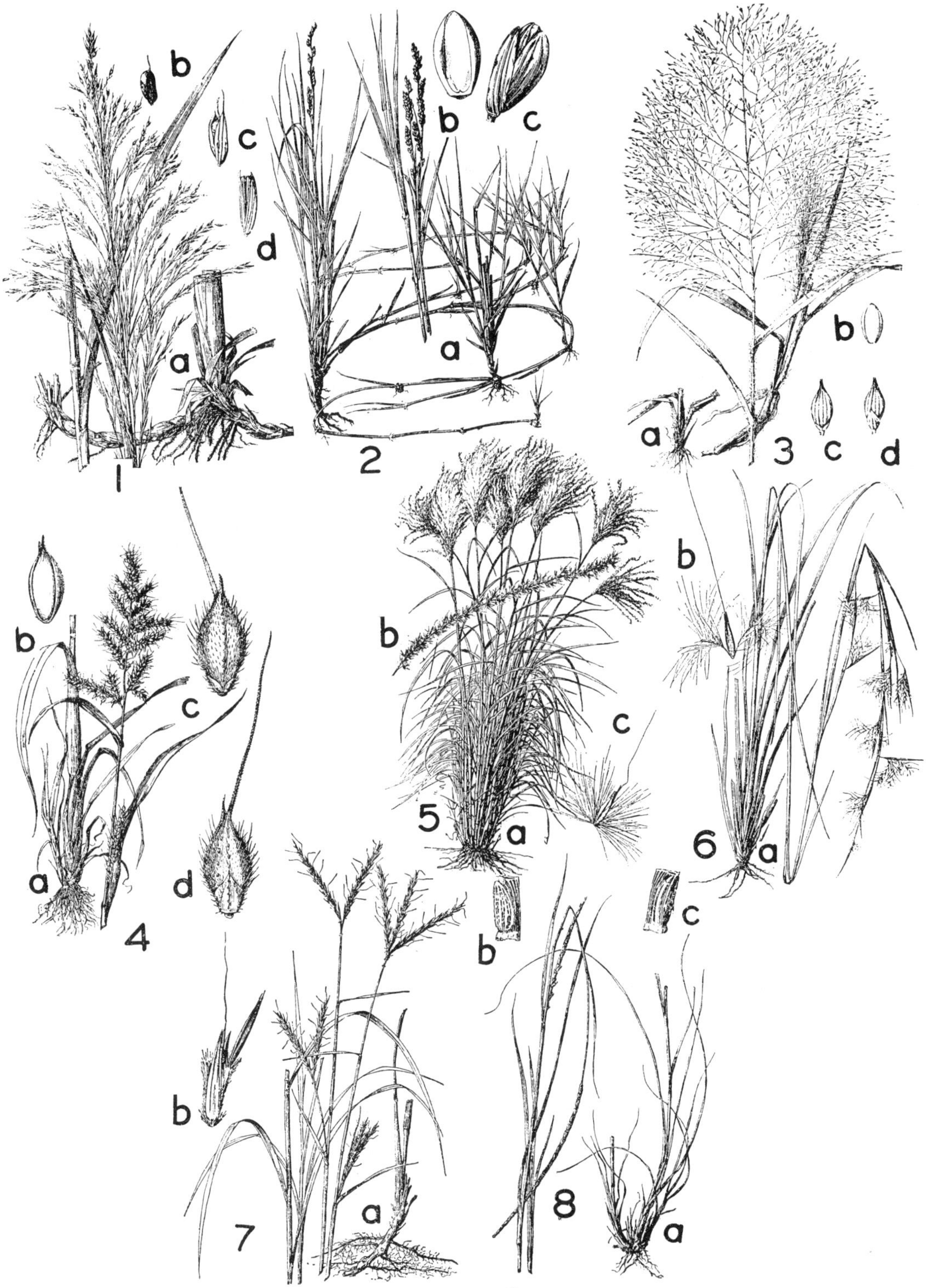


PLATE NO. 62

to Washington, south to Illinois, Missouri, Kansas, Texas, New Mexico, Arizona, and California.

3d. **Echinochloa muricata** var. **microstachya** Wieg. Map 339

*Echinochloa pungens* var. *microstachya* (Wieg.) Fern. & Grisc. [G, Fassett]

*Echinochloa crusgalli* var. *microstachya* (Wieg.) Shinners, Rh. 56: 34. 1954.

Similar situations as var. *muricata*. Scattered over the state, but apparently the rarest of the varieties in the state.

Ranges from Quebec and Ontario to Wyoming, south to New England, Pennsylvania, Michigan, Illinois, Missouri, Texas, and Mexico.

In view of recent changes in rules of nomenclature, the name *Echinochloa muricata* (Beauv.) Fern., as interpreted by Fairbrothers (Rh. 58: 48. 1956) is followed in the present treatment, and supplants the previously used *E. pungens* (Poir.) Rydb.

This species and others of the genus have been reported to be involved as possible causes of hay fever.

The seeds principally of *E. muricata* var. *occidentalis*, var. *microstachya*, other western varieties of *E. muricata* and of *E. crusgalli* var. *zelayensis* and other varieties, serve the Indians of the western United States as food in the form of a starchy flour.

The several varieties, although palatable, and occasionally cut for hay or used for forage, have never become economically important for grazing animals.

## 71. **Setaria** Beauv. Foxtail Grass

Annual or perennial grasses with dense cylindrical, spike-like inflorescences. Spikelets similar to *Panicum*, except that they are surrounded by an involucre of 1–many bristles, which persist after the spikelets fall. The ligule consists of a stiff band of white hairs of varying length. Spikelets may be articulated above, but usually below the glumes.

a. Leaf-sheaths with an edge of fine hairs on some part of the margins; each spikelet with 1–3 bristles attached below its base (caution must be used here not to include bristles from adjacent spikelets) . . *b*

b. Bristles of the inflorescence barbed downwardly (catching on to clothes or rough objects easily); lower length (internode) of inflorescence usually showing between the clusters of spikelets . 3a. S. VERTICILLATA var. VERTICILLATA

b. Bristles of the inflorescence barbed upwardly (not catching on clothing or to rough objects) . . . . *c*

c. Upper surface of leaf-blades with usually many hairs, less frequently with few or no hairs . 5. S. FABERII

c. Upper surface of leaf-blades completely glabrous (rarely with a few hairs) . . . . . . . *d*

d. Top of peduncle (main stalk supporting inflorescence) with backward- or downward-pointing rough hairs (retrorse-scabrous); leaf-blades loosely spreading. . 3b. S. VERTICILLATA var. AMBIGUA

d. Top of peduncle with upward-pointing fine appressed hairs (lying close to and parallel with axis); leaf-blades usually erect or ascending . . . . . . . . . . . . . . . . . *e*

e. Inflorescence narrowly cylindrical, dense, not interrupted, and of nearly same diameter from base to tip, 0.5–2.3 cm. thick, including the bristles; when seeds are mature, the whole spikelet falls together, the seed remaining covered by glumes and lemma; leaves 5–15 mm. wide . . . . . . . . . . . . . . . . . 4. S. VIRIDIS var. VIRIDIS

e. Inflorescence ovoid or broadest in lower half, lobed, or with numerous, short branches, interrupted, or in dwarfed specimens often dense and not interrupted, 1.5–4 cm. thick (sometimes in dwarfed specimens only 1 cm. thick), including the bristles; when seeds are mature, they fall from rest of spikelet, leaving the glumes and lemma to remain on the inflorescence; leaves often 15–30 mm. wide (sometimes narrower and only 6–7 mm. wide in dwarfed specimens) . . . . . . . . . . . . . . . . . . 6. S. ITALICA

a. Leaf-sheaths with glabrous margins; each spikelet with a cluster of 5–20 bristles attached below its base . . . . . . . . . . . . . . . . . . . . . . . . . . . *f*

f. Spikelets mainly 1.9–2.9 mm. (rarely 3) long; perennial plant; culms arising singly or in small tufts from branched underground rhizomes (culm breaks away easily above ground when not carefully collected with rhizome); each spikelet with a cluster of 8–12 bristles attached below its base; leaf-blade (upon drying) with a half twist or loose spiral . . . . . . . 2. S. GENICULATA

f. Spikelets 3–3.5 mm. long; annual plants; culms arising in large tufts from fibrous roots; each spikelet with a cluster of 5–20 bristles attached below its base; leaf-blades straight, without a twist . . . . . . . . . . . . . . . . . . . . . . 1. S. GLAUCA

340 Setaria glauca (Yellow Foxtail)

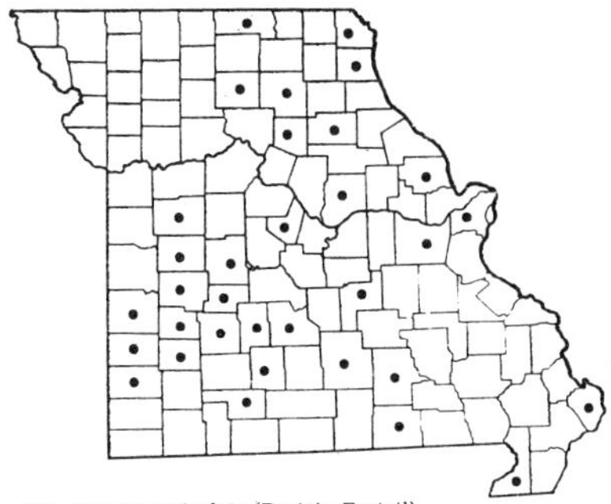

341 Setaria geniculata (Prairie Foxtail)

342 ● Setaria verticillata var. verticillata (Bristly Foxtail)
342 □ Setaria verticillata var. ambigua

1. **Setaria glauca** (L.) Beauv. Yellow Foxtail Map 340

*Setaria lutescens* (Weigel) Hubbard [P & S, Fassett]

Flowers early June–October.

Occurs as a weed in waste and cultivated ground, gravel bars along streams, along roadsides, railroad tracks, in fields, pastures, gardens, and open places.

Throughout Missouri.

Native of Europe and Asia; introduced into North America, ranging from Nova Scotia and Quebec to South Dakota, south to Florida, Texas, and westward to a lesser extent.

For a discussion of the proper Latin name of this plant, see John R. Reeder (Rh. 53: 27–30. 1951).

The bristles are at first green, but eventually turn a tawny, yellow-brown, or golden-brown color, and are $2$–$2\frac{3}{4}$ times the length of the spikelet. They are similar in color to those of the perennial species, *S. geniculata*. The culms and leaf-sheaths are flattened or compressed and keeled, giving the culms a broad, flat appearance, especially near the base where they are 5 mm. or more wide. The leaf-blades have a glaucous- or pale blue-green color on the upper surface. The upper surface of the blades is usually glabrous but occasionally may have a few long, slender hairs. As they dry, they present a half twist or loose spiral.

The seeds of this and the other species of *Setaria* are sometimes prepared into a flour or mealy substance.

2. **Setaria geniculata** (Lam.) Beauv.
Prairie Foxtail Map 341

Flowers July–October.

Occurs in prairies or prairie remnants along railroads, often in sandstone or cherty soil, and occasionally on limestone glades. Most common on prairies in the unglaciated prairie region of southwestern Missouri, scattered elsewhere in prairie openings in the northeastern half of the state and in parts of the Ozark section. Absent from northwestern Missouri, and most of the eastern Ozark section.

Ranges from Argentina north into Central America, Mexico, and the West Indies, and from Florida to Texas, north along the coast to Massachusetts, in the interior north to Pennsylvania, Illinois, Missouri, Kansas, New Mexico, and California.

This is the only perennial and native species of *Setaria* in Missouri. It is easily identified by the short underground branched rhizomes.

3a. **Setaria verticillata** (L.) Beauv. var. **verticillata** Bristly Foxtail, Bur Bristlegrass Map 342

Flowers early June–October.

Occurs along railroad tracks, waste places, gardens, and along streets in some of the larger cities. Rare, known from several counties in the northern half of the state.

Native of the Old World; introduced into the United States, ranging from New Hampshire to North Dakota, south to Virginia, Alabama, Louisiana, and Texas. Also introduced as a weed in gardens and coffee plantations in Guatemala, and in temperate and warm regions of the New and Old World.

In certain parts of Africa the mature inflorescences of this species serve the native inhabitants to keep rats away from stored grain; the adherence of the reversed barbed bristles of mature inflorescences to the fur of these animals acts as a nuisance value preventative.

3b. **Setaria verticillata** var. **ambigua** (Guss.) Parl. Map 342

*Setaria viridis* var. *ambigua* (Guss.) Cosson & Durand [BB]

Introduced along railroads, where known from one recent collection in St. Louis County (Terperance St. freight yard of Mo.-Pac. R. R. [refrigerator car division] along the wooden ramp, south of Rutger

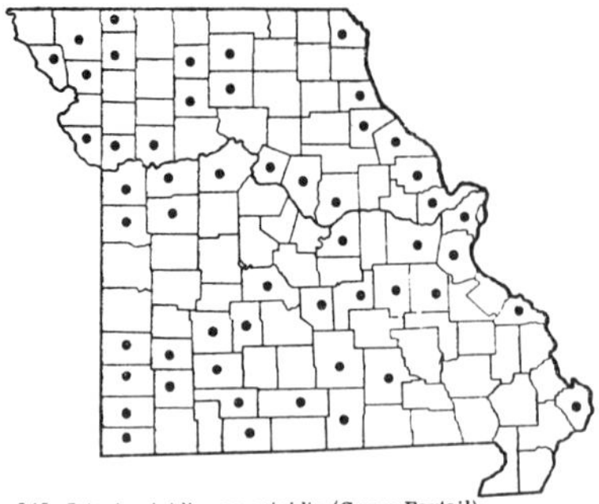

343 Setaria viridis var. viridis (Green Foxtail)

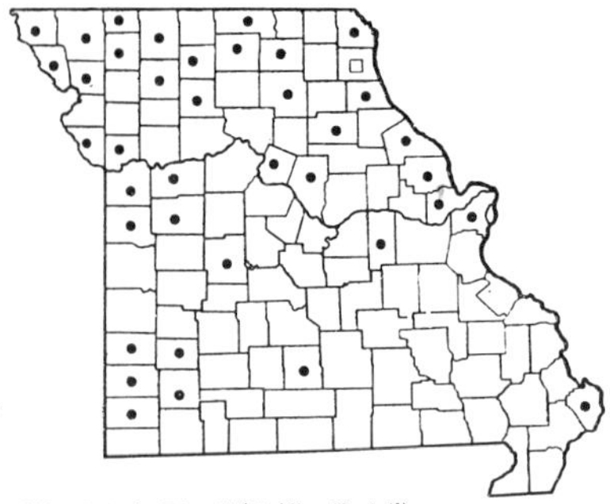

344 ● Setaria Faberii (Nodding Foxtail)
344 □ Setaria Faberii X Setaria viridis

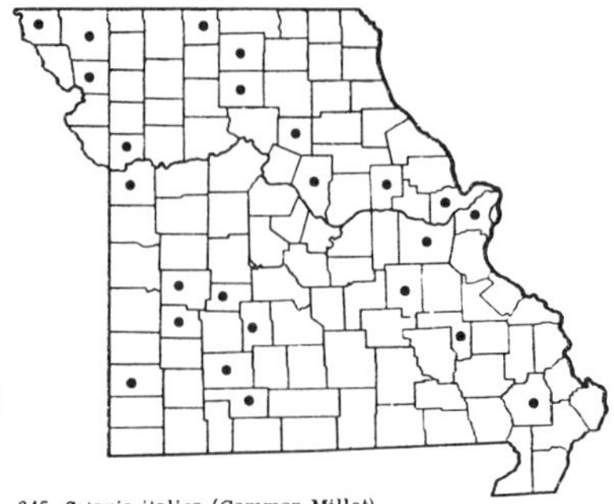

345 Setaria italica (Common Millet)

Street, St. Louis, August 4, 1957, *Muehlenbach 1295*).

The var. *ambigua*, in addition to the upwardly barbed bristles of the inflorescence, differs from var. *verticillata* in having the axis of the inflorescence scabrous (rough to the touch with short stiff projections) instead of hairy as in var. *verticillata*.

The inflorescences of both varieties are of a deep green color and mostly erect and tapering at the tip, and more or less lobed or divided into short rounded dense clusters. The small spikelets (averaging 2 mm. long) are subtended by usually one but rarely 2, bristles.

4. **Setaria viridis** (L.) Beauv. var. **viridis**
Green Foxtail — Map 343
Flowers June–October.

Similar to the habitat of the Yellow Foxtail (*S. glauca*) and frequently associated with it. Occurs throughout Missouri.

Native of Europe; introduced into North America, where naturalized from Newfoundland to British Columbia, south to Florida, California, and Mexico.

Aside from the difference in leaf-sheaths, this species is easily distinguished from *S. glauca*, which it resembles, by the much smaller spikelets 1.9–2.5 mm. long, with the bristles 3–4 instead of 2–2¾ times the length of the spikelets, the much more slender terete or subterete culms, and rather rounded, not compressed, leaf-sheaths, the culms mainly 1–2 mm. wide near the base, and by the deeper green leaf-blades, which are straight, not half-twisted. The spike-like inflorescence may be erect, but tends to nod slightly at the tip, whereas it remains stiffly erect in *S. lutescens*. The color of the bristles in *S. viridis* may be either green or purple, the younger inflorescences frequently being greener.

Although no Missouri specimens have been seen of *S. viridis* var. *Weinmanni* (R & S) Brand, its eventual occurrence in the state is very probable. It differs from *S. viridis* var. *viridis* in the more slender and shorter inflorescences (1–4 cm. long and 5–10 mm. thick), shorter bristles, shorter culms, and generally narrower leaves (to as narrow as 2 mm.). Its known range in North America is from Newfoundland to Ontario, Wisconsin, and Iowa, south to New England, Virginia, and elsewhere.

5. **Setaria Faberii** Herrm.
Nodding Foxtail — Map 344
Flowers July–October.

Introduced in cultivated fields, waste ground along roadsides, railroad tracks, gravel bars along streams, and open places. Rapidly spreading throughout the state.

Native of China; recently introduced into the United States, and known in this country since 1931; now recorded from Massachusetts and Pennsylvania to Illinois and Nebraska, south to North Carolina, Tennessee, and Missouri.

This grass is becoming more common each year, and it is known from all sections of the state except the southeastern quarter, where it has not yet been recorded, except from Mississippi County. It is now spreading along all the highways.

It is much taller and usually more robust in length and width of inflorescence and leaves than *S. viridis* var. *viridis*, which it resembles. The usually curved or nodding spikelike inflorescences, which droop from a point near their bases, and the usually more or less hairy upper surface of the leaf-blades are other characters used to distinguish this species. The spike-

Plate no. 63. **1.** Setaria Faberii, × 3/4. **2.** Setaria italica; *a.* Panicle and leaf, × 3/4; *b.* Floret, × 3 3/4. **3.** Setaria geniculata, × 3/4. **4.** Cenchrus longispinus; *a.* Bur, × 3 3/4; *b, c.* Two views of spikelet, × 3 3/4; *d.* Floret, × 3 3/4. **5.** Setaria viridis, × 3/4. **6.** Setaria verticillata, × 3/4. **7.** Setaria glauca; *a.* Plant, × 3/8; *b.* Floret, × 7 1/2; *c, d.* Two views of spikelet, × 7 1/2.

PLATE NO. 63

lets, moreover, are larger (2.6–3 mm. long) than those of *S. viridis* var. *viridis*. The lower surface of the leaves in *S. Faberii*, while often minutely pubescent, may also be glabrous, while *S. viridis* var. *viridis* has the leaves completely glabrous on both sides. Occasional plants of *S. Faberii* are found which are completely glabrous, as in *S. viridis* (Rh. 61: 220–23. 1959). The leaf-blades of *S. Faberii* are usually of a paler yellow-green as contrasted to a darker green of *S. viridis* var. *viridis*.

Sometimes plants are found which appear to be natural hydrids between *Setaria viridis* and *S. Faberii*. One such collection has been found in Lewis County. More, undoubtedly, will be collected.

6. **Setaria italica** (L.) Beauv.
Common Millet Map 345

This is also known under other common names, such as German, Italian, Hungarian, and Japanese Millet, and Hungarian Grass. *Echinochloa crusgalli* var. *frumentacea*, sometimes called Japanese Millet, is more properly designated Japanese Barnyard Millet. The Common Millet of Europe refers usually to *Panicum miliaceum*, not to *Setaria italica*.

Flowers July–October.

Escaped from cultivation to waste ground, roadsides, and along railroads. Scattered throughout the state.

Native of Europe and Asia; cultivated in the warmer parts of the United States, but escaped throughout the country.

This ranges through a number of variations, differing principally in the color of fruits and bristles and in the degree of lobing of the inflorescence.

Forms with fruit blackish, brownish black, or purplish black are usually called Hungarian Grass (*S. italica* subsp. *nigrofructa* Hubb.). Those with the fruit orange, yellowish to straw-colored or light brown are divided into forms with green bristles, purple bristles, and brown bristles. Further subdivision is carried out on the basis of the degree of lobing of the panicle. Forms of the so-called German Millet occur under the types having orange or yellowish to straw-colored or light brown fruits; in some forms of German Millet, the bristles may be green (*S. italica* subsp. *stramineofructa* Hubb.), in other forms of German Millet the bristles are purple (*S. italica* var. *Hostii* Hubb.).

For a detailed classification of these various subdivisions of the species, the work of F. T. Hubbard (Am. Jour. Bot. 2: 187–88. 1920) should be consulted.

In the United States *S. italica* is usually cultivated as a fodder plant. In parts of Asia the seeds are used as a food plant. At various times in Europe the seeds have been used in puddings and boiled in soups and sauces.

## 72. **Cenchrus** L. Sandbur

**Cenchrus longispinus** (Hack.) Fern. Sandbur
Map 346

*Cenchrus pauciflorus* of many recent authors, not Benth. [BB, Hitchcock, P & S]

Flowers late May–October.

Frequents waste ground, roadsides, railroads, and fields; especially common on sand or in sandy light soils of any kind.

Now widespread over the entire state.

Ranges from New Hampshire to Oregon, south to North Carolina, Kentucky, Missouri, Kansas, and New Mexico.

The spiny globelike burs of this grass easily become detached when mature and lie on the ground where they can cause sudden pain to barefoot children or unsuspecting fingers. These burs attach themselves easily to clothes or other objects with the slightest contact. The spines easily puncture the flesh of grazing animals, causing dangerous wounds. By lodging in the wool of sheep, the burs cause damage to the fleece. In some parts of Africa another species of *Cenchrus* is used by one of the tribes for its edible seeds, from which a cooling drink is made, or as an ingredient or constituent of porridge and other foods.

## Tribe XI. **ANDROPOGONEAE**

Grasses of this tribe have the spikelets occuring in usually 2's, sometimes 3's at each joint of the rachis, one of the spikelets usually stalkless (sessile), perfect and fertile, the other one or two stalked (pedicelled) and either staminate, sterile, or rarely perfect, or lacking, with only the stalk present. The perfect fertile spikelet consists of 1 terminal floret, the fertile lemma of which is often awned, and below this, a staminate or sterile floret represented by 2 or 1 empty awnless glumes. The spikelets are articulated below the glumes. The glumes are firm or hardened, while the sterile and fertile lemmas are thin and transparent.

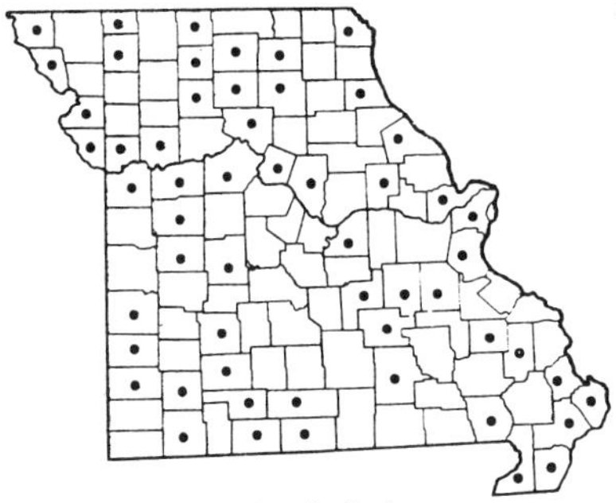

346 Cenchrus longispinus (Sandbur)

347 Miscanthus floridulus (Eulalia)

348 Erianthus strictus (Narrow Plume Grass)

## 73. **Miscanthus** Anderss.

Tall perennial grasses with long leaves and large inflorescences of numerous fan-shaped, plume-like showy, silky hairy racemes. The spikelets are all uniform, perfect, and occur in pairs along the rachis, the latter being continuous and not breaking up into segments.

**Miscanthus floridulus** (Lab.) Warb. Eulalia Map 347

Flowers September–November.

Escaped from cultivation along railroads, where known only from St. Louis County (along right-of-way of Terminal Railroad, between West Florissant Ave. and Broadway, opposite Hussmann Refrigerator Company plant, St. Louis, October 1, 1955, *Muehlenbach 822A*; and Carrie Avenue freight yard of Rock Island Railroad, in the northeastern corner, about 100 m. southeast of the station, St. Louis, November 27, 1955, *Muehlenbach 860*).

Native of eastern Asia; escaped from cultivation as an ornamental grass in the United States, where naturalized from Florida to Mississippi and Texas, north to Massachusetts, New York, Ohio, and Missouri.

Although frequently cultivated in Missouri, this species is known to have become naturalized in the state only from the above stations.

The Missouri material was originally identified and published under the name of *M. sinensis* Anderss. However, as pointed out to the present author by Drs. Fosberg and Van Schaack, that species has narrower leaves 3–5 mm. wide and barely exserted panicles, whereas the more commonly cultivated species with broader leaves and truly exserted panicles is *M. floridulus*.

## 74. **Erianthus** Michx. Plume Grass, Beard Grass

Tall perennial grasses with long leaves and long inflorescences of numerous crowded racemes often covered with silky hairs. One spikelet of each pair is sessile, the other stalked, and the spikelets occur along a jointed rachis which eventually breaks up into distinct segments.

a. Spikelets awnless, or the awn only 3–6 mm. long; cultivated species escaped along railroads or disturbed areas . . . . . . . . . . . . . . . . . . . . . . . . 3. E. RAVENNAE
a. Spikelets conspicuously awned, the awn 10–20 mm. long; native species of wild natural areas . . . *b*
  b. Hairs at base of spikelet much longer than the spikelet it surrounds; awn spiral or twisted at or near the base; plants of dry sandy or rocky open woods or open sandy slopes . . 2. E. ALOPECUROIDES
  b. Hairs at base of spikelet either none or of few hairs many times shorter than the spikelet; awn straight, not twisted; plants of moist or wet soil . . . . . . . . . . . . . 1. E. STRICTUS

1. **Erianthus strictus** Baldw.
Narrow Plume Grass Map 348

Flowers July–October.

Known only from swamps in Dunklin County, southeastern Missouri lowlands (Campbell, October 7, 1910, *Bush 6384*).

Ranges from Florida to Texas, north to Virginia, Tennessee, and southeastern Missouri.

A species of the Atlantic-Gulf Coastal Plain-Mississippi Embayment area found only once in the state. There is reason to believe that it will eventually be found in other sectors of the lowland area of south-

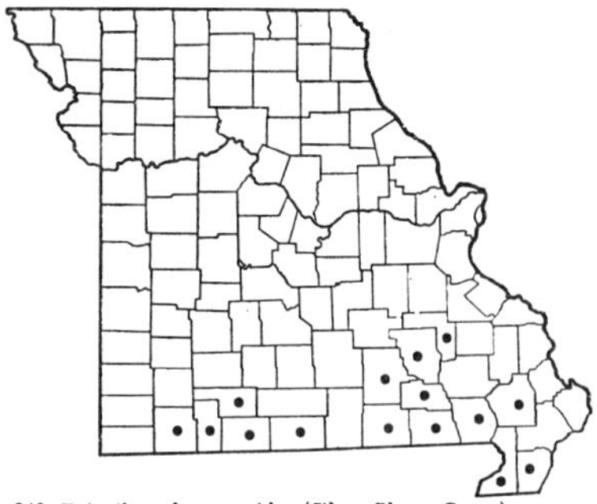

349 Erianthus alopecuroides (Silver Plume Grass)

350 Erianthus Ravennae (Plume Grass)

351 Arthraxon hispidus var. hispidus

eastern Missouri if such places can be located before the present march of industrialization and development has obliterated them.

2. **Erianthus alopecuroides** (L.) Ell.
Silver Plume Grass Map 349
Flowers September–October.

Frequents sandy or cherty, usually dry open woods or open sandy slopes and thickets in acid soils in southern Missouri, east to Stoddard and Pemiscot counties, north to Iron and Reynolds counties, and west to Barry County.

Ranges from Florida to Texas, north to Virginia, New Jersey, Ohio, Indiana, Illinois, Missouri, and Oklahoma.

This is a very showy species, attaining a height of 3 meters. The beautiful large silvery-white to cream-colored beard of the inflorescence is a most striking feature of the plant. It certainly ranks among the showiest and most spectacular of the native Missouri grasses and merits horticultural recognition for its great ornamental value. It may not respond easily to cultivation or transplanting, however, as it seems to require an acid soil. Initial experiments which I carried on to raise the species from seeds have failed.

3. **Erianthus Ravennae** (L.) Beauv.
Plume Grass, Ravenna Grass Map 350
Flowers September–November.

Escaped from cultivation and rarely established. Known only from railroad tracks in St. Louis County (along ditch west of the right-of-way of Burlington Railroad, north of East Grand Avenue, just opposite Cemco Manufacturing & Supply Company's building, October 17, 1954, *Muehlenbach 500*).

Native of southern Europe; cultivated as an ornamental grass in the United States, and escaped and naturalized from Maryland southward and west to Arizona.

This ornamental grass is often cultivated, but rarely escapes and becomes naturalized, the above station being the only one thus far known in the state. The grass attains 3 meters in height. The spikelets are surrounded by a dense silvery beard which lends beauty to this showy species.

## 75. **Arthraxon** Beauv.

Grasses with broad deeply heart-shaped leaf-blades and digitate spikes of 1-flowered, sessile spikelets. An awn arises from the back of the 2-toothed fertile lemma, and the seed is very narrowly linear.

**Arthraxon hispidus** (Thunb.) Makino var. **hispidus** Map 351
*Arthraxon hispidus* (Thunb.) Makino [G]
Flowers September–October.

Known only from St. Louis County (growing about the margins of the northeast portion of the lagoon at the base of 'Art Hill,' Forest Park, St. Louis, September, 1933, *Steyermark*).

Native to eastern Asia; the Missouri station is the only one known in the United States.

From Florida to Arkansas, north to New York is found *Arthraxon hispidus* var. *cryptatherus* (Hack.) Honda, the more commonly collected, although rare variety in the United States. Like *A. hispidus* var. *hispidus*, it is a native of eastern Asia, but differs from *A. hispidus* var. *hispidus* in the minute or only slightly exserted awn, whereas in *A. hispidus* var. *hispidus* the awn is 5–9 mm. long and conspicuously exserted.

With its papillose-hispid leaf-sheaths and broad ovate-lanceolate, heart-shaped leaf-blades ciliate-margined at the base, this grass might superficially be mistaken for *Panicum clandestinum*. However, the creeping culms rooting at their nodes, together with the awned spikelets arranged in digitate spikes, quickly dispel any further resemblance to *Panicum clandestinum*.

This grass was reported originally by the author in Rhodora 36: 313. 1934, and subsequently in Palmer & Steyermark, *Annotated Catalogue of Flowering Plants*, as *Arthraxon hispidus* var. *cryptatherus*, the name by which it was determined at the United States National Herbarium. However, as correctly shown by Dr. Fernald, in the eighth edition of *Gray's Manual*, the long awn would place the Missouri collection with typical *A. hispidus* var. *hispidus* rather than with the minute or short-awned var. *cryptatherus*.

## 76. **Andropogon** L. Beard Grass

Perennial grasses in large clumps with solid stems and long narrow leaf-blades. Inflorescences of solitary, paired, digitate, or panicled racemes. At each joint of the rachis occur spikelets in pairs (of two kinds), one sessile and perfect, the other stalked and staminate, sterile, or not developed. The fertile lemma, which is shorter than the glumes, usually ends in a long awn. The rachis and usually the pedicels (stalks of spikelet) are covered with long silky hairs.

a. Only 1 raceme (unbranched or separated part) attached to the tip of each peduncle (main support of inflorescence) and, even when crowded, appearing at different levels or heights, or arising from different axes . . . . . . 1. A. SCOPARIUS
a. Two or more racemes attached to the tip of each peduncle (main support of inflorescence) and crowded together to appear as a unit arising from the same axis . . . . . . *b*
  b. Inflorescence of a lavender, brown-purple, bronze, yellow-brown, green and purple color, or at least not producing long white, gray-white, or silvery-white silky hairs 5–10 mm. long which equal or are longer than sessile spikelets . . . . . . 2. A. GERARDI
  b. Inflorescence with conspicuous long white, gray-white, or silvery-white silky hairs 5–10 mm. long, which equal or are usually longer than the sessile spikelets . . . . . . *c*
    c. Inflorescence a dense narrow panicle of usually 10 or more short racemes (unbranched lateral units) on an elongated axis . . . . . . 6. A. SACCHAROIDES
    c. Inflorescence of 2–3 racemes arranged digitately (like fingers on a hand or resembling a V-shape) . . . . . . *d*
      d. Mature racemes (unbranched units of inflorescence) mostly straight or closely flowered, rather stiff; sessile spikelet 5–6 mm. long, equaling or longer than the sterile pedicel (stalk) alongside it, broadly lanceolate; leaf-sheaths of lower part of plant not strongly keeled nor alternately overlapping the base of each other in 2 ranks (not equitant) . . . 3. A. TERNARIUS
      d. Mature racemes rather loosely curving or gracefully flexuous (alternately curving in opposite directions); sessile spikelet 3–5 mm. long, narrowly or linear-lanceolate, shorter than the sterile pedicel (stalk) alongside it; leaf-sheaths of lower part of plant compressed, keeled at the back, and alternately overlapping the base of each other in 2 ranks (equitant) . . *e*
        e. No tufts of hair occurring on the culms just below the nodes; awn of fertile sessile spikelet not spiraling or twisted at the base; the shortened leaf-like parts (spathe) at the side or base of each inflorescence not inflated nor strongly overlapping at maturity. 4. A. VIRGINICUS
        e. Conspicuous tufts of white hairs occurring on the culms just below the upper nodes at least (best observed at the base of the leaf-like spathes); awn of fertile sessile spikelet spiraling or twisted at the base; the shortened leaf-like parts (spathe) at the side or base of each inflorescence inflated at maturity or strongly overlapping . . . 5. A. ELLIOTTII

1. **Andropogon scoparius** Michx. Little Bluestem, Broom, Broom Beardgrass Map 352

*Andropogon scoparius* Michx. (typical) [G]
*Andropogon scoparius* var. *scoparius* [BB]
*Andropogon scoparius* var. *genuinus* Fern. & Grisc. [P & S]
*Andropogon scoparius* var. *frequens* Hubbard [G, P & S]
*Andropogon scoparius* var. *divergens* (Anderss.) Hack. [P & S]
*Andropogon scoparius* var. *polycladus* Scribn. & Ball [G]
*Andropogon scoparius* var. *neomexicanus* (Nash) Hitchc. [G]
*Andropogon scoparius* f. *calvescens* Fern. [G]

Flowers September–October.

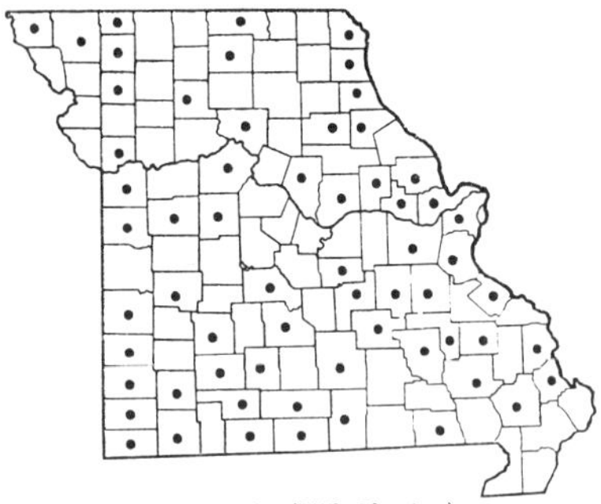

352 Andropogon scoparius (Little Bluestem)

353 • Andropogon Gerardi var. Gerardi (Big Bluestem)
353 □ Andropogon Gerardi var. chrysocomus

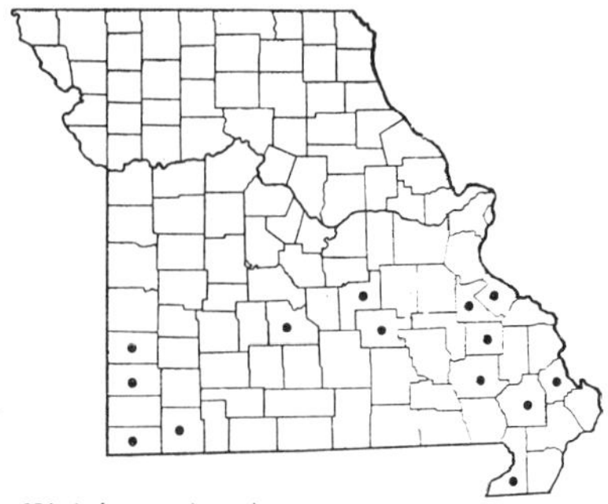

354 Andropogon ternarius

Occurs in prairies, glades, rocky open woods, abandoned fields, sandy open ground, waste places, and along railroads.

Throughout Missouri, except absent in the southeastern lowlands.

Ranges from New Brunswick and Quebec to Alberta, south to Florida, Texas, and Arizona.

In the beginning at flowering time, in a young stage of development, or when moist, the racemes are more or less straight and slender, but at maturity or upon drying, the racemes become curved or arched and expose the grayish-white hairs which stand out on the rachis-joints and sterile pedicels. The ligule is a brownish membrane about 1.5 mm. high with a subtruncate, irregularly broken border often fringed with tiny white hairs.

Several varieties of *Andropogon scoparius* have been described by Hubbard (Rh. 19: 100–105. 1917) and by Fernald & Griscom (Rh. 37: 143–46. 1935). An attempt has been made to fit the abundant Missouri material on hand for study into some of these varieties, but, as will be judged from the above synonymy, the treatment by Fernald and Griscom has not been found to be applicable nor adequate. So little correlation can be found between the bearding on the joints of the rachis and the length of the glumes of the fertile spikelets, or between the number of spikelets in a raceme and the length of the sterile rudiment, that there appear to be no well-defined constant correlations in these characters with the Missouri material. In this respect, the writer is in agreement with Fassett (*Grasses of Wisconsin*, p. 92. 1951) who found Fernald & Griscom's treatment unworkable for Wisconsin. Deam (*Flora of Indiana*, pp. 178–79. 1940, *Grasses of Indiana*, p. 318. 1929) likewise was unable to find adequate criteria for a satisfactory separation of the Indiana material.

Practically all the Missouri material examined has the basal third of the joints of the rachis without any beard and would be interpreted, following Fernald's treatment, as *A. scoparius* var. *scoparius* f. *calvescens*, f. *scoparius*, or var. *polycladus*, but a few might be placed in var. *frequens* or var. *neomexicanus* where bearding is denser. The most obvious variations in the Missouri plants are in glabrous versus pubescent culms and leaves, and in their green versus their silvery-blue or silvery-glaucous appearance. Particularly striking are plants on the 'bald knobs' and limestone glades in southwestern Missouri, whole colonies of which for acres in extent possess silvery-glaucous or blue-glaucous culms and leaves, the latter varying from glabrous to strongly villous. Such plants could be referred to *A. scoparius* var. *scoparius* f. *scoparius*, f. *calvescens*, or var. *polycladus*, with some showing more bearding near the basal third and transitional toward var. *neomexicanus*. One such collection (bald knob, Malva, Taney County, September 17, 1924, *Palmer 26170*) has been identified by Fernald as var. *neomexicanus* in the Gray Herbarium. Shinners (Rh. 56: 37. 1954) notes that this variety in Texas has 'the whole plant . . . pale or glaucous.' A collection from Taney County (*Steyermark 66368*) representing this glaucous phase shows sparse bearding on the lower third of some rachis joints as in var. *neomexicanus*, but most of the joints are free of bearding in the lower third and are thus characteristic of *A. scoparius* f. *calvescens*. The glaucous and green phases in other parts of Missouri often occur together, such as *Steyermark 25957* and *25958* from Lincoln County, *Steyermark 23839* and *23840* from Webster County, *Steyermark 24040* and *24041* from

---

Plate no. 64. **1.** Erianthus alopecuroides, × 3/8. **2.** Erianthus Ravennae, × 3/4. **3.** Erianthus strictus, × 3/8. **4.** Andropogon Elliottii, × 3/4. **5.** Arthraxon hispidus var. cryptatherus, × 3/4. **6.** Andropogon ternarius, × 3/4. **7.** Andropogon scoparius; *a.* Plant, × 3/8; *b.* Pair of spikelets, × 3 3/4. **8.** Andropogon saccharoides, × 3/4.

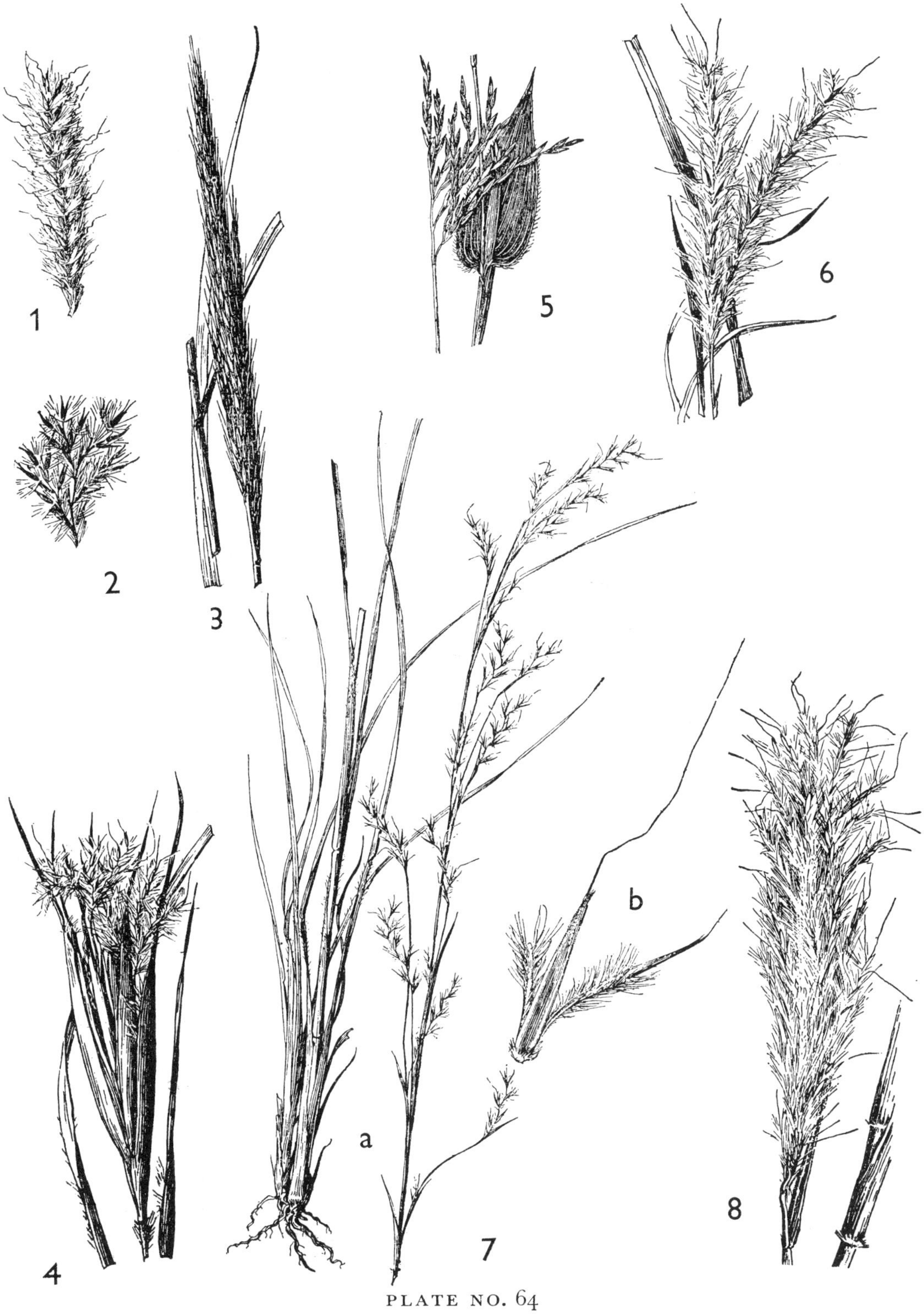


PLATE NO. 64

Polk County, and *Steyermark 25341* and *25342* from Phelps County, with sometimes the green phase and other times the glaucous phase predominating. Since both glabrous or pubescent glaucous plants may be found along with glabrous or pubescent green plants among the populations on the glades and elsewhere, and variations may be found in the amount of bearding on the rachis joints, the applicability of the names var. *scoparius* f. *scoparius* and f. *calvescens*, and var. *neomexicanus*, to fit given plants is uncertain.

*Andropogon scoparius* is a valuable forage grass and provides good grazing when young. It is considered to be of some significance in causing hay fever. In late autumn and through the winter, Little Bluestem is conspicuous over large sections of open landscape, covering bare fields, prairies, and railroad sidings with ornamental clumps of russet-brown or tan dried leaves and stems.

2. **Andropogon Gerardi** Vitman Big Bluestem, Bluejoint Turkeyfoot Map 353

Flowers June–September.

Occurs in prairies, glades, open rocky woods, and along railroads.

Missouri material may be classified into two varieties:

Hairs of rachis and pedicels white, 1–2 mm. long; glumes of sessile spikelet usually scabrous with short rough projections; common type throughout Missouri . . 2a. A. GERARDI var. GERARDI

Hairs of rachis and pedicels creamy to yellowish-white, 2–4 mm. long; glumes of sessile spikelet smoother; southwestern Missouri . 2b. A. GERARDI var. CHRYSOCOMUS

2a. **Andropogon Gerardi** var. **Gerardi** Map 353

*Andropogon provincialis* Lam. [P & S]

*Andropogon Gerardi* Vitman [G]

Throughout Missouri, except absent from the southeastern lowland section, where not known in a wild state nor recorded.

Ranges from Central America, Mexico, and Florida to Texas, north to Maine, Quebec, Ontario, Michigan, Minnesota, Manitoba, Saskatchewan, and Wyoming.

2b. **Andropogon Gerardi** var. **chrysocomus** (Nash) Fern. Map 353

*Andropogon provincialis* var. *chrysocomus* (Nash) Fern. & Griscom [P & S]

Prairies and rocky open or waste ground.

Occurs in western Missouri south of the Missouri River in Jackson, Jasper, Newton (1 mi. east of Christopher, August 18, 1953, *Palmer 56526*), and Stone counties.

Ranges from Quebec to Montana, south to Connecticut, New York, Indiana, Arkansas, Texas, and New Mexico.

Another more western variety, var. *paucipilus* (Nash) Fern., with glabrous racemes and the internodes possessing only a few weak hairs, occurs in Nebraska and Montana. It should be looked for on the loess hills of northwestern Missouri, where it may possibly be found.

Plants either glabrous or pubescent and yellow-green or silvery-glaucous occur throughout the range. Plants vary in height from usually nearly as tall or taller than the average man to individuals with secondary or late-developing culms only 5 dm. or so tall, or up to 2.5 meters. The first inflorescences, the terminal or topmost ones, of the season, usually bear 3 or more (6–12) digitate racemes; the later and lateral ones may often bear only a pair of racemes. In the beginning and during flowering time, the racemes spread apart, but after flowering and at maturity contract and become erect. The ligule is a brownish rather firm membrane 2–3 mm. high gradually rounded to subtruncate at the slightly uneven or somewhat broken margin fringed by a minute border of white hairs. Sometimes behind the ligule at the base of the upper surface of the leaf-blade are found long equal hairs arranged in a row. After frost the plants usually take on a bronze-purple or russet brown with lavender color.

Like *A. scoparius*, this grass sometimes may contribute as a causative in hay fever cases. *Andropogon Gerardi* var. *Gerardi* is of considerable value as a forage grass, and ranks near the top of the list of native grasses for the quality and quantity of its production.

3. **Andropogon ternarius** Michx. Map 354

Flowers July–October.

Prairies, along railroads, and rocky or sandy open woods or glades, especially on sandstone and chert exposures.

Occurs in southern Missouri, north to Ste. Genevieve, Phelps, Laclede, and Barton counties.

Ranges from Florida to Texas, north to New Jersey, Delaware, Maryland, Kentucky, and Missouri.

The ligule is a brownish membrane about 1 mm. high with a subtruncate summit usually fringed with tiny white hairs. A conspicuous upwardly-pointing tuft of silky white hairs surrounds the culm just below each of the upper nodes and spathes. The silky white rather stiff racemes protruding from long peduncles give this species a very distinctive and ornamental

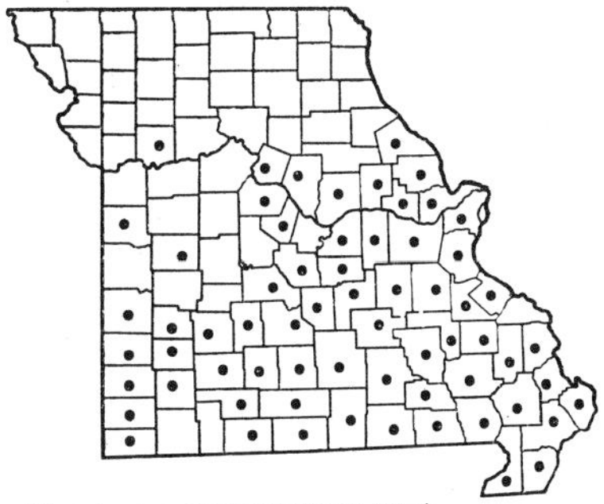

355 Andropogon virginicus (Broom Sedge)

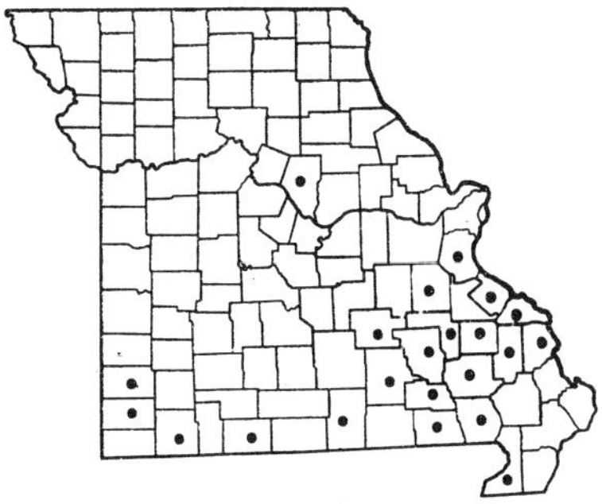

356 Andropogon Elliottii var. Elliottii

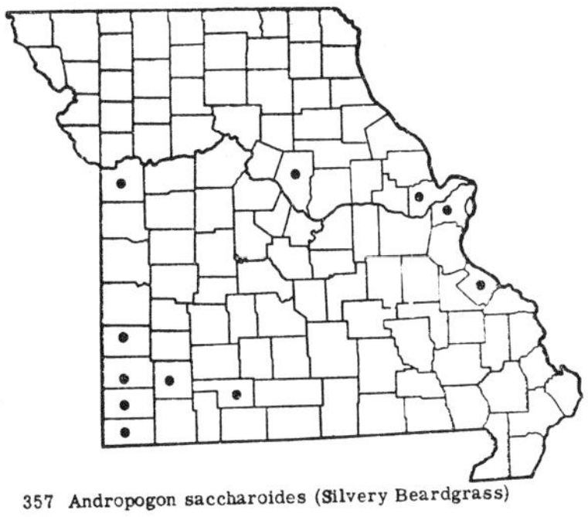

357 Andropogon saccharoides (Silvery Beardgrass)

appearance. The silky white inflorescences continue to remain on the plant well into winter.

The racemes of this species are straight and more rigid usually than in *A. Elliottii* or *A. virginicus* and terminate long erect peduncles. Occasional specimens of *A. ternarius* with more laxly-flowered or less straight racemes may be distinguished by other characters presented in the key.

4. **Andropogon virginicus** L.

Broom Sedge Map 355

*Andropogon virginicus* var. *virginicus* [BB]

*Andropogon virginicus* var. *genuinus* Fern. & Grisc. [P & S]

*Andropogon virginicus* var. *tetrastachyus* (Ell.) Hack. [P & S]

Flowers August–November.

Occurs in prairies, glades of sandstone, chert, or granite, sandy and fallow or abandoned fields, along railroad tracks, rarely in wet open and swampy places. In southern and central Missouri north to Pike, Howard, Ray, and Cass counties; common throughout the Ozarks, southeastern lowland, and unglaciated prairie sections of the state.

Ranges from Florida to Texas and Mexico, north to Massachusetts, New York, Ohio, Indiana, Illinois, Missouri, Kansas, and Oklahoma; also West Indies and Central America.

The ligule is a yellow-brown membrane with a truncate hairy white-fringed border. It is smaller than in *A. Gerardi* and *A. scoparius*, 0.5–1 mm. long.

As in *A. scoparius*, Fernald and Griscom proposed the subdivision of *A. virginicus* into several varieties based upon the branching of the inflorescence, length of racemes, color of leaves, and pubescence of spathes. Missouri material was placed by these authors in *A. virginicus* var. *virginicus* (their var. *genuinus*) and in var. *tetrastachyus* (Ell.) Hack., the latter being separated on the basis of branching instead of simple or nearly simple inflorescence. Again, however, as in the case of *A. scoparius*, a division of the species has not been found workable for the Missouri material. Most of the Missouri specimens can be placed in the simple type of inflorescence, which characterizes *A. virginicus* var. *virginicus*, whereas the var. *tetrastachyus* is less frequently encountered.

The occurrence of this grass usually indicates a definitely acid soil. Great masses of it are found in fallow or abandoned fields and sandy or cherty open soils. Like Little Bluestem (*A. scoparius*), the dried clumps provide a rather striking splash of russet-brown or tan color to the bare or exposed fields of the landscape during the late fall and winter months. Early American housewives of the southern states employed the culms of this grass, using them as makeshift brooms, hence the name broom-sedge. When young, the grass provides considerable forage value for livestock.

5. **Andropogon Elliottii** Chapm. var. **Elliottii** Map 356

*Andropogon Elliottii* var. *projectus* Fern. & Grisc. [BB, P & S]

Flowers September–November.

Occurs most frequently in acid soils derived from sandstone, chert, or granitic rocks, either on rocky glades of these substrata or in rocky open woodland or border of woodland with acid soil, sometimes also in rocky acid prairie and in sterile soils along roadsides, rarely along railroad tracks.

Found in southern and east-central Missouri, and locally north in Boone County (railroad near Providence, Missouri River, October 7, 1953, *Kucera*). Frequent in the eastern Ozark section, but absent from the western Ozarks. Absent mostly from the southeastern lowland section.

Ranges from Florida to Texas, north to New Jersey, Pennsylvania, Kentucky, Ohio, Indiana, Illinois, and Missouri; also in Cuba and British Honduras.

The ligule is 0.5–1 mm. long and similar to that

of *A. virginicus* and *A. ternarius*. As in *A. ternarius*, a conspicuous tuft of long white hairs surrounds the culm just below each of the upper nodes, and is best developed at the base of the leaflike spathes.

Like *A. ternarius* and *A. saccharoides*, *A. Elliottii* is a very beautiful and ornamental species, and lends a distinct yellow-brown charm to the winter landscape with its clumps of dried leaves and culms. The writer has attempted several times to establish this species and *A. ternarius* by transplanting them to acid-soil habitats in his wildflower garden. He has not been successful with these acid-soil-inhabiting species, whereas no difficulty was experienced in transplanting and getting established plants of *A. scoparius* and *A. Gerardi*.

*Andropogon Elliotti* var. *projectus* was described by Fernald and Griscom in 1935 (Rh. 37: 139. 1935), based upon plants in which most of the racemes are exserted on long peduncles from the leaflike spathes. This variety is rejected by Deam (*Fl. Indiana*, p. 179. 1940), however, as merely a phase in the development of the inflorescence; apparently, Fernald himself became convinced of the artificial status of the variety, since he does not recognize it as worthy of separation in the eighth edition of *Gray's Manual*, uniting it with typical *A. Elliottii* var. *Elliottii* by stating in the description (p. 234), 'racemes . . . included within or exserted from the large sheaths on erect peduncles.'

*Andropogon Elliottii* var. *Elliottii* is sometimes found in association with *A. virginicus* and *A. ternarius*, all of them thriving on acid soils.

6. **Andropogon saccharoides** Sw.

Silver Beardgrass Map 357

Flowers July–October.

Occurs in prairies, glades, on rocky ledges, and along railroad tracks, where found scattered in southern and central Missouri north to St. Charles, Boone, and Jackson counties. The grass is native in southwestern Missouri, but is introduced elsewhere in the state.

Ranges from South America, Central America, and the West Indies to Mexico, and from Alabama, Texas, and California, north to Missouri, Kansas, Colorado, and Arizona.

The silvery-white dense, plumelike inflorescences are very showy and ornamental. The ligule is a white membrane 1–1.5 mm. high with a slightly broken subtruncate summit.

## 77. **Sorghum** Adans. Sorghum

(Also spelled *Sorgum*)

Tall grasses with long narrow or broad leaves and large panicles of 1- to 5-jointed short racemes which eventually break apart. Spikelets occur in pairs (one sessile and fertile, the other stalked and sterile or staminate) or in 3's at the ends of the branches (2 of them stalked and staminate or empty).

Plants perennial, creeping with scaly underground rhizomes; leaf-blades usually 10–20 mm. wide; panicles rather open with spreading branches; spikelets broadly lanceolate, disarticulating from the pedicel at maturity . . . . . . . . . . . . . . . . . . . . . . . . . 1. S. HALEPENSE

Plant annual, without any creeping underground rhizomes; leaf-blades up to 75 mm. wide; panicle either open with spreading branches or dense and compact; spikelet at maturity not disarticulating from the pedicel . . . . . . . . . . . . . . . . . 2. SORGHUM (cultivated species and varieties)

1. **Sorghum halepense** (L.) Pers.

Johnson Grass Map 358

Also known as Egyptian Millet and by several other common names.

Flowers June–November.

Very weedy and occupying large areas in low meadows, fertile cultivated ground, along roadsides, railroad tracks, at the base of wooded slopes in thickets, and borders of cultivated fields.

Introduced and established mainly in the southern, central, and western parts of the state, north to St. Charles, Boone, Livingston, Caldwell, and Holt counties; not recorded from the northeastern quarter of the state.

Native of Europe and Asia; introduced into the United States from Turkey about 1830. About 1840 Col. William Johnson introduced it in Alabama. It has spread from Florida and Texas north to Massachusetts, New York, Ohio, Indiana, Illinois, and Iowa; also introduced and a weed in Mexico, the West Indies, Central and South America.

The open loosely spreading inflorescence has a dull purplish or rosy-purple color tinged with green. The culms may reach 3 meters in height. The leaf-blades

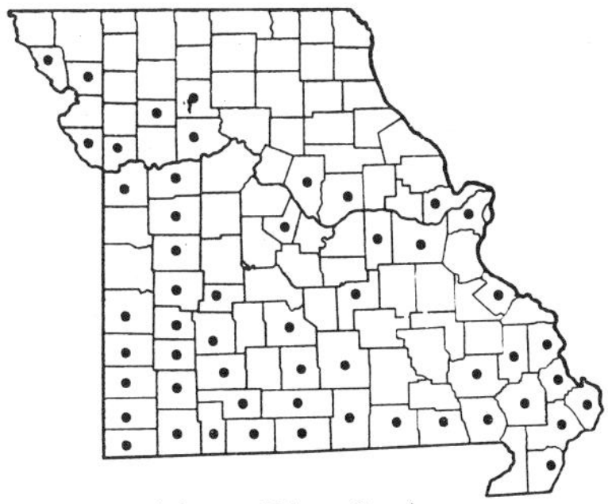

358 Sorghum halepense (Johnson Grass)

359 ‡ Sorghum sudanense (Sudan Grass)
359 □ Sorghum Dochna var. Dochna (Sorgo)
359 x Sorghum Dochna var. technicum (Broom Corn)
359 • Sorghum Durra (Durra)

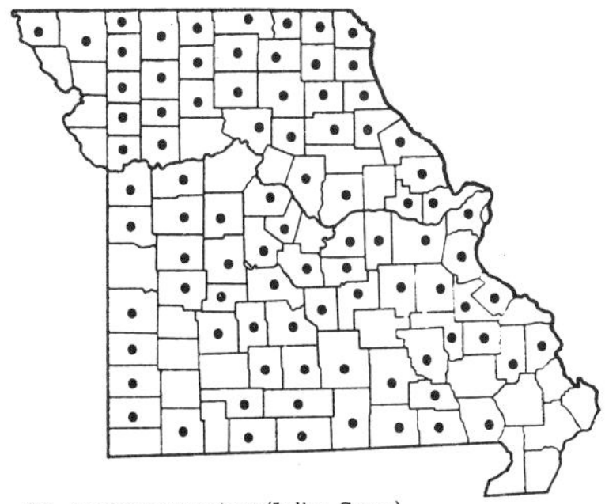

360 Sorghastrum nutans (Indian Grass)

have a conspicuous whitish midnerve, and a hairy-fringed ligule.

Although cultivated for forage, especially in the southern states, it becomes a troublesome weed, forming dense colonies with its creeping rhizomes and covering the ground to the exclusion of other plants. Once having invaded an area, the plant is nearly impossible to eradicate. It can be eradicated by repeated deep plowing, then allowing the ground to lie fallow, followed by cultivating with a different kind of crop. Farmers are being warned not to introduce it in cultivation for hay crop, as it can invade valuable cotton, corn, and potato fields. During its time of flowering it is suspected of causing much hay fever. After a frost, the leaves yield prussic acid, which may fatally poison cattle. A second crop is also likely to develop the poison, especially during a hot dry season. The leaves are often blotched with reddish-purple, sign of a bacterial disease, with which the plant is commonly afflicted.

2. **Sorghum** (cultivated species and varieties) Map 359

Flowers June–October.

Commonly cultivated, and occasionally escaped from cultivation along roadsides and railroad tracks.

Various species or varieties, depending upon personal points of view, are grown in Missouri, either in experimental plots, from which they occasionally escape, or are intensively cultivated as crops.

Some of the most intensive work on the group has been done by Stapf and by Snowden. For more detailed information their works should be consulted (Stapf, *Fl. Trop. Africa* 9, pt. 1: 104–54. 1917; Snowden, J. D. Classification of the Cultivated Sorghums. Kew Bull. 1935: 221–55. 1935: Snowden, J. D. *The Cultivated Races of Sorghum.* 1936). They consider the cultivated *Sorghums* mostly as distinct species, rather than treating them as varieties under *Sorghum vulgare* Pers., as had been done by some previous workers.

The following treatment is based upon the above cited works, and especially that of Snowden, the key presented intended not so much as a definitive contrastive type as ordinarily found in the other portions of this flora, but rather as a general orientation to the main types of *Sorghum* which may be expected to be seen in Missouri.

a. Mature spikelets closed, not exposing the grain . *b*
   b. Culms usually more than 6 mm. in diameter; leaf-blades mainly more than 20 mm. wide; leaf-sheaths longer than the internodes . 2f. Chicken Corn (S. DRUMMONDII)
   b. Culms usually less than 6 mm. in diameter; leaf-blades mainly less than 20 mm. wide; leaf-sheaths shorter than the internodes . 2g. Sudan Grass (S. SUDANENSE)
a. Mature spikelets open and exposing the grain . *c*
   c. Inflorescence a long much-branched loose panicle with elongated branches naked (without flowers) below, arising from a short axis; spikelets usually awned; culm stout, 3–4 meters tall, dry; grown for the branching inflorescence for brooms . . . 2b. Broom Corn (S. DOCHNA var. TECHNICUM)
   c. Inflorescence mostly a dense and compact panicle, sometimes loose and open, erect or recurved (peduncle goosenecked); spikelets with or without awns; culms stout, dry or juicy; grown for the sugar in the culm or forage or seeds as food for livestock and man . . . . . . . . . . . . *d*
      d. Inflorescence a dense or loose panicle, erect or recurved (peduncle goosenecked); culm stout, 1.8–2 meters tall, with an abundant sweet juice; grown for the sugar and syrup in the culm and for forage. . 2a. Sorgho, Sweet Sorghum (S. DOCHNA var. DOCHNA)

d. Inflorescence usually a dense and compact panicle; culm stout, with an acid dry or slightly juicy culm; grown mostly for the seed (grain), but also the culms and leaves grown for forage . . . . *e*

e. Spikelets not awned; seeds mostly black, or red or white; inflorescence oblong or cylindrical, erect; culms with slightly juicy, acid pith. . . 2c. Kafir Corn (S. CAFFRORUM)

e. Spikelets awned or without awns; inflorescence mostly recurved (goose-necked) or erect; culms with dry pith. . . . 2d. Durra (S. DURRA)

e. Spikelets awned or without awns; inflorescence erect or somewhat curving in a flexuous manner; culms with dry pith . . . . . . 2e. Feterita (S. CAUDATUM var. FETERITA)

2a. **Sorghum Dochna** (Forsk.) Snowden var. **Dochna** Sweet Sorghum, Sorgho Map 359
*Sorghum vulgare* var. *saccharatum* (L.) Boerl.
*S. saccharatum* Moench.

Infrequently escaped from cultivation along roadsides and waste ground. Known only from a collection from Jackson County (waste ground, Courtney, September 23, 1933, *Bush 47*).

Sweet Sorghum produces culms with a sweet juicy pith from which is extracted a syrup. Sorghum molasses is a product of this syrup, and is used for flavoring pancakes, as a confectionery and ingredient in the preparation of cakes and cookies, and as a medicine. Sweet Sorghum is also grown for its forage and silage value and is extensively cultivated in Missouri.

2b. **Sorghum Dochna** var. **technicum** (Koern.) Snowden Broom Corn Map 593
*Sorghum vulgare* var. *technicum* (Koern.) Jav.

Very rarely escaped from cultivation, and known as such from St. Louis County (North St. Louis freight yard of Burlington Railroad along the highway which branches off the foot of Carrie Avenue to the north, St. Louis, August 6, 1955, *Muehlenbach 711*).

The elongated panicle branches are used in the manufacture of brushes and brooms.

2c. **Sorghum caffrorum** Beauv. Kafir Corn Map 359
*Sorghum vulgare* var. *caffrorum* (Thunb.) Hubb. and Rehder

This has not yet been recorded as an escape but is widely grown in Missouri as a forage plant for its slightly juicy acid pith, and for the seeds. The Standard Blackhull Kafir with black seeds is the most commonly grown. In other parts of the world these seeds are valued for their flavor and nutritive value, which is similar to corn or maize in its composition and food value.

2d. **Sorghum Durra** Stapf Durra Map 359
*Sorghum vulgare* var. *Durra* (Forsk.) Hubb. & Rehder

This has been found as an escape in Jackson County (Courtney, August 19, 1933, *Bush 12895*) and was identified by Bush as *Sorghum cernuum*.

Grown for its grain. The seeds of the white durra (*Sorghum cernuum* Host var. *cernuum* [Koern.] Snowden) are a favorite poultry feed in California. In India and some parts of southwestern Asia and northern Africa, white durra is widely cultivated.

Milo corn (*Sorghum subglabrescens* Schweinf. & Aschers.) is sometimes considered a form of the durra sorghum. It is widely cultivated in Missouri.

2e. **Sorghum caudatum** Stapf var. **caudatum** Feterita Map 359
*Sorghum vulgare* var. *caudatum* (Hack.) Bailey
*Sorghum caudatum* var. *feterita* Stapf

This is cultivated in Missouri, but has not been recorded as an escape.

2f. **Sorghum Drummondii** (Steud.) Millsp. & Chase Chicken Corn Map 359
*Sorghum vulgare* var. *Drummondii* (Nees) Hitchc.

Sometimes cultivated and reported as a weed in cornfields in southern Indiana. It has an erect, dense panicle with ascending branches and small orange, oval seeds.

2g. **Sorghum sudanense** Stapf Sudan Grass Map 359
*Sorghum vulgare* var. *sudanense* (Piper) Hitchc.

Commonly escaped along roadsides and waste ground, where it is scattered over the state. This is the most commonly encountered as an escape of the cultivated annual species of *Sorghum*. The flowers vary from green to purplish. It is on the increase as a forage plant, producing a crop more certain to thrive and mature in poor dry upland soils than will corn. It is being widely grown in many sections of the state.

Plate no. 65. **1.** Sorghastrum nutans; *a*. Plant, × 3/8; *b*. Spikelet with pedicel and rachis joint, × 3 3/4. **2.** Sorghum halepense; *a*. Plant, × 3/8; *b*, *c*. Two views of terminal raceme, × 3 3/4.

PLATE NO. 65

The above enumerated types of cultivated *Sorghum* are those most likely to be found in Missouri in cultivation or as escapes. Others not mentioned may be cultivated or eventually found as escapes.

Considerable pollen is produced from the various cultivated *Sorghums*, and the cause of some hay fever is sometimes attributed to them. Under certain conditions, many of the cultivated *Sorghums* are able to poison livestock because of prussic acid contained in the plant. It is stated that silage made from *Sorghum* seldom possesses sufficient poison to injure livestock.

Sorghum grain oil is obtained from the seeds of some of the *Sorghums* and found to contain two-thirds as much oil as corn oil, for which it may be used as a substitute for salad oils or as an ingredient in oleomargarine. A wax similar to that of the Carnauba (obtained from the Carnauba palm) is also obtained from the seeds. The culms of one of the African cultivated *Sorghums* is considered to be a source for pulp in papermaking.

## 78. **Sorghastrum** Nash

Tall perennial grasses with long narrow leaf-blades, prominently auricled leaf-sheaths, and elongate panicles. The perfect fertile spikelets are sessile at each joint of the rachis. The racemes are reduced to 2 or 3 joints, the sterile spikelets represented only by a pedicel (stalked part).

**Sorghastrum nutans** (L.) Nash Indian Grass Map 360

Flowers August–September.

Frequents prairies, glades, and rocky open woods; also along railroads. Found in nearly every county of Missouri, except absent from the southeastern lowland section, where not native.

Ranges from Florida to Texas and Mexico, north to Maine, Ontario, Manitoba, North Dakota, and Wyoming.

The stiffish, elongated cartilaginous brown auricles at the base of the leaf-blades are quite characteristic. The culms are quite smooth except for the finely appressed-pubescent nodes. At first the branches of the inflorescence during the flowering period are more or less open and spreading-ascending from the main axis. After flowering, they become contracted and bend in against the main axis of the inflorescence. The fertile spikelet, without the hairs, has a rich chestnut-brown color. The hairs surrounding the fertile spikelet at first are golden brown to fulvous brown, but at late maturity and, upon drying, become grayish. This change in aging accounts for some of the color discrepancies given in descriptions of this species.

## 79. **Manisuris** L. Joint Grass

Glabrous perennial grasses with slender, cylindrical, cartilaginous solitary spikes breaking off from the axis at maturity. The spikelets occur in pairs in cavities at the nodes of the cylindrical jointed axis.

**Manisuris cylindrica** (Michx.) Ktze.
Joint Grass Map 361

Flowers late May–August.

Found in prairies and sandy open ground, where known only from Mississippi, Barton, and Jackson counties.

Ranges from Florida to Texas, north to South Carolina, Missouri, and Oklahoma.

The slender solitary inflorescence resembles the pistillate portion of the inflorescence of the next genus, *Tripsacum.*

## Tribe XII. **TRIPSACEAE**

Grasses of this tribe have pithy stems, spikelets of 1-sexed flowers with the pistillate below, and the staminate toward the tip of the same inflorescence, or the staminate and pistillate flowers in separate inflorescences on the same plant (monoecious).

---

Plate no. 66. **1.** Tripsacum dactyloides; *a.* Plant, × 3/8; *b.* Pair of staminate spikelets with rachis joint, × 3 3/4; *c.* Pistillate spikelet with rachis joint, × 3 3/4. **2.** Zea mays; *a.* 2 branches of staminate inflorescence (tassel), × 3/8; *b.* Pistillate inflorescence (ear), × 3/8; *c.* Staminate spikelet, × 1 1/2; *d.* Single pistillate spikelet soon after flowering, × 3; *e.* Pair of pistillate spikelets attached to rachis (cob) with mature grains, the second glume showing, × 1 1/2.

PLATE NO. 66

361 Manisuris cylindrica (Joint Grass)

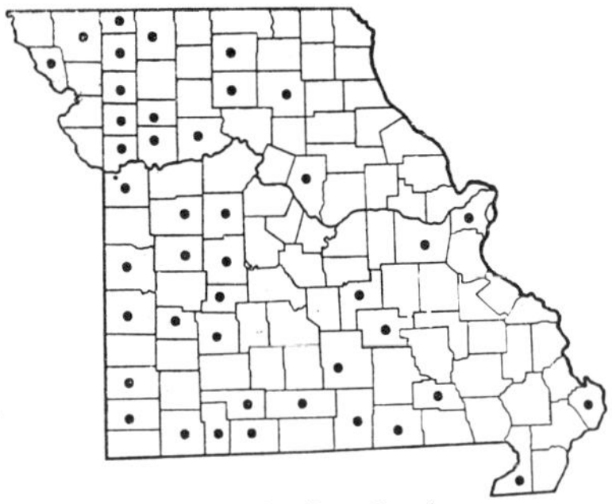

362 Tripsacum dactyloides (Gama Grass)

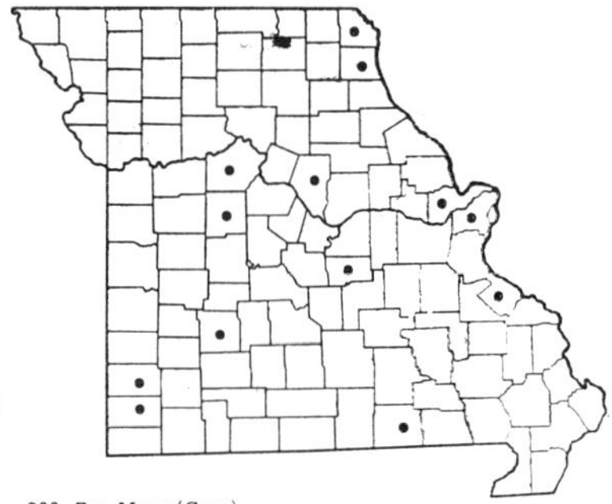

363 Zea Mays (Corn)

## 80. **Tripsacum** L. Gama Grass

Rather tall stout perennials growing in large clumps with thick rhizomes, broad flat leaves, the staminate and pistillate flowers in separate parts of the same many-flowered spikes. The pistillate spikelets are solitary and occur in hollowed portions on opposite sides of the thickened hard joints of the lower part of the rachis; this pistillate portion breaks up at maturity into several 1-seeded joints. The staminate spikelets are 2-flowered and in pairs on one side of a continuous rachis.

**Tripsacum dactyloides** L. Gama Grass Map 362

Flowers May–September.

Frequents prairies, open limestone slopes, borders of woods and thickets, fields, waste, and cultivated ground, and along roadsides and railroads. Scattered over the state.

Ranges from Florida to Texas and Mexico, north to Massachusetts, New York, Michigan, Illinois, Iowa, and Nebraska.

The ligule is a short membrane with a ring of hairs.

At times hundreds of plants of this species may be present in natural prairies located in the southwestern quarter of the state, and occur in large clumps which form conspicuous patches of olive-green leaves and culms.

## 81. **Zea** L. Maize, Indian Corn, Corn

Robust annual grasses with long, spreading, wide leaves. The ligule is membranous with an uneven, broken-edged hairy border. The staminate spikelets are 2-flowered, in pairs on one side of a continuous rachis; the top tassel (a racemose panicle) of the plant is completely staminate. The pistillate inflorescence, borne below in the axils of the leaves, is a short-stemmed spike (ear) enveloped within numerous green spathes (husks).

The pistillate spikes consist of usually 8–30 rows of sessile spikelets surrounding a thick axis (the cob). The pistillate spikelets occur in pairs, with one fertile and one sterile floret, the glumes broad and rounded or indented (emarginate) at the summit; the long slender styles protrude from the summit of the pistillate spike-like silky threads. At maturity the grain is much longer than the glumes.

**Zea Mays** L. Maize, Indian Corn, Corn Map 363

Flowers June–August.

Occasionally found self-seeding itself in rich or rocky woods where unintentionally dropped or carried by man, squirrel, and other animals. Also found as waifs occasionally along roads, railroad tracks, dump heaps, and in waste ground, but not self-seeding nor showing signs of persisting.

Scattered over the state.

Native probably of South America.

The pollen of corn sometimes causes hay fever, but, fortunately, due to the large size of the pollen grains, is only to a slight extent carried around in the atmosphere.

The dried stalks of corn are sometimes used as pulp for making into cardboard.

## Fam. **CYPERACEAE** (Sedge Family)

Plants resembling grasses, but the stems (culms) usually solid or sometimes hollow, often 3-angled, but also round, flattened, or 4- or more-angled. Leaves growing from base or on stem, those on stem usually in 3 ranks and with a sheath usually closed (except at summit); leaves are sometimes absent or inconspicuous because of reduced size. A ligule (scaly membrane) is usually present at the junction of leaf-sheath and leaf-blade. The either perfect or 1-sexed flowers are greatly reduced, each in the axil of a single scale or glume, which overlap into spirally arranged or 2-rowed spikelets or spikes. The perianth is lacking or composed of bristles or scale-like sepals. Stamens 3, 2, or 1. Pistil 1, with 1 style, which is 2–3-cleft, and a superior 1-celled ovary with 1 erect ovule; at maturity this pistil becomes an achene, which may be flattened, 3-sided, or rounded, and is free from the wall (pericarp) of the fruit and not attached to it.

In many instances, it will be noted that satisfactory and certain identification of the sedges is possible only with mature or fruiting plants. It is necessary, therefore, that this be kept in mind when trying to arrive at a particular identification.

Many species of this family shed an abundance of very buoyant pollen, but do not appear to be of any significance in causing hay fever.

a. Ordinary leaves apparently not present on the plant, only the culms (stems) evident . . . . . . . *b*
  b. Inflorescence of 1 spikelet at the top of the stem (culm); achenes crowned with a tubercle. . 3. ELEOCHARIS
  b. Inflorescence of 2 or more spikelets, or, if only of 1 spikelet, this not at very tip of stem; achenes lacking a tubercle. . . . . . . . . . . . . . . . . . . . . . . 6. SCIRPUS
a. Ordinary leaves present, either occurring at base of plant, on culm, or both . . . . . . . . . *c*
  c. Culm (stem) hollow, jointed, terete (round when cut cross-wise), and with many, short leaf-blades; inflorescence arising from the sides of the culm in the axils of the leaf-sheaths. . . . 2. DULICHIUM
  c. Without the above combination of characters . . . . . . . . . . . . . . . . . . *d*
    d. Spikelets not all alike, because the staminate (male) and pistillate (female) flowers are in separate parts of the same inflorescence or in completely separated inflorescences, or the pistillate part is either (1) a pistil surrounded by a tight- or loose-fitting sac (perigynium) or (2) in fruit becomes a miniature white or buff golf-ball-shaped hard achene (fruit) . . . . . *e*
      e. Pistillate part (spikelet) 1-flowered, becoming in fruit a miniature white or buff golf-ball-shaped hard achene (fruit) . . . . . . . . . . . . . . . . . . . . 10. SCLERIA
      e. Pistillate part (spikelet) of more than 1 flower, each pistil surrounded by a tight-or loose-fitting sac (perigynium) . . . . . . . . . . . . . . . . . . . . 11. CAREX
    d. Some or all of the flowers with stamens and pistil in the same flower (perfect), the spikelets appearing to be all alike or essentially so, or at least *none* of the pistillate flowers surrounded by a sac (perigynium) nor in fruit becoming a hard white or buff golf-ball-shaped achene . . . *f*
      f. Bristles, scales, or both present at the base of each flower, sometimes only 1 minute scale present . . . . . . . . . . . . . . . . . . . . . . . . . . *g*
        g. Some bristles (hair-like) at the base of each flower . . . . . . . . . . . . . . *h*
          h. Three bristles alternate with 3 enlarged stalked scales at the base of each flower; some of the leaf-sheaths hairy with evident hairs up to 1.5 mm. long . . . . . . 7. FUIRENA
          h. Only the bristles present at the base of a flower; leaf-sheaths without hairs, at most somewhat roughened (scabrous) . . . . . . . . . . . . . . . . . . . *i*
            i. A triangular or long-beaked extension crowns the top of the achene or ovary from which it is separated by a dividing line; only 1–2 fertile and perfect flowers in a spikelet, located at or near the top, the other flowers sterile, staminate, or imperfect . . . . . . . . . . . . . . . . . . . 9. RHYNCHOSPORA
            i. Achene or ovary tapering directly into the summit without any extra beak or extension present marked by any line of separation; flowers all fertile and perfect and few to many. . . . . . . . . . . . . . . . . . . . . 6. SCIRPUS
        g. No bristles and only 1 minute scale present at base of each flower . . . . 8. HEMICARPHA
      f. No bristles or scales or any other attachment at base of each flower . . . . . . . . . . *j*
        j. A minute tubercle crowns the achene and is separated from it by a dividing line 4. BULBOSTYLIS
        j. No tubercle present on top of achene . . . . . . . . . . . . . . . . . . *k*
          k. Scales of spikelets in 2 ranks, alternating on 2 sides of the axis (rachis) of the spikelet, giving the spikelet a mostly flattened appearance (but immature specimens have a more terete [rounded] or less compressed appearance) . . . . . . . . . . 1. CYPERUS

k. Scales of spikelets spirally arranged, shingle-like (imbricate) in several ranks, presenting a more rounded or cone-shaped appearance . . . . . . . . . . . . . . . . . . . . . . *l*
l. Base of style larger than rest of style . . . . . . . . . . . . . . . . 5. FIMBRISTYLIS
l. Base of style slender, not larger than rest of style . . . . . . . . . . . . . . . . . . . *m*
m. Spikelets more or less cone- or almost ball-shaped, broadest at the base or near the middle, 1½–4 times as long as broad, 1–2.5 mm. wide, well-rounded from all sides; spikelets many-flowered . . . . . . . . . . . . . . . . . . . . . . . . 6. SCIRPUS
m. Spikelets long and narrow (linear-lanceolate or subulate), either conspicuously much longer than broad and 4–15 times longer than broad, of nearly the same width throughout, or, if only 2½–3 times longer than broad, then the spikelets 1–3-flowered . . . . . . . . . 1. CYPERUS

## 1. **Cyperus** L. Umbrella Sedge

Annual or perennial plants with usually triangular culms. Leaves mostly crowded at the base, with 1 or more at the summit of the culm at the base of the umbel-like spikes or heads forming the inflorescence. Spikelets 1–many-flowered, either more or less flattened with the scales 2-ranked, or narrowly cylindrical (terete) or 4-angled. Flowers perfect, without a perianth. Stamens 1–3. Styles 2–3-cleft. Achenes lenticular (flattened with 2 convex surfaces) or triangular.

The references to Kukenthal in this section refer to his recent treatment of the genus (Das Pflanzenreich IV: 20. Heft. 101. Nov. 1956).

a. Scales of spikelets narrowed at tip into a conspicuous outwardly curved awn 1/3–3/8 total length of scale; dried plants or bruised fresh ones giving off an odor of Slippery Elm (*Ulmus rubra*). 5. C. ARISTATUS
a. Without the above combination of characters . . . . . . . . . . . . . . . . . . . . *b*
b. Style split into 2 thread-like divisions; achene or ovary lenticular (rather flattened with 2 convex surfaces or with 2 sides or 2 angles) . . . . . . . . . . . . . . . . . . . . . *c*
c. Flower only 1 in each spikelet; spikelets only slightly longer (2–3 times) than broad. 26. C. TENUIFOLIUS
c. Flowers many (6–35) in each spikelet; spikelets conspicuously longer (4–15 times) than broad . . *d*
d. Scales of spikelet yellow, greenish, yellow-green, or yellow-brown . . . . . . . . . . *e*
e. Spikelets mainly 1–1.9 mm. wide; tips of the scales of spikelets very slightly pointed (mucronulate), prolonged, and projecting slightly from the next scale above. 4. C. POLYSTACHYOS var. TEXENSIS
e. Spikelets mainly 2–3 mm. wide; tips of the scales of spikelets blunt (obtuse), closely appressed to the next scale above. . . . . . . . . . . . . . . . . . . . .
f. Mature achene (seed-like fruit) black, with the cells on the surface (superficial) of a rectangular-oblong shape longer than broad (use magnifying lens of over 15 ×) . 1. C. FLAVESCENS
f. Mature achene brownish or grayish with the cells on the surface 4–5-sided and of about the same diameter throughout (use magnifying lens of over 15 ×) . 3b. C. RIVULARIS var. RIVULARIS f. ELUTUS
d. Scales of spikelet tinged more or less with red-purple, brown-purple, reddish-brown, or mahogany-chestnut-brown . . . . . . . . . . . . . . . . . . . . . . . *g*
g. Style cleft nearly to its base; branches of styles protruding 2–4 mm. beyond scale, usually many of such and remaining attached after flowering; scales of spikelet mainly 2.5–2.7 mm. long, rather loosely overlapping and causing base of achene to be visible in dried specimens . . . . . . . . . . . . . . . . . . . . 2. C. DIANDRUS
g. Style cleft only to about the middle; branches of style protruding mainly 1–1.5 mm. beyond scale, usually few of such and soon falling after flowering or easily falling with the touch of a dissecting needle; scales of spikelet mainly 2–2.4 mm. long, closely overlapping to cause the achenes to be hidden . . 3a. C. RIVULARIS var. RIVULARIS f. RIVULARIS
b. Style split into 3 thread-like divisions; achene or ovary 3-sided (trigonous) . . . . . . . . *h*
h. Scales of spikelet purple-red, reddish-tinged, brown-purple, or reddish-brown . . . . . . . *i*
i. Tip of scale not reaching the base of the next scale above it on the same side, the scales not overlapping, exposing part of the zig-zag joints of the axis of the rachis at flowering time 13. C. ENGELMANNI
i. Tip of each scale reaching the middle or almost the middle of the next scale above it on the same side, the scales prominently overlapping, the axis of the rachis not exposed at flowering time . . . . . . . . . . . . . . . . . . . . . . . . . *j*
j. Scales of spikelet 1–1.5 mm. long; achene 0.8–1 mm. long, grayish-white . 9. C. ERYTHRORHIZOS

j. Scales of spikelet mainly 2–4 (rarely 1.5) mm. long; achene 1–2 mm. long (measure mature ones only), golden brown, gray-brown, pale brown, reddish-brown, or blackish . . . . . . . . . *k*

k. Scales of spikelet mainly 2–2.9 mm. long, conspicuously nerved, loosely overlapping; spikelets 1–2 mm. wide; achenes gray-brown, pale brown, golden-brown, or reddish-brown; annuals without runners at base, or perennials with weak scaly runners (stolons) . . . . . . . . *l*

l. Annuals with numerous fine roots; scales of spikelets dark reddish- or mahogany-brown; spikelets crowded or densely clustered, in age each joint of the rachilla breaking up into separate segments between the scales; scales at extreme tip of spikelet prominently narrowed to give the spikelet a long-pointed or acuminate summit . . . . . . 12. C. FERRUGINESCENS

l. Perennials with slender underground scaly runners (stolons) with sometimes a tuber at the end (careful examination should show at least some sign of a scaly runner, which is usually visible among the finer roots); scales of spikelets with some light reddish-brown, otherwise yellow- to golden-brown; spikelets not crowded or densely massed, with noticeable spacing between spikelets, in age the joints of the rachilla not breaking up into segments, but remaining intact; scales at extreme tip of the spikelet slightly narrowed, but not giving the spikelet a long-pointed or acuminate summit . . . . . . . . . . . 17. C. ESCULENTUS

k. Scales of spikelet mainly 3–4 mm. long (rarely less than 3), faintly nerved or without nerves, closely overlapping and the tips appressed to adjacent scales; spikelets about 2 mm. wide; achenes blackish; perennials sending out firm tough horizontal runners from the base of the plant . . . . . . . . . . . . . . . . . . . . . . . . . . . . *m*

m. Most of the leaf-like bracts of the involucre (at base of inflorescence) longer than the rays (branches) of the inflorescence; runners at the base of plant stout, forming a new horizontal underground stem from their tip; scales of spikelet mainly 3.8–4 mm. long . . 15. C. SETIGERUS

m. Most of the leaf-like bracts of the involucre shorter than or equaling the rays of the inflorescence; runners thread-like, usually ending in little enlargements (tubers); scales of spikelet mainly 2.5–3.5 mm. long . . . . . . . . . . . . 14. C. ROTUNDUS

h. Scales of spikelet mainly brown, yellow, green, straw-color, golden brown, gray-brown, greenish-white, or *combinations* of these shades . . . . . . . . . . . . . . . . . . . . *n*

n. At least some part of the culm (stem) rough to the touch or minutely pubescent . . . . . . . *o*

o. Spikelets flattened in appearance, 1.5–4.5 mm. wide, spreading to ascending (except the lowermost ones of a spike), not crowded and not conspicuously turned down (reflexed); spikelets mainly 5–16-flowered (up to 20), mainly 10–25 mm. long; scales of spikelet 3.5–4.5 mm. long, with conspicuous awns . . . . . . . . . . . . . . . . . . . . *p*

p. Spikelets pinnately or racemosely arranged and ascending; scales with an acuminate tip or the midrib prolonged to 1 mm. long; achenes 2.5–3.5 mm. long . . . 22. C. SCHWEINITZII

p. Spikelets digitately arranged or giving that appearance, the axis very shortened, mainly spreading or somewhat ascending; scales more rounded, blunt, or with a midrib prolonged 0.3–0.5 mm. long; achenes 2.2–2.5 mm. long . . . . . . . . 22a. C. × MESOCHORUS (C. Schweinitzii × C. filiculmis)

o. Spikelets subulate, not flattened, about 1 mm. wide, all except the uppermost conspicuously pointing down or turned down (reflexed) at maturity, thickly crowded in an obovoid or obconic head; spikelets 1–3-flowered, 3–11 mm. long; scales of spikelets not as above. . . . . *q*

q. Rays (branches) of inflorescence rough (scabrous); leaves of involucre (at base of inflorescence) rough; heads (inflorescences) obconic, noticeably narrowed at the base; spikelets sharp-pointed (pungent) . . . . . . . . . . . . . . . . . . . . 21. C. PLUKENETII

q. Rays of inflorescence smooth; leaves of involucre glabrous; heads more or less cylindrical or slightly obovoid; spikelets not sharp-pointed . . . . . . . . 20. C. DIPSACIFORMIS

n. Culm (stem) smooth . . . . . . . . . . . . . . . . . . . . . . . . . . *r*

r. Inflorescence resembling in shape a ball or globe (spherical) or the fingers on a hand (digitate), the spikelets crowded together or all arising on a shortened axis which is 1 cm. or less long, or at least not conspicuously pinnately or racemosely arranged on the axis. . . . . . . . *s*

s. Spikelets distinctly flat, the scales in 2 ranks alternating on 2 sides of the axis (rachis) of the spikelet, or, if not appearing flat, then the achenes linear and only 1–1.2 mm. long; achenes 0.9–1.5 mm. long; spikelets mainly 6–40-flowered; annuals, or, if perennials, without corm-like bases (like small knots or enlargements) . . . . . . . . . . . *t*

t. Scales of spikelets conspicuously 9–13-nerved, mainly 2.5–3 mm. long, striped white and yellow with a prominent green midrib; midrib of scale prolonged as a straight projection; inflorescence resembling the fingers on a hand (digitate), the spikelets spreading from nearly the same point of the shortened axis; spikelets 8–25 mm. long . . 6. C. COMPRESSUS

t. Scales of spikelets 3–5-nerved near midnerve, mainly 1.8–2.1 mm. long, at first rather uniformly pale greenish-white or pale greenish, in age tinged brown; tip of scale slightly curved outward; inflorescence globe-shaped or of compact ball-like masses; spikelets 3–10 mm. long . . . . . *u*

u. Robust perennial, 3–7.5 dm. tall; culm (stem) mainly 2–4 mm. wide in lower half; achene linear, about the same width from one end to the other, 4–6 times as long as wide; scales of spikelet narrowly oblong, closely touching each other, their tips very slightly curved outward or ascending, the numerous spikelets bunched into thick, crowded, compact, often large heads . . . . . . . . . . . . . . . . . . . . . . 7. C. VIRENS

u. Slender annual, mainly 0.5–3.5 dm. tall; culm (stem) mainly 0.5–1.5 mm. wide in lower half; achene oblong or ellipsoid, tapering at each end and broadest near the middle, 2–3 times as long as wide; scales of spikelet ovate, more loosely arranged, showing more space between each flower, the tips more widely curved out, the fewer spikelets grouped into thinner, less compact, smaller heads . . . . . . . . . . . . . . 8. C. ACUMINATUS

s. Spikelets either rather cylindrical with somewhat curved or angled sides and not presenting a flat appearance, or, if appearing flat, then the achenes 1.5–2.5 mm. long or the spikelets 2.5–4 mm. wide; spikelets 2–8-flowered (or up to 12–20 in *C. filiculmis* and *C.* × *mesochorus*); perennials often with hardened corms (like small knots or enlargements) at base . . . . . . . . . . . *v*

v. Spikelets averaging 10–20 to an inflorescence, mainly 2.5–4.5 mm. wide, often appearing flat; scales of spikelets loose or with loose or spreading tips, not with tips closely appressed; spikelets 2–20-(mainly 6–16-)flowered . . . . . . . . . . . . . . . . . . . . . . . . *w*

w. Leafy bracts of inflorescence widely spreading or recurved; culms 0.5–1 mm. in diameter below the inflorescence; scales 1.8–3.5 mm. long, blunt or the uppermost acutely pointed; achenes 1.5–2.3 mm. long; leaves usually less than 2 mm. wide; inflorescence a dense spherical or somewhat globose head, with or without a few short rays . . . . 23. C. FILICULMIS

w. Leafy bracts of inflorescence erect or ascending; culms mainly more than 1 mm. in diameter below the inflorescence; scales 3.5–4 mm. long, with conspicuous mucros or protruding tips 0.3–0.5 mm. long; achenes 2.2–2.5 mm. long; leaves usually 2–3.5 mm. wide; inflorescence with spikelets digitately arranged, some or most of them on elongated, well-developed rays 22a. C. × MESOCHORUS (see also under C. Schweinitzii)

v. Spikelets averaging 20–120 or more to an inflorescence, mainly 1–1.5 mm. wide, with rather rounded, curving sides or 4-angular; scales of spikelet with tips appressed to the adjacent spikelets, closely overlapping; spikelets 2–6-flowered . . . . . . . . . . . . . . . *x*

x. Spikelets 3–6-flowered; spikelets 20–40 in a loose rather globe-shaped (subglobose) to somewhat cylindrical head, with noticeable space between the fewer spikelets; scales of spikelets 2.5–3 mm. long; no hard corm-like developments at base of culm . . 24. C. GLOBULOSUS

x. Spikelets 2–3-flowered; spikelets 80–120 or more, crowded together into a dense globe-shaped head about as long as wide or slightly longer than wide, with little space between the crowded spikelets; scales of spikelet 3.5–4.2 mm. long; hard enlargements (corm-like or tuberous) at base of culm . . . . . . . . . . . . . . . . . . . 25. C. OVULARIS

r. Inflorescence of a broadly cylindrical, obovoid, or ellipsoid shape, with the spikelets arising from a rather elongated axis mostly 1–3 cm. long, or if the axis is shorter than 1 cm., then arranged racemosely (feather- or stepladder-like) (plants normally to be keyed out in this section may sometimes have a more contracted and somewhat ball- or globe-shaped inflorescence as described in the alternate r. section, but should be keyed out here) . . . . . . . . . . . . . . . . . . . . . . .

y. Scales of spikelets 1–1.6 mm. long, obovate, and wider than long when flattened; achenes 1 mm. long, black or blackish; axis (rachilla) of spikelet without any wings . . . . . . . 10. C. IRIA

y. Scales of spikelets 2–6 mm. long, oblong, oblong-lanceolate, or ovate, longer than wide; achenes mainly 1.2–3 mm. long, gray, brown, or black; axis (rachilla) of spikelet with wings. . . . . . *z*

z. Most of the spikelets or at least the lower spikelets turned down (reflexed) or pointing downward, with rounded or curving sides in cross-section, not flattened; scales with 9–11 nerves (including midnerve); achenes mainly 2–3 mm. long; perennials with base of culm (stem) with hard knotty enlargements (corm-like or tuber-like) . . . . . . . . . . . . . . . . *1*

1. Spikelets rather distant from each other on the axis (rachis) of inflorescence, open spaces plainly visible between the spikelets, dull greenish, 10–30 mm. long; achenes 2.5–3 mm. long. 18. C. REFRACTUS

1. Spikelets very crowded and close to each other on the axis (rachis) of inflorescence, scarcely any or no space visible between the spikelets, yellow-brown to brownish, or mixed with gray-green, mainly 7–10 mm. (up to 15) long; achenes 2–2.5 mm. long . . 19. C. LANCASTRIENSIS

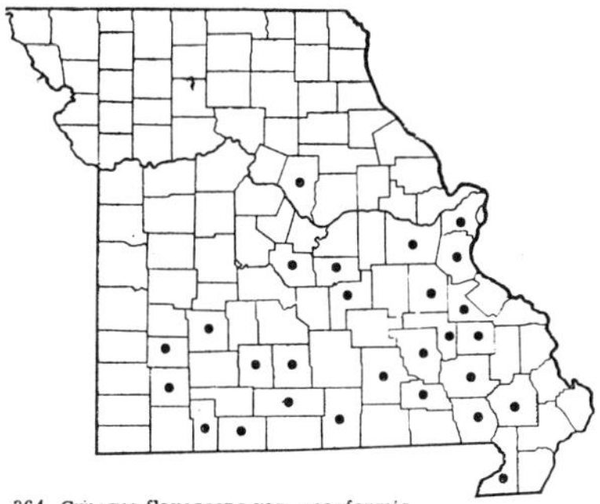

364 Cyperus flavescens var. poaeformis

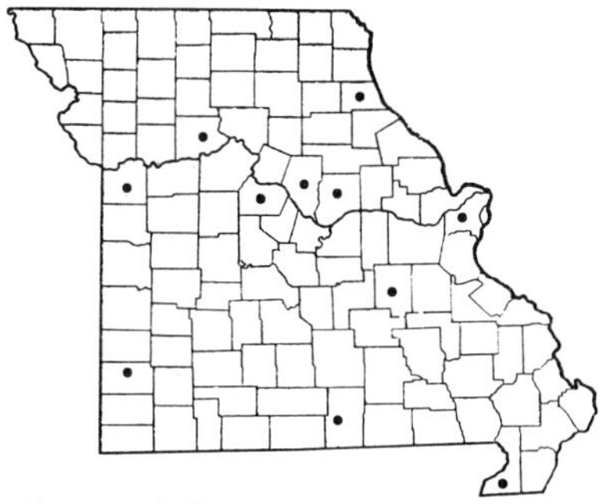

365 Cyperus diandrus

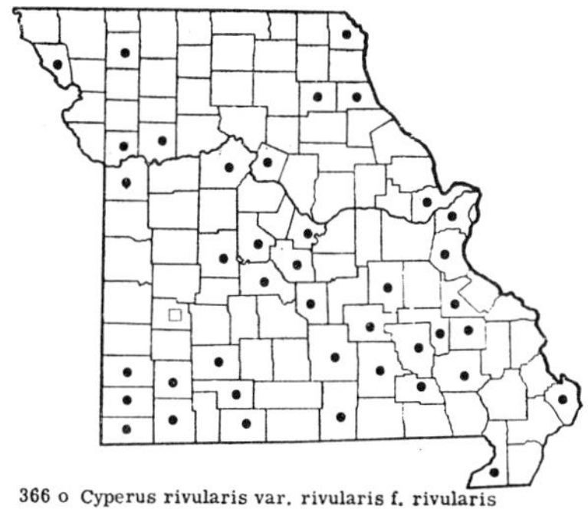

366 o Cyperus rivularis var. rivularis f. rivularis
366 □ Cyperus rivularis var. rivularis f. elutus

z. Most of the spikelets ascending to horizontally spreading, the lowest sometimes turned down, usually more or less flattened in mature specimens, (if not flattened, then not having the other characters given in the alternate z. section above); most of the scales usually with 5–7 nerves (including mid-nerve) or nerveless; achenes mainly 1.2–2 mm. long; annuals, or, if perennials, the base of culm sending out slender or stout stolons (runners), or if hard and enlarged (as in *C. strigosus*), then without the other characters given in the alternate z. section above . . . . . . . . . . . . . . *2*

2. Scales of spikelets rather distant from each other, the tip of a scale not reaching the base of the next scale located above on the same side, the joints of the axis of the spikelet exposed and appearing zig-zag . . . . . . . . . . . . . . . . . . . . 13. C. ENGELMANNI

2. Scales of spikelets close together, overlapping, the tip of a scale reaching up past the base of the next scale located above on the same side, the joints of the axis of the spikelet mostly hidden . . . . . . . . . . . . . . . . . . . . . . . . . . . . . *3*

3. Scales of spikelets chiefly yellow, greenish-yellow, golden-brown, or straw-color; in age the joints of the rachilla do not break up into separate segments, but remain as a whole intact, in place; perennial plants either with a hard enlarged corm-like base or sending out slender or stout stolons (creeping runners) . . . . . . . . . . . . . . . . . . . . . . . . *4*

4. Scales of spikelets mainly 3–4.5 mm. long, conspicuously keeled (with a sharp ridge running down the back or center); base of culm (stem) either with a hardened corm-like enlargement or sending out thick underground stolons (runners) . . . . . . . . . . . . *5*

5. Common plants throughout Missouri; scales of spikelets linear- to oblong-lanceolate, conspicuously nerved, loosely overlapping; achenes brown, linear; base of culm (stem) with a hard enlarged corm-like development . . . . . . . . . 17. C. STRIGOSUS

5. Rare species known only from Jackson County, western Mo.; scales of spikelet elliptic to ovate, faintly nerved or without nerves, closely overlapping and appressed to one another; achenes blackish, oblong; base of culm sending out thick underground stolons (runners) . . . . . . . . . . . . . . . . . . 15. C. SETIGERUS

4. Scales of spikelets mainly 2–2.9 mm. long, faintly or not strongly keeled; base of culm sending out slender underground scaly stolons, sometimes bearing a tuber at the end (careful examination should show at least some sign of a scaly runner, if no tuber is present) . . . . . . . . . . . . . . . . . . . . . . . 16. C. ESCULENTUS

3. Scales of spikelets dull or dark brown; in age each joint of the rachilla breaks up into separate segments between the scales; annual plants with fibrous roots, the base of culm (stem) without stolons (runners) or without a hardened corm-like enlargement . . . . . . . . . . . *6*

6. Scales of spikelet 2–3.5 mm. long, firm or coriaceous (leather-like); achene 1.5–2 mm. long; rare species, known only from the southeastern lowland section . . . . . 11. C. ODORATUS

6. Scales of spikelet 1.5–2.3 mm. long, soft membranaceous; achene 1–1.5 mm. long . . 12. C. FERRUGINESCENS

1. **Cyperus flavescens** L., var. **poaeformis** (Pursh) Fern. Map 364

*Cyperus flavescens* [of P & S, BB]

Flowers July–October.

Moist open sandy or gravelly ground along streams and spring branches, ditches along roadsides and railroad embankments, and wet places in meadows. Southern and central Missouri north to St. Louis and Boone (*Link 328*) counties, west to Polk, Dade, and Lawrence counties.

367 Cyperus polystachyos var. texensis

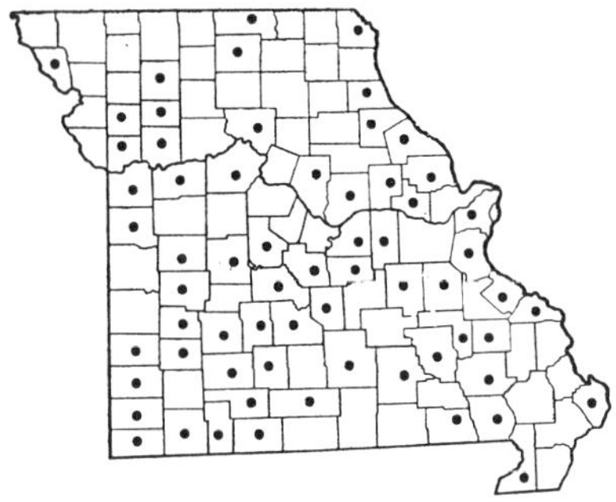

368 Cyperus aristatus

369 Cyperus compressus (Coco Grass)

Ranges from Florida to Texas, north to New York, Pennsylvania, Ohio, Michigan, Illinois, Missouri, and Kansas; also West Indies.

Fernald's interpretation of the species is followed here, var. *flavescens* ranging in Africa and Eurasia, and var. *picens* (Liebm.) Fern. from Mexico to Central America and South America.

The black achenes at maturity have transverse ridges coated with lines of a frosty white substance.

2. **Cyperus diandrus** Torr. Map 365

Flowers June–October.

Occurs in swampy meadows and low ground along streams and sloughs. Scattered in southern and central Missouri northeast to Marion County.

Ranges from Quebec and Maine to Ontario, Minnesota and North Dakota, south to South Carolina, Indiana, and Missouri.

Of the three related annual species (*C. flavescens* var. *poaeformis*, *C. diandrus*, and *C. rivularis*) which resemble one another, this is the most infrequently collected.

3a. **Cyperus rivularis** Kunth var. **rivularis** f. **rivularis** Map 366

*Cyperus rivularis* Kunth [G, BB, P & S]

Flowers July–October.

Occurs in swampy meadows, wet open sandy, gravelly, or muddy bars along streams and spring branches, and moist places on limestone ledges along creeks.

Ranges from Maine and Quebec to Minnesota, Nebraska, and California, south to Georgia, Texas, and Mexico.

3b. **Cyperus rivularis** var. **rivularis** f. **elutus** (C. B. Clarke) Kükenth. Map 366

This form with the scales greenish or only straw-colored, instead of tinged with purplish or purple-brown as in f. *rivularis*, is known from Cedar County (gravel bar by ford along Little Sac River, T34N, R26W, south part of sect. 26, $4\frac{1}{2}$–5 mi. southeast of Stockton, October 17, 1957, *Steyermark 85774*).

Another variety, *C. rivularis* var. *lagunetto* (Steud.) O'Neill, with 2–3 stamens instead of 2 as in var. *rivularis*, is found in Central and South America.

4. **Cyperus polystachyos** Rottb., var. **texensis** (Torr.) Fern. Map 367

*Cyperus Gatesii* Torr. [P & S]

*Cyperus polystachyos* var. *leptostachyus* Boeck. [Kükenthal]

Flowers July–October.

Occurs in wet sands and along sandy spring branches in the Crowley Ridge section of southeastern Missouri, where known only from Stoddard (along spring branch in alder thickets, at junction of flood plain and Crowley Ridge, T25N, R11E, sect. 6, $3\frac{1}{2}$ miles southeast of Bloomfield, August 20, 1954, *Steyermark 76819*) and Dunklin (wet sands, Campbell, September 9, 1910, *Bush 6299*; Milltown, Kennett, August 20, 1897, *Trelease*) counties.

Ranges from Florida to Texas and Mexico, north along the coast to Massachusetts and inland to southeastern Missouri; also in the West Indies, Central and South America, and the Philippines. *Cyperus polystachyos* var. *polystachyos* occurs in the West Indies, Mexico, Central and South America.

5. **Cyperus aristatus** Rottb. Map 368

*Cyperus inflexus* Muhl. [G, P & S]

*Cyperus aristatus* var. *inflexus* (Muhl.) Kük. [Kükenthal]

Flowers late May–October.

Occurs commonly on moist sand, mud, and gravel

Plate no. 67. **1.** Cyperus flavescens var. poaeformis. **2.** Cyperus diandrus. **3.** Cyperus rivularis. **4.** Cyperus polystachyos var. texensis. **5.** Cyperus aristatus. **6.** Cyperus compressus. **7.** Cyperus virens. **8.** Cyperus acuminatus. **9.** Cyperus erythrorhizos.

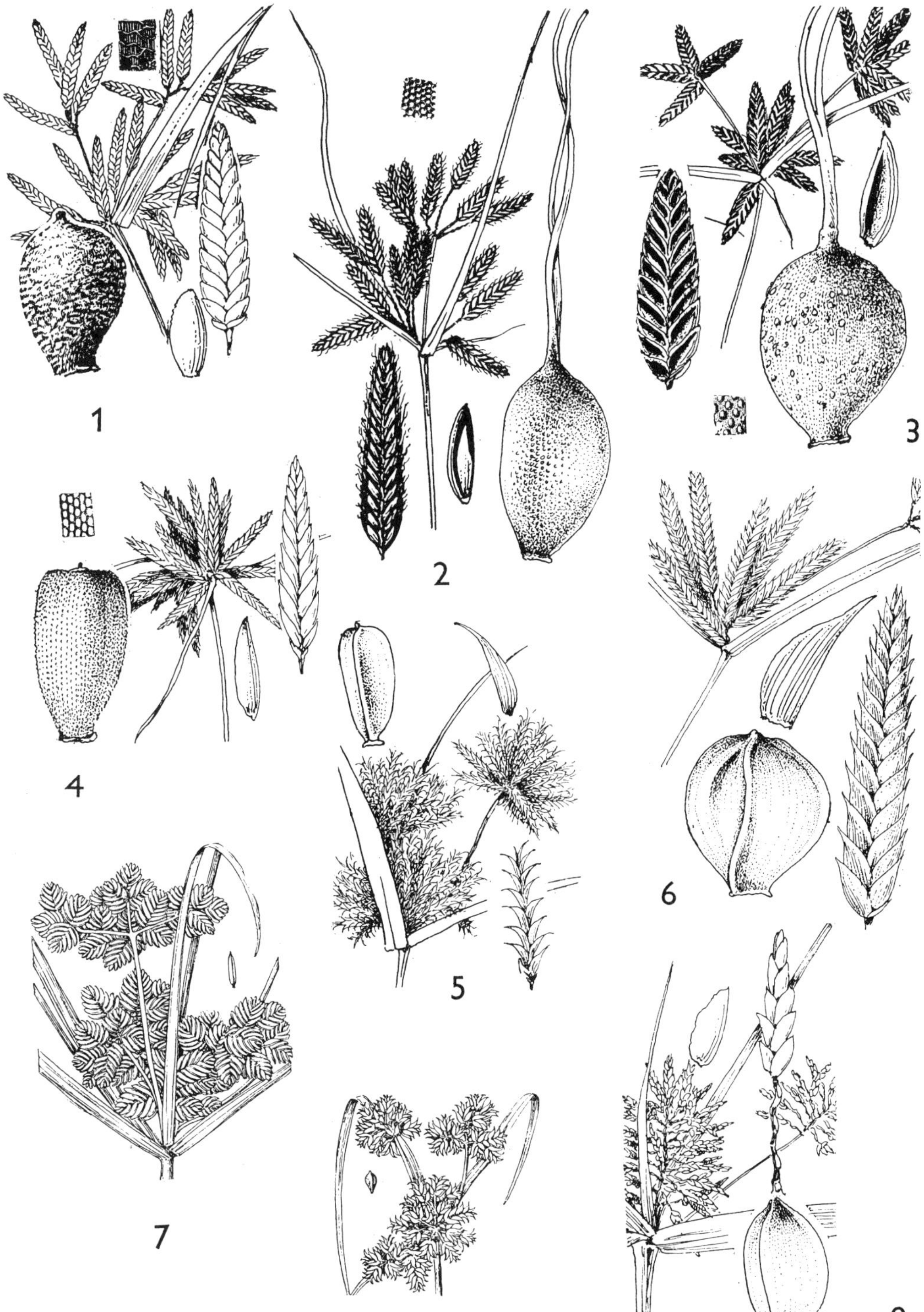


PLATE NO. 67

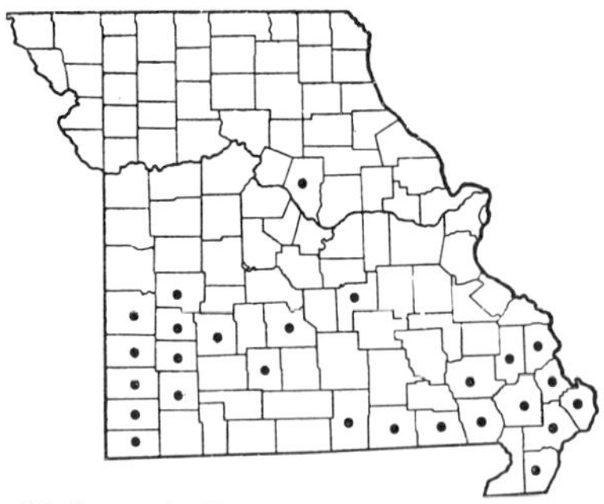

370 Cyperus virens

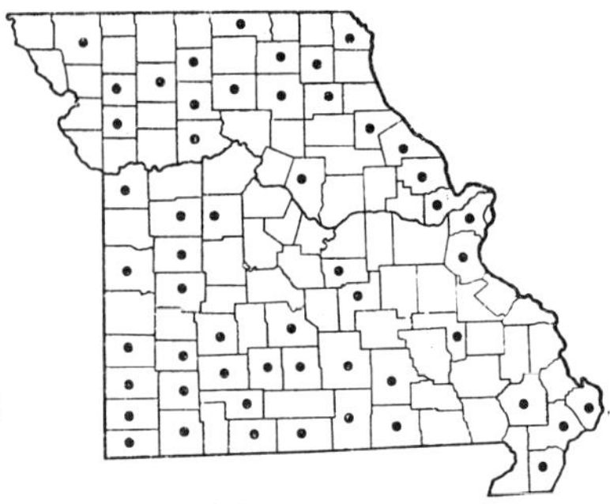

371 Cyperus acuminatus

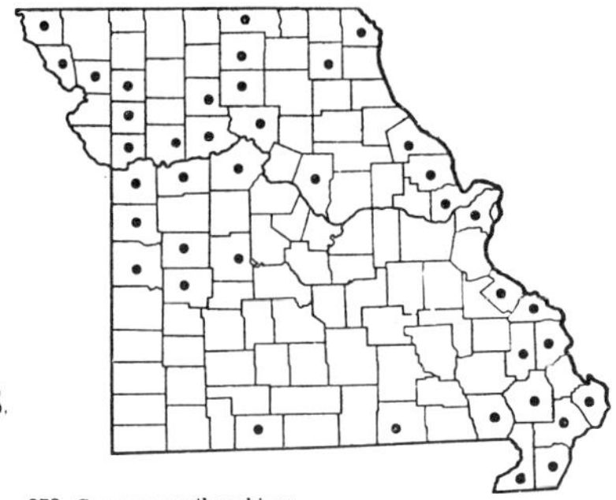

372 Cyperus erythrorhizos

bars along streams, wet depressions in rocky glades, wet places in prairies, on dry openings of ridge of upland slopes, and in alluvial and moist cultivated ground.

Most common in southern and central Missouri, extending north locally to Clark, Putnam, Daviess, and Holt counties. Apparently absent from most of northwestern and extreme southeastern Missouri.

Ranges from South and Central America, the West Indies, and Mexico north in North America to Maine, Quebec, Ontario, Michigan, Minnesota, Manitoba, Saskatchewan, and British Columbia; also Old World tropics.

The odor of slippery elm is a characteristic of dried or freshly bruised plants of this species.

6. **Cyperus compressus** L. Coco Grass Map 369
Flowers August–October.

Known only from wet fields and swampy ground in the southeastern lowland section (Dunklin and Butler counties) and locally north to St. Louis County.

Ranges from South and Central America, the West Indies, and Mexico, into the United States north from Florida to Delaware, west to Texas, and north in the interior to southeast Missouri, introduced elsewhere north in the United States to Minnesota and New York.

7. **Cyperus virens** Michx. Map 370
*Cyperus pseudovegetus* Steud. [BB, P & S]
Flowers late June–October.

Occurs in wet prairies, margins of streams and ponds, swales and swampy ground, along wet ditches and railroad cuts, and sandy openings in upland woods.

Mainly in the southern third of the state north to Cape Girardeau, Phelps, Laclede, St. Clair, and Vernon counties, where most common in the southeastern lowland and unglaciated prairie sections. Recorded locally northward in Boone County.

Ranges from Florida to Texas, north to New Jersey, Indiana, Illinois, Missouri, and Kansas; also in Mexico, the West Indies, Central and South America.

This species averages much more robust with taller culms and denser and usually larger heads of spikelets than *C. acuminatus* Torr. & Hook.

8. **Cyperus acuminatus** Torr. & Hook. Map 371
*Cyperus virens* var. *arenicola* (Boeckeler) Shinners [Shinners]
Flowers late June–October.

Occurs along muddy wet margins of ponds, creeks, spring branches, wet cultivated and alluvial ground, and open moist places. Throughout Missouri, but absent from almost a third of the counties in the state.

Ranges from Florida to Texas, north to Virginia, Ohio, Indiana, Illinois, Minnesota, and North Dakota; also in Washington and Oregon.

9. **Cyperus erythrorhizos** Muhl. Map 372
*Cyperus erythrorhizos* var. *Halei* (Torr.) Kük. [Kükenthal]
Flowers August–October.

Occurs on moist muddy margins of mud flats, banks, and alluvial ground of the larger streams, margins of sloughs, oxbow and other lakes, and rarely in wet woods. Common along the Missouri, Mississippi, Grand, and Osage rivers; elsewhere in the state absent or local; absent from the main part of the Ozark section.

Ranges from Florida to California, north to Massachusetts, New York, Ontario, Michigan, Wisconsin, Minnesota, North Dakota, Wyoming, and Washington.

This is a common inhabitant of the Missouri, Mississippi, and Grand River flood plain, where it frequents the shifting, flooded mud flats associated with other species, such as *Cyperus ferruginescens*, *Rumex maritimus* var. *fueginus*, *Euphorbia serpens*, *Bidens cernua*,

373 Cyperus Iria

374 Cyperus odoratus

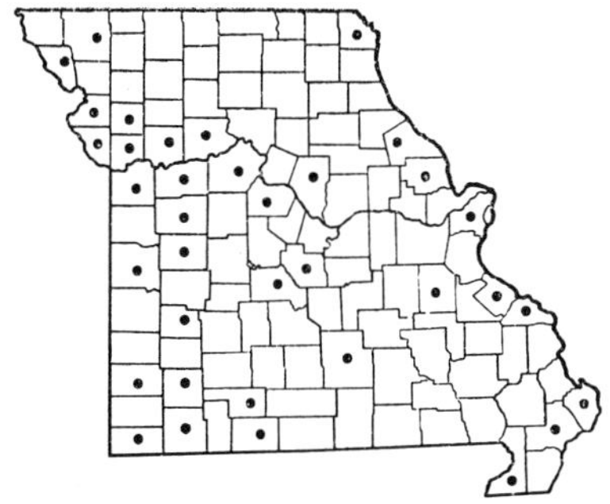

375 Cyperus ferruginescens

and others, which are adapted to this type of environment.

The narrow spikelets, small reddish-brown scales, and small achenes easily distinguish this species from *C. ferruginescens*, *C. odoratus*, or other members of the genus with which it has sometimes been confused.

10. **Cyperus Iria** L. Map 373

Flowers July–October.

Occurs in low ground and damp fallow cultivated ground in the southeastern Missouri lowlands, where known from Butler (¼ mi. south of Ash Hill, October 29, 1956, *Steyermark 83324*) and Pemiscot (3½ mi. southwest of Portageville, August 19, 1955, *Steyermark 79452*) counties.

Native in Eurasia; introduced into the United States, ranging from Florida to Texas, north to Virginia and southeastern Missouri.

11. **Cyperus odoratus** L. Map 374

*Cyperus ferax* Rich. [P & S, BB, Kükenthal]

Flowers August–October.

Occurs on alluvial and wet muddy flats along the Mississippi and St. Francis rivers in Pemiscot and Dunklin (Kennett, September 18, 1893, *Bush*) counties.

Ranges from Mexico, the West Indies, Central and South America, north in the United States to Massachusetts, also in the Old World tropics.

12. **Cyperus ferruginescens** Boeckl. Map 375

*Cyperus odoratus* in part [BB], not L.

*Cyperus ferax* subsp. *speciosus* (Vahl) Kük. [Kükenthal]

*Cyperus ferax* subsp. *speciosus* var. *squarrosus* (Britton) Kük. [Kükenthal]

*Cyperus ferax* subsp. *speciosus* var. *squarrosus* f. *parvus* (Boeckl.) Kük. [Kükenthal]

*Cyperus speciosus* var. *squarrosus* Britton.

Flowers August–October.

Occurs on moist muddy margins of mud flats, banks, and alluvial ground of the larger streams, and margins of sloughs, oxbow and other lakes, and occasionally along railroads. Common along the Mississippi, Missouri, and Osage rivers and tributaries, infrequent or absent elsewhere in the state. Unknown from much of northern Missouri and Ozark section.

Ranges from Connecticut, New York, and Ontario to Oregon, south to Virginia, Alabama, Louisiana, Texas, Arizona, and California.

*Cyperus ferruginescens* is often associated on the mud flats of the larger rivers with *Cyperus erythrorhizos* and other inhabitants of wet open mud. *Cyperus ferruginescens* varies, as do other wet-mud-inhabiting species, subjected to varying flood stages of river, in height of plant, length of leaves of culms and involucre, in length of spikelets, and development of inflorescence. The inflorescence has a characteristic reddish-brown to mahogany-brown or even orange-brown color with a dense cluster of spikelets.

This species has been most often confused with *C. odoratus*, *C. esculentus*, and *C. erythrorhizos*. It is easily distinguished from *C. erythrorhizos* by the larger scales and achenes, and from *C. odoratus* by the smaller golden- or orange-brown, instead of gray or blackish achenes, and by the thinner, smaller, reddish-brown scales, instead of the dull dark or drab brown ones of *C. odoratus*. The characters used in the present key distinguish *C. ferruginescens* from *C. esculentus*.

In his recent treatment of *Cyperus*, Kükenthal (Pflanzenreich IV: 20. Heft 101. p. 620. 1956) includes Missouri or cites specimens from Missouri under *C. ferax* subsp. *speciosus* var. *squarrosus* (St. Louis, *Engelmann*) and *C. ferax* subsp. *speciosus* var. *squarrosus* f. *parvus* (St. Louis, *Riehl 67*). As shown by Fernald (Rh. 37: 150. 1935) *C. parvus* Boeckl. is only a depauperate growth state of the common *C. ferruginescens*. Likewise, Kükenthal's segregated varieties and forms

376 Cyperus rotundus (Nut Grass)

377 Cyperus setigerus

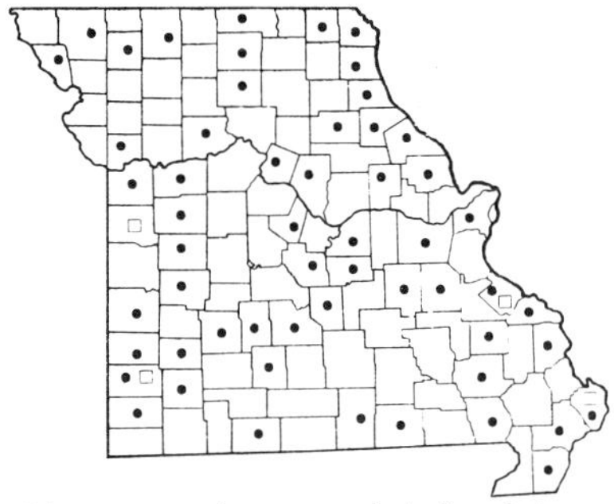

378 • Cyperus esculentus var. esculentus f. esculentus (Yellow Nut Grass)

378 □ Cyperus esculentus var. esculentus f. angustispicatus

above cited are considered in the present work as synonymous with *C. ferruginescens*.

13. **Cyperus Engelmanni** Steud.
See excluded species.

14. **Cyperus rotundus** L. Nut Grass, Coco Grass Map 376

Flowers August–October.

Known only from Dunklin County in the southeastern Missouri lowland section (cultivated field near Senath, April 28, 1942, *Bratton Wallace*).

Native of Eurasia; introduced into the United States, ranging from Florida to Texas and Mexico, north to Virginia, New York, Tennessee, southeastern Missouri, and Arkansas; also on the Pacific Coast, the West Indies, Central and South America.

Although the single Missouri collection is represented only by a specimen in the vegetative state, the ellipsoid tubers and short leafy tufts are characteristic marks of the species. Undoubtedly, additional plants will be found with a more intensive search in the southeastern lowland section of the state.

Kükenthal records this species from Kansas and Missouri (Das Pflanzenreich IV: 20. Heft 101. p. 111. 1956).

The small underground tubers of this are used for food in the same manner as those of *C. esculentus* (which see below).

15. **Cyperus setigerus** Torr. & Hook. Map 377

*Cyperus Hallii* Britt. [P & S]

*Cyperus setigerus* var. *Hallii* (Britton) Kük. [Kükenthal]

Flowers June–September.

Occurs on the sandy alluvial flats of the Missouri River, where known only from Jackson County in western Missouri (sandy bottoms, Sheffield, June 17, 1915, *Bush 7644*).

Ranges from Missouri and Kansas to Texas and New Mexico.

16. **Cyperus esculentus** L.

Yellow Nut Grass Map 378

Flowers mid-June to October.

Occurs in wet open fields and cultivated ground, sand and gravel bars along streams, borders of artificial and natural lakes, swales in prairies, along railroads, and occasionally swales in woodland. Common throughout Missouri.

Ranges from Mexico, the West Indies, Central and South America, north in North America from Florida, Texas, and Arizona to Nova Scotia, Quebec, Ontario, Manitoba, and Washington; also in the Old World.

This species sometimes is thought to cause hay fever, but there has been no substantial evidence for this.

Missouri material may be divided as follows:

Spikelets 0.5–1.5 cm. long, obtuse at tip . . . .
16a. C. ESCULENTUS var. ESCULENTUS f. ESCULENTUS

Spikelets 1.5–3 cm. long, narrowed to an acute tip .
16b. C. ESCULENTUS var. ESCULENTUS f. ANGUSTISPICATUS

16a. **Cyperus esculentus** var. **esculentus** f. **esculentus** Map 378

*Cyperus esculentus* L. [G, BB, P & S]

This is widespread and common throughout the state.

16b. **Cyperus esculentus** var. **esculentus** f. **angustispicatus** (Britt.) Fern. Map 378

*Cyperus esculentus* var. *angustispicatus* Britton [P & S]

*Cyperus esculentus* var. *leptostachyus* Boeckl. [Kükenthal]

Plate no. 68. **1.** Cyperus Iria. **2.** Cyperus odoratus. **3.** Cyperus rotundus. **4.** Cyperus esculentus. **5.** Cyperus strigosus.

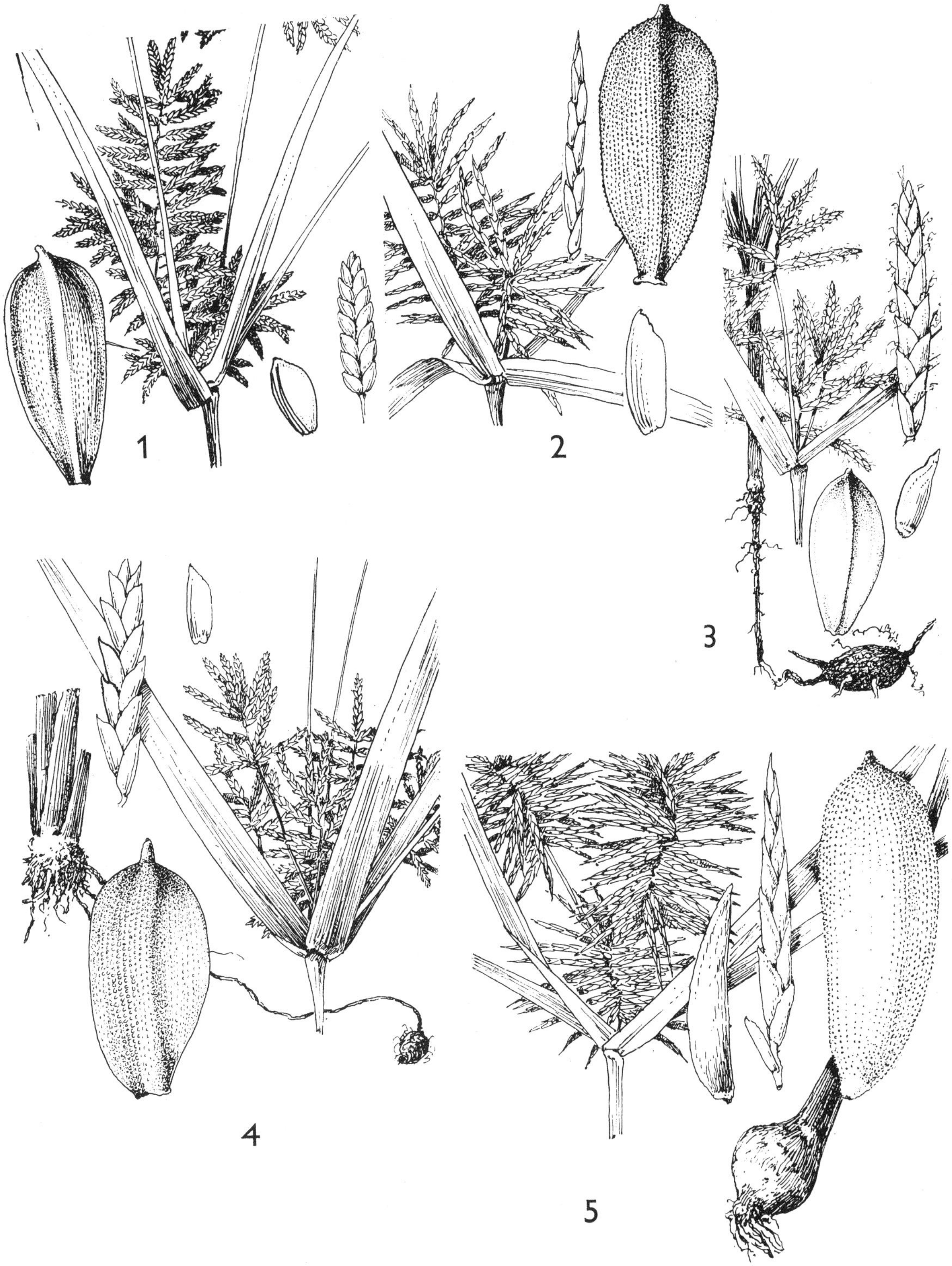


PLATE NO. 68

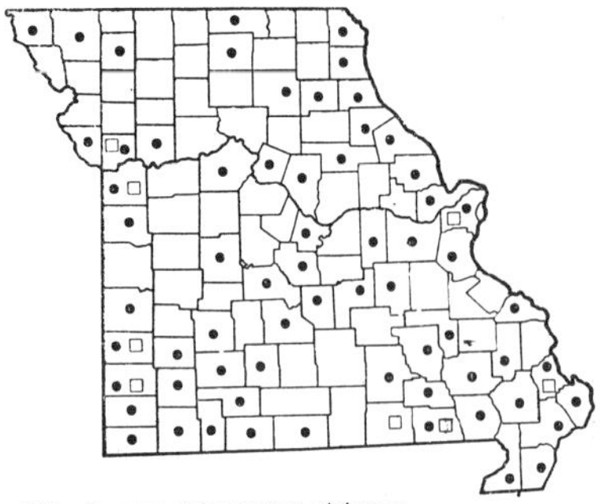

379 • Cyperus strigosus var. strigosus
379 □ Cyperus strigosus var. robustior

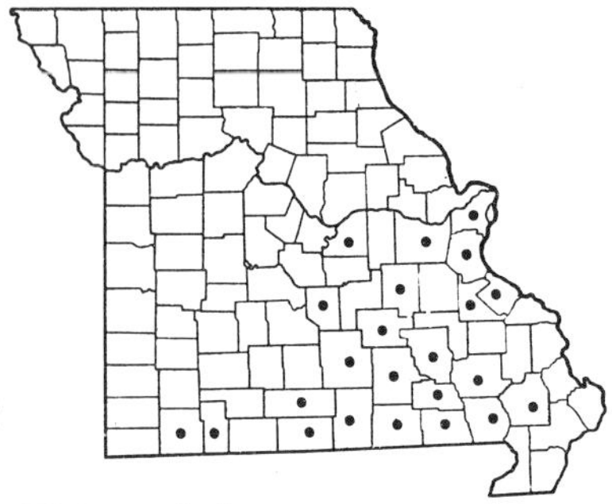

380 Cyperus refractus

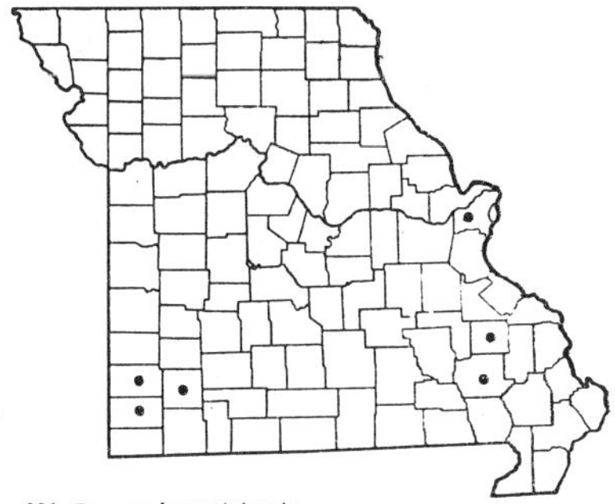

381 Cyperus lancastriensis

Scattered in the state.

Another form, *C. esculentus* var. *esculentus* f. *macrostachyus* (Boeckl.) Fern., differing in having the spikelets 2–3 cm. long, round-tipped, and 2–3 mm. wide, instead of 1.5–2 mm. as in the other two forms, has not been recorded from Missouri, but is to be expected.

A cultivated variety of this species, originating from Africa, *C. esculentus* var. *sativus* Boeckl., known as Chufa, with enlarged tubers at the ends of short underground runners, is sometimes cultivated in the southern United States for the sweetish nutty flesh found in the tubers. The tubers can be eaten raw, boiled, candied, or ground into a flour. In Europe the tubers have found use as a sweetmeat, conserve, refreshing sweet drink, and in historical times as a coffee substitute.

*Cyperus esculentus* is sometimes confused with *C. strigosus*, *C. erythrorhizos*, and *C. ferruginescens*. From *C. erythrorhizos* it differs in the longer yellow- or golden-brown scales, longer achenes, and perennial habit with the scaly runners evident. From *C. strigosus*, in addition to the usual characters of the scaly stolons and smaller scales of the spikelets of *C. esculentus* which are not deciduous from the rachis of the spike, there are additional recognition characters not stressed in the current manuals, and these have been found to be of use as follows: in *C. esculentus* most of the spikelets are separated by a space of 1–2 mm. along the rachis, and the rachis is frequently pubescent, whereas in *C. strigosus* most of the spikelets are much closer together and separated by a space of only 0.2–0.5 mm. along the rachis, which is glabrous; also, the spikelets in *C. esculentus* are relatively fewer per inflorescence than in *C. strigosus*.

17. **Cyperus strigosus** L. Map 379

Flowers mid-June to October.

Occurs in alluvial soils of sand, gravel, and mud along streams, borders of lakes and sloughs, roadside ditches, prairie swales, moist open ground in fallow or cultivated areas, and along railroads. Found throughout Missouri.

Missouri material may be separated into two varieties:

Spikelets 4–14-flowered, 0.5–1.8 cm. long . . . .
17a. C. STRIGOSUS var. STRIGOSUS

Spikelets 12–25-flowered, 2–3 cm. long . . . .
17b. C. STRIGOSUS var. ROBUSTIOR

17a. **Cyperus strigosus** var. **strigosus** Map 379

*Cyperus strigosus* L. [G, BB, P & S]

*Cyperus strigosus* f. *compositus* (Britton) Kük. [Kükenthal]

*Cyperus strigosus* var. *elongatus* (Torr.) Britton

*Cyperus strigosus* var. *capitatus* Britton

This is the common type found throughout the state.

Ranges from Florida to Texas and New Mexico, north to Maine, Quebec, Ontario, Michigan, Wisconsin, Minnesota, and South Dakota; also in the Pacific states.

17b. **Cyperus strigosus** var. **robustior** Britt. Map 379

*Cyperus strigosus* f. *robustior* (Kunth) Kük. [Kükenthal]

Occurs in a few widely scattered localities, but probably of more frequent occurrence.

Ranges from Florida to Mississippi, north to Massachusetts, Connecticut, New York, West Virginia, Indiana, and Missouri.

Of the several named varieties and forms that have been recorded under *C. strigosus*, Fernald and Griscom (Rh. 37: 150–51. 1935) retain var. *robustior* and var. *stenolepis* (Torr.) Kük. as good varieties. Father O'Neill (Rh. 44: 79–81. 1942) favors the maintenance of var. *stenolepis* but would reduce var. *robustior* and other varieties and forms to synonymy. He also argues that a range of intergrading forms can be found

382 Cyperus dipsaciformis

383 Cyperus Plukenetii

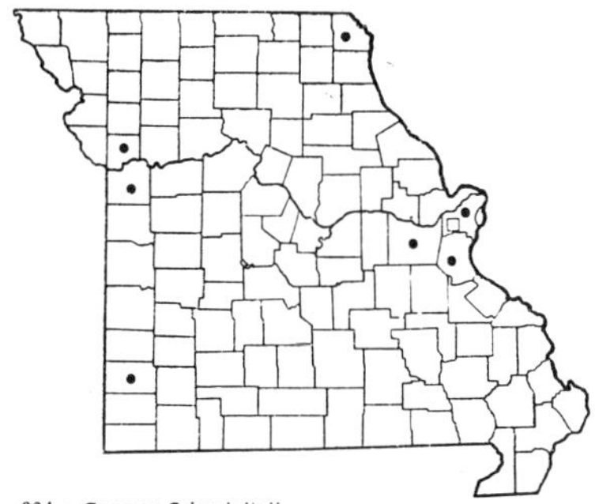

384 • Cyperus Schweinitzii
384 □ Cyperus X mesochorus (C. Schweinitzii X C. filiculmis)

varying from simple inflorescences, where a plant is crowded by others close by, to compound inflorescences (characteristic of f. *compositus*) where a plant is by itself in the open and in better soil. On the basis of the more compound inflorescence, the collections of *Bush* from Oregon Co. in 1892, and *Bush 6175* would be determined as *C. strigosus* f. *compositus*. However, O'Neill believes there is no way to separate such specimens having slightly compound inflorescences (*Engelmann*, St. Louis in 1845) classified by Britton as *C. strigosus* var. *robustior*, from others with more compound inflorescences of f. *compositus*.

In the present treatment the var. *robustior* is being retained tentatively as a geographical variety, according to Fernald and Griscom's studies, until more experimental evidence is forthcoming to prove that these variations are governed by environmental conditions and do not have any genetic value.

18. **Cyperus refractus** Engelm. Map 380

Flowers August–September.

Occurs along gravel bars of streams, dry sandy or sterile open woodland of valleys, and low ground following streams. Southern and east-central Missouri, confined to the Ozark section, north to St. Louis and Osage counties, west to Pulaski, Douglas, and Barry counties.

Ranges from New Jersey to Ohio and Kentucky, south to Georgia, Alabama, Tennessee, Missouri, and Arkansas.

The rather long, slender, rather remote, pale dull green, loosely arranged spikelets, the lowest ones strongly reflexed, mark this species fairly easily. The other Missouri species with strongly reflexed spikelets (*C. lancastriensis, C. Plukenetii, C. dipsaciformis*), all have more numerous spikelets in denser clusters. Of the species having reflexed spikelets, this one is quite characteristic of dry open wooded valleys and sandy or gravelly bars bordering Ozark streams.

19. **Cyperus lancastriensis** Porter Map 381

*Cyperus refractus* var. *lancastriensis* (Porter) Kük. [Kükenthal]

Flowers August–October.

Occurs on rocky open slopes, glades, rocky open woods along streams, swales in prairies, and along railroad cuts through sandstone areas. Known only from Jasper, Newton, and Lawrence counties of southwestern Missouri, and St. Louis, Madison, and Wayne counties in the eastern Ozark section.

Ranges from Georgia to Arkansas and Missouri, north to New Jersey, Pennsylvania, and Ohio.

The short cylindric head of densely clustered spikelets with the lowest ones reflexed usually distinguish this species readily from *C. strigosus*. The spikelets are usually narrower than those of *C. strigosus* and much narrower than those of *C. filiculmis*. Occasional specimens are found (*Palmer 66624* from Newton County) which are somewhat suggestive of hybridity between *C. lancastriensis* and *C. filiculmis*, *C. ovularis*, or *C. strigosus*. Such a specimen has shorter more subglobose inflorescences with somewhat broader (1.5–2 mm.) spikelets, which are 8–12 mm. long and 6–8-flowered. As compared with *C. filiculmis*, *C. strigosus*, and *C. ovularis*, *Cyperus lancastriensis* has a greater number of spikelets to a head than either *C. filiculmis* and *C. strigosus*, and fewer than in *C. ovularis*. The spikelets of *C. lancastriensis* average longer with more flowers than in *C. ovularis*, but shorter than in *C. filiculmis*.

20. **Cyperus dipsaciformis** Fern. Map 382

*Cyperus retrofractus* var. *dipsaciformis* (Fern.) Kük. [BB, Kükenthal]

Flowers July–October.

Occurs in sandy fields and sandy open woods, chiefly on Crowley Ridge of southeastern Missouri, where known only from Mississippi, Stoddard, and Dunklin (sands, Malden, September 8, 1910, *Bush 6269B*) counties.

Ranges from Georgia to southeastern Missouri, north to New Jersey, Maryland, D.C., and Kentucky.

21. **Cyperus Plukenetii** Fern. Map 383

*Cyperus retrofractus* var. *retrofractus* [of BB], not (L.) Torr. as treated by Fernald

*Cyperus retrofractus* [of P & S, Kükenthal], not (L.) Torr.

Flowers July–September.

Occurs in sandy or open oak-hickory woodland in acid soils, where known only from southeastern Missouri in Stoddard, Dunklin (Bernie, August 2, 1895, *Bush 688*; Malden, August 27, 1894, *Bush 6269*), Ripley (Pleasant Grove, August 9, 1899, *Bush 254*), and Wayne counties.

Ranges from Florida to Texas, north to New Jersey, Ohio, and southeastern Missouri.

This species and *C. dipsaciformis* stand out among Missouri species in having obovoid or obconic dense inflorescences narrowed in the lower half with most of the spikelets reflexed, and are not likely to be confused with any other species. However, some difficulty may be encountered in telling the two species apart. The differences brought out in the key to species should clarify the main points of departure.

22. **Cyperus Schweinitzii** Torr. Map 384

Flowers late June–September.

Occurs in dry sandy prairies or low sandy ground along streams and about ponds, and along railroads. Rare and scattered in central, western, and northeastern Missouri, where locally north in Clark County (natural sandy prairie on old river terrace paralleling Des Moines River, T65N, R6W, sect. 9, 1½ mi. southeast of St. Francisville, August 18, 1949, *Steyermark 68859*); absent from the Ozark section.

Ranges from Quebec to Saskatchewan, south to New York, Pennsylvania, Ohio, Indiana, Illinois, Missouri, Oklahoma, Texas, and New Mexico.

*Cyperus Schweinitzii* may be distinguished in general from *C. filiculmis* var. *filiculmis* and var. *macilentus* by (1) the erect or ascending, instead of widely spreading or recurved, involucral bracts, (2) the usually scabrous-rough instead of smooth culm, (3) the culms usually more than 1 mm. in diameter below the inflorescence, instead of 0.5–1 mm., (4) the acuminate, awned, somewhat longer scales, (5) achenes 2.5–3.5 mm. long instead of 1.5–2.3 mm. long, (6) the slightly broader leaves which are mainly 2–6 mm. wide instead of mostly less than 2 mm. wide as in *C. filiculmis*, (7) the prevalence of elongated, well developed rays on some or most of the spikelets, and (8) the pinnately arranged ascending spikelets instead of digitately arranged, spreading spikelets.

It apparently hybridizes with *C. filiculmis* var. *filiculmis* in St. Louis County and in western Missouri (where see discussion under C. × *mesochorus*). The main differences are indicated in the key to species.

22a. **Cyperus × mesochorus** Geise Map 384

*Cyperus Houghtonii* var. *uberior* Kükenthal

*Cyperus Bushii* Britton, as to numerous specimens so identified by him, but not the type (*Bush 619* from Arkansas, which is *C. filiculmis*).

Sister M. Joseph Geise (Am. Midl. Nat. 15: 249–50. 1934) considered this a hybrid between *C. Schweinitzii* and *C. Houghtonii*, whereas Father O'Neill (Rh. 44: 58. 1942) considers it as a form of *C. Schweinitzii* with more densely-clustered heads containing more numerous digitately- instead of more pinnately-arranged spikelets. He indicates (p. 57) that Britton's annotation of a number of specimens in the New York Botanical Garden Herbarium identified as *C. Bushii* Britton are identical with some of the specimens that would be classified as *Cyperus × mesochorus*. Gleason in the *New Ill. Fl.* (vol. 1. p. 252. 1952) interprets these intermediate types as possible hybrids between *C. filiculmis* and *C. Schweinitzii*, which is the viewpoint also taken by Father O'Neill (Rh. 44: 58. 1942).

Some of the specimens of *C. filiculmis* and *C. Schweinitzii* in the western sections of Missouri may show this intermediate charater between *C. Schweinitzii* and *C. filiculmis*, and are to be interpreted as belonging to *C. × mesochorus*.

A collection from St. Louis County which has the characters of this hybrid is *Muehlenbach 1485* (right-of-way of the Wabash Railroad, on a free place opposite the Wabash Fruit Terminal, west of Collins Street, September 7, 1958). Another collection from St. Louis County which suggests this hybrid and is somewhat intermediate between *C. filiculmis* and *C. Schweinitzii* is *Muehlenbach 103* (right-of-way of Wabash R. R., south of the gravel plant at the foot of Branch Street, St. Louis, June 5, 1954).

23. **Cyperus filiculmis** Vahl Map 385

Flowers last of May–October.

Occurs in dry open sandy or rocky woods, dry open ridges, prairies, glades of sandstone, chert, or granite, dry fallow and upland fields, and along railroads, generally in acid soils.

Plate no. 69. **1.** Cyperus refractus. **2.** Cyperus lancastriensis. **3.** Cyperus dipsaciformis. **4.** Cyperus Plukenetii. **5.** Cyperus Schweinitzii.

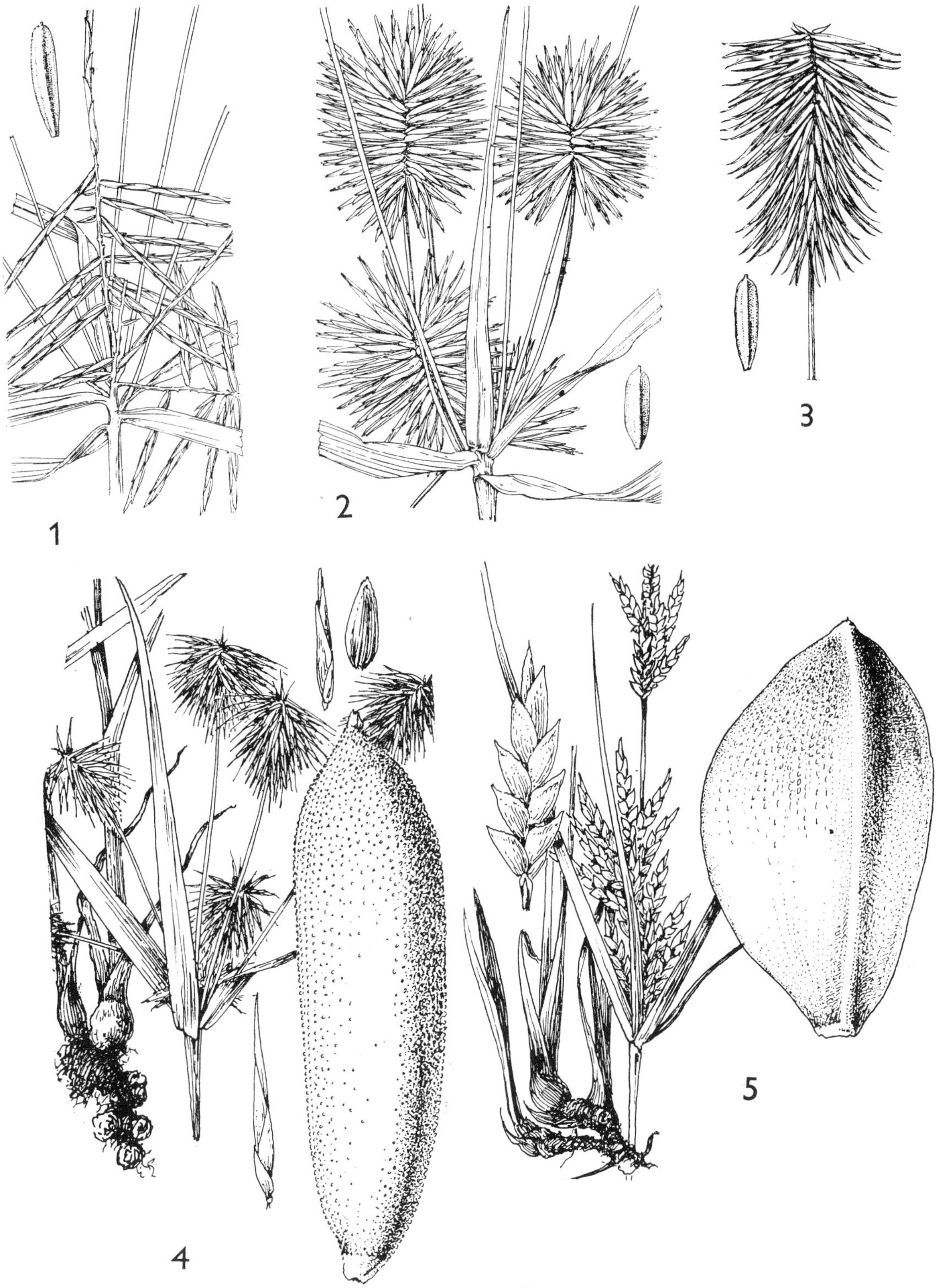


PLATE NO. 69

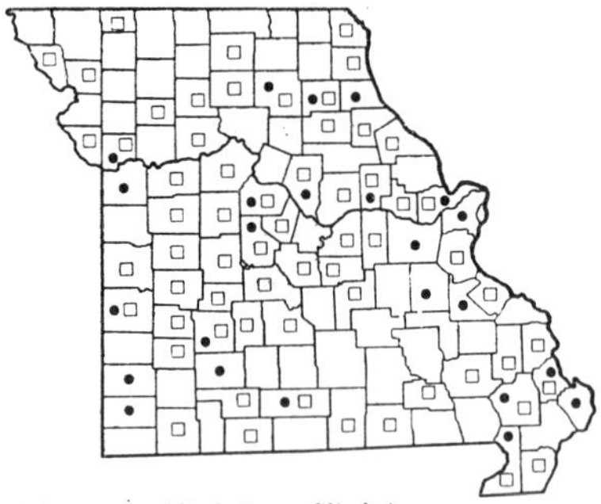

385 • Cyperus filiculmis var. filiculmis
385 □ Cyperus filiculmis var. macilentus

386 Cyperus globulosus

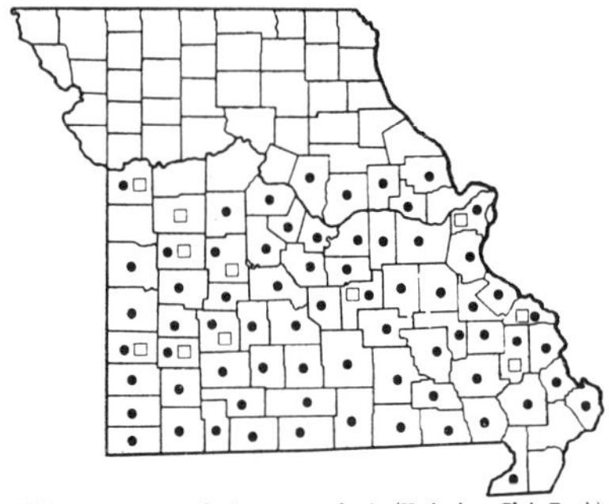

387 • Cyperus ovularis var. ovularis (Hedgehog Club Rush)
387 □ Cyperus ovularis var. sphaericus

Missouri material may be divided into the following varieties:

Scales of spikelets 2.8–3.5 mm. long; spikelets 7–20 mm. long, mainly 6–16-flowered, rather loosely arranged; achenes 1.8–2.3 mm. long
23a. C. FILICULMIS var. FILICULMIS

Scales of spikelets 1.8–2.8 mm. long; spikelets 3–15 mm. long, mainly 2–8-flowered, in a compact usually dense head; achenes 1.5–1.8 mm. long . . . . 23b. C. FILICULMIS var. MACILENTUS

23a. **Cyperus filiculmis** var. **filiculmis** Map 385

*Cyperus filiculmis* Vahl [G, BB, P & S]

*Cyperus Bushii* Britton

*Cyperus Houghtonii* Torr. var. *Bushii* (Britton) Kük. [Kükenthal]

Scattered in southern and central Missouri north to Marion, Shelby, Macon, and Jackson counties.

Ranges from Florida to Texas, north to Massachusetts, New York, Ohio, Michigan, Illinois, Iowa, and Nebraska.

23b. **Cyperus filiculmis** var. **macilentus** Fern. Map 385

This is the more common variety encountered, and is found nearly throughout the state, although relatively more rare in northern Missouri.

Ranges from Maine and Quebec to Minnesota, south to Virginia, Ohio, Indiana, Illinois, and Missouri.

*Cyperus filiculmis* var. *filiculmis* and var. *macilentus* sometimes resembles specimens of *C. compressus*. It is not always possible to know whether the plants are annual or perennial, a character sometimes used in keys to distinguish these species. In the absence of this knowledge, the species can be readily separated on the basis of the achenes, which are only 1.2–1.5 mm. long in *C. compressus*, 1.5–2.3 in the varieties of *C. filiculmis*, scales more separated at maturity in *C. filiculmis* than in *C. compressus*, and more abruptly tipped in *C. filiculmis* instead of having gradually prolonged sharp tips as in *C. compressus*.

The two varieties of *C. filiculmis* inhabit usually quite dry habitats in poor acid soils or on rocks. There is definite intergradation between them, making it difficult to place them in one or the other category. Specimens with few-flowered (3–6) spikelets of *C. filiculmis* var. *macilentus* but with the longer scales and achenes of var. *filiculmis* are frequently found, and other noncorrelated combinations of characters occur, rendering a decision as to proper placement unsatisfactory. The two varieties are tentatively retained for Missouri, pending results of more intensive study.

24. **Cyperus globulosus** Aubl. Map 386

Flowers July–October.

Occurs in open sandy or alluvial ground along the Mississippi and St. Francis rivers and in sandy open woods of Crowley Ridge, where found only in southeastern Missouri in Pemiscot and Dunklin (sands, Malden, September 8, 1910, *Bush 6270*; woods, Campbell, September 7, 1910, *Bush 6222*; Kennett, July 27, 1895, *Bush 683*) counties.

Ranges from Florida to Texas, north to Virginia, Missouri, and Oklahoma; also in Central and South America, the West Indies, and southern Mexico; locally introduced north to New Jersey and Pennsylvania.

25. **Cyperus ovularis** (Michx.) Torr.

Hedgehog Club Rush Map 387

Flowers first of June–September.

---

Plate no. 70. **1.** Cyperus filiculmis var. macilentus. **2.** Cyperus ovularis. **3.** Cyperus globulosus. **4.** Cyperus tenuifolius. **5.** Dulichium arundinaceum.

PLATE NO. 70

Occurs in dry open woods, prairies, glades of sandstone, chert, and granite, in sandy soils along streams, dry fallow or upland fields, and open banks along roadsides and railroads, generally in acid soils.

Two varieties may be distinguished in Missouri:

Inflorescence mostly longer than thick, 11–23 mm. long, 8–18 mm. thick . . . 25a. C. OVULARIS var. OVULARIS

Inflorescence mostly globe-shaped, about as long as thick, 7–12 (–15) mm. long, 7–12 (–15) mm. thick . . . . 25b. C. OVULARIS var. SPHAERICUS

25a. **Cyperus ovularis** var. **ovularis** Map 387

*Cyperus ovularis* (Michx.) Torr. [G, BB]

*Cyperus ovularis* var. *robustus* Boeckl. [P & S]

*Cyperus ovularis* var. *Wolfii* (Wood) Kük. [Kükenthal]

Common in southern and central Missouri north to Lincoln, Boone, and Jackson counties.

Ranges from Florida to Texas, north to New York, New Jersey, Pennsylvania, Ohio, Indiana, Illinois, Missouri, and Kansas.

25b. **Cyperus ovularis** var. **sphaericus** Boeckl. Map 387

Scattered in the range of var. *ovularis*, in southern and central Missouri, but less common. Representative specimens are from Jackson (*Bush 41*), Johnson (*Palmer 66529*), and Polk (*Palmer 66968*) counties.

Ranges from Georgia to Texas, north to Virginia, West Virginia, Ohio, Indiana, Missouri, and Oklahoma.

26. **Cyperus tenuifolius** (Steud.) Dandy Map 388

*Kyllinga pumila* Michx. [P & S]

*Cyperus densicaespitosus* Mattf. & Kükenth. [BB]

Flowers late May–October.

Occurs on gravel bars and muddy banks of streams, moist margins of ponds, wet places in prairies, and wet open woods. Southern and central Missouri north to Clay, Boone, and Marion counties.

Ranges from Mexico, Central and South America, and in North America from Florida and Texas, north to New York, Pennsylvania, Ohio, Indiana, Illinois, Missouri, and Kansas.

*Excluded species*

13. **Cyperus Engelmanni** Steud.

*Cyperus ferax* subsp. *Engelmannii* (Steud.) Kük. [Kükenthal]

The basis for crediting this species to St. Louis, Missouri rests upon specimens collected by *Engelmann* with the data: 'muddy banks of the Cahokia near St. Louis, October, 1845,' and 'on Cahokia Creek near St. Louis, September, 1846.' Unfortunately, Engelmann did not take the trouble to indicate on the labels accompanying these specimens that they were collected in Illinois. All the other specimens of *C. Engelmanni* collected by Engelmann at the same time and place bear data which are more specific, as follows: 'on Cahokia Creek, St. Clair County, Illinois;' 'on Cahokia Creek in the bottom opposite St. Louis, September, 1846;' and 'shady muddy banks of Cahokia Creek opposite St. Louis, October, 1845.'

The dates of October, 1845 and September, 1846 and Cahokia Creek accompany the labels of both the specimens placed in Illinois and those interpreted for Missouri. There is no doubt whatsoever that the data apply to the same locality on Cahokia Creek in St. Clair, Illinois, opposite St. Louis.

With more intensive collecting in the future in the area along the Mississippi River bottoms on the Missouri side, it is expected that *C. Engelmanni* will eventually be found in the state.

**Cyperus filicinus** Vahl

Kükenthal (Das Pflanzenreich IV: 20. Heft. 101. p. 374. 1956) records this species from St. Louis County, citing *C. polystachyos* var. *macrostachyus* Boeckl., in synonymy.

As stated by Fernald (Rh. 41: 531, 1939) and by Father O'Neill (Rh. 44: 84. 1942) *C. filicinus* is a maritime species confined to salt or brackish-water habitats along the Atlantic Coast from Maine to Florida west along the Gulf Coast to Louisiana. As Fernald (*ibid.*) points out, dwarfed plants of *C. polystachyos* var. *texensis* resemble dwarfed extremes of *C. filicinus*, and it is possible that the St. Louis County specimen cited by Kükenthal is based upon such a mistaken resemblance. However, since *C. polystachyos* var. *texensis* is known only from the counties of extreme southeastern Missouri, it is probable that another more common species has been confused with *C. filicinus*.

**Cyperus retrofractus** (L.) Torr.

*Cyperus retrofractus* var. *retrofractus* [BB]

Kükenthal (Das Pflanzenreich IV: 20. Heft 101. pp. 507, 509. 1956) cites *Bush 688* in crediting this species to Missouri. As interpreted in the present flora, however, *Bush 688* from Dunklin County is *C. Plukenetii* Fern., which, following Fernald's treatment, is distinguishable as a different species from *C. retrofractus* (L.) Torr. *Cyperus retrofractus*, following Fernald's interpretation (Rh. 47: 111–12. pl. 878, fig.

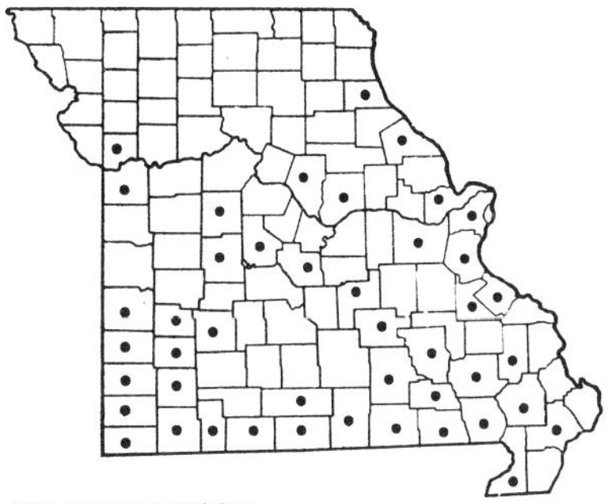

388 Cyperus tenuifolius

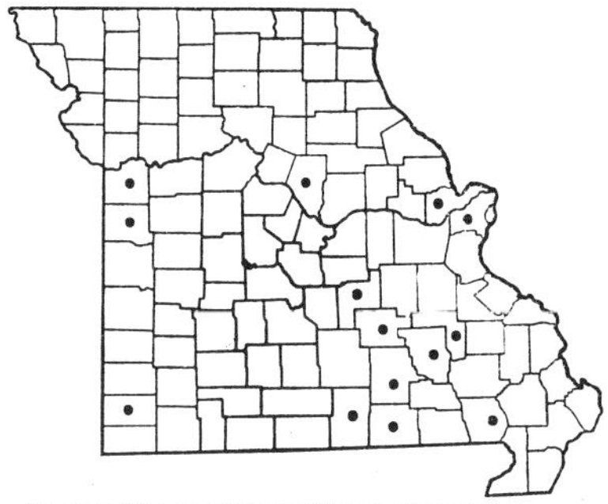

389 Dulichium arundinaceum (Three-way Sedge)

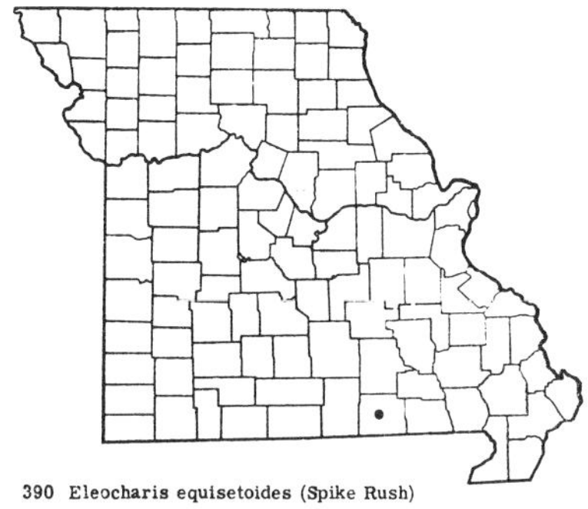

390 Eleocharis equisetoides (Spike Rush)

1–4) of that taxon, is a species with glabrous culms lnd leaves, gla brous rays of the inflorescence, golden brown or drab scales of the spikelets, and achenes 2–2.5 mm. long. The distribution of this species lies to the east of Missouri, and thus far, no true *C. retrofractus* has been collected in the state.

**Cyperus retrorsus** Chapm., var. **Deeringianus** (Britton & Small) Fern. & Griscom
*Cyperus cylindrostachys* Small

Kükenthal (Das Pflanzenreich IV: 20. Heft 101 1956) based his Missouri report on a specimen collected by Tracy from Ocean Springs.

Unfortunately, Ocean Springs is in Mississippi, and all of Tracy's other Ocean Springs collections are known to come from Mississippi, not Missouri. The range of *C. retrorsus* var. *Deeringianus* is in the Coastal states from Mississippi and Florida to Georgia and Virginia.

## 2. **Dulichium** Pers.

Perennials with terete (rounded) culms leafy to the summit, the leaves short and 3-ranked. Spikelets linear, stalkless, in 2 ranks from the leaf-sheaths. Each flower consists of 6–9 downwardly barbed bristles, 3 stamens, and 2-cleft style. Achene beaked with a persistent style.

**Dulichium arundinaceum** (L.) Britt.
Three-way Sedge Map 389
Flowers July–October.

Occurs in the mucky soil and floating mats of sinkhole ponds in the Ozarks, elsewhere in low ground bordering sloughs, ponds, and spring branches. Southern and central Missouri, north to St. Charles, Boone, and Jackson counties. Absent from the southeastern Missouri lowland section. Most abundant in sink-hole ponds of the Ozarks, and locally in southwest Missouri in Newton County (wet rocky ground along bed of spring brook, Haddock Spring, 3 mi. northwest of Wentworth, October 29, 1953, *Palmer 57135*; low wet ground along margin of slough Sagmount, 3 mi. south of Saginaw, July 31, 1953, *Palmer 56325*).

### 3. **Eleocharis** R. Br. Spike Rush

Glabrous annuals or perennials, the culms (stems) with leaf-sheaths at their base. Leaf-blades absent. Flower of usually 6 (3–12) bristles or the bristles not present, 3–1 stamens, and a 2–3-cleft style. Base of style persistent in fruit on summit of achene as a tubercle, either enlarged and conspicuous or not enlarged. Achene 2-sided (lenticular) or 3-sided (trigonous).

a. Culms (stems )4-angled and mainly 2–5 mm. thick . . . . . . . 2. E. QUADRANGULATA var. CRASSIOR
a. Culms flattened, rounded, or angled, but if 4-angled, then only 0.5–1 mm. thick . . . . . . . . *b*
b. Culms marked cross-wise (transversely) by lines (septations) or partitions 1–5 cm. apart . . 1. E. EQUISETOIDES
b. No cross (transverse) lines present on culms . . . . . . . . . . . . . . . . . . *c*
c. Plants usually of limestone glades or prairies, or more rarely openings in upland woods; thickened creeping underground rhizome present, 2–5 mm. thick; culms strongly flattened, conspicuously 2-sided, 1–2 mm. wide; scales of spikelets with long-pointed whitish tips which are often cut into 2 at summit; styles 3-cleft; achenes finely roughened or wrinkled. . . . 12. E. COMPRESSA
c. Without such a combination of characters . . . . . . . . . . . . . . . . . *d*
d. No bristles present at base of any flower or achene . . . . . . . . . . . . . . *e*
e. Styles 2-cleft; achenes 2-sided or with 2 angles . . . . . . . . . . . . . . . *f*
f. At least the scales half-way up the spikelet obtuse or rounded at the summit; 1 sterile scale at the base of spikelet completely encircling the base of spikelet; culms half-way up 0.5–1.5 mm. (rarely up to 2 or more) thick (in dried condition) . . . 11. E. CALVA
f. At least the scales half-way up the spikelet with an acute or acuminate point, rarely obtuse; 2 or 3 sterile scales at very base of spikelet and not encircling the base of spikelet; culms 0.5–5 mm. thick (in dried condition) . . . . . . . . . . . *g*
g. Culms usually soft or flattened; spikelets usually long-tapering to acuminate, 10–30 mm. long; fertile scales rather closely pressed against one other (appressed). 10. E. MACROSTACHYA
g. Culms usually firm, wiry, and terete (cylindrical or in cross-section rounded); spikelets either pointed (acute) or obtuse (blunt) at summit, 5–20 mm. long; fertile scales more loosely ascending . . . . . . . . . . . 9. E. SMALLII
e. Styles 3-cleft; achenes 3-sided or with 3 or more angles . . . . . . . . . . . *h*
h. Plants restricted to salt-impregnated (saline) or alkaline soils or brackish water around salt or sulfur springs, or around old coal strip mines; achenes smooth, without slender longitudinal ribs; top of achene gradually narrowed into the crowning tubercle (persistent base of style) without any constricted area noticeable; culms terete (rounded in cross-section), but marked with fine vertical lines (striae) in dried state . . . . . . . . . . . . . . . . . 3b. E. PARVULA var. ANACHAETA
h. Plants of fresh-water ponds, springs, or sloughs or wet prairies; achenes with slender longitudinal ribs and cross (transverse) ridges or honey-combed (reticulate); top of achene separated from the crowning tubercle by a definite constriction; culms flattened and 2-sided or angled and furrowed . . . . . . . . . . . . . . *i*
i. Plants with firm, strong, tough, dark rhizomes 1–3 mm. thick; achenes honey-combed (reticulate); scales reddish-, purplish-, or dark chestnut brown with white margins . . . . . . . . . . . . . . 13. E. TENUIS var. VERRUCOSA
i. Plants with a slender rhizome or with hair-like tuber-bearing stolons or rhizomes; achenes with longitudinal ribs and transverse ridges; scales either pale green, greenish-white, or buff tinged with pale brown or purplish or purple-lined. . . . *j*
j. Common plant in or bordering fresh-water ponds, springs, or sloughs, and throughout Missouri; culms (stems) less than 0.5 mm. thick, angled (more than 2 angles) and furrowed; spikelets flattened, 0.7–1.5 mm. wide, 3–15-flowered; middle scales of spikelets 1.5–2.5 mm. long; roots bearing hair-like tuberous stolons . . . . . . . . . . . . . . . . . . . 4. E. ACICULARIS
j. Rare plant of wet prairies, and known only from Linn Co.; culms (stems) about 1 mm. (0.6–1.3) thick, flattened and 2-sided; spikelets terete (cylindrical), 2–3 mm. wide, mainly 16–20-flowered; middle scales of spikelets about 3 mm. long; slender creeping rhizomes present . . . . . . . . . 5. E. WOLFII
d. Bristles present at base of each flower or achene . . . . . . . . . . . . . . *k*

k. Plants restricted to salt-impregnated (saline) or alkaline soils or brackish water around salt or sulfur springs, or around old coal strip mines; top of achene gradually narrowed into the crowning tubercle (persistent base of style) with no constricted or separated portion evident . . . 3a. E. PARVULA var. PARVULA

k. Plants of many other kinds of habitats; top of achene separated from the crowning tubercle (persistent base of style) by either a definite constriction or line of separation . . . . . . . . . . *l*

l. Styles 2-cleft; achenes 2-sided or with 2 angles . . . . . . . . . . . . . . . . . . . *m*

m. Annuals with fibrous or fleshy roots and no creeping rhizomes (care must be taken to collect complete specimens with roots and basal parts); top of achene separated from the crowning tubercle by a division line but the tubercle not constricted at its base from the top of the achene . . *n*

n. Scales of spikelet definitely acute (pointed) at tip; spikelets more or less lanceolate and long-tapering or acuminate at summit . . . . . . . . . . . . 8. E. LANCEOLATA

n. Scales of spikelet usually rounded or obtuse (blunt) at summit; spikelets cylindrical to broadly ovoid, rounded or obtuse or slightly pointed (subacute) at summit . . . . . . . *o*

o. Bristles either scarcely reaching to the beginning of the crowning tubercle on the achene or very much shorter; tubercle less than ¼ as long as the achene and noticeably depressed-triangular . . . . . . . . . . . . . . . . . . . . 6. E. ENGELMANNI

o. Bristles usually longer than the combined length of achene and tubercle; tubercle 1/3–1/2 the length of achene, broadly triangular . . . . . . . . . . 7. E. OBTUSA

m. Perennials with creeping or horizontal rhizomes (care must be taken to collect complete specimens with roots and basal parts); top of achene separated from the crowning tubercle (persistent base of style) by a definite constriction . . . . . . . . . . . . . . . . *p*

p. At least the scales half-way up the spikelet obtuse or rounded; 1 sterile scale at base of spikelet completely encircling the base of spikelet; culms 0.5–1.5 mm. thick (in dried condition) . . . . . . . . . . . . . . . . . . . . . . . . 11. E. CALVA

p. At least the scales half-way up the spikelet with an acute or acuminate point, rarely obtuse; 2 or 3 sterile scales at very base of spikelet and not encircling the base of spikelet; culms 0.5–5 mm. thick (in dried condition) . . . . . . . . . . . . . . . . . *q*

q. Culms usually soft or flattened; spikelets usually long-tapering to acuminate, 10–30 mm. long; fertile scales rather closely pressed against one another (appressed). 10. E. MACROSTACHYA

q. Culms usually firm, wiry, and terete (cylindrical or in cross-section rounded); spikelets either pointed (acute) or obtuse (blunt) at summit, 5–20 mm. long; fertile scales more loosely ascending . . . . . . . . . . . . . . . . . . . . . 9. E. SMALLII

l. Styles 3-cleft; achenes with 3 or more angles, not noticeably angled, or only 2-angled . . . . . *r*

r. Culms hair-like or thread-like, mainly 0.1–1 mm. thick, finely angled with ridges and furrows; spikelets chiefly 0.7–2 (up to 2.8) mm. wide; achenes with the surface honey-combed (reticulate) or minutely roughened or wrinkled; perennials creeping and forming mats with either an obvious thickened cord-like rhizome 1–3 mm. thick or a thread-like tuber-bearing minute rhizome . . . . . . . . . . . . . . . . . . . . . . . . . *s*

s. Culms hair-like, usually 0.1–0.4 mm. thick; spikelets mainly 0.7–1.5 mm. wide; scales mainly pale green to greenish-white, only slightly tinged sometimes with brownish or purplish; achenes minutely roughened or wrinkled; rhizome hair-like, delicate, tuber-bearing . . . . . . . . . . . . . . . . . . . . . . 4. E. ACICULARIS

s. Culms of the thickness of coarse thread, 0.3–1 mm. thick; spikelets mainly 1.5–2.5 mm. wide; scales mostly reddish- to purple- or chestnut-brown with broad white or pale margins and tips; achenes with a honey-combed (reticulate) surface; rhizome obvious, thickened, cord-like, 1–3 mm. thick . . . . . . . . . . . . . . 13. E. TENUIS var. VERRUCOSA

r. Culms 0.3–1.5 mm. thick, terete (rounded in cross-section) and rather smooth, without deep furrows; spikelets chiefly 2–5 mm. wide; achenes smooth and shining; annuals with fibrous or fleshy roots . . . . . . . . . . . . . . . . . . . . . . . . . *t*

t. Scales of spikelet definitely acute (pointed) at tip; spikelets more or less lanceolate and long-tapering or acuminate at summit . . . . . . . . . . . . 8. E. LANCEOLATA

t. Scales of spikelets usually rounded or obtuse (blunt) at summit; spikelets cylindrical to broadly ovoid, rounded or obtuse or slightly pointed (subacute) at summit . . . . . . . *u*

u. Bristles either scarcely reaching to the beginning of the crowning tubercle on the achene or very much shorter; tubercle less than ¼ as long as the achene and noticeably depressed-triangular . . . . . . . . . . . . . . . . . . . 6. E. ENGELMANNI

u. Bristles usually longer than the combined length of achene and tubercle; tubercle 1/3–1/2 the length of the achene, broadly triangular . . . . . . . . . . 7. E. OBTUSA

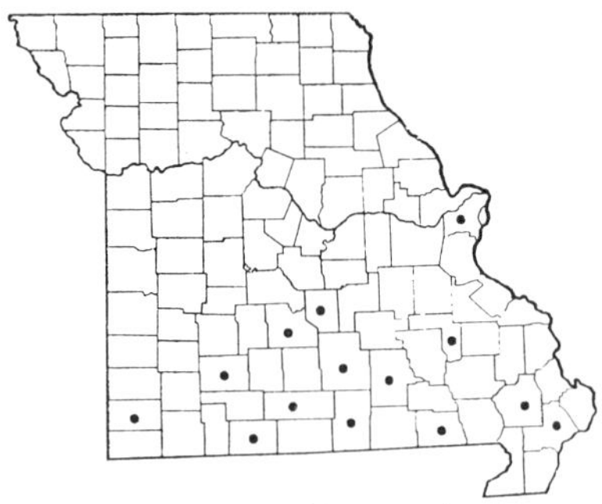

391 Eleocharis quadrangulata var. crassior

392 • Eleocharis parvula var. parvula
392 □ Eleocharis parvula var. anachaeta

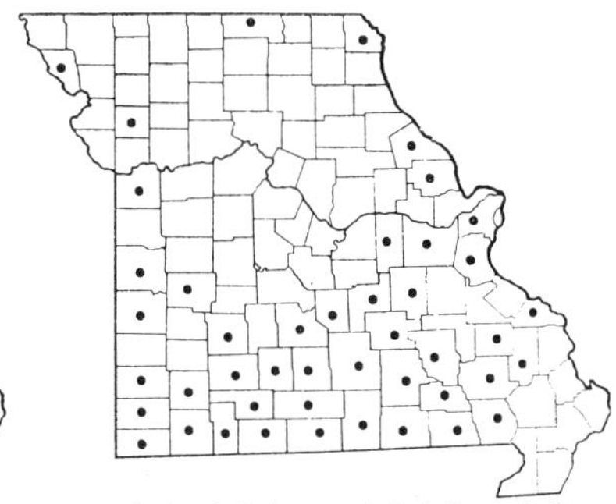

393 Eleocharis acicularis var. acicularis f. acicularis

1. **Eleocharis equisetoides** (Ell.) Torr. Map 390
Flowers July–October.

Known only from a sink-hole pond in Oregon County, southern Missouri (Tupelo Gum Pond, Clark National Forest, T25N, R4W, sect. 4, 9 mi. west of New Liberty, July 24, 1936, *Steyermark 12266*).

Ranges from Florida to Texas, north to North Carolina, Maryland to Massachusetts, New York to Wisconsin, Illinois, and Missouri.

One of the isolated relicts in the sink-hole ponds of the Ozarks.

2. **Eleocharis quadrangulata** (Michx.) R. & S. var. **crassior** Fern. Map 391
*Eleocharis quadrangulata* [of P & S]
Flowers from first part of June–October.

Occurs along muddy margins of ponds and streams in the Ozark section of southern and east-central Missouri northeast to St. Louis County, west to Pulaski, Laclede, Greene, and Newton counties.

Ranges from Florida to Texas, north to Massachusetts, New York, Ontario, Ohio, Michigan, Wisconsin, Missouri, and Oklahoma.

3. **Eleocharis parvula** (R. & S.) Link Map 392
Flowers July–October.

Occurs in marshy ground bordering saline springs and ponds in Ralls, Cooper, and Saline counties, and along alkaline muddy or sandy margins of old coal strip mines in southwest Missouri in Jasper (near Duenweg, June 15, 1952, *Palmer 54312*) and Barton (1 mi. southeast of Burgess, August 27, 1952, *Palmer 54972*; 1½ mi. southeast of Minden Mines, August 19, 1954, *Palmer 58601*) counties.

3a. **Eleocharis parvula** var. **parvula** Map 392

The typical variety with bristles is the commoner variety encountered in the state.

Ranges from Newfoundland to Louisiana, inland locally to New York, Michigan, Missouri, and Minnesota; also in British Columbia to California, the West Indies, South America, Europe, and Africa.

3b. **Eleocharis parvula** var. **anachaeta** (Torr.) Svenson Map 392

This variety which lacks bristles is more rarely encountered in the state.

Ranges from Minnesota and Iowa to Oregon, south to Louisiana, Texas, and Mexico; also in the West Indies and South America.

A rare species restricted to alkaline or saline spots in the state. Around Elk Lick Springs in Saline County, Chouteau Springs in Cooper County, and at a saline lake in the valley of Salt Lick Creek, Ralls County (*Steyermark 66458*), it occurs as a native species, associated frequently with *Zannichellia palustris*, *Scirpus paludosus*, *Typha angustifolia*, and *Diplachne acuminata*, and from a salt lick east of Elk Lick Springs with the isolated inland relict, *Distichlis spicata*. It has been introduced around the old coal strip mines in southwestern Missouri.

4. **Eleocharis acicularis** (L.) R. & S. Map 393
Flowers July–October.

Occurs commonly along margins of natural and artificial ponds, sloughs, and in springs. Often occurring as dense mats in dried-up temporary ponds.

Missouri material may be segregated as follows:

a. Culms sterile (without spikelets at tip)
4b. E. ACICULARIS var. ACICULARIS f. LONGICAULIS
a. Culms fertile (with spikelets at tip) . . . *b*

Plate no. 71. **1.** Eleocharis equisetoides. **2.** Eleocharis quadrangulata var. crassior. **3.** Eleocharis parvula. **4.** Eleocharis acicularis. **5.** Eleocharis Wolfii. **6.** Eleocharis Engelmanni.

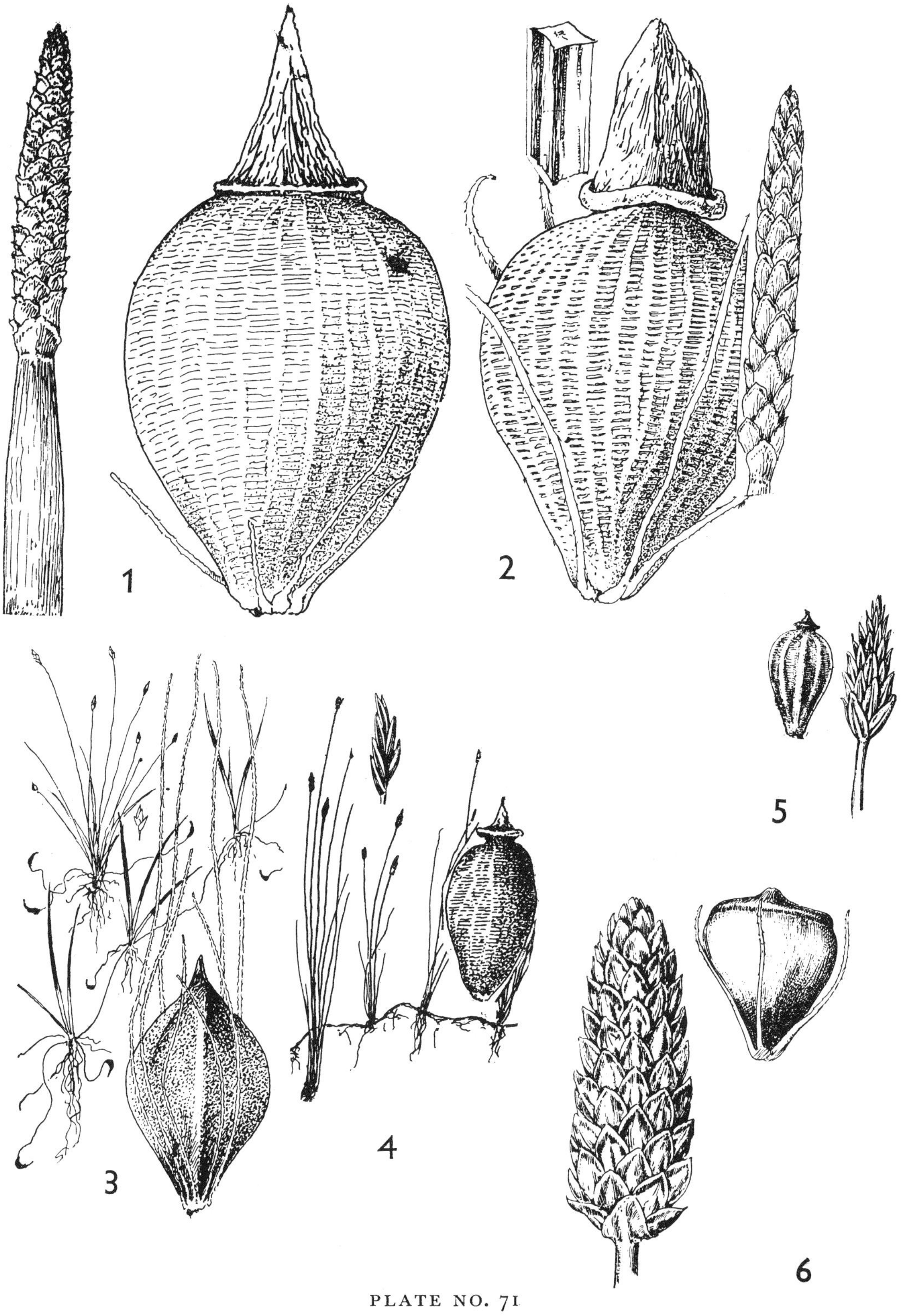

PLATE NO. 71

394 Eleocharis Wolfii

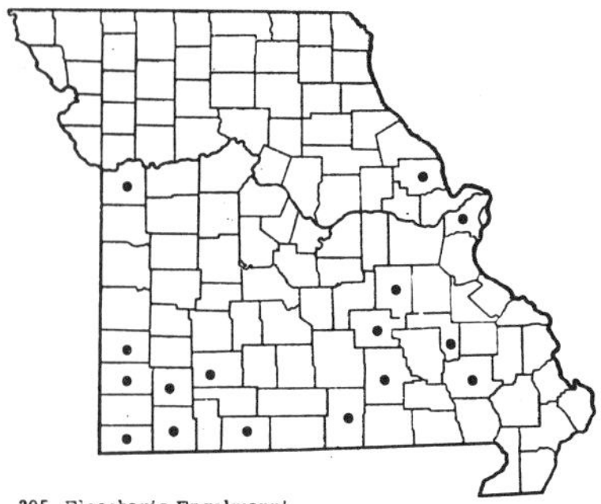

395 Eleocharis Engelmanni

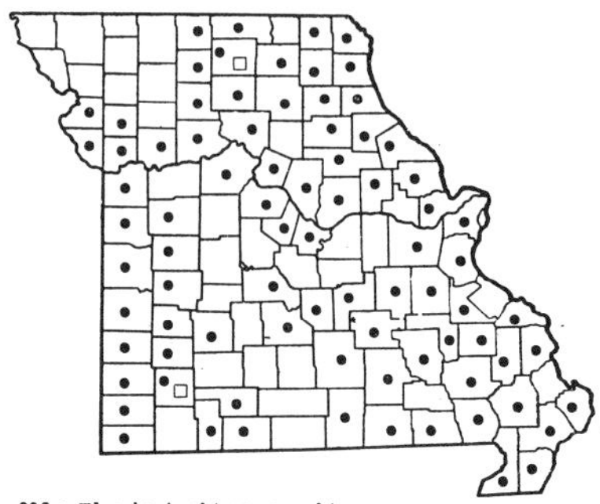

396 • Eleocharis obtusa var. obtusa
396 □ Eleocharis obtusa var. jejuna

b. Culms 0.2–20 cm. long; spikelets mainly 2–6 mm. long, mainly 3–8-flowered, ovate to lanceolate . . . . . 4a. E. ACICULARIS var. ACICULARIS f. ACICULARIS

b. Culms mainly 30–40 cm. long; spikelets mainly 5–12 mm. long, chiefly 9–20-flowered . . . . . . . . 4c. E. ACICULARIS var. ACICULARIS f. FLUITANS

4a. **Eleocharis acicularis** var. **acicularis** f. **acicularis** Map 393

*Eleocharis acicularis* (L.) R. & S. [G, BB, P & S]

This is the common variety encountered in the state, where it occurs throughout the Ozark section and scattered elsewhere northward; absent from large sectors of northern and central Missouri.

Ranges from Labrador to British Columbia, south to Florida, Tennessee, Missouri, Oklahoma, New Mexico, and California; also in Eurasia.

4b. **Eleocharis acicularis** var. **acicularis** f. **longicaulis** (Desmaz.) Hegi Map 393

Scattered throughout the range of the typical variety in the state.

4c. **Eleocharis acicularis** var. **acicularis** f. **fluitans** (Doell.) Svenson Map 393

*Eleocharis acicularis* var. *gracilescens* Svenson [P & S]

Scattered in portions of southern and east-central Missouri, principally in springs and stagnant pools.

Known from Tennessee to Oklahoma and California; also in Europe.

Svenson (Rh. 31: 183–91. 1929) has attempted to subdivide *E. acicularis* into numerous varieties and forms on the basis of assumed ecological and morphological differences. In greater depths of water, it is possible to find plants with the culms more greatly elongated than ordinarily, and these can be classified as f. *longicaulis*, if sterile, or f. *fluitans*, if fertile. Plants occurring in temporary ponds or ponds subject to drying conditions or to changing shore lines often have culms shorter than usual and appear stunted; such plants are common *E. acicularis* var. *acicularis* f. *acicularis*.

Study of material separated by Svenson as var. *gracilescens*, characterized by the elongated culms and linear spikelets, would indicate that this variation cannot be kept separate from f. *fluitans*. The spikelets of these plants, most of the culms of which grow in a greater depth of water, are usually more elongated and have more flowers per spikelet than in the common variety *acicularis* with shorter culms and smaller spikelets. In the specimen of *Palmer 66173* from Newton County, the culms are 40 cm. long, up to 0.5 mm. thick, the spikelets are linear or lanceolate-linear up to 12 mm. long and with flowers from 9–20. The flowers in Palmer's specimen also possess bristles. Actually, although Svenson described var. *gracilescens* as having no bristles, Missouri material is found either with or without bristles. Svenson (N. Am. Fl. 18, part 9: 516. 1957) describes the color of the bristles in *E. acicularis* as 'brownish,' but in fresh or recently dried material the bristles are white or whitish.

As the variations of this species become subjected to more detailed experimental investigations, a more positive statement may be forthcoming as to the relative worth of the various forms that have been described. At present, it would appear that some correlation exists between the relative length of culm and spikelet and the relative depth of water, but that no more than a normal ecological range of variation exists in the environment of this species; and it is, therefore, questionable whether names should be given for such normal variations.

5. **Eleocharis Wolfii** Gray Map 394

Flowers May–July.

Known only from Linn County, northern Missouri (uncommon around swale in bottom prairie along

East Yellow Creek, 6 mi. east of Brookfield, June 5, 1936, *Steyermark 11028*).

Ranges from Indiana to Iowa and Colorado, south to Tennessee, Louisiana, and Oklahoma; also in New York and Saskatchewan.

6. **Eleocharis Engelmanni** Steud. Map 395

*Eleocharis Engelmanni* var. *robusta* Fern. [P & S]

Flowers May–October.

The type locality is St. Louis, Missouri, collected by *Engelmann* in 1845.

Occurs along muddy margins of ponds, sloughs, and, occasionally, streams, rarely along roadside ditches. Scattered in parts of southern and central Missouri north to Lincoln and Jackson counties. A collection from Boone County labeled *E. Engelmanni* (near More's Lake, Columbia, July, 1908, *Daniels*) proves upon examination to be *E. obtusa*.

Ranges from Maine to Saskatchewan and Washington, south to Virginia, Georgia, Arkansas, Texas, Arizona, and California.

A form in which the bristles are absent, f. *detonsa* (Gray) Svenson, has not been found in Missouri, but is known from localities near the state's boundaries, and is eventually expected to occur.

7. **Eleocharis obtusa** (Willd.) Schultes Map 396

Flowers May–October.

Occurs along muddy borders of ponds, sloughs, and streams, temporary pools and roadside ditches. The commonest spike-rush in Missouri, where known from all but the northwestern section of the state.

Missouri material may be classified into the following varieties:

Spikelets 2–16 mm. long, many-flowered (40–150 or more); scales of spikelets pale or yellow-brown with white margins, erect and close against one another; achenes 1–1.5 mm. long, the tubercle 1/3–1/2 the length of the achene . . 7a. E. OBTUSA var. OBTUSA

Spikelets 2–5 mm. long, few-flowered (mainly 6–12, but up to 20); scales of spikelets usually purplish or brown-purple with paler margins, spreading; achenes 0.9–1.3 mm. long, the tubercle nearly 1/2 the length or slightly shorter than 1/2 the length of the achene . . . . 7b. E. OBTUSA var. JEJUNA

7a. **Eleocharis obtusa** (Willd.) Schultes var. **obtusa** Map 396

The common variation in Missouri.

Ranges from Nova Scotia to Minnesota, south to Florida, Alabama, Mississippi, Florida, and Texas; also British Columbia to California, and Hawaii.

7b. **Eleocharis obtusa** var. **jejuna** Fern. Map 396

Known only from two scattered localities, in Sullivan (sandy and muddy shallow bars of East Fork of Medicine Creek, T62N, R22W, sect. 2, 1/4 mi. east of Osgood, September 21, 1955, *Steyermark 79815*) and Lawrence (muddy margins of artificial spring-fed pond, state fish hatchery, Chesapeake, October 17, 1956, *Palmer 64613*) counties.

Known from Nova Scotia and New England to Virginia, and locally in Missouri.

The above-cited collections are the only ones that can be considered to come within the limits of the var. *jejuna*. This variety has slender culms about 0.5 mm. thick and only 1–10 (rarely to 20) cm. tall. Dwarfed specimens of *E. obtusa* var. *obtusa* with culms very slender are sometimes mistaken at first for var. *jejuna*, but have larger heads with more numerous flowers and pale brown scales.

*Eleocharis obtusa* varies considerably in height and thickness of culm, and length, thickness, and shape of spikelet. Svenson (Rh. 31: 216. 1929) considers var. *jejuna* an ecological state of the species growing in muddy places subjected to flooding. He further expresses his belief that this species might be considered as part of a variable series combined with *E. Engelmanni*, *E. lanceolata*, *E. diandra*, and *E. ovata* (N. Am. Fl. 18, part 9: 519. 1957). Until future studies, perhaps along experimental lines, have been conducted with these species, the present treatment is in accord with general botanical opinion.

8. **Eleocharis lanceolata** Fernald Map 397

Flowers June–October.

Occurs in wet sandy open ground, muddy margins of small ponds, and in moist depressions on glades. Known only from southern Missouri in Carter, Jasper, and Barton counties.

Ranges from southern Missouri to Arkansas, Oklahoma, and Texas.

The character of the length of the tubercle, ascribed by authors to this species, as a distinguishing criterion, has been found to break down in the case of some collections. In the collection of *Palmer 65905* from Newton County, the tubercle is mostly only 1/3 instead of 1/2 the height of the achene, thereby breaking down one of the differences often used to separate *E. lanceolata* from *E. obtusa*, as noted in Svenson's treatment in *North America Flora* (p. 518–20).

9. **Eleocharis Smallii** Britt. Map 398

*Eleocharis palustris* [of BB] in part, not (L.) R. & S.

Flowers from the first of June–September.

Occurs along borders of swamps, sloughs, ponds,

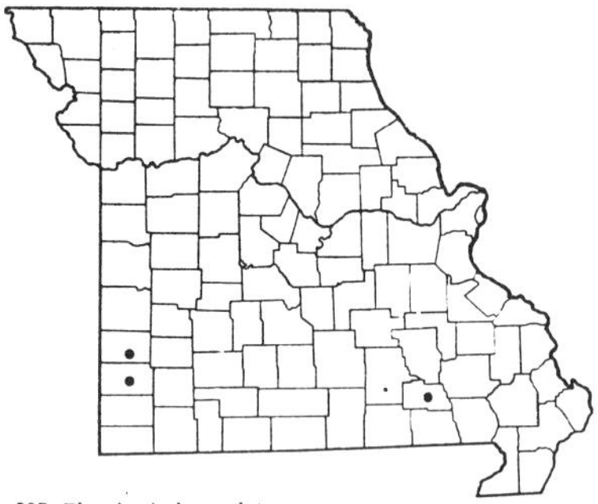

397 Eleocharis lanceolata

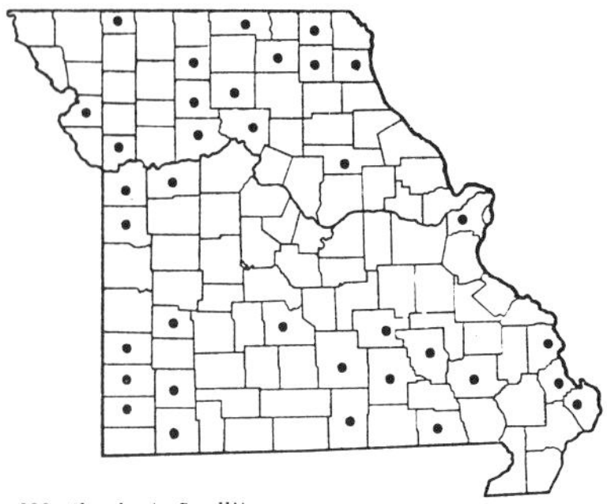

398 Eleocharis Smallii

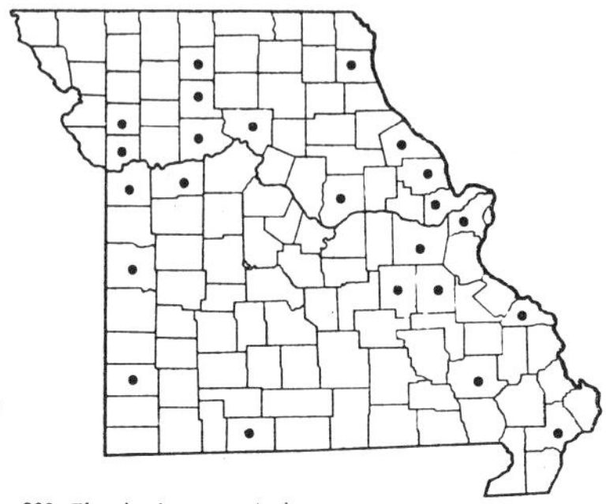

399 Eleocharis macrostachya

streams, roadside ditches, and in wet meadows and prairies. Scattered throughout the state.

Ranges from Newfoundland to Ontario, south to Maryland, Pennsylvania, West Virginia, Indiana, Illinois, and Missouri.

10. **Eleocharis macrostachya** Britt. Map 399

*Eleocharis palustris* [of BB] in part, not (L.) R. & S.

*Eleocharis mamillata* [of P & S], not Lindb. f.

Flowers late May–August.

Occurs along muddy margins of ponds, sloughs, swamps, streams, and in prairie swales. Rather rare and scattered in Missouri, mostly absent in the Ozark section.

Ranges from Minnesota to Saskatchewan and British Columbia, south to Illinois, Louisiana, Texas, Mexico, and California.

11. **Eleocharis calva** Torr. Map 400

*Eleocharis erythropoda* Steud. [Svenson

Flowers from first of June–September.

Occurs along margins of ponds, sloughs, streams, and in wet prairies and meadows.

Throughout Missouri, except apparently absent from the extreme southeastern lowland section and parts of western Missouri.

Ranges from Gaspé Peninsula, Quebec, to Manitoba, south to Virginia, Tennessee, Arkansas, Oklahoma, and New Mexico; also in Hawaii and Asia.

Of the group having prominent creeping rhizomes and growing near the water (*E. Smallii*, *E. macrostachya*, and *E. calva*), this is the most frequently encountered, and the commonest throughout the Ozark section, where it borders most of the small and large streams. It also possesses the most slender culms of the group. As with the other members of the group, bristles may be present or absent.

12. **Eleocharis compressa** Sulliv. Map 401

Flowers early May–July.

Occurs most frequently on limestone glades and wet depressions throughout the Ozark section of southern and central Missouri, infrequently and locally northwest in prairies, along roadside ditches, and openings in upland woods northwest to Nodaway and Atchison counties.

Ranges from Quebec and New York to Saskatchewan, south to Virginia, Georgia, Alabama, Missouri, Texas, and Colorado.

This species is a characteristic one on calcareous exposures, limestone glades, and 'bald knobs' throughout the Ozark region, where it is frequently associated with such other plants as *Leavenworthia uniflora*, *Arenaria patula* f. *Pitcheri*, *Galium virgatum*, and *Ophioglossum Engelmanni*.

It is easily identified usually by the combination of flattened, relatively broad culms, thick creeping rhizomes, and long-pointed, white-tipped scales which are often split in two at the summit; such plants are characteristic of the Ozark area. There are specimens found, however, in western Missouri north of the Missouri River which are not quite typical in certain respects, and may be mistaken for *E. tenuis* var. *verrucosa*. Such plants may have rather more slender culms, which, because of their striations, may even appear angled as in *E. tenuis*, and are somewhat intermediate in appearance between the two species; they may be of hybrid origin.

Svenson likewise indicates (Rh. 34: 217–18. 1932) that difficulty may be experienced in distinguishing *E. compressa* from *E. tenuis* (known at that time as *E. capitata*) at the borders of range of the two species and notes also the narrower culms of *E. compressa* which may be evident in such marginal areas where the ranges of these two species overlap.

Plate no. 72. **1.** Eleocharis obtusa. **2.** Eleocharis lanceolata. **3.** Eleocharis Smallii. **4.** Eleocharis macrostachya. **5.** Eleocharis calva. **6.** Eleocharis compressa.

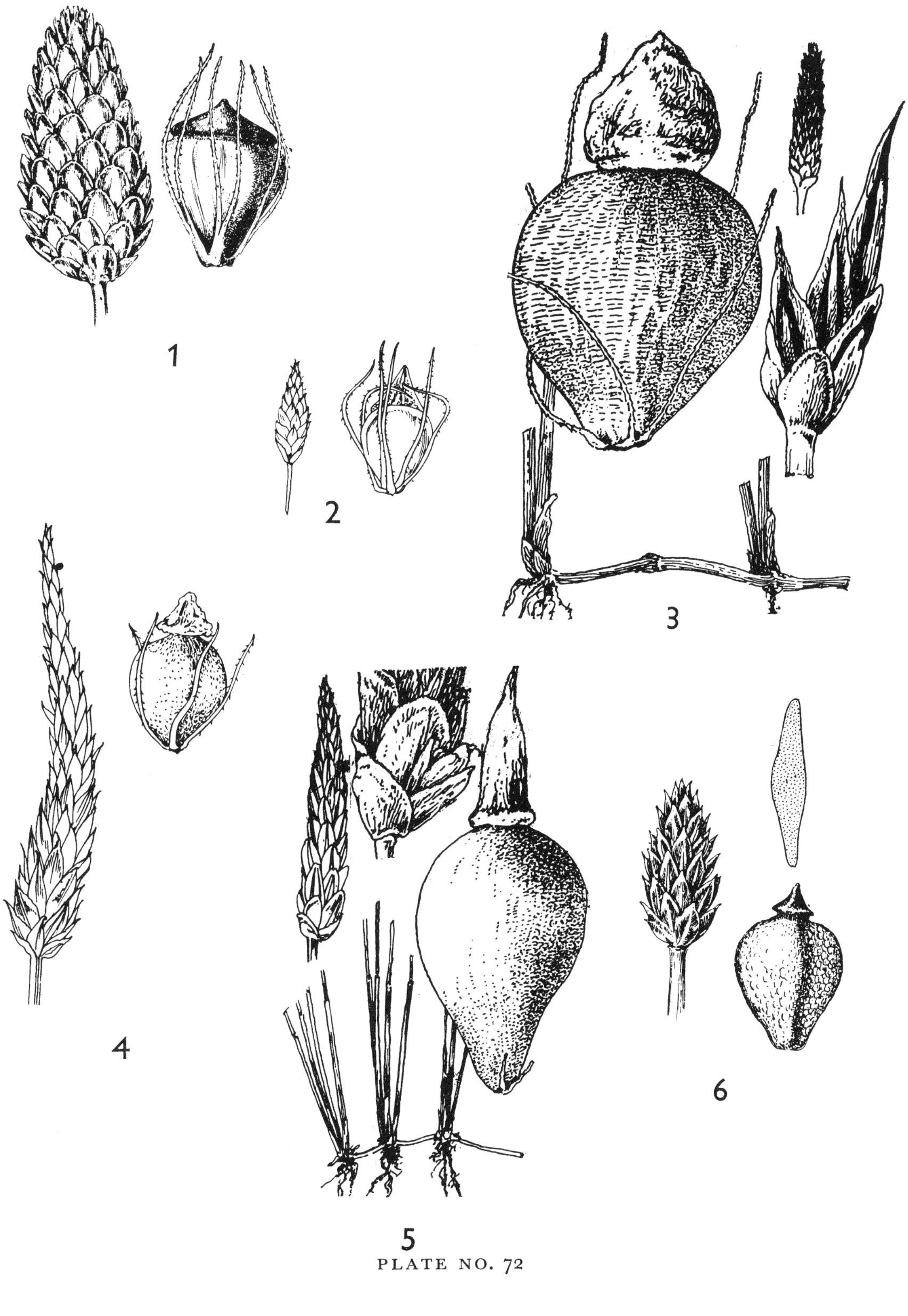


PLATE NO. 72

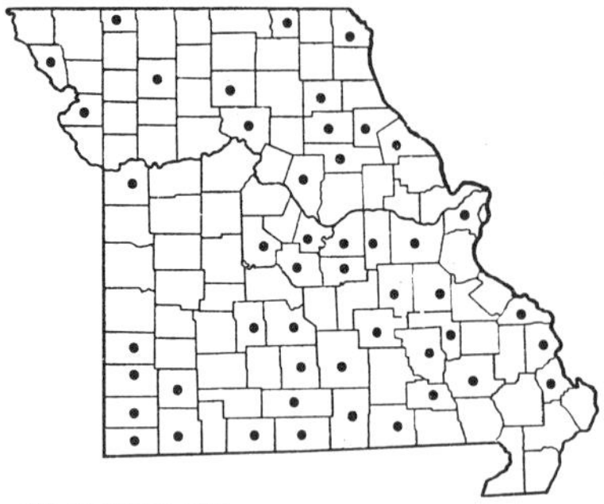

400 Eleocharis calva

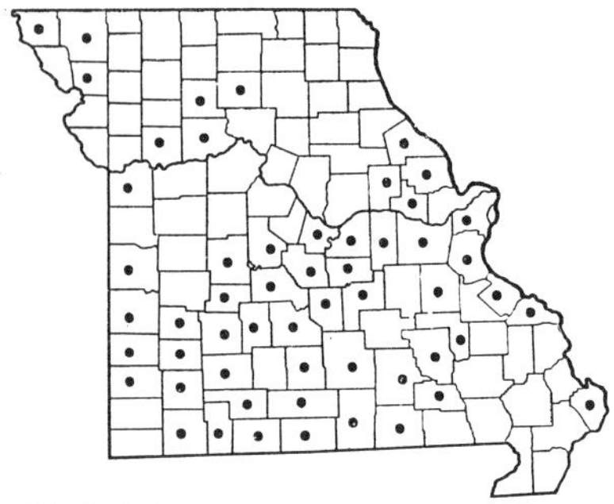

401 Eleocharis compressa

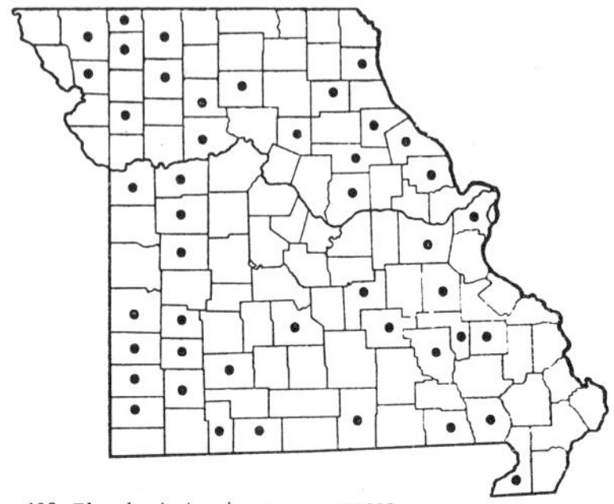

402 Eleocharis tenuis var. verrucosa

13. **Eleocharis tenuis** (Willd.) Schultes var. **verrucosa** Svenson Map 402

*Eleocharis capitata* (L.) R. Br. var. *verrucosa* Svenson [P & S]

Flowers early May–September.

Occurs in wet places in prairies, glades, prairie swales along railroad tracks, roadside openings, along ditches, wet places in woods, and wet ground bordering sloughs. Common throughout Missouri.

Ranges from Indiana and Iowa, south to Kentucky, Illinois, Louisiana, and Texas; also in Virginia.

In general, there is not too much difficulty in distinguishing this species from *E. compressa*, especially in mature fruiting stage, at which time the reticulate achenes of *E. tenuis* var. *verrucosa* easily separate it from the granular-roughened or wrinkled ones of *E. compressa*. However, in younger stages of flowering, difficulty is sometimes experienced, as the two species strongly resemble one another. The scales of *E. tenuis* var. *verrucosa* are often as acute and may have nearly as prominent white margins and summits as in *E. compressa*, the rhizomes are sometimes nearly as thick as in *E. compressa*, although usually more curving and more slender (1–3 mm. thick), and the culms of *E. tenuis* var. *verrucosa* may even appear flattened, after having been dried and pressed. For the most part, however, the culms of *E. tenuis* var. *verrucosa* are found to be more slender (0.3–1 mm. thick) than those of *E. compressa* (mostly 1–2 mm. thick), and, if cross-sectioned, usually show 4 angles with somewhat depressed or concave sides or 5 angles and the 4 to 8 fibrovascular bundles separate; in *E. compressa* the culm is usually broad and flattened with a continuous fibrovascular system. In general, the scales of *E. tenuis* var. *verrucosa* have the greater portion of their area darker maroon and a corresponding smaller area of white margin and apex than in those of *E. compressa*, the scales also being more rounded to obtuse.

Certain collections from western Missouri north of the Missouri River greatly resemble *E. compressa* and seem to merge with that species. Perhaps these are of possible hybrid origin. Svenson notes (Rh. 34: 218. 1932) that the achene in *E. compressa* has a tendency to be more rounded than that of *E. tenuis* with less conspicuous surface markings and the tubercle usually in the form of a low cone.

Some authors have attempted to indicate a difference in the apex of the leaf-sheath, that of *E. tenuis* var. *verrucosa* being cut off at an angle (obliquely) and that of *E. compressa* being cut square across (truncate) at the top. These differences, in the case of much Missouri material examined, do not seem to hold, however, and therefore this character has not been used as a reliable one in distinguishing the two species. Due to the broader culms of *E. compressa*, however, with the summit of the leaf-sheath usually more truncate, there is a somewhat more conspicuous darker margin and color contrast to the culm than in *E. tenuis* var. *verrucosa*.

## 4. **Bulbostylis** (Kunth) C. B. Clarke

Annuals (in the local species) growing in dense tufts with hair-like leaves and culms, the leaf-sheaths finely short-hairy. Stamens 2 or 3. Style 2–3-cleft, the base enlarged and persisting as a tubercle.

**Bulbostylis capillaris** (L.) C. B. Clarke. Map 403

*Stenophyllus capillaris* (L.) Britton [P & S]

*Bulbostylis capillaris* var. *crebra* Fern. [G]

*Bulbostylis capillaris* var. *isopoda* Fern. [G]

Flowers July–October.

Occurs in thin acid soils and on rocks of sandstone,

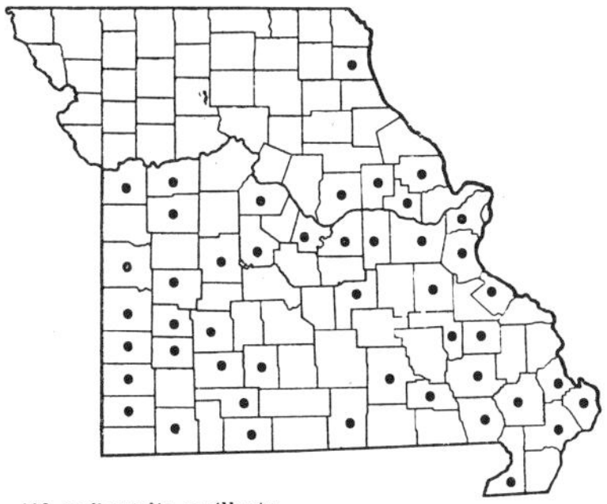

403 Bulbostylis capillaris

404 Fimbristylis caroliniana

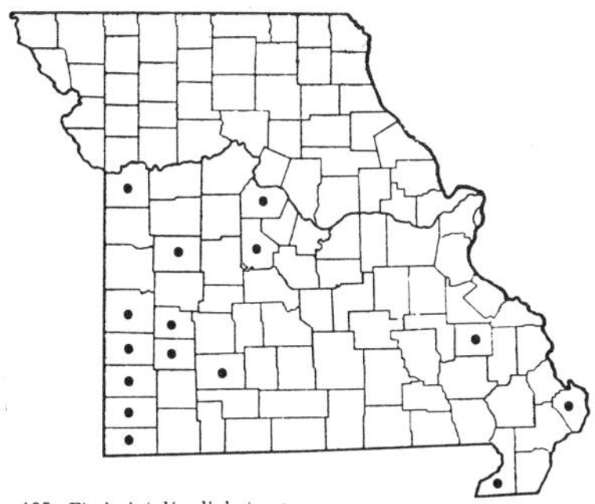

405 Fimbristylis dichotoma

chert, and granite on glades and barrens, and swales in prairies. Southern, central, and eastern Missouri northeast to Lewis County.

Ranges from Maine and Quebec to Minnesota, south to Florida and Texas; also in the Pacific states, northern Mexico, Central America, and Cuba.

Fernald (Rh. 40: 394–95. 1948) has attempted to divide the species into three varieties separated as follows:

a. Crowded sessile spikelets occurring at the base of the leaves . . B. CAPILLARIS var. CAPILLARIS
a. No sessile spikelets occurring at the base of the leaves. . . . . . . . . . . . . *b*
b. Lateral spikelets on unequal pedicels 0.1–1 cm. long . . . B. CAPILLARIS var. CREBRA
b. Lateral spikelets on nearly equal pedicels 0.6–3 cm. long. . B. CAPILLARIS var. ISOPODA

Plants which fit the above classification can be found in the state, but they do not appear to show any constancy in their occurrence, all types of variations being found in a noncorrelated manner and intergrading. It would seem that Missouri material at least cannot fit into Fernald's scheme of classification.

## 5. **Fimbristylis** Vahl

Annuals or perennials with slender narrow leaves. An involucre of 2–3 leaves at the base of the umbellate inflorescence. Flowers consist of 1–3 stamens, 2–3-cleft style, the base of style deciduous, and no bristles or scales.

a. Some hairs present at least on the lowermost scales of a spikelet (often on all the scales of younger spikelets); perennials with a thickened or bulb-like base. . . . . . . . . . . 1. F. CAROLINIANA
a. All the scales of a spikelet glabrous (without hairs); annuals with a soft non-thickened base . . . . *b*
b. Scales of spikelets usually pale green or greenish-white; each spikelet stalkless (sessile), all close together in a compact cluster; spikelets appearing as if hidden in the greatly elongated hair-like leaves and involucre . . . . . . . . . . . . . . . . . . . . . . 3. F. VAHLII
b. Scales of spikelets shades of brown; 1 or more of the spikelets on slender stalks; spikelets raised on elevated culms (stems) usually equalling or longer than the leaves . . . . . . . . . . *c*
c. Spikelets ovoid (egg-shaped), chiefly 2.5–3 mm. wide, chiefly 2–2½ times as long as broad; leaves with fine hairs on margins; style 2-cleft; achene 2-sided (lenticular) with 6–8 conspicuous longitudinal ribs on each surface . . . . . . . . . . . . . 2. F. DICHOTOMA
c. Spikelets narrowly linear- or ovate-lanceolate or narrowly elliptic, 1–1.5 mm. wide, chiefly 3–4 times as long as broad; leaves glabrous; style 3-cleft; achene 3-sided (trigonous), smooth and shining or with a surface faintly and finely honeycombed (reticulate) . . . 4. F. AUTUMNALIS

### 1. **Fimbristylis caroliniana** (Lam.) Fern.

Map 404

*Fimbristylis castanea* (Michx.) Vahl [P & S]
*Fimbristylis castanea* var. *puberula* (Michx.) Britton [P & S]
*Fimbristylis Drummondii* Boeckl. [G]
*Fimbristylis interior* Britton
*Fimbristylis puberula* (Michx.) Vahl

Flowers early May–July.

Occurs on open limestone, chert, or sandstone glades, upland sometimes sandy prairies, and rocky openings in woodland. Southern and central Missouri north to St. Louis, Montgomery, Moniteau, Pettis, and Jackson counties; absent from the southeastern lowland section.

Ranges from Florida to Texas, north to New York,

406 Fimbristylis Vahlii

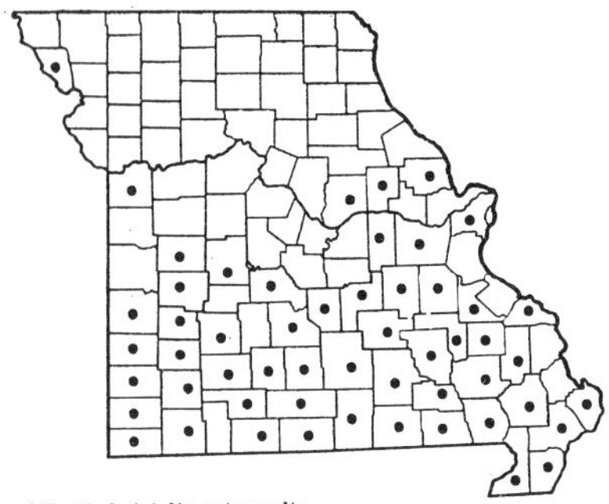

407 Fimbristylis autumnalis

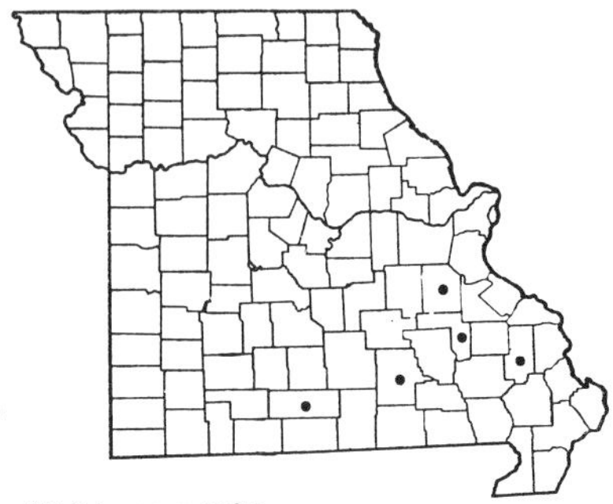

408 Scirpus verecundus

New Jersey, Ontario, Michigan, Illinois, and Missouri.

This species is a characteristic plant of upland prairies and exposed rocky glades, where it withstands extremes of temperature and severe drouth. Fernald (*Gray's Manual*, eighth edition, p. 261. 1950) attempted to separate two variants of this group on the basis of soil differences. The 'acid or subacid habitats' which he attributes to *F. Drummondii* (the plants in Missouri), does not cover the range of soils variations encountered among the Missouri plants as they are more frequently found on alkaline limestone and other calcareous glades, as well as on acidic sandstone or chert rocks.

The spikelets are raised on a slender elongated mostly naked culm often 70 cm. tall, standing higher than the narrow leaves which are densely tufted at the base. The leaves and culms may be glabrous or more or less hairy. The spikelets may be sessile in the early stages of flowering to eventually stalked as the flowering stage develops.

Svenson (*North Am. Fl.* 18, part 9: 551. 1957) has combined this and other related variations under *Fimbristylis spadicea* (L.) Vahl. He is of the opinion that, although these variants differ in certain characters, such as luster and pubescence of scales, spikelet shape and length, and width of leaves, they are all held together by the same type of achene. In the present treatment the name *Fimbristylis caroliniana* is used for those plants having pubescent scales of the spikelets.

2. **Fimbristylis dichotoma** (L.) Vahl Map 405
*Fimbristylis Baldwiniana* (Schultes) Torr. [G]
*Fimbristylis annua* (All.) R. & S. [BB]
*Fimbristylis laxa* Vahl [P & S]
Flowers June–October.

Occurs on sandy prairies, moist sandy open ground, margins of ponds, moist places in glades, and on gravel and sand bars along streams, usually in acid soils. Southern and central Missouri, mainly in the southwestern third of the state, north to Cooper and Jackson counties, locally east to Madison, Mississippi, and Dunklin counties.

Ranges from Florida to Texas and Arizona, north to Pennsylvania, Illinois, Missouri, and Oklahoma; also in Mexico, Central and South America, the West Indies, and in the Old World tropics.

The spikelets have a shape similar to those of *F. caroliniana*, but the plants are generally much shorter, averaging 1–3 dm. tall, the leaves are flat instead of becoming involute (margins rolled in) as in *F. caroliniana*, and the achenes have a different color and markings.

3. **Fimbristylis Vahlii** (Lam.) Link Map 406
Flowers July–October.

Occurs in moist sandy open ground and along sandy margins of ponds, in southern Missouri north to Wayne and Barton counties.

Ranges from Florida to Texas, north to North Carolina and Missouri; also in California, Pennsylvania (introduced), and Central and South America.

The plants occur as small grass-like tufts usually only 3–10 cm. tall with hair-like leaves and involucral bracts, which greatly exceed the small clusters of spikelets. The latter, in contrast to the total height of the leaves, appear almost near the surface of the ground.

4. **Fimbristylis autumnalis** (L.) R. & S. Map 407
*Fimbristylis autumnalis* var. *mucronulata* (Michx.) Fern. [G]
*Fimbristylis mucronulata* (Michx.) Blake [P & S]
Flowers June–October.

Plate no. 73. **1.** Eleocharis tenuis var. verrucosa. **2.** Bulbostylis capillaris. **3.** Fimbristylis caroliniana. **4.** Scirpus verecundus. **5.** Fimbristylis Vahlii. **6.** Fimbristylis dichotoma. **7.** Fimbristylis autumnalis.

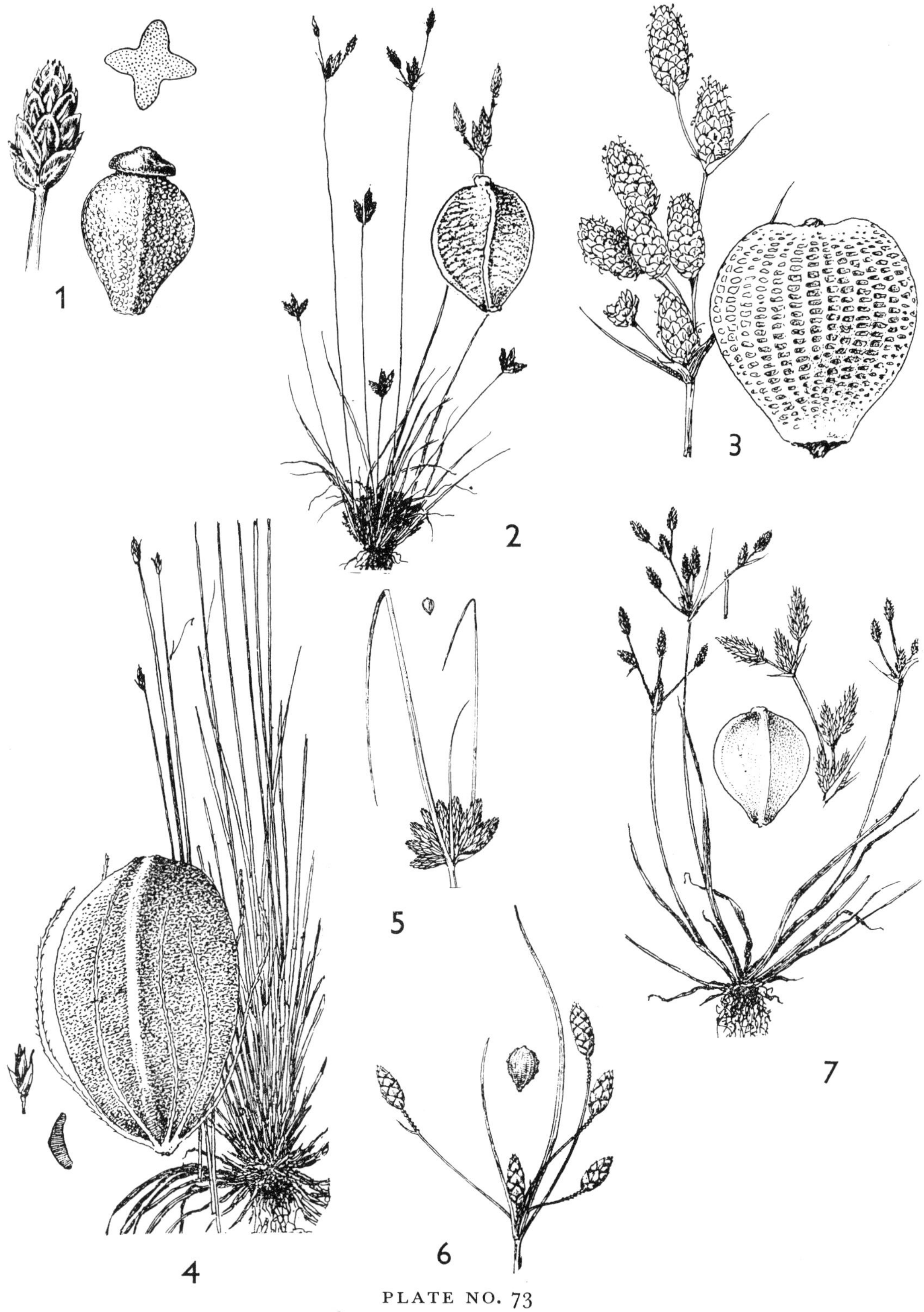


PLATE NO. 73

Occurs in sandy fields, sandy open ground, moist sand and gravel bars, wet muddy shores of streams and ponds, and wet places in prairies and glades. Southern and central Missouri north to Lincoln, Callaway, and Jackson counties, and locally northwest in Holt County.

Ranges from Quebec and Maine to Ontario and Minnesota, south to Florida and Texas, Central America, and Cuba.

Missouri material has been divided into *F. autumnalis* var. *autumnalis* and *F. autumnalis* var. *mucronulata* (Michx.) Fern. with the following key differences emphasizing the main contrasts:

Spikelets ovate-lanceolate or lance-elliptic; scales of spikelets with free slender tips; spikelets simple to compound, 1–50. . . . F. AUTUMNALIS var. AUTUMNALIS

Spikelets linear-cylindric to narrowly fusiform; scales of spikelets with tips more closely appressed; spikelets compound, 5–100 . F. AUTUMNALIS var. MUCRONULATA

These differences do not appear to be correlated well enough to be applicable to collected material, and are interpreted in the present work as modifications occurring as a normal amount of variation influenced partly, no doubt, by changing environmental conditions. Similar conclusions have been drawn or indicated by Deam (*Fl. Indiana* p. 205. 1940) and by Gleason (*New Ill. Fl.* 1: 269. 1952).

## 6. **Scirpus** L. Bulrush

Annuals or perennials with leafy or leafless culms, the leaves sometimes represented by sheaths only. Flowers of 2–3 stamens, a 2–3-cleft style, the base of which is deciduous (falls off), and 1–8 bristles, or these not present.

a. Tip of each culm (stem) ends in only 1 spikelet without any elongated bract . . . . 1. S. VERECUNDUS
a. Tip of each culm (stem) ends in 1 or more leaf-like bracts with 1 or more spikelets . . . . . . . *b*
b. The culm (stem) ends in 1 erect bract (which resembles a continuation of the culm), the inflorescence thus appearing to originate from the side of the culm (stem); culms (stems) without leaves, or leaves inconspicuous, or leafy only near the base . . . . . . . . . . . . . . . *c*
c. All spikelets without or nearly without a stalk, appearing to come directly from the culm or erect involucral bract; culms either 3-angled, or, if terete, 3 mm. or less thick at base, and often less than 1 meter tall, but also taller . . . . . . . . . . . . . . . . . . . . *d*
d. Culms (stems) terete (cylindrical, or with rounded sides, or round or nearly round in cross-section); perennial plants either growing in the water, or soft-based annuals with fibrous roots and slender tufts of culms averaging 5–35 cm. tall . . . . . . . . . . . . . *e*
e. Perennial plants with culms usually submerged and growing in the water; flowers with bristles at base of achene; culms solitary or scattered, with occasional thickened areas (nodulose), usually 30–120 cm. tall . . . . . . . . . . . . 4. S. SUBTERMINALIS
e. Annual plants growing in dry or moist soil; flowers without bristles; culms without thickened areas, in tufts usually 5–35 cm. tall . . . . . . . . . . . . . . . . *f*
f. Culms hair-like, 1 mm. or less thick; style 3-cleft; achene brown, 3-sided; scales of spikelet greenish, prominently keeled; spikelets 1–3 . . . . . . . . 2. S. KOILOLEPIS
f. Culms stouter, usually 1.5–3 mm. thick; style 2-cleft; achene black (when ripe), flattened, 2-sided; scales of spikelet mostly brown with green keel, slightly keeled; spikelets mostly 2–7 (rarely 1) . . . . . . . . . . . . . . . 3. S. HALLII
d. Culms (stems) 3-sided; perennial plants with creeping rhizomes with culms solitary or scattered, averaging 40–150 cm. tall . . . . . . . . . . . . . . . . . . *g*
g. Style 3-cleft (for exceptional 3-cleft style of *S. americanus* var. *polyphyllus*, refer to 2nd g.); achene 3-sided (trigonous), with a long tip (beak) about ¼ length of achene; culms bluntly 3-angled, 0.5–2 mm. thick at summit; scales of spikelets pale or yellow-brown to straw-color, entire, ovate-lanceolate; leaves with scattered thickened areas (nodulose) 5. S. TORREYI
g. Style 2-cleft (rarely 3-cleft in *S. americanus* var. *polyphyllus*); achene 2-sided, with a very short tip (beak) about 1/10–1/9 length of achene; culms sharply 3-angled, mostly 2–10 mm. thick at summit; scales of spikelet reddish- or chestnut-brown, with fringed or irregularly cut margins, ovate to orbicular; leaves without scattered thickened areas . . *h*
h. Common species; upper leaf-blade 2–9 mm. wide at base, sharp-pointed; erect stem-like involucral bract linear, 2–15 cm. long, acute; culm without winged angles, firm, solid, its sides nearly flat or only slightly concave (sunken in middle), mostly 2–6 mm.

thick at the upper leaf-sheath; scales of spikelet prominently and abruptly tipped (mucronate) . . . . . . . . . . . . . . . . . . . 6. S. AMERICANUS

h. Rare species; upper leaf-blade 8–15 mm. wide at base, blunt-tipped; erect stem-like involucral bract narrowly triangular, 1–3.5 cm. long, blunt; culm with winged angles, soft, hollow, its sides deeply concave (as if sunken in the middle), mostly 4–10 mm. thick at the upper leaf-sheath; scales of spikelet inconspicuously tipped . . 7. S. OLNEYI

c. Spikelets separated from the culm (stem) and erect involucral bract by branched or unbranched stalks; culms (stems) terete, or, if 3-sided, 3–20 mm. thick at base, or mostly 1–2 meters tall . . . . . . . . . . . . . . . . . . . . . . . . . . *i*

i. Rare species of sink-hole ponds, known only from Oregon Co.; culm 3-sided; scales of spikelets pale brown or greenish; basal leaf-sheath with a long and developed leaf-blade; stem-like involucral bract 3-angled; main part (body) of achene 4 mm. long . 11. S. ETUBERCULATUS

i. Common species of ponds and sloughs; culm terete or nearly terete (cylindrical or with curved or rounded sides, round in cross-section); scales of spikelets reddish- or chestnut-brown, or pale brown to greenish; basal leaf-sheath lacking a developed leaf-blade; stem-like involucral bract terete (cylindrical and rounded in cross-section); main part (body) of achene 1.7–2.8 mm. long . . . . . . . . . . . . . . . . . . . *j*

j. Spikelets ovoid, usually 4–10 mm. (up to 15) long; scales of spikelets about as long as or only slightly longer than achenes, the margins with fine white hairs; midrib of scales prominently thickened and green, projecting noticeably beyond scale as a tip; achenes 1.5–2.3 mm. long . . . . . . . . . . . . . . . . 8. S. VALIDUS var. CREBER

j. Spikelets more elongated, ranging from linear-cylindrical or narrowly ellipsoidal to lanceolate or narrowly ovoid, mostly 8–23 mm. long (rarely shorter); scales of spikelets much longer than the achenes, either glabrous or with red dots and glandular-hairy toward summit; midrib of scales either not prominent or not projecting noticeably beyond summit of scale; achenes averaging 2–3 mm. long. . . . . . . . . . . . *k*

k. Achene 3-angled (trigonous); style 3-cleft; scales of spikelet glabrous, pale brown, drab, or greenish; bristles shorter than achene; spikelets lanceolate or narrowly ellipsoidal, tapering to a somewhat pointed tip . . . . . . . 10. S. HETEROCHAETUS

k. Achene 2-sided; style usually 2-cleft, rarely 3-cleft; scales of spikelets red-dotted, with a glandular-hairy area toward summit, reddish-brown to dark brown; bristles equaling or slightly shorter than achene; spikelets linear-cylindrical to narrowly ovoid, somewhat rounded or blunt at tip . . . . . . . . . . 9. S. ACUTUS

b. The culm (stem) ends in 2 or more leaf-like spreading bracts, the inflorescence thus terminating the culm; culms with several or many well-developed leaves . . . . . . . . . *l*

l. Culms (stems) sharply 3-angled; spikelets 10–40 mm. long, 5–11 mm. thick; achenes 3–5 mm. long . . . . . . . . . . . . . . . . . . . . . . . . . . . . *m*

m. Common plant of fresh water ponds, sloughs, and swamps; leaves usually more than 6 on a culm (stem); leaves of involucre 3–5; summit of leaf-sheath with a rounded or convex membrane; style 3-cleft; achenes 4–5 mm. long, 3-sided; bristles 6, persistent, equaling or nearly equaling achene . . . . . . . . . . . . 12. S. FLUVIATILIS

m. Rare plant of saline or alkaline soils around salt or sulfur springs, rarely elsewhere; leaves usually 3–6 on a culm (stem); leaves of involucre 1–2 (rarely 3); summit of leaf-sheath with a truncate (cut straight across) or concave (shallowly U-shaped with sloping sides) membrane; style usually 2-cleft; achenes 2.8–3.8 mm. long, mostly 2-sided; bristles 2–6, weak, eventually falling from and shorter than the achene . . . . . . 13. S. PALUDOSUS

l. Culms (stems) obtusely (bluntly) angled; spikelets 2–10 mm. long, 1–4 mm. thick (excluding bristles); achenes 0.7–1.3 mm. long . . . . . . . . . . . . *n*

n. Bristles lacking in flowers . . . . . . . . . . . 14c. S. ATROVIRENS var. GEORGIANUS

n. Bristles present, minute or conspicuous . . . . . . . . . . *o*

o. Bristles in flower and fruit with tiny barbs directed downward (retrorsely) (do not confuse with the smooth filaments which may remain attached at first); stems either solitary or few in a clump from scaly stolons . . . . . . . . . . *p*

p. Common throughout Missouri; 4–9 leaves on each culm (stem) not in a 2-ranked arrangement, averaging 10–20 mm. (rarely 5) wide; bristles somewhat shorter or slightly longer than achene; spikelets pale to dark brown or lead-colored. 14a. S. ATROVIRENS var. ATROVIRENS

p. Known only from swampy meadows in the southeastern Missouri Ozarks; 10–20 or more leaves regularly arranged in 2 ranks (distichous) on each culm (stem),

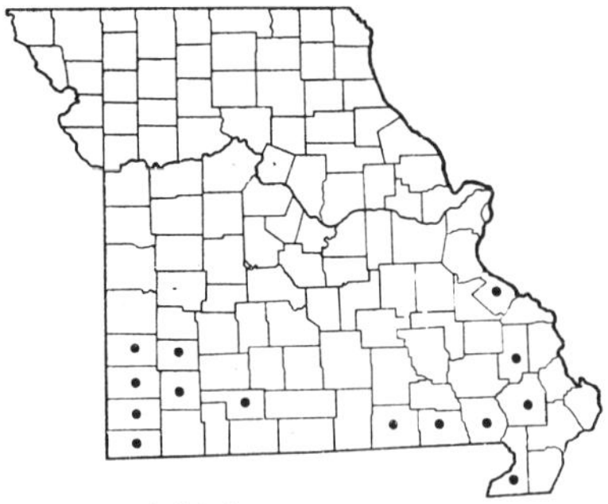

409 Scirpus koilolepis

410 Scirpus Hallii

411 Scirpus subterminalis

averaging 3–8 mm. (rarely 10) wide; bristles about 2 times the length of the achene; spikelets reddish-brown . . . . . . . . . . . . . . . . . . 15. S. POLYPHYLLUS

o. Bristles in flower and fruit smooth, or, if barbed, the barbs few and upwardly ascending (do not confuse with the filaments which may remain attached at first); stems growing from usually large clumps, not producing stolons . . . . . . . . . . . . . . *q*

q. Bristles *at maturity* scarcely longer than the scales of the spikelets; spikelets linear-cylindrical (finger-shaped), 2½–4 times longer than broad; each spikelet alone on a pedicel (stalk) or not clustered with other spikelets . . . . . . . . . . . . *r*

r. Rare plant of southeastern Missouri swamplands; 10–20 leaves regularly arranged in 2 ranks (distichous) on each culm (stem), the internodes short and leaf-sheaths nearly overlapping; lower branches of inflorescence widely spreading; scales of spikelet obtuse (blunt), the midrib not projecting as a conspicuous tip; bristles shorter than the achene . . . . . . . . . . . . . . 16. S. DIVARICATUS

r. Common plant throughout Missouri; usually 5–10 leaves present on each culm (stem) and not in regular distichous (2-ranked) arrangement, the internodes elongated and leaf-sheaths not overlapping; lower and other branches of inflorescence ascending or upwardly spreading; scales of spikelet conspicuously pointed with the green midrib projecting; bristles about twice the length of the achene . . . . . . . . . . . . . . . . . . . . 17. S. LINEATUS

q. Bristles *at maturity* much longer than the scales of the spikelets, conspicuously curled or twisted, giving a somewhat woolly appearance; spikelets ovoid or nearly globe-shaped, nearly as long as broad or at most 1½ times longer than broad; spikelets usually in clusters or grouped together, or, in *S. rubricosus* with the lateral spikelets of each ultimate group on separate stalks . . . . . . . . . . . . . . . . *s*

s. Spikelets all or nearly all in sessile clusters, without stalks for any one spikelet in an ultimate group . . . . . . . . . . . . . . . . . . 18. S. CYPERINUS

s. The lateral spikelet of each ultimate group with its own stalk, the central one of each ultimate group sessile (without a stalk) . . . . . . . . . 19. S. RUBRICOSUS

1. **Scirpus verecundus** Fern. Map 408

*Scirpus planifolius* Muhl. [P & S]

Flowers late April–June.

Occurs on dry rocky wooded slopes, in acid soils of sandstone or chert areas. Infrequently collected, known from only a few Ozark counties north to Washington County and west to Douglas County.

Ranges from Maine to Virginia, west to Ohio, Kentucky, and Missouri.

This plant, growing in characteristic tufts with narrow basal leaves 1–2 mm. wide and slender culms, may be mistaken in the field at first for *Carex artitecta* or some similar species, which it resembles vegetatively, as well as occupying a similar type of dry acid-soil woodland. Careful search in acid soils of dry woodlands should increase its distribution to include a majority of Ozark counties.

2. **Scirpus koilolepis** (Steud.) Gleason Map 409

*Scirpus carinatus* Gray [P & S]

Flowers April–June.

Occurs in sandy open ground along small streams in prairies, in moist sandy soil, and on sandstone and chert glades and barrens. Southern Missouri, where it

Plate no. 74. **1.** Scirpus Hallii. **2.** Scirpus subterminalis. **3.** Scirpus Torreyi.

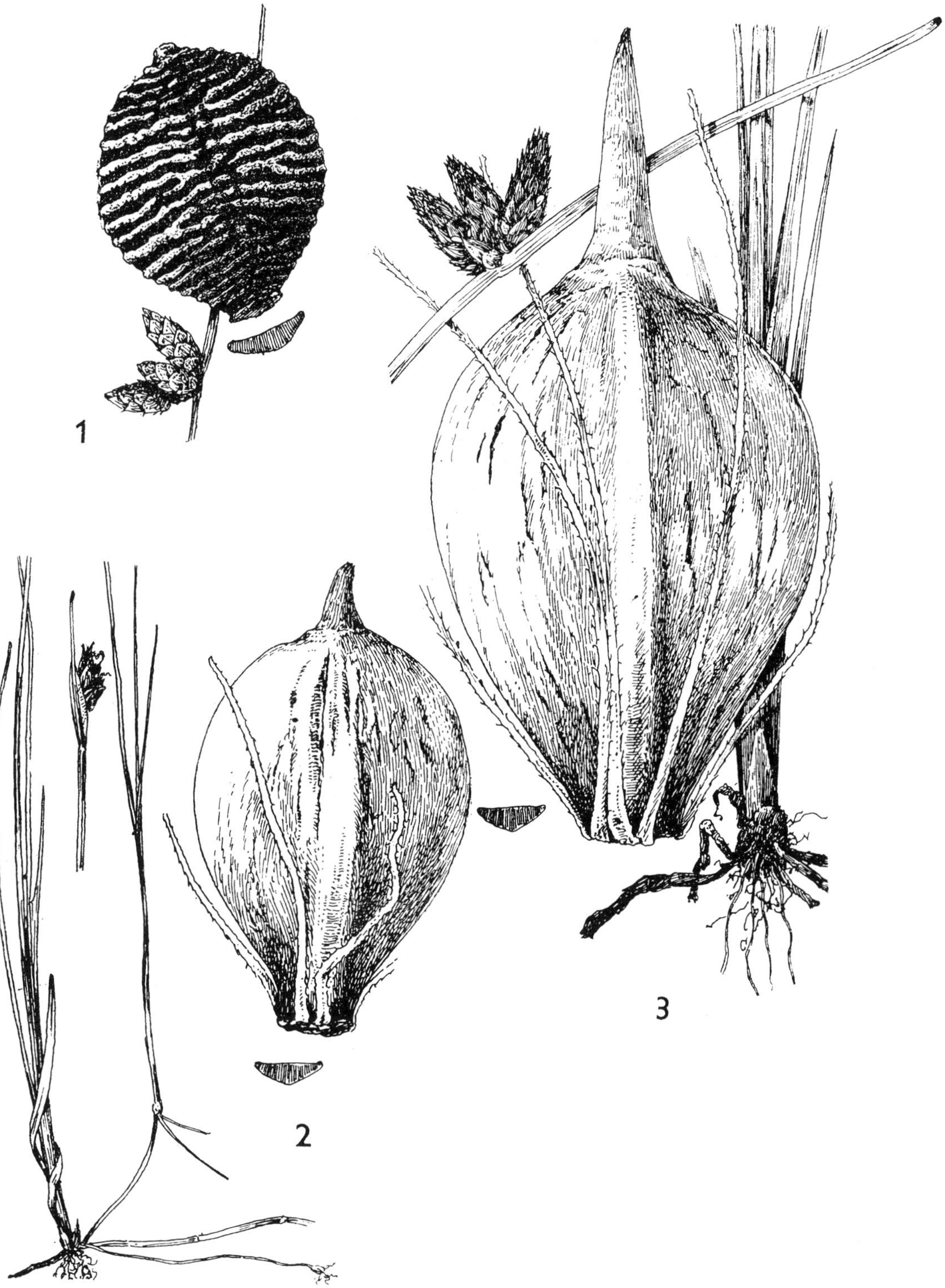


PLATE NO. 74

412 Scirpus Torreyi

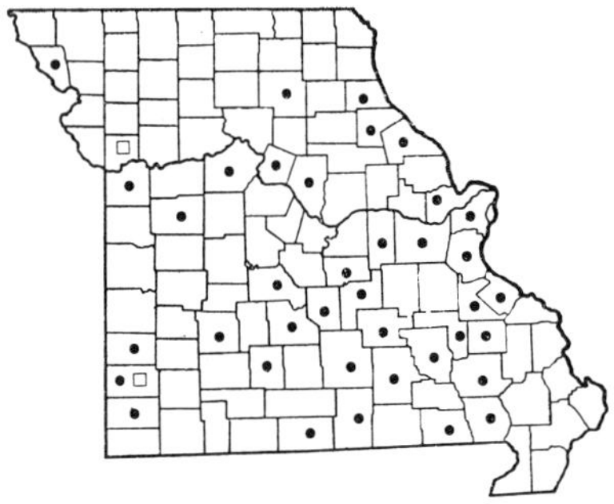

413 • Scirpus americanus var. americanus
413 □ Scirpus americanus var. polyphyllus

414 Scirpus Olneyi

occurs in two distinct areas: (1) southwestern Missouri and (2) southeastern Missouri. It is absent from most of the Ozark section.

Ranges from Alabama to Mexico, north to Tennessee, Missouri, Oklahoma, and California.

The boat-shaped and sharply keeled scales of the spikelets in this species are quite unlike those of any other species of the genus. Plants average about 10 cm. in height, varying from 5 to 18 cm. tall.

3. **Scirpus Hallii** Gray Map 410

Flowers August–October.

Known at present only from a sink-hole pond in Howell County, southern Missouri (pond near Adobosee Pond, T22N, R7W, north part of sect. 1, 9 mi. southeast of West Plains, September 4, 1949, *Steyermark 69122*). Previously known only from a pond in St. Louis County, where now exterminated.

This species occurs as a rare and local plant from Florida and Georgia to Texas and Mexico; Massachusetts, Illinois, and Missouri.

This species was abundant where it occurred at the only known Missouri station. The scales of the spikelet are conspicuously pointed, the outermost one prolonged as a narrow involucral leaf.

4. **Scirpus subterminalis** Torr. Map 411

Flowers July–October.

Known only from Oregon County, southern Missouri (in still water of White's Creek, Irish Wilderness, Clark National Forest, T24N, R2W, sect. 17, 4 mi. south of Wilderness, July 20, 1936, *Steyermark 12054*).

Ranges from Newfoundland to Ontario, south to South Carolina, Georgia, Michigan, Indiana, Illinois, and Missouri; also Alaska to Montana, Idaho, and Oregon.

This unusual aquatic perennial has slender soft rhizomes and threadlike, very long, delicate submerged leaves. The culms likewise, are threadlike. The solitary sessile spikelet is attached to the side of the single, erect, threadlike involucral bract. At the single station known in Missouri it occurred in two feet of still water.

The occurrence of this northern species isolated in a remote section of the southern Ozark region is an example of a relict plant which probably migrated southward to the Ozarks in Pleistocene (Ice Age) times during a relatively cooler period that prevailed, and, following the retreat of the glaciers northward, continued to survive to the present time in a less favorable environment, where this species now just manages to exist, and not spread. Such habitats should be preserved in their natural condition for the important geological and geographical history signified by the occurrence of their relict survivors.

5. **Scirpus Torreyi** Olney Map 412

Flowers June–September.

Known only from a sink-hole pond in Laclede County, southern Missouri (around deep part of margin of upland sink-hole pond along highway 32, sect. 6, 1¼ mi. north of Lynchburg, June 23, 1939, *Steyermark 27146*).

Ranges from New Brunswick to Manitoba, south to New Jersey, Pennsylvania, Kentucky, Indiana, Illinois, Missouri, and South Dakota.

The occurrence of this species at the single Missouri station is another example of the remarkable isolation of several rare sedges known in the state only from sink-hole pond habitats.

6. **Scirpus americanus** Pers. Chairmaker's Rush, Three Square Map 413

Flowers late May–September.

Occurs along gravel bars of streams, sandy and

Plate no. 75. **1.** Scirpus americanus. **2.** Scirpus Olneyi. **3.** Scirpus etuberculatus.

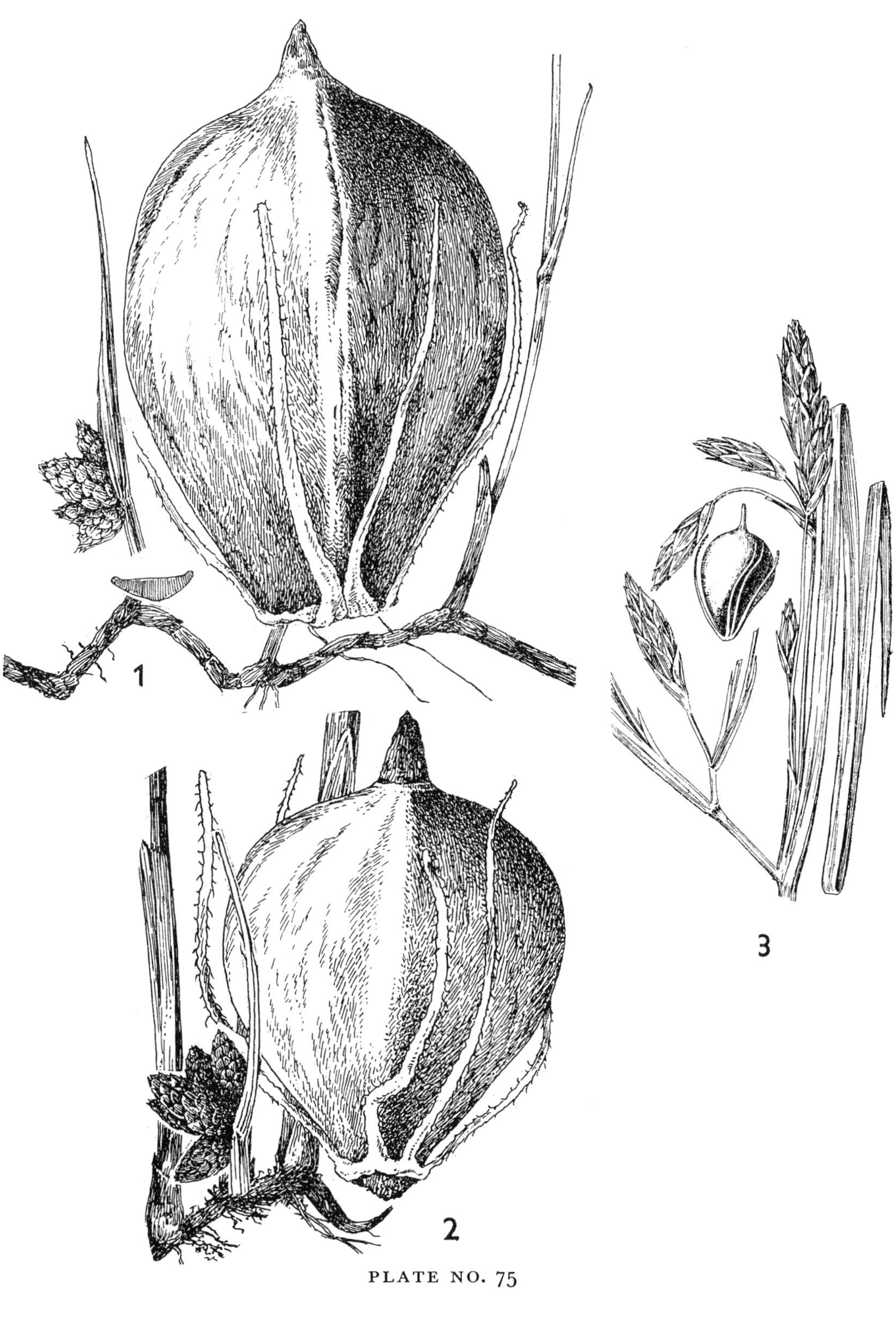


PLATE NO. 75

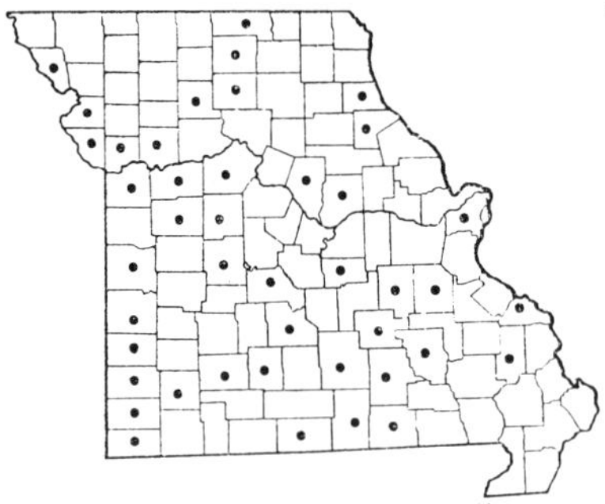

415 Scirpus validus var. creber (Great Bulrush)

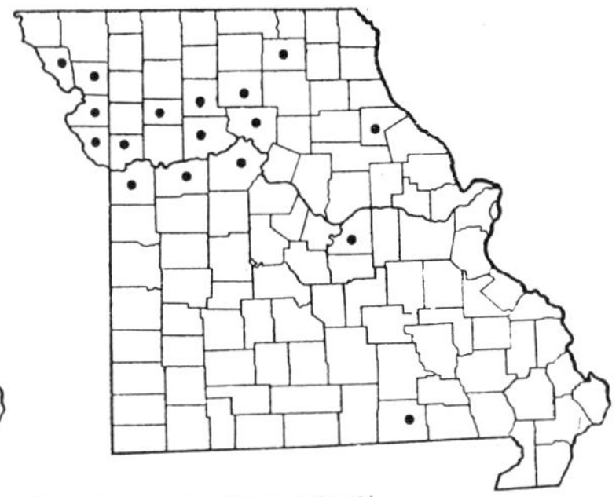

416 Scirpus acutus (Great Bulrush)

417 Scirpus heterochaetus (Great Bulrush)

muddy margins of ponds, sloughs, swamps, ditches, and low wet woods. Frequent in southern and central Missouri north locally to Marion, Macon, and Holt counties; apparently absent from most of northern Missouri.

Ranges from Florida to Texas, north to Newfoundland, Quebec, Ontario, Michigan, Wisconsin, and Nebraska; also Eurasia. Numerous varieties are encountered in western United States, Mexico, Central and South America.

Most of the plants in Missouri have the style normally 2-cleft, ovoid spikelets 6–13 mm. long, averaging $2–2\frac{1}{2}$ times longer than broad, culms averaging 60–100 cm. tall and 3–4 mm. thick, and an involucral bract averaging 4–10 cm. long, and are *S. americanus* var. *americanus*. Plants with the style normally 3-cleft and with unusually long and narrow linear-cylindrical spikelets 15–23 cm. long and 4–5 times longer than broad, short dwarfed culms 15–60 cm. tall and 1–1.5 mm. thick with involucral bract 2–3.5 cm. long have been found in Jasper County, southwestern Missouri (*Steyermark 86153*; *Palmer 67476*) and are distinguished as var. *polyphyllus* (Boeckl.) Beetle. The achenes in this dwarf variation are mostly 2.25–2.5 mm. long, instead of 2.5–3 mm. long as in typical *S. americanus* var. *americanus*.

7. **Scirpus Olneyi** Gray Map 414

Flowers June–September.

Known only from Jefferson County, eastern Missouri (sulphur spring brackish morass, August 10, 1866, *Engelmann*).

Ranges along the coast from Florida to Mexico, north to New Hampshire and Nova Scotia; inland in New York, Ohio, Michigan, Missouri, Idaho, Oregon; also in the West Indies, Central and South America.

This species has hollow culms, an exception to the usual character of Cyperaceae.

8. **Scirpus validus** Vahl var. **creber** Fern.

Great Bulrush Map 415

*Scirpus validus* of auth., not Vahl [P & S, BB]

Flowers late May–September.

Occurs in shallow water and in wet ground of sloughs, ponds, rivers, oxbow lakes of river flood plains, and in wet prairies. Found throughout most of Missouri, the common bulrush of the Ozark section; apparently absent from sectors of northern Missouri.

Ranges from Newfoundland to Alaska, south to New York, Georgia, Tennessee, Missouri, Oklahoma, Texas, New Mexico, and California.

Typical *S. validus* var. *validus* ranges from South and Central America, the West Indies, and Mexico to South Carolina.

This plant, like *S. acutus* and *S. heterochaetus*, the other species with long whiplike culms, usually occurs in large colonies. In some Missouri specimens the style is 3-cleft, although the 2-cleft condition is more frequent.

In this and the following species (*S. acutus*), the dried beaten rootstocks are used by some Indian tribes as meal for bread, while the young tip of the underground rootstock is edible and reputed to serve as a thirst-quencher; the boiled young roots yield a sweet syrup. The achenes are also used for food.

9. **Scirpus acutus** Muhl. Great Bulrush Map 416

Flowers late May–September.

Occurs along the larger rivers (Missouri, Grand and its tributaries) of northwestern and west-central Missouri, south to Jackson, Lafayette, Saline, and Osage counties, locally east to Ralls County and locally south in Oregon County. Absent from practically the entire Ozark sector, as well as the southeastern lowland and southwestern unglaciated prairie sections.

Ranges from Newfoundland to British Columbia,

Plate no. 76. Scirpus validus var. creber.

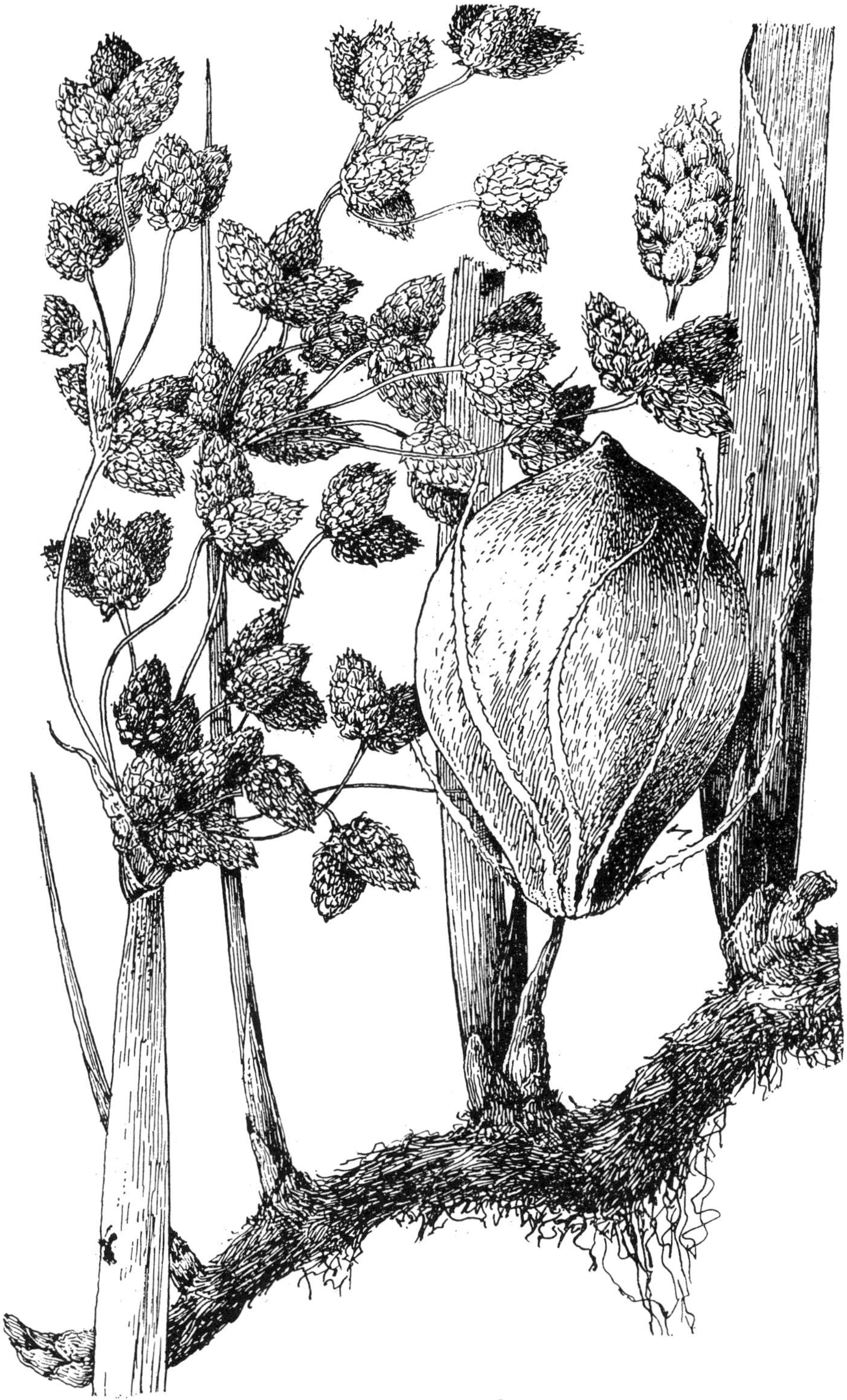

PLATE NO. 76

418 Scirpus etuberculatus

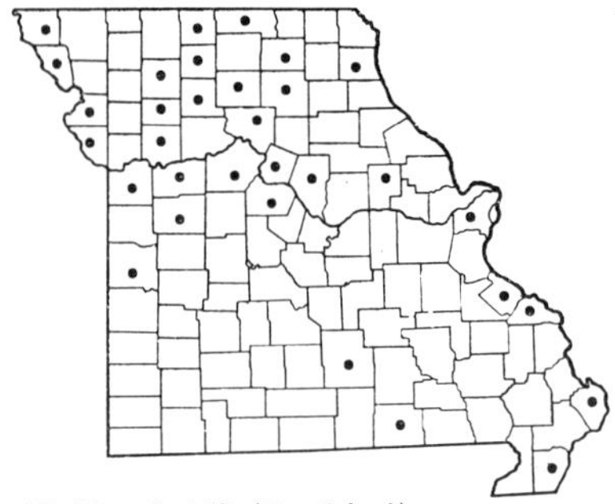

419 Scirpus fluviatilis (River Bulrush)

420 Scirpus paludosus var. paludosus (Bayonet Grass)

south to New Jersey, Pennsylvania, Ohio, Indiana, Illinois, Missouri, Oklahoma, Texas, New Mexico, Arizona, and California.

This species is sometimes confused with *S. validus* var. *creber* and *S. heterochaetus.* From the former it differs in having longer, more linear-cylindrical spikelets and from the latter in 2- instead of 3-sided achenes. The culms (stems) of *S. acutus* are usually harder and firmer, more slender, and of a darker green shade than those of *S. validus* var. *creber.* For this reason, *S. acutus* is often known as Hard-stem Bulrush as contrasted with *S. validus* var. *creber,* the Soft-stem Bulrush.

10. **Scirpus heterochaetus** Chase
Great Bulrush — Map 417
Sometimes called slender bulrush.
Flowers late May–September.

Occurs in shallow water and in wet ground in the border of ponds, sloughs, rivers, and oxbow lakes of river flood plains. Scattered over the state, but probably of wider distribution with more intensive collecting.

Ranges from Quebec to Massachusetts and New York; from Wisconsin and Illinois to North Dakota, south to Kentucky, Missouri, and Oklahoma; and from Idaho and Washington to Oregon.

Some representative Missouri collections are from Laclede (*George Moore*), Texas (*Steyermark 23292*), and St. Charles (*Von Schrenk*) counties.

The spikelets are paler brown or even straw-colored as contrasted with the reddish-or chestnut-brown or dark brown colors found in *S. acutus* and *S. validus* var. *creber.*

11. **Scirpus etuberculatus** (Steud.) Ktze. Map 418
Flowers June–October.

Known only from a natural sink-hole pond in Oregon County, southern Missouri (Tupelo Gum Pond, T25N, R4W, sect. 4, 8 mi. west of New Liberty, October 3, 1948, *Steyermark 66629*).

Ranges from Florida to Louisiana, north to Delaware and Missouri.

This is one of the notable remnant species of the Atlantic and gulf coastal plain flora restricted to sink-hole ponds in the Ozarks, where it, like some other species in the state restricted to such ponds, indicates the former, more widespread, though ancient coastal plain conditions that prevailed before the last Tertiary uplift of the Ozarks. The Ozark sink-hole ponds contain the last survivors of the original pre-Tertiary coastal plain flora that covered the present Ozark area when it was a low-lying peneplain.

12. **Scirpus fluviatilis** (Torr.) Gray
River Bulrush — Map 419
Flowers May–September.

Occurs in shallow water and wet mud of sloughs, swamps, oxbow lakes in river flood plains, and along the larger rivers, where it occupies stretches along the Missouri, Grand and its tributaries, and Mississippi rivers, mostly in northern and central Missouri. Absent from the greater part of the Ozark section.

Ranges from Quebec to Saskatchewan and Washington, south to Delaware, Maryland, Virginia, Pennsylvania, Ohio, Indiana, Illinois, Missouri, Kansas, New Mexico, and California.

This species usually occurs in dense colonies which are frequently all vegetative and without flowers. The leaf-blades and culms are sharp-edged and can easily cut the flesh. The culms are thicker and the leaf-blades wider than in any of the other Missouri species; these characters, plus the dense colonial habit of the culms, leafy to the summit, are unmistakable field marks of identification. The thick creeping rootstocks bear solid cormlike enlargements.

Plate no. 77. **1.** Scirpus acutus f. congestus. **2.** Scirpus acutus f. acutus. **3.** Scirpus heterochaetus.

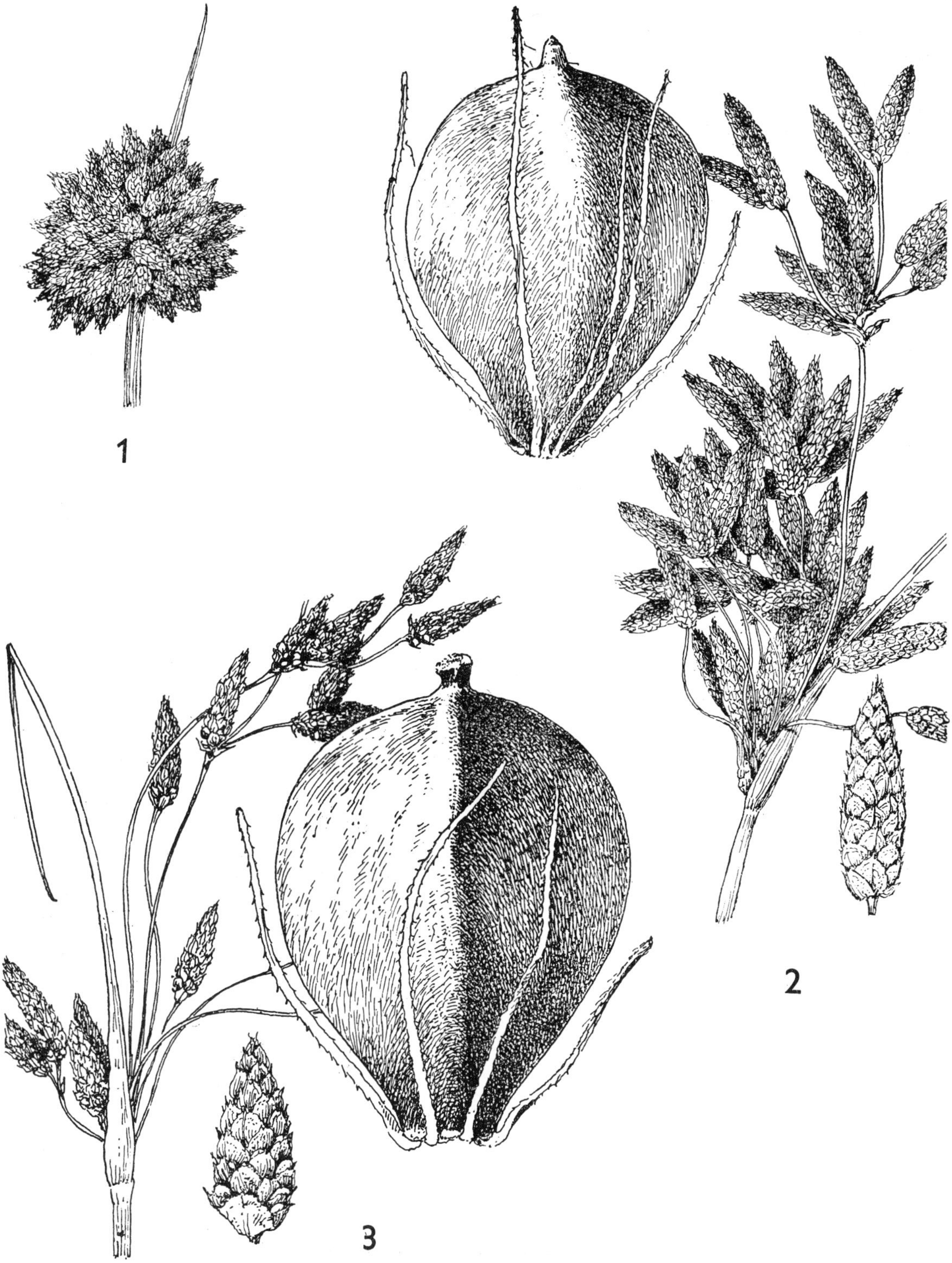


PLATE NO. 77

13. **Scirpus paludosus** Nels. var. **paludosus**
Bayonet Grass, Prairie Bulrush Map 420
*Scirpus campestris* var. *paludosus* (Nels.) Fern.
*Scirpus paludosus* Nels. [G]
*Scirpus maritimus* L., var. *paludosus* (A. Nels.) Gl. [BB]
*Scirpus campestris* Britton, not Roth [P & S]
Flowers June–October.

Occurs in wet salt-impregnated or alkaline ground around salt springs in Saline County (wet swaley ground bordering salt spring and spring branch of Elk Lick Spring, in valley of Heath's Creek, sect. 17, 3 mi. southwest of Ridge Prairie, October 6, 1938, *Steyermark 21587*), and in alkaline sandy mud along an oxbow lake of Missouri River in Holt County (sands along west end of Big Lake State Park, September 2, 1934, *Steyermark 15141*).

Ranges from Minnesota, Iowa, Missouri, Oklahoma, and Texas, west to British Columbia, California, and Mexico.

At the Saline County station this species was associated with other saline to brackish water-inhabiting species, as *Distichlis spicata*, *Eleocharis parvula*, and *Typha angustifolia*.

Although often a robust and tall-growing species elsewhere in its range, at the Missouri stations it is usually less than a meter tall, and in the Holt Co. collection the plants are only 15–25 cm. tall. It most closely resembles *S. fluviatilis*, and, like that species, has the midribs of the spikelet scales prolonged into conspicuous awns.

The enlarged cormlike parts of the rootstock of this species are reported to serve as starchy food for some of the western Indians. The related species, *S. fluviatilis*, probably may, likewise, have similar use.

14. **Scirpus atrovirens** Willd.
Common Bulrush Map 421
Flowers late May–September.

Occurs in wet ground bordering lakes, sloughs, streams, and ditches. Throughout Missouri. So far as records indicate, this is the commonest species of Bulrush in the state.

Missouri material may be divided into the following variations:

a. Scales of the spikelet prolonged at the tip to a long awn; scales narrowly ovate, 2–3 mm. long . . . 14d. S. ATROVIRENS var. PALLIDUS
a. Scales of the spikelet with a short mucro (extension of midrib) at the tip; scales broadly ovate, 1–2 mm. long . . . . . . . . *b*
b. Dried leaf-blades and leaf-sheaths of lower leaves only faintly, if at all, marked with short cross-partitions or thickenings, mainly 5–10 mm. wide; bristles not present or shorter than the achenes . 14c. S. ATROVIRENS var. GEORGIANUS
b. Dried leaf-blades and leaf-sheaths of the lower leaves with conspicuous short cross-mark or thickenings mostly 10–20 mm. wide; bristles slightly longer than the achenes . . . . . . . . . . *c*
c. Bunches of spikelets massed together in 1 or more dense heads; branches (rays) supporting the inflorescence very much shortened . . . . 14b. S. ATROVIRENS var. ATROVIRENS f. SYCHNOCEPHALUS
c. Bunches of spikelets in numerous separate groups; branches (rays) supporting the inflorescence conspicuous, elongated, at least most of them 2–5 cm. long
14a. S. ATROVIRENS var. ATROVIRENS f. ATROVIRENS

14a. **Scirpus atrovirens** var. **atrovirens** f. **atrovirens** Map 421

The common variety throughout Missouri in nearly every county.

Ranges from Quebec to Saskatchewan, south to Georgia, Tennessee, Arkansas, and Kansas.

14b. **Scirpus atrovirens** var. **atrovirens** f. **sychnocephalus** (S. N. Cowles) Blake Map 421

Scattered in the range of var. *atrovirens*. A representative collection is *Sparling* from Livingston County. Probably of more frequent occurrence.

14c. **Scirpus atrovirens** var. **georgianus** (Harper) Fern. Map 421

Known from Iron (open bog near Lopez, September 4, 1926, *Palmer 31550*) and Butler (low woods along Mud Creek, T26N, R7E, sect. 20, 2 mi. northwest of Rombauer, June 30, 1936, *Steyermark 11312*) counties in southeast Missouri.

Ranges from Newfoundland to Ontario and Minnesota, south to Georgia, Tennessee, and Arkansas.

The spikelets in this variety are often smaller (2–3 mm. long instead of 3.5 mm. or more) and in less crowded clusters (usually less than 7 mm. in diameter) than those of *S. atrovirens* var. *atrovirens*.

14d. **Scirpus atrovirens** var. **pallidus** Britton Map 421

*Scirpus pallidus* (Britt.) Fern. [G, P & S]
Known from Jackson and Dekalb counties, western

Plate no. 78. Scirpus fluviatilis.

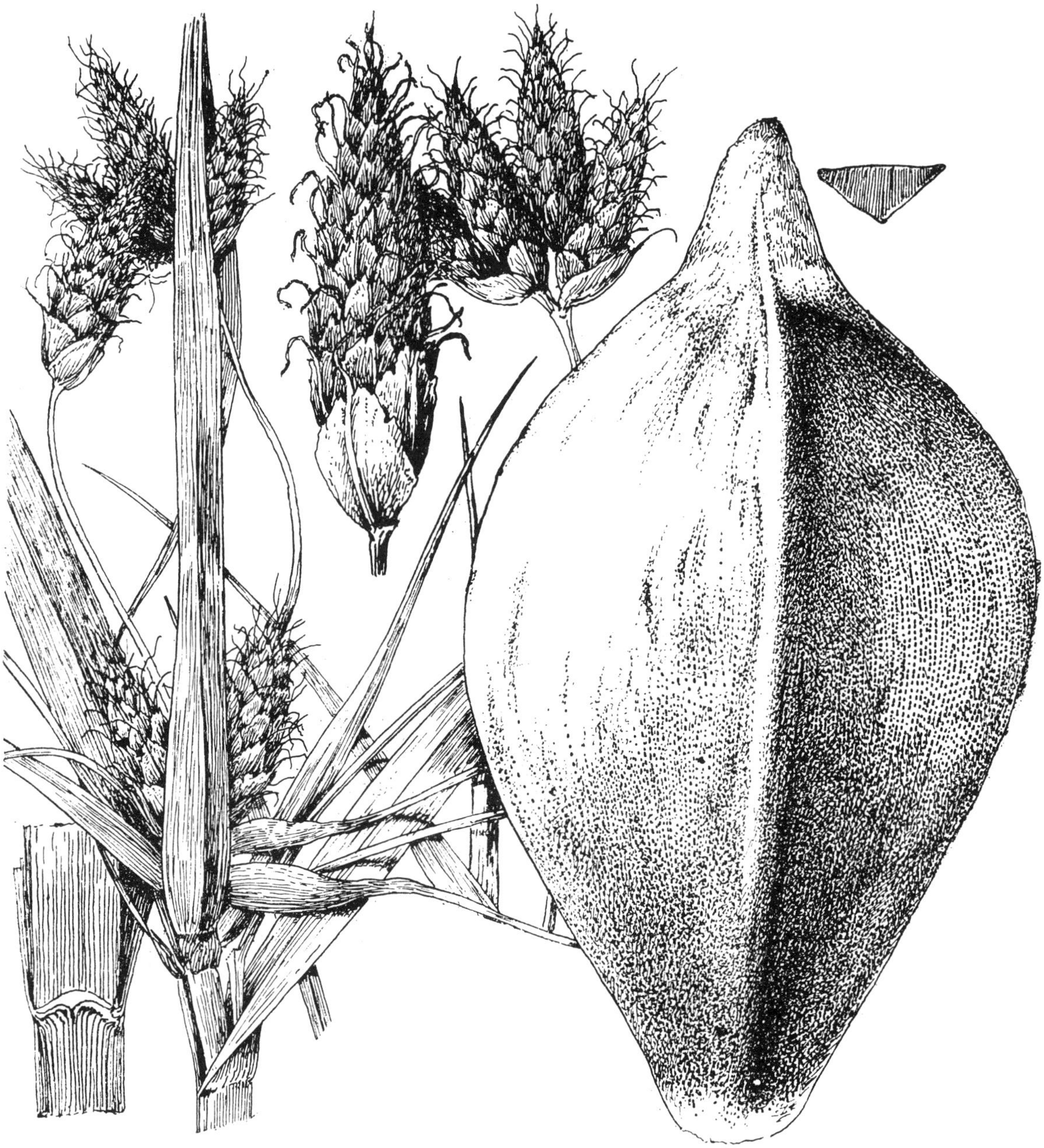

PLATE NO. 78

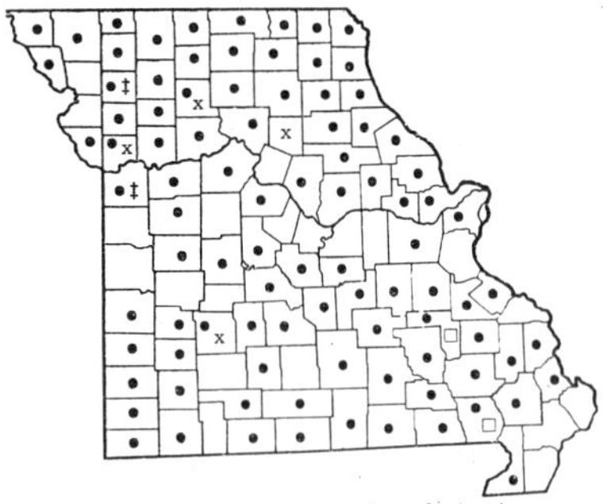

421 • Scirpus atrovirens var. atrovirens f. atrovirens (Common Bulrush)
421 x Scirpus atrovirens var. atrovirens f. sychnocephalus
421 □ Scirpus atrovirens var. georgianus
421 ‡ Scirpus atrovirens var. pallidus

422 Scirpus polyphyllus

423 Scirpus divaricatus

Missouri. The Jasper County collection (*Palmer 2128*), cited in Palmer & Steyermark's *Catalogue* (p. 485) is a misidentified specimen of *S. atrovirens* var. *atrovirens*. The following collections from Jackson County in the Gray Herbarium were referred to this variety by Fernald: *Bush 587*, Courtney, June 22, 1895; *Bush 374*, July 11, 1893. The scales of the spikelets in these specimens are only 1.3–1.6 mm. long and the awns are not particularly long-setulose. These specimens are here referred instead to *S. atrovirens* var. *atrovirens*.

The spikelets of *S. atrovirens* change color materially during their development. In the young bud or early flowering stage, the scales are pale or greenish brown with dull green midribs. In late stages of flowering and fruit the scales turn a dark or chocolate brown with some black marks.

15. **Scirpus polyphyllus** Vahl — Map 422

Flowers June–September.

Occurs in calcareous swampy meadows in the Ozark section of southeastern Missouri, where known only from Bollinger, Stoddard, Madison, and Reynolds (*Steyermark 72002, 72041*) counties.

Ranges from Massachusetts and New York to Ohio, Indiana, and Illinois, south to Georgia, Tennessee, and Missouri.

This rare species is localized in swampy meadows fed by calcareous springs in small valleys bordering the smaller streams. Before such habitats are drained for agriculture or otherwise destroyed by farming or other disturbances, it may eventually be found to have a somewhat greater range of distribution in the eastern section of the Ozarks.

16. **Scirpus divaricatus** Ell. — Map 423

Flowers June–September.

Occurs in swamps and low wet woodland of the southeastern lowland section of the state in Dunklin, Butler, and Ripley counties, and introduced in St. Louis County along a ditch by railroad tracks (in ditch along track of Terminal R. R. Assoc. at Brown Ave. west of Union Blvd., St. Louis, July 1, 1956, *Muehlenbach 980*). Another specimen from the World's Fair was also collected in St. Louis County (August 18, 1892, *Bush 2769*).

Ranges from Florida to Louisiana, north to Virginia and southeastern Missouri.

Young leafy plants and buds are often produced in the inflorescences and leaf-axils of this species.

17. **Scirpus lineatus** Michx. — Map 424

Flowers May–August.

Occurs in wet prairies, wet limestone glades and ledges, roadside ditches, and borders of streams and ponds. Throughout Missouri, except in the extreme northwestern section, where not yet recorded.

Ranges from Maine to Minnesota, south to Florida and Texas, and in Oregon.

In early stages of devolopment the spikelets are scarcely stalked and paler brown, scarcely protruding from the upper leaf-sheaths and involucral bracts, but as they become more mature, they take on a reddish-brown or brick color and are conspicuously exserted on long pedicels of the drooping rays.

18. **Scirpus cyperinus** (L.) Kunth — Map 425

Flowers July–October.

Occurs in swamps, sloughs, and borders of oxbow lakes in river flood plains. Eastern and northern Missouri west to Caldwell County.

The Missouri material falls into the following two categories:

Plate no. 79. **1.** Scirpus paludosus. **2.** Scirpus atrovirens.

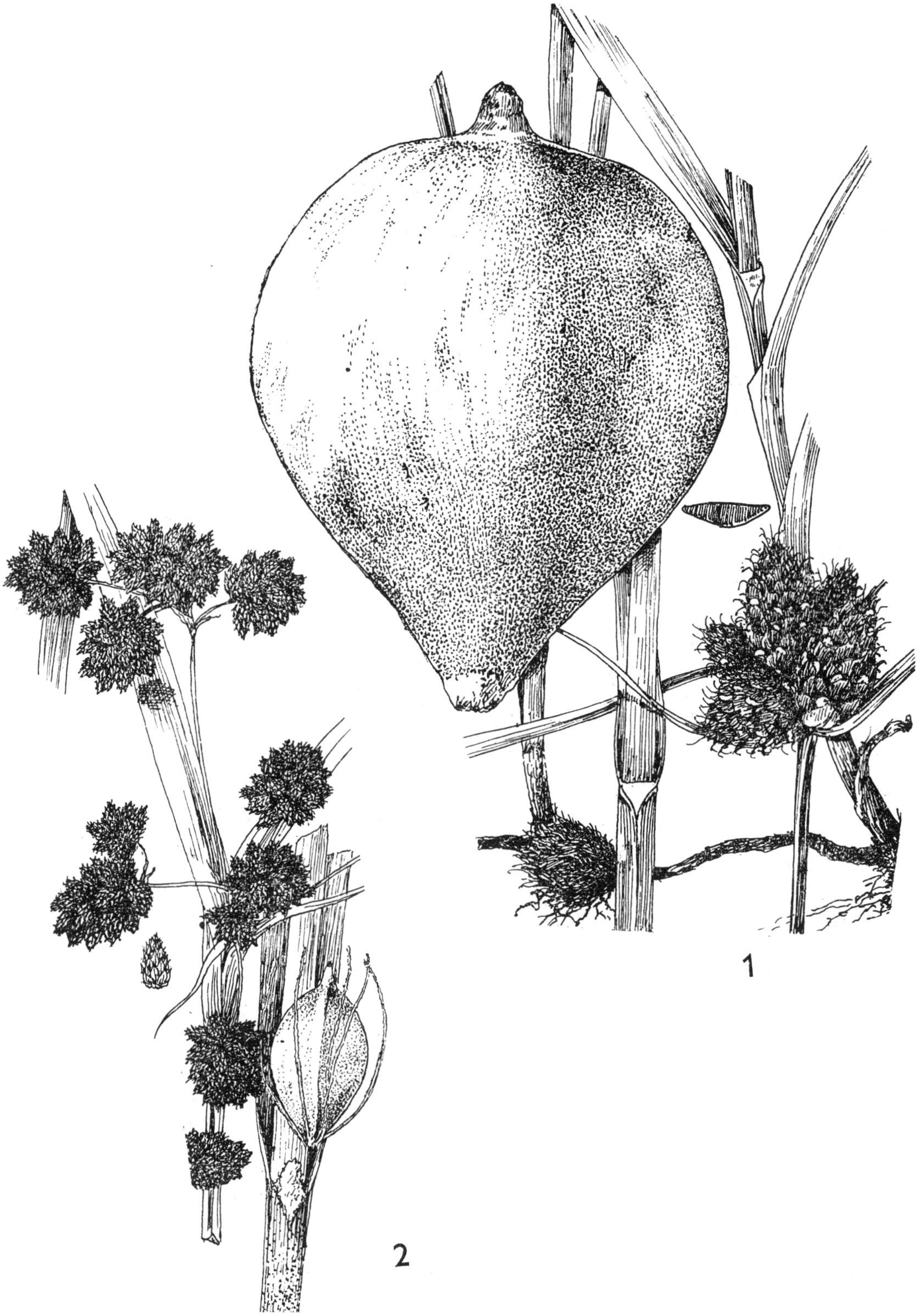


PLATE NO. 79

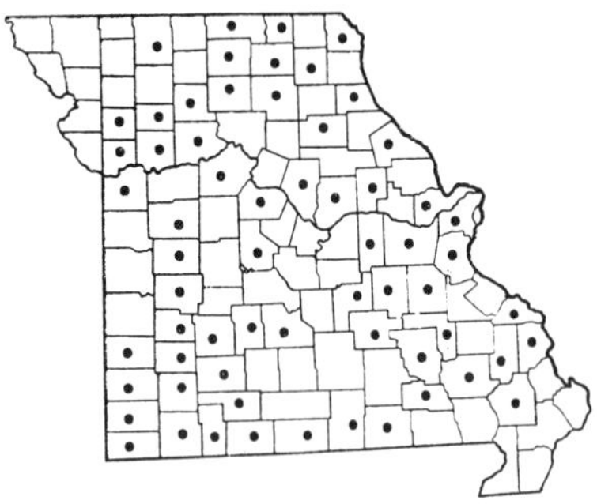

424 Scirpus lineatus

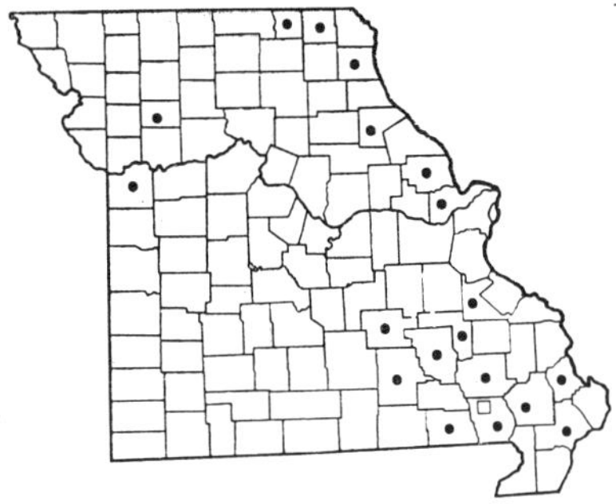

425 ● Scirpus cyperinus var. cyperinus
425 □ Scirpus cyperinus var. pelius

426 Scirpus rubricosus

Involucels (the small colored bracts at the place separating the main rays or branches of the inflorescence from the more slender spikelet-bearing branches) reddish-brown; scales of spikelets reddish-brown; bristles rust-colored . . .
18a. S. CYPERINUS var. CYPERINUS

Involucels blackish or drab, without reddish-brown tinge; scales of spikelets with little or no reddish color; bristles smoky or gray-drab color
18b. S. CYPERINUS var. PELIUS

18a. **Scirpus cyperinus** var. **cyperinus** Map 425

The commoner variety of this species in Missouri.

Ranges from North Carolina to Oklahoma, north to New England, New York, Ohio, Indiana, Illinois, and Iowa.

18b. **Scirpus cyperinus** var. **pelius** Fern. Map 425

Known only from Butler County, southeastern Missouri (swamps, October 16, 1905, *Bush 3707*).

Ranges from Newfoundland to Minnesota, south to Maryland, West Virginia, Ohio, Illinois, Iowa, and Missouri.

19. **Scirpus rubricosus** Fern. Map 426

*Scirpus Eriophorum* Michx.
*Scirpus pedicellatus* [of P & S], not Fern.
*Scirpus cyperinus* [of BB] in part

Flowers July–October.

Occurs in swampy meadows along streams, swamps, sloughs, and borders of oxbow lakes in river flood plains. Northern and eastern Missouri.

Specimens from Putnam and Adair counties, cited in Palmer and Steyermark's *Catalogue* (p. 485) as *S. pedicellatus* Fern., are misidentified specimens of *S. rubricosus*. There is no authentic material of *S. pedicellatus* known from the state.

Gleason (*Ill. Fl.* 1: 276. 1952) has combined *S. Eriophorum*, *S. rubricosus*, *S. pedicellatus*, and *S. atrocinctus* Fern. under *Scirpus cyperinus*, which he regards as a polymorphic species varying in the color of the scales and involucre, length of rays, peduncles, and spikelets, and width of leaves. Until more experimental work is at hand to test the soundness of this viewpoint, the present treatment is maintaining both *S. Eriophorum* and *S. rubricosus* in the light of Fernald's previous detailed studies.

## 7. **Fuirena** Rottb. Umbrella Grass

Perennial plants (in the local species) with leafy culms, more or less pubescent leaf-sheaths, and flat leaf-blades. Spikelets in close clusters with a leaflike bract at their base. Flowers of 3 stamens, a 3-cleft style, and a 3-sided achene surrounded by 3 bristles alternating (in our species) with 3 stalked broadened scales.

**Fuirena simplex** Vahl Map 427

Flowers July–October.

Occurs along gravel bars, wet margins and moist limestone banks of streams, swampy open and wooded ground of spring branches, and muddy margins of oxbow lakes of river flood plains. Except for two localities along the Missouri River in western Missouri in Jackson and Holt (Big Lake State Park, September 1, 1934, *Steyermark 15094*) counties, this species is restricted to the Ozark section of the state, extending north to Ste. Genevieve, Washington, Crawford, Miller, and Camden counties, west to Laclede, Webster, Christian, and Taney counties.

Ranges from northwestern Missouri and Nebraska

Plate no. 80. **1.** Scirpus divaricatus. **2.** Scirpus lineatus. **3.** Scirpus polyphyllus. **4.** Scirpus cyperinus.

PLATE NO. 80

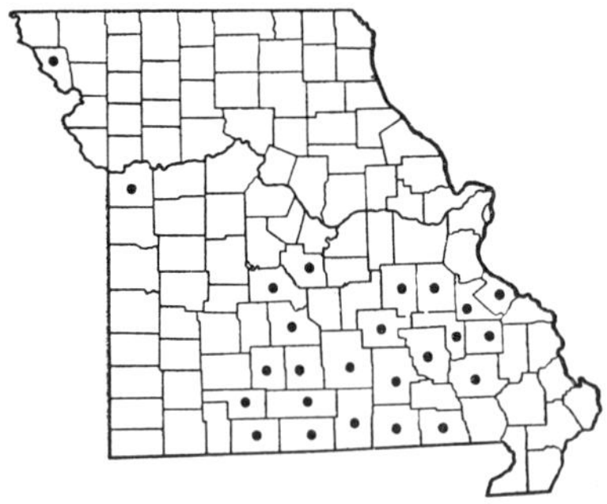

427 Fuirena simplex

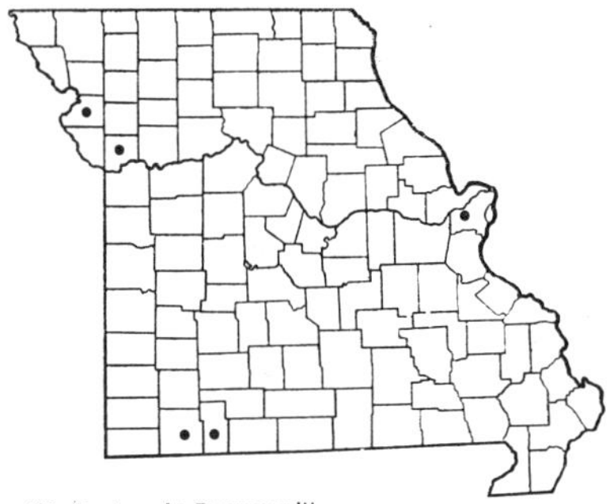

428 Hemicarpha Drummondii

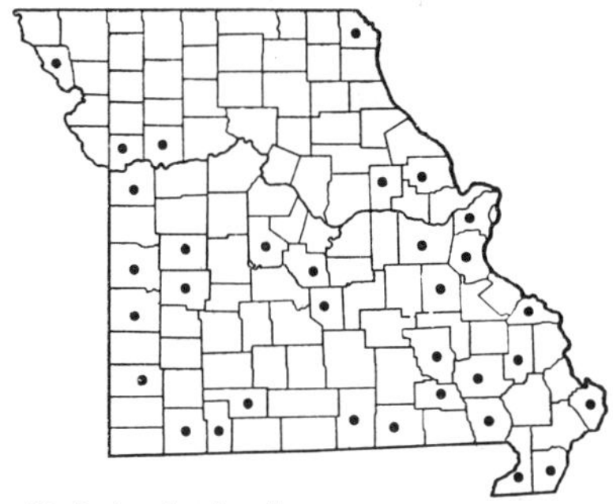

429 Hemicarpha micrantha

south to Oklahoma, Texas, and Mexico; also in Central America and Cuba.

*Excluded species*

**Fuirena squarrosa** Michx.

B. F. Bush (in personal communication) wrote that he had found this species (reported by him as *F. hispida* Ell.) at Anderson, McDonald County. No specimen of this collection has ever been found, nor is the species cited for Missouri in Svenson's recent treatment in *North America Flora* 18: 506. 1957. Therefore, the species is excluded from the present work.

## 8. **Hemicarpha** Nees

Dwarf annual tufted plants, the hairlike culms bearing 2 hair-like leaves at the base, only the upper one with a short leaf-blade. Spikelets very small, 1–3 in a stalkless cluster surpassed by 2 or 3 hairlike involucral bracts. Flowers of 1 stamen, a 2-cleft style and no bristles; in addition to the larger, pointed, spirally overlapping, outer scales, a minute inner scale occurs on the side of the flower next to the axis of the spikelet.

Main outer scales with a long curved awn almost equaling the length of the scale itself; small inner scale equaling and partly inclosing the achene; mature achene obovate, compressed, black . . 1. H. DRUMMONDII

Main outer scales with a short extended tip much shorter than the length of the scale itself; small inner scale much shorter than the achene, often absent; mature achene cylindrical, brown . . 2. H. MICRANTHA

1. **Hemicarpha Drummondii** Nees Map 428

*Hemicarpha aristulata* (Coville) Smyth [G]

*Hemicarpha micrantha* var. *aristulata* Coville [P & S, BB]

Flowers July–October.

Occurs along wet sandy margins of ponds and streams, and in wet depressions of glades. Rare and scattered, in St. Louis, Stone, Barry, Clay, and Buchanan counties.

Ranges from Ohio, Minnesota, Iowa, and Missouri, west to the Pacific states.

The type locality for this species is St. Louis, Missouri.

2. **Hemicarpha micrantha** (Vahl) Pax. Map 429

Flowers July–October.

Occurs on sand bars along streams, muddy borders of ponds and sloughs, and wet depressions in glades of sandstone, chert, or granite. Mostly found in southern and central Missouri, and locally north in Clark and Holt counties.

Ranges from Maine to Ontario and Minnesota, south to Florida, Nebraska, and Kansas; Texas, New Mexico, Arizona, and the Pacific states; also in Mexico, Central and South America, and the West Indies.

## 9. **Rhynchospora** Vahl Beak-rush

Perennial plants (in the local species) with triangular leafy culms. Lower scales of spikelet usually empty, the uppermost ones usually staminate (with stamens only). Stamens 3, when present. Style 2-cleft. Achene globe-shaped with 2 sides or flattened, crowned by a conspicuous beak of the persistent hardened base or portion of the style. Perianth consisting of 6–20 bristles or these absent.

---

Plate no. 81. **1.** Scirpus rubricosus. **2.** Hemicarpha micrantha. **3.** Fuirena simplex.

PLATE NO. 81

430 Rhynchospora macrostachya var. macrostachya (Horned Rush)

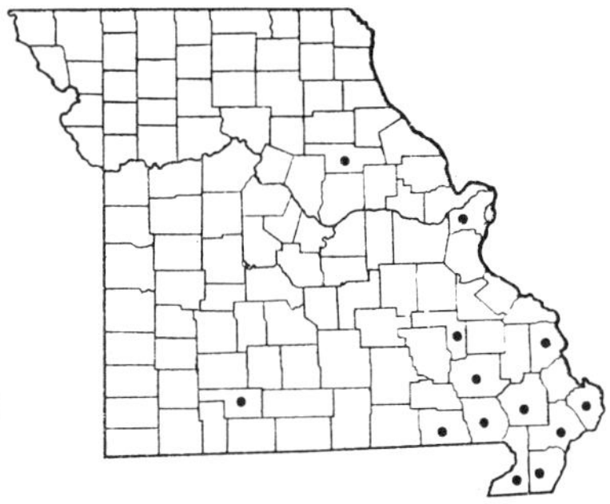

431 Rhynchospora corniculata var. corniculata (Horned Rush)

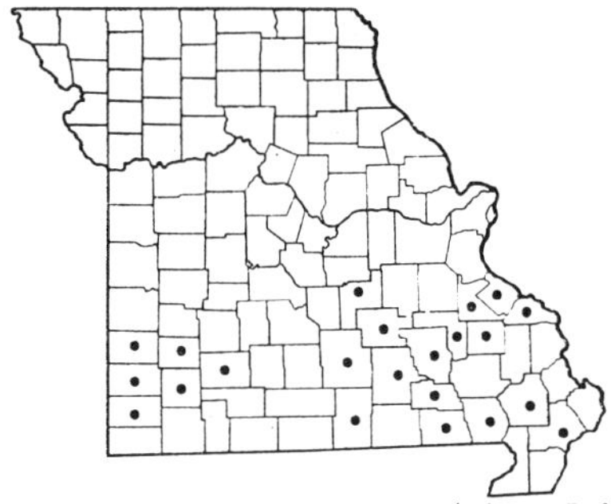

432 Rhynchospora capitellata f. capitellata (False Bog Rush)

a. Leaf-blades and culms (stems) hair-like or very narrowly linear, 0.2–0.4 mm. wide; leaf-blades with margins rolled in (involute); plants of moist limestone ledges and calcareous wet meadows or calcareous swamps . . . . . . . . . . . . . . . . . . . . . . 4. R. CAPILLACEA

a. Leaf-blades 0.5–20 mm. wide; leaf-blades flat; plants not of limestone ledges or calcareous swamps . . *b*

b. Leaf-blades 3.5–20 mm. wide; spikelets (in flower) 10–15 mm. long (in fruit) 15–30 mm. long; achenes 4.4–6 mm. long; crown-like tubercle 14–23 mm. long, longer than the scales . . . . . *c*

c. Bristles 11–14 mm. long, much longer than achenes, 6; inflorescence with stiff ascending branches . . . . . . . . . . . . . . . . . . . . 1. R. MACROSTACHYA

c. Bristles 2–4 mm. long, much shorter than achenes, usually 5 with some of them minute or scarcely evident; inflorescence with wide-spreading branches . . . . . . 2. R. CORNICULATA

b. Leaf-blades 0.5–4 mm. wide; spikelets (in flower and fruit) 2–7 mm. (rarely 10) long; achenes 0.9–3 mm. (rarely 4) long; crown-like tubercle 0.2–2.6 mm. long, mostly shorter than or concealed by the scales . . . . . . . . . . . . . . . . . . . . . . . . . *d*

d. Spikelets much longer (4–5 times) than broad, narrowly elliptical or lance-elliptical; bristles conspicuous, equaling or nearly equaling combined length of achene and tubercle; tubercle about as long as body of achene, long triangular; surface of achenes smooth; teeth or barbs of the bristles pointing downward (retrorsely), or, if rarely pointing upward or smooth, then with smooth achenes and other characters referred to here . . . . . . . . 3. R. CAPITELLATA

d. Spikelets only slightly longer than ($1\frac{1}{2}$–2 times) broad, broadly ovoid to nearly globe-shaped; bristles weak or absent, much shorter than body of achene; tubercle 1/3–1/6 as long as body of achene, low conical or broadly triangular; surface of achenes not smooth, but either minutely honeycombed or with wavy transverse (horizontal) ridges between narrow cells of spaces; teeth or barbs of the bristles pointing upward (antrorsely) . . . . . . . . . . . . *e*

e. Surface of achene minutely honeycombed with cells of about the same diameter, without wavy transverse ridges; tubercle 1/4–1/6 as long as body of achene, depressed conical; midrib of at least the lower scales of spikelets (do not confuse with lowermost bracts) conspicuously projecting into a tip 0.5–1 mm. long . . . . . . . . . . 5. R. HARVEYI

e. Surface of achene with transverse (horizontal) wavy ridges between series of narrow cells or spaces longer than broad; tubercle about 1/3 as long as body of achene, broadly triangular; midrib of the scales of the spikelet (do not confuse with lowermost bracts) scarcely or not at all projecting beyond tip of scale . . . . . . . . 6. R. GLOBULARIS var. RECOGNITA

1. **Rhynchospora macrostachya** Torr. var. **macrostachya** Horned Rush Map 430

Flowers July–October.

Occurs in low swampy woods, around margins of sink-hole ponds, and along small streams in upland prairies. Found in the southern part of the state, principally in the southeastern lowland section and around upland swamps and sink-hole ponds in Oregon and Howell counties, locally west to moist ground in an upland prairie in Barton County, southwestern Missouri (wet margins of small brook, upland prairie, 3 mi. east of Lamar, September 6, 1950, *Palmer 50932*).

Ranges from Florida to Texas, north to Maine, New York, Indiana, Michigan, Missouri, and Kansas.

---

Plate no. 82. **1.** Rhynchospora corniculata. **2.** Rhynchospora capitellata. **3.** Rhynchospora capillacea. **4.** Rhynchospora Harveyi. **5.** Rhynchospora globularis var. recognita. **6.** Scleria triglomerata.

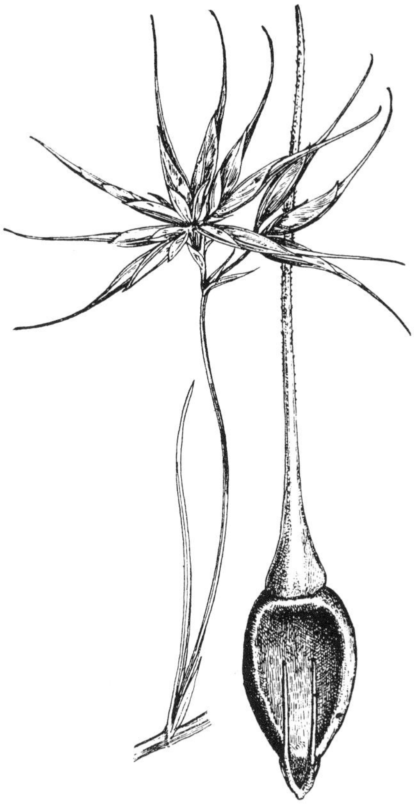

1

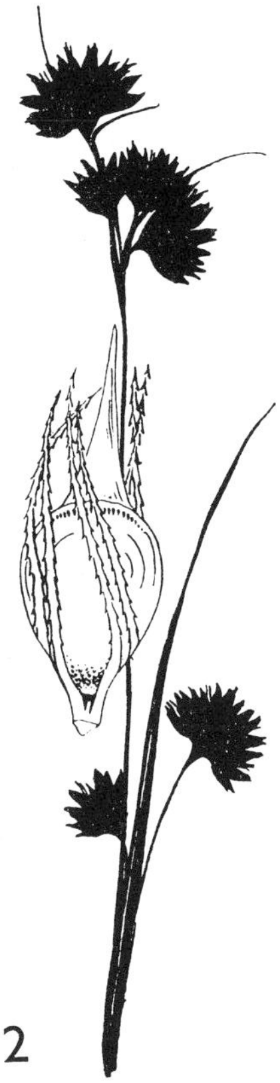

2

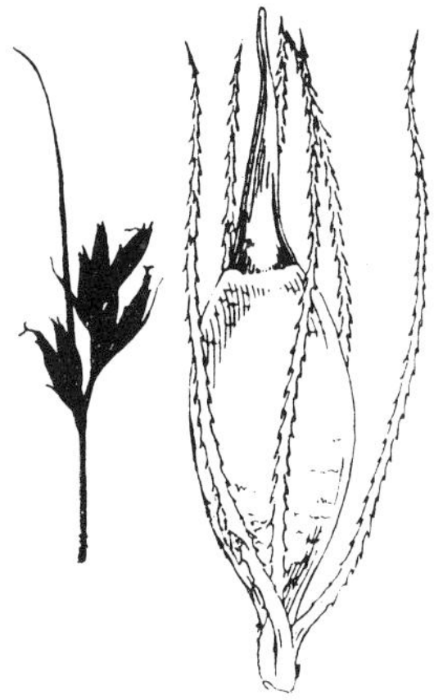

3

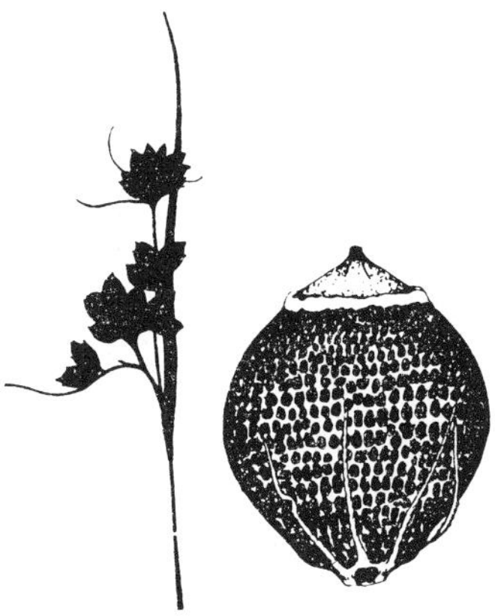

4

5

6

PLATE NO. 82

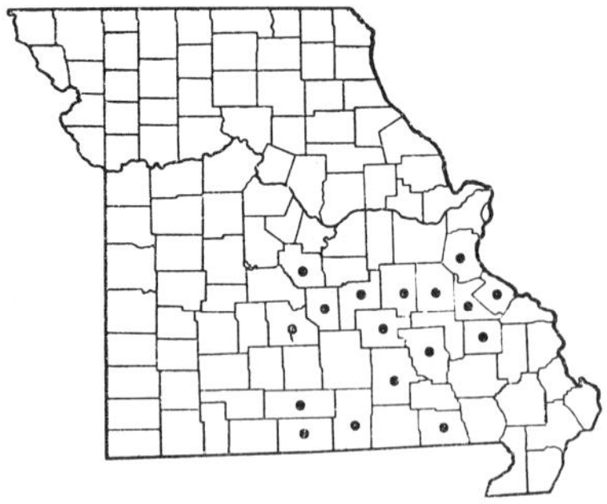

433 Rhynchospora capillacea

434 Rhynchospora Harveyi

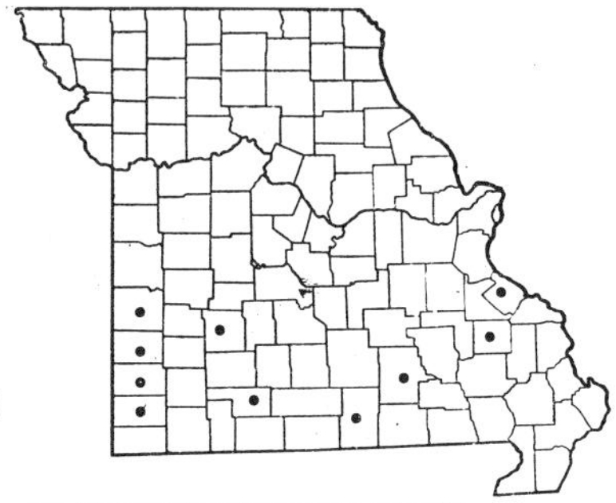

435 Rhynchospora globularis var. recognita

2. **Rhynchospora corniculata** (Lam.) Gray var. **corniculata** Horned Rush Map 431

*Rhynchospora corniculata* (Lam.) Gray [G, BB, P & S]

Flowers June–September.

Occurs in swamps, low wet woods, along ditches, drainage canals, and spring branches. Mostly in southeastern Missouri, where it is a characteristic plant of the southeastern lowland section, northward in eastern Missouri locally in St. Louis and Audrain counties, and at a single location in southwestern Missouri in Christian County (wet sandy margins of spring branch, 2 mi. southwest of Billings, July 13, 1955, *Palmer 60557*).

Ranges from Florida to Louisiana, north to Delaware, Kentucky, and Missouri.

Another variety, var. *interior* Fern., with achenes 4.4–5.3 mm. long and 2.4–2.8 mm. wide, instead of 5–6 mm. long and 2.8–3.3 mm. wide, ranges west to Texas and north to Arkansas and Indiana. A Missouri collection from Ripley County approaches the var. *interior*.

3. **Rhynchospora capitellata** (Michx.) Vahl f. **capitellata** False Bog Rush Map 432

Flowers late May–October.

Occurs along rocky or sandy banks of streams and spring branches, and moist places in upland, or granite substrata. Found in the southern third of the state, north to Ste. Genevieve, St. Francois, Phelps, and Vernon counties.

Ranges from Florida to Texas, Oklahoma, and Missouri, north to Nova Scotia, Quebec, Ontario, and Wisconsin; also in California and Oregon.

Two forms which may be expected in Missouri, but so far have not appeared in collections in the state, are: *Rhynchospora capitellata* f. *controversa* (S. F. Blake) Gale, with the bristles upwardly toothed, and *R. capitellata* f. *discutiens* (C. B. Clarke) Gale, with the bristles smooth. The usual form encountered in Missouri with the downwardly toothed bristles is *R. capitellata* f. *capitellata*.

4. **Rhynchospora capillacea** Torr. Map 433

Flowers early June–September.

Occurs in wet seepy ground of spring-fed calcareous meadows of small valleys or on moist limestone ledges and rocks by small streams. Restricted to the Ozark section, and primarily to the eastern part of the Ozarks north to Jefferson and Miller counties, west to Laclede, Douglas, and Ozark counties.

Ranges from Newfoundland to Saskatchewan, south to New Jersey, Pennsylvania, Virginia, Tennessee, Ohio, Indiana, Illinois, Missouri, and South Dakota.

In Missouri *R. capillacea* is at its southwesternmost limit of range, where it represents a survivor from a time during the Ice Age (Pleistocene) when more northerly-distributed species pushed southward in the wake of the ice sheets and then remained stranded in their present relatively warmer sites following the retreat of the glaciers.

This fine-leaved *Rhynchospora* with delicate slender culms is a characteristic sedge of calcarous swampy meadows of the eastern Ozarks, occurring with other narrow-leaved sedges, such as *Carex interior*, *C. leptalea*, and *Scleria verticillata*.

5. **Rhynchospora Harveyi** Boott Map 434

Flowers June–July.

Occurs on sandstone glades and in swales of upland prairies, where it is known only from the unglaciated prairie section of southwestern Missouri, from St. Clair (ferruginous sandstone glade south of Birdsong, July 13, 1934, *Steyermark 13416*) and Newton (swales in upland prairie, 4 mi. northwest of Diamond, June 17, 1957, *Palmer 65631*) counties.

Ranges from Florida to Texas, north to Virginia, Tennessee, Missouri, and Oklahoma.

The stations in Missouri represent the most northwesterly known ones in the range of the species. *Rhynchospora Harveyi*, *R. capitellata*, and *R. globularis* var. *recognita* occur, in most cases, in soils of an acid reaction.

6. **Rhynchospora globularis** (Chapm.) Small var. **recognita** Gale Map 435

*Rhynchospora cymosa* [of P & S], not Ell.

Flowers late May–October.

Occurs in wet or dry places in upland sandy prairies and meadows, and wet rocky banks. Restricted to the southern third of the state, north to Ste. Genevieve, Shannon, Polk, and Vernon counties, but commonest in the unglaciated prairie region in the southwestern section.

Ranges from Florida to Texas, north to New Jersey, Pennsylvania, Maryland, North Carolina, Tennessee, Ohio, Illinois, Missouri, and Oklahoma.

Sometimes plants of this species may have the flowers infected by a smut, which produces a peculiar condensed appearance of the inflorescence, with only the scales of the spikelets remaining, and the rest of the flowers modified into smut spores. Such a specimen is *Palmer 65881* from Burkhart Prairie in Newton County.

This and other related species of the genus are apparently not browsed by cattle, as the margins and tips of the leaves possess microscopic teeth of silica which renders the leaves unfit for fodder (Gale in Rh. 46: 97. 1944).

## 10. **Scleria** Bergius Nut Rush

Annual or perennial plants with 3-cornered leafy culms (stems). Flowers few, the lower scales empty. Stamens 1–3. Style 3-cleft. Achene bony or enamel-like in hardness, the base surrounded by a simple, scalloped, or lobed disk (hypogynium) or ornamented with rounded or knob-like portions (tubercles).

a. Main body of achene smooth . . . . . . . . . . . . . . . . . . . . . . . . *b*
 b. At the base of the achene are 8–9 small rounded tubercles; leaves thin; rootstock soft . 3. S. OLIGANTHA
 b. The base of the achene does not have 8–9 rounded tubercles but is set in a spongy ring whose surface is covered by minute pebble- or papillae-like grains; leaves hard and firm; rootstock hard, somewhat woody in texture . . . . . . . . . . . . . . . . . . . . . *c*
  c. Surface of the ventral part of leaf-sheath (the side opposite the one that ends in the leaf-blade) covered with a fine dense hairiness, and scarcely contrasted in color with the ligule (thickened summit or appendage in which the ventral sheath ends); achene 2.8–3.3 mm. high, gradually rounded at the tip. . . . . . . . . . . . . . . . . . . . . 2. S. NITIDA
  c. Surface of the ventral part of leaf-sheath glabrous or nearly glabrous (the margins of the sheath usually hairy), contrasting strongly in color with the darker ligule; achene 2–2.5 mm. high, strongly rounded at the tip . . . . . . . . . . . . . . . 1. S. TRIGLOMERATA
a. Main body of achene either warty, roughened with papillae, wrinkled, or honeycombed . . . . . *d*
 d. Surface of achene with hairs in addition to the sculpturing and other ornamentation; lateral (those arising from the side) inflorescences on long, drooping, hair-like peduncles (stalk of inflorescence) . . . . . . . . . . . . . . . . . . . . 6. S. MUHLENBERGII
 d. Surface of achene without hairs, although other ornamentation is present; inflorescences not as above, either sessile (without stalks), or if stalked, the peduncles short and erect or ascending. . . *e*
  e. Clusters of spikelets 2–6 (rarely 1), sessile (stalkless) along an elongated interrupted axis; achene with transverse (horizontal) ridges or wrinkles; no rounded knobs (tubercles) at base of achene; annuals with delicate hair-like culms (stems) and soft leaves . . . . . . . . 7. S. VERTICILLATA
  e. Clusters of spikelets 1–3, close together at the top of the culm (stem), not spaced along an interrupted axis; achene warty or covered with granular or papillose projections; 3, 6, or 9 rounded knobs (tubercles) at base of achene; perennials with stiff or firm culms and hard or firm leaves . . . . . . . . . . . . . . . . . . . . . . . . *f*
   f. Rounded knobs (tubercles) at base of achene 3 (each of the 3 may be lobed into 2 parts); achene 2–3 mm. in diameter . . . . . . . . . . . . . . . . 5. S. CILIATA
   f. Rounded tubercles at base of achene 6 or 9; achene 1.5–2 mm. in diameter . 4. S. PAUCIFLORA

1. **Scleria triglomerata** Michx.
Tall Nut Grass Map 436

Flowers late May–September.

Occurs on dry upland prairies, rocky glades, and rocky open woods. Throughout Missouri. This is the commonest species in the state, although unknown from more than half the counties of the state.

Ranges from Florida to Texas, north to Massachusetts, Vermont, New York, Ontario, Ohio, Michigan, Wisconsin, Minnesota, and Kansas.

The leaf-blades average the broadest (5–9 mm.) and the culms the thickest (2.5–6 mm. at base) of any of the species occurring in Missouri.

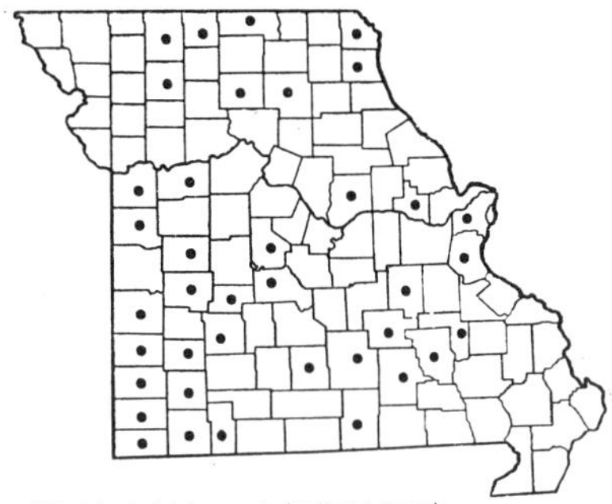

436 Scleria triglomerata (Tall Nut Grass)

437 Scleria nitida

438 Scleria oligantha

2. **Scleria nitida** Willd. Map 437
Flowers May–October.

Occurs on sandy slopes along spring branches and on dry sandstone rock exposures. Known only from southeastern Missouri in Ste. Genevieve (dry exposed bluff tops, Pickle Springs, May 22, 1955, *Mohlenbrock 5244*) and Stoddard (sandy and mossy slopes above and along spring branch in valley at junction of Crowley Ridge and bottomland, on property of Mr. Martin, T25N, R11E, sect. 6, $3\frac{3}{4}$ mi. southeast of Bloomfield, October 17, 1955, *Steyermark 80375*) counties. The locality in Ste. Genevieve County is in the Ozark section of the state, while the Stoddard County station is in the Crowley Ridge section, in a unique botanical area (Steyermark, Rh. 60: 205–8. 1958) where the species is associated with such rarities as *Trisetum pensylvanicum*, *Paspalum setaceum*, *Bartonia paniculata*, *Pyrus melanocarpa*, and *Trichostema setaceum*.

Ranges from Florida to Louisiana, north to New Jersey, Kentucky, and Missouri.

The puberulent-tomentulose leaf-sheaths are sometimes dull purple throughout.

3. **Scleria oligantha** Michx. Map 438
Flowers May–July.

Occurs in rocky, often cherty, open woods and glades, where restricted to the southernmost section of the Ozarks, extending east to Carter and Ripley counties and west to Barry County.

Ranges from Guatemala and Honduras in Central America to Mexico, Florida to Texas, north to Virginia, Kentucky, Indiana, and Missouri.

4. **Scleria pauciflora** Muhl. Map 439
Flowers May–September.

Occurs in acid soils of sandstone, chert, or granite of glades and barrens, open woodland, edges of bluffs, and upland prairies. Restricted to the Ozark section of southern and east-central Missouri southeast of a line extending from Lincoln, Montgomery, Maries, Camden, Hickory, and St. Clair counties to Vernon County, but absent from the southeastern lowland section of the state.

Missouri material may be divided into two varieties:

Culms (stems) glabrous or scarcely hairy; leaves glabrous or sparsely short-hairy . 4a. S. PAUCIFLORA var. PAUCIFLORA

Culms conspicuously hairy; leaves conspicuously hairy, the hairs often 1 mm. long . 4b. S. PAUCIFLORA var. CAROLINIANA

4a. **Scleria pauciflora** var. **pauciflora** Map 439

Ranges from Florida to Texas, north to New Jersey, Pennsylvania, Kentucky, Missouri, and Kansas.

4b. **Scleria pauciflora** var. **caroliniana** (Willd.) Wood Map 439

Ranges from Florida to Tennessee and Missouri, north to New Hampshire, New York, Ohio, and Michigan.

5. **Scleria ciliata** Michx. Map 440
Flowers June–October.

Occurs in acid soils of sandstone, chert, or granite substrata in openings of glades or rocky open woods and on upland sandy prairies.

Missouri material may be divided into the following two varieties:

Leaf-blades 1–2.5 mm. wide, soon turning back the margins (revolute); culms glabrous or slightly hairy; margins of bracts of inflorescence with short hairs; scales of spikelets glabrous or nearly so . . . . . . . 5a. S. CILIATA var. CILIATA

Leaf-blades 3–7 mm. wide, mostly remaining flat;

Plate no. 83. **1.** Scleria oligantha. **2.** Scleria pauciflora. **3.** Scleria Muhlenbergii. **4.** Scleria ciliata var. Elliottii. **5.** Carex Douglasii. **6.** Rhynchospora macrostachya. **7.** Scleria verticillata. **8.** Carex praegracilis.

PLATE NO. 83

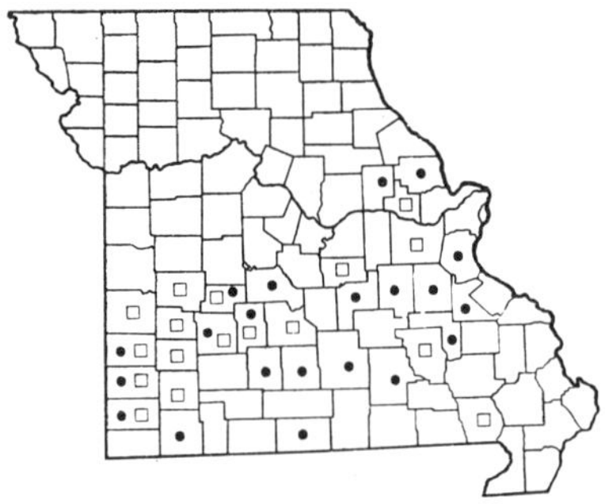

439 • Scleria pauciflora var. pauciflora
439 □ Scleria pauciflora var. caroliniana

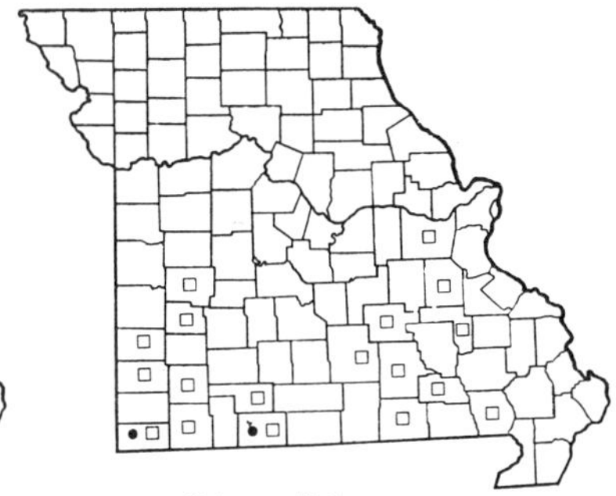

440 • Scleria ciliata var. ciliata
440 □ Scleria ciliata var. Elliottii

441 Scleria Muhlenbergii

culms hairy; margins of bracts of inflorescence conspicuously long-hairy; scales of spikelets hairy . . . . . 5b. S. CILIATA var. ELLIOTTII

5a. **Scleria ciliata** var. **ciliata** Map 440

*Scleria ciliata* Michx. [G, BB, P & S in part]

Known only from southwestern Missouri in Taney (rocky woods, Swan, September 27, 1905, *Bush 3473*) and McDonald (rocky open woods, 1½ mi. SE of Pineville, August 15, 1957, *Palmer 66253*) counties. Both these collections represent the more glabrate extremes of Missouri collections and are therefore referred to the var. *ciliata*.

Ranges from Florida to Texas, north to Virginia and Missouri.

5b. **Scleria ciliata** var. **Elliottii** (Britt.) Fern. Map 440

*Scleria Elliottii* Britton [as synonym of *S. ciliata* var. *ciliata* in P & S]

This is the commoner variety in Missouri, with more pubescent culms, bracts, and scales, and wider leaves. However, transitional types which have the narrower leaf-blades of var. *ciliata* and the pubescent culms of var. *Elliottii* are frequent.

Ranges from Florida to Texas, north to Virginia, Missouri, and Oklahoma.

Achenes are often gray or of a smoky color.

6. **Scleria Muhlenbergii** Steud. Map 441

*Scleria setacea* [of P & S], not Poir.

*Scleria reticularis* var. *pubescens* Britt. [BB]

Flowers August–October.

Known only from Greene County, southwestern Missouri (wet ground, August 16, 1892, *Bush*).

Ranges from Florida to Texas, north to New York, New Jersey, Pennsylvania, Tennessee, Indiana, and Missouri; also in the West Indies, Mexico, and Central America.

This species has not been re-collected since Bush's original collection.

7. **Scleria verticillata** Muhl. Map 442

Flowers July–September.

Occurs in spring-fed calcareous swampy meadows and in moist seepage at base of limestone bluffs. Rare and local in the Ozark region, where known from Jefferson, Reynolds, Camden, and McDonald counties.

Ranges from Connecticut and Ontario to Minnesota, south to Florida and Texas, Mexico, the West Indies, Central and South America.

The leaves of this species, 0.5–2 mm. wide, average narrower than any of the other species in Missouri.

In the calcareous swampy meadows which this species inhabits, it is associated with other calciphiles, such as *Rhynchospora capillacea*, *Juncus subcaudatus*, *Parnassia grandifolia*, *Pedicularis lanceolata*, *Selaginella apoda*, *Solidago patula*, and *Aster puniceus* var. *firmus* f. *lucidulus*.

It is probable that the species will be eventually found at many more stations throughout the Ozarks.

## 11. **Carex** L. Sedge

Perennial plants with usually narrow leaves. Flowers either staminate (male) or pistillate (female), in short head-like or long, slender spikes, with the staminate and pistillate flowers occurring on the same plant in different parts of the same spike, or in separate spikes on the same culm (stem), or rarely on separate plants (dioecious). At the base of the spike is a bract, which is leaf-like and may or may not have a sheath, or is not leaf-like and reduced to a sheath only. Sometimes these spikes appear (a) all *alike* with the staminate flowers at the base of the spike and the pistillate flowers at the top of the same spike (gynecandrous), or with the staminate flowers at the top of the spike and the pistillate flowers at the base of the same spike (androgynous), or the sexes scattered in different parts of the same spike; or sometimes these spikes are (b) all pistillate, or all or part staminate. Each staminate or pistillate flower is in the axil of a scale. The staminate flowers consist of 3 stamens. The pistillate flower consists of 1 pistil with a style of 2 or 3 stigmas enclosed by a loose or tightly-fitting green, yellow, or brown sac (perigynium). The ovary (turning into the mature achene) within the perigynium is either 2-sided (lenticular) or 3-sided (trigonous). Sometimes the achene is nearly rounded.

There are always those who attempt to evaluate all living objects in relation to mankind, and require some explanation as to the possible use or economic value concerned. In the present case of *Carex*, with such an abundance of species and individual plants comprising the flora, it is perhaps surprising that the group as a whole has little or no economic importance. However, the important ecological role performed by the various species is a great one, often overlooked by those who are not satisfied unless they have some immediate explanation for the usefulness of the carices. In Deam's *Flora of Indiana* the thought about the importance of this group is well expressed by the following statement taken from that work (p. 212): 'The species of marshes and muddy borders of lakes which form extensive colonies, and to some extent the less gregarious species, comprise an essential step in the successional stages from open water to the culmination in climax forest or prairie. Much of the fertile soils of our region today would still be barren mudflats were it not for the part played by these sedges in the conversion of the once great boggy areas into a turf, thus enabling less hydrophytic plants to become established and add further to the fertility of the soil.' Furthermore, the achenes of the plants are eaten by various birds, while the leaves are often grazed upon by deer, rabbits, and other wild animals, as well as by cattle. Some species with their large tussocks or mounds of leaves offer nesting sites for some kinds of ground birds, protective cover for certain small birds and fur-bearing animals, and insect habitats. Marsh hay from the northern and eastern states, used for winter mulches and bedding purposes for wintering animals, often includes species of *Carex*. The tough and long leaves of some species may be used for tying loose objects temporarily together.

In order to be certain of one's identification in this group, the largest genus encountered in Missouri, it is necessary to have fully mature pistillate flowers with well developed achenes, and a complete specimen with roots and all. There follow three separate categories which may be used in identification: (1) a summarized listing of species into various groups offered more as a guide which may be useful in the field or for bypassing more technical characters; (2) a key to the sections of the genus in Missouri; and (3) a key to species of the genus.

### **Category** 1.

It may prove of some help and advantage in enabling one to identify various species of *Carex* in the field or to simplify some of the steps in identification by utilizing certain obvious characters to be found associated with a particular habitat, habit, or with easily observable gross features of the plant. With a view towards making available such information for this purpose, the following leads or groupings are offered (species are arranged alphabetically within each grouping):

1. *Leaves glaucous* or *silvery green* (either slightly or greatly)
   *Carex albursina, C. blanda, C. Crawei, C. crus-corvi, C. flaccosperma* var. *flaccosperma, C. flaccosperma* var. *glaucodea, C. granularis, C. hyalinolepis, C. Joori, C. laxiculmis, C. Meadii.*

2. *Leaves* (*leaf-sheaths or blades*) *and/or culms hairy*
   *Carex atherodes, C. Bushii, C. caroliniana, C. complanata* var. *complanata, C. complanata* var. *hirsuta, C. Davisii,*

*C. debilis* (leaf-blades merely scabrous-hirtellous), *C. hirtifolia*, *C. Hitchcockiana* (leaf-sheaths only), *C. oxylepis*, *C. Swanii*, *C. virescens*.

3. *Perigynia hairy*
*Carex abdita*, *C. artitecta*, *C. communis*, *C. Grayii* var. *hispidula*, *C. hirtifolia*, *C. lanuginosa*, *C. nigromarginata*, *C. pensylvanica*, *C. physorhyncha*, *C.* × *subimpressa*, *C. Swanii*, *C. trichocarpa*, *C. umbellata*, *C. virescens*.

4. *Perigynia white or gray-white*
*Carex Buxbaumii*, *C. communis*, *C. Meadii*, *C. nigromarginata*, *C. pensylvanica*.

5. *Plants forming large colonies or beds*
*Carex Buxbaumii*, *C. Emoryi*, *E. Haydenii*, *C. hyalinolepis*, *C. lacustris*, *C. laeviconica*, *C. lanuginosa*, *C.* × *subimpressa*, *C. trichocarpa*, *C. vesicaria* var. *monile*.

6. *Pistillate spikes nodding, curving, or loosely spreading* (at least some of them)
*Carex cherokeensis*, *C. comosa*, *C. crinita*, *C. Davisii*, *C. debilis*, *C. digitalis*, *C. gracillima*, *C. Joori*, *C. hystricina*, *C. laxiculmis*, *C. lurida*, *C. oxylepis*, *C. torta*.

7. *Pistillate flowers 1–9 in a spike*
*Carex abdita*, *C. amphibola* var. *amphibola*, *C. amphibola* var. *globosa*, *C. amphibola* var. *rigida* (or more), *C. amphibola* var. *turgida* (or more), *C. blanda*, *C. Careyana*, *C. communis*, *C. convoluta*, *C. digitalis*, *C. eburnea*, *C. flaccosperma* (or more), *C. gracilescens*, *C. Hitchcockiana*, *C. interior*, *C. intumescens* (or more), *C. Jamesii*, *C. laxiculmis*, *C. leptalea* (or more), *C. nigromarginata*, *C. oligocarpa*, *C. pensylvanica*, *C. physorhyncha*, *C. retroflexa*, *C. rosea*, *C. texensis*, *C. tonsa*, *C. umbellata*.

8. *Pistillate spikes or mixed spikes globose or nearly so or as broad as or broader than long*
*Carex abdita*, *C. albolutescens*, *C. arkansana*, *C. artitecta*, *C. Bicknellii*, *C. brevior*, *C. Bushii*, *C. caroliniana*, *C. cephalophora*, *C. communis*, *C. complanata*, *C. convoluta*, *C. cristatella*, *C. eburnea*, *C. Grayii*, *C. interior*, *C. intumescens*, *C. Jamesii*, *C. Leavenworthii*, *C. mesochorea*, *C. molesta*, *C. Muhlenbergii*, *C. nigromarginata*, *C. normalis*, *C. pensylvanica*, *C. physorhyncha*, *C. projecta*, *C. retroflexa*, *C. rosea*, *C. Sartwellii*, *C. sparganioides*, *C. straminea*, *C. Swanii*, *C. tenera*, *C. texensis*, *C. tonsa*, *C. umbellata*.

9. *Plants of limestone glades*
*Carex austrina*, *C. Crawei*, *C. granularis* var. *Haleana* (or other habitats), *C. Meadii*, *C. microdonta*.

10. *Plants of crevices of limestone bluffs*
*Carex eburnea*, *C. granularis* var. *Haleana*.

11. *Plants of upland natural sink-hole ponds of the Ozarks* (* exclusively)
*Carex alata**, *C. annectens*, *C. comosa*, *C. crinita*, *C.* × *Deamii*, *C. decomposita**, *C. hyalinolepis*, *C. lupulina*, *C. scoparia*, *C. Shortiana*, *C. stipata*, *C. straminea**, *C. tribuloides*.

12. *Plants of calcareous spring-fed swampy meadows of the Ozarks* (* exclusively or nearly so)
*Carex annectens* var. *xanthocarpa*, *C. Buxbaumii*, *C. crinita*, *C. cristatella*, *C. Frankii*, *C. granularis*, *C. hystricina*, *C. interior**, *C. lanuginosa*, *C. leptalea**, *C. lurida*, *C. Shortiana*, *C. squarrosa*, *C. stricta**, *C. suberecta**, *C. trichocarpa**, *C. vulpinoidea*.

13. *Plants in the water of or along the banks of small or large streams* (* characteristic or mostly)
*Carex aggregata*, *C. amphibola*, *C. aquatilis* var. *altior*, *C. blanda*, *C. conjuncta*, *C. crinita*, *C. Davisii*, *C. Emoryi**, *C. Frankii*, *C. granularis*, *C. Haydenii*, *C. Leavenworthii*, *C. laeviconica*, *C. leptalea*, *C. lurida*, *C. normalis*, *C. oklahomensis* *C. squarrosa*, *C. stipata*, *C. torta**, *C. tribuloides*, *C. vulpinoidea*.

14. *Plants of north-facing wooded limestone slopes*
*Carex albursina*, *C. amphibola*, *C. blanda*, *C. Careyana*, *C. cephalophora*, *C. communis*, *C. convoluta*, *C. digitalis*, *C. Hitchcockiana*, *C. Jamesii*, *C. laxiculmis*, *C. oligocarpa*, *C. sparganioides*.

15. *Plants characteristic of rich woods not subject to flooding* (on or at the base of slopes, ravines, and wooded valleys)
*Carex aggregata*, *C. amphibola*, *C. blanda*, *C. Careyana*, *C. conjuncta*, *C. convoluta*, *C. Davisii*, *C. digitalis*, *C. gracilescens*, *C. gravida* var. *Lunelliana*, *C. hirtifolia*, *C. Hitchcockiana*, *C. Jamesii*, *C. laxiculmis*, *C. normalis*, *C. oligocarpa*, *C. oxylepis*, *C. Muhlenbergii*, *C. retroflexa*, *C. rosea*, *C. Shortiana*, *C. sparganioides*, *C. typhina*, *C. virescens*.

16. *Plants of dry acid soils in woodland* (either open woods on upland, along sandstone ledges, sandstone bluffs, or cherty or granite slopes)
*Carex abdita, C. arkansana, C. artitecta, C. austrina, C. cephalophora, C. complanata, C. digitalis, C. flaccosperma* var. *glaucodea, C. gracilescens, C. Muhlenbergii, C. nigromarginata, C. oxylepis, C. pensylvanica, C. retroflexa, C. tonsa, C. virescens.*

17. *Base of plant with purple-red or wine-red or maroon-magenta coloring*
*Carex amphibola* var. *amphibola, C. aquatilis* var. *altior, C. artitecta, C. Bushii, C. Buxbaumii, C. Careyana, C. caroliniana, C. communis, C. complanata, C. conoidea* (often), *C. convoluta* (sometimes), *C. crinita* (frequently), *C. Davisii, C. debilis, C. Emoryi, C. Frankii, C. gigantea, C. gracillima, C. gracilescens* (sometimes to often), *C. Grayii, C. Haydenii, C. hirtifolia, C. Hitchcockiana, C. hystricina, C. intumescens, C. lacustris, C. laeviconica, C. lanuginosa, C. louisianica, C. lupuliformis, C. lupulina* (sometimes), *C. lurida, C. nigromarginata, C. oligocarpa, C. oxylepis, C. pensylvanica, C. physorhyncha, C. Schweinitzii, C. Shortiana, C. stricta* var. *strictior, C.* × *subimpressa, C. Swanii, C. tetanica, C. texensis* (sometimes), *C. tonsa, C. torta, C. trichocarpa, C. typhina* (sometimes), *C. umbellata, C. virescens, C. Woodii.*

18. *Leaf-blades mostly between 0.5–3.5 mm. wide* (leaf-blades which vary to 4.5 mm. wide are not included)
*Carex abdita, C. albolutescens, C. amphibola* var. *amphibola, C. arkansana, C. artitecta, C. brevior, C. Buxbaumii, C. complanata, C. convoluta, C. Crawei, C. Douglasii, C. eburnea, C. festucacea, C. interior, C. Jamesii, C. Leavenworthii, C. leptalea, C. molesta, C. nigromarginata, C. pensylvanica, C. physorhyncha, C. praegracilis, C. retroflexa, C. rosea, C. scoparia, C. straminea, C. suberecta, C. Swanii, C. tenera, C. texensis, C. tonsa, C. umbellata, C. virescens, C. Woodii.*

19. *Leaf-blades averaging 10–30 mm. or more wide* (and with blades never less than 7 mm. wide)
*Carex albursina, C. Careyana, C. lupuliformis, C. stipata* var. *maxima.*

## **Category** 2.

For those who wish to key their plants to sections, the following key, based upon the monograph of *Carex* by Mackenzie (*North American Flora* 18: 1–478), and the treatments of Fernald in *Gray's Manual*, eighth edition, of Kükenthal in Das Pflanzenreich IV. 20. 1909, and of Hermann in Deam's *Flora of Indiana*, is offered.

## **Key to the Sections of Missouri Carices**

Stigmas 2; achenes lenticular (2-sided); spikes usually with both staminate and pistillate flowers (bisexual), the lateral spikelets sessile . . . . . . . . . . . . . . . . . Subgenus I. VIGNEA
Stigmas 3; achenes more or less 3-sided or triangular, or if stigmas 2 and achenes 2-sided, then the lateral spikes peduncled (stalked); spikes mostly either staminate or pistillate (unisexual) . Subgenus II. EUCAREX

### Subgenus I. **Vignea**

a. Terminal or all spikes androgynous (staminate flowers at apex) . . . . . . . . . . . . *b*
b. Culms from long-creeping rootstocks, mostly arising singly or few together . . . . . . . . *c*
c. Beak of perigynium 1/2–5/8 as long as main body of perigynium; upper leaf-sheaths hyaline (transparent) or brown, not finely striate (lined) on the ventral (inner) side . . 1. DIVISAE (p. 330, C. PRAEGRACILIS, C. DOUGLASII)
c. Beak of perigynium 1/4–1/3 as long as main body of perigynium; upper leaf-sheaths green and finely striate on the ventral side . . . . . . . . . . 2. INTERMEDIAE (p. 330, C. SARTWELLII)
b. Culms cespitose with the leafy shoots and culms rather crowded or close together, the rootstocks not long-creeping . . . . . . . . . . . . . . . . . . . . . . . . . . . *d*
d. Spikes generally 10 or fewer (up to 15), in mostly simple close or interrupted heads, often greenish . 3. BRACTEOSAE (p. 330, C. RETROFLEXA, C. CONVOLUTA, C. ROSEA, C. CEPHALOPHORA, C. LEAVENWORTHII, C. ARKANSANA, C. MUHLENBERGII, C. GRAVIDA, C. AGGREGATA, C. SPARGANIOIDES)
d. Spikes numerous, in panicled spike-like heads, at least the lowest lateral branches of the spike bearing 2 or more spikes . . . . . . . . . . . . . . . . . . . . . . *e*
e. Leaf-sheaths tightly enveloping the culm (stem); culms firm and more or less slender; perigynia firm or rigid, flattened or convex on the inner (upper) face, rather closely surrounding the achene, 1.7–4 mm. long . . . . . . . . . . . . . . . . . *f*

f. Perigynia flat on inner face, convex on outer face, thin, yellow, buff, or greenish; inner (ventral) side of leaf-sheath usually roughened or wrinkled transversely; pistillate scales awned; at least the lower bracts of the inflorescence prolonged and setaceous or leaf-like . . . . 4. MULTIFLORAE (p. 335, C. VULPINOIDEA, C. ANNECTENS, C. TRIANGULARIS)

f. Perigynia more or less convex on each side, thick and inflated; inner side of leaf-sheath not transversely roughened or wrinkled; pistillate scales acute to shortly mucronate-tipped; bracts of inflorescence not present or very short . . . . . . . 5. PANICULATAE (p. 336, C. DECOMPOSITA)

e. Leaf-sheaths loose around the culm; culms soft, becoming flattened when pressed; perigynia soft and thin, with a soft spongy or corky base, more or less inflated and loosely surrounding the achene, 4–9 mm. long . . . . . . . . . . . . . . . . . . 6. VULPINAE (p. 338, C. OKLAHOMENSIS, C. STIPATA, C. LAEVIVAGINATA, C. CRUS-CORVI, C. CONJUNCTA)

a. Terminal or all spikes gynecandrous (upper flowers pistillate, lower staminate) or the staminate flowers scattered in the spike . . . . . . . . . . . . . . . . . . . . . . . . *g*

g. Perigynia without winged borders, at most only with thin edges; perigynia thickened or spongy at base . . . . . . . . . . . . . . . . . . . 7. STELLULATAE (p. 340, C. INTERIOR)

g. Perigynia with winged borders; perigynia not thickened nor spongy at base. . . . . . 8. OVALES (p. 340, C. MUSKINGUMENSIS, C. SCOPARIA, C. TRIBULOIDES, C. PROJECTA, C. CRISTATELLA, C. NORMALIS, C. TENERA, C. FESTUCACEA, C. ALBOLUTESCENS, C. SUBERECTA, C. BREVIOR, C. MOLESTA, C. BICKNELLII, C. ALATA, C. STRAMINEA)

## Subgenus II. **Eu-Carex**

a. Style articulated (jointed) with the achene, deciduous (falling) in age from the mature achene . . . *b*

b. Spike solitary, terminal, androgynous (staminate flowers at apex); perigynia without a beak, rounded at the summit, glabrous . . . . . . . . . 9. POLYTRICHOIDEAE (p. 346, C. LEPTALEA)

b. Spikes 2 or more, the perigynia not as above . . . . . . . . . . . . . . . . . . *c*

c. Lower pistillate scales prolonged, leaf- or bract-like; staminate scales with margins united at base, tightly clasping the axis (rachis) . . . . . . 10. PHYLLOSTACHYAE (p. 346, C. JAMESII)

c. Lower pistillate scales not prolonged or leaf- or bract-like; staminate scales loosely ascending, the margins free and not united at base . . . . . . . . . . . . . . . . . *d*

d. Achenes with rounded or convex sides, closely surrounded by the perigynium; lowest bract of inflorescence scale- or bristle-like, without any sheath . 11. MONTANAE (p. 347, C. ARTITECTA, C. NIGROMARGINATA, C. PHYSORHYNCHA, C. PENSYLVANICA, C. COMMUNIS, C. UMBELLATA, C. ABDITA, C. TONSA)

d. Achenes with flat or concave sides; lowest bract of inflorescence with or without a sheath . . *e*

e. Perigynia closely surrounding the achene, tightly filled to the summit by the achene; bracts, when present, with definite sheaths . . . . . . . . . . . . . . . *f*

f. Perigynia glabrous; leaves and culms glabrous . . . . 12. ALBAE (p. 350, C. EBURNEA)

f. Perigynia pubescent; leaves and culms pubescent . 13. TRIQUETRAE (p. 351, C. HIRTIFOLIA)

e. Perigynia not closely surrounding the achene, the summit usually empty and not filled by the achene; bracts with or without sheaths . . . . . . . . . . . . . . . *g*

g. Styles 2; achenes lenticular . . . . . . . . . . . . . . . . . *h*

h. At least the lower pistillate scales with long rough awns mostly 2–4 times longer than perigynia (sometimes only slightly longer); pistillate spikes mostly all peduncled (stalked), usually nodding or arching . 28. CRYPTOCARPAE (p. 372, C. CRINITA)

h. Pistillate scales obtuse to acute, rarely with a short awn, shorter than or somewhat longer than the perigynia; pistillate spikes usually sessile (without stalks), rarely on long peduncles, erect to arching or nodding . 27. ACUTAE (p. 368, C. AQUATILIS var. ALTIOR, C. NEBRASKENSIS, C. HAYDENII, C. EMORYI, C. STRICTA, C. TORTA)

g. Styles 3; achenes 3-sided (trigonous) . . . . . . . . . . . . . . *i*

i. Bract at base of inflorescence without a sheath or scarcely with a sheath (if the sheath prolonged, then the perigynia are plump, pubescent, and with a 2-toothed beak) . . . . . . . . . . . . . . . . . . . . . . . . . *j*

j. Leaves and perigynia glabrous (perigynia minutely scabrous in some species) . . *k*

k. Perigynia without any beak or with a toothless or slightly 2-toothed short-beaked tip . . . . . . . . . . . . . . . . . . . . . *l*

l. Perigynia closely appressed-ascending, flattened, 1–2 mm. broad . . 26. ATRATAE (p. 366, C. BUXBAUMII)

l. Perigynia squarrose (tips turning outward from the center) or spreading, plump or with bulging sides, 1.7–3.5 mm. thick . . . . . . . . . . . . . . . . . . *m*
m. Terminal spike completely staminate; perigynia longer than broad, strongly nerved, 4–5 mm. long . . . . . . . . . . . . . . 25. PENDULINAE (p. 366, C. JOORI)
m. Terminal spike pistillate except at base; perigynia as broad as long, without nerves, 2.5–3 mm. long . . . . . . . 24. SHORTIANAE (p. 366, C. SHORTIANA, C. × DEAMII)
k. Perigynia with a definite beak, minutely scabrous . . . 23. ANOMALAE (p. 366, C. SCABRATA)
j. Leaves or perigynia or both pubescent . . . . . . . . . . . . . . . . . . . . *n*
n. Beak of perigynium without teeth at summit or at most the summit barely notched; leaf-sheaths or blades or both more or less pubescent; plants with long creeping rootstocks 21. VIRESCENTES (p. 363, C. VIRESCENS, C. SWANII, C. COMPLANATA, C. CAROLINIANA, C. BUSHII)
n. Beak of perigynium definitely 2-toothed at summit; leaves glabrous; plants cespitose, without creeping rootstocks . . . . . . . . . . . 22. HIRTAE (p. 364, C. LANUGINOSA)
i. Bract at base of inflorescence with a prolonged closed sheath; perigynia always glabrous . . . . *o*
o. Terminal spike mostly pistillate except at base where staminate . . . . . . 18. GRACILLIMAE (p. 360, C. GRACILLIMA, C. OXYLEPIS, C. DAVISII)
o. Terminal spike usually completely staminate . . . . . . . . . . . . . . . . . . *p*
p. Pistillate spikes short, oblong or linear, erect, or if drooping, the perigynia sharply 3-sided . . *q*
q. Perigynia with few to many strongly raised nerves . . . . . . . . . . . . . *r*
r. Perigynia tapering at base, triangular; perigynia closely surrounding and tight about the achenes . . . . . . . . . . . . . . . . . . . . . . . . . . . *s*
s. Plants with creeping rootstocks and small tufts of leaves; perigynia glaucous or gray-green . . . . . . . 14. PANICEAE (p. 351, C. MEADII, C. TETANICA, C. WOODII)
s. Plants without creeping rootstocks; leaves prominently developed; perigynia green or olive-green . . . . . . . . . . . . 15. LAXIFLORAE (p. 352, C. CAREYANA, C. DIGITALIS, C. LAXICULMIS, C. ALBURSINA, C. BLANDA, C. GRACILESCENS)
r. Perigynia rounded at the base, somewhat rounded in cross-section; perigynia loosely surrounding the achenes . . . . . . . . . . . . . . . 16. GRANULARES (p. 355, C. GRANULARIS, C. CRAWEI, C. MICRODONTA)
q. Perigynia with many fine impressed nerves . . . 17. OLIGOCARPAE (p. 356, C. OLIGOCARPA, C. HITCHCOCKIANA, C. CONOIDEA, C. AMPHIBOLA, C. FLACCOSPERMA)
p. Pistillate spikes elongate, linear to cylindric, on slender peduncles (stalks), at least the lower drooping, nodding, or spreading . . . . . . . . . . . . . . . . . . . . . *t*
t. Base of plant purple or reddish-purple; no green leaf-blades produced from the lower sheaths at the base of the culm; pistillate spikes narrowly linear, 3–4 mm. wide . 19. SYLVATICAE (p. 362, C. DEBILIS)
t. Base of the plant strongly marked with brown; well-developed green leaf-blades produced from the lower sheaths at the base of the culm; pistillate spikes oblong-cylindric, 5–10 mm. wide . . . . . . . . . . . . 20. LONGIROSTRES (p. 363, C. CHEROKEENSIS)
a. Style not articulated but continuous with the achene, persistent to the summit of the achene. . . . *u*
u. Perigynia broadest in the upper half, obovoid or obconic, abruptly contracted and rounded above to a beak . . . . . . . . . 31. SQUARROSAE (p. 378, C. FRANKII, C. SQUARROSA, C. TYPHINA)
u. Perigynia broadest between the base and the middle, gradually tapering to the beak . . . . . *v*
v. Perigynia at maturity subcoriaceous, of firm or thick texture, scarcely inflated; perigynia glabrous or pubescent; leaf-blades and sheaths glabrous or pubescent . . . . . 30. PALUDOSAE (p. 374, C. LACUSTRIS, C. HYALINOLEPIS, C. × SUBIMPRESSA, C. ATHERODES, C. LAEVICONICA, C. TRICHOCARPA)
v. Perigynia at maturity membranaceous, of thin or papery texture, usually conspicuously inflated; perigynia usually glabrous; leaf-blades and sheaths glabrous. . . . . . . . . . *w*
w. Pistillate scales with scabrous awns equaling or longer than the rest of the scale . . . 29. PSEUDO-CYPEREAE (p. 372, C. HYSTRICINA, C. COMOSA, C. LURIDA, C. SCHWEINITZII)
w. Pistillate scales blunt or cuspidate, if awned the awn smooth and less than half as long as the rest of the scale . . . . . . . . . . . . . . . . . . . . . . . . . *x*
x. Staminate spike usually 1 (sometimes 2–5); perigynia 10–20 mm. long; achenes 2.5–6 mm. long, 2–4 mm. wide . . . . . . 33. LUPULINAE (p. 380, C. GRAYII, C. INTUMESCENS, C. LOUISIANICA, C. LUPULINA, C. LUPULIFORMIS, C. GIGANTEA)
x. Staminate spikes usually 2–4 (in the Missouri species); perigynia 5–8 mm. long (in Missouri species); achenes 2–3 mm. long, 1.25–2.5 mm. wide. . . . . . 32. VESICARIAE (p. 380, C. VESICARIA)

The following key to species attempts to utilize characters to be found in the foliage, flowers, and fruit, as well as in the habit and habitat of the plant:

### Category 3.

### Key to species

a. Hairs present on some part of the surface of the leaves (leaf-sheaths or leaf-blades or both), bracts, or culms (stems) (sometimes the hairs are found on the lower leaf-sheaths only) (rough-margined scabrous leaves and culms are placed in the other *a.* section) . . . . . . . . . . . . . . *b*
- b. Perigynia hairy . . . . . . . . . . . . . . . . . . . . . . . . . . . . . *c*
  - c. Perigynia 3.5–5 mm. long; topmost (terminal) flower spike of staminate (male) flowers only; leaf-blades 3–10 mm. wide . . . . . . . . . . . . . . . . . 50. C. HIRTIFOLIA
  - c. Perigynia 1.8–2.5 mm. long; topmost (terminal) flower spike mostly of pistillate (female) flowers except at the base where staminate; leaf-blades 2–4 mm. wide . . . . . . . . . . *d*
    - d. Pistillate spikes linear, narrowed to the base, (11–) 15–40 mm. long, 4–8 times longer than broad; leaves usually shorter than the flowering culms . . . . . . . . 73. C. VIRESCENS
    - d. Pistillate spikes broadly oblong to oblong-globose, rounded or broadened at the base, mainly 4–15 mm. long, mainly 1½–4 times longer than broad; leaves usually longer than the flowering culms . . . . . . . . . . . . . . . . . . . 74. C. SWANII
- b. Perigynia glabrous . . . . . . . . . . . . . . . . . . . . . . . . . . . *e*
  - e. Topmost (terminal) flower spike of staminate (male) flowers only . . . . . . . . . . . *f*
    - f. Styles 2; achenes 2-sided or 2-angled . . . . . . . . . . 87. C. STRICTA var. STRICTIOR
    - f. Styles 3; achenes 3-sided or 3-angled . . . . . . . . . . . . . . . . . . . *g*
      - g. Plant forming large colonies with creeping rhizomes; mature perigynia 7–12 mm. long, ending in a conspicuous 2-pronged beak with teeth 1.6–3 mm. long; pistillate spikes 6–10 mm. wide . . . . . . . . . . . . . . . . . . . . . 96. C. ATHERODES
      - g. Plants in single clumps or tufts; mature perigynia 4.5–7 mm. long, either without beaks or the beaks without conspicuous teeth; pistillate spikes 2–4 mm. wide . . . . . . . *h*
        - h. Only 1–8 flowers in a pistillate (female) spike; plant brown at base (near soil); pistillate spikes erect, 8–25 mm. long; leaf-sheaths with hairs . . . 64. C. HITCHCOCKIANA
        - h. Ten or more flowers in a pistillate spike; plant with purple or red-purple at base; pistillate spikes spreading or nodding, 28–60 mm. long; no hairs on leaf-sheaths 71. C. DEBILIS
  - e. Topmost (terminal) flower spike with pistillate (female) flowers in the upper part and staminate (male) flowers at the base . . . . . . . . . . . . . . . . . . . . . . . *i*
    - i. All spikes 18–45 mm. long, 6–8 times longer than broad, linear to narrowly oblong-cylindrical . . . . . . . . . . . . . . . . . . . . . . . . . . . . *j*
      - j. At least the upper scales of pistillate flowers usually with a long awn or narrowed tip about as long or half as long as the main part of the scale; spikes mainly 5–7 mm. thick; perigynia inflated and loose around the achene, oblong-ovoid, broadest in the lower half; common and throughout Missouri . . . . . . . . . . . . . 70. C. DAVISII
      - j. At least the upper scales of pistillate flowers with a short point or tip much shorter than the main part of the scale; spikes 3–5 mm. thick; perigynia not inflated, close to the achene, narrowly ellipsoid and narrowed below the middle; rare, found only in the southeastern lowland section of Missouri. . . . . . . . . . . . . 69. C. OXYLEPIS
    - i. All spikes, except the topmost (terminal) one, mainly 5–20 mm. long, 1¼–4 times longer than broad, narrowly ovoid or short cylindric . . . . . . . . . . . . . . *k*
      - k. Scales of pistillate flowers 3–6 mm. long, with a prominent awn or projection from the tip and longer than the perigynium; perigynia 3–4 mm. long . . . . . 77. C. BUSHII
      - k. Scales of pistillate flowers 2–3 mm. long, obtuse to short-pointed, and shorter than the perigynium; perigynia 2.2–3.5 mm. long . . . . . . . . . . . . . . . *l*
        - l. Culms, leaf-sheaths, and leaf-blades conspicuously hairy . 75b. C. COMPLANATA var. HIRSUTA
        - l. Culms, leaf-sheaths, and leaf-blades sparsely hairy to nearly or quite hairless . . . . *m*
          - m. Perigynia conspicuously nerved on both faces; perigynia spreading, plump, about equally rounded on outer and inner faces, nearly as thick as wide, short-pointed at the tip; achenes very abruptly bent at tip; common species. . . 76. C. CAROLINIANA
          - m. Perigynia usually without nerves on the upper face, only weakly nerved on

the lower face; perigynia more appressed or closely ascending, somewhat flattened with the inner face flattish, much wider than thick, rounded at tip; achenes only slightly bent at tip; rare . . . . . . . . . . . . . . . . . 75a. C. COMPLANATA var. COMPLANATA

a. Hairs absent on leaves, bracts, and culms (rough or scabrous margins or surfaces are keyed out in this section) . . . . . . . . . . . . . . . . . . . . . . . . . . . . *n*

n. Surface of perigynia hairy or with a minute rough puberulence (rough-toothed or serrulate beaks of glabrous perigynia are not included here) . . . . . . . . . . . . . . . . . . *o*

o. Pistillate spikes 20–40 mm. thick, forming a ball- or globe-shape; perigynia 12–20 mm. long 103b. C. GRAYII var. HISPIDULA

o. Pistillate spikes and perigynia smaller . . . . . . . . . . . . . . . . . . . *p*

p. Most or all of the flowering culms much shorter than and/or hidden by the leaf-blades (do not confuse leaf-like bracts of inflorescence with ordinary leaf-blades; on late-season plants, remains of old fruiting stalks [peduncles] may be found among the leaf-bases); flowering culms mostly 0.1–2 dm. tall . . . . . . . . . . . . . . . . . . . . . . *q*

q. Only 1 peduncle (stalk supporting inflorescence) arising from each culm; most of peduncles conspicuous and elongated, 5–20 cm. long, mostly all similar and producing both staminate and pistillate flowers; rare species of southern and southeastern Missouri 42. C. NIGROMARGINATA

q. 2 to 4 peduncles of different lengths arising from each culm; peduncles not alike, 1 elongated with staminate or staminate and pistillate flowers, the other 1–3 very short (1–5 cm. long) and pistillate only and hidden in the densely tufted bases, or rarely a few of the pistillate peduncles elongated (5–20 cm. long); common species . . . . . . . *r*

r. Beak of perigynium 0.9–1.7 mm. long, about 3/4 as long as the main body of perigynium; scales of pistillate flowers with usually long-pointed (acuminate) tips 46. C. UMBELLATA

r. Beak of perigynium 0.5–1 mm. long, about 1/2 as long as the main body of perigynium; scales of pistillate flowers with usually short-pointed (acute) tips . . . . 47. C. ABDITA

p. Without the above combination of characters; most or all of the flowering culms longer than and not hidden by the leaf-blades (do not confuse leaf-like bracts of inflorescence with ordinary leaf-blades); or culms 2–12 dm. tall . . . . . . . . . . . . . . . . *s*

s. Topmost (terminal) flower spike with pistillate (female) flowers in the upper part and staminate (male) flowers at the base . . . . . . . . . . . . . . . . . . . *t*

t. Scales of pistillate flowers 3–6 mm. long, with a prominent awn or point projecting and longer than the perigynium; perigynia 3–4 mm. long . . . . . . 77. C. BUSHII

t. Scales of pistillate flowers 2–3 mm. long, obtuse to short-pointed, and shorter than the perigynium; perigynia 2.2–3.5 mm. long. . . . . . . . . . . . . . *u*

u. Perigynia conspicuously nerved on both faces; perigynia spreading, plump, about equally rounded on outer and inner faces, nearly as thick as wide; common species . . . . . . . . . . . . . . . . . . . . 76. C. CAROLINIANA

u. Perigynia usually without nerves on the upper face, only weakly nerved on the lower face; perigynia more appressed or closely ascending, somewhat flattened, much wider than thick, with the inner face flattish; rare . . . 75a. C. COMPLANATA var. COMPLANATA

s. Topmost (terminal) flower spike all staminate (male) flowers . . . . . . . . . . *v*

v. Perigynia with a minute roughened (scabrous) puberulence; no purple-red or wine-purple color at base of plant; beak of perigynium curved, nearly as long as main body of perigynium. . . . . . . . . . . . . . . . . . 79. C. SCABRATA

v. Without the above combination; perigynia with short or long, sparse or dense hairs; purple-red or wine-purple color at base of plant; beak of perigynia, if present, straight . . . . . . . . . . . . . . . . . . . . . . . . . *w*

w. Plants of wet or swampy ground, forming large colonies with creeping rhizomes; pistillate spikes 10–90 mm. long, much longer (3–6 times) than thick; culms mostly 5–12 dm. tall. . . . . . . . . . . . . . . . . . . . . . *x*

x. Perigynia 2.5–5 mm. long; pistillate spike 3–8 mm. thick; no raised lines or ribs showing on surface of perigynia . . . . . . . . . . . 78. C. LANUGINOSA

x. Perigynia 5–10 mm. long; pistillate spike 8–15 mm. thick; faint or prominent raised lines or ribs showing on surface of perigynia . . . . . . . . . . *y*

y. Ribs strong or prominent on perigynia; perigynia ending in 2 conspicuous teeth 0.7–2.7 mm. long . . . . . . . . . . . . 98. C. TRICHOCARPA

y. Ribs weak or faint on perigynia; perigynia ending in 2 short teeth 0.5–1 mm. long . . . 95a. C. × SUBIMPRESSA

w. Plants of dry ground or dry rocky places, forming tufts or clumps; pistillate spikes mainly 3–9 mm. long, either as thick as long or only slightly longer ($1\frac{1}{2}$–2 times) than thick; culms mostly 0.5–5 dm. tall . . . . . . . . . . . . . . . . . . . . . . . . . . . . . . *z*

z. Main body of perigynium about as long as wide, at maturity not or very slightly 3-angled . . . *1*

1. Plant with elongated horizontal scaly stolons at base (plants must be dug up completely to show this); base of leafy shoots conspicuously fibrillose; ligule at summit of leaf-sheath short, much wider than long; leaf-blades 1–3 mm. wide; plants of acid soils of northern and central Missouri . . . . . . . . . . . . . . . . . . 44. C. PENSYLVANICA

1. Plant without long horizontal stolons; base of leafy shoots faintly or not fibrillose; ligule at summit of leaf-sheath prominent, longer than wide; leaf-blades 3–7 mm. wide; plants of wooded limestone slopes of southwestern Missouri . . . . . . . . . 45. C. COMMUNIS

z. Main body of perigynium longer than wide, at maturity noticeably 3-angled . . . . . . . *2*

2. Leaves mainly 3–7 mm. wide; ligule at summit of leaf-sheath prominent, longer than wide; perigynia about 1.5 mm. thick . . . . . . . . . . . . . . . 45. C. COMMUNIS

2. Leaves 0.5–3 mm. wide; ligule at summit of leaf-sheath, short, wider than long; perigynia 0.5–1.3 mm. thick . . . . . . . . . . . . . . . . . . . . . . . . *3*

3. Base of plant with elongated horizontal scaly stolons; staminate spike usually with a short stalk; perigynia whitish-green . . . . . . . . . . 43. C. PHYSORHYNCHA

3. Base of plant without elongated horizontal stolons; staminate spike usually sessile (without a stalk); perigynia olive green or dull green . . . . . . . . . . . . *4*

4. Common plant throughout Missouri; scales of staminate spike often dark or purplish or purple-red; pistillate spikes usually not bunched together; staminate spike 3–15 mm. long; scales of staminate spike obtuse (blunt) or short-pointed, closely appressed; flowering culms upright, rather stiff . . . . . . . . . . . . 41. C. ARTITECTA

4. Rare plant of southeastern Missouri lowlands; scales of staminate spike often green or green and brown; pistillate spikes, at least the upper ones, bunched together; staminate spike 3–8 mm. long; scales of staminate spike pointed, ascending or loosely spreading; flowering culms curving, arching, or spreading, weak . 42. C. NIGROMARGINATA

n. Perigynia glabrous or nearly so (roughened, toothed, or serrulate margins of beaks are included here) . . *5*

5. Leaves thread-like, 0.5–1.5 mm. wide (mainly 1 mm. or less); growing in the crevices and ledges of limestone bluffs; 2–4 thread-like peduncles appearing from the top of the main peduncle; perigynia 1.5–2 mm. long; stigmas 3 and achenes 3-sided . . . . . . . . . . 49. C. EBURNEA

5. Without the above combination of characters . . . . . . . . . . . . . . . . . . . *6*

6. Lowermost broadest basal leaves (overwintering or last years's) mostly 15–40 mm. wide (rarely 10), bluish- or gray-green; no purple or red-purple color at base of plant; plants mainly of north-facing or east-facing wooded limestone slopes; stigmas 3 and achenes 3-sided; perigynia with rounded angles at least below the middle. . . . . . . . . . . . 57. C. ALBURSINA

6. Without the above combination of characters . . . . . . . . . . . . . . . . . . *7*

7. Pistillate spikes globose or nearly globose (ball-like or spherical); perigynia 10–20 mm. long; 1 staminate (male) spike at the top of the inflorescence . . . . . . . . . . . . . . *8*

8. Found throughout the state, including the southeastern section; perigynia cuneate (tapering inward or narrowed) at the base, not shining; staminate spike often sessile (without a stalk) . . . . . . . . . . . . . . . . . . . . . 103. C. GRAYII

8. Found only in southeastern Missouri lowland section; perigynia rounded at base, shining; staminate spike usually with a long peduncle (stalk) . . . . . . . 104. C. INTUMESCENS

7. Without the above combination of characters . . . . . . . . . . . . . . . . . . *9*

9. Lowest scale of pistillate flower prolonged, green, and leaf-like, 15–50 mm. long (do not confuse this with a leaf-like bract at base of a flowering spike); each culm with only 1 spike with 2–4 perigynia; all flowering culms much shorter than the dense clump of dark green leaves; styles 3; body of perigynium rounded, globe-shaped, abruptly tapering to a long beak; plants of rich low woods or wooded slopes . . . . . . . 40. C. JAMESII

9. Without such a combination of characters; no leaf-like pistillate scales, all reduced in size and less than 10 mm. long (the leaf-like bract at base of the pistillate inflorescence may be greatly prolonged but should not be confused here with the lowest pistillate scale) . . . . . . . . . . . . . . . . . . . . . . . . . . *10*

10. Flowering culms 1–10 cm. tall; most or all of the flowering culms much shorter than the leaf-blade, the pistillate spikes hidden in the densely tufted leaf-bases (remains of

old fruiting stalks [peduncles] may be found among the leaf-bases; do not confuse leaf-like bracts of inflorescence with ordinary leaf-bases); flowering peduncles of a culm not alike, 1 elongated with staminate or staminate and pistillate flowers, the other 1–3 very short and pistillate only . . . . . . . . . . . . . . . . . . . . . . . . . 48. C. TONSA

10. Without the above combination of characters . . . . . . . . . . . . . . . . . . *11*

11. Each culm with only 1 spike, the male flowers at the top; leaves all thread-like, 0.5–1.3 mm. wide; plants of limey open swampy meadows or at the base of moist shaded limestone river bluffs; perigynia beakless, 2.5–3.5 mm. long; styles 3 . . . . . . . . . 39. C. LEPTALEA

11. Without the above combination of characters . . . . . . . . . . . . . . . . . *12*

12. Styles 3; achenes (inside perigynia) 3-sided or 3-angled . . . . . . . . . . . *13*

13. Topmost (terminal) flower spike with pistillate (female) flowers in the upper part and staminate (male) flowers at the base . . . . . . . . . . . . . . *14*

14. Mature pistillate spikes 10–22 mm. thick; mature perigynia 3.5–8.5 mm. long; beak at the summit of perigynium 2–3.5 mm. long . . . . . . . . . . *15*

15. Beaks of mature perigynia widely to nearly horizontally spreading; style strongly curved near its base (look inside perigynium); scales of pistillate flowers with an awn or sharp point at tip . . . . . . . . 100. C. SQUARROSA

15. Beaks of mature perigynia upwardly pointing or ascending; style nearly or quite straight (look inside perigynium); scales of pistillate flowers mainly blunt or only a short point at tip. . . . . . . . . . . . . 101. C. TYPHINA

14. Mature pistillate spikes 2–8 mm. thick; mature perigynia 2–4 mm. long; perigynium either without a beak or the beak at most 1.5 mm. long . . . . . . *16*

16. Some or all of the pistillate spikes drooping, nodding, or spreading at maturity on thread-like peduncles (stalks) nearly equaling the length of the spikes . . . *16a*

16a. Pistillate spikes narrowly linear, 10–12 times longer than thick, 2–3 mm. thick . . . . . . . . . . . . . . . . 68. C. GRACILLIMA

16a. Pistillate spikes cylindric, oblong-cylindrical or linear, 4–7 times longer than thick, 5–7 mm. thick . . . . . . . . . . . . 70. C. DAVISII

16. Pistillate spikes erect to ascending at maturity, either sessile (no stalk) or at least the upper ones nearly sessile or with a short stalk . . . . . . . . . *17*

17. Only 2–4 leaves present at base of culm (stem); leaves pale green and glaucous (bluish-green); perigynia whitish . . . . . . . 82. C. BUXBAUMII

17. Leaves at base of culm more numerous; leaves olive or dark green, not glaucous; perigynia olive green to dark brown . . . . . . . . . . *18*

18. Pistillate spikes mostly 3½–7 times as long as thick; leaf-blades 4–10 mm. wide; perigynia without nerves on the faces, the tip or upper half of mature perigynium spreading outward (squarrose) . . . . . . . *19*

19. Perigynium with a beak 1–1.5 mm. long; pistillate spikes 7–8 mm. thick; rarely found . . . . . . . . . . . 80a. C. × DEAMII

19. Perigynium beakless or with only a minute beak less than 0.5 mm. long; pistillate spikes mostly 3.5–6 (up to 7) mm. thick; common throughout Missouri . . . . . . . . . . . 80. C. SHORTIANA

18. Pistillate spikes mostly 1½–3 times as long as thick; leaf-blades 1.5–5 mm. wide; perigynia with 1 or more nerves on at least one of the faces, the tip or upper half continuing in the same direction as the rest of the perigynium . . . . . . . . . . . . . . . . . . *20*

20. Perigynia usually without nerves on the upper face, only weakly nerved on the lower face; perigynia closely ascending or appressed, somewhat flattened, much wider than thick, with the inner face flattish; rare . . . 75a. C. COMPLANATA var. COMPLANATA

20. Perigynia conspicuously nerved on both faces; perigynia spreading, plump, about equally rounded on outer and inner faces, nearly as thick as wide; common species . . . 76. C. CAROLINIANA

13. Topmost (terminal) flower spike with all the flowers staminate (male) . . . . . *21*

21. Perigynia minutely roughened (scabrous); beak of perigynium curved, nearly as long as the body of perigynium . . . . . . . . . . . . 79. C. SCABRATA

21. Without the above combination; perigynia glabrous or smooth; beaks of perigynia straight or bent . . . . . . . . . . . . . . . . . . . . . *22*

22. Beak of perigynium abruptly bent or turned to one side . . . . . . . . *23*

23. Perigynia rounded or at least not long tapering at the base, rounded in cross-section, loosely surrounding the achenes, dark brown or olive green . . . . . . . . . . 60. C. GRANULARIS
23. Perigynia tapering and conspicuously narrowed at the base, 3-sided, closely surrounding the achenes, whitish, pale green, gray-green, or pale brown . . . . . . . . . . . . *24*
24. Pistillate scales usually purplish brown on each side of midrib; plants with underground elongated rootstocks; leaves gray-green or glaucous, stiffish, most of them at base of plant; plants of limestone glades and prairies . . . . . . . . . . . . . . . 51. C. MEADII
24. Without the above combination of characters; pistillate scales pale green or greenish-white; plants growing in tufts, without creeping rootstocks; leaves mostly dark or dull green, soft, scattered on the plant; plants of dry or wet woodland . . . . . . . . . . . . *25*
25. Dark purplish-red, reddish-brown, mahogany-red, or similar color at base of plant; staminate spike on usually a long peduncle (stalk) 1–4 cm. long standing well above the uppermost pistillate spike; scales of staminate spike usually tinged brown or purplish; beak of perigynium snout-like, long and narrow . . . . 59. C. GRACILESCENS
25. Green or pale to dull brown color at base of plant; staminate spike either sessile (stalkless) or on a very short peduncle, very close to and not much, if any, longer than the uppermost pistillate spike; scales of staminate spike usually greenish-white to whitish; beak of perigynium short and broad . . . . . . . . . . . 58. C. BLANDA
22. Beak of perigynium, if present, straight or slightly curved, not abruptly bent or turned to one side . *26*
26. Mature perigynia 10–20 mm. long . . . . . . . . . . . . . . . . . . . . *27*
27. Perigynia widely or nearly horizontally spreading at maturity, the beak 2–3 times as long as the inflated body of the perigynium; achene broader than long, nearly straight across (subtruncate) at summit . . . . . . . . . . . . . . . . . . . 108. C. GIGANTEA
27. Perigynia either closely overlapping and appressed to one another or spreading-ascending at maturity, but not horizontally widely spreading, the beak shorter or at most somewhat longer than the inflated body of the perigynium; achene longer than broad or as broad as long, tapering or narrowed above to the summit . . . . . . . . . . . . *28*
28. Pistillate spikes short-cylindric to broadly oblong-cylindric, mainly $1\frac{1}{2}$–2 times longer than thick; achene (inside perigynium) much longer than broad, the angles not prominently knobbed, the sides flat or nearly so . . . . . . . . . . . *29*
29. Common species throughout Missouri; culms (stems) bunched together in tufts; perigynia closely overlapping, appressed-ascending, 13–20 mm. long; scales of pistillate spikes long-tapering or awned at tip; leaf-blades mainly 5–15 mm. wide . . . . . . . . . . . . . . . . . . . . 106. C. LUPULINA
29. Found only in the southeastern lowland section; culms not bunched, arising solitary or a few from a creeping elongated rootstock; perigynia spreading-ascending, loosely arranged, 10–14 mm. long; scales of pistillate spikes blunt (obtuse) to short-pointed (acute); leaf-blades 2–6 mm. wide . . 105. C. LOUISIANICA
28. Pistillate spikes long-cylindric, mainly 2–4 times longer than thick; achenes about as broad as long, the angles prominently knobbed, the sides deeply sunken in center (concave) . . . . . . . . . . . . . . . . . . . 107. C. LUPULIFORMIS
26. Mature perigynia 2–9 mm. long (or, if 10 mm. or more, the pistillate spikes either 2–4 mm. thick or drooping or on slender thread-like peduncles) . . . . . . . . . . . . . *30*
30. Perigynium ends in a prominent 2-toothed or 2-pronged beak, the teeth of the beak 0.2–2.2 mm. long . . . . . . . . . . . . . . . . . . . . . . *31*
31. Pistillate spikes very slender, linear-cylindric, 10–15 times longer than thick, 2–4 mm. thick, the lower drooping or spreading; perigynia only faintly nerved at the base, otherwise nerveless; base of achene raised on a stalk 0.5–1.5 mm. long . . 71. C. DEBILIS
31. Pistillate spikes thick-cylindric to shortly oblong, mostly 2–5 times longer than thick, mainly 4–20 mm. (rarely 3) thick, erect to drooping; perigynia usually many-nerved or ribbed, rarely nerveless; base of achene without a stalk . . . . . . . . . *32*
32. Plants of limestone glades or upland limey soils; pistillate spikes 10–23 mm. long, 3–7.5 mm. thick; low plants with creeping rootstocks, the culms mainly 1–2 dm. tall; leaves glaucous, 3–15 cm. long, the short basal ones 3–10 cm. long; perigynia 3–4.5 mm. long . . . . . . . . . . . . . . . 62. C. MICRODONTA
32. Without the above combination of characters; plants of wet or swampy ground of valleys and lowlands; pistillate spikes mainly 25–105 mm. (rarely 10) long, mainly 4–13 mm. thick; taller plants, the mature ones usually 3–15 dm. tall; basal and other leaves generally much longer, 30–100 cm. long; perigynia mainly 4–9 mm. long . *33*

33. Main body of perigynium (excluding its beak) broadest in the upper half; all scales of pistillate spike with a long awn much longer than the length of the perigynium; perigynia 3.5–5 mm. long . 99. C. FRANKII
33. Main body of perigynium broadest in the lower half; scales of pistillate spike mostly shorter than or equaling the length of the perigynia (in some cases slightly exceeding); perigynia mainly 5–9 mm. long . . . . . . . . . . . . . . . . . . . . . . *34*
34. Staminate spike only 1 on each flowering culm; scales of pistillate flowers tipped with a rough (scabrous) awn which is equal to or longer than the main body of the scale . . . *35*
35. No red-purple or wine-purple color at base of plant; most of the mature perigynia, or at least the lower ones, pointing downward (reflexed); teeth of the beak 1.2–2.2 mm. long, outwardly arching . . . . . . . . . . . . . . . . . . 91. C. COMOSA
35. Red-purple or wine-purple color at base of plant; mature perigynia ascending or spreading, not reflexed; teeth of the beak 0.1–1 mm. long (averaging 0.5 mm. long). . *36*
36. Scales of staminate spike with a short slender tip or a short abrupt cusp at tip; all pistillate spikes erect or ascending, narrowed or tapering at base; perigynia erect or strongly ascending; culms (stems) not in clumps, but 1 or few from long-creeping underground rhizomes . . . . . . . . . . 93. C. SCHWEINITZII
36. Scales of staminate spike tipped with a rough (scabrous) awn; at least the lower mature pistillate spikes drooping or spreading, not narrowed or tapering at base, but about as thick at base as higher up; perigynia spreading or spreading-ascending; culms growing in clumps. . . . . . . . . . . . . . . . *37*
37. Perigynia with 12–20 nerves; teeth of perigynia spreading or ascending; mature perigynia 1.5–2 mm. thick, only barely inflated . . . . 90. C. HYSTRICINA
37. Perigynia with 7–10 nerves; teeth of perigynia erect or very close to one another; mature perigynia mostly 2.5–3 mm. (2–4) thick, conspicuously inflated . . . . . . . . . . . . . . . . . . . 92. C. LURIDA
34. Staminate spikes 2–6 on each flowering culm; scales of pistillate spike smooth or rough-tipped, with or without an awn . . . . . . . . . . . . . . . . . . *38*
38. Plants growing in large colonies from creeping underground rhizomes; perigynia of thick or firm texture, either smooth and without nerves or with many (12–25) conspicuous, delicately impressed or barely elevated nerves . . . . . . . . . *39*
39. Rarely with wine-red or wine-purple at base of plant; leaves glaucous or pale bluish-green; mature perigynia smooth and without nerves or with fine, impressed or barely elevated nerves . . . . . . . . . . . . . . 95. C. HYALINOLEPIS
39. Purple-red or wine-purple commonly at base of plant; leaves grass-green; mature perigynia with conspicuous strong ribs or nerves . . . . . . . . . . . . *40*
40. Teeth at end of perigynia 1–2.2 mm. long; the inner band of the middle and upper leaf-sheath conspicuously feather-veined (pinnate) at the summit; leaf-blades mainly 2–6 mm. (–8) wide . . . . . . . . 97. C. LAEVICONICA
40. Teeth at end of perigynia averaging 0.5 mm. (0.3–1) long; no feather-veined (pinnate) summit on the inner band of the leaf-sheath; leaf-blades mainly 8–15 mm. (rarely only 6) wide . . . . . . . . . . 94. C. LACUSTRIS
38. Plants growing in dense clumps, not in large colonies; mature perigynia thin or papery, 7–10-nerved . . . . . . . . . . . . . . . 102. C. VESICARIA var. MONILE
30. Tip of perigynium ends abruptly and cut off straight across or at an angle (obliquely), either without teeth or with only a slight notch without conspicuous projections . . . . . . . . *41*
41. At least the lower mature pistillate spikes spreading, nodding, or drooping from very slender peduncles (stalks supporting inflorescences) . . . . . . . . . . . . . . . . . . *42*
42. Pistillate spikes many-flowered (10–60 or more); perigynia tapering into an obvious or prominent beak 1–2 mm. long . . . . . . . . . . . . . . . . . . . *43*
43. Red-purple or wine-purple color at base of plant; pistillate spikes very slender, linear-cylindric, 10–15 times longer than thick, 2–4 mm. thick; perigynia 6–10 mm. long, only faintly nerved at the base, otherwise nerveless; base of achene raised on a stalk 0.5–1.5 mm. long . . . . . . . . . . . . . . . . . . . 71. C. DEBILIS
43. Shades of brown at base of plant; pistillate spikes thick-cylindric, 3–5 times longer than thick, 5–10 mm. thick; perigynia 4–6 mm. long, strongly nerved; base of achene without a stalk . . . . . . . . . . . . . . . . . . . . . *44*
44. Scales of pistillate spike with a rough (scabrous) awn projecting at the tip; mature perigynia squarrose (outwardly spreading or curving from the center), broadest at or above the middle . . . . . . . . . . . . . . . . 81. C. JOORI

44. Scales of pistillate spike with a smooth sharp-pointed (acuminate) tip; mature perigynia erect or straight, not curving outward or squarrose, broadest in the lower half . 72. C. CHEROKEENSIS

42. Pistillate spikes mainly with only 3–9 flowers; perigynia without any beak or the beak, if present, less than 1 mm. long . . . . . . . . . . . . . . . . . . . . . . . *45*

45. Leaf-blades 1.5–5 mm. wide, dull or dark green, not glaucous or gray-green; pistillate spikes without any staminate (male) flowers at the base . . . . . . . . . . 55. C. DIGITALIS

45. Leaf-blades 6–15 mm. wide, glaucous or gray-green; pistillate spikes with 1–2 staminate (male) flowers or empty scales at base . . . . . . . . . . . . . . 56. C. LAXICULMIS

41. Mature pistillate spikes erect or ascending . . . . . . . . . . . . . . . . . . . . *46*

46. Lowermost basal leaves mainly 10–17 mm. wide (less often 8), forming an evergreen spreading rosette; base of leaf-tufts, culms, and sheaths of bracts of the infloresence dark red- or maroon-purple . . . . . . . . . . . . . . . . . . . . . . . . 54. C. CAREYANA

46. Without the above combination of characters . . . . . . . . . . . . . . . . *47*

47. Pistillate spikes with only 1–9 perigynia . . . . . . . . . . . . . . . . *48*

48. Plants of limestone glades and prairies, rarely in wet meadows, with horizontally creeping underground slender rootstocks or superficial stolons; perigynia glaucous (whitish) or pale greenish, with several strong ribs at maturity; scales of pistillate spikes with usually purplish-brown margins . . . . . . . . . . . . . . *49*

49. Common species throughout Missouri on limestone glades and prairies; with horizontally creeping underground slender rootstocks; leaves glaucous, gray- or blue-green, stiffish; lower leaf-sheaths bearing leaf-blades; lower pistillate spike 5–10 mm. thick; perigynia usually in 6 rows in the spike, mostly overlapping, 2–2.5 mm. wide . . . . . . . . . . . . . . . . . . . 51. C. MEADII

49. Rare species, known only from swampy meadows in Howell County; with superficial stolons at the base of plant; leaves green, thin; lower leaf-sheaths mostly without leaf-blades; lower pistillate spikes 3–5 mm. thick; perigynia alternate-flowered, in 2 rows in the spike, at least the lower not overlapping, 1.5–2 mm. wide . . . . . . . . . . . . . . . . . . . . . 53. C. WOODII

48. Plants growing in clumps or bunches, without horizontally creeping rhizomes or stolons; perigynia dull or dark green or brown, not glaucous, either with many fine lines (striate) or impressed nerves, but not with strong elevated ribs; scales of pistillate spikes mostly pale green and white to pale or dull brown, rarely tinged with red . . . *50*

50. Perigynia more or less narrowed or tapering from the middle to the base. . . . *51*

51. Scales of pistillate spike acute or only short-pointed, smooth; perigynia with distinct elevated nerves; peduncles (supporting stalk) of at least the lower pistillate spikes much elongated, longer than their own length as well as their leaf-sheath; base of plant dull green, drab, or brownish . . . 55. C. DIGITALIS

51. Most of the scales of the pistillate spike with long rough (scabrous) awns; perigynia with many impressed nerves, producing a vertically wrinkled or lined surface; peduncles of the lower pistillate spikes slightly longer than the leaf-sheath; base of plant dark wine-red purple or brown-purple, sometimes brown or greenish-white . . . . . . . . . . . . . . . . *52*

52. Base of plant always dark purplish-red or reddish-purple; leaf-blades mainly 2–3 mm. (–5) wide; perigynia very loosely alternately scattered, the tip of the lowest perigynia in the spikelets barely or not reaching the base of the next higher perigynia on the same side of the rachis; perigynia abruptly constricted at the tip into a short beak, definitely 3-angled and nearly completely filled to the top by the achene; lowest leaf-like bract (at base of lowest pistillate spike) usually shorter than the uppermost spikes; lowest leaves of plant usually much shorter than the culms; staminate spike conspicuous, usually on a long peduncle (stalk) often 2–3 or more cm. long and projecting usually 2–5 cm. above the uppermost pistillate spike . . . . . . . . . . . . . . 63. C. OLIGOCARPA

52. Base of plant variable in color, some with greenish-white or brown, others with dark brown-purple or reddish-purple; leaf-blades mainly 3–10 mm. wide; perigynia more or less closely overlapping, in at least the uppermost pistillate spike with the tip of the perigynium usually reaching to and above the base of the next higher perigynium on the same side of the rachis; perigynia usually more gradually tapered, curving, or rounded

at the tip (sometimes more abruptly constricted as in *C. oligocarpa*, in which case use other key characters), the sides mostly curved or rounded (sometimes more definitely angled, in which case use other key characters), mainly empty inside above and not filled by the achene, much longer than the achene; lowest leaf-like bract usually equaling or longer than the uppermost spikes; lowest leaves of plant nearly equaling or only slightly shorter than the culms; staminate spike nearly or quite sessile, usually inconspicuous and projecting usually only 0.5–2 cm. beyond uppermost pistillate spike . . . . 66. C. AMPHIBOLA

50. Perigynia round or broadened at the base . . . . . . . . . . . . . . . *53*

53. Leaves firm, more or less glaucous (silvery-white) or gray-green, or pale green; sheaths of the bracts at the base of the spikes enlarged and loose at summit; scales of pistillate spike smooth-tipped with a long point or short awn, usually shorter than or at most equaling the length of the mature perigynia . . . . . . . . . . . . 67. C. FLACCOSPERMA

53. Leaves thin, grass-green or dull or dark green, not glaucous; sheaths of the bracts at the base of spikes tight and narrowly cylindrical, not enlarged at summit; scales of pistillate spikes with a conspicuous long and rough (scabrous) awn, usually longer than the perigynia . . . . . . . . . . . . . . . . . . . . . . 66. C. AMPHIBOLA

47. Pistillate spikes usually with 10–60 perigynia . . . . . . . . . . . . . . . . *54*

54. Leaf-like bract of the lowest pistillate spike without a sheath or nearly without one; base of plant conspicuously chestnut brown; scales of pistillate spike with a rough (scabrous) awn; mature perigynia squarrose (outwardly spreading or curving from the center), broadest at or above the middle . . . . . . . . . . . . . . . . . . . . . . . 81. C. JOORI

54. Without the above combination of characters; leaf-like bract of the lowest pistillate spike with an elongated tubular closed sheath; base of plant silvery, green, brown, purple, or reddish-tinged . . . . . . . . . . . . . . . . . . . . . . . . . . *55*

55. Staminate spike very conspicuous, usually on a prominent long peduncle (stalk) much longer than the uppermost pistillate spike; plants of open limestone glades, and wet or dry prairies, rarely of wet meadows . . . . . . . . . . . . . . . . . . . *56*

56. Perigynia glaucous (whitish) or pale greenish, with several strong ribs at maturity, broadest in the upper half, narrowed toward the base; scales of pistillate spikes usually with purplish-brown margins; perigynia closely enveloping the achene . . . . . . *57*

57. Rare plant of swampy meadows in Howell County; with superficial stolons at the base of plant; lower leaf-sheaths mostly without leaf-blades; perigynia alternate-flowered, in 2 rows in the spike, at least the lower not overlapping . . 53. C. WOODII

57. Common plants of limestone glades and prairies (except for the rare *C. tetanica*); with horizontally creeping underground slender rootstocks; lower leaf-sheaths bearing leaf-blades; perigynia in 3 or 6 rows in the spike, more crowded and overlapping . . . . . . . . . . . . . . . . . . . . . . . . *58*

58. Common plant of limestone glades and prairies; leaves glaucous, gray- or blue-green, stiffish; lower pistillate spike 5–10 mm. thick; perigynia usually in 6 rows in the spike, 2–2.5 mm. wide, more or less spreading and inflated when mature, with an abruptly narrowed tip, somewhat turned or curved to one side . . . . . . . . . . . . . . . . . . . . 51. C. MEADII

58. Rare, known only from dry prairie in Atchison County, northwestern Missouri; leaves green, rather thin or membranous; lower pistillate spike 3–4.5 mm. thick; perigynia usually in 3 rows in the spike, 1.5–2.2 mm. wide, appressed or straight and not inflated when mature; without any curvature at the tip . 52. C. TETANICA

56. Perigynia dull or dark green or brown, not glaucous, with many vertically lined impressed nerves or nearly nerveless or few-nerved, broadest at the base or in the lower half, rounded at the base; scales of pistillate spikes either pale green and white or pale or dull brown with white or green, rarely tinged purplish or reddish; perigynia loosely enveloping the achene . . . . . . . . . . . . . . . . . . *59*

59. Plants of limestone glades in the Ozark region; culms solitary, arising from horizontally creeping underground rootstocks; leaf-blades glaucous, gray- or blue-green, stiffish; perigynia with a few raised nerves, often resinous-dotted . 61. C. CRAWEI

59. Rare plant of dry prairie in Callaway and Phelps counties; culms several in a dense clump; leaf-blades deep green, not stiffish; perigynia with many vertically lined impressed nerves, not resinous-dotted . . . . . . . . 65. C. CONOIDEA

55. Staminate spikes either sessile or shortly stalked, close to, shorter, and almost hidden by uppermost pistillate spike (in some variations of *C. amphibola* the staminate spike is more prominently raised on a longer stalk exceeding the uppermost pistillate spike); shade-loving plants typically of rich, moist, dry or rocky woodlands, or on moist shaded bluffs and banks along streams . . *60*

60. Scales of pistillate spikes usually only 1/3–1/2 length of perigynia, either short-pointed or with a short inconspicuous smooth awn 1 mm. or less long; leaves often or usually glaucous (whitish or silvery) and firm or thickish . . . . . . . . . . . . . . . . *61*

61. Perigynia 2–4 mm. long, with strong elevated nerves, the tip of the mature perigynium bent over or curving outward (squarrose), or straight in one variety; pistillate spikes mainly 3–5 mm. thick; sheaths of leaf-like bracts tight and not enlarged at the summit . . . . . . . . . . . . . . . . . . . . 60. C. GRANULARIS

61. Perigynia mostly 3–6 mm. long, with fine slender nerves, the tip of the perigynium straight; pistillate spikes 5–8 mm. thick; sheaths of leaf-like bracts enlarged and loosened at the summit . . . . . . . . . . . . . . . . 67. C. FLACCOSPERMA

60. Most of the scales of pistillate spikes usually equal or longer than the perigynia, with a conspicuous rough (scabrous) awn usually 1–4 mm. long (scales sometimes shorter than perigynium, but still prominently rough-awned); leaves usually dull or dark green, more rarely glaucous . . . . . . . . . . . . . . . . . . 66. C. AMPHIBOLA

12. Styles 2; achenes (inside perigynia) 2-sided or 2-angled . . . . . . . . . . . . . *62*

62. Uppermost spike completely staminate (male); lower or lowermost spikes completely pistillate (female) or some of the spikes all staminate and some all pistillate . . . . . . . . *63*

63. Inflorescence a more or less continuous cluster 15–60 mm. long, of very short crowded spikes, each of the spikes only slightly longer than broad and only 4–12 mm. long, often not distinguishable from one another . . . . . . . . . . . . . . . . . . *64*

64. Upper leaf-sheaths green and finely lined (striate) on the inner (ventral) side; beak of perigynium 1/4–1/3 as long as main body of perigynium . . . . . . 3. C. SARTWELLII

64. Upper leaf-sheaths transparent (hyaline) or brown, not finely lined on the inner (ventral) side; beak of perigynium 1/2–5/8 as long as main body of perigynium . . . *65*

65. Culm smooth below the spikes . . . . . . . . . . . . . 2. C. DOUGLASII

65. Culm rough below the spikes . . . . . . . . . . . . . 1. C. PRAEGRACILIS

63. Inflorescence of distinctly separated spikes which are several to many times (3–20) longer than broad . . . . . . . . . . . . . . . . . . . . . . . . *66*

66. At least the lower scales of pistillate spikes with a long bristle-like awn mostly 2–4 times as long as the perigynia (sometimes only slightly longer than perigynia); pistillate spikes usually nodding or arching; tip of perigynium not 2-toothed; face of perigynium without 2–4 ribs . . . . . . . . . . . . . . . . . . . . 89. C. CRINITA

66. Without the above combination of characters; scales of pistillate spikes blunt (obtuse) or acute to sharp pointed (cuspidate), rarely with a short awn, usually shorter than or somewhat longer than the perigynia; pistillate spikes erect to drooping or arching; tip of perigynium either entire or 2-toothed; face of perigynium nerveless or with 2–4 strong ribs . . . . . . . . . . . . . . . . . . . . . . . . *67*

67. Beak of perigynia with 2 small teeth at the summit; in addition to the rib on each side, the perigynia have 2–4 strong nerves on each face . . . . 84. C. NEBRASKENSIS

67. Beak of perigynia entire or cut off flat on the top or at least without 2 small teeth at the summit; faces of perigynia completely nerveless or with 1 or 2 faint nerves if any . . . . . . . . . . . . . . . . . . . . . . . . *68*

68. At least the lowest mature pistillate spikes widely spreading, arching, or drooping; tip of mature perigynium twisted or bent and rather conspicuous when dry . . . . . . . . . . . . . . . . . . . 88. C. TORTA

68. Pistillate spikes erect or definitely ascending; tip of mature perigynium when dry not twisted, but more or less flat and straight . . . . . . . . . *69*

69. Leaves usually glaucous (silvery or gray mixed with green); plants usually in small tufts of few culms; flowering or fruiting (fertile) culms surrounded at the base by the dried-up leaves of the previous year; flowering or fruiting culms shorter than or at most as long as the leaves; lowest leaves of the current year possessing leaf-blades (phyllopodic); mature perigynia mainly 2.7–3.1 mm. long . . . . . . . . 83. C. AQUATILIS var. ALTIOR

69. Leaves grass- or dark green, not glaucous; plants in dense large clumps, beds, or patches; flowering or fruiting (fertile) culms arising laterally (at

the side), not surrounded at the base by the dried-up leaves of the previous year; flowering or fruiting culms usually longer or taller than the clump of leaves; lowest leaves of the current year with sheaths only, without leaf-blades (aphyllopodic); mature perigynia mainly 2–2.7 mm. long . *70*

70. Plants of the wet limey meadows of the Ozark section; at least some of the lowest leaf-sheaths consisting of thread-like or fiber-like parts on the inner (ventral) side. . 87. C. STRICTA var. STRICTIOR

70. Plants along banks of streams throughout Missouri, or prairies or along railroads of northern Missouri; lowest leaf-sheaths without thread-like or fiber-like parts . . . . . *71*

71. Plants mostly of prairie habitats along railroads of northern Missouri; mature perigynia inflated, each side convex (bulging at center); scales of pistillate spike loose, somewhat spreading away from the perigynia at maturity, sharp-pointed or slender-tipped; ligule at summit of leaf-sheath usually slightly longer than wide; pistillate spikes 10–40 mm. (averaging 12–27) long . . . . . . . . . . . 85. C. HAYDENII

71. Plants most commonly found along banks of streams throughout Missouri; mature perigynia not inflated, rather closely enveloping the achene; scales of pistillate spike appressed-ascending, close to or pressed against the perigynia, obtuse (blunt) to acute; ligule at summit of leaf-sheath usually wider than long; pistillate spikes 20–100 mm. long . . . . . . . . . . . . . . . . . . . . . . . 86. C. EMORYI

62. All the spikes alike or nearly so and/or with both staminate (male) and pistillate (female) flowers in the same spike (where stamens have fallen or disappeared from mature or old spikes, the location of the staminate flowers may usually be detected by the empty scales at base or tip of spike) . . . *72*

72. Plants with long black or brown creeping rootstocks; the culms (stems) arising solitary or few together . . . . . . . . . . . . . . . . . . . . . . . . . . . *73*

73. Upper leaf-sheaths green and finely lined (striate) on the inner (ventral) side; beak of perigynium 1/4–1/3 as long as main body of perigynium . . . . . . . . 3. C. SARTWELLII

73. Upper leaf-sheaths transparent (hyaline) or brown, not finely lined on the inner (ventral) side; beak of perigynium 1/2–5/8 as long as main body of perigynium . . . . . . . *74*

74. Culm smooth below the spikes . . . . . . . . . . . . . . 2. C. DOUGLASII

74. Culm rough below the spikes . . . . . . . . . . . . . . 1. C. PRAEGRACILIS

72. Plants growing in close tufts or clumps, the leafy shoots and those with flowers or fruits arising from a non-creeping or compact base . . . . . . . . . . . . . . . . . . . *75*

75. Scales of the pistillate flowers (located by the perigynia) rounded or obtuse (blunt) at tip, not tapering or slender-pointed . . . . . . . . . . . . . . . . . . . *76*

76. Plants usually of wooded situations, throughout Missouri; lower part or base of mature perigynium corky, spongy, or soft-thickened as contrasted with the firmer upper part; some or all of the spikes with the staminate (male) flowers at the top, pistillate (female) flowers at the base and middle of the same spike (where stamens have fallen or disappeared from mature or old spikes, the location of the staminate flowers may usually be detected by the empty or closely overlapping or bunched scales at the tip or base of the individual spikes) . . . . . . . . . . . . . . . . . . . . . *77*

77. Stigmas long, slender, straight, not twisted; perigynium more or less gradually tapering into a beak . . . . . . . . . . . . . . . . . . . 6. C. ROSEA

77. Stigmas shorter, stout, twisted; perigynium more abruptly contracted or narrowed into a beak . . . . . . . . . . . . . . . . . . . 5. C. CONVOLUTA

76. Plants typically of swampy limey meadows or on dripping limestone ledges at the base of bluffs, only in the Ozark section; mature perigynia firm throughout, not corky nor soft spongy-thickened at base; at least the topmost (terminal) spike with the staminate (male) flowers at the base . . . . . . . . . . . . . 23. C. INTERIOR

75. Scales of the pistillate flowers (with the perigynium) short- to long-pointed or awned at the tip, at least tapering or narrowed at the tip . . . . . . . . . . . . . . *78*

78. Staminate (male) flowers located at the tip of some or all of the spikes (where stamens have fallen or disappeared from mature or old spikes, the location of the staminate flowers may usually be detected by the empty or closely overlapping or bunched scales at the top of the individual spikes) . . . . . . . . . . . . . . . . . . *79*

79. Lower part or base of mature perigynium corky, spongy, or soft-thickened as contrasted with the firmer upper part (test the perigynium with a dissecting needle or pin) . . . . . . . . . . . . . . . . . . . . . . . *80*

80. Margins of the perigynia, including their beak, smooth (use 10 × magnifying lens) . . . . . . . . . . . . . . . . . . . . . . . *81*

81. Common throughout the Ozark section, usually in dry or rocky woodland or on ledges of bluffs; lower 1/2 to 2/3 of mature perigynia spongy and distinctly nerved or with vertical lines . . . . . . . . . . . . . . . . 4a. C. RETROFLEXA var. RETROFLEXA

81. Known only from wet woodland in Dunklin County, southeastern Missouri; only the lowest 1/4 or 1/5 part of the mature perigynia spongy, not nerved on the inner side . 4b. C. RETROFLEXA var. TEXENSIS

80. Upper margins of body or beak of perigynium minutely roughened or serrulate (saw-toothed) (use magnifying lens of 10 ×) . . . . . . . . . . . . . . . . . . *82*

82. Leaf-blades 1–3 (–4) mm. wide; lowest bract (at base) of inflorescence greatly prolonged and leaf-like, 50–200 mm. long, 1–2 mm. wide; inflorescence 6–8 mm. thick, simple and unbranched, all the spikes arising directly from the main axis (rachis) with 1 spike at each node of the axis; leaf-sheaths tight and closely enveloping the culms; culms firm and slender . . . . . . . . . . . . . . . . . . . . . . 9. C. ARKANSANA

82. Leaf-blades mainly 3–17 mm. (rarely 2.5) wide; inflorescence without a lowest bract or lowest bract not conspicuous or elongated; inflorescence 7–40 mm. thick, compound, with 2 or more spikes placed on at least the lowest lateral (side) branch from the main axis; leaf-sheaths loose, not closely enveloping the culms, the inner thinner (ventral) side (the thinner side without vertical ribs or stripes) either with tiny brown- or purple-red dots or cross-wrinkled (seen especially when old or dried); culms rather soft and flattened . . . *83*

83. Beak of perigynium shorter than the length of the main body of the perigynium; the flat inner face of the perigynium not nerved or with short nerves only at base . . . *84*

84. Inner (ventral) side (the thinner side without vertical ribs or stripes) of leaf-sheath cross-wrinkled; leaf-blades mainly 5–10 mm. wide; scales of pistillate flowers whitish . . . . . . . . . . . . . . . . . . . 22. C. CONJUNCTA

84. Inner thinner (ventral) side of leaf-sheath not cross-wrinkled; leaf-blades mainly 2.5–4.5 (–5) mm. wide; scales of pistillate flowers brownish or tinged brown . 18. C. OKLAHOMENSIS

83. Beak of perigynium as long as or longer than length of the main body of perigynium; the flat inner face of the perigynium either strongly nerved, or if not nerved, the perigynium very long-beaked or of a lanceolate shape and tapering into the beak . . . . . . . . . . . . . . . . . . . . . . . . . *85*

85. Beak of perigynium 2–3 times the length of the main body of perigynium; mature perigynia abruptly enlarged at base and noticeably wider there, not nerved or faintly nerved on the flat inner side; inner (ventral) side of leaf-sheath red-purple dotted but not cross-wrinkled . . . . . . . . . . . . 21. C. CRUS-CORVI

85. Beak of perigynium about as long as or 1½ (rarely 2) times length of main body of perigynium; mature perigynium with the base continuous with the rest of the body, tapering gradually, not abruptly enlarged at base, strongly nerved on the flat inner side; inner thinner (ventral) side of the leaf-sheath either cross-wrinkled, or (as in *C. laevivaginata*) not cross-wrinkled . . . . . . . . . . . . . . . . . *86*

86. Leaf-sheaths with a thin summit which is prolonged as a loose membranous projection beyond the base of the leaf-blade; inner thinner (ventral) side of the leaf-sheath usually cross-wrinkled . . . . . . . . . . . 19. C. STIPATA

86. Leaf-sheaths with a thick-margined summit which is sunken in the middle (concave) or nearly straight-edged (subtruncate); inner thinner (ventral) side of leaf-sheath not or rarely cross-wrinkled . . . . . . . 20. C. LAEVIVAGINATA

79. Perigynia not corky, spongy, or softer-thickened at the base, but firmer throughout from base upward . . . . . . . . . . . . . . . . . . . . . . . . *87*

87. Plants of sink-hole ponds; inflorescence compound, with conspicuous side branches, some of the lower branches up to 4 cm. long; perigynium abruptly contracted into a beak, broadest above the middle and tapering at the base, 2–2.6 mm. long, nearly as broad as long; 1 or both sides of perigynium bulged out in the middle, olive-brown to black . . . . . 17. C. DECOMPOSITA

87. Plants not of sink-hole ponds, but if so, then without the above combination; inflorescence without conspicuous side branches or these not usually elongated; perigynia broadest at the base or at the middle; perigynium flat on at least one side, greenish, straw-colored, yellow, or brown . . . . . . . . . . . . . . . . . . . . . . . . *88*

88. Leaf-sheaths loose, mottled or speckled with green and white and here and there marked with dark green or brown transverse or cross lines or thickenings (septate-nodulose) between the vertical or long ribs or lines of the dorsal (back) side . . . . . . . . *89*

89. Spikes crowded and dense in a more or less uninterrupted ovoid or cylindric head, 1–4 cm. long, the spikes all touching or with little space between the lowest spike and the others; perigynia pale brown or drab; beak of perigynium 1/4–1/3 the length of the ovate or suborbicular body of the perigynium; culms rough and minutely toothed on the angles . . . . . 11. C. GRAVIDA

89. Spikes in a more elongated interrupted cylindrical inflorescence, 2.5–15 cm. long, at least the lower spikes distant and well-separated from the others, or, if all spikes close together, then the culms usually smooth on the angles; perigynia green; beak of perigynium 1/3–1/2 the length of the ovate body of the perigynium . . . . . . . . . . . . . . . . . 90

90. Inflorescence 2–10 cm. long, the lower spikes of inflorescence usually distant and well-separated, sometimes as much as 2–3 cm. from those above; summit of inner thinner side of leaf-sheaths with a truncate (as if cut straight across) edge, the inner thinner side of lower leaf-sheaths usually cross-wrinkled and delicate; flowering culms minutely toothed on angles; leaf-blades 4.5–10 mm. long; stigmas short and stout. . . . 13. C. SPARGANIOIDES

90. Inflorescence 2–5 cm. long, all the spikes close together; summit of inner thinner side of leaf-sheaths with a concave or sunken center, the inner side of the lower leaf-sheaths usually firm and not cross-wrinkled; flowering culms smooth on the angles; leaf-blades 3–6 mm. wide; stigmas long and slender . . . . . . . . . . 12. C. AGGREGATA

88. Leaf-sheaths tight, closely enveloping the culms, not marked with dark green or brown transverse or cross lines or thickenings (septate-nodulose) between the vertical or long ribs or lines of the dorsal (back) side . . . . . . . . . . . . . . . . . . . . . . . . 91

91. Inflorescence a dense, more or less solid, single, unbranched, head-like mass of spikes, mostly 4–25 mm. long, usually without any interrupted areas or spaces between the spikes . . . . 92

92. Perigynia widest at the broadly rounded or truncate (as if cut straight across) base; scales of pistillate flowers short-pointed; leaf-sheaths not thickened at summit . 8. C. LEAVENWORTHII

92. Perigynia widest near or below the middle, more or less tapering to the base; scales of pistillate flowers long-pointed or awned; leaf-sheaths thickened at summit. . . . . . 93

93. Perigynia 1.5–1.9 mm. wide, 2–3.3 mm. long; scales much shorter than the bodies of the perigynia and mostly concealed by the perigynia. . . . . . 7a. C. CEPHALOPHORA var. CEPHALOPHORA

93. Perigynia 2–2.5 mm. wide, 3–3.5 mm. long; scales longer than to slightly shorter than the bodies of the perigynia, the body of the scale nearly or quite as long as the body of the perigynium. . . . . . . . . . 7b. C. CEPHALOPHORA var. MESOCHOREA

91. Inflorescence more elongated and more interrupted, varying from linear-oblong to oblong, or cylindric to narrowly ovoid, mostly 15–100 mm. long, usually with interrupted areas or spaces between the spikes . . . . . . . . . . . . . . . . . . . . . 94

94. Inner (thinner) side of leaf-sheath usually cross-wrinkled, at least in the mature stage of growth; perigynia 1.7–3.5 mm. long; inflorescence compound, with 2 or more spikes placed on at least the lowest lateral (side) branch from the main axis (the lateral branches are often closely pressed against the main axis, making it necessary to examine them carefully) . 95

95. Known only from swampy areas of southeastern Missouri; perigynia 2.5–3 mm. wide, suborbicular (nearly round) to reniform (kidney-shaped) . . . . 16. C. TRIANGULARIS

95. Throughout Missouri; perigynia 1–2.4 mm. wide, usually longer than broad, ovate to broadly ovate-lanceolate . . . . . . . . . . . . . . . . . 96

96. Beak of perigynium about equaling main body of perigynium; perigynium 1–1.8 mm. wide; leaves usually longer than the flowering or fruiting culms . . . . 14. C. VULPINOIDEA

96. Beak of perigynium much shorter than the main body of perigynium; perigynium 1.5–2.4 mm. wide; leaves usually shorter than the flowering or fruiting culms 15. C. ANNECTENS and var. XANTHOCARPA

94. Leaf-sheath usually not cross-wrinkled; perigynia 3–4.5 mm. long; inflorescence simple, all the spikes arising directly from the main axis (rachis) with 1 spike at each node of the axis, not on any lateral (side) branches from the axis. . . . . . . . . . . 97

97. Perigynia 3–3.5 (–4) mm. long, 2–2.5 mm. wide, usually spreading at maturity; scales shorter or equal to and narrower than the main body of the perigynium, short-awned or pointed; bracts of inflorescence narrow at the base, not broadly dilated . . 10. C. MUHLENBERGII

97. Perigynia 3.5–4.5 mm. long, 2.5–3 mm. wide, mostly ascending at maturity; at least the lower scales longer and broader than the perigynia, long-awned; bracts of inflorescence broad and dilated at the base . . . . 10c. C. MUHLENBERGII var. AUSTRALIS

78. Staminate (male) flowers located at the base of some or all of the spikes (where stamens have fallen or disappeared from mature or old spikes, the location of the staminate flowers may usually be detected by the empty or closely overlapping or bunched scales at the base of the individual spikes) . . . . . . . . . . . . . . . . . . . . . . . . . . . . *98*

98. Spikes 15–25 mm. long, fusiform and narrowed at both ends, 3–4 times as long as wide; perigynia 6.5–10 mm. long, lanceolate . . . . . . . . . . . . 24. C. MUSKINGUMENSIS

98. Without the above combination; spikes either shorter or, if as long, of a different shape; perigynia either shorter and/or of a different shape . . . . . . . . . . . . . . *99*

99. Perigynia lanceolate or ovate-lanceolate, 3–4 times as long as wide . . . . . . . . . *100*

100. Sterile leafy culms abundant and strongly developed with spreading leaf-blades, these not clustered at the tip; main leaf-blades 2.5–8 mm. wide; mature perigynia 1–1.5 mm. wide, the wing rather abruptly narrowed just below the middle part; leaf-sheaths loose . *101*

101. Perigynia erect with strongly ascending, closely appressed tips; spikes crowded and overlapping . . . . . . . . . . . . . . . . . 26. C. TRIBULOIDES

101. Perigynia with spreading, slightly recurved, or loosely ascending tips, not closely appressed to one another; spikes scattered and separated in an elongated inflorescence . . . . . . . . . . . . . . . . . 27. C. PROJECTA

100. Sterile leafy culms few with ascending leafblades clustered at the tip; main leaf-blades 1–3 mm. wide; mature perigynia 1.2–2.6 mm. wide, the wing continuous to the base; leaf-sheaths closely enveloping the culms . . . . . . . . . . 25. C. SCOPARIA

99. Perigynia ovate or obovate or suborbicular, nearly as broad as long or, when longer than broad, then $1\frac{1}{4}$–$2\frac{1}{2}$ times as long as wide . . . . . . . . . . . . . . . . *102*

102. Tips of mature perigynia widely spreading or recurved; spikes globose (ball-shaped), usually crowded; sterile leafy culms abundant and strongly developed with spreading leaf-blades, these not clustered at the tip; main leaf-blades 3–7 mm. wide; leaf-sheaths loose; wing of perigynium abruptly narrowed below the middle . . . 28. C. CRISTATELLA

102. Without the above combination of characters . . . . . . . . . . . . . *103*

103. Mature perigynia less than 2 mm. wide and mostly less than 4 mm. long . . . *104*

104. Main body of perigynium broadest in the upper half or toward the summit, nearly orbicular (rounded) to broadly obovate . . . . 32. C. ALBOLUTESCENS

104. Main body of perigynium broadest in the lower half at the base or near the middle, not obovate . . . . . . . . . . . . . . . . *105*

105. Sterile leafy culms present and frequent with leaves clustered at summit; principal leaf-blades 4–7, mainly 3.5–6.5 mm. (rarely 2) wide; leaf-sheaths loosely enveloping the culms; spikes mostly close together and crowded (except scattered and distant in f. *perlonga*) 29. C. NORMALIS

105. No sterile leafy culms present; principal leaf-blades 3–5, mainly 0.5–3.5 mm. wide; leaf-sheaths tight, closely enveloping the culms; spikes mostly scattered and distant or well-separated from one another . *106*

106. Body of perigynium suborbicular (nearly round), rounded above and rather abruptly continuing into the beak, usually 3/4 as wide as long or nearly as wide as long; spikes conspicuously tapering at the base, 6–10 mm. long; leaf-blades up to 3.5 mm. wide; inflorescence arching to nearly straight, but not nodding 31. C. FESTUCACEA

106. Body of perigynium narrowly ovate, gradually tapering into the beak, about 1/2 as wide as long; spikes mostly rounded or slightly tapering at the base, 4–8 mm. long; leaf-blades up to 2.5 mm. wide; inflorescence usually nodding from above the lowest spike 30. C. TENERA

103. Mature perigynia more than 2 mm. wide and usually more than 4 mm. long (sometimes only 3.5 mm. long) . . . . . . . . . . . . . . . . *107*

107. Main body of perigynium broadest in the upper half or toward the summit, broadly obovate to nearly orbicular. . . . . . . . . . *108*

108. Plants of sink-hole ponds, and of swamps in southeastern Missouri, forming large dense clumps; scales of pistillate flowers with a slender or short birstle-like tip; perigynia 2.5–4 mm. wide; achenes constricted to a slender stalk-like base; leaf-blades 2.5–5.5 mm. wide . 37. C. ALATA

442 Scleria verticillata

443 Carex praegracilis

444 Carex Douglasii

108. Plants of other habitats, forming small tufts or clumps; scales of pistillate flowers short- to slender-pointed (acute to acuminate); perigynia 1.5–2.5 (–3.5) mm. wide; achenes without a constricted stalk-like base or with a slight stalked base; leaf-blades 2–3.5 (–4) mm. wide *109*

109. Perigynia pale green to dull brown, nerved on both faces, 1.5–2.5 mm. wide; scales of pistillate flowers pale, whitish or white with green. . . . . . . . 32. C. ALBOLUTESCENS

109. Perigynia brown, or brown and green, not nerved on the inner (ventral) face or rarely faintly nerved, 2.5–3.5 mm. wide; scales of pistillate flowers usually rusty- or yellow-brown with green midnerve . . . . . . . . . . . . . . . . . . 34. C. BREVIOR

107. Main body of perigynium broadest in the lower half at the base or near the middle, not obovate . *110*

110. Mature perigynia with erect or appressed tip . . . . . . . . . . . . . . . *111*

111. Spikes rounded or obtuse at summit . . . . . . . . . . . . . 34. C. BREVIOR

111. Spikes tapering to a rather pointed summit . . . . . . . . . . . . . . . *112*

112. Plants of swampy limey meadows and at the base of wet limestone bluffs along creeks, only in the Ozark section; heads tawny or rusty-brown; perigynia of firm texture, obviously thicker than their scales and well expanded across and covering the achene, the achene not showing its impression, the inner face of the perigynium nerveless or only faintly nerved . . . . . . . . . . . . 33. C. SUBERECTA

112. Plants of prairies and meadows and other open habitats, throughout Missouri; heads straw-colored to yellow-brown; perigynia of very thin texture, of about the same as that of their scales, only barely stretched over the achene within, the achene easily showing its outline, the inner face of the perigynium finely nerved . 25. C. SCOPARIA

110. Mature perigynia with spreading or ascending loose or free tips . . . . . . . . . . *113*

113. Mature perigynium flat, thin and nearly transparent, mainly 5.5–7.5 mm. long, 2.7–4.8 mm. wide, buff- or straw-colored, nerved on both faces; pistillate scales pale brown or straw-colored . . . . . . . . . . . . . . . . . . . 36. C. BICKNELLII

113. Without the above combination of characters; mature perigynia usually plano-convex (one side flat, the other side convex or slightly raised in the middle), thicker, firmer, subcoriaceous (leathery or hard), not transparent, pale green, brown, or green and brown; pistillate scales pale brown, yellow-brown with green, or gray-green . . . . *114*

114. Scales of pistillate flowers tipped with a long point or awn; leaf-sheaths green, strongly nerved on the thinner inner (ventral) side nearly to the summit; plants of sink-hole ponds in the Ozark section . . . . . . . . . . 38. C. STRAMINEA

114. Scales of pistillate flowers short- or slender-pointed or blunt, not awned; leaf-sheaths mostly white and transparent on the thinner inner (ventral) side; plants found throughout Missouri in prairie swales, along railroad tracks, meadows, and other open habitats . . . . . . . . . . . . . . . . . . . . . *115*

115. Inflorescence mostly compact, of crowded, pale green, whitish-green, or drab spikes; spikes broadly rounded at top and bottom; perigynia nerved on both sides, more gradually tapering with a long slope from the body into the relatively longer beak; scales of pistillate flowers rather blunt or dull-pointed, reaching only to base of the beak of perigynium . . . . . . . 35. C. MOLESTA

115. Inflorescence of a more elongated type with the spikes more separated from one another, brown or brown and green; spikes usually more narrowed or tapering at the bottom, conic to slightly rounded at the top; perigynium

445 Carex Sartwellii

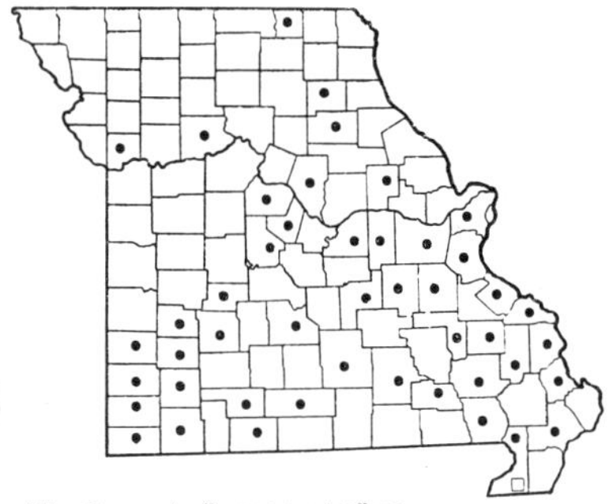

446 ● Carex retroflexa var. retroflexa
446 □ Carex retroflexa var. texensis

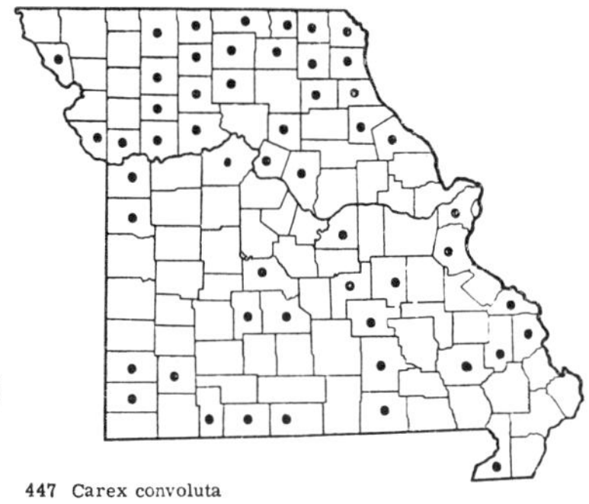

447 Carex convoluta

nerveless or nearly so on the inner (ventral) face, more abruptly changing from the rounded upper side of the body on a shorter slope into the relatively shorter beak; scales of pistillate flowers usually slender-pointed (acuminate) and almost as long as the beak of perigynium . . . . . . . . 34. C. BREVIOR

The following sequence of species indicates the sections or species-groups to which they belong, and the arrangement follows essentially that of Mackenzie (*North American Flora* 18: 1–478. 1931–35).

## Subgenus I. **Vignea**

### Section 1. **Divisae**

1. **Carex praegracilis** W. Boott Map 443

Flowers June–July.

Occurs in dry open ground and along railroads.

Known only from St. Louis (right-of-way of the Manufacturers' Railway, Arsenal Street, St. Louis, on the banks of the Mississippi River, June 7, 1959, *Muehlenbach 1522*) and Jackson (Sheffield, *Bush 4798*) counties.

Ranges from Manitoba, Michigan, Iowa, Missouri, and Oklahoma, west to Yukon and California and south to northern Mexico.

2. **Carex Douglasii** Boott Map 444

Flowers early May–late June.

Occurs along railroad tracks in St. Louis County, where collected several times at the Carrie Avenue Freight Yard in St. Louis by Dr. Victor Muehlenbach (May 8, 1955, *Muehlenbach 571*; June 24, 1956, *Muehlenbach 971*; June 1, 1957, *Muehlenbach 1178*).

Ranges from Missouri, Iowa, and Manitoba, west to British Columbia and south to Colorado and California.

### Section 2. **Intermediae**

(included under section *Arenariae* in *Gray's Manual*)

3. **Carex Sartwellii** Dewey Map 445

Flowers June–July.

Known only from low meadows in Jackson County, western Missouri.

Ranges from Quebec to British Columbia, south to New York, Ohio, Indiana, Illinois, Missouri, Nebraska, and Colorado.

### Section 3. **Bracteosae**

4. **Carex retroflexa** Muhl. Map 446

The species occurs in two varieties as indicated above in the key to species.

4a. **Carex retroflexa** var. **retroflexa** Map 446

*Carex retroflexa* Muhl. [G, P & S]

Flowers late April–June.

Frequents dry open, rocky, or sandy woods, wooded sandstone ledges of bluffs, cherty slopes, rich ravine slopes, wooded valleys, and rarely in grassy open places. Most frequently in acid soils. Throughout the Ozark section and locally northeast in northern

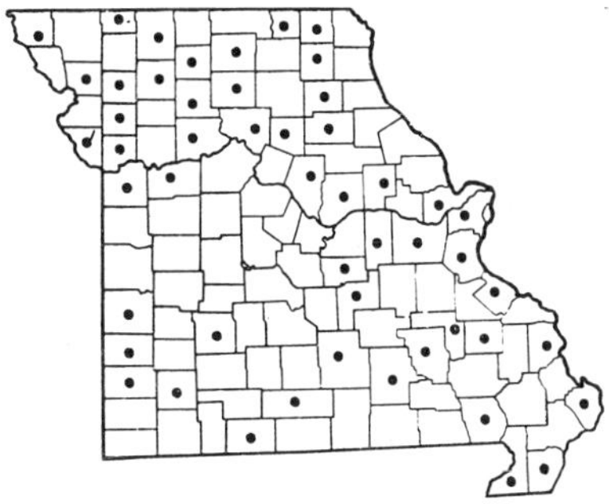

448 Carex rosea

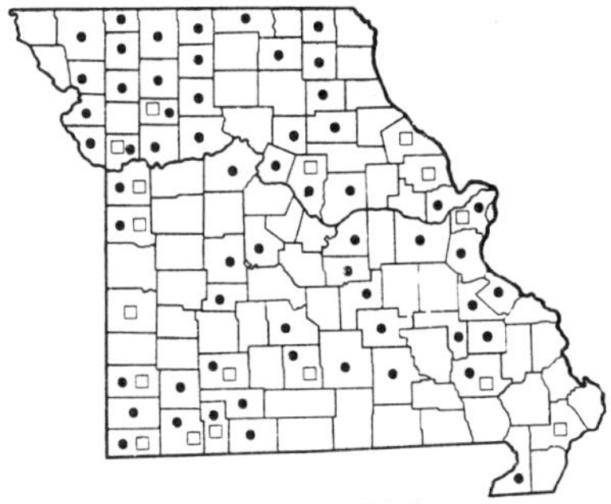

449 • Carex cephalophora var. cephalophora
449 □ Carex cephalophora var. mesochorea

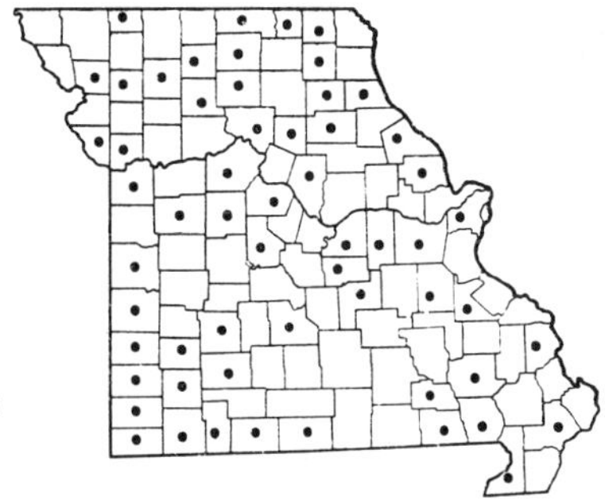

450 Carex Leavenworthii

Missouri in Monroe, Shelby, and Schuyler counties.

Ranges from Florida to Texas, north to New Hampshire, Vermont, New York, Ohio, Michigan, Illinois, Missouri, and Kansas.

4b. **Carex retroflexa** var. **texensis** (Torr.) Fern. Map 446

*Carex texensis* (Torr.) Bailey [G, P & S]

Flowers May–June.

Known only from the southeastern lowland section in Dunklin County (wet places, May 24, 1893, *Eggert*). A collection from Cape Girardeau County, previously identified as *C. texensis*, is now placed in *C. convoluta*.

Ranges from Georgia to Texas, north to New Jersey, Maryland, Ohio, Kentucky, and southeastern Missouri.

The small differences separating the two variations appear to justify their being maintained as varieties rather than as species.

5. **Carex convoluta** Mackenz. Map 447

Flowers mid-May–late June.

Frequents rich wooded ravines, slopes, and valleys. Throughout Missouri, but not recorded from some sections.

Ranges from Nova Scotia to Manitoba, south to South Carolina, Alabama, Arkansas, and Kansas.

6. **Carex rosea** Schkuhr Map 448

Flowers April–July.

Frequents rich wooded ravines, slopes, and valleys. Throughout Missouri, but not recorded from some sections.

Ranges from Nova Scotia and Quebec to North Dakota, south to Georgia, Alabama, and Louisiana.

Except for the chief difference in the styles, this species cannot be readily distinguished from *C. convoluta*, and by Gleason (*New Illustrated Flora* 1: 309. 1952) the two are united, along with *C. radiata*, under *C. rosea*. In the present work the two species are being retained as separate, pending further observations and experimental evidence on the significance of straight versus twisted styles.

7. **Carex cephalophora** Muhl. Map 449

The species is represented in Missouri, by two varieties, as indicated in the key to species.

7a. **Carex cephalophora** var. **cephalophora** Map 449

*Carex cephalophora* Muhl. [G, P & S]

Flowers May–August.

Frequents dry or rocky wooded slopes, ridges, uplands, and openings of woods, mostly in cherty or sandy soils, pastured fields, and rarely at the base of bluffs along stream banks. Most frequently in acid soils. Throughout Missouri.

Ranges from Florida to Texas, north to Maine, Quebec, Ontario, Michigan, and Manitoba.

7b. **Carex cephalophora** var. **mesochorea** (Mackenz.) Gl. Map 449

*Carex mesochorea* Mackenz. [G, P & S]

Flowers May–August.

Occurs in dry upland prairies, pastured fields, openings on dry chert or limestone wooded slopes, open fields, and rocky open valleys. Scattered in southern and central Missouri north to Pike, Boone, and Caldwell counties.

Ranges from Massachusetts to Wisconsin, south to Virginia, Tennessee, Missouri, and Texas.

Separated from var. *cephalophora* only by the larger perigynia, generally taller culms, and coarser habit of growth.

8. **Carex Leavenworthii** Dew. Map 450

*Carex cephalophora* var. *angustifolia* Boott [Shinners]

Flowers April–June.

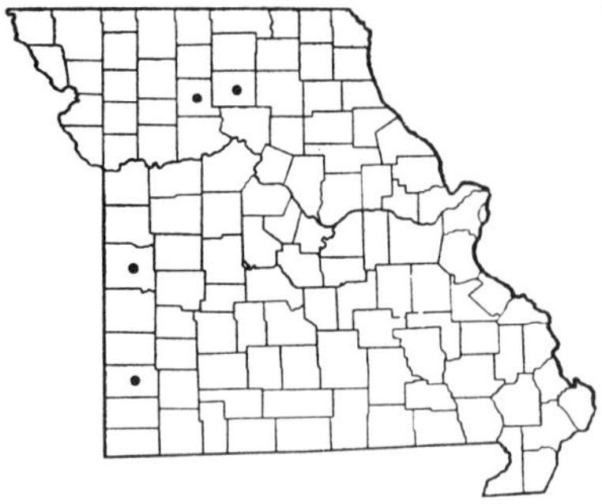

451 Carex arkansana

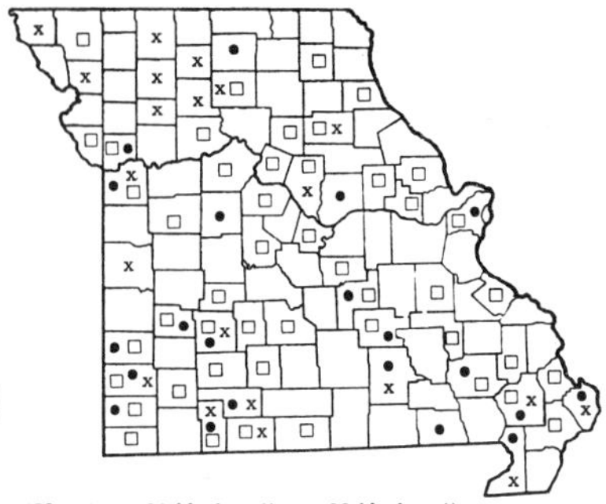

452 • Carex Muhlenbergii var. Muhlenbergii
452 □ Carex Muhlenbergii var. enervis
452 x Carex Muhlenbergii var. australis

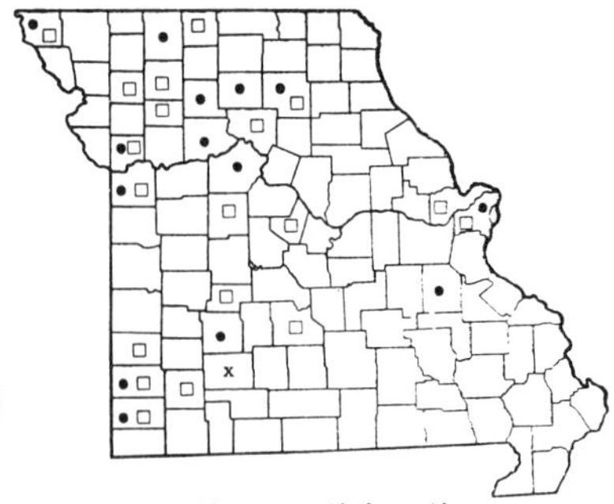

453 • Carex gravida var. gravida f. gravida
453 x Carex gravida var. gravida f. laxifolia
453 □ Carex gravida var. Lunelliana

Occurs in a variety of situations, from rich and low wet woodland, wet wooded valleys, along streams, and alluvial open ground in valleys, to prairie swales, wet meadows in open valleys, and open or dry woods and ridges in cherty, sandstone, limestone, or granitic soils. Throughout Missouri.

Ranges from Florida to Texas, north to New Jersey, Pennsylvania, Ontario, Michigan, Illinois, Iowa, and Kansas.

This species is easily distinguished from the two varieties of *C. cephalophora* above by the broad-based perigynia. Both varieties of *C. cephalophora* and *C. Leavenworthii* have a small congested usually single head of spikes.

9. **Carex arkansana** Bailey Map 451

Flowers May–June.

Occurs in wet river bottom prairies, prairie swales, and openings in oak woods. Rare, known only in western Missouri in Linn, Livingston, Bates, and Jasper counties. In the wet bottom prairies of the Grand River and tributaries in Linn and Livingston counties this species reaches its northeastern known limits of geographical distribution.

Ranges from northeastern Texas to western Arkansas northeast to western Missouri.

This species is readily recognized by the very long leafy bracts of the inflorescence, being much longer than in any other species of this group.

10. **Carex Muhlenbergii** Schk. Map 452

Missouri material is represented by three varieties:

a. Perigynia strongly nerved on both faces . . . 10a. C. MUHLENBERGII var. MUHLENBERGII
a. Perigynia nerveless on the inner (ventral) face or with only short nerves at the base . . . *b*
  b. Perigynia 3–3.5 mm. long, 2–2.5 mm. wide, usually spreading at maturity; scales shorter or equal to and narrower than the main body of the perigynia, short-awned or pointed; bracts of inflorescence narrow at the base, not broadly dilated . . . 10b. C. MUHLENBERGII var. ENERVIS
  b. Perigynia 3.5–4.5 mm. long, 2.5–3 mm. wide, mostly ascending at maturity; at least the lower scales longer and broader than the perigynia, long-awned; bracts of inflorescence broad and dilated at the base 10c. C. MUHLENBERGII var. AUSTRALIS

10a. **Carex Muhlenbergii** var. **Muhlenbergii** Map 452

*Carex Muhlenbergii* Schk. [G, P & S]

Flowers mid-May–July.

Frequents usually dry rocky or open woods, usually in acid soils of chert, sandstone, or granite, but also in limestone areas; also in sandy open ground, along railroads, and occasionally in rich woods or low wet woodland. Throughout Missouri, less common than var. *enervis*, into which it overlaps.

Ranges from Florida to Texas, north to Maine, Quebec, Ontario, Ohio, Michigan, Wisconsin, and Minnesota.

10b. **Carex Muhlenbergii** var. **enervis** Boott Map 452

Flowers mid-May–June.

Frequents the same habitats as var. *Muhlenbergii*, but commoner in the state.

Ranges from Florida to Texas, north to Maine, Ontario, Michigan, Illinois, Iowa, and Nebraska.

Plate no. 84. **1.** Carex Sartwellii. **2.** Carex retroflexa. **3.** Carex convoluta. **4.** Carex cephalophora. **5.** Carex Leavenworthii. **6.** Carex arkansana. **7.** Carex gravida. **8.** Carex sparganioides. **9.** Carex Muhlenbergii var. Muhlenbergii. **10.** Carex Muhlenbergii var. australis. **11.** Carex aggregata.

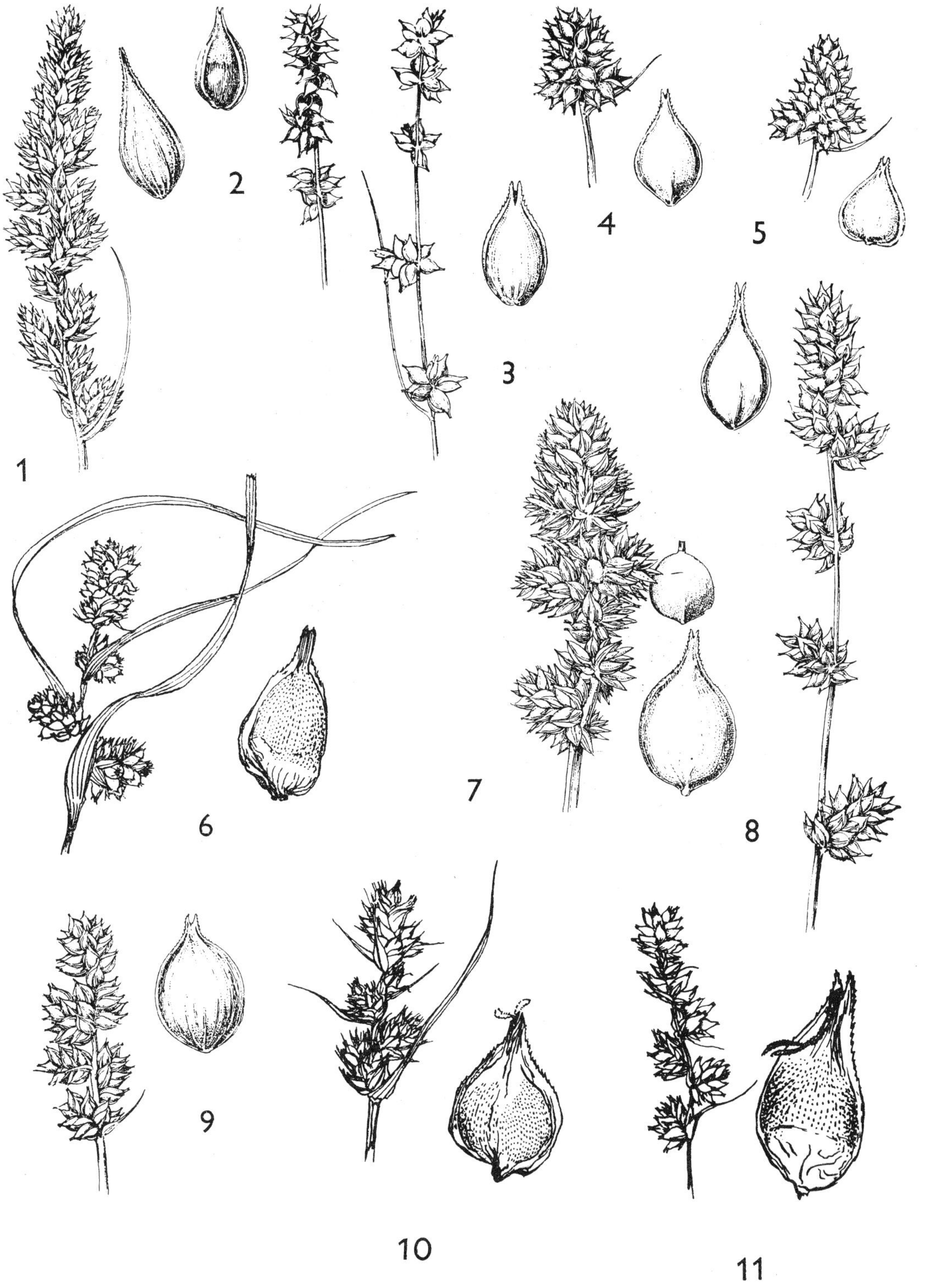


PLATE NO. 84

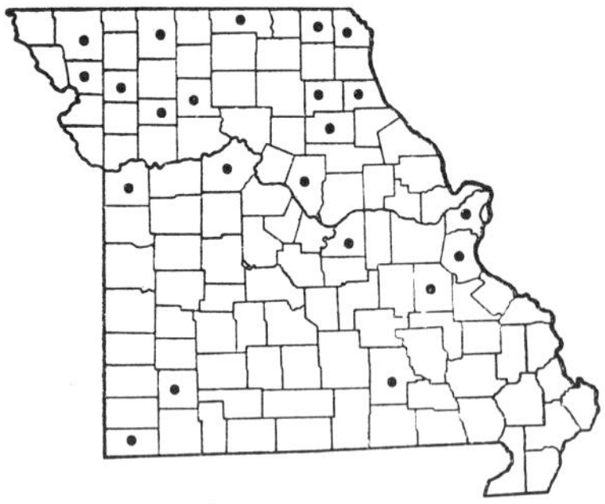

454 Carex aggregata

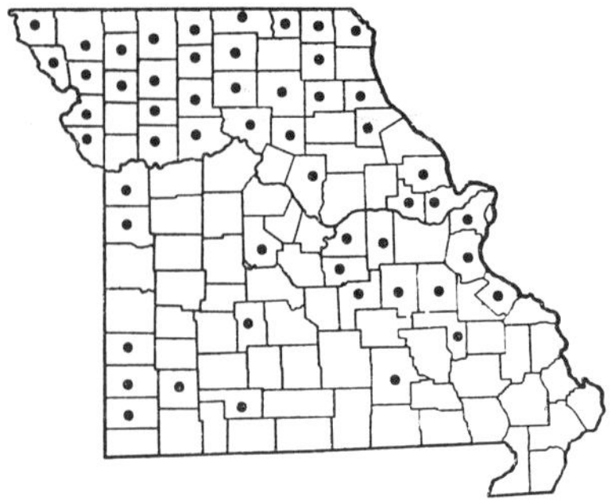

455 Carex sparganioides

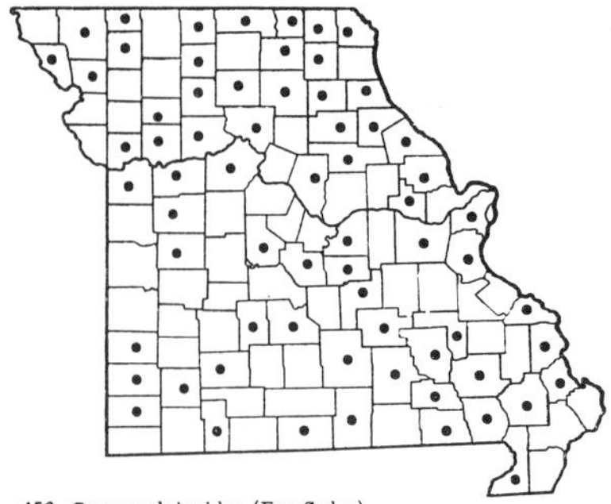

456 Carex vulpinoidea (Fox Sedge)

10c. **Carex Muhlenbergii** var. **australis** Olney Map 452

*Carex austrina* (Small) Mackenz. [G, P & S]

*Carex Muhlenbergii* var. *austrina* Small

Flowers late April–July.

Occurs in dry upland woods, openings on dry chert or limestone wooded slopes, but mostly in dry upland, cherty limestone glades, sandy prairies, roadside open banks, and railroads. Mostly in western and southern Missouri.

Ranges from Arkansas and Texas, north to Missouri and Kansas.

Intergrading plants between var. *australis* and var. *Muhlenbergii* and var. *enervis* would appear to justify treating var. *australis* as a well-marked variety rather than as a species.

11. **Carex gravida** Bailey Map 453

Missouri material is represented by 3 variations:

a. Perigynia ovate, 2–3 mm. wide, gradually tapering into a beak about 1 mm. long, faintly nerved or without nerves on the outer lower (dorsal) face; scales of pistillate flowers mostly with long awns or long slender points . . *b*
  b. Inflorescence 1–3 cm. long . 11a. C. GRAVIDA var. GRAVIDA f. GRAVIDA
  b. Inflorescence 3–5 cm. long . 11b. C. GRAVIDA var. GRAVIDA f. LAXIFOLIA

a. Perigynia broadly ovate to suborbicular, up to 3.8 mm. wide, abruptly passing into a shorter beak about 0.5 mm. long, usually more strongly and definitely nerved on the outer lower (dorsal) face; scales of pistillate flowers usually only short slender-pointed (acuminate) or short-awned . . . . . . . 11c. C. GRAVIDA var. LUNELLIANA

11a. **Carex gravida** var. **gravida** f. **gravida** Map 453

*Carex gravida* Bailey [G, P & S]

Flowers early May–July.

Occurs in prairie swales, dry open banks, and open places in wooded sections. Scattered in northern, western, and central Missouri; apparently absent from most of the Ozark section.

Ranges from Ontario and North Dakota to Wyoming, south to Ohio, Indiana, Illinois, Missouri, and Kansas.

11b. **Carex gravida** var. **gravida** f. **laxifolia** (Bailey) Kükenth. Map 453

*Carex gravida* f. *laxifolia* [G]

Known only from a single collection in Greene County (*Palmer 62263*).

11c. **Carex gravida** var. **Lunelliana** (Mackenz.) Herm. Map 453

*Carex Lunelliana* Mackenz. [P & S]

Flowers May–June.

More common in Missouri than var. *gravida*, occurring in dry upland prairies and prairie openings, open grassy slopes and pastures, and less commonly in rich wooded slopes or wooded valleys. Scattered in northern, central, and western Missouri; apparently absent from the Ozark and southeastern lowland sections of the state.

Ranges from Indiana, Michigan, Illinois, Iowa, and Kansas, south to Arkansas, Oklahoma, Texas, and New Mexico.

12. **Carex aggregata** Mackenz. Map 454

*Carex sparganioides* Muhl. var. *aggregata* (Mackenz.) Gl. [BB]

*Carex cephaloidea* [of P & S], not Dewey

Flowers early May–early June.

Frequents wooded valleys, rich wooded slopes and ravines, low wet woodland, along streams, alluvial thickets and fields, and in wet river bottom prairies. Scattered over the state.

Ranges from New York to Iowa, south to D. C., Kentucky, Missouri, and Oklahoma.

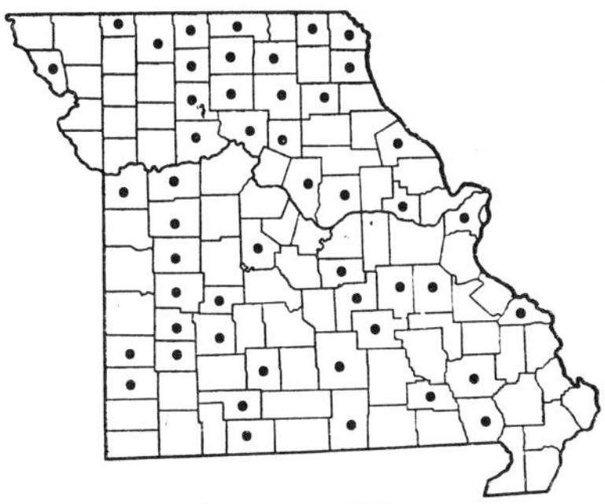

457 ○ Carex annectens var. annectens
457 ● Carex annectens var. xanthocarpa

458 Carex triangularis

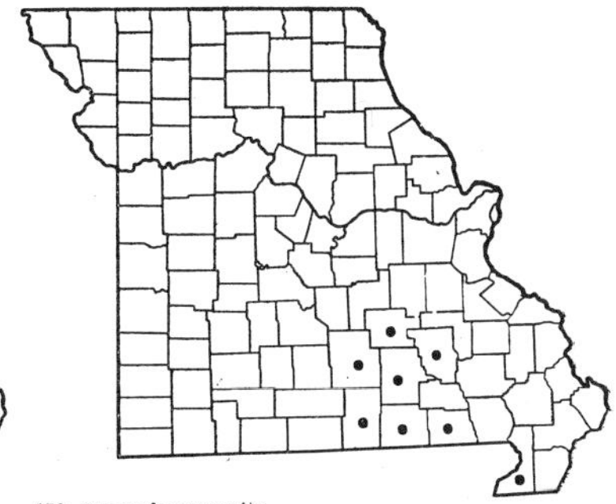

459 Carex decomposita

The status of this as a species distinct from *C. sparganioides* is doubtful. In their extremes the two can be readily distinguished. But some plants (such as *Palmer 59972* from Christian County and *Palmer 59577* and *Palmer 60085* from McDonald County) may show certain characters of *C. sparganioides* (inner ventral side of leaf-sheaths prominently cross-wrinkled) and other characters of *C. aggregata* (leaf-blades 3–6 mm. wide, concave summit of firm inner band of leaf-sheath). Additional collections may show other types of intergradation, casting considerable doubt as to the validity of *C. aggregata* as a true species. Pending further and more detailed studies of plants so identified, *C. aggregata* is being retained in the specific category.

Some specimens from St. Louis and Jackson counties, previously erroneously placed under *C. cephaloidea* in Palmer and Steyermark's *Annotated Catalogue*, are now referred to *C. aggregata*.

13. **Carex sparganioides** Muhl. Map 455

*Carex sparganioides* var. *sparganioides* [BB]

Flowers late April–July.

Usually found in rich woodland of valleys, ravines, slopes (especially north-facing limestone or sandstone), moss-covered boulders, more rarely in river bottom prairie or wet open valleys. Throughout Missouri, but apparently absent from the southeastern lowland section.

Ranges from New Hampshire and Quebec to Ontario, Minnesota, and South Dakota, south to Virginia, Tennessee, Missouri, and Kansas.

Usually abundantly distinct in its relatively broad leaf-blades, well-separated lower spikes of the elongated inflorescence, and other characters indicated in the key, but, as stated above under *C. aggregata*, intergrading and overlapping specimens are found, making it sometimes difficult to place collections with absolute certainty.

The perigynia in this species vary from 3–4.5 mm. long. The pistillate scales are usually acuminate both in this species and in *C. aggregata*. Although separated vegetatively from *C. Muhlenbergii* var. *Muhlenbergii* and var. *enervis* by the loose leaf-sheaths which are conspicuously mottled and cross-marked on the back side, the perigynia of *C. sparganioides* are longer (3–4.5 mm. as contrasted with 3–3.5) with longer beaks than those of *C. Muhlenbergii* var. *Muhlenbergii* and var. *enervis*.

## Section 4. **Multiflorae**

14. **Carex vulpinoidea** Michx. Fox Sedge Map 456

Flowers May–July.

Frequents prairie swales, wet river bottom prairies, wet open valleys, margins of ponds, streams, and spring branches, limey open or wooded swamps, and low wet woods. Throughout Missouri, and one of the most abundant species in the state.

Ranges from Newfoundland to British Columbia, south to Florida, Texas, Arizona, and Oregon.

Although many plants show the typical cross-wrinkled inner (ventral) side of the leaf-sheath, other plants may exhibit only a purple-red-dotted leaf-sheath. Although the bracts of the inflorescence are usually bristlelike or setaceous, the lowest one is sometimes rather leaflike (foliaceous) rather than setaceous.

15. **Carex annectens** Bickn. Map 457

Two varieties are usually distinguished:

Perigynia 1.7–2.4 mm. wide, mostly more than 2 mm. wide, usually clearly nerved on the outer lower (dorsal) side, and with a rather conspicuous beak . . . 15a. C. ANNECTENS var. ANNECTENS

460 Carex oklahomensis

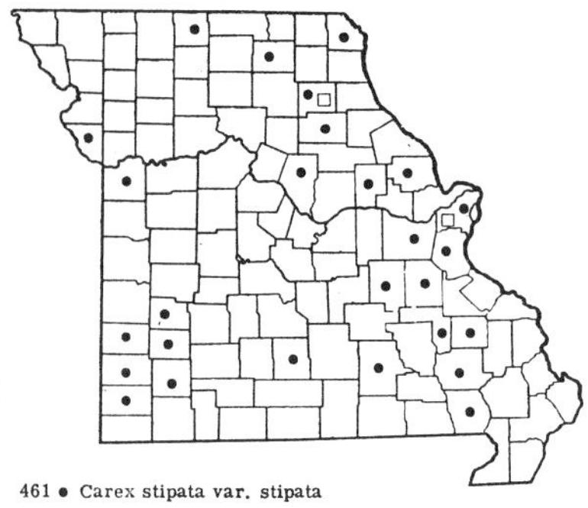

461 • Carex stipata var. stipata
461 □ Carex stipata var. maxima

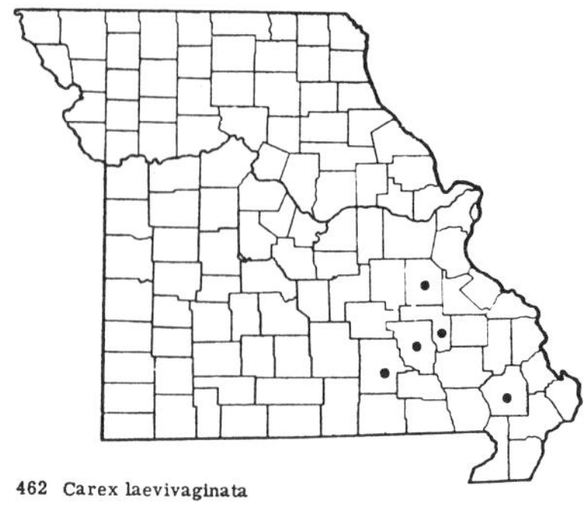

462 Carex laevivaginata

Perigynia 1.4–1.8 mm. wide, mostly less than 2 mm. wide, without nerves or only faintly nerved on the outer lower (dorsal) side, and with a very small beak . 15b. C. ANNECTENS var. XANTHOCARPA

15a. **Carex annectens** var. **annectens** Map 457

*Carex annectens* Bickn. [G, P & S in part]

Occurs in dry and wet prairies, open wet ground in valleys, borders of sink-hole or other ponds, and swampy limey meadows.

Less common in Missouri than var. *xanthocarpa*.

Ranges from Maine to Wisconsin, south to Florida and Texas.

15b. **Carex annectens** var. **xanthocarpa** (Bickn.) Wieg. Map 457

*Carex annectens* [of P & S in part], not Bickn.
*Carex brachyglossa* Mackenz.

Frequents the same situations as var. *annectens*, but the commoner variety in Missouri.

Ranges from Maine and Quebec to Wisconsin and Iowa, south to Virginia, Ohio, Illinois, Missouri, and Kansas.

Most of the material encountered in Missouri pertains to var. *xanthocarpa*, which intergrades freely with var. *annectens*. The two varieties are not separated on the distribution map. The perigynia of *C. annectens* are appressed-ascending in the spikes, whereas in *C. vulpinoidea* and *C. triangularis*, they spread out widely at maturity.

16. **Carex triangularis** Boeckl. Map 458

Flowers early May–June.

Known only from swamps and swampy ground in southeastern Missouri from Wayne (swamps, May 12, June 5, 1893, Williamsville, *Eggert*) and Dunklin (May 17, 1892, *Bush*) counties.

Ranges from Mississippi to Texas, north to southeastern Missouri and Oklahoma.

## Section 5. **Paniculatae**

17. **Carex decomposita** Muhl. Map 459

Flowers late April–July.

Known only from sink-hole ponds in the southeastern Ozarks and in the lowland swamps of Dunklin County, southeastern Missouri.

Ranges from Florida to Louisiana, north to Virginia, New York, Ohio, Indiana, Michigan, and Missouri.

A very distinct species, found in greatest abundance in the natural sink-hole ponds in the upland oak-hickory woods of the southeastern Ozarks, where it is associated with *C. alata*, *C. comosa*, *C. lupulina*, and *C. tribuloides*. Like *C. alata* and *C. comosa*, it forms dense large tussocks or leafy mounds in the water. Its distinctive habitat and habit, together with its compound inflorescence of elongated thick lower branches, and thick swollen perigynia broadest in the upper half, readily prevent this species from being confused with other sedges.

---

Plate no. 85. **1.** Carex vulpinoidea. **2.** Carex annectens var. xanthocarpa. **3.** Carex triangularis. **4.** Carex decomposita. **5.** Carex oklahomensis. **6.** Carex stipata. **7.** Carex laevivaginata. **8.** Carex crus-corvi.

PLATE NO. 85

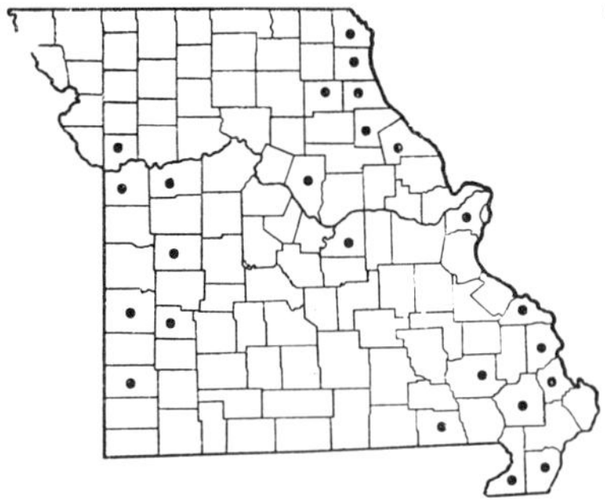

463 Carex crus-corvi

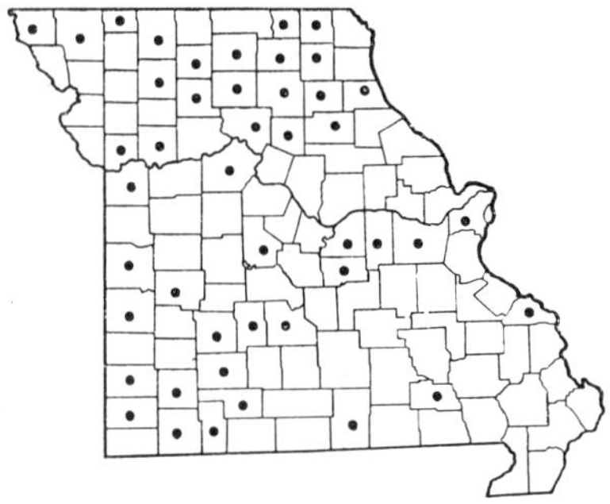

464 Carex conjuncta

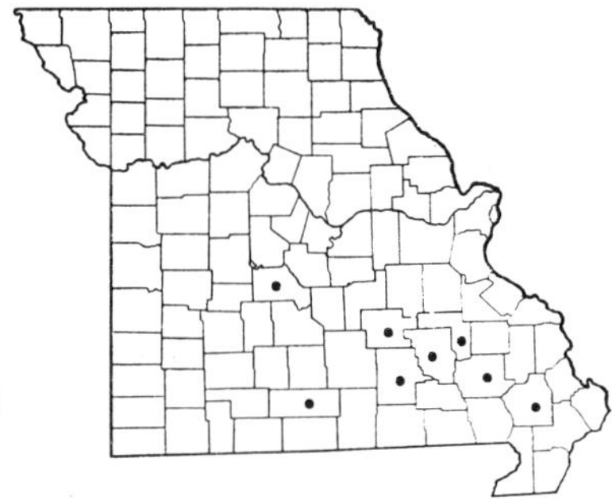

465 Carex interior

## Section 6. **Vulpinae**

18. **Carex oklahomensis** Mackenz. Map 460

*Carex stipata* var. *oklahomensis* (Mackenz.) Gl. [BB]

Flowers May–July.

Occurs in low river bottom prairies, wet places in upland prairies, swampy open meadows around upland ponds, wet sandy ground bordering sloughs and streams, and swales in low open woods. Known only from Laclede, Christian (*Palmer 59918*), Jasper (*Palmer 1962, 2174, 2442,* and *25299*), and Newton (*Palmer 54200*) counties of southwestern Missouri.

Ranges from southwest Missouri and Arkansas, south to Oklahoma and Texas.

Some Missouri material previously referred to this species has been found to be erroneously identified, and is now referred to *C. stipata* var. *stipata* and *C. conjuncta*.

19. **Carex stipata** Muhl. Map 461

Two varieties may be recognized in Missouri:

Perigynia 4–5 mm. long, the beak about equaling the length of the main body; leaf-blades 4–8 mm. wide . . . . . 19a. C. STIPATA var. STIPATA

Perigynia 5–6 mm. long, the beak longer than the main body; leaf-blades 8–17 mm. wide . 19b. C. STIPATA var. MAXIMA

19a. **Carex stipata** var. **stipata** Map 461

*Carex stipata* Muhl. [G, P & S]

Flowers May–July.

Occurs along streams, spring branches, wet river bottom prairies, wet open ground in valleys, and low wet woods. Scattered throughout the state.

Ranges from Labrador to Alaska, south to North Carolina, Tennessee, Missouri, Kansas, New Mexico, and California.

19b. **Carex stipata** var. **maxima** Chapm. Map 461

*Carex uberior* (Mohr) Mackenz.

Occurs in wet low ground in eastern Missouri, where known from St. Louis (Allenton, July 30, 1900, *Letterman*) and Shelby (open alluvial meadow along east side of Tiger Fork, 1½ mi. north of Burksville, June 5, 1951, *Steyermark 71719*) counties.

Ranges from Florida to Texas, north to New Jersey, Pennsylvania, Indiana, and Missouri.

The two varieties intergrade and are not separated on the distribution map of the species.

20. **Carex laevivaginata** (Kükenth.) Mackenz. Map 462

Flowers May–June.

Occurs in swampy meadows along valleys of small streams and in low wet woods in the southeastern quarter of the state.

Ranges from Maine to Minnesota, south to Florida, Alabama, and Missouri.

21. **Carex crus-corvi** Shuttlw. Map 463

Flowers late May–July.

Occurs in wooded swamps and swales, and alluvial wooded flood plains, where especially common along the Mississippi and Missouri rivers and other large streams tributary to those waters; also in wet river bottom prairies. Mainly found in eastern and central Missouri, and in the western part of the state mostly south of the Missouri River; absent from northwestern Missouri and most of the Ozark section.

The foliage of *C. crus-corvi* varies from glaucous (gray-green or bluish green) to deep green. The species is easily recognized by the compound much-

Plate no. 86. **1.** Carex interior. **2.** Carex muskingumensis. **3.** Carex scoparia. **4.** Carex tribuloides. **5.** Carex normalis. **6.** Carex tenera. **7.** Carex projecta. **8.** Carex cristatella. **9.** Carex festucacea.

PLATE NO. 86

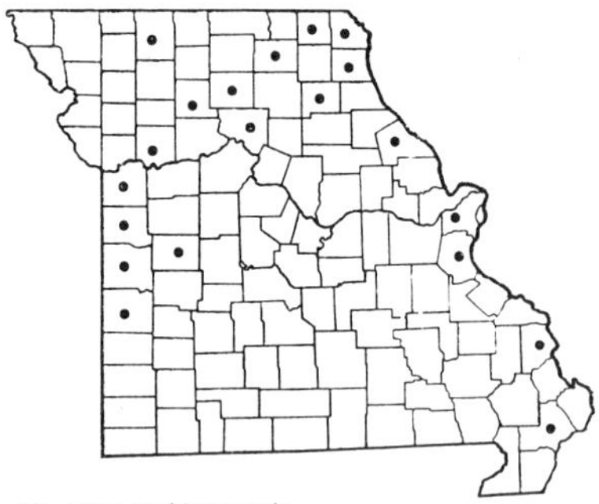

466 Carex muskingumensis

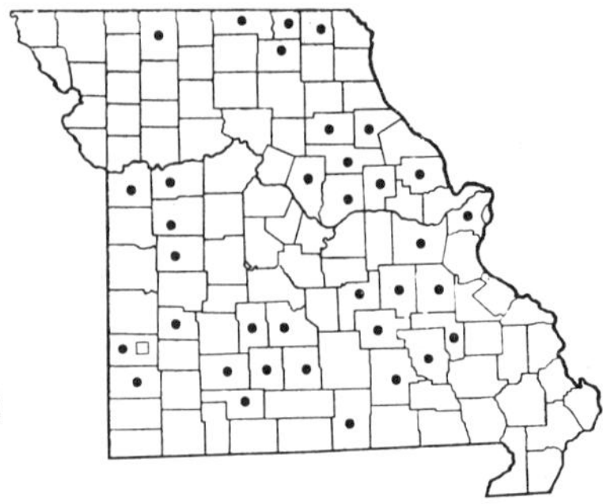

467 • Carex scoparia var. scoparia f. scoparia
467 □ Carex scoparia var. scoparia f. moniliformis

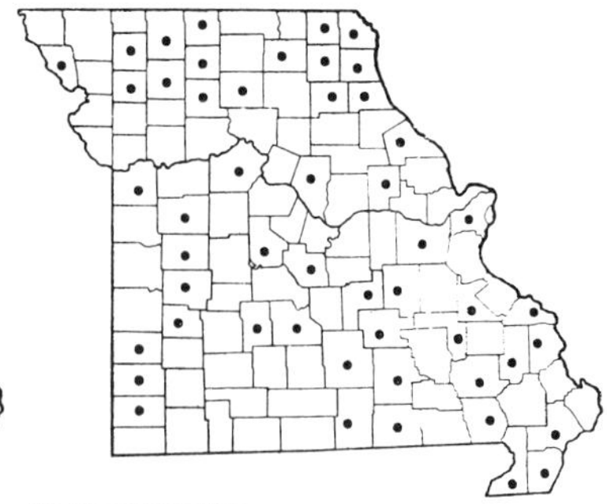

468 Carex tribuloides

branched inflorescence with elongated lower branches, the whole up to 25 cm. long and 6 cm. thick, and by the very elongated beak of the perigynium.

22. **Carex conjuncta** Boott Map 464
Flowers early May–late June.

Occurs in wet river bottom prairies, prairie swales, wet open valleys, alluvial thickets, along streams, but most commonly in low wet, alluvial, or rich woods. Common in northern, central, and western Missouri; apparently rare or absent in most sections of the Ozarks and altogether absent from the southeastern lowland section.

Ranges from New York to South Dakota, south to Virginia, Tennessee, Missouri, and Kansas.

## Section 7. **Stellulatae**

23. **Carex interior** Bailey Map 465
Flowers early April–June.

Frequents swampy, calcareous, spring-fed meadows and occasionally found on dripping limestone ledges at the base of bluffs. Mainly in the southeastern Ozark section and locally west in Camden and Douglas counties.

Ranges from Labrador to British Columbia, south to Delaware, Pennsylvania, West Virginia, Ohio, Indiana, Illinois, Missouri, Kansas, New Mexico, California, and northern Mexico.

This well-marked species of *Carex* with very narrow leaf-blades (1–3 mm. wide) and few-flowered small clusters of widely spreading or recurving perigynia is a characteristic inhabitant of swampy, spring-fed, calcareous meadows, where it is associated with *C. leptalea*, *C. suberecta*, and *C. stricta*.

It is another one of the boreal North American species at one of its southern limits in the swampy calcareous meadows of the Ozarks, where it probably represents a relict survivor of an Ice Age (Pleistocene) flora.

## Section 8. **Ovales**

24. **Carex muskingumensis** Schwein. Map 466
Flowers late May–October.

Frequents wooded swamps and alluvial flood plain woods of the Mississippi River in eastern Missouri and of other large streams (Fabius, Wyaconda, Salt, Chariton, Locust, Grand, Thompson, South Grand, and Marais des Cygnes) of northern and west-central Missouri. Absent from the Ozark region, southeastern lowland, and most of the unglaciated prairie sections.

Ranges from Ontario to Manitoba, south to Kentucky, Missouri, and Kansas.

This well-marked species is easily recognized by the narrowly elliptical, long and pointed, pale brown spikes containing long, narrow, thin perigynia, together with the numerous leafy sterile shoots with conspicuously spreading leaf-blades.

25. **Carex scoparia** Schkuhr Map 467
Represented in Missouri by two variations:

Spikes crowded or overlapping one another in an erect or slightly arching inflorescence . . .
25a. C. SCOPARIA var. SCOPARIA f. SCOPARIA

All spikes, or at least the lower ones, distant from one another in a more or less stretched out, flexuous (zigzag curving) type of inflorescence
25b. C. SCOPARIA var. SCOPARIA f. MONILIFORMIS

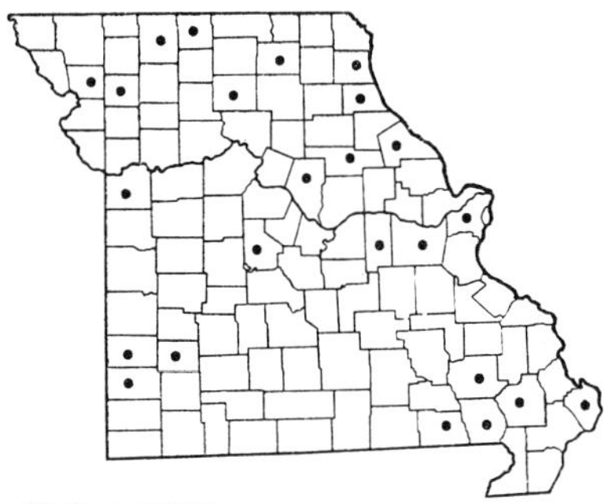

469 Carex projecta

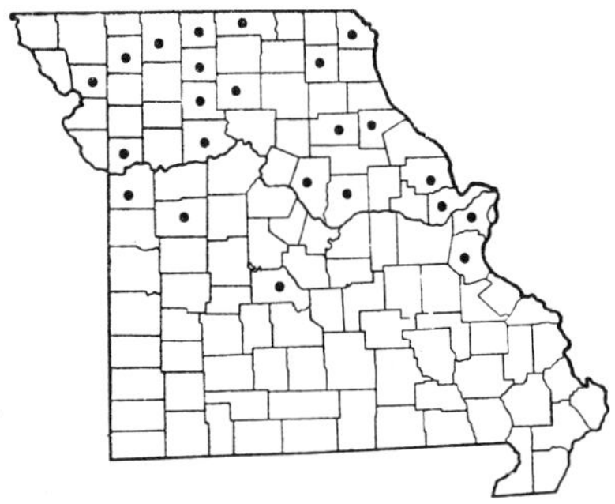

470 Carex cristatella

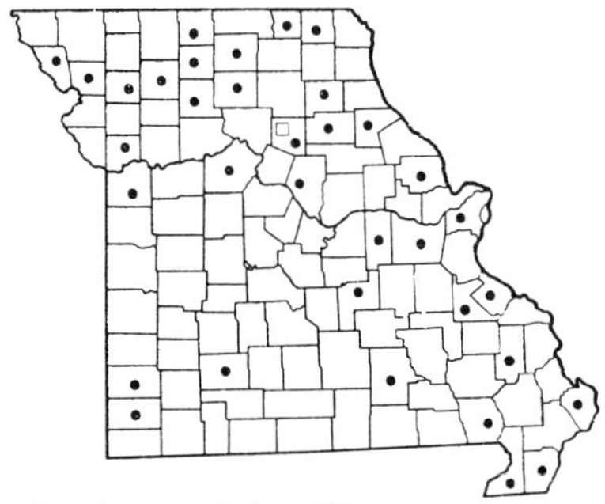

471 • Carex normalis f. normalis
471 □ Carex normalis f. perlonga

25a. **Carex scoparia** var. **scoparia** f. **scoparia** Map 467

*Carex scoparia* Schkuhr [G, BB, P & S]

Flowers late May–July.

Frequent in wet river bottom prairies, wet open valleys, prairie swales, upland prairies, margins of sink-hole ponds, and roadside ditches. Throughout Missouri, and the common variation encountered.

Ranges from Newfoundland to British Columbia, south to South Carolina, Tennessee, Arkansas, New Mexico, and Oregon.

Sometimes mistaken for *C. tribuloides*, but easily distinguished vegetatively from that species by the much narrower leaf-blades and culms, tightly fitting leaf-sheaths, and few ascending leaf-blades bunched near the top of the sterile culms; the spikes in *C. scoparia* usually taper to a point, whereas they are blunt or obtuse in *C. tribuloides*.

25b. **Carex scoparia** var. **scoparia** f. **moniliformis** (Tuckerm.) Kükenth. Map 467

*Carex scoparia* f. *moniliformis* (Tuckerm.) Kükenth. [G]

*Carex scoparia* var. *moniliformis* Tuckerm. [BB]

Represented by a single collection from southwestern Missouri in Barton County (sandy soil, upland prairie, along small stream, ½ mi. west of Duvall, May 26, 1952, *Palmer 53945*).

Ranges from Newfoundland to British Columbia, south to Virginia, New York, and Missouri.

The inflorescence in this form varies from 3.5–6 cm. long, that in f. *scoparia* from 1.5–5 cm. long.

26. **Carex tribuloides** Wahlenb. Map 468

*Carex tribuloides* var. *sangamonensis* Clokey

Flowers May–August.

Frequent in swampy or low wet woods of alluvial soils, borders of streams and spring branches, sink-hole and other ponds, wet river bottom prairies, wet open valleys, and low meadows. Throughout Missouri, and one of the commonest species in the state.

Ranges from Quebec, Ontario, Michigan, Wisconsin, Minnesota, and Nebraska, south to Florida, Louisiana, and Oklahoma.

The gray-green or pale brown blunt spikes are somewhat variable in size, shape, and in the length of the perigynia, but are readily spotted by the thin and erect or appressed perigynia with the wing abruptly tapering below the middle, and by the numerous sterile culms bearing many spreading leaves.

Plants having more slender culms, narrower leaves, and narrow perigynia 3–4 mm. long have been separated as *C. tribuloides* var. *sangamonensis*, but do not appear to merit recognition. Plants answering this description are two *Daniels* collections from wet meadow near dam, Columbia, Boone County, July 3, 1903, and specimens from St. Louis and Jackson counties.

27. **Carex projecta** Mackenz. Map 469

Flowers May–October.

Occurs in wet river bottom prairies, prairie swales, wet open valleys, swampy ground in bald cypress forest, and low wet woods or wooded alluvial soils of flood plain. Scattered throughout Missouri, but probably more common.

Ranges from Newfoundland to Manitoba, south to D. C., West Virginia, Ohio, Indiana, Illinois, and Missouri.

Very similar to *C. tribuloides*, from which it is mainly distinguished by the more loosely ascending to spreading tips of the perigynia and by the more usually elongated, loosely arranged, more scattered heads of the inflorescence.

28. **Carex cristatella** Britt. Map 470

Flowers late May–July.

Frequents wet river bottom prairies, swampy

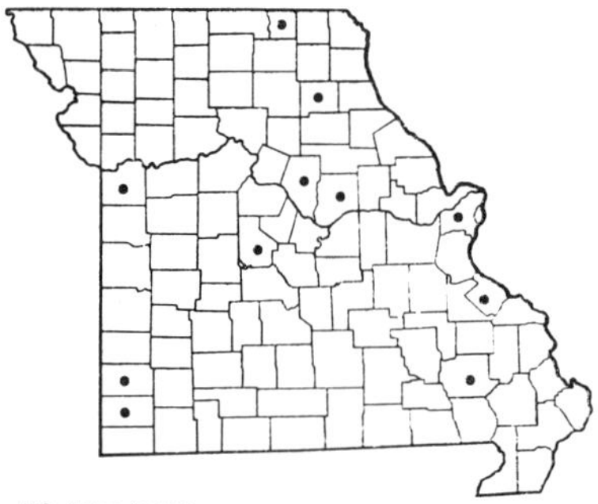

472 Carex tenera

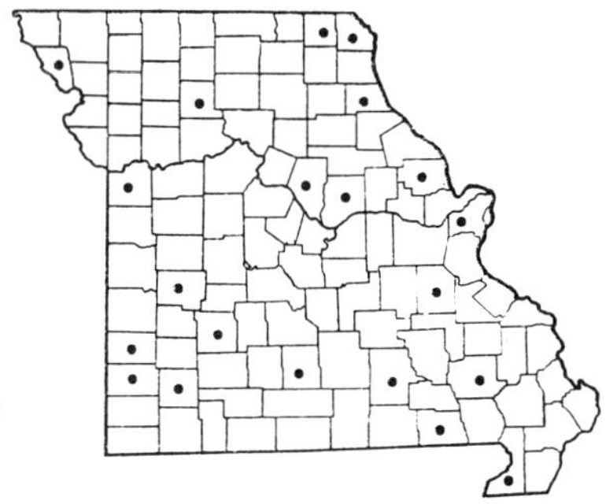

473 Carex festucacea

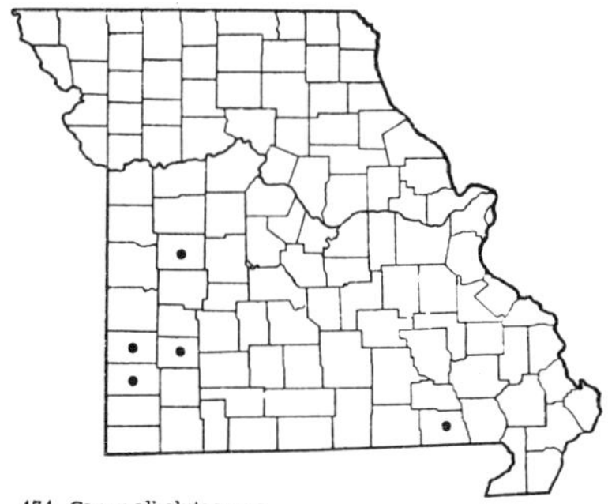

474 Carex albolutescens

meadows, alluvial soils in wet open valleys, thickets, borders of natural ponds, swamps, sloughs, and ditches, along spring branches, along railroads, and in rich wooded valleys. Northern and central Missouri south to Jefferson, Callaway, Johnson, and Jackson counties, and locally south in Camden Co.; absent from the Ozark region, the southeastern lowland and southwestern unglaciated prairie sections.

Ranges from Quebec to North Dakota, south to Virginia, Kentucky, Missouri, and Nebraska.

This species is usually readily identified in the field by the crowded nearly globose spikes with the perigynia widely spreading in all directions in a series of little rosettes, and, vegetatively, by the many widely spreading leaves produced from the numerous sterile culms. Some authors consider it to have probably originated as a hybrid between *C. tribuloides* and *C. Bebbii*.

29. **Carex normalis** Mackenz. Map 471

Represented in Missouri by two variations:

Spikes more or less crowded and usually overlapping in a rather straight or erect inflorescence 1.5–4 cm. long . . 29a. C. NORMALIS f. NORMALIS

Most or all the spikes rather remote and separated from one another and not overlapping in a flexuous (alternately zigzag curving) arching inflorescence 3–7 cm. long . . 29b. C. NORMALIS f. PERLONGA

29a. **Carex normalis** f. **normalis** Map 471

*Carex normalis* Mackenz. [G, BB, P & S]

Flowers early May–July.

Frequents low wet woods, wooded swamps or alluvial soils, rich wooded valleys, dry and open wooded slopes, open alluvial soils, prairie swales, low river bottom prairies, along small streams, and along railroads. Scattered over the state, but most common in northern, central, and eastern Missouri; absent from most parts of the Ozark section.

Ranges from Maine to Manitoba, south to North Carolina, Tennessee, Missouri, and Oklahoma.

29b. **Carex normalis** f. **perlonga** Fern. Map 471

Known from Randolph County, central Missouri. Scattered throughout the range of f. *normalis*.

30. **Carex tenera** Dewey Map 472

Flowers late April–June.

Occurs in upland prairies, fields, and grassy open places, wet low woods bordering ditches, occasionally in dry open woods. Scattered in Missouri, and infrequently collected.

Ranges from Quebec to Alberta, south to North Carolina, Ohio, Indiana, Illinois, Missouri, South Dakota, and Montana.

This species, with small spikes of relatively few flowers and small perigynia, has been found in less than a dozen counties of the state. Perhaps future collecting will reveal its greater abundance and frequency. The absence of sterile leafy culms in this species and in *C. festucacea* is a distinctive character separating the two species from others of the *Ovales* section in Missouri.

31. **Carex festucacea** Schkuhr Map 473

Flowers May–July.

Occurs in prairie swales, upland dry prairies, and low or flat wet woods. Scattered over most sections of the state, but not common.

Ranges from Massachusetts, New York, Ohio, Michigan, Illinois, and Iowa, south to Georgia, Louisiana, and Oklahoma.

Like *C. tenera*, this species, although well-marked by the small spikes of relatively small perigynia, has been collected from less than twenty counties in the state. *Carex festucacea* is very close to and difficult to distinguish from *C. tenera*, the main distinction being in the shape of the body of the perigynium, which in *C. tenera* is narrowly ovate and half as wide as long,

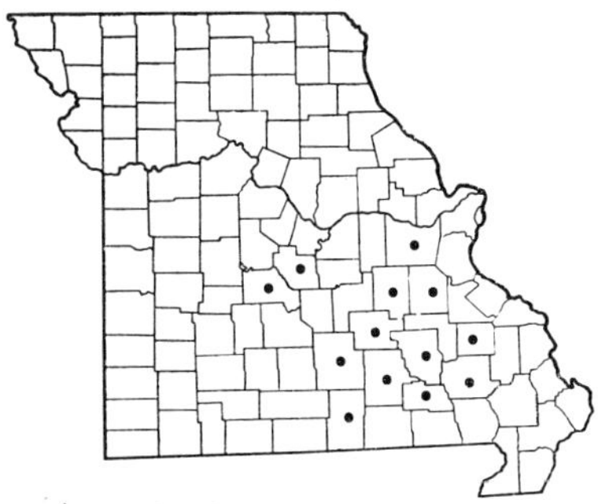

475 Carex suberecta

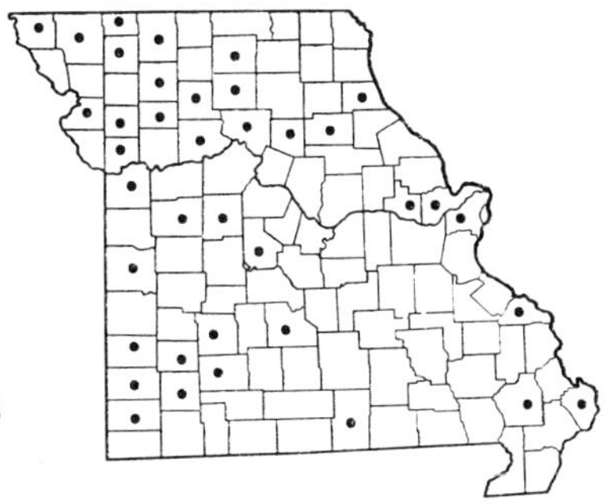

476 Carex brevior

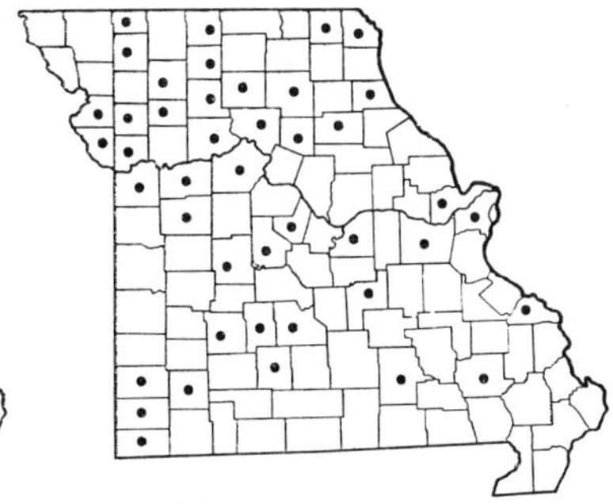

477 Carex molesta

while in *C. festucacea* it is suborbicular to broadly ovate and from nearly 3/4 as wide as long to nearly as wide as long. Other differences, as given in the key, indicate additional items of divergence of these two species.

32. **Carex albolutescens** Schwein. Map 474
*Carex straminea* according to Mackenzie [P & S], not *C. straminea* Willd.
Flowers May–July.

Occurs in moist depressions and swales of upland prairies, low wooded swamps, open places, and along shaded banks of streams. Infrequently collected, and known from less than six counties in southern Missouri north to Henry County; future collecting should reveal a greater frequency of occurrence.

Ranges from Florida to Texas, north near the coast to Nova Scotia, and inland to Pennsylvania, Michigan, Indiana, Illinois, and Missouri.

This species is stated to hybridize with *C. tribuloides*. Its small spikes with relatively small perigynia resemble most closely those of *C. tenera* and *C. festucacea*, and sometimes may also be mistaken for plants of *C. brevior* with small spikes and short perigynia less than 4 mm. long. From these other three species it is mainly distinguished by the body of the perigynia being broadest above the summit of the achene in the upper third and by the nerved inner face of the perigynium.

33. **Carex suberecta** (Olney) Britt. Map 475
Flowers late April–August.

Frequents calcareous (limey), spring-fed, swampy meadows and moist limestone ledges along small creeks in the eastern Ozark section west to Miller, Camden, Texas, and Howell counties.

Ranges from Ontario to Minnesota, south to Virginia, Ohio, Indiana, Illinois, and Missouri.

A characteristic sedge of spring-fed, calcareous, swampy meadows, where it is associated with *C. leptalea*, *C. stricta*, and *C. interior*, and other calciphiles. In most parts of its range, *C. suberecta* is a species of glaciated terrain, but in Missouri it is in strictly unglaciated stations, and is probably a relict survivor of a more boreal flora that took refuge or migrated southward during one of the advances of the recent Ice Age (Pleistocene). The habitat and rather pointed, somewhat crowded, rusty-brown or tawny spikes, with closely appressed perigynia, should help identify this species in the field.

34. **Carex brevior** (Dew.) Mackenz. Map 476
Flowers May–June.

Most frequent in dry upland prairies and prairie swales along railroads, open grassy slopes and prairie openings, wet river bottom prairies, and wet open valleys. Mostly absent from the Ozark section, but elsewhere common, especially in northern, central, and western Missouri.

Ranges from Maine to British Columbia, south to D. C., Tennessee, Arkansas, Texas, New Mexico, and Oregon.

Somewhat variable in the shape and size of the perigynia, which vary from broadly ovate to suborbicular, and 3.5–5.5 mm. long by 2.5–3.5 mm. broad. Spikes with short perigynia less than 4 mm. long are distinguished from *C. festucacea* by having the beak of the perigynia less than half the length of the body instead of as in *C. festucacea* half or more the length of the body, and the perigynia more than 2 mm. wide, while from *C. tenera* such plants may be distinguished by the broader perigynia and the suborbicular instead of narrowly ovate body of the perigynia.

35. **Carex molesta** Mackenz. Map 477
Flowers May–early July.

Usually found in dry or wet prairies of upland or river bottom, open thickets in valleys, pastures, grassy slopes, and dry open woodland or low wooded alluvial soils of flood plain. Throughout Missouri, except not recorded in the southeastern lowland section.

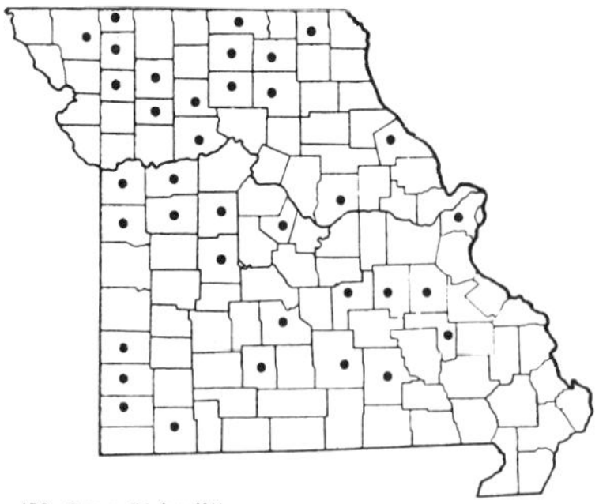

478 Carex Bicknellii

479 Carex alata

480 Carex straminea

Ranges from Massachusetts and Vermont to Saskatchewan, south to D. C., Tennessee, Arkansas, Kansas, and Colorado.

Sometimes mistaken for *C. brevior*, from which it is mainly distinguished by the gray-green or pale green spikes rounded at both ends, the perigynia nerved on both faces, shorter, blunter pistillate scales, and the more gradually tapering beak of the perigynia. Specimens of *C. molesta* are sometimes confused with *C. Bicknellii* having ovoid heads rounded at the base, but may be distinguished by the relatively shorter as well as thicker perigynia 3 mm. or less broad.

Although apparently a recognizable species in Missouri and in most other parts of its range, Gleason (*New Ill. Fl.* 1: 325. 1952) indicated that it may represent only a hybrid with *C. normalis*. However, as shown by Mr. Richard J. Eaton (Rh. 59: 33–36. 1957), it appears to be a recognizable species exhibiting occasional intermediate transitions to *C. brevior* at the northeast margin of its range, as indicated by Dr. Hermann, who considers it quite distinct from *C. brevior* in the midwestern states and east to Pennsylvania and western New York.

36. **Carex Bicknellii** Britt. Map 478

Flowers April–July.

Most frequently found in both dry upland prairie and wet river bottom prairie, also in openings on dry wooded chert or limestone slopes. Throughout Missouri, except apparently absent from the extreme southeastern lowland section.

Ranges from Maine to Saskatchewan, south to Delaware, Pennsylvania, Ohio, Indiana, Illinois, Arkansas, Oklahoma, and New Mexico.

This is a fairly tall-growing *Carex*, attaining a height of 1 meter with the flowering culms, which greatly exceed the narrow leaves. The thin, transparent, large perigynia, mostly 5.5–7.5 mm. long by 2.7–4.8 mm. wide, many-nerved on each face, and the silvery brown or straw-buff light-colored scales and perigynia are marks of recognition. The spikes vary from elongate and tapering at the base to ovoid or subglobose with rounded bases. Specimens with the latter type of rounded bases are sometimes confused with *C. molesta*, from which they may be distinguished by the thinner, flatter, and more translucent perigynia over 3 mm. wide.

37. **Carex alata** T. & G. Map 479

Flowers late April–June.

In sink-hole ponds of the southeastern Ozarks in Reynolds, Ripley, Oregon, and Howell counties, and occasionally in open swamps of ditches in the southeastern lowland section in Scott County.

Ranges from Florida to Texas, north near the coast to Massachusetts, New York, and Pennsylvania, inland from New York to Michigan, Indiana, and Missouri.

This is a characteristic species of sink-hole ponds in the Ozarks, where it forms large tussocks or mounds of culms and leaves, often in the crotches of *Cephalanthus* bushes. In such places it is often associated with *Carex comosa*, *C. decomposita*, *Glyceria acutiflora*, *Rosa palustris*, *Viola lanceolata*, *Galium tinctorium*, and *Bidens discoidea*, some of which are also found growing in the tussocks of the *Carex* or among the crotches of *Cephalanthus* bushes.

38. **Carex straminea** Willd. Map 480

*Carex straminea* var. *straminea* [BB]

*Carex Richii* Mackenz.

*Carex hormathodes* var. *Richii* Fern.

Flowers early May–June.

Known only from a sink-hole pond in the southern

Plate no. 87. **1.** Carex albolutescens. **2.** Carex suberecta. **3.** Carex brevior. **4.** Carex alata. **5.** Carex molesta. **6.** Carex Bicknellii. **7.** Carex straminea. **8.** Carex leptalea. **9.** Carex Jamesii.

PLATE NO. 87

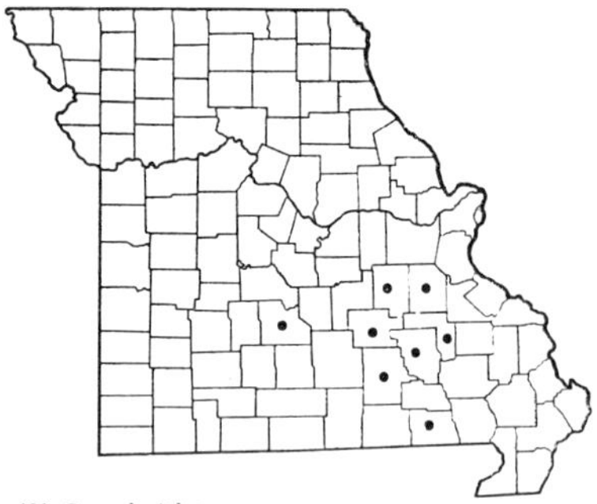

481 Carex leptalea

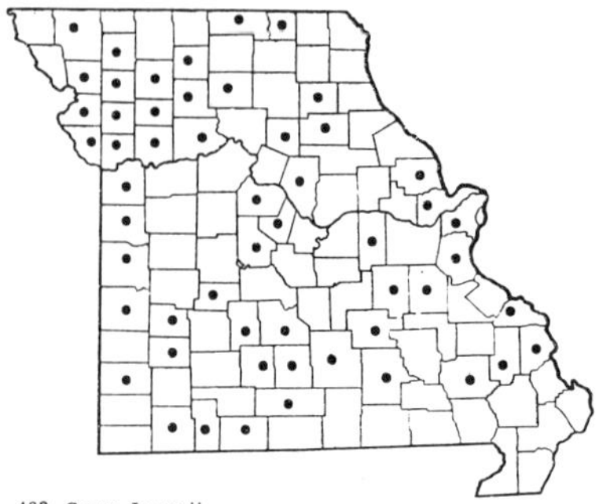

482 Carex Jamesii

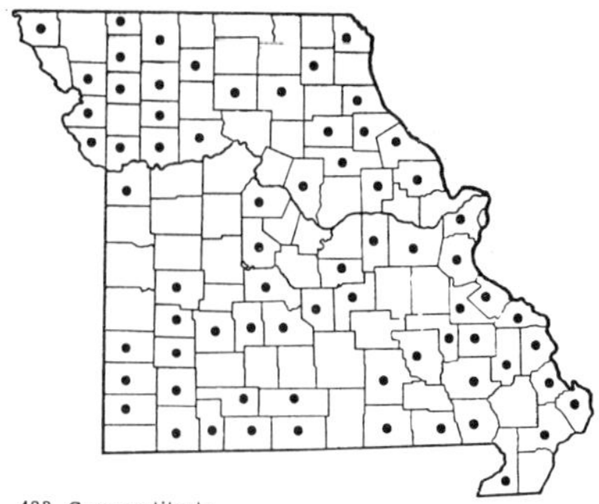

483 Carex artitecta

Ozarks in Shannon County (T31N, R6W, sect. 32, 7 mi. south of Cedar Grove, May 3, 1947, *Steyermark 64267*; also same locality, June 26, 1951, *Steyermark 71854, 71855*).

Ranges from Massachusetts to Michigan, south to D. C., Indiana, and Missouri.

In this species the culms attain 1 meter in height, exceeding the leaves. The pale brown spikes are mostly distant and well-separated from one another in an arched, curving or nodding, flexuous, elongated inflorescence, the perigynia are round-ovate to suborbicular, abruptly narrowed to a pronounced beak, and the spikes are long-tapering at the base.

## Subgenus II. **Eucarex**

### Section 9. **Polytrichoideae**

39. **Carex leptalea** Wahlenb. Map 481

Flowers May–July.

Occurs in swampy, calcareous, spring-fed meadows and at the base of moist shaded limestone ledges of bluffs along streams and spring branches. Mainly concentrated in a section of the southeastern Ozarks and locally west to Laclede County.

Ranges from Labrador to Alaska, south to North Carolina, Tennessee, Missouri, North Dakota, Colorado, and California.

This is another one of the boreal North American species, which apparently migrated or was forced southward into the Ozarks during one of the early advances of the Pleistocene ice, and has since remained in its present sites as a relict survivor of that period. It is an easily recognizable species with its fine thread-like leaves and single terminal spike of few-flowered, erect-appressed, beakless perigynia.

### Section 10. **Phyllostachyae**

40. **Carex Jamesii** Schwein. Map 482

Flowers late April–July.

Frequents rich woods, occurring in wet and dry situations, on rocky (mostly calcareous) or nonrocky slopes, at the base of bluffs, in flat lowland, and wet swampy forest; usually in neutral to alkaline soils. Common throughout the state, except absent in the southeastern lowland section.

Ranges from New York and Ontario to Michigan and Iowa, south to Virginia, Tennessee, Missouri, and Kansas.

This species is easily recognized in the field by the low dark green clumps of short, soft leaves which mostly exceed the inconspicuous, few-flowered spikes with subglobose, slender-beaked perigynia.

With its clusters of dark green, grasslike leaves, *C. Jamesii* superficially resembles the familiar lily turf (*Ophiopogon japonicus* and *Liriope spicata*), often used for borders or ground plants in conservatories of the northern states or as outdoor turf and pot plants in the southern states.

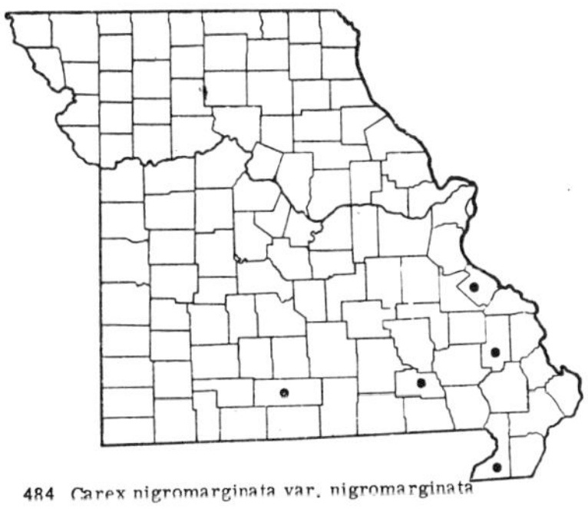

484 Carex nigromarginata var. nigromarginata

485 Carex physorhyncha

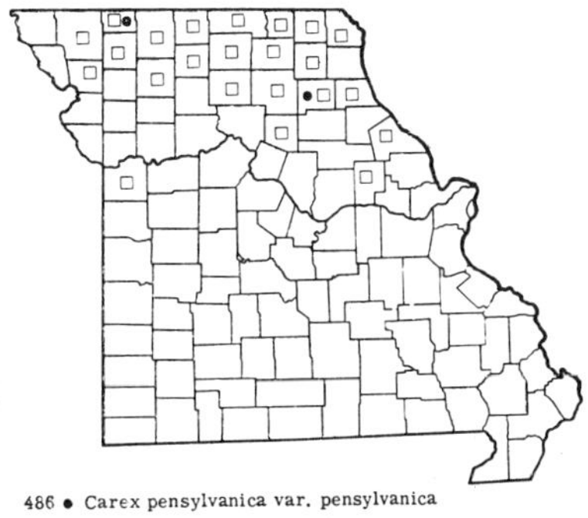

486 • Carex pensylvanica var. pensylvanica
486 □ Carex pensylvanica var. digyna

## Section 11. **Montanae**

41. **Carex artitecta** Mackenz. Map 483

*Carex nigromarginata* Schw. var. *Muhlenbergii* (Gray) Gl. in part [BB]

Flowers March–May.

Frequents mostly acid, dry soils of chert, sandstone, and granite, but also on limestone, and on wooded slopes, ridges, uplands, sandstone ledges, openings in woodland, and occasionally on sandy alluvium of low flat woodlands. Throughout Missouri.

Ranges from Maine and Quebec to Ontario, Michigan, Illinois, and Iowa, south to South Carolina, Tennessee, Kansas, and Texas.

This is one of the earliest-flowering species of *Carex*, along with *C. umbellata* and *C. pensylvanica*, to appear in the spring. *Carex artitecta* may be recognized in the field by its nonstoloniferous dense tufts of long slender leaves, which are purplish-red at the base and mostly shorter than the elongated slender culms. The fact that it is one of the first of the *Carex* species to flower, its relative abundance, and its usual affiliation for dry acid wooded soils are additional aids in its field identification.

42. **Carex nigromarginata** Schw. var. **nigromarginata** Map 484

*Carex nigromarginata* var. *nigromarginata* [BB]

Missouri material is referred to typical var. *nigromarginata*, the var. *floridana* (Schwein.) Kükenth. being a local Atlantic Coastal Plain variety.

Flowers April–May.

Occurs on acid soils of wooded sandstone slopes, crevices of sandstone or chert outcrops, and on gravelly wooded soils of Crowley Ridge. Rare and known from a few counties of the southern Ozark section (Ste. Genevieve, Bollinger, Wayne, and Douglas counties) and on Crowley Ridge in Dunklin County (Campbell, *Bush 6397*; *Bush* and *Palmer 3580*).

Ranges from Florida to Louisiana, north to Connecticut, New York, Pennsylvania, Indiana, and Missouri.

43. **Carex physorhyncha** Liebm. Map 485

*Carex nigromarginata* var. *Muhlenbergii* (Gray) Gl. in part [BB]

Flowers late April–June.

Known only from two collections in the lowlands of southeastern Missouri, from Mississippi (sandy open ground in Charleston Cemetery, 2 mi. west of Charleston, May 2, 1936, *Steyermark 10268*) and Dunklin (May 17, 1892, *Bush*) counties.

Ranges from Florida to Texas and Mexico, north to Virginia, Missouri, and Oklahoma.

The Dunklin County specimen in the herbarium of the University of Missouri was originally labeled *C. novae-angliae* by Bush, and later identified by Mackenzie as *C. Emmonsii* Dew. The latter species was later treated by Mackenzie under *C. albicans* Willd. (*N. Am. Fl.* 18: 190. 1935), but *C. albicans* is now considered synonymous with *C. Emmonsii*.

44. **Carex pensylvanica** Lam. Map 486

This species, including its varieties, sheds considerable pollen, which is occasionally suspected of causing some hay fever.

Missouri material may be divided into two varieties:

Mature perigynia 1.3–1.8 mm. wide, the body obtusely 3-angled in cross-section; beak of perigynium 1/6–2/5 (averaging 1/4) as long as the body . . 44a. C. PENSYLVANICA var. PENSYLVANICA

Mature perigynia 1.5–2.2 mm. wide, the body scarcely angled and round or more or less circular in cross-section; beak of perigynium 1/3–1/2 as long as the body . 44b. C. PENSYLVANICA var. DIGYNA

487 Carex communis

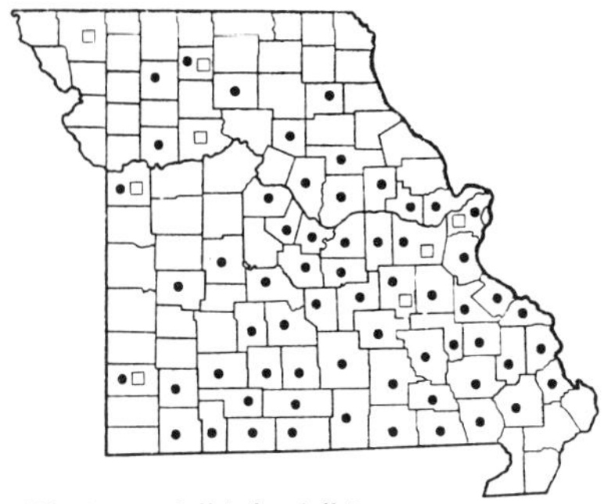

488 • Carex umbellata f. umbellata
488 □ Carex umbellata f. vicina

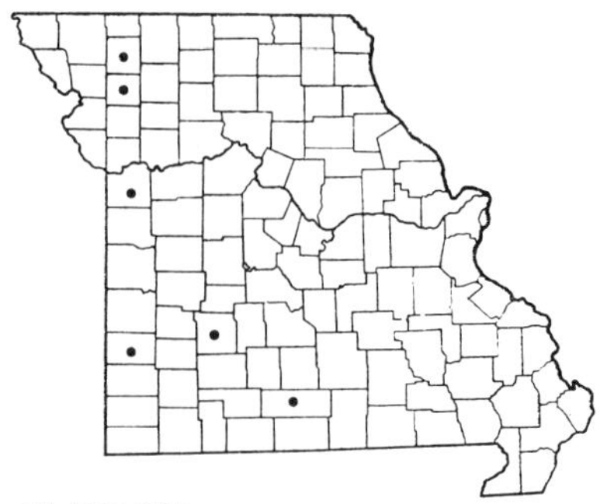

489 Carex abdita

44a. **Carex pensylvanica** var. **pensylvanica** Map 486

*Carex pensylvanica* Lam. [G]

Flowers early April–June.

Occurs in dry acid soils of upland wooded slopes and ridges. Known in northern Missouri from Shelby and Worth counties. This is the less common of the two varieties in the state.

Ranges from Quebec to Ontario and North Dakota, south to South Carolina, Tennessee, and Missouri.

44b. **Carex pensylvanica** var. **digyna** Boeckl. Map 486

*Carex heliophila* Mackenz. [BB]

Flowers early April–June.

Occurs in dry acid soils of upland wooded slopes and ridges. Common in northern Missouri south to Pike, Montgomery, Randolph, and Jackson counties.

Ranges from Ontario to Alberta, south to New York, Indiana, Illinois, Missouri, Kansas, and New Mexico.

Sometimes treated as a distinct species by some authors, but it seems preferable to consider var. *digyna* as a well-marked western variety of *C. pensylvanica* with which it intergrades in the width of the perigynia and length of the beak.

Closely resembling *C. artitecta* in the field with similar leaves and purplish-red color at the base of the plants, but with leafless horizontal stolons from the base, and perigynia with a more or less globose or subglobose body. The achenes in *C. pensylvanica* also more often have a more whitish or paler gray-white color than those of *C. artitecta.*

45. **Carex communis** Bailey Map 487

Flowers late April–June

Known from rich woods and wooded limestone slopes from Livingston County in northern Missouri and two counties in southwestern Missouri (Taney Co.: wooded limestone slopes below Table Rock along White River, April 25, 1936, *Steyermark 10116*; Barry Co.: half-way up wooded limestone slopes along White River, east of Smith Ford, south of Shell Knob, April 26, 1936, *Steyermark 10229*).

Ranges from Quebec to Ontario and Minnesota, south to Georgia, Kentucky, and Arkansas.

The stations noted above along White River have now been destroyed by the impoundment of the Table Rock Dam. Other stations from wooded limestone slopes along streams in southwestern Missouri should be carefully searched for new localities where it is expected the species may exist.

46. **Carex umbellata** Schkuhr Map 488

Represented in Missouri by two variations:

Culms all crowded among the bases of the leaves . . . 46a. C. UMBELLATA f. UMBELLATA

At least some of the culms elongated, 3–20 cm. tall . . . . . . . 46b. C. UMBELLATA f. VICINA

46a. **Carex umbellata** f. **umbellata** Map 488

*Carex umbellata* Schkuhr [G, BB, P & S]

Flowers late March–May.

Occurs in acid soils of dry upland woods, often on cherty, sandstone, or granitic soils, on slopes, crests of hills, ridges, and upland areas, also on sandstone outcrops and in dry prairie, prairie openings, open grassy slopes, and upland pastures. The commoner variation in Missouri, where it occurs throughout the Ozark region, and north locally in northern Missouri to Shelby, Linn, Grundy, and Daviess counties; absent

Plate no. 88. **1.** Carex artitecta. **2.** Carex nigromarginata. **3.** Carex physorhyncha. **4.** Carex pensylvanica. **5.** Carex communis. **6.** Carex umbellata. **7.** Carex abdita. **8.** Carex tonsa. **9.** Carex eburnea. **10.** Carex hirtifolia. **11.** Carex Meadii. **12.** Carex tetanica.

PLATE NO. 88

490 Carex tonsa

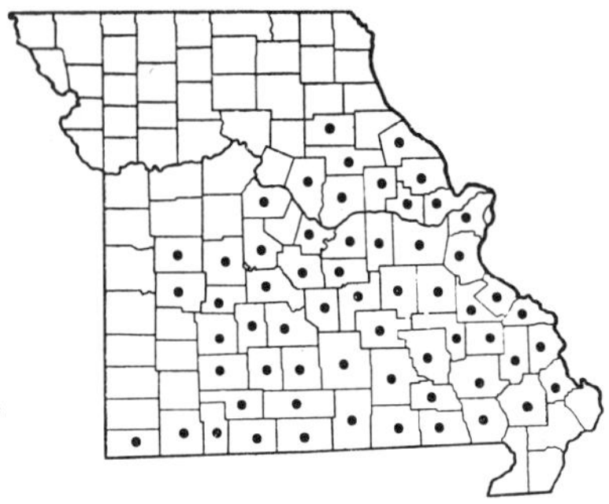

491 Carex eburnea

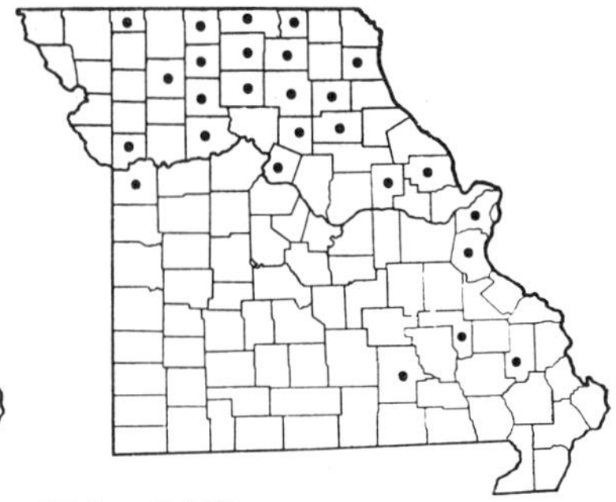

492 Carex hirtifolia

from the lowland section of southeastern Missouri.

Ranges from Newfoundland to Ontario and Minnesota, south to Virginia, Indiana, Illinois, and Missouri.

Like *C. artitecta* and *C. pensylvanica*, this species is one of the earliest of *Carex* species to flower in the spring. It is often overlooked, because of its resemblance to a bunch of grass and because the flowering culms are usually hidden among the leaves. In this respect it is like the closely related *C. abdita* and *C. tonsa*.

46b. **Carex umbellata** f. **vicina** (Dew.) Wieg. Map 488

Scattered throughout the range of the typical form and often occurring with it.

47. **Carex abdita** Bickn. Map 489

*Carex umbellata* var. *brevirostris* Boott
*Carex microrhyncha* Mackenz.
*Carex umbellata* in part [of P & S], not Schkuhr
Flowers late March–May.

Occurs in rocky open ground, prairies, prairie openings, and dry open woods. Found in western Missouri east to Gentry, Polk, and Douglas counties. Future collecting should reveal a greater frequency of occurrence of this species.

Ranges from Newfoundland and Saskatchewan, south to Virginia, Tennessee, Illinois, and Missouri.

This differs from *C. umbellata* mainly in the smaller perigynia with shorter beaks, and in the shorter, less pointed tips of the pistillate scales.

48. **Carex tonsa** (Fern.) Bickn. Map 490

*Carex umbellata* var. *tonsa* Fern.
Flowers April–June.

Known only from the southeastern Ozarks in Ste. Genevieve County (sandstone wooded slopes of La Motte sandstone outcrops and bluffs along River aux Vases, T36N, R7E, sect. 26 and 27, $3\frac{1}{2}$–4 mi. northwest of Coffman, May 11, 1957, *Steyermark 84407*).

Ranges from Quebec and Alberta, south to Virginia, Indiana, Wisconsin, Missouri, and Minnesota.

Differing from *C. umbellata* and *C. abdita* in the glabrous instead of pubescent perigynia with longer beaks, and in the firmer, more scabrous and broader leaves.

By some authors the last three species are united under *C. umbellata*. However, the various characters by which *C. abdita*, *C. tonsa*, and *C. umbellata* are distinguished appear to be fairly constant in the main, and, pending additional experimentation, it would seem preferable to retain them in separate specific rank.

## Section 12. **Albae**

49. **Carex eburnea** Boott Map 491

Flowers April–August.

Frequents the crevices of limestone bluffs, rarely found otherwise on bluffs of St. Peter sandstone or on the ground at the base of limestone talus. Common throughout the entire Ozark section, north to Pike, Monroe, and Boone counties, west to Cooper, Henry, St. Clair, and McDonald counties.

Ranges from Newfoundland to Alaska, south to Virginia, Alabama, Arkansas, and Texas.

By its distinctive habitat alone, this species can be easily identified. This, together with its tufts of numerous, fine, threadlike leaves, few-flowered spikes with pale pistillate scales, and small olive-green to black perigynia, make it further readily recognized in the field.

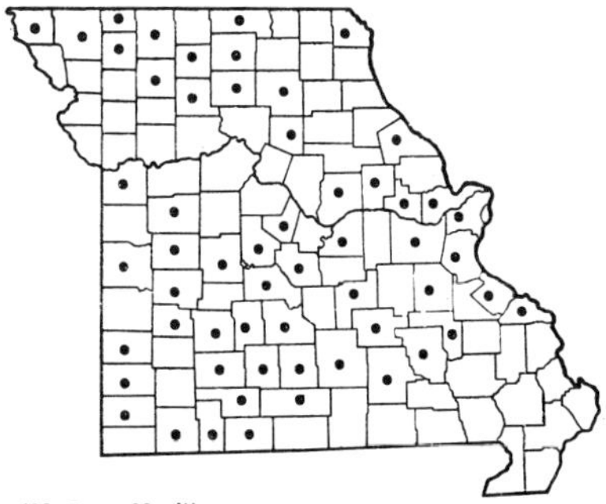

493 Carex Meadii

494 Carex tetanica

495 Carex Woodii

## Section 13. **Triquetrae**

50. **Carex hirtifolia** Mackenz. Map 492

Flowers late April–late June.

Frequents rich wooded areas with or without rock outcrops, on slopes, in ravine bottoms, and flat level wooded valleys. Known only from northern, central, and eastern Missouri; absent from the greater part of the Ozarks, southeastern lowland, and unglaciated prairie sections.

Ranges from New Brunswick and Quebec to Ontario and Minnesota, south to Maryland, Kentucky, Missouri, and Kansas.

The soft hairy leaves and culms, together with the habitat in rich woodland, easily identify this *Carex* in the field.

## Section 14. **Paniceae**

51. **Carex Meadii** Dew. Map 493

Flowers late March–June.

Occurs principally on limestone glades and dry upland prairies or prairie openings, also in wet meadows in valleys, and along railroads. Throughout Missouri, except absent in the southeastern lowland section; commonly found throughout the Ozark section on limestone or cherty limestone glades and on top of limestone bluffs.

Ranges from New Jersey, Pennsylvania, Ontario, Michigan, Wisconsin, Minnesota, Manitoba, and Saskatchewan, south to Georgia and Texas.

This is usually easily recognized in the field by the combination of limestone glade or prairie habitat, elongated underground rootstocks, usually glaucous or gray-green stiffish leaves, purplish-brown pistillate scales, glaucous or gray-green perigynia, and usually slender club-shaped staminate spike conspicuously raised on a long stalk. The only other species for which it can be mistaken are *C. Crawei*, which also occurs on limestone glades, and *C. tetanica*, known only from the loess hills of Atchison County. The main differences lie in the pistillate spikes and perigynia: in *C. Crawei* there are usually 3 or 4 pistillate spikes scattered along the length of the culm, some of the lower spikes being on distinct stalks, the dull green to brown perigynia, which taper to a short straight beak, are few-nerved, broadest at or below the middle and often resinous-dotted; in *C. Meadii* there are frequently only 1 or 2 sometimes sessile or stalkless pistillate spikes, the glaucous to pale green or greenish-white perigynia, which taper to the base and have a bent or curved beak, are strongly several-ribbed, broadest in the upper half, and not resinous-dotted.

As compared with *C. Meadii*, *C. tetanica* has the leaves thinner and green without being glaucous, the pistillate spikes are more slender with fewer rows of perigynia, which are closer together, straight, not as wide, less inflated, and more gradually tapering to the tip. However, plants which are intermediate and have overlapping combinations of characters occur and are difficult to place, as likewise noted for Indiana material by Hermann (Deam, *Flora of Indiana*, p. 245. 1940).

52. **Carex tetanica** Schkuhr Map 494

*Carex tetanica* var. *tetanica* [BB]

Flowers May–late June.

Known only from the loess hills of northwestern Missouri in Atchison County (Watson, June 1, 1894, *Bush 696*).

Ranges from Massachusetts to Alberta, south to Virginia, Pennsylvania, Ohio, Indiana, Illinois, northwestern Missouri, and South Dakota.

496 Carex Careyana

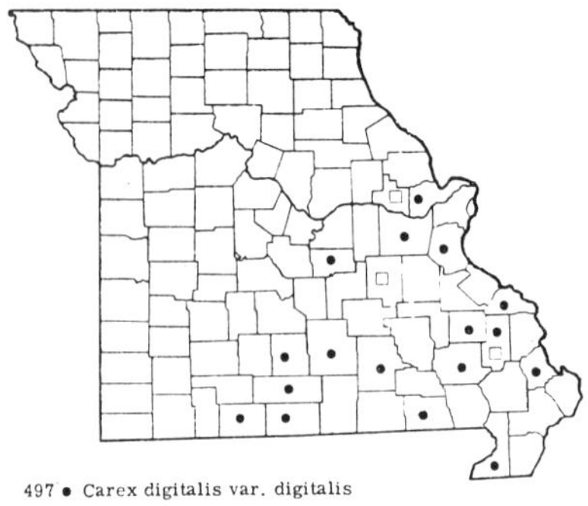

497 • Carex digitalis var. digitalis
497 □ Carex digitalis var. macropoda

498 Carex laxiculmis

Specimens from limestone glades in the White River section of southwestern Missouri, identified by their collectors as *C. tetanica*, have been referred by the present author to *C. Crawei*. *Carex tetanica* differs from *C. Crawei* mainly in having the perigynia glaucous, greenish-white, or gray-green, broadest in the upper half, tapering at the base, and strongly several-ribbed, and in the thinner, greener, not glaucous leaves.

53. **Carex Woodii** Dew. Map 495
*Carex tetanica* var. *Woodii* (Dew.) Bailey [BB, P & S in part]

Flowers late April–June.

Known only from the southern Ozarks in Howell County (swampy meadow along spring branch along Bennett Bayou, sect. 30, 5 mi. south of Caulfield, April 29, 1938, *Steyermark 5237*).

Ranges from Connecticut and New York to Manitoba, south to D. C., Ohio, Indiana, Illinois, and Missouri.

The only known station thus far encountered in the state is from a moist, calcareous, spring-fed meadow, quite unlike the woodland, although calcareous, habitats ascribed for the species in other parts of its range.

## Section 15. **Laxiflorae**

54. **Carex Careyana** Torr. Map 496
Flowers May–June.

Frequents rich woods, especially north-facing wooded limestone slopes, also in ravines, wooded valleys, and at the base of slopes. Restricted to the Ozark section in unglaciated terrain north to Lincoln, Callaway, and Boone counties, and west to Dallas, Christian, and Stone counties.

Ranges from New York to Ontario and Michigan, south to Virginia, Ohio, Indiana, Illinois, and Missouri.

This species develops rosettes of basal evergreen leaves which are prominently many-nerved and dark reddish-purple at the very base. The leaves, which are 8–17 mm. wide, may be distinguished from *C. albursina*, which also has persistent evergreen leaves, by the reddish-purple color at the base, by the darker green, not glaucous upper leaf surface, and by the relatively narrower (8–17 mm. wide as contrasted with usually 15–40 mm.) leaf-blades.

55. **Carex digitalis** Willd. Map 497
Flowers May–July.

Missouri material is represented by 2 variations within the species:

Upper pistillate spikes short-stalked or nearly stalkless; staminate spike on stalks 0.2–3 cm. (rarely 4.5) long; staminate spike of the inflorescence usually rising above the middle bract of the inflorescence; culms 0.5–5 dm. tall . . . 55a. C. DIGITALIS var. DIGITALIS

All the pistillate spikes on long stalks; staminate spike on stalks 5–20 cm. long; inflorescence more frequently shorter than the leaf-like middle bract; culms 3–6 dm. tall . 55b. C. DIGITALIS var. MACROPODA

55a. **Carex digitalis** var. **digitalis** Map 497
*Carex digitalis* Willd. [G, BB, P & S]

Occurs usually on north- or northwest-facing dry wooded slopes in acid soils, often of chert, sand, or sandstone, sometimes in crevices of shaded sandstone ledges or bluffs, also in ravines or at the base of

Plate no. 89. **1.** Carex Careyana. **2.** Carex digitalis. **3.** Carex laxiculmis. **4.** Carex albursina. **5.** Carex blanda. **6.** Carex gracilescens. **7.** Carex granularis.

PLATE NO. 89

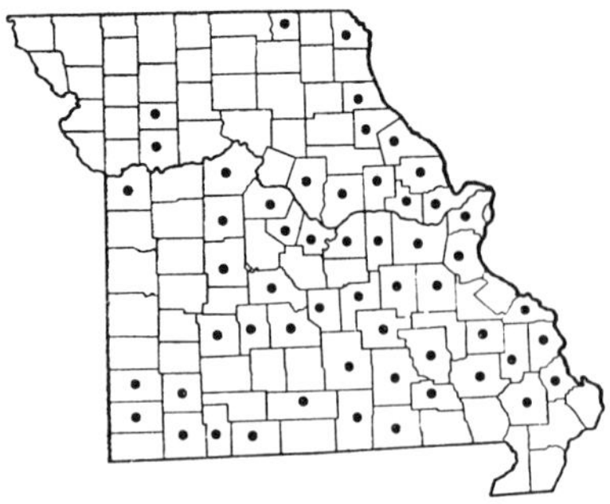

499 Carex albursina

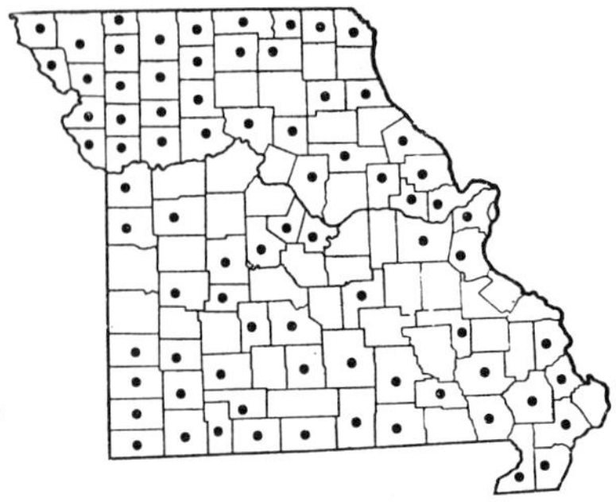

500 Carex blanda

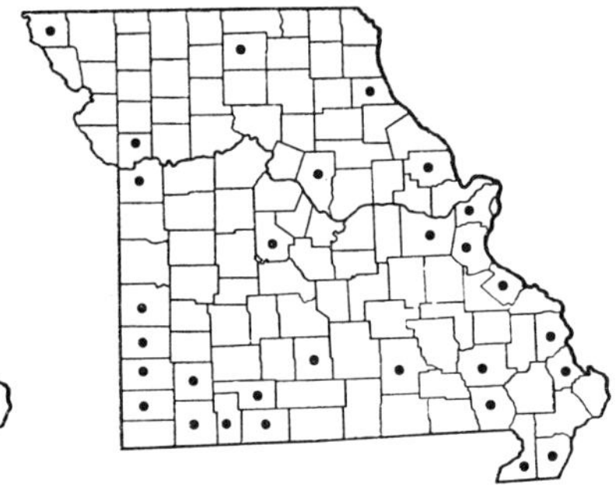

501 Carex gracilescens

slopes in limestone areas, and also along wooded moist margins of upland sink-hole ponds.

This, the commoner variety in the state, is found in the Ozark section, occurring north to St. Charles, Franklin, and Maries counties, and west to Wright and Taney counties, and in the southeastern lowland section on Crowley Ridge in Dunklin County.

Ranges from Maine to Ontario and Wisconsin, south to Florida, Mississippi, and Missouri.

55b. **Carex digitalis** var. **macropoda** Fern. Map 497

In similar habitats to var. *digitalis*.

Scattered within the range of var. *digitalis* in Warren (*Steyermark 73327*), Bollinger (*Steyermark 69613 a*), and Crawford counties.

Ranges from Virginia to Missouri, south to Alabama and Louisiana.

Some authors are inclined not to recognize the above variety *macropoda* because of intergrading and noncorrelating characters. Until more studies have been conducted throughout the range of the plants, they are, for the present, being retained as recognizable, intergrading varieties.

56. **Carex laxiculmis** Schwein. Map 498

Flowers May–June.

Frequents rich wooded areas, especially north-facing limestone slopes, ravines, and wooded valleys. Rare, known only from Bollinger, Madison, Ripley, Oregon, and Shannon counties in the southeastern Ozark section.

Ranges from Maine to Ontario and Wisconsin, south to North Carolina, Tennessee, and Missouri.

The specimen from Shannon County, previously included by Palmer & Steyermark in their *Annotated Catalogue* as *C. laxiculmis* var. *copulata* (Bailey) Fern. is now referred to *C. laxiculmis*. *Carex laxiculmis* var. *copulata* is treated by Fernald in the eighth edition of *Gray's Manual* as *C.* × *copulata* (Bailey) Mack., considered to be a hybrid between *C. digitalis* var. *digitalis* and *C. laxiculmis*. It differs from *C. laxiculmis* in having culms with sharp instead of blunt angles, sheaths of the bracts of the inflorescence minutely serrulate-toothed, instead of entire, and perigynia rounded, blunt, or round-tapering at the tip and abruptly short-beaked.

57. **Carex albursina** Sheldon Map 499

*Carex laxiflora* Lam. var. *latifolia* Boott [BB]

Flowers late April–late June.

Usually found on north-facing wooded, rich or limestone slopes, throughout the Ozarks, and locally north to Clark, Schuyler, Boone, Saline, Caldwell, and Jackson counties; absent from the southeastern lowland section and from parts of western Missouri.

Ranges from Quebec and Vermont to Minnesota, south to Virginia, Tennessee, and Arkansas.

The habitat and broad, gray-green, lilylike leaves, which are the largest of any of the species in Missouri, serve to identify this sedge easily in the field.

58. **Carex blanda** Dew. Map 500

*Carex laxiflora* var. *blanda* (Dewey) Boott [BB]

*Carex anceps* of P & S, not Muhl.

Flowers April–July.

Found in dry or wet woodland of all types, low wet woods, ravines, rich wooded slopes with or without rocks, alluvial thickets, and along rivers. Found throughout the state, and of expected occurrence in every county.

Ranges from Quebec to Ontario and North Dakota, south to Alabama and Texas.

This species varies in its foliage from dull green to glaucous, with leaf-blades 4–12 mm. wide. It is the most abundant and commonest species of *Carex* in Missouri, as it is also in Indiana and some other states. After eventually becoming acquainted with it, it becomes an easily recognized, if somewhat variable, species.

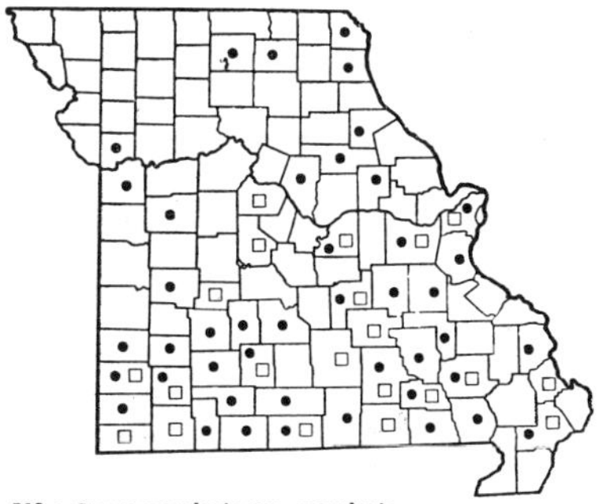

502 • Carex granularis var. granularis
502 □ Carex granularis var. Haleana

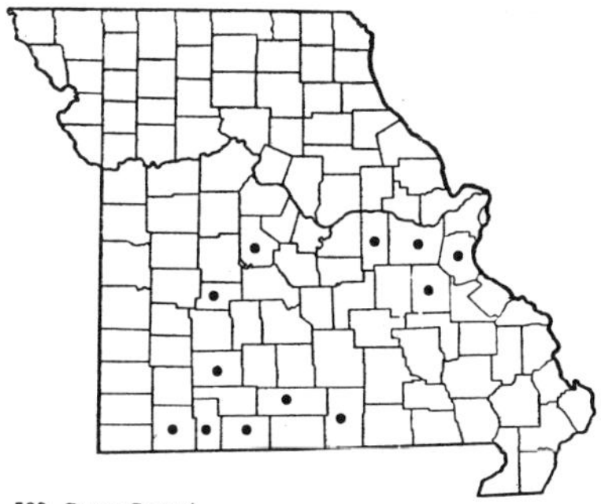

503 Carex Crawei

504 Carex microdonta

By Gleason (*New Ill. Fl.* 1: 336. 1952) this species is considered, along with *C. albursina*, *C. gracilescens*, and other species, as varieties of *C. laxiflora*. The group has long been subjected to various interpretations. In the present treatment the works of Fernald and Hermann, which maintain these as separate species, are being followed.

59. **Carex gracilescens** Steud. Map 501

*Carex laxiflora* var. *gracillima* (Boott) Robins. & Fern. [BB]

*Carex laxiflora* in part [P & S], not Lam.

Flowers late April–June.

Most frequently occurring in rich usually dry woods, with or without rocks, on or at the base of slopes, ravines, and valleys. Scattered over the state, and much less common than *C. blanda*; rare in the northern half of the state.

Ranges from Vermont and Quebec to Ontario and Wisconsin, south to Alabama and Texas.

With repeated experience this species can be readily distinguished from *C. blanda* by the usually long-stalked, brown to purplish staminate spike, the more strongly curved and slender snoutlike beak of the perigynia, and the dark purplish- or mahogany-red or purplish-brown color at the base of the culms.

## Section 16. **Granulares**

60. **Carex granularis** Muhl. Map 502

Flowers late April–late July.

Two varieties are represented in Missouri:

Perigynia 2.5–4 mm. long, 1.5–2.5 mm. wide, broadly ovoid to subglobose, becoming squarrose at maturity with spreading or outward-turning tapering tips . 60a. C. GRANULARIS var. GRANULARIS

Perigynia mainly 2–2.8 (–3) mm. long, 1–1.5 mm. wide, narrowly ovoid or oblong-ellipsoid, more or less erect-ascending and appressed at maturity, with rounded and abruptly minutely beaked tips not outwardly spreading or turning . . . . 60b. C. GRANULARIS var. HALEANA

60a. **Carex granularis** var. **granularis** Map 502

Frequent in wet woods and wooded swamps or alluvial flood plain, along streams, spring branches, spring-fed calcareous swampy meadows, moist limestone ledges of bluffs, river bottom prairies, and prairie swales. Common throughout the Ozark and unglaciated prairie sections and scattered in northern Missouri; apparently absent from the northwestern portion of the state.

Ranges from Vermont to Ontario and Minnesota, south to Florida, Arkansas, and Kansas.

60b. **Carex granularis** var. **Haleana** (Olney) Porter Map 502

Most often occurring in moist crevices of limestone bluffs, on rocky limestone glades, along spring branches, in swampy woods or wooded flood plain, and less commonly in wet open prairies. Less common than var. *granularis* in Missouri, where it is restricted to the Ozark section north to St. Louis, Osage, and Cooper counties, west to Hickory and Jasper counties.

This is sometimes treated as a distinct species by some authors. Both varieties often have glaucous or nearly glaucous leaves.

61. **Carex Crawei** Dewey Map 503

Flowers April–June.

Found on limestone glades, but rather rare, with two concentrations of occurrence, one in the northeastern sector of the Ozarks, the other in the southwestern sector, with an isolated station along the northern Ozark border in Morgan County.

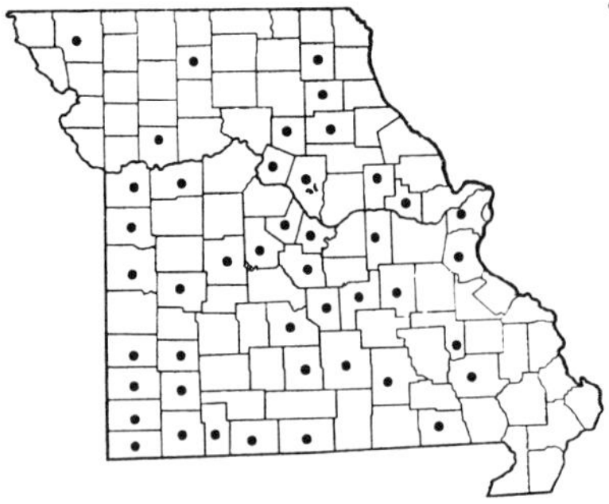

505 Carex oligocarpa

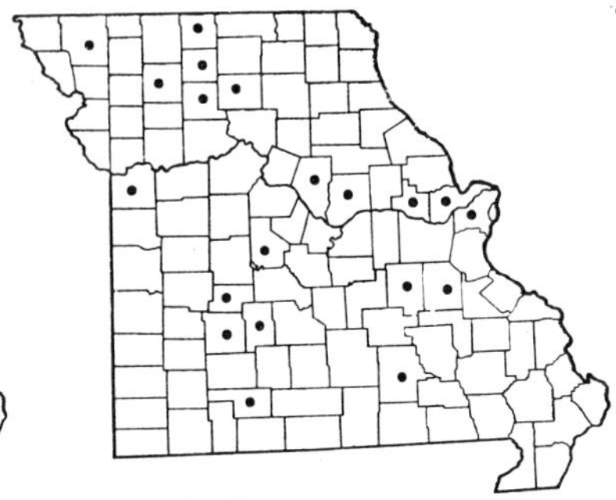

506 Carex Hitchcockiana

507 Carex conoidea

Ranges from Quebec to Alberta, south to New Jersey, Alabama, Tennessee, Missouri, Kansas, Wyoming, and Washington.

This species is frequently confused with *C. Meadii* and *C. tetanica* (see under *C. Meadii* and *C. tetanica* for their separation).

62. **Carex microdonta** Torr. & Hook. Map 504

Flowers April–July.

Occurs on limestone glades, dry prairies, and wet calcareous open soils. Rare, and known only from Osage (*Steyermark 27682*), Clay, Vernon, and Bates (*Palmer 53869*) counties in western and central Missouri.

Ranges from Missouri and Oklahoma, south to Mississippi and Texas.

Section 17. **Oligocarpae**
(including section *Griseae*)

63. **Carex oligocarpa** Schkuhr Map 505

Flowers late April–July.

Commonly found on north-facing, well-drained, wooded limestone slopes and on rich wooded slopes and ravines, also in alluvial flood plain forest, ledges along limestone bluffs, and rarely on open grassy slopes. Absent from the southeastern lowland section, elsewhere widespread in Missouri, especially in limestone wooded sections of the Ozarks.

Ranges from Vermont, Quebec, Ontario, and Michigan to Iowa, south to Florida and Texas.

Frequently confused with the varieties of *C. amphibola* having narrow leaves and slender pistillate spikes (for discussion of which see under that species). The base of the culms in *C. oligocarpa* has a dark purplish-red color.

64. **Carex Hitchcockiana** Dew. Map 506

Flowers May–June.

Occurs in rich woods, with or without rock, often on north-facing wooded limestone slopes, and in rich ravines or forested valleys. Rather rare and scattered in parts of northern Missouri and the Ozark region; apparently absent from the southeastern lowland and unglaciated prairie sections.

Ranges from Quebec and Vermont to Ontario and Wisconsin, south to Virginia, Tennessee, and Missouri.

The habitat, purple-red base of the plants, and pubescent leaf-sheaths are field characters which should aid in identifying this distinct species. The pubescence on the sheaths is short but evident. The species is often associated with other rich woods-inhabiting species of *Carex*, such as *C. hirtifolia*, *C. Jamesii*, *C. oligocarpa*, *C. albursina*, *C. sparganioides*, *C. blanda*, and *C. Careyana*.

65. **Carex conoidea** Schkuhr Map 507

Flowers May–June.

Known only from upland virgin prairies in Callaway County, central Missouri (virgin prairie [Tucker Prairie], west of Kingdom City, June 7, 1940, *W. B. Drew 104*) and Phelps County, central Ozarks (virgin prairie along creek, T36N, R7W, sect. 17, 2 mi. southwest of Elk Prairie, about 10 mi. southeast of Rolla, May 7, 1951, *Steyermark 71201*).

Ranges from Newfoundland to Ontario and Minnesota, south to Delaware, Ohio, Indiana, Illinois, and Missouri.

This species is distinguished from *C. Crawei* and *C.*

Plate no. 90. **1.** Carex Crawei. **2.** Carex oligocarpa. **3.** Carex Hitchcockiana. **4.** Carex conoidea. **5.** Carex amphibola var. rigida. **6.** Carex flaccosperma var. flaccosperma. **7.** Carex flaccosperma var. glaucodea. **8.** Carex gracillima.

PLATE NO. 90

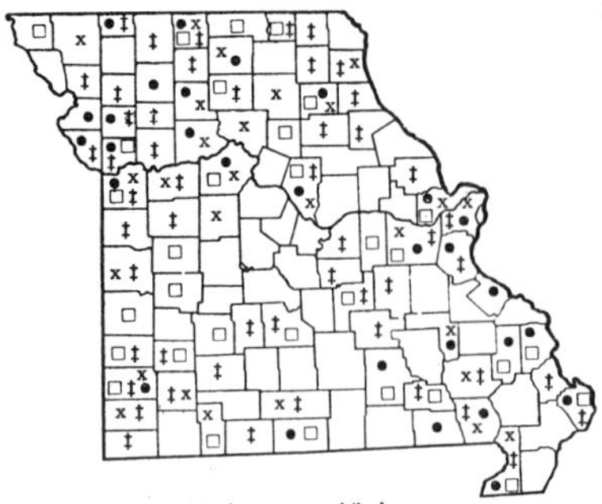

508 • Carex amphibola var. amphibola
508 x Carex amphibola var. rigida
508 ‡ Carex amphibola var. turgida
508 □ Carex amphibola var. globosa

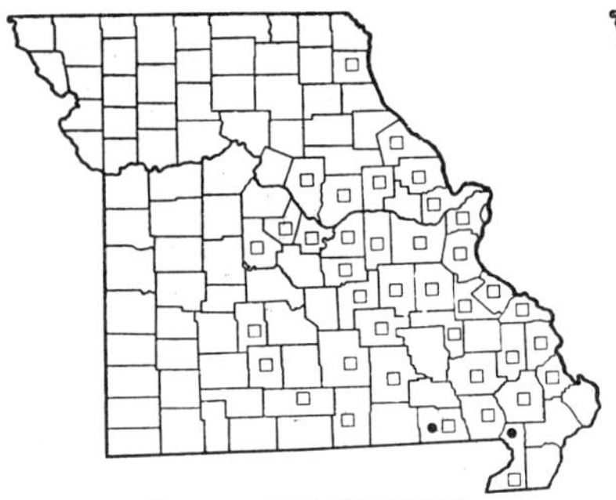

509 • Carex flaccosperma var. flaccosperma
509 □ Carex flaccosperma var. glaucodea

510 Carex gracillima

*Meadii*, which it resembles, in the deep green instead of glaucous leaves, the relatively long- and rough-awned pistillate scales which equal or exceed the perigynia, and the bunched culms.

66. **Carex amphibola** Steud. Map 508

Flowers late April–July.

This species is usually either divided into *C. grisea* and *C. amphibola*, or is split up into 4 separate rather poorly marked varieties of *C. amphibola*. Either procedure has its difficulties, but, since it does not seem that two distinct species can be represented in the mass of variation found in Missouri, the course followed in the present work is to maintain the division into the 4 named varieties, the main distinctions of which are outlined in the following key:

a. Base of plant strongly tinged purplish- or wine-red; leaves 1.5–4 mm. wide; mature perigynia rather tightly surrounding the achene, scarcely or little inflated; pistillate spikes often widely separated, the lower ones near the base of the plant . . 66a. C. AMPHIBOLA var. AMPHIBOLA

a. Base of plant usually brown or green; leaves mainly 4–10 mm. (or as little as 2) wide; mature perigynia mostly loose around the achene and more or less definitely inflated; pistillate spikes more approximate, the lowest not near the base of the plant . . . . . *b*

b. Perigynia somewhat narrowed or tapering at both ends, obtusely angled, somewhat inflated; pistillate spikes 3–6 mm. thick; staminate spike usually elevated conspicuously above the nearest pistillate spike. . 66b. C. AMPHIBOLA var. RIGIDA

b. Perigynia more rounded at both ends, with rounded angles, definitely more inflated; pistillate spikes 5–8 mm. thick; staminate spike usually partly surpassed or hidden by the upper pistillate spike . . . . . *c*

c. Main leaf-blades 4–10 mm. wide; larger pistillate spikes 7–20-flowered, 1–3 cm. long; perigynia 4–5.5 mm. long, more or less oblong, definitely longer than thick 66c. C. AMPHIBOLA var. TURGIDA

c. Main leaf-blades 2–5 mm. wide; larger pistillate spikes 3–7-flowered, 0.7–2 cm. long; perigynia 4–4.5 mm. long, nearly globose to obovoid-oblong, slightly longer than thick or nearly as thick as long . . 66d. C. AMPHIBOLA var. GLOBOSA

66a. **Carex amphibola** var. **amphibola** Map 508

*Carex amphibola* Steud. [BB, P & S, G, Deam]

Found in rich woods, with or without rock, often on north-facing wooded limestone slopes, or in ravines, wooded valleys, low wet or alluvial forested areas, and along streams. Scattered in Missouri, apparently more common in the northern and eastern sectors.

Ranges from Florida to Texas, north to Delaware, Pennsylvania, Tennessee, and Missouri.

The separation of this variety from *C. oligocarpa* often offers great difficulties, as the two overlap in such characters as few-flowered pistillate spikes with cuspidate or awned scales, relatively tight perigynia narrowed or tapering at both ends, narrow leaves, purplish-red base of culms, and elevated staminate spikes. In fact, the only criterion for placing such overlapping types in one or the other species is on the basis of the more definitely angled and beaked perigynium nearly filled by the achene in *C. oligocarpa* contrasted with a more obscurely or obtusely angled, scarcely or not beaked perigynium empty above in *C. amphibola* var. *amphibola*. Since both *C. amphibola* var. *amphibola* and var. *rigida* have the perigynium somewhat tapering to base and summit, the character of the tapering base characteristic of *C. oligocarpa* cannot be used as a mark of separation between these three. Likewise, the few-flowered pistillate spikes and narrow leaf-blades of *C. amphibola* var. *globosa*, characters also found in *C. oligocarpa*, cannot be used in separating these two.

66b. **Carex amphibola** var. **rigida** (Bailey) Fern. Map 508

*Carex grisea* var. *rigida* Bailey [P & S]

*Carex grisea* Wahl. in part [BB, Deam]

In similar situations as var. *amphibola.*

Scattered over the state.

Ranges from Florida to Texas, north to Massachusetts, New York, Indiana, Illinois, and Missouri.

The pistillate spikes in this variety are usually 5–20-flowered and 1–2.5 cm. long, and the leaf-blades mainly 3–7.5 mm. wide. As this variety has the perigynia somewhat tapering at the base and summit, similar to the condition in *C. oligocarpa,* the main points for separating it from that species are the relatively broader leaves, and more crowded, usually more numerous, slightly more inflated perigynia of the pistillate spikes.

Some of the earlier students (Bailey, Mackenzie) have been puzzled by this variety, and Fernald (Rh. 44: 314. 1942) considers it intermediate between the var. *amphibola* and var. *turgida.*

66c. **Carex amphibola** var. **turgida** Fern. Map 508

*Carex grisea* Wahlenb. in part [BB, P & S, Deam]

Usually found in wet, low or alluvial woodland, along streams, and in rich woods, ravines, and valleys. Throughout Missouri, where it appears to be the commonest variation of the species.

Ranges from New Brunswick to Ontario and Minnesota, south to Georgia, Alabama, Louisiana, and Texas.

In this variety the plants are much more robust, from 2–8 dm. tall, the leaf-blades are broader (4–10 mm. wide) than in the other varieties, and the pistillate spikes are thicker and longer with more perigynia. In fact, var. *turgida* stands at the opposite extreme of variation from that shown by *C. amphibola* var. *amphibola,* and in their extremes these two varieties can be readily distinguished. It is the intergradation between them and the other varieties which makes it difficult to recognize sharp distinctions in the group of *C. amphibola.*

66d. **Carex amphibola** var. **globosa** Bailey Map 508

*Carex grisea* var. *globosa* Bailey [P & S]

*Carex grisea* Wahlenb. in part [BB, Deam]

In similar habitats to var. *amphibola* and var. *turgida.*

Scattered throughout Missouri.

Ranges from Louisiana and Texas, north to Missouri.

Fernald has discussed at some length the perplexities encountered in the variation and nomenclature of *C. amphibola* and *C. grisea* (Rh. 44: 311–15. 1942). In view of the statement made by him that there is considerable doubt as to the true identity of what is being called *C. amphibola* on account of the mixed material contained in the original specimen, together with the doubtful application of the name *C. grisea,* it can be seen that there would be little justification in assigning two doubtful specific names to the mass of Missouri material. So far as attempting to divide this material into the two species, *C. grisea* and *C. amphibola,* such a course of procedure inevitably leads to the hopeless attempt to place plants having purple-based culms of *C. amphibola* with leaf-blades of 4–10 mm. wide, supposedly characteristic of *C. grisea.* The more conservative treatment as here presented reflects the uncertainties of the present status of an admittedly confused group, which needs considerably more study.

The very fact that there is difficulty in distinguishing some of the variations of *C. amphibola* (such as *C. amphibola* var. *amphibola*) from *C. oligocarpa* would seem to justify the placing of these two species into the same section, as was done by Fernald and likewise followed in the present work, rather than placing *C. oligocarpa* and *C. amphibola* in separate sections *Oligocarpae* and *Griseae* respectively, as was done in the works of Gleason and Deam.

67. **Carex flaccosperma** Dew. Map 509

Missouri material is represented by two varieties:

Foliage green to somewhat glaucous (gray- or silvery green); perigynia 4–6 mm. long; swampy ground in extreme southeastern Missouri . . . 67a. C. FLACCOSPERMA var. FLACCOSPERMA

Foliage definitely glaucous; perigynia 3–5 mm. long; throughout the Ozark section in dry, open or rocky woodland, often on ridges and upland slopes, or in valley woods along streams. . . 67b. C. FLACCOSPERMA var. GLAUCODEA

67a. **Carex flaccosperma** var. **flaccosperma** Map 509

*Carex flaccosperma* Dew. [G, BB, P & S]

Flowers May–early June.

In low wet woodland or swampy ground, in the lowland section of southeastern Missouri, where known only from Dunklin (moist ground, May 17 and May 20, 1892, *Bush*) and Ripley (depression in wooded swamp surrounded by sandy wooded knolls,

511 Carex oxylepis

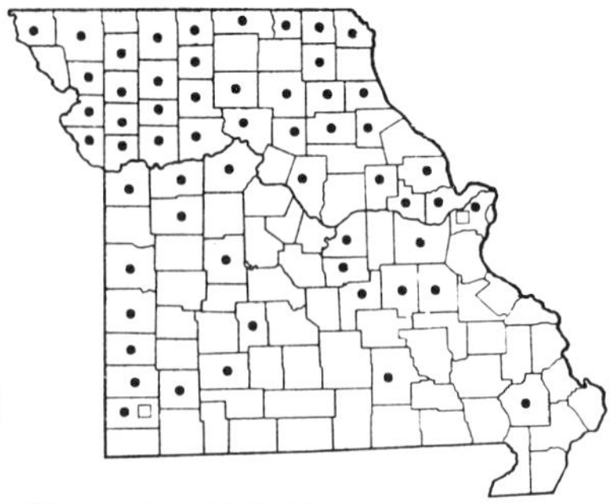

512 ● Carex Davisii f. Davisii
512 □ Carex Davisii f. glabrescens

513 Carex debilis var. debilis

T22N, R4E, southeast sect. 35, $4\frac{3}{4}$ mi. south of Naylor, May 28, 1951, *Steyermark 71239*) counties.

Ranges from Florida to Texas, north to Virginia, Tennessee, and southeastern Missouri.

At the station in Ripley County, this sedge was associated with the rare and local shrub, *Lindera melissaefolium*, and with *Carex debilis* var. *debilis*, *C. gigantea*, *C. typhina*, *C. crus-corvi*, *C. caroliniana*, and other species of the genus, as well as with other plants characteristic of the flora of the Atlantic coastal plain and Mississippi River Embayment of the gulf coastal plain.

67b. **Carex flaccosperma** var. **glaucodea** (Tuckerm.) Kükenth. Map 509

*Carex glaucodea* Tuckerm. [BB, P & S]

Flowers late April–early July.

Frequents usually acid soils of sandstone, chert, or granite wooded slopes, upland ridges, ravines, in usually dry situations, sometimes in wooded valleys along streams. Throughout the Ozark section west to Boone, Moniteau, Morgan, Dallas, Webster, and Douglas counties, north to Pike County, and locally in northeast Missouri in Lewis County.

Ranges from Alabama to Louisiana, north to Massachusetts, New York, Ontario, Ohio, Indiana, Illinois, and Missouri.

An easily recognized *Carex*, because of the thick glaucous leaves, which last throughout the winter months. This sedge is accepted by some authors as a distinct species. The overlapping range of the length of perigynia and the sometimes glaucous foliage of var. *flaccosperma* appear justifiable grounds for including the two under one species.

## Section 18. **Gracillimae**

68. **Carex gracillima** Schwein. Map 510

Flowers late May–late June.

Occurs in low woods along streams. Very rare, and known only from St. Louis (Fox Creek bottoms, Allenton, June 10, 1887, *Eggert*) and Greene (Springfield, June 16, 1889, *S. Weller* in Drury College Herbarium) counties.

Ranges from Newfoundland to Manitoba, south to Virginia, Tennessee, and Missouri.

69. **Carex oxylepis** Torr. & Hook. Map 511

Flowers late April–July.

Occurs in low rich damp woodland or in acid soils of dry sandy wooded knolls, and wet sandy ground. Known only from the southeastern lowland section in either low ground or on knolls of Crowley Ridge, in New Madrid (*Steyermark 69659*), Pemiscot (*Steyermark 82256*), Dunklin (May 24, 1893, *Eggert*; May 17, 1892, *Bush*; *Steyermark 26580A*), and Butler counties.

Ranges from Florida to Texas, north to Virginia, Tennessee, and southeastern Missouri.

This species grows in either rich lowland wooded areas or on dry sandy wooded slopes and knolls. At the Pemiscot County station, where it occurred in a virgin wooded section 3 mi. southwest of Portageville, it was associated with the rare and local *Polymnia laevigata*, and with such species of *Carex* as *C. Grayii*, *C. Frankii*, *C. tribuloides*, and *C. louisianica*. At the New Madrid County locality it occurred on dry sandy acid soils with *Quercus stellata*, *Q. falcata*, *Ulmus alata*, *Panicum Boscii* var. *molle*, *Carex artitecta*, *Galium pilosum*, and *Aster patens*.

Plate no. 91. **1.** Carex oxylepis. **2.** Carex Davisii. **3.** Carex debilis var. debilis. **4.** Carex cherokeensis. **5.** Carex virescens. **6.** Carex Swanii. **7.** Carex complanata.

PLATE NO. 91

514 Carex cherokeensis

515 Carex virescens

516 Carex Swanii

70. **Carex Davisii** Schw. & Torr. Map 512

Flowers May–July.

Occurs in low or wet woods, wooded alluvial soils of river flood plain, along streams, open fields and valleys, and on wooded rich slopes in ravines or valleys. Throughout northern, western, and central Missouri, less common in the Ozark section and absent in many parts of that section; apparently absent from the lowlands of southeastern Missouri.

Ranges from Vermont to Michigan and Minnesota, south to Maryland, Tennessee, Missouri, Oklahoma, and Texas.

Two variations are encountered in Missouri:

Leaf-sheaths and usually the lower surface of the leaf-blades more or less hairy . . 70a. C. DAVISII f. DAVISII

No hairs present on leaf-sheaths or leaf-blades, the plant completely glabrous. . 70b. C. DAVISII f. GLABRESCENS

70a. **Carex Davisii** f. **Davisii** Map 512

*Carex Davisii* Schw. & Torr. [G, BB, P & S]

This is the commoner variation found throughout the state.

70b. **Carex Davisii** f. **glabrescens** Kükenth. Map 512

Known only from Newton (thickets along brook, George Washington Carver National Monument, May 2, 1959, *Palmer 68561*; moist ground, thickets along brook, George Washington Carver National Monument, June 29, 1959, *Palmer 68562*) and St. Louis (*Eggert*) counties.

This species is common throughout the northern half of the state. During the growth of the plant, various modifications in the development of the pistillate spikes and perigynia produce corresponding changes of gross appearance. At first the pistillate spikes are erect, later they may become spreading or nodding. Likewise, the perigynia at first are pale green, but later turn to brown or yellow-brown. The pubescent leaf-sheaths and usually the pubescent lower surface of the leaf-blades are distinctive marks of this species.

Section 19. **Sylvaticae**

71. **Carex debilis** Michx. Map 513

Flowers late May–August.

Two varieties are represented in Missouri:

Perigynia 6–10 mm. long, narrowly lanceolate, broadest below the middle, gradually tapering to the tip; scales of pistillate flowers usually pale, whitish, or greenish-white. . 71a. C. DEBILIS var. DEBILIS

Perigynia 4.5–7 mm. long, broadly or ovate-lanceolate, broadest at or near the middle, more abruptly tapering to the tip; scales of pistillate flowers usually tinged with brown 71b. C. DEBILIS var. RUDGEI

71a. **Carex debilis** var. **debilis** Map 513

Known from low swampy woods in the southeastern Missouri lowlands in Stoddard (T25N, R11E, sect. 6, $3\frac{1}{2}$ mi. southeast of Bloomfield, August 20, 1954, *Steyermark 76827*) and Ripley (T22N, R4E, southeast sect. 35, $4\frac{3}{4}$ mi. south of Naylor, May 28, 1951, *Steyermark 71228*) counties.

Ranges from Florida to Texas, north along the coast to Massachusetts and inland north to Indiana and southeastern Missouri.

At the Ripley County station this sedge was growing with the rare and local *Lindera melissaefolium* in swampy depressions surrounded by sandy wooded knolls. At the Stoddard County station it was associated with *Commelina virginica*, *Osmunda cinnamomea*, and *Ilex opaca*.

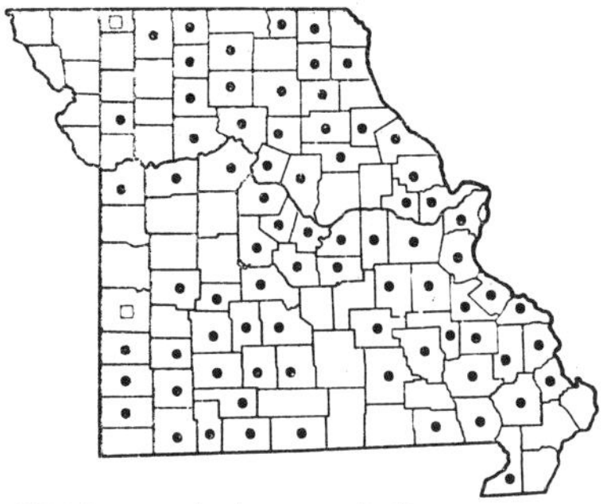

517 □ Carex complanata var. complanata
517 • Carex complanata var. hirsuta

518 Carex caroliniana

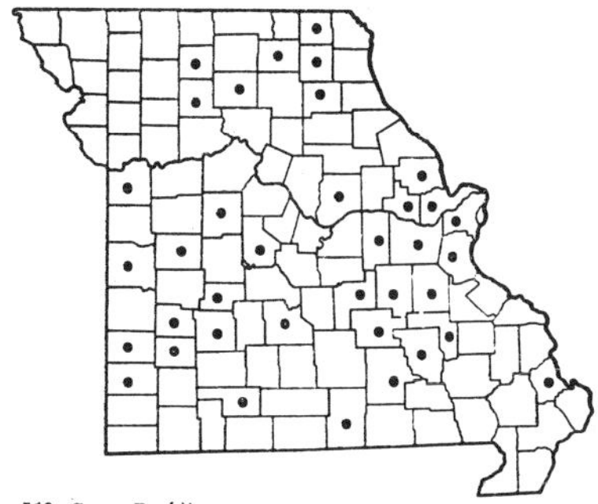

519 Carex Bushii

71b. **Carex debilis** var. **Rudgei** Bailey Map 513

Recorded from Missouri.

Ranges from Newfoundland to Ontario, Wisconsin, and Minnesota, south to Virginia, Tennessee, and Missouri.

Section 20. **Longirostres**

72. **Carex cherokeensis** Schw. Map 514

Flowers April–May.

Known only from low wet woods in the lowlands of southeastern Missouri in Butler County (May 1, 1905, *Bush 2658*).

Ranges from Florida to Texas, north to Georgia, Tennessee, southeastern Missouri, and Oklahoma.

Section 21. **Virescentes**

73. **Carex virescens** Muhl. Map 515

Flowers mid-May–July.

Occurs usually in acid soils of wooded sandstone ledges, north- or northwest-facing wooded slopes of chert or sandstone, more rarely at the base of wooded limestone slopes, and also in moist ground either in sandy moist soils along spring branches, or in low wet woods in the valleys of streams. Known only from a few stations in the southeastern Ozark section west locally to Texas County, and on Crowley Ridge in Stoddard County adjacent to the southeastern lowland section.

Ranges from Maine to Quebec, Michigan, and Indiana, south to Georgia, Tennessee, and Missouri.

74. **Carex Swanii** (Fern.) Mackenz. Map 516

*Carex virescens* [of P & S], not Muhl.

Flowers mid-May–July.

Occurs in wet woods and swampy low ground. Known only from the lowlands of southeastern Missouri in Scott, Dunklin, Butler, and Ripley counties.

Ranges from Nova Scotia to Ontario and Wisconsin, south to North Carolina, Tennessee, and Arkansas.

This species is found in low moist woods, whereas *C. virescens* is usually found on drier slopes and ledges in better-drained sites.

75. **Carex complanata** Torr. & Hook. Map 517

Missouri material is represented by 2 varieties, whose differences are indicated in the key to species.

75a. **Carex complanata** var. **complanata** Map 517

Flowers May–early July.

Occurs on dry upland open wooded slopes in western Missouri in Vernon and Worth (*Steyermark 65565*) counties.

Ranges from Florida to Texas, north near the coast to New Jersey, inland north to Tennessee and Missouri.

75b. **Carex complanata** var. **hirsuta** (Bailey) Gl. Map 517

*Carex hirsutella* Mackenz. [G, P & S]

Flowers May–July.

Frequents dry acid soils of open wooded slopes, ridges, and upland, mostly in cherty, sandstone, or granitic soils of the Ozarks or in sterile sandy or clay

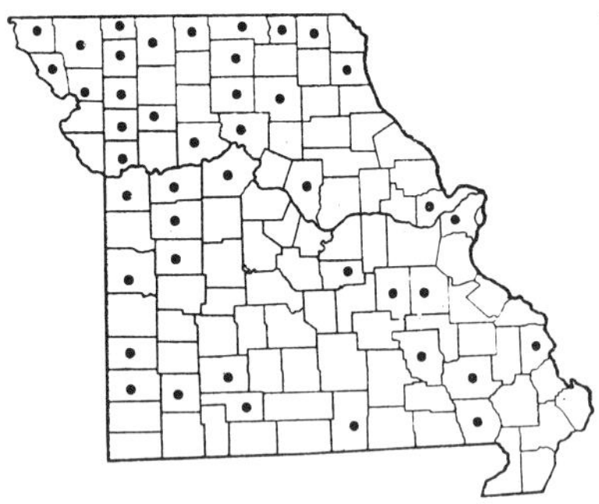

520 Carex lanuginosa

521 Carex scabrata

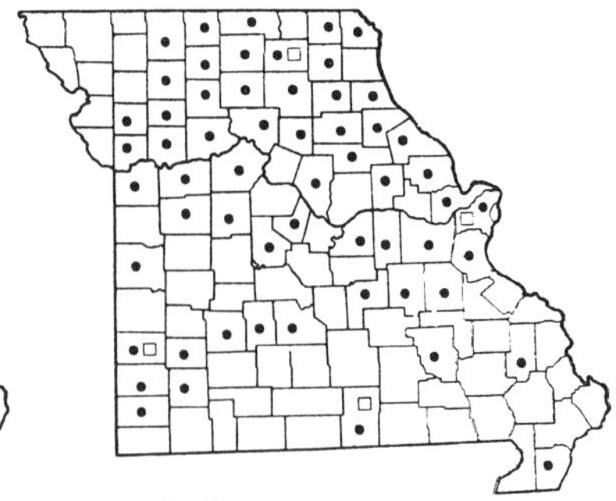

522 • Carex Shortiana
522 □ Carex X Deamii

soils elsewhere. Common in most parts of the state, but absent from the extreme northwestern counties and from a few in the extreme western and southeastern lowland sections.

Ranges from Maine and Quebec to Ontario, Michigan and Iowa, south to Alabama, Arkansas, and Texas.

This is one of the commonest and most frequently encountered species of *Carex* in Missouri on the drier upland and acid soils of open woodland, where it is associated frequently with *Carex artitecta*, *C. umbellata*, and *Danthonia spicata.*

The differences separating this from *C. complanata* var. *complanata* do not seem sufficiently well-marked or sharp enough to maintain the two as separate species, and I agree with Gleason's interpretation of this species in his *New Illustrated Flora* 1: 348. 1952.

76. **Carex caroliniana** Schw. Map 518

Flowers May–June.

Frequents low wet woods or wooded swamps and river flood plain forest, also in prairie openings, prairie swales, open grassy slopes, and upland pastures and prairies. Scattered in southern and central Missouri, north to St. Louis, Cooper, and Jackson counties.

Ranges from Georgia to Texas, north to New Jersey, Pennsylvania, Ohio, Indiana, Missouri, and Oklahoma.

77. **Carex Bushii** Mackenz. Map 519

Flowers early May–early July.

Frequently found in dry upland prairies, prairie openings, open grassy slopes, dry fields and pastures, wet river bottom prairie, wet open swales in meadows and valleys, along railroads, borders of ponds, open swamps and ditches, openings in dry upland wooded areas, upland sandstone outcrops, rarely on rich woodland slopes and ravines. Frequent in most parts of Missouri, but absent from the northwestern and southeastern lowland sections.

Ranges from Massachusetts and New York to Michigan and Iowa, south to Virginia, Mississippi, Arkansas, and Texas.

## Section 22. **Hirtae**

78. **Carex lanuginosa** Michx. Map 520

*Carex lasiocarpa* Ehrh. var. *latifolia* (Boeckl.) Gl. [BB]

Flowers late April–July.

Occurs most commonly in wet meadows and river bottom prairies of open valleys and alluvial soils, also in swales of upland prairies, calcareous spring-fed swampy meadows, and more rarely in wooded swamps. Found throughout the state, but most common in the northern half; absent from the southeastern lowland section and from some parts of the Ozarks.

Ranges from Quebec to British Columbia, south to Virginia, Tennessee, Arkansas, Oklahoma, Texas, New Mexico, Arizona, and California.

Plate no. 92. **1.** Carex scabrata. **2.** Carex Shortiana. **3.** Carex Joori. **4.** Carex Buxbaumii.

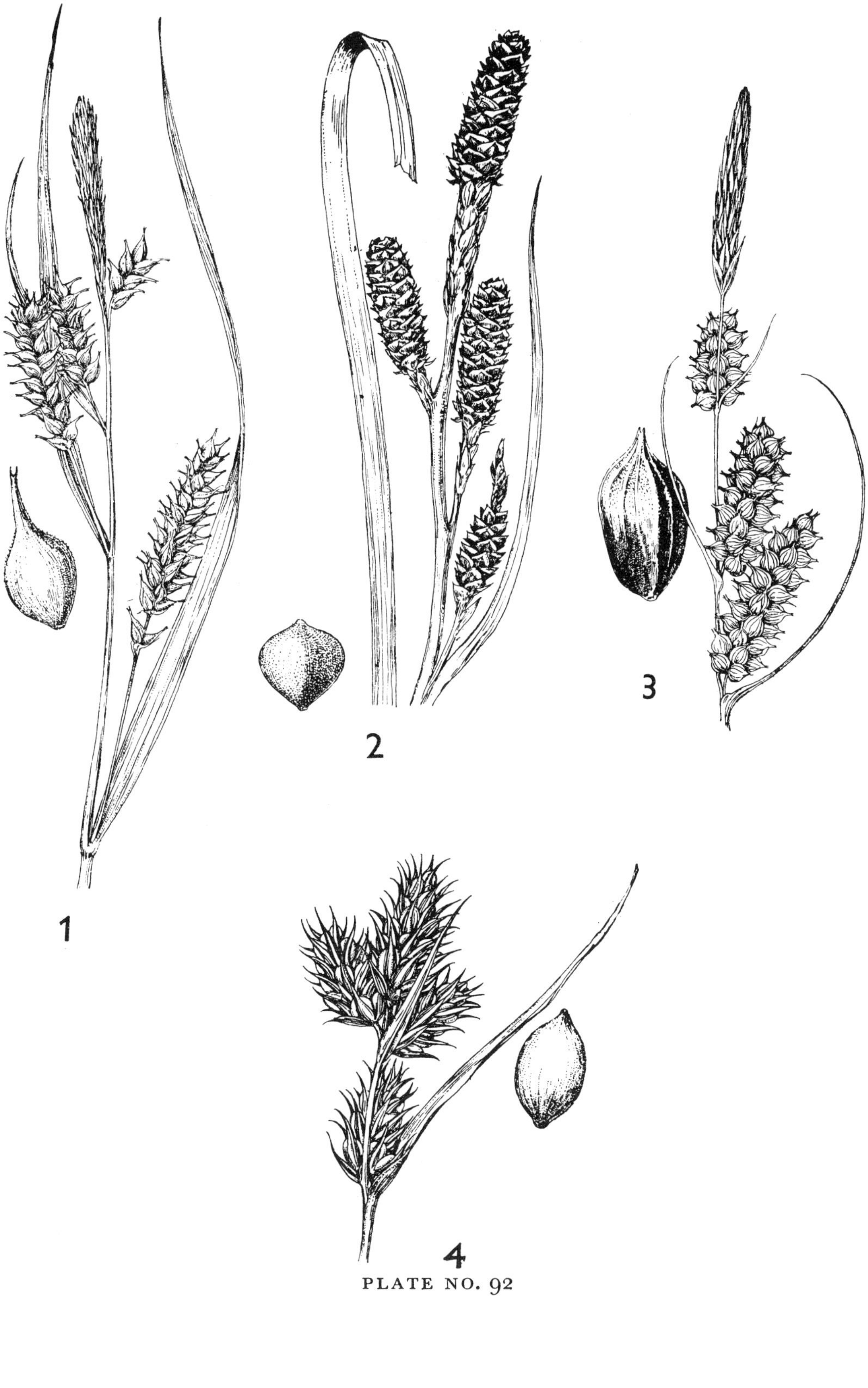


PLATE NO. 92

523 Carex Joori

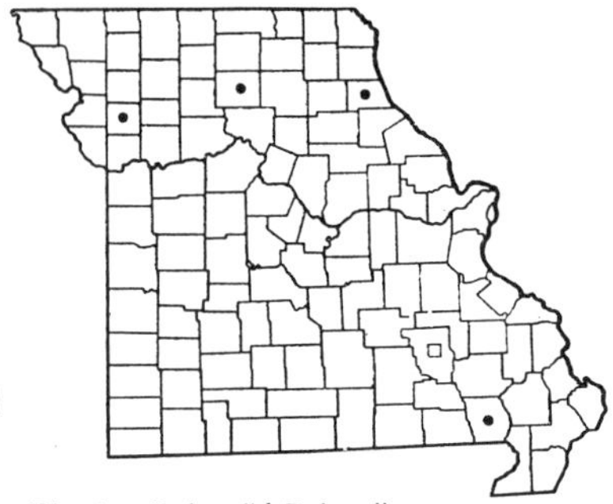

524 ● Carex Buxbaumii f. Buxbaumii
524 □ Carex Buxbaumii f. dilutior

525 Carex aquatilis var. altior

## Section 23. **Anomalae**

79. **Carex scabrata** Schwein. Map 521

Flowers June–August.

Known only from a single collection from southwestern Missouri in Greene County (Springfield, July 3, 1889, *S. Weller* in the herbarium of Drury College).

Ranges from Nova Scotia and Quebec to Ontario and Michigan, south to South Carolina, Ohio, Tennessee, and Missouri.

## Section 24. **Shortianae**

80. **Carex Shortiana** Dewey Map 522

Flowers early May–early July.

Frequents prairie swales, wet river bottom prairies, wet open valleys, borders of ponds, open swamps, and ditches, calcareous spring-fed swampy meadows, wooded swamps and flood plain forests, rich woods usually at the base of slopes, or in ravines and wooded valleys. Common in most parts of the state; apparently absent from extreme northwestern Missouri, most of the lowlands of the southeastern section, and from most parts of the southern Ozarks.

Ranges from Pennsylvania, Ontario to Iowa, south to West Virginia, Tennessee, Missouri, and Oklahoma.

80a. **Carex** × **Deamii** F. J. Herm. Map 522

*Carex Shortiana* Dew. × *Carex typhina* Michx. and possibly *Carex squarrosa* L.

Flowers late May–September.

Occurs in low wet or alluvial woods and wet margins of upland sink-hole ponds. Found in Missouri at four widely scattered stations in Adair (*Steyermark 79705*), St. Louis (*Letterman*), Howell (*Steyermark 78724*), and Barton (*Palmer 53821*) counties. Known elsewhere only from Indiana.

Originally described from Indiana as a hybrid between *C. Shortiana* and *C. typhina*. At the Missouri stations encountered it has been found growing with *C. Shortiana* and *C. typhina* as well as with *C. squarrosa*, or with *C. Shortiana* and *C. squarrosa*. Probably future exploration will reveal many more stations for this hybrid which has thicker spikes (7–8 mm. thick) and longer beaks (1–1.5 mm. long) of the perigynia than those of *C. Shortiana*, the parent it most resembles. It will be interesting to learn whether *C. typhina* or *C. squarrosa* proves to be the commoner of the other one of the presumed parents of *C.* × *Deamii*.

## Section 25. **Pendulinae**

81. **Carex Joori** Bailey Map 523

Flowers August–October.

Known only from low wet woods in the southeastern Missouri lowlands from Dunklin County (Campbell, September 12, 1910, *Bush 6327* and *6327A*).

Ranges from Florida to Texas, north to Virginia and Maryland, and in the interior to Tennessee and southeastern Missouri.

---

Plate no. 93. **1.** Carex aquatilis var. altior. **2.** Carex nebraskensis.

1 2

PLATE NO. 93

526 Carex nebraskensis

527 Carex Haydenii

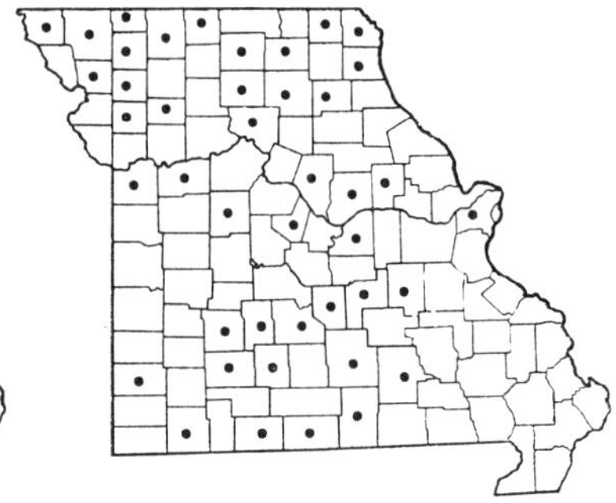

528 Carex Emoryi

## Section 26. **Atratae**

82. **Carex Buxbaumii** Wahl. Map 524
Flowers late April–July.
Two variations occur in Missouri:

Scales of pistillate flowers purple-black, brown, or blackish-brown . 82a. C. BUXBAUMII f. BUXBAUMII
Scales of pistillate flowers whitish or pale brown 82b. C. BUXBAUMII f. DILUTIOR

82a. **Carex Buxbaumii** f. **Buxbaumii** Map 524
*Carex Buxbaumii* Wahl. [G, BB, P & S]

Occurs in prairie swales of upland prairie along railroads, in wet river bottom prairie and alluvial meadows, and swampy woods. Rare and scattered in northern Missouri, where known from Marion (prairie between Monroe City and Hunnewell, May 28, 1934, *Steyermark 8315*), Linn (bottom prairie, valley of Locust Creek, Pershing State Park, June 26, 1941, *Steyermark 40432*), and Clinton counties, and in the lowlands of southeastern Missouri where known from Butler County (swamps, May 1, 1905, *Bush 2663*).

This species forms large colonies and clumps from the long horizontal stolons. The whitish, nearly nerveless perigynia, narrow pale green or glaucous leaves, and dark red-purple base of the plant are additional recognition characters of this distinct species. While it bears a superficial resemblance to *C. Bushii*, the two are immediately distinguished in the field by the completely glabrous leaf-sheaths and blades of *C. Buxbaumii*.

82b. **Carex Buxbaumii** f. **dilutior** Kükenth. Map 524

Known only from a calcareous, spring-fed, swampy meadow in the southeastern Ozark region from Reynolds County (along West Fork of Black River, T33N, R3W, sect. 23 and northeast sect. 26, 3–3½ mi. northwest of Greeley, July 6, 1951, *Steyermark 71984*).

Dr. Hermann believes that the pale color of the pistillate scales is the result of bleaching with age. If this is true, it is probable that no definite name should be given this form (Rh. 57: 312. 1955).

## Section 27. **Acutae**

83. **Carex aquatilis** Wahlenb. var. **altior** (Rydb.) Fern. Map 525
*Carex aquatilis* Wahl. in part [BB]
*Carex substricta* (Kükenth.) Mackenz.
*Carex aquatilis* var. *substricta* Kükenth. [Deam]
Flowers June–August.

Known only in the southeastern Ozark section from Carter County (alluvial banks of Ten Mile Creek, T26N, R3E, sect. 36, July 13, 1936, *Steyermark 11735*).

Ranges from Newfoundland and Quebec to British Columbia, south to New Jersey, New York, Ohio, Indiana, Wisconsin, Missouri, Nebraska, Colorado, and Oregon.

This is apparently one of the Pleistocene relict species of a more boreal flora which has survived in an isolated sector of the Ozarks.

84. **Carex nebraskensis** Dewey Map 526
Flowers May–July.

Introduced along railroad tracks in St. Louis County (north St. Louis, freight yard of Burlington

Plate no. 94. **1.** Carex lanuginosa. **2.** Carex hyalinolepis.

PLATE NO. 94

Railroad, north of Carrie Avenue, along the fifth track counted from east, June 15, 1957, *Muehlenbach 1199*).

Ranges from western Kansas and South Dakota to British Columbia, south to New Mexico and California; introduced in Missouri.

Dr. F. J. Hermann, *Carex* authority, has identified this specimen.

The plants have long creeping rootstocks with large blackish-brown scales. The leaves on the Missouri material are 3 dm. long, 3.5–5 mm. long, and are longer than the flowering culms which are 2.1–2.5 dm. tall. The pistillate spikes are on short to long slender peduncles which may attain 5 cm. in length. The lowest peduncles are sometimes drooping, but most are erect. The 2–5 pistillate spikes are cylindrical, 2–2.5 cm. long, 6–7 mm. thick in Missouri material. The pistillate scales are lanceolate, brown-purple or dark brown with a pale green or buff-nerved stripe down the middle, tapering into a slender, somewhat rigid cusp or awn 1–1.5 mm. long. The pistillate scales equal to slightly exceed the perigynia which are straw-colored or buff, sometimes speckled with maroon-red near the summit, and strongly nerved on the faces. In the Missouri specimen the perigynia are 2.7–3 mm. long, 1.6–1.8 mm. wide (perigynia up to 3.5 mm. long are described in Rydberg's *Flora of the Prairies and Plains*). The achenes have a hooked or bent style.

This species differs from *C. crinita*, with which it might at first be confused, by the mostly erect pistillate spikes, the bidentate beak of the strongly ribbed perigynia, and the uniformly short-awned or short-cuspidate instead of usually long-awned perigynia.

85. **Carex Haydenii** Dew. Map 527

Flowers May–July.

Occurs in wet river bottom prairies and along streams. Infrequent in northern and central Missouri, mostly in the northwestern third of the state.

Ranges from New Brunswick and Ontario to Minnesota, south to New Jersey, Pennsylvania, Ohio, Indiana, Illinois, Missouri, and Nebraska.

Often confused with *C. Emoryi*, a much more common species in the state.

86. **Carex Emoryi** Dew. Map 528

*Carex stricta* [of P & S], not Lam.

*Carex stricta* var. *elongata* (Boeckl.) Gl. [BB]

Flowers late April–June.

Commonly occurring along banks of streams in the Ozark section or in wet river bottom prairies and valleys in northern and central Missouri. Widespread in the state, most common in northern Missouri and the Ozark section; apparently absent from the southeastern lowland section, and rare in the unglaciated prairie region.

Ranges from Florida to Texas and New Mexico, north to New Jersey, New York, Ohio, Indiana, Wisconsin, Minnesota, Manitoba, and Colorado.

Along the alluvial banks and among the rocky limestone boulders at the base of bluffs of Ozark streams, this species is conspicuous with its large colonies and masses of long, narrow, dark green leaves. The perigynia start falling from the spikes after the middle of July, so that fruiting material for identification must be collected usually before this time. Once, however, acquaintance has been made in recognizing this species in fruiting condition, it becomes easy to identify it in repeated encounters by habitat, habit, and foliage characters.

Specimens included by Palmer and Steyermark in their *Annotated Catalogue* under *Carex stricta* Lam. are referred to *C. Emoryi* in the present work. The specimen from Greene County (Springfield, June 4, 1889, *S. Weller* in the herbarium of Drury College) was labelled *C. stricta* var. *angustata* Gray.

87. **Carex stricta** Lam. var. **strictior** (Dew.)

Carey Tussock Sedge Map 529

*Carex stricta* var. *stricta* in part [BB]

Flowers May–July.

Known only from calcareous, spring-fed, swampy meadows of the southeastern Ozarks in Maries (valley of Little Tavern Creek, along highway 42, T40N, R11W, W sect. 33, 11 mi. south of Meta, July 6, 1952, *Steyermark 73686*) and Reynolds (along Bee Fork, T32N, R2W, sect. 23, 4½ mi. southeast of Bunker, May 29, 1951, *Steyermark 71303*; and July 7, 1951, *Steyermark 72032*) counties.

Ranges from Quebec to Ontario and Minnesota, south to North Carolina, Tennessee, and Missouri.

Like several other more northern species of *Carex*, this is another one of the elements of the relict Pleistocene flora which has managed to survive in the swampy meadows of the Ozark region.

At the Maries County locality this sedge was growing with and under the tall culms of the rare and local *C. trichocarpa*.

The previous records for *C. stricta* in Missouri, originally included in Palmer and Steyermark's *Annotated Catalogue*, are now referred to *C. Emoryi*.

Plate no. 95. **1.** Carex Haydenii; After Gleason, The New York Botanical Garden. **2.** Carex stricta var. strictior; After Gleason, The New York Botanical Garden.

PLATE NO. 95

529 Carex stricta var. strictior

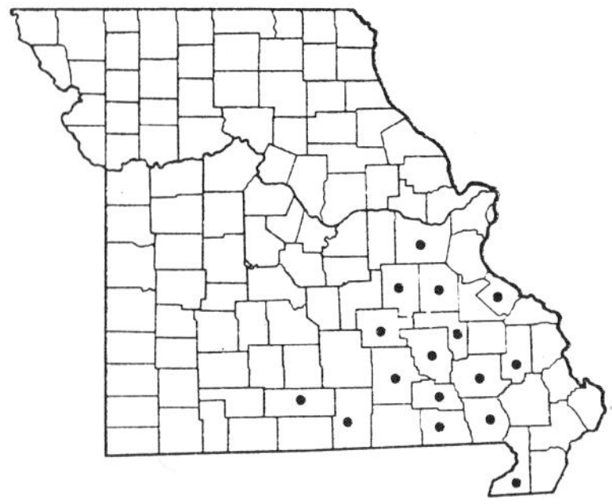

530 Carex torta

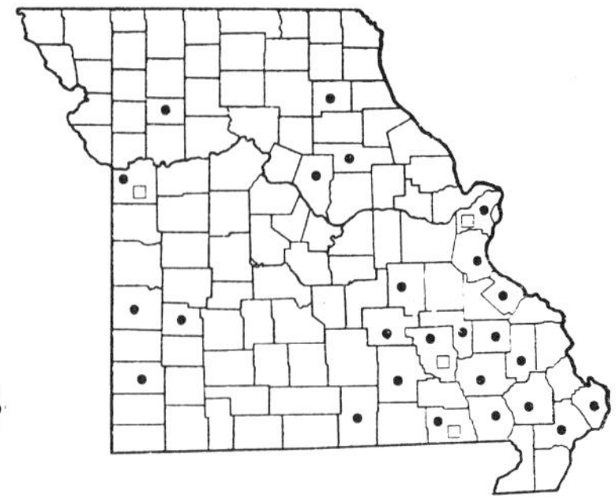

531 ● Carex crinita var. crinita

531 □ Carex crinita var. brevicrinis

The formation of dense cushions or tussocks of long leaves is very characteristic of the varieties of *C. stricta*.

88. **Carex torta** Boott Map 530

Flowers mid-April–July.

Frequents banks of small running streams and spring branches. Found throughout the eastern Ozark section west to Dent, Douglas, and Howell counties.

Ranges from Quebec and Nova Scotia to Minnesota, south to Georgia, Tennessee, and Arkansas.

## Section 28. **Cryptocarpae**

89. **Carex crinita** Lam. Fringed Sedge Map 531

Flowers May–July.

Frequents spring branches, fast-running streams, spring-fed calcareous swampy meadows, margins of ponds, wet river bottom prairies, prairie swales, and sloughs. Two varieties are represented in Missouri:

Pistillate spikes rarely with staminate flowers at summit of spike; awns at the tip of the lower scales 2–4 times as long as the perigynia, upper scales longer than perigynia; perigynia 2–3 (–3.5) mm. long, 1–2 mm. thick . 89a. C. CRINITA var. CRINITA

Pistillate spikes frequently with staminate flowers at summit; awns at the tip of the lower scales equal to or 2 times as long as the perigynia, upper scales about equaling or slightly longer than perigynia; perigynia 3–4 mm. long, 2–3 mm. thick. 89b. C. CRINITA var. BREVICRINIS

89a. **Carex crinita** var. **crinita** Map 531

*Carex crinita* Lam. [G, P & S]

The commoner variety in the state, in southern and central Missouri north to Shelby, Boone, and Caldwell counties.

Ranges from Newfoundland to Manitoba, south to Georgia, Tennessee, and Missouri.

89b. **Carex crinita** var. **brevicrinis** Fern. Map 531

*Carex crinita* var. *crinita* in part [BB]

*Carex crinita* var. *gynandra* [of P & S], not (Schwein.) Schwein. & Torr.

Scattered in southern and central Missouri.

Ranges from Florida to Texas, north near or along the Coastal Plain to Massachusetts, inland north to Kentucky, Tennessee, and Missouri.

The above varieties grade into one another. They are not recognized as distinct by Gleason.

## Section 29. **Pseudo-Cypereae**

90. **Carex hystricina** Muhl. Map 532

Flowers May–July.

Most commonly occurring in spring-fed, calcareous, swampy meadows and wet open places. Rare in southern and central Missouri, north to St. Louis, Saline, and Jackson counties.

Ranges from Quebec to Alberta and Washington, south to Virginia, Tennessee, Missouri, Oklahoma, Texas, New Mexico, Arizona, and California.

91. **Carex comosa** Boott Map 533

Flowers July–late August.

Plate no. 96. **1.** Carex torta. **2.** Carex crinita.

PLATE NO. 96

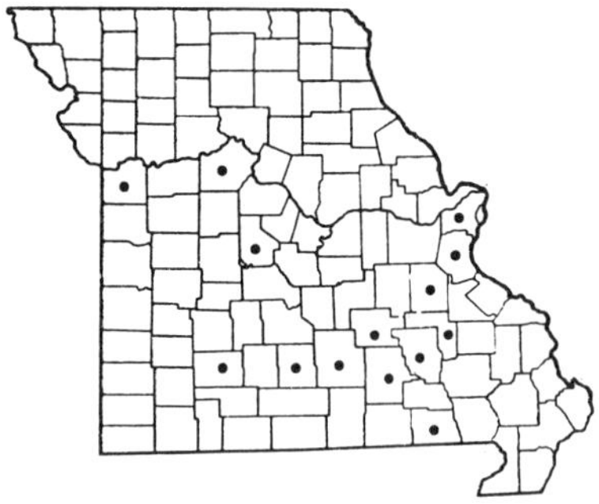

532 Carex hystricina

533 Carex comosa

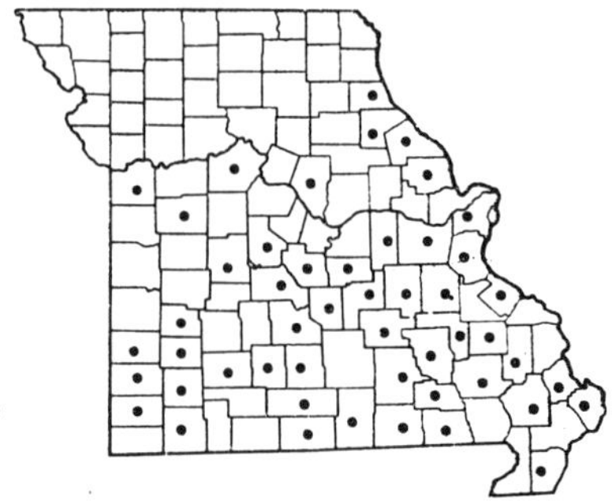

534 Carex lurida

Occurs in upland sink-hole ponds in the southeastern Ozarks in Shannon and Oregon counties, in swampy ground in the southeastern lowland section in Butler and Dunklin counties, and along margins of oxbow lakes along the Missouri River in west-central Missouri in Platte County (Iatan Pond, ½ mile north of Iatan, September 5, 1934, *Steyermark 15244*).

Ranges from Florida to Louisiana, north to Nova Scotia, Maine, Quebec, Ontario, Michigan, Wisconsin, Minnesota, and Nebraska; also from Idaho and Washington to California.

92. **Carex lurida** Wahlenb. Map 534

Flowers early May–September.

Occurs commonly along spring branches, small streams, borders of ponds in valleys, spring-fed calcareous swampy meadows, prairie swales, wet river bottom prairie, and wet open places in valleys. Common throughout southern and central Missouri north to Jackson, Saline, and Boone counties, northeast to Marion County.

Ranges from Florida to Texas and Mexico, north to Nova Scotia, Maine, Quebec, Ontario, and Minnesota.

This is placed in section *Vesicariae* in the works of Gleason and Deam.

93. **Carex Schweinitzii** Dew. Map 535

Flowers May–July.

Known only from three collections, all from St. Louis County (June, 1883, *Letterman 21* in New York Botanical Garden Herbarium; west of engine house of Baden freight yard of Mo.-Kan.-Top. R. R., St. Louis, May 4, 1957, *Muehlenbach 1136*; freight yard Carrie Ave. of Rock Island R. R. on large free place south of Carrie Ave., St. Louis, May 7, 1955, *Muehlenbach & Kramer 548*).

Ranges from Vermont to Ontario and Michigan, south to New Jersey, North Carolina, Tennessee, and Missouri.

This is one of the rarest carices in Missouri. The *Muehlenbach 1136* collection was found among bushes in the low area in the north part of St. Louis adjacent to the Mississippi River.

*Carex Schweinitzii* is placed in section *Vesicariae* in the work of Gleason.

## Section 30. **Paludosae**

94. **Carex lacustris** Willd. Map 536

*Carex lacustris* var. *lacustris* [BB]

*Carex riparia* Curtis var. *lacustris* (Willd.) Kükenth.

Flowers May–July.

Occurs in wet river bottom prairies, prairie swales, and wet alluvial soils.

Known only from northern Missouri in Mercer, Linn, and Caldwell counties.

Ranges from Quebec to Manitoba, south to Virginia, Ohio, Indiana, Illinois, Missouri, and South Dakota; also Idaho.

Previous records attributed to this species from other counties in Missouri are now referred to *C. hyalinolepis*.

95. **Carex hyalinolepis** Steud. Map 537

*Carex lacustris* var. *laxiflora* Dew. [BB]

Flowers late April–July.

Frequents bottom prairies, wet meadows, swampy woods, open wet swales in alluvial flood plain soils especially along the larger rivers, borders of sloughs, open swamps, ditches, oxbow lakes in the valleys of

Plate no. 97. **1.** Carex hystricina. **2.** Carex comosa. **3.** Carex lurida. **4.** Carex Schweinitzii. **5.** Carex lacustris.

PLATE NO. 97

535 Carex Schweinitzii

536 Carex lacustris

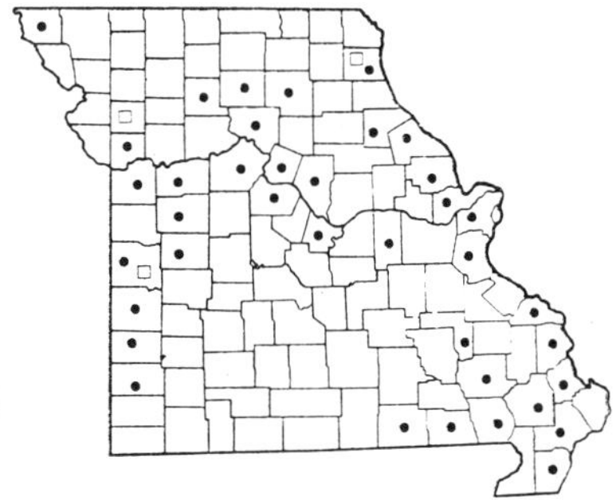

537 • Carex hyalinolepis
537 □ Carex X subimpressa

rivers, and rarely along margins of upland sink-hole ponds in the Ozarks. Found mostly in eastern, central, and western Missouri, following principally the Mississippi, Missouri, St. Francis, Grand, and Osage rivers and some of their main tributaries; absent from most of the Ozark section and the tiers of counties in extreme northern Missouri; in the lowlands of southeastern Missouri it often inhabits bald cypress (*Taxodium*) swamps.

Ranges from Florida to Texas, north to New Jersey, Pennsylvania, Ontario, Ohio, Michigan, Wisconsin, Iowa, and Nebraska.

This species forms dense stands of glaucous or silvery-green leaves and culms, and grows frequently in depressions in flood plains or sloughs where it remains with its culms in water for indefinite prolonged periods of time. It often frequents the wet alluvial soils where forests of silver maple and cottonwood dominate, becoming a dominant ground cover in such areas, and also is found in thickets with buttonbush.

95a. **Carex × subimpressa** Clokey Map 537
*Carex lanuginosa × Carex hyalinolepis*
Flowers May–June.

Occurs in wet meadows and swampy ground of alluvial flood plain. Known from northern and central Missouri in Bates, Clinton (open swales of meadow in draw along highway C, T54N, R31W, sect. 6, $2\frac{3}{4}$ mi. southeast of Plattsburg, May 23, 1957, *Steyermark 84652*), and Lewis (swamp in alluvial bottoms of Mississippi River along highway 61, 2 mi. north of Canton, May 14, 1939, *Steyermark 26489*) counties.

Known only from Indiana, Illinois, and Missouri.

This is a supposed hybrid between *C. lanuginosa* and *C. hyalinolepis*. It has a more robust appearance than *C. lanuginosa*, but is less robust than *C. hyalinolepis*. The grass-green leaves are similar in color to those of *C. lanuginosa*. At the Lewis County station, only *C. lanuginosa* and *C. vesicaria* were in the immediate neighborhood of the hybrid plant.

96. **Carex atherodes** Spreng. Map 538
Flowers May–July.

Occurs in rich swampy woods and in sandy wet swales in northern and west-central Missouri, where it is known from Mercer (rich swampy woods, $4\frac{1}{2}$ mi. northeast of Saline, May 27, 1934, *Steyermark 9505*) and Jackson (sandy swales, Courtney, May 27, 1916, *Bush 7725* and *7725a*) counties.

Ranges from Ontario to Yukon, south to New York, Ohio, Indiana, Wisconsin, Missouri, Nebraska, Colorado, Utah, and Oregon; also in Maine.

97. **Carex laeviconica** Dew. Map 539
Flowers late April–July.

Occurs in wooded swamps and in wet soils of alluvial flood plain forests, also along streams, in wet low bottom prairies, wet open valleys, and prairie swales. Infrequent in northern and central Missouri, south from Clark to Livingston, Cooper, and Jackson counties.

Ranges from Manitoba and Saskatchewan, south to Illinois, Missouri, Kansas, and Montana.

98. **Carex trichocarpa** Muhl. Map 540
Flowers June–August.

Found in spring-fed, calcareous, swampy meadows in valleys along small streams in the eastern Ozarks in Maries (valley of Little Tavern Creek, along highway 42, T40N, R11W, west part sect. 33, 11 mi. south of

Plate no. 98. **1.** Carex vesicaria var. monile. **2.** Carex atherodes. **3.** Carex trichocarpa. **4.** Carex laeviconica. **5.** Carex Grayii. **6.** Carex intumescens. **7.** Carex Frankii.

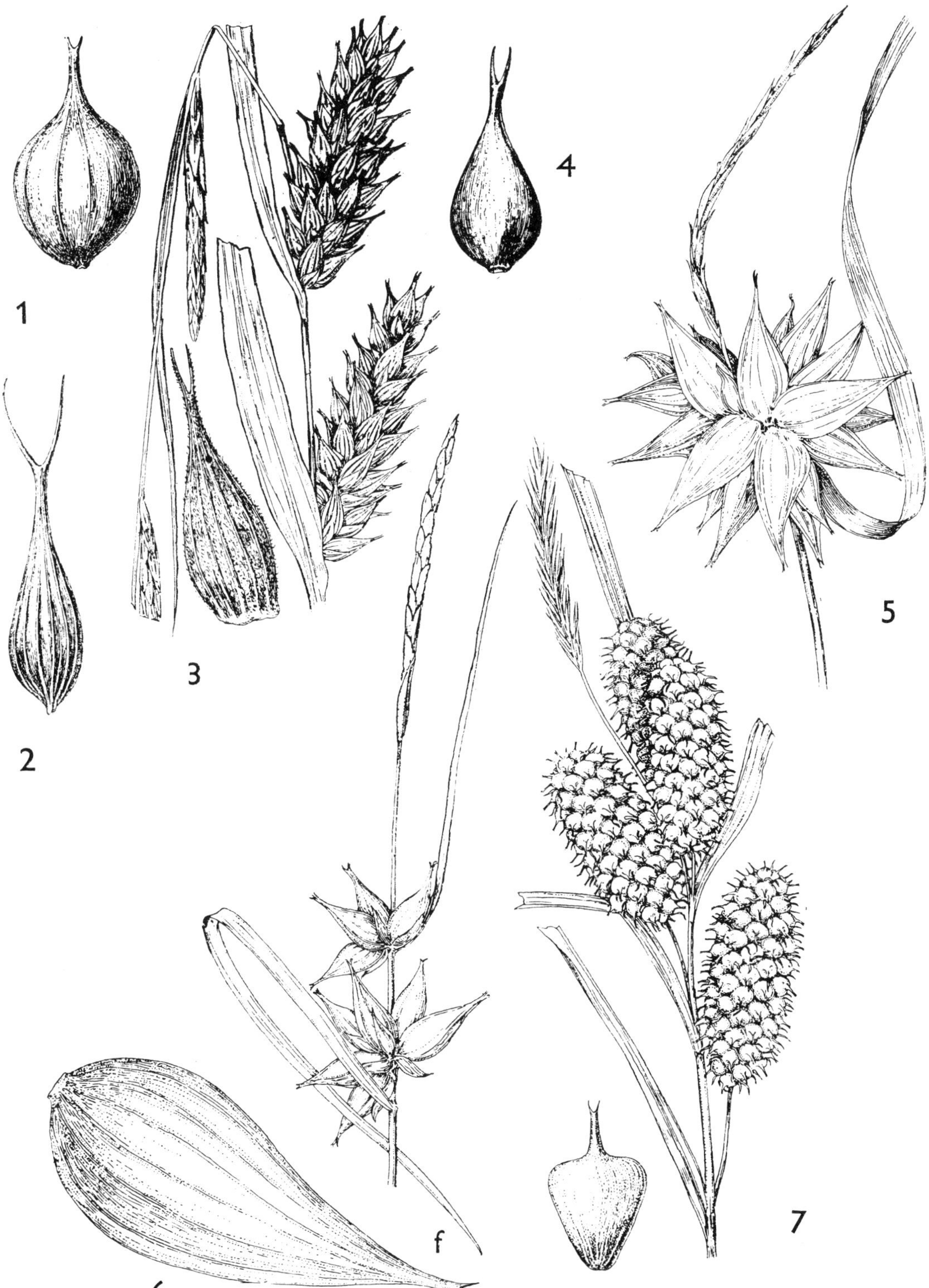


PLATE NO. 98

538 Carex atherodes

539 Carex laeviconica

540 Carex trichocarpa

Meta, July 6, 1952, *Steyermark 73685*) and Reynolds (along West Fork of Black River, T33N, R3W, sect. 23 and NE sect. 26, 3–3½ mi. northwest of Greeley, July 6, 1951, *Steyermark 71977*) counties.

Ranges from Quebec and Vermont to Ontario and Minnesota, south to Delaware, Ohio, Indiana, Illinois, and Missouri.

*Carex trichocarpa* attains heights of 1.3 meters, and forms large dense leafy colonies. It is the tallest-growing and most stately species of *Carex* in Missouri.

## Section 31. **Squarrosae**

99. **Carex Frankii** Kunth Map 541

Flowers late May–September.

Occurs in moist soil or gravel bars along small streams and spring branches, margins of ponds, spring-fed calcareous swampy meadows, low wet woods or alluvial flood plain forests, wet low river bottom prairie, prairie swales, wet open valleys, and meadows. Abundant and widespread in the state, especially common in the Ozark section, but absent in the northwestern and extreme north-central sections of the state.

Ranges from Georgia to Texas, north to New York, Pennsylvania, Ohio, Michigan, Illinois, Missouri, and Kansas.

Easily recognized by the long leaflike bracts, which are 2–4 times as long as the inflorescence, by the very long rough awns of the pistillate scales, and by the widely spreading perigynia which are broadest near the summit of their bodies.

100. **Carex squarrosa** L. Map 542

Flowers late April–September.

Most commonly found in wooded swamps, low wet woods, or forested alluvial flood plain, sometimes also in open wet swales in river valleys and prairies, and low swampy meadows. Common in the eastern half of the state and in parts of western Missouri, following generally the larger streams; apparently absent from the extreme northwestern, southwestern, and parts of west-central Missouri.

Ranges from Connecticut to Ontario, Minnesota, and Nebraska, south to North Carolina, Tennessee, and Arkansas.

The spikes are broadly elliptical-oblong or short-cylindric. A misleading illustration in *Gray's Manual*, 8th edition, p. 371, shows the spikes globose or subglobose, a shape not seen in the average Missouri *C. squarrosa*.

101. **Carex typhina** Michx. Map 543

Flowers late April–September.

Frequents swampy or low wet woods, forested alluvial flood plain, low rich wooded valleys, wet river bottom prairies, prairie swales, and wet open valleys. Mostly found in the eastern half of the state, where it follows generally the larger streams and their tributaries, locally west to Jackson, Vernon, and Barton counties south of the Missouri River, and north of the Missouri River west to Mercer, Grundy, Linn, Chariton, and Boone counties; apparently absent from most of the Ozark, unglaciated prairie, and northwestern sectors of the state.

Plate no. 99. **1.** Carex typhina; After Gleason, The New York Botanical Garden. **2.** Carex squarrosa; After Gleason, The New York Botanical Garden.

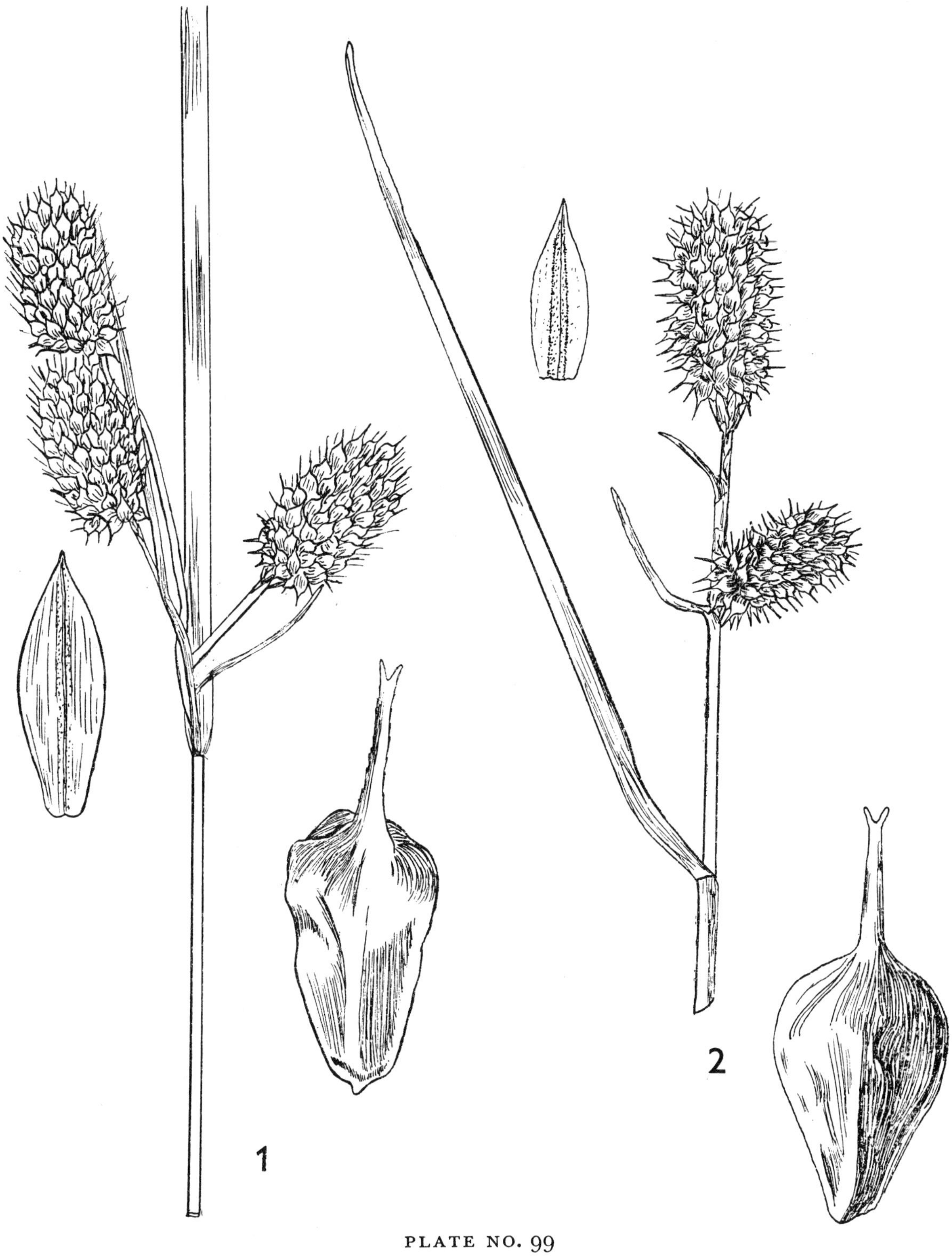


PLATE NO. 99

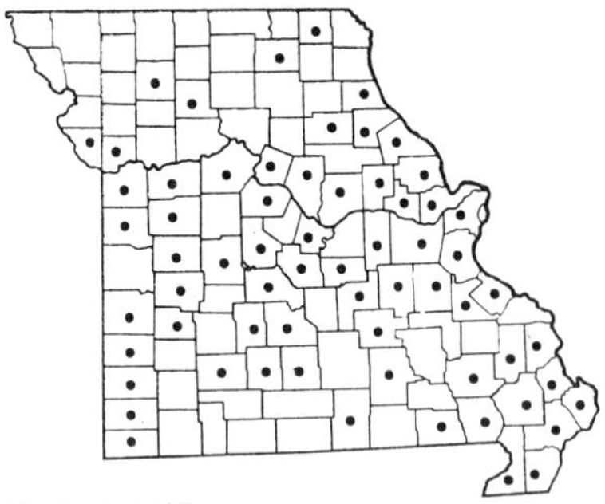

541 Carex Frankii

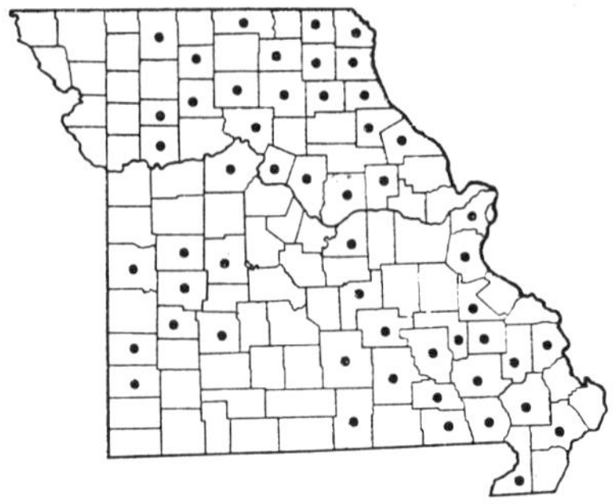

542 Carex squarrosa

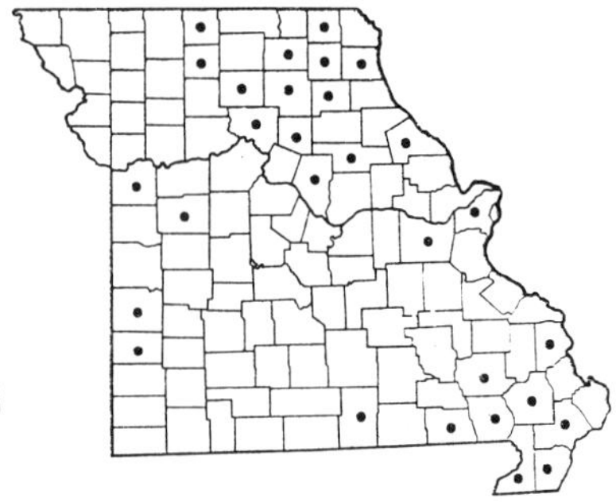

543 Carex typhina

## Section 32. **Vesicariae**

102. **Carex vesicaria** L. var. **monile** (Tuckerm.) Fern. Map 544

*Carex vesicaria* in part [BB, P & S], not L.

Flowers May–August.

Most commonly found in low wet river bottom prairie, prairie swales, swampy open ground in valleys, and swampy woods or forested alluvial flood plain. Rare in northern and central Missouri, south to St. Louis, Saline, and Jackson counties.

Ranges from Quebec and Ontario, south to Delaware, Pennsylvania, Ohio, Indiana, Illinois, and Missouri.

The more northern ranging typical *C. vesicaria* var. *vesicaria* has somewhat longer perigynia (7–9 mm. long instead of 5–8 mm.), with more slender, ovoid-conic bodies 1/3–1/2 as thick as long, tapering more gradually to a beak, and the leaf-blades average broader (4–7 mm. instead of 2–5 mm. wide).

Like *C. trichocarpa*, this species attains considerable height, up to 1 meter tall, and is one of the tallest of the *Carex* species in Missouri. It usually forms dense tussocks or clumps. The long leafy bracts of the inflorescence usually greatly extend above the inflorescence.

## Section 33. **Lupulinae**

103. **Carex Grayii** Carey Map 545

Flowers May–October.

Most commonly occurs in swampy or low wet woods and forested alluvial flood plain, where it is found along some of the larger rivers and their tributaries.

Missouri material may be divided into 2 varieties:

Perigynia glabrous (without hairs) . 103a. C. GRAYII var. GRAYII

Perigynia with short hairs . . . 103b. C. GRAYII var. HISPIDULA

103a. **Carex Grayii** var. **Grayii** Map 545

*Carex Grayii* Carey [G]

*Carex Grayii* in part [BB]

*Carex Asa-Grayii* Bailey in part [P & S]

Found in parts of eastern, northern, and central Missouri along the larger streams and tributaries; absent from the Ozark section. Less common than var. *hispidula*.

Ranges from Quebec to Iowa, south to New Jersey, Pennsylvania, Ohio, Indiana, Illinois, and Arkansas.

103b. **Carex Grayii** var. **hispidula** Gray Map 545

*Carex Asa-Grayii* Bailey in part [P & S]

*Carex Grayii* in part [BB]

The commoner variety in Missouri and more frequent in some of the counties of the western sector south of the Missouri River.

Ranges from Connecticut and New York to Ohio, Michigan, Indiana, and Illinois, south to Georgia, Mississippi, and Missouri.

There is considerable doubt as to whether var. *hispidula* should merit recognized varietal status, and some authors include it with typical *C. Grayii*. According to Dr. Hermann (Deam's *Flora of Indiana*, p. 270. 1940), observations made by J. K. Underwood of the University of Tennessee Agricultural Experiment Station indicate that the glabrous and pubescent perigynia are variations or seasonal fluctuations which may be found even on the same plant. However, until careful studies have been made throughout the range of *C. Grayii*, it seems preferable to give tentative recognized status to var. *hispidula*.

Plate no. 100. **1.** Carex louisianica. **2.** Carex lupulina. **3.** Carex lupuliformis. **4.** Carex gigantea.

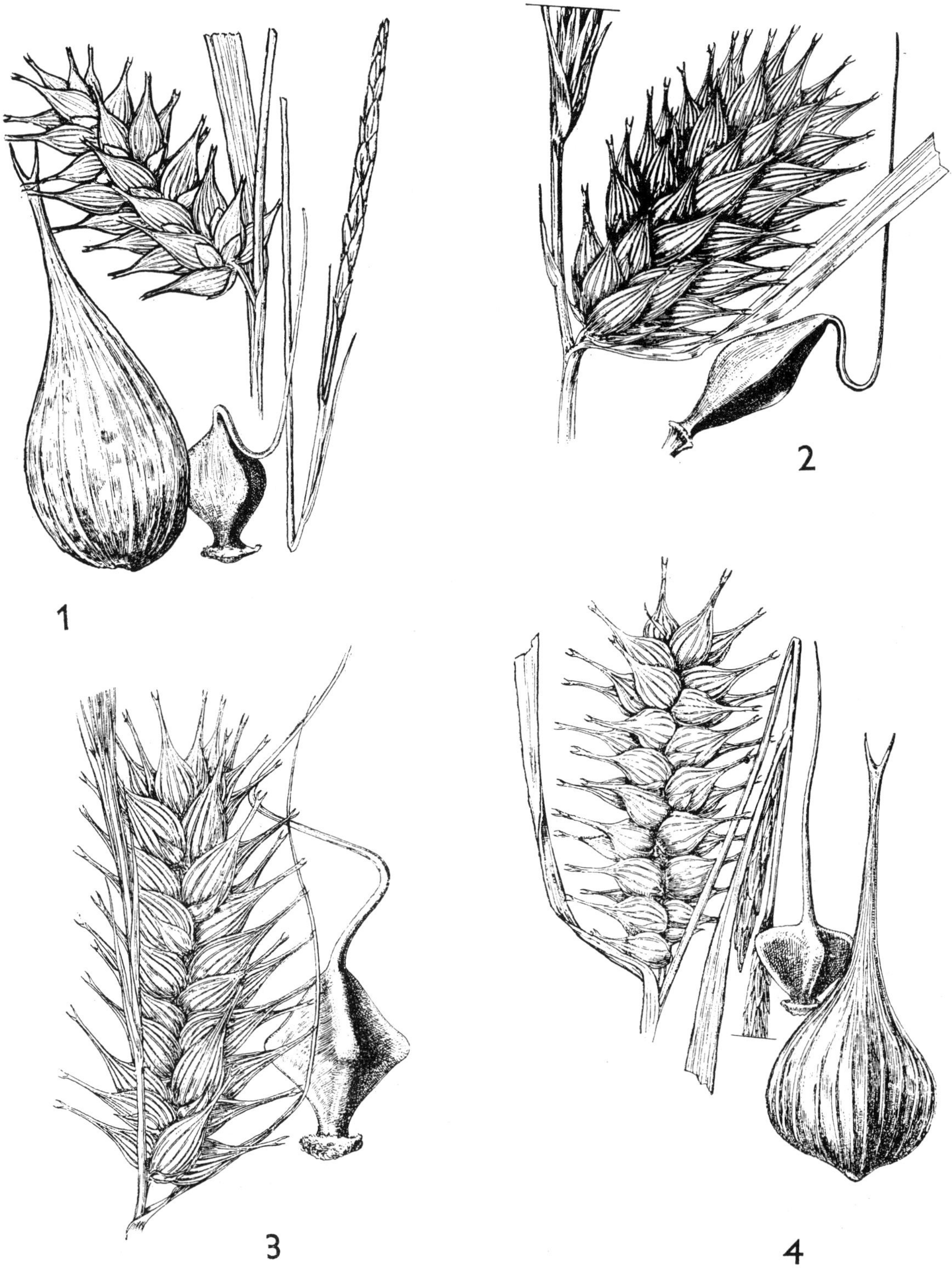


PLATE NO. 100

544 Carex vesicaria var. monile

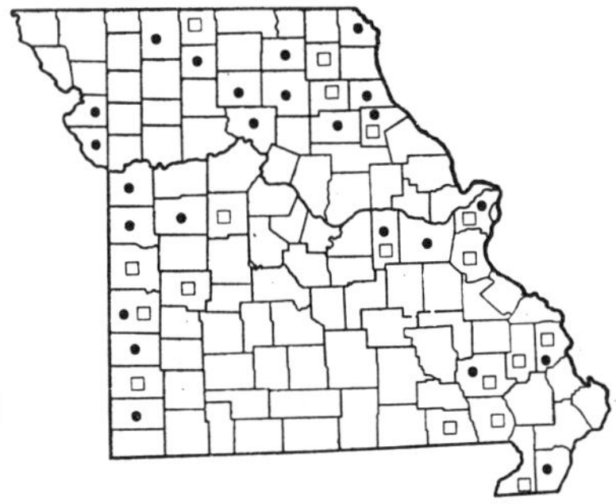

545 • Carex Grayii var. Grayii
545 □ Carex Grayii var. hispidula

546 Carex intumescens var. intumescens

Although the usual shape of the pistillate spikes is of a globose type, occasional plants are found (*Palmer 2196* from Quaker Mill, Jasper Co., for example) in which the spikes are longer than broad.

104. **Carex intumescens** Rudge var. **intumescens** Map 546

*Carex intumescens* Rudge [G, BB in part, P & S]

Flowers late April–July.

Occurs in low wet woods and wooded swamps or forested alluvial flood plain. Known only from the lowlands of southeastern Missouri north to Cape Girardeau and Wayne counties, and west to Ripley County, where it inhabits margins of upland sink-hole ponds (*Steyermark 66114*).

Ranges from Florida to Texas, north to Nova Scotia, Maine, Quebec, New York, Ontario, Michigan, and Wisconsin.

The more northern var. *Fernaldii* Bailey has narrower perigynia (3.5–5 mm. instead of 5–8 mm. thick) which are 1/4–1/3 instead of nearly 1/2 as broad as long.

105. **Carex louisianica** Bailey Map 547

Flowers late April–August.

Occurs in low wet woods and wooded swamps or forested alluvial flood plain. Known from the lowlands of southeastern Missouri north to Cape Girardeau and Wayne counties, and west to Butler County, locally north in St. Louis County.

Ranges from Florida to Texas, north to New Jersey, Kentucky, Indiana, and Missouri.

106. **Carex lupulina** Muhl. Hop Sedge Map 548

Flowers early May–September.

Occurs in low wet woods, forested alluvial flood plain, borders of upland sink-hole and other ponds, open wet swales in valleys, wet low river bottom prairies, prairie swales, and along ditches. Widespread and common in Missouri, but not recorded from some of the western Ozark and southwestern counties, and apparently absent from extreme northwestern Missouri.

Missouri material is represented by 2 varieties:

Pistillate spikes without a peduncle (stalk) or nearly so, mostly close to one another . . . 106a. C. LUPULINA var. LUPULINA

Pistillate spikes usually with a definite peduncle, rather distant and well apart from one another 106b. C. LUPULINA var. PEDUNCULATA

106a. **Carex lupulina** var. **lupulina** Map 548

*Carex lupulina* Muhl. [G, P & S in part, BB in part]

This is the more common variety in Missouri.

Ranges from Florida to Texas, north to Nova Scotia, Maine, Quebec, New York, Ontario, Michigan, Wisconsin, and Minnesota.

106b. **Carex lupulina** var. **pedunculata** Gray Map 548

*Carex lupulina* Muhl. in part [BB, P & S]

This variety is scattered in parts of southern and central Missouri northeast to Marion County, and is less common than var. *lupulina*.

Ranges from Quebec to Ontario, south to Georgia, Illinois, and Missouri.

This variety is not recognized by some authors (Gleason, Deam), and only intensive future studies will reveal the relative significance of the peduncled pistillate spikes.

107. **Carex lupuliformis** Sartwell Map 549

Flowers early July–September.

Occurs in swamps, low wet woods and margins of ponds in river bottom or lowland areas. Rare and scattered in eastern and northern Missouri, where known from Putnam, Livingston, Montgomery, St. Louis, Stoddard, Butler, and Ripley counties.

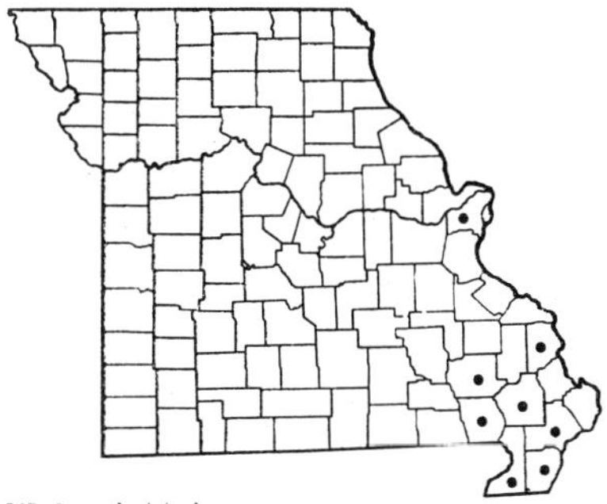

547 Carex louisianica

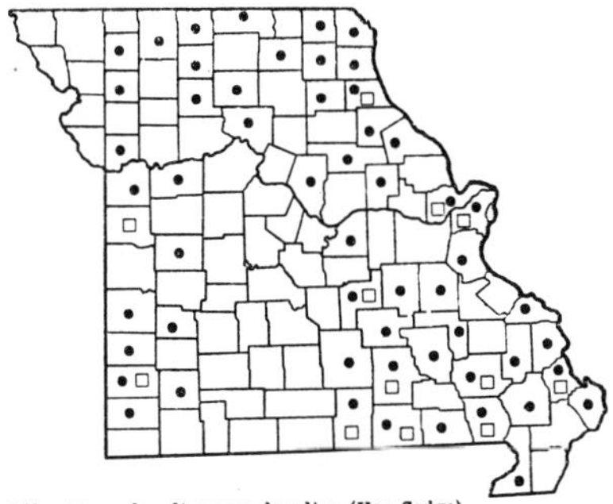

548 • Carex lupulina var. lupulina (Hop Sedge)
548 □ Carex lupulina var. pedunculata

549 Carex lupuliformis

Ranges from Vermont to Minnesota, south to Virginia, Kentucky, Louisiana, and Texas.

108. **Carex gigantea** Rudge Map 550

Flowers late May–September.

Occurs in low wet woods, wooded swamps, or forested alluvial flood plain. Known only from the lowlands of southeastern Missouri north to Scott, Stoddard, and Wayne counties, west to Ripley County.

Ranges from Florida to Texas, north to Delaware, Indiana, Kentucky, and southeastern Missouri.

*Excluded species*

**Carex cephaloidea** Dew.

Included in Palmer and Steyermark's *Annotated Catalogue*, the collections from St. Louis (Prospect Hill, St. Louis, *Glatfelter*; Crescent, *Eggert*) and Jackson counties, upon which the species was included, are now referred to *C. aggregata* Mackenz.

**Carex hormathodes** Fern.

The Jasper County collection, which formed the basis for the inclusion of this species in Palmer and Steyermark's *Annotated Catalogue* is now referred to *C. Bicknellii.*

**Carex laxiflora** Lam.

*Carex anceps* Muhl., included by Palmer and Steyermark in their *Annotated Catalogue* on the basis of several collections from western Missouri, is a synonym of *C. laxiflora* Lam. The records constituting the basis for this report are now referred to *C. blanda* Dew.

## Order **ARALES**

### Fam. **Araceae** (Arum Family)

a. Leaf-blades sword-shaped, iris-like, long and narrow; leaves and rootstocks pleasantly fragrant when crushed . . . . . 3. ACORUS
a. Leaf-blades not sword-shaped, but of a broader type; leaves and underground parts not fragrant . . . *b*
  b. Leaves simple, the blades sagittate (arrowhead-shaped) . . . . . 2. PELTANDRA
  b. Leaves not simple, divided into 3 or 5–15 leaflets (divisions) . . . . . 1. ARISAEMA

### 1. **Arisaema** Mart. Jack-in-the-Pulpit, Green Dragon

Leaf-divisions (leaflets) 3 . . . . . 1. A. ATRORUBENS
Leaf-divisions (leaflets) 5–15 . . . . . 2. A. DRACONTIUM

1. **Arisaema atrorubens** (Ait.) Blume

Jack-in-the-Pulpit, Indian Turnip Map 551

Flowers April–June.

Missouri material may be divided into 3 forms:

a. Hoodlike flap of flower green only . . . 1c. A. ATRORUBENS f. VIRIDE
a. Hoodlike flap of flower entirely purple or purple to bronze with pale or greenish stripes . *b*
  b. Hoodlike flap entirely purple without any pale stripes. 1a. A. ATRORUBENS f. ATRORUBENS
  b. Hoodlike flap purple or bronze with pale or greenish stripes . . 1b. A. ATRORUBENS f. ZEBRINUM

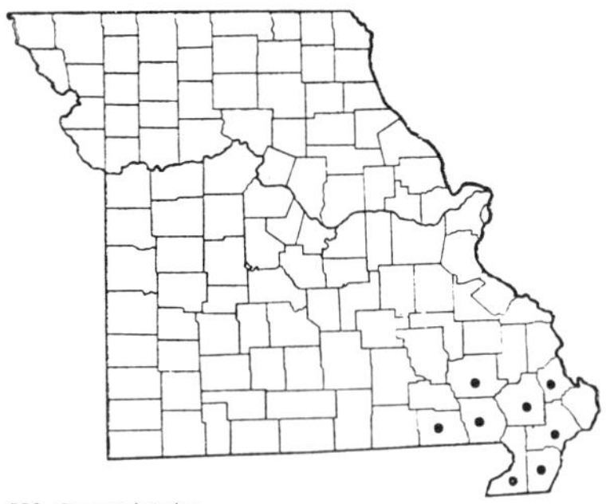

550 Carex gigantea

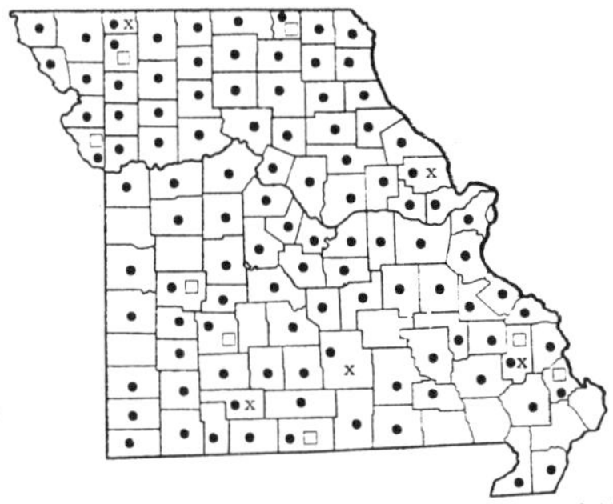

551 • Arisaema atrorubens f. atrorubens (Jack-in-the-Pulpit)
551 □ Arisaema atrorubens f. viride
551 x Arisaema atrorubens f. zebrinum

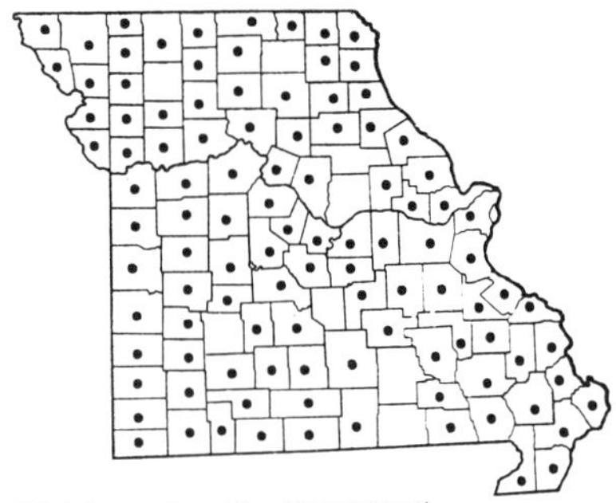

552 Arisaema Dracontium (Green Dragon)

1a. **Arisaema atrorubens** f. **atrorubens** Map 551

*Arisaema triphyllum* [of P & S], not (L.) Schott
*Arisaema triphyllum* var. *triphyllum* [BB]

Occurs in rich, level or hilly, rocky or nonrocky wooded areas or in low bottom thickets.

Throughout Missouri, doubtless in every county.

Ranges from New Brunswick and Quebec to Manitoba, south to South Carolina, Tennessee, Arkansas, and Kansas.

1b. **Arisaema atrorubens** f. **zebrinum** (Sims) Fern. Map 551

Throughout the range of *A. atrorubens*.

1c. **Arisaema atrorubens** f. **viride** (Engler) Fern. Map 551

Throughout the range of *A. atrorubens*.

At first green, the fruiting head of fleshy, berrylike fruits finally becomes scarlet or orange-red, adding a striking color touch to the woodland floor in late summer and autumn.

The fresh rootstocks and leaves containing crystals of calcium oxalate may cause poisoning if large amounts are eaten raw, and a few people develop a dermatitis after touching the plant. After being thoroughly dried (but not boiled), the round, solid underground starchy corm can be grated and made into a type of flour, and was so used by the Indians in eastern North America.

2. **Arisaema Dracontium** (L.) Schott
Green Dragon Map 552

Flowers April–June.

Occurs in moist or rich, rocky or nonrocky woodland, usually in level ground, at the base of slopes, in ravines, or flood plain forest, mostly in wetter ground than that of the preceding species.

Occurs throughout Missouri, doubtless in every county.

Ranges from New Hampshire and Quebec to Ontario, Michigan, and Wisconsin, south to Florida and Texas.

The rounded solid starchy corm may be dried and used for a flour in the same manner as the preceding species. The berrylike fruits are of a similar color to those of the Indian Turnip.

## 2. **Peltandra** Raf. Arrow Arum

**Peltandra virginica** (L.) Schott & Endl.
Arrow Arum Map 553

Flowers April–June.

Occurs in wet, mucky ground bordering sloughs and oxbow lakes of river bottoms.

Missouri material is represented by 2 forms:

Lobes at base of leaf-blade 3–7.5 cm. broad; main upper part of leaf-blade 8–18 cm. broad . . . P. VIRGINICA f. VIRGINICA

Lobes at base of leaf-blade 1–3.5 cm. broad; main upper part of leaf-blade 3–12 cm. broad . . . P. VIRGINICA f. HASTIFOLIA

**Peltandra virginica** f. **virginica** Map 553

Very rare, known only from St. Charles, St. Louis, Jefferson, Cape Girardeau, and Wayne counties in eastern Missouri.

Ranges from Maine and Quebec to New York, Ontario, and Michigan, south to Florida and Texas.

Plate no. 101. **1.** Arisaema Dracontium, × 2/7; Details from Small, The New York Botanical Garden. **2.** Arisaema atrorubens, × 2/7; Details from Small, The New York Botanical Garden. **3.** Peltandra virginica, × 2/7; Details from Small, The New York Botanical Garden. **4.** Acorus Calamus, × 2/7; Details from Small, The New York Botanical Garden.

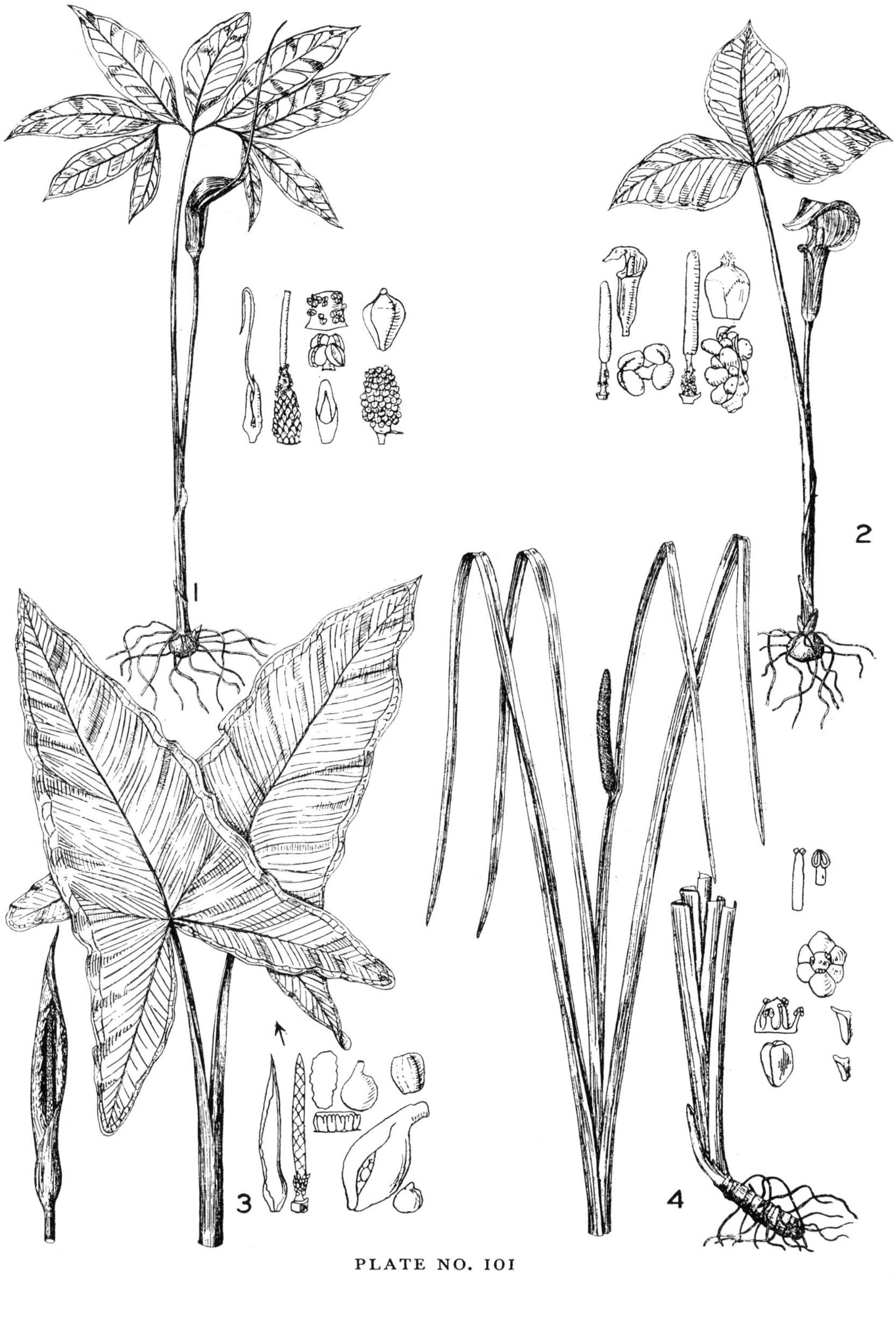


PLATE NO. 101

553 • Peltandra virginica f. virginica (Arrow Arum)
553 □ Peltandra virginica f. hastifolia

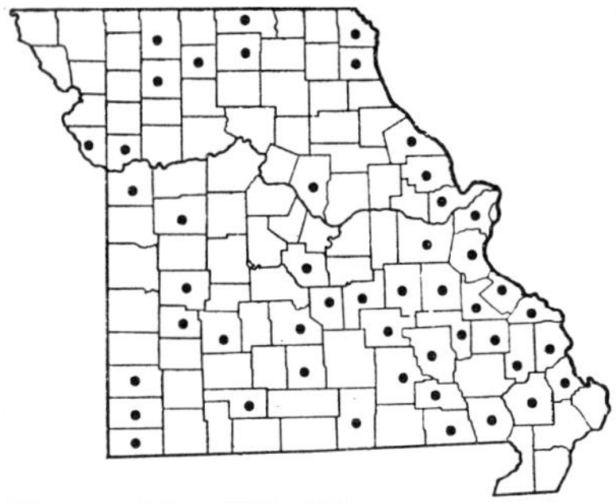

554 Acorus Calamus (Sweet Flag)

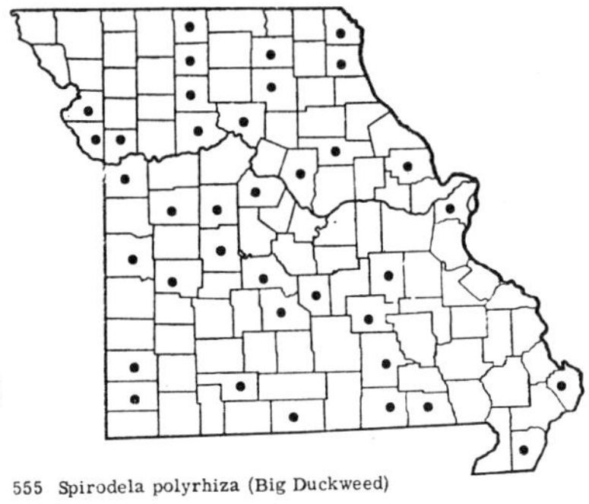

555 Spirodela polyrhiza (Big Duckweed)

**Peltandra virginica** f. **hastifolia** Blake Map 553

Known only from Dunklin County, southeastern Missouri.

Other leaf variations have been described and may eventually be found in the state.

In contrast to Indian Turnip or Green Dragon, the root is very large and perpendicular, and the green calla-lilylike spathe with paler margins surrounds the mass of tiny flowers which occur throughout the elongated spadix instead of just at its base as in *Arisaema*. In the fruiting stage the berries are green or brownish in a thick fleshy head surrounded at the base by the thick spathe with a beaklike summit. Some of the North American Indians are reported to have eaten the berries, which, after boiling, are said to be somewhat sweet. A type of bread can be made from the seeds. The roots are stated to be poisonous when raw, but, ofter being cooked, can be used as a starchy vegetable, similar to potatoes in flavor.

## **Acorus** L. Sweet Flag, Calamus

**Acorus Calamus** L. Sweet Flag Map 554

Flowers May–July.

Occurs in wet meadows, swales of prairies, along spring branches, open wet ground, and along sloughs and ponds. Found throughout the state, although unrecorded from more than half the counties of the state.

Native of Europe; introduced in North America, where found from Nova Scotia to Montana and Oregon, south to Florida, Texas, and Colorado.

The plant usually occurs in large colonies, occupying a continuous area. Sometimes entire stands are found with no flowers or fruit present.

The drug Calamus is made from the fragrant rootstock, in a powdered form being employed for toilet powders and in distilled form as an oil in perfumeries. It is also employed in flavorings and medicine. The spicy-peppery roots also are used as a candy, after having been boiled and sugared. The tender interior of the tip of the young leafy shoots in spring may be used fresh as a cold salad substitute.

## Fam. **Lemnaceae** (Duckweed Family)

a. No roots present . . . . . . . . . . *b*
  b. Plants 6–8 mm. long, many times longer than broad, linear, thin . . . . . 4. WOLFFIELLA
  b. Plants at most 1.5 mm. long, only slightly longer than broad, or nearly as broad as long, oblong to ovoid, thickish . . . . . . . . 3. WOLFFIA
a. Plants with 1 or more roots . . . . . . . . . . *c*
  c. Roots 2 to several to a plant; plant purplish-red underneath . . . . . 1. SPIRODELA
  c. Roots 1 to a plant; plant green underneath . . . . . . . 2. LEMNA

Plate no. 102. **1.** Spirodela polyrhiza; *a.* $\times 3/4$; *b.* $\times 3/4$; *c.* $\times 4\,1/2$. **2.** Spirodela oligorhiza; *a.* $\times 3/4$; *b.* $\times 3\,3/4$; *c.* $\times 4\,1/2$. **3.** Lemna trisulca; *a.* $\times 3/4$; *b.* $\times 3/4$; *c.* $\times 4$.

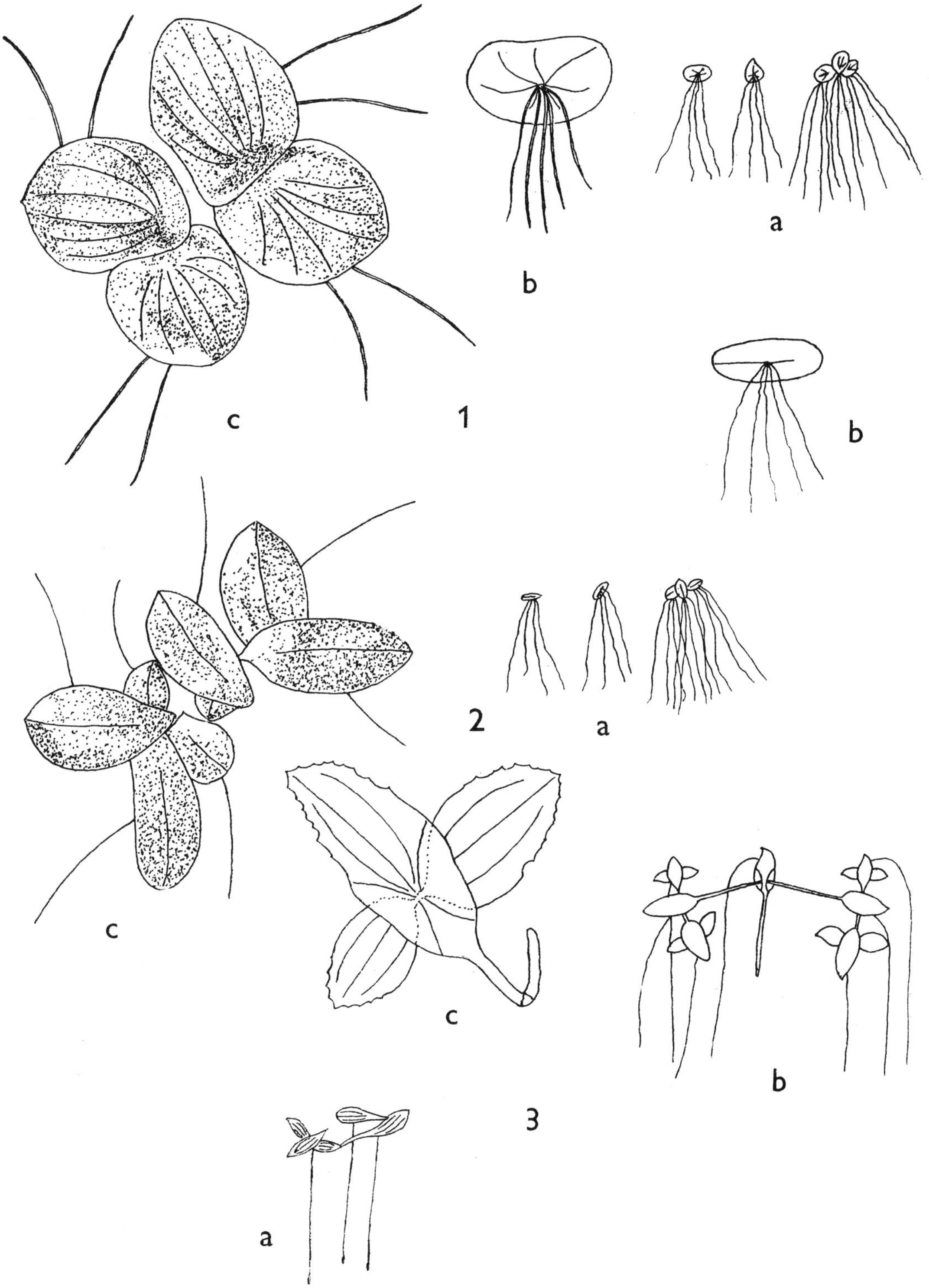


PLATE NO. 102

556 Spirodela oligorhiza

557 Lemna trisulca (Star Duckweed)

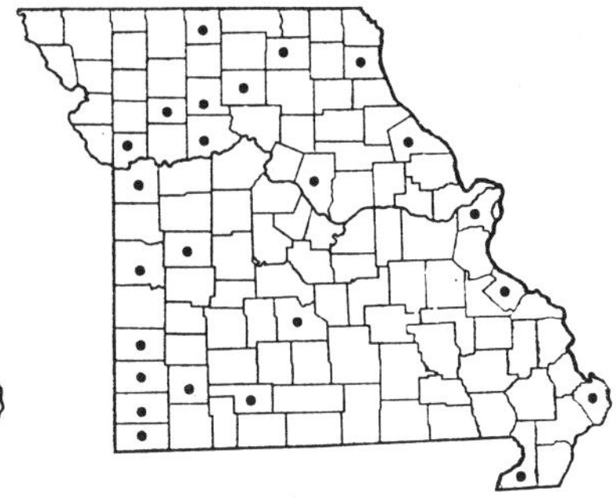
558 Lemna minor (Lesser Duckweed)

### 1. **Spirodela** Schleiden Big Duckweed

Plant more or less circular or broadly obovate, 3–8 mm. long, with 5–11 nerves and 6–18 roots. 1. S. POLYRHIZA
Plant oblong or narrowly obovate, 2–4 mm. long, with 4–6 nerves and 2–3 (rarely –6) roots . 2. S. OLIGORHIZA

1. **Spirodela polyrhiza** (L.) Schleid.
Big Duckweed Map 555

Occurs in sloughs, artificial and natural ponds, oxbow lakes of river bottoms, and sluggish margins of streams. Throughout Missouri.

Ranges from Nova Scotia, Ontario, Minnesota, Nebraska, and British Columbia, south to Florida, Texas, California, Mexico; also Central and South America, the West Indies, Europe, and Asia.

Recorded as being a good food for wildfowl.

2. **Spirodela oligorhiza** (Kurtz) Hegelm.
Map 556

Known only from a slow-flowing stream in the wooded section of Swope Park, Kansas City, in Jackson County, where originally collected by Dr. A. C. Saeger on November 2, 1930, and from a goldfish pond in southwestern Missouri by Dr. F. H. Woods in September, 1932 (Bull. Torr. Bot. Club 61: 233–36. 1934).

Known in the United States only from Missouri. Ranges in warm and temperate areas of Australia, India, Java, Japan, and New Zealand.

### 2. **Lemna** L. Duckweed

With the exception of *Lemna trisulca*, a species usually submerged in the water and with long narrow stalked fronds, the remainder of the species are very difficult to separate from one another and usually require examination with either a high-powered or binocular microscope.

a. Plants usually submerged, connected with one another by their narrowed and long-stalked bases, the green leafy joints 4–10 mm. long . . . . . . . . . . . . . . . . . . 1. L. TRISULCA
a. Plants usually floating on the surface (at least during the spring, summer, and fall months), rounded at their base or not long-stalked, the green leafy joints 1.5–5 mm. long . . . . . . . . . . *b*
  b. The green joints of plant more or less symmetrical (the right half similar in shape and size to the left half) . . . . . . . . . . . . . . . . . . . . . . . . . . . *c*
    c. Both upper and lower surface of green joints convex (sloping or curved to a higher center); joints with 3 weak nerves, with thickened margins . . . . . . . . . . . 2. L. MINOR
    c. Lower surface of green joints flat; joints with 1 weak nerve, with thin margins. . . . 3. L. MINIMA
  b. The green joints of plant not symmetrical (right and left halves not completely similar), one side or end shorter or more curved than the other . . . . . . . . . . . . . . . . . . *d*
    d. Joints with 3 faint or recognizable nerves, nearly as broad as long or at most $1\frac{3}{4}$ times as long as broad, 1.2–2.5 mm. wide, with curved or rounded sides, obovate or oblong-obovate 4. L. PERPUSILLA
    d. Joints with 1 faint nerve or nerveless, 2–3 times as long as broad, mainly 0.5–1.5 mm. wide, with nearly parallel sides, narrowly oblong or narrowly elliptic . . . . . . . . 5. L. VALDIVIANA

559 Lemna minima (Least Duckweed)

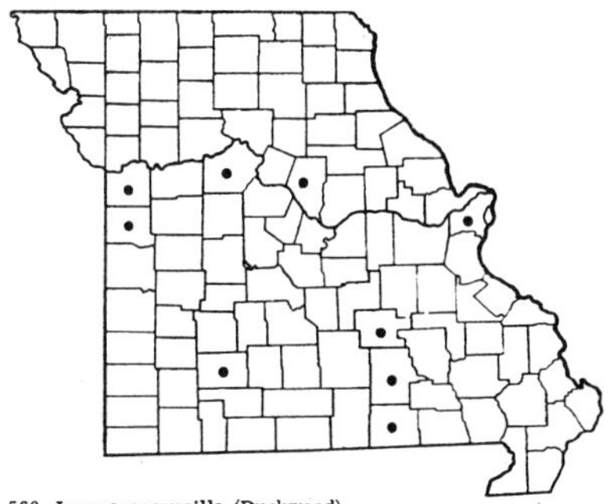

560 Lemna perpusilla (Duckweed)

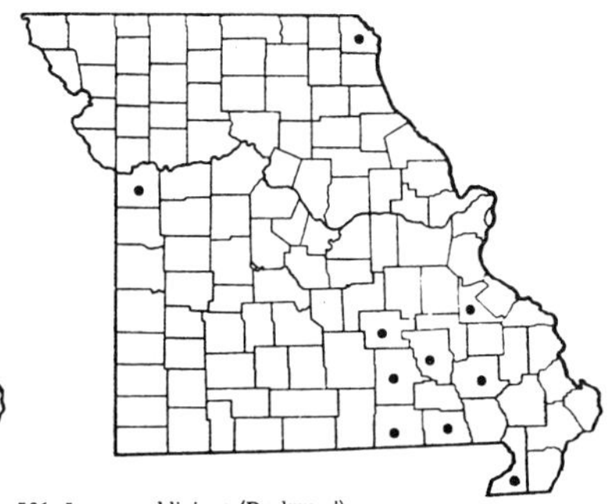

561 Lemna valdiviana (Duckweed)

1. **Lemna trisulca** L. Star Duckweed Map 557

Also called Ivy-leaved Duckweed and Submerged Duckweed.

Rare, known only from a few cold fresh-water springs in the southeastern Ozarks (Ripley, Oregon, and Shannon counties), and from cold spring-fed ponds or lakes in Jackson and Mercer counties of northern and western Missouri.

Ranges from Nova Scotia and Ontario to British Columbia, south to Florida, Texas, California, and Mexico; also in the Old World.

Occasionally plants are found lacking roots. The plants form large tangled masses in the water and are of usually a dark blue-green color, as contrasted with a pale to grass-green shade of the other species of Lemna. They are often an important wildfowl food as well as eaten by small aquatic animals.

2. **Lemna minor** L. Lesser Duckweed Map 558

Found in sloughs, ponds, and slow-flowing streams. Scattered throughout Missouri.

Ranges from South America, Central America, and Mexico, north throughout the United States north to Labrador, Ontario, and British Columbia; also in the Old World.

This species is an important food of wildfowl, and is eaten also by muskrats and small aquatic animals.

3. **Lemna minima** Philippi Least Duckweed Map 559

Known only from upland natural sink-hole ponds in Oregon (Tupelo Gum Pond, Irish Wilderness, sect. 4, May 23, 1938, *Steyermark 5365*) and Howell (between Moody and South Fork, sect. 7, 2 mi. south of South Fork, April 28, 1938, *Steyermark 5222*) counties in the southern Ozark region.

Ranges from South America, Central America, and Mexico, north to Ohio, Indiana, Missouri, Minnesota, Wyoming, and California.

The above-cited specimens were determined by Lawrence E. Hicks, a student of the *Lemnaceae.*

4. **Lemna perpusilla** Torr. Duckweed Map 560

*Lemna perpusilla* var. *trinervis* Austin [P & S]

Occurs in sloughs, ponds, and stagnant water.

Scattered in parts of southern and central Missouri.

Ranges from Florida, Arkansas, and Oklahoma, north to Massachusetts, New York, Ohio, Indiana, Illinois, Wisconsin, Minnesota, and North Dakota. Also in South America.

The tip of the root in this species is pointed, whereas in *L. minor* it is rounded. The root-sheath in *L. perpusilla* is furnished with lateral wing appendages, and in *L. valdiviana* the appendages are absent.

This species is eaten by ducks.

5. **Lemna valdiviana** Phil. Duckweed Map 561

Occurs in sloughs, ponds, and stagnant water, sometimes in organic debris along the margins of sink-hole ponds. Scattered in the state, mostly in the southeastern quarter, locally north in Clark County and west in Jackson County.

Ranges from South and Central America, the West Indies, and Mexico, north into the United States to Massachusetts, New York, Ohio, Indiana, Michigan, Wisconsin, Wyoming, Nevada, and Oregon.

This species is recorded as being eaten by ducks.

### 3. **Wolffia** Horkel Water-meal

The species of this genus are considered to be the smallest flowering plants known and are considered to provide good food and cover for fish. As in *Lemna*, the species are very difficult to distinguish and must be studied under a high-powered or binocular microscope.

562 Wolffia columbiana (Water-meal)

563 Wolffia punctata (Water-meal)

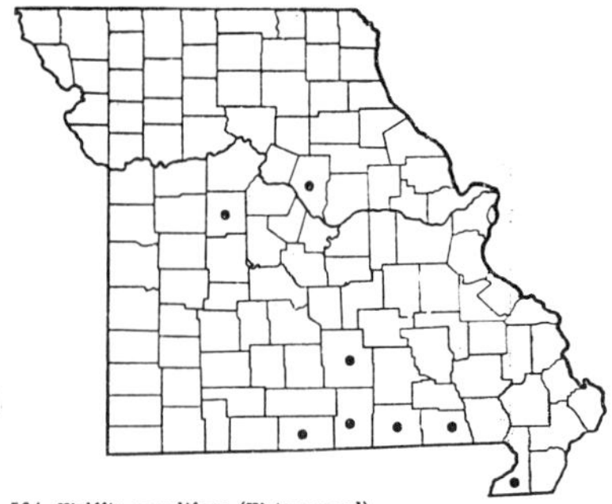

564 Wolffia papulifera (Water-meal)

a. Plants without dots, the upper surface convex (sloping or curved to a higher center); shape of plant mainly globose or nearly circular in outline; upper surface of plant chiefly below the surface of the water . . . . . . . . . . . . . . . . . . . . . . . . 1. W. COLUMBIANA
a. Plants with numerous minute brown dots, the upper surface not convex; shape of plant mainly ovate or oblong to ellipsoid; upper surface of plant floating chiefly above the surface of the water . . . *b*
 b. Upper surface of plant nearly flat, raised at one end into a tiny papilla (bump) . . . 2. W. PUNCTATA
 b. Upper surface of plant with a tiny pointed papilla (bump) near the center . . . 3. W. PAPULIFERA

1. **Wolffia columbiana** Karst. Water-meal Map 562

Occurs in still waters of ponds and sloughs. Scattered in parts of central and southwestern Missouri.

Ranges from South and Central America to Mexico, north to Massachusetts, Quebec, New York, Ontario, Michigan, Wisconsin, and Minnesota.

Plants of this species measure 0.3–1 mm. long and nearly as wide. They are recorded as being eaten by ducks and muskrats.

2. **Wolffia punctata** Griseb. Water-meal Map 563

Known only from the cool waters of a spring-fed lake of Yancy Mills Spring in Phelps County of the Ozark section.

Ranges from the West Indies, and from Florida and Texas, north to Connecticut, New York, Ontario, Michigan, and Minnesota.

Eaten by ducks and muskrat.

3. **Wolffia papulifera** Thompson Water-meal Map 564

Occurs in sloughs, and in stagnant waters of ponds and slow-flowing streams, often in organic floating debris of sink-hole ponds. Scattered in southern and central Missouri, frequent in sink-hole ponds of the southern Ozarks.

Ranges from Virginia to Illinois, Kentucky, Tennessee, Missouri, Arkansas, and Kansas.

## 4. **Wolffiella** Hegelm.

**Wolffiella floridana** (Smith) Thompson Map 565

Known only from the swamps of the southeastern lowland section of the state in Dunklin County (Kennett, April 24, 1897, *Trelease*; St. Francis River, August 6, 1895, *Widmann*).

Ranges from Mexico, Florida and Texas, north to Massachusetts, Ontario, Michigan, Wisconsin, and Missouri.

Plants occur either singly or in star-shaped pale green bodies, often forming tangled masses, usually just the base of the plant showing above the surface of the water. They are eaten by wildfowl.

Plate no. 103. **1.** Lemna minor; *a.* × $^{3}/_{4}$; *b.* × 3. **2.** Lemna minima; *a.* × $^{3}/_{4}$; *b.* × 3. **3.** Lemna perpusilla; *a.* × $^{3}/_{4}$; *b.* × 3. **4.** Lemna valdiviana; *a.* × $^{3}/_{4}$; *b.* × $4^{1}/_{2}$. **5.** Wolffia columbiana; *a.* × $^{3}/_{4}$; *b.* top view, × 15; *c.* side view, × 15. **6.** Wolffia punctata; *a.* × $^{3}/_{4}$; *b.* top view, × 15; *c.* side view, × 15. **7.** Wolffia papulifera; *a.* × $^{3}/_{4}$; *b.* top view, × $7^{1}/_{2}$; *c.* side view, × $18^{3}/_{4}$. **8.** Wolffiella floridana; *a.* × $^{3}/_{4}$; *b.* cluster of leaves in natural position in water, × $3^{3}/_{4}$; *c.* single leaf, × $3^{3}/_{4}$; *d.* cluster of leaves, top view, × $3^{3}/_{4}$.

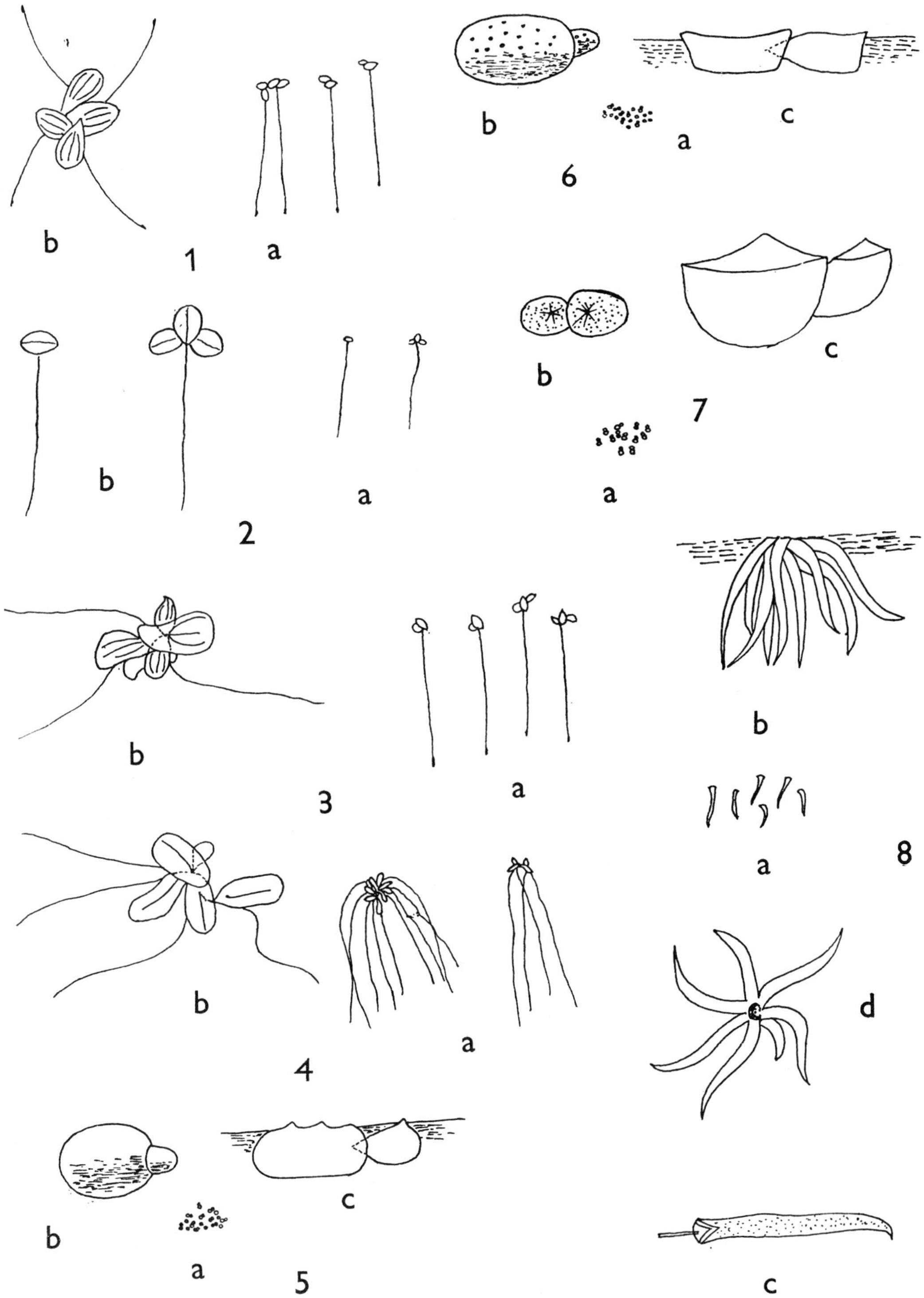


PLATE NO. 103

565 Wolffiella floridana

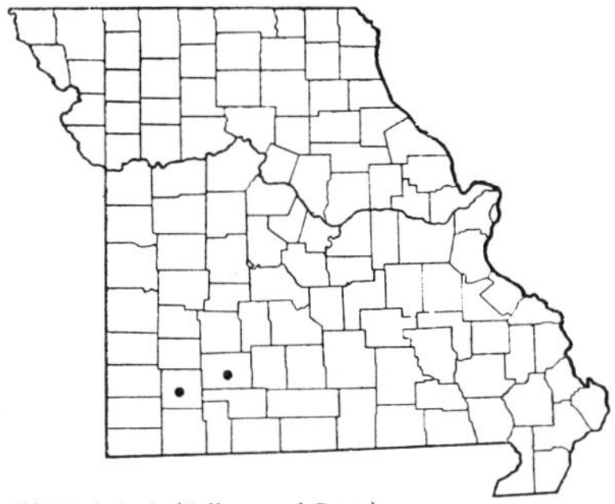

566 Xyris torta (Yellow-eyed Grass)

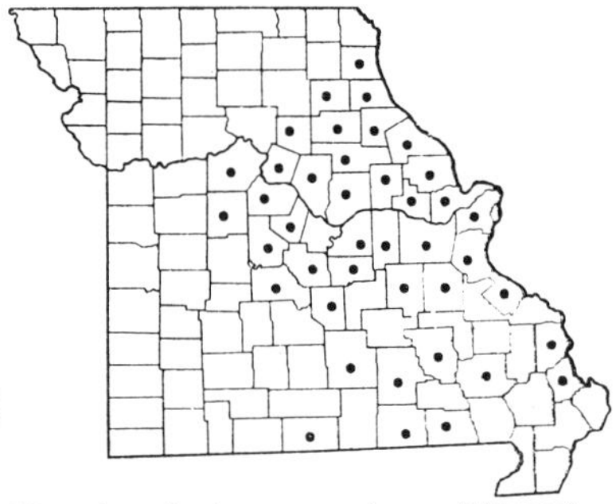

567 Tradescantia subaspera var. subaspera (Spiderwort)

## Fam. **XYRIDACEAE** (Yellow-eyed Grass Family)

### **Xyris** L. Yellow-eyed Grass

**Xyris torta** Sm. Yellow-eyed Grass Map 566
*Xyris flexuosa* [of P & S], not Muhl.

Flowers July–September.

Very rare and local in Missouri, where known only from two localities in the southwestern sector of the state, in Greene (Asher, July 16, 1887, *Blankinship* in Drury College herbarium; Goose Pond, *Bush*) and Lawrence (wet sandy soil in deep railway cut through Pennsylvanian sandstone, 2 mi. southeast of Aurora, July 13, 1953, *Palmer 56161* [in bud]; same locality, along ditches, August 25, 1953, *Palmer 56626* [in flower]) counties.

Ranges from Georgia to Arkansas and Texas, north to New Hampshire, New York, Ohio, Minnesota, and Iowa.

The flowers are yellow and have 3 sepals, 3 petals with claws, 3 stamens, a 3-cleft style, and a 1-celled ovary with many ovules or seeds on parietal placentae. The flowers are arranged in the axils of the overlapping scales, the whole resembling a tiny pine cone, at the tip of a long green scape (leafless stem bearing the flowers). The grasslike leaves are erect and arise from the thickened bulbous base of the plant.

Mallard ducks are known to feed on the seeds of this and other species of the genus.

## Fam. **COMMELINACEAE** (Spiderwort Family)

The 3 petals equal, all alike; stamens 6, all alike; flowers not surrounded by a folded spathe . 1. Tradescantia

One of the petals slightly or much smaller than the other 2; 3 of the 5 or 6 stamens smaller and different; flowers surrounded by a folded spathe . . . . . . . . . . . . . . . . . . 2. Commelina

### 1. **Tradescantia** L. Spiderwort

The following key will separate the majority of plants encountered. It should be pointed out, however, that the species hybridize or cross with one another, so that combinations of characters may be found on a few specimens which do not appear to fit those given in the key.

a. Dwarf plants with short stems 2–10 cm. tall and crowded leaves . . . . . . . . . . . . *b*

  b. Sepals or pedicels or both bearing some gland-tipped hairs (tiny swellings or knobs at the tip) (use magnifying lens of 10 ×) . . . . . . . . . . . . . . . . . . . . . . . . *c*

    c. Plants of prairies, meadows, railroads, and open roadsides of the northern 2/3 of the state 7. T. bracteata

    c. Plants of woodland in the southeastern or southwestern Ozark region . . . . . . . . . *d*

      d. Leaf-blades 5–12 mm. wide, hairy, dull or grass-green without any glaucous covering (smooth gray or silvery coating which can be rubbed off); petals deep purple, deep blue or deep rose; southeastern Ozark region . . . . . . . . . . . . . . . . . 8. T. longipes

      d. Leaf-blades 15–50 mm. wide, hairless except for minutely hairy margins, more or less glaucous and silvery or gray-green; petals of light shades of pink or lavender-blue, or white; southwestern Ozark region . . . . . . . . . . . . . . . . . 2. T. ozarkana

b. Sepals or pedicels with no gland-tipped hairs (use magnifying lens of 10 ×) . . . . . . . . . *e*
  e. Blades of the upper leaves much broader than their sheaths . . . . . . . . 3. T. ERNESTIANA
  e. Blades of the upper leaves narrower than or about as broad as their sheaths . . . . . . . *f*
    f. Stems and leaves with many conspicuous hairs; leaves usually edged with pink or purple; sepals petal-like, thin with usually purplish tips or margins, not inflated; southwestern Missouri plants. . . . . . . . . . . . . . . . . . . . . . 5. T. THARPII
    f. Stems and leaves mainly glabrous (without hairs) or with a few sparse hairs at most; leaves not edged with purple; sepals green, more or less inflated; plants of the eastern half of Missouri . . . . . . . . . . . . . . . . . . . . 4. T. VIRGINIANA

a. Plants averaging 15–70 cm. tall . . . . . . . . . . . . . . . . . . . . . . . *g*
  g. Sepals glabrous, or if hairs are present, they are at the very tip and do not bear glands at tip 6. T. OHIENSIS (the hybrid between T. OHIENSIS and T. SUBASPERA may be keyed here)
  g. Sepals with numerous hairs from base to tip . . . . . . . . . . . . . . . . . . *h*
    h. Blades of at least the upper leaves much broader than their sheaths . . . . . . . . . *i*
      i. Plants of the eastern half of Missouri west to the central part of the state, west to Saline, Pettis, Camden, and Ozark counties; pedicels (flower-stalks) 10–17 mm. long; stem more or less conspicuously zigzag or flexuous; flowers pale to deep blue; flowering mostly from the first week of June through September . . . . . . . . . . . . . 1. T. SUBASPERA
      i. Plants of southwestern Missouri only, east to Douglas and Ozark counties; pedicels 20–32 mm. long; stem more or less straight; flowers white, pale pink, pale to deep purple, lavender-blue, rose-red, or deep blue; flowering mostly in April and May . . . . . . . . . . *j*
        j. Sepals 4–10 mm. long, not inflated, with some gland-tipped hairs (with a minute swelling or knob at tip); leaf-blades more or less glaucous (with a whitish coating which can be rubbed off) and silvery or gray-green; petals of light shades of pink or lavender-blue or white . . . . . . . . . . . . . . . . . . . . . . 2. T. OZARKANA
        j. Sepals 10–14 mm. long, more or less inflated, without any gland-tipped hairs; leaf-blades deep green, not glaucous; petals usually deep blue, pale to deep purple or rose-red . . . . . . . . . . . . . . . . . . . . . . 3. T. ERNESTIANA
    h. Blades of the upper leaves narrower than to about as broad as their sheaths . . . . . . . *k*
      k. Sepals with gland-tipped hairs; plants of the prairies, meadows, railroads, and open road-sides . . . . . . . . . . . . . . . . . . . . . . . . 7. T. BRACTEATA
      k. Sepals without any gland-tipped hairs; plants of the woodland or ledges along bluffs or sandstone exposures . . . . . . . . . . . . . . . . . . . 4. T. VIRGINIANA

1. **Tradescantia subaspera** Ker. var. **subaspera** Map 567

*Tradescantia subaspera* Ker. [G, P & S]

Flowers first week of June–September.

Occurs in rich woods, shaded bluffs, usually of limestone areas, and in wooded valleys along streams. Eastern and middle Missouri north to Lewis and Shelby counties, west to Saline, Pettis, Morgan, Camden, Texas, and Ozark counties; absent from the southeastern lowland section. Another variety, var. *montana* (Shuttlw.) Anderson & Woodson, occurs in some of the southeastern states.

Ranges from West Virginia to Illinois, south to Tennessee, and Missouri.

This hybridizes with no. 6 (*T. ohiensis*), producing a plant with the broader leaf-blades and somewhat zigzag stems of *T. subaspera*, the silvery glaucous character of the leaf of *T. ohiensis*, the mostly glabrous or hairy-tipped sepals of *T. ohiensis*, and the cymes of flowers terminal and from the upper nodes of the stem, a character of *T. subaspera*. The hybrid is known from Monroe, Boone, and Washington counties.

*Tradescantia subaspera* is a good plant for wildflower gardens, flowering throughout the summer and fall months in shaded areas. It favors a rich loamy soil.

2. **Tradescantia ozarkana** Anderson & Woodson Map 568

Flowers April–May.

Occurs in rich rocky woods, steep wooded slopes, and moist ledges along wooded bluffs, in limestone areas. Occurs along the White River and tributaries of southwestern Missouri from Ozark County west to Barry County.

Ranges from southwest Missouri and northwestern Arkansas to eastern Oklahoma.

*Tradescantia ozarkana* varies in color of petals from white to pink, rose, or lavender. This species was formerly abundant along the White River and tributaries, but the impoundments of the various dams which have been erected have destroyed millions of plants.

This species hybridizes with no. 3 (*T. Ernestiana*) in Barry County at Eagle Rock near Roaring River State

568 Tradescantia ozarkana (Spiderwort)

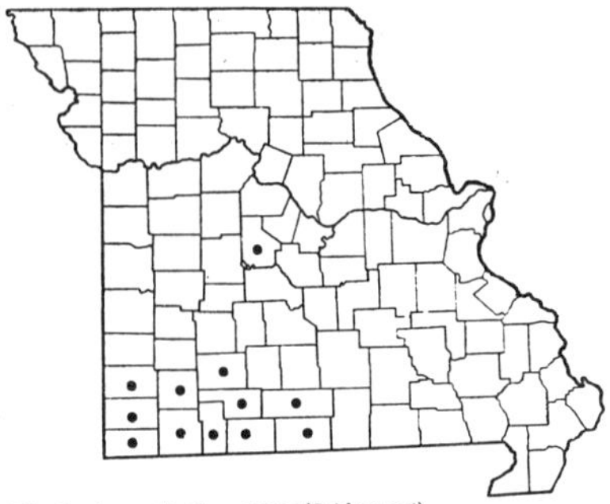

569 Tradescantia Ernestiana (Spiderwort)

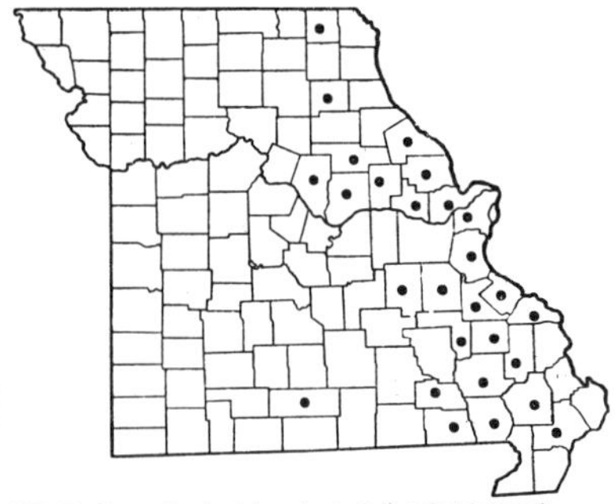

570 Tradescantia virginiana f. virginiana (Spiderwort)

Park, and with no. 6 (*T. ohiensis*) in Ozark (*Steyermark 27091*) and Taney (*Steyermark 5267* and *5268*) counties. The latter hybrid is a broad-leaved, glaucous plant with often glabrous or only hairy-tipped sepals.

The stems vary from glabrous (the usual condition) to very pubescent (*Steyermark 5293* from Taney County).

The plant does well in wildflower gardens when provided with rich, well-drained, shaded soils.

3. **Tradescantia Ernestiana** Anderson & Woodson f. **Ernestiana** Map 569

Flowers April–May.

Occurs in valleys, ravines, rich or rocky slopes of woodland, and wooded ledges of bluffs, where it is found in southwestern Missouri east to Douglas and Ozark counties, north to Greene and Jasper counties; introduced around Syracuse in Morgan County (herbarium Wm. Jewell College).

Ranges from southwestern Missouri to northwestern Arkansas, and eastern Oklahoma.

As noted above, this species crosses in nature with no. 2 (*T. ozarkana*). The colors of the petals of *T. Ernestiana* run to deeper shades of blue, rose-red, and deep purple than those in *T. ozarkana*. A white-flowered form, f. *alba* Waterfall, has not yet been recorded from Missouri. The species does very well in wildflower gardens in rich, shaded soils.

4. **Tradescantia virginiana** L. f. **virginiana** Map 570

*Tradescantia virginiana* L. [BB, G, P & S]

Flowers late April–June.

Occurs in rocky or nonrocky open wooded slopes or valleys, often on the crests or upper slopes in acid soils overlying sandstone, chert, or granite, and along moist shaded ledges of wooded bluffs of sandstone and granitic rocks. Found in the eastern half of Missouri west to Scotland, Shelby, Boone, Crawford, Reynolds, Carter, and Ripley counties, and locally southwest in Douglas County. The Jackson County record (barrens, Dodson, May 10, 1896, *Mackenzie* in the herbarium of New York Botanical Garden), cited in the work on Tradescantia by Anderson and Woodson (Contr. Arn. Arb. 9: 65. 1935), and the record from Clinton County cited in Palmer and Steyermark's *Annotated Catalogue* should be deleted for this species and transferred to *T. bracteata* instead.

Ranges from Connecticut to Wisconsin, south to Georgia, Tennessee, and Missouri; escaped northeast to Maine.

The usual color of this species is a deep purple, blue, or deep blue-purple or violet, occasionally a deep rose-red. A white-flowered form, f. *albiflora* Britt., has not yet been recorded from Missouri, but is to be expected. The leaves are either glabrous or have some scattered hairs on the surface and margins of the blades, and vary from narrow- to broad-leaved types.

Hybrids between this species and no. 6 (*T. ohiensis*) have been found in Mississippi (*Steyermark 10338*) and Butler counties, producing plants in habit somewhat intermediate with low stature only 3 dm. tall, subglaucous leaves with the upper ones recurved, and the sepals varying from glabrous to bearded at the tip to pubescent throughout.

Dwarf plants with more crowded leaves and short stems less than 1 dm. tall are sometimes puzzling and confused with *T. longipes* or *T. Tharpii*. The nonglandular hairs of the inflated green calyx and more glabrous stems and leaves distinguish this species from *T. longipes*, and the shorter pedicels, inflated green

---

Plate no. 104. **1.** Xyris torta, × 2/7; *a.* Head of bracts, × 2; After Gleason, details from Small, The New York Botanical Garden. **2.** Tradescantia subaspera, × 2/7; Details from Small, The New York Botanical Garden. **3.** Tradescantia ozarkana, × 1; Details from Small, The New York Botanical Garden. **4.** Tradescantia Ernestiana, × 2/7. **5.** Tradescantia virginiana, × 2/7.

PLATE NO. 104

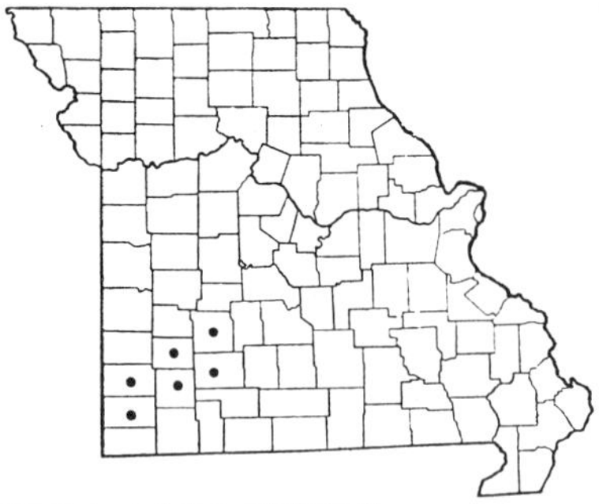

571 Tradescantia Tharpii (Spiderwort)

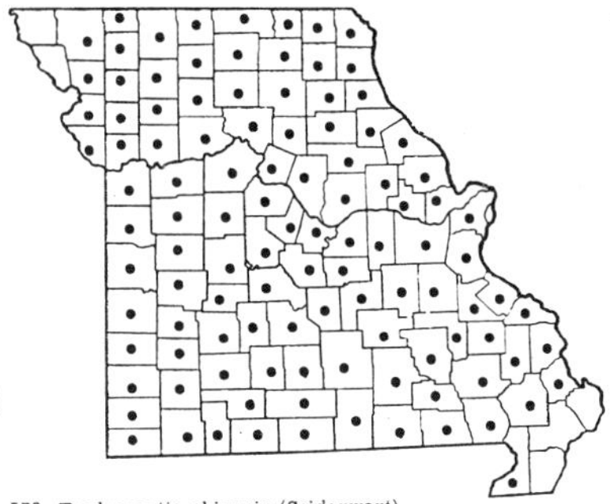

572 Tradescantia ohiensis (Spiderwort)

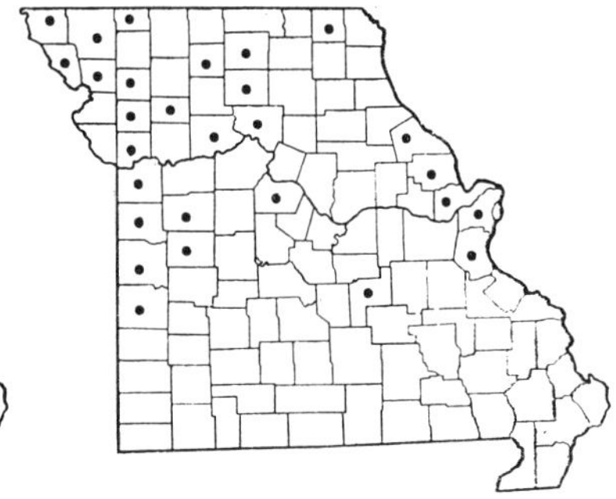

573 Tradescantia bracteata (Spiderwort)

sepals, and mainly glabrous stems and leaves distinguish it from *T. Tharpii.*

*Tradescantia virginiana* favors acid soils similar to those of *T. longipes*. Cultivated plants have been derived from *T. virginiana* and hybrids between it and other species. The fleshy juicy leaves and stems of this and other species of *Tradescantia* are sometimes eaten as a salad or cooked green vegetable.

5. **Tradescantia Tharpii** Anderson & Woodson Map 571

Flowers April–May.

Occurs usually on rocky prairies of sandstone or chert, or on sandstone, limestone, or chert glades and open slopes, and along railroads in southwestern Missouri in the unglaciated prairie section north and east to Polk and Greene counties.

Ranges from southwestern Missouri and Kansas to Oklahoma and Texas.

The sepals, as in *T. longipes*, are somewhat thin and petal-like with a purple or rose hue, and not inflated and completely green, as in *T. virginiana* or *T. Ernestiana*. The color of the petals varies from deep rose to purple or purplish-blue.

This species hybridizes with no. 6 (*T. ohiensis*) in Lawrence (*Steyermark 22224*; *Palmer 53722*) and Jasper (*Palmer 30130, 48587-A*) counties.

6. **Tradescantia ohiensis** Raf. Map 572

*Tradescantia canaliculata* Raf. [P & S]

Flowers early May–July.

Commonly occurs in prairies, meadows, thickets, and along railroads and roadsides, more rarely in open woodland. Found throughout Missouri; not recorded from a couple of counties in extreme northwestern Missouri, and three counties in the extreme southeastern section.

Ranges from Massachusetts to Minnesota and Nebraska, south to Florida and Texas.

The color of the petals is normally a clear blue, but rose and purplish-lavender colors occasionally occur. White-flowered forms are rarely noted. The petals vary in size, in some individuals being much larger than normal.

This species has been recorded as hybridizing with all the other species in Missouri except *T. Ernestiana* and *T. bracteata*, and eventually hybrids will probably be found with these latter two as well.

The leaf-sheaths vary from glabrous to hairy, and the sepals from completely glabrous to tipped with nonglandular hairs.

This species is glaucous with a silvery-green color on all parts and easily recognized by this character together with the glabrous stems, leaves, and sepals.

7. **Tradescantia bracteata** Small Map 573

Flowers May–July.

Occurs usually in open exposed soils of prairies, meadows, along railroads and roadsides in northern and central Missouri south to Jefferson, Cooper, Henry, and Vernon counties, and locally south in Phelps County.

Ranges from Michigan to Montana, south to Indiana, Illinois, Missouri, and Kansas.

The color range of this species is quite varied, ranging from usually bright rose to purple, less frequently blue, lavender, violet, white, and blends of combinations of these colors. Plants of this species do well in garden beds exposed to full sun, and increase rapidly from the roots. Unlike most of the other species of Missouri, which retain their foliage for a relatively long period after flowering, this species quickly disappears after the short blooming season. The leaves vary from narrow- to broad-leaved types.

8. **Tradescantia longipes** Anderson & Woodson

Wild Crocus Map 574

Flowers April–May.

Occurs on slopes or level valleys of rocky open woodland in acid soils associated with sandstone, chert,

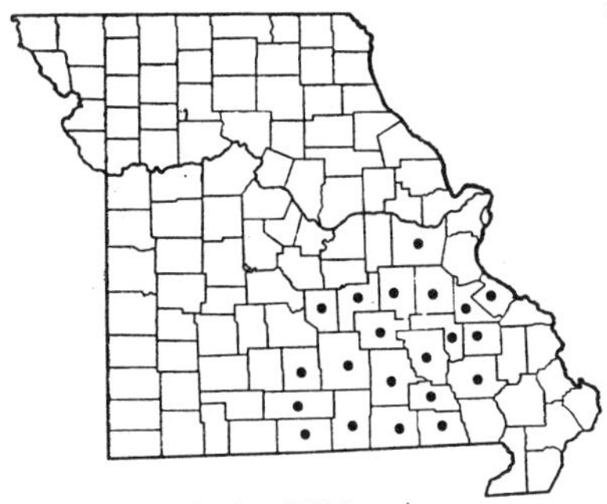

574 Tradescantia longipes (Wild Crocus)

574 A ‡ Tradescantia Ernestiana X ozarkana
574 A † Tradescantia ohiensis X ozarkana
574 A * Tradescantia ohiensis X longipes
574 A ✢ Tradescantia ohiensis X subaspera
574 A □ Tradescantia ohiensis X virginiana
574 A ✣ Tradescantia ohiensis X Tharpii

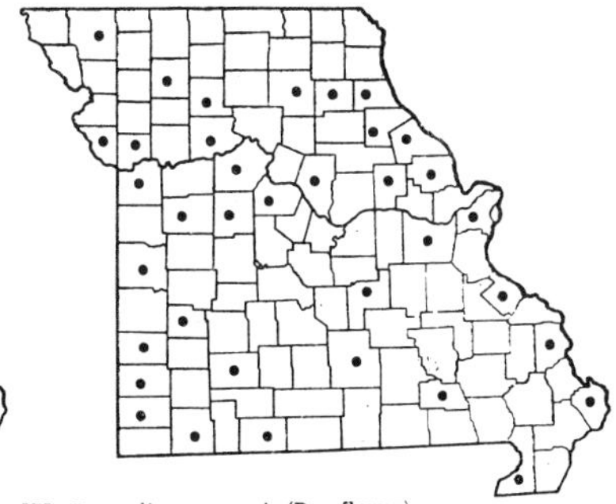

575 Commelina communis (Day-flower)

granite or porphyritic rock. Restricted to the Ozark section of the state, extending north to Franklin, Phelps, and Pulaski counties, west to Wright, Douglas, and Ozark counties. It occurs commonly on dry upland slopes and crests of hills and ridges where chert or Roubidoux sandstone predominates, and is frequently found in pine-oak woodland. Not known outside of Missouri, but to be expected in northern Arkansas.

Sometimes known as wild crocus in the St. Francois hills section of the Ozarks (St. Francois, Madison, Iron, and Reynolds counties), the flowers of this species appear to arise just above the ground level and are usually deep purple, violet, or deep rose, occasionally pink or bluish. It is a desirable type of plant for the rock-garden, but requires an acid soil free of lime rock. The flowers have a pleasant fragrance, resembling that of lily-of-the-valley.

*Tradescantia longipes* hybridizes with *T. ohiensis*, records having been collected from Iron and Wayne counties, producing a plant with longer stems, and longer, somewhat glaucous leaves, combined with thin somewhat rose-colored sepals.

The pedicels are longer than those of *T. Tharpii*, the other dwarf species which it resembles, and the stem does not elongate as much in fruit as in *T. Tharpii.*

## 2. **Commelina** L. Day-flower

a. The folded green spathe that surrounds the flowers is open from one end to the flower, the margins being free all the way to the base; plants with slender roots . . . . . . . . . . . . . . . *b*

b. The smallest 1 of the 3 petals white; the larger 2 blue petals with the main part (blade) 8–15 mm. long, ovate; anthers 6; leaf-sheaths rarely with long hairs along the margins at the summit; flower stalks (peduncles) 10–50 mm. long, not hair-like . . . . . . . . . . . . 1. C. COMMUNIS

b. All 3 petals blue; the larger 2 blue petals with the main part (blade) 2.5–8 mm. long, bean-shaped; anthers 5; leaf-sheaths with long hairs along the margins at the summit; flower stalks (peduncles) 5–20 mm. long, hair-like . . . . . . . . . . . . . . . . . . . . . . . . *c*

c. The green spathe 25–35 mm. long; stems erect or spreading . . . . . . . . 2. C. CAROLINIANA

c. The green spathe 10–25 mm. long; stems creeping . . . . . . . . . . . 3. C. DIFFUSA

a. The folded green spathe has its margins united at the base (the end nearest the flower stalk); plants with thick roots . . . . . . . . . . . . . . . . . . . . . . . . . . . . *d*

d. Leaf-sheaths not prolonged at the top, with rusty-colored bristles on the margins; all 3 petals blue, nearly equal; plants of low moist woodland . . . . . . . . . . . . . . 4. C. VIRGINICA

d. Leaf-sheaths prolonged at the top into a spreading rounded loose piece with pale white hairs on the margins; 2 of the petals blue, 1 white and much smaller; plants of gravel and sand bars, stream banks, rocky hills and bluffs, and rocky woods . . . . . . . . . . . 5. C. ERECTA

### 1. **Commelina communis** L. Day-flower

Map 575

Flowers May–October.

Occurs in cultivated and waste ground, moist alluvial soils, and in low woods and thickets. Throughout Missouri.

Native of Asia; introduced into the United States, ranging from Massachusetts to Wisconsin,

576 Commelina caroliniana (Day-flower)

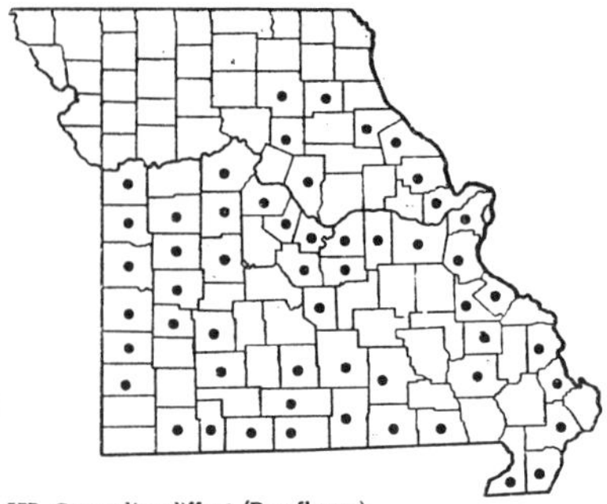

577 Commelina diffusa (Day-flower)

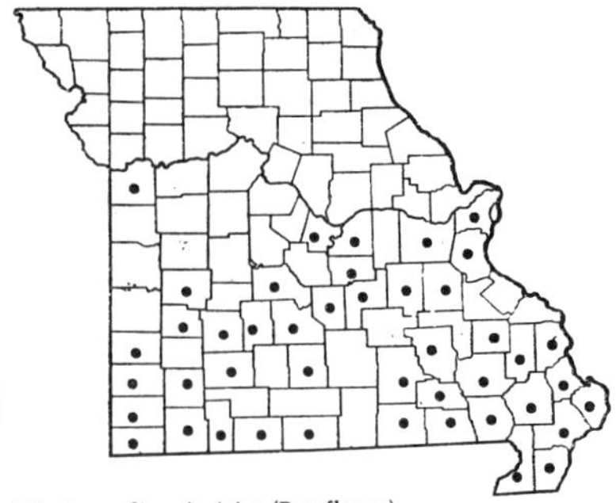

578 Commelina virginica (Day-flower)

outh to Alabama, Arkansas, Kansas, and Nebraska.

Two varieties, var. *communis* and var. *ludens* (Miq.) Clarke, have been distinguished as follows:

var. *communis*. Main part (blade) of larger petals 10–15 mm. long, pale violet-blue; lobed sterile anthers all yellow; leaf-sheaths glabrous at the summit and on the margins.

var. *ludens*. Main part (blade) of larger petals 8–10 mm. long, deep violet-blue; lobed sterile anthers with a madder-brown center; leaf-sheaths usually with short hairs on the summit and margins.

Apparently both varieties occur in the state, but their proper status remains to be more carefully evaluated.

This is the common day-flower often encountered around dwelling places, alleys, back yards, and as a weed in gardens. In the East Indies the plants are reported to be cooked and eaten as a green vegetable.

2. **Commelina caroliniana** Walt. Day-flower Map 576

Flowers August–September.

Introduced in Jackson County, west-central Missouri (Sheffield, August 30, 1905, *Bush 3307*; Sheffield, September 14, 1905, *Bush 3332*).

Ranges from South Carolina to Florida west to Mississippi; introduced in Missouri.

The specimens collected by Bush were originally identified by him as *C. nudiflora*.The spathes on the specimens are elongate, acuminate, and their margins are free to the base. They are clearly not *C. elegans* HBK., a species having closed spathes with a short apex, but were filed under that species at the Gray Herbarium.

3. **Commeliha diffusa** Burm. Day-flower Map 577

*Commelina longicaulis* Jacq. [P & S]

Flowers early July–October.

Occurs in moist sandy or muddy alluvial soils of river bottom, in low fields, valleys, and moist open cultivated ground, sometimes weedy. Common in all of southern and central Missouri north to Shelby, Macon, Saline, and Jackson counties.

Ranges from South and Central America, the West Indies, and Mexico, north in the United States to Virginia, Ohio, Indiana, Illinois, Missouri, and Kansas; introduced in Massachusetts; also in the Old World tropics.

The slender stems creep along the moist ground, rooting at the joints (nodes). The black seeds are 2–3 mm. long.

This species, likewise, is reported to be cooked and eaten as a green vegetable.

4. **Commelina virginica** L. Day-flower Map 578

Flowers July–September.

Occurs in wet low woodland in valleys along streams, and borders of swamps and sloughs. Found throughout the Ozark and southeastern lowland section, extending north to St. Louis, Franklin, Osage, Cole, Camden, St. Clair, and Barton counties, and locally northwest in Jackson County.

Ranges from Florida to Texas, north to New Jersey, Pennsylvania, Kentucky, Illinois, Missouri, and Kansas.

This species is the most robust of the native species of *Commelina* with stout erect stems up to 1.2 meters tall, and spiderwort- or lilylike dark green leaf-blades 2.5–5 cm. broad and up to 20 cm. long. The spathes

Plate no. 105. **1.** Tradescantia Tharpii, $\times 2/7$. **2.** Commelina erecta var. erecta, $\times 2/7$. **3.** Tradescantia ohiensis, $\times 2/7$. **4.** Tradescantia bracteata, $\times 2/7$. **5.** Tradescantia longipes, $\times 2/7$. **6.** Commelina diffusa, $\times 2/7$; *a.* Fruit, $\times 1\,3/7$; *b.* X-section of fruit; *c.* Seed, highly magnified; After Britton and Brown, details from Small, The New York Botanical Garden. **7.** Commelina virginica; *a.* Fruit, $\times 1\,3/7$; After Britton and Brown, The New York Botanical Garden.

PLATE NO. 105

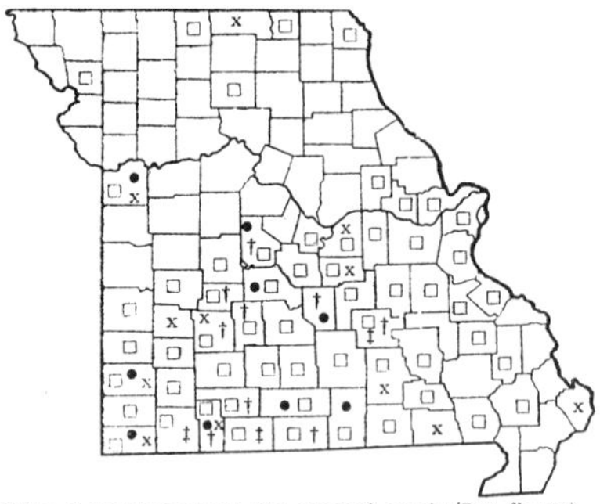

579 • Commelina erecta var. erecta f. erecta (Day-flower)
579 † Commelina erecta var. erecta f. intercursa
579 □ Commelina erecta var. angustifolia f. angustifolia
579 x Commelina erecta var. angustifolia f. crispa
579 ‡ Commelina erecta var. Deamiana

580 Eichhornia crassipes (Water Hyacinth)

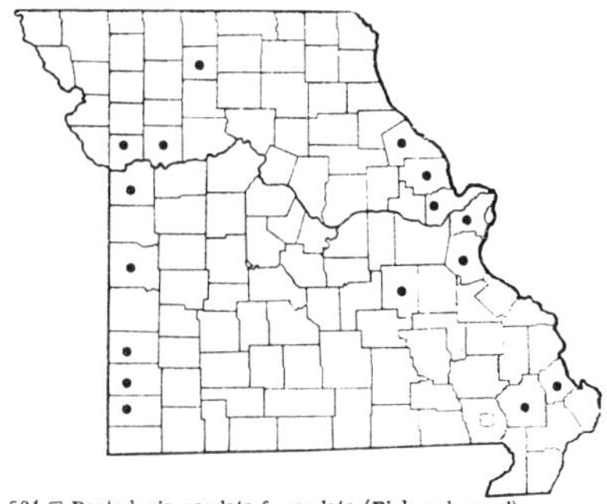

581 □ Pontederia cordata f. cordata (Pickerel-weed)
581 • Pontederia cordata f. latifolia

are crowded near the summit of the stem and surround the large blue flowers, the blades of which are 10 mm. or more long. It is a handsome species.

5. **Commelina erecta** L. Day-flower Map 579
Flowers last of May–October.

Occurs commonly on sand and gravel bars, alluvial banks of streams, rocky wooded slopes, and exposed bluffs and glades. Common throughout the Ozark region in southern and central Missouri, locally north in northern Missouri.

The following varieties are represented in Missouri material:

a. The larger developed main leaves principally 20–40 mm. (sometimes 15) wide, lanceolate to ovate-lanceolate; mature spathes 22–36 mm. long . . . . . . . . . . . . *b*
  b. Spathes nearly glabrous or with short hairs
    5a. C. ERECTA var. ERECTA f. ERECTA
  b. Spathes with rather conspicuous white hairs
    5b. C. ERECTA var. ERECTA f. INTERCURSA
a. The larger developed main leaves principally 4–20 mm. wide, linear to narrowly lanceolate; mature spathes 10–30 mm. long . . . . *c*
  c. Mature spathes mostly 10–20 mm. (sometimes up to 25) long; the longest leaves 4–10 cm. long . . . . . . . . . *d*
    d. Spathes covered with short dense hairs
      5c. C. ERECTA var. ANGUSTIFOLIA f. ANGUSTIFOLIA
    d. Spathes with long white hairs at and near the base . . . . 5d. C. ERECTA var. ANGUSTIFOLIA f. CRISPA
  c. Mature spathes mostly 25–30 mm. long; the longest leaves 7–15 cm. long . . .
    5e. C. ERECTA var. DEAMIANA

The above arrangement, modified slightly from Fernald's original treatment (Rh. 42: 435–41. 1940), will account for the variations to be encountered. Although some of the extremes of variation seem distinct enough, considerable overlapping between characters occurs. This intergradation and lack of correlation often break down the lines of separation between the various varieties and forms, casting considerable doubt as to their acceptability.

5a. **Commelina erecta** var. **erecta** f. **erecta** Map 579
*Commelina erecta* L. [G, P & S]
Scattered in southern and central Missouri.

Ranges from New York to Kansas, south to Florida and Texas.

This form and f. *intercursa* appear to be less common in Missouri than var. *angustifolia* f. *angustifolia* and f. *crispa*.

5b. **Commelina erecta** var. **erecta** f. **intercursa** Fern. Map 579
*Commelina erecta* f. *intercursa* Fern. [G]
Scattered in southern and central Missouri.
Scattered within the range of f. *erecta*.

5c. **Commelina erecta** var. **angustifolia** (Michx.) Fern. f. **angustifolia** Map 579
*Commelina erecta* var. *angustifolia* (Michx.) Fern. [G, P & S]

This is the commonest of the variations of *C. erecta* in Missouri, according to the known records. It and f. *crispa* are more commonly found on exposed places of glades, bluffs, and sand and gravel bars.

Ranges from Central America and Mexico, north into the United States to Delaware, West Virginia, Illinois, Missouri, and Nebraska.

A form with all white petals (f. *albina* Fern.) occurs, but has not yet been found in Missouri.

5d. **Commelina erecta** var. **angustifolia** f. **crispa** (Wooton) Fern. Map 579

*Commelina erecta* var. *crispa* (Wooton) Palmer & Steyerm. [P & S]

Less common than var. *angustifolia* f. *angustifolia* in the state.

Found throughout the range of f. *angustifolia*.

The fleshy roots of *C. erecta* var. *angustifolia* f. *angustifolia* and f. *crispa*, and other varieties and forms of *C. erecta* contain considerable starchy carbohydrates and have been suggested as being edible as a type of cooked vegetable.

5e. **Commelina erecta** var. **Deamiana** Fern. Map 579

This variety has been collected in Missouri thus far only in Dent, Taney, and Barry counties.

Ranges from Michigan and Indiana to Kansas and Texas.

## Fam. PONTEDERIACEAE (Pickerel-weed Family)

a. Plants normally floating on the surface of water, the roots hanging below the surface; some or all the petioles of the leaves inflated, swollen around the middle and tenpin-shaped . . . . . . 1. EICHHORNIA

a. Plants normally and usually rooted to mud, sand, or gravel in or out of water; petioles of leaves not inflated or tenpin-shaped . . . . . . . . . . . . . . . . . . . . . . . . . *b*

b. Plants 50–100 cm. tall; leaf-blades 5–15 cm. wide, mostly 10–20 cm. long; flowers numerous in a long dense spike; stamens 6; flowers irregular, 2-lipped . . . . . . . . . . 2. PONTEDERIA

b. Plants 6–25 cm. tall, or, if taller, the leaves are grass-like; leaf-blades 0.2–6 cm. wide, mostly 2–6 cm. long, but if longer, then the leaves are grass-like; flowers 1–16, loosely-flowered; stamens 3; flowers regular, about equally 6-parted . . . . . . . . . . . . . 3. HETERANTHERA

### 1. Eichhornia Kunth Water Hyacinth

**Eichhornia crassipes** (Mart.) Solms

Water Hyacinth Map 580

Flowers July–August.

Known only from drainage canals of the southeastern lowland section from Dunklin and Butler (Neeleyville, July 12, 1930, *Kellogg 15142*) counties.

Ranges from South and Central America, the West Indies, Mexico, north in the United States from Florida and Texas to Virginia and Missouri. Introduced in Europe and Asia.

The plant itself is not considered to have any direct food value for wildfowl, but indirectly attracts minute animal life which in turn is eaten by fish and waterfowl.

In Florida and other sections of the United States, and in Europe, it has clogged up waterways and interfered seriously with navigation and water transportation to the extent of having become a serious pest. The young leaf-blades, petioles, and inflorescences are cooked and eaten by the inhabitants of the Dutch East Indies, where it was introduced.

### 2. Pontederia L. Pickerel-weed

**Pontederia cordata** L. Pickerel-weed Map 581

Flowers first week of June–October.

Occurs in sloughs, open swamps and oxbow ponds of river bottoms, and other ponds. Scattered in eastern and western sections of the state, and locally northward in Grundy County.

Ranges from Nova Scotia to Ontario and Minnesota, south to Florida, Missouri, and Oklahoma.

Missouri material is represented by the following 2 categories:

Leaves narrowly triangular-ovate, tapering with straight sides from the base to the apex. . . . a. P. CORDATA f. CORDATA

Leaves broadly ovate, gradually curved from the broad base to the blunt summit . . b. P. CORDATA f. LATIFOLIA

a. **Pontederia cordata** f. **cordata** Map 581

This is the less frequently encountered form in Missouri.

b. **Pontederia cordata** f. **latifolia** (Farw.) House Map 581

This is the common form in Missouri. The most northern station in the state for the form is in Livingston County (Gooseneck Lake, in bottoms of Thomp-

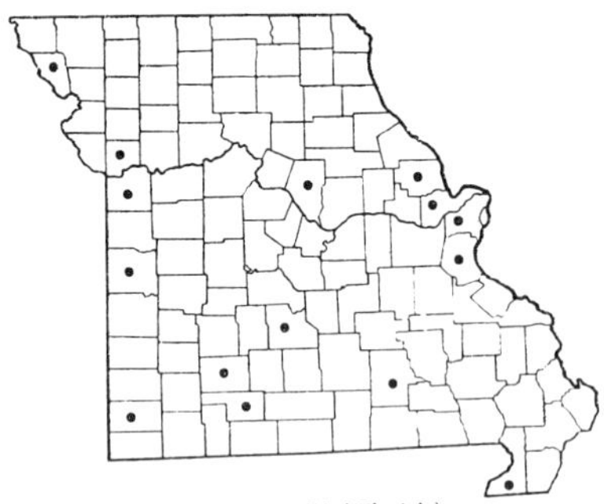

582 Heteranthera reniformis (Mud Plantain)

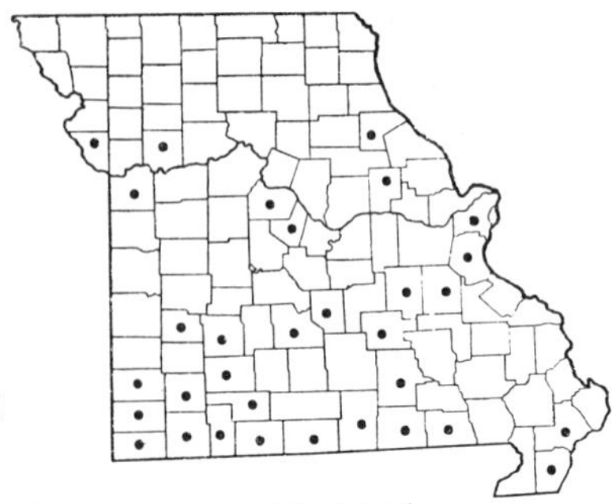

583 Heteranthera limosa (Mud Plantain)

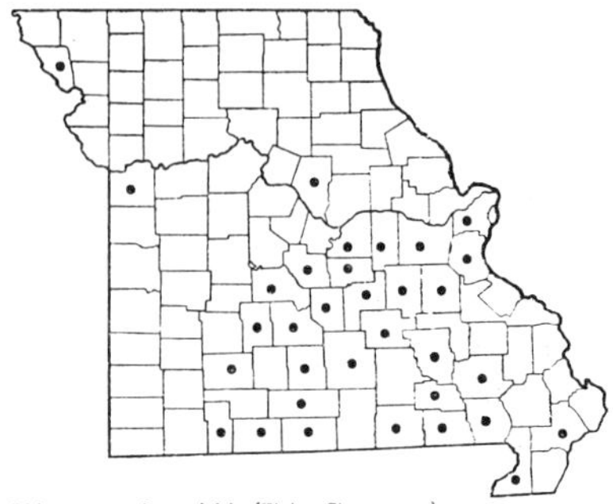

584 Heteranthera dubia (Water Star-grass)

son River, T60N, R24W, northwest sect. 34, 6½ mi. south of Trenton, August 10, 1952, *Steyermark 74216*). This is a natural lake which has many other interesting aquatic plants, such as *Myriophyllum heterophyllum, Utricularia vulgaris, Sparganium androcladum, S. eurycarpum,* and *Anacharis Nuttallii,* some of which are not known elsewhere in northern Missouri.

Narrowly triangular-leaved (f. *angustifolia*) and submerged, linear-leaved (f. *taenia*) forms also have been described, but have thus far not been recorded from the state.

Gleason has identified the Missouri material as *P. lanceolata* Nutt., a tropical American species which extends north into the United States to Virginia and Texas. Based upon the presence of a persistently pubescent perianth with short glandular hairs, all the Missouri material examined would key to *P. lanceolata,* according to Gleason's concept. However, the Missouri material appears not to differ in respect to the pubescence of the perianth from the widely distributed *P. cordata* to which species it is assigned in the present flora.

The ripe raw fruits are edible. It has been recommended that the fleshy stems and young leaves, raw or cooked, may be tried as a vegetable, since it is closely related to the edible southeastern Asiatic *Monochoria.* The seeds are eaten by ducks and muskrats.

## 3. **Heteranthera** R. & P. Mud Plantain

a. Leaves grass-like, linear; flowers yellow, solitary . . . . . . . . . . . . . . . . . 3. H. DUBIA
a. Leaves broad, of an oval, oblong, round, or round-bean shape; flowers blue or white, 1–16 in the inflorescence . . . . . . . . . . . . . . . . . . . . . . . . . . . . . *b*
  b. Leaf-blades longer than broad, narrowed to an acute, obtuse, or slightly heart-shaped base; flower only 1 in each inflorescence; tube of flower 20–35 mm. long . . . . . . . . . 2. H. LIMOSA
  b. Leaf-blades about as broad as long with a deeply heart-shaped base; flowers 2–16 in the inflorescence; tube of flower 6–10 mm. long . . . . . . . . . . . . . . 1. H. RENIFORMIS

### 1. **Heteranthera reniformis** R. & P.

Mud Plantain Map 582

*Heteranthera peduncularis* Benth. [BB]

Flowers July–October.

Occurs along sloughs, ditches of valleys and swampy depressions, and margins of artificial and natural ponds, especially of oxbow lakes in river bottoms. In southern and central Missouri, northwest along the Missouri River to Holt County.

Ranges from South and Central America, the West Indies and Mexico, and in the United States from Florida to Texas, north to Connecticut, New York, Kentucky, Indiana, Illinois, Missouri, and Nebraska.

Sometimes plants may become detached from the muddy soil and float to the surface. Most of them remain rooted, however. The rich green leaves are very shiny.

Gleason (*New Illustrated Flora,* vol. 1: 385. 1952) has assigned specimens from western Missouri (ponds, Floyd, Jackson Co., July 11, 1921, *Bush 9418A*) to

---

Plate no. 106. **1.** Commelina communis, × 2/5; Details from Small, The New York Botanical Garden. **2.** Heteranthera dubia, × 2/5; After Gleason, details from Small, The New York Botanical Garden. **3.** Pontederia cordata, × 2/5; *a.* Flower, × 2/5; After Gleason, details from Small, The New York Botanical Garden. **4.** Eichhornia crassipes, × 3/10; After Gleason, details from Small, The New York Botanical Garden. **5.** Heteranthera limosa, × 2/5; After Gleason, The New York Botanical Garden. **6.** Heteranthera reniformis, × 2/5; After Gleason, details from Small, The New York Botanical Garden.

PLATE NO. 106

*H. peduncularis* Benth., which is differentiated from *H. reniformis* as follows:

Spathe and the leaf next below it separated by a conspicuous internode; base of the short spike included by the spathe. . . . H. RENIFORMIS
Spathe and the leaf next below it contiguous; base of the spike exserted from the spathe . . H. PEDUNCULARIS

I have not been able to separate the Missouri material on the above stated differences, and have assigned all the specimens to *H. reniformis*. Actually, the amount of exsertion of the base of the spike from the spathe and development of a conspicuous internode below the spathe and the leaf next below it appear to be growth phases of the plant, rather than real criteria for the separation of species.

2. **Heteranthera limosa** (Sw.) Willd.
Mud Plantain Map 583
Flowers June–September.

Occurs along sloughs, ditches, in artificial or natural ponds, and margins of slow streams. Southern and central Missouri north to Ralls, Cooper, Ray, and Buchanan counties.

Ranges from South and Central America, the West Indies, and Mexico, and in the United States from Florida to New Mexico, north to Kentucky, Illinois, Minnesota, Nebraska, and Colorado.

Plants are sometimes submerged with only the flowers protruding above the water surface.

3. **Heteranthera dubia** (Jacq.) MacM.
Water Star-grass Map 584
*Zosterella dubia* (Jacq.) Small [BB]
Flowers June–September.

Occurs in streams, springs and spring branches, and sloughs. Common throughout most of the streams in the Ozarks, where it extends north to St. Louis, Franklin, Gasconade, and Osage counties, westward to Camden, Dallas, Greene, and Stone counties, and locally outside the Ozarks in Boone, Jackson, and Holt counties.

Ranges from Florida to Texas and Mexico, north to Maine, New Hampshire, Quebec, Ontario, Minnesota, Idaho, and Washington.

This is a common species found in most of the Ozark streams, where it roots in the muddy gravel or sand beds near the shallower margins and banks; it favors mud substrata and is most frequently found on mud. It grows in both shallow or deep, and swift or still waters, often forming long leafy strands. The leaves may attain a length of 15 cm. Unlike some aquatic plants, it has a wide temperature tolerance, since it is found in the cold waters (54–60° F.) of Ozark spring branches and in the warmer waters of Ozark streams. Where the water is shallower or swifter, the plants may produce much shorter stems with shorter internodes, forming denser and more circular patches. Plants which get stranded on the mud outside of the water develop short and somewhat stiffish leaves and stems and have been named *H. dubia* f. *terrestris* (Farw.) Vict. As this is undoubtedly a mere ecological or temporary condition correlated with fluctuating water levels, it would not appear to merit formal recognition.

Wildfowl sometimes eat this species. The leaves have an olive green or dull green color. When the plants are found without flowers, they may be distinguished from similar looking species of *Potamogeton* by the complete absence of a midnerve in the leaves of *Heteranthera*.

## Order **LILIALES**

### Fam. **JUNCACEAE** (Rush Family)

Plants resembling grasses or sedges, but with small regular greenish or brownish flowers with 3 narrow sepals, 3 similar petals, 6 or 3 stamens, a superior ovary with 1 cell or 3 cells, and numerous ovules and seeds. No ligule, as occurs in the grasses, is found at the junction of the leaf-blade and leaf-sheath. However, an auricle (ear-like extension or membrane) may be present at the summit of the leaf-sheath, and is often an important character in distinguishing some of the species.

The pollen in this family is not known to cause hay fever, as it is in some of the grasses.

Plants without any hairs; ovules or seeds usually many . . . . . . . . . . . . . . 1. JUNCUS
Plants with hairs; ovules or seeds 3 . . . . . . . . . . . . . . . . . . . . . 2. LUZULA

#### 1. **Juncus** L. Rush

The stems (culms) are either hollow or are solid with pith. It is necessary that plants be collected with mature fruits and with the complete underground parts present in order to reach a satisfactory identification.

a. Inflorescence (flower-clusters) appearing to come from the side of the stem which continues upward as the involucral bract . . . . . *b*

b. Common throughout southern and central Missouri; culms (stems) close together in dense clumps, marked with vertical lines or ribs (seen best when dry); stamens 3 . . . . . 1. J. EFFUSUS

b. Known only from west-central Missouri in Jackson and Lafayette counties; culms well separated, arising in a single row from a creeping rootstock, with a smooth unfurrowed surface, or the surface not vertically lined or ribbed; stamens 6 . . . . . 2. J. BALTICUS var. LITTORALIS

a. Inflorescence (flower-clusters) at the very tip of the stem . . . . . *c*

c. Leaves without any cross partitions which show up as darker or harder places at regular intervals . . *d*

d. Flowers crowded together in 2–12-flowered heads with only a bract at the base of each pedicel and no pair of bracteoles at the base of the flower; leaves 1–7 mm. wide . . . . . *e*

e. Culms (stems) stout, 3–5 mm. thick at base, 6–13 dm. tall; leaves at the base mainly 4–7 mm. wide; heads mainly 20–100 to a stem; stamens exserted (protruding above the sepals and petals) in fruit or equaling the sepals, persistent . . . . . 10. J. BIFLORUS

e. Culms (stems) slender, mainly 1–2 mm. (up to 3) thick at base, 1.5–6 dm. (up to 8) tall; leaves at the base mainly 1–5 mm. wide; heads mainly 2–20 (up to 30) to a stem; stamens not exserted in fruit, shorter than sepals, not persisting but shriveling . . . 9. J. MARGINATUS

d. Flowers occurring either singly or scattered on the branches of the inflorescence, or, if more or less clustered, each flower has 2 bracteoles at its base and 1 bractlet at the base of the pedicel; leaves 0.5–1.5 mm. wide . . . . . *f*

f. Inflorescence taking up 1/3–3/4 or more the height of the entire plant; plants annual with a soft base; leaf-sheaths without any auricles . . . . . 3. J. BUFONIUS

f. Inflorescence mainly 1/5 or less the height of the entire plant (rarely 1/4 the height); plants perennials with a firm hard base; leaf-sheaths with auricles (ear-like extensions or loose membranes) at summit . . . . . *g*

g. Auricles at summit of leaf-sheath white and very thin, 1–3.5 mm. long, loose, like a tiny flap protruding at the summit of the sheath . . . . . 6. J. TENUIS

g. Auricles at summit of leaf-sheath short and rounded, not loose or protruding as a tiny flap . . *h*

h. Involucral bract (leaf at base of inflorescence) shorter than the inflorescence; auricle thin and membranous; flowers chiefly arranged along the inner side of the branches which curve in near the tips; leaves usually less than 1/3 the height of the culms (stems); perianth 2.5–3.5 mm. long . . . . . 5. J. SECUNDUS

h. Involucral bract longer than the inflorescence; auricle either firm or rigid like cartilage; flowers somewhat clustered, crowded, or, if loosely branched and scattered, not arranged just along the inner side of the branches; leaves usually about 1/2 the height of the culms (stems); perianth mainly 3.6–6 mm. (sometimes 3) long. . . . . *i*

i. Auricles yellow and glossy, rigid, like cartilage . . . . . 8. J. DUDLEYI

i. Auricles pale brown or gray-brown, greenish or straw-colored, firm but not rigid or glossy . . . . . *j*

j. The 2 small bracteoles at the base of each flower sharp-pointed or tapering to a conspicuous point; capsule 1-celled or imperfectly 3-celled; sepals and petals erect and appressed to capsule or slightly spreading, 3–4 mm. long . 7. J. INTERIOR

j. The 2 small bracteoles at the base of each flower blunt or slightly pointed at tip; capsule completely 3-celled; sepals and petals spreading-ascending, mostly 4–6 mm. long. . . . . . 4. J. KANSANUS

c. Leaves with cross partitions which show up as darker or harder places at regular intervals . . . . *k*

k. Sepals and petals more or less obtuse or blunt at tip; known only from Jackson County, west-central Missouri . . . . . 22. J. ALPINUS var. FUSCESCENS

k. Sepals and petals tapering to a narrow or slender-pointed tip . . . . . *l*

l. Fruiting capsule twice or nearly twice as long as the sepals and petals; fruiting capsule lanceolate-linear; flowers only 2–7 in each head . . . . . 21. J. DIFFUSISSIMUS

l. Without the above combination of characters; fruiting capsule either shorter or somewhat longer than the sepals and petals, but never nearly twice as long. . . . . *m*

m. Capsule conspicuously shorter than the sepals and petals, only 1/2–2/3 their length 15. J. BRACHYCARPUS

m. Capsule equaling or longer than the sepals and petals, or at least only slightly shorter than the sepals and petals . . . . . *n*

n. Seeds with tail-like tips or appendages at each end (examine with magnifying lens 10 × or binocular microscope) . . . . . *o*

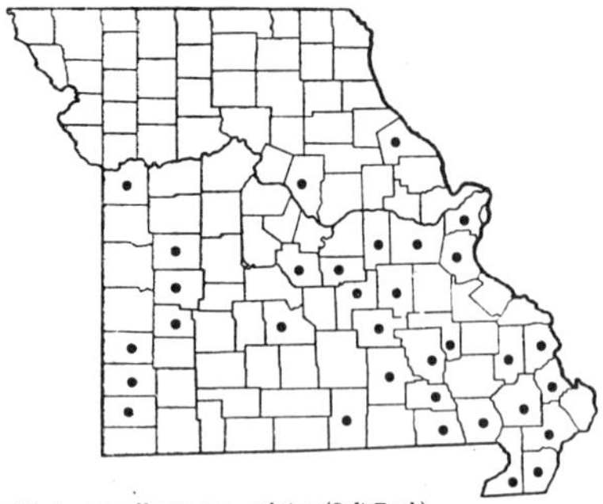

585 Juncus effusus var. solutus (Soft Rush)

586 Juncus balticus var. littoralis

587 Juncus bufonius var. bufonius (Toad Rush)

o. Seeds 1.3–1.9 mm. long, the tails 2/3 to almost as long as the main body of the seed; capsule abruptly short-tipped and somewhat rounded at the summit, 3.8–5 mm. long . . . . . . . . . . . . . . . . . . . . . . . . 11. J. CANADENSIS

o. Seeds 0.7–1 mm. long, the tails much shorter and only about 1/3–1/4 as long as the main body of the seed; capsule tapering to the summit, 2–4 mm. long. 12. J. SUBCAUDATUS

n. Seeds without tail-like appendages, either blunt or dark-pointed at the ends (examine with magnifying lens 10 × or binocular microscope) . . . . . . . . . . *p*

p. Mature heads few-flowered with only 2–7 flowers (rarely to 10), 3–6 mm. thick; sepals and petals mainly 1.9–2.5 mm. long . . . . . . . . . . . . . *q*

q. Heads of flowers 3–35; leaves 0.5–1 mm. thick; capsule conspicuously longer than the sepals and petals and tapering to an acute point . . . . 19. J. DEBILIS

q. Heads of flowers mainly 200 or more; leaves 3–5 mm. thick; capsule about the same length as the sepals and petals and somewhat blunt or rounded at the tip 20. J. NODATUS

p. Mature heads usually many-flowered with usually 10–100 flowers (rarely only 5–10), usually 6–15 mm. thick (rarely 5); sepals and petals mainly 3–6 mm. long (rarely 2.5–3) . . . . . . . . . . . . . . . . . . . . . . . . *r*

r. Leaves more or less flattened, 2–4 mm. broad. . . . . . . . . 13. J. VALIDUS

r. Leaves terete (circular or round when cut crosswise), 1–3 mm. broad . . . . . *s*

s. All the roots fibrous, slender, and thread-like, no thickened or slender creeping rootstocks or tubers present; heads of flowers hemispherical (half spherical or half circular in shape). . . . . . . . . . . . 18. J. ACUMINATUS

s. Thickened or slender creeping rootstocks or tuberous enlargements present; heads of flowers spherical, globe- or ball-shaped . . . . . . . . . . *t*

t. Stamens 3; capsule distinctly longer than the sepals and petals; underground rootstock thick and whitish, not bearing tuberous enlargements 14. J. SCIRPOIDES

t. Stamens 6; capsule shorter than or equaling the sepals and petals, or at most only slightly longer than the sepals and petals; underground rootstock bearing tuberous enlargements . . . . . . . . . . . . . *u*

u. Common throughout Missouri; sepals somewhat longer than the petals; flowers mainly 4–5 mm. long; plants usually 4–10 dm. tall . 17. J. TORREYI

u. Known only from Jackson County, west-central Missouri; sepals either equaling or shorter than the petals; flowers mainly 3–4 mm. long; plants usually 1–4 dm. tall . . . . . . . . . . 16. J. NODOSUS

1. **Juncus effusus** var. **solutus** Fern. & Wieg.
Soft Rush Map 585
Flowers May–September.

Occurs along margins of streams, sloughs, and ponds, in wet depressions of open fields, pastures, prairies, and roadside ditches. Southern and central Missouri north to Pike, Boone, and Jackson counties.

Ranges from Central America and Mexico, north in the United States from Florida to Texas, north to Newfoundland, Quebec, and Ontario.

Plate no. 107. **1.** Juncus effusus var. solutus. **2.** Juncus balticus var. littoralis.

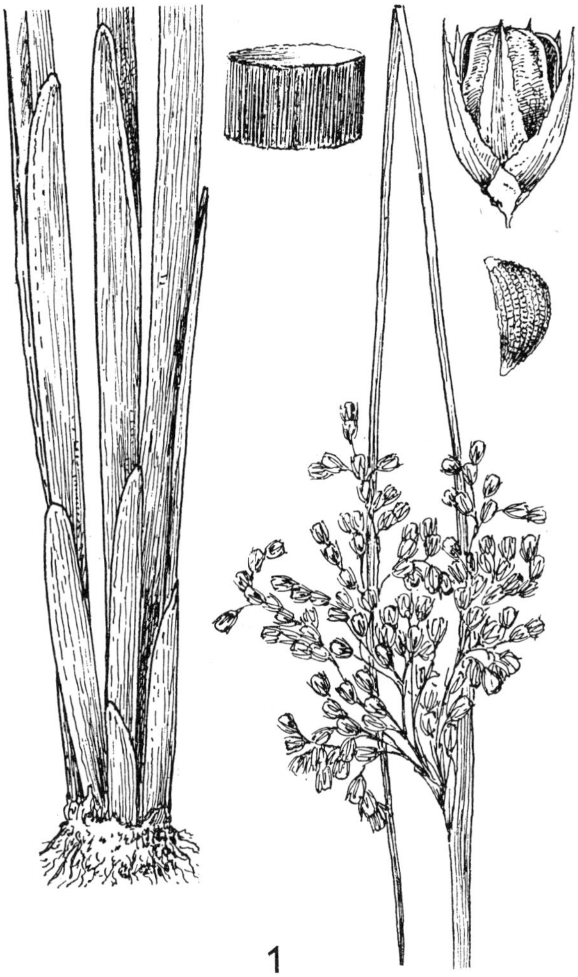

1

2

PLATE NO. 107

588 Juncus kansanus

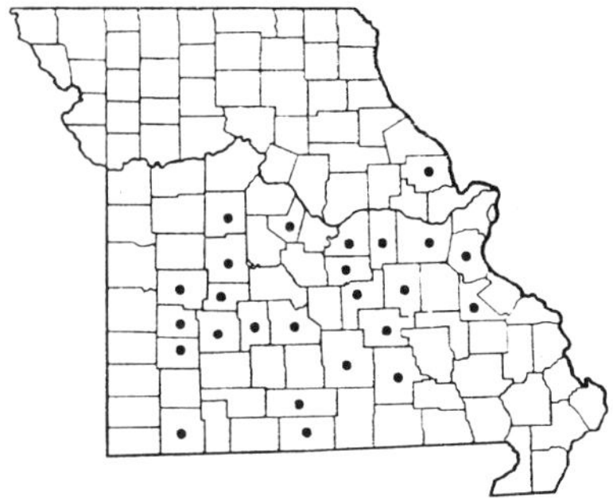

589 Juncus secundus

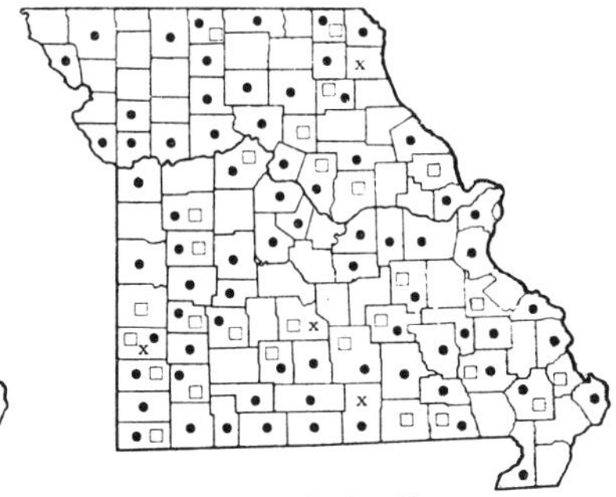

590 • Juncus tenuis f. tenuis (Path Rush)
590 □ Juncus tenuis f. anthelatus
590 x Juncus tenuis f. discretiflorus

This species serves occasionally as the food of muskrat. In Guatemala, the stems are used for weaving mats.

2. **Juncus balticus** Willd. var. **littoralis** Engelm. Map 586

Flowers May–September.

Occurs in prairie swales along railroads and in sand bars and mud flats along the Missouri River. Known only from west-central Missouri in Jackson and Johnson (wet low place in prairie along highway 58 and railroad, T46N, R28W, northeast ¼ sect. 5, 3 mi. northwest of Holden, May 18, 1949, *Steyermark 67897*).

Ranges from Labrador to British Columbia, south to New York, Pennsylvania, Ohio, Indiana, Illinois, Missouri, and Kansas.

3. **Juncus bufonius** L. var. **bufonius**
Toad Rush Map 587

*Juncus bufonius* L. [G, P & S]

Flowers late May–November.

Occurs in sandy and gravelly ground along the Missouri River and in waste places on moist shady paths in open upland woods of Crowley Ridge. Found along the Missouri River in Clay and Jackson counties in west-central Missouri and on Crowley Ridge in Dunklin County (May 26, 1939, *Steyermark 26592*).

Ranges throughout North America; also in the Old World.

4. **Juncus kansanus** Herm. Map 588

Flowers June–August.

Occurs on dry rocky slopes in Pike County, east-central Missouri (Eolia, June 25, 1917, *Davis 8824*) and on upland sandstone glades and sandy prairies of the unglaciated prairie section of southwestern Missouri in Henry, St. Clair, and Dade counties.

Known only from Missouri and Kansas.

5. **Juncus secundus** Beauv. Map 589

Flowers late May–October.

Occurs on dry upland prairies, sandy or rocky open ground, and along railroads, in acid soils. Southern and central Missouri north to Lincoln, Moniteau, and Pettis counties, and west to St. Clair, Dade, and Barry counties.

Ranges from Maine to Ontario and Indiana, south to Georgia, Tennessee, Missouri, and Arkansas.

6. **Juncus tenuis** Willd. Path Rush Map 590

Flowers late May–September.

Missouri material is represented by the following varieties and forms, the key adapted from the work of Dr. Hermann:

a. Flowers clustered 2–6 together at the tips of the branches . . . . 6a. J. TENUIS f. TENUIS
a. Flowers scattered along the axis of the branch . . *b*
  b. Ultimate flowering branches rarely over 4 cm. long; sepals about same length as petals; capsule averaging 3/4 length of the sepals . . . . 6b. J. TENUIS f. ANTHELATUS
  b. Ultimate flowering branches up to 7 cm. long; sepals usually longer than the petals; capsule averaging 1/2 length of the sepals 6c. J. TENUIS f. DISCRETIFLORUS

6a. **Juncus tenuis** f. **tenuis** Map 590

*Juncus tenuis* Willd. [G]

*Juncus macer* S. F. Gray [P & S]

Occurs commonly in paths and along roadsides, open woods, fields, pastures, wet open ground, along streams, and waste places. Throughout Missouri, and doubtless in every county.

Plate no. 108. **1.** Juncus bufonius. **2.** Juncus secundus. **3.** Juncus tenuis. **4.** Juncus interior. **5.** Juncus Dudleyi.

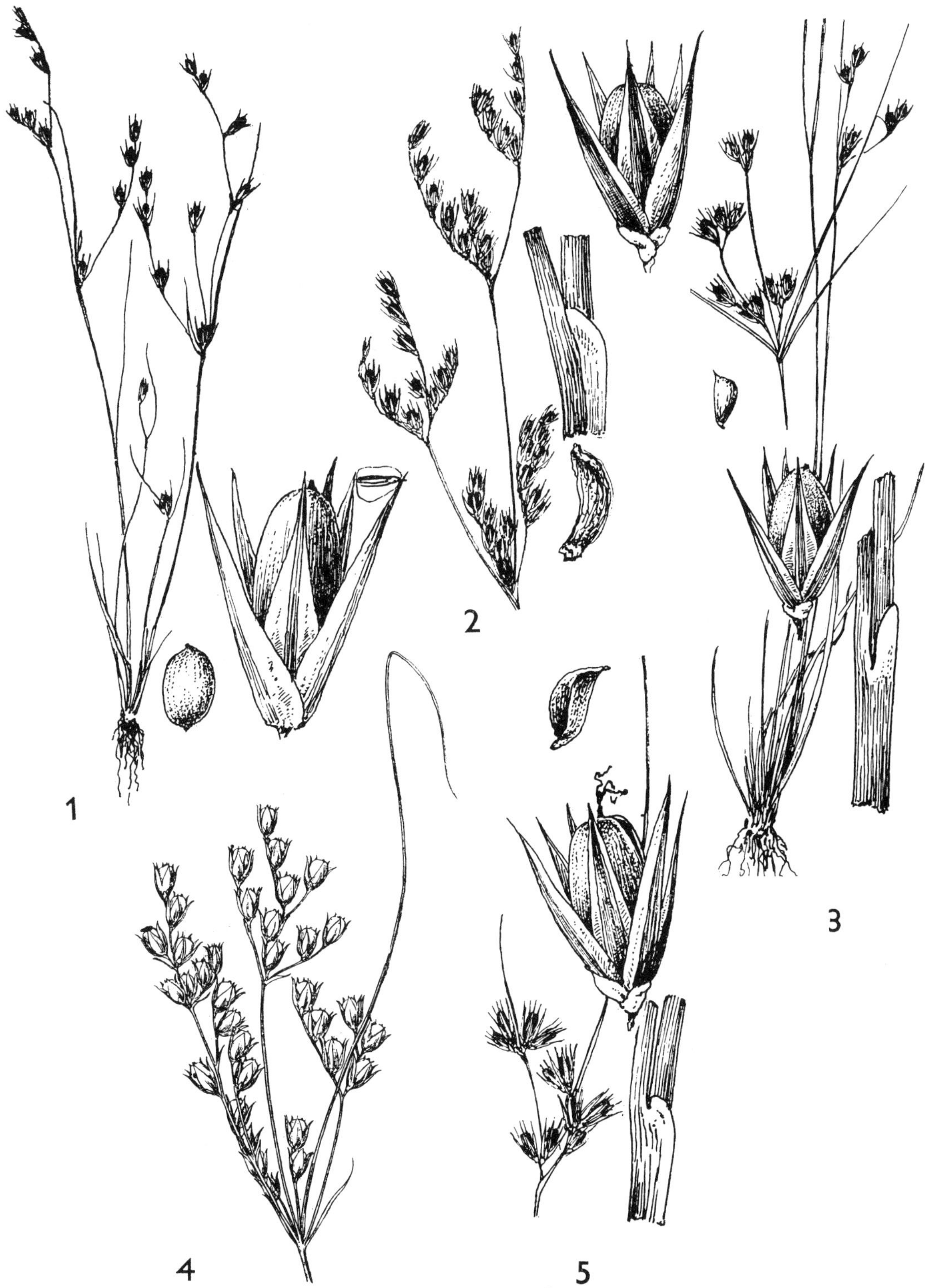


PLATE NO. 108

591 Juncus interior

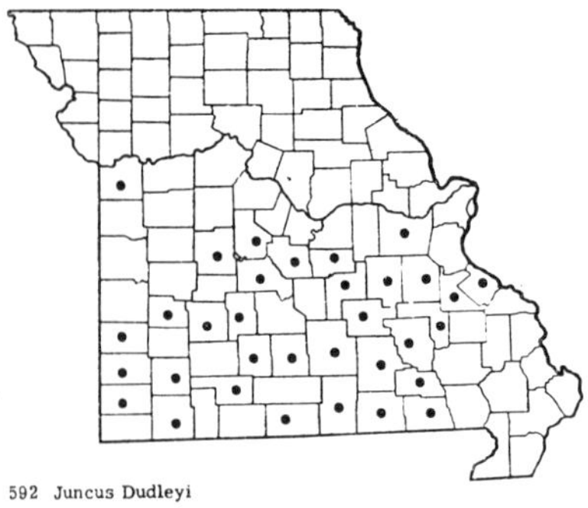

592 Juncus Dudleyi

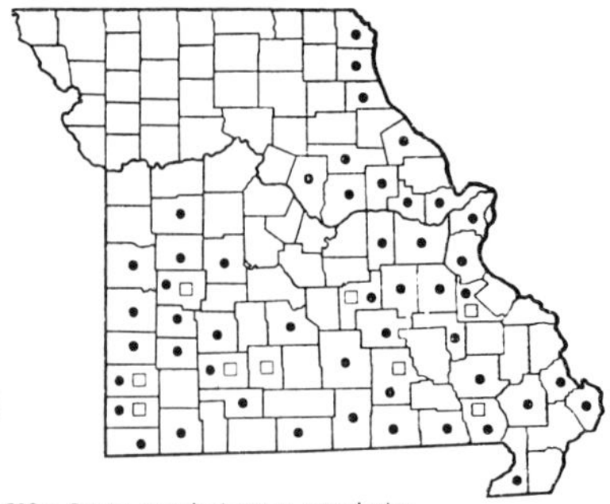

593 ● Juncus marginatus var. marginatus
593 □ Juncus marginatus var. setosus

Ranges throughout North America, Mexico, Central and South America, and in Europe, North Africa, and Australia.

The seeds and stems are eaten by various wildfowl. Deer also browse the plants. Other species of juncus are known to furnish seed for various song and marsh birds and wildfowl.

6b. **Juncus tenuis** f. **anthelatus** (Wieg.) Herm. Map 590

*Juncus tenuis* var. *anthelatus* Wieg. [G]

*Juncus macer* var. *anthelatus* (Wieg.) Fern. [P & S]

Scattered throughout the range of f. *tenuis* and somewhat less common. As in Indiana, this form is usually found in wetter places bordering ponds and low fields, and less frequently on paths or dry roadsides.

6c. **Juncus tenuis** f. **discretiflorus** (Herm.) Fern. Map 590

*Juncus macer* f. *discretiflorus* Herm.

This is the least common of the forms of *J. tenuis* in Missouri. It is known thus far from Lewis, Howell, Laclede (swampy meadow and pond in depression along road, T33–34N, R16–17W, sect. 36, 8 mi. southwest of Lebanon, July 5, 1947, *Steyermark 64701*), and Barton counties.

Known from Guatemala and a few states in the United States (Ohio, Indiana, Kentucky, Illinois, and Missouri, and elsewhere).

7. **Juncus interior** Wieg. Map 591

Flowers May–August.

Occurs in upland prairies, prairie swales, meadows, limestone, sandstone, or chert glades, fallow fields, and openings. Throughout Missouri, except not recorded from the southeastern lowland section.

Ranges from Ontario, Michigan, and Wisconsin to Wyoming, south to Indiana, Illinois, Missouri, Arkansas, Kansas, Texas, and New Mexico.

8. **Juncus Dudleyi** Wieg. Map 592

Flowers May–September.

Occurs along calcareous spring branches, borders of streams and ponds, swampy calcareous and other meadows, wet places in prairies, and moist ground. Restricted, except for a station in Jackson County, to the Ozark region, extending there to Franklin, Maries, Morgan, and Pettis counties.

Ranges from Newfoundland to British Columbia, south to Maryland, Tennessee, Missouri, Oklahoma, Texas, New Mexico, Arizona, and Mexico.

The inflorescence is somewhat compact, the flowers chiefly in small clusters, and the sepals and petals usually spreading-ascending and mainly 4–6 mm. long. This species is characteristic of calcareous spring branches and wet meadows of small valleys in the Ozark region.

9. **Juncus marginatus** Rostk. Map 593

Flowers May–September.

Missouri material may be divided into two categories:

Petals blunt or somewhat rounded at tip with a short abrupt point, ovate to oblong . . . . 9a. J. MARGINATUS var. MARGINATUS

Petals with a long and slender tip, lanceolate . 9b. J. MARGINATUS var. SETOSUS

9a. **Juncus marginatus** var. **marginatus** Map 593

Occurs in prairies, meadows, sandstone glades and open sandy ground, limestone glades, and sterile upland fields. Throughout the Ozark region in southern

Plate no. 109. **1.** Juncus marginatus. **2.** Juncus canadensis. **3.** Juncus biflorus. **4.** Juncus subcaudatus.

PLATE NO. 109

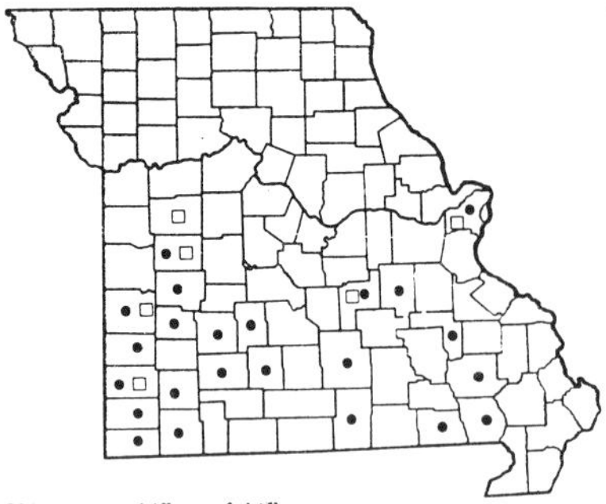

594 • Juncus biflorus f. biflorus
594 □ Juncus biflorus f. adinus

595 Juncus canadensis var. canadensis

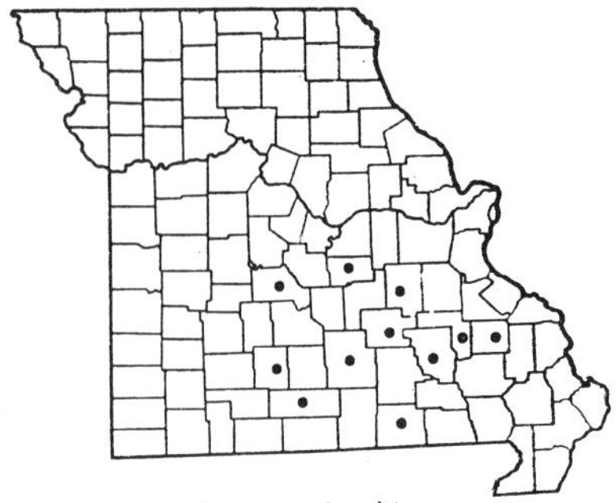

596 Juncus subcaudatus var. subcaudatus

and central Missouri, north along the eastern border of the state to Clark County.

Ranges from Florida to Texas, north to Nova Scotia, Maine, New York, Ontario, Michigan, Missouri, and Kansas.

This is the commoner of the two varieties in the state.

9b. **Juncus marginatus** var. **setosus** Coville Map 593

Occurs in similar situations as var. *marginatus*. Known from several counties in the southern third of the state.

Ranges from Central America and Mexico, and in the United States from Florida to Texas, north to Wisconsin, Minnesota, and Nebraska.

10. **Juncus biflorus** Ell. Map 594

Flowers May–September.

Occurs in wet meadows, prairie swales, and sandy open ground, usually in acid soils.

Ranges from Florida to Texas, north to Massachusetts, Pennsylvania, Ohio, Michigan, Illinois, Missouri, and Oklahoma.

Missouri material is represented by the following categories:

Flower-clusters mainly 2–3-flowered (up to 6), scattered or well separated in a more or less loose inflorescence . . . 10a. J. BIFLORUS f. BIFLORUS

Flower-clusters mainly 3–6-flowered, close together or crowded in a compact inflorescence . 10b. J. BIFLORUS f. ADINUS

10a. **Juncus biflorus** f. **biflorus** Map 594

Found in southern and central Missouri, where restricted to the Ozark and the unglaciated prairie sections.

This is the form more commonly encountered in the state.

10b. **Juncus biflorus** f. **adinus** Fern. & Grisc. Map 594

Scattered in portions of the range of *biflorus*.

Both *J. biflorus* and *J. marginatus* may occur together, although *J. biflorus* usually is found in wetter places. It is much more robust than *J. marginatus* with a thicker, more branched, and more knotty underground stem.

11. **Juncus canadensis** J. Gay var. **canadensis** Map 595

*Juncus canadensis* J. Gay [G, P & S]

Flowers July–October.

Known only from southern Missouri in Howell (August 13, 1892, *Bush 2598* World's Fair Collection in University of Missouri herbarium) and Bollinger (swampy meadow bordering alder thicket along Wash Creek in narrow valley, T32N, R8E, west part sect. 26, 1½ mi. east of Marquand, September 17, 1954, *Steyermark 77565*) counties.

Ranges from Quebec, Ontario, and Minnesota, south to Georgia, Tennessee, and Louisiana.

12. **Juncus subcaudatus** (Engelm.) Coville & Blake var. **subcaudatus** Map 596

*Juncus subcaudatus* (Engelm.) Coville & Blake [G, P & S]

Flowers July–October.

Occurs usually in calcareous, spring-fed, swampy meadows and along spring branches in the Ozark section north to Crawford, Maries, and Camden counties and southwest to Webster and Douglas counties.

Ranges from Georgia and Missouri, north to Massachusetts, New York, Pennsylvania, and Ohio.

Another variety, var. *planisepalus* Fern., occurs in Nova Scotia.

13. **Juncus validus** Coville Map 597

Flowers July–August.

Occurs in prairie swales and swampy ground.

597 Juncus validus

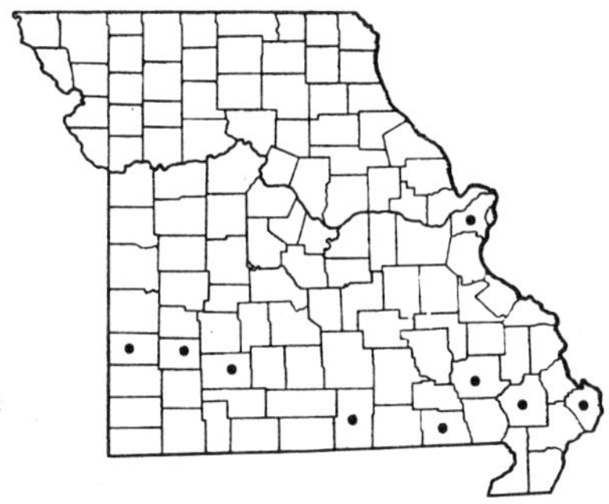

598 Juncus scirpoides var. scirpoides

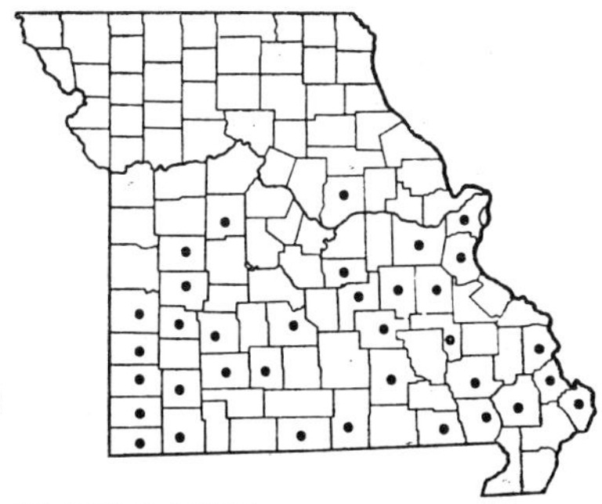

599 Juncus brachycarpus

Known only from the unglaciated prairie section in Vernon, Barton, and Jasper counties, and from a wet prairie in Howell County in the Ozark section.

Ranges from Florida to Texas, north to Georgia, Missouri, and Oklahoma.

This is a mostly tall robust species with a large inflorescence of widely spreading branches which support spherical many-flowered heads 10–15 mm. in diameter. The petals and sepals are mainly 4–6 mm. long and much shorter than the narrowly lanceolate slender-tipped capsule.

14. **Juncus scirpoides** Lam. var. **scirpoides** Map 598

Flowers June–October.

Occurs in wet sandy ground, prairie swales, and open swampy places, in acid soils. Known from southern Missouri, and locally northeast to St. Louis County.

Ranges from Florida to Texas, north to New York, Pennsylvania, Indiana, Michigan, Illinois, Missouri, and Oklahoma.

Another variety, var. *meridionalis* Buchenau, occurs from Florida north to Long Island, New York.

This species resembles *J. brachycarpus*, from which it may be distinguished by the exserted capsule much exceeding the petals and sepals.

15. **Juncus brachycarpus** Engelm. Map 599

Flowers May–September.

Occurs in wet depressions and draws in prairies, wet meadows, and sandy margins of small streams. Southern and central Missouri, north to St. Louis, Franklin, Callaway, Pettis, Henry, and Vernon counties.

Ranges from Georgia to Texas, north to Massachusetts, Ohio, Ontario, Michigan, Indiana, Illinois, Missouri, and Oklahoma.

16. **Juncus nodosus** L. Knotted Rush Map 600

Flowers July–August.

Occurs on sand bars, sandy banks, and swales along the Missouri River in Jackson County, west-central Missouri (swales, Courtney, July 26, 1904, *Bush 2138*; sandy banks, July 18, 1912, *Bush 6848*; low sandy ground, July 14, 1889, *Bush 1*).

Ranges from Newfoundland to Alaska, south to Virginia, Ohio, Indiana, Illinois, Missouri, Nebraska, Texas, and New Mexico.

A specimen collected from Jasper County (Webb City, *Palmer 609*) and labelled *J. nodosus* is *J. Torreyi*.

17. **Juncus Torreyi** Coville Map 601

Flowers July–October.

Occurs along wet sandy or muddy shores of ponds, sloughs, and small streams, in open ditches, and in swales and wet depressions of prairies and meadows. Throughout Missouri, except not recorded in the southeastern lowland section.

Ranges from Massachusetts and New York to Saskatchewan and Washington, south to Alabama, Missouri, Texas, California, and northern Mexico.

Small-headed forms of this have sometimes been mistaken for *J. nodosus*. *Juncus Torreyi*, and the related *J. nodosus*, *J. acuminatus*, and *J. subcaudatus*, sometimes are found with the inflorescences changed into leafy hornlike clusters of galls.

18. **Juncus acuminatus** Michx.

Knotty-leaved Rush Map 602

Flowers May–August.

Occurs in calcareous, spring-fed meadows, along spring branches, margins of ponds, sloughs, and small streams, wet depressions in prairies and ditches, and open wet ground. Common throughout the Ozark region and in most parts of southern and central Missouri, more locally northward in the state north to Clark, Schuyler, Grundy, and Clinton counties.

Ranges from Georgia to Texas and Mexico, north to Nova Scotia, Maine, Ontario, Michigan, Wisconsin, and Minnesota.

600 Juncus nodosus (Knotted Rush)

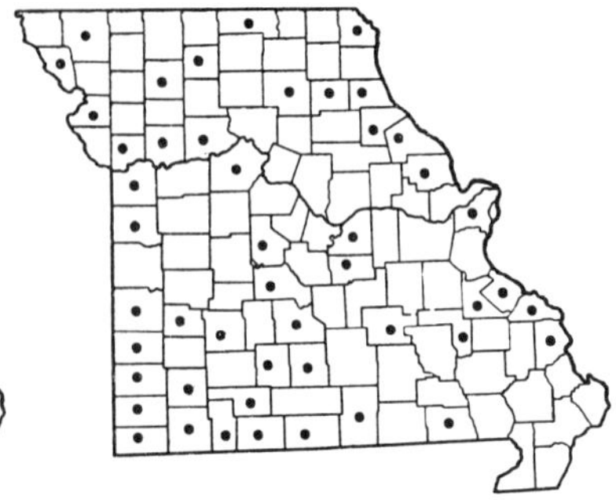

601 Juncus Torreyi

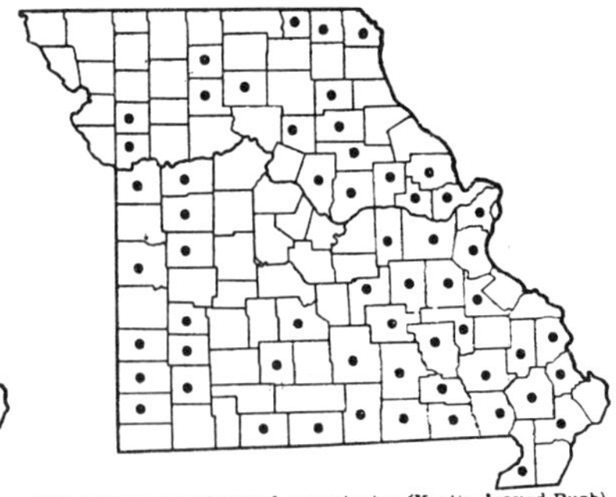

602 Juncus acuminatus f. acuminatus (Knotty-leaved Rush)

All the Missouri material has the capsule short-pointed at the summit and is typical *J. acuminatus* f. *acuminatus*. A form with the capsule rounded or ending somewhat abruptly at the summit is known as f. *obtusatus* Herm., but has not been found thus far in the state.

In autumn the fruiting heads of this species often produce leafy offshoots and branches, which are capable of producing new plants directly. This habit is also found in some of the other species, such as *J. debilis*.

19. **Juncus debilis** Gray Map 603

Flowers June–August.

Known only from Ripley County in the southeastern Ozark section (still water of Little Barren Creek, T25N, R1W, west part sect. 28, 1½ mi. southeast of Handy P. O., July 22, 1951, *Steyermark 72221*).

Ranges from Florida to Mississippi, north to Rhode Island, New York, Connecticut, Kentucky, and southeastern Missouri.

In the Steyermark collection the culms are all very weak and slender and nearly all submerged. The inflorescence developed leafy offshoots and branches in this collection.

20. **Juncus nodatus** Coville Map 604

Flowers May–August.

Occurs in wet sandy open ground, wet depressions in upland prairies, borders of ponds, sloughs, slow streams, and ditches, and low wet woods. Found in southeastern Missouri in the lowland section and in the southernmost Ozark counties west to Howell County, and in the extreme western border counties south of the Missouri River, and locally northeast in St. Louis County.

Ranges from Mississippi to Texas, north to Indiana, Illinois, Missouri, and Kansas.

21. **Juncus diffusissimus** Buckl. Map 605

Flowers late May–July.

Occurs in calcareous, spring-fed, swampy meadows, wet rocky banks of small streams, sandy, gravelly, or muddy margins of streams, margins of ponds and sloughs, and wet open places. Restricted to the Ozark and unglaciated prairie sections, extending north to Cape Girardeau, Iron, Crawford, Miller, Benton, Johnson, and Vernon counties.

Ranges from Georgia to Texas, north to Virginia, Indiana, Missouri, and Kansas.

This species is easily recognized by the very long and narrow capsules which are much longer than the short sepals and petals. The inflorescence is widely spreading and much branched, comprising about one-third the height of the plant, and has many small heads of flowers. The plants are usually small, as compared with most of the other species in Missouri, averaging 2–3 dm. in height, but sometimes attaining as much as 5 or more dm.

22. **Juncus alpinus** Vill. var. **fuscescens** Fern. Map 606

Flowers July–October.

Known only from sandbars and sandy swales along the Missouri River in Jackson County, west-central Missouri (common along river, Courtney, October 19, 1897, *Bush 274*; sandy bottoms, October 6, 1913, *Bush 7093*; July 26, 1904, *Bush 2137*; sands, August 24, 1910, *Bush 6166*; sandbars, Courtney, October 24, 1897, *Mackenzie 946*).

Ranges from Quebec to Saskatchewan, south to New York, Ohio, Indiana, Illinois, and Missouri.

---

Plate no. 110. **1.** Juncus Torreyi. **2.** Juncus scirpoides. **3.** Juncus brachycarpus. **4.** Juncus nodosus. **5.** Juncus acuminatus.

PLATE NO. 110

603 Juncus debilis

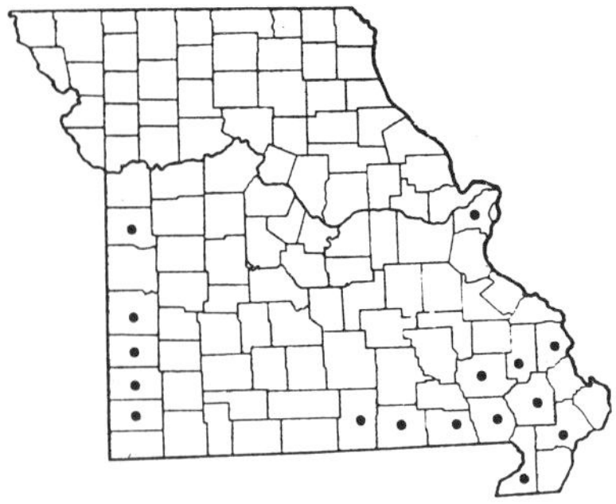

604 Juncus nodatus

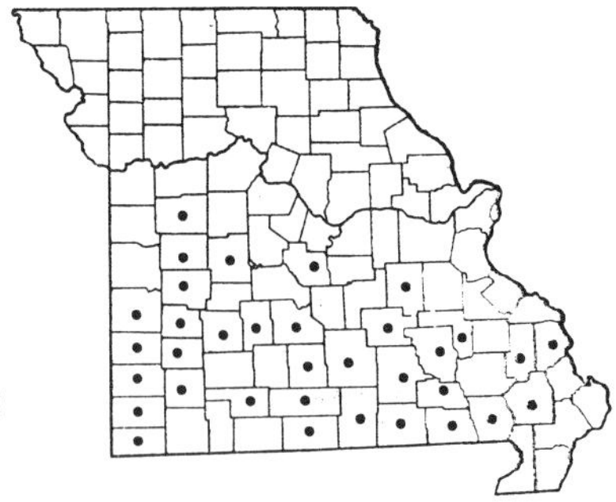

605 Juncus diffusissimus

## 2. **Luzula** DC. Wood Rush

**Luzula bulbosa** (Wood) Rydb. Wood Rush
Map 607

*Luzula campestris* (L.) DC. var. *bulbosa* A. Wood [P & S]

*Luzula campestris* (L.) DC. [BB]

*Luzula multiflora* var. *bulbosa* (Wood) Hermann [Deam]

Flowers April–June.

Occurs most commonly in acid soils of dry open woodland, usually in oak-hickory or pine-oak-hickory types, on slopes, upland ridges, or in low level valleys; also occurs on chert, granite, or sandstone glades and dry rocky prairies. Restricted to the Ozark and unglaciated prairie sections of the state, extending north to St. Charles, Warren, Montgomery, Callaway, Boone, Moniteau, Morgan, Henry, and Vernon counties.

Ranges from Florida to Texas, north to Massachusetts, Pennsylvania, Indiana, Illinois, Missouri, and Kansas.

Plate no. 111. **1.** Juncus debilis. **2.** Juncus nodatus. **3.** Juncus diffusissimus. **4.** Juncus alpinus var. fuscescens. **5.** Luzula bulbosa.

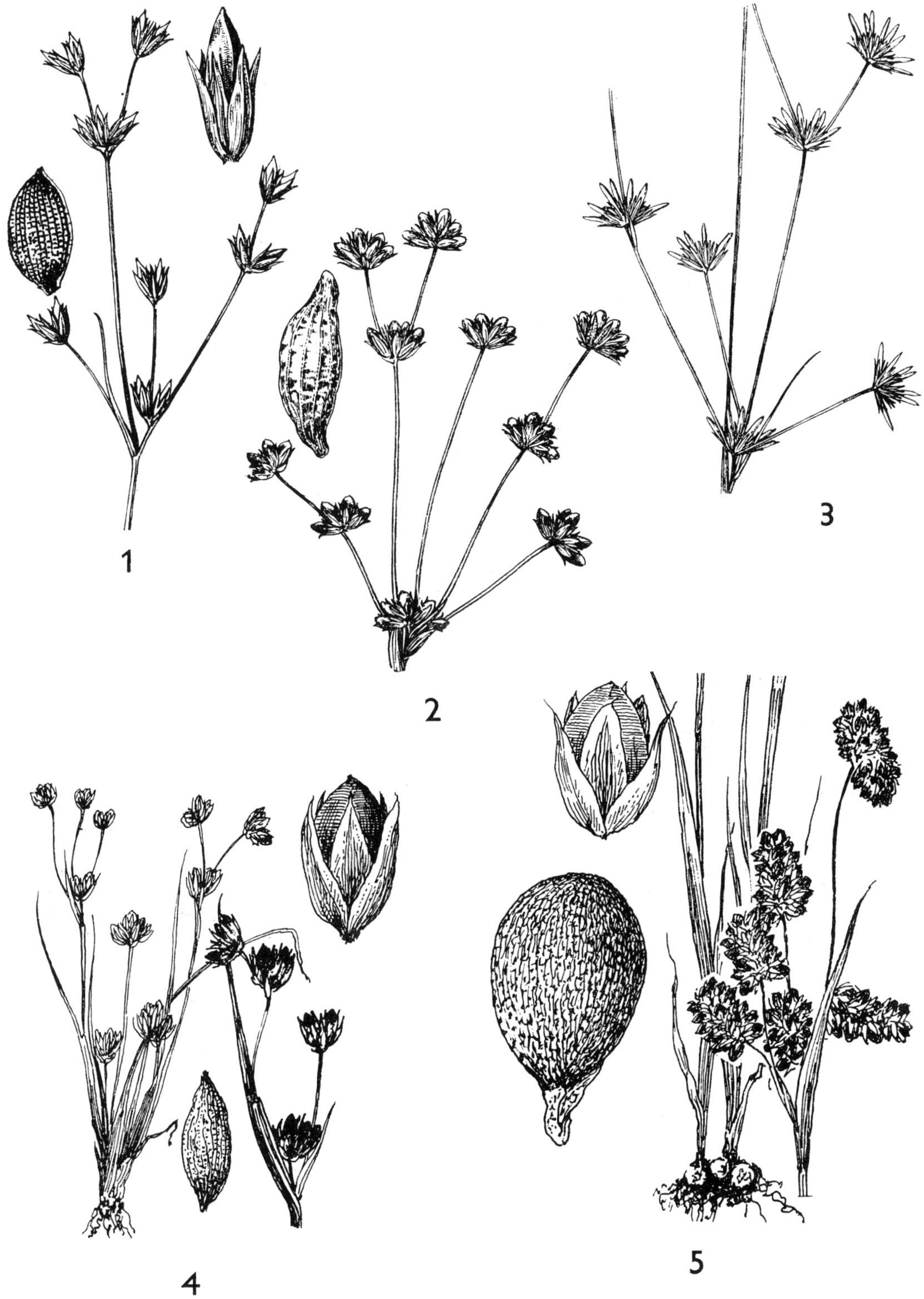


PLATE NO. III

## Fam. **LILIACEAE** (Lily Family)

The treatment of this family follows the traditional one maintained by Fernald in the eighth ed. of *Gray's Manual* and by Gleason in the *New Britton and Brown Illustrated Flora.*

Some botanists have followed a system proposed by Hutchinson in his *Families of Flowering Plants* (vol. 2: 129 and following pages. 1934) to include in the Liliaceae only those plants in which the inflorescence is not subtended by bracts or a spathe. This would exclude *Allium* and *Nothoscordum* among those genera found in Missouri. Also some botanists now exclude *Yucca* from the Liliaceae, placing it and *Agave* in a separate family, Agavaceae. *Smilax* is sometimes removed to a separate family, Smilacaceae.

Evidently, then, at the present time there is no general agreement. While admitting some justification for the splitting up of the Liliaceae, and the realignment of some of the genera into the Amaryllidaceae, the present treatment follows the Engler-Prantl arrangement, at present adopted by many botanists.

a. Leaves absent at time of flowering or fruiting. . . . . . . . . . . . . . 7. ALLIUM TRICOCCUM
a. Leaves present at time of flowering or fruiting . . . . . . . . . . . . . . . . . *b*
b. Leaf-like parts (really branchlets) thread-like, the true leaves reduced to tiny scales . . 17. ASPARAGUS
b. Leaves not thread-like . . . . . . . . . . . . . . . . . . . . . . *c*
c. The main leaves arising at the base or crowded near the very base of the plant, a leafless or nearly leafless naked stem (scape) supporting the flower, flowers, or fruits (do not judge any reduced green leaf-like bracts as leaves here) . . . . . . . . . . . . . . . . . *d*
d. Only 2 leaves present at flowering or fruiting time (only 1 when sterile) on 1 plant; plant with only 1 flower, 1 bud, or 1 fruit . . . . . . . . . . . . . . . 11. ERYTHRONIUM
d. Without the above combination; usually 3 or more leaves present at flowering or fruiting time, or, if 2 leaves present, the flowers blue; or plant with 2 or more flowers, buds or fruits . . *e*
e. Lowest leaves near ground very broad, 6 cm. or more broad; leaves at base of plant and those on stem broadest at or above the middle, oblanceolate to elliptical; axis of inflorescence somewhat hairy with short fine hairs . . . . . . . . . . 4. VERATRUM
e. All leaves linear, ribbon- or grass-shaped, at most 4 cm. broad . . . . . . . . . *f*
f. Leaves with thread-like fibers separating loose from the margins (more prominent during spring and fall months); leaves thick and tough, hard to tear; plants rising from a strong woody base; inflorescence 30–200 cm. long; flowers white, large, 3.5–7 cm. long, 2–3 cm. wide; fruit a large capsule 2–7 cm. long; seeds large, black, 6–13 mm. long 16. YUCCA
f. Without the above combination of characters . . . . . . . . . . . . . . *g*
g. Flower orange or yellow, 10–13 cm. long, 10–12 cm. wide; roots a cluster of thickened tubers; flowering stem (scape) 50–200 cm. tall . . . . 9. HEMEROCALLIS
g. Without the above combination of characters . . . . . . . . . . . . . *h*
h. Flowers or fruits arranged in an umbel (flower stalks arising from about the same place at the same level) . . . . . . . . . . . . . . . . . *i*
i. All parts of plant with an onion or garlic odor; flowers usually some shade of pink, rose, lavender, or rarely white; flowers or fruits usually 8–50 or more in one umbel, or sometimes changed to little bulblets instead of flowers; ovules or seeds 1–2 in each cell of ovary or fruit . . . . . . . 7. ALLIUM
i. No part of plant with an onion or garlic odor; flowers white to greenish-white, usually 3–8 (sometimes up to 12) in one umbel, never changed to little bulblets; ovules or seeds 4–10 in each cell of ovary or fruit . 8. NOTHOSCORDUM
h. Flowers or fruits arranged in a raceme, or panicle, or solitary, the flower- or fruit-stalks arising along the elongated axis at different levels . . . . . . . *j*
j. Perianth of flower almost globe-shaped or short tube-shaped, its segments united at least to the middle or more; flowers longer than their stalks (pedicels), in a dense raceme with little space between the flowers; low plants at most 2.5 dm. tall when in flower . . . . . . . . . 15. MUSCARI
j. Perianth with its parts divided or separate nearly or all the way to the base; flowers often shorter than, but also longer than their stalks in looser racemes or panicles, with much space at least between the lower flowers; low or tall plants, sometimes up to 15 dm. tall . . . . . . . . . . . . . . *k*
k. Open flowers and buds or fruits only 1–12 to an inflorescence . . . . . *l*

l. Perianth of flower deep blue; pedicels (the individual flower-stalks) mostly 2–9 mm. long; bracts at base of pedicels very short, 2–4 mm. long . . . . . . . . . . . 13. SCILLA
l. Perianth of flower white with a green stripe on the outside; pedicels 10–60 mm. long; bracts at base of pedicels 10–40 mm. long . . . . . . . . . . . . . . 14. ORNITHOGALUM
k. Open flowers and buds or fruits mainly 20 or more to an inflorescence . . . . . . . . . *m*
m. Style 1; perianth usually shades of lavender or blue; fruit showing 1 beak or point at summit 12. CAMASSIA
m. Styles 3; perianth white, greenish-yellow, or green-and-purple; fruit showing 3 beaks or points at summit . . . . . . . . . . . . . . . . . . . . . . . . . . *n*
n. Bracts at the base of the pedicels (the individual flower-stalks) 10–20 mm. long; perianth 6–12 mm. long; 1–2 dark or colored spots (glands) at or above base of each perianth segment . . . . . . . . . . . . . . . . . . . . . . 3. ZIGADENUS
n. Bracts at the base of the pedicels small and inconspicuous, about 2 mm. long; no spots or glands present on perianth segments . . . . . . . . . . . . . . . . . *o*
o. Pedicels (stalks) of flowers none or at most 2 mm. long at flowering time, much shorter than the flowers; perianth segments of flower narrowly pointed (acuminate) at tip; mature fruit with 3 short beaks at tip . . . . . . . . . . . 2. STENANTHIUM
o. Pedicels (stalks) of flowers 15 mm. or more long, much longer than the flowers; perianth segments of flower rounded at summit; mature fruit with conspicuous 3 horn-like lobes 1. AMIANTHIUM
c. The main leaves scattered alternately on the stem or in 1 or more whorls (circles) of 3 or more . . . *p*
p. Leaves in 1, 2, or more whorls (circles) of 3 or more leaves . . . . . . . . . . . . . *q*
q. Leaves in 1 whorl of 3 leaves . . . . . . . . . . . . . . . . . . 21. TRILLIUM
q. Leaves in 2 or more whorls . . . . . . . . . . . . . . . . . . . . . *r*
r. Leaves in 2 whorls . . . . . . . . . . . . . . . . . . . . 20. MEDEOLA
r. Leaves in 6–11 whorls . . . . . . . . . . . . . . . 10. LILIUM MICHIGANENSE
p. Leaves scattered alternately along the stem . . . . . . . . . . . . . . . . . *s*
s. Axis of inflorescence or upper part of stem more or less hairy or downy with some type of hairiness . . . . . . . . . . . . . . . . . . . . . . . . . . . . *t*
t. Flowers large and showy, 75–100 cm. long, the perianth segments 12–25 mm. wide; dark small bulblets in the upper part of the stem at the base of the leaves . . 10. LILIUM TIGRINUM
t. Flowers much smaller, the perianth segments only 5–13 mm. long, 2–5 mm. wide; no bulblets present at the base of the leaves . . . . . . . . . . . . . . . . . . *u*
u. Plants of open prairies, meadows, and along railroads; all the leaves broadly linear, broadly ribbon- or grass-like, nearly the same width throughout, 1.5–3 cm. wide; 2 dark glands on the inside of the base of each perianth segment . . . . . . . . 5. MELANTHIUM
u. Plants of wooded slopes; lowest leaves near ground very broad, 6 cm. or more broad; leaves at base of plant and on stem oblanceolate to elliptical, broadest at or above the middle; no glands at the base of the perianth segments . . . . . . . 4. VERATRUM
s. Axis of inflorescence and stem glabrous (without hairs) . . . . . . . . . . . . . *v*
v. Base of leaf completely encircling or growing around the stem . . 6. UVULARIA GRANDIFLORA
v. Base of leaf not encircling the stem . . . . . . . . . . . . . . . . . . . *w*
w. Leaves narrowed to a short or long petiole (stalk of leaf); climbing or tough- or woody-stemmed plants with or without tendrils or spines, or, if soft-stemmed herbs, the leaves with a network of veins connecting the main nerves . . . . . . . . . . 22. SMILAX
w. Leaves lacking a petiole, attached directly to main stem; soft-stemmed herbs, not climbing, not woody, not spiny and without tendrils; or leaves all parallel-veined, lacking a network of connecting veins . . . . . . . . . . . . . . . . *x*
x. Mature plant forking or branching at the 1st or 2nd leaf; flower and fruit solitary, the flower pale yellow or cream-colored and 12–25 mm. long, the fruit a sharply 3-angled capsule . . . . . . . . . . . . . . . . . . 6. UVULARIA SESSILIFOLIA
x. Without the above combination; mature plant with a simple unbranched stem; flowers and fruits not as above . . . . . . . . . . . . . . . . . . . *y*
y. Flowers or fruits produced from the sides of the stem or hanging below the leaves 19. POLYGONATUM
y. Flowers or fruits at the top of the stem . . . . . . . . . . . . . . . *z*
z. All parts of plant with an onion or garlic odor; flowers or fruits arranged in an umbel (flower stalks arising from about the same place at the same level), sometimes the flowers changed to little bulblets . . . . . . . . . 7. ALLIUM

606. Juncus alpinus var. fuscescens

607 Luzula bulbosa (Wood Rush)

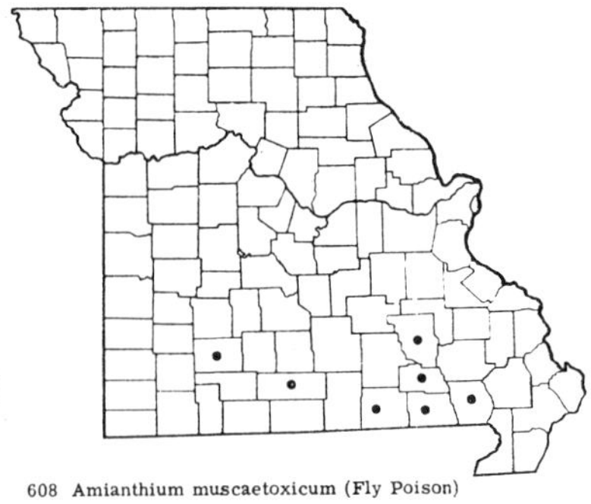

608 Amianthium muscaetoxicum (Fly Poison)

z. No part of plant with an onion or garlic odor; flowers or fruits arranged in a raceme or panicle, the flower- or fruit-stalks arising along the elongated axis at different levels . . . . . . . . . . . . *1*
 1. All the leaves linear, ribbon- or grass-like, 4–15 mm. wide, about the same width from one end to the other, 25–40 times or more longer than broad; fruit a capsule . . . . . . 2. STENANTHIUM
 1. Leaves oblong, oval, or lanceolate, the larger ones 10–90 mm. wide, broader near the middle or base, 3–7 times longer than broad; fruit a berry . . . . . . . . . . . . . 18. SMILACINA

## 1. **Amianthium** Gray Fly Poison

**Amianthium muscaetoxicum** (Walt.) Gray
Fly Poison Map 608
Flowers May–July.

In acid soils of sandy or cherty ground of low wooded slopes or in level ground of small valleys. Southern Missouri in the Ozark region, east to Butler County, west to Greene County, and north to Reynolds and Greene counties.

Ranges from Florida to Mississippi, Arkansas, and Oklahoma, north to southern New York, Pennsylvania, West Virginia, Kentucky, and Missouri.

The leaves and underground bulb are poisonous, especially so to cattle and sheep.

The flowers are white at first, soon turning yellow-green or eventually green with purplish color, and on one inflorescence the various stages of flowering show a striking color variation, the older lower flowers having already turned green, while the freshly open flowers higher up towards the top are white. The leaves spread outward and arch downward.

## 2. **Stenanthium** Gray

**Stenanthium gramineum** (Ker) Morong var. **gramineum** Featherbells Map 609
*Stenanthium gramineum* (Ker) Morong [G, P & S]
*Stenanthium gramineum* f. *robustum* (Wats.) Palm. & Steyerm. [of P & S], not *Stenanthium gramineum* var. *robustum* (S. Wats.) Fern. [G]
Flowers mid-June–September.

Occurs in moist rocky woods, rich wooded slopes, along and at the base of wooded bluffs, open banks along railroads, and occasionally in open thickets; most frequent in acid soils. Throughout Missouri, but commonest in the Ozark section; locally north in Audrain, Linn, and Harrison counties.

The whitish or greenish-white flowers are produced during summer on an elongated panicle usually 4–8 dm. tall with drooping or spreading lateral branches. The plant itself attains a height of 1.5 meters, the lowest section of the stem being 4–10 mm. thick. The long, ribbonlike, pale green leaves are crowded toward the base in the lower third of the plant. The

---

Plate no. 112. **1.** Amianthium muscaetoxicum, × $^2/_7$; *a.* Flower, × $^6/_7$; Details from Small, The New York Botanical Garden. **2.** Zigadenus Nuttallii, × $^2/_7$; Details from Small, The New York Botanical Garden. **3.** Stenanthium gramineum var. gramineum; *a.* Inflorescence, × $^1/_{14}$; *b.* Leaf, × $^2/_7$; *c.* Portion of inflorescence, × $^2/_7$; *d.* Single flower, × $1^1/_7$; Details from Small, The New York Botanical Garden. **4.** Zigadenus elegans, × $^2/_7$; Details from Small, The New York Botanical Garden. **5.** Veratrum Woodii; *a.* Terminal part of inflorescence, × $^2/_7$; *b.* Lower portion of inflorescence, × $^2/_7$; *c.* Single basal leaf, × $^2/_7$; Details from Small, The New York Botanical Garden. **6.** Melanthium virginicum; *a.* Single basal leaf, × $^2/_7$; *b.* Portion of inflorescence, × $^2/_7$; *c.* Single flower, × $^4/_7$; After Gleason, details from Small, The New York Botanical Garden.

PLATE NO. 112

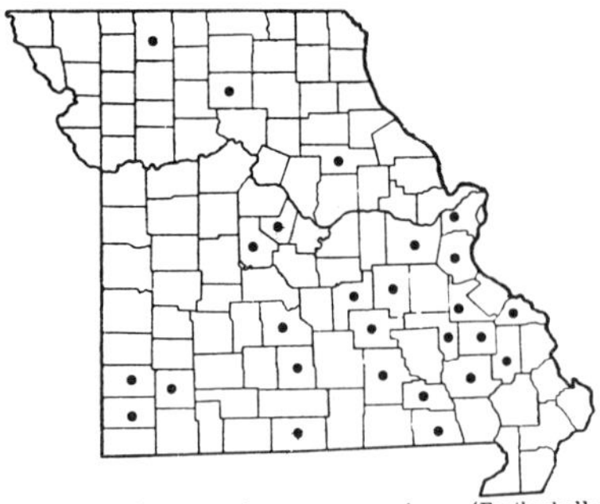

609 Stenanthium gramineum var. gramineum (Featherbells)

610 Zigadenus elegans (White Camas)

611 Zigadenus Nuttallii (Death Camas)

leaves, which in Missouri material vary mainly from 7–15 mm. wide, have a somewhat corrugated surface because of the prominently raised ribs. Another variety, var. *robustum* (Wats.) Fern., not known in Missouri, has the larger leaves mainly 10–30 mm. wide, the ribs more depressed and sunken in the leaf tissue, crowded up to the main branches of the panicle, and the lowest section of the dried stem 7–15 mm. thick. Plants which have been transplanted to the author's wildflower garden are still living after ten years, but show an irregularity in flowering, some years producing only a leafy basal rosette, other years flowering. New plants have been produced from extensions sent out underground by the bulbous base of the plant.

## 3. **Zigadenus** Michx. Death Camas

a. Plants of wet crevices of north-facing limestone bluffs, flowering in late June–early August; perianth segments 8–12 mm. long; gland at base of perianth segment 2-lobed; mature fruiting capsule 13–22 mm. long; bracteoles 5–13 mm. long; pedicels 4–30 mm. long . . . . . . . . 1. Z. ELEGANS
a. Plants of dry limestone open woods and glades, flowering in May; perianth segments 6–8 mm. long; gland at base of perianth segment not lobed; mature fruiting capsule 13–16 mm. long; pedicels 10–26 mm. long in flower, 17–45 mm. long in fruit . . . . . . . . . . . . . 2. Z. NUTTALLII

### 1. **Zigadenus elegans** Pursh White Camas

Map 610

Flowers late June–early August.

Occurs in moist crevices of north-facing limestone bluffs. Very rare and known only from the southern Ozark region at Jam-up Bluff in Shannon County, along the Gasconade River in Pulaski County (by Peterson Cave, sect. 8, 5 mi. northwest of Waynesville, August 24, 1937, *Steyermark 25263*), and along Indian Creek in Douglas County (Holy Cliff and vicinity, T26–27N, R11W, sect. 4 and 33, 3½ mi. northeast of Topaz, July 19, 1937, *Steyermark 23345*).

Ranges from Alaska to Arizona and New Mexico, east to Manitoba, Minnesota, Iowa, and Missouri.

This is another one of the remarkable northern species of plants apparently isolated in the Ozarks after the retreat of one of the stages of Pleistocene ice and now surviving at one of its southernmost outposts as a relict. At the Jam-up locality this species occurs with an assemblage of other remarkable northern relict species, such as *Galium boreale* var. *hyssopifolium* and *Campanula rotundifolia*.

An alkaloid, zygadenine, contained in this and other species of the genus, is the source of a poison which may produce harmful effects on cattle, horses, and sheep, especially when the fresh leaves, stems, or flowers are eaten. The seeds are considered quite poisonous also. Poisoning may occur among children if the bulbs are eaten.

### 2. **Zigadenus Nuttallii** Gray Death Camas

Map 611

Flowers May.

Known only from Oregon County in the southern Ozark region, where it occurs at 'The Narrows' on a dry narrow limestone open wooded ridge or glade above a bluff above Blue Spring along Fredericks Fork, west of Calm (May 11, 1935, *Steyermark 18920*; July 13, 1933, *Palmer & Steyermark 41667*).

Ranges from Tennessee, Missouri, and Kansas to Oklahoma and Texas.

At the Missouri locality it occurs with *Arenaria patula* and *Melica nitens*.

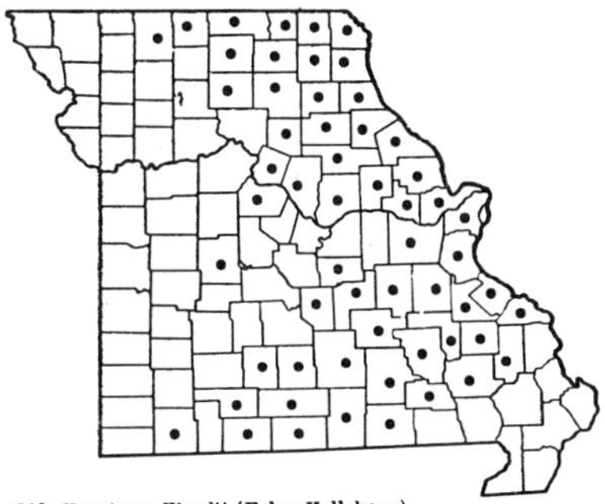

612 Veratrum Woodii (False Hellebore)

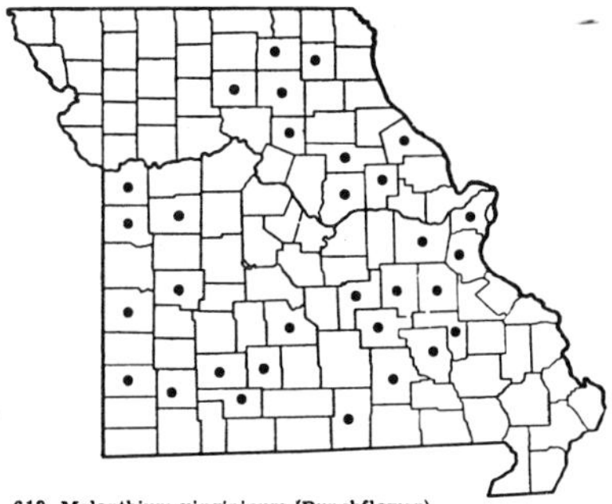

613 Melanthium virginicum (Bunchflower)

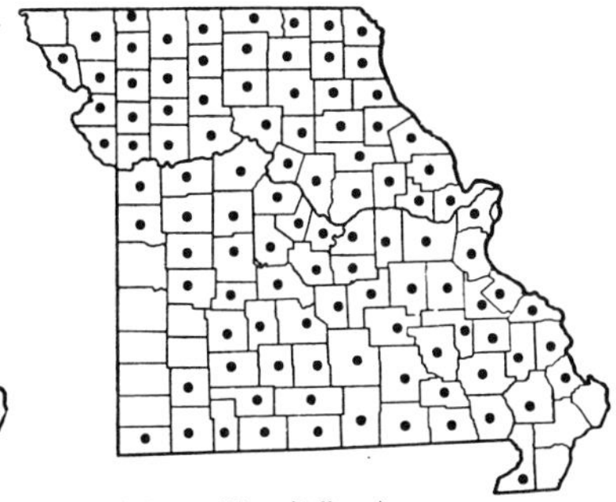

614 Uvularia grandiflora (Bellwort)

## 4. **Veratrum** L. False Hellebore

**Veratrum Woodii** Robbins False Hellebore
Map 612

Flowers July–September.

Occurs usually on north- or east-facing wooded limestone slopes in the Ozark region, and on rich wooded slopes in northern Missouri. Found in the Ozark section of the state northward in the eastern half of Missouri, west to Harrison, Linn, and Howard counties, and south of the Missouri River west to Cooper, Benton, Pulaski, Webster, Christian, and Barry counties.

Ranges from Ohio and Iowa, south to Missouri and Oklahoma.

The plant is commonly found in the eastern half of the state, but is most often seen in the sterile leafy condition, attracting much attention by its broad, handsome, plaited basal leaves. These root leaves emerge in early spring. Flowering appears to be very irregular, some seasons passing without any flowering taking place. At various localities where flowering was observed, only a small percentage of the plants bore flowering stalks. On other occasions a greater percentage of plants had flowers. On still other occasions no plants were observed in flower. Plants in flower attain a height of 1.3–1.8 meters. The perianth of the flowers is dark maroon, fading to dull greenish with purple.

At the author's wildflower garden where two plants have been growing for eleven years, one has flowered twice during this time, the first time waiting seven years before flowering, the second time flowering three years later. The second plant has flowered only once, this during the same year as the first plant.

Several alkaloids are found in related species which are a source of poison. In Europe a species of *Veratrum* is used as an insecticide. The green shoots and seeds, when eaten, sometimes poison chickens, and an overdose of drugs containing *Veratrum* may cause poisoning symptoms. Ordinarily, most grazing animals avoid the plant because of its sharp burning taste.

## 5. **Melanthium** L. Bunchflower

**Melanthium virginicum** L. Bunchflower
Map 613

Flowers about June 10–early August.

Occurs in prairies, wet meadows, thickets, and wet ground below bluffs. Scattered throughout southern and central Missouri, and north of the Missouri River in the eastern half of the state west to Adair and Linn counties.

Ranges from Florida to Texas, north to New York, Ohio, Indiana, Illinois, and Iowa.

Horses have been reported being poisoned after having eaten hay containing seeds of this species.

The broadly ovoid to ovoid-pyramidal inflorescence is very showy with cream-colored flowers which turn to green and purplish shades with age. The plant itself usually attains a height of 1–1.5 meters.

## 6. **Uvularia** L. Bellwort

Base of leaf completely encircling or growing around the stem . . . . . . . . . . 1. U. GRANDIFLORA
Base of leaf not encircling nor growing around the stem . . . . . . . . . . . 2. U. SESSILIFOLIA

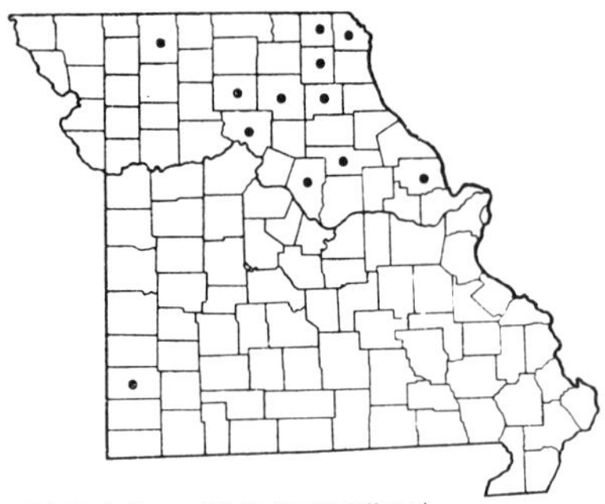

615 Uvularia sessilifolia (Small Bellwort)

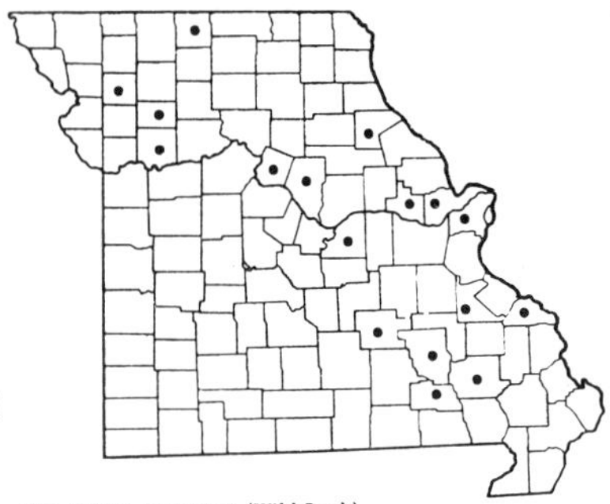

616 Allium tricoccum (Wild Leek)

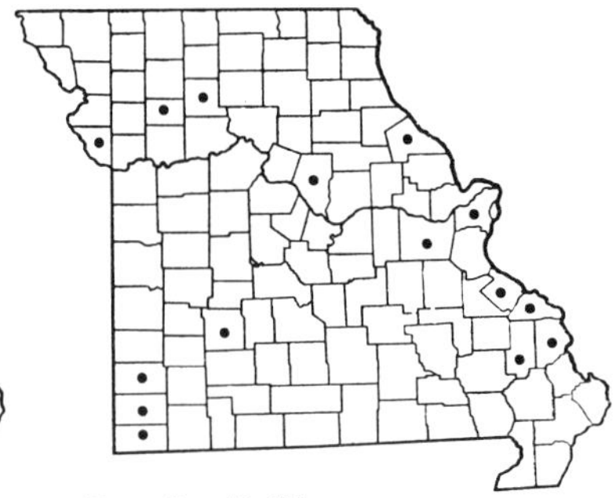

617 Allium sativum (Garlic)

1. **Uvularia grandiflora** Sm. Bellwort Map 614

Flowers April–May.

Occurs in rich or rocky soil of woodland, on steep slopes or level alluvial valleys. Throughout Missouri, except absent from some of the counties of the unglaciated southwestern prairie section and southeastern lowland section.

Ranges from Quebec to North Dakota, south to Georgia, Arkansas, and Oklahoma.

The starchy roots may be cooked as a vegetable.

2. **Uvularia sessilifolia** L. Small Bellwort Map 615

Flowers April–May.

Occurs in rich or alluvial soils in mostly level ground of low woods along streams, occasionally along north-facing wooded bluffs. Scattered in the northeastern and north-central counties of the state south to Lincoln, Audrain, and Boone counties, west to Harrison, Linn, and Chariton counties, and locally in Jasper County, southwestern Missouri (rocky wooded north-facing hillside along bluff of Turkey Creek, 1/8 mi. east of 'Castle Rock,' 1 mi. north of Joplin, April 29, 1949, *Palmer 48613*). The isolated occurrence of the species in Jasper County, the only station south of the Missouri River in the state, may be regarded as a probable relict survivor from Pleistocene times, but it is difficult to explain why it has not survived in other southern sectors of the Ozarks.

Ranges from New Brunswick and Ontario to North Dakota, south to Georgia, Alabama, Tennessee, Missouri, and Arkansas.

The young shoots of this species are reported to be eaten, when cooked, as a substitute for asparagus, and it is supposed that the young shoots of *U. grandiflora* can be similarly used.

## 7. **Allium** L. Onion, Garlic, Leek

a. Leaves absent at flowering time; leaves found during spring and with elliptical or elliptic-lanceolate leaf-blades 3–10 cm. wide with a petiole . . . . . . . . . . . . . . . 1. A. TRICOCCUM

a. Leaves present at flowering time; leaves linear or ribbon-like, without a petiole . . . . . . . . *b*

  b. Stems bearing leaves 1/3 to 1/2 way up the stem . . . . . . . . . . . . . . . . . . *c*

    c. Leaf-blades 5–15 mm. wide, flat, ribbon-like; base of stem often dark red-purple; perianth of flowers greenish, whitish, or pink, its segments lanceolate and long-pointed (acuminate) . . 2. A. SATIVUM

    c. Leaf-blades 1.5–3 mm. wide (in dried condition), hollow, terete (like a quill of a feather or porcupine) or nearly terete (convex on one side, plane or concave on other side); base of stem usually bluish- or gray-green; perianth of flowers pink, lavender, rose-purplish, or occasionally white or greenish, its segments obtuse to somewhat short-pointed (acutish). . . . 3. A. VINEALE

  b. All the leaves at the base or nearly so . . . . . . . . . . . . . . . . . . . . . *d*

    d. Leaves terete (like the quill of a feather or porcupine) or nearly so, hollow . . . . . . . *e*

      e. Stem enlarged or inflated somewhere below the middle; leaves spreading or drooping; bulb globose or ball-shaped, much thicker than the stem; pedicels (the individual flower stalks) much longer than the greenish white to lilac flowers . . . . . . . . . 4. A. CEPA

      e. Stem not inflated, of the same slender diameter throughout; leaves erect; bulb oblong-ovoid, scarcely thicker than the stem; pedicels shorter than the pink or rose-purple flowers 5. A. SCHOENOPRASUM

    d. Leaves flat, not hollow . . . . . . . . . . . . . . . . . . . . . . . . . *f*

f. The umbels of flowers mostly producing little bulblets, the actual flowers and fruits few or none . . . . . . . . . . . . . . . . . . . . . . 6. A. CANADENSE
f. The umbels of flowers producing flowers and fruits only . . . . . . . . . . . . . *g*
g. Flowering mainly late April–late June; outer coats of the bulb with a criss-cross mat or network of coarse hair-like fibers; sepals and petals longer than the stamens; no horns or triangular projections produced from any of the 3 valves or cells of the ovary or fruit 7. A. MUTABILE
g. Flowering mainly July–November; outer coats of the bulb only vertically nerved and membrane-like, not with any network of coarse fibers; sepals and petals shorter than the eventually exserted stamens; each of the 3 valves or cells of the ovary or fruit producing 2 horns or triangular projections at the summit (6 all together) . . . . . . *h*
h. The main supporting stem (scape) of the inflorescence arched or hooked at summit; the umbel of flowers drooping or nodding . . . . . . . . . . . . 8. A. CERNUUM
h. The main supporting stem (scape) of the inflorescence erect to the tip, not arched or hooked at summit; the umbel of flowers erect . . . . . . . . 9. A. STELLATUM

1. **Allium tricoccum** Ait. Wild Leek Map 616
Flowers mid-June–July.

Occurs mainly on rich soils of north-facing wooded slopes or limestone bluffs. Northern, central, and eastern Ozark region west to Osage, Dent, Reynolds, and Carter counties.

Ranges from New Brunswick and Quebec to Minnesota, south to Georgia, Tennessee, Illinois, and Missouri.

The leaves appear in earliest spring, resembling tulip leaves in size and shape. They disappear at the time of flowering. In late May the inflorescence is still in bud stage, and not until later in June do the whitish flowers actually open. The capsule ripens later in the summer, exposing the 3 steel gray, bulletlike seeds.

Most of the Missouri plants of this species have the upper surface of the leaf-blade a rich olive- or grass-green, the lower surface paler green, not shining on either side and tapering into a wine purple or reddish-purple petiole; the broadly elliptical, elliptic-oblong or elliptical-lanceolate blades are quite broad, averaging mainly 4–9 cm. wide. The flowering scapes of such plants are also purple or red-purple. This is in contrast to the plants of other sections of the country, especially in the northern and eastern states, where the tendency is towards specimens with narrower leaf-blades 2–4 cm. wide, darker and deeper green and shining on both sides and without any purple or reddish-purple at the base. The scapes in these plants are usually pale green. When grown together, the plants with purple in the petiole appear above the ground a week to ten days earlier than those with no purple color. These two races have been discussed in a paper by Clarence R. Hanes and Marion Ownbey (Rh. 48: 61–63. 1946), but no names have been given as yet to these plants which are being studied by Drs. Ownbey and Aase. Plants from the Missouri Ozarks brought north to the author's wildflower garden have borne out the observations made by Dr. Ownbey, and can easily be distinguished from the other plants of *A. tricoccum* native to northern Illinois by the earlier appearance of the leaves, their broader, darker green color with purple petioles and scapes.

The young leaves may be used as a seasoning in soups or cooked into an onion soup. They also may be cut up in salads.

2. **Allium sativum** L. Garlic Map 617
Flowers late May–July.

Occurs in waste ground, along roadsides and railroads, fields, meadows, thickets, and grassy places. Scattered in various parts of Missouri and probably more common than the records indicate.

Native to Eurasia; introduced into the United States where it ranges from New York to Indiana, south to Kentucky, Tennessee, and Missouri.

The ribbonlike gray- or silvery-green leaves are smooth and handsome in early spring when the species is seen in best foliage condition. The stems are usually dark purple-red at the base. Usually, many plants are found growing together.

It can become a bad pest if allowed to become established in cultivated areas.

3. **Allium vineale** L. Field Garlic Map 618
Flowers May–July.

Occurs in fields, meadows, thickets, borders of woods and streams, along railroads, and waste ground.

Missouri material is represented by two variations:

Umbels with bulblets mixed with normal stalked flowers . . . . . . 3a. A. VINEALE f. VINEALE
Umbels consisting entirely of bulblets. 3b. A. VINEALE f. COMPACTUM

3a. **Allium vineale** L. f. **vineale** Map 618

Common over the southern and central portions of the state and locally north to Marion and Living-

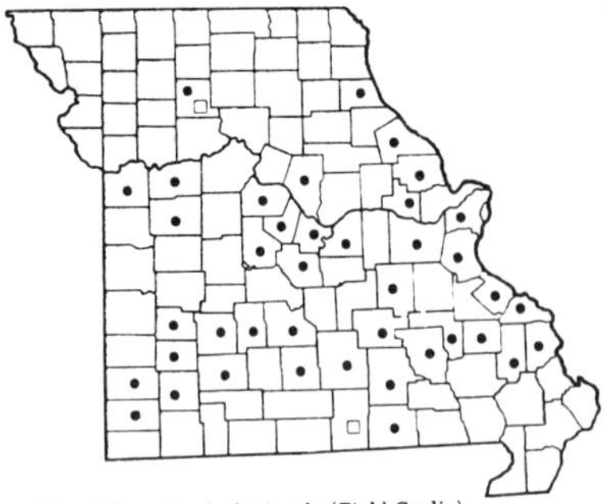

618 • Allium vineale f. vineale (Field Garlic)
618 □ Allium vineale f. compactum

619 Allium Cepa var. viviparum (Onion)

620 Allium Schoenoprasum var. Schoenoprasum (Chives)

ston counties. This is the commoner of the forms encountered.

Native of Europe; introduced into the United States, where it ranges from Georgia to Arkansas, north to Massachusetts, New York, Ohio, Michigan, Illinois, Missouri, and Kansas.

The leaves are bluish-green. They may be terete or subterete with one side rounded and convex and the other side plane to concave or sunken.

This species is responsible for contaminating milk and other dairy products with a garlic flavor in the late winter and early spring months. Not only is the flavor transmitted by the actual eating of the plant by cows, but by inhaling the volatile oil emanating from the fresh plants, the same effect is produced. It is among the strongest-flavored of the strong-flavored species, and produces a burning sensation if tasted.

It is also a bad weed in wheat-producing areas, and, when bulblets become harvested with the wheat, they must be removed before the wheat is acceptable for milling. It has been suggested that this species may be used to advantage as a spring pasture for meat-producing beef and sheep.

*Allium vineale* is sometimes confused with *A. oleraceum* L., Wild Garlic, which has not been recorded thus far from Missouri. *Allium oleraceum* has the leaf-blades flattened toward the summit instead of nearly terete and hollow, the bulbs are simple and not crowded or with older or smaller secondary side bulbs, the filaments of the flowers are slender and without appendages, whereas in *A. vineale* they are dilated at the base, with the upper half 3-cuspidate or divided into three linear divisions or appendages; finally, in *A. oleraceum* there are two leafy, elongated, linear bracts forming the spathe which persist below the umbel, whereas in *A. vineale* there is only one thin transparent bract which quickly falls away from the umbel.

3b. **Allium vineale** f. **compactum** (Thuill.) Aschers. Map 618

Known only from Howell County and Livingston County (along ditch, partially shaded, T58N, R24W, sect. 14, 3.3 mi. north and 1.2 mi. west of Chillicothe, June 15, 1950, *Sparling 57*).

4. **Allium Cepa** L. Onion Map 619

Flowers late May–July.

Native of southwestern Asia. Commonly cultivated in the United States and rarely escaped to waste ground.

*Allium Cepa* var. *Cepa* was listed by Daniels in his *Flora of Columbia* [as *Allium Cepa*], but no specimen is extant.

*Allium Cepa* var. *viviparum* Metz, the Top Onion, has been found in St. Louis County (St. Louis, on the free place between the signal box at Davis Street Junction and the grain elevator on Mo. Pacific Railroad, May 20, 1957, *Muehlenbach 1165*). It differs from the ordinary onion in having a small undeveloped bulb and the umbel of the inflorescence developing bulblets. These latter are used for propagating purposes. In *Allium Cepa* var. *Cepa* the underground bulb is large and well-developed, recognized when mature as the familiar common onion, and the large umbel is filled with a dense cluster of numerous flowers borne on upwardly curving pedicels.

There have been occasional reports of horses being

Plate no. 113. **1.** Uvularia grandiflora, $\times\ ^{2}/_{7}$; *a.* Flowering plant; *b.* Fruiting plant; Details from Small, The New York Botanical Garden. **2.** Allium tricoccum; *a.* Fruit, $\times\ ^{6}/_{7}$; After Britton and Brown, The New York Botanical Garden. **3.** Uvularia sessilifolia, $\times\ ^{2}/_{7}$; *a.* Flowering plant; *b.* Fruiting plant. **4.** Allium canadense, $\times\ ^{2}/_{7}$. **5.** Allium sativum, $\times\ ^{2}/_{7}$. **6.** Allium vineale, $\times\ ^{2}/_{7}$; *a.* Flowering plant, $\times\ ^{2}/_{7}$; *b.* Head with bulblets instead of normal flowers, $\times\ ^{2}/_{7}$; *c.* Single flower, $\times\ 3^{3}/_{7}$; Details from Small, The New York Botanical Garden. **7.** Allium Cepa var. viviparum, $\times\ ^{2}/_{7}$; *a.* Base of plant; *b.* Middle portion of fruiting stem; *c.* Upper part of fruiting stem with fruits; Details from Small, The New York Botanical Garden.

PLATE NO. 113

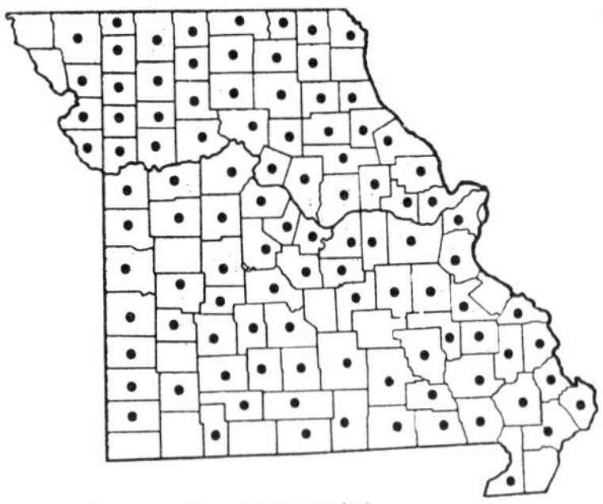

621 Allium canadense (Wild Garlic)

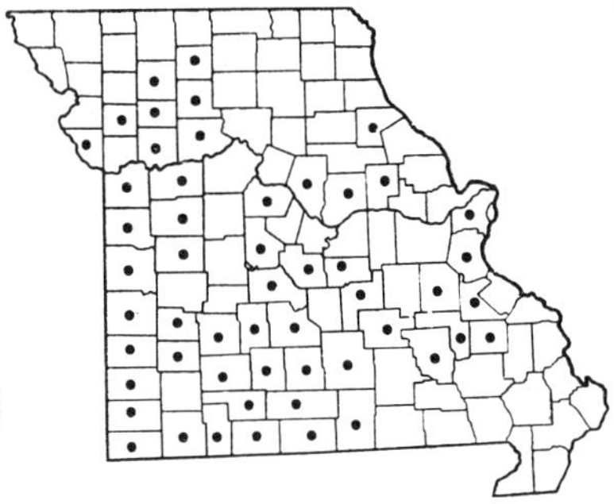

622 Allium mutabile

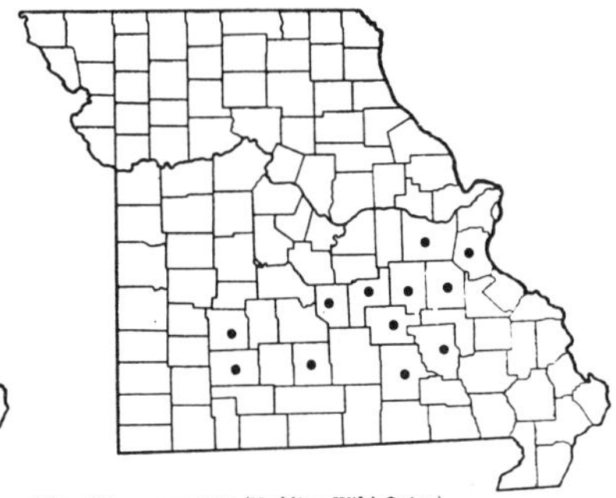

623 Allium cernuum (Nodding Wild Onion)

poisoned from eating *A. Cepa*, but such poisoning appears to have resulted from the eating of partly decayed bulbs.

5. **Allium Schoenoprasum** var. **Schoenoprasum** (L.) Hartm. Chives Map 620
Flowers June–August.

Known only as an escape in Lawrence and in St. Louis (St. Louis, June 30, 1910, *Earl E. Sherff*) counties.

Native of Europe. Commonly cultivated, but rarely escaped to waste ground.

The stem is only 2–5 dm. tall, the plants occurring in dense clumps from crowded tiny bulbs. The numerous leaves nearly equal or sometimes are longer than the stems. The many pink to rose-colored flowers are borne in a dense hemispherical umbel.

This is a well-known plant used for seasoning cheese, soups, relishes, and salads.

6. **Allium canadense** L. Wild Garlic Map 621
*Allium canadense* L. var. *canadense* [Ownbey & Aase]
Flowers early May–July.

Occurs in moist open woodland, usually in level ground of valleys along streams, alluvial banks and slopes, thickets, meadows, edge of bluffs, prairies, open fields, along roadsides, railroads, and waste ground. Throughout Missouri and doubtless in every county.

Ranges from Florida to Texas, north to New Brunswick, Quebec, Ontario, Wisconsin, and Minnesota.

The flowers, which are usually pink or lilac and pink, rarely whitish, are few to an umbel.

Ownbey and Aase (Res. Stud. State College Wash. Monogr. Supp. 1, vol. 23, no. 4: 1–106. 1955) have attempted to group into one species, *A. canadense*, with several varieties, all the native species in Missouri having coarsely anastomosing or reticulate criss-cross fibers of the outer coats of the bulb. They maintain *A. canadense* var. *canadense* as an asexual variety, with most or all of the flowering pedicels replaced by bulblets, for the plants which have previously ordinarily passed as *A. canadense*. However, since the Missouri material falls clearly into what many other botanists would consider distinct and easily recognizable species, I am treating *A. canadense* in the category of a species rather than as a variety. Seen not only in the field, but grown in the garden, the Missouri plants do not intergrade, but retain the clearly marked characters which distinguish them. Shinners, likewise, in his *Spring Flora of the Dallas-Fort Worth Area, Texas* (pp. 93–94. 1958) has preferred the status of species rather than variety for a similar situation existing among the Texas constituents of the same group related to *Allium canadense*.

Like *Allium vineale*, *A. canadense* is also responsible for giving an onion flavor to milk and other dairy products in the late winter and early spring months when it is eaten by cows grazing the valley woodlands and meadows.

Unlike the strong-scented *A. vineale*, however, the bulbs of *A. canadense* are considered to have a rather sweet flavor and are edible, while the bulblets of the inflorescence can be pickled and eaten as such. The entire plant, before it reaches the flowering stage, can be boiled and made into a soup, while the young fresh leaves can be cut up to season salads or as a garnish for flavoring meats, cheese, and spaghetti dishes.

7. **Allium mutabile** Michx. Map 622
*Allium canadense* var. *mobilense* (Regel) Ownbey [Ownbey & Aase]
*Allium microscordion* Small
*Allium canadense* var. *lavendulare* (Bates) Ownbey [Ownbey & Aase]
*Allium acetabulum* (Raf.) Shinners var. *lavendulare* (Bates) Shinners
Flowers late April–late June.

Occurs chiefly on limestone glades and prairies, sandstone exposures, upland cherty openings above bluffs, and along railroads throughout central and southern Missouri in the Ozark and unglaciated prairie sections, northward in glaciated prairie mainly in the western sector to Grundy, Daviess, Clinton, and Platte counties, and locally east in Ralls County.

Ranges from Florida to Texas, north to North Carolina, Missouri, and Nebraska.

This is an attractive plant with numerous flowers on elongated pedicels. The flowers are usually some shade of rose, pink, or lilac, but also vary to white. Plants here referred to *A. mutabile* are treated by Ownbey and Aase (loc. cit. pp. 29, 45–50, 51–55) as two varieties. Normally, the plants are robust and average 3–5 dm. tall (varying from 2–6) with rather thickened, rigid pedicels. Infrequently, in some of the southwestern counties, occasional plants are found which are more slender, usually 1–2.5 dm. tall (varying to 5), and with very slender, threadlike pedicels. Ownbey and Aase consider the taller and more robust plants with stouter pedicels as *Allium canadense* var. *lavendulare*, while the shorter plants with slender, filiform pedicels are treated as *A. canadense* var. *mobilense*. They assign the var. *mobilense* to collections from Howell, Taney, Wright, Jasper, and McDonald, all in southwestern Missouri. The var. *lavendulare* they consider to be the widespread type. Both varieties are maintained by them as sexually developed varieties having all the flowering pedicels bearing flowers and fruits as contrasted to the bulblet-bearing, mostly asexual var. *canadense*. However, as stated under *A. canadense*, the Missouri material appears to be clearly recognizable as two ordinarily distinct species, which can be distinguished both in the field and in the garden.

The slender, shorter plants with more filiform pedicels referred by Ownbey and Aase to var. *mobilense* are considered in this flora as intermediates and possibly hybrids between *A. mutabile* and *A. canadense*. The specimen from McDonald County (barrens, Noel, May 4, 1909, *Bush 5613*), labelled *A. Helleri* by Bush and previously catalogued as *Allium Nuttallii* Wats. [= *A. Drummondii* Regel] by Palmer and Steyermark (*Annot. Cat. Fl. Pl.* p. 501. 1935) is one of these smaller individuals, which Ownbey and Aase would place under *A. canadense* var. *mobilense*.

8. **Allium cernuum** Roth
Nodding Wild Onion — Map 623
Flowers July–September.

Occurs mostly along rocky wooded ledges of bluffs and shaded slopes. Ozark region of southern and east-central Missouri, north to Jefferson, Franklin, and Pulaski counties, west to Polk and Greene counties.

Ranges from New York to Minnesota and British Columbia, south to Georgia, Tennessee, Missouri, Texas, and Arizona.

Plants with white petals and sepals occur infrequently (*Steyermark 19331* from Pulaski County).

Dr. Ownbey has referred all the Missouri specimens to *Allium stellatum*, but those plants having the summit of the peduncle hooked with drooping umbels I have placed in *A. cernuum*.

9. **Allium stellatum** Fraser Wild Onion — Map 624
Flowers July–November.

Occurs on limestone glades and exposed ledges of bluffs. Restricted to the unglaciated Ozark area in southern and central Missouri north to Lincoln, Boone, Pettis, and Henry counties.

Ranges from Ohio and Ontario to Saskatchewan, south to Illinois, Missouri, Arkansas, Kansas, and Texas; introduced in Indiana.

This species has showy rose to pink flowers and does well in cultivation when grown in a sunny exposure with a limey rocky soil. Flowering as it does late in the season it is a welcome addition to a rock garden.

*Excluded species*

**Allium Drummondii** Regel
*Allium Nuttallii* Wats. [P & S]

Recorded by Palmer and Steyermark (*Annot. Cat. Fl. Pl.* p. 501. 1935) from McDonald County, the collection (*Bush 5613*) is here referred to *A. mutabile* Michx. as a small slender specimen of that species. It is placed by Ownbey and Aase under *A. canadense* var. *mobilense* (Regel) Ownbey.

## 8. **Nothoscordum** Kunth False Garlic

**Nothoscordum bivalve** (L.) Britton
False Garlic, Crow Poison — Map 625
Flowers late March–May and October–November.

Occurs on rocky glades of limestone, chert, granite, or sandstone, sandy open alluvial soils along streams, rocky exposed ledges, upland open wooded areas, and on upland or river bottom prairies. Southern and central Missouri, locally north to Ralls, Audrain, Linn, and Livingston counties; absent from most of the southeastern lowland section where found only on Crowley Ridge.

Ranges from Florida to Texas and Mexico, north to Virginia, Ohio, Indiana, Illinois, Missouri, and Nebraska.

Plants often flower again in fall, following late rains and warm temperatures.

624 Allium stellatum (Wild Onion)

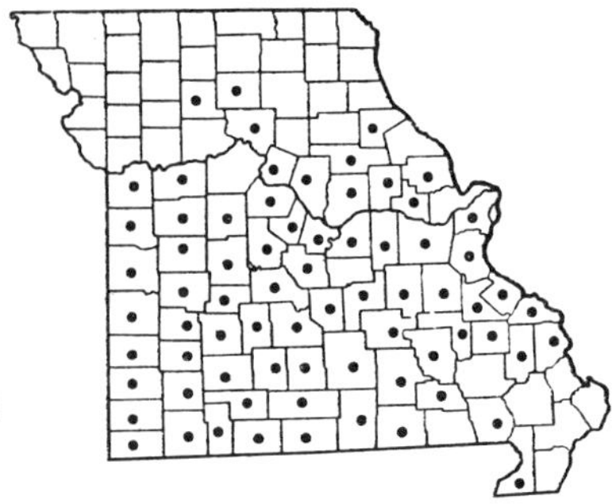

625 Nothoscordum bivalve (False Garlic)

626 ● Hemerocallis fulva var. fulva (Orange Day Lily)
626 □ Hemerocallis fulva var. Kwanso (Double Orange Day Lily)

## 9. **Hemerocallis** L. Day Lily

**Hemerocallis fulva** L. Orange Day Lily Map 626

Flowers late May–August.

Commonly cultivated, and often escaped and well established on moist wooded banks along streams, alluvial thickets, fields, roadsides, and railroads. Scattered over the state.

Native to Eurasia; introduced into North America, where it has spread from New Brunswick to Ontario, south to North Carolina, Tennessee, Missouri, Kansas, and Nebraska.

The plant spreads only by its tuberous roots in North America, as it is not known to produce seed naturally.

Two variations are encountered in Missouri:

Flowers with normal 6 perianth segments . . H. FULVA var. FULVA

Flowers double, with more than 6 perianth segments to a flower . . . H. FULVA var. KWANSO

**Hemerocallis fulva** L. var. **fulva** Map 626

This is the common variation encountered with large orange-brick red flowers, usually with an apricot-colored line down the middle of each petal. Plants usually form large clumps.

**Hemerocallis fulva** var. **Kwanso** Regel Map 626

This double-flowered variation is known thus far as an escape only in St. Louis County (St. Louis, Carrie Avenue freight yard of Rock Island R. R. in northeast corner of the yard, July 4, 1957, *Muehlenbach 1257*).

This has been found naturalized in Virginia. It is introduced from Asia.

Var. *Kwanso* is a less double-flowered variation than var. *flore pleno*.

Another species of *Hemerocallis*, *H. flava* L., the yellow day lily or lemon lily, is commonly cultivated and sometimes becomes established away from cultivation, but no definite records are known in Missouri of its having become naturalized.

For many years *Hemerocallis fulva* and *H. flava* had been the only day lilies in cultivation. With the pioneer work by Dr. A. B. Stout, followed by other workers, breeding of day lilies and development of new strains have increased the numbers of available horticultural types now cultivated into the thousands. None of these, however, have been found escaped from cultivation, only *H. fulva* var. *fulva* and var. *Kwanso* having been recorded.

The plant may be prepared for food in various ways. Fully grown flower buds or open flowers, browned in oil or butter in a batter of eggs, flour, and milk, are considered as a tasty dish. Flowers soaked in water are added in final stages of cooking to soups, meats, and noodles to impart a pleasant flavor as well as substance to the dish. The tuberous roots, boiled in salted water, are reported to possess the flavor of corn and salsify.

---

Plate no. 114. **1.** Allium mutabile, × $^{2}/_{7}$. **2.** Allium stellatum, × $^{2}/_{7}$; *a.* Fruit, × $^{6}/_{7}$; After Gleason, The New York Botanical Garden. **3.** Allium cernuum, × $^{2}/_{7}$; *a.* Flower, × $^{4}/_{7}$. **4.** Hemerocallis fulva; *a.* Leaf, × $^{2}/_{7}$; *b.* Lower part of plant with basal leaves, × $^{2}/_{21}$; *c.* Flowering portion, × $^{2}/_{7}$; Details from Small, The New York Botanical Garden. **5.** Lilium michiganense, × $^{2}/_{7}$; Details from Small, The New York Botanical Garden. **6.** Erythronium albidum var. albidum, × $^{2}/_{7}$; Details from Small, The New York Botanical Garden. **7.** Erythronium albidum var. mesochoreum, × $^{2}/_{7}$; Details from Small, The New York Botanical Garden. **8.** Nothoscordum bivalve, × $^{2}/_{7}$.

PLATE NO. 114

627 Lilium michiganense (Michigan Lily)

628 Lilium tigrinum (Tiger Lily)

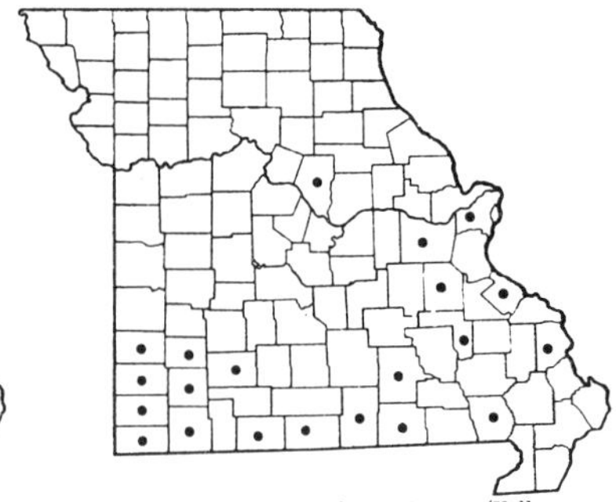

629 Erythronium americanum f. americanum (Yellow Adder's Tongue)

## 10. **Lilium** L. Lily

Leaves in 6–11 whorls (circles) on the stem . . . . . . . . . . . . . . 1. L. MICHIGANENSE
Leaves scattered alternately along the stem . . . . . . . . . . . . . . 2. L. TIGRINUM

1. **Lilium michiganense** Farw. Michigan Lily Map 627

*Lilium canadense* L. ssp. *michiganense* (Farwell) Boivin & Cody

*Lilium canadense* ssp. *michiganense* f. *uniflorum* (Farwell) Boivin & Cody

*Lilium michiganense* f. *uniflorum* (Farwell) Wherry

*Lilium michiganense* f. *peramoenum* (Farwell) Stoker

*Lilium superbum* [of BB], not L.

Flowers mainly June 20–July 12.

Occurs in prairie swales, swampy meadows, low woods in valleys, low wooded slopes, and prairie remnants along railroads. Throughout Missouri, except apparently absent in the southeastern lowland sector.

Ranges from Michigan and Ontario to Manitoba, south to Tennessee, Arkansas, and Kansas.

The number of flowers per plant varies from one, described as var. *uniflorum* Farwell to several. Vigorous plants with more numerous leaves, 8–16 to a whorl, and with large and numerous flowers, have been described as *L. peramoenum* Farwell. In the present work I have considered such differences in numbers of flowers and leaves as varying responses to the amount of moisture in the soil as well as age and vigor of the individual plant, and as not meriting recognition.

The final taxonomic status of *Lilium michiganense*, the Turk's Cap Lily of Missouri, is still open to argument. Competent botanists have considered it as a distinct species (Mill, Deam, Wherry, Fernald), as the same as or synonymous with *Lilium superbum* L. (Gleason, Hull), and as a subspecies of *L. canadense* L. (Boivin and Cody). In general appearance of having strongly recurved sepals and petals, *L. michiganense* appears more closely related to *L. superbum*, yet in various characters, such as the ribbing of the sepals and petals, shorter anthers, occurrence of tiny projections (spicules) on the leaf margins and veins of the lower leaf surface, circumneutral to subacid soil preference, time of flowering in early summer, the tendency toward an umbellate type of inflorescence, and the morphology of the karyotypes of the somatic chromosomes, *L. michiganense* is definitely more closely related to *L. canadense*.

Although *L. michiganense* may occasionally overlap into *L. superbum* in anther length, relative smoothness of leaf margin and veins on lower surface, ribbing of the sepals and petals, and type of inflorescence, yet in the main it shows fairly stable characters by which it can be recognized both in the field and herbarium, and is treated in the present work as a distinct species. For further details on arguments pro and con as regards *L. michiganense* the reader is referred to Rhodora 44: 220–27, 453–56. 1942; Rhodora 46: 86–87. 1944; Rhodora 58: 14–20. 1956; Amer. Jour. Bot. 34: 9–26. 1947; and Proc. Iowa Acad. Sci. 31: 265. 1926.

This species does well in cultivation and in wildflower gardens in either sun or shade, and multiplies over the years.

The fleshy bulbs of this and other species of the genus were eaten by North American Indians. After being cooked, they possess a somewhat sweet flavor and starchy consistency.

2. **Lilium tigrinum** L. Tiger Lily Map 628

Flowers July–September.

This is a well-known cultivated species, which rarely is found in Missouri as an escape from gardens. It is definitely known in Missouri to have become naturalized or adventive in St. Louis and Christian (2 mi. southwest of Billings, July 13, 1955, *Palmer 60565*) counties.

Native of eastern Asia; introduced into the United States, ranging there as an escaped plant from New England to North Dakota, south to Virginia and Missouri.

## 11. **Erythronium** L. Dog-tooth Violet

a. Flowers yellow, either partly or entirely so; a glandular spot is present on each margin of the petals near the base; sepals and petals with strongly forking side (lateral) veins which arch toward the margins (seen best on dried specimens); stigmas erect, slightly if at all separated, usually shorter than 2 mm.; leaves usually conspicuously mottled brown and green on upper surface (but also unmottled), with a solid rich green color on lower surface without any silvery glaucous or pale brown-lavender color or mottling . . . . . . . . . . . . . . . . . . 1. E. AMERICANUM

a. Flowers white, pinkish-white, lilac-white, or bluish-white with greenish and yellow at base; no glandular spot present on sepals or petals; sepals and petals without a rounded or subcordate base, the veins mainly parallel or only the outermost side ones slightly forking and arching (seen best on dried specimens); stigmas spreading and curving outward, separated, mostly 2–3 mm. long; leaves less mottled or not at all mottled on upper surface, the lower surface gray- or silvery-green (glaucous) with or without a pale brown-lavender color or mottled effect. . . . . . . . . . . *b*

b. Plants growing chiefly in low or moist woods along streams, valleys, on slopes and wooded bluffs; sepals and petals usually recurved or conspicuously spreading, with chiefly white predominant, often tinged with pinkish, lavender, or bluish; leaf-blades 1.3–4 cm. broad, flat, elliptical-lanceolate to ovate-lanceolate, somewhat or more or less mottled on both surfaces; colonies of numerous sterile 1-leaved plants present and predominating, the flowering plants not numerous; bulbs sending out slender long runners with a new bulb forming at the tip of the runner . . 2a. E. ALBIDUM var. ALBIDUM

b. Plants growing mostly on glades, prairies, or dry open woods; sepals and petals more ascending and remaining closer together or often only slightly spreading, not strongly recurved, with lavender or bluish-lilac predominating; leaf-blades mainly 0.5–2 cm. broad, slightly folded with margins upcurved, mostly linear-lanceolate to narrowly lanceolate, not mottled or very rarely mottled above, not mottled below; flowering plants generally more scattered but more numerous, the sterile 1-leaved types less numerous and appearing mostly later than the flowering individuals; bulbs generally without runners, a new bulb forming at the base of the old one. . . 2b. E. ALBIDUM var. MESOCHOREUM

1. **Erythronium americanum** Ker Yellow Adder's Tongue, Yellow Dog-tooth Violet

Map 629

Flowers late March–May.

Occurs in low rich woods along streams or low wooded slopes and along bluffs. Uncommon, mainly in southern Missouri, north in central Missouri to St. Louis, Franklin, and Boone counties.

Ranges from New Brunswick to Ontario and Minnesota, south to Florida, Alabama, Tennessee, Arkansas, and Oklahoma.

The plants occurring in Missouri belong to *E. americanum* f. *americanum* with the flowers mainly yellow. The leaves are usually mottled with rich brown, but unmottled leaves also occur. The three outer sepals are usually yellow within and yellow with pale brown down the center on the outside, less commonly are they greenish within, brownish at the tip, and greenish-brown on the outside, while the three inner petals are usually yellow and somewhat dull orange mottled within toward the base, in age becoming more orange-brown. Flowers with a chestnut-brown color, known as f. *castaneum* L. B. Sm., have not been recorded from the state.

Large colonies are usually found consisting chiefly of 1-leaved sterile plants, the number of flowering plants with 2 leaves being comparatively small. About four years are required following seed germination to flower production. Also, bulbs from flowering plants which have been transplanted usually appear slow in recovering from the change in transplanting and may require more than a year before they again produce flowers. They do well in wildflower gardens in rich, loose, often sandy, shaded soils.

In early spring the leaves are sometimes used as a cooked vegetable, and the bulbs of this and other species of the genus are reputed to be edible either raw or cooked. The bulbs are also reported to have emetic properties.

2a. **Erythronium albidum** Nutt. var. **albidum**
White Dog-tooth Violet. Also called White Trout Lily, White Fawn Lily, Map 630
White Adder's Tongue.
*Erythronium albidum* var. *coloratum* Sterns [Shinners]
*Erythronium albidum* Nutt. [G, P & S]
Flowers late March–May.

Occurs chiefly in low or moist woods, often alluvial soils, along streams, in valleys, on wooded slopes and bluffs. Common throughout Missouri and doubtless in every county.

Ranges from Ontario to Minnesota, south to Georgia, Kentucky, Arkansas, and Oklahoma.

This species reacts to transplanting similarly to *E. americanum*, the bulbs requiring a year or two to regain the setback temporarily caused by transplanting. Once established, however, they thrive in wildflower gardens.

Flowers vary in color from pinkish-white to bluish-white. *Erythronium albidum* var. *coloratum* Sterns (Bull. Torr. Club 15: 111. 1888) is described as a pink dog-tooth violet with the sepals white and 'more or less suffused with rose-purple, varying to bright red' and with the 'leaves more strongly mottled with green and brown than in the type.' Since the mottled leaves and habitat in 'shaded woods near streams' are more characteristic of typical var. *albidum* than of var. *mesochoreum*, the description would seem to apply more generally as a color variation of var. *albidum*, the interpretation regarded in the present flora, rather than as synonymous with var. *mesochoreum*, as interpreted by Shinners.

Sterile plants or colonies of individuals of *E. albidum* may be distinguished from similar plants of *E. americanum* by noting the lower surface of the living leaves, those of *E. americanum* being rich green without any trace of mottling, or of any glaucous-silvery color mixed with pale brown-lavender, whereas in *E. albidum* the lower surface is glaucous-silvery with a pale brown-lavender mottling showing.

2b. **Erythronium albidum** var. **mesochoreum** (Knerr) Rickett Prairie Dog-tooth Violet
Map 630
*Erythronium mesochoreum* Knerr [P & S]
Flowers late March–May.

Occurs in prairies, rocky glades, and open woods.

Ranges from Iowa and Nebraska, south to Missouri, Oklahoma, and Texas.

It is questionable whether this should be retained as a variety or as a species. Rickett (Rh. 39: 101–5. 1937) has produced evidence to show intergradation and overlapping of characters between *E. mesochoreum* and *E. albidum*. While such intergradation does occur and specimens are found which are difficult to place definitely, my own field experience has been that the majority of plants can be readily distinguished and separated in the living or growing condition. We may have here actually, instead of a variety, an example of an incipient or young species which has not become absolutely stabilized, with intermediate types occurring as hybrids or having hybrid origin where the ranges of the two taxa (categories) overlap. The same problem is encountered in various other members of the native flora. Only future detailed investigations of these two *Erythronium* will reveal a more definite answer as to how distinct they are.

In addition to the differences presented in the key, I have observed another one which should be subjected to future study and detailed observation. This apparent difference was noted in the fruiting stage of the plant, where it was found that in *E. albidum* var. *mesochoreum* the fruiting peduncle bends so strongly downward that it carries the fruiting capsule nearly or quite to the ground. In *E. albidum* var. *albidum*, on the other hand, the fruiting peduncle remains rather spreading or slightly drooping, but does not arch so strongly as in var. *mesochoreum*.

## 12. **Camassia** Lindl. Wild Hyacinth

Flowering early April–late May; flowers pale blue or pale lilac to whitish; inflorescence chiefly 7–50-(less frequently 50–85-) flowered, 4–18 (rarely to 32) cm. long; petals and sepals 10–15 mm. long, 3–4 mm. wide; style 4.5–7 mm. long, whitish to pale bluish; filaments 6–8 mm. long, pale lilac, pale blue to whitish; capsule nearly globose, as broad as or broader than long, 6–11 mm. long, 5–12 mm. wide; pedicels spreading-ascending or widely spreading in fruit, the capsules spreading away from main axis of inflorescence; stem with flowers or fruit mainly 3.2–5.5 (up to 7.5) dm. tall, the leaves 1/2 height or nearly equaling the height; bracts 0–2 on flower-stem (scape) below lowest flower-stalk, at most 8 cm. below the lowest flower-stalk, straight or slightly curved after flowering; petals and sepals spreading after flowering, their upper portions not joining after flowering . . . . . . 1. C. SCILLOIDES
Flowering early May–late June; flowers lavender to pale purple; inflorescence chiefly 50–100-flowered (where flowers have disappeared, the flower-stalks remaining in place indicate location of flower), 15–40 cm. long; petals and sepals 7–9 mm. long, 1.5–3 mm. wide; style 3.5–4 mm. long, lavender;

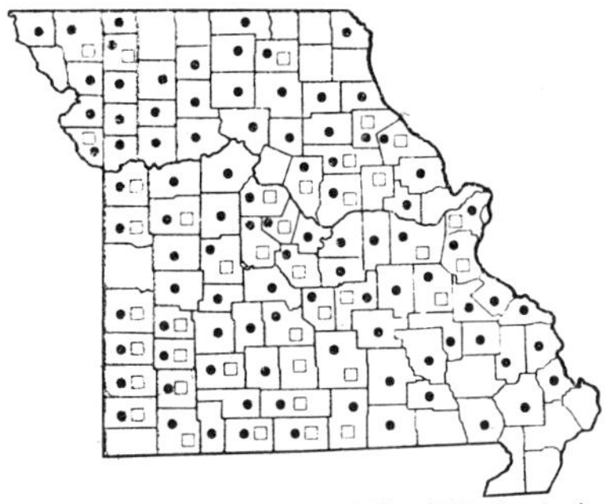

630 • Erythronium albidum var. albidum (White Dog-tooth Violet)
630 □ Erythronium albidum var. mesochoreum (Prairie Dog-tooth Violet)

631 • Camassia scilloides f. scilloides (Wild Hyacinth)
631 □ Camassia scilloides f. Petersenii
631 x Camassia scilloides f. variegata

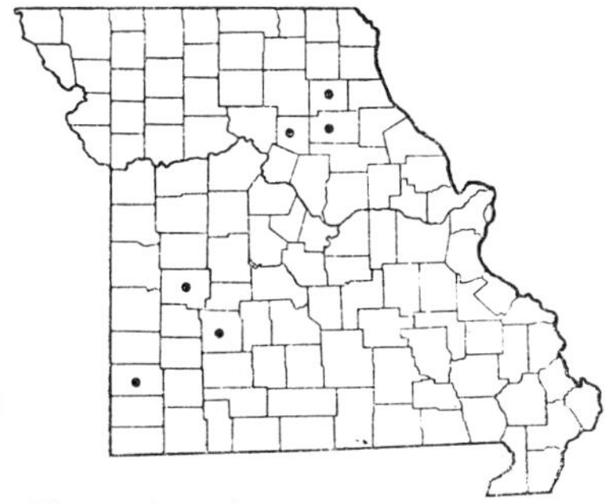

632 Camassia angusta

filaments 4–6 mm. long, lavender; capsule ovoid-oblong, larger than broad, 7–13 mm. long, 6–10 mm. wide; pedicels usually strongly erect-ascending or upcurved in fruit, the capsules appearing more closely appressed and near main axis of inflorescence; stems with flower or fruit mainly 7.5–11 dm. tall, the leaves occupying 3/8–1/2 this height; bracts 7–21 on flower-stem (scape) below base of lowest flower-stalk, 17–40 cm. below base of lowest flower-stalk, more or less sigmoid, curved, or twisted during or after flowering; upper portion of petals and sepals sometimes joining or united above the ovary after flowering . . . . . . . . . . . . . . . . . . . . . . 2. C. ANGUSTA

1. **Camassia scilloides** (Raf.) Cory
Wild Hyacinth — Map 631
*Camassia hyacinthina* (Raf.) Palmer & Steyerm. [P & S]

Flowers early April–middle May.

Occurs on prairies, rocky glades, mostly of limestone, low rich woods along streams, and open woods and rocky open slopes. Common throughout southern and central Missouri, and in the northern portion of the state north to Marion, Shelby, Schuyler, Linn, and Daviess counties; absent from the lowland section of southeastern Missouri.

Missouri material is represented by three variations:

a. Leaves with creamy-white along the borders
1c. C. SCILLOIDES f. VARIEGATA
a. Leaves green throughout . . . . . . . *b*
b. Leaves mainly 4–15 mm. broad . . .
1a. C. SCILLOIDES f. SCILLOIDES
b. Leaves mainly 15–40 mm. broad . . .
1b. C. SCILLOIDES f. PETERSENII

1a. **Camassia scilloides** (Raf.) Cory f. **scilloides** — Map 631
*Camassia scilloides* (Raf.) Cory [G, BB]
*Camassia hyacinthina* (Raf.) Palmer & Steyermark [P & S]

This is the common type found in the state.

Ranges from Alabama to Texas, north to Pennsylvania, Ontario, Michigan, Wisconsin, Iowa, and Kansas.

1b. **Camassia scilloides** f. **Petersenii** Steyerm. — Map 631

Known only from Pike County (*Steyermark 28620*) and St. Louis County (limestone slopes, Osage Hills, April 29, 1936, *Petersen*; same locality, April 23, 1937, *Steyermark 22499*).

Recorded thus far only from Missouri.

The scape in this form attains a thickness of 10 mm. and the roundish bulb attains a diameter of 5 cm.

1c. **Camassia scilloides** f. **variegata** Steyerm. — Map 631

Known only from Christian County in southwestern Missouri (northwest-facing slopes, Pedelo Creek, T27N, R19W, sect. 6, 4½ mi. northeast of Sparta, April 25, 1941, *Steyermark 28324*).

Recorded thus far only from the type locality in Missouri. The bulbs of a western species of Wild Hyacinth, *C. Quamash*, are recorded as being eaten by the Indians in the western United States.

Most plants have lilac to bluish petals and filaments, but occasionally white predominates on the petals and filaments.

2. **Camassia angusta** (Engelm. & Gray) Blankinship — Map 632

Flowers early May–late June.

Occurs in prairies and rocky open ground in Monroe (*Steyermark 85377*), Shelby (*Steyermark 64553*), Randolph (*Naylor* and *McVeigh*), Polk (*Palmer 62666* and *62725*), St. Clair (*Steyermark 83080*), and Jasper (*Palmer 25278*) counties.

633 Scilla sibirica (Squill)

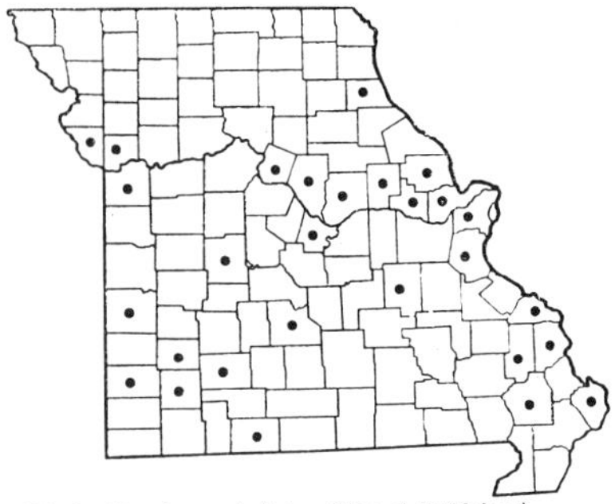

634 Ornithogalum umbellatum (Star-of-Bethlehem)

635 Ornithogalum nutans (Star-of-Bethlehem)

Ranges from Illinois (Mason Co.) to Missouri, south to Kansas, Oklahoma, and Texas.

Gould in his treatment of *Camassia* (Am. Midl. Nat. 28: 712–42. 1942) considered *C. angusta* as possibly a distinct subspecies of *C. scilloides.* Basing his opinion on slender, small-flowered plants of Kansas, Oklahoma, and Texas, he remarks, 'These plants generally have small flowers whose perianth segments spread for a short time and then are connivent over the ovary. In most cases, fertilization apparently is not accomplished, and the flowers are deciduous without further development of the ovary.' In the Illinois and Missouri material, however, while many flowers are deciduous, many others are fertilized and develop mature fruit.

I have been familiar with this plant in the field since 1947, when Mr. Bill Bauer of Kimmswick, Missouri, first called it to my attention, and have had plants growing under observation in my wildflower garden which were transplanted from Missouri and also sent to me from Illinois by Mr. Virginius Chase.

Study and observation of these plants for more than ten years in the field and herbarium have convinced me that they are specifically distinct from the common earlier-blooming *C. scilloides,* as indicated by the main differences given in the key.

Besides the much later blooming period, this species differs in the more robust, taller plants, with a much longer raceme of more numerous flowers of a lavender to purple instead of pale bluish to whitish color, smaller sepals and petals, shorter lavender to purple styles and filaments, 7–21 conspicuous bracts occurring on the main scape 17–40 cm. below the base of the inflorescence and lowest pedicels, more erect-ascending fruiting pedicels with the longer than broad fruiting capsules more closely appressed to the main axis of the inflorescence, proportionately shorter leaves as compared with the height of the flowering stem, and more rounded, globose bulbs. Plants of *C. angusta* start flowering at a time when most of the plants of *C. scilloides* have finished flowering and already have begun to form fruits. This is true even in the same general latitude or locality where the two may be present. The incurving-erect fruiting pedicels with the capsules nearly appressed to the axis of this species are similar in that respect to those of the western *C. Quamash.* The latter species differs from *C. angusta* in its longer (12–35 mm.) and broader (2–8 mm.) sepals and petals, in the sepals and petals always converging together over the capsule and persisting on it after flowering has occurred, in the relatively smaller number of flowers to the inflorescence, and longer capsules.

## 13. **Scilla** L. Squill

**Scilla sibirica** Andr. Squill Map 633

Flowers in March and April.

Cultivated in gardens and rarely escaped to roadsides, vacant grassy areas, and waste places in Pike and Boone counties.

Native of Eurasia.

The plant from Pike County was previously referred erroneously by Palmer and Steyermark (*Annot. Cat. Fl. Pl.* p. 502. 1935) to *Scilla biflora* L. (= *S. bifolia* L.), a species as yet not known as naturalized in the state. This error was later corrected in Steyermark's *Spring Flora of Missouri* (p. 72).

---

Plate no. 115. **1.** Ornithogalum umbellatum, × 2/7. **2.** Erythronium americanum, × 2/7. **3.** Scilla sibirica, × 2/7. **4.** Camassia scilloides; *a.* Camassia scilloides f. scilloides, × 2/7; *b.* Camassia scilloides f. Petersenii; Leaf, × 2/7; Details from Small, The New York Botanical Garden. **5.** Yucca Smalliana; *a.* Habit of plant, × 1/14; *b.* Two flowers, × 2/7; *c.* Single leaf, × 2/7. **6.** Muscari botryoides, × 2/7; Details from Small, The New York Botanical Garden. **7.** Ornithogalum nutans, × 2/7; *a.* Stamen, × 4/7; Details from Small, The New York Botanical Garden.

PLATE NO. 115

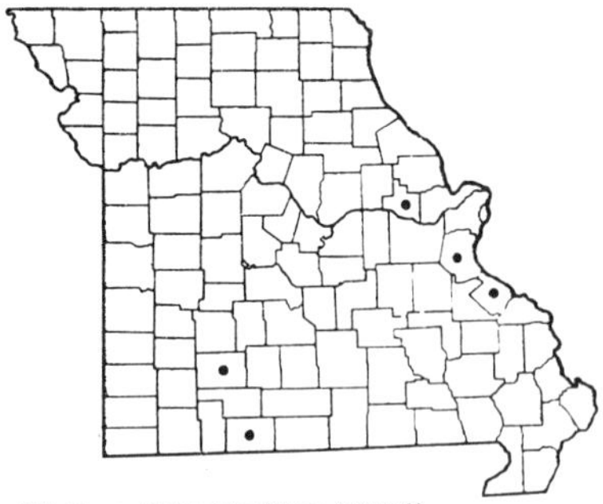

636 Muscari botryoides (Grape Hyacinth)

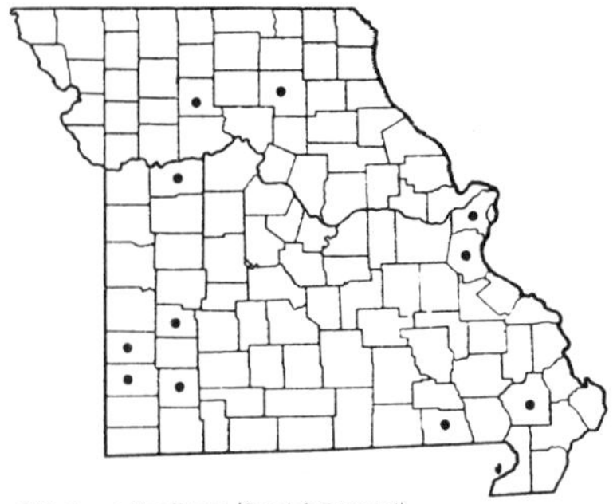

637 Yucca Smalliana (Spanish Bayonet)

638 □ Yucca glauca var. glauca (Soapweed)
638 • Yucca glauca var. mollis (Soapweed)

## 14. **Ornithogalum** L. Star-of-Bethlehem

Perianth of flower 1–2 cm. long; flowers ascending, 3–7 with pedicels (flower-stalks) 2–6 cm. long. . 1. O. UMBELLATUM
Perianth of flower 2.5–3.5 cm. long; flowers eventually nodding, 8–20 (rarely less) with pedicels (flower-stalks) 1 cm. or less long . . . . . . . . . . . . . . . . . . . . . . 2. O. NUTANS

1. **Ornithogalum umbellatum** L.
Star-of-Bethlehem Map 634
Flowers April–June.

Commonly cultivated and frequently escaped and becoming naturalized in open fields, grassy areas, along railroad tracks, roadsides, rocky grazed slopes, shaded banks, and alluvial soils in wooded valleys along streams. Frequent in parts of southern and central Missouri, northeast to Marion County.

Native of Europe; introduced and naturalized in North America from Newfoundland to Ontario and Nebraska, south to Mississippi, Missouri, and Kansas.

Cattle and other stock can be poisoned from eating any part of the plant, either fresh, or dried in hay. Another species, *O. thyrsoides* Jacq., native of South Africa and now widely sold by florists in the United States under the name *Chincherinchee*, is poisonous when eaten and should be kept out of reach of children or domestic animals.

2. **Ornithogalum nutans** L.
Star-of-Bethlehem Map 635
Flowers April–May.

Cultivated in gardens, but rarely escaped. Known naturalized in Missouri only from fields and lawns in Boone County (lawn, Columbia, April 20, 1929, *Rickett*).

Native of Europe; introduced into the United States, ranging as a naturalized plant from New York to D. C., Pennsylvania, and Missouri.

## 15. **Muscari** Mill. Grape Hyacinth

**Muscari botryoides** (L.) Mill.
Grape Hyacinth Map 636
Flowers April–May.

Commonly cultivated in gardens, and occasionally escaped in grassy open ground, low fields, and waste ground in Warren, Jefferson, Ste. Genevieve, Greene, and Taney counties.

Native of Europe; introduced in the United States, ranging there as a naturalized plant from New England to Ontario and Minnesota, south to Virginia, Missouri, and Kansas.

Sometimes this species becomes so well established as to form patches of solid blue, with thousands of plants massed together. Such an observation was made by the author in a fertile valley in Ste. Genevieve County in the early spring of 1958.

## 16. **Yucca** L. Bear Grass, Spanish Bayonet

a. Flowering or fruiting axis with several long conspicuous side (lateral) branches, the inflorescence a branched panicle; flowering and fruiting axis and branches covered with short hairs; leaves shorter than the lowest side branches of the flowering or fruiting portion; main leaves 2–4 cm. broad. 1. Y. SMALLIANA

a. Flowering or fruiting axis simple and unbranched, or with a few short branches at the base of the axis, the inflorescence a raceme-like panicle; leaves longer than the lower part of the flowering or fruiting portion; main leaves mostly 0.4–2.5 cm. broad . . . . . . . . . . . . . . . *b*
b. Plants of the loess hills of northwestern Missouri; leaves stiff and rigid, 0.4–1.2 cm. broad; seeds 10–13 mm. long; petals and sepals short-pointed (acute) . . . . . . 2a. Y. GLAUCA var. GLAUCA
b. Plants of the southernmost Ozark counties; leaves rather soft and flexible, mostly 1–2.5 cm. broad; seeds 8–10 mm. long; petals and sepals mostly blunt (obtuse) at tip . . 2b. Y. GLAUCA var. MOLLIS

1. **Yucca Smalliana** Fern. Spanish Bayonet, Adam's Needle Map 637
*Yucca filamentosa* [of BB, P & S and most authors], not L.
Flowers May–August.

Commonly cultivated in gardens, and occasionally escaped and established away from cultivation in thickets, along roadsides, open banks, railroads, and open woods. Scattered in Missouri.

Native of the United States from Florida to Louisiana, north to North Carolina and Tennessee.

According to Fernald (Rh. 46: 5–9. 1944), the Yucca which Linnaeus described as *Y. filamentosa* is a plant of the coastal plain ranging from Georgia to New Jersey.

The fleshy succulent flowers of this and the other species of Yucca may be used fresh in salads in combination with other vegetables, or may be cooked as a vegetable, dipped in eggs and fried, or mixed with meats, as is usually done with those of *Y. elephantipes* Regel, a species of Central America and Mexico.

The flattened black seeds packed closely together in vertical tiers resemble those of the Dutchman's Pipe Vine (*Aristolochia tomentosa*).

2a. **Yucca glauca** Nutt. var. **glauca** Soapweed Map 638
*Yucca angustifolia* Pursh [Shinners]
Flowers May–July.

Occurs only on open, dry exposed slopes of loess hills in Holt and Atchison counties, northwestern Missouri.

Ranges from Iowa and North Dakota to Montana, south to northwestern Missouri, Texas, New Mexico, and Arizona.

The roots of this species contain a soapy substance, which may be employed as a soap substitute. The leaves are sometimes made into stable brooms.

Some of the Indians of the western United States eat the fresh fruiting capsules or the dried roasted ones, ground into meal. A fermented beverage is reported to be made by them from the fruits.

2b. **Yucca glauca** var. **mollis** Engelm. Map 638
*Yucca arkansana* Trel. [BB, P & S]
Flowers May–June.

Occurs on rocky wooded slopes and banks along streams in limestone soils, and infrequently on dry rocky limestone glades. Known only from the southernmost counties bordering Arkansas, from Oregon County on the east to Barry County on the west.

Ranges from southern Missouri and Oklahoma to Arkansas and Texas.

The flowers are greenish-white in general, but the sepals show a tinge of purplish-pink over their whole surface with a broad greenish-white margin. The petals are somewhat longer and broader than the sepals.

## 17. **Asparagus** L. Asparagus

**Asparagus officinalis** L. Asparagus Map 639
Flowers May–June.

Commonly cultivated and often escaped along railroad tracks, roadsides, fields, thickets, and waste ground.

Native of Europe; introduced into the United States and Canada where naturalized throughout the United States north to Quebec and Ontario.

The small greenish-yellow flowers on slender stalks along the branches are followed in late summer and fall by scarlet or orange-red globular berries containing a few rounded seeds. The small brownish scales on the stem are the true leaves, whereas the fine, green, threadlike, leafy-appearing clusters in the axils of the scales are actually divisions of branchlets which function as leaves, but instead have the structure of a stem or branch.

Wild escaped plants are less tender than the cultivated, but, nevertheless, when gathered as young sprouts just out of the ground, make a suitable morsel for a meal.

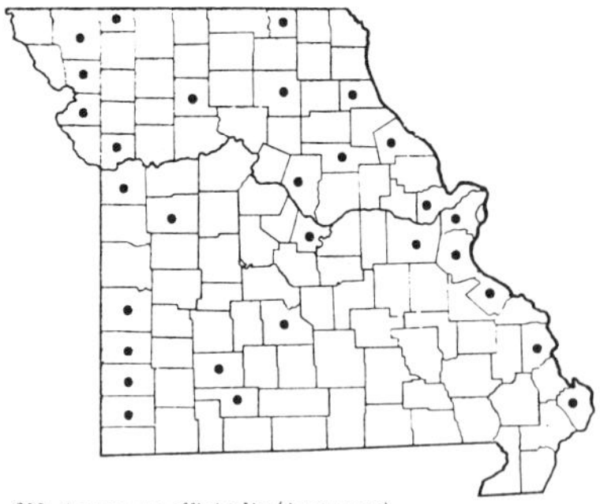

639 Asparagus officinalis (Asparagus)

640 • Smilacina racemosa var. racemosa (False Solomon's Seal)

640 x Smilacina racemosa var. cylindrata (False Solomon's Seal)

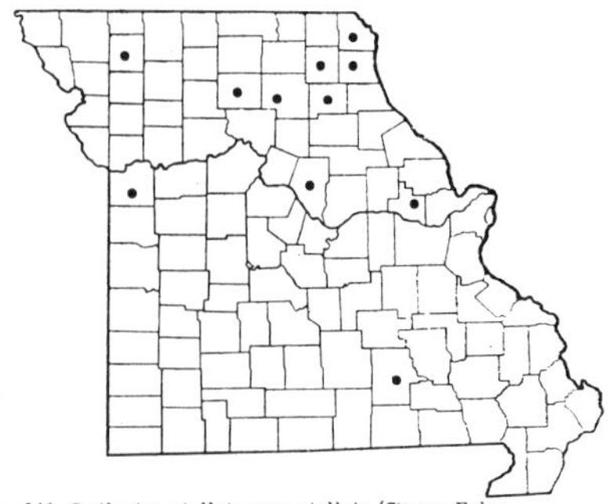

641 Smilacina stellata var. stellata (Starry False Solomon's Seal)

## 18. **Smilacina** Desf. False Solomon's Seal

Flowers numerous, 40 or more on short pedicels (the individual flower-stalks), in a branched panicle; mature berries red, numerous; leaves with 3 or 5 conspicuous ribs on both lower and upper surface and dark or grass-green, not covered with a pale or gray-bluish (glaucous) coating; sepals and petals 1.5–3 mm. long . . . . . . . . . . . . . . . . . . . . . . . . . . **1.** S. RACEMOSA

Flowers few, less than 15, on long pedicels about 4–6 mm. long, in a simple unbranched raceme; mature berries green or black with black stripes; leaves with less conspicuous ribs on both surfaces and with a pale or gray-bluish (glaucous) coating; sepals and petals 4–6 mm. long . . . . 2. S. STELLATA

1. **Smilacina racemosa** (L.) Desf. False Solomon's Seal, False Spikenard Map 640

Flowers May–June.

Occurs in rich nonrocky or rocky woodland. Throughout Missouri.

As treated by Fernald, Missouri material has been divided into the following two variations:

Inflorescence ovoid to pyramidal, 3–10 cm. in diameter, 7–17 cm. long, 3/8–3/4 as broad as long, its longer branches 2–6 cm. long and with 8–24 flowers; peduncle (main supporting stem of inflorescence) usually less than half the length of the inflorescence; larger leaf-blades 3.5–9.5 cm. broad, 10–25 cm. long; main stem 5–10 dm. high . . 1a. S. RACEMOSA var. RACEMOSA

Inflorescence nearly cylindrical, 1.5–3 cm. in diameter, 4.5–8.5 cm. (rarely to 13 cm.) long, 1/4–3/8 as broad as long, its longer branches 1–2.5 cm. long and with 6–10 flowers; peduncle half as long as or longer than the length of the inflorescence; larger leaf-blades 3.5–6 cm. broad, 8.5–17 cm. long; main stem shorter, at most up to 7.5 dm. high . 1b. S. RACEMOSA var. CYLINDRATA

1a. **Smilacina racemosa** (L.) Desf. var. **racemosa** Map 640

Scattered throughout Missouri.

Ranges from Quebec to British Columbia, south to Virginia, North Carolina, Tennessee, and Missouri.

1b. **Smilacina racemosa** var. **cylindrata** Fern. Map 640

Scattered over the state, somewhat more common.

Ranges from Georgia to Arizona, north to New Hampshire, New York, Ontario, Ohio, Michigan, Wisconsin, Kansas, and Colorado.

These two variations need to be studied more carefully throughout their ranges before any final statement can be made as to their relative merit. Until those studies are forthcoming, the varieties are being maintained in all fairness to Fernald's observations.

The cooked young shoots are used as a substitute for asparagus, and the starchy rootstocks were eaten, after having been soaked first in lye and then parboiled like potatoes, by some Indian tribes. The fresh berries have a flavor said to suggest that of bitter molasses, but possess a cathartic action.

2. **Smilacina stellata** (L.) Desf. var. **stellata** Starry False Solomon's Seal Map 641

Plate no. 116. **1.** Yucca glauca var. glauca, × 2/21; *a.* Portion of inflorescence, × 2/7; Details from Small, The New York Botanical Garden. **2.** Yucca glauca var. mollis, × 2/7; Details from Small, The New York Botanical Garden. **3.** Asparagus officinalis, × 2/7; Details from Small, The New York Botanical Garden. **4.** Smilacina stellata, × 2/7. **5.** Smilacina racemosa, × 2/7; Details from Small, The New York Botanical Garden.

PLATE NO. 116

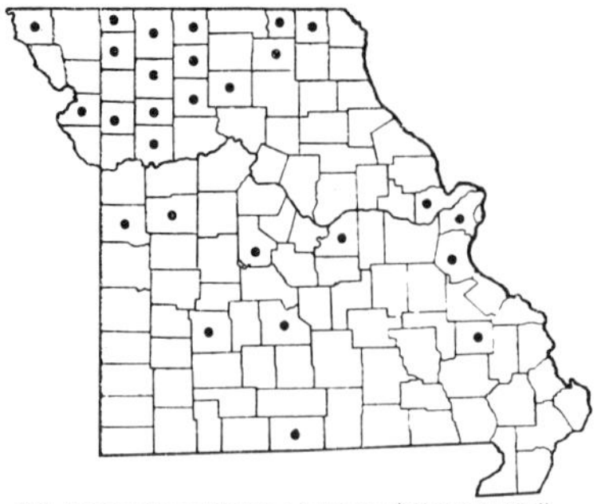

642 Polygonatum biflorum f. biflorum (Solomon's Seal)

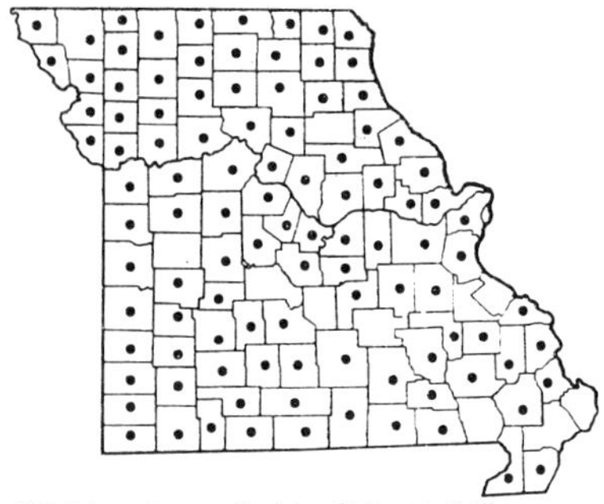

643 Polygonatum canaliculatum (Solomon's Seal)

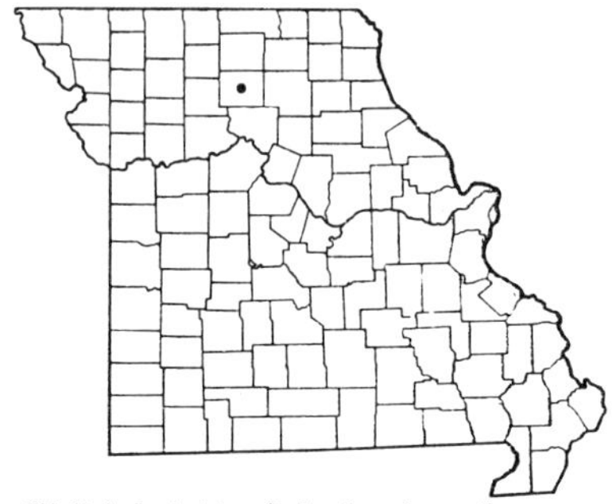

644 Medeola virginiana (Indian Cucumber-root)

Flowers May–June.

Occurs on wooded alluvial banks and on rich wooded slopes, mostly in alluvial low valleys along streams of northern and central Missouri south to Warren, Boone, and Jackson counties, and locally south in the Ozarks in Shannon County along Jack's Fork of Current River in the vicinity of Jam-up Bluff.

Ranges from Newfoundland to British Columbia, south to Virginia, Ohio, Indiana, Illinois, Missouri, Kansas, New Mexico, Arizona, and California. A var. *crassa* Vict. is distributed from Labrador to Maine and Connecticut.

Before reaching the flowering stage, the young stems and leaves may be picked and cooked as 'green.'

## 19. **Polygonatum** Mill. Solomon's Seal

Flowers usually 1–3 (rarely to 5) hanging from each peduncle (main stalk supporting inflorescence); peduncles terete (round in cross-section) or nearly so; perianth of flower 10–17 mm. long, the lobes 3–4 mm. long; topmost small leaf-blade with 20–66 nerves; leaf-blades sessile (without a stalk) or nearly so, the main nerves 1–5 (up to 9) and becoming faint or lost below the tip; stem 2–9 dm. high, slender . . . . . . . . . . . . . . . . . . . . . . . . . . . . 1. P. BIFLORUM

Flowers usually 2–10 hanging from each peduncle; peduncles flattened; perianth of flower 17–22 mm. long, the lobes 5–6.5 mm. long; topmost small leaf-blade with 58–112 nerves; leaf-blades narrowed to a somewhat stalked base, the main nerves 7–19 and mostly conspicuous to the tip; stem 6–20 dm. tall 2. P. CANALICULATUM

1. **Polygonatum biflorum** (Walt.) Ell. f. **biflorum**
Solomon's Seal Map 642
*Polygonatum canaliculatum* in part [of P & S], not (Muhl.) Pursh
*Polygonatum biflorum* (Walt.) Ell. [G, BB]
Flowers May–June.

Occurs in rich low woods and wooded slopes. Throughout Missouri.

Ranges from Florida to Texas, north to Connecticut, New York, Ontario, Michigan, Illinois, Iowa, and Nebraska.

For relative merit of this species, see discussion under *P. canaliculatum.*

2. **Polygonatum canaliculatum** (Muhl.) Pursh Map 643
*Polygonatum canaliculatum* var. *ovatum* (Farwell) Palmer & Steyerm. [P & S]
Flowers May–June.

Occurs in rich or rocky woods, alluvial thickets and stream banks, along railroads, and open places. Throughout Missouri, and more common than the preceding species.

Ranges from New Hampshire to Manitoba, south to South Carolina, Tennessee, Missouri, and Oklahoma.

The young leafy shoots may be cooked and eaten as a substitute for asparagus, while the dried rootstocks are reputed to be used as a bread. The rootstocks resemble those of the poisonous May-apple (*Podophyllum peltatum*), but have large circular scars, whereas these circular scars are absent in May-apple.

*Polygonatum canaliculatum* and *P. biflorum*, while apparently often recognizable as distinct, are frequently difficult to separate from one another because of overlapping and intergrading characters. Plants occur combining 1–3-flowered peduncles with a longer perianth and leaves characteristic of *P. canaliculatum*, or combining short-perianth forms with leaves and other characters of *P. canaliculatum*, as well as other types of noncorrelation of characters. Therefore, there is considerable doubt in the author's mind as to whether these two Polygonatums can be maintained as separate species. They may be no more than ecological variations or even growth forms of the same species with at most recognition in the category of variety or form.

Although the group has been monographed by several students, there is much need for more detailed field and breeding investigations of the species in this genus, especially in the *P. canaliculatum-biflorum* complex.

## 20. **Medeola** L. Indian Cucumber-root

**Medeola virginiana** L. Map 644

Flowers May–June.

Known from a collection from Linn County, northern Missouri (near bank of stream, May 22, 1860, *Broadhead* in University of Missouri herbarium).

Ranges from Quebec, Ontario, and Minnesota, south to Florida, Alabama, and Louisiana.

Numerous localities have been searched for in likely looking spots in Linn County and other northern Missouri counties in an effort to relocate this species which Broadhead presumably collected in 1860. It is to be hoped that it will eventually be found at some place in the state. The small 3–9 flowers have greenish-yellow, recurved perianth-segments, protruding above the upper tier of 3–5 ovate leaves. The fleshy globose berries turn dark purple, and in the fall the leaves become flushed with purple.

The whorled leaves resemble somewhat those of the orchid, *Isotria verticillata*.

The succulent thickened rootstock may be eaten raw, resembling somewhat the taste of cucumber, or used in salad form with vinegar and dressing.

## 21. **Trillium** L. Trillium, Wake Robin

a. Flower sessile (arising directly from top of stem and leaf clusters) . . . . . . . . . . . . . *b*

b. Mature leaves narrowed to a petiole (leaf-stalk); sepals turned down (reflexed); petals abruptly narrowed at base into a distinct claw . . . . . . . . . . . . . . . . 3. T. RECURVATUM

b. Mature leaves sessile, without a petiole; sepals spreading to ascending; petals without a claw or narrowed to a claw-like base . . . . . . . . . . . . . . . . . . . . . . . . *c*

c. Stamens about $\frac{1}{2}$–$\frac{5}{8}$ length of petals; petals 15–36 mm. long, 2.8–5 times longer than broad, broadest in the lower half or near the middle, narrowly to broadly elliptical, without a claw-like base, the narrowed base at most 2 mm. long; connective of the anther prolonged to a broadly triangular-lanceolate tip, somewhat pointed, flattened, 2–3.5 mm. long . . . 1. T. SESSILE

c. Stamens mostly between $\frac{1}{4}$–$\frac{1}{3}$ (infrequently $\frac{3}{8}$) length of petals; petals 36–75 mm. long, 5–8 times longer than broad, broadest in the upper half, oblanceolate to linear-oblanceolate to nearly linear, conspicuously narrowed in the lower one-third to one-fifth to a claw-like base 7–15 mm. long; connective of the anther prolonged into a rounded thick tip mostly 0.5–1.5 mm. long . . . . . . . . . . . . . . . . . . . . . . . . . 2. T. VIRIDE

a. Flower on a peduncle (stalk) which connects the flower to the top of the stem . . . . . . . . *d*

d. Leaves very large, 7–20 cm. broad, 7–20 cm. long, broadly rhombic (shaped roughly like a baseball diamond), abruptly long-pointed (acuminate) at tip; peduncle (flower-stalk) 3–12 cm. long, usually spreading horizontally or turned down while in full flower; main part of style lacking or not evident; ovary and fruit with 6 sharp angles . . . . . . . . . 4. T. FLEXIPES

d. Leaves small or medium-sized, 1–3 cm. broad, 1.5–8.5 cm. long, oblong to oblong-lanceolate to ovate, blunt (obtuse) at tip; peduncle (flower-stalk) 1–3 cm. long, erect and standing above the leaves while in full flower (in *T. nivale* turning down in fruit); main part of style 2 mm. or more long, evident; ovary and fruit with 3 rounded angles . . . . . . . . . . . . . . *e*

e. Mature leaves narrowed to a definite petiole (leaf-stalk); stems mostly 3–10 cm. (rarely 15) tall; peduncle turned down and recurved in fruit; petals white at all stages; plants mainly of alkaline soils of north-facing limestone slopes mainly north of the Missouri River . . 5. T. NIVALE

e. Mature leaves without a petiole (leaf-stalk); stems mostly 10–30 cm. tall; peduncle remaining erect in fruit; petals partially or entirely pink or rose, at least on outer or lower surface, white elsewhere; plants of acid soils of upland oak-hickory or pine woods in shallow draws. 6. T. PUSILLUM var. OZARKANUM

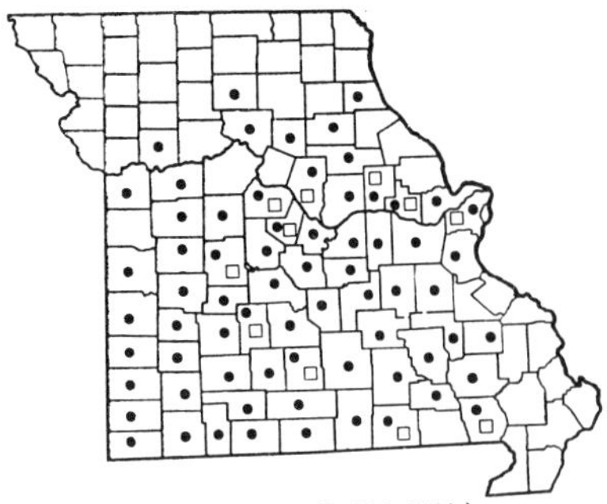

645 • Trillium sessile f. sessile (Wake Robin)
645 □ Trillium sessile f. viridiflorum (Wake Robin)

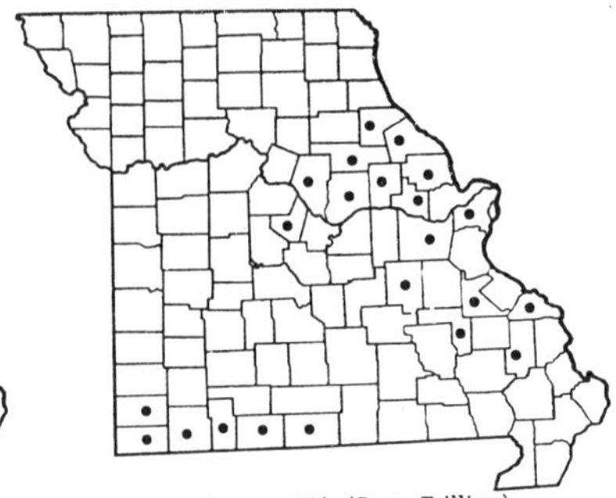

646 Trillium viride var. viride (Green Trillium)

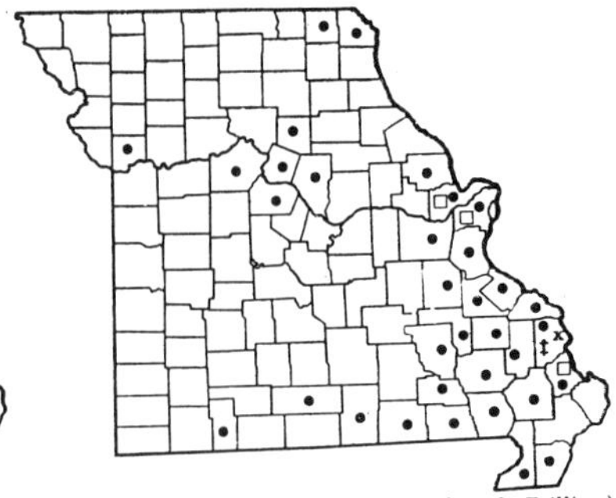

647 • Trillium recurvatum f. recurvatum (Purple Trillium)
647 □ Trillium recurvatum f. Shayi
647 ‡ Trillium recurvatum f. petaloideum
647 x Trillium recurvatum f. foliosum

1. **Trillium sessile** L. Wake Robin Map 645
Flowers early April–early June.

Occurs in either low rich bottom woods of valleys or on wooded slopes, in rocky or nonrocky soils.

Ranges from Georgia to Arkansas, north to New York, Pennsylvania, Ohio, Indiana, Illinois, and Missouri.

Missouri material may be divided into the following two variations:

Petals brown-purple, maroon, or brick-colored throughout or mixed with dull green . 1a. T. SESSILE f. SESSILE

Petals mainly greenish-yellow, brownish-yellow, yellow, or greenish . 1b. T. SESSILE f. VIRIDIFLORUM

1a. **Trillium sessile** L. f. **sessile** Map 645
*Trillium sessile* L. [G, BB, P & S]

Common throughout southern and central Missouri, and ranging northward to Marion, Monroe, Randolph, Linn, Ray, and Jackson counties; absent from the lowland counties of southeastern Missouri. The commoner of the two color variations in the state. The anthers and filaments are usually maroon or purplish-brown in this form.

1b. **Trillium sessile** f. **viridiflorum** Beyer Map 645
*Trillium sessile* f. *luteum* (Muhl.) Peattie [P & S, Deam], not *T. luteum* (Muhl.) Harbison, nor *T. sessile* var. *luteum* Muhl.

Scattered in the range of f. *sessile* and often occurring with it. The anthers of this form are usually green and the filaments yellow.

*Trillium sessile* is the commonest species of the genus in the state. Some of the standard manuals disagree on the matter of mottling of the leaves in *T. sessile*. Actually, both mottled and unmottled leaves are found over the range in the state, with unmottled-leaved individuals more common.

The relatively shorter petals broadest in the lower half with a very short narrowed base and the stamens about one-half the length of the petals will distinguish this species and its forms from the next one, *T. viride*.

The unfolding leaves of young plants of this and other species of the genus may be cooked as a green vegetable. The roots are reported to be emetic.

2. **Trillium viride** Beck var. **viride**
Green Trillium Map 646
*Trillium viride* Beck [G, P & S]
*Trillium viridescens* Nutt. [P & S]
*Trillium luteum* [of G in part], not (Muhl.) Harbison
*Trillium sessile* var. *Nuttallii* Watson
Flowers April–May.

Occurs in rich soils of valleys along streams, and on rich or rocky wooded slopes, generally in limestone areas. Found in two areas, (1) extreme southwestern Missouri along the White River and tributaries, and (2) eastern Missouri from Bollinger and Reynolds counties to Ralls and Audrain counties.

Ranges from southern Illinois, Missouri, and eastern Kansas, south to northern Arkansas and eastern Oklahoma.

This species may be readily distinguished from *T. sessile* on the basis of the characters presented in the key. While there is variation in the shape, size, and coloring of the petals and in the amount of tapering of the leaf-tip, prolonged field study and growing of

Plate no. 117. **1.** Polygonatum canaliculatum, × $^2/_7$; Details from Small, The New York Botanical Garden. **2.** Medeola virginiana, × $^2/_7$; Details from Small, The New York Botanical Garden. **3.** Trillium viride var. viride, × $^2/_7$. **4.** Trillium flexipes, × $^2/_7$. **5.** Trillium recurvatum, × $^2/_7$; Details from Small, The New York Botanical Garden. **6.** Trillium sessile, × $^2/_7$.

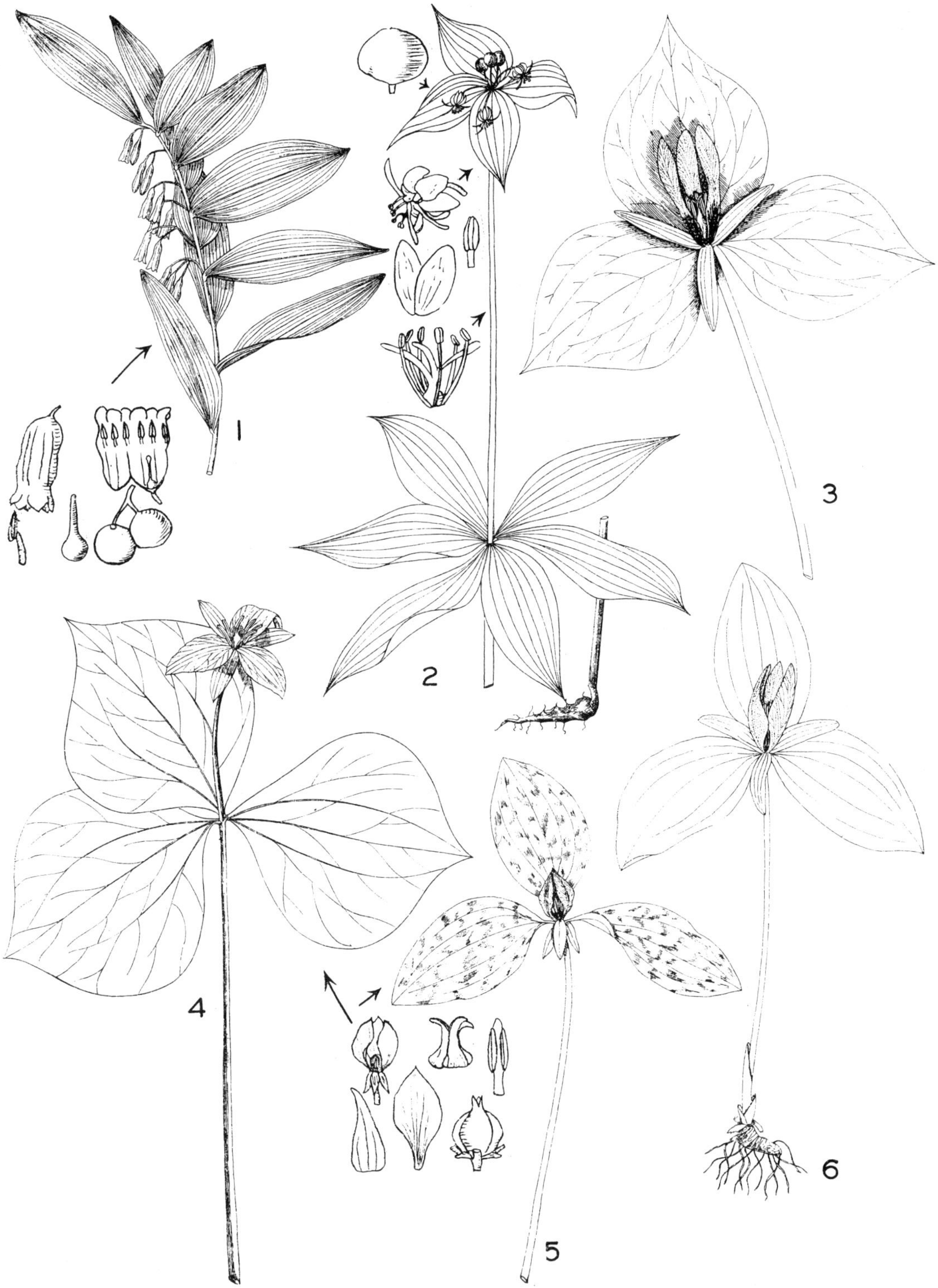


PLATE NO. 117

*T. viride* var. *viride* has convinced the author of its distinctness from *T. sessile.* The usual coloring of the petals is greenish or greenish-yellow to yellow over the main portion with maroon, purple, or purplish-red in the narrowed clawlike base or lower part. Rarely are found plants (*Steyermark 84448* from Lincoln County) in which the petals are dull maroon throughout. The petals vary from usually oblanceolate to almost linear and taper to a narrow clawlike base 2–5 mm. wide; sometimes the base is not so distinctly narrowed. The filament and connective are usually dark maroon or purple, and the anthers range from yellow to purplish.

Study of *T. viridescens* in the field and herbarium shows that it cannot be separated even as a form or variety from *T. viride* var. *viride.* While *T. viridescens* has sharp-pointed (acuminate) leaves, and *T. viride* var. *viride* has them mostly short-pointed (acute), yet plants of *T. viride* often vary within a given locality from acute to acuminate, and cannot be distinguished from equally acuminate leaf tips of *T. viridescens.*

As regards the matter of hairs in *T. viride* var. *viride* being present at the summit of the stem and at the base of the main nerves of the lower side of the leaves, this character has been found after extensive field and herbarium studies to be unsuitable for differentiating *T. viride* var. *viride* from *T. sessile* for the following reasons: (1) these hairlike excrescences are often not too evident or well-developed, and (2) close examination of typical plants of *T. sessile* reveals similar, although on the average shorter, excrescences. Sometimes the hairlike projections on *T. viride* var. *viride* are well-developed and conspicuous, but this is by no means a typical condition to be relied upon.

Gleason (*New Ill. Fl.* 1: 432. 1952) has given the most realistic presentation of the close relationship of *T. viride* var. *viride* with *T. luteum* (Muhl.) Harbison and *T. Hugeri* Small [= *T. cuneatum* Raf.] by treating these last taxa as one variety, var. *luteum* (Muhl.) Gl. For a number of years I have had growing in my wildflower garden plants of *T. luteum* (Muhl.) Harbison [= *viride* var. *luteum*] from eastern Tennessee. The petals, which are greenish-yellow to yellow throughout, lack the maroon or purple base of *T. viride* var. *viride* and are broader and less contracted in the lower half than var. *viride.* The leaves of *T. viride* var. *luteum* with their stronger mottling of pale green and gray-white are in contrast to a duller and darker type of mottling in var. *viride.* The summit of the stem and the nerves of the lower leaf surface are practically glabrous in *T. viride* var. *luteum,* but there are intergrading forms which show some hairlike projections. For plants in Missouri referred by *Gray's Manual* to *T. luteum,* see under excluded species following treatment of this genus.

3. **Trillium recurvatum** Beck Purple Trillium Map 647

Also called Bloody Butcher and Purple Wake Robin. Flowers early April–May.

Occurs usually in rich woods in alluvial soils of valleys and along streams, and at the base of wooded slopes and rocky bluffs, occasionally in open low grassy areas bordering fields and roadsides. Most common in eastern Missouri, with scattered stations westward in central Missouri to Saline and Clay counties, and in the southernmost counties bordering the Arkansas line west to Stone County.

Ranges from Ohio, Michigan, Minnesota, and Iowa, south to Alabama, Mississippi, and Arkansas.

Missouri material is represented by the following variations:

a. In addition to the 3 normal leaves, the sepals, petals, stamens, and carpels are changed to green leaf-like parts . . . 3d. T. RECURVATUM f. FOLIOSUM
a. Plants with normal leaves and with colored sepals, petals, stamens, and carpels . . . *b*
  b. Petals yellow or yellow-green; stamens yellow or yellow-green . 3b. T. RECURVATUM f. SHAYI
  b. Petals maroon, purplish, or red-brown; stamens dark purple or maroon . . . *c*
    c. The 3 sepals green with or without purple edging, thicker than the petals . 3a. T. RECURVATUM f. RECURVATUM
    c. The 3 sepals maroon or purplish, about as thin as the petals . 3c. T. RECURVATUM f. PETALOIDEUM

3a. **Trillium recurvatum** Beck f. **recurvatum** Map 647

*Trillium recurvatum* Beck [G, BB, P & S]

The common form encountered in the state.

3b. **Trillium recurvatum** f. **Shayi** Palmer and Steyerm. Map 647

Known in Missouri only from St. Louis (Monarch Hill, near Monarch, April 29, 1935, *W. F. Shay,* type collection in Missouri Botanical Garden Herbarium), St. Charles (Weldon Springs, August Busch Wildlife area, spring, 1957, *Bruce Dowling* in University of Missouri herbarium), and Scott counties.

Recently recorded from southern Illinois.

A very handsome form which should be more widely grown.

648 Trillium flexipes f. flexipes (White Trillium)

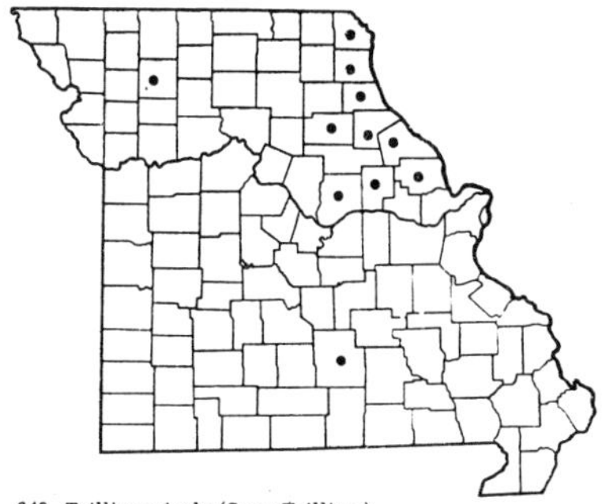

649 Trillium nivale (Snow Trillium)

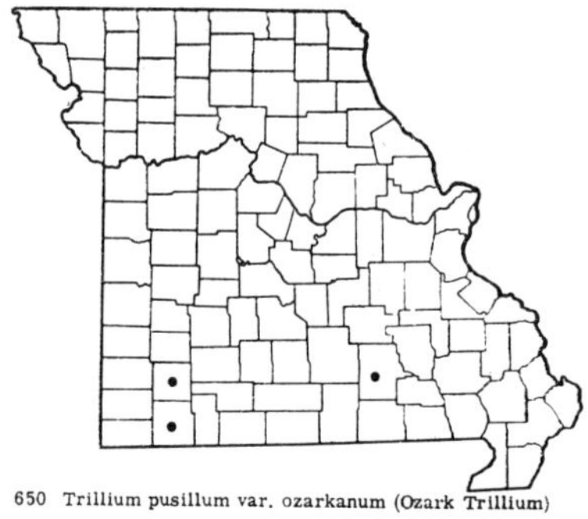

650 Trillium pusillum var. ozarkanum (Ozark Trillium)

3c. **Trillium recurvatum** f. **petaloideum** Steyerm. Map 647

Known only from Cape Girardeau County (Houck Woods, Bloomfield Road, April 6, 1949, *Wayne Adams*, type collection in Southeast Missouri State College herbarium).

3d. **Trillium recurvatum** f. **foliosum** Steyerm. Map 647

Known only from Cape Girardeau County (Houck Woods, Bloomfield Road, April 6, 1949, *Wayne Adams*, type collection in Southeast Missouri State College herbarium).

When the leaves of *Trillium recurvatum* are first unfolding, the narrowed petiolar base of the leaf-blade is not evident. However, fully grown leaves show a definite petiole.

4. **Trillium flexipes** Raf. f. **flexipes** White Trillium, White Wake Robin Map 648
*Trillium flexipes* Raf. [G]
*Trillium Gleasoni* Fern. [BB, P & S]
Flowers April–late May.

Occurs usually in rich woods on slopes or bluffs or in valleys of ravines, and along streams. Eastern Missouri north from Stoddard County to Ralls County, west to Wayne, St. Francois, Franklin, Cole, and Boone counties; absent from most of the Ozark and southeastern lowland sections.

Ranges from Michigan to Minnesota, south to Ohio, Kentucky, Tennessee, and Missouri; also in New York and Maryland.

This is the tallest of the species of *Trillium* in Missouri. The showy white petals occur at the tip of the usually horizontally spreading or reflexed peduncle. Occasionally the peduncle is nearly upwardly ascending. The fruit, when mature, is a rose-red color during summer. A form, f. *Walpolei* (Farw.) Fern., with maroon or purple petals, filaments, and ovary is known, but has not been recorded for Missouri.

5. **Trillium nivale** Riddell Snow Trillium Map 649

Flowers from about March 12–April 16.

Occurs usually on limestone or cherty limestone soils of steep eroded forested slopes bordering streams, mostly in forests of Sugar Maple associated with White Oak and Northern Red Oak. Found mainly in northeastern and east-central Missouri from Lincoln and Callaway counties north to Clark County, locally west in northwestern Missouri to Daviess County (*Steyermark 84807*) and locally south along Jack's Fork of Current River in Texas County.

Ranges from Pennsylvania to Minnesota, south to Kentucky and Missouri.

Upon drying in a plant press the leaves and petals shrink considerably.

This is the smallest species of *Trillium* in the state. Where it is found, it usually occurs in abundance, with hundreds or thousands of individual plants.

During the middle of March, when it starts flowering, only a few other plants, such as *Hepatica nobilis* var. *acuta* (*acutiloba*), *H. nobilis* var. *obtusa* (*americana*), and *Erigenia bulbosa*, are in bloom, so that the large white petals stand out conspicuously against the dark earth background. Slightly later, *Dicentra Cucullaria*, *Sanguinaria canadensis*, *Isopyrum biternatum*, *Erythronium albidum*, *Dentaria laciniata*, and *Lindera Benzoin* are in bloom with the *Trillium*. On slopes where the soil is not bare or is grown over by other vegetation, this *Trillium* is absent, apparently favoring the more eroded places where the soil is more exposed and has scattered limestone fragments visible.

It transplants easily and does well in a wildflower garden if provided with some shade and a limey soil with good drainage. It increases from year to year by seed or by underground shoots from the short rhizome.

By late July or early August the plants above ground shrivel and disappear.

6. **Trillium pusillum** Michx. var. **ozarkanum** (Palmer & Steyerm.) Steyerm. Ozark Wake Robin
Map 650

*Trillium ozarkanum* Palmer & Steyerm. [P & S]
*Trillium pusillum* [of G and BB in part], not Michx.
Flowers April 13–early May.

Occurs in acid soils of shallow draws in thin cherty-flinty soils of oak-hickory, oak-pine, or oak-chestnut woodland in the southern Missouri Ozarks, where known from Shannon (west side of highway 80, T27N, R5W, sect. 34, 2 mi. south of Birch Tree, May 3, 1947, *Steyermark 64262*; same locality, 1946, *Bill Bauer*), Lawrence (*Palmer*), and Barry (3 mi. south of Cassville along highway 112 along road to Roaring River State Park, April 20, 1935, *Steyermark 18628*, type collection in Missouri Botanical Garden Herbarium; same locality, April 13, 1930, *Cora Shoop*) counties.

Known only from southern Missouri and northwestern Arkansas.

The leaves are membranaceous and soft in texture, and dull grass green on both sides. The petals assume a flat-horizontal, spreading, or arched-recurving position, but do not become erect or ascending. They vary in color from pure white within or tinged with varying amounts of pink or are deep madder- or rose-purple; on the lower or outer side they are usually tinged with varying amounts of rose-purple or pink. Usually it was noted in the field that the petals which had the deepest dark-purple-rose color were the narrowest ones of all.

Transplanted specimens from Missouri to the author's wildflower garden have grown and increased over a period of twelve years. The upper surface of the petals is white and the lower surface is pinkish; in age the lower surface turns a deeper pink, but the petals as a whole remain a more subdued or paler shade of pale rose or pink and have not attained the deeper richer colors seen in the southern Missouri Ozarks.

The flowers have a fragrance somewhat resembling that of the Dandelion (*Taraxacum*).

Typical *T. pusillum* var. *pusillum* occurs in South Carolina and var. *virginianum* in Virginia. The var. *ozarkanum* differs from these in its longer peduncles, more strongly nerved, normally 5- instead of 3-nerved leaves and sepals, averaging broader, and petals averaging longer and broader. The Ozark variety inhabits dryish woodland, the other varieties a damp woodland.

*Excluded Species*

**Trillium viride** var. **luteum** (Muhl.) Gl.

*Trillium luteum* (Muhl.) Harbison [G]

Fernald in the eighth edition of *Gray's Manual* (p. 445) credits Missouri in the range of *T. luteum*. Missouri specimens placed in the *T. luteum* folder at Gray Herbarium were originally identified by W. A. Anderson in 1931 as *T. Hugeri* Small, as *T. Underwoodii* Small, by Fernald in 1943 as *T. cuneatum* Raf., and by S. J. Smith in 1943 as *T. viride* Beck. I agree with Dr. Smith's identifications. Apparently, specimens which appeared glabrous at the summit of the stem and on the nerves of the lower side of the leaves were identified by Anderson and Fernald as identical with specimens from the more eastern states. Since there appears to be no difference between such specimens and others which I have placed in *T. viride* var. *viride*, the Gray Herbarium material from Missouri should be referred to var. *viride*, whereas the material from the eastern and southeastern states may be identified as *T. viride* var. *luteum*, *T. cuneatum*, and *T. Hugeri*. For further discussion on the variability in *T. viride* var. *viride* and the weakness of the pubescence character previously relied on, the reader is referred to the author's discussion under *T. viride* var. *viride*.

## 22. **Smilax** L. Catbrier, Greenbrier

a. No part of the stem with prickles or thorns; low or climbing soft-stemmed herbaceous plants, dying back to the ground each year, and no part of the plant woody . . . . . . . . . . . . . *b*
  b. Plant mostly erect, mainly 1.5–6 dm. tall (rarely up to 10 dm.), not climbing, without tendrils or with only weak tendrils in the upper part of stem; some or all of the peduncles (main flower-bearing stalks) arising from the axils of the bracts on the lower part of the stem below the lowest leaf . . . . . . . . . . . . . . . . . . . . . . . . 3. S. ECIRRHATA
  b. Plant eventually climbing by tendrils from the axils of most of the middle and upper leaves; peduncles arising from the axils of the normally developed leaves . . . . . . . . . . . *c*

Plate no. 118. **1.** Trillium nivale, × 2/7; *a.* Fruiting plant, × 2/7. **2.** Trillium pusillum var. ozarkanum, × 2/7. **3.** Smilax herbacea var. herbacea, × 2/7. **4.** Smilax glauca; *a.* Smilax glauca; *b.* Smilax glauca. **5.** Smilax ecirrhata, × 2/7; *a.* Staminate flower, × 1 1/7.

PLATE NO. 118

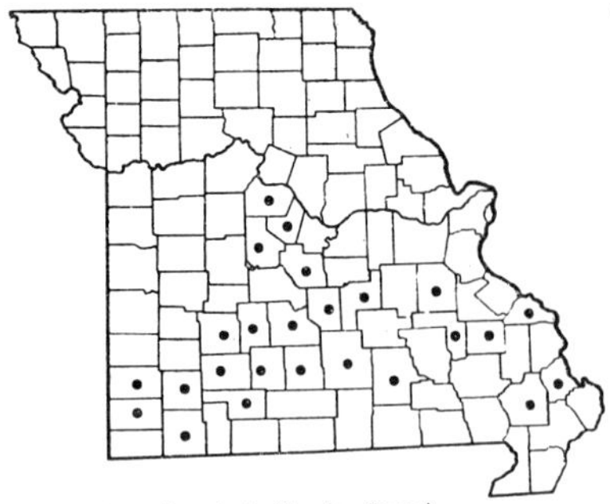

651 Smilax pulverulenta (Carrion-flower)

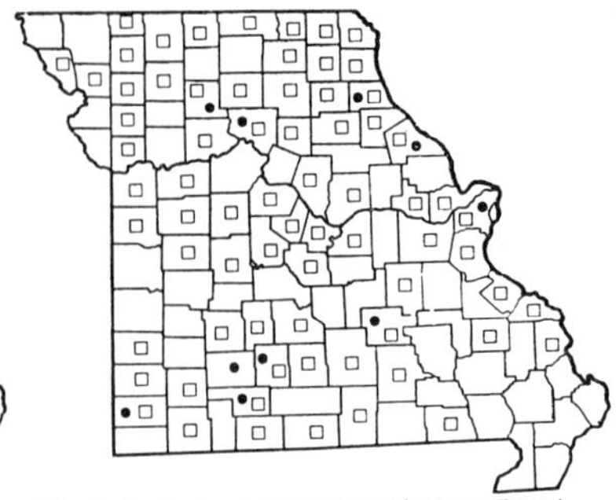

652 • Smilax herbacea var. herbacea (Carrion-flower)
652 □ Smilax herbacea var. lasioneura (Carrion-flower)

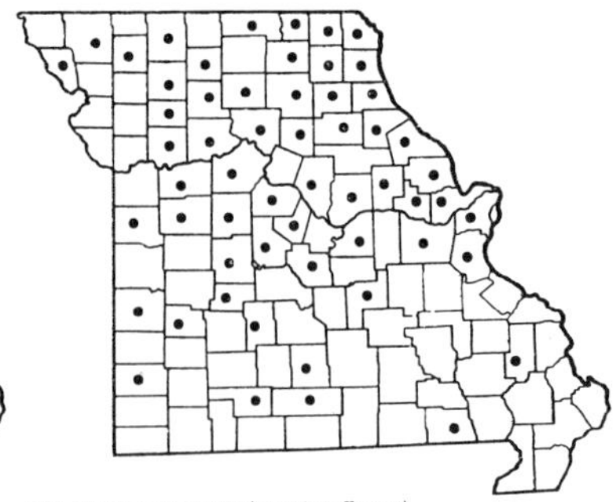

653 Smilax ecirrhata (Carrion-flower)

c. Lower surface of leaf rich grass green like the upper surface, both surfaces glossy; leaf-blades thin and membranaceous, ending in a long-pointed (acuminate) tip, the margins minutely irregularly toothed and with projecting small hairs; inflorescence containing 10–35 flowers; fruit black . . . . . . . . . . . . . . . . . . . . . . 1. S. PULVERULENTA

c. Lower surface of leaf paler than upper surface, and mostly gray-green or pale green, neither surface glossy; leaf-blades mostly obtuse (blunt) or acute (short-pointed), less frequently shortly acuminate, the margins usually not irregularly toothed nor with projecting hairs; inflorescence containing 20–100 or more flowers; fruit bluish-gray or covered with a bloom which can be rubbed off . . . . . . . . . . . . . . . . . . . . . . . . *d*

d. Lower surface of leaves glabrous (without hairs); bracts on the stem below the regular leaves usually appressed (fitting close to the stem) . . . . . 2a. S. HERBACEA var. HERBACEA

d. Lower surface of leaves more or less hairy with rather short or minute hairs; bracts on the stem below the regular leaves usually loose or spreading . . . 2b. S. HERBACEA var. LASIONEURA

a. Some part of the stem, but usually the lowest part, bearing prickles or thorns; climbing plants with stem woody at least toward the base, tough otherwise and not dying back to the ground . . . . . *e*

e. Lower surface of leaves whitened, bluish-gray, or silvery . . . . . . . . . . . . 4. S. GLAUCA

e. Lower surface of leaves about the same green color as the upper surface . . . . . . . . . . *f*

f. Leaf-margin much more thickened than rest of leaf-blade; at least the older stems more or less stellate-pubescent (with tufts of small star-shaped clustered hairs); tooth-like projections and irregularities of leaf-margins, when present, colored; leaves sometimes fiddle-shaped (panduriform), otherwise ovate; veins and veinlets thickened and conspicuous in a network over the leaf surface . . . . . . . . . . . . . . . . . . . . . 6. S. BONA-NOX

f. Leaf-margin not thicker than rest of leaf-blade; stem glabrous throughout; tooth-like projections and prickles of leaf-margins colorless; leaves never fiddle-shaped; veins and veinlets slender forming a delicate network which is not thickened conspicuously . . . . . . . . *g*

g. Plants found only in southeastern Missouri; prickles of stem pale or with a dark tip, flattened and with a broad base; peduncle (main stem bearing the flowering or fruiting cluster) 3–15 mm. long, mostly shorter than or about equaling the leaf-stalk (petiole) of the leaf from whose axil it appears; fruit blue-black, covered with a grayish-blue coating (glaucous); seeds 2–3 in each berry . . . . . . . . . . . . . . . . . . 7. S. ROTUNDIFOLIA

g. Found throughout the state; mature prickles black throughout (green or brown on young stems), terete (round in cross-section) and bristle-like; peduncle 15–65 mm. long, much longer than the leaf-stalk (petiole) from whose axil it arises; fruit black; seed 1 in each berry . . . . . . . . . . . . . . . . . . . 5. S. TAMNOIDES var. HISPIDA

1. **Smilax pulverulenta** Michx. Carrion-flower
Map 651

Flowers April–May.

Occurs in rich or rocky woods in the Ozark section of the state north to Washington, Osage, and Cooper counties, west to Polk and Jasper counties.

Ranges from Georgia, Tennessee, and Arkansas, north to New York, Pennsylvania, West Virginia, Ohio, Indiana, Illinois, and Missouri. The perianth is often described as being 4–6 mm. long, but in the dried specimen is frequently only 2.5–3 mm. long.

This appears to be a species quite distinct from *S. herbacea*, as indicated by the thinner, long-acuminate leaves, which are shining and rich green on both sides, erose-ciliate on the margins, and by the relatively fewer-flowered inflorescence with black fruits.

2a. **Smilax herbacea** L. var. **herbacea**
Carrion-flower Map 652
*Smilax herbacea* L. [G, P & S]
Flowers May–June.

Occurs in rich or rocky woods and thickets. Scattered and much less common in Missouri than the var. *lasioneura*.

Ranges from New Brunswick and Quebec, south to Georgia, Tennessee, and Missouri.

2b. **Smilax herbacea** var. **asioneura** (Hook.) A. DC. Carrion-flower Map 652
*Smilax lasioneura* Hook. [G]
Flowers May–June.

Occurs in rich or rocky woods and thickets. Common throughout Missouri, except apparently absent in the southeastern lowland section.

Ranges from Ontario and Ohio to Minnesota, Saskatchewan and Montana, south to Georgia, Arkansas, Oklahoma, and Colorado.

The leaves are usually gray, rarely green beneath and the petioles vary from 1.2–9 cm. in length. The berries usually have a bluish-black color and are coated with a grayish 'bloom' (glaucous).

The two varieties of *S. herbacea* here enumerated are frequently found side by side or in the same location. They do not appear to merit specific segregation, as treated by Fernald in the eighth edition of *Gray's Manual*. Miss Bernice M. Speese, who has studied the group most recently, treats *S. lasioneura* in the varietal category.

The young tender shoots of these two varieties, as well as of *S. pulverulenta*, may be cooked in early spring and eaten as asparagus.

3. **Smilax ecirrhata** (Engelm.) Wats.
Carrion-flower Map 653
Flowers May–June.

Occurs in rich woods in ravines, at the base of bluffs, and along streams. Common throughout northern and central Missouri, and scattered southward, becoming rare or absent in most of the Ozark section; absent also from the southeastern lowland section.

Ranges from Ontario to Minnesota and South Dakota, south to Tennessee and Missouri.

The lower surface of the leaves is gray-hairy in much the same manner as in *S. herbacea* var. *lasioneura*. The mature leaf-blades of *S. ecirrhata*, however, are much broader, becoming broadly ovate to nearly round, and with a heart-shaped to rounded base. They also average larger, attaining a length of 18 cm. and a width of 12 cm. The inflorescences in *S. ecirrhata* are also fewer-flowered in both the male (less than 25 flowers) and female (less than 20 flowers) plants than those of *S. herbacea*, in which the inflorescences are usually 30–110-flowered in both male and female plants.

Some descriptions of *S. ecirrhata* give the berry of this species as being 3-seeded. Observations of Missouri and other specimens indicate a berry with 1–6 seeds.

Although some students of the genus seem inclined to combine this species with *S. herbacea* as a variety, the two are mainly quite distinct in their habit of growth and in other characters presented above and in the key. Usually the stems of *S. ecirrhata* remain low and erect.

4. **Smilax glauca** Walt. Greenbrier, Catbrier
Map 654
Flowers May–June.

Occurs in acid soils of chert, sandstone, or granite in rocky woodland, wooded valleys and moist ground along streams, open ground in cultivated or fallow fields, along fence rows, open banks, and roadsides, often becoming a persistent weed.

Missouri material is represented by two variations:

Lower surface of leaves covered with minute hairs or projections (use magnifying lens of 10 ×) . 4a. S. GLAUCA var. GLAUCA
Lower surface of leaves glabrous (hairless) or smooth (use magnifying lens of 10 ×) . . . 4b. S. GLAUCA var. LEUROPHYLLA

4a. **Smilax glauca** var. **glauca** Map 654
*Smilax glauca* Walt. [G]
*Smilax glauca* var. *genuina* Blake [P & S]

Known from Cape Girardeau, Madison, Carter, and Dunklin counties.

Ranges from Florida to Texas, north to New Jersey, West Virginia, Ohio, Indiana, Illinois, and Missouri.

4b. **Smilax glauca** var. **leurophylla** Blake
Map 654

This is the commoner variety in Missouri, found throughout the southeastern lowland section north into the southeastern Ozark section to Ste. Genevieve and St. Francois counties, west to Iron, Reynolds, Carter, and Oregon counties.

*Smilax glauca* is a handsome vine, becoming more or less evergreen; the leaves which remain on the stem through the winter have a greenish coppery upper surface. The dark green upper surface of the leaves is sometimes blotched or marked with white or gray throughout the growing season. The bluish-gray or

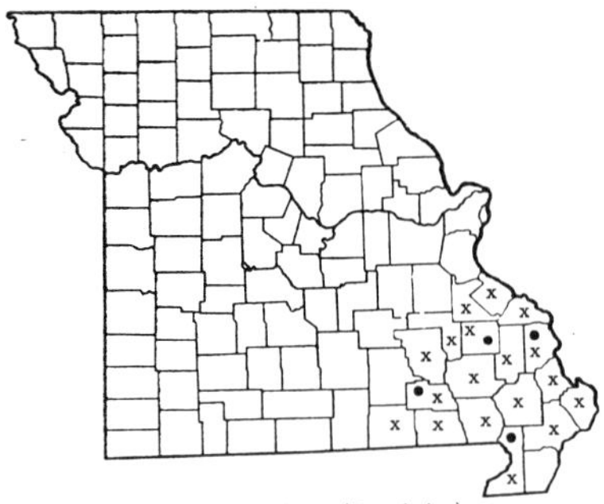

654 • Smilax glauca var. glauca (Greenbrier)
654 x Smilax glauca var. leurophylla (Greenbrier)

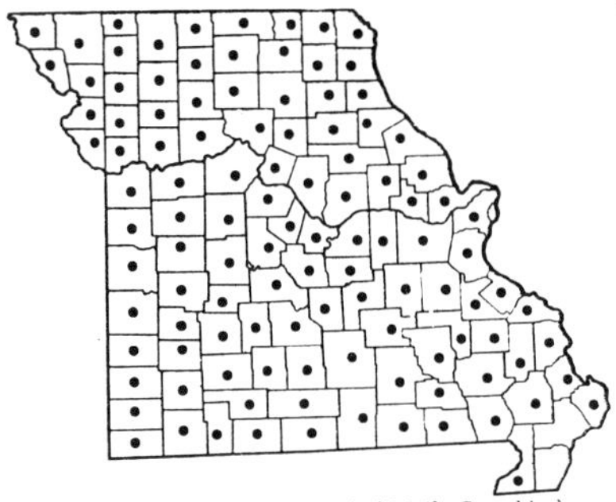

655 Smilax tamnoides var. hispida (Bristly Greenbier)

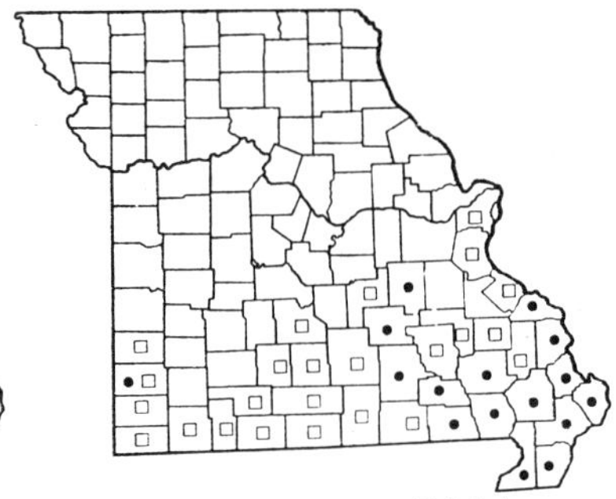

656 • Smilax Bona-nox var. Bona-nox (Catbrier)
656 □ Smilax Bona-nox var. hederaefolia (Catbrier)

whitish lower surface of the leaf-blades imparts a striking color contrast to the leaves of this species, which are definitely whiter beneath than those of any other species in Missouri, while the stems are definitely more slender than those of the other spiny-stemmed species of Missouri.

An amber-colored sweet jelly is obtained from the slender rootstocks, which have been cut up, boiled, and had sugar added in equal bulk. This jelly can be mixed with water to make a sweet drink.

5. **Smilax tamnoides** L. var. **hispida** (Muhl.) Fern. Bristly Greenbrier, Catbrier Map 655
*Smilax hispida* Muhl. [BB, P & S]
Flowers May–June.

Occurs in low woods in valleys and thickets, along banks of streams, and on rich wooded slopes. The most common species of woody greenbrier in Missouri, where known from practically every county in the state.

Ranges from New York and Ontario to Minnesota and South Dakota, south to Georgia, Alabama, Arkansas, and Texas.

The older stems of this catbrier are covered with numerous, crowded spines, presenting a very bristly appearance.

*Smilax tamnoides* var. *hispida* has the leaves ovate, elliptic, or somewhat round, whereas *S. tamnoides* var. *tamnoides*, occurring mostly along the Coastal Plain, and not in Missouri, has some or all of the leaves fiddle-shaped (panduriform).

6. **Smilax Bona-nox** L. Catbrier, Greenbrier, Bullbrier Map 656
Flowers May–June.

Occurs in rocky open woods, low ground in valleys and along streams, limestone glades and outcrops, in thickets, and fields. Common throughout the Ozark region north to St. Louis, Crawford, Phelps, Laclede, Webster, Christian, and Jasper counties.

Missouri material may be divided into two variations:

Leaves of flowering or fruiting stems usually mottled with white with bristlelike hairs on margins; leaf-blades mostly fiddle-shaped (panduriform) with usually short rounded lobes, the main part of the blade narrowly triangular to oblong and 1.5–4 cm. broad at the base; low plant with slender stems, rarely high-climbing . . . 6a. S. BONA-NOX var. BONA-NOX

Leaves of flowering or fruiting stems green (except rarely mottled with white on young basal sprouts), without bristlelike hairs on margins or with a few weak easily falling hairs; leaf-blades mostly broadly ovate with a heart-shaped (cordate) base, 3–8 cm. broad; plants with stouter stems, often high-climbing . . . 6b. S. BONA-NOX var. HEDERAEFOLIA

6a. **Smilax Bona-nox** L. var. **Bona-nox** Map 656
*Smilax Bona-nox* L. [G, BB, P & S]

Scattered throughout the range in the Ozark and southeastern lowland sections.

Ranges from Mexico, and from Florida to Texas, north to Maryland, Kentucky, Indiana, Illinois, Missouri, and Kansas.

6b. **Smilax Bona-nox** L. var. **hederaefolia** (Beyrich) Fern. Map 656

---

Plate no. 119. **1.** Smilax rotundifolia, $\times\ ^{2}/_{7}$; *a.* Female flower, $\times\ 1^{5}/_{7}$; *b.* Male flower $\times\ 1^{5}/_{7}$. **2.** Smilax tamnoides var. hispida, $\times\ ^{2}/_{7}$; *a.* Male flower, $\times\ 1^{5}/_{7}$; *b.* Leaf and part of stem, $\times\ ^{2}/_{7}$. **3.** Smilax Bona-nox var. Bona-nox, $\times\ ^{2}/_{7}$. **4.** Dioscorea villosa, $\times\ ^{2}/_{7}$; *a.* Female inflorescence, $\times\ ^{2}/_{7}$; *b.* Male inflorescence, $\times\ ^{2}/_{7}$; *c.* Fruit, $\times\ ^{2}/_{7}$; Details from Small, The New York Botanical Garden. **5.** Dioscorea quaternata, $\times\ ^{2}/_{7}$; *a.* Male inflorescence, $\times\ ^{2}/_{7}$; *b.* Fruit, $\times\ ^{2}/_{7}$; Details from Small, The New York Botanical Garden.

PLATE NO. 119

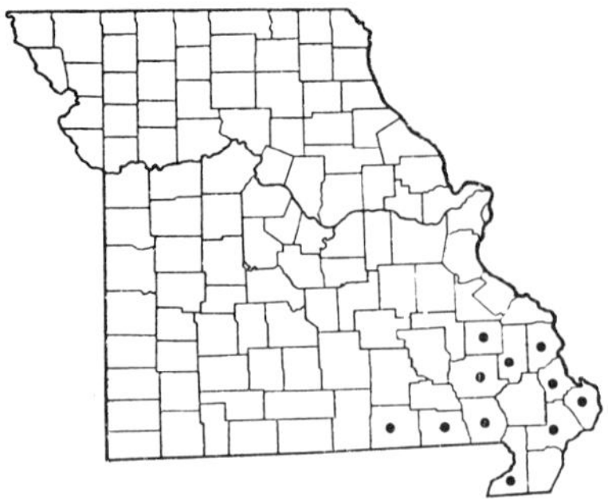
657 Smilax rotundifolia (Catbrier)

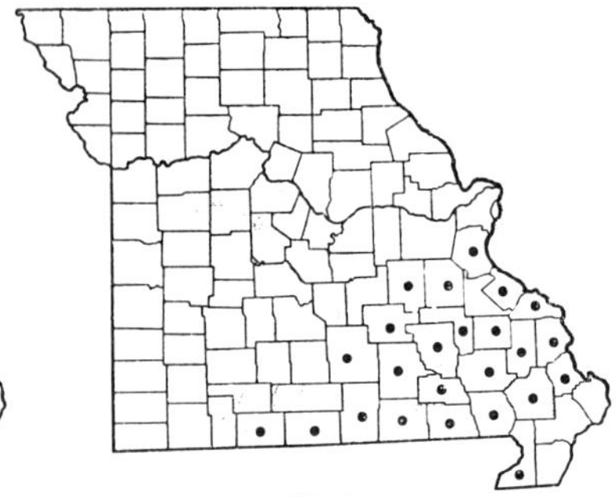
658 Dioscorea quaternata (Yam)

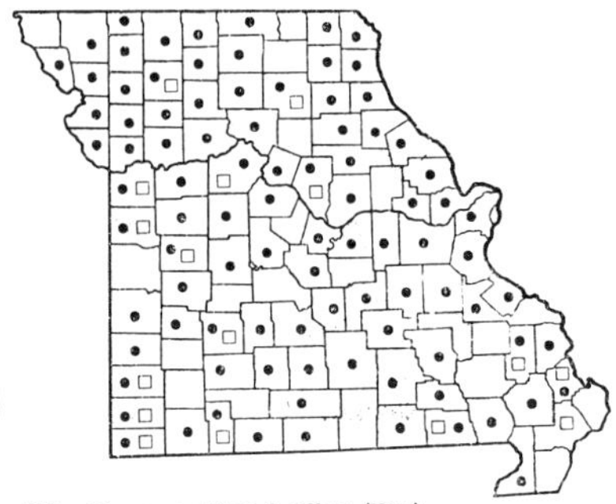
659 • Dioscorea villosa f. villosa (Yam)
659 □ Dioscorea villosa f. glabrifolia (Yam)

This is the more common variety encountered in Missouri (representative collections found in the Gray Herbarium at Harvard University are *Steyermark 7749* from St. Louis County, *Steyermark 7716* from Jefferson County, *Palmer 22848* from Stone County, *Palmer 32939* from Ozark County, and *Palmer 22317, 22727, 22745* from Jasper County).

Ranges from Florida to Texas, north to Delaware and Massachusetts, Tennessee, Illinois, Missouri, and Kansas.

It is possible that future detailed studies on the variability in the leaves of *S. Bona-nox* may show that the above varieties are correlated with stages of maturity of the plant, as noted in the leaf variations found in some of the tropical aroids, such as *Syngonium podophyllum*. In the present treatment both varieties are accepted, pending further study of living plants.

The root of this and other prickly-stemmed species in Missouri may be used in the preparation of a root-beer-like (sarsaparilla) drink, in which molasses and sassafras are some of the added ingredients.

Where *S. Bona-nox* is plentiful, it often impedes walking, the tangled prickly stems tearing or catching the clothes and making it necessary to detour around these thickets. It is especially troublesome on the dry rocky Ozark slopes and in the wet thickets and forests of the southeastern lowlands.

7. **Smilax rotundifolia** L. Greenbrier, Catbrier
Map 657

*Smilax rotundifolia* var. *quadrangularis* (Muhl.) Wood [P & S]

Flowers late April–May.

Occurs in mostly alluvial soils of low ground along streams or in valleys, and in wet woodland and thickets. Southeastern Missouri, mainly in the southeastern lowland section northeast to Cape Girardeau and Bollinger counties, and extending into a limited section of the southeastern Ozarks northwest to Madison County, and west to Oregon County. A specimen from Pike County (Norris Farm at Eolia) collected by Reverend John Davis is undoubtedly taken from a cultivated plant.

Ranges from Florida to Texas, north to Nova Scotia, Maine, New York, Ontario, Ohio, Illinois, Missouri, and Oklahoma.

The tender young or new shoots may be eaten raw or cooked either for use as a boiled vegetable like asparagus or for a salad, said to suggest the flavor of Alligator Pear (*Persea*). The tender shoots of the other prickly-stemmed species should likewise be tried. As in *S. glauca*, the boiled rootstocks, to which sugar has been added in equal bulk, makes a brown sweet jelly. As with *S. Bona-nox*, the root may be used in the preparation of a rootbeer-like drink.

## **DIOSCOREACEAE** (Yam Family)

### **Dioscorea** L. Yam

a. At least the lower leaves in whorls (circles) of 4–7; dried underground rootstocks 10–15 mm. thick, very irregularly contorted or with many short, knob-like or stubby lateral branches; seeds 17–18 mm. broad . . . . . . . . . . . . . . . . . . . . . . . . . . 1. D. QUATERNATA

a. The lowest leaves in whorls of 3 (rarely 4) or all the leaves alternate; dried underground rootstock 5–8 (up to 10) mm. thick, comparatively straight and linear with few or no lateral branches (the older rootstocks with some lateral branches); seeds 7–12 mm. broad . . . . . . . . 2. D. VILLOSA

1. **Dioscorea quaternata** (Walt.) J. F. Gmel. Yam Map 658

*Dioscorea quaternata* var. *glauca* (Muhl.) Fern. *Dioscorea glauca* Muhl.

Flowers late April–early June.

Occurs in rich or rocky woods along bluffs, on talus slopes, and in thickets. Eastern and southern Ozark region, where found north to St. Louis and Crawford counties, and west to Texas and Taney counties.

Ranges from Pennsylvania to Indiana, Illinois, and Missouri, south to Florida, Louisiana, and Oklahoma.

Capsules in Missouri material vary from 2–3 cm. long. The leaves vary from green on both sides to paler green or silvery-gray beneath. Such paler-leaved types were at one time separated as var. *glauca*.

2. **Dioscorea villosa** L. Yam Map 659

Flowers late April–June.

Occurs in rich or rocky woods and thickets, in ravines, on slopes, at the base of bluffs, and in valleys.

Missouri material is represented by two variations:

Nerves on the lower side of the leaf-blades with short hairs . . . . . . 2a. D. VILLOSA f. VILLOSA
Nerves on the lower side of the leaf-blades glabrous (without hairs) . . . . 2b. D. VILLOSA f. GLABRIFOLIA

2a. **Dioscorea villosa** L. f. **villosa** Map 659

Throughout Missouri and probably in every county. This is the commoner form in the state.

Ranges from Massachusetts to Minnesota, south to Virginia, Ohio, Tennessee, Arkansas, and Texas.

2b. **Dioscorea villosa** f. **glabrifolia** (Bartlett) Fern. Map 659

*Dioscorea villosa* var. *glabrifolia* (Bartl.) Stone [P & S]

Scattered over the state throughout the range of f. *villosa*.

In Missouri material of *D. villosa* the larger leaves may attain a length of 14 cm.

## **AMARYLLIDACEAE** (Amaryllis Family)

As in the Liliaceae, the present treatment follows that of Fernald in the eighth edition of *Gray's Manual* and of Gleason in the *New Britton & Brown Illustrated Flora*. Some botanists, following Hutchinson, as indicated in the Liliaceae section of this flora, separate the Liliaceae from the Amaryllidaceae on the basis of the type of inflorescence and presence of bracts instead of on the traditional superior *vs.* inferior ovary. As such, the Amaryllidaceae includes those plants having the inflorescence in the form of an umbel subtended by an involucre of one or more usually papery bracts or spathes; sometimes the inflorescence is reduced to a single flower, but bracts or a spathe are present. Such a treatment would include *Allium* and *Nothoscordum* in the Amaryllidaceae instead of Liliaceae. Moreover, another family, Agavaceae, is recognized by some botanists to include 1) *Agave*, traditionally placed in the Amaryllidaceae, and 2) *Yucca*, traditionally placed in the Liliaceae.

In the present work the usual arrangement, following the Engler-Prantl system, adopted by many botanists, is maintained, so *Agave* is to be found in the Amaryllidaceae, and *Yucca*, *Allium*, and *Nothoscordum* have been kept in the Liliaceae.

a. Outside of flower and flower-stalks (pedicels) hairy; foliage often hairy . . . . . . . . 4. HYPOXIS
a. All parts of the plant glabrous (without hairs) . . . . . . . . . . . . . . . . . . . . *b*
  b. Leaves mostly broadest in the lower half, conspicuously narrowed to a pointed tip, mostly lanceolate; margins (edges) of leaves usually roughened with minute teeth or projections; tall plants, the flowering stem 1–2 meters high, without a bulb; flowers or fruit arranged in a long narrow spike . . . . . . . . . . . . . . . . . . . . . . . . . 5. AGAVE
  b. Leaves of about the same width throughout with more or less parallel sides, usually linear or strap-shaped, only gradually narrowed to a short or rounded tip; margins of leaves smooth, without fine teeth; smaller plants, at most 0.7 meter high, arising from a bulb; flowers or fruit solitary or 2 or more in an umbel . . . . . . . . . . . . . . . . . . . . . . *c*
    c. Flowers or fruit sessile (without a separate stalk or pedicel), attached directly to the main flower-stem (scape); summer-flowering (July–August) native plants of the lowlands of southeastern Missouri; flower erect with entirely white perianth and crown . . . . 2. HYMENOCALLIS
    c. Flowers or fruit on separate stalks or pedicels before being connected with the main flower-stem (scape); spring-flowering (April–May) garden plants escaped from cultivation, throughout Missouri; flowers nodding or loosely spreading . . . . . . . . . . . . . . . . *d*
      d. Perianth white with green tips to the segments; flower with a bell-shaped perianth, without

660 Narcissus poeticus (Poet's Narcissus)

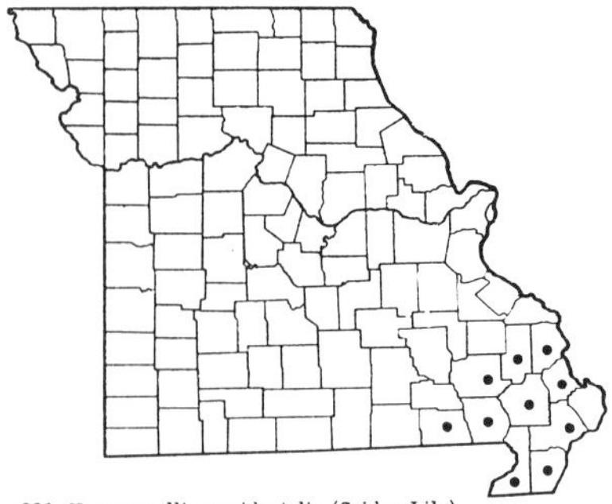

661 Hymenocallis occidentalis (Spider Lily)

662 Leucojum aestivum (Snowflake)

a slender tube or crown; flowers 2–several hanging from the main flower-stem (scape) . 3. LEUCOJUM

d. Perianth white with a red- or buff-margined yellow cup, or completely yellow, or yellow and white; flower with a slender tube and cup-like crown; flowers usually 1, rarely 2 on the main flower-stem (scape) . . . . . . . . . . . . . . . . . . 1. NARCISSUS

### 1. **Narcissus** L. Narcissus

Flower yellow, the crown about as long as or longer than the spreading perianth-segments . . . 1. N. PSEUDO-NARCISSUS

Flower with a shallow cup-like red- or buff-margined yellow crown which is much shorter than the white spreading perianth-segments . . . . . . . . . . . . . . . . . . 2. N. POETICUS

1. **Narcissus Pseudo-Narcissus** L. Daffodil Map 660

Flowers February–April.

Commonly planted in gardens throughout Missouri. Occasionally escaping into grassy open or shaded banks and thickets.

Native of Europe.

2. **Narcissus poeticus** L. Poets' Narcissus Map 660

Flowers April–May.

Commonly planted throughout Missouri, and occasionally escaping from gardens and cemeteries to roadsides, open woods, meadows, and vacant ground. Known as definitely established in Stoddard County (base of sandy wooded slopes across from Triplett cemetery, T25N, R11E, northwest ¼ sect. 6, 3½ mi. southeast of Bloomfield, April 25, 1955, *Steyermark 78305*) and probably in many other counties.

Native of Europe.

This is the last of the *Narcissus* group to flower in Missouri.

### 2. **Hymenocallis** Salisb. Spider Lily

**Hymenocallis occidentalis** (Le Conte) Kunth Spider Lily Map 661

Flowers July–late August. Flowering times have been recorded on July 23, 29, August 2 and August 20–21.

Occurs in mucky soil of bald cypress swamps, and swampy woodland in level low ground, often persisting in open fields of cut-over land. Restricted to the southeastern lowland section of the state, where it extends north to Cape Girardeau, Bollinger, and Wayne counties, and west to Ripley County.

Ranges from Georgia to Alabama, north to Kentucky, southern Indiana, southern Illinois, and southeastern Missouri.

The beautiful, showy white flowers are very fragrant, the fragrance lasting during the evening. The species does well in cultivation. It requires a long growing season and succeeds better in the southern portion of the United States. The author has transplanted it to his wildflower preserve northwest of Chicago, where the much shorter growing season curtails the time required by this species for certain flowering. The flower buds at this northern garden often appear too late or are killed by early frosts. Only twice in the past twelve years have flowers developed and matured.

The leaves usually persist during the period of development of the fruit which begins late in August

The flowering scape is compressed and 2-sided. The plant is often found in woodland dominated by *Populus heterophylla*, *Acer rubrum* var. *Drummondii*, *Cornus foemina*, *Quercus lyrata*, *Q. Michauxii*, *Fraxinus tomentosa*, and *Cephalanthus occidentalis*, associated with such plants as *Iris fulva*, *Carex gigantea*, *Scutellaria lateriflora*, *Lysimachia radicans*, and *Bignonia capreolata*.

## 3. **Leucojum** L. Snowflake

**Leucojum aestivum** L. Snowflake Map 662

Flowers April–May.

Commonly planted in gardens, and rarely escaped from cultivation. Known only from St. Charles County (Weldon Springs, Busch Wildlife Area, April, 1957, *Bruce Dowling*, in University of Missouri Herbarium).

Native of Europe. Established and naturalized also in Maine, New York, Delaware, Missouri and elsewhere.

The flowers in the Missouri specimen cited vary from 1.1 cm. in anthesis to 2.5 cm. in post-anthesis with the perianth segments 1.5 cm. long in post-anthesis. The specimen seems intermediate between typical *L. aestivum* var. *aestivum* with flowers about 2.5 cm. long and var. *pulchellum* with flowers only 1.2 cm. long.

## 4. **Hypoxis** L. Star Grass

**Hypoxis hirsuta** (L.) Coville Yellow Star Grass Map 663

Flowers mostly April–May, occasionally later until October.

Occurs in prairies, meadows, rocky glades, and dry open woods, in acid soils. Common throughout Missouri, except absent in most of the southeastern lowland section.

Ranges from Florida to Texas, north to Maine, New York, Ohio, Indiana, Wisconsin, Manitoba, and North Dakota.

The following 2 variations occur in Missouri:

Main flower-stem (scape), flower-stalks (pedicels), outside of sepals and petals, and fruit sparsely or short-hairy, becoming nearly hairless (glabrate) a. H. HIRSUTA f. HIRSUTA

Main flower-stem (scape), flower-stalks (pedicels), outside of sepals and petals, and fruit densely or noticeably hairy with crowded hairs 3–4 mm. long b. H. HIRSUTA f. VILLOSISSIMA

a. **Hypoxis hirsuta** f. **hirsuta** Map 663

This is the rare form in the state and is known only from St. Louis and Marion counties.

b. **Hypoxis hirsuta** f. **villosissima** Fern. Map 663

This is the common form in Missouri.

*Excluded Species*

**Hypoxis leptocarpa** Engelm.

*Hypoxis hirsuta* var. *leptocarpa* (Engelm.) Brackett [P & S]

This was recorded by Palmer and Steyermark in their *Annotated Catalogue* based on a specimen from Ste. Genevieve County, southeastern Missouri (low woods along Jonca Creek, May 30, 1933, *Steyermark 8800*). It shows a sparsely pubescent calyx, but generally pubescent pedicels and peduncles, and is best placed with *H. hirsuta* f. *villosissima*.

## 5. **Agave** L. American Aloe

**Agave virginica** L. False Aloe Map 665

Also called Rattlesnake-master, a name used likewise for *Eryngium yuccifolium*, a member of the Umbelliferae or Carrot Family.

Flowers late June–August.

Occurs on rocky open glades, usually of limestone, and in rocky chert or sandy open woods in mostly dry upland ridges or flats, occasionally in low ground along streams; usually in alkaline soils on limestone, but also found in acid leached sandy and cherty soils. Restricted to the Ozark section of southern and east-central Missouri, where it extends north to St. Louis, Franklin, Phelps, Pulaski, Laclede, Greene, and Jasper counties.

Ranges from Florida to Texas, north to North Carolina, West Virginia, Ohio, Indiana, Illinois, and Missouri.

Missouri material is represented by two variations:

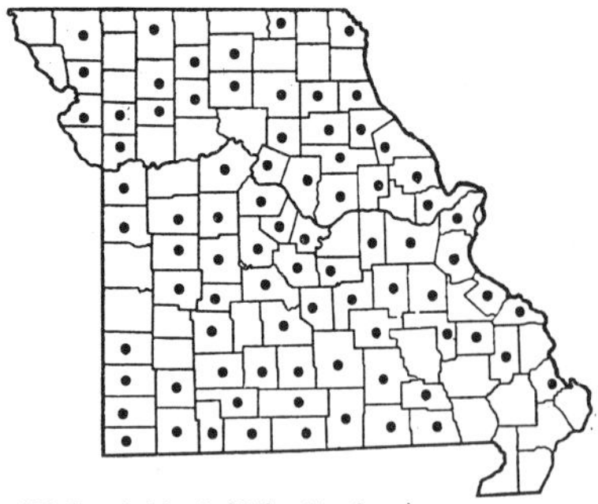

663 Hypoxis hirsuta (Yellow Star Grass)

• • •
664 Excluded species

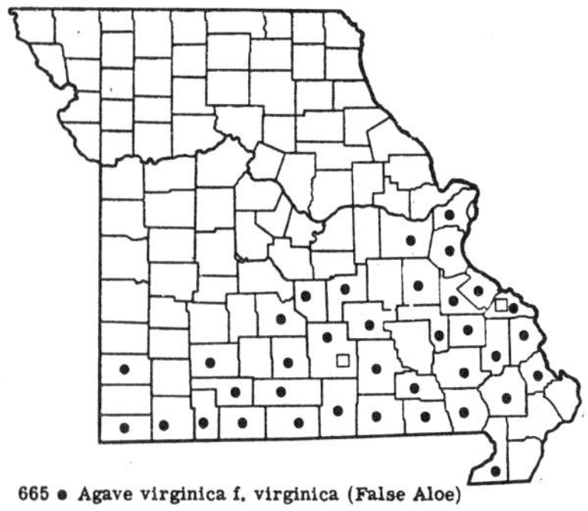

665 • Agave virginica f. virginica (False Aloe)
665 □ Agave virginica f. tigrina

666 • Iris cristata (Crested Iris)
666 □ Iris cristata var. alba

Leaves green throughout or mostly so . . . A. VIRGINICA f. VIRGINICA
Leaves with some purplish spots or mottling . A. VIRGINICA f. TIGRINA

**Agave virginica** L. f. **virginica** Map 665
*Agave virginica* L. [G, BB, P & S]
*Agave lata* Shinners [Shinners]
This is the common type encountered in Missouri.

**Agave virginica** f. **tigrina** (Engelm.) Palmer & Steyerm. Map 665

This form occurs infrequently within the Missouri range, and is definitely known only from Perry and Texas counties.

Shinners separates (*Spring Flora of Dallas-Fort Worth Area, Texas*, p. 92, 1958) *A. virginica* from *A. lata* as follows:

Leaves 12–18 cm. long, 2–7 cm. wide; perianth (including ovary) 2.6–3.5 cm. long; anthers 1.3–1.5 cm. long . . . . . . . . A. LATA
Leaves 12–30 cm. long, 1–4.5 cm. wide; perianth (including ovary) 2–2.3 cm. long; anthers 0.8–1 cm. long. . . . . . . . . . A. VIRGINICA

On the basis of measurements of both living and dried material, Missouri material combines the characters of *A. virginica* and *A. lata* to such an extent that no correlation can be found, broad-leaved individuals with smaller flowers and narrow-leaved individuals with larger flowers occurring throughout the range. It is possible, therefore, to find a more or less complete series of overlapping in the characters given by Shinners, and, although specimens answering in part to the description of *A. lata* are found in Missouri, they cannot be accepted, at least for Missouri, as significant variations.

The flowers are fragrant, possessing a scent reminiscent of Easter Lily. The perianth is mainly greenish-yellow tinged on the outside of the lanceolate erect lobes with pale madder-brown or brick color. The filaments are yellow and the anthers gray or straw-colored.

## IRIDACEAE (Iris Family)

a. Flowers with the sepals and petals much different in shape, size, and position, the 3 petals usually upwardly pointing or spreading, the 3 sepals drooping or downward spreading; the 3 style branches large, broad, and petal-like, hiding the stamens; leaves mostly 8–30 mm. broad . . . . . . 1. IRIS
a. Flowers with all 6 divisions regularly spread outward and more or less similar in shape, size, and position; the style branches or style not large, broad, nor petal-like; leaves mainly 1–11 mm. wide, or when wider, the roots and rhizome orange, the fruiting heads resembling a blackberry, and the perianth orange-red or salmon with brown-purple spots. . . . . . . . . . . . . . . . *b*
  b. Plant with iris-like leaves 16–25 mm. broad; roots and rhizome orange; perianth orange-red or salmon with brown-purple spots; fruit blackberry-like with fleshy black seeds. . . . . 3. BELAMCANDA
  b. Leaves not iris-like, much narrower, 1–11 mm. broad; roots or underground parts not orange; perianth blue, purple, white, or rarely yellow; fruit not as above . . . . . . . . . . . . *c*

Plate no. 120. **1.** Narcissus poeticus, × 2/7. **2.** Hypoxis hirsuta, × 2/7; Details from Small, The New York Botanical Garden. **3.** Narcissus Pseudonarcissus, × 2/7; After Gleason, The New York Botanical Garden. **4.** Hymenocallis occidentalis, × 2/7; Details from Small, The New York Botanical Garden. **5.** Leucojum aestivum, × 2/7; After Gleason, The New York Botanical Garden. **6.** Agave virginica, × 2/7; *a*, *b*. Leaves of two shapes; *c*. Portion of inflorescence; Details from Small, The New York Botanical Garden. **7.** Iris cristata, × 2/7. **8.** Iris pumila, × 2/7.

PLATE NO. 120

c. Stem flattened, 2-edged, arising from slender stringy or thread-like fibrous roots; leaves usually 5–many per clump, not conspicuously marked with ribs and furrows running lengthwise; flowers 10–20 mm. across, the individual segments 5–10 mm. long; mature fruiting capsule 2–6 mm. long . . . . . . . . . . . . . . . . . . . . 4. SISYRINCHIUM

c. Stem terete (round in cross-section), arising from a bulb; leaves usually 1–4 per plant, conspicuously marked with ribs and furrows running lengthwise; flowers 40–70 mm. across, the individual segments 20–35 mm. long; mature fruiting capsules 10–20 mm. long . . 2. NEMASTYLIS

## 1. **Iris** L. Iris, Flag

a. The 3 spreading to drooping sepals with a conspicuous strip or beard of long hairs down the middle . . *b*

b. Dwarf plants, the combined stems and leaves less than 25 cm. tall at flowering time; flowering in March–April . . . . . . . . . . . . . . . . . . . . . . 2. I. PUMILA

b. Plants of more moderate height, the combined stems and leaves 30–90 cm. tall at flowering time; flowering usually in late April–June . . . . . . . . . . . . . . . . . . *c*

c. Bracts (spathe-valves) which enclose buds and remain at base of open flowers mainly green on lower portion, scarious (thin, dry, and transparent) along margin and summit . . 3. I. GERMANICA

c. Bracts (spathe-valves) which enclose buds and remain at base of open flowers completely silvery-scarious (thin, dry, and transparent) throughout . . . . . . . . . 4. I. PALLIDA

a. The 3 spreading to drooping sepals without a strip or beard of long hairs down the middle. . . . . *d*

d. Plants (with the leaves) usually less than 20 cm. tall while in flower, without a definite upright stem; rootstocks slender with alternate constrictions and thickenings and creeping close to the surface of the ground; flowering mainly in late April to early May; the 3 sepals with a fringed crest . . . . . . . . . . . . . . . . . . . . . . . . 1. I. CRISTATA

d. Plants (with the leaves) 20–100 cm. tall while in flower, with a conspicuous upright stem; rootstocks thickened uniformly and not creeping close to the surface of the ground; flowering from late April–July; the 3 sepals without any crest . . . . . . . . . . . . . . . . . *e*

e. Flowers orange, brick-brown, or copper-colored, or deep yellow . . . . . . . . . . . *f*

f. Flowers deep yellow; sepals and petals of quite different lengths; ovary and capsule bluntly 3-angled; plants introduced from cultivation . . . . . . . . . . . 6. I. PSEUDACORUS

f. Flowers orange, brick-brown, or copper-colored; sepals and petals of nearly the same length, the petals slightly smaller; ovary and capsule 6-angled; native plants of the lowlands of southeastern Missouri . . . . . . . . . . . . . . . . . . . 7. I. FULVA

e. Flowers some shade of blue, lavender, or purple . . . . . . . . . . . . . . . . *g*

g. Flowers in the lower part of the plant, appearing hidden among the leaves, the leaves much longer than the flowers and flower-stem; stem (bearing the flowers or fruit) short, somewhat zigzag, mostly 10–35 cm. high; ovary and capsule 6-angled; rare plants mostly of low rich woods, at the base of wooded slopes, or bordering wooded sloughs and streams. . 8. I. BREVICAULIS

g. Flowers at the top of an elevated stem, plainly visible, most of leaves either shorter, or equal, or slightly longer than the flowering stem; stems (bearing the flowers or fruit) elongated, straight, mostly 40–100 cm. high; ovary and capsule bluntly 3-angled; common plants chiefly of open swampy or wet ground of meadows, river bottom prairies, swales along railroads, and other open wet areas, throughout Missouri . . 5. I. VIRGINICA var. SHREVEI

1. **Iris cristata** Ait. Crested Iris, Dwarf Wild Iris Map 666

Flowers April–early May.

Occurs on rocky wooded slopes and along bluffs, in cherty or limestone soils, and along sandy banks of streams. Restricted to the southeastern Ozark section in Shannon, Oregon, Reynolds, Carter, Ripley, Wayne, and Butler counties, and locally east in Scott County (*Steyermark 76890*).

Ranges from D. C. to Indiana and Missouri, south to North Carolina, Alabama, Mississippi, Arkansas, and Oklahoma.

The alternately constricted and enlarged parts of the creeping rhizome are characteristic of this species. It thrives in shaded rock gardens, preferring well-drained slopes. In the author's wildflower garden it has multiplied by the hundreds. The usual color of the flowers is lilac to lavender, the sepals having a

Plate no. 121. **1.** Iris pallida, × 2/7. **2.** Iris virginica var. Shrevei, × 2/7. **3.** Iris fulva, × 2/7. **4.** Iris brevicaulis, × 2/7. **5.** Iris Pseudacorus, × 2/7; *a.* Fruit, × 2/7; Details from Small, The New York Botanical Garden.

PLATE NO. 121

667 Iris pumila (Dwarf Iris)

668 Iris germanica (German Iris)

669 Iris pallida (Flag)

yellow or whitish crest with a white area bordered below with violet or deep purple. The petal-like style branches are pale lilac with deep lavender or purple forking or lobed tips. A rare form with the flowers all white except for the yellow crest is known as var. *alba* Dykes. It has been found in Missouri by Mr. Bill Bauer on property adjoining Big Spring State Park, Carter County, in 1937, but has not been recorded from any other locality in the state.

2. **Iris pumila** L. Dwarf Iris — Map 667

Flowers late March–April.

Commonly planted in rock and other gardens throughout Missouri, but rarely escaped from cultivation. Known definitely as established in Washington and Boone (waste fields west of Columbia, April 16, 1902, *Paul Hayhurst*, in University of Missouri Herbarium) counties.

Native to Europe and Asia Minor.

The usual color of the flower is a dark reddish-purple with a bluish or yellow beard. Various other colors may occur, including one with yellow flowers (var. *flaviflora* Fuss).

3. **Iris germanica** L. German Iris, Blue Flag — Map 668

Flowers April–June.

Commonly grown as a garden flower throughout the state, but rarely naturalized.

Known definitely as an escape from cultivation in St. Louis (Burlington R. R., north of E. Grand Ave., St. Louis, April 30, 1955, *Muehlenbach 530*), Lawrence (*Palmer 55490*), and Butler (*Huron Smith 629*) counties.

Native to Europe.

Numerous garden varieties have originated from this species, which is one of the sources of orris-root used as an ingredient in perfumes. The flowers are usually lilac to purple, varying to white, the sepals usually a dark purple with a yellow beard.

The Muehlenbach specimen has been interpreted by Dr. Edgar Anderson as probably representing a hybrid between *I. pallida* and *I. variegata* L. The latter species is considered to be the probable ancestor of much of the *I. germanica* group.

4. **Iris pallida** Lam. Flag — Map 669

Flowers April–June.

Like *I. germanica*, commonly grown as a garden flower throughout Missouri. Rarely escaping and established in Washington, Boone, and Vernon counties.

Native of Europe and Asia.

The flowers are purple, varying to white with a yellow beard. This species and *I. germanica* are often confused. From the rootstocks of this species is obtained an essential oil, which has an odor of violets (*Viola*) and often used in a tincture of orris-root to impart a fragrance to violet perfume, while the dried root is the basis of violet face powder.

5. **Iris virginica** L. var. **Shrevei** (Small) E. Anders. Southern Blue Flag — Map 670

*Iris Shrevei* Small [BB]

*Iris virginica* [of P & S], not L.

Flowers late May–early July.

Occurs mostly in open swampy or wet ground of meadows, river bottom prairies, swales along railroads, ditches, and borders of ponds and sloughs of river valleys. Common throughout northern and central Missouri and scattered elsewhere in the state; not recorded from a large sector of western Missouri south of the Missouri River.

Typical *I. virginica* var. *virginica* has capsules which average shorter, 3–7 cm. instead of 5–10 cm. long, and they are ovoid or ellipsoid instead of long-cylindric as in var. *Shrevei*.

Some persons are reported as getting dermatitis from handling the rootstocks and other parts of this species. The rootstock possesses an irritant substance which, when eaten, can cause complications in the digestive tract, leading to mild cases of poisoning.

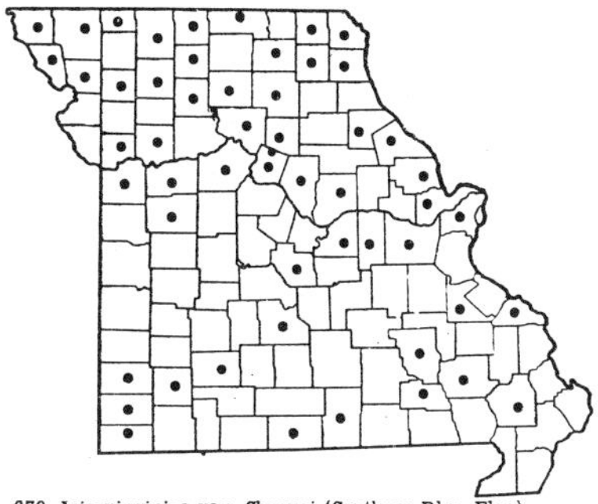

670 Iris virginica var. Shrevei (Southern Blue Flag)

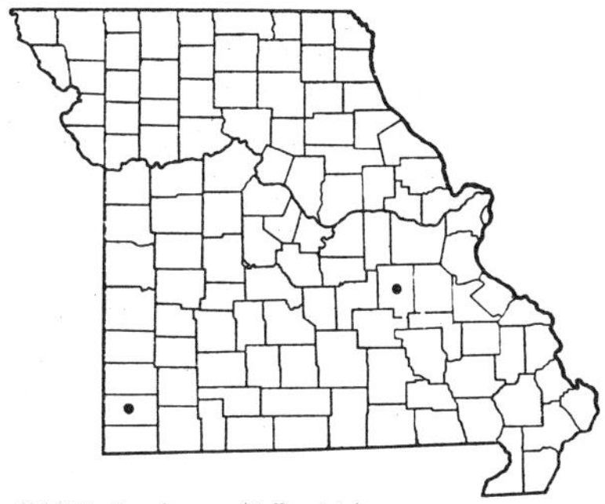

671 Iris Pseudacorus (Yellow Iris)

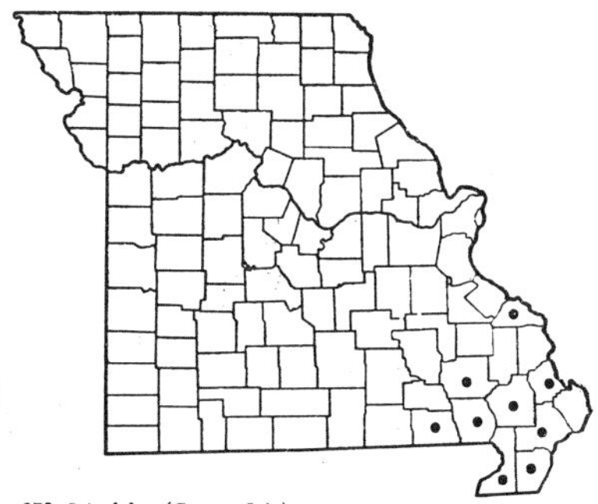

672 Iris fulva (Copper Iris)

Grazing animals do not eat much of this plant because of its sharp taste.

6. **Iris Pseudacorus** L. Yellow Iris, Yellow Flag
Map 671

Flowers May–July.

Planted in gardens, often around pools or low wet ground, and rarely escaping or becoming naturalized in Missouri. Known definitely established only in Crawford (low woods along creek below Blue Spring, 2 mi. southeast of Bourbon, August 12, 1939, *Steyermark 27908*) and Newton (along rocky margins of Hickory Creek, Neosho, May 19, 1954, *Palmer 57567*) counties, where it grows in wet ground and low moist woods.

Native of Europe.

This is a handsome and striking species with tall erect leaves a meter or more high. The tall stems bearing the bright yellow flowers may equal or are shorter than the leaves.

7. **Iris fulva** Ker Copper Iris, Red Iris Map 672

Flowers late April–June.

Occurs in bald cypress swamps, borders of sloughs, bayous, and ponds, along open ditches and drainage canals, and swampy woodland. Restricted to the lowland section of southeastern Missouri, extending north to Perry County and west to Wayne and Ripley counties.

Ranges from Georgia to Louisiana, north to southern Illinois and southeastern Missouri.

This iris is a striking species when in flower, often growing nearly submerged in the water during the flowering season. The soil often becomes dried up later on in the year.

At the author's wildflower preserve this species has prospered and produced blossoms numerous times in the fifteen years it has been growing at this locality, situated four hundred miles north of its usual range.

8. **Iris brevicaulis** Raf. Short-stemmed Iris
Map 673

*Iris foliosa* Mack. & Bush [P & S]

Flowers late May–early July.

Occurs usually in woodland habitats, of low rich valley woods bordering sloughs or streams or at the base of wooded slopes in ravines. Rare in the state, where known from portions of eastern, northern, and central Missouri; absent from most of the Ozark region and from the southwestern unglaciated prairie section.

Ranges from Alabama to Texas, north to Ontario, Ohio, Indiana, Illinois, Missouri, and Kansas.

This species may be somewhat more common than appears at present evident from the known records, since it often grows vegetatively and does not produce flowers. It is, therefore, probable that it has been overlooked by collectors when not in flower.

The flowers have a deeper blue or blue-purple than those of the other native species in the state. The yellow crest area on the sepals is prominently surrounded by a broad white zone which gives way to the remainder of the blue-purple expanded portion of the sepal. The expanded style branches are greenish white with a conspicuous lavender stripe down the middle ending in the lavender somewhat ovate crests.

## 2. **Nemastylis** Nutt. Nemastylis

Dr. R. C. Foster of Harvard University has identified the Missouri material and has furnished some of the data in the following key:

Flowering April–May; stems of leaves well-developed, mainly 5–11 mm. wide; filaments free or only

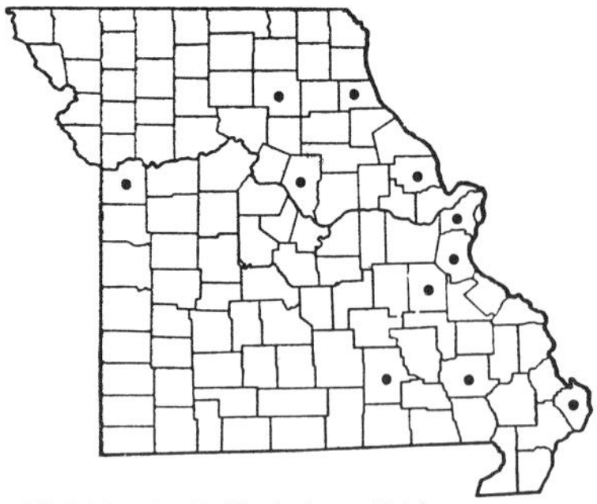

673 Iris brevicaulis (Short-stemmed Iris)

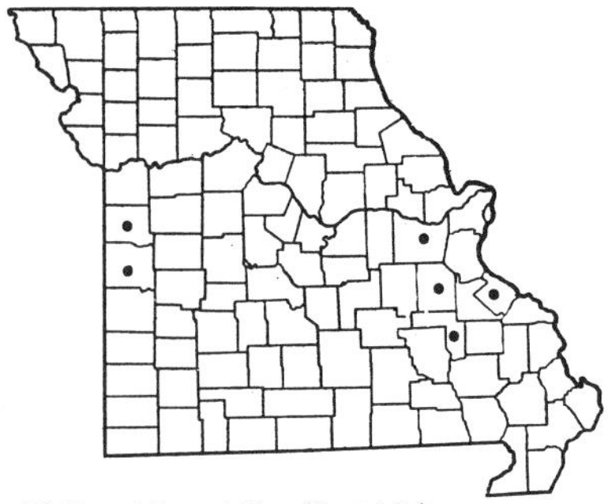

674 Nemastylis geminiflora (Nemastylis)

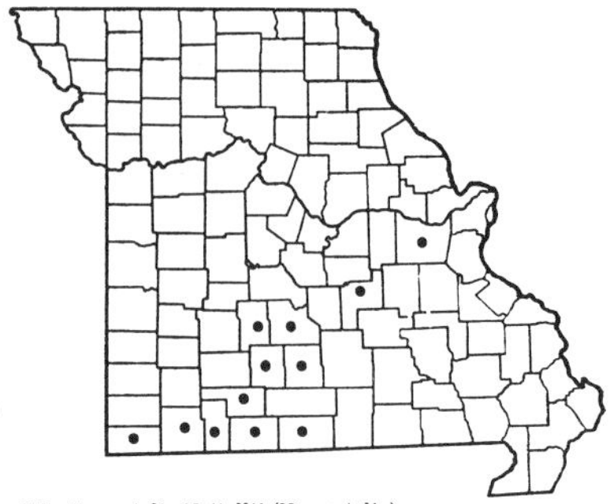

675 Nemastylis Nuttallii (Nemastylis)

slightly joined (connate) at base; flowers blue to lilac-blue, 4–7 cm. broad, opening in the morning and closing before sundown . . . . . . . . . . . . . . . . . . . . . . 1. N. GEMINIFLORA
Flowering June–July; stems of leaves reduced, or, if developed, narrow, 1–4 (–6) mm. wide; filaments completely or partially united; flowers deep lavender or pale purple, smaller, 2–3.5 cm. broad (smaller in dried specimens), opening from 5–7 in afternoon and staying open in the evening. . . . 2. N. NUTTALLII

1. **Nemastylis geminiflora** Nutt Map 674
*Nemastylis acuta* (Bart.) Herb. [P & S, BB]
Flowers April 20–May 15.

Occurs on limestone glades and calcareous prairies, mostly in the eastern section of the Ozarks in Ste. Genevieve, Washington, Franklin, and Iron counties, and locally west to Cass County.

Ranges from Louisiana and Texas, north to Tennessee, Missouri, and Kansas.

2. **Nemastylis Nuttallii** Pickering Map 675
*Nemastylis coelestina* (Bartr. ex Willd.) Nutt. as to name only;
*Nemastylis acuta* of some authors in part [P & S], not Herb.
Flowers mainly June 1–July (flowering dates June 1, June 21, 22, 24, and early July).

Occurs in prairies, limestone glades near *Juniperus* trees, cherty open woods near limestone slopes above bluffs along streams, cherty slopes just above contact with limestone glades, and post oak-white oak-*Andropogon* covered slopes along bluffs, mostly in the southwestern sector of the Ozarks from McDonald County east to Ozark County and north to Franklin (*Steyermark 41039*), Phelps (*Steyermark 70906, 71752*), Laclede, and Dallas counties.

Ranges from Missouri to Arkansas and Oklahoma.

This species has smaller, more lavender to purple flowers, and blooms later in the season (June 1, June 21, 22, 24, and July flowering dates) than *N. geminiflora*.

According to the author's field observations, this species may occur more frequently in more acid soils than does *N. geminiflora*. The latter species in the eastern Ozarks is on calcareous open glades, whereas *N. Nuttallii* has been collected more frequently in cherty soils in grassy oak chert ridges near the zone of contact of the chert with the limestone. It is also found on the limestone glade proper frequently with red cedar and with other limestone glade plants, such as *Penstemon Cobaea* var. *purpureus*, *Parthenium hispidum*, and *Baptisia minor*, so that it may occur in both situations. At the virgin prairie in Phelps County, where it was common, the soil is slightly acid. More field work needs to be done regarding the soil differences of these species.

## 3. **Belamcanda** Adans. Blackberry Lily

**Belamcanda chinensis** (L.) DC. Map 676
Flowers early July–August.

Occurs in rocky open woods, thickets, limestone glades, limestone ledges and pockets of bluffs. Throughout Missouri, where it has escaped from cultivation in old gardens and become established.

Plate no. 122. **1.** Sisyrinchium Bermudiana, × 2/7; *a.* Fruiting plant, × 2/7. **2.** Sisyrinchium atlanticum, × 2/7; Details from Small, The New York Botanical Garden. **3.** Sisyrinchium campestre, × 2/7; Details from Small, The New York Botanical Garden. **4.** Nemastylis geminiflora, × 2/7; Details from Small, The New York Botanical Garden. **5.** Belamcanda chinensis, × 2/7; *a.* Fruit, × 2/7; After Gleason, details from Small, The New York Botanical Garden. **6.** Sisyrinchium albidum, × 2/7.

PLATE NO. 122

676 Belamcanda chinensis (Blackberry Lily)

677 Sisyrinchium albidum (Blue-eyed Grass)

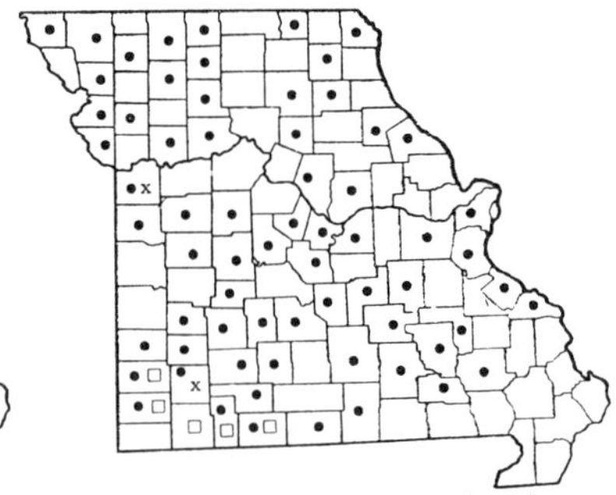

678 • Sisyrinchium campestre f. campestre (Prairie Blue-eyed Grass)
678 □ Sisyrinchium campestre f. kansanum
678 x Sisyrinchium campestre f. flaviflorum

Native of Asia; introduced into the United States, where it ranges from Connecticut to Nebraska, south to Georgia and Texas.

The leaves are very similar to those of iris, and the entire plant when not in flower greatly resembles that of the vesper iris (*Iris dichotoma* Pall.), both having an elongated stem up to a meter tall with fan-shaped leaves crowded well up the stem. When in leaf only, the orange roots and rhizomes of this species will distinguish Belamcanda from any *Iris* species encountered in Missouri.

### 4. **Sisyrinchium** L. Blue-eyed Grass

a. Stems branched above; the flowers or fruits and the 2 thinner spathes which surround them are located at the top of a long peduncle which arises from the side of a leaf-like bract . . . . . . . . *b*
  b. Stems and leaves silvery- or blue-green (glaucous), not blackened in drying; stems narrow-margined, 1–3 mm. wide; peduncles thread-like, delicate; perianth violet-blue; inner bract of spathe 1–1.5 cm. long; capsules 3–4.5 mm. long . . . . . . . . . . . . 4. S. ATLANTICUM
  b. Stems and leaves deep green or dark green, blackened in drying; stems broadly and conspicuously winged, 2–6 mm. wide; peduncles flat and winged; perianth pale blue to violet; inner bract of spathe 1.5–2 cm. long; capsule 4–6 mm. long . . . . . . . . . . . . 3. S. BERMUDIANA
a. Stem simple and unbranched; the flowers or fruits and the 1 or 2 thinner spathes which surround them arise directly from the top of the stem with a much longer green bract surrounding them. . . *c*
  c. Flowers yellow . . . . . . . . . . . . . . . . . 2. S. CAMPESTRE f. FLAVIFLORUM
  c. Flowers white, lilac, purplish, or blue . . . . . . . . . . . . . . . . . . . *d*
    d. The longer green bract encloses 2 thinner spathes and inner bracts which surround the flowers . . . . . . . . . . . . . . . . . . . . . . . . . 1. S. ALBIDUM
    d. The longer green bract encloses only 1 thinner spathe which surrounds the flower . 2. S. CAMPESTRE

1. **Sisyrinchium albidum** Raf.
Blue-eyed Grass Map 677
Flowers May–June.

Occurs in meadows, prairies, open grassy swales along railroads, and open woods. Known only in eastern Missouri from Scott County north to St. Louis and Franklin counties, and locally northwest into Randolph County.

Ranges from Georgia and Alabama to Louisiana, north to North Carolina, Ontario, Ohio, Michigan, Wisconsin, and Missouri; also recorded northeastward to New York.

The flowers vary from white to pale violet, and the pale green or bluish-green leaves and stems vary from 1–3 mm. wide.

2. **Sisyrinchium campestre** Bickn.
Prairie Blue-eyed Grass Map 678
Flowers April–June.

Occurs in prairies, meadows, grassy places along railroads, rocky open woods, and rocky glades. Throughout Missouri.

Ranges from Wisconsin to Manitoba, south to Illinois, Louisiana, and Texas.

Missouri material may be divided into the following 3 variations:

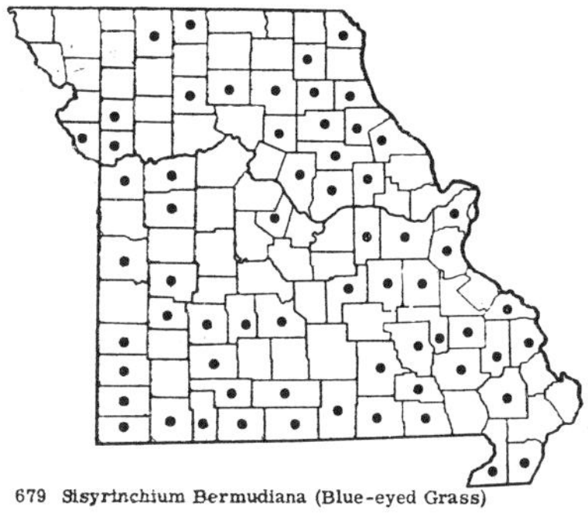

679 Sisyrinchium Bermudiana (Blue-eyed Grass)

680 Sisyrinchium atlanticum

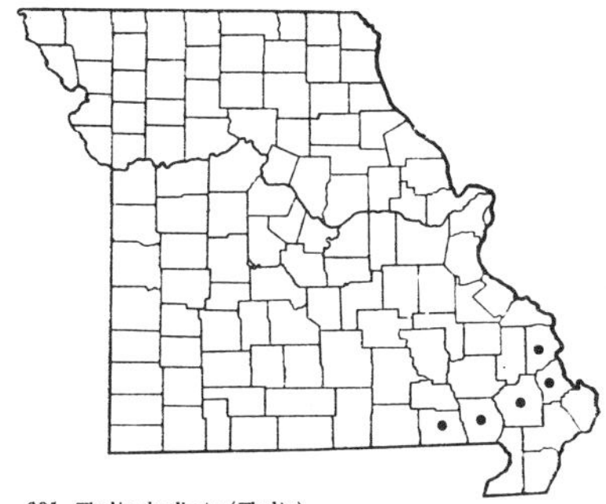

681 Thalia dealbata (Thalia)

a. Flowers yellow . 2c. S. CAMPESTRE f. FLAVIFLORUM
a. Flowers white, lilac, purplish, or blue . . . *b*
 b. Flowers white . 2b. S. CAMPESTRE f. KANSANUM
 b. Flowers bluish or lilac . . 2a. S. CAMPESTRE f. CAMPESTRE

2a. **Sisyrinchium campestre** f. **campestre** Map 678

*Sisyrinchium campestre* Bickn. [G, BB, P & S]

This is the common type encountered in the state.

2b. **Sisyrinchium campestre** f. **kansanum** (Bickn.) Steyerm. Map 678

*Sisyrinchium kansanum* (Bicknell) Alexander [BB]; *Sisyrinchium campestre* var. *kansanum* Bicknell [P & S].

This white-flowered form is known from several counties in southwestern Missouri, but is not known elsewhere in the state.

In Gleason's *New Illustrated Flora* (1: 451. 1952) it is treated as a distinct species, distinguished by the glabrous instead of scabridulous (minutely rough-hairy) spathe-values. I have not been able to use this character in distinguishing the southwestern Missouri material from the other *S. campestre* in the state, and have relegated it to a form under *S. campestre*.

2c. **Sisyrinchium campestre** f. **flaviflorum** (Bicknell) Steyermark Map 678

Known originally only from post oak woods in Jackson County, west-central Missouri, where collected by Bush. The only other known collection is one recently made in Lawrence County, southwestern Missouri (swales in upland prairie, 5 mi. east of Sarcoxie, May 10, 1951, *Palmer 51785*).

3. **Sisyrinchium Bermudiana** L. em. Fernald Blue-eyed Grass Map 679

*Sisyrinchium angustifolium* Mill. [G]
*Sisyrinchium graminoides* Bicknell [BB, P & S]
Flowers May–July.

Occurs mostly in woodland, frequently in wooded valleys in moist soil along streams or on wooded slopes and thickets, less frequently in more open situations. Found throughout Missouri, but absent from most of northwestern Missouri and some northern and central counties.

Ranges from Florida to Texas, north to Newfoundland, Quebec, Ontario, Ohio, Indiana, Illinois, Missouri, and Kansas.

Dr. Shinners (Rh. 59: 159. 1957) has recently shown that the correct name for this species is *S. Bermudiana*.

4. **Sisyrinchium atlanticum** Bickn. Blue-eyed Grass Map 680

Flowers late April–June 19.

Occurs in acid soils of low sandy prairies, wet prairie swales of upland, and wet upland open meadows bordering sink-hole ponds. Rare and scattered in the southeastern third of the state, in Scott, Mississippi (*Steyermark 10294*), Phelps (*Steyermark 71633*), and Howell (*Steyermark 40028*) counties.

Ranges from Florida to Mississippi, north to Nova Scotia, Maine, New Hampshire, Vermont, and southeastern New York, and in the interior in Ohio, Indiana Michigan, Illinois, and Missouri.

This species is typically one along or near the coastal plain. The stations where it is found in Missouri are from localities whose flora shows ancient connections with the Atlantic and gulf coastal plain vegetation.

The conspicuously forking stems with long slender peduncles and the very glaucous leaves and stems are notable features of this species.

## Order **SCITAMINALES**

### Fam. **MARANTACEAE** (Arrowroot Family)

### **Thalia** L.

**Thalia dealbata** Roscoe Thalia Map 681

Flowers July 7–August 25.

Occurs along drainage canals, open swamps, and natural lakes in the lowland section of southeastern Missouri, where it is known from Ripley, Butler, Stoddard, Scott, and Cape Girardeau counties.

Ranges from Florida to Texas, north to South Carolina and southeastern Missouri.

This is often grown as an ornamental plant for outdoor pools and pond margins in St. Louis and elsewhere in the state, as well as a greenhouse plant. Its large ovate-lanceolate, long-petioled leaves are covered with a white powdery coating which covers all parts of the plant. The flowers are blue and purple and arranged in an open panicle.

Earlier collectors in Missouri had found this plant only in swamps around Poplar Bluff and Neeleyville, in Butler County. More intensive collecting in the state has revealed a wider distribution in the southeastern lowland section where it reaches the known northwestern limits of natural range. Its natural habitat is in swamps and lakes associated with *Zizaniopsis miliacea*, *Ludwigia glandulosa*, *Hydrolea ovata*, *Polygonum hydropiperoides* var. *opelousanum*, and *Hypericum Walteri*. Such natural lakes occur in those sections of the southeastern lowland forest where Bald Cypress, Swamp Tupelo, Swamp Cottonwood, and Swamp Red Maple are abundant.

## Order **ORCHIDALES**

### **ORCHIDACEAE** (Orchid Family)

a. No leaves present at flowering time . . . . . . . . . . . . . . . . . . . . . . . *b*

b. Flowers all or nearly all white or creamy-white (some green or yellow may be present on the lip), in a more or less spirally twisted inflorescence; flowers sessile (without a stalk); lip (largest lower division) without ridges . . . . . . . . . . . . . . . . . . . . . . 9. SPIRANTHES

b. Flower of other colors, not spirally twisted; flower on short or long stalks; lip (largest lower division) with 1 or more ridges . . . . . . . . . . . . . . . . . . . . . . *c*

c. Sepals about 20 mm. long; lip about 15 mm. long, with 5–7 conspicuous ridges; plants blooming July–late August; pollen masses 8 . . . . . . . . . . . . 15. HEXALECTRIS

c. Sepals 3–15 mm. long; lip 2.5–12 mm. long, with 1–3 low ridges; plants blooming February-June or July–October; pollen masses 4 . . . . . . . . . . . . . . . . . . *d*

d. Sepals 10–15 mm. long; lip 10–12 mm. long; stem originating from a solid round tuber (corm) . . . . . . . . . . . . . . . . . . . . . . . . 14. APLECTRUM

d. Sepals 3–8 mm. long; lip 2.5–6 mm. long; stem originating from a mass of coral-like or shortly branched fleshy roots . . . . . . . . . . . . . . . . 11. CORALLORHIZA

a. Leaves present at flowering time . . . . . . . . . . . . . . . . . . . . . . . *e*

e. Only 1 well-developed leaf present, either on the stem or at the base of the plant. . . . . . . . *f*

f. Leaf present only in autumn, winter, and early spring, disappearing before or just after flowering; leaf dark green, 4 cm. or more wide . . . . . . . . . . 14. APLECTRUM

f. Leaf present during and after flowering; leaf pale or grass green, 0.2–3 cm. wide . . . . . *g*

g. Leaf only $1\frac{1}{4}$–2 times as long as broad, 3–6 cm. long; flowers minute, greenish; sepals 1.3–1.5 (–2) mm. long, the lip (largest petal) 2–2.5 mm. long; lateral petals linear, thread-like, 1 mm. long . . . . . . . . . . . . . . . . . . . . . . 12. MALAXIS

g. Leaf mostly 3–20 times as long as broad, 3–40 cm. long; flowers rose-colored, lilac or pink with sepals 15–22 mm. long and lip (largest petal) 15–20 mm. long, or the flowers greenish with white or yellow and with sepals 2–5 mm. long and lip 3–5 mm. long; lateral petals elliptic, lanceolate, or broadly ovate, 2–22 mm. long . . . . . . . . . . . . . *h*

h. Flowers greenish with white or yellow; sepals 2–5 mm. long; lip 3–5 mm. long . . 3. HABENARIA CLAVELLATA

h. Flowers mainly rose, pink, or lavender; sepals 15–22 mm. long; lip 15–20 mm. long . . . *i*

i. Leaf grass-like, linear, mainly 2–20 mm. wide, mainly 12–30 cm. long; sepals ending in an acute point . . . . . . . . . . . . . . . . . . . 7. CALOPOGON

i. Leaf lanceolate to oblong-elliptic or obovate, mainly 10–30 mm. wide, 2–9 cm. long; sepals blunt or rounded . . . . . . . . . . . . . . . . . . 4. POGONIA

e. 2 or more well-developed leaves present, either on the stem or at the base of the plant . . . . . *j*

j. 5–6 (rarely up to 10) leaves arranged in a whorl (circle) at the top of the stem. . . . 5. ISOTRIA

j. Leaves not as above, but either at the very base of the plant, or arranged alternately along the stem, or at the base as well as on the stem . . . . . . . . . . . . . . . . *k*

k. Leaves only at the base of the plant . . . . . . . . . . . . . . . . . . *l*

l. Only 2 leaves at the base of the plant . . . . . . . . . . . . . . . . *m*

m. Leaves 0.3–1.5 cm. wide; flowers mostly white or cream-colored, or with some yellow or green on the lip (largest lower division), in a more or less spirally twisted inflorescence; flowers sessile (without stalks) . . . . . . . . . . 9. SPIRANTHES

m. Leaves 2–7 cm. wide; flowers of other combinations, either lavender and white, or purple-brown or lavender with green or yellow-green; flowers on short or long stalks (pedicels) . . . . . . . . . . . . . . . . . . . . . . . *n*

n. Upper half of flower purple or lavender, lower half (lip) white; leaves very dark green; flower-stem arising from a cluster of long fleshy roots . . . . . 2. ORCHIS

n. Flowers either mostly brown-lavender or purple-brown with some parts greenish or flowers yellowish-green throughout; leaves olive- or grass green; flower-stem swollen at the base, arising from a rounded bulbous base . . . . . . 13. LIPARIS

l. Several leaves arranged in a rosette at the base. . . . . . . . . . . . . . *o*

o. Upper leaf-surface with dark green background marked with gray-white nerves and cross veins . . . . . . . . . . . . . . . . . . . . 10. GOODYERA

o. Upper leaf surface plain green without a network of white venation . . . 9. SPIRANTHES

k. Leaves mainly on stem, alternately arranged, or both on the stem and at the base of the plant . . . . . . . . . . . . . . . . . . . . . . . . . . *p*

p. All the leaves of stem very small and ovate, 6–18 mm. long, 5–8 mm. wide; flowers white and lavender-pink . . . . . . . . . . . . . . . . . . 6. TRIPHORA

p. Without the above combination or lacking the above characters . . . . . . . . . *q*

q. Flowers numerous in a more or less spirally twisted inflorescence, white or creamy white, the lip with yellow, green, or white; lip (largest lower petal) 3–15 mm. long; sepals 3–14 mm. long; flowers sessile (without stalks) . . . . . . . 9. SPIRANTHES

q. Without the above combination or lacking the above characters . . . . . . . . . *r*

r. Flowers 1–3 to a stem . . . . . . . . . . . . . . . . . . . . . *s*

s. Flowers with a large showy inflated balloon-like pouch (the lip) which is 18–50 mm. long; stem or leaves or both hairy . . . . . . . . . 1. CYPRIPEDIUM

s. Without the above combination; lip not inflated or a balloon-like pouch, 17–22 mm. long; stem and leaves glabrous (without hairs) . . . . . . 4. POGONIA

r. Flowers 5–many to a stem . . . . . . . . . . . . . . . . . . . . *t*

t. Flowers green and purple, without any spur (downward prolonged back part of lip); lip slightly inflated or sac-like at its base; sepals and petals ending in an acute or sharp point; leaves spreading out more or less horizontally, shining, strongly nerved . . . . . . . . . . . . . . . . . . . . 8. EPIPACTIS

t. Without the above combination or lacking the above characters; flower with a definite spur; lip not inflated or sac-like; sepals often blunt or rounded; leaves mainly erect or ascending . . . . . . . . . . . . 3. HABENARIA

## 1. **Cypripedium** L. Lady-slipper

a. Sepals and lateral petals white, blunt at tip, shorter than the lip; lip rose or rose-purplish with white; plants of moist crevices of limestone bluffs along streams or in wet calcareous meadows in the Ozarks, or in rich moist woods of northern Missouri . . . . . . . . . . . . . . . . 3. C. REGINAE

a. Sepals and lateral petals purplish-brown, lavender and green or greenish-yellow, slender- or long-pointed at tip, the sepals longer than the lip, the lateral petals equaling or longer than the lip; lip yellow or white with purple spots or lines toward the inside; plants of acid soils, most frequently along the upper half of steep north- or east-facing wooded oak-hickory or pine-oak slopes, less frequently on lower wooded slopes or moist prairies . . . . . . . . . . . . . . . . . . *b*

b. Lip white, with purple dots or stripes inside, 1.8–2.3 cm. long; lateral petals 2–3 cm. long; upper sepal 2–3 cm. long; leaves 1–4 cm. wide, usually 3–4 on a stem; very rare in Missouri. 2. C. CANDIDUM

b. Lip yellow, 2–5 cm. long; lateral petals 2.5–9 cm. long; upper sepal 2.5–7 cm. long; leaves

682 o Cypripedium Calceolus var. parviflorum (Small Yellow Lady-slipper)

682 x Cypripedium Calceolus var. pubescens (Large Yellow Lady-slipper)

683 Cypripedium candidum (Small White Lady-slipper)

684 Cypripedium reginae (Showy Lady-slipper)

mostly 4–12 cm. (less frequently 2 cm.) wide, mostly 4–6 (rarely 3) on a stem; throughout Missouri except absent in southeastern lowland section . . . . . . . . . . . . . . . . . *c*

c. The common type in the eastern half of the state; lip large, 3–5 cm. long; lateral petals 5–9 cm. long; sepals usually greenish-yellow with purple lines; leaves 3–4 (rarely 5) on a stem. . . 1b. C. CALCEOLUS var. PUBESCENS

c. The common type in the western half of the state, and east in the Ozarks to Washington, Dent, Shannon, and Oregon counties; lip smaller, mainly 2–3 cm. (rarely to 4) long; lateral petals 3.5–5 cm. long; sepals dull madder purple; leaves usually 4–6 on a stem . . . 1a. C. CALCEOLUS var. PARVIFLORUM

1a. **Cypripedium Calceolus** L. var. **parviflorum** (Salisb.) Fern. Small Yellow Lady-slipper Map 682

*Cypripedium parviflorum* Salisb. [P & S]

*Cypripedium Calceolus* var. *pubescens* [of BB], not (Willd.) Correll

Flowers late April–early June.

Occurs most frequently in acid soils along the upper half of steep north- or east-facing wooded oak-hickory or pine-oak slopes along streams or valleys and in ravines, less frequently on the lower slopes. Northwestern, central, and western Missouri, and locally east in Randolph, Ralls, Boone, Callaway, Crawford, Dent, Shannon, and Oregon counties.

Ranges from Newfoundland to British Columbia, south to Georgia, Tennessee, Ohio, Indiana, Illinois, Missouri, Texas, New Mexico, Utah, and Washington.

Gleason in the *New Illustrated Flora* combines this variety with var. *pubescens*.

1b. **Cypripedium Calceolus** var. **pubescens** (Willd.) Correll Large Yellow Lady-slipper Map 682

*Cypripedium parviflorum* var. *pubescens* (Willd.) Knight [P & S]

*Cypripedium Calceolus* in part [BB], not L.

Flowers late April–early June.

Similar habitats as var. *parviflorum*. Northern, central, and eastern Missouri, west to Gentry, De Kalb, Carroll, Jackson, Moniteau, Texas, and Greene counties.

Ranges from Nova Scotia and Maine to Minnesota, south to Georgia, Alabama, Tennessee, and Missouri.

The foliage turns a pale yellow-green in autumn.

As indicated by the general geographical distribution, var. *pubescens* occurs mainly in the southeastern and eastern half of Missouri, and in the eastern half of the United States, whereas var. *parviflorum* is of more western and northern range, both in Missouri and in the United States. There is intergradation between the two varieties and they are sometimes difficult to distinguish. Both are closely related to *C. Calceolus* var. *Calceolus* of Europe and Asia, and are now considered as only American varieties of this species (see Correll, Bot. Mus. Harv. Univ. Leaflets 7: 14. 1938, and Fernald, Rh. 48: 4. 1946). In their extremes the vars. *parviflorum* and *pubescens* may be rather readily distinguished, and their differences were carefully brought out by Mr. Albert M. Fuller (Bull. Pub. Mus. Milwaukee 14: 64–70. 1933. The Orchids.). The habitat differences usually indicated for the two

---

Plate no. 123. **1.** Thalia dealbata, × 2/7; After Gleason, details from Small, The New York Botanical Garden. **2.** Cypripedium Calceolus var. parviflorum, × 2/7. **3.** Cypripedium Calceolus var. pubescens, × 2/7. **4.** Cypripedium candidum, × 2/7. **5.** Orchis spectabilis, × 2/7; Details from Small, The New York Botanical Garden. **6.** Cypripedium reginae, × 2/7; Details from Small, The New York Botanical Garden.

PLATE NO. 123

varieties are not at all marked in Missouri, as both of them occupy similar woodland habitats in similar acid soils. The calcareous bogs, mossy swamp, and other wet habitats attributed to var. *parviflorum*, certainly do not apply to Missouri plants encountered.

Handling the stems, leaves, or flowers of both varieties may cause a dermatitis condition to those sensitive to the plant. The roots of these and other species of lady-slipper have been dug for their supposed medicinal value, and the plants greatly diminished in numbers thereby. As their presumed curative properties have never been fully substantiated and other drugs have been found more suitable, it is hoped that the plants will be protected against future exploitation and disturbance. In general both varieties are more abundant and occur in greater numbers with more individuals to a stand in northern Missouri than in the Ozark region.

Of the native lady-slippers, both the yellow types do well in wildflower gardens, especially in shaded acid well-drained situations provided with oak leaf mulch.

2. **Cypripedium candidum** Muhl.
Small White Lady-slipper Map 683
Flowers late April–early June.

Occurs in wet prairies or prairie swales, and low wooded slopes bordering streams and valleys. Very rare, and known at the present time only from a colony discovered in 1947 by the author in Nodaway County, northwestern Missouri (*Steyermark 64321*). Other counties from which specimens have been collected are in St. Louis (prairies, St. Louis, May, 1838, *Riehl*), Shannon (Monteer, May 15, 1894, *Bush 545*), Atchison, and Andrew. These stations, so far as known, have all been eliminated, and the plants are not now extant there. The Nodaway County locality is a west-facing slope bordering the One Hundred and Two River. Here, on May 18, 1947, were found fifteen plants growing in a clayey loam.

Ranges from New York to North Dakota, south to New Jersey, Pennsylvania, Kentucky, and Missouri.

The flowers are fragrant and the leaves are grass-green in this species. The stems and leaves are less pubescent, and often more nearly glabrous, than in the yellow lady-slippers.

I have grown this species in my wildflower preserve for twelve years, where it has prospered and even increased in numbers of stems per plant. In the more northern sectors of its range it frequents open wet sandy, but calcareous prairies and often forms plants with numerous stems having as many as 15 stems and 15 flowers to a single clump.

3. **Cypripedium reginae** Walt. Showy Lady-slipper Map 684
Flowers May 11–June 5.

Occurs in the Ozarks in moist crevices and seepage of the lower half of mainly north-facing limestone bluffs bordering canyons and narrow cool valleys along small streams and in wet calcareous meadows, where it is found in several Ozark counties (Iron, Shannon, Texas, Howell, Douglas, and Ozark), and locally north in Putnam County, where it has been collected near Livonia from rich moist woods, according to Bush who collected it there. The author has visited the Livonia area several times in unsuccessful attempts to relocate the Bush collecting locality.

Ranges from Newfoundland to Manitoba and North Dakota, south to Georgia, Tennessee, and Missouri. A form in which the lip is entirely white, known as f. *albolabium* Fern. & Schub., has been described, but is not known in Missouri.

This is one of the most beautiful of native wild flowers. Fortunately, for its own protection and continued existence, it usually grows in remote and sometimes inaccessible situations on bluffs out of reach of unscrupulous persons. It is mainly a plant of alkaline or limestone wet soils or rocks, but can be grown in wildflower gardens in somewhat drier situations, although it attains its best growth under moist conditions.

The stems and foliage of this species are more densely covered with hairs than in the other native Missouri species. These hairs have glands which contain a substance often irritating to persons handling the plants, and causing a type of dermatitis similar to that of poison ivy.

## 2. **Orchis** L. Orchis

**Orchis spectabilis** L. Showy Orchis. Map 685
Flowers late April–June.

Occurs in rich woods, either on slopes of bluffs or ravines as well as in rich level soils in valleys or low woods of river bottoms. It has been found in northern, central, and Ozark sections of the state, but is absent from the unglaciated prairie region of the southwestern portion of the state, from some areas in the

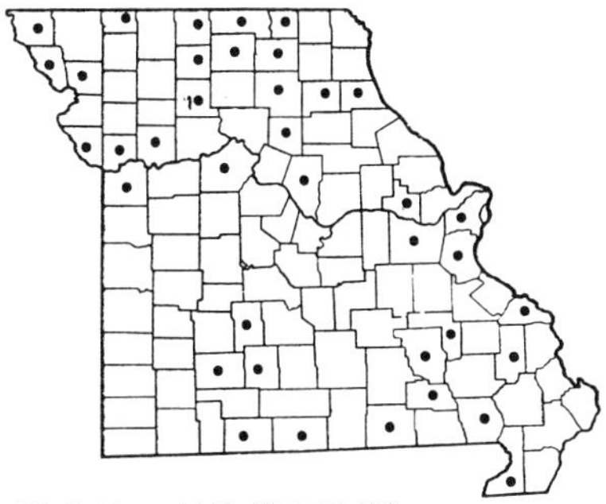

685 Orchis spectabilis (Showy Orchis)

686 Habenaria clavellata var. clavellata (Green Wood Orchid)

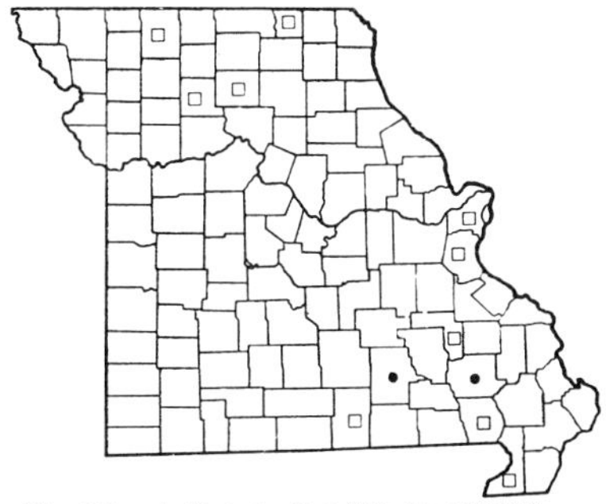

687 • Habenaria flava var. flava (Pale Green Orchid)
687 □ Habenaria flava var. herbiola (Pale Green Orchid)

Ozarks, the southeastern lowlands, and glaciated prairie sections.

Ranges from Quebec and Ontario, south to Georgia, Alabama, Tennessee, Missouri, and Kansas.

This species will grow well in a wildflower garden when provided with cool shade and rich loose soil covered with leaf-mold. The flowers are delicately fragrant.

### 3. **Habenaria** Willd. Rein-orchis, Fringed Orchis

a. The lip (largest 3rd petal) fringed, deeply lobed or cut, or toothed all along the margins . . . . . *b*
- b. Flowers orange; lip not 3-parted nor 3-lobed, the main body of the lip oblong . . . 3. H. CILIARIS
- b. Flowers white, rose, lavender, purple, yellow-green, or yellow-white with bronze; lip 3-parted or 3-lobed . . . . . *c*
  - c. Flowers white or yellow-green or greenish-white with bronze; lip deeply fringed, the hair-like fringed portions as long as or longer than the main body of the lobe itself; largest leaves 0.5–3.5 cm. wide . . . . . *d*
    - d. Flowers completely white or greenish-white, the sepals and lateral petals 8–10 mm. long; lip 15–20 mm. long; spur 20–40 mm. long; lateral petals with toothed margins . 5. H. LEUCOPHAEA
    - d. Flowers mainly yellow-green or green with white or bronze, the sepals and lateral petals 4–6 mm. long; lip 7–15 mm. long; spur 10–17 mm. long; lateral petals with entire (not toothed) margins . . . . . 4. H. LACERA
  - c. Flowers rose-purple, lavender, or lilac-pink; lip shallowly toothed to deeply fringed, if deeply fringed, then the fringed portions not as long as the main body of the lobe itself; largest leaves 2–9 cm. wide . . . . . *e*
    - e. Lip shallowly toothed; the teeth 1 mm. or less long; lip 18–30 mm. broad . . 7. H. PERAMOENA
    - e. Lip deeply fringed or cut, the fringed or cut portions 2–5 mm. long; lip 6–16 mm. broad 6. H. PSYCHODES

a. The lip (largest 3rd petal) with mostly an entire (toothless) margin or with some teeth at tip or along margins, but not lobed, nor fringed, nor with fine teeth all along margins. . . . . *f*
- f. Only 1 well-developed leaf on stem (rarely 2); lip with 3 shallow rounded teeth at tip; bracts shorter than the flowers . . . . . 1. H. CLAVELLATA var. CLAVELLATA
- f. Mostly 2–5 leaves (rarely 1) on stem; lip with a few teeth on the margins; bracts equaling or much longer than the flowers . . . . . 2. H. FLAVA

1. **Habenaria clavellata** (Michx.) Spreng. var. **clavellata** Green Wood Orchid Map 686
Flowers July–August.

Occurs in swampy woodlands and spring branches of Crowley Ridge in the southeastern lowland section of Missouri, where known from Dunklin (Kennett, July 21, 1900, *Dimmitt*) and Stoddard (wooded spring branch in Crowley Ridge, T25N, R10E, northeast $\frac{1}{4}$ sect. 1, $3\frac{1}{2}$–$3\frac{3}{4}$ mi. southeast of Bloomfield, October 17, 1955, *Steyermark 80385*; spring branch in valley of Crowley Ridge, T25N, R10E, southeast $\frac{1}{4}$ sect. 11, 4–$4\frac{1}{4}$ mi. south of Bloomfield, October 17, 1955, *Steyermark 80427*) counties.

Ranges from Florida to Texas, north to Massachusetts, New Brunswick, New York, Ohio, Indiana, Wisconsin, and Minnesota.

In Missouri this orchid is found in seepage and wet soils of spring branches of Crowley Ridge, near its

688 Habenaria ciliaris (Yellow Fringed Orchid)

689 Habenaria lacera var. lacera (Ragged Orchid)

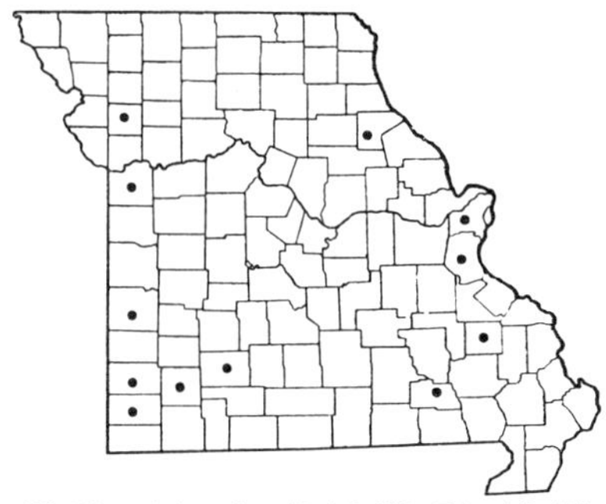

690 Habenaria leucophaea (Prairie White Fringed Orchid)

junction with the swampy level lowlands proper. In such situations it is associated with the rare American Holly (*Ilex opaca*).

2. **Habenaria flava** (L.) R.Br.
Pale Green Orchid — Map 687
Flowers May 20–September.

Occurs in low or level woods in valleys bordering streams and swamps, and in swales of river bottom prairies.

Missouri material is represented by 2 varieties:

Inflorescence loosely-flowered; most of floral bracts shorter than the flowers, the lowest bracts equaling or slightly longer than the flowers; larger leaves of stem usually 2, less frequently 1 or 3 . . . . . . . 2a. H. FLAVA var. FLAVA
Inflorescence compact; lowest floral bracts much longer than the flowers; the larger leaves of stem 2–5 . . . . . . . 2b. H. FLAVA var. HERBIOLA

2a. **Habenaria flava** var. **flava** — Map 687

Known only from the southeastern Ozark section in Wayne and Shannon counties.

Ranges from Florida to Texas, north to Maryland, Kentucky, Indiana, Illinois, and Missouri; also in Nova Scotia.

2b. **Habenaria flava** var. **herbiola** (R.Br.) Ames & Correll — Map 687

Occurs in northern and eastern Missouri west in the Ozark region to Howell County. It is the commoner variety encountered in the state.

Ranges from New Brunswick to Ontario, south to North Carolina, Tennessee, Ohio, Indiana, Illinois, Wisconsin, and Missouri.

The sepals and 2 lateral petals in *H. flava* are mostly greenish, varying to greenish-yellow and greenish-white.

3. **Habenaria ciliaris** (L.) R.Br.
Yellow Fringed Orchid — Map 688
Flowers July 25–October 13.

Occurs in acid soils of sandy wet springy ground along spring branches, wet wooded margins of upland sink-hole ponds, and in moist pine woods. Known only from Iron, Ripley (Pleasant Grove, *Bush*; margin of Cupola Pond, T25N, R1W, sect. 29, $1\frac{1}{2}$ mi. southeast of Handy, August 19, 1951, *Steyermark 72493*), and Stoddard (base of Crowley Ridge, near junction with lo·vland, T25N, R11E, northwest $\frac{1}{4}$ sect. 6, $3\frac{1}{2}$ mi. southeast of Bloomfield, August 20, 1954, *Steyermark 76790*).

Ranges from Florida to Texas, north to Massachusetts, Ontario, Michigan, Wisconsin, and Missouri.

This is one of the most beautiful as well as one of the rarest of orchids in Missouri. This orchid does not do well in a wildflower garden, and should remain undisturbed in its native haunts. Areas where it occurs should be rigidly protected against vandalism. The Stoddard County station, which the author discovered, has now been purchased by the Missouri Chapter of Nature Conservancy for protection to be kept

Plate no. 124. **1.** Habenaria flava, $\times {}^2/_7$; *a.* Fruit, $\times {}^2/_7$; After Britton and Brown, The New York Botanical Garden. **2.** Habenaria ciliaris, $\times {}^2/_7$; *a.* Flower, $\times {}^4/_7$; After Gleason, The New York Botanical Garden. **3.** Habenaria psychodes, $\times {}^2/_7$; After Gleason, The New York Botanical Garden. **4.** Habenaria lacera, $\times {}^2/_7$; *a.* Flower, $\times 1{}^5/_7$. **5.** Habenaria leucophaea, $\times {}^2/_7$; After Gleason, details from Small, The New York Botanical Garden. **6.** Habenaria clavellata, $\times {}^2/_7$. **7.** Habenaria peramoena, $\times {}^4/_7$; After Britton and Brown, The New York Botanical Garden. **8.** Pogonia ophioglossoides, $\times {}^2/_7$; After Gleason, details from Small, The New York Botanical Garden. **9.** Isotria medeoloides, $\times {}^2/_7$; *a.* Fruiting plant, $\times {}^2/_7$. **10.** Isotria verticillata, $\times {}^2/_7$; *a.* Fruiting plant, $\times {}^2/_7$. **11.** Triphora trianthophora, $\times {}^4/_7$; Details from Small, The New York Botanical Garden. **12.** Calopogon pulchellus, $\times {}^2/_7$; Details from Small, The New York Botanical Garden.

PLATE NO. 124

691 Habenaria psychodes (Small Purple Fringed Orchid)

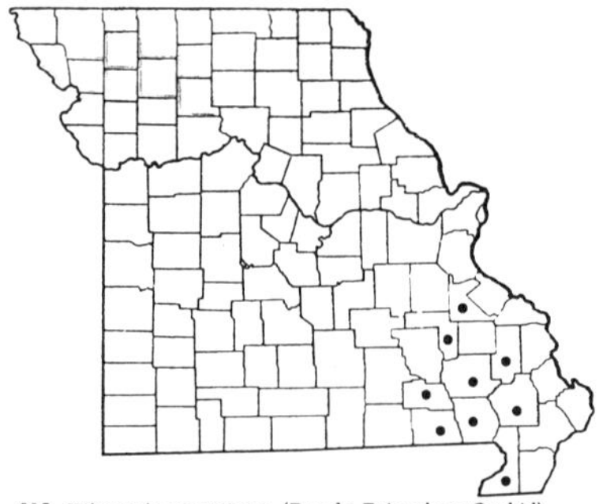

692 Habenaria peramoena (Purple Fringeless Orchid)

693 Pogonia ophioglossoides (Snake-mouth)

forever in its natural state. The orange or orange-yellow flowers are arranged in a compact, many-flowered inflorescence 5–15 cm. long and 4–8 cm. thick. The inflorescence, which starts to bloom in Missouri normally around the early to middle part of August, generally continues in bloom for a period of a month, especially in woodland situations.

4. **Habenaria lacera** (Michx.) Lodd. var. **lacera**
Ragged Orchid Map 689
Flowers May–early August.

Occurs in moist meadows and swales and along small streams in upland prairies, also in upland flat wet woods bordering sink-hole ponds. Southern and east-central Missouri, north to St. Louis, Audrain, and Boone counties, west to Hickory, Barton, Jasper, and Newton counties.

Ranges from the Magdalen Islands to Ontario and Minnesota, south to Florida, Alabama, Mississippi, and Texas.

Another variety, var. *terrae-novae* Fern., with smaller flowers, occurs from Newfoundland to Nova Scotia.

5. **Habenaria leucophaea** (Nutt.) Gray
Prairie White Fringed Orchid Map 690
Flowers June 10–July. Most of the flowering dates in Missouri occur between June 10 and July 4.

Occurs in swales of upland prairies, low moist river bottom prairies, and spring-fed, calcareous meadows along small valleys in southern and central Missouri locally north to Ralls and Clinton counties.

Ranges from Nova Scotia to Maine, Ontario to North Dakota, south to New York, Ohio, Indiana, Illinois, Louisiana, and Kansas.

This very showy white-flowered fringed orchid is found most frequently in alkaline moist soils, such as calcareous spring-fed meadows in small valleys of the Ozarks, as in Carter County, or in wet river bottom prairies at the base of limestone bluffs, as in Ralls County.

6. **Habenaria psychodes** (L.) Spreng.
Small Purple Fringed Orchid Map 691
Flowers July–August.

Ranges from Newfoundland, Quebec, and Ontario, south to Georgia, Tennessee, Ohio, Indiana, Illinois, and Iowa.

This species is accepted tentatively on the basis of a flowering specimen taken by the author on July 23, 1949, from a plant seen in the wildflower garden of Mr. Bill Bauer. This garden at that time was located in Old Orchard in Webster Groves in St. Louis County, and reputed to have been found by Bauer in the southern Ozarks in Ozark County in the Caney Mountain Refuge along Caney Creek, T23N, R13W, sect. 8, 5½ mi. north of Gainesville. A visit was made with Mr. Bauer the next day, July 24, to this station in order to verify the record and to study the environment of the orchid. A search was made around a wet wooded area bordering a spring and spring branch where Mr. Bauer said the orchid was seen, but a recently fallen tree at this locality had obstructed the site and destroyed all herbaceous ground vegetation. The result was that no evidence could be found for the occurrence of a single plant of this species.

For the record, the collection from Mr. Bauer's garden, *Steyermark 68585*, is the only extant one that may afford a clue to a later rediscovery of this species from the Ozark County locality or another station.

7. **Habenaria peramoena** Gray
Purple Fringeless Orchid Map 692
Flowers late June–August. Most of the flowering dates recorded in Missouri are between July 4 and July 24.

Occurs in low wet woods along streams, wooded margins of lakes, rocky moist wooded ground around springs, and less frequently in low swampy meadows along creeks. Known in Missouri only from the south-eastern portion of the state north to St. Francois

694 Isotria verticillata (Large Whorled Pogonia)

695 Isotria medeoloides (Small Whorled Pogonia)

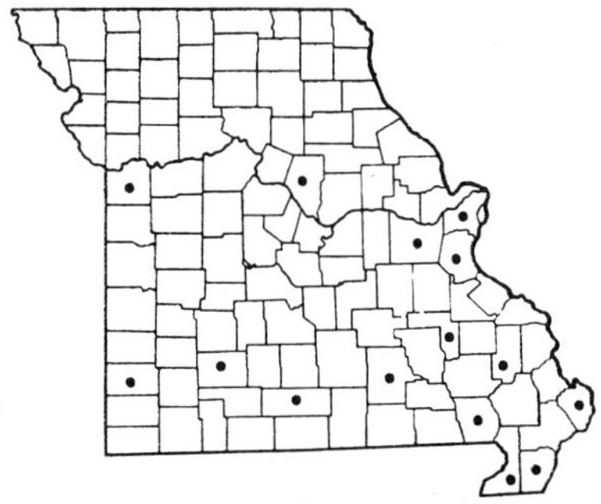

696 Triphora trianthophora (Nodding Pogonia)

County and west to Iron, Carter, and Ripley counties

Ranges from New Jersey and Pennsylvania to Ohio, Illinois, and southeast Missouri, south to North Carolina, Alabama, and Tennessee.

This is a handsome orchid with showy rose-purple flowers in a large inflorescence 6–18 cm. long and 4–7 cm. thick.

This and the other species of *Habenaria* do not succeed in wildflower gardens and die out after a year or two. It is hoped that they will be protected, undisturbed, and allowed to perpetuate themselves wherever found.

## 4. **Pogonia** Juss. Snake-mouth

**Pogonia ophioglossoides** (L.) Ker
Snake-mouth Map 693

Flowers early June–July. Flowering dates in Missouri are June 4, July 4–10.

This species was discovered in 1948 by Mr. Bill Bauer from a calcareous swampy meadow fed by a spring in Reynolds County. The data are: swampy meadow bordering spring branch tributary to Middle Fork of Black River, along highway 21, T32N, R1E, sect. 15, 1.8 mi. southwest of bridge over Middle Fork, about 3 mi. northeast of Centerville, July 4, 1948, *Bill Bauer*. It was subsequently collected by the author from the same locality, *Steyermark 71735* on June 17, 1951. Typical *P. ophioglossoides* var. *ophioglossoides*, which is the one found in Missouri, ranges from Newfoundland west to Ontario, south coastwise to Florida to Texas, inland to Pennsylvania, Tennessee, Indiana, Illinois, Missouri, and Minnesota. Another variety, var. *brachypogon* Fern., with practically no beard on the lip, is found in Nova Scotia.

It is hoped that other stations will eventually be found in Missouri for this rare orchid, as the permanence of the present known station is endangered because of disturbance by hogs and other livestock. even though the swamp has been fenced.

## 5. **Isotria** Raf. Whorled Pogonia

Flower-stalk (peduncle) shorter than the ovary; fruiting capsule 1.3–2.5 cm. long, its stalk (peduncle) 0.5–2 cm. long; sepals as long as the petals or at most 1½ times the length of the petals, 1.5–2.5 cm. long . . . 2. I. MEDEOLOIDES

Flower-stalk (peduncle) longer than the ovary; fruiting capsule 2.5–3.5 cm. long, its stalk (peduncle) 2.5–6 cm. long; sepals 2–3 times as long as the petals, 3.5–6 cm. long . . . . . . . . 1. I. VERTICILLATA

1. **Isotria verticillata** (Willd.) Raf.
Large Whorled Pogonia Map 694

Flowers May.

Occurs in acid soils of woodland on slopes in deep ravines or on level ground. Known only from southeastern Missouri in Butler (vicinity of Poplar Bluff, August 3, 1897, *Savage & Stull 1362* in herb. Univ. of Iowa and Chi. Nat. Hist. Museum) and Ste. Genevieve (along Pickle Creek, about ¼ mi. down valley from Pickle Spring, T36N, R7E, about 6 mi. southeast of Sprott, June 7, 1952, *Steyermark 73313*) counties.

Ranges from Florida to Texas, north to Maine, New York, Ohio, Michigan, and Missouri.

The Ste. Genevieve County record of this very rare and local orchid was first discovered by Mr. Oscar Petersen of St. Louis County. In the pine-oak covered slopes of the ravine where it grows, bluffs and outcrops of LaMotte sandstone occur, producing a fine sandy acid soil in which these plants prosper. Over a hundred plants of this orchid were seen at this locality, occuring in colonies, and associated with *Rhododendron roseum, Vaccinium vacillans*, and other acid-soil plants. Over a period of ten years' observation at this locality, annual variations occurred in the relative abundance of plants; some years the plants were plentiful, other years they were scarce. Some years also witnessed little or no flowering.

The sepals in this species are linear, colored with dull brown-purple above and yellow-green below; the lateral petals are yellow-green, and the lip is green on the main portion with a larger white lobe and smaller white and purple-striped lateral lobes.

2. **Isotria medeoloides** (Pursh) Raf.
Small Whorled Pogonia Map 695
*Isotria affinis* (Aust.) Rydb. [P & S]
Flowers May.

Known only from southeastern Missouri in Bollinger County (wooded limestone hill, near Glen Allen, May, 1897, *Colton Russell*).

Ranges from Maine and Vermont south to North Carolina; and Missouri.

This is considered the most local and the rarest native orchid of the eastern and central United States.

The author has visited the vicinity of Glen Allen several times, as well as other favorable adjacent areas in the county in the hope of rediscovering the station for this species, but he has never met with success. Russell's collection is the only known Missouri record.

The sepals and lateral petals are greenish-yellow, while the lip has a pale green color.

## 6. Triphora Nutt. Nodding Pogonia

**Triphora trianthophora** (Sw.) Rydb.
Nodding Pogonia Map 696
*Pogonia trianthophora* (Sw.) BSP. [P & S]
Flowers July 28–October 5.

Rich woods, along slopes, in level ground of valleys along creeks in the Ozarks, and in level ground of rich alluvial forest in the lowlands of southeastern Missouri, sometimes on rotting logs. Southern and central Missouri, north to St. Louis, Boone, and Jackson counties.

Ranges from Florida to Texas, north locally to Maine, New York, Ohio, Michigan, Wisconsin, and Iowa.

This delicate orchid has a lavender and green stem about 8–10 cm. tall, but may attain a height of 30 cm., arising from a small tuber which may send out slender tuber-tipped stolons to give rise to new plants, thus developing a small or large colony. The 3 or 4 short ovate leaves are scattered on the stem which bears at its tip usually 2 or 3 flowers. These have pinkish-lavender sepals, pink and white lateral petals, and a white lip marked with 3 green ridges.

## 7. Calopogon R. Br. Grass Pink

**Calopogon pulchellus** (Salisb.) R.Br. var. **pulchellus** Grass Pink Map 697
*Calopogon pulchellus* (Salisb.) R.Br. [G, BB, P & S]
Flowers late May (May 24)–early July.

Occurs in moist upland prairies, wet calcareous meadows in spring-fed valleys of small creeks, and sometimes in moist open woods. Southern Missouri north to Reynolds, Shannon, and Henry counties.

Ranges from Florida to Texas, north to Newfoundland, Quebec, Ontario, and Minnesota. A broader-leaved variety, var. *latifolius* (St. John) Fern., with leaves equaling or longer than the short flower-stem, occurs in the Magdalen Islands and Nova Scotia.

## 8. Epipactis Sw. Helleborine

**Epipactis Helleborine** (L.) Crantz.
Helleborine Map 698
*Epipactis latifolia* (L.) All. [P & S]
Flowers late June–September.

Known only from southwestern Missouri in Jasper County (moist open ground at foot of north-facing limestone bluff along Turkey Creek, at 'Castle Rock,' 1 mi. NE of Joplin, July 5, 1928, *Palmer 34965*, in herb. Botanical Museum, Harvard University).

Native of Europe; introduced into the United

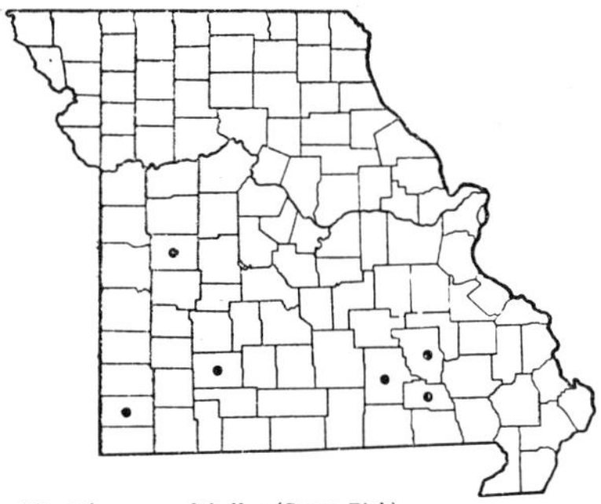

697 Calopogon pulchellus (Grass Pink)

698 Epipactis Helleborine (Helleborine)

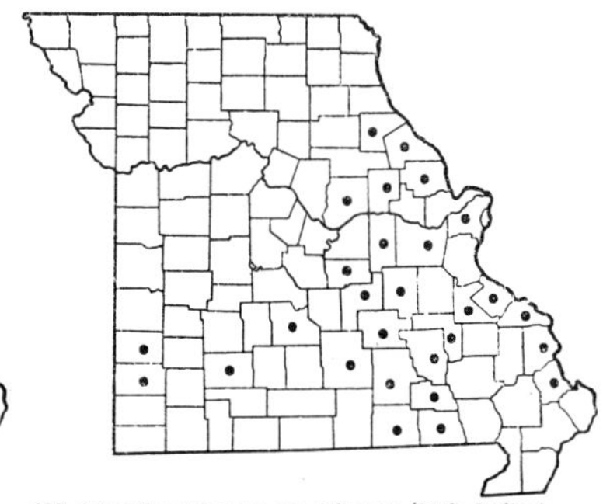

699 Spiranthes tuberosa var. tuberosa (Little Ladies' Tresses)

States and Canada, ranging from Quebec and Ontario to Wisconsin and Montana, south to D. C., Illinois, and Missouri.

The habit of the plant with its broad spreading leaves reminds one of *Smilacina racemosa* which it somewhat resembles superficially. The sepals and lateral petals are green with dull purple, while the lip is similar but darker on the inner side.

This orchid, used for its supposed medicinal value, was introduced into the United States, first having been noted around Syracuse, New York. It often grows in undisturbed woodland habitats, often appearing as if it were a member of the native flora, and prefers a calcareous soil, according to most of the stations where the orchid is known.

## 9. **Spiranthes** Richard  Ladies' Tresses

a. Flowering late May–mid-June; lip and ovary glabrous (without hairs); lip orange or orange-yellow; axis of inflorescence mainly glabrous or minutely puberulent; leaves present at flowering time, conspicuous at base of plant . . . . . . . . . . . . . . . . . . . . 6. S. LUCIDA

a. Without the above combination of characters; flowering late July–late November; lip minutely or conspicuously pubescent (with hairs) on the outside; ovary glabrous to pubescent; lip white, or with green or yellow in center; axis of inflorescence pubescent or glabrous; leaves absent or present at flowering time . . . . . . . . . . . . . . . . . . . . . . . . . . . . *b*

  b. Flowers or fruits ascending to the tip in 1 spiral or twisted series around the flowering axis. . . . *c*

    c. Axis of inflorescence, ovaries and bracts of flowers glabrous (without hairs); leaves usually absent, dried, or wilted at flowering time, when present all from the base, only 1–5 cm. long, and ovate or elliptic-oblong; lip white or white with a green center; flowers 3–5 mm. long. . . *d*

      d. Root 1, vertical; lip white; leaves always absent at flowering time . . . . . 1. S. TUBEROSA

      d. Roots 2 or more, spreading; lip with a green center; leaves rarely, but sometimes, present at flowering time . . . . . . . . . . . . . . . . . . . . . 2. S. GRACILIS

    c. Axis of inflorescence, ovaries and bracts of flowers hairy (pubescent); leaves present at flowering time, occurring at the base and on the lower part of stem, the larger ones 6–20 cm. long, linear or linear-lanceolate; lip yellow; flowers 6–11 mm. long. . . . . . . 3. S. VERNALIS

  b. Flowers or fruits ascending to the tip in 2 or more spiral or twisted series around the flowering axis . . *e*

    e. Very common, and in open meadows, prairies, and glades; flowers 7–10 mm. long; inflorescence stout, mostly 15–25 mm. across (thick); none of stem-leaves conspicuous at flowering time; root (basal) leaves often absent at flowering time . . . . . . . . . . 4. S. CERNUA

    e. Very rare, and in low wet or rich woods; flowers 4–5 mm. long; inflorescence slender, mostly 10–14 mm. across (thick); lower stem-leaves conspicuous and well-developed at flowering time; root (basal) leaves usually present at flowering time . . . . . . . . . . . . 5. S. OVALIS

1. **Spiranthes tuberosa** var. **tuberosa** Raf.
Little Ladies' Tresses  Map 699
*Spiranthes tuberosa* Raf. [*G*, BB in part]
*Spiranthes Beckii* of auth. [P & S], not Lindl.

Flowers August 21–October.

Occurs in acid soils usually in dry, mossy or rocky (chert, sandstone, or granite) upland oak-hickory or pine-oak woodland, on ridges, crests of hills, leached

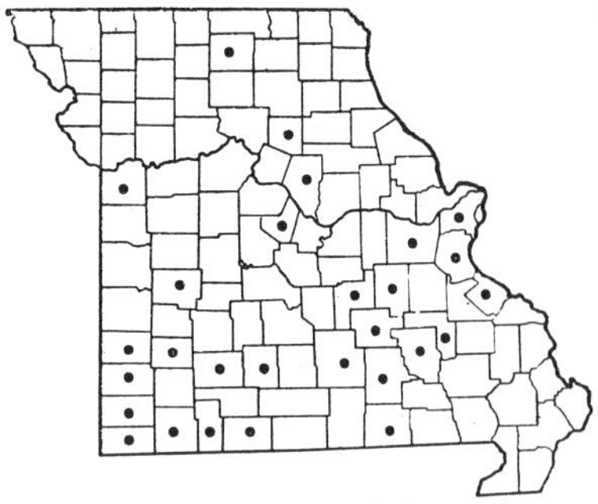

700 Spiranthes gracilis (Slender Ladies' Tresses)

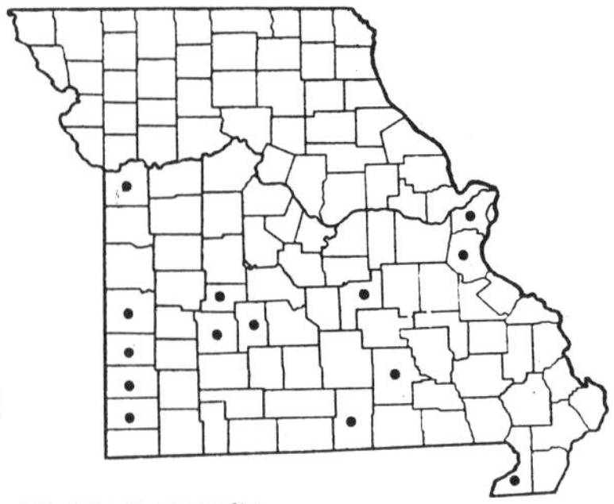

701 Spiranthes vernalis

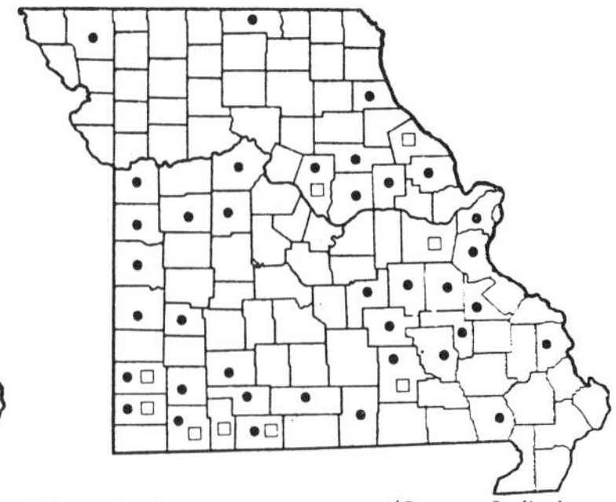

702 ● Spiranthes cernua var. cernua (Common Ladies' Tresses)
702 □ Spiranthes cernua var. ochroleuca (Common Ladies' Tresses)

upper slopes above bluffs and escarpments, and sometimes in dry prairie or meadows. Ozark and east-central sections of the state north to Ralls County and west along a line from Callaway, Maries, Laclede, and Greene to Barton County.

Ranges from Georgia to Mississippi, north to New Jersey, Ohio, Indiana, and Missouri.

Another variety, var. *Grayi* (Ames) Fern., occurs from Massachusetts to North Carolina.

This is a very slender and delicate-stemmed species, attaining a height usually of 1.5–2 dm., less frequently taller. The stem just below the inflorescence is usually less than 1 mm. thick.

The length of the perianth is given as 2–3 mm. in the eighth edition of *Gray's Manual*, but Missouri material usually measures 4–4.5 mm.

2. **Spiranthes gracilis** (Bigel.) Beck
Slender Ladies' Tresses Map 700
Flowers August 20–October.

Occurs in similar situations to those of *S. tuberosa*, in acid soils usually in dry, mossy, or rocky (chert, sandstone, or granite) upland oak-hickory or pine-oak woodland, on ridges, crests of hills, and leached upper slopes above bluffs and escarpments, and less frequently, in upland prairies. Southern and central Missouri, mostly in the Ozark section north to St. Louis, Randolph, and Jackson counties, and locally north in Sullivan County.

Ranges from Florida to Texas, north to Maine, New York, Ohio, Indiana, Wisconsin, Missouri, and Oklahoma.

Like *S. tuberosa*, this is a very slender- and delicate-stemmed species with a stem diameter of usually less than 1 mm. just below the inflorescence. The stems are somewhat taller than in *S. tuberosa*, varying from usually 2.5–4 dm. tall, less frequently taller.

3. **Spiranthes vernalis** Engelm. & Gray Map 701
Flowers July 11–early September.

Occurs in upland prairies and swampy meadows, usually in acid soils. Infrequent and scattered in southern and central Missouri north to St. Louis, Phelps, Hickory, and Jackson counties.

Ranges from Florida to Texas, north to Massachusetts, New York, Ohio, Tennessee, Missouri, and Kansas.

This species of Ladies' Tresses grows taller than any other Missouri species, usually attaining a height of 5–10 dm. The hairs on the outer surface of the lip are more conspicuous after the flowers are dry, in the fresh condition appearing as numerous papillae or projections.

In one swampy meadow in Howell County this species occurred with *Viola sagittata*, *Rhexia virginica*, and *Lysimachia lanceolata*, while in a dry section of a large natural prairie in Phelps County it was blooming on August 18 with *Asclepias stenophylla*, *Hieracium longipilum*, *Linum medium* var. *texanum*, *Sabatia angularis*, *Buchnera americana*, and *Silene regia*.

4. **Spiranthes cernua** (L.) Richard Common Ladies' Tresses, Nodding Ladies' Tresses Map 702
Flowers August 1–November 23.

Occurs on rocky limestone glades, upland dry prairies, wet meadows and thickets. Common in the southern and central sections of the state, known in northern Missouri in Marion, Putnam, and Nodaway

Plate no. 125. **1.** Epipactis Helleborine, × 2/5. **2.** Goodyera pubescens, × 2/5; *a.* Flower, × 2 4/5; After Gleason, details from Small, The New York Botanical Garden. **3.** Spiranthes cernua, × 2/5; Details from Small, The New York Botanical Garden. **4.** Spiranthes gracilis, × 2/5; *a.* Flower × 2 4/5. **5.** Spiranthes lucida, × 2/5. **6.** Spiranthes tuberosa, × 2/5.

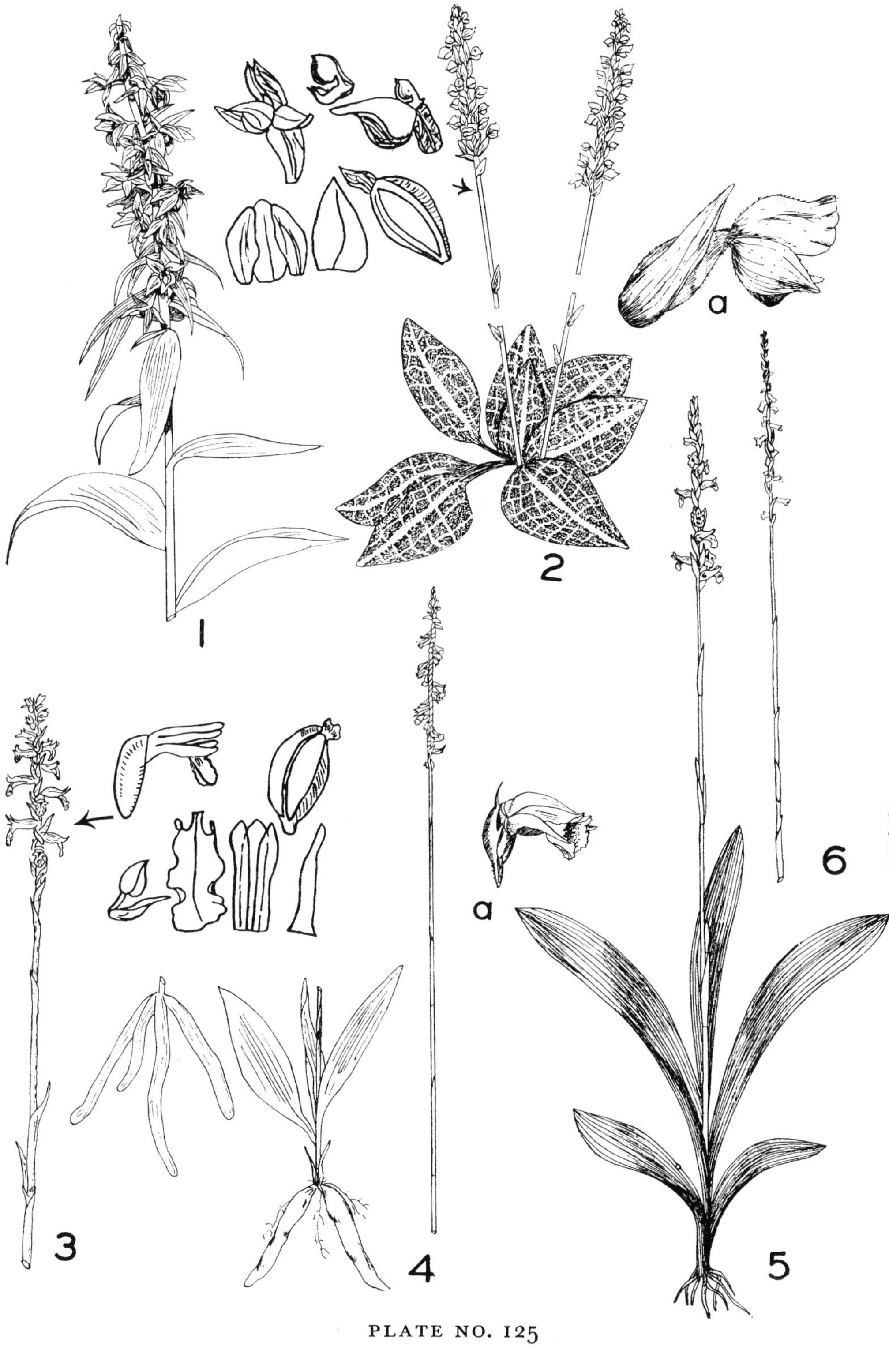


PLATE NO. 125

703 Spiranthes ovalis

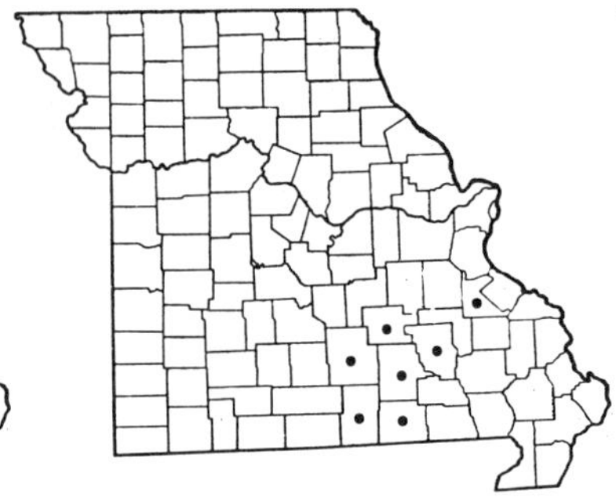

704 Spiranthes lucida

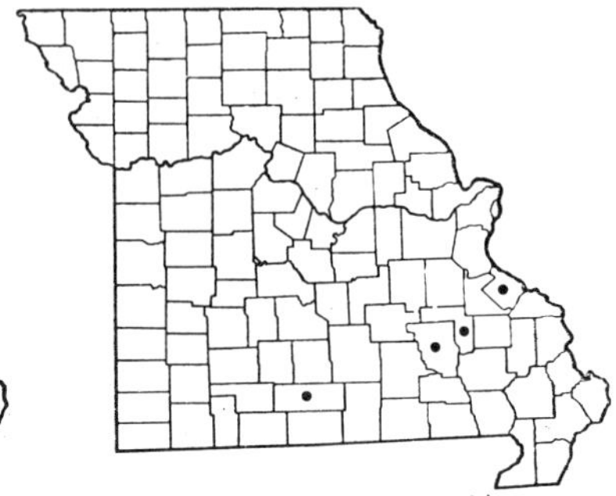

705 Goodyera pubescens (Rattlesnake Plantain)

counties; absent from the lowlands of southeastern Missouri.

Missouri material is represented by 2 poorly separated, difficult-to-distinguish varieties:

Bracts, at least the lower ones, shorter than the corolla; flowers white; seeds with 2 or more embryos . . . . . . 4a. S. CERNUA var. CERNUA

Bracts longer than the corolla; flowers whitish, greenish, or strawcolored; seeds with 1 embryo 4b. S. CERNUA var. OCHROLEUCA

4a. **Spiranthes cernua** var. **cernua** Map 702

*Spiranthes cernua* (L.) Richard [G, BB, P & S]

Flowers August 1–November 23.

This is the commoner of the two varieties in Missouri.

Ranges from Florida to Texas, north to Nova Scotia, Ontario, Michigan, Wisconsin, Minnesota, and South Dakota.

The flowers are very fragrant, often reminiscent of vanilla extract flavor as well as that of lily-of-the-valley, but sometimes flowers are not fragrant. This orchid is really the only common one in the state, and sometimes hundreds of them may be counted in a single meadow, prairie, or glade. They are among the last wild flowers to be seen in bloom in late autumn. The usual height is from 1.5–3.5 dm., but taller plants are found up to 6 or more dm. tall.

4b. **Spiranthes cernua** var. **ochroleuca** (Rydb.) Ames Map 702

Flowers mainly September 17–October 10.

Scattered in southern and east-central Missouri north to Pike, Franklin, and Boone counties.

Range is the same as var. *cernua*, but less common.

This variety is not easily separated from var. *cernua*, into which it intergrades. Apparently the best distinguishing difference between the two varieties is the number of embryos in the seed, a difficult character to use. The variety *ochroleuca* is being maintained as separate, as it has been adopted by various orchid authorities.

5. **Spiranthes ovalis** Lindl. Map 703

Flowers September 12–October 3.

Occurs in low or rich moist woodland. Rare and scattered in the state, where known only from St. Louis and Jackson counties in central Missouri, and Mississippi County in the southeastern lowland section of the state.

Ranges from Florida to Texas, north to Virginia, Kentucky, Indiana, and Missouri.

In the Three States Timber Tract adjacent to Big Oak State Park in Mississippi County, the only locality where the author has seen this rare orchid, it occurred in moist soil of deep tall forest where Bald Cypress (*Taxodium*) and a rich assemblage of other kinds of forest trees abounded.

6. **Spiranthes lucida** (H. H. Eat.) Ames Map 704

Flowers late May–mid-June. Flowering dates in Missouri are May 25, May 27, May 29, May 31, June 4.

Occurs usually in alkaline soils in swampy, calcareous, spring-fed meadows, moist places at the base of or along the ledges of limestone or sandstone rocks or small bluffs bordering small creeks. Restricted to the southeastern Ozark section, where known from St. Francois, Reynolds, Dent, Shannon, Oregon, Texas, and Howell counties.

---

Plate no. 126. **1.** Corallorhiza trifida var. verna, × 2/5; *a.* Flower, front view, × 2 4/5; *b.* Flower, side view, × 2 4/5. **2.** Corallorhiza Wisteriana, × 2/5; *a.* Flower, side view, × 2 4/5; *b.* Flower, front view, × 7. **3.** Corallorhiza odontorhiza, × 2/5; *a.* Flower, side view, × 5. **4.** Malaxis unifolia, × 1; *a.* Flower, × 20; Details from Small, The New York Botanical Garden. **5.** Liparis Loeselii, × 2/5; *a.* Flower, side view, × 1 3/5; *b.* Flower, front view, × 1 3/5.

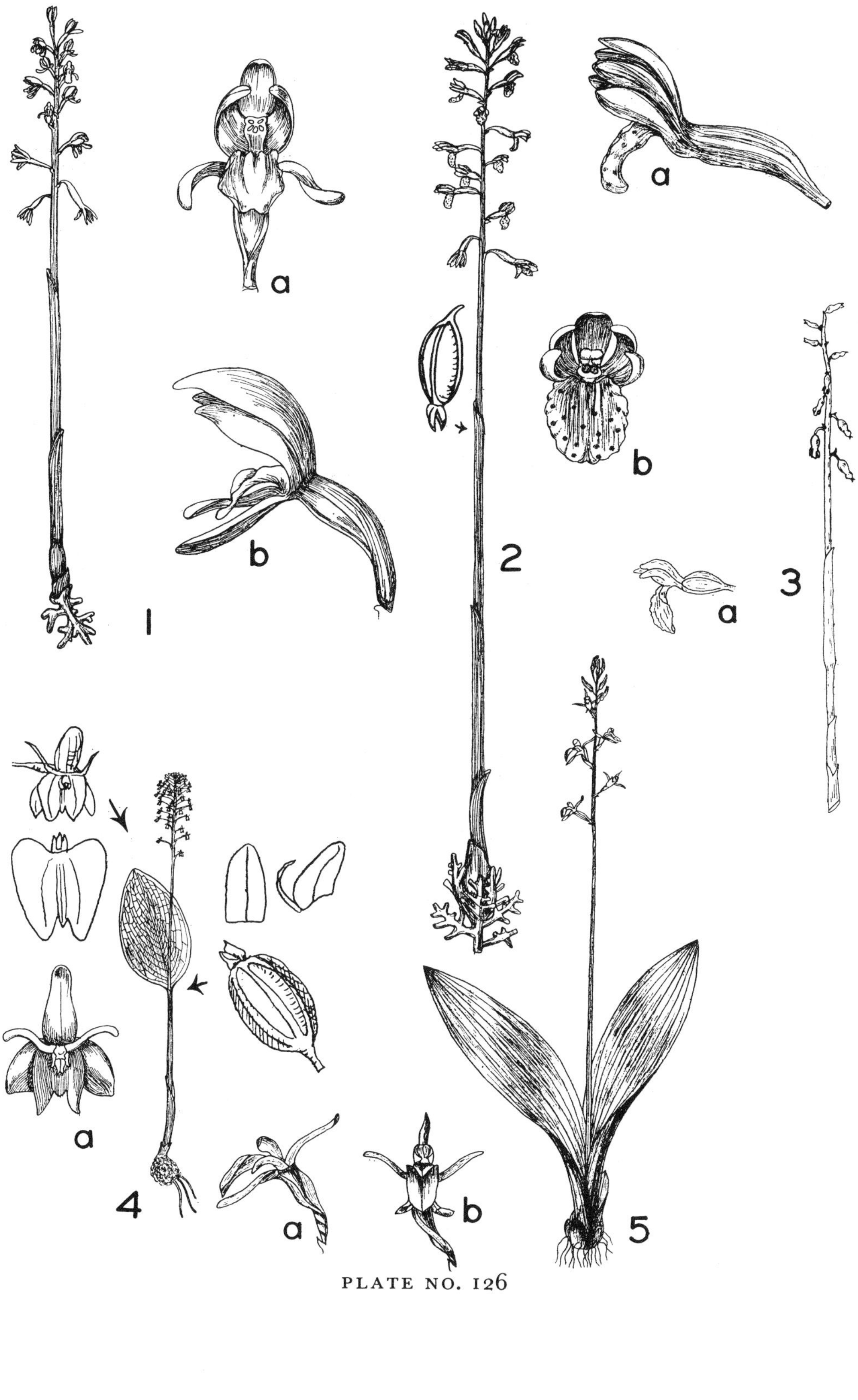


PLATE NO. 126

Ranges from New Brunswick to Ontario and Minnesota, south to North Carolina, Ohio, Michigan, Indiana, Illinois, and Missouri.

This is a low-growing species usually only 0.5–1.5 dm. tall, rarely up to 3 dm. The leaves, which are rich grass-green and shining, measure as much as 15 cm. long and 2.3 cm. wide, are of an oblanceolate or oblong shape, and occur mainly at or near the base of the plant.

In a wet springy meadow in Reynolds County where this species was found, it was associated with *Carex leptalea*, *Cardamine bulbosa*, *Castilleja coccinea*, and *Senecio aureus*. Along Plattin Creek in St. Francois County, where it grew in moist places at the base of shallow limestone bluffs, it occurred with *Selaginella apoda*, *Rhynchospora capillacea*, *Juncus Dudleyi*, and other species of calcareous moist places.

## 10. **Goodyera** R.Br. Rattlesnake Plantain

**Goodyera pubescens** (Willd.) R.Br.
Rattlesnake Plantain Map 705

Flowers late July–mid-September. Flowers mainly August 15–September 15.

Occurs in acid soils overlying sandstone or chert in pine or pine-oak-hickory woodland, at the bottom of deep ravines or small canyons, and along north-facing wooded slopes of sandstone or chert-covered hills from the base to three-quarters distance up. Rare and scattered in the southern Ozarks, where known only from Ste. Genevieve (*Steyermark 63863*), Iron, Reynolds (*Steyermark 73115, 76419*), and Douglas (*Steyermark 23638, 85969*) counties.

Ranges from Maine and Quebec to Ontario, south to Florida, Alabama, Tennessee, and Missouri.

This easily recognized orchid with the decorative white-veined, dark green leaves may eventually be found to have a wider distribution in the Ozark region, when most of the hills in the Ozark counties have been thoroughly searched. In the few stations where it is now known, it is found on well-shaded north-facing slopes in acid soils which are mossy-covered and not completely taken over by ground vegetation. *Hepatica nobilis* var. *obtusa* (*americana*), *Cypripedium Calceolus* var. *pubescens*, *Carex digitalis*, *Thelypteris hexagonoptera*, *Liparis liliifolia*, *Rhododendron roseum*, and *Solidago arguta*, are some of the herbaceous species which have been found associated with Rattlesnake Plantain.

## 11. **Corallorhiza** Chatelain Coral Root

a. Blooming July–October; sepals and petals of the flower 3–4 mm. long. . . . . . . 3. C. ODONTORHIZA
a. Blooming February–June; sepals and petals of the flower 4.5–8 mm. long. . . . . . . . . . . *b*
  b. Common throughout southern and central Missouri; stem purple, brownish, or purple-red; lip not lobed, white dotted pale or dull purple; sepals and petals 6–8 mm. long, greenish-yellow prominently marked with purple-brown or purple . . . . . . . . . . . . 2. C. WISTERIANA
  b. Known only from Lawrence County, southwestern Missouri; stem yellowish or greenish-yellow; lip with a shallow lobe on each side at the base, white, not spotted or with very few reddish dots; sepals and petals 4.5–6 mm. long, mostly yellow-green, not spotted . . . . 1. C. TRIFIDA var. VERNA

1. **Corallorhiza trifida** Chatelain var. **verna** (Nutt.) Fern. Early Coral Root, Pale Coral Root Map 706

Flowers late April–May.

Known only from Lawrence County in southwestern Missouri (rocky woods along creek near Red Oak, May 1, 1925, *Palmer 26962*).

Ranges from Newfoundland to British Columbia, south to Georgia, Tennessee, Ohio, Indiana, Wisconsin, Missouri, South Dakota, Colorado, and Oregon.

Typical *C. trifida* var. *trifida* with red-spotted or brown-tinged perianth, ranges farther north and occurs also in Eurasia.

2. **Corallorhiza Wisteriana** Conrad
Coral Root Map 707

Flowers February–June.

Occurs in rich or rocky open woods on slopes, crests of hills, and valley and ravine bottoms. Common in southern and central Missouri north to Pike, Boone, Saline, Lafayette, and Platte counties.

Ranges from Florida to Texas, north to New Jersey, Pennsylvania, West Virginia, Ohio, Indiana, Illinois, Missouri, and reported from South Dakota.

Mr. Bill Bauer has grown this species successfully in his wildflower garden, where it spread over a period of several years. He reports the earliest flowering took

706 Corallorhiza trifida var. verna (Early Coral Root)

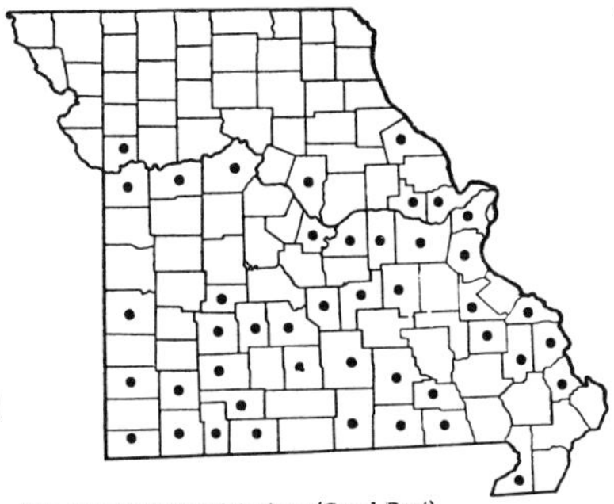

707 Corallorhiza Wisteriana (Coral Root)

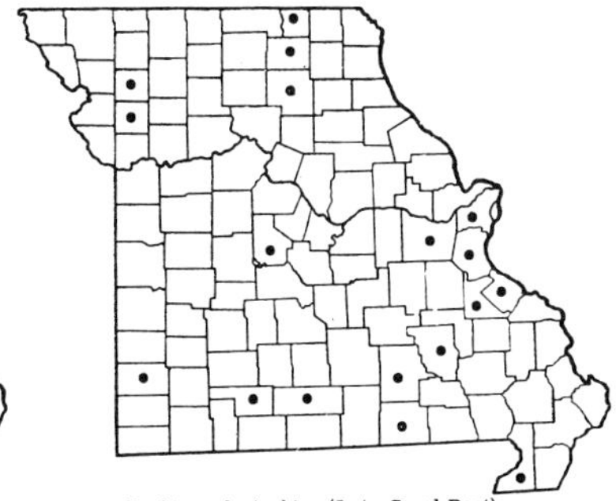

708 Corallorhiza odontorhiza (Late Coral Root)

place in February. The fruiting capsules are 8–12 mm. long.

3. **Corallorhiza odontorhiza** (Willd.) Nutt. f. **odontorhiza** Late Coral Root Map 708
*Corallorhiza odontorhiza* (Willd.) Nutt.
Flowers August–October.

Occurs in wooded rocky or nonrocky areas, on slopes, upland ridges, and in low valleys, often in acid soils. Scattered over the state but less common than *C. Wisteriana*.

Ranges from Maine to Minnesota, south to Georgia, Alabama, Mississippi, and Missouri.

The sepals and lateral petals in typical *C. odontorhiza* f. *odontorhiza* are greenish with purple to brown with maroon or dull dark purple, and the lip is white with purple-magenta margins and dots. The small capsules are 5–8 mm. long. In *Corallorhiza odontorhiza* f. *flavida* Wherry the lip is unspotted and the flower-stem, sheaths, sepals, and lateral petals are yellow without purplish markings.

## 12. **Malaxis** Sw. Adder's Mouth

**Malaxis unifolia** Michx. f. **unifolia**
Green Adder's Mouth Map 709
*Malaxis unifolia* Michx. [G, BB, P & S]
Flowers May–June.

Occurs in dry rich woodland, on upland slopes or ridges or in ravine bottoms and valley floors bordering streams. In acid or near-neutral soils. Rare and scattered in the Ozark section north to St. Louis and Franklin counties and west to Dent, Greene, and Howell counties.

Ranges from Newfoundland to Saskatchewan, south to Florida, Louisiana, and Texas.

This diminutive orchid seldom grows over 20 cm. tall, usually varying from 6–15 cm.; some specimens rarely attain over 20 cm. The single oval or broadly elliptical leaf without a stalk about halfway up the flowering stem is characteristic of typical *M. unifolia* f. *unifolia* which cannot be confused with any other orchid in the state. A rare form occurs with two leaves known as f. *bifolia* Mousley; this has not been found in Missouri.

## 13. **Liparis** Richard Twayblade

a. Plants of dry wooded slopes or rocky ledges, common over the state; flowers mainly brownish-purple, the lip 7–12 mm. long . . . . . . . . . . . . . . . . . . . 1. L. LILIIFOLIA
a. Plants of wet swampy thickets and wet meadows, known only from Shannon County; flowers yellow-green, the lip 4–5 mm. long. . . . . . . . . . . . . . . . . . . . 2. L. LOESELII

1. **Liparis liliifolia** (L.) Richard f. **liliifolia**
Large Twayblade, Lily Twayblade Map 710
*Liparis liliifolia* (L.) Richard [G, BB, P & S]
Flowers May–June.

Occurs in acid soils in rocky or nonrocky woods, frequently on leached crests or upper slopes of hills or on sandstone ledges along creeks. Common in the Ozark region and again in the northernmost tiers of counties, but apparently absent from a large area of the state between these two areas.

Ranges from New Hampshire to Minnesota, south to Georgia, Alabama, Tennessee, and Missouri.

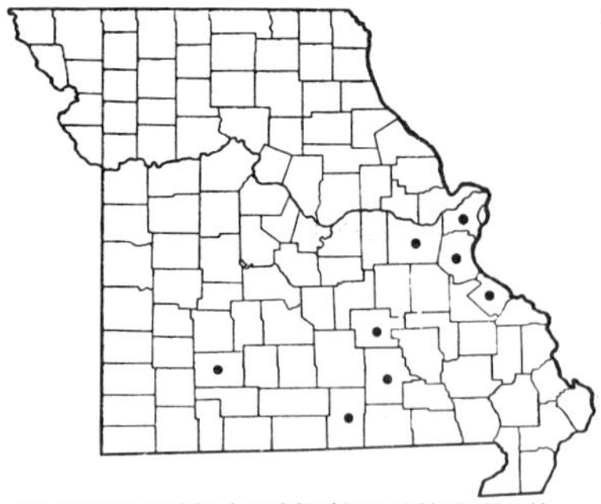

709 Malaxis unifolia f. unifolia (Green Adder's Mouth)

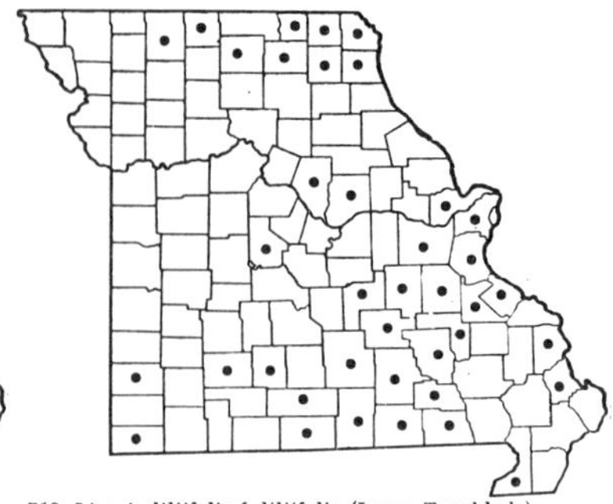

710 Liparis liliifolia f. liliifolia (Large Twayblade)

711 Liparis Loeselii (Loesel's Twayblade)

In typical *L. liliifolia* f. *liliifolia* the flowers are mainly brownish-purple with some pale green, greenish-white, or yellow-green on the sepals and lateral petals, but in f. *viridiflora* Wadmond, not known in Missouri, the flowers are pale green throughout. The leaves are rich green and shining, ovate to broadly elliptic, generally 5–15 cm. long and 3–9 cm. wide.

New plants are formed by offshoots from the old bulb, several plants eventually growing up as a small colony.

2. **Liparis Loeselii** (L.) Richard
Loesel's Twayblade Map 711
Flowers May–June.

Occurs in a calcareous swampy meadow and alder thicket in Shannon County in the southeastern Ozark section (valley of Big Creek, T31N, R2W, sect. 5, September 26, 1936, *Steyermark 20123*; July 3, 1949, *Steyermark 68380*).

Ranges from Quebec to Saskatchewan, south to Alabama, Ohio, Indiana, Illinois, Missouri, and North Dakota.

This rare orchid has been searched for in similar swampy meadows in other sections of the Ozarks without success. This species is another one of the isolated relict survivors which has been stranded in the swampy meadows in this section of Missouri following one of the advances and retreats of an ice sheet during Pleistocene times.

## 14. **Aplectrum** (Nutt.) Torr. Adam-and-Eve, Putty Root

**Aplectrum hyemale** (Muhl.) Torr. f. **hyemale**
Adam-and-Eve, Putty Root Map 712
*Aplectrum hyemale* (Muhl.) Torr. [G, BB, P & S]
Flowers May–June.

Occurs in low rich woods of ravine bottoms or valleys along streams, or on low wooded slopes. Central and eastern Missouri north to St. Charles, Warren, Boone, and Clay counties, locally in southwest Missouri in Greene County.

Ranges from Quebec and Vermont to Saskatchewan, south to Georgia, Tennessee, and Arkansas.

Typical *A. hyemale* f. *hyemale* has the flowers purple-green or dull purple, but in f. *pallidum* House, not known in Missouri, they are pale yellowish throughout.

This orchid is easily identified in late fall, winter, and early spring by the single, large, ascending or erect basal leaf with a blade 7–20 cm. long and 4 or more cm. wide. The coloring of the blade is dark blue-green above and conspicuously ridged or plaited and often rose-purple beneath. Sooner or later in spring the leaf disappears, either before, during, or shortly after the flower-stem appears. The flower-stem then remains all summer and early fall without a leaf, but towards the beginning of October a new leaf starts up again.

If grown in a rich soil with humus, this orchid will last indefinitely in a wildflower garden. The author has had several plants for twelve years at his garden, and has learned of similar results at the Babson estate near St. Charles, Illinois.

---

Plate no. 127. **1.** Aplectrum hyemale, × 4/5; *a.* Flower, × 1 1/5. **2.** Liparis liliifolia, × 2/5. **3.** Hexalectris spicata, × 2/5; Details from Small, The New York Botanical Garden.

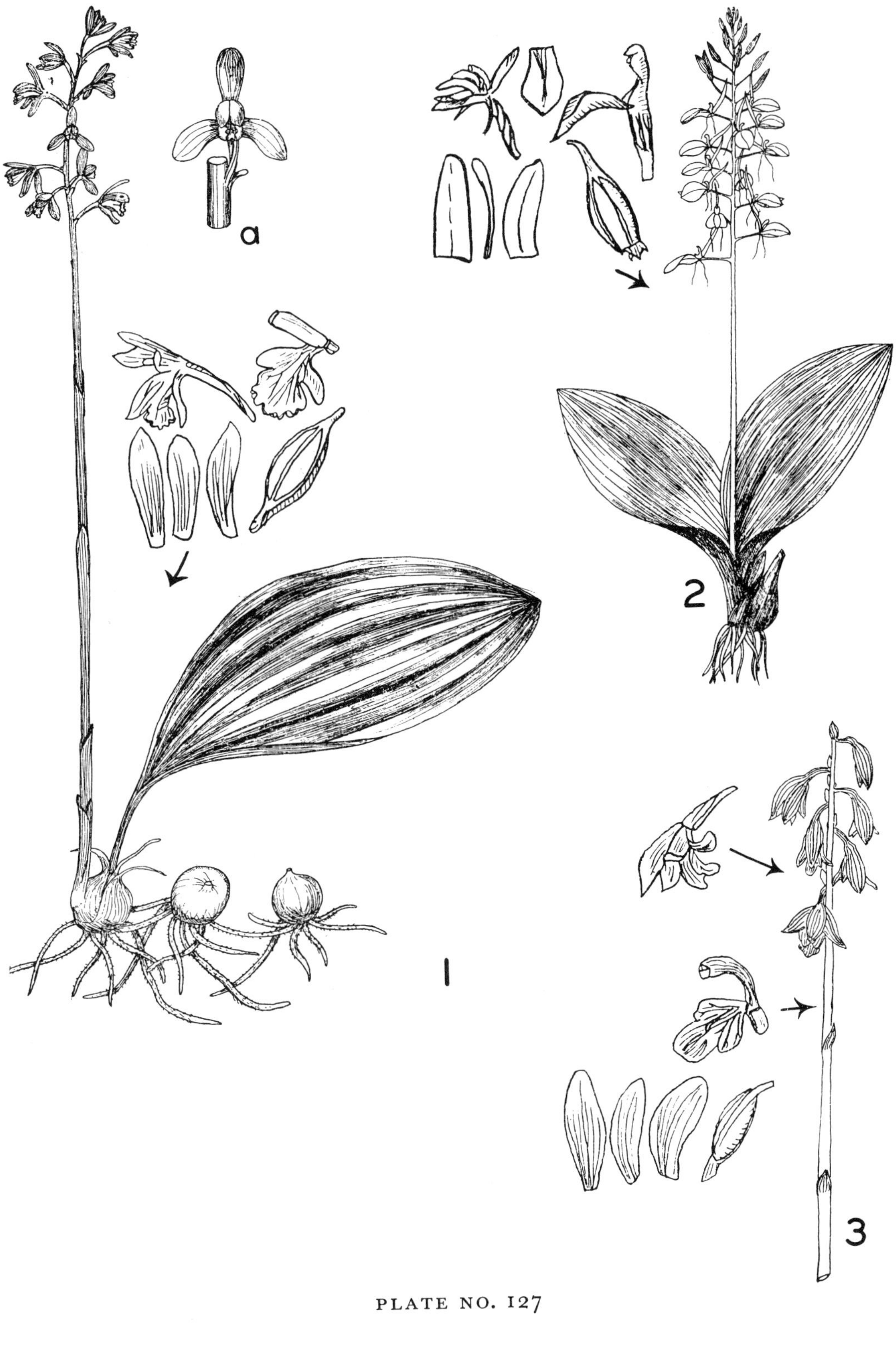


PLATE NO. 127

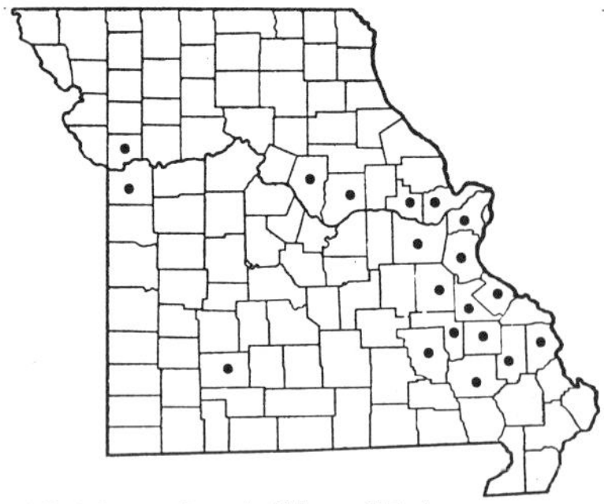

712 Aplectrum hyemale (Adam-and-Eve)

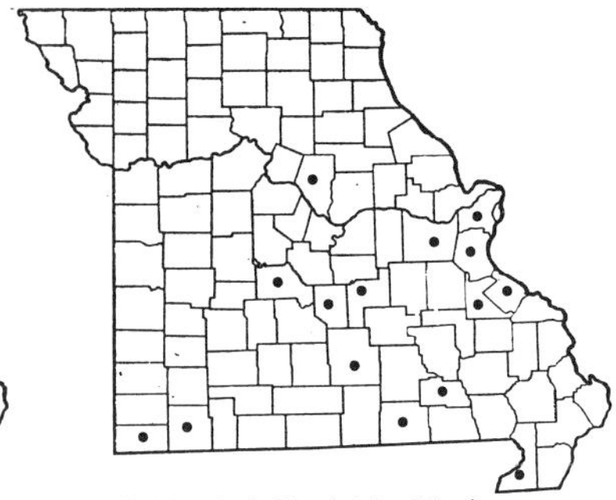

713 Hexalectris spicata (Crested Coral Root)

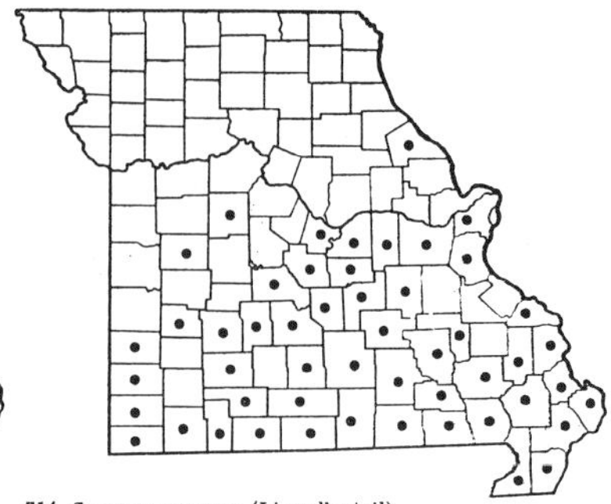

714 Saururus cernuus (Lizard's-tail)

## 15. **Hexalectris** Raf. Crested Coral Root

**Hexalectris spicata** (Walt.) Bernh.
Crested Coral Root Map 713

Flowers July 15–August 26 (flowering dates in Missouri: July 15, 20, 21, 23, 28, August 7, August 26).

Occurs in dry, usually rocky woods of limestone substrata, along or at the base of bluffs, or sometimes bordering limestone openings of glades near red cedar trees. Restricted to the Ozark section of the state north to St. Louis, Franklin, and Boone counties, west to Camden and McDonald counties, and on Crowley Ridge in Dunklin County.

Ranges from Mexico, and Florida to Arizona, north to Maryland, West Virginia, Ohio, Indiana, and Missouri.

This summer-flowering orchid has larger flowers than the other saprophytic Coral Root orchid (*Corallorhiza*). The showy lip is white and 3-lobed with orchid-colored or rose-purple stripes; the sepals and lateral petals are yellowish-brown or salmon-brown with brownish-purple stripes and have somewhat recurved or spreading tips. The sepals are slightly longer than the petals. Each of the very short-stalked flowers has an ovate bract at the base of the stalk, similar in color detail to the sepals and lateral petals.

## Subclass 2. **DICOTYLEDONEAE,** Dicotyledons

Embryo with usually two cotyledons. Woody (vascular) bundles usually united into a ring surrounding a central pith. Parts of flowers mostly in 4's or 5's, but many other combinations may occur. Leaves mostly with the veins forming a network.

## Order **PIPERALES**

### Fam. **SAURURACEAE** (Lizard's-Tail Family)

**Saururus cernuus** L. Lizard's-tail Map 714

Flowers May–September.

Occurs in swampy woods, borders of sloughs, spring branches, and slow streams. Southern and central Missouri north to Pike, Osage, Cole, Pettis, Henry, and Barton counties.

Ranges from Florida to Texas, north to Rhode Island, Connecticut, Quebec, Ontario, Michigan, Illinois, Missouri, and Kansas.

## Order **SALICALES**

### Fam. **SALICACEAE** (Willow Family)

Leaves several to many times as long as broad; bud scale 1 to a bud; flowering or fruiting catkins erect, ascending, or spreading, not drooping or very rarely drooping; stamens 1–10 to each flower in the catkin . . . . . . 1. SALIX
Leaves triangular or ovate, about as broad as long; bud scales several to a bud; flowering or fruiting catkins arching or hanging down; stamens 8–30 or more to each flower in the catkin . . . . 2. POPULUS

### 1. **Salix** L. Willow

The species of this genus are difficult to identify unless leaves, both male and female flowers, and fruits are collected. Since the male and female flowers are found on different plants, it is necessary to collect specimens from both types in order to have a complete representation of the species. A tree or shrub first collected in flower should be revisited later for mature leaves and, if possible, fruit. The species often hybridize.

Willow pollen is of only minor importance in hay fever. Although the willows are primarily insect-pollinated, they are also wind-pollinated to some extent.

The following keys are based on (1) leaves and twig characters (Key I and II), and on (2) flowers and fruits (Key III and IV).

#### I. *Key based mainly on leaf and twig characters*

a. Leaves green on both sides (those at tips of shoots sometimes covered with silvery-silky hairs) . . . . *b*
  b. Buds sticky; petioles with glands just below base of leaf-blade; introduced tree . . . 4. S. FRAGILIS
  b. Buds not sticky; petioles without glands; mostly native trees or shrubs . . . . . . . . . . *c*
    c. Teeth on margins of leaves remotely and unequally spaced, only 3–12 to an inch (2.5 cm.); leaves with very short stalks (petioles) at most 3 mm. long or nearly stalkless . . . 7. S. INTERIOR
    c. Teeth on margins of leaves closely and equally spaced, mostly 13–25 to an inch (2.5 cm.); stalks (petioles) of leaves 4–14 mm. long . . . . . . . . . . *d*
      d. Large tree; leaves narrowly lanceolate with conspicuous long tapering tip, 5–10 times as long as broad; branchlets brittle at base; catkins appearing with the leaves . . . 1. S. NIGRA
      d. Small or tall shrubs; leaves broader, of oblong-lanceolate to ovate shape, gradually tapering at tip (acuminate), but not conspicuously long-tapering, mostly 3–4 times as long as broad; branchlets not brittle at base; catkins appearing before or with the leaves. . . . 8. S. RIGIDA
a. Leaves silvery, gray-white, pale bluish-green, or glaucous (whitish coating) on lower side . . . . . *e*
  e. Margins of leaves entire (without teeth), shallowly wavy-edged (undulate-crenate), or with only 2–11 broad low teeth to an inch (2.5 cm.) . . . . . . . . . . *f*
    f. Lower side of leaf more or less hairy; twigs often hairy; low shrub mainly 0.3–2 m. tall, in prairies, dry open woods, rocky washes, slopes and thickets . . . . . . . . 11. S. HUMILIS
    f. Lower side of leaf glabrous; mature twigs glabrous (without hairs) or nearly so; plants of other kinds of habitats, mostly very wet . . . . . . . . . . *g*
      g. Some of the leaves opposite or nearly opposite, thin, purple-tinged; escaped from cultivation . . . . . . . . . . 14. S. PURPUREA
      g. All the leaves alternate, thick, not purple-tinged; native shrub . . . . . . 10. S. DISCOLOR
  e. Margins of leaves finely and distinctly toothed, the teeth more crowded, 12–25 to an inch (2.5 cm.) . . *h*
  h. Lower side (surface or veins or both) of some or all of the leaves more or less pubescent (with hairs) . . *i*
    i. Youngest twigs or topmost part of twigs or branchlets hairy . . . . . . . . . *j*
      j. Leaves 5–9 times as long as broad, oblong-lanceolate or narrowly lanceolate; scales (bracts) of catkins yellow to white, falling before the capsule matures; plants of rocky stream beds and gravel bars and rocky banks mainly of the Ozark region of southern and central Missouri . . . . . . . . . . 2. S. CAROLINIANA
      j. Leaves $2\frac{1}{2}$–5 times as long as broad, oblong-lanceolate or oblanceolate to ovate-oblong; scales (bracts) of catkins mostly brown or darkened at the tip, remaining attached to the catkin until it shrivels; plants mainly of other habitats . . . . . . . . . . *k*

k. Stipules at the base of the leaf-stalk on the younger shoots conspicuous, semi-ovate to nearly circular or somewhat bean-shaped . . . . . . . . . . 8. S. RIGIDA f. MOLLIS
k. Stipules of younger shoots all small and not conspicuous . . . . . . . . . . . *l*
l. Lower surface of leaves often silky-hairy with hairs pressed against surface; leaves narrowed or tapering at base of blade; capsules or ovaries frequently somewhat silky-hairy; branchlets brittle, breaking easily . . . . . . . . . . . 8. S. × MYRICOIDES
l. Lower surface of leaves with soft more or less spreading hairs; leaves more rounded or curved at base of blade; capsules or ovaries glabrous (without hairs); branchlets hard to break . . . . . . . . . . . . . . . . . . . . 9. S. ERIOCEPHALA
i. Youngest twigs or topmost part of twigs or branchlets glabrous (without hairs) or nearly so . . *m*
m. Stipules conspicuous (0.5–1.5 cm. long) ovate, and mostly persistent on the young or new leafy shoots . . . . . . . . . . . . . . . . . . . . . . 2. S. CAROLINIANA
m. Stipules either none or minute or lanceolate, small and inconspicuous, or not persisting, but soon falling . . . . . . . . . . . . . . . . . . . . . . . . *n*
n. Leaves not usually turning black when dried; planted tree, spread from cultivation; capsules or ovary glabrous (without hairs) . . . . . . . . . . . . . 6. S. ALBA
n. Leaves turning black when dried; native shrubs; capsules or ovary finely silky-hairy. . . *o*
o. Growing in calcareous swampy meadows, around springs, and spring branches in the eastern Ozark region; teeth occur along margins of leaves from base to tip and teeth not tipped with glands; stipules of young shoots lanceolate, eventually falling; twigs or branchlets easy to break, brittle . . . . . . . . . . . . 13. S. SERICEA
o. Growing in low ground along streams in northeastern Missouri; teeth do not occur at base of leaf, and teeth which are present are tipped with glands; stipules none or very tiny and quickly disappear; twigs or branchlets hard to break or tear . 12. S. GRACILIS var. TEXTORIS
h. Lower side of leaves glabrous (without hairs) or mainly so . . . . . . . . . . . . . *p*
p. Youngest twigs or topmost part of twigs or branchlets hairy . . . . . . . . 2. S. CAROLINIANA
p. Youngest twigs or topmost part of twigs glabrous (not hairy) . . . . . . . . . . . *q*
q. Strong leafy sprouts with conspicuous semi-ovate, nearly circular, or somewhat bean-shaped stipules occurring at the base of the leaf-stalks . . . . . . . . . . . . 8. S. RIGIDA
q. Strong leafy sprouts with no stipules or stipules very tiny and quickly disappearing . . . . *r*
r. Shrub; scales (bracts) of catkins pale brown to yellow, remaining attached until catkins dry up . . . . . . . . . . . . . . . . . . . . . 12. S. GRACILIS var. TEXTORIS
r. Tree; scales (bracts) of catkins yellow to white, falling before the capsule is mature . . . . *s*
s. Tiny glands are not present at the summit of the leaf-stalk (petiole); leaf-stalk is twisted; native tree found along the Missouri, Mississippi, and large streams of eastern, central, and northern Missouri . . . . . . . . . . . 3. S. AMYGDALOIDES
s. Tiny glands occur at the summit of the leaf-stalk (petiole); leaf-stalk not twisted; planted trees spread from cultivation . . . . . . . . . . . . . . . . . . *t*
t. Branchlets of previous year drooping or hanging down . . . . . . 5. S. BABYLONICA
t. Branchlets not drooping or hanging down . . . . . . . . . . . . . . *u*
u. Branchlets hard to break, tough; leaf-stalk (petiole) of young leaves not sticky; teeth averaging 7–11 per cm. of leaf-margin . . . . . . . . . . 6. S. ALBA
u. Branchlets easily broken, brittle; leaf-stalk (petiole) of young leaves sticky toward the tip; teeth averaging 3–7 per cm. of leaf-margin . . . . 4. S. FRAGILIS

II. *Key based on floral, fruit, vegetative, and habitat characters*

a. Catkins expanding with or after the leaves (see also *S. petiolaris* keyed out in other *a.*); ovary or capsule glabrous (without hairs), or rarely thinly silky in *S. interior* (*S. petiolaris* with pubescent capsules should be keyed in other *a.*) . . . . . . . . . . . . . . . . . . . . . *b*
b. Scales of catkins dark brown, remaining attached to catkin and persisting until catkin dries up; filaments in male flowers glabrous; male flowers with 1 basal gland . . . . . . . . . . *c*
c. Growing along the Missouri and Mississippi and large streams of northern, central, and eastern Missouri; fruiting capsules 7–10 mm. long; style 0.6–1.2 mm. long; main twigs hairy; stipules of strong leafy shoots small and inconspicuous; fruiting catkins 6–10 cm. long . 9. S. ERIOCEPHALA
c. Along streams, spring branches, and wet meadows and thickets throughout Missouri; fruiting

capsules 4–7 mm. long; style 0.2–0.5 mm. long; main twigs glabrous or hairy; stipules of strong leafy shoots large and conspicuous; fruiting catkins 2.5–6 cm. long . . . . . . . . 8. S. RIGIDA

b. Scales of catkins yellow to whitish, falling and disappearing before capsule ripens; filaments in male flowers hairy at least at the base; male flowers with 2 basal glands . . . . . . . . *d*

d. Flowers in more or less definite whorls, circles, or tiers from base to tip of catkin; 3–12 stamens to each male flower . . . . . . . . . . . . . . . . . . . . . . . . . . . *e*

e. Branchlets or twigs mostly yellowish, tough, hard to break; stipules of strong leafy shoots none or minute; pistillate flowers with 2 glands . . . . . . . . . . 3. S. AMYGDALOIDES

e. Older branchlets or twigs mostly red- or purple-brown, brittle at base, easy to break; stipules of strong leafy shoots conspicuous, persistent; pistillate flowers with 1 gland . . . . *f*

f. Growing along gravel bars and rocky banks of streams mainly in the Ozark section of southern and central Missouri; leaves whitened or silvery on lower surface; branchlets chiefly hairy; stipules broadly ovate . . . . . . . . . . . . . . 2. S. CAROLINIANA

f. Growing along more muddy sections of streams, sloughs, and ponds and throughout the state; leaves green both sides; branchlets at first hairy, later glabrate; stipules semi-cordate, acute, serrate with glands on tips of teeth . . . . . . . . . . . 1. S. NIGRA

d. Flowers spirally arranged along the axis of the catkin, not in whorls or tiers; 1–2 stamens to each male flower . . . . . . . . . . . . . . . . . . . . . . . . . . . *g*

g. Native shrub or small tree, spreading by underground stems and forming colonies along mud flats and sand bars along streams, sloughs, and ponds; catkins frequently clustered at the tips of crowded branchlets; leaves nearly stalkless, the leaf-stalks (petioles) at most 3 mm. long; 3–12 rather distantly spaced teeth to each inch (2.5 cm.) of leaf-margin . . . . . . . . . . . . . . . . . . . . . . . 7. S. INTERIOR

g. Planted, introduced large tree, not forming colonies; catkins at the tips of scattered short leafy side branches; leaves with a leaf-stalk (petiole) 5–15 mm. long; 15–25 closely spaced teeth to each inch (2.5 cm.) of leaf-margin . . . . . . . . . . . . . . . *h*

h. Young leaf-stalks (petioles) without any glands nor sticky at the summit; female flowers with 1 gland; branchlets tough, not easily broken, hairy or glabrous (without hairs) . 6. S. ALBA

h. Young leaf-stalks (petioles) with glands or sticky at the summit; female flowers with 2 glands; branchlets brittle at base, easily broken, usually quite glabrous . . . . . . *i*

i. Branchlets drooping or hanging down; mature leaves 2–12 cm. long, 0.5–2 cm. broad; capsules 1–2 mm. long, almost or quite sessile (without a stalk) . . . 5. S. BABYLONICA

i. Branchlets not drooping nor hanging down; mature leaves 10–15 cm. long, 2.5–4 cm. broad; capsules 4.5–5.5 mm. long, on stalks (pedicels) 0.5–1 mm. long . . 4. S. FRAGILIS

a. Catkins expanding before the leaves unfold; ovary or capsule hairy . . . . . . . . . . . *j*

j. Leaves nearly opposite on the branchlets; filaments of male flowers united to the summit; male flowers with 2 glands; ovary and capsule in female flowers sessile (without a stalk) . 14. S. PURPUREA

j. Leaves alternately arranged on the branchlets; filaments of male flowers free; male flowers with 1 gland; ovary and capsule in female flowers on definite stalks . . . . . . . . . . . *k*

k. Plants known only from extreme northeastern Missouri; fruiting catkins 2–3 cm. thick; female flowers with a definite elongated style 0.5–0.7 mm. long; capsules 7–12 mm. long; stipules of strong leafy shoots large, conspicuous, 1 cm. long, and persistent. . . . . . . . 10. S. DISCOLOR

k. Without the above combination of characters; fruiting catkins 0.5–2 cm. thick; female flowers with no style; capsules 3–9 mm. long; stipules of strong leafy or fertile shoots none or minute or lanceolate and quickly falling. . . . . . . . . . . . . . . . . . . . . . . *l*

l. Low shrub mainly 0.3–2 m. tall, in prairies, dry open woods, rocky washes, slopes, and thickets throughout Missouri; fruiting capsules mainly 7–9 mm. long; male flowers with glabrous filaments; leaves not turning black when dried; twigs and young branchlets mainly gray- or white-hairy covering a pale brown to yellow surface; leaves thickish. . . . . . . . . . . . . . . . . . . . . . . . . 11. S. HUMILIS

l. Shrub mainly 2–4 m. tall, in swamps and wet meadows and wet ground along streams, either in the eastern Ozark section or in northeastern Missouri; fruiting capsules mainly 3–7 mm. (rarely to 9) long; male flowers with filaments glabrous or hairy at base; leaves turning black when dried; twigs and young branchlets mainly glabrous, mainly dark brown or purplish, or also green to olive brown; leaves thinner and more membranaceous . . . . *m*

m. Plants of swampy meadows, springs, spring branches, and open swamps of the eastern Ozark region; scales of catkins dark brown to black; capsules 2–5 mm. long, broadly rounded above; leaves evenly and finely toothed along margins from base to tip; twigs and branchlets brittle at base, easily broken; stipules of strong leafy shoots lanceolate . 13. S. SERICEA

m. Plants along streams of northeastern Missouri; scales of catkins pale brown to yellow; capsules 5–7 (rarely 9) mm. long, tapering above to a blunt tip; leaves usually without teeth at base; twigs and branchlets tough, hard to break; stipules none or minute . . . . 12. S. GRACILIS var. TEXTORIS

### III. *Key based principally on male flowers*

a. Two glands at base of each male flower . . . . . . . . . . . . . . . . . . . . *b*
b. Stamens 3–12 on each flower . . . . . . . . . . . . . . . . . . . . *c*
c. Stipules not present or very small; branchlets yellowish, tough, hard to break . 3. S. AMYGDALOIDES
c. Stipules conspicuous, persisting; branchlets with red or purple, brittle at base, easy to break . . *d*
d. Young unfolding leaves green both sides; large trees growing along more muddy sections of streams, sloughs, and ponds throughout the state; branchlets at first hairy, later glabrate. . . . . . . . . . . . . . . . . . . . . . . . 1. S. NIGRA
d. Young unfolding leaves whitened or glaucous beneath; small trees or shrubs growing along gravel bars and rocky banks of streams mainly in the Ozark section of southern and central Missouri; branchlets chiefly hairy . . . . . . . . . . . . . . . 2. S. CAROLINIANA
b. Stamens 1 or 2 on each flower . . . . . . . . . . . . . . . . . . . . . . *e*
e. Stamen 1, the 2 filaments completely united; leaves mostly almost opposite or nearly so . 14. S. PURPUREA
e. Stamens 2, the filaments free and distinct; leaves alternate . . . . . . . . . . . *f*
f. Native shrub or small tree, spreading by underground stems and forming colonies along mud flats and sand bars along streams, sloughs, and ponds; catkins frequently clustered at the tips of crowded branchlets; leaves nearly stalkless, the leaf-stalk (petioles) at most 3 mm. long; 3–12 rather distantly spaced teeth to each inch (2.5 cm.) of leaf-margin; filaments densely long-hairy in lower half . . . . . . . . . . . . . . . . . . 7. S. INTERIOR
f. Planted, introduced large tree, not forming colonies; catkins at the tips of scattered short leafy side branches; leaves with a leaf-stalk (petiole) 5–15 mm. long; 15–25 closely spaced teeth to each inch (2.5 cm.) of leaf-margin . . . . . . . . . . . . . . . . . *g*
g. Young leaf-stalks (petioles) without any glands nor sticky at the summit; branchlets tough, not easily broken, hairy or glabrous (without hairs) . . . . . . . . 6. S. ALBA
g. Young leaf-stalks (petioles) with glands or sticky at the summit; branchlets brittle at base, easily broken, usually quite glabrous . . . . . . . . . . . . . . . . . *h*
h. Branchlets drooping or hanging down; mature leaves 2–12 cm. long, 0.5–2 cm. broad; staminate catkins 6–8 mm. thick, 1–4 cm. long . . . . . . . . 5. S. BABYLONICA
h. Branchlets not drooping nor hanging down; mature leaves 10–15 cm. long, 2.5–4 cm. broad; staminate catkins 9–10 mm. thick, 3–5 cm. long . . . . . . . 4. S. FRAGILIS
a. One gland at base of each male flower . . . . . . . . . . . . . . . . . . *i*
i. Filaments hairy at base . . . . . . . . . . . . . . . . . . . . . . . *j*
j. Staminate catkins 1.5–2.5 cm. thick; found in extreme northeastern Missouri . . . 10. S. DISCOLOR
j. Staminate catkins 0.5–1.3 cm. thick; found in eastern Ozarks and northeastern Missouri. . . . *k*
k. Staminate catkins 5–8 mm. thick; bracts of catkin blackish; branches dark brown or purple, brittle at base, easy to break; found in swampy meadows and swamps of the eastern Ozarks . . . . . . . . . . . . . . . . . . . . . . . 13. S. SERICEA
k. Staminate catkins 10–13 mm. thick; bracts of catkin pale brown with black tip; branches yellow, greenish, or reddish-purple, tough, hard to break; found along streams in northeastern Missouri . . . . . . . . . . . . . . . . 12. S. GRACILIS var. TEXTORIS
i. Filaments mainly glabrous . . . . . . . . . . . . . . . . . . . . . . . *l*
l. Catkins opening before unfolding of leaves; low shrubs 0.3–2 m. tall, in prairies, dry open woods, rocky washes and slopes . . . . . . . . . . . . . . . . . 11. S. HUMILIS
l. Catkins opening together with the unfolding leaves . . . . . . . . . . . . . . . *m*
m. Main twigs hairy; stipules of strong leafy shoots small and inconspicuous; growing along the Missouri, Mississippi, and larger streams of northern, central, and eastern Missouri. . . . . . . . . . . . . . . . . . . . . . . 9. S. ERIOCEPHALA
m. Main twigs glabrous or hairy; stipules of strong leafy shoots large and conspicuous; growing along streams, spring branches, wet meadows, and thickets throughout Missouri . . 8. S. RIGIDA

### IV. *Key based primarily on female flowers*

a. Leaves mostly opposite or nearly so; catkins frequently in 2's at the nodes (joints of axis), 5–6 mm. thick; capsule sessile, 2–3 mm. long . . . . . . . . . . . . . . . 14. S. PURPUREA

a. Without the above combination of characters; leaves alternate; catkins occurring singly along the branches or at the tips of the branchlets . . . . . . . . . . . . . . . . . . . . *b*

b. Flowers in more or less definite whorls, circles, or tiers from base to tip of catkin . . . . . . . *c*

c. Leaves green both sides; branchlets brittle at base . . . . . . . . . . . . . 1. S. NIGRA

c. Leaves whitish or silvery-gray on lower surface . . . . . . . . . . . . . . . . *d*

d. Petioles 3–7 mm. long; stipules large, persistent; branchlets reddish or purple, brittle at base, hard to break; small trees or shrubs of gravel bars and rocky banks of streams mostly in the Ozark section of southern and central Missouri. . . . . . . . . . 2. S. CAROLINIANA

d. Petioles 10–30 mm. long; stipules small, quickly falling; large trees along alluvial muddy soils of the Missouri, Mississippi, and large streams of eastern, central, and northern Missouri . . . . . . . . . . . . . . . . . . . . . 3. S. AMYGDALOIDES

b. Flowers and capsules spirally arranged around the axis of the catkin . . . . . . . . . . *e*

e. Scales of catkin yellow, white, or pale brown, falling before the capsule ripens. . . . . . . *f*

f. Native shrub or small tree, spreading by underground stems and forming colonies along mud flats and sand bars along streams, sloughs, and ponds; catkins frequently clustered at the tips of crowded branchlets; leaves nearly stalkless, the leaf-stalks (petioles) at most 3 mm. long; 3–12 rather distantly spaced teeth to each inch (2.5 cm.) of leaf-margin . 7. S. INTERIOR

f. Planted, introduced large tree, not forming colonies; catkins at the tips of scattered short leafy side branches; leaves with a leaf-stalk (petiole) 5–15 mm. long; 15–25 closely spaced teeth to each inch (2.5 cm.) of leaf-margin . . . . . . . . . . . . . . . . . . *g*

g. Young leaf-stalks (petioles) without any glands nor sticky at summit; female flowers with 1 gland; capsule 3–5 mm. long; branchlets tough, not easily broken, hairy or glabrous (without hairs) . . . . . . . . . . . . . . . . . . . . 6. S. ALBA

g. Young leaf-stalks (petioles) with glands or sticky at the summit; female flowers with 2 glands; capsule 1–2 mm. long or about 5 mm. long; branchlets brittle at base, easily broken, usually quite glabrous . . . . . . . . . . . . . . . . . . *h*

h. Branchlets drooping or hanging down; capsules 1–2 mm. long, almost or quite sessile; mature leaves 2–12 cm. long, 0.5–2 cm. broad . . . . . . . 5. S. BABYLONICA

h. Branchlets not drooping nor hanging down; capsules 4.5–5.5 mm. long, on slender stalks (pedicels) 0.5–1 mm. long; mature leaves 10–15 cm. long, 2.5–4 cm. broad . 4. S. FRAGILIS

e. Scales of catkins dark brown to blackish, remaining attached to catkin until catkin dries up or capsules open . . . . . . . . . . . . . . . . . . . . . . . . *i*

i. Capsule, ovary, and flower-stalk glabrous; catkins flowering with or after the leaves unfold . . *j*

j. Main twigs hairy; stipules of strong leafy shoots small and inconspicuous; capsules 7–10 mm. long; fruiting catkins 6–10 cm. long; growing along the Missouri, Mississippi, and large streams of northern, central, and eastern Missouri . . . . . . 9. S. ERIOCEPHALA

j. Main twigs glabrous or hairy; stipules of strong leafy shoots large and conspicuous; capsules 4–7 mm. long; fruiting catkins 2.5–6 cm. long; growing along streams, spring branches, wet meadows, and thickets throughout Missouri . . . . . . . . 8. S. RIGIDA

i. Capsule, ovary, and flower-stalk hairy; catkins flowering before the leaves unfold . . . . . *k*

k. Style definite, elongate, 0.5–0.7 mm. long; capsules 7–12 mm. long; fruiting catkins 2–3 cm. thick; known only from extreme northeastern Missouri . . . . . . 10. S. DISCOLOR

k. Style not evident; capsules 3–9 mm. long; fruiting catkins 0.5–2 cm. thick . . . . . . . *l*

l. Low shrub mainly 0.3–2 m. tall, in prairie, dry open woods, rocky washes, slopes, and thickets throughout Missouri; fruiting capsules 7–9 mm. long; twigs and young branchlets mainly gray- or white-hairy covering a pale brown to yellow surface . 11. S. HUMILIS

l. Shrub mainly 2–4 m. tall, in swamps, wet meadows, and wet ground along streams, either in the eastern Ozark section or in northeastern Missouri; fruiting capsules mainly 3–7 mm. (rarely to 9) long; twigs and young branchlets mainly glabrous, mainly dark brown to purplish, or also green to olive-brown . . . . . . . . . *m*

m. Plants of swampy meadows and open swamps of the eastern Ozark region; scales of catkins dark brown to black; capsules 2–5 mm. long, broadly rounded above; branchlets brittle at base, easily broken . . . . . . . . . . . 13. S. SERICEA

m. Plants along streams of northeastern Missouri; scales of catkins pale brown to yellow; capsules 5–7 (rarely 9) mm. long, tapering above to a blunt tip; twigs and branchlets tough, hard to break . . . . . . . 12. S. GRACILIS var. TEXTORIS

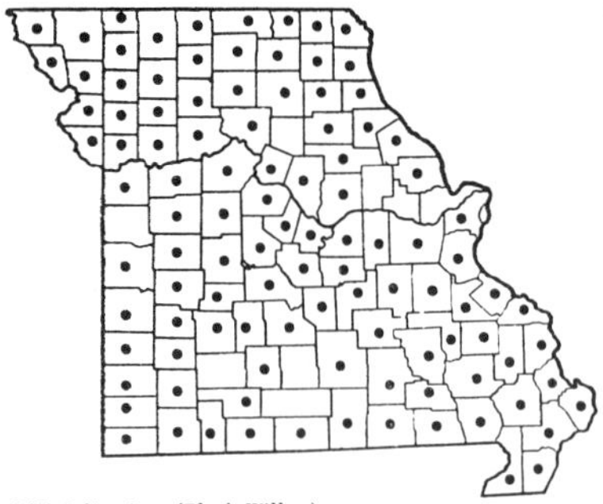

715 Salix nigra (Black Willow)

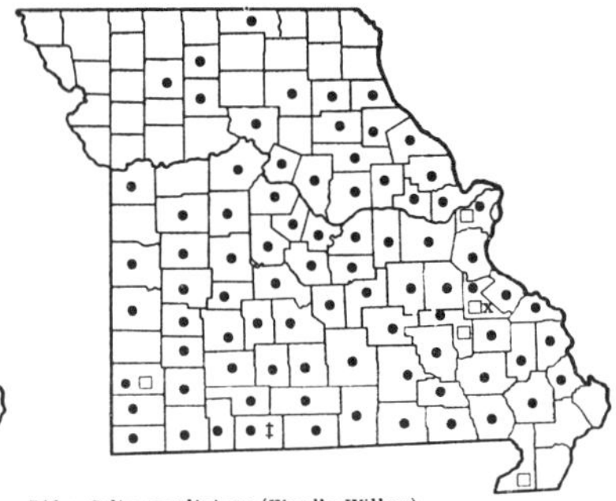

716 • Salix caroliniana (Ward's Willow)
716 □ Salix caroliniana X Salix nigra
716 x Salix caroliniana X Salix sericea
716 ‡ Salix caroliniana X Salix rigida

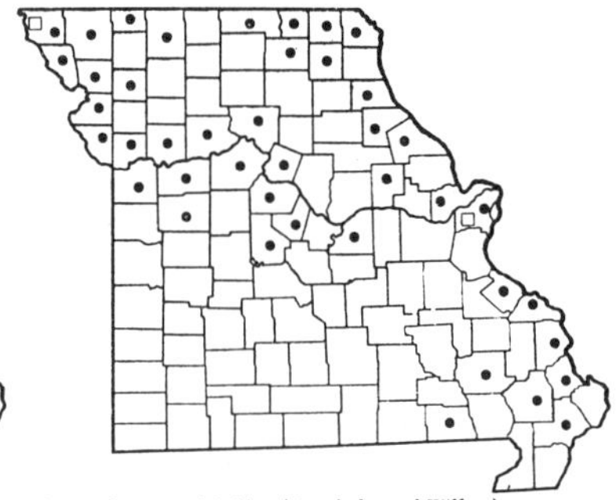

717 • Salix amygdaloides (Peach-leaved Willow)
717 □ Salix X Glatfelteri (Salix amygdaloides X Salix nigra)

1. **Salix nigra** Marsh. Black Willow Map 715
Flowers April–May.

Occurs along borders of streams, sloughs, and ponds generally in muddy alluvial soils. Common throughout Missouri and doubtless in every county of the state.

Ranges from New Brunswick to North Dakota, south to Florida and Texas.

Along the streams in the Ozark region this willow is found, not on the immediate gravel bed next to the stream occupied by the pioneering Ward's Willow, *Salix caroliniana*, but in the deeper alluvial soils of mud and silt which have been built up in later stages of formation of river banks. This finer deeper silt occurs back of or behind the zone occupied by *Salix caroliniana*. Outside of the Ozark region, Black Willow is found along the muddy margins and flats bordering all the streams.

This is the largest and tallest of any native species of willow in the United States. Used as mattresses along the larger rivers for reinforcing levees and protecting them from washing, it helps control erosion of river banks.

The wood does not warp or splinter, and is used in the manufacture of toys, packing cases, and some kinds of furniture where strength is not an item of importance, since the wood is one of the softest and lightest of all native eastern hardwood trees. Together with other species of willow, the flexible strips of young branches are made into wicker baskets and ornamental pieces. Small amounts of the wood go into pulping for the papermaking business in the United States. The bitter bark of the root has been used as an ingredient for blood medicines during spring. This is one of the species I have observed beaver using for dam construction.

2. **Salix caroliniana** Michx. Ward's Willow Map 716

*Salix longipes* Shuttlew. var. *Wardii* (Bebb) Schneider [P & S]

Flowers April–May.

Occurs along gravel bars, sandy gravel beds, and rocky banks of streams throughout the Ozark region of southern and central Missouri and northward to Marion, Shelby, Macon, Putnam, Grundy, Daviess, and Jackson counties.

Ranges from Florida to Texas, north to Maryland, West Virginia, Pennsylvania, Indiana, Illinois, Missouri, and Kansas; also in Cuba.

This attractive willow, with white lower leaf-surfaces, is abundant along all the Ozark streams where it pioneers as one of the first woody species to occupy the outer edges of gravel bars and rocks in the stream. It is one of the species used in the Ozarks for making into wickerwork for baskets, furniture, and ornamental pieces.

This species hybridizes with *S. nigra*, *S. sericea*, and *S. rigida* in Missouri. Presumed crosses have been recorded between *S. caroliniana* and *S. nigra* scattered in southern and central Missouri, between *S. caroliniana* and *S. sericea* from St. Francois County, and between *S. caroliniana* and *S. rigida* from Taney County.

3. **Salix amygdaloides** Anders.
Peach-leaved Willow Map 717

---

Plate no. 128. **1.** Saururus cernuus, × 2/7. **2.** Salix nigra, × 2/7; *a.* Male inflorescence, × 2/7; *b.* Female inflorescence, × 2/7. **3.** Salix caroliniana, × 2/7; *a.* Male inflorescence, × 2/7. **4.** Salix babylonica, × 2/7; *a.* Female inflorescence, × 2/7; *b.* Male inflorescence, × 2/7. **5.** Salix amygdaloides, × 2/7; *a.* Male inflorescence, × 2/7; *b.* Female inflorescence, × 2/7. **6.** Salix fragilis, × 2/7; *a.* Female inflorescence, × 2/7; *b.* Male inflorescence, × 2/7; Details from Small, The New York Botanical Garden.

PLATE NO. 128

718 Salix fragilis (Crack Willow)

719 Salix babylonica (Weeping Willow)

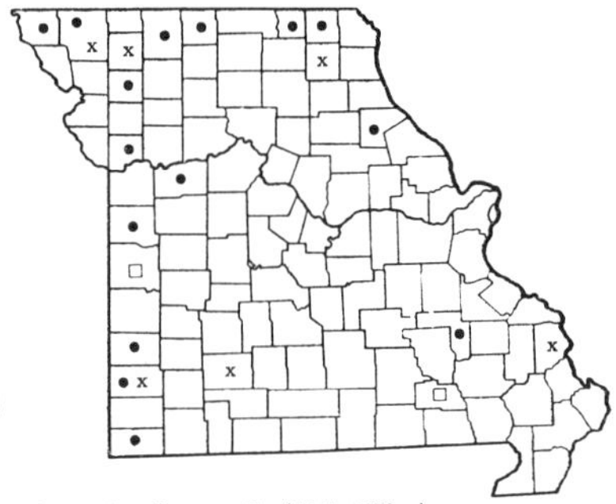

720 ● Salix alba var. alba (White Willow)
720 □ Salix alba var. calva
720 x Salix alba var. vitellina

Flowers April–June.

Occurs along alluvial muddy banks and low woods bordering the Missouri, Mississippi, and other large rivers of northern, central, and eastern Missouri, south of the Missouri River west to Ripley, Wayne, Ste. Genevieve, St. Louis, Osage, Morgan, Johnson, and Jackson counties.

Ranges from Vermont and Quebec to British Columbia, south to Massachusetts, New York, Ohio, Indiana, Illinois, Missouri, the Rocky Mountain states, south to Texas and New Mexico, and Washington.

The slightly drooping branches give this species the most 'weeping' effect of any of the Missouri species, but not as great as the cultivated Weeping Willow (*S. babylonica*), a native of the Old World. The trees attain a height of nearly 70 feet and are of some importance in acting as a natural agent in reinforcing soils and controlling them against washing along the Missouri and Mississippi rivers and their tributaries. In Europe the split younger branches are woven into baskets.

A hybrid between this willow and the Black Willow (*S. nigra*), known as *Salix* × *Glatfelteri* Schneid., occurs in St. Louis County.

4. **Salix fragilis** L. Crack Willow Map 718
Flowers April–June.

Occurs along streams, in swamps, and wet woods. Infrequent in the state, where known from Pike, St. Louis, Wayne, Dekalb, and Atchison counties.

Native of Europe; planted and escaped from cultivation in the United States and Canada, where naturalized from Newfoundland to Ontario, south to Virginia, Kentucky, Illinois, and Missouri.

This willow attains a height of about 80 feet. During early colonial times it and *S. nigra* were used for powdered charcoal as an ingredient of the black gunpowder then popular. Also called Brittle Willow because of the twigs, which are very brittle at their base.

5. **Salix babylonica** L. Weeping Willow Map 719
Flowers April–May.

Commonly planted as an ornamental tree, but rarely escaped or naturalized in Missouri away from cultivation in wet ground and meadows, as in Dent and Jackson counties.

Native of China and introduced from Europe into the United States. True *S. babylonica* is hardy only in the southern third of the United States. It hybridizes with *S. alba* and *S. fragilis*.

This tree may attain a height of 30 feet. With its long branches which hang nearly to the ground, it is beautiful the year round. In fall the leaves turn to greenish yellow or pale yellow.

This species is commonly confused with other cultivated 'weeping' willows, such as the Golden Weeping Willow (*S. alba* var. *tristis*), Niobe or Wisconsin Weeping Willow (*S.* × *blanda*), and Thurlow Weeping Willow (*S.* × *elegantissima*). The four may be distinguished by their branches and leaves, *S. alba* var. *tristis* having yellow or golden branches with the leaves somewhat silky-hairy on the lower surface, and on the upper as well, in younger leaves; *S. babylonica* with mostly olive-brown or red-brown, sometimes yellow, branches and glabrous narrowly lanceolate leaves 8–15 mm. wide; *S.* × *elegantissima* and *S.* × *blanda*, similar to *S. babylonica*, but with broader leaves 15–22 mm. wide, the veins of the leaves in *S.* × *elegantissima* spreading at an angle of more than 45°. The branches of *S.* × *elegantissima* also are about twice as long as those of *S.* × *blanda* and the leaves in *S.* × *blanda* are shining above, those in *S.* × *elegantissima* not shining above.

6. **Salix alba** L. White Willow Map 720
Flowers April–May.

Often planted as an ornamental tree and sometimes escaped and naturalized in wet ground along streams and ponds.

Native of Europe and introduced into the United States and Canada.

Three of the varieties of *S. alba* which have escaped from cultivation may be distinguished as follows:

a. Branchlets olive-green to brown, silky-hairy; leaves white-silky hairy on lower surface . . 6a. S. ALBA var. ALBA
a. Branchlets brown or yellow, becoming glabrous or nearly so (without hairs); leaves becoming glabrous on lower surface . . . . *b*
  b. Branchlets brown . . 6b. S. ALBA var. CALVA G. F. W. Mey.
  b. Branchlets yellow . 6c. S. ALBA var. VITELLINA (L.) Stoker

All three varieties have sparingly spread from cultivation.

Another variety, the strongly weeping variation of *S. alba*, known as the Golden Weeping Willow, *S. alba* var. *tristis* Gaud., has long bright yellow branches which hang down, much as in *S. babylonica*, *S.* × *blanda*, and *S.* × *elegantissima*. The last two are considered to be of hybrid origin between *S. fragilis* and *S. babylonica*. *Salix alba* var. *tristis* is much hardier farther north in the United States and Canada and commonly planted in those regions, but has not been reported as escaped from cultivation in Missouri.

*Salix alba*, like *S. fragilis*, became introduced in early colonial times in North America, and was used similarly for making into charcoal for gunpowder purposes and for medicine. The medicine was used as a substitute for quinine and for fevers. The bark of this and other large willows, such as *S. fragilis* and *S. nigra*, has been used for tanning purposes, and the wood used in the making of charcoal. This species and *S. fragilis* are commonly planted around farmyards and pastures in Missouri and elsewhere, where it serves as hedges and for posts, while the younger stems are used for fishweirs and whistles.

The leaves turn a pale yellow in autumn.

7. **Salix interior** Rowlee Sandbar Willow Map 721

Flowers May–June.

Occurs on sand bars, mud flats, and alluvial muddy banks of streams, oxbow lakes and ponds of river bottoms and flood plains throughout Missouri, except absent apparently in the lowlands of the southeastern section.

Two variations occur in the state:

Leaves, especially the youngest, rather permanently and densely silky-hairy, the older ones mostly 10–14 mm. (–18) wide, and 3–5 cm. (–7) long. 7b. S. INTERIOR var. INTERIOR f. WHEELERI
Leaves, especially the youngest, less densely silky-hairy, the older ones mainly 5–12 mm. (–15 on strong shoots) wide, and 5–14 cm. long . . . 7a. S. INTERIOR var. INTERIOR f. INTERIOR

7a. **Salix interior** var. **interior** f. **interior** Map 721

*Salix interior* Rowlee [G, BB, P & S]

This is the commoner form encountered in the state.

Ranges from Quebec to Alaska, south to Virginia, West Virginia, Ohio, Kentucky, Illinois, Arkansas, Louisiana, Texas, and the Great Plains states; mainly absent from New England and Coastal Plain.

7b. **Salix interior** var. **interior** f. **Wheeleri** (Rowlee) Rouleau Map 721

*Salix interior* var. *Wheeleri* Rowlee [BB, P & S]

This is known from several counties in the state, and is less commonly collected.

Of similar range to var. *interior* f. *interior*, but more common around the Great Lakes area and sandhills of Nebraska and Wyoming, and North Dakota.

The Sandbar Willow attains a height of about 20 feet and is generally found in dense stands or thickets of rather impenetrable growth. It is associated with Silver Maple (*Acer saccharinum*) and Cottonwood (*Populus deltoides*) on the river flats of the Missouri and Mississippi rivers for stretches of many miles on the alluvial silty soils deposited by these rivers. In such places this willow has pioneered to become established, and these areas are poor botanically with only a small variety of plant life, such as *Aster lateriflorus* or *Carex hyalinolepis*. Sandbar Willow makes a good soil binder on these river soils and prevents much washing and erosion which might otherwise take place.

8. **Salix rigida** Muhl. Map 722

Flowers April–May.

Occurs in wet ground along streams, spring branches, and in wet meadows. Throughout the state, except absent from the lowland section of southeastern Missouri.

The following variations may be distinguished:

a. Leaf margin with few or no teeth . 8c. S. RIGIDA var. RIGIDA f. SUBINTEGRA
a. Leaf margin with many teeth scattered from base to tip . . . . . . . . . . *b*
  b. Young twigs and branchlets densely hairy; lower side of leaves, especially the mid-

721 • Salix interior var. interior f. interior (Sandbar Willow)
721 □ Salix interior var. interior f. Wheeleri

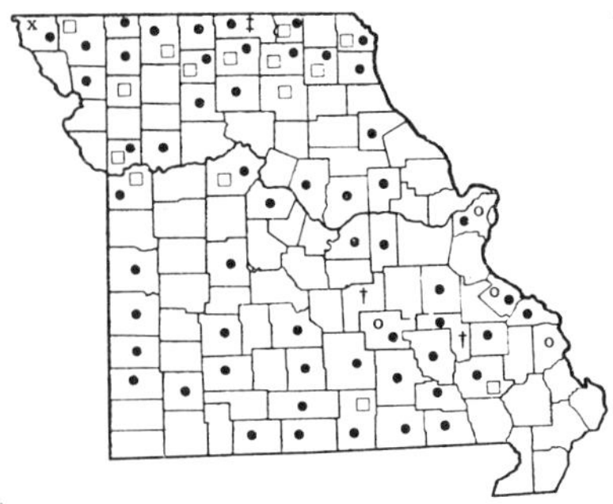

722 • Salix rigida var. rigida f. rigida
722 □ Salix rigida var. angustata
722 ‡ Salix rigida X Salix humilis
722 x Salix rigida X Salix nigra
722 † Salix rigida var. rigida f. subintegra
722 o Salix X myricoides (Salix rigida X Salix sericea)

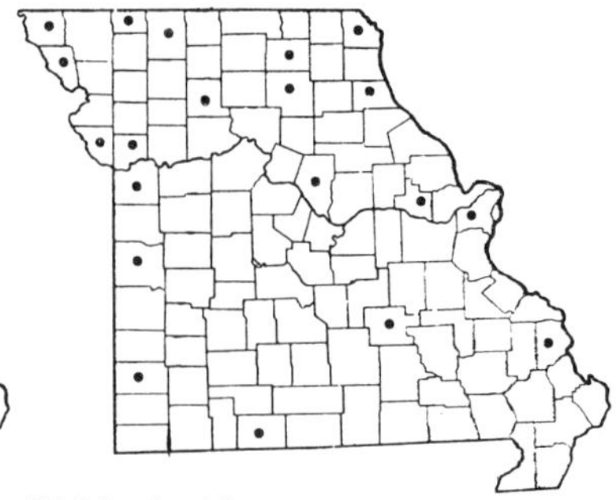

723 Salix eriocephala

nerve, hairy; leaf-stalks (petioles) hairy . 8b. S. RIGIDA var. RIGIDA f. MOLLIS

b. Young twigs and branchlets mainly glabrous (without hairs) or nearly so; lower side of leaves glabrous or becoming so; leaf-stalks (petioles) glabrous or essentially so . *c*
 c. Leaves narrowly lanceolate, gradually tapering to the base, mainly 4–9 times as long as broad, the larger ones 9–22 mm. broad . 8d. S. RIGIDA var. ANGUSTATA
 c. Leaves oblong-lanceolate, broader in the middle and more abruptly pointed, mostly broadly rounded to nearly heart-shaped at the base, mainly 2½–5 times as long as broad, the larger ones 15–45 mm. broad . . . . . . 8a. S. RIGIDA var. RIGIDA f. RIGIDA

8a. **Salix rigida** var. **rigida** f. **rigida** Map 722

*Salix rigida* Muhl. [G, BB in part]; *Salix cordata* Muhl. and auth., not Michx. [P & S]

This is the commonly encountered variation in Missouri.

Ranges from Newfoundland to Ontario, south to North Carolina, Mississippi, Arkansas, and Kansas.

8b. **Salix rigida** var. **rigida** f. **mollis** (Palmer & Steyerm.) Fern. Map 722

*Salix cordata* f. *mollis* Palmer & Steyerm.

Scattered throughout the state.

8c. **Salix rigida** var. **rigida** f. **subintegra** (Palmer & Steyerm.) Steyerm. Map 722

*Salix cordata* f. *subintegra* Palmer & Steyerm. in Ann. Mo. Bot. Gard. 25: 770. 1938.

Known only from Iron and Phelps (shore of lake at Yancy Mills Spring, November 28, 1936, *Steyermark 20910*, holotype in Missouri Botanical Garden Herbarium) counties.

8d. **Salix rigida** var. **angustata** (Pursh) Fern. Map 722

*Salix cordata* var. *angustata* (Pursh) Anderss. [P & S]

Mainly found in northern and west-central Missouri, and south locally in Wayne and Howell counties.

Ranges from Quebec to Ontario, south to Georgia, Alabama, and Missouri.

*Salix rigida* hybridizes with several other species of willow. Those which are known in Missouri are hybrids between *S. rigida* and *S. nigra* from Atchison Co.; *S. rigida* and *S. caroliniana* from Taney Co.; *S. rigida* × *S. humilis* from Putnam Co.; and *S.* × *myricoides* (*S. rigida* × *S. sericea*) from St. Louis, Ste. Genevieve, Cape Girardeau, and Dent counties.

9. **Salix eriocephala** Michx. Map 723

*Salix missouriensis* Bebb [P & S]
*Salix rigida* var. *vestita* Anderss. [BB]

Flowers March–April.

Occurs in alluvial soils of silt and mud following the Missouri and Mississippi and other large rivers of northern, central, and eastern Missouri, mainly north of the Missouri River, and south locally to Cape Girardeau and Bates counties.

Plate no. 129. **1.** Salix alba, × 2/7; *a.* Female inflorescence, × 2/7. **2.** Salix rigida var. rigida, × 2/7; *a.* Male inflorescence, × 2/7; *b.* Salix rigida var. angustata, × 2/7. **3.** Salix interior, × 2/7; *a.* Male flower-branch; *b.* Female flower-branch. **4.** Salix eriocephala, × 2/7; *a.* Leaf with stipule, × 2/7; *b.* Male flower-branch, × 2/7. **5.** Salix discolor, × 2/7; *a.* Female flower-branch, × 2/7; *b.* Male flower-branch, × 2/7. **6.** Salix humilis var. hyporhysa, × 2/7; *a.* Male flower-branch, × 2/7.

PLATE NO. 129

724 Salix discolor var. discolor (Pussy Willow)

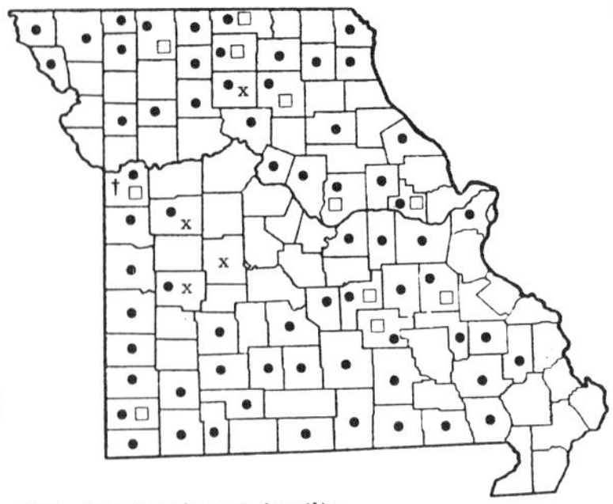

725 x Salix humilis var. humilis
725 • Salix humilis var. hyporhysa
725 □ Salix humilis var. microphylla
725 † Salix humilis X Salix eriocephala

726 Salix gracilis var. textoris

Ranges from Indiana to Minnesota and South Dakota, south to Kentucky, Missouri, and Nebraska.

This species may become a large tree up to 50 ft. tall. The stipules are smaller and less conspicuous than in *S. rigida*, but the fruiting catkins, capsules, and styles are longer, the leaves are often broader, and the branchlets and lower leaf surface much more pubescent than in *S. rigida*. By some botanists *S. eriocephala* is sometimes combined with *S. rigida* or treated as only a variety of that species, but it seems amply distinct and recognizable from that species, not only on the basis of the characters mentioned above, but also on the habit and habitat; in Missouri *S. eriocephala* is usually a medium- to large-sized tree along the larger streams, whereas *S. rigida* is usually a shrub found along smaller streams and wet meadows. The catkins of *S. eriocephala* also tend to open somewhat earlier than the unfolding leaves, whereas in *S. rigida* they open as the leaves unfold.

A hybrid between *S. eriocephala* and *S. humilis* has been recorded from Jackson County.

As in *S. amygdaloides* and *S. interior*, this species plays an important role in holding back the soil along the river banks and preventing soil erosion. Also the wood, because of its relative permanence when contacting the soil, is used for fence posts.

10. **Salix discolor** Muhl. var. **discolor**
Pussy Willow Map 724
*Salix discolor* Muhl. [G, BB, P & S]
Flowers February–April.

Known only from Clark County, northeastern Missouri (August 30, 1892, *Bush*, from World's Fair specimen 2388 in University of Missouri Herbarium). In a letter dated October 21, 1934, Bush wrote the author that this specimen was collected 'just north of Dumas on the river bank some 3–4 mi. north of the town.'

Ranges from Labrador to Alberta, south to Maryland, Kentucky, Missouri, South Dakota, and Montana.

The branchlets are glabrous or, if at first hairy, eventually become glabrous in this variety of *S. discolor*, whereas in var. *latifolia* they are more permanently pubescent, as are the bud-scales.

The soft silky hairs clothing the bracts of the catkins give this species much more of a 'Pussy Willow' effect than in most of the other native species, except possibly *S. sericea* and *S. humilis*. The 'Pussy Willow' sold in florists' shops in early spring and planted as an ornamental shrub or small tree is a species from Europe and Asia, known as Goat Willow, *S. caprea* L. It has much larger fuzzy catkins than *S. discolor*, but only sparingly escapes from cultivation, and thus far has not been recorded as an escape in Missouri.

11. **Salix humilis** Marsh. Prairie Willow
Map 725
Flowers March–May.

Occurs in prairies, open woods, rocky draws or washes, rocky slopes, and thickets.

Missouri material is represented by three variations:

a. Staminate catkins 7–8 mm. thick, 5–12 mm. long; fruiting catkins 10–25 mm. long; leaves 5–15 mm. (up to 17) broad, mostly 15–50 mm. (up to 70) long, nearly sessile (without a stalk) to very short-stalked; shrub 2.5–10 dm. tall . . . 11c. S. HUMILIS var. MICROPHYLLA
a. Staminate catkins 10–23 mm. thick, 10–30 mm. long; fruiting catkins 20–50 mm. (up to 80) long; leaves 7–30 mm. broad, mostly 50–100 mm. (rarely 40) long, short-stalked; shrub 10–30 dm. tall . . . . . . . . . . *b*
b. Lower side of leaves definitely covered with hairs . . . . 11a. S. HUMILIS var. HUMILIS
b. Lower side of leaves only slightly hairy or

becoming glabrous (without hairs) or nearly so . . 11b. S. HUMILIS var. HYPORHYSA

11a. **Salix humilis** var. **humilis** Map 725

This is scattered over the state, but is less common than var. *hyporhysa.*

Ranges from Quebec to Minnesota, south to North Carolina, Kentucky, Louisiana, and Kansas.

Some hairy-leaved Missouri specimens have been referred to var. *keweenawensis* Farw. in Gleason's *New Illustrated Flora* (vol. 2: 22. 1952), but are placed in the present treatment under var. *humilis.*

11b. **Salix humilis** var. **hyporhysa** Fern. Map 725

*Salix humilis* var. *rigidiuscula* of auth. [BB, P & S], not Anderss.

Throughout Missouri. This is the commonest variation in the state.

Ranges from Florida to Texas, north near the Coastal Plain to Connecticut and Pennsylvania, inland north to West Virginia, Ohio, Michigan, Wisconsin, Iowa, and Oklahoma.

11c. **Salix humilis** var. **microphylla** (Anderss.) Fern. Dwarf Gray Willow Map 725

*Salix tristis* Ait. [BB, P & S]

Scattered in portions of northern and central Missouri, south to Washington, Dent, and Phelps counties.

Ranges from Maine to Minnesota, south to Florida, Mississippi, Louisiana, and Oklahoma.

Although var. *microphylla* is treated by some botanists as a distinct species, it seems better regarded as a dwarf variety of *S. humilis*, smaller in all respects but intergrading with the other varieties of that species.

*Salix humilis* and varieties can be readily recognized in the field generally by its low growing habit and dry open habitat in prairies, rocky open woods, and similar situations quite in contrast to the wet swampy habitats selected in general by the other species of willow in Missouri.

Presumed hybrid plants have been recorded between this species and *S. rigida* from Putnam County, and between *S. humilis* and *S. eriocephala* from Jackson County.

12. **Salix gracilis** Anderss. var. **textoris** Fern. Map 726

*Salix petiolaris* of some auth. [P & S, BB], not Sm.

Flowers April–June.

Known only from low wet ground in mud or sandy gravel along streams in northeastern Missouri, from Clark (*Davis 1500*), Marion, Ralls (*Steyermark 25769*), and Pike counties.

Ranges from Quebec to Manitoba, south to New Jersey, Pennsylvania, Ohio, Indiana, Illinois, Iowa, Missouri, and Nebraska.

This is a low growing shrub, similar in this respect to *S. humilis* and varieties, but with green to olive-brown, slender flexible branches. The narrow leaves, linear to narrowly lanceolate, are entire or have short teeth, and are 2.5–10 cm. long by 3–20 mm. wide.

13. **Salix sericea** Marsh. Silky Willow Map 727

Flowers March–May.

Occurs in usually calcareous swampy meadows, swampy ground around springs, and spring branches. Eastern Ozark region north to St. Louis, Washington, and Crawford counties, and west to Texas County.

Two variations are encountered in Missouri:

Lower surface of mature leaves silky-hairy . . 13a. S. SERICEA f. SERICEA

Lower surface of mature leaves glabrous (without hairs) . . . . . 13b. S. SERICEA f. GLABRA

13a. **Salix sericea** f. **sericea** Map 727

This is the common variation encountered.

Ranges from Quebec to Wisconsin and Iowa, south to South Carolina, Tennessee, and Missouri.

13b. **Salix sericea** f. **glabra** Palmer & Steyermark Map 727

Known only from Dent (north prong of Meramec River, between Stone Hill and Indian Trail State Park, August 4, 1936, *Steyermark 12548*, holotype in herbarium of Missouri Botanical Garden) and Reynolds (along Brushy Creek, 1 mi. north of Moses Store, August 13, 1936, *Steyermark 12918*) counties in the central-eastern Ozarks.

*Salix sericea* is very distinctive in the field, growing as a densely rounded shrub 2–3 meters tall with dark brown to purplish-brown slender twigs and narrowly lanceolate, finely and regularly toothed leaves, which are dark green on upper surface and gray-silvery on lower surface with a fine silky hairiness of appressed hairs. In the calcareous, swampy, spring-fed meadows of the Ozarks it is associated with many herbaceous species of a more northern geographical range, whose presence in the southeastern Ozark area indicates a survival of a relict type left over from and related to a time of prehistoric Pleistocene ice advances.

14. **Salix purpurea** L. Basket Willow Map 728

Also called Purple Osier.

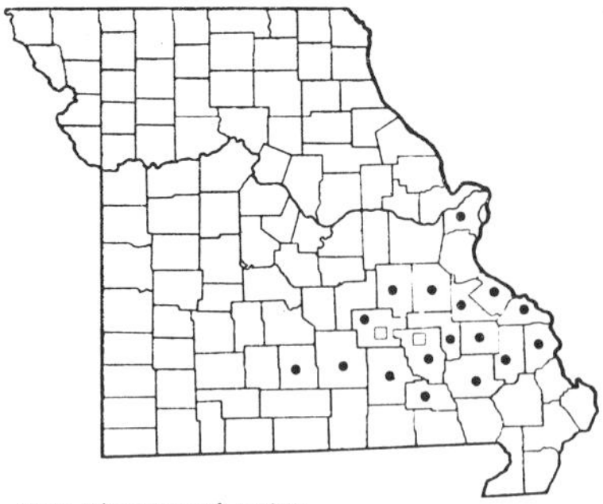

727 • Salix sericea f. sericea
727 □ Salix sericea f. glabra

728 Salix purpurea (Basket Willow)

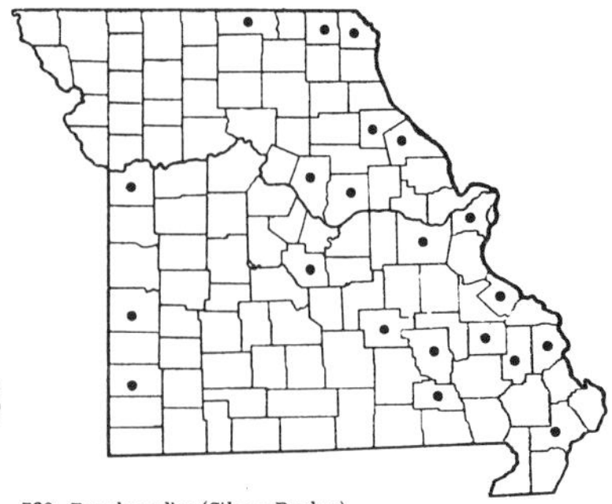

729 Populus alba (Silver Poplar)

Flowers April–May.

Known only from Ralls County, northeastern Missouri (Oakwood, September 17, 1912, *Davis 364*; Oakwood, May 12, 1913, *Davis 250*).

Native of Europe; introduced into the United States and Canada, where ranging from Newfoundland to Ontario and Wisconsin, south to Virginia, West Virginia, Ohio, Illinois, and Missouri.

Several other specimens from Missouri have been misidentified as this species, but pertain elsewhere.

*Salix purpurea* was brought to this country from Europe in early colonial times and grown for basket-making. The leaves are often purple-tinged and covered with a silvery or gray color; they are entire or finely toothed.

## 2. **Populus** L. Poplar, Aspen

This genus of trees has wind-pollinated flowers. The pollen is very light and is easily carried by the wind, thus making the poplars eligible for hay fever consideration. While some of the species in the southern and western states are of some importance, those in the central and eastern United States seem to have little importance in hay fever cases.

Instances are reported where the bitter bast substance of the bark has been dried and ground into a flour as an emergency food. The same use has been reported for willows.

### I. *Key based mainly on leaves and twigs*

a. Young twigs and lower surface of fully grown leaves covered with a white felt of hairs; at least the leaves at end of branches maple-like with 3–5 lobes . . . . . . . . . . . . . . . . 1. P. ALBA

a. Twigs at maturity not covered with a white felt of hairs (some white hairiness may be present on young twigs in the beginning stages of growth); fully grown leaves not covered with a white felt of hairs, not 3–5-lobed . . . . . . . . . . . . . . . . . . . . . . . . . . . . *b*

b. Leaves coarsely toothed with large or broad teeth, these usually less than 15 on each edge . . . . *c*

c. Teeth of leaves somewhat triangular and short-pointed; mature leaf-blades mainly 6–12 cm. (rarely 4) long, 4–9 cm. broad, eventually glabrous (without hairs) or nearly so; at least the topmost bud hairy . . . . . . . . . . . . . . . . . . 2. P. GRANDIDENTATA

c. Teeth of leaves mostly blunt or rounded at their tips; mature leaf-blades averaging smaller, mainly 2–6 cm. long, 2–4 (–7) cm. broad, densely hairy to nearly glabrous; buds not hairy . 3. P. TREMULA

b. Leaves with closely spaced teeth, these usually 15–50 or more on each edge (rarely less than 15) . . *d*

d. Pith of stems orange; tip of leaf usually blunt or round (rarely short-pointed); base of leaf usually heart-shaped or prominently rounded with no space between the rounded portions (sinus deep and closed); leaf-stalks (petioles) terete (rounded in cross section); trees of the swamps of southeastern Missouri . . . . . . . . . . . . . . . . 5. P. HETEROPHYLLA

d. Pith of stems white; tip of leaf short- to long pointed; base of leaf tapering or narrowed or rather straight-edged, if rounded, at least wi h an open space (sinus) between the rounded portions; leaf-stalks (petioles) flattened or compressed toward the summit; trees of other habitats or in other parts of Missouri . . . . . . . . . . . . . . . . . . . *e*

e. Leaves whitened, definitely paler, and somewhat hairy on lower side; leaf-stalks (petioles) terete or nearly so (rounded in cross section) in upper part . . . . . 6. P. × GILEADENSIS

e. Leaves usually equally green on both sides, but if pale or grayish on the lower side, then not hairy; leaf-stalks (petioles) flattened laterally (from the sides) in upper part toward leaf-blade . . . . . . . . . . . . . . . . . . . . . . . . . . *f*

f. Teeth along edge (margin) of leaves rather blunt or rounded at tip; leaves usually broadest near the middle or above the base, often as long as broad or longer than broad, as well as shorter than broad, broadly ovate or somewhat round in shape, mostly less than 7 cm. wide; margin of leaves without a thinner translucent border (not permitting light to pass through); trees of northern Missouri only . . . . 4. P. TREMULOIDES

f. Teeth along edge (margin) of leaves ending in a more prominent or narrowed or curved point; leaves usually broadest at the very base and usually much broader than long (rarely longer than broad), usually more broadly triangular (deltoid) or somewhat rhombic (somewhat broadly baseball-diamond-shaped), mostly more than 7 cm. broad; margin of leaves with a thin translucent border (permitting light to pass through); trees found in all parts of Missouri . . . . . . . . . . . . . . . . . . . *g*

g. Tree with a narrow, erect, slender, columnar shape, the branches all conspicuously ascending or erect; no glands at base of leaf-blade; no tiny hairs on edges of leaf-blade. 7. P. NIGRA var. ITALICA

g. Trees with more widely spreading branches and a more open crown, not narrow and slender-looking; glands usually present at base of leaf-blade (sometimes absent in *P. × canadensis*); tiny hairs present on edges of leaf-blade . . . . . . . . . . *h*

h. Common native tree along streams and in low ground; tree with a broad open crown and spreading branches; leaves broadly triangular, usually as broad as long, the base mostly straight-edged (truncate) or somewhat heart-shaped; margins of leaves with numerous hairs; glands 2–4, always present at base of leaf-blade at junction with leaf-stalk . . . . . . . . . . . . . . . 9. P. DELTOIDES

h. Planted tree, sometimes escaping from cultivation and becoming naturalized; tree with a narrower crown, the branches more ascending; leaves triangular-ovate, the base more curved, narrowed, or cuneate; margins of leaves with fewer and more scattered hairs; glands 1–2, either present or often absent at base of leaf-blade. 8. P. × CANADENSIS

## II. *Key based on male and female flowers, fruit, and vegetative characters*

a. Scales of catkins either with short triangular teeth on margin or with 3–7 narrowly triangular or linear-lanceolate, deeply cut lobes or segments, these with long hairs at their tips; stigmas thread-like, terete; stamens 6–12 to each flower of the male catkin; topmost overwintering buds 3–10 mm. long . . *b*

b. Scales of the male and female flowers with short teeth, not deeply cut; young twigs covered with a white hairy felt; topmost bud covered with fine soft downy hairs; mature leaves white-hairy on lower side . . . . . . . . . . . . . . . . . . . . . . 1. P. ALBA

b. Scales of the male and female flowers with 3–7 narrowly triangular or linear-lanceolate deeply cut lobes or segments; twigs and winter-buds glabrous or hairy, but not soft white-downy; mature leaves eventually glabrous, young leaves glabrous or white-velvety on lower side. . . . . *c*

c. Winter buds hairy; branches somewhat gray-hairy at first; unfolding leaves thick and with white-velvety hairs . . . . . . . . . . . . . . . . 2. P. GRANDIDENTATA

c. Winter buds glabrous; young branches glabrous or nearly so; unfolding leaves thin and glabrous or soon becoming so . . . . . . . . . . . . . . . . . . . *d*

d. Leaves irregularly wavy-toothed with the coarse teeth usually less than 12 along either half; tip of leaf-blade usually blunt . . . . . . . . . . . . . . . . 3. P. TREMULA

d. Leaves finely and regularly toothed with the small teeth mainly 15–40 along either half; tip of leaf-blade usually short-pointed . . . . . . . . . . . . . 4. P. TREMULOIDES

a. Scales of catkins fringed with 9 to many hair-like segments, these without any hairs; stigmas with broadly expanded or dilated lobes; stamens mainly 12–60 to each flower of the male catkin; topmost overwintering bud 10–25 mm. long . . . . . . . . . . . . . . . . . *e*

e. Pith orange; topmost bud on the branch not sticky (viscid); stigmas raised up on a definite slender

730 Populus grandidentata (Large-toothed Aspen)

731 Populus tremula (European Aspen)

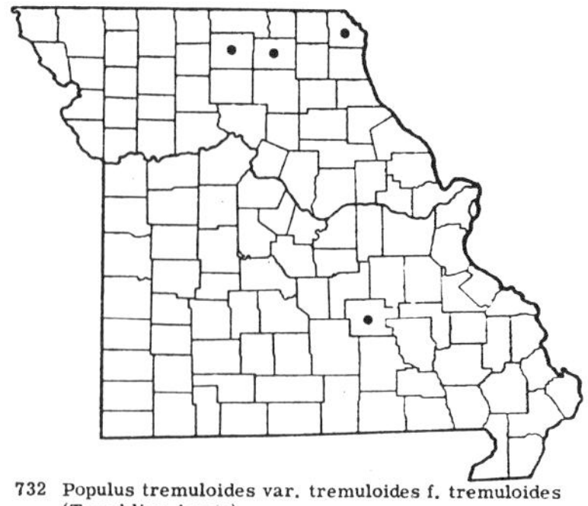
732 Populus tremuloides var. tremuloides f. tremuloides (Trembling Aspen)

style; unfolding leaves with a white-hairy felt; trees of swamps of the southeastern Missouri lowlands . . . . . . . . . . . . . . . . . . . . . . 5. P. HETEROPHYLLA

e. Pith white; topmost bud on the branch sticky (viscid); stigmas sessile (with a stalk) or nearly so; unfolding leaves without a white-hairy felt; trees of other habitats or in other parts of Missouri . . *f*

f. Stigmas 3–4; stamens 40–60; common native tree along streams and in low ground throughout Missouri . . . . . . . . . . . . . . . . . . . . . . . 9. P. DELTOIDES

f. Stigmas 2; stamens 12–30; planted trees rarely escaping from cultivation and becoming naturalized. . . . . . . . . . . . . . . . . . . . . . . . . *g*

g. Branchlets or twigs hairy; young leaves white on lower surface, hairy, the base heart-shaped. 6. P. × GILEADENSIS

g. Branchlets or twigs glabrous (without hairs); young leaves glabrous, or if somewhat hairy, the base truncate (as if cut straight across or straight-edged) or curved, narrowed, or cuneate, rarely slightly heart-shaped. . . . . . . . . . . . . . . . . *h*

h. Tree with a narrow, erect, slender, columnar shape, the branches all conspicuously ascending or erect; no glands at base of leaf-blade; no tiny hairs on edges of leaf-blades. 7. P. NIGRA var. ITALICA

h. Tree with a more elliptical or narrowly ovate shape, the branches ascending but the crown more open; glands either 1–2 at the base of the leaf-blade at its junction with the leaf-stalk (petiole) or no glands present; scattered hairs occur on edges of leaf-blades. 8. P. × CANADENSIS

1. **Populus alba** L. Silver Poplar
Also called White Poplar. Map 729
Flowers March–May.

At one time commonly planted and now well established along streams, fence rows, and roadsides. Found throughout the state, and probably naturalized in every county.

Native of Europe.

This species spreads by underground suckers, forming thickets, and apparently can grow in most kinds of soils and varied situations. It is not commonly planted now as formerly.

2. **Populus grandidentata** Michx.
Large-toothed Aspen Map 730
Flowers March–May.

Known only from Clark County in northeastern Missouri (August 30, 1892, *Bush*, world's fair specimen 2401 in University of Missouri Herbarium).

Ranges from Quebec to Ontario and Minnesota, south to North Carolina, Tennessee, and Iowa; introduced in northeast Missouri.

According to correspondence with Bush (letters of September 15, 1933, and October 21, 1934), he stated that he collected this specimen 'just east of the depot... [at Dumas]... in a field on edge of woods and I decided years ago that the trees were adventive there' and 'escaped from cultivation.' On the basis of Bush's letters and since no trees other than these have been seen in Missouri, I am treating the species as introduced in the state.

In the fall the foliage turns to pale yellow. The

Plate no. 130. **1.** Salix humilis var. microphylla, $\times {}^2/_7$; *a.* Female flower-branch, $\times {}^2/_7$; *b.* Male flower-branch, $\times {}^2/_7$. **2.** Salix gracilis var. textoris, $\times {}^2/_7$; *a.* Male flower-branch, $\times {}^2/_7$. **3.** Populus alba, $\times {}^2/_7$. **4.** Salix sericea, $\times {}^2/_7$; *a.* Female flower-branch, $\times {}^2/_7$; *b.* Male flower-branch, $\times {}^2/_7$. **5.** Salix purpurea, $\times {}^2/_7$; *a.* Male flower-branch, $\times {}^2/_7$. **6.** Populus grandidentata, $\times {}^2/_7$. **7.** Populus tremula, $\times {}^2/_7$; Details from Small, The New York Botanical Garden. **8.** Populus tremuloides, $\times {}^2/_7$; *a.* Female flower-branch, $\times {}^2/_7$; *b, c.* Male catkins, $\times {}^2/_7$.

PLATE NO. 130

wood is used for pulp in papermaking. The sapwood is white, and the heartwood a pale brown. Some radio cabinets and bedroom furniture are occasionally made from a fancy grain which rarely appears in the wood of this species.

3. **Populus tremula** L. European Aspen Map 731

Flowers March–May.

Known only from Washington County in the eastern Ozark section (near Palmer, T36N, R1W, sect. 13, August 18, 1936, *Steyermark 12997*).

Native of Europe, Asia, and North Africa.

The trees were plentiful where this specimen was found and were well established at this station, where they have escaped from cultivation.

The wood of this species is used for pulp in papermaking.

4. **Populus tremuloides** Michx. var. **tremuloides** f. **tremuloides** Trembling Aspen Map 732

Also called Quaking Aspen.

Flowers March–April.

Occurs in thickets and borders of upland woods. Known only from northern Missouri in Clark, Adair, and Sullivan counties; also in Dent County, but not now extant at the Dent County station.

Ranges from Labrador to Alaska, south to Virginia, Tennessee, Missouri, Alberta, New Mexico, and Lower California.

In 1941 the author collected this species in Clark Co. (from upland clay till slopes, 2½ miles southwest of Athens, T67N, R7W, sect. 26, *Steyermark 28640*), where a half dozen large trees were growing in a semi-grazed area. It has also been reported as having been seen in the same county northeast of St. Patricks by Mr. Frederick Dunlap. The species is quite rare in the state and no collections more recent than the above date are known. Bush in a letter of September 15, 1933, writes that '*Populus tremuloides* occurs clear across the state in the northern counties,' but there are no herbarium specimens to support this statement. It is quite probable after still more intensive botanical exploration has been carried out in the northern section of Missouri, that new stations for this species will turn up.

This species was found by Frederick Dunlap, at one time state forester, in Dent County, situated in the Ozarks (along old road to Rolla ¼ mi. west of Salem, March 30, 1934 [flowers] and June 16, 1933 [leaves]), but the trees died in the summer of 1934. A clump of ten stems occurred at this place in a slight depression on the south side of the old road leading west from Salem toward Rolla and about ¼ mi. from the railroad crossing. Dunlap reported (in letter of December 31, 1936 to Dr. W. J. Robbins in the University of Missouri Herbarium) that this 'location was rather low and inclined to be wet, being in the head of a small valley, and not the sort of place where one would plant trees or build a house. Moreover, it is, as Bush says, not a tree which is planted. I am of the opinion its occurrence is natural. In his list of trees of Greene County, Professor Shepard included this species from that county, but in reviewing his list, Bush states "certainly not native in Greene County. I have never seen this in cultivation anywhere. Nurserymen often call *P. alba* 'Aspen' and 'American aspen,' and I suspect Shepard must have mistaken this for *P. tremuloides*." A number of other species of trees in Shepard's list were based upon misidentifications, and Mr. Dunlap's observations are undoubtedly correct (Bush, Am. Midl. Nat. 15: 581–85. 1934; Shepard, E. M. A List of the Native Trees of the [Greene] County. Mo. Geol. Surv. 12: 46–47. 1898).

*Populus tremuloides* is one of the most naturally widely distributed species of trees in North America. It germinates well in fire-scarred areas, rapidly building up a forest on burnt over soil. The foliage of this aspen assumes a pale yellow color in autumn. It is the species chiefly used in North America for pulpwood in papermaking.

5. **Populus heterophylla** L.
Swamp Cottonwood Map 733

Flowers March–April.

Occurs in inundated wooded swamps throughout the lowlands of southeastern Missouri north to Cape Girardeau and Bollinger counties and west to Wayne and Ripley counties.

Ranges from Florida to Louisiana, north near the Coastal Plain to Connecticut and Pennsylvania, and inland north to Ohio, Michigan, Illinois, and Missouri.

At first the branchlets are white-hairy, later become glabrous. The winter buds have some hairiness at their base. The large leaves are white-hairy at first, but later become glabrous.

The wood is used for pulp in papermaking, and for crates and wooden boxes. It also has been used for interior finish. Nearly half of excelsior material is derived from this species.

6. **Populus** × **gileadensis** Rouleau
Balm-of-Gilead Map 734

*Populus balsamifera* × *P. deltoides* var. *missouriensis*

*Populus candicans* [of BB, Steyerm.], not Ait.

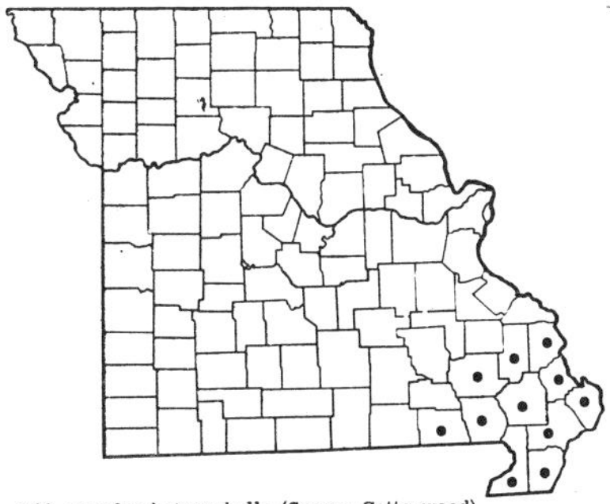

733 Populus heterophylla (Swamp Cottonwood)

734 Populus X gileadensis (Balm-of-Gilead)

735 Populus nigra var. italica (Lombardy Poplar)

Flowers March–April.

Known only from a tree escaped from cultivation in Taney County, southwestern Missouri.

Escaped and spread from cultivation from Newfoundland to Ontario, southward.

The origin of this tree is doubtful. Whether it is of hybrid origin between *P. balsamifera* and *P. deltoides* var. *missouriensis* or represents a sterile clone of *P. balsamifera* var. *subcordata*, reproducing by sprouts, remains doubtful. The name used here designates its status as of hybrid origin.

7. **Populus nigra** L. var. **italica** Muenchh.
Lombardy Poplar — Map 735
Flowers March–May.

Commonly cultivated throughout Missouri, but rarely escaped and naturalized, as far as known, only in Jasper County, southwestern Missouri (open ground, border of woods, 4 mi. northwest of Webb City, September 27, 1952, *Palmer 55194*).

This tree is known as a staminate tree only and is believed to have originated as a freak or sport from typical *P. nigra* var. *nigra* in Italy, and since propagated through cuttings. As the trees are subject to a canker disease and are short-lived, they are recommended for planting only as a quick-growing hedge or screen for temporary puposes.

8. **Populus** × **eugenei** Simon-Louis
Eugene Poplar — Map 736
Also called Carolina Poplar.
*Populus deltoides* × *Populus nigra* var. *italica*
*Populus canadensis* var. *eugenei* (Simon-Louis) Schelle [P & S]

Formerly this tree was widely planted, but has rarely escaped from cultivation in Missouri, where it is known to have become established only in Jasper County (open alluvial ground, about old mines, valley of Center Creek, 1 mi. south of Oronogo, October 18, 1953, *Palmer 57074*), southwestern Missouri.

Although extensively planted in the past, this cottonwood, like other cultivated species of *Populus*, is now no longer recommended by nurserymen, and is prohibited by law in some sections of the United States, because of the widely spreading roots which clog up drains and water pipe systems or lift sidewalks and streets. Also, like other species of poplar, it has weak wood, the twigs and limbs breaking away to litter the streets.

This tree is believed to be of hydrid origin between *P. deltoides* and *P. nigra* and possibly includes various hybrid clones which have originated in different places.

It is similar to *P. deltoides*, but the tree has a more upright growth with the branches more ascending. It is sometimes separated as a variety from *P.* × *canadensis* Moench, a hybrid between *P. deltoides* and *nigra*, by its more pyramidal habit of growth and shorter staminate catkins, but by other authors is placed under the synonymy of *P.* × *canadensis* (Little, *Check List of Trees of the U.S.*, p. 281. 1953).

9. **Populus deltoides** Marsh. Cottonwood
Map 737
Flowers March–May.

Occurs in alluvial soils along streams and in low wet ground. Throughout the state, and doubtless occurring in every county.

Ranges from Quebec to Manitoba, south to Florida and Texas.

Missouri material is represented by the following variations:

a. Leaf-stalks somewhat hairy when young; veins of lower side of leaf-blade somewhat hairy until late in season . . . . 9b. P. DELTOIDES var. DELTOIDES f. PILOSA

a. Leaf-stalks and leaf-blades usually glabrous (without hairs) from the beginning or soon becoming so . . . . . . . . . . *b*

736 Populus X eugenei (Eugene Poplar)

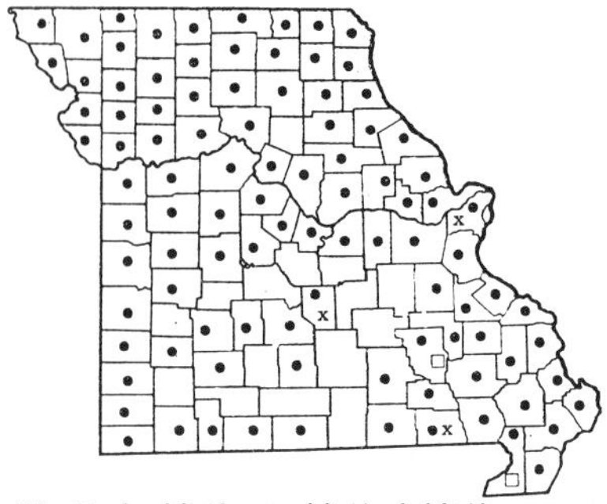

737 • Populus deltoides var. deltoides f. deltoides (Cottonwood)
737 □ Populus deltoides var. deltoides f. pilosa
737 x Populus deltoides var. missouriensis (Southern Cottonwood)

738 Leitneria floridana (Corkwood)

b. Leaf-blades broader than long or as broad as long . . 9a. P. DELTOIDES var. DELTOIDES f. DELTOIDES
b. Leaf-blades usually longer than broad. . 9c. P. DELTOIDES var. MISSOURIENSIS

9a. **Populus deltoides** var. **deltoides** f. **deltoides** Map 737

*Populus deltoides* Marsh. [G, BB, P & S]

This is the widely distributed and common type found throughout the state.

9b. **Populus deltoides** var. **deltoides** f. **pilosa** (Sarg.) Sudw. Map 737

*Populus deltoides* f. *pilosa* (Sarg.) Sudw. [P & S]

Known only from Dunklin and Reynolds (in back of gravel bar along Middle Fork of Black River, just west of Lesterville, May 21, 1939, *Steyermark 26530*) counties.

9c. **Populus deltoides** var. **missouriensis** (Henry) Rehd. Southern Cottonwood Map 737

Known from St. Louis (Bridgeton, *Fritchey*), Ripley (southwest of Grandin, *Steyermark 11826*), Wayne (2½ mi. south of Shook, T 27 N, R 7 E, sect. 8, *Steyermark 6566*; Sam Baker State Park, *Steyermark 1874*), and Pulaski (2 mi. southeast of Hazel Run, *Steyermark 25094*) counties.

More frequent southward in the range of *P. deltoides*, especially from Vermont to Ohio and Missouri, south to Florida and Mississippi.

The foliage of the cottonwood at first turns greenish yellow and eventually a dull yellow color. They are rapid-growing but short-lived trees, attaining heights of 150 feet and living only 75–100 years. The leaves may be as small as 4 cm. long by 4 cm. broad, but usually are much larger.

The wood is of very light weight and used in pulp for papermaking. Formerly cottonwood was much employed for making into packing boxes and crates of all sizes, excelsior, posts, and barrel staves. The thick bark was reported to be used by children of early pioneer families for whittling out toys.

*Excluded species*

**Populus Sargentii** Dode Great Plains Cottonwood

This species was reported as extending into Missouri in Rydberg's *Flora of the Prairies and Plains*, p. 247. 1932. The report may have been based upon a specimen collected in Jackson County, western Missouri (bottoms, Courtney, June 25, 1915), but this specimen has glabrous bud scales and the leaves are toothed as in typical *P. deltoides*, to which species the specimen is referred in the present flora. No other records are extant to support evidence of its occurrence in the state, but as it is known as far east as Oklahoma, Kansas, and Nebraska, it should be looked for in westernmost Missouri. It differs from *P. deltoides* in having hairy-margined bud-scales and very coarsely dentate leaves.

---

Plate no. 131. **1.** Populus heterophylla, × 2/7. **2.** Populus × gileadensis, × 2/7. **3.** Populus × eugenei, × 2/7. **4.** Populus deltoides var. deltoides, × 2/7; *a.* Male catkins, × 2/7; *b.* Female catkins, × 2/7. **5.** Leitneria floridana, × 2/7; *a.* Male catkins, × 2/7; *b.* Female inflorescences, × 2/7; Details from Small, The New York Botanical Garden. **6.** Juglans nigra, × 2/7; *a.* Fruits; *b.* Seed; *c.* Leaf; *d.* Flowering branch.

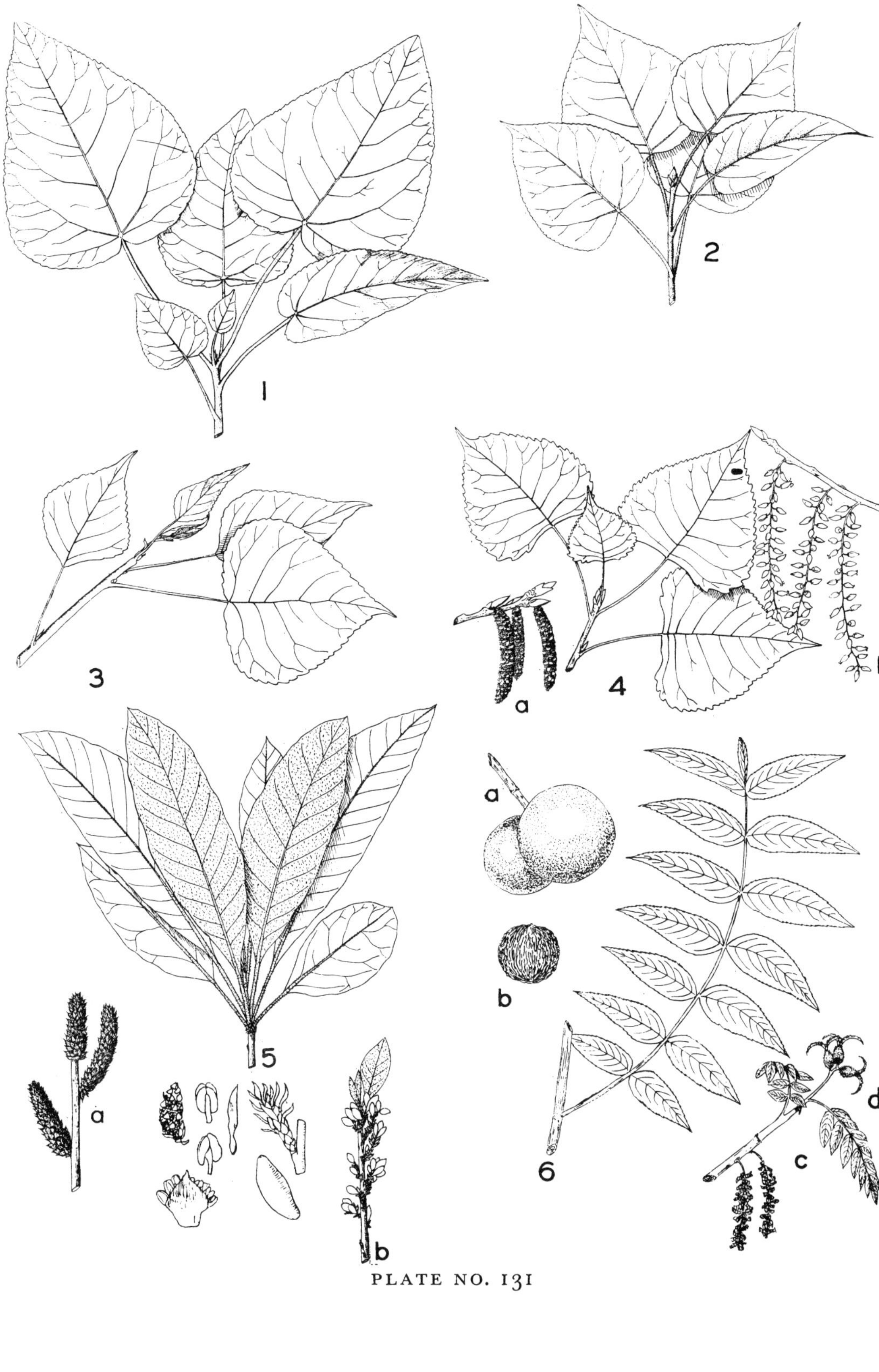


PLATE NO. 131

## Fam. **LEITNERIACEAE** (Corkwood Family)

**Leitneria floridana** Chapm. Corkwood Map 738
Flowers March–April.

Occurs in wooded or open swamps and wet thickets along roadsides in the lowlands of southeastern Missouri, where known from Dunklin, Butler, and Ripley counties.

Ranges from Florida to Texas, north to Georgia, Arkansas, and southeastern Missouri.

This shrub or small tree attains a height of about 7.5 meters (25 feet), but often is only a little over a meter (3¼ feet) tall. Plants are either male or female and form thickets by root suckers. The wood, which is lighter than cork, has the lightest weight of any woody plant native to the United States, weighing only 13 pounds per cubic foot, but is not as light as the Central American Balsawood (*Ochroma*). It is sometimes used for fishing-net floats and bottle stoppers.

The rather leathery, dark olive to dull green leaves, reminding one of some kinds of similar foliage of tropical and subtropical plants, are 7–15 cm. long, and more or less crowded near the top of the new season's growth of the slender stem, which is slightly enlarged at the base and attains a diameter of not over 10–12.5 cm. (4–5 inches), and usually much less so.

The plant is quite hardy far north of its natural known geographical limits. Plants originally transplanted from Missouri to a pond in the author's wildflower garden northwest of Chicago have spread into a sizeable thicket in ten years. Moreover, plants succeed in dry soils, as can be testified by their growth at the Missouri Botanical Garden in St. Louis and at the Morton Arboretum west of Chicago. The leaves remain green until late in the autumn, and are still green well after most other woody plants have either turned color or dropped their leaves.

## Fam. **JUGLANDACEAE** (Walnut Family)

Pith of twigs or branchlets, when cut lengthwise, separates into chambers or stepladder-like tiers; leaves usually with 11–19 (sometimes only 7) divisions, the middle divisions larger than those at the tip; male catkins not in groups of 3's, without stalks and appearing on branchlets of the previous year's growth; stamens glabrous, 8–40 on each flower of the male catkin; husk of fruit not splitting open, the shell furrowed . . . . . . . . . . . . . . . . . . . . . . . . . . . . 1. JUGLANS

Pith of twigs continuous and solid brown; leaves usually with 5–11 (sometimes up to 17) divisions, the topmost divisions generally largest; male catkins in groups of 3's, long- or short-stalked, appearing on the new season's growth; stamens hairy, 3–8 on each flower of the male catkin; husk of fruit splitting open, the shell smooth or slightly net-veined . . . . . . . . . . . . . . . . . 2. CARYA

### 1. **Juglans** L. Walnut

a. Pith of branches light brown, the brown part much thicker than the separating partitions; bark dark brown or black with rough ridges; upper part of leaf-scar of the previous year's leaves without any mat of hairs; fruit usually somewhat globe-shaped and about as broad as long (or rarely in f. *oblonga* longer than broad), the husk not sticky . . . . . . . . . . . . . . . . . . . . 1. J. NIGRA

a. Pith of branches dark brown, the brown part about as thick as the separating partitions; bark gray with smooth ridges; upper part of leaf-scar of the previous year's leaves with a mustache-like mat of hairs; fruit longer than broad in the native tree, husk sticky, or globe-shaped or ovoid in the cultivated tree . . . . . . . . . . . . . . . . . . . . . . . . . . . . . . . *b*

  b. Native tree; fruit longer than broad, 7–10 cm. long, occurring 2–5 together; nut with 4 prominent and 4 less prominent ridges . . . . . . . . . . . . . . . . . . . . 2. J. CINEREA

  b. (See excluded species); cultivated tree; fruit globe-shaped or ovoid, 5–6 cm. long occurring with as many as 20 in a long drooping raceme; nut rough, 2-edged . . . . . . . . 3. J. AILANTIFOLIA

1. **Juglans nigra** L. Walnut Map 739
Flowers April–late May.

Occurs usually in rich woods at the base of slopes or bluffs, in valleys along streams, and in open and upland woods; sometimes planted or persisting in old fields and pastures. Throughout Missouri and doubtless in every county.

Ranges from Massachusetts to Minnesota, South Dakota, and Nebraska, south to Florida and Texas.

Two variations are found in Missouri:

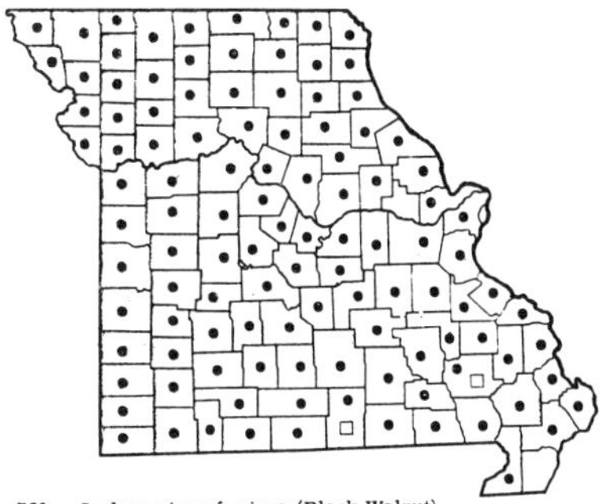

739 ● Juglans nigra f. nigra (Black Walnut)
739 □ Juglans nigra f. oblonga

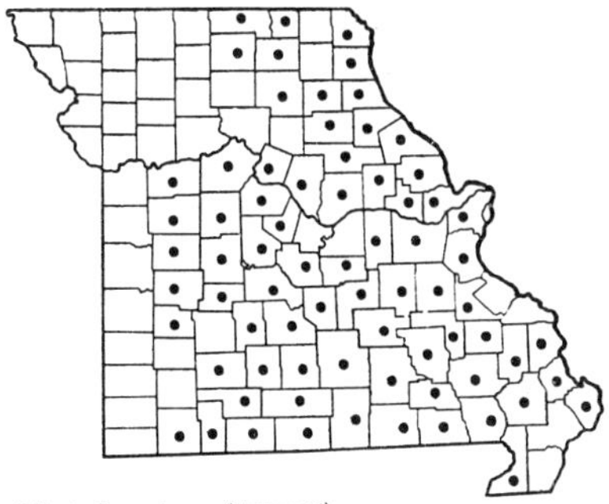

740 Juglans cinerea (Butternut)

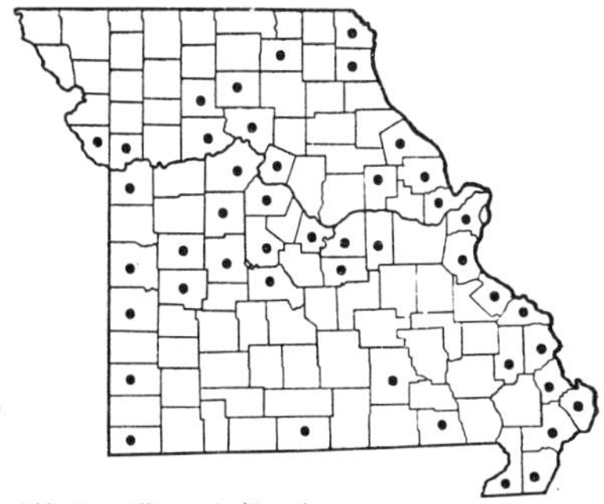

741 Carya illinoensis (Pecan)

Common type with fruit globe-shaped, spherical, about as broad as long . . . 1a. J. NIGRA f. NIGRA
Rare type with fruit ellipsoid and longer than broad . . . . . . . 1b. J. NIGRA f. OBLONGA

1a. **Juglans nigra** f. **nigra** Map 739

This is the form ordinarily encountered.

1b. **Juglans nigra** f. **oblonga** (Marsh.) Fern. Map 739

Known only from Howell (Pinebrook Farms on highway 14, Siloam Springs, 1948, *John T. Woodruff*) and Wayne (along road bordering swampy meadow in valley of Stanley Creek, T27N, R7E, sect. 18, 6–7 mi. northeast of Wappapello, September 11, 1949, *Steyermark 69265*) counties.

Known also in Virginia.

The foliage of the Black Walnut which turns to a light yellow shade in autumn serves as one of the hosts to the fall webworm (*Hyphantria cunea*), which forms webby masses over the ends of branches. Although much pollen is shed by this species, it does not appear to enter into hay fever cases, although it is sometimes suspected. The kernel of the nut is edible and walnut nutting parties are among autumn activities. Besides the obvious use of the plain nuts in a fresh or dried state, they were prepared in several ways as food by the Indians of the eastern United States in soups and gravies, and as butter. The early colonists pickled the young fruits, and the practice is occasionally continued today in various parts of the United States. The boiled sap of Black Walnut, tapped in early spring, was also used by the Indians as a sweet syrup.

Commercially, Black Walnut is one of the most important timber trees. It is the source of the best cabinet wood available in the United States, and is highly prized for its beautiful grain used in furniture and interior finish. For its durability it was used in building ships and boats. It is also the source of much of the wood used in gunstocks, and at one time supplied much wood for railroad ties and rafters. Formerly, many rail fences were made from walnut. Missouri, Illinois, Indiana, and Ohio continue to furnish the bulk of the commercial supply of this species.

2. **Juglans cinerea** L. Butternut Map 740

Flowers April–late May.

Occurs in rich woods along the base of slopes or bluffs, and along streams. Common in the Ozarks and eastern half of the state, north of the Missouri River ranging west to Putnam, Sullivan, Macon, Monroe, Audrain, and Howard counties, south of the Missouri River ranging west to Lafayette, Johnson, Henry, St. Clair, Cedar, Greene, Christian, and Barry counties.

Ranges from New Brunswick to North Dakota, south to Georgia and Arkansas.

Like Black Walnut, the nut of this species has a sweet edible kernel and has been used similarly by the Indians; the sap, likewise, was boiled by them for a sweet syrup. Also, like Black Walnut, the wood is highly valued for its grain and durability, furnishing wood for cabinetmaking and interior finish.

Small amounts are used as pulp when mixed with other hardwoods for papermaking. The bark of the root is sometimes employed in medicine as a liver stimulant, while the early settlers in this country prepared an orange and yellow dye from the bark and green fruits of this species and of the Black Walnut to color their homespun cloth.

*Excluded species*

3. **Juglans ailantifolia** Carr. Japanese Walnut

*Juglans Sieboldiana* Maxim., 1872 not *J. Sieboldiana* Goeppert, 1854 (fossil)

*Juglans cordiformis* Maxim. var. *ailantifolia* Rehd.

A specimen from a cultivated tree in Palmyra, Marion County, growing 3 blocks south of the hotel along highway 61 was collected on July 20, 1929 by Frederick Dunlap, former state forester, and deposited in the herbarium of the University of Missouri. As the species has not been found as an escape from cultivation, it is not included as an introduced member of the Missouri flora. It is native to Japan and cultivated in North America where it is hardy as far north as Ontario. The Heartnut (*J. ailantifolia* var. *cordiformis* [Makino] Rehder) is similar, but with heart-shaped, smooth, and thinner-shelled nuts.

The Persian or English Walnut (*Juglans regia* L.) is planted in Missouri, but is not known as an escape from cultivation. It is an important nut tree in California, and is distinguished from the other walnuts by the 7–9 nearly entire, instead of toothed, leaflets.

## 2. **Carya** Nutt. Hickory

### I. *Key based on fruit, flowers, and vegetative characters*

a. Scales of bud (best seen on uppermost bud) 4–6, in pairs, valvate (edges touching but not overlapping); leaflets (divisions of 1 leaf) 5–17, often with one side wider or more curved than the other side (falcate); at least some of the male catkins arising from separate side buds located near the top of the branches of the previous year's growth; nuts thin-shelled; sutures on fruit (where the fruit splits) with definite wings or keels; scars of the bud-scales separate, rather broad, not forming a ring . . *b*

b. Nut longer than broad; fruit longer than broad, 3.5–8 cm. long, nearly terete with usually rounded or curved sides or rarely compressed; clusters of male catkins without stalks (sessile) or nearly without stalks; leaflets mainly 9–17; lobes of the seed entire or with a slight notch at the end; winter buds with tiny clusters of yellow hairs . . . . . . . . . . . . . . . . *c*

c. Leaflets usually 11–17, rarely 9; nut cylindric, ovoid-oblong, mostly terete with rounded or curved sides; nut sweet; common in many parts of southern, central, and eastern Missouri 1. C. ILLINOENSIS

c. Leaflets usually 9–13, rarely 7 or 15; nut compressed, oblong; nut bitter; rare, known only from the lowland swamps of Dunklin County, southeastern Missouri . . . . 2. C. × LECONTEI

b. Nut as broad as or broader than long; fruit either longer than broad or nearly as broad as long, 2–3.5 cm. long, strongly or somewhat flattened or compressed, more or less angled; clusters of male catkins with definite stalks (peduncles); leaflets mainly 5–13 (rarely 15); lobes of seed deeply 2-lobed; winter buds without yellow hairs, but with yellow scales which either remain attached or fall off . . . . . . . . . . . . . . . . . . . . . . . . . . . . . *d*

d. Found only in lowland swamps of southeastern Missouri; winter buds dark brown or reddish-brown, with yellow glands which disappear eventually; leaflets 7–13 (rarely 15), the lower surface glabrous or nearly so when mature and without rust-colored hairs; stamens 3–6 to each flower of the male catkin; husk of fruit winged to the base and splitting to the base when mature; shell of nut furrowed or wrinkled, reddish-brown . . . . . . . . 3. C. AQUATICA

d. Found throughout the state in low woods or rich slopes; winter buds mustard-yellow or orange-yellow, the yellow glands remaining permanently attached; leaflets mostly 7–9, rarely 5 or 11, the lower surface, even when mature, with pubescence and with rusty-colored scales; stamens 4 to each flower of the male catkin; husk of fruit not winged near the base and splitting only to just below middle when mature; shell of nut rather smooth, gray. . 4. C. CORDIFORMIS

a. Scales of bud (best seen on uppermost bud) 6–12, overlapping (imbricate); leaflets (divisions of 1 leaf) 3-9, usually not falcate or only barely so; clusters of male catkins all arising from the bud-scales located at the lowest part of the leafy shoot of the present season during which it was collected; nut usually thick-shelled; sutures on fruit (where the fruit splits) with no wings or keels or at most very slightly keeled; scars of bud-scales coming together, very narrow, forming a definite ring . . . . . *e*

e. Uppermost (terminal) bud large, mainly 12–30 mm. long, with 10–12 scales; fruit 3.5–7 cm. long with a husk 4–12 mm. thick; nut usually noticeably angled; branchlets usually stout . . . . *f*

f. Leaflets usually 5, or, if rarely 7, then the teeth provided with persistent dense hairy tufts just below the tip of each tooth, or the petioles, young branchlets, and lower surface of leaves not covered with curly hairs, or the fruit splitting to the base, or the shell of the nut only 1–2 mm. thick . . . . . . . . . . . . . . . . . . . . . . . . . 5. C. OVATA

f. Leaflets usually 7–9, or if rarely 5, then the shell of the nut 3–6 mm. thick, or the young branchlets, petioles, and lower surface of leaves covered rather densely with curly hairs, or the fruit not splitting to the base, or the teeth of leaves lacking dense hairy tufts just below their tips . . . . . . . . . . . . . . . . . . . . . . . . . . . . . g

g. Trees usually of dry or high ground; bark of trees always tight, not shaggy; 1st year branchlets brown or reddish-brown; leaf-stalks and axis of leaves, lower surface of leaves, and young branchlets covered with dense, rather persistent, curly hairiness; shell of nut 5–6 mm. thick; husk of fruit not splitting to the base; outer bud-scales of uppermost bud falling in autumn to expose the inner silky bud-scales . . . . . . . . . 7. C. TOMENTOSA

g. Trees of low or alluvial ground; bark of old trees loosening, shaggy; 1st year branchlets brown-orange or tan; leaf-stalks and axis of leaves, lower surface of leaves, and young branchlets either glabrous (without hairs) or slightly short-hairy or scurfy, but not covered with a dense persistent curly hairiness; shell of nut 3–6 mm. thick; husk of fruit splitting to the base; outer bud-scales of uppermost bud remaining attached to bud from autumn until the following spring . . . . . . . . . . . . . . . . . . . . 6. C. LACINIOSA

e. Uppermost (terminal) bud smaller or medium-sized, mainly 6–12 mm. long, with 6–10 scales; fruit 1.5–4 cm. long with a husk 1.5–4 mm. (rarely 5) thick; nut either rounded or angled; branchlets usually slender . . . . . . . . . . . . . . . . . . . . . . . . . . h

h. Outer bud-scales, young branchlets, young leaves, and male catkins with a rusty pubescence or scurfiness; nut dark reddish-brown, the shell with a conspicuous pale network; bark dark. 10. C. TEXANA

h. Outer bud-scales and young branchlets glabrous (without hairs) or with a pale hairiness; nut pale brown, the shell lacking a conspicuous network; bark gray . . . . . . . . . . . . i

i. Trees of southeastern Missouri only; leaflets most commonly 5, rarely 7; bark of old trees tight, furrowed, and ridged, not separating into small or long plates or small scales; husk of fruit not splitting or splitting only above the middle, dark brown, shining; nut smooth; shell of nut about 1.5 mm. thick; female flowers and bracts covered with a fine gray short hairiness; male catkins 5–8 cm. long, their bracts with noticeable hairs. . . . 8. C. GLABRA

i. Trees of southern, central, and eastern Missouri: leaflets most commonly 7, rarely 5; bark of trunk a few feet above the ground usually slightly scaly, on mature trees separating into longer plates; husk of fruit eventually splitting all the way to base, light brown, dull, warty; nut angled; shell of nut thinner, less than 1.5 mm. thick; female flowers and bracts covered with a yellow scurfiness; male catkins 8–17 cm. long, their bracts minutely hairy (puberulent) . . . . . . . . . . . . . . . . . . . . . . . . 9. C. OVALIS

## II. *Alternate key based on foliage and buds*

The following key is adapted from the treatment by Dr. Wayne Manning (Rh. 52: 196–99. 1950), whose work on the genus is the most complete and detailed to date (*Carya* × *Lecontei*, considered a hybrid between *C. illinoensis* and *C. aquatica*, is not keyed out here, but in other key).

a. Scales of bud 4–6, in pairs, valvate (edges touching but not overlapping); buds often yellow or orange-yellow, rather flattened; leaflets 7–17, never only 5, either always or else sometimes more curved and wider on one side than the other (falcate) . . . . . . . . . . . . . . . b

b. Leaflets mostly 7–9, rarely 5 or 11, sometimes or rarely falcate; terminal leaflet mostly sessile (without a stalk), rarely short-stalked . . . . . . . . . . . . . . . . 4. C. CORDIFORMIS

b. Leaflets mostly 11–17, rarely 7–9, strongly falcate; terminal leaflet stalked. . . . . . . . . . c

c. Teeth of leaflet usually prominent; side veins of leaflets not crowded, branching rather freely near the tips, these branches usually curving downward and passing to the center of the tooth . . . . . . . . . . . . . . . . . . . . . . . . . 1. C. ILLINOENSIS

c. Teeth of leaflets usually quite inconspicuous; side veins of leaflets crowded (usually about 30 pairs), about 1 to each tooth, rarely branching at the tip, many of them ending at the sinus (space between teeth) . . . . . . . . . . . . . . . . . . . . 3. C. AQUATICA

a. Scales of buds 6–12, overlapping (imbricate); mature buds brown or gray, sometimes covered with yellow glands, plump, not flattened but more or less curved or rounded on all sides; leaflets mainly 5–9, not falcate . . . . . . . . . . . . . . . . . . . . . . . . . . d

d. Margins of leaflets strongly hairy when young, each tooth with a persistent dense tuft of hairs on one or both sides of the tooth just below its tip (many of the tufts disappear on some of the leaflets towards autumn); leaflets typically 5 (rarely 7); branchlets of the present year (youngest) usually hairy; topmost bud 10–27 mm. long, the outermost scales persistent and remaining attached . . . . . . . . . . . . . . . . . . . . . . . . 5. C. OVATA

d. Margins of leaflets sometimes hairy when young with groups (fascicles) of hairs, but teeth without dense tufts of hairs just below the tip; leaflets 5–9; leaflets and leaf-axis (rachis) glabrous or glabrate (mostly without hairs) on those species having only 5 leaflets, or with the margins not hairy, or with the twigs slender and glabrous, or with the outermost bud-scales falling off in autumn . . . . . . . . . . . . . . . . . . . . . . . . . . . . . . *e*

e. Branchlets stout; lower surface of leaflets densely hairy, most of the hairs in groups (fascicles); mature topmost bud 10–27 mm. long; branchlets of the present year hairy, at least early in season, or else branchlets light orange or buff in autumn; leaflets 7–9, never 5 alone, the margins usually hairy, especially early in the season; outer bud-scales either falling or persistent . . *f*

f. Twigs or branchlets tan, buff, or pale orange at least on one side, gray on the other; hairs on axis of leaf (rachis) not in groups (fascicles), usually straight, the axis of the leaf (rachis) frequently almost glabrous; bark shaggy; outer leathery bud-scales of topmost bud persistent; leaf-stalks (petioles) often remaining attached in winter . . . . . . 6. C. LACINIOSA

f. Twigs or branchlets brown; hairs on axis of leaf (rachis) in separated groups (fascicles), curly; axis of leaf (rachis) mostly hairy; bark deeply ridged, not shaggy; outer bud-scales of topmost bud falling off, leaving the light grayish-brown silky-hairy inner bud-scales exposed; leaf-stalks (petioles) usually fallen from the branches by winter . . . . . 7. C. TOMENTOSA

e. Branchlets slender; lower surface and side veins of lower surface of leaflets mostly glabrous (sometimes hairy in *C. texana* var. *villosa*); mature topmost bud short, 5–12 mm. long; branchlets of the present year glabrous, in autumn reddish- or chestnut-brown; leaflets 5 alone on some trees, 5–9 in others, the margins usually glabrous (without hairs) or glabrate; outer bud-scales falling early . . . . . . . . . . . . . . . . . . . . . . . . . . . . . *g*

g. Outer bud-scales, young branchlets, and young leaves with a rusty hairiness or scurfiness; young leaflets with numerous reddish or rusty-colored scales; twigs and axis of leaf (rachis) sometimes with hairs in groups (fascicles) . . . . . . . . . . . . . 10. C. TEXANA

g. Outer bud-scales and young branchlets glabrous or with a pale hairiness; young leaflets without reddish or rusty-colored, but with scattered dark scales; twigs not hairy; axis of leaf (rachis) glabrous, or if hairy, the hairs crowded, not in fascicles . . . . . . . . . *h*

h. Leaflets usually 7, or 5 and 7, usually ovate to obovate or oblanceolate, thickish; leaf-stalk (petiole) frequently red; topmost buds ovate, blunt or acute, the tips of the outer bud-scales often hairy; bark of trunk scaly or shaggy in older trees . . . . . 9. C. OVALIS

h. Leaflets usually 5, rarely 5 and 7, the lateral usually lanceolate or oblong-elliptic, rather thin; topmost buds usually lanceolate, acuminate, the tips of the outer bud-scales glabrous; bark of trunk tight, not scaly . . . . . . . . . . . . 8. C. GLABRA

1. **Carya illinoensis** (Wang.) K. Koch Pecan
Map 741

*Carya Pecan* (Marsh.) Engl. & Graebn. [P & S]

Flowers April–May.

Occurs on alluvial soils of river bottom flood plain in woodland and margins of river bottom prairie; often growing alone in open fields where other trees have been cut away; sometimes cultivated, forming open groves. Common along the counties in eastern and central Missouri bordering the Mississippi, the Missouri, Osage, South Grand, and Grand rivers, scattered elsewhere along a few other rivers; generally absent from most of the Ozark section and much of northwestern and extreme northern Missouri.

Ranges from Alabama to Texas, north to Ohio, Indiana, Illinois, and Iowa; also in Mexico.

Pecan is the state tree of Texas. It is the most important hickory from the standpoint of hay fever, and often is ranked next to ragweed in its importance. The bark on Pecan is deeply and irregularly divided into narrow ridges and is not loose.

Unlike other hickories, which are noted for their strength and tenacity, the wood of the Pecan is rather weak and poor in comparison, and for that reason was used chiefly for fuel, although occasionally it was used also in the manufacture of wagon and agricultural implements. The nuts, however, produced by the tree, give to it the honor of being the best of the native nut-bearing trees of North America. Nuts from the wild trees are much smaller than those from cultivated and improved varieties, of which well over a hundred horticultural types are known, including the so-called 'paper-shelled.'

Georgia is the leading pecan-growing state, but

Plate no. 132. **1.** Juglans cinerea, × 2/7; *a.* Flowering branch; *b.* Fruit; *c.* Seed; Details from Small, The New York Botanical Garden. **2.** Carya illinoensis, × 2/7; *a.* Flowering branch; *b.* Seed; *c.* Fruit; Details from Small, The New York Botanical Garden. **3.** Carya aquatica, × 2/7; *a.* Fruiting branch; *b.* Leaf. **4.** Carya cordiformis f. cordiformis, fruiting branch, × 2/7; *a.* Carya cordiformis f. latifolia, leaf, × 2/7.

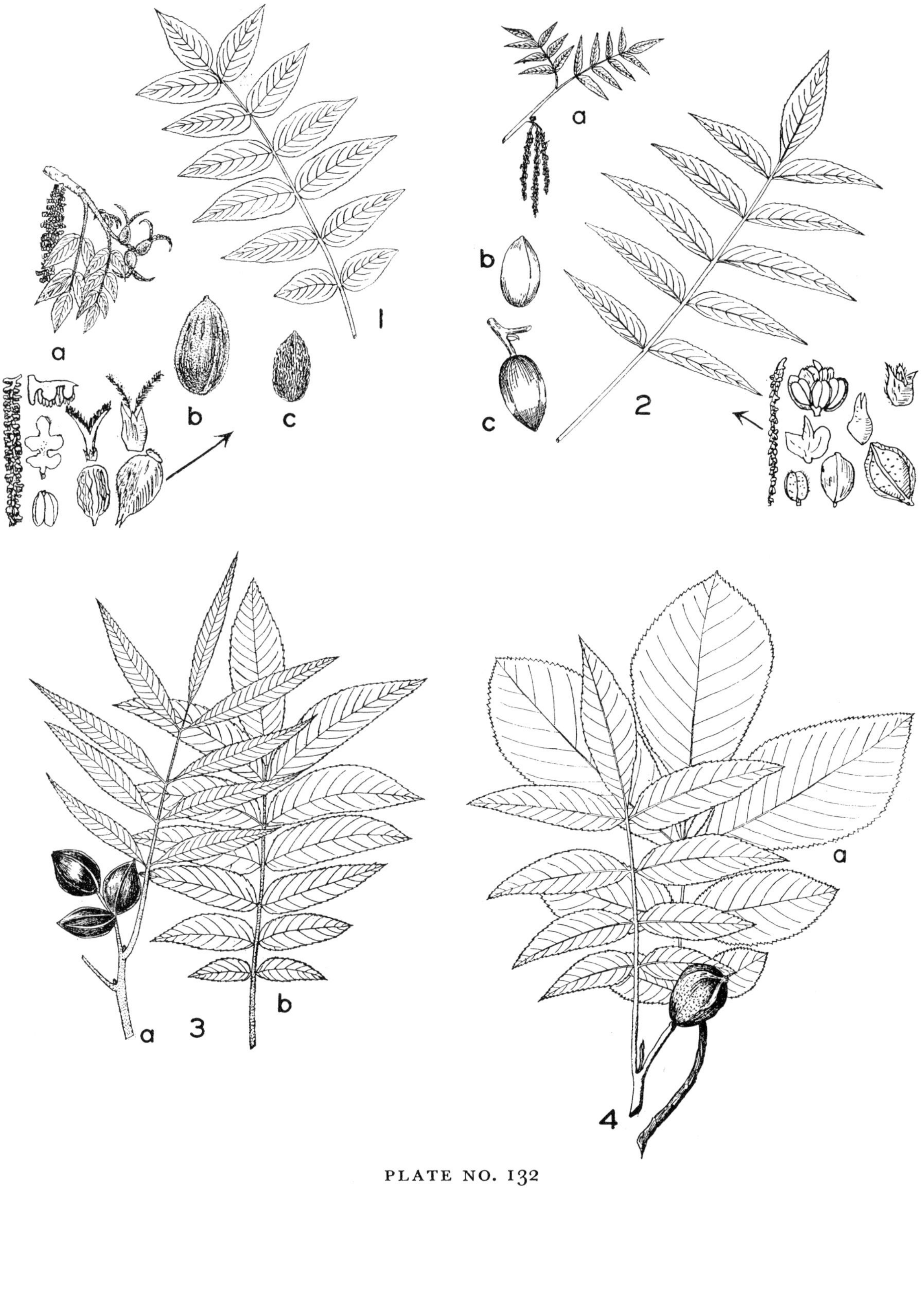


PLATE NO. 132

742 Carya X Lecontei (Bitternut Hickory)

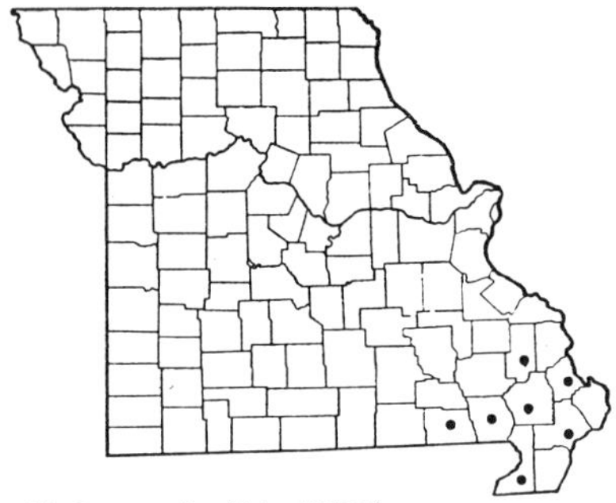

743 Carya aquatica (Water Hickory)

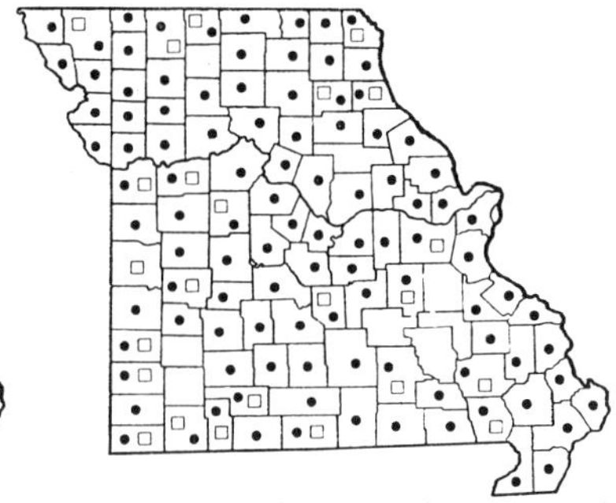

744 • Carya cordiformis f. cordiformis (Bitternut Hickory)
744 □ Carya cordiformis f. latifolia

pecan orchards do well in the Pacific states of California, Oregon, and Washington, and are grown especially in the southern states. There are a number of pecan orchards in Missouri, especially in the southeastern section of the state. Pecans are reputed to have a higher percentage of fat (over 70 per cent) than that of any other vegetable. The shells themselves yield a tannin and have been employed as mulches for covering roses, etc.

2. **Carya** × **Lecontei** Little
Bitternut Hickory Map 742
*Carya aquatica* × *Carya illinoensis*
*Carya texana* (Le Conte) C. DC., 1862 and auth. [*Steyerm.*], not *C. texana* Buckl., 1860
Flowers April–May.

Known only from the lowlands of southeastern Missouri in Dunklin County (near Campbell, September 4, 1935, *Kellogg 27036*).

Ranges from Missouri and Arkansas, south to Mississippi, Louisiana, and Texas.

Mr. E. J. Palmer (Jour. Arn. Arb. 18: 133–35. 1937) has shown that this should be considered a hybrid between *C. aquatica* × *C. illinoensis*, and it has been recognized as such by the U.S. Forest Service. It combines the bitter seeds and compressed fruit of *C. aquatica*, and the sessile male catkins, yellow, hairy winter buds, and elongate nut, which is longer than broad, of *C. illinoensis*.

The bark is roughened with platelike scales which are rather closely appressed to the main trunk. The winter buds are covered with pale yellow hairs. The fruit is oblong or oblong-obovoid, more or less covered with yellow scurfiness, and 3.5–5 cm. long. The nut has a thin papery shell, is oblong-ovoid, compressed, and obscurely 4-angled.

3. **Carya aquatica** (Michx. f.) Nutt.
Water Hickory Map 743
Flowers April–May.

Occurs in swamps and low wet woods in the lowlands of southeastern Missouri north to Bollinger and west to Ripley County.

Ranges from Florida to Texas, north to Virginia, Illinois, and Missouri.

The bark separates into long and thin plates or scales. The compressed nut resembles the shape of the next species, *C. cordiformis*, but, in addition to the differences mentioned in the key, is more or less 2-celled at the base, whereas in *C. cordiformis* the nut is 4-celled at the base and lower half.

4. **Carya cordiformis** (Wang.) K. Koch
Bitternut Hickory, Pignut Hickory Map 744
Flowers April–May.

Occurs in alluvial and rich soils, either in low woods and river bottoms along streams, sloughs, and ponds, or at the base of rich slopes and bluffs. Throughout Missouri.

Ranges from Florida to Texas, north to New Hampshire, Quebec, Ontario, Michigan, Wisconsin, Minnesota, and Nebraska.

Missouri material may be divided into two variations:

Leaflets mainly 2–4 (rarely 4.5) cm. broad . . 4a. C. CORDIFORMIS f. CORDIFORMIS
Leaflets mainly broader, 5–10 cm. broad. . . 4b. C. CORDIFORMIS f. LATIFOLIA

4a. **Carya cordiformis** f. **cordiformis** Map 744

Throughout Missouri and doubtless in every county; this is the commoner form encountered.

4b. **Carya cordiformis** f. **latifolia** (Sarg.) Steyerm. Map 744
*Carya cordiformis* var. *latifolia* Sarg. Trees & Shrubs 2: 206. 1913
Scattered throughout the range in Missouri.

The bright mustard-yellow or orange-yellow buds

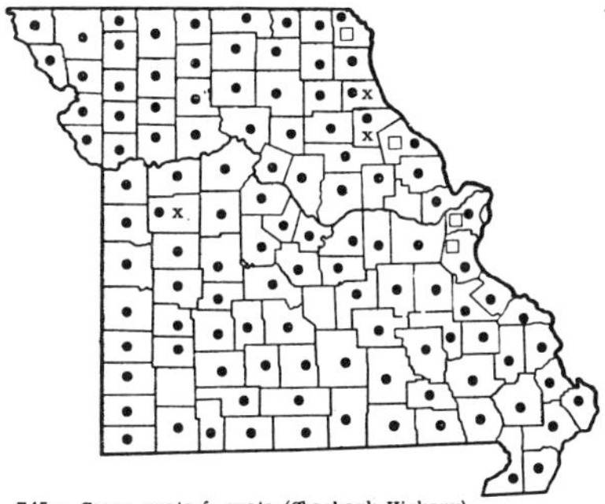

745 • Carya ovata f. ovata (Shagbark Hickory)
745 x Carya ovata f. ellipsoidalis
745 □ Carya ovata f. fraxinifolia

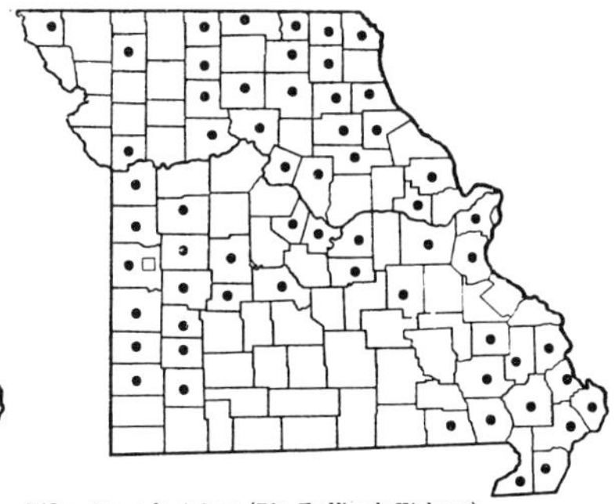

746 • Carya laciniosa (Big Shellbark Hickory)
746 □ Carya X Nussbaumerii (Carya laciniosa X Carya illinoensis)

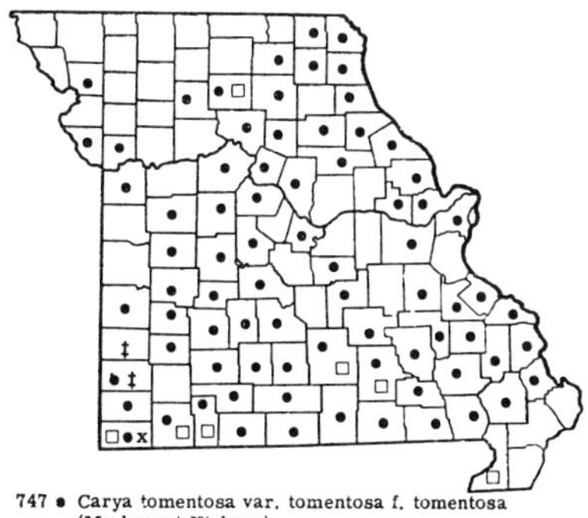

747 • Carya tomentosa var. tomentosa f. tomentosa (Mockernut Hickory)
747 x Carya tomentosa var. tomentosa f. ovoidea
747 ‡ Carya tomentosa var. tomentosa f. ficoidea
747 □ Carva tomentosa var. subcoriacea

at all seasons of the year easily distinguish this species. The bark is shallowly and irregularly furrowed, tight, and light gray. The kernel, as in *C. aquatica* and *C.* × *Lecontei* is bitter. The wood is very strong and tough, and has been used for fuel, hoops, and ox-yokes. It is stated that the shaven bark has been used by manufacturers under the name of 'yellow-bud' hickory for the making of chairs with split-bottoms.

It is reported that the early settlers obtained from the nut an oil used in their lamps as well as in the role of a rheumatism medicine.

5. **Carya ovata** (Mill.) K. Koch
Shagbark Hickory — Map 745
Flowers April–May.

Occurs in lowland or upland woods, in either valleys along streams, or on slopes, limestone hills, and dry upland areas.

Ranges from Florida to Texas, north to Maine, Quebec, Ontario, Michigan, Wisconsin, Minnesota, and Nebraska.

Missouri material is represented by the following variations, of which some may be distinguished only in fruit:

a. Larger leaflets 2.5–5 cm. wide; fruit smaller than in other forms, 3 cm. long; nut 2.3–3 cm. long, 1.6–2.4 cm. broad, 1–1.6 cm. thick, not depressed at apex; husk of fruit 4–5 mm. thick. . . . 5d. C. OVATA f. FRAXINIFOLIA
a. Larger leaflets 5–11 cm. wide; fruit larger, 3.5–6 cm. long . . . . . . . . . . *b*
  b. Nuts ellipsoidal, much longer than broad, abruptly narrowed and long-pointed at one end . . . . . 5c. C. OVATA f. ELLIPSOIDALIS
  b. Nuts top-shaped, obovoid, nearly globe-shaped or somewhat broadly bean-shaped, only slightly longer than broad or broader than long, at most short-pointed . . . *c*
    c. Fruit 3.5–6 cm. long; nut obovoid, ovate, or oval, slightly compressed or flattened, 2–3 cm. long, 1.5–2.5 cm. broad, more or less pointed at tip . 5a. C. OVATA f. OVATA
    c. Fruit smaller; nut subglobose or nearly rounded, or obcordate, strongly compressed, 1.5–2 cm. long, 1–1.5 cm. broad, rarely pointed at tip . . . 5b. C. OVATA f. NUTTALLII

5a. **Carya ovata** f. **ovata.** — Map 745
*Carya ovata* (Mill.) K. Koch [G, BB, P & S]

This is the commonest form found and occurs in doubtless every county of the state.

5b. **Carya ovata** f. **Nuttallii** (Sarg.) Steyerm. — Map 745
*Carya ovata* var. *Nuttallii* Sarg. Trees and Shrubs 2: 207. 1913

Scattered in Missouri with the f. *ovata*.

5c. **Carya ovata** f. **ellipsoidalis** (Sarg.) Steyerm. — Map 745
*Carya ovata* var. *ellipsoidalis* Sarg. Bot. Gaz. 66: 235. 1918 [P & S]

Known definitely from Marion (near Hannibal, September 5, 1913, *Davis 2071*, holotype) and Ralls (Oakwood, September 13, 1913, *Davis 2132*) counties, and reported but not collected by Palmer from Johnson County.

5d. **Carya ovata** f. **fraxinifolia** (Sarg.) Steyerm. — Map 745
*Carya ovata* var. *fraxinifolia* Sarg. Trees and Shrubs 2: 207. 1913 [G]

Known from Clark, Pike, St. Louis, and Jefferson counties, and probably more widespread over the state.

The bark of older trees of *C. ovata* separates into

long thin strips a foot or more long; that of young trees is tight and nonscaly, but the bark begins to scale after a growth of 1–2 dm. in diameter has been reached.

Much more field work and experimental studies need to be carried out on the variation of nuts in this species before any final opinion can be rendered about the various forms treated. Some authors (Gleason) have not recognized the variations as sufficiently significant to possess nomenclatural value. Others (Fernald, Deam) have maintained them as varieties, even though Deam admits that 'no dependence can be placed upon... size or shape of nuts' (*Trees of Indiana*, p. 71. 2nd ed. 1932). As the variations in leaf width, and shape and size of fruit and nut appear to be characters which may appear here and there throughout the range of the species, and as extremes of variation, it appears preferable here to treat them all as forms, not varieties.

The leaves of this species become dull gold or deep yellow in the fall, changing color at the beginning of autumn. This is one of the principal species of *Carya*, other than *C. illinoensis*, whose pollen is known to be responsible for causing hay fever. The flesh of the nut is sweet and, after the Pecan, is usually accorded second place in taste quality among the sweet-fleshed hickories. As with the walnuts, the sap of this and other species of hickory can be boiled for its sweet-yielding syrup or sugar, much as in the Sugar Maples. A particularly good sugar is reputed to come from the Shagbark Hickory. The Indians used the oily portion of the boiled liquid in their cookery.

The wood of this species is very strong and tough. It generally has fewer knots and badly formed spots than any of the other species, and has been employed in the manufacture of ax handles, skis, various agricultural implements, barrel hoops, wagons, carriages, and baskets. This species of hickory is widely used for fuel, ranking next to *Robinia Pseudo-acacia* (Black Locust) in value, while the green wood of this and other hickory species is greatly esteemed and used for barbecuing various meats. It burns long with little or no smoke, similar to the wood of ash, and thus is unexcelled for outdoor cooking. The straight young saplings often are made into walking sticks. Early colonists and pioneers employed the inner bark as a yellow dye for their homespun cloth.

6. **Carya laciniosa** (Michx.) Loud.
Big Shellbark Hickory, Kingnut Map 746
Flowers April–May.

Occurs in low alluvial woods of valleys along streams and in river flood plain, often associated with *Carya cordiformis*, *C. Pecan*, *Quercus palustris*, and other trees favoring wet woodland. Found mostly along the larger streams in eastern, northern, central, and western Missouri, common in the tall swamp forests of the southeastern lowland section, but absent from a large part of the Ozark region.

Ranges from New York and Ontario to Iowa and Nebraska, south to North Carolina, Georgia, Alabama, Louisiana, Kansas, and Oklahoma.

The leaflets normally are 7 or 9, but occasionally may be 5. The upper leaflets average larger than those of the Shagbark Hickory (*C. ovata*), varying from 1.2–2.3 dm. long and 7–12 cm. broad, the terminal leaflet sometimes attaining a length of 2.3 dm. The fruit is only slightly longer than that of *C. ovata*, becoming 6.5 cm. long, but the nut is much larger, being 3–6 cm. long instead of 2–3 cm. long as in *C. ovata*. The shell of the nut in *C. laciniosa* is yellow-brown to red-brown; in *C. ovata* it is pale to whitish-brown.

Those who have tasted the sweet kernels of *C. laciniosa* often prefer it to the much-esteemed *C. ovata* Both are of excellent quality. The hard nuts are rendered more easily cracked by first soaking them and then allowing them to dry, the heat splitting the shell sufficiently to make it easier to crack them open. The various high qualities possessed by the wood of *C. ovata* apply equally well to that of *C. laciniosa*.

6a. **Carya × Nussbaumerii** Sarg. Map 746
*Carya illinoensis × laciniosa*

This hybrid is known from the vicinity of Rockville, Bates County, western Mo. It is found also in Illinois, Indiana, and Iowa.

It has leaves similar to those of *C. laciniosa*, slender branchlets and shape of fruit of *C. illinoensis*, white or nearly white nuts, and fruit which lacks wings on the sutures.

7. **Carya tomentosa** Nutt. Mockernut Hickory
Also called White Hickory. Map 747
Flowers April–May.

Occurs mostly in dry upland woods on high ridges or slopes, frequently in acid soils overlying sandstone, chert, flint, or granite, occasionally in low or alluvial woods along streams. Throughout Missouri, except absent from the extreme southeastern lowland corner

Plate no. 133. **1.** Carya ovata, × 2/7; *a.* Terminal bud; *b.* Nut; *c.* Fruit; *d.* Flowering branch. **2.** Carya laciniosa, × 2/7; *a.* Fruit, × 2/7. **3.** Carya tomentosa, × 2/7; *a.* Nut; *b.* Fruit. **4.** Carya glabra, × 2/7; *a.* Nut.

PLATE NO. 133

and most of the extreme northern and northwestern Missouri sectors.

Ranges from Florida to Texas, north to Massachusetts, New Hampshire, New York, Ontario, Michigan, Illinois, Iowa, and Nebraska.

Missouri material is represented by the following variations, which can be distinguished only by mature fruit:

a. Dried fruit oblong, longer than broad, 5 cm. long; nut oblong to ovate, 4.5 cm. long, strongly angled, pointed at both ends or sometimes more rounded at base; leaflets thick, very hairy on lower surface . . . 7d. C. TOMENTOSA var. SUBCORIACEA

a. Dried fruit usually globe-shaped or round, rarely ovoid, obovoid, or shortly ellipsoid, broader than long or slightly longer than broad, 3–3.5 cm. long; nut somewhat round or globe-shaped, rarely ellipsoid, somewhat longer than broad or broader than long, 2.5–3.5 cm. long, short- to long-pointed at top and rounded or pointed at the base, 4–6-angled; leaflets thinner, less hairy on lower surface . *b*

  b. Fruit with a long-pointed (long acuminate) summit . . . . . . . 7b. C. TOMENTOSA var. TOMENTOSA var. OVOIDEA

  b. Fruit short-pointed (acute) or gradually narrowed at summit. . . . . . . *c*

    c. Fruit obovoid with a prominent stalk-like base. 7c. C. TOMENTOSA var. TOMENTOSA f. FICOIDEA

    c. Fruit ellipsoidal or obovoid, gradually narrowed at the base, not with any stalk-like base. 7a. C. TOMENTOSA var. TOMENTOSA f. TOMENTOSA

7a. **Carya tomentosa** var. **tomentosa** f. **tomentosa** Map 747

*Carya tomentosa* Nutt. [G, BB, P & S]

This is the most commonly encountered form in Missouri.

7b. **Carya tomentosa** var. **tomentosa** f. **ovoidea** (Sarg.) Palmer & Steyerm. Map 747

*Carya tomentosa* f. *ovoidea* (Sarg.) Palmer & Steyerm. [P & S]

Known only from McDonald County (Noel, September 5, 1913, *Palmer 4119*, holotype in herb. Arnold Arboretum; same locality, October 23, 1910, *Palmer 3287*), southwestern Missouri.

7c. **Carya tomentosa** var. **tomentosa** f. **ficoidea** (Sarg.) Palmer & Steyerm. Map 747

*Carya tomentosa* f. *ficoidea* (Sarg.) Palmer & Steyerm. [P & S]

Known only from Barton and Jasper (sandy soil in Mt. Hope Cemetery, Webb City, October 8, 1911, *Palmer & Sargent 3493*, holotype in herb. Arnold Arboretum) counties, southwestern Missouri.

7d. **Carya tomentosa** var. **subcoriacea** Sarg. Map 747

Known mostly from southernmost Missouri, and locally in Linn County.

Ranges from Missouri and Arkansas to Oklahoma and Texas.

Some authors (Gleason and Fernald) do not consider the variations treated above as more than 'merely phases in the normal variation of the species.' However, Deam, who had first-hand knowledge of var. *subcoriacea*, considered it as a distinct and separable variety. Until more detailed field and experimental studies have been completed on the variations found in the fruits and nuts of this species, it would seem preferable to maintain most of the variations as forms occurring here and there in the range of the common variation, and one of them, var. *subcoriacea*, as a variety with a marked southern geographical range.

The bark in *C. tomentosa* is tight, shallowly furrowed, usually thick and densely covered with lichens. In certain river bottom areas the bark may be thinner and lighter colored.

This species has unusually fragrant resinous foliage, although all the hickories possess a resinous fragrance. The wood has been shown to be the strongest of any of the species of hickory, and is highly prized for making into splints and rustic furniture. The name White Hickory is derived from the whitish sapwood, which in *C. tomentosa* is much whiter than usual.

8. **Carya glabra** (Mill.) Sweet var. **glabra**

Pignut Hickory Map 748

Also called Broom Hickory; in Indiana called Black Hickory.

Flowers April–May.

Occurs in usually acid sandy or gravelly soils of dry or upland wooded ridges. Southeastern Missouri, where it is found on Crowley Ridge in Dunklin, Stoddard (top of Crowley Ridge, north of Ardeola school, near Ardeola, November 8, 1936, *Steyermark 20772*), and Scott (top of ridge above east-facing sandstone slopes along Mississippi River, T29N, R14E, sect. 13, 2 mi. north of Commerce, November 7, 1936, *Steyermark 20746*) counties, and at the edge of the Ozark escarpment in Cape Girardeau County. The specimen from Stoddard County (*Steyermark*

748 Carya glabra var. glabra (Pignut Hickory)

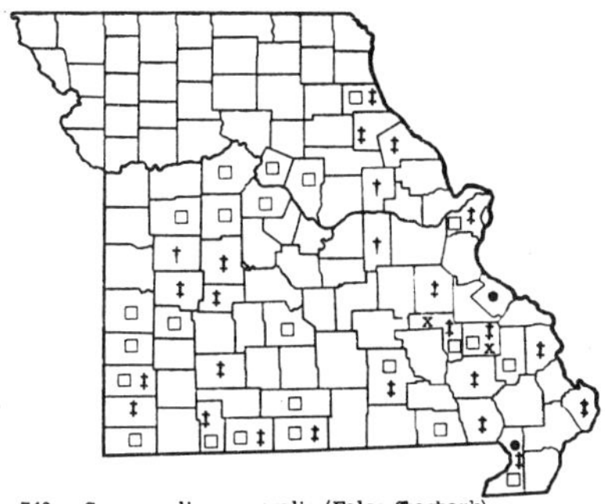

749 • Carya ovalis var. ovalis (False Shagbark)
749 □ Carya ovalis var. obcordata f. obcordata
749 † Carya ovalis var. obcordata f. vestita
749 ‡ Carya ovalis var. obovalis
749 x Carya ovalis var. odorata

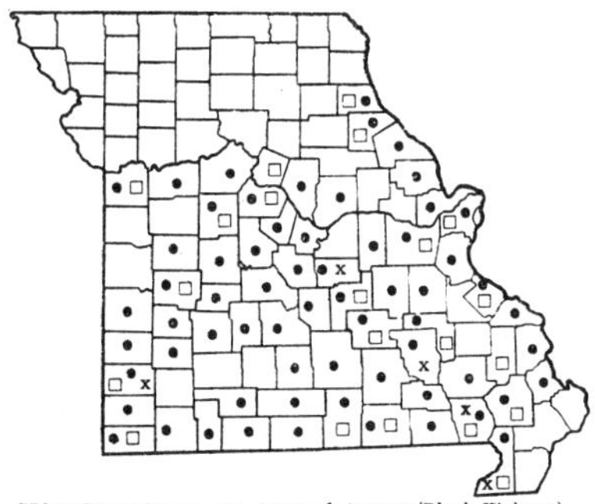

750 • Carya texana var. texana f. texana (Black Hickory)
750 x Carya texana var. texana f. glabra
750 □ Carya texana var. villosa

*20772*) has fruits approaching the larger size of var. *megacarpa* Sarg.

Ranges from Massachusetts and Vermont to Ontario, south to Georgia, Alabama, Mississippi, Illinois, Missouri, and Arkansas.

This species and the following, *C. ovalis*, are combined by some botanists. As there appear to be justifiable grounds for maintaining the two as distinct, a view held by Dr. Wayne Manning, current authority on the genus, both hickories are accorded full specific rank in the present work.

The bark of *C. glabra* is close and tight, shallow to deeply furrowed, and generally dark. Similar uses are found for the wood of this hickory, as for that of other species, such as for tool handles, manufacture of agricultural implements, wagons, chairs and other furniture, skis, sulkies for trotting horses, and fuel. Early settlers split strips of the small saplings of this species to make into brooms, hence the name Broom Hickory in some sections of the eastern states.

This species is considered to be one of those of the genus responsible for causing hay fever.

9. **Carya ovalis** (Wang.) Sarg.
False Shagbark — Map 749

Also called Sweet Pignut, Red Hickory, Small-fruited Hickory.

Flowers April–May.

Occurs mostly in dry or upland, rocky or nonrocky woods, usually in acid soils derived from or overlying chert, flint, sandstone, or granite. Throughout the Ozark region of southern and central Missouri northeast to an unglaciated section of Marion County.

*Carya ovalis* is united with *C. glabra* by the United States Forest Service, but is held to be distinct from that species by other botanists, including Deam, E. J. Palmer, and the current authority on the genus, Dr. Wayne Manning. While many trees are found with fruits intermediate between the two types of hickory, and certain separation of *C. glabra* from *C. ovalis* depends upon collection of mature fruit in late autumn, extreme trees of the two types appear distinct and they are being maintained in the present treatment as species following Dr. Manning's studies (see Rh. 50: 191–92. 1948).

The following variations are found in Missouri, but can be distinguished mainly by mature fruits:

a. Lower surface of leaflets conspicuously covered with glands and rather sticky; fruit small, 1.3–1.5 cm. thick, splitting to the base along the sutures which are distinctly broadly winged
9e. C. OVALIS var. ODORATA

a. Lower surface of leaflets not conspicuously covered with glands; fruit larger, 1.5–3 cm. thick, splitting to the base along the sutures which are not winged or at most with a narrow wing or keel . . . . . . . . . . . *b*

b. Fruit broadest in upper half, narrowed at the base (obovoid) 9d. C. OVALIS var. OBOVALIS

b. Fruit either as broad as long and more or less globe-shaped and round, or broadest in the middle and tapering or narrowed at each end (ellipsoid) . . . . . . . *c*

c. Nut narrowed at the summit to a short point, ovoid to ellipsoid . . 9a. *C.* OVALIS var. OVALIS

c. Nut rounded and broadest above the middle to a rounded, slightly or definitely heart-shaped summit . . . . . *d*

d. Branches of the 1st year (youngest) rather short-hairy, those of older growth of 2nd and 3rd year more or less hairy . 9c. C. OVALIS var. OBCORDATA f. VESTITA

d. Branches more or less glabrous . .
9b. C. OVALIS var. OBCORDATA f. OBCORDATA

9a. **Carya ovalis** var. **ovalis** Map 749

*Carya ovalis* (Wang.) Sarg. [G, P & S]

*Carya glabra* var. *glabra* in part, of [Little], not *C. glabra* (Mill.) Sweet

Known in southeastern Missouri from Ste. Genevieve (wooded upland slopes overlying LaMotte sandstone near Pickle Creek, 3 mi. west of Pickle, February 3, 1934, *Steyermark 7755*) and Dunklin counties.

Ranges from Massachusetts to Wisconsin and Iowa, south to Georgia, Alabama, Mississippi, and Missouri.

9b. **Carya ovalis** var. **obcordata** (Muhl. & Willd.) Sarg. f. **obcordata** Map 749

*Carya glabra* var. *glabra* in part, of [Little], not *C. glabra* (Mill.) Sweet

Generally common in the Ozark region of southern and central Missouri northeast to Marion County.

Ranges from Massachusetts to Wisconsin and Iowa, south to Georgia, Alabama, Mississippi, and Missouri.

9c. **Carya ovalis** var. **obcordata** f. **vestita** Sarg. Map 749

Known from Montgomery, Gasconage, and Henry counties.

Recorded from Indiana and Missouri.

9d. **Carya ovalis** var. **obovalis** Sarg. Map 749

*Carya glabra* var. *glabra* in part, of [Little], not *C. glabra* (Mill.) Sweet

Generally common in the Ozark region of southern and central Missouri northeast to Marion County.

Ranges from Massachusetts to Indiana and Missouri, south to Georgia, Alabama, and Arkansas.

9e. **Carya ovalis** var. **odorata** (Marsh.) Sarg. Map 749

*Carya glabra* var. *glabra* in part, of [Little], not *C. glabra* (Mill.) Sweet

Known only from Madison and Iron counties in the southeastern Ozark section.

Ranges from Massachusetts to Ontario, south to Georgia, Mississippi, and Missouri.

The bark of *C. ovalis* and its varieties is more or less scaly as in Shagbark Hickory (*C. ovata*), but this does not flake off in plates as thin or large as in that species. On the lower 5–12 feet of trunk of older trees, the bark is tight, close, and slightly ridged. As in other members of the genus, the wood of this hickory is used in similar ways for tool handles, manufacture of agricultural implements, wagons, and for fuel.

Since this species is closely related to *C. glabra*, which is believed to be one of those responsible for hay fever, it is likely that *C. ovalis* also may cause some cases.

10. **Carya texana** Buckl. Black Hickory Map 750

Also called Ozark Pignut Hickory.

Flowers April–May.

Occurs in dry or rocky upland woods, usually in acid soils derived from or overlying chert, flint, sandstone, or granite.

Common in the Ozarks throughout southern and central Missouri and northeast in unglaciated territory to Marion County.

Ranges from Louisiana and Texas, north to Indiana, Illinois, Missouri, and Oklahoma.

Missouri material may be divided into the following variations:

a. Young branchlets, veins and midrib of lower surface of fully grown leaflets, as well as lower surface itself, glabrous (without hairs) or nearly so to the end of the season. 10b. C. TEXANA var. TEXANA f. GLABRA

a. Young branchlets and lower surface or midrib or veins of lower surface of fully grown leaflets more or less hairy to the end of the season . *b*

b. Midrib or veins of lower surface and/or lower surface of fully grown leaflets covered with rusty or tawny hairs; leaf-stalk (petiole) and axis of leaf (rachis) glabrous or nearly so at maturity, rarely with hairs . 10a. C. TEXANA var. TEXANA f. TEXANA

b. At least the midrib of lower surface and usually the complete lower surface of mature leaflets densely covered with hairs; leaf-stalk (petiole) and axis of leaf (rachis) usually quite hairy . 10c. C. TEXANA var. VILLOSA

10a. **Carya texana** var. **texana** f. **texana** Map 750

*Carya texana* Buckl., in part [G, Little], not *C. texana* C. DC.

*Carya Buckleyi* var. *arkansana* [BB, P & S, Steyerm.]

*Carya texana* var. *arkansana* (Sarg.) Little

This is the commonest variation encountered in Missouri.

10b. **Carya texana** var. **texana** f. **glabra** (Palmer & Steyerm.) Steyerm. Map 750

*Carya Buckleyi* var. *arkansana* f. *glabra* Palmer & Steyerm., Ann. Mo. Bot. Gard. 25: 770–71. 1938. [Steyerm.]

Scattered within the range of var. *texana* f. *texana* in Missouri in Dunklin, Butler (*Steyermark 11461*, holotype in Mo. Bot. Gard. Herb.), Reynolds, Maries, and Jasper counties.

10c. **Carya texana** var. **villosa** (Sarg.) Little
Map 750

*Carya Buckleyi* var. *villosa* (Sarg.) Sarg. [BB, P & S, Steyerm.]

*Carya texana* Buckl. in part [Little], not *C. texana* C. DC.

Of frequent occurrence in the Ozark region of southern and central Missouri northeast to Marion County, but less common than typical *C. texana* var. *texana* f. *texana*.

Known only from Missouri, but to be expected in adjacent states bordering the range of var. *villosa*.

The bark of *C. texana* and varieties is tight, thick, and deeply furrowed. The wood is hard and brittle and often used for fuel. The foliage has a beautiful yellow color in autumn. It is often attacked by the fall webworm (*Hyphantria cunea*) which forms webby masses over the ends of branches.

On the drier upland ridges, on level upland, and on upper slopes of hills where chert, sandstone, or granite impregnate the soil, this species is abundant, occurring with various species of oak, such as *Quercus alba*, *Q. velutina*, *Q. coccinea*, *Q. marilandica*, *Q. stellata*, and *Q. falcata* (in the southern part of the state), pine, other hickories, such as *C. tomentosa*, *C. ovata*, and *C. ovalis*, flowering dogwood (*Cornus florida*), and shrubs, such as Deerberry (*Vaccinium stamineum*), Tree Huckleberry (*Vaccinium arboreum*), and Lowbush Blueberry (*Vaccinium vacillans*).

## Order **FAGALES**

### Fam. **CORYLACEAE** (Hazel Family)

(*Betulaceae* Birch Family [BB, P & S, Steyerm.])

#### I. *Key based on bark and mature leaves*

a. Lower surface of leaves whitish or gray-white; bark peeling off in reddish-brown papery strips, exposing a pale smooth layer . . . . . . . . . . . . . . . . . . . . 4. BETULA
a. Lower surface of leaves green or pale green, not white or gray-white; bark smooth or rough, not peeling . . . . . . . . . . . . . . . . . . . . . . . . . . . *b*
  b. Leaf-blades with a blunt or rounded tip and rounded teeth . . . . . . . . . . 5. ALNUS
  b. Leaf-blades with a pointed (acute or acuminate) tip or pointed teeth, or with both . . . . . . *c*
    c. Leaf-blade usually with 5–8 veins on each side of midrib. . . . . . . . . . . 1. CORYLUS
    c. Leaf-blade usually with 9–15 veins on each side of midrib . . . . . . . . . . . *d*
      d. Bark smooth, twisty, blue-gray; lower surface of fully grown leaves glabrous (without hairs), except for some hairy tufts in the axils of the veins; none of the lower large lateral nerves forked . . . . . . . . . . . . . . . . . . . . 3. CARPINUS
      d. Bark scaly, light brown or brownish-gray; lower surface of fully grown leaves mostly hairy; some of the lower large lateral nerves forked . . . . . . . . . . . . 2. OSTRYA

#### II. *Key based on flowers, fruits, and immature leaves*

a. Female catkin resembling a small pine cone, its bracts woody and long-persisting; base of catkin or main supporting stalk of flowering portion glabrous (without hairs); flowering catkins appear before the leaves . . . . . . . . . . . . . . . . . . . . . . 5. ALNUS
a. Female catkin not as above; base of catkin or main supporting stalk of flower portion densely hairy . . *b*
  b. Flowering catkins appear usually before the leaves; flowering female catkin very small, few-flowered, and inconspicuous, 5 mm. long or less, with red or rose-colored short hair-like stigmas; mature fruit a nut 10–15 mm. long enclosed in a leafy involucre . . . . . . . . 1. CORYLUS
  b. Flowering catkins appear usually with the leaves; flowering female catkin conspicuous, many-flowered, 10 mm. or more long, lacking the red stigmas; mature fruit 3–5 mm. long . . . . . . *c*
    c. Female catkins closely-flowered, their scales densely overlapping, showing no space between; anthers glabrous (without hairs); male flowers with a calyx; female flowers without a calyx; nut winged. . . . . . . . . . . . . . . . . . . . . . . . . 4. BETULA
    c. Female catkins loosely-flowered, their scales loosely overlapping with much space exposed in between; anthers hairy at tip (use magnifying lens); male flowers without a calyx; female flowers with a calyx; nut wingless . . . . . . . . . . . . . . . . . . . *d*

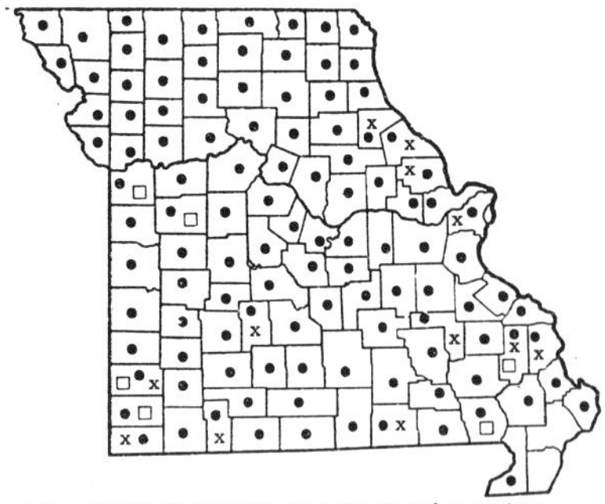

751 ● Corylus americana var. americana f. americana (Hazelnut)
751 x Corylus americana var. americana f. missouriensis
751 □ Corylus americana var. indehiscens

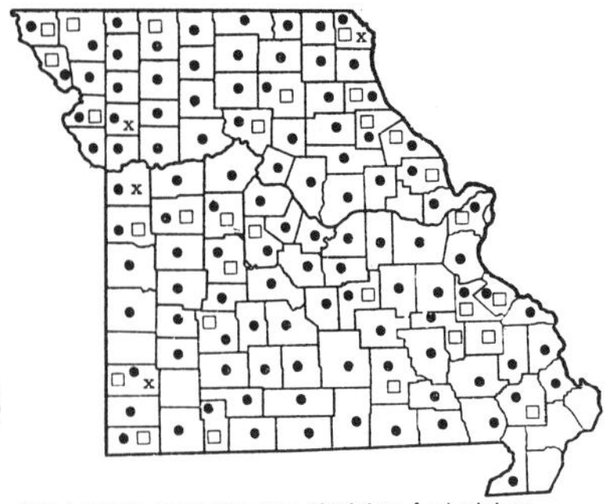

752 ● Ostrya virginiana var. virginiana f. virginiana (Hop Hornbeam)
752 x Ostrya virginiana var. virginiana f. glandulosa
752 □ Ostrya virginiana var. lasia

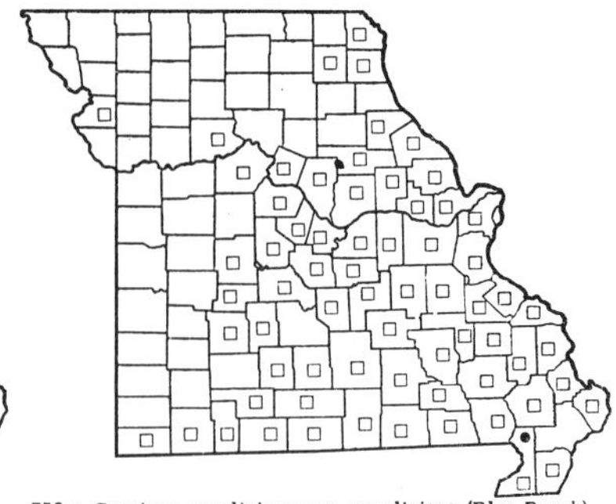

753 ● Carpinus caroliniana var. caroliniana (Blue Beech)
753 □ Carpinus caroliniana var. virginiana

d. Each bract of female or fruiting catkins 2–3-lobed, flat, open, and attached to or subtending the ovary or nut; bracts of flowering female catkins glabrous or sparsely hairy; male catkins enclosed in bud-scales in winter . . . . . . . . . . . . . . . . . . 3. CARPINUS

d. Each bract of female or fruiting catkins unlobed, bladdery, closed, surrounding an ovary or nut; bracts of flowering female catkins densely hairy; male catkins naked in winter, not enclosed in bud-scales . . . . . . . . . . . . . . . . . . . . . . 2. OSTRYA

## 1. **Corylus** L. Hazelnut

**Corylus americana** Walt. Hazelnut Map 751

a. Bracts of fruit united on one side to their summit and partly united on the other side, so that fruit does not appear open. c. C. AMERICANA var. INDEHISCENS

a. Bracts of fruit two, distinct and open on both sides to the nut exposing the fruit . . . . *b*

b. Branches, leaf-stalks (petioles), and involucres around flower or fruits with gland-tipped hairs . . . . . a. C. AMERICANA var. AMERICANA f. AMERICANA

b. Branches, leaf-stalks, and involucres around flowers or fruits without gland-tipped hairs b. C. AMERICANA var. AMERICANA f. MISSOURIENSIS

a. **Corylus americana** var. **americana** f. **americana** Map 751

*Corylus americana* Walt. [G, BB, P & S]

Flowers February–April.

Occurs in dry or moist thickets, woodland, and border of woodland, in valleys and upland. Throughout Missouri and doubtless in every county.

Ranges from Maine to Saskatchewan, south to Georgia, Arkansas, and Oklahoma.

b. **Corylus americana** var. **americana** f. **missouriensis** (A. DC.) Fern. Map 751

*Corylus americana* f. *missouriensis* [G, P & S, Steyerm.]

Frequent in the state, but records of its occurrence are less common than f. *americana*.

Occurs in the range of the typical form.

c. **Corylus americana** var. **indehiscens** Palmer & Steyerm. Map 751

Known from Bollinger, Butler, Jackson, Johnson, Jasper, and Newton counties, but to be expected in other sections of the state.

Ranges from North Carolina to Missouri, Arkansas, and Oklahoma.

Material at one time identified as *Corylus cornuta* Marsh. is now referred to this variety.

The nuts of *C. americana*, if they can be collected before eaten first by blue jays and other birds, are very good and similar to the European filbert. A meal ground from the nuts can be made into a bread of similar type to that made from filberts. The foliage is quite striking in autumn, becoming quite varied. It may be dull orange with brick-red and dull green at

---

Plate no. 134. **1.** Carya ovalis, $\times\ ^{2}/_{7}$; *a.* Fruit, $\times\ ^{2}/_{7}$. **2.** Carya texana var. texana, $\times\ ^{2}/_{7}$; *a.* Fruit, $\times\ ^{2}/_{7}$. **3.** Corylus americana, $\times\ ^{2}/_{7}$; *a.* Flowering branch, $\times\ ^{2}/_{7}$; Details from Small, The New York Botanical Garden. **4.** Ostrya virginiana, $\times\ ^{2}/_{7}$; *a.* Flowering branch; Details from Small, The New York Botanical Garden. **5.** Carpinus caroliniana; *a.* Carpinus caroliniana var. virginiana, flowering branch, $\times\ ^{2}/_{7}$; *b.* Carpinus caroliniana var. virginiana, fruit, $\times\ 1^{3}/_{7}$; *c.* Carpinus caroliniana var. caroliniana, leaf $\times\ ^{2}/_{7}$; *d.* Carpinus caroliniana var. virginiana, fruiting branch, $\times\ ^{2}/_{7}$; Details from Small, The New York Botanical Garden.

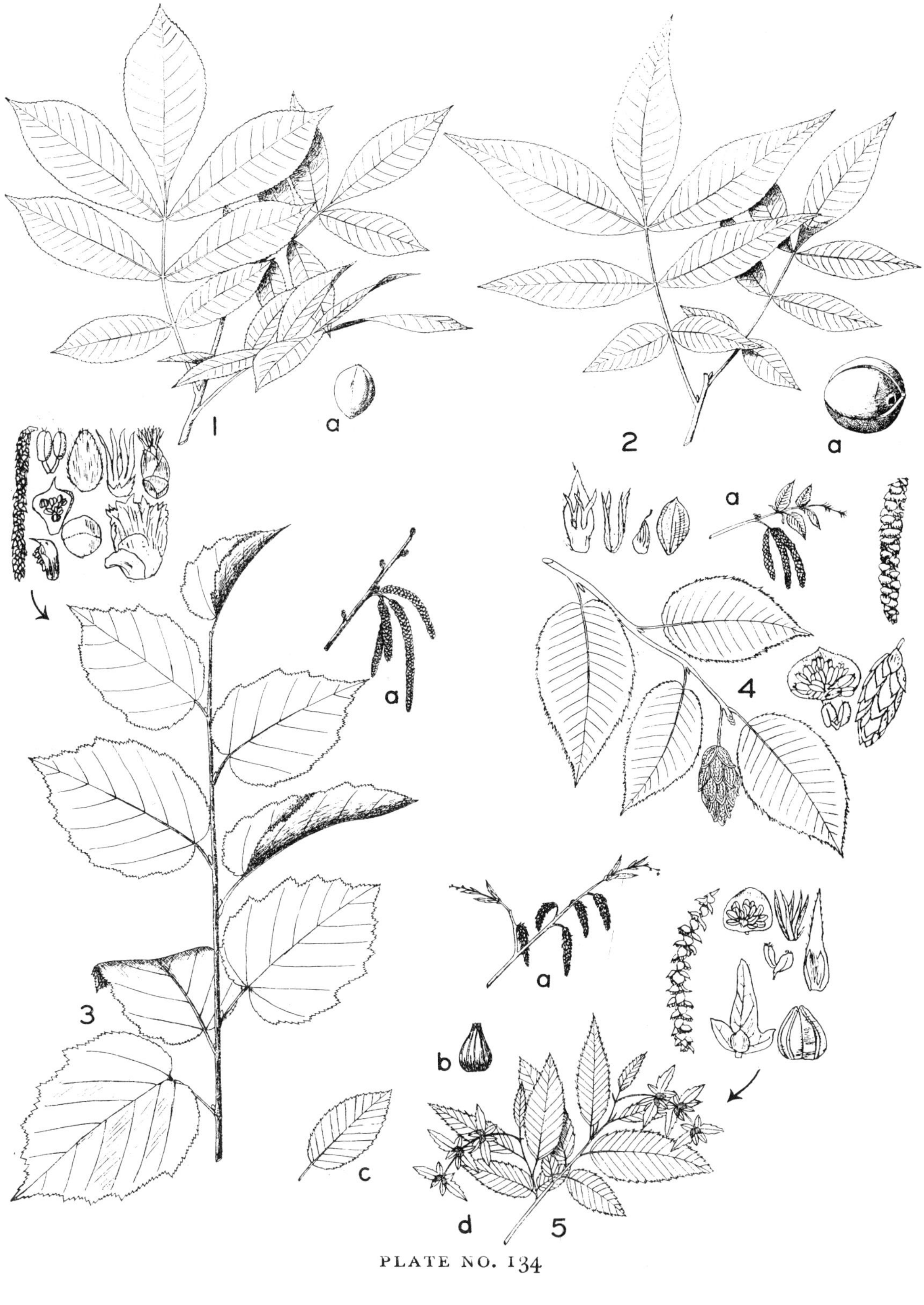


PLATE NO. 134

the same time or dull purple-red or with combinations of rose, orange, yellow, and pale green.

*Excluded Species*

**Corylus cornuta** Marsh. Beaked Hazelnut

This was included by Palmer & Steyermark in their *Annotated Catalogue* on the basis of a specimen from Butler County. This is now found to be referable to *C. americana* var. *indehiscens* as an extreme form of that variety. Another specimen (*Palmer 16422*) from Bollinger County, labeled *C. rostrata* [= *C. cornuta*], is also referable to *C. americana* var. *indehiscens*.

## 2. **Ostrya** Scop. Hop Hornbeam, Ironwood

**Ostrya virginiana** (Mill.) K. Koch

Hop Hornbeam Map 752

More frequently called Ironwood in certain sections.

Flowers April–May.

Occurs in usually dry soil on rocky slopes, along bluffs, borders of upland woods, and rarely along streams. Throughout Missouri, and apparently absent from some of the extreme southeastern lowland counties.

Ranges from Nova Scotia to Manitoba, south to Florida, Tennessee, Arkansas, Texas, and northern Mexico.

Missouri material is represented by the following variations:

a. Young or new branchlets with stalked glands . . . b. O. VIRGINIANA var. VIRGINIANA f. GLANDULOSA
a. Young or new branchlets without stalked glands . . . . . . . . . . . . . b
 b. Young or new branchlets glabrous (without hairs) or very sparsely hairy . . . a. O. VIRGINIANA var. VIRGINIANA f. VIRGINIANA
 b. Young or new branchlets densely hairy . c. O. VIRGINIANA var. LASIA

a. **Ostrya virginiana** var. **virginiana** f. **virginiana** Map 752

*Ostrya virginiana* (Mill.) K. Koch [G, BB, P & S]

This is the most common form encountered in the state.

b. **Ostrya virginiana** var. **virginiana** f. **glandulosa** (Spach) Macbr. Map 752

*Ostrya virginiana* f. *glandulosa* (Spach) Macbr. [G, Steyerm.]

*Ostrya virginiana* var. *glandulosa* (Spach) Sarg. [P & S]

Scattered within the range in Missouri in Clark, Clinton, Jackson, and Jasper counties.

c. **Ostrya virginiana** var. **lasia** Fern. Map 752

Frequent throughout the state, but less common than var. *virginiana*.

Ranges from Florida to Texas, north to Massachusetts, Tennessee, Illinois, Iowa, and South Dakota.

The wood of Hop Hornbeam ranks among the hardest and strongest of woods known in eastern North America. It is harder than oak, hickory, locust, or persimmon, and surpassed only by flowering dogwood. The limited size and diameter of the tree, however, makes it useful only for such small-sized articles as tool handles, ax handles, mallets, and for fence posts.

## 3. **Carpinus** L. Hornbeam, Blue Beech

**Carpinus caroliniana** Walt. Blue Beech Map 753

Flowers April–May.

Occurs on north-facing bluffs, in rich woods at the base of bluffs, rocky slopes along streams, ravine bottoms, low wooded valleys, and moist woodland.

Two variations may be recognized:

Rare type, known only from Dunklin County; leaves oblong, acute, 1–3.5 (rarely 4) cm. broad, 2.5–8 cm. long, with broad low teeth rarely 1 mm. high; fruiting bracts blunt or pointed, entire (without teeth) or with 1–2 blunt teeth . . . a. C. CAROLINIANA var. CAROLINIANA
Common type, throughout eastern Mo., the Ozarks, and elsewhere; leaves oval or narrowly, abruptly long-tipped, 2.5–6 cm. broad, 5–12 cm. long, with sharp slender teeth up to 1–3 mm. high; fruiting bracts acute, usually with 1–5 sharp teeth . . b. C. CAROLINIANA var. VIRGINIANA

a. **Carpinus caroliniana** var. **caroliniana** Map 753

*Carpinus caroliniana* Walt. [G, Steyerm., BB in part, P & S in part]

Known only from low woods in Dunklin County, southeastern Missouri.

Ranges from Florida to Texas, north to Maryland, Tennessee, Illinois, and southeastern Missouri.

754 Betula nigra (River Brich)

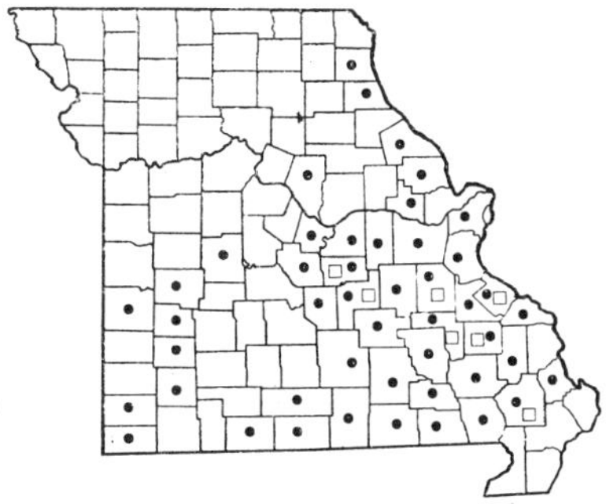

755 ● Alnus serrulata var. serrulata f. serrulata (Alder)
755 □ Alnus serrulata var. serrulata f. novaboracensis

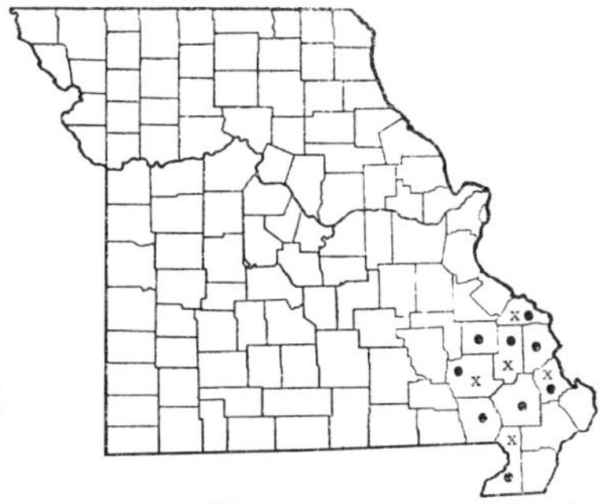

756 ● Fagus grandifolia var. caroliniana f. caroliniana (Beech)
756 x Fagus grandifolia var. caroliniana f. mollis

b. **Carpinus caroliniana** var. **virginiana** (Marsh.) Fern. Map 753

*Carpinus caroliniana* in part [of BB, P & S], not Walt.

Eastern Missouri, Ozark region west to Cooper, Morgan, Benton, Hickory, Polk, and McDonald counties, and northwest along the Missouri River locally to Buchanan County.

Ranges from New England to Ontario and Minnesota, south to North Carolina, Tennessee, and Arkansas.

In fall the foliage turns an orange russet with brick-red flushed with greenish, or is dull yellow to pale greenish-yellow. The sapwood is nearly white, and is used for levers, tool handles, and other small articles. The blue-gray fluted stems have a sculptured appearance, resembling twisted muscles. The nuts may be eaten, but, considering their small size, a considerable number would have to be gathered to be of much food value.

## 4. **Betula** L. Birch

**Betula nigra** L. River Birch Map 754

Also called Red Birch.

Flowers April–May.

Occurs in alluvial ground along streams and borders of gravel beds. Nearly throughout Missouri, absent in the extreme northwestern and west-central sectors.

Ranges from Florida to Texas, north to New Hampshire, New York, Pennsylvania, West Virginia, Ohio, Michigan, Wisconsin, Minnesota, and Kansas.

The foliage turns a pale to rich yellow in autumn. Large quantities of pollen are shed by this and other species of birch and constitute one of the important tree pollens in causing hay fever. The wood is used for pulp in papermaking and employed principally in the manufacture of furniture and woodenware. It is used for shuttles, but not to as much extent as *Cornus florida*, the Flowering Dogwood. At one time it was used for ox yokes and wooden shoes. Along Missouri streams this species germinates along the silty sands and gravels washed with mud deposited by the river and is one of the pioneer species in building up a new forest along stream bank alluvium. Following close behind the initial stages started on pure gravel beds by Vernal Witch Hazel (*Hamamelis vernalis*) and Ward's Willow (*Salix caroliniana*), River Birch, Sycamore, Green Ash, and Silver Maple continue the forest succession as finer sandy alluvium is built up along the river bed and banks.

## 5. **Alnus** B. Ehrh. Alder

**Alnus serrulata** (Ait.) Willd. Common Alder

Also called Smooth Alder, Tag Alder. Map 755

Flowers March–April.

Occurs along stream banks, springs, spring branches, and in wet spring-fed meadows. Ozark region north along the Mississippi River to Lewis County.

Ranges from Florida to Louisiana, north to Nova Scotia, Maine, New Hampshire, Vermont, New York, West Virginia, Ohio, Indiana, Illinois, Missouri, and Oklahoma. Missouri material is represented by the following variations:

Lower surface of fully grown leaves glabrous or nearly so. . . a. A. SERRULATA var. SERRULATA f. SERRULATA

Lower surface of fully grown leaves densely and permanently short-hairy, velvety to touch . . b. A. SERRULATA var. SERRULATA f. NOVABORACENSIS

a. **Alnus serrulata** var. **serrulata** f. **serrulata** Map 755

*Alnus rugosa* of auth. in part [P & S, Steyerm.], not (DuRoi) Spreng.

*Alnus serrulata* (Ait.) Willd. [G, BB]

This is the common type found in the state.

b. **Alnus serrulata** var. **serrulata** f. **novaboracensis** (Britt.) Fern. Map 755

*Alnus serrulata* f. *novaboracensis* (Britt.) Fern. [G]

Occasionally found in parts of the Ozark region. Found throughout the range of var. *serrulata*.

The foliage remains dark green for well into autumn after many of the other species have turned color. The leaves stay dark green, finally becoming tinged with dull brown. The presence of this alder indicates boggy ground, usually the location of a spring or spring-fed stream. Associated with this habitat are to be found many rare and unusual species in Missouri.

## Fam. FAGACEAE (Beech Family)

a. Buds long, slender, and pointed, at least 4 times as long as broad; leaves as thin as tracing or onion-skin paper; male flowers in rounded head-like or globe-shaped masses; nuts sharply triangular . 1. FAGUS

a. Buds less than 4 times as long as broad, not long, slender, and pointed; leaves thicker and firmer; male flowers in long narrow catkins; nuts rounded or flattened on one side, not sharply triangular. . *b*

b. Male catkins upright or spreading, pleasantly fragrant; young female flower and fruit covered with prickles; female flowers 2–4 in each involucre; nuts flattened on one side or compressed; fruit a prickly bur . . . . . . . . . . . . . . . . . . . . . . . . 2. CASTANEA

b. Male catkins drooping or hanging, not fragrant; young female flower not prickly; female flower solitary; nuts with curved or rounded sides; fruit an acorn . . . . . . . . . . . 3. QUERCUS

### 1. **Fagus** L. Beech

**Fagus grandifolia** Ehrh. Beech Map 756

Flowers April–May.

Occurs in usually acid soils overlying sands, gravels, or clays of Crowley Ridge or on sandstone formations, in rich woods of ravines, slopes, and small valleys bordering streams and spring branches. Southeastern Missouri, mostly on Crowley Ridge in Dunklin, Butler, Stoddard, and Scott counties, and northeast on St. Peter sandstone substrata to Perry County.

Missouri material may be referred to the following variations:

Lower surface of fully grown leaves glabrous (without hairs) . a. F. GRANDIFOLIA var. CAROLINIANA f. CAROLINIANA

Lower surface of fully grown leaves soft-hairy . b. F. GRANDIFOLIA var. CAROLINIANA f. MOLLIS

a. **Fagus grandifolia** var. **caroliniana** (Loud.) Fern. & Rehd. f. **caroliniana** Map 756

*Fagus grandifolia* var. *caroliniana* (Loud.) Fern. [G, BB, P & S, Steyerm.]

This is the more commonly encountered variation in Missouri.

Ranges from Florida to Texas, north to Massachusetts, Ohio, Indiana, Illinois, and Missouri.

b. **Fagus grandifolia** var. **caroliniana** f. **mollis** Fern. & Rehd. Map 756

Scattered throughout the range of f. *caroliniana* in Missouri.

Found throughout the range of var. *caroliniana* f. *caroliniana*.

The pollen of the Beech plays only an insignificant role in causing hay fever. The foliage turns a clear yellow in autumn contrasting with the beautiful smooth gray bark. The sweet edible nuts vary in abundance, about every third year reaching a maximum of harvest. They are also a favorite food of hogs

Plate no. 135. **1.** Betula nigra; Details from Small, The New York Botanical Garden. **2.** Alnus serrulata var. serrulata, × 2/7; *a.* Fruiting branch; *b.* Flowering branch; Details from Small, The New York Botanical Garden. **3.** Fagus grandifolia; *a.* Fagus grandifolia var. caroliniana, flowering branch, × 2/7; *b.* Fagus grandifolia var. caroliniana, fruit, × 2/7; *c.* Fagus grandifolia var. caroliniana, leaf, × 2/7; *d.* Fagus grandifolia var. grandifolia, leaf, × 2/7; Details from Small, The New York Botanical Garden. **4.** Castanea ozarkensis, × 2/7. **5.** Castanea dentata, × 2/7; *a.* Fruiting branch; *b.* Nut; Details from Small, The New York Botanical Garden.

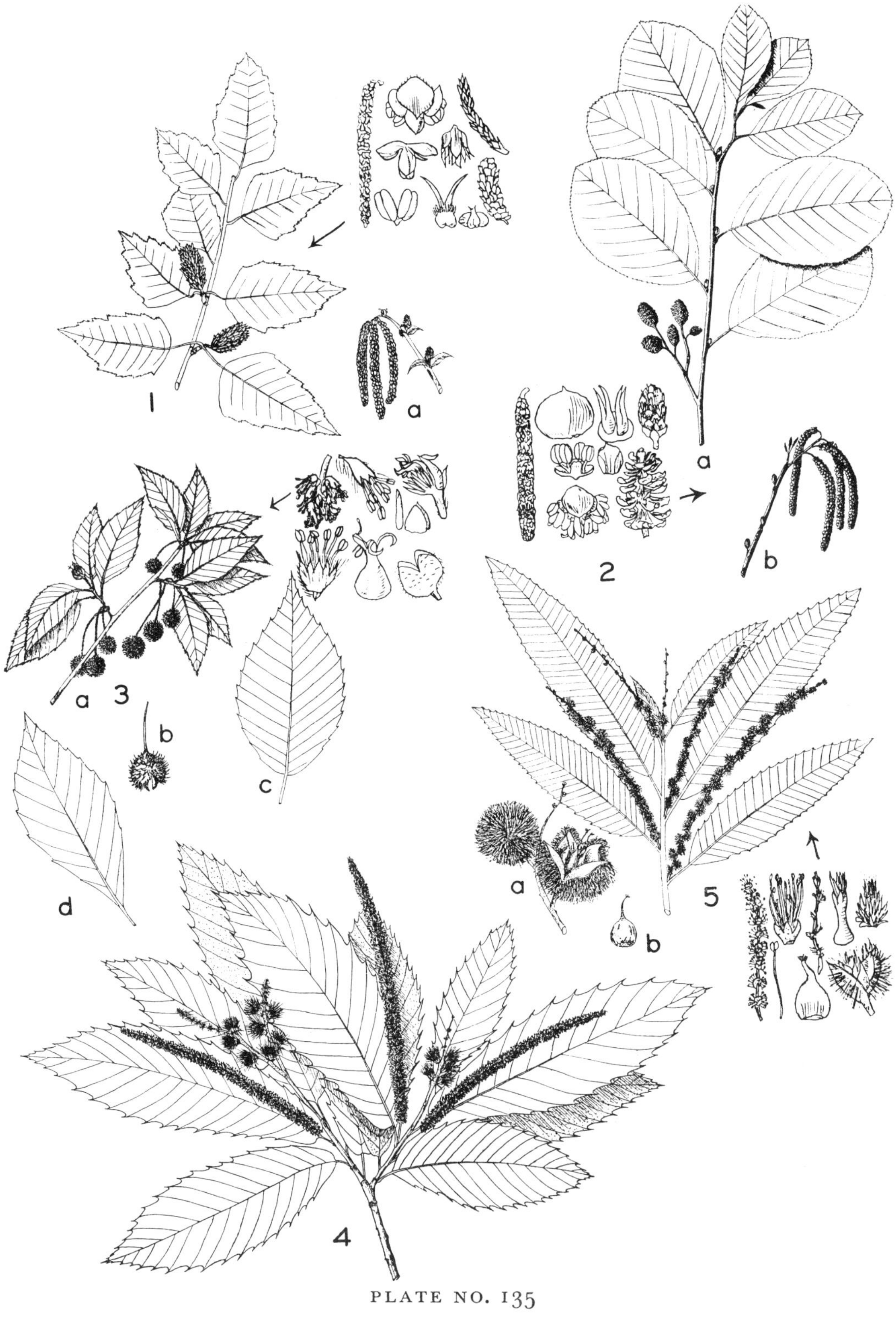


PLATE NO. 135

in those areas where open range is permitted. The young unfolding leaves may be cooked as a vegetable. Beech oil, extracted from the nuts of the European Beech, *F. sylvatica,* serves as a table oil in France and other countries and is said to be similar in delicacy to olive oil. Roasted beech nuts are sometimes used as a coffee substitute.

The wood is not very durable. A small amount is used as pulp for papermaking. At one time the wood was largely used for fuel. Although beech wood enters into the manufacture of some furniture, such as chairs, and in flooring, most of the wood is employed for tool handles, shoe-lasts, railroad ties, boxes, barrels, and such articles as clothespins, picnic plates, and spoons.

Although one of the most handsome and stately trees in North America, native beech is rarely used as a shade or ornamental tree, because it not only is difficult to transplant, but responds poorly to cultivation.

*Excluded variety*

**Fagus grandifolia** Ehrh. var. **grandifolia**
*Fagus grandifolia* Ehrh. [G, P & S, Steyerm.]

This variety, distinguished from var. *caroliniana* by its thinner leaves, which are more narrowed or cuneate at the base, more coarsely and distinctly toothed, and with the prickles of the mature fruit longer, 4–7 mm., instead of 3 mm. or less long, occurs from Nova Scotia and New Brunswick to Ontario and Minnesota, south to Virginia and West Virginia.

Records in Missouri previously referred to this variety are now included under var. *caroliniana.*

## 2. **Castanea** Mill. Chestnut

Lower surface of fully grown leaves glabrous (without hairs) and pale green; spines of fruit glabrous (without hairs); each fruit with usually 2 or 3 nuts, flattened on 1 or 2 sides; cultivated or escaped from cultivation . . . . . . . . . . . . . . . . . . . . . . . 1. C. DENTATA
Lower surface of fully grown leaves minutely hairy and yellowish-cream or whitish; spines of fruit with hairs; each fruit with only 1 nut not flattened; native tree of southwest Missouri . . . 2. C. OZARKENSIS

1. **Castanea dentata** (Marsh.) Borkh.
Chestnut Map 757
Flowers late May–July.

At one time ranged from Maine and Massachusetts to New York to Ohio, Michigan, Indiana, and southern Illinois, south to Georgia, Alabama, and Mississippi. Apparently now extinct in Florida.

Formerly more planted on farms, woodlots, and around buildings. Sometimes trees planted away from cultivation have persisted in wild or semiwild areas, and specimens from these trees have been collected in Crawford (*Steyermark 63074*) and Howell (*Steyermark 23586*) counties. The Crawford County trees, originally four in number, grew in a dry open pastured slope in a small valley, and were not near any dwellings. Sixty feet or more tall with a spread of a hundred feet, they produced an abundance of fruit, and at the time of the author's original visit in 1936, and again in 1945, showed no sign of the chestnut blight disease. On his last visit to this station, on October 28–29, 1945, along a branch of Crooked Creek near the Dent County line, T35N, R4W, sect. 35, 2½ miles northeast of Sligo, only one of the trees was still surviving. Seeds from the healthy trees taken to the author's wild flower preserve in northern Illinois and germinated there, beginning in 1946, flowered for the first time eleven years later although they were only 5 feet tall at that time.

The Howell County record, based on *Steyermark 23586*, from upland woods on top of a dry ridge bordering a field, T27N, R10W, sect. 4, 5 mi. southwest of Willow Springs, July 23, 1937, was collected from one of three large trees occurring with pine, Scarlet Oak (*Quercus coccinea*), Black Oak (*Q. velutina*), and other oaks and hickories, and appearing almost as if native.

Trees which have been planted in various parts of the state are known from the following records collected from the cultivated tree: Nodaway Co. (Northwest Missouri State College at Maryville, *Singleton 118*), Clay Co. (on Rust farm by spring near stream, at Liberty, *Hurst 70*), and Clark Co.

Although nearly wiped out by the chestnut blight, many sections where it originally thrived as tall forest trees still have living trees which send up stump sprouts which survive for a number of years until killed by the disease or occasionally grow to larger specimens which seem to resist the disease. Many dead trees continue to supply a market for chestnut rail fence poles and posts. The wood is used as pulp in papermaking and is also used for railroad ties, cheaper types of furniture, and interior trimming for houses. The bark has been much used in tanning.

Home owners and horticulturists are now planting

757 Castanea dentata (Chestnut)

758 Castanea ozarkensis (Ozark Chinquapin)

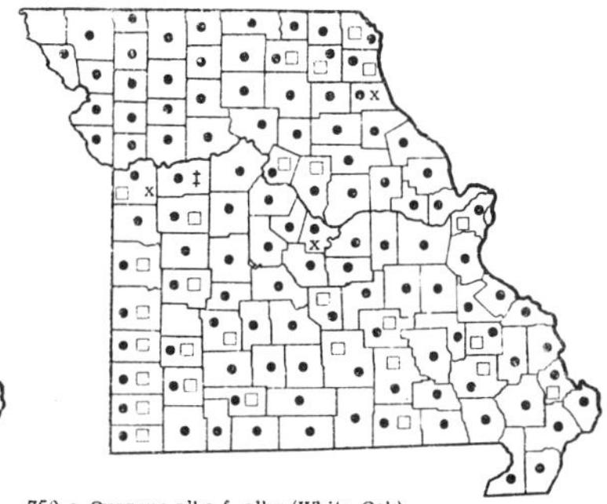

759 • Quercus alba f. alba (White Oak)
759 □ Quercus alba f. latiloba
759 x Quercus X Bebbiana (Quercus alba X macrocarpa)
759 ‡ Quercus X Jackiana (Quercus alba X bicolor)

other kinds of chestnuts for nut and ornamental trees, such as the Chinese Chestnut (*C. mollissima*) and hybrid races which show more or less immunity to this disease.

2. **Castanea ozarkensis** Ashe Ozark Chinquapin
Also called Ozark Chestnut. Map 758
Flowers late May–June.

Occurs in acid soils of dry upland ridges in oak-hickory forest, in southwestern Missouri, in Stone, Barry, McDonald, and Newton (*Palmer 56575*, along Indian Creek, 3½ mi. southeast of Christopher) counties; locally east in Howell County on upland low slopes along small draw in abandoned thicket, 1 mi. southeast of Moody, August 11, 1934, *Steyermark 14398*).

Ranges from Missouri and Arkansas to Oklahoma. Louisiana and Mississippi are sometimes mentioned as in the range of this species, but specimens examined from those states have proved to be *C. pumila* (L.) Mill.

This tree reaches a height of about sixty-five feet in its Missouri habitat, where it usually occupies the upper reaches of ravines and narrow crests of stony chert or flint ridges where the soil is acid and noncalcareous and with a good accumulation of leaf mold. Here it is associated with Sour Gum (*Nyssa sylvatica* var. *caroliniana*), Flowering Dogwood, White Oak, Blackjack Oak, Post Oak, Black Oak, Ozark Hickory (*Carya texana*), Mockernut, and Shagbark Hickory. On the forest floor at its base grow Blueberry (*Vaccinium vacillans* var. *crinitum*), Deerberry (*V. stamineum* and var. *neglectum*), and Farkleberry (*V. arboreum* var. *glaucescens*), along with many other kinds of acid-soil plants.

The leaves are similar to, and may be at first mistaken for those of, *Quercus Muehlenbergii*, the Chinquapin Oak, but the margins are more sharply toothed and the leaf has a rich yellow- or grass-green on its upper surface instead of the blue-green or dull green of the oak. The Chinquapin Oak also occurs in limestone and calcareous soils.

The mature nuts of the Ozark Chinquapin, like those of the eastern *C. dentata*, are sweet and valued as food. They are smaller than the Eurasian species, *C. sativa*, but have a good flavor. The nut of the Eurasian species has served as an important bread source, often being substituted for potatoes or wheat; it has also been cooked as a starchy vegetable dish or used as a flour for soups. The French, especially, use chestnuts to a greater extent in their cookery than people of other countries, even to the degree of extracting sugar from the nuts. *Marrons glacés*, sometimes seen on restaurant menus or sold in grocery stores, are chestnuts which have been first boiled, afterwards placed in sugar, and then dried. The eastern American Chestnut, *C. dentata*, was used by the early colonists in their soups and as starchy vegetables, and the American Indian included the nuts in their corn dough. The nuts were even roasted as a substitute for coffee and chocolate. Although the nuts of the Ozark species have not received any especial attention nor are known to have been used for the several types of dishes mentioned above, it is probable that they could serve in similar preparations if given an opportunity.

So far as known, *C. ozarkensis* has not suffered from attack by the fungus parasite, *Endothia parasitica*, responsible for the chestnut blight disease, but the blight is recorded as occurring in Missouri on various planted trees of the eastern *C. dentata* (see Maneval, W. E. A list of Missouri fungi. Univ. Mo. Stud. 12, no. 3, pp. 3, 100. 1937).

### 3. **Quercus** L. Oak

In number of species oaks outrank any other group of trees in the forests of Missouri and are also the most important timber trees in the state. The wood of various species is much used for railroad ties, furniture, such as chairs and tables, office desks, interior paneling and veneer, flooring, barrels, wagons, agricultural implements, baskets and novelties, fuel, and charcoal. The bark is rich in tannin and is used for tanning leather.

The wind-blown pollen is very light and abundantly produced by all species. Among trees, more cases of hay fever have been attributed to oaks than to any other kinds of trees. If oak leaves are browsed by cattle to the exclusion of other kinds of food, it is believed that toxic effects result, most cases of poisoning occurring in early spring when other natural growth is less plentiful. The tannic acid itself, found in the leaves, is not believed to cause the poisoning, but rather other substances, which are contained, are believed responsible for any serious trouble which may develop.

In identifying oaks, it must be kept in mind that as many characters as possible, leaves, twigs, winter buds, and fruit, must be taken into consideration. Use of one character alone may lead to difficulties.

Of the three main groups of oaks in Missouri (1) the White Oak group matures its fruit the first and same year the young acorns develop on the newest shoots, the leaves have rounded lobes or teeth without bristly or pin-point tips, the inside of the acorn shell is smooth, the kernel is sweet, and the bark is pale gray; (2) the Black Oak group matures its fruit at the end of the second growing season, fully grown fruit being found only on last year's twigs or the ones previous to the season in which collected, the leaves have bristle or pin-point tips, the inside of the acorn shell is silky-hairy, the kernel is bitter, and the bark is usually blackish or dark gray; (3) the Willow Oak group, to which belong Shingle and Willow Oak, in which the leaves are entire (without teeth or lobes), and the fruit ripens the second year.

Many species of oaks cross with one another, and the following key does not include hybrids. These are mentioned after the description of a species and in a separate list at the end. Since the leaves are best developed after May 1, greater use of the following key will be after that date. If no fruit is available on the branches at the time of collection, it is often possible to find old fruits of last year lying on the ground somewhere around the base of the tree.

a. No teeth or lobes on margins of leaves (rarely a seedling or young sprout may show some teeth) . . *b*
 b. Leaves broadest around the middle or of about the same width throughout . . . . . . . . . *c*
  c. Lower surface of leaves hairy; leaves mainly $2\frac{1}{2}$–4 times as long as broad, averaging 3–6 cm. wide, rarely narrower; leaf-stalk (petiole) usually hairy; common, in most counties of Missouri 8. Q. IMBRICARIA
  c. Lower surface of leaves mostly glabrous (without hairs) or with some hairiness along the mid-nerve or in the axils of the nerves; leaves mainly 5–8 times as long as broad, usually 0.6–2.5 cm. (rarely to 3) wide; leaf-stalk (petiole) mainly glabrous (without hairs); swamps and low ground of southeast Missouri . . . . . . . . . . . . . . . . . . . . . 9. Q. PHELLOS
 b. Leaves broadest towards the top, enlarged or broadened upward . . . . . . . . . . . *d*
  d. Lower surface of leaves hairy; leaves only slightly, if at all, longer than broad, sometimes as wide as long, usually blunt, rounded, or somewhat heart-shaped at base, rarely straight or tapering; lower surface of leaves yellow-brown or yellow-green; cup of fruit more or less top-shaped or inversely conical, 1.5–2 cm. broad . . . . . . . . . . 11. Q. MARILANDICA
  d. Lower surface of leaves without hairs (glabrous) or nearly without (glabrate); leaves usually $1\frac{1}{2}$–2 times as long as broad, tapering to a narrowed, acute, or triangular base; lower surface of leaves pale green; cup of fruit saucer-shaped, 1–1.5 cm. broad . . . . . . 10. Q. NIGRA
a. Teeth or lobes appearing along the margins or near the top of the leaves, or leaves deeply cut. . . . *e*
 e. Lobes or teeth of mature leaves lacking bristle-like points, the mid-nerve and side nerves remaining included in leaf tissue (teeth of leaves pointed with short tip in *Q. prinoides* var. *acuminata* and some other species, but not bristle-tipped) . . . . . . . . . . . . . . . . . *f*
  f. Lower surface of fully grown leaves without hairs (rarely some hairs along main nerves or in their axils) . . . . . . . . . . . . . . . . . . . . . . . . . . . *g*
   g. Lower surface of fully grown leaves whitened; leaves usually with 6–10 (rarely 4) lobes; acorn 1.8–3 cm. long; acorn cup with warty or corky scales; common tree throughout Missouri. 1. Q. ALBA
   g. Lower surface of fully grown leaves dull or slightly paler green, not whitened; leaves with 2–5 lobes; acorn 0.8–1.6 cm. long; cup with flat unthickened scales; tree of lowlands of southeast Missouri . . . . . . . . . . . . . . . . . . . . 10. Q. NIGRA

f. Lower surface of fully grown leaves with hairs more or less covering the surface . . . . . . . *h*

h. Margins of leaves wavy or with 3–13 rounded or pointed teeth along each margin, but not lobed; the spaces (sinuses) between the teeth shallow, running less than a third the distance to the main mid-nerve; main side (lateral) nerves of the lower surface of leaves more or less regularly spaced and equally prominent . . . . . . . . . . . . . . . . . . . . *i*

i. Some or most of the side nerves of the lower surface of leaves not ending in the teeth; stalks (peduncles) of fruits 2–6 cm. long, longer than the leaf-stalks (petioles) . . . . 5. Q. BICOLOR

i. All side nerves of the lower surface of leaves ending in the teeth; stalks (peduncles) of fruits none or shorter than the leaf-stalks (petioles), at most 1 cm. long . . . . . . . . . . *j*

j. Trees found only in southeastern Missouri north to Cape Girardeau, Bollinger, Madison, and Iron counties; hairs on lower surface of leaves simple and erect or few-branched (use magnifying lens), the leaf surface beneath more or less velvety to the touch; cup of fruit 2.5–3 cm. broad with scales free to the base . . . . . . . . . 6. Q. MICHAUXII

j. Trees in other parts of Missouri, or also in southeastern Missouri; hairs on lower surface of leaves star-shaped (stellate), the rays spreading out from the center (use magnifying lens), the leaf surface beneath not velvety to the touch; cup of fruit 1–1.5 (rarely 2) cm. broad, only the tips of the scales free or distinct . . . . . . . . . . . . . *k*

k. Usually low shrub 1–3 m. tall; teeth of leaf mostly 3–8 on each side, sometimes more typically blunt but also pointed or acute; leaves mainly 4–10 cm. long (up to 13 cm.); leaf-stalks (petioles) mainly 5–15 mm. long; western half of Missouri east to Putnam, Sullivan, Macon, Howard, Benton, Dallas, Greene, and Stone counties . . 7a. Q. PRINOIDES var. PRINOIDES

k. Usually medium to large tree, 20–30 m. tall, but becoming shrubby and only 2 meters tall; teeth of leaf mainly 8–13 on each side, sometimes 6–7, typically sharp, pointed, and prominent, but varying to blunt, low, and less prominent; leaves mainly 10–20 cm. long; leaf-stalks (petioles) (10)–15–30 mm. long; throughout Missouri in practically every county . . . . . . . . . . . 7b. Q. PRINOIDES var. ACUMINATA

h. Margins of leaves with 1 or more (up to 5) deep lobes projecting from each margin, at least 1 of the spaces (sinuses) prominent and extending more than one-third the distance to the main mid-nerve; main side (lateral) nerves of the lower surface of leaves not equally spaced and not equally conspicuous . . . . . . . . . . . . . . . . . . . . . . . *l*

l. Twigs and branchlets rather densely hairy throughout the season; leaf blades usually with 3–5 main lobes; hairs on lower surface of leaves erect and with only few branches (use magnifying lens); lower scales of cup of acorn flat or somewhat concave; mature nut usually less than 12 mm. in diameter . . . . . . . . . . . . . . 2. Q. STELLATA

l. Twigs and branchlets chiefly glabrous (without hairs) at maturity; leaf blades usually with 5–9 main lobes; hairs on lower surface of leaves horizontally spreading and with many branches (use magnifying lens); lower scales of cup of acorn with some thickening or enlargement in the upper part; mature nut more than 12 mm. in diameter . . . . . . *m*

m. The deepest space (sinus) of the leaf about half-way up each side divides a generally broader and more shallowly lobed upper portion from a narrower and more deeply lobed lower portion; leaf-blades mainly 10–25 cm. long; upper (inner) scales of cup of acorn prolonged into slender tips, forming a fringe; throughout Missouri, practically in every county . . . . . . . . . . . . . . . . . . . . 4. Q. MACROCARPA

m. The spaces (sinuses) of the leaf about half-way up scarcely or not deeper than the other spaces; leaf-blades mainly 10–20 cm. long; upper (inner) scales of cup of acorn acute (short-pointed) or blunt, not prolonged, but sometimes forming a ragged rim around the top not to be mistaken as prolonged awned scales; southeastern Missouri north along Mississippi River to St. Louis County and locally in Taney County . . 3. Q. LYRATA

e. Lobes or teeth of mature leaves or summit of leaves with bristle-like points projecting (main nerves and side nerves projecting from leaf tissue) . . . . . . . . . . . . . . . . . *n*

n. Lower surface of fully grown leaves more or less hairy on some part of surface or over the whole surface . . . . . . . . . . . . . . . . . . . . . . . . . . . *o*

o. Lower surface of fully grown leaves with a yellowish, grayish, or whitish hairiness; the individual star-shaped (stellate) hairs so minute that they are barely visible under a 10× magnifying lens; leaves conspicuously drooping or hanging from the branches; scales of cup of acorn with a reddish-brown dark border; nut enclosed by cup about $\frac{1}{3}$ its length; trees of extreme southern and southeastern Missouri . . . . . . . . . . . . . . . . 12. Q. FALCATA

o. Lower surface of fully grown leaves with a rusty or brownish hairiness; the individual star-shaped (stellate) hairs large enough to be seen under a 10 × magnifying lens; leaves mostly stiffly erect, ascending, or spreading from the branches; scales of cup of acorn without a dark border; nut enclosed by cup about ½ its length; trees of wide range mostly throughout Missouri . . . . . *p*

p. Leaves mostly fan-shaped or more broadened around top, usually with 3, sometimes 5, broad lobes, other lobes none or poorly developed; fully grown twigs with scattered scurfy hairiness; buds covered with rusty-colored hairs; leaf-stalks (petioles) mostly 10–20 mm. (as little as 4 mm. long or up to 25 mm.); acorns 10–18 (rarely 20) mm. long; cup of acorn 13–20 mm. broad . . . . . . . . . . . . . . . . . . . . . 11. Q. MARILANDICA

p. Leaves usually narrower at the top, broadest around the middle, with usually 5–9 long and narrow or short and broad lobes separated by prominent spaces, the longest lobes along the sides, not at the top; fully grown twigs (except in f. *missouriensis*) mostly without hairs; buds covered with gray or buff hairs; leaf-stalks (petioles) mostly 25–80 mm. long; acorns 15–25 mm. long; cup of acorn 20–25 mm. broad . . . . . . . . . . . . 13. Q. VELUTINA

n. Lower surface of fully grown leaves without hairs, except for small tufts in the main axils where the side nerves join the mid-nerve . . . . . . . . . . . . . . . . . . . . . . . *q*

q. Leaves fan-shaped or broadened toward the summit with a generally 3-lobed upper half . . . . *r*

r. Leaves only slightly, if at all, longer than broad, sometimes as wide as long, usually blunt, rounded, or somewhat heart-shaped at base, rarely straight or tapering; lower surface of leaf yellow-brown or yellow-green; leaf-stalks (petioles) mainly 5–15 mm. long; cup of fruit more or less top-shaped or inversely conical, 1.5–2 cm. broad . . . . . . . . 11. Q. MARILANDICA

r. Leaves usually 1½–2 times as long as broad, tapering to a narrowed, acute, or triangular base; lower surface of leaf pale green; leaf-stalks (petioles) mainly less than 5 mm. long; cup of fruit saucer-shaped, 1–1.5 cm. broad . . . . . . . . . . . . . . . . . . 10. Q. NIGRA

q. Leaves not fan-shaped or broadened toward the summit, but narrower at the summit and broadest at the middle or toward the base with 5–11 long and narrow or short and broad lobes separated by spaces, the longest lobes near the middle of the leaf . . . . . . . . . . . . . . *s*

s. The longest lateral (side) lobes of the leaf shorter than or about equaling to slightly longer than the width of the solid green undivided central part of the leaf . . . . . . . . . *t*

t. Main midrib on upper surface of leaf usually glabrous (without hairs); leaf-stalks (petioles) glabrous (without hairs) or nearly so; topmost (terminal) winter buds glabrous, chestnut-brown, 3–7 mm. long, not strongly angled; cup of acorn mostly saucer-shaped, with tight, slightly hairy scales; leaf-blades dull, not shining above, rather bluish-green, thin; lower surface of fully grown leaves seldom with conspicuous tufts of hairs at junction of midrib and main side nerves . . . . . . . . . . . . . . . . . . . . . 18. Q. RUBRA

t. Main midrib on upper surface of leaf hairy; leaf-stalks (petioles) partly or completely hairy; topmost (terminal) winter buds gray-hairy, 6–12 mm. long, strongly angled; cup of acorn top-shaped or inversely conical or cup-like, with loosely overlapping hairy scales; leaf-blades shining above, dark green, thick; lower surface of fully grown leaves always with conspicuous tufts of hairs at junction of midrib and main side nerves . . . . 13. Q. VELUTINA

s. The longest lateral (side) lobes of the leaf 2–6 times as long as the width of the solid green undivided central part of the blade . . . . . . . . . . . . . . . . . . . *u*

u. Main midrib on upper surface of leaf hairy; leaf-stalks (petioles) partly or completely hairy; upper scales of acorn cup loosely overlapping (imbricated), thin, forming a loose fringe; topmost (terminal) buds gray-hairy, 6–12 mm. long, strongly angled . 13. Q. VELUTINA

u. Main midrib on upper surface of leaf usually glabrous (without hairs); leaf-stalks (petioles) glabrous (without hairs) or nearly so; upper scales of acorn cup tight, appressed to one another, thickish, not forming a loose fringe; topmost (terminal) buds 1–7 mm. long, glabrous or hairy . . . . . . . . . . . . . . . . . . . . . . . *v*

v. Acorn cup 3–5 mm. high, 1–1.5 cm. broad; nut 10–13 mm. long; fully grown buds very small, 1–5 mm. long; base of many leaves often tapering or often V-shaped. . 17. Q. PALUSTRIS

v. Acorn cup higher or broader than noted in alternate v. above; nut 12–26 mm. long; fully grown buds 4–7 mm. long; base of leaves either truncate (straight) or more or less tapering or V-shaped . . . . . . . . . . . . . . . . . . . . . . *w*

w. Scales of cup without hairs, shining, brown or chestnut-colored; trees of the south-eastern Ozarks west to Douglas and Howell counties . . . . . . 15. Q. COCCINEA

w. Scales of cup hairy with a silky or appressed hairiness, dull, ashy-gray; trees of northern, or central and southern, or southernmost Missouri . . . . . . . . . *x*

x. Mature leaves 15–20 cm. long; leaf-stalks (petioles) 4–6 cm. long; cup of acorn saucer-shaped to top-shaped, rounded or flattened at base with a shallow rim, usually enclosing $\frac{1}{4}$–$\frac{1}{3}$ of nut, 2–3 cm. broad; fully mature buds completely glabrous (without hair) . . . . . . . . . . 16. Q. SHUMARDII

x. Mature leaves 7–16 cm. long; leaf-stalks (petioles) 2–5 cm. long; cup of acorn top-shaped to deep cup-shaped, sloping or stalked at base, usually enclosing $\frac{1}{3}$–$\frac{5}{8}$ of nut, 1–1.8 cm. broad; fully mature buds noticeably hairy or at least slightly hairy on margins of upper scales . . . . . . . . . *y*

y. Acorn 12–20 mm. long, usually longer than wide and narrowed at both ends; cup gradually sloping to base; inner bark yellow or orange; shape of tree mostly broadly oblong, not pyramidal; leaves generally broadest at the middle, oblong-obovate, either abruptly narrowed or truncate (straight) at base; known only from Harrison County, northern Missouri. . . 14. Q. ELLIPSOIDALIS

y. Acorn 20–28 mm. long, oblong-ovoid, scarcely longer than wide, rounded at the ends; cup enclosing $\frac{1}{3}$–$\frac{1}{2}$ of the acorn, abruptly contracted to a stalk-like base 2–7 mm. long; inner bark not orange nor yellow; shape of tree pyramidal; leaves mostly broadest in upper half, obovate or rhombic-obovate, usually abruptly or narrowly tapering or V-shaped at the base; known only from Butler County, southeastern Missouri . . . . . . . . . . . . . . 19. Q. NUTTALLII

1. **Quercus alba** L. White Oak Map 759
Flowers early April–May.

Occurs most frequently on dry upland slopes and ridges and also in low ground of valleys and ravine bottoms. Throughout Missouri, doubtless in every county.

Ranges from Florida to Texas, north to Maine, Quebec, Ontario and Minnesota.

Two variations may be distinguished in Missouri material:

Leaves cut 1/2–3/4 distance inwardly toward the central mid-nerve, the lobes long and narrow, mostly 8–20 mm. (rarely to 30) broad . 1a. Q. ALBA f. ALBA

Leaves cut less than half-way to the central mid-nerve, the lobes relatively shorter and broader, mostly 15–30 (up to 40) broad . . 1b. Q. ALBA f. LATILOBA

1a. **Quercus alba** f. **alba** White Oak Map 759
This is the common type encountered in Missouri.

1b. **Quercus alba** f. **latiloba** (Sarg.) Palmer & Steyerm. Map 759
Scattered through the range of the commoner type in Missouri, north to Clark, Knox, and Adair counties; not recorded from the northwestern quarter of the state.

*Quercus alba* is the state tree of Maryland, and is often considered the most valuable species of tree in the United States from the standpoint of the general all-around purposes to which the wood is applied. At one time it was much used in general construction work, but is being replaced by cheaper woods. Early settlers in North America employed the durable wood for their ships, homes, barns, bridges, and mills. Today it is more often used for furniture, interior panelling, railroad ties, agricultural implements, and barrels. It has always been widely used for fuel purposes. Up until recently, the wood was also used for wagons and the bark used for tanning.

This species and other members of the White Oak group have sweet acorns rich in starch and oil, which are eaten by squirrels, hogs, and various other animals. The American Indians prepared a type of pasty bread from some of the species, after first leaching out the bitter qualities of the raw acorns. The foliage of white oak turns a dull rose-red or rose-purplish in fall.

This oak is known to hybridize with several species, and in Missouri three hybrids involving this species are known:

**Quercus** × **Bebbiana** Schneid. (*Q. alba* × *macrocarpa*) is known from Marion, Jackson, and Cole counties. A tree in Forest Park, in St. Louis County measures 7 feet, 8 inches in circumference, has a height of 62 feet and a spread of 77 feet, and is claimed to represent the largest tree recorded for this hybrid oak.

**Quercus** × **Fernowi** Trel. (*Q. alba* × *stellata*) is known from St. Louis and Dade (*Palmer 54717*) counties.

**Quercus** × **Jackiana** Schneid. (*Q. alba* × *bicolor*) is known from Lafayette Co.

2. **Quercus stellata** Wang. Post Oak Map 760
Flowers April–May.

Occurs mostly in dry or rocky upland woods and glades, and sometimes along small streams. Common throughout Missouri; not recorded from a few extreme northwestern counties.

Missouri material is represented by two principal variations:

Lower surface of leaves noticeably hairy; leaves mainly 10–20 cm. long; main side lobes of leaves horizontally spreading, large, broadened toward summit, narrowed or constricted at base, truncate

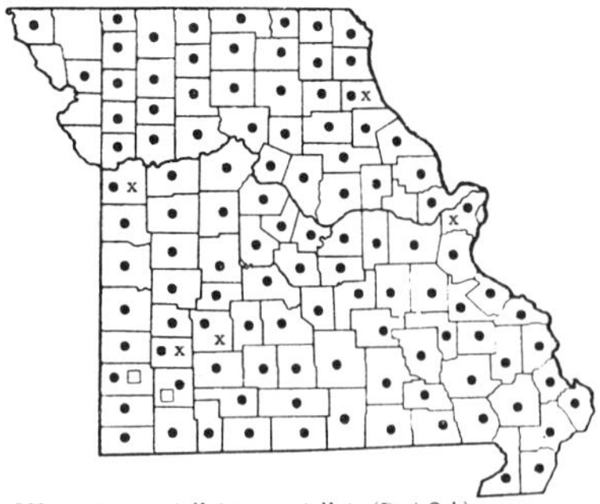

760 • Quercus stellata var. stellata (Post Oak)
760 □ Quercus stellata var. Margaretta
760 x Quercus Fernowii (Quercus alba X stellata)

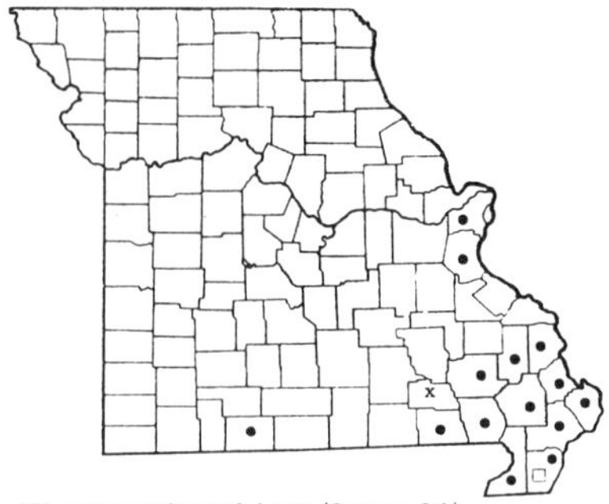

761 • Quercus lyrata f. lyrata (Overcup Oak)
761 □ Quercus lyrata f. viridis
761 x Quercus X Sterrettii (Quercus stellata X lyrata)

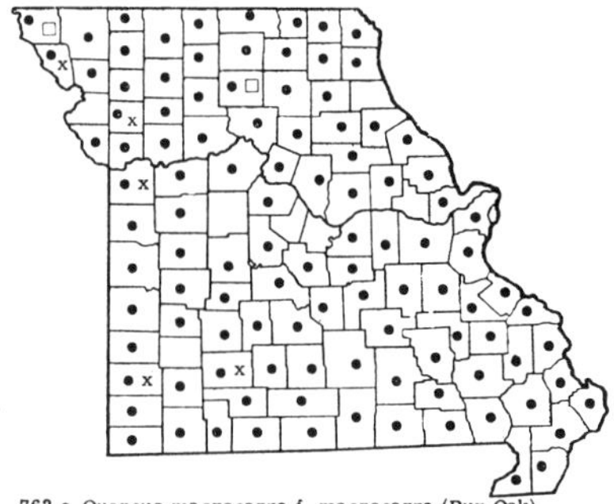

762 • Quercus macrocarpa f. macrocarpa (Bur Oak)
762 □ Quercus macrocarpa f. olivaeformis
762 x Quercus X Deamii (Quercus macrocarpa X prinoides var. acuminata)

(as if cut straight across) or emarginate (shallowly notched or sunken in the center) at the summit; common type . . 2a. Q. STELLATA var. STELLATA

Lower surface of leaves nearly without hairs or only sparsely hairy; leaves shorter, 4–10 cm. long; lobes ascending or spreading, entire (without notches or toothing), not constricted or narrowed at the base; rare and known only from southwestern Missouri. 2b. Q. STELLATA var. MARGARETTA

2a. **Quercus stellata** var. **stellata** Map 760

The common variety, throughout the state, except not recorded from a few extreme northwestern counties.

Ranges from Florida to Texas, north to Massachusetts, New York, Pennsylvania, Ohio, Indiana, Illinois, Iowa, and Kansas.

2b. **Quercus stellata** var. **margaretta** (Ashe) Sarg. Map 760

*Quercus margaretta* Ashe [BB]

Known only from Jasper (near Prosperity) and Lawrence (open woods with limestone outcrops above bluff, ½ mi. southeast of Sycamore School, Lincoln Township, September 20, 1950, *Palmer 51154*) counties.

Ranges from Florida to Texas, north to Virginia, locally to Massachusetts, Missouri, and Oklahoma.

Various other varieties have been described but are not usually recognized.

The foliage of Post Oak in fall turns to an orange-brown or chestnut-brown. The wood is somewhat tougher than White Oak and is chiefly used for railroad ties, but also for fencing, fuel, and construction. It is often associated with Black Jack Oak (*Q. marilandica*) on the poorer and more acid upland soils and rocks, frequently in cherty, sandstone, and granite areas.

*Quercus stellata* hybridizes with a few other species, of which three others are involved in Missouri:

**Quercus** × **Fernowi** Trel. (*Q. alba* × *stellata*), referred to under *Q. alba*.

**Quercus** × **stelloides** Palmer (*Q. prinoides* var. *prinoides* × *stellata*), known from Putnam and Jackson counties.

**Quercus Sterrettii** Trel. (*Q. lyrata* × *stellata*), known from a questionable sterile specimen from Carter County (along Current River, near Van Buren, May, 1945, *Bill Bauer*, in herb. Chi. Nat. Hist. Mus.).

3. **Quercus lyrata** Walt. Overcup Oak Map 761

Flowers early April–May.

Occurs in alluvial or wet ground in low or swampy woods in the valleys of streams and bayous, most frequently encountered in southeastern Missouri, extending north along the Mississippi and Meramec river bottoms and flood plain forests to St. Louis County, and a local stand formerly along Long Creek in Taney County, southwestern Missouri.

Ranges from Florida to Texas, north to New Jersey along the coastal plain, inland north to Indiana Illinois, Iowa, and Missouri.

Missouri material is represented by two variations:

Lower surface of leaves whitened, the hairs usually dense . . . . . . 3a. Q. LYRATA f. LYRATA

Lower surface of leaves green, the hairs scattered, not dense . . . . . . 3b. Q. LYRATA f. VIRIDIS

3a. **Quercus lyrata** f. **lyrata** Map 761

*Quercus lyrata* Walt. [G, BB, P & S]

This is less common than the other variation.

3b. **Quercus lyrata** f. **viridis** Trel. Map 761

Apparently f. *viridis* is the more common form in Missouri.

A remarkable isolated stand of this oak occupied a natural swamp in an old creek meander of Long Creek, tributary to White River, T21N, R22W, sect. 3, 1½ mi. north-northwest of Oasis (Cedar Valley), 9 mi. southwest of Branson, in Taney County (*Steyermark 80729*), and collections by *Moore & Iltis* and *Palmer*), and was dominant at this locality. This relict area, which also contained a remnant flora of such other coastal plain and Mississippi Embayment species as *Trachelospermum difforme*, was in existence until 1956, when the impounded waters of the Table Rock Dam on White River turned the station into part of the new lake bottom.

The foliage of this species turns in autumn an orange mixed with brown and green, or sometimes more red and yellow are combined with the orange. The wood is employed similarly to that of White Oak.

*Quercus lyrata* hybridizes with a few other species, of which the following two are involved as hybrids in Missouri:

**Quercus × humidicola** Palmer (*Q. bicolor × lyrata*), known from Dunklin County (Campbell, October 6, 1910, *Bush 6365*, holotype).

**Quercus × Sterrettii** Trel. (*Q. lyrata × stellata*), referred to previously under *Q. stellata*.

4. **Quercus macrocarpa** Michx. Bur Oak

Map 762

Also called Mossy Cup Oak.

Flowers early April–May.

Occurs usually in low woods or slopes in valleys along streams, especially in the Ozark section, but in glaciated northern Missouri often in upland woods and on higher slopes, as well as in valleys. Throughout Missouri, and doubtless in every county.

Ranges from New Brunswick, Maine, and Quebec to Manitoba, south to Maryland, Alabama, Louisiana, Arkansas, Oklahoma, and Texas.

Missouri material is represented by two variations:

The broad upper portion of leaf with relatively short and broad, oblong or ovate, blunt lobes; cup of acorn covering 2/3 of the acorn; acorn flat at base, oblong to ovoid, about 2/3 as thick as long

4a. Q. MACROCARPA f. MACROCARPA

The broad upper portion of leaf with relatively narrower and longer lobes; cup of acorn covering only 1/3–1/2 of the acorn; acorn strongly rounded at base, oval, about 1/2 as thick as long . . .

4b. Q. MACROCARPA f. OLIVAEFORMIS

4a. **Quercus macrocarpa** f. **macrocarpa**

Map 762

*Quercus macrocarpa* Michx. [G, P & S, Steyerm.]
*Quercus macrocarpa* var. *macrocarpa* [BB]

This is the common form encountered in Missouri.

The fruit is quite variable, both in size of acorn, cup, and length and fringing of scales of the cup.

4b. **Quercus macrocarpa** f. **olivaeformis** (Michx. f.) Trel.

Map 762

*Quercus macrocarpa* var. *olivaeformis* (Michx. f.) Gray [BB, P & S, Steyerm.]

Known only from Atchison (*Palmer* from Watson, September 4, 1920) and Linn counties in northern Missouri.

The branches of younger trees of the Bur Oak generally develop corky wings which are mostly absent on the older trees. The autumnal coloring of this oak is one of the least colorful of the genus, turning to a dull yellow or greenish-yellow and brown combination. Among the species of oaks, the trunk of this one attains the greatest diameter in Missouri, the Big Oak State Park in Mississippi County possessing the largest tree known in the state. Only the base of the tree now remains to testify to its once former great size, as it was severely damaged a number of years ago by storms. Its circumference 4½ feet above the ground measures 21 feet 4 inches, and before the damage occurred, the tree had a height of 143 feet, and a spread of 114 feet. This tree is estimated to have been at least 400 years old. The tap root of the Bur Oak is comparatively short in comparison with the widespreading branches above ground.

The wood of Bur Oak is used similarly to that of White Oak, and has been used in general construction work, furniture, barrels, shipbuilding, agricultural implements, railroad ties, fences, and for fuel.

*Quercus macrocarpa* hybridizes with a few other species, of which the following three are involved as hybrids in Missouri:

**Quercus × Bebbiana** Schneid. (*Q. alba × macrocarpa*), previously accounted for under *Q. alba*.

**Quercus × Deamii** Trel. (*Q. macrocarpa × Q. prinoides* var. *acuminata* [*Muehlenbergii*]) Deam Oak
*Quercus × fallax* Palmer [G]
*Quercus × Hillii* [of BB, P & S, Steyerm., Deam], not Trel.

Known from Holt, Clinton, Jackson, and Greene counties.

The name *Quercus × Deamii* has been misapplied. The reader is referred to Dr. H. H. Bartlett's discussion of this hybrid (Rh. 53: 249–64. 1951) for more details. He has submitted evidence to show that it represents a cross between *Q. macrocarpa*, the Bur Oak, and *Q. prinoides* var. *acuminata* [*Muehlenbergii*], the Chestnut Oak, instead of, as originally thought and recorded in the literature, a cross between *Q. alba*, the White Oak,

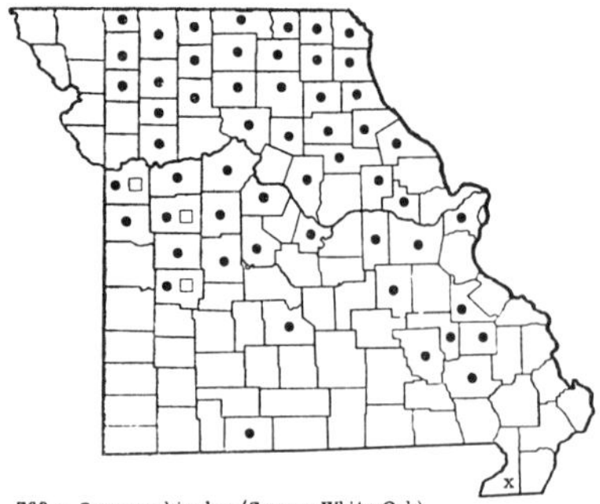

763 • Quercus bicolor (Swamp White Oak)
763 □ Quercus X Hillii (Quercus bicolor X macrocarpa)
763 x Quercus X humidicola (Quercus bicolor X lyrata)

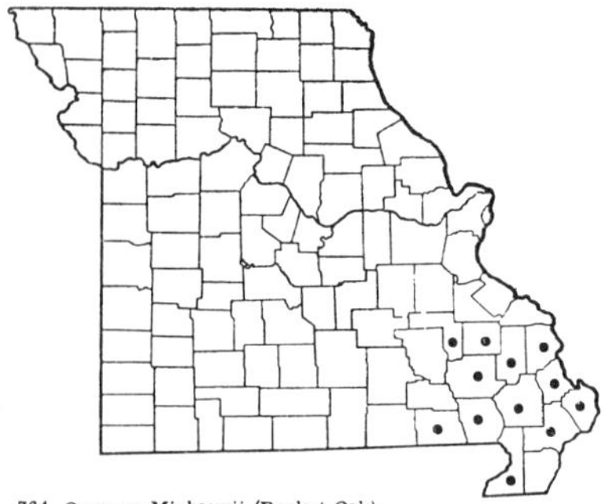

764 Quercus Michauxii (Basket Oak)

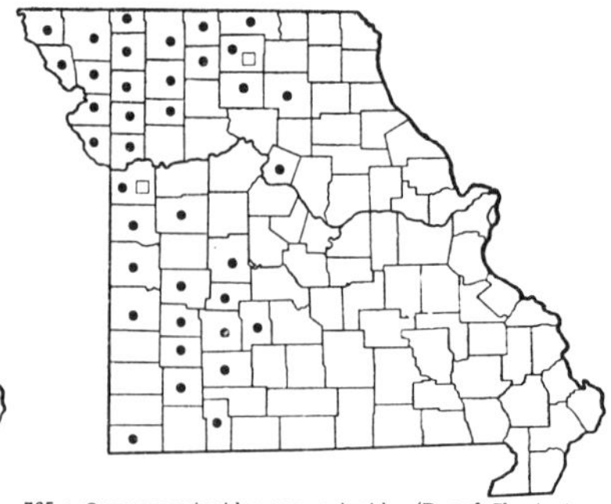

765 • Quercus prinoides var. prinoides (Dwarf Chestnut Oak)
765 □ Quercus X stelloides (Quercus prinoides var. prinoides X stellata)

and *Q. prinoides* var. *acuminata* [*Muehlenbergii*]. Bartlett based his conclusions on progeny tests. Mr. E. J. Palmer believes, however, that the Deam Oak is a cross between *Q. alba* and *Q. bicolor*, judging by a tree grown at Arnold Arboretum from an acorn sent by Deam from the original tree.

**Quercus** × **Schuettei** Trel. (*Q. bicolor* × *macrocarpa*)
*Quercus* × *Hillii* Trel. [G]

Known only from St. Clair County (low alluvial woods along Monegaw Creek near Monegaw Springs, October 3, 1936, *Steyermark 20216*).

5. **Quercus bicolor** Willd. Swamp White Oak
Map 763

Flowers early April–May.

Occurs in alluvial and rich soils of low woods in valleys and on slopes, in wet ground bordering swamps and oxbow lakes of flood plain and stream meanders, and along streams. Common in northern, central, and most of eastern Missouri, south to Wayne, Reynolds, Crawford, Laclede, St. Clair, and Cass counties, locally in Taney County along White River; absent from much of the unglaciated prairie and Ozark sections, and from the southeastern lowland area.

Ranges from Maine and Quebec to Minnesota and Nebraska, south to Georgia, Kentucky, Tennessee, Arkansas, and Oklahoma.

This species has longer stalks (peduncles) supporting the fruit than that of any other native species of oak in Missouri, and, among the species with which it might be confused, only the cultivated English Oak (*Q. robur* L.) has as long or longer peduncles (from 3–8 cm. long). The base of the leaf of *Q. bicolor* is rounded or wedge-shaped, while that of *Q. robur* is auricled with rounded projections. The silvery-white lower surface of the leaves is quite conspicuous in *Q. bicolor*, and, when the wind blows through the foliage, this paler side of the leaf is quickly visible.

The wood of this species is not distinguished commercially from that of White Oak, and is employed for construction, furniture, interior finish, shipbuilding, barrels, railroad ties, and fuel.

*Quercus bicolor* is known to hybridize with a few other species, involving the following three hybrids in Missouri:

**Quercus** × **humidicola** Palmer (*Q. bicolor* × *lyrata*), previously referred to under *Q. lyrata*.

**Quercus** × **Jackiana** Schneid. (*Q. alba* × *bicolor*), previously referred to under *Q. alba*.

**Quercus** × **Schuettei** Trel. (*Q. macrocarpa* × *bicolor*), previously referred to under *Q. macrocarpa*.

6. **Quercus Michauxii** Nutt. Basket Oak
Map 764

Also called Cow Oak, Swamp Chestnut Oak

*Quercus Prinus* of some authors [P & S, Steyerm., Deam], not L.

Flowers early April–May.

Wet low woods in depressions and valleys, along bayous and slow streams principally in the lowland section of southeastern Missouri, north to Cape Girardeau, Bollinger, Madison, and Iron counties, and west to Wayne and Ripley counties.

Ranges from Florida to Texas, north to New Jersey, Indiana, Illinois, and Missouri.

This species has very large acorns, compared with some other oaks in Missouri, varying from 2.5–3.5 cm. long and 2–3 cm. thick, with cups 2–3 cm. broad. The tree is a handsome one with large rounded summit. The bark is conspicuous because of its silvery white or light gray color, and forms a striking contrast in autumn when the foliage turns a crimson or dark red color. Because the wood naturally splits into long, narrow, tough sections, splints from it were widely used in the southern states for baskets for carrying cotton, and, when woven, make satisfactory chair bottoms. Other uses for which it has served are agricultural implements, wheels and axles, other parts of wagons, railroad ties, barrel staves, furniture, cant-hooks used in lumbering, fences, and fuel. *Quercus Michauxii* is also a source of tannin.

The name Cow Oak originates from the fact that cows frequently eat the acorns of this species in the southern states. The kernels contain less bitterness than is found in any other of the oak group, and are sufficiently sweet to eat raw.

Although the species is known to hybridize with White Oak (*Q. alba*) and Bur Oak (*Q. macrocarpa*), no hybrids have been found in Missouri involving these species and *Q. Michauxii*.

7. **Quercus prinoides** Willd. Map 765

This has usually been treated as distinct from *Q. Muehlenbergii*. However, during the past two decades much more opportunity to observe these oaks in the field has made it evident that they intergrade and are not readily separated as species. They seem more naturally maintained as varieties, and are so treated in the present work, following Gleason's similar treatment in the *New Britton and Brown Illustrated Flora*. Differences between the two have been presented in the general key to *Quercus*.

7a. **Quercus prinoides** Willd. var. **prinoides**
Dwarf Chestnut Oak Map 765
Also called Scrub Oak, Dwarf Chinquapin Oak
*Quercus prinoides* Willd. [G, Deam, P & S, Steyerm.]
Flowers April–May.

Occurs in dry exposed soils in open woods, glades, along bluffs, open banks along roadsides, thickets, and borders of upland woods and prairies. Western half of Missouri, mainly in the glaciated and unglaciated prairie sections, east to Putnam, Sullivan, Macon, Howard, Benton, Hickory, Dallas, Greene, and Stone counties. Previous records from St. Louis, Shannon, and Howell counties are now referred to *Q. prinoides* var. *acuminata*.

Ranges from Maine to Minnesota and Nebraska, south to Virginia, Alabama, Tennessee, Arkansas, Oklahoma, and Texas.

*Quercus prinoides* var. *prinoides* is very distinct in its extreme form as a low spindly-branched bush frequently forming patches only a few feet high with slender twiggy branches, small leaves mostly 5–10 cm. long with 3–7 low blunt teeth on each margin, and small acorn cups 1–1.5 cm. broad covered with small tubercled scales. Unfortunately, many plants are found which approach or grade into *Q. prinoides* var. *acuminata* (*Q. Muehlenbergii*), not only as small trees, but with longer leaves with 8–10 teeth which may be low and blunt, or prominent and as pointed as in characteristic *Q. prinoides* var. *acuminata*.

In Missouri *Q. prinoides* var. *prinoides* hybridizes with only one species, *Q. stellata*:

**Quercus × stelloides** Palmer (*Q. prinoides* var. *prinoides* × *Q. stellata*), known from Sullivan and Jackson counties, and previously referred to under *Q. stellata*.

Another hybrid involving this species and White Oak (*Q. alba*) has not been recorded from Missouri.

7b. **Quercus prinoides** var. **acuminata** (Michx.) Gl. Chestnut Oak Map 765a
Also called Yellow Oak, Chinquapin Oak
Flowers April–May.

Most frequently found in limestone rocky areas on bluffs or borders of glades and upland woods, and also in alluvial ground in valleys along streams or at the base of limestone bluffs or slopes. Throughout Missouri, absent only in two counties in extreme southeastern Missouri.

Two forms are encountered in Missouri:

Leaves oblong or lanceolate, mostly 2.5–8.5 cm. broad, typically with rather prominent sharply pointed teeth . . 7b. Q. PRINOIDES var. ACUMINATA f. ACUMINATA
Leaves obovate to obovate-oblong, mostly 9–15 cm. broad, typically with less prominent, shorter and broader teeth . 7c. Q. PRINOIDES var. ACUMINATA f. ALEXANDERI

7b. **Quercus prinoides** var. **acuminata** f. **acuminata** Map 765a
*Quercus prinoides* var. *acuminata* (Michx.) Gl. [BB]
*Quercus Muehlenbergii* Engelm. [G, P & S, Steyerm., Deam]

This is the commonly encountered form of var. *acuminata* throughout Missouri.

Ranges from Florida to Texas, north to Vermont, New York, Ontario, Michigan, Wisconsin, Minnesota, and Nebraska. Also in New Mexico and Mexico.

7c. **Quercus prinoides** var. **acuminata** f. **Alexanderi** (Britt.) Steyerm. Map 765a
*Quercus Alexanderi* Britton, Man. Fl. North. States Canada 336. 1901
*Quercus Muhlenbergii* f. *Alexanderi* (Britt.) Trel. [G, P & S, Steyerm.]
*Quercus acuminata* var. *Alexanderi* (Britton) Farwell, Mich. Acad. Sci. Rpt. 6: 206. 1904.

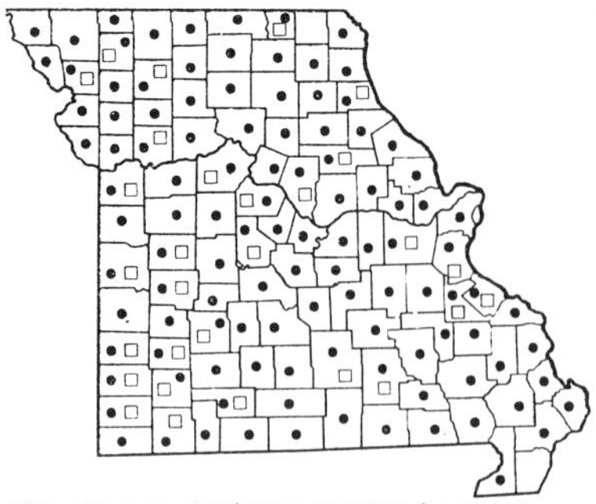

765a ● Quercus prinoides var. acuminata f. acuminata (Chestnut Oak)
765a □ Quercus prinoides var. acuminata f. Alexanderi

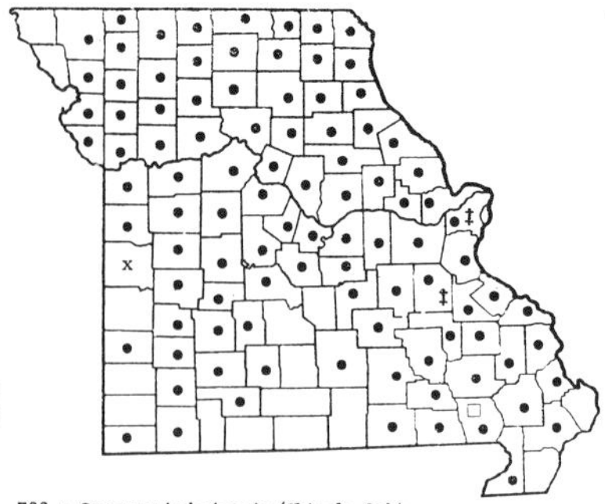

766 ● Quercus imbricaria (Shingle Oak)
766 □ Quercus X anceps (Quercus falcata X imbricaria)
766 ‡ Quercus X tridentata (Quercus imbricaria X marilandica)
766 x Quercus X Egglestonii (Quercus imbricaria X Shumardii var. Schneckii)

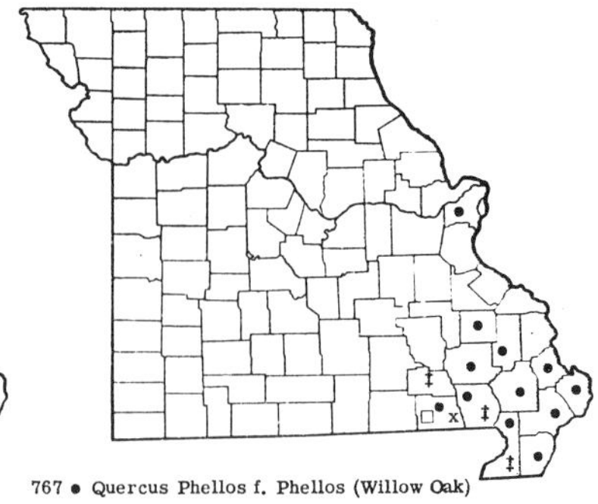

767 ● Quercus Phellos f. Phellos (Willow Oak)
767 □ Quercus X filialis (Quercus Phellos X velutina)
767 ‡ Quercus X ludoviciana (Quercus falcata X Phellos)
767 x Quercus X moultonensis (Quercus Phellos X Shumardii var. Shumardii)

Generally occupying low ground in bottom woods, river flood plain, or low slopes along streams. Scattered throughout the range of var. *acuminata* in Missouri.

In many parts of western Missouri especially where *Q. prinoides* var. *prinoides* occurs, dwarf plants of *Q. prinoides* var. *acuminata* are difficult to distinguish as such from var. *prinoides*. The leaves of var. *acuminata* are generally larger, 10–20 cm. long, with the margins of the leaf-blades with usually 8–13 teeth, but there may be only 7 and these vary from blunt or rounded and low to prominent and more sharply pointed, and when such specimens have smaller leaf-blades, there are no other characters to separate them from *Q. prinoides* var. *prinoides*. Frequently, fruiting specimens only 2–3 feet tall of var. *acuminata* occur (*Steyermark 72315* from Carter Co.; *Steyermark 69988b* from Nodaway Co.; *Steyermark 70200* from Sullivan Co.; *Steyermark 70101* from Buchanan Co.; *Steyermark 72821* from Clinton Co.; and many others) with the slender twigs of var. *prinoides* and with little or no difference in size of acorns or cups as compared with those of var. *prinoides*. While *Q. prinoides* var. *prinoides* may frequently have somewhat smaller acorns and acorn cups with somewhat more tuberculate scales than those of var. *acuminata*, such differences certainly do not hold and are found to be extremely inconstant. Such frequent series of intergradations seen in the field and collected for deposit in the herbarium have made it seem hopeless to recognize the two oaks as more than extreme varieties of one variable species. If they may be called incipient species, they have not diverged sufficiently to merit ordinary specific recognition.

In autumn the foliage of *Q. prinoides* var. *acuminata* turns an orange or brilliant red color. The kernel of the acorns of this variety are quite sweet and good to eat. The hard durable wood is used for railroad ties, fences, wheels, barrels, and fuel.

Hybrids are known between this variety and other oaks. In Missouri only *Quercus* × *Deamii* Trel. (*Q. macrocarpa* × *prinoides* var. *acuminata* [*Muehlenbergii*]) is the hybrid found involving var. *acuminata*. It was discussed under *Q. macrocarpa*.

8. **Quercus imbricaria** Michx. Shingle Oak
Also called Laurel Oak. Map 766
Flowers early April–May.

Occurs in a variety of habitats, varying from upland and dry woods, borders of prairies, slopes, ravine bottoms, valleys, and bottom woods along streams. Common in all parts of Missouri with the exception of some of the extreme southern and southwestern counties, where apparently absent.

Ranges from New Jersey, Pennsylvania, to Wisconsin, Iowa and Nebraska, south to South Carolina, Tennessee, Mississippi, Arkansas, and Kansas. Reported as introduced in Massachusetts and New York.

This oak holds its green color when most of the other oaks have turned other colors. Eventually the foliage in late autumn turns orange or russet brown,

---

Plate no. 136. **1.** Quercus alba f. alba, × 2/7; *a*. Quercus alba f. latiloba, × 2/7. **2.** Quercus stellata var. stellata, × 2/7. **3.** Quercus lyrata, × 2/7; *a*. Flowering branch, × 2/7. **4.** Quercus macrocarpa, × 2/7, showing two types of leaves. **5.** Quercus bicolor, × 2/7, showing two types of leaves. **6.** Quercus nigra var. nigra, × 2/7. **7.** Quercus imbricaria, × 2/7. **8.** Quercus Phellos, × 2/7. **9.** Quercus Michauxii, × 2/7. **10.** Quercus prinoides var. prinoides, × 2/7. **11.** Quercus prinoides var. acuminata, × 2/7; *a*. Quercus prinoides var. acuminata f. acuminata; *b*. Quercus prinoides var. acuminata f. Alexanderi. **12.** Quercus velutina, × 2/7; *a, b*. Quercus velutina f. velutina; *c*. Quercus velutina f. missouriensis. **13.** Quercus marilandica. **14.** Quercus falcata, × 2/7; *a, b*. Quercus falcata var. falcata; *c*. Quercus falcata var. pagodaefolia.

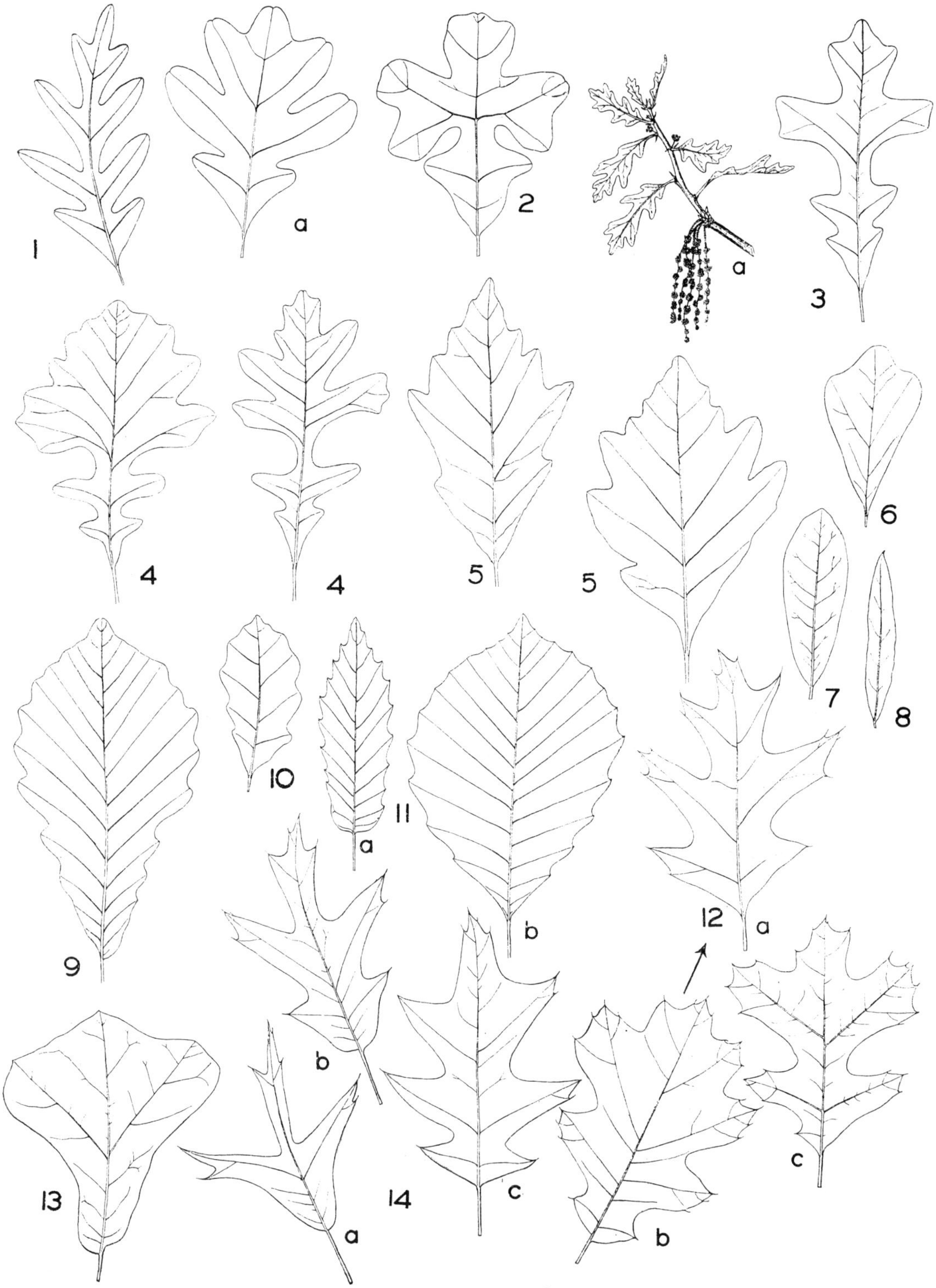


PLATE NO. 136

and sometimes red with green or brown. The leaves, which are somewhat like those of species of Mountain Laurel (*Rhododendron maximum* and *catawbiense*) in size and shape, although thinner, persist for some time into the winter. The tree can be used in ornamental plantings for hedges, screens, and windbreaks.

The wood of this species is considered relatively inferior as oak goes, and is only occasionally used in construction. However, the early settlers of Illinois used this oak especially for shingles, and in this capacity the wood is still used in Tennessee and some other states. The species is often subject to leaf spot and other fungus diseases which sometimes discolor leaves or defoliate trees.

*Quercus imbricaria* is known to hybridize with no less than six other species, all in the black oak group of which the following six are involved in Missouri:

**Quercus** × **anceps** Palmer (*Q. falcata* × *imbricaria*), known only from Butler County (upland woods, Poplar Bluff, October 5, 1921, *Palmer 20674*).

**Quercus** × **Egglestonii** Trel. (*Q. imbricaria* × *Shumardii*); *Quercus* × *Shirlingii* Bush ex Palmer & Steyerm. [P & S, Steyerm.], known only from Jackson County (Oak Grove, October 22, 1927, *Bush 11637*, holotype).

**Quercus** × **exacta** Trel. (*Q. imbricaria* × *palustris*), known from De Kalb, Sullivan, Shelby, Boone, and Jefferson counties.

**Quercus** × **leana** Nutt. (*Q. imbricaria* × *velutina*), known in a number of widely scattered counties throughout Missouri. In Kansas City, Jackson County, a tree of this hybrid has been measured as 9 feet, 3 inches in circumference at 4½ feet above the ground, with a height of 68 feet and a spread of 62 feet. It is at present recorded as the largest tree of this hybrid known.

**Quercus** × **runcinata** (A. DC.) Engelm. (*Q. imbricaria* × *rubra* [*borealis* var. *maxima*]), known from St. Louis (Allenton, *Letterman 2, 5, 6, 36*), Carter (Van Buren, *Palmer 19497*), and Jackson (near Independence, *Bush 623, 628, 10347-A*; Kansas City, *Laughlin*) counties.

**Quercus** × **tridentata** (A. DC.) Engelm. (*Q. imbricaria* × *marilandica*) St. Louis Oak

Known definitely in Missouri only from Cape Girardeau County (Cape Girardeau, June 22, 1920, *Palmer 18012*). A record attributed supposedly to St. Louis County (6 mi. east of St. Louis, autumn, 1849, *George Engelmann*) should be judged instead as occurring in the vicinity of East St. Louis, Illinois.

9. **Quercus Phellos** L. f. **Phellos**
Willow Oak — Map 767
*Quercus Phellos* L. [G, BB, P & S]

Flowers early April–May.

Occurs in wet or low woods bordering swamps, slow streams, bayous, and drainage canals in the lowlands of southeastern Missouri north to Scott, Bollinger, and Madison counties, west to Wayne and Ripley counties; locally north in St. Louis County where reported by Murtfeldt, but no extant tree is known in that county.

Ranges from Florida to Texas, north on coastal plain to New York, New Jersey, Pennsylvania, and inland to Illinois, Missouri, and Oklahoma.

A form of *Q. Phellos*, in which the lower surface of the leaves is permanently silky hairy with whitish hairs, known as f. *intonsa* Fern., is not recorded from Missouri.

The foliage, among the last of the oaks to turn color, eventually becomes pale yellow or dull gold in autumn. It makes a good shade tree and adorns the streets of many southern towns, along with evergreen live oak (*Q. virginiana*) and with laurel oak (*Q. laurifolia*). The wood of this species at one time served for rims or felloes of wheels and was also employed for railroad car timber. Today, sold on the market as red oak, it supplies the market for barrels, interior finish, stairs, posts, railings and supports for railings, church seats and pulpits, and for other kinds of items where the surface is subjected to wear and use.

Eight hybrids are recorded, involving *Quercus Phellos* as one of the parents. The following four are known from Missouri:

**Quercus** × **filialis** Little (*Q. Phellos* × *velutina*)
*Quercus* × *inaequalis* Palmer & Steyerm. [P & S, Steyerm.]

Known only from Butler County (Poplar Bluff, September 11, 1919, *Palmer 16342*, holotype of *Q.* × *inaequalis*).

**Quercus** × **heterophylla** Michx. f. (*Q. Phellos* × *rubra* [*borealis* var. *maxima*]) Bartram Oak

Known only from Wayne (northwest of Wappapello, September 2, 1938, *Steyermark 6428*) and Dunklin (Crowley Ridge, August 7, 1932, *Kellogg 26039*) counties.

**Quercus** × **ludoviciana** Sarg. (*Q. falcata* × *Phellos*)
*Quercus subfalcata* Trel.

Known from southeastern Missouri from Carter, Butler, and Dunklin (Campbell, July 28, 1895, *Bush 610-B*) counties.

**Quercus** × **moultonensis** (*Q. Phellos* × *Shumardii* var. *Shumardii*).

Known only from Ripley County (along Current River in low woods near Current View, August 7, 1934, *Steyermark 14334*).

768 Quercus nigra var. nigra f. nigra (Water Oak)

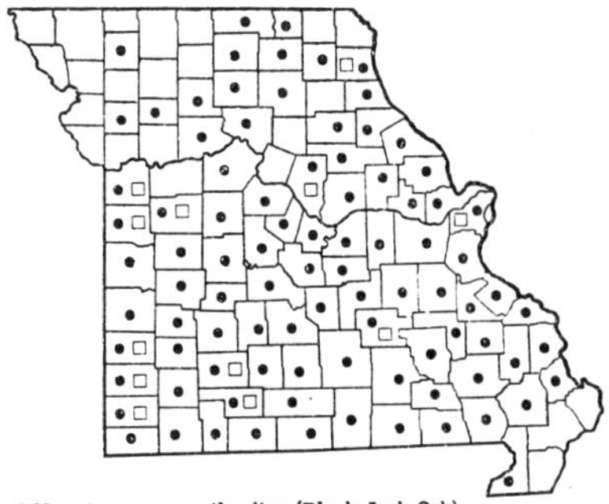

769 • Quercus marilandica (Black Jack Oak)
769 □ Quercus X Bushii (Quercus marilandica X velutina)

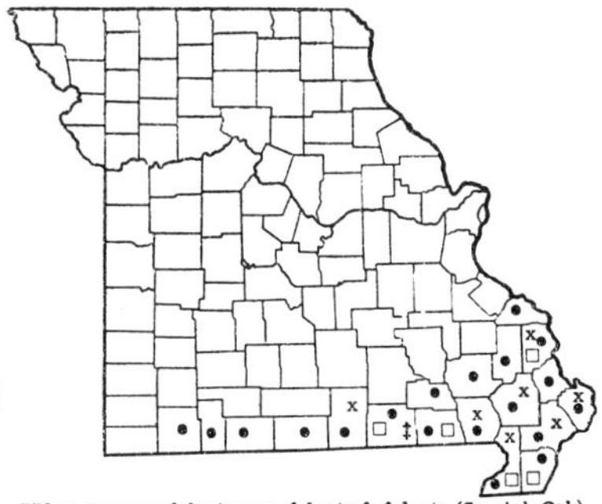

770 • Quercus falcata var. falcata f. falcata (Spanish Oak)
770 ‡ Quercus falcata var. falcata f. angustior
770 □ Quercus falcata var. falcata f. triloba
770 x Quercus falcata var. pagodaefolia (Cherry Bark Oak)

10. **Quercus nigra** L. var. **nigra** f. **nigra**
Water Oak — Map 768
Flowers early April–May.

Occurs in low wet woodland bordering swamps, bayous, or slow streams of the lowland section of southeastern Missouri, where it is found in Pemiscot, Dunklin, Butler, and Ripley counties.

Ranges from Florida to Texas, north on the coastal plain to Delaware, and inland north to Kentucky, Missouri, and Oklahoma.

The leaves vary considerably in shape and size. A form of *Q. nigra* var. *nigra* in which the upper part of the leaf is deeply cleft into 3 narrow oblong lobes is known as f. *tridentifera* (Sarg.) Trel., but has not been recorded from Missouri, nor has the var. *heterophylla* (Ait.) Ashe, in which some or all of the leaves are deeply cut almost to the midrib with 2 narrow long lobes below the ones of the broad summit.

In the southeastern states the tree is often seen as a shade tree. It is occasionally used in southeastern Missouri as such. The wood does not seem to be much used except for fuel. Seven hybrids involving *Q. nigra* as one of the parent species are recorded, but none has been recorded from Missouri.

11. **Quercus marilandica** Muenchh.
Black Jack Oak — Map 769
Flowers early April–May.

Occurs in acid soils on dry or rocky upland, often level woods or slopes, sometimes on rocky glades of sandstone, chert, or granite. Commonly found in southern and central Missouri, especially throughout the Ozark and unglaciated prairie sections, ranging northward on poor and acid soils to the Iowa border, extending west in northern Missouri to Putnam, Sullivan, Livingston, and Clinton counties, and locally northwest in Gentry County.

Ranges from Florida to Texas, north to New York, Pennsylvania, Ohio, Michigan, Indiana, Illinois, Iowa, and Nebraska.

The leaves are rather thicker and more rigid than in most of the other native Missouri species. In autumn they turn brilliant combinations varying from crimson and mixed orange, red, and brown, to crimson or carmine with russet-brown. The leaves are rather variable in shape, but in general have a broadly dilated fan-shape with usually three main broad shallow lobes around the summit. A specimen from Putnam County (*Palmer 27929*) is a peculiarly lobed, small-leaved form with additional lateral lobes below the usual dilated fan-shaped upper portion, and may be considered as an exceptional leaf variant of the species.

The wood is largely used for fuel and in the manufacture of charcoal. Many sections of the eastern Ozarks at one time supplied the iron smelters of that region with charcoal derived from this species. Such settlements as Sligo and St. James had smelter furnaces until about 1915.

Of the eight hybrids recorded involving *Quercus marilandica* as one of the parent species, the following two are known to occur in Missouri:

**Quercus** × **Bushii** Sarg. (*Q. marilandica* × *velutina*)

This hybrid is known from nearly a dozen counties scattered through southern, central, and eastern Missouri northeast to Lewis County. The largest tree recorded for this hybrid exists in Forest Park, St. Louis County. It has a circumference of 8 feet, 7 inches at 4½ feet above the ground, is 67 feet tall, and has a spread of 49 feet.

**Quercus** × **tridentata** (A. DC.) Engelm. (*Q. imbricaria* × *marilandica*)

This has been previously referred to under *Q. imbricaria*.

12. **Quercus falcata** Michx. Spanish Oak Map 770

Also known as Southern Red Oak, Cherry Bark Oak

Flowers early April–May.

Very variable in the shape of the leaves and their lobing. The following variations are found in Missouri:

a. Leaves uncut except toward summit into 3 short blunt or rather blunt lobes, the middle lobe about equal to the other lobes in length . 12c. Q. FALCATA var. FALCATA f. TRILOBA
a. Leaves cut into 3–5 long narrow lobes or into 5–11 short and broad or long and narrow lobes, the terminal lobe usually longer than the other lobes . . . . . . . . . . *b*
 b. The 5–11 lobes of the leaf more or less regular and equal-sided, scarcely curved on one side or sickle-shaped, with some of the lateral (side) lobes usually cut less than halfway to the midrib and spreading at nearly right angles to main length of leaf-blade . . 12d. Q. FALCATA var. PAGODAEFOLIA
 b. The 3–5 (rarely 7) lobes of the leaf unequal, the terminal lobe or 2 of the side lobes greatly prolonged and often unequal-sided, curved, or sickle-shaped, and usually cut more than halfway to the midrib . . . *c*
  c. Leaves ovate or obovate, the lateral lobes 4–8 cm. long, 1.5–2 cm. broad, 2–3 times as long as broad . 12a. Q. FALCATA var. FALCATA f. FALCATA
  c. Leaves oblanceolate or oblong, the shorter lateral lobes 1–3.5 cm. long, 1.5–2 cm. broad, about as long as broad or only 1½ times as long as broad . . 12b. Q. FALCATA var. FALCATA f. ANGUSTIOR

12a. **Quercus falcata** Michx. var. **falcata** f. **falcata**
Spanish Oak, Southern Red Oak Map 770
*Quercus falcata* Michx. [G, P & S, Steyerm., Deam]

Occurs in acid soils of chert or sandstone on rocky or clayey upland ridges and hills, or in valley or river bottom woods. Southern Missouri, throughout the lowlands of southeastern Missouri north to Perry County and west along the southern border counties to Barry County. This is the commonest variation in Missouri.

Ranges from Florida to Texas, north to New Jersey, Pennsylvania, Ohio, Indiana, Illinois, Missouri, and Oklahoma.

12b. **Quercus falcata** var. **falcata** f. **angustior** Palmer & Steyerm. Map 770
*Quercus falcata* f. *angustior* Palmer & Steyerm. [Steyerm.]

Known only from Oregon County (along Eleven Points River, ½ mi. north of McCormack Hollow, 3 mi. north of Greer, July 27, 1936, *Steyermark 12318*, holotype in Mo. Bot. Gard. Herbarium).

12c. **Quercus falcata** var. **falcata** f. **triloba** (Michx.) Palmer & Steyerm. Map 770
*Quercus falcata* f. *triloba* (Michx.) Palmer & Steyerm. [P & S, Steyerm., Deam]
*Quercus falcata* var. *triloba* (Michx.) Nutt. [G]

Found in southeastern Missouri north to Cape Girardeau County and west to Oregon County.

12d. **Quercus falcata** var. **pagodaefolia** Ell.
Cherry Bark Oak Map 770
*Quercus falcata* var. *leucophylla* (Ashe) Palmer & Steyerm. [P & S, Steyerm., Deam]

Occurs in low bottom or valley woods in the southeastern lowland section, north to Cape Girardeau County and west to Butler County.

Ranges from Florida to Louisiana, north to New Jersey, Kentucky, Indiana, Illinois, and southeastern Missouri.

This variety receives the name Cherry Bark, because the scaly reddish-tinged bark resembles somewhat that of a black cherry.

It does not appear possible to maintain var. *leucophylla* as a separate variety, the character of whitish versus brownish pubescence, and diversity in types of leaves on upper and lower branches not appearing to be consistent or reliable differences.

Much more field work and experimental studies are needed before a final judgment can be drawn on the relative values of variation in leaf lobing and shape of the many variations found in *Q. falcata*.

The foliage of *Q. falcata* turns a dull orange and brown color in autumn. The tree is sometimes seen as a shade specimen in some of the towns of southeastern Missouri. The wood is used as pulp for paper making and fuel, and the bark is used for tanning leather. The wood of var. *pagodaefolia* is often used for telegraph poles, furniture, crates, and boxes, and is considered superior to that of var. *falcata*.

Eight hybrids have been recorded involving *Q. falcata* as one of the parent species. In Missouri two of these are known, *Quercus* × *ludoviciana* Sarg. (*Q. falcata* × *Phellos*), already referred to previously under *Q. Phellos*, and *Quercus* × *anceps* Palmer (*Q. falcata* × *imbricaria*), already referred to under *Q. imbricaria*.

13. **Quercus velutina** Lam. Black Oak Map 771

Also called Yellow-barked Oak

Flowers early April–May.

Occurs often in acid thin soils on rocky or dry sandy or upland slopes and ridges, rocky glades of sandstone, chert, or granite, in rich or well-drained ground along streams, and borders of woods and fields. Found throughout Missouri, and doubtless in every county.

Ranges from Florida to Texas, north to Maine, New York, Ontario, Michigan, Illinois, Minnesota, and Nebraska.

The leaves of this species vary considerably in lobing and pubescence. Missouri material is divided into the following variations:

a. Fully grown twigs, leaf-stalks (petioles), and lower surface of fully grown leaves more or less covered with hairs . . . 13b. Q. VELUTINA f. MISSOURIENSIS

a. Fully grown twigs mostly glabrous (without hairs), the young ones hairy; fully grown leaf-stalks and lower surface of leaves mostly glabrous except for remains of hairs along some of main nerves and their axils . 13a. Q. VELUTINA f. VELUTINA

13a. **Quercus velutina** f. **velutina** Map 771

*Quercus velutina* Lam. [G, BB, P & S, Steyerm.]
*Quercus velutina* f. *macrophylla* (Dippel) Trel. [G]
*Quercus velutina* f. *dilaniata* Trel. [G]
*Quercus velutina* f. *pagodaeformis* Trel. [G]
*Quercus leiodermis* Ashe [P & S]

This form, including the various leaf variations, is the more commonly encountered one in Missouri. The various forms included above in the synonymy of f. *velutina* are not recognized principally because differences in leaf lobing are often to be found on one and the same tree, from upper or lower branches, or represent various stages of maturity. Typical *Q. velutina* f. *velutina* is considered to be a rather rare form of *Q. velutina*, but is encountered frequently on young shoots, in which the leaves are obovate and sinuate or very shallow-lobed; *Q. velutina* f. *macrophylla* has the leaves cut less than halfway to the midrib with the toothed broadly oblong lobes many times broader than the narrow spaces (sinuses); *Q. velutina* f. *dilaniata* has the leaves cut from 1/2–3/4 distance to the midrib, the oblong lobes are often narrower than the broad spaces (sinuses) and much toothed at their summit; *Q. velutina* f. *pagodaeformis* has deeply lobed leaves and the long-pointed lobes are scarcely or not at all toothed. All these forms occur in Missouri, the commonest one being f. *dilaniata*, with f. *macrophylla* also frequent. Leaves from juvenile or immature shoots have leaf-shapes and lobing found in both typical f. *velutina* and f. *macrophylla*.

13b. **Quercus velutina** f. **missouriensis** (Sarg). Trel. Map 771

*Quercus velutina* var. *missouriensis* Sarg. [P & S, Steyerm.]

Scattered throughout the state.

In autumn Black Oak usually changes to a bright yellow color, but also may be dull orange to red brown. The inner bark, which is yellow, yields a yellow dye known as *quercitron* and contains the substance *quercitrin*. The yellow extract is used in cotton and wool printing producing a yellow color said to be comparable to that obtained from fustic (the dye extracted from the tropical American tree, *Chlorophora tinctoria*). One of the glucosides found in *quercitrin* is being studied because of its vitamin possibilities. The bark contains much tannic acid, and, like some other species of oaks, is used in tanning.

The wood, which is judged inferior to red oak, is sold nevertheless as red oak and is used in similar ways. Besides its use as fuel, it is employed for railroad cars, flooring, barrels, and rough lumber.

Ten hybrids have already been described which involve *Quercus velutina* as one of the parent species, and in Missouri the following five are known to occur:

**Quercus** × **Bushii** Sarg. (*Q. marilandica* × *velutina*), already referred to previously under *Q. marilandica*.

**Quercus** × **filialis** Little (*Q. Phellos* × *velutina*), referred to previously under *Q. Phellos*.

**Quercus** × **Hawkinsiae** Sudw. (*Q. rubra* [*borealis* var. *maxima*] × *velutina*)
*Quercus* × *Porteri* Trel.

Known from Clark (Dumas, *Palmer 21904*, *Bush 9528*), Dunklin (Crowley Ridge, *Kellogg 26043*), and Jackson (Holmes Park, Kansas City, *Bush 9944*; Swope Park, Kansas City, *Laughlin*) counties.

**Quercus** × **leana** Nutt. (*Q. imbricaria* × *velutina*), previously accounted for under *Q. imbricaria*.

**Quercus** × **vaga** Palmer & Steyerm. (*Q. palustris* × *velutina*).

Known only from Nodaway County (Maryville, June 13, 1924, *Palmer 25421*, holotype in Arnold Arboretum Herbarium).

14. **Quercus ellipsoidalis** E. J. Hill
Jack Oak Map 772
Also called Northern Pin Oak, Hill's Oak
Flowers May–June.

Known only from dry upland bordering woods in acid soil near the Iowa border in northern Missouri in Harrison County (near Eagleville, July 6, 1933, *Palmer & Steyermark 41358*).

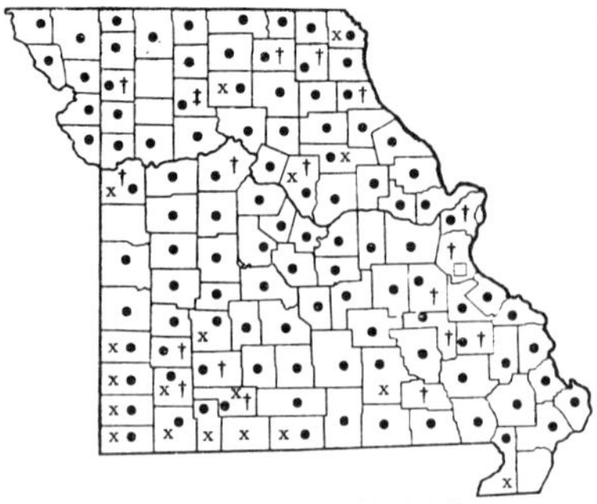

771 • Quercus velutina f. velutina (Black Oak)
771 ‡ Quercus velutina f. macrophylla
771 □ Quercus velutina f. dilaniata
771 x Quercus velutina f. missouriensis
771 † Quercus X Leana (Quercus imbricaria X velutina)

772 Quercus ellipsoidalis (Jack Oak)

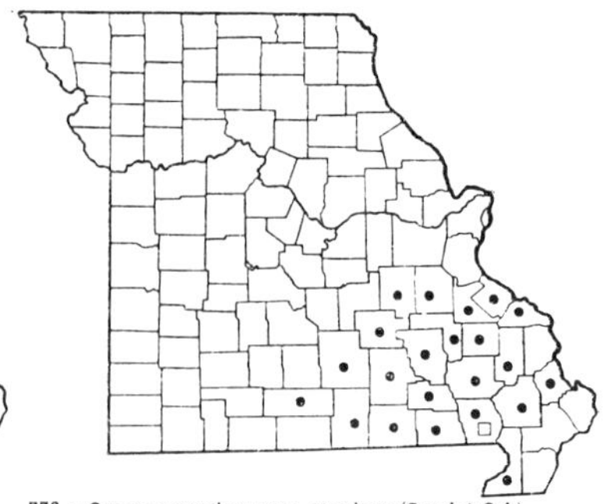

773 • Quercus coccinea var. coccinea (Scarlet Oak)
773 □ Quercus coccinea var. tuberculata

Ranges from southern Michigan to southern Manitoba, south to northern Ohio, northern Indiana, northern Illinois, Iowa, northern Missouri, and southeastern North Dakota.

The species has been divided into two forms based on the shape and size of the acorn:

Acorn distinctly longer than thick, ellipsoid-cylindric, narrowed at both ends, 15–20 mm. long
Q. ELLIPSOIDALIS f. ELLIPSOIDALIS
Acorn somewhat longer than thick, oblong-ovoid to subglobose, rounded at base and at summit, 12–16 mm. long . . . Q. ELLIPSOIDALIS f. DEPRESSA

Since both types and all gradations of acorns may be found on the same tree, as observed repeatedly on trees in northern Illinois, no importance should be given to this range of normal variation. The leaves vary from deeply lobed with the cutting more than halfway to the midrib to less than halfway, and with the base of the leaves wedge-shaped to truncate. The opportunity to live in an area in northern Illinois during the past twenty years where this species is the dominant tree around the author's residence has afforded him unlimited and frequent observations of winter buds. Although winter buds are sometimes reported as only 1–3 mm. long (Wadmond, S. C., Trans. Acad. Sci. 28: 203. 1933), they are, when fully grown, 4–7 mm. long. Rather than having the upper scales 'inconspicuously ciliate-pubescent or nearly glabrous' under which division of the key this species would fall in Palmer's key to the species and varieties of red oaks (Am. Midl. Nat. 27: 739, 1942), the winter bud-scales in *Q. ellipsoidalis* are covered with a gray silky or appressed type of fine pubescence which is general over most of the surface. The cups and acorns vary in size.

In having a yellow inner bark *Quercus ellipsoidalis* may be more easily mistaken for *Q. velutina* than any other species. The best criterion for separating the two lies in the loose overlapping or fringe development of the upper scales of the acorn cup in *Q. velutina* as contrasted with the tightly appressed ones of *Q. ellipsoidalis*, and in the longer more angled, more densely pubescent terminal winter buds of *Q. velutina*, which are 6–12 mm. long, as contrasted with 4–7 mm. long in *Q. ellipsoidalis*. Also, it can be stated that the leaves of *Q. velutina* are much thicker and more shining on the upper surface than those of *Q. ellipsoidalis*, but this can be judged only when one is sufficiently well acquainted with the two species, so this leaf contrast at best is only relative. The leaves of *Q. ellipsoidalis*, in fact, are about as thin as in *Q. Shumardii* and nearly as thin as in *Q. coccinea*.

The foliage of *Q. ellipsoidalis* takes on brilliant colors in autumn, sometimes scarlet or crimson-red hues. In early spring the young leaves are deep purple-red or blood red, and the older ones become green and yellow with some rose. The lower branches of the tree have a habit of dying, with the spreading to drooping dead limbs remaining indefinitely attached and giving the tree an untidy appearance. The species also appears to be one of the weaker oaks and is subject to windfall and storms which are known to topple over large trees of this species. Trees which have been blown down during wind storms usually reveal a root system which appears to be rather weak and superficial, and a trunk with a heart wood which had begun to rot near the base after the tree had reached a certain age.

An effort should be made throughout the counties in northern Missouri to locate more stands of this interesting species. No timber value is known for *Q. ellipsoidalis*. Bronze grackles, and gray and fox squirrels are fond of the acorns of this and other species of *Quercus*.

A hybrid is recorded between this species and *Q. velutina* (as *Q.* × *palaeolithicola* Trel.) but has not been found in Missouri.

15. **Quercus coccinea** Muenchh. Scarlet Oak
Map 773

Flowers April–May.

Occurs in acid soils associated with sandstone, chert, flint, or granite rocks, either on narrow ridges, ravine slopes, or in upland woods bordering head-waters of tributary streams. Restricted to the eastern half of the Ozark region ranging there north to Ste. Genevieve, Washington, and Crawford counties, west to Dent, Texas, Howell, and eastern Douglas County, where it reaches its southwesternmost station in the sandstone hills following the drainage of Indian Creek, northeast and south of Topaz.

Ranges from Maine, New York, Ontario, Ohio, southern Indiana, southern Illinois, and southern Missouri, south to Georgia, Alabama, northern Mississippi, and northeastern Arkansas.

The species is sometimes divided into two varieties on the basis of fruiting characters. Much more field work and experimental studies need to be carried out in connection with the variation of the fruits of this species. Until more exhaustive work has been done, the var. *tuberculata*, which has not been recognized either in *Gray's Manual*, eighth edition, or in Gleason's *New Britton and Brown*, is being maintained in the present work.

Acorns mainly 20–25 mm. long; scales of cup not thickened with any tubercle (wartlike enlargement); cup 20–23 mm. wide; common type . 15a. Q. COCCINEA var. COCCINEA

Acorns mainly 25–30 mm. long; lower scales of cup with a conspicuous thickened tubercle (wartlike enlargement); cup often 25–30 mm. wide; rare, known only from Butler County . 15b. Q. COCCINEA var. TUBERCULATA

15a. **Quercus coccinea** var. **coccinea** Map 773

*Quercus coccinea* Muenchh. [G, BB, P & S, Steyerm.]

*Quercus* × *Richteri* Baenitz [P & S, Steyerm.]

This is the usual fruiting variation encountered.

15b. **Quercus coccinea** var. **tuberculata** Sarg.
Map 773

Known only from Butler County in southeastern Missouri (gravelly hills near Poplar Bluff, October 5, 1921, *Palmer 20678*; same locality, November 14, 1935, *Palmer 29638*).

Ranges from Massachusetts to Indiana and Missouri, south to Alabama and Tennessee.

Among the most brilliantly red trees in autumn this species of oak ranks near the top in Missouri, along with Sour Gum and Red Maple as rivals. In early autumn Scarlet Oak is still mostly green, but in late autumn about the first part of November or the very last of October, it attains its most brilliant aspects of deep red. It is sometimes planted for its handsome foliage, but transplants with great difficulty.

The bark of mature trees of Scarlet Oak resembles that of Northern Red Oak (*Q. rubra*) in having a smooth pale gray or gray-brown surface interrupted by darker shallow vertical furrows. The bark is usually given a pale mottled gray and greenish-white appearance by the pale lichens which attach over much of the surface. The inner bark is reddish. The conspicuous acorn cup with nearly glabrous, closely set, tightly appressed, glossy scales, and glabrous lower portion of the buds are reliable characters that serve to separate Scarlet Oak from other species with which it might be mistaken. The winter buds are more pubescent than in either Northern Red Oak (*Q. rubra*) or Shumard Oak (*Q. Shumardii*), but less pubescent than in Black Oak (*Q. velutina*), the pubescence on the buds of *Q. coccinea* limited to the upper section and only sparsely so. The leaves of Scarlet Oak are quite thin and have a texture similar to thin parchment paper, and are green on both sides. They are not shiny as in either Black Oak (*Q. velutina*) or Shumard Oak (*Q. Shumardii*). They are thinner than those of either Black Oak or Shumard Oak, more deeply cut than either those of Black Oak or Northern Red Oak (*Q. rubra*) and as deeply cut as in Pin Oak (*Q. palustris*), Jack Oak (*Q. ellipsoidalis*), and Shumard Oak (*Q. Shumardii*). The axillary tufts of hairs on the lower leaf surface, which are so conspicuous in the Black Oak and Shumard Oak, are either not present or else not conspicuous in Scarlet Oak.

On the narrow ridge tops and heads of tributary ravines in the Ozarks it is usually associated with other species of oak, sour gum, pine, flowering dogwood, and red maple.

The wood of this species is used principally for wagons, barrels, chairs, boats, and agricultural implements, and is usually marketed as black oak.

No hybrids are known in Missouri involving this species. Palmer (Jour. Arn. Arb. 29: 38. 1948) concurs with Croizat (Torreya 36: 139–42. 1936) that *Q.* × *Richteri* is synonymous with *Q. coccinea* and not a hybrid between *Q. rubra* [*borealis* var. *maxima*] and *palustris* as formerly maintained. Specimens from Osage and Benton counties, cited from Missouri previously as *Q.* × *Richteri*, are still to be considered as hybrids between *Q. rubra* and *palustris*, but no formal name has been assigned to this hybrid up to the present time.

16. **Quercus Shumardii** Buckl. Map 774

Flowers early April–May.

The species is represented by two variations, based principally on the shape of the acorn cup:

Cup of acorn flattish, shallow, saucer-shaped, 20–30 mm. broad, enclosing about 1/4 of the acorn; trees usually of low or alluvial woods, but also in upland woods . . . 16a. Q. SHUMARDII var. SHUMARDII

Cup of acorn deeper, strongly rounded or convex below, bowl-shaped, 20 mm. or less broad, enclosing 1/3–2/5 of the acorn; trees of drier, upland or rocky woods, glades, or edges of bluffs . 16b. Q. SHUMARDII var. SCHNECKII

16a. **Quercus Shumardii** var. **Shumardii**

Shumard Oak Map 774

Also called Water Oak, Spotted Oak, and Red Oak

*Quercus Shumardii* Buckl. [G, P & S, Steyerm.]

Occurs frequently in the valleys and along banks of the larger Ozark streams usually in low and alluvial ground, but also found in upland drier situations. Southern and central Missouri north to Ralls, Saline, and Jackson counties.

Ranges from Florida to Texas, north to Maryland, Pennsylvania, West Virginia, Ohio, Indiana, Illinois, Missouri, Kansas, and Oklahoma.

16b. **Quercus Shumardii** var. **Schneckii** (Britt.) Sarg. Schneck Oak Map 774

Usually frequents drier and rockier situations on the upland, on glades and edges of bluffs, and in rocky woods. Southern and central Missouri, where common throughout the Ozark and unglaciated prairie sections, north to Pike, Monroe, Boone, Saline, and Clay counties; absent from most of the lowland section of southeastern Missouri.

Ranges from Ohio, Indiana, Missouri, and Kansas, south to Tennessee, Arkansas, and Oklahoma.

This variety is not recognized by the United States Forest Service, it being combined with *Q. Shumardii* var. *Shumardii*. While it is true that intermediates and intergrading forms occur and cannot be identified positively as one or the other variety, it can also be maintained that frequent field contact with these varieties serves to emphasize more their differences. Until additional new evidence is at hand to indicate that var. *Shumardii* and var. *Schneckii* are not distinct varieties, the present treatment maintains them along with the judgment of Deam, Palmer, Fernald, Gleason, and the author's own field knowledge of them. Palmer (Am. Midl. Nat. 27: 739. 1942) also emphasizes differences between var. *Shumardii* and var. *Schneckii* in leaf variation: in var. *Schneckii* the leaves are essentially uniform on upper and lower branches, whereas in var. *Shumardii* those on the upper branches of old trees are more often deeply divided than the ones on the lower branches. Since these differences are frequently not observed in the field nor are to be judged on herbarium material, it is usually impossible to rely on these vegetative criteria, and it can only be recommended here that such observations be encouraged for future appraisal of the distinctness of the varieties.

These two varieties are easily distinguished from Northern Red Oak (*Q. rubra*) by the gray, glabrous, acute buds and more deeply cut leaves which are shiny on the upper surface. The bark of the *Q. Shumardii* varieties is dark with broken furrows, while that of Northern Red Oak is paler gray and the furrows more continuous. The completely glabrous gray buds distinguish the *Q. Shumardii* group from Jack Oak (*Q. ellipsoidalis*), Scarlet Oak (*Q. coccinea*), and Black Oak (*Q. velutina*). The bark is not rough as in *Q. velutina* or *Q. ellipsoidalis*, and the inner bark is reddish to gray, not yellow. For leaf distinctions between *Q. Shumardii* and *Q. Nuttallii*, refer to remarks under the latter species.

The wood of the *Q. Shumardii* group is usually classified by lumber merchants as red oak and accordingly given uses similar to red oak (*Q. rubra*). The trees have been found to do well when planted along streets and highways.

Six hybrids have been recorded which involve either one or the other of these varieties as parents, and in Missouri the following three have been found:

**Quercus × Egglestonii** Trel. (*Q. imbricaria × Shumardii* var. *Schneckii*), already referred to previously under *Q. imbricaria*,

**Quercus × moultonensis** (*Q. Phellos × Shumardii* var. *Shumardii*), already referred to previously under *Q. Phellos*, and

Plate no. 137. **1.** Quercus ellipsoidalis, × 2/7. **2.** Quercus coccinea, × 2/7. **3.** Quercus Shumardii var. Schneckii, × 2/7. **4.** Quercus palustris, × 2/7. **5.** Quercus rubra var. borealis, × 2/7; Details from Small, The New York Botanical Garden. **6.** Quercus Nuttallii, × 2/7; **7.** Quercus × Leana, × 2/7. **8.** Quercus × Bushii, × 2/7. **9.** Quercus × exacta, × 2/7. **10.** Quercus × Hawkinsiae, × 2/7. **11.** Quercus heterophylla, × 2/7; *a*, *b*. Two types of leaves of this hybrid. **12.** Ulmus americana, × 2/7; *a*. Flowering branch; *b*. Fruiting branch.

PLATE NO. 137

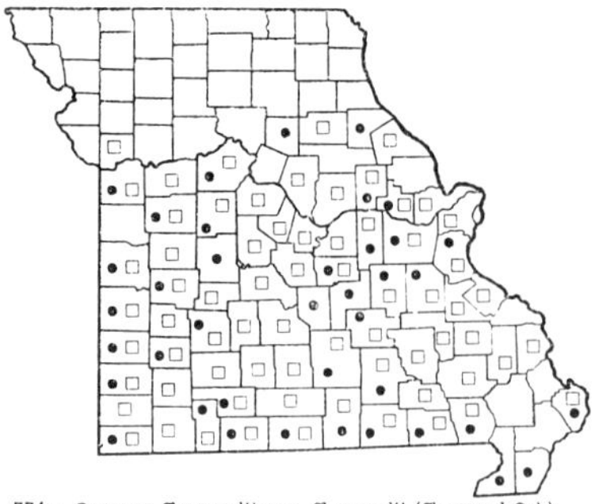

774 • Quercus Shumardii var. Shumardii (Shumard Oak)
774 □ Quercus Shumardii var. Schneckii (Schneck Oak)

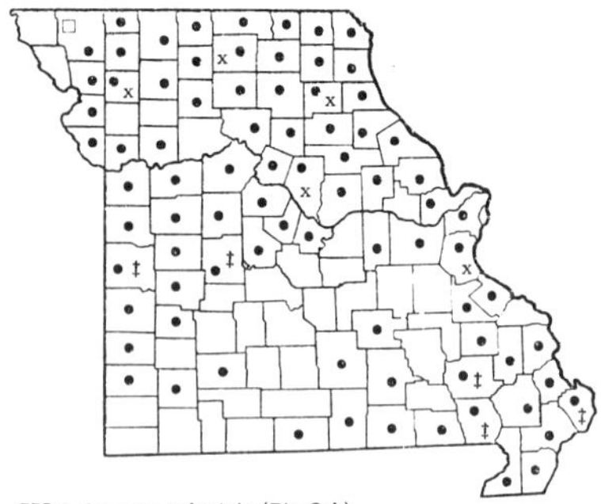

775 • Quercus palustris (Pin Oak)
775 □ Quercus X vaga (Quercus palustris X velutina)
775 ‡ Quercus X mutabilis (Quercus palustris X Shumardii var. Schneckii)
775 x Quercus X exacta (Quercus imbricaria X palustris)

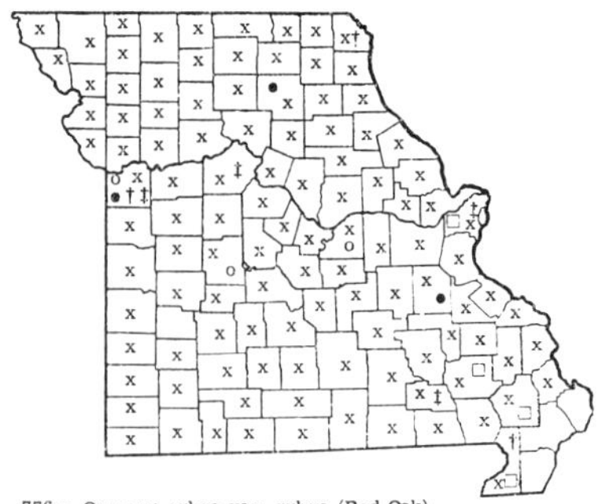

776 x Quercus rubra var. rubra (Red Oak)
776 • Quercus rubra var. borealis
776 ‡ Quercus X runcinata (Quercus rubra var. rubra X imbricaria)
776 □ Quercus X heterophylla (Quercus rubra var. rubra X Phellos)
776 † Quercus X Hawkinsiae (Quercus rubra var. rubra X velutina)
776 o Quercus rubra var. rubra X palustris

**Quercus** × **mutabilis** Palmer & Steyerm. (*Q. palustris* × *Shumardii* var. *Schneckii*).

Known from Mississippi, Wayne (*Palmer 19423*), Butler (*Bush 3770*), Benton (*Palmer 26358*), and Bates (Monteith Junction, September 10, 1924, *Palmer 26069*, holotype) counties.

17. **Quercus palustris** Muenchh. Pin Oak Map 775

Also called Turkey Oak, Swamp Oak, and Water Oak in some portions of Missouri.

Flowers early April–May.

Occurs in alluvial ground in valleys and low ground of flood plain forests, bordering streams, oxbow lake meanders, sloughs, and river bottom meadows, sometimes on dry upland as relicts around old sink-hole ponds in the Ozarks. Throughout Missouri, but absent from most sectors of the main Ozark region.

Ranges from South Carolina to Louisiana and Oklahoma, north to Massachusetts, Pennsylvania, Ontario, Michigan, Illinois, Iowa, and Kansas.

For distinctions between the leaves of this species and *Q. Nuttallii*, refer to the remarks under that species.

The foliage of this species in autumn is quite colorful, turning from green tinged with carmine, or reddish-green, to scarlet. Especially in open or sunlit places, the tree develops a pyramidal shape with more or less horizontal, long, slender middle branches and drooping lower branches, giving it a distinctive and graceful appearance. Because of this, it is frequently planted as an ornamental tree. The campus at Washington University in St. Louis has beautiful specimens of this species, and many parks and streets in Missouri and elsewhere show off the landscaping qualities of the Pin Oak to good advantage. In nature the lowest drooping branches of the Pin Oak are usually much crowded and shaded by branches of other trees in a lowland forest, and eventually die. They remain attached to the tree for some time, but the pinlike or spurlike stumps of the branches which have broken away remain and give the name Pin Oak. Where the tree is allowed to develop in a situation where it is not crowded nor shaded, these lower branches continue to grow and add to the beauty of the tree.

The wood is considered inferior to that of Red Oak (*Q. rubra*) because it is usually full of knots produced from the numerous branches of the tree, but is used in construction, rough siding as clapboard for buildings, and for shingles.

Of the four hybrids involving this species, three of them are known from Missouri:

**Quercus** × **exacta** Trel. (*Q. imbricaria* × *palustris*), previously referred to under *Q. imbricaria.*

**Quercus** × **mutabilis** Palmer & Steyerm. (*Q. palustris* × *Shumardii* var. *Schneckii*), previously referred to under *Q. Shumardii.*

**Quercus** × **vaga** Palmer & Steyerm. (*Q. palustris* × *velutina*), previously referred to under *Q. velutina.*

18. **Quercus rubra** L. Red Oak Map 776

Also called Northern Red Oak

Flowers April–May.

Two variations, based chiefly on the fruit, occur in Missouri:

Common type found throughout Missouri; cup of acorn shallow, saucer-shaped, flat on the

bottom, covering about 1/4 of cup, 18–30 mm. broad . . . . . . 18a. Q. RUBRA var. RUBRA
Rarely encountered, known only from Jackson and Macon counties; cup of acorn deeper, enclosing about 1/3 of acorn, bowl-shaped (hemispherical) or top-shaped (turbinate), 15–20 mm. (up to 25) broad . . 18b. Q. RUBRA var. BOREALIS

18a. **Quercus rubra** var. **rubra** Map 776

*Quercus rubra* L. [G]

*Quercus borealis* Michx. f. var. *maxima* [BB, P & S, Steyerm., Deam]

Occurs in well drained soils in rich ravines, north- and east-facing wooded bluffs, near heads of tributary streams and along streams, both in upland and lowland, on slopes or at the base of bluffs, in limestone, sandstone, chert, or granitic soil areas. Throughout Missouri, where known from essentially every county except the three most southeastern ones. It does not occur in soil which is subject to periodic inundation and flooding by streams, such as endured by Pin Oak.

Ranges from Nova Scotia, Quebec, Ontario, Michigan, Wisconsin, Minnesota, and Nebraska, south to Georgia, Alabama, Mississippi, Louisiana, Arkansas, and Oklahoma.

18b. **Quercus rubra** var. **borealis** (Michx. f.) Farw. Map 776

*Quercus borealis* Michx. f. var. *borealis* [BB]

*Quercus borealis* Michx. f. [P & S, Steyerm., Deam]

Known only from Macon and Jackson (woods along creek, Greenwood, September 10, 1921, *Palmer 20450*) counties in northern and west-central Missouri.

Ranges from Nova Scotia, Quebec, and Ontario, south to Maine, New York, Pennsylvania, North Carolina, Michigan, Wisconsin, Iowa, and Missouri.

The United States Forest Service unites the two varieties, but the two, although intergrading at the margins of their ranges, may be separated by the differences given in the key. The var. *borealis* is a more northern-ranging variety.

The foliage of this tree in autumn turns to orange or russet-brown with green. It makes a desirable ornamental tree, grows rapidly, and is recommended as one of the best of trees for planting along streets. Fortunately, unlike some other species of oaks, *Q. rubra* usually transplants easily. The inner bark, which is reddish, is responsible for the name of this and other related members of the group.

Commercially, red oak includes this species combined with other members of the clan, but the quality of the wood of the true red oak (*Q. rubra* and var. *borealis*) is considered superior for the most part to that of the other kinds, but not as good as white oak. It is much used, however, in construction work for clapboards and rough lumber, barrels, railroad ties, as well as for furniture and interior finish of homes and offices.

Five hybrids have been recorded involving this oak and its variety as one of the parents. The following four are known to occur in Missouri:

**Quercus** × **Hawkinsiae** Sudw. (*Q. rubra* [*borealis* var. *maxima*] × *velutina*), previously referred to under *Q. velutina*.

**Quercus** × **heterophylla** Michx. f. (*Q. Phellos* × *rubra* [*borealis* var. *maxima*]), previously referred to under *Q. Phellos*.

**Quercus** × **runcinata** (A.DC.) Engelm. (*Q. imbricaria* × *Q. rubra* [*borealis* var. *maxima*]), previously referred to under *Q. imbricaria*.

**Quercus rubra** [*borealis* var. *maxima*] × **palustris**, referred to under discussion of *Q.* × *Richteri* Baenitz. Specimens representing the above cross are known from Osage and Benton counties.

19. **Quercus Nuttallii** Palmer Nuttall's Oak Map 777

Flowers April–May.

Occurs in low or wet woods, where known in the state only from New Madrid (Paw Paw Junction, September 3, 1897, *Bush 224*) and Butler (October 19, 1905, *Bush 3770*) counties in southeastern Missouri. Trees have been seen on the Joe Adamson place, T26N, R7E, sect. 27, 2 mi. north of Rombauer, along Mud Creek, where Mr. Adamson calls it 'Black Cherry-bark Oak,' a name usually applied to *Q. falcata* var. *pagodaefolia*. Here it has been observed by Mr. Albert Chandler, amateur botanist of Kirkwood, Missouri.

Ranges from Alabama to Texas, north to western Tennessee, southeastern Missouri, Arkansas, and southeastern Oklahoma.

Until described by Mr. E. J. Palmer in 1927 (Jour. Arn. Arb. 8: 52. 1927), this oak had not been distinguished. It is considered a commercially important species, known as red oak in the lumber industry, where it is employed in much the same manner as other red oak lumber.

This oak is sometimes confused with Pin Oak (*Q. palustris*) on account of the similar leaves and dead lower branches of the tree, but is easily distinguished in fruit by the deeper acorn cups which are narrowed into a stalklike base and the larger acorns and winter buds. The fruit of *Q. Nuttallii* also distinguishes it from *Q. Shumardii* var. *Shumardii* and var. *Schneckii*. By leaves alone, *Q. Nuttallii* may be distinguished, according to Dr. Dyal (Rh. 38: 60. 1936) from *Q. palustris* and *Q. Shumardii* var. *Shumardii* and var. *Schneckii* as follows: in *Q. Nuttallii* the lobes of the leaf are usually long-tapering and the main lobes have no small bristles at the sides less than 5 mm. from the very tip; in *Q. palustris* and *Q. Shumardii*, on the other hand, the leaf lobes are rather blunt and some of the main lobes, at least, have 1–2 small bristles at the sides less than 5 mm. from the very tip.

A hybrid involving this species and *Q. Shumardii* var. *Shumardii* has been recorded, but up to the present time has not been found in Missouri.

## HYBRID OAKS RECORDED FROM MISSOURI

For the geographical distribution of these in the state, refer to either of the parent species indicated in the cross.

1. **Quercus × anceps** Palmer (*Q. falcata × imbricaria*)
2. **Quercus × Bebbiana** Schneid. (*Q. alba × macrocarpa*)
3. **Quercus × Bushii** Sarg. (*Q. marilandica × velutina*)
4. **Quercus × Deamii** Trel. (*Q. macrocarpa × prinoides* var. *acuminata* [*Muehlenbergii*])
5. **Quercus × Egglestonii** Trel. (*Q. imbricaria × Q. Shumardii* var. *Schneckii*)
6. **Quercus × exacta** Trel. (*Q. imbricaria × palustris*)
7. **Quercus × Fernowii** Trel. (*Q. alba × stellata*)
8. **Quercus × filialis** Little (*Q. Phellos × velutina*)
9. **Quercus × Hawkinsiae** Sudw. (*Q. rubra* [*borealis* var. *maxima*] × *velutina*)
10. **Quercus × heterophylla** Michx. f. (*Q. Phellos × rubra* [*borealis* var. *maxima*])
11. **Quercus × humidicola** Palmer (*Q. bicolor × lyrata*)
12. **Quercus × Jackiana** Schneid. (*Q. alba × bicolor*)
13. **Quercus × Leana** Nutt. (*Q. imbricaria × velutina*)
14. **Quercus × ludoviciana** Sarg. (*Q. falcata × Phellos*)
15. **Quercus × moultonensis** Ashe (*Q. Phellos × Q. Shumardii* var. *Shumardii*)
16. **Quercus × mutabilis** Palmer & Steyerm. (*Q. palustris × Q. Shumardii* var. *Schneckii*)
17. **Quercus rubra** [*borealis* var. *maxima*] **× palustris**
18. **Quercus × runcinata** (A.DC.) Engelm. (*Q. imbricaria × rubra* [*borealis* var. *maxima*])
19. **Quercus × Schuettei** (*Q. bicolor × macrocarpa*)
20. **Quercus × stelloides** Palmer (*Q. prinoides* var. *prinoides × stellata*)
21. **Quercus × Sterrettii** (*Q. lyrata × stellata*)
22. **Quercus × tridentata** (A.DC.) Engelm. (*Q. imbricaria × marilandica*)
23. **Quercus × vaga** Palmer & Steyerm. (*Q. palustris × velutina*)

## HYBRID OAKS NOT RECORDED FROM MISSOURI

The following list of hybrids is given below to indicate the possible discoveries that may be expected as a result of future collecting in the state. Both parent species are present in Missouri in all the hybrids listed below and the nearest state to Missouri from which the hybrid is known is indicated wherever possible after the name. (Numbers 18–22 are possible combinations, but have so far not been supported by sufficient fruiting material.)

1. **Quercus × Beadlei** Trel. (*Q. alba × Michauxii*). Indiana
2. **Quercus × Byarsii** Sudw. (*Q. macrocarpa × Michauxii*). Tennessee
3. **Quercus × Demarei** Ashe (*Q. nigra × velutina*). Arkansas
4. **Quercus × Faxonii** Trel. (*Q. alba × prinoides* var. *prinoides*). Michigan
5. **Quercus × Garlandensis** Palmer (*Q. falcata × nigra*). Arkansas
6. **Quercus guadalupensis** Sarg. (*Q. macrocarpa × stellata*). Texas
7. **Quercus × Joorii** Trel. (*Q. falcata × Shumardii* var. *Shumardii*). Texas
8. **Quercus × neopalmeri** Sudw. (*Q. nigra × Q. Shumardii* var. *Shumardii*). Arkansas
9. **Quercus × palaeolithicola** Trel. (*Q. ellipsoidalis × velutina*). Iowa, Illinois
10. **Quercus × Rudkinii** Britton (*Q. marilandica × Phellos*). Arkansas

11. **Quercus** × **Schochiana** Dieck (*Q. palustris* × *Phellos*). Arkansas
12. **Quercus** × **sterilis** Trel. (*Q. marilandica* × *nigra*). Georgia
13. **Quercus** × **Willdenowiana** (Dippel) Zabel (*Q. falcata* × *velutina*). Arkansas
14. **Quercus alba** × **prinoides** var. **acuminata** [*Muehlenbergii*]
15. **Quercus alba** × **lyrata**
16. **Quercus coccinea** × **palustris**
17. **Quercus falcata** × **marilandica**
18. **Quercus falcata** × **nigra**
19. **Quercus Nuttallii** × **Shumardii** var. **Shumardii**
20. **Quercus Phellos** × **imbricaria**
21. **Quercus rubra** [*borealis* var. *maxima*] × **ellipsoidalis**
22. **Quercus rubra** [*borealis* var. *maxima*] × **marilandica**

## Order **URTICALES**

### Fam. **ULMACEAE** (Elm Family)

a. Leaves with 4–6 main nerves on each side of midrib, the lowest pair arising from the base of blade, forming a 3-nerved or V-nerved part with the midrib; margins of leaves without teeth or with teeth only along part of the margin; branches with chambered pith; all the flowers appearing on the newest leafy branches; sepals separate, not united; fruit fleshy, a 1-seeded round drupe with thin pulp and tough skin, ripening in late summer or fall, but well-developed when the leaves are fully grown . . . . . 3. CELTIS

a. Leaves with usually 6–20 main nerves on each side of midrib, the lowest pair arising above the base of blade, all the nerves at nearly the same angle parallel to one another; margins of leaves toothed rather regularly from base to summit; branches with solid pith; some of the flowers appearing on branches of the previous year; sepals united forming a calyx tube with lobes; fruit dry, either a flat winged fruit (samara) or a wingless bur-like nut, ripening and falling before June . . . . . *b*

b. Flowers appearing before the leaves; fruit flat with a broad wing; margins of leaves with most of teeth usually double-toothed (each tooth cut into or broken); bark rough or ridged; trunk single . . . . . 1. ULMUS

b. Flowers appearing with the leaves; fruit a wingless bur-like nut; margins of leaves with most of teeth mostly simply-toothed (without any break or cut on each tooth); bark thin and scaly; trunk forking usually near base . . . . . 2. PLANERA

### 1. **Ulmus** L. Elm

Species of this genus produce pollen which has been shown to be irritating and to cause some cases of hay fever.

a. Upper surface of leaves rough, like sandpaper, to the touch, with erect stiff hairs; bud-scales conspicuously covered with dense orange-brown or rusty-colored hairs; inner bark slippery when chewed . . . . . 4. U. RUBRA

a. Upper surface of leaves more or less smooth to the touch, without hairs or these not stiff or erect; bud-scales either without hairs or the hairs either inconspicuous or not orange-brown or rusty-colored; inner bark not slippery when chewed . . . . . *b*

b. Cultivated tree or shrub, sometimes escaped from cultivation; leaves nearly equal-sided at the base; margins of leaves simply toothed (without any break or cut on each tooth); lower surface of leaves glabrous (without hairs); flowers nearly without stalks (sessile), not drooping; margins of fruits without hairs projecting . . . . . 5. U. PUMILA

b. Without the above combination of characters: native trees; leaves mostly unequal-sided at the base, one side longer or wider or more curved than other; margins of leaves usually double-toothed (each tooth slightly cut into or broken on one side); lower surface of leaves hairy or sometimes glabrous (without hairs); flowers with long stalks, soon drooping; margins of fruits with dense hairs projecting . . . . . *c*

c. None of the branches developing corky wings or woody outgrowths; flowers occurring in clusters or groups, their stalks originating from the same point; fruits glabrous on the side which covers the seed; lower surface of leaves hairy or becoming glabrous . . . 1. U. AMERICANA

c. Some of the young or old branches, or both, developing corky wings or woody outgrowths;

777 Quercus Nuttallii (Nuttall's Oak)

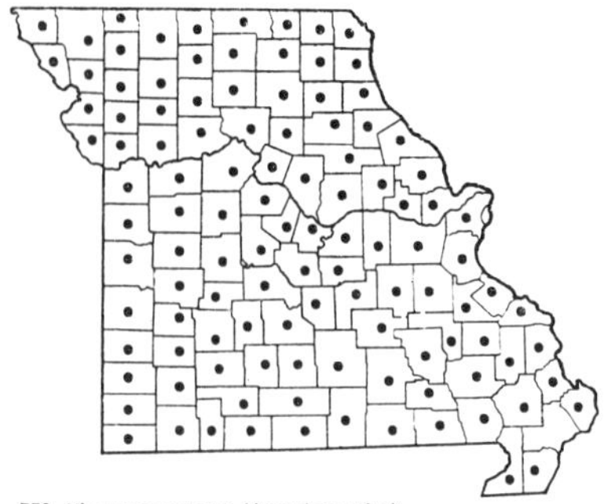

778 Ulmus americana (American Elm)

779 Ulmus Thomasi (Rock Elm)

flowers in racemes becoming longer than broad; fruits usually hairy on the side which covers the seed; lower surface of leaves generally remaining hairy . . . . . . . . . . . . *d*

d. Common in the Ozark region, often in dry and rocky places, but also in lowlands or along streams; leaf-stalks (petioles) usually 1–3 mm. (rarely to 5) long; base of leaf-stalk hairy on upper and lower sides; buds glabrous (without hairs) or nearly so; young branchlets often developing corky wings; fully grown leaves mostly 4–8 cm. (rarely 9) long, mostly 1–3 cm. (rarely 4) broad; fruits 3–5 mm. broad . . . . . . . . . . . . 3. U. ALATA

d. Rare, and chiefly outside of the Ozark area in northern, central, and extreme southeastern Missouri, along streams and wet woods; leaf-stalks (petioles) 3–10 mm. long; base of leaf-stalks glabrous (without hairs) on lower side; buds rather hairy; corky growth only on older branches; fully grown leaves mostly 8–15 cm. long, 3–9 cm. broad; fruits 9–15 mm. broad . . . . . . . . . . . . . . . . . . . . . . . 2. U. THOMASI

1. **Ulmus americana** L. American Elm Map 778

Also called White Elm

Flowers February–April.

Occurs usually along streams and low moist ground. Throughout Missouri and doubtless in every county.

Ranges from Newfoundland to Saskatchewan, south to Florida and Texas.

This species varies considerably in the pubescence of the young branchlets and upper surface of leaves.

Fernald (Rh. 47: 132–33. 1945) has attempted to account for these variations found by recognizing four forms as follows:

Leaves smooth or smoothish above
  Young branchlets pubescent. f. PENDULA (Ait.) Fern.
  Young branchlets glabrous. . f. LAEVIOR Fern.
Leaves harshly scabrous above
  Young branchlets pubescent . f. ALBA (Ait.) Fern.
  Young branchlets glabrous . f. INTERCEDENS Fern.

All four of these variations occur throughout Missouri.

Since it has never been determined to which of these the original Linnean specimen applies, it is at present not possible to designate any one form as typical f. *americana* as represented by the actual type specimen of Linnaeus. In the present work, no subdivisions of the species are being maintained, pending future realignment of the names used by Fernald.

In autumn the foliage turns yellow and eventually orange brown. The wood has been used in many ways, especially for wagon hubs, agricultural implements, furniture, barrels, basket handles, flooring, shipbuilding, boxes, and chopping bowls. It is also used for the manufacture of fiber for roofing felt. The bark was at one time made into ropes and ox whips, and the Indians used the bark for construction of canoes. The wood is unsatisfactory, however, for posts or firewood because of its quick-rotting and water holding properties.

The vaselike form of the tree has made American Elm one of the most prized ornamental and shade trees in the United States. Since 1930, when the first outbreak of Dutch Elm disease was recorded in the United States, ever increasing damage from this combination fungus and beetle scourge together with the phloem necrosis has taken its toll of this graceful species. Some trees have been killed in Missouri, but in general the diseases at the present time have affected a relatively small percentage of elm in the state.

2. **Ulmus Thomasi** Sarg. Rock Elm Map 779

Also called Cork Elm

Flowers March–April.

Occurs in low wet or alluvial ground along streams and in flood plain forest. Scattered in portions of

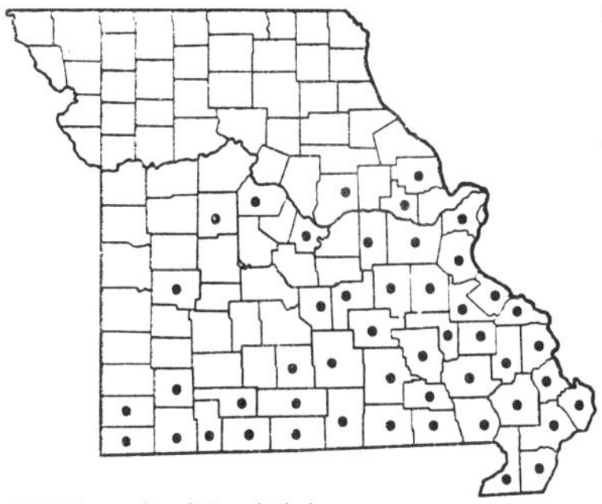

780 Ulmus alata (Winged Elm)

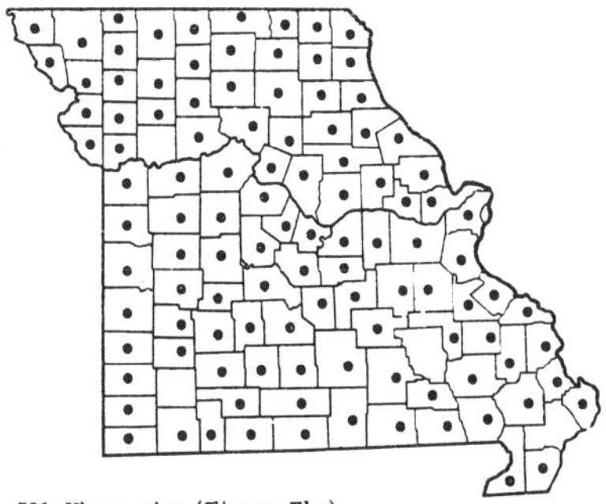

781 Ulmus rubra (Slippery Elm)

782 Ulmus pumila (Siberian Elm)

northern, central, and southeastern Missouri; absent from the Ozark and unglaciated prairie sections.

Ranges from Vermont and Quebec to Michigan, Minnesota, and South Dakota, south to New York, Ohio, Tennessee, Missouri, and Kansas.

The corky growth is limited to the older branches, and is sometimes poorly developed on the trees. The wood is tough, solid, and flexible, and these qualities have in the past given it some preference for such items as railroad ties, hubs of wagon wheels, various agricultural implements, framework of chairs, stock for other furniture, barrels, and ax handles. It is used as plywood in boxes and trunks.

3. **Ulmus alata** Michx. Winged Elm Map 780

Also called Wahoo (a name also applied to *Euonymus atropurpurea*).

Flowers February–March.

Occurs in dry rocky upland woods, borders of rocky glades, and in low ground of valleys, ravine bottoms, and along streams, generally in acid soils. Found in the Ozark region of southern and east-central Missouri, north to Lincoln, Callaway, Cooper, and Pettis counties, west to St. Clair, Lawrence, and Newton counties.

A specimen in the herbarium of William Jewell College from Clay County (river banks and ridges, woods along river, south of Liberty, May 6, 1948, *Shirley Landers 83*), is doubtfully identified as this species, and is not included on the distribution map. More material needs to be seen from this station before the status of the collection can be properly evaluated.

Ranges from Florida to Texas, north to Virginia. Kentucky, Indiana, Illinois, and Missouri.

This distinct species is common throughout most of the Ozark forests in either lowland or upland situations. It varies from a dwarf scrubby tree only several feet tall to one of fairly majestic proportions, up to 100 feet tall. The foliage turns a dull green and yellow in autumn.

The wood is difficult to split, but is used for such articles as hubs of wagon wheels, tool handles, and table legs. The inner bark has been employed as a rope to fasten together the covers of cotton bales.

4. **Ulmus rubra** Muhl. Slippery Elm Map 781

*Ulmus fulva* Michx. [P & S, Steyerm.]

Flowers February–April.

Occurs in dry upland or rocky woods, open wooded bluffs and along streams, generally commonest in limestone areas. Throughout Missouri, and doubtless in every county.

Ranges from Florida to Texas, north to Maine, Quebec, Ontario, Minnesota, and North Dakota.

The twigs are generally rough-pubescent, and usually more so than any of the other species of Missouri. The mucilaginous inner bark is sometimes chewed. It was used by some of the American Indian tribes as a food cooked with fat. It has been referred to as an emergency food, and the Kiowa Indians employ the bark as a tea. This bark has also been used in medicine as a demulcent in the treatment of fevers and inflammations.

The wood is easy to split, is strong and durable, and these qualities have seen it used for fence posts, railroad ties, window sills, wagon hubs, and agricultural implements.

5. **Ulmus pumila** L. Siberian Elm Map 782

Also called Dwarf Elm

Flowers March–April.

Often planted, and occasionally escaped from cultivation in St. Louis (along overpass of Kingshighway between Manchester Ave. and right-of-way of Mo. Pac. R.R., St. Louis, August 31, 1958, *Muehlenbach 1475*), Douglas, Newton (rocky waste ground along Hickory Creek, Neosho, October 19, 1953, *Palmer 57208*; *57218*), and McDonald (near Beaver Brook Spring, Anderson, October 19, 1954, *Palmer 52940*) counties.

Native of Asia; naturalized and escaped from cultivation from Minnesota to Kansas.

This tree is often collected in sterile condition. It is frequently planted in the central and western United States for shade, screen, windbreak, and shelter. It has sometimes been mistaken for and confused with the Chinese Elm, *Ulmus parvifolia* Jacq., a species which flowers in autumn, whereas *U. pumila* blossoms in the spring.

*Excluded species*

**Ulmus serotina** Sarg. September Elm, Red Elm

This species occurs in adjacent southern Illinois and northern Arkansas. It has thus far not been found in Missouri. False reports of the occurrence of this species in Missouri have been taken from the following mislabeled collection: The label on the specimen reads 'top of high rocky bluff of White River, opposite Cotter, in Marion Co., October 2, 1937, *Palmer 43881*,' and has 'Flora of Missouri' at the top of the label, but the collector forgot to replace 'Missouri' with 'Arkansas,' Cotter, Marion Co., being, of course, in Arkansas.

## 2. **Planera** J. F. Gmel Water Elm

**Planera aquatica** (Walt.) J. F. Gmel.
Water Elm Map 783
Flowers March–April.

Occurs in swamps and low wet woodland of the southeastern lowland section, north to Mississippi and Wayne counties and west to Ripley County.

Ranges from Florida to Texas, north to North Carolina, Kentucky, Illinois, and Missouri.

The sexes are separate, the staminate or male flowers occurring in small clusters of last year's twigs, the pistillate on short slender stalks from the leaf-axils of the present season. The thin outside of the 1-seeded fruit has slender short rough appendages 1–3 mm. long.

The gray or light brown bark separates into large scales which reveal a reddish-brown inner bark.

## 3. **Celtis** L. Hackberry, Sugarberry

This genus is greatly in need of a complete revision. The limits of the species, as presently interpreted, leave unsolved many problems of variability and intergradations. The following treatment is based upon the work of Fernald, Deam, and Sargent.

a. Leaves of both fruiting branches and leafy shoots with 10–40 teeth all around the margins except at very base, usually abruptly long-acuminate (long-pointed); mature fruits 8–11 mm. long, usually with a short beak, quite wrinkled in drying, purple-black to dark brown or orange-red; fruiting stalks (pedicels) usually much longer than the leaf-stalks (petioles); stone of fruit 7–9 mm. long . . . 1. C. OCCIDENTALIS

a. Leaves of fruiting branches without teeth or with few teeth on one or both margins (if many-toothed then uniformly green on both surfaces and only gradually long-pointed); leaves of vegetative (leafy) branches similar or with teeth nearly all around the margins; mature fruits 5–8 mm. long, without or nearly without a beak, not much wrinkled in drying, orange to cherry-red or brownish; fruiting stalks either shorter or slightly longer than leaf-stalks; stone of fruit 4.5–7 mm. long . . . . . . . . *b*

  b. Shrub or small tree growing usually in dry, rocky, or exposed situations; leaves of fruiting branches $\frac{1}{2}$–$\frac{3}{4}$ as broad as long, 2–8 cm. long, ovate, with either blunt, acute, or short-acuminate (short slender tip) summit, without or nearly without teeth, gray-green on both surfaces or darker above . . . 2. C. TENUIFOLIA

  b. Usually medium to large trees of low wet bottomland along streams and in valleys, but also small trees on glades and bluffs; leaves of fruiting branches less than half as broad as long, 4–10 cm. long, narrowly lanceolate to lance-ovate or lance-oblong, tapering to a long slender tip, which is often falcate (curved to one side), either without teeth or conspicuously toothed, pale or dull green on both surfaces . . . . . . . . . . . . . . . . . . . . 3. C. LAEVIGATA

---

Plate no. 138. **1.** Ulmus Thomasi, × 2/7; *a.* Fruiting branch; *b.* Flowering branch. **2.** Ulmus alata, × 2/7; *a.* Fruiting branch; *b.* Flowering branch; Details from Small, The New York Botanical Garden. **3.** Ulmus rubra, × 2/7; *a.* Flowering branch; *b.* Fruiting branch. **4.** Planera aquatica, × 2/7; *a.* Flowering branch; *b.* Fruiting branch; Details from Small, The New York Botanical Garden.

PLATE NO. 138

783 Planera aquatica (Water Elm)

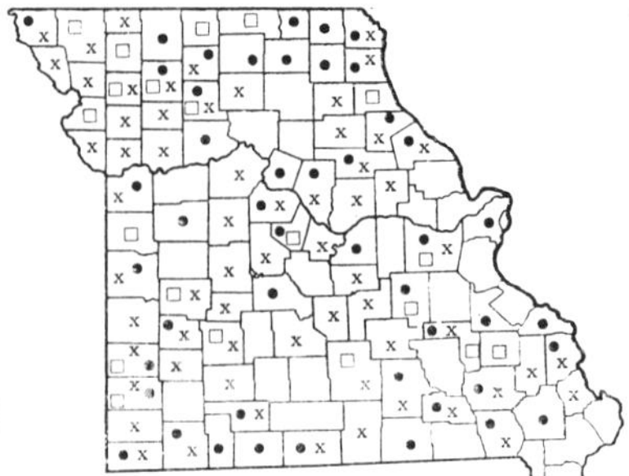

784 • Celtis occidentalis var. occidentalis (Hackberry)
784 □ Celtis occidentalis var. pumila
784 x Celtis occidentalis var. canina

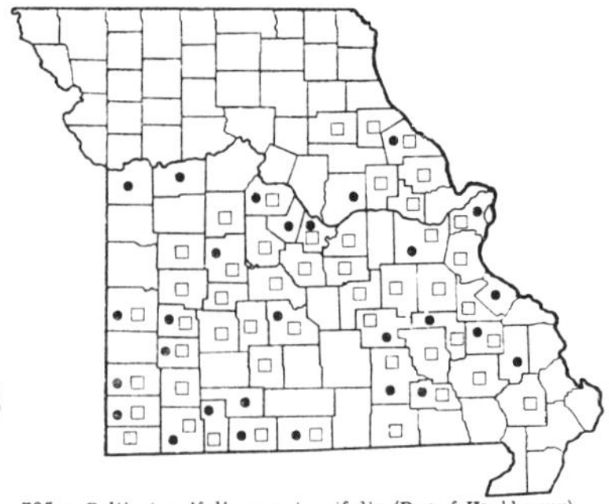

785 • Celtis tenuifolia var. tenuifolia (Dwarf Hackberry)
785 □ Celtis tenuifolia var. georgiana

1. **Celtis occidentalis** L. Hackberry Map 784
Called Nettle Tree in some sections of Missouri. Flowers April–May.

Occurs in low woods along streams and upland slopes. Throughout the state.

The variations of this species are ill-defined and as yet not satisfactorily worked out. Fernald's interpretation of the names (Rh. 50: 155–59. 1948) has been adopted in the present treatment. Intensive future work will be necessary to clarify further the present unsolved problems.

a. Leaves rather thick (coriaceous), distinctly rough (scabrous) to the touch; fruit orange-red to dark brown, more or less spherical (globe-shaped), on stalks 3–15 mm. long . . 1a. C. OCCIDENTALIS var. OCCIDENTALIS

a. Leaves rather thin or membranaceous, smooth or nearly so to the touch; fruit blackish-purple or blackish or dark brown, ellipsoid-ovoid to ellipsoid-obovoid, on stalks mainly 8–35 mm. long . . . . . . . . . . . . . *b*

b. Sides of leaves strongly unequal, one side of the fully grown main leaves much broader and more strongly rounded, curved, or heart-shaped at base; leaves usually more than half as broad as long; the larger leaves of fruiting branches chiefly 3–9 cm. broad at the base . 1b. C. OCCIDENTALIS var. PUMILA

b. Sides of leaves nearly equal, the side which is slightly broader tapers gradually to a more wedge-shaped, straighter-sided, or only slightly rounded base; leaves usually less than half as broad as long; the larger leaves of fruiting branches chiefly 1.5–5 cm. broad at the base . . 1c. C. OCCIDENTALIS var. CANINA

1a. **Celtis occidentalis** var. **occidentalis** Map 784

*Celtis occidentalis* var. *crassifolia* (Lam.) Gray [BB, P & S, Steyerm., Deam]

*Celtis occidentalis* L. [G]

Throughout Missouri and very common.

Ranges from Massachusetts to Idaho, south to Florida, Tennessee, Arkansas, and Oklahoma.

1b. **Celtis occidentalis** var. **pumila** (Pursh) Gray Map 784

*Celtis occidentalis* of auth. [P & S, Steyerm., Deam], not L.

*Celtis pumila* Pursh, not auth.

Throughout Missouri, apparently less common than the two other variations.

Ranges from Quebec to North Dakota, south to Georgia, Alabama, Arkansas, and Oklahoma.

1c. **Celtis occidentalis** var. **canina** (Raf.) Sarg. Map 784

Throughout Missouri, apparently the most common variation in the state.

Ranges from Quebec to Utah, south to Georgia, Tennessee, Missouri, and Oklahoma.

The foliage of *C. occidentalis* turns pale yellow in autumn. Some authorities consider this species the most important one of the genus, so far as investigated, in causing some cases of hay fever. On account of being commonly infected by a 'witches'-broom,' this species is not recommended as a street tree. The pale yellow or yellowish-white wood has good bending qualities, for which it is sometimes used for barrel hoops. It is also used for making into cabinets, cheap furniture, fence posts, boxes, and crates.

A powder obtained from the dried mashed-up stones of the fruit is reported to be used as an article of seasoning by Indians of the Dakotas. Mature frost-ripened fruits are edible, their thin pulp faintly suggesting the taste of date fruits, although much less sweet.

2. **Celtis tenuifolia** Nutt. Dwarf Hackberry Map 785

Flowers April–May.

Missouri material is represented by the following two variations:

a. Upper leaf-surface smooth; lower leaf-surface glabrous or sparsely hairy; leaves rather thin; leaf-stalks (petioles), fruiting-stalks (pedicels), and branchlets glabrous or sparsely hairy . 2a. C. TENUIFOLIA var. TENUIFOLIA

a. Upper leaf-surface harshly scabrous (rough); lower leaf-surface densely or more conspicuously hairy; leaves thick and rather stiff; leaf-stalks, fruiting-stalks, and branchlets more densely hairy . 2b. C. TENUIFOLIA var. GEORGIANA

2a. **Celtis tenuifolia** var. **tenuifolia** Map 785

*Celtis tenuifolia* Nutt. [G, BB in part]

*Celtis pumila* of auth. [P & S, Steyerm., Deam], not Pursh

*Celtis pumila* var. *Deamii* Sarg. [P & S, Steyerm., Deam]

Occurs in rocky open woods, borders of glades and along bluffs. Southern and central Missouri north to Pike, Pettis, and Jackson counties; less common than the var. *georgiana* in Missouri.

Ranges from Florida to Louisiana, north to Pennsylvania, Virginia, Indiana, and Missouri.

2b. **Celtis tenuifolia** var. **georgiana** (Small) Fern. & Schub. Map 785

*Celtis pumila* var. *georgiana* (Small) Sarg. [P & S, Steyerm.]

Occurs in rocky open woods, glades, and on bluffs. Common throughout the Ozark section north to Ralls, Monroe, Cooper, Lafayette, and Vernon counties; absent from the lowlands of southeastern Missouri.

Ranges from Georgia to Louisiana and Oklahoma, north to Virginia, Kentucky, Indiana, Missouri, and Kansas.

Specimens from Iron and St. Francois counties, previously treated by Palmer and Steyermark as *C. pumila* var. *Deamii* in their *Annotated Catalogue* are referred in the present work to *C. tenuifolia* var. *tenuifolia* and var. *georgiana*. Deam (*Trees of Indiana*, 2nd revised ed. 160. 1932, and *Fl. Ind.* 394. 1940) considered this variety as synonymous with typical *C. pumila* = *C. tenuifolia* var. *tenuifolia*. While *Palmer 19510* and *18119* from Iron Co., both labeled by Palmer *C. pumila* var. *Deamii* in the Gray Herbarium material, may be referred to *C. tenuifolia* var. *tenuifolia* because of the nearly glabrous, thinner leaves which are smooth above, nevertheless *Palmer 18051* from St. Francois Co. and *Palmer 18095* from Iron Co., also in the Gray Herbarium and identified there as *C. pumila* var. *Deamii*, have the petioles and branchlets more pubescent and may be referred rather to *C. tenuifolia* var. *georgiana*.

This species varies from a dwarf stunted shrub a few feet tall to a tree up to 8 meters. It usually is found in the most exposed rocky sections on top of limestone bluffs and escarpments or bordering limestone glades, where it is frequently associated with *Bumelia lanuginosa* var. *oblongifolia*, *Viburnum rufidulum*, *Diospyros virginiana*, *Ulmus alata*, and *Rhus aromatica* var. *serotina*, forming with these one of the earlier stages in forest succession over a limestone glade area.

The fruit varies from orange to light cherry red in late summer to a darker cherry-red in fall.

3. **Celtis laevigata** Willd. Sugarberry Map 786

Also known as Southern Hackberry

Flowers April–May.

Missouri material is quite variable and represented by the following variations:

a. Leaf-stalks usually hairy; leaves thick (coriaceous), without teeth on the margins; upper leaf-surface rough (scabrous) to the touch . 3c. C. LAEVIGATA var. TEXANA

a. Leaf-stalks usually glabrous (without hairs); leaves thin and membranaceous or nearly so, with or without teeth on the margins; upper leaf-surface smooth or sparsely hairy . . . *b*

b. Margins of leaves without teeth . . . 3a. C. LAEVIGATA var. LAEVIGATA

b. Margins of leaves with numerous teeth . 3b. C. LAEVIGATA var. SMALLII

3a. **Celtis laevigata** var. **laevigata** Map 786

*Celtis laevigata* Willd. [G, P & S, Steyerm.]

*Celtis occidentalis* var. *integrifolia* Nutt.

Occurs in low wet or alluvial soils along streams and flood plains, rarely on higher rocky ground in southern and central Missouri mostly outside the Ozark region, north to Marion County on the east and Andrew County on the west. At the Andrew County station (*Steyermark 84519*), it occupies the upper exposed portion of rocky west-facing limestone slopes bordering the valley of One Hundred and Two River, T59N, R35W, SW $\frac{1}{4}$ sect. 7, SE $\frac{1}{4}$ sect. 12, 3 mi. east of Savannah.

Ranges from Florida to Texas and Mexico, north to Virginia, Kentucky, Indiana, Illinois, Missouri, and Oklahoma.

3b. **Celtis laevigata** var. **Smallii** (Beadle) Sarg. Map 786

Occurs in low wet or alluvial soils along streams

and flood plains. Known only from Marion, Pemiscot, and Jackson counties.

Ranges from Florida to Louisiana, north to Virginia, Kentucky, Illinois, and Missouri.

3c. **Celtis laevigata** var. **texana** Sarg. Map 786

*Celtis reticulata* [of Steyerm.], not Torr.

Occurs in rocky open woods, glades, and bluffs. Common throughout the Ozark region in southern and central Missouri north to Pike, Montgomery, Callaway, Boone, Pettis, and Jackson counties.

Ranges from Illinois, Missouri, and Kansas, southwest to Texas and New Mexico.

An especially thick- and reticulate-leaved specimen from St. Clair Co. (dominant on limestone glade on headlands of Osage River, 3½ mi. north of Iconium, T39N, R24W, sect. 10, August 8, 1937, *Steyermark 24397*) was originally identified as *C. reticulata* Torr. and accepted by Palmer and Steyermark (Ann. Mo. Bot. Gard. 25: 780. 1938) and by Steyermark (*Spring Flora*, p. 148. 1940). Despite the unusually rugose-reticulate leaves, however, a re-examination of the specimen would seem to place it better in *C. laevigata* var. *texana*.

These variations grade into one another, and are not distinct. Trees with the scabrous leaves of var. *texana* but the thinner leaves of var. *laevigata*, and other similar non-correlations occur. Moreover, there is frequently great difficulty in distinguishing the thick scabrous, more or less entire leaves of *C. laevigata* var. *texana* from similar leaves of *C. tenuifolia* var. *georgiana*, especially since the two varieties may occupy the same type of rocky exposed habitat. Much field work remains to be done to clarify the problems of intergradation and variation not only within *C. laevigata*, but between the var. *texana* and *C. tenuifolia* and var. *georgiana*. As a lowland tree with thin, relatively narrow, and entire leaves, *C. laevigata* var. *laevigata* is fairly constant. As it occupies drier and more exposed situations as var. *texana* it becomes less clearly marked. Deam (*Fl. Ind.* pp. 393–94. 1940) believes that the thickening of the leaves in *C. leavigata* may be due to growing in an open situation.

Birds eat the fruit of this and other species of *Celtis*, and in autumn the mature fruits secrete a sweet sticky substance attractive to mealy bugs. The wood is used for similar purposes as that of *C. occidentalis*. Since this species is quite resistant to the witches'-broom infection, it is recommended as a street tree in the southern states, where it may be seen in wide use.

## MORACEAE (Mulberry Family)

a. Leaf compound, divided into 3–7 separate, narrow, toothed leaflets; erect herb . . . . . . 4. CANNABIS
a. Leaf simple, unlobed or deeply lobed, but not divided into separate leaflets; tree, shrub, or vine . . . *b*
  b. No teeth on leaf-margin; stout thorns 1–2 cm. long usually present on branches . . . . 3. MACLURA
  b. Teeth present on leaf-margins; no thorns on branches . . . . . . . . . . . . . . *c*
    c. Stems vining, sprawling, or spreading along the ground; plants with watery juice; rough hairs on stem and leaf-stalks (petioles) pointed downward (with reference to tip of stem). 5. HUMULUS
    c. Trees or shrubs with milky juice; branches and leaf-stalks when hairy without rough downward-pointing hairs . . . . . . . . . . . . . . . . . . . . . . . . *d*
      d. Midrib of upper leaf-surface with noticeably longer hairs as compared with other hairs; lower surface of leaf soft-hairy with dense long hairs; leaf-stalks (petioles) 3–10 cm. long, they and young branches with long hairs; 2–3 scales to each bud; female catkins and fruit ball-shaped (globose); bark of tree smooth . . . . . . . . . . . 2. BROUSSONETIA
      d. Midrib of upper leaf-surface without noticeably longer hairs as compared with other hairs; lower surface of leaf smooth or with short hairs; leaf-stalks (petioles) 1.5–3 cm. long, they and young branches either without hairs or with short hairs closely pressed to the stem; 3–6 scales to each bud; female catkins and fruit short-cylindric; bark of tree scaly or furrowed . . . . . . . . . . . . . . . . . . . . . . 1. MORUS

---

Plate no. 139. **1.** Celtis occidentalis, × 2/7; Details from Small, The New York Botanical Garden. **2.** Celtis tenuifolia var. tenuifolia, × 2/7; Details from Small, The New York Botanical Garden. **3.** Celtis laevigata, × 2/7; *a.* Flowering branch. **4.** Morus rubra, × 2/7; *a.* Branch with male inflorescences; *b.* Fruit; *c.* Branch with female inflorescence; Details from Small, The New York Botanical Garden. **5.** Broussonetia papyrifera, × 2/7; *a.* Branch with male inflorescences; Details from Small, The New York Botanical Garden. **6.** Morus alba, × 2/7; *a*, *b*. Two types of leaves; Details from Small, The New York Botanical Garden.

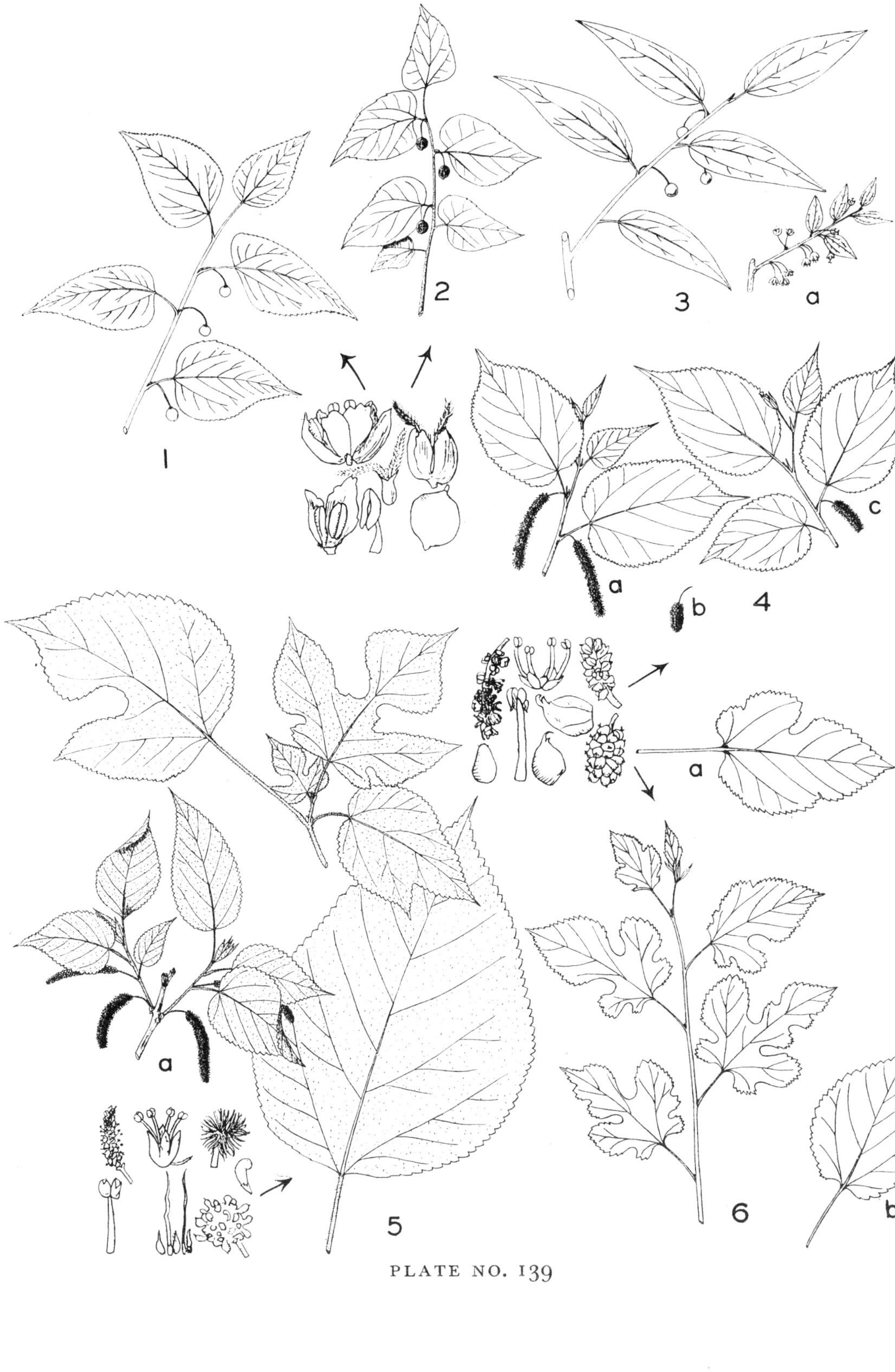


PLATE NO. 139

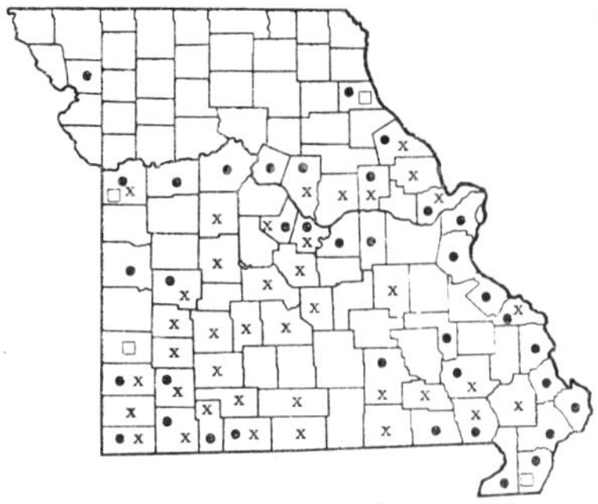

786 • Celtis laevigata var. laevigata (Sugarberry)
786 □ Celtis laevigata var. Smallii
786 x Celtis laevigata var. texana

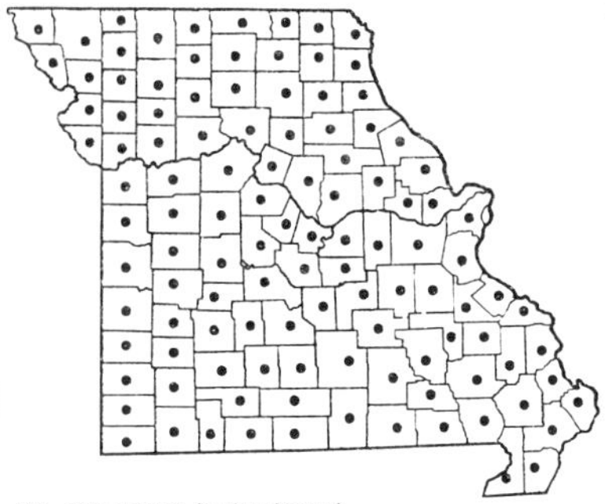

787 Morus rubra (Red Mulberry)

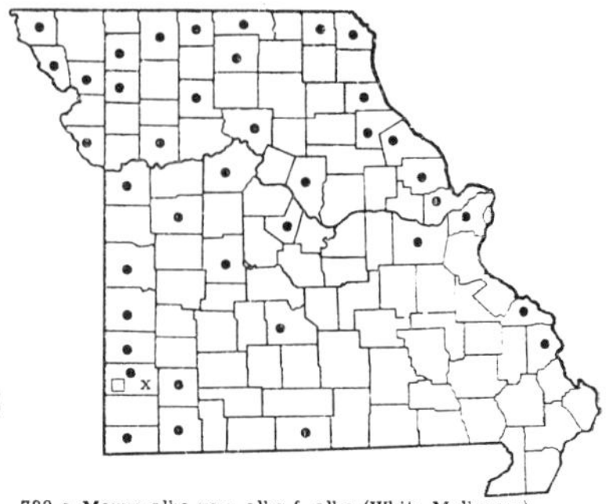

788 • Morus alba var. alba f. alba (White Mulberry)
788 x Morus alba var. alba f. Skeletoniana
788 □ Morus alba var. alba f. macrophylla

## 1. **Morus** L. Mulberry

Occasional cases of hay fever have been traced to species of this genus which shed large quantities of pollen.

Lower surface of leaf hairy between the main nerves . . . . . . . . . . . . . 1. M. RUBRA
Lower surface of leaf without hairs, or, if hairs are present, they are along midrib or main nerves or as small tufts in the axils of main nerves . . . . . . . . . . . . . . . . . . . . 2. M. ALBA

1. **Morus rubra** L. Red Mulberry Map 787
Flowers April–May.

Occurs in rich woods, or open rocky places, either in lowland or upland, along roadsides, fields, and pastures. Known from every county in Missouri.

Ranges from Florida to Texas, north to Vermont, New York, Ontario, Minnesota, and South Dakota.

The foliage turns pale or dull green to greenish yellow in autumn. It is quite variable, ranging from unlobed to deeply lobed, and from rough-hairy to glabrous on the upper surface. Leaves with glabrous upper surfaces are often mistaken for those of *M. alba*.

The sweet reddish to dark purple fruits attract many birds, especially robins and cedar waxwings, and squirrels. They are made into jellies, jams, pies, as well as served for refreshing drinks, or are frequently eaten just as fresh fruit.

The wood is light in weight and quite durable, used for fence posts, barrels, boat building, and farm tools. As in various tropical Central and South American countries, where the Indians employed the bark of certain genera of the Mulbery family for cloth, so some of the Indian women of Mississippi and Louisiana at one time used the bark of this tree as an outer robe.

2. **Morus alba** L. White Mulberry Map 788
Flowers April–May.

Escaped from cultivation and commonly found in low wet ground along streams and stream banks, and along fence rows and pastures. Scattered throughout Missouri, most frequently found naturalized in northern and central Missouri, with few records known from the Ozark region.

Native of eastern Asia; naturalized in various parts of the United States from New York southward.

The species is extremely variable and many forms are in cultivation. Only the following are known to have escaped from cultivation in Missouri:

a. Leaves deeply and rather regularly divided into narrow long-pointed lobes . 2c. M. ALBA var. ALBA f. MACROPHYLLA
a. Leaves unlobed or lobed, but not regularly cut into narrow long-pointed lobes . . . *b*
b. Leaves chiefly 15–22 cm. long, not lobed 2b. M. ALBA var. ALBA f. SKELETONIANA
b. Leaves mainly 6–15 cm. (up to 18) long, lobed or unlobed . . . . 2a. M. ALBA var. ALBA f. ALBA

2a. **Morus alba** var. **alba** f. **alba** Map 788

This is the common form found naturalized in Missouri.

2b. **Morus alba** var. **alba** f. **skeletoniana** (Schneider) Rehder Map 788
*Morus alba* var. *laciniata* Beissn. [Bailey]

Known only from Jasper County, southwestern Missouri (open ground along roadway, Webb City, October 4, 1954, *Palmer 59051*; October 18, 1954, *Palmer 59163*).

789 Broussonetia papyrifera (Paper Mulberry)

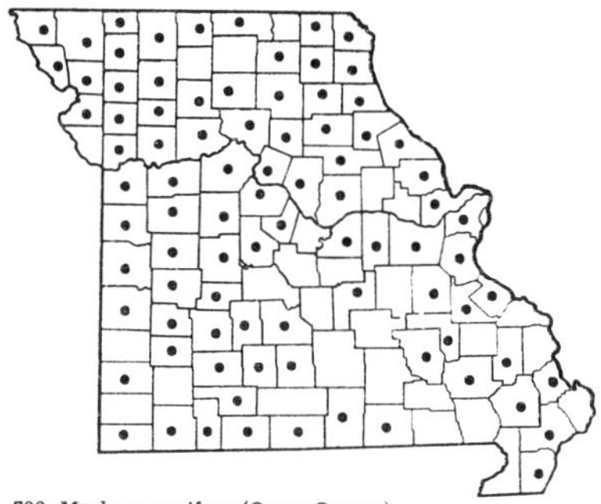

790 Maclura pomifera (Osage Orange)

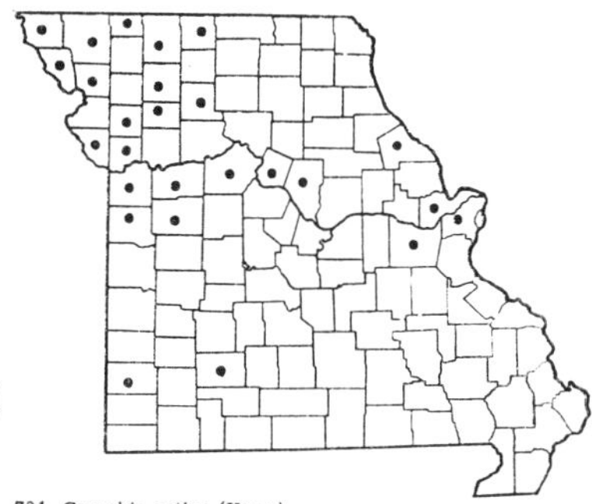

791 Cannabis sativa (Hemp)

2c. **Morus alba** var. **alba** f. **macrophylla** (Loud.) Schneider — Map 788

*Morus alba* f. *macrophylla* (Loud.) Schneider

*Morus alba* var. *macrophylla* Loddiges ex Loud.

Known only from Lawrence County, southwestern Missouri (along rocky bluffs of Clear Creek, Pierce City, August 22, 1950, *Palmer 50832*).

Some of the leaves on this specimen are only 13–15 cm. long.

Found commonly cultivated, but without records of their having become naturalized, are *M. alba* var. *pendula* Dipp., the Weeping Mulberry, the Texas form of which originated at the Texas Nursery of Carthage, Missouri, in 1883, with long hanging branches and usually small lobed leaves; and *M. alba* var. *tatarica* (L.) Loudon, or Russian Mulberry, a small, bushy-headed tree with small usually lobed leaves 4–8 cm. long and a small red or purple, rarely whitish, fruit about 1 cm. long. In *M. alba* var. *alba* f. *alba* the fruit is somewhat longer, 1–2.5 cm. long and mostly whitish or pinkish.

The fruit is an attractive morsel to birds and people alike. *Morus alba* is the species upon whose leaves the caterpillar of the silkworm moth feeds, and, upon whose production the silk of commerce depends. The young leafy shoots of this species are used in Asia as a cooked vegetable.

## 2. **Broussonetia** L'Her. Paper Mulberry

**Broussonetia papyrifera** (L.) Vent.

Paper Mulberry — Map 789

Flowers April–May.

This is commonly planted as a shade tree in the lowland section of southeastern Missouri and occasionally is found naturalized there in Stoddard, Butler, and Dunklin counties, and also in St. Louis and Johnson counties. Probably more widely naturalized in the southern part of the state.

Native of Asia; introduced in the United States where naturalized from New York to Missouri and southward.

The irregularly lobed leaves greatly resemble those of *Morus*, but may be distinguished by the longer softer pubescence on the lower surface and the longer hairs of the midrib of the upper surface. In the Far East the bark has been used for centuries in paper-making, and the Pacific island natives make a cloth from the inner bark, known as Tapa cloth. A kind of glue is made from the sap by the Polynesian islanders.

The pollen of this tree is known to be responsible for some hay fever cases.

## 3. **Maclura** Nutt. Osage Orange

**Maclura pomifera** (Raf.) Schneid.

Osage Orange — Map 790

Known also as Bois-d'Arc or Bodark, Bowwood, and Hedge Apple.

Flowers May–June.

Occurs commonly in low alluvial woods in valleys, along streams, borders of woods, pastures, fence rows, and in thickets. Throughout Missouri, doubtless in every county.

The tree is native to Arkansas, Oklahoma, and Texas, and perhaps Louisiana, but is believed to be introduced and naturalized in Missouri and elsewhere north to southern New England, New York, Ohio, Indiana, Illinois, and Iowa.

This tree was formerly widely planted in Missouri and other central states as a living fence, for which it is admirably suited by virtue of its stout thorns, and as a windbreak, but is now being replaced in many sections by living fences of *Rosa multiflora*.

The wood is exceptionally hard and the most durable of any of the woods used for fence posts at which it excels. Formerly, it supplied wood for wheel-stock and railroad ties. The bark was at one time used for tanning leather, and the orange roots and inner bark were used as a yellow dye. The Indian tribes of Arkansas, Oklahoma, and adjacent states at one time used the strong wood for bows and tomahawks.

This tree, called by the French Bois d'Arc, or Bowwood, is the one responsible for the name Ozark applied to the Missouri and Arkansas hills, the land of the Bois d'Arc or Bodark or Bowwood. The foliage remains green long after most trees have turned color, but eventually in late autumn turns an orange-lemon, light yellow, or greenish-yellow color. The orange-shaped fruits, which start falling in October, are at first green, turning yellow. They are sometimes eaten by squirrels, opossums, and other animals. The fruit is reputed to have insect-repellent properties, for which it is recommended to place a few in the basement and in closets, or in rooms infested with roaches or waterbugs.

Occasional cases of hay fever are reported as caused by the pollen of this wind-pollinated tree, while some cases of dermatitis are occasionally caused by contact with the milky sap contained in the leaves, branches, and fruits.

## 4. **Cannabis** L. Hemp, Marijuana

**Cannabis sativa** L. Hemp, Marijuana Map 791

Flowers June–October.

Occurs most commonly in rich fertile alluvial ground of river flood plain and valley bottoms frequently near streams, also in open and waste ground. Common in the northwest quarter of the state, where frequent in the Missouri River bottoms, and scattered elsewhere in central and western Missouri, south to St. Louis, Franklin, Boone, Howard, Saline, and Cass counties, locally south in Greene and Jasper counties; absent from the Ozark and southeastern lowland sections. Large sections of vacant lots in St. Louis, Kansas City, and other large towns, are sometimes seen populated with this plant, especially in low ground.

Native of Asia; introduced into North America, where naturalized from Quebec to British Columbia and southward.

This plant has the male and female flowers separated on distinct plants. These plants, when found, usually occur in large stands, with an aspect similar to that of Giant Ragweed or Horseweed (*Ambrosia trifida*). In fact, both are annuals, often 6 or more feet tall, and occupy similar habitats along river flood plains. Both also cause hay fever, *Cannabis* pollen being considered an important cause in those regions, such as Nebraska, where it is abundant and constitutes 17 per cent of the pollen contained in the air in late summer and fall.

The strong fiber of the plants is made into many kinds of paper, one of its early uses. At present most of the commercial use of the fiber is for rope, sacks, twine, cloth, caulking material, packing, and cordage. Hemp seeds are often found in bird seed mixtures for wild bird food, and in parts of Europe the parched seeds are sometimes eaten by native folk in cakes or fried. The upper leaves and flower parts and seeds from the female plant, when dried, are the source of a drug originally employed in medicine for relieving pains and treating nervous ailments. This same drug became used as a narcotic in liquid form or smoked as *hashish*. It entered the cigarette trade as *marijuana*, and because the smoking of it in excess produces serious effects on its victims, it is prohibited from use. In the United States it can be grown under special government permit only for certain uses. The seeds yield a drying oil used for soaps, paints, and varnishes, and as an illuminating oil.

---

Plate no. 140. **1.** Maclura pomifera, × 2/7; *a.* Male flowering branch; *b.* Female flowering branch; *c.* Leafy branch; *d.* Fruit with leaves; Details from Small, The New York Botanical Garden. **2.** Cannabis sativa; *a.* Leaf, × 2/7; *b.* Upper portion of plant, × 2/7; *c.* Male flower, × 1 1/7; *d.* Fruit, × 3 1/7; Details from Small, The New York Botanical Garden. **3.** Humulus japonica, leaf, × 2/7; *a.* Bract, × 4/7; After Gleason, details from Small, The New York Botanical Garden. **4.** Humulus Lupulus; *a.* Pistillate spike, × 2/7; *b.* Bract, × 4/7; After Gleason, details from Small, The New York Botanical Garden. **5.** Urtica urens, × 2/7. **6.** Urtica chamaedryoides, × 2/7; Details from Small, The New York Botanical Garden. **7.** Urtica dioica var. procera, × 2/7.

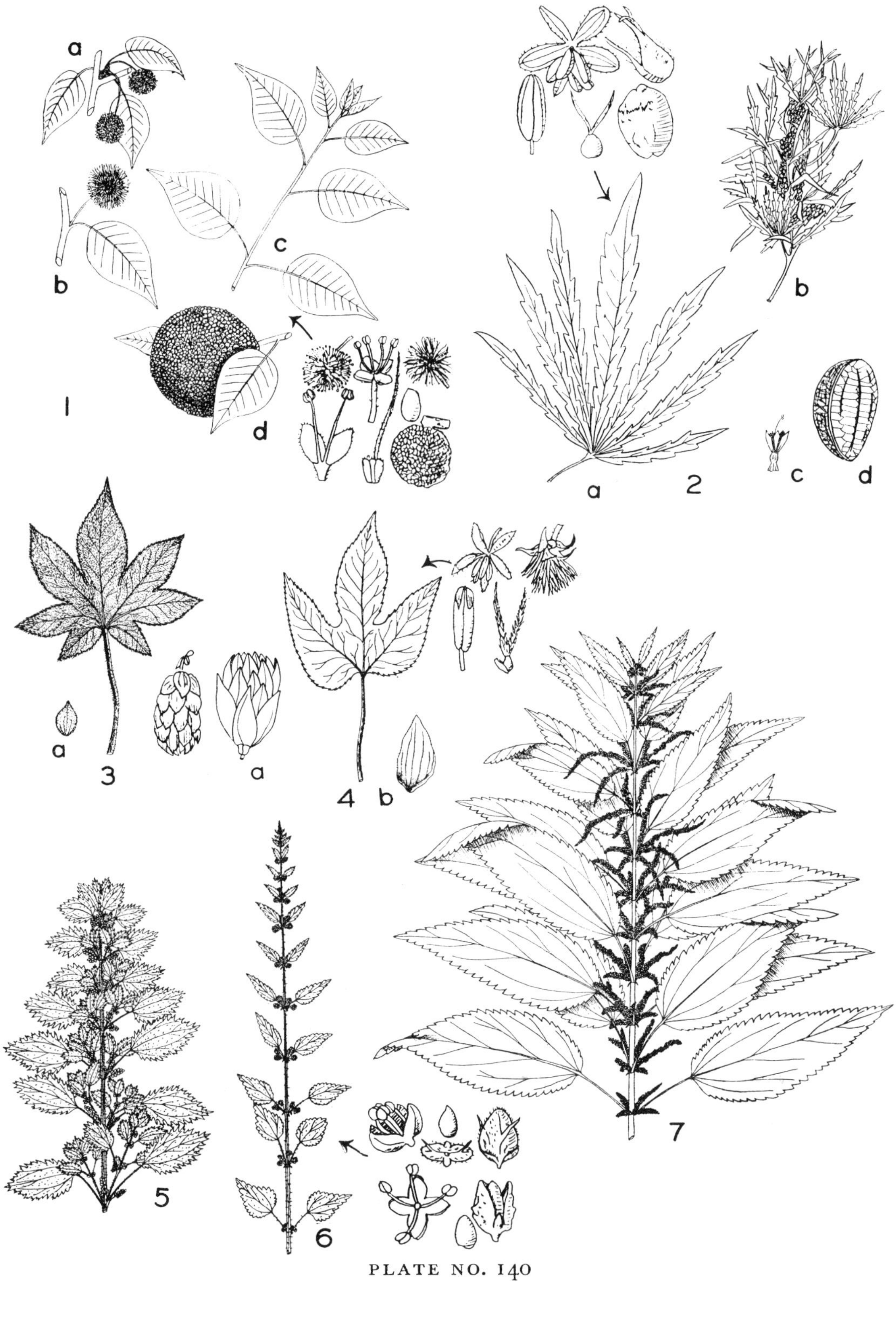


PLATE NO. 140

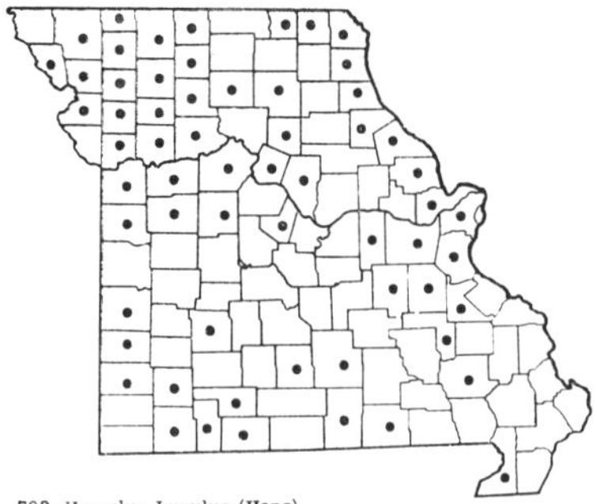

792 Humulus Lupulus (Hops)

793 Humulus japonicus (Japanese Hops)

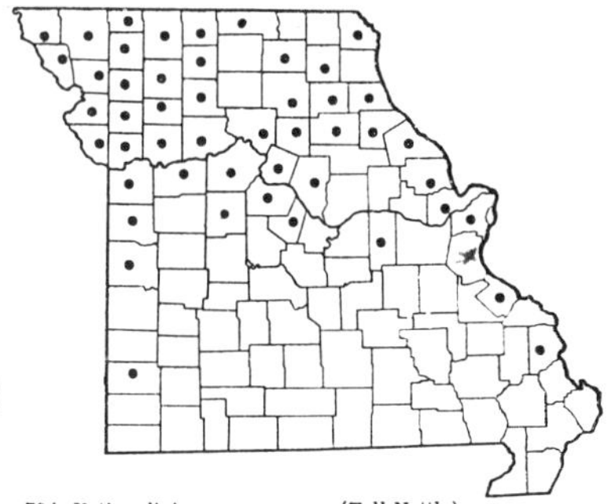

794 Urtica dioica var. procera (Tall Nettle)

## 5. **Humulus** L. Hops, Hop

The wind-blown pollen from the male plants of species of *Humulus* is sometimes thought to be responsible or cases of hay fever.

The large main leaves with usually 3 lobes (but sometimes all the leaves on the young stems or on the upper part of stem without any lobes); leaf-stalk (petiole) mainly shorter than or equaling the leaf-blade; bracts of the female flowers not hairy on the margins, not green, and with little glands at the base; upper bracts of female inflorescence blunt; bracts of female inflorescence in fruit becoming broad, rounded and covering the seed-like fruits; anthers of the male flowers without glands . . . 1. H. LUPULUS

The large main leaves with usually 5–7 lobes; leaf-stalk (petiole) chiefly longer than the leaf-blade; bracts of the female flowers conspicuously long-hairy on the margins (ciliate), green, and usually lacking little glands at the base; all bracts of the female inflorescence conspicuously long-pointed (acuminate); bracts of female inflorescence in fruit narrow, long-pointed, and narrower than the seed-like fruits; anthers of the male flowers with glands . . . . . . . . . . . . . . . . . 2. H. JAPONICUS

1. **Humulus Lupulus** L. Hops Map 792
*Humulus americanus* Nutt. [Bailey, Deam]
Flowers July–August.

Occurs in fertile, open, and waste ground, fence-rows, shaded thickets and wooded slopes on bluffs or at the base of bluffs, and along railroads. Found throughout Missouri in most counties, but not recorded from most of the lowland counties of southeastern Missouri.

Widely distributed from Nova Scotia to Manitoba, Montana, and California, south to North Carolina, Ohio, Kentucky, Illinois, Missouri, Arkansas, Oklahoma, New Mexico, and Arizona. Also in Europe and Asia.

The American plant has been considered by Bailey and some authors as a distinct species, *H. americanus* Nutt., with a western variety, but the characters used for separating it from *H. Lupulus*, mainly referring to the apex of the leaf lobes, the fineness of their toothing, and the relative amount of glands present on the lower surface of the leaves, do not appear to be reliable criteria.

This species is widely known for its use in the brewing industry, which employs the aromatic glandular female inflorescences to give beer its bitter taste and to prevent decay from bacterial processes during stages of formation. Hops also play a role medicinally, being used to relieve pain and induce sleep. The discarded remains of the hops are sometimes used as a mulch for rose-beds and shrubs. The very young shoots can be cooked as a vegetable, and are claimed to have unusual flavor.

In some rural communities, before the common use of commercial yeast, hops were used for breadmaking: a handful of hops was boiled in a pint of water. To this was added potatoes which were boiled until tender, then mashed and thickened with corn meal. The whole mass was then flattened and cut into squares to dry. Such squares were stored as yeast cakes and were ready in 3 or 4 days for use in breadmaking.

2. **Humulus japonicus** Sieb. & Zucc.
Japanese Hops Map 793
*Humulus scandens* (Lour.) Merr. [Bailey]
Flowers July–October.

Occurs in open ground, along railroads, and waste places. Scattered and rare, as an escape from cultivation; known only in St. Louis, Schuyler, Jackson, and Jasper counties.

Native of eastern Asia; introduced into the United States, where naturalized from New England to Michigan, south to Virginia and Missouri.

## URTICACEAE (Nettle Family)

a. No teeth on leaf-margins . . . . . . . . . . . . . . . . . . . . . 5. PARIETARIA
a. Teeth present on leaf-margins . . . . . . . . . . . . . . . . . . . . . b
b. Leaves alternate . . . . . . . . . . . . . . . . . . . . . 2. LAPORTEA
b. Leaves opposite . . . . . . . . . . . . . . . . . . . . . . . . c
c. Plant completely without hairs . . . . . . . . . . . . . . . . . . . 4. PILEA
c. Plant with hairs on some part of stem, leaf, flower, or fruit . . . . . . . . . . . . d
d. Plants with stinging hairs; female flowers with 4 separate sepals; inflorescence of loosely interrupted unbranched clusters shorter than the leaf-stalks (petioles), or, if longer than the leaf-stalks, the flower-clusters are in forked, branched, or loosely-flowered inflorescences with broad intervals of space between the flower-clusters; stigma of female flowers globe-shaped, without a definite style . . . . . . . . . . . . . . . . . . 1. URTICA
d. Plants lacking stinging hairs; female flowers with the sepals united into a tubular or cup-shaped calyx; inflorescence a simple elongated spike of densely crowded clusters, the whole inflorescence longer than the leaf-stalks (petiole); style of female flowers long and thread-like with the stigmatic portion along one side . . . . . . . . . . . . 3. BOEHMERIA

### 1. **Urtica** L. Nettle

a. Stem without spreading hairs; main leaves of the stem 7.5–15 cm. long, more than twice as long as broad; tall perennials with tough stems, mainly 9–21 dm. tall; stipules erect, 5–15 mm. long; flower-clusters branched, usually longer than the leaf-stalks (petioles) . . . . . 1. U. DIOICA var. PROCERA
a. Stem with noticeable spreading hairs; main leaves of the stem 1.5–7.5 cm. long, as broad as long to about twice as long as broad; annuals with soft or slender stems, 1–8 dm. tall; stipules spreading or turning down, 1–3 mm. long; flower-clusters unbranched, usually shorter than the leaf-stalks (petioles) . . . . . . . . . . . . . . . . . . . . . . . . . . . b
b. All leaves about the same size from base to tip of plant; margins of leaves saw-toothed, the teeth sharp-pointed, triangular, straight-sided . . . . . . . . . . . . . . . . . 2. U. URENS
b. Upper leaves noticeably smaller than middle or lower ones; margins of leaves wavy-toothed (crenate-serrate), the teeth blunt, the sides curved (convex) . . . . . . . . 3. U. CHAMAEDRYOIDES

1. **Urtica dioica** L. var. **procera** (Muhl.) Wedd. Tall Nettle Map 794

*Urtica procera* Muhl. [G, P & S, Steyerm.]

Flowers May–October.

Occurs commonly in low alluvial bottoms of river valley and flood plain, bordering streams, and in cultivated and waste ground. Northern and central Missouri mainly along the larger streams and south to Cape Girardeau, Gasconade, Moniteau, Pettis, Bates, and Jasper counties.

Ranges from Newfoundland to British Columbia, south to North Carolina, Louisiana, and New Mexico.

The upper portion of the stem is short hairy. Many of the Missouri specimens have the lower surface of the leaf-blades glabrous or nearly so and the pairs of teeth often range between 13–23, which, according to Fernald's treatment in the eighth edition of *Gray's Manual*, would place the Missouri material in *U. gracilis* Ait., a species of more northerly range. However, as Dr. Hermann has demonstrated (Am. Midl. Nat. 35: 773-78. 1946), Fernald's separation of the perennial species of *Urtica* in the United States east of the Rocky Mountains into four separate species is based largely on assumed and quantitative differences which are not constant and frequently cannot be correlated. The present treatment is in agreement with Dr. Hermann's conclusions.

Young shoots of this species, when boiled, are reported to serve as a cooked vegetable, somewhat similar to spinach.

2. **Urtica urens** L. Dwarf Nettle Map 795

Also called Burning Nettle.

Flowers May–September.

Known only from Clark and Jackson (waste ground, Courtney, May 13, 1930, *Bush 11707, 11707A*) counties.

Native of Europe and Asia; introduced in North America, where ranging from Newfoundland to British Columbia, south to New York, Pennsylvania, West Virginia, Illinois, Missouri, and California.

The stiff, brittle hairs of this and other species of the genus, when coming into contact with the skin, produce an itching for a short time. The young leaves of this and other species of nettles may be boiled and eaten as a vegetable.

3. **Urtica chamaedryoides** Pursh Nettle Map 796

Flowers April–September.

795 Urtica urens (Dwarf Nettle)

796 Urtica chamaedryoides (Nettle)

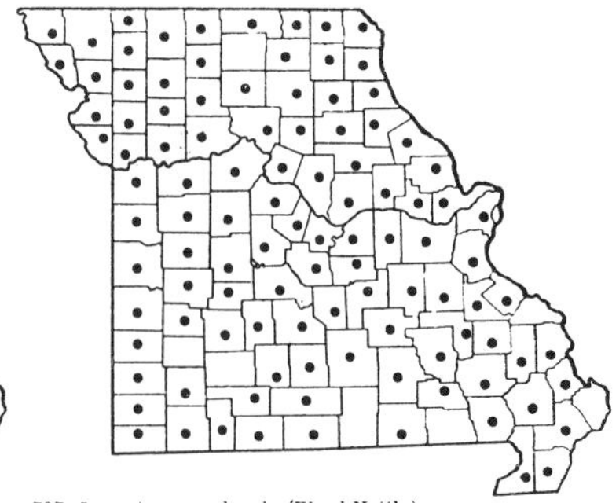

797 Laportea canadensis (Wood Nettle)

Occurs in low rich woods in alluvial ground and along slopes bordering streams, or in open low ground. Southwestern Missouri from McDonald to Taney counties, and locally east in Pemiscot County, southeastern Missouri. The stations along White River, previously occupied by this species, have now been destroyed by the impoundment of Table Rock Dam.

Ranges from Florida to Texas and Mexico, north to West Virginia, Kentucky, Missouri, and Oklahoma; locally north in Massachusetts.

## 2. **Laportea** Gaud. Wood Nettle

**Laportea canadensis** (L.) Gaud.
Wood Nettle Map 797

Flowers May 30–August.

Occurs in wet or rich low woodland, in valleys and along streams, where it usually forms extensive stands. Throughout Missouri, and doubtless in every county.

Ranges from Nova Scotia and Quebec to Manitoba, south to Florida, Alabama, Mississippi, Missouri, and Oklahoma.

This nettle is, by far, the one of greatest concern and nuisance to anyone tramping the wooded valleys of Missouri during summer and autumn. As it usually occurs in extensive stands over large areas, it is difficult to avoid or detour around, and thick pants are strongly recommended. The stinging hairs, which break off at only slight contact, produce an intense itching sensation which lasts for several minutes. It does not compare in severeness, however, with some of the really irritating hairs of tropical american Urticaceae (*Urera* spp.) or Euphorbiaceae (*Cnidoscolus*).

## 3. **Boehmeria** Jacq. False Nettle

**Boehmeria cylindrica** (L.) Sw. False Nettle
Map 798

Flowers June–October.

Occurs in swampy low spring-fed meadows, along spring branches and borders of small streams, and low wet woods.

Missouri material is represented by two variations:

Leaf-stalks (petioles) of the middle leaves usually (20–) 25–90 mm. long; tip of leaf long acuminate; leaves not folded, flat, spreading to ascending; stem mostly glabrous; fruiting achenes without purple marks, without hairs on the main body; upper surface of leaves smooth to slightly rough (scabrous); lower surface of leaves mostly without hairs or only slightly hairy . . . B. CYLINDRICA var. CYLINDRICA

Leaf-stalks (petioles) of the middle leaves 3–20 mm. (rarely 25) long; tip of leaf shorter-pointed; some of leaves more or less folded and drooping; stem generally more pubescent; fruiting achenes marked with purple, with hooked hairs on body and wings; upper surface of leaves usually rough (scabrous); lower surface of leaves usually hairy
B. CYLINDRICA var. DRUMMONDIANA

---

Plate no. 141. **1.** Laportea canadensis, $\times {}^{2}/_{7}$. **2.** Boehmeria cylindrica var. cylindrica, $\times {}^{2}/_{7}$; After Gleason, The New York Botanical Garden. **3.** Pilea pumila, $\times {}^{2}/_{7}$; After Gleason, The New York Botanical Garden. **4.** Parietaria pensylvanica var. pensylvanica, $\times {}^{2}/_{7}$. **5.** Phoradendron flavescens, $\times {}^{2}/_{7}$; *a.* Inflorescence, $\times 1{}^{1}/_{7}$; *b.* Fruiting branch, $\times {}^{2}/_{7}$. **6.** Asarum canadense, $\times {}^{2}/_{7}$; *a.* Flowers with spreading and recurved perianth segments. **7.** Comandra Richardsiana, $\times {}^{2}/_{7}$; *a.* Flower, $\times {}^{1}/_{7}$. **8.** Aristolochia Serpentaria, $\times {}^{2}/_{7}$; *a.* Flower, side and front view, $\times {}^{2}/_{7}$; Details from Small, The New York Botanical Garden.

PLATE NO. 141

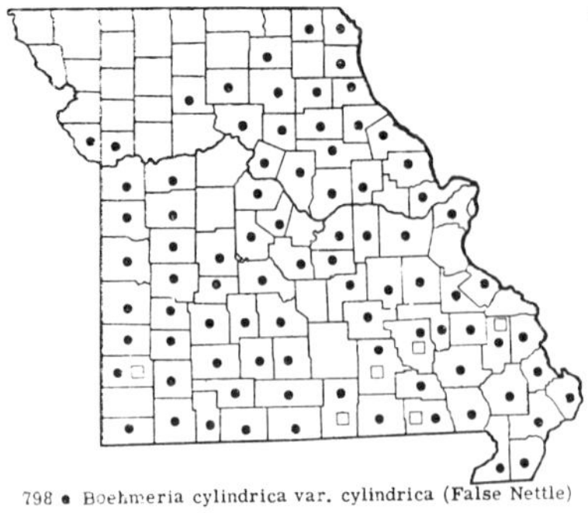

798 • Boehmeria cylindrica var. cylindrica (False Nettle)
798 □ Boehmeria cylindrica var. Drummondiana

799 Pilea pumila (Clearweed)

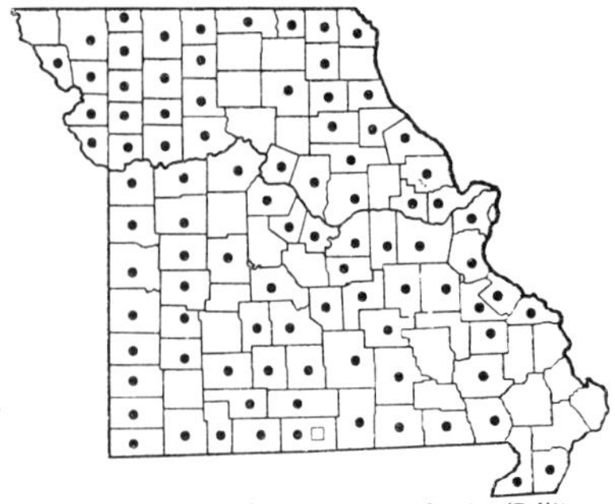

800 • Parietaria pensylvanica var. pensylvanica (Pellitory
800 □ Parietaria pensylvanica var. obtusa

**Boehmeria cylindrica** var. **cylindrica** Map 798

This is the common variation in the state, occurring throughout southern, central, and eastern Missouri north to Clark, Scotland, Adair, Linn, Livingston, Clay, and Platte counties.

Ranges from Florida to Texas, north to Maine, Quebec, Ontario, and Minnesota.

**Boehmeria cylindrica** var. **Drummondiana** Wedd. Map 798

This variety is limited to moist thickets, spring branches and swampy open meadows fed by spring branches issuing from calcareous rocks in the Ozark section of southern Missouri, north to Bollinger, Reynolds, Shannon, Howell, and Jasper counties. Probably of wider distribution within the Ozarks.

Ranges from Florida to Texas, north to Massachusetts, New York, Michigan, Illinois, Missouri, and Nebraska.

This variety is found in generally more exposed situations than var. *cylindrica* and frequents calcareous swampy habitats where the soil is quite alkaline. Deam (*Fl. Ind.* p. 401. 1940) likewise noted an affinity of the var. *Drummondiana* for alkaline soils in Indiana. While the varieties show various stages of intergradation, for the most part they appear to be satisfactorily separated by the characters indicated in the key.

## 4. **Pilea** Lindl. Clearweed

**Pilea pumila** (L.) Gray Clearweed Map 799

*Pilea pumila* var. *Deamii* (Lunell) Fern. [G]

Flowers July–October.

Occurs in rich low or alluvial ground in wooded valleys, and wet woodland along streams and ponds. Throughout Missouri and doubtless in every county.

Ranges from Quebec to Ontario and Minnesota, south to Florida, Louisiana, and Texas.

Both var. *pumila* and var. *Deamii* occur in Missouri, but the numerous intermediate specimens found between the two variations and lack of correlation and inconstancy of characters, as reported by Deam in Indiana (*Fl. Ind.* pp. 399–400. 1940), apply similarly to all the material examined in Missouri.

Another species, *P. fontana* (Lunell) Rydb., which has not yet been found in Missouri, but approaches the state's borders, is distinguished from *P. pumila* by the blackish instead of green, purplish, or straw-colored fruits with whitish margins, pericarp (outer covering) purplish within, instead of whitish or pale brown, relatively shorter petioles (generally 1/5–1/2 length of leaf-blade), and fewer teeth (generally 4–9) on each side of the leaf-blade.

The clear, translucent plants of Clearweed are often used in biology classroom experiments to demonstrate the passage of liquids upward from the lower part of the stem to the petioles and leaf-blades. The colored liquid used, often red eosin, is clearly visible through the plant's watery cells. The plant has been suggested, because of its succulence, as a possible source for greens as a cooked vegetable, but no records indicate its definite use for such purposes.

## 5. **Parietaria** L. Pellitory

**Parietaria pensylvanica** Muhl. Pellitory Map 800

Flowers May–October.

Occurs in open or rocky, often dry woodland, along bluffs, on glades, in thickets, and in moist open

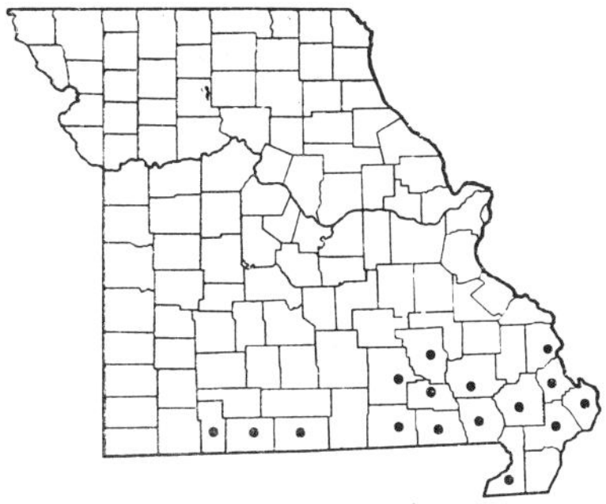

801 Phoradendron flavescens (Mistletoe)

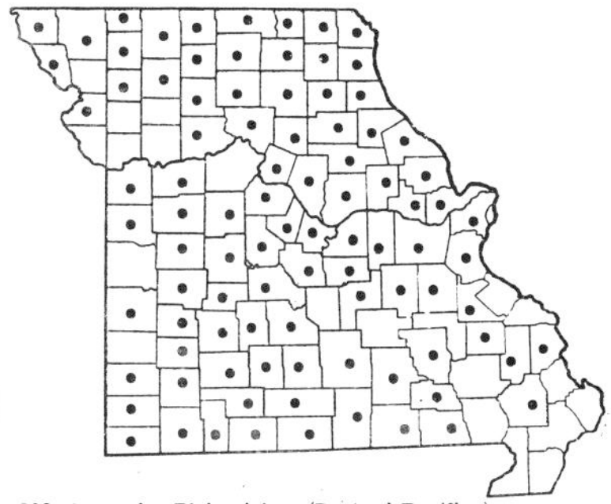

802 Comandra Richardsiana (Bastard Toadflax)

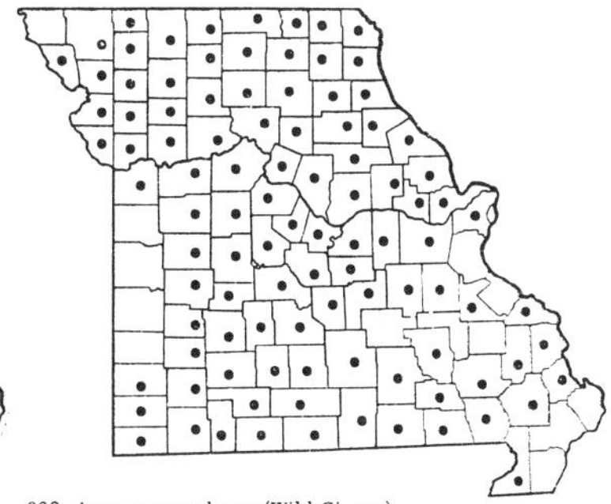

803 Asarum canadense (Wild Ginger)

ground; frequents limestone areas, but also in other types of soils. Throughout Missouri, but absent from most of the lowland counties of the extreme southeastern section.

Missouri material is represented by two variations:

Bracts which are at base of flowers of well-developed plants 2–3 times length of fully grown calyx; stem with a minute hairiness of mostly curled or hooked hairs or these mixed with a few straight and longer hairs; middle and upper leaf-blades oblong-lanceolate, long-pointed, up to 7 cm. long . . . . . . . . . . . . . a. P. PENSYLVANICA var. PENSYLVANICA

Bracts which are at base of flowers of well-developed plants about as long as or shorter than the length of fully grown calyx; stem with a denser hairiness of a few to many straight and long hairs mixed with short spreading and hooked ones; middle and upper leaf-blades broadly ovate or ovate-oblong, often more blunt at tip, up to 4 cm. long . . . b. P. PENSYLVANICA var. OBTUSA

a. **Parietaria pensylvanica** var. **pensylvanica** Map 800

This is the common variation, occurring throughout the state as indicated above.

Ranges from Florida to Texas and Mexico, north to Maine, Quebec, Ontario, Minnesota, Montana, and British Columbia.

b. **Parietaria pensylvanica** var. **obtusa** (Rydb.) Shinners Map 800

*Parietaria obtusa* Rydb. [G, P & S]

Known only from Ozark County, southern Missouri (shaded ledges, dolomite bluffs of North Fork of White River, near Tecumseh, October 8, 1927, *Palmer 32949*).

Ranges from Missouri and Rocky Mountain region, south to Texas and Mexico.

I agree with Shinners' interpretation that the intergradation between *P. obtusa* and *P. pensylvanica* makes it impossible to maintain *P. obtusa* as more than a variety (Field & Lab. 18: 42. 1950).

## Order SANTALALES

### Fam. LORANTHACEAE (Mistletoe Family)

**Phoradendron flavescens** (Pursh) Nutt. Mistletoe Map 801

*Phoradendron serotinum* (Raf.) M. C. Johnston var. *serotinum* [Shinners]

Flowers March–August; fruits September–January.

Occurs in low woods in valleys and along streams, in southern Missouri, northeast in the southeastern lowland section to Cape Girardeau and Wayne counties, north and west along the St. Francis, Black, Eleven Points, and Current Rivers to Reynolds, Shannon, and Oregon counties, and then west in the White River and North Fork section of the southwestern part of the state in Ozark, Taney, and Stone counties; in the latter three counties now nearly exterminated because of the impoundments from Norfork, Bull Shoals, and Table Rock dams.

Ranges from Florida to Texas, north to New Jersey, Pennsylvania, Ohio, Indiana, Illinois, Missouri, and Kansas.

This well-known Christmas symbol is a parasite on only certain deciduous trees in Missouri, of which Sycamore is the commonest host. Other trees in the state on which it is found are American Elm, Swamp Tupelo (*Nyssa aquatica*), Tupelo (*Nyssa sylvatica* var. *caroliniana*), and River Birch. It has been found in other states on various other kinds of trees.

The sticky whitish berries, which mature in the fall and winter, are eaten and distributed by birds, but considered poisonous if eaten by human beings.

Johnston (Southwest. Nat. 2: 45–47. 1957) considers the glabrous plants of this species as *P. serotinum* var. *serotinum*.

## Fam. SANTALACEAE (Sandalwood Family)

**Comandra Richardsiana** Fern.
Bastard Toadflax Map 802
Flowers May–July.

Occurs in dry or rocky open woods, usually in the upland on ridges and upper slopes of hills, along bluffs, in rocky glades, and prairies, most frequently in acid soils. Throughout Missouri where common; apparently absent from most of the southeastern lowland section of the state.

Ranges from Newfoundland and Labrador to Manitoba, south to Maine, Vermont, New York, Kentucky, Tennessee, Arkansas, and Kansas.

The horizontal underground rootstock which is just below the surface of the ground sends out slender suckers which become attached to and parasitic on roots of other plants. The nearly ripe fruit of this species may be eaten for its rather sweet flavor.

## Order ARISTOLOCHIALES

## Fam. ARISTOLOCHIACEAE (Birthwort Family)

Stemless plants, the leaf-stalk appearing to rise directly from the ground; flower symmetrical with long-pointed calyx-lobes; stamens 12; seeds ovoid, thick, with curved or rounded sides . . . . . . 1. ASARUM
Plants with definite stems, either herbs or vining or trailing plants; flower not symmetrical, but with a long strongly curved or bent calyx-tube with blunt or rounded lobes; stamens 6; seeds flattened, horizontally arranged . . . . . . . . . . . . . . . . . . . . 2. ARISTOLOCHIA

### 1. Asarum L. Wild Ginger

**Asarum canadense** L. Wild Ginger Map 803
*Asarum canadense* var. *canadense* [BB]
*Asarum canadense* var. *reflexum* (Bickn.) Robins. [G, BB, P & S, Steyerm.]
*Asarum canadense* var. *acuminatum* Ashe [G, P & S, Steyerm.]
*Asarum canadense* var. *ambiguum* (Bickn.) Farw. [G]
Flowers April–May.

Occurs on rich wooded slopes, in valleys, ravine bottoms, and at the base of bluffs. Throughout Missouri, except absent from a few extreme western and southeastern counties.

Ranges from Quebec and New Brunswick to Ontario and Minnesota, south to North Carolina, Alabama, Tennessee, Arkansas, and Kansas.

This species has been divided into no less than four species or varieties, based principally on the length, position, and shape of the calyx-lobes. By Fernald in the eighth edition of *Gray's Manual* these are recognized as four varieties, but by Gleason in the *New Britton & Brown Ill. Fl.* as only two varieties with the admission by Gleason that the two varieties intergrade under the name var. *ambiguum* Bickn. While it is possible to separate Missouri material into all four categories, of which var. *acuminatum*, with the longest calyx-lobes, is the most commonly encountered, the ascending, spreading, or reflexed position, and relative length of the calyx-lobes appear to vary too much with the age and maturity of the plant to justify any formal recognition of this variability.

The fresh or dried rootstock, which is quite aromatic, is sometimes used as seasoning. Cut up into short lengths, then boiled and cooked in sufficient sugar, the rootstocks serve as a substitute for preserved ginger. Some persons are reported to be affected by a dermatitis after having touched the leaves.

### 2. Aristolochia L. Birthwort

Short erect herb, growing only to 4.5 dm. tall (rarely taller); lower surface of leaves not soft-hairy. . 1. A. SERPENTARIA
Climbing up trees or trailing along the ground with long stems; lower surface of leaves soft-hairy. . 2. A. TOMENTOSA

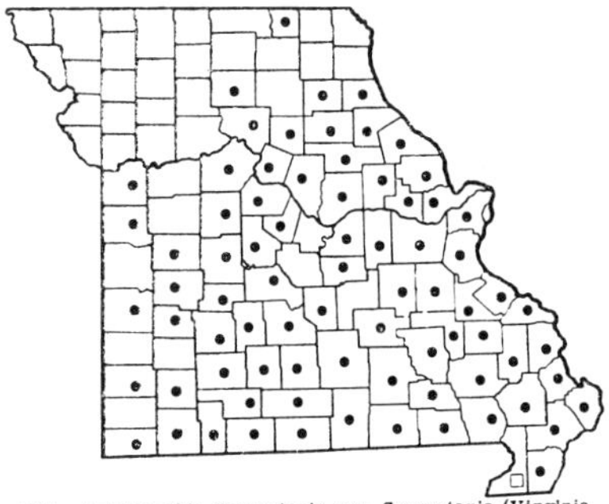

804 • Aristolochia Serpentaria var. Serpentaria (Virginia Snakeroot)
804 □ Aristolochia Serpentaria var. hastata

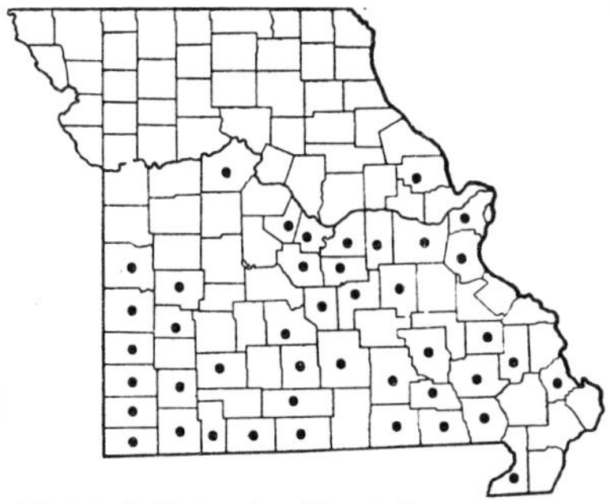

805 Aristolochia tomentosa (Pipe-vine)

806 Eriogonum longifolium (Umbrella Plant)

1. **Aristolochia Serpentaria** L.
Virginia Snakeroot Map 804

Flowers late May–July.

Occurs in rich or rocky lowland or upland woods and thickets, generally on somewhat dry or well-drained slopes and ravines.

Two variations are represented in Missouri material:

Common type with ovate or ovate-oblong leaf-blades 2–9 cm. broad . . . 1a. A. SERPENTARIA var. SERPENTARIA

Rare type with narrowly triangular to lanceolate leaf-blades 0.5–2.5 cm. broad . 1b. A. SERPENTARIA var. HASTATA

1a. **Aristolochia Serpentaria** var. **Serpentaria** Map 804

Common throughout southern and central Missouri, especially in the Ozark region, north to Marion, Shelby, Schuyler, Linn, Saline, and Jackson counties.

Ranges from Florida to Texas, north to Connecticut, New York, Ohio, Indiana, Illinois, Missouri, and Kansas.

1b. **Aristolochia Serpentaria** var. **hastata** (Nutt.) Duchartre Map 804

Known only from Dunklin and Dent counties in the southeastern one-quarter of the state.

Ranges from Florida to Texas, north to Virginia and Missouri.

Virginia Snakeroot usually occurs as scattered individual plants within an area. At one time the fragrant rootstocks, which resemble the odor of turpentine or menthol, were more frequently dug and used in medicine as a bitter tonic.

2. **Aristolochia tomentosa** Sims Pipe-vine Map 805

Also known as Dutchman's Pipe

Flowers May–June.

Occurs in low alluvial woodland along streams, often along the stream banks, rarely in open ground. Rather frequent over most of the Ozark section in southern and central Missouri, north to Lincoln, Gasconade, Osage, Cole, Moniteau, Saline, St. Clair, and Bates counties; not recorded from some of the eastern and western Ozark counties.

Ranges from Florida to Texas, north to North Carolina, Indiana, Illinois, Missouri, and Kansas.

An unusual habitat for this species occurred in Scott County (sandy prairie along north side of road H, T27N, R14E, sect. 27, 3 mi. south of Blodgett, June 6, 1957, *Steyermark 85108*), where the plant trailed along the ground in an open sandy prairie. At first, it was believed to be referable to a different species, but was not found to be distinguishable from *A. tomentosa*.

The prominently 6-ridged capsule contains numerous flat seeds arranged in tiers somewhat resembling those of *Yucca* plants.

A related species, *A. durior* Hill, native to the eastern United States, and differing from *A. tomentosa* by the nearly glabrous stems, leaves, and flowers, is cultivated, but has not been recorded from Missouri as definitely naturalized. A specimen of this species at one time examined in the former herbarium of the Hannibal-La Grange College indicated 'Marion Co.' on the label, and was collected without doubt from a cultivated plant for the college collection.

## Order **POLYGONALES**

### Fam. **POLYGONACEAE** (Buckwheat Family)

a. Vine, woody at least in lower part of stem, the stem tough and bearing tendrils at tips of branches; the mature sepals are joined to the flower- or fruit-stalk (pedicel) by a wing-like growth . . 6. BRUNNICHIA

a. No tendrils present on the plants, which are either erect or sprawling herbs or soft-stemmed, vining or sprawling plants, or, if woody on the stems, then the leaves are very narrow and hair-like; the mature sepals not joined to the flower- or fruit-stalk (pedicel) by any wing . . . . . . . . *b*

b. Lower surface of leaves covered with a dense white felt mat of closely crowded hairs; base of plant, stem, and outside of flowers covered with a dense white hairiness; leaves without stipules 1. ERIOGONUM

b. Without the above characters; lower surface of leaves not covered with a white mat of hairs; stem, and outside of flowers not covered with any dense white hairiness; leaves with stipules in the form of sheaths (ochreae) . . . . . . . . . . . . . . . . . . . . . . . *c*

c. Leaves very narrow, almost hair-like or needle-like; plants with tough woody branches. 5. POLYGONELLA

c. Leaves not almost hair-like nor needle-like, or if nearly so, then no part of plant is woody. . . *d*

d. Sepals 6; the inner 3 sepals longer and enlarged in fruit; usually a cluster of leaves present at base of plant . . . . . . . . . . . . . . . . . . . 2. RUMEX

d. Sepals usually 5 or 4, but, if 6, then the flowers not occurring in many-flowered inflorescences; the sepals nearly equal in length in fruit or the inner sepals smaller; usually no cluster of leaves present at base of plant. . . . . . . . . . . . . . . . . . *e*

e. Flowers at the top of the stem in a broad corymb-like inflorescence broader than long or as broad as long, white, fragrant; leaves triangular-heart shaped; mature achene 3-sided, longer than and protruding from the soon withering calyx . . . . 4. FAGOPYRUM

e. Flowers either in short clusters in the axils of the leaves or in longer than broad, loose or dense spikes, racemes, or panicles, white, pink, rose, or greenish; leaves narrow and linear to heart-shaped or arrow-shaped; mature achene 2-sided or 3-sided, enclosed within the surrounding calyx, or, if longer than and protruding from the persisting calyx, then the leaves narrow or the flowers in small clusters in the axils of the leaves . 3. POLYGONUM

### 1. **Eriogonum** Michx. Umbrella Plant

**Eriogonum longifolium** Nutt. Map 806

Flowers July–October.

Occurs on limestone glades and exposed edges of limestone bluffs bordering streams in southern Missouri, where known only from Oregon (1890, *Blankinship*), Ozark (west-facing bald knob along North Fork of White River, T21N, R12W, sect. 2, between Pumphrey Ford and White Ferry, 1½ mi. south of Udall, September 4, 1955, *Steyermark 79628*), and Taney (limestone glade on top of southwest-facing limestone bluffs along White River, T21N, R19W, sect. 7, 2 mi. south of Groom, 4–5 mi. south of Cedar Creek, April 30, 1938, *Steyermark 5564*, with sterile basal rosettes) counties.

Ranges from Missouri and Kansas to Oklahoma and Texas.

This is one of the rarest plants in Missouri, originally collected in 1890 in Oregon County by J. W. Blankinship without any locality being indicated by him. Basal rosettes of the plant were found by the present author in 1938 in Ozark and Taney counties and transplanted to his wildflower garden. Not until 1955 did he find flowering plants at the Ozark County station. The flowers, which are yellow-green, are in clusters subtended by a gray-white hairy cup-shaped involucre. The inflorescence is loosely corymbosely branched at the summit of the single gray-white hairy tall stem which varies from 1.5–1.65 meters (5–5½ feet) tall, and has on it only 3 or 4 scattered leaves. The plant forms a basal rosette of long, narrow leaves which are dull olive-green on the upper side and have a dense, velvety, gray-white hairiness below. This rosette persists over winter. I have had the original Missouri plant in my botanical preserve since 1938 and it produced flowers for the first time during August and September of 1958.

---

Plate no. 142. **1.** Aristolochia tomentosa, × $^2/_7$. **2.** Eriogonum longifolium; *a*. Inflorescence, × $^2/_7$; *b*. Basal leaf, × $^2/_7$; *c*. Single flower, × $2^4/_7$; Details from Small, The New York Botanical Garden. **3.** Rumex altissimus, × $^2/_7$; *a*. Fruit, × $2^4/_7$. **4.** Rumex acetosella, × $^2/_7$; Details from Small, The New York Botanical Garden. **5.** Rumex hastatulus, × $^2/_7$. **6.** Rumex mexicanus, × $^2/_7$; *a*. Fruit, × $1^3/_7$; Details from Small, The New York Botanical Garden. **7.** Rumex verticillatus, × $^2/_7$; *a*. Fruit, × $1^3/_7$.

PLATE NO. 142

## 2. **Rumex** L. Dock, Sorrel

The flowers of this genus are wind-pollinated and some of the species are known to cause hay fever.

The seeds of some species were used by some of the American Indian tribes for making into a meal for a breadstuff.

a. At least the lowest leaves arrowhead-shaped (hastate or sagittate) with basal lobes; leaves very sour to the taste; plants dioecious (male and female flowers on different plants) . . . . . . . . . . *b*

b. Female sepals not enlarged in fruit; achene protruding from the female sepals; perennial plant with slender horizontal creeping rootstocks . . . . . . . . . . . . . . 1. R. ACETOSELLA

b. Female sepals enlarged and prominently veined in fruit; achene shorter than and enclosed by mature female sepals; annual plants with a more or less vertical tap root . . . . 2. R. HASTATULUS

a. Leaves without basal lobes at base, either narrowed, rounded, more or less straight-sided, or heart-shaped at base; leaves not or scarcely sour to the taste; plants with most of flowers perfect or at least with both sexes on the same plant . . . . . . . . . . . . . . . . . . . . *c*

c. Fruiting sepals (valves) with conspicuous long bristles, or spiny, coarse, or sharp teeth . . . . . *d*

d. Plants glabrous (without hairs) . . . . . . . . . . . . . . . . 9. R. STENOPHYLLUS

d. Plants with minute hairs either on leaf-stalk, lower surface of at least the lower leaves, or on stem . . . . . . . . . . . . . . . . . . . . . . . . . . . . *e*

e. Lower leaf-blades lanceolate, mostly less than 2 cm. broad, often tapering or narrowed at the base (also rounded, truncate, or heart-shaped); bristles of fruiting sepals (valves) $1\frac{1}{2}$–2 times as long as width of mature sepal; low annual plants with hollow stems; plants of Missouri and Mississippi river muddy banks, mud flats, and sand bars . . . 13. R. MARITIMUS var. FUEGINUS

e. Lower leaf-blades broadly oblong to broadly ovate, truncate (as if cut horizontally) or mostly heart-shaped at the base; teeth of fruiting sepals (valves) less than width of mature sepal; tall perennial plants with firm stems; plants of fields, roadsides, waste ground, about dwellings throughout Missouri . . . . . . . . . . . . . . . . . . . . *f*

f. 3 grain-like tubercles (enlarged outgrowths) present, 1 appearing from center of each fruiting sepal (valve); lower leaf-blades 1.5–6 cm. broad; inflorescence interrupted, all the clusters of flowers separated from one another by relatively wide space intervals; flower or fruiting pedicels (stalks) shorter or at most as long as the flowering or fruiting sepals; rarely collected, known only from St. Louis County . . . . . . 11. R. PULCHER

f. Only 1 well-developed grain-like tubercle (enlarged outgrowth) appearing from center of one of the fruiting sepals (valves); lower leaf-blades 5–15 cm. broad; upper clusters of flowers close together, not conspicuously separated from one another; flower or fruiting pedicels (stalks) much longer than the sepals; common throughout Missouri 12. R. OBTUSIFOLIUS

c. Fruiting sepals (valves) either with smooth, wavy, or slightly bluntly toothed margins . . . . . *g*

g. Only 1 developed grain-like tubercle (enlarged outgrowth) present, appearing from center of one of the fruiting sepals (valves) or sometimes not even 1 developed . . . . . . . . *h*

h. Larger lower leaf-blades 10–18 cm. wide; grain-like tubercle much less than $\frac{1}{2}$ the length of the fruiting sepal; fruiting sepals (valves) 6–9 mm. long, 8–9 mm. broad . . . 6. R. PATIENTIA

h. Larger leaf-blades less than 10 cm. wide; largest grain-like tubercle at least $\frac{1}{2}$ the length of the fruiting sepal; fruiting sepals (valves) 4–6 mm. long, 4–6 mm. broad . . . . . . . *i*

i. Leaves conspicuously wavy or wrinkled on margins; grain-like tubercle about $\frac{2}{3}$ as wide as long . . . . . . . . . . . . . . . . . . . . . . . . 8. R. CRISPUS

i. Leaves flat and smooth, without wrinkled or wavy margins; grain-like tubercle $\frac{1}{2}$ as wide as long or narrower . . . . . . . . . . . . . . . . 5. R. ALTISSIMUS

g. 3 well-developed grain-like tubercles (enlarged outgrowths) present, 1 appearing from center of each of the fruiting sepals (valves) . . . . . . . . . . . . . . . . . . . *j*

j. Side-veins of leaves spreading almost at right angles to midrib; pedicels (flower- or fruit-stalks) without a noticeable swollen joint . . . . . . . . . . . . 7. R. ORBICULATUS

j. Side-veins of leaves ascending at acute angles to midrib; pedicels (flower- or fruit-stalks) with a noticeable swollen joint . . . . . . . . . . . . . . . . . . . . . *k*

k. Fruiting pedicels (stalks) straight, conspicuously turned downward, some of them 2–5 times as long as the fruiting sepals and 6–14 mm. long; plants usually growing at some stage in water, or in very wet places . . . . . . . . . . . . 3. R. VERTICILLATUS

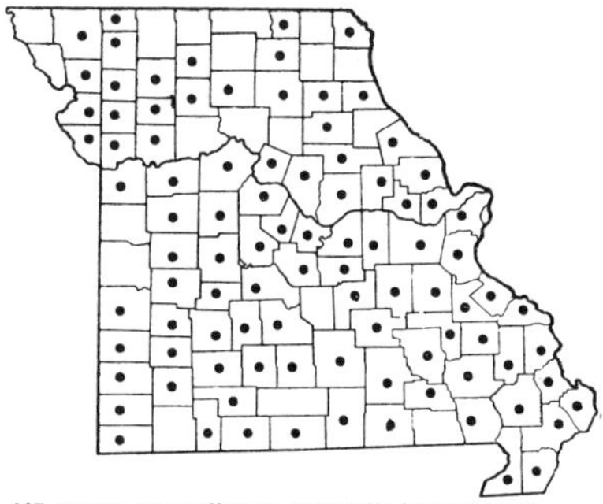

807 Rumex Acetosella var. Acetosella (Sheep Sorrel)

808 Rumex hastatulus (Wild Sorrel)

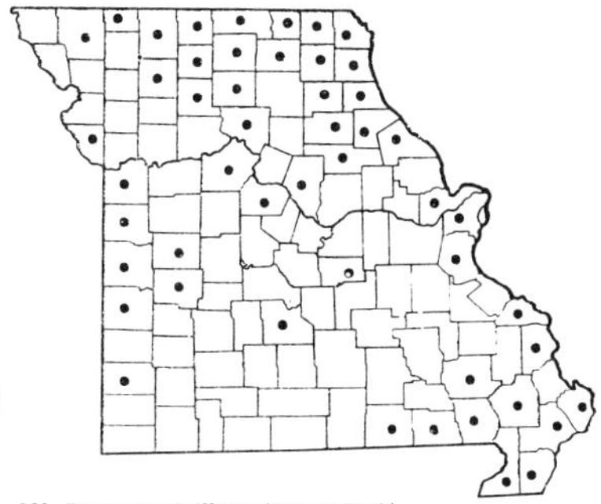

809 Rumex verticillatus (Swamp Dock)

k. Fruiting pedicels (stalks) curved, shorter than to mostly less than twice as long as fruiting sepals, and mostly 1–4 mm. long; plants of dry or moist ground, but not normally in water . . . . . . . . . . . . . . . . . . . . . . . . *l*

l. Mature fruiting sepals (valves) about 1.5 mm. broad, only slightly wider than the grain-like tubercle in the center; clusters of flowers separated from one another with noticeable gaps between them; middle and lower clusters of flowers conspicuously subtended by leaves; branches of the inflorescence spreading-ascending. 10. R. CONGLOMERATUS

l. Mature fruiting sepals (valves) 3–6 mm. broad, much broader than the grain-like tubercle in the center; clusters of flowers much more crowded and closer; most of flower-clusters without subtending leaves (some leaves occasionally present among lower clusters); branches of the inflorescence erect or strongly ascending. . . . . . *m*

m. Leaves conspicuously wavy or wrinkled on margins, dull dark green; stems usually lacking side branches (rarely producing small leaves at the sides); grain-like tubercle about $\frac{2}{3}$ as wide as long . . . . . . . . . . . . 8. R. CRISPUS

m. Leaves flat and smooth, without wrinkled or wavy margins, pale or light green; stems usually with side branches or with tufts of leaves from the sides; grain-like tubercle $\frac{1}{2}$ as wide or narrower. . . . . . . . . . . . . . . . . . . . *n*

n. Rare; fruiting sepals (valves) triangular-ovate, slightly pointed (subacute); larger leaves 1.5–3 cm. broad; 3 well-developed grain-like tubercles present, 1 in the center of each fruiting sepal (valve) . . . . . . . . . 4. R. MEXICANUS

n. Common; fruiting sepals (valves) round-ovate or broadly ovate, rounded at tip; larger leaves 2.5–5.5 cm. broad; if 3 grain-like tubercles are present, only 1 is well-developed, the other 2 just slightly developed or poorly so . 5. R. ALTISSIMUS

1. **Rumex Acetosella** L. var. **Acetosella**

Sheep Sorrel Map 807

Also called Red Sorrel, Common Sorrel, Field Sorrel, and Sorrel.

Flowers May–September; fruits June–October.

Occurs in sour soils of fallow fields, open waste ground, along railroads, sandstone, chert, or granite glades and outcrops. Throughout Missouri, and doubtless in every county.

Native of Europe; introduced into North America, where naturalized throughout the United States and Canada.

So far as records indicate, the Missouri material all falls into var. *Acetosella*, in which the terminal lobe of the leaves is lanceolate or oblanceolate, somewhat acute, and 2–12 mm. broad, as opposed to var. *pyrenaeus* (Pourret) Timbal-Lagrave, in which the terminal lobe of the basal leaves is obovate or elliptic, blunt or rounded at tip, and 12–30 mm. broad. There is also a form of var. *Acetosella*, f. *integrifolius* (Wallr.) G. Beck, in which all or most of the basal leaves lack lobes. Since these variations are found in adjacent states and may eventually be found in Missouri, they are here mentioned with their salient characters indicated.

Although indicative of acid soils, this species will also do well on neutral or slightly alkaline soil low in nitrate plant food.

Some individuals are sensitive to the plants, receiving a dermatitis through contact. This species sheds large amounts of pollen in May and June, and is considered as a cause of hay fever.

The sour leaves are a good thirst-quencher. The plant may be prepared for food in a number of ways, as a cooked vegetable, an ingredient in salads, as a seasoning in various dishes, as a soup, and as an acid beverage.

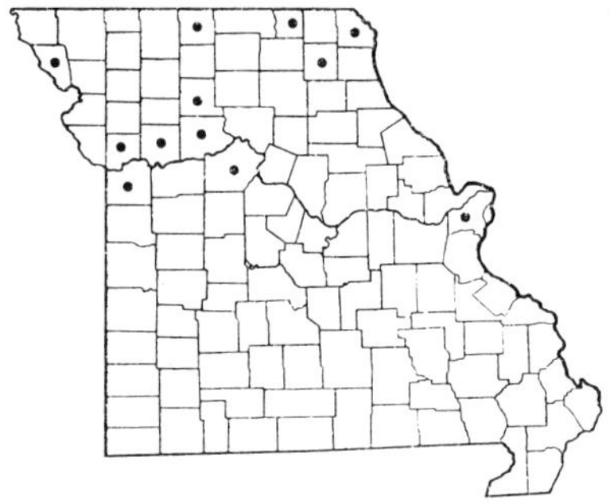
810 Rumex mexicanus

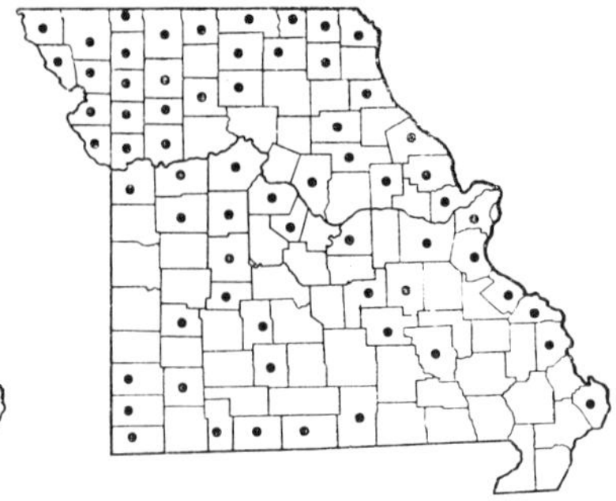
811 Rumex altissimus (Pale Dock)

812 Rumex Patientia (Patience Dock)

2. **Rumex hastatulus** Baldw. Wild Sorrel Map 808

Flowers April–May; fruits May–August.

Occurs in sandy open ground, along railroads, sandy or cherty glades, and prairies. Southeastern Missouri north along the Mississippi River counties to St. Louis County, and locally west in Newton County.

Ranges from Florida to Texas, north near the coast locally to Massachusetts, and inland north to southern Illinois, Missouri, and Kansas.

The enlarged fruiting sepals, much more conspicuous than in *R. Acetosella*, become straw-colored to pink when mature.

3. **Rumex verticillatus** L. Swamp Dock Map 809

Flowers April–June; fruits June–September.

Occurs in swamps, borders of sloughs, oxbow-lake meanders, and low wet woods. Throughout Missouri, but not recorded from many of the Ozark counties.

Ranges from Florida to Texas, north to Quebec, Ontario, Michigan, Wisconsin, and Minnesota.

This is the most aquatic of the common native species, the only other aquatic among the native species of Missouri being *R. orbiculatus*, which is very rare. *Rumex verticillatus* usually occurs in dense stands. It is common in bald cypress swamps of southeastern Missouri, and in around borders of natural ponds and sloughs elesewhere in the state. Its leaf-blades are smooth and flat, as in *R. mexicanus* and *R. altissimus*, and numerous sterile or basal leaves stand stiffly and prominently above the water.

This species, like many other docks, may be cooked as greens.

4. **Rumex mexicanus** Meisn. Map 810

*Rumex triangulivalvis* (Danser) Rech. f. [Steyerm.]

Flowers April–May; fruits June–September.

Occurs along borders of swamps, low ground along streams, sloughs, and ponds. Scattered in northern and central Missouri, mostly in the western sectors of those areas, south to St. Louis, Saline, and Jackson counties.

Ranges from Newfoundland and Quebec to British Columbia, south to New York, Pennsylvania, Ohio, Indiana, Illinois, Missouri, Texas, New Mexico, and California; also in Mexico and Guatemala.

The young leafy shoots are eaten as a cooked vegetable.

5. **Rumex altissimus** Wood Pale Dock Map 811

Flowers April–May; fruits May–August.

Occurs in alluvial ground along streams, sloughs, and ponds, often in flood plains, fields, thickets, waste ground in towns and cities, and along railroads. Throughout Missouri.

Ranges from Georgia to Texas, north to New York, Ohio, Michigan, Wisconsin, Minnesota, Nebraska, and Colorado; introduced in New England, New Jersey, and Virginia.

This is one of the commonest species of docks in Missouri. Alone, or in combination with Sour Dock (*R. crispus*), the young shoots in spring may be cooked as tasty greens.

6. **Rumex Patientia** L. Patience Dock Map 812

*Rumex Patientia* var. *kurdicus* Boiss. [P & S, Steyerm.]

Flowers April–June; fruits June–July.

Occurs in waste ground and along railroads. Known only from St. Louis, Jackson, Clay, and Buchanan counties.

Native of Europe and Asia; introduced into the

---

Plate no. 143. **1.** Rumex Patientia, $\times\ ^{2}/_{7}$. **2.** Rumex orbiculatus, $\times\ ^{2}/_{7}$; *a.* Fruiting branch, $\times\ ^{2}/_{7}$; *b.* Fruit, $\times\ 1^{5}/_{7}$. **3.** Rumex crispus, $\times\ ^{2}/_{7}$. **4.** Rumex pulcher, $\times\ ^{2}/_{7}$; *a.* Fruit, $\times\ 1^{3}/_{7}$; After Gleason, The New York Botanical Garden. **5.** Rumex conglomeratus, $\times\ ^{2}/_{7}$; *a.* Fruiting branch, $\times\ ^{2}/_{7}$; *b.* Fruit, $\times\ 2^{6}/_{7}$.

PLATE NO. 143

813 Rumex orbiculatus (Water Dock)

814 • Rumex crispus (Sour Dock)
814 □ Rumex crispus X Rumex obtusifolius

814A Rumex stenophyllus

United States, where naturalized from New England to Minnesota and Idaho, south to New York, Pennsylvania, Missouri, Oklahoma, and Utah.

The leaves are either flat or scarcely wavy on the margins, the main ones being rounded at the base. They are used as greens and are sometimes prepared into a thick soup.

7. **Rumex orbiculatus** Gray Water Dock Map 813

Known also as Great Water Dock.

*Rumex Brittanica* of auth. [P & S, Steyerm.], not L.

Flowers April–May; fruits June–September.

Occurs in moist alluvial ground along the Missouri River, where known only from St. Louis County (Natural Bridge Road, St. Louis, right-of-way of Mo. Pac. R.R., west of Knox Ave., August 14, 1955, *Muehlenbach 725*).

Ranges from Newfoundland and Quebec to North Dakota, south to New Jersey, Pennsylvania, Ohio, Indiana, Illinois, Missouri, and Nebraska.

The leaves are flat and entire or finely wavy-toothed on the margins.

8. **Rumex crispus** L. Sour Dock Map 814

Also known as Yellow Dock.

*Rumex elongatus* Guss. [P & S, Steyerm.]

Flowers April–May; fruits June–July.

Occurs in fields, garden soils, waste ground, along railroad tracks, and sometimes in low woods bordering streams, ponds, or sloughs. Throughout Missouri, and doubtless in every county.

Native of Europe; introduced into North America, where naturalized throughout the United States and Canada.

This is one of the species commonly used in the spring, either alone, or in combination with *R. altissimus*, wild lettuce (*Lactuca canadensis*), Sow Thistle (*Sonchus* spp.), Dandelion (*Taraxacum* spp.), and Pokeweed (*Phytolacca americana*).

Handling the plant occasionally causes dermatitis in some persons. The root was used at one time for medicine and sold as Yellow Dock.

A rare variant, f. *unicallosus* Peterm., of *R. crispus* with only one well developed grain-like tubercle instead of 3 on the fruiting valve, has not been found in Missouri, but may be expected.

8a. **Rumex crispus** × **obtusifolius** Map 814

Hybrid plants between these species of Rumex have been found in St. Louis, Jefferson, Carroll, and Buchanan counties.

9. **Rumex stenophyllus** Ledeb. Map 814A

*Rumex alluvius* Gates & McGregor

Flowers May–June; fruits June–August.

Occurs in alluvial ground along the larger streams. Known only from St. Charles (east and west of new highway 61 bridge, August 5, 1939, *Bauer 519*; Hamburg, August 6, 1939, *Bauer 578*), Clay (alluvial deposit, Missouri River valley, 2 mi. southeast of Missouri City, June 15, 1959, *McGregor 14395*), and Buchanan (along margin of Singleton Lake, T56N, R36W, sect. 25, 1½–2 mi. northwest of Halls, August 20, 1950, *Steyermark 70087*) counties.

Native of Eurasia; introduced in North America, where known from Manitoba, Wyoming, Colorado, North Dakota, South Dakota, Minnesota, Nebraska, Kansas, and Missouri.

This species resembles a hybrid between *R. crispus* and *R. obtusifolius*, but the fruiting sepals bear 3 grain-like tubercles, 1 on each valve. Besides the difference in number of tubercles present on the fruiting valves, *R. stenophyllus* differs from the true hybrid *R. crispus* × *obtusifolius* in the shape of the leaves and in being fertile instead of sterile, as in the hybrid.

815 Rumex conglomeratus

816 Rumex pulcher (Fiddle Dock)

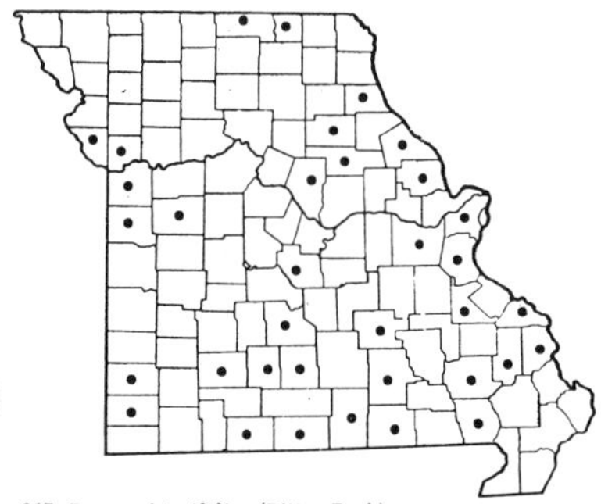

817 Rumex obtusifolius (Bitter Dock)

10. **Rumex conglomeratus** Murr. Map 815

*Rumex sanguineus* of some auth. [P & S], not L.

Flowers April–May; fruits June–July.

Known only from waste places in St. Louis County.

Native of Europe; introduced into the United States, where locally established from Virginia to South Carolina, Ohio, Indiana, Michigan, Illinois, Missouri, and the Pacific states.

The leaves are slightly wavy-crisped on the margins.

11. **Rumex pulcher** L. Fiddle Dock Map 816

Flowers April–July; fruits June–July.

Occurs along railroad tracks, known only from St. Louis County (St. Louis, along Mo. Pac. R.R., Plum Street, June 17, 1956, *Muehlenbach 957, 1261*; along Wabash R.R., north of Palm Street, June 28, 1958, *Muehlenbach 1422*).

Native of Europe; introduced in the United States, where naturalized from Florida to Texas and California, north locally to New York, Missouri, Oklahoma, and Oregon; also in Mexico.

The leaves are rather firm and thickish with a flat and entire or somewhat wavy margin.

12. **Rumex obtusifolius** L. Bitter Dock Map 817

Also called Blunt-leaved Dock, Broad-leaved Dock.

Flowers April–June; fruits June–September.

Occurs in fields, garden soils, waste ground, about dwellings, and rather damp soil of woodland and along streams. Throughout Missouri, but not recorded from most of northern sections of western Missouri.

Native of Europe; introduced into North America, where naturalized from Newfoundland and Quebec to British Columbia, south to Florida, Texas, and Arizona.

The lowest leaves at the base are reddish-veined and have a rounded or blunt tip, while the upper ones are acutely pointed; the leaves are wavy-crisped on the margins.

13. **Rumex maritimus** L. var. **fueginus** (Phil.) Dusen Golden Dock Map 818

Flowers May–July; fruits July–October.

Occurs on wet alluvial mud banks and flats and sand bars of the Missouri and Mississippi rivers. Found from Clark County along the Mississippi River in northeastern Missouri south along that river to Perry County, and from Holt County along the Missouri River in northwestern Missouri south and east along that river to Jackson, Saline, Boone, Franklin, St. Charles, and St. Louis counties. Probably occurring in all the counties along those two rivers.

Ranges from Quebec to British Columbia, south along the coast to New Jersey, inland south to New York, Ontario, Illinois, Arkansas, New Mexico, Arizona, and California. Also in South America.

The leaves are quite thick and fleshy and rather crisped on the margins. The flower clusters usually extend nearly to the base of the plant and are frequently interrupted with leaves. At late maturity the fruiting masses turn a reddish-brown or copper color.

### 3. **Polygonum** L. Knotweed, Smartweed

a. Stem armed with prickles . . . . . . . . . . . . . . . . . . . . . . . . . . . . *b*
b. Rare, known only from southeastern Missouri; lobes at base of leaves horizontally or wide-spreading, nearly at right angles to midrib; achenes 2-sided (lenticular) . . . . 20. P. ARIFOLIUM
b. Common, in many parts of Missouri; lobes at base of leaves pointing in the opposite direction from that of the tip of leaf; achenes 3-sided . . . . . . . . . . . . . 19. P. SAGITTATUM
a. Stems not armed with prickles . . . . . . . . . . . . . . . . . . . . . . . . . . *c*
c. Leaves broadly ovate, and most of them usually conspicuously heart-shaped or broadly rounded at base, rarely tapering or truncate at base . . . . . . . . . . . . . . . . . *d*
d. Leaves truncate (as if cut horizontally across) or cuneate (wedge-shaped) at base; plants dioecious, the male and female flowers separated on different plants . . . . 24. P. CUSPIDATUM
d. Leaves usually more or less heart-shaped at base or with strongly curved or rounded base; plants with stamens and pistils on the same flower . . . . . . . . . . . . . . . . *e*
e. Plants growing in the water with floating leaves; leaves glabrous (without hairs) on both sides; stems glabrous . . . . . . . . . . . . . . . . . . . . . . . . . *f*
f. Rare; inflorescence less than 4 cm. long, ellipsoid, narrowly ovoid, or thick cylindric; main stalk (peduncle) of inflorescence glabrous (without hairs); ocreolae (small sheathing stipules subtending flowers along axis of inflorescence) glabrous . . . . 7. P. AMPHIBIUM var. STIPULACEUM f. NATANS
f. Common; inflorescence 4–18 cm. long, narrowly cylindric; main stalk (peduncle) of inflorescence hairy; ocreolae hairy . . . . . 8. P. COCCINEUM var. COCCINEUM f. NATANS
e. Plants not growing in the water, without floating leaves; some part of leaves or stems or both with minute or conspicuous hairs . . . . . . . . . . . . . . . . . . *g*
g. Stems not twining, trailing, nor sprawling over the ground, 6–30 dm. tall . . . . . . *h*
h. Flowers greenish or greenish white, few and separated in a long narrow inflorescence; styles persistent and protruding in fruit as hooked or recurved beaks . . 21. P. VIRGINIANUM
h. Flowers rose to rose-red, many in a dense cylindric inflorescence; styles withering or falling in fruit, not hooked . . . . . . . . . . . . . . . . . . . . *i*
i. Stem usually 1.2–3 m. (4–10 feet) tall, with soft spreading hairs; lower leaf-surface soft hairy; inflorescence arching or drooping . . . . . . . . . 13. P. ORIENTALE
i. Stem much shorter, mostly 0.6–0.9 m. (2–3 feet) tall, with hairs usually parallel to surface or lacking; lower leaf surface with hairs parallel to surface or lacking; inflorescence straight, erect or ascending . . . . . . . . . . 8. P. COCCINEUM
g. Stems twining, trailing, or sprawling over the ground, if erect only a few dm. tall . . . . *j*
j. Outer divisions of calyx keeled, but not winged, in fruit; achene dull, minutely roughened; fruiting stalks (pedicels) of flowers 3.5–4.5 mm. long . . 22. P. CONVOLVULUS
j. Outer divisions of calyx strongly winged in fruit; achene shining, smooth; fruiting stalks (pedicels) of flowers as much as 4–10 mm. long . . . . . . . . . . . *k*
k. Mature fruit, measured from the joint, 10–15 mm. long (the mature calyx 8–10 mm. long); wings of fruiting calyx 1.5–3 mm. broad at summit; achenes 3.5–6 mm. long, 2.5–4 mm. broad; common . . . . . . . . 23a. P. SCANDENS var. SCANDENS
k. Mature fruit, measured from the joint, 7–10 mm. long (the mature calyx 4–6 mm. long); wings of fruiting calyx 0.25–1 mm. broad; achene 2–3.5 mm. long, 1.5–2 mm. broad; rare . . . . . . . . . . . . 23b. P. SCANDENS var. CRISTATUM
c. Leaves either narrowly linear, lanceolate, or oblong, either tapering or narrowed to rounded at base, but not ovate with a heart-shaped or broadly rounded base . . . . . . . . . . *l*
l. All the flowers appearing at the base of the leaves in short clusters in their axils, sometimes only 1 or 2 in a cluster; leaves jointed at the base to the leaf-stalk (petiole) . . . . . . . . *m*
m. Stem with angled sides; leaves with 2 furrows, 1 on each side of midrib; margins of leaves with minute rough projections (difficult to see if leaf is folded in); plants of sandstone, chert, or granite glades and outcrops . . . . . . . . . . . . . . . . . 6. P. TENUE
m. Stem with rounded or curved sides, terete (rounded in cross-section); leaves not furrowed; margins of leaves entire (smooth-edged) or very faintly irregular-edged with some projections; plants of prairies, fields, waste ground, along railroads, moist and alluvial ground, and around houses and farmlots . . . . . . . . . . . . . . . . . . . . *n*
n. Leaves oval, oblong, or elliptic, blunt or rounded at tip, $1\frac{1}{2}$–3 times as long as broad, 3–25 mm. (mainly 10–25) broad . . . . . . . . . . . . . . . . . . *o*

o. Leaves yellow-green; divisions of calyx yellow-green with yellow-green to white margins; mature flower-stalks (pedicels) protruding beyond and longer than the ocreae (sheathing stipules); achene mainly 3–3.5 mm. (sometimes 2.5) long . . . . . . . . . 3. P. ERECTUM

o. Leaves blue-green or silvery-green; divisions of calyx blue-green or green with rose- or purple-tinged margins or some of them with white margins; mature flower-stalks (pedicels) mostly included within and shorter than the ocreae (sheathing stipules); achene 2–2.5 mm. long. . . *p*

p. Fruiting calyx 3.5–4 mm. long, lobed to above the middle, strongly constricted below the narrowed tip; margins of inner divisions of calyx white; achene dull yellow-brown or olive; 3 outer divisions of calyx hooded (cucullate) near tip, somewhat ridged in the center, noticeably longer than the inner two divisions . . . . . . . . . . 4. P. ACHOREUM

p. Fruiting calyx 2–3 mm. long, lobed to below the middle, not constricted below any narrowed tip; margins of calyx divisions mainly rose, pink, or purple, sometimes white; achene dark brown; 3 outer divisions of calyx flat, not hooded nor with a ridge, usually equaling or shorter than the inner two divisions . 5. P. AVICULARE var. VEGETUM and var. LITTORALE

n. Leaves narrowly linear, oblong-linear, narrowly lanceolate or oblanceolate, acute or blunt at tip, mainly $3\frac{1}{2}$–9 times as long as broad, mainly 1–5 mm. (up to 10) broad . . . . . . . . *q*

q. Stems mainly sprawling, lying on the ground or loosely ascending, generally less than 1.5 dm. tall; 3 outer divisions of calyx flat, not hooded nor with a ridge, usually equaling or shorter than the inner two divisions; calyx divisions lobed to just below middle; common weed of waste ground, lawns, dooryard, farmlot, salt licks and borders of saline springs . . . . 5. P. AVICULARE var. AVICULARE

q. Stems erect with usually strongly ascending branches, mainly 3–20 dm. (rarely 1.5–3) tall; 3 outer divisions of calyx hooded (cucullate) near tip, somewhat ridged in the center, noticeably longer than the inner two divisions; calyx divisions lobed nearly to the base; plants of prairies and open ground. . . . . . . . . . . . . . . . . . . . . . . . . *r*

r. Leaves yellow-green; divisions of calyx yellow-green with yellow margins; mature calyx divisions 3–4 mm. long; achene 2.2–6.5 mm. long; flower-stalks (pedicels) usually longer than the calyx, the flowers usually protruding beyond the ocreae (sheathing stipules) . 2. P. RAMOSISSIMUM

r. Leaves bluish green; divisions of calyx green with white or pink margins; mature calyx divisions 1.3–1.8 mm. long; achene 2 mm. long; flower-stalks (pedicels) shorter than the calyx, the flowers scarcely or not protruding beyond the ocreae (sheathing stipules) . . 1. P. PROLIFICUM

l. Flowers several to many in dense or loose elongated inflorescences arising at the ends of the stems and branches; leaves not jointed to the leaf-stalk (petiole) . . . . . . . . . . . . . . . *s*

s. Sheathing stipules (ocreae) surrounding stem lacking a fringe of bristles at the top . . . . . . *t*

t. Main flower-stalk (peduncle) and axis of inflorescence with hairs lying along and parallel to surface; perennial plants with creeping rhizomes and stolons; the sheathing stipules (ocreae) not prolonged on one side into an ovate extension, scarcely or not oblique (unequal-sided), shaped like a funnel or ice cream cone . . . . . . . . . . . . . . 9. P. DENSIFLORUM

t. Main flower-stalk (peduncle) and axis of inflorescence either completely glabrous (without hairs) or with stalked or sessile glands; annuals without creeping rhizomes and stolons; the sheathing stipules (ocreae) prolonged on one side into an ovate extension, noticeably oblique (lop-sided) . . . . . . . . . . . . . . . . . . . . . . . . . . . . *u*

u. Main flower-stalk (peduncle) and axis of inflorescence either completely glabrous or with sessile (without stalks) glands; inflorescences mostly arching or drooping at tip, 3–9 mm. thick; achenes 1.8–2.2 mm. long, 1.6–2 mm. broad; mature calyx constricted toward tip to form a beak of sepals over the achene . . . . . . . . . . 12. P. LAPATHIFOLIUM

u. Main flower-stalk (peduncle) and axis of inflorescence with stalked glands (or completely glabrous in *P. pensylvanicum* var. *eglandulosum*, a rarely collected variety); inflorescence erect or ascending, straight, 10–15 mm. thick; achenes 2.5–3.5 mm. long, 2.2–3.5 mm. broad; mature calyx not constricted toward tip to form a beak of sepals over the achene . . . . . *v*

v. Common type encountered with the flowers uniform, the styles and stamens mainly included within the calyx; one or both faces of the achene depressed or sunken in the center . . . . . . . . . . . . . . . . . . . . . 11. P. PENSYLVANICUM

v. Rare type encountered with two types of flowers, some plants with styles long protruding from calyx and stamens included, other plants with styles included and stamens long protruding from calyx; achenes with both faces convex (curved outward with highest part in center) . . . . . . . . . . . . . . . . . . 10. P. BICORNE

818 Rumex maritimus var. fueginus (Golden Dock)

819 Polygonum prolificum

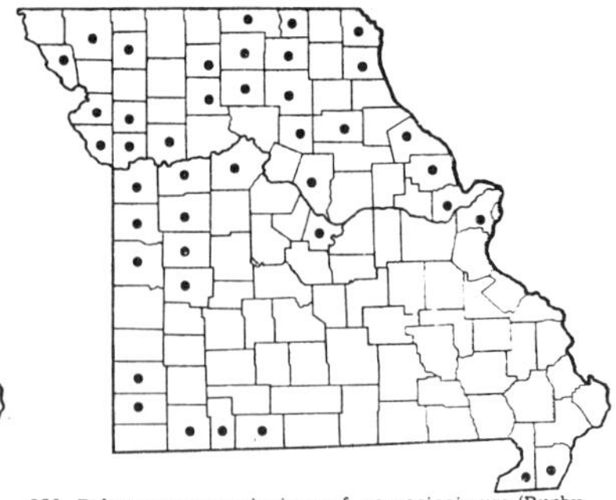
820 Polygonum ramosissimum f. ramosissimum (Bushy Knotweed)

s. Sheathing stipule (ocrea) surrounding stem provided with a fringe of bristles at the top . . . . . . *w*

w. Calyx divisions with pale or dark dots . . . . . . . . . . . . . . . . . . . . . . . . *x*

x. Achenes dull; calyx greenish or with purple tips; internodes 2–4 cm. long; inflorescence drooping or nodding at tip . . . . . . . . . . . . . 14. P. HYDROPIPER and varieties

x. Achenes shining; calyx white; internodes 3–8 cm. long; inflorescence slightly or not at all drooping . . . . . . . . . . . . . . . 17 ~~16~~. P. PUNCTATUM and varieties

w. Calyx divisions lacking dots . . . . . . . . . . . . . . . . . . . . . . . . . . . *y*

y. Ocreolae (stipules at base of flower clusters along axis of inflorescence) with or without short hairs at the summit margins, the hairs when present at most 1 mm. long and shorter than the flowers; most of the achenes 2-sided (sometimes 3-sided); inflorescence dense, 7–12 mm. thick; mature sepals with prominent reticulation (network) at the base . . . . 16 ~~17 16~~. P. PERSICARIA

y. Ocreolae with some of the hairs at the summit margins 2–3.5 mm. long; achenes 3-sided; inflorescence with looser, less crowded flowers, mostly less than 7 mm. thick; mature sepals lacking any reticulation at their base . . . . . . . . . . . . . . . . . . . . . *z*

z. Commonly encountered throughout the state; native plants of swamps, streams, sloughs, ponds, and wet soils, often growing in the water; achenes 2–3 mm. long; the longer flowering or fruiting spikes mainly 3–8 cm. (sometimes 2) long; perennial plants with firm elongated rootstocks, often with basal offshoots in autumn; inflorescence pink, white, greenish-white, dull purple, or pale rose . . . . . . . . . . . 18. P. HYDROPIPEROIDES and varieties

z. Rarely encountered, known only from St. Louis County; introduced plants of waste ground, lawns, and streets; achenes 2–2.4 mm. long; the longer flowering or fruiting spikes 1–4 cm. long; annual plants with fibrous roots lacking elongated rootstocks or basal offshoots in autumn; inflorescence deep rose to rose-purplish . . . . 15. P. CESPITOSUM var. LONGISETUM

1. **Polygonum prolificum** (Small) Robins. Map 819

Flowers July–November.

Occurs in open places in flood plain or alluvial soils near streams. Known only from Clay and Jackson counties in west-central Missouri.

Ranges from Maine to Virginia, and inland from Minnesota to Saskatchewan and Washington, south to Arkansas, Oklahoma, and New Mexico.

2. **Polygonum ramosissimum** Michx. f. **ramosissimum** Bushy Knotweed Map 820

*Polygonum ramosissimum* Michx. [G, P & S, Steyerm.]

*Polygonum exsertum* Small [G, P & S]

*Polygonum triangulum* Bicknell [P & S]

Flowers July–October.

Occurs in upland and low river bottom prairies, and in moist alluvial and open ground. Northern, central, and western Missouri, south to St. Louis, Cole, and Taney counties, locally in the southeastern lowland section in Pemiscot and Dunklin counties; absent from all the Ozark section except where found in the White River counties.

This species shows great variation in the length and shape of the fruit and in the extent of its protrusion from the calyx. This varies with the season of the year and on the same plant. As noted by Gleason (*New Ill. Fl.* vol. 2: 74. 1952), a broadly

Plate no. 144. **1.** Rumex obtusifolius, × 2/7; *a.* Fruiting branch, × 2/7; *b.* Fruit, × 2 2/7. **2.** Rumex maritimus var. fueginus, × 2/7. **3.** Polygonum ramosissimum, × 2/7; *a.* Fruit, × 3 1/7. **4.** Polygonum erectum, × 2/7. **5.** Polygonum aviculare var. aviculare, × 2/7; Details from Small, The New York Botanical Garden. **6.** Polygonum achoreum, × 2/7.

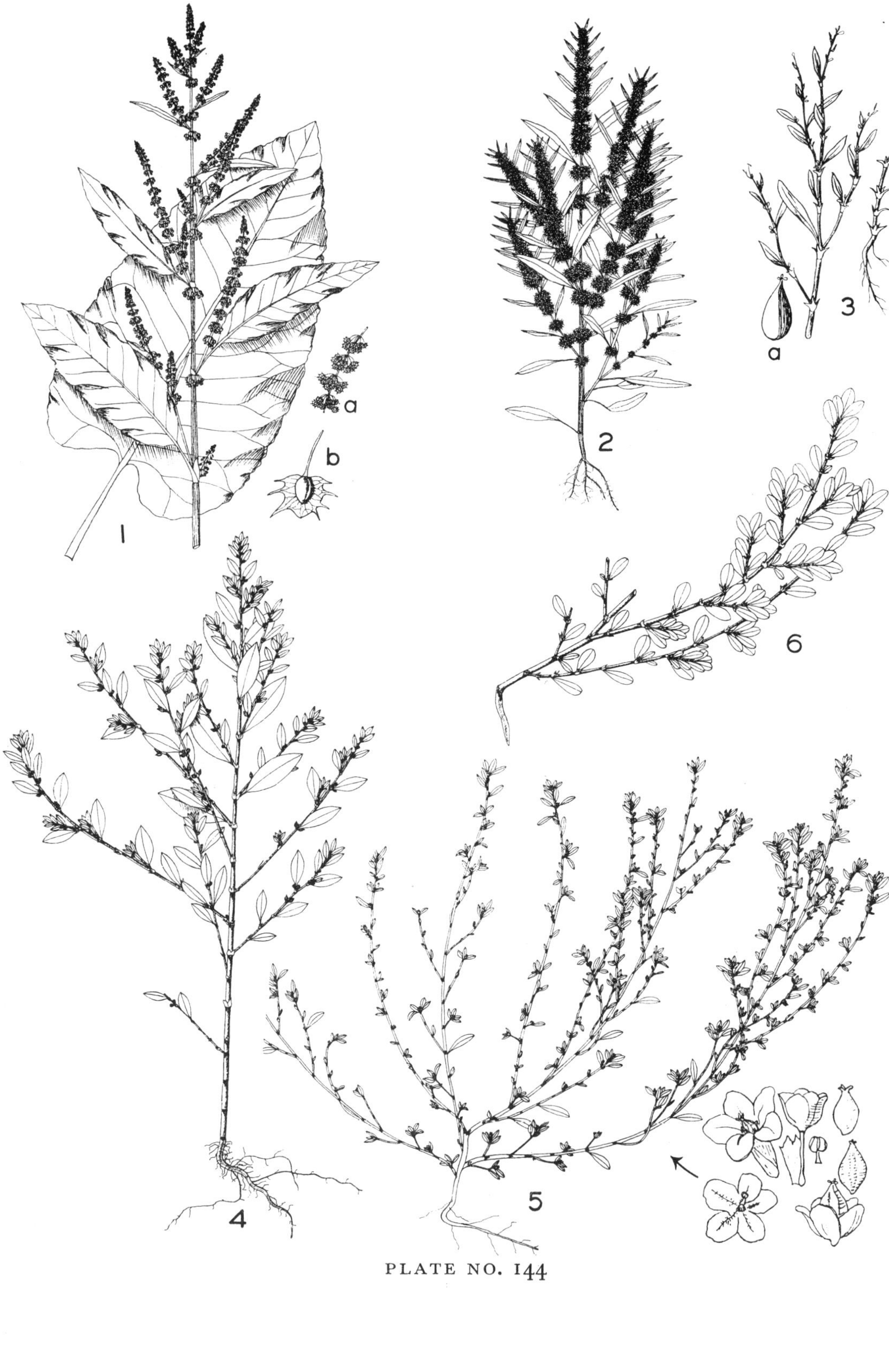


PLATE NO. 144

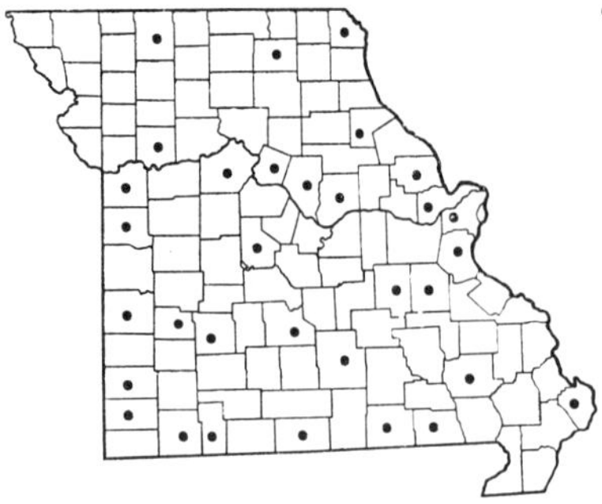

821 Polygonum erectum

822 Polygonum achoreum

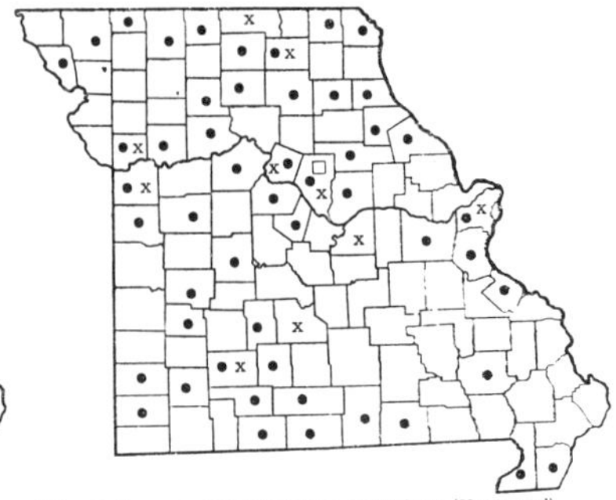

823 • Polygonum aviculare var. aviculare (Knotweed)
823 □ Polygonum aviculare var. vegetum
823 x Polygonum aviculare var. littorale

ovoid type of achene, scarcely broader than long and included in the calyx, may occur earlier in the season and at the older flowering nodes. An extreme form of this type of fruit, described as *P. triangulum* Bicknell from Atherton, Jackson County, Missouri, grades into typical *P. ramosissimum*. On the other hand, the narrowly ovoid to lanceolate type of achene, conspicuously protruding beyond the length of the calyx, is encountered later in the season or at the upper flowering nodes, and this longer type of achene, represented by *P. exsertum* Small, is 4–6.5 mm. long, whereas the smaller, broader, and included type of achene is 2.2–3.5. Both types of achenes are found on the same plant and at the same time, as noted by Gleason, and study of Missouri material is in full agreement with his conclusions. Fernald (eighth ed. *Gray's Manual*, p. 578) noted that the achenes of *P. ramosissimum* were both included and exserted, with the latter up to 4 mm. long.

Another form of *P. ramosissimum*, f. *atlanticum* Robins., with pink-margined calyx divisions, has not been recorded from Missouri.

3. **Polygonum erectum** L. Map 821

Flowers late May–October.

Occurs in low damp woodland, alluvial banks of streams and ponds, and in roadside thickets. Throughout Missouri, but represented by few collections from northern Missouri.

Ranges from Quebec and Vermont to Wisconsin and Iowa, south to Georgia, Tennessee, Missouri, and Kansas; also in the Pacific states.

This simulates *P. ramosissimum* in the yellow-green color of leaves and calyx, but has broader and elliptical leaves.

As noted by Gleason (*New Ill. Fl.* vol. 2: 75. 1952), the achenes may be either shining, included, and 2.5 mm. long, or dull, protruding, and 3–3.5 mm. long.

4. **Polygonum achoreum** Blake Map 822

Flowers mid-May–September.

Known only from waste ground in Jackson County, west-central Missouri (introduced, Sheffield, July 19, 1906, *Bush 4060* in Gray Herbarium).

Ranges from Newfoundland to Alaska, south to New York, Ontario, Michigan, Wisconsin, Missouri, South Dakota, Montana, and Colorado.

The plant resembles *P. erectum* in habit, but is more compact and the leaves and calyx are bluish-green. The ocreae are persistent and not torn into narrow fibrous shreds, as are those of *P. aviculare* and *P. erectum*.

## Section **Avicularia** Knotweed

5. **Polygonum aviculare** L. Knotweed Map 823

Also known by Knotgrass, Dooryard Weed, Hogweed, and many other names.

Flowers May–November.

This is a highly variable species, different in habit of growth, and in size, shape, and texture of leaves. The following variations occur in Missouri:

a. Leaves 1.5–5 mm. broad, usually linear or narrowly lanceolate and $3\frac{1}{2}$–9 times as long as broad . . 5a. P. AVICULARE var. AVICULARE

a. Leaves 4–15 mm. broad, oblong-lanceolate or oblong to oval or spatulate, mostly $1\frac{1}{2}$–3 times as long as broad . . . . . . . . . *b*

b. Leaves rather thin, pale green, acute (pointed) at tip, slightly wavy on margins, not prominently veined; stems erect or ascending . . 5b. P. AVICULARE var. VEGETUM

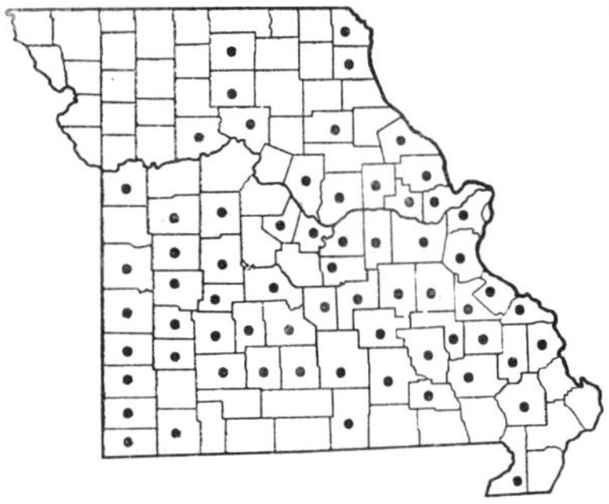

824 Polygonum tenue

825 Polygonum amphibium var. stipulaceum f. fluitans (Water Smartweed)

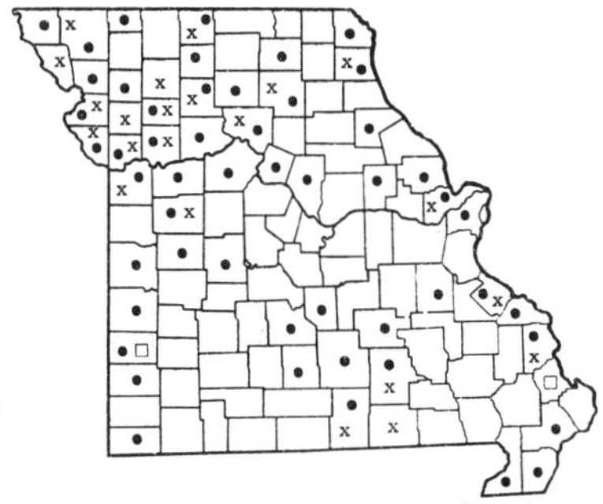

826 • Polygonum coccineum var. coccineum f. coccineum (Water Smartweed)

826 □ Polygonum coccineum var. coccineum f. natans

826 x Polygonum coccineum var. pratincola

b. Leaves rather thick and firm or subcoriaceous, bluish- or gray-green, blunt or rounded at tip, the side veins rather prominent; stems usually spreading along the ground . 5c. P. AVICULARE var. LITTORALE

5a. **Polygonum aviculare** var. **aviculare** Map 823

*Polygonum aviculare* var. *angustissimum* Meisn. [P & S, Steyerm.]

*Polygonum neglectum* Bess. [Deam]

*Polygonum avicalare* L. [G, BB, P & S, Steyerm.]

Occurs in waste ground, along streets, lawns and gardens about dwellings, pastures, fields, and borders of salt licks and saline springs, often in clayey and hard-packed soils. Throughout Missouri, and doubtless in every county.

Ranges throughout the United States and Canada; also in Europe and Asia. Some authors consider this variation as a European native introduced in North America, others as representing both native North American and introduced Eurasian strains. It would appear that both types are present in Missouri. Certainly the fleshy-leaved, prostrate types found around salt licks and salt marshes of central Missouri in Saline, Howard, and Cooper counties show every indication of being part of the native flora of those distinctive areas.

5b. **Polygonum aviculare** var. **vegetum** Ledeb. Map 823

Occurs in similar situations to those occupied by var. *aviculare*, except absent from the saline soil areas. Probably of wide occurrence throughout the state, but only rarely collected. Represented by a specimen from Boone Co. (north shore of Brushwood Lake, July 4, 1933, *Drouet 637*).

Ranges throughout North America.

5c. **Polygonum aviculare** var. **littorale** (Link) W. D. J. Koch Map 823

*Polygonum buxiforme* Small [P & S, Steyerm.]

Occurs in fields and waste ground, but is less common than var. *aviculare*. Scattered in Missouri. Representative collections are from St. Louis (Sept., 1845, *Engelmann*) and Jackson (Courtney, September 22, 1928, *Bush 11647*; September 5, 1913, *Bush 7069*) counties.

Ranges from Newfoundland to Alaska, south along the coast to South Carolina, and inland to New York, Tennessee, Arkansas, Oklahoma, Texas, and New Mexico.

The hard fruits of *Polygonum aviculare* var. *aviculare* and var. *littorale* have been used by various Indian tribes as an ingredient for their pinole meal or flour. These knotweeds are well-known as common weeds, but are frequently mistaken for grass by the uninformed. They persist during winter as densely leafy patches growing close to the ground among the grass, or are often present in poultry yards where they survive, better than grass, the constant tramping of chickens. The leaves are eaten by poultry and cattle. The house sparrow and other birds eat the achenes.

6. **Polygonum tenue** Michx. Map 824

Flowers June–October.

Occurs in usually dry acid soils of sandstone, chert, or granitic glades, rocky prairies, and gravel bars along streams. Common throughout the Ozark and unglaciated prairie sections of southern and central Missouri, and locally in the northeastern half of the state in Clark, Lewis, Sullivan, Linn, and Chariton counties.

Ranges from Georgia, Alabama, Arkansas, Oklahoma, and Texas, north to Maine and Minnesota.

This rather spare-looking plant has slender erect stems and branches with erect narrowly lanceolate or linear pointed leaves. It is frequently associated on dry acid soils in exposed situations with *Oenothera linifolia*, *Hypericum gentianoides*, *Bulbostylis capillaris*, and *Crotonopsis elliptica*.

7. **Polygonum amphibium** L. var. **stipulaceum** (Colem.) Fern. f. **fluitans** (Eat.) Fern. Map 825
Water Smartweed
*Polygonum natans* (Michx.) Eat. [P & S]
*Polygonum natans* f. *natans* [BB]
Flowers June–September.

Growing in water, known only from Greene County, southwestern Missouri.

Ranges from Labrador to Alaska, south to New Jersey, Pennsylvania, Ohio, Indiana, Illinois, Missouri, Nebraska, Colorado, and California.

8. **Polygonum coccineum** Muhl. Map 826
Water Smartweed
Also known as Shoestring Smartweed.
Flowers June–October.

Occurs usually in or bordering water of swamps, sloughs, oxbow lakes of river bottoms, streams, river bottom prairies or meadows, and along railroad tracks. Common throughout Missouri, although not recorded from many of the Ozark counties.

This is a highly variable species whose several modified forms appear to be responses to environmental changes correlated with a fluctuating water depth.

Missouri material may be divided into the following variations:

- a. Growing in water with the floating leaves and branches glabrous (without hairs); upper leaves heart-shaped (cordate). 8b. P. COCCINEUM var. COCCINEUM f. NATANS
- a. Growing on land or out of water with the upper leaves at most rounded at base; stems often hairy; leaves often hairy . . . . . *b*
  - b. Leaf-stalks attached near the base of the ocreae (sheath-like stipule); upper portions of stem and leaves glabrous or with finely appressed hairs; inflorescences up to 8 cm. long . 8a. P. COCCINEUM var. COCCINEUM f. COCCINEUM
  - b. Leaf-stalks usually attached near the middle of the ocreae (sheath-like stipule); upper portions of stems and leaves densely hairy; inflorescences up to 18 cm. long . 8c. P. COCCINEUM var. PRATINCOLA

8a. **Polygonum coccineum** var. **coccineum** f. **coccineum** Map 826
*Polygonum coccineum* Muhl. [G, BB, P & S]

This variety appears to be the commonest in the state, where widely distributed, and probably in every county.

Ranges from Quebec to Washington, south to North Carolina, Kentucky, Arkansas, Oklahoma, Texas, and California.

8b. **Polygonum coccineum** var. **coccineum** f. **natans** (Wieg.) Stanf. Map 826
*Polygonum coccineum* f. *natans* (Wieg.) Stanf. [G]

Known only from Scott County, southeastern Missouri (swamp along road H, T28N, R14E, sect. 29, 3½ mi. north of Blodgett, June 6, 1957, *Steyermark 85128*). At this locality the stems were submerged in 2 feet of water and the leaves were floating. Plants were growing with *Scirpus heterochaetus*, *Eleocharis Smallii*, *Juncus acuminatus*, *Carex alata*, *Ranunculus sceleratus*, *Sparganium eurycarpum*, and *Sium suave*.

Occurs throughout the range of var. *coccineum*.

8c. **Polygonum coccineum** var. **pratincola** (Greene) Stanford Map 826

This is less common than typical var. *coccinea*, but is probably throughout the same area.

Ranges from Indiana to North Dakota, south to Arkansas, Oklahoma, Texas, and Mexico.

*Polygonum coccineum* and varieties, which some botanists prefer to treat as a single variable species, generally occupies large stands where it occurs, and frequently produces leafy stems only. Also in the drouth years few flowering spikes are found. These are of a deep rose-red color. The creeping black rootstocks, somewhat resembling black shoelaces, give the plant its common name in some localities. The achenes are eaten by various species of wildfowl.

9. **Polygonum densiflorum** Meisn. Map 827
*Polygonum portoricense* Bertero
Flowers August–October.

Occurs in low wet woodland and swampy ground in southeastern Missouri, where known only from Scott (swamps, August 31, 1894, *Eggert*), Dunklin (Kennett, September 18, 1893, *Bush*; E. Bertig, St. Francis River, September 28, 1897, *Trelease 776*), and Carter counties.

Ranges from South and Central America, the West

Plate no. 145. **1.** Polygonum densiflorum, × 2/7. **2.** Polygonum aviculare var. littorale, × 2/7. **3.** Polygonum tenue, × 2/7. **4.** Polygonum coccineum, × 2/7. **5.** Polygonum bicorne, × 2. **6.** Polygonum pensylvanicum, × 2/7. **7.** Polygonum lapathifolium, × 2/7. **8.** Polygonum Persicaria, × 2/7. **9.** Polygonum Hydropiper, × 2/7.

PLATE NO. 145

827 Polygonum densiflorum

828 Polygonum bicorne

829 • Polygonum pensylvanicum var. pensylvanicum (Pinkweed)

829 □ Polygonum pensylvanicum var. laevigatum f. laevigatum

829 ‡ Polygonum pensylvanicum var. laevigatum f. albineum

829 x Polygonum pensylvanicum var. eglandulosum

Indies, and Florida to Texas, north to New Jersey and Missouri.

10. **Polygonum bicorne** Raf. Map 828

*Polygonum longistylum* Small [G, BB, P & S]

Flowers July–October.

Occurs in swampy or low wet ground near streams, and along railroads. Rare, known only in southern and central Missouri, from St. Louis (along Mo. Pac. R.R., August 21, 1954, *Muehlenbach 359*), Dunklin (September 14, 1893, *Bush 117A*), Clay, Jackson (swamps, Atherton, August 1, 1902, *Mackenzie 79*; September 7, 1895, *Bush 515*), and Jasper counties.

Ranges from Louisiana to New Mexico, north to Illinois, Missouri, and Kansas.

Dr. Shinners (Rh. 59: 265–67. 1957) has recently produced evidence to show that the earlier and correct name for this species is *P. bicorne*. The flowers are pale rose to pink, and the plant as a whole resembles the next species, *P. pensylvanicum*.

11. **Polygonum pensylvanicum** L. Pinkweed Map 829

Flowers May–October.

Occurs in wet ground bordering swamps, sloughs, streams, oxbow lakes of flood plain and other ponds, gravel bars, waste ground, and along roadsides and railroads.

Missouri material is represented by the following variations:

a. Lower surface of leaves with hairs lying along the surface . . . . 11a. P. PENSYLVANICUM var. PENSYLVANICUM f. PENSYLVANICUM
a. Lower surface of leaves glabrous or soon becoming glabrous. . . . . . . . . . *b*
  b. Main flower-stalk (peduncle) and axis completely glabrous (without hairs) . . . 11d. P. PENSYLVANICUM var. EGLANDULOSUM
  b. Main flower-stalk (peduncle) and axis with stalked glands . . . . . . . . . *c*
    c. Flowers rose or pink . 11b. P. PENSYLVANICUM var. LAEVIGATUM f. LAEVIGATUM
    c. Flowers white. . 11c. P. PENSYLVANICUM var. LAEVIGATUM f. ALBINEUM

11a. **Polygonum pensylvanicum** var. **pensylvanicum** Map 829

*Polygonum pensylvanicum* L. [G, P & S, Steyerm.]

This is the commoner variety in Missouri, and occurs throughout the state, doubtless in every county.

Ranges from Nova Scotia to Ontario and Minnesota, south to Florida and Texas.

11b. **Polygonum pensylvanicum** var. **laevigatum** Fern. f. **laevigatum** Map 829

*Polygonum pensylvanicum* var. *laevigatum* [G, BB, P & S, Steyerm.]

This variety is recorded from less counties than var. *pensylvanicum*, but doubtless occurs in most of them; it has not been recorded from most of the Ozark section nor the southeastern lowland area of the state.

Ranges from Quebec to Minnesota, South Dakota, and Colorado, south to Virginia, North Carolina, Tennessee, Missouri, and Oklahoma.

11c. **Polygonum pensylvanicum** var. **laevigatum** f. **albineum** Farw. Map 829

This white-flowered form is known from Knox (open swale bordering ditch along highway 6, 4.6 mi. east of Edina, September 19, 1955, *Steyermark 79656*) and Pike (barnyard, Washington University experimental farm, west of Clarksville, May 24, 1948, *Etter 154*) counties.

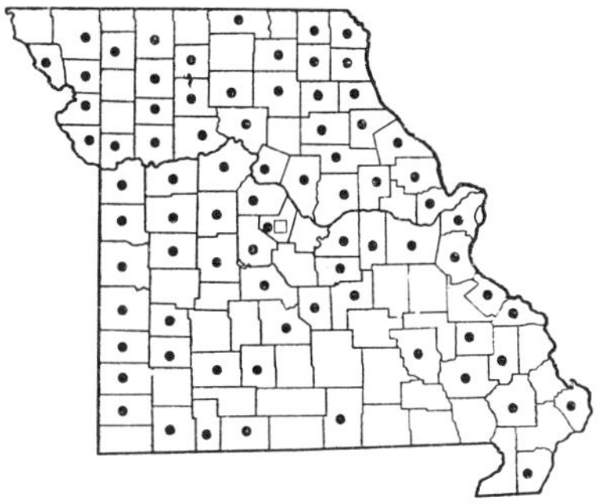

830 ● Polygonum lapathifolium var. lapathifolium
830 □ Polygonum lapathifolium X Polygonum pensylvanicum

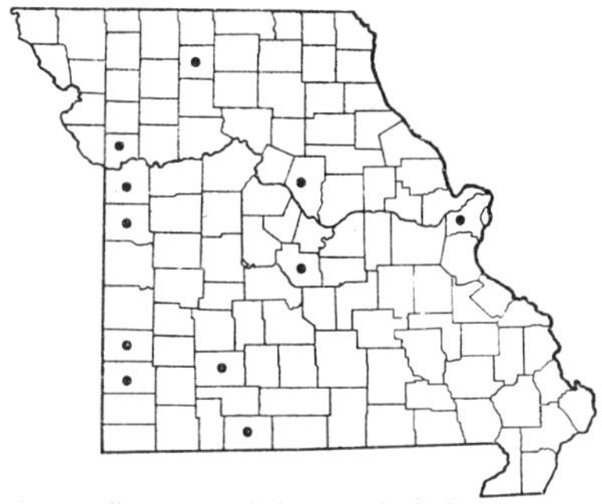

831 Polygonum orientale (Prince's Feather)

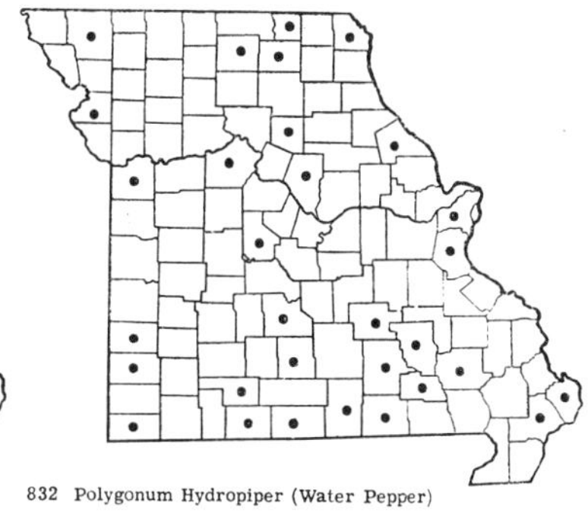

832 Polygonum Hydropiper (Water Pepper)

The achenes of *P. pensylvanicum* are a source of food for wildfowl and upland game birds, and the plant itself is browsed by deer and muskrat. In areas which have become recently impounded by high dams, this species, along with other members of the *Persicaria* group, take over and dominate newly available muddy shores. The fragrant flowers are much visited by bees.

11d. **Polygonum pensylvanicum** var. **eglandulosum** J. C. Myers — Map 829

Known only from Greene County, southwestern Missouri (in mud along banks of a small stream near the east end of Fellows Lake, July 23, 1958, *Paul L. Redfearn, Jr. 3796*, in herb. Southwest Mo. State College).

Ranges in northern Ohio and Missouri, but probably much more common.

12. **Polygonum lapathifolium** L. var. **lapathifolium** — Map 830

*Polygonum lapathifolium* L. [G, P & S]

*Polygonum lapathifolium* var. *nodosum* (Raf.) Weinm. [BB]

Flowers early June–November.

Occurs most frequently in low wet alluvial ground of mud flats and gravel bars along streams, borders of sloughs and ponds, in cultivated fields, waste ground, and along railroads. Throughout Missouri, but not recorded from some of the Ozark counties.

Ranges from Newfoundland to British Columbia, south throughout the United States and Mexico.

Other varieties have not been found in Missouri. This species and its varieties is one of the species of the genus most frequently eaten by various kinds of wildfowl. The flowers are either white, pink, or greenish, or combinations of these, the base of the calyx generally being greenish with white, straw-colored, or pinkish above. The slenderly cylindrical spikes are usually arching or nodding at the tip, but are sometimes more erect.

12a. **Polygonum lapathifolium** × **pensylvanicum** — Map 830

A plant intermediate between the two species and thought to represent a hybrid has been collected in Moniteau County.

13. **Polygonum orientale** L. Prince's Feather — Map 831

Flowers June–October.

Occurs in waste and cultivated ground, where it has escaped from cultivation. Known from scattered counties over the state, but uncommonly found.

Native of Asia; introduced into North America where naturalized.

The large spikes of brightly rose-colored flowers droop from densely hairy peduncles, and are an attractive feature of this tall, stout herb.

14. **Polygonum Hydropiper** L. Water Pepper — Map 832

Known also as Common Smartweed, Pepper Plant, and several other local names.

*Polygonum Hydropiper* var. *projectum* Stanf. [P & S]

Flowers late May–November.

Occurs in fields, pastures, moist ground bordering sloughs, streams, ponds, and wet often open logged woodland. Scattered throughout Missouri, and probably in every county.

Ranges from Quebec to British Columbia, south to Florida, Texas, and California.

The slender spike with greenish or white-bordered, mostly 4-parted flowers tapers to a slender arching tip. The glabrous annual stems are usually reddish or copper-red. The leaves are intensely burning to the taste, and cause a smarting sensation and irritation if the juice from the plant comes into contact with the

832A Polygonum cespitosum var. longisetum

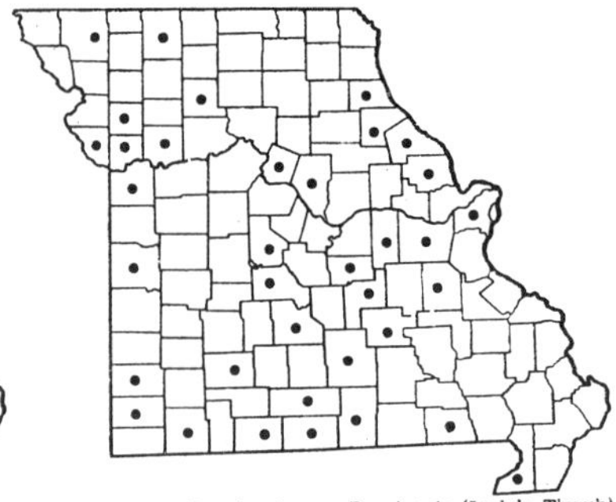

833 Polygonum Persicaria var. Persicaria (Lady's Thumb)

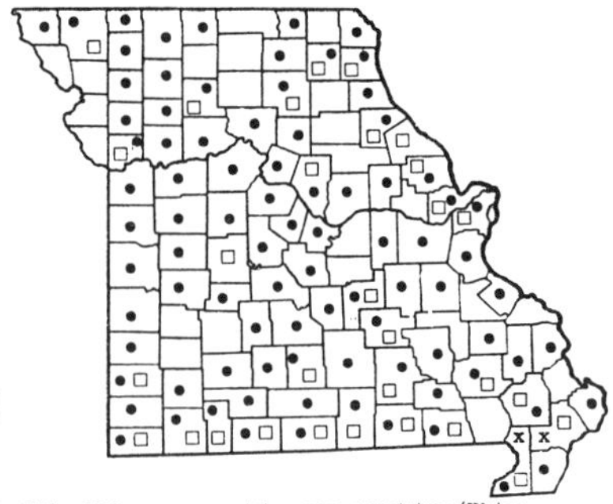

834 • Polygonum punctatum var. punctatum (Water Smartweed)
834 □ Polygonum punctatum var. confertiflorum
834 x Polygonum punctatum var. majus

eyes or nostrils. Some cases in Europe have been reported of poisoning of stock traced to eating the plants. Rare cases of dermatitis from handling this species have been reported.

15. **Polygonum cespitosum** Blume var. **longisetum** (De Bruyn) Stewart Map 832 A
*Polygonum caespitosum* Blume [BB]
*Polygonum longisetum* De Bruyn
Flowers June–October.

Occurs in waste ground, lawns, along streets and roadsides, where known only from St. Louis County (St. Louis, along sidewalk, 2300 Tower Grove Avenue, outside the grounds of Missouri Botanical Garden, October 17, 1959, *Steyermark 86179*; along grassy border of sidewalk, St. Louis, 2020 Tower Grove Avenue, October 17, 1959, *Isely & Pohl*).

Native of Asia; introduced and naturalized in the United States from Virginia and Kentucky to Missouri, north to Massachusetts and Illinois.

This species somewhat resembles *P. Persicaria*, but differs in the smaller, 3-sided achenes, mature sepals lacking the prominently reticulated bases of *P. Persicaria*, and generally more slender inflorescences with longer cilia (hairs) on the margins of the ocreolae. From *P. hydropiperoides* it differs in the smaller achenes, annual habit, differently shaped leaves, shorter inflorescences, and generally darker rose to rose-purple instead of pink, white, greenish-white or dull purplish sepals.

## Section **Persicaria** Smartweed

16. **Polygonum Persicaria** L. var. **Persicaria**
Lady's Thumb Map 833
*Polygonum Persicaria* L. [G, BB, P & S]
Flowers May–October.

Occurs in waste and cultivated ground, along railroads, wet meadows, and alluvial soils along streams. Scattered over the state, but recorded from fewer of the northern counties.

Native of Europe; introduced into North America, where naturalized throughout the United States and Canada.

This nearly glabrous annual usually has pink or dull rose-purplish and green, rarely white, flowers. Often the upper sides of the leaves are marked with a purplish or brownish patch near the middle. This species is reported to have similar properties of poisoning as those given for *P. Hydropiper*.

This is one of the species of *Polygonum* from which smartweed honey is derived. Other species are *P. pensylvanicum*, *P. longistylum*, and *P. hydropiperoides*. The achenes are frequently eaten by wildfowl and upland game birds.

17. **Polygonum punctatum** Ell.
Water Smartweed Map 834
Flowers July–November.

Occurs in wet ground bordering streams, sloughs, and ponds, low damp places in woodland, and swampy places.

Missouri material may be divided into the following variations:

a. Plants annual, the stems from a short tap-root, rooting at few or no nodes and scarcely prostrate at base; achenes usually both 3-sided and 2-sided on the same plant . 17b. P. PUNCTATUM var. CONFERTIFLORUM

a. Plants perennial, the stems prostrate at base and rooting at the nodes; achenes all 3-sided . *b*

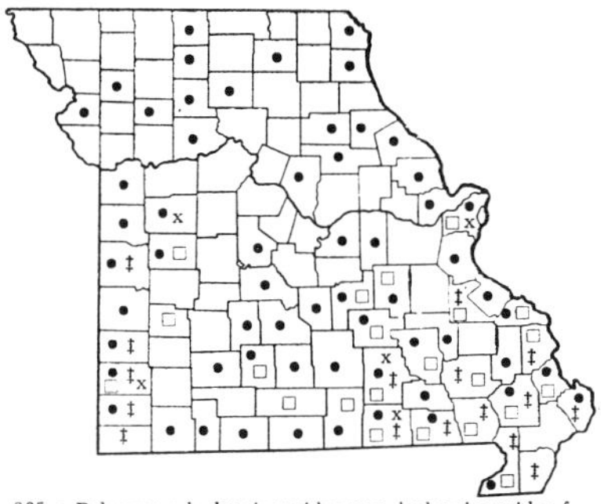

835 • Polygonum hydropiperoides var. hydropiperoides f. hydropiperoides (Wild Water Pepper)
835 x Polygonum hydropiperoides var. hydropiperoides f. strigosum
835 ‡ Polygonum hydropiperoides var. opelousanum
835 □ Polygonum hydropiperoides var. setaceum

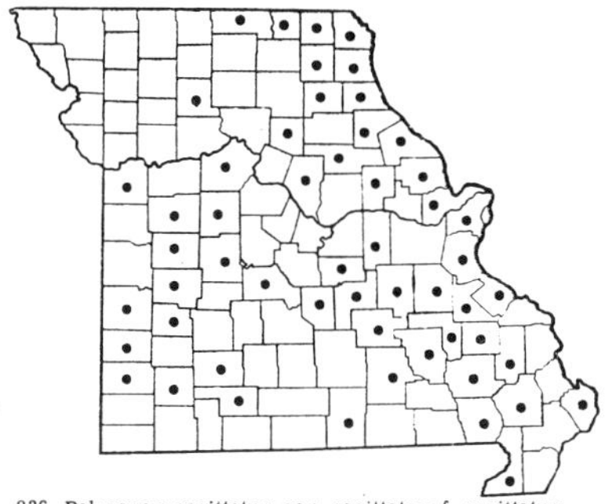

836 Polygonum sagittatum var. sagittatum f. sagittatum (Arrow-leaved Tear-thumb)

837 Polygonum arifolium var. pubescens (Halberd-leaved Tear-thumb)

b. Main leaves 1–2 cm. broad; fruiting calyx 3–3.5 mm. long; branches of the inflorescence from the axil of a blade-bearing leaf; stems much branched above the rooting base; common throughout Missouri 17a. P. PUNCTATUM var. PUNCTATUM

b. Main leaves 2–4 cm. wide; fruiting calyx 3.5–5 mm. long; branches of the inflorescence from the axil of an ocrea not bearing a leafy blade; stems simple or nearly so above the rooting base; rare, known only from Dunklin Co., southeastern Mo. . . 17c. P. PUNCTATUM var. MAJUS

17a. **Polygonum punctatum** var. **punctatum** Map 834

*Polygonum punctatum* Ell. [G, P & S]

This variety is common throughout Missouri and probably occurs in every county.

Ranges from the West Indies, Florida and Texas, north to New York, Iowa, and Kansas.

17b. **Polygonum punctatum** var. **confertiflorum** (Meisn.) Fassett Map 834

*Polygonum punctatum* var. *leptostachyum* (Meisn.) Small and other auth. [G, BB, P & S], not *P. acre* var. *leptostachyum* Meisn., a tropical American variety north to Florida and Texas.

This variety is widespread in Missouri, but apparently somewhat less common than var. *punctatum*. It probably occurs in most counties of the state.

Ranges from Quebec to British Columbia, south to Virginia, Tennessee, Arkansas, Oklahoma, Colorado, Idaho, and Oregon.

17c. **Polygonum punctatum** var. **majus** (Meisn.) Fassett Map 834

*Polygonum punctatum* var. *robustius* Small

*Polygonum robustius* (Small) Fern. [G, BB, P & S]

Known only from southeastern Missouri in New Madrid (Pawpaw Junction, September 4, 1897, *Bush 391*) and Dunklin (Kennett, August 22, 1894, *Bush 557*) counties.

Ranges from South and Central America, and the West Indies, and from Florida to Nova Scotia, locally inland to Indiana and Missouri.

As Fassett has shown in his paper on *Polygonum punctatum* (Brittonia 6: 369–93. 1949), the occurrence of so many intermediate specimens makes it impossible to maintain *P. robustius* as a species distinct from *P. punctatum* and varieties, although in certain parts of their range, as in Nova Scotia, they appear more distinct with little or no intergradation.

Occasionally, the stems of *P. punctatum* and varieties are purple-red simulating those of *P. Hydropiper*. The spikes have less of a tendency to droop as much at the tip as those to *P. Hydropiper*.

The achenes of *P. punctatum* and varieties serve as desirable food for wildfowl and upland game birds, and the plant is browsed by deer and muskrat. The leaves possess a sharp burning quality similar to those of *P. Hydropiper*, and have the same type of poisoning and irritant effects as that species.

18. **Polygonum hydropiperoides** Michx.
Wild Water Pepper Map 835
Flowers June–November.

Occurs in and bordering water of swamps, natural and made ponds, along streams and spring branches, and low wet woodland.

Missouri material may be divided into the following varieties:

a. Hairs of the ocreae (sheath-like stipules) spreading (do not confuse with the hairs projecting from the margins at the summit) . 18d. P. HYDROPIPEROIDES var. SETACEUM

a. Hairs of the ocreae (sheath-like stipules) erect and appressed (parallel to or lying along surface) . . . . . . . . . . . . *b*

b. Mature achene slightly protruding from calyx, averaging 1.9 mm. long (1.5–2.5 mm.), 1.5–2 mm. wide (4/5 as wide); leaves narrowly lanceolate to nearly linear, 0.4–1.5 cm. broad . . . 18c. P. HYDROPIPEROIDES var. OPELOUSANUM

b. Mature achene included within the calyx, averaging 2.2 mm. long and 1.7–2 mm. wide (2/3 as wide); leaves narrowly to broadly lanceolate, 0.5–2.5 cm. broad . . . . *c*

c. Leaves glabrous to minutely strigose (with hairs lying pressed to surface or paralleling surface) on one or both surfaces . . . 18a. P. HYDROPIPEROIDES var. HYDROPIPEROIDES f. HYDROPIPEROIDES

c. Leaves densely long-strigose . . . . 18b. P. HYDROPIPEROIDES var. HYDROPIPEROIDES f. STRIGOSUM

18a. **Polygonum hydropiperoides** var. **hydropiperoides** f. **hydropiperoides** Map 835

*Polygonum hydropiperoides* Michx. [G, P & S]

*Polygonum hydropiperoides* var. *Bushianum* Stanford [G]

*Polygonum hydropiperoides* var. *euronotorum* Fern. [G]

Occurs throughout Missouri and is the commonest variation found in the state.

Ranges from Nova Scotia to British Columbia, south to Florida, Texas, and California.

18b. **Polygonum hydropiperoides** var. **hydropiperoides** f. **strigosum.** (Small) Stanford Map 835

*Polygonum hydropiperoides* f. *strigosum* (Small) Stanford [G, P & S]

Scattered in southern and central Missouri.

Ranges from Quebec to Pennsylvania and Missouri.

18c. **Polygonum hydropiperoides** var. **opelousanum** (Riddell) Stone Map 835

*Polygonum opelousanum* Riddell [G, P & S]

Occurs in southeastern Missouri northeast to Cape Girardeau and St. Francois counties, west to Oregon County, and reappears in southwestern Missouri from McDonald County north to Bates County.

Ranges from Florida to Texas and Mexico north on the coastal plain to Massachusetts and New York, inland north to Tennessee, Illinois, Missouri, and Oklahoma.

18d. **Polygonum hydropiperoides** var. **setaceum** (Baldw.) Gl. Map 835

*Polygonum setaceum* Baldw. [G, P & S]

*Polygonum setaceum* var. *interjectum* Fern. [G]

*Polygonum setaceum* var. *tonsum* Fern.

*Polygonum hydropiperoides* Michx. var. *hydropiperoides* [of Shinners]

Occurs in southern and east-central Missouri, north to St. Louis, Phelps, Henry, and Jasper counties.

Ranges from Florida to Texas, north to Massachusetts, New York, Michigan, Missouri, and Oklahoma.

Gleason's interpretation (*New Ill. Fl.* vol. 2: 82. 1952) of this group has been followed in the present treatment and is believed to be the most general satisfactory alignment of the species available at the present time. The present author has collected and studied a large series of specimens of the *P. hydropiperoides – P. setaceum – P. opelousanum* alliance during the past twenty-five years, but in attempting to separate the plants according to the outline proposed by Fernald in the eighth edition of *Gray's Manual* (p. 576, pp. 586–87), I have found numerous intergradations and intermediates at all levels of comparison. In *P. hydropiperoides* it was not found possible to maintain var. *Bushianum* apart from var. *hydropiperoides* on account of intergradation in length of cilia on the ocreae, and many plants with long cilia in Missouri possessed narrow leaves that would have fit the description for var. *euronotorum* Fern., supposedly a variety confined to Virginia and South Carolina. Similarly with *P. setaceum* no constancy could be found between the pubescence of the leaves and ocrea of the three variations indicated by Fernald. Moreover, habit and habitat characters given by Fernald, supposed to distinguish *P. opelousanum* and *P. setaceum,* were found inadequate and unreliable guides, and, as noted by Gleason, it is not possible to distinguish *P. opelousanum* from *P. hydropiperoides* except at maturity.

Continued maintenance of these three entities as distinct species, in the light of such intergradation and lack of definition, would only lead to further confusion. Until a completely new and badly needed revision of

Plate no. 146. **1.** Polygonum punctatum, × $^{2}/_{7}$; Details from Small, The New York Botanical Garden. **2.** Polygonum hydropiperoides var. hydropiperoides, × $^{2}/_{7}$. **3.** Polygonum sagittatum, × $^{2}/_{7}$; Details from Small, The New York Botanical Garden. **4.** Polygonum virginianum, × $^{2}/_{7}$; *a.* Fruit, × $1^{5}/_{7}$; Details from Small, The New York Botanical Garden. **5.** Fagopyrum sagittatum, × $^{2}/_{7}$; Details from Small, The New York Botanical Garden. **6.** Polygonum arifolium var. pubescens, × $^{2}/_{7}$. **7.** Polygonum scandens var. scandens, × $^{2}/_{7}$; Details from Small, The New York Botanical Garden. **8.** Polygonum Convolvulus var. Convolvulus, × $^{2}/_{7}$.

PLATE NO. 146

this alliance is forthcoming, Gleason's interpretation may well be taken as a guide to the future approach of study of these plants.

The flowers of *P. hydropiperoides* var. *hydropiperoides* are either pink or roseate, and rarely white, those of var. *setaceum* usually white, and of var. *opelousanum* greenish-white or green and purplish. *Polygonum hydropiperoides* var. *hydropiperoides* usually is found in dense stands, in shallow water. It also occurs in a purely vegetative state submerged in the cold water of springs and spring branches in the Ozark region. In such situations, and in fast-running spring water, the submerged stems and leaves bend horizontally, paralleling the stream bed, with their olive-green leaves often flushed with purple and red shades waving in the current. Frequently these submersed leaves are bronze-purple on the lower surface and olive-green on the upper surface, or the leaves may be tinged copper or reddish in some plants, orange and green in others. Achenes of *P. hydropiperoides* frequently are eaten by various kinds of wildfowl.

## Section **Echinocaulon** Tear-thumb

19. **Polygonum sagittatum** L. var. **sagittatum** f. **sagittatum** Arrow-leaved Tear-thumb Map 836

*Polygonum sagittatum* L. [G, BB, P & S]

Flowers June–October.

Occurs in swampy places around springs, spring branches, sloughs, and streams. Common over a large section of eastern, southern, and central Missouri, and locally northwest to Livingston County; not recorded from a number of Ozark counties.

Ranges from Florida to Texas, north to Newfoundland, Quebec, Ontario, and Saskatchewan.

A nearly smooth-stemmed variation, var. *gracilenta* Fern., occurs in Virginia, and a form of var. *sagittatum*, f. *chloranthum* Fern., with green flowers and reduced ciliation of leaves, occurs from Maine to New York.

20. **Polygonum arifolium** L. var. **pubescens** (Keller) Fern. Halberd-leaved Tear-thumb Map 837

*Polygonum arifolium* of some auth. [P & S], not L.

Flowers July–October.

Occurs in wet sandy swales of spring branches. Known only from Crowley Ridge in Stoddard County in southeastern Missouri (swampy sandy ground of spring branch at base of Crowley Ridge, T25N, R11E, northwest ¼ sect. 6, 3½ mi. southeast of Bloomfield, August 20, 1954, *Steyermark 76781*; T25N, R10E, southeast ¼ sect. 11, 4–4¼ mi. south of Bloomfield, October 18, 1955, *Steyermark 80440*; without locality in Stoddard Co., *Bush*).

Ranges from New Brunswick to Ontario and Minnesota, south to New Jersey, Pennsylvania, Ohio, Indiana, and Missouri.

Typical *P. arifolium* var. *arifolium* has larger achenes 4–4.2 mm. instead of 3–3.5 mm. broad and 3–3.2 mm. instead of 2.2–2.6 mm. thick with a slight point, sometimes projecting from the center of the sides of the achene.

This rare *Polygonum* in Missouri is associated at one of its Stoddard County stations with such other relict eastern species as *Bartonia paniculata*, *Trisetum pensylvanicum*, and *Pyrus melanocarpa*, accompanied by such other rarities as *Habenaria ciliaris*, *Eupatorium fistolosum*, and *Ilex opaca*.

## Section **Tovara** Virginia Knotweed

21. **Polygonum virginianum** L.
Virginia Knotweed Map 838

Flowers July–October.

Occurs in rich woodland. Throughout Missouri.

Two variations occur in Missouri:

Leaves usually hairy on lower surface, hairy on upper surface; leaves thick and firm . . . . 21a. P. VIRGINIANUM var. VIRGINIANUM

Leaves glabrous (without hairs) or nearly so; leaves thin and membranous . 21b. P. VIRGINIANUM var. GLABERRIMUM

21a. **Polygonum virginianum** var. **virginianum** Map 838

*Polygonum virginianum* L. [BB, P & S]

*Tovara virginiana* (L.) Raf. [G]

This is the common variation in the state and doubtless occurs in every county.

Ranges from Quebec and New Hampshire to Ontario and Minnesota, south to Florida and Texas.

21b. **Polygonum virginianum** var. **glaberrimum** (Fern.) Steyerm. Map 838

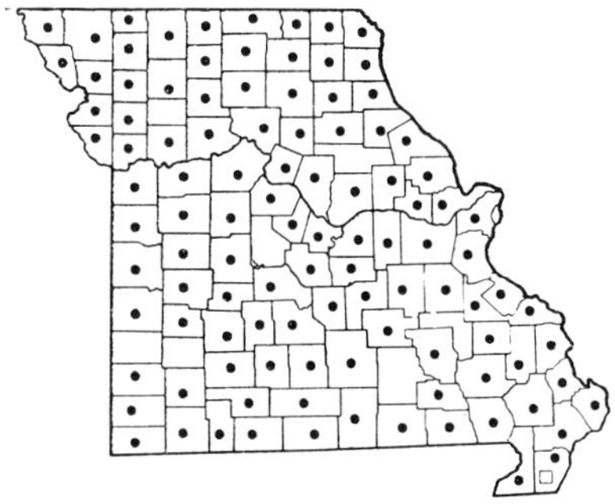

838 • Polygonum virginianum var. virginianum (Virginia Knotweed)
838 □ Polygonum virginianum var. glaberrimum

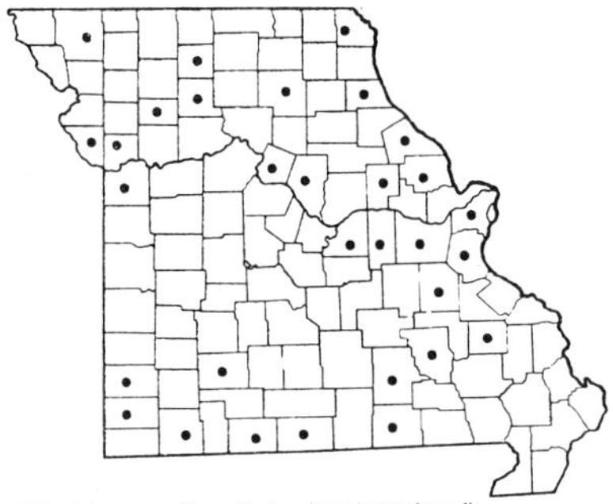

839 Polygonum Convolvulus (Black Bindweed)

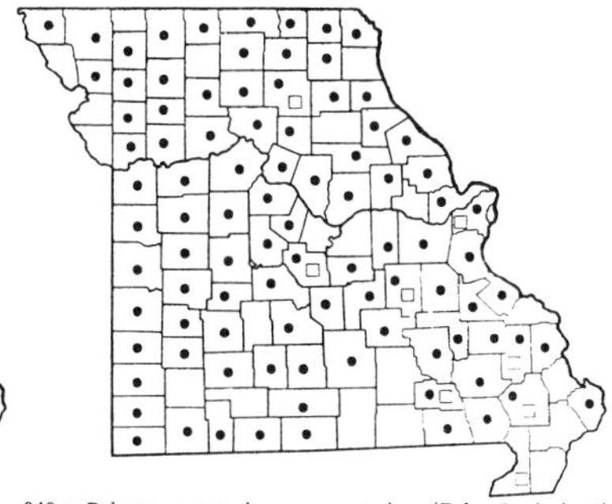

840 • Polygonum scandens var. scandens (False Buckwheat)
840 □ Polygonum scandens var. cristatum

*Tovara virginiana* var. *glaberrima* Fern. [G]

Known only from Pemiscot County in southeastern Missouri (virgin woods on level ground on property of Son Rone, T20N, R12E, sect. 12, 3 mi. south-southwest of Portageville, July 29, 1956, *Steyermark 82232*).

Ranges from Virginia to Florida and Missouri.

22. **Polygonum Convolvulus** L. var. **Convolvulus**
Black Bindweed Map 839

Flowers May–November.

Occurs along roadsides, railroads, in cultivated fields and waste ground. Scattered throughout Missouri, and probably eventually to be found in most of the counties.

Native of Europe; introduced into North America, where naturalized over most of the United States and Canada.

Another variation, var. *subalatum* Lej. & Court., differs in the fruiting calyx 5–8 mm. instead of 4–5 mm. long, and with 3 developed winged keels.

Some primitive peoples are reported to use the achenes which are mashed into a meal or flour said to possess somewhat the qualities of buckwheat flour.

Section **Tiniaria** Climbing Buckwheat

23. **Polygonum scandens** L. False Buckwheat
Map 840

Flowers July–November.

Occurs in moist thickets, alluvial woods and flood plain forest, about lakes and sloughs, streams, along rocky bluffs, and on dry wooded slopes.

23a. **Polygonum scandens** var. **scandens**
Map 840

*Polygonum scandens* L. [G, P & S]

This is the common variation encountered.

Throughout Missouri, and doubtless in every county.

Ranges from Florida to Texas, north to Nova Scotia, Quebec, Ontario, and Manitoba.

23b. **Polygonum scandens** var. **cristatum**
(Engelm. & Gray) Gl. Map 840

*Polygonum cristatum* Engelm. & Gray [G]

*Polygonum scandens* var. *dumetorum* (L.) Gl. [BB]

*Polygonum dumetorum* of some auth., not L.

This is the less common variation in Missouri, and is known in scattered sections of the state in Macon (*Steyermark 77373*), Miller, Crawford, Bollinger (*Steyermark 77735*), and Stoddard (*Steyermark 76845*) counties.

Ranges from Florida to Texas, north to Massachusetts, Vermont, New York, Indiana, Illinois, Minnesota, and Kansas.

Gleason's interpretation of the *Polygonum scandens* – *P. cristatum* alliance is followed in the present treatment, rather than Fernald's since the variations in the size of the achene, the length of the fruiting calyx, and the relative width of the wing appear to intergrade too freely to permit recognition of these entities as distinct species.

The achenes of *P. scandens* var. *scandens*, as with those of *P. Convolvulus*, have been used in making meal.

24. **Polygonum cuspidatum** Sieb. & Zucc.
Japanese Knotweed Map 841

Also known as Mexican Bamboo.

Flowers August–September.

Escaped into waste ground, fence rows, and along roadsides and railroads. Scattered in a few sections of

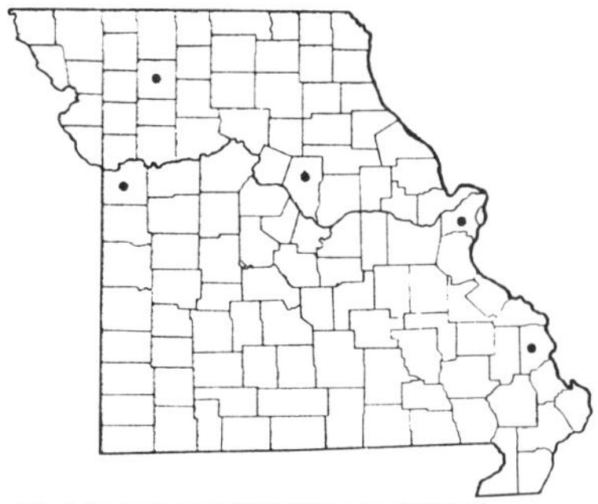

841 Polygonum cuspidatum (Japanese Knotweed)

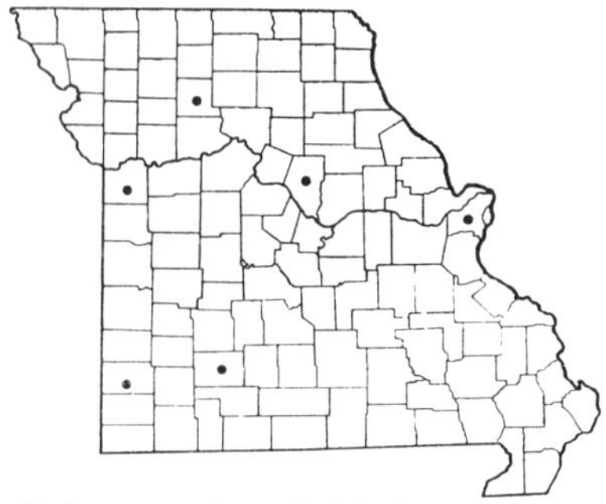

842 Fagopyrum sagittatum (Buckwheat)

843 Polygonella americana (Jointweed)

the state, in Daviess, Jackson, Boone, St. Louis, and Cape Girardeau counties.

Native of Japan; introduced into North America, where naturalized from Newfoundland to Ontario and Minnesota, south to Maryland and Missouri.

The young underground rootstocks and young leafy stems can be cooked as a palatable vegetable substitute for asparagus, or treated as a salad.

## 4. **Fagopyrum** Mill. Buckwheat

**Fagopyrum sagittatum** Gilib. Buckwheat
Map 842

*Fagopyrum esculentum* Moench. [BB, P & S, Steyerm.]

Flowers May–September.

Escaped into fields, pastures, cultivated and waste ground, along roadsides, and railroads. Rarely persisting and known as an occasional waif from a few scattered counties in the state.

Native of Asia; introduced into North America.

The well-known buckwheat flour is produced from the achenes of this plant. Some people are allergic to buckwheat, developing a rash after eating dishes prepared from the flour. Some persons are also sensitive to the leaves, developing a dermatitis after contact. Stock of all kinds having an uncolored skin may be poisoned if they are directly exposed to the sun after eating buckwheat. The fragrant flowers are the source of buckwheat honey, the plants are often grown as a green manure to enrich the soil, and the discarded dried hulls or remains of the fruits are used as a summer mulch for roses and other flower beds.

## 5. **Polygonella** Michx. Jointweed

**Polygonella americana** (Fisch. & Mey.) Small
Jointweed Map 843

Flowers July–October.

Occurs in dry sandy ground, mainly on Crowley Ridge in the southeastern lowland section of the state in Scott, Stoddard, and Dunklin counties, and locally in Iron County in the southeastern Ozarks.

Ranges from southeastern Missouri to Georgia and Texas.

This plant with its small, crowded, needlelike, gray-green leaves has the appearance of a small spreading heather or juniper. The branches usually are low and spreading, mostly 3–9 dm. tall, and at their tips develop racemes of showy white to pink-tinged flowers, which continue to bloom through summer and fall. This plant does satisfactorily in a rock garden if provided with acid sand and good drainage. A plant has prospered at the author's botanical preserve since 1954. It can also be easily grown from seed and should make a desirable horticultural accessory.

---

Plate no. 147. **1.** Polygonella americana; *a.* Flowering portion of plant, $\times\ ^2/_5$; *b.* Leafy portion of plant, $\times\ ^2/_5$; *c.* Flower, side and top views, $\times\ 2^2/_5$; Details from Small, The New York Botanical Garden. **2.** Brunnichia cirrhosa, $\times\ ^2/_5$; *a.* Fruit $\times\ ^4/_5$; Details from Small, The New York Botanical Garden. **3.** Cycloloma atriplicifolium, $\times\ ^2/_5$; *a.* Fruit, $\times\ 1^3/_5$; *b.* Female flower, $\times$ 6; Details from Small, The New York Botanical Garden. **4.** Kochia scoparia, $\times\ ^2/_5$; *a.* Fruiting calyx, $\times\ 3^1/_5$. **5.** Chenopodium Pumilio, $\times\ ^2/_5$; Details from Small The New York Botanical Garden. **6.** Chenopodium Botrys, $\times\ ^2/_5$; Details from Small, The New York Botanical Garden

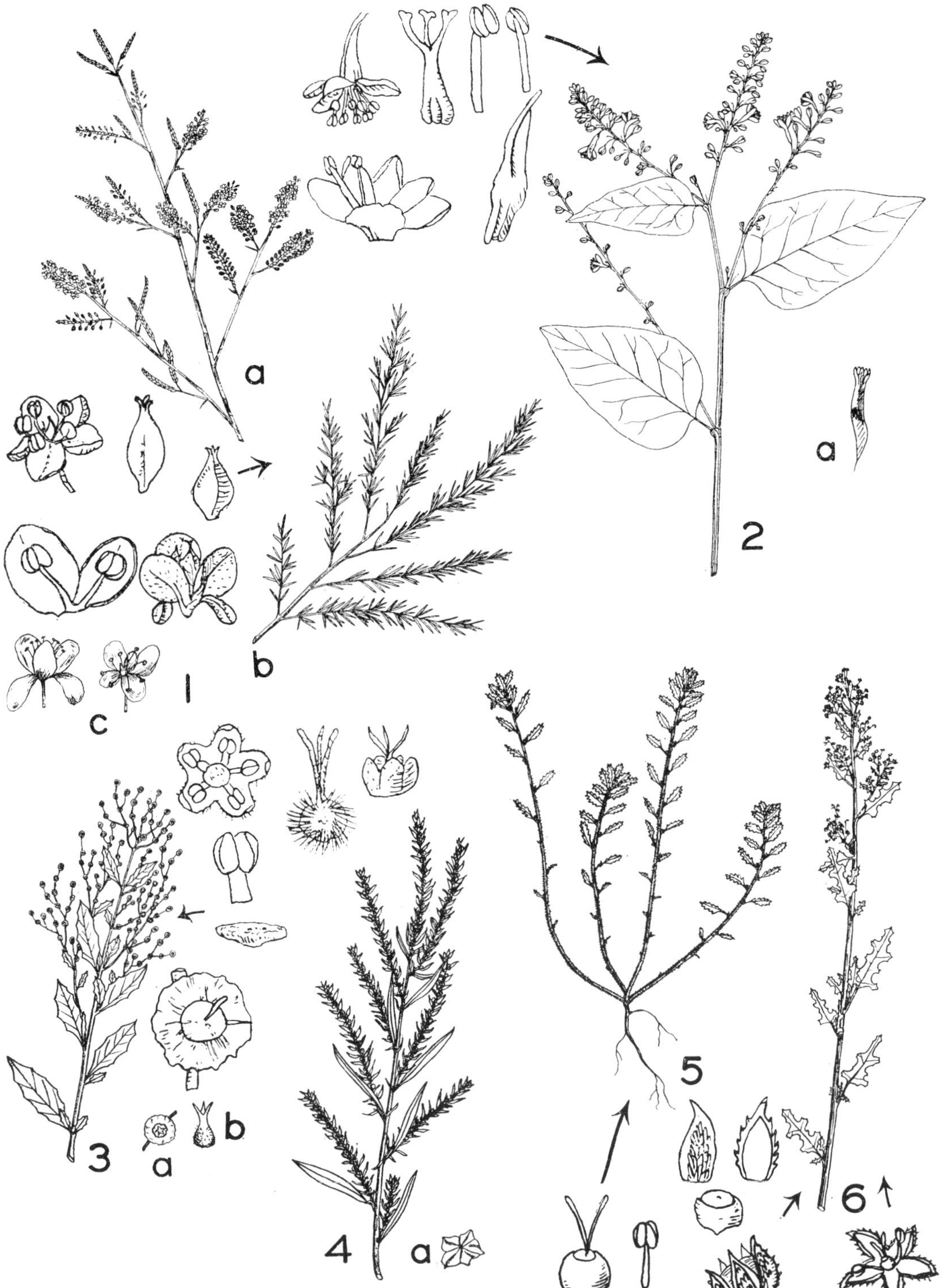


PLATE NO. 147

### 6. **Brunnichia** Banks

**Brunnichia cirrhosa** Gaertn. Ladies' Eardrops. Map 844

Also known as Buckwheat Vine.

Flowers May–August; fruits August–October.

Occurs in swamps, low wet woods, and moist alluvial thickets in the lowland section of southeastern Missouri, north to Bollinger and Wayne counties and west to Ripley County.

Ranges from Florida to Texas, north to South Carolina, Kentucky, Illinois, Missouri, and Oklahoma.

The common name is derived from the appearance of the long narrow winged fruit to ear decorations. This woody vine often climbs to the tops of tall trees in favorable situations. The flowers are greenish.

## Order **CARYOPHYLLALES**

## Fam. **CHENOPODIACEAE** (Goosefoot Family)

### I. *Key based chiefly on vegetative characters*

a. Leaves or flower-bracts ending in a more or less prickly or spiny stiff tip . . . . . . . . 9. SALSOLA
a. Neither leaves nor flower-bracts ending in a prickly or spiny stiff tip . . . . . . . . . . . *b*
  b. Star-shaped hairs (stellate pubescence) present on lower surface of leaves, on calyx of female flowers, or on some part of plant (do not confuse plants with scales or scurfy or mealy covering, which do not belong here) . . . . . . . . . . . . . . . . . . . . . . . . *c*
    c. Leaves narrowly linear with sides parallel or nearly so, sessile (without stalks) . . 7. CORISPERMUM
    c. Leaves lanceolate to elliptic-ovate, usually broader in the middle or below the middle, short- or long-stalked . . . . . . . . . . . . . . . . . . . . . 5. AXYRIS
  b. Plants hairy or without hairs, but if hairy the hairs not star-shaped (scurfy, scaly, or mealy coverings, if present, belong here) . . . . . . . . . . . . . . . . . . . . *d*
    d. Main leaves either thread-like, narrowly linear, or narrowly lanceolate, with their sides parallel or nearly so, nearly of equal width throughout or slightly broader in the middle, or if broader and dilated at the base then many times (8–15) as long as broad; no teeth, lobes, or irregularities on margins of leaves . . . . . . . . . . . . . . . . . . . *e*
      e. Hairs present on some part of plant . . . . . . . . . . . . . . . . 2. KOCHIA
      e. Hairs absent . . . . . . . . . . . . . . . . . . . . . . . . . *f*
        f. No mealy, scaly, or scurfy covering on leaves, stems, or flowers . . . . . . . . . . *g*
          g. Leaves thick and fleshy, nearly terete (mostly rounded in cross-section) or flattened above, not nerved; calyx 5-parted; fruit surrounded or enclosed by the calyx . 8. SUAEDA
          g. Leaves not fleshy, flattened with upper and lower sides, 1-nerved; calyx of 1 small sepal, or if rarely with 3 sepals, these minute and thin-transparent; fruit free from calyx and exposed and protruding above it . . . . . . . . . . . 7. CORISPERMUM
        f. Mealy, scaly, or scurfy covering, this sometimes white or gray, on leaves, stems, or flowers . . . . . . . . . . . . . . . . . . . . . . . . . . . *h*
          h. Upper part of stem and lower surface of leaves densely white- or gray-mealy . . 3. CHENOPODIUM
          h. Stem and leaves dark or dull green throughout . . . 4. ATRIPLEX PATULA var. LITTORALIS
    d. Main leaves oblong, elliptical, broadly lanceolate, ovate to triangular, usually broader at the middle or in the lower half, mostly only 1½–5 times as long as broad, or if linear, thread-like, or narrow, at least some teeth or irregularities are present; teeth, lobes, or irregularities on margins of leaves . . . . . . . . . . . . . . . . . . . . . . . . . *i*
      i. Plants aromatic, the stems and leaves or flowers with glandular hairs or sessile glands. . 3. CHENOPODIUM
      i. Plants not aromatic, no part of plant with glandular hairs or glands (plants with mealy, scurfy, or scaly covering belong here) . . . . . . . . . . . . . . . . . *j*
        j. Upper part of stem and young leaves with white cobwebby hairs . . . . 1. CYCLOLOMA
        j. No white cobwebby hairs on any part of stem or leaves . . . . . . . . . . . *k*
          k. Sepal only 1 in each flower; stamen 1 . . . . . . . . . . . . 6. MONOLEPIS

k. Calyx 2–5-parted in all flowers or at least in the staminate flowers; stamens mostly 3–5 (rarely 1–2) . . . . . *l*
  l. Stamens and pistils separated from each other on different flowers of the same plant (monoecious) or on different plants (dioecious); female flowers mostly without any calyx; fruit hidden in a pair of rounded, triangular, or baseball-diamond-shaped little bracts . . . . . 4. Atriplex
  l. Stamens and pistils in the same flowers, flowers all perfect; fruit surrounded by the calyx, without any bracts around it . . . . . *m*
    m. Fruiting calyx with a continuous wing around it; plants frequently of open sandy places . . . . . 1. Cycloloma
    m. Fruiting calyx without any wing; plants more often found in mud, on loamy, clayey, or fertile organic soils in woodland, along bluffs, bottomland, waste ground, and about dwellings . . . . . 3. Chenopodium

## II. *Key based mainly on flower and fruit characters*

a. Calyx of usually only 1 sepal in each flower (if rarely 3 sepals, these minute and thin-transparent); fruit exposed, free, not surrounded or enclosed by bracts or by the calyx . . . . . *b*
  b. Main leaves linear, with mostly parallel sides and of about equal width throughout; no leaves with any teeth or irregularities on margin, sessile (without a stalk); sepals 1–3, minute and thin-transparent; fruit usually with a thin or winged margin . . . . . 7. Corispermum
  b. Main leaves somewhat arrowhead-shaped, at least the upper ones broadest near the base, or if narrowed at the base and linear or not arrowhead-shaped, at least with 1 or more teeth or irregularities on margins; main leaves with at least 1 large tooth on each side, petioled (with a stalk); sepal 1; fruit without any wing . . . . . 6. Monolepis
a. Calyx 2–5-parted in all flowers or at least in the staminate flowers; fruit either surrounded and enclosed by the calyx or hidden by 2 bracts . . . . . *c*
  c. Stamens and pistils separated from each other on different flowers of the same plant (monoecious) or on different plants (dioecious); female flowers mostly without any calyx; fruit hidden in a pair of rounded, triangular, or baseball-diamond-shaped little bracts . . . . . 4. Atriplex
  c. Without the above combination of characters . . . . . *d*
    d. Leaves with teeth, lobes, or irregularities on the margins; leaves ovate to triangular, oblong, elliptical, or broadly lanceolate, usually broader at the middle or in the lower half, mostly only $1\frac{1}{2}$–5 times as long as broad . . . . . *e*
      e. Fruiting calyx with a continuous wing around it; plants frequently of open sandy places 1. Cycloloma
      e. Fruiting calyx without any wing; plants more often found in mud, on loamy, clayey, or fertile organic soils in woodland, bottomland, along bluffs, waste ground, and about dwellings . . . . . 3. Chenopodium
    d. Leaves without teeth, lobes, or irregularities on margins; leaves either thread-like, narrowly linear, or lanceolate, their sides parallel or nearly so, nearly of equal width throughout or slightly broader in the middle, many times (8–15) as long as broad, or, if elliptic-ovate then the flowers unisexual with male and female flowers separate . . . . . *f*
      f. Fruiting calyx or fruit partly or wholly winged, the fruit either 2-winged at its summit or the fruiting calyx-lobes themselves developing wings or transverse keels . . . . . *g*
        g. Fruit 2-winged at its summit, the calyx not winged; lower surface of leaves and calyx of female flowers with star-shaped hairs (stellate pubescent); stamens and pistils in separate flowers on the same plant (monoecious), the staminate flowers raised high above rest of inflorescence; fully grown female bracts and calyx conspicuous, white . 5. Axyris
        g. Fruit not winged, but each of the 5 calyx-lobes are winged or transversely keeled; hairs, when present, simple, not star-shaped; stamens and pistils in the same flower, perfect; calyx and bracts not white . . . . . *h*
          h. Leaves and bracts with bristly or spiny tips . . . . . 9. Salsola
          h. Leaves and bracts soft, lacking bristly or spiny tips . . . . . 2. Kochia
      f. Fruiting calyx and fruit without any winged part . . . . . 8. Suaeda

844 Brunnichia cirrhosa (Ladies' Eardrops)

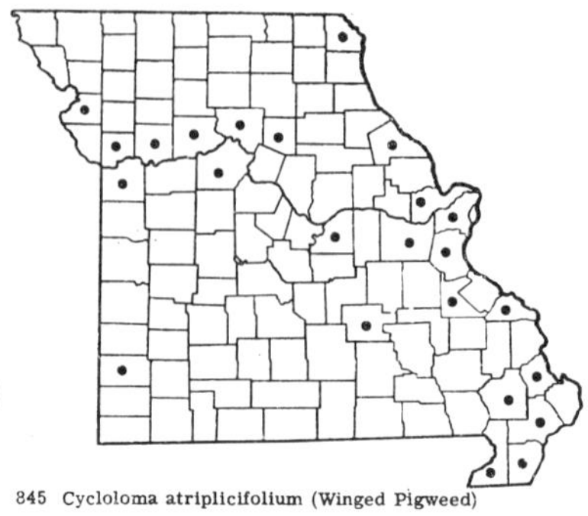

845 Cycloloma atriplicifolium (Winged Pigweed)

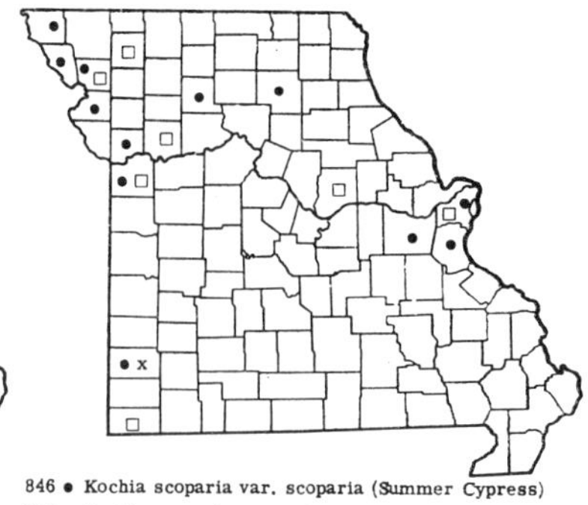

846 • Kochia scoparia var. scoparia (Summer Cypress)
846 x Kochia scoparia var. pubescens
846 □ Kochia scoparia var. culta (Firebush)

## 1. **Cycloloma** Moq. Winged Pigweed

**Cycloloma atriplicifolium** (Spreng.) Coult.
Winged Pigweed — Map 845
Flowers June–October.

Occurs in sand flats along the Missouri and Mississippi rivers, sand bars along a few Ozark streams, sand ridges, sandy fields and sandy open ground, occasionally along railroads and sandy roadsides. Found along the counties bordering the Mississippi and Missouri rivers in eastern and central Missouri, and scattered elsewhere in southern Missouri in Phelps, Dent, St. Francois, and Jasper counties; on sand of Crowley Ridge in southeastern Missouri in Scott, Stoddard, and Dunklin counties.

Ranges from Indiana to Manitoba, south to Illinois, Arkansas, Texas, Mexico, Arizona, and California, occasionally spread east to Quebec, New England, and New Jersey.

The plant is a characteristic sand-inhabiting species. Late in autumn the fruit and the rest of the plant may turn a reddish-purple.

## 2. **Kochia** Roth

**Kochia scoparia** (L.) Schrad. Summer Cypress, Fire Bush — Map 846
Flowers July–October.

Missouri material is represented by the following variations:

a. Leaves narrowly linear, thread-like, mostly 1–2 mm. broad, inconspicuously hairy; shape of plant ovoid or globular, branches close together; plant turning purplish-red or deep rose-red . . . . c. K. SCOPARIA var. CULTA

a. Leaves linear-lanceolate or oblanceolate, 3–8 mm. broad, more conspicuously hairy, the hairs up to 2 mm. long; shape of plant pyramidal, branches loosely spreading, not close together; plant remaining green . . . . *b*

- b. Stem downy, conspicuously densely hairy with sordid or brownish-white hairs; leaves and flowers conspicuously hairy; rare type . . . b. K. SCOPARIA var. PUBESCENS
- b. Stem sparsely hairy to nearly glabrous; leaves and flowers sparsely or moderately hairy to nearly glabrous; common type . a. K. SCOPARIA var. SCOPARIA

a. **Kochia scoparia** var. **scoparia** — Map 846
*Kochia scoparia* (L.) Roth [G, BB, P & S]
*Kochia prostrata* (L.) Schrad. [P & S]
*Kochia alata* Bates, Am. Bot. 24: 52. 1918.

Occurs in waste ground, along railroads and roadsides, where escaped from cultivation. Mostly in some of the counties of northern and central Missouri, especially so along the northwestern section bordering the Missouri River, and locally south in Jasper County.

Native of Eurasia; introduced in the western United States, from where it has spread eastward to New England and southward.

b. **Kochia scoparia** var. **pubescens** Fenzl — Map 846
*Kochia scoparia* var. *subvillosa* Moq.
*Kochia scoparia* var. *soongorica* Moq.
*Kochia Sieversiana* (Pall.) C. A. Mey.

Known only from Jasper County, southwestern Missouri (rich waste ground, Stockyard Switch, Joplin, August 10, 1949, *Palmer 49750*; September 26, 1949, *Palmer 49835-A*; August 12, 1951, *Palmer 52979*; October 15, 1952, *Palmer 55345*).

c. **Kochia scoparia** var. **culta** Farw. Firebush
Map 846

*Kochia scoparia* var. *trichophila* (Stapf) Bailey
*Kochia trichophylla* Stapf

Planted commonly in gardens as an ornamental. Rarely escaping and becoming naturalized in fields and along railroads, from a few scattered localities. The plants turn purplish red to red in autumn.

*Kochia scoparia* is highly variable. The leaves vary in length, width, and pubescence. *Kochia alata* appears to be one of these variants. The var. *pubescens* is well-marked as an extreme variation in pubescence.

Large quantities of pollen shed during the summer months by this species account for quite a number of hay fever cases in Iowa, South Dakota, Wyoming, Colorado, Wisconsin, and Minnesota. In Missouri it is too rare to be of any importance in this respect. In Japan and China the plant is grown chiefly for the seeds used as a cereal or in bread. There the young tips of the plant are cooked as a vegetable, and the entire plant, when dried, used as a broom for sweeping.

## 3. **Chenopodium** L. Goosefoot, Pigweed

Some of the species of this genus can be identified only with mature seeds. The genus has been studied by several botanists in the past four to five decades beginning with Standley in 1916 (N. Am. Fl. 21: 1–93). In 1929 Aellen published on the genus (Fedde Rep. Spec. Nov. Regn. Veg. 26: 31–64, 119–60), and in 1943 Aellen and Just further elaborated (Am. Midl. Nat. 30: 47–76. 1943). The latest work on the North American species by H. A. Wahl (Bartonia 27: 1–46. 1954) is considered by its author as a preliminary study. As many changes in names and in their interpretations have resulted from the studies of these various authors, the synonymy cited below the presently acceptable names indicates the cross-references to these various authors in order to clarify their work and to enable the student to refer to a work which he may have previously followed. The present treatment follows more or less the recent work of Wahl, as indicated above.

a. Plants aromatic or strongly scented, with simple or glandular hairs or with resinous sessile glands on some parts . . . . . . . . . . . . . . . . . . . . . . . . *b*

b. Leaves without hairs; stems chiefly hairless, and without glands, the hairs, if present, not spreading; calyx-lobes smooth or minutely puberulent (short-haired), but without conspicuous yellowish glands or stalked glandular hairs; leaf-blades usually 4–15 or more cm. long, the lower ones up to 15 or more cm. . . . . . . . . . . . . . . . . . . 3. C. AMBROSIOIDES

b. Leaves with obvious hairs; stems glandular and hairy, the hairs spreading; calyx-lobes with either conspicuous yellowish sessile glands or with stalked glandular hairs; leaf-blades small, usually less than 4.5 cm. long, the lower ones longer . . . . . . . . . . . . . . . . *c*

c. Calyx-lobes with yellowish sessile glands, hairs if present without glands; leaves coarsely and rather evenly wavy-toothed, the teeth simple, blunt and mainly 1–4 mm. long; flowers in small glomerules (clusters) of the leaf axils; stems ascending mainly from a spreading or nearly prostrate position . . . . . . . . . . . . . . . . . . . . 1. C. PUMILIO

c. Calyx-lobes with short-stalked glandular hairs; leaves more deeply and irregularly wavy-lobed (sinuate-pinnatifid), the lobes angled or lobed, the larger ones 3–6 mm. long; flowers in forked spreading short cymes; stems mainly erect . . . . . . . . . . . . . 2. C. BOTRYS

a. Plants glabrous (without hairs) or parts of the plant mealy, but not with hairs or yellowish glands; plants not aromatic . . . . . . . . . . . . . . . . . . . . . . . . *d*

d. Lower surface of leaves densely white-mealy . . . . . . . . . . . . . . . . *e*

e. Main leaves linear, oblong, narrowly lanceolate, oblong-elliptical, or ovate-lanceolate, 3–many times as long as broad, mostly less than 1.5 cm. broad (occasionally to 2 cm. near base); no teeth or lobes on leaves . . . . . . . . . . . . . . . . . . . . . . *f*

f. Leaves 3-nerved or pinnately veined, narrowly lanceolate or broader, mostly broader in the middle or near the base or not of the same width throughout, the sides mostly not parallel; pericarp (covering of fruit) easily removed from seed; seeds 1–1.2 mm. broad; scattered in waste ground and along railroads . . . . . . . . . . . 7. C. DESICCATUM

f. Leaves 1-nerved, linear or nearly so with sides mostly parallel to one another, of about the same width throughout; pericarp (covering of fruit) firmly attached to seed, removed with difficulty; seeds 1.2–1.5 mm. broad; plants of rocky and sandy places, glades and bluffs . . . . . . . . . . . . . . . . . . . . . . 8. C. PALLESCENS

e. Main leaves triangular, triangular-rhombic, ovate, or lanceolate to ovate-lanceolate, 1–3 times as long as broad, mostly more than 2 cm. broad; 1, 2, or more teeth or lobes on leaves . . . . *g*

g. Leaves mostly 1–3 cm. long, wavy-margined (sinuate); calyx-lobes often 3 or 4 (or 5 in some horizontal flowers which are terminal in the clusters); calyx-lobes flat or rounded with the midrib flat or only slightly thickened or ridged, not definitely keeled; seeds 0.6 mm. broad, mostly vertical; pericarp (covering of fruit) free from the seed, easily removed; calyx-lobes only partially covering fruit . . . . . . . . . . . . . 5. C. GLAUCUM

g. Without the above combination of characters; leaves mostly larger; calyx-lobes 5; calyx-lobes either conspicuously or somewhat keeled in the center; seeds 0.9–1.5 mm. broad, horizontal; pericarp (covering of fruit) usually firmly attached to the seed and removed with difficulty, or free from the seed in *C. desiccatum*; calyx-lobes mostly completely covering the fruit, or only partially covering in *C. strictum* var. *glaucophyllum* . . . . . . . . . . *h*

h. Surface of seeds either smooth or with other markings, but not honeycombed . . . . . *i*

i. Pericarp (covering of fruit) free from the seed, easily removed. . . . . . . . . . *j*

j. Stems erect, up to 8 dm. tall; leaves lanceolate to ovate-lanceolate, mostly 3–8 times as long as broad, with only basal lobes . . . . . . . . 7. C. DESICCATUM

j. Stems low and spreading, up to 4–5 dm. tall; leaves triangular-ovate or rounded-ovate, about as broad as long, often wavy-toothed or lobed . . . . . . . . . . 9. C. INCANUM

i. Pericarp (covering of fruit) firmly attached to the seed, removed with difficulty . . . . . . *k*

k. Seed 1.5–2 mm. broad; surface of seed with a honeycombed appearance . . 18. C. BUSHIANUM

k. Seed 0.9–1.5 mm. broad; surface of seed more or less smooth . . . . . . . . . . . *l*

l. Seeds mainly 1.1–1.5 mm. broad; leaves often very mealy . . . . . . . . 14. C. ALBUM

l. Seeds 0.9–1.2 mm. broad; leaves often greener . . . . . . . . . . . . . . . . *m*

m. Calyx-lobes only partially covering fruit; lower leaves with short teeth; middle leaves oblong, smooth-edged; flower clusters on erect strict branches; flowering May–September . . . . . . . . . . . . 16. C. STRICTUM var. GLAUCOPHYLLUM

m. Calyx-lobes mostly covering the fruit; lower and middle leaves coarsely toothed, at most $1\frac{1}{2}$ times as long as broad; flower clusters on arching branches; flowering September–October. . . . . . . . . . . . . . . . . . 15. C. MISSOURIENSE

h. Surface of seeds with a honeycombed appearance (alveolate) . . . . . . . . . . . . *n*

n. Seed 1.5–2 mm. broad; calyx-lobes slightly keeled along center; base of style little if at all prominent; pericarp not showing a yellow area at base of style; leaves rather thin to membranaceous. . . . . . . . . . . . . . . . . . . . . . . 18. C. BUSHIANUM

n. Seed 1–1.5 mm. broad; calyx-lobes very prominently keeled along the center, the keel often equaling half the width of the calyx-lobe; base of style often prominent; pericarp showing a yellow area around base of style; leaves rather thick . . . . . . . . . 17. C. BERLANDIERI

d. Lower surface of leaves various shades of green or even tawny or reddish-purple, but not white . . . . *o*

o. Calyx mainly 3-lobed (or 4–5-lobed in horizontal flowers which are terminal in the glomerules); seeds mainly all vertical (except for some of the terminal flowers of the glomerules with horizontal seeds) . . . . . . . . . . . . . . . . . . . . . . . . . . . . *p*

p. Mature flowers in dense heads usually 5–10 mm. in diameter; calyx very fleshy and bright red in fruit, becoming berry-like; seeds all erect or vertical, with an acute or narrow margin, about 1.5 mm. wide . . . . . . . . . . . . . . . . . . . 4. C. CAPITATUM

p. Mature flower clusters smaller; calyx in fruit not becoming berry-like; seeds mostly vertical but with a single horizontal terminal flower in the glomerules, with rounded margins, 0.6–0.8 mm. wide . . . . . . . . . . . . . . . . . . . . . . . . 6. C. RUBRUM

o. Calyx 5-lobed; seeds all horizontal (very exceptionally a few vertical) . . . . . . . . . . *q*

q. Surface of seeds with a honeycombed appearance (alveolate) . . . . . . . . . . . *r*

r. Seed 1.5–2 mm. broad; calyx-lobes only slightly keeled along center; base of style little if at all prominent; pericarp not showing a yellow area at base of style; leaves rather thin to membranaceous. . . . . . . . . . . . . . . . . . . . 18. C. BUSHIANUM

r. Seed 1–1.5 mm. broad; calyx-lobes very prominently keeled along the center, the keel often equaling half the width of the calyx-lobe; base of style often prominent; pericarp showing a yellow area around base of style; leaves rather thick . . . . . . 17. C. BERLANDIERI

q. Surface of seeds either smooth or with other markings, but not honeycombed . . . . . . . . *s*

s. Seed 1.5–2.5 mm. broad, either smooth or with indistinct and irregular grooves or wrinkles, easily removed from pericarp (covering of fruit); leaves thin, rounded or heart-shaped at base, long-pointed (acuminate) with large long-pointed teeth; leaf-blades large, mainly 7–17 cm. long and 4–12 cm. wide; often growing along shaded ledges of bluffs or at their base in woodland, or ravines . . . . . . . . . . . . . . 11. C. GIGANTOSPERMUM

s. Without the above combination of characters; seed smaller, 0.8–1.5 mm. broad, either more or less smooth or with little dots or ridges, easy or hard to remove from pericarp (covering of fruit); leaves thin to thick, with either blunt or long-pointed teeth; leaf-blades usually smaller; either woodland plants or in waste and open ground or other habitats . . . . . . . . . . . . . . . . . . . . . . . . . . . . . *t*

t. Pericarp (covering of fruit) not firmly attached to the seed, easy to remove. . . . . . *u*

u. Common woodland plant, usually in shaded ground; usually delicate, slender-stemmed plant with thin green leaves and mostly without any mealiness; calyx-lobes flat or rounded, faintly or not at all keeled, only partially covering the mature fruit; seed 1–1.5 mm. broad . . . . . . . . . . . . . . . 10. C. STANDLEYANUM

u. Rare plant, introduced in open waste ground in Jackson Co., western Mo.; low spreading plants with thick leaves and conspicuous mealiness; calyx-lobes with a definite keel in the center, completely covering the mature fruit; seed 0.9–1.1 mm. broad. . . . . . . . . . . . . . . . . . . . . 9. C. INCANUM

t. Pericarp (covering of fruit) firmly attached to the seed, difficult to remove. . . . . . *v*

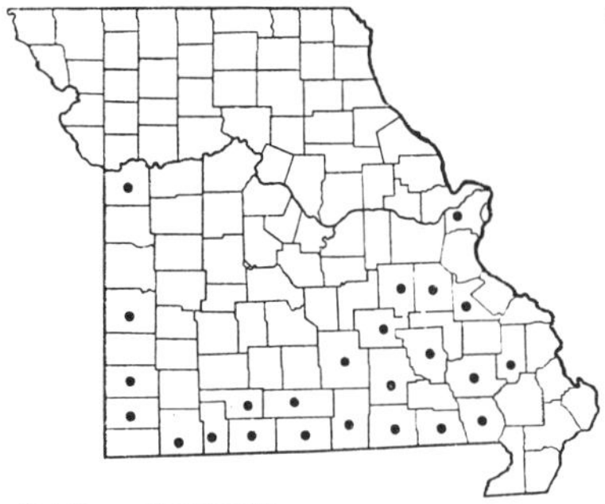

847 Chenopodium Pumilio

848 Chenopodium Botrys (Jerusalem Oak)

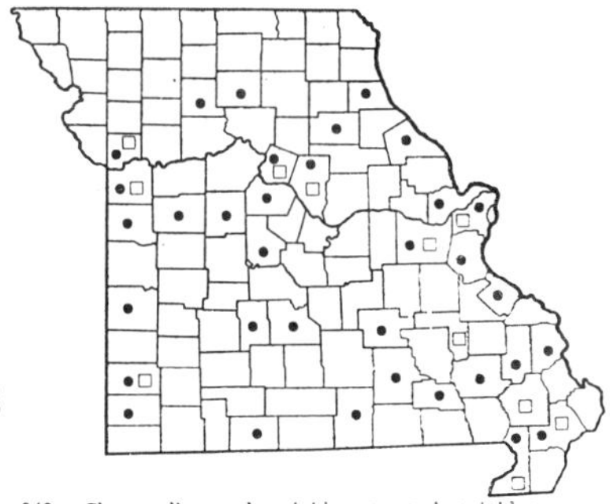

849 • Chenopodium ambrosioides var. ambrosioides (Mexican Tea)

849 □ Chenopodium ambrosioides var. anthelminticum

v. Stigmas short (0.1 mm. or less) and stout; leaves mostly triangular and conspicuously marked with long-pointed (acuminate) teeth; leaf-blades mostly rather thin and shining above . . . . . . . . *w*

w. Calyx-lobes definitely keeled in the center, mealy; seeds 1.2–1.5 mm. broad, dull, with fine ridges or dots over the surface, the margins narrowed or acute; inflorescences shorter than the leaves . . . . . . . . . . . . . . . . . . . . . . . . . . . 12. C. MURALE

w. Calyx-lobes rounded or flat, not keeled, not mealy; seeds 1–1.2 mm. broad, shining, smooth, the margins rounded; inflorescences usually longer than the leaves . . . . . . . . 13. C. URBICUM

v. Stigmas long or short but not stout; leaves variously shaped and toothed; leaf-blades thick to membranaceous, not shining above . . . . . . . . . . . . . . . . . . . . . *x*

x. Seeds mainly 1.1–1.5 mm. broad; leaves often very mealy . . . . . . . . . . 14. C. ALBUM

x. Seeds 0.9–1.2 mm. broad; leaves often greener . . . . . . . . . . . . . . . . . . *y*

y. Calyx-lobes only partially covering fruit; lower leaves with short teeth; middle leaves oblong, smooth-edged; flower clusters on erect strict branches; flowering May–September . 16. C. STRICTUM var. GLAUCOPHYLLUM

y. Calyx-lobes mostly covering the fruit; lower and middle leaves coarsely toothed, at most $1\frac{1}{2}$ times as long as broad; flower clusters on arching branches; flowering September–October . . . . . . . . . . . . . . . . . . . . . 15. C. MISSOURIENSE

1. **Chenopodium Pumilio** R. Br. Map 847

*Chenopodium incanum* (S. Wats.) Heller [G, P & S]

Flowers August–October.

Occurs in rocky, sandy, or gravelly ground, in fields, waste ground, along railroads, and about dwellings. Southern Ozark region, where common, northeast to St. Louis County and northwest locally in Jackson County.

Native of Australia; introduced into the United States where naturalized in Massachusetts, New York, New Jersey, Pennsylvania, D. C., and from Missouri to Texas and California.

2. **Chenopodium Botrys** L. Jerusalem Oak Map 848

Also known as Feather Geranium.

Flowers July–October.

Occurs in waste ground and along railroads. Infrequent and scattered in southern and central Missouri north to St. Louis and Jackson counties.

Native of Eurasia; introduced into North America, where naturalized from Quebec to Washington, south to Virginia, Kentucky, Missouri, and California.

This species may cause some cases of hay fever, but is not common enough in Missouri to be of any present concern.

3. **Chenopodium ambrosioides** L.

Mexican Tea, Wormseed Map 849

Flowers August–November.

Occurs in rich waste ground, on dumps, fields, along railroads, and about dwellings. Scattered throughout the state, but not recorded from the northernmost counties.

Missouri material may be divided into two varieties:

Inflorescences interrupted with leaves or leafy bracts; calyx-lobes slightly keeled. . . . . 3a. C. AMBROSIOIDES var. AMBROSIOIDES

Inflorescences mostly without leaves or leafy bracts; calyx-lobes not keeled . 3b. C. AMBROSIOIDES var. ANTHELMINTICUM

3a. **Chenopodium ambrosioides** var. **ambrosioides** Map 849

*Chenopodium ambrosioides* ssp. *eu-ambrosioides* Aellen [Aellen, Aellen & Just]

850 Chenopodium capitatum (Strawberry Blite)

851 ● Chenopodium glaucum var. glaucum (Oak-leaved Goosefoot)
851 □ Chenopodium glaucum var. salinum

852 Chenopodium rubrum var. rubrum (Coast Blite)

This is the more commonly collected variety in the state, recorded north to Marion, Linn, and Livingston counties; probably in most of the counties in the state.

Native of tropical South and Central America, Mexico, and the West Indies; naturalized in the United States and Canada, ranging north to Maine, New York, Ontario, Wisconsin, Iowa, and California.

3b. **Chenopodium ambrosioides** var. **anthelminticum** (L.) Gray — Map 849

*Chenopodium ambrosioides* ssp. *eu-ambrosioides* var. *anthelminticum* (L.) Aellen [Aellen, Aellen & Just]

Apparently less frequently found in Missouri, where recorded north to St. Louis, Boone, Howard, and Jackson counties.

Native of tropical America; naturalized in the United States, in some sections more common than var. *ambrosioides*.

Aellen has divided *C. ambrosioides* into 5 subspecies with a number of varieties and numerous forms. The two variations in Missouri have been treated in the present flora as varieties and await further studies by other workers.

This species is sometimes grown for the oil which is used as an efficient agent for getting rid of intestinal parasites, such as hook-worm. For this purpose it is frequently used in the southern United States, Mexico, and parts of Central and South America. In some parts of Guatemala it is used for flavoring food, as a poultice, and as a narcotic for inducing sleep. Stock usually do not eat the plant. A few cases have been recorded of poisoning from taking too large quantities of the oil. The large amount of pollen shed sometimes is considered important in causing cases of hay fever. Although many species of *Chenopodium* are cooked as green vegetable, this one should not be used because of its possible poisonous effects from the oil contained.

4. **Chenopodium capitatum** (L.) Aschers. Strawberry Blite — Map 850

Also known as Strawberry Spinach.

Flowers May–August.

Occurs in waste ground in Pike County, northeastern Missouri.

Ranges from Nova Scotia and Quebec to Alaska, south to New Jersey, Pennsylvania, Michigan, Indiana, Illinois, Wisconsin, Missouri, Minnesota, New Mexico, and California; also in Eurasia.

Where it is more common farther north, the fleshy fruits are sometimes eaten, raw or cooked, and the plant cooked as a spinach substitute.

5. **Chenopodium glaucum** L. Oak-leaved Goosefoot — Map 851

Flowers May–October.

Occurs in alluvial muddy soils along the Missouri River, around ponds, open waste ground, and along railroads.

Missouri material is represented by two variations:

Tip of leaf and teeth of margins obtuse or blunt; flower-clusters mostly without bracts . 5a. C. GLAUCUM var. GLAUCUM

Tip of leaf and teeth of margins acute; flower-clusters mostly with bracts. . . 5b. C. GLAUCUM var. SALINUM

5a. **Cheopodium glaucum** var. **glaucum** — Map 851

*Chenopodium glaucum* L. [G, BB, P & S, Steyerm.]

*Chenopodium glaucum* ssp. *eu-glaucum* Aellen [Aellen & Just]

Known only from St. Louis, Jackson, Dekalb, and Holt counties in central and western Missouri.

Native of Europe; naturalized in North America, ranging from Quebec to Saskatchewan, south to Virginia, Ohio, Indiana, Illinois, Missouri, Nebraska, Colorado, Wyoming, Utah, New Mexico, and California.

In some areas where this plant abounds, it is re-

853 • Chenopodium desiccatum var. desiccatum
853 □ Chenopodium desiccatum var. leptophylloides

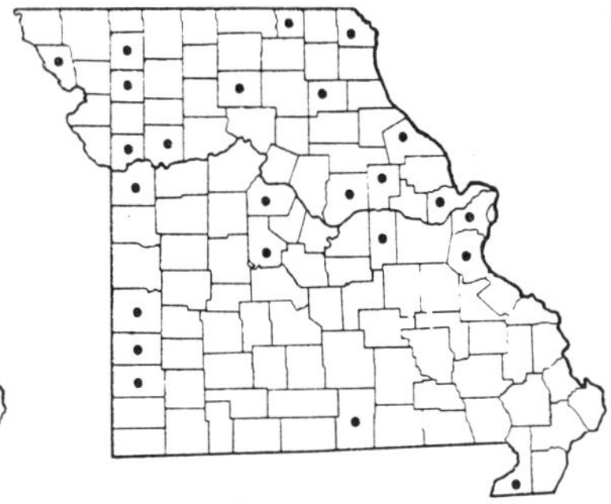

854 Chenopodium pallescens

855 Chenopodium incanum

garded as an important cause of hay fever, but in Missouri is too rare to be of any present concern.

5b. **Chenopodium glaucum** var. **salinum** (Standley) Boivin Map 851

*Chenopodium salinum* Standl. [G, BB]

*Chenopodium glaucum* ssp. *salinum* (Standley) Aellen [Aellen & Just]

Known from St. Louis County, east-central Missouri (St. Louis, freight yard (Burlington) in northern part south of switch 11, June 30, 1956, *Muehlenbach 975*). The plant, determined by Wahl, is one of two specimens under this number collected by Dr. Muehlenbach, the other specimen being determined as typical var. *glaucum*.

Ranges from Manitoba to Iowa and Missouri, west to Oregon and Arizona.

6. **Chenopodium rubrum** L. var. **rubrum** Coast Blite Map 852

*Chenopodium rubrum* L. [G, BB, Aellen, Aellen & Just]

Flowers July–November.

Known only from Jackson County, west-central Missouri (July 15, 1892, *Bush*, World's Fair specimen in Univ. of Mo. Herb.).

Ranges from Newfoundland to Washington, south to New Jersey, New York, Missouri, Nebraska, New Mexico, Arizona, and California; also in Europe.

7. **Chenopodium desiccatum** A. Nelson Map 853

Flowers early June–October.

Occurs in waste and open ground and along railroads.

Missouri material is represented by two variations:

Leaves not lobed, mostly oblong-elliptic, rather thick; calyx-lobes more green than yellow, blunt or obtuse, more prominently keeled in the center; plants varying from low and spreading to erect-spreading . . . C. DESICCATUM var. DESICCATUM

Leaves lobed or not lobed, lanceolate to ovate-lanceolate; calyx-lobes more yellowish than green, acute or short-pointed, less prominently keeled in the center; plants erect . C. DESICCATUM var. LEPTOPHYLLOIDES

7a. **Chenopodium desiccatum** var. **desiccatum** Map 853

*Chenopodium pratericola* var. *oblongifolium* (S. Wats.) Wahl (Wahl in Bartonia 27: 19. 1954)

*Chenopodium leptophyllum* of auth., in part, not Nutt. [G, BB, P & S]

*Chenopodium desiccatum* A. Nels.

*Chenopodium pratericola* ssp. *desiccatum* (A. Nelson) Aellen [Aellen, Aellen & Just]

*Chenopodium leptophyllum* var. *oblongifolium* S. Wats.

Occurs in waste ground. Known from Jackson County. Less common than the next variety, var. *leptophylloides*.

Ranges from Missouri and Nebraska west to Idaho and Arizona.

7b. **Chenopodium desiccatum** var. **leptophylloides** (Murr.) Wahl Map 853

*Chenopodium pratericola* ssp. *eu-pratericola* Aellen var. *leptophylloides* (Murr.) Aellen [Aellen & Just]

*Chenopodium pratericola* Rydb. var. *pratericola* (Wahl in Bartonia 27: 18. 1954)

*Chenopodium pratericola* Rydb. [P & S]

Plate no. 148. **1.** Chenopodium ambrosioides, × 2/5; Details from Small, The New York Botanical Garden. **2.** Chenopodium glaucum, × 2/5; Details from Small, The New York Botanical Garden. **3.** Chenopodium desiccatum, × 2/5; *a.* Flower, × 2 4/5. **4.** Chenopodium gigantospermum, × 2/5; *a.* Flower, × 2 4/5; *b.* Flower, × 2 4/5. **5.** Chenopodium album var. album, × 2/5.

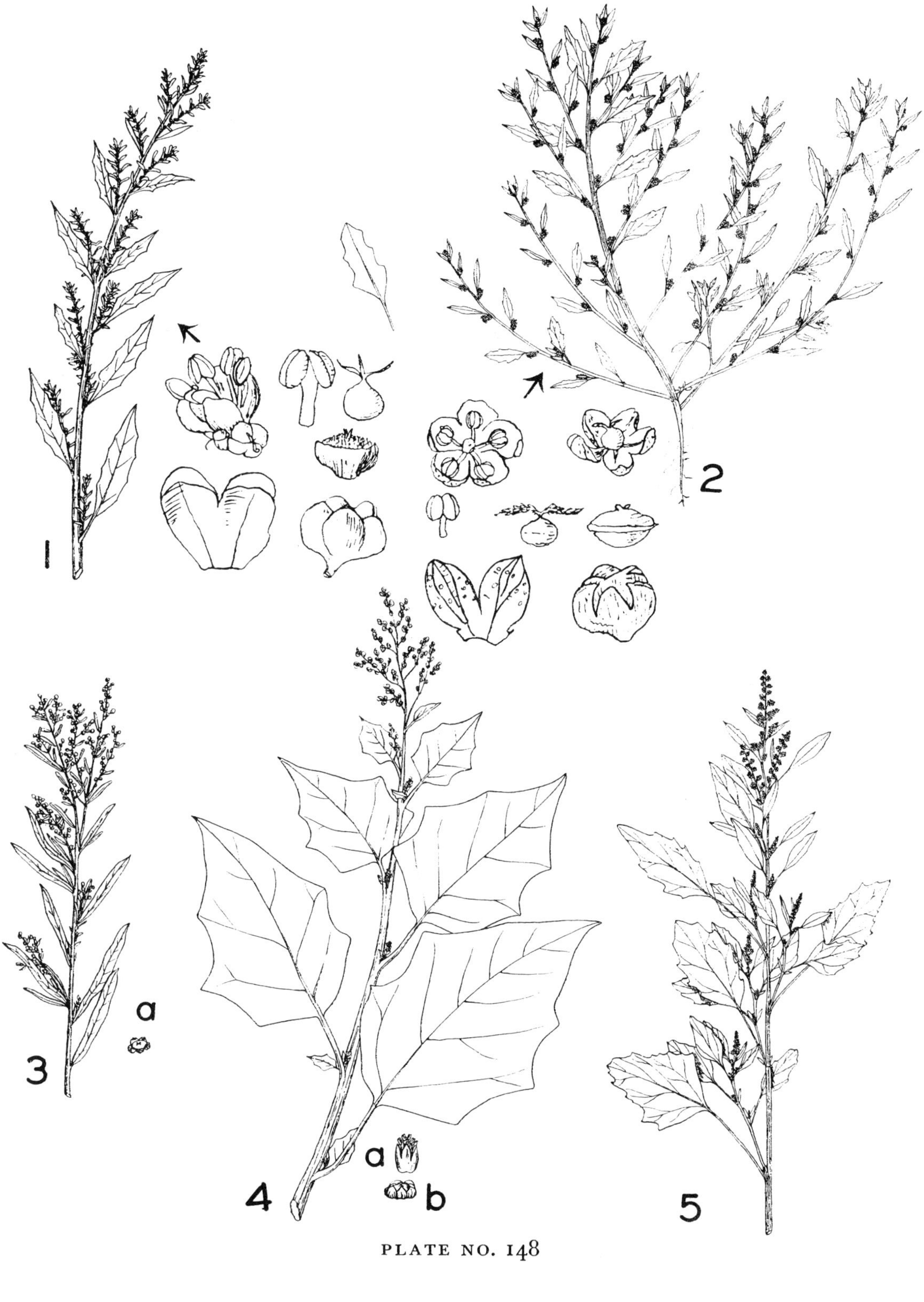


PLATE NO. 148

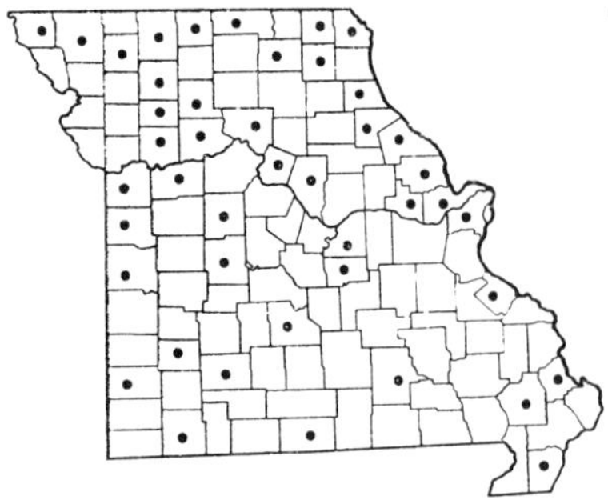

856 Chenopodium Standleyanum

857 Chenopodium gigantospermum (Maple-leaved Goosefoot)

858 Chenopodium murale (Nettle-leaved Goosefoot)

*Chenopodium leptophyllum* of auth., in part, not Nutt. [G, BB, P & S in part]
*Chenopodium leptophyllum* var. *leptophylloides* (Murr.) Thellung & Aellen [P & S in part]
*Chenopodium pratericola* var. *Thellungianum* Aellen

Occurs in waste ground and along railroads. Scattered in St. Louis (*Muehlenbach 61, 222*), and Jackson (Kansas City, *Bush 9424*; Courtney, *Bush 9021, 9034*) counties.

Ranges from New Jersey, Indiana, Illinois, Missouri, Nebraska, North Dakota, Montana, Idaho, and Washington, south to Virginia, Louisiana, Texas, New Mexico, Arizona, and California; also in Mexico and South America.

Material properly placed in *C. pallescens* has been confused with this species and with *C. leptophyllum* Nutt. Typical *C. leptophyllum* Nutt. is not known in Missouri.

8. **Chenopodium pallescens** Standley Map 854
*Chenopodium leptophyllum* of auth. in part, not Nutt. [G, BB, P & S]
*Chenopodium leptophyllum* var. *leptophylloides* of some auth., not (Murr.) Thellung & Aellen [P & S]
*Chenopodium subglabrum* [of BB in part], not (Wats.) A. Nels.

Flowers early June–October.

Occurs on rocky or sandy open ground, along edges of bluffs, quarries, limestone outcrops and glades, waste ground, and along railroads.

Scattered throughout northern, central, and western Missouri, south to Jefferson, Gasconade, Morgan, and Jasper counties, and locally south in Dunklin and Howell counties.

Ranges from Indiana, Missouri, and Oklahoma to Texas and New Mexico.

This species has been confused previously by various authors with *C. leptophyllum* of auth., not Nutt. (= *C. desiccatum* and variety *leptophylloides*). Most of the specimens previously cited in Palmer & Steyermark's *Annotated Catalogue* as *C. leptophyllum* and var. *leptophylloides*, and specimens, labeled as such in various herbaria, should be referred to *C. pallescens*, characterized by the thick, linear 1-nerved leaves without any teeth or lobes and with the pericarp firmly attached to the seed.

9. **Chenopodium incanum** (S. Wats.) Heller Map 855

Flowers June–October.

Known only from waste ground in Jackson County, west-central Missouri (Courtney, June 18, 1920, *Bush 8984*).

Ranges from Missouri, Nebraska, South Dakota, Colorado, Nevada, and California, south to Oklahoma, Texas, New Mexico, and Arizona; also in Mexico.

10. **Chenopodium Standleyanum** Aellen Map 856
*Chenopodium Boscianum* Moq. of auth. [G], not as to type.

Flowers June–October.

Occurs in woodland, shaded thickets, dry or moist soil, rocky or rich ground, and shaded ledges of bluffs. Throughout Missouri; probably in every county.

Ranges from Florida to Texas, north to Connecticut, Quebec, Ontario, Minnesota, and North Dakota.

This is an easily recognizable species, with its combination of relatively thin, mostly dull green, often entire leaves, small interrupted slender inflorescences, rounded calyx-lobes which do not cover the fruit, mostly delicate, slender habit, and woodland habitat.

11. **Chenopodium gigantospermum** Aellen
Maple-leaved Goosefoot Map 857
*Chenopodium gigantospermum* var. *Standleyanum* Aellen [P & S, Aellen & Just]
*Chenopodium hybridum* of Am. auth. [BB], not L.
*Chenopodium hybridum* var. *gigantospermum* (Aellen) Rouleau [G]

859 Chenopodium urbicum (City Goosefoot)

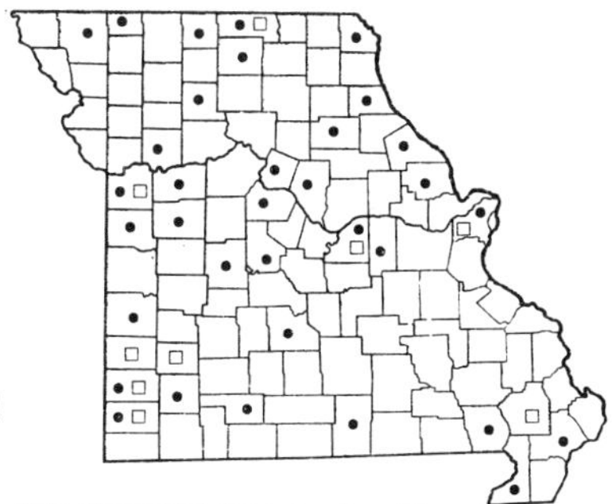

860 • Chenopodium album var. album (Pigweed)
860 □ Chenopodium album var. lanceolatum

861 Chenopodium missouriense

*Chenopodium hybridum* var. *Standleyanum* (Aellen) Fern. [G]

Flowers June–October.

Occurs along shaded ledges of bluffs, at the base of rocky wooded slopes and bluffs, especially of limestone, and in rich or fertile open soils or in woodland. Common in most parts of the state; not recorded from the southeastern and most of the northeastern sectors, but of expected occurrence in most counties.

Ranges from Quebec to British Columbia, south to Virginia, Kentucky, Missouri, Oklahoma, Texas, New Mexico, and California.

This is another easily recognized species in the genus by virtue of the large, thin, rich green leaves, which taper to a long acuminate apex and are rounded or subcordate at the base, have conspicuous acuminate large teeth or lobes, and have large, leafless clusters of flowers with relatively large seeds.

By some authors this is considered as only a variety of the European *C. hybridum* L. Besides the larger seeds of the American plants, Wahl (Bartonia 27: 30. 1954) contrasts the two in such other respects as the thick, black, strongly reticulate, attached pericarp of *C. hybridum* instead of the thinner, greenish-transparent, less reticulate pericarp of *C. gigantospermum* which is usually more separable and easily removed from the seed, although attached also, by the more rounded margin of the seed in *C. hybridum*, by the broad sepals with contrasting green center and pale margins covering more of the fruit in *C. hybridum* instead of the narrow sepals with less contrast in color of *C. gigantospermum*, and by differences in the inflorescence and base of leaves.

12. **Chenopodium murale** L.

Nettle-leaved Goosefoot — Map 858

Flowers June–November.

Occurs in waste ground and about dwellings. Scattered in central and southern Missouri, north to St. Louis, Warren, Boone, and Jackson counties. Undoubtedly of wider occurrence when the state has been more thoroughly collected.

Native of Europe; naturalized in North America, from Quebec to British Columbia, south throughout the United States.

13. **Chenopodium urbicum** L. City Goosefoot

Map 859

*Chenopodium urbicum* var. *intermedium* (Mert. & Koch) Koch [P & S]

Flowers July–October.

Occurs in waste ground, dumps, and about dwellings. Rarely collected, known only from St. Louis, Boone, and Greene counties.

Native of Europe; naturalized in North America from Quebec to Ontario and Wisconsin, south to Maryland, West Virginia, Indiana, Illinois, and Missouri; also naturalized on the Pacific Coast, and South America.

Both this species and *C. murale* are similar to *C gigantospermum* in having rich green, rather thin leaves with sharply pointed teeth. The leaves, however, are generally smaller and more broadly triangular than those of *C. gigantospermum* and wedge-shaped or truncate instead of rounded to subcordate.

14. **Chenopodium album** L.

Pigweed, Lamb's Quarters — Map 860

Flowers May–October.

Occurs in waste and cultivated ground and along railroads. Throughout Missouri, and probably in every county.

A very variable species, the following variations of which are recognized in Missouri based upon Dr. Wahl's studies:

Plants with erect branches; flower-clusters mostly in contact with each other, without conspicuous separated intervals between; calyx-lobes more

frequently pointed and with yellow margins; all the individual flowers sessile (without stalks); leaves usually broadly ovate and prominently toothed, varying to lanceolate nontoothed bracts
14a. C. ALBUM var. ALBUM

Plants with spreading branches; flower-clusters conspicuously separated from one another; calyx-lobes more frequently blunt (obtuse) and with white margins; some of the individual flowers on stalks; leaves narrowly ovate to lanceolate, those of the inflorescence without teeth . . 14b. C. ALBUM var. LANCEOLATUM

14a. **Chenopodium album** var. **album** Map 860

*Chenopodium album* L. [G, BB, P & S, Steyerm.]

*Chenopodium album* ssp. *eu-album* (Ludwig) Aellen var. *polymorphum* Aellen [Aellen & Just]

Throughout Missouri.

Throughout the United States and Canada; also in Eurasia. Both native North American and introduced Old World races of this species are interpreted as *C. album* var. *album.*

According to Wahl (Bartonia 27: 34. 1954), var. *album* is a plant of cultivated places, in usually disturbed soils, such as gardens and cornfields.

14b. **Chenopodium album** var. **lanceolatum** (Muhl.) Coss & Germ. Map 860

*Chenopodium album* f. *lanceolatum* (Muhl.) Aellen [P & S, Aellen & Just]

*Chenopodium album* f. *viride* (L.) Aellen [P & S, Aellen & Just]

*Chenopodium album* ssp. *eu-album* var. *polymorphum* f. *glomerulosum* (Reichenb.) Aellen [Aellen & Just]

*Chenopodium album* ssp. *eu-album* var. *polymorphum* f. *cymigerum* (Koch) Aellen [Aellen & Just]

Scattered throughout Missouri, and probably more common after thorough collecting has been done.

According to Wahl (above), var. *lanceolatum* is a plant of both cultivated and undisturbed soil, often persisting in waste ground, and in cities and towns. Future studies of the respective habitats of var. *album* and var. *lanceolatum* in Missouri may throw additional light on this phase of the subject.

The present author believes that Dr. Wahl's conservative treatment of the variations of *C. album* is more justified than that of Aellen which recognizes additional subspecies, varieties, and forms.

This species, and those following in this treatment, make excellent spring greens, the tender young shoots and leaves being cooked alone as a substitute for spinach or in combination with such other greens as *Amaranthus*, *Lactuca*, *Sonchus*, *Rumex*, and *Phytolacca*. American Indians prepared a type of bread and porridge from the richly nutritious seeds of various species of this genus, of which *C. album* is one. In the Andean countries of South America, particularly in Ecuador, Peru, and Bolivia, *C. Quinoa* or Quinoa produces large quantities of nutritious seeds consumed by the Indians of those countries in soups, bread, and cakes. A fermented drink is also prepared by them from these seeds.

In its earlier period of growth, while the stems are still soft, this and the other similar and related following species of *Chenopodium* may be plowed under as a green manure to enrich the soil, or fed to poultry and hogs as green food. The abundant pollen produced by this and other species similar in appearance is considered to be responsible for causing hay fever in some areas where the species are abundant. In fall the foliage of this and related species turns bronze to purplish-red.

15. **Chenopodium missouriense** Aellen Map 861

*Chenopodium missouriense* var. *Bushianum* Aellen [P & S, Aellen & Just]

*Chenopodium paganum* Standl. [Standl.]

*Chenopodium album* of some auth. [BB in part], not L.

*Chenopodium Berlandieri* of some auth. [G in part], not Moq.

Flowers September–October.

Occurs in waste ground and along railroads. Scattered in Missouri, but doubtless in most of the counties.

Ranges from Florida to Texas, north to New England, Pennsylvania, Ohio, Michigan, Indiana, Illinois, Wisconsin, and Minnesota.

The var. *Bushianum* Aellen, distinguished by Aellen on the basis of the mealy lower surface of the leaves, does not merit distinction, since both mealy and non-mealy leaves may be present on the same plant, or various degrees of mealiness may be found on the same plant.

The strictly autumnal flowering of this species is considered an easy mark of recognition to separate it from *C. album*, which begins to flower in May.

16. **Chenopodium strictum** Roth var. **glaucophyllum** (Aellen) Wahl Map 862

*Chenopodium glaucophyllum* Aellen [P & S]

*Chenopodium strictum* ssp. *glaucophyllum* (Aellen) Aellen [Aellen & Just]

Flowers July–September.

Occurs in waste ground. Known only from western Missouri in Jasper (Webb City, August 6, 1920, *Bush 9061A*) and Jackson (Sheffield, Courtney, *Bush 8065*, *8065A*, *8095*, *8108*, *8110*, *8111*, *8195A*, *8546B*, and others by Bush) counties.

862 Chenopodium strictum var. glaucophyllum

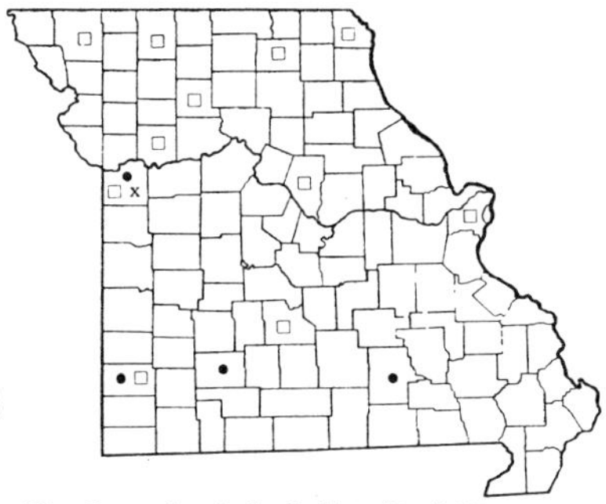

863 • Chenopodium Berlandieri var. Boscianum
863 □ Chenopodium Berlandieri var. Zschackei
863 x Chenopodium X variabile var. Murrii

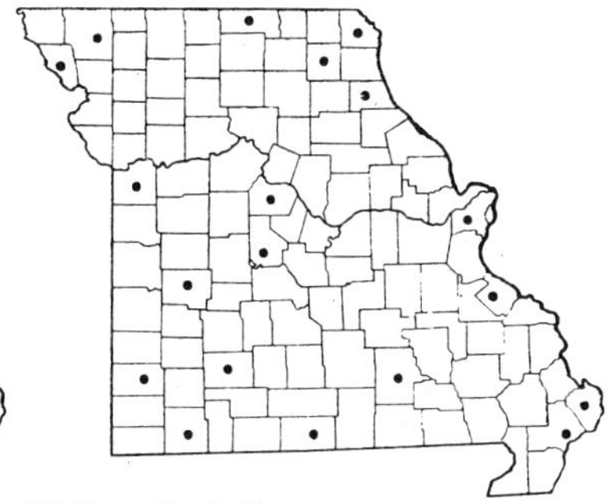

864 Chenopodium Bushianum

Ranges from Quebec to Manitoba south to Connecticut, New York, Pennsylvania, Michigan, Illinois, Missouri, Arkansas, Nebraska, and California.

*Chenopodium strictum* var. *strictum* is Asiatic and differs in the more prominently serrate oblong-ovate leaves 3 or more times as long as broad instead of only twice as long as broad as in var. *glaucophyllum*. The small seeds combined with the shallowly serrate, oblong lower leaves are considered the best marks of distinction of *C. strictum* var. *glaucophyllum*.

17. **Chenopodium Berlandieri** Moq. Map 863

Flowers June–October.

Occurs in waste and cultivated ground and along railroad tracks.

Many subspecies (8), varieties (6), and forms (9) have been published by Aellen under this species and are to be found categorized in their respective details in Aellen and Just's work (Am. Midl. Nat. 30: 71–73. 1943). The following key and treatment of Missouri material is in accordance with Dr. Wahl's paper (Bartonia 27: 40–44. 1954):

a. Leaves membranaceous, 2–4 cm. long, rhombic-triangular, conspicuously dentate or the upper ones becoming entire; upper portions of the inflorescence becoming bractless; seeds 1–1.3 mm. in diameter . . 17a. C. BERLANDIERI var. BOSCIANUM

a. Leaves more often thicker, but also thin, variable in size and shape but larger than 4 cm. long; sepals strongly keeled; inflorescence leafy; seeds 1.2–1.5 mm. in diameter . . . 17b. C. BERLANDIERI var. ZSCHACKEI

17a. **Chenopodium Berlandieri** var. **Boscianum** (Moq.) Wahl Map 863

*Chenopodium Berlandieri* ssp. *Boscianum* (Moq.) Aellen [P & S, Aellen & Just]

*Chenopodium Boscianum* Moq., in part, not of Am. auth.

Occurs in waste ground. Known from Jackson, Jasper, Greene, and Shannon counties in southern and central Missouri.

Ranges from Florida to Texas, north to Virginia and Missouri.

17b. **Chenopodium Berlandieri** var. **Zschackei** (Murr.) Murr. Map 863

*Chenopodium Berlandieri* ssp. *Zschackei* (Murr.) Zobel [Aellen & Just]

*Chenopodium Berlandieri* Moq. in part [G]

*Chenopodium album* in part [BB], not L.

*Chenopodium Berlandieri* ssp. *Zschackei* var. *foetens* (Ludwig) Aellen [P & S, Aellen & Just]

*Chenopodium Berlandieri* ssp. *Zschackei* var. *farinosum* (Ludwig) Aellen [P & S, Aellen & Just]

*Chenopodium Berlandieri* ssp. *Zschackei* var. *typicum* (Ludwig) Aellen [P & S, Aellen & Just]

*Chenopodium Berlandieri* ssp. *Zschackei* var. *typicum* f. *angustius* (Ludwig) Aellen [P & S, Aellen & Just]

*Chenopodium Berlandieri* ssp. *Zschackei* var. *typicum* f. *latifolium* (Ludwig) Aellen [P & S, Aellen & Just]

*Chenopodium Berlandieri* ssp. *Zschackei* var. *typicum* f. *pedunculare* Aellen [P & S, Aellen & Just]

*Chenopodium Berlandieri* ssp. *Zschackei* var. *typicum* f. *neglectum* Aellen [P & S, Aellen & Just]

*Chenopodium dakoticum* Standley [P & S, Standley]

Concurring with the preliminary conclusions reached by Dr. Wahl, it will be noted that numerous varieties and forms treated as such by Aellen and by Aellen and Just are considered in the present flora as variations within var. *Zschackei* not meriting recognition.

By Aellen and Just var. *typicum* is considered a slightly mealy to nearly glabrous variation, while var. *foetens* and var. *farinosum* (syn. *C. dakoticum*) are abundantly white-mealy; var. *foetens* is distinguished by having short, obtuse, slightly 3-lobed leaves often with a rank odor, whereas var. *farinosum* has shorter, ovate-elliptic, dentate leaves without an odor. Under var. *typicum* forms involving Missouri material are based upon narrower leaves with a longer tip (f. *angustius*), broader leaves often somewhat rounded at tip (f. *latifolium*), terminal inflorescence large and cymose (f. *pedunculare*), and inflorescence short, axillary, and cymose (f. *neglectum*). All these varieties and forms are based upon specimens collected by Bush from Jackson County.

The var. *Zschackei* occurs in waste ground scattered in various parts of Missouri, and of expected occurrence in nearly every county of the state.

Ranges from Ontario to James Bay and Mackenzie Basin, south to Virginia, Pennsylvania, Illinois, Arkansas, Oklahoma, Texas, New Mexico, Arizona, and California.

17c. A supposed hybrid between *Chenopodium album* and *C. Berlandieri* var. *Zschackei* known as **Chenopodium × variabile** var. **Murrii** Aellen has been collected by Bush in Jackson County, west-central Missouri.

18. **Chenopodium Bushianum** Aellen Map 864

*Chenopodium album* of some auth. in part [BB], not L.

*Chenopodium paganum* of Am. auth. [G, Standley, P & S], not Reichenb.

Flowers August–October.

Occurs in sandy fields and alluvial ground along rivers, waste places, wooded slopes, and dry open or shaded ground. Scattered throughout Missouri, and of expected occurrence in nearly every county of the state.

Ranges from Quebec to Alaska, south to Virginia and Missouri.

The reticulate large seeds and large thin leaves are characters which separate this species from related ones. The leaves have a paler green color than those of *C. album* or *C. missouriense*, and the usually lead-grey, drooping irregular inflorescences, on plants which are often leafless by autumn, further serve as aids in identifying *C. Bushianum*.

## 4. **Atriplex** Orach

a. Branches and leaves mainly opposite in the lower half of the plant; leaves mainly green or only slightly gray-green, mostly without teeth (entire) or slightly toothed or with spreading lobes at base; stem mainly green, pale green, or slightly gray-green; fruiting bracts which surround the fruit are usually without teeth (entire) or only minutely toothed, soft or thin and easily separating from each other, their margins free to below the middle . . . . . . . . . . . . . . . . . . . . . *b*
  - b. Larger leaf-blades mainly 8–20 cm. (less frequently 3.5–7 cm.) long; fruiting bracts which surround the fruit 10–15 mm. broad, with conspicuous reticulate venation (net-like); pistillate (pistil-bearing) flowers with a 3–5-lobed calyx; rarely encountered, known only from St. Louis and Jackson counties, central Missouri . . . . . . . . . . . . . . . . 1. A. HORTENSIS
  - b. Larger leaf-blades 2.5–8 cm. long; fruiting bracts which surround the fruit 2–6 mm. broad, without prominent venation; pistillate (pistil-bearing) flowers without any calyx; scattered in northern, central, and western sections of Missouri . . . . . . . . . . . . 2. A. PATULA

a. Branches and leaves alternate in the lower half of the plant; leaves gray-silvery or whitish-gray, mostly toothed; at least the young stems gray-silvery or whitish-gray, the older becoming straw-colored; fruiting bracts which surround the fruit are toothed, hard and firm, their margins united to at least the middle or above the middle or free only in the upper half . . . . . . . . . . . . . . . . . *c*
  - c. Fruiting bracts surrounding the fruit dentate (merely toothed) along the sides, broadly ovate to rhombic, with a broad or spreading base, broadest in the lower half, united to the middle, free in the upper half. . . . . . . . . . . . . . . . . . . . . . . . . . 3. A. ROSEA
  - c. Fruiting bracts surrounding the fruit lacerate or deeply toothed across the summit, rounded and somewhat wedge-shaped with a slightly tapering base, broadest at the middle or upper half, united to above the middle, free only in the upper third . . . . . . . . . . 3. A. ARGENTEA

1. **Atriplex hortensis** L. Garden Orach Map 864A

Flowers August–October.

Occurs in waste and fallow ground. Known only from St. Louis (introduced, nearly spontaneous, St. Louis, August, 1839, *Engelmann*) and Jackson (introduced, Sheffield, June 10, 1907, *Bush 4799*) counties, central Missouri.

Native of Asia; introduced in North America, from Quebec and New England to Montana, south to New York, Kentucky, Missouri, Colorado, and Utah.

This species is sometimes grown as a garden vegatable and used like spinach for greens. The foliage is normally green, but crimson-leaved plants (var. *atrosanguinea* Hort.), sometimes over 1 meter tall, are

864A Atriplex hortensis

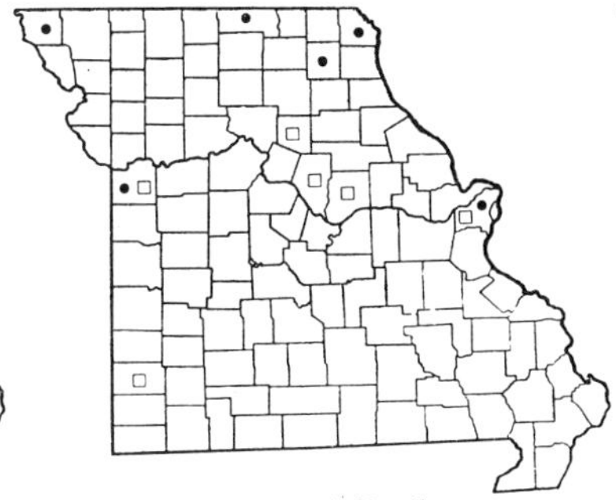

865 • Atriplex patula var. patula (Orach)
865 □ Atriplex patula var. hastata

866 Atriplex rosea (Red Scale)

cultivated as ornamentals. The small flowers are often purplish.

2. **Atriplex patula** L. Orach, Spear Scale Map 865

Flowers early June–November.

Occurs in cultivated or waste ground, dumps, and along railroads.

Two variations are represented in Missouri material:

Main leaves triangular or arrowhead-shaped with spreading lobes, often toothed; fruiting bracts truncate (as if cut straight across) or broadly rounded at base . 2b. A. PATULA var. HASTATA

Main leaves oblong or lanceolate, mostly without teeth or with a few low teeth; fruiting bracts slightly narrowed or narrowly rounded at base 2a. A. PATULA var. PATULA

2a. **Atriplex patula** var. **patula** Map 865

*Atriplex patula* L. [G, P & S]

Scattered in northern and central Missouri.

Ranges from Newfoundland to British Columbia, south to Florida, Alabama, Missouri, and California; also in Eurasia.

2b. **Atriplex patula** var. **hastata** (L.) Gray Map 865

Scattered in central and western Missouri.

Ranges from Newfoundland to British Columbia, south to South Carolina, Ohio, Indiana, Illinois, Missouri, Texas, and California.

In the western states *Atriplex patula* is one of the species causing hay fever during the summer months, and is considered as an especially important cause of hay fever in parts of California. The young growing tips and leaves may be cooked for greens, and are considered superior to *Chenopodium album* and related species.

3. **Atriplex rosea** L. Red Scale Map 866

Also known as Red Orach.

Flowers August–October.

Occurs in salt-impregnated soils bordering saline or sulfur springs and in waste ground and along railroads. Known only in central Missouri from St. Louis, Howard (alkaline ground along Salt Creek and adjacent salt springs at Boonslick, 3 mi. west of Boonsborough, August 23, 1934, *Steyermark 14785*), and Jackson counties. In Howard County this species is found around a saline spring.

Native of Eurasia; introduced into the United States, where naturalized from New York to Wisconsin, south to Florida, Missouri, and Kansas; in the western states from Wyoming to Washington, south to California; also Mexico and the West Indies.

In some of the western states this *Atriplex* is considered as one of the important hay fever contributing species.

4. **Atriplex argentea** Nutt. Silver Scale Map 867

Also known as Saltbrush.

Flowers June–October.

Occurs in waste ground and along railroad tracks. Known only from St. Louis (along railroad near Lead Works, 5500 Manchester Ave., St. Louis, September 22, 1939, *Petersen*; North St. Louis, Burlington freight yard, June 30, 1956, *Muehlenbach 974*) and Jackson (Courtney, July 16, 1907, *Bush 4813*) counties.

Ranges from Manitoba and Saskatchewan to Montana, south to Kansas, Oklahoma, Texas, New Mexico, and California; introduced eastward to Michigan, Ohio, and Missouri.

The upper leaves are very short-petioled to nearly sessile. The species is considered an important cause of hay fever in the western states.

867 Atriplex argentea (Silver Scale)

868 Axyris amaranthoides (Russian Pigweed)

869 Monolepis Nuttalliana (Poverty Weed)

## 5. **Axyris** L.

**Axyris amaranthoides** L. Russian Pigweed Map 868

Flowers May–October.

Known only from Jackson County (May, 1918, *Bush 8308*), west-central Missouri, where introduced in waste ground. Not found since the early collection by Bush.

Native of Siberia; introduced into North America, ranging from Quebec to Alberta, south to Massachusetts, Missouri, Nebraska, North Dakota, and westward.

## 6. **Monolepis** Schrad.

**Monolepis Nuttalliana** [R. & S.] Greene Poverty Weed Map 869

Flowers late May–October.

Occurs in waste ground and along railroad tracks. Known only from St. Louis (Bremen Ave. freight yard of Terminal Railroad Association, June 2, 1956, *Muehlenbach 907*) and Jackson (along railroad, Courtney, May 18, 1918, *Bush 8276*) counties.

Ranges from Minnesota and Manitoba, south to Missouri, Oklahoma, Texas, New Mexico, and California; also in Mexico, and introduced in New England.

This is usually a low-growing plant from 0.4–2 dm. tall, rarely becoming 4–5 dm. The stems and leaves are mainly glabrous or with sparse, scattered mealy granules. The leaves are usually narrow and petioled, and the small sessile flowers are in little clusters in the axils of the upper and middle leaves. The plant at first superficially resembles a dwarfed *Kochia scoparia* or *Salsola Kali* var. *tenuifolia* until the toothed leaves and floral characters are observed.

## 7. **Corispermum** L. Bugseed

a. Fruit lacking a wing or practically so . . . . . . . . . . . 3. C. ORIENTALE var. EMARGINATUM
a. Fruit with a definite wing along the margin . . . . . . . . . . . . . . . . . . . b
b. Fruit 3.5–4.5 mm. long; spikes of flowers or fruits dense and continuous except at very base; bracts of flowers and fruits conspicuously overlapping, at least the middle and upper ones hiding the main axis and longer and broader than the fruits, and the lower bracts broader than the fruits 1. C. HYSSOPIFOLIUM

---

Plate no. 149. **1.** Atriplex patula var. hastata, × $^2/_5$; *a.* Male flower, × $2^4/_5$; *b.* Female flower with bracts, × $2^4/_5$. **2.** Atriplex argentea, × $^2/_5$; *a.* Female flower with bracts, × $2^4/_5$; After Gleason, details from Small, The New York Botanical Garden. **3.** Atriplex hortensis; *a, b.* Two types of leaves, × $^2/_5$; *c.* Fruiting bract, × $2^4/_5$; After Gleason, details from Small, the New York Botanical Garden. **4.** Axyris amaranthoides, × $^2/_5$; After Gleason, The New York Botanical Garden. **5.** Monolepis Nuttalliana, × $^2/_5$. **6.** Corispermum hyssopifolium, × $^2/_5$; *a.* Flower, × 2; *b.* Fruit, × 2; After Gleason, The New York Botanical Garden. **7.** Suaeda depressa, × $^2/_5$; *a.* Fruit, × 8; Details from Small, The New York Botanical Garden. **8.** Salsola Kali var. tenuifolia, × $^2/_5$; *a, b.* Two types of leaves, × $^2/_5$; *c.* Flower, × $1^1/_5$; Details from Small, The New York Botanical Garden. **9.** Celosia argentea var. cristata, × $^2/_5$; Details from Small, The New York Botanical Garden.

PLATE NO. 149

870 Corispermum hyssopifolium (Bugseed)

871 Corispermum nitidum

872 Corispermum orientale var. emarginatum

b. Fruit 2–3 mm. long; spikes of flowers or fruits slender and rather interrupted; bracts of flowers and fruits not overlapping, and more separated, exposing the main axis, the upper bracts only slightly if at all longer than the fruits, the lower bracts much narrower than the fruit. . 2. C. NITIDUM

1. **Corispermum hyssopifolium** L. Bugseed Map 870

Flowers August–October.

Occurs on sand dunes and flats along the Missouri and Mississippi rivers in western, central, and eastern Missouri. Records are from only Clay, Jackson, Ray, St. Louis, and Perry counties, but doubtless to be found in sandy situations along all the counties bordering the Missouri and Mississippi rivers.

Ranges from Quebec to Washington, south to Indiana, Illinois, Missouri, Oklahoma, New Mexico, and Arizona. Introduced along railroads eastward to New York. Also in Mexico and Eurasia.

2. **Corispermum nitidum** Kit. Map 871

Flowers August–October.

Occurs in sandy and waste ground. Known only from Clark (introduced, Medill, August 23, 1920, *Bush 9153*) and Jackson (sands, Courtney, August 27, 1905, *Bush 3284*) counties.

Ranges from Indiana to Idaho, south to Missouri, Kansas, Texas, and Arizona; also in Europe.

3. **Corispermum orientale** Lam. var. **emarginatum** (Rydb.) Macbr. Map 872

*Corispermum villosum* Rydb.

Flowers August–October.

Known only from sandy open places in Jackson County, west-central Missouri (sandy bottoms, Courtney, September 4, 1933, *Bush 12964*).

Ranges from Alberta to Colorado and Nevada east to Missouri.

## 8. **Suaeda** Forsk. Sea Blite

**Suaeda depressa** (Pursh) Wats. Sea Blite Map 873

Flowers July–October.

Known only from Jackson County, west-central Missouri (July 15, 1892, *Bush*, World's Fair specimen in Univ. of Mo. Herb.).

Ranges from Minnesota and Manitoba to Montana, south to Missouri, Kansas, Texas, Nevada, and California; introduced in Missouri.

The fleshy tender leaves and branches of this and other species of the genus are sometimes cooked as a vegetable in those areas where the plants abound.

## 9. **Salsola** L. Saltwort, Russian Thistle

**Salsola Kali** L. Saltwort, Russian Thistle Map 874

Flowers July–October.

Occurs in dry, sandy, rocky, or open ground, waste places, along roadsides, and railroads.

Missouri material is represented by two variations:

Main leaves usually 1–2 mm. broad, tapering gradually upward from a broader base, up to 3 cm. long, all stiff and spiny-pointed; fruiting calyx 6–10 mm. wide; rarely found in Missouri a. S. KALI var. KALI

Main leaves about 1 mm. broad, threadlike, usually 3–7 cm. long, especially those on the

873 Suaeda depressa (Sea Blite)

874 • Salsola Kali var. Kali (Saltwort)
874 □ Salsola Kali var. tenuifolia (Russian Thistle)

875 Celosia argentea var. cristata (Cockscomb)

younger and leafy stems, fruiting calyx 3–6 mm. wide; the more commonly encountered type found . . . . . b. S. KALI var. TENUIFOLIA

a. **Salsola Kali** var. **Kali** Saltwort Map 874
*Salsola Kali* L. [G, P & S]

This is rarely found in Missouri, and known only from Jackson and Jasper counties.

Ranges from Newfoundland and Quebec, south along the coast to Georgia and Louisiana, rarely inland; also in Eurasia.

b. **Salsola Kali** var. **tenuifolia** Tausch.
Russian Thistle Map 874

The more commonly found variation occurs along sand bars and flats of the Missouri and Mississippi rivers, especially more common in the northwestern and west-central sectors, south locally to Jasper, Lawrence, and St. Francois counties.

Native of Asia; introduced into the United States, where ranging from Quebec to British Columbia, south to North Carolina, West Virginia, Ohio, Indiana, Illinois, Missouri, Kansas, Texas, and California.

The flowers are at first greenish, but the fruiting calyx becomes pinkish. *Salsola Kali* var. *tenuifolia* is considered to cause more cases of hay fever than any other member of the Chenopodiaceae, and, where it is more common than the ragweeds in some of the prairie regions and western states, it is thought to be the main cause of many cases of hay fever in those areas.

Although the plant has been much and justifiably maligned as a weed, it makes nourishing food for livestock when put up and cured for hay, and the young plants are eaten by grazing animals. In years of drouth in some of the western states it has carried over livestock in otherwise impoverished soils, and also has helped to build up loose soils in windblown sections of the western states.

## Fam. **AMARANTHACEAE** (Amaranth Family)

The genus *Acnida*, formerly maintained as a distinct genus, is now combined with *Amaranthus*, according to the recent studies of Dr. Sauer (Madroño 13: 5–46. 1955), whose work is here followed.

a. Leaves alternate . . . . . . . . . . . . . . . . . . . . . . . . . . . . *b*
 b. Flowers white, pink, red, or orange, in a broad and flattened fasciated (enlarged and grown together) mass; flowers perfect (stamens and pistil in the same flower); ovary with 2 or more ovules or seeds; filaments of stamens united at the base . . . . . . . . . . . . 1. CELOSIA
 b. Flowers green, straw-colored, or rarely shades of red, in usually long and narrow simple or branched spikes or panicles or short axillary clusters; stamens and pistils usually separated on different flowers on the same (monoecious) or different (dioecious) plants, or polygamous (both perfect and unisexual flowers present); ovary with 1 ovule or seed; filaments of stamens separate and distinct . . . . . . . . . . . . . . . . . . . . . . . 2. AMARANTHUS
a. Leaves opposite . . . . . . . . . . . . . . . . . . . . . . . . . . . . *c*
 c. Low sprawling plant, often lying on the ground; leaves 1–3 cm. long; stem and leaves covered with tiny star-shaped hairs (stellate pubescence), the uppermost leaves gray-white with a dense pubescence; flowers along the stem in the axils of the leaves . . . . . . . . . . 3. TIDESTROMIA
 c. Stems mainly and eventually erect; leaves mainly 3–12 cm. long; stem and leaves glabrous or

hairy, but, when hairy, the hairs simple, not star-shaped; flowers in inflorescences at the ends of the stems . . . . . . . . . . . . . . . . . . . . . . . . . . . . . *d*

d. Leaves with definite petioles (stalk); leaf-blade thin, ovate, the lower surface glabrous or sparsely hairy; flowers in a branching panicle . . . . . . . . . . . . . . 4. IRESINE

d. Leaves sessile (without a stalk) or nearly sessile; leaves thickish, linear, narrowly lanceolate, or elliptical-lanceolate, the lower surface densely woolly- or silky-hairy; flowers in an unbranched narrow, slender or thick spike . . . . . . . . . . . . . . 5. FROELICHIA

## 1. **Celosia** L. Cockscomb

**Celosia argentea** L. var. **cristata** (L.) Ktze.
Cockscomb Map 875
Flowers July–October.

Commonly cultivated as an ornamental annual, and rarely escaped along railroads and waste ground, where known in Missouri only from St. Louis County (Mo. Pac. R.R., north of Dover St., St. Louis, September 14, 1957, *Muehlenbach 1364*; Mo.-Kan.-Tex. R.R., Baden freight yard, St. Louis, September 1, 1957, *Muehlenbach 1334*).

Native of tropical America; introduced into the United States, and generally cultivated and introduced in temperate and tropical regions.

So far as known this plant does not persist away from cultivation.

## 2. **Amaranthus** L.

(*Acnida* L. [G, BB, P & S])

The reader is referred to Dr. Jonathan Sauer's paper, 'Revision of the dioecious amaranths' (Madroño 13: 5–46. 1955).

a. Stout spines present at the base of most of the leaves . . . . . . . . . . . . 5. A. SPINOSUS

a. No spines present on the plant . . . . . . . . . . . . . . . . . . . . . . . . . *b*

b. Male and female flowers in separate flowers on different plants (dioecious), the plants either all male (staminate) or all female (pistillate) . . . . . . . . . . . . . . . . . . . . . *c*

c. Calyx absent in pistillate flowers or at most consisting of 1 or 2 sepals . . . . . . . . . *d*

d. Sepals completely lacking in pistillate flowers; outer sepals of male flowers without heavy midribs, nearly equaling the inner sepals, blunt or acute; sepals of the male (staminate) flowers acute (short-pointed) or with short subulate tip 0.3 mm. or less; mature fruit splitting open irregularly or not splitting open, longer than the subtending bracts; bracts 1–1.5 mm. long, without heavy midribs . . . . . . . . . . . . . 1. A. TUBERCULATUS

d. 1 to 2 sepals present in pistillate flowers; outer sepals of male flowers with heavy midribs, longer than the inner sepals, cuspidate (abruptly tipped with a sharp or firm point); sepals of the male (staminate) flowers mostly long-pointed (long-acuminate) to a long slender tip; mature fruit circumscissile (splitting by a transverse circular line separating upper and lower section); bracts 1.5–2 mm. long, with heavy midribs . . . . . . . 2. A. TAMARISCINUS

c. Calyx present in pistillate flowers, consisting of 5 sepals . . . . . . . . . . . . . . *e*

e. Bracts 4–6 mm. long, 2–3 times as long as the flowers, with scarious (thin) margins $\frac{1}{3}$–$\frac{1}{2}$ or more of their length; outer sepals conspicuously longer than inner sepals; main leaves tapering to the tip, of an ovate or lance-oblong shape; longest outer sepals short- to long-pointed at tip . . . . . . . . . . . . . . . . . . . . . . . 3. A. PALMERI

e. Bracts 1.5–2.5 mm. long, usually shorter than or at most slightly longer than the flowers, with the scarious (thin) margins almost all the way to the tip; outer sepals scarcely or longer than inner sepals; main leaves rounded or blunt at tip, oblong-lanceolate to oblanceolate; sepals obtuse (blunt) or retuse (indented at tip) . . . . . . . . . . . 4. A. TORREYI

b. Male and female flowers in separate flowers on the same plant (monoecious) . . . . . . . . *f*

f. Flowers in short clusters along the stem in the axils at the base of the leaves; plants usually growing low with spreading or sprawling branches or lying along the ground; stamens 2–3 . . . *g*

g. Seeds 1.3–1.7 mm. broad; the often purplish stems usually lying on the ground (prostrate); bracts ovate-oblong, more or less equal in length to the calyx and fruit or only slightly exceeding them . . . . . . . . . . . . . . . . . . . . 11. A. GRAECIZANS

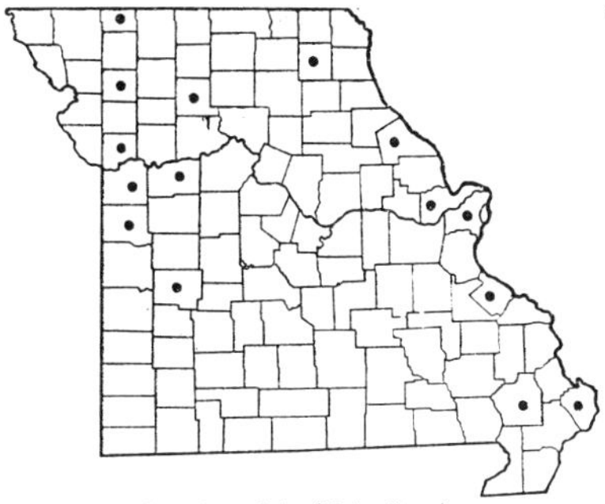
876 Amaranthus tuberculatus (Water Hemp)

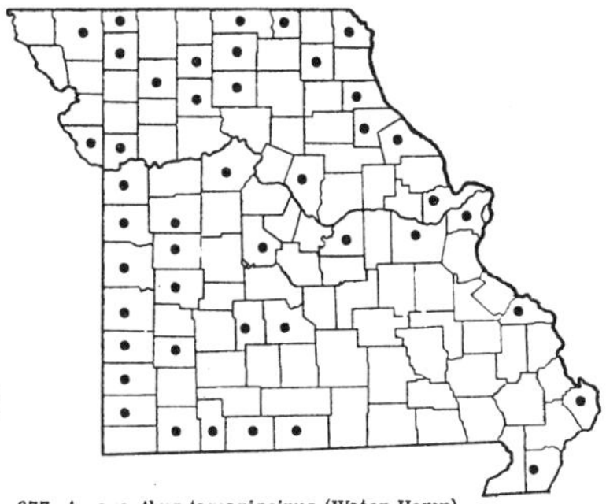
877 Amaranthus tamariscinus (Water Hemp)

878 Amaranthus Palmeri

g. Seeds 0.7–1 mm. broad; the usually whitish stems more commonly erect with spreading to ascending branches; bracts subulate (narrow with a slender point), much longer (2–3 times) than the calyx and fruit . . . . . . . . . . . . . . . . . 10. A. ALBUS

f. Flowers mainly in long and slender simple or branched spikes or panicles at the ends of the stems and branches and also in the leaf axils; plants with mainly erect stems; stamens 5, or 3 in *A. Powellii* . . . . . . . . . . . . . . . . . . . . . . . . . . . *h*

h. Sepals of pistillate flowers 1.5–2.3 mm. long; main bracts of pistillate flowers equaling, somewhat longer than, or up to twice as long as the sepals, 2–3.5 mm. long, not rigid. . . . *i*

i. At least the outer sepals of the pistillate flowers acute, 1.5–2 mm. long; bracts of pistillate flowers green or tinged with red, (rarely red); calyx equal to or somewhat longer than the fruit; lateral (side) branches (spikes) of the inflorescence erect or spreading-ascending . . . . . . . . . . . . . . . . . . . . . . 7. A. HYBRIDUS

i. All the sepals of the pistillate flowers blunt (obtuse) or rounded at tip, 1.5 mm. long; bracts of pistillate flowers red or purple, sometimes green; calyx shorter than the fruit; lateral (side) branches (spikes) of the inflorescence widely spreading . . . . 6. A. CAUDATUS

h. Sepals of pistillate flowers 3–4 mm. long; main bracts of pistillate flowers 2–3 times as long as the sepals, 5–9 mm. long, rigid . . . . . . . . . . . . . . . . *j*

j. Sepals of pistillate flowers rounded or truncate (as if cut straight across) or emarginate (shallowly notched) at tip; stamens 5; lateral (side) branches of inflorescences 1–5 cm. long; abundant throughout Missouri. . . . . . . . . . . . . . 8. A. RETROFLEXUS

j. Sepals of pistillate flowers acute or short-pointed at tip; stamens 3; lateral (side) branches of inflorescences 4–12 cm. long; rarely found, known only from St. Louis County . . . . . . . . . . . . . . . . . . . . . . 9. A. POWELLII

1. **Amaranthus tuberculatus** (Moq.) Sauer

Water Hemp Map 876

*Acnida altissima* Riddell [G, BB]

*Acnida altissima* var. *subnuda* (S. Wats.) Fern. [G]

*Acnida subnuda* (Wats.) Standl. [BB]

*Acnida altissima* var. *prostrata* (Uline & Bray) Fern. [G]

*Acnida tuberculata* Moq. [P & S]

*Acnida tuberculata* var. *subnuda* Wats. [P & S]

*Acnida tuberculata* var. *prostrata* (Uline & Bray) Robinson [P & S]

Flowers August–October.

Occurs on sand bars and mud flats along the Missouri and Mississippi rivers, and along the larger streams and their oxbow lakes. Scattered in northern, central, eastern, and western Missouri; absent from the Ozark section.

Ranges from New York, Quebec, Ontario, Wisconsin, to North Dakota, south to Alabama, Tennessee, Mississippi, Louisiana, Arkansas, Missouri, and Nebraska; introduced in New England.

The variations in leaf shape, habit of plant, and arrangement of the pistillate spikes are considered best treated under the one name.

This species is less common in the state than the next one, *A. tamariscinus*. This species is believed to have some association with causing hay fever, as is definitely the case in the next species, *A. tamariscinus*.

2. **Amaranthus tamariscinus** Nutt.

Water Hemp Map 877

*Acnida tamariscina* (Nutt.) Wood [G, BB, P & S]

Flowers July–October.

Occurs in moist alluvial soils of mud flats and on banks of streams, sloughs, oxbow lakes, and cultivated fields. Common along the larger streams in northern,

879 Amaranthus Torreyi

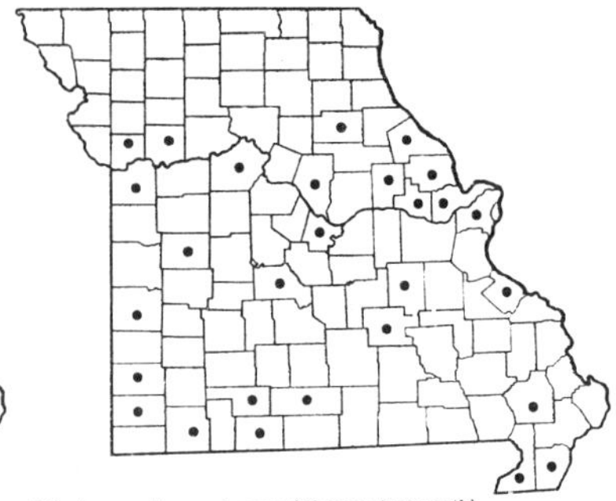

880 Amaranthus spinosus (Thorny Amaranth)

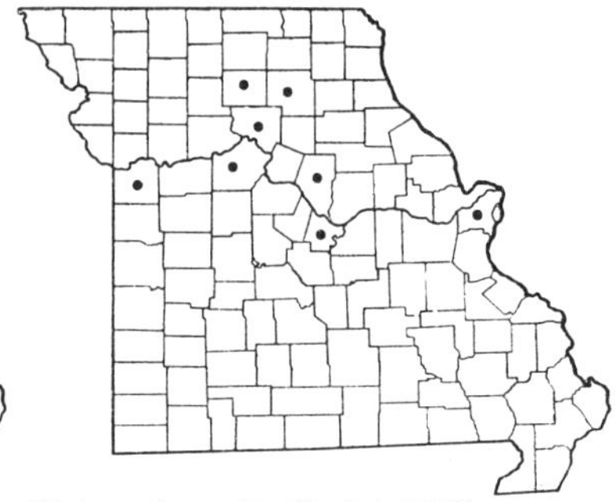

881 Amaranthus caudatus (Purple Amaranth)

central, western, and eastern Missouri; absent from most of the Ozark region.

Ranges from Indiana, Wisconsin, Iowa, Minnesota, North and South Dakota, and Idaho, south to Alabama, Tennessee, Mississippi, Louisiana, Texas, and New Mexico; introduced in the eastern states of Maine, Massachusetts, New Jersey, and Delaware.

This is much more common in Missouri than *A. tuberculatus*. In sections of its range where the species is abundant (parts of Oklahoma, Nebraska, Iowa, South Dakota, and Missouri), this species has been found to be important in causing cases of hay fever in late summer, contributing to 35 per cent of cases in some areas.

3. **Amaranthus Palmeri** S. Wats. Map 878

Flowers August–October.

Occurs in waste ground and along railroads. Known only from central and western Missouri in St. Louis (St. Louis, Rock Island R.R., south of Carrie Avenue, August 14, 1954, *Muehlenbach 331*), Jackson, and Jasper (waste ground along railway grade, Stockyard Switch, Joplin, August 10, 1949, *Palmer 49748*) counties.

Ranges from Oklahoma to California, south to Texas and Mexico; introduced in Kansas, Missouri, Illinois, Massachusetts, New York, Pennsylvania, and elsewhere in the East.

In part of its range, where abundant, this species is believed to be an important cause of hay fever.

4. **Amaranthus Torreyi** (Gray) Benth. Map 879

*Amaranthus arenicola* I. M. Johnston

Flowers late June–October.

Occurs in waste ground and along railroads. Known only from central Missouri in Jackson and St. Louis (St. Louis, Burlington R.R. north of E. Grand Ave., July 5, 1954, *Muehlenbach 228*) counties.

Ranges from Iowa to Colorado, south to Texas and New Mexico; introduced in Missouri, Illinois, Michigan, Pennsylvania, New Jersey, D.C., and Virginia.

5. **Amaranthus spinosus** L.

Thorny Amaranth Map 880

Also called Careless Weed.

Flowers early June–October.

Occurs in waste and cultivated ground. Southern and central Missouri north to Pike, Monroe, Saline, Ray, and Clay counties; not recorded from northern Missouri, but doubtless of expected occurrence in that section.

Probably native of tropical America and naturalized in the United States from Florida to Texas, north to Maine, Ontario, and Minnesota; also naturalized in the Old World tropics.

In some tropical American countries the young leaves and shoots are cooked as a vegetable.

6. **Amaranthus caudatus** L. Purple Amaranth Map 881

Also called Love-Lies-Bleeding; Tassel Flower.

*Amaranthus cruentus* L. [G]

*Amaranthus paniculatus* L. [P & S]

*Amaranthus hybridus* in part [BB], not L.

Flowers August–October.

Cultivated as a garden annual, rarely escaping along roadsides, dumps, and waste ground in Missouri, where known from a few counties in the northern and central part of the state.

Probably native of tropical America; introduced into North America, where naturalized from Quebec and throughout most of the United States; also naturalized in the Old World.

This species does not appear to persist or maintain itself, so that one finds only an occasional waif here and there escaped from gardens. The differences between *A. caudatus* and *A. cruentus* do not appear to be distinct enough for separating them as species, and it seems preferable to keep all the red and purple forms under one name. In Mexico the seeds of this plant are mashed up and mixed with water as a type of mush.

7. **Amaranthus hybridus** L. Green Amaranth
Map 882

Also known as Prince's Feather, Pigweed, Wild Beet, Careless Weed, Slender Pigweed.

*Amaranthus hybridus* var. *hypochondriacus* (L.) Robinson [P & S]

Flowers August–October.

Occurs as a weed in cultivated fields, pastures, farm lots, around dwellings, gardens, along railroads, and waste places. Throughout Missouri, and doubtless in every county, but only recorded from a relatively small number of counties.

Probably native of tropical America; naturalized in many parts of the Old World, and in the New World found throughout the temperate and tropical portions, in North America ranging from Quebec, Massachusetts, Ontario, and Michigan to Iowa, south throughout the United States.

The var. *hypochondriacus*, sometimes known as Prince's Feather, with usually copper-red or red-tinged inflorescences, is cultivated and escaped in St. Louis and Boone (city dump, Columbia, August 18, 1933, *Drouet 983*; ballast in East Highlands, Columbia, August 9, 1933, *Drouet 921*) counties, but is usually not considered varietally distinct.

*Amaranthus hybridus* makes a very palatable cooked vegetable and spinach substitute when the young plants or shoot tips are gathered. They can be eaten alone or mixed with *Chenopodium*, *Rumex*, *Lactuca*, and *Phytolacca*. The nutritious seeds can be made into a bread, for which purpose they were employed by some southwestern tribes of American Indians. Pigs are fond of eating this and the next species, accounting for one of the common names of the plant.

8. **Amaranthus retroflexus** L.
Rough Green Amaranth Map 883

Also called Rough Green Pigweed, Careless Weed, Wild Beet.

Flowers July–October.

Occurs in cultivated fields, farm lots, along roadsides, railroads, waste places, gardens, and occasionally in dry open places on top of bluffs and prairies. Throughout Missouri; actual records are relatively

Plate no. 150. **1.** Amaranthus tuberculatus, female plant, × $^2/_5$; Details from Small, The New York Botanical Garden. **2.** Amaranthus Powellii, × $^2/_5$. **3.** Amaranthus spinosus, × $^2/_5$. **4.** Amaranthus graecizans, × $^2/_5$. **5.** Amaranthus retroflexus, × $^2/_5$; Details from Small, The New York Botanical Garden.

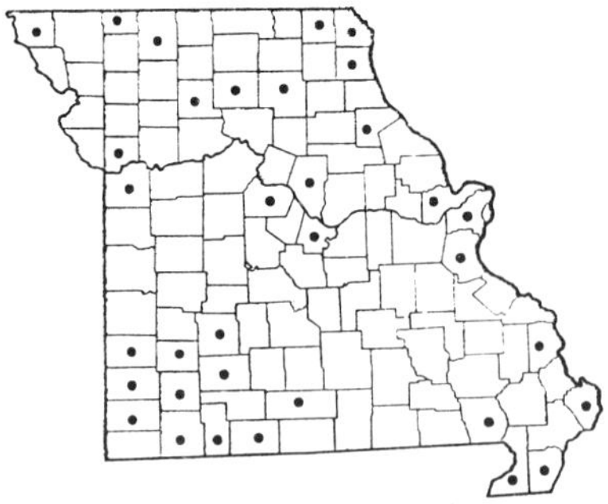

882 Amaranthus hybridus (Green Amaranth)

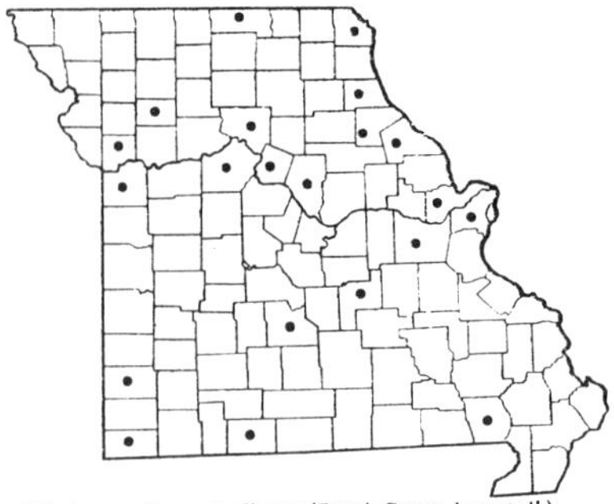

883 Amaranthus retroflexus (Rough Green Amaranth)

884 Amaranthus Powellii

few, but doubtless to be found in most counties.

Native of tropical America; naturalized in North America and throughout the United States; naturalized also in many parts of the Old World.

As in the case of *A. hybridus*, the young plants of this species make an excellent cooked vegetable. Normally *A. retroflexus* is eaten by livestock without having any harmful effects, but some reported cases of poisoning, producing a bloated condition, have been attributed to the eating by cattle and horses of this or other plants containing a high percentage of stored nitrates, particularly where plants had been growing on a soil rich in fertilizer and nitrates or in years of drouth.

Late summer cases of hay fever are partly caused by this species in areas where it is abundant. The foliage in autumn may turn a brownish red.

9. **Amaranthus Powellii** S. Wats. Map 884

Flowers late June–October.

Occurs along railroads and waste ground, where known only from St. Louis County (St. Louis, Terminal R.R. Association between W. Florissant Ave. and Broadway, opposite Hussmann Refrigerator Co., August 28, 1955, *Muehlenbach 726*; Bremen Ave. freight yard of Terminal R.R. Association, July 4, 1958, *Muehlenbach 1435*).

Ranges from South Dakota to Oregon, south to Texas and California; introduced in Missouri, and from Maine to Pennsylvania.

10. **Amaranthus albus** L. Tumbleweed Map 885

Flowers July–October.

Occurs in cultivated and waste ground, along roadsides and railroads, sandy, gravelly, or muddy banks and flats along streams. Throughout Missouri.

Two variations are represented in Missouri material:

Plant glabrous (without hairs) . . . 10a. A. ALBUS var. ALBUS

Plant finely hairy . . 10b. A. ALBUS var. PUBESCENS

10a. **Amaranthus albus** var. **albus** Map 885

*Amaranthus albus* L. [G, BB]

*Amaranthus graecizans* of auth. [P & S], not L.

This is the commonly encountered variation in the state. Actually recorded from relatively few counties, but doubtless to be found in every county.

Ranges from southern Canada, south throughout the United States; introduced into Europe.

10b. **Amaranthus albus** var. **pubescens** (Uline & Bray) Fern. Map 885

Known from Jackson and Barton (open ground along roadside near river, near Bushnell, July 8, 1952, *Palmer 54468*; sandy open ground along Horse Creek, 5 mi. northeast of Milford, August 23, 1952, *Palmer 54893*) counties in western Missouri.

Ranges from Colorado, Nevada, and New Mexico, east to Missouri.

This is a very common tumbleweed. After maturing, the plants, like Russian Thistle (*Salsola Kali* var. *tenuifolia*), are blown around, accumulating along fence rows, roadsides, and ditches, and becoming a nuisance in road clearance or even injurious to fences.

11. **Amaranthus graecizans** L. Tumbleweed Map 886

Also called Prostrate Amaranth, Spreading Pigweed.

*Amaranthus blitoides* S. Wats. [P & S]

Flowers early June–October.

---

Plate no. 151. **1.** Iresine rhizomatosa, × 2/5; After Gleason, details from Small, The New York Botanical Garden. **2.** Tidestromia lanuginosa, × 2/5; After Gleason, The New York Botanical Garden. **3.** Froelichia floridana var. campestris, × 2/5. **4.** Froelichia gracilis, × 2/5.

PLATE NO. 151

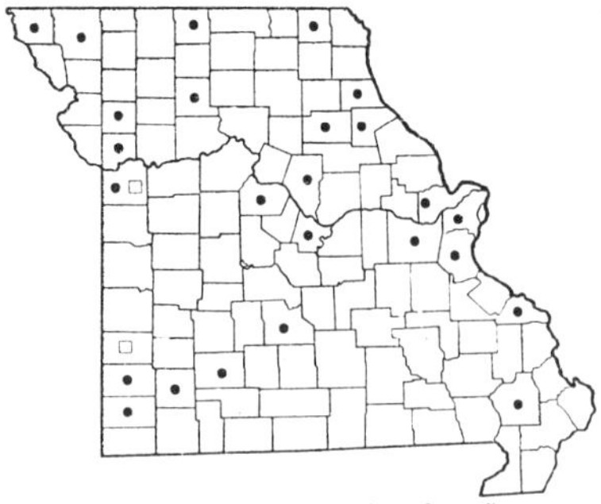

885 • Amaranthus albus var. albus (Tumbleweed)
885 □ Amaranthus albus var. pubescens

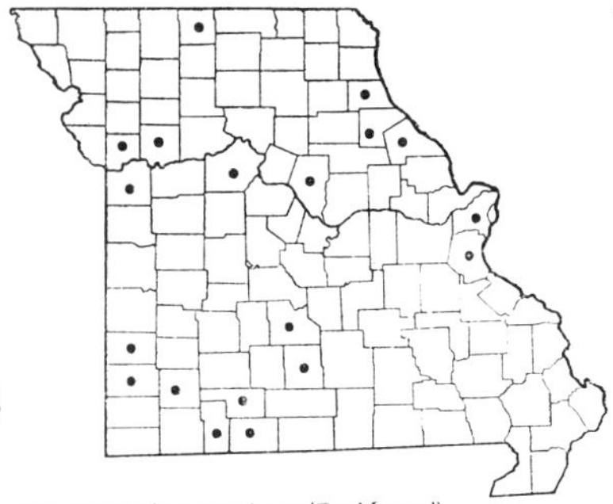

886 Amaranthus graecizans (Tumbleweed)

887 Tidestromia lanuginosa

Occurs in waste and cultivated ground, along roadsides, railroads, and sandy or gravelly places along streams. Scattered throughout Missouri, although actually recorded from but few counties, but doubtless to be found in most of them.

Native of the western states from North Dakota to British Columbia, south to Texas and California; established throughout most of the eastern United States and northeast to Quebec.

This species is more prostrate in habit than *A. albus*. It is a good source of green food for poultry and hogs.

## 3. **Tidestromia** Standl.

**Tidestromia lanuginosa** (Nutt.) Standl Map 887

*Cladothrix lanuginosa* Nutt. [P & S]

Flowers July–October.

Occurs in waste ground and along railroads, where known only from west-central Missouri in Jackson County (Kansas City, September 26, 1921, *Bush 9695, 9695A, 9695B*; Courtney, July 16, 1916, *Bush 8523*; September 10, 1918, *Bush 8783, 8783A*; August 1, 1918, *Bush 8161*).

Ranges from South Dakota to Colorado and Nevada, south to Texas and Mexico; introduced in Missouri and Illinois.

This prostrate or sprawling annual with many forking stems is covered with a dense coating of gray stellate pubescence. It somewhat resembles *Glinus lotoides* in habit of growth and appearance of leaves and flowers.

## 4. **Iresine** P. Br. Bloodleaf

**Iresine rhizomatosa** Standl. Bloodleaf Map 888

Also known as Juba's Bush.

*Iresine paniculata* (L.) Ktze. [P & S]

Flowers August–October.

Occurs in low wet woods and thickets near streams. Southern and central Missouri, north to St. Louis, Franklin, Gasconade, Osage, Camden, Benton, St. Clair, and Vernon counties.

Ranges from Alabama to Texas, north to Virginia and Maryland, Indiana, Illinois, Missouri, and Kansas.

Although the flowers themselves are quite minute, the silvery-white color of the bracts and calyx makes the elongated, pyramidal panicle quite showy in the deep shade in which the plants grow.

## 5. **Froelichia** Moench. Cottonweed

Woolly fruiting calyx with toothed or irregular margins (remove the wool to see clearly); flower spikes 10–13 mm. thick; leaves elliptic-lanceolate or oblanceolate; larger leaf-blades 5–25 mm. broad; stem stout, up to 7 mm. thick and up to 2 m. tall, either unbranched or with a few erect branches . . . 1. F. FLORIDANA var. CAMPESTRIS

Woolly fruiting calyx with slender spine-like, nearly distinct projections on the margins (remove the

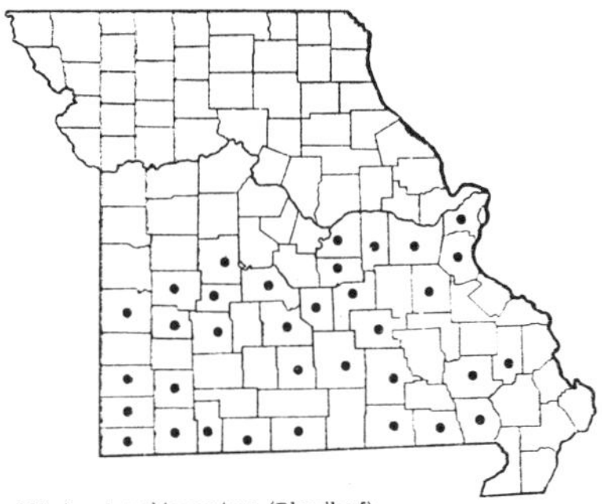

888 Iresine rhizomatosa (Bloodleaf)

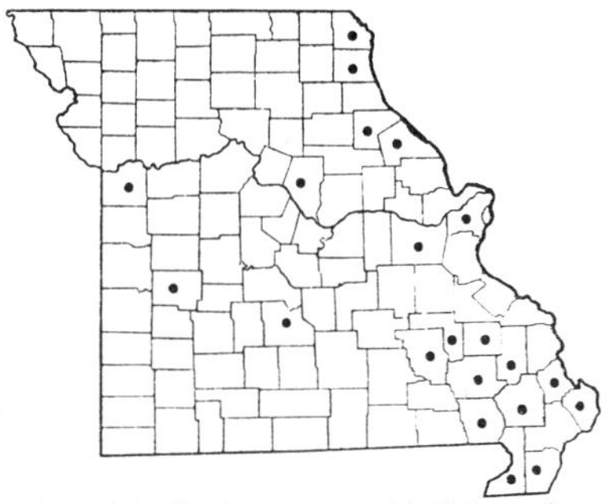

889 Froelichia floridana var. campestris (Cottonweed)

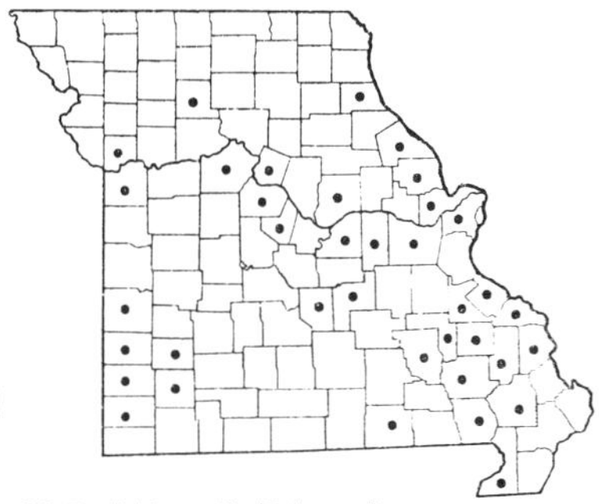

890 Froelichia gracilis (Cottonweed)

wool to see clearly); flower spikes 7–8 mm. thick; leaves linear to narrowly lanceolate; larger leaf-blades mainly 3–9 mm. (rarely up to 15 mm.) broad; stem slender, 1–2 mm. thick, 1.5–4.5 dm. (up to 6 dm.) tall, commonly with spreading branches at the base . . . . . . . . . . . 2. F. GRACILIS

1. **Froelichia floridana** (Nutt.) Moq. var. **campestris** (Small) Fern. Cottonweed Map 889
*Froelichia floridana* in part [BB, P & S in part]
Flowers May–September.

Occurs in sandy open ground and along railroads, frequently on sand and gravel bars along streams. Eastern Missouri mainly, west in central Missouri to Jackson County and in southern Missouri to Laclede and St. Clair counties.

Ranges from Indiana to Minnesota and South Dakota, south to Arkansas, Oklahoma, Texas, and New Mexico; introduced in Delaware and Maryland.

This species is much more robust and luxuriant-looking than the next, *F. gracilis*. From typical var. *floridana* of the southeastern states, var. *campestris* differs in the leaves rounded at the tip and broadest above the middle.

2. **Froelichia gracilis** (Hook.) Moq.
Cottonweed Map 890
Flowers May–September.

Occurs in sandy open ground and along railroads, and on sand and gravel bars along streams. Southern and central Missouri north to Marion and Livingston counties; frequent along some of the sandy and gravelly flats bordering Ozark streams.

Ranges from Indiana to Iowa and Colorado, south to Arkansas, Texas, New Mexico, and Mexico; introduced eastward to New York, New Jersey, and Virginia.

A collection by E. J. Palmer from Jasper County (*Palmer 56454* from Joplin, August 11, 1953) is somewhat intermediate between *F. gracilis* and *F. floridana* var. *campestris*, with the fruiting calyx characteristic of *F. gracilis* but the larger and usual form of the leaves of *F. floridana* var. *campestris*; it may be indicative of hybrid origin.

## Fam. NYCTAGINACEAE (Four-o'clock Family)

Flowers arising from a green calyx-like 5-lobed involucre; flowers conspicuous, the pink to purple portion (the calyx) 7–10 mm. long . . . . . . . . . . . . . . . . . . . . 1. MIRABILIS
Flowers not arising from a lobed involucre, but accompanied at their base by minute distinct bracts; flowers minute, the pink to white portion (calyx) 1–1.5 mm. long . . . . . . . . 2. BOERHAAVIA

### 1. Mirabilis L. Four-o'clock

a. Leaves (except the uppermost) with definite petioles (stalk) 10–30 mm. long; main leaf-blades ovate or triangular-ovate, broadest at the base . . . . . . . . . . . . . . . . . 1. M. NYCTAGINEA
a. Leaves all sessile (without stalks) or with a very short petiole; leaf-blades linear, lanceolate, or oblong, nearly the same width throughout or only slightly broader at the base . . . . . . . . . . . *b*
b. Leaves linear, usually at most 5 mm. wide, 15–30 times as long as wide; stems glabrous (without any hairs) below the inflorescence; angles of fruit smooth . . . . . . . . . . 4. M. LINEARIS
b. Leaves narrowly lanceolate or narrowly oblong, the main ones usually 10–40 mm. wide (or

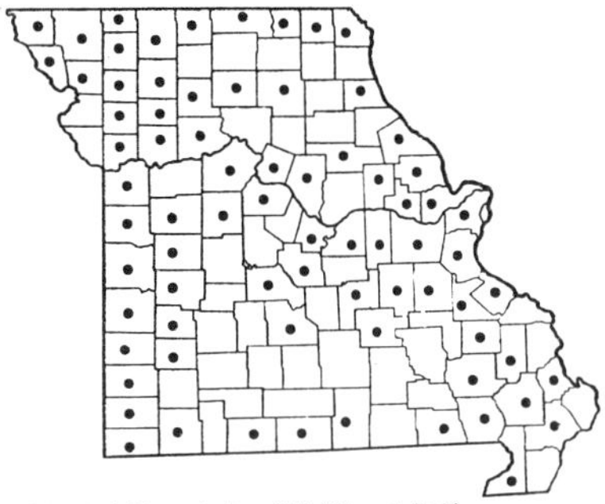

891 Mirabilis nyctaginea (Wild Four o'clock)

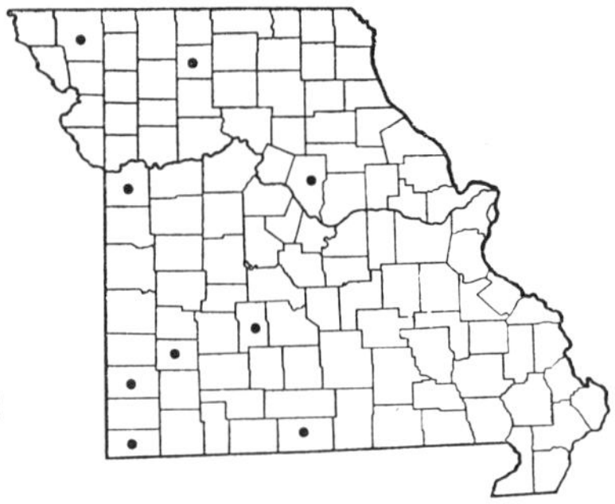

892 Mirabilis hirsuta

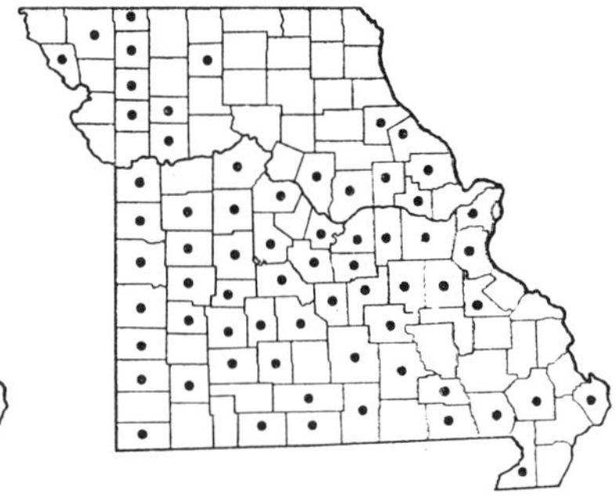

893 Mirabilis albida

rarely 4–9 mm. wide), 2–12 times as long as wide; stems either noticeably hairy with spreading hairs or inconspicuously hairy with 2 lines of short hairs . . . . . . . . . . . . . . . . c

c. Stem noticeably hairy at least in the lower part with hairs 1–2.5 mm. long; leaf-blades 1–4 times as long as wide, broadly wedge-shaped at base; angles of fruit smooth . . . 2. M. HIRSUTA

c. Stem inconspicuously hairy with 2 lines of short incurved hairs about 0.5 mm. long; leaf-blades 4–12 times as long as wide, tapering at base; angles of fruit roughened with little projections (tuberculate) . . . . . . . . . . . . . . . . . . . . . . . 3. M. ALBIDA

1. **Mirabilis nyctaginea** (Michx.) Mac M.
Wild Four-o'clock Map 891
Also known as Heart-leaved Umbrella-wort.
*Oxybaphus nyctagineus* (Michx.) Sweet [P & S]; *Oxybaphus floribundus* Chois. [P & S]
Flowers May–October.

Occurs mostly along railroad tracks, also in open ground, prairies, and waste places. Throughout Missouri, in most counties.

Ranges from Wisconsin and Manitoba to Montana and Colorado, south to Alabama, Louisiana, and Texas; introduced east to Massachusetts and Virginia.

2. **Mirabilis hirsuta** (Pursh) Mac M. Map 892
*Oxybaphus hirsutus* (Pursh) Sweet [P & S]
*Oxybaphus hirsutus* var. *integrifolius* Chois. [P & S]
Flowers May–October.

Occurs in dry open ground, glades, and along roadsides. Scattered in western Missouri east to Grundy, Boone, Dallas, and Ozark counties.

Ranges from Wisconsin to Saskatchewan and Wyoming, south to Missouri, Oklahoma, Texas, New Mexico, and Arizona; introduced eastward to New England.

3. **Mirabilis albida** (Walt.) Heimerl Map 893
*Oxybaphus albidus* (Walt.) Sweet [P & S]
*Allionia decumbens* Nutt. [Rydberg]
*Allionia bracteata* Rydb. [Rydberg]
Flowers May–October.

Occurs along edges of bluffs, limestone glades, prairies, and open ground, rarely in rocky open woods. Western, southern, and central Missouri, northeast to Ralls County; absent from the northeastern sector of the state.

Ranges from Georgia to Texas, north to South Carolina, Tennessee, Iowa, and Kansas; rarely introduced northeastward.

This species varies somewhat in width of leaves and amount of pubescence on the stems. The stems are silvery-gray or glaucous and have hairy lines, but appear glabrous.

4. **Mirabilis linearis** (Pursh) Heimerl Map 894
*Oxybaphus linearis* (Pursh) Robinson [P & S]
Flowers May–October.

Occurs on rocky prairies, limestone glades, edges of limestone escarpments, and open wooded rocky limestone hills. Scattered and rare in western Missouri in Jackson, Jasper, Lawrence, and Miller counties.

Ranges from Minnesota to Montana, south to Missouri, Oklahoma, Arizona, and Mexico; rarely introduced northeastward to New England.

Many narrowly lanceolate-leaved collections of *M. albida* have been misidentified as this species. Actually, *M. linearis* is the rarest of the species in Missouri.

---

Plate no. 152. **1.** Mirabilis nyctaginea, × 2/7. **2.** Mirabilis hirsuta, × 2/7; Details from Small, The New York Botanical Garden. **3.** Mirabilis albida, × 2/7; Details from Small, The New York Botanical Garden. **4.** Boerhaavia erecta, × 2/7; *a.* Fruit, × 3 3/7; Details rom Small, The New York Botanical Garden. **5.** Mirabilis linearis, × 2/7. **6.** Phytolacca americana, × 2/7; Details from Small, The New York Botanical Garden.

PLATE NO. 152

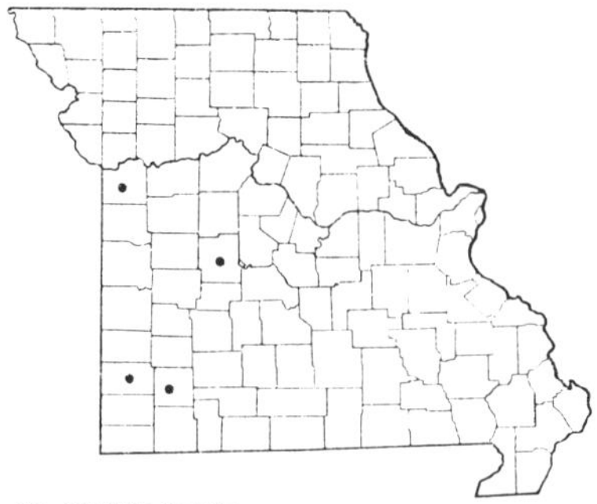
894 Mirabilis linearis

895 Boerhaavia erecta

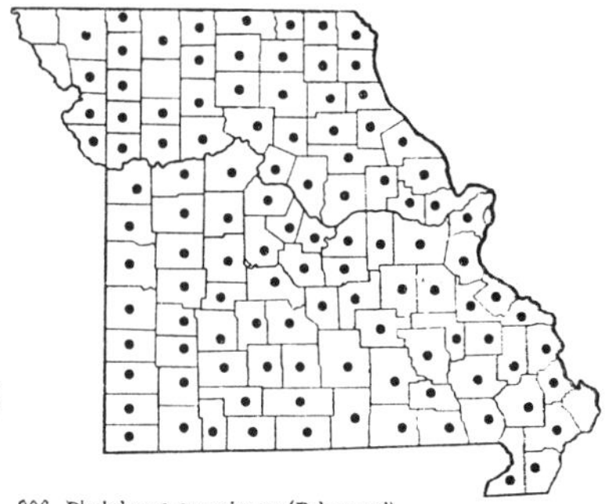
896 Phytolacca americana (Pokeweed)

## 2. **Boerhaavia** L.

**Boerhaavia erecta** L. Map 895

*Boerhaavia erecta* f. *subepunctata* Heimerl

Flowers July–October.

Occurs along railroads. Known only from St. Louis County (St. Louis, Bremen Ave. freight yard of the Terminal Railroad Assoc., September 1, 1958, *Muehlenbach 1481*).

Native of tropical America, ranging from South to Central America, the West Indies, and Mexico, and extending in the southern United States from Florida to Texas and Arizona, north to South Carolina, Missouri, and Arkansas.

This plant has more or less procumbent stems. The leaves are cooked and eaten like spinach in some parts of the American tropics.

## Fam. **PHYTOLACCACEAE** (Pokeweed Family)

**Phytolacca americana** L. Pokeweed Map 896

Known also as Poke.

Flowers late May–October.

Occurs in rich soils in waste ground, farm lots, about dwellings, fields, along roadsides and railroads, thickets, and borders of woods. Throughout the state and doubtless in every county.

Ranges from Quebec, Maine, New York, and Ontario to Minnesota, south to Florida and Texas. Also in Mexico; introduced in Europe.

The roots and purple rind of the fully grown stem are poisonous, but the tender young leaves and shoots make a tasty cooked vegetable, prepared alone or mixed with other wild greens. Brought indoors before winter, the root, placed in earth in a cellar, can be forced into sending up succulent sprouts, which are cut when young and serve as an asparagus substitute during the winter and early spring months. The root has been used in medicine for treatment of skin diseases, and for garget (swelling of the throat) in cattle. The berries are actually not poisonous, as sometimes stated, but the seeds are suspected of having poisonous properties as children have been poisoned after eating the berries. The berries are eaten by birds in great quantity. The purple juice of the berries is used for coloring foods, such as frostings and candies, and beverages. The dried berries, cut up and put into whiskey, have been used as a remedy for rheumatism. The juice of the fresh berries is also used as a red ink.

## Fam. **AIZOACEAE** (Carpet-weed Family)

The genus *Geocarpon*, previously placed in this family, has been transferred to Fam. **Caryophyllaceae** (Bull. Torr. Club 77: 268–73. 1950), and is to be found in the present flora in its newly placed position.

a. Leaves opposite, in more or less unequal pairs . . . . . . . . . . . . . . . . 1. TRIANTHEMA
a. Leaves in whorls (circles) of 3–8, unequal in size and shape . . . . . . . . . . . . . . . . *b*

---

Plate no. 153. **1.** Trianthema Portulacastrum, × $^{2}/_{5}$; After Gleason, details from Small, The New York Botanical Garden. **2.** Glinus lotoides, × $1^{3}/_{5}$; After Gleason, details from Small, The New York Botanical Garden. **3.** Mollugo verticillata, × $^{2}/_{5}$.

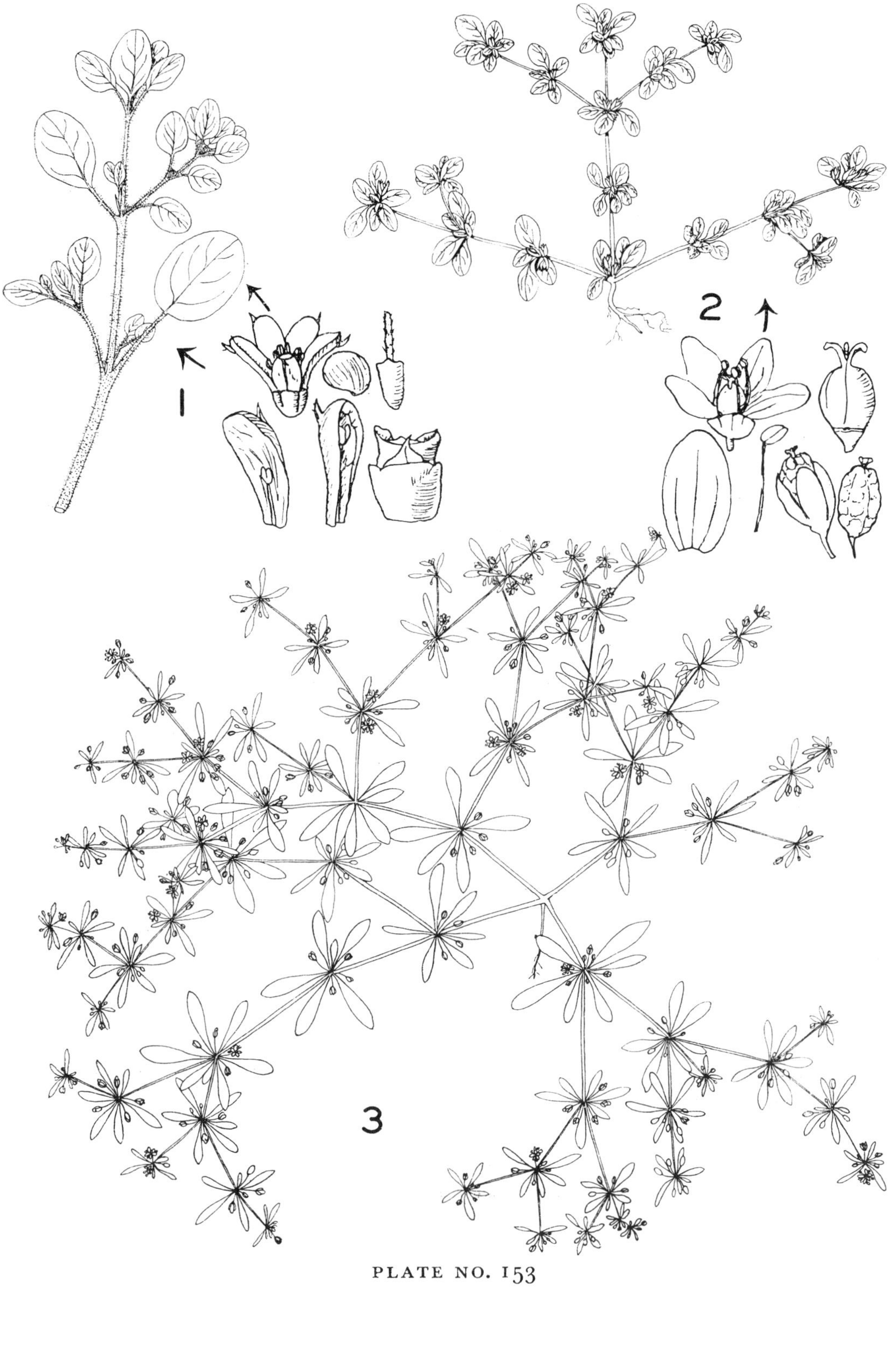


PLATE NO. 153

897 Trianthema Portulacastrum (Sea Purslane)

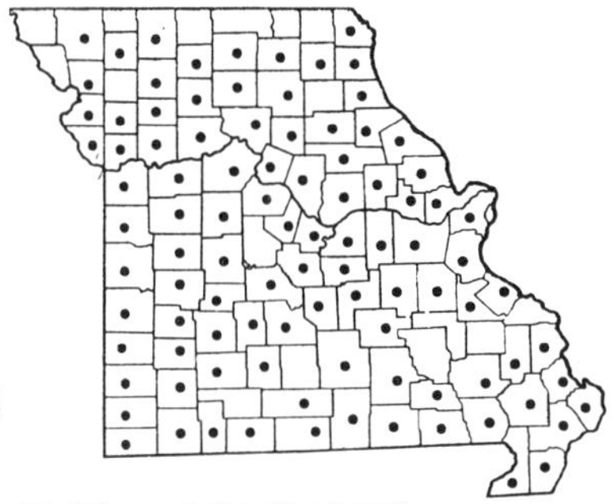

898 Mollugo verticillata (Carpet-weed)

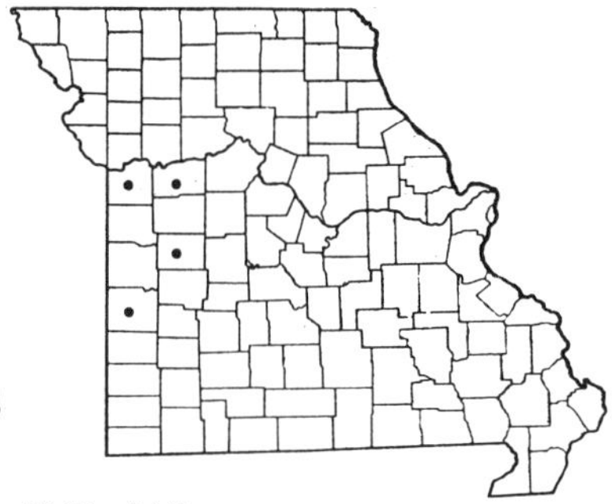

899 Glinus lotoides

b. Plants glabrous (without hairs); leaves narrow, spatulate, the larger ones usually 3–5 (rarely to 10) mm. wide and 4–8 times as long as broad; flowers on slender stalks (pedicels) . . . 2. MOLLUGO

b. Plants hairy with star-shaped hairs (stellate-pubescent); leaves round-obovate, the larger ones 8–20 mm. (rarely only 5) wide and $1\frac{1}{2}$–2 times as long as broad; flowers sessile (without stalks) 3. GLINUS

## 1. **Trianthema** L. Sea Purslane

**Trianthema Portulacastrum** L. Sea Purslane Map 897

Flowers May–October.

Occurs in waste ground, where introduced. Known only from west-central and southwestern Missouri in Jackson and Jasper (rich waste ground along Stockyard Switch, Joplin, September 26, 1949, *Palmer 49835*) counties.

Throughout the tropical regions of both the Old and New World; in the United States ranging from Florida to Texas and California, north to New Jersey, Missouri and Oklahoma.

## 2. **Mollugo** L. Carpet-weed

**Mollugo verticillata** L. Carpet-weed Map 898

Flowers May–November.

Occurs as a weed in waste and cultivated ground and along railroads. Throughout Missouri and doubtless in every county.

Found in the Old World and in temperate and tropical North America, Central and South America and the West Indies; naturalized in North America from Florida to Texas, north to Nova Scotia, Quebec, Ontario, and Washington.

## 3. **Glinus** L.

**Glinus lotoides** L. Map 899

Flowers July–September.

Occurs in wet muddy alluvial ground along rivers and oxbow lakes of stream meanders. Known only from western Missouri in Jackson, Lafayette, Henry, and Vernon counties.

Native of tropical Africa and southern Europe; introduced into the United States, where ranging from Louisiana to California, north to Missouri and Oklahoma.

The stems are widely spreading, procumbent to ascending and repeatedly forked. With their short hairiness of star-shaped hairs and leaf-shape, the plants closely resemble the amaranthaceous *Tidestromia lanuginosa*, from which they are best distinguished by the many ovules in a 3–5-celled ovary in *Glinus* as opposed to the 1-celled ovary with 1 ovule in *Tidestromia*. In both genera the calyx consists of 5 distinct sepals or nearly so.

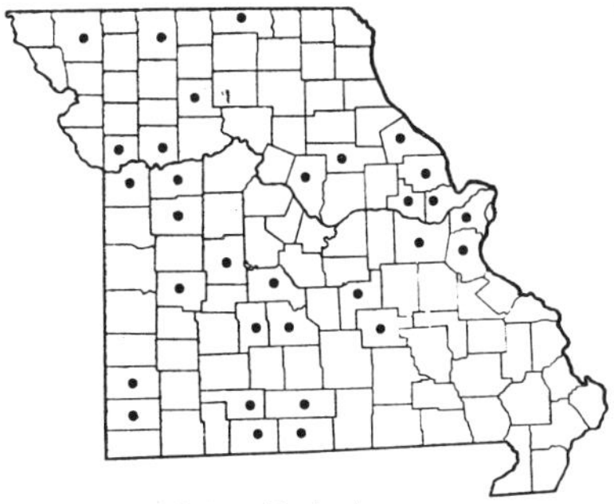

900 Portulaca oleracea (Purslane)

901 Portulaca neglecta (Purslane)

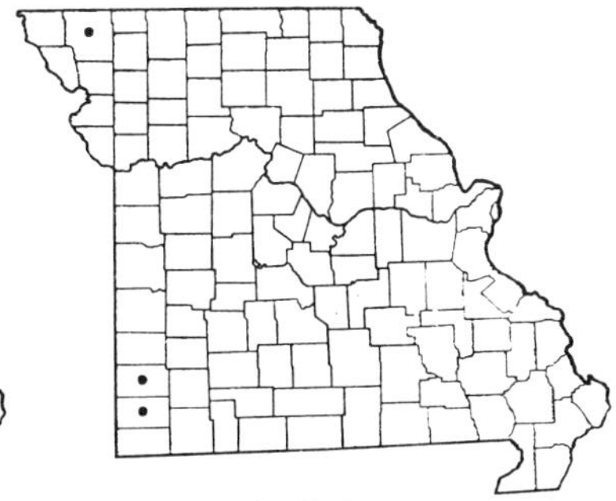

902 Portulaca retusa (Purslane)

## Fam. **PORTULACACEAE** (Purslane Family)

a. Only 2 leaves generally present on each flowering stem; leaf-blades 4–15 cm. long . . . 2. CLAYTONIA
a. Several to many leaves present with each flowering stem; leaf-blades 0.6–5 cm. long . . . . . . *b*
 b. Leaves scattered along the flowering stem from near base to near tip of stem; flowers sessile (without stalks) or nearly so . . . . . . . . . . . . . . . . . . . . 1. PORTULACA
 b. Leaves in a tuft or rosette at the base of the plant; flowers stalked, supported by a long, erect naked stem (scape) which is much longer than the leaves . . . . . . . . . . . . . 3. TALINUM

### 1. **Portulaca** L. Purslane

a. Plants completely glabrous (without hairs); leaves flat, broadest in the upper half, the larger ones 8–16 mm. wide; petals yellow . . . . . . . . . . . . . . . . . . . . . . . . . . *b*
 b. Calyx-lobes obtuse; seeds covered with rather sharp-pointed conical projections (tubercles); leaves often retuse (with a shallow notch at summit) . . . . . . . . . . . . . 3. P. RETUSA
 b. Calyx-lobes pointed; seeds covered with blunt or rounded projections (tubercles); leaves commonly rounded or truncate (as if cut straight across) at summit, rarely retuse (with a shallow notch at tip) . . . . . . . . . . . . . . . . . . . . . . . . . . *c*
  c. Commonly found; stems lying on ground (prostrate); main leaves 1–3 cm. long; stamens 7–12 1. P. OLERACEA
  c. Rare, western Mo. plant; stems ascending; main leaves 2.5–5 cm. long; stamens 12–18 . 2. P. NEGLECTA
a. Plants with hairy tufts in the axils of the leaves; leaves cylindrical or nearly so with curved or rounded sides, about the same width throughout, about 2 mm. wide; petals red, rose-purple, pink, white, orange, or yellow . . . . . . . . . . . . . . . . . . . . . . . . . *d*
 d. Flowers red or rose-purple, less than 1 cm. wide; petals 5–7.5 mm. long; stamens 10–30; native species of rocky limestone, sandstone, or chert glades . . . . . . . . . . . . 4. P. MUNDULA
 d. Flowers red, orange, white, or rich yellow, 2–4 cm. wide; petals 15–30 mm. long; stamens about 40; not native, escaped from gardens in waste ground and along railroads . . . . 5. P. GRANDIFLORA

1. **Portulaca oleracea** L. Purslane Map 900

Also called Pusley or Pursley.

Flowers June–November.

Occurs in cultivated and waste ground, rocky escarpments along bluffs, and open places. Scattered throughout Missouri, and doubtless in every county, although represented by relatively few county collections.

Native of Europe; introduced in North America, where found from southern Canada south throughout the United States. Also naturalized in temperate and tropical regions of the New and Old World.

The plant may be eaten in various ways, the stems and leaves either cooked as a vegetable alone or in combination with turnip tops or mustard greens or with bread-crumbs and eggs, or used as a fresh green in salad form, or pickled. Some of the Indians of the southwestern United States are reported using the seeds in a kind of bread or mush. The plants are one of the favorite hog foods. They also help in building up soils.

2. **Portulaca neglecta** Mack. & Bush Purslane Map 901

Flowers July–September.

Occurs on limestone glades, prairies, and rich alluvial bottoms. Western Missouri, where known from Caldwell, Jackson, Cass, Johnson, and Jasper counties.

Ranges from Missouri to Arkansas.

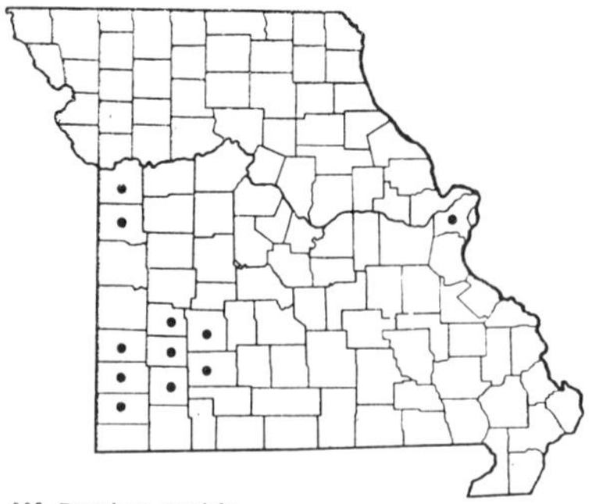

903 Portulaca mundula

904 Portulaca grandiflora (Portulaca)

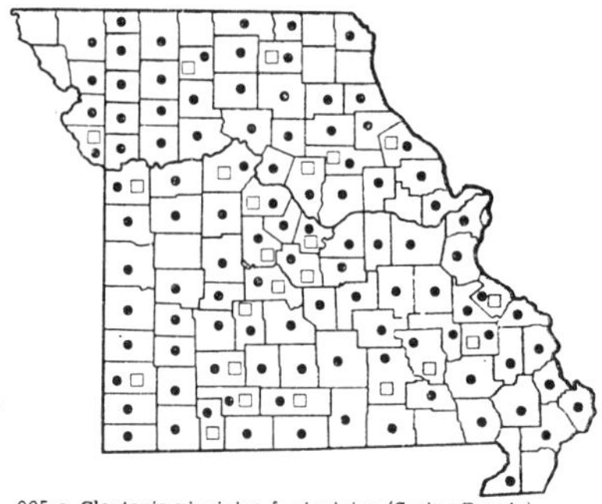

905 • Claytonia virginica f. virginica (Spring Beauty)
905 □ Claytonia virginica f. robusta

Gleason (*New Ill. Fl.* 2: 114. 1952) has indicated that this may be only a luxuriant form of *P. oleracea*, and considers it as of doubtful specific rank, and Shinners (*Spring Flora*, p. 137) has relegated it to synonymy under *P. oleracea*. However, until more detailed field investigations can be carried out in the future, the species is being retained as such, based upon the characters as given in the key.

3. **Portulaca retusa** Engelm. Purslane Map 902
Flowers July–September.

Occurs on limestone and chert glades, where known only from southwestern Missouri in Jasper and Newton (chert barrens, near Reding's Mill, October 7, 1927, *Palmer 33158*) counties.

Ranges from Missouri to Utah, south to Arkansas, Texas, and Arizona.

4. **Portulaca mundula** Johnston Purslane Map 903

*Portulaca parvula* of auth. [G, BB, P & S, Steyerm.], not Gray

Flowers last of May–October.

Occurs on sandstone, chert, and limestone glades and edges of rocky exposed bluffs and escarpments in western Missouri south of the Missouri River from Jackson County south to Newton County, east to Polk and Greene counties, and exceptionally introduced along railroads in east-central Missouri in St. Louis County (St. Louis, along Mo.-Kan.-Tex. R.R., Baden Freight Yard, October 7, 1956, *Muehlenbach 1116*).

Ranges from Missouri and Kansas to Oklahoma, Texas, and New Mexico.

On limestone glades and outcrops on the edge of southwest-facing limestone bluffs this species is often associated with *Sphaeralcea angusta*, *Mentzelia oligosperma*, *Ophioglossum Engelmanni*, and other calcareous-inhabiting types. On sandstone exposures and glades it occurs with a different selection of associates, such as *Sedum Nuttallianum*, *Oenothera linifolia*, *Linaria canadensis* var. *texana*, *Krigia virginica*, *Talinum parviflorum*, *Selaginella rupestris*, and *Crotonopsis elliptica*, species all favoring acid soils.

Missouri material has been confused previously with true *P. parvula* Gray. Dr. I. M. Johnston (Jour. Arn. Arb. 29: 195–96. 1948) has shown that the Missouri specimens constitute an undescribed species, characterized by the red-purplish or purple-red obovate petals often as much as 6–7.5 mm. long, 3–4.5 mm. wide, retuse at the summit, by the lower part of the capsule 2.5–3.5 mm. in diameter, by the shortly stipitate capsule, by the mature capsule having the upper part smooth and hemispherical with hairy calyx and stamens cohering, by the 10–15 (rarely 30) stamens, and by the black stellate-tuberculate seeds 0.3–0.5 mm. in diameter.

Contrasted with *P. mundula*, typical *P. parvula* Gray of Oklahoma, Texas, Arizona, and New Mexico has yellow, orange, or bronzy petals 2–2.5 mm. long, small, conspicuously stipitate fruit, which, after dehiscence and shedding of the seeds, is only 1.5–2 mm. in diameter at the saucer-shaped base. The well-developed stipe is 1–1.5 mm. long, and the seeds are 3–3.5 mm. long, at first brownish, becoming black when mature and covered with minute, crowded, flattened stellate surface markings.

5. **Portulaca grandiflora** Hook.
Portulaca, Rose Moss Map 904

---

Plate no. 154. **1.** Portulaca oleracea, × 2/5 (Scribner's). **2.** Portulaca mundula, × 2/5. **3.** Portulaca grandiflora, × 2/5; Details from Small, The New York Botanical Garden. **4.** Claytonia virginica, × 2/5. **5.** Talinum parviflorum, × 2/5. **6.** Talinum calycinum, × 2/5; Details for 5 and 6 from Small, The New York Botanical Garden.

PLATE NO. 154

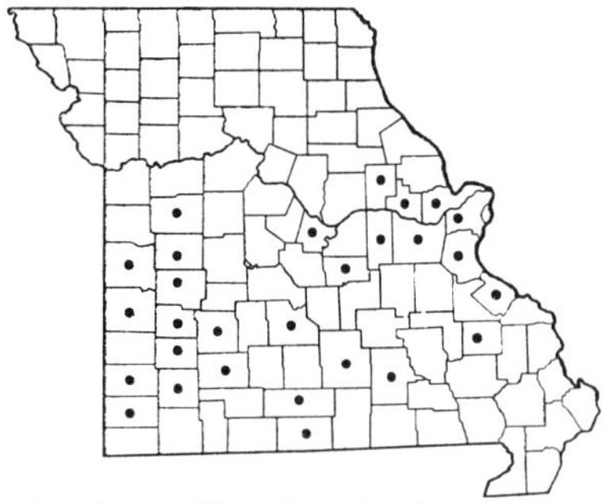

906 Talinum parviflorum (Fame Flower)

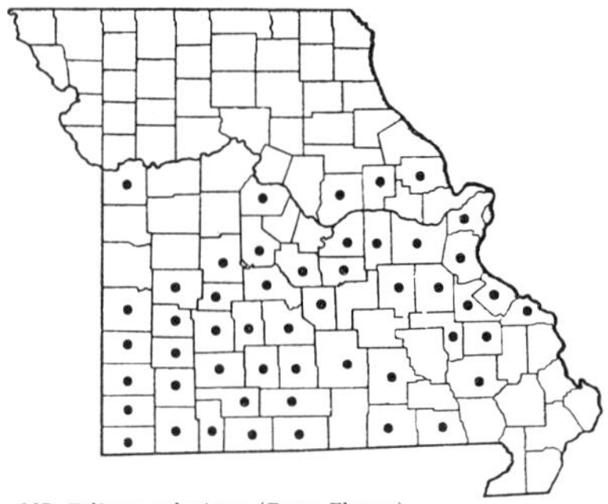

907 Talinum calycinum (Fame Flower)

908 Scleranthus annuus (Knawel)

Flowers July–September.

Commonly cultivated, but rarely escaping along roadsides and along railroads. Known only from Boone (roadside, Columbia, July, 1903, Daniels) and St. Louis (St. Louis, Mo. Pac. R.R. under the loading ramp of the Cornel Seed Co., August 23, 1958, *Muehlenbach 1468*).

Native of Argentina.

## 2. **Claytonia** L. Spring Beauty

**Claytonia virginica** L. Spring Beauty Map 905

Flowers February–May.

Occurs commonly in dry upland open woods on level ground or hillsides, moist alluvial wooded or open valleys, prairies, and rocky ledges. Throughout Missouri.

Ranges from Nova Scotia, Quebec, and Ontario to Minnesota, south to Georgia, Alabama, Mississippi, Louisiana, and Texas.

Two variations are represented in Missouri material:

Leaves 2–10 mm. wide, linear or linear-oblanceolate . . . . . . a. C. VIRGINICA f. VIRGINICA
Leaves 10–33 mm. wide, rather broad in the middle, lanceolate or elliptical b. C. VIRGINICA f. ROBUSTA

a. **Claytonia virginica** f. **virginica** Map 905

*Claytonia virginica* L. [G, BB, P & S, Steyerm.]

This is the commoner form found doubtless in every county of the state.

b. **Claytonia virginica** f. **robusta** (Somes) Palmer & Steyerm. Map 905

Scattered throughout the state and not uncommon. This form has been confused at various times with the more eastern and northern species, *C. caroliniana* Michx., of which no authentic specimens have been found in the state.

The young plants of Spring Beauty, in regions of abundance, may be eaten as a cooked vegetable, while the thick, round, underground corms, when boiled in salted water, resemble the taste of chestnuts.

## 3. **Talinum** Adans. Fame Flower, Rock Pink

Petals 3–7 mm. long, pale or dull pink; stamens 4–8 (mainly 5–6); capsules 3–5 mm. high . 1. T. PARVIFLORUM
Petals 10–15 mm. long, deep rose-pink or rose-red; stamens 25–45; capsules 6–8 mm. high 2. T. CALYCINUM

1. **Talinum parviflorum** Nutt.
Fame Flower, Rock Pink Map 906

Flowers May–August.

Occurs on more or less level sandstone and chert glades and rocky outcrops above bluffs. Southern and central Missouri, in the Ozark and unglaciated prairie sections, north to St. Louis, St. Charles, Warren, Montgomery, Cole, Johnson, and Bates counties; absent from the southeastern lowland area.

Ranges from Minnesota to South Dakota and Colorado, south to Missouri, Arkansas, Texas, Arizona, and Mexico.

This species has shorter leaves and shorter, more delicate peduncles than the following one.

2. **Talinum calycinum** Engelm.
Fame Flower, Rock Pink Map 907

Flowers May–August.

Occurs on sandstone, chert, granite, or limestone glades and outcrops of exposed bluff escarpments on more or less level rocks. Known only from the Ozark and unglaciated prairie sections of southern and central Missouri, north to Lincoln, Montgomery, Callaway, Cooper, Benton, St. Clair, and Vernon counties, and locally in Jackson County.

Ranges from Missouri, Kansas, and Oklahoma to Arkansas and Texas.

This species does well in rock gardens if provided with sufficient sunlight and drainage. It does not appear particular as to the type of rock on which it grows. The author has had plants which have maintained and seeded themselves during the past fifteen years in northern Illinois. The flowers open around noon and close after a few hours. Plants often occur in abundance, and in some places, where they occur on sandstone, their bright rose-red flowers mingle colorfully with the yellow flowers and silvery green foliage of *Sedum Nuttallianum*, the delicate white sprays of *Arenaria patula*, and the purple flowers of *Specularia leptocarpa*.

## Fam. **CARYOPHYLLACEAE** (Pink Family)

a. Transparent papery-like stipules at the base of the leaves . . . . . *b*
  b. Leaves in 2's opposite one another, elliptic to oblanceolate, 0.5–3 cm. long; petals normally absent; native plants of woodland and rocky open places . . . . . 3. Paronychia
  b. Leaves appearing in whorls of 6–8, filiform, 2–5 cm. long; petals present, white; introduced plant of waste and cultivated ground and along railroads . . . . . 4. Spergula
a. No stipules at the base of the leaves . . . . . *c*
  c. Calyx of separate sepals, these not united at the base. . . . . *d*
    d. Inflorescence umbellate, the 3–8 flowers on slender stalks (pedicels) of about equal length and arising at the tip of the main flower stalk (peduncle) . . . . . 9. Holosteum
    d. Inflorescence not as above, the flowers arranged otherwise . . . . . *e*
      e. Leaves all thread- or hair-like, 0.5–1 mm. broad . . . . . *f*
        f. Styles normally 3; stems erect; petals 5, conspicuous, about 2 times the length of the sepals; sepals 5 . . . . . 6. Arenaria
        f. Styles 4 or 5; stems mostly ascending or spreading; petals 0 or 4–5, inconspicuous, shorter, equaling, or slightly longer than sepals; sepals 4 or 5 . . . . . 5. Sagina
      e. Leaves broader. . . . . *g*
        g. Styles 5; stems either with noticeable spreading hairs or these not occurring in definite lines; fruit much longer than broad, cylindrical, with usually 10 hair-like teeth at the summit . . . . . 8. Cerastium
        g. Styles 3; stems either glabrous (without hairs) or with very short inconspicuous hairs occurring in definite lines or pressed close to the stem; fruit ovoid or ellipsoid, with no teeth at summit or less than 10 teeth. . . . . *h*
          h. The white petals without a notch or at most with only a shallow notch at the tip . 6. Arenaria
          h. The white petals deeply notched or split nearly halfway or more to the base . 7. Stellaria
  c. Calyx with the sepals united into a short or long tube . . . . . *i*
    i. Flowers all green, petals normally absent; calyx 3–5.5 mm. long . . . . . *j*
      j. Stems 1–4 cm. tall or long, more or less erect; larger leaves about 5 mm. long, narrowly oblong or ovate-oblong, obtuse or blunt at tip; style none; stigmas 3; fruit many-seeded . 2. Geocarpon
      j. Stems up to 15 cm. tall or long, sometimes longer, the main part spreading or reclining; larger leaves 5–15 mm. long, thread- or hair-like, slender-pointed at tip; styles 2; fruit 1-seeded. . . . . 1. Scleranthus
    i. Flowers with green, whitish, or pink calyx and white, rose, pink, or purple petals with petals normally present (absent in *Silene antirrhina* f. *apetala*); calyx 5–25 mm. long . . . . . *k*
      k. 1–3 pairs of scales or leaf-like bracts are present at base of calyx; calyx-tube 30–40-nerved 14. Dianthus
      k. No scales or leaf-like bracts at base of calyx; calyx-tube faintly nerved or with 5–10–20 nerves (in species found in Missouri) . . . . . *l*
        l. The elongated calyx-lobes longer than and extending beyond the purple-rose petals; calyx-lobes longer than the calyx-tube . . . . . 10. Agrostemma
        l. The calyx-lobes shorter than the petals or not extending beyond them; calyx-lobes much shorter than the calyx-tube, less than half as long as the tube. . . . . *m*

m. Some or all flowers male (staminate) or female (pistillate), the sexes of the flowers usually separated on different plants (dioecious), or sometimes mixed with both unisexual and bisexual or perfect flowers (polygamo-dioecious) . . . . . . 11. LYCHNIS
m. Most of the flowers perfect with both stamens and pistils in the same flower . . . . . *n*
  n. Styles 5; each of the 5 valves of the capsule with 2 teeth at the summit . . 11. LYCHNIS
  n. Styles 2 or 3; capsule with 3, 4, or 6 teeth at the summit . . . . . . . . . *o*
    o. Styles 2; capsule with 4 teeth at the summit, 4-valved; calyx scarcely nerved or 5-nerved . . . . . . . . . . . . . . . . . . . 13. SAPONARIA
    o. Styles 3; capsule with 3 or 6 teeth at the summit, 3- or 6-valved; calyx 5-10- or 20-nerved . . . . . . . . . . . . . . . . . . . . 12. SILENE

## 1. **Scleranthus** L. Knawel

**Scleranthus annuus** L. Knawel Map 908

Flowers March–October.

Occurs in moist depressions of upland sandstone glades, where known only from Greene County, southwestern Missouri (on northeast side of railroad tracks and highway 123, 0.2 mi. northwest of crossroad, and 1 mi. north of junction of road AC with highway 123, T30N, R23W, SE¼ of SE¼ sect. 8, 1 mi. [by air] southwest of Cave Spring, 5 mi. southeast of Walnut Grove, May 3, 1958, *Steyermark*, *Voigt*, and *Mohlenbrock 86081*).

Native of Europe; introduced into North America, where ranging from Quebec and Ontario to Minnesota, south to Florida, Mississippi, and Missouri, and on the Pacific Coast.

This annual was found growing in a wild environment, associated with *Geocarpon minimum*, *Saxifraga texana*, and *Isoetes Butleri*. The spreading stems with subulate leaves superficially resemble *Polypremum procumbens* of the Loganiaceae.

## 2. **Geocarpon** Mackenzie

**Geocarpon minimum** Mackenzie Map 909

Flowers early April–May.

Occurs on upland sandstone glades in moist depressions where a small trickle of drainage seeps through or near the plant. Known only from the unglaciated prairie section of southwestern Missouri, in Jasper, Greene, Polk, and St. Clair counties.

Known only from southwestern Missouri and southeastern Arkansas (Drew and Bradley counties).

Until 1957 this plant was thought to be the rarest one in Missouri, and one of the rarest in the United States, having been known only from the originally collected locality north of Alba in Jasper County, southwestern Missouri. The genus was described in 1914 by Mackenzie (Torreya 14: 67–68) from specimens collected by E. J. Palmer, and was known to occur nowhere else in the world. In 1957 the present author found a second station in St. Clair County, 57 miles by air northeast of the original station. In 1958, accompanied by Drs. Voigt and Mohlenbrock of Southern Illinois University, two new stations were found in Polk and Greene counties, and during the same spring of 1958 Dr. Dwight M. Moore of Arkansas Agricultural and Mechanical College collected the plant in Drew and Bradley counties of southeastern Arkansas for the first occurrence of the genus and species outside of Missouri. There is every reason to suppose that many more stations will be found on sandstone exposures in the unglaciated section of southwestern Missouri and perhaps from other states adjacent to Missouri.

The plants are often dull wine-purplish throughout or at maturity may turn an inconspicuous dull gray-green color. The perianth is pale green with the lower part frequently tinged with wine purplish color. The plant is quite diminutive and possesses a succulent aspect. For further details the reader is referred to Bull. Torr. Club 77: 268–73. 1950; Bull. Torr. Club 85: 124–27. 1958.

---

Plate no. 155. **1.** Paronychia fastigiata var. paleacea, × 3/4; *a.* Flower with bract and stipule, × 3. **2.** Scleranthus annuus. × 3/4; *a.* Flower, × 3 3/4; After Britton and Brown, details from Small, The New York Botanical Garden. **3.** Geocarpon minimum; *a.* Plant, × 3/4; *b.* Portion of flowering branch, × 3 3/4.

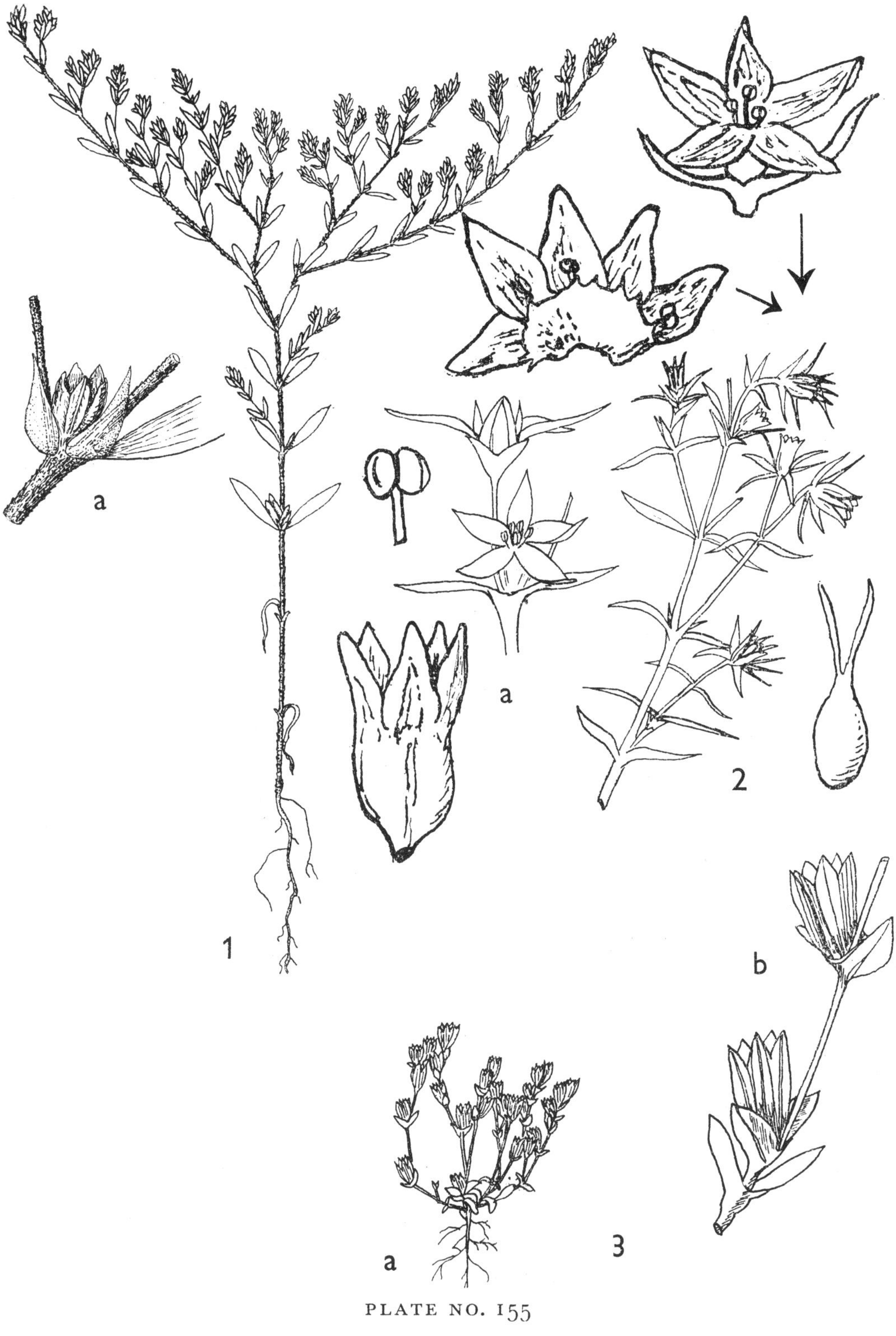


PLATE NO. 155

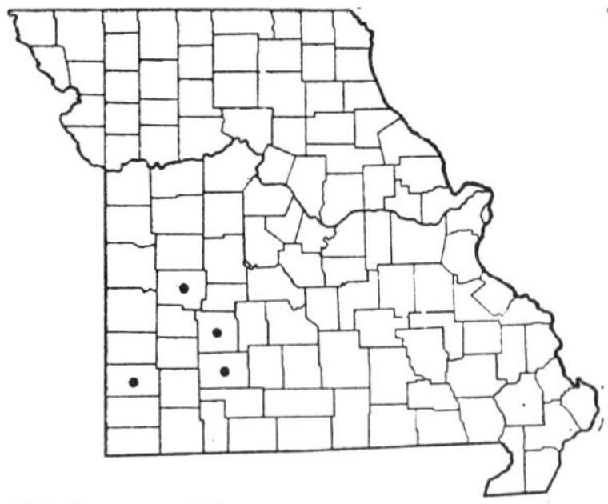

909 Geocarpon minimum

910 Paronychia canadensis (Forked Chickweed)

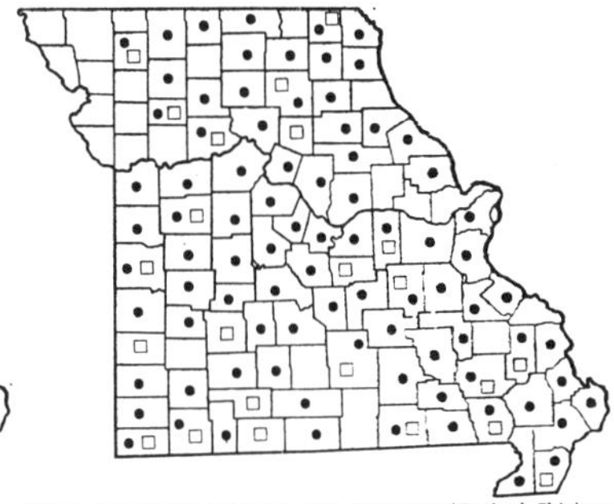

911 ● Paronychia fastigiata var. fastigiata (Forked Chickweed)
911 □ Paronychia fastigiata var. paleacea

## 3. **Paronychia** Mill. Whitlow, Forked Chickweed

This genus is sometimes placed in a separate family, *Illecebraceae* [P & S, Steyerm.], but in the present treatment this family is combined with the *Caryophyllaceae*.

Stem glabrous (without hairs); calyx 1–1.5 mm. long with flat non-ribbed sepals; sepals shorter than the mature fruit, blunt at tip . . . . . . . . . . . . . . . . . . . . . 1. P. CANADENSIS
Stem or branches minutely hairy; calyx 2–3 mm. long with ribs or strong nerves on the sepals; sepals usually as long as or longer than the mature fruit, with a minute pointed tip. . . . . . 2. P. FASTIGIATA

1. **Paronychia canadensis** (L.) Wood
Forked Chickweed Map 910
*Anychia canadensis* (L.) BSP. [P & S]
Flowers late May–October.

Occurs in dry open woods and shaded ground. Common throughout Missouri; apparently absent only from the lowland section of southeastern Missouri.

Ranges from New Hampshire to Ontario and Minnesota, south to Georgia, Alabama, Tennessee, Arkansas, and Kansas.

2. **Paronychia fastigiata** (Raf.) Fern.
Forked Chickweed Map 911
Flowers late May–October.

Occurs in dry open woods, rocky open woods, and sandstone, chert, or granite glades, mostly in acid soils.

Two variations are encountered in Missouri material:

The small, thin, transparent stipular bracts accompanying the flowers are mainly shorter than or at most equal to the calyx . . 2a. P. FASTIGIATA var. FASTIGIATA
The small, thin, transparent stipular bracts accompanying the flowers are either equal to or longer than the calyx . . . . 2b. P. FASTIGIATA var. PALEACEA

2a. **Paronychia fastigiata** var. **fastigiata**
Map 911
*Paronychia fastigiata* (Raf.) Fern. [G]
*Anychia polygonoides* Raf. in part [P & S]

Nearly throughout Missouri; absent only from the counties of extreme northwestern Missouri. This is the commoner of the two variations in the state.

Ranges from Massachusetts to Minnesota, south to Florida and Texas.

2b. **Paronychia fastigiata** var. **paleacea** Fern.
Map 911
*Anychia polygonoides* in part [of P & S], not Raf.

Scattered throughout the state, but less common than var. *fastigiata*.

These variations intergrade and are often placed in one or the other rather arbitrarily. Usually plants of the var. *paleacea* are greenish and are found in less exposed situations, whereas in var. *fastigiata* the plants are copper-reddish or brick-colored, have more crowded flowers, and occur in more exposed or drier and sunnier situations, as pointed out by Deam (*Fl. Ind.* p. 443).

---

Plate no. 156. **1.** Paronychia canadensis, × 2/5; Details from Small, The New York Botanical Garden. **2.** Sagina decumbens, × 2/5; Detail from Small, The New York Botanical Garden. **3.** Spergula arvensis, × 2/5; Details from Small, The New York Botanical Graden. **4.** Arenaria serpyllifolia, × 2/5. **5.** Arenaria lateriflora, × 2/5. **6.** Arenaria stricta, × 2/5

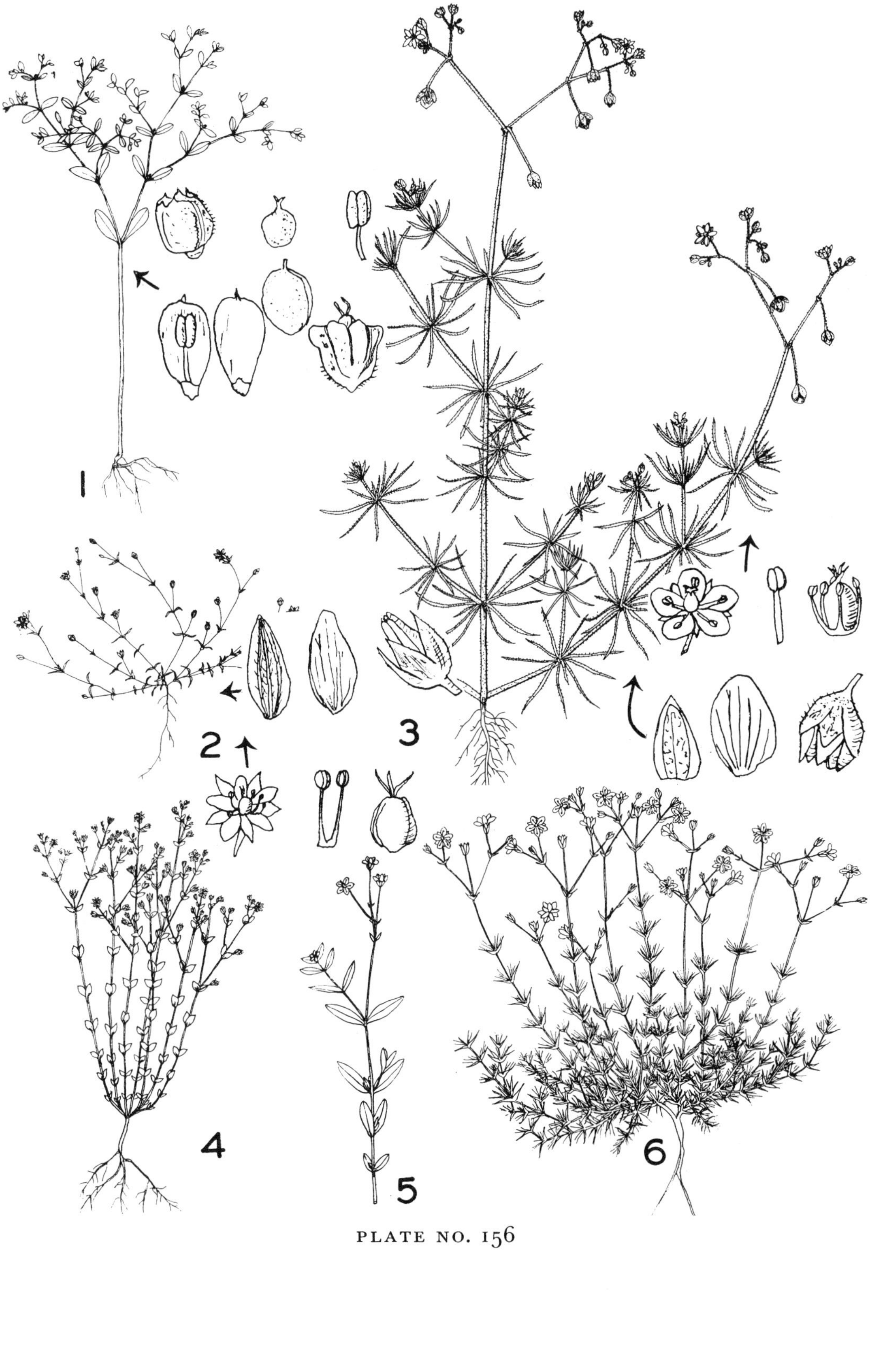


PLATE NO. 156

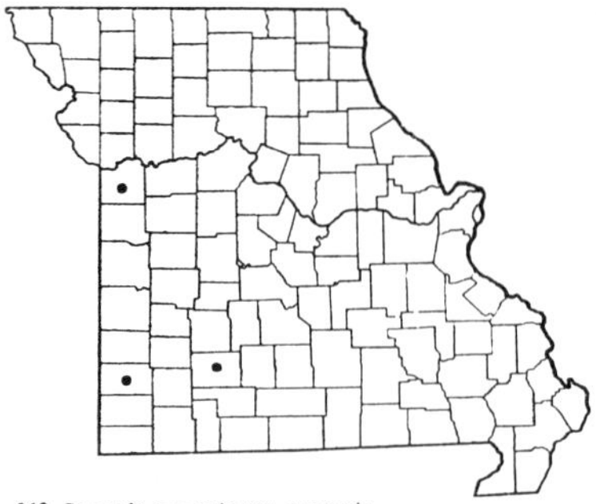

912 Spergula arvensis var. arvensis

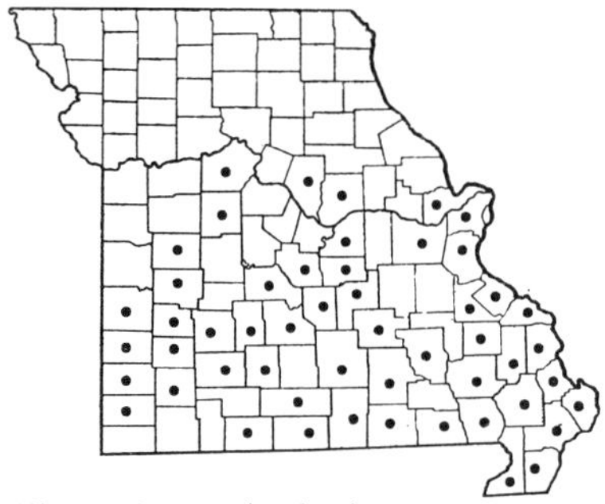

913 Sagina decumbens (Pearlwort)

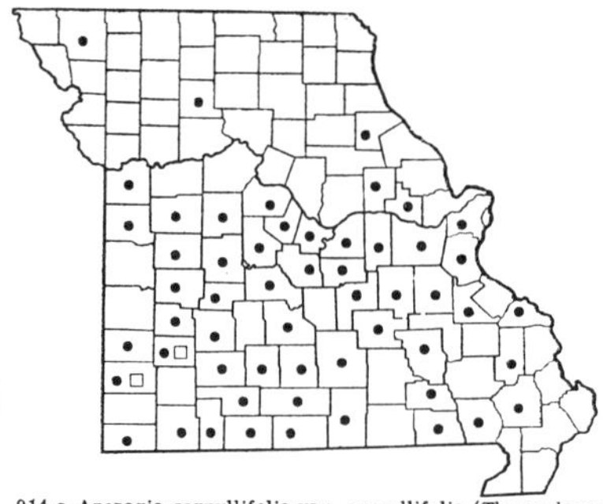

914 • Arenaria serpyllifolia var. serpyllifolia (Thyme-leaved Sandwort)
914 □ Arenaria serpyllifolia var. tenuior

## 4. **Spergula** L. Spurrey

**Spergula arvensis** L. var. **arvensis**
Corn Spurrey Map 912
*Spergula arvensis* L. [G, P & S, Steyerm.]
Flowers May–October.

Occurs in waste and cultivated ground and along railroads. Known only from Greene, Jasper, and Jackson counties, but doubtless much commoner in the state.

Native of Europe; introduced into North America, where ranging from Newfoundland to Alaska, south to Florida, Missouri, and California.

## 5. **Sagina** L. Pearlwort

**Sagina decumbens** (Ell.) T. & G. Pearlwort Map 913
Flowers March–August.

Occurs in fields, pastures, sandy or rocky open ground, sandy open wooded ridges, and waste places. Southern and central Missouri north to St. Louis, St. Charles, Franklin, Callaway, Boone, Saline, Henry, and Vernon counties.

Ranges from Florida to Texas, north to Massachusetts, Vermont, New York, Indiana, Illinois, Missouri, and Kansas.

The typical variation has petals which equal or are shorter than the sepals, or are entirely lacking in var. *Smithii* (Gray) Wats. So far as known the variety without petals has not been recorded from Missouri.

## 6. **Arenaria** L. Sandwort

a. Leaves linear, thread-like or needle-like, 0.5–1.5 mm. wide; mature fruit opening by 3 valves . . . . *b*
  b. Leaves occurring in 2's opposite each other on the stem, without additional leaves clustered in their axils; soft annuals . . . . . . . . . . . . . . . . . . . . . . . 4. A. PATULA
  b. Leaves with more than 2 clustered together and with tufts of smaller leaves in the main axils, stiff, needle-like; perennials . . . . . . . . . . . . . . . . . . . . . 3. A. STRICTA
a. Leaves broadened, expanded to ovate, oval, oblong, or lanceolate blades, mostly 3–15 mm. wide; mature fruit opening by 6 teeth or valves . . . . . . . . . . . . . . . . . . . . *c*
  c. Leaf-blades usually blunt (obtuse) at tip, the larger ones 1–3.5 cm. long; known only from woodlands of northeastern Missouri . . . . . . . . . . . . . . . . 2. A. LATERIFLORA
  c. Leaf-blades pointed (acute) at tip, 0.2–0.8 cm. long; common throughout the southern half of the state in fields, glades, open places, sandy or rocky soil, and scattered north of the Missouri River . . . . . . . . . . . . . . . . . . . . . . . . 1. A. SERPYLLIFOLIA

1. **Arenaria serpyllifolia** L.
Thyme-leaved Sandwort Map 914
Flowers April–August.

Occurs in sandy or rocky soil, glades, fields, ledges of bluffs, along railroads and roadsides, and open ground. Common throughout southern and central Missouri, but scattered northward in Ralls, Livingston, and Nodaway counties. Probably to be found in

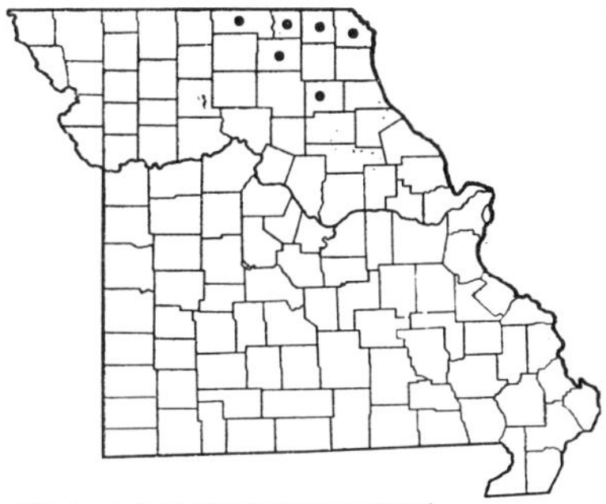

915 Arenaria lateriflora (Grove Sandwort)

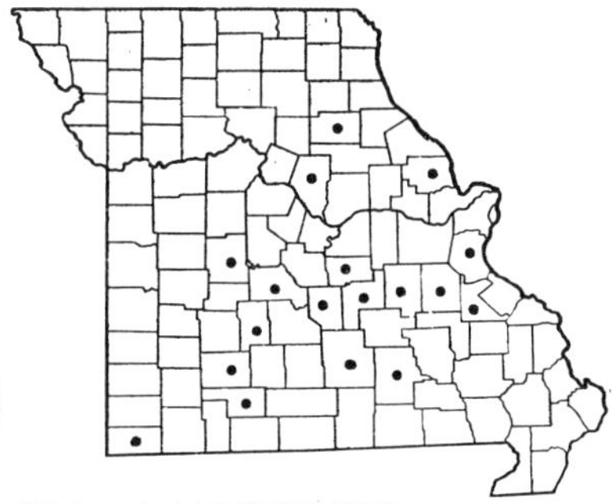

916 Arenaria stricta (Rock Sandwort)

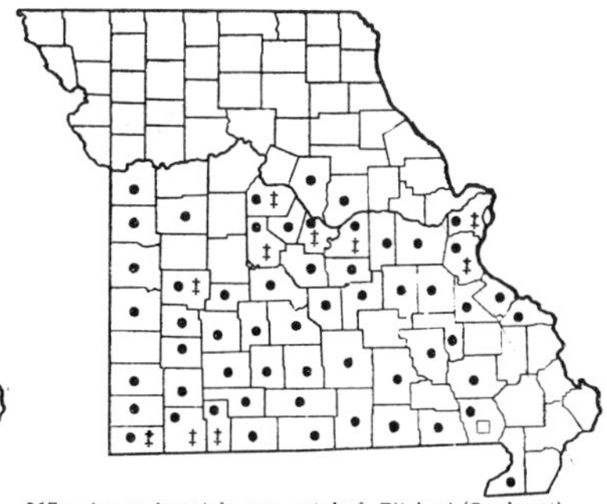

917 ● Arenaria patula var. patula f. Pitcheri (Sandwort)
917 ‡ Arenaria patula var. patula f. media
917 □ Arenaria patula var. robusta

most of the counties north of the Missouri River.

Two variations are known in Missouri:

Fruiting calyx 3–4 mm. long, 2–3 mm. broad at base; mature fruit ovoid; leaves ovate . . . 1a. A. SERPYLLIFOLIA var. SERPYLLIFOLIA
Fruiting calyx 2–3 mm. long, 1.2–2 mm. broad at base; mature fruit narrowly ovoid to nearly cylindric; leaves lance-ovate to lanceolate . . 1b. A. SERPYLLIFOLIA var. TENUIOR

1a. **Arenaria serpyllifolia** var. **serpyllifolia** Map 914

*Arenaria serpyllifolia* L. [G, P & S, Steyerm.]

This is the common variation encountered.

Native of Europe; introduced into North America, where ranging from Quebec to British Columbia, south to Florida, Texas, and California.

1b. **Arenaria serpyllifolia** var. **tenuior** Mert. & Koch Map 914

Known only from southwestern Missouri in Jasper (1 mi. north of Joplin, June 8, 1951, *Palmer 49133*) and Dade (along highway near South Greenfield, June 10, 1951, *Palmer 52137*) counties.

Native of Europe; introduced and naturalized in North America from Quebec to Michigan, south to Georgia and Missouri.

2. **Arenaria lateriflora** L. Grove Sandwort Map 915

Flowers May–August.

Occurs mostly in acid soils of upland woodland of dissected oak-hickory-clad hills, crests of ravines, rarely in low woods. Known only from northeastern Missouri, in Clark, Scotland, Schuyler, Putnam, Adair and Shelby counties.

Ranges from Labrador to Alaska, south to New Jersey, Pennsylvania, Ohio, Illinois, Missouri, South Dakota, and New Mexico and California; also in Eurasia.

The black seeds of this species are provided with a conspicuous appendage, distinguishing it from most other species of the genus.

3. **Arenaria stricta** Michx. Rock Sandwort Map 916

*Arenaria stricta* subsp. *stricta* [BB]

*Arenaria stricta* var. *texana* Robins. [G, P & S, Steyerm.]

*Arenaria stricta* subsp. *texana* (Robins.) Maguire [BB]

Flowers May–July.

Occurs on bluff escarpments, ledges, and rocky glades of limestone strata. Southern and east-central Missouri, north and east to Benton, Boone, Monroe, and Lincoln counties, in unglaciated or apparently unglaciated territory.

Ranges from New Hampshire and Quebec to Ontario and Nebraska, south to South Carolina, Kentucky, Arkansas, and Texas.

The species has been divided into subspecies or varieties, the Missouri material passing mainly as var. *texana*, distinguished from typical var. *stricta* by having leaves 5–20 mm. instead of 15–30 mm. long, by the flowering stems mainly leafy only in the lower third rather than leafy for more than half their length, and by the somewhat shorter flowering stems 5–20 cm. instead of 10–40 cm. tall.

Missouri material could be classified in either variety, with most of it passing as var. *texana*. The differences, however, do not stand out in Missouri specimens, and the variations in leaf length, stiffness, and concentration of leaves in the lower third of the stem seem to be related to the occurrence in exposed dry habitats in which the var. *texana* often occurs. In shadier bluff crevices and ledges the leaves are longer and the stem has a leafier appearance for more than half its length.

This species is a handsome gem for the limestone rock garden, but it is difficult to transplant, because the long tough root resents disturbance.

4. **Arenaria patula** Michx. Sandwort Map 917
Flowers April–June.

Occurs on sandstone, chert, and limestone glades and rocky sections in prairies. Throughout the Ozark region of southern and central Missouri north to St. Louis, Franklin, Gasconade, Callaway, Boone, Cooper, Johnson, and Jackson counties.

Missouri material is represented by the following variations:

a. Leaves mostly 2–3.2 mm. broad; sepals 3-nerved; seeds 0.7–0.9 mm. wide . 4c. A. PATULA var. ROBUSTA
a. Leaves mostly 0.5–1.5 mm. broad; sepals 5-nerved; seeds 0.5–0.7 mm. wide . . . . *b*
  b. Pedicels (stalk of flower or fruit) and sepals glabrous. 4a. A. PATULA var. PATULA f. PITCHERI
  b. Pedicels and sepals with some glandular hairs . . 4b. A. PATULA var. PATULA f. MEDIA

4a. **Arenaria patula** var. **patula** f. **Pitcheri** (Nutt.) Steyerm. Map 917
*Arenaria patula* f. *Pitcheri* (Nutt.) Steyerm. [G]
*Arenaria patula* var. *patula* in part [BB]

This is the more commonly encountered variation in Missouri, with the stems, leaves, pedicels, and sepals glabrous or nearly so.

Ranges from Ohio and Indiana to Minnesota, south to Virginia, Alabama, Arkansas, Kansas, Oklahoma, and Texas.

4b. **Arenaria patula** var. **patula** f. **media** Steyerm. Map 917
*Arenaria patula* f. *media* Steyerm. [G]
Scattered throughout the range in Missouri.

4c. **Arenaria patula** var. **robusta** (Steyerm.) Maguire Map 917
*Arenaria patula* f. *robusta* Steyerm. [G]

Known only from open sandy ground in open woods in Butler County, southeastern Missouri (Gillis Bluff, sect. 8 and 17, 5 mi. southwest of Quilin, May 27, 1939, *Steyermark 26652*).

Ranges from Kentucky, Tennessee, and Missouri, south to Arkansas, Kansas, and Texas.

Some intergradation occurs between var. *robusta* and var. *patula* f. *media*, and frequent intergradation occurs between the forms of *A. patula* var. *patula*. Apparently, typical *A. patula* var. *patula* f. *patula*, with the stems and leaves with glandular hairs, does not occur in Missouri, or at least no specimens have thus far been recorded.

Except for var. *robusta*, which is isolated in sandy open woods on Crowley Ridge in Butler County, all the other variations of *A. patula* are on rocky glades of usually limestone strata in the Ozark section, although plants occur also on sandstone and chert glades and barrens. While in flower, the numerous dainty white blossoms carpet the glades so thickly that they suggest natural rock gardens. The species has a very brief flowering season, beginning in April and ending in June at which time the seed is ripened and the plant dies. Seeds germinate the following autumn and the plant lives over the winter as a winter annual, recognized by threadlike little leaves on the ground.

## 7. **Stellaria** L. Chickweed, Starwort

a. Leaves ovate to elliptic or elliptic-oblong, noticeably broadest around the middle or near the base, $1\frac{1}{2}$–$4\frac{1}{2}$ times as long as broad, mainly 8–45 mm. wide; stems with short hairs, often appearing in 2 vertical lines . . . . . . . . . . . . . . . . . . . . . . . . . . . . *b*
  b. Lower and middle leaves of plant with a stalk (petiole); leaf-blades ovate, 5–40 mm. long; sepals longer than the petals; common species . . . . . . . . . . . . . . . . 1. S. MEDIA
  b. Most of leaves sessile (without a stalk); leaf-blades elliptical to elliptic-oblong, 20–100 mm. long; sepals shorter than the petals; rare species, known only from Franklin County . . . . . 2. S. PUBERA
a. Leaves narrow, linear to oblanceolate, mainly of approximately the same width throughout, 6–10 or more times as long as broad, 1.5–7 mm. wide; stems glabrous (without hairs) . . . . . . . . *c*
  c. Flowering branch arising from the very top of the leafy stem; sepals 4–5.5 mm. long; leaves often broadest at the base; rare introduced plant known from open grassy lawns in Boone County. 3. S. GRAMINEA
  c. Flowering branch arising at the side of the leafy stem below the very tip; sepals 2.5–4 mm. long; leaves often broadest at or above the middle; native species of low wet woods, wooded swamps, and open wet meadows in eastern Missouri . . . . . . . . . . . . 4. S. LONGIFOLIA

Plate no. 157. **1.** Arenaria patula, × 2/5; Details from Small, The New York Botanical Garden. **2.** Stellaria media, × 2/5; Details from Small, The New York Botanical Garden. **3.** Stellaria pubera var. pubera, × 2/5. **4.** Stellaria graminea, × 2/5. **5.** Stellaria longifolia, × 2/5.

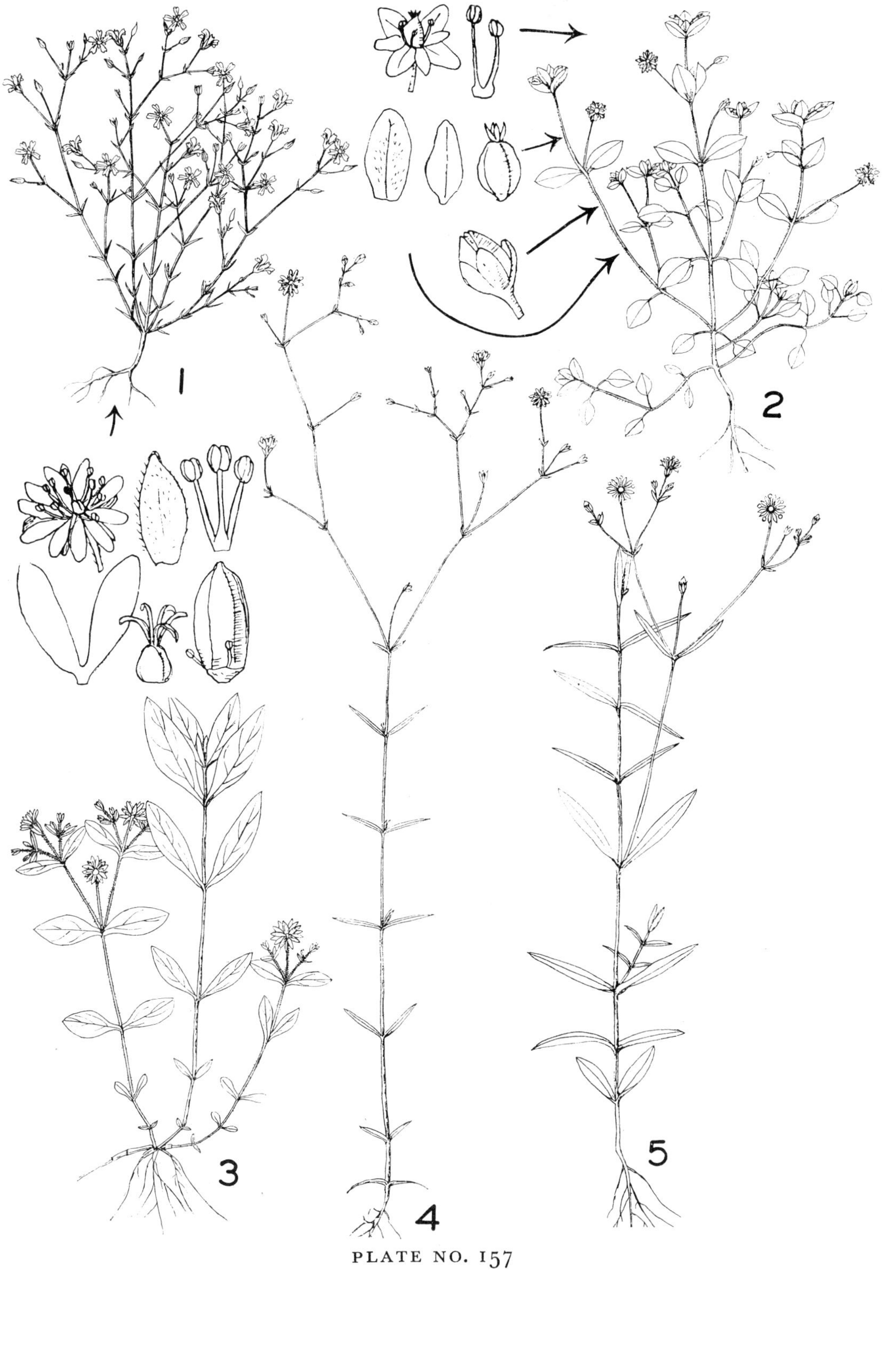


PLATE NO. 157

918 Stellaria media var. media (Common Chickweed)

919 Stellaria pubera var. pubera (Great Chickweed)

920 Stellaria graminea (Common Stitchwort)

1. **Stellaria media** (L.) Cyrillo var. **media**
Common Chickweed Map 918
Flowering January through December.

Occurs as a weed in lawns, gardens, and about dwellings, waste ground, roadsides, along railroads, sometimes in low woodland and valleys. Common in southern and central Missouri, and doubtless occurring throughout northern Missouri, but recorded there only from Marion, Caldwell, and Andrew counties.

Native of Europe; introduced and naturalized throughout North America; also naturalized in Guatemala in Central America and South America.

A var. *glaberrima* G. Beck, differing from var. *media* in being completely glabrous throughout, has not yet been found in Missouri.

In Europe the plant is sometimes cooked as a vegetable as a spinach substitute, but only the younger growing tips are recommended for use. The small seeds are sometimes found in seed used for cage birds. Cases of poisoning are reported where young lambs have eaten quantities of the plant. However, hogs and rabbits eat the plants, and domestic fowl and wild birds are fond of the seed.

2. **Stellaria pubera** Michx. var. **pubera**
Great Chickweed Map 919
*Stellaria pubera* var. *silvatica* [of P & S, Steyerm.], not (Béguinot) Weath.
Flowers late March–May.

Known only from Franklin County, east-central Missouri (Gray Summit, April 20, 1927, *Kellogg* in herb. Mo. Bot. Gard.).

Ranges from New Jersey to Indiana, Illinois, and Missouri, south to Florida.

The leaves of the Missouri specimen are sessile and the plant should be referred, therefore, to var. *pubera* instead of to var. *silvatica*, as treated in Palmer and Steyermark's *Annotated Catalogue*.

The Missouri record is probably based upon a planted specimen originally grown at the Missouri Botanical Garden's Arboretum at Gray Summit in Franklin County. Native Missouri plants as well as some not native were introduced there, and Kellogg's specimen gives no detailed data concerning its origin. Unfortunately, no other records are available at the Arboretum that would indicate the introduction of this species there. The plant thrives in shaded wildflower gardens and spreads over the years, often forming dense carpets over the soil. It prefers rich loose humus. The white starlike flowers, produced in early spring, are attractive against the dark green leafy background.

3. **Stellaria graminea** L. Common Stitchwort
Map 920
Flowers May–October.

Known only from grassy open ground in Boone County, central Missouri, where introduced (University campus, Columbia, May 31, 1929, *Rickett*, in herb. Univ. Mo.).

Native of Europe; introduced and naturalized in North America from Newfoundland and Quebec to Ontario and Minnesota, south to North Carolina, West Virginia, Ohio, Indiana, Illinois, and Missouri; also Washington.

Broad-leaved plants are sometimes treated as var. *latifolia* Peterm., but the Missouri specimen appears referable to typical *S. graminea*.

4. **Stellaria longifolia** Muhl.
Long-leaved Stitchwort Map 921
Flowers May–July.

Occurs in low wet woods, wooded swamps, and wet meadows. Known only from eastern Missouri, where found from Butler and Scott counties north to St. Louis County and locally in northeastern Missouri in Clark County.

Ranges from Labrador and Newfoundland to Alaska, south to South Carolina, Kentucky, Missouri, Louisiana, Nebraska, New Mexico, and Arizona. Also in Europe and Asia.

## 8. **Cerastium** L. Mouse-ear Chickweed

The European and South American species have been recently studied by Dr. W. Möschl of Austria.

a. Petals normally absent . . . . . 5b. C. VISCOSUM f. APETALUM
a. Petals (except in mature fruiting plants) normally present . . . . . *b*
b. Petals showy, 2–3 times as long as sepals, 10–15 mm. long; plants perennial with overwintering leafy offshoots or branches attached at base . . . . . 2. C. ARVENSE var. VILLOSUM
b. Petals shorter than or equaling or somewhat longer than the sepals, only 4–6 mm. long; mostly annuals with a delicate root-system and no overwintering leafy offshoots at the base, or, if perennials as in *C. vulgatum*, the petals only 4–6 mm. long . . . . . *c*
c. Perennial plants with overwintering leafy offshoots or branches attached at base; leaves thick; stems often purplish at least at base; upper bracts of inflorescence with broad scarious margins (thin, whitish, or transparent); stamens 10 . . . . . 1. C. VULGATUM
c. Annual plants with a delicate root system and without the above combination of characters; if stems purplish, then the bracts are green throughout and leaf-like in texture, and not scarious-margined, or the stamens are 4–5, or the plants are annuals; stamens 4, 5, or 10 . . . . . *d*
d. Stamens 4 or 5 . . . . . *e*
e. Bracts of the inflorescence green throughout and similar to the leaves in texture; stamens mainly 4, sometimes 5 . . . . . 6. C. TETRANDUM
e. Bracts of the inflorescence with thin, whitish, transparent, or colorless summit or margin, not green throughout; stamens mainly 5. . . . . . *f*
f. Upper half of bracts scarious (thinner, transparent, colorless, or whitish); sepals with broad scarious margins and tips, the scarious or white part occupying about $\frac{1}{2}$ of the sepal width; petals shallowly notched, none of the veins branched . 8. C. SEMIDECANDRUM
f. Only the narrow margin and summit of the bracts and sepals scarious, the scarious or white part of the sepal occupying only about $\frac{1}{5}$ its width; petals deeply 2-cleft about $\frac{1}{4}$ way from the tip, at least the lateral veins with a simple fork above 9. C. PUMILUM
d. Stamens 10 . . . . . *g*
g. Bracts of inflorescence ending in long tufts of hairs; entire plant covered with long hairs, the hairs in life giving a grayish-white hairy appearance to the plant; filaments of the stamens and claws of the petals long-ciliate (hairy on margins) . . . 7. C. BRACHYPETALUM
g. Without the above combination of characters; bracts not as above nor hairs so long as to give a gray-hairiness in life; filaments of stamens and claws of petals glabrous. . . *h*
h. Stalks (pedicels) of fruits 2–5 mm. long, shorter than or as long as the calyx; inflorescence compact, the flowers bunched close together, even at maturity the terminal clusters close together; leaves firm and thickish; leaves at the middle of the stem ovate or oblong-obovate, usually blunt to rounded at tip, $1\frac{1}{2}$–3 times as long as broad; sepals all sharply pointed at tip; sepals mainly covered with long hairs both gland-tipped and non-glandular . . . . . 5. C. VISCOSUM
h. Stalks (pedicels) of fruits 4–55 mm. long, equaling to 12 times as long as the calyx; inflorescence open and loosely flowered; leaves thin and membranous; leaves at the middle of the stem lanceolate, oblong-lanceolate, or of an elongated type, acute or acutish (rather short-pointed) at tip, mainly 4–9 times as long as broad; at least the outer sepals rather blunt at tip; sepals mainly covered with short gland-tipped hairs . . . . . *i*
i. At least the lower stalks of the fruit much longer than the capsules, 10–55 mm. long; pedicels (stalks) hooked or curved at tip; leaves mainly acute or short-pointed at tip, the well-developed ones 4–5 times as long as the calyx . . 3. C. NUTANS
i. Stalks of the fruit shorter than or but slightly longer than the capsules, mainly 4–10 mm. long; pedicels (stalks) not hooked nor curved at tip; leaves mainly obtuse or blunt at tip, the well-developed ones not more than twice as long as the calyx . . . . . 4. C. BRACHYPODUM

1. **Cerastium vulgatum** L. Common Mouse-ear Chickweed Map 922

Flowers April–November.

Occurs in fallow and cultivated fields, pastures, moist meadows, farmlots, on lawns about dwellings, along roadsides and railroads, and open and grazed woodland. Throughout Missouri, and doubtless in every county, but not recorded from a number of northern Missouri counties and elsewhere.

Native of Eurasia; introduced and naturalized

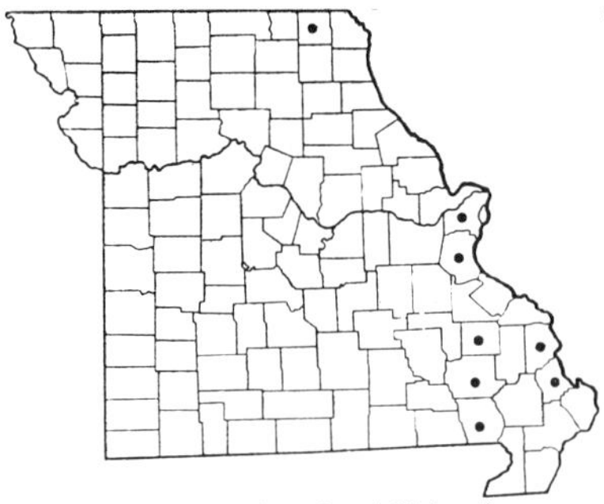

921 Stellaria longifolia (Long-leaved Stitchwort)

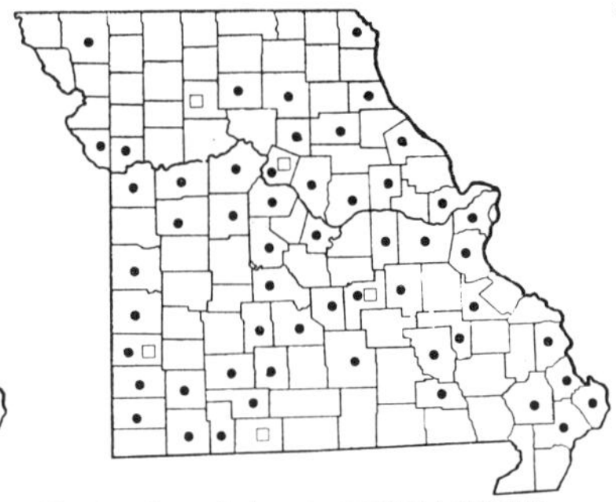

922 • Cerastium vulgatum var. vulgatum f. vulgatum (Common Mouse-ear Chickweed)
922 □ Cerastium vulgatum var. vulgatum f. glandulosum

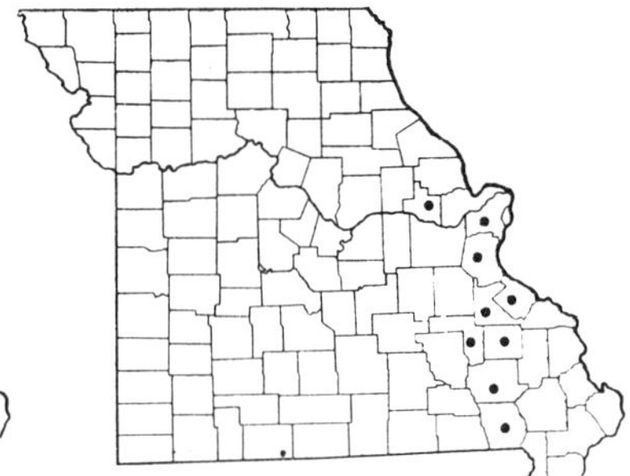

923 Cerastium arvense var. villosum f. oblongifolium

throughout temperate North America.

Missouri material is represented by the following variations:

Stems covered with hairs which are not gland-tipped . 1a. C. VULGATUM var. VULGATUM f. VULGATUM
Stems covered with gland-tipped hairs . . .
1b. C. VULGATUM var. VULGATUM f. EGLANULOSUM

1a. **Cerastium vulgatum** var. **vulgatum** f. **vulgatum** Map 922

*Cerastium vulgatum* L. [G]

*Cerastium vulgatum* var. *hirsutum* Fries. [P & S, Steyerm., BB]

This is the more commonly encountered variation in Missouri.

1b. **Cerastium vulgatum** var. **vulgatum** f. **glandulosum.** (Boenn.) Druce Map 922

*Cerastium vulgatum* f. *glandulosum* (Boenn.) Druce [G, BB]

Scattered throughout the range of the var. *vulgatum*, but less frequent.

2. **Cerastium arvense** L. var. **villosum** (Muhl.) Hollick & Britt. f. **oblongifolium** (Torr.) Pennell Map 923

*Cerastium arvense* var. *oblongifolium* (Torr.) Hollick & Britt. [P & S, Steyerm., BB]

Flowers April–June.

Occurs along rocky stream banks, on moist open or wooded ledges and outcrops of sandstone, granite, and limestone rock, and on limestone talus and rocky slopes of bluffs. Eastern Missouri from Butler County north to St. Louis and Warren counties.

Ranges from Ontario to Idaho, south to Virginia, Tennessee, and Missouri.

This form of *C. arvense* is recognized by the glabrous or nearly glabrous upper surface of the green, mostly lanceolate leaves, which are 3–10 mm. broad, and up to 6.5 cm. long. This is a very ornamental plant with large, attractive white flowers. It does well in wild-flower rock gardens, on either siliceous or limestone rocks.

3. **Cerastium nutans** Raf. Nodding Chickweed Map 924

*Cerastium nutans* Raf. var. *nutans* [BB]

Flowers March–June.

Occurs in low ground of valleys in fallow and cultivated fields, moist meadows and pastures, open or rocky woodland, and ledges along streams. Throughout Missouri, doubtless in every county.

Ranges from Quebec to British Columbia, south to Florida, Alabama, Arkansas, Texas, New Mexico, and Arizona.

Petals are occasionally wanting in cleistogamous flowers of this species.

4. **Cerastium brachypodum** (Engelm.) Robins. Map 925

*Cerastium nutans* var. *brachypodum* Engelm. [BB]

Flowers March–June.

Occurs in open woods, rocky slopes, moist meadows, cultivated and fallow fields, waste ground, and along roadsides and waste ground. Occurs commonly throughout the southern half of the state, north to St. Charles, Warren, Boone, Carroll, and Jackson counties.

Ranges from Guatemala to Mexico, and Georgia to Arizona, north to Virginia, Tennessee, Illinois, North Dakota, Alberta, and Washington.

In general this species seems sufficiently distinct from the preceding one, *C. nutans*. In the main, the stems are shorter and the leaves are smaller, usually not exceeding 3 cm. long. Some specimens collected from the northern part of the state, previously determined as *C. brachypodum*, are now referred to *C. nutans*. Throughout the southern half of the state the plants

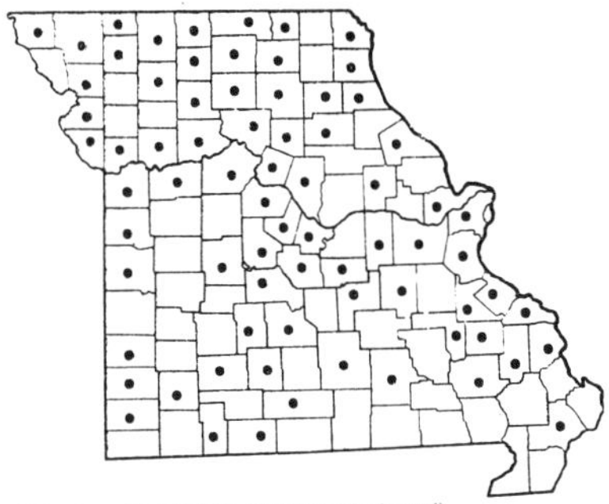

924 Cerastium nutans (Nodding Chickweed)

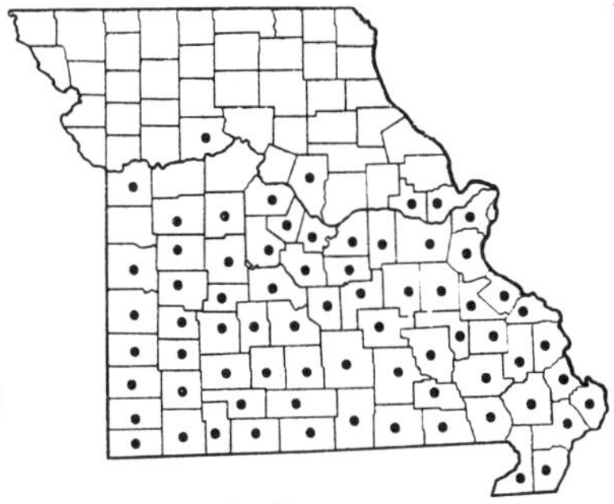

925 Cerastium brachypodum

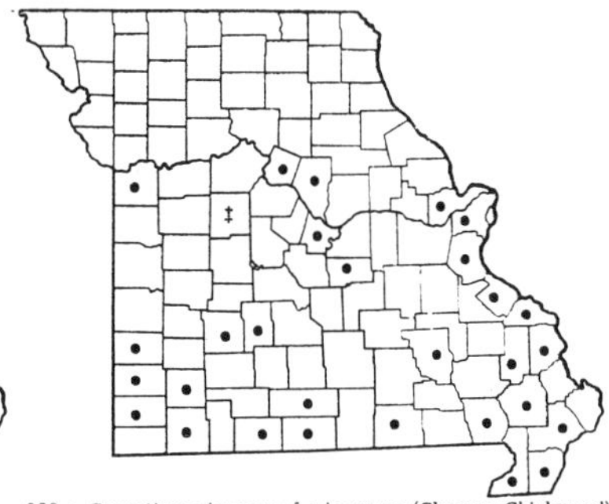

926 • Cerastium viscosum f. viscosum (Clammy Chickweed)
926 ‡ Cerastium viscosum f. apetalum

appear more uniformly constant and clearly recognizable.

5. **Cerastium viscosum** L. Clammy Chickweed. Map 926

*Cerastium glomeratum* Thuillier [Shinners]

Flowers March–July.

Occurs in fallow and cultivated fields, pastures, along roadsides and waste ground, wet rocky and open woodland, and open rocky places.

Native of Europe; introduced and naturalized in North America from Florida to Texas and California, north to Newfoundland, Ontario, and British Columbia.

Missouri material is represented by the following variations:

Petals present, somewhat shorter than the sepals or about equaling them . . . 5a. C. VISCOSUM f. VISCOSUM

Petals either completely absent or so minute as to be very inconspicuous . . . 5b. C. VISCOSUM f. APETALUM

5a. **Cerastium viscosum** f. **viscosum** Map 926

This is the more commonly encountered form in the state. Known only from southern and central Missouri north to St. Charles, Boone, Howard, and Jackson counties; to be expected in the northern half of the state, but not recorded as yet from that section.

5b. **Cerastium viscosum** f. **apetalum** (Dumort.) Mert. & Koch Map 926

Known only from Pettis County, west-central Missouri (open grassy ground on top of ridge on cherty limestone upper slopes above northwest-facing wooded bluffs along Flat Creek, T45N, R21W, sect. 24, $2\frac{1}{2}$ mi. south of Sedalia, May 20, 1949, *Steyermark 67973*). Undoubtedly to be found in most parts of the range of typical f. *viscosum*.

Most of the small annual species of *Cerastium*, such as *C. viscosum*, and others treated in the present work, may be used as cooked vegetables if gathered very early in spring while still in the leafy and nonflowering stage of growth.

*Cerastium viscosum* usually grows as a small plant with erect or ascending stems with closely crowded flowers at the tip. The stems and sepals are usually covered with gland-tipped hairs, while the thickish leaves are mainly hirsute and nonglandular. In some specimens the stems have mostly spreading, nonglandular hairs.

6. **Cerastium tetrandrum** Curtis Map 927

Flowers March–May.

Known only from Crawford County, eastern Missouri (along highway 8, south of Scotia, between highway to Scotia and Huzzah Creek, April 28, 1957, *Steyermark 83932, 83934*).

Native of Europe; introduced into the United States, where now recorded from Virginia and Missouri.

Since this species is just becoming established along highways and is not fully described in the standard manuals, the following short description is given to aid further in the identification of the plant: Stems 3–20 cm. tall, slender, with short gland-tipped hairs. Middle and upper leaves ovate, obtuse to acute, 7–17 mm. long, 3–6 mm. broad, the lower ones elliptic to elliptic-obovate, narrowed at base and more pointed, densely pubescent with rather stiff mostly nonglandular hairs. Pedicels 4–13 mm. long, covered with short gland-tipped hairs equaling or longer than the calyx. Sepals are scarious-margined, 4–4.5 mm. long, and covered with mostly gland-tipped hairs. The claws of the petals are glabrous, and the 4 or 5 stamens are also glabrous. Capsule 7–8 mm. long, slightly curved, not quite twice the length of the calyx, at maturity reaching 2–5 times the length of the pedicels.

The plants resemble *C. viscosum* in general habit,

927 Cerastium tetrandrum

928 Cerastium brachypetalum

929 Cerastium pumilum

but the cymose inflorescence is more open and loose with a dichotomous or forked branching, the leaves average smaller, the stems more slender, and the pedicels longer, especially in fruit. This species eventually will probably be found in all the counties of southern and central Missouri.

7. **Cerastium brachypetalum** Pers. Map 928
Flowers April–May.

Occurs along highways and pastured hills, where now known in four counties in the southern third of the state. First collected in 1954 by Harry Ahles in Ste. Genevieve County (pastured hillside, 1 mi. north of Zell, April 30, 1954, *Ahles 7930*, in herb. Univ. of Ill.), and more recently collected in Miller (fallow field along highway 52, 3 mi. northwest of Tuscumbia, May 6, 1957, *Steyermark 83988*), Phelps (along highway 66, just southwest of junction of highway 63 and 66, northwest of Rolla, April 21, 1957, *Steyermark 83643*), and Wayne (along highway 34, 2 mi. west of Piedmont, May 4, 1958, *Steyermark, Voigt*, and *Mohlenbrock 86097*) counties.

Native of Europe; introduced into the United States, where at present known from Virginia and Missouri.

This is another one of the annual species of Mouse-ear Chickweed which has recently become introduced in the state along highways. It is predicted that the species will eventually be found at least throughout all the southern and central Missouri counties.

In the field the plant has a characteristic gray-hairy appearance. It usually occurs in dense masses so that at a distance the spreading gray-white hairs, which densely cover the stems, leaves, pedicels, bracts, and sepals, stand out conspicuously. Superficially, the plants resemble *C. vulgatum*, but differ in their slender roots with annual habit, without persistent leafy offshoots, ciliate filaments of stamens and claws of the petals, and the conspicuous long tufts of hairs on the nonscarious margined bracts of the inflorescence. The stems are usually 10–25 cm. tall, the leaves are elliptical-oblong to ovate-oblong, obtuse or slightly acute, 7–18 mm. long, and 4–7 mm. broad. The pedicels are 4–17 mm. long, and at maturity are 2–5 times the length of the calyx. The filaments of the 10 stamens and claws of the petals are ciliate.

All parts of the plant are usually covered with long nonglandular hairs. Most of the Missouri material seen appears to belong to this category and is known as **Cerastium brachypetalum** f. **brachypetalum**

*Cerastium brachypetalum* var. *strigosum* (Fries.) Aschers. & Graebn. Syn. 5: 680. 1918; *Cerastium brachypetalum* subsp. *strigosum* (Fries.) Lousing; *Cerastium brachypetalum* var. *eglandulosum* Fenzl in Ledeb. Fl. Ross. 1: 404. 1842.

The form with glandular hairs on the sepals and stems is known from Ste. Genevieve County (*Ahles 7930*) and is known as **Cerastium brachypetalum** f. **glandulosum** Koch, Syn. Fl. Germ. & Helv. 1: 121. 1835.

*Cerastium brachypetalum* subsp. *tauricum* (Sprengel) Murbeck

8. **Cerastium semidecandrum** L.

Excluded species. Easily confused with *C. pumilum*, but doubtless to be found eventually in the state.

A specimen from Bates County (waste places, Amoret, May 15, 1936, *Bush 15493* in Mo. Bot. Gard. Herb.) labeled *C. semidecandrum* is a misidentified *C. vulgatum*. It has a perennial habit, mostly 10 stamens, and other characteristics of *C. vulgatum*. See the discussion under the next species, *C. pumilum*.

Plate no. 158. **1.** Cerastium vulgatum, × 2/5; Details from Small, The New York Botanical Garden. **2.** Cerastium arvense var. villosum f. oblongifolium, × 2/5; Details from Small, The New York Botanical Garden. **3.** Cerastium nutans, × 2/5. **4.** Cerastium viscosum, × 2/5. **5.** Holosteum umbellatum, × 2/5; Details from Small, The New York Botanical Garden.

PLATE NO. 158

9. **Cerastium pumilum** Curtis Map 929

Flowers April–May.

Occurs along highways in southern Missouri in Cape Girardeau (along highway 34, just west of junction with road U near Bollinger Co. line, May 4, 1958, *Steyermark, Voigt,* and *Mohlenbrock 86106*) and Dallas (along highway 43, 1 mi. north of Buffalo, May 2, 1958, *Steyermark, Voigt,* and *Mohlenbrock 86064*) counties. Doubtless to be eventually expected in most of the counties of southern and central Missouri.

Native of Europe; introduced into the United States, where known at present from New Jersey and Missouri.

Like *C. tetrandrum* and *C. brachypetalum,* this species has just been recently introduced in the state, but will doubtless eventually be found as a weed in most of the counties of the southern and central part of the state. The differences by which it is separated from *C. semidecandrum* L., another European species, seem slight, and perhaps eventually *C. pumilum* may be found synonymous with *C. semidecandrum* in Missouri. The main differences between the two are based largely on the simple or forked veins of the petals and relative proportion of scarious to non-scarious surface of the bracts and sepals as stated in the keys. The two Missouri collections cited are intermediate between *C. semidecandrum* and *C. pumilum,* combining conspicuously margined bracts and sepals, characteristic of *C. semidecandrum,* with simply forked veins, more characteristic of *C. pumilum.* In the Missouri collections cited, the midvein is simple and unbranched, while the lateral veins are simply forked in the upper part. It is upon the basis of the forked lateral veins that the cited collections are placed with *C. pumilum.* There would also appear to be some difference of opinion and interpretation as to the relative conspicuousness of the scarious portion of the bracts and sepals in the Missouri material. Hegi's illustration of the two species (*Ill. Fl.* Band III, teil 1: 365–66. *fig. 601, e–p, q–t*) shows the broad scarious margin of the sepals in *C. semidecandrum* occupying one-half or more of the sepal, whereas in *C. pumilum* the narrow scarious margin occupies only about one-fifth of the sepal. In the Missouri material the outer sepals have a broad scarious margin, like that of *C. semidecandrum,* whereas the inner sepals have a narrow scarious margin of *C. pumilum.*

The sepals in these collections are 3.5–5 mm. long. The stems, pedicels, and sepals are covered with glandular hairs. The leaves vary from oblong or elliptic to broadly ovate.

## 9. **Holosteum** L. Jagged Chickweed

**Holosteum umbellatum** L. Jagged Chickweed Map 930

Flowers March–May.

Occurs along roadsides and railroads. Recently introduced in the state (known since 1950) and now spreading rapidly. Established in southern and central Missouri north to St. Louis and Boone counties. First recorded from Washington County in 1950 (one mile south of Caledonia, April 22, 1950, *Ahles*).

Native of Europe; introduced and naturalized in the United States from Massachusetts and Pennsylvania, south to Georgia, Kentucky, Ohio, and Missouri.

After a brief flowering the plants quickly become dried and tawny brown and disappear. Like other early spring annuals, they often occur in dense masses at the time of flowering. Eventually this species undoubtedly should spread over the greater portion of the state, as detailed highway collecting will probably reveal.

## 10. **Agrostemma** L. Corn Cockle

**Agrostemma Githago** L. Corn Cockle Map 931

Flowers late May–September.

Occurs in fallow and cultivated (usually grain) fields, waste ground, and along roadsides and railroads. Known in most parts of southern and central Missouri, but recorded northward from only Clark, Livingston, and Caldwell counties. Undoubtedly of expected occurrence in all the counties.

Native of Europe; introduced and naturalized throughout most of North America.

The seeds of this hairy annual are poisonous, and poultry and stock fed with wheat screenings containing Corn Cockle seeds are sometimes known to have died afterwards. When ground with wheat flour, it may cause poisoning, if a large enough quantity becomes included, as well as give the flour a poor taste. The seeds have been used in medicine for gastritis and paralytic cases, according to reports in herbals.

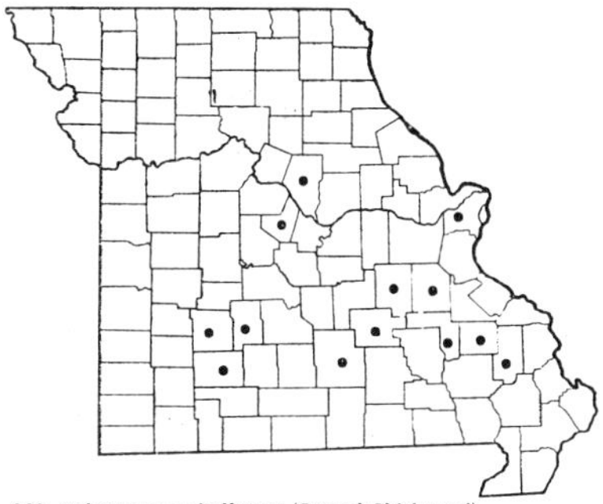
930 Holosteum umbellatum (Jagged Chickweed)

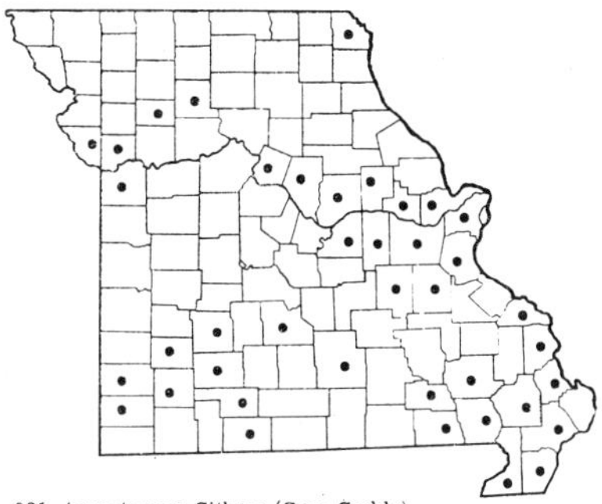
931 Agrostemma Githago (Corn Cockle)

932 Lychnis dioica (Red Campion)

## 11. **Lychnis** L. Campion

Petals usually rose-colored, rose-purple, or pink, rarely white, opening in the morning; teeth of the calyx short-pointed, triangular-lanceolate; mature capsule nearly as long as broad, the teeth recurved when the capsule opens . . . . . . . . . . . . . . . . . . . . . . . . . . 1. L. DIOICA
Petals usually white, rarely pinkish, opening in the evening and early morning; teeth of the calyx slenderly tapering, lanceolate-linear; mature capsule conical-ovoid, tapering and narrowed at tip, the teeth either erect or slightly spreading when the capsule opens . . . . . . . . . . . 2. L. ALBA

1. **Lychnis dioica** L. Red Campion Map 932
Flowers late May–September.

Known only from Jackson County, west-central Missouri.

Native of Eurasia; introduced and naturalized from Newfoundland to Ontario, south to Delaware, Pennsylvania, Ohio, Illinois, and Missouri.

2. **Lychnis alba** Mill. White Campion Map 933
Flowers late May–September.

Occurs in fallow and cultivated fields, waste ground, and along roadsides and railroads. Scattered throughout Missouri, and to be expected in every county.

Native of Europe; introduced and naturalized in North America from Quebec to British Columbia, south to North Carolina, Ohio, Indiana, Illinois, Missouri, Kansas, Colorado, Utah, and California.

The flowers are quite fragrant. This species and *Silene noctiflora* are both night-flowering plants with fragrant white flowers. While they are frequently confused, the essential differences between them are as follows:

Styles 5; flowers mostly unisexual, either entirely with stamens or entirely with pistils, and these on different plants, rarely some perfect flowers scattered among the unisexual ones; plants biennial or short-lived perennial with coarsely hairy stems with rather conspicuous hairs; calyx with 20 nerves, of which 10 are strong and 10 faint; lobes of calyx triangular-lanceolate, acute, 3–5 mm. long; teeth of capsule 10, each of the main 5 teeth split in 2 . LYCHNIS ALBA
Styles 3; flowers perfect, with the stamens and pistils in the same flowers, smaller than in *Lychnis alba*; plants annual with hairs of stem short and not conspicuous; calyx with 10 nerves; lobes of calyx linear-lanceolate, slenderly pointed (subulate), 4–8 mm. long; teeth of capsule 6
SILENE NOCTIFLORA

## 12. **Silene** L. Catchfly, Campion

a. Most of leaves, or at least those at the middle of the stem, in whorls of 4's; petals fringed . 8. S. STELLATA
a. All the leaves in 2's opposite each other; petals not fringed, but notched or cut, or sometimes minute, inconspicuous, or absent. . . . . . . . . . . . . . . . . . . . . . . . . . . . . *b*
b. All parts of stem glabrous (without hairs); calyx glabrous or with only minute hairiness at base . . *c*
c. Leaves on the stem narrow, linear or linear-elliptic to oblanceolate, 1–12 mm. broad; some part of the stem (at least the upper part) with a dark sticky area . . . . . . 1. S. ANTIRRHINA
c. Leaves on the stem broader, from lanceolate to lanceolate-oblong or ovate, 8–50 mm. broad; stem completely smooth, without any sticky area . . . . . . . . . . . . . . . . *d*
d. Stems and leaves green; flowers few on a plant, solitary on slender, spreading pedicels (stalks) arising from the upper leaf axils; expanded blades of petals about 8 mm. long, each with a

small scale crowning the base; calyx with veins scarcely showing between the nerves . 7. S. NIVEA

d. Stems and leaves gray-silvery green, glaucous (with a coating which can be rubbed off); flowers rather numerous, more closely grouped in a bracted inflorescence; expanded blades of petals 2–6 mm. long, without any scale crowning the base; calyx with the veins prominent between the nerves . . . . . . . . . . . . . . . . . . . . . . . . e

e. Calyx becoming 10–13 mm. in fruit, firm, not inflated or very slightly inflated in fruit but the summit of the calyx in contact with and lying against the protruding capsule; nerves of calyx not or scarcely connected by the veins; calyx not constricted at the summit . . . . . . . . . . . . . . . . . . . . . . 5. S. CSEREI

e. Calyx becoming 20 mm. in fruit, papery, much inflated in fruit, not at all in contact with the included capsule; nerves of calyx connected with one another their entire length by veins; calyx constricted at the summit . . . . . . . . . . 6. S. CUCUBALUS

b. At least the lower part of the stem hairy with short or long hairs; calyx hairy (or if glabrous in *S. antirrhina* the stems have a dark sticky area) . . . . . . . . . . . . . . . . . . f

f. Some part of the stem with a dark sticky area (except in *S. antirrhina* f. *Deaneana*); at least the upper portion of stem glabrous (without hairs) . . . . . . . . . . . 1. S. ANTIRRHINA

f. Stem completely lacking dark sticky areas; stems completely hairy . . . . . . . . . . g

g. Most of the petal or all of it white; expanded part of petals 5–12 mm. long . . . . . . h

h. Calyx with about 30 prominent green ribs, conical and tapering from a broad base. 2. S. CONOIDEA

h. Calyx with 5–10 prominent ribs, not conic nor tapering from a broad base . . . . . . i

i. Calyx mainly 15–23 mm. long; lobes of calyx mainly 5–9 mm. long; leaves ovate-lanceolate, the lower spatulate or elliptic-oblanceolate, 10–50 mm. wide; flowers opening at night . . . . . . . . . . . . . . . . . . 9. S. NOCTIFLORA

i. Calyx mainly 6–16 mm. long; lobes of calyx 3–4 mm. long; leaves lanceolate, linear-oblong to spatulate, mainly 3–18 mm. wide (rarely to 35 mm.); flowers opening during the day . . . . . . . . . . . . . . . . . . . . . . . . j

j. Inflorescence forking, the flowers nodding; leaves with an acutely pointed, tapering tip; petals deeply cut more than half way; ribs on calyx covered by a comb-like arrangement of hairs; fruiting calyx 10–16 mm. long . . . . . . 4. S. DICHOTOMA

j. Inflorescence simple and 1-sided, raceme-like, the flowers ascending; leaves with a rounded or blunt broadened tip; petals at most with only a slight notch at the tip; calyx covered with hairs, lacking any regular pattern of arrangement; fruiting calyx 7–10 mm. long . . . . . . . . . . . . . . . . . 3. S. GALLICA

g. All of the petal red, rose-red, or deep pink; expanded part of petals 8–22 mm. long . . . . k

k. Leaves of the stem in 10–30 pairs, thickish, lanceolate to ovate, most of them rounded at base or clasping; flowers mainly June–October; stems mainly 80–120 cm. tall; plants of prairies and rocky open places . . . . . . . . . . . . . . . . 12. S. REGIA

k. Leaves of the stem in 2–6 pairs, thin, oblanceolate to oblong-spatulate, the middle and lower ones tapering or narrowed at the base and not clasping; flowers mainly April–May; stems mainly 8–60 cm. tall; plants mainly of open woodland and wooded rocky ledges, sometimes in exposed cuts along roadsides . . . . . . . . . . . . . . . . . l

l. Flowers red or scarlet, the petals usually split at the tip and with linear-oblong blades 20–25 mm. long; plants usually 20–60 cm. tall . . . . . . . . . . . 11. S. VIRGINICA

l. Flowers deep pink or rose-pink, the summit of the petals either smooth-edged or with slightly wavy edges, the blade of the petals wedge-shaped (cuneate-obovate) and 8–15 mm. long; plants usually 8–20 cm. tall . . . . . 10. S. CAROLINIANA var. WHERRYI

l. Flowers deep pink or rose, the petals split at the tip and with linear-oblong blades; plants 8–20 cm. tall . . . . . . . . 11a. S. VIRGINICA × S. CAROLINIANA var. WHERRYI

1. **Silene antirrhina** L. Sleepy Catchfly Map 934
Flowers April–September.

Occurs along roadsides and railroads, pastures, rocky ledges of bluffs, rocky open glades, fallow and cultivated fields, waste places, and alluvial or open woodland. Throughout Missouri.

Ranges from Quebec to British Columbia, south to Florida, California, and Mexico; also in South

---

Plate no. 159. **1.** Agrostemma Githago, × 2/7; Details from Small, The New York Botanical Garden. **2.** Lychnis alba, × 2/7; Details from Small, The New York Botanical Garden. **3.** Silene antirrhina, × 2/7. **4.** Silene gallica, × 2/7. **5.** Silene Cserei, × 2/7. **6.** Silene conoidea, × 2/7; After Gleason, The New York Botanical Garden. **7.** Silene Cucubalus, × 2/7. **8.** Silene dichotoma, × 2/7.

PLATE NO. 159

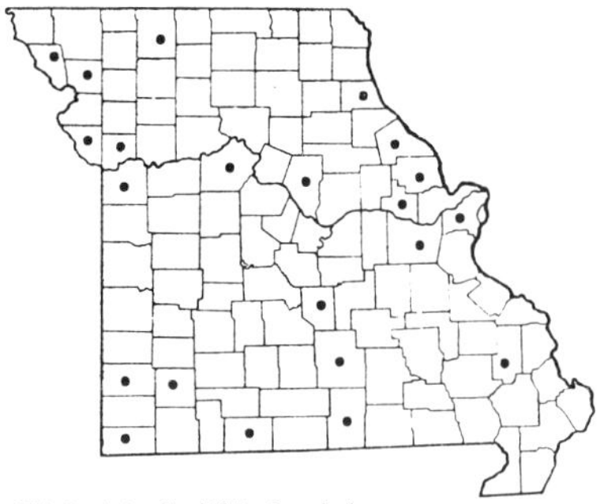

933 Lychnis alba (White Campion)

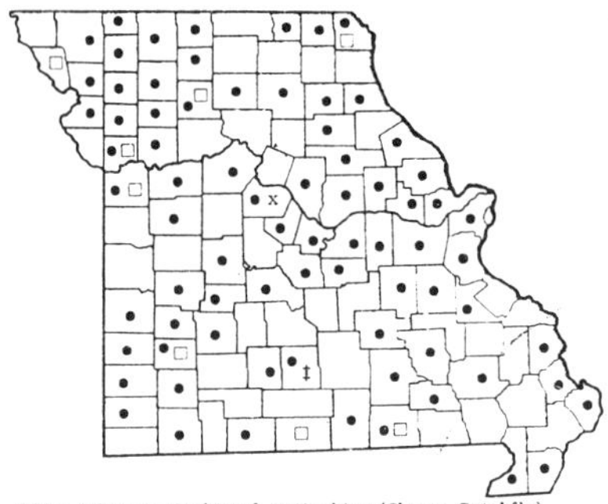

934 • Silene antirrhina f. antirrhina (Sleepy Catchfly)
934 ‡ Silene antirrhina f. bicolor
934 □ Silene antirrhina f. apetala
934 x Silene antirrhina f. Deaneana

934A Silene conoidea

America, and introduced in Europe.

Missouri material is represented by the following variations:

a. No sticky (glutinous) area present on any part of stem or branches of inflorescence . . . 1d. S. ANTIRRHINA f. DEANEANA
a. A sticky (glutinous) area present on some part of stem or branches of inflorescence . . . *b*
  b. Petals normally absent . . 1c. S. ANTIRRHINA f. APETALA
  b. Petals normally present, although small or inconspicuous . . . . . . . . *c*
    c. Petals pink, rose, or purplish . . . 1a. S. ANTIRRHINA f. ANTIRRHINA
    c. Petals white above, pink or rose or purplish below . 1b. S. ANTIRRHINA f. BICOLOR

1a. **Silene antirrhina** f. **antirrhina** Map 934
*Silene antirrhina* L. [G, BB, P & S, Steyerm.]

This is the commonly encountered form in the state. Doubtless to be found in every county of the state.

1b. **Silene antirrhina** f. **bicolor** Farw. Map 934

Known only from Wright County (fallow field by cemetery, by Webster County line, along headwaters of Bryant Creek, T28N, R16W, sect. 22 and 27, 1–1½ mi. [by air] southwest of Cedar Gap, April 26, 1955, *Steyermark 78373*), southern Missouri, but doubtless to be found throughout the state.

1c. **Silene antirrhina** f. **apetala** Farw. Map 934
*Silene antirrhina* var. *divaricata* Robins. [BB, P & S, Steyerm.]

Scattered throughout the range of f. *antirrhina* in the state, but less common.

1d. **Silene antirrhina** f. **Deaneana** Fern. Map 934

Known only from Cooper Co., central Missouri (along west side of road by field along Moniteau Creek, T46N, R16W, W half sect. 19 and E hal sect. 24, 4½–5 mi. [by air] southeast of Pisgah, 6 mi. southwest of Prairie Home, April 27, 1955, *Steyermark 78400*). Doubtless to be found throughout the state when better collected.

The sticky areas of *S. antirrhina* sometimes trap ants at the time of flowering. Those plants of the species found in Missouri in woodland areas may be native to the state, others encountered as weeds in fields, or along roadsides and railroads may have originated from an introduced strain not native to the state. This is a problem which future studies may work out more clearly.

2. **Silene conoidea** L. Map 934a

Flowers May–July.

Occurs along railroads. Known only from St. Louis County (St. Louis, Baden freight yard of Mo.-Kan.-Tex. R.R. between the two most western sidings, June 14, 1959, *Muehlenbach 1538*).

Native of Europe; introduced in the United States along the Pacific Coast, and locally in Delaware and Missouri.

This species closely resembles *S. conica* L., from which it differs in the larger fruiting calyx (20–30 mm. instead of 17 mm. or less), inflated in fruit, blades of the petals 8–12 mm. long instead of 3–6 mm., and seeds 1.2–1.5 mm. instead of 0.6–0.9 mm. wide.

3. **Silene gallica** L. f. **gallica** Map 935
*Silene gallica* L. [G, BB, P & S, Steyerm.]

Flowers April–September.

Occurs in fallow fields and waste ground. Known only from Cape Girardeau (*Steyermark 4813*) and Jackson (*Bush 2056, 3023*) counties.

Native of Europe; introduced and naturalized in North America from Nova Scotia to Michigan, south

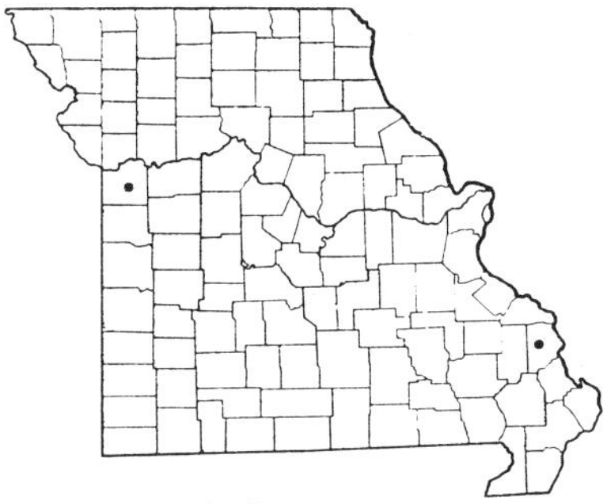

935 Silene gallica f. gallica

936 Silene dichotoma (Forking Catchfly)

937 Silene Cserei

to New Jersey, Illinois, Missouri, and Texas, and in the Pacific states; also naturalized in Central and South America.

The ordinary f. *gallica* has the petals white; in f. *quinquevulnera* (L.) Mert. & Koch, not known in Missouri, the petals are purple or dark brownish-red.

4. **Silene dichotoma** Ehrh. Forking Catchfly Map 936

Flowers June–September.

Occurs along railroads. Known only from St. Louis County (Burlington Railroad, between East Grand Ave. and Ferry St. along the siding, St. Louis, July 9, 1955, *Muehlenbach 672*).

Native of Europe; introduced and naturalized in North America from Quebec to Montana, south to Virginia, Missouri, Texas, and California.

5. **Silene Cserei** Baumg. Map 937

Flowers May–September.

Occurs in waste ground and along railroads. Known from St. Louis, Livingston, and Grundy counties.

Native of Europe; introduced and naturalized in the United States from New York to Montana, south to New Jersey, Pennsylvania, Ohio, Indiana, Illinois, and Missouri.

6. **Silene Cucubalus** Wibel var. **Cucubalus** Bladder Campion Map 938

*Silene latifolia* (Mill.) Britten & Rendle [P & S, Steyerm.]

Flowers May–August.

Occurs in fields, waste ground, and along railroads. Known from St. Louis, Taney, Jackson, and Platte counties.

Native of Europe; introduced and naturalized in North America from Newfoundland to British Columbia, south to Virginia, Tennessee, Missouri, Kansas, Colorado, and Oregon.

The young leafy shoots just emerging above the ground are sometimes gathered and used as a cooked vegetable or made into a thick soup.

7. **Silene nivea** (Nutt.) Otth Snowy Campion Map 939

Flowers early June–August.

Occurs in alluvial soils of wooded valleys and flood plain forest, wooded stream banks, and occasionally found along railroads and waste ground. Rather rare and scattered in northern and central Missouri south to Franklin, Callaway, Boone, and Caldwell counties.

Ranges from Pennsylvania to Minnesota and South Dakota, south to Maryland, Virginia, Tennessee, Missouri, and Nebraska.

The stems are rather slender and weak, sometimes half-sprawling among other herbaceous plants. The leaves are thin, resembling those of *Phlox paniculata* in shape. The solitary flowers are on slender spreading pedicels, appearing from the upper axils of the leaves. The rather showy white petals are 2-lobed at the summit.

8. **Silene stellata** (L.) Ait. f. Starry Campion Map 940

*Silene stellata* var. *scabrella* (Nieuwland) Palmer & Steyerm. [BB]

*Silene stellata* var. *stellata* [BB]

Flowers June–September.

Occurs in usually dry upland woods or wooded slopes, rarely in clearings or along railroads. Throughout Missouri and doubtless in every county.

Ranges from Massachusetts to Minnesota, south to Georgia, Alabama, Arkansas, Oklahoma, and Texas.

By some authors the Missouri material would be placed in var. *scabrella*, distinguished from typical var. *stellata* by the densely puberulent instead of glabrous calyx and pedicels, the slightly stouter pedicels, and the calyx 10 mm. instead of 8 mm. long.

938 Silene Cucubalus var. Cucubalus (Bladder Campion)

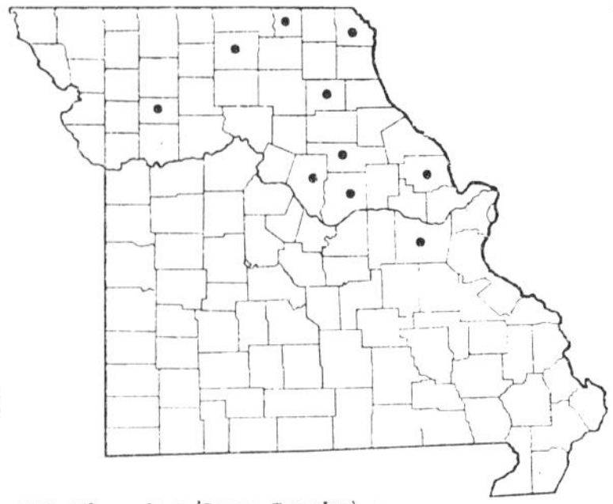

939 Silene nivea (Snowy Campion)

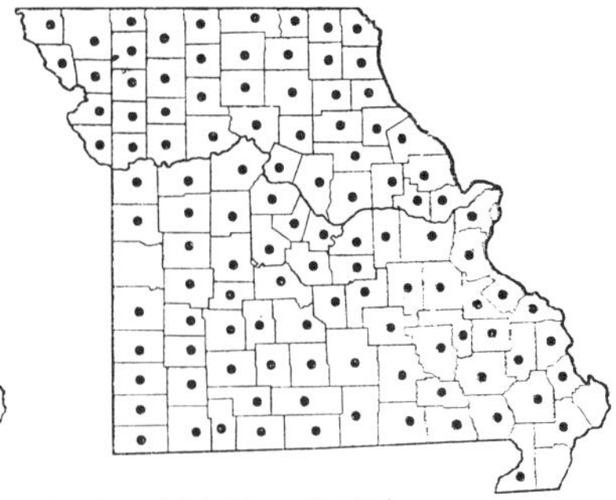

940 Silene stellata (Starry Campion)

9. **Silene noctiflora** L. Night-flowering Catchfly Map 941

Flowers May–September.

Occurs in waste and cultivated ground, along roadsides and railroads.

Scattered in central and southwestern Missouri, in St. Louis, Franklin, Boone, Saline, and Jasper counties.

Native of Europe; introduced and naturalized in south Canada and much of the United States south to Florida, Missouri, Utah, and Washington.

This has been mistaken for *Lychnis alba*, a frequently encountered species and much more common in Missouri than *Silene noctiflora*. The main distinguishing characters between these species were indicated under *Lychnis alba*.

10. **Silene caroliniana** Walt. var. **Wherryi** (Small) Fern. Wild Pink Map 942

*Silene Wherryi* Small [P & S, Steyerm.]

*Silene caroliniana* subsp. *Wherryi* (Small) Clausen [BB]

Flowers April–May.

Occurs in acid soils of rocky upland woods in chert and Roubidoux sandstone soils, and on ledges and bluffs of Roubidoux sandstone, chert, or flint outcrops. Known only from a limited section of the Ozark region bounded by Crawford, Dent, Shannon, Phelps, Pulaski, Laclede, Texas, and Douglas counties.

Ranges in Ohio, Kentucky, Alabama, and Missouri.

This is one of the most beautiful wildflowers in Missouri. It favors acid soils in rocky upland woodland. The ascending or somewhat spreading stems arise from a rosette of pale green leaves. Sometimes a single plant may produce 50–100 showy rosy-pink flowers. The plant may be successfully grown in a wildflower rock garden if provided with ample drainage, acid rocks, and some shade. Plants are often found with their roots entangled only in a loose mass of chert and flint rocks. If the roots are not disturbed in transplanting, and an environment created similar to the one the plant came from, the chances for success are good in naturalizing this plant. Several plants have prospered during the past fifteen years in the author's wildflower garden in northern Illinois.

11. **Silene virginica** L. Fire Pink Map 943

Flowers April–June.

Occurs in rocky woods, ledges, and wooded slopes in sandstone, chert, granite, or limestone soils. Restricted to the Ozark section of southern and east-central Missouri, north to Jefferson, Washington, Crawford, Miller, Laclede, Greene, Lawrence, and Newton counties, and in the Ozark area of Montgomery County.

Ranges from New York and Ontario to Minnesota, south to Georgia, Alabama, Arkansas, and Oklahoma.

This showy-flowered species is common in the Ozark section and adapts itself easily to wildflower gardens, preferring a semi-shaded, well-drained situation. It does well on shaded rock ledges.

11a. **Silene virginica** × **Silene caroliniana** var. **Wherryi** Map 943

This hybrid is known only from Shannon County (upland chert woods, 3 mi. south of Dent-Shannon Co. line, in Shannon Co., May 11, 1935, *Steyermark 18923* and *18924*). Both parent species were present, represented by *S. virginica* (*Steyermark 18921*) and *S. caroliniana* var. *Wherryi* (*Steyermark 18922*). Number *18923* had the shape of the petals of *S. virginica*, but an

Plate no. 160. **1.** Silene nivea, × 2/7. **2.** Silene stellata, × 2/7; Details from Small, The New York Botanical Garden. **3.** Silene noctiflora, × 2/7. **4.** Silene caroliniana var. Wherryi, × 2/7. **5.** Silene virginica, × 2/7. **6.** Saponaria officinalis, × 2/7; After Britton and Brown, details from Small, The New York Botanical Garden. **7.** Silene regia, × 2/7.

PLATE NO. 160

941 Silene noctiflora (Night-flowering Catchfly)

942 Silene caroliniana var. Wherryi (Wild Pink)

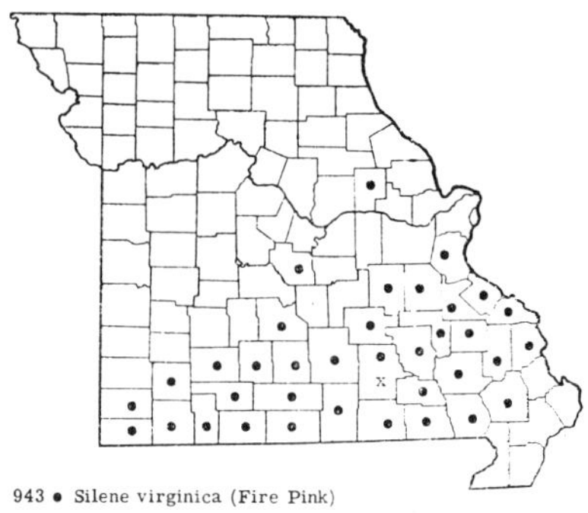

943 • Silene virginica (Fire Pink)
943 x Silene virginica X Silene caroliniana var. Wherryi

intermediate color between pink and red, with the pink predominating. Collection number *18924* had the red petals of *S. virginica*, but the stature of the plant and shape of the petals of *S. caroliniana* var. *Wherryi*.

12. **Silene regia** Sims Royal Catchfly Map 944

Flowers late May (May 30, 1879, St. Louis Co., *Letterman*) – October.

Occurs on rocky prairies, rocky open woods, thickets, and borders of rocky glades. Restricted to the Ozark section of southern and east-central Missouri, north to St. Louis, Franklin, Maries, Cole, Laclede, Polk, Cedar, and Jasper counties.

Ranges from Georgia to Louisiana and Oklahoma, north to Ohio, Indiana, Illinois, and Missouri.

The striking crimson-scarlet or bright red flowers are among the reddest of any wildflowers in Missouri, rivaling or surpassing those of Cardinal Flower (*Lobelia cardinalis*) in intensity. The author has grown the species in his wildflower garden in northern Illinois from seed, as well as having successfully transplanted individuals which have maintained themselves and seeded new plants during the past fifteen years. The species has a long flowering season and produces flowers well up to frost in northern Illinois.

## 13. **Saponaria** L.

Leaves narrowed at the base; calyx not winged, cylindrical with curved or rounded sides, 20-nerved; flowers in dense clusters at the top and sides of the stem; perennial with dull or dark green leaves, not glaucous (no bloom that can be rubbed off) . . . . . . . . . . . . . . . 1. S. OFFICINALIS

Leaves rounded and clasping at the base; calyx conspicuously 5-angled and in winged fruit, ovoid, 5-ribbed; flowers in a loosely branched inflorescence at the top of the stem; annual with pale or blue-green leaves with a glaucous 'bloom' which can be rubbed off. . . . . . . . . . . 2. S. VACCARIA

1. **Saponaria officinalis** L. Bouncing Bet Map 945

Also known as Soapwort.

Flowers June–October.

Commonly found on gravel and sand bars along streams, roadsides, railroads, fallow fields, sandy or open waste ground, and rarely in wooded areas. Throughout Missouri, doubtless in every county.

Native of Europe; introduced and naturalized throughout North America.

This species was formerly cultivated as a garden plant, but has become well established as a weed. The deliciously spicy-fragrant pink-lilac to whitish flowers are either single or double. A soapy green lather can be made from the plant by rubbing the leaves and stems in some water. The juice of the plant is reputed to be slightly poisonous as in the following species, *S. Vaccaria*.

2. **Saponaria Vaccaria** L. Cow Herb Map 946

Also known as Cow Soapwort, Cow Cockle.

*Vaccaria segetalis* (Neck.) Garcke [BB]

*Vaccaria vulgaris* Host [Small]

*Vaccaria pyramidata* Moench [Royal Hort. Dict., Shinners]

Flowers late May–September.

Occurs along railroads and roadsides. Scattered in southern, central, and eastern Missouri; not recorded from the northwestern third of the state, but to be expected throughout most of the counties.

Native of Europe; introduced and naturalized throughout the greater part of North America.

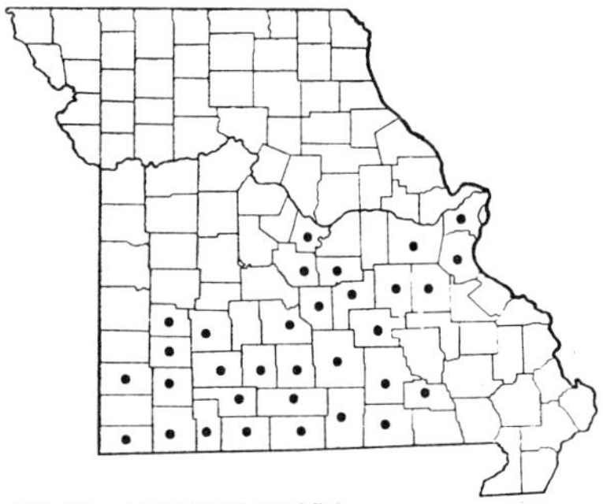
944 Silene regia (Royal Catchfly)

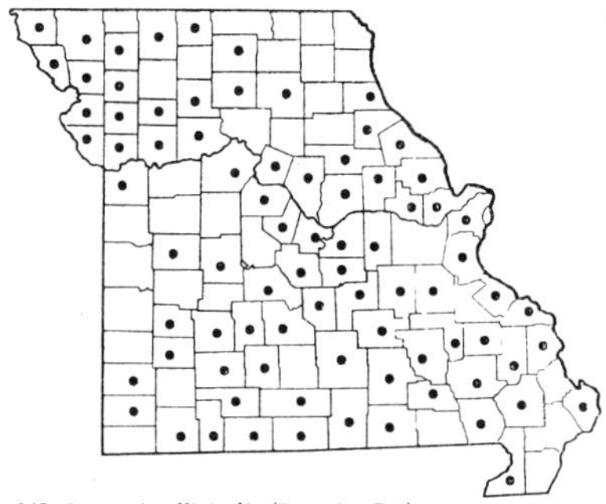
945 Saponaria officinalis (Bouncing Bet)

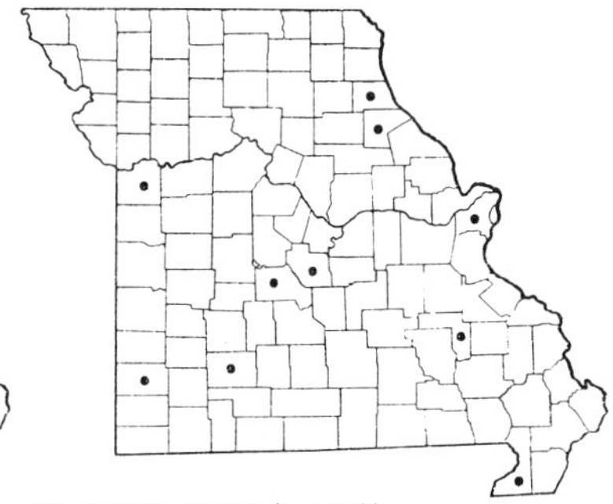
946 Saponaria Vaccaria (Cow Herb)

The flowers vary from deep pink to pale pink, the latter color sometimes recognized as var. *rosea* Hort., and white-flowered individuals as var. *alba* Hort. The seeds contain a toxic substance which sometimes poisons poultry, rabbits, and other animals in screening of grain containing some of the seeds.

Other species of *Saponaria* are cultivated as garden plants. A trailing, much branched, pink-flowered species with small leaves, *S. ocymoides* L., is frequently found in gardens. There is no definite record of its having become established outside of gardens. A specimen from Platte County in the herbarium of Park College (Weatherby Lake, on dam, 1½ mi. north of Parkville, May 9, 1957, *Kent Drew 91*) was collected, so far as known, from a planted clump.

## 14. **Dianthus** L. Pink, Carnation

All the species of *Dianthus* found in Missouri are of European origin. Some species, such as *D. plumarius* L., not here included, may eventually be found as escapes from cultivation.

a. Leaves 10–18 mm. broad, lanceolate to elliptic-oblong . . . . . 2. D. BARBATUS
a. Leaves 1.5–8 mm. broad, linear or linear-lanceolate . . . . . *b*
  b. Calyx tube with 5 to 15 nerves . . . . . 4. D. PROLIFER
  b. Calyx tube with 20–40 nerves . . . . . *c*
    c. Each flower solitary on its own single long stalk of 1–4 cm. length; leaves of stem 1–4 cm. long; stem glabrous or with short hairs; calyx with 30–40 nerves; low perennial with trailing leafy shoots, forming loose mats, 1–4 dm. tall. . . . . . 1. D. DELTOIDES
    c. Inflorescence consisting of few or several closely crowded flowers in clusters, the flowers without stalks or nearly so; leaves of stem 2–8 cm. long; stem more noticeably hairy; calyx with 20–25 nerves; annual or biennial plant with erect stem 2–8 dm. tall, and without any trailing leafy shoots, not forming loose mats . . . . . 3. D. ARMERIA

1. **Dianthus deltoides** L. Maiden Pink Map 947

Flowers May–August.

Known only from Franklin County, east-central Missouri, where escaped from gardens.

Native of Europe; naturalized in the United States from New Hampshire to Michigan, south to New Jersey, Illinois, and Missouri.

2. **Dianthus barbatus** L. Sweet William Map 948

Flowers June–August.

Known only from Jasper County, southwestern Missouri, where escaped from gardens (waste ground near Webb City, June 16, 1923, *Palmer 23314* in Palmer herb.).

Native of Europe; naturalized in North America from Quebec to North Dakota, south to Delaware and Missouri.

3. **Dianthus Armeria** L. Deptford Pink Map 949

Flowers late May–October.

Occurs in open woodland, along roadsides, railroads, waste ground, fields, and pastures. Throughout Missouri; not recorded from a number of northern Missouri counties, but doubtless to be found there.

Native of Europe; introduced and naturalized in North America from Quebec and Ontario, south to Georgia, Kentucky, Missouri, and Oklahoma.

4. **Dianthus prolifer** L. Map 950

*Tunica prolifera* (L.) Scop. [BB]

Flowers May–October.

947 Dianthus deltoides (Maiden Pink)

948 Dianthus barbatus (Sweet William)

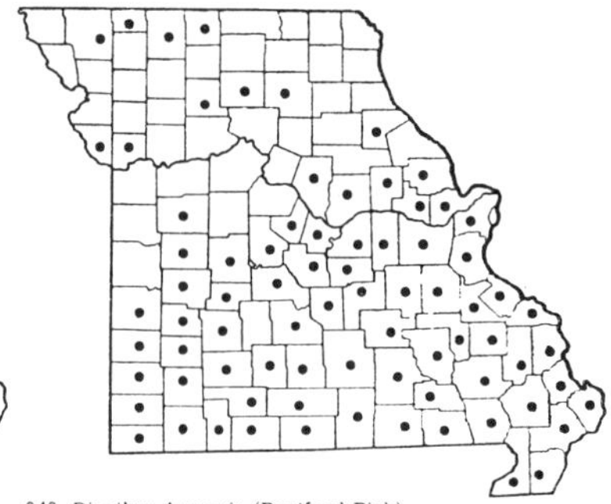

949 Dianthus Armeria (Deptford Pink)

Known only in Stone County, southwestern Missouri (dry open rocky places near pasture above gravel bar along Big Creek, about $\frac{1}{2}$ mi. northwest of junction with White River, T22N, R24W, NE sect. 18, NW sect. 17, $5\frac{1}{2}$ mi. southeast of Shell Knob, July 13, 1956, *Steyermark 81924*).

Native of Europe; introduced and naturalized in the United States from New York to South Carolina, Kentucky, Ohio, Missouri, Oklahoma, and California.

The station from which this plant was collected is now destroyed and part of the lake bottom of the impoundment of Table Rock Dam on White River. Presumably, careful search will reveal its presence in other, possibly southern, sections of the state.

## Order **RANALES**

### Fam. **CERATOPHYLLACEAE** (Hornwort Family)

#### **Ceratophyllum** L. Hornwort

For a detailed discussion, the reader is referred to the most recent revision of this genus by Norman C. Fassett (North American Ceratophyllum, in Comunicaciones del Instituto de Investigaciones Cientificas. Universidad de El Salvador no. 2: 25–45. March, 1953).

Leaves forked 1–2 times, or simple and not forked; teeth on the leaf-divisions arising from a broad green base connected with the leaf; fruit not winged, with 1 long spine at one end and 2 spines at the other basal end: common type found in Missouri . . . . . . . . . . . . . . . . 1. C. DEMERSUM

Leaves forked 2–4 times; teeth on the leaf-divisions arising directly from the main leaf-division with little or no green base connecting the teeth to the leaf-division; fruit narrowly winged, sometimes also with spines along the margins; rarer type found . . . . . . . . . . . . . . 2. C. ECHINATUM

1. **Ceratophyllum demersum** L. Coontail, Hornwort — Map 951

Flowers July–October.

Occurs in sluggish streams, sloughs, artificial and natural ponds, oxbow lakes of river bottom, and in slow water of spring branches. Mainly in southern and central Missouri north to St. Charles, Boone, Macon, Saline, Carroll, Clay, Platte, and Buchanan counties: absent or not recorded from most of northern Missouri.

Ranges from Quebec to British Columbia, and in most of the United States, south to Mexico, Central and South America, and the West Indies; also in Europe and Asia.

The seeds of this species and, sometimes, the foliage, constitute a food item of importance to wildfowl. The large branching masses of leaves and stems afford a good shelter for young fish and small invertebrate animals, and serve as a habitat for many aquatic insects eaten by fish.

---

Plate no. 161. **1.** Saponaria Vaccaria, $\times {}^2/_5$; Details from Small, The New York Botanical Garden. **2.** Dianthus deltoides, $\times {}^2/_5$. **3.** Dianthus Armeria, $\times {}^2/_5$; Details from Small, The New York Botanical Garden. **4.** Dianthus barbatus, $\times {}^2/_5$. **5.** Dianthus prolifer, $\times {}^2/_5$; After Gleason, The New York Botanical Garden. **6.** Ceratophyllum demersum, $\times {}^2/_5$; Details from Small, The New York Botanical Garden.

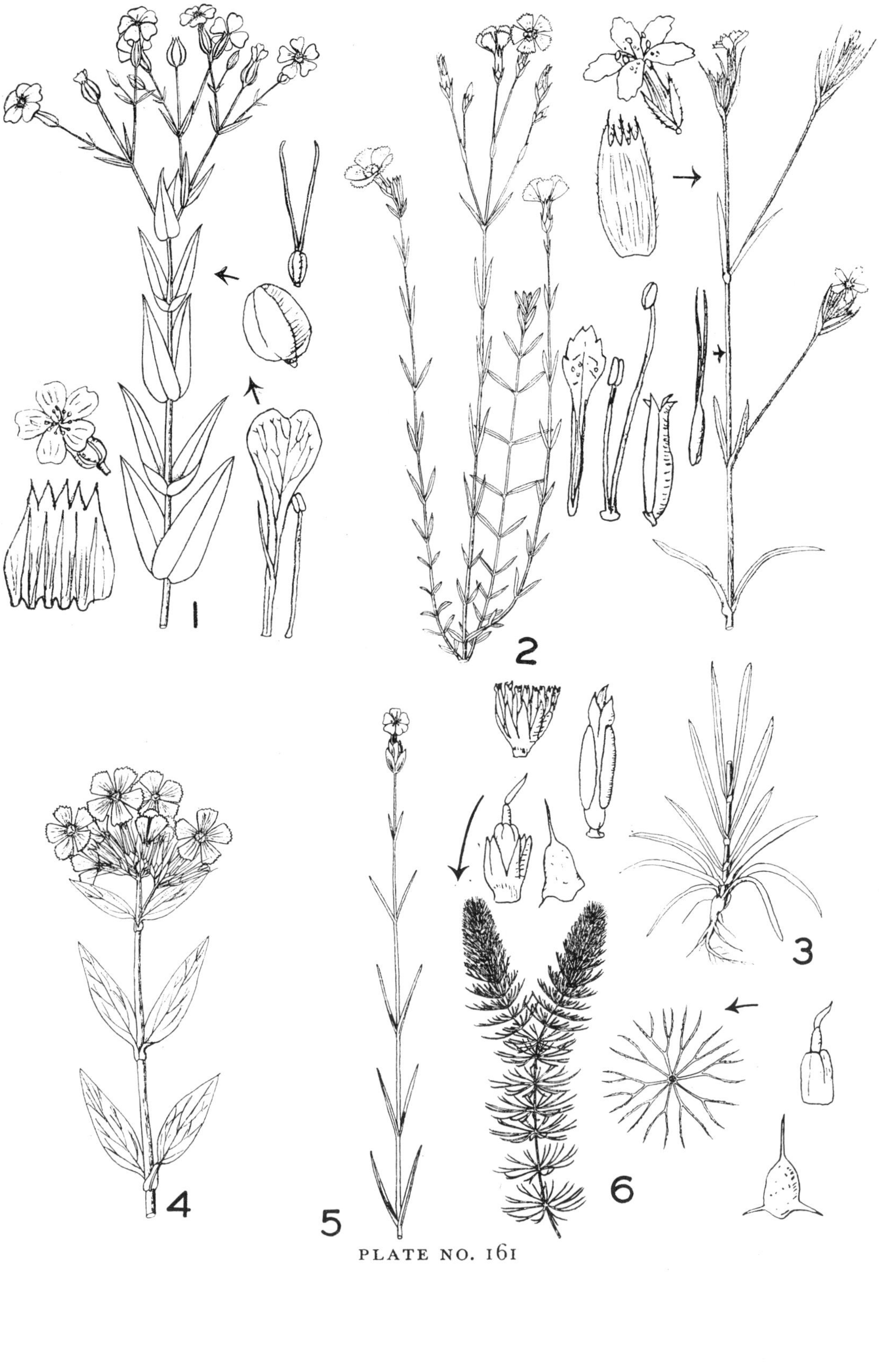


PLATE NO. 161

950 Dianthus prolifer

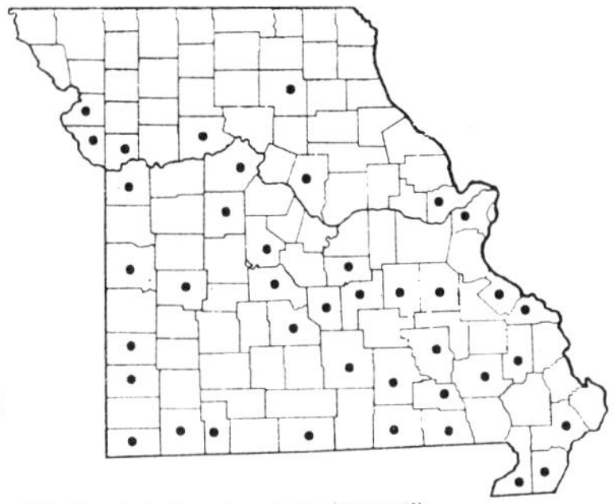

951 Ceratophyllum demersum (Coontail)

952 Ceratophyllum echinatum (Coontail)

The plants vary considerably in aspect, having soft or stiff and hard foliage, dark green or sometimes dull bronze, or lighter green tinged with reddish-brown. The leaves vary from slender to thickish and from short leaves only 3–5 mm. long to longer ones 7–22 mm. long. It is one of the more characteristic plants found in the fresh water springs of the Ozarks, ranking ninth in frequency of species encountered (Steyermark, Bot. Ser. Field Mus. Nat. Hist. 9, no. 6: 493. 1941). In these springs or spring branches it favors the deeper and more quiet waters with a muddy substratum.

*Ceratophyllum demersum* may sometimes be confused with the superficially similar *Myriophyllum*, *Chara*, and *Nitella*. From *Myriophyllum* it may be distinguished by the forking and branching leaves with teeth or irregularities on the margins of the leaf-divisions as contrasted with the pinnately parted leaves of *Myriophyllum* with no teeth or irregularities on the leaf-divisions. Both *Chara* and *Nitella* pertain to the Characeae of the algae, whose leaflike divisions are only elongated narrow cells unlike the multicellular leaf tissue of a leaf segment of *Ceratophyllum* or other spermatophytes.

2. **Ceratophyllum echinatum** Gray Coontail, Hornwort Map 952
Flowers July–October.

Occurs in natural oxbow lakes of river bottom, upland sink-hole ponds in the Ozarks, and in sluggish waters of slow streams and ponds. Known only from five counties in southern and east-central Missouri northeast to St. Charles County. The following two collections have been seen by the present author and verified by Dr. Fassett: St. Charles County (Merry Rundo Lake, 5 mi. north of St. Peters, October 5, 1933, *Steyermark 8864* in herb. Univ. of Mo.); Oregon County (in 2–3 feet of water, Red Brush Pond, T24N, R2W, sect. 16, 4 mi. south of Wilderness, July 20, 1936, *Steyermark 12072*).

Ranges from Florida to Texas and Mexico, north to Maine, Quebec, New York, Ohio, Michigan, Wisconsin, Missouri, and Kansas.

This species, according to Dr. Fassett's studies, does not show the variability as exhibited by *C. demersum*. The leaves are more uniform in length, 17–21 mm. long with slender terminal segments 0.1–0.2 mm. wide.

## Fam. **NYMPHAEACEAE** (Water Lily Family)

a. All the submerged leaves deeply and finely dissected . . . . . 5. CABOMBA
a. None of the leaves dissected. . . . . . *b*
  b. Leaves split or cleft at one end, the leaf-stalk (petiole) joining the leaf-blade at the base of the cleft (sinus) . . . . . *c*
    c. Most of the main nerves of the leaf-blade unbranched or forking only near the leaf-margin and spreading directly from the main midrib; flowers yellow, with small, thick, stamen-like petals 6–9 mm. long; the 5, 6, or more sepals are yellow or tinged red or green, and roundish; floating or aerial leaves oval . . . . . 1. NUPHAR
    c. Main nerves of the leaf-blade much branched and forked and at least half of them originating directly from the top of the petiole (main leaf-stalk); flowers white, pink, or rose, with showy colored petals mostly 45–90 mm. long, and not stamen-like; the 4 sepals are green or purplish tinged on outside, and oblong-lanceolate; floating leaves round or nearly so . . . 2. NYMPHAEA
  b. Leaves entire, not split nor cleft at any place, the leaf-stalk (petiole) joining the leaf-blade approximately at the middle of the blade. . . . . *d*

d. Leaves small, floating, mainly 0.2–1 dm. across, oval; stems, buds, and lower surface of leaves covered with a jelly-like, slimy coating; petals and sepals purple, 16 mm. or less long . 4. BRASENIA

d. Leaves very large, aerial (standing above the water), mainly 3–7 dm. across, circular; stems and leaves not covered with any jelly-like, slimy coating; petals yellow or rose, they and the sepals 60–100 mm. or more long . . . . . . . . . . . . . . . . . . . 3. NELUMBO

### 1. **Nuphar** Sm. Yellow Pond Lily, Spatterdock, Yellow Water Lily

For a more detailed elaboration of this genus, the reader is referred to Ernest O. Beal's Taxonomic Revision of the Genus *Nuphar* Sm. of North America and Europe (Jour. Elisha Mitchell Sci. Soc. 72, no. 2: 317–46. 1956), which treatment has been followed in the present flora.

**Nuphar luteum** (L) Sibth. & Sm. Yellow Pond Lily, Spatterdock, Yellow Water Lily Map 953
Flowers May–October.

Missouri material is represented by the following two variations:

Inner part of the sepals green to yellow (rarely tinged with red); fruit green (rarely tinged with red); leaves either standing erect above the surface of the water or floating . . . a. N. LUTEUM subsp. MACROPHYLLUM

Inner part of the sepals conspicuously tinged with red; fruit more or less tinged with red; leaves mostly floating . . . . . b. N. LUTEUM subsp. OZARKANUM

a. **Nuphar luteum** subsp. **macrophyllum** (Small) Beal Map 953
*Nuphar advena* (Ait.) Ait. f. [G, BB, Steyerm.]
*Nymphozanthus advena* (Ait.) Fern. [P & S]

Occurs along borders of streams, ponds, and sloughs. Southern and east-central Missouri, north to Lincoln, Franklin, Phelps, Pulaski, Laclede, St. Clair, and Bates counties.

Ranges in Cuba, northeastern Mexico, and from Florida to Texas, north to Maine, New Hampshire, New York, Ontario, Michigan, Wisconsin, Missouri, and Kansas.

This subspecies can usually be recognized by its ovate leaves with spreading lobes, terete to more or less flattened petioles, and yellow-green sepals, but grades into the next subspecies, subsp. *ozarkanum*. The leaves vary from erect and glabrous to floating and pubescent on the lower surface.

b. **Nuphar luteum** subsp. **ozarkanum** (Miller & Standley) Beal Map 953
*Nuphar ozarkanum* (Mill. & Standl.) Standl. [G, BB, Steyerm.]
*Nymphozanthus ozarkanus* (Mill. & Standl.) Palmer & Steyerm. [P & S]

Occurs along borders of streams and sloughs in the Ozark section north to Washington, Crawford, Pulaski, Miller, and Benton counties, west to Greene and Taney counties, and east to Iron, Wayne, and Butler counties.

Ranges from Missouri Ozarks to Arkansas and eastern Oklahoma.

The lobes of the leaves in subsp. *ozarkanum* are usually approximate and closer together than in subsp. *macrophyllum*, although this is not constant. The leaves are oblong-ovate to orbicular, while the petioles are terete. The seeds vary from 15–48 per fruit.

By previous workers the two subspecies treated above have always been held to be distinct species, but Dr. Beal's studies show that the shape of leaves, presence of red color, and number of seeds, previously used as characters for separating these entities, are scarcely reliable for use in distinguishing them even as subspecies.

The starchy rootstocks have been used by some of the Indian tribes as a boiled or roasted vegetable. The seeds, after the removal of the hard shell, of a related northwestern subspecies, are sometimes eaten in parched form flavored with salt and placed in cream, or the kernels ground into a flour for breadstuff. Both subspecies are of food value to wild animals, the stem and leaves being eaten by deer and muskrat, the roots by beaver, and the seeds by wildfowl. The lower surface of the leaves is a habitat for aquatic insects and other invertebrates.

The underground rootstock in these two subspecies of *Nuphar* is very thick and spongy, at least 5 cm. thick, and bears prominent circular scars of the remains of the flower-stalks and semicircular scars of the remains of the leaf-stalks.

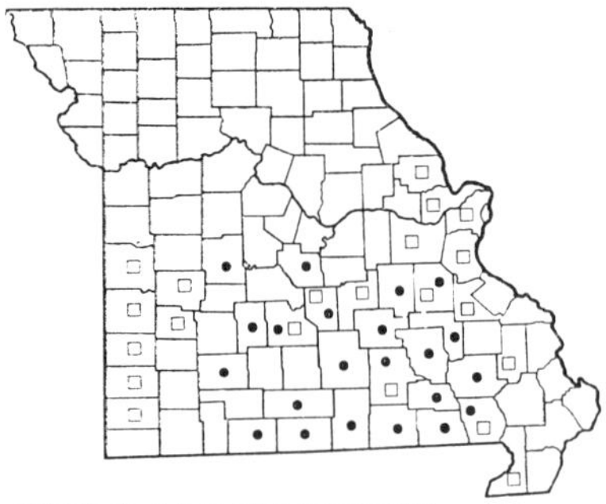

953 □ Nuphar luteum subsp. macrophyllum
953 • Nuphar luteum subsp. ozarkanum

954 • Nymphaea odorata var. odorata f. odorata (Fragrant Water Lily)
954 x Nymphaea odorata var. odorata f. rubra
954 □ Nymphaea odorata var. gigantea

955 Nymphaea tuberosa (Tuberous Water Lily)

## 2. **Nymphaea** L. Water Lily

a. Petals, or at least the outer ones, rose-colored or pink . . . . 1b. N. ODORATA var. ODORATA f. RUBRA
a. Petals white . . . . . . . . . . . . . . . . . . . . . . . . . . . . . *b*
  b. Flowers very fragrant, mostly 5–12 cm. across (10–20 cm. in var. *gigantea*); petals with tapering ovate obtuse tip; lower surface of leaves purple or purple-red; petiole (leaf-stalk) not striped; inner filaments narrower than anthers . . . . . . . . 1a. N. ODORATA var. ODORATA f. ODORATA
  b. Flowers not fragrant or with a faint scent, 10–25 cm. across; petals with a broadly rounded summit, obovate or spatulate; lower surface of leaves mainly green; petiole (leaf-stalk) green, striped above with brown; filaments broader than anthers . . . . . . . . . . 2. N. TUBEROSA

1. **Nymphaea odorata** Ait. Fragrant Water Lily Map 954

Flowers late May–September.

Occurs in ponds and sloughs.

The following variations are found in Missouri:

a. Petals, at least the outer ones, rose-colored or pink . . 1b. N. ODORATA var. ODORATA f. RUBRA
a. Petals white . . . . . . . . . . . *b*
  b. Flowers mostly 5–12 cm. across; sepals 2.8–8 cm. long, 1–2.5 cm. broad; leaves not turning upward at the edge . . . . 1a. N. ODORATA var. ODORATA f. ODORATA
  b. Flowers 10–20 cm. across; sepals 5–10 cm. long, 2.4–3.6 cm. broad; leaves (when growing) turning upward at the edge . . 1c. N. ODORATA var. GIGANTEA

1a. **Nymphaea odorata** var. **odorata** f. **odorata** Map 954

*Nymphaea odorata* Ait. [G, P & S, Steyerm.]

*Nymphaea odorata* var. *odorata* [BB]

Scattered and rare in southern and central Missouri north to St. Louis, Boone, Ray, and Platte counties. Ranges from Newfoundland to Manitoba, south to Florida, Louisiana, and Texas.

1b. **Nymphaea odorata** var. **odorata** f. **rubra** Guillon Map 954

*Nymphaea odorata* f. *rubra* Guillon [G]

Known only from Macon County in northern Missouri (Ethel Lake, T59N, R17W, south part sect. 25 and north part sect. 36, $\frac{3}{4}$ mi. west of Ethel, September 14, 1954, *Steyermark 77311*).

This was introduced from plants purchased from a Chicago company fifteen years ago and is well established in the lake.

1c. **Nymphaea odorata** var. **gigantea** Tricker Map 954

*Nymphaea odorata* var. *villosa* Caspary [Shinners]

Known from Ray County, west-central Missouri.

Ranges from Florida to Texas and Mexico, north to Massachusetts, Illinois, Missouri, and Oklahoma.

As in *Nuphar lutea*, the underground rootstock is thick and spongy, at least 5 cm. thick, and bears prominent semi-circular and circular scars.

The flower buds of *N. odorata* are sometimes cooked as a vegetable by the northern Ojibwa Indians.

2. **Nymphaea tuberosa** Paine Tuberous Water Lily Map 955

Plate no. 162. **1.** Nuphar luteum subsp. macrophyllum, × 2/7. **2.** Nuphar luteum subsp. ozarkanum, × 2/7. **3.** Nelumbo lutea. × 2/7; After Britton and Brown, The New York Botanical Garden. **4.** Nymphaea odorata var. odorata, × 2/7. **5.** Cabomba caroliniana, × 2/7. **6.** Brasenia Schreberi, × 2/7. Details from Small, The New York Botanical Garden.

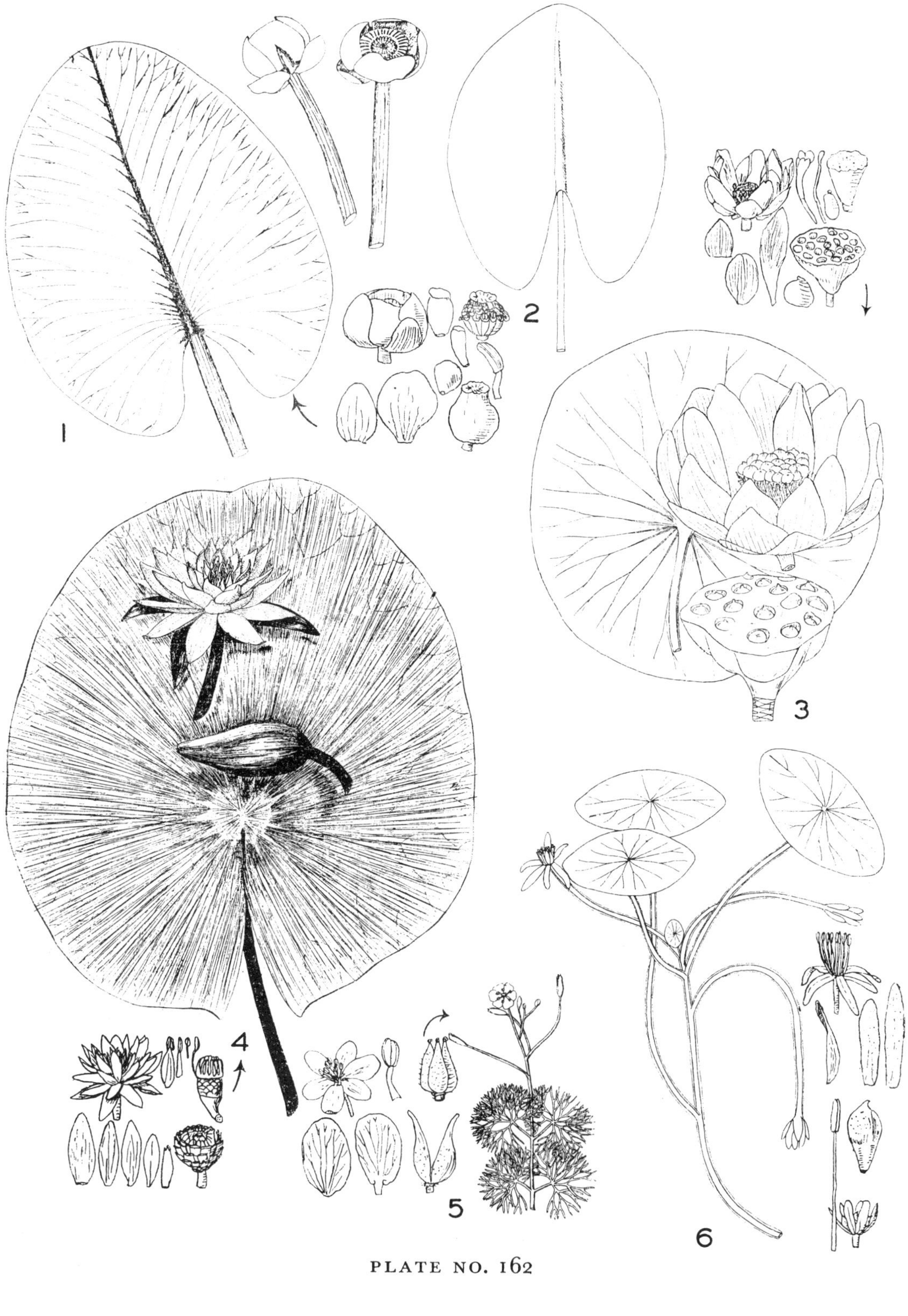


PLATE NO. 162

956 Nelumbo lutea (American Lotus)

957 Nelumbo nucifera (Sacred Lotus)

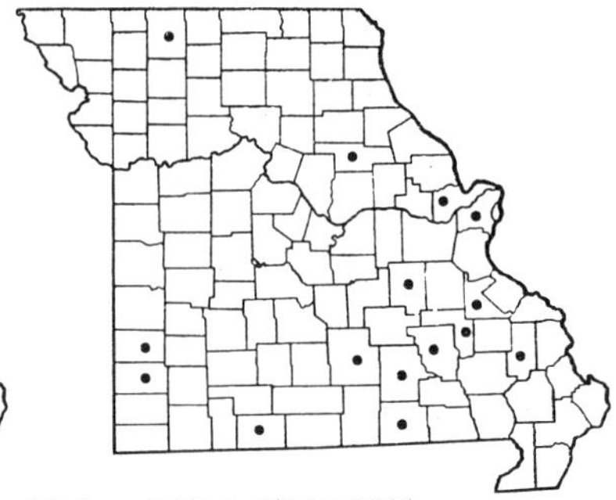

958 Brasenia Schreberi (Water-shield)

Flowers June–October.

Occurs in ponds and slow streams.

Known only from eastern Missouri in St. Louis (Natural Bridge road, June 17, 1893, *Eggert*) and Dunklin (St. Francis River, Bertig, September 28, 1897, *Trelease 11*) counties.

The St. Louis County record, so far as known, is now exterminated, and there is no collection of the species since the one by Trelease in 1897.

Ranges from Quebec to Ontario, Minnesota, and Nebraska, south to Maryland, Ohio, Indiana, Illinois, Missouri, and Arkansas.

The dormant tuberous branches of the rootstock of some of the African species of *Nymphaea* are starchy and are boiled or roasted, while the kernels are cooked whole or ground into a flour. The brown tubers of the rootstock of *N. tuberosa* probably possess similar human food value. The rootstocks of *N. tuberosa* and *N. odorata* are eaten by beavers, muskrat, and deer, while the seeds are eaten by various wildfowl. The leaves provide shade and shelter for fish.

## 3. **Nelumbo** Adans.

Flowers pale yellow; petiole (leaf-stalk) and peduncle (flower-stalk) rather smooth and glabrous . 1. N. LUTEA
Flowers pink or rose-colored; petiole (leaf-stalk) and peduncle (flower-stalk) rather rough . . 2. N. NUCIFERA

1. **Nelumbo lutea** (Willd.) Pers. American Lotus Map 956

Also called Yanquapin, Water Chinquapin, Lotus Lily.

Flowers late June (June 25)–September.

Occurs in oxbow lakes and ponds in river flood plain, upland sink-hole ponds, artificial lakes and ponds, and sloughs. Throughout Missouri, commonest and in greatest abundance in oxbow lakes in the flood plain of the Mississippi and Missouri rivers and their tributaries, especially the Grand, Chariton, Mussel Fork, Thompson, Osage, Mariais des Cygnes, and South Grand rivers.

Ranges from Florida to Texas, north to Massachusetts, New York, Ontario, Minnesota, and Iowa.

For an interesting account of the species, the reader is referred to the article by Thomas F. Hall and Wm. T. Penfound in Am. Midl. Nat. 31: 744–58. 1944.

The American Lotus was extensively used by the American Indian for food. The starchy tuberous enlargements of the rootstock, produced in autumn, when baked, make a palatable dish, as do the young leaf-stalks and young leaves when cooked as a vegetable. The seeds have a chestnutlike flavor if eaten in the earlier stage before the shell gets hard, while the kernel of the ripe seeds, after having been removed from the hard shell by parching, is cooked or ground up for breadstuff.

The roots are eaten by beaver and the seeds are eaten by some wildfowl.

The seeds possess unusual powers of longevity and dormancy, records showing seeds germinating following a dormant period of at least 200 years. The late Mr. John Kellogg of the Missouri Botanical Garden related to this author that he had seen seeds of Yanquapin, which had previously occupied some of the Fox River bottoms near Allenton, germinate 40 years later when the abandoned fields containing the seeds were newly plowed for cultivation.

The underground rootstock is 1 cm. or less thick in contrast to the much thicker (5 cm. and more) rootstocks of *Nuphar* and *Nymphaea*. The large blue-green leaves possess the character, as in Touch-me-not (*Impatiens capensis* and *pallida*) and a few other native plants, of shedding water when held below the surface of the water. The plants are quite ornamental, and the large leaves afford shelter and habitats for wildlife, and attract marsh birds and wildfowl. The large leaves act as a shelter and shade for fish.

2. **Nelumbo nucifera** Gaertn. Sacred Lotus Map 957

Also known as East Indian Lotus.

Flowers late June–August.

Introduced into lakes, where established in Iron (Iron Mountain Lake, July 23, 1932, *Kellogg 26126*) and Platte (Basswood Lakes, 4 mi. east of Platte City, July 12, 1956, *Radloff*) counties.

Native of Asia and Australia.

The seeds and rootstocks are used for food in southern China, where the species is cultivated.

### 4. **Brasenia** Schreb. Water-shield

**Brasenia Schreberi** Gmel. Water-shield Map 958

Flowers May–September.

Occurs in natural and artificial ponds and sloughs. Scattered in Missouri, mainly in the southern third of the state, north locally in Harrison County.

Ranges from Central America, the West Indies, and Mexico, north from Florida to Texas, north to Nova Scotia, Ontario, Minnesota, Nebraska, Manitoba, British Columbia, and Oregon; also in Asia, Africa, and Australia.

The gelatinous covering of the stems, young buds, and lower leaf surface readily identifies this species. All the leaves are floating in this plant. It is of food value to ducks, and the tuberous roots are reported to be eaten by Indians in California. The young leaves and petioles, before the gelatinous covering is prominent, are also reported as eaten as a salad by some of the Japanese. In its natural haunts the leaves afford shade and shelter for fish.

### 5. **Cabomba** Aubl. Fanwort, Carolina Water-shield

For a recent revision of the genus, the reader is referred to Norman C. Fassett, A Monograph of Cabomba (Castanea 18: 116–28. 1953).

**Cabomba caroliniana** Gray Fanwort Map 959

Flowers May–September.

Occurs in swamps, sloughs, and introduced occasionally into ponds. Rare and scattered in eastern Missouri from St. Charles County south; most common in the lowlands of southeastern Missouri and along swamps in the flood plain near the junction of the Missouri and Mississippi rivers.

Ranges from Florida to Texas, north to Massachusetts, New York, Kentucky, Michigan, Illinois, and Missouri; considered to be introduced in the Atlantic states north of Virginia.

The flowers are rather showy, 1.2–1.8 cm. broad, white with yellow spots at the base, rising above the water on long stalks. The plant is considered to be of slight food value to water birds, but is a valuable food producer and cover for fish. The decorative dark green fan-shaped leaves are often seen in fresh water aquaria as an ornamental aquatic plant.

## Fam. **MAGNOLIACEAE** (Magnolia Family)

Leaves longer than broad, not toothed or lobed, with only 1 tip . . . . . . . . . . . 1. MAGNOLIA
Leaves about as broad as long, with 4–6 lobes, the tip spreading into 2 points . . . . 2. LIRIODENDRON

### 1. **Magnolia** L. Magnolia

**Magnolia acuminata** L. var. **acuminata**
Cucumber Tree Map 960

*Magnolia acuminata* var. *ozarkensis* Ashe [P & S, Steyerm.]

*Magnolia acuminata* L. [G, BB, P & S, Steyerm.]

Flowers April–May.

Occurs in low woods in valleys of streams in the southwestern Ozarks, at the base of wooded limestone

959 Cabomba caroliniana (Fanwort)

960 Magnolia acuminata var. acuminata (Cucumber Tree)

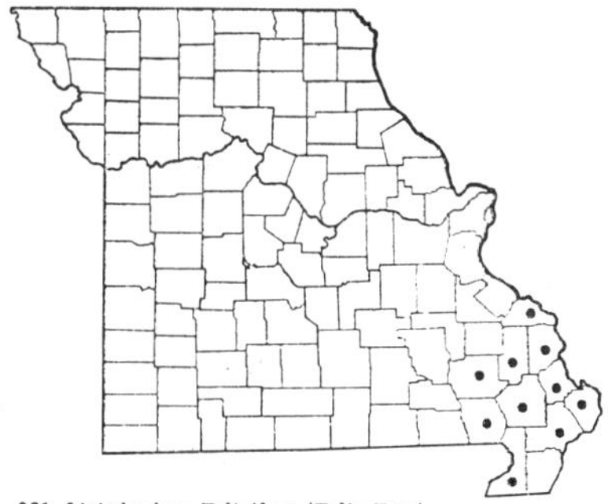

961 Liriodendron Tulipifera (Tulip Tree)

and gravel bluffs and thickets along the Mississippi River, and in gravelly or clayey soil of low wooded hills of Crowley Ridge in southeastern Missouri. Restricted to extreme southern Missouri, in two areas: (1) southeastern Missouri along Crowley Ridge in Dunklin, Butler, Stoddard, and Scott counties and northeast along the Mississippi River to Cape Girardeau County; (2) southwestern Missouri, along the White River and North Fork of White River in Ozark, Taney, and Barry counties. The latter stations have now been exterminated by the creation of the Bull Shoals and Table Rock dams, since the trees had occupied the low valleys of these areas.

Ranges from Georgia to Louisiana and Oklahoma, north to New York, Ontario, Ohio, Indiana, Illinois, and Missouri.

*Magnolia acuminata* var. *ozarkensis*, separated from typical *M. acuminata* var. *acuminata* on the basis of having the lower surface of the leaf-blades green and completely glabrous or slightly hairy on the veins instead of pale green and pubescent on the lower surface, does not appear to be sufficiently constant in these characters to merit recognition as formerly was accorded it and as later field studies seem to bear out.

The flowers, 6–8 cm. long, have six petals, of which the outer three are glaucous-green tinged with yellow, the inner three yellow in the upper half and green in the lower half; the styles are yellow and the receptacle green. The fragrance of the flowers resembles that of daffodils. In autumn the foliage takes on a greenish-yellow or dull yellow color. The seeds of the cucumberlike green cones of the tree are quite bitter and are not eaten by birds and fur-bearing animals, but the green cones were at one time eaten by early pioneers after the bitter properties had been extracted. After being mashed up and placed in whiskey, the green cones are sometimes used to make a mildly bitter medicine.

The tree is used in the lumber industry for flooring, cabinetmaking, interior finish, barrels, boxes, and crates, and in mixture with other hardwoods the wood is pulped for paper making.

In the Trail-of-Tears State Park in Cape Girardeau County there are many magnificent specimens of the tree, and it is sometimes cultivated as an ornamental shade tree because of its rapid growth. It can be grown throughout Missouri.

### *Excluded Species*

**Magnolia tripetala** L. Umbrella Tree

There are no valid records for the occurrence of this species in Missouri. It has been reported from Missouri in the eighth edition of *Gray's Manual* (p. 676), as well as in the previous edition of that manual. This report may possibly have been based upon a specimen labeled by Eggert on May 9, 1898 from 'Normandy?' recopied to read 'Normandy.' As Eggert collected *Magnolia tripetala* on the same day (May 9, 1898) in Franklin Co., Tennessee, his specimen from Normandy undoubtedly referred to Normandy in Bedford Co., Tennessee, which is close to Franklin Co. and not Normandy in St. Louis County, Missouri. It would have been possible in 1898 for Eggert to have been in Franklin and Bedford counties, Tennessee, on the same day, but the slow transportation and poor roads would have prevented his being in Franklin Co., Tennessee and St. Louis Co., Missouri on the same day. Eggert, who often collected in St. Louis County, Missouri, occasionally neglected, as have other collectors in other places, to cross out 'Missouri' from the top of the label, which he ordinarily used, when collecting in other states.

The nearest stations to Missouri for this species are in Pope County, Arkansas, about 75 miles south of the Missouri border.

### 2. **Liriodendron** L. Tulip Tree

**Liriodendron Tulipifera** L. Tulip Tree, Yellow Poplar Map 961

Also known by other names, such as Whitewood, Canoewood, White Poplar, Poplar, and Hickory Poplar.

Flowers May–June.

Occurs in rich woods of ravines, in upland woods and along streams of Crowley Ridge, and at the base of wooded bluffs along the Mississippi River. Southeastern Missouri north to Bollinger and Perry counties and west to Wayne and Butler counties. Reported from Madison, Ripley, and Oregon counties by Transeau (Ecol. 16: 430, fig. 15. July, 1935), but the records have not been substantiated by herbarium specimens. Reported also as escaped from cultivation in St. Louis County.

Ranges from Florida and Louisiana, north to Massachusetts, Vermont, New York, Ontario, Michigan, Indiana, southern Illinois, and Missouri.

In southeastern Missouri this is often associated with beech and cucumber tree, as well as with sugar maple, white oak and other species. It is a tree of the richer and well-drained soils. It is widely planted in St. Louis and other cities and towns of Missouri. The flowers, growing erect at the tip of the leafy branches, are often missed when the tree is in bloom, but are quite showy. They are slightly fragrant and the 6 petals of the tulip-shaped, cuplike flower (the state flower of Indiana) are greenish-yellow with a bright orange color inside near the base. The foliage turns yellow to orange-yellow in autumn.

The tulip tree grows taller than any other hardwood tree in North America, reaching a height of 200 feet in some areas of the eastern United States. The wood, which is one of the lightest North American woods known, and the softest of the hardwoods, is used for many purposes, including insulating material, exterior and interior finish of houses, boatbuilding, shingles, brooms, woodenware, hat blocks, crates, boxes, and millwork. Mixed with other hardwoods, it is also used as pulp for papermaking. Early pioneers also used it for canoes, furniture, and general construction. The bitter inner bark of the roots was formerly used in medicine as a tonic and stimulant, and an alkaloidal substance, extracted from the bark, was used as a heart stimulant.

## Fam. **ANNONACEAE** (Custard Apple Family)

### **ASIMINA** Adans. Pawpaw

**Asimina triloba** (L.) Dunal Pawpaw Map 962

Flowers March–May.

Occurs in low bottom woods and wooded slopes along streams, ravines, and at the base of bluffs. Throughout Missouri; absent from the extreme northern counties, and found north to Clark, Shelby, Macon, Linn, Grundy, Daviess, De Kalb, Andrew, and Atchison counties.

Ranges from Florida to Texas, north to New Jersey, New York, Ontario, Michigan, Illinois, southern Iowa, and southeastern Nebraska.

The tree sends up suckers from the roots, forming groves or thickets. The foliage remains green long after other trees have changed colors in autumn, but eventually turns greenish-yellow to yellow before falling late in autumn.

The sweet fruits, which ripen during September, long before touched by any frost in Missouri, resemble small stout bananas about 7–12.5 cm. ($2\frac{3}{4}$–5″) long. They are good to eat when just beginning to feel soft to one's touch. At first the thin covering of the fruit is pale green, but eventually turns brown to black when over-ripe. The flesh of the fruit, at first creamy white, becomes pale yellow, resembling egg-custard, and has a sweet flavor. The fruits are eaten raw or may be made into an ice cream, sherbet, pudding, pie, jelly, or combined with 'Jello,' eggs, and cornstarch for a dessert. Some people are allergic to the fruits and develop a dermatitis after handling or eating them. The trees are difficult to transplant, but are easily grown from seed. Plants, started from seed at the author's botanical preserve in northern Illinois, did not begin to flower until after twelve years' growth, but farther south in Missouri should start flowering much earlier. The large, long, drooping dark green leaves are quite ornamental in appearance.

A strong alkaloid contained in the seeds has a depressing effect on the brains of animals, although the fruit is relished by such fur-bearers as opossum, raccoon, and gray squirrels. The bark, which is difficult to tear, strips off into long pieces, and was used by some early pioneer fishermen for stringing fish.

The inner bark was woven into a cloth by Indians of Louisiana. In this connection it is interesting to note that the similar difficult-to-tear properties of the bark of some tropical Venezuelan trees of this family serve the Indians of that country as substitutes for rope and head-strap carriers or tumplines.

Fam. **RANUNCULACEAE** (Crowfoot Family)

a. Plants mainly submerged in the water with finely divided foliage . . . . . . . . . 9. RANUNCULUS
a. Plants mainly growing out of the water, or on rocks, or in soil . . . . . . . . . . . . . *b*
  b. Some or all of the leaf-blades, main leaf-stalks (petioles) of the leaves, or leaf-like bracts of the involucre opposite each other on the stem in 2's or in whorls (circles) of 3 or more . . . . . . *c*
    c. Stems climbing, vining, or sprawling . . . . . . . . . . . . . . . . 15. CLEMATIS
    c. Stems not climbing, vining, nor sprawling . . . . . . . . . . . . . . . . . *d*
      d. Leaves opposite each other on the stem in 2's . . . . . . . . . . . . . 15. CLEMATIS
      d. Leaf-like bracts of the involucre (which resemble leaves) in whorls (circles) of 3 or more . . . *e*
        e. Lobes of leaf-like bracts sharply toothed . . . . . . . . . . . . . 12. ANEMONE
        e. Lobes of leaf-like bracts rounded . . . . . . . . . . . . . . . 14. ANEMONELLA
  b. Leaf-blades or the main leaf-stalks (petioles) of the leaves alternate on the stem, or arising from the base of the plant, or both; herbaceous plants . . . . . . . . . . . . . . . . . *f*
    f. All the leaves grass-like, linear-spatulate, and arising at the base of the plant; flowers greenish, on a long and slender mousetail-shaped column (receptacle) 10–15 or more times as long as broad . . . . . . . . . . . . . . . . . . . . . . . . 11. MYOSURUS
    f. Plants without the above combination of characters . . . . . . . . . . . . . . . *g*
      g. Flowers with 1 or more spurs (slender tubes or extensions projecting) . . . . . . . . *h*
        h. Flowers both red or pink and yellow (rarely completely pale yellow), with 5 spurs; leaves ternately compound, the basal ones with usually 3 long-stalked compound divisions . 6. AQUILEGIA
        h. Flowers usually blue, purple, white, or mixtures of these, sometimes all pink or rose, with only 1 spur; leaves palmately divided or cut . . . . . . . . . . 7. DELPHINIUM
      g. Flowers lacking spurs . . . . . . . . . . . . . . . . . . . . . . . . *i*
        i. Flowers yellow . . . . . . . . . . . . . . . . . . . . . . . . . *j*
          j. Petals not bearing a nectar-pit or scale at the base; petals 6–9; flower solitary, 1 on a stem; leaves dissected into slender segments, sessile on an erect terrestrial stem; pistils hairy . . . . . . . . . . . . . . . . . . . . . . . 10. ADONIS
          j. Petals bearing a nectar-pit or scale at the base; petals mostly 5, usually smooth-edged; flowers 1 or more; leaves, when finely dissected, occurring in the water; pistils usually glabrous . . . . . . . . . . . . . . . . . . . . . . 9. RANUNCULUS
        i. Flowers of other colors, not yellow, or yellow with purple at base . . . . . . . . *k*
          k. Leaves lobed only, not split or divided to the base, nor compound . . . . . . . *l*
            l. Leaves with 3 (rarely 4) smooth-edged (entire) lobes; leaves all arising at the base of the plant . . . . . . . . . . . . . . . . . . . . . 13. HEPATICA
            l. Leaves with 5–11 toothed lobes; leaves on the stem as well as at the base of plant. . . *m*
              m. Leaves 2 on the stem, near the summit; flower 1 on a plant; fruit red, fleshy; rootstock yellow . . . . . . . . . . . . . . . . . 1. HYDRASTIS
              m. Leaves 3 or more on the stem, scattered; flowers several or many on a plant; fruit green to brown, dry; rootstock not yellow . . . . . . 5. TRAUTVETTERIA
          k. Leaves split or divided to the base, or compound . . . . . . . . . . . . *n*
            n. Petals pale yellow with purple at base; leaves scattered and alternate on stem, dissected into numerous narrowly linear segments . . . . . . . . . 10. ADONIS
            n. Flowers white, pink, rose, lavender, purplish, or greenish; leaves, if finely dissected, arising mainly from the base of the plant . . . . . . . . . . . . . . *o*
              o. Petals or petal-like parts less than 5 mm. long; flowers many, crowded on a simple or branching axis . . . . . . . . . . . . . . . . . . *p*
                p. Flowers perfect (both stamens and pistils in the same flower), decidedly white, in close clusters; leaflets sharply toothed . . . . . . . . . . . *q*

Plate no. 163. **1.** Magnolia acuminata, × 2/7; *a.* Fruit, × 2/7. **2.** Liriodendron Tulipifera, × 2/7; *a.* Fruit, × 2/7; *b.* Terminal bud covered by stipules, × 2/7. **3.** Asimina triloba, × 2/7. **4.** Hydrastis canadensis, × 2/7. **5.** Cimicifuga racemosa var. racemosa, × 2/7. Details from Small, The New York Botanical Garden.

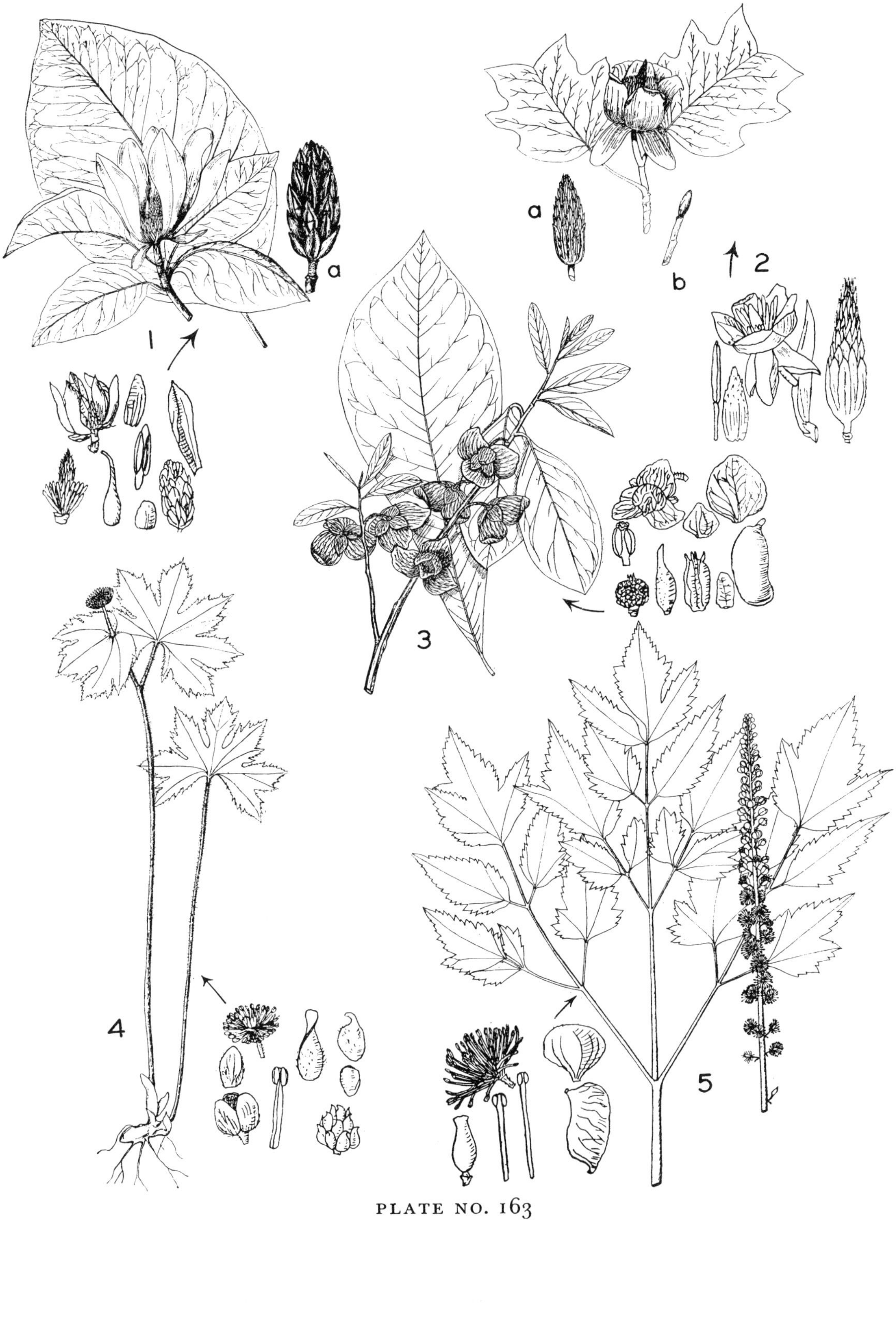


PLATE NO. 163

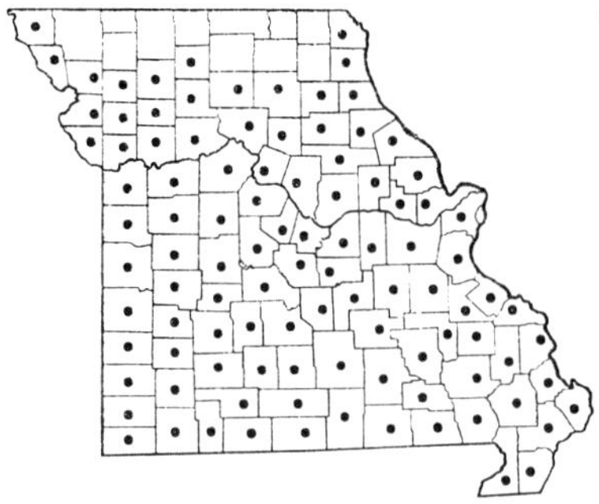

962 Asimina triloba (Pawpaw)

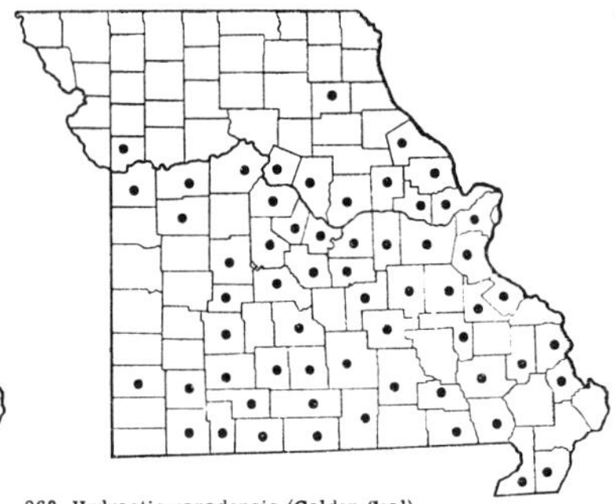

963 Hydrastis canadensis (Golden Seal)

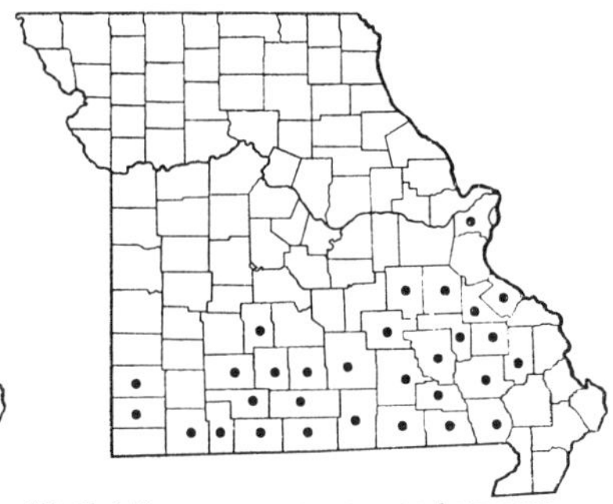

964 Cimicifuga racemosa var. racemosa f. racemosa (Black Cohosh)

q. Flowers in a very long, narrow cylindrical spray 15–90 cm. long; plants 1–2.6 m. tall; fruit dry greenish-brown follicles . . . . . . 2. CIMICIFUGA

q. Flowers in a short dense ovate cluster 3–8 cm. long; plants 0.4–0.8 m. tall; fruit white, fleshy, berry-like . . . . . . . . . . . 3. ACTAEA

p. Flowers unisexual (either male or female), greenish, whitish or purplish, in loose scattered clusters; leaflets with rounded or blunt teeth . . . 4. THALICTRUM

o. Petal-like parts always obvious, 5 mm. or more long; flowers solitary, 1 or few on the stem . . . . . . . . . . . . . . . . . . . . . *r*

r. Lobes of leaflets or leaf-like bracts of involucre sharply toothed; leaf-like bracts of involucre opposite or in a whorl (circle) . . . . . . 12. ANEMONE

r. Lobes of leaflets or leaf-like bracts of involucre rounded; leaves alternate or leaf-like bracts of involucre in a whorl . . . . . . . . . . . . *s*

s. Leaves on the stem alternate; roots string-like (sometimes with tuber-like thickenings); plant of low or moist ground . . . . . . . 8. ISOPYRUM

s. Leaf-like bracts of involucre which resemble the stem-leaves arranged in a whorl (circle); roots a cluster of thickened tubers; plants of dry, rocky, or upland woods . . . . . . . . . . . 14. ANEMONELLA

## 1. **Hydrastis** Ellis Golden Seal

**Hydrastis canadensis** L. Golden Seal Map 963

Flowers April–May.

Occurs in rich woods, on slopes, in ravines, and on the floor of valley woods and thickets. Not uncommon throughout the Ozark region and central Missouri north to Pike, Shelby, Boone, Howard, Saline, Lafayette, and Clay counties; mostly absent from the unglaciated prairie region of southwestern Missouri.

Ranges from Vermont to Minnesota and Nebraska, south to Georgia, Alabama, Arkansas, Kansas, and Oklahoma.

The roots and sometimes the leaves are used in medicine. The plant is sometimes encountered in large colonies or as scattered clumps. The red berries are very attractive against the leafy background.

## 2. **Cimicifuga** L.

**Cimicifuga racemosa** (L.) Nutt. var. **racemosa** f. **racemosa** Black Cohosh, Black Snakeroot Map 964

*Cimicifuga racemosa* (L.) Nutt. [G, P & S, Steyerm.]

Flowers late May–August.

Occurs in rocky woods and along the base of bluffs. Restricted to the Ozark region of southern and east-central Missouri north to St. Louis, Washington, Crawford, Texas, Dallas, Greene, and Jasper counties.

Ranges from Massachusetts to New York to Ontario, Ohio, and Indiana, south to Georgia, Tennessee, and Missouri.

The roots have been used in medicine.

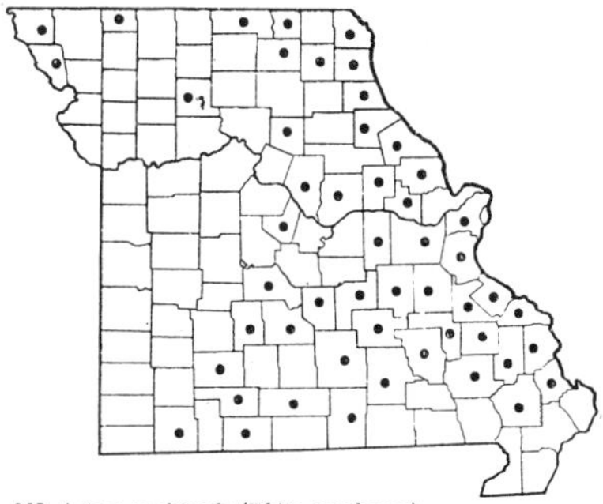

965 Actaea pachypoda (White Baneberry)

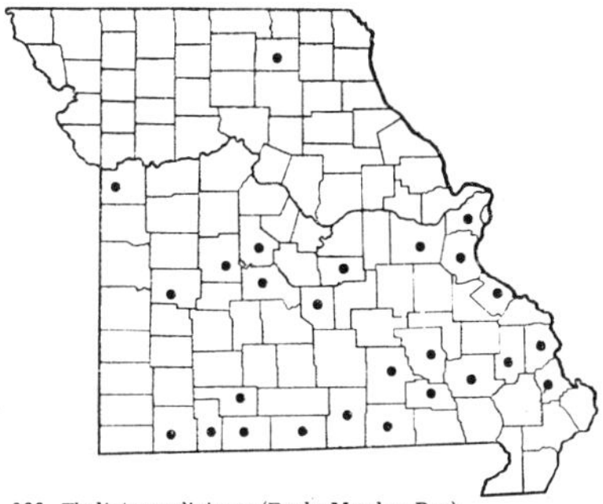

966 Thalictrum dioicum (Early Meadow Rue)

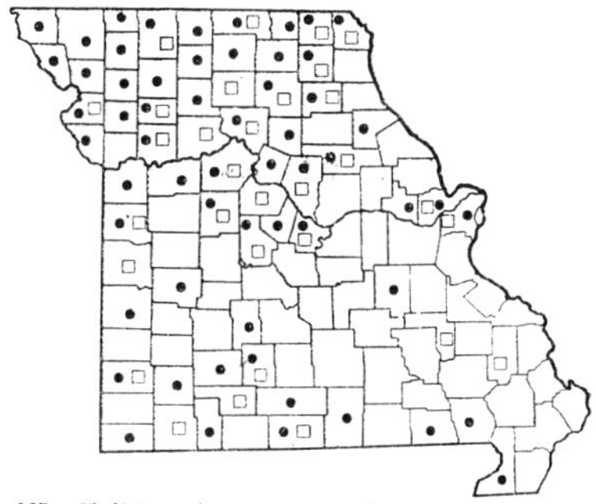

967 ● Thalictrum dasycarpum var. dasycarpum
967 □ Thalictrum dasycarpum var. hypoglaucum

## 3. **Actaea** L. Baneberry

**Actaea pachypoda** Ell. White Baneberry
Map 965

Also known as Doll's Eyes.

*Actaea brachypoda* Ell. [P & S]

*Actaea alba* (L.) Mill. [BB]

Flowers May–June.

Occurs in deep rich woods and ravines, often on north-facing slopes and at the base of bluffs. Eastern half of Missouri west in the Ozark region to Moniteau, Camden, Dallas, Greene, Christian, and Barry counties, and in northern Missouri west to Livingston, Holt, and Atchison counties.

Ranges from Quebec to Ontario, Minnesota, and Manitoba, south to Georgia, Alabama, Louisiana, and Oklahoma.

The rhizomes of this species have medicinal properties. The fruit of the related *A. rubra* (Ait.) Willd. is poisonous, while the rootstock of that species has strong purgative and emetic properties. Cases of poisoning to children have been ascribed to a related European species, *A. spicata*, from eating the fruit, and this species is reported as having injurious effects on the heart.

## 4. **Thalictrum** L. Meadow Rue

a. Middle and upper stem-leaves with a long petiole (main leaf-stalk) attaching the compound leaf to the main stem; flowering in April and early May; plants generally 3–6 dm. tall; leaflets (leaf-divisions) thin, usually not longer than broad . . . . . . . . . . . . . . . . 1. T. DIOICUM

a. Middle and upper stem-leaves sessile, the three main divisions of the compound leaf directly attached to the stem without a common petiole; flowering after mid-May; plants generally 6–20 dm. tall; leaflets (leaf-divisions) thicker, usually averaging longer than broad . . . . . . . . . . . *b*

- b. Lower leaf-surface or veins or pistils more or less dotted with minute, round sessile glands or gland-tipped, short-stalked hairs . . . . . . . . . . . . . 3a. T. REVOLUTUM f. REVOLUTUM
- b. Lower leaf-surface or veins or pistils either glabrous (without hairs) or with fine hairs lacking glands on tip . . . . . . . . . . . . . . . . . . . . . . . . . . . . *c*
  - c. Lower leaf-surface with fine hairs . . . . . . . . . 2a. T. DASYCARPUM var. DASYCARPUM
  - c. Lower leaf-surface glabrous (without hairs) . . . . . . . . . . . . . . . *d*
    - d. Leaf-divisions rather thick, with revolute (rolled under) margins, the veins on lower surface prominent and forming a conspicuous network. . . . . . . . 3b. T. REVOLUTUM f. GLABRUM
    - d. Leaf-divisions comparatively thin, with flat margins, the veins on lower surface not prominent nor forming a conspicuous network . . . . . . 2b. T. DASYCARPUM var. HYPOGLAUCUM

1. **Thalictrum dioicum** L. Early Meadow Rue
Map 966

Also called Quicksilver Weed.

Flowers early April–early May.

Occurs usually on rich north-facing wooded slopes and ledges, mostly in the vicinity of limestone rock outcrops. Mainly in the Ozark section of southern and central Missouri north to St. Louis, Franklin, Maries, Morgan, Benton, and St. Clair counties, north locally in Jackson and Adair counties.

Ranges from Georgia, Kentucky, Alabama, and Missouri, north to Quebec, Ontario, Minnesota,

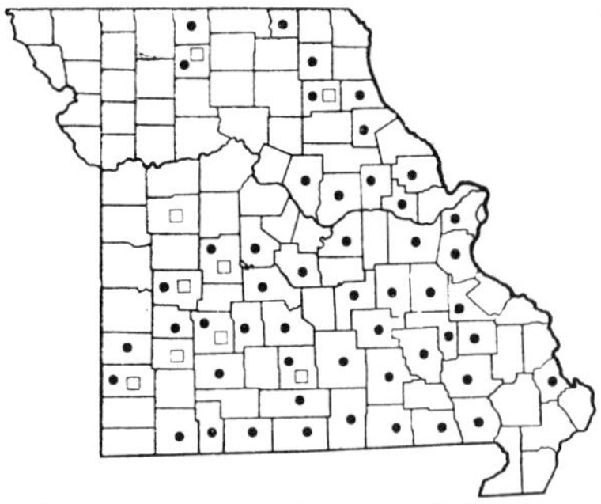

968 • Thalictrum revolutum f. revolutum (Wax-leaved Meadow Rue)
968 □ Thalictrum revolutum f. glabrum

969 Trautvetteria caroliniensis (False Bugbane)

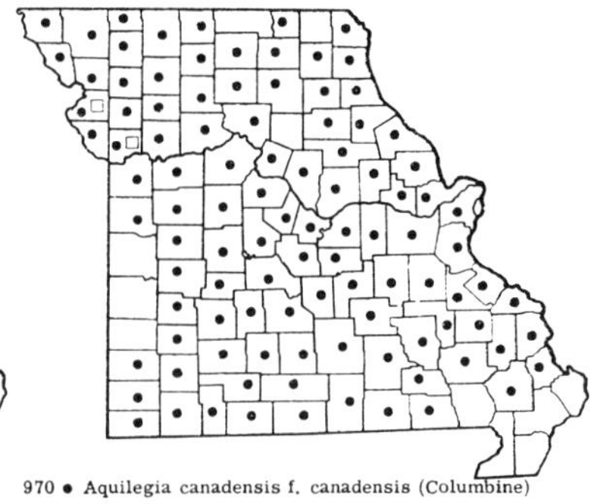

970 • Aquilegia canadensis f. canadensis (Columbine)
970 □ Aquilegia canadensis f. flaviflora

North Dakota, and Manitoba.

The gray-green fernlike foliage is quite decorative. As it persists during the often dry summer and autumn months, it makes a good ground cover or permanent green foliage understory in the shaded wildflower garden.

2a. **Thalictrum dasycarpum** Fisch. & Lall. var. **dasycarpum** Purple Meadow Rue Map 967
*Thalictrum dasycarpum* Fisch. & Lall. [G, BB, P & S, Steyerm.]
Flowers late May–July.

Occurs in low damp ground in wooded valleys and ravines, moist banks of streams, and moist thickets. Throughout Missouri, where most abundant in the northern, central, and western parts; rare or absent from most of the Ozark section.

Ranges from Ontario to Manitoba and Alberta, south to Ohio, Indiana, Illinois, Missouri, Kansas, Oklahoma, New Mexico, and Arizona.

2b. **Thalictrum dasycarpum** var. **hypoglaucum** (Rydb.) Boivin Map 967
*Thalictrum dasycarpum* f. *hypoglaucum* (Rydb.) Steyerm. [Steyerm.]

Occurs in similar situations and with a similar geographical range within the state as var. *dasycarpum*, but somewhat less common.

In addition to usually thinner leaves, var. *hypoglaucum* has generally longer filaments (4–7 mm. instead of about 4 mm. long) and longer anthers (2.2–3.2 mm. instead of 1.5–2.5 mm. long) than var. *dasycarpum*.

Ranges from Minnesota and South Dakota to British Columbia, south to Louisiana, Texas, and Arizona.

3a. **Thalictrum revolutum** DC. f. **revolutum**
Wax-leaved Meadow Rue, Waxy Meadow Rue Map 968
Also known as Purple Meadow Rue.
*Thalictrum revolutum* DC. [G, BB, P & S, Steyerm.]

Occurs in prairies and open ground along roadsides, thickets and openings in woodland. Mostly in the Ozark section of southern and central Missouri and in the eastern half of northern Missouri west to Mercer and Grundy counties.

Ranges from Massachusetts to Ontario, south to Florida, Alabama, Tennessee, and Missouri.

3b. **Thalictrum revolutum** f. **glabrum** Pennell Map 968

Scattered in the range of f. *revolutum* and less common.

## 5. **Trautvetteria** Fisch. & Mey. False Bugbane

**Trautvetteria caroliniensis** (Walt.) Vail
False Bugbane Map 969
Flowers around June 1–August.

Occurs in moist, shaded crevices of north-facing limestone bluffs along streams. Known only from bluffs along Jack's Fork and Current River in Shan-

Plate no. 164. **1.** Actaea pachypoda, × 2/7; *a.* Inflorescence; *b.* Fruiting branch; Details from Small, The New York Botanical Garden. **2.** Trautvetteria caroliniensis, × 2/7. **3.** Thalictrum dioicum, × 2/7. **4.** Thalictrum dasycarpum, × 2/7; *a.* Male plant; *b.* Female plant; Details from Small, The New York Botanical Garden. **5.** Aquilegia canadensis, × 2/7; Details from Small, The New York Botanical Garden.

PLATE NO. 164

non County, southern Missouri.

Ranges from Pennsylvania to Kentucky and Georgia; southern Indiana, western Illinois, and southern Missouri.

This is one of the rarest wildflowers of Missouri. It is found at several stations, however, at each of which it is locally somewhat plentiful. One is at Jam-Up Bluff along Jack's Fork of Current River (July 17, 1932, *Steyermark 7119*; July 27, 1898, *Bush 178*), the others along Current River near and north of Eminence. At these stations it is associated with *Parnassia grandifolia*, *Adiantum Capillus-Veneris*, and sometimes *Campanula rotundifolia*, *Galium boreale* var. *hyssopifolium*, and *Zigadenus elegans*. Its corymbiform clusters of white flowers are quite showy. This species has prospered in the author's wildflower garden in northern Illinois during the past fifteen years, where it was started from seed obtained from Missouri specimens. It thrives there in a calcareous rich soil in low, slightly damp woodland.

## 6. **Aquilegia** L. Columbine

**Aquilegia canadensis** L. Columbine Map 970

Called Wild Honeysuckle in some sections of the state.

Flowers April–July.

Occurs in the Ozark region usually on limestone bluffs and ledges, elsewhere in rich woods or on steep wooded slopes, occasionally along railroads, and openings in logged areas. Throughout Missouri, where found in all but a few of the extreme southeastern lowland counties and in some of the southwestern counties by the Kansas line.

Ranges from Nova Scotia to Saskatchewan, south to Florida and Texas.

Missouri material may be divided into the following two variations:

Flowers yellow with red or pink . . . . . 6a. A. CANADENSIS f. CANADENSIS

Flowers entirely yellow or pale yellow . . . 6b. A. CANADENSIS f. FLAVIFLORA

6a. **Aquilegia canadensis** L. f. **canadensis** Map 970

*Aquilegia canadensis* L. [G, BB, P & S, Steyerm.]

*Aquilegia canadensis* var. *coccinea* (Small) Munz [G]

*Aquilegia canadensis* var. *latiuscula* (Greene) Munz [G]

Throughout Missouri.

6b. **Aquilegia canadensis** f. **flaviflora** (Tenney) Britt. Map 970

*Aquilegia canadensis* var. *flaviflora* (Tenney) Britt. [P & S]

Known only from Buchanan (*Broadhead*) and Clay (wooded ravine slopes with limestone outcrops along Dry Creek, Lautenschlager's Farm, T51N, R31W, sect. 13, 3–3½ mi. west-southwest of Missouri City, May 17, 1949, *Steyermark 67740*) counties in northwestern and west-central Missouri. At the Clay County station five plants were seen. One of them was transplanted to the author's wildflower garden in northern Illinois where it is still growing. Seed from another plant produced a lone one of the resulting progeny with all the flowers pale yellow, while the others had normal red-and-yellow flowers. The lone descendant with the pale yellow flowers comes up each year true to color and has paler green leaves than the normal-colored type, as well as having a yellow-green or pale green stem in contrast to a purplish-red stem of the normal-colored plants.

The forms having the flowers salmon-pink, f. *Phippenii* (J. Robins.) R. Hoffm., or white, f. *albiflora* House, have not yet been recorded from Missouri.

This species was divided by Dr. Munz (Gent. Herb. 7: 116–23. 1946) into several varieties based upon variation in size and shape of sepals, length of spurs, and the extent of lobing of the leaflets and degree of division of the basal leaves. According to his interpretation and that of Fernald (eighth edition of *Gray's Manual*) who followed, var. *coccinea* and var. *latiuscula* are the variations found in Missouri, with apparently var. *coccinea* the commoner one encountered having longer sepals 15–20 mm. and oblong-ovate, instead of 10–14 mm. long and ovate or ovate-lanceolate, and with stout instead of slender spurs. Study of Missouri material over a period of years has forced the present author to conclude that the differences suggested by Dr. Munz are not sufficiently correlated or distinct to merit their recognition, and for this reason they are eliminated from consideration in the present work. This is the conclusion, likewise, reached by Dr. Gleason (*New Ill. Fl.* 2: 164. 1952). Plants growing on limestone bluffs and in their crevices are usually shorter-stemmed than those growing away from such habitats, and in some of the habitats in northern Missouri where plants are often found in soil free from rocks the stems are two to three times as tall with a more luxurious appearance. Correlations in sepal, spur, and leaf variations do not appear, however, to be constant nor clearly defined.

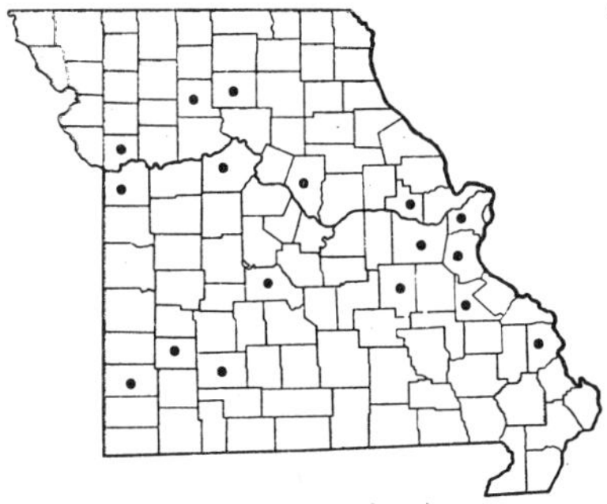

971 Delphinium Ajacis (Rocket Larkspur)

972 Delphinium tricorne f. tricorne (Dwarf Larkspur)

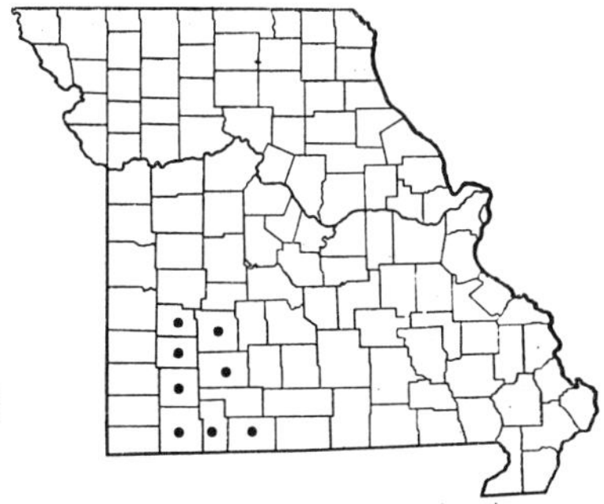

973 Delphinium Treleasei (Trelease's Larkspur)

## 7. **Delphinium** L. Larkspur

a. Pistil or fruit (follicle) 1 to each flower; annuals escaped from cultivation. . . . . . 1. D. AJACIS
a. Pistils or fruits (follicles) 3 or 5 to each flower; native perennials . . . . . . . . . . . . *b*
  b. Follicles (fruits) spreading at maturity; stems averaging shorter than the other species, mainly 2–3.5 dm. tall (up to 6–9 dm.); roots short, thickened, tuber-like, fleshy; inflorescence mainly 4–10 cm. long (up to 20 cm.); lower petals (do not confuse with the larger colored petal-like sepals) split into 2; seeds smooth and unwrinkled. . . . . . . . . . . . . 2. D. TRICORNE
  b. Follicles (fruits) erect and parallel to one another at maturity; stems tall, mainly 7–20 dm. tall (and as short as 5 dm.); roots elongate, not conspicuously thickened, rather firm; inflorescence mainly 10–30 cm. long; lower petals (do not confuse with the larger colored petal-like sepals) not split into 2; seeds rough or covered with scales or wrinkled even when otherwise smooth. . . . . *c*
    c. Stem glabrous (without hairs) below the inflorescence; seeds not covered with scales, wrinkled but otherwise smooth; rare species found only in southwestern or southern Missouri . . . . . *d*
      d. Spring-flowering (May–June); flowers deep blue or blue-purple; leaf-divisions narrow and linear, 3–8 mm. broad; lower stalks (pedicels) of flowers or fruits mainly 2.5–8 cm. long . 3. D. TRELEASEI
      d. Summer-flowering (July–August); flowers purple, pale blue or whitish with lavender; leaf-divisions broad, lanceolate or wedge-shaped, mostly 10–25 mm. broad; lower stalks (pedicels) of flowers or fruits chiefly 1–2.5 cm. long . . . . . . . . . . 4. D. EXALTATUM
    c. Stem hairy below the inflorescence; seeds covered with scales, roughened; common species throughout the Ozarks and northern and central Missouri . . . . . . . . . . . . . *e*
      e. Flowers mainly a deep blue or violet color, or mixed blue, lavender, and lilac; seeds covered with appressed scales, slightly winged on the margins; plants chiefly of limestone glades or rocky lime or chert openings of the Ozark and Ozark border region (apparently hybridizes with the following species in paler flowers with more white predominating) . . . . . . . . . . . . . . . . . . . . . . 5. D. CAROLINIANUM
      e. Flowers mainly white, greenish-white, or mixtures of white with lilac or blue; seeds densely covered with projecting scales, not winged on margins; plants chiefly of prairies, railroad tracks, rocky openings, and open places in northern and western Missouri in the glaciated and unglaciated prairie sections of the state (apparently hybridizes with the preceding species with bluer or more lavender-tinged plants) . . . . . . . . . . . . 6. D. VIRESCENS

1. **Delphinium Ajacis** L. Rocket Larkspur
Map 971

Flowers May–September.

Commonly planted as a garden flower and escaped from cultivation along streets in towns, along roadsides and railroad tracks, fields, farmlots, and waste ground. Scattered throughout Missouri.

Native of Europe; introduced and naturalized in North America from Nova Scotia to Minnesota and Montana, south to South Carolina and Texas.

Flowers of this species have blue, purple, lavender rose, pink, or white flowers.

2. **Delphinium tricorne** Michx. f. **tricorne**
Dwarf Larkspur Map 972

Flowers April–early June.

Occurs in rich or rocky woods, on slopes, valley and ravine bottoms, ledges along bluffs, and banks of streams. Throughout Missouri; absent from the lowland counties of extreme southeastern Missouri, but to be expected otherwise in every county.

Ranges from Pennsylvania to Minnesota and Nebraska, south to Georgia, Alabama, Arkansas, and Oklahoma.

The flowers are mainly blue or violet or shades of these mixed with white in typical *D. tricorne* f. *tricorne*. In f. *albiflora* Millsp. they are white throughout. The closest approach to an all-white corolla is the collection of *Drouet 2854* from Boone County (along lower Brushy Creek, Federal Wildlife area, east of Ashland, May 4, 1936 in Univ. Mo. herb.). However, since the collection has the note 'almost white corolla form,' it cannot be considered typical f. *albiflora*.

Plants of this species are poisonous to stock, and are probably the commonest cause of cattle poisoning in early spring when the green plants are grazed.

3. **Delphinium Treleasei** Bush Trelease's Larkspur Map 973

Flowers May–June.

Occurs on limestone glades and bald knobs in the White River region and on rocky open limestone exposures and glades elsewhere. Known only from southwestern Missouri circumscribed in a small area including Polk, Cedar, Dade, Greene, Lawrence, Taney, Stone, and Barry counties.

Known only from Missouri; probably in adjacent Arkansas.

This is a handsomely colored plant with deep blue or blue-purple flowers. There is some variation in size of flowers (2.5–3.7 cm. long), in the amount of yellow and brown spotting on the sepals, and the amount of yellow beard present on the petals.

The species has done well far north of its known natural range in the author's wildflower garden in northern Illinois. As long as it is provided with a sunny, open, well-drained, preferably limestone rocky soil free from competition with grasses, there should be no difficulty in getting this attractive species well established.

4. **Delphinium exaltatum** Ait. Tall Larkspur Map 974

Flowers July–August.

Occurs on limestone or cherty limestone wooded, usually steep slopes of north- or west-facing exposures of hills bordering Current River, Jack's Fork of Current River and tributary streams in Shannon County and South Fork of Spring River in Howell County. Known only from southern Missouri in Shannon (dry rocky open hillside above Current River, 0.5 mi. below Round Spring, August 8, 1942, *Hubricht B 2266*; $\frac{1}{2}$ way up west-facing wooded limestone slopes just bordering and below limestone glade along highway 19, T30N, R4W, SW $\frac{1}{4}$ sect. 20, 0.5 mi. southwest of Round Spring State Park, above valley of Spring Valley, July 24, 1955, *Steyermark 78915*; $\frac{1}{4}$–$\frac{1}{2}$ way up open west-facing wooded limestone slopes, along Jack's Fork of Current River along south and east side of river, T28N, R5W, sect. 3, 8$\frac{1}{2}$–9 mi. [by air] southwest of Eminence, July 25, 1955, *Steyermark 78921*) and Howell ($\frac{1}{2}$–$\frac{2}{3}$ way up eroded limestone to cherty limestone steep wooded slopes along South Fork of Spring River, T22N, R8W, sect. 15, 10 mi. [by air] south of West Plains, 5 mi. [by air] northwest of Lanton, June 26, 1955, *Steyermark 78760* [vegetative condition only]) counties.

Ranges from Pennsylvania to North Carolina and Alabama, west to Ohio, Tennessee, and Missouri.

At all the above-cited stations the plants were locally frequent to common. At one of the Shannon County stations on west-facing, rather dry slopes, the species was associated with *Agrimonia rostellata*, *Polygala Senega* var. *latifolia*, *Silphium Gatesii*, and *Liatris squarrosa*. At the Howell County station on north-facing slopes it was growing with *Viburnum Rafinesquianum* var. *affine*, *Dirca palustris*, *Solidago arguta*, *Gerardia flava* var. *macrantha*, *Veronicastrum virginicum* f. *villosum*, *Trillium recurvatum*, *Panicum Boscii*, and *Asclepias quadrifolia*.

Plants grown at the author's botanical preserve in northern Illinois from seed collected in southern Ohio have done well in a wooded situation during the past ten years, and have spread through seeding. In such shaded situations the corolla becomes paler lavender.

There is a specimen in the Drury College Herbarium from Bethesda, Greene County, collected May 19, 1887, by J. W. Blankinship, which may be this species, but the early date of flowering suggests a confused label, as the species does not flower until July.

5. **Delphinium carolinianum** Walt. Carolina Larkspur Map 975

Also called Wild Blue Larkspur.

Flowers May–June.

Occurs on usually limestone and cherty limestone glades, prairies, and rocky openings in woods, also along railroads in prairie habitats.

This species is quite variable in color of flower and pubescence of stem, as well as vigor of plant, and size

Plate no. 165. **1.** Delphinium Ajacis, × 2/5. **2.** Delphinium Treleasei, × 2/5. **3.** Delphinium tricorne, × 2/5. **4.** Delphinium exaltatum, × 2/5.

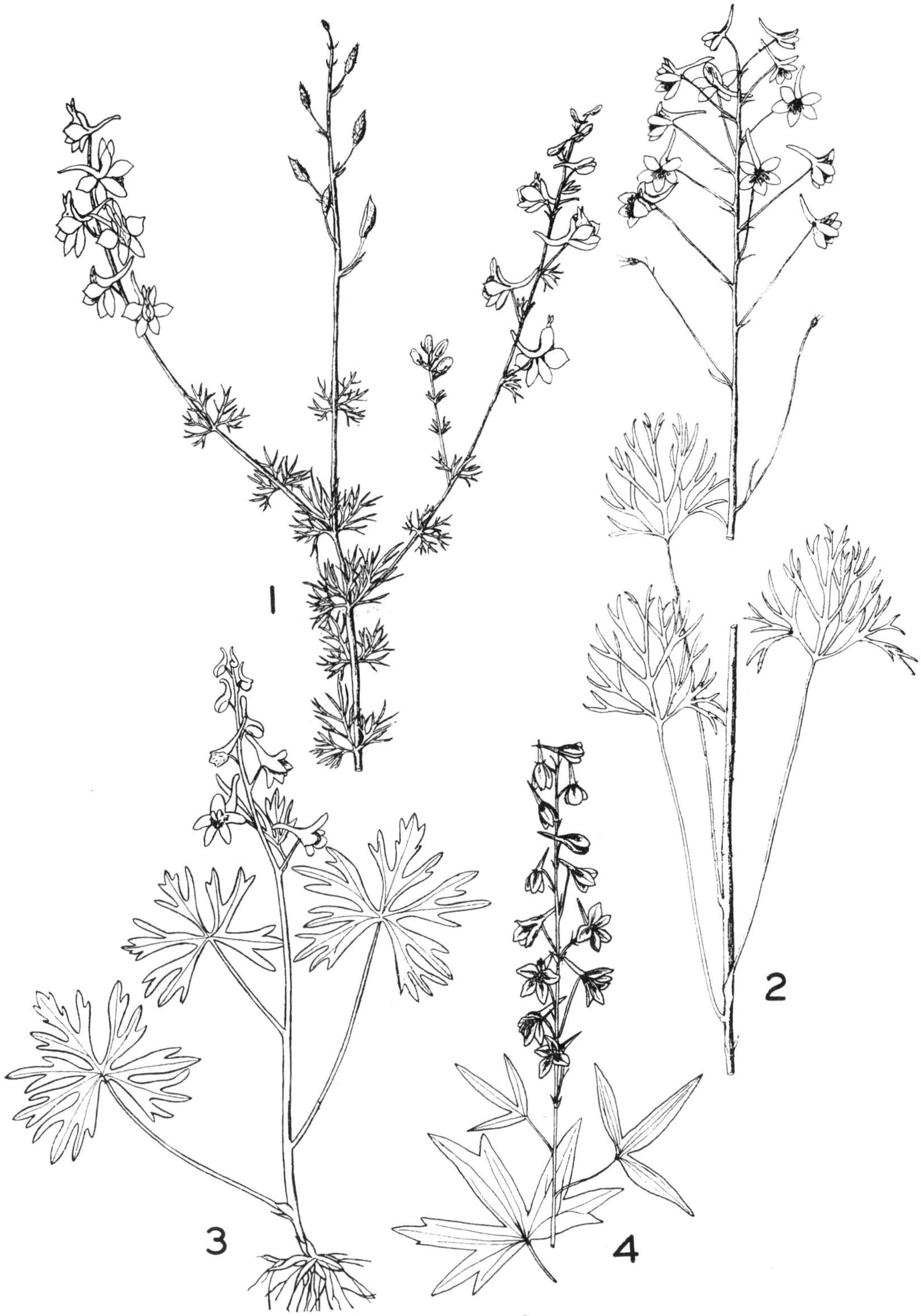


PLATE NO. 165

974 Delphinium exaltatum (Tall Larkspur)

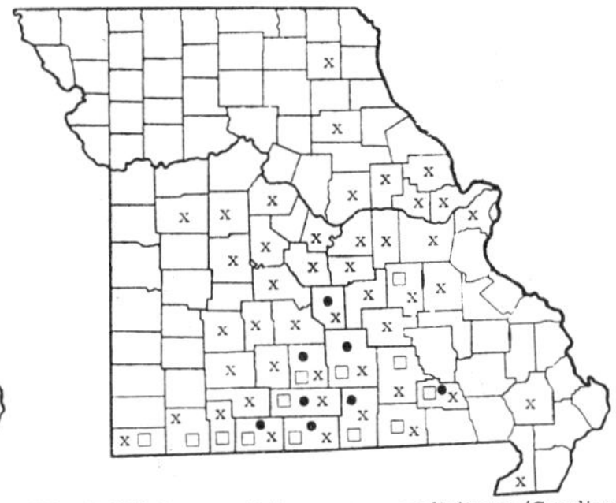

975 • Delphinium carolinianum var. carolinianum (Carolina Larkspur)
975 □ Delphinium carolinianum var. Nortonianum
975 x Delphinium carolinianum var. crispum

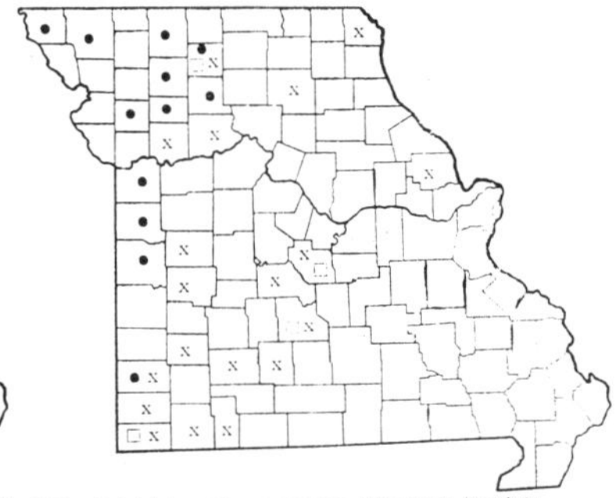

976 • Delphinium virescens var. virescens (Prairie Larkspur)
976 □ Delphinium virescens var. Penardi
976 x possible hybrids between Delphinium virescens and D. carolinianum var. crispum

of leaves. The following variations may be recognized in Missouri:

a. None of the hairs on stem with glands at their tips . . . 5c. D. CAROLINIANUM var. CRISPUM
a. Some of the hairs on upper part of stem with glands at their tips . . . . . . . . *b*
  b. Hairs on stem rather short, not conspicuous 5a. D. CAROLINIANUM var. CAROLINIANUM
  b. Hairs on stem long, conspicuous, bunching together in soft tufts . . 5b. D. CAROLINIANUM var. NORTONIANUM

5a. **Delphinium carolinianum** Walt. var. **carolinianum** Map 975

*Delphinium carolinianum* Walt. [G, BB, Steyerm.]

*Delphinium azureum* Michx. [P & S]

Scattered in the Ozark section of the state.

Ranges from Florida to Texas, north to Virginia, Missouri, Oklahoma.

5b. **Delphinium carolinianum** var. **Nortonianum** (Mackenz. & Bush) Perry Map 975

*Delphinium azureum* var. *Nortonianum* (Mack. & Bush) Palmer & Steyerm. [P & S]

*Delphinium Nortonianum* Mack. & Bush

Known from several counties in the Ozark section, but uncommon. Usually in more acid soils of chert in rocky open woods or bordering oak-hickory or pine-oak woods, also in prairies along railroads, less common on glades.

Known only from Missouri, Arkansas, and Oklahoma.

5c. **Delphinium carolinianum** var. **crispum** Perry Map 975

Occurs usually on limestone glades, rocky openings, and prairies, and along railroads. This is the commonest of the variations in Missouri and is found throughout the Ozark region for the most part, and the region bordering it on the north to Lincoln, Montgomery, Callaway, Cooper, Pettis, and Johnson counties, locally northward in northeastern Missouri to Monroe and Knox counties.

Ranges from Ohio to Illinois and Missouri, south to Arkansas and Oklahoma.

Ewan (Univ. Colo. Stud. 2: 55–244. 1945) in his revision of *Delphinium* combines the above variations of *D. carolinianum,* believing that the varying types of pubescence and glandularity do not merit recognition. His view may eventually have to be followed, but, pending detailed field and experimental investigations of the variations involved, the present treatment recognizes such variation. Plants lacking glandular hairs and having pale blue or whitish mixtures show intermediate combinations between this and the next species and may represent hybrids.

6. **Delphinium virescens** Nutt. Prairie Larkspur Map 976

Flowers May–July.

Occurs in prairies, along railroad tracks, and sometimes on rocky exposures and openings in woods.

Two variations are recognized in this species:

Only the upper part of stem with gland-tipped hairs . . . . . 6a. D. VIRESCENS var. VIRESCENS
Only the lower part of stem with gland-tipped hairs . . . . . 6b. D. VIRESCENS var. PENARDI

6a. **Delphinium virescens** var. **virescens** Map 976

*Delphinium virescens* Nutt. [G, BB, Steyerm.]

*Delphinium Penardi* in part [of P & S], not Huth

This is the more common variation of the species in the state and is found in western Missouri south to Jasper County east to Harrison, Grundy, and Livingston counties.

Ranges from Wisconsin to Manitoba, south to Missouri, Oklahoma, and Texas.

6b. **Delphinium virescens** var. **Penardi** (Huth) Perry Map 976

*Delphinium Penardi* Huth [P & S in part]

Ranges from Nebraska and Colorado, south to Missouri, Oklahoma, and Texas.

Of var. *Penardi*, if recognizable, only a few specimens have been seen of characteristic material from Missouri. Missouri specimens at the Gray Herbarium labeled var. *Penardi* (*Bush 5744* and *5744A* from Noel, McDonald County, and *Steyermark 19225* from Miller County) do not show any glandular hairs, and should be referred to the glandless type referred to below. In the Bush specimens cited, the hairs on the lower part of the stem are longer and more yellow, but not glandular, while in the Steyermark material the stem is crisp-pubescent with short hairs, no glands, and a pale bluish flower.

Ewan (above) combines the variety *Penardi* with variety *virescens*, recognizing the one species as variable in glandularity on the respective upper and lower parts of the stem. Fassett, however, finds that all the Wisconsin material belongs to var. *virescens* (Trans. Wisc. Acad. Sci. 38: 197. 1946), but states that the species is very variable where its range overlaps that of *D. caroliniana*, as noted by Steyermark in the *Spring Flora of Missouri*, p. 203. Throughout the range inhabited in Missouri by *D. virescens* and often more common than either glandular variation of *D. virescens* occur plants having the whitish flowers of *D. virescens* but the hairs of the stem throughout glandless. These may be of hybrid origin between *D. carolinianum* var. *crispum* and *D. virescens*, since the plants are most frequent in the range of overlap between *D. virescens* and *D. carolinianum* var. *crispum*.

The length of the spur, sometimes used as a key character for separating *D. carolinianum* (spur 1½ times length of upper sepal) from *D. virescens* (spur 1½–2 times length of upper sepal) has not been found a reliable one, and has therefore not been used. Actually, the two taxa, *D. carolinianum* and *D. virescens* are sometimes so difficult to distinguish, especially where they occur in contact at the margins of their ranges, as to appear as species of questionable status. Future studies may show that only one variable species is involved rather than two.

## 8. **Isopyrum** L. False Rue Anemone

**Isopyrum biternatum** (Raf.) T. & G. False Rue Anemone Map 977

Flowers March–May.

Occurs in moist, alluvial, or low rich woods of valleys, ravine bottoms, and river flood plains. Throughout Missouri, except absent from the lowland counties of extreme southeastern Missouri.

Ranges from Ontario to Minnesota, south to Florida, Alabama, Arkansas, Oklahoma, and Texas.

## 9. **Ranunculus** L. Buttercup, Crowfoot

a. Petals mainly white; water plants with most of the stem and leaves growing under water; leaf-divisions thread-like with rounded sides, not flattened; seed-like fruits with transverse ridges . 1. R. LONGIROSTRIS

a. Petals entirely yellow; plants of land or water; leaves not thread-like, or, if finely divided, with flat, ribbon-like leaf-divisions; seed-like fruits not with transverse ridges . . . . . . . . . . . . *b*

b. Most of plant submerged in water; stem-leaves finely dissected into many narrow linear divisions 2 mm. or less wide . . . . . . . . . . . . . . . . . . . . . . 8. R. FLABELLARIS

b. Land plants, or, if growing in or near water, the leaf-divisions more than 2 mm. wide. . . . . . . *c*

c. All the leaves on the stem as well as at the base of the plant unlobed and undivided, the margins of the leaves merely toothed or smooth (entire) . . . . . . . . . . . . . *d*

d. Leaves chiefly round, heart-shaped, lima-bean shaped, or ovate with crenate or toothed margins, the teeth rounded; summit of leaf rounded; seed-like achene with longitudinal nerves; plants of mud banks and shores of the Missouri River and lakes of northwestern and west-central Missouri . . . . . . . . . . . . . . . . . . 4. R. CYMBALARIA

d. Leaves, especially the stem-leaves, lanceolate to linear or narrowly elliptic, or the basal ones oblong, oval, or ovate, with sparsely or finely toothed or smooth (entire) margins, the teeth pointed; most of leaves with narrowed or pointed summit; seed-like achenes without longitudinal nerves; plants of swamps, margins of ponds and sloughs, and wet places in southern and east-central Missouri . . . . . . . . . . . . . . . . . . . *e*

e. Petals 5–10, conspicuous, 3–9 mm. long, about twice as long as the sepals; stamens 10–30 . . . 2. R. LAXICAULIS

e. Petals 1–3, inconspicuous, 1–1.5 mm. long, about as long as or shorter than the sepals; stamens 3–10 . . . 3. R. PUSILLUS

c. Some or all of the leaves deeply lobed, divided, or dissected . . . *f*

f. One or more of the basal leaves not lobed or divided but merely crenate or scallop-edged with rounded or wavy teeth or projections; stem-leaves usually different . . . *g*

g. Petals conspicuous, mostly 5–8 mm. long, about twice as long as the sepals . . 7. R. HARVEYI

g. Petals inconspicuous, 1.5–3.5 (–4) mm. long, about equal to or shorter than the sepals . . *h*

h. Plant completely glabrous (without hair); receptacle (flower axis or support) mainly hairy (take off the pistils to see this); achenes shining; plants mostly of moist or low ground . . . 5. R. ABORTIVUS

h. At least the base of the stem or leaf-stalks (petioles) with some hairs; receptacle glabrous (without hairs); achenes dull; plants mostly of high, dry, or rocky ground, but also in low or alluvial soils . . . 6. R. MICRANTHUS

f. All the leaves more or less lobed or deeply divided or dissected (if an occasional basal leaf is only slightly lobed, then the petals conspicuous and 7–15 mm. long) . . . *i*

i. Plants normally growing in water, but sometimes stranded with stems creeping on wet mud of ponds, sloughs, and swamps; all the leaf divisions without stalks; seed-like achenes with a conspicuous corky keel thickened below the middle. 8. R. FLABELLARIS f. RIPARIUS

i. Without the above combination of characters . . . *j*

j. Petals small, 1–5 mm. long, shorter than or at most equaling the length of the petals . . *k*

k. Plant completely glabrous (without hairs); stems hollow; seed-like achenes without a beak or essentially so . . . 9. R. SCELERATUS

k. Stems or leaves or both more or less hairy; stems not hollow; seed-like achenes with a recurved or a flat triangular beak . . . *l*

l. Sides of seed-like achenes with minute hairs or spines, the beak flat and triangular; leaf-blades chiefly 1.5–2 cm. long, 1–2.5 (–3) cm. broad; petals 1–2 mm. long; dwarf annuals with spreading stems mainly 0.2–1.5 (–3) dm. tall; stems several and branched; receptacle glabrous . . . 17. R. PARVIFLORUS

l. Sides of seed-like achenes glabrous (without hairs), minutely pitted, the beak hooked or recurved; leaf-blades 2–9 cm. long, 3–12 cm. broad; petals 2.5–3.5 mm. long, rarely longer; erect perennials mainly 2–7.5 dm. (as low as 1.5) tall; stems usually 1 and slightly or not at all branched; receptacle hispid . . . 10. R. RECURVATUS

j. Petals conspicuous, 5–16 mm. long, definitely longer than the sepals . . . *m*

m. Noticeable bristles or prickles 2 mm. or more long spreading or curving from the seed-like fruits (achenes) . . . 19. R. ARVENSIS

m. No noticeable bristles or prickles present on the achenes, their sides smooth or barely roughened . . . *n*

n. None of the divisions or lobes of the leaves stalked, all sessile (without stalks) and connected to one another by leaf-tissue . . . 11. R. ACRIS

n. At least one of the divisions of the main leaves stalked . . . *o*

o. Most of the ultimate and smallest lobes or divisions of the leaves narrow, linear or narrowly oblong, and blunt, chiefly 2–8 mm. broad, and without teeth along margins (entire); leaves pinnately divided or cleft; leaves and stems gray-green or gray from the abundant silky-appressed hairiness; roots fleshy-thickened, short, tuber-like; petals mostly narrowly oblong and $\frac{1}{4}$–$\frac{2}{5}$ as wide as long . . . 16. R. FASCICULARIS

o. Main lobes or divisions of the larger and principal leaves broad and ovate, wedge-shaped, or rhombic (somewhat baseball-diamond-shaped), usually acute or acuminate at tip, often 10–40 mm. or more broad, and prominently toothed along margins; leaves palmately divided; leaves and stems with a definite green (yellow-, grass-, or dark-green) appearance; roots not short and tuber-like, but long and stringy, fibrous; petals mostly proportionately

Plate no. 166. **1.** Delphinium carolinianum, × $^{2}/_{5}$; Details from Small, The New York Botanical Garden. **2.** Delphinium virescens, × $^{2}/_{5}$; Details from Small, The New York Botanical Garden. **3.** Isopyrum biternatum, × $^{2}/_{5}$; Details from Small, The New York Botanical Garden. **4.** Ranunculus longirostris, × $^{2}/_{5}$.

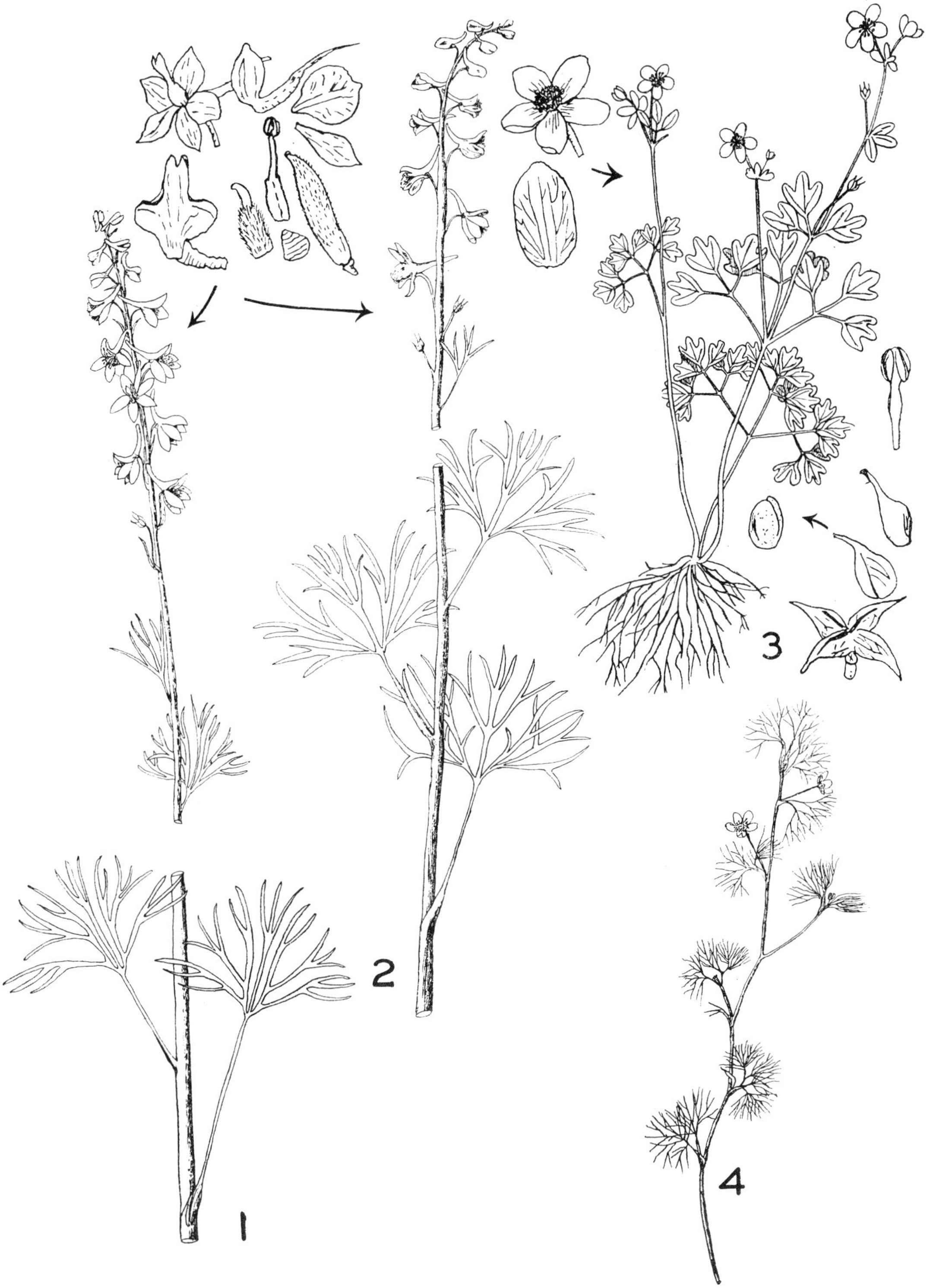


PLATE NO. 166

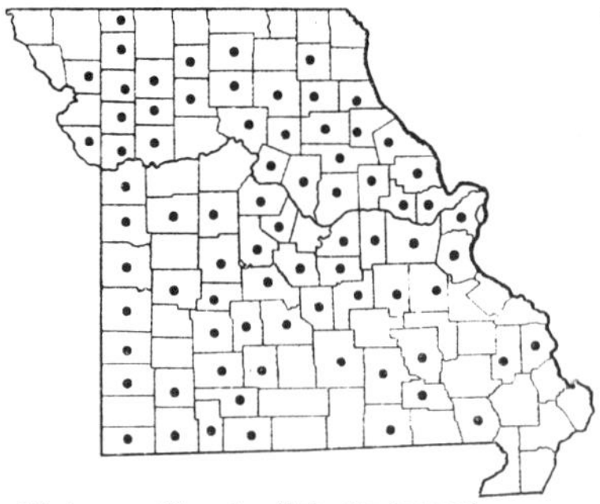

977 Isopyrum biternatum (False Rue Anemone)

978 Ranunculus longirostris (White Water Crowfoot)

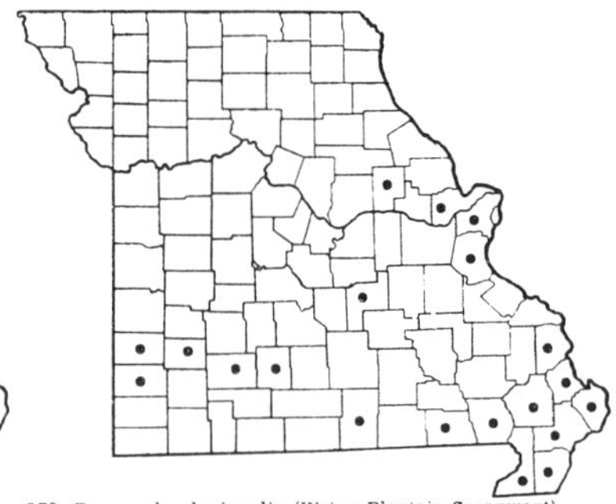

979 Ranunculus laxicaulis (Water Plantain Spearwort)

broader, often obovate or broadly oblong-obovate and $\frac{1}{2}$–$\frac{7}{8}$ as wide as long, or if narrower in proportion, at least having the other characters given. . . . *p*

p. At least the lower leaves with each of the 3 main leaf-divisions of a leaf on an individual stalk (petiolule) before joining the main leaf-stalk (petiole); stems either creeping or the older ones finally lying along the ground and rooting from the lower nodes or tips (but younger or incompletely collected plants may not show this rooting or creeping character). . . . . *q*

q. Common native plant of moist or low woodland and ravines; style straight or slightly curved, the stigma portion only at the tip of the long (1.5–2 mm.) style and 0.2–0.5 mm. long; beak of mature seed-like achene 1.5–3 mm. long; mature achene flattened on the sides, broadly keeled; dilated (stipular) base of the leaf-stalk of the main leaves lavender or brown, prominent, with the summit often free or loose and rounded, projecting upward or outward, 6–20 mm. long 14. R. SEPTENTRIONALIS

q. Rare introduced and escaped garden plant on lawns, grassy places, or in the shade of trees near houses; style ending in a small hook, the stigma portion 0.4–1 mm. long and nearly covering the inner (upper) side of the short (0.5–1 mm.) style; beak of mature seed-like achene 0.7–1.4 mm. long; mature achene plump, scarcely keeled; stipular base of leaf-stalk similar, but usually more prominent, 20–40 mm. long . . . . . . . . . . . . . . . . . . . . 13. R. REPENS

p. The 2 lateral divisions or leaflets of some or most of the main leaves sessile (not stalked), the leaf tissue directly connected to the top of the main leaf-stalk (petiole) or nearly sessile; stems remaining erect or ascending, not creeping or rooting at the lower nodes or tips even at maturity . . . . . . . . . . . . . . . . . . . . . *r*

r. Native woodland plant of mostly acid, dry or rocky soils of the southern half of Missouri; style 1.5–2 mm. long, the stigma portion 0.2–0.5 mm. long, only at the tip of the long style and deciduous in fruit; beak of mature seed-like achene 1.3–2 mm. long; sepals spreading, or in aged flowers turning down (reflexed) . . . 15. R. HISPIDUS

r. Introduced plants of open pastures, wet low fields, along ditches, and waste ground; style 0.3–1 mm. long, nearly covering the one side of the short style and persistent in fruit; beak of mature seed-like achene 0.4–0.8 mm. long; sepals quickly turned down (reflexed) in flower. . . *s*

s. Perennial plant with the base enlarged, thickened, bulb-like, and 10–35 mm. thick; beak of mature seed-like achene recurved, 0.4–0.8 mm. long; sides of the achene smooth; petals 8–14 mm. long, 7–12 mm. wide . . . . . . . . . . . . . 12. R. BULBOSUS

s. Annual plant not thickened with any bulb-like enlargement; beak of mature seed-like achene curved at tip, 0.2–0.4 mm. long; sides of the achene with nipple-like projections (papillae) or sometimes smooth; petals 5–9 mm. (–12) long, 4–7 mm. wide . . . 18. R. SARDOUS

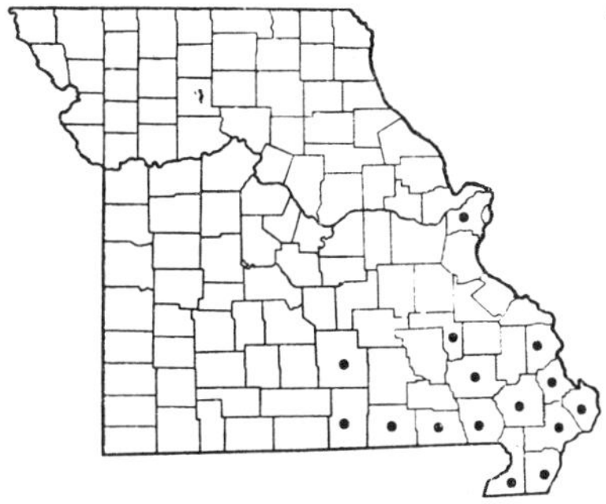

980 Ranunculus pusillus (Low Spearwort)

981 Ranunculus Cymbalaria var. Cymbalaria f. hebecaulis (Seaside Crowfoot)

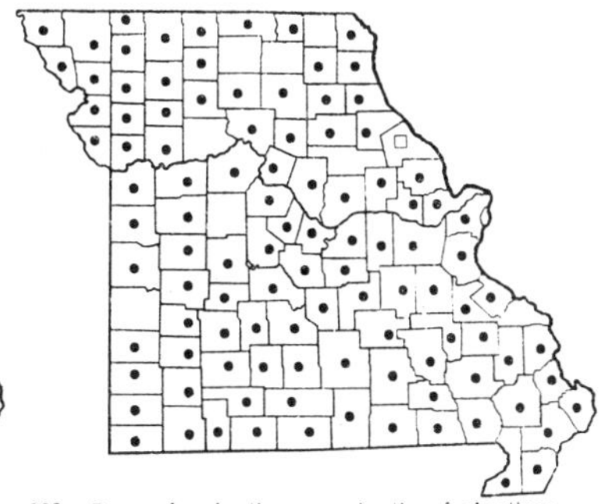

982 • Ranunculus abortivus var. abortivus f. abortivus (Small-flowered Crowfoot)

982 □ Ranunculus abortivus var. indivisus

1. **Ranunculus longirostris** Godron
White Water Crowfoot Map 978
*Ranunculus aquatilis* L. var. *capillaceus* [of P & S], not DC.
*Ranunculus circinatus* Sibth. [BB]
Flowers late May–September.

Occurs in coldwater springs and spring branches, spring-fed ponds, streams, and sloughs. Southern and central Missouri, north to St. Louis, Crawford, Phelps, Pulaski, Laclede, Dallas, and Jasper counties, and locally in Jackson County; mainly in the vicinity of the larger springs of the Ozark region of southern Missouri.

Ranges from Quebec to Oregon, south to Delaware, Pennsylvania, Indiana, Tennessee, Missouri, Kansas, Oklahoma, Texas, New Mexico, and Arizona.

The seedlike achenes as well as the foliage of this species are eaten by wildfowl, and the plant serves as a source of food for trout.

This plant is often found only in vegetative condition in swiftly running spring water but may be distinguished in this stage by its alternate compound leaves from Hornwort (*Ceratophyllum*) and Milfoil (*Myriophyllum*), both of which have the compound leaves in whorls of 4, 5, or 6. *Ceratophyllum*, furthermore, has the margins of the leaf-divisions with irregular teeth or projections, wheras the margins of the leaf-divisions in *R. longirostris* are smooth-edged or entire. Where the water is more stagnant, as in sloughs and stream-margins, plants in flowering condition are more frequent.

2. **Ranunculus laxicaulis** (T & G) Darby
Water Plantain Spearwort Map 979
*Ranunculus oblongifolius* of authors [P & S, Steyerm.] not Ell.
*Ranunculus texensis* Engelm. apud. Engelm. & Gray [Benson]
Flowers May–June.

Occurs in swampy open ground, low depressions and pools in upland prairies, and borders of sloughs and ditches. Southern and east-central Missouri, north to St. Louis, St. Charles, Montgomery, Phelps, Webster, Greene, and Barton counties.

Ranges from Florida to Texas, north to Connecticut, Indiana, Illinois, Missouri, and Kansas.

3. **Ranunculus pusillus** Poir. Low Spearwort
Map 980
*Ranunculus oblongifolius* Ell., not of most authors
*Ranunculus pusillus* var. *typicus* Benson [Benson]
Flowers May–June.

Occurs in swamps, borders of ditches, natural upland sink-hole ponds, and low wet woods. Southeastern Missouri west to Texas and Howell counties, north to Cape Girardeau and Iron counties, and locally in St. Louis County, where exterminated.

Ranges from Florida to Texas, north to New York, Ohio, Indiana, Missouri, and Oklahoma; California.

Benson (Am. Midl. Nat. 40: 196–97. 1948) treats the Missouri material as typical *R. pusillus* var. *pusillus*, differentiating it from a more southern var. *angustifolius* (Engelm.) Benson on the basis of rounded or truncate instead of an acute tip of the lower leaves.

4. **Ranunculus Cymbalaria** Pursh var. **Cymbalaria** f. **hebecaulis** Fern. Seaside Crowfoot
Map 981
*Ranunculus Cymbalaria* f. *hebecaulis* Fern. [G, P & S]
*Ranunculus Cymbalaria* Pursh [BB]
*Ranunculus Cymbalaria* var. *typicus* Benson [Benson]
Flowers May–October.

Occurs along sand bars, sandy moist open ground, and mud flats along the Missouri River and backwater oxbow lakes of this river. Known only from northwestern and west-central Missouri in Jackson, Clay, and Holt counties.

Ranges from Greenland and Labrador to Alaska, south to New Jersey, New York, Illinois, Arkansas,

Kansas, Texas, New Mexico, Utah, Idaho, and Washington.

Benson recognizes two other varieties. Typical *R. Cymbalaria* var. *Cymbalaria* f. *Cymbalaria*, glabrous throughout, is not known in Missouri.

5. **Ranunculus abortivus** L.

Small-flowered Crowfoot Map 982

Also known as Kidneyleaf Crowfoot, Smooth-leaved Crowfoot.

Flowers March–June.

Occurs on moist banks of streams, moist ledges of bluffs, low moist ground in valleys, ravines, woodland, fields, along ditches, banks, roadsides, railroads, and waste ground. Throughout Missouri, and doubtless in every county.

Ranges from Florida to Texas, north to Maine, New York, Ohio, Michigan, Wisconsin, Minnesota, and Saskatchewan.

Several variations have been described, of which only the following two are known from Missouri:

All the leaves on the stem as well as all the basal and root leaves simple and undivided; rare type encountered . . . 5b. R. ABORTIVUS var. INDIVISUS

Uppermost leaves of the stem but often all of those on the stem divided to the base into narrow divisions; basal leaves simple or some of them divided; common type encountered . . . . 5a. R. ABORTIVUS var. ABORTIVUS f. ABORTIVUS

5a. **Ranunculus abortivus** var. **abortivus** f. **abortivus** Map 982

*Ranunculus abortivus* L. [G, P & S, Steyerm.]

*Ranunculus abortivus* var. *abortivus* [BB]

The common variation found throughout the state.

5b. **Ranunculus abortivus** var. **indivisus** Fern. Map 982

Known only from Pike County, northeastern Missouri (grassy dry lower part between portions of Ben Watts Knob, T52N, R1W, sect. 13, 4 mi. east of Eolia, April 19, 1941, *Steyermark 28556*), but probably more common.

Recorded also from Virginia.

Fassett's study of *Ranunculus abortivus* (Am. Midl. Nat. 27: 512–22. 1942) indicates that the var. *eucyclus* Fern., supposedly characterized by having some of the basal leaves with a narrow or closed sinus, cannot be maintained, and is best regarded as comprising individual plants having some of the leaves large and well-developed as a result of certain environmental conditions.

This species is considered to be somewhat poisonous, but loses this property when dried, and is therefore not harmful in hay fed to cattle.

6. **Ranunculus micranthus** Nutt. Map 983

Flowers March–May.

Occurs in open or rocky woodland, moist shaded ledges of bluffs, dry or moist soils on lowland or upland.

The following two variations are found in Missouri:

Basal leaves somewhat to strongly heart-shaped (cordate) at base with 11–19 broad and blunt, squarish, flat-topped teeth . . 6a. R. MICRANTHUS var. MICRANTHUS

Basal leaves with a wedge-shaped, rounded, or somewhat truncate (as if cut straight across) base with 5–13 more prolonged ovate or triangular teeth . . . 6b. R. MICRANTHUS var. DELITESCENS

6a. **Ranunculus micranthus** var. **micranthus** Map 983

*Ranunculus micranthus* Nutt. [G, P & S, Steyerm., BB in part]

This is the commoner variation encountered in the state, in southern and central Missouri, north to Pike, Monroe, Adair, Boone, Cooper, and Clay counties.

Ranges from Maryland to Illinois and Missouri south to Georgia, Alabama, Arkansas, and Oklahoma.

6b. **Ranunculus micranthus** var. **delitescens** (Greene) Fern. Map 983

*Ranunculus micranthus* [of BB in part], not Nutt.

Scattered in the range of var. *micranthus* in Missouri, but less common.

Ranges from Massachusetts, New York, Pennsylvania, and Ohio to Illinois and Missouri, south to North Carolina and Arkansas; also South Dakota.

Benson does not maintain the var. *delitescens* in his treatise on North American Ranunculus (Am. Midl. Nat. 40: 161. 1948), but Fassett's preliminary mass collection studies (Am. Midl. Nat. 27: 519. 1942) showed that there might be some justification in retaining var. *delitescens* as valid. Pending future, more detailed investigation of these variations, the present work has retained the two types found in Missouri.

---

Plate no. 167. **1.** Ranunculus laxicaulis, × 2/5; Details from Small, The New York Botanical Garden. **2.** Ranunculus Cymbalaria, × 2/5. **3.** Ranunculus Harveyi, × 2/5. **4.** Ranunculus pusillus, × 2/5. **5.** Ranunculus abortivus, × 2/5. **6.** Ranunculus micranthus, × 2/5.

PLATE NO. 167

983 • Ranunculus micranthus var. micranthus
983 □ Ranunculus micranthus var. delitescens

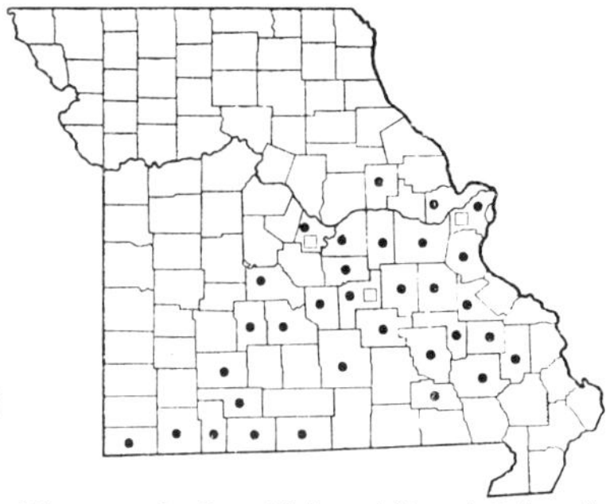

984 • Ranunculus Harveyi f. Harveyi (Harvey's Buttercup)
984 □ Ranunculus Harveyi f. pilosus

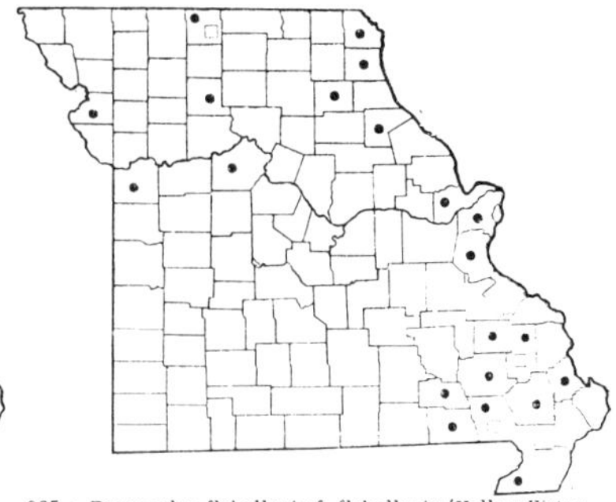

985 • Ranunculus flabellaris f. flabellaris (Yellow Water Crowfoot)
985 □ Ranunculus flabellaris f. riparius

7. **Ranunculus Harveyi** (Gray) Britt.
Harvey's Buttercup Map 984
Flowers late March–May.

Occurs in acid soils associated with chert, sandstone, and granite rocks on rocky wooded slopes, ledges of bluffs and outcrops, upland ridges, and rocky open ground.

Two variations are encountered in Missouri material:

Stems and leaves mainly glabrous (without hairs) . . . 7a. R. HARVEYI f. HARVEYI
Stems and leaves mostly hairy . . . . . 7b. R. HARVEYI f. PILOSUS

7a. **Ranunculus Harveyi** f. **Harveyi** Map 984
*Ranunculus Harveyi* var. *typicus* Benson [Benson]
*Ranunculus Harveyi* (Gray) Britt. [G, BB, P & S, Steyerm.]

Restricted to the Ozark section in southern and east-central Missouri north to St. Louis, St. Charles, Montgomery, Osage, Cole, Camden, Dallas, and Greene counties to McDonald County.

Known only from Alabama, southern Illinois, Missouri, Arkansas, and eastern Oklahoma.

7b. **Ranunculus Harveyi** f. **pilosus** (Benke) Palmer & Steyerm. Map 984

Scattered within the range of f. *Harveyi*.

The basal leaves of *R. Harveyi* are heart-shaped or nearly so at the base as in *R. abortivus* var. *abortivus* f. *abortivus* and are either simple and undivided or mixed with some deeply divided ones. Benson does not recognize f. *pilosus* in his work (loc. cit. p. 159), whereas Fassett (loc. cit. p. 519–20, 522) would retain the pilose variation as a form, as treated in the present work.

Benson treats plants from Lower California as *R. Harveyi* var. *australis* (Brand.) Benson. In the present author's opinion Brandegee's *R. australis* is considered as of specific rank.

8. **Ranunculus flabellaris** Raf. Yellow Water Crowfoot Map 985
Flowers early May–June.

Occurs in swamps, bayous, and ponds in bald cypress (*Taxodium*) woods, sloughs, and oxbow lakes of river flood plain of the larger streams, and on muddy banks and in shallow water of slow streams. Along the valleys of the Mississippi, Missouri, and other large streams of northern, central, and eastern Missouri; more common in the lowlands of southeastern Missouri.

Ranges from Maine to British Columbia, south to North Carolina, Louisiana, Kansas, Oklahoma, Utah, and California.

Two variations occur in Missouri:

Plants growing submerged in the water, the leaves finely cut into many very narrow divisions and glabrous . 8a. R. FLABELLARIS f. FLABELLARIS
Plants growing on soil out of water, the leaves with broader divisions and either glabrous or hairy . . . . . 8b. R. FLABELLARIS f. RIPARIUS

8a. **Ranunculus flabellaris** f. **flabellaris** Map 985
*Ranunculus flabellaris* Raf. [G, BB, Steyerm.]
*Ranunculus delphinifolius* Torr. [P & S]

This is the more usual form found in the state.

8b. **Ranunculus flabellaris** f. **riparius** Fern. Map 985
*Ranunculus delphinifolius* var. *terrestris* (Gray) Farwell

This is a stranded or shore form which occurs

Plate no. 168. **1.** Ranunculus flabellaris. **2.** Ranunculus sceleratus, × 2/5. **3.** Ranunculus recurvatus, × 2/5; *a.* Fruit, × 4. **4.** Ranunculus acris, × 2/5.

PLATE NO. 168

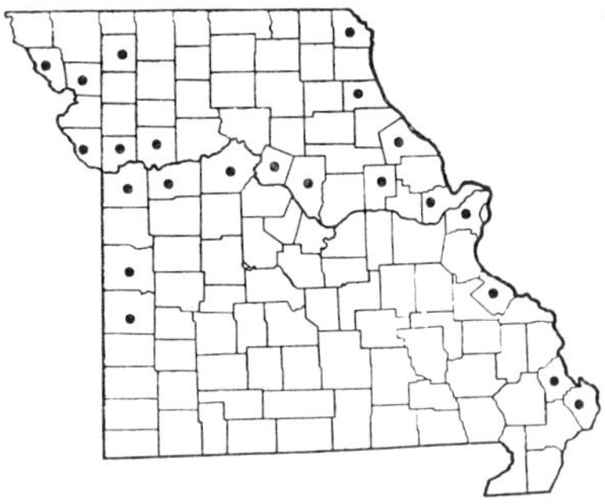

986 Ranunculus sceleratus var. sceleratus f. sceleratus (Cursed Crowfoot)

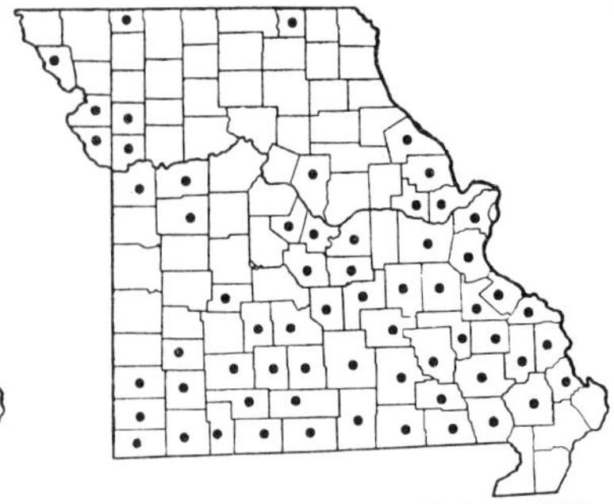

987 Ranunculus recurvatus var. recurvatus f. recurvatus (Hooked Crowfoot)

988 Ranunculus acris var. acris (Tall Buttercup)

when the water dries up and grades into the aquatic phase. It is found with the usual submerged type.

While the two leaf types contrast greatly according to occurrence in water or on land, the land form is to be considered merely a normal phase in the varying aquatic environment of the plant. Thus, the retention of the name f. *riparius* is, in this case, one of convenience only, as a means of distinguishing, by some arbitrary way, i.e., a name, between the two forms occupying different habitats.

The seedlike achenes and leaves are eaten by wildfowl.

9. **Ranunculus sceleratus** L. var. **sceleratus** f. **sceleratus** Cursed Crowfoot Map 986
*Ranunculus sceleratus* L. [G, BB, P & S, Steyerm.]
*Ranunculus sceleratus* var. *typicus* Benson [Benson]
Flowers May–August.

Occurs in moist alluvial ground on mud flats and sand bars principally of the Missouri, Mississippi, and other rivers, in swampy depressions and ditches of lowlands and valleys, and muddy wet margins of oxbow lakes and sloughs of river flood plains. Northern, central, and eastern Missouri, locally south in western Missouri to Bates and Vernon counties, and in eastern Missouri south to Scott and Mississippi counties.

Ranges from Newfoundland to Alaska, south to Florida, Louisiana, Arkansas, Oklahoma, Texas, New Mexico, and California; also in Eurasia.

The plant contains a substance which poisons cows and other cattle which eat it. Milk from cows which have eaten the plant is bitter or contains a reddish color. Human contact with the plant may irritate the skin in the form of blisters in sensitive individuals. In Europe the plant, after being boiled, is sometimes eaten as a green vegetable when the boiled water has been poured off to get rid of the poisonous substance.

An aquatic form, in which the petioles are elongated to 5 dm. long in deep water and the leaf-blades are floating, is known as *R. sceleratus* var. *sceleratus* f. *natans* Glück, but this has not been found as yet in Missouri.

10. **Ranunculus recurvatus** Poir. var. **recurvatus** f. **recurvatus** Hooked Crowfoot Map 987
*Ranunculus recurvatus* var. *typicus* Benson [Benson]
*Ranunculus recurvatus* Poir. [G, BB, P & S, Steyerm.]
Flowers May–July.

Occurs in usually moist places along spring branches, margins of small streams, wet mucky borders of ponds and springs, moist ledges of bluffs, and rich or rocky open woods. Throughout the state, but most common in the Ozark section of southern and east-central Missouri, and scattered elsewhere, locally northward in Schuyler, Worth, and Holt counties.

Ranges from Newfoundland to Manitoba, south to Florida, Alabama, Mississippi, Louisiana, and Texas.

In this variation the stems are more or less hairy with spreading hairs, as contrasted to var. *recurvatus* f. *laevicaulis* Harger, with glabrous stems, and to var. *adpressipilis* Weath., in which the hairs of the stem are short, scattered, and appressed closely to the surface. Neither of the latter two variations has been found in Missouri. Benson includes under *R. recurvatus* a var. *tropicus* (Griseb.) Fawc. & Rendle from the West Indies, by some botanists treated as a distinct species, *R. cubensis* Griseb., or as a variety of *R. repens* L.

11. **Ranunculus acris** L. var. **acris**
Tall Buttercup Map 988
Known also as Meadow Buttercup.
*Ranunculus acris* L. [G, BB, P & S, Steyerm.]
*Ranunculus acris* L. var. *typicus* G. Beck [Benson]
Flowers May–August.

Occurs along railroads, where known only from St. Louis and Jackson (Sheffield, May 26, 1896, *Bush*

989 Ranunculus bulbosus var. bulbosus (Bulbous Buttercup)

990 • Ranunculus repens var. repens (Creeping Buttercup)
990 □ Ranunculus repens var. pleniflorus

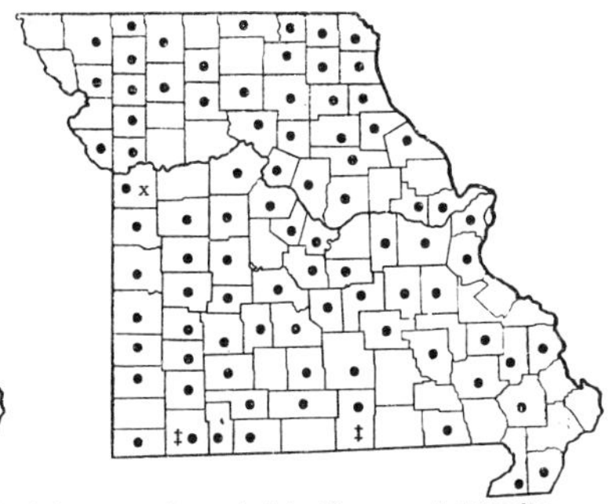

991 • Ranunculus septentrionalis var. septentrionalis (Swamp Buttercup)
991 x Ranunculus septentrionalis var. caricetorum
991 ‡ Ranunculus septentrionalis var. pterocarpus

*654*; Courtney, June 9, 1905, *Bush 2989*; June 17, 1900, *Bush 794*; Lake City, May 20, 1921, *Bush 9357*; Sugar Creek, June 2, 1906, *Bush 3908*) counties.

Native of Europe; introduced and naturalized in North America from Labrador and Newfoundland to Ontario and Alaska, south to North Carolina, West Virginia, Ohio, Indiana, Illinois, Missouri, Kansas, and Oregon.

This species is common in meadows and pastures of the northeastern United States, where it is the commonest cause of poisoning from among the buttercups. As in the case of *R. sceleratus*, milk from cattle which have eaten the plant becomes bitter or tainted, and this persists in butter. Since the plant is a rare introduction along the railroads in the state, and shows no tendency to spread, it would appear to be of little concern to cattlemen.

12. **Ranunculus bulbosus** L. var. **bulbosus**
Bulbous Buttercup Map 989
*Ranunculus bulbosus* L. [G, BB, Steyerm.]
Flowers late April–June.

Occurs in wet open pastures, orchards, and grassy open places. Known only from Boone (University Orchard, Columbia, May 14, 1936, *Rickett 1313*) and Dunklin (wet open upland pasture, Campbell, May 27, 1939, *Steyermark 26666*) counties.

Native of Europe; introduced and naturalized in North America from Newfoundland to Ontario and British Columbia, south to Georgia and Louisiana and California.

The *Steyermark 26666* collection has small petals 5–6 mm. long and 5 mm. wide and the non-bulbous base of *R. Sardous*; but has the smooth achenes of *R. bulbosus*.

The pubescence of the stem in var. *bulbosus* varies from silky soft hairy to nearly glabrous, whereas in var. *valdepubens* (Jord.) Briq. it is very abundant with a grayish-white appearance. In var. *dissectus* Babey the leaflets of the basal leaves are more finely cut into narrow oblong or linear divisions.

The bulbous thickened base and roots may be eaten as a starchy vegetable or relish, after being boiled. They have a pungent quality, resembling radish.

13. **Ranunculus repens** L. Creeping Buttercup
Map 990

Flowers May–September.

Missouri material is represented by two variations:

Petals 5–9; stamens numerous; base of the middle leaflet of the basal leaves wedge-shaped (cuneate) to nearly truncate (as if cut straight across) . . . . . . 13a. R. REPENS var. REPENS

Petals very numerous; base of middle leaflet of basal leaves rounded or somewhat heart-shaped (cordate) . . 13b. R. REPENS var. PLENIFLORUS

13a. **Ranunculus repens** var. **repens** Map 990
*Ranunculus repens* L. [G, BB, P & S, Steyerm.]
*Ranunculus repens* var. *typicus* G. Beck [Benson]

Known only from St. Louis County (large colony around shade of trees, in vacant lot, University City, 2 blocks east of Jackson Avenue, on Bolson Avenue, May 13, 1936, *Steyermark 10398*). This locality is now destroyed.

13b. **Ranunculus repens** var. **pleniflorus** Fern.
Map 990
*Ranunculus repens* var. *florepleno* of auth., not DC.

Known only from Clay County (woods, 325 Morse Avenue, Liberty, 1948, *Shirley Landers 38* in herb. Wm. Jewell College).

Escaped from gardens, and naturalized from Vermont and New York to Ohio and Missouri, south to North Carolina; also from Alaska, and Idaho to

Washington, south to Utah and California.

In young plants of *R. repens* the creeping stems may not be evident and are difficult to distinguish from young plants of *R. septentrionalis*, similarly lacking creeping or prostrate stems. In such cases the differences in the pistils, as presented in the key, may be depended on. The leaves of *R. repens* are also often mottled with white and are usually thicker, darker green, and in var. *repens* the hairs of the stems and leaf-stalks are appressed. In other varieties of *R. repens*, which have not been recorded from Missouri, the stems and leaf-stalks may be glabrous (var. *glabratus* DC.), or hairy with widely spreading hairs (var. *villosus* Lamotte), the creeping or trailing stems may be completely absent (var. *erectus* DC.), or the lobes and teeth of the leaves lanceolate to linear and acuminate (var. *linearilobus* DC.).

14. **Ranunculus septentrionalis** Poir.
Swamp Buttercup — Map 991

Also known as Marsh Buttercup.

Flowers April–June.

Occurs in low or alluvial ground along streams, in valleys and ravines, at the base of bluffs, and in rich or moist woodland slopes.

The following variations occur in Missouri material:

a. Hairs, when present, retrorse-hispid (directed down or toward the roots) on the lower part of stem and leaf-stalks . 14b. R. SEPTENTRIONALIS var. CARICETORUM

a. Hairs, when present, spreading outward or upward (away from roots), or plants nearly without any hairs . . . . . . . . . *b*

- b. Common type; main body of mature seed-like achenes 3–4.5 mm. long with a wing or keel down the center 0.5 mm. wide or less; mature achenes in a fruiting head 15–30 or more; sepals 5–10 mm. long, pubescent to glabrous, spreading or becoming reflexed (turned down) in older flowers; leaves definitely hairy to nearly glabrous . . . . 14a. R. SEPTENTRIONALIS var. SEPTENTRIONALIS
- b. Uncommon type; main body of mature seed-like achenes 3.7–5 mm. long with a wing or keel down the center 0.5–1 mm. wide; mature achenes in a fruiting head 5–20; sepals 3.5–5 mm. long, glabrous, promptly reflexed in early flower; leaves smoothish. . . . 14c. R. SEPTENTRIONALIS var. PTEROCARPUS

14a. **Ranunculus septentrionalis** var. **septentrionalis** — Map 991

*Ranunculus septentrionalis* Poir. [G, P & S, Steyerm.]

Throughout Missouri, and doubtless in every county. This is the common variation found in the state.

Ranges from Quebec to Manitoba, south to Maryland, Kentucky, Arkansas, and Kansas.

14b. **Ranunculus septentrionalis** var. **caricetorum** (Greene) Fern. — Map 991

*Ranunculus caricetorum* Greene [P & S]

*Ranunculus sicaeformis* Mack. & Bush [Rydberg]

Known only from Jackson County, west-central Missouri (low prairies, Buckner, May 30, 1898, *Mackenzie 95*).

Ranges from Maryland and Ohio to Minnesota, Iowa, and Missouri.

Many Missouri specimens are densely pubescent with conspicuous hairs, but the hairs are spreading horizontally or upward and not downward as in true var. *caricetorum*.

14c. **Ranunculus septentrionalis** var. **pterocarpus** Benson — Map 991

*Ranunculus carolinianus* DC. [G, BB, Benson, Shinners]

*Ranunculus palmatus* of most auth., not Ell. [Small]

The following two collections, in the herbarium of Missouri Botanical Garden, were identified as *R. carolinianus* by Dr. Lyman Benson: Barry Co. (alluvial ground in woods at base of east-facing limestone slopes at Prentiss Ford along White River, between Viola and Shell Knob, April 26, 1936, *Steyermark 10208*); Howell Co. (wet swale in valley along tributary of Wyatt Creek, T21N, R7W, sect. 32, 2½ mi. east of Lanton, April 27, 1938, *Steyermark 5145*). Both specimens had been previously identified by the collector as *R. septentrionalis*.

Ranges from Florida to Texas, north to Maryland, West Virginia, Indiana, Illinois, Missouri, and Nebraska.

*Ranunculus septentrionalis* is highly variable in the pubescence of its stems, petioles, and leaves. The pubescence ranges from spreading to ascending, retrorse, appressed, or may be nearly absent.

It is extremely doubtful whether the plants recognized as *R. septentrionalis* var. *pterocarpus* or *R. carolinianus* can be maintained in any rank separate from

Plate no. 169. **1.** Ranunculus bulbosus, × 2/7. **2.** Ranunculus repens var. repens, × 2/7. **3.** Ranunculus septentrionalis var. septentrionalis, × 2/7. **4.** Ranunculus hispidus var. hispidus, × 2/7. **5.** Ranunculus fascicularis, × 2/7. **6.** Ranunculus parviflorus, × 2/7. **7.** Ranunculus Sardous, × 2/7.

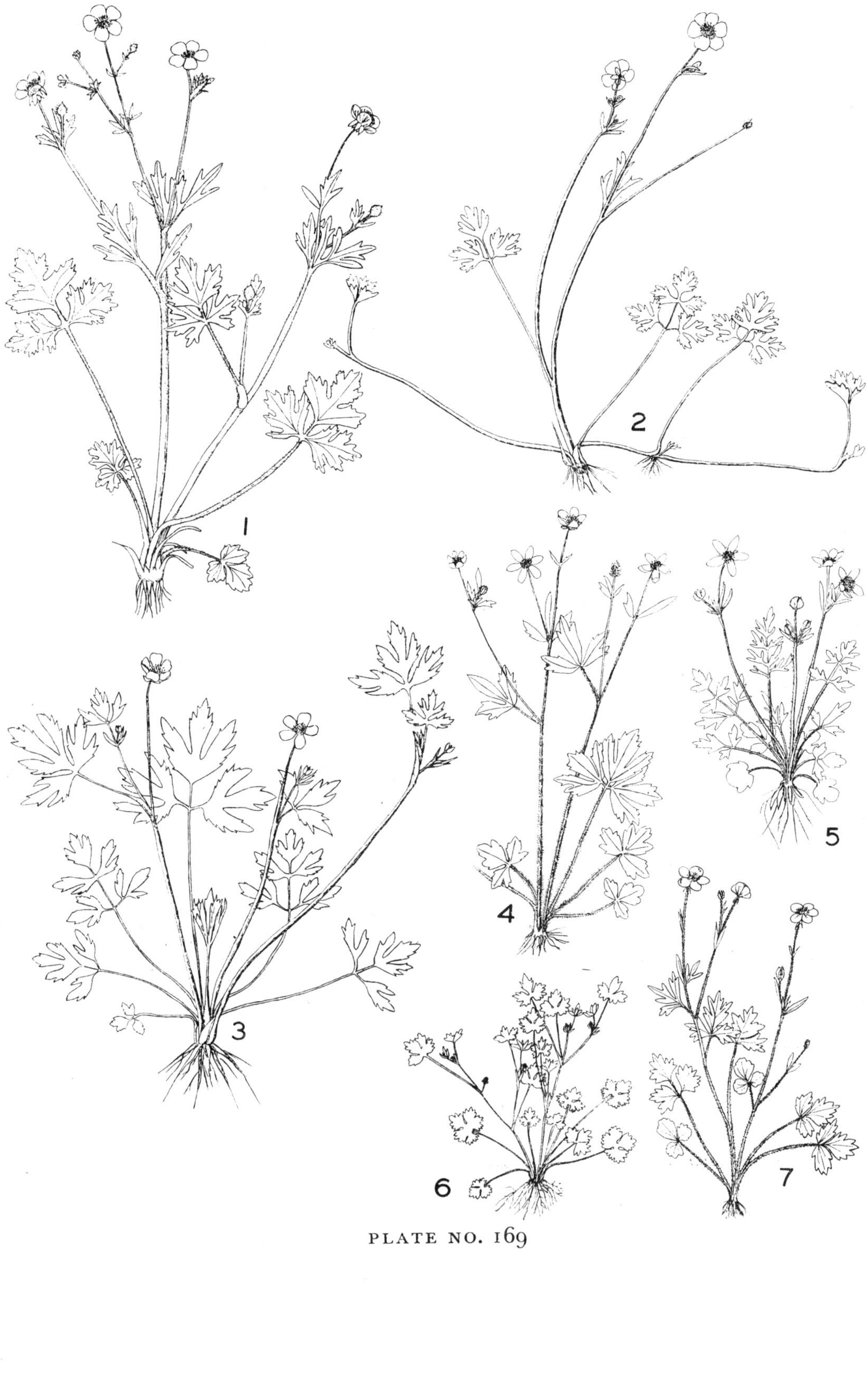


PLATE NO. 169

*R. septentrionalis* var. *septentrionalis*, and in the present flora the var. *pterocarpus* is kept up only tentatively, pending future field and experimental work. Although Fernald stresses as marks of distinction for *R. carolinianus* (Rh. 41: 543–44. 1939) short and promptly reflexed, glabrous sepals, relatively few and large achenes with a broad marginal wing, there is little evidence that such characters are really significant in Missouri material studied. The achene length between *R. septentrionalis* var. *septentrionalis* and var. *pterocarpus* grades from one to another imperceptibly, while the width of the marginal wing of the achene varies with the age of the achene. Likewise, the presumed shorter glabrous sepals of var. *pterocarpus* or *R. carolinianus* can be matched by sepals of var. *septentrionalis* which are 5 mm. long or glabrous. Specimens, such as *Steyermark 83857* from Dent County, have petals and sepals as short as *R. carolinianus* but the leaves and stems are noticeably pubescent. Since the sepals of older flowers of var. *septentrionalis* are as reflexed as those of var. *pterocarpus*, it is not possible to use this character for separation in such flowers. The matter of relatively few achenes in a fruiting head of var. *pterocarpus* or *R. carolinianus* is likewise a criterion of extreme doubt. This is believed to be too variable in plants to have real significance. In plants of otherwise typical *R. septentrionalis* var. *septentrionalis* fruiting heads are frequently found having only 7 achenes (*Steyermark 83890* from Crawford Co.) or less than 20 achenes. The smoothish or nearly glabrous leaves supposedly characteristic of *R. carolinianus* can be matched by similar nearly glabrous plants of *R. septentrionalis*. Benson (Am. Midl. Nat. 40: 82–83. 1948) characterizes the plants of *R. carolinianus* as 'subglabrous,' but Fernald (*Gray's Manual*, eighth ed., p. 653) describes the stem as varying from 'subglabrous, pilose or spreading-hirsute.'

Benson (loc. cit. p. 82–83) admits difficulty in assigning specific rank to *R. carolinianus* and hesitates, in the light of insufficient study of material seen, to give full weight to the value of the fruiting characters of *R. carolinianus*. He further admits the intergradation between what he considers *R. carolinianus* and non-fruiting specimens of both *R. hispidus* and *R. septentrionalis*. In the *New Ill. Fl.* Gleason inadvertently accepts and describes both *R. carolinianus* (p. 178) and *R. septentrionalis* var. *pterocarpus* (p. 177), although one is synonymous with the other.

The present author believes that plants which have been identified as *R. carolinianus* or *R. septentrionalis* var. *pterocarpus* may be referred either to *R. septentrionalis* or to *R. hispidus* depending upon to what extent are shown the characters usually associated with those species. If there is a tendency after flowering for the leafy stems to be trailing or rooting and for the 3 leaflets to have petiolules, the plants can be assigned to *R. septentrionalis*. If no rooting or trailing stems are developed and the earliest basal leaves are smaller than the later ones with the lateral leaflets sessile or nearly so, then the plants can be identified as *R. hispidus*.

15. **Ranunculus hispidus** Michx. Hispid Buttercup
Map 992

Flowers late March–early June.

Occurs in rich or rocky woods, on slopes, crests of hills and ridges, in valleys and ravines, mainly in drier and more acid soils of chert, sandstone, and granite than occupied by *R. septentrionalis*.

Missouri material is represented by the following variations:

Hairs of stems and leaf-stalks usually widely spreading; body of mature seedlike achenes 3–3.5 mm. long, 2.5–3.3 mm. broad . . . . 15a. R. HISPIDUS var. HISPIDUS

Hairs of stems and leaf-stalks mostly appressed (hairs lying parallel to or along length of stem); body of mature achenes 2–2.5 mm. long, 1.8–2.3 mm. broad . . 15b. R. HISPIDUS var. MARILANDICUS

15a. **Ranunculus hispidus** var. **hispidus**
Map 992

*Ranunculus hispidus* Michx. [G, P & S, Steyerm.]
*Ranunculus hispidus* var. *eurylobus* Benson [G, BB, Benson]

Common in the Ozark region of southern and east-central Missouri, and locally northeast to Pike and Marion counties, northwest in west-central Missouri to Jackson and Platte counties; absent from the lowlands of southeastern Missouri and from the unglaciated prairie section of southwestern Missouri.

Ranges from Georgia to Arkansas and Oklahoma, north to New York, Ohio, Indiana, Illinois, and Missouri.

15b. **Ranunculus hispidus** var. **marilandicus** (Poir.) Benson
Map 992

*Ranunculus hispidus* var. *falsus* Fern. [G, BB, P & S, Steyerm.]
*Ranunculus palmatus* Ell., not of most auth.

Occasionally found scattered in the range of var. *hispidus* in Missouri, but much less common.

Ranges from Massachusetts and Vermont to Ontario, Ohio, and Missouri, south to Florida, Georgia, Alabama, Tennessee, and Arkansas.

Benson characterizes *R. hispidus* var. *hispidus* as

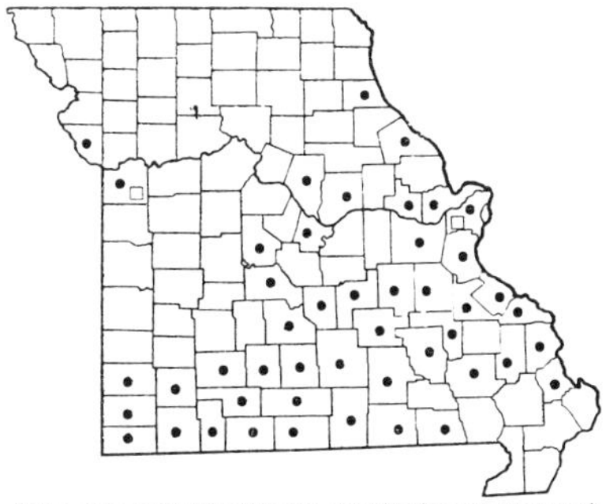

992 • Ranunculus hispidus var. hispidus (Hispid Buttercup)
992 □ Ranunculus hispidus var. marilandicus

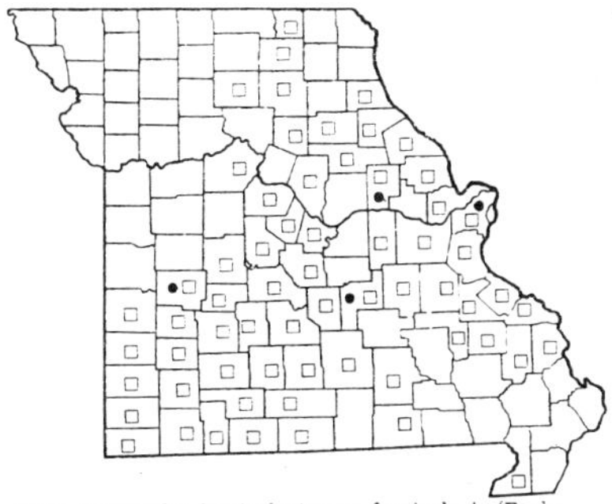

993 • Ranunculus fascicularis var. fascicularis (Early Buttercup)
993 □ Ranunculus fascicularis var. apricus

994 Ranunculus parviflorus (Small-flowered Crowfoot)

having the basal leaf-blades nearly always completely 3-parted or pinnately compound, as contrasted to the mostly simple 3-lobed or 3-parted leaves of var. *eurylobus*. Benson (loc. cit. p. 86) cites a Missouri collection from McDonald County (2 mi. west of Mountain, *Steyermark 4497*) as probably referable to var. *eurylobus*. Actually, study of Missouri material shows that most of the specimens have smaller earliest basal leaves only 3-lobed or somewhat 3-parted with the lobes or sessile terminal leaflet broadly rhombic or suborbicular and merely shallowly toothed or obscurely lobed, and could, therefore, be referred to var. *eurylobus*. However, this type of basal leaf, while common in Missouri material, occurs with plants showing more deeply parted basal leaves. The two types intergrade and this does not appear to be a character of any significance. The var. *marilandicus* is being retained for the present, pending future field and experimental studies of the group.

As indicated under the discussion of *R. septentrionalis*, some plants, identified by some authors with *R. carolinianus* or *R. septentrionalis* var. *pterocarpus*, should be referred to *R. hispidus*. Plants having relatively short sepals and short or narrow petals have been called *R. carolinianus*, when actually they are *R. hispidus*. *Steyermark 84329* from Perry County, for example, has sepals only 4–5 mm. long and the petals 7–9 mm. long (characteristic presumably of *R. carolinianus*), but has the hairy sepals, leaves, and long, spreading hairs on the petioles and stems, as well as the erect, nonrooting habit of *R. hispidus*. Actually, *R. hispidus* has sepals varying mainly from 4–6.5 mm. long, instead of 5.5–7.5 mm. as given in the eighth edition of *Gray's Manual*. Benson, likewise, (loc. cit. p. 84) gives the length of *R. hispidus* as 4–6 mm. In drying, it should be taken into account that the petals and sepals shrink somewhat both in length and width, which may be the basis for some of the differences of measurement existing in the current treatments of the same species.

Usually this species may be distinguished from nonrooting or nontrailing early stages of *R. septentrionalis* var. *septentrionalis* by the less dilated and narrower stipular base of the leaf-stalk. This stipular base, moreover, does not extend upward or outward at the summit or project as a rounded free or loose part as in *R. septentrionalis*. However, other vegetative characters, such as the sessile or nearly sessile lateral leaflets of the main leaves combined with the smaller usually 3-lobed earliest basal leaves, and the nonrooting and erect or ascending stems of the plant in all stages further distinguish *R. hispidus* from *R. septentrionalis*. The sepals in *R. hispidus* are usually 4–6.5 mm. long and ovate, those of *R. septentrionalis* var. *septentrionalis* vary from 5–10 mm. long and from ovate to ovate-lanceolate. The petals of *R. septentrionalis* var. *septentrionalis* average slightly broader than those of *R. hispidus*, varying from 4.5–13 mm. broad in *R. septentrionalis* var. *septentrionalis* and 4–11 mm. broad in *R. hispidus*. The habitat of *R. hispidus* is usually drier, rockier, upland, and more acid soils as compared with the low or moist or alluvial woodland, valley and ravine floor or base of bluffs of the habitat of *R. septentrionalis* var. *septentrionalis*.

The main characters used to distinguish *R. hispidus* from *R. septentrionalis* var. *pterocarpus* (*R. carolinianus*), if that doubtful taxon can be maintained, are the sessile lateral leaflets of the main leaves of *R. hispidus*, the nonrooting or nontrailing stems in all stages of growth, the smaller achenes (2–3.5 mm. long of *R. hispidus* vs. 3.7–5 mm. long of *R. septentrionalis* var. *pterocarpus*), and the relatively narrower keel or wing of the achene (0.2–0.5 mm. broad in *R. hispidus* vs. 0.5–1 mm. broad in *R. septentrionalis* var. *pterocarpus*).

16. **Ranunculus fascicularis** Muhl.
Early Buttercup — Map 993
Also called Prairie Buttercup.
Flowers March–May.

995 Ranunculus Sardous

995A Ranunculus arvensis

996 Adonis aestivalis var. citrina (Summer Adonis)

Occurs in dry or upland open woods of acid soils associated with chert, sandstone, or granite substrata, rocky glades, and prairies. Common throughout the eastern half of the state and the Ozark section of southern and central Missouri, north of the Missouri River west to Schuyler, Adair, Linn, Randolph, and Boone counties, and south of the Missouri River north to Saline, Benton, St. Clair, and Vernon counties. Absent from the lowlands of southeastern Missouri.

Two variations appear which may be separated as follows:

Main leaf divisions of the later developed leaves usually entire (smooth-edged), shallowly lobed or with only a few teeth towards the tip; leaf segments usually broadest above the middle, oblanceolate to narrowly elliptic . . . . .
16b. R. FASCICULARIS var. APRICUS

Main leaf divisions of the later developed leaves deeply 3–7-parted and again angularly toothed; leaf segments with linear or oblong obtuse lobes
16a. R. FASCICULARIS var. FASCICULARIS

16a. **Ranunculus fascicularis** var. **fascicularis** Map 993

*Ranunculus fascicularis* Muhl. [G, BB, P & S]

This is the rare variation found in the state, where known from St. Louis (*F. C. Prince*) and Montgomery (*Drouet 1363*) counties.

Ranges from New Hampshire to Ontario and Minnesota, south to Georgia, Tennessee, Mississippi, Missouri, Kansas, and Oklahoma.

16b. **Ranunculus fascicularis** var. **apricus** (Greene) Fern. Map 993

This is the common variation found in the state, the range indicated at the beginning.

Ranges from Kentucky, Illinois, Missouri, and Kansas, south to Alabama, Mississippi, Louisiana, and Texas. Specimens from Michigan to Iowa may belong here also.

The petals of *R. fascicularis* are relatively narrower in proportion to their length than the other large-flowered species, ranging from 7–15 mm. long and 2–7 mm. wide, or about $\frac{1}{4}$–$\frac{2}{5}$ as wide as long.

Plants are often difficult to place where occurring near the range of overlap of the two variations, especially in the zone from Michigan to Iowa. Most of the Missouri material appears referable to var. *apricus*, if that variety can be maintained in the future.

17. **Ranunculus parviflorus** L.
Small-flowered Crowfoot Map 994
Flowers late March–June.

Occurs in open and waste ground, often in rocky pastures and dry soils. Southern Missouri in Stoddard, Dunklin, Butler, Ripley, and Taney counties.

Native of Europe; introduced and naturalized in North America from Florida to Texas, north to New York, Kentucky, Missouri, and Oklahoma; also in California and the West Indies.

18. **Ranunculus Sardous** Crantz Map 995
Flowers May–August.

Known only from St. Louis County (St. Louis, *Sherff 210*, in Gray Herb.).

Native of Europe; introduced and naturalized in North America in New Brunswick, New York, Pennsylvania, New Jersey, Virginia, North Carolina, South Carolina, Georgia, Alabama, Illinois, Missouri, Oregon and California.

This species is common in adjacent southern Illinois, where it occurs in moist low ground of fields, meadows, along ditches, and wet depressions. It is to be expected in the southeastern sections of the state, especially around Cape Girardeau. It is a showy annual with many yellow flowers terminating the erect stems.

19. **Ranunculus arvensis** L. Map 995A
Flowers May–June.

Occurs along railroads. Known only from St. Louis County (St. Louis, Baden freight yard of Mo.-Kan.-Tex. R. R., 'between the two most western sidings, June 14, 1959, *Muehlenbach 1537*).

Native of Europe; introduced and naturalized in the United States from New Jersey and Ohio to Missouri, west to Utah.

*Excluded Species*

**Ranunculus ambigens** Wats.

*Ranunculus laxicaulis* of auth. [P & S in part], not (T & G) Darby

*Ranunculus obtusiusculus* of recent auth., not Raf.

Two specimens previously identified by Palmer and Steyermark in their *Annotated Catalogue* as *R. laxicaulis* (= *R. ambigens*) are now referred to true *R. laxicaulis* (T & G) Darby (*R. texensis* Engelm., *R. oblongifolius* of auth., not Ell.). These specimens are from Montgomery and Greene (Springfield, June 11, 1887, *Blankinship* in Drury College Herb.) counties.

**Ranunculus Beckii** G. Don

Benson (loc. cit. p. 249) places this under 'Epithets of Undetermined Application' and assigns it possibly to *R. fascicularis* or var. *apricus*. The collection was based upon a specimen collected by Beck from 'Native of Missouri in plains.'

**Ranunculus rhomboideus** Goldie

The occurrence of this species in Missouri is based upon a questionable specimen in the Schweinitz Herbarium in the Acad. of Nat. Sci. of Phila. collected questionably from Missouri at Ft. Gratiot by a questioned collector, Thenor. Since both the locality and collector are questioned as authentic, the species is excluded from the flora of the state.

## 10. **Adonis** L. Pheasant's Eye

**Adonis aestivalis** L. var. **citrina** Hoffm.

Summer Adonis Map 996

Flowers early May–June.

Occurs in cultivated fields and open ground. Known from east-central Missouri in St. Louis (in wheat field near Gumbo, May 7, 1949, *Bauer 201*) and Jefferson (Festus, May 20, 1946, *Bauer*) counties.

Native to Europe; occasionally planted as a garden ornamental and rarely escaped.

This is an annual plant with 6–9 petals 10–17 mm. long, about twice the length of the sepals, and pale yellow with purple at the base. The flowers are solitary and terminal on the usually simple erect stems 2.5–9 dm. tall.

Cases of poisoning have been reported from Europe of horses eating hay containing *Adonis aestivalis* and other species of the genus.

## 11. **Myosurus** L. Mousetail

**Myosurus minimus** L. Mousetail Map 997

Flowers March–May.

Occurs in moist fallow or cultivated fields and pastures, low alluvial ground and valleys, and waste places. Nearly throughout the state, north to Pike, Macon, Linn, Grundy, Gentry, and Nodaway counties; absent from the extreme northeastern sector.

Ranges from Florida to Texas, north to Virginia, Indiana, Illinois, Wisconsin, Minnesota, Ontario; also in British Columbia, California, and Mexico; also in Eurasia.

This is a winter annual, starting from seed in the fall and surviving the winter as a young seedling, then flowering in early spring and developing seed immediately thereafter. It occurs commonly with other plants showing similar phases of growth, such as *Androsace occidentalis*, *Viola Kitaibeliana* var. *Rafinesquii*, *Houstonia minima*, *Draba brachycarpa*, *D. verna*, *Cardamine parviflora* var. *arenicola*, and *Arabis virginica*.

## 12. **Anemone** L. Anemone

a. Flowers with all the stamens petal-like . . . . . . . . . . . . . 3. A. VIRGINIANA f. PLENA

a. Flowers with all the stamens normal and bearing anthers . . . . . . . . . . . . . *b*

b. Petal-like sepals 10–20, white usually tinged with lavender, lilac, pink, or rose; main leaves and involucre (green leaves at base of flower-stalk) cut into narrow divisions; only 1 flower or fruiting head on a stem; roots thickened, tuber-like . . . . . . . . . . . . . . 1. A. CAROLINIANA

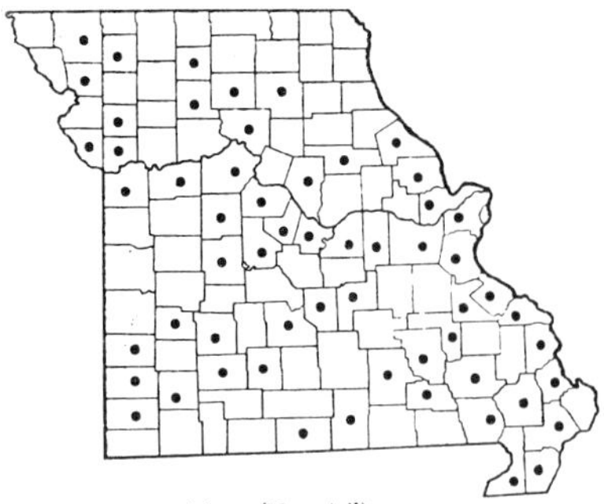

997 Myosurus minimus (Mousetail)

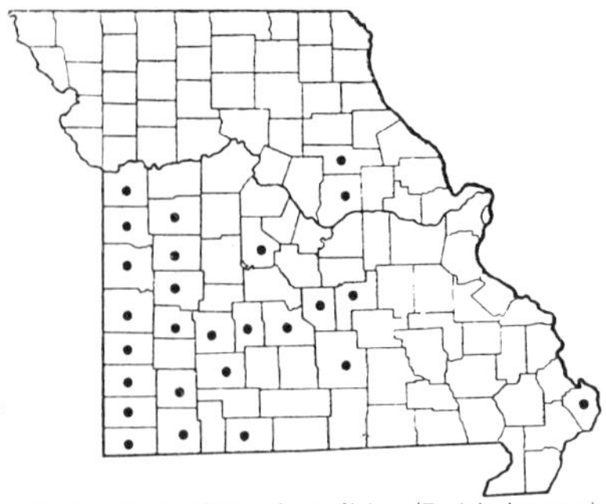

998 Anemone caroliniana f. caroliniana (Prairie Anemone)

999 Anemone cylindrica (Thimbleweed)

b. Petal-like sepals 4–8, white, creamy, or greenish-white; main leaves and involucre cut into larger and broader divisions; 2 or more flowers appearing on a stem; roots long and slender, not thickened . . . . . . . . . . . . . . . . . . . . . . . . . . . . . . . *c*

c. Leaves (involucre) at base of flower-stalk not stalked (sessile); flower-stalks equaling or less than twice as long as the leaf-divisions at their base; fruiting heads globe-shaped or spherical, about as broad as long; pistils or fruiting achenes somewhat hairy with straight hairs or glabrous, the hairs not hiding the surface of the fruit. . . . . . . . . . . . . 4. A. CANADENSIS

c. Leaves (involucre) at base of flower-stalk on definite stalks (petioles); flower-stalks at least twice as long as leaf-divisions at their base; fruiting heads cylindric or ovate-oblong, longer than broad; pistils or fruiting achenes long-hairy, at maturity the achenes densely woolly, their surface hidden by the dense wool . . . . . . . . . . . . . . . . . . . . . . *d*

d. Leaf-divisions narrowly wedge-shaped (cuneate-obovate), toothed only above the middle, narrowed or straight-sided at the base; leaves of the involucre 5–9 (rarely 3) with 3 of the leaves large and the others small; main flower-stalks (peduncles) lacking any leaves; plant covered with a gray silky hairiness; fruiting heads more than twice as long as broad, 2–4 cm. long, 0.6–1 cm. thick . . . . . . . . . . . . . . . . . . 2. A. CYLINDRICA

d. Leaf-divisions ovate, toothed to below the middle, usually curved toward the base; leaves of the involucre 2–3; some of the main flower-stalks (peduncles) bearing some leaves (involucels); plant greener, not covered with gray silky hairiness; fruiting heads less than twice as long as broad, 1–2.5 (–3) cm. long, 1.2–1.5 cm. thick . . . . . . 3. A. VIRGINIANA

1. **Anemone caroliniana** Walt. f. **caroliniana**
Prairie Anemone Map 998
Also called Carolina Anemone.
*Anemone caroliniana* Walt. [G, BB, P & S, Steyerm.]
Flowers late March–early May.

Occurs in open rocky prairies and glades in acid soils associated usually with chert or sandstone rocks, meadows, waste ground, and along roadsides. Almost restricted to the unglaciated prairie section of the southwestern third of the state north to Jackson County, east to Morgan, Phelps, Texas, and Taney counties, northeast locally in a part of the glaciated prairie section in Callaway and Audrain counties, and in sandy ground in Mississippi County, southeastern Missouri (*Steyermark 80936*). Introduced in Taney, Phelps, Pulaski, and Texas counties, but native elsewhere.

Ranges from Florida to Texas, north to North Carolina, Indiana, Wisconsin, Minnesota, and South Dakota.

The deep violet colored form of the species, *A. caroliniana* f. *violacea* Clute, has not been recorded from Missouri. Most of the stations in Missouri are from acid soils. It is noted, however, that in other portions of its range the species is sometimes found in calcareous gravels.

2. **Anemone cylindrica** Gray Thimbleweed
Map 999

Flowers May–July.

Occurs on dry open loess hills and loess prairies and dry open wooded hills of northwestern Missouri

Plate no. 170. **1.** Adonis aestivalis var. citrina, × 2/5; Details from Small, The New York Botanical Garden. **2.** Myosurus minimus, × 4/5; Details from Small, The New York Botanical Garden. **3.** Anemone caroliniana, × 2/5. **4.** Anemone cylindrica, × 2/5. **5.** Anemone virginiana, × 2/5; Details from Small, The New York Botanical Garden. **6.** Anemone canadensis, × 2/5; Details from Small, The New York Botanical Garden.

PLATE NO. 170

1000 • Anemone virginiana f. virginiana (Thimbleweed)
1000 □ Anemone virginiana f. leucosepala
1000 x Anemone virginiana f. plena

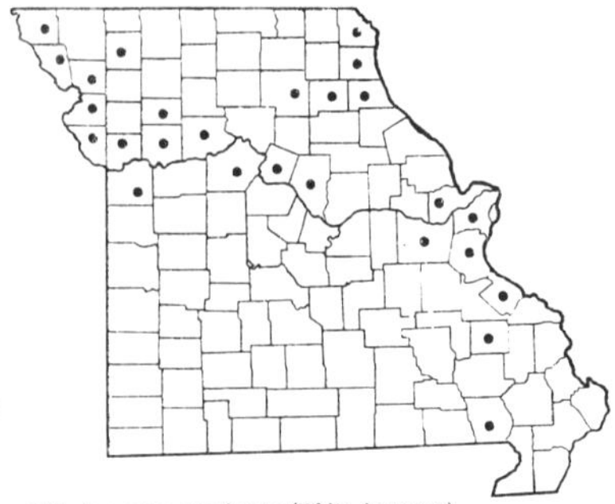

1001 Anemone canadensis (White Anemone)

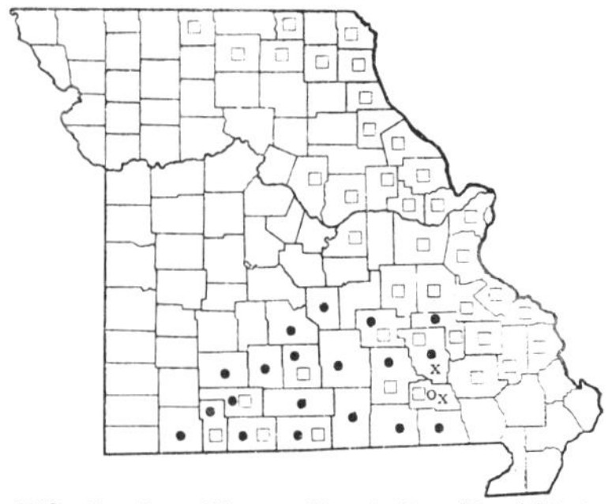

1002 • Hepatica nobilis var. obtusa f. obtusa (Round-lobed Hepatica)
1002 □ Hepatica nobilis var. acuta f. acuta (Sharp-lobed Hepatica)
1002 x Hepatica nobilis var. obtusa f. obtusa X Hepatica nobilis var. acuta f. acuta

and west-central Missouri in Atchison, Nodaway, Holt, and Jackson counties.

Ranges from Maine to British Columbia, south to New Jersey, Pennsylvania, Ohio, Indiana, Illinois, Missouri, Kansas, New Mexico, and Arizona.

This is the rarest species in Missouri.

3. **Anemone virginiana** L. Thimbleweed
Map 1000

Flowers late April–August.

Occurs in rocky or dry open woods and prairies. Throughout Missouri.

Ranges from Quebec and Maine to North Dakota, south to Georgia, Alabama, Tennessee, Arkansas, and Oklahoma.

The following variations are in Missouri:

a. Flowers with all the stamens petal-like . . . 3c. A. VIRGINIANA f. PLENA
a. Flowers with all the stamens normal and bearing anthers . . . . . . . . . . *b*
  b. Petal-like sepals greenish or greenish-yellow and leathery . . . . . 3a. A. VIRGINIANA f. VIRGINIANA
  b. Petal-like sepals white, thinner and more petaloid . 3b. A. VIRGINIANA f. LEUCOSEPALA

3a. **Anemone virginiana** f. **virginiana** Map 1000

*Anemone virginiana* L. [G, BB, P & S, Steyerm.]

This is the common variation found throughout the state.

3b. **Anemone virginiana** f. **leucosepala** Fern.
Map 1000

Less common and scattered.

3c. **Anemone virginiana** f. **plena**
Palmer & Steyermark Map 1000

Known only from St. Louis County (Rockwoods Reservation, 1956, *George E. Moore*, holotype in Chi. Nat. Hist. Mus. Herb.).

4. **Anemone canadensis** L. White Anemone
Map 1001

Also called Meadow Anemone.

Flowers May–July.

Occurs in low moist alluvial ground of river flood plain, low meadows and thickets, low ground along railroads, sometimes in low woods or wooded thickets at the foot of river bluffs, and along levees. Chiefly along the Missouri and Mississippi rivers and a few of their tributaries, in northern, central, and eastern Missouri; absent from the main part of the Ozark and unglaciated prairie sections of the state.

Ranges from Quebec to British Columbia, south to Maryland, West Virginia, Ohio, Indiana, Illinois, Missouri, Kansas, and New Mexico.

This is a very showy-flowered species, usually occurring in large colonies. The plants spread from slender underground rhizomes, and show a tendency to spread when introduced into perennial gardens.

## 13. **Hepatica** Mill. Hepatica, Liverleaf

Lobes of leaf and of sepal-like green bracts (involucre) just below flower rounded . a. H. NOBILIS var. OBTUSA
Lobes of leaf and of sepal-like green bracts (involucre) just below flower pointed or acute at apex. . b. H. NOBILIS var. ACUTA

a. **Hepatica nobilis** Schreb. var. **obtusa** (Pursh) Steyermark f. **obtusa** Round-lobed Hepatica, Round-lobed Liverleaf Map 1002

*Hepatica triloba α obtusa* Pursh, Fl. Am. Syst. 2: 391. 1814

*Hepatica triloba* var. *americana* DC. Sept. 1 : 216. 1817

*Hepatica americana* (DC.) Ker in Edwards' Bot. Reg. 5. *t. 387*. 1819 [G, BB, P & S, Steyerm.]

Flowers March–April, sometimes a second flowering November–December.

Occurs on rich or rocky wooded slopes and ravine bottoms, and along mossy banks and ledges, usually in acid soils associated with chert, sandstone, or granite rocks, but also with limestone. Southern Missouri in the Ozark section north to Iron, Dent, Pulaski, and Laclede counties, and east to Iron, Carter, and Ripley counties.

Ranges from Quebec and Nova Scotia to Minnesota and Manitoba, south to Florida, Alabama, and Missouri.

The usual color of the flowers is lavender to bluish. The white-flowered form occurs scattered throughout the range and is known as:

**Hepatica nobilis** var. **obtusa** f. **candida** (Fern.) Steyermark

*Hepatica americana* f. *candida* Fern. [G, P & S, Steyerm.]

Weatherby (Rh. 27: 131–32. 1925) grew plants having various color phases in his garden, and reported that the white-flowered forms were constant, whereas the other colors varied from one year to the next. More experimental work needs to be carried out along these lines.

b. **Hepatica nobilis** Schreb. var. **acuta** (Pursh) Steyermark f. **acuta** Sharp-lobed Hepatica, Sharp-lobed Liverleaf Map 1002

*Hepatica triloba β acuta* Pursh, Fl. Am. Sept. 2: 391. 1814

*Hepatica acuta* (Pursh) Britton, Ann. N.Y. Acad. Sci. 6: 234. 1891

*Hepatica triloba* var. *acutiloba* (DC.) Warne in Am. Ent. & Bot. 2: 313. 1870

*Hepatica acutiloba* DC. Prodr. 1: 22. 1824 [G, BB, P & S, Steyerm.]

*Anemone acutiloba* (DC.) Laws. Rev. Canad. Ranunc. 30. 1870.

Flowers February–April, sometimes a second

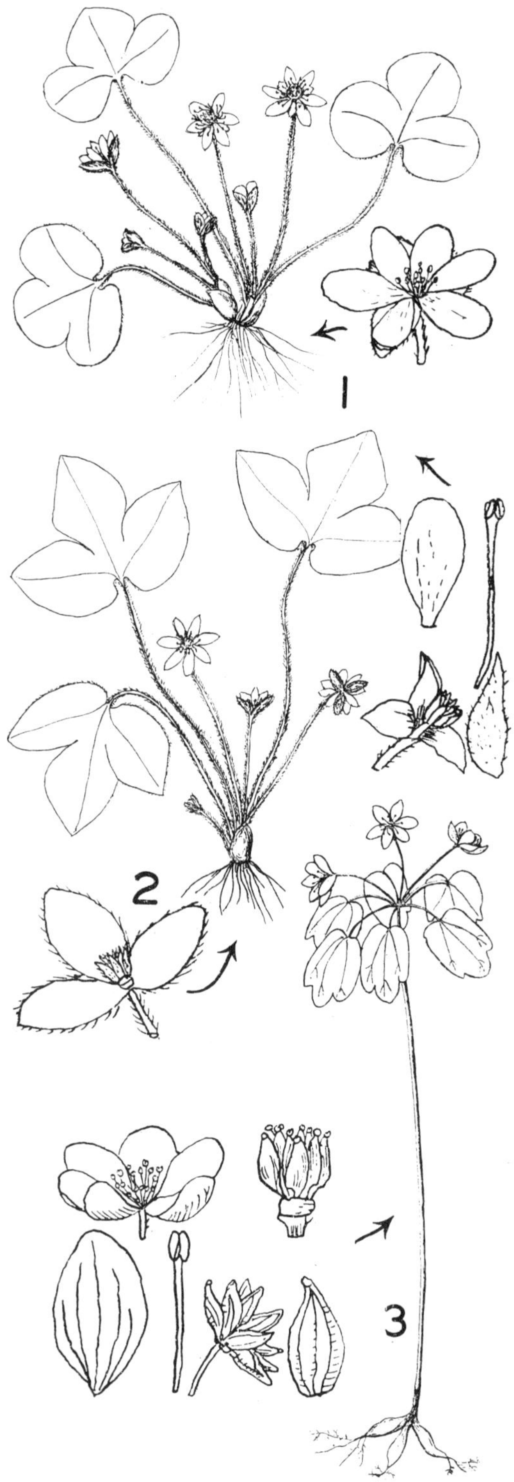


Plate no. 171. **1.** Hepatica nobilis var. obtusa f. obtusa, × ²/₅. **2.** Hepatica nobilis var. acuta f. acuta, × ²/₅. **3.** Anemonella thalictroides, × ²/₅; Details from Small, The New York Botanical Garden.

flowering November–December.

Occurs on rich, usually north-facing wooded slopes of well-drained soil, mostly around limestone ledges, bluffs, and outcrops. Eastern and southern Missouri; west in northern Missouri to Mercer and Sullivan counties, and west in southern Missouri to Wright, Christian, Taney, and Stone counties.

Ranges from Quebec and Maine to Minnesota, south to Georgia, Alabama, and Missouri.

The usual color of the flowers is lavender or bluish. Both white and rosy or pink forms are found scattered in Missouri throughout the range. The white-flowered one is known as:

**Hepatica nobilis** var. **acuta** f. **albiflora** (R. Hoffm.) Steyermark

*Hepatica acutiloba* DC. f. *albiflora* R. Hoffm. [G, P & S, Steyerm.]

The rosy or pink form is known as:

**Hepatica nobilis** var. **acuta** f. **rosea** (R. Hoffm.) Steyermark

*Hepatica acutiloba* f. *rosea* R. Hoffm. [G, P & S, Steyerm.]

This variety hybridizes with var. *obtusa* where the two come together at the margins of overlap of their range in the state. Such hybrids are known from Reynolds (steep north-facing wooded slopes along West Fork of Black River, T32N, R2W, sect. 1, southeast of West Fork P. O., 7½ mi. [by air] northeast of Bunker, April 26, 1952, *Steyermark 73114*) and Carter counties, and probably occur elsewhere. At the Reynolds County locality the uppermost acid chert slopes were occupied by *H. nobilis* var. *obtusa*, while the lower and middle limestone slopes were taken over by var. *acuta*. The individual hybrid plants occurred in the zone between the two varieties.

Such hybrid plants show an intermediate condition in the lobes of the leaves between rounded and pointed. The American plants of *Hepatica* are only slightly different from the European species, *H. nobilis* Schreb. Previously considered two separate species, *H. americana* and *H. acutiloba*, differing only in the matter of type of apex of the leaf lobes, they would appear more logically regarded as American varieties of the European species, in much the same manner as the American Yellow Lady Slipper orchids are considered to be varieties of the European *Cypripedium Calceolus* L. Gleason (*New Ill. Fl.* 2: 183. 1952) indicates likewise that the difference between *H. americana* and the European *H. nobilis* is 'slight and scarcely warrants specific segregation.' In the European plants there are also obtuse- and acute-lobed types.

## 14. **Anemonella** Spach. Rue Anemone

**Anemonella thalictroides** (L.) Spach

Rue Anemone Map 1003

Flowers late March–June.

Occurs in dry open or rocky woods, usually in acid soils on upland slopes and ridges. Absent from the northwest sector and the lowlands of extreme southeastern Missouri, otherwise throughout the state; in northern Missouri west to Mercer, Grundy, Livingston, and Carroll counties.

Ranges from Maine to Minnesota, south to Florida, Alabama, Mississippi, Arkansas, and Oklahoma.

The following variations occur in Missouri:

a. Stamens and pistils are all petal-like, rosy-lavender, pink, or white . . . . . . . c. A. THALICTROIDES f. FAVILLIANA

a. Stamens and pistils all normal, not petal-like . *b*

b. The petal-like sepals are white, rose, or lavender shades . . . a. A. THALICTROIDES f. THALICTROIDES

b. The petal-like sepals are green and leaflike b. A. THALICTROIDES f. CHLORANTHA

a. **Anemonella thalictroides** f. **thalictroides** Map 1003

*Anemonella thalictroides* (L.) Spach [G, BB, P & S, Steyerm.]

This is the widely distributed and common variation found.

b. **Anemonella thalictroides** f. **chlorantha** Fassett Map 1003

Known only from Polk County in the western Ozarks (damp limestone cliff, Brighton, April 24, 1937, *Fassett 18606*, holotype in U. of Wis. Herb.).

c. **Anemonella thalictroides** f. **Favilliana** Bergseng Map 1003

This 'double-flowered' form is known only from Putnam County, northern Missouri, where the author saw it growing in the garden of Mrs. Guernsey Furgason, who stated she collected it on dry open wooded slopes of Solly Mountain, along Dog Branch, tributary to Chariton River, T66N, R16W, sect. 17, just south of Mapleton, 4 mi. south of Livonia, May 26, 1956. An herbarium specimen was made from the upper part of the plant.

This is a beautiful form of Rue Anemone, resembling a miniature tuberous begonia, marigold, or zinnia in the buttonlike condensed flowers. The author has had plants of this form growing for the past nineteen years in his woodland, where they were originally found native. Plants can be increased through propagation of the fleshy tuberous roots.

Rue Anemone is one of the longest-flowering of the species in the spring flora, the petal-like sepals of the flowers persisting from one to three months.

Plants with rose-colored petal-like sepals have been named *A. thalictroides* f. *rosea* (Millsp.) Moldenke (Boissiera 7: 1. 1943).

## 15. **Clematis** L. Clematis, Leather Flower, Virgin's Bower

a. Stem erect, not climbing, short, mostly 1.5–6 dm. tall; leaves all simple, not compound, sessile (without leaf-stalk) or nearly so . . . . . . . . . . . . . . . 7. C. FREMONTII var. RIEHLII

a. Stem climbing, sprawling, or trailing, very elongated; at least some of the leaves compound, divided into 3–5 leaflets, on conspicuous petioles (leaf-stalks) . . . . . . . . . . . . . . . *b*

 b. Flowers white or whitish, in a compound many-flowered inflorescence, small, the petal-like sepals 0.6–1.8 cm. long, the petal-like sepals all widely spreading; flowering time mostly July–October . . *c*

  c. Leaflets usually toothed, mostly 3 (more rarely 5) to the compound leaf; petal-like sepals 0.6–1.2 cm. long; anthers 0.6–1.5 mm. long . . . . . . . . . . . . 1. C. VIRGINIANA

  c. Leaflets mainly entire (smooth-edged), sometimes wavy-margined or roundly lobed, usually 5 to the compound leaf; petal-like sepals 1–1.7 cm. long; anthers 2–4 mm. long . 2. C. DIOSCOREIFOLIA

 b. Flowers purple, bluish, brown-lavender, or lavender and green, solitary on long stalks (peduncles) or 3 together at the most, large, 1.5–4.5 cm. long, the petal-like sepals bell-shaped, mostly united except at the tips; flowering time mostly April–July . . . . . . . . . . . . . . . . *d*

  d. Lower surface of leaflets noticeably silvery-gray or whitened (glaucous); leaflets rounded or blunt at tip, thickened, coriaceous (leathery and firm); tail-like styles conspicuously feathery with spreading hairs (plumose) in the fruiting stage; stems and leaflets glabrous (without hairs) . . . . . . . . . . . . . . . . . . . . . 4. C. VERSICOLOR

  d. Without the above combination of characters . . . . . . . . . . . . . . . . *e*

   e. Tail-like styles conspicuously feathery with spreading hairs (plumose) in the fruiting stage; petal-like sepals joined together for nearly their entire length, free only near tip; rare species in the southern Ozark region . . . . . . . . . . . . . . . . . 3. C. VIORNA

   e. Tail-like styles at most covered with short erect hairs and sometimes silky hairy or nearly hairless above the middle; petal-like sepals free in the half towards the tip, this half strongly recurved; plants common throughout the state or else mainly confined to the lowlands of southeastern Missouri . . . . . . . . . . . . . . . . . . . . . . *f*

    f. Petal-like sepals mostly 3–4.5 cm. long, the upper half spreading and broadened above the middle with thin broad wavy margins; leaflets glabrous, thin, not noticeably reticulated (with a network of conspicuous veins); plants mainly of the swamps of southeastern Missouri . . . . . . . . . . . . . . . . . . . 6. C. CRISPA

    f. Petal-like sepals 1.5–2.5 cm. long, the upper half recurved but with a narrow margin which is not wavy-edged; leaflets usually with some hairs on at least nerves or veins of lower surface, rather thick and firm, noticeably reticulated (with a network of conspicuous veins); common plants throughout Missouri, but rarely found in the swamp section of southeastern Missouri . . . . . . . . . . . . . . 5. C. PITCHERI

### 1. **Clematis virginiana** L. Virgin's Bower

Map 1004

Flowers July–September.

Occurs in moist or low ground of woodland and thickets bordering streams, ponds, and fence rows. Scattered throughout Missouri.

Ranges from Nova Scotia and Quebec to Manitoba, south to Georgia, Alabama, Mississippi, Louisiana, and Oklahoma.

Missouri material varies as follows:

Lower surface of leaflets glabrous (without hairs) or sparsely hairy . 1a. C. VIRGINIANA f. VIRGINIANA

Lower surface of leaflets densely hairy with hairs persisting in fruit . . . . . 1b. C. VIRGINIANA f. MISSOURIENSIS

### 1a. **Clematis virginiana** f. **virginiana** Map 1004

*Clematis virginiana* L. [G, P & S]

*Clematis virginiana* var. *virginiana* [BB]

Mostly recorded from the eastern and northern halves of the state.

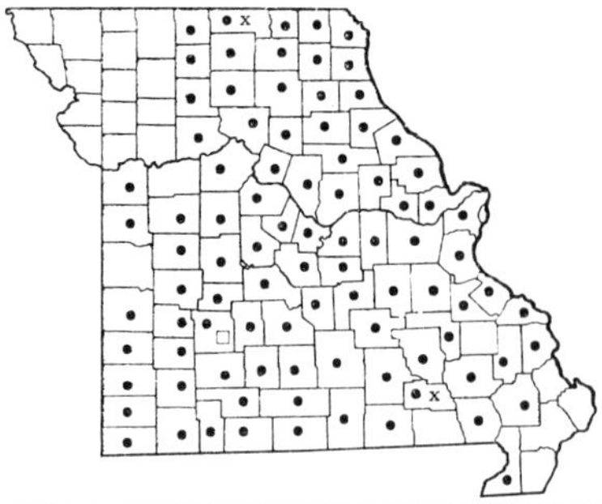

1003 • Anemonella thalictroides f. thalictroides (Rue Anemone)
1003 □ Anemonella thalictroides f. chlorantha
1003 x Anemonella thalictroides f. Favilliana

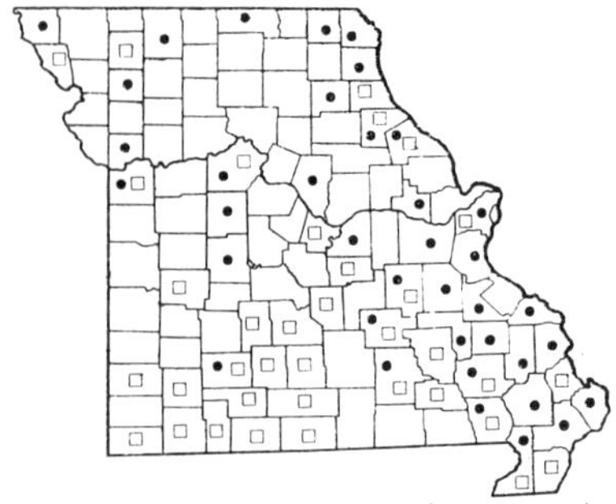

1004 • Clematis virginiana f. virginiana (Virgin's Bower)
1004 □ Clematis virginiana f. missouriensis

1005 Clematis dioscoreifolia var. dioscoreifolia (Virgin's Bower)

1b. **Clematis virginiana** f. **missouriensis** (Rydb.) Fern. Map 1004

*Clematis virginiana* var. *missouriensis* (Rydb.) Palmer and Steyerm. [BB, P & S]

Mostly occurring in the southern half of the state, and locally in northern Missouri.

Some individuals, upon contact with the leaves, develop a type of dermatitis.

Gleason (*New Ill. Fl.* 2: 184. 1952) has segregated these two variations as varieties instead of forms, further distinguishing the typical form by the thickened rounded margins of the achenes as contrasted with no thickened rounded margins for f. *missouriensis*. I have not been able to correlate these differences with the density of leaf pubescence.

2. **Clematis dioscoreifolia** Lévl. & Vaniot var. **dioscoreifolia** Virgin's Bower Map 1005

*Clematis dioscoreifolia* Lévl. & Vaniot [G, BB]

*Clematis paniculata* Thunb. var. *dioscoreifolia* (Lévl. & Vaniot) Rehder

Flowers July–October.

Occurs along railroads, roadsides, and waste places. Known from St. Louis (south of Frisco R. R. right-of-way on the River des Peres sewer slope, opposite the emporium of Finer Iron and Metal Company, St. Louis, October 9, 1954, *Muehlenbach 487*; September 10, 1955, *Muehlenbach 770*), Jasper (along roadside, Webb City, August 24, 1952, *Palmer 54920*), Lawrence (waste ground, Aurora, August 15, 1950, *Palmer 50751*), and Newton counties.

Native of Japan; commonly cultivated as an ornamental for the fragrance and abundance of white flowers, and escaped from cultivation sparingly from Massachusetts to Florida, Missouri, and Oklahoma.

Another variation, var. *robusta* (Carr.) Rehd. differs in the thinner leaflets, which are ovate instead of round-ovate or suborbicular, and less strongly cordate at the base.

3. **Clematis Viorna** L. var. **Viorna** Leather Flower Map 1006

*Clematis Viorna* L. [G, BB, P & S, Steyerm.]

Flowers May–July.

Occurs in rocky woods and around limestone bluffs. Southern Missouri, in Bollinger, Wayne, and Taney counties.

Ranges from Georgia to Texas, north to Pennsylvania, Ohio, Indiana, southern Illinois, and Missouri.

The leaflets are rather thin and membranaceous, and usually somewhat hairy on the lower surface. They are not noticeably reticulate, as in *C. Pitcheri* and *C. versicolor*, and are green on both sides.

4. **Clematis versicolor** Small Leather Flower Map 1007

Flowers early May–June.

Occurs in rocky open woods, and ledges of bluffs usually of limestone. Southern Ozark region from McDonald County east to Oregon County and north to Texas and Greene counties.

Ranges from Kentucky and Tennessee to Missouri, Arkansas, and Oklahoma.

The flower, which is blue-lavender with greenish tips, resembles that of both *C. Pitcheri* and *C. Viorna* in general coloring. The stems are nearly cylindrical with curved sides, not 6-angled as in *C. Viorna*. The nodes of the stems may be slightly hairy sometimes, as they are usually in *C. Viorna*. The vegetative parts of the plants are otherwise glabrous. The leaves are much firmer and thicker and more conspicuously reticulated than those of *C. Viorna*, and have the lower

---

Plate no. 172. **1.** Clematis virginiana, × 2/5; Details from Small, The New York Botanical Garden. **2.** Clematis dioscoreifolia, × 2/5; Details from Small, The New York Botanical Garden. **3.** Clematis Viorna, × 2/5.

PLATE NO. 172

1006 Clematis Viorna var. Viorna (Leather Flower)

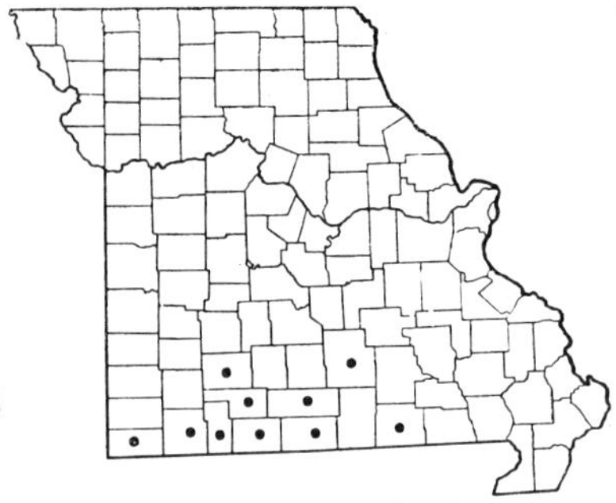

1007 Clematis versicolor (Leather Flower)

1008 Clematis Pitcheri (Leather Flower)

surface usually conspicuously whitened or glaucous, a character lacking from the other species closely related. The leaflets in *C. Viorna* are always acutely pointed, whereas in *C. versicolor* they are always rounded or blunt at the tip, a character which likewise usually distinguishes it from *C. Pitcheri,* and always from *C. crispa.*

5. **Clematis Pitcheri** T. & G. Leather Flower Map 1008

Also called Bluebell.

Flowers May–late September.

Occurs in low or rocky woodland, ledges of bluffs, thickets, level ground, or on slopes. Throughout Missouri, except apparently absent from most of the lowland counties of extreme southeastern Missouri.

Ranges from Indiana to Iowa and Nebraska, south to Tennessee, Arkansas, Oklahoma, and Texas.

This is the commonest species of the Leather Flower group in the state. The leaves vary somewhat in shape, size, and texture, but are usually rather firm, green on both sides, and more or less prominently reticulated.

6. **Clematis crispa** L. Swamp Leather Flower Map 1009

Also called Blue Jasmine.

Flowers April–July.

Occurs in swampy low woods, borders of swamps, and along fence rows in low open ground. Known only from the swamps of southeastern Missouri from Cape Girardeau County on the north to Ripley County on the west, and locally in Ozark County. A specimen from Newton County (*Palmer 14369*) previously identified as *C. crispa* is here referred to *C. Pitcheri.*

This is a much showier-flowered species than any of the other Leather Flowers, and has more bluish color than the flowers of either *C. Pitcheri, C. Viorna,* or *C. versicolor.* A purple-blue or blue-lilac color pervades the outside and outer margin of the petal-like sepals, while white normally occurs around the middle section on the inside. The flower buds are long-pointed and longer-tapering than those of the other species.

7. **Clematis Fremontii** S. Wats. var. **Riehlii** Erickson Fremont's Leather Flower Map 1010
*Clematis Fremontii* [of P & S, Steyerm.], not S. Wats.

Flowers April–May.

Occurs on limestone glades of the eastern Ozark region in St. Louis, Franklin, Jefferson, Ste. Genevieve, St. Francois, and Washington counties, and locally in Ozark County, southern Missouri (north side of road BB, T24N, R12W, sect. 25, $3\frac{1}{2}$ mi. south of Dora, August 19, 1956, *Steyermark 82365*; *Bauer*); a collection is also known from Shannon County (April, 1948) from a plant sent to Bill Bauer by a friend, but has no definite locality data.

Known only from Missouri and west-central Kansas.

Typical var. *Fremontii* from Nebraska and Kansas differs chiefly in having thicker, often coarsely toothed, often broader and more orbicular to rounded ovate leaves, and shorter peduncles 0.5–4 cm. long; in var. *Riehlii* the peduncles may lengthen to 4–6.5 cm. long.

For a discussion of the ecological and taxonomic interest of this species, the reader is referred to Erickson's paper on the genus (Ann. Mo. Bot. Gard. 30: 1–62. 1943) and on the dolomite glades of east-central Missouri in connection with this species (Erickson, R. O., L. G. Brenner, and Joseph Wraight in Ann. Mo. Bot. Gard. 29: 89–101. 1942) as well as to Fernald's later discussion of variation (Rh. 45: 410–12. 1943) within the group.

Plate no. 173. **1.** Clematis pitcheri, $\times\ ^2/_7$. **2.** Clematis crispa, $\times\ ^2/_7$. **3.** Clematis versicolor, $\times\ ^2/_7$. **4.** Clematis Fremontii var. Riehlii, $\times\ ^2/_7$; Details from Small, The New York Botanical Garden.

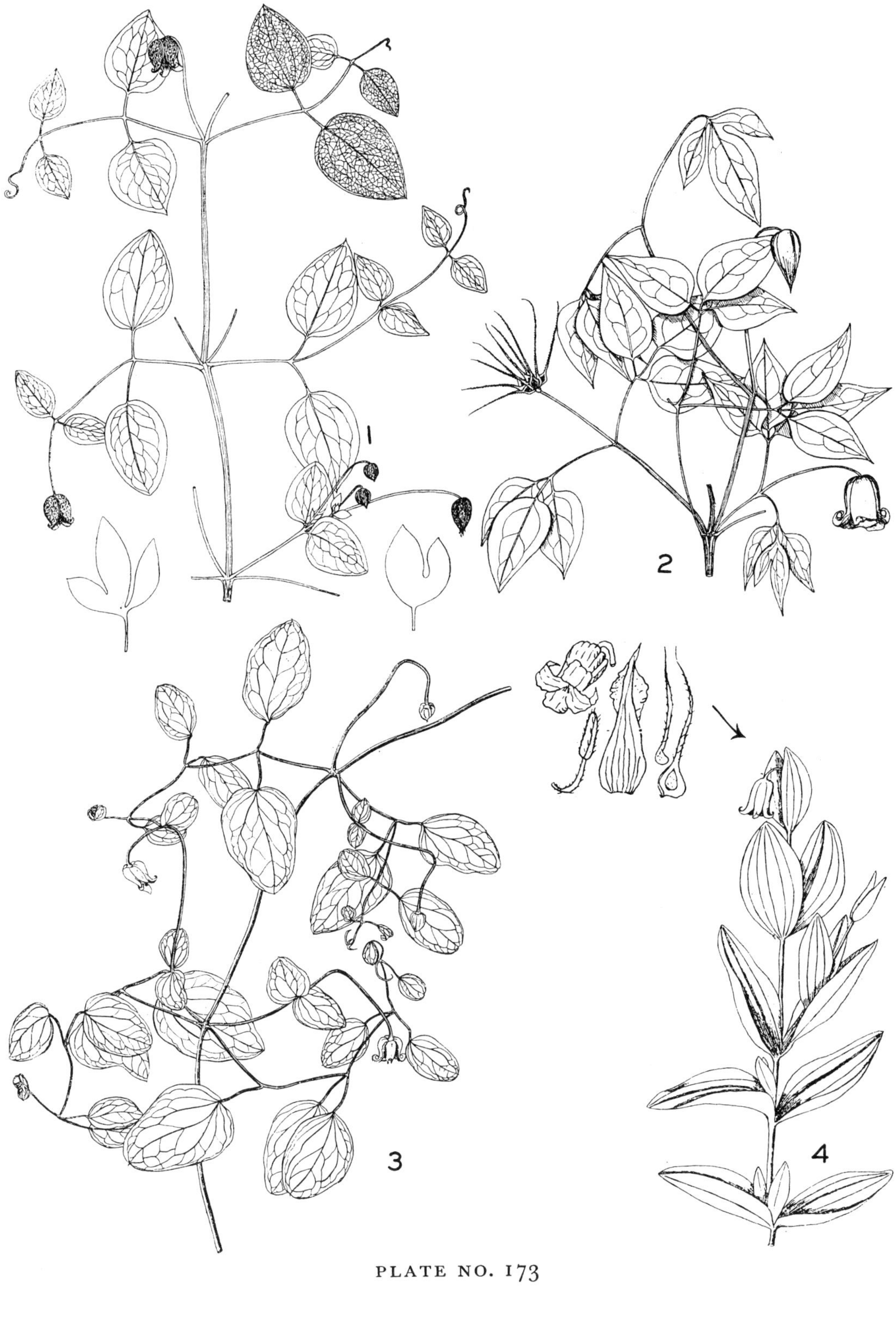


PLATE NO. 173

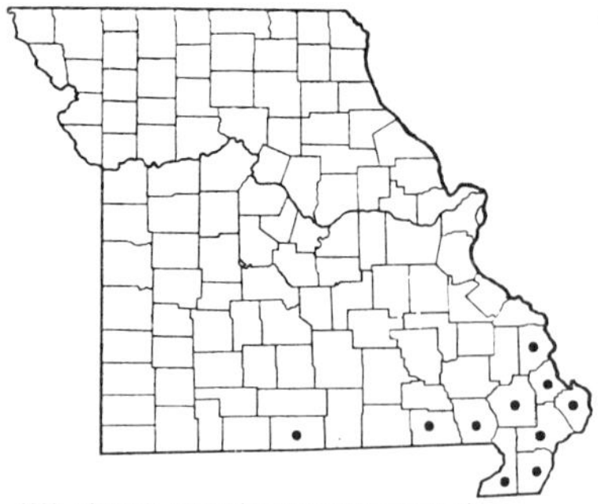

1009 Clematis crispa (Swamp Leather Flower)

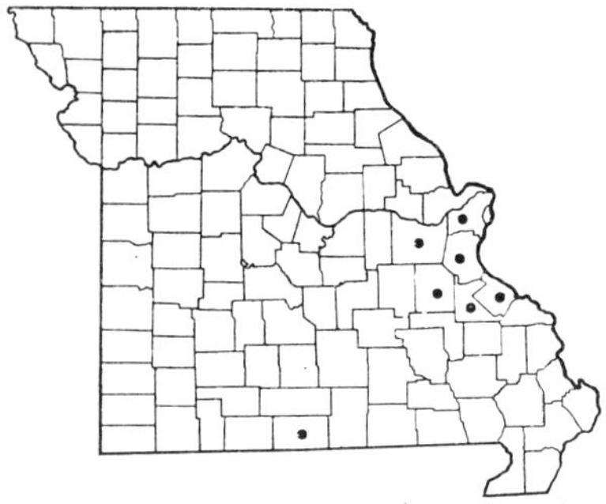

1010 Clematis Fremontii var. Riehlii (Fremont's Leather Flower)

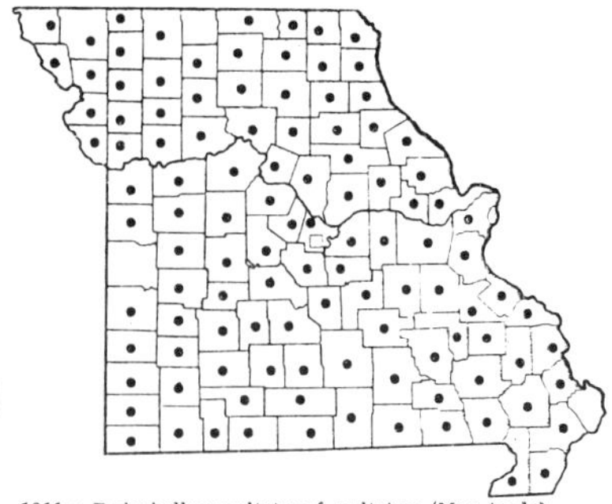

1011 • Podophyllum peltatum f. peltatum (May Apple)
1011 □ Podophyllum peltatum f. Deamii

*Excluded Species*

**Clematis glaucophylla** Small

Cited from Missouri by Fernald (p. 665 of eighth ed. *Gray's Man.*), but no specimens have been seen either at the Gray Herbarium or elsewhere. It is found in Garland County, Arkansas, some considerable distance south of the Missouri border. It is not cited from Missouri by Erickson in his monograph, and should be excluded from the flora of the state until a definite authentic record is known.

**Clematis ligusticifolia** Nutt.

Reported as occurring in Missouri by Fernald (p. 664 of eighth ed. *Gray's Man.*). There is a specimen so labeled in the Drury College Herbarium at Springfield, Mo., collected by E. M. Shepard in Greene County in 1880. While the leaves are 5-foliolate and firm in texture, the specimen appears rather to be a 5-foliolate variant of *C. virginiana* f. *missouriensis*; *C. ligusticifolia* is a more western species and the evidence for including it in the state flora on the above basis is inconclusive and open to serious doubt.

**Clematis verticillaris** DC.

In the University of Missouri Herbarium is a plant correctly identified as this species with a label containing 'Dunklin Co., November 1856' on one part of the label and 'Montana' written along the right hand side. Undoubtedly, Montana is the correct state from which the collection was taken, rather than Missouri, as Montana and other western and northern states are included in the usual geographical range of this species. The collection should be disregarded for Missouri.

## Fam. **BERBERIDACEAE** (Barberry Family)

a. Shrubs usually with prickles; wood yellow; leaves either smooth-edged (entire) or with fine teeth along margins; flowers yellow; fruit (when mature) a red berry . . . . . . . . . . 3. BERBERIS
a. Soft-stemmed herb; leaves deeply lobed or divided into many leaflets; flowers white, pink, or yellowish-green; fruit (when mature) deep blue, greenish, yellow, or rarely maroon-purple . . . . . *b*
  b. Leaf deeply lobed, umbrella-like, attached at the middle to the main stalk; flowers white or pink, large, 1 on each plant; fruit usually pale green or yellow, rarely maroon-purple . 1. PODOPHYLLUM
  b. Leaf compound, divided into many leaflets; flowers yellowish-green, small, several on each plant; mature fruit (actually the seed) deep blue . . . . . . . . . . . 2. CAULOPHYLLUM

### 1. **Podophyllum** L. May Apple

**Podophyllum peltatum** L. May Apple Map 1011

Also known as Mandrake.

Flowers late March–May; fruit ripe July–August.

Occurs in low moist or dry open woods and thickets, occasionally along railroads. Throughout Missouri, doubtless in every county.

Ranges from Quebec to Ontario and Minnesota, south to Florida and Texas.

The following variations are known in Missouri:

Common type; petals white; ovary yellow-green; fruit yellow-green or pale yellow; seeds yellow; fruiting stalk (peduncle) yellow-green . . . a. P. PELTATUM f. PELTATUM

Rare type; petals pink or pinkish; ovary deep

wine-purple; fruit maroon-purple; seeds dark purple-brown; fruiting stalk (peduncle) purplish
b. P. PELTATUM f. DEAMII

a. **Podophyllum peltatum** f. **peltatum** Map 1011
*Podophyllum peltatum* L. [G, BB, P & S, Steyerm.]
The common type throughout Missouri.

b. **Podophyllum peltatum** f. **Deamii** Raymond Map 1011

Known only from Cole County, central Missouri (wooded slopes above small tributary at south end of Hough Park, a few hundred yards southwest of golf course hole no. 12, ½ mi. south of limits of Jefferson City, T44N, R11W, west-central part sect. 19, July 17, 1950, *Steyermark 69987*; same locality, May 6, 1951, *Steyermark 71155*).

Known only from Pennsylvania, Indiana, Illinois, and Missouri.

The plants occur in acid soil. In addition to the differences noted in the key, this form also has stem, leaf-stalk, and flower-stalk flecked with lavender, and the tip of the rhizome bud a dark wine-purple color. Plants in the author's wildflower preserve have been growing there since 1950, when originally transplanted there from the Missouri station. The single plant has now formed a large colony a couple of yards in diameter.

Many years ago the late B. F. Bush wrote the author that he had found *Podophyllum peltatum roseum* (a rose-colored form) but had not distributed the collections. Although he had promised to send living plants the following spring, none ever arrived. Mrs. Fred Glenn wrote me that she had found the pink-flowered May Apple on their property near Peculiar, Cass County, but I have seen no specimens of the plant. Undoubtedly, this f. *Deamii* will be found eventually in other sections of Missouri. A form having orange-colored fruit (f. *Biltmoreanum* Steyerm.) is known from Illinois and should be expected to occur in Missouri and elsewhere.

May Apple is bitter and therefore avoided by grazing animals. The rootstocks have been used in medicine for their cathartic properties. Both the rootstocks and the leaves of the plant are poisonous, but the ripe fruit when pale yellow is edible. It may be eaten raw, having a slightly sweet, mildly acid quality, or may be prepared into a jelly, marmalade, or preserves slightly reminiscent of the tropical guava jelly or paste. A refreshing drink may also be prepared from the juice of the fruit when mixed with lemonade or other fruit juices, or mixed with sugar.

Some persons develop a dermatitis after handling the rootstocks. The rootstocks have been the object of some study in recent years as a possible remedy in treating certain afflictions, such as cancer, but no positive results have appeared from such studies.

## 2. **Caulophyllum** Michx. Blue Cohosh

**Caulophyllum thalictroides** (L.) Michx.
Blue Cohosh Map 1012

Flowers late March–May.

Occurs in rich woods of valleys, ravines, north-facing wooded slopes and bluffs, and moist banks at the base of bluffs. Throughout the eastern half of the state, scattered westward locally to Atchison, Clay, Jackson, and Jasper counties.

Ranges from New Brunswick to Manitoba, south to South Carolina, Alabama, Tennessee, and Missouri; also in eastern Asia.

The unusual ultramarine blue spherical fruits are actually naked seeds surrounded by their fleshy coat. They are generally mature in summer, and are considered as poisonous if eaten by children especially.

Like the May Apple, this plant is quite bitter, and is therefore avoided by grazing animals. The thickened rootstocks have been used in medicine. Some persons are susceptible to handling the plant and develop a dermatitis.

## 3. **Berberis** L. Barberry

a. Leaves smooth-edged (entire); none of the prickles forking, all simple . . . . . . . 3. B. THUNBERGII
a. Leaves with some teeth; some or most prickles forking or 3-pronged . . . . . . . . . . . . *b*
b. Teeth of mature leaves ending in a bristle; 2-year-old branchlets grayish; usually 13–20 teeth on each side of leaf; petals notched at tip; shrubs 1–3 m. high, escaped from cultivation . 2. B. VULGARIS
b. Teeth of mature leaves not ending in a bristle; 2-year-old branchlets orange- or ruddy-brown; usually 2–11 teeth on each side of leaf; petals smooth-edged (entire); shrubs mostly less than 1 m., usually 0.3–0.9 dm. tall, native in southern Missouri . . . . . . . . . . . 1. B. CANADENSIS

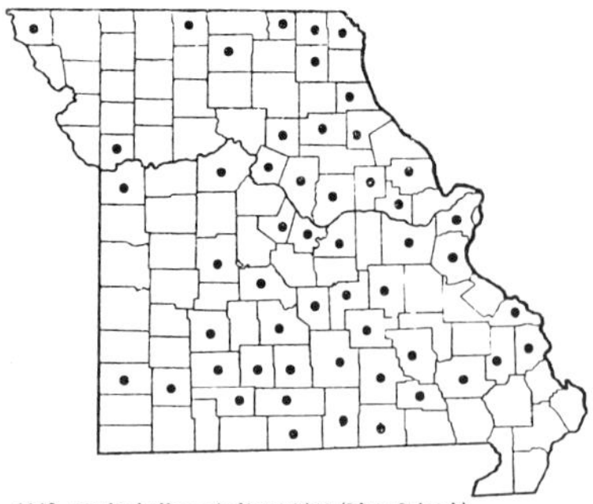
1012 Caulophyllum thalictroides (Blue Cohosh)

1013 Berberis canadensis (American Barberry)

1014 Berberis vulgaris (Common Barberry)

1. **Berberis canadensis** Mill. American Barberry Map 1013

Flowers May; fruit June–July.

Occurs on rocky wooded, usually north-facing upper ledges of bluffs along streams, and on exposed upper portions of bluffs, on limestone, or sandstone. Known only from Shannon, Texas, and Howell counties in the southern Ozark section.

Ranges from West Virginia and Virginia to Georgia and Alabama, and in Indiana and Missouri.

This is the only native species of Barberry in Missouri, and should be rigidly protected where it is found. It appears to be restricted to the edges of limestone bluffs where the soil is leached out, or near the contact between the chert or Roubidoux sandstone with the limestone. In Howell County it is found at 'County Hollow' by the bluffs of Eleven Points River, about 5 mi. northeast of Peace Valley (*Steyermark 14502*), in Texas County between Barn Hollow Canyon and highway 17 along Jack's Fork of Current River (*Steyermark 19086*), and in Shannon County from the top of Jam-up Bluff along Jack's Fork of Current River and another bluff at 'Bear Hole Bluff' about ½ mile west of Jam-up Bluff, where it occurs on the top of that bluff (*Steyermark 26890*), associated or near other rare relict species for Missouri, such as *Galium boreale* var. *hyssopifolium*, *Campanula rotundifolia*, *Trautvetteria caroliniensis*, and *Zigademus glaucus*.

2. **Berberis vulgaris** L. Common Barberry Map 1014

Flowers May–June.

Escaped from cultivation sparingly into pastures, open woods, and occasionally along rocky spring branches. Known as escaped from plantings only from Putnam, Clark, Marion, and Phelps counties.

Native of Europe; introduced and naturalized in North America from Nova Scotia to Minnesota, south to Delaware, Pennsylvania, Indiana, Iowa, and Missouri.

Formerly more common in Missouri but mostly exterminated from previous localities as a result of the barberry eradication campaign carried on at one time by the United States Department of Agriculture to get rid of the black stem rust of wheat, oats, rye, barley, and various wild and cultivated grasses. This rust had one stage of its life cycle on the cereals and other grasses, and the alternate stage occurred on the European Barberry. By exterminating the alternate stage of the Barberry, the rust was not able to survive on the grass host.

The ripe fruits of this species may be used in the making of preserves, jellies, jams, and pies, while the juice from the fruit is extracted for a syrup to serve as a refreshing drink. The fruit gives a certain sour or tart taste to preserved fruits and can be used like cranberries as an acid sauce or cooked in soups to lend a special flavor.

3. **Berberis Thunbergii** DC. Japanese Barberry Map 1015

Flowers April–May.

Commonly cultivated as a hedge plant, but rarely escaped into waste ground and open woods. Known only from Marion (pasture 1 mi. north of entrance of Riverview Park, Hannibal, October 11, 1933, *Drouet 1249*; escaped in open woods throughout the city of Hannibal) and Lincoln (grazed wooded valley in thin soil below sandstone outcrops along Sandy Creek, T50N, R2E, sect. 34, 2 mi. northwest of Foley, October 8, 1956, *Steyermark 83032*) counties.

Native of Japan; introduced and naturalized from

Plate no. 174. **1.** Caulophyllum thalictroides, × 2/5; Details from Small, The New York Botanical Garden. **2.** Berberis vulgaris, × 2/5. **3.** Berberis Thunbergii, × 2/5. **4.** Berberis canadensis, × 2/5; *a, b.* Types of leaves; Details from Small, The New York Botanical Garden.

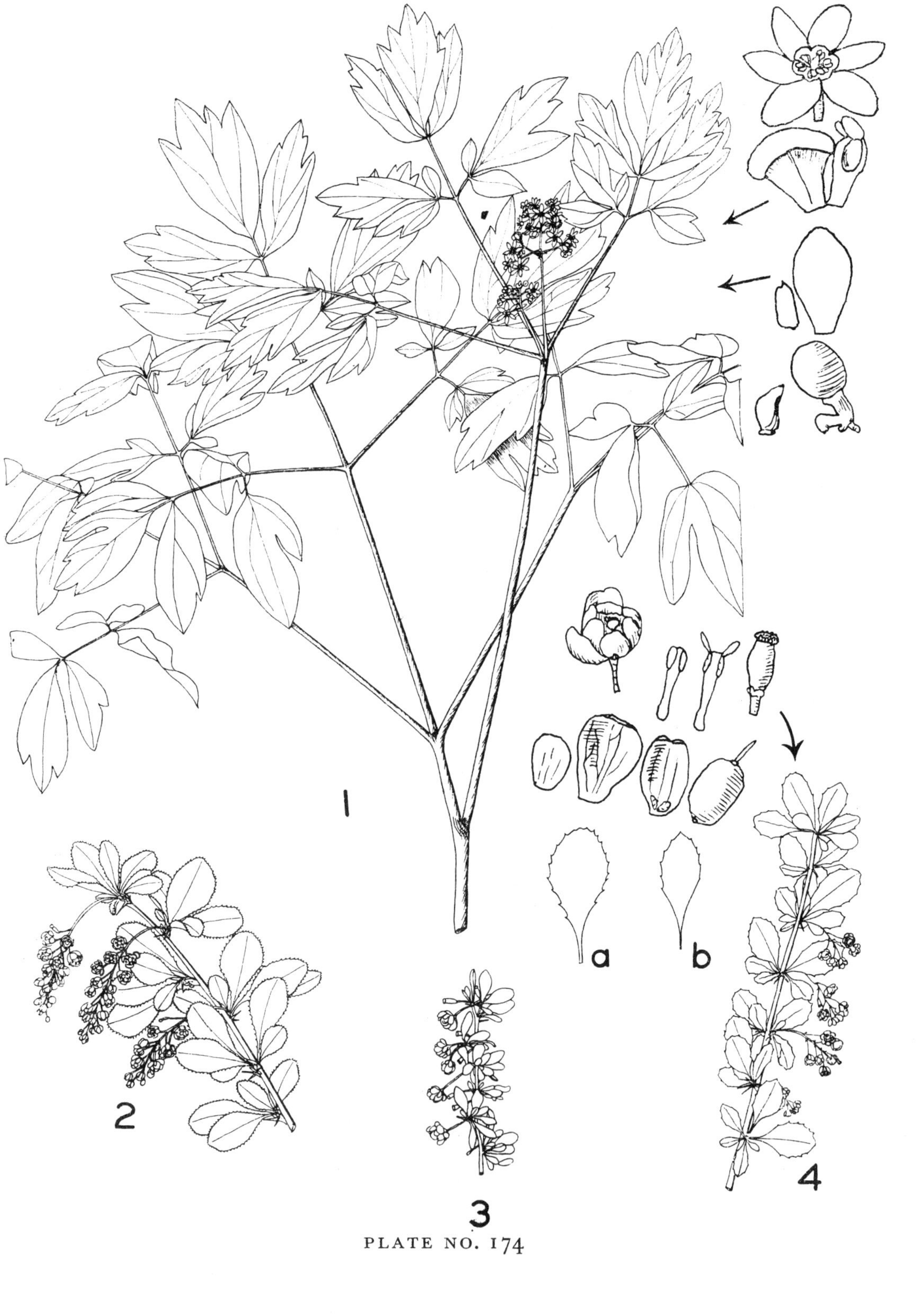


PLATE NO. 174

1015 Berberis Thunbergii (Japanese Barberry)

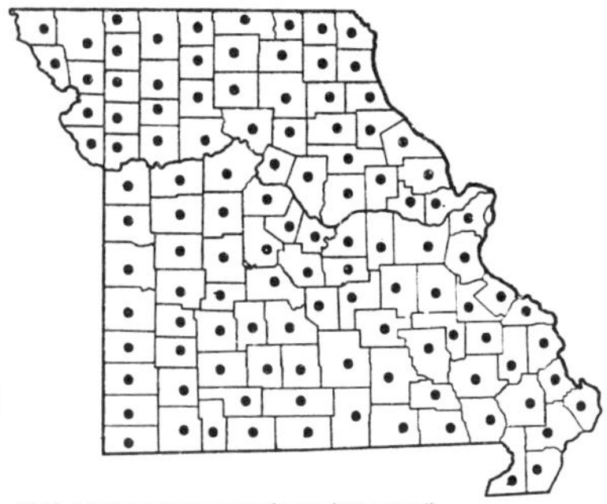

1016 Menispermum canadense (Moonseed)

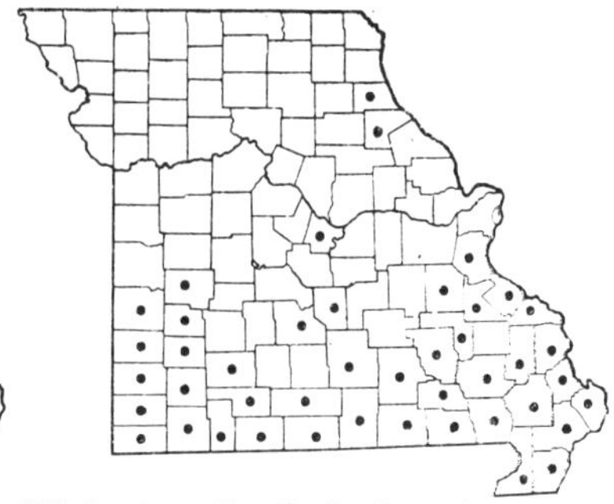

1017 Cocculus carolinus (Carolina Moonseed)

Nova Scotia to Michigan, south to North Carolina and Missouri.

*Excluded Genus and Species*

**Jeffersonia diphylla** (L.) Pers. Twin-leaf

The late B. F. Bush wrote the author that his specimen of this species from Missouri was contained in the collections he made between 1899–1902 and that these collections were stolen. He stated definitely that he had found this plant 'in the creek bottom about a mile above the mouth of Billin Creek on the east bank' near Swan in Taney County. Attempts have been made to relocate this plant, but without success.

As there are no definite specimens to corroborate Bush's report, the species is excluded from the present flora. The nearest stations to the Missouri border are found in southern Illinois, so that search for the plant should be directed in some of the counties bordering the Mississippi River as well as a renewed careful search in the area of Swan, Taney County, where Bush indicated.

## Fam. **MENISPERMACEAE** (Moonseed Family)

a. Usually some of leaf-blades mostly longer than broad, ovate or triangular, usually smooth-edged (entire) and without lobes or with wavy lobes or 3–5-lobed, densely hairy on lower surface; fruit red; stamens of the male (staminate) flowers 6; petals 6; sepals 6 . . . . . . . . . 2. COCCULUS

a. Leaf-blades usually as broad as long or broader than long, usually 3–5–7-angled or lobed, more rarely without lobes, only sparsely hairy on lower surface or nerves or nearly glabrous; fruit bluish-black or black; stamens of the male (staminate) flowers 12–24; petals 0 or 6–8; sepals 4–8 . . . . . *b*

b. Top of leaf-stalk (petiole) joining the leaf on the lower side of the leaf-blade near the lower margin of the blade; leaves shallowly 3–7-angled or lobed; lower surface of leaf-blade pale gray or silvery-gray; leaf-blade firm and thickish; petals 6–8; sepals 4–8; stamens 12–24 in the male (staminate) flowers, represented by 6 rudiments in the female (pistillate) flowers . . 1. MENISPERMUM

b. Top of leaf-stalk (petiole) joining the leaf directly at the edge of the angle or crotch at the base of the leaf-blade; leaves deeply 3–7-lobed with long-pointed lobes; lower surface of leaf-blade green, scarcely paler than the upper surface; leaf-blades thin; petals 0; sepals 6; stamens 12 in the male (staminate) flowers, 9 in the female (pistillate) flowers . . . . . . . 3. CALYCOCARPUM

### 1. **Menispermum** L. Moonseed

**Menispermum canadense** L. Moonseed Map 1016

Flowers May–July.

Occurs in low moist woods and thickets in ravines and along streams, in valleys, along fence rows, and at the base of bluffs. Throughout Missouri, where it

---

Plate no. 175. **1.** Menispermum canadense, $\times {}^{2}/_{7}$; *a, b, c.* Various leaf shapes. **2.** Podophyllum peltatum, $\times {}^{2}/_{7}$. **3.** Lindera Benzoin var. Benzoin, $\times {}^{2}/_{7}$; *a.* Male flower, $\times 1{}^{5}/_{7}$; *b.* Male flower-clusters, $\times {}^{2}/_{7}$. **4.** Cocculus carolinus, $\times {}^{2}/_{7}$; *a.* Branch with male inflorescence and leaves, $\times {}^{2}/_{7}$; *b.* Male flower $\times 1{}^{1}/_{7}$; *c.* Female flower $\times 1{}^{1}/_{7}$; *d.* Another leaf type, $\times {}^{2}/_{7}$; After Britton and Brown, The New York Botanical Garden. **5.** Sassafras albidum, $\times {}^{2}/_{7}$; *a.* Branch with male flowers, $\times {}^{2}/_{7}$; *b.* Male flower, $\times 2$. **6.** Calycocarpum Lyoni, $\times {}^{2}/_{7}$; All details from Small, The New York Botanical Garden.

PLATE NO. 175

is known from every county.

Ranges from Quebec to Manitoba, south to Georgia, Alabama, Arkansas, Oklahoma, and Texas.

The rootstock was at one time used in medicine. The fruit, about 1 cm. long, is bluish-black and covered with a bloom (glaucous coating). It occurs in small grapelike clusters, but is not to be eaten and has been responsible for several cases of poisoning. This vine, which twines from left to right, is difficult to eradicate when once introduced into the home grounds.

## 2. **Cocculus** DC. Carolina Moonseed

**Cocculus carolinus** (L.) DC. Carolina Moonseed
Map 1017

Also called Carolina Snailseed, Fishberry.

Flowers July–August.

Occurs in rocky open woods, limestone glades, alluvial ground and thickets bordering streams and ponds, fence rows, and roadsides. Southern Missouri, north to Jefferson, Washington, Pulaski, Cole, Greene, St. Clair, and Vernon counties, and introduced in northeastern Missouri in Ralls and Marion (Hannibal, August, 1921, *Mrs. Wm. C. Parker*) counties; the specimens from Ralls and Marion counties undoubtedly were taken from cultivated or introduced plants; common in the southeastern lowland counties.

Ranges from Florida to Texas, north to Virginia, Kentucky, Illinois, Missouri, and Kansas.

The red fleshy fruits, 5–8 mm. long, are very showy. The seed has a $\frac{3}{4}$ moon-shaped appearance, sunken along the sides. The author has kept plants growing in his botanical preserve in northern Illinois since 1948, but they have not flowered during this time.

## 3. **Calycocarpum** Nutt. Cupseed

**Calycocarpum Lyoni** (Pursh) Gray Cupseed
Map 1018

Flowers May–June.

Occurs in low alluvial rich ground in wooded valleys and along banks of streams, in thickets, and borders of fields. Southern and east-central Missouri, restricted to the Ozark and southeastern lowland sections, north to St. Louis, Franklin, Osage, Cole, Miller, St. Clair, and Jasper counties.

Ranges from Florida to Louisiana, north to Kentucky, Illinois, Missouri, Kansas, and Oklahoma.

This vine climbs fairly high on trees. The fruit, which is 20–25 mm. long, is blackish and has a seed which is scooped out on one side like a cup.

# Fam. **LAURACEAE** (Laurel Family)

At least some of the leaves with 1–3 lobes (rarely 5-lobed), with 3 main nerves arising near the base of the leaf-blade; young twigs yellow-green; flowers and fruits appearing only at the tips of the branches, in loosely stalked clusters; flowers usually appearing with the young leaves . . . . . . . 1. SASSAFRAS

None of the leaves lobed, all simple, pinnately veined with 5–7 pairs of nerves along the midvein; young twigs some shade of brown; flowers and fruits appearing along the sides as well as at the tips of the twigs, in nearly sessile (not stalked) round clusters; flowers appearing before the leaves come out 2. LINDERA

## 1. **Sassafras** Nees Sassafras

**Sassafras albidum** (Nutt.) Nees Sassafras
Map 1019

Flowers early April–May.

Occurs in dry acid soils on the border of woods or prairies, borders of rocky chert, sandstone, or granite glades, wooded alluvial soils in valleys, roadsides, railroads, fence rows, fallow fields, and thickets.

Two variations occur in Missouri:

Lower surface of leaves glabrous (without hairs) or nearly so from the beginning . . a. S. ALBIDUM var. ALBIDUM

Lower surface of at least some of the fully grown leaves more or less hairy, densely hairy when young . . . . . b. S. ALBIDUM var. MOLLE

a. **Sassafras albidum** var. **albidum** Map 1019

Sometimes known as White Sassafras.

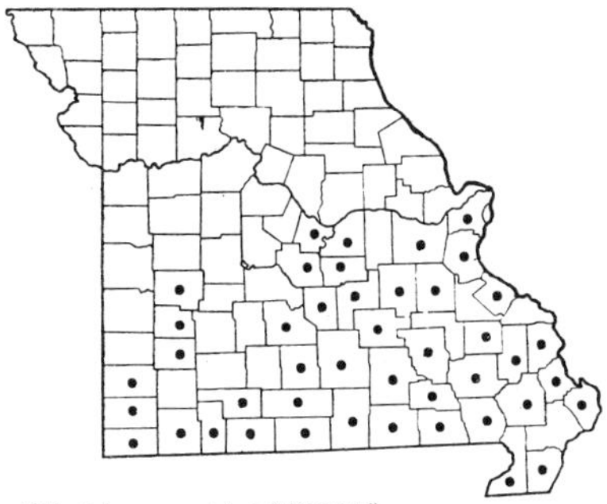

1018 Calycocarpum Lyoni (Cupseed)

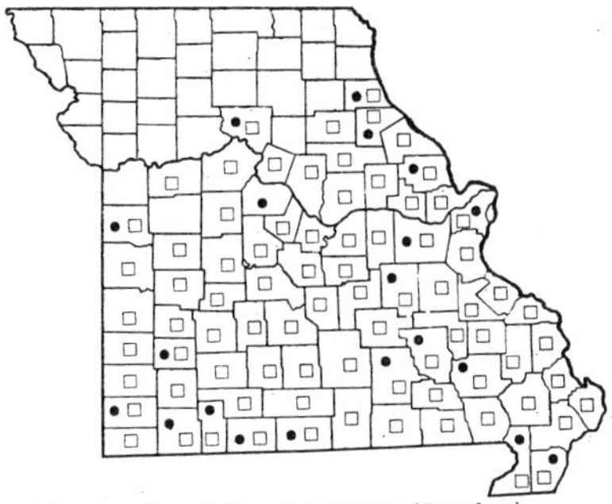

1019 • Sassafras albidum var. albidum (Sassafras)
1019 □ Sassafras albidum var. molle

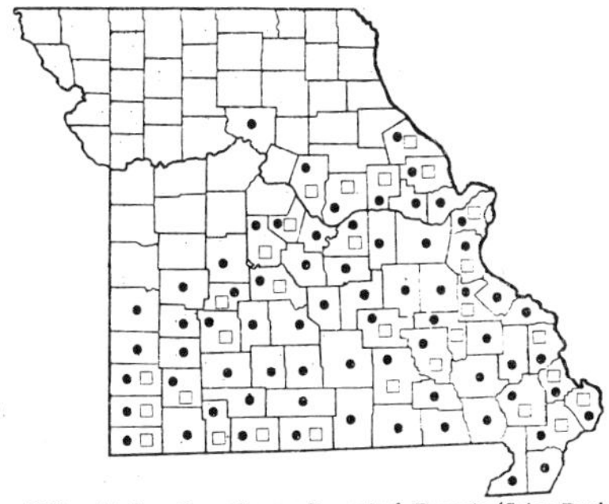

1020 • Lindera Benzoin var. Benzoin f. Benzoin (Spice Bush)
1020 □ Lindera Benzoin var. pubescens

*Sassafras albidum* (Nutt.) Nees [G, BB, Steyerm.]
*Sassafras officinale* Nees & Eberm. var. *albidum* (Nutt.) Blake [P & S]

This is much less common in Missouri than the more pubescent var. *molle*, and is scattered in southern and east-central Missouri north to Marion, Chariton, and Cass counties.

Ranges from Virginia to Arkansas, north to Maine, New Hampshire, Vermont, New York, Ohio, Michigan, Illinois, and Missouri.

b. **Sassafras albidum** var. **molle** (Raf.) Fern. Map 1019

Sometimes known as Red Sassafras.

*Sassafras officinale* Nees & Eberm. [P & S]

This is the commoner variation encountered throughout southern, central, and part of northeastern Missouri, ranging north to Marion, Monroe, Randolph, Chariton, Saline, Lafayette, and Cass counties.

Ranges from Florida to Texas, north to Maine, New Hampshire, New York, Ohio, Indiana, Illinois, Iowa, Missouri, and Kansas.

The twigs of this variety are usually more pubescent than those of var. *albidum*, but are also glabrous.

Sassafras usually is found in colonies, sending up root suckers freely. The shape of the tree and its branching is similar throughout the colony, since all the individual trees have arisen from an original parent plant. The leaves of sassafras begin to turn color fairly early in the season, beginning sometimes in late August to show some orange color, eventually changing from pale orange, yellow, or salmon to orange-red.

The twigs are spicy-fragrant and are frequently chewed or nibbled for their pleasant spicy taste. The bark from the root is often gathered for sassafras tea, an aromatic orange-brown drink made by boiling pieces of the root-bark and often used as well as a spring tonic of uncertain therapeutic value. The mucilaginous young leaves and pith of the stems, after being dried and powdered, are used as a thickener in gumbo soup in the southern states under the name of 'filet' or 'filé,' and also as a condiment.

An oil obtained from the roots and bark is used as a flavoring in root beer and other carbonated beverages, tooth pastes and powders, candies, perfume for soap, and in medicines. The orange or yellow dye from the bark has been used for dyeing cloth. The dark blackish-blue fruits, with a glaucous coating, are eaten by birds.

The wood is used for fence posts, rails, and crossties because of its durability, also in construction of small boats, and to a limited extent in flooring and articles of furniture manufacture. In some of the southern states it was traditional to make bedsteads from sassafras wood, chicken houses from the smaller pole wood, and sassafras sticks to stir the kettle in which soap was being prepared.

Sassafras is a desirable ornamental tree, but is very difficult to transplant. It can be grown easily from the seed, however. Trees are usually of either one sex or the other.

## 2. **Lindera** Thunb. Spice Bush

a. Rare shrub, known only from lowlands in Ripley County, southeastern Missouri; leaf-blades drooping, thin, membranaceous, about as green on the lower as on the upper surface, rounded or obtuse at the curved base; lower surface of leaf-blade with conspicuous elevated venation between the main lateral nerves; lowest 2 pairs of lateral nerves not parallel to those higher up, ascending at a markedly different angle; fruiting pedicels (stalks) stout, 9–12 mm. long, conspicuously enlarged at the summit, 2.5–3 mm. wide; mature fruit (in dried state) 10–11.5 mm. long, 7–8 mm. wide; calyx-segments of male (staminate) flowers 1–1.25 mm. wide; pedicels (stalks) of male flowers slightly hairy; pedicels of female (pistillate) flowers 2.5 mm. long; fruiting stalks persistent from previous year and lasting until time of flowering . . . . . . . . . . . . . 2. L. MELISSAEFOLIUM

a. Common shrub of southern and east-central Missouri; leaf-blades erect or ascending, rather firm and thickish, pale or grayish (subglaucous) on lower surface, narrowed to an acute or wedge-shaped base; lower surface of leaf-blade without conspicuous venation between the main lateral nerves; lateral nerves mostly all parallel and spreading at about the same angle; fruiting pedicels slender, 3–4 mm. long, not conspicuously enlarged at summit, 1–1.5 mm. wide at summit; mature fruit (in dried state) 8–10 mm. long, 5–7 mm. wide; calyx-segments of male flowers 1.5–2 mm. wide; pedicels of male flowers glabrous (without hairs); pedicels of female flowers 1–1.5 mm. long; fruiting stalks deciduous, not persistent to the next flowering season . . . . . . . . . . . . . 1. L. BENZOIN

1. **Lindera Benzoin** (L.) Blume Spice Bush Map 1020

Flowers March–May.

Occurs in low or moist woodland and thickets along streams, in valleys, ravine bottoms, base of bluffs, along spring branches, and seepage of wooded slopes. Southern and east-central Missouri[6] north to Pike, Montgomery, Callaway, Boone, Chariton, Moniteau, Morgan, Benton, St. Clair, and Vernon counties.

Two variations are found in Missouri:

Lower surface of leaf-blades glabrous (without hairs) or nearly so; leaf-stalk (petiole) and bud glabrous or nearly so; branches glabrous . . 1a. L. BENZOIN var. BENZOIN

At least the nerves on the lower surface of the leaf-blades more or less hairy; leaf-stalk (petiole) and bud more or less hairy; branches often more hairy . . . . . 1b. L. BENZOIN var. PUBESCENS

1a. **Lindera Benzoin** var. **Benzoin** f. **Benzoin** Map 1020

*Lindera Benzoin* (L.) Blume [G]

*Lindera Benzoin* var. *Benzoin* [BB]

*Benzoin aestivale* of Nees [P & S, Steyerm.], not *Laurus aestivalis* L.

Ranges from Maine to Ontario, Michigan, and Illinois, south to North Carolina, Kentucky, Missouri, Kansas, and Oklahoma.

1b. **Lindera Benzoin** var. **pubescens** (Palmer & Steyerm.) Rehd. Map 1020

*Benzoin aestivale* var. *pubescens* Palmer & Steyerm. [P & S, Steyerm.]

Nearly as common in Missouri as the glabrous variety.

Ranges from Florida to Texas, north to New Jersey, Pennsylvania, Kentucky, Indiana, Michigan, Illinois, Iowa, and Oklahoma.

The shrubs of Spice Bush are usually sexually distinct, having either all male (staminate) or all female (pistillate) flowers. The staminate plants have larger and showier masses of yellow flowers than the pistillate plants. The shrubs vary from 1.6–4.5 meters (about 5–14 feet) tall and are bushy and much-branched. Their foliage persists as dark green well into fall, eventually turning greenish-yellow.

The spicy aromatic twigs, leaves, and fruits can be used to make a pleasant fragrant tea. During the Revolutionary War the fruit was dried and powdered and used as a substitute for allspice.

The bright red fruits of the pistillate shrubs and the bright masses of yellow flowers appearing before the leaves in early spring, especially showy on the staminate shrubs, make the Spice Bush a plant of ornamental value, particularly in shaded areas or for naturalizing in shrub borders or woodland. They are easily transplanted or grown from seed.

2. **Lindera melissaefolium** (Walt.) Blume Pond Berry Map 1021

Flowers late March–early April.

Occurs in swamps, where known only from Ripley County, southeastern Missouri (wooded depression, T22N, R4E, sect. 35, 4 mi. south of Naylor, October 19, 1948, *Steyermark 66947*; same locality, March 29, 1949, *Steyermark 67084* and *67089* [staminate plants], and *67090* [pistillate plants]).

Ranges from Florida to Louisiana, north to North Carolina and Alabama; locally northwest in southeastern Missouri.

This is one of the rarest shrubs in the United States, and at the single locality in Missouri where it was discovered it is known as Pondberry (See Steyermark, Rh. 51: 153–62. 1949 for discussion of this species). It is a lower less-branched shrub than *L. Benzoin*, varying in height from 0.6–2 meters (approximately 2–6½ feet) tall, and the leaves have a more drooping habit than those of Spice Bush. It grows in the water of swamps and swampy depressions, and at the Missouri station these depressions are surrounded by

1021 Lindera melissaefolium (Pond Berry)

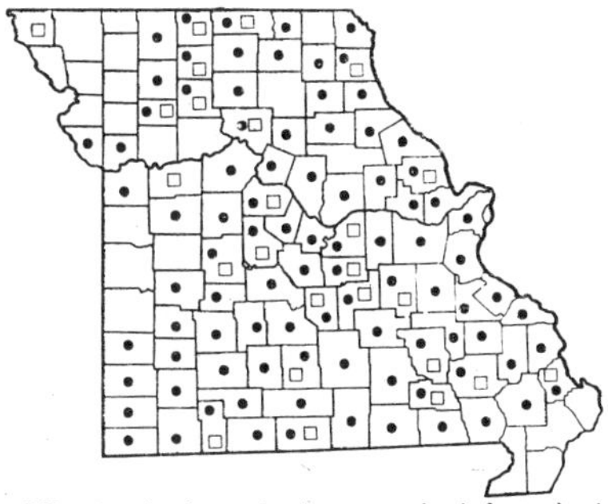

1022 • Sanguinaria canadensis var. canadensis f. canadensis (Bloodroot)

1022 □ Sanguinaria canadensis var. rotundifolia

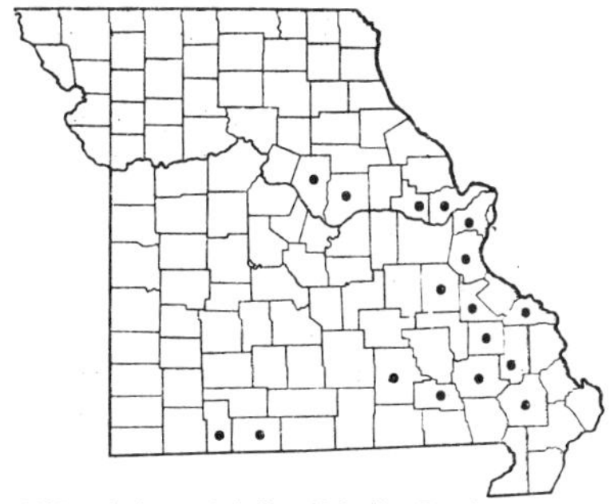

1023 Stylophorum diphyllum (Celandine Poppy)

sandy knolls. On the drier adjacent knolls ordinary Spice Bush (*L. Benzoin*) occurs.

Search for the species in other localities in the southeastern lowland section of Missouri has failed thus far to reveal any new stations, but it should eventually be found elsewhere in that area and in adjacent Arkansas. The scarlet fruits, larger than in *L. Benzoin*, are quite showy in the late fall, and are used locally as ammunition for 'pop guns,' tubular contrivances constructed from twigs of Elderberry (*Sambucus canadensis*) in this section of the state.

Plants grown from seed at the author's botanical preserve in northern Illinois die back near the tips each winter, but renew growth each spring. They are gradually dying out, however, after ten years, apparently needing a more acid soil and a less rigorous winter climate.

## Order **PAPAVERALES**

### Fam. **PAPAVERACEAE** (Poppy Family)

a. The 1 or 2 leaves arising only from the base of the plant; petals usually 8 or more (rarely 6); juice of plant orange-red . . . . . . . . . . . . . . . . . . . . . . . . . . . . 1. SANGUINARIA

a. Leaves attached to the main stem as well as arising at the base of the plant; petals 4–6; juice of plant yellow or white . . . . . . . . . . . . . . . . . . . . . . . . . . . . . *b*

  b. Leaves or stems prickly . . . . . . . . . . . . . . . . . . . . . . . 5. ARGEMONE

  b. Leaves and stems not prickly . . . . . . . . . . . . . . . . . . . . . . . . *c*

    c. Petals red, salmon, or bluish-white with a purple center; juice of plants not yellow . . 4. PAPAVER

    c. Petals yellow or orange; juice of plants yellow or orange-yellow. . . . . . . . . . . *d*

      d. Native woodland plant; petals 18–30 mm. long; sepals hairy; fruit (capsule) elliptic-oblong, hairy . . . . . . . . . . . . . . . . . . . . . 2. STYLOPHORUM

      d. Rare introduced plant of waste ground, known only from Jackson County, west-central Missouri; petals 8–13 mm. long; sepals glabrous (without hairs); fruit (capsule) linear or slenderly cylindric, glabrous . . . . . . . . . . . . . . . . . 3. CHELIDONIUM

### 1. **Sanguinaria** L. Bloodroot

**Sanguinaria canadensis** L. Bloodroot Map 1022

Also known as Red Puccoon.

Flowers March–April.

Occurs in rich or rocky wooded slopes, at the base of bluffs, in ravines, and valley bottoms, most common and abundant around limestone outcrops. Throughout Missouri, but apparently absent or unrecorded from most of the extreme northwestern portion and some of the western unglaciated prairie and southeastern lowland counties.

Two variations are known in Missouri:

Leaf with 5–9 lobes, the upper lobes of about equal width but narrower than the 2 lowest lobes; margins of the broader lowest lobes and summits of the other lobes coarsely or irregularly bluntly toothed . . . . . . . . . a. S. CANADENSIS var. CANADENSIS f. CANADENSIS

Leaf with no lobes or with mostly 3 lobes, the

middle one rather small as compared with the others (if 5-lobed, then 3 are shallowly lobed and narrower than the broader other 2); margins of the lobes slightly wavy or entire (without teeth) . . . b. S. CANADENSIS var. ROTUNDIFOLIA

a. **Sanguinaria canadensis** var. **canadensis** f. **canadensis** Map 1022

*Sanguinaria canadensis* L. [G, BB, P & S, Steyerm.]

This is a common variation in Missouri.

Ranges from Quebec to Manitoba, south to New England, Pennsylvania, Kentucky, Illinois, Missouri, and Kansas.

b. **Sanguinaria canadensis** var. **rotundifolia** (Greene) Fedde Map 1022

This is a common variation in Missouri and in some places commoner than the var. *canadensis*.

Ranges from Florida to Texas, north to New York, New Jersey, Pennsylvania, Ohio, Indiana, and Wisconsin.

A final evaluation of the significance of the variation in leaf lobing found in Bloodroot must be left for future field and experimental studies. In some areas of Missouri the var. *rotundifolia* is the common and dominant variation, but in other areas both variations are found, and no correlation appears evident.

The 'double' Bloodroot with numerous white petals is f. *multiplex* (E. H. Wilson) Weath., and a form with pink petals is f. *Colbyorum* Benke, but neither has been recorded in a wild state in Missouri.

The rootstock has been used in medicine for its emetic, purgative, and stimulant properties, and the juice was employed as a stain by American Indians for war paint and for dyeing the quills of their arrows. The bitter rootstock is poisonous as a drug if taken in excess dosage. It has been observed by Torrey (Rh. 44: 13–15. 1942) that ants help in the dissemination of seeds of bloodroot, the ants apparently being attracted to the seeds by a transparent gelatinous crest (caruncle) attached to the seed.

## 2. **Stylophorum** Nutt. Celandine Poppy

**Stylophorum diphyllum** (Michx.) Nutt. Celandine Poppy Map 1023

Also called Wood Poppy.

Flowers April–June.

Occurs in rich soil of woodland in low ground at the base of bluffs, along streams, or in ravine bottoms. Eastern and east-central Missouri from Stoddard and Shannon counties north to St. Charles, Warren, Callaway, and Boone counties, and locally southwest in the White River region of southwestern Missouri in Taney and Stone counties, the latter localities now exterminated by the dams on White River.

Ranges from Pennsylvania to Wisconsin, south to Virginia, Tennessee, and Missouri.

This is a handsome wild flower, sometimes occurring in the same woodland with Virginia Bluebells (*Mertensia virginica*) where its orange blossoms contrast strikingly with the blue of *Mertensia*. The whitened or silvery-gray lower surface of the leaves is also striking. The plant succeeds well in shaded wild flower gardens and eventually its abundant self-seeding produces a large stand. It makes a good ground cover for shade.

## 3. **Chelidonium** L. Celandine

**Chelidonium majus** L. Celandine Map 1024

Flowers April–August.

Known only as an introduction in waste ground in Jackson County, west-central Missouri.

Native of Europe; introduced and naturalized in North America from Quebec to Ontario, south to Georgia, Tennessee, and Missouri.

The plant was used at one time in medicine.

A cut-leaved variety, var. *laciniatum* Koch, is sometimes cultivated.

---

Plate no. 176. **1.** Sanguinaria canadensis var. canadensis, × 1; Details from Small, The New York Botanical Garden. **2.** Sanguinaria canadensis var. rotundifolia, × $^2/_7$. **3.** Stylophorum diphyllum, × $^2/_7$; Details from Small, The New York Botanical Garden. **4.** Chelidonium majus, × $^2/_7$; Details from Small, The New York Botanical Garden. **5.** Papaver somniferum, × $^2/_7$; Details from Small, The New York Botanical Garden. **6.** Papaver Rhoeas, × $^4/_7$, fruit. **7.** Papaver dubium; *a.* Fruit, × $^4/_7$. **8.** Argemone albiflora, × $^2/_7$; Details from Small, The New York Botanical Garden. **9.** Dicentra Cucullaria, × $^2/_7$. **10.** Dicentra canadensis, × $^2/_7$; Details from Small, The New York Botanical Garden.

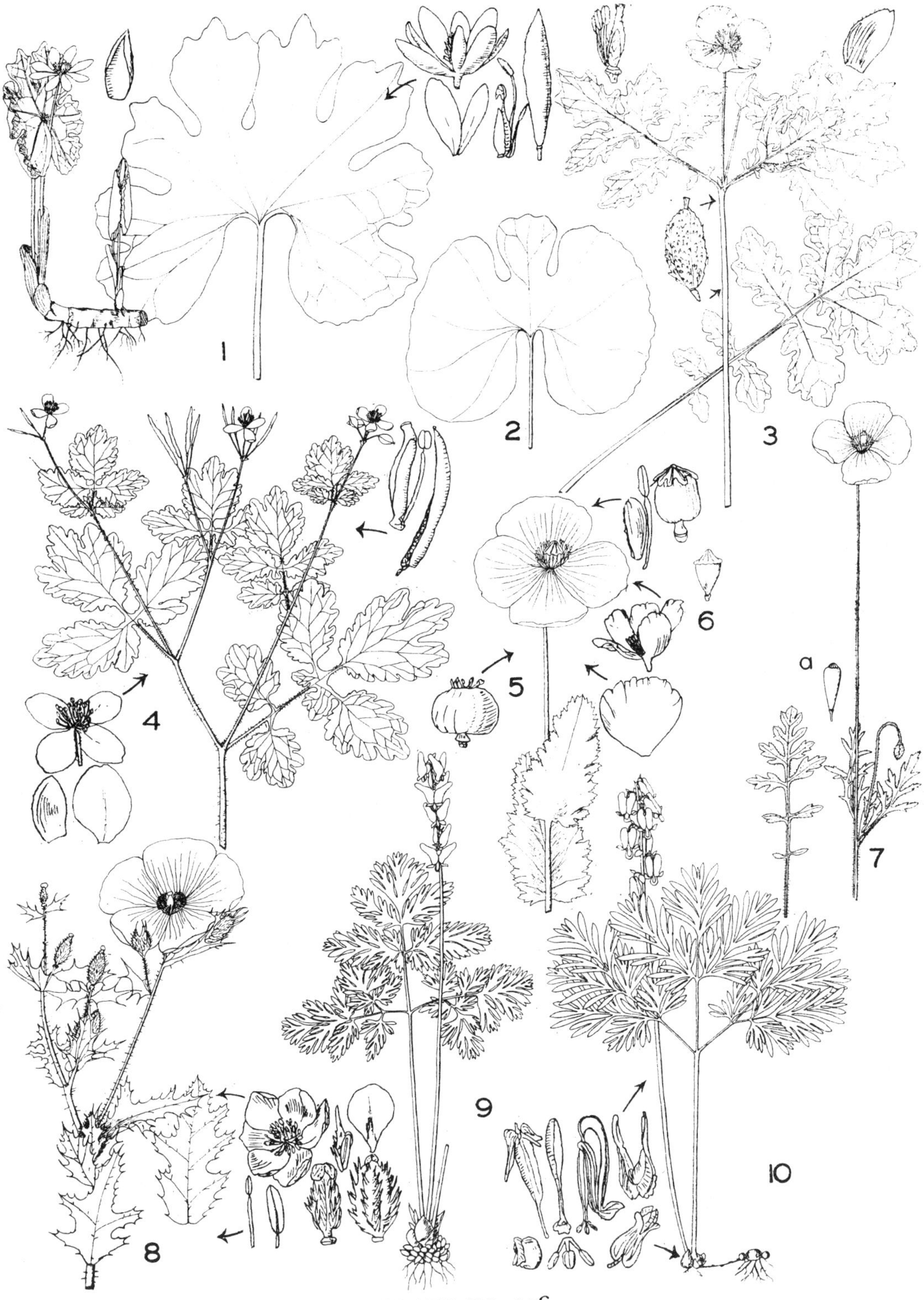


PLATE NO. 176

1024 Chelidonium majus (Celandine)

1025 Papaver somniferum (Common Poppy)

1026 Papaver dubium (Blind Eyes)

## 4. **Papaver** L. Poppy

a. Plants glabrous (hairless), covered with a white or gray-silvery coating (glaucous); base of leaves clasping the stem; leaves wavy-margined or coarsely toothed; petals bluish-white with a purple center . . . . . . . . . . . . . . . . . . . . . . . . . 1. P. SOMNIFERUM
a. Plants hairy, green, not glaucous; base of leaves not clasping; leaves deeply cut or pinnately divided, petals crimson-red or salmon-red, varying to purple or white . . . . . . . . . . . . . . . *b*
 b. Seed-pod club-shaped or obovoid, noticeably longer than broad, narrowest at the base . 2. P. DUBIUM
 b. Seed-pod somewhat globe-shaped, only slightly longer than broad or as broad as long, rounded at the base. . . . . . . . . . . . . . . . . . . . . . . . 3. P. RHOEAS

1. **Papaver somniferum** L. Common Poppy Map 1025

Flowers May–September.

Cultivated and sometimes escaped. Known only from Greene and Jackson (waste places, Courtney, June 5, 1918, *Bush 8344*) counties.

Native of Eurasia; introduced and naturalized in North America from Newfoundland to North Dakota, southward.

Opium is obtained from the dried milky juice of the unripened seed pod. Some cases are reported in Europe of poisoning in cattle which have eaten the unripe capsules. Morphine and codeine are two of the alkaloids of opium used for their narcotic and sedative action in relieving pains and inducing sleep. The seeds are commonly used by the bakeries for rolls and bread and as a bird seed known as 'Maw Seed.' An oil obtained from the seeds is used in pressed form as food and as a salad oil. It is also used as an illuminating oil and for soap and paints.

2. **Papaver dubium** L. Blind Eyes Map 1026

Flowers May–July.

Cultivated and occasionally escaped from gardens to waste ground in Adair, Cooper, Morgan, and Jackson counties.

Native of Europe; introduced and naturalized from Massachusetts to North Carolina, Tennessee, and Missouri.

The petals are red or rarely white. This and the next species are questionably distinct.

3. **Papaver Rhoeas** L. Corn Poppy, Field Poppy, Shirley Poppy Map 1027

Flowers May–October.

Commonly cultivated and escaped from gardens to waste ground and along roadsides. Scattered over the state.

Native of Europe; introduced and naturalized from Nova Scotia to North Dakota, south to Virginia, Missouri, and Kansas.

## 5. **Argemone** L. Prickly Poppy

The following key and treatment of the taxa is based upon the most recent study of the genus by Dr. Gerald B. Ownbey (Mem. Torr. Bot. Club 21: 1–159. 1958).

a. Petals bright or pale yellow . . . . . . . . . . . . . . . . . . . . . 1. A. MEXICANA
a. Petals white . . . . . . . . . . . . . . . . . . . . . . . . . . *b*
 b. Lower leaves of the stem cut in about ⅔ the distance to the midrib; sepal horns 3–5 mm. long;

1027 Papaver Rhoeas (Corn Poppy)

1028 Argemone mexicana (Prickly Poppy)

1029 • Argemone albiflora subsp. albiflora
1029 □ Argemone albiflora subsp. texana

capsules of 3–7 carpels, oblong to oblong-elliptic, the spines all rather stout, even-sized and widely spaced . . . . . . . . . . . . . . . . . . . . 2a. A. ALBIFLORA subsp. ALBIFLORA

b. Lower leaves of the stem cut in about $\frac{4}{5}$ the distance to the midrib; sepal horns 4–6 (rarely to 10) mm. long; capsules of 3–5 carpels, narrowly elliptical, the widely spaced large spines interspersed with uneven-sized smaller spines and prickles . . . . . 2b. A. ALBIFLORA subsp. TEXANA

1. **Argemone mexicana** L. Map 1028

Flowers June–August.

Occurs in rocky or open ground and along railroads. Known only in southern and central Missouri from St. Louis (St. Louis, *Kellogg*), Franklin (Pacific, *Kellogg 1754*), and Wayne (Piedmont, *Trelease*) counties.

Native of the West Indies and probably parts of Central America and Florida; introduced in Mexico, and in the United States introduced from Georgia to Texas, north to Massachusetts, New York, Pennsylvania, Illinois, Missouri, Kansas, and Nebraska.

The seeds contain about 36 per cent oil, which has purgative and emetic properties, and in Mexico has been used in making soap. The latex of the plant is sometimes used by Guatemalan Indians to relieve eye afflictions.

2a. **Argemone albiflora** Hornem. subsp. **albiflora** Map 1029

*A. intermedia* [G] as to Mo. specimens, not *A. intermedia* Sweet.

Flowers June–September.

Introduced in waste ground and along roadsides and railroads. Scattered in southern and central Missouri in St. Louis (Allenton, *Letterman*), Dunklin (Campbell, *Bush 3*), Howell (Carson, *Eggert*), and Jackson (Sheffield, *Bush 9480*), counties.

Native possibly in northern Florida, introduced elsewhere in the United States from Florida to Mississippi, north to Connecticut, Illinois, and Missouri.

2b. **Argemone albiflora** Hornem. subsp. **texana** G. B. Ownbey Map 1029

*Argemone alba* of auth. [BB, Steyerm.], not *A. alba* Lestib.

*Argemone intermedia* of auth. [G as to Missouri specimen], not *A. intermedia* Sweet

Flowers May–August.

Introduced in waste ground and along railroads. Known only from southeastern Missouri in Butler County (Poplar Bluff, *Dewart*; Neelyville, *Kellogg 15186*).

Known from Missouri, Arkansas, and Texas.

The plant is reported to be responsible for some cases of poisoning when eaten by grazing animals, and the prickles can set up an irritation.

## Fam. FUMARIACEAE (Fumitory Family)

All leaves arising at the base of the plant, not attached to the flowering stem; flowers white, tinged with yellow or pink, with 2 spurs or sacs projecting from the ends; small tubers or bulb-like portions attached to plant below the ground . . . . . . . . . . . . . . . . . . . . 1. DICENTRA

Leaves attached to the flowering stem; flowers yellow or orange with 1 spur at the tip; normal thread-like roots present . . . . . . . . . . . . . . . . . . . . 2. CORYDALIS

1030 Dicentra Cucullaria f. Cucullaria (Dutchman's Breeches)

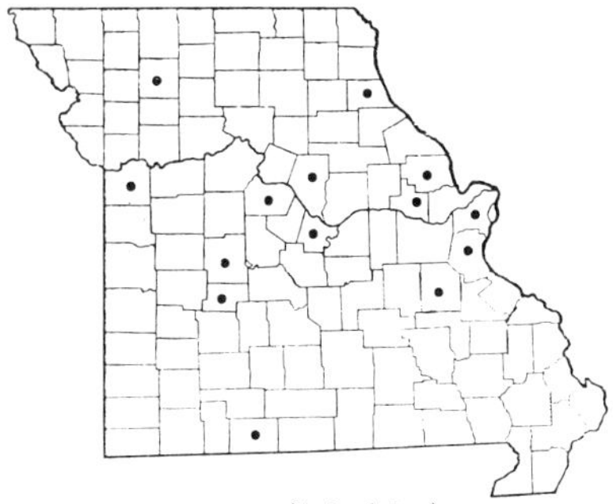

1031 Dicentra canadensis (Squirrel Corn)

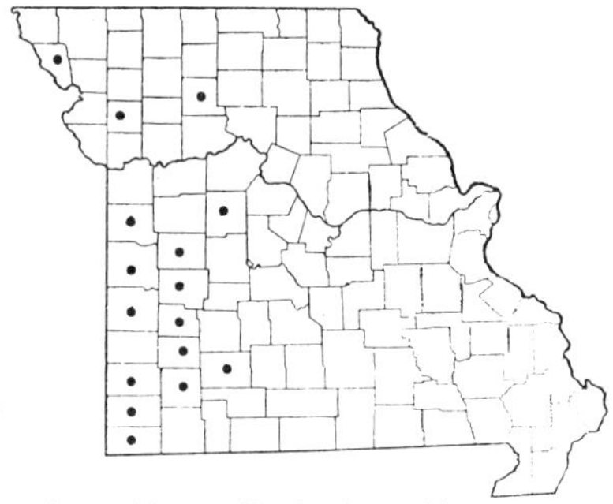

1032 Corydalis crystallina (Mealy Corydalis)

## 1. **Dicentra** Bernh.

Spurs prominent, triangular, pointing in different directions, widely spreading; flowers not fragrant; stem and leaves arising from a pink or white loosely scaly bulb, the scaly parts pointed; lower surface of leaves pale greenish or slightly glaucous (grayish with a 'bloom'), the tips of the segments acutely pointed . . . . . . . . . . . . . . . . . . . . . . . . . . . . . . 1. D. CUCULLARIA

Flowers heart-shaped, the spurs short and rounded and pointing more or less in the same direction; flowers fragrant; stem and leaves arising from a horizontal rhizome bearing round, yellow pea- or corn-like grains; lower surface of leaves conspicuously whitened or glaucous, the tips of the segments more or less rounded but mucronate (with a projection at tip) . . . . . . . . . 2. D. CANADENSIS

1. **Dicentra Cucullaria** (L.) Bernh. f. **Cucullaria**
Dutchman's Breeches Map 1030
*Dicentra Cucullaria* (L.) Bernh. [G, BB, P & S, Steyerm.]
Flowers late March–May.

Occurs in rich or rocky woods, valleys, ravine bottoms, rocky slopes at the base of bluffs, and ledges along streams. Throughout Missouri, and doubtless in every county except the extreme southeastern lowland counties.

Ranges from Quebec to North Dakota, south to Georgia, Alabama, Arkansas, Kansas, and Oklahoma.

The usual color of the flower is white tipped with cream or yellow, sometimes tinged with pale pink, but in f. *purpuritincta* E. H. Eames, not recorded from Missouri, the calyx is deep purple and the corolla is pink with deep orange at the junction of the spurs.

The scaly bulb and the rest of the plant contain an alkaloidal substance poisonous to cattle.

2. **Dicentra canadensis** (Goldie) Walp.
Squirrel Corn Map 1031
Flowers early April–May.

Occurs usually on rich north-facing slopes of wooded bluffs or at their base, rich ravines, and rich slopes along streams. Scattered and rare, mainly in the northern half of the state, south to Washington, Benton, and Hickory counties, and locally south in Taney County along White River, but now destroyed by the Bull Shoals Dam on White River.

Ranges from Quebec and Nova Scotia to Minnesota, south to North Carolina, Tennessee, and Missouri.

The flowers partake the fragrance of hyacinths, and are greenish-white tinged with creamy-yellow, rose, or lavender.

As in *D. Cucullaria* a toxic alkaloid is contained in the plant and is poisonous to grazing animals, but not as strongly toxic as in *D. Cucullaria.*

Large stands of the plants are often found with few or no flowers, in contrast to *D. Cucullaria* which usually bears flowering stems on most plants. After being transplanted to wildflower gardens this species generally requires several years to reestablish itself: only leafy stems appear for a number of years.

*Excluded Species*

**Dicentra spectabilis** (L.) DC. Bleeding Heart

This commonly cultivated plant with rose or white flowers occurs throughout Missouri in gardens, but has not been definitely recorded as an escape. A specimen exists in the University of Missouri Herbarium collected by C. A. Benson from the town lot at Laclede, Linn County, on June 4, 1930, but it is not known under what circumstances the plant was found. It has been thought best to exclude this uncertain record, pending possible future collections of the plant as definitely naturalized.

1033 Corydalis flavula (Pale Corydalis)

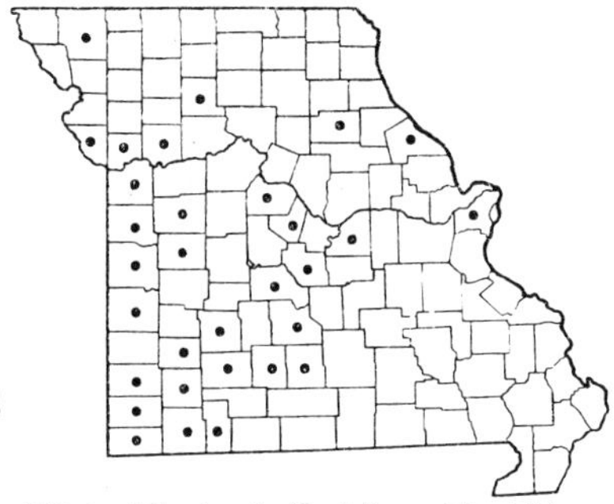

1034 Corydalis micrantha (Small-flowered Corydalis)

1035 Corydalis Halei

## 2. **Corydalis** Medic. Corydalis

a. Ovary of flower or fruit covered with a gray mealiness from the many short thick whitish inflated hairs (vesicles) . . . . . . . . . . . . . . . . . . . . . . . . 1. C. CRYSTALLINA
a. Ovary of flower or fruit smooth and glabrous, not covered with any mealiness . . . . . . . . . *b*
  b. Spur of corolla only 1.5–2 mm. long, usually incurved in the direction of the flower-stalk (pedicel); fruiting pedicels (stalks) 10–15 mm. (rarely 6) long with widely spreading or drooping fruit (capsule); outer petal with a prominent 3–4-toothed crest . . . . . . . . . . . 2. C. FLAVULA
  b. Spur of corolla longer, 2.5–6 mm. or more long, straightish or upwardly curved, not incurved in the direction of the pedicel; fruiting pedicels (stalks) shorter, with erect or spreading fruits; outer petal with a non-toothed crest or a keel. . . . . . . . . . . . . . . . . . . . *c*
    c. Corolla 6–15 mm. long (including the spur); outer petals near their summit with a prominent crest or wing along the center; spur $\frac{1}{4}$–$\frac{1}{3}$ as long as the corolla (including spur) . . . . . . . *d*
      d. Topmost inflorescence usually standing above the foliage, eventually elongated, prolonged and remotely flowered; intervals of 15–25 mm. separating the lower flowers of inflorescence; fruiting capsule 15–25 mm. long, arched-ascending, with constricted portions at intervals (torulose); plants of southeastern Missouri . . . . . . . . . . . . 4. C. HALEI
      d. Topmost inflorescence usually surrounded by the foliage, rarely overtopping it, compactly flowered; intervals of 2–12 mm. separating the lower flowers of the inflorescence; fruiting capsule 6–15 mm. long, erect, with scarcely any constricted portions at intervals; plants of the northern and western half of Missouri . . . . . . . . . . . 3. C. MICRANTHA
    c. Corolla 10–20 mm. long (including the spur); outer petals at their summit with only a hollow keel along the center, not crested; spur nearly $\frac{1}{2}$–$\frac{2}{3}$ as long as the corolla (including spur) . . . . *e*
      e. Capsules drooping or recurving; seeds with rounded margins . . . . . . . . . 5. C. AUREA
      e. Capsules erect or strongly ascending; seeds with acute margins . . . . . . . 6. C. MONTANA

1. **Corydalis crystallina** Engelm.
Mealy Corydalis — Map 1032
Flowers April–June.

Occurs in barrens, rocky glades, prairies, and open ground on usually chert or sandstone strata. Western Missouri east to Livingston, Pettis, Henry, St. Clair, Cedar, Dade, Greene, and Lawrence counties.

Ranges from Iowa to Arkansas, Oklahoma, and Texas.

The foliage is very silvery-glaucous.

2. **Corydalis flavula** (Raf.) DC. Pale Corydalis
Map 1033
Also known as Yellow Fumewort.
Flowers April–May.

Occurs in rich low woods along streams, ravines, moist rocky banks, and ledges along bluffs. Southern and central Missouri north to Marion, Boone, Howard, Chariton, Livingston, Clay, and Platte counties.

Ranges from Connecticut to Ontario and Minnesota, south to Virginia, Tennessee, Louisiana, Kansas, and Oklahoma.

The foliage is dull gray green, but not silvery-glaucous as in the other species.

3. **Corydalis micrantha** (Engelm.) Gray
Small-flowered Corydalis — Map 1034
Also called Slender Fumewort.
*Corydalis micrantha* subsp. *micrantha* (Engelm.) G. B. Ownbey
*Corydalis micrantha* var. *micrantha* [Shinners]
*Corydalis micrantha* var. *pachysiliquosa* Fedde: originally found at Independence, Jackson Co., *Bush 23*.

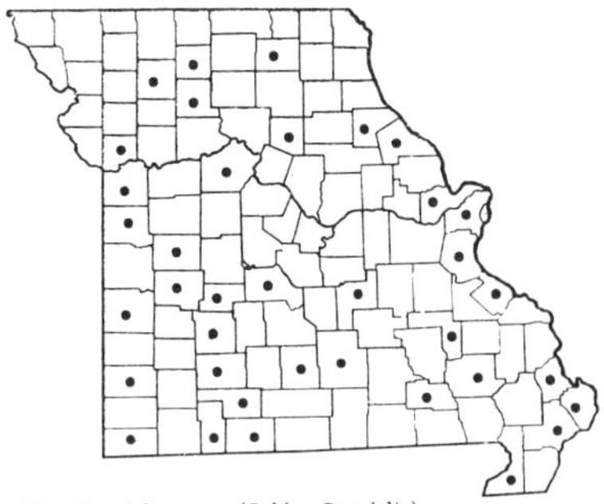

1036 Corydalis aurea (Golden Corydalis)

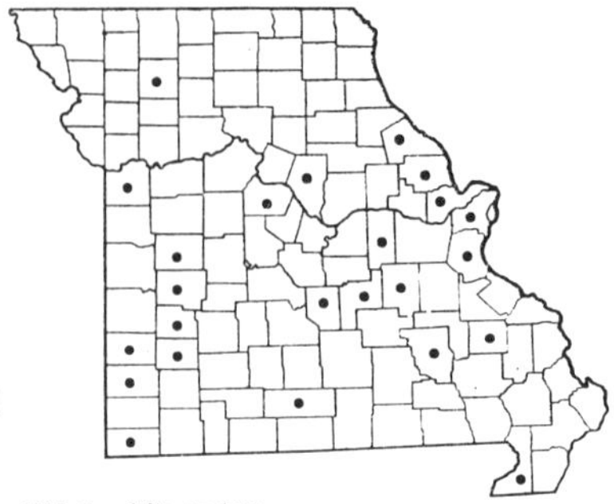

1037 Corydalis montana

1038 Brassica hirta (White Mustard)

Flowers May–June.

Occurs in rocky open woods, ledges along bluffs, rocky glades, and open ground, sometimes on shaded banks. Northern and western half of Missouri, rare in northern Missouri in Nodaway, Livingston, Monroe, and Pike counties, and east south of the Missouri River to New Madrid, Osage, Miller, Camden, Laclede, Wright, and Stone counties; absent from the eastern and central part of the Ozark area.

Ranges from Illinois to Minnesota and Nebraska, south to Tennessee, Arkansas, and Oklahoma.

The foliage is very silvery (glaucous) green.

4. **Corydalis Halei** (Small) Fern. & Schub. Map 1035

*Corydalis micrantha* subsp. *australis* (Chapm.) G. B. Ownbey

*Corydalis micrantha* var. *australis* (Chapm.) Shinners [Shinners]

*Corydalis aurea* var. *australis* Chapm.

Flowers April–May.

Occurs in low open and sandy ground in the lowlands of southeastern Missouri in Scott (sandy ground along railroad, along highway 60, just west of city limits of Sikeston, April 6, 1956, *Steyermark 80855*) and Stoddard counties.

Ranges from Florida to Texas, north to Virginia and inland to southeastern Missouri.

The foliage is very silvery (glaucous) green.

5. **Corydalis aurea** Willd. Golden Corydalis Map 1036

*Corydalis micrantha* subsp. *australis* [of Ownbey], not *C. aurea* var. *australis* Chapm.

*Corydalis micrantha* var. *australis* [of Shinners], not *C. aurea* var. *australis* Chapm.

Flowers March–June.

Occurs in prairies, glades, old fields, roadsides, rocky woods, and ledges. Throughout Missouri.

Ranges from Quebec to Alaska, south to West Virginia, Ohio, Illinois, Missouri, Texas, New Mexico, and California.

The foliage is silvery green.

6. **Corydalis montana** Engelm. Map 1037

*Corydalis campestris* (Britt.) Buchholz & Palmer [G, P & S]

*Corydalis micrantha* subsp. *australis* [of Ownbey], not *C. aurea* var. *australis* Chapm.

*Corydalis micrantha* var. *australis* [of Shinners], not *C. aurea* var. *australis* Chapm.

*Corydalis aurea* var. *occidentalis* Engelm. [G, P & S]

Flowers March–June.

Occurs in rocky woods and open ground. Scattered throughout Missouri.

Ranges from Illinois to Montana, south to Missouri, Arkansas, Oklahoma, Texas, and Mexico.

The foliage is silvery green. Some alkaloids have been found in this and related species (*C. aurea*) which have the same toxic properties as those mentioned under *Dicentra Cucullaria* in connection with poisoning by grazing animals. Sheep are poisoned from eating certain species of the genus found in the western United States.

---

Plate no. 177. **1.** Corydalis crystallina, × 2/5; *a.* Fruit, × 1 1/5. **2.** Corydalis flavula, × 2/5. **3.** Corydalis micrantha, × 2/5; Details from Small, The New York Botanical Garden. **4.** Corydalis aurea, × 2/5; Details from Small, The New York Botanical Garden. **5.** Brassica hirta, × 2/5. **6.** Brassica nigra, × 2/5; Details from Small, The New York Botanical Garden.

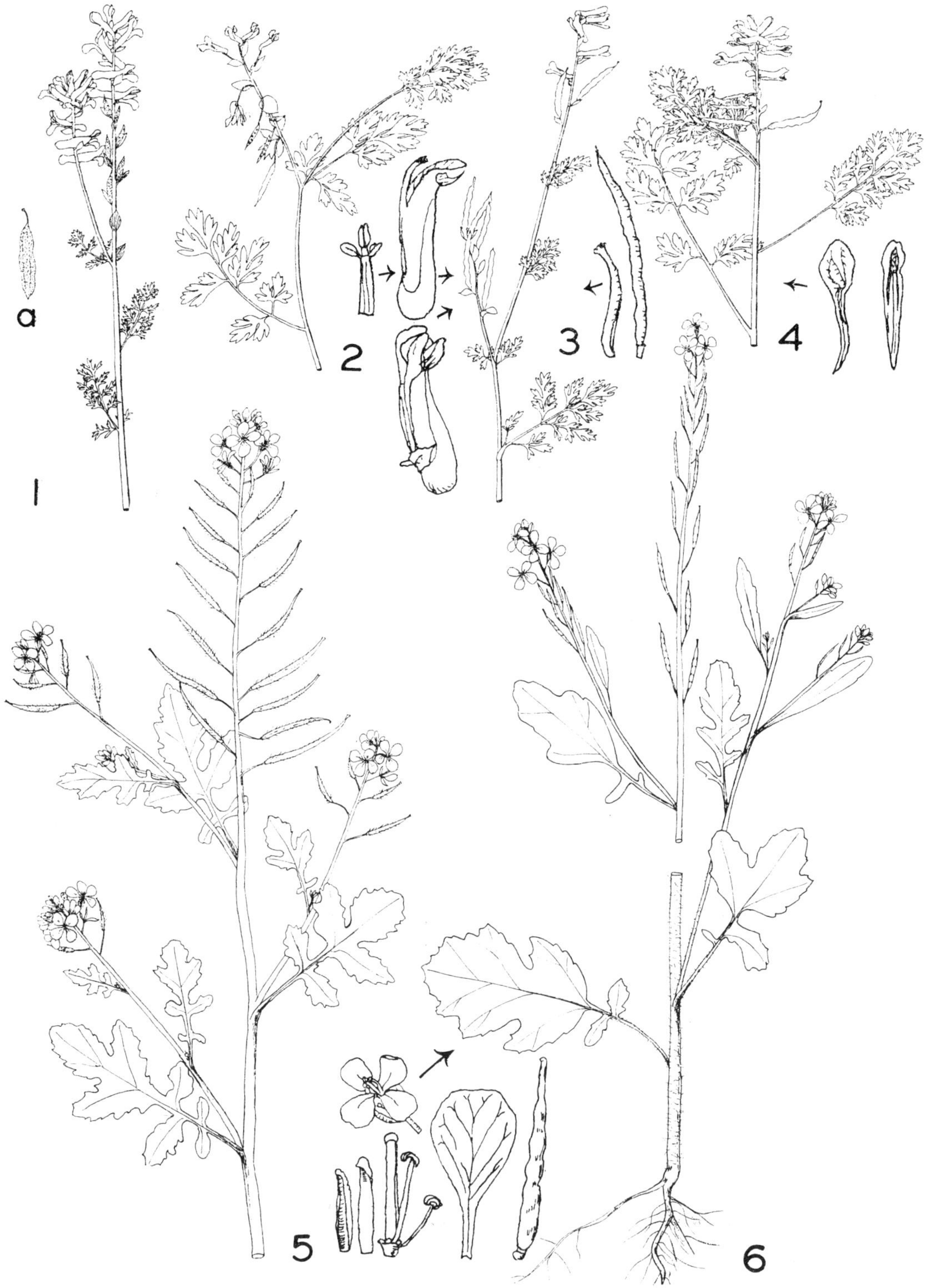


PLATE NO. 177

## Fam. **CRUCIFERAE** (Mustard Family)

a. Leaves all arising apparently from the base of the plant, none on stem . . . . . . . . . . . *b*
b. Leaves deeply cut or parted . . . . . . . . . . . . . . . . . . 19. LEAVENWORTHIA
b. Leaves without teeth (entire) or scarcely toothed . . . . . . . . . . . . . . 14. DRABA
a. Leaves on stem as well as at base of plant. . . . . . . . . . . . . . . . . . . . *c*
c. Leaves of the stem in a whorl of 3, each one of the 3 leaves deeply parted or divided into 3–5 palmately arranged divisions (radiating from a common center like the fingers of a hand) 18. DENTARIA
c. Leaves alternately arranged on the stem, simple or, if lobed or divided, the divisions pinnately arranged (feather-like) . . . . . . . . . . . . . . . . . . . . . . . . *d*
d. Some of the upper or middle leaves, or all of the leaves on the stem entire (without teeth) . . . *e*
e. Base of leaves of stem arrowhead-shaped or clasping or with 2 rounded or pointed ear-like lobes . . . . . . . . . . . . . . . . . . . . . . . . . *f*
f. Base of uppermost leaves completely surrounding the stem or branch, the uppermost leaves ovate or almost circular in outline, the lowest leaves finely dissected . . . . 6. LEPIDIUM PERFOLIATUM
f. Without the above combination . . . . . . . . . . . . . . . . . . . . *g*
g. Stems and leaves glabrous (without hairs), or at most with a few hairs at the very base of the stem or on the lowermost leaves . . . . . . . . . . . . . . . . *h*
h. Ovary or fruit as broad as long or only slightly longer than broad; margins of fruit expanded into a wing; ovules or seeds 2–8 in each cell of the ovary or fruit 9. THLASPI
h. Ovary slender, linear; fruit 10 or more times as long as broad; no wing on fruit; ovules or seeds several to many in each cell of the ovary or fruit . . . . . . . . *i*
i. Petals white in living plants (yellow flowers often fade to white in drying); basal lobes of leaves tapering to an acute or pointed tip. . . . . . . 21. ARABIS
i. Petals some shade of yellow (sometimes very pale or creamy-yellow) in living plants (yellow flowers often fade to white in drying); basal lobes of leaves rounded . . . . . . . . . . . . . . . . . . . . . . . *j*
j. Petals pale yellow or creamy-yellow; mature fruits mainly 8–15 cm. (sometimes 5) long; fruit not constricted into a prominent beak . . . 5. CONRINGIA
j. Petals rich or bright yellow; mature fruits 3–7 cm. long; fruit constricted above the middle into a prominent beak . . . . . . . . . . . 1. BRASSICA
g. Stems hairy on some part, or more or less throughout its length; leaves more or less hairy. . . . . . . . . . . . . . . . . . . . . . . . . . *k*
k. Rosette or basal leaves usually present and usually deeply lobed and cut; ovary or fruit triangular or wedge-shaped with the narrowest part at the base . 10. CAPSELLA
k. Rosette or basal leaves often absent, or, if present, not lobed or deeply cut; ovary or fruit of other shapes, but not as above . . . . . . . . . . . . . . . *l*
l. Ovary slender, linear; fruit 10 or more times as long as broad; ovules or seeds several to many in each cell of the ovary or fruit. . . . . . . . . 21. ARABIS
l. Ovary or fruit as broad as long or only somewhat longer than broad; ovule 1 in each cell of the ovary or fruit . . . . . . . . . . . . . . . . *m*
m. Petals white in living plants; hairs simple, not branched; ovule 1 in each cell of the ovary . . . . . . . . . . . . . . . . . . . . . . *n*
n. Petals not showy, 1–2.2 mm. long; stem noticeably hairy with short spreading hairs; fruit with a heart-shaped notch at summit, 5–6 mm. long when mature, the top half with a wing; annual or biennial plants 6. LEPIDIUM CAMPESTRE
n. Petals showy, 3–4 mm. long; some or all of stem with a fine downy inconspicuous hairiness; fruit not notched at summit, tipped by the prominent style, 3–4 mm. long, not winged; perennial plants. . . 8. CARDARIA
m. Petals pale yellow or creamy-yellow in living plants (often fading to white on drying); hairs branched or 2-pronged; ovules 2–12 in each cell of the ovary . . . . . . . . . . . . . . . . . . . . . . . . *o*
o. Petals 4–6 mm. long; fruit smooth, without a honeycombed or net-like surface, obovoid and tapering to a narrow base; ovules 4–12 in each cell of the ovary; seeds usually 4 in each cell of the fruit . . . 31. CAMELINA

o. Petals 2–3 mm. long; fruit with a net-like or honeycombed surface, rather globe-shaped, about as long as broad and rounded at the base; ovules 2 in each cell of the fruit; fruit with 1 seed . 11. NESLIA

e. Base of leaves of stem not arrowhead-shaped, not clasping, nor with 2 lobes . . . . . . . . . . *p*

p. Plants growing in the water of swamps and sloughs; lower leaves and those growing in the water deeply dissected and finely cut . . . . . . . . . . . . . . . . 16. ARMORACIA AQUATICA

p. Without the above combination of characters; if growing in water, then the lower leaves not finely or deeply cut . . . . . . . . . . . . . . . . . . . . . . . . . . . *q*

q. Ovary or fruit nearly round in outline or about as broad as long, or only slightly longer than broad . . . . . . . . . . . . . . . . . . . . . . . . . . . . . *r*

r. Petals yellow, showy, 5–9 mm. long; plant covered with star-shaped (stellate) hairs or scales; fruit globe-shaped, inflated, covered with hairs or scales; ovules 2–10 in each cell; plants of the southern half of Missouri . . . . . . . . . . . . . 15. LESQUERELLA

r. Petals, if present, white in living plants, often absent, at most 1.5–2 mm. long; plant either glabrous (without hairs) or some part of stem or branches with minute, simple, unbranched hairs; fruit flattened or compressed, glabrous; ovule 1 in each cell; found throughout the state. . . . . . . . . . . . . . . . . . . . . . . . . 6. LEPIDIUM

q. Ovary or fruit longer than broad, 2–many times as long as broad . . . . . . . . . . . *s*

s. Petals with purple veins, the petal either white, yellowish, or lavender; fruit glabrous (without hairs), spongy with a prominent thick or wide and flattened beak 5–20 mm. long, either not opening (indehiscent) nor splitting lengthwise and with swollen ribbed corky portions alternating with narrower constricted portions and with 4–10 seeds or not constricted or ribbed and with 2–4 seeds, or opening and splitting lengthwise and with several seeds . . . . . . . . . . . . . . . . . . . . . . . . . *t*

t. Pedicels (stalks) of flowers or fruits 8–20 mm. long; fruit more or less cylindrical and terete (more or less rounded in cross-section), not opening or splitting lengthwise and either with or without swollen ribbed corky portions alternating with narrower constricted portions with 2–10 seeds; ovules 2–10; beak of fruit conic and thickened or flattened and triangular . . . . . . . . . . . . . . . . . 4. RAPHANUS

t. Pedicels (stalks) of flowers 0.5–3 mm. long, of fruits up to 6 mm. long; fruit somewhat 4-sided, splitting open lengthwise, without constricted portions, usually with more numerous seeds; ovules numerous; beak of fruit flattened, triangular-lanceolate . . 3. ERUCA

s. Without the above combination of characters . . . . . . . . . . . . . . . . . *u*

u. Petals rose-lavender or tinged with pink or purplish, rarely all white; some of the stem hairy with simple, unbranched hairs; some of the root- or basal leaves more or less round; plants of the northeastern section of the state . . . . 17. CARDAMINE DOUGLASSII

u. Petals white, yellow, creamy-yellow, or orange; hairs on stems or leaves, when present, simple or branched or star-shaped; without the above combination of characters . . . *v*

v. Stems glabrous (without hairs) and fruit or ovary 10-30 or more times as long as broad. . . . . . . . . . . . . . . . . . . . . . . . . . . *w*

w. Petals white in living plants; most of middle and upper leaves of stem without teeth or with few teeth; mature fruit with no beak or an inconspicuous short beak less than 5 mm. long; plants of bluff crevices or of low or wet ground . . . . *x*

x. Plants of crevices of limestone or sandstone bluffs or rocky slopes; petals 3.5–8 mm. long; leaves of stem linear to spatulate, very narrow, 1.5–5 mm. wide; basal leaves deeply divided or wavy-toothed; base of stem usually hairy; valves of fruit 1-nerved in the middle . . . . . . . . . . . . . 21. ARABIS LYRATA

x. Plants of low or wet woods, swamps, or meadows; petals 7–16 mm. long; leaves of stem linear-oblong, oblong or ovate, mainly 4–30 mm. wide; basal leaves scarcely toothed or toothless; stem glabrous throughout (without hairs); valves of fruit without a distinct nerve. . . . . . . . . . . . . 17. CARDAMINE

w. Petals yellow in living plants; at least the lower and basal leaves lobed or deeply cut at their base; mature fruit with a conspicuous beak more than 5 mm. long; plants of fields, roadsides, along railroads, or waste ground . . . . . . . 1. BRASSICA

v. Without the above combination of characters; stems hairy, or if glabrous (without hairs) the fruit or ovary only 3–7 times as long as broad . . . . . . . . . . *y*

y. Petals pale yellow, creamy-yellow, or orange; hairs of stem and leaves 2-pronged, attached near their middle and lying flat and pressed along the stem and leaves (hairs may appear simple but under 10× magnification show their 2-pronged character) . . . . . . . . . . . . . . . . . . . 28. ERYSIMUM

y. Petals, when present, white; some or all hairs of stem and leaves either simple, branched, or star-shaped (stellate), but not as described above under *Erysimum* . . . . . . . . . . . . *z*
z. Fruit or ovary 2–7 times as long as broad . . . . . . . . . . . . . . . . . . . *1*
1. Stems shorter than 3 dm.; a rosette or cluster of leaves usually present at base of plant; petals, when present, not 2-parted; seeds not winged on the margins; ovules or seeds 6–50 in each cell of ovary or fruit; commonly found. . . . . . . . . . . . . . . 14. DRABA
1. Stems 3–10 dm. tall; no rosette or cluster of leaves present at base of plant; petals conspicuously 2-parted; margin of seeds extending into a wing; ovules or seeds 2–6 in each cell of ovary or fruit; rare. . . . . . . . . . . . . . . . . . . 13. BERTEROA
z. Fruit or ovary 10–30 or more times as long as broad . . . . . . . . . . . . . . *2*
2. Plants 0.5–4.5 dm. tall; petals 2–3 mm. long; leaves of rosette small, less than 1 cm. wide and 1–5 cm. long; at most 6 leaves on stem below branches of inflorescence . 30. ARABIDOPSIS
2. Plants mainly 3–10 dm. (rarely 3) tall; petals 3–5 mm. long; leaves of rosette larger, mainly 1.5–4 cm. wide and 4–13 cm. long; usually 8 or more leaves present on stem below branches of inflorescence . . . . . . . . . . . . . . . . . . 21. ARABIS
d. Some or all of the leaves on stem shallowly or deeply toothed, or deeply lobed or dissected or divided into separate leaflets . . . . . . . . . . . . . . . . . . . . . . . . . . *3*
3. Base of leaves of stem arrowhead-shaped or clasping or with 2 rounded or pointed ear-like lobes . . *4*
4. Lobes at base of leaves scarcely evident, not clasping or arrowhead-shaped, but mere auricles (ear-like flaps) 0.5–3 mm. long; petals, if present, yellow; fruit $1\frac{1}{2}$–6 times as long as broad . 23. RORIPPA
4. Lobes at base of leaves conspicuous or evident, clasping the stem or arrowhead-shaped, mainly 2–5 mm. or more long; petals white, or, if yellow, the fruit or ovary 10 or more times as long as broad . . . . . . . . . . . . . . . . . . . . . . . . . . . . *5*
5. Stems and leaves glabrous (without hairs) or at most with a few hairs at the very base of the stem or on the lowermost leaves . . . . . . . . . . . . . . . . . . *6*
6. Ovary or fruit as broad as long or only slightly longer than broad; margins of fruit expanded into a wing; ovules or seeds 2–8 in each cell of the ovary or fruit. . . 9. THLASPI
6. Ovary slender, linear; fruit 10 or more times as long as broad; no wing on fruit; ovules or seeds several to many in each cell of the ovary or fruit . . . . . . . . . . *7*
7. Petals white in living plants (yellow flowers often fade to white in drying) . . 21. ARABIS
7. Petals yellow in living plants (yellow flowers often fade to white in drying) . . . . . *8*
8. Some of middle or upper leaves of stem coarsely toothed, angled, or deeply lobed; beak of the fruit not conspicuous, only 1–3 mm. long . . . . . . . 24. BARBAREA
8. Middle and upper leaves not coarsely toothed, angled, or deeply lobed; beak of the fruit conspicuous, 8–20 mm. long . . . . . . . . . . . . . . 1. BRASSICA
5. Stems hairy on some part, or more or less throughout its length; leaves more or less hairy . . *9*
9. Rosette or basal leaves usually present and deeply lobed and cut; ovary or fruit triangular or wedge-shaped with the narrowest part at the base . . . . . . . . . 10. CAPSELLA
9. Rosette or basal leaves often absent, or, if present, not lobed or deeply cut; ovary or fruit of other shapes, but not as above . . . . . . . . . . . . . . . . *10*
10. Ovary slender, linear; fruit 10 or more times as long as broad; ovules or seeds several to many in each cell of the ovary or fruit . . . . . . . . . . 21. ARABIS
10. Ovary or fruit as broad as long or only somewhat longer than broad; ovule 1 in each cell of the ovary or fruit . . . . . . . . . . . . . . . . . . *11*
11. Petals not showy, 1–2.2 mm. long; stem noticeably hairy with short spreading hairs; fruit with a heart-shaped notch at summit, 5–6 mm. long when mature, the top half with a wing; annual or biennial plants . . . 6. LEPIDIUM CAMPESTRE
11. Petals showy, 3–4 mm. long; some or all of stem with a fine downy inconspicuous hairiness; fruit not notched at summit, tipped by the prominent style, 3–4 mm. long when mature, not winged; perennial plants . . . . . 8. CARDARIA
3. Base of leaves of stem not arrowhead-shaped, not clasping, nor with 2 prominent pointed or rounded lobes . . . . . . . . . . . . . . . . . . . . . . . . . *12*
12. At least the middle and upper leaves on stem more or less shallowly toothed, but not deeply lobed nor dissected nor divided into separate leaflets . . . . . . . . . . . . . *13*
13. Stalked glands projecting from stems, fruits, and stalks (pedicels) of flowers and fruits; petals blue-lavender or purplish . . . . . . . . . . . . . . . 26. CHORISPORA
13. Without the above combination of characters . . . . . . . . . . . . . . . *14*
14. Plants growing in the water of swamps and sloughs; lower leaves and those growing in the water deeply dissected and finely cut . . . . . . . 16. ARMORACIA AQUATICA

14. Without the above combination of characters; if growing in water, then the lower leaves not finely or deeply cut . . . . . . . . . . . . . . . . . . . . . . . . *15*
  15. Ovary or fruit about as broad as long or very slightly longer than broad . . . . . . . *16*
    16. Petals yellow; hairs on plant star-shaped (stellate); fruit covered with star-shaped hairs 15. LESQUERELLA
    16. Petals, if present, white, sometimes absent; plant glabrous (without hairs) or the hairs simple and unbranched; fruit glabrous . . . . . . . . . . . . . . . . . . *17*
      17. Petals inconspicuous, either lacking or at most 2 mm. long; ovule or seed 1 in each cell of ovary or fruit; fruit flattened . . . . . . . . . . . . . . 6. LEPIDIUM
      17. Petals conspicuous, 6–8 mm. long; ovules or seeds many in each cell of the ovary or fruit; fruit inflated . . . . . . . . . . . . . . . . . 16. ARMORACIA
  15. Ovary or fruit 2–many times as long as broad . . . . . . . . . . . . . . . . . *18*
    18. Petals some shade of yellow or orange in living plants (fading often to white when dried) . *19*
      19. Hairs of stem and leaves 2-pronged, attached near their middle, lying flat and pressed along the stem and leaves (hairs may appear simple but under 10 × magnification show their 2-pronged character) . . . . . . . . . . . . . . 28. ERYSIMUM
      19. Hairs absent or, if present, simple and unbranched and not lying flat . . . . . . . *20*
        20. Petals absent or up to 2 mm. long; beak of fruit absent or inconspicuous, at most 1 mm. long; fruit 1½–6 times as long as broad . . . . . . . . . 23. RORIPPA
        20. Petals 6–20 mm. long; beak of fruit well-developed and conspicuous, 2.5–20 mm. long; fruit (including beak) 9–15 times as long as broad . . . . . . . . . *21*
          21. Mature fruit with regular corky enlarged sections alternating with narrow constricted sections, the enlarged portions prominently ribbed and containing the seeds; fruit glabrous (without hairs); petals conspicuously veined, the veins often purplish or sometimes of other color combinations; petals 10–20 mm. long . . . . . . . . . . . . . . . . . . . . . 4. RAPHANUS
          21. Mature fruit without alternating enlarged and constricted sections, or, if with enlarged and constricted sections, then the fruit hairy; fruit usually glabrous, except as noted; petals all one color, the veins of the same color and less conspicuous; petals 6–12 mm. long . . . . . . . . . . . 1. BRASSICA
    18. Petals white, lavender, rose, or purple in living plants . . . . . . . . . . . . . *22*
      22. Petals lavender, purple, or rose . . . . . . . . . . . . . . . . . . . . *23*
        23. Ovary or fruit with only 2–4 ovules or seeds; mature fruit spongy, 8–10 mm. thick, with a prominent beak about 1 cm. long; taproot greatly thickened . 4. RAPHANUS SATIVUS
        23. Ovary or fruit with many ovules or seeds; mature fruit not spongy, 3 mm. or less wide, a beak, if present, less than 0.5 cm. long; all roots long and slender, stringy . *24*
          24. Stems 0.7–3 dm. tall; teeth of leaves rather blunt or coarse, or, if somewhat pointed, then few (1–4) on each margin to a leaf; fruit with a definite beak 17. CARDAMINE DOUGLASSII
          24. Stems 3–10 dm. tall; teeth of leaves sharp-pointed, either numerous and close together or 8–25 or more on each margin; fruit without a definite beak. . . *25*
            25. Stems and leaves usually glabrous (without hairs); petals 10–13 mm. long; mature fruit 1.8–3 cm. long; native plant of low or moist woodland 25. IODANTHUS
            25. Stem or leaves more or less hairy with simple and branched hairs; petals 18–25 mm. long; mature fruit 5–14 cm. long; escaped garden plant along roads, railroads, or open ground . . . . . . . . . . . . 27. HESPERIS
      22. Petals white . . . . . . . . . . . . . . . . . . . . . . . . . . *26*
        26. Hairs on plant star-shaped (stellate); fruit or ovary 2–7 times as long as broad; leaves mostly grouped at the base of the plant, or, if regularly scattered on the stem, the fruit only 1–17 mm. long. . . . . . . . . . . . . 14. DRABA
        26. Plants without hairs or hairs simple or branched, but not star-shaped (stellate); fruit or ovary 10–30 times as long as broad; leaves regularly scattered along the stem with fruits 1.5–10 cm. long . . . . . . . . . . . . . . . . *27*
          27. Petals 7–12 mm. long; fruit 1.5–2.5 cm. long, with a definite beak 1–3 mm. long; plant glabrous throughout (no hairs present) . . . . 17. CARDAMINE BULBOSA
          27. Petals 2–8 mm. long; fruit 1.5–10 cm. long, without any beak; simple and branched hairs usually present on rosette leaves, or at base of stem, or elsewhere, or, if completely glabrous, the fruits eventually recurving . . . 21. ARABIS

12. Most or all leaves on stem deeply lobed, dissected, or divided into separate leaflets . . . . . . *28*
28. Ovary and fruit broader than long, as broad as long, or only slightly longer than broad; 1 ovule or seed in each cell of ovary or fruit; petals white, less than 2 mm. long . . . . . *29*
29. Stems erect or ascending; flowers and fruits in inflorescences at the tip of the main stem and branches; fruit smooth, flattened . . . . . . . . . . . . . . . . 6. LEPIDIUM
29. Stems low-spreading and depressed; inflorescences arising from the sides of the stems and branches; fruit roughened or wrinkled, rounded on both sides . . . . . . . 7. CORONOPUS
28. Without the above combination of characters; ovary and fruit $1\frac{1}{2}$–many times longer than broad with 2–many ovules or seeds in each cell of ovary or fruit; petals yellow, orange-yellow, or, if white, usually larger or even lacking . . . . . . . . . . . . . . . . . . *30*
30. Petals 10–20 mm. long, with purple veins, the petal either white, yellowish, or lavender; hair or hairs sometimes prominent near tip of sepals, or absent; fruit glabrous (without hairs), spongy with a prominent thick or wide and flattened beak 5–20 mm. long, neither opening (indehiscent) nor splitting lengthwise and with swollen ribbed corky portions alternating with narrower constricted portions and with 4–10 seeds, or not constricted or ribbed and with 2–4 seeds, or opening and splitting lengthwise and with several seeds. . *31*
31. Pedicels (stalks) of flowers or fruits 8–20 mm. long; hair or hairs usually prominent near tip of sepals, especially in bud; fruit more or less cylindrical and terete (more or less rounded in cross-section), not opening or splitting lengthwise and either with or without swollen, ribbed, corky portions alternating with narrower constricted portions with 2–10 seeds; ovules 2–10; beak of fruit conic and thickened or flattened and triangular-elongated . . . . . . . . . . . . . . . . . 4. RAPHANUS
31. Pedicels (stalks) of flowers 0.5–3 mm. long, of fruits up to 6 mm. long; hairs mostly absent from sepals; fruit somewhat 4-sided, splitting open lengthwise, without constricted portions, and usually more numerous seeds; ovules numerous; beak of fruit flattened, triangular-lanceolate . . . . . . . . . . . . . . . . . 3. ERUCA
30. Without the above combination of characters; petals lacking, or up to 12 mm. long, and either all white, yellow, or orange-yellow, the veins not purple nor of a different color from the rest of the petal; fruit either splitting open lengthwise along the thickened sutures and with many ovules in the ovary or seeds in the fruit, or, if the fruit has alternately enlarged and constricted portions, then bristly with spreading hairs . . . . . . . . . . *32*
32. Stems and/or leaves covered with glandular or star-shaped (stellate) hairs. 32. DESCURAINIA
32. Stems and leaves either glabrous (without hairs) or the hairs simple and unbranched . *33*
33. Petals white in living plants (flowers with yellow petals sometimes fading white when dried should be keyed out in the alternate 33) . . . . . . . . . . *34*
34. Plants normally growing in the cold water of fresh-water springs and spring-fed branches; leaves with a peppery taste; petals 4–5 mm. long, about twice as long as sepals . . . . . . . . . . . . . . . . . 22. NASTURTIUM
34. Plants either of dry or wet ground, but usually out of the water and not normally growing in the cold water of springs and spring-fed streams; petals 1.5–4 mm. long, less than twice as long as sepals . . . . . . . . . . *35*
35. Lower leaves with 5–14 divisions on each side of leaf; fruiting portion of plant occupying $\frac{2}{3}$–$\frac{3}{4}$ length of plant; fruit not or scarcely beaked; seeds or ovules with narrow or winged margins . . . . . . . . . 20. SIBARA
35. Lower leaves with 1–6 divisions on each side of leaf; fruiting portion of plant occupying usually $\frac{1}{3}$–$\frac{1}{5}$ length of plant, sometimes up to $\frac{3}{4}$ length of plant; fruit with a short beak; seeds or ovules without a winged margin . 17. CARDAMINE
33. Petals some shade of yellow in living plants (often fading to white when dried) . . *36*
36. Base of all the pedicels (stalks) of the flowers and fruits, or all but the uppermost ones, accompanied by a deeply cut leaf-like bract . . . . . . . *37*
37. Stem and leaves completely glabrous (without hairs); petals deep or orange-yellow, 8–10 mm. long; fruit much flattened, 5–8 mm. wide at maturity, at most 4 times as long as broad, tipped by a long slender style 5 mm. long; ovules or seeds 6–8 in each cell; plants of sandstone and chert glades, sandy fields and rocky prairies of southwestern Missouri . . 12. SELENIA
37. Stem and leaves more or less hairy; petals pale yellow turning white when dried, 5–7 mm. long; fruit somewhat 4-angled, 1.5–2 mm. wide at maturity, 12–20 times as long as broad, tipped by a style 2–3 mm. long; ovules or seeds many in each cell; plants of railroads and waste ground . 2. ERUCASTRUM

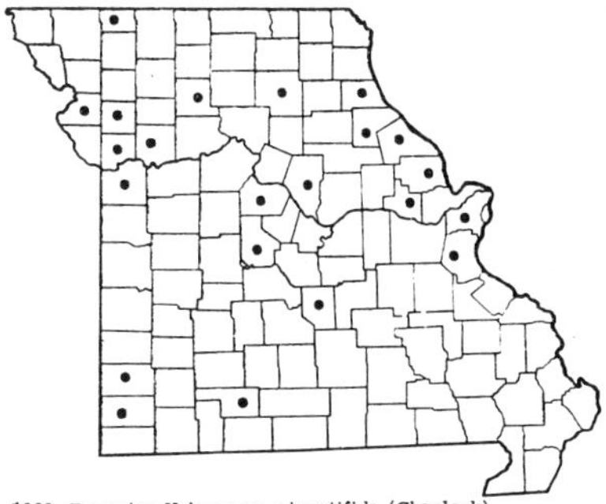

1039 Brassica Kaber var. pinnatifida (Charlock)

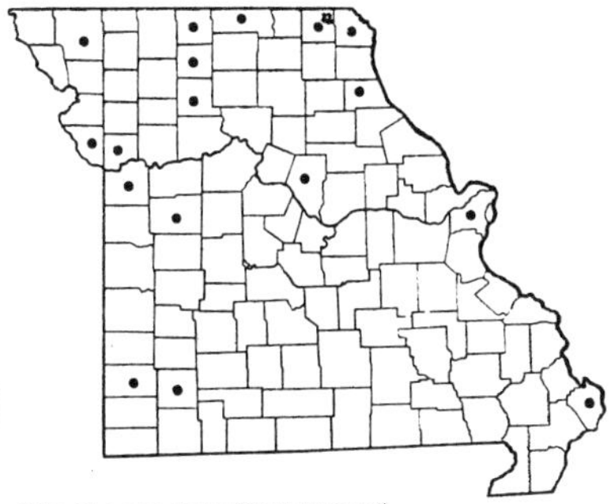

1040 Brassica nigra (Black Mustard)

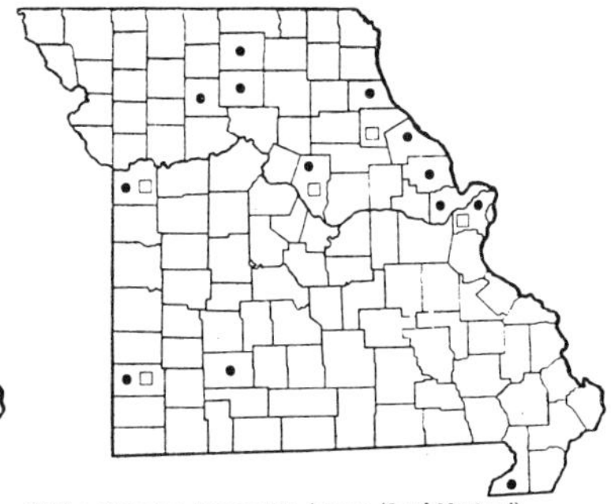

1041 • Brassica juncea var. juncea (Leaf Mustard)
1041 □ Brassica juncea var. crispifolia

36. No leaf-like bracts at base of the pedicels (stalks) of the flowers or fruits (except rarely at base of lowest pedicel) . . . . . *38*
  38. Ovary and fruit with conspicuous beak, the beak in the fruit 2.5–20 mm. long; petals 6–12 mm. long; sepals erect at time of flowering . . . . . 1. BRASSICA
  38. Ovary and fruit without a conspicuous or scarcely any beak, the style in fruit 0.5–3 mm. long; petals none or up to 8 mm. long; sepals somewhat spreading at time of flowering . . . . . *39*
    39. Ovary long and slender, each side 1–3-nerved; each valve of fruit 1–3-nerved; seeds in 1 row in each cell of fruit; fruit 15–30 times or more as long as broad; style short and thick or none, not definite . . . . . 29. SISYMBRIUM
    39. Ovary relatively short and broad to short-cylindric, each side without nerves; each valve of fruit without nerves; seeds in 2 rows in each cell of the fruit; fruit mainly $1\frac{1}{2}$–7 times as long as broad; style short but slender and definite, 0.5–3 mm. long . . . . . 23. RORIPPA

## 1. **Brassica** L. Mustard

a. Bases of upper and often middle leaves clasping or partly surrounding the stem . . . . . *b*
  b. Root branched, not fleshy and thickened . . . . . 5. B. RAPA
  b. Root thickened, fleshy . . . . . 6. B. NAPUS ssp. NAPOBRASSICA
a. Base of upper and middle leaves not clasping nor partly surrounding the stem . . . . . *c*
  c. Plant glabrous (without hairs), often glaucous (with a whitish coating which can be rubbed off) 4. B. JUNCEA
  c. Some parts of the plant, usually the lower part of stem, more or less hairy . . . . . *d*
    d. Ovary of fruit hairy; pedicels of mature fruits about 10 mm. long; seeds pale yellow, usually 3 or less in each cell, 1.5–2 mm. thick; teeth and lobes of leaves usually blunt to rounded . 1. B. HIRTA
    d. Ovary or fruit glabrous (without hairs); pedicels of mature fruits 3–7 mm. long; seeds dark brown to black, 5–50 in each cell, about 1 mm. thick; teeth and lobes of leaves often more acutely pointed . . . . . *e*
      e. Upper leaves as well as the others on the stem on their own petioles (stalks); stalks of mature fruits 3–4 mm. long; mature fruit erect, remaining close to main axis of inflorescence, 1–2 cm. long; beak of fruit short, slender, cylindrical, 1–4 mm. long; seeds usually 9–50 in each cell of the fruit . . . . . 3. B. NIGRA
      e. At least the upper leaves of stem sessile (without stalks) or nearly so; stalks of mature fruits 3–7 mm. long; mature fruit ascending or spreading-ascending away from main axis, 2.5–4.5 cm. long when mature; beak of fruit prominent, flattened, 4-angled, about as wide as lower part of fruit, 10–15 mm. long when mature; seeds 5–10 in each cell of the fruit; ovules similarly relatively few in ovary . . . . . 2. B. KABER var. PINNATIFIDA

1. **Brassica hirta** Moench White Mustard Map 1038

*Brassica alba* [of P & S, Steyerm.], not Gilib.

Flowers April–July.

Occurs in fields, waste ground, and along railroads. Rare and scattered in the state, where known from St. Louis (*Muehlenbach 562*), Jasper, Boone, and Marion counties.

Native of Eurasia; introduced and naturalized from Prince Edward Island to British Columbia, south locally throughout the United States.

The plant is cultivated both for mustard made

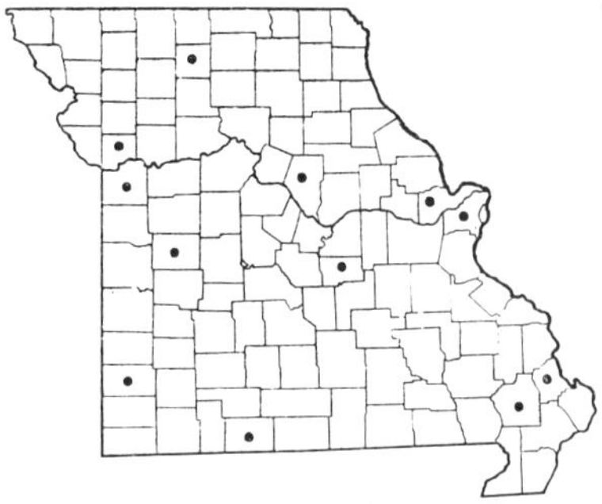

1042 Brassica Rapa (Field Mustard)

1043 Brassica Napus var. Napobrassica (Rutabaga)

1044 Erucastrum gallicum

from the seeds and as a vegetable. Oil from the seeds is used in medicine as a counter-irritant and as a condiment, and in industry for lubrication and illuminating fuel.

2. **Brassica Kaber** (DC.) L. C. Wheeler var. **pinnatifida** (Stokes) L. C. Wheeler Charlock Map 1039

*Brassica arvensis* (L.) Rabenh. [P & S, Steyerm.], not L.

*Brassica Kaber* in part [BB]

Flowers April–July.

Occurs in fields, waste ground, along roadsides and railroads. Scattered throughout Missouri, and probably to be expected in every county.

Native of Eurasia; introduced and naturalized throughout North America.

3. **Brassica nigra** (L.) Koch Black Mustard Map 1040

Flowers April–November.

Occurs in fields, waste ground, and along roadsides. Scattered in Missouri and less common than the preceding species.

Native of Eurasia; introduced and naturalized in various parts of North America.

The seeds of this species provide the chief source of the table mustard of commerce, and are often mixed with those of *B. hirta*, the White Mustard. The seeds also are used in medicine as a counter-irritant, and oil from the seeds is used in making of some soaps.

4. **Brassica juncea** (L.) Coss. Leaf Mustard Map 1041

Known also as Indian Mustard, Chinese Mustard.

Flowers April–September.

Two variations occur in Missouri:

Margins of leaves not curled, crisped, nor finely fringed . . . . . 4a. B. JUNCEA var. JUNCEA

Margins of leaves curled, crisped, or finely fringed . . . . 4b. B. JUNCEA var. CRISPIFOLIA

4a. **Brassica juncea** var. **juncea** Map 1041

*Brassica juncea* (L.) Coss. [G, P & S, Steyerm.]

Occurs in fields, waste ground, along roadsides and railroads. Scattered throughout Missouri.

Native of Eurasia; introduced and naturalized from Nova Scotia to Saskatchewan, south to Florida, New Mexico, and California.

This variety is cultivated as Chinese Mustard as a leafy green vegetable. The leaves may be used fresh in salads or cooked as a green vegetable. The seeds are used in some parts of Europe and Asia as a spice and in cooking, while oil from the seeds is sometimes used for the body.

4b. **Brassica juncea** var. **crispifolia** Bailey Curled Mustard Map 1041

*Brassica japonica* [P & S, Steyerm.], not Thunb.

*Brassica juncea* var. *japonica* (Thunb.) Bailey [BB]

Occurs in fields, waste ground, along roadsides and railroads. Rare and scattered in Missouri in St. Louis, Ralls, Boone, Jackson, and Jasper counties.

Native of Eurasia; introduced and naturalized in various parts of North America.

5. **Brassica Rapa** L. Field Mustard Map 1042

Also known as Bird's Rape.

*Brassica campestris* L. [of BB, P & S, Steyerm.]

Flowers April–September.

---

Plate no. 178. **1.** Brassica Kaber var. pinnatifida, × 2/7; Details from Small, The New York Botanical Garden. **2.** Brassica juncea var. crispifolia, × 2/7. **3.** Brassica Rapa, × 2/7. **4.** Conringia orientalis, × 2/7; Details from Small, The New York Botanical Garden. **5.** Brassica juncea var. juncea, × 2/7. **6.** Erucastrum gallicum, × 2/7. **7.** Eruca sativa, × 2/7; Details from Small, The New York Botanical Garden. **8.** Raphanus Raphanistrum, × 2/7; After Britton and Brown, details from Small, The New York Botanical Garden.

PLATE NO. 178

1045 Eruca sativa (Garden Rocket)

1046 Raphanus Raphanistrum (Wild Radish)

1047 Raphanus sativus (Radish)

Occurs in fields, waste ground, along roadsides and railroads. Frequent throughout Missouri and probably to be found in every county.

Native of Eurasia; introduced and naturalized throughout much of North America.

Young plants or shoots of this and other species may be cooked as greens. Certain races of this species are cultivated in some countries for the oil produced from the seeds, used as a lubricating oil for machinery, soap preparations, greasing loaves of bread, oiling woolen goods, and for illuminating purposes.

6. **Brassica Napus** L. var. **Napobrassica** (L.) Peterm. Rutabaga Map 1043

*Brassica Napobrassica* Mill. [G]

*Brassica Rapa* [of BB in part], not L.

*Brassica oleracea* L. var. *Napobrassica* L.

Flowers April–September.

Introduced along railroads, where known only from St. Louis County (St. Louis, Mo. Pac. R. R. tracks, sidings of the Ozark elevator, May 5, 1956, *Muehlenbach 879*).

Native of Eurasia; introduced and sparingly naturalized in North America.

Probably more common as an escaped plant.

Other species or varieties are to be expected along railroads or roadsides. One of these, *B. Napus* L., the turnip, (*B. Rapa* of authors, not L.) resembles *B. Rapa* L. (*B. campestris*) but is green instead of glaucous, completely glabrous, with a thickened tuberlike base of the stem, and smaller flowers.

## 2. **Erucastrum** Presl.

**Erucastrum gallicum** (Willd.) O. E. Schulz Map 1044

*Erucastrum Pollichii* Schimper & Spenner [P & S]

Flowers May–September.

Occurs in waste ground and along railroads. Known only from St. Louis (St. Louis, Carrie Ave. freight yard of Terminal Railroad Assoc., June 28, 1958, *Muehlenbach 1427*) and Jackson (Courtney, along railroad, June 26, 1918, *Bush 8440*) counties.

Native of Europe; introduced and naturalized from Newfoundland to British Columbia, south to West Virginia, Kentucky, Indiana, Illinois, Missouri, and Texas.

## 3. **Eruca** Mill. Garden Rocket

**Eruca sativa** Mill. Garden Rocket Map 1045

*Eruca vesicaria* (L.) Cav. [BB]

Flowers May–October.

Occurs along railroads. Known only from Clark (Dumas, July 7, 1909, *Bush 5895*) and Jackson (Courtney, June 28, *1909*, *Bush 5854*) counties.

Native of Europe; introduced and naturalized from Ontario to North Dakota and Washington, south to New Jersey, Pennsylvania, Illinois, Missouri, Texas, New Mexico, and California.

The young leaves are used for their peppery taste in salads.

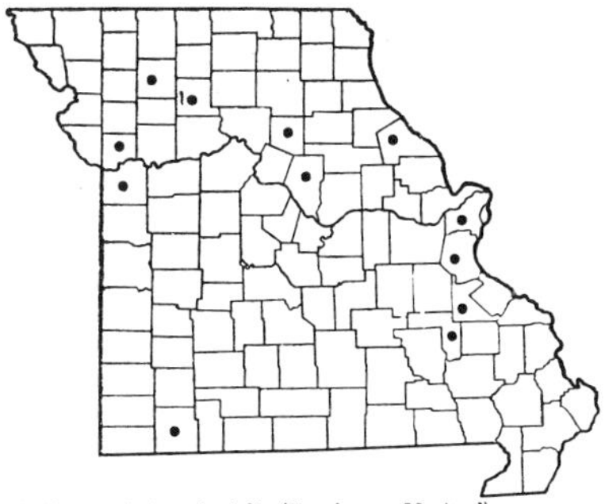

1048 Conringia orientalis (Hare's-ear Mustard)

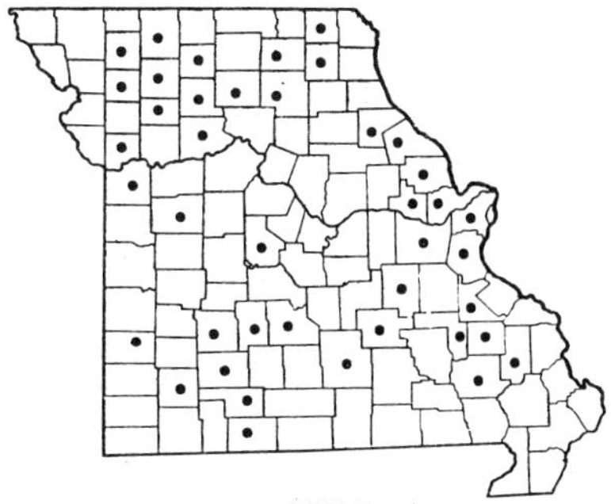

1049 Lepidium campestre (Field Cress)

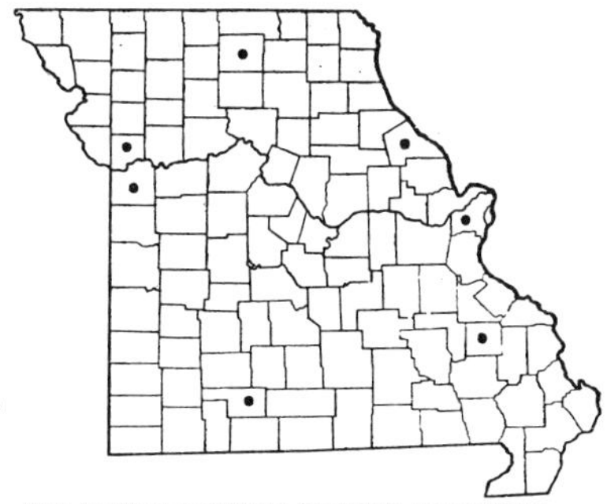

1050 Lepidium perfoliatum (Perfoliate Pepper Grass)

## 4. **Raphanus** L. Radish

Flowers pale yellow in living plants (drying white); fruit 4–10-seeded, conspicuously ribbed on the enlarged corky portions which alternate with the narrow constricted section, 4–6 mm. thick . 1. R. RAPHANISTRUM

Flowers pink-lavender or purplish to white; fruit 2–4-seeded, not ribbed, 6–10 mm. thick, not constricted between the seeds . . . . . . . . . . . . . . . . . . . . . . . 2. R. SATIVUS

1. **Raphanus Raphanistrum** L. Wild Radish Map 1046

Also called Jointed Charlock.

Flowers May–November.

Occurs along railroads. Known only from St. Louis County (North St. Louis, Burlington R. R. freight yard, just south of Humboldt Ave., May 30, 1955, *Muehlenbach 620*; Baden freight yard, Mo.-Kan.-Texas R. R., July 5, 1958, *Muehlenbach 1452*).

Native of Europe; introduced and naturalized from Newfoundland to British Columbia, south to Virginia, Kentucky, Indiana, Illinois, Missouri, Texas, and California.

2. **Raphanus sativus** L. Radish Map 1047

Flowers May–August.

Commonly cultivated and infrequently escaping along railroads, fields, and waste places. Known from St. Louis, Boone, Jackson, and Jasper counties.

## 5. **Conringia** Adans. Hare's-ear Mustard

**Conringia orientalis** (L.) Dumort.

Hare's-ear Mustard Map 1048

Flowers May–August.

Occurs along railroads, roadsides, and waste ground. Scattered in Missouri, commonest in the northern and eastern halves.

Native of Europe; introduced and naturalized from Quebec to Manitoba and Oregon, south to Florida, Texas, and California.

## 6. **Lepidium** L. Pepper Grass

a. Stems densely hairy with spreading noticeable hairs . . . . . . . . . . . . 1. L. CAMPESTRE

a. Stems either glabrous (without hairs) or with inconspicuous minute hairiness . . . . . . . . *b*

b. Base of upper leaves broadly clasping or surrounding the stem . . . . . . . . 2. L. PERFOLIATUM

b. Base of upper leaves tapering or narrowed to the base, not clasping the stem . . . . . . . . *c*

c. Petals present, up to 2 mm. long, equaling or longer than the sepals; fruits 2.5–4 mm. long; cotyledons of seed with their edges against the radicle (accumbent) . . . . . 3. L. VIRGINICUM

c. Petals absent or shorter than the sepals; fruits 2–2.5 mm. long; cotyledons of seed with the back of one of them against the radicle (incumbent) . . . . . . . . . . 4. L. DENSIFLORUM

1. **Lepidium campestre** (L.) R. Br.

Field Cress Map 1049

Also known as Cow Cress or Field Pepper Grass.

Flowers April–June.

1051 Lepidium virginicum var. virginicum (Pepper Grass)

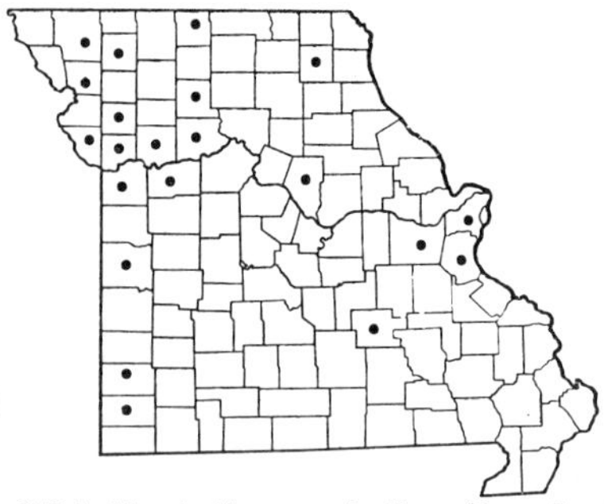

1052 Lepidium densiflorum var. densiflorum (Pepper Grass)

1053 Coronopus didymus (Wart Cress)

Occurs along roads and railroads, fields, and waste ground, usually in large colonies. Throughout Missouri, and probably to be expected from all counties, but many not yet represented.

Native of Europe; introduced and naturalized from Nova Scotia and Ontario to North Dakota, south to North Carolina, Illinois, Missouri, Kansas, and Oklahoma; also in the Pacific states.

The plant has a gray-green appearance from the covering of hairs on the stems and leaves. The fruiting pedicels are widely spreading, further adding to the easy recognition of the plant.

2. **Lepidium perfoliatum** L. Perfoliate Pepper Grass Map 1050

Flowers April–June.

Occurs along railroads, roadsides, and waste ground, sometimes along dry creek beds. Rare and scattered in Missouri.

Native of Europe; introduced and naturalized from New England to Michigan, Iowa, Nebraska, and Utah, south to Pennsylvania, Illinois, Missouri, Oklahoma, and California; commoner in the western states.

The contrasting dissected lower leaves and the more rounded clasping upper ones give this species a very distinctive appearance.

3. **Lepidium virginicum** L. var. **virginicum** Pepper Grass Map 1051

*Lepidium virginicum* L. [G, P & S, Steyerm.]

Flowers February–November.

Occurs in fields, pastures, rocky glades, prairies, rocky outcrops, along roadsides, railroads, and open waste ground. Throughout Missouri, and doubtless in every county.

Ranges from Newfoundland to South Dakota, south to Florida and Texas.

Several other varieties of the species are found in the western states, differing in having incumbent or oblique cotyledons.

The peppery seed pods are relished by canaries and other seed-eating cage birds. These or the seeds also furnish tasty seasoning for soups and salads, and the seeds can be mixed with vinegar and salt for a meat dressing. The young spring shoots can be eaten raw sprinkled in salads as a substitute for water cress.

4. **Lepidium densiflorum** Schrad. var. **densiflorum** Pepper Grass Map 1052

*Lepidium densiflorum* Schrad. [G, BB, P & S, Steyerm.]

*Lepidium densiflorum* var. *typicum* Thell. [Hitchc.]

*Lepidium neglectum* Thell.

*Lepidium apetalum* of auth., not Willd.

Flowers April–November.

Occurs in fields, pastures, rocky glades, prairies, rocky outcrops, along roadsides, railroads, and open waste ground. Scattered over the state, but probably to be found in most all the counties.

Ranges from Quebec to Montana and Oregon, south to Florida, Texas, and California.

The other varieties of this species are more western and have fruits averaging 3 mm. or slightly more long, more oblong-obovate rather than obovate-rounded, and with more conspicuously flattened pedicels (see Hitchcock in Madrono 3: 278. 1936).

A collection which Dr. Hitchcock considers as a possible hybrid between *L. virginicum* var. *virginicum* and *L. densiflorum* var. *densiflorum* has been collected in Jackson County (Courtney, June 12, 1900, *Bush 781*). Considering the frequency of both species in the state,

Plate no. 179. **1.** Lepidium campestre, × $^2/_5$. **2.** Lepidium perfoliatum, × $^2/_5$. **3.** Lepidium virginicum var. virginicum, × $^2/_5$; Details from Small, The New York Botanical Garden. **4.** Coronopus didymus, × $^2/_5$; Details from Small, The New York Botanical Garden.

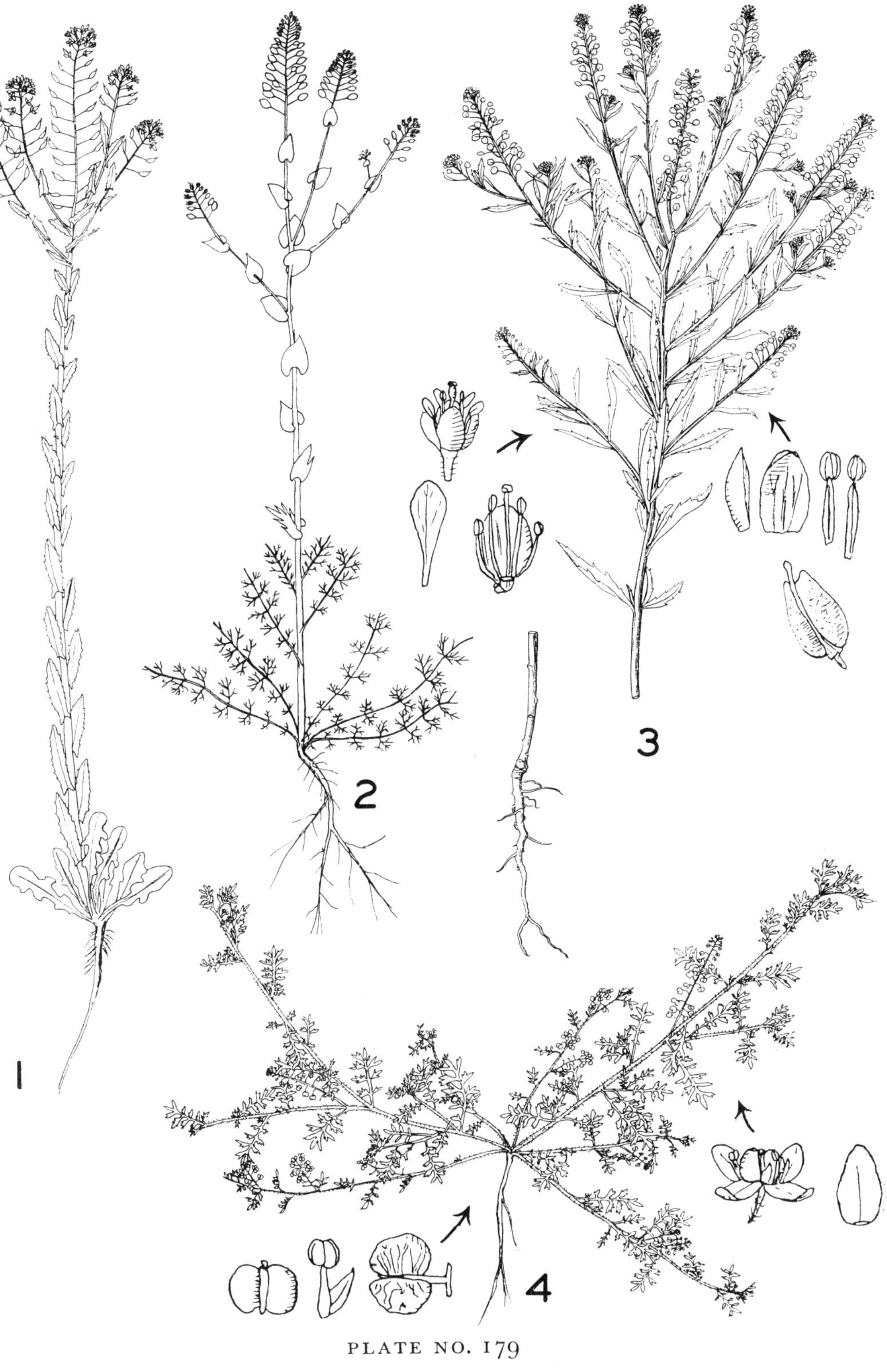


PLATE NO. 179

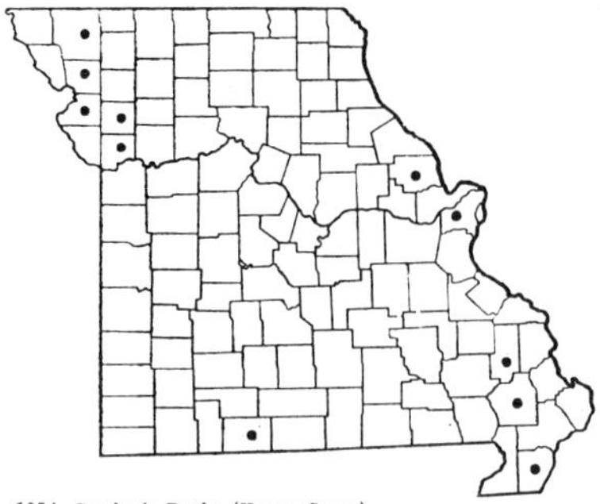
1054 Cardaria Draba (Hoary Cress)

1055 Cardaria pubescens var. elongata (White Top)

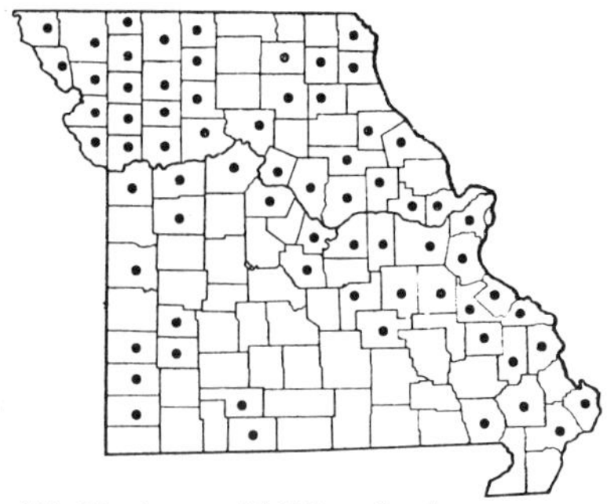
1056 Thlaspi arvense (Field Penny Cress)

it is probable that more collections of such hybrid origin will come to light with future collecting in the state.

*Excluded Species*

**Lepidium ruderale** L.

Specimens collected by Bush from Cedar (*Bush 15086*) and St. Clair (*Bush 15076, 15091*) counties and distributed under this name show flowers having short petals about equaling the sepals. *Lepidium ruderale* is a species in which the petals are absent and Bush's specimens should be referred to *L. virginicum* var. *virginicum*.

## 7. **Coronopus** Trew Wart Cress

**Coronopus didymus** (L.) Sm. Wart Cress Map 1053

Flowers April–October.

Occurs in waste ground and open places. Known only from Dunklin and Butler counties in southeastern Missouri.

Native of Europe; introduced and naturalized from Florida to Texas, north to Newfoundland, Quebec, Ohio, and Missouri.

## 8. **Cardaria** Desv.

Fruit glabrous, broadly heart-shaped or somewhat triangular with a heart-shaped base; inflorescences crowded into a broad flat-topped corymb, the summits of the individual racemes reaching about the same height; mature fruiting stalks 10–15 mm. long . . . . . . . . . . . . . . . . 1. C. DRABA
Fruit hairy, more or less globular; inflorescence of more separated distinct racemes not crowded into a flat-topped corymb; mature fruiting stalks 3–5 mm. long . . . . . . . . . . . . 2. C. PUBESCENS

1. **Cardaria Draba** (L.) Desv. Hoary Cress Map 1054

*Lepidium Draba* L. [P & S, Steyerm.]

Flowers April–June.

Occurs along roadsides, railroads, and open waste ground. Scattered in parts of eastern and western Missouri.

Native of Europe; introduced and naturalized from Nova Scotia to British Columbia, south to Florida, and California, but commonest in the western states.

This is a showy flowered perennial with fragrant white flowers in a broad corymbose crowded inflorescence.

2. **Cardaria pubescens** (C. A. Mey.) Rollins var. **elongata** Rollins White Top Map 1055

*Hymenophysa pubescens* C. A. Meyer [BB]

Flowers May–September.

Occurs along railroads. Known only from St. Louis

Plate no. 180. **1.** Cardaria Draba, × 2/5. **2.** Cardaria pubescens var. elongata, × 2/5; After Britton and Brown, The New York Botanical Garden. **3.** Thlaspi perfoliatum, × 2/5; Details from Small, The New York Botanical Garden. **4.** Thlaspi arvense, × 1; Details from Small, The New York Botanical Garden. **5.** Capsella Bursa-pastoris, × 2/5. **6.** Neslia paniculata, × 2/5; After Britton and Brown, The New York Botanical Garden.

PLATE NO. 180

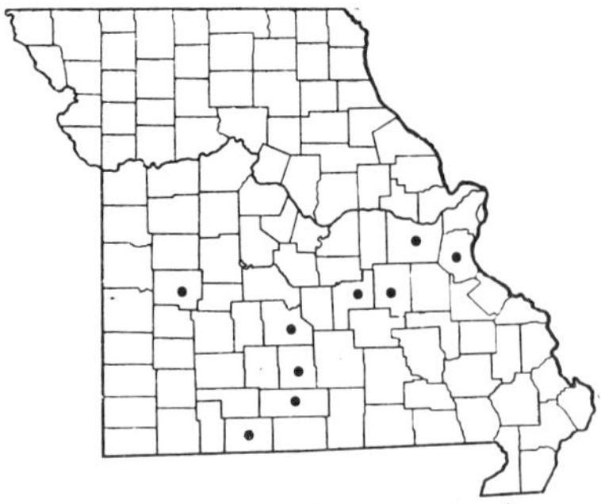

1057 Thlaspi perfoliatum (Perfoliate Penny Cress)

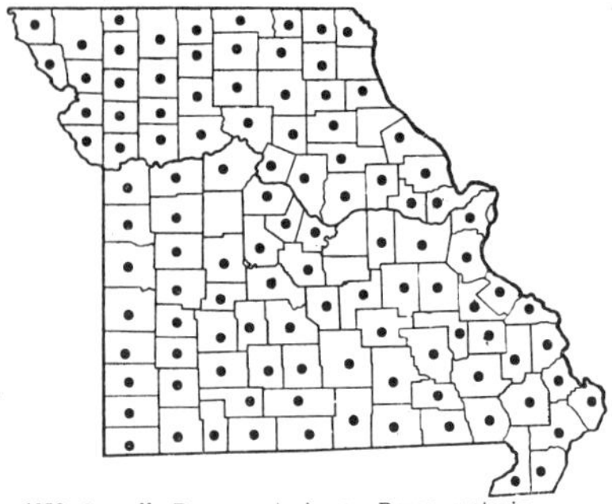

1058 Capsella Bursa-pastoris var. Bursa-pastoris (Shepherd's Purse)

1059 Neslia paniculata (Ball Mustard)

County (St. Louis, freight yard of Terminal Railroad, between Victor and Rutgers Streets, May 23, 1954, *Muehlenbach 62*).

Native of Asia; introduced and naturalized from Washington to California, east locally to Missouri, Michigan, and Pennsylvania.

## 9. **Thlaspi** L. Penny Cress

Middle and upper leaves of stem oblong to lanceolate, usually rather regularly toothed, the lobes at base of leaves slender and narrowly or acutely pointed; fruits 10–18 mm. long . . . . . . 1. T. ARVENSE
Middle and upper leaves broadly ovate or ovate-oblong, conspicuously wider at the base, usually with only a few teeth or more frequently without teeth, the lobes at base of leaf more rounded or blunt; fruit 4–6 mm. long . . . . . . . . . . . . . . . . . . . . . . . 2. T. PERFOLIATUM

1. **Thlaspi arvense** L. Field Penny Cress
Map 1056

Flowers April–June.

Occurs along roadsides, railroads, fields, meadows, and waste ground. Throughout Missouri, becoming common along roadsides. Probably to be found in every county of the state, but still unrecorded from many Ozark counties.

Native of Europe; introduced and naturalized from Greenland and Labrador to Alaska, south to Florida, Arkansas, Oklahoma, Texas, and California.

The peppery seeds are sometimes used as a substitute season for mustard, while the young leaves may be eaten as a salad, having a combination mustard-onion flavor.

2. **Thlaspi perfoliatum** L. Perfoliate Penny Cress
Map 1057

Flowers March–May.

Occurs along roadsides, rocky open fields, low valleys in open ground near streams, and waste and cultivated ground. Known only from southern and east-central Missouri, but becoming increasingly more common along highways.

Native of Europe; introduced and naturalized from Ontario and New York to Maryland, Virginia, Kentucky, and Missouri.

## 10. **Capsella** Medic. Shepherd's Purse

**Capsella Bursa-pastoris** (L.) Medic. var.
**Bursa-pastoris** Shepherd's Purse Map 1058

*Capsella Bursa-pastoris* (L.) Medic. [G, BB, P & S, Steyerm.]

Flowers January–December.

Occurs in fields, pastures, cultivated ground, lawns, along roadsides and railroads. Throughout Missouri and doubtless in every county.

Native of Europe; introduced and naturalized throughout North America.

The seeds are reported to be sometimes ground into a meal by some of the California Indians, while the young leaves may be used as a salad green or cooked as a green vegetable. The young leaves are much relished by poultry. The leaves, together with those of various other mustards, dandelion, wild lettuce, and poke, are sometimes used as a tea for medicinal purposes, especially for stopping various kinds of hemorrhages.

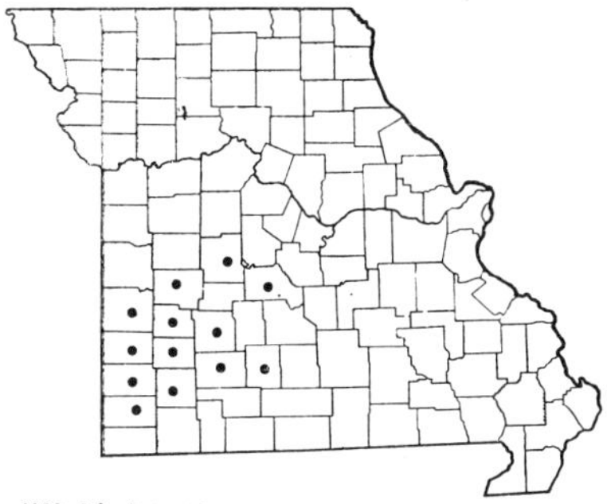

1060 Selenia aurea

1061 Berteroa incana

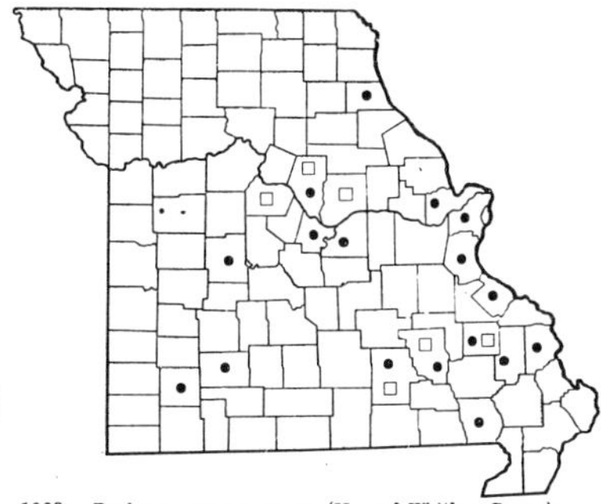

1062 ● Draba verna var. verna (Vernal Whitlow Grass)
1062 □ Draba verna var. Boerhaavii

## 11. **Neslia** Desv. Ball Mustard

**Neslia paniculata** (L.) Desv. Ball Mustard
Map 1059

Flowers May–September.

Occurs along railroads. Known only from St. Louis (St. Louis, Carrie Avenue freight yard of Terminal R. R. Assoc., June 1, 1957, *Muehlenbach 1183*; Wabash Fruit Terminal, May 26, 1957, *Muehlenbach 1153*) and Christian (Billings, May 17, 1955, *Palmer 59952*) counties.

Native of Europe; introduced and naturalized from Newfoundland to British Columbia, south to New Jersey, Pennsylvania, Ohio, Indiana, Illinois, Missouri, and South Dakota.

## 12. **Selenia** Nutt.

**Selenia aurea** Nutt. Map 1060

Flowers April–May.

Occurs in acid soils of cherty and sandstone glades and rocky prairies, sandy fields, and waste ground. Known only from the unglaciated prairie section of southwestern Missouri north to Benton, St. Clair, and Vernon counties, east to Camden and Webster counties.

Ranges from southwest Missouri to Kansas, Arkansas, Oklahoma, and Texas.

This is a very showy annual with the leaves deeply and regularly cut into narrow segments. The stems ordinarily are 0.5–2 dm. tall.

## 13. **Berteroa** DC.

**Berteroa incana** (L.) DC. Hoary Alyssum
Map 1061

Flowers late May–September.

Occurs along railroads and waste ground. Known from St. Louis (St. Louis, Burlington R. R. north of E. Grand Ave., May 30, 1955, *Muehlenbach 616*), Phelps, and Greene counties.

Native of Europe; introduced and naturalized from Nova Scotia to Montana, south to New Jersey, West Virginia, Ohio, Indiana, Illinois, Missouri, Kansas, and Oklahoma; also in the Pacific states.

The plant is easily recognized by the gray stems and foliage and small white flowers.

## 14. **Draba** L. Whitlow Grass

a. Leaves scattered up and down the length of the stem . . . . . . . . . . . . . . . . . *b*

b. Fruits or ovaries glabrous (without hairs); stems usually several or branched; inflorescences elongated at maturity, the stems with elongated or leafy branches; commonly occurring and collected . . . . . . . . . . . . . . . . . . . . . . . . . . . 2. D. BRACHYCARPA

b. Fruits or ovaries minutely hairy; stems often more unbranched and solitary; inflorescences short

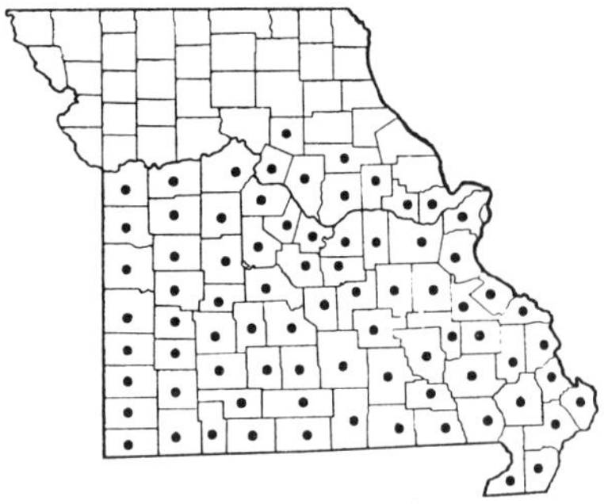

1063 Draba brachycarpa (Whitlow Grass)

1064 Draba aprica (Whitlow Grass)

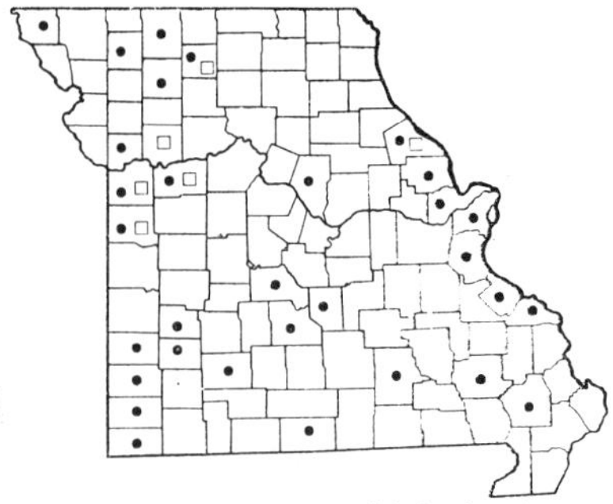

1065 • Draba reptans var. reptans (Whitlow Grass)
1065 □ Draba reptans var. micrantha

and abbreviated, arising from the axils of the middle and upper leaves; rare species of the southeastern Ozarks . . . . . . . . . . . . . . . . . . . . . 3. D. APRICA

a. All the leaves near or at the base of the plant . . . . . . . . . . . . . . . . . . . *c*

c. Leaves in a rosette at base of plant; no leaves present on flowering stems (scapes); petals deeply split into 2 narrow lobes; fruits shorter than their stalks (pedicels) . . . . . . . . . 1. D. VERNA

c. One or more leaves on the flowering stems above the base of the plant; petals not split into 2 narrow lobes; fruits mostly longer than their stalks (pedicels) . . . . . . . . . . . . . . *d*

d. Leaves with some teeth; axis of inflorescence and stalks (pedicels) of flowers and fruits hairy. 5. D. CUNEIFOLIA

d. Leaves entire (without teeth); axis of inflorescence and stalks of flowers and fruits glabrous (without hairs) . . . . . . . . . . . . . . . . . . . . . 4. D. REPTANS

1. **Draba verna** L. Vernal Whitlow Grass Map 1062

Flowers February–April.

Occurs in grassy or rocky open places, lawns, pastures, roadsides, fallow or cultivated fields, in valleys or on upland. Southern, central, and northeastern Missouri.

Two variations occur in Missouri:

Fruits $2\frac{1}{2}$–4 times as long as broad, 4–10 mm. long, 1.5–2.3 mm. broad, with 40–60 seeds . . 1a. D. VERNA var. VERNA

Fruits nearly as broad as long or at most 2 times as long as broad, 2.5–6 mm. long, 2–4 mm. broad, with fewer seeds . . . . . 1b. D. VERNA var. BOERHAAVII

1a. **Draba verna** var. **verna** Map 1062

*Draba verna* L. [G, BB, P & S, Steyerm.]

Southern, central, and northeastern Missouri north to Marion, Boone, and Benton counties. Probably to be found in most of the counties of the state, especially throughout the southern half.

Ranges from Massachusetts to North Carolina, west to Missouri.

1b. **Draba verna** var. **Boerhaavii** Van Hall Map 1062

*Draba verna* var. *aestivalis* Lejeune [BB]

Southern and central Missouri north to Callaway and Boone counties. Probably to be found over most of the state.

Ranges from Vermont to Ontario and Minnesota, south to Georgia, Alabama, Tennessee, and Missouri.

*Draba verna* has been used in the treatment of whitlow disease, an inflammation of the hoof in animals, and the common name is derived from this association.

2. **Draba brachycarpa** Nutt. Whitlow Grass Map 1063

Flowers March–April.

Occurs in fallow or cultivated fields, lawns, grazed pastures, rocky glades and prairies, along roadsides, railroads, in open or wooded valleys or on the upland. Common throughout all of southern and central Missouri north to St. Charles, Warren, Montgomery, Audrain, Randolph, Saline, Lafayette, and Jackson counties. To be expected in many counties north of this area.

Plate no. 181. **1.** Selenia aurea, $\times {}^{2}/_{5}$. **2.** Draba verna var. verna, $\times 1{}^{1}/_{5}$. **3.** Draba aprica, $\times {}^{2}/_{5}$; *a.* Fruit, $\times 4$. **4.** Draba brachycarpa, $\times {}^{2}/_{5}$. **5.** Draba reptans, $\times {}^{2}/_{5}$; Details from Small, The New York Botanical Garden. **6.** Draba cuneifolia, $\times {}^{2}/_{5}$; Details from Small, The New York Botanical Garden. **7.** Lesquerella filiformis, $\times {}^{2}/_{5}$; Details from Small, The New York Botanical Garden.

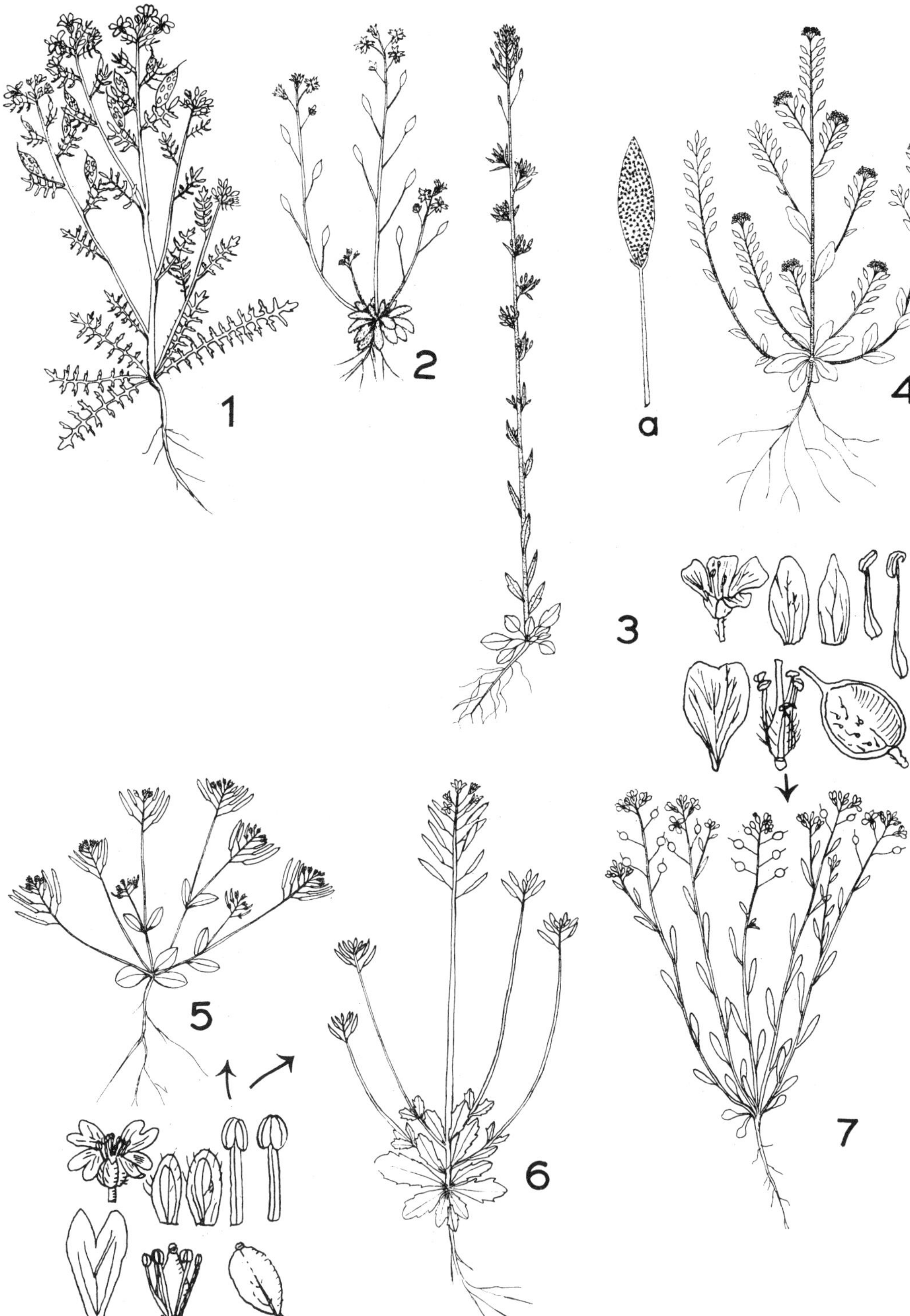


PLATE NO. 181

1066 Draba cuneifolia var. cuneifolia (Whitlow Grass)

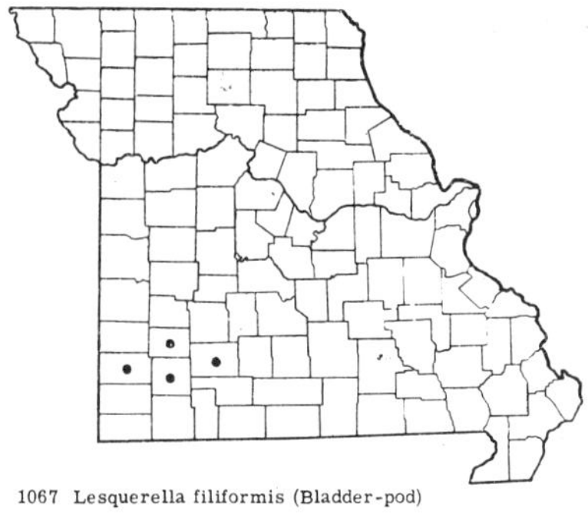

1067 Lesquerella filiformis (Bladder-pod)

1068 Lesquerella gracilis var. gracilis

Ranges from Florida to Texas, north to Virginia, Indiana, Illinois, Missouri, and Kansas.

The flowers may have petals or be without them (apetalous), and may even develop seeds without opening the flower (cleistogamous). Plants vary from 4–20 cm. tall.

3. **Draba aprica** Beadle Whitlow Grass
Map 1064

*Draba brachycarpa* var. *fastigiata* Nutt.

Flowers April–May.

Occurs in low rocky or alluvial woodland in valleys of the southeastern Ozark section. Known only from Madison (St. Francis 'shut-ins,' 14 mi. south of Fredericktown, April 27, 1930, *Steyermark 1750*) and Reynolds (alluvial low woods along Black River, between the mouth of Cave Spring Hollow and Logslide Bluff, T29N, R2E, sect. 13 and 24, 6 mi. northwest of Piedmont, April 30, 1939, *Steyermark 22097*) counties. The latter station has been exterminated by impoundment from Clearwater Dam.

Ranges from Georgia to Arkansas and southeast Missouri. This is a very rare species and is known from only a few stations in the United States (see Rh. 42: 32–33. 1940 and Rh. 36: 361–63. 1934).

4. **Draba reptans** (Lam.) Fern. Whitlow Grass
Map 1065

Flowers February–May.

Occurs on rocky open glades and ledges, often of sandstone, chert, or granitic rocks, sandy and fallow fields, pastures, roadsides, and waste ground.

Two variations occur in Missouri:

Fruits glabrous (without hairs) . . . . . . 4a. D. REPTANS var. REPTANS

Fruits minutely hairy . . . . 4b. D. REPTANS var. MICRANTHA

4a. **Draba reptans** var. **reptans** Map 1065

*Draba reptans* (Lam.) Fern. [G, P & S, Steyerm.]

Throughout Missouri. This is the commoner variety encountered in the state.

Ranges from Georgia to New Mexico, north to Massachusetts, New York, Pennsylvania, Ontario, Michigan, Wisconsin, Minnesota, North Dakota, Colorado, and Washington.

4b. **Draba reptans** var. **micrantha** (Nutt.) Fern.
Map 1065

Scattered in the northern half of the state.

Ranges from Illinois and Wisconsin westward to the Pacific.

5. **Draba cuneifolia** Nutt. var. **cuneifolia**
Whitlow Grass Map 1066

*Draba cuneifolia* Nutt. [G, BB, P & S, Steyerm.]

Flowers February–May.

Occurs along the edges and top of bluff escarpments, usually of limestone, but also of sandstone or other rocks, rocky open glades and ledges, and in rocky open woods, most frequently in limestone areas. Restricted to the unglaciated prairie and Ozark sections of the state north to Ralls, Boone, Saline, and Jackson counties; absent from the southeastern lowland counties.

This species is a good one for rock gardens if planted from seed, as it is a winter-annual. In most specimens the leaves are near the base of the plant, but occasional ones seen in southern Illinois have 1–2 leaves higher on the stem.

## 15. **Lesquerella** Wats. Bladder-pod

Plant with a gray-silvery appearance due to the dense covering of star-shaped scales; stems slender, wire-like, 1–2 dm. tall; ovules 2 in each cell of ovary or seeds 2 in each cell of fruit; fruit directly attached (sessile) to its stalk (pedicel); limestone glade plants of southwestern Missouri . . . 1. L. FILIFORMIS

Plant with a greener appearance due to the sparser covering of star-shaped scales; stems stouter, less wire-like, 1–5 dm. tall; ovules 8–12 in each cell of ovary or seeds 8–12 in each cell of fruit; fruit with a short constricted stalk (stipe) 1–2 mm. long between its seed-bearing portion and the fruiting pedicel (stalk) to which it is attached; railroad introductions in the southern and central portions of the state 2. L. GRACILIS

1. **Lesquerella filiformis** Rollins Map 1067
*Lesquerella angustifolia* of auth. [G, BB, P & S, Steyerm.], not (Nutt.) S. Wats.
Flowers April–May.

Occurs on limestone glades, barrens, and outcrops, and in rocky open woods. Restricted to the southwestern unglaciated prairie section, where known from Greene, Dade, Lawrence, and Jasper counties.

Known only from southwestern Missouri in its presently recorded range.

This handsome plant has showy yellow flowers which contrast against the silvery stems and foliage. The plants usually occur in masses and make desirable additions to the rock garden. Being annuals, they should be started from seed on a gritty limestone, and eventually may seed themselves. Dr. Rollins (Rh. 58: 199–202. 1956) has recently clarified the status of this plant, previously confused by Payson and later authors with *L. angustifolia* (Nutt.) S. Wats. of Oklahoma and Texas.

2. **Lesquerella gracilis** (Hook.) S. Wats. var. **gracilis** Map 1068
Flowers April–June.

Occurs along railroads, where introduced in southeastern and west-central Missouri, in Bollinger (Glen Allen, May, 1900, *Colton Russell*) and Jackson (Courtney, May 24, 1894, *Bush 44*) counties. The latter collection was originally identified by Bush as *Camelina sativa*. Its correct identification has been verified by Dr. Rollins.

Ranges from Iowa and Nebraska to Kansas, Oklahoma, and Texas.

Specimens from Greene, Dade, Lawrence, and Jasper counties were erroneously cited under *L. gracilis* in Palmer and Steyermark's *Annotated Catalogue*, and these should have been cited under *L. angustifolia* of that catalogue (= *L. filiformis* Rollins), whereas the Greene County specimen cited in that catalogue under *L. angustifolia* should have been placed with *L. gracilis* (Hook.) Wats. of the *Catalogue*.

## 16. **Armoracia** Gaertn., Mey. & Scherb.

Plant growing in the water; basal and lower leaves finely dissected into hair-like divisions . . 1. A. AQUATICA

Plant normally growing out of the water; none of the leaves finely dissected, the lower with crenately toothed (scalloped or wavy-edged) margins . . . . . . . . . . . . . . . . . . . 2. A. RUSTICANA

1. **Armoracia aquatica** (Eat.) Wieg. Lake Cress Map 1069
*Rorippa aquatica* (Eat.) Palmer & Steyerm. [P & S, Steyerm.]
*Neobeckia aquatica* (Eat.) Greene [Jones *et al.*]
Flowers May–August.

Occurs in bald cypress swamps, other kinds of wooded swamps and sloughs, and slow streams. Rare and scattered in southern and eastern Missouri; mainly in the southeastern part of the state and locally in Jasper County.

Ranges from Quebec, Ontario, and Minnesota, south to Florida and Texas.

The dissected leaves somewhat resemble and are sometimes mistaken for those of Milfoil (*Myriophyllum*), Hornwort (*Ceratophyllum*), or White or Yellow Crowfoot (*Ranunculus longirostris* and *R. flabellaris*).

2. **Armoracia rusticana** (Lam.) Gaertn. Horseradish Map 1070
*Armoracia lapathifolia* Gilib. [G]
*Rorippa Armoracia* (L.) Hitchc. [P & S, Steyerm.]
*Cochlearia Armoracia* L. [Royal Hort. Dict.]
Flowers May–July.

Escaped from cultivation in fields, along roadsides, fence rows, and waste or cultivated ground. Scattered in the state.

Native of Europe; introduced and naturalized from Quebec to British Columbia, south to Florida and Texas.

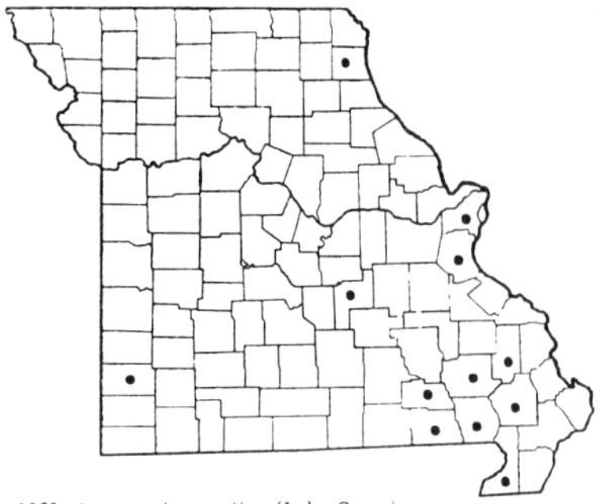

1069 Armoracia aquatica (Lake Cress)

1070 Armoracia rusticana (Horseradish)

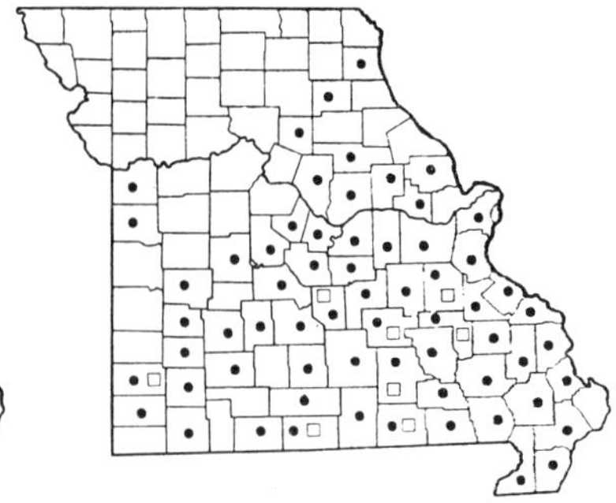

1071 • Cardamine bulbosa f. bulbosa (Spring Cress)
1071 □ Cardamine bulbosa f. fontinalis

The large basal leaves with their crenately toothed margins are quite conspicuous and characteristic of the plant. The large vertical taproot provides the horseradish of commerce. It is reputed to have beneficial effects on digestion and in the prevention of scurvy.

## 17. **Cardamine** L. Bitter Cress

a. Leaves divided to the midrib into 2–12 or more divisions; petals small, 1.5–4 mm. long . . . . . . *b*
 b. Base of petiole (stalk) of stem-leaves hairy; leaves at base of plant persistent, numerous, thickish, often hairy, their petioles (stalks) hairy; leaves on stem 2–5; rarely collected species . . 3. C. HIRSUTA
 b. Base of petiole of stem-leaves glabrous (without hairs); leaves at base of plant absent or few, rather thinner, glabrous (without hairs), their petioles glabrous; leaves on stem usually 4–10; commonly collected species . . . . . . . . . . . . . . . . . . . . . . . . . . *c*
  c. Leaf-divisions along the sides of the leaves oblanceolate to linear, mostly of the same narrow width throughout, their bases ending at the main axis (rachis) of the leaf and not connecting with one another or inconspicuously connecting by continuous green leaf tissue; topmost leaflet of the stem-leaves scarcely or only slightly wider than the leaflets along the side . . 5. C. PARVIFLORA var. ARENICOLA
  c. Leaf-divisions along the sides of the leaves broader in the middle or upper part and obovate, linear-oblanceolate, oval or oblong, their bases not ending at the main axis (rachis) of the leaf but conspicuously connecting with one another by continuous green leaf tissue; topmost leaflet of the stem-leaves usually wider than the leaflets along the side . . . 4. C. PENSYLVANICA
a. Leaves smooth-edged (entire) or with few coarse, blunt or pointed teeth, but not divided to the midrib . . . . . . . . . . . . . . . . . . . . . . . . . . . . *d*
 d. Stems usually completely glabrous (without hairs); leaves on stem mostly 4–8, sometimes more; petals white . . . . . . . . . . . . . . . . . . . . . . . . 1. C. BULBOSA
 d. Stems more or less hairy, at least in the lower half; leaves on stem mostly 3–5; petals pale lilac or rose-lavender . . . . . . . . . . . . . . . . . . . . . . 2. C. DOUGLASSII

1. **Cardamine bulbosa** (Schreb.) BSP.
Spring Cress Map 1071
Flowers late March–June.

Occurs in low wet woodland, wet meadows, margins of spring branches and spring-fed streams, and moist ledges and slopes of bluffs. Southern, central, and northeastern Missouri.

Missouri material is represented by the following variations:

Plants growing in the water of springs and spring branches; basal leaves more or less orbicular, more or less heart-shaped (cordate) at the base; stem not rising from a tuberous-thickened, but from a fibrous base . . 1b. C. BULBOSA f. FONTINALIS

Plate no. 182. **1.** Armoracia aquatica, × 2/5; Details from Small, The New York Botanical Garden. **2.** Armoracia rusticana, × 2/5; Details from Small, The New York Botanical Garden. **3.** Cardamine bulbosa, × 2/5; *a.* Flower × 4/5. **4.** Cardamine Douglassii, × 2/5; Details from Small, The New York Botanical Garden. **5.** Cardamine hirsuta, × 2/5; After Gleason, The New York Botanical Garden.

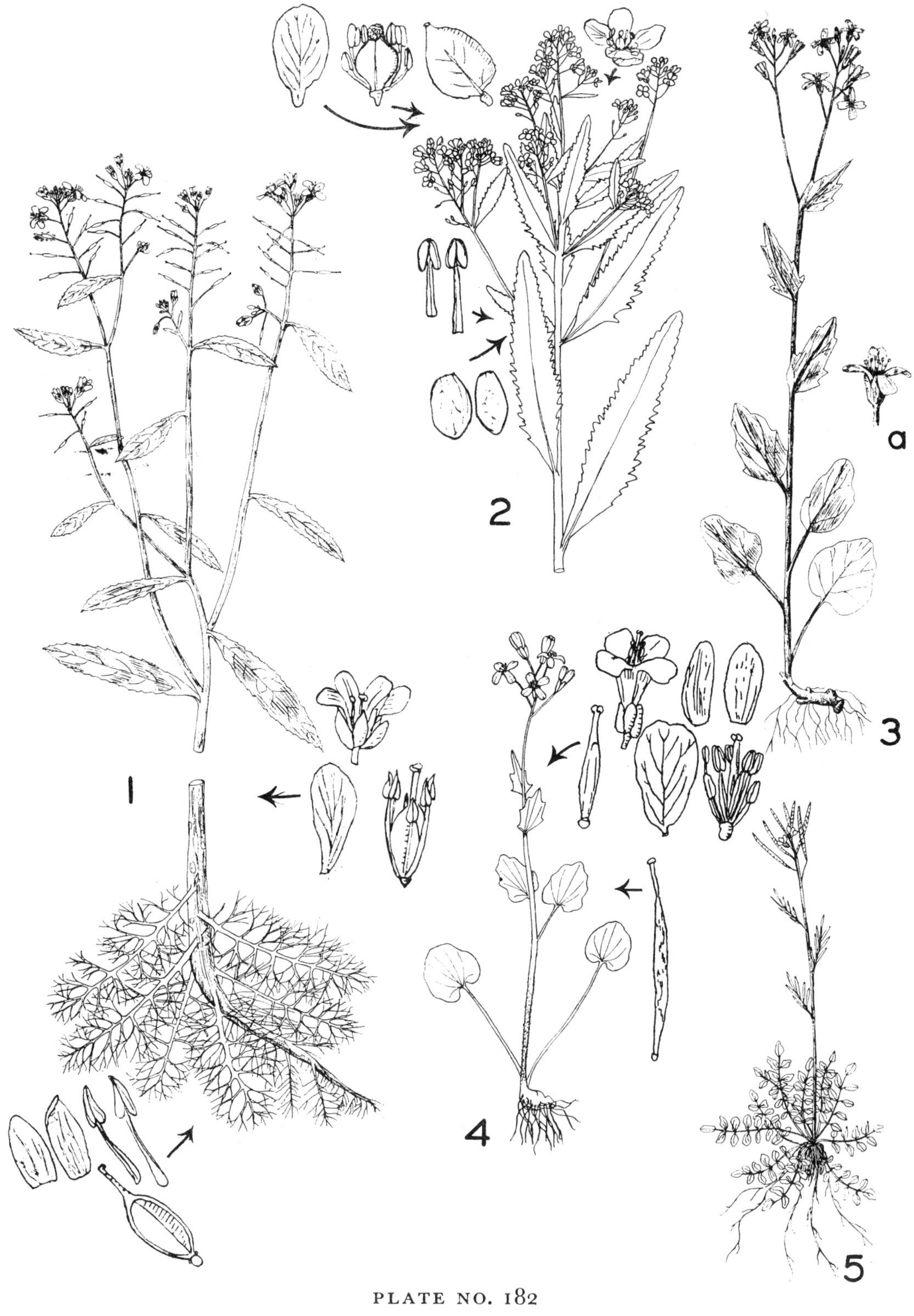


PLATE NO. 182

1072 Cardamine Douglassii (Purple Cress)

1073 Cardamine hirsuta

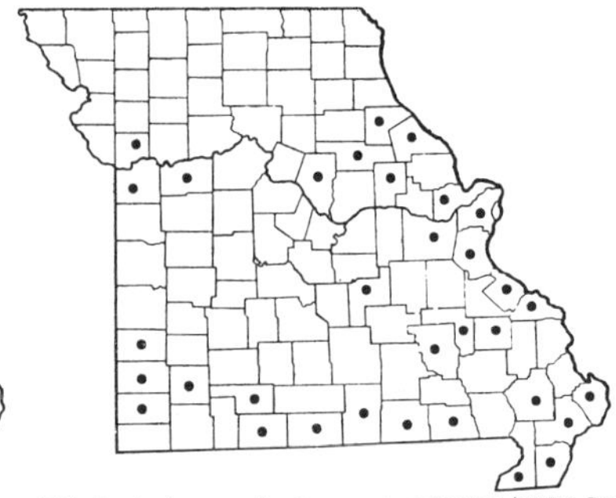

1074 Cardamine pensylvanica var. pensylvanica (Bitter Cress)

Plants growing out of water; basal leaves less rounded, often more oblong or ovate, not heart-shaped at the base; stem arising mostly from a tuberous-thickened base . 1a. C. BULBOSA f. BULBOSA

1a. **Cardamine bulbosa** f. **bulbosa** Map 1071
*Cardamine bulbosa* (Schreb.) BSP. [G, BB, P & S, Steyerm.]

Southern, central, and northeastern Missouri, north to Lewis, Shelby, Randolph, and Jackson counties. This is the common variation in Missouri.

Ranges from Florida to Texas, north to New Hampshire, Vermont, Quebec, Ontario, Michigan, Wisconsin, Minnesota, and South Dakota.

1b. **Cardamine bulbosa** f. **fontinalis** Palmer & Steyerm. Map 1071

Occurs in Ozark springs and spring branches of southern Missouri north to Washington, Dent, Pulaski, and Jasper counties (the holotype is *Steyermark 12414* from Pulltight Spring, Shannon County, in the Mo. Bot. Gard. herb.).

Known only from Missouri.

This aquatic or semiaquatic form of *C. bulbosa* somewhat resembles and has been mistaken for *C. rotundifolia*, which does not occur in Missouri.

Stems of *C. bulbosa* may occasionally show some hairs, but in the majority of specimens examined, they are completely or practically glabrous.

The thickened base of the stem has a flavor similar to that of horseradish, and, when grated and mixed with vinegar, may be used as a horseradish substitute. The young leaves and stems also have a horseradish flavor and may be cut up and used for salad.

2. **Cardamine Douglassii** (Torr.) Britt. Purple Cress Map 1072

Also called Northern Bitter Cress.

Flowers March–April.

Occurs in low or swampy woodland. Northeastern and eastern Missouri west to Adair, Macon, and Chariton counties north of the Missouri River, and locally south of the Missouri River only in Cape Girardeau County.

Ranges from Connecticut, New York, and Ontario, to Wisconsin, south to Virginia, Kentucky, Tennessee, and Missouri.

This is an attractive early-flowering species, usually growing lower (usually 10–25 cm. tall) and flowering one to two weeks earlier in Missouri than *C. bulbosa*. This is in accord with Deam's observations (*Fl. Ind.* p. 498. 1940).

3. **Cardamine hirsuta** L. Map 1073

Flowers March–April.

Known only from southwestern Missouri in Webster County (opening in valley along Finley Creek, T28N, R17W, sect. 27, 3½ mi. south of Seymour, April 17, 1955, *Steyermark 78382*).

Native of Europe; introduced and naturalized from New York to Georgia, Alabama, and Kentucky west to Illinois and Missouri.

4. **Cardamine pensylvanica** Muhl. var. **pensylvanica** Bitter-Cress Map 1074
*Cardamine pensylvanica* Muhl. [G, BB, P & S, Steyerm.]

Flowers March–July.

Occurs in low wet woods, along spring branches and streams, and at the base of moist bluffs. Southern and central Missouri north to Ralls, Audrain, Boone, and Clay counties. Probably throughout northern Missouri when more intensively collected.

Ranges from Labrador to British Columbia, south to Florida, Texas, and California.

Although the stems are described in some manuals as being hispid at the base, they are usually completely glabrous when growing as ordinary terrestrial plants in Missouri. The species is quite variable in texture and shape of leaflets, and is often mistaken for and

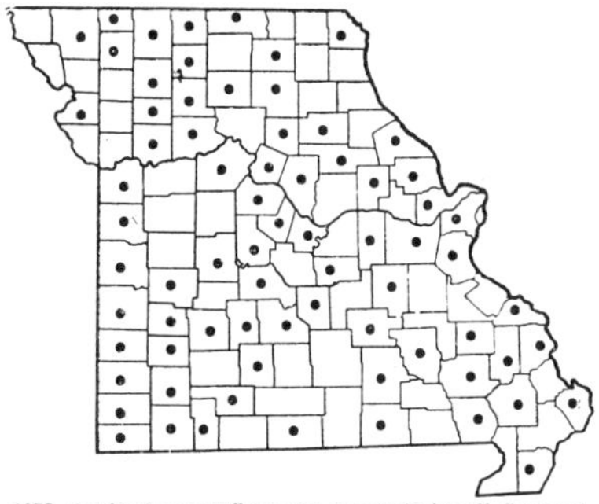

1075 Cardamine parviflora var. arenicola (Small-flowered Bitter Cress)

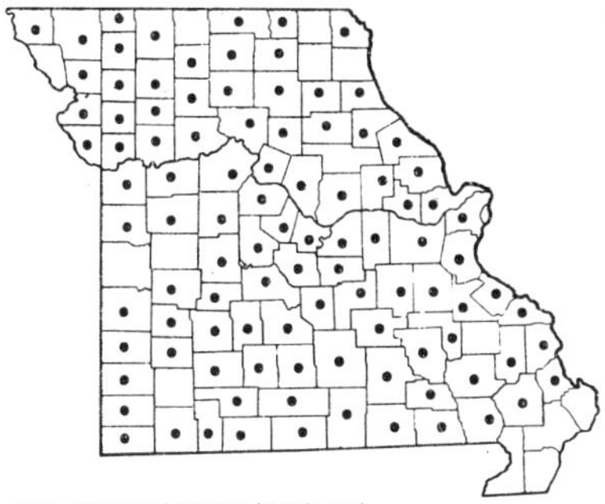

1076 Dentaria laciniata (Toothwort)

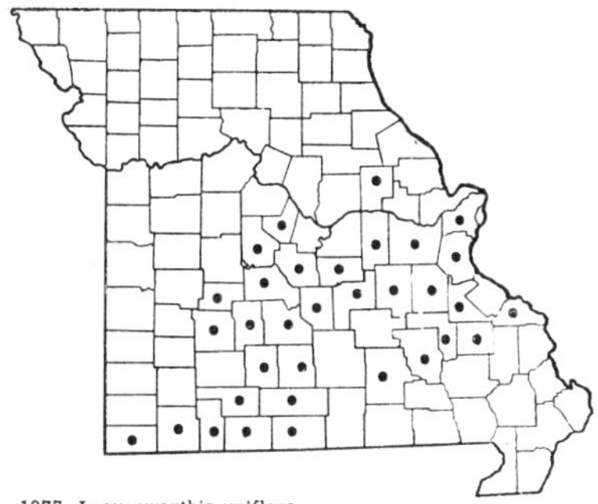

1077 Leavenworthia uniflora

difficult to separate from the next species. In general this species selects the low valleys and bottomland soils, whereas the next species inhabits generally drier upland habitats.

The fresh green stems and leaves may be eaten in salads or as an ingredient of salads as in Water Cress, but lack the soft tenderness of that plant.

5. **Cardamine parviflora** L. var. **arenicola** (Britt.) O. E. Schulz Small-flowered Bitter Cress Map 1075

Flowers March–July.

Occurs in thin, more or less acid soils of open dry, upland or rocky woodland, on slopes and ridges, wet depressions in glades and rocky ledges, and fallow and cultivated fields. Throughout Missouri, and doubtless in every county.

Ranges from Florida to Texas, north to Newfoundland, Quebec, Ontario, Wisconsin, and Minnesota; and from Oregon to British Columbia.

This varies in size and luxuriance from short delicate simple stems 5–8 cm. tall to more luxuriant plants having taller and more branched stems.

## 18. **Dentaria** L. Toothwort

**Dentaria laciniata** Muhl. Toothwort Map 1076

*Dentaria laciniata* var. *integra* (Schulz) Fern. [P & S, Steyerm.]

*Dentaria laciniata* var. *latifolia* Farwell [Steyerm.]

Flowers March–May.

Occurs in rich woods, either in low valley forest or on slopes of ravines and bluffs. Throughout Missouri and doubtless in every county except possibly those from the extreme southeastern lowlands.

Ranges from Vermont and Quebec to Minnesota and Nebraska, south to Florida, Alabama, Louisiana, and Oklahoma.

This species varies in the width and amount of toothing of the leaf-segments, as well as in the number and arrangement of the main leaves. These variations appear to represent inconstant manifestations of variability and have not been accorded recognition by recent students of various floras.

The fleshy rootstock may be eaten in salads. It has a radishlike flavor. When grated and mixed with vinegar, it can be used as a horseradish substitute.

## 19. **Leavenworthia** Torr.

**Leavenworthia uniflora** (Michx.) Britt. Map 1077

Flowers March–April.

Occurs on limestone glades, barrens, and 'bald knobs.' Restricted to the Ozark region of southern and central Missouri, north to St. Louis, Franklin, Montgomery, Moniteau, Morgan and Hickory counties to McDonald County; absent from a number of southern and southeastern Ozark counties.

Ranges from Alabama to Arkansas, north to Ohio, Indiana, and Missouri.

This low winter-annual, 5–15 cm. tall, has a brief season, quickly disappearing from the rocky dry glades after maturing the seed in late spring. The dark green rosette of deeply cut leaves are very attractive with the showy white flowers.

## 20. **Sibara** Greene

**Sibara virginica** (L.) Rollins Map 1078
*Arabis virginica* (L.) Poir. [BB, P & S, Steyerm.]

Flowers March–May, occasionally October–November.

Occurs in open or wooded alluvial ground, fallow fields, low pastures, along roadsides, and waste ground. Throughout southern and central Missouri and north locally to Ralls, Linn, De Kalb, Gentry, and Andrew counties.

Ranges from Florida to Texas and Mexico, north to Virginia, Ohio, Indiana, Illinois, Missouri, Kansas; also California.

This species is often found in fallow fields with such species as *Myosurus minimus*, *Valerianella radiata*, *Myosotis verna*, *Poa Chapmaniana*, and other annuals. It closely resembles *Cardamine parviflora* var. *arenicola*, but is easily distinguished by such obvious differences as stems hairy at their base, the more numerous leaf-divisions of a leaf, and the broader fruits.

## 21. **Arabis** L. Rock Cress

a. All the leaves of stem narrowed at the base, or not clasping or not partly surrounding the stem by their leaf-bases . . . . . *b*
- b. Plant 0.7–3.5 dm. tall; fruits 1–4.5 cm. long, ascending when mature; leaves of stem linear to spatulate, 1.5–5 mm. broad . . . . . 1. A. LYRATA
- b. Plant 3–10 dm. tall; fruits 5–10 cm. long, widely spreading or recurved or pointing downward when mature; leaves of stem oblong-lanceolate, elliptic, or linear, 5–25 mm. broad . 6. A. CANADENSIS

a. Some or all of the leaves of the stem clasping or partly surrounding the stem by their leaf-bases . . . . *c*
- c. Stems and leaves on the stem (not the rosette leaves) glabrous (without hairs) . . . . . *d*
  - d. Leaves on lower part of stem sharply toothed or irregularly to deeply cut; petals 6–8 mm. long, about twice the length of the sepals; leaves dark or grass green without a glaucous (grayish) coating which can be rubbed off . . . . . 4. A. MISSOURIENSIS
  - d. Leaves on lower part of stem toothed to smooth-edged (entire); petals 3–5 mm. long, about equal to or slightly longer than the sepals; leaves bluish- or gray-green with a glaucous (grayish) coating which can be rubbed off . . . . . 3. A. LAEVIGATA
- c. Stems and/or leaves on the stem hairy . . . . . *e*
  - e. Petals 2–4 mm. long; stalks (pedicels) of flowers 0.8–2 mm. long, of fruits 1–4 mm. long; lower side of basal leaves with star-shaped hairs; usually several curved stems arising from the base of the plant; seeds wingless . . . . . 5. A. SHORTII
  - e. Petals 4–8 mm. long; stalks (pedicels) of flowers 3–9 mm. long, of fruits 6–10 mm. long; lower side of basal leaves with simple hairs; usually 1 stiff erect stem arising from the base of the plant; seeds winged . . . . . *f*
    - f. Petals 6–8 mm. long; fruits 6–9 cm. long, 1.7–2 mm. wide, outwardly curving or recurved; rosette leaves deeply and irregularly cut (laciniate) to pinnatifid with a large terminal lobe 4. A. MISSOURIENSIS
    - f. Petals 4–6 mm. long; fruits 1.5–5 cm. long, 0.7–1 mm. wide, stiffly erect; rosette leaves somewhat toothed or slightly wavy-edged or without teeth (entire) . . . . . 2. A. HIRSUTA

1. **Arabis lyrata** L. Rock Cress Map 1079

Flowers April–May.

Occurs in rocky crevices of limestone or sandstone bluffs and on rocky wooded slopes along bluffs. Eastern Missouri north to Knox County and west to Boone, Phelps, Texas, and Shannon counties.

Ranges from Vermont to Ontario and Minnesota, south to Georgia, Tennessee, and Missouri; also in Alberta.

Two variations occur in Missouri:

Fruits 2–4.5 cm. long; petals 5–8 mm. long; style 0.65–1.25 mm. long; common type . 1a. A. LYRATA var. LYRATA f. LYRATA

Fruits 1–2 cm. long; petals 3.5–5 (–6) mm. long; style less than 0.5 mm. long; rare type . . . 1b. A. LYRATA var. LYRATA f. PARVISILIQUA

Plate no. 183. **1.** Cardamine pensylvanica var. pensylvanica, × 2/5. **2.** Cardamine parviflora var. arenicola, × 2/5; After Gleason, The New York Botanical Garden. **3.** Dentaria laciniata, × 2/5; Details from Small, The New York Botanical Garden. **4.** Leavenworthia uniflora, × 2/5; Details from Small, The New York Botanical Garden. **5.** Sibara virginica, × 2/5. **6.** Arabis lyrata var. lyrata, × 2/5.

PLATE NO. 183

1078 Sibara virginica

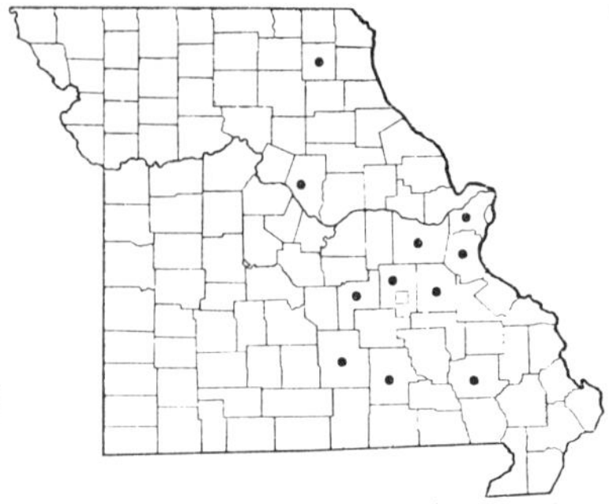

1079 • Arabis lyrata var. lyrata f. lyrata (Rock Cress)
1079 □ Arabis lyrata var. lyrata f. parvisiliqua

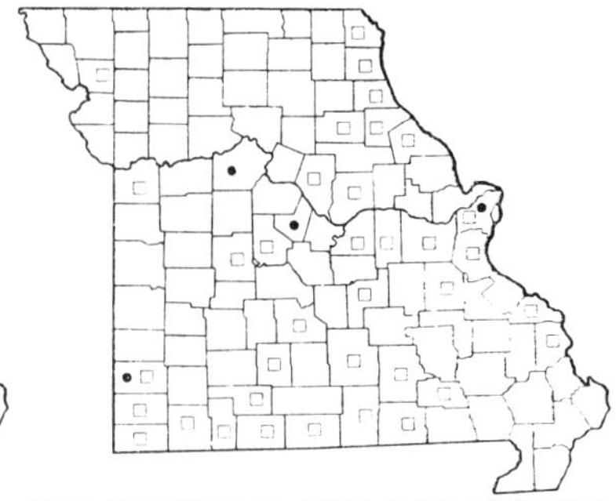

1080 • Arabis hirsuta var. pycnocarpa (Hairy Rock Cress)
1080 □ Arabis hirsuta var. adpressipilis

1a. **Arabis lyrata** var. **lyrata** f. **lyrata** Map 1079
*Arabis lyrata* L. [G, BB, P & S, Steyerm.]
*Arabis lyrata* var. *typica* [Hopkins.]

This is the common variety in Missouri and has the range indicated for the state.

1b. **Arabis lyrata** var. **lyrata** f. **parvisiliqua** Hopkins Map 1079
*Arabis lyrata* f. *parvisiliqua* [G]

Known only from Crawford County (sandy banks of Meramec River, April, 1882, *Eggert*).

2. **Arabis hirsuta** (L.) Scop. Hairy Rock Cress Map 1080

Flowers May–June.

Occurs on rocky ledges and bluffs, usually of limestone exposures.

Two variations occur in Missouri:

Hairs of stem spreading, simple, not forked . . . 2a. A. HIRSUTA var. PYCNOCARPA
Hairs of stem closely pressed against the stem and forked . . . 2b. A. HIRSUTA var. ADPRESSIPILIS

2a. **Arabis hirsuta** var. **pycnocarpa** (M. Hopkins) Rollins Map 1080
*Arabis hirsuta* of auth. [P & S] not (L.) Scop.
*Arabis pycnocarpa* M. Hopkins [Steyerm.]
*Arabis pycnocarpa* var. *typica* [Hopkins]

Southern and central Missouri, where known from St. Louis, Moniteau, Saline, and Jasper counties. This is the most uncommon of the variations as far as records go.

Ranges from Quebec to Alaska, south to Pennsylvania, Georgia, Arkansas, Kansas, New Mexico, Arizona, and California.

2b. **Arabis hirsuta** var. **adpressipilis** (M. Hopkins) Rollins Map 1080
*Arabis pycnocarpa* var. *adpressipilis* Hopkins [Steyerm., Hopkins]

Scattered throughout Missouri, but apparently absent from most of the northwestern third of the state, locally northwest in Andrew County. This is the commoner variation in Missouri.

Ranges from Ontario and Minnesota, south to West Virginia, Indiana, Illinois, and Arkansas.

3. **Arabis laevigata** (Muhl.) Poir. var. **laevigata** Smooth Rock Cress Map 1081
*Arabis laevigata* (Muhl.) Poir. [G, BB, P & S, Steyerm.]

Flowers April–June.

Occurs on ledges of and at the base of rocky bluffs, alluvial ground and low woods along streams, and rocky wooded hills. Southern, central, and eastern Missouri, and north locally to Mercer County.

Ranges from Quebec to Minnesota and Colorado, south to Georgia, Alabama, Arkansas, and Oklahoma.

Another variety, var. *Burkii* Porter, has leaves which are not clasping at the base, but has not been recorded from Missouri.

The leaves of *A. laevigata* are somewhat variable, the shape varying from oblong-lanceolate to linear, the margins of the leaves on the stem varying from entire to conspicuously toothed. The rosette leaves of this biennial plant are slightly hairy the first year, but glabrous the second year and coarsely to irregularly toothed. They are generally wine-purple on the lower surface. The fruits vary from numerous, crowded, and elongated on luxuriant plants to few, remote, and shorter on dwarfed or depauperate specimens.

---

Plate no. 184. **1.** Arabis hirsuta, $\times {}^2/_7$. **2.** Arabis laevigata var. laevigata, $\times {}^2/_7$; Details from Small, The New York Botanical Garden. **3.** Arabis missouriensis, $\times {}^2/_7$. **4.** Arabis canadensis, $\times {}^2/_7$. **5.** Arabis Shortii, $\times {}^2/_7$. **6.** Nasturtium officinale, $\times {}^2/_7$; Details from Small, The New York Botanical Garden.

PLATE NO. 184

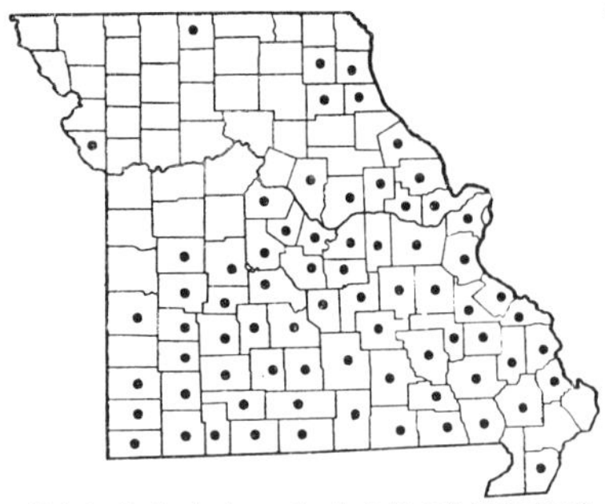

1081 Arabis laevigata var. laevigata (Smooth Rock Cress)

1082 ● Arabis missouriensis var. missouriensis (Rock Cress)
1082 □ Arabis missouriensis var. Deamii

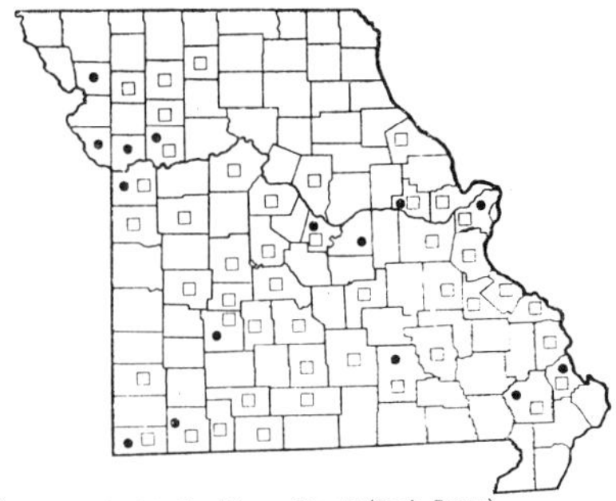

1083 ● Arabis Shortii var. Shortii (Rock Cress)
1083 □ Arabis Shortii var. phalacrocarpa

4. **Arabis missouriensis** Greene Rock Cress
Map 1082

Flowers April–June.

Occurs in generally acid soils associated with chert, sandstone, or granite, on rocky wooded slopes, upland ridges, crests of hills, and sand hills of Crowley Ridge. Restricted to the Ozark section of southern and east-central Missouri north to St. Charles, Franklin, Phelps, Pulaski, Laclede, and Jasper counties, and southeast on sand hills of Crowley Ridge in Stoddard County.

Two variations occur in Missouri:

Plants completely glabrous (without hairs) . . . 4a. A. MISSOURIENSIS var. MISSOURIENSIS
Stems and leaves more or less hairy . . . . 4b. A. MISSOURIENSIS var. DEAMII

4a. **Arabis missouriensis** var. **missouriensis**
Map 1082

*Arabis missouriensis* Greene [G]

*Arabis viridis* Harger [P & S in part, Steyerm.]

*Arabis viridis* var. *viridis* [BB]

This is the common variation in Missouri and has the range indicated above.

Ranges from Maine to New York, south to Pennsylvania; also in Georgia and Michigan; Missouri to Arkansas and Oklahoma.

The original description for this species was based upon a Missouri specimen (*Bush 31*, from Monteer, Shannon County, holotype in Notre Dame herb.).

4b. **Arabis missouriensis** var. **Deamii** (Hopkins) Hopkins
Map 1082

*Arabis viridis* var. *Deamii* Hopkins [BB, Steyerm.]

Known only from Stoddard (sand hills, Crowley Ridge, T25N, R11E, NW¼ sect. 6, 3½ mi. southeast of Bloomfield, April 25, 1955, *Steyermark 78279*) and Dent (wooded bluffs of Creek, Howe's Mill, *Palmer 34950*) counties.

Known from Wisconsin, Indiana, and Missouri.

*Arabis missouriensis* is easily distinguished from *A. laevigata* by its green, nonglaucous leaves and stem, the more deeply toothed, incised, or pinnatifid rosette leaves, erect to closely ascending, more or less overlapping leaves of the stem, and by the occurrence on more acid soils in drier locations.

5. **Arabis Shortii** (Fern.) Gl. Rock Cress
Map 1083

Flowers April–June.

Occurs in rocky or alluvial soils along stream banks, rich or low woods bordering streams, and at the base of and on rocky ledges of wooded bluffs. Scattered throughout southern, central, and western Missouri; apparently absent or not collected from most of the northeastern third of the state.

Two variations occur in Missouri:

Fruits and ovaries finely hairy with minute, star-shaped (stellate) hairs . 5a. A. SHORTII var. SHORTII
Fruits and ovaries glabrous (without hairs) . . 5b. A. SHORTII var. PHALACROCARPA

5a. **Arabis Shortii** var. **Shortii** Map 1083

*Arabis perstellata* Braun var. *Shortii* Fern. [G]

*Arabis dentata* (Torr.) T & G. [P & S in part, Hopkins], not *A. dentata* Clairv.

Occurs in southern, central, and western Missouri.

Ranges from New York to Minnesota and South Dakota, south to Virginia, Tennessee, Arkansas, and Oklahoma.

5b. **Arabis Shortii** var. **phalacrocarpa** (M. Hopkins) Steyerm.
Map 1083

*Arabis dentata* var. *phalacrocarpa* M. Hopkins, Rh. 39: 169. 1937. [Steyerm., Hopkins]

*Arabis perstellata* var. *phalacrocarpa* (M. Hopkins) Fern. Rh. 48: 208. 1946 [G]

*Arabis dentata* in part [of P & S], not *A. dentata* (Torr.) T & G.

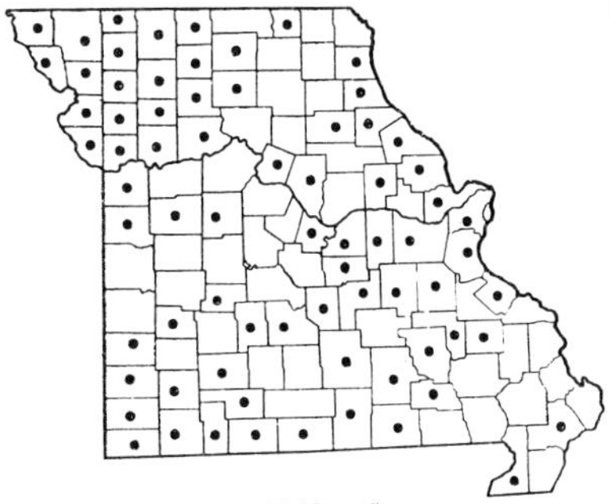

1084 Arabis canadensis (Sickle-pod)

1085 Nasturtium officinale var. officinale (Water Cress)

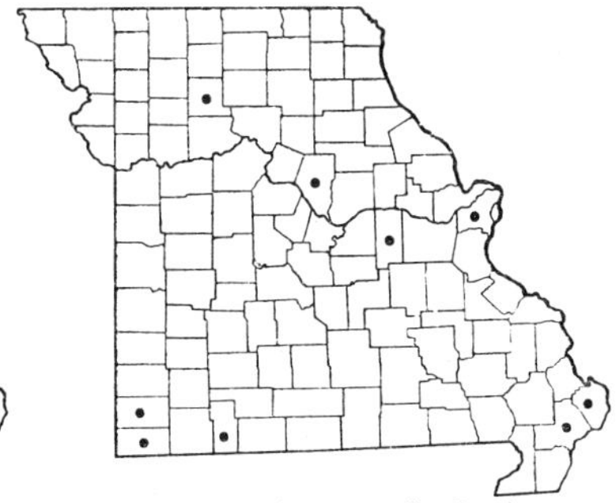

1086 Rorippa sylvestris (Creeping Yellow Cress)

Occurs in southern, central, and western Missouri, northeast to Pike County.

Known only from Iowa, Missouri, Arkansas, and Oklahoma.

6. **Arabis canadensis** L. Sickle-pod Map 1084

Flowers April–June.

Occurs in rocky or open, usually dry woodland, on slopes or upland ridges, rocky ledges of dry wooded bluffs, and open ground. Throughout Missouri, and probably in every county of the state.

Ranges from Maine to Ontario and Minnesota to Nebraska, south to Georgia, Alabama, Arkansas, Oklahoma, and Texas.

## 22. **Nasturtium** R. Br. Water Cress

**Nasturtium officinale** R. Br. Water Cress Map 1085

Flowers April–October.

Occurs in and around the water of cold springs and spring-fed rivulets and streams. Restricted to southern and central Missouri north to St. Louis, Boone, and Clay counties; absent from the counties of most of the unglaciated prairie section between Morgan, Benton, Dade, and Jasper counties to Jackson County.

All the known stations for Water Cress are included on one map without variations being indicated.

The following two principal variations are represented by Missouri material:

Leaflet at the summit of each leaf with a round or oval shape; leaflets along the side of the leaf rounded at their base, oval, elliptical, or obovate; common variation . . . . . a. N. OFFICINALE var. OFFICINALE

All the leaflets of the leaf with a nearly oblong shape; less common variation . . b. N. OFFICINALE var. SIIFOLIUM

a. **Nasturtium officinale** var. **officinale** Map 1085

*Nasturtium officinale* R. Br. [G, BB]

*Rorippa Nasturtium-aquaticum* (L.) Schinz & Thell. [P & S, Steyerm.]

This is the common variation found in the state.

Ranges from southern Canada throughout the United States; also in Eurasia.

b. **Nasturtium officinale** var. **siifolium** (Reichenb.) Koch Map 1085

This variation is found throughout the range in Missouri.

Ranges from Massachusetts to Iowa and westward, south to North Carolina, Texas, and California.

Another variation, var. **microphyllum** (Boenn.) Thell., differs from both of the others indicated by the longer (usually 1.5–2.7 cm. long) and narrower (1–1.5 mm. broad) fruits, somewhat longer pedicels (0.8–2.5 cm. instead of 0.6–1.5 cm. long), and longer, more slender styles (0.5–2 mm. long) of the fruit, those of the other two varieties being somewhat thicker and 0.1–1 mm. long. This variety has not been recorded from Missouri.

Although Water Cress is stated by most manuals to be introduced in North America, all the Missouri stations seen by the author appear to be indigenous native plants associated with cold fresh water springs and, so far as known, do not suggest having been introduced. The springs are often located in remote sections of the Ozark region and invariably the Water Cress is found, seemingly, as a native plant associated with other native aquatic species, such as *Sparganium americanum*, *Anacharis Nuttallii*, *Najas guadalupensis*, *Ranunculus longirostris*, *Potamogeton foliosus*, *P. illinoensis*,

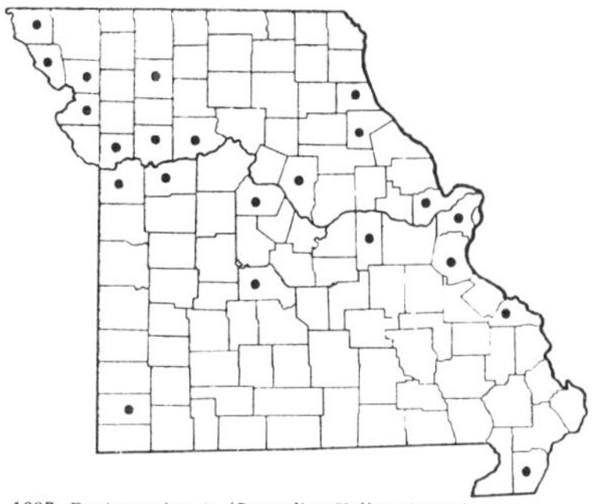

1087 Rorippa sinuata (Spreading Yellow Cress)

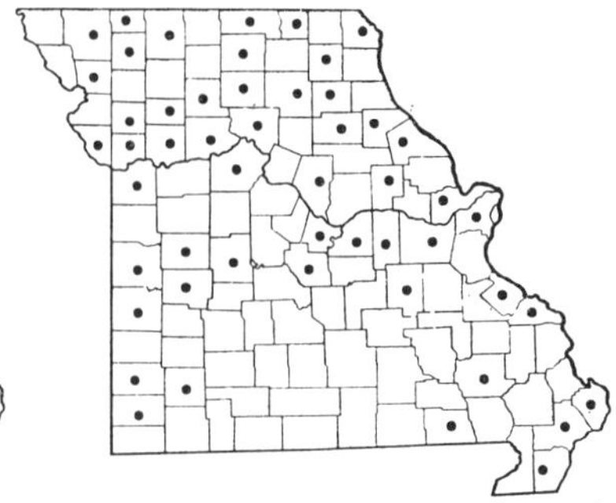

1088 Rorippa sessiliflora (Sessile-flowered Yellow Cress)

1089 Rorippa obtusa var. obtusa (Blunt-leaved Yellow Cress)

*P. amplifolius*, *Lemna trisulca*, *Ceratophyllum demersum*, and *Myriophyllum heterophyllum*. Until evidence can be produced to the contrary, I am regarding the Water Cress found in Ozark springs as a native indigenous North American plant.

This is commonly used for salads and for garnishing meats and other dishes where a peppery or pungent flavor is desired. The leaves are eaten by ducks, muskrats, and deer, and the plants serve as shelter for small aquatic life.

## 23. **Rorippa** Scop. Yellow Cress, Marsh Cress

a. Petals 4–8 mm. long, showy and longer than the sepals; perennial plants with creeping underground stems . . . *b*
- b. Main leaves of the stem pinnately divided (cut in all the way to the midrib); divisions of leaves toothed or cut; no auricles (rounded or ear-like lobes or projections) protruding from the base of the leaves . . . 1. R. SYLVESTRIS
- b. Main leaves of the stem deeply cut but not to the midrib (pinnatifid); divisions of leaves smooth-edged (entire) or with few teeth; auricles (rounded or ear-like lobes or projections) extending from the base of at least the middle leaves of the stem . . . 2. R. SINUATA

a. Petals none or up to 2 mm. long, shorter than or only equaling the sepals; annual or biennial plants without a creeping underground stem . . . *c*
- c. Fully grown stalks (pedicels) of fruits 3–7 mm. long; fruits shorter than to at most 2 times as long as their stalks (pedicels) . . . 5. R. ISLANDICA
- c. Fully grown stalks (pedicels) of fruits 3 mm. or less long; fruits 2–10 times as long as their stalks (pedicels) . . . *d*
  - d. Main leaves of the stem pinnately parted or divided (cut all the way to the midrib); stamens 6; style evident, about 1 mm. long on the fruits; fruits 2–4 times as long as their stalks (pedicels) 4. R. OBTUSA
  - d. Main leaves of the stem mostly shallowly toothed or wavy-edged, sometimes coarsely toothed or one or two of the lower teeth more prominent; stamens 4; style practically absent, at most 0.5 mm. in fruit; fruits 6–10 times as long as their stalks (pedicels) . . . 3. R. SESSILIFLORA

1. **Rorippa sylvestris** (L.) Bess.
Creeping Yellow Cress — Map 1086
Flowers May–September.

Occurs in moist low ground, along stream banks, railroads, wet fields, and meadows. Rare and scattered in Missouri north to Livingston County.

Native of Europe; introduced into North America where ranging from Newfoundland to Ontario and North Dakota, south to North Carolina, Kentucky, and Louisiana.

2. **Rorippa sinuata** (Nutt.) Hitchc.
Spreading Yellow Cress — Map 1087
Flowers April–July.

Occurs in moist open ground in river flood plain, alluvial banks of the Missouri, Mississippi, and other large streams, and along railroads. Scattered in Missouri, mostly in the northern and eastern halves of the state following the larger rivers and locally in the southwestern part of the state in Newton County.

Ranges from Ontario to Montana and Washington,

---

Plate no. 185. **1.** Rorippa sylvestris, × 2/5. **2.** Rorippa sessiliflora, × 2/5. **3.** Rorippa sinuata, × 2/5. **4.** Rorippa obtusa, × 2/5. **5.** Rorippa islandica var. Fernaldii, × 2/5; Details from Small, The New York Botanical Garden.

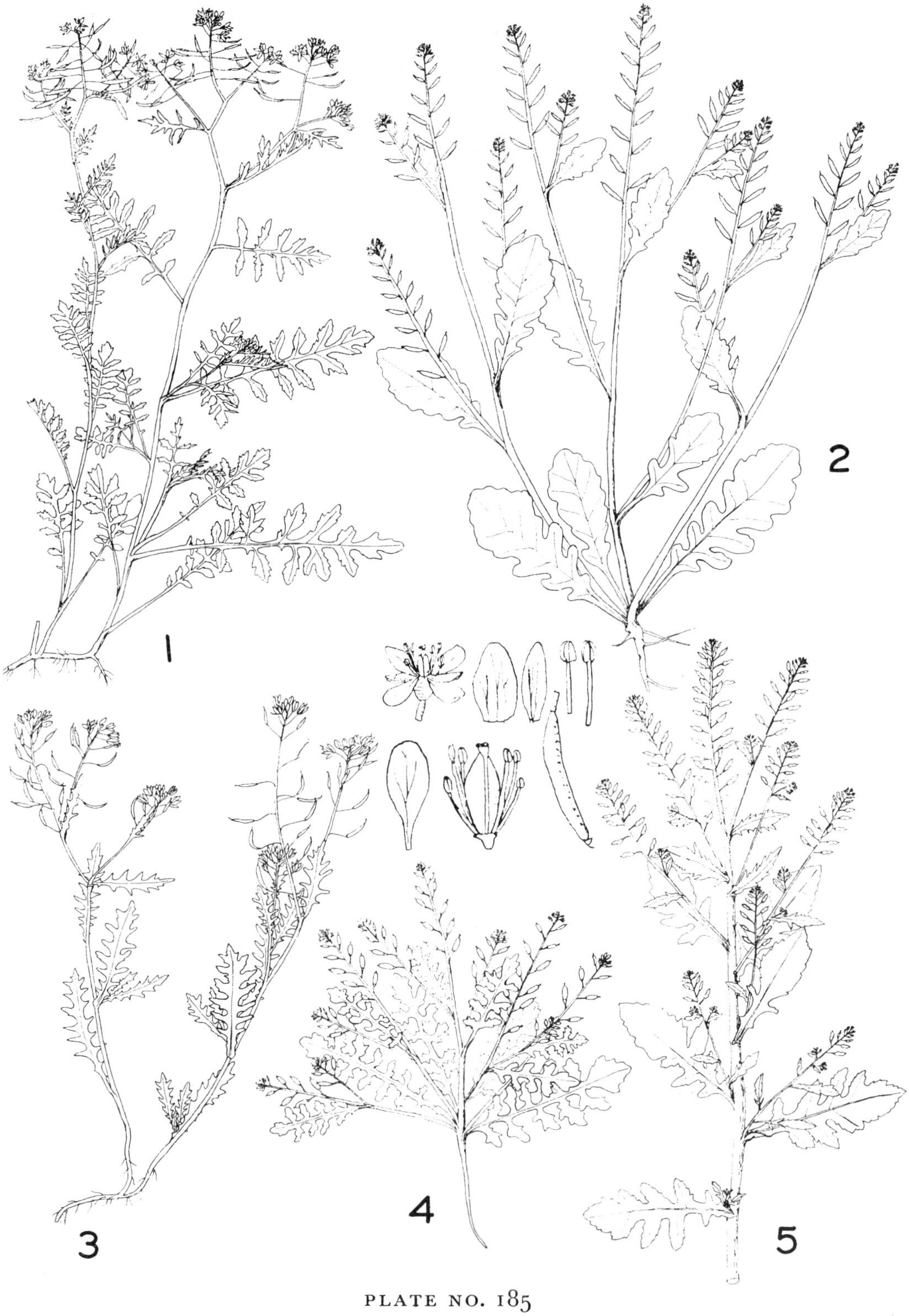


PLATE NO. 185

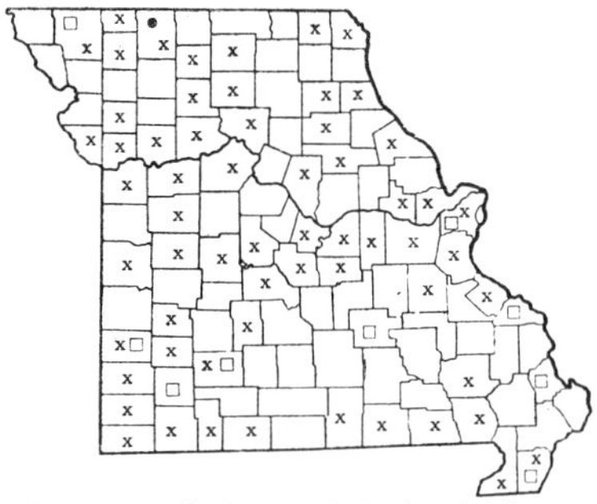

1090 • Rorippa islandica var. islandica (Marsh Yellow Cress)
1090 x Rorippa islandica var. Fernaldiana
1090 □ Rorippa islandica var. hispida

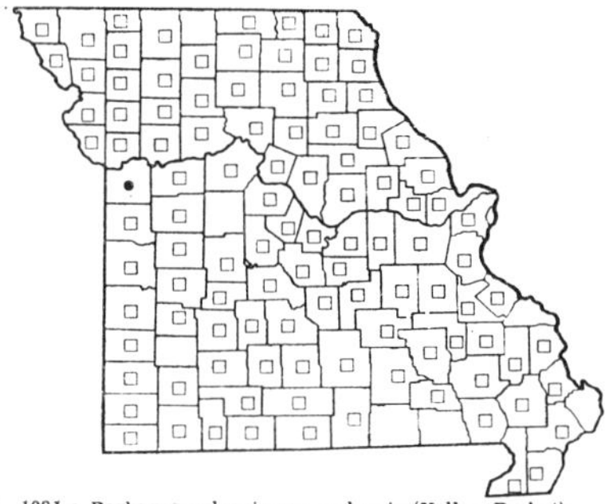

1091 • Barbarea vulgaris var. vulgaris (Yellow Rocket)
1091 □ Barbarea vulgaris var. arcuata

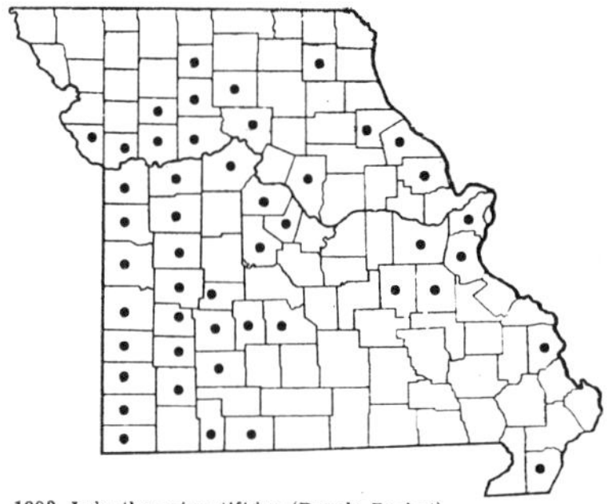

1092 Iodanthus pinnatifidus (Purple Rocket)

south to Michigan, Illinois, Arkansas, Oklahoma, Texas, Arizona, and California.

3. **Rorippa sessiliflora** (Nutt.) Hitchc.
Sessile-flowered Cress — Map 1088
Flowers April–October.

Occurs in wet muddy soils on the margins of sloughs, streams, and oxbow lakes in river flood plain, low swampy open woodland of alluvial ground, and along railroads. Throughout Missouri except in the Ozarks, where absent from the main part of that area.

Ranges from Indiana to Wisconsin and Minnesota to Nebraska, south to Florida and Texas; also in Virginia.

4. **Rorippa obtusa** (Nutt.) Britt. var. **obtusa**
Blunt-leaved Yellow Cress — Map 1089
Flowers May–September.

Occurs in wet soils bordering sloughs and banks of the Missouri and Mississippi rivers, and along railroad tracks. Rare, and known only from St. Louis, Saline, Jackson, and Clay counties in central Missouri.

Ranges from British Columbia to California, east to Ohio, West Virginia, Illinois, Missouri, and Texas.

5. **Rorippa islandica** (Oeder) Borbas
Marsh Yellow Cress — Map 1090
Also known as Yellow Water Cress.
Flowers May–October.

Occurs in wet muddy alluvial soils bordering streams, sloughs, ponds, especially oxbow lakes of river bottoms, wet woodland, along railroads, low fallow fields, and along ditches. This is the most common species in Missouri, represented by the following varieties:

a. Upper as well as lower leaves pinnate (cut to the midrib) or deeply pinnatifid (rather deeply cut towards the midrib) . . . .
5a. R. ISLANDICA var. ISLANDICA
a. Upper leaves coarsely or shallowly toothed to only slightly toothed or nearly entire (without teeth), not deeply parted nor pinnate . *b*
b. Base of stem or lower leaves usually hairy; fruits 2–5.5 mm. long, 1.7–4 mm. thick, broadly ovoid to ovoid-oblong . . . .
5c. R. ISLANDICA var. HISPIDA
b. Stem and leaves usually glabrous (without hairs), sometimes hairy on lower part of stem; fruits 3–9 mm. long, 1–2.5 mm. thick, nearly cylindric or narrowly ellipsoidal .
5b. R. ISLANDICA var. FERNALDIANA

5a. **Rorippa islandica** var. **islandica** — Map 1090
*Rorippa islandica* (Oeder) Borbas [G]

Known from Harrison County in northern Missouri (edge of slough along Wolf Creek along road, T67N, R28W, southwest corner of sect. 33, southeast corner of sect. 32, 6 mi. northeast of Hatfield, June 9, 1957, *Steyermark 85292*).

Ranges from Quebec to Michigan, south to New Jersey, Pennsylvania, and Missouri; also in Greenland and Eurasia.

5b. **Rorippa islandica** var. **Fernaldiana**
Butt. & Abbe — Map 1090
*Rorippa hispida* var. *glabrata* of auth. [P & S, Steyerm.], not Lunell

This is the commonest variety in the state, and is found more or less generally distributed except absent from some of the south-central Ozark counties.

Ranges from Labrador to Alaska, south to Florida, Texas, and California.

5c. **Rorippa islandica** var. **hispida** (Desv.)
Butt. & Abbe — Map 1090
*Rorippa hispida* (Desv.) Britton [P & S, Steyerm.]

Scattered in Missouri and less common than var. *Fernaldiana.*

Ranges from Newfoundland to British Columbia, south to Florida, Louisiana, Texas, New Mexico, and Oregon.

1093 Chorispora tenella

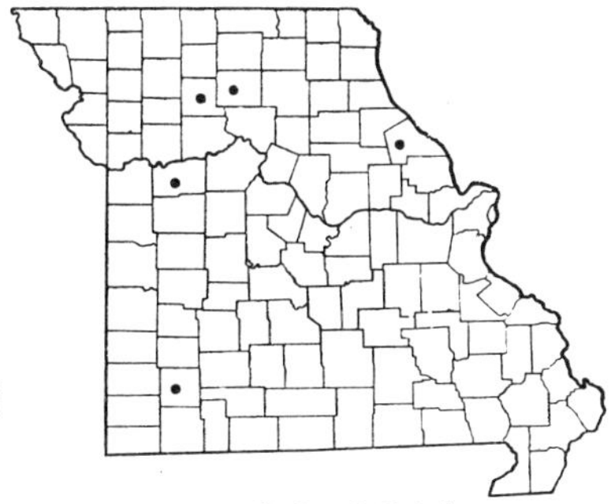

1094 Hesperis matronalis (Dame's Rocket)

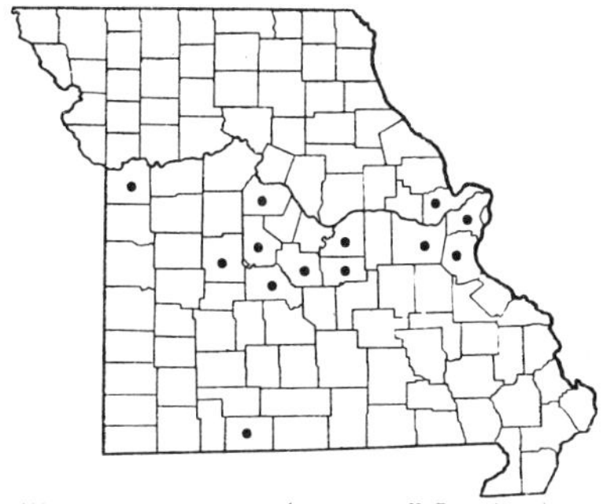

1095 Erysimum capitatum (Western Wall-flower)

## 24. **Barbarea** R. Br. Winter Cress

**Barbarea vulgaris** R. Br. Yellow Rocket, Winter Cress Map 1091

Flowers April–June.

Occurs in fallow and cultivated fields, wet meadows, pastures, alluvial ground near streams, and along roadsides and railroads. Throughout Missouri and doubtless in every county.

Two variations are represented in Missouri material:

Fruits erect or conspicuously ascending, pressed close to the main axis and overlapping; pedicels ascending to erect or appressed; rare type . . a. B. VULGARIS var. VULGARIS

Fruits spreading outward or away from main axis, or loosely ascending; pedicels spreading . b. B. VULGARIS var. ARCUATA

a. **Barbarea vulgaris** var. **vulgaris** Map 1091

*Barbarea vulgaris* R. Br. [G]

*Barbarea vulgaris* var. *longisiliquosa* Carion [P & S, Steyerm.]

*Barbarea stricta* of auth. in part, not Andrz.

Known only from Jackson County, west-central Missouri.

Native of Europe; introduced and naturalized in North America from Newfoundland to Ontario, south to Virginia, Ohio, Indiana, Illinois, Missouri, and Kansas.

b. **Barbarea vulgaris** var. **arcuata** (Opiz) Fries Map 1091

*Barbarea vulgaris* of some auth. [P & S, Steyerm.], not R. Br.

This is the common variation found in the state.

Native of Europe; introduced and naturalized in North America from Quebec to Minnesota, south to Virginia, Kentucky, Arkansas, and Oklahoma.

Other varieties are known but have not been collected in the state.

The young foliage and tender young stems of *Barbarea vulgaris* can be cooked as a green vegetable and resemble dandelion somewhat in flavor.

## 25. **Iodanthus** T. & G. Purple Rocket

**Iodanthus pinnatifidus** (Michx.) Steud. Purple Rocket Map 1092

Flowers May–June.

Occurs in moist alluvial soils of valley and river flood plain forests, and low wet woods. Widely distributed over the state, but absent or not known from most of northwestern Missouri and from much of the Ozark and southeastern lowland sections; more frequent in western Missouri south of the Missouri River.

Ranges from West Virginia and Pennsylvania to Illinois and Minnesota, south to Alabama, Ohio, Indiana, Arkansas, Oklahoma, and Texas.

## 26. **Chorispora** R. Br.

**Chorispora tenella** (Pall.) DC. Map 1093

Flowers April–June.

Occurs along railroads. Known only from St. Louis County, east-central Missouri (St. Louis, Terperance Street freight yard of the Mo. Pac. R. R., June 7, 1958, *Muehlenbach 1412*).

Native of Asia; introduced and naturalized in the United States, ranging from Massachusetts to Iowa, Nebraska, Missouri, Oklahoma, and west to Washington.

The lower leaves of this plant are usually deeply incised with the segments directed backward (runcinate), while the middle and upper ones are wavy-toothed with a few low teeth.

## 27. **Hesperis** L. Dame's Rocket

**Hesperis matronalis** L. Dame's Rocket
Map 1094

Flowers May–June.

Occurs along roadsides in either shaded or open ground and waste places. Scattered in the state, where known from Pike, Linn, Livingston, Lafayette, and Lawrence counties.

Native of Eurasia; introduced into North America as a garden plant and naturalized from Quebec and Nova Scotia to Michigan and Iowa, south to Georgia, Kentucky, Illinois, and Missouri.

Formerly more widely used as a garden plant for its fragrant rose-lavender to purplish, pink, or white flowers. The white-flowered race, *alba*, has been collected in Pike County (shaded bank below driveway along entrance road, just south of bridge, at experimental farm, near Clarksville, May 17, 1948, *Etter 119*).

## 28. **Erysimum** L.

a. Flowers very showy, with orange, orange-yellow, or copper-colored petals 15–30 mm. long; plants of rocky limestone bluffs, glades, and outcrops . . . . . 1. E. CAPITATUM
a. Flowers less conspicuous, with pale or bright yellow petals 3.5–10 mm. long; plants of roadsides, railroad tracks, and waste ground . . . . . *b*
  b. Petals 3–5.5 mm. long; fruits 1–3 cm. long; flower- and fruit-stalks hair-like, very slender, much narrower than thickness of fruit; lower fruiting stalks 10–15 mm. long . . . . 2. E. CHEIRANTHOIDES
  b. Petals 6–10 mm. long; fruits 2–12 cm. long; flower- and fruit-stalks thickened, their diameter half as thick or about as thick as that of the fruit; fruiting stalks 2–9 mm. long . . . . . *c*
    c. Mature fruits 1.5–5 cm. long, erect or stiffly ascending; leaves mainly entire (without teeth) or with remote low shallow toothing; ovary, fruit, and stem gray-hairy; perennial plants with stiff erect stems . . . . . 4. E. INCONSPICUUM
    c. Mature fruits 5–12 cm. long, widely spreading; leaves wavy-toothed to more deeply toothed or nearly entire; ovary and fruit paler green, often nearly glabrous, with few hairs, or with numerous hairs; stem pale green with scattered or many hairs; annual plants with delicate roots and erect or often widely branching stem (young earliest-flowering plants have simple erect stems) . . . . . 3. E. REPANDUM

1. **Erysimum capitatum** (Dougl.) Greene
Western Wall-flower Map 1095
*Erysimum arkansanum* Nutt. [G]
*Erysimum asperum* of some auth. [BB, P & S, Steyerm.], not (Nutt.) DC.

Flowers May–July.

Occurs on limestone bluffs, glades, and rocky open ground. Scattered in central Missouri along the Missouri, Meramec, Osage, and Gasconade rivers and tributaries north to St. Charles, Cooper, and Jackson counties, locally in Taney County, southwestern Missouri along bluffs and open ground above bluffs of White River.

Ranges from Ohio to Washington, south to Illinois, Missouri, Texas, New Mexico, and California.

This is a very showy biennial. It can be grown from seed, but does not transplant successfully. The present name adopted is in accordance with recent studies by Dr. George Rossbach (Madroño 14: 261–67. 1958).

2. **Erysimum cheiranthoides** L.
Wormseed Mustard Map 1096

Plate no. 186. **1.** Barbarea vulgaris var. arcuata, × 2/7; Details from Small, The New York Botanical Garden. **2.** Iodanthus pinnatifidus, × 2/7; Details from Small, The New York Botanical Garden. **3.** Chorispora tenella, × 2/7. **4.** Hesperis matronalis, × 2/7; Details from Small, The New York Botanical Garden. **5.** Erysimum capitatum, × 2/7; Details from Small, The New York Botanical Garden. **6.** Erysimum cheiranthoides, × 2/7. **7.** Erysimum repandum, × 2/7. **8.** Sisymbrium officinale, × 2/7; Details from Small, The New York Botanical Garden. **9.** Sisymbrium altissimum, × 2/7; Details from Small, The New York Botanical Garden.

PLATE NO. 186

1096 Erysimum cheiranthoides (Wormseed Mustard)

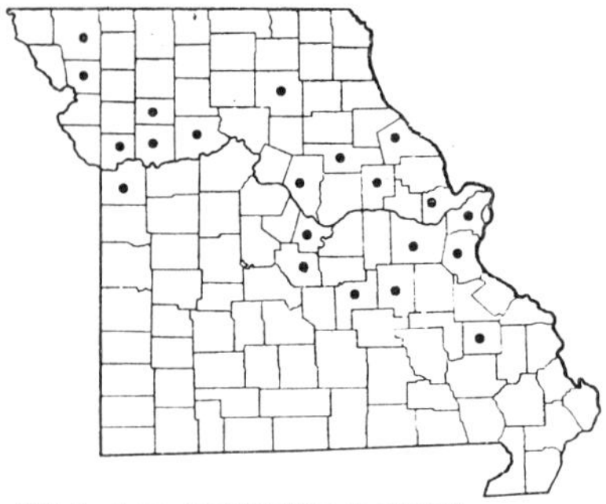
1097 Eyrsimum repandum (Treacle Mustard)

1098 Erysimum inconspicuum

Flowers May–September.

Occurs along roadsides, railroads, and waste ground.

Ranges in North America from Newfoundland to Alaska and British Columbia, south to North Carolina, Tennessee, Arkansas, Colorado, and Oregon; also in Eurasia, from which region it is believed by some botanists to have been introduced into North America.

3. **Erysimum repandum** L. Treacle Mustard
Map 1097

Flowers April–June.

Occurs along roadsides, railroads, and waste ground.

Native of Europe; introduced into the United States, ranging from Massachusetts to Washington, south to Florida, Alabama, Arkansas, Texas, Utah, Arizona, and California.

This species, like some other crucifers, when young and seen in early stages of growth, is quite different from older and mature plants. Young early spring-flowering plants are often dwarfed and have simple, erect stems in contrast to the later, often widely branching and taller stems of older or more mature plants. Plants often occur in dense masses along the roadsides in early spring lending a pale sulfur-yellow color to the highway border.

4. **Erysimum inconspicuum** (S. Wats.) MacM.
Map 1098

*Erysimum parviflorum* Nutt., not Pers.

Flowers May–June.

Occurs along roadsides and waste ground. Known only from western Missouri in Jackson and Livingston (T58N, R23W, sect. 8, Chillicothe, May 12, 1951, *Sparling 766*) counties.

Ranges from Alaska and British Columbia to Nevada and Colorado, east to Ontario, Michigan, Minnesota, North Dakota, and South Dakota; introduced south to Missouri and Kansas, east to New England and Nova Scotia.

## 29. **Sisymbrium** L.

a. Upper leaves of the stem finely divided into slender, linear, thread-like divisions; fruits 5–10 cm. long . . . 2. S. ALTISSIMUM

a. Upper leaves of the stem either coarsely cut with lanceolate, triangular-ovate, or triangular lobes, or sometimes nearly without teeth; fruits 1–3.5 cm. long . . . . . . . . . . . . . . . . *b*

b. Petals very small, 3 mm. long; fruits 1–2 cm. long, erect and closely placed next to the main axis of the stem and inflorescence; pedicels (stalks) of flowers and fruits erect, 2–3 mm. long; stem mostly hairy at base or lower part of plant . . . . . . . . . . . . . 1. S. OFFICINALE

b. Petals more conspicuous, 5–6.5 mm. long, longer than the sepals; fruits 2–3.5 cm. long, widely spreading or ascending, not erect along the axis; pedicels (stalks) of flowers and fruit spreading to ascending, 7–15 mm. long; most of stem rather thickly covered with hairs . . . . . . 3. S. LOESELII

1. **Sisymbrium officinale** (L.) Scop.
Hedge Mustard Map 1099

Flowers May–October.

Occurs in fields, pastures, barnyards, farmlots, along roadsides and railroads.

Missouri material is represented by two variations:

Fruits hairy . . 1a. S. OFFICINALE var. OFFICINALE

Fruits glabrous (without hairs) . 1b. S. OFFICINALE var. LEIOCARPUM

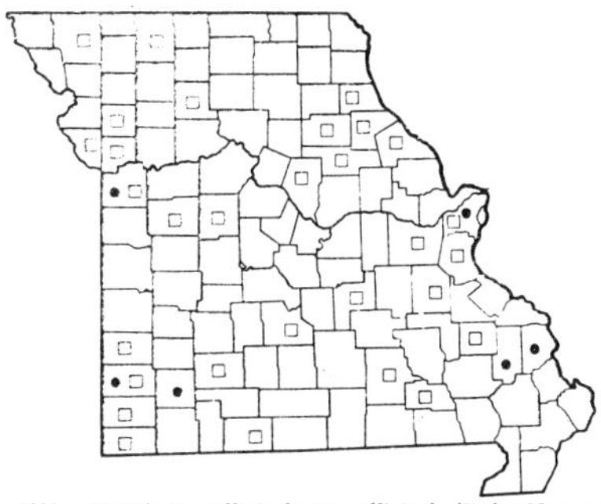

1099 ● Sisymbrium officinale var. officinale (Hedge Mustard)
1099 □ Sisymbrium officinale var. leiocarpum

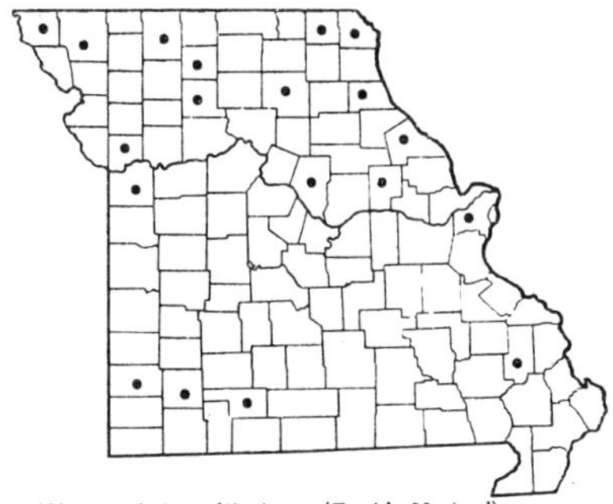

1100 Sisymbrium altissimum (Tumble Mustard)

1101 Sisymbrium Loeselii

1a. **Sisymbrium officinale** var. **officinale** Map 1099

*Sisymbrium officinale* (L.) Scop. [G, P & S, Steyerm.]

Scattered in a few counties in the state. This is the rarer of the two variations.

Native of Europe; introduced and naturalized in Canada and the United States less commonly than the following variety.

1b. **Sisymbrium officinale** var. **leiocarpum** DC. Map 1099

Throughout Missouri, this is the commoner variation encountered.

Native of Europe; introduced and naturalized from Nova Scotia to British Columbia, south to Florida and California.

The young plants can be cooked as a green vegetable, and the young leaves can be used in salads.

2. **Sisymbrium altissimum** L. Tumble Mustard Map 1100

Flowers May–August.

Occurs in fields, pastures, along roadsides, railroads, and sandy or rocky open ground. Throughout Missouri, and probably to be found in most of the counties.

Native of Europe; introduced and naturalized in North America from Quebec to Ontario and British Columbia, south to Florida, Texas, and California.

3. **Sisymbrium Loeselii** L. Map 1101

Flowers May-October.

Occurs along railroads, where known only from St. Louis (Terperance St. freight yard of Terminal Railroad, June 19, 1954, *Muehlenbach 183*; south of Humboldt Ave., May 8, 1955, *Muehlenbach 570*) and Christian (waste open ground along railway, Billings, June 24, 1954, *Palmer 57901*) counties.

Native of Europe; introduced and naturalized from Connecticut to North Dakota, Idaho, and Washington, south to Indiana, Illinois, Missouri, and Nebraska.

## 30. **Arabidopsis** Heyn. Mouse-ear Cress

**Arabidopsis Thaliana** (L.) Heyn.

Mouse-ear Cress Map 1102

Flowers April–May.

Occurs in sandy or rocky open ground in fallow fields, pastures, waste ground, and along roadsides. Southern and east-central Missouri north to St. Charles and Callaway counties, west to Christian and Taney counties.

Native of Europe; introduced and naturalized from Georgia, Arkansas, Oklahoma, and Utah, north to Massachusetts, Ontario, Michigan, Wisconsin, and Washington.

## 31. **Camelina** Crantz False Flax

Common type found; stems and leaves noticeably hairy, the simple hairs much longer than the abundant shorter star-shaped (stellate) ones; mature fruits 4–6.5 mm. (up to 7) long and 4–5 mm. wide . . . 1. C. MICROCARPA

Rare type found; stems nearly glabrous (without hairs) to sparsely hairy, the few simple hairs slightly or not longer than the few star-shaped (stellate) ones; mature fruits mainly 7–9 mm. long and 6–7 mm. wide . . . . . . . . . . . . . . . . . . . . . . . . . 2. C. SATIVA

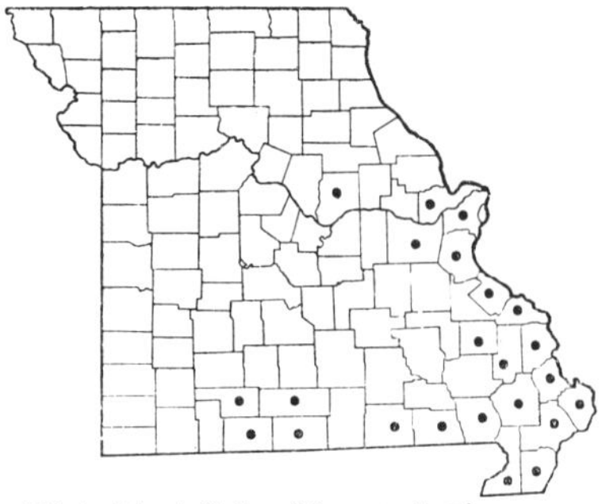

1102 Arabidopsis Thaliana (Mouse-ear Cress)

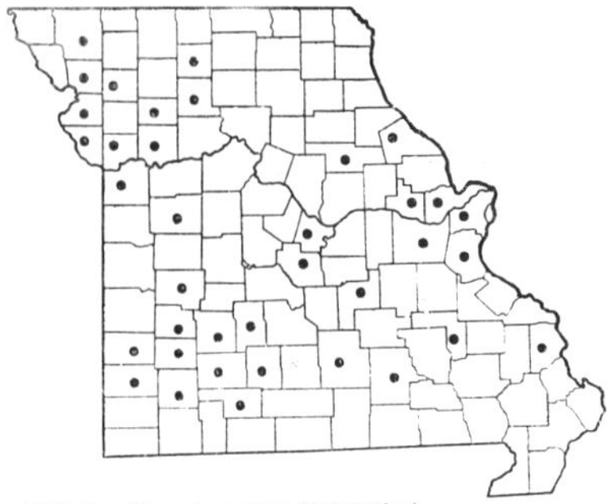

1103 Camelina microcarpa (False Flax)

1104 Camelina sativa (False Flax)

1. **Camelina microcarpa** Andrz. False Flax Map 1103

Flowers April–September.

Occurs mainly along railroads and roadsides. Scattered throughout the state, and probably in every county when railroad collecting has become more thorough.

Native of Europe; introduced and naturalized in North America from Newfoundland to British Columbia, south to Virginia, Tennessee, Missouri, Oklahoma, Texas, and California.

This species sometimes is difficult to distinguish from the next and intergrades in length of fruits, some Missouri specimens having fruits 7 mm. long with brown seeds 1 mm. long. In addition to intergradation in fruit and seed length, the length of the pedicels, generally averaging longer (12–30 mm.) in *C. sativa*, has been found too variable to rely on as a character for separation of the two species. Perhaps future investigations may reveal that only one species can be maintained.

2. **Camelina sativa** (L.) Crantz False Flax Map 1104

Flowers April–August.

Occurs along railroads and waste ground, where known only from St. Louis and Jackson counties in central Missouri.

Native of Europe; introduced and naturalized in North America from Quebec to British Columbia, south to South Carolina, Illinois, Missouri, Kansas, Colorado, and California.

## 32. **Descurainia** Webb & Berth. Tansy Mustard

a. Common type encountered; stem and leaves with a green appearance, most of the hairs on the stems and/or leaves tipped with glands . . . . . . . . . . . . . 1b. D. PINNATA var. BRACHYCARPA

a. More rarely encountered; stem and leaves with a grayish appearance, most of the hairs on the stems and leaves not gland-tipped . . . . . . . . . . . . . . . . . . . . . . . *b*

b. Fruits 5–10 mm. long, 1–2 mm. wide, $3\frac{1}{2}$–7 times as long as broad; seeds in 2 rows in each compartment of fruit; fruiting stalks widely or nearly horizontally spreading . 1a. D. PINNATA var. PINNATA

b. Fruits mainly 15–30 mm. (rarely 10) long, 0.7–1 mm. wide, mostly 15–30 times as long as broad; seeds in only 1 row in each compartment of fruit; fruiting stalks upwardly spreading to ascending 2. D. SOPHIA

1a. **Descurainia pinnata** (Walt.) Britt. var. **pinnata** Tansy Mustard Map 1105

*Descurainia pinnata* (Walt.) Britt. [G]

Flowers March–May.

Occurs in sandy or rocky open woodland. Southern Missouri, where known only from New Madrid (sandy ground of burnt-over wooded knoll on Sikeston Ridge, T25N, R14E, sect. 34, 4 mi. southeast of Sikeston, April 6, 1956, *Steyermark 80858*) and Lawrence (rocky hillside woods, 1 mi. southeast of Bonham Siding, T25N, R24W, sect. 25, near southwest corner of county, April 21, 1953, *Palmer 55387*) counties.

Ranges from Florida to Texas, north to Virginia and in the interior to Missouri.

---

Plate no. 187. **1.** Sisymbrium Loeselii, $\times\ ^2/_5$. **2.** Camelina microcarpa, $\times\ ^2/_5$; Details from Small, The New York Botanical Garden. **3.** Descurainia pinnata var. brachycarpa, $\times\ ^2/_5$. **4.** Arabidopsis Thaliana, $\times\ ^2/_5$; Details from Small, The New York Botanical Garden. **5.** Descurainia Sophia, $\times\ ^2/_5$; Details from Small, The New York Botanical Garden.

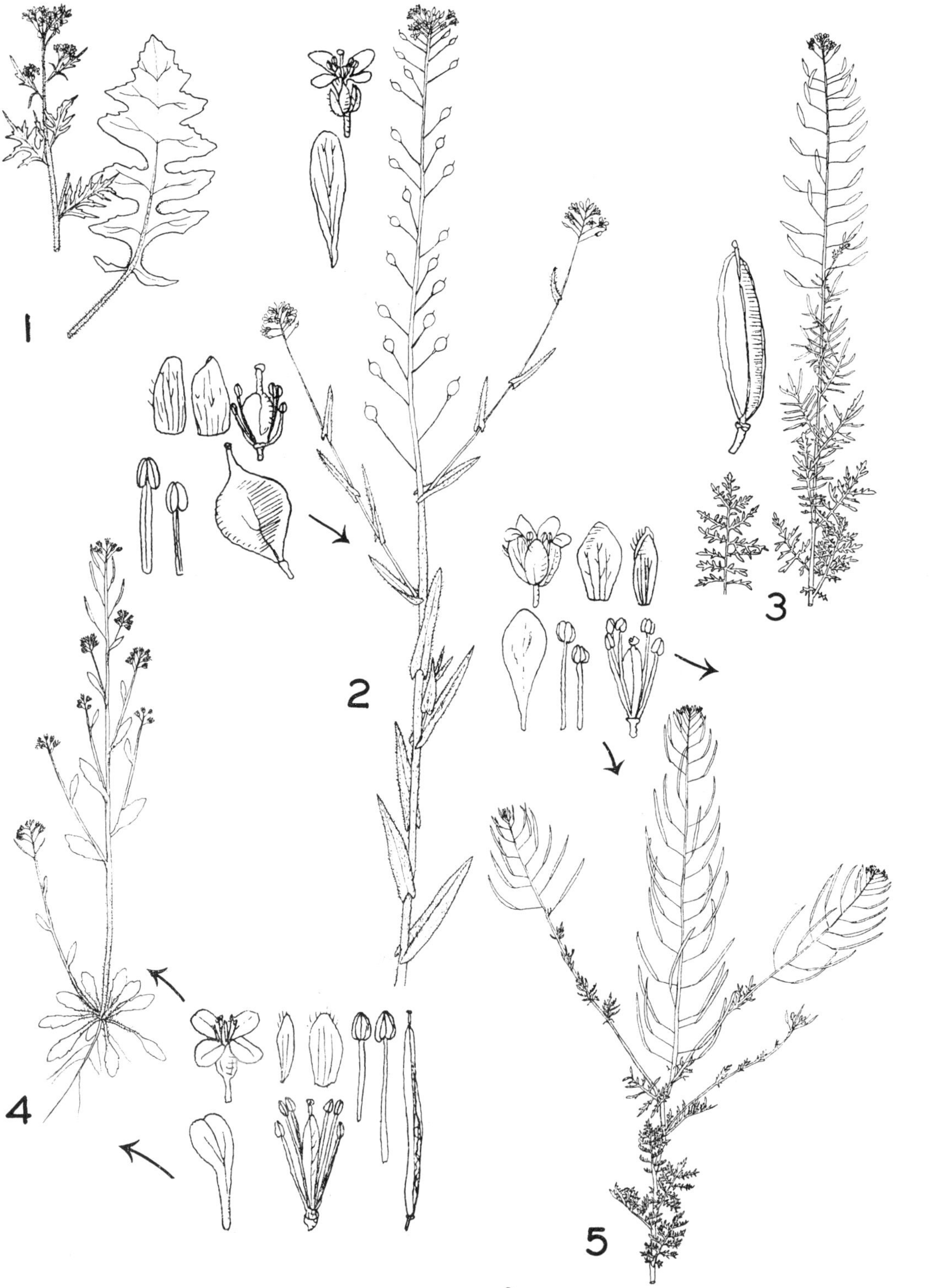


PLATE NO. 187

1105 □ Descurainia pinnata var. pinnata (Tansy Mustard)
1105 • Descurainia pinnata var. brachycarpa

1106 Descurainia Sophia (Tansy Mustard)

1107 Cleome serrulata (Rockly Mountain Bee Plant)

1b. **Descurainia pinnata** var. **brachycarpa** (Richards.) Fern. Tansy Mustard Map 1105
*Descurainia intermedia* (Rydb.) Daniels [P & S]
*Descurainia pinnata* in part of some auth. [Steyerm.], not (Walt.) Britt.
Flowers March–July.

Occurs on rocky open ledges, bluffs, glades, prairies, sand and gravel bars along streams, open woods, along roadsides, railroads, and waste ground. Common throughout Missouri, and probably in every county.

Ranges from Quebec to Mackenzie, south to West Virginia, Tennessee, Arkansas, Oklahoma, Texas, and Colorado.

A western variety of *D. pinnata* in New Mexico has been reported to poison cattle, especially when eaten during spring in the flowering stage.

The flowers are stated by some manuals (see *Gray's Man.*, eighth ed., p. 711) to be whitish, but actually the flowers are always pale yellow to yellow.

2. **Descurainia Sophia** (L.) Webb
Tansy Mustard Map 1106
Flowers May–July.

Occurs in open ground along roadsides, alluvial open soils, along railroads, and waste places. Rare and scattered in Missouri throughout the state.

Native to Europe; introduced and naturalized in North America, from Quebec to Washington, south to Delaware, Pennsylvania, Illinois, Missouri, Kansas, Oklahoma, Colorado, Utah, and California.

*Excluded Species*

**Isatis tinctoria** L. Dyer's Woad

A specimen in the herbarium of the School-of-Ozarks collected in Taney County was taken from a plant found near a garden, and cannot be accepted as an established or naturalized member of the flora of the state.

## Fam. **CAPPARIDACEAE** (Caper Family)

Stamens 6; fruit on a long slender special stalk (stipe) above the summit of the regular pedicel (stalk of flower or fruit); petals with a notch at their summit . . . . . . . . . . . . . . . . 1. CLEOME
Stamens 8 or more; fruit set directly on top of its regular pedicel (stalk) or with a short special stalk between it and the pedicel; petals without any notch at summit . . . . . . . . . . . . 2. POLANISIA

### 1. **Cleome** L.

Leaves divided into 3 leaflets; stem glabrous (without hairs) or very sparsely hairy when young. 1. C. SERRULATA
Leaves divided into 5–7 leaflets; stem conspicuously sticky-hairy . . . . . . . . . . 2. C. SPINOSA

1. **Cleome serrulata** Pursh
Rocky Mountain Bee Plant Map 1107
Also known as Stinking Clover.
Flowers June–September.

Occurs on the loess hills of Atchison County, northwestern Missouri, and introduced along railroads and waste ground in St. Louis and Jackson counties, central Missouri.

Ranges from Manitoba and Saskatchewan to Washington, southern Illinois, Missouri, Kansas, Oklahoma, New Mexico, and Arizona.

In the Rocky Mountain states this is an important bee plant.

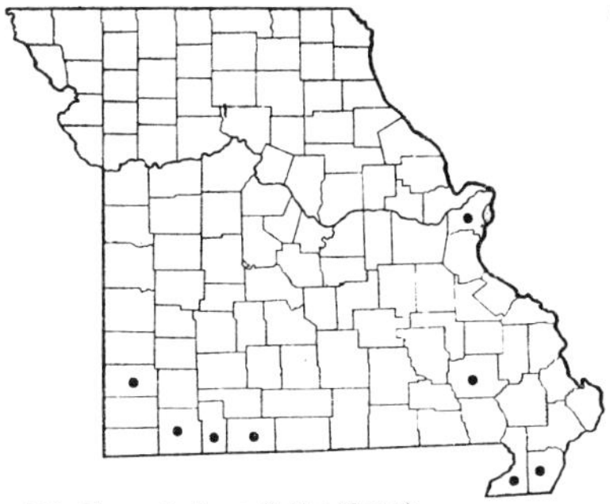

1108 Cleome Houtteana (Spider Flower)

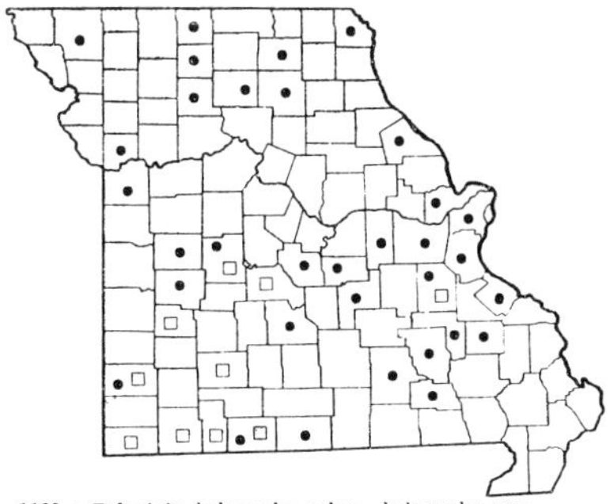

1109 • Polanisia dodecandra subsp. dodecandra var. dodecandra (Clammy-weed)
1109 □ Polanisia dodecandra subsp. dodecandra var. trachysperma

1110 Reseda Luteola

2. **Cleome Houtteana** Schlecht. Spider Flower Map 1108

*Cleome spinosa* Hort., not auth. [G, BB, P & S]

Flowers June–September.

Commonly cultivated as an annual garden plant and rarely escaping to dump heaps, sandy gravel bars and clay banks along streams, waste ground, roadsides, and railroads. Scattered in southern and central Missouri northeast to St. Louis County.

Native of South America; introduced and naturalized in the United States from Florida to Texas, north to Massachusetts, New York, Ohio, Indiana, Missouri, and Oklahoma.

The leaflets vary from finely toothed to entire (without teeth).

## 2. **Polanisia** Raf. Clammy-weed

The following treatment is in accord with the most recent studies of the genus by Dr. Hugh H. Iltis (Brittonia 10: 33–58. 1958).

Longest petals 3.5–6.5 (up to 8) mm. long; longest stamens 4–10 (up to 14) mm. long, scarcely longer than the petals; common species of gravel bars, railroads, and open places throughout Missouri . . a. P. DODECANDRA subsp. DODECANDRA var. DODECANDRA

Longest petals mostly 8–13 (rarely 7 or up to 16) mm. long; longest stamens mainly 12–30 (rarely 9) mm. long, usually much longer than the petals; rare species mostly of the southwestern portion of the state on rocky glades and bluffs . . . . . . b. P. DODECANDRA subsp. DODECANDRA var. TRACHYSPERMA

a. **Polanisia dodecandra** (L.) DC. subsp. **dodecandra** var. **dodecandra** Map 1109

*Polanisia graveolens* Raf. [G, BB, P & S, Steyerm.]

Flowers late May–October.

Occurs on sand and gravel bars along streams, rocky open and waste ground, along railroads and roadsides. Throughout Missouri.

Ranges from Quebec, Vermont, Minnesota, and North Dakota to Manitoba and Alberta, south to Maryland, West Virginia, Ohio, Indiana, Tennessee, Arkansas, Oklahoma, and Texas.

b. **Polanisia dodecandra** subsp. **dodecandra** var. **trachysperma** (Torr. & Gray) Iltis Map 1109

*Polanisia trachysperma* Torr. & Gray [G, BB, P & S, Steyerm.]

Flowers June–September.

Occurs on rocky open glades, bluffs, and open ground. Mostly in southwestern Missouri north to Benton and Camden counties, east to Greene and Taney counties, and locally east in Washington County.

Ranges from Indiana, Illinois, Wisconsin, and Minnesota to Manitoba, Alberta, and Washington, south to Missouri, Oklahoma, Texas, Colorado, New Mexico, California, and northern Mexico; introduced east to Connecticut.

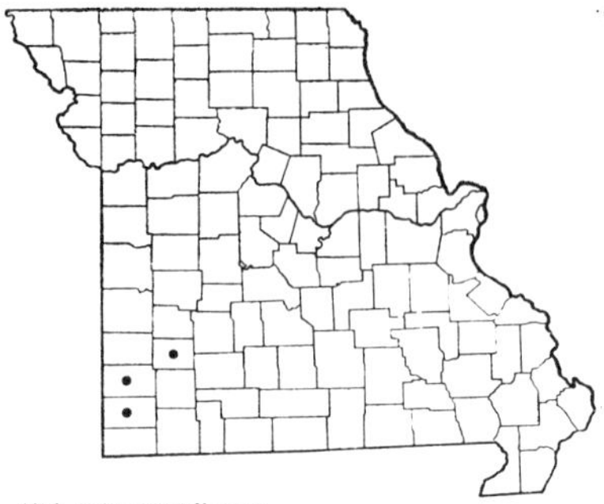

1111 Sedum Nuttallianum

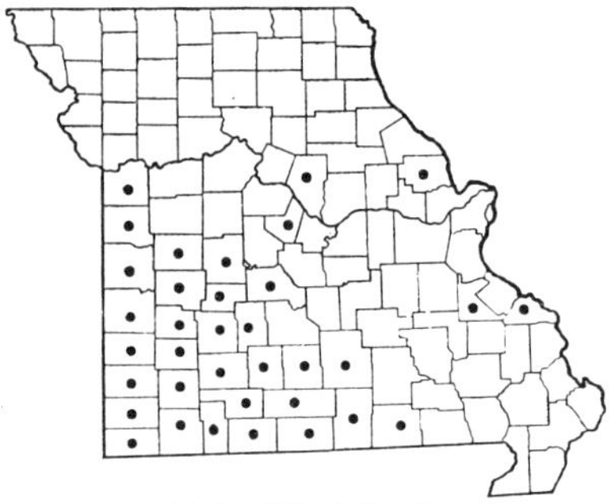

1112 Sedum pulchellum (Widow's Cross)

1113 Sedum sarmentosum

## Fam. **RESEDACEAE** (Mignonette Family)

### **Reseda** L. Mignonette

**Reseda Luteola** L. Dyer's Rocket Map 1110

Also called Yellow Weed or Weld.

Flowers May–October.

Occurs in waste ground. Known only from Laclede County in the western Ozark region (June, 1937, *George Moore*, in Chi. Nat. Hist. Mus. Herb.).

Native of Europe; introduced and naturalized locally in the United States from Massachusetts to Pennsylvania and in Missouri.

This plant was at one time cultivated as a source of yellow dye, obtained from the leaves.

## Order **ROSALES**

## Fam. **CRASSULACEAE** (Stonecrop Family)

The genus *Penthorum* has been placed in this family by various authors. Other authors now place the genus in the *Saxifragaceae*, to which family it is referred in the present flora.

### **Sedum** L. Stonecrop, Orpine

a. Leaves mainly 15–35 mm. wide, broadly oblong, elliptic, or obovate; stems mostly 3–6 dm. tall . 5. S. TELEPHIUM

a. Leaves mainly 1–12 mm. wide, varying from linear, oblong, or oblong-lanceolate in some species to obovate with a wedge-shaped base in another species; stems, when upright, mainly less than 2 dm. tall . . . . . . . . . . . . . . . . . . . . . . . . . . . . . *b*

b. Most of leaves, and all the middle and lower ones, opposite in 2's or in whorls of 3's. . . . . . *c*

c. Flowers white; leaves obovate, wedge-shaped or narrowed at base and usually broadly rounded at the summit; flowering branches upright, leafy stems creeping to ascending . 4. S. TERNATUM

c. Flowers yellow; leaves oblong-lanceolate to broadly lanceolate, acutely pointed at tip; flowering and leafy stems mainly creeping . . . . . . . . . . . . 3. S. SARMENTOSUM

b. All the leaves alternately scattered on the stem, although close together . . . . . . . . . *d*

d. Petals yellow; leaves mainly 4–6 mm. (up to 12 mm.) long, oblong; plants of chert and sandstone glades and ledges of southwestern Missouri only . . . . . . . 1. S. NUTTALLIANUM

d. Petals white, pink, or rose-purple; leaves mainly 10–25 mm. (rarely 5 on young plants) long, linear or linear-spatulate on mature stems; plants of limestone, chert, and sandstone glades of southern and central Missouri north to Lincoln, Boone, and Jackson counties . 2. S. PULCHELLUM

---

Plate no. 188. **1.** Cleome serrulata, × $^{2}/_{5}$; After Gleason, details from Small, The New York Botanical Garden. **2.** Polanisia dodecandra subsp. dodecandra var. dodecandra, × $^{2}/_{5}$. **3.** Reseda Luteola, × $^{2}/_{5}$; After Britton and Brown, The New York Botanical Garden. **4.** Sedum Nuttallianum, × $^{2}/_{5}$. **5.** Sedum sarmentosum, × $^{2}/_{5}$; After Gleason, The New York Botanical Garden. **6.** Cleome Houtteana, × $^{2}/_{5}$; *a.* Fruit, × $3^{3}/_{5}$; After Britton and Brown, details from Small, The New York Botanical Garden. **7.** Sedum pulchellum, × $^{2}/_{5}$; Details from Small, The New York Botanical Garden.

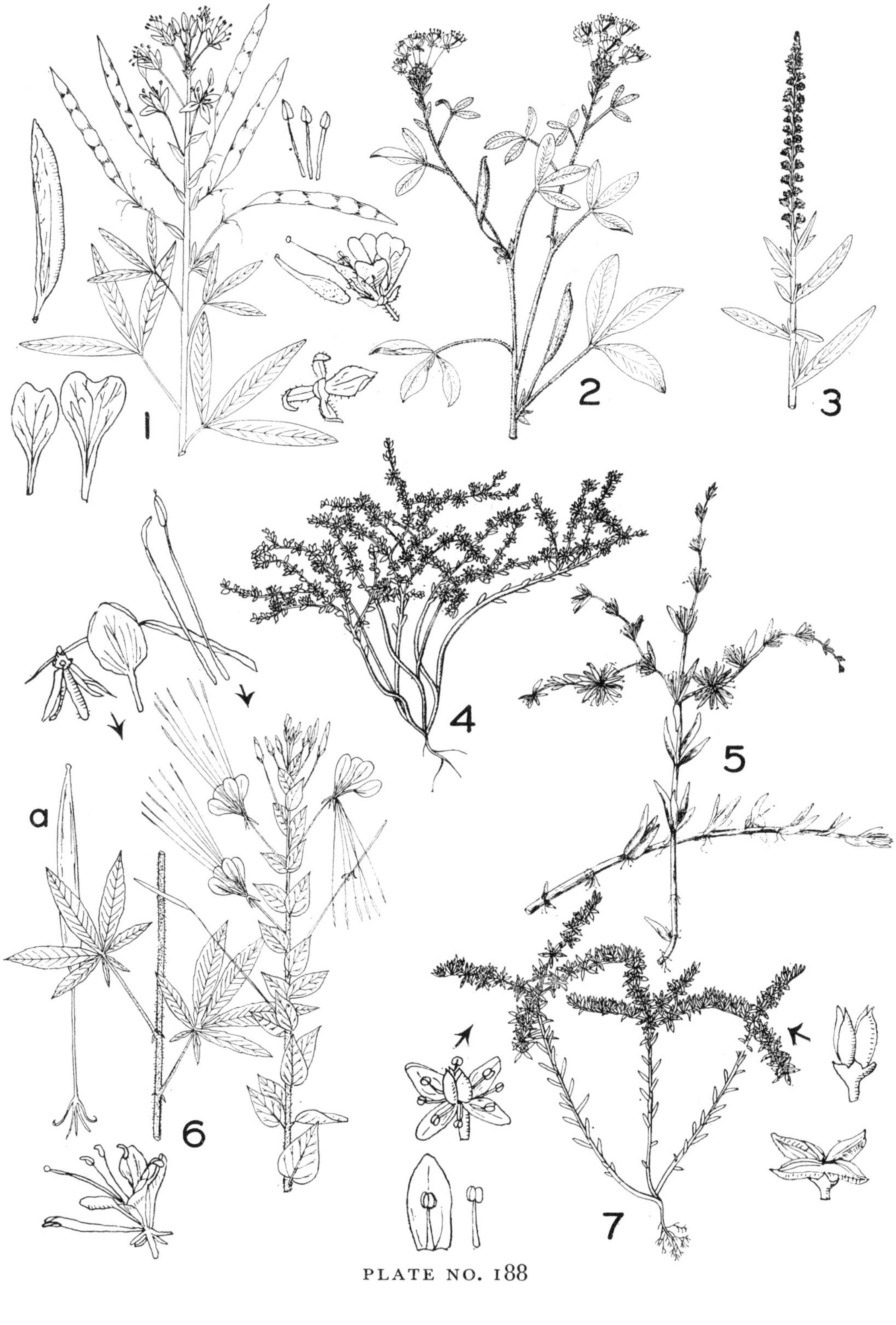


PLATE NO. 188

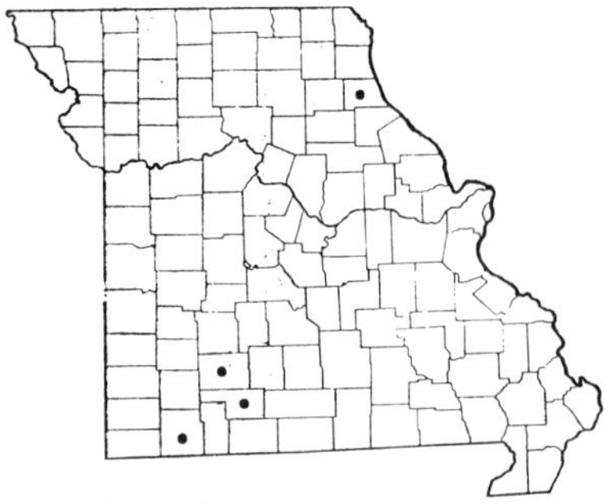
1114 Sedum ternatum

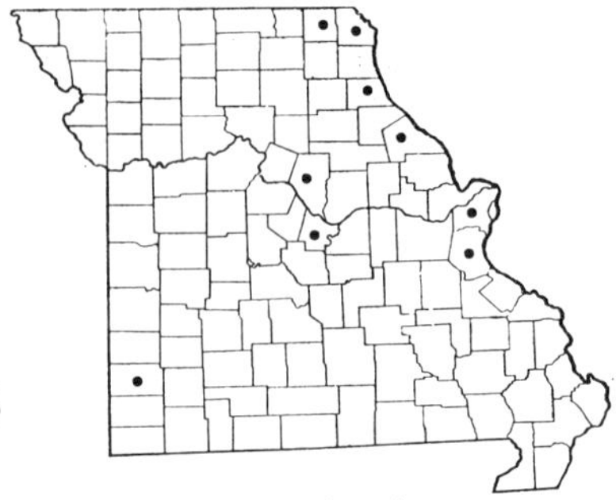
1115 Sedum Telephium (Live-forever)

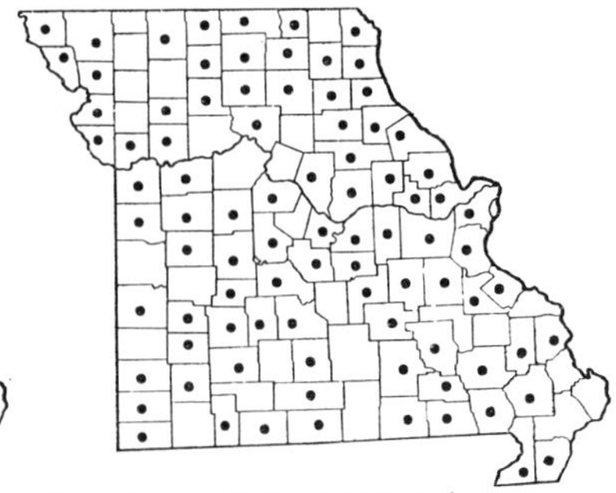
1116 Penthorum sedoides (Ditch Stonecrop)

1. **Sedum Nuttallianum** Raf. Map 1111
Flowers May–June.

Occurs on sandstone and chert glades and ledges. Known only from the unglaciated prairie section of the state in Dade, Jasper, and Newton counties.

Ranges from Missouri and Arkansas to Oklahoma and Texas.

This annual grows in dense small clumps, often forming considerable patches on the rocks. The fleshy silvery green leaves are at first close together on the stem, eventually becoming more separated as the stem elongates. The small yellow flowers are very attractive. The species often grows in association with *Talinum calycinum* and *T. parviflorum*, *Selaginella rupestris*, *Crotonopsis elliptica*, *Linaria canadensis*, *Saxifraga texana*, *Selenia aurea*, *Scirpus koilolepis*, *Portulaca mundula*, and other acid-soil species.

2. **Sedum pulchellum** Michx. Widow's Cross Map 1112
Flowers May–July.

Occurs chiefly on limestone glades, ledges, and outcrops, either in full sun or in wooded sites, also on chert and sandstone glades and outcrops. Mainly in the western Ozark and unglaciated prairie sections of the state, locally north and east in the Ozark and Ozark border areas to Oregon, Perry, St. Francois, Lincoln, and Boone counties; absent from most of the eastern and northern Ozark counties and from all the lowland counties of southeastern Missouri.

Ranges from Virginia to southern Illinois, Missouri, Kansas, south to Georgia, Alabama, Arkansas, Oklahoma, and Texas.

This winter annual germinates in late fall, producing a small rosette of gray-green, flattened, broad, oblong-spatulate or spoon-shaped leaves. These overwinter and early in spring flowering stems appear with narrowly linear, cylindrical, terete leaves.

3. **Sedum sarmentosum** Bunge Map 1113
Flowers June–September.

Commonly grown in gardens and escaped from cultivation in southwestern Missouri in Newton (moist mossy rocks, Haddock Spring, November 3, 1953, *Palmer 57195*) and Jasper (moist limestone ledge at Carter's Spring, 3 mi. northwest of Webb City) counties.

Native of eastern Asia; introduced and naturalized in the United States from New Jersey and Pennsylvania to Ohio and Missouri.

4. **Sedum ternatum** Michx. Map 1114
Flowers April–June.

Occurs on moist mossy, shaded ledges around springs at the base of bluffs and damp low woods along streams, where apparently native in Marion County (base of wooded slopes of shale bluffs along Clear Creek, T56N, R5W, sect. 11, 3 mi. northwest of Hannibal, May 15, 1938, *Steyermark 5323*), northeastern Missouri, and in Barry and Christian (base of northwest-facing limestone bluffs on mossy boulders along Bull Creek, T25N, R20W, south part sect. 8 and NW¼ sect. 17, 9 mi. [by air] southwest of Chadwick, 6 mi. [by air] southeast of Christian Center, October 18, 1957, *Steyermark 85853*; also *Palmer*) counties, southwestern Missouri. Elsewhere introduced from plantings in Greene County (Springfield) and Marion County (old farm yard, south of Hannibal, May 16, 1921, *Davis 3514*).

Ranges from New York to Michigan, Illinois, and Missouri, south to Georgia, Tennessee, and Arkansas.

5. **Sedum Telephium** L. Live-forever Map 1115
*Sedum triphyllum* (Haw.) S. F. Gray [P & S]
*Sedum purpureum* (L.) Link [G]
*Sedum Telephium* var. *purpureum* L.
Flowers July–October.

Commonly cultivated and sometimes escaped from cemeteries and gardens to roadsides and waste ground. Scattered in Missouri.

Native of Eurasia; introduced and naturalized in North America from Newfoundland to Minnesota, south to Virginia, Pennsylvania, Indiana, and Missouri.

This species is quite variable, plants with comparatively narrow, shallowly toothed leaves and spreading petals referred to ssp. *Fabaria* (Koch) Schinz & Keller, and those with broader, more coarsely toothed leaves and downwardly curved petals referred to ssp. *purpureum* (Link) Schinz & Keller.

The tender young leaves and stems may be used as a fresh salad, or, when gathered as late as July, cooked as a green vegetable. The fleshy tuberous roots are sometimes pickled or used as a relish.

## Fam. **SAXIFRAGACEAE** (Saxifrage Family)

The genus *Penthorum,* sometimes classified in the *Crassulaceae,* is placed in the present flora in the Saxifragaceae.

Some authors split this family into other families: Parnassiaceae (Parnassia), Penthoraceae (Penthorum), Saxifragaceae (Mitella, Heuchera, Saxifraga, Sullivantia), Hydrangeaceae (Hydrangea, Philadelphus), Iteaceae (Itea), and Grossulariaceae (Ribes).

a. Soft-stemmed herbaceous plants . . . . . . . . . . . . . . . . . . . . . . *b*
- b. Flowering stem with 6 or more prominent, well-developed leaves; ovary 5–7-celled; fruit with 5 or 7 spreading portions (follicles) eventually splitting open by a top-like cap. . . . 1. PENTHORUM
- b. Flowering stem either with all the leaves at its base or bearing at most 1 or 2 prominent leaves, any other leaves present being very small or inconspicuous; ovary 1–3-(rarely 4) celled; fruit not splitting open by a many-lobed cap . . . . . . . . . . . . . . . . . . *c*
  - c. Leaves on the flowering stem 2, opposite each other and conspicuous; petals deeply fringed, suggesting snowflakes . . . . . . . . . . . . . . . . . 5. MITELLA
  - c. Leaves either all at the base of the flowering stem or only 1 conspicuous leaf on flowering stem, any others present being small or inconspicuous . . . . . . . . . . . . *d*
    - d. Flowering stem bearing only 1 flower or 1 fruit; flowering stem with 1 conspicuous rounded leaf present just below the middle of the stem; all leaves without teeth; plant completely glabrous (without hairs); petals 15–22 mm. long . . . . . . . . . . 6. PARNASSIA
    - d. Flowering stem bearing more than 1 flower or fruit; flowering stem with the leaves all clustered at the base or with 1 just above the base, any other leaves present being inconspicuous or small bracts subtending the flower branches; leaves with some teeth, or, if lacking teeth, the plant possessing some hairs on leaf, stem, or in the flowering portion; petals much smaller, at most 5.5 mm. long and the complete calyx and petals less than 10 mm. long . . . *e*
      - e. Stamens 10; basal leaves usually longer than broad, gradually narrowed to a short stalk which is scarcely developed; only short teeth along the margins of leaves . . 3. SAXIFRAGA
      - e. Stamens 5; basal leaves about as broad as or broader than long, somewhat circular or heart-shaped at base, resembling the leaf of a cultivated Geranium (Pelargonium); lobes or deeper indentations along the margins of leaves . . . . . . . . . . . . *f*
        - f. Leaf-stalks (petioles) glabrous (without hairs); leaf-blades thin, membranous, mostly glabrous on both upper and lower surface; ovary 2-celled; calyx-tube united with only the base of the ovary; seeds strongly winged . . . . . . . . . 2. SULLIVANTIA
        - f. Leaf-stalks hairy, or if hairless or nearly so in *H. puberula* f. *glabrata* then with rather thick leaf blades hairy on upper surface; leaf-blades thickish, usually hairy on both surfaces (except glabrous on lower surface in *H. puberula* f. *glabrata*); ovary 1-celled; calyx-tube united with the lower half of the ovary; seeds not winged. . . 4. HEUCHERA

a. Shrubs or woody plants . . . . . . . . . . . . . . . . . . . . . . . . *g*
- g. Leaves deeply lobed or irregularly and deeply cut, unequally toothed, about as broad as long; fruit a fleshy berry . . . . . . . . . . . . . . . . . . . . 10. RIBES
- g. Leaves regularly and shallowly toothed, not lobed or deeply cut, usually longer than broad; fruit a dry capsule . . . . . . . . . . . . . . . . . . . . . . *h*
  - h. Leaves alternate . . . . . . . . . . . . . . . . . . . . 9. ITEA
  - h. Leaves opposite, in pairs . . . . . . . . . . . . . . . . . . . . *i*
    - i. Leaves without teeth or with 1–5 teeth per 2.5 cm. (1 inch); all the petals large and showy, 10–20 mm. long; stamens 20–40; fruit 3–4-celled . . . . . . . . . . 7. PHILADELPHUS
    - i. Leaves with 6–11 teeth per 2.5 cm. (1 inch); petals small, at most 3 mm. long, the flowers along the outer edge of the inflorescence sometimes with enlarged sepals and sterile; stamens 8 or 10; fruit 2-celled in the lower portion . . . . . . . . . . . . 8. HYDRANGEA

1117 Sullivantia renifolia

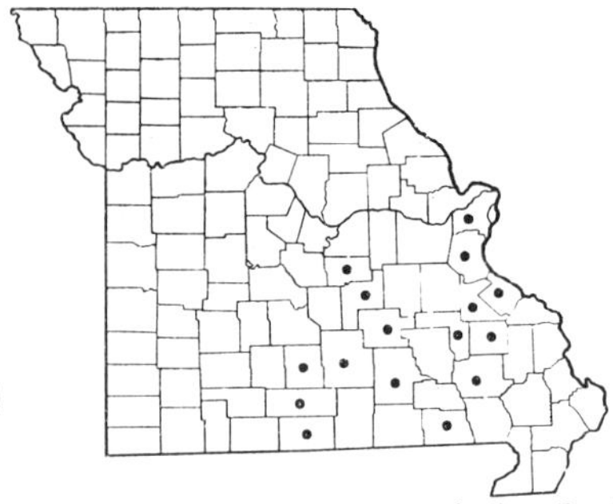

1118 Saxifraga virginiensis f. virginiensis (Early Saxifrage)

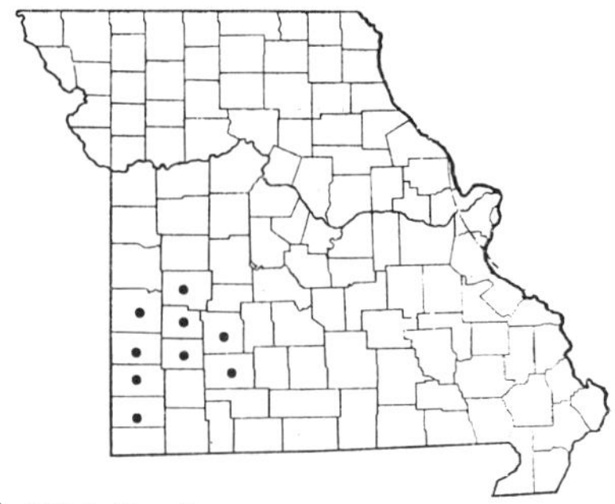

1119 Saxifraga texana

## 1. **Penthorum** L. Ditch Stonecrop

**Penthorum sedoides** L. Ditch Stonecrop Map 1116

Flowers July–October.

Occurs in wet soils bordering sloughs, ditches, swamps, oxbow lakes of river flood plains, streams, low wet woodland, fallow wet fields, and sometimes along railroads; frequently partly submerged or standing in water. Throughout Missouri and doubtless in every county.

Ranges from Florida to Texas, north to Quebec, Maine, Ontario, Michigan, Wisconsin, Minnesota, and Nebraska.

The fruit turns reddish at maturity.

## 2. **Sullivantia** T. & G.

**Sullivantia renifolia** Rosend. Map 1117

Flowers May–August.

Occurs on moist shaded, north-facing bluffs of limestone or St. Peter sandstone along streams. Rare and scattered in east-central and south-central Missouri in the Ozark and Ozark border counties. Known only from Warren, Jefferson, Franklin, Shannon, Howell, and Douglas counties.

Ranges from southwestern Wisconsin and southeastern Minnesota to northeastern Iowa and northwestern Illinois; and Missouri.

In its range outside the Ozarks, this species is limited to the so-called 'Driftless Area,' which was not covered by the last advance (Wisconsin) of the Pleistocene ice. The plants in Missouri represent relict survivors of a species which had a wider geographical distribution before the last advance of the ice sheet, and which is now maintaining itself on secluded favorable moist bluffs in the Ozarks and in the 'Driftless Area.'

## 3. **Saxifraga** L. Saxifrage

a. Petals greenish; leaves mainly 10–30 cm. long; flowering stems mainly 3.5–7 dm. (up to 9) tall; calyx-lobes turned down (reflexed). . . . . . . . . . . . . . . 3. S. PENSYLVANICA var. FORBESII

a. Petals white; leaves mainly 1.5–5 cm. (up to 8) long; flowering stems mainly 0.5–2.5 dm. (up to 3.5) tall; calyx-lobes ascending to spreading . . . . . . . . . . . . . . . . . . . . . *b*

b. Plants of southwestern Missouri east to Polk and Greene counties; inflorescence compact, the axis not elongating or loosening, the flowers remaining close together; main flower-stem sparsely short-hairy with few or no glandular hairs; flower-stalks (pedicels) glabrous or glabrate (nearly without hairs); carpels usually 3 or 4; petals 2–3.5 mm. long, usually only slightly longer than the calyx-lobes; calyx-lobes broadly rounded, rose- or wine-colored on thinner margins . . 2. S. TEXANA

b. Plants of the eastern Ozark region west to Maries, Phelps, Wright, Douglas, and Ozark counties; inflorescence eventually loosening, the axis becoming lax and branching, the flowers eventually becoming separated and distant from one another; main flower-stem and flower-stalks (pedicels) more or less glandular-hairy; carpels mainly 2; petals 3.5–6 mm. long, mostly 2–3 times as long as the calyx lobes; calyx lobes obtuse or mostly acutely pointed and mucronate, green throughout, without colored thinner margins . . . . . . . . . . . . . . . . . . . . 1. S. VIRGINIENSIS

1. **Saxifraga virginiensis** Michx. f. **virginiensis**
Early Saxifrage Map 1118
*Saxifraga virginiensis* Michx. [G, BB, P & S, Steyerm.].
Flowers February–June.

Occurs on moist wooded ledges, bluffs, and rocky glades of usually sandstone or granitic bluffs, rarely on limestone. Restricted to the Ozark region of south-central and east-central Missouri north to St. Louis and Maries counties and west to Phelps, Wright, Douglas, and Ozark counties.

Ranges from Quebec to Ontario and Minnesota, south to Georgia, Tennessee, and Missouri.

The leaves are usually dark green above and rose or purple beneath on the earliest winter leaves, becoming paler green on the later developed ones. The leaves vary from toothed with a scalloped edge or with shallow broad teeth.

A number of variations have been described as forms, but none except typical f. *virginiensis* has been thus far recorded from Missouri.

This saxifrage can be grown successfully in wild flower rock gardens if given sufficient shade, drainage, and moisture; usually sand rocks and fine sandy soil suit its requirements.

2. **Saxfraga texana** Buckl. Map 1119
Flowers April–May.

Occurs in acid rocky sandstone or chert glades and rocky prairies. Restricted to the unglaciated prairie region of southwestern Missouri north to St. Clair and Vernon counties and east to Polk and Greene counties.

Ranges from Missouri and southeast Kansas to Arkansas, Oklahoma, and Texas.

This species often occurs with those of similar habitat requirements on sandstone or chert glades having a similar restricted range both within and outside of Missouri. Such species as *Selenia aurea* and *Sedum Nuttallianum* are examples of this type of association. The rare and local *Geocarpon minimum* is likewise associated with *Saxifraga texana* where it occurs in Missouri.

3. **Saxifraga pensylvanica** L. var. **Forbesii** (Vasey) Engl. & Irmsch. Swamp Saxifrage Map 1120
Also known as Pennsylvania Saxifrage.

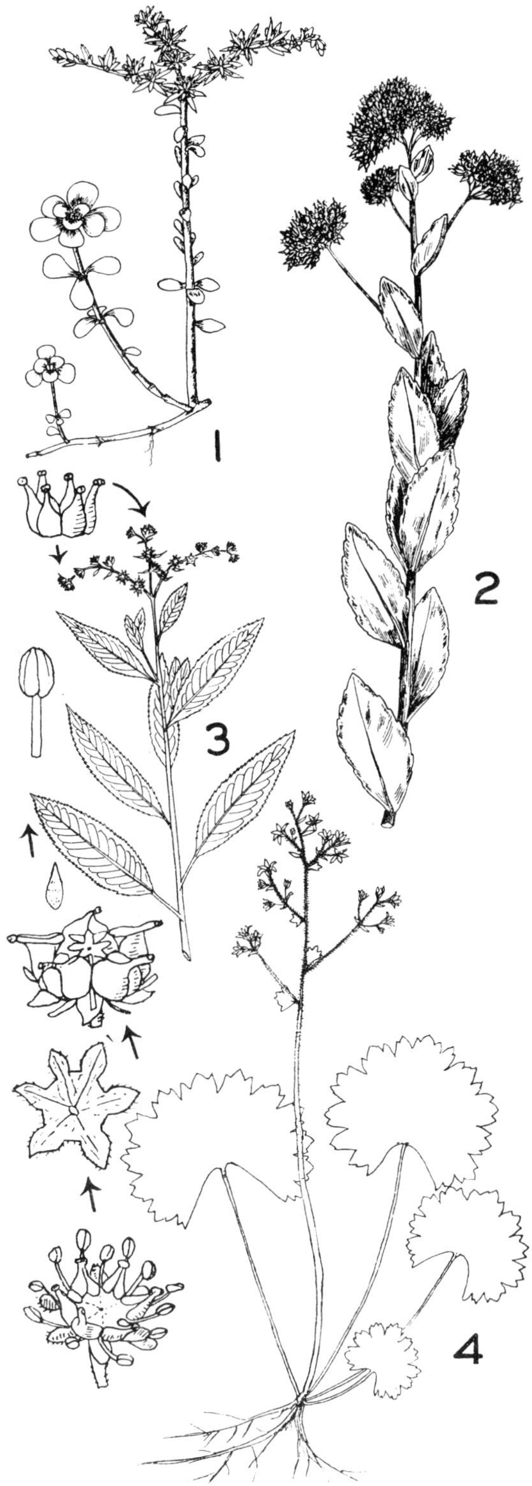


Plate no. 189. **1.** Sedum ternatum, × 2/5. **2.** Sedum Telephium, × 2/5; After Gleason, The New York Botanical Garden. **3.** Penthorum sedoides, × 2/5; After Gleason, details from Small, The New York Botanical Garden. **4.** Sullivantia renifolia, × 2/5.

*Saxifraga Forbesii* Vasey [BB, Johnson, Burns]
*Saxifraga pensylvanica* in part [G], not L.
*Saxifraga Forbesii* f. *bracteosa* Burns
Flowers April–June.

Occurs on moist, usually north-facing sandstone bluffs of St. Peter sandstone and occasionally on Roubidoux and other sandstones. Restricted to a small section of east-central Missouri north to Lincoln, Montgomery, and Boone counties, south to St. Francois, Washington, and Crawford counties, west to Boone and Crawford counties.

Ranges in the 'Driftless Area' of northwestern Illinois, southwestern Wisconsin, and southeastern Minnesota; eastern Missouri and southern and north-central Illinois.

This saxifrage is treated as a distinct species by some authors and merged with *S. pensylvanica* by others. The chief character used to separate it as a species by Burns (Am. Midl. Nat. 28: 127–60. 1942) from *S. pensylvanica* is the position and degree of cohesion of the two follicles: in *S. pensylvanica* the follicles are stated to remain united to or beyond the middle and the body of the follicle is erect with the styles erect to recurved; in *S. pensylvanica* var. *Forbesii* the follicles are nearly or quite separate from each other to their base and are outwardly spreading with the styles always recurved. My own examination of herbarium material, however, does not bear out these conclusions. Jones recognizes the two taxa as distinct species and in his *Flora of Illinois* (2nd ed. pp. 155–56. 1950) attempts to differentiate them as follows:

'2. Leaves pilose beneath; petals white, longer than the elliptical sepals; filaments filiform; moist shaded sandstone cliffs . . . S. FORBESII
2. Leaves glabrous or nearly so beneath; petals greenish, equalling the deltoid sepals; filaments subulate; meadows . . S. PENSYLVANICA'

The above differences do not correlate well in Missouri material, which has green petals and leaves pilose beneath. Actually, I have never observed any of the *S. pensylvanica* group that had any except greenish petals, although on herbarium specimens the petals often appear pale in the dried state. Many specimens of *S. pensylvanica* vary from glabrous to pilose on the midrib and other parts beneath. Burns in his treatment distinguishes f. *bracteosa* (Am. Midl. Nat. 28: 159. 1942) from *S. Forbesii* on the basis of longer, foliaceous lower bracts of the inflorescence, those of f. *bracteosa* being lance-ovate, 5–9 cm. long and 2–4 cm. wide, and of *S. Forbesii* f. *Forbesii* being linear and up to 1.5 cm. long. Specimens collected from Allenton, St. Louis County by Letterman (June 15, 1875 and June 20, 1884 [holotype]) and Kellogg (June 19, 1884), and from Jefferson County by Eggert (Silica, May 5, 1898 and May 11, 1887) are cited by Burns under his f. *bracteosa*. Missouri specimens which I have observed in the field show great variation in width and length of the lower bracts of the inflorescence, and I do not believe a separate name is justified for those individuals showing extremes of the bracts.

It may be that the plants of moist sandstone bluffs known as *S. Forbesii* or *S. pensylvanica* var. *Forbesii* represent ecotypes (ecological variations more or less fixed to certain habitats), but whether or not the character of follicle cohesion and divergence is a real or assumed one remains the subject of much more detailed field work, breeding, and transplanting investigations. My own observations would indicate that the follicle cohesion and divergence is not a real difference, but merely a matter of age and maturity of the follicle. In view of the present uncertain status of this saxifrage, it is here maintained in varietal category, as indicating some divergence from typical *S. pensylvanica* var. *pensylvanica*, instead of as a distinct species.

The young unfolding leaves are sometimes used in green salads.

## 4. **Heuchera** L. Alum Root

a. Leaf-stalks (petioles) without hairs or nearly so or only sparsely hairy; lower surface of leaf-blades without hairs or nearly so . . . . . . . . . . . . . . . . . . 2b. H. PUBERULA f. GLABRATA
a. Leaf-stalks either conspicuously hairy with long or spreading hairs or covered with minute hairs; lower surface of leaf-blade more or less hairy . . . . . . . . . . . . . . . . . . *b*

---

Plate no. 190. **1.** Saxifraga virginiensis, × 2/5. **2.** Saxifraga texana × 2/5. **3.** Heuchera villosa var. macrorhiza; *a.* Flower in vertical section, × 2 2/5; Details after Rosendahl; *b.* Basal leaf, × 2/5; *c.* Inflorescence, × 2/5; **4.** Saxifraga pensylvanica var. Forbesii, × 2/5; Details from Small, The New York Botanical Garden. **5.** Heuchera puberula, × 2/5 *a.* Flower in vertical section, × 2 2/5; Details after Small, The New York Botanical Garden, and Rosendahl. **6.** Heuchera missouriensis; *a, b, c.* Bracts of inflorescence, × 2 4/5; (After Rosendahl); *d.* Basal leaf, × 2/5; *e.* Part of dissected flower, × 4; (After Rosendahl); *f.* Flower at anthesis, × 4; (After Rosendahl); Details after Rosendahl.

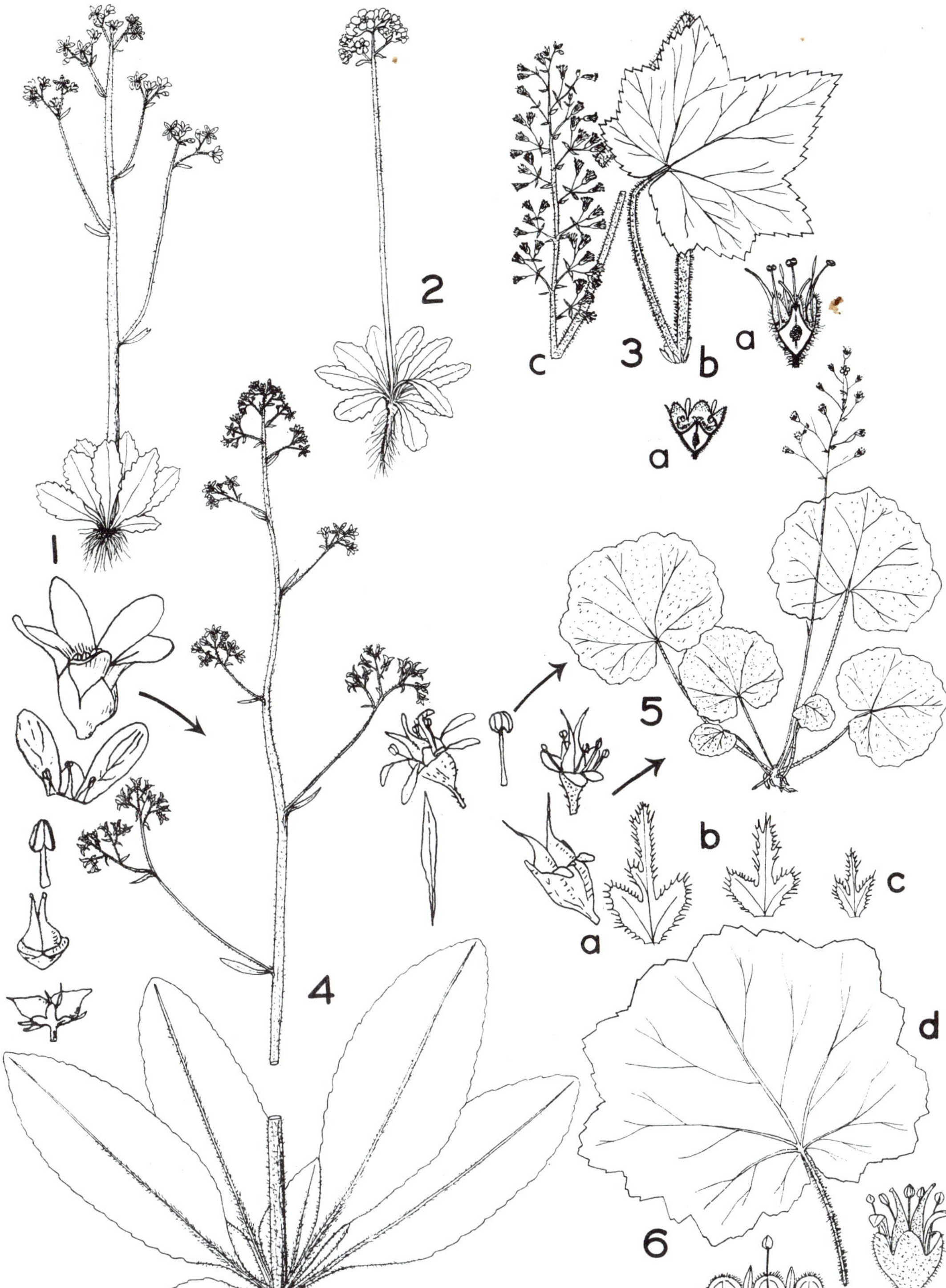


PLATE NO. 190

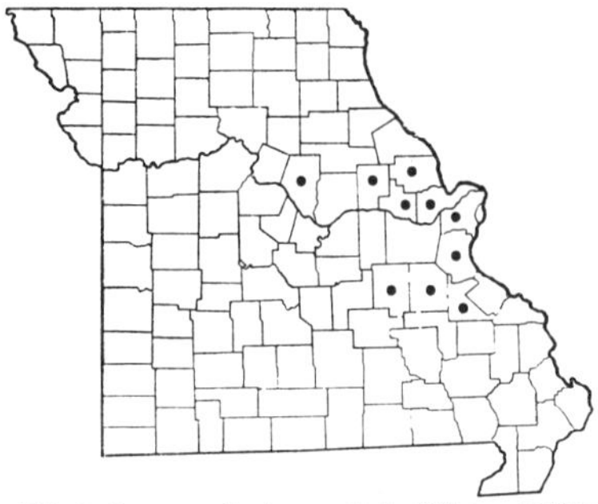

1120 Saxifraga pensylvanica var. Forbesii (Swamp Saxifrage)

1121 Heuchera villosa var. macrorhiza

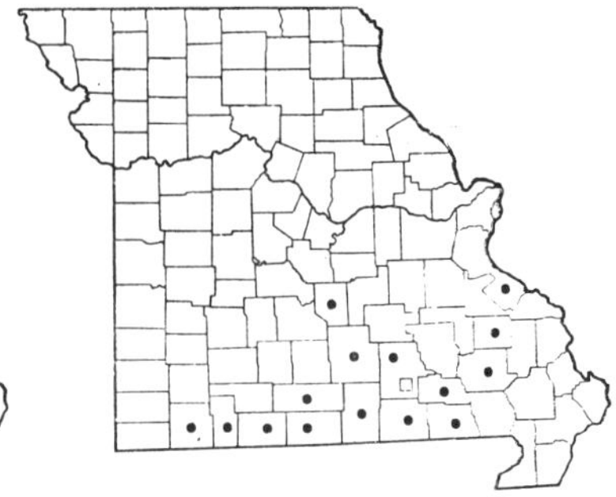

1122 • Heuchera puberula f. puberula
1122 □ Heuchera puberula f. glabrata

b. Outside of flower covered with hairs which are not gland-tipped . . . . . . . . . . . . *c*

c. Leaf-stalks and flowering stems and branches covered with short, minute gland-tipped hairs; seeds smooth or with little ridges; flowering usually from July–November . . . . . . . 2a. H. PUBERULA f. PUBERULA

c. Leaf-stalks and flowering stems with rather long and conspicuous hairs which are not gland-tipped; seeds covered with tiny pointed projections; flowering usually from June–August. . 1. H. VILLOSA var. MACRORHIZA

b. Outside of flower covered with short or long gland-tipped hairs . . . . . . . . . . . *d*

d. Petals white, about 1.5 mm. long; green outer part of flower (calyx) 2–2.5 mm. long (measured from base of tube to tip of calyx-lobes); capsule 2.5–3 mm. long; calyx covered with rather long spreading gland-tipped hairs; flowering in July and August . . . . . . 3. H. MISSOURIENSIS

d. Petals greenish or copper-colored or tinged with reddish color, 1–4 mm. long; green outer part of flower (calyx) 3–10 mm. long (measured from base of tube to tip of calyx-lobes); capsule 5–10 mm. or more long; calyx covered with minute short gland-tipped hairs; flowering from April–July . . . . . . . . . . . . . . . . . . . . . . . . . . . . *e*

e. The green calyx of the flower 3–4.5 mm. long; calyx-tube 1–1.6 mm. long, the upper part of it which is not attached to the ovary shorter than the lower part which is attached to the ovary; petals 1–3 mm. long; plants usually growing along ledges of bluffs or in their crevices, also in rocky woodland . . . . . . . . . . . . . . . . 4. H. AMERICANA

e. The green calyx of the flower 5–10 mm. long; calyx-tube 4–5 mm. long, the upper part of it which is not attached to the ovary longer than the lower part which is attached to the ovary; petals 3–4 mm. long; plants more commonly found on prairies, railroads, open banks, and woodland slopes . . . . . . . . . . . . . . . . 5. H. RICHARDSONII

1. **Heuchera villosa** Michx. var. **macrorhiza** (Small) Rosend., Butt. & Lak. Map 1121

*Heuchera macrorhiza* Small [P & S]

*Heuchera villosa* [of P & S], not Michx.

Flowers June–August and occasionally to November.

Known only from specimens grown in the garden of Mr. Albert Chandler of Kirkwood, Missouri, who stated that he collected it originally in Iron County in 1928 on limestone bluffs between Annapolis and Sabula (represented by *Steyermark 7714*, *7714A*, and *7715* in herb. Univ. Minn. and Mo. Bot. Gard.). Another specimen (*Steyermark 7715A*), also from Mr. Chandler's garden, was said to have been collected by him several miles away from the locality indicated for No. *7714* and *7714A*.

Ranges from West Virginia to Indiana and Missouri, south to Georgia, Alabama, Kentucky, and Tennessee.

Attempts to rediscover this plant at the original station have failed. The author found plants of a *Heuchera*, which he identified as *H. americana* var. *hirsuticaulis* (*Steyermark 65982*) on west-facing limestone outcrops of small bluffs along highway 49 near Big Creek, T31N, R3E, sect. 3, ½ mi. south of Sabula in Iron County. This was the station probably visited by Mr. Chandler, and the only likely habitat available between Annapolis and Sabula for a rock-ledge or bluff-inhabiting type favored by *H. villosa* var. *macrorhiza*. In view of these unfavorable results, it seems possible that the origin of the recorded station by Mr. Chandler had been confused at the time of introduction of the plant in Mr. Chandler's garden. More intensive exploration of limestone bluffs in Iron and

1123 Heuchera missouriensis

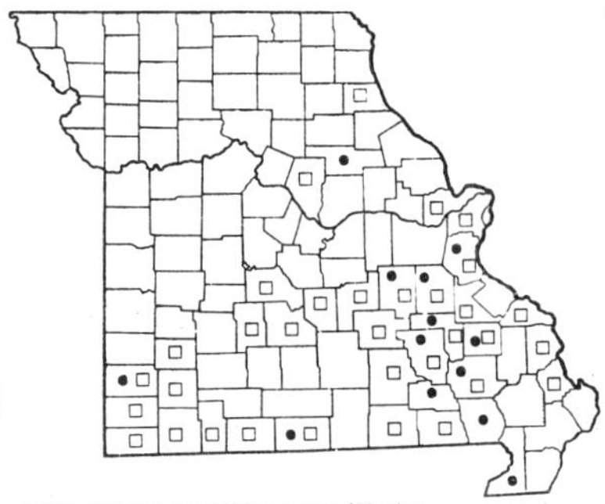

1124 • Heuchera americana var. interior
1124 □ Heuchera americana var. hirsuticaulis

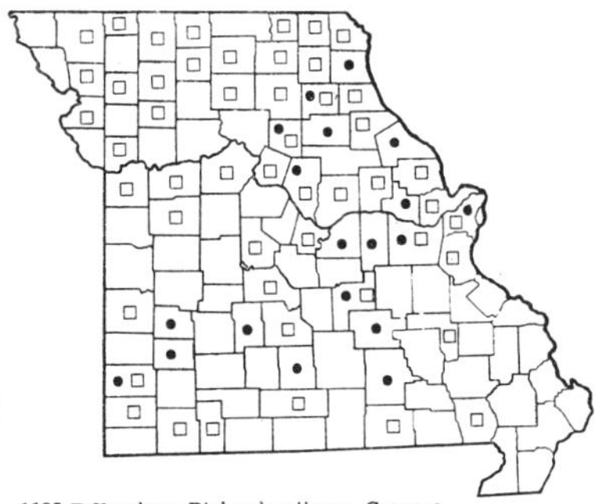

1125 □ Heuchera Richardsonii var. Grayana
1125 • Heuchera Richardsonii var. affinis

adjacent counties may eventually reveal the presence of this plant, or it may have to be eliminated from the state flora.

2. **Heuchera puberula** Mackenz. & Bush Map 1122

Flowers from July but mainly September–November.

Occurs on usually north- or east-facing shaded limestone bluffs, usually on the drier part where there is less seepage. Southern Ozark section north to Ste. Genevieve, Madison, Shannon, Pulaski, and Douglas counties to Barry County.

Ranges from Missouri to Arkansas; also western Kentucky.

Two variations are represented by Missouri material:

Common variation; leaf-stalks (petioles) and flowering stem (scape) densely covered with minute gland-tipped hairs; lower surface of leaf-blade densely soft-hairy and usually purplish . 2a. H. PUBERULA f. PUBERULA

Rare variation; leaf-stalks and flowering stem without hairs or nearly so, or sparsely hairy; lower surface of leaf-blade nearly without any hairs (glabrate) and pale green 2b. H. PUBERULA f. GLABRATA

2a. **Heuchera puberula** f. **puberula** Map 1122

*Heuchera puberula* Mackenz. & Bush [G, BB, P & S]

2b. **Heuchera puberula** f. **glabrata** Steyerm. Map 1122

Known only from Shannon County, in the southeastern Ozarks (north-facing base of limestone bluffs, at Jam-Up Bluff along Jack's Fork of Current River, T27N, R6W, sect. 5, 4 mi. north of Teresita, 6 mi. northwest of Monteer, October 2, 1948, *Steyermark 66615*, holotype in herb. Chi. Nat. Hist. Mus.).

This species occurs principally in the Missouri Ozarks, its Arkansas and Kentucky records being based on one known station in those states. In Missouri it is often associated with the sedge, *Carex eburnea*, and although receiving shade and some moisture, it favors the drier part of the bluffs, rather than the moist or dripping seepage favored on limestone bluffs by *Parnassia grandifolia* and *Sullivantia renifolia*.

3. **Heuchera missouriensis** Rosendahl Map 1123

Flowers July–August.

Occurs in crevices of limestone bluffs, where known only from the southeastern Ozarks in Madison (along St. Francis River, between mouth of Captain Creek and bridge over highway to Jewett, November 15, 1936, *Steyermark 20980*) and Wayne (Hall's Bluff along St. Francis River, 4 mi. south of Kime, 6 mi. southeast of Greenville, south of Davidson's Blue Spring, T27N, R6E, sect. 5, July 9, 1936, *Steyermark 11542*; September 1, 1938, *Steyermark 6342* [holotype]; October 21, 1948, *Steyermark 66966*) counties.

Known only from Missouri.

This species, an endemic of the Missouri Ozarks like *Tradescantia longipes* and others, was described by Dr. C. O. Rosendahl in 1951 (Rh. 53: 105–9.), and is the most recently described new species from Missouri. It was reported previously by the present author on the basis of Dr. Rosendahl's original determination first as *H. puberula* × *americana* var. *hirsuticaulis* (Rh. 40: 253. 1938) and later as *H. parviflora* Bartl. var. *Rugelii* (Shuttlw.) Rosend., Butt. & Lak. (Rh. 42: 99. 1940). With the more complete material available to him, Dr. Rosendahl recognized it as a distinct species.

It was feared that the species had been exterminated by the Wappapello Dam from its Wayne County station, but when the author revisited that place in 1948, it was still surviving, although newly exposed to unaccustomed exposure and sunlight as a result of felling of the adjacent forest. Only a few plants were found on the upper ledges of Hall's Bluff, most of them occurring on the middle and lower portions. It will be interesting to observe how this Missouri endemic survives the future in view of these changes.

4. **Heuchera americana** L. Map 1124

Flowers April–June.

Occurs mainly along ledges and crevices of bluffs, also in rocky open woodland.

Two variations are represented by Missouri material:

Calyx during period of anthesis (when stamens are present) 3–3.5 mm. long; calyx nearly symmetrical, one side only slightly longer than the other side; petals 1–1.5 mm. long. . . . . 4a. H. AMERICANA var. INTERIOR

Calyx during period of anthesis 4–4.5 mm. long; calyx not symmetrical, one side rather noticeably longer than the other side; petals 2–3 mm. long. 4b. H. AMERICANA var. HIRSUTICAULIS

4a. **Heuchera americana** var. **interior** Rosend., Butt. & Lak. Map 1124

*Heuchera americana* [of P & S], not L.

*Heuchera hirsuticaulis* in part [of BB], not (Wheelock) Rydb.

Scattered in southern and central Missouri, north to Audrain Co.; mostly in the eastern Ozarks.

Ranges from Indiana and Illinois to Missouri, south to Tennessee and Arkansas.

4b. **Heuchera americana** var. **hirsuticaulis** (Wheelock) Rosend., Butt. & Lak. Map 1124

*Heuchera hirsuticaulis* (Wheelock) Rydb. [P & S, BB in part]

Southern, central, and northeastern Missouri, north to Marion and Boone counties north of the Missouri River, and west to Camden, Dallas, Dade, and Jasper counties. This is the commoner variation of *H. americana* in Missouri.

Ranges from Indiana to Illinois and Missouri.

Gleason (*New Ill. Fl.* 2: 269. 1952) includes both the Missouri varieties under *H. hirsuticaulis*, retaining that as a species distinct from *H. americana*, the latter not known in Missouri. The varieties given below under *H. Richardsonii* are also treated by Gleason in the *H. hirsuticaulis* group. See discussion under the following species.

5. **Heuchera Richardsonii** R. Br. Map 1125

Flowers April–July.

Occurs on prairies, along railroads, open banks, open woodland slopes, and upland.

Two variations are represented in Missouri material:

Newly flowering calyx 6–10 mm. long; calyx with one side markedly longer than the other side . 5a. H. RICHARDSONII var. GRAYANA

Newly flowering calyx 5–7 mm. long; calyx with one side somewhat longer than the other side . 5b. H. RICHARDSONII var. AFFINIS

5a. **Heuchera Richardsonii** var. **Grayana** Rosend., Butt. & Lak. Map 1125

*Heuchera hirsuticaulis* in part [of BB], not (Wheelock) Rydb.

Throughout Missouri, except apparently absent from the lowland counties of southeastern Missouri. This is the commonest *Heuchera* in Missouri.

Ranges from Indiana, Michigan, and Wisconsin, south to Illinois and Missouri.

5b. **Heuchera Richardsonii** var. **affinis** Rosend., Butt. & Lak. Map 1125

*Heuchera hirsuticaulis* in part [of BB], not (Wheelock) Rydb.

Southern, central, and northeastern Missouri northeast to Lewis County, west to Randolph, Boone, Cedar, and Jasper counties.

Ranges from Michigan and Wisconsin, south to Indiana, Illinois, and Missouri.

As indicated under *H. americana*, Gleason treats the above two variations of *H. Richardsonii*, along with the two Missouri variations of *H. americana*, as one variable species, *H. hirsuticaulis* (Wheelock) Rydb. There is some justification for following this concept, since the relative amount of regularity and irregularity (obliqueness or asymmetry) of the calyx-tube varies, as does also its relative length. Specimens are frequently encountered which intergrade and it is difficult to place them with certainty. While it is admitted that Gleason's attitude represents a practical course of procedure, and may eventually be favored or adopted, the present author of this flora, in deference to the many years of work on this genus by its monographers, prefers to retain their conclusions concerning these taxa (also retained by such able botanists as Deam, Jones, and Fernald) until future investigations add further proof or disproof to the present facts known.

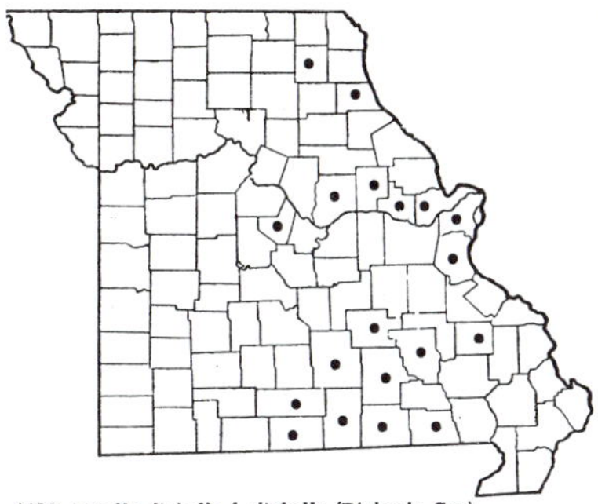

1126 Mitella diphylla f. diphylla (Bishop's Cap)

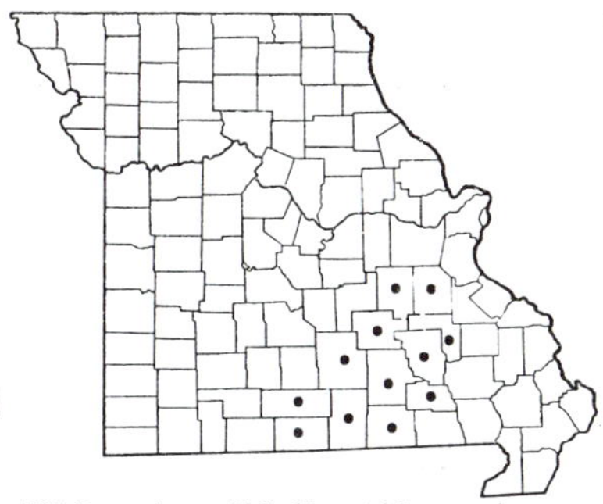

1127 Parnassia grandifolia (Grass-of-Parnassus)

1128 Philadelphus pubescens var. verrucosus (Mock Orange)

## 5. **Mitella** Miterwort, Bishop's Cap

**Mitella diphylla** L. f. **diphylla**
Miterwort, Bishop's Cap Map 1126
*Mitella diphylla* L. [G, BB, P & S, Steyerm.]
Flowers April–June.

Occurs on north-facing wooded limestone or sandstone bluffs or rocky slopes, in mossy pockets, moist ledges, and among humus-covered talus. Eastern half of Missouri west to Knox, Moniteau, Texas, Douglas, and Ozark counties; absent from the lowland counties of southeastern Missouri and from most of the northeastern counties.

Ranges from Quebec and New Hampshire to Ontario and Minnesota, south to South Carolina, Alabama, Tennessee, Mississippi, and Missouri.

Forms not recorded from Missouri have been described with the leaves on the stem in threes (f. *triphylla* Rosend.) or the paired leaves petioled (f. *oppositifolia* [Rydb.] Rosend.).

## 6. **Parnassia** L. Grass-of-Parnassus

**Parnassia grandifolia** DC. Grass-of-Parnassus
Map 1127
Flowers August–October.

Occurs in swampy open meadows in small valleys fed by calcareous spring water, moist limestone ledges along streams, and moist crevices at the base of north-facing limestone bluffs. Restricted to the eastern and south-central Ozark area north to Washington and Crawford counties and west to Texas, Douglas, and Ozark counties.

Ranges from Florida to Texas, north to Virginia, West Virginia, Tennessee, and Missouri.

The lower larger leaves in many Missouri specimens are only 3 cm. long, while averaging 5–10 cm. in other portions of the range. The anthers in some Missouri plants are 2 mm. long and the petals sometimes shorter than the 15–22 mm. given for them elsewhere in their range.

This species brightens the swampy meadows in autumn with the open white flowers, often flowering at the same time as *Spiranthes cernua*, *Pedicularis lanceolata*, *Solidago Riddellii*, *Aster puniceus* var. *lucidulus* f. *firmus*, *Rudbeckia palustris*, and *Lobelia siphilitica*. Along the moist limestone ledges of small streams it is likely to be encountered with *Gentiana quinquefolia* var. *occidentalis* and *Selaginella apoda*.

## 7. **Philadelphus** L. Mock Orange

**Philadelphus pubescens** L. var. **verrucosus**
(Schrad.) Hu Mock Orange Map 1128
*Philadelphus pubescens* L. [G, Steyerm.]
Flowers late April–May.

Occurs on north-facing steep, wooded limestone bluffs, where known only from southwestern Missouri in McDonald County ($\frac{3}{4}$ mi. southeast of Southwest City, May 31, 1938, *Steyermark 5573*; same locality, September 25, 1947, *Steyermark 65196*).

Ranges from Tennessee, Illinois, and Missouri, south to Arkansas and Oklahoma.

Typical var. *pubescens* has ovate leaves rounded at the base, whereas var. *verrucosus* has elliptic or ovate-elliptic leaves acute or obtuse at base.

This shrub occurs near the Oklahoma-Arkansas border, and is the dominant understory where it occurs. It was associated at this station with *Bumelia lanuginosa* var. *oblongifolia*, *Cacalia Muhlenbergii*, and *Polymnia canadensis* var. *radiata*. In the autumn the leaves turn greenish yellow.

*Philadelphus pubescens* is a handsome shrub when in full bloom and can be easily grown from seed or transplantings. It is the most commonly cultivated species. Plants raised from seed collected at the Missouri station of var. *verrucosus* have prospered in the author's botanical preserve in northern Illinois, where they produce an abundance of flowers and seeds several hundred miles north of their natural northern limit of geographical range. They are not bothered by rabbits, a distinction that can be claimed for few shrubs observed and grown on the author's property. Cultivated specimens of the var. *verrucosus* are represented by herbarium material collected at the Missouri Botanical Garden in St. Louis (May, 1902, *Kellogg*) and of the typical variety *pubescens* in St. Louis (July 13, 1910, *Sherff*).

## 8. **Hydrangea** L. Hydrangea

**Hydrangea arborescens** L. Wild Hydrangea Map 1129

Also known as Seven-bark.

Flowers late May–July.

Occurs in rich or rocky wooded slopes, talus, at the base of bluffs, along streams, and in ravines.

Missouri material is represented by the following variations:

a. Lower surface of leaf-blades noticeably hairy, even the finer veins short-hairy . . . . *b*
- b. Most or all of the flowers large and showy with an expanded 3- or 4-lobed calyx, not seed-producing . . . . f. H. ARBORESCENS var. DEAMII f. ACARPA
- b. All or all but the flowers along the outer edge of the inflorescence small, perfect, and seed-producing . . . . e. H. ARBORESCENS var. DEAMII f. DEAMII

a. Lower surface of leaf-blades glabrous (without hairs) or nearly so, sparsely hairy only along the midrib or main veins . . . . . . *c*
- c. All the flowers large and showy with an expanded 3- or 4-lobed calyx, not seed-producing . . . . . . . . . *d*
  - d. Main leaf-blades broadly ovate to nearly orbicular with a broadly rounded or cordate base . . . b. H. ARBORESCENS var. ARBORESCENS f. GRANDIFLORA
  - d. Main leaf-blades narrowly ovate to narrowly oblong or lanceolate-elliptic, tapering or narrowed or gradually rounded at the base . d. H. ARBORESCENS var. OBLONGA f. STERILIS
- c. All or all but the flowers along the outer edge of the inflorescence small, perfect, and seed-producing . . . . . . . . *e*
  - e. Main leaf-blades broadly ovate to nearly orbicular with a broadly rounded or cordate base . . . a. H. ARBORESCENS var. ARBORESCENS f. ARBORESCENS
  - e. Main leaf-blades narrowly ovate to narrowly oblong or lanceolate-elliptic, tapering or narrowed or gradually rounded at the base . . . . c. H. ARBORESCENS var. OBLONGA f. OBLONGA

a. **Hydrangea arborescens** var. **arborescens** f. **arborescens** Map 1129

*Hydrangea arborescens* L. [G, BB, P & S, Steyerm.]

*Hydrangea arborescens* subsp. *arborescens* [McClintock]

This is the commonest variation in Missouri, occurring throughout the Ozark region in southern and central Missouri north to St. Charles, Pike, Callaway, and Saline counties, west to Miller, Saline, Hickory, and Newton counties.

Ranges from Georgia to Oklahoma, north to New York, Ohio, Indiana, Illinois, and Missouri.

b. **Hydrangea arborescens** var. **arborescens** f. **grandiflora** (E. G. Hill) Rehd. Map 1129

*Hydrangea arborescens* f. *grandiflora* (E. G. Hill) Rehder [G, P & S, Steyerm.]

*Hydrangea arborescens* subsp. *arborescens* [McClintock]

Known only from Osage (around rocks in woods below base of southwest-facing limestone bluffs along Osage River, T43N, R11W, sect. 34, 1¾ mi. [by air] west of Folk, 7 mi. southwest of Westphalia, August 15, 1955, *Steyermark 79257*) and Stoddard (steep shaded banks, north side of Crowley Ridge, near Messler, July 1, 1928, *Palmer 34868*) counties.

c. **Hydrangea arborescens** var. **oblonga** T. & G. f. **oblonga** Map 1129

Plate no. 191. **1.** Heuchera americana var. hirsuticaulis, × 2/7. **2.** Mitella diphylla, × 2/7. **3.** Parnassia grandifolia, × 2/7; After Britton and Brown, The New York Botanical Garden. **4.** Philadelphus pubescens var. verrucosus, × 2/7. **5.** Hydrangea arborescens, × 2/7; All details from Small, The New York Botanical Garden.

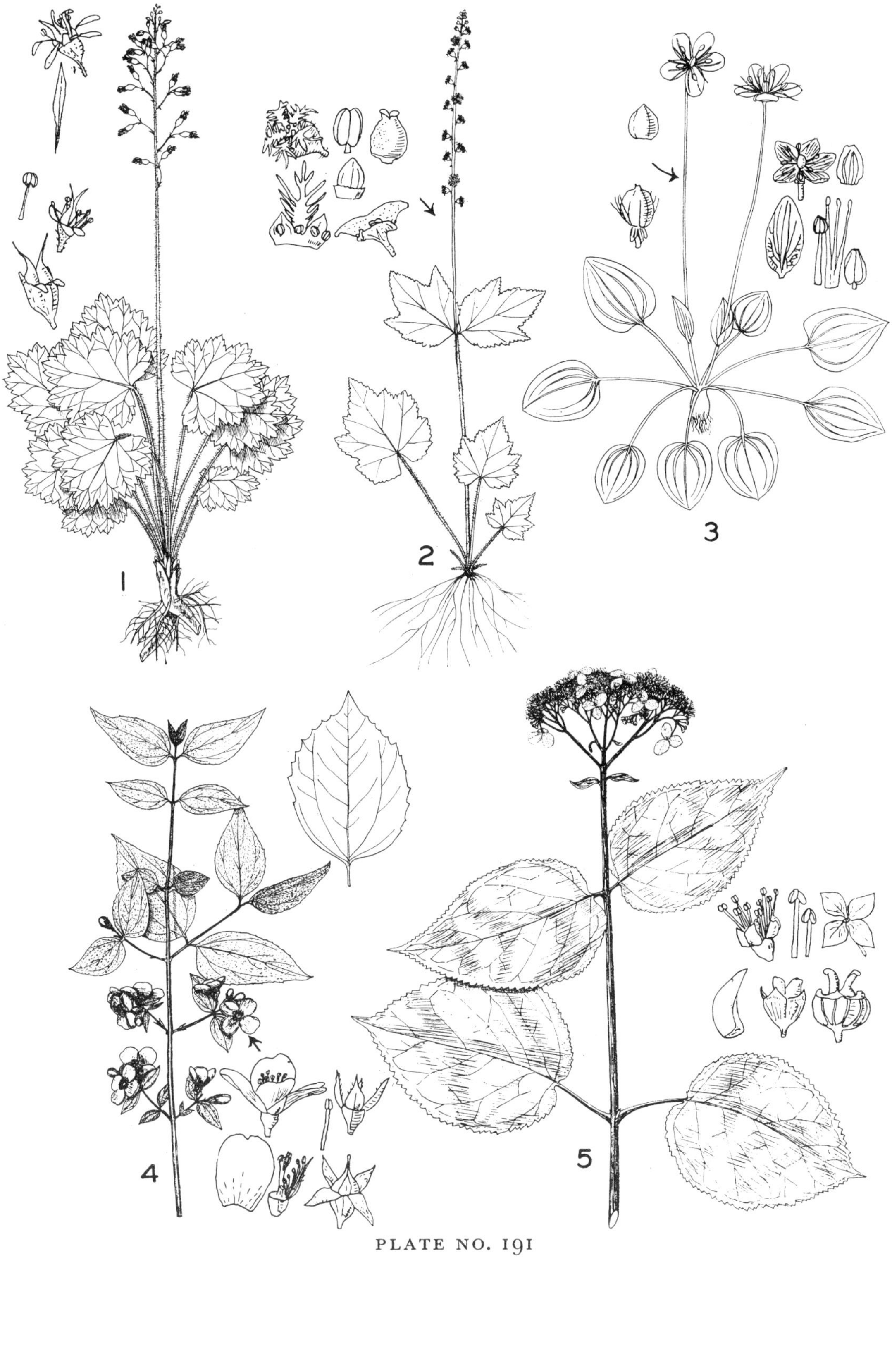


PLATE NO. 191

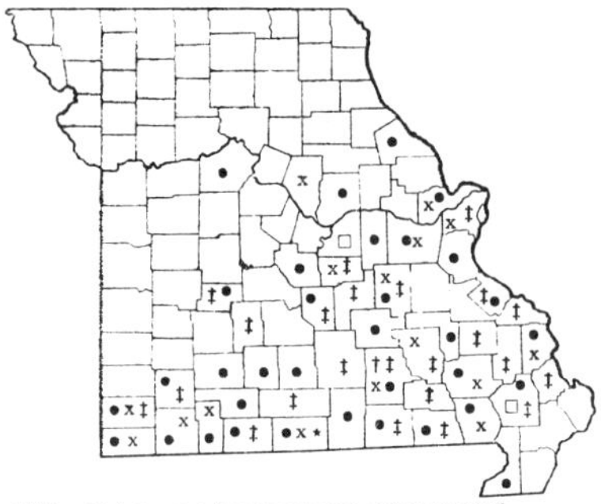

1129 ● Hydrangea arborescens var. arborescens f. arborescens (Wild Hydrangea)
1129 □ Hydrangea aborescens var. arborescens f. grandiflora
1129 x Hydrangea arborescens var. oblonga f. oblonga
1129 • Hydrangea arborescens var. oblonga f. sterilis
1129 ‡ Hydrangea arborescens var. Deamii f. Deamii
1129 † Hydrangea arborescens var. Deamii f. acarpa

1130 Itea virginica

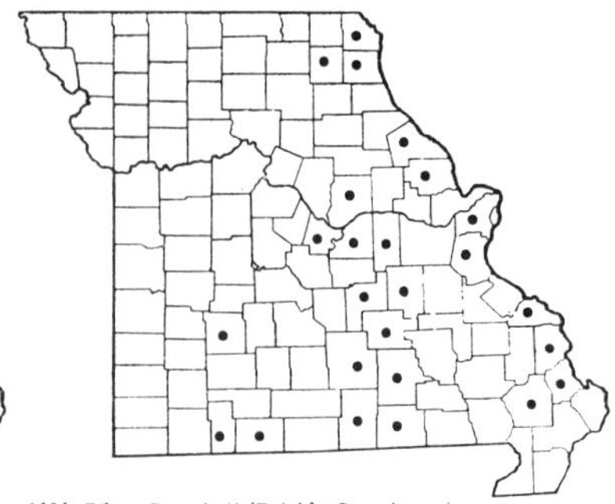

1131 Ribes Cynosbati (Prickly Gooseberry)

*Hydrangea arborescens* var. *oblonga* T. & G. [G, P & S, Steyerm.]

*Hydrangea arborescens* subsp. *arborescens* [McClintock]

Ozark region of southern and central Missouri, north to St. Charles and Boone counties, west to Maries and Newton counties.

Ranges from Florida to Louisiana and Oklahoma, north to New York, Ohio, Indiana, Illinois, and Missouri.

d. **Hydrangea arborescens** var. **oblonga** f. **sterilis** (T. & G.) St. John Map 1129

*Hydrangea arborescens* subsp. *arborescens* [McClintock]

Known only from Ozark County (shaded limestone bluffs of Little North Fork of White River near Pontiac, June 27, 1928, *Palmer 34783*).

e. **Hydrangea arborescens** var. **Deamii** St. John f. **Deamii** Map 1129

*Hydrangea arborescens* var. *Deamii* St. John [G, P & S, Steyerm.]

*Hydrangea arborescens* subsp. *discolor* (Seringe) McClintock [McClintock]

Ozark region of southern and east-central Missouri north to St. Louis, Crawford, Maries, and Hickory counties tô Newton County.

Ranges from Georgia to Oklahoma, north to West Virginia, Indiana, Illinois, and Missouri.

f. **Hydrangea arborescens** var. **Deamii** f. **acarpa** (T. & G.) St. John Map 1129

*Hydrangea arborescens* subsp. *discolor* (Seringe) McClintock [McClintock]

Known only from Shannon County (Monteer, August 6, 1910, *Bush 6116*, holotype of *H. cinerea* Small var. *sterilis* Rehd.).

McClintock (Proc. Cal. Acad. Sci. 4th Ser., 29: 147–255. 1957 and Jour. Arn. Arb. 37: 373–74. 1956) treats the above variations as two subspecies, *H. arborescens* subsp. *arborescens* and *H. arborescens* subsp. *discolor*. Under subsp. *arborescens* are included typical var. *arborescens* and var. *oblonga* with their respective showy-flowered, sterile forms included in synonymy. Under subsp. *discolor* are placed both the showy-flowered, sterile form and the regular fertile-flowered type of var. *Deamii*.

The above reduction to two subspecies without due recognition of the conspicuous sterile forms, which are of horticultural significance, is not followed in the present treatment. Deam (*Fl. Ind.* p. 520. 1940) concluded, after growing plants of var. *oblonga* together with typical var. *arborescens* that the leaf shape was a constant character and not subject to environmental variation. While admittedly there are intermediate specimens and intergrading forms (Rh. 23: 204. 1921), it would seem that the variations are recognizable, with differences in leaf shape and lack or presence of sterile flowers maintaining themselves under cultivation or after being transplanted.

The foliage turns a pale greenish-yellow in the fall. Some cases of poisoning among cattle are believed to have resulted after the leaves, which contain a glucoside yielding prussic acid, have been eaten.

## 9. **Itea** L.

**Itea virginica** L. f. **virginica** Virginia Willow
Map 1130

Flowers May–June.

Occurs in swamps, low wet woods, and along sandy spring branches. Restricted to the lowlands and base of Crowley Ridge in the southeastern Missouri counties, north to Cape Girardeau and Bollinger counties, west to Stoddard and Butler counties.

Ranges from Florida to Texas, north to New Jersey, Pennsylvania, Kentucky, Illinois, Missouri, and Oklahoma.

A form, f. *abbreviata* Fern., with the inflorescence 1–2.2 cm. instead of the usual 5–20 cm. long, and with shorter leaves 1–2.5 cm. instead of 5–8 cm. or more long, sometimes occurs, but has not been recorded from Missouri.

## 10. **Ribes** L. Currant, Gooseberry

a. No spines or prickles on any stem or branch; flowers golden-yellow; leaves pale green or grass green . . . 3. R. ODORATUM
a. Spines or prickles present on some of the stems or branches; flowers whitish-green or yellow-green; leaves dark green. . . . *b*
b. Ovary and fruit covered by slender prickles; flowers yellow-green; calyx-lobes much shorter than the calyx-tube. . . . 1. R. CYNOSBATI
b. Ovary and fruit without prickles; flowers whitish-green; calyx-lobes longer than or at least equaling the calyx-tube . . . 2. R. MISSOURIENSE

1. **Ribes Cynosbati** L. var. **Cynosbati** f. **Cynosbati**
Prickly Gooseberry Map 1131

Also known as Dogberry.

Flowers April–June; fruits July–September.

Occurs on north-facing shaded bluffs and wooded ledges. Scattered in the Ozark section of southern and east-central Missouri and north in eastern Missouri to Lewis, Knox, and Clark counties.

Ranges from New Brunswick, Maine, and Quebec to Manitoba, south to North Carolina, Alabama, Kentucky, Arkansas, and Oklahoma.

Spineless forms (f. *inerme* Rehd.) and varieties with mostly glabrous or nearly glabrous leaves (var. *glabratum* Fern.) and very bristly fruiting branches (var. *atrox* Fern.) occur, but have not been recorded from Missouri.

The fruits may be cooked and eaten or put up as preserves.

2. **Ribes missouriense** Nutt. Missouri Gooseberry
Map 1132

Also called Wild Gooseberry.

Flowers April–May; fruits June–September.

Occurs in rocky or open dry woods, thickets, upland or lowland woodland borders, and grazed or cut-over areas. Throughout Missouri, except absent from the southeastern lowland counties.

Two varieties occur in Missouri:

Leaves hairy on both upper and lower surface; hairs of the petioles fringed . 2a. R. MISSOURIENSE var. MISSOURIENSE
Leaves mostly without hairs on upper and lower surface; hairs of the petioles mostly simple . . 2b. R. MISSOURIENSE var. OZARKANUM

2a. **Ribes missouriense** var. **missouriense**
Map 1132

This is the common variation, occurring throughout the state.

Ranges from Connecticut to Minnesota and South Dakota, south to Tennessee, Arkansas, and Kansas.

2b. **Ribes missouriense** var. **ozarkanum** Fassett
Map 1132

Southwestern Ozark section, where known in Howell, Cedar, Greene, Christian, and Taney counties.

Ranges from Missouri to Arkansas.

This gooseberry is the common one in the state and often dominates grazed-over woodland and pastures.

The insipid fruit can be eaten when ripe, and like other kinds, can be cooked or put up as preserves. However, it does not compare favorably with some of the other wild or cultivated species.

3. **Ribes odoratum** Wendland f.
Golden Currant Map 1133

Also called Flowering Currant, Missouri Currant, and Buffalo Currant.

Flowers April–June; fruits June–August.

Occurs on exposed high rocky limestone bluffs along White River and tributaries in Taney, Stone,

1132 • Ribes missouriense var. missouriense (Missouri Gooseberry)
1132 □ Ribes missouriense var. ozarkanum

1133 Ribes odoratum (Golden Currant)

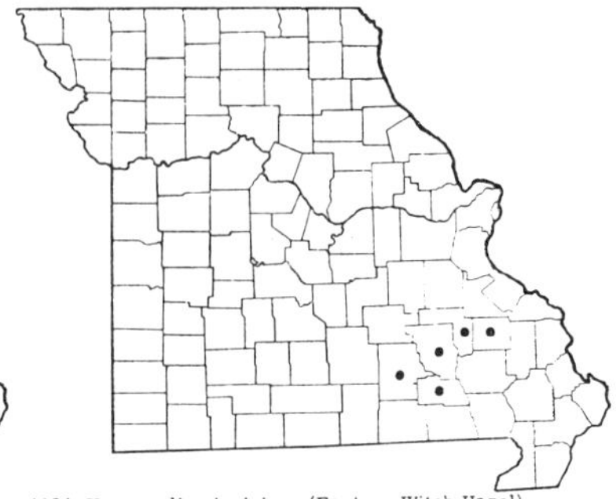

1134 Hamamelis virginiana (Eastern Witch Hazel)

and Barry counties, along Jack's Fork of Current River in Shannon County (west-facing bluffs, T29N, R4W, W½ sect. 25, 1 mi. northeast of Eminence, June 12, 1950, *Steyermark 69942*; south-facing bluffs, T29N, R3W, north part sect. 20, 4½ mi. northeast of Eminence, *Steyermark 69943*), and introduced along railroads in St. Louis County (Mo. Pac. R. R. north of Dover St., along upper terrace of Mississippi River, September 18, 1957, *Muehlenbach 1365*).

Ranges from Minnesota to South Dakota, south to Missouri, Arkansas, Oklahoma, and Texas; widely cultivated and escaped from cultivation in the eastern states.

The flowers are very fragrant, resembling a combination spiciness of cloves and carnation, or suggesting also *Viburnum Carlesii* scent. The fragrance is sometimes evident only in the afternoon.

The berries are generally black, but a yellow-fruited form, f. *xanthocarpum* Rehd., occasionally occurs. This has not been recorded from Missouri.

## Fam. **HAMAMELIDACEAE** (Witch Hazel Family)

Large trees with a 5–7-pointed star-shaped leaf; flowers and fruits in dense globe-shaped heads; flowers greenish . . . . . . . . . . . . . . . . . . . . . . . . . . 2. LIQUIDAMBAR

Shrubs with wavy-toothed, unlobed leaves; flowers in small clusters scattered along the branches; flowers yellow, salmon, or reddish . . . . . . . . . . . . . . . . . . . . . . 1. HAMAMELIS

### 1. **Hamamelis** L. Witch Hazel

Flowering from September to the end of December; usually no part of flower with an orange or reddish tinge; adult leaves mostly fallen or mature at flowering time; twigs glabrous or slightly hairy; petals 15–20 mm. long, yellow; calyx-lobes brownish-yellow, yellowish-green, or greenish on inside; growing mostly on wooded north-facing hillsides, sometimes in valleys along streams . . . . . . 1. H. VIRGINIANA

Flowering from January to April; some part of the flower usually tinged with orange, reddish, or salmon; adult leaves usually developing at flowering time; twigs densely hairy or woolly; petals 10–15 mm. long, yellow with reddish or orange at base or reddish throughout; calyx-lobes reddish, orange, or salmon on inside; growing mostly in gravel or rocky stream-beds, or at the base of rocky slopes along streams, rarely on wooded hillsides . . . . . . . . . . . . . . . . . . . 2. H. VERNALIS

1. **Hamamelis virginiana** L.
Eastern Witch Hazel Map 1134
Also known as American Witch Hazel.
Flowers September–December.
Occurs in rich woods on north-facing slopes or in wooded valleys along streams. Rare, and restricted to the southeastern Ozarks in Madison, Iron, Reynolds, Shannon, and Carter counties.

Ranges from Quebec to Minnesota, south to Georgia, Tennessee, and Missouri.

---

Plate no. 192. **1.** Itea virginica, × 2/5; Details from Small, The New York Botanical Garden. **2.** Ribes Cynosbati, × 2/5; *a.* Flower, × 2 4/5. **3.** Ribes missouriense var. missouriense, × 2/5. **4.** Ribes odoratum, × 2/5; Details from Small, The New York Botanical Garden. **5.** Hamamelis virginiana, × 2/5; Details from Small, The New York Botanical Garden. **6.** Hamamelis vernalis, × 2/5; *a, b.* Two types of leaves; *c.* Fruit.

PLATE NO. 192

The extract of commercial witch hazel, commonly employed as a toilet water and after-shaving lotion, is obtained from the branches of this shrub. The bark and leaves are used in medicine for their astringent properties. In pioneer days it was traditional to employ the forked branch of a witch hazel to search for a source of water or for mineral deposits, much as Geiger counters are used today to hunt uranium deposits.

The foliage in autumn turns a pale to rich yellow or yellow tinged with orange. The flowers have a delicate fragrance. The flowers can withstand considerable freezing weather, and after fertilization, develop slowly into the fruit. The process of fruit formation requires nearly a year for complete development. It is remarkable that the ovary, following fertilization in the late autumn, can survive subzero weather for long periods during the winter and still continue the process of growth into a mature fruit the following year. The fruit, when ripe, splits open suddenly, throwing the black seeds as much as 30 feet away.

2. **Hamamelis vernalis** Sarg.
Vernal Witch Hazel Map 1135
Also called Ozark Witch Hazel.
Flowers January–April.

Occurs in gravel and rocky stream-beds, at the base of rocky slopes along streams, and rarely on wooded hillsides in rocky draws. Ozark region of southern and east-central Missouri, north to Ste. Genevieve, Washington, Franklin, Phelps, Pulaski, Laclede, and Dallas counties, west to Webster, Christian, Stone, Barry, and McDonald counties.

Ranges from Missouri and Arkansas to Oklahoma; doubtfully in Louisiana.

The following variations occur in Missouri:

a. Petals, inside of calyx, and stamens entirely dark red . . . . 2c. H. VERNALIS f. CARNEA
a. Petals, inside of calyx, and stamens orange, salmon, orange with yellow, or other combinations, not all dark red . . . . . . *b*
b. Lower surface of leaves glabrous (without hairs) or slightly hairy . . 2a. H. VERNALIS f. VERNALIS
b. Lower surface of leaves densely velvety-hairy . . . 2b. H. VERNALIS f. TOMENTELLA

2a. **Hamamelis vernalis** f. **vernalis** Map 1135
*Hamamelis vernalis* Sarg. [G, BB, P & S, Steyerm.]
This is the common variation found.

2b. **Hamamelis vernalis** f. **carnea** Rehd. Map 1135

Occasionally found in the range of f. *vernalis*, in Madison and Iron counties.

2c. **Hamamelis vernalis** f. **tomentella** Rehd. Map 1135

Occasionally found in the range of f. *vernalis*, in Taney, Stone, and Barry counties.

The flowers are quite fragrant, and are the first of any woody species to be found in bloom in Missouri. The foliage turns pale yellow in the autumn. Occasional overlapping occurs in habitats between this species and *H. virginiana*. Such a locality is found along the 'Johnson shut-ins' in Wilderness State Park along the East Fork of Black River in Reynolds County. At this place the *H. virginiana* is commonest on the adjacent rocky wooded slopes, descending the wooded valley to the edges of the rocky stream-bed, while *H. vernalis* keeps to the rockiest portions of the stream bar and never occurs on the adjacent wooded slopes. There apparently is no opportunity for hybridization to take place, since the time of flowering is quite different for the two species (see Steyermark in Rh. 36: 97–100. 1934).

This species spreads by a suckering habit and forms extensive stands along most of the gravel beds in the Ozarks. It makes a handsome shrub in cultivation and transplants readily, enjoying full sun and well-drained rocky situations.

## 2. **Liquidambar** L. Sweet Gum

**Liquidambar Stryraciflua** L. Sweet Gum Map 1136
Also called Red Gum, Bilsted, or Gum.
Flowers April–May.

Occurs in low woods along streams in the southeastern Ozarks and wet or flat rich woodland in the lowland counties of southeastern Missouri, rarely introduced along railroads. Restricted to the southeastern section of the state, throughout the lowland counties north along the border of the southeastern Ozarks to Perry, Madison, Wayne, Ripley, and Oregon counties; introduced along railroads in St. Louis County (St. Louis, Terminal Railroad Association, west of Broadway at the 2nd gate of the Hussman Refrigerator Plant, September 28, 1957, *Muehlenbach 1386*).

Ranges from Central America and Mexico into the United States from Florida to Texas, north to

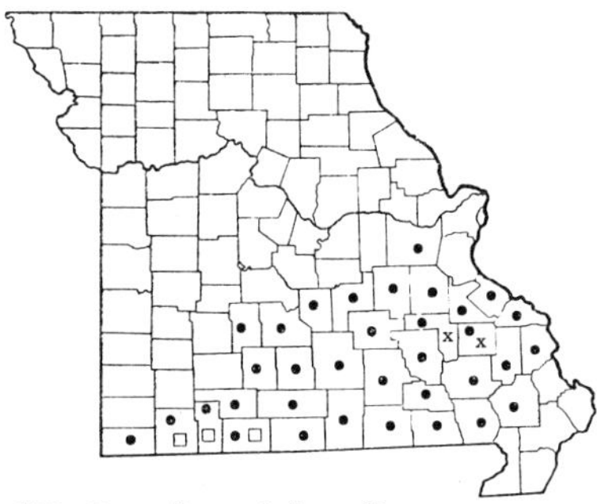

1135 • Hamamelis vernalis f. vernalis
1135 x Hamamelis vernalis f. carnea
1135 □ Hamamelis vernalis f. tomentella

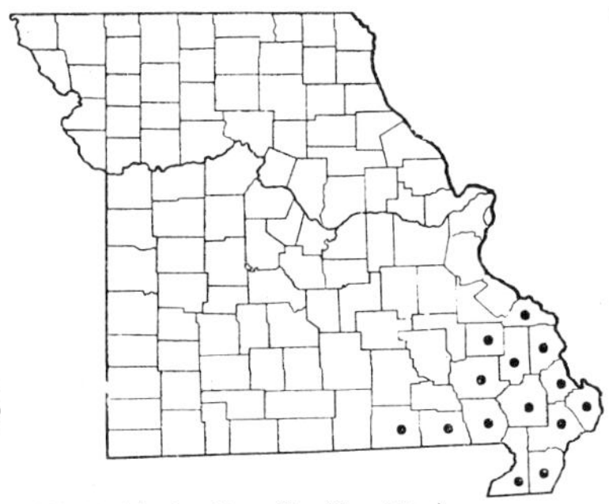

1136 Liquidambar Styraciflua (Sweet Gum)

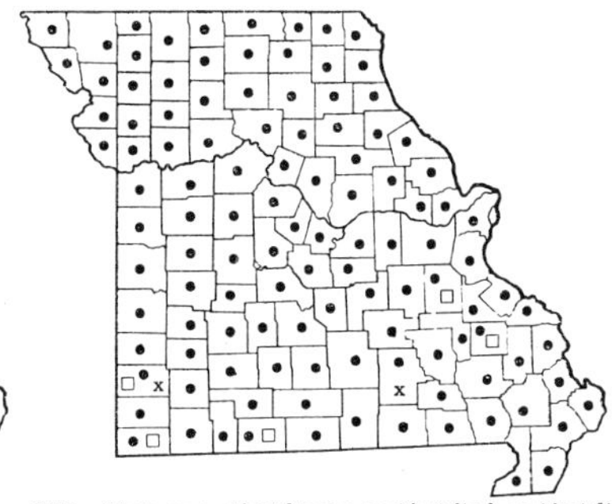

1137 • Platanus occidentalis var. occidentalis f. occidentalis
1137 x Platanus occidentalis var. occidentalis f. attenuata
1137 □ Platanus occidentalis var. glabrata

Connecticut, New York, Ohio, Indiana, Illinois, Missouri, and Oklahoma.

The leaves have a pleasant fragrance when bruised, and in autumn give a magnificent display of varied coloring, some of the leaves remaining green and yellow while others are rose-red, orange and purple, the foliage eventually becoming purple and crimson or deep red. In some parts of Central America a decoction of the leaves is used for bathing.

The fragrant but rather bitter resinous sap of the tree hardens into a gum which is sometimes chewed, and has been used in domestic medicine for treating catarrh, coughs, dysentery, sores and wounds of people and domestic animals, and as a base in various salves. This resinous substance is also employed in flavoring tobacco, and used in soaps, perfumes and cosmetics, adhesives and lacquers. In some countries it is used as an incense in the temples and houses.

The wood was at one time extensively cut for use in making tobacco boxes and crates, but is now more commonly employed in the lumber industry for cabinet wood, furniture, veneers, interior finish, barrels, and wooden dishes. In the lumber trade it is sometimes called Satin Walnut or used as an imitation Circassian Walnut. In the southern states it is now generally used for pulpwood in papermaking.

The largest known tree in Missouri as well as in the United States, according to Mr. Kendall Laughlin, is located in New Madrid, New Madrid County, on the property of Laura Dawson, and measures 16 feet, 11 inches in diameter (measured 54 inches above the ground) with a height of 112 feet.

## Fam. **PLATANACEAE** (Plane Tree Family)

### **Platanus** L. Sycamore, Plane Tree

**Platanus occidentalis** L. Sycamore, Plane Tree
Map 1137

Also called Buttonwood.

Flowers April–June.

Occurs usually in valleys along streams, either on gravel bars or in alluvial soils of low or wet woodland, occasionally on moist places on upland slopes, bluffs, and ridges. Throughout Missouri, occurring in every county.

The following variations are found in Missouri:

a. Leaves with 3 main, long-pointed lobes which usually lack teeth (entire), the 2 lateral (side) lobes sometimes with 1 or 2 smaller lobes . c. P. OCCIDENTALIS var. GLABRATA

a. Leaves with 3–5 main lobes furnished with shallow teeth . . . . . . . . . . *b*

b. Base of leaf truncate (as if cut straight across) or cordate (heart-shaped); common type . . . . . . . a. P. OCCIDENTALIS var. OCCIDENTALIS f. OCCIDENTALIS

b. Base of leaf conspicuously long-tapered or narrowed to its junction with the leaf-stalk; rare type . . . . . b. P. OCCIDENTALIS var. OCCIDENTALIS f. ATTENUATA

a. **Platanus occidentalis** var. **occidentalis** f. **occidentalis** Map 1137

*Platanus occidentalis* L. [G, BB, P & S, Steyerm.]

This is the commonest variation of the species in the state. Found in every county.

Ranges from Maine and Ontario to Michigan, Minnesota, and Nebraska, south to Florida and Texas.

b. **Platanus occidentalis** var. **occidentalis** f. **attenuata** Sarg. Map 1137

*Platanus occidentalis* f. *attenuata* Sarg. [G, P & S, Steyerm.]

Known from Shannon and Jasper counties in southern Missouri, but to be expected over the greater part of its range.

c. **Platanus occidentalis** var. **glabrata** (Fern.) Sarg. Map 1137

Known from a few counties in the southern half of the state.

Ranges from Iowa and Missouri to Texas and Mexico.

The Sycamore is the largest deciduous tree in the United States, attaining a greater trunk diameter than any other species. It is not the tallest, however. The largest specimen recorded in Missouri is 26 feet, 10 inches in circumference (d.b.h.), located in what was formerly a virgin timber tract in Mississippi County (Bull. Mo. Bot. Gard. 20: 151. *pl. 48, figs. 1 and 2.* 1932). In a list furnished the author by Kendall Laughlin, the largest sycamore known to him measured 16 feet, 2 inches in circumference (d.b.h.), and 109 feet tall, and is located in Kansas City, Jackson County. The largest tree recorded in the United States was located near Worthington, Indiana, and had a circumference of 43 feet, 3 inches at 5 feet above the ground (Deam, *Trees of Ind.* 2nd ed. p. 182. 1932), although a trunk measured by Francois Michaux in 1802 in Ohio was recorded to be 47 feet in circumference 4 feet above the ground.

The foliage in autumn may be green mixed with an orange-brown or pale yellow and green, or eventually changes to yellow or orange-brown. The flowers are wind-pollinated and shed large quantities of pollen from the staminate flowers. It is reported that sycamore pollen is exceeded only by oaks in its importance in being a cause of some cases of hay fever.

The trunk, which is often hollow in large trees, was frequently employed by early American settlers for storing grain and smoking meat. At one time it was the chief wood used for tobacco boxes and ox-yokes, and later became widely employed for butcher blocks, since it practically does not split when subjected to constant chopping. Other uses for which it serves are for furniture, interior finish, veneer, piling, boxes and crates, wooden ware, barrels, and handles for brushes and other articles. Small amounts are used with other hardwoods for pulp in papermaking. It sometimes enters the lumber industry as Lacewood, because of a mottled shiny appearance in the quarter-sawed state. The sweet sap of the tree has been used by some Indian tribes for making a sugar and syrup.

It is recorded that the now extinct Carolina paroquet relished the fruit of this tree as its favorite food. The fruit which hangs on the tree during the winter is also eaten by many other species of birds, including chickadees and juncos.

---

Plate no. 193. **1.** Liquidambar Styraciflua, × 2/5; Details from Small, The New York Botanical Garden. **2.** Platanus occidentalis, × 2/5; Details from Small, The New York Botanical Garden. **3.** Physocarpus opulifolius var. intermedius, × 2/5; Details from Small, the New York Botanical Garden. **4.** Spiraea tomentosa var. rosea, × 2/5. **5.** Spiraea prunifolia var. plena, × 2/5; *a.* Flowering branch; *b.* Leafy branch; After Gleason, The New York Botanical Garden. **6.** Spiraea alba, × 2/5; Details from Small, The New York Botanical Garden.

PLATE NO. 193

## Fam. **ROSACEAE** (Rose Family)

Some authors divide this family into Malaceae, Amygdalaceae, and Rosaceae, the first including Crataegus, Amelanchier, Pyrus, Malus among Missouri genera, Amygdalaceae to include Prunus, and the remainder of the Missouri genera in Rosaceae.

a. Trees, shrubs, low or tall woody-stemmed plants, or plants with spines or prickles . . . . . . . . *b*
  b. Leaves divided into 3 or more separate divisions (leaflets) . . . . . . . . . . . . . . . *c*
    c. Petals usually white, rarely pink or rose-tinged; leaves palmately compound (the divisions starting from the same point at the summit of the leaf-stalk) with 3–5 leaflets; fruit usually black when ripe, rarely white, made up of a cluster of fleshy 1-seeded little drupes massed upon a conical or convex receptacle . . . . . . . . . . . . . . . . . . 14. Rubus
    c. Petals usually rose-colored or pink, rarely white; leaves pinnately compound (the divisions arranged feather-like with an odd one at the tip) with 3–11 leaflets; fruit red or scarlet, consisting of an apple- or pear-shaped fleshy outer calyx-tube enclosing bony seed-like achenes . 17. Rosa
  b. Leaves either toothed or with lobes or deep cuts, not divided into separate leaflets . . . . . . *d*
    d. Stems arching, ascending, or trailing, but not upright, with or without a few prickles or bristles; fruit usually black when ripe, made up of a cluster of fleshy 1-seeded little drupes massed upon a conical or convex receptacle . . . . . . . . . . . . . . 14. Rubus
    d. Without the above combination of characters . . . . . . . . . . . . . . . . . . *e*
      e. Ovary or ovaries superior, not concealed within the calyx-tube (hypanthium) which is saucer-shaped to globe-shaped; fruit superior, usually with the remains of the calyx persisting at its base . . . . . . . . . . . . . . . . . . . . . . . *f*
      f. Pistil or ovary 1 in each flower . . . . . . . . . . . . . . . . 18. Prunus
      f. Pistils or ovaries 2–5 in each flower . . . . . . . . . . . . . . . . . . *g*
        g. Petals lacking; calyx-lobes prominent, petal-like, green, white to greenish-white, leafy, 7–12 mm. long; one expanded flower 1.8–2.5 cm. across; fruit consisting of fleshy indehiscent drupe-like achenes . . . . . . . . . . . . . 15. Neviusia
        g. Petals present, white, pink, or rose-colored; calyx-lobes small, green, 1–3 mm. long; one expanded flower less than 1 cm. across; fruit consisting of dry dehiscent inflated or non-inflated follicles . . . . . . . . . . . . . . . . . . . . . *h*
          h. Some of the leaves usually lobed, 3 or more principal veins arising at the base (palmately veined); stipules usually present at the base of young leaves; fruit inflated, 7–10 mm. long; bark peeling off, shreddy; plants usually along rocky banks and bluffs of streams throughout the Ozark section and eastern Missouri 1. Physocarpus
          h. All the leaves merely toothed, not lobed, the principal veins usually feather-like (pinnate) in arrangement; stipules absent; fruit not inflated, 2.5–3 mm. long; bark not peeling or shreddy; plants of river bottom meadows and alluvial bottoms of northern and western Missouri, southeastern Missouri lowlands, or introduced cultivated plants escaped and naturalized . . . . . . . . . . . 2. Spiraea
      e. Ovary or ovaries inferior, concealed within the calyx-tube (hypanthium); fruit inferior with the seeds or achenes within the hypanthium and the calyx-lobes or their remains at the top . . . . . . . . . . . . . . . . . . . . . . . . . . *i*
      i. Thorns present on at least some of the branches . . . . . . . . . . . . . . *j*
        j. Styles distinct; ovules 1 in each cell; mature carpels hard, stony, seed-like; stamens 5–20 (rarely 25) . . . . . . . . . . . . . . . . . . . . . 7. Crataegus
        j. Styles joined together at their bases (free or separate to base in *P. communis*); ovules 2 in each cell; mature carpels leathery or papery, opening easily; stamens 15–50 . 5. Pyrus
      i. Thorns absent on the branches . . . . . . . . . . . . . . . . . . . *k*
        k. Flowers in racemes mostly longer than broad; petals usually at least twice as long as broad; ovary with 6–10 cells; ovule 1 in each cell; fruit apparently 10-celled, usually with 10 seeds; stamens usually 20 . . . . . . . . . . . . . 6. Amelanchier
        k. Flowers not in racemes, usually broader than long; petals slightly longer than broad; ovary with 2–5 cells; ovules 2 in each cell; fruit 2–5-celled, each carpel with 1 or 2 seeds; stamens 15–50 . . . . . . . . . . . . . . . . . . . . 5. Pyrus

a. Soft-stemmed herbaceous plants or stems dying to the ground each year . . . . . . . . . . *l*
  l. All or most all the leaves appearing to be at the base of the plant . . . . . . . . . . *m*
    m. Petals yellow; at least the middle (terminal) one of the 3 leaflets as broad as long; leaflets with

unequal, irregular teeth, somewhat lobed; main side nerves of a leaflet 2–4 on each side of midrib, connected with the other nerves by secondary nerves; all the leaflets wedge-shaped (cuneate) and narrowed to the base; ovaries 2–10; achenes on a dry, slightly enlarged and not brightly-colored receptacle; very rarely found . . . . . . . . . . . . . . . . 10. WALDSTEINIA

m. Petals white; all leaflets longer than broad; leaflets with all the teeth uniform and more or less equal, no lobes present; main side nerves of a leaflet 5–7 or more on each side of midrib, mostly not connected by secondary nerves; lower side at base of outer two leaflets (the lateral ones) curved or rounded; ovaries many, more than 50; achenes on a juicy, greatly enlarged, red receptacle; commonly encountered . . . . . . . . . . . . . . . . . . . 8. FRAGARIA

l. Some leaves on the flowering stem as well as at the base of the plant . . . . . . . . . . . *n*

n. All the leaves divided into 3 leaflets (2 large stipules at the base of the 3 leaflets appear in *Gillenia* and may be mistaken for 2 extra leaflets) . . . . . . . . . . . . . . . . . . . . *o*

o. No green bractlets present at the base of the calyx nor alternating with the calyx-lobes; petals long and narrow, 10–22 mm. long, pointed at the tip, much longer than the calyx-lobes; ovaries 2–10; fruit a dehiscent follicle . . . . . . . . . . . . . . . . . 4. GILLENIA

o. Green bractlets present at the base of the calyx or alternating with the calyx-lobes; petals short and broad, at most 9 mm. long, mostly rounded or blunt at tip, shorter than or equal to somewhat longer than the calyx-lobes; ovaries numerous, more than 10; fruit an indehiscent achene or drupe . . . . . . . . . . . . . . . . . . . . . . . . . . . *p*

p. Petals white; style long and slender, bent near the middle, persisting after the petals have fallen, and at maturity hooked at the summit . . . . . . . . . . . . . 13. GEUM

p. Petals yellow, yellow white, or creamy; styles short and inconspicuous, falling after the flowering period . . . . . . . . . . . . . . . . . . . . . . . . . *q*

q. The green bractlets much larger than the calyx-lobes with which they alternate; flowers all solitary, arising from the leafy joints along the stem; each of the 3 leaflets with individual stalks (petiolules) before joining the main leaf-stalk (petiole); receptacle spongy, red, and greatly enlarged at maturity, bearing the numerous achenes on its surface . . . . . . . . . . . . . . . . . . . . . 9. DUCHESNEA

q. The green bractlets about equal in size to the calyx-lobes with which they alternate; flowers arranged in a leafy or paniculate cyme, usually 2 or more in an inflorescence; receptacle dry, not red, scarcely enlarged . . . . . . . . . . . 11. POTENTILLA

n. Some of the leaves divided into 5 or more leaflets . . . . . . . . . . . . . . . . *r*

r. Leaves palmately compound (the base of the leaflets arising from approximately the same point) . . . . . . . . . . . . . . . . . . . . . . . . . . . . s

s. Style long and slender, bent near the middle, persisting after the petals have fallen, and at maturity hooked at the summit . . . . . . . . . . . . . . . . 13. GEUM

s. Style short and inconspicuous, falling after the flowering period . . . . . . . 11. POTENTILLA

r. At least the lowest leaves or some of the leaves of the stem pinnately compound (the leaflets arranged feather-like with an odd one at the top) or ternately 2 or 3 times pinnately divided . . . *t*

t. Stipules absent; leaves ternately 2 or 3 times pinnately divided; flowers dioecious, the male (staminate) and female (pistillate) on separate plants; fruit consisting of dehiscent follicles 3. ARUNCUS

t. Stipules present at the base of the leaf; leaves simply once pinnately divided; flowers perfect, the stamens and pistils in the same flower; fruit consisting of indehiscent achenes . . . *u*

u. No green bractlets present at the base of the calyx nor alternating with the calyx-lobes; pistils 1–15 . . . . . . . . . . . . . . . . . . . . . . . . . *v*

v. Petals yellow; summit of calyx-tube with hooked bristles; pistils 1–4 . . . . 16. ARGIMONIA

v. Petals pink; calyx not bristly; pistils 5–15 . . . . . . . . . . 12. FILIPENDULA

u. Green bractlets present at the base of the calyx or alternating with the calyx-lobes; pistils numerous, 30–160 . . . . . . . . . . . . . . . . . . . . . *w*

w. Styles short and inconspicuous, falling after the flowering period. . . . . 11. POTENTILLA

w. Styles long and slender, bent near the middle, persisting after the petals have fallen, and at maturity hooked at the summit . . . . . . . . . . . . . 13. GEUM

## 1. **Physocarpus** Maxim. Ninebark

**Physocarpus opulifolius** (L.) Maxim. var. **intermedius** (Rydb.) Robins. Ninebark Map 1138

Flowers May–June.

Occurs on gravel bars, rocky banks, and bluffs

1138 Physocarpus opulifolius var. intermedius (Ninebark)

1139 Spiraea prunifolia var. plena

1140 Spiraea alba (Meadow-sweet)

along streams and moist thickets. Eastern, central, and southern Missouri, north to Jackson, Saline, Boone, Shelby, and Clark counties.

Ranges from New York to Minnesota and Colorado, south to Indiana, Illinois, Arkansas, and Oklahoma.

This variety differs from var. *opulifolius* in having the fruits permanently hairy instead of glabrous.

The shrub is often cultivated. The foliage turns a yellow-green in autumn.

## 2. **Spiraea** L. Spiraea

a. Flowers commonly 'double,' the stamens and pistils mostly changed to petals; flowers 2–6 in umbel-like inflorescences arising along the sides of the branches of the previous season; flowers usually appearing before the leaves; leaves elliptic-oblong, less than 2 times as long as broad . . . . 1. S. PRUNIFOLIA var. PLENA

a. Without the above combination of characters; flowers single, the usual stamens and pistils present; flowers numerous in longer than wide panicles arising from the end of the branches of the new season; flowers appearing after the leaves are fully developed; leaves 1½–6 times as long as broad. . . *b*

b. Flowers white; leaves glabrous (without hairs) or nearly so on the lower surface . . . . 2. S. ALBA

b. Flowers rose-colored or pink; leaves more or less densely hairy on the lower surface . . . . . . *c*

c. Fruit (follicle) hairy; leaves obtuse (blunt) or rounded at their summits; native plant of Dunklin County, southeastern Missouri. . . . . . . . . . . . . . . 4. S. TOMENTOSA

c. Fruit (follicle) glabrous (without hairs); leaves acutely pointed at their summits; cultivated foreign plants, escaped along roadsides in various sections of Missouri . . . . . 3. S. DOUGLASII

1. **Spiraea prunifolia** Sieb. & Zucc. var. **plena** Schneid. Map 1139

*Spiraea prunifolia* var. *flore-pleno* Hort.

Flowers April–May.

Commonly cultivated in old gardens and cemeteries, and occasionally escaped to wooded sections, fence rows, and roadsides in Lewis (fence rows above ravines bordering Wyaconda River, T62N, R6W, sect. 18, 7 mi. northeast of Canton, July 27, 1952, *Steyermark 74161*), Cole (escaped from old cemetery in woods on ridge tops and established as a shrub forming thickets on north-facing, steep, wooded bluffs along Missouri River, T44N, R10W, sect. 16, just north and northwest of Osage City, May 20, 1950, *Steyermark 69747*), and Jefferson (along highway 110, 4 mi. from junction of highways 67 and 110, east of DeSoto, April 4, 1956, *Steyermark 80796*) counties.

Native of eastern Asia.

Plants usually occur in colonies and bear flowers on simple long branches. The leaves, which follow the flowers, are elliptic to ovate-oblong and finely toothed, and pointed at each end.

2. **Spiraea alba** DuRoi Meadow-sweet Map 1140

Flowers June 12–August.

Occurs in wet river bottom prairies, wet prairies along railroads, alluvial soils bordering oxbow lakes of rivers in valley terrain, and open ground along streams. Rare in northern Missouri in Shelby, Linn, Grundy, and Harrison counties, and locally in western Missouri south of the Missouri River in St. Clair County (lake along Fowler Branch, T37N, R28W, sect. 6, ½ mi. south of Osage River, 3 mi. west of Taberville, September 27, 1938, *Steyermark 9587*). At the last locality it was growing in an alluvial flat around shade of Buttonbush (*Cephalanthus*).

Ranges from Quebec and Vermont to Saskatchewan, south to Virginia, North Carolina, Ohio, In-

1141 Spiraea Douglasii

1142 Spiraea tomentosa var. rosea (Steeple Bush)

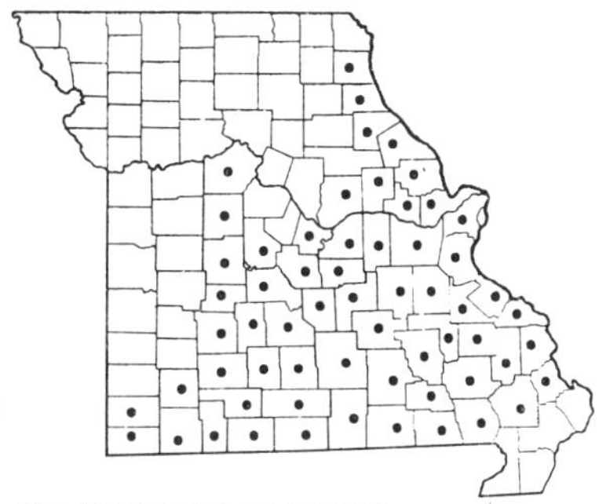

1143 Aruncus dioicus var. pubescens

diana, Illinois, Missouri, and South Dakota.

This species occurs in thickets about 3–4 feet tall, usually growing in full sun. It is reported as occurring also in Livingston County.

3. **Spiraea Douglasii** Hook. Map 1141
Flowers June–August.

Commonly cultivated, and sometimes escaping along roadsides in Cape Girardeau County (meadow 1 mi. north of campus of Southeast Missouri State College, June 25, 1953, *Donald Rhodes 145*).

Native of western North America and introduced into cultivation.

This collection has the lower surface of the leaves grayish pubescent and approaches *S. Menziesii* Hook. f. *eriophylla* Zab., which is sometimes placed varietally under *S. Douglasii*.

4. **Spiraea tomentosa** L. var. **rosea** (Raf.) Fern.
Hardhack, Steeple Bush Map 1142
Flowers June–August.

Known only from Crowley Ridge in Dunklin County, southeastern Missouri (common, Malden, July 22, 1895, *Bush 108*); not collected in the state since.

Ranges from Ontario to Manitoba, south to Virginia, Georgia, Tennessee, and Arkansas.

In the Missouri material the leaves are shallowly and coarsely serrate with few teeth along the margins. This is an ornamental plant, but is difficult to grow unless provided with a moist acid soil.

## 3. **Aruncus** Adans. Goat's Beard

**Aruncus dioicus** (Walt.) Fern. var. **pubescens** (Rydb.) Fern. Goat's Beard Map 1143
*Aruncus sylvester* of auth. in part [P & S], not Kostel.
*Aruncus allegheniensis* Rydb. var. *pubescens* (Rydb.) Fern. [Steyerm.]
*Aruncus dioicus* in part [BB], not (Walt.) Fern.
Flowers May–July.

Occurs along bluffs, base of wooded slopes, and in rocky or moist woodland. Ozark region of southern and central Missouri west to Saline, Pettis, Benton, Hickory, Polk, Greene, Lawrence, and Newton counties, north to Callaway and the other Ozark border counties, northeast to Lewis County.

Ranges from Kentucky, Illinois, and Iowa, south to Arkansas and Oklahoma.

This plant is easily transplanted or raised from seed and becomes well established in shaded borders and wild gardens. The creamy-white flowers are borne in elongated showy inflorescences. In autumn the foliage becomes a pale yellow-green or greenish-yellow.

## 4. **Gillenia** Moench Indian Physic

Stipules leaf-like, ovate, coarsely or deeply toothed; lower surface of leaflets more or less covered with glands; petals white, 10–13 mm. long; common in southern, central, and eastern Missouri . . 1. G. STIPULATA

Stipules narrow, inconspicuous, less than 5 mm. wide, smooth-edged or with slight toothing; lower surface of leaflets without glands; petals pink, 12–22 mm. long; rare, known only from Lawrence Co., southwestern Missouri . . . . . . . . . . . . . . . . . . . . . . . 2. G. TRIFOLIATA

1. **Gillenia stipulata** (Muhl.) Trel.
Indian Physic Map 1144

Also called American Ipecac.
Flowers May–July.

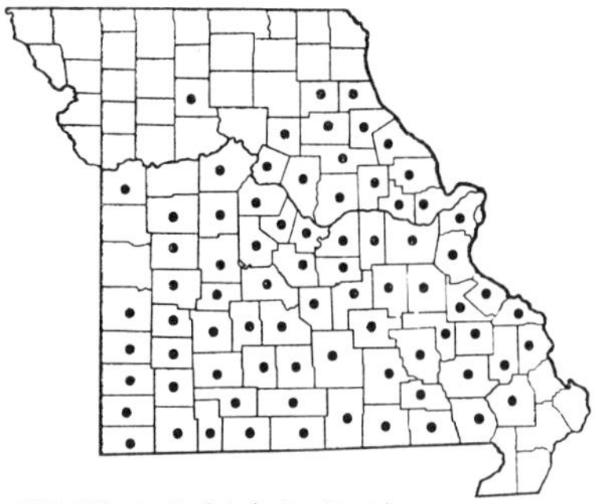
1144 Gillenia stipulata (Indian Physic)

1145 Gillenia trifoliata (Indian Physic)

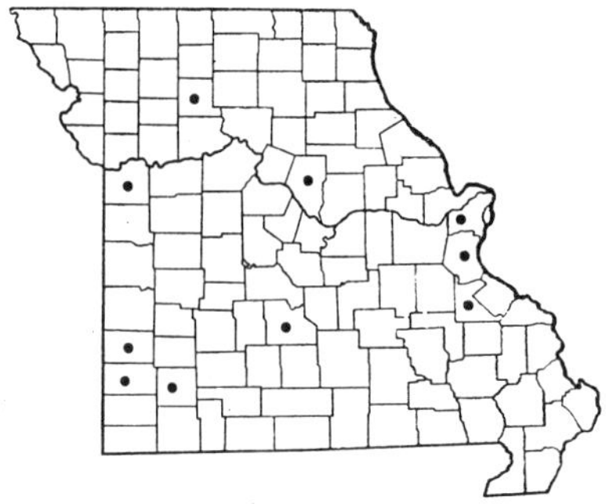
1146 Pyrus communis (Pear)

Occurs in usually acid soils of dry, rocky, or upland and open woods. Southern, central, and northeastern Missouri, north to Jackson, Livingston, Randolph, Shelby, and Marion counties.

Ranges from Georgia and Alabama to Texas, north to New York, Ohio, Indiana, Illinois, Missouri, and Kansas.

This has showy pointed petals protruding from the bell-shaped calyx. It has established and seeded itself in the author's wildflower preserve in northern Illinois in an acid rocky soil where it receives some sunlight during the day.

2. **Gillenia trifoliata** (L.) Moench
Indian Physic, Bowman's Root Map 1145
Flowers May–June.

Known only in southwestern Missouri from Lawrence County (open rocky bank, border of woods, along right-of-way of the Mo. Pac. R. R., ½ mi. southwest of Bonham Siding, T25N, R24W, NW corner sect. 26, May 14, 1951, *Palmer 59134*; May 14, 1955, *Palmer 59873*).

Ranges from New York to Ontario and Michigan, south to Georgia, Alabama, Kentucky, and Missouri.

This species has larger petals than *G. stipulata* and comes into flower a couple of weeks earlier.

## 5. **Pyrus** L.

*Aronia* Medic [BB]

*Malus* Mill. [P & S, Steyerm.]

The present treatment retains the genus *Pyrus* in the sense employed by Fernald in the eighth edition of *Gray's Manual* (p. 757), which recognizes as subgenera, *Malus* (Apples), *Aronia* (Chokeberries), *Sorbus* (Mountain Ash), and *Pirophorum* (Pears) instead of as genera, as sometimes maintained by Lawrence (Gent. Herb. 8: 40–43. 1949) and other authors. Gleason in the *New Ill. Fl.* (2: 334. 1952) combines the pears and apples under *Pyrus*, while separating out *Aronia* and *Sorbus* as distinct genera.

a. Midrib of the upper surface of the leaves with glands present; petals chiefly less than 10 mm. long and 7 mm. broad; fruit small, black, 7–10 mm. in diameter. . . . . . . . . 6. P. MELANOCARPA

a. Midrib of the upper surface of the leaves without any glands; petals 10–20 mm. long, 8–15 mm. broad; fruit green, yellowish-green, or red, 8–30 mm. or more in diameter . . . . . . . . . *b*

b. Petals white; flower-clusters compound with a short stout central column from which flower-stalks arise; anthers reddish; winter-buds mostly glabrous; a cushion-like thickening (disk) partially or almost completely closes the summit of the green cup-shaped (receptacle) part of flower; styles free or separate to their base; grit cells present in flesh of fruit (observed as a ring when fruit is cut cross-wise); fruit pear-shaped (pyriform) . . . . . . . . . . 1. P. COMMUNIS

b. Petals usually rosy or pink, fading to white, or rarely white; flower-clusters mainly simple, without a columnar central stalk; anthers yellow or red; winter-buds hairy; green cup-shaped (receptacle) part of flower without any cushion-like thickening at the summit obstructing the opening; bases

---

Plate no. 194. **1.** Aruncus dioicus var. pubescens, × $^{2}/_{5}$; Details from Small, The New York Botanical Garden. **2.** Pyrus communis, × $^{2}/_{5}$; Details from Small, The New York Botanical Garden. **3.** Gillenia stipulata, × $^{2}/_{5}$; Details from Small, The New York Botanical Garden. **4.** Pyrus Malus, × $^{2}/_{5}$. **5.** Pyrus coronaria var. coronaria, × $^{2}/_{5}$.

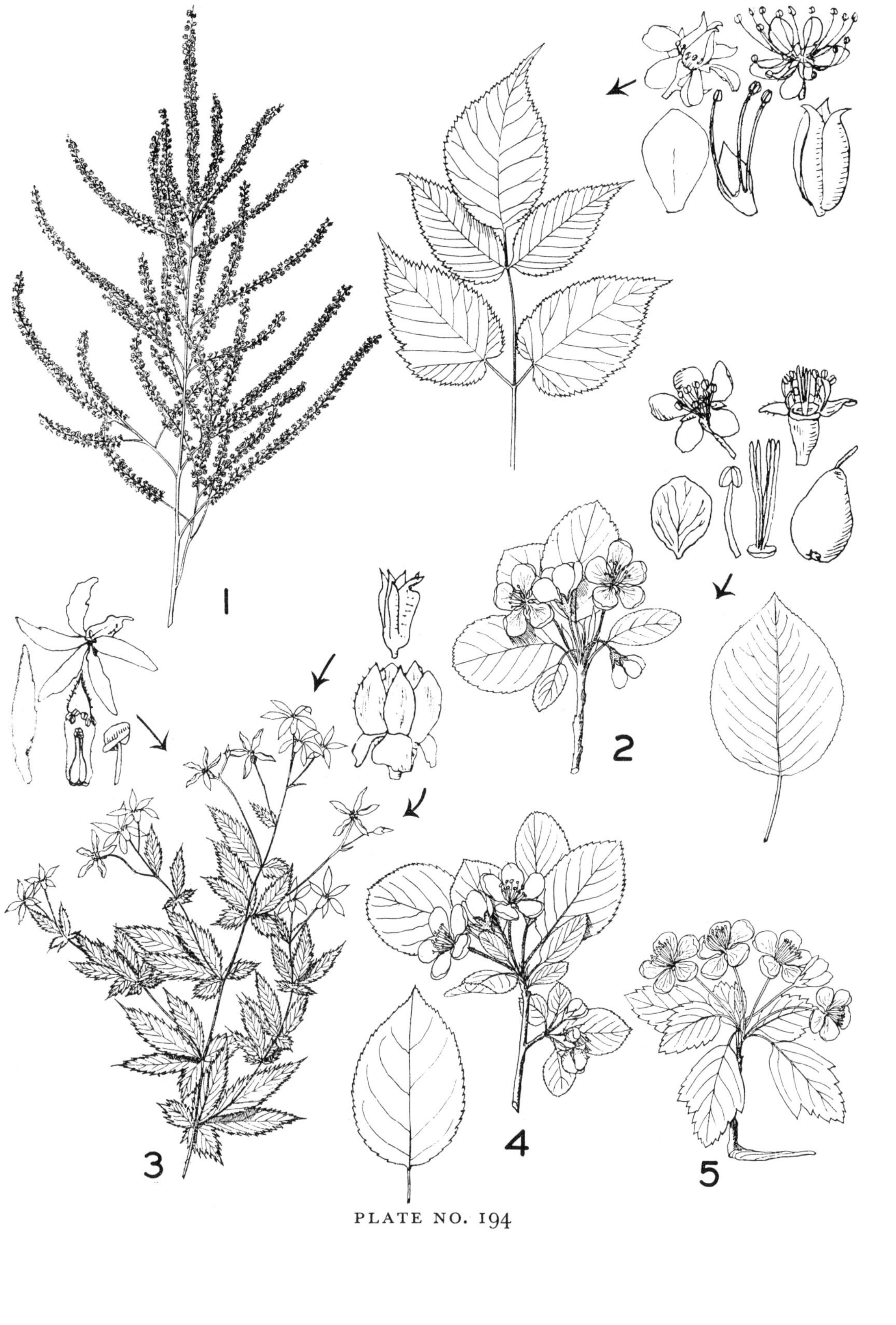


PLATE NO. 194

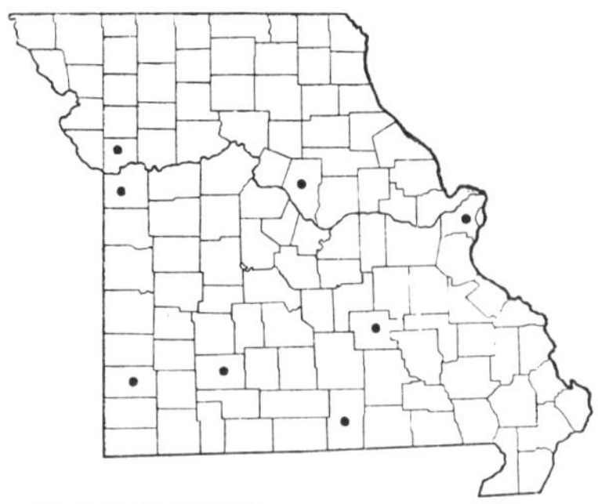

1147 Pyrus Malus (Apple)

1148 Pyrus angustifolia var. spinosa (Wild Crab)

1149 • Pyrus coronaria var. coronaria (Wild Crab)
1149 □ Pyrus coronaria var. lancifolia

of styles joined together; no grit cells present in flesh of fruit; fruit more or less round or globe-shaped, apple-like. . . . . . . . . . . . . . . . . . . . . . . . . . . . . *c*

c. Anthers yellow; leaves crenate-serrate with rounded or obtuse teeth; leaves similar in shape and toothing on both vegetative and fruiting branches; flower-stalks (pedicels) 1.4–2 mm. in diam. at time of flowering; calyx-lobes 2.5–4.5 mm. wide at the base only. . . . . 2. P. MALUS

c. Anthers red; leaves more or less sharply toothed, those on the vegetative (sterile) branches more coarsely cut, toothed or lobed than those of fruiting branches; leaves and teeth of leaves of both vegetative and fruiting branches variable in size, not similar; flower-stalks (pedicels) more slender, 1 mm. or less in diam. at time of flowering; calyx-lobes 1–2 mm. wide at the base only . . . . . . . . . . . . . . . . . . . . . . . . . . . . *d*

d. Green cup-shaped calyx-like receptacle and calyx-lobes densely hairy on the outside . . . . *e*

e. Common type found; fruit 2.5–3.5 cm. in diameter . . . . . . . . . . 5. P. IOENSIS

e. Rare type found; fruit 5 or more cm. in diameter . . . . . . . . . 5a. P. × SOULARDI

d. Green cup-shaped calyx-like receptacle and calyx-lobes glabrous (without hairs) or sparsely hairy . . . . . . . . . . . . . . . . . . . . . . . . . . *f*

f. Leaves of flower- or fruit-bearing branches rounded, blunt, or shortly mucronate at tip, with an oblong-lanceolate to narrowly elliptic shape . . . . . . . . 3. P. ANGUSTIFOLIA

f. Leaves of flower- or fruit-bearing branches usually acutely tipped, of a broadly ovate, oval, or broadly lanceolate shape . . . . . . . . . . . . . . 4. P. CORONARIA

**Pyrus communis** L. Pear Map 1146

Flowers April–May.

Commonly planted, and sometimes escaping to fence rows or persisting in the wild state in abandoned farm lots. Scattered in the state.

Native of Eurasia.

The leaves resemble those of *Prunus Mahaleb*. When not in flower, *Pyrus communis* may be distinguished from *Prunus Mahaleb* by its spiny-tipped branches and absence of glands between the teeth of the leaves.

2. **Pyrus Malus** L. Apple Map 1147

*Malus pumila* Mill. [P & S, Steyerm.]

Flowers April–May.

Commonly planted. Occasionally escaping along fence rows and banks of streams, or persisting in the wild state in abandoned farm lots. Scattered in the state.

Native of Eurasia.

**Pyrus angustifolia** Ait. var. **spinosa** (Rehd.) Bailey Wild Crab Map 1148

*Malus ioensis* var. *spinosa* Rehd. [P & S, Steyerm.]

*Pyrus angustifolia* in part [BB], not Ait.

Flowers April–May.

Occurs in lowland or upland woods and thickets, on Crowley Ridge in southeastern Missouri in Dunklin County (low woods, Campbell, April 22, 1912, *Bush 6630*, holotype, and *6631*; same locality, October 6, 1912, *Bush 6890*, in Gray Herbarium; Campbell, *Palmer 3537* in Mo. Bot. Gard. herb.).

Ranges from Florida to Louisiana, north to New Jersey, Kentucky, and Missouri.

Palmer has identified two sterile collections, deposited in the Gray Herbarium, as this variety (*Steyermark 10295* from Scott Co., and *Palmer & Steyermark 41713* from Oregon Co.), but both have definitely acute to acuminate leaves and are to be referred to one of the variations of *P. ioensis*. Typical *P. angustifolia* var. *angustifolia*, with branchlets and pedicels glabrous instead of densely hairy as in var. *spinosa*, is reported from Missouri in the eighth edition of *Gray's Manual* (p. 759), but no specimens other than pertaining to var. *spinosa* have been found.

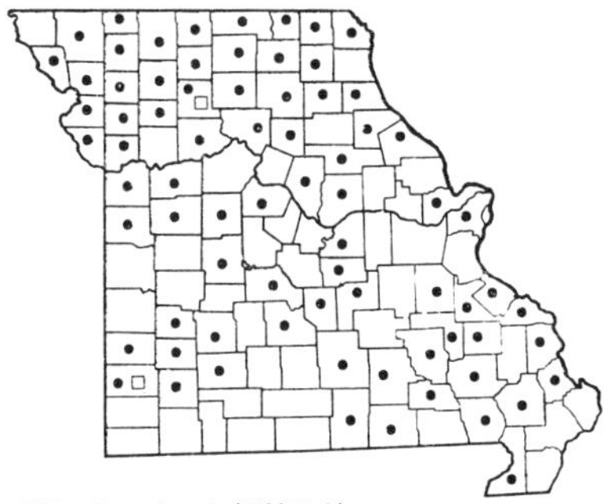

1150 • Pyrus ioensis (Wild Crab)
1150 □ Pyrus X Soulardi (Soulard Crab)

1151 Pyrus melanocarpa (Black Chokeberry)

1152 Amelanchier humilis var. humilis (Low Shadbush)

4. **Pyrus coronaria** L. Wild Crab Map 1149
Flowers early April–May.

Occurs in low open or upland woods, thickets along streams, and prairie openings.

Two variations are represented in Missouri material:

Leaves of the flower- or fruit-bearing branches broadly ovate or roundish, $\frac{5}{8}-\frac{4}{5}$ as broad as long. 4a. P. CORONARIA var. CORONARIA

Leaves of the flower- or fruit-bearing branches broadly lanceolate to lance-ovate, acutely pointed to acuminate at tip, $\frac{1}{3}-\frac{1}{2}$ as broad as long. . 4b. P. CORONARIA var. LANCIFOLIA

4a. **Pyrus coronaria** var. **coronaria** Map 1149
*Pyrus coronaria* L. [G, BB in part, P & S, Steyerm.]

Scattered in southern and central Missouri in Clay, Jackson (*Bush 6426, 3074*), Saline, Texas, Madison (*Palmer 30286*), and Butler (*Palmer 20696*) counties.

Ranges from New York and Ontario to Wisconsin, south to North Carolina, Tennessee, and Missouri.

4b. **Pyrus coronaria** var. **lancifolia** (Rehd.) Fern. Map 1149
*Malus lancifolia* Rehder [P & S, Steyerm.]
*Malus bracteata* Rehder [P & S, Steyerm.]

Scattered in southeastern, east-central, and central Missouri, where known in Dunklin (upland woods, Campbell, April 22, 1922, *Bush 6631A*, holotype of *Malus bracteata; Bush 6618*), Bollinger (*Palmer 39096*), Iron (*Palmer 19512*), St. Francois (*Palmer 31519*), Wayne (*Palmer 6150*), St. Louis (*Letterman*, April, 1883), Boone (*Palmer 35948*), and Jackson (*Bush 10047*; rich woods, Courtney, April 29, 1906, *Bush 3869*, holotype of *Malus lancifolia*) counties.

Ranges from Pennsylvania to Illinois, south to North Carolina, Tennessee, and Missouri. The fruits of these varieties contain an abundance of pectin and may be prepared into a jelly or marmalade.

The fruits of this and related species of wild crab were sometimes made into a cider drink by some of the early colonists in the United States.

It seems preferable to regard this as one variable species in the lobing, shape, and pubescence of leaves, characters which have been used as criteria for separating *Malus lancifolia*, *M. bracteata*, and *M. coronaria*.

5. **Pyrus ioensis** (Wood) Bailey Wild Crab Map 1150
*Malus ioensis* (Wood) Bailey [P & S, Steyerm.]
*Malus ioensis* var. *Palmeri* Rehd. [P & S, Steyerm.]
*Malus ioensis* var. *Bushii* Rehd. [P & S, Steyerm.]
Flowers April–May.

Occurs in open woods, thickets, along streams, borders of woods, and pastures. Throughout Missouri.

Ranges from Wisconsin, Minnesota, and Nebraska, south to Indiana, Illinois, Arkansas, and Oklahoma.

Leaf variations ranging from extremes of little pubescence and slight lobing (*P. ioensis* var. *Bushii*) to those with more rounded tips and usually crenately-serrate (*P. ioensis* var. *Palmeri*) are sometimes recognized, but do not appear to be constant, and the same trees may show a range of variations from one extreme to the other.

In autumn the leaves turn to a dull rose color blended with yellow and dull green. A form having large, double-rose-colored flowers, blooming later than typical *P. ioensis*, is cultivated and sold extensively as Bechtel's Crab (*P. ioensis* f. *plena* Rehd.).

The flowers vary in color from deep rich rose-pink or pale pink to nearly white, and are usually quite fragrant. The fragrance varies in intensity and according to temperature extremes. The fruit which contains an abundance of pectin, is easily made into a jelly or marmalade. The plant, because of its long deep, running, suckering stem, is very difficult to transplant, but is easily propagated from seed.

5a. **Pyrus × Soulardi** Bailey Soulard Crab
Map 1150

× *Malus Soulardi* (Bailey) Britton [P & S, Steyerm.]

Flowers April–May.

A natural hybrid between the apple (*P. Malus*) and wild crab (*P. ioensis*) with leaves similar to those of *P. ioensis*, but with fruit much larger (5–7 cm. in diameter) than *P. ioensis*, and retaining the tartness and general inedible qualities of *P. ioensis*.

Known only from Livingston and Jasper counties.

6. **Pyrus melanocarpa** (Michx.) Willd.
Black Chokeberry Map 1151

*Aronia melanocarpa* (Michx.) Ell. [BB]

Flowers April–May.

Known only from sandy wet or boggy ground along spring branches at the base of Crowley Ridge in Stoddard County, southeastern Missouri (in midst of dense alder thickets along spring branch at junction of flood plain and Crowley Ridge, east and southeast of Pleasant Valley Church, T25N, R11E, sect. 6, 3½ mi. southeast of Bloomfield, August 20, 1954, *Steyermark 76825*; along sandy spring-fed branch, T25N, R11E, NW¼ sect. 6, 3½ mi. southeast of Bloomfield, April 25, 1955, *Steyermark 78277*).

Ranges from Newfoundland to Ontario and Minnesota, south to South Carolina, Georgia, Tennessee, and Missouri.

About twenty-five plants, ranging from 3–9 dm. tall, were found along the moist sandy soil of the spring branch. Here, associated with such eastern species as *Ilex opaca*, *Bartonia paniculata*, *Trisetum pensylvanicum*, and *Trichostema setaceum*, this shrub is found along with these species at the western known limits of its natural range (Rh. 60: 205–8. 1958). It is the most recent shrub to be discovered in the state (in 1954).

*Pyrus melanocarpa* is often cultivated as an ornamental shrub for its showy clusters of white flowers. The purple-black berries have a puckery quality, but, with their abundant pectin contents, make up into a good jelly, dark reddish-purple in color.

## 6. **Amelanchier** Medic. Shadbush, Service Berry

Dwarf shrub, mainly 3–15 dm. (1–4½ ft.) tall, growing in colonies from suckering stems; petals 7–10 mm. long; leaves rounded, blunt (obtuse), or with merely a short abrupt projection at tip; teeth of leaves coarse, 3–5 per cm. (⅜″); inflorescence upright; summit of ovary densely hairy; cultivated shrub, sometimes escaping along roadsides . . . . . . . . . . . . . . . . . . . 1. A. HUMILIS

Small to medium tree or shrub, mainly 2–8 meters (6′8″–26′2″) or more tall, not suckering; petals 10–14 mm. long; leaves definitely acutely pointed or long tapering at tip; teeth of leaves fine, 4–10 per cm. (⅜″); inflorescence nodding or drooping; summit of ovary and young fruit glabrous (without hairs) or nearly so; native trees . . . . . . . . . . . . . . . . . . . 2. A. ARBOREA

1. **Amelanchier humilis** Wieg. var. **humilis**
Low Shadbush Map 1152

*Amelanchier humilis* Wieg. [G, P & S, Steyerm.]

*Amelanchier spicata* [of BB, Jones] in part, not K. Koch

Flowers April–May.

Occurs along open banks and roadsides in the unglaciated prairie section of southwestern Missouri in Pettis (open gravelly bank, open prairie near Sedalia, May 9, 1926, *Palmer 30009*) and Lawrence (prairie banks near house, near Hoberg, April 20, 1935, *Steyermark 18635*).

Ranges from Quebec to Ontario, south to Vermont, Pennsylvania, Ohio, Michigan, Wisconsin, Minnesota, and South Dakota; naturalized in Missouri.

This dwarf suckering shrub does well in cultivation, but is uncommonly cultivated. The fruit, at first reddish, turns purplish-black with a 'bloom,' but is often quickly eaten by birds before maturing fully.

2. **Amelanchier arborea** (Michx. f.) Fern.
Shadbush Map 1153

Also known as Service Berry, Sarvice Berry, Sarviss Tree, June Berry, Shadblow, Sugar Plum.

*Amelanchier canadensis* of auth. [P & S, Steyerm.], not (L.) Medic.

*Amelanchier canadensis* f. *nuda* Palmer & Steyerm.

Flowers March–May.

Occurs in open or rocky woods, steep wooded slopes, and bluffs. Throughout Missouri, but absent

---

Plate no. 195. **1.** Pyrus coronaria var. lancifolia, × 2/5; *a.* Leafy branch; *b.* Flowering branch; *c.* Flowering branch; Details from Small, The New York Botanical Garden. **2.** Pyrus ioensis, × 2/5; *a.* Flowering branch; *b, c, d.* Leaf variations. **3.** Pyrus melanocarpa, × 2/5. **4.** Amelanchier arborea, × 2/5. **5.** Amelanchier humilis var. humilis, × 2/5; Details from Small, The New York Botanical Garden.

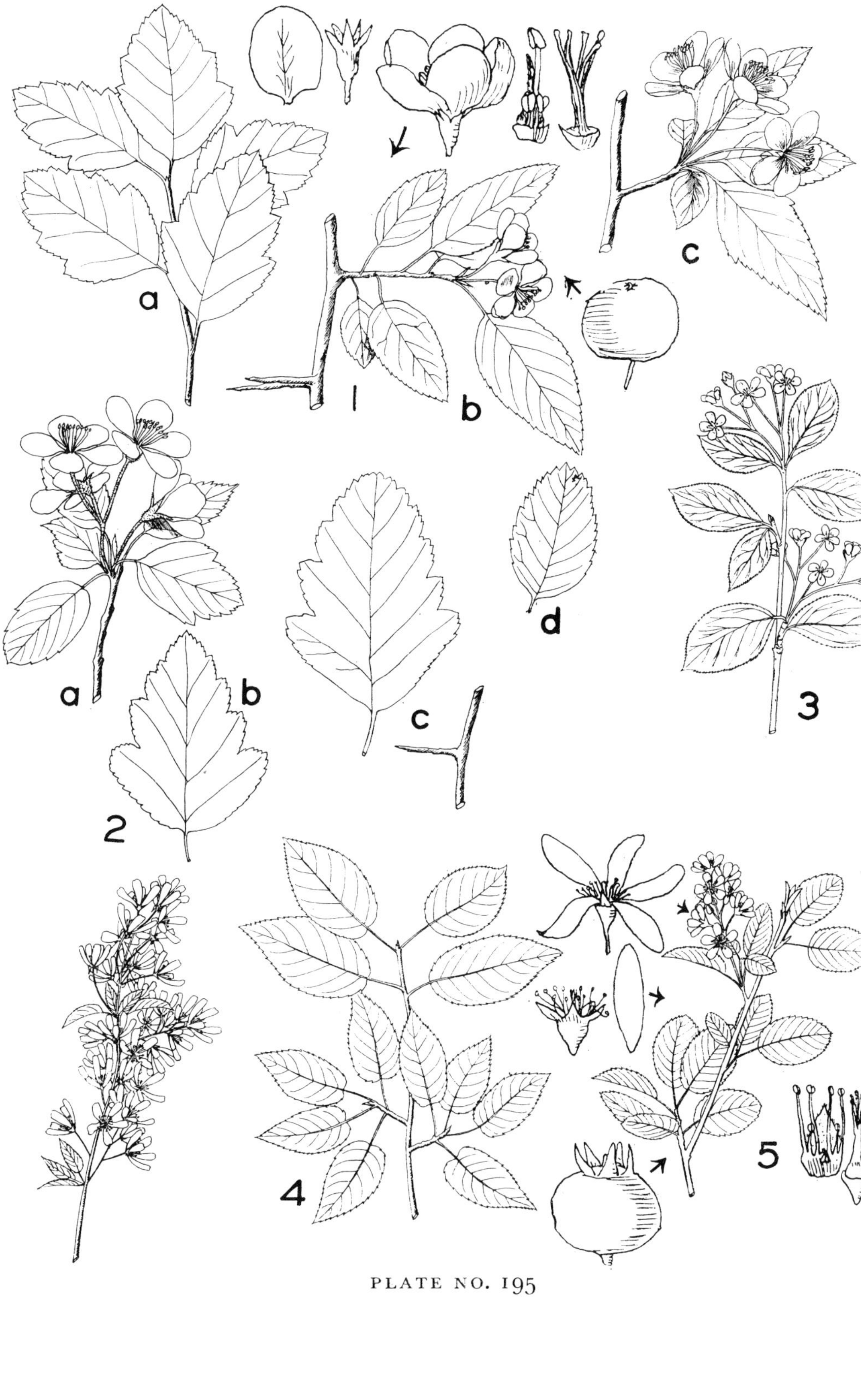


PLATE NO. 195

from the extreme southeastern lowland counties and some of northwestern Missouri.

Ranges from Florida to Louisiana and Oklahoma, north to Maine, New Brunswick, Quebec, Ontario, Michigan, and Minnesota.

The leaves at first, as they unfold and in the early stages of growth, are densely hairy on the lower leaf surface and on the petioles. As they mature, however, much of the pubescence disappears, and sometimes the lower surface is nearly or completely glabrous. Trees having the glabrous leaves were described as f. *nuda*, but the glabrous condition appears to be a matter of aging and inconstant over the years of growth. The showy white flowers are among the first of the early spring trees and shrubs to light up the woodland, and appear before the leaves unfold.

The fruits, which ripen in June and July, at first rosy or reddish, eventually turn wine-purplish when ripe. They have a rather weak taste when eaten raw, but, when cooked for pies, puddings, or muffins, add a pleasant flavor. The fruit was used by the Indians in breadmaking, being prepared into a paste and then dried and mixed with corn meal. Birds usually devour the fruits before they fully ripen.

The foliage is very colorful in autumn, turning a pale orange or gold blended with russet or rose and green. The wood is sometimes used for the making of tool handles and other small articles. It ranks with persimmon as the heaviest wood among North American trees, and as fifth in hardness.

The word Sarviss is said to be a modified form of *Sorbus*, the name applied to a fruit known to the Romans and resembling that of *Amelanchier*.

## 7. **Crataegus** L. Hawthorn, Red Haw

By Ernest J. Palmer

*Key to the series*

a. Veins of the larger leaves running to the sinuses as well as to the points of the lobes; flowers 0.6–1.5 cm. wide; fruit 4–8 mm. thick . . . . . . . . . . *b*

- b. Leaves mostly ovate, incised and often trilobate, acuminate at apex, rounded or cordate at base; flowers 1–1.3 cm. wide in many-flowered glabrous corymbs, opening usually from the middle of May to early June; fruit bright red with deciduous calyx exposing ends of nutlets, persistent until late in the season . . . . . . . . . . 1. CORDATAE
- b. Leaves spatulate or broadly ovate, nearly entire or deeply incised, acute or rounded at apex; flowers 0.7–1.2 cm. wide in many-flowered glabrous or pubescent corymbs, opening usually from the end of April to the middle of May; fruit red or greenish red, with persistent calyx, falling early in the season . . . . . . . . . . 2. MICROCARPAE

a. Veins of the leaves running only to the points of the lobes; flowers 1–2.6 cm. wide; fruit 5–16 mm. thick . . . . . . . . . . *c*

- c. Flowers usually opening from late April to the middle of May; nutlets 1–5, not pitted on the ventral surface . . . . . . . . . . *d*
  - d. Leaves narrowed or attenuate at the base . . . . . . . . . . *e*
    - e. Leaves relatively small, mostly 0.8–2 cm. wide or slightly wider at the ends of shoots, oblong or oblong-obovate, serrate or crenate, sometimes slightly lobed; flowers single or rarely 2–3 in simple clusters; stamens 20–25; sepals foliaceous, pectinately glandular-serrate, persistent in fruit; slender shrubs 1–1.5 m. tall . . . . . . . . . . 3. PARVIFOLIAE
    - e. Leaves larger, 2–6 cm. wide or sometimes much wider at the ends of shoots; flowers in simple or compound corymbs or cymes; stamens 5–20; sepals entire or glandular-serrate, not foliaceous arborescent shrubs or trees . . . . . . . . . . *f*
      - f. Leaves of flowering branchlets ovate, rhombic or rarely obovate, narrowed at base, or sometimes rounded or truncate at the ends of shoots . . . . . . . . . . *g*
        - g. Foliage and inflorescence conspicuously glandular; flowers in nearly simple, mostly 3–8-flowered corymbs; fruit dull red, orange-red or russet, with thin dry flesh; usually arborescent shrubs . . . . . . . . . . 4. INTRICATAE
        - g. Foliage and inflorescence eglandular or sometimes with a few deciduous glands; flowers in simple or compound corymbs; fruit red or dull orange; becoming trees, but sometimes flowering as stout shrubs . . . . . . . . . . *h*
          - h. Leaves thin or firm, glabrous except for small axillary tufts of tomentum, or rarely

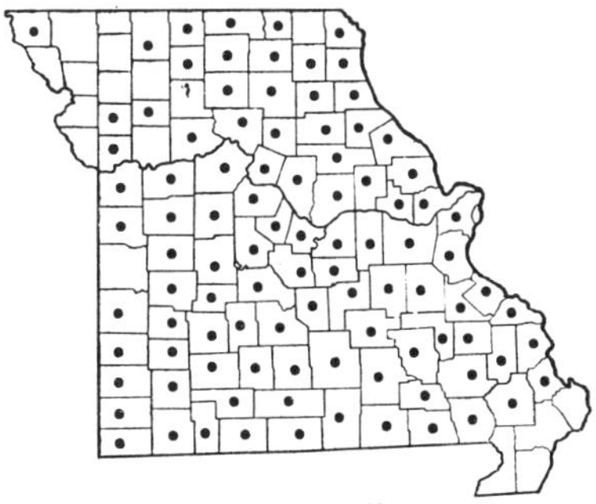

1153 Amelanchier arborea (Shadbush)

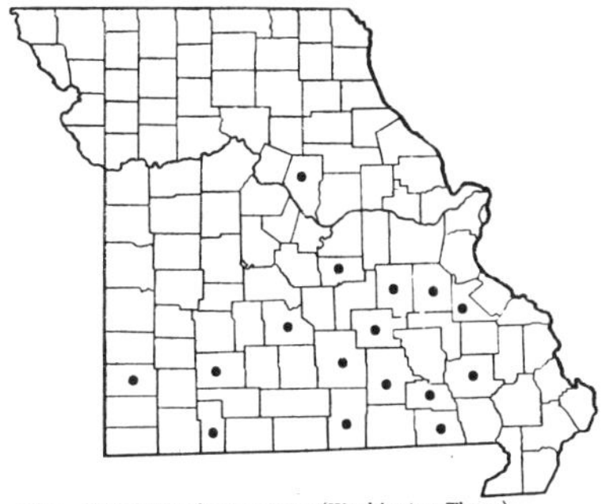

1154 Crataegus Phaenopyrum (Washington Thorn)

1155 Crataegus spathulata

slightly villous beneath; flowers in many-flowered compound corymbs; stamens about 20; fruit 0.5–1 cm. thick; nutlets 3–5; trees up to 10–12 m. tall, with thin scaly bark over reddish inner bark, growing in wet or alluvial ground . . . 5. VIRIDES

h. Leaves firm or thick, glabrous or slightly pubescent, without conspicuous axillary tufts of tomentum; flowers in few-flowered compact nearly simple corymbs; stamens 10 or 20; fruit 0.6–1.2 cm. thick; nutlets 2–5; small trees or sometimes stout shrubs, with rough bark, growing in drier ground . . . . . 6. ROTUNDIFOLIAE

f. Leaves of flowering branchlets mostly obovate or oblong-obovate, seldom over 2–3 cm. wide; terminal shoot leaves ovate, elliptic or rarely suborbicular, sometimes slightly lobed; flowers mostly in compound corymbs; stamens 5–20; nutlets 1–5 . . . . . . *i*

i. Leaves firm to subcoriaceous, often glossy above, the veins inconspicuous or rarely slightly impressed above, unlobed or sometimes obscurely lobed at the ends of shoots; flowers 1–1.7 cm. wide; fruit remaining hard and dry or rarely becoming mellow late in the season, green, dull red, dull orange-red at maturity; nutlets 1–2 or rarely more . . . . . . . . . . . . . . . . . . . . . . 7. CRUS-GALLI

i. Leaves firm, dull green, with the veins distinctly impressed above at maturity, often slightly lobed especially at the ends of shoots; flowers 1.2–2.3 cm. wide; fruit red or dark red, usually becoming mellow; nutlets 3–5 or rarely only 2 . . . . 8. PUNCTATAE

d. Leaves obtuse, rounded or cordate at the base . . . . . . . . . . . . . . . . . . *j*

j. Foliage and inflorescence glabrous or rarely slightly pubescent; terminal shoot leaves truncate to cordate at the base, seldom over 5–6 cm. wide, more or less lobed; flowers 1.5–2 cm. wide, in few-flowered, usually nearly simple corymbs; sepals entire, finely serrate or rarely glandular-serrate; fruit remaining hard and dry, often pruinose, sometimes 5-angled; nutlets 2–5 . . . . . . . . . . . . . . . . . . 9. PRUINOSAE

j. Foliage and inflorescence glabrous, villous or tomentose; leaves of flowering branchlets mostly rounded to subcordate at base (except in form of no. 32); terminal shoot leaves truncate to cordate at base, up to 7–8 cm. wide, more or less lobed; flowers mostly 2–2.6 cm. wide, in compound many-flowered corymbs; sepals glandular-serrate; fruit relatively large, becoming mellow or succulent, highly flavored and edible; nutlets 3–5, often 5 . . . *k*

k. Foliage and inflorescence villous or densely tomentose, at least while young; flowers 2–2.3 cm. wide; anthers small, pale yellow or red; fruit 1–1.8 cm. thick, pubescent while young, usually ripening in August or September, or sometimes later, scarlet, orange or crimson; nutlets 3–5 . . . . . . . . . . . . . . . . . . . . 10. MOLLES

k. Foliage and inflorescence glabrous; flowers 2.4–2.6 cm. wide; anthers large, rose-colored; fruit ripening in September or October, bright crimson, often angular . . . 11. DILATATAE

c. Flowers opening in May or early June; nutlets 2–5, usually 2–3, pitted on the ventral surfaces . . 12. MACRACANTHAE

*Key to the species*

Series 1. **Cordatae**

Only one species in Missouri . . . . . . . . . . . . . . . . . . 1. C. PHAENOPYRUM

1156 Crataegus Marshallii (Parsley Haw)

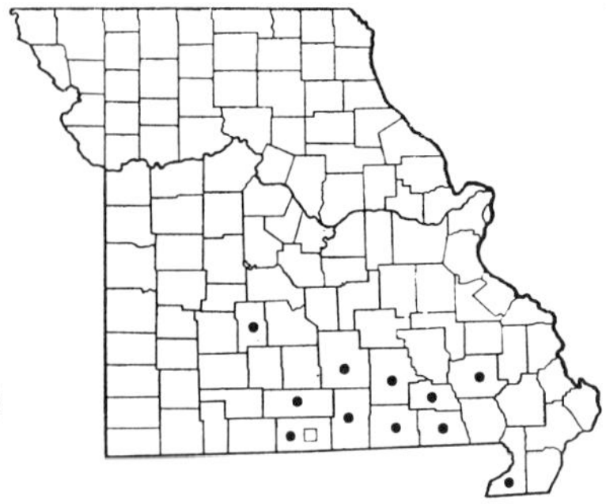

1157 • Crataegus uniflora
1157 □ Crataegus uniflora X Engelmanni var. Engelmanni

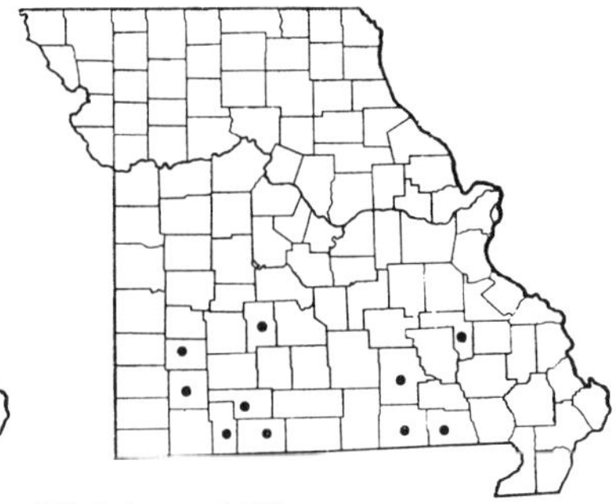

1158 Crataegus neobushii

1. **Crataegus Phaenopyrum** (Ehrh.) Medic. Washington Thorn Map 1154

Flowers mid-May–early June.

Scattered in open ground, thickets and borders of woods. Central and southern Missouri, mostly in the Ozark region and introduced in Boone and Jasper counties.

Ranges from Pennsylvania to Florida, west to Missouri and Arkansas.

Often planted as an ornamental tree, distinctive in its foliage and bright red long-persistent fruit; and valuable as bird food.

Series 2. **Microcarpae**

Leaves spatulate or obovate, narrowed or attenuate at base, with shallow rounded lobes, glabrous; fruit subglobose . . . . . . . . . . . . . . . . . . . . . . . . 2. C. SPATHULATA

Leaves broadly ovate, truncate to subcordate at base, deeply incised and lobed, pubescent; fruit oblong 3. C. MARSHALLII

2. **Crataegus spathulata** Michx. Map 1155

Flowers May.

Moist or fertile ground, thickets, and open woods. Rare in the White River region of southern Missouri and northwest to Jasper County.

Ranges from Virginia to Florida, west to southern Missouri and eastern Texas.

3. **Crataegus Marshallii** Egglest. Parsley Haw Map 1156

Flowers May.

Low wet or alluvial ground. Rare in southeastern lowlands. Ranges from Virginia to Florida, west along the Coastal Plain to eastern Texas.

Ornamental and distinctive in its foliage; sometimes planted in parks and along roads; would probably make a good hedge plant.

Series 3. **Parvifoliae**

Only one species in Missouri . . . . . . . . . . . . . . . . . . . . . 4. C. UNIFLORA

4. **Crataegus uniflora** Muenchh. Map 1157

Flowers May.

Open woods, sandy or rocky ground with acid or subacid soil. Southern Ozark region.

Ranges from New York and Pennsylvania to Florida, west to southern Missouri and eastern Texas.

Series 4. **Intricatae**

a. Foliage and inflorescence glabrous or essentially so . . . . . . . . . . . . . . . . . . . *b*

b. Leaves of flowering branchlets mostly ovate, broadest below the middle, sharply lobed; sepals glandular-serrate or pectinate; fruit orange-red . . . . . . . . . . . . . 5. C. NEOBUSHII

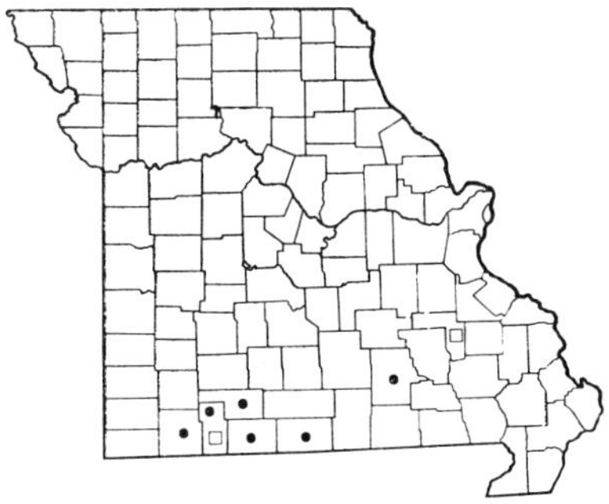

1159 • Crataegus padifolia var. padifolia
1159 □ Crataegus padifolia var. incarnata

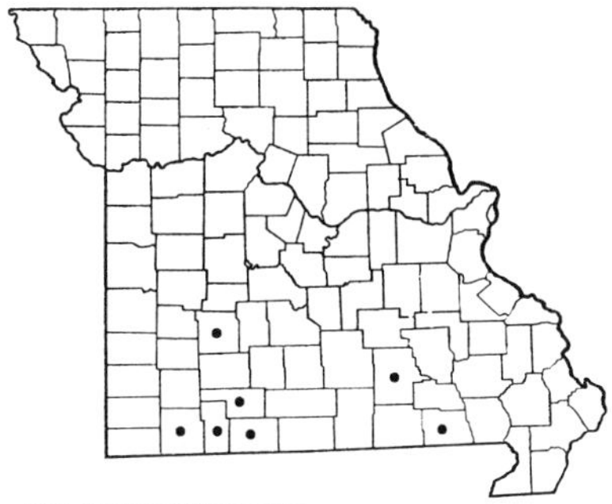

1160 Crataegus biltmoreana

1161 Crataegus Harveyana

b. Leaves of flowering branchlets mostly oval or elliptic, broadest about the middle, unlobed or with shallow lobes; sepals entire or finely serrate; fruit dull red or dull orange-red . . 6. C. PADIFOLIA

a. Foliage and leaves pubescent . . . . . . . . . . . . . . . . . . . . . . . . . . *c*

c. Leaves mostly ovate or oval, pointed at apex, narrowed or rounded, or at the ends of shoots sometimes truncate at base; flowers mostly 3–7, in nearly simple corymbs; stamens about 10; anthers white or pale yellow; fruit 0.9–1.4 cm. thick; nutlets 3–5 . . . . . . . 7. C. BILTMOREANA

c. Leaves ovate to deltoid, acuminate at apex, narrowed to rounded at base; flowers mostly 6–12, in compound corymbs; stamens about 20; anthers deep pink; fruit about 1 cm. thick; nutlets usually 3 . . . . . . . . . . . . . . . . . . . . . 8. C. HARVEYANA

5. **Crataegus neobushii** Sarg. Map 1158

Flowers May.

In rocky or well drained ground, thickets, and open woods. Scattered in southern Ozark region.

Ranges from western Pennsylvania to southern Illinois, southern Missouri and Arkansas.

6. **Crataegus padifolia** Sarg. Map 1159

Two varieties are recognized:

Leaves nearly unlobed; fruit dull orange-red, 0.9–1.2 cm. thick, remaining hard and dry; nutlets 2–3 . . . . . . 6a. C. PADIFOLIA var. PADIFOLIA

Leaves often slightly lobed, at least at the ends of shoots; fruit dull red, 1–1.4 cm. thick, becoming mellow or succulent; nutlets 3–4 . . 6b. C. PADIFOLIA var. INCARNATA

6a. **Crataegus padifolia** var. **padifolia** Map 1159

Flowers late April–May.

Open upland woods and thickets. Southern Ozark region, mainly along White River.

Ranges in Missouri and northern Arkansas.

6b. **Crataegus padifolia** var. **incarnata** Sarg. Map 1159

Flowers about mid-April.

Rocky upland woods. Known only from Stone and Iron counties.

Ranges from southern Missouri and northern Arkansas.

7. **Crataegus biltmoreana** Beadle Map 1160

Flowers late April–May.

Rocky open woods and hillsides. Scattered in southern Ozark region.

Ranges from Vermont to North Carolina, west to Indiana, Missouri, and Arkansas.

8. **Crataegus Harveyana** Sarg. Map 1161

Flowers May.

Rocky open woods and hillsides. Known in Missouri only from McDonald County, but should be looked for in adjoining counties.

Ranges from southern Missouri and northern Arkansas.

Series 5. **Virides**

a. Foliage and inflorescence glabrous except for small axillary tufts of tomentum sometimes found on the under surface of the leaves; fruit bright red or rarely orange-red . . . . . . . . . . . *b*

b. Leaves thin, scarcely glossy above; flowers 1.2–1.5 cm. wide; fruits 5–7 mm. thick; nutlets 3–5, usually 5 . . . . . . . . . . . . . . . . . . . . . . 9. C. VIRIDIS

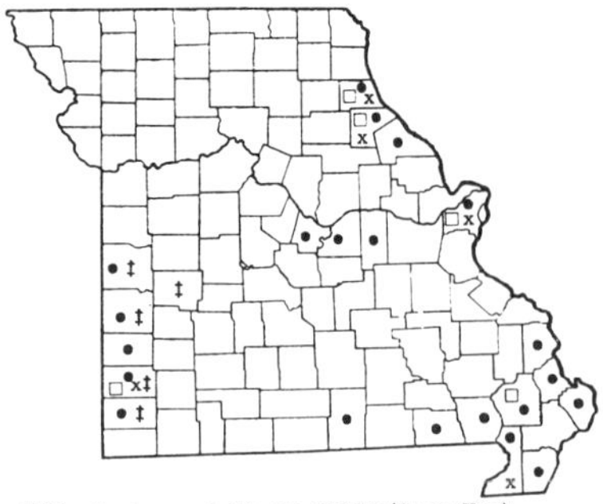

1162 • Crataegus viridis var. viridis (Green Haw)
1162 □ Crataegus viridis var. ovata
1162 x Crataegus viridis var. lanceolata
1162 ‡ Crataegus viridis var. lutescens

1163 Crataegus nitida

1164 Crataegus atrorubens

b. Leaves firm or thick, decidedly glossy above; flowers 1.5–1.7 cm. wide; fruit 8–10 mm. thick; nutlets 3–5. . . . . . . . . . . . . . . . . . . . . . . . . 10. C. NITIDA
a. Foliage and inflorescence slightly villous, at least while young; fruit orange-red or crimson . 11. C. ATRORUBENS

9. **Crataegus viridis** L. Green Haw  Map 1162
Four varieties are recognized in Missouri:

a. Leaves not serrate near base; fruit with thin flesh becoming soft when ripe; nutlets normally 5 . . . . . . . . . . . . . *b*
 b. Leaves variable and sometimes asymmetric, on flowering branchlets mostly obovate to rhombic; terminal shoot leaves ovate or broadly-ovate, coarsely serrate and usually deeply cut or laciniate toward the base . 9a. C. VIRIDIS var. VIRIDIS
 b. Leaves less variable and mostly symmetric; terminal shoot leaves often sharply lobed but not laciniate toward the base . . . *c*
  c. Leaves of flowering branchlets mostly ovate or oblong-ovate, often obtuse or rounded at apex; terminal shoot leaves broadly ovate to suborbicular . . . 9b. C. VIRIDIS var. OVATA
  c. Leaves of flowering branchlets mostly lance-ovate, distinctly longer than broad usually acute or acuminate at apex; terminal shoot leaves similar but larger and relatively broader . . 9c. C. VIRIDIS var. LANCEOLATA
a. Leaves sharply serrate nearly to base; fruit remaining hard and dry; nutlets 3–5 . . . 9d. C. VIRIDIS var. LUTESCENS

9a. **Crataegus viridis** var. **viridis** L.  Map 1162
Flowers May.

Low wet or alluvial ground, or sometimes on slopes with seepage water. Southern and eastern Missouri, mostly along the larger streams, north to Marion County, and also in southwestern border counties from Bates to Newton counties.

Ranges from Virginia to Florida, west to southern Illinois, eastern Oklahoma, and Texas.

Often becomes a large tree for the genus, up to 10–12 m. tall, and with trunk 4 dm. in diameter.

9b. **Crataegus viridis** var. **ovata** (Sarg.) Palmer  Map 1162

Flowers May.

In similar situations to var. *viridis*. Scattered, mostly along the Mississippi River, and also in Jasper County.

Ranges from North Carolina to Missouri and Arkansas.

9c. **Crataegus viridis** var. **lanceolata** (Sarg.) Palmer  Map 1162

Flowers May.

In similar situations to var. *viridis*. Scattered in eastern Missouri, and also in Jasper County.

Ranges from Illinois to Alabama, west to Missouri, Arkansas, and Louisiana.

9d. **Crataegus viridis** var. **lutescens** (Sarg.) Palmer  Map 1162

Plate no. 196. **1.** Crataegus Phaenopyrum, × 2/5. **2.** Crataegus spathulata, × 2/5. **3.** Crataegus Marshallii, × 2/5. **4.** Crataegus uniflora, × 2/5. **5.** Crataegus padifolia, × 2/5. **6.** Crataegus viridis, × 2/5. **7.** Crataegus nitida, × 2/5. **8.** Crataegus Margaretta, × 2/5. **9.** Crataegus crus-galli, × 2/5. **10.** Crataegus acutifolia, × 2/5. **11.** Crataegus regalis, × 2/5. **12.** Crataegus Palmeri, × 2/5. **13.** Crataegus fecunda, × 2/5. **14.** Crataegus Engelmanni, × 2/5. **15.** Crataegus verruculosa, × 2/5. **16.** Crataegus Lettermani, × 2/5. **17.** Crataegus Kelloggii, × 2/5.

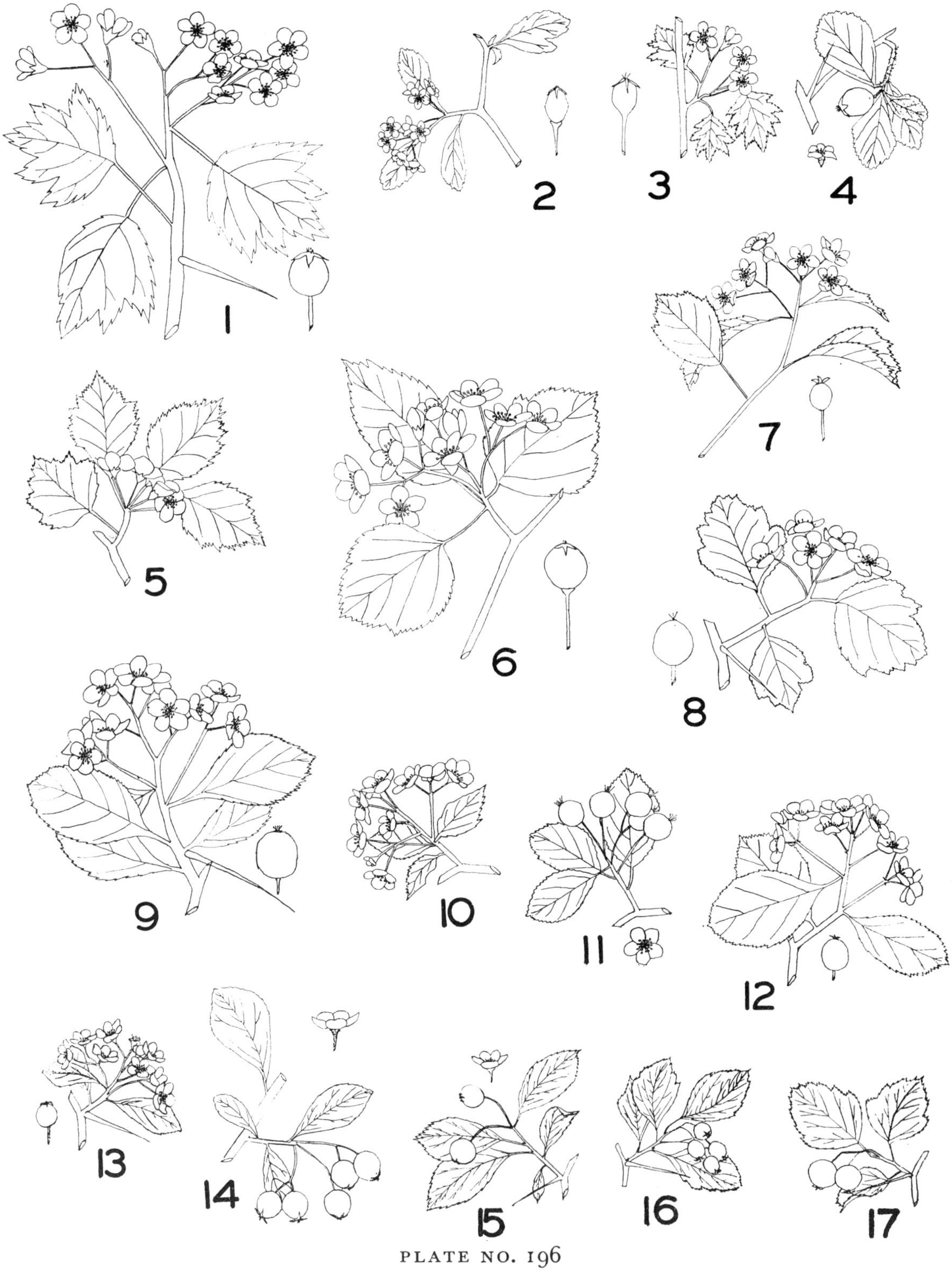


PLATE NO. 196

Flowers May.

Open woods and thickets along streams. Southwestern Missouri, from Bates and St. Clair to Newton counties.

Ranges from Missouri to southeastern Kansas and northeastern Oklahoma.

10. **Crataegus nitida** (Engelm.) Sarg. Map 1163

Flowers May.

Open woods, along streams. Scattered along Mississippi River, and also in Newton County.

Ranges from Ohio and Illinois to Missouri and Arkansas.

11. **Crataegus atrorubens** Ashe Map 1164

Flowers May.

Moist alluvial ground. Known only from St. Louis County.

Series 6. **Rotundifoliae**

Leaves extremely variable but mostly oval or broadly obovate or sometimes suborbicular at the ends of shoots, generally obtuse or rounded at apex; flowers 0.8–1.5 cm. wide, mostly 6–10 in corymb; stamens about 20; nutlets 2–4, usually 3 . . . . . . . . . . . . . . . . 12. C. MARGARETTA

Leaves more uniform, mostly elliptic or oblong-ovate, acute at apex; flowers 1.5–2 cm. wide, mostly 3–6 in corymb; stamens about 10; nutlets 3–5 . . . . . . . . . . . . . . . . 13. C. SICCA

12. **Crataegus Margaretta** Ashe Map 1165

Two varieties are recognized in Missouri:

Flowers 1.2–1.5 cm. wide; fruit 0.9–1.2 cm. thick 12a. C. MARGARETTA var. MARGARETTA

Flowers 0.8–1 cm. wide; fruit 6–8 mm. thick . 12b. C. MARGARETTA var. BROWNEI

12a. **Crataegus Margaretta** var. **Margaretta** Map 1165

Flowers late April–early May.

Thickets and rocky open ground. Eastern Missouri, from Putnam and Clark to Madison counties, and also in several southwestern counties.

Ranges from southern Ontario and Michigan to Pennsylvania, Ohio, and Missouri.

12b. **Crataegus Margaretta** var. **Brownei** (Britt.) Sarg. Map 1165

Flowers late April–May.

Rocky open woods and thickets. Scattered, northeast Missouri, and also in Greene County.

Ranges from Virginia and Pennsylvania to Indiana and Missouri.

13. **Crataegus sicca** Sarg. Map 1166

Two varieties are recognized:

Foliage and inflorescence slightly villous; flowers 1.5–2 cm. wide . . . . 13a. C. SICCA var. SICCA

Foliage and inflorescence glabrous; flowers 1–1.5 cm. wide. . . . 13b. C. SICCA var. GLABRIFOLIA

13a. **Crataegus sicca** var. **sicca** Map 1166

Flowers late April–early May.

Dry limestone hills or rocky ground along streams. Scattered in Ozark region.

Ranges in southern Missouri.

13b. **Crataegus sicca** var. **glabrifolia** (Sarg.) Palmer Map 1166

Flowers late April–May.

Rocky open woods. Known only from Pike County.

Ranges from Ohio to southern Missouri.

Series 7. **Crus-galli**

a. Foliage and inflorescence glabrous or essentially so; leaves glossy above except sometimes in shade . . *b*
  b. Leaves of flowering branchlets mostly obovate, oblong-obovate or cuneate, seldom more than 2 cm. wide, entire except for serrate margins; terminal shoot leaves similar but larger, often oblong-obovate, unlobed or rarely obscurely lobed . . . . . . . . . . . . . . . . *c*
    c. Leaves thick or subcoriaceous, pointed to rounded at apex, sharply serrate; stamens about 10 . . *d*
      d. Leaves of flowering branchlets mostly obovate or cuneate, symmetric; nutlets 1–3, usually 1–2 . . . . . . . . . . . . . . . . . . . . . 14. C. CRUS-GALLI
      d. Leaves of flowering branchlets mostly oblong-obovate, often slightly asymmetric; nutlets usually 3 . . . . . . . . . . . . . . . . . . . . . 15. C. TANTULA
    c. Leaves firm but thinner, mostly acute at apex; stamens 10 or 20; nutlets 2–3 . . . . . 16. C. PYRACANTHOIDES var. ARBOREA

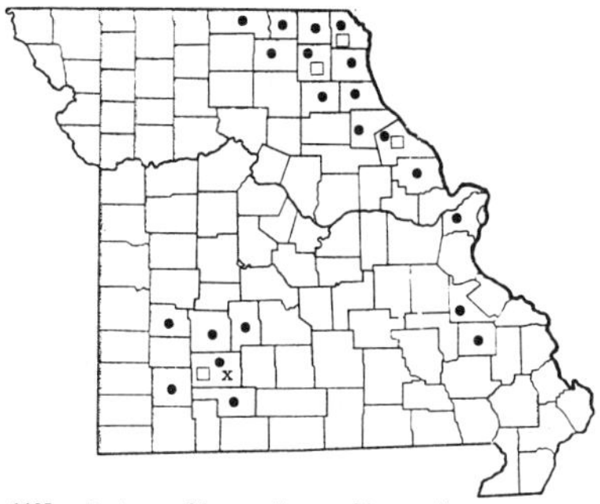

1165 • Crataegus Margaretta var. Margaretta
1165 □ Crataegus Margaretta var. Brownei
1165 x Crataegus Margaretta var. Margaretta X mollis

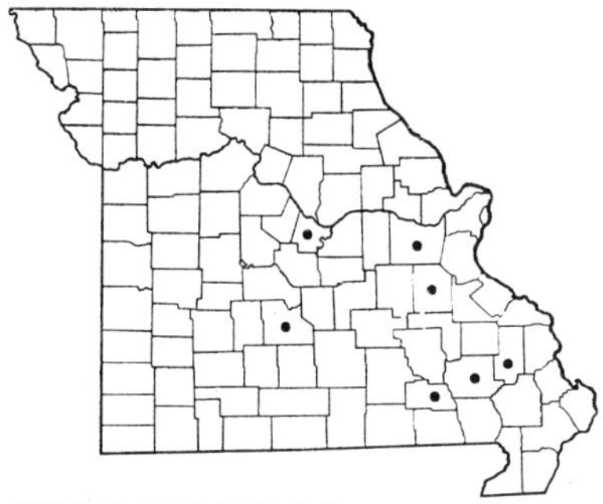

1166 Crataegus sicca var. sicca

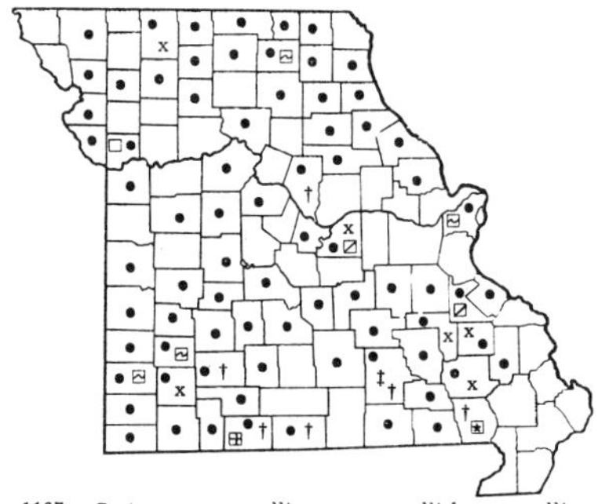

1167 • Crataegus crus-galli var. crus-galli f. crus-galli (Cockspur Thorn)
1167 ⊞ Crataegus crus-galli var. crus-galli f. truncata
1167 † Crataegus crus-galli var. macra
1167 □ Crataegus crus-galli var. pyracanthifolia
1167 ‡ Crataegus crus-galli var. exigua
1167 ⧄ Crataegus crus-galli var. pachyphylla
1167 ⊟ Crataegus crus-galli var. barrettiana
1167 x Crataegus crus-galli var. leptophylla
1167 ⊠ Crataegus crus-galli var. bellica

b. Leaves of flowering branchlets mostly oblong-obovate to elliptic, 1.5–2.5 cm. wide; terminal shoot leaves broadly obovate or suborbicular, often slightly lobed . . . . . . . . . . . . *e*
- e. Veins of the leaves not noticeably impressed above; branchlets brown or orange-red at the end of the first season . . . . . . . . . . . . . . . . . . . . . . . . . *f*
  - f. Leaves sharply or finely serrate, or more coarsely serrate at the ends of shoots, but teeth not enlarged and lobe-like . . . . . . . . . . . . . . . . . . . . *g*
    - g. Terminal shoot leaves often suborbicular, 2–3 times as broad as those of the flowering branchlets . . . . . . . . . . . . . . . . . . . . . . . . *h*
      - h. Terminal shoot leaves often obscurely lobed above the middle, pointed or acuminate; anthers pale yellow; fruit ellipsoidal or rarely subglobose; nutlets 2–3 . 17. C. ACUTIFOLIA
      - h. Terminal shoot leaves usually entire except for serrate margins, mostly rounded or obtuse at apex; anthers red; fruit subglobose; nutlets 3–5 . 18. C. REVERCHONI var. DISCOLOR
    - g. Terminal shoot leaves mostly broadly ovate or oblong-elliptic, seldom more than twice as broad as those of the flowering branchlets . . . . . . . . . . . . . . *i*
      - i. Leaves pointed or acuminate at apex, sharply and deeply serrate; lobes of the terminal leaves, if present, usually above the middle; nutlets 2–3 . . . 19a. C. REGALIS var. REGALIS
      - i. Leaves usually obtusely pointed or rounded at apex; lobes of the terminal leaves, if present, usually near the base; nutlets usually 3. . . . . . . . . . . 20. C. PALMERI
  - f. Leaves coarsely serrate or dentate, those at the ends of shoots usually slightly lobed and with some of the teeth enlarged and lobe-like . . . . . . . . . . . 21. C. PERMIXTA
- e. Veins of the leaves impressed above at maturity; branchlets yellowish or pale olive-green at the end of the first season; nutlets 1–3, usually 2 . . . . . . . . . . 22. C. HANNIBALENSIS

a. Foliage and inflorescence more or less pubescent at least while young (except in rare forms of numbers 24 and 26) . . . . . . . . . . . . . . . . . . . . . . . . . . *j*
- j. Leaves relatively small, 0.8–3 cm. wide, mostly obovate, oblong-obovate or spatulate, reticulately veined at maturity . . . . . . . . . . . . . . . . . . . . . 24. C. ENGELMANNI
- j. Leaved larger, 1.5–5 cm. wide, mostly broadly obovate or oblong-obovate, not reticulately veined at maturity . . . . . . . . . . . . . . . . . . . . . . . . . . . *k*
  - k. Foliage and inflorescence sparsely villous, becoming glabrous or nearly so at maturity . . . . *l*
    - l. Leaves finely serrate above the middle; terminal shoot leaves 4–4.5 cm. wide; fruit 1.4–1.6 cm. thick, glabrous. . . . . . . . . . . . . . . . . . . 23. C. VALLICOLA
    - l. Leaves sharply and deeply serrate nearly to base; terminal shoot leaves up to 6–8 cm. wide; fruit 1–1.4 cm. thick, slightly pubescent while young . . . . . . . . . . 25. C. FECUNDA
  - k. Foliage and inflorescence more or less villous (except in rare form of number 26); leaves villous at least along the veins beneath at maturity . . . . . . . . . . . . . . *m*
    - m. Leaves of flowering branchlets mostly oblong-obovate, finely serrate; fruit oblong or obovoid, 6–8 mm thick; nutlets 2–3, usually 2. . . . . . . 19b. C. REGALIS var. PARADOXA

1168 Crataegus tantula

1169 Crataegus pyracanthoides var. arborea

1170 Crataegus acutifolia var. acutifolia

m. Leaves of flowering branchlets rhombic, oblong-elliptic or oblong-obovate, sharply serrate; terminal shoot leaves similar but larger and more coarsely serrate; fruit subglobose, 8–12 mm. thick; nutlets 2–3, usually 3 . . . . . . . . . . . . . . . . . . . . . . 26. C. DANIELSII

14. **Crataegus crus-galli** L. Cockspur Thorn Map 1167

Eight varieties and two forms are recognized in Missouri:

a. Foliage and inflorescence glabrous or essentially so; leaves scattered at the ends of branchlets . . . . . . . . . . *b*
b. Stamens about 10; nutlets 1–3 . . . . *c*
c. Leaves pointed to rounded at apex. . *d*
d. Leaves thick or subcoriaceous, glossy above (except sometimes in shade) . *e*
e. Terminal shoot leaves seldom over 3–3.5 cm. wide, the veins not impressed above . . . . . . . *f*
f. Terminal shoot leaves unlobed or rarely obscurely lobed; nutlets 1–3 . . . . . . . . *g*
g. Leaves of flowering branchlets mostly obovate, 1.5–2.5 cm. wide. . . . . . *h*
h. Fruit subglobose, often slightly 5-angled, 0.8–1 cm. thick; nutlets 1–2 . 14a. C. CRUS-GALLI var. CRUS-GALLI f. CRUS-GALLI
h. Fruit subglobose or oval, 5–8 mm. thick; nutlets 1–3 . 14c. C. CRUS-GALLI var. MACRA
g. Leaves of flowering branchlets obovate or lance-obovate, 1–1.5 cm. wide . . 14d. C. CRUS-GALLI var. PYRACANTHIFOLIA
f. Terminal shoot leaves often slightly incised or lobed; nutlets usually 1 . 14e. C. CRUS-GALLI var. EXIGUA
e. Terminal shoot leaves 4–5 cm. wide, the veins slightly impressed above at maturity . 14f. C. CRUS-GALLI var. PACHYPHYLLA
d. Leaves firm or moderately thick, not glossy above, the veins slightly impressed at maturity . 14g. C. CRUS-GALLI var. BARRETTIANA
c. Leaves truncate or rounded at apex . 14b. C. CRUS-GALLI var. CRUS-GALLI f. TRUNCATA
b. Stamens about 20; nutlets 3–4, or rarely 5 14h. C. CRUS-GALLI var. LEPTOPHYLLA
a. Foliage and inflorescence usually slightly villous while young; leaves in rosette-like clusters at the ends of branchlets. . . . 14i. C. CRUS-GALLI var. BELLICA

14a. **Crataegus crus-galli** var. **crus-galli** f. **crus-galli** Map 1167

Flowers May–early June.

Thickets and rocky pastures. Found throughout the state and probably in every county.

Ranges from Quebec to South Carolina, west to southern Minnesota, eastern Kansas and eastern Texas.

14b. **Crataegus crus-galli** var. **crus-galli** f. **truncata** (Sarg.) Palmer Map 1167

Recorded only from Taney County, but may occur elsewhere.

14c. **Crataegus crus-galli** var. **macra** (Beadle) Palmer Map 1167

Scattered, mostly in southern Missouri, north to Boone County.

Ranges from Georgia to Missouri and Arkansas.

14d. **Crataegus crus-galli** var. **pyracanthifolia** Ait. Map 1167

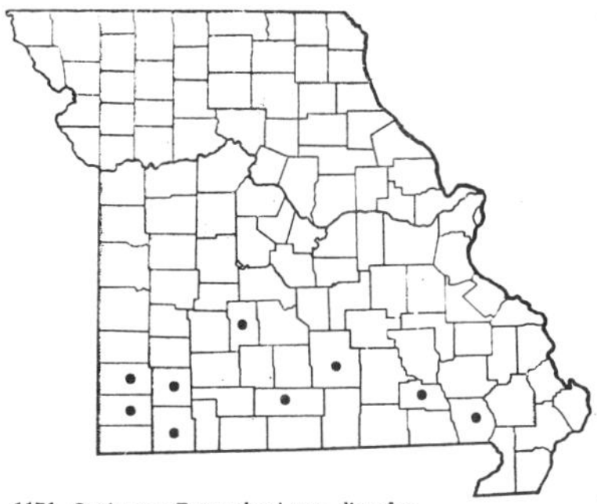

1171 Crataegus Reverchoni var. discolor

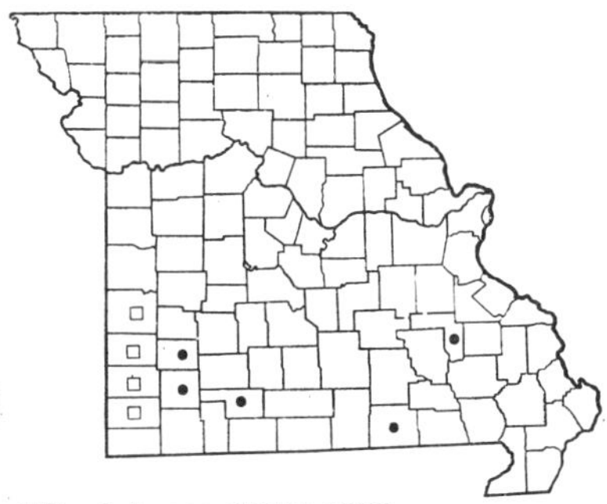

1172 ● Crataegus regalis var. regalis
1172 □ Crataegus regalis var. paradoxa

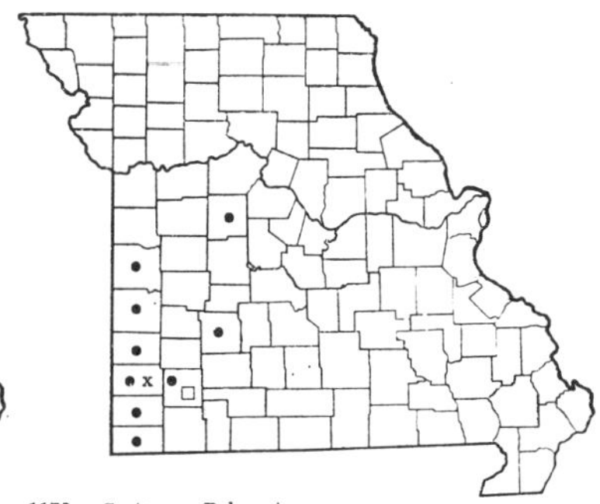

1173 ● Crataegus Palmeri
1173 x Crataegus Palmeri X viridis
1173 □ Crataegus Palmeri X collina

Recorded only from Clay County, but probably occurs elsewhere.

Range same as typical var., but scattered.

14e. **Crataegus crus-galli** var. **exigua** (Sarg.) Egglest. Map 1167

Recorded only from Shannon County, but probably occurs elsewhere.

14f. **Crataegus crus-galli** var. **pachyphylla** (Sarg.) Palmer Map 1167

Scattered in eastern and central Missouri.

Ranges from Ohio and Indiana to Missouri.

14g. **Crataegus crus-galli** var. **barrettiana** (Sarg.) Palmer Map 1167

Widely scattered in Missouri, but uncommon.

Ranges from Ohio and Illinois to Missouri.

14h. **Crataegus crus-galli** var. **leptophylla** (Sarg.) Palmer Map 1167

Widely scattered in Missouri.

Ranges from Ohio to Missouri and Arkansas.

14i. **Crataegus crus-galli** var. **bellica** (Sarg.) Palmer Map 1167

Local in southeastern Missouri; recorded only from Butler County.

Ranges from Missouri and Arkansas to eastern Texas.

15. **Crataegus tantula** Sarg. Map 1168

Flowers May.

Thickets and rocky uplands. Local in southwestern Missouri, Jasper, Newton, and Lawrence counties.

Not known elsewhere.

16. **Crataegus pyracanthoides** Beadle var. **arborea** (Beadle) Palmer Map 1169

Flowers late April–May.

Low hills or fertile ground along streams. Southeastern Missouri.

Ranges from South Carolina to Florida, west to Indiana, Missouri, and eastern Texas.

17. **Crataegus acutifolia** Sarg. var. **acutifolia** Map 1170

Flowers May.

Open woods, moist or fertile soil along streams. Scattered, mostly in eastern and northern Missouri.

Ranges from Indiana and Illinois to Missouri.

18. **Crataegus Reverchoni** Sarg. var. **discolor** (Sarg.) Palmer Map 1171

Flowers May.

Thickets and open woods along small streams. Southern Missouri from Jasper and Dallas to Butler counties. The typical var. is not known in Missouri.

Ranges from southern Missouri Arkansas, and southeast Kansas.

19. **Crataegus regalis** Beadle Map 1172

Two varieties are recognized:

Foliage and inflorescence glabrous; flowers 1.2–1.4 cm. wide . . 19a. C. REGALIS var. REGALIS

Foliage and inflorescence slightly villous; flowers about 1.5 cm. wide . . . . . 19b. C. REGALIS var. PARADOXA

19a. **Crataegus regalis** var. **regalis** Map 1172

Flowers May.

Thickets and open woods along small streams. Scattered in Ozark region.

Ranges from North Carolina to southern Indiana, Missouri, and Arkansas.

19b. **Crataegus regalis** var. **paradoxa** (Sarg.) Palmer Map 1172

1174 Crataegus permixta

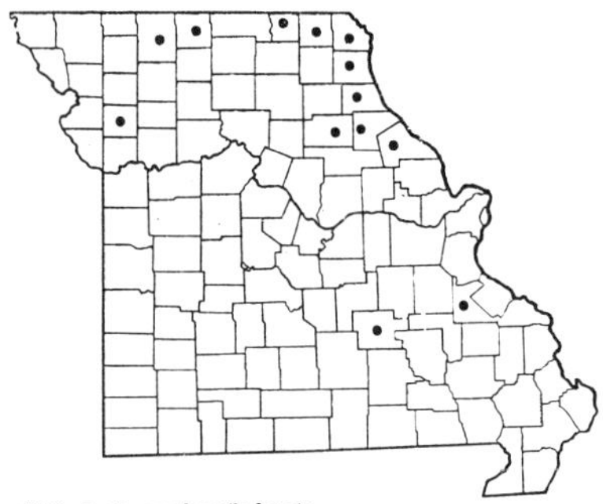

1175 Crataegus hannibalensis

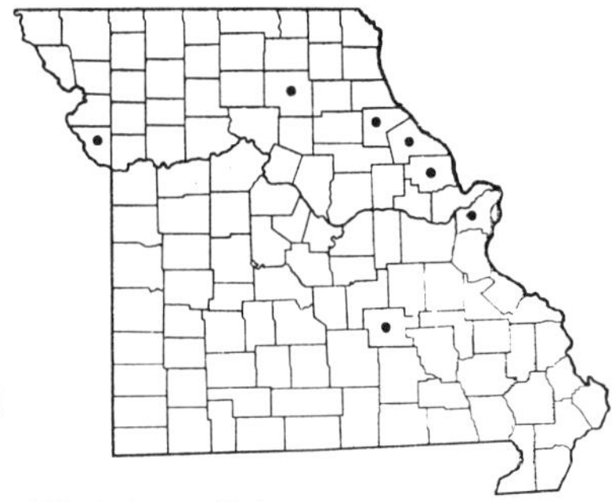

1176 Crataegus vallicola

Flowers May.

Thickets and rocky open woods. Southwestern Missouri, from Vernon to Newton counties.

Ranges from southwestern Missouri to southeastern Kansas and Arkansas.

20. **Crataegus Palmeri** Sarg. Map 1173

Flowers May.

Thickets and open woods along small streams. Southwestern Missouri and north to Pettis County.

Ranges from Missouri to southeast Kansas, northeast Oklahoma and Arkansas.

21. **Crataegus permixta** Palmer Map 1174

Flowers May.

Open uplands. Northeast Missouri, Marion to Pike County.

Ranges from northwestern Illinois to Missouri.

22. **Crataegus hannibalensis** Palmer Map 1175

Flowers May.

Thickets and open woods along bluffs and banks of streams. Northeast Missouri and scattered west to Clinton and south to Dent County.

Ranges from Ohio and Illinois to southeast Iowa and Missouri.

23. **Crataegus vallicola** Sarg. Map 1176

Flowers late April–May.

Open woods, in fertile soil. Scattered mostly in eastern Missouri.

Ranges from Ohio to Missouri.

24. **Crataegus Engelmanni** Sarg. Map 1177

Two varieties and two forms are recognized in Missouri:

a. Foliage and inflorescence pubescent or villous . *b*
 b. Leaves slightly pubescent along the veins beneath at maturity; stamens about 10; anthers pink or rarely white . . . . 24a. C. ENGELMANNI var. ENGELMANNI f. ENGELMANNI
 b. Leaves and flowering corymbs densely pilose-pubescent; stamens 10 or rarely 20; anthers pale yellow or rarely pink . . . 24c. C. ENGELMANNI var. SINISTRA
a. Foliage and inflorescence glabrous or essentially so . . . . . 24b. C. ENGELMANNI var. ENGELMANNI f. NUDA

24a. **Crataegus Engelmanni** var. **Engelmanni** f. **Engelmanni** Map 1177

Flowers May.

Thickets and rocky open woods. Southern Missouri, Ozark region.

Ranges from Illinois to Mississippi, Missouri and eastern Oklahoma.

24b. **Crataegus Engelmanni** var. **Engelmanni** f. **nuda** Palmer Map 1177

With the typical form. Scattered in southern Missouri. Not known elsewhere.

24c. **Crataegus Engelmanni** var. **sinistra** (Beadle) Palmer Map 1177

Flowers May.

Thickets and open woods. Scattered in southern Missouri.

Ranges from Tennessee to Missouri and Arkansas.

25. **Crataegus fecunda** Sarg. Map 1178

Flowers May,

---

Plate no. 197. **1.** Crataegus neobushii, × 2/5; *a.* Flowering branch; *b.* Fruiting branch. **2.** Crataegus biltmoreana, × 2/5; *a.* Fruit, × 4/5; *b.* Flower, × 4/5; *c.* Leaves, × 2/5. **3.** Crataegus sicca, × 2/5; *a, b, c.* Leaves. **4.** Crataegus permixta, × 2/5; *a, b, c.* Leaves. **5.** Crataegus pyracanthoides var. arborea. **6.** Crataegus collina, × 2/5; *a.* Flower, × 4/5; *b.* Fruit, × 4/5. **7.** Crataegus Reverchonii var. discolor, × 2/5. **8.** Crataegus incaedua, × 2/5. **9.** Crataegus noelensis, × 2/5. **10.** Crataegus dispessa, × 2/5; *a.* Flower, × 4/5.

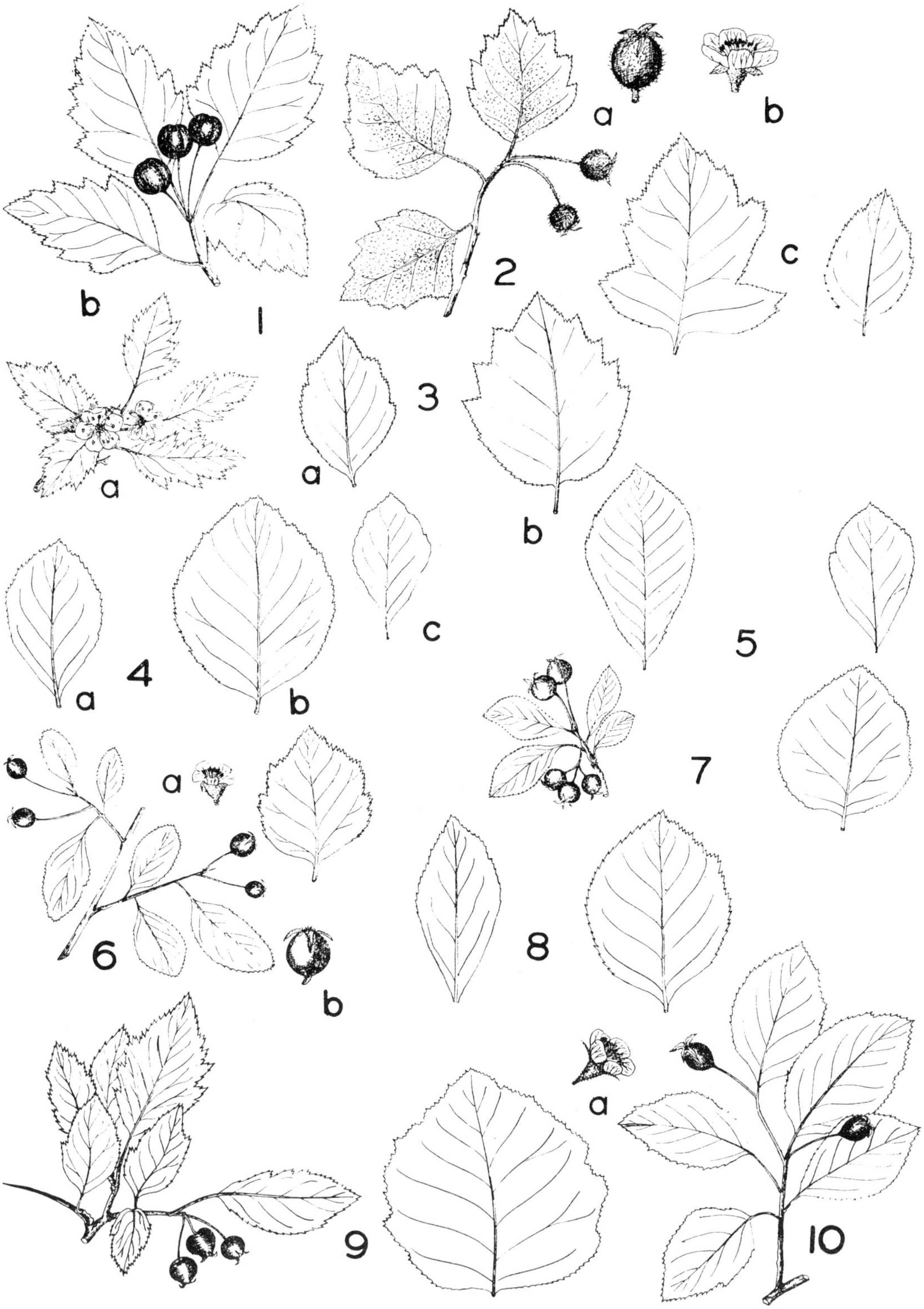


PLATE NO. 197

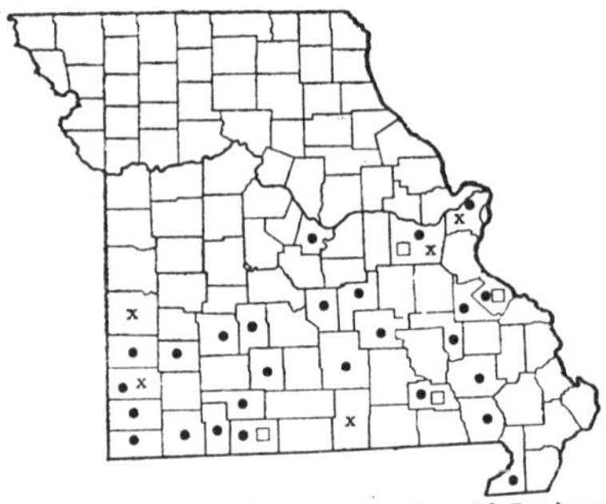

1177 • Crataegus Engelmanni var. Engelmanni f. Engelmanni
1177 x Crataegus Engelmanni var. Engelmanni f. nuda
1177 □ Crataegus Engelmanni var. sinistra

1178 Crataegus fecunda

1179 • Crataegus Danielsii f. Danielsii
1179 x Crataegus Danielsii f. glabra

Open woods, moist or fertile ground along streams. Scattered in southeast Missouri and north to Boone County.

Ranges from western Illinois and Missouri.

26. **Crataegus Danielsii** Palmer Map 1179

Two variations are recognized:

Foliage and inflorescence slightly villous . . . 26a. C. Danielsii f. Danielsii

Foliage and inflorescence glabrous . . . . 26b. C. Danielsii f. glabra

26a. **Crataegus Danielsii** f. **Danielsii** Map 1179

Limestone glades and hillsides. Boone County, Missouri. Not known elsewhere.

26b. **Crataegus Danielsii** f. **glabra** Palmer Map 1179

With the typical form.

Series 8. **Punctatae**

a. Leaves of flowering branchlets mostly obovate or oblong-obovate, broadest above the middle or near the middle; nutlets 2–5 . . . . . . . . . . . . . . . . . . . . . . . . . . . . . . *b*
  b. Leaves not lustrous above; nutlets 2–5, usually more than 2. . . . . . . . . . . . . . . . *c*
    c. Leaves of flowering branchlets pointed to rounded at apex; flowers 1.5–2 cm. wide. 27. C. collina
    c. Leaves of flowering branchlets pointed to short-acuminate at apex; flowers 1.8–2.3 cm. wide 28. C. verruculosa
  b. Leaves somewhat lustrous above at maturity; nutlets 2–3, usually 2 . . . . . . 29. C. incaedua
a. Leaves of flowering branchlets mostly oblong-obovate to elliptic, broadest about the middle; nutlets 4–5 . . . . . . . . . . . . . . . . . . . . . . . . . . . . . . *d*
  d. Terminal shoot leaves usually lobed; nutlets usually 5; flowers in May . . . . 30. C. Lettermani
  d. Terminal shoot leaves unlobed or obscurely lobed; nutlets 4–5; flowers in late April or early May 31. C. hirtiflora

27. **Crataegus collina** Chapm. Map 1180

Flowers late April–May.

Thickets and open woods along small streams.

Five varieties are recognized in Missouri:

a. Young branchlets, foliage and inflorescence more or less villous . . . . . . . . *b*
  b. Leaves firm, dull above; stamens 5–15 . *c*
    c. Leaves of flowering branchlets mostly obovate, elliptic or lance-elliptic, unlobed or obscurely lobed. . . . . *d*
      d. Leaves of flowering branchlets mostly 1.5–3 cm. wide; flowers 1.5–2 cm. wide; nutlets 3–5, usually 5. . . 27a. C. collina var. collina
      d. Leaves of flowering branchlets mostly 1.5–2.5 cm. wide; flowers 1.2–1.5 cm. wide; nutlets 2–4, usually 3 . . . 27b. C. collina var. succincta
    c. Leaves of flowering branchlets mostly obovate or oblong-obovate with shallow rounded lobes above the middle; stamens 5–10, usually 5 . . . . . 27c. C. collina var. secta
  b. Leaves thinnish, slightly lustrous above; stamens about 20 . . . 27d. C. collina var. sordida
a. Young branchlets, foliage and inflorescence glabrous or essentially so . . 27e. C. collina var. collicola

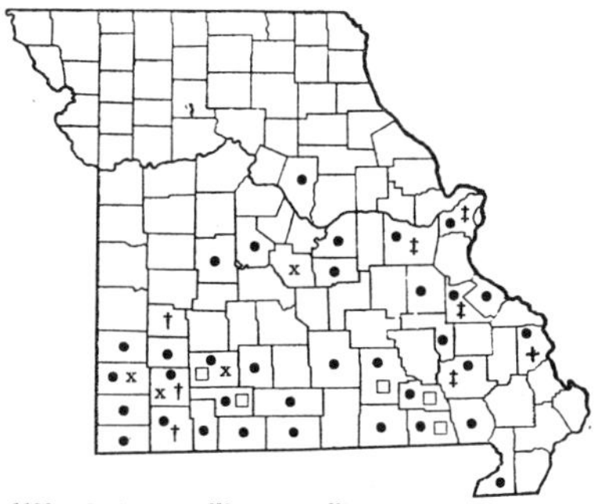

1180 • Crataegus collina var. collina
1180 ‡ Crataegus collina var. succincta
1180 x Crataegus collina var. secta
1180 □ Crataegus collina var. sordida
1180 † Crataegus collina var. collicola
1180 + Crataegus collina X viridis

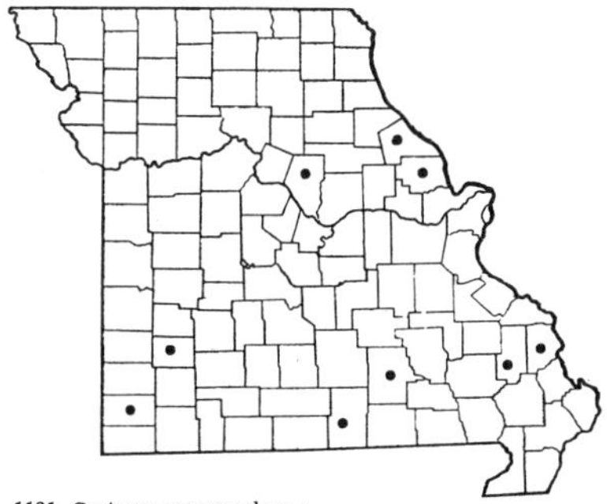

1181 Crataegus verruculosa

1182 Crataegus incaedua

27a. **Crataegus collina** var. **collina** Map 1180

Southern Missouri, north to St. Louis and Boone County.

Ranges from Virginia to South Carolina and Georgia, west to Indiana, eastern Oklahoma, and Arkansas.

27b. **Crataegus collina** var. **succincta** (Sarg.) Palmer Map 1180

Scattered in eastern Missouri, from St. Louis to Wayne County.

Ranges in southern Missouri and Arkansas.

27c. **Crataegus collina** var. **secta** (Sarg.) Palmer Map 1180

Southwestern Missouri, north to Miller County.

Ranges in southern Missouri and northern Arkansas.

27d. **Crataegus collina** var. **sordida** (Sarg.) Palmer Map 1180

Southern Missouri, from Ripley to Greene counties, and not known elsewhere.

27e. **Crataegus collina** var. **collicola** (Ashe) Palmer Map 1180

Southwestern Missouri, known only from Lawrence and Barry counties, but probably occurring elsewhere.

Ranges from Virginia and North Carolina to Kentucky and Missouri.

28. **Crataegus verruculosa** Sarg. Map 1181

Flowers late April–May.

Thickets and fertile open ground along streams. Scattered in central and southern Missouri from Pike and Boone counties to Newton County.

Ranges from Kentucky to Missouri and Arkansas.

29. **Crataegus incaedua** Sarg. Map 1182

Flowers late April or May.

Thickets and rocky open woods. Scattered in southern Missouri.

Ranges from southern Indiana to Missouri and Arkansas.

30. **Crataegus Lettermani** Sarg. Map 1183

Flowers late April–May.

Alluvial woods along streams. Rare and scattered from St. Louis to Jasper County.

Not known elsewhere. It may be a hybrid between some variety of *C. collina* and *C. mollis*.

31. **Crataegus hirtiflora** Sarg. Map 1184

Flowers April or May.

Open woods along stream, Taney County.

Known only from the type locality. This is probably also a hybrid of similar origin to the last species, though differing in several characters.

## Series 9. **Molles**

a. Leaves yellowish-green or bright green, firm to subcoriaceous, more or less pubescent beneath at maturity; anthers small, pale yellow or pink . . . . . . . . . . . . . . . . . . . . . *b*

b. Leaves mostly broadly ovate, deltoid or suborbicular; terminal shoot leaves often as broad as long or broader; stamens about 20 . . . . . . . . . . . . . . . . . . . . . . *c*

1183 Crataegus Lettermani

1184 Crataegus hirtiflora

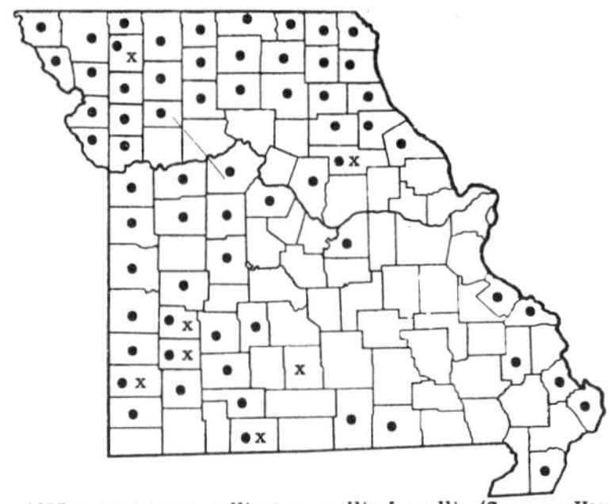

1185 • Crataegus mollis var. mollis f. mollis (Summer Haw)
1185 x Crataegus mollis var. mollis f. dumetosa

c. Leaves relatively large, sometimes 8–10 cm. wide at the ends of shoots, rounded to cordate at base; flowers 2–2.3 cm. wide; fruit becoming mellow or succulent, edible . . . . 32. C. MOLLIS

c. Leaves smaller, mostly ovate or oblong-ovate, abruptly narrowed, rounded or truncate at base; terminal shoot leaves sometimes suborbicular, up to 6–7 cm. wide; flowers 1.4–1.6 cm. wide; fruit remaining dry and hard . . . . . . . . . . . . . . . 33. C. KELLOGGII

b. Leaves mostly elliptic or oblong-ovate, distinctly longer than broad except sometimes at the ends of shoots; stamens 10 or 20 . . . . . . . . . . . . . . . . . . . . . . . . *d*

d. Leaves mostly oval or oblong-ovate; terminal shoot leaves usually ovate, rounded at base; flowers in compact several-flowered corymbs; stamens about 10 . . . . . . . . . . . *e*

e. Terminal shoot leaves up to 6–7 cm. wide; flowers 2–2.3 cm. wide; anthers pink or rarely yellow . . . . . . . . . . . . . . . . . . . . . . 34. C. NOELENSIS

e. Terminal shoot leaves seldom over 4–5 cm. wide, acute or narrowed at base; flowers about 1 cm. wide; anthers pale yellow . . . . . . . . . . . . . . . 35. C. LATEBROSA

d. Leaves mostly oblong-obovate or elliptic, broadest about the middle; terminal shoot leaves similar but larger, narrowed at base; flowers mostly 5–10, in compound pubescent corymbs; stamens about 20 . . . . . . . . . . . . . . . . . . . . . 36. C. DISPESSA

a. Leaves bluish-green, subcoriaceous, closely velutinous beneath at maturity; stamens about 20; anthers large, pink or rose-color; fruit remaining hard until late in the season, often slightly angled. . 37. C. LANUGINOSA

32. **Crataegus mollis** (T. & G.) Scheele
Summer Haw, Turkey Apple Map 1185
Flowers April.

Open woods, usually in fertile soil along small streams. Throughout most of the state, but commonest in northern Missouri and apparently absent from some of the Ozark counties.

Ranges from southern Ontario and Michigan to Alabama, west to Minnesota, eastern Kansas and eastern Oklahoma.

This species is quite variable in the shape and size of leaves, in the shape, size, color and time of ripening of the fruit. Only a single variety and a form have been recognized in Missouri.

*Crataegus mollis* var. *mollis* forma *dumetosa* (Sarg.) Palmer differs in the relatively narrower leaves, narrowed or cuneate at base. It is found occasionally with the typical form from Indiana to northeast Oklahoma.

*Crataegus mollis* is the state flower of Missouri, under the name Hawthorn. It is a handsome tree, distinguished for its large early flowers and its large bright red edible fruit. The fruit has been used for jelly and preserves; and it has great possibilities in cultivation through selection and spraying.

33. **Crataegus Kelloggii** Sarg. Map 1186
Flowers late April–early May.

Low open woods along streams. Rare and scattered in Missouri.

Ranges from Indiana to Missouri. Probably originated as a hybrid between *C. Margaretta* and *C. mollis.*

34. **Crataegus noelensis** Sarg. Map 1187
Flowers April.

Open woods, usually in fertile soil along streams. Southwestern Missouri from Henry and Benton to Taney and McDonald counties.

Ranges in Missouri and northern Arkansas.

35. **Crataegus latebrosa** Sarg. Map 1188
Flowers April–early May.

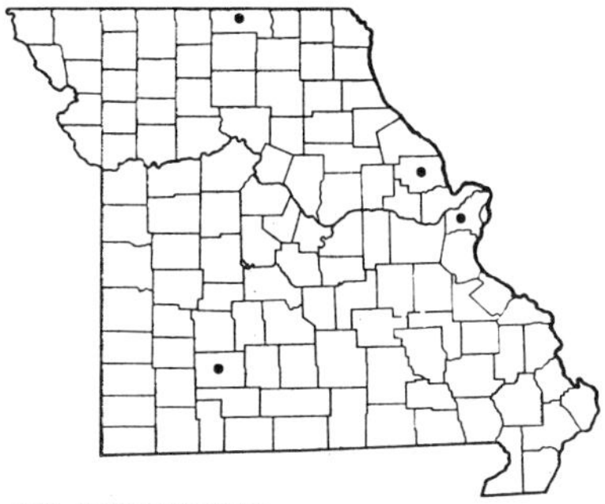

1186 Crataegus Kelloggii

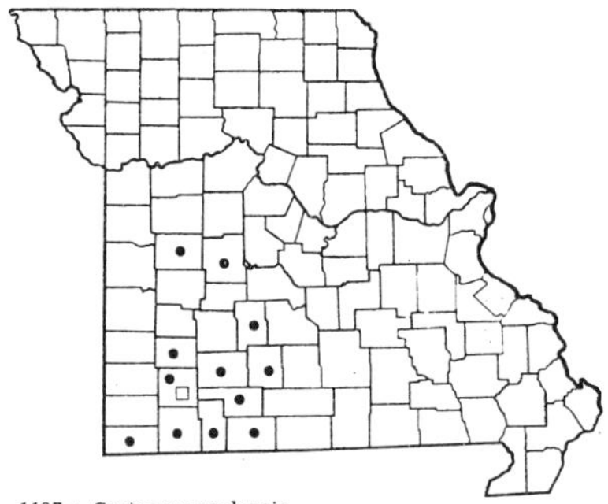

1187 • Crataegus noelensis
1187 □ Crataegus noelensis X collina

1188 Crataegus latebrosa

Open woods along stream. McDonald County.

Known only from type locality. Probably a hybrid between a variety of *C. collina* and *C. noelensis.*

36. **Crataegus dispessa** Ashe Map 1189

Flowers late April–May.

Open woods, usually in fertile soil along streams. Scattered in southern Missouri.

Ranges from southern Illinois to Missouri and northeastern Oklahoma.

37. **Crataegus lanuginosa** Sarg. Map 1190

Flowers April–early May.

Rocky open woods and thickets. Southwestern Missouri, from Greene to Jasper and Barry counties.

Ranges from Missouri to southeastern Kansas and Arkansas.

Series 10. **Pruinosae**

a. Foliage and inflorescence glabrous or essentially so; stamens 20 or rarely 10 . . . . . . . . . . *b*
  b. Leaves of flowering branchlets ovate or oblong-ovate, abruptly narrowed or rounded at base; terminal shoot leaves similar but larger, rounded to truncate at base; fruit sometimes 5-angled. . . *c*
    c. Leaves mostly ovate, usually sharply lobed, firm or thick; terminal shoot leaves seldom over 5–6 cm. wide; flowers mostly 5–10, in slightly branched open corymbs; fruit firm pruinose . 38. C. PRUINOSA
    c. Leaves ovate to oblong-ovate, thin but firm; terminal shoot leaves up to 7–8 cm. wide; flowers 3–6, in compact corymbs; fruit not pruinose . . . . . . . . . . 39. C. DISJUNCTA
  b. Leaves of flowering branchlets ovate or deltoid, rounded, truncate or subcordate at base; terminal shoot leaves sometimes as broad as long, subcordate or cordate at base . . . . . . . . . . *d*
  d. Leaves of flowering branchlets oblong-ovate or narrowly deltoid, sharply lobed, the terminal lobe elongated; terminal shoot leaves broadly ovate or deltoid, up to 4–5 cm. wide . 40. C. GATTINGERI
  d. Leaves of flowering branchlets ovate or oblong-ovate, relatively broad, more or less lobed, the terminal lobe not noticeably elongated; terminal shoot leaves up to 6–8 cm. wide. . . . . . . . *e*
    e. Flowers 2–2.2 cm. wide; fruit subglobose or depressed-globose, 1.5–1.8 cm. thick, bright red or orange-red . . . . . . . . . . . . . . . . . . . . . 41. C. PLATYCARPA
    e. Flowers 1.6–2 cm. wide; fruit subglobose, sometimes slightly 5-angled, 0.9–1.6 cm. thick, dull red or green at maturity . . . . . . . . . . . . . . . . . . . . . . *f*
      f. Leaves of flowering branchlets seldom over 4 cm. wide; terminal shoot leaves up to 5–6 cm. wide; fruit with a prominent elevated calyx; nutlets 3–5 . . 42a. C. MACKENZII var. MACKENZII
      f. Leaves of flowering branchlets 4–5 cm. wide; terminal shoot leaves 5–7 cm. wide; fruit with slightly elevated calyx . . . . . . . . . . . . . . . . . 43. C. RUGOSA
a. Foliage and inflorescence villous or pubescent, at least while young; stamens about 20 . . . . . . *g*
  g. Leaves bluish green, sparsely short-pilose above while young and more or less pubescent along the veins beneath throughout the season; flowers 1.7–1.8 cm. wide, 5–7 in compact corymbs; anthers pink; fruit about 1 cm. thick . . . . . . . . . . . 42c. C. MACKENZII var. ASPERA
  g. Leaves yellowish green, copiously short-pilose above while young and slightly pubescent along the veins beneath throughout the season; flowers 2–2.4 cm. wide, mostly 7–10 in compound villous corymbs; fruit 1–1.3 cm. thick . . . . . . . . . . . . . . 44. C. LOCUPLES

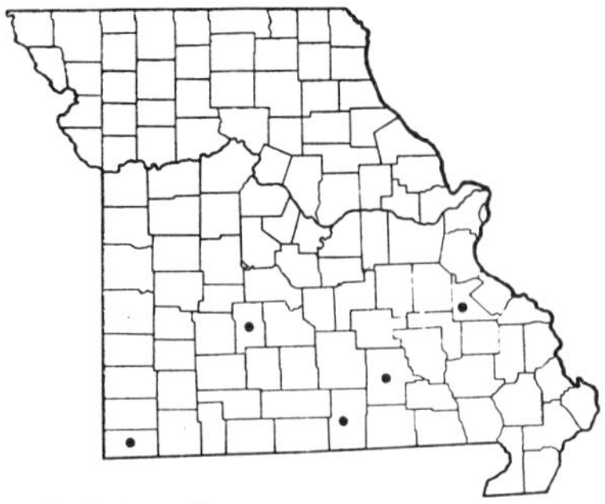

1189 Crataegus dispessa

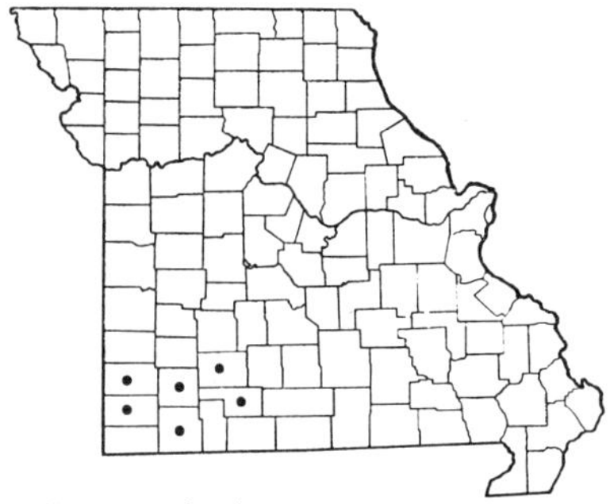

1190 Crataegus lanuginosa

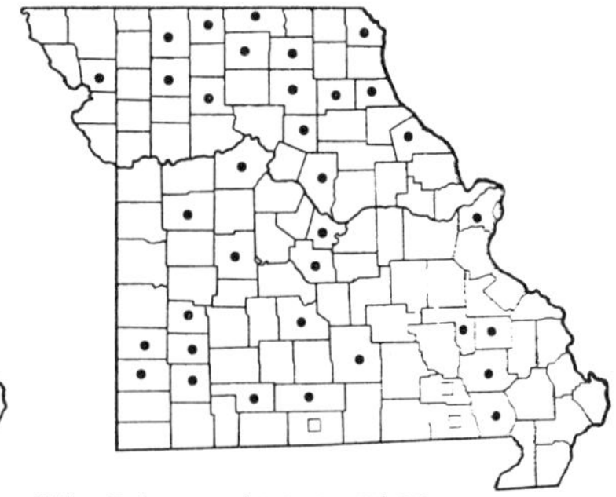

1191 • Crataegus pruinosa var. pruinosa
1191 □ Crataegus pruinosa var. brachypoda

38. **Crataegus pruinosa** (Wendl.) K. Koch Map 1191

Flowers late April–May.

Rocky open woods and thickets.

Two varieties are recognized in Missouri:

Flowers 1.7–2 cm. wide; stamens about 20; anthers pink or rarely creamy white; fruiting calyx large and elevated . . . . C. PRUINOSA var. PRUINOSA

Flowers 1.5–1.7 cm. wide; stamens 10 or 20; anthers creamy white; fruiting calyx nearly sessile . . . . C. PRUINOSA var. BRACHYPODA

38a. **Crataegus pruinosa** var. **pruinosa** Map 1191

Scattered throughout Missouri.

Ranges from Newfoundland, New England to North Carolina, Wisconsin, Kentucky, and northern Arkansas.

38b. **Crataegus pruinosa** var. **brachypoda** (Sarg.) Palmer Map 1191

Southeastern Missouri, Carter and Ripley counties.

Ranges from Kentucky to southern Missouri and northern Arkansas.

39. **Crataegus disjuncta** Sarg. Map 1192

Flowers late April–May.

Thickets and open woods along small streams. Scattered in southern Missouri, from St. Louis to Stoddard and west to Jasper counties.

Ranges from Ohio and Kentucky to Missouri.

40. **Crataegus Gattingeri** Ashe Map 1193

Flowers late April–May.

Thickets and rocky open woods. Southern Missouri, mostly in granitic region of eastern Ozarks, and also in Newton County.

Ranges from Pennsylvania and Ohio to Tennessee, Missouri, and Arkansas.

41. **Crataegus platycarpa** Sarg. Map 1194

Flowers late April–early May.

Rocky open woods. Eastern Missouri, from Lincoln to Dent and Butler counties.

Ranges from southern Illinois to Missouri and Arkansas.

42. **Crataegus Mackenzii** Sarg. Map 1195

Flowers late April–May.

Rocky open woods and thickets.

Three varieties are recognized in Missouri:

a. Foliage and inflorescence glabrous or essentially so . . . . . . . . . . . *b*
  b. Leaves broadly ovate or deltoid, rounded to truncate or at the ends of shoots subcordate at base; fruit 1–1.6 cm. thick . 42a. C. MACKENZII var. MACKENZII
  b. Leaves mostly ovate, abruptly narrowed, rounded or truncate at base; fruit 0.9–1.2 cm. thick . 42b. C. MACKENZII var. BRACTEATA
a. Foliage and inflorescence more or less villous 42c. C. MACKENZII var. ASPERA

42a. **Crataegus Mackenzii** var. **Mackenzii** Map 1195

Scattered throughout Missouri, but commonest in southern and central counties.

Ranges from Kentucky to southeastern Iowa, Missouri, eastern Oklahoma, and Arkansas.

Plate no. 198. **1.** Crataegus lanuginosa, × $^{2}/_{5}$; *a.* Flower, × $^{4}/_{5}$. **2.** Crataegus disjuncta, × $^{2}/_{5}$. **3.** Crataegus Gattingeri, × $^{2}/_{5}$; *a.* Flower, × $^{4}/_{5}$. **4.** Crataegus Mackenzii, × $^{2}/_{5}$. **5.** Crataegus coccinioides, × $^{2}/_{5}$. **6.** Crataegus succulenta, × $^{2}/_{5}$. **7.** Crataegus Calpodendron, × $^{2}/_{5}$; *a.* Flower, × $^{4}/_{5}$; *b.* Fruit, × $1^{1}/_{5}$. **8.** Crataegus Vailiae, × $^{2}/_{5}$. **9.** Crataegus mollis, × $^{2}/_{5}$. **10.** Crataegus pruinosa, × $^{2}/_{5}$; All details from Small, The New York Botanical Garden.

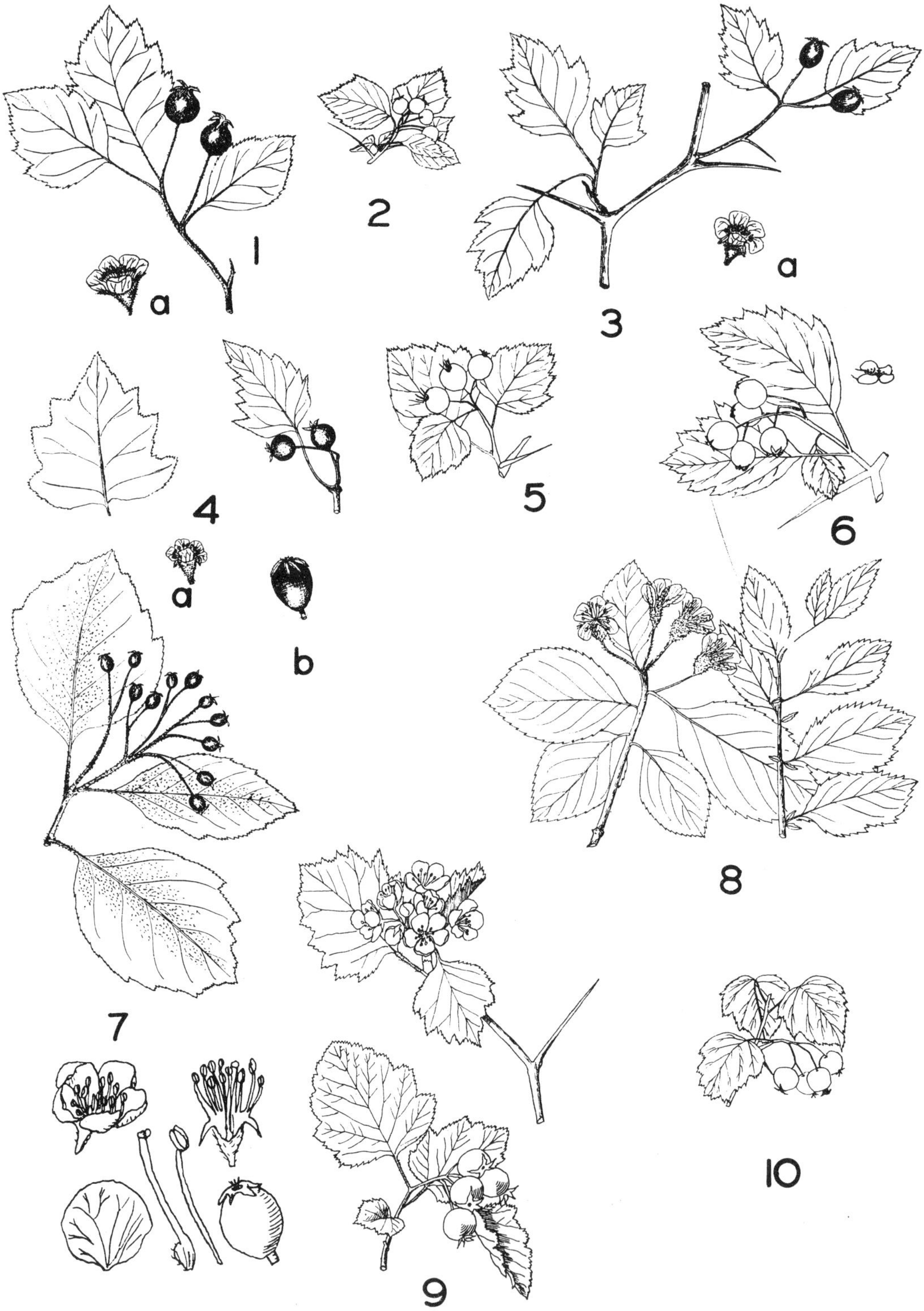


PLATE NO. 198

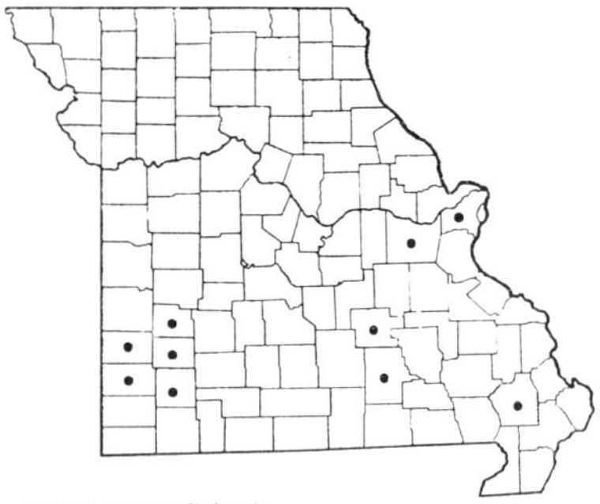

1192 Crataegus disjuncta

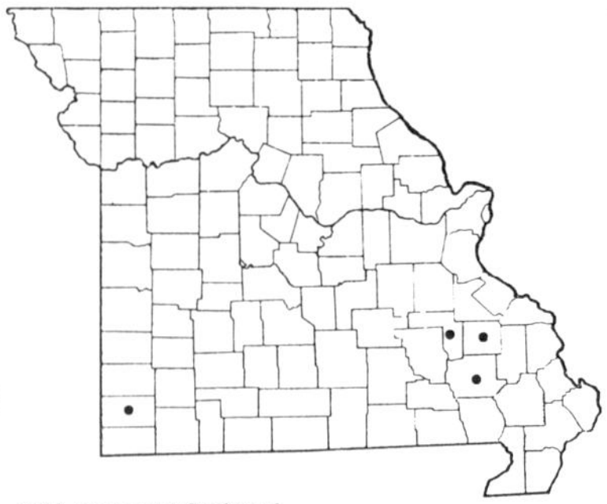

1193 Crataegus Gattingeri

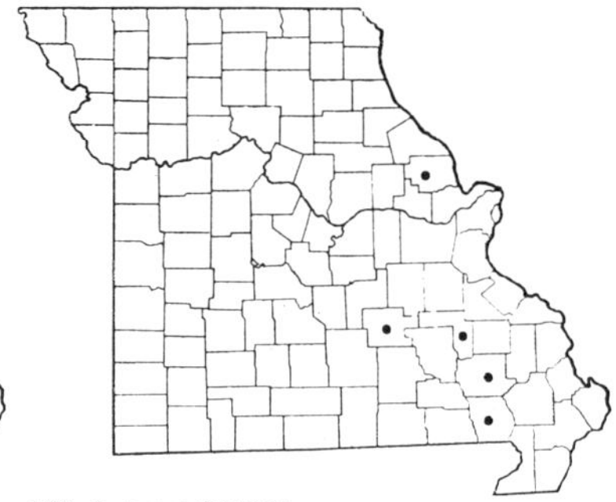

1194 Crataegus platycarpa

42b. **Crataegus Mackenzii** var. **bracteata** (Sarg.) Palmer — Map 1195

Southern Ozark region, north to Boone and Johnson counties.

Ranges from Ohio to Missouri, southeast Kansas, northeast Oklahoma, and Arkansas.

42c. **Crataegus Mackenzii** var. **aspera** (Sarg.) Palmer — Map 1195

Southern Missouri, from Jasper to Ripley County.

Ranges from southern Missouri and northern Arkansas.

43. **Crataegus rugosa** Ashe — Map 1196

Flowers late April–May.

Dry open woods, thickets and rocky fields. Scattered in northern and central Missouri.

Ranges from Pennsylvania to South Carolina, west to Indiana, southeastern Iowa, and Missouri.

44. **Crataegus locuples** Sarg. — Map 1197

Flowers late April–May.

Open woods, usually in fertile ground along streams. Rare and scattered in Ozark region, from St. Louis to Dent and Carter counties.

Ranges from Ohio and Kentucky to Missouri.

## Series 11. **Dilatatae**

Only one species in Missouri . . . . . . . . . . . . . . . . . . . . 45. C. COCCINIOIDES

45. **Crataegus coccinioides** Ashe — Map 1198

Flowers late April–May.

Thickets and open woods, mostly on limestone hills. Southwestern Missouri, from St. Clair to Newton, Barry, and Christian counties, also in St. Louis and Franklin counties.

Ranges from southern Illinois to Missouri, southeastern Kansas and northern Arkansas.

## Series 12. **Macracanthae**

a. Leaves of flowering branchlets mostly ovate or oblong-ovate, pointed to rounded at apex; terminal shoot leaves similar but larger and more deeply lobed . . . . . . . . . . . . . . . *b*
  b. Leaves thin but firm, or if subcoriaceous, with the veins distinctly impressed above; flowers mostly 5–15, in compound corymbs; nutlets 2–4, often 2 . . . . . . . . . . 46. C. SUCCULENTA
  b. Leaves firm, dull yellowish green, short villous above while young, and persistently pubescent at least along the veins beneath; terminal shoot leaves up to 7–9 cm. wide; flowers mainly 15 or more; nutlets 2–3, often 2 . . . . . . . . . . . . . . . . . . 47. C. CALPODENDRON
a. Leaves of flowering branchlets mostly oval or elliptic, acute or acuminate at apex; terminal shoot leaves broadly ovate to suborbicular, unlobed or slightly lobed, up to 8 cm. wide. . . . . . . . . *c*
  c. Leaves thin but firm, dull green; terminal shoot leaves broadly ovate or oblong-ovate, sometimes slightly lobed; flowers 3–13, in nearly simple or compact compound corymbs; nutlets 2–3. . . . . *d*
    d. Terminal shoot leaves similar to those of the flowering branchlets except large, relatively wider and more coarsely serrate, 4–6 cm. wide, in many-flowered compact branching corymbs; nutlets usually 2 . . . . . . . . . . . . . . . . . . . . . . 48. C. NUDA

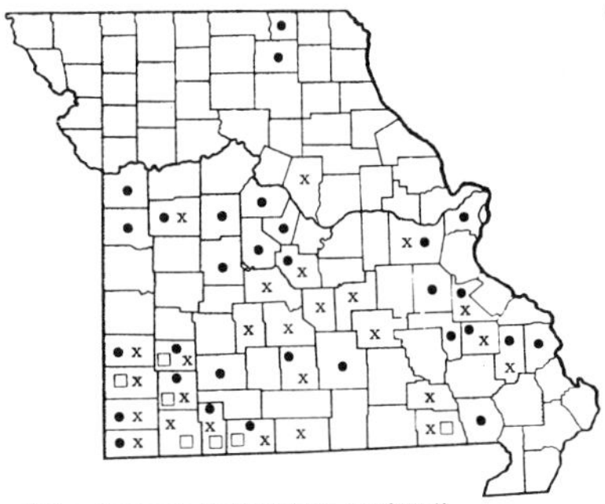

1195 • Crataegus Mackenzii var. Mackenzii
1195 x Crataegus Mackenzii var. bracteata
1195 □ Crataegus Mackenzii var. aspera

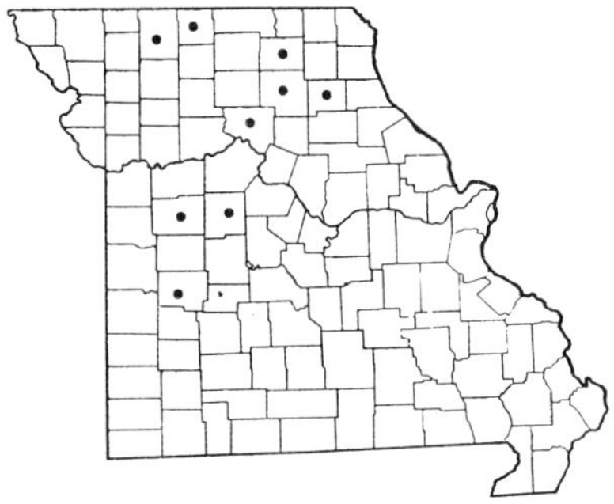

1196 Crataegus rugosa

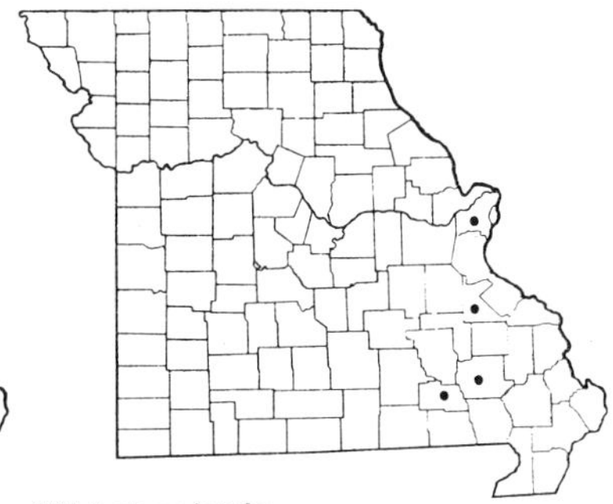

1197 Crataegus locuples

d. Terminal shoot leaves broadly oval to suborbicular, up to 6 cm. wide; flowers 1.6–1.8 cm. wide, in many-flowered slightly villous corymbs; nutlets usually 3 . . . . . . . . 49. C. SIMULATA

c. Leaves subcoriaceous, the veins slightly impressed above, unlobed or sometimes slightly lobed especially at the ends of shoots; flowers mostly 2–6, in simple tomentose corymbs; nutlets usually 3 . . . . . . . . . . . . . . . . . . . . . 50. C. VAILIAE

46. **Crataegus succulenta** Link Map 1199

Flowers May–early June.

Rocky woods and bluffs.

Three varieties are recognized in Missouri:

a. Leaves green both sides or only slightly paler beneath, glabrous or nearly so at maturity . *b*
  b. Stamens about 20; fruit 7–12 mm. thick, becoming succulent . . 46a. C. SUCCULENTA var. SUCCULENTA
  b. Stamens about 10 or rarely more; fruit 6–8 mm. thick, remaining hard and dry or slightly mellow at maturity. . . . . 46b. C. SUCCULENTA var. NEOFLUVIALIS

a. Leaves dark green above, much paler and permanently pubescent beneath . . . . 46c. C. SUCCULENTA var. PERTOMENTOSA

46a. **Crataegus succulenta** var. **succulenta** Map 1199

Rare and scattered throughout Missouri.

Ranges from southeastern Canada, New England and New York west to Iowa and Missouri.

46b. **Crataegus succulenta** var. **neofluvialis** (Ashe) Palmer Map 1199

Known only from Montgomery and Christian counties.

Ranges from southern Ontario, New England and New York to North Carolina, west to Missouri.

46c. **Crataegus succulenta** var. **pertomentosa** (Ashe) Palmer Map 1199

Scattered throughout Missouri, but commonest in northern Missouri.

Ranges from Illinois to Iowa, Missouri, and eastern Kansas.

47. **Crataegus Calpodendron** (Ehrh.) Medic. Map 1200

Flowers May–early June.

Rocky open woods and bluffs.

Five varieties are recognized in Missouri:

a. Leaves with usually 5–6 pairs of primary veins, slightly impressed above at maturity . . . *b*
  b. Leaves firm, more or less pubescent beneath at maturity; stamens about 20 . . *c*
    c. Fruit usually oblong or obovoid, becoming soft and succulent in October . *d*
      d. Leaves serrate with shallow teeth and broad shallow often irregular lobes; fruit 7–9 mm. thick . . . 47a. C. CALPODENDRON var. CALPODENDRON
      d. Leaves deeply serrate with sharp narrow teeth; lobes small, sharp and spinulose; fruit 6–7 mm. thick . . 47b. C. CALPODENDRON var. HISPIDULA
    c. Fruit subglobose or short-oblong, remaining hard and dry or only becoming mellow in November . . . . . 47c. C. CALPODENDRON var. GLOBOSA
  b. Leaves thick, velvety-pubescent beneath at maturity; stamens about 10 . . . . 47d. C. CALPODENDRON var. MOLLICULA

a. Leaves with usually 6–8 pairs of primary veins, distinctly impressed above at maturity; fruit subglobose, 6–8 mm. thick, becoming succulent . 47e. C. CALPODENDRON var. MICROCARPA

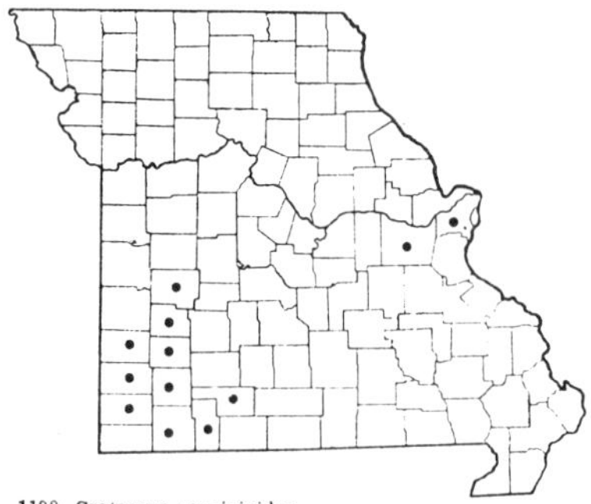

1198 Crataegus coccinioides

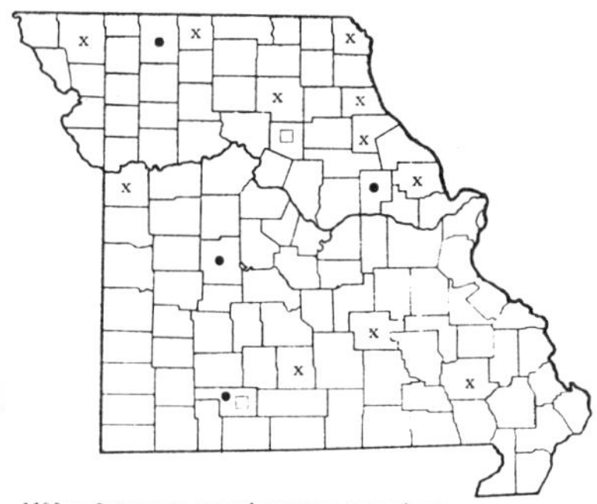

1199 • Crataegus succulenta var. succulenta
1199 □ Crataegus succulenta var. neofluvialis
1199 x Crataegus succulenta var. pertomentosa

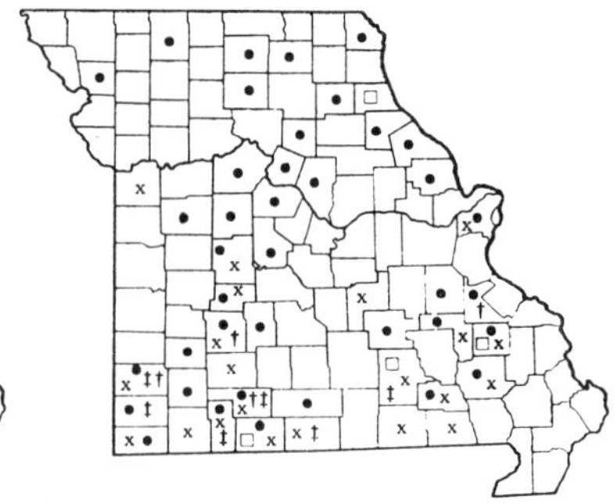

1200 • Crataegus Calpodendron var. Calpodendron
1200 † Crataegus Calpodendron var. hispidula
1200 x Crataegus Calpodendron var. globosa
1200 ‡ Crataegus Calpodendron var. mollicula
1200 □ Crataegus Calpodendron var. microcarpa

47a. **Crataegus Calpodendron** var. **Calpodendron** Map 1200

Scattered throughout Missouri except in southeastern lowlands.

Ranges from southern Ontario and New York to Georgia and Alabama, west to Minnesota and Missouri.

47b. **Crataegus Calpodendron** var. **hispidula** (Sarg.) Palmer Map 1200

Scattered in southern Missouri from Iron to Jasper counties.

Ranges in Missouri, southeastern Kansas and northern Arkansas.

47c. **Crataegus Calpodendron** var. **globosa** (Sarg.) Palmer Map 1200

Scattered in southern Missouri north to St. Louis and Jackson counties.

Ranges from Kentucky to eastern Kansas, Arkansas, and eastern Texas.

47d. **Crataegus Calpodendron** var. **mollicula** (Sarg.) Palmer Map 1200

Southwestern Missouri, Jasper to Shannon counties.

Ranges in Missouri and Arkansas.

47e. **Crataegus Calpodendron** var. **microcarpa** (Chapm.) Palmer Map 1200

Rare and scattered in Missouri.

Ranges from Virginia and North Carolina to Missouri.

48. **Crataegus nuda** Sarg. Map 1201

Flowers late April–May.

Rocky open woods. Taney County, Missouri.

Not known except from type locality. Perhaps a hybrid between *C. crus-galli* and a var. of *C. succulenta.*

49. **Crataegus simulata** Sarg. Map 1202

Flowers May.

Moist open woods along streams. Local, Jasper and Lawrence counties.

Not known elsewhere. Perhaps a hybrid between *C. Calpodendron* and *C. Palmeri.*

50. **Crataegus Vailiae** Britt. Map 1203

Known in Missouri only from Shannon (near Monteer, October 6, 1920, *Palmer 19329*) and Ripley counties.

Ranges from Virginia and North Carolina to southern Missouri.

Rare, and probably a hybrid between *C. Calpodendron* and *C. uniflora.*

Besides the species and other taxa listed above, some of which may be of hybrid origin, as suggested, several other unnamed supposed hybrids have been found in Missouri as follows:

*Crataegus collina* × *C. noelensis* Lawrence Co. Map 1187
*Crataegus collina* × *C. Palmeri* Lawrence Co. Map 1173
*Crataegus collina* × *C. viridis* Cape Girardeau Co. Map 1180
*Crataegus Palmeri* × *C. viridis* Jasper Co. Map 1173
*Crataegus Engelmannii* × *C. uniflora* Ozark Co. Map 1157
*Crataegus Margaretta* × *C. mollis* Greene Co. Map 1165

These and other hybrids can be recognized by the characters they show that are intermediate between those of the supposed parent species. But such identifications can usually only be hypothetical, and hybrids of the same parentage can be expected to differ considerably from each other in appearance and characters.

1201 Crataegus nuda

1202 Crataegus simulata

1203 Crataegus Vailiae

## 8. **Fragaria** L. Strawberry

a. Hairs on main flower stem (scape) and sometimes the leaf-stalks (petioles) pressed against the surface lengthwise, not spreading; topmost tooth of the leaflets higher than the ones on either side of it; petals mainly 4–7 mm. long; leaves thin, membranous; fully grown inflorescence often longer than the leaves; inflorescence appearing raceme-like, the main branches of the few flowers unequal and the leading one elongated as the axis; calyx-lobes spreading to turned downward from the young fruit; seed-like achenes scattered over the surface of the ripe, pulpy receptacle, but not set in pits below the surface . . . . . . . . . . . . . . . . . . . . . . 1. F. VESCA var. AMERICANA

a. Hairs on main flower-stem (scape) and leaf-stalks (petioles) spreading; topmost tooth of the leaflets usually shorter and smaller than the ones on either side of it; petals mainly 7–14 mm. long; fully grown inflorescence usually shorter than the leaves; inflorescence appearing flat and umbel-like, the main branches of the more numerous 5–10 flowers more or less equal; calyx-lobes pressed upon or surrounding the young fruit; seed-like achenes set in pits below the surface of the ripe, pulpy receptacle . . . . . . . . . . . . . . . . . . . . . . . . . . . . *b*

b. Common native plants; petals 7–10 mm. long; leaves firm, not noticeably thick, the leaf surface not conspicuously furrowed with deeply set veins and veinlets, the lower surface pale or gray green; fruit 1–1.5 cm. in diameter; leaf-stalks (petioles) 1–2 mm. thick; plants mostly 5–25 cm. tall . . . . . . . . . . . . . . . . . . . . . . . . 2. F. VIRGINIANA

b. Garden plants rarely escaped from cultivation; petals mainly 10–14 mm. long; leaves thick, the leaf-surface conspicuously furrowed with deeply set veins and veinlets, the lower surface silvery white or reddish; fruit mainly 2–3 cm. in diameter; leaf-stalks (petioles) 2–3 mm. thick; plants mostly 20–40 cm. tall. . . . . . . . . . . . . . . . . . . . . 3. F. × ANANASSA

1. **Fragaria vesca** L. var. **americana** Porter
Wild Strawberry Map 1204
Flowers April–May.

Occurs on rich, usually north-facing, steep wooded slopes and bluffs along streams. Rare in the northeastern and east-central section of the state. In their *Annotated Catalogue* (p. 565) Palmer and Steyermark record this strawberry from Jasper County, but the specimen is now referred to *F. virginiana* var. *illinoensis.*

Ranges from Quebec and Newfoundland to Manitoba and South Dakota, south to Virginia, Ohio, Indiana, Illinois, Missouri, Nebraska, and New Mexico.

Typical *F. vesca* var. *vesca* with the pubescence spreading on the petioles and main flower-stem (scape) is not known from Missouri.

2. **Fragaria virginiana** Duchesne
Wild Strawberry Map 1205
Flowers April–May.

Occurs on open slopes, prairies, along railroads, openings in woodland, and cleared places. Throughout Missouri.

Two variations are represented in Missouri material:

Hairs on the flower-stalks (pedicels) pressed lengthwise on the stalk or strongly ascending; more rarely encountered . . . 2a. F. VIRGINIANA var. VIRGINIANA f. VIRGINIANA

Hairs on the flower-stalks (pedicels) more or less widely spreading; commonly encountered . . 2b. F. VIRGINIANA var. ILLINOENSIS

2a. **Fragaria virginiana** var. **virginiana** f. **virginiana** Map 1205

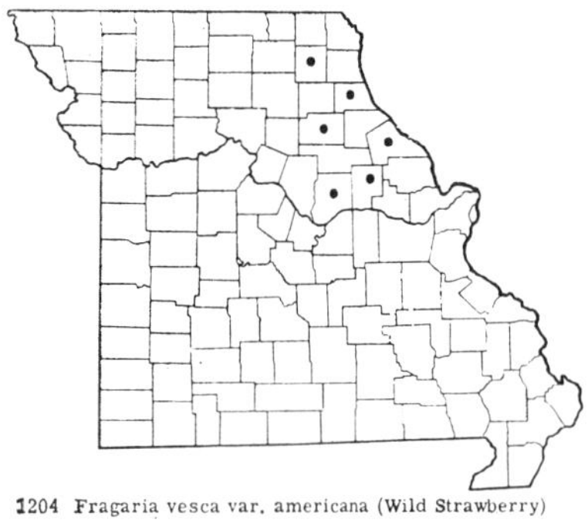

1204 Fragaria vesca var. americana (Wild Strawberry)

1205 • Fragaria virginiana var. virginiana f. virginiana (Wild Strawberry)
1205 □ Fragaria virginiana var. illinoensis

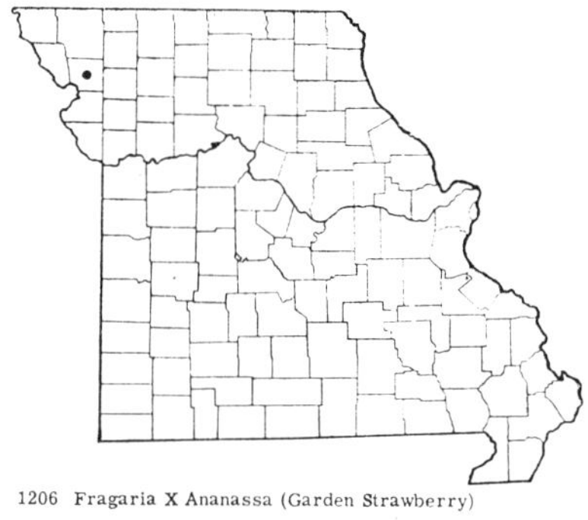

1206 Fragaria X Ananassa (Garden Strawberry)

*Fragaria virginiana* Duchesne [G, P & S, Steyerm.]
*Fragaria virginiana* var. *virginiana* [BB]

Rare in Missouri, where known from Callaway, Greene, and Jackson counties.

Ranges from Newfoundland and Quebec to Alberta, south to Georgia, Tennessee, Missouri, and Oklahoma.

2b. **Fragaria virginiana** var. **illinoensis** (Prince) Gray — Map 1205

This is the common type throughout Missouri. Although it has not been recorded from the extreme southeastern sector of the state, it is probably to be found in every county.

Ranges from New York to Minnesota, south to Alabama and Louisiana.

The ripe berries, although smaller than those of the cultivated type, are considerably sweeter, and are used fresh, served on short-cakes or with ice cream or made into jam. The cooked leaves can be used as tea.

3. **Fragaria** × **Ananassa** Duchesne
Garden Strawberry — Map 1206
*Fragaria chiloensis* Duchesne var. *Ananassa* (Duchesne) Hort. ex Bailey [BB, P & S, Steyerm.]
*Fragaria chiloensis* × *virginiana* Duchesne [Jones et al.]

Flowers April–May.

Commonly cultivated throughout the state, but rarely escaping and naturalizing. Known only from Andrew County in northwestern Missouri (prairies near Savannah, May 16, 1929, *Palmer 35746*).

The common cultivated strawberry is considered to be of hybrid origin between *F. chiloensis* and *F. virginiana*, where it was first developed in France.

## 9. **Duchesnea** Sm. Indian Strawberry

**Duchesnea indica** (Andr.) Focke
Indian Strawberry — Map 1207

Flowers April–June.

Occurs in open woods, prairies, and waste ground. Rare and scattered in southern and central Missouri in Dunklin, Shannon, Dent, Cole, and Jackson counties.

Native of Asia; introduced and naturalized in the United States from Florida to Oklahoma, north to Connecticut, New York, Ohio, Indiana, and Iowa.

The red fruit resembles strawberries, but is not juicy and is poor in taste.

## 10. **Waldsteinia** Willd. Barren Strawberry

**Waldsteinia fragarioides** (Michx.) Tratt.
Barren Strawberry — Map 1208

Flowers April 6–May.

Occurs on shaded, moist humus of steep wooded

---

Plate no. 199. **1.** Fragaria vesca var. americana, × 2/7. **2.** Fragaria virginiana var. illinoensis, × 2/7; *a*. Fruiting cluster; Details from Small, The New York Botanical Garden. **3.** Fragaria × Ananassa, × 2/7; Details from Small, The New York Botanical Garden. **4.** Duchesnea indica, × 2/7; Details from Small, The New York Botanical Garden. **5.** Waldsteinia fragarioides, × 2/7; Details from Small, The New York Botanical Garden. **6.** Potentilla simplex, × 2/7.

PLATE NO. 199

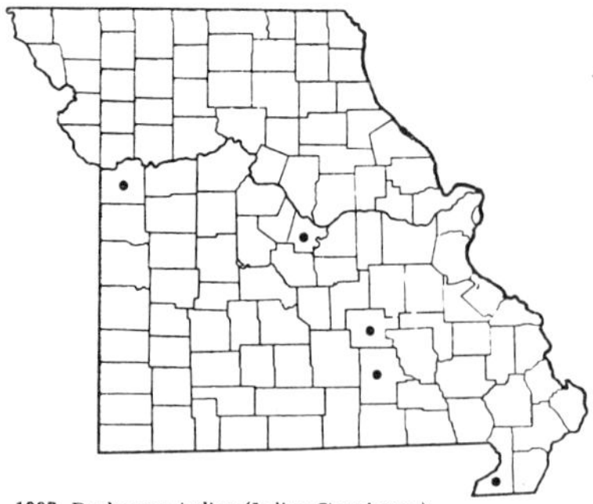

1207 Duchesnea indica (Indian Strawberry)

1208 Waldsteinia fragarioides (Barren Strawberry)

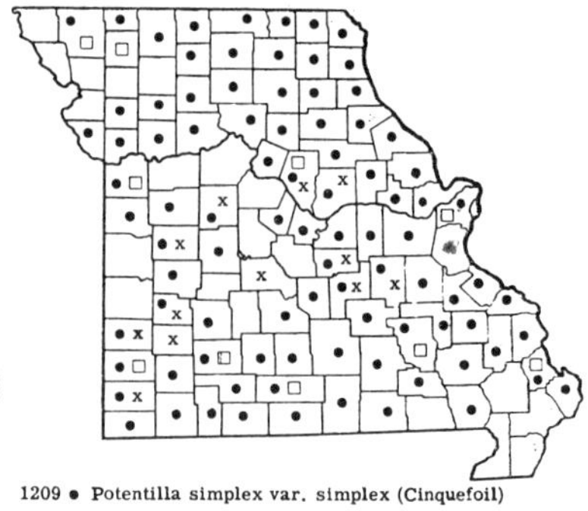

1209 • Potentilla simplex var. simplex (Cinquefoil)
1209 □ Potentilla simplex var. calvescens
1209 x Potentilla simplex var. argyrisma

north-facing slopes and ledges near the base of or on the upper portions of bluffs of Roubidoux sandstone or cherty limestone exposures. Very rare and known only from the south-central Ozarks in Texas (Roubidoux sandstone and chert near the top of slopes above limestone bluffs along Jack's Fork of Current River, 5 mi. southeast of Arroll, April 13, 1935, *Steyermark 18597*; 1 mile downstream from Harlow Ford, 4½ mi. southeast of Arroll, August 16, 1934, *Steyermark 14605*), Dallas (cherty steep slopes at edge of and above steep north-facing wooded bluffs along Niangua River, T34N, R18W, sect. 8, 1½ mi. south of Windyville, June 2, 1951, *Steyermark 71475*), and Douglas (acid humus soil in upland pine-oak wooded slopes around Roubidoux sandstone outcrops associated with *Vaccinium vacillans*, east of Bryant Creek, southeast of Sweden, April 6, 1935, *Steyermark 18528*; lower cherty limestone ledges of steep north-facing bluffs along Little Indian Creek, T27N, R11W, SW¼ sect. 24 and SE¼ sect. 23, 7 mi. [by air] southeast of Ann, October 19, 1957, *Steyermark 85971*) counties.

Ranges from Maine and Quebec to Ontario and Minnesota, south to Georgia, Tennessee, Indiana, and Missouri.

This low-growing herb has attractive yellow flowers with dark green wedge-shaped, strawberrylike, evergreen leaves. Plants are usually found in small colonies as a result of the creeping underground stems. At some of the Missouri stations several hundred plants are found in a small area occupying only several yards in length. Considering the long period in time during which this species has undoubtedly occupied the Ozark region, since at least before the last glacial advance, it is surprising that the colonies of plants are not more extensive in area occupied. Apparently a relict plant from a more northerly distributed range, it has survived in only a few choice situations where it has been able merely to survive to a limited degree where competition from other species is not too severe.

The plant makes a handsome one for the woodland wildflower or shaded rock garden where it spreads over moss-covered rocky slopes having good drainage. It grows well in humus soils or loose, rocky situations, preferably in acid soils.

## 11. **Potentilla** L. Cinquefoil, Five-finger

a. Flowers solitary and remote, arising on long leafless flower-stalks (pedicels) from the axils of the leaves; stems either trailing or some of them arching or eventually rooting at the tip, or the early stems ascending to erect . . . . . . . . . . . . . . . . . . . . . . . . *b*

b. First flower arises from the first node (joint) located above the first well-developed internode; stem at flowering time creeping or lying along the ground; middle leaflet of the leaves obovate-wedge-shaped, the length usually less than twice the width; found only in the Ozark region. . . 2. P. CANADENSIS var. VILLOSISSIMA

b. First flower arises from the second node (joint) located above the second well-developed internode; stem at flowering time ascending to erect, eventually arching or rooting at the tips; middle leaflet of the middle and upper leaves narrower, oblanceolate or oblong-elliptic, the length usually more than twice the width; throughout the state . . . . . . . . . . 1. P. SIMPLEX

a. Flowers few to several in an inflorescence, developed mostly in the upper half of the stem, occasionally only 1 flower developed in young plants; stems mainly erect . . . . . . . . . . . . . *c*

c. No more than 3 leaflets on all the leaves . . . . . . . . . . . . . . . . . . . *d*
 d. Petals equaling or somewhat shorter than the calyx-lobes, 3–5 mm. long; stamens 15–20; seed-like achenes with the ribs running lengthwise, 0.8–1.3 mm. long; common throughout the state . . . . . . . . . . . . . . . . . . . . . . 3. P. NORVEGICA
 d. Petals much shorter than the calyx-lobes, 1.5–2 mm. long; stamens 10–15; seed-like achenes smooth, without ribs, 0.5–0.8 mm. long; rare . . . . . . . . 4b. P. RIVALIS var. MILLEGRANA
c. Five or more leaflets on some of the leaves . . . . . . . . . . . . . . . . . . *e*
 e. At least the lower leaves pinnately compound, with a single leaflet at the tip, the others attached along the sides of the leaf axis. . . . . . . . . . . . . . . . . . . *f*
  f. Petals creamy-white or yellowish-white, 5–7 mm. long; stamens 25–30; hairs tipped with glands on at least the flowering branches; topmost leaflet of the lower leaves mostly 3–4 cm. wide; plants of dry prairies and rocky open woods. . . . . . . . . . . 8. P. ARGUTA
  f. Petals yellow, smaller, 2–5 mm. long; stamens 5–20; hairs lacking glands; topmost leaflet of the lower leaves 1.5 cm. or less wide; plants mainly of low alluvial soils in valleys and along streams . . . . . . . . . . . . . . . . . . . . . . . *g*
   g. All the leaves pinnately compound with a single leaflet at the tip, the others attached along the sides; leaflets 7–11 on all the leaves . . . . . . . . . . . 5. P. PARADOXA
   g. Only the lower leaves pinnately compound with a single leaflet at the tip; leaflets 5–7 on the lower leaves . . . . . . . . . . . . . . . . . . . . . . *h*
    h. Ovary or seed-like achenes with a corky swelling on one side; achenes with ribs running lengthwise; leaflets of the lower leaves rather separated and remote . 6. P. NICOLLETII
    h. Ovary or seed-like achene without a corky swelling on one side; achenes smooth, without ribs; leaflets of the lower leaves close together . . . 4a. P. RIVALIS var. RIVALIS
 e. At least the basal and lower leaves digitately compound, the leaflets all attached at or near the tip of the leaf-stalk (petiole) . . . . . . . . . . . . . . . . . . . . *i*
  i. Petals showy, 8–10 mm. long, equaling or longer than the calyx-lobes; stamens 30; fully grown calyx 10–15 mm. long; inflorescence with leafy parts mostly or only at the base; seed-like achenes with a honeycomb-like surface . . . . . . . . . . . . . 7. P. RECTA
  i. Petals not showy, 1.5–2 mm. long, much shorter than the calyx-lobes; stamens 5; fully grown calyx 5–8 mm. long; inflorescence with leaves scattered throughout; seed-like achenes smooth, without any markings on surface . . . . . 4c. P. RIVALIS var. PENTANDRA

1. **Potentilla simplex** Michx. Cinquefoil, Five-finger Map 1209

Flowers April–June.

Occurs in rocky or open woods, prairies, fallow fields, and along railroads.

Missouri material is represented by the following 3 variations:

a. Lower surface of leaflets with a dense silvery-silky (sericeous) hairiness . . . 1c. P. SIMPLEX var. ARGYRISMA
a. Lower surface of leaflets green or grayish-green, and more or less hairy, but without a silvery-silky hairiness . . . . . . . . *b*
 b. Hairs on stems and leaf-stalks (petioles) more or less widely spreading . . . . 1a. P. SIMPLEX var. SIMPLEX
 b. Hairs on stems and leaf-stalks (petioles) appressed (pressed lengthwise along the surface) or hairs nearly or quite absent . 1b. P. SIMPLEX var. CALVESCENS

1a. **Potentilla simplex** var. **simplex** Map 1209

*Potentilla simplex* Michx. [G, P & S, Steyerm., BB in part]

This is the commonest variation in Missouri and is doubtless found in every county.

Ranges from Nova Scotia, Quebec, and Ontario to Minnesota, south to North Carolina, Tennessee, Missouri, and Oklahoma.

1b. **Potentilla simplex** var. **calvescens** Fern. Map 1209

*Potentilla simplex* in part [BB]

Scattered throughout Missouri, but less common than the var. *simplex*.

Ranges from Newfoundland to Minnesota, south to South Carolina, Illinois, Missouri, and Oklahoma.

1c. **Potentilla simplex** var. **argyrisma** Fern. Map 1209

Southern half of Missouri north to Pettis, Boone, and Callaway counties, and east to Crawford County; of expected occurrence in more of the counties of the unglaciated prairie section.

Ranges from Pennsylvania and Indiana to Tennessee, Illinois, Missouri, Arkansas, and Kansas.

Gleason (*New Illus. Fl.* 2: 293. 1952) regards these variations as mere intergrading forms. Pending future and more detailed field and experimental studies of

1210 Potentilla canadensis var. villosissima (Cinquefoil)

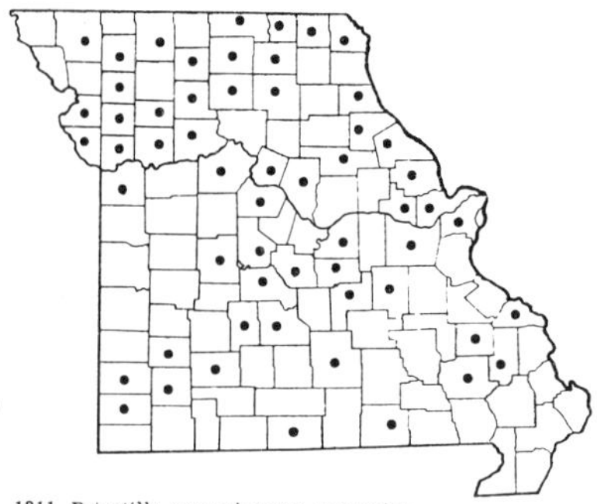

1211 Potentilla norvegica var. norvegica

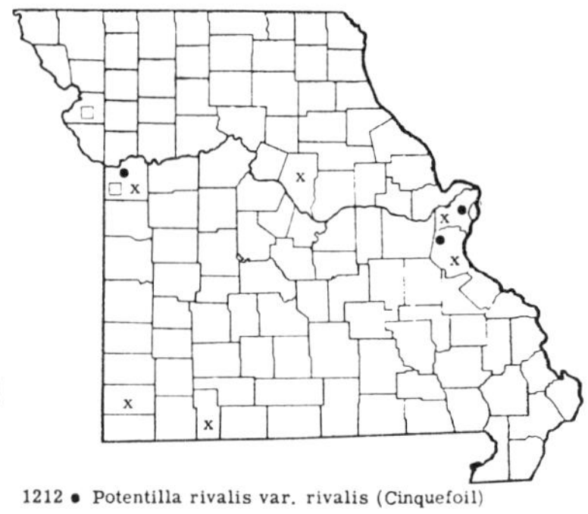

1212 • Potentilla rivalis var. rivalis (Cinquefoil)
1212 x Potentilla rivalis var. millegrana
1212 □ Potentilla rivalis var. pentandra

these variations, they are being retained in the present flora as varieties.

2. **Potentilla canadensis** L. var. **villosissima** Fern. Cinquefoil, Five-finger Map 1210

Flowers April–May.

Occurs in acid soils of dry rocky open woods on uplands or in valleys around chert and sandstone strata. Restricted to the Ozark region east to Washington, Reynolds, and Ripley counties, north to Washington, Crawford and Phelps counties, and west to Texas, Douglas, and Ozark counties.

Ranges from Maryland to Ohio, south and west to Georgia, Tennessee, and Missouri.

This is a good ground cover plant for rocky sterile acid soils and rock gardens having a sandstone, chert, or granite base. For the past twelve years the author has had this plant growing in northern Illinois in his wildflower garden where it has formed a dense carpet. In early spring its yellow flowers, protruding above the interlacing mat of stems and leaves, are very attractive.

3. **Potentilla norvegica** L. var. **norvegica** Map 1211

*Potentilla norvegica* var. *hirsuta* (Michx.) Lehm. [P & S, Steyerm.]

*Potentilla norvegica* L. [G, BB]

Flowers May–October.

Occurs in fallow and cultivated fields, meadows, pastures, prairies, farm lots, open woodland, in open valleys along streams, along roadsides, railroads, and waste ground. Throughout Missouri, and probably in every county.

Ranges from Greenland and Labrador to Alaska, south to D. C., Indiana, Illinois, Missouri, Oklahoma, Texas, New Mexico, and California; also in Mexico and Eurasia.

The inflorescence is a few-flowered leafy cyme, but in the early stages of development the plant may show only a solitary flower.

A variation with glabrous stems and nearly glabrous leaves is known as var. *labradorica* (Lehm.) Fern.

4a. **Potentilla rivalis** Nutt. var. **rivalis** Cinquefoil Map 1212

*Potentilla rivalis* Nutt. [G, BB in part, P & S, Steyerm.]

Flowers May–August.

Occurs in open ground and wet open banks along the Missouri and Mississippi rivers in central Missouri in St. Louis, Jefferson, and Jackson counties.

Ranges from Manitoba to British Columbia, south to Missouri, Kansas, and California; also in Mexico.

In var. *rivalis* the lower leaves are pinnately compound with 5–7 leaflets, or only 3 leaflets may be present with the terminal one 3-parted.

4b. **Potentilla rivalis** var. **millegrana** (Engelm.) Wats. Map 1212

*Potentilla millegrana* Engelm. [G]

*Potentilla rivalis* in part [BB]

Flowers May–August.

Occurs in wet open soils along the Missouri and Mississippi rivers, waste ground, and pastured soils. Scattered in southern and central Missouri, north to St. Louis, Boone, and Jackson counties.

---

Plate no. 200. **1.** Potentilla canadensis var. villosissima, × 2/7; Details from Small, The New York Botanical Garden. **2.** Potentilla norvegica var. norvegica, × 2/7. **3.** Potentilla paradoxa, × 2/7. **4.** Potentilla rivalis var. rivalis, × 2/7; Details from Small, The New York Botanical Garden. **5.** Potentilla recta, × 2/7. **6.** Potentilla rivalis var. millegrana, × 2/7. **7.** Potentilla arguta, × 2/7. **8.** Filipendula rubra, × 2/7; Details from Small, The New York Botanical Garden.

PLATE NO. 200

1213 Potentilla paradoxa (Bushy Cinquefoil)

1214 Potentilla Nicolletii (Cinquefoil)

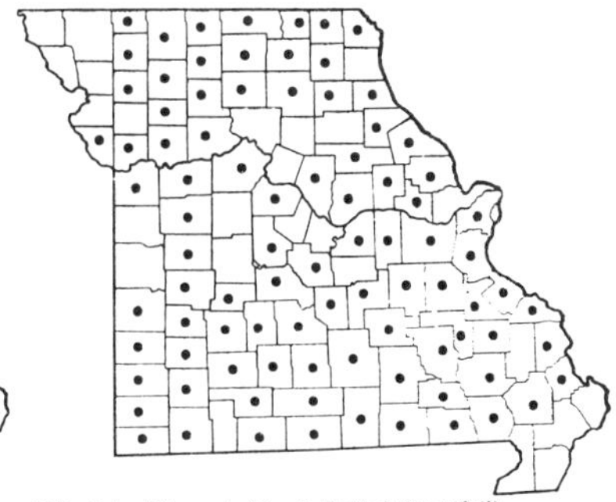

1215 Potentilla recta (Rough-fruited Cinquefoil)

Ranges from Manitoba to Washington, south to Illinois, Missouri, Kansas, Oklahoma, New Mexico, and California.

In this variety all the leaves have 3 leaflets.

4c. **Potentilla rivalis** var. **pentandra** (Engelm.) Wats. Map 1212

*Potentilla pentandra* Engelm. [G]

*Potentilla rivalis* in part [BB]

Flowers May–August.

Occurs in waste and alluvial ground. West-central Missouri in Jackson (sandy bottoms, Courtney, July 7, 1933, *Bush 12674A*) and Buchanan (lake in Missouri River bottoms, T56N, R36W, southwest sect. 10, 2 mi. northeast of Kenmoor, August 20, 1950, *Steyermark 70110*) counties.

Ranges from Minnesota to Alberta, south to Arkansas and Oklahoma.

The basal and lower leaves in var. *pentandra* have either 5 leaflets or only 3 with the lower leaflets deeply divided.

The above variations have been treated by Fernald (eighth edition, *Gray's Manual*) as separate species, and by Gleason (*New Illus. Fl.*) as part of a highly variable species often producing seed without fertilization.

5. **Potentilla paradoxa** Nutt. Bushy Cinquefoil Map 1213

Flowers May–September.

Occurs in wet, alluvial muddy soils of river banks and flood plains along the Missouri and Mississippi rivers, and occasionally in low ground along railroads. Northwestern, central, and eastern Missouri following the Missouri River from Holt County to St. Louis County and along the Mississippi River to Perry County.

Ranges from Ontario to British Columbia, south to New York, Pennsylvania, Ohio, Illinois, Missouri, Kansas, and New Mexico. Also in eastern Asia.

6. **Potentilla Nicolletii** (S. Wats.) Sheld. Cinquefoil Map 1214

*Potentilla paradoxa* in part [BB], not (S. Wats.) Sheld.

Flowers May–August.

Occurs on mud flats and sandy ground along the Missouri and Mississippi rivers from Jackson County on the Missouri River to Perry County on the Mississippi River.

Ranges from Minnesota and North Dakota to Missouri and Kansas.

The lower leaves of this species are pinnately compound with 5–7 leaflets, while the middle and upper ones are digitately compound with 3 leaflets. Gleason (*New Illus. Fl.* 2: 295. 1952) considers this as a synonym of *P. paradoxa*.

7. **Potentilla recta** L. Rough-fruited Cinquefoil Map 1215

Flowers May–August.

Occurs in fields, meadows, pastures, waste ground, along roadsides and railroads. Throughout Missouri and doubtless in every county.

Native of Europe; introduced and naturalized in North America from Newfoundland to Ontario and Minnesota, south to North Carolina, Tennessee, Arkansas, Oklahoma, and Texas.

8. **Potentilla arguta** Pursh Tall Cinquefoil Map 1216

Flowers late May–August.

Occurs on prairies, glades, prairie remnants along railroads, and rocky open woods. Rare and scattered over the state in the Ozark and glaciated prairie sections; absent from the unglaciated prairie section, the southeastern lowlands, and most of the eastern Ozarks.

Ranges from Quebec to Mackenzie, south to D. C., West Virginia, Ohio, Indiana, Illinois, Missouri, Oklahoma, Colorado, and Arizona.

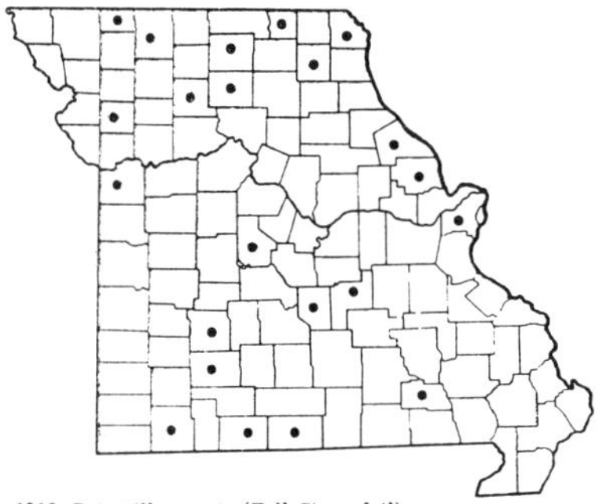

1216 Potentilla arguta (Tall Cinquefoil)

1217 Filipendula rubra (Queen of the Prairie)

1218 Geum vernum (Early Water Avens)

## 12. **Filipendula** Adans.

**Filipendula rubra** (Hill) Robins.
Queen of the Prairie Map 1217
Flowers late June–August.

Known only from a swampy calcareous meadow in the southeastern Ozarks in Reynolds County (along West Fork of Black River on property of D. C. Miner, T33N, R3W, sect. 23 and northeast sect. 26, 3–3½ mi. northwest of Greeley, July 6, 1951, *Steyermark 71974*).

Ranges from Pennsylvania to Michigan and Iowa, south to Georgia, Kentucky, Illinois, and Missouri.

Frequently cultivated as a garden perennial and escaped and naturalized in some of the eastern states from New England to Nova Scotia.

The stems are generally 1–2.5 meters tall bearing large clusters of numerous pale pink flowers; the leaves are fragrant.

This is another relict species, in Missouri restricted like many others of more northern and eastern affinity, to the swampy calcareous meadows of the southeastern Ozark section, where it now survives. At the only known station for it in the state it is associated with the rare *Carex trichocarpa, Parnassia grandifolia, Pedicularis lanceolata, Phlox maculata, Melanthium virginicum, Fuirena simplex,* and *Juncus Dudleyi.*

## 13. **Geum** L. Avens

a. Petals yellow to cream-color, 1.5–2 mm. long; lateral divisions of some of basal leaves 0.5–2 cm. long, 0.5–1.5 cm. wide; upper division of style glabrous (without hairs); the cluster or head of fruiting seed-like achenes raised on a stalk above the calyx; calyx lacking tiny bractlets between the calyx-lobes . . . . . . . . . . . . . . . . . . . . . . . . . . 1. G. VERNUM

a. Petals white, 2–9 mm. long; lateral divisions of the basal leaves large, 2–6 cm. long, 2–6 cm. wide; upper division of style hairy above the joint; the cluster or head of fruiting seed-like achenes not stalked; calyx with small bractlets alternating between the calyx-lobes . . . . . . . . . . . *b*

b. Common species throughout Missouri; petals equaling or longer than the calyx, 5–9 mm. long; stalks bearing the flowers and fruiting heads covered with a short hairiness and with or without scattered longer hairs; remove the pistils or fruiting achenes from the center of the flower- or fruit-cluster and note the brisly hairs covering the receptacle (supporting base of flower); lower part of stem glabrous or sparsely hairy with hairs 1 mm. long . . . . . . . . . 2. G. CANADENSE

b. Rarer species of northern and central Missouri only; petals mostly shorter than the calyx-lobes, 2–5 mm. long; stalks bearing the flowers and fruiting heads with many conspicuous hairs widely or downward spreading; remove the pistils or fruiting achenes from the center of the flower- or fruit-cluster and note the hairless or nearly hairless condition of the receptacle (supporting base of flower); lower part of stem conspicuously hairy with spreading hairs about 2 mm. long. . 3. G. LACINIATUM var. TRICHOCARPUM

1. **Geum vernum** (Raf.) T. & G.
Early Water Avens Map 1218
Also called Spring Avens.
Flowers April–June.

Occurs in low moist woods and wet ground along streams, wooded valleys at base of bluffs, and low thickets. Common in the southern and eastern half of the state, but known only from Gentry County in the northwestern third of Missouri.

Ranges from New York to Ontario and Michigan,

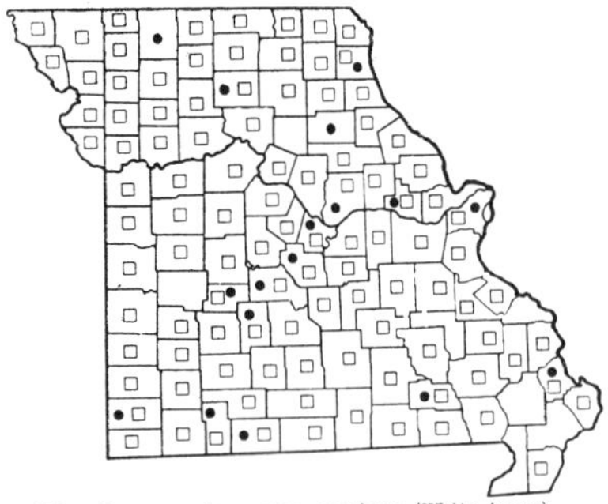

1219 • Geum canadense var. canadense (White Avens)
1219 □ Geum canadense var. camporum

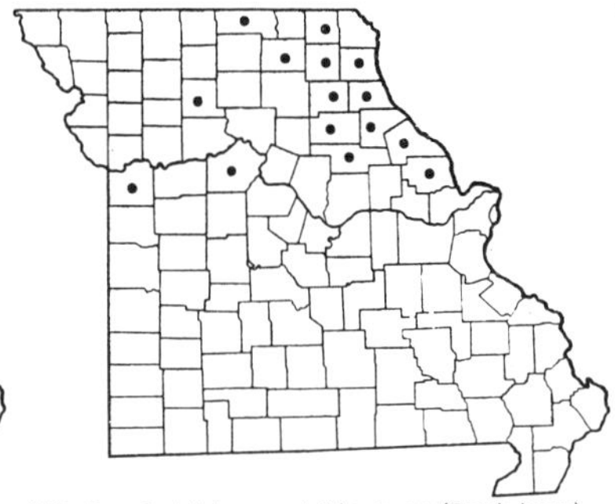

1220 Geum laciniatum var. trichocarpum (Rough Avens)

1221 Rubus idaeus var. strigosus (Red Raspberry)

south to Maryland, West Virginia, Tennessee, Arkansas, Kansas, and Oklahoma.

2. **Geum canadense** Jacq. White Avens Map 1219
Also called Red Root.
Flowers late May–October.
Occurs in rich or rocky woods on hillsides, in valleys along streams, in ravines, and sandy thickets.
Two variations are represented in Missouri material:

Pistils or fruiting achenes 30–60 in a head; mature seedlike achenes 2.5–3 mm. long with broadly ovate to obovate bodies; upper surface of stem-leaves mainly without hairs . . 2a. G. CANADENSE var. CANADENSE

Pistils or fruiting achenes 60–150 in a head; mature seedlike achene 3–4.5 mm. long with narrowly obovate to wedge-shaped bodies; upper surface of stem-leaves with many hairs lying parallel to the leaf-surface . . . 2b. G. CANADENSE var. CAMPORUM

2a. **Geum canadense** var. **canadense** Map 1219
*Geum canadense* Jacq. [G, P & S, Steyerm. in part, BB in part]
This is the rarer variation which is scattered over the state.
Ranges from New Brunswick to Minnesota, south to South Carolina, Tennessee, Missouri, and Kansas.

2b. **Geum canadense** var. **camporum** (Rydb.) Fern. & Weath. Map 1219
*Geum canadense* in part [BB]
This is the common variation occurring throughout Missouri and doubtless in every county.
Ranges from Quebec to North Dakota, south to Alabama, Oklahoma, and Texas.
Gleason (*New Illus. Fl.* 2: 817. 1952) does not recognize any variations for this species as distinguished by Fernald and Weatherby. The two variations above indicated in the present flora intergrade, but, for the most part, are recognizable.

3. **Geum laciniatum** Murr. var. **trichocarpum** Fern. Rough Avens Map 1220
Flowers May–July.
Occurs in wet alluvial soils bordering ditches, swamps, natural ponds and sloughs, and in rich moist woodland near streams. Northern and central Missouri, mostly in the northeastern sector of the state, south to Lincoln, Audrain, Saline, and Jackson counties.
Ranges from Nova Scotia to Ontario and Minnesota, south to North Carolina, Illinois Missouri, and Kansas.

---

Plate no. 201. **1.** Geum vernum, × 2/7; Details from Small, The New York Botanical Garden. **2.** Geum canadense var. camporum, × 2/7; Details from Small, The New York Botanical Garden. **3.** Geum laciniatum var. trichocarpum, × 2/7. **4.** Rubus ideaeus var. strigosus, × 2/7. **5.** Rubus laciniatus, × 2/7. **6.** Rubus occidentalis, × 2/7. **7.** Rubus bifrons, × 2/7.

PLATE NO. 201

## 14. **Rubus** L. Bramble

This genus includes the Blackberry, Raspberry, and Dewberry, familiar to everyone. The species, often difficult to separate from one another, have received different interpretations by each one of the botanists who has studied the genus.

In order to identify plants of Rubus, it is necessary to collect both the vegetative shoots of the first year's growth, bearing usually 3–5 leaflets, as well as the flowering and fruiting canes or stems which represent the second year's growth and which have only 3 leaflets. Notes should always be taken in the field on the habit of growth of the plant, whether trailing, arching, ascending, or erect, and the height of the plant.

Members of this group are perennial, but the individual canes are biennial, those bearing flowers and fruits representing the second year's growth which dies at the end of the season, whereas the vegetative canes represent the growth of the first year, persisting into the following season when they then bear flowers and fruits.

In the present state of our limited knowledge concerning variations among the native blackberries, and pending future field and experimental studies on the genus, it is believed that the following treatment will account for the large majority of specimens collected in Missouri. Some of the treatment is based upon Gleason's studies in the *New Illus. Fl.* 2: 305–16. 1952.

a. Leaflets deeply cut or dissected to the midrib into several to many smaller and narrower divisions; calyx armed with prickles; petals 3-lobed . . . . . . . . . . . . . . 5. R. LACINIATUS
a. Leaflets finely to coarsely toothed; calyx hairy or bristly, but not armed with prickles; petals not 3-lobed . . . . . . . . . . . . . . . . . . . . . . . . *b*
  b. Lower surface of leaflets white or whitened . . . . . . . . . . . . . . . . . . *c*
    c. Flowers or fruits 2–7 in a short umbel-, corymb-, or raceme-like inflorescence; fruit separating when ripe from the central green receptacle, the latter remaining attached to the stalk of the fruit . . . . . . . . . . . . . . . . . . . . . . . . *d*
      d. Some of stems whitened with a coating which can be rubbed off (glaucous); fruit black when ripe; stalks of flowers or fruits bearing prickles or spines; stems bearing prickles or spines; when 5 leaflets are present on 1st year vegetative canes, they are digitately compound, all arising from the summit of the main leaf-stalk (petiole); common species, throughout Missouri . . . . . . . . . . . . . . . . 2. R. OCCIDENTALIS
      d. Stems usually not whitened with a glaucous coating; fruit red when ripe; stalks of flowers or fruits bearing bristles and glands but no spines or prickles; stems bearing bristles but no prickles or spines; when 5 leaflets are present on 1st year vegetative canes, they are pinnately compound, one pair of leaflets arising on the main leaf-stalk (petiole) below the other 3; rare species, known only from northwestern Missouri from Atchison and Holt to Jackson counties . . . . . . . . . . . . . . . . . . . . 1. R. IDAEUS var. STRIGOSUS
    c. Flowers or fruits in a pyramidal panicle; fruit remaining attached to the central green receptacle, even when ripe . . . . . . . . . . . . . . . . . . . . . *e*
      e. First year's vegetative canes (primocanes) strongly angled, conspicuously hairy, especially near tips; prickles of inflorescence curved and flat, sickle-shaped; inflorescence about equally thick throughout its length; topmost (terminal) one of the 5 leaflets of the leaves of the 1st year's canes (primocanes) coarsely toothed and broadly ovate with a broad rounded base 4. R. PROCERUS
      e. First year's vegetative canes (primocanes) nearly terete (rounded in cross-section) with curved sides, slightly hairy or glabrous (without hairs); prickles of inflorescence straight; inflorescence pyramid-shaped, broadest at base; topmost (terminal) one of the 5 leaflets of the leaves of the 1st year's canes (primocanes) finely toothed, usually narrowed at base 3. R. BIFRONS
  b. Lower surface of leaflets green or grayish-green, but not white or whitened . . . . . . . *f*
    f. Canes of 1st year and usually of 2nd year and leaf-stalks (petioles) bearing purplish-red bristles or hairs . . . . . . . . . . . . . . . . . . . . . . 6. R. TRIVIALIS
    f. Canes and leaf-stalks smooth, spiny, or hairy, but without purplish-red bristles . . . . . . *g*
      g. Main stems trailing, sprawling, low-arching, or forming domes over the ground, the tips or branches of the first year's vegetative canes rooting at the tip . . . . . . . . . . *h*
        h. Main stems of both vegetative first year's and flowering or fruiting second year's growth trailing, sprawling or low-arching over the ground; flowering shoots appearing to rise from the sides of the trailing cane . . . . . . . . . . . . . . . . .
          i. Flower-stalks (pedicels) with gland-tipped hairs . . . . . . . . . . 9. R. INVISUS

i. Flower-stalks (pedicels) without gland-tipped hairs . . . . . . . . . . . . . . . . *j*
j. Terminal (topmost) one of the 3 leaflets of the flowering cane ovate, broadest near the base or below the middle, the sides curved or rounded . . . . . . . . . . 7. R. FLAGELLARIS
j. Terminal (topmost) one of the 3 leaflets of the flowering cane oblong to oblanceolate or obovate, broadest at or above the middle, narrowed toward the wedge-shaped (cuneate) base, with the sides straight or nearly so below the middle . . . . . . . . 8. R. ENSLENII
h. Main stems high-arching or forming domes to 1 meter high above the surface of the ground, but the tips of the branches of the vegetative first year's canes reaching the ground and rooting at the tips. . . . . . . . . . . . . . . . . . . . . . . . . . . . 10. R. MISSOURICUS
g. Main stems erect to ascending with arching tips, but not with rooting or trailing tips . . . . . . *k*
k. Numerous gland-tipped hairs present on the stalks of the flowers and fruits, and often on other parts of inflorescence and younger parts of axis of vegetative cane of first year's growth. . . . . *l*
l. The main developed inflorescences 2–4 times as long as thick, rather elongated and cylindrical; topmost (terminal) leaflet of leaves of the first year's canes rarely more than ⅝ as wide as long, ovate-oblong, obovate-oblong, or narrowly ovate . . . . . . . . . . 11. R. ALLEGHENIENSIS
l. The main developed inflorescences as long as thick or at most 2 times as long as thick, broadest at the summit or with the flowers mostly crowded at the summit; topmost (terminal) leaflet of leaves of the first year's canes ¾ as wide to as wide as long, broadly ovate to nearly round in shape . . . . . . . . . . . . . . . . . . . . . . . . . 12. R. ORARIUS
k. Gland-tipped hairs mostly absent on the stalks of the flowers and fruits and the young growth (a few inconspicuous ones may be present) . . . . . . . . . . . . . . . . . . *m*
m. The upper 3 leaflets of fully grown leaves of the vegetative first year's canes (primocanes) 2–3 times as long as broad with rather straight or nearly straight margins from below the middle to the base; flowers of medium size, the petals narrow, about 5–7 mm. broad . 13. R. ARGUTUS
m. The upper 3 leaflets of fully grown leaves of the vegetative first year's canes (primocanes) usually less than twice as long as broad; flowers medium to large, with broad rounded petals 9–15 mm. broad . . . . . . . . . . . . . . . . . . . . . . . . . . *n*
n. Main fully developed inflorescences short-cylindrical racemes not broadened across the top, more or less of equal width throughout and with an elongated axis . . . . . 14. R. MOLLIOR
n. Main fully developed inflorescences broadened across the top with a short axis, the racemes more or less corymbose. . . . . . . . . . . . . . . . . . . . . . . . *o*
o. Leaf-like bracts 7–12 in the inflorescence, a large bract at the base of each or nearly each flower- or fruit-stalk; topmost (terminal) of the 5 leaflets of the first year's vegetative canes (primocanes) widest toward the base, broadly ovate; main inflorescence short and rather hidden in the subtending foliage, of a corymbiform type with the lowest pedicels elongated . . . . . . . . . . . . . . . . . . . . . 15. R. PENSILVANICUS
o. Leaf-like bracts fewer in the inflorescence, at the base of the lower and middle flower- or fruit-stalks only; topmost (terminal) of the 5 leaflets of the first year's vegetative canes (primocanes) widest near the middle, elliptic-oblong, ovate-oblong, or oblong-obovate; main inflorescences standing above the subtending foliage, of a short-raceme type . . 16. R. OSTRYIFOLIUS

1. **Rubus idaeus** L. var. **strigosus** (Michx.) Maxim. f. **strigosus** Red Raspberry Map 1221
*Rubus strigosus* Michx. [BB]
*Rubus idaeus* var. *strigosus* (Michx.) Maxim. [G, P & S, Steyerm.]
Flowers May–June; fruit June–August.

Occurs on wooded slopes of loess hills in northwestern Missouri in Atchison and Holt counties, south to Jackson County, west-central Missouri.

Ranges from Labrador to British Columbia, south to Virginia, Pennsylvania, Ohio, Indiana, Illinois, Missouri, Nebraska, Wyoming, and Arizona.

Forms of this variety without bristles (f. *tonsus* Fern.), and with amber-white instead of red fruits (f. *albus* [Bailey] Fern.) are known, but have not been collected in Missouri. Of the native species, this is the rarest and least known in the state.

The cultivated Red Raspberry (*R. idaeus* L. var. *idaeus*) is well known for its delectable fruit. The leaves of the wild red raspberry are sometimes used for tea.

2. **Rubus occidentalis** L. f. **occidentalis** Black Raspberry Map 1222
*Rubus occidentalis* L. [G, BB, P & S, Steyerm.]
Flowers April–June; fruit June–July.

Occurs in open woods, along bluffs, and in thickets. Throughout Missouri; absent from most of the lowland section of southeastern Missouri, but elsewhere doubtless in every county.

Ranges from Quebec to North Dakota and Colo-

1222 Rubus occidentalis f. occidentalis (Black Raspberry)

1223 Rubus bifrons

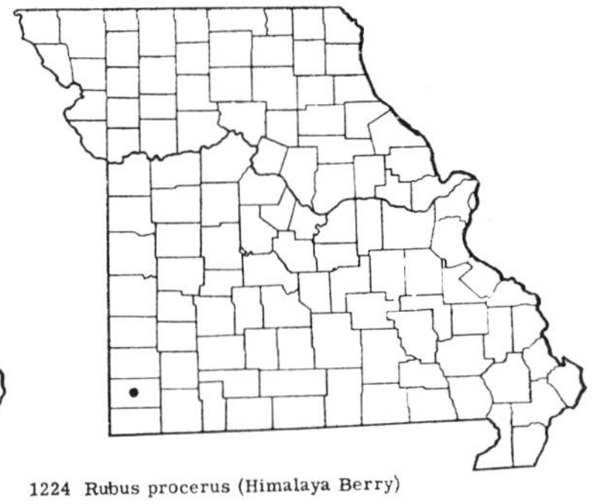

1224 Rubus procerus (Himalaya Berry)

rado, south to Georgia, Arkansas, and Oklahoma.

The fruit is ordinarily purplish-black. A yellow-amber to whitish-fruited form (f. *pallidus* [Bailey] Robins.) has been described, but, thus far, has not been found in Missouri.

Improved horticultural varieties have originated from this species and are commonly cultivated. The fruit of the wild black raspberry is very sweet, in some years, depending upon the amount of rainfall, larger and more abundant than in other years. As with other members of the genus, the fruit is eaten fresh, cooked into pies, preserves, syrup for drinks, and the leaves may be used as a substitute for tea.

3. **Rubus bifrons** Vest Map 1223
*Rubus procerus* of some auth. [P & S, Steyerm.], not P. J. Muell.
Flowers May–June; fruit July–August.

Occurs along rocky stream beds, banks, and open ground. Known only from southwestern Missouri, in Christian, and Lawrence (2 mi. south of Verona, June 4, 1953, *Palmer 55746*) counties.

Native of Europe; introduced and naturalized in the United States from Florida to Louisiana, north to Rhode Island, Tennessee, Missouri, and Oklahoma.

Palmer's specimen from Lawrence County has the flower-bearing canes arching, 2–2.5 m. long with pale pink petals on the flowers.

This species has been confused in Missouri and elsewhere previously with *R. procerus*, the Himalaya Berry. The flowers are white or rosy-tinged.

4. **Rubus procerus** P. J. Muell.
Himalaya Berry Map 1224
Flowers May–June.

Known only from southwestern Missouri in Newton County (roadside bank, 1 mi. south of Granby, May 12, 1957, *Palmer 65071*).

Native of Europe; introduced and naturalized in the United States from Delaware to Virginia, and locally in Missouri.

The flowers are white or rosy-tinged.

Most of the Missouri material, previously identified as *R. procerus*, is now referred to *R. bifrons*. However, the collection by *Palmer* (*65071*), cited above, is placed in *R. procerus* because of the strongly angled, pubescent primocanes, broadly ovate, round-based rather coarsely toothed terminal leaflet of the primocane leaves, and the somewhat curved flat prickles in the inflorescence.

5. **Rubus laciniatus** Willd. Cut-leaved Blackberry Map 1225
Flowers May–June; fruit July–August.

Known only from southwest Missouri in Barry County (farm, Seligman, June 10, 1931, *W. W. Farnham & Drouet* in herb. U. of Mo.).

Native of Europe; introduced into cultivation and naturalized from Massachusetts to Michigan and southward; commonly naturalized in the Pacific northwestern states.

The petals are usually rose-colored. The specimen cited above was collected from a cultivated plant, but is included in the present flora because of the tendency of the species to escape from cultivation elsewhere where planted. Bailey has determined the Missouri collection, the leaflets of which are definitely gray underneath. Bailey in his key (Gent. Herb. 5: 838–39. 1945) states that the foliage of this species is 'green or greenish, or at least not cano- or griseo-tomentose underneath,' whereas Fernald (eighth ed., *Gray's Man.*, p. 824) states that the foliage is 'greenish or but slightly grayish beneath.'

The species is often grown for its decorative cut leaves and is planted in some areas to cover eroded spots. The large, globe-shaped, edible and very delicious fruit is represented by various horticultural races, such as Atlantic, Black Diamond, Pan American, Starr, and Wonder.

1225 Rubus laciniatus (Cut-leaved Blackberry)

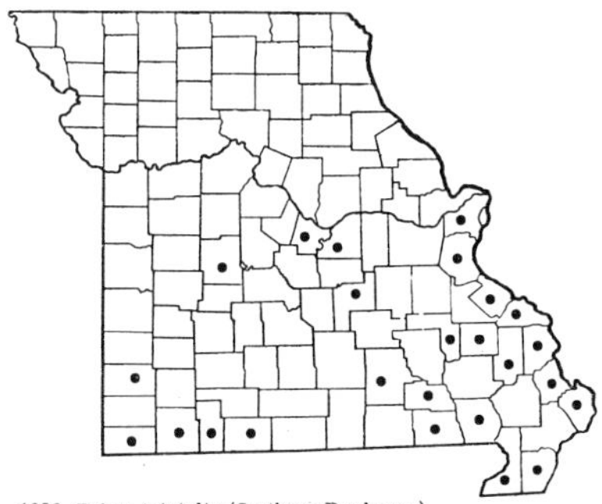

1226 Rubus trivialis (Southern Dewberry)

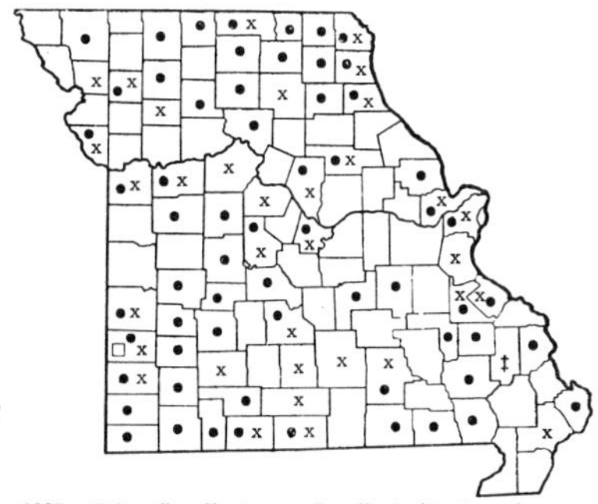

1227 • Rubus flagellaris var. flagellaris (Dewberry)
1227 x Rubus flagellaris var. occidualis f. occidualis
1227 ‡ Rubus flagellaris var. occidualis f. roseus
1227 □ Rubus flagellaris var. occidualis f. roseo-plenus

6. **Rubus trivialis** Michx. Southern Dewberry Map 1226

Flowers April–May; fruit late May–June.

Occurs in low or alluvial woods, moist thickets, and wooded banks of streams. Southern and central Missouri, north to St. Louis, Osage, Cole, Benton, and Jasper counties; mostly in the counties of southeastern Missouri, and those bordering the Mississippi and White rivers.

Ranges from Florida to Texas, north to Maryland, Illinois, Missouri, and Oklahoma.

This species is easily recognized by the slender reddish-purple glandular hairs and bristles on the trailing stems. The leaves are more or less evergreen.

7. **Rubus flagellaris** Willd. Dewberry Map 1227

Flowers April–June; fruits June–August.

Occurs in rocky open woods, thickets, prairies, along roadsides, and railroad embankments.

Ranges from Nova Scotia and Quebec to Minnesota, south to Florida and Texas.

Throughout Missouri and probably in every county.

Missouri material may be divided into the following 4 variations:

a. Lower surface of leaflets of flowering, fruiting, or leafy canes glabrous (without hairs) or the hairs on the veins only and either short or appressed to the surface . 7a. R. FLAGELLARIS var. FLAGELLARIS
a. Lower surface of leaflets of flowering, fruiting, or leafy canes hairy, soft or velvety to the touch *b*
b. Petals white; common type. . . . . 7b. R. FLAGELLARIS var. OCCIDUALIS f. OCCIDUALIS
b. Petals pink or pink-tinged; rare type . . *c*
c. Petals mostly 5, the flowers normally 'single'. . . . . 7c. R. FLAGELLARIS var. OCCIDUALIS f. ROSEUS
c. Petals many, other parts of the flowers turned to petals, the flowers 'double' . 7d. R. FLAGELLARIS var. OCCIDUALIS f. ROSEO-PLENUS

7a. **Rubus flagellaris** var. **flagellaris** Map 1227

*Rubus flagellaris* Willd. [G, BB, P & S, Steyerm.]

*Rubus Baileyanus* Britt. [G]

This is common throughout Missouri.

Ranges from Maine and Quebec to Ontario and Minnesota, south to Florida and Texas.

7b. **Rubus flagellaris** var. **occidualis** Bailey f. **occidualis** Map 1227

*Rubus flagellaris* var. *occidualis* Bailey [P & S, Steyerm.]

*Rubus occidualis* Bailey [G]

*Rubus flagellaris* in part [BB], not Willd.

Common throughout Missouri.

Ranges from Illinois and Iowa, south to Missouri, Kansas, and Oklahoma.

7c. **Rubus flagellaris** var. **occidualis** f. **roseus** Steyerm. Map 1227

Known only from southeastern Missouri in Bollinger County (low woods in valley of creek tributary to Castor River, T29N, R8E, sect. 5, 3¾ mi. northwest of Buchanan P. O., May 8, 1957, *Steyermark 84212*).

Occurs only in Missouri.

7d. **Rubus flagellaris** var. **occidualis** f. **roseo-plenus** (Palmer & Steyerm.) Steyerm. Map 1227

*Rubus flagellaris* f. *roseo-plenus* Palmer & Steyerm., Brittonia 10: 114. 1958.

Known only from southwestern Missouri in Barton County (open bank along railway grade, ½ mi. southeast of Liberal, May 3, 1955, *Palmer 59655*, holotype in Chi. Nat. Hist. Mus. Herb.).

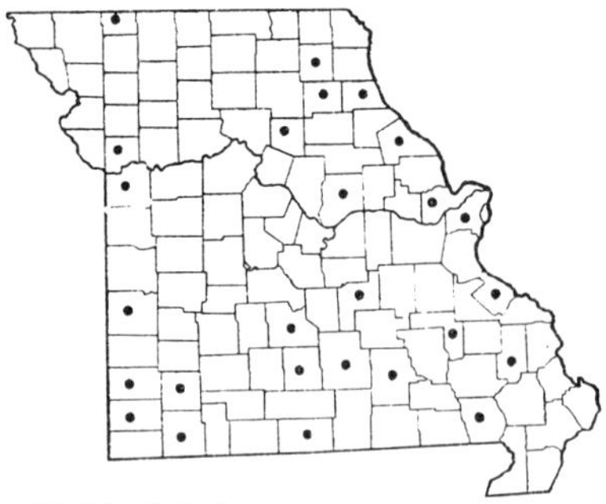
1228 Rubus Enslenii

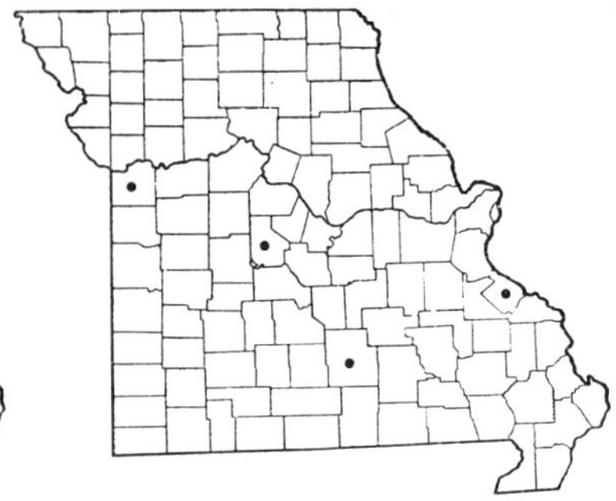
1229 Rubus invisus

1230 Rubus missouricus (Prickly Groundberry)

Occurs only in Missouri.

*Rubus flagellaris* and its varieties comprise the commonest dewberry in Missouri.

The variations of *Rubus flagellaris* have been maintained by Bailey and by Fernald as separate species, and by Gleason as a single collective species. Differences in interpretation of one character, such as pubescence of the lower surface of the leaflets, by different workers, has led to lack of agreement as to proper placement of certain names. *Rubus Baileyanus* Britt., for example, is described by Bailey (Gent. Herb. 2: 324. 1932) as having the leaflets 'soft-pubescent underneath,' whereas both Gleason (*New Illus. Fl.* 2: 310. 1952) and Fernald (eighth ed., *Gray's Man.* 831. 1950) refer to the leaflets as glabrous beneath. Again, Bailey (Gent. Herb. 2: 317. 1932) places *R. aboriginum* Rydb. in the synonymy of *R. flagellaris*, yet considers it a pubescent phase (*loc. cit.* p. 322) of that species having soft-pubescent lower surfaces of the leaves, but Gleason interprets *R. aboriginum* as part of the *R. Enslenii* complex with the lower surface of the leaflets glabrous or nearly so (Gleason, *loc. cit.* p. 310). Although Bailey considers *R. aboriginum* as a phase of *R. flagellaris* having soft-pubescent lower leaf surfaces, he describes *R. flagellaris* var. *occidualis* as another variation with the leaves softly pubescent beneath. If the decision is made in the future that *R. aboriginum* and *R. flagellaris* var. *occidualis* are synonymous, then the former name, which is older, will have to replace var. *occidualis*, if placed as a variety under *R. flagellaris*. On the other hand, if *R. aboriginum* is correctly interpretated as part of the *R. Enslenii* instead of *R. flagellaris* complex, then the var. *occidualis* can be maintained as such.

The fruit is commonly eaten, but the plants bear less prolifically than the high-bush blackberries.

8. **Rubus Enslenii** Tratt. Map 1228

*Rubus frustratus* Bailey
*Rubus mundus* Bailey
*Rubus aboriginum* Rydb.
*Rubus nefrens* Bailey [G, P & S, Steyerm.]

Flowers April–June; fruits June–August.

Occurs in fallow fields, prairies, open banks, and along roadsides. Scattered throughout Missouri; apparently less common then *R. flagellaris* and varieties.

Ranges from Alabama to Mississippi, north to Massachusetts, Pennsylvania, Ohio, Michigan, and Wisconsin.

Fernald (eighth ed., *Gray's Man.* p. 827, 831. 1950) considers *R. Enslenii* as distinct from *R. nefrens* on the basis of its more slender, more flexible canes 1–3 mm. thick at base instead of 3–6 mm. thick, thinner membranaceous leaflets, and fewer-flowered (1–3) inflorescences. Gleason (*New Illus. Fl.* 2: 310. 1952), on the other hand, combines *R. nefrens*, *R. aboriginum*, and many other taxa, recognized by Bailey as species, with *R. Enslenii*. Until future studies reveal new relationships or other details not now evident, the present treatment of this species follows that of Gleason.

9. **Rubus invisus** (Bailey) Britt. Map 1229

*Rubus Rosagnetis* Bailey [G]
*Rubus Deamii* Bailey [G, P & S, Steyerm.]
*Rubus flagellaris* var. *invisus* Bailey [Bailey]
*Rubus arundelanus* [of BB in part], not Blanchard

Flowers April–June; fruits June–August.

Occurs in dry rocky woods and on open rocky exposures. Rare and scattered in southern and central Missouri, in Ste. Genevieve (dry, exposed La Motte sandstone boulders near Chimney Rocks near River aux Vases, 5 mi. east of Pickle, June 4, 1933, *Steyer-*

Plate no. 202. **1.** Rubus trivialis, × 2/7. **2.** Rubus flagellaris var. flagellaris, × 2/7. **3.** Rubus Enslenii, × 2/7. **4.** Rubus invisus, × 2/7. **5.** Rubus missouricus, × 2/7. **6.** Rubus allegheniensis var. allegheniensis, × 2/7. **7.** Rubus orarius, × 2/7; All details from Small, The New York Botanical Garden.

PLATE NO. 202

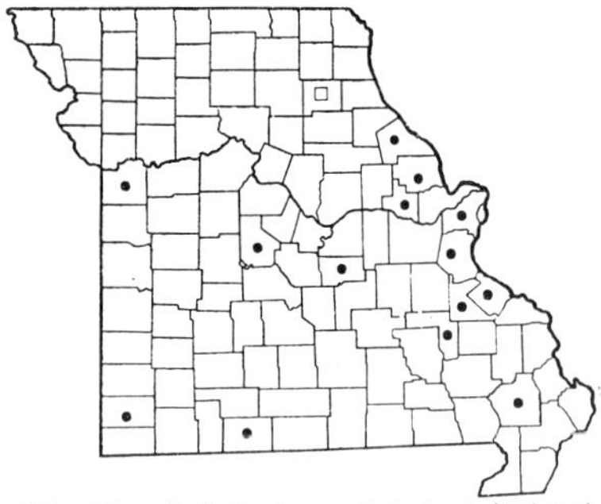

1231 • Rubus allegheniensis var. allegheniensis (High-bush Blackberry)
1231 □ Rubus allegheniensis var. plausus

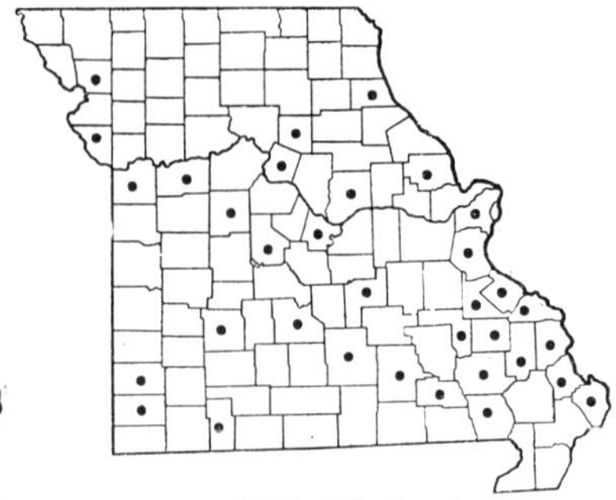

1232 Rubus orarius (High-bush Blackberry)

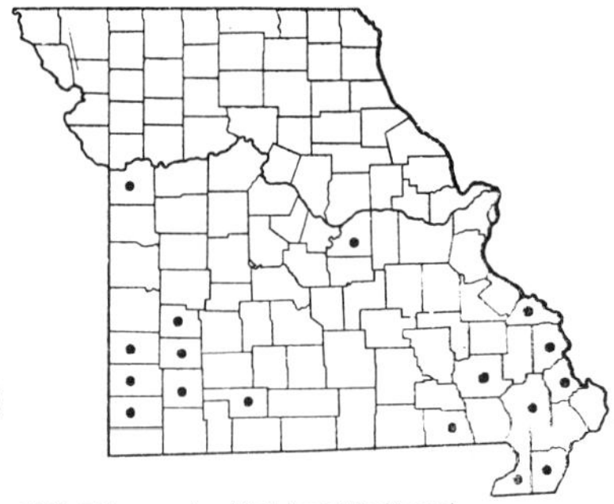

1233 Rubus argutus (High-bush Blackberry)

*mark 8541*), Texas, Morgan (near mouth of Proctor Creek, July 7, 1934, *Steyermark 13167*), and Jackson counties.

Ranges from New York, Ohio, and Indiana, south to Virginia, Kentucky, Tennessee, and Missouri.

In the present treatment *Rubus invisus* is interpreted as a dewberry type of the *Procumbentes* section having the pedicels glandular-hairy, as are also the young canes. *Rubus Deamii* and *R. Rosagnetis* do not appear to differ sufficiently from *R. invisus* to be maintained separately. Gleason, however, places *R. Deamii* in the *R. arundelanus* complex having 'inflorescences of a racemiform type' with most of the few to several flowers 'subtended by a pair of stipules only' (*loc. cit.* p. 306, 308), whereas *R. Rosagnetis* is placed in the *R. Enslenii* group with 'inflorescence of an ascendate type' with the 1 to few or several flowers 'subtended by simple or 3-foliolate leaves' (loc. cit. p. 306, 310), and *R. invisus* is placed in *R. flagellaris* next to *R. Enslenii.* Bailey, Deam (following Bailey), and Fernald place *R. Deamii* and *R. Rosagnetis* as members of the *Procumbentes* section of the dewberries, most closely related to *R. flagellaris*, *R. Enslenii*, and *R. invisus*, and I agree with this interpretation rather than that of Gleason.

10. **Rubus missouricus** Bailey
Prickly Groundberry Map 1230
*Rubus hispidus* [of BB in part], not L.

Flowers April–June; fruit June–August.

Occurs in prairies, thickets, and open ground. Rare and scattered in southern and central Missouri in Lincoln (Silex, *Davis*), Stone (Galena, *Palmer*), and Jackson (prairies, Lake City, June 7, 1932, *Bush 12446* [holotype], *12447*, *12448*, *12449*; May 23, 1932, *Bush 12427*, *12428*) counties.

Known only from Missouri.

Fernald (eighth ed., *Gray's Man.* 840. 1950) includes this species in his key under 'Pedicels glabrous or only inconspicuously pilose,' but the holotype collection (*Bush 12446*) and other Bush collections cited have the pedicels and rachis of the inflorescence densely appressed-pubescent.

The leaflets of the primocanes are mostly 5, gradually tapering to long acuminate tips; those of the flowering and fruiting canes are blunt or somewhat acute at the tip and shallowly toothed. The 10- or more-flowered inflorescences are broader than long and are of a cymose-corymbiform type. The primocanes, up to about 1 m. high, become recurved, their tips eventually touching and rooting; they have scattered, thin, subulate prickles 4–5 mm. long.

Bailey considers this species the most western representative of the section *Jacentes*, whereas Fernald has redefined this section and called it *Tholiformes*, characterizing it on the basis of the arching or doming habit of growth exhibited by the members of the section.

11. **Rubus allegheniensis** Porter
High-bush Blackberry Map 1231

Flowers April–June; fruit June–August.

Occurs in rocky open woods, along bluffs, thickets, and open valleys.

Missouri material is represented by two variations:

Stalks of flowers and fruits and axis of inflorescence without prickles; leaf-stalks (petioles) and stalks of leaflets without prickles or remotely armed . 11a. R. ALLEGHENIENSIS var. ALLEGHENIENSIS
Some of the stalks of flowers and fruits and axis of inflorescence bearing prickles; leaf-stalks (petioles) and stalks of leaflets with strong prickles. . 11b. R. ALLEGHENIENSIS var. PLAUSUS

11a. **Rubus allegheniensis** var. **allegheniensis**
Map 1231
*Rubus allegheniensis* Porter [G, BB, P & S, Steyerm.]

Southern and central Missouri north to Pike and

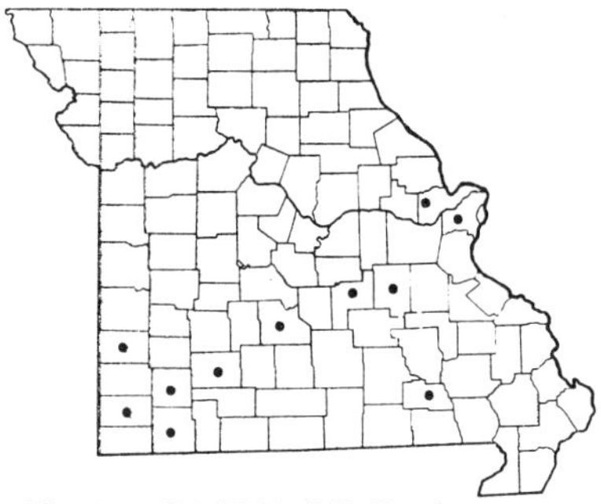
1234 Rubus mollior (High-bush Blackberry)

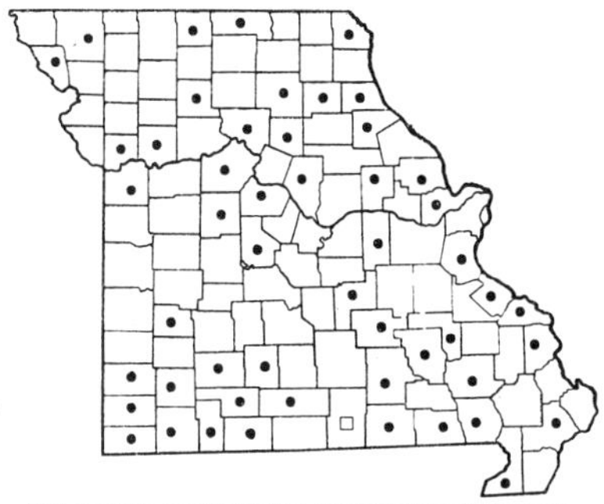
1235 • Rubus pensilvanicus f. pensilvanicus (High-bush Blackberry)
1235 □ Rubus pensilvanicus f. albinus

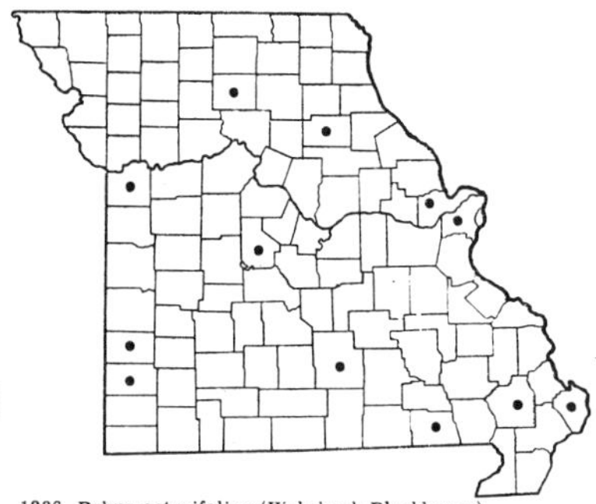
1236 Rubus ostryifolius (High-bush Blackberry)

Jackson counties. This is the commoner variation in the state.

Ranges from New Brunswick and Quebec to Minnesota, south to North Carolina, Tennessee, and Missouri.

11b. **Rubus allegheniensis** var. **plausus** Bailey Map 1231

Known only from Shelby County, northeastern Missouri (open ground of valley between ravines tributary to wooded slopes along north side of North River, 'Miller's Hills,' T29N, R9W, west part sect. 2, 2½ mi. southeast of Burksville, June 5, 1951, *Steyermark 71705*).

Gleason excludes Missouri from the range of *R. allegheniensis*, whereas Fernald includes the state.

12. **Rubus orarius** Blanchard
High-bush Blackberry Map 1232
*Rubus alumnus* Bailey [G, P & S, Steyerm.]; originally described from Jackson County, western Missouri.
*Rubus impos* Bailey

Flowers April–June; fruits June–August.

Occurs in thickets, open woods, prairies, along roadsides, and open ground.

Rather common throughout Missouri, but scarcely recorded from most of the northern sector of the state.

Ranges from Quebec and Maine to Minnesota, south to Virginia, Kentucky, and Missouri.

This species is one of those most commonly sought for its fruits, and is the source of several cultivated varieties.

Fernald (eighth ed., *Gray's Man.* 861. 1950) places *R. orarius* in synonymy with *R. pensilvanicus*, a species lacking the glandular-tipped hairs characteristic of the *Allegheniense*s section. However, he places *R. impos* in synonymy (loc. cit. p. 855) with *R. alumnus*, both of which are characterized by the presence of glandular hairs in the inflorescence. In the present treatment I am following Gleason's usage of the name *R. orarius* as the oldest acceptable one for the group of species involving *R. alumnus* and *R. impos*.

13. **Rubus argutus** Link High-bush Blackberry Map 1233
*Rubus Blakei* Bailey [G]

Flowers April–June; fruits June–August.

Occurs in low and wet woods and thickets. Southern and central Missouri north to Perry, Osage, and Jackson counties.

Ranges from Georgia and Alabama to Arkansas, north to Massachusetts, Pennsylvania, Kentucky, Illinois, and Missouri.

The fruits of this species are smaller and less desirable, having a more seedy and more acid quality than those of the preceding or following species.

14. **Rubus mollior** Bailey High-bush Blackberry Map 1234

Flowers April–June; fruits June–August.

Occurs in shallow draws in upland, thickets and dry open woods. Southern and east-central Missouri, north to St. Charles, Phelps, Laclede, and Barton counties.

Ranges from Missouri and Kansas, south to Arkansas and Oklahoma.

The type of this species was described from Washington County, northwestern Arkansas. The labelled Missouri specimens in the Gray Herbarium consist mostly of sterile primocane leaves collected by Dr. Wm. Blanchard.

Further study of this species may prove that it is not distinct from *R. argutus*.

15. **Rubus pensilvanicus** Poir.
High-bush Blackberry Map 1235
*Rubus summotus* Bailey

*Rubus pratensis* Bailey [BB]; described from prairies, Cooper Co., Missouri, *Bush 14826* (holotype)
*Rubus laudatus* Berger [G, P & S, Steyerm.]
*Rubus abactus* Bailey [G, P & S, Steyerm.]
*Rubus ozarkensis* Bailey [G]; described from Arkansas originally, specimens cited from Missouri by Bailey (Gent. Herb. 5: 818, fig. 379. 1945)
*Rubus pubifolius* Bailey [G]; described from near Eagle Rock, Barry Co., Missouri, *Kellogg 26047* (holotype)
*Rubus frondosus* Bigel. [BB, P & S, Steyerm.]
*Rubus recurvans* Blanchard [BB, G]
*Rubus heterophyllus* of Am. auth. [P & S, Steyerm.] not Willd.

Flowers April–June; fruit June–August.

Occurs in open, rich or rocky wooded hills, thickets, meadows, pastures, prairie openings, and along fence rows and roadsides. Common throughout Missouri.

Ranges from Newfoundland to Ontario and Minnesota, south to Virginia, Alabama, Tennessee, Arkansas, and Oklahoma.

The fruit of *R. pensilvanicus* is usually sweet and juicy and varies from 1–3 cm. long. As with other members of the genus, the abundance of local rains in any one season is often responsible for the size and quality of the fruit. Among the native blackberries this species and the following, *R. ostryifolius*, yield the best-tasting fruit. The Bundy Blackberry, derived from *R. laudatus*, considered here synonymous with *R. pensilvanicus*, originated at Piedmont, Wayne County, Missouri, where it was developed by T. B. Bundy in 1905. The Bundy is considered to be a seedling of Early Harvest, the original plant of which, according to Hedrick, was found 'growing wild in Illinois some time previous to 1880' (Bailey, Gent. Herb. 5: 710. 1945).

Two variations in the fruit are represented in Missouri material:

Ripe fruit black or purple-black . . . . .
15a. R. PENSILVANICUS f. PENSILVANICUS
Ripe fruit white or whitish . . . . . . .
15b. R. PENSILVANICUS f. ALBINUS

15a. **Rubus pensilvanicus** f. **pensilvanicus** Map 1235

*Rubus pensilvanicus* Poir. [G, BB]
Throughout Missouri.

15b. **Rubus pensilvanicus** f. **albinus** Palmer & Steyerm. Map 1235

Known only from Howell County, southern Missouri (open thickets around oak-hickory clumps in upland on slopes, property of Buford Skaggs, T24N, R7W, sect. 6, 6 mi. northeast of West Plains, 5 mi. south southeast of White Church, July 23, 1955, *Steyermark 78868*, holotype in herb. Chi. Nat. Hist. Mus.; same locality, May 28, 1957, *Steyermark 85084, 85087*).

This albino form is known thus far only from Missouri, where it has been growing in the wild state on the property of the Buford Skaggs family and under their observation during the past thirty years, but only by chance circumstances brought to the attention of the present author.

The flowering canes of *R. pensilvanicus* are often arching or spreading and ascending, while the vegetative first-year primocanes are usually erect in habit. They vary from somewhat less than a meter tall to nearly 3 meters tall.

Some of the species listed above as synonyms of *R. pensilvanicus* by Fernald (eighth ed., *Gray's Man.*, p. 861) are treated by Gleason (*New Illus. Fl.* 2: 312. 1952) as synonyms under *R. ostryifolius* (*R. laudatus, R. pubifolius, R. ozarkensis, R. abactus*). It is possible also that future studies may show that both *R. pensilvanicus* and *R. ostryifolius* will have to be united. Specimens from Missouri labeled *R. recurvans* in the Gray Herbarium consist of sterile primocane leaves only collected from Phelps County (*Blanchard 157*) identified as such by Dr. Wm. Blanchard, and a specimen from Jackson County (*Bush 7566*), the determination of which is questioned by Fernald as doubtful.

16. **Rubus ostryifolius** Rydb. High-bush Blackberry Map 1236

*Rubus ablatus* Bailey; described from prairies, Morgan Co., Missouri, *Bush 14793* (holotype)
*Rubus jugosus* Bailey [G]
*Rubus virilis* Bailey [BB]; described from Licking, Texas Co., Missouri, *Kellogg 26037* (holotype)
*Rubus sertatus* Bailey [BB]; described from rich bottom woods in Grain Valley, Jackson Co., Missouri, *Bush 10502* (holotype)
*Rubus Bushii* Bailey [BB, G]; described from rocky prairies, Webb City, Jasper Co., *Bush 32422*
*Rubus Schneckii* Bailey [BB]; described from southern Illinois originally, specimens cited from Missouri by Bailey (Gent. Herb. 5: 698, fig. 314. 1945)
*Rubus Kelloggii* Bailey [BB]; described from along railroad, Dexter, Stoddard Co., Missouri, *Kellogg 26021* (holotype); Licking, Texas Co., *Kellogg 26038*
*Rubus abactus* Bailey [BB, P & S, Steyerm.]
*Rubus ozarkensis* Bailey [BB]; described from northwestern Arkansas (holotype), and Missouri (Noel, McDonald Co., *Palmer 45346*; Alba, Jasper Co.,

*Palmer 30121*; Clarkson, Lawrence Co., *Palmer 45329*; and Cedar Spring, Cedar Co., *Steyermark 27394*)

*Rubus pubifolius* Bailey [BB]; described from near Eagle Rock, Barry Co., Missouri, *Kellogg 26047* (holotype); Oasis, Taney Co., *Kellogg 26150*; Swan, Taney Co., *Kellogg 26042*

*Rubus laudatus* Berger [BB]

Flowers April–June; fruits June–August.

Occurs in woodland, prairies, thickets, pastures, fields, along fencerows and roadsides. Scattered throughout Missouri and probably as common as *R. pensilvanicus*.

Ranges from Maine to Michigan and Minnesota, south to Virginia, Kentucky, Arkansas, and Oklahoma.

I am following Gleason's concept of this species (*New Illus. Fl.* 2: 312–13. 1952) in the present flora. Bailey at one time had considered this as a distinct widespread blackberry, but in 1945 (Gent. Herb. 5: 726–29) changed his concept of *R. ostryifolius* as being a shade form localized in New Jersey. The numerous names cited as synonymous with *R. ostryifolius* include those so considered by Gleason, with whose conclusions the present author is in accord. An attempt to separate these taxa as distinct species has failed, the supposed differences between them being based upon criteria which do not appear to be constant or significant. It will be noted that some of the species considered by Gleason as synonymous with *R. ostryifolius* (*R. abactus*, *R. laudatus*, *R. pubifolius*, *R. ozarkensis*) are placed by Fernald as synonyms under *R. pensilvanicus*. Whether or not *R. ostryifolius* will continue to be maintained as distinct from *R. pensilvanicus* will depend upon results of future studies.

## 15. **Neviusia** A. Gray

**Neviusia alabamensis** A. Gray Snow Wreath
Map 1237

Flowers April.

Known only from Butler County in southeastern Missouri (sandy loam, about 8 mi. west of Poplar Bluff, 1918, *J. C. Th. Uphof*).

Occurs in Alabama, Arkansas, and Missouri.

This shrub was originally excluded from Palmer & Steyermark's *Annotated Catalogue*, because it was at that time (1935) believed that the reports for the Missouri occurrence by Uphof (Das Vorkommen von *Neviusia alabamensis* Gray im Suden von Missouri. Mitt. Deutsch. Dendrol. Gesells. 282–83. 1921; Am. Jour. Bot. 9: 7. 1922) were based upon misidentified specimens probably of a *Physocarpus opulifolius* var. *intermedius*. Moreover, no specimens were available for examination, they having been destroyed. Later, however, specimens were collected in 1925 by Demaree from Conway County, Arkansas, and again in 1955 from the same locality by Dwight M. Moore (Rh. 58: 187–91. 1956). These newly recorded stations in central Arkansas indicated the occurrence of the shrub outside Alabama, and, together with the accurate illustration of it reproduced by Uphof in his original publication (*loc. cit.* 1921), left no doubt that the shrub had certainly been collected in that portion of southeastern Missouri from which Uphof reported it. The range is essentially similar to that of *Draba aprica*, known from Georgia, Arkansas, and southeastern Missouri, and a number of other rare species. Search was made by Dr. Moore in 1955 in Butler County, Missouri, and the present author accompanied him in the spring of 1956 in an attempt to relocate the *Neviusia*, but the efforts were unsuccessful. Continued search by the present author has thus far failed to rediscover the old or new stations in Missouri. It is expected, however, that with continued diligent exploration, its rediscovery will eventually be made.

The shrub is 9–18 dm. (3–6 feet) tall and usually grows in a clump or small colony increasing each year by creeping underground root suckers, much as in *Kerria* to which *Neviusia* is most closely related. When in flower, *Neviusia* is quite showy with feathery clusters of numerous white stamens surrounded by 5 spreading to reflexed, green, white or greenish-white, petal-like calyx lobes. Petals are absent. There are 1–4 flowers arising at the tip of naked pedicels from the ends of the shoots. The fruit consists of 2–4 drupelike, small ovoid achenes. The leaves resemble those of *Carpinus caroliniana*, the American Hornbeam, as well as unlobed leaves of *Physocarpus opulifolius* var. *intermedius*, and are thin and doubly serrate. The foliage remains green well into late November long after other shrubs and trees have turned their autumnal colors. Eventually the leaves disappear with little color change. Shrubs grown at the author's botanical preserve in northern Illinois have prospered since 1946, flowering each spring and increasing in area by root suckering. Plants should be protected at least during the winter against rabbits, which are fond of the tender shoots. They succeed in a rich loamy, well-drained soil in shaded or woodland situations.

1237 Neviusia alabamensis (Snow Wreath)

1238 Agrimonia gryposepala

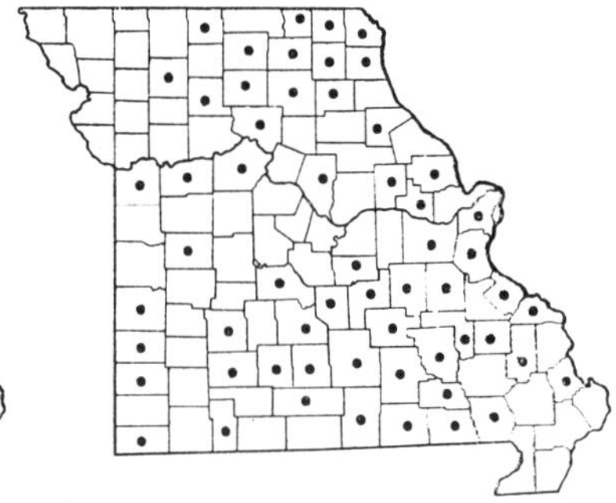
1239 Agrimonia parviflora

## 16. **Agrimonia** L. Agrimony

a. Plants of moist or wet ground of open swamps, wet prairies, and meadows; main well-developed leaf-divisions (leaflets) 11–17 (excluding the small leaf-divisions between the main divisions). 2. A. PARVIFLORA

a. Plants of dry usually wooded areas; main well-developed leaf-divisions (leaflets) 5–9 (excluding the small leaf-divisions between the main divisions) . . . . . . . . . . . . . . . . . *b*

b. Lower surface of leaf-divisions more or less noticeably hairy, but not dotted with any resinous glands . . . . . . . . . . . . . . . . . . . . . . . . . 4. A. PUBESCENS

b. Lower surface of leaf-divisions faintly or plainly dotted with few to many resinous glands, the rest of the leaf surface with or without hairs . . . . . . . . . . . . . . . . . . *c*

c. Axis of inflorescence with scattered, spreading, noticeable hairs; mature fruit 6–8 mm. long, 5–6 mm. in diameter, top-shaped or inversely conical, narrowed or tapering to the base, strongly grooved; rare species of the northern third of Missouri . . . . . . . 1. A. GRYPOSEPALA

c. Axis of inflorescence bearing minute glands but lacking noticeable or spreading hairs; mature fruit 3.5–4.5 mm. long, 2.5–3 mm. in diameter, hemispherical, rounded at base, faintly or not at all grooved; common species of all but the northwest and north-central sections of the state 3. A. ROSTELLATA

1. **Agrimonia gryposepala** Wallr. Map 1238

Flowers July–August.

Occurs in rich or rocky woodland and thickets. Northern third of Missouri, where known only from Clark, Macon, Putnam, and Clinton counties.

Ranges from New Brunswick, Maine, and Quebec to Ontario, Michigan and North Dakota, south to North Carolina, Tennessee, Missouri, and Kansas; also from British Columbia to California and New Mexico.

2. **Agrimonia parviflora** Ait. Map 1239

Flowers August–September.

Occurs in usually wet open ground along streams or spring branches in prairies, swamps, boggy ground, swampy meadows, roadside ditches, and moist thickets. Throughout Missouri, except not recorded in some of the northwestern and extreme southeastern lowland counties.

Ranges from Florida to Texas, north to Connecticut, New York, Ontario, Ohio, Michigan, Illinois, Iowa, and Nebraska; also in Mexico and the West Indies.

The stems are densely hairy with spreading hairs, usually more conspicuously so than *A. pubescens* or the other species in Missouri. The leaflets are also narrower, of a lanceolate type, and more acuminate at the tip than in the other Missouri species, and the lower surface of the middle and upper leaves is abundantly gland-dotted.

3. **Agrimonia rostellata** Wallr. Map 1240

Flowers July–September.

Occurs generally on slopes, ridges, or upland of

---

Plate no. 203. **1.** Rubus argutus, $\times {}^{2}/_{7}$. **2.** Rubus pensylvanicus, $\times {}^{2}/_{7}$. **3.** Rubus ostryifolius, $\times {}^{2}/_{7}$. **4.** Neviusia alabamensis, $\times {}^{2}/_{7}$; Details from Small, The New York Botanical Garden. **5.** Agrimonia gryposepala, $\times {}^{2}/_{7}$; *a.* Fruit, $\times 1{}^{1}/_{7}$; After Britton and Brown, The New York Botanical Garden. **6.** Agrimonia pubescens, $\times {}^{2}/_{7}$; After Britton and Brown, details from Small, The New York Botanical Garden. **7.** Rosa setigera, $\times {}^{2}/_{7}$; After Britton and Brown, The New York Botanical Garden. **8.** Agrimonia parviflora, $\times {}^{2}/_{7}$; *a.* Fruit, $\times 1{}^{1}/_{7}$; After Britton and Brown, details from Small, The New York Botanical Garden. **9.** Rosa Eglanteria, $\times {}^{2}/_{7}$. **10.** Rosa $\times$ centifolia, $\times {}^{2}/_{7}$.

PLATE NO. 203

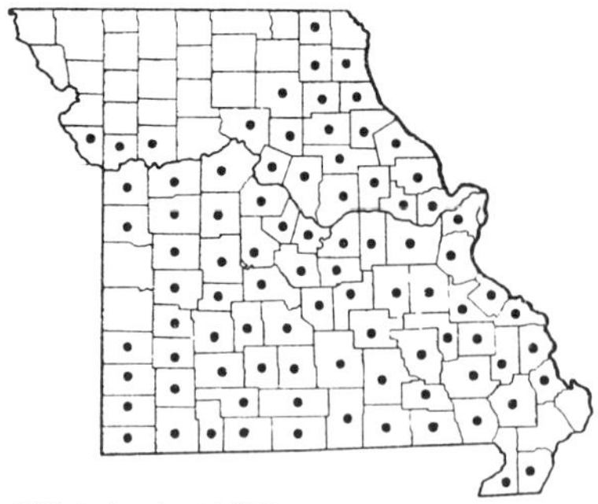

1240 Agrimonia rostellata

1241 Agrimonia pubescens

1242 Rosa multiflora (Multiflora Rose)

dry open or rocky woodland. Southern, central, and northeastern Missouri, south of a line from Scotland, Macon, Chariton, and Ray counties to Platte County.

Ranges from Georgia and Louisiana to Oklahoma, north to Massachusetts, New York, Ohio, Indiana, Illinois, Missouri, and Kansas.

4. **Agrimonia pubescens** Wallr. Map 1241

Flowers July–September.

Occurs in dry open or rocky woodland, either in lowland or upland situations, sometimes in low woods along streams and thickets. Throughout Missouri and probably in every county.

Ranges from Georgia to Oklahoma, north to Massachusetts, New York, Ontario, Michigan, Illinois, Minnesota, and Nebraska.

The pubescence of the stems is usually ascending or curved in toward the stem. The fruiting hypanthium is as long as or longer than broad.

Missouri records reported in Palmer & Steyermark's *Annotated Catalogue* (pp. 568–69) as *A. microcarpa* Wallr. and *A. platycarpa* Wallr. were misidentified and should be referred to *A. pubescens* Wallr.

*Excluded Species*

**Agrimonia microcarpa** Wallr.

*Agrimonia platycarpa* Wallr. [P & S]

Missouri specimens reported in Palmer & Steyermark's *Annotated Catalogue* (pp. 568–69) as belonging to *A. platycarpa* and *A. microcarpa* have been found to be misidentified *A. pubescens* and *A. rostellata*, and should be referred to those species.

## 17. **Rosa** L. Rose

The numerous ovaries are enclosed by the green cup-shaped calyx-tube (receptacle) of the flower, on the bottom or sides of which receptacle they are inserted. This receptacle is also referred to as the hypanthium and in fruit becomes fleshy, red, scarlet, or sometimes purplish-black, the hairy ovaries becoming bony seed-like achenes.

Many of the species of this genus are difficult to identify, and whenever possible mature fruiting specimens should be collected later in the season from the same plants from which the flowers were obtained. Notes should also be recorded on the shape of the fruit and position of the calyx-lobes in fruit.

Roses have often been blamed for causing hay fever or 'rose fever,' but, although it has been shown that they can cause hay fever, actually they rarely ever do. Most modern roses in cultivation are of the double type, producing only a very small amount of pollen. Most single-flowered roses, wild species, and *Rosa rugosa* produce more abundant quantities of pollen.

a. Stems entirely smooth, without prickles or thorns or practically without them . . . . . . . . . *b*
  b. Stems more or less straight and upright, usually 4.5–9 dm. (1½–3 feet) tall; plants of northeastern Missouri . . . . . . . . . . . . . . . . . . . . . . . . . 14. R. BLANDA
  b. Stems long and spreading, whip-like, downward-arching or climbing, commonly 1.8 meters (6 feet) or more long; plants of southern, central, and east-central Missouri . . . . . . . . *c*
    c. Lower surface of leaflets glabrous (without hairs) or with only a few hairs on the nerves. . 2. R. SETIGERA var. SETIGERA f. INERMIS
    c. Lower surface of leaflets more or less hairy. . . . . 2. R. SETIGERA var. TOMENTOSA f. SERENA
a. Stems with few to many prickles or thorns . . . . . . . . . . . . . . . . . . . *d*
  d. Upper or lower surface, or both surfaces, of leaflets bearing little, round, rust-colored resinous glands . . . . . . . . . . . . . . . . . . . . . . . . . . . . . . *e*
    e. Both upper and lower surfaces of leaflets abundantly gland-dotted; leaflets quite fragrant when bruised; styles hairy; calyx-lobes persisting and long-lasting until ripening of the fruit . 6. R. EGLANTERIA
    e. Upper surface of leaflets very sparsely gland-dotted or nearly or quite without glands; foliage slightly or scarcely fragrant when bruised; styles glabrous (without hairs) or nearly glabrous; calyx-lobes promptly falling after flowering, not long-lasting into fruit . . . . 7. R. MICRANTHA
  d. Neither surface of leaflets with resinous glands, but with or without hairs . . . . . . . . . *f*
    f. Stipules at base of leaf-stalks (petioles) deeply dissected, fringed, and comb-like (fimbriate-pectinate); flowers small, 1.5–2.8 cm. across . . . . . . . . . . . . 1. R. MULTIFLORA
    f. Stipules not as above; flowers medium to large, 2–8 cm. across . . . . . . . . . . . *g*
      g. Surface of stems and branches covered with hairs as well as thorns and bristles; most of the thorns or their bases hairy . . . . . . . . . . . . . . . . . 12. R. RUGOSA
      g. Surface of stems and branches without hairs; thorns or prickles when present not hairy . . . *h*
        h. Styles united and protruding noticeably from the opening or mouth of the cup- or vase-shaped receptacle; leaves on old stems mainly with 3 leaflets, on new stems 3 or 5; stems arching, recurved, climbing, or sprawling . . . . . . . . . . . . . 2. R. SETIGERA
        h. Styles free, not protruding or only shortly protruding from the mouth of the receptacle, usually forming a dense head-like brush in the mouth and closing it, if rarely protruding, then the styles united; leaves on old and new stems mainly with 5–11 leaflets, or if 3–5, then the stems erect, ascending, or forming a bush . . . . . . . . . . . . *i*
          i. Flower-stalk (pedicel) without a bract at the base; flower usually solitary; leaflets 7–11, mostly 0.5–1.5 cm. long . . . . . . . . . . . . . 15. R. SPINOSISSIMA
          i. Without the above combination of characters; flower-stalk (pedicel) usually with a bract at the base; flowers either solitary or several arranged in a corymbose type of inflorescence (more or less flat-topped or convex), if solitary the flower-stalk usually with a bract at the base . . . . . . . . . . . . . . . . . . . . *j*
            j. Outer calyx-lobes or some of them pinnately cut into several narrow segments; mouth of the receptacle (through which the styles protrude) about 1 mm. in diameter; styles shortly or long protruding from mouth of hypanthium; chiefly introduced cultivated shrubs . . . . . . . . . . . . . . . . . . *k*
              k. Receptacle smooth; flower-stalks (pedicels) usually glabrous; leaves of flowering branchlets mostly with 5–9 leaflets . . . . . . . . . . . . . . . . *l*
                l. Lower surface of leaflets with glands, and with or without hairs. . . . 8a. R. CANINA var. CANINA

l. Lower surface of leaflets without glands, and with hairs . . . . 8b. R. CANINA var. DUMETORUM

k. Receptacle with hairs and/or glands; flower-stalks (pedicels) always bearing glands or gland-tipped hairs; leaves of flowering branchlets mostly with 3 or 5 leaflets. . . . . . . . . . . *m*

m. Flower usually nodding, double, solitary; prickles mostly straight; calyx-lobes persisting and spreading after flowering; plant sterile, not forming fruit . . . . . . 4. R. × CENTIFOLIA

m. Flowers erect, usually single or semi-double, rarely double, usually 2–5 together in an inflorescence, or solitary; prickles or thorns hooked or curved; calyx-lobes reflexed, falling away from receptacle after flowering; plant fertile, forming fruit . . . . . . . . . . . . . *n*

n. Teeth of leaves tipped with a rounded gland; leaflets rather thickish, rough, rigid, and leather-like in texture; fruiting receptacle globe-shaped or hemispherical; prickles unequal in size and shape . . . . . . . . . . . . . . . . . . . . . 3. R. GALLICA

n. Teeth of leaves simply toothed without a gland at tip; leaflets thinner, rather soft and flexible; fruiting receptacle obovate, elongated-pear-shaped, 2.5 cm. long; prickles more or less equal and uniform in size and shape. . . . . . . . . . . . 5. R. × DAMASCENA

j. All the calyx-lobes entire (smooth-edged) or with simple shallow teeth, or some of them with 1–4 narrow, elongated lobes arising from the upper or lower half; mouth of the receptacle (through which the styles protrude) 2–4 mm. in diameter; styles not protruding, forming a dense head-like brush or disk in the mouth of the receptacle and closing it; native species . . . . . . . . . . *o*

o. Calyx-lobes remaining attached to receptacle long after flowering period, more or less persisting in fruit; receptacle and flower-stalk (pedicel) usually smooth from the beginning of flowering period; pistils and seed-like achenes attached to the inside walls as well as on the bottom of the receptacle or 'hip' . . . . . . . . . . . . . . . . . . . . . . . . . *p*

p. Prickles or thorns present only at base of stem or with a few scattered above, but ending well below the flowering stem; leaflets usually 5 or 7, sometimes 9; flowers arising from the side branches of the previous season's stems; plants found only in northeastern Missouri . 14. R. BLANDA

p. Without the above combination; prickles or thorns well distributed and usually numerous on the stem and flowering branches; leaflets chiefly 9–11; flowers arising at the tips of the new shoots as well as from the side branches of the previous season's stems; plants of northern and western Missouri . . . . . . . . . . . . . . . 13. R. ARKANSANA var. SUFFULTA

o. Calyx-lobes usually disappearing in fruit, falling from the receptacle after the flowering period; at least some of the receptacles and flower-stalks (pedicels) during the flowering period with gland-tipped hairs, often disappearing in fruit or with scars or stubs of bases of the hairs remaining on surface of receptacle; pistils and seed-like achenes on the bottom only of the receptacle or 'hip' . . *q*

q. Large shrubs of swamps or wet ground, mostly 1.5–2.5 m. (5–8½ ft. approximately) tall; leaves finely toothed, 12–25 teeth above the middle of each margin, the teeth near the widest part of the leaf 0.3–1.2 mm. high; stipules more or less convolute (rolled up longitudinally with sides curving in) at least on the shoots; plants found only in southeastern Missouri west to Reynolds, Carter, and Ripley counties and north to Cape Girardeau, Bollinger, Madison, and Iron counties . . . . . . . . . . . . . . . . . . . . . . 9. R. PALUSTRIS

q. Plants mainly of dry or rocky ground, usually less than 9 dm. (3 ft.) tall in *R. carolina*, but up to 2 m. (6½ ft.) tall in *R. virginiana*; leaves coarsely toothed, usually 5–15 teeth above the middle of each margin, the teeth near the widest part of the leaf 0.6–2 mm. high; stipules flat; plants occurring throughout all parts of the state . . . . . . . . . . . . . . . . *r*

r. Common species; infrastipular prickles (those at base of or just below the stipules) slenderly needle-like and straight, terete (rounded in cross-section or sides rounded); dilated portion of base of larger thorns usually less than half the length of the thorn; prickles usually numerous between nodes of the stem; flowers usually arising from the new growth of the season; the lower part of the stipule which is connected with the main petiole (stalk) of the leaf is 0.5–1.5 cm. long; the combined width of the pair of stipules which are connected with the main petiole of the leaf is 0.5–2 (–3) mm. . . . . . . . . . 10. R. CAROLINA

r. Rare species, mainly northern Missouri; infrastipular prickles stout, flattened and broadened near the base, often curved or hooked, sometimes straight; dilated portion of base of larger thorns usually longer than half the length of the thorn; prickles usually absent between nodes of the stem; flowers usually arising from the branches of the old wood of the previous season; the lower part of the stipule which is connected with the main petiole of the leaf is 1–4 cm. long; the combined width of the pair of stipules which are connected with the main petiole of the leaf is 3–10 mm. . . . . . . . . . . . . 11. R. VIRGINIANA

1. **Rosa multiflora** Thunb. Japanese Rose, Multiflora Rose Map 1242
Flowers May–June.

Commonly planted along fence rows and roadsides; escaping into roadside thickets, pastures, and open ground. Naturalized in a few scattered counties, but of expected more frequent occurrence.

Native of Japan; introduced and naturalized from New England and New York south and west to Missouri, Oklahoma, and Texas.

This species is now frequently planted in the midwestern states as a so-called living fence to serve as a barrier against trespassers or for keeping in stock, as well as to provide shelter and cover for birds and other wildlife. In many sections it has replaced or been substituted for Osage Orange or Hedge Apple (*Maclura pomifera*). When fully developed, it forms large clumps of arching canes 5–10 feet or more tall. When in flower, the numerous, small, white flowers are very showy on the arching or drooping branches and resemble blackberry bushes at a distance.

2. **Rosa setigera** Michx. Prairie Rose, Climbing Rose Map 1243
Flowers late May–July.

Occurs in moist ground and rocky places along streams and spring branches, moist thickets, low open woodland, pastures, prairie thickets, clearings, fence rows, and along roadsides.

The following variations are represented in Missouri material:

a. Stems entirely smooth, without prickles or thorns or practically without them . . . *b*
  b. Lower surface of leaflets glabrous (without hairs) or with only a few hairs on the nerves
    2b. R. SETIGERA var. SETIGERA f. INERMIS
  b. Lower surface of leaflets more or less hairy
    2e. R. SETIGERA var. TOMENTOSA f. SERENA
a. Stems with conspicuous prickles or thorns . *c*
  c. Lower surface of leaves glabrous (without hairs) or with only a few hairs on the nerves
    2a. R. SETIGERA var. SETIGERA f. SETIGERA
  c. Lower surface of leaves more or less hairy . *d*
    d. Common type; petals pink or rose-colored, sometimes fading to whitish, 2–3 cm. long, the flowers 4–8 cm. broad
      2c. R. SETIGERA var. TOMENTOSA f. TOMENTOSA
    d. Rare type; petals white from the beginning, smaller, 1–2 cm. long, the flowers 3–4 cm. broad . 2d. R. SETIGERA var. TOMENTOSA f. ALBA

2a. **Rosa setigera** var. **setigera** f. **setigera** Map 1243
*Rosa setigera* Michx. [G, BB in part, P & S, Steyerm.]

This form is rare and scattered in southern and central Missouri.

Ranges from Florida to Texas, north to New York, Ohio, Indiana, Illinois, Missouri, and Kansas; also naturalized in New England and Michigan.

2b. **Rosa setigera** var. **setigera** f. **inermis** Palmer & Steyerm. Map 1243
*Rosa setigera* f. *inermis* Palmer & Steyerm. [G, P & S, Steyerm.]

Scattered in the Ozark sector of the state north to Crawford County, but less common than the other thornless form (var. *tomentosa* f. *serena*).

Scattered in parts of the range of var. *setigera*.

2c. **Rosa setigera** var. **tomentosa** f. **tomentosa** T. & G. Map 1243
*Rosa setigera* var. *tomentosa* T. & G. [G, BB, P & S, Steyerm.]

This is the commonest variation of *R. setigera* in Missouri and occurs throughout the state, doubtless in every county.

Ranges from Georgia to Texas, north to Ontario, Indiana, Illinois, and Nebraska; naturalized in New England.

2d. **Rosa setigera** var. **tomentosa** f. **alba** Steyermark Map 1243
*Rosa setigera* f. *alba* Steyerm. Rh. 54: 254. 1952.

Known only from Reynolds (roadside gravelly bank above swampy meadow along Bee Fork, T32N, R2W, sect. 22, on property of Mr. Reese, 4 mi. southeast of Bunker, July 7, 1951, *Steyermark 72011*, holotype in Chi. Nat. Hist. Mus. herb.), Howell (*Steyermark 78797*), and Callaway (thickets along spring branch, Tucker Prairie, along south side of highway 40, west of Kingdom City, July 21, 1958, *Steyermark 86124*) counties.

Known thus far only from Missouri.

The leaves are usually paler green in this albino form.

2e. **Rosa setigera var. tomentosa** f. **serena** (Palmer & Steyerm.) Fern. Map 1243
*Rosa setigera* var. *serena* Palmer & Steyerm. [P & S, Steyerm.]

Common throughout the Ozark region and northeast locally to Pike and Ralls counties.

Scattered in the range of var. *tomentosa*.

*Rosa setigera* is easily grown and makes a desirable rose for shady spots and borders where it is too shady

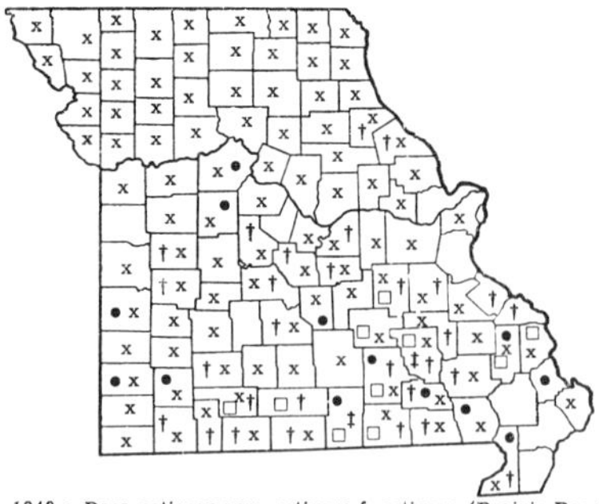

1243 • Rosa setigera var. setigera f. setigera (Prairie Rose)
1243 □ Rosa setigera var. setigera f. inermis
1243 x Rosa setigera var. tomentosa f. tomentosa
1243 ‡ Rosa setigera var. tomentosa f. alba
1243 † Rosa setigera var. tomentosa f. serena

1244 • Rosa gallica var. gallica (French Rose)
1244 □ Rosa gallica var. officinalis (Red Rose of Lancaster)

1245 Rosa X centifolia (Cabbage Rose)

for other roses. It also does well in full sun. It is a relatively later bloomer as compared with other native species. The foliage often turns a deep rose-red in late fall.

3. **Rosa gallica** L. French Rose, The Rose of Provins Map 1244
Flowers May–July.

Occasionally planted as an 'old-fashioned' or 'species' rose, and rarely escaping into thickets and along railroads.

Native of Europe; naturalized in the United States north to New England.

Missouri material is represented by two variations:

Flowers not double, the parts of the flower normally developed with stamens and pistils . . 3a. R. GALLICA var. GALLICA

Flowers double, the stamens and pistils mostly changed to petals. . . 3b. R. GALLICA var. OFFICINALIS

3a. **Rosa gallica** var. **gallica** Map 1244
*Rosa gallica* L. [G, BB]

Known only from Iron (along railway embankment, Arcadia, May 18, 1923, *Palmer 22674*) and Clark (thickets along small stream near Dumas, September 7, 1922, *Palmer 21876*) counties.

3b. **Rosa gallica** var. **officinalis** Thory
Red Rose of Lancaster Map 1244

Known only from Boone County (persisting and spreading in abandoned farmyards, federal wildlife area east of Ashland, May 19, 1936, *Drouet 2959*, in U. of Mo. Herb.).

The flowers of *R. gallica* are usually solitary or 3–4 together, and are pink to crimson to purple. A variety of this species with striped red and white petals (var. *versicolor* Thory), known as Rosa Mundi, is more commonly grown. Purple-colored variations (Cardinal deRichelieu and President de Sẹze) are sometimes cultivated as old-fashioned unusual roses. The rose hips of this species are ornamental and may be boiled for rose-hip tea.

4. **Rosa × centifolia** L. Cabbage Rose Map 1245
Flowers May–July.

Grown as an old-fashioned rose and rarely escaping and persistent. Known only from Clark County.

This was considered at one time to be a wild species from the Caucasus Mountain area, but is now believed to have been derived from garden origin as a hybrid between four wild species (*R. rubra*, *R. phoenicia*, *R. moschata*, and *R. canina*).

An unusual variation, known as Moss Rose (*R. × centifolia* var. *muscosa* [Ait.] Sér.), is sometimes grown as an old-fashioned type; it has very glandular hairy pedicels and calyx, and the calyx-lobes are finely divided. The so-called Crested Moss (var. *cristata* Prév.) is a similar type with the calyx-lobes crested with a ruffled fringe or crest. Another unusual variation is Variegata di Bologna, with magenta stripes on the white and pink-tinged petals.

5. **Rosa × damascena** Mill. Damask Rose Map 1246

Flowers May–July.

Planted as an 'old-fashioned' or 'species' rose and rarely escaping and established along roads and thickets. Known only from Clark (upland clay slopes along road, T67N, R7W, sect. 25, 1½ mi. southwest of Athens, 1¼ mi. east of highway 81, May 30, 1941, *Steyermark 28641*), Grundy and Dunklin (along road in sect. 11, east of Glennonville, on Crowley Ridge, May 26, 1939, *Steyermark 26591*, in herb. Arnold Arboretum) counties.

1246 Rosa X damascena (Damask Rose)

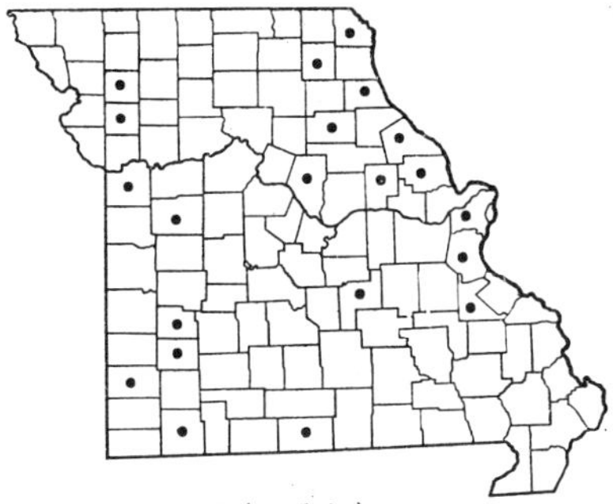

1247 Rosa Eglanteria (Sweetbrier)

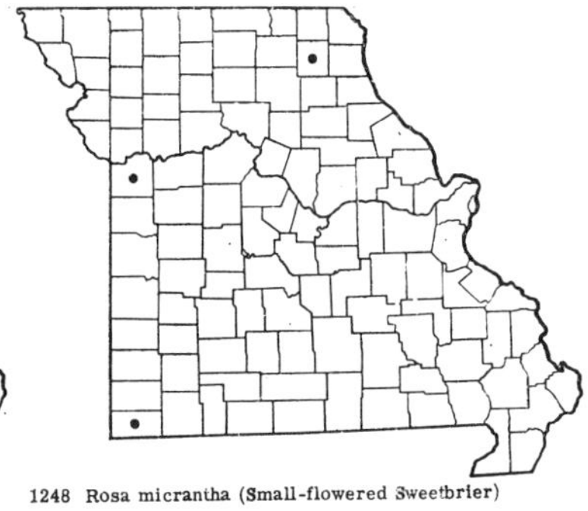

1248 Rosa micrantha (Small-flowered Sweetbrier)

This rose is now believed to have been derived through hybrid origin involving *R. gallica* as one of the parents. The hybrid perpetual class of roses can be traced in part to an ancestry involving *R.* × *damascena*.

The Damask Rose is one of the sources for the 'Otto (attar) of roses,' the perfume obtained through steam distillation from the blossoms of this rose and *R.* × *alba* L., produced in Bulgaria.

The flowers, often 6–12 grouped in a corymbose inflorescence, are usually double, and very fragrant, varying from usually deep rose, or pink to red, bluish-white, or with both pink and white flowers in the York and Lancaster Rose (*R.* × *damascena* var. *versicolor* West). A number of named horticultural varieties are in cultivation.

6. **Rosa Eglanteria** L. Sweetbrier, Eglantine Map 1247

Flowers May–July.

Planted as 'old-fashioned' or 'species' rose, and formerly more commonly grown. Escaped and established in pastures, sandy or open ground in valley bottoms, and in waste ground. Scattered throughout Missouri.

Native of Europe; introduced and naturalized in many sections of North America.

Of the various foreign species found as escapes or established from cultivation, this one is the most commonly encountered. It is usually a large bush up to 3.5 meters tall, but may also grow somewhat sprawling. The thorns are stout and strongly hooked. The flowers, which are single, are pink, and 3–5 cm. across, occurring either singly or few to several together.

7. **Rosa micrantha** Sm. Small-flowered Sweetbrier Map 1248

Flowers May–July.

Sometimes found along roadsides, open ground, and waste places. Scattered in Knox, Jackson (roadsides, Lees Summit, June 6, 1897, *Bush 298*, in herb. Arnold Arboretum), and McDonald (Southwest City) counties.

Native of Europe; introduced and naturalized in North America from Anticosti to Ontario and Wisconsin, south to North Carolina, Kentucky, Missouri, and Texas.

The flowers, 2–3 cm. across, are somewhat smaller than those of *R. Eglanteria*, and paler pink varying to white.

8. **Rosa canina** L. Dog Rose, Dog Briar Map 1249

Flowers May–July.

Persisting from cultivation in old gardens in abandoned farmyards and pastures, along roadsides, and thickets along streams.

Missouri material is represented by the following two variations previously keyed out:

8a. **Rosa canina** var. **canina** Map 1249

*Rosa canina* L. [G, BB, P & S, Steyerm.]

Known only from Dade, Lawrence, and Newton counties, southwestern Missouri.

Native of Europe; introduced and naturalized in North America, from Nova Scotia to New York, south to Virginia, Tennessee, and Missouri.

8b. **Rosa canina** var. **dumetorum** Baker Map 1249

*Rosa dumetorum* Thuill. [Bailey Cyclop.]

*Rosa corymbifera* Borkh. [Royal Hort. Soc. Dict.]

*Rosa canina* var. *dumetorum* Mert. & Koch [Hegi]

Known only from Boone (persisting in abandoned farmyards, federal wildlife area east of Ashland, May 26, 1936, *Drouet 3019* in Univ. of Mo. herb.) and Worth (established along road, upland clay slopes, T65N, R32W, sect. 17, 1¾ mi. northeast of Oxford, June 8, 1957, *Steyermark 85233A*) counties.

Native of Europe; introduced and naturalized in

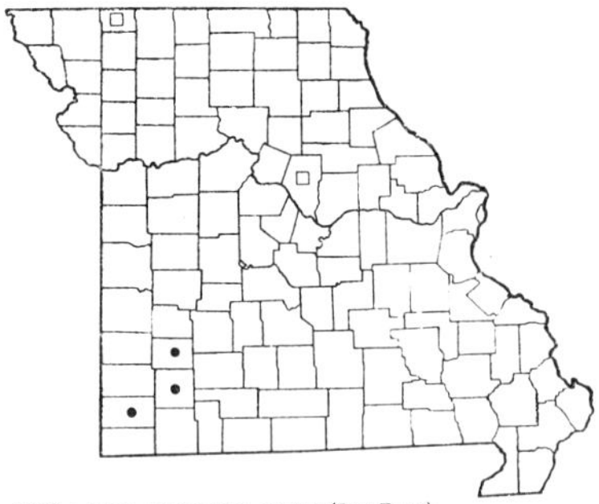

1249 • Rosa canina var. canina (Dog Rose)
1249 □ Rosa canina var. dumetorum

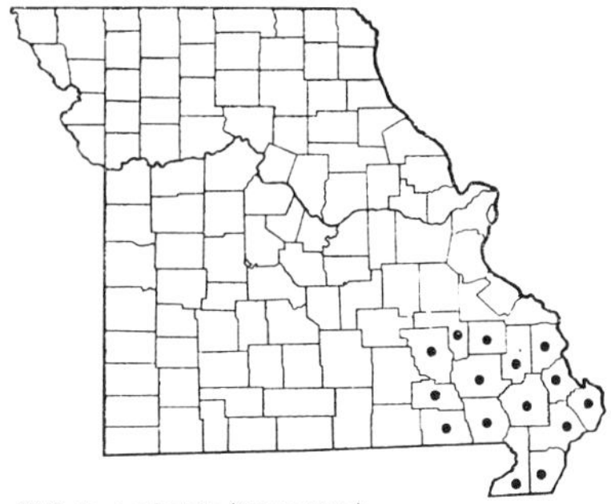

1250 Rosa palustris (Swamp Rose)

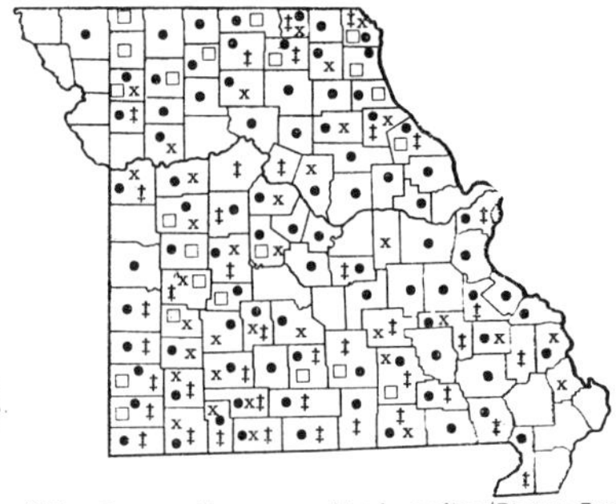

1251 • Rosa carolina var. carolina f. carolina (Pasture Rose)
1251 ‡ Rosa carolina var. carolina f. glandulosa
1251 x Rosa carolina var. villosa
1251 □ Rosa carolina var. grandiflora

North America from Quebec to Kentucky and Missouri.

The petals in *R. canina* var. *canina* and var. *dumetorum* are pink, rose, or white. The flowers are fragrant and 3.5–5 cm. across. The plants are sometimes only 7.5–9 dm. (2½–3 ft.) tall, but may reach a height of 3 meters (about 10 feet). Typical var. *canina* is often employed as stock for grafting purposes.

9. **Rosa palustris** Marsh. Swamp Rose Map 1250
*Rosa palustris* var. *dasistema* (Raf.) Palmer & Steyerm. [P & S]
Flowers late May–July.

Occurs in swamps, low wet woods, borders of upland wooded sink-hole ponds, streams, bayous, and drainage ditches. Restricted to southeastern Missouri, from the lowland counties north to Cape Girardeau, Bollinger, Madison, and Iron counties, and west to Reynolds, Carter, and Ripley counties.

Ranges from Florida to Arkansas, north to Nova Scotia, Quebec, Ontario, Michigan, Wisconsin, and Minnesota.

The fruiting receptacle may be smooth or covered with glandular hairs.

10. **Rosa carolina** L. Pasture Rose Map 1251
Flowers May–June.

Occurs in open woods, glades, prairies, thickets, clearings, and along railroads and roadsides. Throughout Missouri.

This is a complex and variable species, as here interpreted, consisting of numerous minor forms and variations of which the following named taxa are represented in Missouri material:

a. Flowers 4–7 cm. across; leaflets obovate or broadly oval . 10d. R. CAROLINA var. GRANDIFLORA
a. Flowers 3.5–5.5 cm. across; leaflets ovate-lanceolate to narrowly obovate . . . . *b*
b. Lower surface of leaflets soft-hairy . . . 10c. R. CAROLINA var. VILLOSA
b. Lower surface of leaflets glabrous (without hairs) or sparsely hairy . . . . . . . . *c*
c. Axis of leaf glabrous (without hairs); tip of teeth of the leaflets without glands . 10a. R. CAROLINA var. CAROLINA f. CAROLINA
c. Axis of leaf with stalked glands; tip of teeth of the leaflets with glands. . . 10b. R. CAROLINA var. CAROLINA f. GLANDULOSA

10a. **Rosa carolina** L. var. **carolina** f. **carolina** Map 1251
*Rosa carolina* L. [G, BB, P & S, Steyerm.]

This is the commonest variation in Missouri, occurring throughout the state, and doubtless in every county except possibly actually absent from most of the extreme lowland counties of southeastern Missouri.

Ranges from Florida to Texas, north to Nova Scotia, Ontario, Michigan, Wisconsin, Minnesota, and Nebraska.

Of all the roses, this particular variation is the commonest in Missouri of any species encountered.

10b. **Rosa carolina** var. **carolina** f. **glandulosa** (Crépin) Fern. Map 1251
*Rosa carolina* f. *glandulosa* (Crépin) Fern. [G]
*Rosa carolina* var. *glandulosa* (Crépin) Farw. [P & S, Steyerm.]
*Rosa subserrulata* Rydb. [P & S, Steyerm.]; origi-

Plate no. 204. **1.** Rosa canina var. canina, × 2/5. **2.** Rosa palustris, × 2/5. **3.** Rosa carolina var. carolina, × 2/5. **4.** Rosa rugosa, × 2/5. **5.** Rosa virginiana, × 2/5. **6.** Rosa arkansana var. suffulta, × 2/5. All details from Small, The New York Botanical Garden.

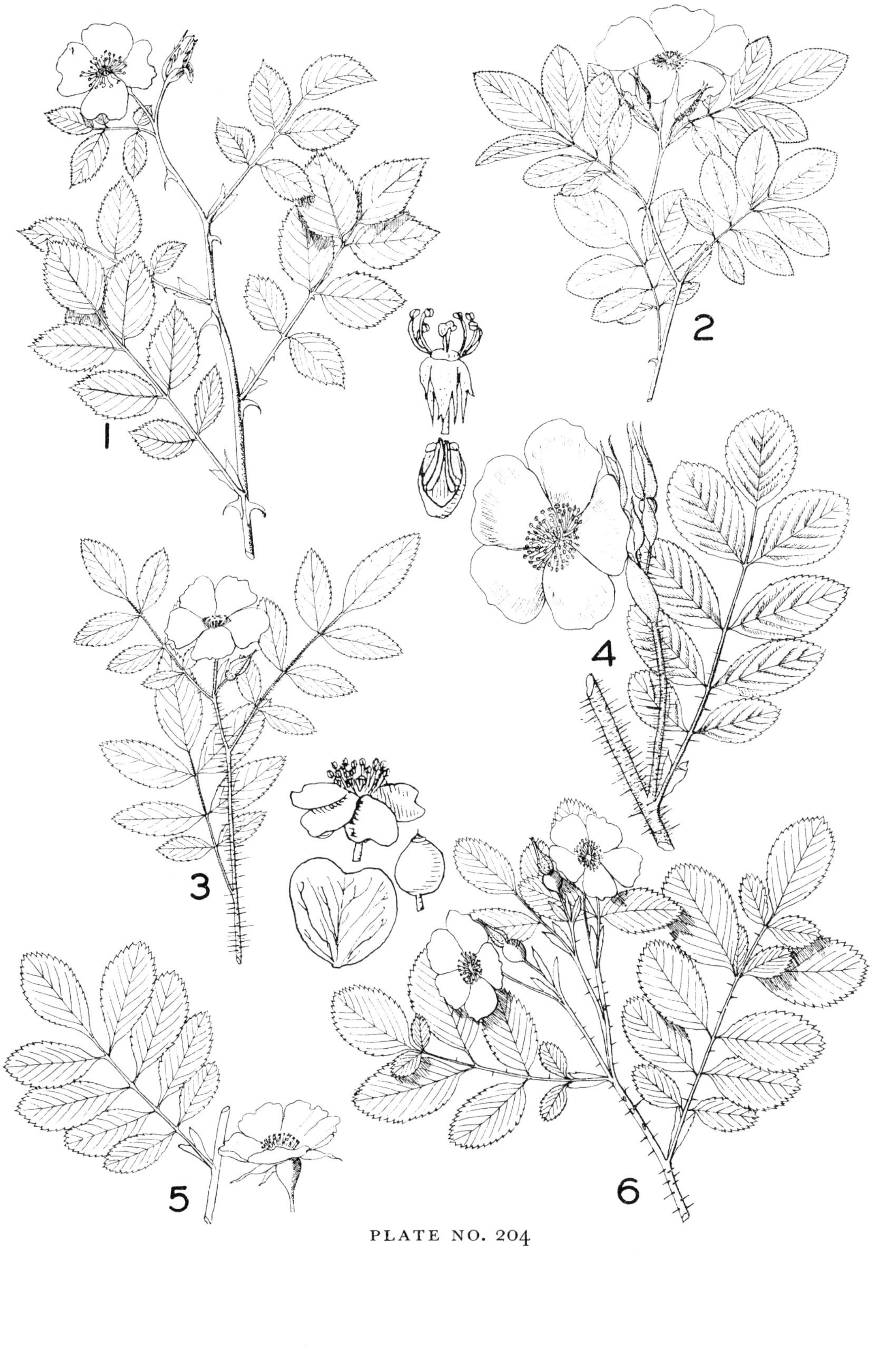

PLATE NO. 204

1252 Rosa virginiana var. virginiana

1253 Rosa rugosa (Japanese Rose)

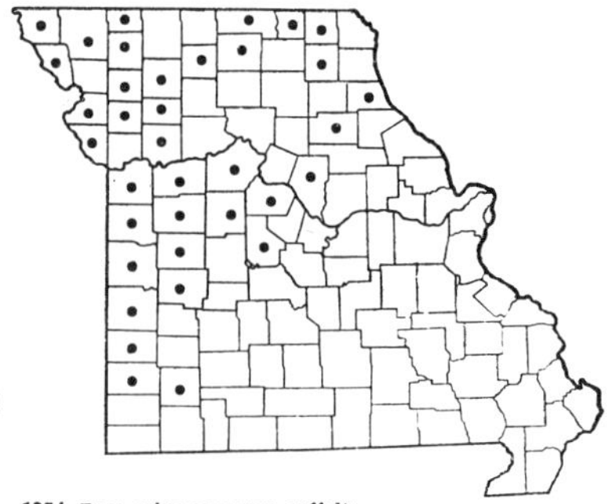
1254 Rosa arkansana var. suffulta

nally described from Swan, Taney County, Missouri, *Bush 42* in herb. N. Y. Bot. Gard.

Occurring mostly in southern and central Missouri in the Ozark section, but also known in the northeastern section of the state.

Throughout the range of var. *carolina*.

10c. **Rosa carolina** var. **villosa** (Best.) Rehd. Map 1251

*Rosa Palmeri* Rydb. [P & S, Steyerm.]; originally described from Carthage, Jasper County, Missouri, *Palmer 3428* in Gray Herb.

*Rosa carolina* var. *Lyoni* (Pursh) Palmer & Steyerm. [P & S]

Throughout Missouri.

Ranges from New Hampshire to Minnesota, south to Georgia, Louisiana, and Oklahoma.

10d. **Rosa carolina** var. **grandiflora** (Baker) Rehd. Map 1251

*Rosa obovata* Raf.

*Rosa carolina* var. *obovata* (Raf.) Deam [Deam]

Scattered in Missouri in most sectors of the state.

Ranges from Maine to Ontario, Wisconsin and Iowa, south to Georgia, Tennessee, and Arkansas.

There is great variation in the above varieties and forms in the density and abundance of prickles, height of plant, size and shape of leaflets, and amount and abundance of hairs and glands. Upon such differences a number of species have been described, but these differences can be placed into the single species, *R. carolina*. The receptacle varies from smooth to abundantly glandular.

In fall the foliage turns scarlet and red hues. The aerial stems arise from long underground stems and often die back from year to year.

11. **Rosa virginiana** Mill. var. **virginiana** Map 1252

*Rosa Aucuparia* Rydb. [P & S, Steyerm.]; originally described from Dumas, Clark County, Missouri, *Bush 5866*.

*Rosa petiolata* Rydb. [P & S, Steyerm.]; originally described from Clark County, Missouri, August 27, 1892, *Bush* in herb. Mo. Bot. Gard.

Open or rich woods and thickets.

Occurs in northern and central Missouri south to Jackson, Chariton, and Macon counties.

Ranges from Newfoundland to Ontario, south to Virginia, North Carolina, Alabama, Tennessee, and Missouri.

By some authors (Gleason, *New Illus. Fl.* 2: 326. 1952) this species is considered to be limited to the eastern states only. But, as has been pointed out elsewhere (Rh. 56: 71, 74–76. 1954), both *R. Aucuparia* and *R. petiolata*, the latter described from Clark County (*Bush*, August 27, 1892), may be judged as identical with what is passing in the eastern United States as *R. virginiana* on the basis of thorn and other characters of that species.

This species, like *R. carolina* and *R. arkansana* var. *suffulta*, is quite variable in such characters as height of plant, abundance of prickles and thorns, size and shape of leaflets, and shape and glandularity of the receptacle. As noted in a previous publication (Rh. 56: 74–76. 1954) the receptacle varies from globose or short ellipsoid and rounded at the base to decidedly pear-shaped or ellipsoid and narrowed or long-tapering at base. Plants are usually robust and may attain 2 m. in height.

12. **Rosa rugosa** Thunb. Japanese Rose, Rugosa Rose Map 1253

Flowers May–September.

Commonly planted in gardens and borders, but rarely found as an escape in Missouri. Known only as naturalized in Greene County (along roads, fields, etc., Springfield, May 8, 1930, *Martha Montague*, in herb. Southwest Missouri State College).

Native of eastern Asia; introduced and naturalized

from Quebec to Minnesota, south to New Jersey, the lower Great Lakes, and Missouri.

This very hardy rose is the most adaptable of any to the rigors of sea breezes and salt sprays of the ocean and is one of the few rose species to persist in such an environment. The flowers are red, white, or purple, and usually large, 6–8 cm. across, occurring either solitary or few together.

A number of horticultural varieties of *R. rugosa* are commonly grown, such as Ruskin, Sarah Van Fleet, Grootendorst (pink, white, and red forms), and a trailing white form. It is one of the thorniest of the rose species, numerous prickles and bristles covering the stems from base to tip.

A tea, prepared from the ripe 'hips' of this rose, is very pleasant. Other species of Rosa, likewise yield a pleasant-tasting drink. The petals of *R. rugosa* and of other species are sometimes candied or prepared as a salad, and the pulp of the fruit can be made into jelly.

13. **Rosa arkansana** Porter var. **suffulta** (Greene) Cockerell Map 1254

*Rosa suffulta* f. *alba* Rehder

*Rosa suffulta* var. *valida* Erlanson [P & S, Steyerm.]

*Rosa suffulta* Greene [BB]

*Rosa conjuncta* Rydb. [G, P & S, Steyerm.]; described originally from Atchison Co., Missouri, *Bush 101*

*Rosa carolina* × *suffulta* [BB]

*Rosa polyanthema* Lunell [P & S, Steyerm.]

*Rosa Bushii* Rydb. [P & S, Steyerm.]

*Rosa relicta* Erlanson [P & S, Steyerm.]; *Bush 11336, 11337* from Wellington, Lafayette County cited in original description.

*Rosa suffulta* var. *relicta* (Erlanson) Deam [Deam]

Flowers May–July.

Occurs in prairies, open banks, loess hills, bluffs, thickets, along roads, and railroads. Prairie region of glaciated northern Missouri and unglaciated western Missouri south of the Missouri River.

Ranges from New York to Alberta, south to D. C., Indiana, Wisconsin, Missouri, Kansas, Texas, and New Mexico.

This species is as variable as *R. carolina* and *R. virginiana*, especially in height of plant, vigor and robustness of stems and foliage, shape and size of leaflets, abundance of prickles, amount of pubescence, and variation in shape of receptacle. This has already been discussed in a previous publication (in Rh. 56: 72–74, 76–78. 1954) on the identity of *R. polyanthema*, *R. relicta*, and *R. conjuncta* with *R. arkansana* var. *suffulta*. The soft hairy covering on the leaf-stalks, axis of leaf, and lower surface of leaf-blades, glandless stipules, lack of or poorly differentiated infrastipular prickles, and persistent calyx-lobes in fruit are recognition characters for *R. arkansana* var. *suffulta*.

*Rosa rudiuscula* Greene is intermediate between *R. carolina* var. *caroliniana* f. *caroliniana* and *R. arkansana* var. *suffulta*, combining in various degrees the characters of the two species, and has been produced experimentally by Mrs. Erlanson as a cross between *R. carolina* and *R. arkansana* var. *suffulta* (Genetics 16: 75–96. 1931). This situation was discussed in a previous publication by the present author (Rh. 56: 78–79. 1954). It seems preferable to refer *R. rudiuscula* to the status of excluded species found at the end of this treatment.

The ordinary form of *R. arkansana* var. *suffulta* f. *suffulta* has pink or rose-colored petals. A form with white petals has been collected in Bates County (shale barrens along Miami Creek, T41N, R33W, sect. 13, 2 mi. southeast of Merwin, June 2, 1938, *Steyermark 5708* )and is known as *Rosa arkansana* var. *suffulta* f. *alba* (Rehder) Steyerm.

14. **Rosa blanda** Ait. f. **blanda** Map 1255

Flowers May–July.

Occurs in thickets, open woods, along open banks, and roadsides. Northeastern and north-central Missouri, where rare and found only in Clark, Schuyler, Putnam, Adair, and Linn counties.

Ranges from Quebec to Manitoba, south to New York, Pennsylvania, Ohio, Indiana, Illinois, Missouri, and Nebraska.

The plants resemble *R. arkansana* var. *suffulta* except that they are unarmed or bear few prickles mostly near the base of the stem, have fewer leaflets generally, and carry the bloom on the stems of the previous or second year's growth.

15. **Rosa spinosissima** L. var. **spinosissima**

Scotch Rose Map 1256

Flowers May–July.

Planted and rarely escaped from cultivation to waste ground or roadsides. Known only from central Missouri in Boone County.

Native of Europe; introduced and naturalized in a few widely scattered parts of North America.

The leaflets of this rose are 7–13 and ovate to orbicular. They are quite small, 0.5–2 cm. long. The flowers are white, yellow, or pink, and small (2–4 cm. broad) as compared with some other rose species. They are quite fragrant. *Rosa spinosissima* var. *altaica* Rehd. is a popular old-fashioned variety with somewhat larger white petals, smooth pedicels, and fewer

1255 Rosa blanda f. blanda

1256 Rosa spinosissima var. spinosissima (Scotch Rose)

1257 Prunus spinosa (Blackthorn)

bristles. Another popular and commonly grown old-fashioned rose descended from *R. spinosissima* is the Harisons Yellow Rose (*R. Harisonii* Rivers), a hybrid between *R. spinosissima* and the Austrian Rose (*R. foetida* Herrmann).

*Excluded Species*

**Rosa coriifolia** Fries

A specimen in the Gray Herbarium (*Bush 10190* from along streams, Southwest City, McDonald County, August 29, 1923) so labeled should be referred to *R. micrantha.* It has the leaflets glandular underneath and pubescent above.

**Rosa Woodsii** Lindl. var. **Fendleri** (Crépin) Rydb.

This rose is recorded from Missouri in the eighth edition of *Gray's Manual* (p. 873). An examination of the specimens in the Gray Herbarium so annotated by Fernald seems to indicate better their identity with both *R. carolina* var. *carolina* f. *glandulosa* (*Bush 113* and *124* from Montier, Shannon County, June 30, 1894, with glandular hypanthium and glandular spreading calyx-lobes, which are not in a state of mature fruit) and *R. arkansana* var. *suffulta* (*Bush 112* from Independence, Jackson County, June 7, 1894, with glabrous hypanthium in flower and pubescent lower surface of leaflets). Neither of these collections show the infrastipular prickles more clearly differentiated from the prickles of the internodes than in many Missouri collections of *R. arkansana* var. *suffulta* or *R. carolina* var. *carolina* f. *glandulosa.*

**Rosa rudiuscula** Greene

Originally described from a collection from Jackson County, Missouri (Little Blue, *Bush 208*), but the name is based upon a specimen intermediate between *R. carolina* var. *carolina* f. *carolina* and *R. arkansana* var. *suffulta,* and apparently of hybrid origin between these two taxa. As indicated in a previous publication (Rh. 56: 78–79. 1954), the name should be abandoned in the interests of clear taxonomy.

## 18. **Prunus** L. Plum, Cherry, Almond, Apricot

### I. *Key to Flower-bearing Specimens*

a. Flowers appearing before the leaves or before they have half opened . . . . . . . . . . . . . . *b*
  b. Petals pink to rose color; flowers without stalks or nearly so; ovary velvety-hairy. . . 7. P. PERSICA
  b. Petals white; flowers on definite elongated stalks; ovary glabrous . . . . . . . . . . *c*
    c. Flowers appearing solitary or 2 together at most; young branchlets hairy . . . . 1. P. SPINOSA
    c. Without this combination of characters; flowers appearing together in groups of 3 or more; young branchlets glabrous or hairy . . . . . . . . . . . . . . . . . . *d*
      d. No glands present or essentially none on calyx-lobes . . . . . . . . . . . . *e*
        e. Petals 3–5 mm. long; suckering shrubs forming colonies usually less than 15 feet tall; branchlets usually ending in a sharply narrowed or spine-like tip. . . . 6. P. ANGUSTIFOLIA
        e. Petals 8–15 mm. long; trees 20 or more tall, either solitary or in colonies; branchlets not ending in spine-like tips . . . . . . . . . . . . . . . . . . . *f*
          f. Flower-stalks and calyx-tube glabrous (without hairs); branchlets glabrous (without hairs); trees usually occurring in thickets. . . . . . . . . . . 3. P. AMERICANA
          f. Flower-stalks and calyx-tube more or less hairy; branchlets finely short-hairy; trees usually occurring singly or solitary . . . . . . . . . . . . . 2. P. MEXICANA

d. Glands definitely present on margins of calyx-lobes . . . . . . . . . . . . . . *g*

g. Petals 10–15 mm. long; calyx-lobes glabrous (without hairs) throughout; inflorescence with leafy green bracts at base . . . . . . . . . . . . . . . . . 8. P. CERASUS

g. Petals 4–7 mm. long; some hairiness present on some part of calyx-lobes; inflorescence without leafy green bracts at base . . . . . . . . . . . . . . . . . . . . *h*

h. Flowers opening and developed before the leaves expand, on short lateral spurs; leaves rolled longitudinally (convolute) in bud . . . . . . . . . 5. P. MUNSONIANA

h. Flowers opening and developing during the expansion of the leaves, when the leaves are about half developed, on one-year old slender branches; leaves folded together lengthwise (conduplicate) in bud . . . . . . . . . . . . . 4. P. HORTULANA

a. Flowers appearing when the leaves are half grown or after the leaves have fully developed . . . . . *i*

i. Flowers numerous, 15–30 in elongated racemes definitely longer than broad; glands usually present on margins of calyx-lobes . . . . . . . . . . . . . . . . . . . . . . *j*

j. Low shrub or small tree occurring usually on north-facing wooded slopes, mainly in northern and central Missouri; calyx-lobes blunt (obtuse), the glands conspicuous on the margins. . 11. P. VIRGINIANA

j. Large trees occurring throughout Missouri; calyx-lobes acutely pointed, the glands sparse on the margins of the calyx-lobes or not present . . . . . . . . . . . . 10. P. SEROTINA

i. Flowers fewer, 1–12 in short umbels, few-flowered broad racemes, or even solitary or 2 together; glands never present on margins of calyx-lobes . . . . . . . . . . . . . . . . *k*

k. Flowers only solitary or at most occurring 2 together; leaf-stalks without small green leaf-like bracts at base . . . . . . . . . . . . . . . . . . . . . . . . 1. P. SPINOSA

k. Flowers 4–12 in a corymb-like raceme; leaf-stalks with small green leaf-like bracts at base . 9. P. MAHALEB

II. *Key to Leaf- or Fruit-bearing Specimens*

a. Leaves broadly ovate to rather round, not much, if any longer than broad, the base heart-shaped to broadly rounded . . . . . . . . . . . . . . . . . . . . . . . 9. P. MAHALEB

a. Leaves lanceolate, oblong, elliptic, obovate, or various combinations of these, obviously longer than broad, not heart-shaped or broadly rounded at base . . . . . . . . . . . . . . . . . . *b*

b. A colored, usually brownish-red or brownish (sometimes pale) gland at tip or appearing as if from the side of most of the teeth along the leaf-margin and most evident on younger leaves (old leaves may lose the glands and have only a tiny scar indicating where the gland occurred) . . . . . . *c*

c. Mature leaf-blades mainly 0.5–2 cm. broad, mainly 2–5 (up to 8) cm. long; old calyx-lobes hairy but lacking glands on the margins . . . . . . . . . . . . 6. P. ANGUSTIFOLIA

c. Mature leaf-blades mainly 2.5–5.5 cm. broad, mainly 5–11 cm. long; old calyx-lobes with glands or teeth on the margins . . . . . . . . . . . . . . . . . . . . . . *d*

d. Stone of fruit spherical or globe-shaped; teeth of leaves relatively coarse and prominent, at least 1 mm. long; leaf-stalk without any hairs, with undeveloped glands; leaf-like bracts conspicuous at base of fruit-stalks or old flower-clusters . . . . . . . . . 8. P. CERASUS

d. Stone of fruit more or less 2-sided or compressed, longer than broad; teeth of leaves relatively fine or shorter, 0.5–0.75 mm. long; leaf-stalk with 1, 2, or more conspicuous glands near their summit; leaf-like bracts not developed at base of fruit- or flower-clusters . . . . . . . *e*

e. Gland arising from the very tip of each tooth of the leaf; teeth of leaves pointed, conspicuous, spreading away from the margin, or ascending; fully grown leaves flat with each half spread out, not folded . . . . . . . . . . . . . . 4. P. HORTULANA

e. Gland arising from that end of each tooth facing the indented lower side of the margin, teeth of leaves low or not pointed, not conspicuous or spreading; fully grown leaves more or less folded lengthwise, trough-like . . . . . . . . . . . . 5. P. MUNSONIANA

b. No gland arising from teeth of leaves . . . . . . . . . . . . . . . . . . . . . *f*

f. Leaf-blades obtuse (blunt) or nearly so at tip, mainly 2–4.5 cm. long, less than 2 cm. wide; spiny or thorny shrub or tree 10–15 ft. tall; fruit 10–15 mm. in diam., blue turning black. . 1. P. SPINOSA

f. Leaf-blades short- to long-pointed at tip, mainly 4–15 cm. long, mainly 2–6 cm. wide; non-spiny or non-thorny shrubs or trees 10–100 ft. tall; fruit either small, 6–10 mm. in diam. and dark red to black or if large and 2–3 or more cm. in diam., then red, gray-blue or purple, or yellow-green with orange or red . . . . . . . . . . . . . . . . . . . . . . *g*

g. Buds strongly hairy; fruit or old ovary without a stalk, velvety-hairy; stone deeply pitted and sculptured; leaves trough-like, the halves more or less folded lengthwise, conspicuously drooping . . . . . . . . . . . . . . . . . . . . . . . . 7. P. PERSICA

g. Buds glabrous (without hairs) or with a few hairs; fruit or old ovary on a definite elongated stalk, glabrous; stone not sculptured; leaves with both halves spread out flat, erect to spreading or somewhat drooping, but not conspicuously drooping . . . . . . . . . *h*

h. Fruits and old inflorescences in elongated racemes longer than broad, with 15–30 fruits or fruiting or flowering stalks to each inflorescence, appearing on branchlets of the new year's growth; mature fruits 6–10 mm. in diam.; bracts occurring at base of fruit-stalks or old flower-stalks . . . . . . . . . . . . . . . . . . . . . . . . *i*

i. Teeth of leaf sharp or conspicuously pointed, spreading upward or ascending, but not curved inward; calyx-lobes disappearing in fruit; leaf-blades rather thin and membranaceous, dull green, rather more abruptly and shorter pointed at tip, ovate or obovate; low shrub or small tree occurring usually on north-facing wooded slopes, mainly in northern and central Missouri . . . . . . . . . . . 11. P. VIRGINIANA

i. Teeth of leaf blunt, short and curved in along the margin; calyx-lobes persisting in fruit; leaf-blades thick, firm, dark green and rather shiny, gradually long tapering to the longer tip, lanceolate-oblong to broadly oblong; tree becoming 100 ft. or more tall, occurring throughout Missouri . . . . . . . . . . 10. P. SEROTINA

h. Fruits or old inflorescences with 1 or at most few to an umbel- or corymb-like cluster, appearing on branchlets of the preceding year; mature fruits 20–30 mm. in diam.; no bracts occurring at base of fruit-stalks or old flower-stalks . . . . . . . . . . . *j*

j. Leaf-stalks hairy all around; branchlets more or less hairy; lower surface of mature leaf-blades hairy; fruit eventually turning gray-blue or gray-lavender. . . 2. P. MEXICANA

j. Leaf-stalk glabrous (without hairs) on lower side; branchlets glabrous (without hairs); lower surface of mature leaf-blade glabrous (without hairs) or sparsely hairy on the main nerves; fruit eventually turning red . . . . . . . . 3. P. AMERICANA

1. **Prunus spinosa** L. Blackthorn, Sloe Map 1257
Flowers April–May.

Occasionally grown as an ornamental shrub and rarely escaped from cultivation. Known only from Jasper County, southwestern Missouri (thickets in limestone hills near Carthage, September 12, 1924, *Palmer 26115*; same locality, October 6, 1925, *Palmer 29053*).

Native of Eurasia and Africa; introduced and naturalized in a few places in North America.

This is a thorny shrub to small tree with small white flowers. Its abundant early flowering, ornamental blue-black fruits, and dense foliage are the especial merits of this species as an introduction for the home grounds.

2. **Prunus mexicana** S. Wats. Big Tree Plum, Wild Plum Map 1258
*Prunus americana* var. *lanata* of auth. [G, BB], not Sudw.
*Prunus lanata* of auth. [P & S, Steyerm., Deam], not (Sudw.) Mackenz. & Bush
*Prunus arkansana* Sarg.
*Prunus americana* var. *mollis* of auth., not T. & G.
Flowers April–May; fruits July–September.

Occurs in rocky or open woodland and thickets. Throughout Missouri and probably in every county.

Ranges from Indiana to Iowa, south to Tennessee, Arkansas, Texas, and Mexico.

Dr. Shinners (Rh. 58: 330–31. 1956) has pointed out that the plants previously passing as *P. lanata* or *P. americana* var. *lanata* are to be called *P. mexicana*. Since it does not seem possible to separate the common pubescent-leaved plants of Missouri, previously referred to *P. lanata*, from the more southwestern *P. mexicana*, they are here treated as synonymous under *P. mexicana*.

Usually this species is a solitary, nonthicket forming tree, but I have also recorded it as forming colonies (*Steyermark 82771* from Pike County, *Steyermark 40407* from Linn County, *Steyermark 68826* from St. Clair County, and others). In the eighth edition of *Gray's Manual* (p. 877) the color of the fruit is given as 'red to yellow' for *P. americana* and not further differentiated for *P. americana* var. *lanata* (= *P. mexicana* as here treated). Actually, the fruit of *P. mexicana* eventually turns to shades of blue, lavender, or rose covered with a gray-glaucous bloom, although at first when unripe it is yellowish or glaucous-green.

From this species have arisen such horticultural varieties as Quaker, Van Buren, and Wolf.

Plate no. 205. **1.** Rosa blanda, × 2/5. **2.** Rosa spinosissima, × 2/5. **3.** Prunus mexicana, × 2/5. **4.** Prunus spinosa, × 2/5; *a*. Flower, × 1 1/5. **5.** Prunus americana, × 2/5. **6.** Prunus hortulana, × 2/5; Details from Small, The New York Botanical Garden. **7.** Prunus Munsoniana, × 2/5.

PLATE NO. 205

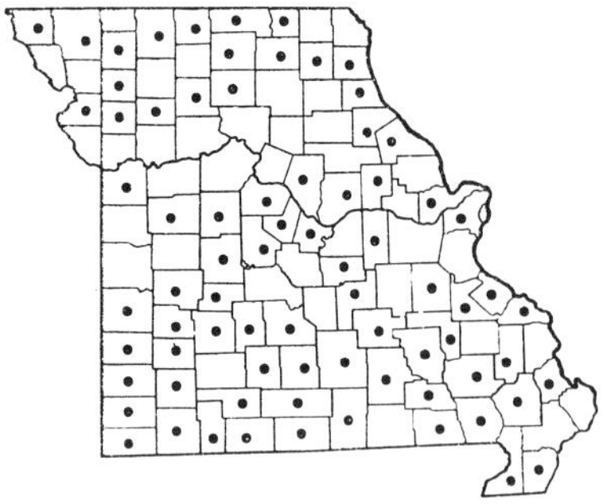

1258 Prunus mexicana (Big Tree Plum)

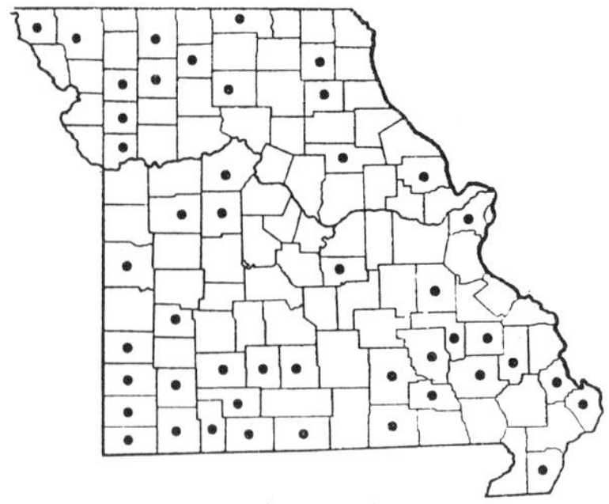

1259 Prunus americana (Wild Plum)

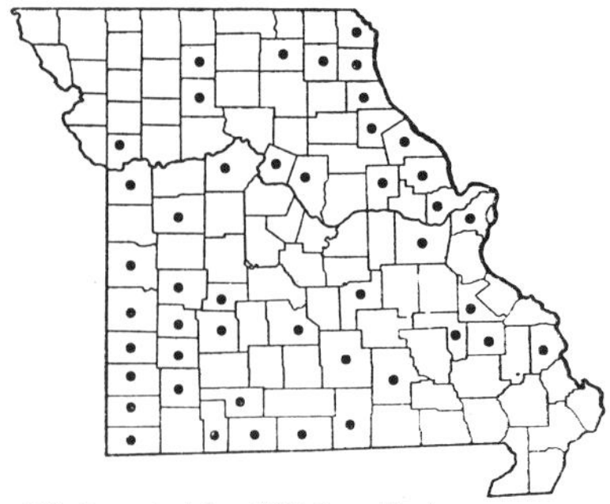

1260 Prunus hortulana (Wild Goose Plum)

3. **Prunus americana** Marsh. Wild Plum
Map 1259

Flowers April–May; fruits July–September.

Occurs in woodland, pastures, and thickets. Throughout Missouri.

Ranges from Florida to New Mexico and Arizona, north to Massachusetts, New York, Ontario, Michigan, Wisconsin, Minnesota, Manitoba, Wyoming, and Utah.

This species generally occurs in colonies or thickets by sending up root suckers and shoots. Occasionally, single solitary trees are encountered, but are usually surrounded by a number of smaller root sprouts. The fruit is sweet and red when ripe, and can be eaten fresh, or made into pies, jellies, or preserves.

This is one of the first shrubs or small trees to bloom in the woodland, its showy clusters of fragrant white flowers, appearing before the leaves unfold and while the woods are mostly bare of foliage, providing a bright note of early spring.

From this species of wild plum many horticultural varieties have been derived, including Blackhawk, De Soto, Forest Garden, Hawkeye, and Rollingstone.

4. **Prunus hortulana** Bailey Wild Goose Plum, Hortulan Plum Map 1260

*Prunus Palmeri* Sarg. Trees and Shrubs 2: 247. pl. 192. 1913.

*Prunus hortulana* var. *pubens* Sarg. [P & S, Steyerm.]

Flowers March–May; fruits July–October.

Occurs in open woodland, borders of woods, along streams, and thickets. Scattered in Missouri; not recorded from the extreme southeastern or northwestern sectors of the state.

Ranges from Indiana to Iowa, south to Alabama, Tennessee, Arkansas, and Oklahoma.

This is usually a small tree which generally does not sucker from the root to form thickets, although I have recorded it as forming thickets occasionally (*Steyermark 22932* from Taney County, *Steyermark 66457* from Ralls County, *Steyermark 27388* from Cedar County, and other collections). The fruit eventually turns red when ripe and can be used for pies, jellies, and preserves.

Occasionally, specimens are found which appear to be hybrids between this species and *P. mexicana.* Such plants, known as *Prunus Palmeri* Sarg., are represented by collections from Vernon (*Steyermark 9806*) and Jasper (rich rocky hillside in limestone soil near Carterville, April 11, and June 15, 1909, April 14, 1910, *Palmer 6* (holotype); Smithfield, June 13, 1909, *Palmer 13*) counties in western Missouri.

This species has given rise to a number of horticultural varieties, such as the Miner Plum (var. *Mineri* Bailey), Golden Beauty, Kanawha, and Wayland.

5. **Prunus Munsoniana** Wight & Hedrick Wild Goose Plum Map 1261

Flowers March–May; fruits June–August.

Occurs in thickets, prairies, borders of streams, borders of woodland, and waste ground. Scattered throughout Missouri, but not recorded from some of the counties in the extreme northwestern and southeastern sectors.

Ranges from Ohio and Kentucky, to Kansas, south to Louisiana, Oklahoma, and Texas.

This is a small tree with suckering habit, forming rather dense thickets. The bright red or yellow fruits, which are early ripening, can be made into delicious pies, jellies, and preserves. It is the species from which such varieties as Wild Goose, Newman, and Robinson have originated. The origin of the name 'Wild Goose' is derived from the discovery of a seed of this species found in the craw of a wild goose which had been shot by a Captain Means of Nashville, Tennessee. From this seed, which was planted, grew a plum which was later developed by nurserymen into a superior strain.

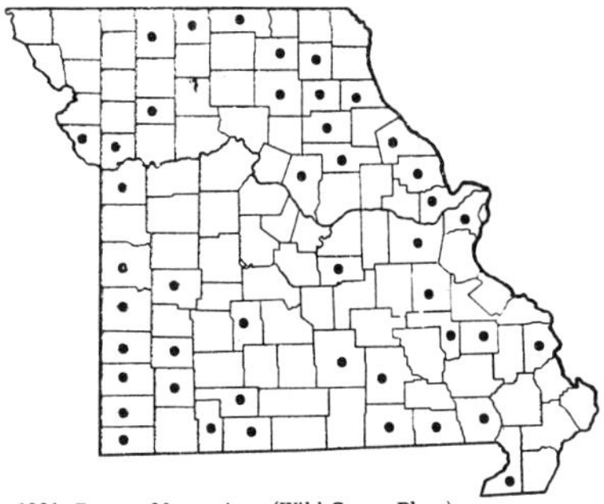

1261 Prunus Munsoniana (Wild Goose Plum)

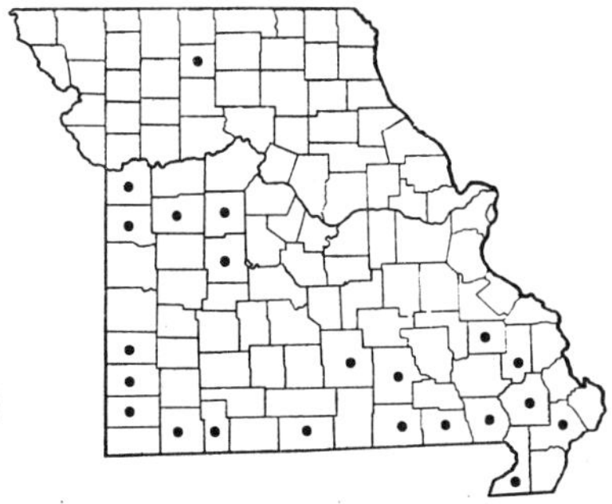

1262 Prunus angustifolia var. angustifolia (Chickasaw Plum)

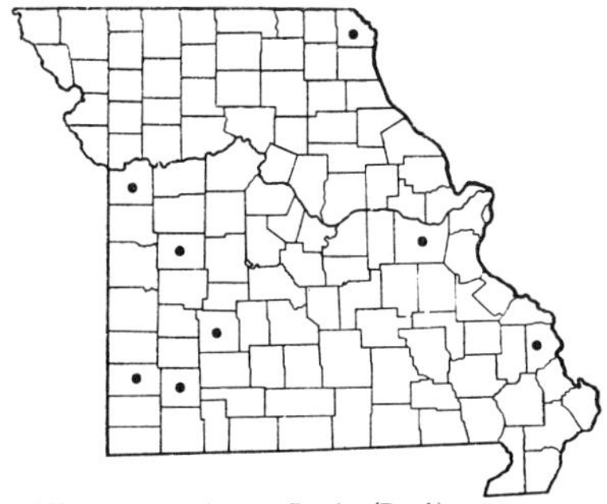

1263 Prunus Persica var. Persica (Peach)

6. **Prunus angustifolia** Marsh var. **angustifolia**
Chickasaw Plum Map 1262
*Prunus angustifolia* Marsh. [G, BB, P & S, Steyerm.]
*Prunus angustifolia* var. *varians* Wight & Hedrick [P & S, Steyerm.]
Flowers March–April; fruit June–July.

Occurs in thickets, pastures, fields, along fence rows, roadsides, and prairie streams. Scattered in southern and central Missouri, and north locally in Grundy County.

Ranges from Florida to Texas, north to New Jersey, West Virginia, Indiana, Missouri, and Kansas.

This is a thicket-forming species which has a small nearly globular, red or yellow cherrylike fruit usually 12–18 mm. in diameter. As with the other wild species, the fruit can be made into pies, jellies, and preserves. A horticultural variety known as Caddo Chief has originated from *P. angustifolia*.

*Prunus angustifolia* var. *varians* Wight & Hedrick, separated by some authors on the basis of its somewhat longer leaves and pedicels, and a stone somewhat more pointed at the apex, is considered in the present treatment as synonymous with *P. angustifolia* var. *angustifolia*, because of the intergradation of the characters used for separation.

7. **Prunus Persica** (L.) Batsch var. **Persica** Peach
Map 1263
Flowers March–April; fruit July–October.

Commonly planted and escaped from cultivation to thickets, fence rows, and roadsides. Scattered throughout Missouri, and probably to be found as an established plant in most counties.

Native of China; introduced and naturalized in the United States from New England to Michigan southward.

Many horticultural races have been derived from this wild species, including showy red double-flowered types with drooping branches, purple-leaved and dwarfed variations. A smooth-skinned variation, the Nectarine (var. *Nectarina* Maxim.), is well-known. Various hybrid strains have been produced by crosses with the Almond (*P. Amygdalus* Batsch).

8. **Prunus Cerasus** L. var. **Cerasus** Sour Cherry
Map 1264
Flowers April–May.

Commonly planted and escaped from cultivation to fence rows, thickets, borders of woodlands and prairies, and along roadsides. Scattered in Missouri.

Native of Eurasia; introduced and naturalized in North America from Prince Edward Island to Michigan and southward.

The sour, red, globular fruits are commonly used in pies. A number of horticultural races are known. Few fruit trees are suited for shady situations, but the Sour Cherry is one which may be grown in such places.

9. **Prunus Mahaleb** L. var. **Mahaleb** Mahaleb, Perfumed Cherry Map 1265
Flowers April–May; fruit July.

Occurs as an escape from cultivation in thickets, along roadsides, and in woodland. Scattered in southern, central, and northeastern Missouri north to Jackson, Moniteau, and Marion counties.

Native of Europe; introduced and naturalized in North America from New England to Ontario, south to Delaware, Indiana, and Missouri.

This species is often encountered without flowers or fruits, and in this state is sometimes confused with *Pyrus communis*. It may be distinguished from *Pyrus communis* by the glands present between the teeth of the leaves and by the nonspinescent branches. The flowers are very fragrant, 1.5–2 cm. broad, with petals 5–8 mm. long, and appear with or after the developing leaves.

*Prunus Mahaleb* is important in horticulture as it is used as an understock for grafting cherries. The aromatic bark is used in the manufacture of stems for pipes.

1264 Prunus Cerasus var. Cerasus (Sour Cherry)

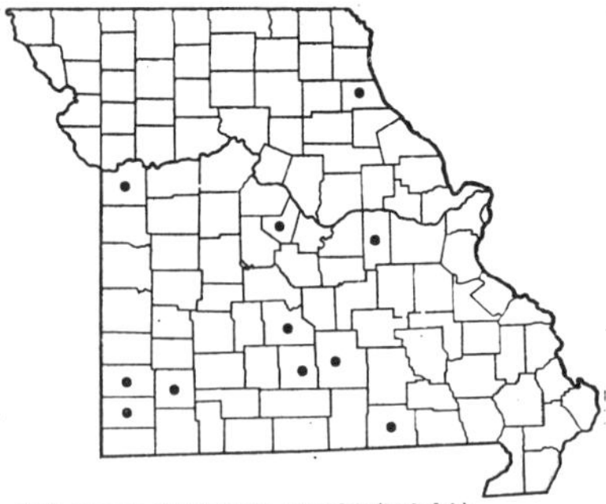
1265 Prunus Mahaleb var. Mahaleb (Mahaleb)

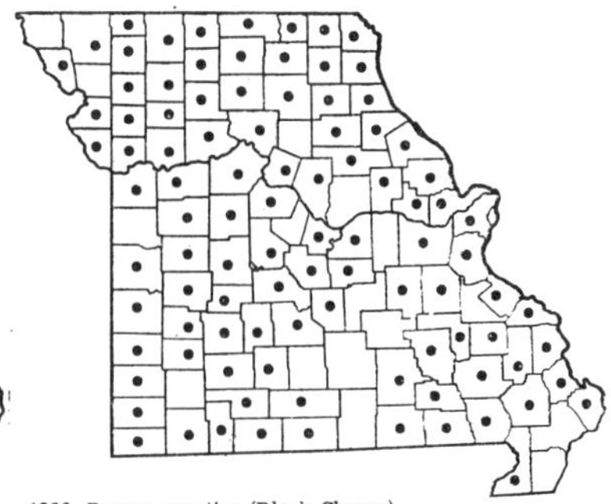
1266 Prunus serotina (Black Cherry)

10. **Prunus serotina** Ehrh. Black Cherry Map 1266

Also known as Rum Cherry.

Flowers April–May; fruits July–August.

Occurs in low or upland woods, and along streams. Common throughout Missouri and doubtless in every county.

Ranges from Mexico and Florida to Texas, north to Nova Scotia, Quebec, Ontario, Minnesota, and North Dakota.

The foliage of this species is most attractive in autumn, when the leaves combine a pale yellow with dull rose-green combination; some trees eventually turn to all yellow or yellow and rose. The wood is occasionally used as pulpwood for papermaking, but is principally employed for furniture, cabinet wood, interior finish, paneling, show counters, cases, bars, weighing apparatus, scientific instruments, handles, musical instruments, and caskets. The wood is reddish-brown and much resembles mahogany.

The hydrocyanic or prussic acid, contained in the leaves, is the cause of some cases of poisoning in cattle which have consumed such leaves. Although both fresh and wilted leaves are known to be poisonous, the wilted ones are considered the more dangerous.

The bitter aromatic bark is used for its astringent properties in cough medicines, expectorants, and for sore throats. The fruit, at first dark red eventually turning purple-black, is sometimes made into a jelly when combined with the juice of sour apple, and the jelly stated to resemble guava in quality. Early New England settlers made a drink, known as 'Cherry Bounce,' from the fruit; the fruits were placed in brandy and produced a bitter cordial. At present the fruit is occasionally used for flavoring alcoholic liqueurs. The fruits are commonly eaten and dispersed by birds. The species eventually dies out in heavily shaded areas, as it is not shade-enduring, although it inhabits forested areas where it gets sufficient light.

11. **Prunus virginiana** L. Choke Cherry Map 1267

Flowers April–May; fruits August–September.

Occurs on rich, mostly north-facing wooded slopes and bluffs, and ravines, rarely on borders of woods, thickets, fence rows, ditches, and roadsides. Most frequent in northern and central Missouri north of the Missouri River, occurring locally south of that river in Osage, Moniteau, Cooper, Saline, Texas, Laclede, Benton, and Barton counties.

Ranges from Newfoundland to Saskatchewan, south to Virginia, Georgia, North Carolina, Tennessee, Missouri, Kansas, and Oklahoma.

Missouri material is represented by the following 2 variations:

Lower surface of leaf-blades glabrous (without hairs) or with mere tufts of hairs in the axils of the nerves . . . . . . . . 11a. P. VIRGINIANA var. VIRGINIANA f. VIRGINIANA

Lower surface of leaf-blades more or less hairy . 11b. P. VIRGINIANA var. VIRGINIANA f. DEAMII

11a. **Prunus virginiana** var. **virginiana** f. **virginiana** Map 1267

*Prunus virginiana* L. [G, BB, P & S, Steyerm.]

This is the more commonly found form in Missouri.

11b. **Prunus virginiana** var. **virginiana** f. **Deamii** G. N. Jones Map 1267

*Prunus virginiana* f. *Deamii* G. N. Jones [G]

Commonly found within the range in Missouri.

The f. *Deamii* is sometimes characterized as having the young shoots and the axis of the inflorescence pubescent, in addition to the pubescent lower surface

Plate no. 206. **1.** Prunus angustifolia, × $^{2}/_{5}$. **2.** Prunus Persica, × $^{2}/_{5}$. **3.** Prunus Cerasus var. Cerasus, × $^{2}/_{5}$. **4.** Prunus Mahaleb, × $^{2}/_{5}$. **5.** Prunus virginiana, × $^{2}/_{5}$. **6.** Prunus serotina, × $^{2}/_{5}$; *a.* Portion of inflorescence, × $^{4}/_{5}$.

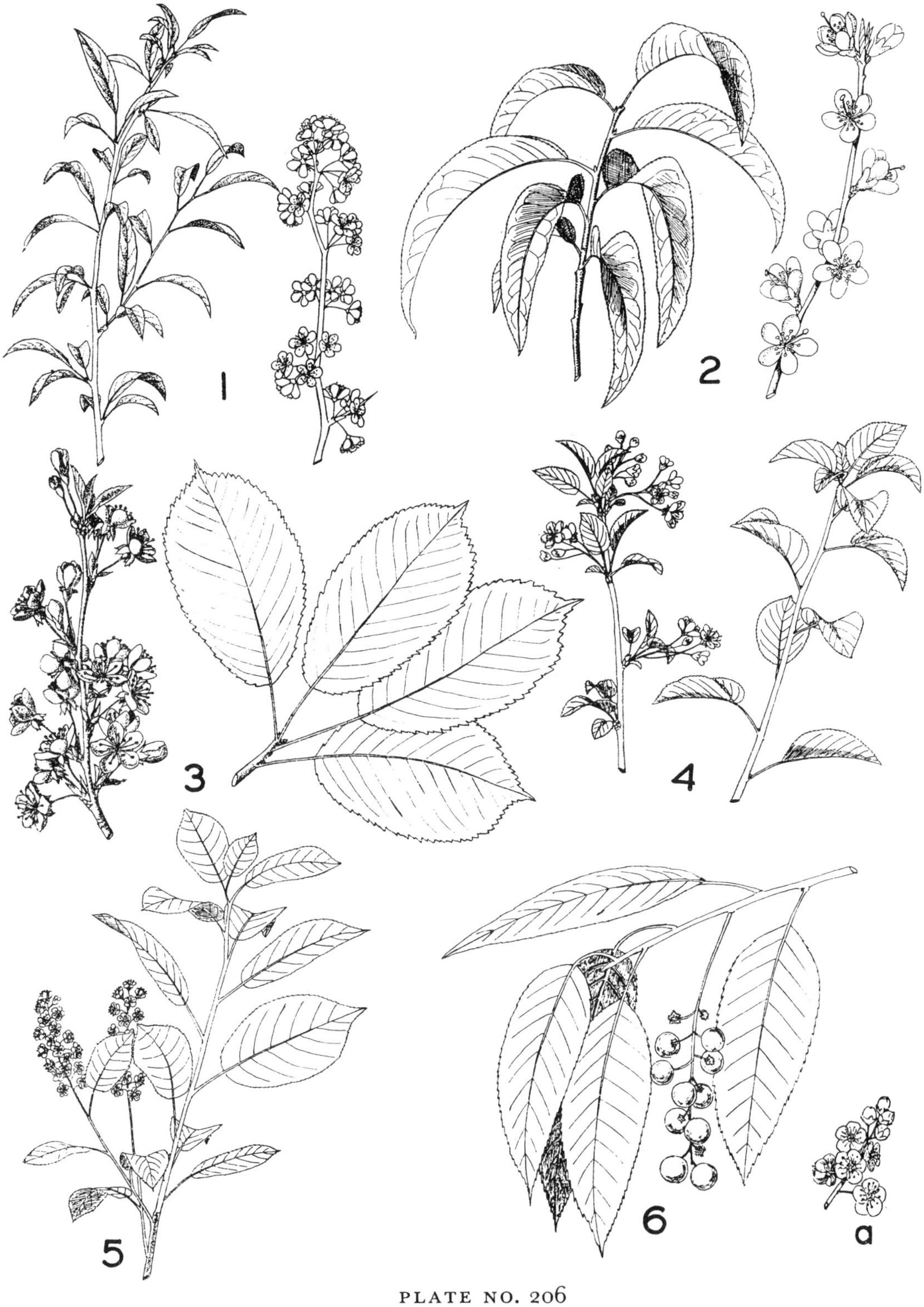


PLATE NO. 206

of the leaf-blades, but in the Missouri material there does not appear to be any correlation with regard to the pubescent axis of the inflorescence and the pubescent lower leaf surface. Most of the material in Missouri referred to f. *Deamii* possesses a glabrous rachis as in typical f. *virginiana* but with the lower leaf surface variously pubescent.

Other varieties of *Prunus virginiana* are encountered in the western states.

The fruit is usually deep red to red-purple when ripe, but in f. *leucocarpa* (S. Wats.) Haynie is amber-colored to whitish. This pale-fruited form has not been recorded from Missouri. The fruits are greatly relished by birds.

The bark of this cherry is not aromatic, whereas the inner bark of *P. serotina* is aromatic. The flowers of the Choke Cherry also appear a couple of weeks earlier in the season than those of *P. serotina.*

Mixed with apple juice, the juice of the Choke Cherry can be made into a tasty jelly.

The wilted leaves of *P. virginiana*, as in *P. serotina*, are poisonous if eaten by cattle.

*Excluded Species*

A number of cultivated species of *Prunus* may possibly be established or naturalized in various parts of the state, but no authentic record for them exists at the present time. The following are of such unverified occurrence:

**Prunus domestica** L. Common Plum, European Plum

Listed by Daniels in his *Flora of Columbia* as spontaneous in Boone County, but no specimens are extant.

**Prunus domestica** var. **insititia** (L.) Bailey
Damson Plum, Bullace
*Prunus insititia* L.
Commonly cultivated.

**Prunus Avium** L. Sweet Cherry
Cultivated.

## Fam. **LEGUMINOSAE** (Pea Family)

(*Mimosaceae, Caesalpiniaceae, Papilionaceae* [BB])

a. Leaves absent at flowering time . . . . . . . . . . . . . . . . . . . . . . . *b*
  b. Trees or shrubs covered with rose, pink, magenta, or white flowers; throughout Missouri . 8. CERCIS
  b. Woody vines with blue-purple or lavender flowers; swamps of southeast Missouri or cultivated 23. WISTERIA
a. Leaves present at time of flowering . . . . . . . . . . . . . . . . . . . . *c*
  c. Stems, branches, or bark with thorns or bristles . . . . . . . . . . . . . . *d*
    d. Low-spreading, non-woody herb; stems and leaf-stalks (petioles) bearing numerous curved bristles or thorns; leaflets 1.5–2.5 mm. broad . . . . . . . . . . . . . 3. SCHRANKIA
    d. Trees or shrubby plants; thorns on stem few, none, or widely spaced, none on leaf-stalks (petioles); leaflets 5–30 mm. broad . . . . . . . . . . . . . . . . . . *e*
      e. Compound leaves with an even number of leaflets, or with a pair of leaflets at the tip; leaflets with minutely wavy (faintly crenulate) margins . . . . . . . . . 5. GLEDITSIA
      e. Compound leaves with odd number of leaflets, or with 1 leaflet at the tip; leaflets with smooth (entire) margins . . . . . . . . . . . . . . . . . . 24. ROBINIA
  c. No thorns or bristles on stems, branches, or bark . . . . . . . . . . . . . . *f*
    f. All the leaves simple and undivided. . . . . . . . . . . . . . . . . . . *g*
      g. Small trees or shrubs with heart-shaped large or broad, more or less roundish leaves; flowers rose, pink, magenta, or white . . . . . . . . . . . . . . . 8. CERCIS
      g. Small herbs with non-cordate, obovate, elliptic, oblong, or lanceolate leaves; flowers yellow . . . . . . . . . . . . . . . . . . . . . . . . . 11. CROTALARIA
    f. Some or all of the leaves divided into 2 or more leaflets . . . . . . . . . . . . *h*
      h. Trees or shrubs or woody vines or plants with woody stems or bases . . . . . . . . *i*
        i. Compound leaves with an even number of leaflets, or with a pair of leaflets at the tip . . . *j*
          j. Foliage, stems, and calyx bearing many black glands . . . . . . 6. HOFFMANSEGGIA
          j. Lacking black glands . . . . . . . . . . . . . . . . . . . . . *k*
            k. Low shrub, less than 1 meter (3′ 4″) tall; leaflets very small, 3–6 mm. long, 18–40 pairs on each leaf . . . . . . . . . . . . . . . . . 1. ACACIA
            k. Trees; leaflets 15–40 mm. long, 6–14 pairs on each leaf . . . . . . . . . . *l*
              l. Individual leaflets short- to long-pointed at tip, the larger ones 2–5 cm. wide when fully grown, smooth-edged (entire) . . . . . . . . . 4. GYMNOCLADUS
              l. Individual leaflets mostly blunt (obtuse) to rounded at tip, less than 1.5 cm. wide, with minutely wavy (faintly crenulate) margins . . . . . 5. GLEDITSIA
        i. Compound leaves with an odd number of leaflets, or with 1 leaflet at the tip. . . . . . *m*
          m. Woody vines, high-climbing or trailing . . . . . . . . . . . . . . . . *n*
            n. Leaves divided into 3 leaflets . . . . . . . . . . . . . . 46. PUERARIA
            n. Leaves divided into 9–15 leaflets . . . . . . . . . . . . . 23. WISTERIA
          m. Trees or shrubs . . . . . . . . . . . . . . . . . . . . . . . . *o*
            o. Lower surface of leaflets dotted with glands, or when glands are not evident, the lower surface is densely covered with hairs . . . . . . . . . . 20. AMORPHA
            o. Lower surface of leaflets not gland-dotted nor covered with hairs . . . . . . . *p*
              p. Leaflets 2.5–9 cm. wide, only 7–11 to a leaf; some or all of the leaflets abruptly narrowed or pointed at tip; topmost (terminal) leaflet noticeably broadest at or above the middle; calyx-lobes broadly rounded at tip; rare tree known only from the White River region of southwestern Missouri. . . . . 10. CLADRASTIS
              p. Leaflets 0.8–3.5 cm. wide, 7–19 to a leaf; leaflets mostly broadly rounded at tip; topmost (terminal) leaflet not noticeably broadest at or above middle; lower calyx-lobes short- to long-pointed at tip; common tree throughout Missouri . . . . . . . . . . . . . . . . . . . . . . . . 24. ROBINIA
      h. Upright, spreading, or creeping herbs or vines without woody stems. . . . . . . . . *q*
        q. All leaves divided into 2 leaflets . . . . . . . . . . . . . . . 35. LATHYRUS
        q. Some or all of the leaves divided into 3 or more leaflets . . . . . . . . . . *r*
          r. Some or all of the leaves divided into 3 leaflets . . . . . . . . . . . . *s*
            s. Leaflets more or less dotted with glands . . . . . . . . . . . . . . *t*
              t. Only 1 flower produced from the peduncle (main support of inflorescence); flowers pink-red with yellow or white; fruit not wrinkled, containing 4–7 seeds. 16. LOTUS

t. Many flowers produced on the peduncle (main support of inflorescence); flowers purple, bluish, or lavender, rarely all white; fruit wrinkled, containing 1 seed . . . . . . . . . 17. PSORALEA

s. Leaflets not dotted with glands . . . . . . . . . . . . . . . . . . . . . . . *u*

u. Leaflets with minute teeth on margins . . . . . . . . . . . . . . . . . . . . *v*

v. Middle leaflet sessile (without a stalk) or nearly so . . . . . . . . . . . 12. TRIFOLIUM

v. Middle leaflet on a distinct stalk (petiolule) . . . . . . . . . . . . . . . . *w*

w. Flowers in long narrow racemes, not crowded, well separated; stipules (at base of leaf-stalk) bristle-like, very narrow and subulate; only 1 seed in the fruit . . . . . . 14. MELILOTUS

w. Flowers in short, rounded or cylindrical-shaped heads or in 2–3-flowered clusters, but crowded or close together; stipules (at base of leaf-stalk) lanceolate or ovate; several seeds in the fruit . . . . . . . . . . . . . . . . . . . . . . . . . . *x*

x. Stipules (at base of leaf-stalk) smooth-edged (entire), connected about half their length to the base of the leaf-stalk; fruit mostly straight; petals more or less persisting in fruit 12. TRIFOLIUM

x. Stipules (at base of leaf-stalk) mostly with teeth or irregularities along the margins, if entire, the stem 4-angled, connected for less than half their length to the base of the leaf-stalk; fruit usually curved, coiled, or twisted; petals falling after flowering . . 13. MEDICAGO

u. Leaflets with entire (toothless) margins . . . . . . . . . . . . . . . . . . . *y*

y. Flowers 2–6 cm. long . . . . . . . . . . . . . . . . . . . . . . . . . *z*

z. Flowers 4–6 cm. long, only 1–2 in the axils of the leaves; all 10 stamens separate from one another; middle leaflet on a much longer stalk (petiolule) than the other 2 leaflets; fruit linear-oblong, flattish . . . . . . . . . . . . . . . . . . . . 37. CLITORIA

z. Flowers 2–3 cm. long, several to many in a raceme at the end of the main stem or side branches; the stamens united by their filaments at base; middle leaflet either sessile (stalkless) or with a stalk no longer than the stalks of the other 2 leaflets; fruit globe-shaped to cylindrical, inflated . . . . . . . . . . . . . . . . . . . 9. BAPTISIA

y. Flowers 1.5 cm. or less long . . . . . . . . . . . . . . . . . . . . . . *1*

1. Only 1 flower produced from the peduncle (main support of inflorescence) . . . . 16. LOTUS

1. Two or more flowers produced from the peduncle (main support of inflorescence) . . . . . *2*

2. No scale-like outgrowth (stipel) appears at base of any of the 3 leaflets; petals yellow, orange, or creamy-colored . . . . . . . . . . . . . . . . . . . . . *3*

3. Stipule partly united to the base of the leaf-stalk and forming a tube around the stem 32. STYLOSANTHES

3. Stipule entirely free from the leaf-stalk and stem . . . . . . . . . . . . *4*

4. Stem erect; calyx equally 6-lobed; flowers in dense heads of spikes, creamy-colored . . . . . . . . . . . . . . . . . . . 30. LESPEDEZA

4. Stem trailing or twining; calyx unequally 4-parted with the upper lobe 2-cleft; flowers in rather few-flowered racemes, orange or deep yellow . . . 47. RHYNCHOSIA

2. Each of the 3 leaflets with a scale-like outgrowth (stipel) at its base (either at the base of its blade or its stalk); petals blue, purple, pink, white, or various combinations of these colors . . . . . . . . . . . . . . . . . . . . . . . . *5*

5. Calyx with an upper and lower lip, not equally or unequally 4–5-lobed . . . . . . *6*

6. Corolla with keel spirally coiled . . . . . . . . . . . . . 39. PHASEOLUS

6. Keel of corolla at most merely arched or curved in, not spirally coiled . . . . . *7*

7. Rare introduced plant; fruit 20–35 cm. long; valves of the fruit continuous, not breaking into joints; axis of inflorescence conspicuously nodose (marked by knots, thickenings, or knobby places) . . . . . . . . . . 43. CANAVALIA

7. Commonly encountered native plant; fruit at most 10 cm. long; valves of the fruit 1–several jointed; axis of inflorescence not nodose . . . . . 29. DESMODIUM

5. Calyx equally or unequally 4–5-lobed, but not 2-lipped with an upper and lower lip . . *8*

8. Calyx 4-lobed or 4-parted . . . . . . . . . . . . . . . . . . . *9*

9. Leaflets mainly 10–20 cm. long; plants covering large areas with coarse, long trailing, twining, or climbing stems with rusty hairiness; the bracts of inflorescence conspicuously longer than the unexpanded flowers; introduced, commonly planted along roads . . . . . . . . . . . . . . . 46. PUERARIA

9. Leaflets 2–9 (–10) cm. long; plants slender, not greatly extended in area covered by twining or trailing, sometimes not twining; the bracts of the inflorescence lacking or inconspicuous; native plants . . . . . . . . *10*

10. Calyx without a pair of little bracts (bracteoles) at the base; calyx more or less equally toothed . . . . . . . . . . . . . . . . . . . . . . . . 44. AMPHICARPA
10. Calyx with a pair of little bracts (bracteoles) at the base; calyx unequally and irregularly lobed . *11*
11. Style with a beard running lengthwise along the upper surface; keel of the corolla strongly upwardly or inwardly curved . . . . . . . . . . . . . . . . 41. STROPHOSTYLES
11. Style without a beard; keel of the corolla nearly straight or only scarcely incurved. 45. GALACTIA
8. Calyx 5-toothed or 5-lobed . . . . . . . . . . . . . . . . . . . . . . . . . *12*
12. Fruit with only 1 seed; calyx about equally 5-toothed . . . . . . . . . 30. LESPEDEZA
12. Fruit with several to many seeds; calyx unequally 5-toothed with some of the teeth definitely shorter . . . . . . . . . . . . . . . . . . . . . . . . . . *13*
13. Stems vining, climbing, or long-trailing, in the native species, but erect or bushy in the introduced species; keel of corolla spirally coiled; native or introduced species . 39. PHASEOLUS
13. Stems erect and bushy-branched to slightly twining; keel of corolla merely arched or curved, but not spirally coiled; introduced cultivated plants . . . . . . . . . . *14*
14. Stems brown-hairy; fruit covered with conspicuous brown hairs; erect bushy plant; style without a beard . . . . . . . . . . . . . . . . . . . . 42. GLYCINE
14. Stems glabrous (without hairs) or nearly glabrous; fruit glabrous (without hairs) or nearly so; stem trailing or running; style with a beard running lengthwise on the upper side . . . . . . . . . . . . . . . . . . . . . . . 40. VIGNA
1. Leaves divided into 4 or more leaflets . . . . . . . . . . . . . . . . . . . . . *15*
15. At least the lower surface of leaflets dotted with glands, or foliage, stems, and calyx bearing many black glands . . . . . . . . . . . . . . . . . . . . . . . . . *16*
16. Leaves palmately divided, all the 5 or 6 leaflets arising at the tip of the main leaf-stalk (petiole) . . . . . . . . . . . . . . . . . . . . . . . . . . 17. PSORALEA
16. Leaves pinnately compound, the leaflets arising from the sides of the leaf axis, one above the other. . . . . . . . . . . . . . . . . . . . . . . . . . . *17*
17. Leaves twice pinnately compound (the main pinnae or divisions again divided into smaller divisions); black glands present on stems, leaves, calyx, and fruits . 6. HOFFMANSEGGIA
17. Leaves once pinnately compound (the leaf divisions arising directly from the central axis of the leaf); black glands, when present, not on stems, but on other parts; waxy or brown glands often present on stems, leaves, calyx, and fruit . . . . . . . . . . *18*
18. Upper part of stem abundantly covered with glands, the distance between the glands being less than the diameter of the stem; inflorescences arising as axillary spikes below the tip of the stem from the axils of the leaves; flowers yellow; fruit covered with hooked prickles . . . . . . . . . . . . . . . . . 7. GLYCYRRHIZA
18. If glands are present on upper part of stem, they are so sparse that the distance between them is greater than the diameter of the stem; inflorescences usually arising from the very tip of the stem, or at the tips of lateral branches; flowers purple, lavender, rose, or white; fruit not covered with prickles . . . . . . . . . . *19*
19. Each flower in the inflorescence with only 1 petal surrounding the stamens and style; fruit protruding from the mature calyx, 2–5 times as long as the calyx; calyx-teeth short, triangular, more or less equal . . . . . . . . 20. AMORPHA
19. Each flower in the inflorescence with 5 petals; fruit not protruding from the mature calyx; calyx-teeth slender and elongated . . . . . . . . . . . *20*
20. Plants mainly of loess mounds of northwestern Missouri and of alluvial soils along the Missouri and Mississippi rivers and lakes; calyx and bracts with numerous conspicuous glands; stamens 10 or 9; corolla somewhat papilionaceous (resembling the flower of sweet pea); the keel- and wing-petals attached to the middle of the bundle of filaments; claws of petals $\frac{1}{4}$–$\frac{1}{2}$ as long as the blades . . . . . . . . . . . . . . . . . 18. DALEA
20. Common plants of prairies, glades, and rocky woods throughout Missouri; calyx and bracts with few or less conspicuous glands; stamens 5; corolla not appearing papilionaceous or scarcely so; the keel- and wing-petals attached at the summit of the bundle of filaments; claws of petals shorter than $\frac{1}{4}$ as long as the blades . . . . . . . . . . 19. PETALOSTEMON
15. Lower surface of leaflets not gland-dotted . . . . . . . . . . . . . . . . . 21
21. Compound leaves with an even number of leaflets, or with a pair of leaflets at the tip . . . *22*
22. At least some of the leaflets ending in tendrils . . . . . . . . . . . . . . *23*
23. Stipules larger than the lower leaflets; calyx-lobes or teeth more or less leafy . 36. PISUM

23. Stipules smaller than the lower leaflets; calyx-lobes or teeth not leafy . . . . . *24*
24. Style hair-like, with a bearded tuft of hairs only at the tip; wings of corolla joined to the keel. . . . . . . . . . . . . . . . . . . . . . . . . 34. VICIA
24. Style somewhat enlarged and flattened toward the tip, bearded only along the upper (inner) side; wings of corolla not joined to the keel, free or nearly free from it. . 35. LATHYRUS
22. None of the leaflets ending in tendrils . . . . . . . . . . . . . . . . . . . *25*
25. Leaflets 4; lower part of stipules joined to the base of the leaf-stalk . . . . . 33. ARACHIS
25. Leaflets 4–many; stipules free, not joined to the base of the leaf-stalk . . . . . . . *26*
26. Leaf-stalk (petiole) bearing 1 or more glands . . . . . . . . . . . . . . *27*
27. Leaves twice pinnately compound (the main pinnae or divisions again divided into smaller divisions); flowers greenish-white, symmetrical (regular); stamens much longer than and protruding far beyond the perianth . . . . . . . 2. DESMANTHUS
27. Leaves once pinnately compound (the leaf divisions arising directly from the central axis of the leaf); flowers yellow, more or less irregular with stamens or petals somewhat equal; stamens shorter than or equaling the petals . . . 7. CASSIA
26. No glands present on leaf-stalk (petiole) . . . . . . . . . . . . . . . *28*
28. Stem more or less hairy; ultimate smallest leaflets 3–6 mm. long; leaves twice pinnately compound (the main pinnae or divisions again divided into smaller divisions); flowers symmetrical (regular), in globose heads, creamy- or yellowish-white to salmon-pink; stamens numerous, much longer than and protruding far beyond the perianth. . . . . . . . . . . . . . . . . . 1. ACACIA
28. Stem and rest of plant smooth and glabrous (without hairs); individual ultimate leaflets 20–30 mm. long; leaves once pinnately compound (the leaf divisions arising directly from the central axis of the leaf); flowers irregular, pea-like in a raceme, yellow often spotted with purple; stamens 10, shorter than the rest of the flower. 22. SESBANIA
21. Compound leaves with an odd number of leaflets, or with 1 leaflet at the tip . . . . . . . *29*
29. Stems twining or climbing; largest leaflets 20–50 mm. wide, 40–85 mm. long, rarely smaller; flowers brown-purple . . . . . . . . . . . . . . . . . . . . . . 38. APIOS
29. Stems not twining or climbing; leaflets 2–15 mm. wide, 10–40 mm. long; flowers yellow, rose, pink, purple, or white, or various combinations of these . . . . . . . . . . . *30*
30. Leaflets always 5; leaves without stipules, the lowest pair of leaflets resembling stipules. 16. LOTUS
30. Main leaflets of the stem 5–29, if sometimes only 5, the leaves then on a definite leaf-stalk; leaves with stipules (sometimes obscure or covered with hairs in *Tephrosia*). . . . . . *31*
31. All leaves arising at ground level at the base of the plant, none elevated on the main stem . . . . . . . . . . . . . . . . . . . . . . . . . *32*
32. Petals yellow, 8–11 mm. long; peduncle (main stalk of inflorescence) shorter than or barely longer than the leaves, up to 10 cm. long . . . 25. ASTRAGALUS LOTIFLORUS
32. Petals purple to bluish, 15–20 mm. long; peduncle (main stalk of inflorescence) much longer than the leaves, up to 25 cm. long . . . . . . . . . 26. OXYTROPIS
31. At least some leaves arising above ground level or attached to the stem . . . . . *33*
33. Inflorescence or main stalk (peduncle) of inflorescence terminating the main or leafy stem, none of the peduncles arising from the sides of the stem. . . . . . *34*
34. Flowers yellow; calyx enclosing the mature fruit; leaflets mainly 5–11 (sometimes 13 or the lower with 1–3); terminal leaflet of some of the leaves often larger than the others; flower or fruit-clusters with an involucre of 3-cleft green bracts at the base . . . . . . . . . . . . . . . . . 15. ANTHYLLIS
34. Flowers creamy-yellowish and rose-lavender or pinkish; mature fruit prominently protruding, leaving the small calyx at its base; leaflets mainly 15–25 (or up to 39 or as few as 9); flowers or fruit clusters without any involucre of bracts . . . . . . . . . . . . . . . . . . . 21. TEPHROSIA
33. Some or all of the main stalks (peduncle) of the inflorescence arising from the sides of the stem . . . . . . . . . . . . . . . . . . . . . *35*
35. Flowers in umbels, the flower-stalks all originating from the tip of the peduncle (main stalk of inflorescence); fruit with 3–7 joints . . . . . . 28. CORONILLA
35. Flowers in short to elongated spikes or racemes, the flowers arising at different levels on the axis of the inflorescence; fruit not 3–7-jointed . . . . . . *36*
36. Commonly encountered native plants; ovary with few to many ovules; fruit with few to many seeds; fruit not spiny or prominently honey-combed, 10–30 mm. long; calyx-lobes usually shorter than or as long as the calyx-

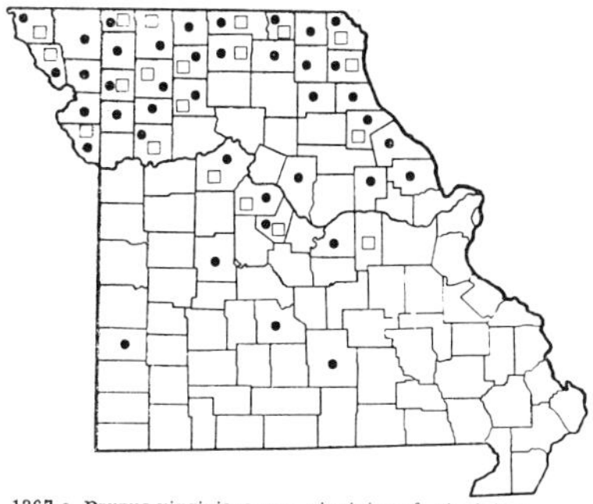

1267 • Prunus virginiana var. virginiana f. virginiana (Choke Cherry)
1267 □ Prunus virginiana var. virginiana f. Deamii

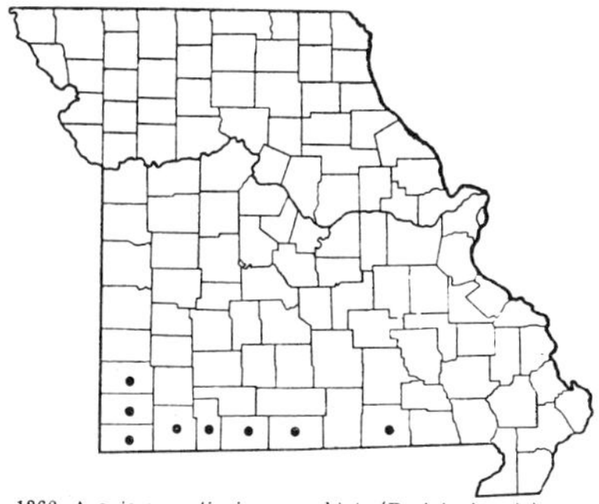

1268 Acacia angustissima var. hirta (Prairie Acacia)

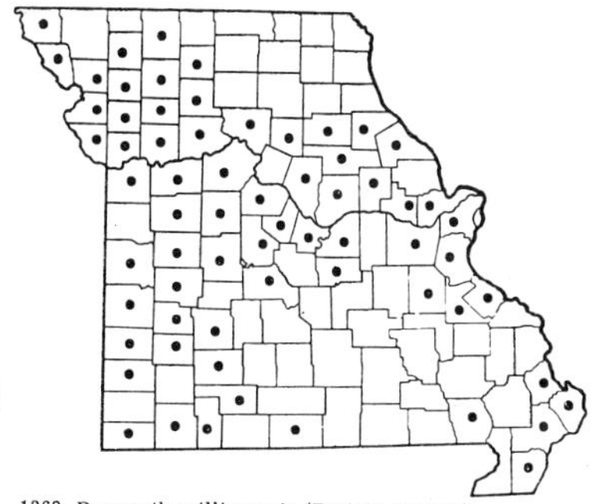

1269 Desmanthus illinoensis (Prairie Mimosa)

tube; petals purple, lavender, yellow, cream-color, greenish-white or some combination of these colors. . . . . . . . . . . . . . . . . . . . . . . . . . 25. ASTRAGALUS

36. Rare introduced, not native plant; ovary with 1–2 ovules; fruit 1-seeded; fruit with spines or prominently honeycombed, 5–8 mm. long; calyx-lobes longer than the calyx-tube; petals rose-colored or pink. . . . . . . . . . . . . . . . . . . . . . . . . 31. ONOBRYCHIS

The Leguminosae are treated by some botanists as 3 separate families (Mimosaceae, Caesalpiniaceae, and Fabaceae) and by other botanists as 3 subfamilies (Mimosoideae, Caesalpinioideae, and Papilionoideae). The treatment followed in the present flora is to treat these units as subfamilies, and the genera are arranged under their respective subfamilies. The following key summarizes the principal differences between these subfamilies as applicable to the Missouri flora:

a. Flowers regular, hypogynous; stamens much longer than the perianth; corolla valvate (edges only touching) in bud stage; stamens 5–many . . . . . . . . . . . . . . . . I. MIMOSOIDEAE (Acacia, Desmanthus, Schrankia)

a. Flowers more or less irregular, sometimes nearly regular, hypogynous or perigynous; stamens usually shorter than or equaling the perianth; corolla imbricate (edges overlapping) in bud stage; stamens 10 or fewer . . . . . . . . . . . . . . . . . . . . . . . . . . . . . *b*

b. Flowers with the upper or odd petal within and inclosed by the other two lateral petals; flowers perigynous; stamens usually distinct and separate from one another; corolla somewhat irregular or nearly regular . . . . . . . . . . . . . . . . . . . . . II. CAESALPINIOIDEAE (Gymnocladus, Gleditsia, Cassia, Cercis)

b. Flowers with the upper or odd petal on the outside, larger than and inclosing the other petals; flowers hypogynous; some or most of the stamens usually united by their filaments, rarely all free and distinct; corolla usually irregular and papilionaceous or pea-like with an upper larger petal (standard), 2 lateral petals (wings), and 2 lower petals often clinging together at their forward edges to form the keel . . . . . . . . . . . . . . . . . . . . III. PAPILIONOIDEAE (remainder of genera treated)

## Subfam. I. **MIMOSOIDEAE**

### 1. **Acacia** Mill. Acacia

**Acacia angustissima** (Mill.) Ktze. var. **hirta** (Nutt.) Robinson Prairie Acacia Map 1268

*Acacia angustissima* of auth. [P & S], not (Mill.) Ktze.

Flowers late June–October.

Occurs on rocky glades, open hillsides, and exposed ledges along bluffs, mostly on open limestone and cherty limestone exposures. Southwestern extreme of the state, from Jasper to McDonald County east along the southern border to Oregon County.

Ranges from Texas and Arkansas, north to Kansas and Missouri.

Palmer (Rh. 55: 160. 1953) reported typical *A. angustissima* with glabrous stems from Barry Co. (Roaring River State Park, July 25, 1951, *Palmer 52795*), but Wiggins has shown that all the material in Missouri is *A. angustissima* var. *hirta*, varying from glabrous to hirsute stems.

The Missouri stations are the most northeasterly known in North America for this genus, known elsewhere in North America in the southwestern United States and Mexico, southward into Central and South America, and elsewhere in Australia and Africa. This slender-stemmed, almost herb-like shrub is usually less than a meter tall with numerous (18–40 pairs) leaflets included in 8–14 pairs of pinnae. Plants strongly resemble *Desmanthus illinoensis* but have no glands on the petioles, the stamens are numerous instead of 5, and the fruit is about twice as wide as in *Desmanthus illinoensis*.

The seeds are stated to have a high protein content and are, therefore, good food for wildlife, and the foliage is palatable and nutritious for livestock. The bark of the Mexican variety of this Acacia is reported to be used in Mexico in fermented form for alcoholic drinks.

The so-called Mimosa or Silk Tree (*Albizzia Julibrissin* Durazzini), commonly cultivated in Missouri for its fragrant clusters of pink or lilac and white flowers, lacy fernlike foliage, and broad, flat-topped spreading branches, is a close relative of *Acacia* and *Mimosa*.

## 2. **Desmanthus** Willd.

Commonly encountered; about 30–50 flowers to one head; peduncles (main stalk of flower- or fruit-cluster) 2.5–7.5 cm. long; fruits many together, strongly curved, 1.5–2.5 cm. long, 4.5–6 mm. wide, in dense heads; stem usually erect; pinnae 6–15 (usually 9–15) pairs . . . . . . . 1. D. ILLINOENSIS

Rarely found; about 5–12 flowers to one head; peduncles mainly 1–2 cm. long; fruits 3–8, straight, 3–7 cm. long, 1.5–4 mm. wide, finger-like or palmately arranged; stems usually prostrate or lying along the ground; pinnae mostly 5–8 pairs . . . . . . . . . . . . . . . . . . 2. D. LEPTOLOBUS

1. **Desmanthus illinoensis** (Michx.) MacM.
Prairie Mimosa Map 1269
Also called Illinois Mimosa, False Sensitive Plant.
Flowers mid-June–August.

Occurs in rocky open ground, rocky wooded slopes, thickets, glades, prairies, and open ground along roadsides and railroads. Common in western, central, east-central, and extreme southeastern Missouri; apparently absent or not recorded from most of the Ozark and northeastern sectors of the state.

Ranges from Alabama to Texas, north to Ohio, Indiana, Illinois, Minnesota, North Dakota, and Colorado.

The stems are usually 1–2 meters tall, and may attain a height of 2.5 meters. The leaves of this species and of *D. leptolobus* are not as sensitive to the touch as are those of *Schrankia uncinata*. This species is listed by some range botanists as the most important native legume from the standpoint of its very high protein content and value as a nutritious plant for all kinds of livestock. It is quite drouth-resistant and adapts itself to a wide variety of soils and climates.

2. **Desmanthus leptolobus** T. & G. Map 1270
Flowers June–August.

Introduced along railroads in St. Louis (station post ½ mi. east of Allenton on Frisco R. R., July 10, 1898, *Letterman*), Cape Girardeau (Delta, September 8, 1926, *Palmer 31640*), and Boone (along Mo.-Kan.-Top. R. R. in valley of Perche Creek in Missouri River bottoms, T47N, R13W, southeast ¼ sect. 20, just southeast of Providence, July 10, 1958, *Steyermark & Kucera 86118*).

Ranges from Texas and Oklahoma to Kansas, and introduced in Missouri.

The stems are usually more prostrate or lying along the ground and the leaves have a more bluish-green color than those of *D. illinoensis*.

## 3. **Schrankia** Willd. Sensitive Brier

**Schrankia uncinata** Willd. Sensitive Brier
Map 1271
Also called Bashful Brier.
*Schrankia Nuttallii* (DC.) Standley [G, BB, Steyerm.]
Flowers late May–September.

Occurs in rocky open glades, open woods and

Plate no. 207. **1.** Acacia angustissima var. hirta, × 2/5; Details from Small, The New York Botanical Garden. **2.** Desmanthus leptolobus, × 2/5. **3.** Desmanthus illinoensis, × 2/5; Details from Small, The New York Botanical Garden. **4.** Schrankia uncinata, × 2/5; Details from Small, The New York Botanical Garden. **5.** Gymnocladus dioica, × 2/5; *a.* Portion of inflorescence, × 2/5; *b.* Fruit, × 2/5; *c.* Leaf, × 1/5; Details from Small, The New York Botanical Garden.

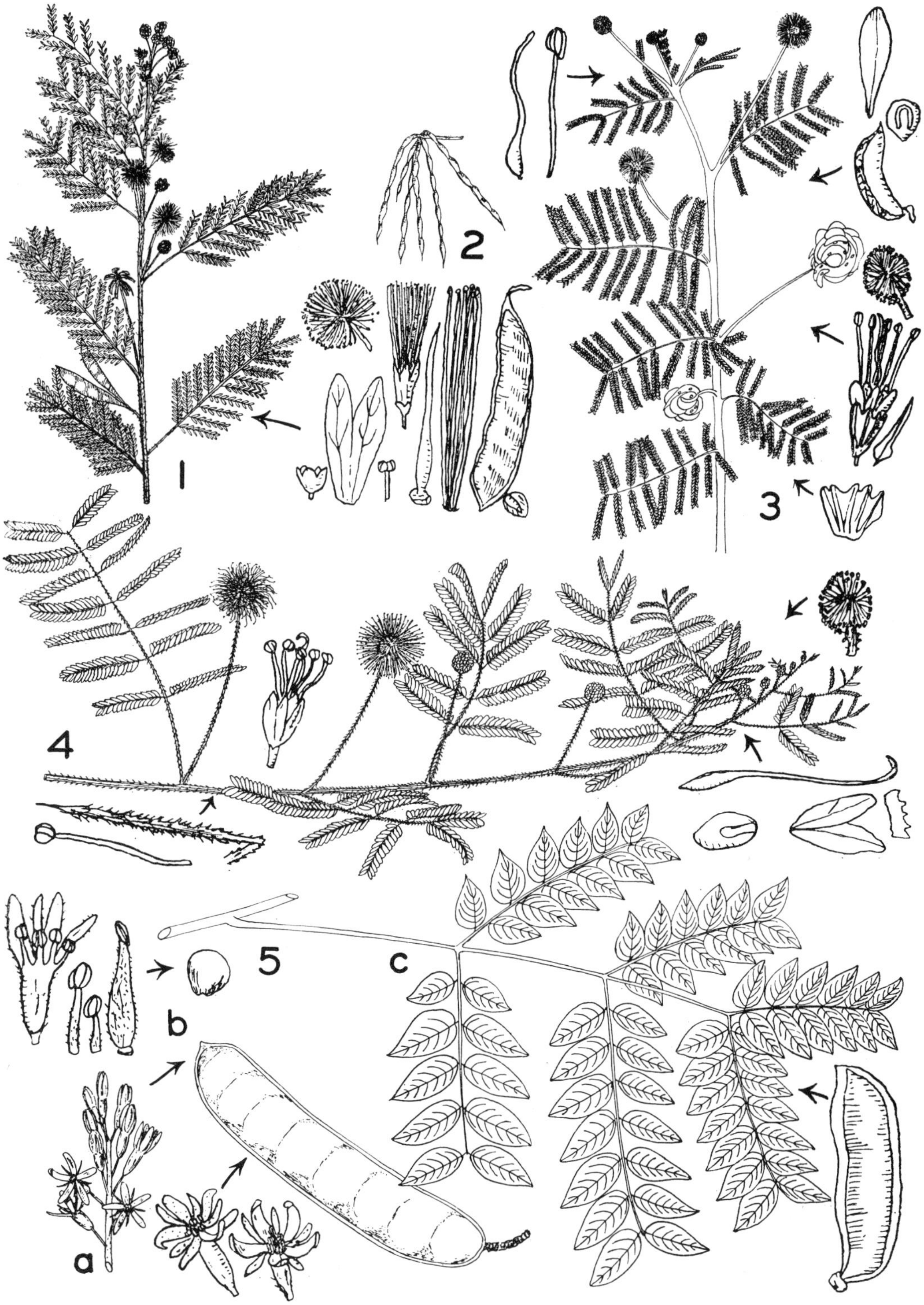


PLATE NO. 207

1270 Desmanthus leptolobus

1271 Schrankia uncinata (Sensitive Brier)

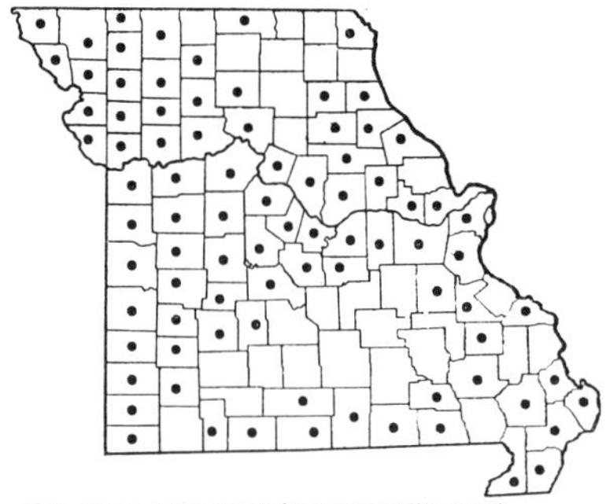

1272 Gymnocladus dioica (Kentucky Coffee Tree)

thickets, rocky prairies and glades, and roadside banks. Common nearly throughout Missouri, except absent from the extreme northwestern and southeastern sectors.

Ranges from Alabama to Texas, north to North Carolina, Illinois, Missouri, Iowa, Nebraska, and South Dakota.

The heads of lavender-rose flowers are very showy. The leaves are the most sensitive to touch of any of the native Missouri species, responding immediately to any contact by folding the leaflets and lowering the petiole. This species does not transplant successfully and the author's efforts to germinate the seeds have always met with failure. *Desmanthus illinoensis*, on the other hand, is both easily transplanted and germinates readily from seed.

The leaves are eaten by wild turkeys and by all classes of livestock. The plant is very nutritious, containing 25–45 per cent protein in the young growth. The seeds are sometimes employed in certain medicines for their laxative effects.

## Subfam. II. **CAESALPINIOIDEAE**

### 4. **Gymnocladus** Lam. Kentucky Coffee Tree

**Gymnocladus dioica** (L.) K. Koch
Kentucky Coffee Tree Map 1272
Flowers May–June.

Occurs in low or rich woods, at the base of bluffs, and along streams. Throughout Missouri, but not recorded from a number of Ozark and northeastern Missouri counties.

Ranges from New York and Ontario to Wisconsin, South Dakota, and Nebraska, south to Kentucky, Tennessee, Arkansas, and Oklahoma; naturalized in Alabama, West Virginia, Virginia, and Delaware.

This tree occurs sometimes in small colonies of rather widely separated individuals, resulting from the habit of the species of sending up root suckers at some distance from the parent tree. In the autumn, this is one of the first trees to lose its foliage, which turns yellow and falls quickly.

Some of the early settlers in Kentucky and Tennessee tried using the seeds as a coffee-substitute, but their usage for this purpose never became popular. Certain Indian tribes were reported to have roasted the seeds for food. The wood does not rot easily in contact with the soil, and has thus been used for fence posts. At one time its wood furnished an attractive cabinet wood for the pioneer settlers.

The twice pinnately compound leaves of this tree are the largest of any of the native species, expanding sometimes to a length of three feet (9 dm.).

### 5. **Gleditsia** L. Honey Locust

Fruit usually with only 1 seed (rarely with 2 or 3 seeds), 3–5 cm. long, $1\frac{1}{2}$–$2\frac{1}{2}$ times as long as broad; trees found only in the swamps of southeastern Missouri and north to east-central Missouri in St. Charles County . . . . . . . . . . . . . . . . . . . . . . . . . . . . . 2. G. AQUATICA

Fruit with many seeds, mainly 15–45 cm. long, many times longer than broad; trees occurring in every county in Missouri . . . . . . . . . . . . . . . . . . . . . . . . . 1. G. TRIACANTHOS

1273 • Gleditsia triacanthos f. triacanthos (Honey Locust)
1273 □ Gleditsia triacanthos f. inermis

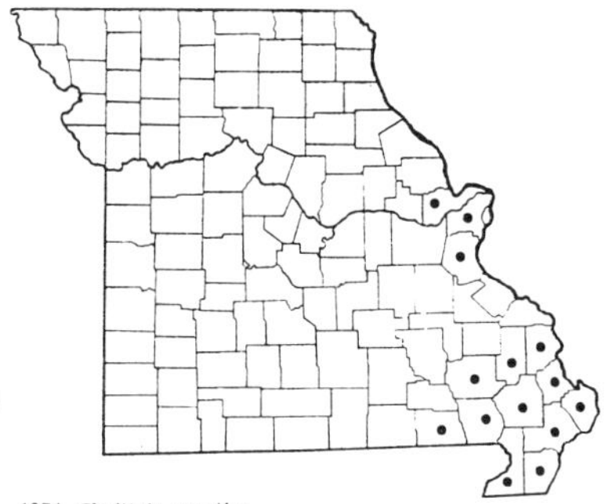

1274 Gleditsia aquatica

1275 Hoffmanseggia Jamesii

1. **Gleditsia triacanthos** L. Honey Locust Map 1273

Also known as Honey Shucks, and Sweet Locust. Flowers May–June.

Occurs usually in low or rich woodland along streams and their valleys, but also on dry or upland slopes and pastures. Common throughout Missouri where known from every county.

Ranges from New York and Pennsylvania to Minnesota and South Dakota, south to Florida and Texas.

Missouri material is represented by two variations:

Branches or bark with conspicuous thorns . . 1a. G. TRIACANTHOS f. TRIACANTHOS
Branches and bark without thorns . . . . 1b. G. TRIACANTHOS f. INERMIS

1a. **Gleditsia triacanthos** f. **triacanthos** Map 1273

*Gleditsia triacanthos* (L.) [G, BB, P & S, Steyerm.]

This is the common variation in the state found in every county.

1b. **Gleditsia triacanthos** f. **inermis** (Pursh) Schneid. Thornless Honey Locust Map 1273

*Gleditsia triacanthos* var. *inermis* Pursh [P & S]

Scattered throughout the state, where it occurs in low or alluvial ground near streams.

A strain of the thornless variation sold as 'Moraine' locust is thornless as in ordinary f. *inermis*, but does not produce fruit, whereas fruit is developed in ordinary f. *inermis*. The 'Moraine' locust is being planted as a substitute for the Dutch Elm Disease-prone American Elm. Another variation, known as 'Sunburst,' with the newer shoots and young foliage of the tips of the branchlets yellow or yellowish instead of green, has also become popular.

The leaves of Honey Locust turn to a rich yellow in autumn. The foliage appears late in the spring and falls early in the autumn, permitting a maximum amount of light to reach the lawn grass in the immediate vicinity of the tree. It is considered an excellent tree for withstanding adverse city conditions.

The thin jellylike pulp surrounding the seeds is sweet and provides a somewhat pleasant diversion to taste when one is tramping outdoors. The long twisted pods which fall to the ground are eaten by cattle. The thorns tied to sticks or poles sometimes serve as sharp weapons used by country folk in hunting bullfrogs at night. The thorns have also served some of the early pioneers and mountain folk as large needles or pins for tying woolen sacks together and for carding wool.

From the standpoint of strength, Honey Locust ranks second (after Black Locust) among native North American woods. Because of its great durability, it serves well for railway ties and fence posts, but the tree rarely enters the lumber market compared with many other species. It is stated that the Cherokee Indians of Tennessee made bows from this wood.

The common name, Honey Locust, usually applied to this tree, is confused with *Robinia Pseudo-Acacia*, both trees being called Honey Locust. Country folk especially have applied the name Honey Locust to *Robinia Pseudo-Acacia*, because of the fragrant flowers of that species, whereas the flowers of *Gleditsia* are greenish and not particularly fragrant. Most botanical authorities, foresters, and horticulturists follow the general usage, however, of Black Locust for *Robinia Pseudo-Acacia* and Honey Locust for *Gleditsia triacanthos*.

2. **Gleditsia aquatica** Marsh. Water Locust Map 1274

Flowers in May.

Occurs in bald cypress and other swamps, borders of bayous and sloughs, and low wet woods. Lowland counties of southeastern Missouri west to Ripley and Wayne counties and north along the Mississippi River locally to Jefferson, St. Louis, and St. Charles (Mis-

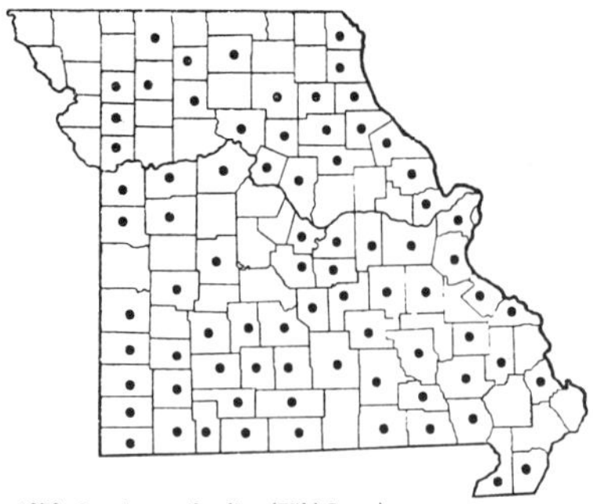

1276 Cassia marilandica (Wild Senna)

1277 Cassia occidentalis (Coffee Senna)

1278 Cassia Tora (Sicklepod)

souri Point, June 24, 1887, *Eggert*) counties.

Ranges from Florida to Texas, north to North Carolina, Indiana, Illinois, and Missouri.

This is a smaller tree than Honey Locust, from 15–20 m. tall (50–60 feet) as compared with a 25–50 m. (75–160 feet) height attained by *G. triacanthos*.

Because of the great durability of the wood against rot in contact with wet soil, it has been used for fence posts. The excellent polish taken by this species qualifies it for use as a cabinet wood.

## 6. **Hoffmanseggia** Cav.

**Hoffmanseggia Jamesii** T. & G. Map 1275

Flowers June–September.

Found as a waif on dumps and waste ground, where introduced. Known only from Jasper County in southwestern Missouri (Joplin, 1901, *Thom*, in Univ. Mo. Herb.).

Ranges from Kansas and Oklahoma to Texas and California; introduced in Missouri.

## 7. **Cassia** L. Senna

a. Leaflets 2–5 mm. broad, 6–20 mm. long; flowers solitary or in small clusters along the sides of the stem in the axils of the leaves; stipules persistent . . . . . . *b*
- b. Flowers small with petals 3–8 mm. long; flower-stalks (pedicels) 2–4 mm. long; stamens 5; mature fruit 2.5–4 cm. long; gland on petiole stalked . . . . . . 5. C. NICTITANS
- b. Flowers showy with petals 10–20 mm. long; flower-stalks (pedicels) 10–25 mm. long; stamens 10; mature fruit 4–7 cm. long; gland on petiole not stalked . . . . . . 4. C. FASCICULATA

a. Leaflets 6–35 mm. broad, 20–80 mm. long; flowers in racemes clustered toward top of stem; stipules not persisting, disappearing . . . . . . *c*
- c. Leaflets only 4 or 6, the uppermost pair broadly rounded or obtuse at summit; gland present between the lowest pair of leaflets; fruit 4-sided . . . . . . 3. C. TORA
- c. Leaflets mostly 8–20, slightly pointed (acutish) to definitely longer-pointed (acuminate) at the tip; gland present near base of the leaf-stalk (petiole); fruit flat or rather flat . . . . . . *d*
  - d. Leaflets usually 8–12, definitely long-pointed (acuminate) at the tip, the largest 25–40 mm. wide . . . . . . 2. C. OCCIDENTALIS
  - d. Leaflets mainly 10–20, slightly pointed (acutish) with a short mucro (projection) at the tip, the largest 10–25 mm. wide . . . . . . 1. C. MARILANDICA

1. **Cassia marilandica** L. Wild Senna Map 1276

*Cassia Medsgeri* Shafer [P & S]

Flowers July–August.

Occurs in open or rocky woods, alluvial thickets, along rocky branches, base of slopes and bluffs, alluvial woodland soils, and wet meadows in valleys. Common throughout Missouri.

Ranges from Pennsylvania to Iowa and Kansas, south to Florida and Texas.

2. **Cassia occidentalis** L. Coffee Senna Map 1277

Also called Coffee Weed.

Flowers August–September.

Occurs in low alluvial open ground, along roadsides, and railroads, where introduced in New Madrid, Dunklin, Ripley, and Jackson counties.

Native of tropical areas in South and Central America and Mexico; naturalized in the Old World tropics; introduced and naturalized in the United States from Florida to Texas, north to Virginia, Indiana, Illinois, Iowa, and Kansas.

The fruits vary from glabrous to sparsely pubescent.

The roasted and ground seeds were used at one time in parts of the southern United States as a coffee substitute, and are still so used in some parts of Central America. The young leaves and undeveloped fruits are cooked as a vegetable in parts of the Dutch East Indies, and in some regions of Central America the leaves furnish the ingredients of an ointment used as a remedy for ringworm and other skin afflictions.

3. **Cassia Tora** L. Sicklepod — Map 1278

Also called Coffee Weed.

Flowers July–September.

Occurs in sandy alluvial soils and rocky open ground along streams, and open or waste ground. Scattered in southern and central Missouri north to St. Louis (prairie, St. Louis, August, 1838, *Riehl 87*) and Jackson (waste ground, Sheffield, September 26, 1921, *Bush 9689*) counties.

Ranges from South and Central America, the West Indies, and Mexico, and in the United States from Florida to Texas, north to Pennsylvania, Indiana, Michigan, Illinois, Missouri, and Kansas; also naturalized in the Old World tropics.

The seeds are sometimes used in parts of Mexico and Central America as a coffee substitute, and are stated to be used sometimes in Europe for adulterating coffee. In some parts of the Dutch East Indies the young shoots of the plants are cooked as a vegetable, while the cooked and roasted seeds in those regions are said to be eaten with rice. In India the seeds are sometimes employed as a mordant in dyeing cloth blue. As in various Old World species of *Cassia*, the leaves are believed to possess the purgative properties ascribed to the official drug senna.

4. **Cassia fasciculata** Michx. Partridge Pea — Map 1279

Flowers June–October.

The following variations are encountered in Missouri:

a. Petals white . . . . . 4b. C. FASCICULATA var. FASCICULATA f. JENSENI
a. Petals yellow. . . . . . . . . . *b*
  b. Hairs on stem short, ascending, or incurved 4a. C. FASCICULATA var. FASCICULATA f. FASCICULATA
  b. Hairs on stem prominent, spreading, often 2 mm. long . . . . . . . . . . *c*
    c. Commonly encountered; stems upright; leaflets 18–36 . . . 4c. C. FASCICULATA var. ROBUSTA
    c. Rarely found, known only from Washington County; stems sprawling or widely spreading; leaflets 12–20 . . 4d. C. FASCICULATA var. DEPRESSA

4a. **Cassia fasciculata** var. **fasciculata** f. **fasciculata** — Map 1279

*Cassia fasciculata* Michx. [G, P & S]

*Cassia fasciculata* var. *fasciculata* [BB]

Occurs in prairies, glades, fallow fields, along roadsides and railroads, and open thickets. Throughout Missouri and doubtless in every county.

Ranges from Florida to Texas, north to Massachusetts, New York, Ontario, Ohio, Indiana, Wisconsin, Minnesota, and South Dakota.

4b. **Cassia fasciculata** var. **fasciculata** f. **Jenseni** Palmer and Steyerm. — Map 1279

*Cassia fasciculata* f. *Jenseni* Palmer & Steyerm. [G, P & S]; originally described from plants found in Franklin County, Missouri.

Known only from Franklin (Gray Summit, August 20, 1928, *L. P. Jensen*, holotype in Mo. Bot. Gard. Herb.) and Jefferson counties in east-central Missouri.

Occurs, so far as known, in Missouri.

4c. **Cassia fasciculata** var. **robusta** (Pollard) Macbr. — Map 1279

Occurs in wet river bottom prairies, alluvial open soils of valleys, and sandy upland fields. Scattered in southern, central and extreme eastern Missouri along the Mississippi River to Clark County, and along the Missouri River to Boone and Jackson counties.

Ranges from Florida to Louisiana, north to Virginia, Ohio, Indiana, Illinois, and Missouri.

4d. **Cassia fasciculata** var. **depressa** (Pollard) Macbr. — Map 1279

Known only from Washington County, eastern Missouri.

Ranges from Florida to Missouri.

The true status of var. *depressa* is yet to be determined. No specimens have been found in Missouri since the original collection. It is possible that the Missouri collection represents an abnormal, depauperate state of var. *robusta*.

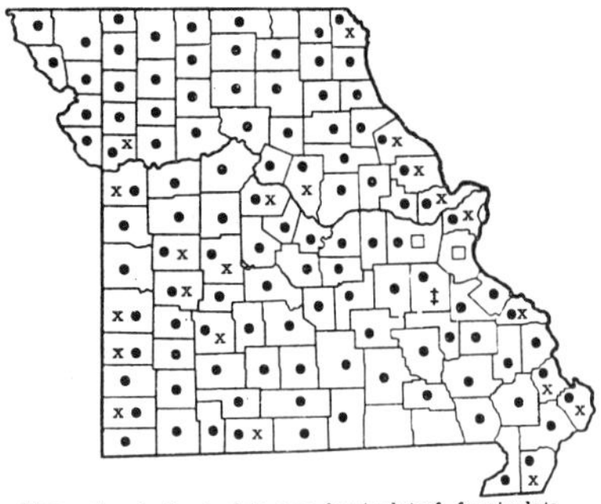

1279 • Cassia fasciculata var. fasciculata f. fasciculata (Partridge Pea)
1279 □ Cassia fasciculata var. fasciculata f. Jenseni
1279 x Cassia fasciculata var. robusta
1279 ‡ Cassia fasciculata var. depressa

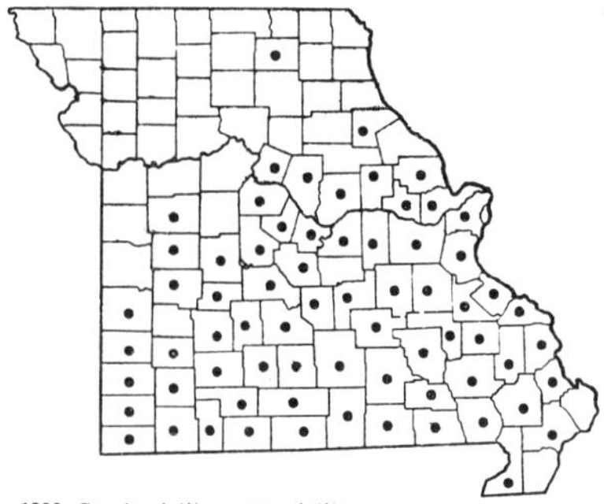

1280 Cassia nictitans var. nictitans

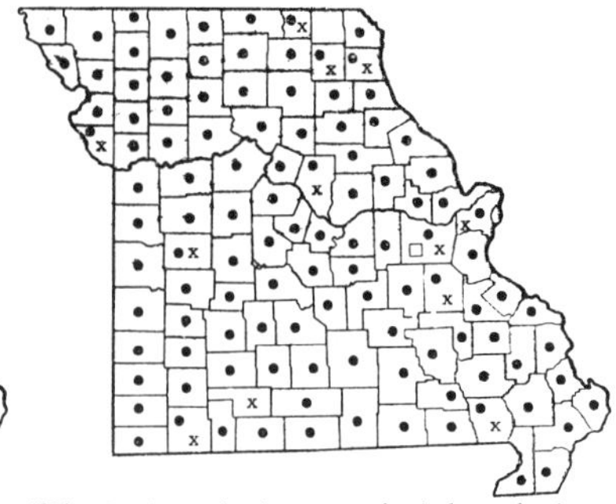

1281 • Cercis canadensis var. canadensis f. canadensis (Redbud)
1281 x Cercis canadensis var. canadensis f. glabrifolia
1281 □ Cercis canadensis var. canadensis f. alba

*Cassia fasciculata* is a showy annual wild flower and may be raised easily from seed. However, it can become a serious and aggressive plant, once it gets started, as the seeds are extremely viable and large stands of the plant are produced, each year becoming larger in extent. The ripe seed pods, upon splitting open, throw the seeds some distance from the parent plant, thus increasing the original area occupied. The seeds are eaten by wild turkey and are much relished by pheasants in the northern states.

5. **Cassia nictitans** L. var. **nictitans**
Sensitive Pea Map 1280
Also called Wild Sensitive Plant.
*Cassia nictitans* L. [G, P & S]
Flowers July–September.

Occurs in acid soils of rocky open woods, upland slopes, ridges, bluffs, prairies, fallow and rocky fields, and open thickets. Southern and central Missouri, where common throughout the Ozark region north to Lincoln, Montgomery, Callaway, Boone, Howard, Cooper, Johnson, and Vernon counties, and locally in northeast Missouri in Ralls and Adair counties.

Ranges from Florida to Texas, north to Massachusetts, Vermont, New York, Ohio, Indiana, Illinois, Missouri, and Kansas.

The leaflets of this species and of *C. fasciculata* are sensitive to the touch and fold against one another eventually. They are not as sensitive, however, as those of *Schrankia uncinata*.

The fruits of var. *nictitans* are finely pubescent with short, incurved appressed hairs. Other varieties having the fruits glabrous (var. *leiocarpa* Fern.) and fruits with spreading hairs (var. *hebecarpa* Fern.) are not known from Missouri. The seeds are commonly eaten by wild turkey and bob-white.

## 8. **Cercis** L. Redbud, Judas Tree

**Cercis canadensis** L. var. **canadensis**
Eastern Redbud Map 1281
Flowers late March–early May.

Occurs in open woodland, borders of woods, thickets, limestone glades and openings, and along rocky streams and bluffs. Throughout Missouri, and known from every county.

Three variations are represented among Missouri material:

a. Petals white . C. CANADENSIS var. CANADENSIS f. ALBA
a. Petals rose-purple or lavender-rose . . . *b*
 b. Lower surface of fully grown leaves more or less hairy . . . . . . . a. C. CANADENSIS var. CANADENSIS f. CANADENSIS
 b. Lower surface of fully grown leaves glabrous (without hairs) . . . . b. C. CANADENSIS var. CANADENSIS f. GLABRIFOLIA

Plate no. 208. **1.** Gleditsia triacanthos, × $^{2}/_{7}$; *a.* Flowering branch; *b.* Fruit. **2.** Gleditsia aquatica, × $^{2}/_{7}$; *a.* Flowering branch; *b.* Leaf and spine; *c.* Fruiting branch; Details from Small, The New York Botanical Garden. **3.** Hoffmanseggia Jamesii, × $^{2}/_{7}$. **4.** Cassia marilandica, × $^{2}/_{7}$; *a.* Fruit, × $^{2}/_{7}$. **5.** Cassia occidentalis, × $^{4}/_{21}$; Details from Small, The New York Botanical Garden. **6.** Cassia Tora, × $^{6}/_{35}$; Details from Small, The New York Botanical Garden. **7.** Cassia fasciculata, × $^{2}/_{7}$; Details from Small, The New York Botanical Garden. **8.** Cassia nictitans, × $^{4}/_{21}$. **9.** Cercis canadensis, × $^{2}/_{7}$; *a.* Flowering branch; *b.* Fruiting branch; Details from Small, The New York Botanical Garden.

PLATE NO. 208

a. **Cercis canadensis** var. **canadensis** f. **canadensis** Map 1281

*Cercis canadensis* L. [G, BB, P & S, Steyerm.]

*Cercis canadensis* var. *canadensis* [Shinners]

This is the commonest form encountered in Missouri, occurring in every county.

Ranges from Florida to Texas and northeastern Mexico, north to Connecticut, New York, Pennsylvania, Ontario, Michigan, Illinois, Iowa, and Nebraska.

b. **Cercis canadensis** var. **canadensis** f. **glabrifolia** Fern. Map 1281

*Cercis canadensis* f. *glabrifolia* Fern. [G, Steyerm.]

Scattered throughout Missouri, but less common than f. *canadensis*.

Scattered throughout the range of typical f. *canadensis*.

c. **Cercis canadensis** var. **canadensis** f. **alba** Rehd. White Redbud Map 1281

*Cercis canadensis* f. *alba* Rehd. [G, Steyerm.]; originally described from a plant found in Franklin County, Missouri. See Mitt. Deutsch. Dendr. Ges. 1907: 72. 1907, and Mo. Bot. Gard. Bull. 10: 110. 1922.

Known in the wild state in Missouri only from Franklin County, eastern Missouri (Alt Farm, Pacific, April 5, 1938, *Ed. A. Alt, Jr.*). See Mo. Bot. Gard. Bull. 26: 82. 1938.

This white-flowered form, so far as known, occurs in a wild state only in Franklin County, Missouri, where originally introduced into cultivation. It is becoming increasingly more popular in home plantings.

The Eastern Redbud is the state tree of Oklahoma. It is very common throughout Missouri and brightens the early spring landscape with its rosy-lavender flowers. It always begins to bloom before the Flowering Dogwood (*Cornus florida*), but delayed flowering and a long blooming period may often carry it through into Flowering Dogwood time in the Ozarks (which usually occurs from mid-April to mid-May) so that the two are frequently seen at the same time in flower. Usually, however, much of the Redbud flowering has been completed by the time Flowering Dogwood is at its best. The foliage turns a yellowish color or pale greenish yellow in autumn.

Typical var. *canadensis* has rather thin leaf-blades, dull green above, about as long as broad or slightly longer than broad and the summit somewhat or definitely pointed, as contrasted with var. *texensis* (S. Wats.) Hopkins and var. *mexicana* (Rose) Hopkins, in each of which varieties the leaf-blades are thicker, more glaucous above, usually broader than long with a more reniform shape, and the summit more rounded or blunt or even depressed in the middle. These other varieties occur in Oklahoma, Texas, and Mexico.

Redbud in Missouri is commonest on limestone soils. It is easily grown from seed or transplanted and prefers sunny situations.

The flowers of Redbud have a sour flavor and are stated to have been used in salads and pickled dishes. Like some other members of the pea family, the flowers and young fruits can be cooked and make a tasty morsel when fried.

## Subfam. III. **PAPILIONOIDEAE**

### 9. **Baptisia** Vent. False Indigo

a. Leaves, or stems, or ovaries, or young fruits hairy; at least the lowest flower-stalks (pedicels) with persistent and prominent leaf-like, reticulate-veined bracts 1–3 cm. long at their base; flowers or fruits borne on a more or less horizontal or declined axis; flowers cream-colored . . 2. B. LEUCOPHAEA

a. Leaves, stems, ovaries, and young fruits glabrous (without hairs); bracts at base of flower-stalks (pedicels) falling early, not plainly reticulate, or less than 1 cm. long; flowers or fruits arising from an erect axis; flowers white, blue-purple, or clear yellow . . . . . . . . . . . . . . . . *b*

b. Flowers blue-purple; stipules at base of leaves broad, persisting and longer than the leaf-stalks; fruit tipped with a beak 8–20 mm. long . . . . . . . . . . . 1. B. AUSTRALIS var. MINOR

b. Flowers white or clear yellow; stipules at base of leaves narrow, soon falling (deciduous) and shorter than the leaf-stalks (petioles); fruit tipped with a beak 3–8 mm. long . . . . . . . . *c*

c. Commonly found; flowers with white petals; mature fruit drooping, much longer than broad,

Plate no. 209. **1.** Baptisia australis var. minor, × 2/7; *a.* Inflorescence. **2.** Baptisia leucantha, × 2/7. **3.** Baptisia leucophaea, × 2/7; *a.* Fruiting branch. All details from Small, The New York Botanical Garden.

PLATE NO. 209

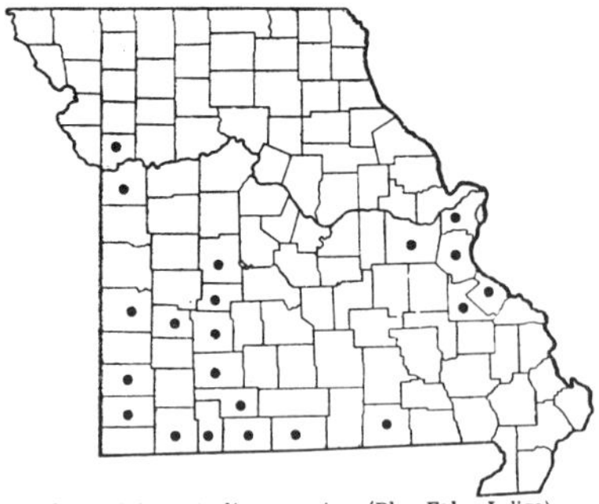

1282 Baptisia australis var. minor (Blue False Indigo)

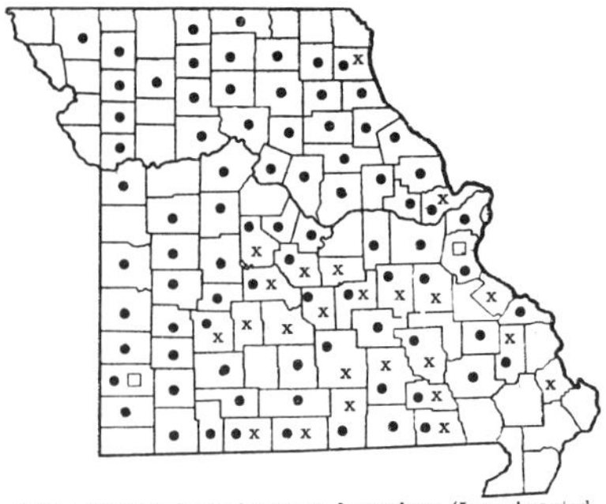

1283 • Baptisia leucophaea var. leucophaea (Long-bracted Wild Indigo)
1283 x Baptisia leucophaea var. glabrescens
1283 □ Baptisia X bicolor (Baptisia leucophaea X australis var. minor)

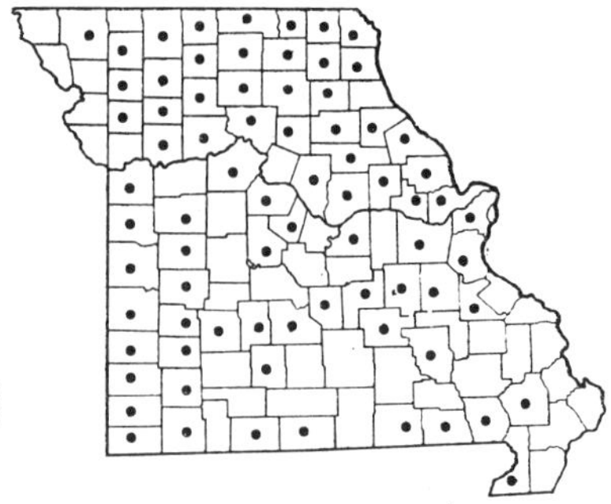

1284 Baptisia leucantha (White Wild Indigo)

ellipsoid-oblong, the main body of the fruit 2.5–4 cm. long; some of the leaf-stalks (petioles) 5–12 mm. long . . . . . . . . . . . . . . . . . . . . . 3. B. LEUCANTHA

c. Rarely found; flowers with clear yellow petals; mature fruit erect, slightly longer than broad or nearly as broad as long, almost globe-shaped or broadly ovoid-oblong, the main body of the fruit 1.2–2 cm. long; leaf-stalks (petioles) 2–4 mm. long . . . . . . . 4. B. SPHAEROCARPA

1. **Baptisia australis** (L.) R. Br. var. **minor** (Lehm.) Fern. Blue False Indigo Map 1282
*Baptisia minor* Lehm. [G, Larisey]
*Baptisia vespertina* Rydb. [P & S, Steyerm.]
Flowers early May–June.

Occurs on limestone glades and prairies. Southern and central Missouri north to St. Louis, Franklin, Benton, and Clay counties; absent from most of the Ozark region, concentrated in the northeastern sector of the Ozarks and the unglaciated prairie region, with an isolated occurrence in Oregon County.

This very showy-flowered plant does well in cultivation and can be raised easily from seed. The author has grown both *B. australis* var. *australis* and var. *minor* in his northern Illinois botanical preserve. While these differ in a number of details, the differences appear to be no more than varietal rather than specific. The var. *minor* has generally shorter stems with more spreading branches, smaller leaflets 1.5–4 instead of 4–8 cm. long, and stipe of the fruit about twice the length of the calyx instead of scarcely or not longer than the calyx. The flowers in both varieties are of about the same length; although a greater length is often ascribed to var. *minor* (2.7–3.5 cm. being given for var. *minor* by some authors versus 2–2.7 cm. for var. *australis*), actually many specimens of var. *minor* can be found having flowers 2.3–3 cm. long, and I have not been able to separate the two on floral length differences. Also, although the petioles are stated by some authors to be 5–20 mm. long in var. *australis* versus 1–4 mm. long in var. *minor*, I have found many specimens of var. *minor* with petioles up to 7 mm. long. The various degrees of intergradation, then, in corolla length, petiole length, and height of plant leave principally the differences in leaflet length and relative length of fruiting stipe to calyx as the chief criteria for distinction between these taxa.

A hybrid between *B. australis* var. *minor* and *B. leucophaea* has been described and is known as *Baptisia* × *bicolor* Greenman & Larisey. The stems and leaves in this hybrid are more or less pubescent with the large persistent bracts, pubescent ovary, and long pedicels of *B. leucophaea* and a two-toned flower with blue-purple standard and yellow to cream-colored wings and keel. It is known in Missouri from Jefferson (Dittmer, May 4, 1930, *Kellogg 15216*) and Jasper (Webb City, May 4, 1902, *Palmer 127*, holotype in Mo. Bot. Gard. herb.; same locality, May 10, 1931, *Palmer 39282*) counties. Outside Missouri it is found in Kansas and Oklahoma.

Country children often use the ripe fruits of this and the other species of *Baptisia* as rattles.

2. **Baptisia leucophaea** Nutt.
Long-bracted Wild Indigo Map 1283
Flowers late April–June.

Occurs in rocky dry open woods, prairies, and open ground, usually in acid soils.

Two variations are represented by Missouri material:

Stems hairy; leaves hairy . . 2a. B. LEUCOPHAEA var. LEUCOPHAEA

Stems glabrous (without hairs); leaves glabrous 2b. B. LEUCOPHAEA var. GLABRESCENS

2a. **Baptisia leucophaea** var. **leucophaea**

Map 1283

*Baptisia leucophaea* Nutt. [G, BB, P & S, Steyerm.]

This is the commoner variation and occurs throughout Missouri, except absent in the lowland counties of the southeastern part of the state.

Ranges from Arkansas to Texas, north to Michigan, Wisconsin, Minnesota, and Nebraska.

2b. **Baptisia leucophaea** var. **glabrescens** Larisey

Map 1283

*Baptisia leucophaea* var. *laevicaulis* Gray [P & S, Steyerm.]

Occurs in rocky open woods, usually in acid soils, and prairies. Ozark region of Missouri west to Morgan, Camden, Polk, and Taney counties, north to St. Charles, Maries, and Morgan counties, and locally in Lewis County, northeastern Missouri.

Ranges from Louisiana and Texas, north to Illinois, Wisconsin, Missouri, and Oklahoma.

3. **Baptisia leucantha** T. & G.

White Wild Indigo Map 1284

Flowers May–July.

Occurs on prairies, glades, rocky open slopes, alluvial soils along streams, bottom prairies, wet meadows, valleys, and along railroads. Throughout Missouri, probably in every county.

Ranges from Mississippi to Texas, north to Ontario, Ohio, Michigan, Wisconsin, Minnesota, and Nebraska.

This species grows taller (up to 1.5 m. high) than any of the other Missouri species. The stems are smooth and covered with a grayish-white (glaucous) covering. It is easily grown from seed. It is known to hybridize with *B. tinctoria*, a species which does not occur in Missouri.

This and the other species in Missouri, if eaten in early spring in sufficient quantity, are capable of causing poisoning among horses and cattle.

4. **Baptisia sphaerocarpa** Nutt.

Yellow Wild Indigo Map 1285

Also known as Round-fruited Yellow Wild Indigo

*Baptisia viridis* Larisey [BB]

Flowers May–June.

Introduced in prairie and cleared places along railroads in southern and central Missouri in St. Louis, Ralls, Randolph, Greene, and Barry counties.

Ranges from Louisiana to Texas, north to Missouri and Oklahoma.

Dr. Larisey separated *B. viridis* from *B. sphaerocarpa* (Ann. Mo. Bot. Gard. 27: 139, 196–98. 1940) on the basis of smaller leaflets (2.5–5 cm.) of *B. viridis* versus 4.5–8 cm. long of *B. sphaerocarpa* and geniculate branches of *B. viridis* versus straight branches of *B. sphaerocarpa.* Although the leaflets are somewhat smaller in *B. viridis,* the branches are as straight as in ordinary specimens of *B. sphaerocarpa.* There do not appear to be other differences between the two. The author has grown *B. sphaerocarpa* from Missouri material at his wildflower garden in northern Illinois for the past ten years and it has multiplied.

Two hybrid plants between this species and *B. australis* var. *australis* originated spontaneously in the author's wildflower graden in the spring of 1958. The plants were intermediate in every respect between the parent species, possessing the smaller height and leaflets of *B. sphaerocarpa,* and two-toned flowers with a yellow keel and brown and dull lavender wings and standard. The brown and dull lavender coloring was a blending of the yellow and purple colors. The fruiting pods had the shape more of *B. australis* var. *australis,* but were shorter than in that species. Upon drying, the foliage remained green as in *B. sphaerocarpa.*

Larisey has also recorded hybrids involving *B. viridis* and *B. leucantha* (*B.* × *fragilis* Larisey) which is synonymous with *B.* × *sulphurea* Engelm. (*B. sphaerocarpa* and *B. leucantha*). This has not been recorded from Missouri, but is found in Louisiana, Texas, and Oklahoma. Another hybrid, involving *B. sphaerocarpa* and *B. leucophaea,* has been found in Arkansas and Oklahoma.

*Baptisia sphaerocarpa* is exceptional among the Missouri species of the genus in having the leaves on the smaller branches with only 1 or 2 leaflets.

## 10. **Cladrastis** Raf. Yellow Wood

**Cladrastis lutea** (Michx. f.) K. Koch

Yellow Wood Map 1286

Flowers May–June.

Occurs along rich wooded limestone slopes and bluffs of the White River and tributaries of southwestern Missouri, in Taney, Stone, and Barry counties.

Ranges from Alabama, Georgia, North Carolina, and Kentucky to Tennessee, Indiana, Illinois, and Missouri.

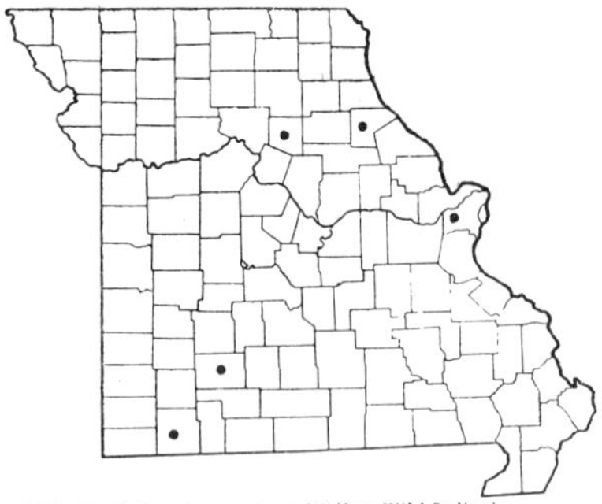

1285 Baptisia sphaerocarpa (Yellow Wild Indigo)

1286 Cladrastis lutea (Yellow Wood)

1287 Crotalaria sagittalis var. sagittalis (Rattlebox)

Thousands of these handsome and rare trees have been exterminated from their original stations in Missouri by dams created on various sections of the White River in the past ten years.

Before synthetic and commercial dyes became available, a yellow coloring was extracted from the wood by some of the early settlers and used in dyeing their homespun cloth yellow. The wood was early used for gunstocks. The pendent sprays of white flowers are very showy, somewhat resembling those of the Black Locust (*Robinia Pseudo-Acacia*).

## 11. **Crotalaria** L. Rattlebox

Common species; low plant 1–4 dm. tall, the stems with usually spreading or loose hairs; calyx hairy; flowers small, 7–10 mm. long, the corolla shorter than or barely equaling calyx; leaves 3–7 cm. long . . . 1. C. SAGITTALIS

Rarely collected species; tall plant 5–20 dm. tall, the stems with hairs paralleling and pressed to the surface; calyx glabrous; flowers large, 15–25 mm. long, the corolla much longer than the calyx; larger main leaves 6–20 cm. long . . . . . . . . . . . . . . . . . . . . 2. C. SPECTABILIS

1. **Crotalaria sagittalis** L. var. **sagittalis**
Rattlebox Map 1287
Flowers May–September.

Occurs in prairies, glades, open wooded slopes, sandy or rocky open ground, fallow fields, and along railroads. Common in southern and central Missouri, and extending northward to Clark, Macon, Linn, Livingston, Caldwell, and Clay counties, and locally northwest in Nodaway County.

Ranges from South and Central America and Mexico to the West Indies, and in the United States from Florida to Texas, north to Massachusetts, Vermont, New York, Ohio, Michigan, Wisconsin, Minnesota, and South Dakota.

Cases of fatal poisoning have been reported for horses which have eaten the fresh or dried plants contained in hay.

2. **Crotalaria spectabilis** Roth Map 1288
*Crotalaria Retzii* Hitchc.
Flowers August–November.

Occurs in sandy fields, open ground, and along roadsides and railroads. Southeastern and central Missouri, in Scott, Mississippi, Dunklin, Boone, and Saline (Napton, October 14, 1937, *J. M. Barrett*) counties.

Native of the Old World tropics; introduced and naturalized in the United States from Florida to Texas north to Virginia and Missouri.

This species has been known in Missouri since 1937, when it was first collected by Mr. John H. Kellogg (See Rh. 40: 361–63. 1938) near Charleston, Mississippi County, and between Kennett and Holcomb, Dunklin County. It has been used by the farmers of that section of the state as a cover crop and silage

Plate no. 210. **1.** Baptisia sphaerocarpa, × 2/7; *a.* Fruiting branch. **2.** Cladrastis lutea, × 2/7; Details from Small, The New York Botanical Garden. **3.** Trifolium pratense, × 2/7. **4.** Trifolium incarnatum, × 2/7; *a.* Flower, × 1 3/7. **5.** Crotalaria sagittalis, × 2/7; Details from Small, The New York Botanical Garden. **6.** Crotalaria spectabilis, × 2/7; After Gleason, details from Small, The New York Botanical Garden. **7.** Trifolium arvense, × 2/7; Details from Small, The New York Botanical Garden; *a.* Flower, × 1 5/7. **8.** Trifolium resupinatum, × 2/7; *a.* Leaf, × 1 1/7; *b.* Fruit, × 3 5/7; *c.* Flower, × 2.

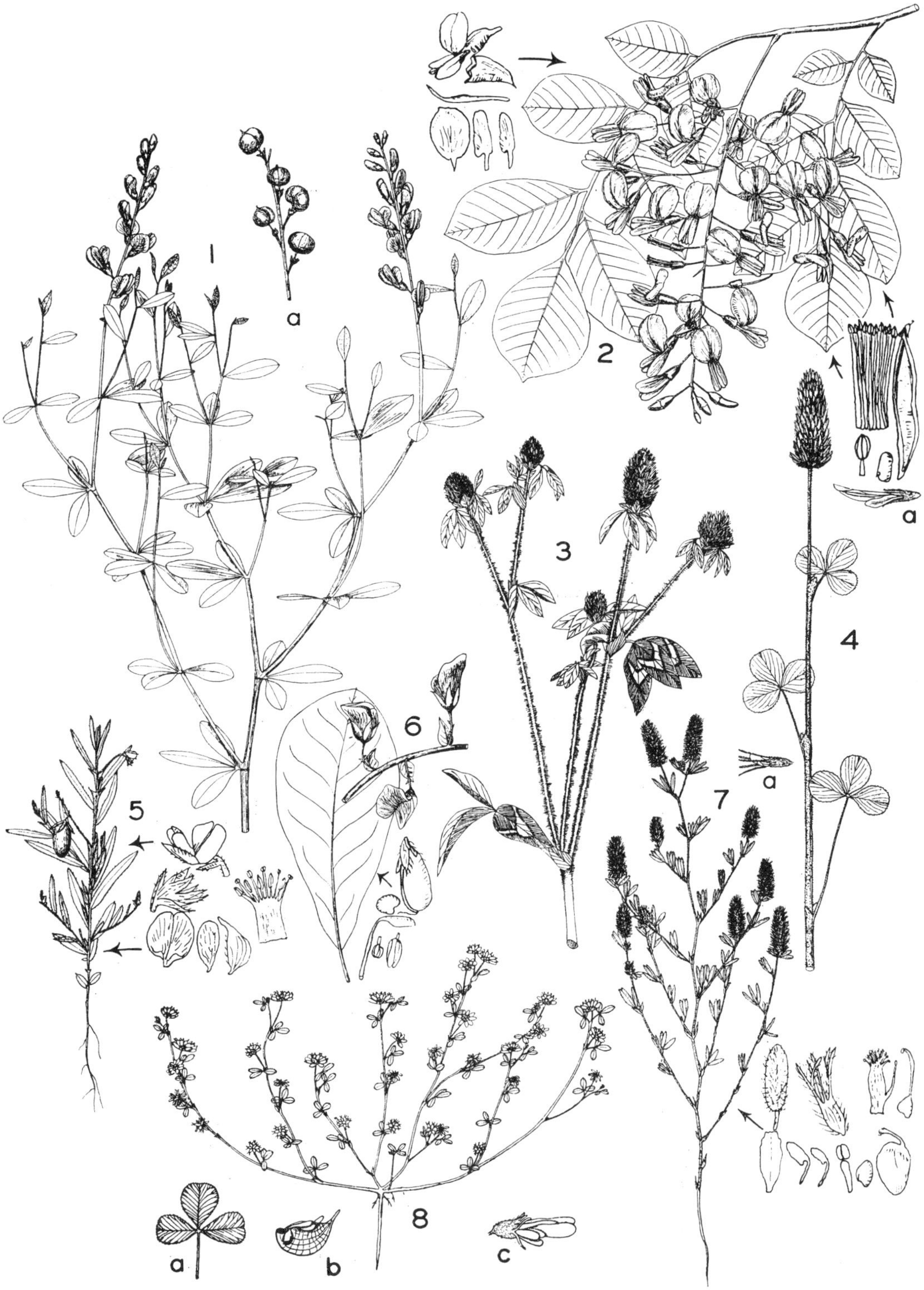


PLATE NO. 210

plant. However, the plant is known to be poisonous when fed to domestic animals, so that its future value will probably be limited to its use as an ornamental subject grown for its showy yellow flowers. It has met with success as a garden annual in the St. Louis area.

## 12. **Trifolium** L. Clover

The author has drawn upon two recent studies of the genus for various items noted in the following key and in discussion of the species: Isely, D. The Leguminosae of the North-Central United States. I. Loteae and Trifolieae. Iowa State College Jour. Sci. 25: 439–82. 1951; Hermann, F. J. A Botanical Synopsis of the Cultivated Clovers (Trifolium). USDA Monogr. 22: 1–45. 1953.

a. Middle (terminal) one of the three leaflets on a stalk much longer than the stalks of the other leaflets . . *b*
 b. Leaf-stalks (petioles) usually shorter than the leaflets; flower-clusters 5–8 mm. thick with 5–15 flowers; flowers 2.5–3 mm. long, the standard (uppermost petal) scarcely or faintly marked with grooves or ridges; rarely collected species . . . . . . . . . . . . . . 12. T. DUBIUM

 b. Leaf-stalks (petioles) of larger leaves of lower and middle part of stem usually longer than the leaflets; flower-clusters 8–15 mm. thick with many (15–40) flowers; flowers 3–4 mm. long, the older ones with the standard distinctly marked with grooves and ridges; commonly found species 11. T. CAMPESTRE
a. Middle (terminal) one of the three leaflets either without a stalk or on a short stalk not longer than the stalks of the other leaflets . . . . . . . . . . . . . . . . . . . . . . . . . *c*
 c. Flowers yellow; fruit distinctly stalked in the calyx; rare species . . . . . . . 10. T. AGRARIUM
 c. Flowers not yellow (either white, pink, rose, red, purple, crimson, or some combination of these), turning brown in age; includes common species . . . . . . . . . . . . . . . . . *d*
  d. Stems definitely hairy. . . . . . . . . . . . . . . . . . . . . . . . *e*
   e. The individual flowers of a head on definite stalks 2 mm. or more long . . . . . . . . *f*
    f. Flowers 4.5–6 mm. long; flower-clusters small, 10–17 mm. thick; teeth of the calyx $1\frac{1}{2}$–2 times as long as the calyx-tube; leaflets 3–15 mm. long; stem about 1–1.5 mm. in diameter; rarely collected . . . . . . . . . . . . . . . 7. T. CAROLINIANUM
    f. Flowers 7–13 mm. long; flower-clusters larger, 25–45 mm. thick; teeth of the calyx 2–3 times as long as the calyx-tube; leaflets mostly 20–30 mm. long; stem usually 1.5–3 mm. in diameter; commonly collected . . . . . . . . . . . . . . 9. T. REFLEXUM
   e. The individual flowers of a head either without a stalk or on a very short stalk 1 mm. or less long. . . . . . . . . . . . . . . . . . . . . . . . . . . *g*
    g. Leaf-stalks (petioles) of at least the lower and middle leaves shorter than the leaflets; leaflets narrowly oblong to narrowly oblanceolate, 3–6 times as long as broad, mostly 2–6 mm. (up to 10) wide; flowers only 3–7 mm. long, the corolla shorter than the calyx-teeth; flower-clusters with a gray or pink silky covering of hairs . . . . . . 3. T. ARVENSE
    g. Leaf-stalks (petioles) of the lower and middle leaves longer than the leaflets; leaflets oval or obovate, 1–2 times as long as broad, mostly 10–35 mm. (down to 5) wide; flowers 12–20 mm. long, slightly longer or longer than the calyx-teeth; flower-clusters without a gray or pink silky covering . . . . . . . . . . . . . . . . . . . . . *h*
     h. Stipules (at base of leaf-stalk) not ending in a bristle; petals scarlet; flower-heads longer than broad, cylindrical or narrowly ovoid, on definite peduncles (flower-stem) not bearing any leaves or stipules . . . . . . . . . . . . . . 2. T. INCARNATUM
     h. Stipules (at base of leaf-stalk) ending in a bristle; petals rose-purple, rose, or white: flower-heads globe-shaped, as broad as long or broader than long, on a short peduncle (flower-stem) or none at all and bearing a pair of leaves at the base . . . 1. T. PRATENSE
  d. Stems glabrous (without hairs) or essentially so . . . . . . . . . . . . . . . . *i*
   i. The individual flowers of a head either without a stalk or on a very short stalk 1 mm. or less long; calyx short-hairy on one side, becoming greatly enlarged and inflated after flowering; nerves of calyx-tube conspicuously enlarged with a honeycomb effect; stipules (at base of leaf-stalk) only 1–1.5 mm. wide in the free portion not united with the leaf-stalk 4. T. RESUPINATUM
   i. The individual flowers of a head on definite stalks 2 mm. or more long; calyx completely glabrous (without hairs) or minutely hairy, not becoming enlarged or inflated, or only slightly so after flowering; nerves of calyx-tube not enlarged or honeycombed; stipules

1288 Crotalaria spectabilis

1289 • Trifolium pratense var. pratense f. pratense (Red Clover)

1289 □ Trifolium pratense var. pratense f. leucochraceum

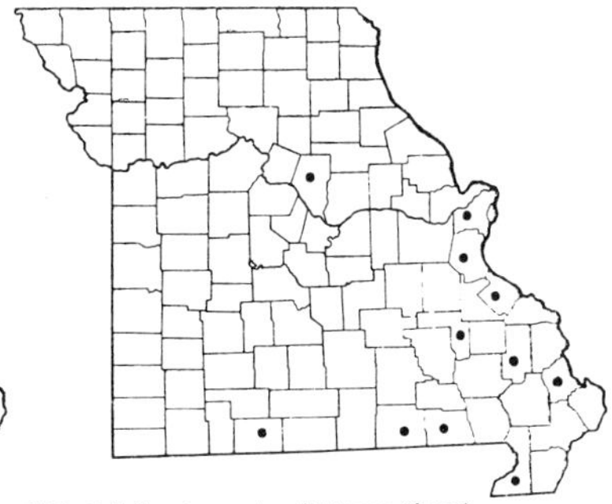

1290 Trifolium incarnatum (Crimson Clover)

(at base of leaf-stalk) 2–10 mm. wide in the free portion not united with the leaf-stalk . . . . *j*

j. Stems widely creeping or with long basal runners; stems usually rooting at the nodes; all the leaves on long stalks (petioles) . . . . . . . . . . . . . . . . . *k*

k. Commonly found plant; stipules (at base of leaf-stalk) usually less than 1 cm. long, pale and thin; calyx-teeth equaling or shorter than the calyx-tube; peduncle (flower-stem) rising from the ground level from creeping stems lying along the ground . 5. T. REPENS

k. Rarely found; stipules (at base of leaf-stalk) 1–2 cm. long, green and leaf-like; calyx-teeth $1\frac{1}{2}$–3 times as long as the calyx-tube; peduncle (flower-stem) rising from ascending stems above ground level . . . . . . . . . . . 8. T. STOLONIFERUM

j. Stems erect or ascending; stems not rooting at the nodes; upper leaves of the stem on usually shorter leaf-stalks (petioles) than the leaves at or near the base of the plant . . . *l*

l. Calyx 3–5 mm. long, the calyx-teeth 2–3 mm. long and about equaling to about $1\frac{1}{2}$ times as long as calyx-tube; stipules 3–8 times as long as wide . . . . 6. T. HYBRIDUM

l. Calyx 6–8 mm. long, the calyx-teeth 3–4 mm. long and 2–3 times as long as calyx-tube; stipules $1\frac{1}{2}$–$2\frac{1}{2}$ times as long as wide . . . . . . . . . . . 9. T. REFLEXUM

1. **Trifolium pratense** L. Red Clover Map 1289

Flowers May–September.

Occurs in fields, meadows, waste and cultivated ground, along roadsides and railroads. Throughout Missouri.

Native of Europe; introduced and naturalized in North America, from Labrador to British Columbia southward throughout the United States.

The following variations are represented in Missouri material:

a. Flowers creamy-white . . . 1b. T. PRATENSE var. PRATENSE f. LEUCOCHRACEUM

a. Flowers rose-colored or rose-lavender . . . *b*

b. Larger leaflets 1–3 cm. long, 0.5–1.5 cm. wide; heads of flowers 1.2–3 cm. long; stems 1–5 dm. tall . . . 1a. T. PRATENSE var. PRATENSE f. PRATENSE

b. Larger leaflets 3–7 cm. long, 1.5–3.5 cm. wide; heads of flowers 3–4 cm. long; stems 4–8 dm. tall . 1c. T. PRATENSE var. SATIVUM

1a. **Trifolium pratense** var. **pratense** f. **pratense** Map 1289

*Trifolium pratense* L. [G, BB, P & S, Steyerm.]

This is less common than the var. *sativum*, but is found throughout the state.

1b. **Trifolium pratense** var. **pratense** f. **leucochraceum** Aschers. & Prantl. Map 1289

Known only from Ozark County, southern Missouri (along Caney Creek, Caney Mountain Refuge, T23N, R13W, sect. 8, $5\frac{1}{2}$ mi. north of Gainesville, July 23, 1949, *Steyermark 68576*), but doubtless to be found in many other parts of the state.

1c. **Trifolium pratense** var. **sativum** Schreb. Cultivated Red Clover Map 1289

This is the commoner variation found in the state and is the one usually sown for pasture and fodder. It is much coarser than var. *pratense* with the stems from 4–8 dm. tall. No attempt has been made on the distribution map to separate var. *pratense* f. *pratense* and var. *sativum*. A white-flowered form of var. *sativum* is known as f. *flavicans* (Vis.) Hayek, but has not yet been recorded from Missouri.

After alfalfa, red clover is considered to be the most important forage legume crop in the United States. It is often grown along with timothy and

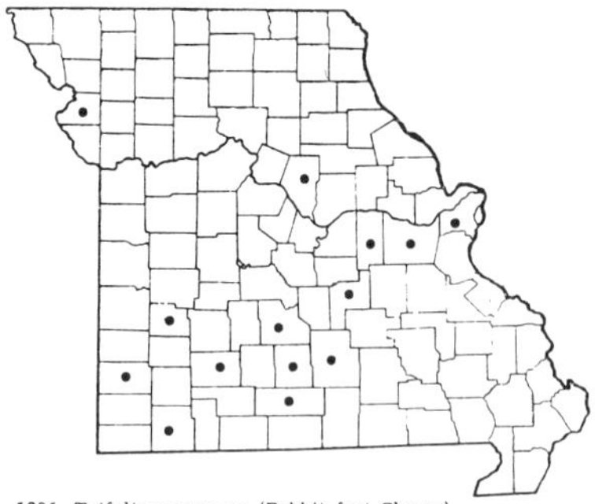

1291 Trifolium arvense (Rabbit-foot Clover)

1292 Trifolium resupinatum (Persian Clover)

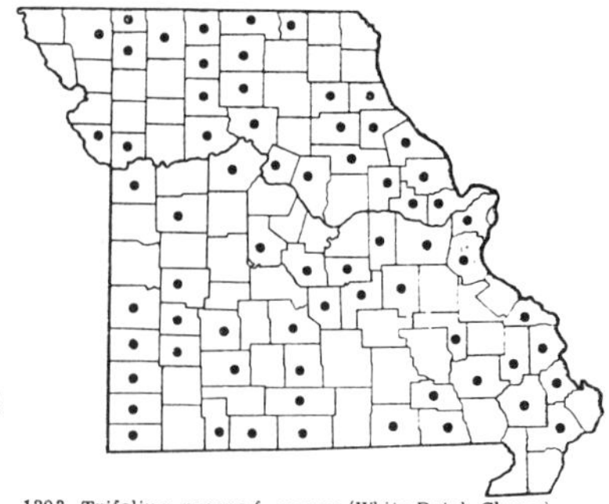

1293 Trifolium repens f. repens (White Dutch Clover)

other forage crops, or by itself. Most of the crop is grown in the north-central, northeastern, and Pacific northwest states. Two chief forms of Red Clover, Medium Red and American Mammoth, are grown in this country.

2. **Trifolium incarnatum** L. Crimson Clover Map 1290

*Trifolium incarnatum* var. *elatius* Gibelli & Belli [G]

Flowers May–July.

Occurs in fields and waste ground. Scattered and rare in southern and central Missouri west to Boone and Taney counties and north to St. Louis and Boone counties.

Native of Europe; introduced and naturalized in the United States from Florida to Texas, north to Virginia, Maine, southern Illinois, Missouri, and Oklahoma.

This is planted for early pasture, hay, and soil improvement in the southern United States, and most commonly grown from New Jersey south to Georgia, west to Mississippi and Kentucky, and to a lesser extent in the Pacific northwest states. It is not hardy in the northern states.

3. **Trifolium arvense** L. Rabbit-foot Clover Map 1291

Flowers May–October.

Occurs in waste ground, pastures, fallow fields, along roadsides and railroads. Infrequent and scattered in southern and central Missouri north to St. Louis, Boone, and Buchanan counties.

Native of Europe; introduced and naturalized in North America from Quebec and Ontario to Minnesota, south to Florida and Texas, and from British Columbia south to Oregon; also in Africa, Asia, and Hawaii.

A white-flowered form, known as f. *albiflorum* Sylven, has not been recorded from Missouri up to the present time.

4. **Trifolium resupinatum** L. Persian Clover Map 1292

Also called Reversed Clover.

Flowers May–September.

Occurs in grassy lawns and along railroads. Known only from St. Louis and Johnson (lawns along sidewalk, Warrensburg, July 5, 1930, *C. G. Schmitt*, in Univ. of Mo. Herb.) counties.

Native of Europe, Africa, and Asia; introduced and naturalized in the United States from Massachusetts to Illinois, Wisconsin, and South Dakota, south to Alabama, Kansas, Oklahoma, and Texas.

The pinkish-purple flowers are 4–7 mm. long, and twisted in such a way that the standard appears ventral.

In the southern United States this clover is a minor forage crop. It is regarded as having a high food value for hay and pasture and for its use as green manure. It is considered of value for early spring pasture.

5. **Trifolium repens** L. White Clover Map 1293

Flowers May–October.

Occurs in lawns, fields, pastures, waste ground, along roadsides and railroads. Common throughout the state and doubtless in every county.

Native of Europe and Asia; introduced and naturalized throughout North America.

The two main variations found in Missouri are differentiated as follows:

Flowers 6–10 mm. long; heads of flowers 1.5–2 cm. in diameter; leaflets 1–3 cm. long, 0.75–2.5 cm. wide. . . . . . . 5a. T. REPENS f. REPENS
Flowers about 10 mm. long; heads of flowers 3 cm. in diameter; leaflets 4–15 cm. long, 3–12.5 cm. wide. . . . . . 5b. T. REPENS f. GIGANTEUM

5a. **Trifolium repens** f. **repens**

White Dutch Clover Map 1293

*Trifolium repens* L. [G, BB, P & S, Steyerm.]

This is the common form found in Missouri.

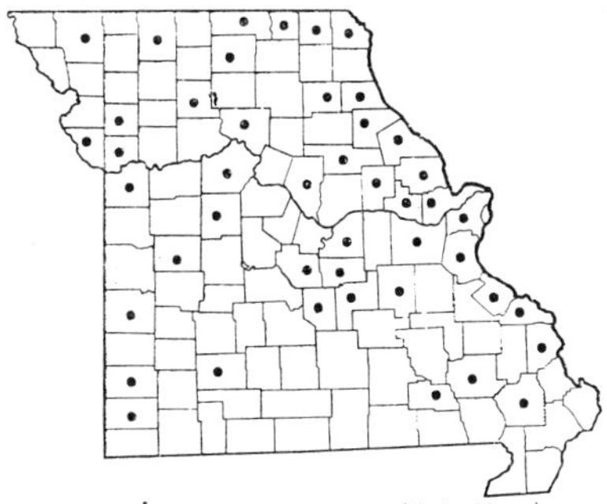

1294 Trifolium hybridum var. pratense (Alsike Clover)

1295 Trifolium carolinianum

1296 Trifolium stoloniferum (Running Buffalo Clover)

5b. **Trifolium repens** f. **giganteum** Lagr.-Foss.
Ladino Clover Map 1293

This is a giant form which originated from Lodi, Italy, and was introduced about 1900 into the northeastern states. It has been seen in various parts of Missouri in cultivation.

Various other strains of White Clover are in cultivation, differing from one another in size. The English White Clovers are the smallest and lowest in size, while the common White or White Dutch Clover is intermediate in size. The Ladino Clover is the largest.

White Clover is commonly used in mixtures for sowing pastures and also used in lawn seed. It is grown for its palatable hay and silage value. It is also employed for soil improvement, erosion control, and food for wildlife. Clover honey is largely derived from this species and Red Clover.

The seeds and dried flowers of this species can be made into a type of bread said to be nutritious and edible. A tea can also be prepared from an infusion of the dried heads of flowers and is reputed to have some medicinal value.

6. **Trifolium hybridum** L. var. **pratense** Rabenh.
Alsike Clover Map 1294

*Trifolium hybridum* var. *elegans* (Savi) Boiss. [G]

*Trifolium hybridum* of auth. in part [BB, P & S, Steyerm.]

Flowers May–October.

Occurs in fields, pastures, waste ground, and along roadsides and railroads. Common throughout Missouri and doubtless in every county.

Native of Europe and Asia; introduced and naturalized throughout North America.

Typical *T. hybridum* var. *hybridum* is cultivated in North America, but has spread southward only into the northern United States from Newfoundland to British Columbia, and is not represented in Missouri material. It is larger in all respects than var. *pratense* with leaflets 1–6 cm. long (1–3.2 cm. long in var. *pratense*), 1–3 cm. wide (0.7–2.5 cm. wide in var. *pratense*), heads 30–50-flowered (about 30-flowered in var. *pratense*), 2–3.5 cm. in diameter (1.25–2.5 cm. in diameter in var. *pratense*), and flowers 6–11 mm. long (5–7 mm. long in var. *pratense*).

Alsike Clover is sown as a pasture and fodder plant primarily as a substitute for red clover in wet or sour soils. It is especially grown in the north-central, eastern, and Pacific northwest states. It is also a source of honey for clover-honey.

Some individuals get a dermatitis from handling this clover, and cases of photosensitization in domestic animals are reported to be caused by this species in Europe.

7. **Trifolium carolinianum** Michx. Map 1295

Flowers April–June.

Occurs in glades, rocky prairies, and along railroads. Known only from Jasper (limestone barrens near Webb City, May 5, 1920, *Palmer 17355*) and Newton counties in southwestern Missouri, and introduced along railroads in St. Louis County (*Muehlenbach 195*).

Ranges from Florida to Texas, north to South Carolina, Missouri, and Kansas.

The flowers are purplish.

8. **Trifolium stoloniferum** Eat.
Running Buffalo Clover Map 1296

Flowers May–August.

Occurs in open woodland and prairies.

Rare and scattered in southern and central Missouri in Wayne, Jasper, St. Louis, and Boone counties.

Ranges from West Virginia to South Dakota, south to Ohio, Kentucky, Missouri, Kansas, and Oklahoma.

The flowers are usually white tinged with lavender.

This species may be confused with *T. reflexum* var. *glabrum*, both species having the calyx teeth with or without hairs either in the sinuses or at the summit. The

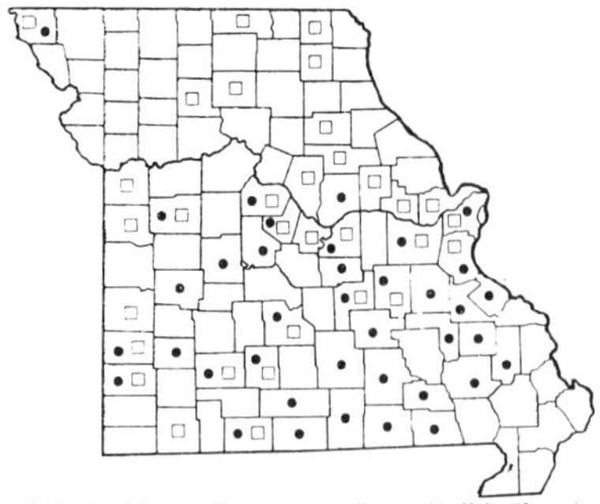

1297 • Trifolium reflexum var. reflexum (Buffalo Clover)
1297 □ Trifolium reflexum var. glabrum

1298 Trifolium agrarium (Yellow Clover)

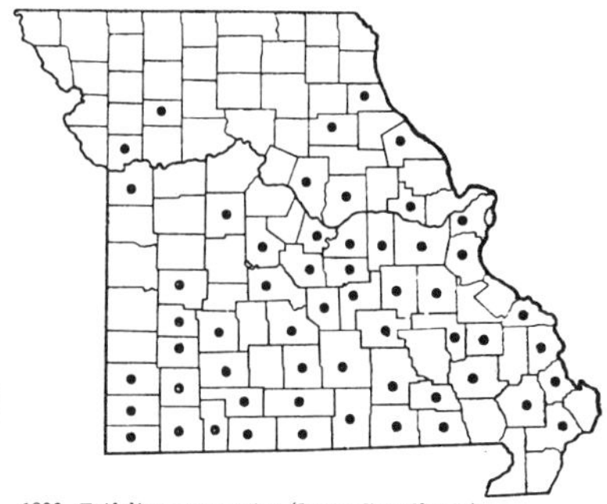

1299 Trifolium campestre (Large Hop Clover)

two may be distinguished by the differences presented in the key.

9. **Trifolium reflexum** L. Buffalo Clover
Map 1297

Flowers May–August.

Occurs in rocky open woods, glades, fallow fields, and prairies, usually in acid soils.

The two variations in Missouri are distinguished as follows:

Stem sparsely to densely hairy; calyx-tube or teeth, or both, more or less hairy (rarely calyx glabrous) . . . 9a. T. REFLEXUM var. REFLEXUM
Stem entirely glabrous; calyx entirely glabrous . 9b. T. REFLEXUM var. GLABRUM

9a. **Trifolium reflexum** var. **reflexum** Map 1297
*Trifolium reflexum* L. [G, P & S, Steyerm.]

Occurs mainly in the Ozark section of southern and central Missouri north to St. Louis, Callaway, Boone, Cooper, Johnson, and Barton counties, and locally northwest in Atchison County.

Ranges from Florida to Texas, north to Virginia, Kentucky, Illinois, Missouri, and Kansas.

9b. **Trifolium reflexum** var. **glabrum** Lojacono
Map 1297
*Trifolium reflexum* f. *glabrum* (Loj.) Isely [Isely]

Occurs mainly outside the Ozark region in northern, central, and western Missouri north and west of a line drawn from Jefferson, Franklin, Crawford, Phelps, Laclede, Webster, and Greene counties to Taney County.

Ranges from Alabama to Oklahoma, north to New York, Ontario, Indiana, Illinois, Iowa, and South Dakota.

It is true, as Dr. Isely maintains (Iowa State College Jour. Sci. 25: 475. 1951), that intergradations occur among plants of *T. reflexum* ranging from stems and calyces completely pubescent to those with stems and calyces completely glabrous. Nevertheless, study of Missouri material shows that the completely glabrous plants are more or less restricted to the southern half of the state, especially in the Ozark region, whereas the plants having stems and calyces more or less pubescent are found mostly outside the Ozark section and well into the glaciated northern part of the state. Since the glabrous or glabrate type also has a more northerly range in the United States, it would appear to merit recognition as a good geographical variety rather than only a mere form, in which case no marked geographical range would be exhibited, which is not the situation here.

The flowers of Buffalo Clover have white or pink wings and keel and purplish-red or rose-red standard. After flowering, the corollas turn brownish but persist, the pedicels recurving or turning down in age and bearing the dried, brown papery corollas.

10. **Trifolium agrarium** L. Yellow Clover
Map 1298

Also called Hop Clover, Field Hop Clover.

Flowers June–September.

Occurs in fields and waste ground.

Known only from St. Charles (O'Fallon, June 20, 1903, *E. K. Harris* in Univ. Mo. Herb.) and Christian counties.

Native of Europe and Asia; introduced and natu-

Plate no. 211. **1.** Trifolium repens, × 2/7. **2.** Trifolium carolinianum, × 2/7; *a.* Flower, × 1 3/7. **3.** Trifolium hybridum var. pratense, × 2/7. **4.** Trifolium stoloniferum, × 2/7. **5.** Trifolium campestre, × 2/7. **6.** Trifolium reflexum, × 2/7. **7.** Trifolium dubium, × 2/7; *a.* Leaf, × 1 1/7; *b.* Flower, × 5 5/7. **8.** Arachis hypogaea; Details from Small, The New York Botanical Garden. **9.** Canavalia ensiformis; Details from Small, The New York Botanical Garden.

PLATE NO. 211

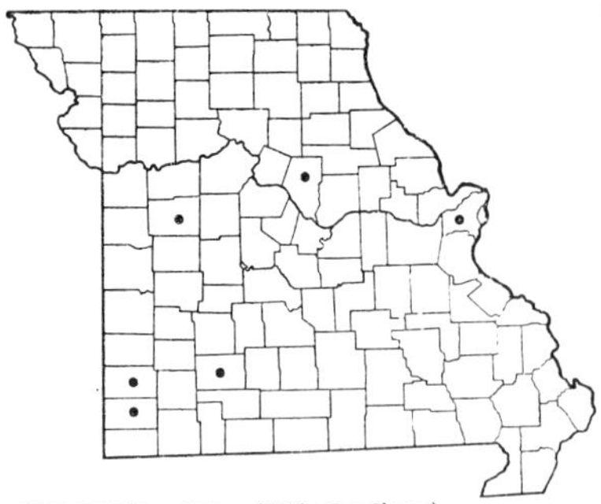

1300 Trifolium dubium (Little Hop Clover)

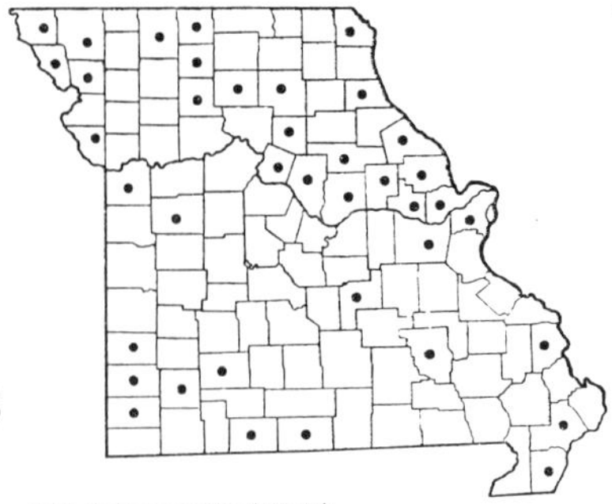

1301 Medicago sativa (Alfalfa)

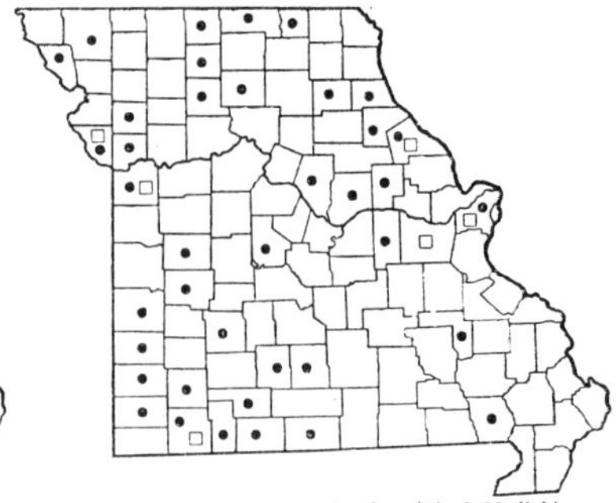

1302 ● Medicago lupulina var. lupulina (Black Medick)
1302 □ Medicago lupulina var. glandulosa

ralized in North America from Newfoundland to British Columbia, south to South Carolina and Missouri.

This species has yellow flowers.

11. **Trifolium campestre** Schreb.

Large Hop Clover Map 1299

Also called Low Hop Clover, Bishop Clover.

*Trifolium procumbens* of Am. authors and of L. Fl. Suec. ed. 2, 261. 1755 [G, BB, P & S, Steyerm.], not L. Sp. Pl. 1753 which is largely *T. dubium.*

Flowers May–September.

Occurs in fallow fields, pastures, rocky open ground, open woodland, along roadsides and railroads. Throughout southern and central Missouri extending north to Marion, Monroe, Boone, Caldwell, and Clay counties.

Native of Europe, Africa, and Asia; introduced and naturalized in North America from Quebec to North Dakota, south to Georgia, Mississippi, Arkansas, and Kansas, and on the Pacific coast.

This species has yellow flowers.

Dr. Hermann (USDA Monogr. 22: 9–10. 1953) has submitted evidence pointing to the use of *T. campestre* as the acceptable name for what has been passing as *T. procumbens.*

This clover is sown for its value as forage and for soil improvement in the southeastern and Pacific states. It provides forage in late winter and early spring in many sections where grown.

12. **Trifolium dubium** Sibth. Little Hop Clover Map 1300

Also called Small Hop Clover.

Flowers April–October.

Occurs in waste ground, border of wet low woodland, gravelly open places in valleys, limestone glades, rocky open woods, roadsides, and railroads. Rare and scattered in southern and central Missouri in Greene, Jasper, Newton, Dunklin, St. Louis, Boone, and Johnson counties.

Native of Europe and Asia; introduced and naturalized in North America, from Florida to Texas, north to Nova Scotia, New York, Ontario, Michigan, Wisconsin, and British Columbia.

This is grown to some extent in the southern United States, where it is employed for early spring pasturage. It is commonly planted in mixtures with White Clover. This species, like *T. agrarium* and *T. campestre*, has yellow flowers.

## 13. **Medicago** L. Medick

a. Flowers purple or blue-lavender, rarely white, 7–12 mm. long; plants perennial with upright or ascending stems; leaflets usually more than twice as long as broad, narrowly oblong to oblanceolate . . . 1. M. SATIVA

a. Flowers yellow, smaller, 2–5 mm. long; plants usually annual with low spreading, sprawling, to ascending stems; leaflets usually less than twice as long as broad, nearly round or broadly obovate to wedge-shaped or obcordate . . . . . . . . . . . . . . . . . . . . . . . . . . *b*

b. Stipules (at base of leaf-stalks) shallowly toothed or even entire (without teeth); fruit without spines. . . . . . . . . . . . . . . . . . . . . . . . . . 2. M. LUPULINA

b. Stipules (at base of leaf-stalks) deeply cut and lacerate (jagged-edged); fruit usually with spines or prickles . . . . . . . . . . . . . . . . . . . . . . . . . . *c*

c. Upper surface of leaflet frequently with a dark (brown or purple-red) blotch in the center; middle (terminal) one of the three leaflets with a stalk only slightly longer than the stalks of the other leaflets; base of peduncle (stem supporting inflorescence) with loose spreading hairs;

stipules of upper leaves cut less than half-way to the base . . . . . . . . . 3. M. ARABICA
c. Upper surface of leaflet without a dark spot in center; middle (terminal) one of the three leaflets with a stalk 2–4 mm. long, noticeably longer than the stalks of the other leaflets; base of peduncle glabrous (without hairs) or with a few short straight hairs parallel to length of peduncle; stipules of upper leaves cut in more than half-way to the base . . . . . .
4. M. POLYMORPHA var. VULGARIS

1. **Medicago sativa** L. f. **sativa** Alfalfa Map 1301

Flowers May–September.

Commonly cultivated and escaped to roadsides, waste ground, and railroads. Throughout Missouri and doubtless in every county.

Native of the Old World, probably in Asia; also known in Africa; introduced and naturalized throughout the United States and in much of Canada.

A form with white flowers, known as f. *alba* Benke, has not yet been recorded for Missouri. There is variation in size and color of flowers, habit of plant, and degree of coiling of the pods.

Alfalfa is the number 1 legume forage crop in the United States grown for hay and pasturage. It is also considered the most valuable legume for soil improvement and is used in silage and feed concentrates. It is one of three commercial sources for the extraction of Carotine, a vitamin A synthetic product, and in the form of a meal is added to many of the prepared baby foods and cage bird foods as a source of vitamin A. Most of the seed production of it is in the western states, but it is widely used throughout the north-central and eastern United States. Due to the greater amount of humidity and high summer temperature, it is scarcely cultivated in the southeastern states.

Cases are on record of persons who have suffered hay fever from sensitivity to alfalfa pollen.

2. **Medicago lupulina** L. Black Medick Map 1302

Flowers February–December.

Occurs in grassy lawns, fields, waste ground, along roadsides and railroads. Throughout Missouri.

Native of Eurasia and Africa, introduced and naturalized throughout the United States and much of Canada; also naturalized in Mexico, the West Indies, Guatemala, South America, and Hawaii.

Two variations occur in Missouri:

Peduncles (stalk supporting inflorescence) and fruits hairy or nearly without hairs (glabrate) .
2a. M. LUPULINA var. LUPULINA
Peduncles and fruits with stalked glands . . .
2b. M. LUPULINA var. GLANDULOSA

2a. **Medicago lupulina** var. **lupulina** Map 1302

*Medicago lupulina* L. [G, BB, P & S, Steyerm.]

This is the commoner variation found throughout Missouri.

2b. **Medicago lupulina** var. **glandulosa** Neilr. Map 1302

Scattered in parts of the state, and known at present from Pike, St. Louis, Franklin, Barry, Jackson, and Platte counties, but probably to be expected in every county.

I am in agreement with Shinners (Rh. 58: 1–2. 1956) that this glandular variation probably deserves no more recognition than form rather than variety.

Black Medick is frequently mistaken for two of the Yellow Hop Clovers, *Trifolium dubium* and especially *T. campestre*. From them it may be distinguished as follows:

Stems angled or square; stipules fused to petiole for less than half their length; flower clusters tightly flowered, mostly ovoid or higher than broad; flowers not reflexed; petals falling in fruit; fruit black, not enclosed within the perianth
MEDICAGO LUPULINA
Stems rounded or terete; stipules fused to petiole for half or more their length; flower clusters loosely flowered, broader than wide, spherical or becoming dome-shaped or higher than broad; flowers reflexed after blossoming; petals persistent into fruiting stage; fruit brown, enclosed within the persistent perianth . . . TRIFOLIUM DUBIUM and T. CAMPESTRE

Black Medick is usually an annual. A perennial strain is known as *M. lupulina* var. *cupaniana* (Guss.) Boiss. In Hawaii Black Medick is highly valued for forage, but in the continental United States it has not assumedimportance and is used only to a slight extent as a cover crop and for winter pasturage.

3. **Medicago arabica** (L.) Huds. Spotted Medick Map 1303

Also known as Spotted Bur Clover.

Flowers April–October.

Known only from New Madrid County, southeastern Missouri (open ground and roadways, 1½ mi. south of Sikeston, T25N, R14E, sect. 6, May 10, 1955, *Dan Neely*; open fields of Sikeston, Expt. Farm of Univ. of Mo., May 10, 1955, *Kucera*, in Univ. Mo. Herb.).

Native of the Old World; introduced and naturalized in North America from Mexico, Florida to Texas north to New Brunswick, New England,

1303 Medicago arabica (Spotted Medick)

1304 Medicago polymorpha var. vulgaris f. vulgaris (California Bur Clover)

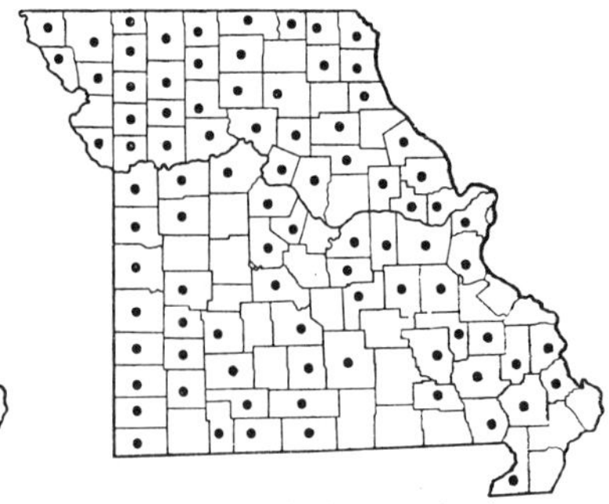

1305 Melilotus officinalis (Yellow Sweet Clover)

Illinois, Missouri, Oklahoma, and Washington; reported from North Dakota and sometimes cultivated in Nebraska.

This species is mostly found in the southern United States.

4. **Medicago polymorpha** L. var. **vulgaris** (Benth.) Shinners f. **vulgaris** Rh. 58: 310. 1956.

California Bur Clover Map 1304

Also called Bur Clover, Toothed Bur Clover.

*Medicago denticulata* Willd.

*Medicago denticulata* Willd. var. *vulgaris* Benth.

*Medicago polymorpha* var. *ciliaris* L. f. *ciliaris*

*Medicago polymorpha* var. *ciliaris* (Ser.) Shinners [Shinners in Rh. 58: 9. 1956]

*Medicago hispida* var. *denticulata* (Willd.) Urban [Isely]

*Medicago hispida* var. *oligogyra* f. *denticulata* (Willd.) Urban

Flowers April–October.

Occurs along railroads, where known only from St. Louis (Wabash R. R. along the most western track north of Palm Street, St. Louis, May 30, 1958, *Muehlenbach 1404*) and Jackson (June 15, 1891, *Bush*) counties.

Native of Eurasia; introduced and naturalized in North America from Florida and Texas, north to Quebec, Michigan, Nebraska, Montana, and British Columbia; also naturalized in Mexico, Guatemala, and South America.

This species has been recently studied by Shinners (Rh. 58: 1–13, 310. 1956), whose nomenclature is followed in the present flora. Numerous subspecies, varieties, and forms are known, of which the var. *vulgaris* f. *vulgaris* is here considered applicable to the commoner variation of plants having fruits with well developed prickles or spines and 2–3½ coils. Spineless forms [*M. polymorpha* var. *vulgaris* f. *tuberculata* (Godron) Shinners] and those with spines shorter than the radius of the spirals of the fruit [*M. polymorpha* var. *vulgaris* f. *apiculata* (Willd.) Shinners] have not been recorded from Missouri.

In California and some of the southwestern states this Bur Clover is highly esteemed for early spring pasturage, hay, green manure, and as a cover crop. As with *M. arabica* this species does best in areas where the winters are mild and moist.

## 14. **Melilotus** Mill. Sweet Clover

Flowers yellow, 5–7 mm. long; fruit conspicuously cross-ribbed and honeycombed; uppermost petal (standard) about as long as the wings . . . . . . . . . . . . . . . . . . 1. M. OFFICINALIS

Flowers white, 4–5.5 mm. long; fruit honeycombed-nerved (reticulate) or nearly smooth; uppermost petal (standard) longer than the wings . . . . . . . . . . . . . . . . . . . 2. M. ALBUS

Plate no. 212. **1.** Medicago sativa, × 2/7; *a.* Flower, × 2; Details from Small, The New York Botanical Garden. **2.** Medicago lupulina, × 2/7; Details from Small, The New York Botanical Garden. **3.** Medicago polymorpha var. vulgaris, × 2/7; *a.* Flower, × 1 3/7; *b.* Fruit, × 5; Details from Small, The New York Botanical Garden. **4.** Melilotus officinalis, × 2/7; Details from Small, The New York Botanical Garden. **5.** Melilotus albus, × 2/7; Details from Small, The New York Botanical Garden. **6.** Lotus Purshianus, × 2/7; Details from Small, The New York Botanical Garden. **7.** Anthyllis Vulneraria, × 2/7; *a.* Bract, × 6/7; After Gleason, The New York Botanical Garden. **8.** Psoralea psoralioides var. eglandulosa, × 2/7; Details from Small, The New York Botanical Garden. **9.** Lotus corniculatus, × 2/7; *a.* Fruiting cluster, × 2/7; Details from Small, The New York Botanical Garden. **10.** Psoralea Onobrychis, × 2/7.

PLATE NO. 212

1306 Melilotus albus (White Sweet Clover)

1307 Anthyllis Vulneraria (Lady's Fingers)

1308 Lotus corniculatus (Bird's-foot Trefoil)

1. **Melilotus officinalis** (L.) Lam.
Yellow Sweet Clover Map 1305

Also known as Yellow Melilot.

Flowers May–October.

Occurs along roadsides, railroads, fields, and waste places. Throughout Missouri, doubtless in every county.

Native probably of Asia; also found in Europe, Australia, and introduced in South America; introduced and naturalized in North America from Quebec to British Columbia, southward through much of the United States.

This species is highly regarded as a bee plant for honey production. It is also used as a green manure, for hay, and pasture. The coumarin contained in this clover may become toxic in improperly cured or spoiled hay when fed to domestic animals, but varieties of the plant lacking coumarin have been developed recently. For pasturage this legume is thought to cause less bloat among livestock than some other legumes. Its chief advantage over alfalfa and red clover lies in its ability to withstand dry sterile soils, and in this role it plays an important part, especially in the southeastern United States, in reclaiming poor or impoverished soils and at the same time being used for hay. It matures earlier and is a better seed producer than *M. alba*, the White Sweet Clover, and establishes a stand better under unfavorable, dry conditions than does *M. alba*.

2. **Melilotus albus** Desr. White Sweet Clover
Map 1306

Also known as White Melilot.

*Melilotus alba* Desr. [G, P & S, Steyerm.]

Flowers May–October.

Occurs along roadsides, railroads, fields, and waste ground. Throughout Missouri and doubtless in every county.

Native probably of Asia; also found in Europe, Australia, and introduced in South America; introduced and naturalized throughout the United States and much of Canada; also in Mexico and the West Indies.

Like Yellow Sweet Clover, this species is valued as a bee plant for honey production. It has similar value to *M. officinalis* as a green manure, and for grazing and fodder. It is stated to produce a somewhat higher yield than Yellow Sweet Clover and is, therefore, planted probably more often for hay or pasturage. This is also a bad weed in areas where not desired, and invades recently cleared or disturbed open ground. Along railroads where prairie vegetation thrives and persists, the presence of White Sweet Clover, along with the Yellow Sweet Clover, is an indication that the native flora has been disturbed. Where no disturbance has occurred, the *Melilotis* is absent. In the young growing stages, the leaves of alfalfa resemble those of both *Melilotus albus* and *officinalis*, but, when crushed, are without fragrance, whereas the crushed leaves of *Melilotus* are fragrant.

## 15. **Anthyllis** L.

**Anthyllis Vulneraria** L. Lady's Fingers
Map 1307

Flowers June–July.

Occurs in fields and waste ground. Recorded from Missouri in *Gray's Manual*, eighth edition, but the only specimen seen was one based upon a plant cultivated at the Missouri Botanical Garden, in St. Louis, collected by Mr. John H. Kellogg.

Native of Europe; introduced and naturalized in North America from Quebec and Ontario, south to New England, Pennsylvania, Ohio, Michigan, Missouri, and North Dakota.

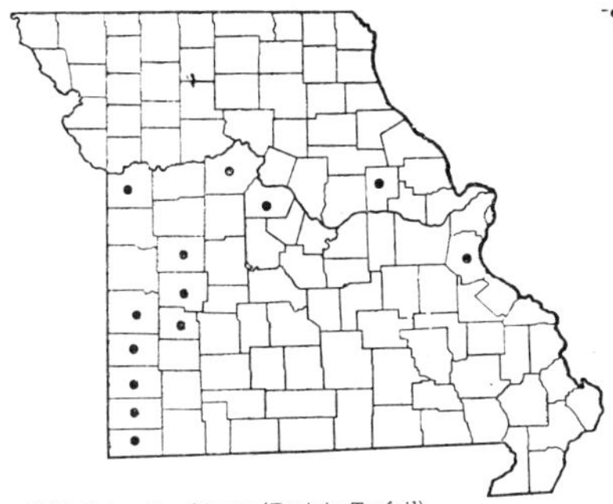
1309 Lotus Purshianus (Prairie Trefoil)

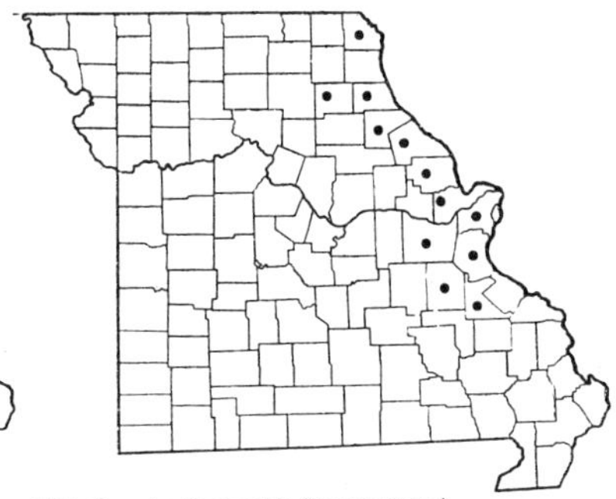
1310 Psoralea Onobrychis (French Grass)

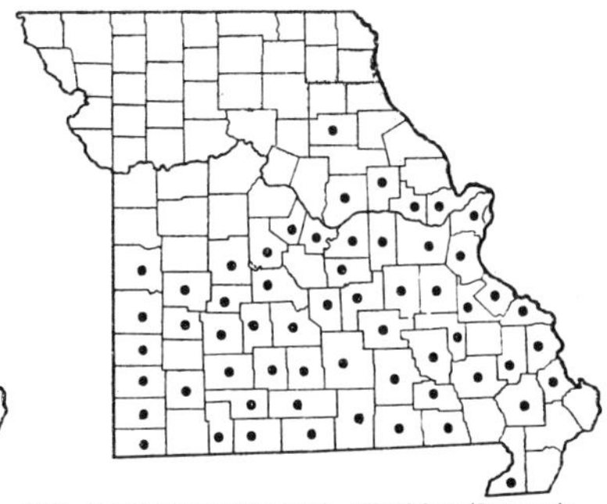
1311 Psoralea psoralioides var. eglandulosa (Sampson's Snakeroot)

This is sometimes grown as a forage plant in Europe and western Asia, but it is little used in North America. It is sometimes grown as an ornamental plant for its heads of showy, yellow flowers.

## 16. **Lotus** L. Trefoil

Flowers yellow; leaflets 5, the lowest pair resembling stipules; flowers several together in an umbel . . 1. L. CORNICULATUS

Flowers pink and red mixed with yellow and whitish; leaflets 3; flowers usually solitary with 1 on each stalk rising from the upper leaf axils . . . . . . . . . . . . . . . . . 2. L. PURSHIANUS

1. **Lotus corniculatus** L. Bird's-foot Trefoil Map 1308

Also known as Ground Honeysuckle.

Flowers May–September.

Occurs in fields, along roadsides, and in waste ground. Known only in central and southwestern Missouri in Boone, Clay, and Newton ($1\frac{1}{2}$ mi. southeast of Joplin, July 26, 1957, *Palmer 66172*) counties.

Native of Europe; also occurs in Asia, Africa, Australia, and New Zealand; introduced and naturalized in North America, from Newfoundland to Minnesota, south to Virginia, Ohio, and Missouri; also on the Pacific Coast.

This legume is sometimes cultivated for its pasturage value, succeeding on poor and dry soils where alfalfa and clover (*Trifolium*) do not flourish. It is believed likely that this species may have future importance as a pasture legume in this country.

2. **Lotus Purshianus** Clements and Clements Prairie Trefoil Map 1309

*Lotus americanus* (Nutt.) Bisch. [G, BB, P & S, Steyerm.], not *L. americanus* Vell.

Flowers late May–September.

Occurs in prairies, glades, open grassy limestone slopes, top of bluffs, along roadsides and railroads. Mainly in the unglaciated prairie region of western Missouri south of the Missouri River, and east in central Missouri to Callaway and Jefferson counties.

Ranges from Manitoba and Minnesota to British Columbia, south to Arkansas, Oklahoma, Texas, New Mexico, Arizona, and California.

## 17. **Psoralea** L. Scurf-pea

a. Middle (terminal) one of the three leaflets on a conspicuous stalk (petiolule) 2–several times longer than the stalks (petiolules) of the other two leaflets . . . . . . . . . . . . . . . . . *b*

b. Leaflets 2–6 cm. broad, long-pointed, lance-ovate; peduncles (stems bearing inflorescences) shorter than the leaves at their base; calyx, fruit, and bracts of the inflorescence without glandular dots . . . . . . . . . . . . . . . . . . . . . . . . . . . 1. P. ONOBRYCHIS

b. Leaflets 0.5–1.5 cm. broad, blunt or obtuse at tip, narrowly oblong to lanceolate; peduncles much longer than the leaves at their base; calyx, fruit, and bracts of the inflorescence often with glandular dots . . . . . . . . . . . . . . . . . . . . 2. P. PSORALIOIDES var. EGLANDULOSA

a. All the leaflets either sessile (without petiolules) or on stalks (petiolules) of about equal length

(palmately compound or arising at about the same level from the top of the main petiole [stalk] of the leaf) . . . . . . . . . . . . . . . . . . . . . . . . . . . . . . . . *c*
c. Stems covered with conspicuous spreading hairs; flowers showy, 16–20 mm. long . . . 5. P. ESCULENTA
c. Stems either covered with a minute hairiness to nearly smooth or with a silvery silky type of hairiness, but not with conspicuous spreading hairs; flowers smaller, 5–10 mm. long . . . . . . *d*
d. Common plants, occurring nearly throughout Missouri; stem covered with a minute hairiness or nearly smooth; calyx 2–3 mm. long . . . . . . . . . . . . . . 3. P. TENUIFLORA
d. Rare plants, known only from the loess hills of northwestern Missouri and in Pike and Jackson counties; stem and leaves covered with a white silky hairiness; calyx (from base of tube to end of longest lobe) 8–10 mm. long . . . . . . . . . . . . . . . . . 4. P. ARGOPHYLLA

1. **Psoralea Onobrychis** Nutt. French Grass Map 1310

Also called Sainfoin Psoralea.

Flowers May–September.

Occurs on rich wooded slopes, alluvial ground in low woodland, river banks, low open thickets in valleys, and low open ground. Eastern Missouri from Clark County south to St. Francois County, west to Shelby, Franklin, and Washington counties.

Ranges from Ohio to Illinois and Iowa, south to West Virginia, Kentucky, Tennessee, and Missouri.

2. **Psoralea psoralioides** (Walt.) Cory var. **eglandulosa** (Ell.) Freeman Sampson's Snakeroot Map 1311

*Psoralea pedunculata* [of P & S], not (Mill.) Vail

Flowers May–July.

Occurs in acid soils associated with chert, sandstone, or granitic rocks in rocky open woods, open rocky exposures and glades, and prairies. Southern and central Missouri where restricted to the Ozark and unglaciated prairie regions, north to St. Charles, Warren, Montgomery, Callaway, Cole, Moniteau, Morgan, Benton, St. Clair, and Bates counties; reported from Monroe County.

Ranges from Georgia to Texas, north to Virginia, Ohio, Indiana, Illinois, Missouri, Kansas, and Oklahoma.

3. **Psoralea tenuiflora** Pursh Scurfy Pea Map 1312

Also known as Few-flower Psoralea, Wild Alfalfa.

Flowers May–September.

Occurs on rocky open glades, prairies, and in open woods, usually in limestone soils.

The following variations occur in Missouri:

a. Petals white . . . . . . . 3b. P. TENUIFLORA var. TENUIFLORA f. ALBA
a. Petals blue-lavender . . . . . . . . *b*
b. Common type encountered; flowers mostly 1 or 2 at a node, about 5 mm. long; calyx 2–2.5 mm. long . . . . 3a. P. TENUIFLORA var. TENUIFLORA f. TENUIFLORA
b. Rare type found only in west-central and southwestern Missouri; flowers usually 2–4 at a node, 6–7 mm. long; calyx 3 mm. long 3c. P. TENUIFLORA var. FLORIBUNDA

3a. **Psoralea tenuiflora** var. **tenuiflora** f. **tenuiflora** Map 1312

*Psoralea tenuiflora* Pursh [G, BB, P & S, Steyerm.]

This is the commonest variation in Missouri, occurring throughout northern, central, and western Missouri, but mainly absent from most of the central, eastern, and southern Ozark region and from the lowland counties of southeastern Missouri.

Ranges from Indiana to Nebraska, south to Missouri, Texas, New Mexico, and Arizona.

3b. **Psoralea tenuiflora** var. **tenuiflora** f. **alba** Steyerm. Map 1312

*Psoralea tenuiflora* f. *alba* Steyerm. [G]

Known only from Benton County in the northwestern Ozark section (limestone glade on top of east-facing bluffs along Osage River, T40N, R23W, sect. 13, 2½ mi. west of Warsaw, June 3, 1938, *Steyermark 5779*).

So far as known, occurring only in Missouri.

3c. **Psoralea tenuiflora** var. **floribunda** (Nutt.) Rydb. Map 1312

Known only from west-central and southwestern Missouri in Jackson (Sheffield, June 29, 1894, *Bush 215*; Independence, June 4, 1894, *Bush 210*), and Bates (Amoret, May 15, 1936, *Bush 15504*), and Jasper (3 mi. west of Neck City, May 7, 1911, *Palmer 3367*) counties.

Ranges from Illinois to Minnesota and Nebraska, south to Arkansas and Texas.

Intergradation occurs between typical var. *tenuiflora* and var. *floribunda*, some specimens (*Kellogg 327* from Jerome, Phelps County) having the more numerous clusters of flowers of var. *floribunda* but with the shorter dimensions of var. *tenuiflora*. Although many specimens are labeled var. *floribunda*, that variety appears to be very rare and confined to the western

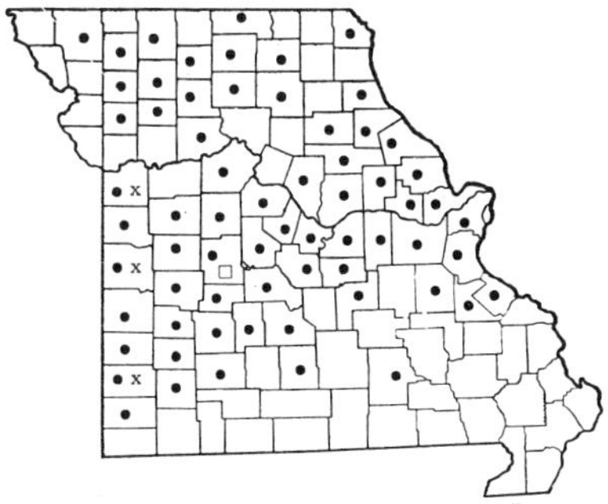

1312 • Psoralea tenuiflora var. tenuiflora f. tenuiflora (Scurfy Pea)
1312 □ Psoralea tenuiflora var. tenuiflora f. alba
1312 x Psoralea tenuiflora var. floribunda

1313 Psoralea argophylla (Silvery Psoralea)

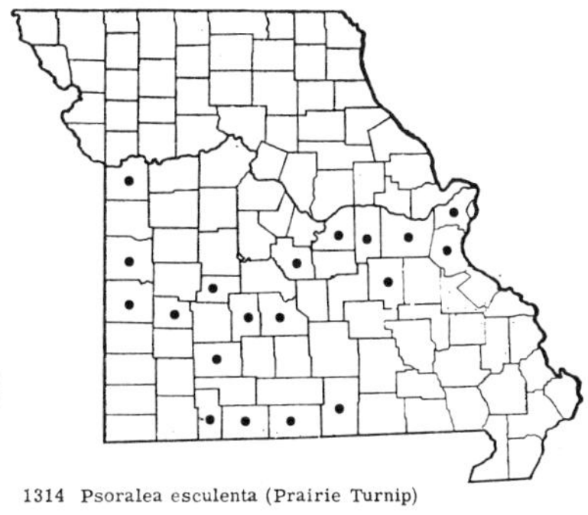

1314 Psoralea esculenta (Prairie Turnip)

borders of the state. Although the leaflets are described as 5–8 mm. broad, they attain a width of 17 mm. on some Missouri specimens.

*Psoralea tenuiflora* is considered to be mildly poisonous to cattle, but since it is seldom eaten in quantity, it is of not much importance in this respect.

4. **Psoralea argophylla** Pursh Silvery Psoralea
Map 1313

Flowers June–August.

Occurs on loess hills, rocky bluffs, prairie areas, and open valleys. In northwestern Missouri confined to the loess hills of Atchison and Holt counties and to open river valleys in Nodaway County (thickets in valley of 102 River, T65N, R33W, sect. 22, 2½ mi. southeast of Pickering, June 19, 1938, *Steyermark 5837*); also in Jackson County, and locally in east-central Missouri in Ralls County (rocky bluffs, Oakwood, September 15, 1916, *Davis 3593*).

Ranges from Wisconsin to Saskatchewan, south to Missouri, Oklahoma, Wyoming, Colorado, and New Mexico.

This is a poisonous plant if eaten by stock, but is rarely eaten. Children are reported to have been poisoned from eating the seeds.

The leaflets vary from 3–5.

5. **Psoralea esculenta** Pursh Prairie Turnip
Map 1314

Also known as Indian Bread-root, Wild Potato, Pomme de Prairie.

Flowers late April–July.

Occurs on rocky open limestone glades, prairies, and rocky open hillsides. Rare and local in southern and central Missouri, where scattered in portions of the northern and western Ozarks and unglaciated prairie region north to St. Louis, Franklin, Gasconade, Osage, Miller, Hickory, and Jackson counties.

Ranges from Manitoba and Wisconsin to Alberta, south to Missouri, Texas, Colorado, and New Mexico.

The starchy root in the raw state has a turniplike flavor. By the Indians and early travelers of the western states the root was cooked as a substitute for potatoes, and is stated to be nutritious and of good flavor.

*Excluded Species*

**Psoralea cuspidata** Pursh

A sheet in the Missouri Botanical Garden Herbarium determined as this species (rocky prairies, 3 mi. west of Neck City, Jasper County, May 7, 1911, *Palmer 3367*) is a misidentified specimen of *P. tenuiflora* var. *floribunda*. The broad leaflets of this specimen are found in both *P. cuspidata* and *P. tenuiflora*, but the leaflets do not possess the narrowed or cuspidate apex characteristic of *P. cuspidata*. The specimen is in a young bud stage and well matches other immature flowering specimens of *P. tenuiflora* var. *floribunda*.

## 18. **Dalea** Willd.

Leaflets usually 19–35 (as few as 13); spikes densely flowered, the flowers overlapping one another; bracts of inflorescence quickly falling from inflorescence; petals rose-colored, pale lilac, or whitish . 1. D. ALOPECUROIDES

Leaflets 5–11 (up to 13); spikes loosely flowered, open with axis of inflorescence visible, the flowers rather widely separated; bracts of inflorescence persistent; petals white and yellowish . . 2. D. ENNEANDRA

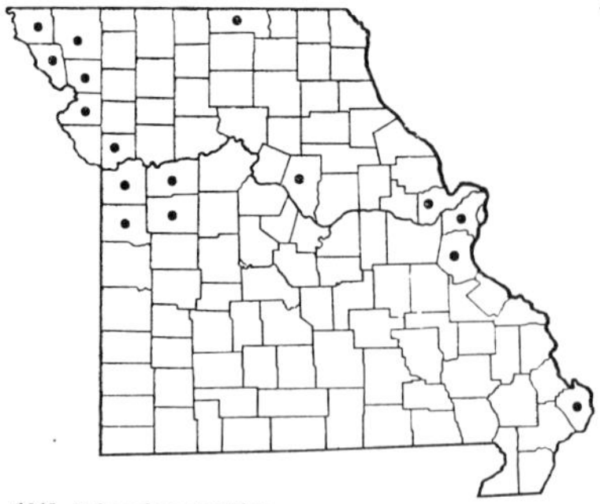
1315 Dalea alopecuroides

1316 Dalea enneandra

1317 Petalostemon multiflorum

1. **Dalea alopecuroides** Willd. Map 1315

Flowers July–September.

Occurs in sandy and wet alluvial open ground along rivers and borders of oxbow lakes, loess hills, and along railroads and waste ground. Mainly following the Missouri River counties from northwestern Missouri in Nodaway and Atchison counties south to Jackson and Bates counties, thence east to Boone and St. Charles counties, and south along the Mississippi River to St. Louis, Jefferson, and Mississippi counties.

Ranges from Indiana, Illinois, Wisconsin, and Minnesota to South Dakota, south to Missouri, Colorado, and New Mexico; introduced in Alabama.

The flowers vary from rose to pale lilac or whitish. The plant is an annual.

2. **Dalea enneandra** Nutt. Map 1316

*Dalea laxiflora* Pursh [Shinners, Field and Lab. 17: 86–87. 1949]

Flowers June–August.

Occurs only on loess hills of Atchison and Holt counties, in extreme northwestern Missouri.

Ranges from Iowa to North Dakota and Colorado, south to Missouri, Oklahoma, Texas, and New Mexico.

The plant is a perennial. The petals of the corolla have a pale yellow keel, and white wings and standard.

## 19. **Petalostemon** Michx. Prairie Clover

(Dalea in part [Shinners])

The genus is often spelled *Petalostemum.* The present treatment follows the spelling *Petalostemon* as adopted in the *Nomina Generica Conservanda.*

This genus is usually maintained as distinct from *Dalea* by most of the previous botanists who have studied this family. Shinners has submitted some evidence to indicate that the two genera might be united under *Dalea* (Field and Lab. 17: 81–89. 1949), such species as *Petalostemon multiflorum* combining certain characters of *Dalea* and of *Petalostemum.* Isely, however, in his most recent studies (Iowa State College. Jour. Sci. 33: 29–30. 1958) maintains both genera as distinct. Until more definite agreement has been reached on this matter, pending monographic work on the two genera, both genera are retained in the present flora in their orthodox sense. I am obliged to Dr. Isely for allowing me the use of certain data obtained from studies on this genus made by him and Dr. Welch.

a. Petals violet or rose-purple; calyx-tube densely velvety hairy; bracts of inflorescence densely hairy on outer surface . . . . . . . . . . . . . . . . . . . . . . . . 3. P. PURPUREUM

Plate no. 213. **1.** Psoralea tenuiflora var. tenuiflora, × 2/7; Details from Small, The New York Botanical Garden. **2.** Dalea enneandra, × 2/7; *a.* Calyx, × 1 3/7; *b.* Staminal cluster, × 1 3/7; After Britton and Brown, The New York Botanical Garden. **3.** Psoralea esculenta, × 2/7. **4.** Dalea alopecuroides, × 2/7; *a.* Calyx with fruit, × 1 3/7; After Britton and Brown, The New York Botanical Garden. **5.** Petalostemon candidum, × 2/7; Details from Small, The New York Botanical Garden. **6.** Onobrychis viciaefolia, × 2/7; *a.* Flower, × 6; After Gleason, The New York Botanical Garden. **7.** Petalostemon purpureum, × 2/7. **8.** Amorpha canescens, × 2/7; Details from Small, The New York Botanical Garden. **9.** Amorpha fruticosa var. angustifolia, × 2/7; *a.* Leaf of Amorpha fruticosa var. tennesseensis, × 2/7; *b.* Leaf of Amorpha fruticosa var. oblongifolia, × 2/7.

PLATE NO. 213

1318 Petalostemon candidum (White Prairie Clover)

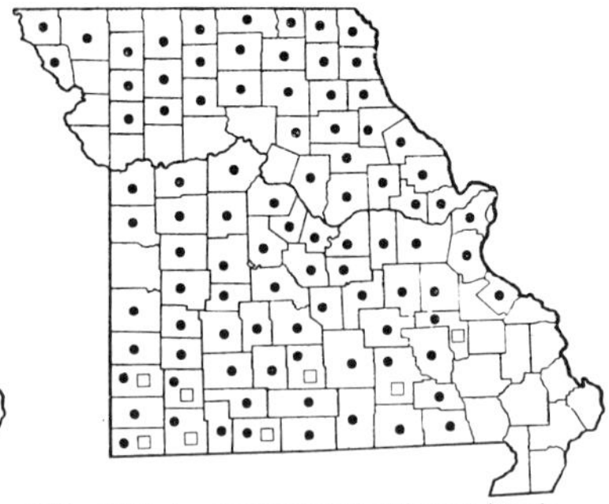

1319 • Petalostemon purpureum f. purpureum (Purple Prairie Clover)
1319 □ Petalostemon purpureum f. pubescens

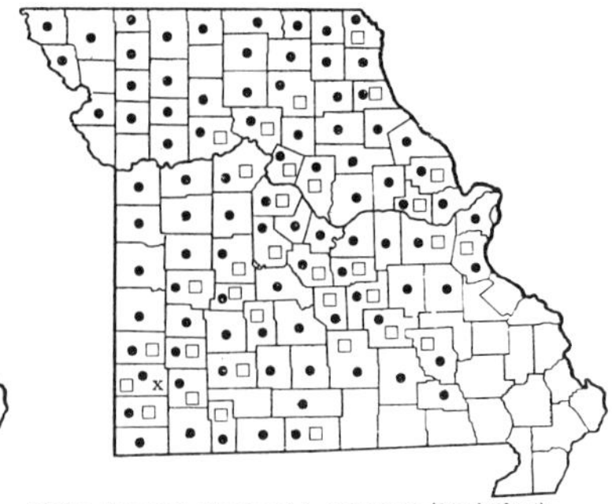

1320 • Amorpha canescens f. canescens (Lead Plant)
1320 □ Amorpha canescens f. glabrata
1320 x Amorpha X notha (Amorpha canescens X fruticosa)

a. Petals white; calyx-tube glabrous (without hairs) or nearly so; bracts of inflorescence glabrous or nearly so on the outer surface . . . . . . . . . . . . . . . . . . . . . . *b*

b. Fully-developed flower-spikes cylindrical, 1.4–7 cm. in flowering stage, 2–5 times as long as thick; bract of flower about equaling or longer than the calyx; common species . . . . . 2. P. CANDIDUM

b. Fully-developed flower-spikes globose to shortly ovoid, 0.7–1.6 cm. long in flowering stage; bract of flower $\frac{1}{3}$–$\frac{2}{3}$ as long as the calyx; rare species, known only from Jackson County . 1. P. MULTIFLORUM

1. **Petalostemon multiflorum** Nutt. Map 1317

*Dalea multiflora* (Nutt.) Shinners [Shinners]

Flowers June–September.

Known from west-central Missouri in Jackson County (common, Sheffield, August 13, 1900, *Bush 826*; same locality, September 6, 1896, *Mackenzie*).

Ranges from Iowa and Nebraska, south to Arkansas and Texas.

2. **Petalostemon candidum** (Willd.) Michx.
White Prairie Clover Map 1318
*Dalea candida* Willd. [Shinners]

Flowers late May–July.

Occurs in prairies, rocky open glades, along railroads, and rocky or open woodland. Throughout Missouri except absent from the lowlands of southeastern Missouri.

Ranges from Indiana to Minnesota and Saskatchewan, south to Tennessee, Alabama, Mississippi, Louisiana, and Texas.

Fassett has identified a number of Missouri specimens as *P. oligophyllum* (Torr.) Rydb. = *P. occidentale* (Gray) Fern., but study of these specimens does not show that they differ in any respect from ordinary *P. candidum* having dense uninterrupted spikes, elongate floral bracts which are longer than the calyces, and glabrous calyx-tubes. According to recent studies by Drs. Isely and Welch, true *P. occidentale* differs from *P. candidum* by having the spikes more loosely flowered and somewhat flexuous at maturity, finally exposing the floral axis, attachment scars of flowers separated by intervening areas of stem surface, ribs of the calyx winglike or flangelike, instead of rounded, and surface of calyx-tube usually minutely diffusely pubescent.

Shinners at first interpreted *Petalostemon oligophyllum* and *P. occidentale* (Gray) Fern. as *Dalea oligophylla* (Torr.) Shinners (Field and Lab. 17: 82–83. 1949), but in his *Spring Flora of the Dallas-Fort Worth Area* (p. 211) treats *Dalea oligophylla* and *Petalostemon occidentale* as var. *oligophylla* (Torr.) Shinners under *Dalea candida*. Study of these closely related taxa by Drs. Isely and Welch appears to indicate that *Petalostemon multiflorum*, *P. occidentale* (*P. oligophyllum*), and *P. candidum* can be separated as distinct species.

3. **Petalostemon purpureum** (Vent.) Rydb.
Purple Prairie Clover Map 1319

Flowers last of May–September.

Occurs in prairies, rocky open glades, along railroads, and rocky or open woods. Throughout Missouri, except absent from the lowland counties of the southeastern section.

Ranges from Indiana to Saskatchewan and Montana, south to Tennessee, Arkansas, Texas, and New Mexico; also in Alabama and introduced east to New York.

Two variations are found in Missouri:

Stems and leaves glabrous (without hairs) or with a few hairs . . 3a. P. PURPUREUM f. PURPUREUM

Stems and leaves densely hairy . . . . . 3b. P. PURPUREUM f. PUBESCENS

3a. **Petalostemon purpureum** f. **purpureum**
Map 1319

*Petalostemon purpureum* (Vent.) Rydb. [G, BB, P & S, Steyerm.]
This is the common variation in the state.

3b. **Petalostemon purpureum** f. **pubescens** (Gray) Fassett Map 1319

*Petalostemon purpureum* var. *pubescens* Gray [P & S]
This form is scattered in southern Missouri, where known from Iron, Shannon, Wright, Taney, Barry, Lawrence, Jasper, and McDonald counties.

Some of the specimens cited above under f. *pubescens* were identified at the Gray Herbarium by Fernald as *P. pulcherrimum* Heller, and are the basis for the Missouri report of that species in *Gray's Manual*, eighth edition (p. 901). These specimens so identified are *Bush 711* from McDonald Co., *Bush 151* and *51* from Taney Co., and *Lansing 3176* from Wright Co. A study of this material does not show any essential differences from *P. purpureum* f. *pubescens*. Similarly, Fassett has identified some of the above material as *P. pulcherrimum* as well as the following specimens: (*Bush 8445* from Jackson Co.; *Palmer 18112* from Iron Co.; *Davis 4422* from Lincoln Co.; *Bush* from Shannon Co.; *Weller* from Lawrence Co.; *Fritchey*, *Eggert*, and *Letterman* from St. Louis Co.). All of this material should be identified as *P. purpureum* f. *pubescens* or f. *purpureum* and referred to those taxa. So far as can be judged at present, none of the Missouri material identified as *P. pulcherrimum* can be separated from *P. purpureum*, and it is doubtful whether *P. pulcherrimum* can be retained as a distinct species.

Purple Prairie Clover is a nutritious legume with especially high protein content in the young foliage and stems, and is eaten by various kinds of livestock. It is ranked as one of the three important native legumes for agriculture. It is stated that a tea made from the leaves of Purple Prairie Clover produces a constipating effect.

*Excluded Species*

**Petalostemon occidentale** (Gray) Fern.

*Petalostemon oligophyllum* (Torr.) Rydb.
*Dalea oligophylla* (Torr.) Shinners, Field and Lab. 17: 82–83. 1949
*Dalea candida* var. *oligophylla* (Torr.) Shinners [Shinners]

As stated under *Petalostemon candidum*, Missouri material identified by Fassett as *P. oligophyllum* is referred to *P. candidum* in the present treatment.

**Petalostemon pulcherrimum** Heller

*Dalea Helleri* Shinners [Shinners]

As indicated under *Petalostemon purpureum*, Missouri material identified as *P. pulcherrimum* by Fassett and by Fernald, the latter's identifications forming the basis of the Missouri report in *Gray's Manual*, eighth edition, is referred in the present flora chiefly to *P. purpureum* f. *pubescens*, the remainder of the specimens being referred to typical *P. purpureum* f. *purpureum*. According to Dr. Isely, *P. pulcherrimum* does not occur in Missouri, but some populations of *P. purpureum* in the southern part of the state, which reflect previous genetic influence of *P. pulcherrimum*, are the basis for the Missouri reports of *P. pulcherrimum*.

## 20. **Amorpha** L.

a. Low plants mostly less than 1 m. tall of dry, high, or rocky ground; leaves nearly sessile (without a stalk), the petioles (leaf-stalk) mostly shorter than the width of the lowest leaflets; leaflets rarely more than 17 mm. long; calyx-lobes all lanceolate, either $\frac{2}{3}$ as long as the calyx-tube or equaling or longer than the calyx-tube . . . . . b
 b. Plant glabrous (without hairs) or nearly so; fruit glabrous (without hairs), resin-dotted; fruit longer than the style and beak; rarely found, known only in southwestern Missouri . 2. A. BRACHYCARPA
 b. Plant densely to sparsely hairy; fruit white-hairy; fruit about equaling the style and beak; common species, throughout Missouri . . . . . 1. A. CANESCENS
a. Taller plants, mainly 1–4 m. tall of low or moist ground and along streams and wet rocky banks; leaves distinctly stalked (petioled), the stalks (petioles) longer than the width of the lowest leaflets; leaflets usually 20–40 mm. long; 4 of the calyx-lobes triangular or nearly round and not more than $\frac{1}{4}$ as long as the calyx-tube, the fifth calyx-lobe longer and narrower. . . . . 3. A. FRUTICOSA

1. **Amorpha canescens** Pursh Lead Plant Map 1320

Flowers May–August.
Occurs in prairies, glades, and rocky open woods. Nearly throughout Missouri, but apparently absent from the southeastern Ozark and lowlands areas.

Ranges from Michigan to Saskatchewan, south to Indiana, Illinois, Arkansas, Texas, and New Mexico.

1321 Amorpha brachycarpa

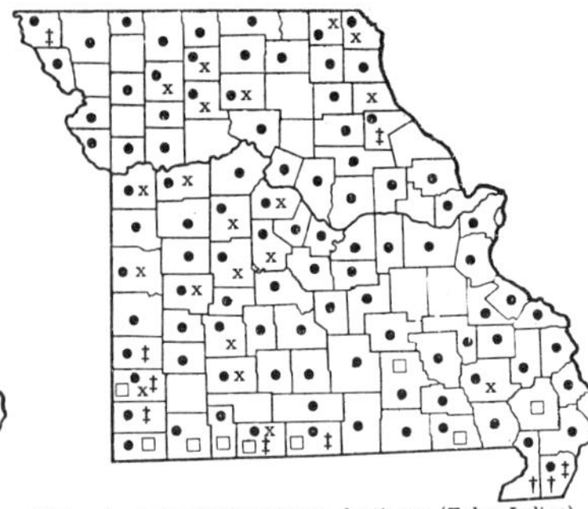

1322 • Amorpha fruticosa var. fruticosa (False Indigo)
1322 □ Amorpha fruticosa var. tennesseensis
1322 ‡ Amorpha fruticosa var. oblongifolia
1322 x Amorpha fruticosa var. angustifolia
1322 † Amorpha fruticosa var. croceolanata

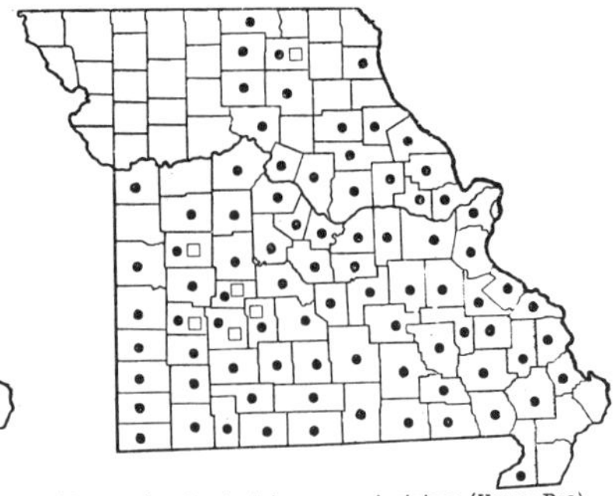

1323 • Tephrosia virginiana var. virginiana (Hoary Pea)
1323 □ Tephrosia virginiana var. holosericea

Missouri material is represented by two variations:

Lower surface of leaflets covered with a gray-white or white hairiness . . . 1a. A. CANESCENS f. CANESCENS

Lower surface of leaflets greener, with scattered hairs . . . . . 1b. A. CANESCENS f. GLABRATA

1a. **Amorpha canescens** f. **canescens** Map 1320

*Amorpha canescens* Pursh [G, BB, P & S, Steyerm.]

This is the common variation encountered in the state.

1b. **Amorpha canescens** f. **glabrata** (Gray) Fassett Map 1320

*Amorpha canescens* var. *glabrata* Gray [P & S]

This variation, although less common, commonly occurs in southern, central, and eastern Missouri; it is apparently absent from the southeastern, northwestern, and north-central sections of the state.

The Lead Plant is a palatable legume and is, therefore, considered to be one of the important native legumes on the range. It is sometimes grown as an ornamental plant, and adapts itself well to sunny dry situations. It is quite showy with the purple flower spikes and grayish-white foliage.

2. **Amorpha brachycarpa** Palmer Map 1321

Flowers May–August.

Occurs in rocky open woods and glades.

Known only from Stone (Galena, September 27, 1920, *Palmer 19197*, holotype in Arn. Arb. Herb.; October 1, 1921, *Palmer 20649*; May 24, 1923, *Palmer 22853*) and Lawrence (open limestone bank, border of woods, near Bonham Siding, October 13, 1950, *Palmer 51355*) counties, southwestern Missouri.

Limited, so far as known, to southwestern Missouri.

3. **Amorpha fruticosa** L. False Indigo Map 1322

Also known as Indigo Bush.

Flowers May–June.

Occurs in moist ground in thickets along streams, rocky banks, borders of ponds, and low open wet woods. Throughout Missouri.

The following variations, which are intergrading, are known in Missouri (see also the hybrid between *Amorpha canescens* and *A. fruticosa* at end of treatment):

a. Lower surface of leaflets softly hairy; calyx, branches, and lower surface of leaflets with tawny or orange-brown hairs when young . 3e. A. FRUTICOSA var. CROCEOLANATA

a. Leaflets minutely hairy with grayish hairs or glabrous (without hairs); calyx, branches, and lower surface of leaflets grayish-hairy to nearly glabrous (without hairs) . . . . . . *b*

b. Leaflets chiefly elliptic or obovate, acute or narrowed at base; hairs appressed (lying parallel to surface, not spreading) . . . 3d. A. FRUTICOSA var. ANGUSTIFOLIA

b. Leaflets chiefly ovate or oblong, rounded or obtuse at base; hairs of young growth spreading . . . . . . . . . . . *c*

c. Leaflets 13–25, averaging 1½–2 times as long as broad . . . . 3a. A. FRUTICOSA var. FRUTICOSA

c. Leaflets 21–35 or more, averaging 2–3 times as long as broad . . . . . *d*

d. Leaflets mostly 1–2 cm. long, more or less hairy; fruits straight or nearly straight. . . . . 3b. A. FRUTICOSA var. TENNESSEENSIS

d. Leaflets mostly 2–5 cm. long, sparsely hairy to nearly glabrous; fruits curved 3c. A. FRUTICOSA var. OBLONGIFOLIA

3a. **Amorpha fruticosa** var. **fruticosa** Map 1322

*Amorpha fruticosa* L. [G, BB, P & S, Steyerm.]

Common throughout the state and probably in every county.

Ranges from Florida to Louisiana, north to Pennsylvania, Ohio, Michigan, Wisconsin, and Kansas.

3b. **Amorpha fruticosa** var. **tennesseensis** (Shuttlew.) Palmer Map 1322

Southern Missouri, north to Stoddard, Shannon, and Jasper counties.

Ranges from Florida to Texas, north to Kentucky, Illinois, Missouri, and Kansas.

3c. **Amorpha fruticosa** var. **oblongifolia** Palmer Map 1322

Scattered in the state.

Known from Missouri and Arkansas.

3d. **Amorpha fruticosa** var. **angustifolia** Pursh Map 1322

Scattered over the state.

Ranges from Texas to Arizona and Mexico, north to Wisconsin, Minnesota, Manitoba, and Saskatchewan.

3e. **Amorpha fruticosa** var. **croceolanata** (P. W. Wats.) Schneid. Map 1322

*Amorpha croceolanata* P. W. Wats. [BB, P & S, Steyerm.]

Known only from Dunklin (Malden, May 25, 1926, *Palmer 30317*) and Pemiscot (thickets in low rich wet woods along North Deering Ditch lateral, T18N, R10E, sect. 31, 3 mi. southwest of Deering, April 26, 1938, *Steyermark 5131*) counties in southeastern Missouri.

Ranges from Florida to Louisiana, north to Georgia, Kentucky, Illinois, and Missouri.

*Amorpha fruticosa* and varieties lend themselves well to ornamental plantings.

A plant presumed to be a hybrid between *Amorpha canescens* and *Amorpha fruticosa* has been described by Palmer (Rh. 55: 157–59. 1953.) and is known as **Amorpha** × **notha** Palmer.

Known only from Jasper County, southwestern Missouri (low rocky bank between upland prairie and the alluvial valley of Center Creek, about 1 mi. north of Webb City, May 27, 1951, *Palmer 52047*, holotype; same locality, August 9, 1951, *Palmer 52930*).

This plant resembles *A. canescens* in the dense gray pubescence, of the young branches and foliage, as well as in the pubescent inflorescence and fruit. However, in the size, shape, and arrangement of the leaflets, and in the form of the fruit with its curved back and short erect beak and in the character of the sepals it resembles *A. fruticosa* more closely. The plant is 1–1.5 m. tall. The leaves are 5–10 cm. long, 2–4.5 cm. wide, with 19–27 leaflets 10–16 mm. long and 4–6 mm. wide. The lateral leaflets are oblong, rounded at base and apex, while the terminal leaflets are obovate and cuneate at the base but truncate and emarginate at the apex; they are densely gray-hairy beneath and green and sparsely hairy above. The entire panicle and the calyx are densely gray-hairy. The fruit is 4–5 mm. long and pubescent between the large black glandular dots.

## 21. **Tephrosia** Pers. Hoary Pea

**Tephrosia virginiana** (L.) Pers. Goat's Rue Map 1323

Also known as Catgut, Hoary Pea.

Flowers May–August.

Occurs in acid soils associated with usually chert, sandstone, or granitic rocks, in rocky open woods, rocky glades, and prairies. Throughout southern, central, and eastern Missouri, less common or rare in northeastern Missouri, south and east of a line drawn from Putnam, Sullivan, Linn, Chariton, and Saline counties, to Jackson County.

Two variations are represented by Missouri material:

Upper surface of the leaflets glabrous . . . a. T. VIRGINIANA var. VIRGINIANA

Upper surface of the leaflets hairy . . . . b. T. VIRGINIANA var. HOLOSERICEA

a. **Tephrosia virginiana** var. **virginiana** Map 1323

*Tephrosia virginiana* (L.) Pers. [G, P & S, Steyerm.]

This is the commoner variation in Missouri.

Ranges from Florida to Texas, north to New Hampshire, Massachusetts, New York, Ontario, Michigan, Wisconsin, Iowa, Missouri, and Kansas.

b. **Tephrosia virginiana** var. **holosericea** (Nutt.) T. & G. Map 1323

This is less common and is known from the counties of the unglaciated prairie region, the Ozark border counties, and some of the counties in northeastern Missouri.

Ranges from Massachusetts, New York, Pennsylvania, Ohio, Michigan, Wisconsin, and Minnesota, south to Maryland, South Carolina, Georgia, Alabama, Arkansas, Texas, and Oklahoma.

There is much intergradation between these two variations which Dr. Carroll E. Wood, Jr. (Rh. 51: 267–76. 1949) prefers to regard as one variable species without varietal recognition being accorded names. The most extreme pubescence is exhibited by specimens in Dallas, Polk, Hickory, Henry, and St. Clair counties in the unglaciated prairie region in which the leaflets are very silky pubescent on both sides, giving a gray-white effect. There is a tendency, as Dr. Wood indicates (p. 275), for the specimens with longer and denser hairs to predominate in the northwestern portion of the range of the species, and it is upon the basis of this tendency of geographical divergence that the two varieties are being retained in the present flora.

The plant acts as a poison to fish, but is not poisonous to mammals. The Indians used the plant as a fish poison, and some of the tropical American, African, and Australian species of the genus are used today for that purpose. The Indians of eastern North America used the roots of this species as a vermifuge as well. A percentage of rotenone has been found in the xylem but not in the bark of the roots of this species (Sievers, A. F., Russell, G. A., *et al.* in USDA Tech. Bull. 595. 40 pp. 1938), and varies considerably among individuals, with a greater amount of rotenone being shown from plants in the southern part of the range in Florida, Georgia, Louisiana, Texas, and Oklahoma.

The roots harbor nitrogen-fixing bacteria and are therefore valuable for soil-building. The plant is a good soil binder for dry, sterile, and eroded soils. It is eaten by all classes of livestock and is considered nutritious and palatable. The seeds are eaten by bobwhite and wild turkey.

The plant is quite showy when in flower with its contrasting creamy yellow standard and pink or rose-lavender wings and keel. The plant does not transplant successfully because of the long tap root and possibly because of sensitive soil mycorrhizal relationships. Of the more than one thousand species of native Missouri plants I have grown on my botanical preserve, this species and *Schrankia uncinata* are two which have failed to grow either from seed or through transplantings. It requires a strongly acid soil.

## 22. **Sesbania** Scop.

**Sesbania exaltata** (Raf.) Cory Map 1324

*Sesbania macrocarpa* Muhl. [P & S]

Flowers July–October.

Occurs in low sandy fields, sand bars of streams, alluvial ground along sloughs and borders of oxbow lakes, and along roadsides and railroads. Lowlands of southeastern Missouri, low ground along the White River in Taney, Stone, and Barry counties, and locally north in St. Clair, Moniteau, Cole, and St. Louis counties.

Ranges from Alabama to Texas, north to Illinois, Missouri, and Oklahoma; introduced in Pennsylvania and New York.

## 23. **Wisteria** Nutt. Wisteria

**Wisteria frutescens** (L.) Poir. var. **macrostachya** (Nutt.) Torr. & Gray Wisteria Map 1325

*Wisteria macrostachya* Nutt. [G, BB, P & S, Steyerm.]

Flowers April–May.

Occurs in low swampy woods and along the borders of swamps, bayous, and slow streams; occasionally cultivated and escaped from cultivation. Southeastern Missouri in the lowland counties north and west to Scott, Stoddard, Wayne, and Ripley counties, and escaped from cultivation in Adair County (Univ. of Mo. Herb.) and Crawford County (*Steyermark 27905*).

Ranges from Louisiana and Texas, north to Kentucky, Illinois, Missouri, and Oklahoma.

The fresh flowers may be eaten as a salad or cooked in batter as fritters.

This Wisteria is usually separated as a species distinct from *W. frutescens*. Fernald, however, in the eighth edition of *Gray's Manual* (p. 903) states that *W. macrostachya* is 'perhaps better treated as a var. of

---

Plate no. 214. **1.** Amorpha fruticosa var. croceolanata, × 2/7. **2.** Robinia Pseudo-acacia, × 2/7; Details from Small, The New York Botanical Garden. **3.** Wisteria frutescens var. macrostachya, × 2/7. **4.** Tephrosia virginiana, × 2/7; Details from Small, The New York Botanical Garden. **5.** Sesbania exaltata, × 2/7; *a.* Fruit, × 2/7; After Gleason, details from Small, The New York Botanical Garden. **6.** Wisteria frutescens var. macrostachya, × 2/7; variation in inflorescence. **7.** Robinia hispida, × 2/7; Details from Small, The New York Botanical Garden.

PLATE NO. 214

1324 Sesbania exaltata

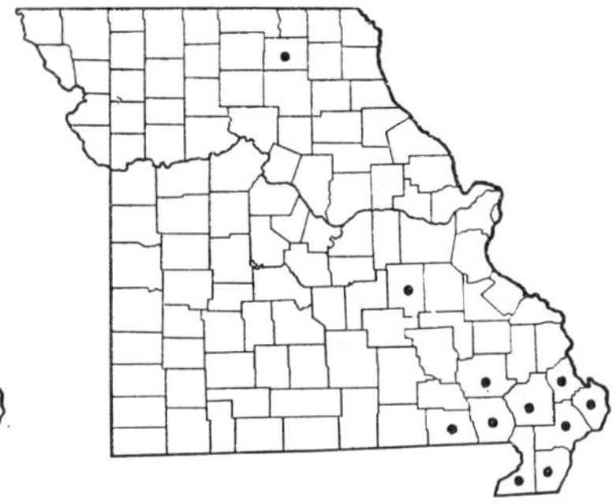

1325 Wisteria frutescens var. macrostachya (Wisteria)

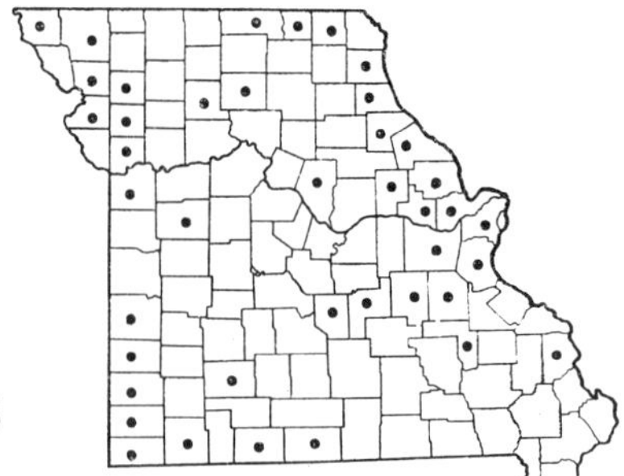

1326 Robinia Pseudo-acacia (Black Locust)

no. 1 [W. frutescens],' and a study of material of both species when applied to Missouri material seems to bear this out. Much of the Missouri material of *W. macrostachya* has the racemes of flowers only 7–12 cm. long, while according to the treatment in some of the manuals, such short racemes are characteristic of *W. frutescens*. In other words, short racemes are to be found frequently in *W. macrostachya* as well as in *W. frutescens*. The abundance of club-shaped glands on the calyx and pedicels in *W. macrostachya* as contrasted with their lack or scarcity in *W. frutescens* is the only character left for separation of these two taxa in *Gray's Manual*, eighth edition, whereas in Gleason's *New Illustrated Flora*, the two are separated by 'upper lip of the calyx about a third as long as the tube' in *W. frutescens* as contrasted with 'upper lip of the calyx nearly or quite as long as the tube' in *W. macrostachya.* Deam (*Flora of Indiana*) presents these and other differences, some of which are overlapping (such as length of leaves and fruits) but Indiana material was based upon only one collection of *W. macrostachya*, and the differences stated were not gleaned from a large suite of specimens for comparative study.

*Excluded Species*

**Wisteria frutescens** (L.) Poir. var. **frutescens**
Wisteria

This species, partly discussed above, was included in Palmer and Steyermark's *Annotated Catalogue* and in Steyermark's *Spring Flora* on the basis of specimens escaped from cultivation in Adair, Daviess, and Crawford counties. The Adair County specimen has numerous glands on the calyx and pedicels and should be referred to *W. frutescens* var. *macrostachya*, while the other specimens mentioned have not persisted as escapes from cultivation. Typical *W. frutescens* var. *frutescens* should, therefore, be eliminated from the flora of the state as based upon present records.

## 24. **Robinia** L. Locust

Flowers white with a yellow blotch on the uppermost petal (standard); twigs glabrous (without hairs); common tree . . . . . . . . . . . . . . . . . . . . . . . . 1. R. PSEUDO-ACACIA
Flowers rose or rose-purple; twigs bristly; rare shrub escaped . . . . . . . . . . . 2. R. HISPIDA

1. **Robinia Pseudo-Acacia** L. Black Locust
Map 1326

Also called Honey Locust by many country folk, and Yellow Locust.

Flowers May–June.

Occurs in dry or rocky upland woods, along streams, in pastures, thickets, and waste ground. Throughout Missouri, but native only in the Ozark section of the southern and east-central sections of the state, introduced elsewhere and doubtless in every county.

Native in the United States from Georgia to Louisiana and Arkansas, north to Pennsylvania, West Virginia, southern Indiana, southern Illinois, southern Missouri, and eastern Oklahoma; elsewhere planted and naturalized from Maine to Nova Scotia, Quebec, and Ontario, westward; also in Oregon and other western states.

The twigs are zigzag and without a terminal bud. The foliage turns a greenish-yellow in late fall. The leaflets have a habit, as in some other legumes, of drooping on their stalks and folding together in the evening hours. The drooping racemes of wisterialike flowers are all white except for a blotch of yellow on the uppermost petal (standard). The very fragrant flowers are much visited by bees and are an important source of honey in some sections. As with a number of other members of the legume family, the flowers may be eaten; they are fried or cooked as a vegetable. A pleasant drink is also made from an infusion of the flowers. The roots have a sweet, licoricelike flavor, but they and the bark are reported to be somewhat poisonous and children have been reported poisoned from chewing the inner bark. Reports of cases of poisoning are also at hand of chickens which have eaten the leaves and of livestock which have fed on the bark and young shoots of the locust. The seeds are a favorite food of pheasants in the northern states.

A number of forms of Black Locust are in cultivation, some of them having unarmed branches (f. *inermis* [Mirb.] Rehder and f. *Bessoniana* [Nichols.] Voss), pale rose-colored flowers (f. *Decaisneana* [Carr.] Voss), or other special characteristics. A variety, known as Shipmast Locust (var. *rectissima* Raber), seldom produces seed and has more durable wood than any of the other varieties known. It has been used in reforestation plantings largely.

The Black Locust is known to have the strongest wood of any species in North America, outside of the American tropics, and is listed as the seventh hardest of any in North America. It is also the stiffest and shrinks least in drying among the major hardwood species. It is also known to be the most durable, and this makes it desirable for fence posts, railroad ties, tree nails, rungs for ladders, teeth for rakes, tool handles, and policemen's clubs. The fuel value of this tree is stated to be higher than that of any other American tree, a cord of it nearly equaling a ton of anthracite coal. Black Locust has been used extensively for reforestation purposes, especially in covering newly exposed ground, such as mine excavations and dumps, and eroded areas. It has been used extensively in some prairie states as a shelter belt tree. Efforts to use the tree in reforestation plantings have sometimes met with failure due to the susceptibility of the species to attacks by borers, bag worms, and leaf miners, but borer-resistant strains are gradually being developed.

The parent tree establishes colonies all around it by sending out long underground root suckers which may extend a long distance before rising above the ground level. The tree may be easily propagated from these root suckers, as well as from seed. Since the tree may become troublesome because of these suckers, it is well to avoid planting in wildflower preserves or other wild areas where it is desirable not to disturb the natural flora.

2. **Robinia hispida** L. var. **hispida**

Bristly Locust — Map 1327

Also called Rose Acacia.

Flowers May–June.

Escaped from cultivation to wooded slopes, borders of woods, and open thickets along roadsides. Known only from Platte (upper edge of wooded slopes, just west of old Union Mill distillery, along Platte River, T54N, R33W, sect. 6, 1–1¾ mi. northwest and west northwest of Edgerton, May 15, 1949, *Steyermark 67530*) and Howell (open ground along roadside, T23N, R8W, northeast ¼ sect. 16, 3½ mi. south of West Plains, June 25, 1955, *Steyermark 78699*) counties.

Native from Virginia and Kentucky, south to Georgia and Alabama; escaped from cultivation and naturalized north to New York and Minnesota and west to Missouri, Kansas, Oklahoma, and Texas.

This is a very showy-flowered shrub, sometimes only 0.5 meters tall (1 foot, 8 inches), but attaining a height of nearly 3 meters.

## 25. **Astragalus** L. Milk Vetch

a. Plants mainly 6–15 dm. (2–5 ft.) tall (rarely less); inflorescences 5–12 (up to 20) cm. long; flowers 30–70 in an inflorescence . . . . . 3. A. CANADENSIS

a. Plants shorter, 0.5–5 dm. tall; inflorescences 1–5 (–7) cm. long; flowers 3–25 in an inflorescence . . . *b*

- b. Flowers cream-colored or yellowish . . . . . *c*
    - c. Plants known only from the loess hills of northwestern Missouri in Atchison and Holt counties; flowers small, yellowish to cream-color, 8–11 mm. long; stems nearly lacking or up to 1 dm. tall; leaflets 7–15 (–21); fruit leathery, dry, dehiscent (splitting open) . . . . 2. A. LOTIFLORUS
    - c. Plants found throughout southern and central Missouri north to Lincoln, Boone, Saline, and Jackson counties; flowers showy, creamy-colored with lilac or blue at the tip of the pointed keel petals, 15–25 mm. long; stems 1–4 dm. tall in flowering stage, taller or longer in fruiting stage; leaflets mainly 17–33 (sometimes 15); fruit fleshy, 'plum-like,' indehiscent (not opening) or opening late in the season . . . . . 5. A. MEXICANUS var. TRICHOCALYX
- b. Flowers mostly purplish, lilac, or bluish-lavender . . . . . *d*

1327 Robinia hispida var. hispida (Bristly Locust)

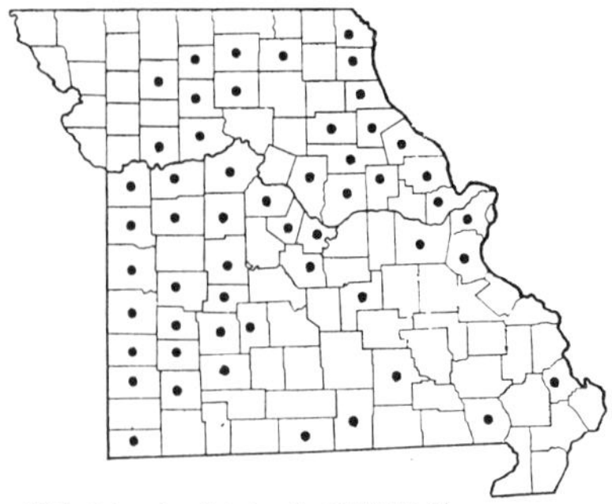

1328 Astragalus distortus (Bent Milk Vetch)

1329 Astragalus lotiflorus (Low Milk Vetch)

d. Stems glabrous (without hairs) or nearly so or sparsely hairy with remote hairs; calyx-tube 2–3 mm. long; flowers 9–11 mm. long; leaflets green, glabrous or nearly so, or remotely hairy beneath, usually blunt or indented at tip; fruit leathery, dry, dehiscent (splitting open) . 1. A. DISTORTUS

d. At least the upper part of the stems covered with a fine gray-whitish hairiness of hairs appressed (lying parallel to the surface); calyx-tube 7–10 mm. long; flowers 15–25 mm. long; leaflets grayish-green, appressed-hairy at least on lower surface (hairs pressed along surface), often acute or pointed at tip; fruit fleshy, 'plum-like,' indehiscent (not opening) or opening late in the season . . . . . . . . . . . . . . . . . . . . . 4. A. CARYOCARPUS

1. **Astragalus distortus** T. & G.
Bent Milk Vetch Map 1328
*Astragalus distortus* var. *Engelmannii* (Sheldon) M. E. Jones [Shinners]
Flowers April–June.

Occurs on limestone, sandstone, and chert glades, edges of limestone escarpments, rocky prairies, shaley open slopes, and along roadsides. Throughout Missouri, except apparently absent in the extreme northwestern counties and in much of the Ozark region.

Ranges from Illinois and Iowa to Kansas, south to Mississippi, Louisiana, and Texas; and in Maryland, West Virginia, and Virginia.

Shinners employs *A. distortus* var. *distortus* for plants having white-petaled flowers, and var. *Engelmannii* for those with lavender or violet-blue petals.

This is a very ornamental species with many rounded clusters of purplish flowers. It is usually found in areas of limestone or shale. Although an excellent subject for the rock garden, because of its low size, it is very difficult to establish through transplanting and seed germination often yields poor results.

2. **Astragalus lotiflorus** Hook. Low Milk Vetch
Map 1329
Flowers April–May.

Occurs on loess hills. Known only from Atchison and Holt counties, extreme northwestern Missouri.

Ranges from British Columbia to New Mexico, east to Manitoba, Minnesota, Iowa, Missouri, Oklahoma, and Texas.

This is one of the species of the Great Plains flora which is known in Missouri only from the loess hills along the Missouri River, associated with such species of similar range as *Psoralea argophylla*, *Oxytropis Lambertii*, *Buchloe dactyloides*, *Gaura coccinea*, *Oenothera serrulata*, *Lactuca pulchella*, and *Yucca glauca*.

3. **Astragalus canadensis** L. var. **canadensis**
Rattle Weed Map 1330
*Astragalus canadensis* L. [G, P & S, Steyerm.]
Flowers May–August.

Occurs in rocky or open lowland or upland woods, thickets, pastures, fields, and prairies. Throughout Missouri, and apparently absent only from most of the southeastern lowland counties.

Ranges from Quebec and Vermont to Hudson Bay and British Columbia, south to Virginia, West Virginia, Arkansas, Texas, and Colorado.

This is the commonest species of *Astragalus* in Missouri.

4. **Astragalus caryocarpus** Ker Ground Plum
Map 1331

---

Plate no. 215. **1.** Astragalus distortus, $\times {}^{2}/_{7}$; *a.* Fruiting cluster, $\times {}^{2}/_{7}$; Details from Small, The New York Botanical Garden. **2.** Astragalus lotiflorus, $\times {}^{2}/_{7}$; *a.* Fruit, $\times {}^{6}/_{7}$. **3.** Astragalus canadensis, $\times {}^{2}/_{7}$; *a.* Flower, $\times {}^{4}/_{7}$. **4.** Astragalus caryocarpus, $\times {}^{2}/_{7}$; *a.* Flower, $\times {}^{4}/_{7}$. **5.** Astragalus mexicanus var. trichocalyx, $\times {}^{2}/_{7}$. **6.** Coronilla varia, $\times {}^{2}/_{7}$. **7.** Oxytropis Lambertii, $\times {}^{2}/_{7}$. **8.** Glycyrrhiza lepidota var. lepidota, $\times {}^{2}/_{7}$; *a.* Fruit, $\times {}^{4}/_{7}$.

PLATE NO. 215

1330 Astragalus canadensis var. canadensis (Rattle Weed)

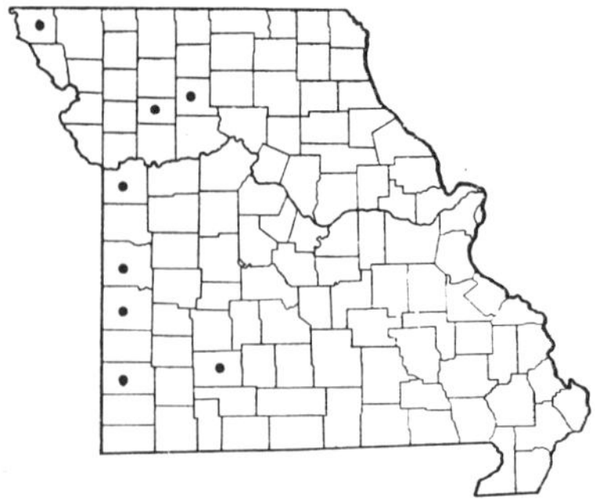

1331 Astragalus caryocarpus (Ground Plum)

1332 Astragalus mexicanus var. trichocalyx (Ground Plum)

*Astragalus crassicarpus* Nutt. var. *crassicarpus* [BB]
*Astragalus carnosus* Pursh [Shinners]
Flowers April–May.

Occurs on rocky open wooded limestone slopes, limestone glades, prairies, loess hills, and shaley open mounds. Western Missouri east to Livingston, Jackson, and Greene counties.

Ranges from Manitoba to Saskatchewan and Montana, south to Arkansas, Oklahoma, Texas, and New Mexico.

The unripe green fruits, resembling small plums, are eaten either raw or cooked.

5. **Astragalus mexicanus** A. DC. var. **trichocalyx** (Nutt.) Fern. Ground Plum Map 1332
*Astragalus trichocalyx* Nutt. [Shinners]
*Astragalus mexicanus* of auth. [P & S, Steyerm.], not A. DC.
*Astragalus crassicarpus* Nutt. var. *trichocalyx* (Nutt.) Barneby [BB]
Flowers March–May.

Occurs in rocky open woods, ledges along bluffs, rocky prairies, and glades. Southern and central Missouri north to Lincoln, Montgomery, Callaway, Boone, Cooper, Saline, and Jackson counties.

Ranges from Texas and Louisiana, north to Missouri and Illinois.

As in *A. caryocarpus*, the unripe green fruits, which resemble small plums, may be eaten either raw or cooked.

*Excluded Species*

The following species have been reported for Missouri in some earlier published works, but are not substantiated by any specimens and are, undoubtedly, based upon erroneous determinations or refer to 'Missouri Territory' on the old labels. In the Botany of King's expedition (*U. S. Geol. Expl.* of the *Fortieth Parallel*, vol. 5. 1871) by Sereno Watson, several species of *Astragalus* are cited for 'Missouri.' These have not been found in the state of Missouri, and are species which occur far to the west of the present state of Missouri in what was at the time of that report 'Missouri Territory.' They should be eliminated, therefore, from future references.

**Astragalus racemosus** Pursh

Recorded for Missouri in the seventh edition of *Gray's Manual* (p. 516) and in the first edition of Britton and Brown. No evidence exists for the occurrence of this species in Missouri.

**Astragalus tennesseensis** Gray

Recorded for Missouri in the first edition of Britton and Brown and reported in the seventh edition of *Gray's Manual* (p. 515). No evidence exists for the occurrence of this species in Missouri.

**Astragalus gracilis** Nutt.

*Astragalus parviflorus* (Pursh) MacM., not Lam.

Recorded for Missouri in the seventh edition of *Gray's Manual* as *A. parviflorus* and in the first edition of Britton and Brown, but no evidence exists for the occurrence of this species from the state. Missouri is recorded (p. 438) for section *Microlobi* (includes *A. gracilis* and *A. microlobus*) in the Botany of King's expedition.

**Astragalus missouriensis** Nutt.

Reported for Missouri in the first edition of Britton and Brown, but no specimens exist to support the report.

1333 Oxytropis Lambertii (Loco Weed)

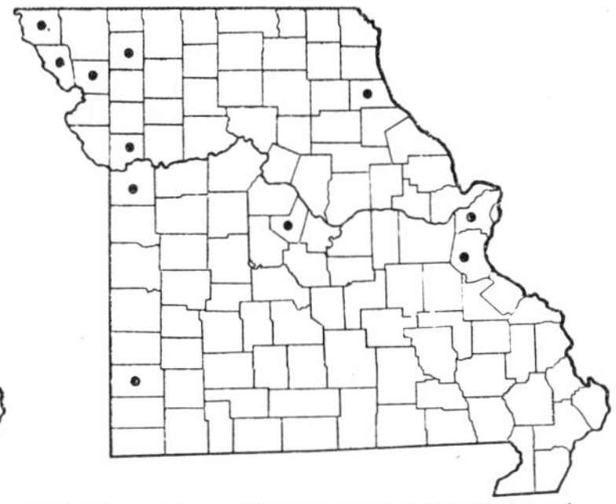

1334 Glycyrrhiza lepidota var. lepidota (Wild Licorice)

1335 Coronilla varia (Crown Vetch)

## 26. **Oxytropis** DC.

**Oxytropis Lambertii** Nutt. Loco Weed Map 1333

*Oxytropis plattensis* [of Steyerm.], not Nutt.

*Astragalus Lambertii* (Pursh) Sprengel var. *Lambertii*

Flowers May–June.

Occurs on loess hills. Known only from Atchison and Holt counties, northwestern Missouri.

Ranges from Minnesota and Manitoba to Montana, south to Missouri, Oklahoma, and Texas.

This is one of the species known to poison grazing animals. When there is sufficient forage available, however, the loco weeds are left alone. A number of other species of *Oxytropis* and *Astragalus* are known to poison livestock if they feed upon either the green or dried parts of the plants, but these harmful species are not ones found in Missouri.

## 27. **Glycyrrhiza** L. Licorice

**Glycyrrhiza lepidota** (Nutt.) Pursh var. **lepidota**

Wild Licorice Map 1334

Flowers May–August.

Occurs in prairies, on loess hills, open waste ground, and along railroads. Native in northwestern Missouri, and scattered elsewhere in northern, central, and western portions of the state.

Ranges from Ontario and Wisconsin to Washington, south to Missouri, Arkansas, Texas, New Mexico, California, and Mexico; introduced eastward in Illinois, New York, and New England.

The sweet, licorice-flavored root was eaten raw or in a baked form by Indians. It is sometimes chewed by country folk for its pleasant taste. The species is related to the Licorice of commerce (*G. glabra* L.), from the root of which is obtained an extract used in medicine as a laxative and expectorant, and in brewing, confectioneries, and flavoring tobacco.

## 28. **Coronilla** L. Crown Vetch

**Coronilla varia** L. Crown Vetch Map 1335

Flowers May–August.

Occurs in waste ground and along roadsides. Known only from St. Louis, Taney, and Jasper counties.

Native of Europe, Asia, and Africa; introduced and naturalized in the United States from Maine to South Dakota, south to Virginia, West Virginia, Kentucky, Illinois, and Missouri.

The seed is considered to be poisonous. This perennial plant forms large clumps of spreading stems with showy umbellate clusters of rose-colored flowers. It does well as a cultivated plant, making an attractive addition to the perennial border.

## 29. **Desmodium** Desv. Tick Trefoil, Beggar's Ticks, Tick Clover

In recent years this genus has received special study by Dr. Bernice Schubert and Dr. Duane Isely. The following key is based partly on the present author's studies and partly on those of Dr. Schubert (Rh. 52: 135–55. 1950; *Gray's Manual* eighth ed., pp. 915–23. 1950); Dr. Isely (Iowa State Coll. Jour. Sci. 30: 33–118. 1955; Am. Midl. Nat. 49: 920–33. 1953), and Dr. Fassett (*The Leguminous Plants of Wisconsin*, pp. 80–98. 1939).

a. Branch bearing the flowers arising at ground level from the base of the plant. . . . . 1. D. NUDIFLORUM
a. Flowers at the top of the stem or on branches arising from the sides of the stem or much above the base of the plant . . . . . *b*
 b. Leaves closely clustered or nearly in a whorl (in a circle around the stem) arising about midway on the flowering stem . . . . . 2. D. GLUTINOSUM
 b. Leaves scattered on the flowering stem, not giving a whorled appearance . . . . . *c*
  c. Vegetative stems trailing or prostrate over the ground . . . . . *d*
   d. Common Ozark species; especially the middle (terminal) one of the three leaflets of an orbicular or rounded shape, nearly as broad as, or broader than long, the other two (lateral) leaflets similarly usually orbicular or nearly so; hairs of the leaves usually straight or curved, none (in Missouri material) hooked . . . . . 4. D. ROTUNDIFOLIUM
   d. Rare species in southern or southeastern Missouri; middle (terminal) one of the three leaflets of an ovate or rhombic-ovate shape, longer than broad, the other two (lateral) leaflets either similarly longer than broad or sometimes as broad as long; some or many of the hairs hooked . . . . . *e*
    e. Stipules (at base of main leaf-stalks) with a heart-shaped or somewhat heart-shaped base more or less clasping the stem, 6–12 mm. long, minutely hairy on outer surface; leaflets thick, conspicuously reticulate-nerved, with minute hooked hairs on upper and lower surface; flowers creamy or whitish, on stalks 12–20 mm. long; joints of fruit 7–10 mm. long, 5–8 mm. wide, more or less twisted, glabrous (without hairs) or nearly so on the surfaces, with hooked hairs on the margins . . . . . 5. D. OCHROLEUCUM
    e. Stipules (at base of main leaf-stalks) not clasping the stem nor with a heart-shaped base, 4.5–8 mm. long, the margins finely hairy (ciliate); leaflets thinner, not conspicuously reticulate-nerved, with minute hooked hairs chiefly on the midrib and main side nerves of the upper surface rather than on the surface itself; flowers purplish, on stalks up to 9 mm. long; joints of fruit 6–8 mm. long, 4–5 mm. wide, not twisted, with minute hooked hairs on the surfaces as well as on the margins. . . . . 17. D. HUMIFUSUM
  c. Vegetative stems upright . . . . . *f*
   f. Stipels (stipule-like outgrowths at base of leaflets) practically absent; flowers white; stem low, mainly 1.5–4.5 dm. (up to 5.5) tall; calyx-lobes much shorter than calyx-tube; fruit raised on a stipe (stalk above the calyx) 5–9 mm. long at least more than 3 times the length of the calyx. . . . . 3. D. PAUCIFLORUM
   f. Stipels (stipule-like outgrowths at base of leaflets) persistent or noticeable; flowers rose, pink, purple, green, or white; stems taller, usually 5–20 dm. tall; calyx-lobes equaling or longer than calyx-tube; fruit either without a stipe (no stalk above calyx) or raised on a short stipe usually only slightly longer than the calyx . . . . . *g*
    g. Lower surface of leaflets glaucous (strongly whitened or silvery-white) and mostly glabrous (without hairs) or with at most only a few hairs; stem mainly without hairs, presenting a smooth surface . . . . . 18. D. LAEVIGATUM
    g. Without the above combination of characters; lower surface of leaflets not glaucous, but if pale, then either covered with hairs or the leaflets chiefly 1.5–2.5 cm. long and 1–1.5 cm. wide, or the stems hairy; stems hairy or without hairs . . . . . *h*
     h. Leaves without stalks (petioles) or the stalks 3 mm. or less long; leaflets linear-oblong or linear-lanceolate, mostly 4–8 times as long as broad . . . . . 8. D. SESSILIFOLIUM
     h. At least leaves from the middle part of stem on petioles 3–100 mm. long; leaflets of various shapes, mostly $1\frac{1}{2}$–$2\frac{1}{2}$ times as long as broad, but when 4–8 times as long as broad (in *D. paniculatum*), then the petioles of the middle leaves are 10–55 mm. long . . *i*
      i. Leaflets of the principal leaves 1–3 cm. long, 0.5–1.7 cm. broad, slightly longer than broad to at most $1\frac{1}{2}$–2 times as long as broad . . . . . *j*
       j. Upper surface of leaflets hairy; leaves from the middle part of stem on stalks (petioles) mostly 4–10 mm. long, from the lower part of stem on stalks 5–15

(–20) mm. long, the stalks shorter than the lateral leaflets; stalks of flowers 4–9 mm. long, rather stiff; stems usually conspicuously hairy . . . . 10. D. CILIARE

j. Upper surface of leaflets glabrous (without hairs) or essentially glabrous; leaves from the middle part of stem on stalks 15–25 mm. long, from the lower part of stem on stalks 20–42 mm. long, the stalks about equaling the length of the lateral leaflets; stalks of flowers mostly 10–19 (rarely 8) mm. long, slender and often curved; stems hairless or sparsely hairy . . . . . . . 11. D. MARILANDICUM

i. Leaflets of the principal leaves mainly 3.5–14 cm. long, mainly 2–10 cm. wide, or, if only 1 cm. wide, 3–8 times as long as broad . . . . . . . . . . . . . . . . . . . . . . . . *k*

k. Hairs on at least lower surface of leaflets hooked; the stipules (at base of leaf-stalk) 3–5 mm. wide, somewhat clasping the stem at their base . . . . . . . . . . . . . . . . . . . . *l*

l. Axis of inflorescence with minute gland-tipped or hooked hairs; stem usually unbranched with a solitary inflorescence at the top; leaflets rather thick with the the nerves on the lower surface joined into a conspicuous network; petals white mainly (rarely pinkish); mature joints of fruit rounded or curved on both margins, 4–6 (rarely 7) mm. long, 3.5–4.5 mm. wide . . 7. D. ILLINOENSE

l. Axis of inflorescence with spreading hairs longer than diameter of axis; stem usually much branched with several inflorescences; leaflets thinner, the nerves on the lower surface joined into only a faint network; petals pink to white, drying blue-green; mature joints of fruit angled or acutish on the lower (ventral) margin, 6–13 mm. long, 4–7 mm. wide . 6. D. CANESCENS

k. Hairs if present on lower surface of leaflets not hooked; the stipules (at base of leaf-stalk) 0.5–3 mm. wide at the base, not clasping the stem at their base . . . . . . . . . . . . . . *m*

m. Middle (terminal) one of the three leaflets usually ending in a long point (long-acuminate); stipules (at base of leaf-stalk) conspicuous, 10–20 mm. long . . . . . . . 13. D. CUSPIDATUM

m. Without the above combination of characters; middle (terminal) one of the three leaflets blunt (obtuse), rounded, or short-pointed (acute or acutish) at tip; stipules (at base of leaf-stalk) narrow and subulate to lanceolate or ovate, mainly 2–10 mm. long . . . . . . . . . . *n*

n. Flowers small, 3–6 mm. long; calyx 2–3 mm. long; mature joints of fruits mostly 1–2 (or 3); floral bracts 2–4 mm. long. . . . . . . . . . . . . . . . . . . 9. D. RIGIDUM

n. Without the above combination of characters; flowers 5–13 mm. long; calyx mainly 3–7 mm. (rarely to 2 mm. in *D. paniculatum*) long; mature joints of fruits principally 4–6 (rarely 3); floral bracts 2–10 mm. long . . . . . . . . . . . . . . . . . . . . *o*

o. Flowers large and showy, mainly 9–13 mm. long; floral bracts attached to inflorescence before flowering conspicuous, 4–10 mm. long; stalks (petioles) of middle and upper leaves 5–10 mm. long (the lower on longer petioles up to 25 mm. long); combined length of petiole and stalk (petiolule) of middle (terminal) one of the three leaflets of middle and lower leaves mostly $\frac{1}{5}$–$\frac{1}{3}$ the length of middle leaflet; joints of mature fruit rounded on the lower (ventral) margin, 5–7 mm. long . . . . . . . . . 12. D. CANADENSE

o. Without the above combination of characters; flowers smaller, 5–8 mm. long; floral bracts attached to inflorescence before flowering small, not conspicuous, 2–3 mm. long; stalks (petioles) of middle and upper leaves mainly 12–50 mm. long, or, if shorter then having the other characters here noted; combined length of petiole and stalk (petiolule) of middle (terminal) one of the three leaflets of middle and upper leaves sometimes longer than or nearly as long as length of the middle leaflet (in *D. paniculatum*), sometimes only $\frac{1}{4}$–$\frac{1}{2}$ the length of middle leaflet (as in *D. viridiflorum* and *D. Nuttallii*); joints of mature fruit angled or rounded (in *D. Nuttallii*) on the lower (ventral) margin, 3–8 mm. long . . . . . . . . . . . . . . . . . . . . . . . . . . . *p*

p. Stalks (petioles) of middle and upper leaves less than 10 mm. long . 16. D. PANICULATUM

p. Stalks (petioles) of middle and upper leaves longer than 10 mm. . . . . . . . . *q*

q. Hairs often absent or very sparse, or if present on lower surface of leaflets either appressed (lying parallel to leaf surface) or spreading, but the lower surface not velvety-hairy; floral bracts long-triangular or lanceolate, tapering to a long-pointed tip, sparsely hairy to nearly glabrous; stipules (at base of leaf-stalk) not reddish, subulate, or hair-like, tapering to a long pointed tip, 8–10 or more times as long as broad . . . . . . . . . . . . . . . 16. D. PANICULATUM

q. Lower surface of leaflets velvety-hairy, the hairiness thick, soft, or dense; floral bracts broad at their base, triangular but not long-pointed, $1\frac{1}{2}$–$2\frac{1}{2}$ times as long as broad, hairy; stipules usually reddish, ovate-triangular, $1\frac{1}{2}$–$2\frac{1}{2}$ times as long as broad . . . *r*

r. Mature fruit nearly straight, the 4–5 joints 5–8 mm. long and usually obtusely angled on the lower (ventral) margin; leaflets broadly ovate, the middle

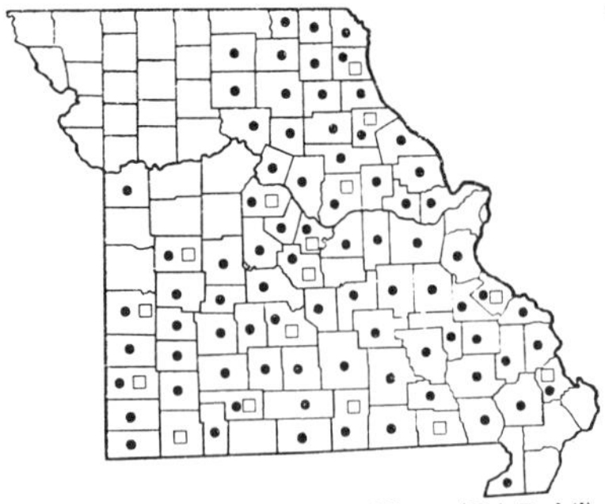

1336 • Desmodium nudiflorum f. nudiflorum (Tick Trefoil)
1336 □ Desmodium nudiflorum f. foliolatum

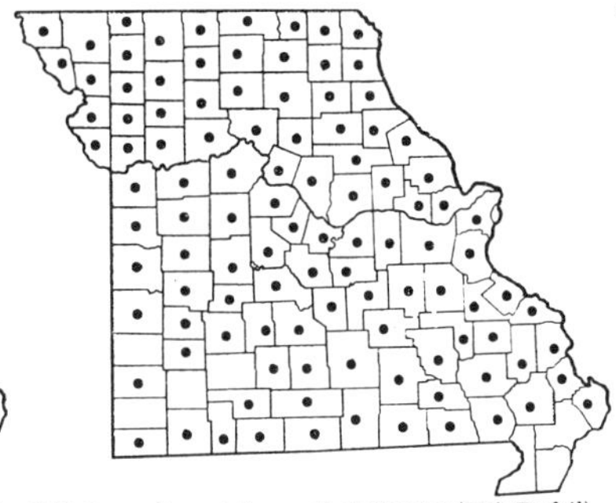

1337 Desmodium glutinosum f. glutinosum (Tick Trefoil)

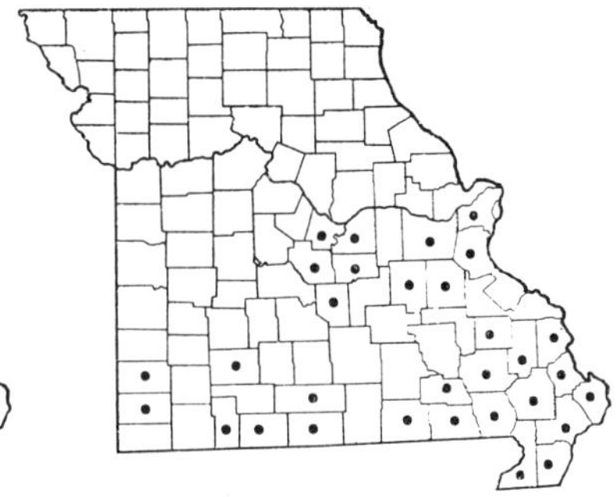

1338 Desmodium pauciflorum (Tick Trefoil)

(terminal) one of the three leaflets often over 10 cm. long, more than half as long as wide, tending to be broadest at or near the base . . . 14. D. VIRIDIFLORUM

r. Mature fruit usually curved inwards, the 3–4 joints 3.5–5 mm. long, rounded on both margins; leaflets ovate to elliptic, rarely up to 10 cm. long, tending to be broadest nearer the middle . 15. D. NUTTALLII

1. **Desmodium nudiflorum** (L.) DC.
Tick Trefoil Map 1335
Flowers late June–August.

Occurs in acid soils of chert, sandstone, or granite in open or rocky dry woodland. Southern, central, and eastern Missouri, east of a line drawn from Schuyler, Sullivan, Linn, Chariton, Howard, Cooper, Benton, and Henry counties to Vernon County; also locally in Jackson County (Courtney, September 18, 1922, *Bush 9917*; same locality, July 16, 1915, *Bush 7684*); absent from northwest Missouri and three southeastern lowland counties.

Ranges from Maine to Quebec and Minnesota, south to Florida, Mississippi, Louisiana, and Texas.

Two variations are found in Missouri:

No leaves on the flowering stem . . . . .
1a. D. NUDIFLORUM f. NUDIFLORUM
Leaves present on the flowering stem . . .
1b. D. NUDIFLORUM f. FOLIOLATUM

1a. **Desmodium nudiflorum** f. **nudiflorum**
Map 1336
*Desmodium nudiflorum* (L.) DC. [G, BB, P & S, Steyerm.]

This is the common variation found in Missouri.

1b. **Desmodium nudiflorum** f. **foliolatum**
(Farwell) Fassett Map 1336

Scattered throughout the range of f. *nudiflorum* in Missouri.

The seeds of this species are a favorite food of the wild turkey, and the leaves are browsed by white-tailed deer.

2. **Desmodium glutinosum** (Muhl.) Wood
f. **glutinosum** Tick Trefoil Map 1337
Also called Beggar's Lice.
*Desmodium acuminatum* (Michx.) DC. [P & S]
Flowers early June–August.

Occurs in rocky open or rich woods, in lowland or upland, and along wooded bluffs. Throughout Missouri, doubtless in every county.

Ranges from Maine to Quebec, Minnesota, and North Dakota, south to Florida, Texas, and Mexico.

The seeds of this species are eaten by wild turkey and bobwhite, and the leaves are browsed by white-tailed deer.

3. **Desmodium pauciflorum** (Nutt.) DC.
Tick Trefoil Map 1338
Also called Beggar's Lice.
Flowers June–late September.

Occurs in rich or alluvial woods, in valleys and ravine bottoms, generally in the vicinity of streams and at base of bluffs. Ozark region of southern and east-central Missouri north to St. Louis, Franklin, Osage, Cole, Miller, Greene, and Jasper counties. Some specimens from Boone County, previously identified as this species, are misdetermined *D. nudiflorum* and *D. cuspidatum*.

Ranges from New York to Ohio, Indiana, Illinois, Missouri, and Kansas, south to Florida, Alabama, Louisiana, and Texas.

Plate no. 216. **1.** Desmodium nudiflorum; *a.* Habit of plant, × 1/6; *b.* Fruit, × 1 1/2; *c.* Leaf, × 1/2. **2.** Desmodium rotundifolium, × 1/2. **3.** Desmodium ochroleucum; *a.* Portion of stem and leaves, × 3/8; *b.* Fruit, × 1; Details from Small, The New York Botanical Garden.

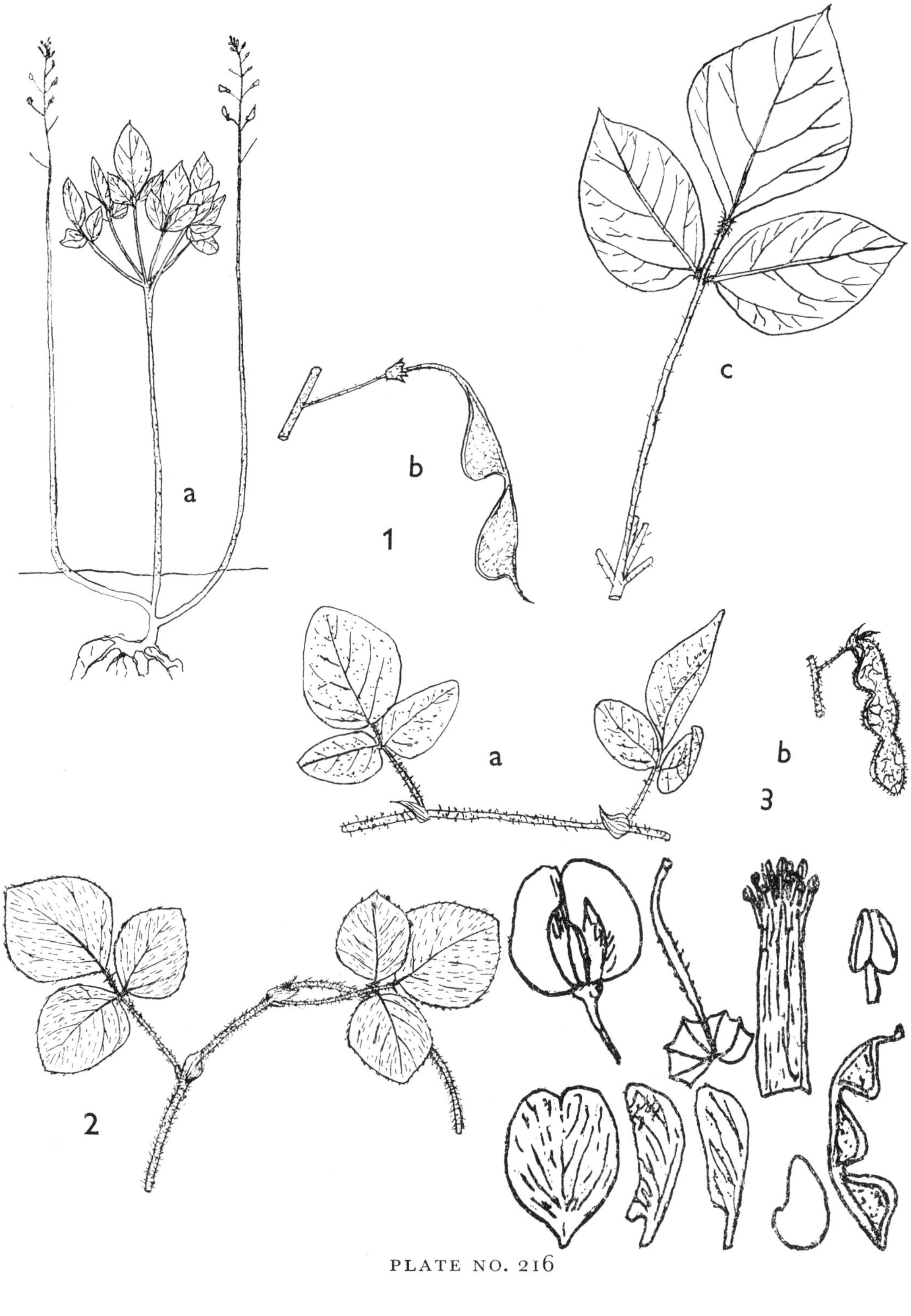


PLATE NO. 216

1339 Desmodium rotundifolium (Tick Trefoil)

1340 Desmodium ochroleucum (Tick Trefoil)

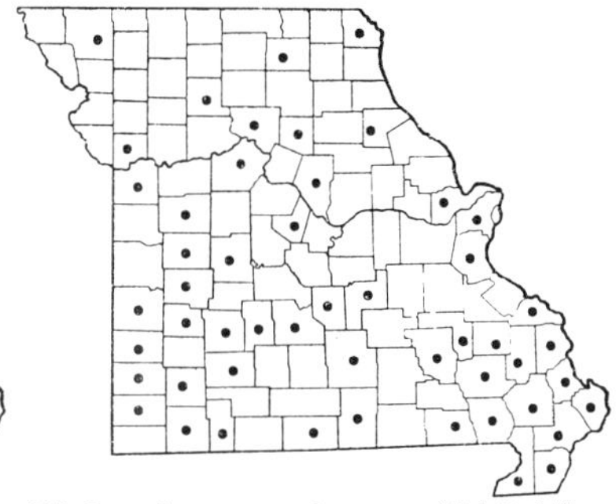

1341 Desmodium canescens f. canescens (Tick Trefoil)

4. **Desmodium rotundifolium** DC.
Tick Trefoil Map 1339
Flowers July–September.

Occurs in acid soils of chert, sandstone, or granite in rocky open dry woodland, usually on crests of hills and tops of ridges. Ozark region of southern and central Missouri south and east of a line drawn from Lincoln, Montgomery, Callaway, Boone, Moniteau, Morgan, Benton, St. Clair, and Cedar counties to Jasper County.

Ranges from Florida to Texas, north to Massachusetts, Vermont, New York, Ontario, Michigan, Illinois, Missouri, and Oklahoma.

This is a common species of the dry upland oak-hickory and pine-oak woodland. The flowers and fruit of this species have been recorded as eaten by wild turkey.

A form of this species having the main stem hairy but the branches glabrous is known as f. *glabratum* (Gray) Schub., but has not been recorded for Missouri.

5. **Desmodium ochroleucum** M. A. Curtis
Tick Trefoil Map 1340
Flowers in August–September.

Occurs in sandy and low open woods. Known only from Dunklin (Malden, August 27, 1894, *Bush 78*; September 12, 1893, *Bush*), Stoddard, and Shannon counties.

Ranges from Georgia to Missouri, north to Delaware and Tennessee.

6. **Desmodium canescens** (L.) DC. f. **canescens**
Tick Trefoil Map 1341
Also called Beggar's Lice.
*Desmodium canescens* (L.) DC. [G, BB, P & S, Steyerm.]
*Desmodium canescens* var. *hirsutum* (Hook.) Rob. [P & S]
Flowers July–September.

Occurs in dry open woods, thickets, open ground in valleys and upland, alluvial ground along streams, roadsides, and railroads.

Ranges from Massachusetts to Ontario, Wisconsin, Iowa, and Nebraska, south to Florida, Alabama, Mississippi, Louisiana, and Texas.

A white-flowered form, f. *albinum* Fern., is known, but has not been recorded from Missouri. The flowers and fruits of this species have been recorded as food of wild turkey, and white-tailed deer are known to eat the leaves of this species.

7. **Desmodium illinoense** Gray Tick Trefoil
Map 1342
Also called Beggar's Lice.
Flowers early June–September.

Occurs in prairies, glades, prairie remnants along railroads, and rarely in rocky open woods. Northern, central, and western Missouri, absent from Ozark region except for isolated occurrences in prairie areas in Crawford, Phelps, and Wright counties.

Ranges from Ohio and Michigan to Minnesota, South Dakota, and Nebraska, south to Indiana, Illinois, Missouri, Oklahoma, and Texas.

The bracts of the inflorescence of this species are prominently ribbed with ciliate margins. This, together with the prominently reticulate lower surface of the leaflets, the longer petioles, and the usually unbranched solitary inflorescence readily distinguishes it from *D. canadense*, another prairie species often associated with it. The upper surface of the leaflets is rather sticky to the touch.

Plate no. 217. **1.** Desmodium glutinosum; *a.* Fruit, × $1^1/_2$; *b.* Habit of plant, × $^3/_8$; *c.* Leaf, × $^9/_{16}$. **2.** Desmodium sessilifolium; *a.* Fruit, × $2^1/_4$; *b.* Leaf, × $^1/_2$. **3.** Desmodium canescens; *a.* Portion of inflorescence, × $^1/_2$; *b.* Leaf, × $^1/_2$; *c.* Fruit, × 1.

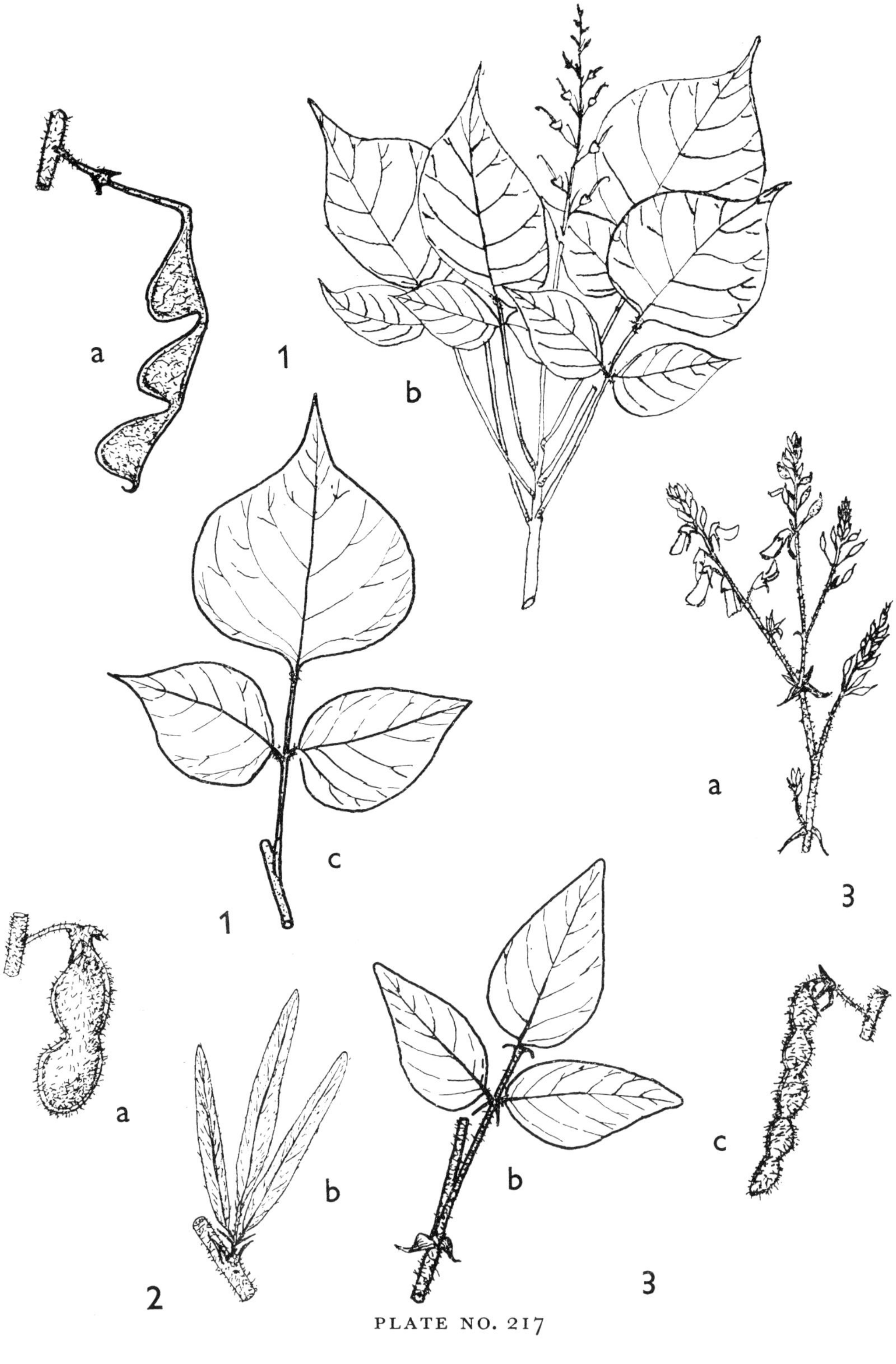


PLATE NO. 217

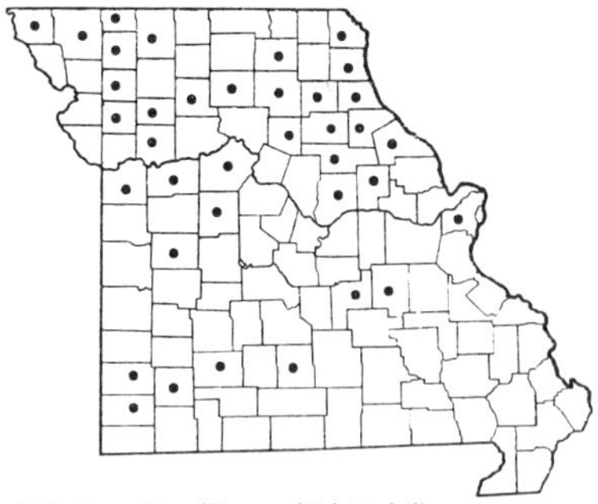

1342 Desmodium illinoense (Tick Trefoil)

1343 Desmodium sessilifolium (Tick Trefoil)

1344 Desmodium rigidum (Tick Trefoil)

8. **Desmodium sessilifolium** (Torr.) T. & G.
Tick Trefoil Map 1343
Also called Beggar's Lice.
Flowers early June–September.

Occurs in prairies, glades, rocky open woods, open banks and roadsides, and prairie remnants along railroads, mostly in dry soils. Southern, central, and eastern Missouri west to Sullivan, Linn, Chariton, Saline, and Jackson counties; absent from a large part of the southeastern Ozark section.

Ranges from Massachusetts to Michigan, Illinois, Missouri, and Kansas, south to South Carolina, Georgia, Alabama, Mississippi, Louisiana, and Texas.

The fruits of this species are eaten by wild turkey, bobwhite, and other birds. The plant is eaten by all classes of livestock as a palatable and nutritious forage species, and is valuable for its protein.

9. **Desmodium rigidum** (Ell.) DC.
Tick Trefoil Map 1344
Also called Beggar's Lice.
Flowers July–September.

Occurs in dry rocky woods, in acid soils of chert, sandstone, or granite, and in glades and prairies. Ozark and unglaciated prairie regions of Missouri, north to Ralls, Montgomery, Callaway, Boone, Pettis, Johnson, and Jackson counties.

Ranges from Massachusetts and New York to Ohio, Michigan, Indiana, Illinois, Missouri, and Kansas, south to South Carolina, Florida, Louisiana, and Texas.

This species varies considerably in size and shape of the leaflets, and length of petioles. Since these proportions are somewhat different from those stated in some of the manuals, the following measurements are shown by Missouri material: petioles 0.5–2 cm. or more long; leaf-blades 4.5–7.5 cm. long, 1.8–3.3 cm. broad. The leaflets vary from narrowly to broadly ovate to lanceolate, rounded to truncate at the base, usually obtuse at the apex, but also sometimes slightly acute. It is often associated with *D. ciliare*, *D. marilandicum*, and other species of *Desmodium* of acid soils of rocky open woods or prairies.

10. **Desmodium ciliare** (Muhl.) DC. var. **ciliare**
Tick Trefoil Map 1345
Also called Beggar's Lice.
*Desmodium obtusum* of auth. [P & S], not (Muhl.) DC.
Flowers August–October.

Occurs in dry acid soils and on rocks of chert, sandstone, or granite in rocky open woods, rocky glades, and open ground. Ozark and unglaciated prairie regions of southern and central Missouri north to St. Charles, Warren, Callaway, Boone, Cooper, Pettis, Johnson, and Bates counties, and locally in Randolph County.

Ranges from Florida to Texas and Mexico, north to Massachusetts, New York, Ohio, Michigan, Indiana, Illinois, Missouri, and Kansas; also in the West Indies and Central America (British Honduras).

The seeds are eaten by wild turkey and bobwhite.

A variation with the leaflets lanceolate to lance-ovate, known as var. *lancifolium* Fern. & Schub., is found in Virginia.

11. **Desmodium marilandicum** (L.) DC.
Tick Trefoil Map 1346
Also known as Beggar's Lice.
Flowers July–September.

Occurs in dry acid soils of chert, sandstone, or granite in rocky or open woodland, usually on ridges, uplands, and upper slopes, thickets, and occasionally in open places. Southern, central, and northeastern

Plate no. 218. **1.** Desmodium pauciflorum; *a.* Leaf, × $^1/_2$, *b.* Habit of plant, × $^1/_4$. **2.** Desmodium illinoense; *a.* Leaf, × $^1/_2$; *b.* Inflorescence, × $^1/_2$; *c.* Fruit, × 1. **3.** Desmodium rigidum; *a.* Leaf, × $^1/_2$.

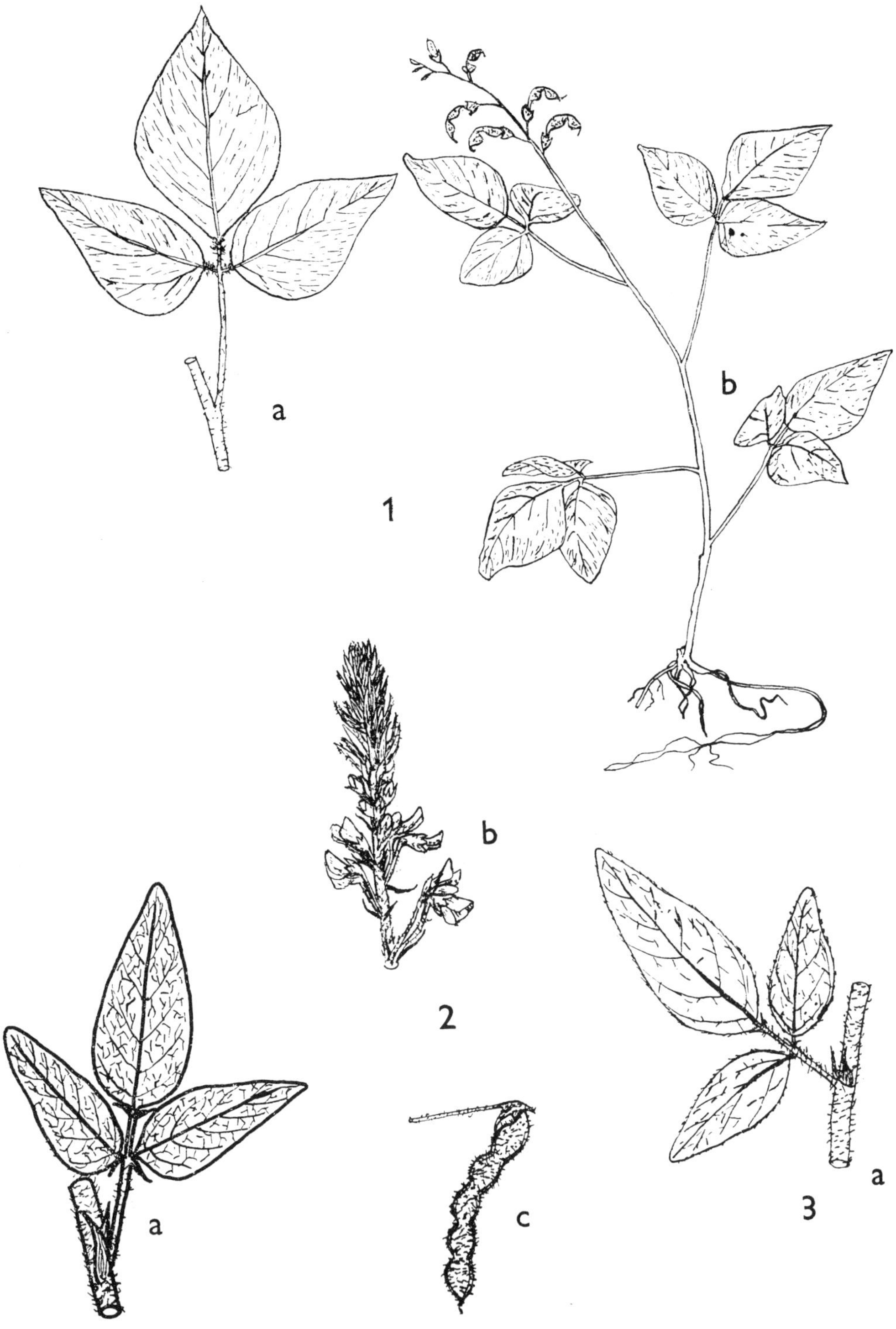


PLATE NO. 218

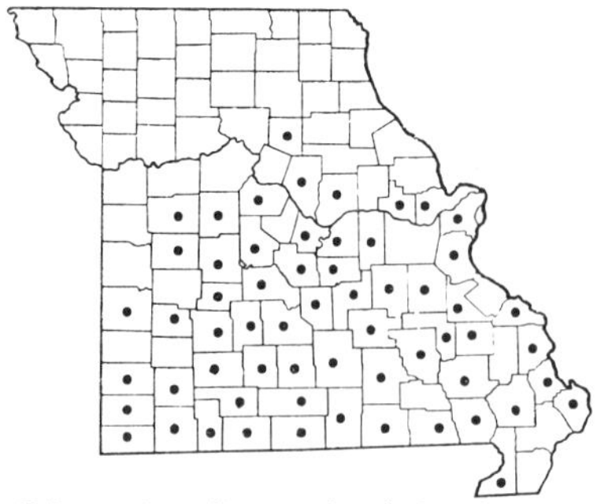

1345 Desmodium ciliare var. ciliare (Tick Trefoil)

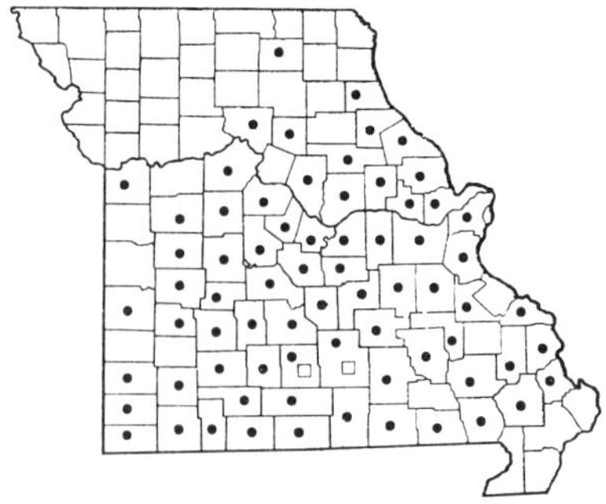

1346 • Desmodium marilandicum (Tick Trefoil)
1346 □ Desmodium ciliare X marilandicum

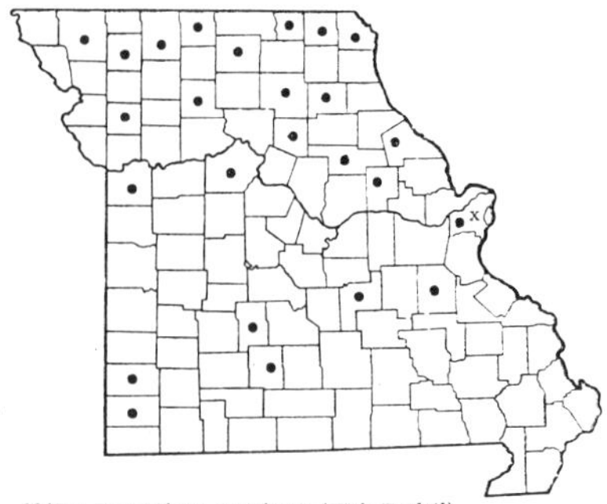

1347 • Desmodium canadense (Tick Trefoil)
1347 x Desmodium canadense X paniculatum var. Dillenii

Missouri, north to Marion, Adair, Chariton, Saline, and Jackson counties; absent from the lowlands of southeastern Missouri, but on Crowley Ridge in Scott and Stoddard counties.

Ranges from Massachusetts to Ontario, Michigan, Illinois, Missouri, and Kansas, south to Georgia, Texas, and Oklahoma.

Plants which appear to represent hybrids between *D. ciliare* and *D. marilandicum* have been collected in Texas and Wright counties.

The leaves are browsed by white-tailed deer, and the seeds are eaten by wild turkey and bobwhite.

12. **Desmodium canadense** (L.) DC.
Tick Trefoil Map 1347
Also called Beggar's Lice.
Flowers July–September.

Occurs in prairies, wet meadows in valleys, along spring branches, and open thickets. Northern and central Missouri in glaciated prairie, and locally south of the Missouri River in prairie remnants in Crawford, Phelps, Dallas, Webster, Jasper, and Newton counties.

Ranges from Quebec to North Dakota and Saskatchewan, south to Virginia, West Virginia, Ohio, Indiana, Illinois, Missouri, and Oklahoma.

13. **Desmodium cuspidatum** (Muhl.) Loud.
Tick Trefoil Map 1348
Also called Beggar's Lice.
*Desmodium bracteosum* (Michx.) DC.
Flowers July–September.

Occurs in dry or rocky woodland and thickets, along bluffs or at base of slopes, ravines, valleys, often in limestone soils. Throughout Missouri, except absent from the lowland counties of southeastern Missouri.

Two varieties are encountered in Missouri material:

Stems, leaves, and bracts of flowers glabrous or nearly so. . 13a. D. CUSPIDATUM var. CUSPIDATUM
Stems and leaves with long straight spreading and shorter hooked hairs; bracts of flowers with longer hairs on margins and minute hairs on outer surface . . . 13b. D. CUSPIDATUM var. LONGIFOLIUM

13a. **Desmodium cuspidatum** var. **cuspidatum** Map 1348
*Desmodium cuspidatum* (Muhl.) Loud. [G]
*Desmodium grandiflorum* (Walt.) DC., not of auth. generally
*Desmodium bracteosum* (Michx.) DC. [P & S]

This is the common variation in Missouri throughout its occurrence in the state.

Ranges from New Hampshire and Vermont to Michigan, and Wisconsin, south to Florida, Arkansas, and Texas.

13b. **Desmodium cuspidatum** var. **longifolium** (T. & G.) Schub. Map 1348
*Desmodium bracteosum* var. *longifolium* (T. & G.) Robins. [P & S]

Scattered mainly in the northern and central portions of Missouri south to Phelps and Jasper counties.

Ranges from Ohio, Michigan, Indiana, Illinois, Wisconsin, Minnesota, Iowa, and Nebraska, south to Alabama, Louisiana, and Kansas.

The large flowers, conspicuous stipules and floral bracts, long-petioled leaves with large, long-acuminate terminal leaflets, and long fruits with large joints

---

Plate no. 219. **1.** Desmodium ciliare; *a.* Leaf, × 1; *b.* Fruit, × 2. **2.** Desmodium marilandicum; *a.* Leaf, × 1; *b.* Fruit, × $1^1/_4$. **3.** Desmodium cuspidatum; *a.* Leaf, × $^1/_2$; *b.* Inflorescence, × $1^1/_2$; *c.* Fruit, × $^3/_4$. **4.** Desmodium canadense; *a.* Fruit, × $1^1/_4$; *b.* Leaf, × $^1/_4$.

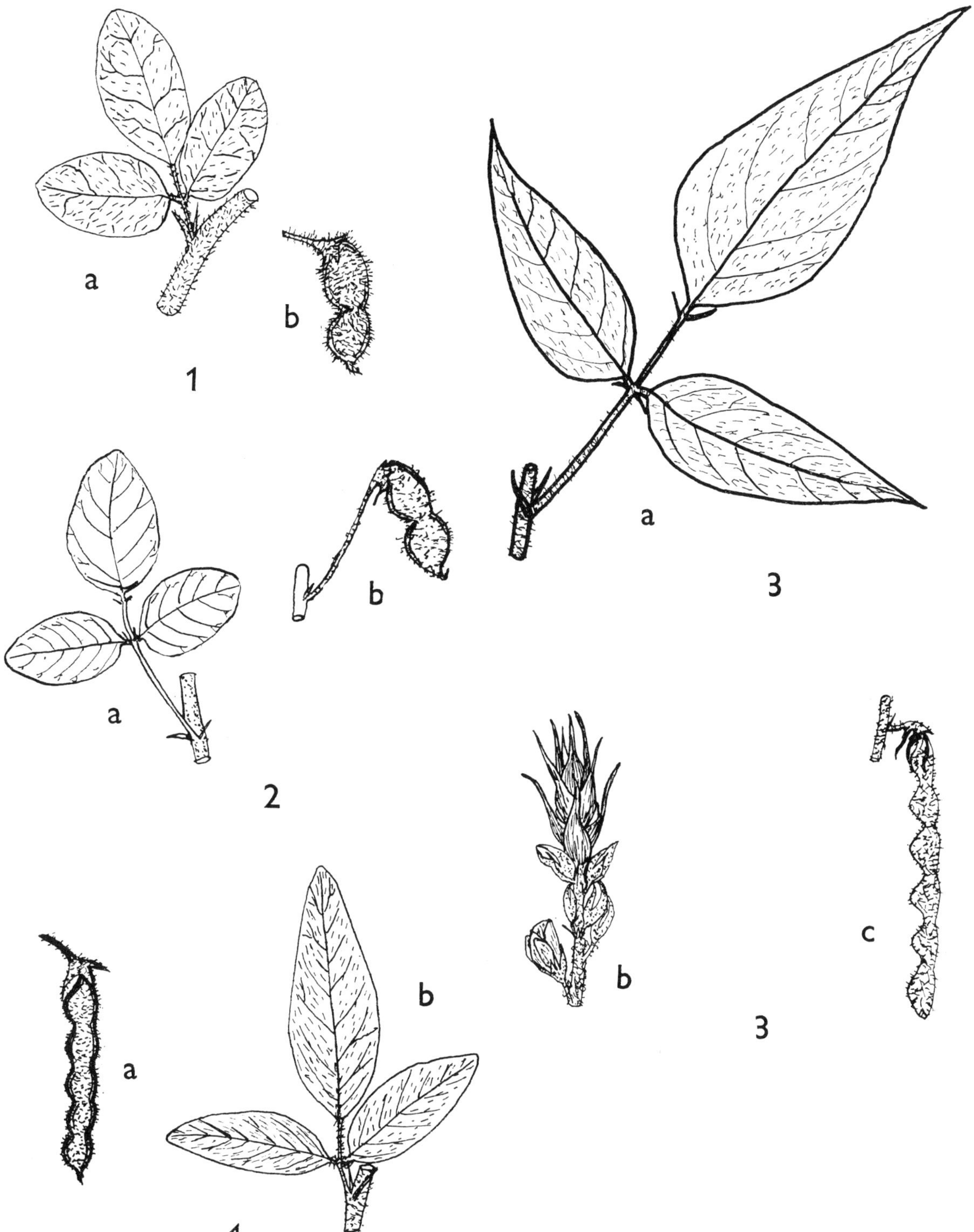


PLATE NO. 219

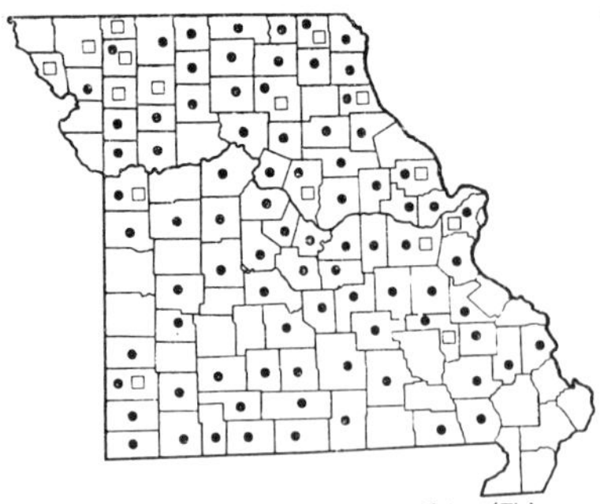

1348 • Desmodium cuspidatum var. cuspidatum (Tick Trefoil)
1348 □ Desmodium cuspidatum var. longifolium

1349 Desmodium viridiflorum

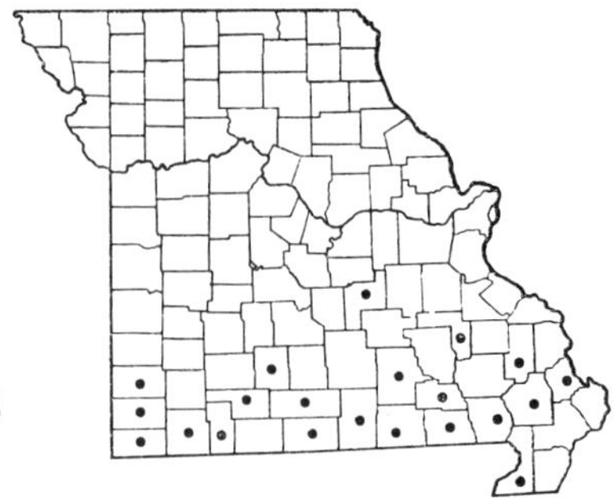

1350 Desmodium Nuttallii (Tick Trefoil)

are distinguishing marks of this species. The leaves are browsed by white-tailed deer.

14. **Desmodium viridiflorum** (L.) DC.
Tick Trefoil Map 1349
Also called Beggar's Lice.
Flowers July–September.

Occurs in sandy open woods. Known only from Crowley Ridge, southeastern Missouri, in Stoddard and Dunklin counties.

Ranges from Florida to Texas, north to Delaware, Tennessee, Missouri, and Oklahoma.

Future studies may show that this and the following species, *D. Nuttallii*, may have to be combined under *D. viridiflorum*, as already has been done by Gleason in the *New Illus. Fl.* (vol.2: 431. 1952). The chief difference between these two taxa is in the curvature of the lower margin of the fruiting joints.

From rather pubescent-leaved plants of *D. paniculatum* var. *Dillenii*, this species and *D. Nuttallii* are best separated by the shape of the stipules and floral bracts, as indicated in the key.

15. **Desmodium Nuttallii** (Schindl.) Schub.
Tick Trefoil Map 1350
Also called Beggar's Lice.
*Desmodium viridiflorum* of auth. in part [BB, P & S], not (L.) DC.
Flowers August–September.

Occurs in rocky open woods, rocky glades, and prairies, chiefly in acid soils of chert, sandstone, and granite. Southern Missouri north to Scott, Bollinger, Phelps, Webster, Christian, and Jasper counties.

Ranges from New York to Ohio, Indiana, Illinois, and Missouri, south to Florida, Alabama, Louisiana, and Texas.

The range of this species, as shown on the map, includes all the stations previously cited under *D. viridiflorum* in Palmer and Steyermark's *Annotated Catalogue*. The status of *D. Nuttallii* as a species distinct from *D. viridiflorum* is highly questionable, and it is being retained in the present flora, largely upon the conclusions reached by Dr. Duane Isely (Am. Midl. Nat. 49: 927–33. 1953; Iowa State College Jour. Sci. 30: 62–64, 72. 1955), and partly, because as a recently recognized species, it must await sufficiently detailed, future field and herbarium studies for proper evaluation. Differences between the two taxa are brought out in the key to the species in the present flora.

16. **Desmodium paniculatum** (L.) DC.
Tick Trefoil Map 1351
Flowers July–September.

This is a variable species. Two variations may be recognized in Missouri:

Middle (terminal) one of the three leaflets usually 3–8 times as long as broad, mostly linear-lanceolate to lanceolate; lower surface of the leaflets with hairs, when present, usually appressed (lying parallel to the surface or pressed against it); stems nearly glabrous (without hairs) to shortly hairy with mainly hooked hairs . . 16a. D. PANICULATUM var. PANICULATUM

Middle (terminal) one of the three leaflets mostly $1\frac{1}{2}$–$2\frac{1}{2}$ (up to 3) times as long as broad; lower surface of leaflets hairy with most of the hairs spreading or loose; stems hairy with both spreading and shorter hooked hairs . 16b. D. PANICULATUM var. DILLENII

16a. **Desmodium paniculatum** var. **paniculatum**
Map 1351

Plate no. 220. **1.** Desmodium viridiflorum; *a.* Fruit, $\times 1^1/_4$; *b.* Leaf, $\times {}^1/_2$. **2.** Desmodium Nuttallii; *a.* Fruit, $\times 1^1/_4$; *b.* Leaf, $\times {}^1/_2$. **3.** Desmodium paniculatum; *a.* Leaf, $\times {}^1/_2$; *b.* Portion of inflorescence, $\times 1^1/_2$. **4.** Desmodium laevigatum; *a.* Portion of inflorescence, $\times 1^1/_2$; *b.* Fruit, $\times 1^1/_2$; *c.* Leaf, $\times {}^1/_4$.

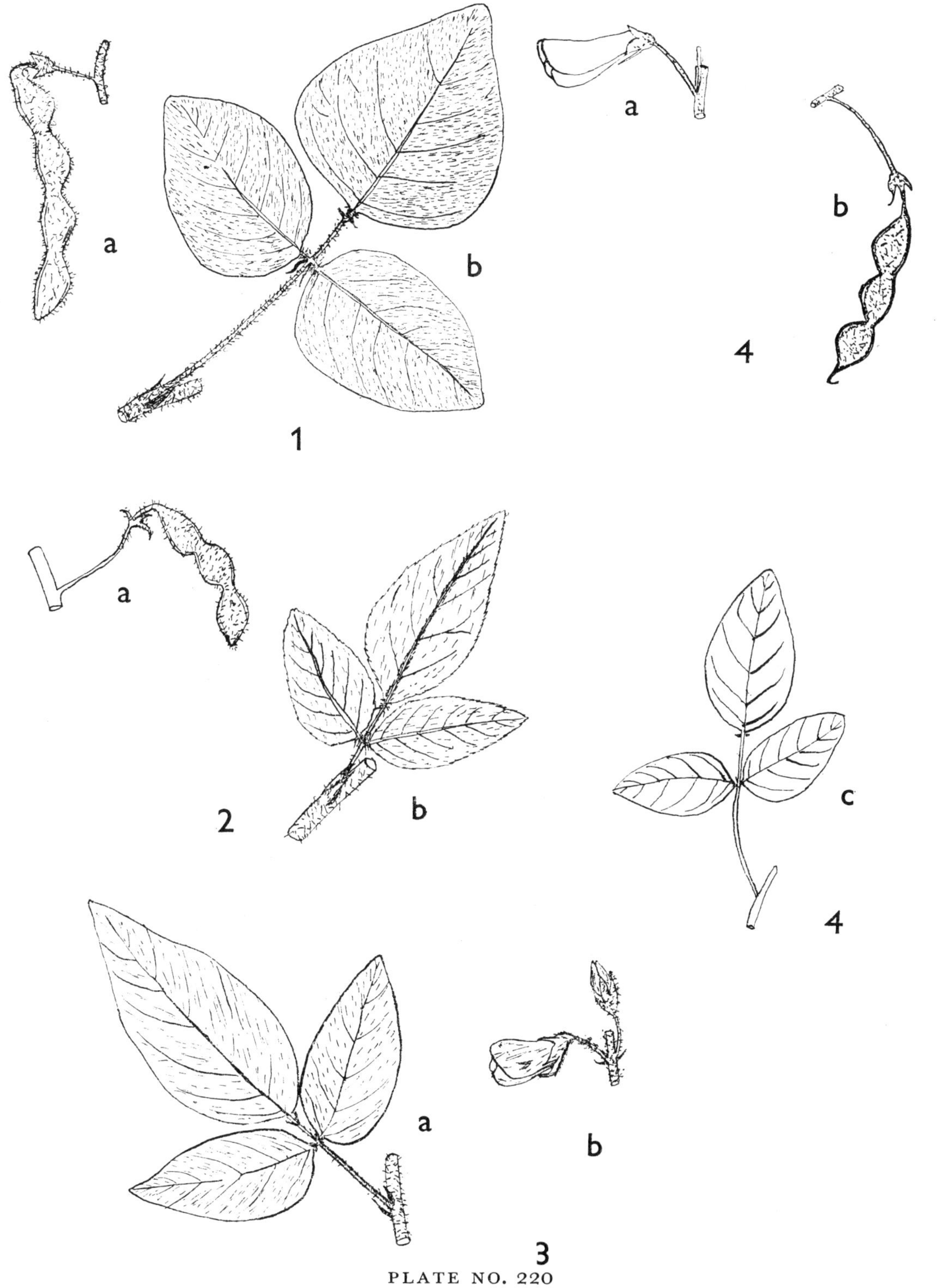


PLATE NO. 220

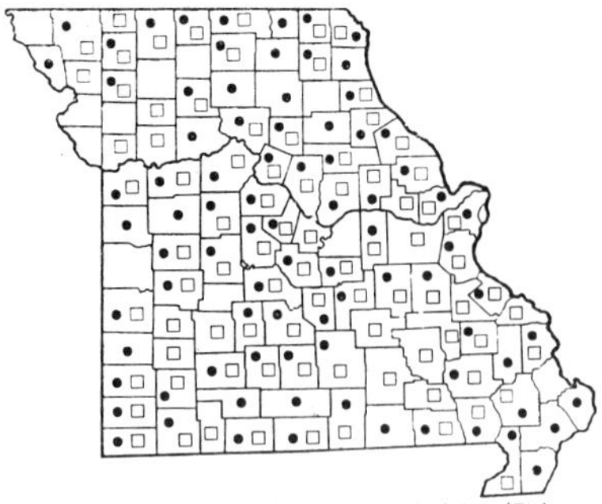

1351 ● Desmodium paniculatum var. paniculatum (Tick Trefoil)
1351 □ Desmodium paniculatum var. Dillenii

1352 Desmodium humifusum (Tick Trefoil)

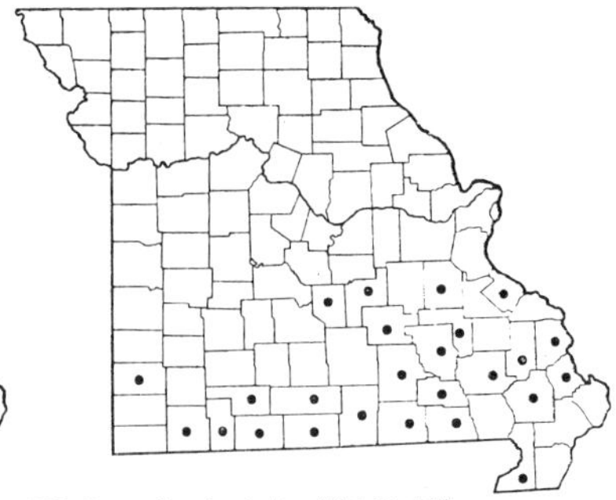

1353 Desmodium laevigatum (Tick Trefoil)

*Desmodium paniculatum* (L.) DC. [G, BB, P & S]
*Desmodium paniculatum* var. *angustifolium* T. & G. [P & S]

Occurs in thickets, dry upland woods, rich woods, open and alluvial ground along streams, borders of ponds and swamps, fallow fields, along roadsides, and railroads. Throughout Missouri, doubtless in every county.

Ranges from Maine to Ontario, Michigan, Iowa, and Nebraska, south to Florida and Texas.

16b. **Desmodium paniculatum** var. **Dillenii** (Darl.) Isely Map 1351

*Desmodium Dillenii* Darl. [P & S]
*Desmodium perplexum* Schub. [G]
*Desmodium glabellum* of auth. in part [G, as to Missouri specimens and range], not (Michx.) DC.

Occurs in rocky open woods, ravines, upland slopes and ridges, upland glades, prairies, and dry thickets. Throughout Missouri.

Ranges from Maine to Wisconsin and Minnesota, south to Florida and Texas.

This group has been studied independently by Dr. Schubert and Dr. Isely, each of whom has reached different conclusions, Dr. Schubert segregating *D. Dillenii* into *D. perplexum* and *D. glabellum*, Dr. Isely placing *D. Dillenii* under *D. paniculatum* as a variety of that species. Although most of the Missouri material plainly falls into what was passing previously as either *D. paniculatum* or *D. Dillenii*, there is intergradation between these two categories of plants in shape and size of leaflets, length of petiole, and abundance and type of pubescence on the stems and leaves. It has not been found possible, however, to arrange the Missouri material, formerly identified as *D. Dillenii*, into the categories, *D. perplexum* and *D. glabellum*, as distinguished by Dr. Schubert in her treatment in *Gray's Manual*, eighth edition. On the other hand, as treated by Dr. Isely (Am. Midl. Nat. 49: 920–27. 1953; Iowa State College Jour. Sci. 30: 64–68. 1955) the *D. Dillenii* complex appears to fit more realistically into the present situation as a variety of *D. paniculatum*, and in the present flora his point of view is followed. Missouri collections, previously identified as *D. Dillenii* and determined as *D. glabellum* by Dr. Schubert, are: Howard Co. (*Steyermark 26324*), Jefferson Co. (*C. H. Thompson*), Reynolds Co. (*Steyermark 19771*), Wright Co. (*Steyermark 25089*), Barry Co. (*Bush 15146*), and Jasper Co. (*Palmer 2696A*). Apparent hybrid plants have been collected between *D. paniculatum* var. *Dillenii* and *D. canadense* from St. Louis County.

The color of the flowers varies in this *D. paniculatum* group from lavender or rose-purple to blue- or greenish-lavender, to white. As the flowers mature or wilt, the flower color usually becomes a dull blue-green. The seeds are eaten by bobwhite and wild turkey.

17. **Desmodium humifusum** (Muhl.) Beck

Tick Trefoil Map 1352

Also called Beggar's Lice.
Flowers July–September.

Occurs in rocky open woods in acid soils of chert or sandstone. Known only from southern Missouri, in Shannon (sandy woods, Monteer, August 6, 1910, *Bush 6118* and *6118A*), Pulaski (dry cherty slopes above northwest- and north-facing wooded limestone bluffs along Gasconade River west of Portuguese Point, T37N, R11W, sect. 36 and 27, 6–7 mi. [by air] south southwest of Dixon, September 10, 1956, *Steyermark 82538*), and Jasper (rocky woods, Carthage, September 11, 1910, *Palmer 3161*) counties.

Ranges from New England to Pennsylvania, Maryland, and Delaware; and Missouri.

The Missouri specimens have been determined by Dr. Schubert, and the Pulaski County material has been matched with authentic specimens. The Missouri stations represent an unusual separation in range, but the Missouri material appears to differ in no respect from typical *D. humifusum*. The peculiar distribution recalls somewhat that of *Trillium pusillum* and var. *ozarkanum*. All the Missouri specimens have a trailing or procumbent habit and agree in all other respects with typical *D. humifusum*.

In order that future collectors may be on the alert for this rare species, the following characters, in addition to those given in the key to species, are provided: stems sparsely to densely hairy with both long spreading straight hairs mixed with shorter, hooked ones; petioles 2.8–4.8 cm. long, with densely spreading pubescence; terminal leaflet 4.7–6.6 cm. wide, 3–5 cm. broad, ovate to rhombic-ovate, the hairs appressed on both surfaces with occasional short hooked hairs on the midrib and lateral nerves of the upper surface; lateral leaflets similar in shape or nearly roundish, with a truncate base, 3.6–6 cm. long, 2–4 cm. broad; floral bracts falling early from the inflorescence; flower-stalks with short hooked hairs; corolla up to 9.5 mm. long; joints of fruit 3–4, the upper margin more or less straight or rhombic and angled.

18. **Desmodium laevigatum** (Nutt.) DC.
Tick Trefoil Map 1353
Also called Beggar's Lice.
Flowers July–September.
Occurs in rocky open woodland in acid soils of chert, sandstone, or granite. Southern Ozark region and Crowley Ridge in southeastern Missouri, north to Ste. Genevieve, Washington, Phelps, Pulaski, Douglas, Christian, and Jasper counties. Reports of this species north of the Missouri River are based upon misidentifications with other species.

Ranges from New York and Pennsylvania to Indiana, Illinois, and Missouri, south to Florida and Texas.

Measurements of this species as given in some of the manuals should be modified to the following to include Missouri material: lateral leaflets 3.2–10 cm. long, 2.5–5.5 cm. broad; pedicels 7–19 mm. long.

This species is easily recognized by its very glaucous or blue-silvery lower leaf-surface which is usually glabrous, although some hairs may be present both on the upper and lower surfaces and on the midrib and veins of the lower side of the leaflet. The leaflets are characteristically firm and stiffish with rather revolute margins. The flowers are rather large and showy, deep rose to purplish, and 8–10 mm. long; their slender pedicels, 7–19 mm. long, are covered with short hooked hairs. The stems are usually glabrous but may also rarely bear sparse, short hairs. The fruit rises on a mostly hairless stipe 6–6.5 mm. long and has 2–5 joints 5–7 mm. long and 3.5–4 mm. wide, angled on both the upper and lower margins.

The leaves of this species are browsed by deer.

## 30. **Lespedeza** Michx. Bush Clover

This genus has received special attention in recent years from Dr. Isely (Iowa State College Jour. Sci. 30: 76–110. 1955) and Dr. Fassett (*The Leguminous Plants of Wisconsin*, pp. 98–109. 1939). Some data in the following key have been drawn from these works.

a. Stipules (at base of leaves or leaf-stalks) brownish, conspicuous and broad, persisting, glabrous (without hairs), as long as or longer than the leaf-stalks; plants annual with weak roots . . . . . . *b*
b. Leaves appearing nearly stalkless, the stalks (petioles) on main stems 1–3 mm. (up to 5) long; hairs on stems pointing toward base of plant; margins of leaves not conspicuously hairy; calyx covering ½–⅘ of the fruit . . . . . . 14. L. STRIATA
b. Principal leaves of the main stem on stalks (petioles) 4–10 mm. long; hairs on stems pointing toward tip of plant; margins and lower midrib of young leaflets conspicuously hairy; calyx covering ⅓–½ of the fruit . . . . . . 15. L. STIPULACEA
a. Stipules barely evident, needle-shaped or very narrow or hair-like, soon falling off or withering, more or less covered with hairs, mostly shorter than the leaf-stalks; plants perennial with a strong main root . . . . . . *c*
c. Plants with the stems trailing or lying upon the ground . . . . . . *d*
d. Hairs of stem and usually of leaf-stalks spreading; inflorescences with usually 6–10 flowers . 7. L. PROCUMBENS
d. Hairs, when present, of stem and leaf-stalks, appressed (lying flat upon or parallel to surface); inflorescences with usually 4–6 flowers . . . . . . 1. L. REPENS
c. Plants with the stems upright or ascending or upwardly spreading, but not lying upon the ground or trailing . . . . . . *e*

e. Hairs of stem spreading (at right or acute angles from stem surface) . . . . . . . . . . *f*

f. Peduncles (stalk bearing inflorescences) longer than the subtending leaves which immediately arise at their base . . . . . . . . . . . . . . . . . . . . . . *g*

g. Flowers yellowish-white; leaflets rounded-obovate to nearly orbicular, $1\frac{1}{4}$–2 times as long as broad; most of the leaf-stalks (petioles) much shorter than the length of the middle (terminal) one of the three leaflets ($\frac{1}{8}$–$\frac{1}{3}$ length), apetalous (normally lacking petals) flowers few or none; common . . . . . . . . . . . . . 12. L. HIRTA

g. Petal-bearing flowers purplish or rose-lavender; leaflets elliptical to oblong, 2–3 times as long as broad; most of the leaf-stalks (petioles) about equaling length of or about $\frac{1}{2}$ the length of the middle (terminal) one of the three leaflets; apetalous flowers usually numerous, located in stalkless or nearly stalkless clusters in the axils of the leaves next to the stem; rare . . . . . . . . . . . . . . . . . . . . . . *h*

h. Calyx of petal-bearing flowers about equaling length of fruit, 6–8 mm. long . 8. L. MANNIANA

h. Calyx of petal-bearing flowers shorter than the fruit, 4–6 mm. long . . 7. L. NUTTALLII

f. Peduncle (stalk bearing inflorescences) shorter than the subtending leaves which immediately rise at their base . . . . . . . . . . . . . . . . . . . . . . *i*

i. Calyx of petal-bearing flowers shorter than the fruit, 3–5 mm. long; stipules without evident nervation . . . . . . . . . . . . . . . . . . . . 10. L. STUEVEI

i. Calyx of petal-bearing flowers equaling or longer than the fruit, 6–10 mm. long; stipules 3-nerved . . . . . . . . . . . . . . . . . . . . . . . . *j*

j. Flowers creamy-yellow or yellowish-white; larger main leaflets 3–5 cm. long; leaf-stalks (petioles) mostly 2–5 mm. long; commonly found . . . . . . . 11. L. CAPITATA

j. Flowers purplish or rose; larger main leaflets 1.5–3.5 (rarely 5) cm. long; leaf-stalks (petioles) mainly 4–13 mm. long; rarely found . . . . . . . . . . 9. L. SIMULATA

e. Hairs of stem none or appressed (appearing pressed against or lying parallel to surface of stem) . . *k*

k. Hairs of stem occurring principally in lines on the conspicuous ridges; petal-bearing flowers yellowish-white with a purple-red spot on upper petal (standard), in 2–3-flowered clusters in the axils of the leaves next to the stem . . . . . . . . . . . 13. L. CUNEATA

k. Hairs of stem not occurring in lines, the stems not conspicuously ridged or lined; petal-bearing flowers purplish or rose-lavender, in 4–many-flowered racemes either on peduncles (stalk bearing inflorescence) or stalkless . . . . . . . . . . . . . . . . *l*

l. Petal-bearing flowers on long slender peduncles (stalks bearing inflorescences), most of them 2–4 times longer than the subtending leaves which arise immediately at their base . . *m*

m. Flowers all of one kind and large and showy, 12–17 mm. long; inflorescence long and curving, many-flowered, up to 20 cm. long; fruit 7–10 mm. long; leaf-stalks (petioles) 3–7 cm. long; escaped from cultivation . . . . . . . . . . . . 4. L. THUNBERGII

m. Flowers of 2 kinds, the petal-bearing ones 6–9 mm. long; inflorescence short, at most 5 cm. long; fruit 4–6 mm. long; leaf-stalks (petioles) of the larger leaves 1.5–4 cm. long; common native species . . . . . . . . . . . . . . . 3. L. VIOLACEA

l. Petal-bearing flowers on stouter peduncles (stalks bearing inflorescences), some or all of which are shorter than the subtending leaves which arise immediately at their base . . . *n*

n. Most of leaflets narrowly oblong or linear, 2–7 mm. broad, most of them usually 4–6 times as long as broad; petal-bearing inflorescences mostly without peduncles (stalks) or nearly without peduncles . . . . . . . . . . . . . . . . 5. L. VIRGINICA

n. Leaflets oval or broadly elliptic, mainly 6–20 mm. broad, $1\frac{1}{2}$–3 times as long as broad; petal-bearing inflorescences often with short peduncles . . . . . . 6. L. INTERMEDIA

1. **Lespedeza repens** (L.) Bart. Map 1354

Flowers June–September.

Occurs in usually acid soils of chert, sandstone, or granite in rocky dry open woods on upland slopes and ridges, more rarely in valleys and open meadows of river bottoms.

Southern and central Missouri in the Ozark and unglaciated prairie region north to Jefferson, Franklin, Gasconade, Cole, Moniteau, Benton, Henry, and Cass counties, and locally northeast in Pike County. Records previously ascribed to Saline, Boone, and Lincoln counties were misidentified and are referred

Plate no. 221. **1.** Lespedeza repens, × 1. **2.** Lespedeza procumbens, × 1. **3.** Lespedeza violacea; *a.* Portion of fruiting plant, × 1$^1/_2$; *b.* Portion of Flowering plant, × 2. **4.** Lespedeza intermedia, × $^1/_2$; Details from Small, The New York Botanical Garden.

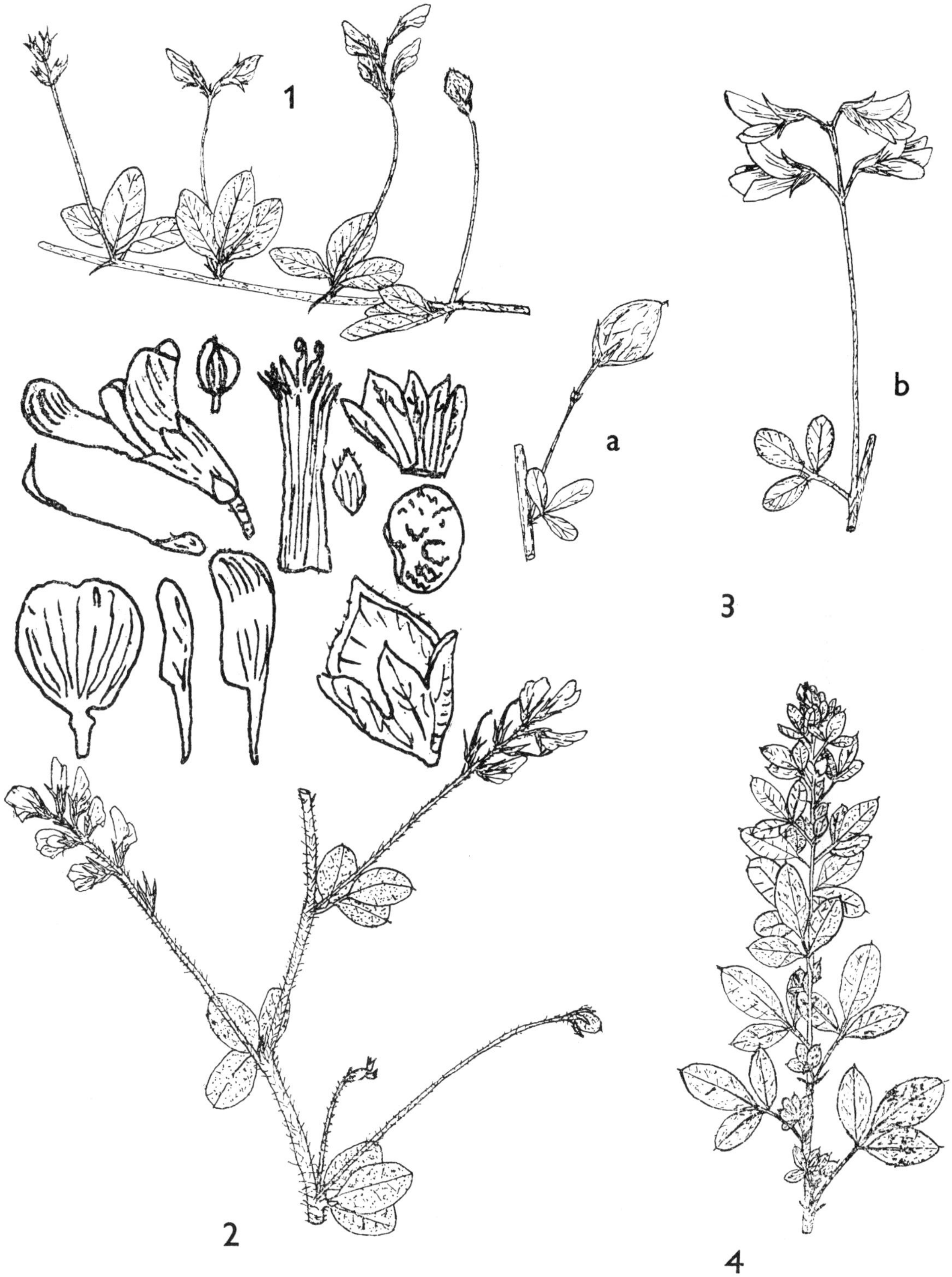


PLATE NO. 221

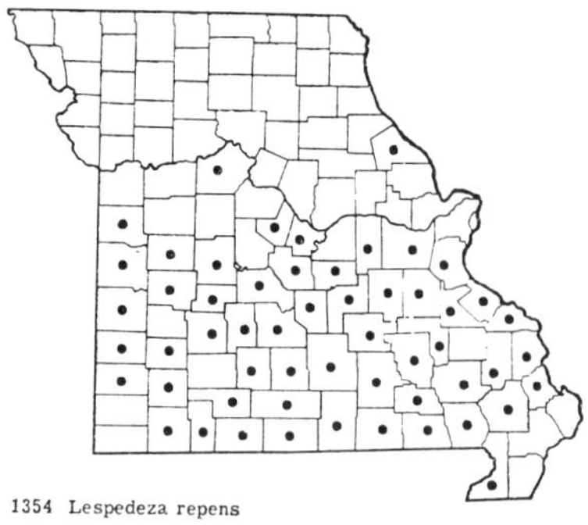

1354 Lespedeza repens

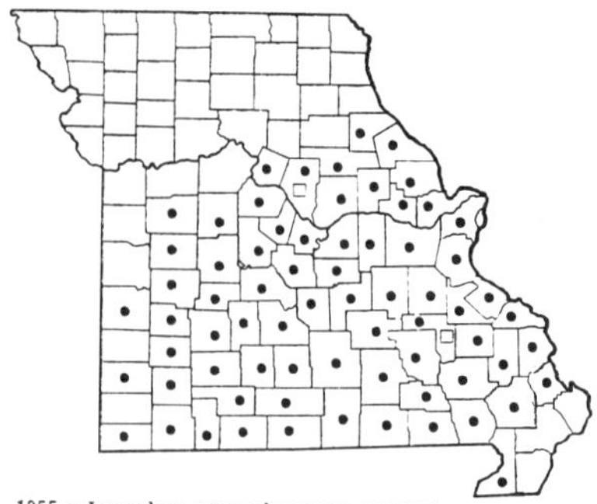

1355 • Lespedeza procumbens var. procumbens
1355 □ Lespedeza procumbens var. elliptica

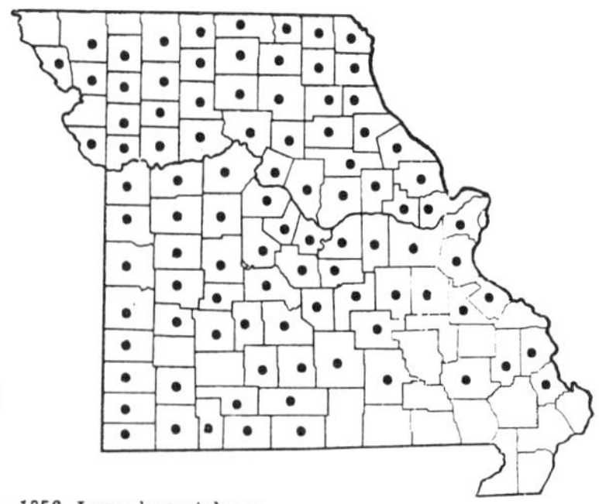

1356 Lespedeza violacea

in the present flora to *L. procumbens* (*Steyermark 8123* from Lincoln Co.) and *L. violacea* (*Daniels, Favor* from Boone Co.).

Ranges from Florida to Texas, north to Connecticut, New York, Ohio, Indiana, southwestern Wisconsin, southeastern Iowa, Missouri, and Kansas.

This species is sometimes confused with *L. procumbens* and *L. violacea*. From *L. procumbens* it is easily distinguished by the appressed pubescence, while from *L. violacea* it mainly is distinguished by the prostrate habit of growth, the short petioled (0.5–1.5 cm.) leaves with the larger leaflets of the main stem only 1–2 cm. long, as contrasted with petioles 2–4 cm. long and larger leaflets 2–4 cm. long in *L. violacea*. The leaflets in *L. repens* are usually more narrowed and cuneate at the base, whereas in *L. violacea* they are more rounded at the base and more oblong in shape.

The leaves of *L. repens* are eaten by white-tailed deer, and the seeds form one of the foods taken by wild turkey and bobwhite.

2. **Lespedeza procumbens** Michx. Map 1355

Flowers August–October.

Occurs in dry rocky woods in acid soils of chert, sandstone, or granite, usually on upland slopes and ridges. Ozark and unglaciated prairie region of southern and central Missouri, northeast to Ralls County, thence south and east of a line from Audrain, Boone, Howard, Pettis, and Johnson counties to Vernon County.

Ranges from Florida to Texas, north to New Hampshire, Massachusetts, New York, Ohio, Indiana, Wisconsin, Iowa, and Kansas.

Two variations are known from Missouri:

Leaflets 2 times or less as long as broad, oval, 1.2–3.5 cm. long, 0.7–2 cm. broad . . . . . 2a. L. PROCUMBENS var. PROCUMBENS

Leaflets about 4 times as long as broad, narrowly elliptic-oblong, the larger ones 1.8–3.5 cm. long, 4–9 mm. broad . . . . . 2b. L. PROCUMBENS var. ELLIPTICA

2a. **Lespedeza procumbens** var. **procumbens** Map 1355

*Lespedeza procumbens* Michx. [G, BB, P & S]

This is the common variation in Missouri.

Ranges from Florida to Texas, north to New Hampshire, Massachusetts, New York, Ohio, Indiana, Wisconsin, Iowa, and Kansas.

2b. **Lespedeza procumbens** var. **elliptica** Blake Map 1355

Occurs in Boone (clay hillside, 1½ mi. west of Marion Chapel, Brown's Station, September 16, 1933, *Drouet 1161*) and Iron (rocky open woods near Middlebrook, September 4, 1926, *Palmer 31529*) counties. The Iron County specimen was originally identified as *L. Manniana*.

Ranges from Alabama to Missouri, north to Massachusetts and Indiana.

The leaves of *L. procumbens* are browsed by white-tailed deer and wild turkey and the fruits are the food of wild turkey and bobwhite.

The two variations grade into one another.

3. **Lespedeza violacea** (L.) Pers. Map 1356

Flowers July–September.

Occurs in rocky or dry open, usually upland woods, thickets, and rocky prairies. Throughout Missouri, except absent from the lowlands of extreme southeastern Missouri.

Ranges from Florida to Texas, north to New Hampshire, Vermont, New York, Ohio, Michigan, Wisconsin, and Kansas.

The plant is eaten by white-tailed deer.

This species apparently hybridizes with *L. virginica* to produce what is known as *L. acuticarpa* Mackenz.

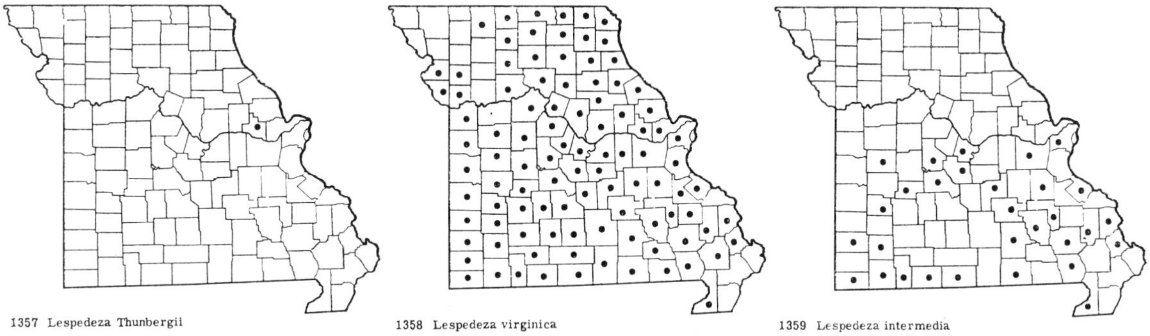

1357 Lespedeza Thunbergii 1358 Lespedeza virginica 1359 Lespedeza intermedia

& Bush, which has narrower and hairier leaves than typical *L. violacea* and possesses both short and elongated peduncles. Plants collected in late fall sometimes have most or all the fruits axillary (close to the stem) and are confused with *L. intermedia*. They may be distinguished as *L. violacea* by the remnants of the elongated peduncles of the petal-bearing flowers and by more distantly separated leaves of the stem.

4. **Lespedeza Thunbergii** (DC.) Nakai Map 1357
Flowers August–October.

Cultivated, and rarely escaped in open ground; known only from Warren County (Massa's Creek, T47N, R4W, sect. 25, 3 mi. south of Jonesburg, September 30, 1951, *Steyermark 73020*).

Native of eastern Asia; naturalized in the United States from Massachusetts southward and westward.

This is a showy-flowered cultivated species with deep rose-purple flowers in showy racemes.

5. **Lespedeza virginica** (L.) Britt. Map 1358
Flowers late May–September.

Occurs in rocky or dry open woods, sandy and gravelly margins of streams, thickets, prairies, bluffs, glades, roadsides, and railroads. Throughout Missouri, except not recorded from extreme northwestern Missouri.

Ranges from Florida to Texas, north to New Hampshire, Massachusetts, New York, Ontario, Michigan, Wisconsin, and Kansas.

The flowers vary from deep rose-purple to white with pale lilac.

The plant known as *L. acuticarpa* Mackenz. & Bush is believed to represent a hybrid between *L. virginica* and *L. violacea*. It possesses an aspect of *L. virginica* except for the broader leaflets and slightly exserted peduncles of a few of the inflorescences, but also has characteristics of *L. violacea* as indicated above for that species (see excluded species).

This species is eaten by white-tailed deer, and its fruits are the food of the bobwhite, wild turkey, and other wildlife. It provides protein food for all classes of livestock, providing for them a palatable and nutritious forage plant where it occurs in prairies and open wooded sections subjected to grazing.

6. **Lespedeza intermedia** (S. Wats.) Britt. f. **intermedia** Map 1359
Flowers July–September.

Occurs in rocky open woods, usually in acid soils of chert, sandstone, or granite, and thickets.

Ozark and unglaciated prairie region of southern and central Missouri north to St. Louis, Franklin, Cole, Morgan, Henry, and Jasper counties.

Ranges from Florida to Texas, north to Maine, New Hampshire, Vermont, New York, Ontario, Michigan, Wisconsin, and Kansas.

This species varies in pubescence of the stem, width, size, and degree of crowding of the leaves on the main stem. Forms with spreading hairs on the stem are placed in f. *Hahnii* (Blake) M. Hopkins, but have not been recorded from Missouri. Typical f. *intermedia* has the appressed type of hairs on the stem varying from sparsely to moderately abundant.

The leaves are eaten by white-tailed deer and the fruits are the food of wild turkey and bobwhite.

7. **Lespedeza Nuttallii** Darl. Map 1360
*Lespedeza Nuttallii* var. *Nuttallii* [BB]
Flowers August–October.

Occurs in dry rocky woods and thickets.

Rare and scattered in the Ozark region north to Iron, Phelps, Miller, Lawrence, and Jasper counties.

Ranges from New Hampshire to Michigan, south to South Carolina, Tennessee, Missouri, and Kansas.

This species varies somewhat in the degree of exsertion of the racemes. It is believed to hybridize with some of the other species, such as *L. procumbens*, *L. violacea*, and *L. virginica*.

1360 Lespedeza Nuttallii

1361 Lespedeza Manniana

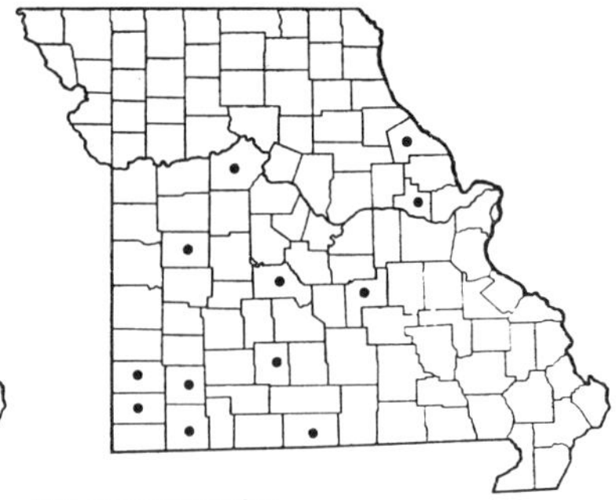

1362 Lespedeza simulata

8. **Lespedeza Manniana** Mackenz. & Bush Map 1361

*Lespedeza Nuttallii* Darl. var. *Manniana* (Mackenz. & Bush) Gl. [BB]

Flowers July–October.

Occurs in rocky open woods and glades.

Scattered and rare in southern, central, and eastern Missouri in Lewis, St. Louis, Phelps, Jackson, and Newton counties.

Ranges from Michigan to Arkansas and Texas.

This Lespedeza combines the calyx characters of *L. hirta* or *L. capitata* with the purplish corollas of *L. virginica*-like plants, or of plants resembling *L. Stuevei*, *L. intermedia*, or *L. Nuttallii*, and is believed to be of hybrid origin.

9. **Lespedeza simulata** Mackenz. & Bush Map 1362

Flowers July–September.

Occurs in dry open woods, prairies, sandy open places along streams, and limestone glades. Southern and central Missouri north to Pike, Saline, and Henry counties.

Known from Missouri, Tennessee, Pennsylvania, and New Jersey.

This Lespedeza is believed most likely to represent a hybrid between *L. capitata* and *L. virginica*, *L. intermedia*, or *L. Stuevei* var. *angustifolia*. The calyx and infloresence is like that of *L. capitata* but the flowers are purple, while the silvery-hairy leaves resemble those of *L. virginica* or *L. Stuevei* var. *angustifolia*, or sometimes of *L. intermedia*.

10. **Lespedeza Stuevei** Nutt. Map 1363

Flowers July–September.

Occurs in dry open woods, thickets, prairies, sandy soils, fallow fields, limestone glades, and river bottom meadows. Southern and central Missouri, north to Ste. Genevieve, Maries, Camden, Johnson, and Vernon counties.

Two variations are encountered in Missouri material:

Leaflets elliptical or ovate-oblong, $1\frac{1}{2}$–2 times as long as broad . . 10a. L. STUEVEI var. STUEVEI

Leaflets narrowly elliptical to linear or linear-oblong . . . 10b. L. STUEVEI var. ANGUSTIFOLIA

10a. **Lespedeza Stuevei** var. **Stuevei** Map 1363

*Lespedeza Stuevei* Nutt. [G, BB, P & S]

This is the common variation in Missouri.

Ranges from Alabama to Texas, north to Massachusetts, Vermont, New York, Ohio, Indiana, Illinois, Missouri, and Kansas.

10b. **Lespedeza Stuevei** var. **angustifolia** Britt. Map 1363

*Lespedeza neglecta* (Britton) Mackenz. & Bush

*Lespedeza Stuevei* var. *neglecta* Britton

*Lespedeza Stuevei* f. *angustifolia* (Britton) M. Hopkins

Rare and known in southern and central Missouri from St. Louis, Iron, Dunklin (September 12, 1893, *Bush 31*), Pulaski, and Johnson counties.

Ranges from Georgia to Texas, north to New Jersey, Indiana, and Missouri.

This narrow-leaved variant is believed to represent a hybrid between *L. Stuevei* and *L. virginica*.

Typical *L. Stuevei* var. *Stuevei*, according to Dr. Isely's studies (Iowa State College Jour. Sci. 30: 104. 1955), is best developed and predominant in the Ozark region of Missouri and Arkansas, exhibiting here plants with broader and hairier leaves with the pubescence on both leaf surfaces, and with conspicuously hairy fruits.

Plate no. 222. **1.** Lespedeza virginica; *a.* Plant, upper portion, $\times\ ^{3}/_{4}$; *b.* Leaf, $\times\ 1^{1}/_{2}$; *c.* Portion of inflorescence, $\times$ 1. **2.** Lespedeza Nuttallii; *a.* Leaf $\times\ ^{3}/_{4}$; *b.* Fruit, $\times\ 3^{3}/_{4}$; *c.* Upper portion of flowering plant, $\times\ ^{3}/_{4}$. **3.** Lespedeza Stuevei; *a.* Leaf, $\times$ 1; *b.* Fruit, $\times\ 3^{3}/_{4}$.

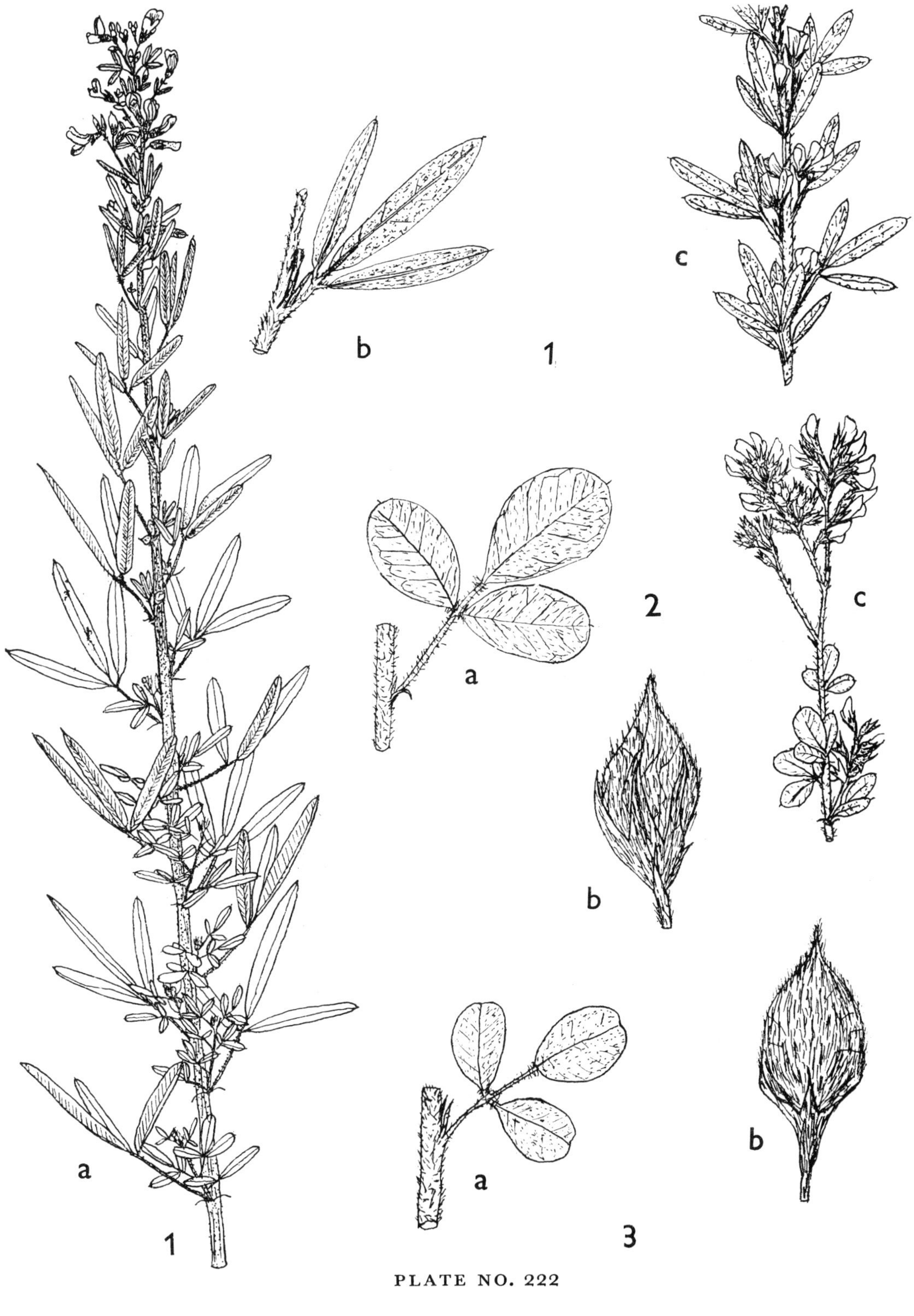


PLATE NO. 222

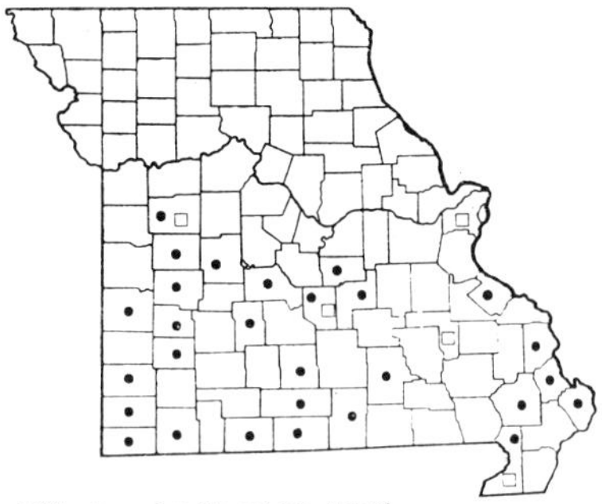

1363 • Lespedeza Stuevei var. Stuevei
1363 □ Lespedeza Stuevei var. angustifolia

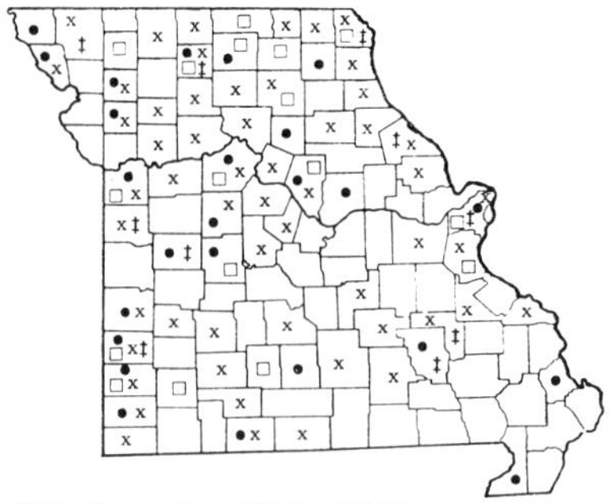

1364 • Lespedeza capitata var. capitata
1364 x Lespedeza capitata var. vulgaris
1364 □ Lespedeza capitata var. stenophylla f. stenophylla
1364 ‡ Lespedeza capitata var. stenophylla f. argentea

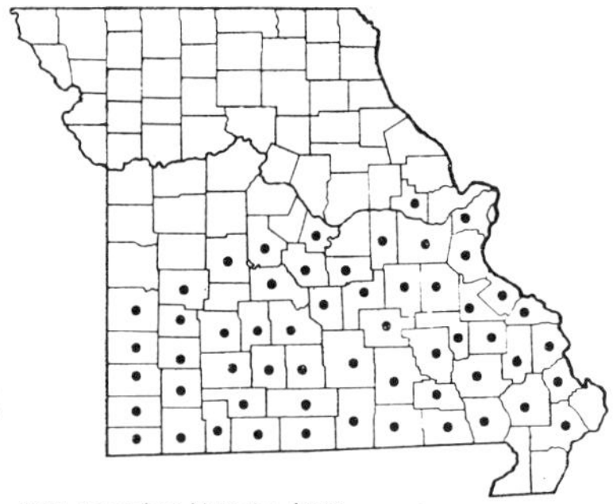

1365 Lespedeza hirta var. hirta

11. **Lespedeza capitata** Michx. Map 1364
Flowers July–October.

Occurs in upland and river bottom prairies, glades, loess mounds, sandy bottoms along streams, openings in upland woods, thickets, and along roadsides and railroads.

The following variations may be recognized for Missouri:

a. Upper surface of leaflets green with few or no hairs . . . . . . . . . . . . . . *b*
  b. Leaflets linear-lanceolate . 11c. L. CAPITATA var. STENOPHYLLA f. STENOPHYLLA
  b. Leaflets oblong or narrowly elliptical. . 11b. L. CAPITATA var. VULGARIS
a. Upper surface of leaflets grayish with a silky (sericeous) hairiness. . . . . . . . . *c*
  c. Leaflets linear-lanceolate . 11d. L. CAPITATA var. STENOPHYLLA f. ARGENTEA
  c. Leaflets oblong or narrowly elliptical . . 11a. L. CAPITATA var. CAPITATA

11a. **Lespedeza capitata** var. **capitata** Map 1364
*Lespedeza capitata* Michx. [G]
*Lespedeza capitata* var. *sericea* H. & A. [P & S]
Scattered throughout Missouri.

Ranges from Florida to Texas, north to New England, New York, Pennsylvania, Tennessee, Wisconsin, Minnesota, and Nebraska.

11b. **Lespedeza capitata** var. **vulgaris** T. & G. Map 1364
*Lespedeza capitata* of auth. [P & S], not Michx.
Common throughout Missouri.

Ranges from Maine and Quebec to Minnesota and Nebraska, south to North Carolina and Missouri.

11c. **Lespedeza capitata** var. **stenophylla** Bissel & Fern. f. **stenophylla** Map 1364
*Lespedeza capitata* var. *stenophylla* Bissel & Fern. [G]
*Lespedeza capitata* var. *longifolia* of auth. in part [P & S], not *L. longifolia* DC., the basonym of var. *longifolia* (DC.) T. & G.
Scattered throughout Missouri.

Ranges from Massachusetts to Wisconsin, south to New Jersey and Missouri.

11d. **Lespedeza capitata** var. **stenophylla** f. **argentea** Fern. Map 1364
*Lespedeza capitata* var. *longifolia* of auth. in part [P & S], not *L. longifolia* DC., the basonym of var. *longifolia* (DC.) T. & G.
Scattered throughout Missouri.

Range as in *L. capitata* var. *stenophylla* f. *stenophylla*.

*Lespedeza capitata* var. *stenophylla* f. *argentea* and *L. capitata* var. *capitata*, with conspicuous gray appressed hairiness on the upper leaf-surface, and dense silvery silky hairiness on the lower leaf surface are usually prevailing variations encountered in the driest and most exposed habitats in Missouri, such as glades and prairies.

Intergradation, however, occurs between all the variations encountered in the state. Especially noticeable is the intergradation between var. *vulgaris* and var. *stenophylla*, with many collections having leaflets almost narrow enough to place with var. *stenophylla* or with leaflets intermediate between lanceolate or linear-lanceolate of var. *stenophylla* and narrowly elliptic of var. *vulgaris*. In their extremes, these and other variations are quite distinct, but their occurrence together in the same habitat and lack of correlation of the length of peduncles with their subtending leaves and leaf shape and pubescence, have led some botanists (Isely, Iowa State College Jour. Sci. 30: 86–90. 1955 and Fassett, *Leguminous Plants of Wisconsin* pp.

Plate no. 223. **1.** Lespedeza hirta; *a.* Leaf, × 1; *b.* Fruit, × 3; *c.* Flowering top, × 1. **2.** Lespedeza Thunbergii, × 3/4.

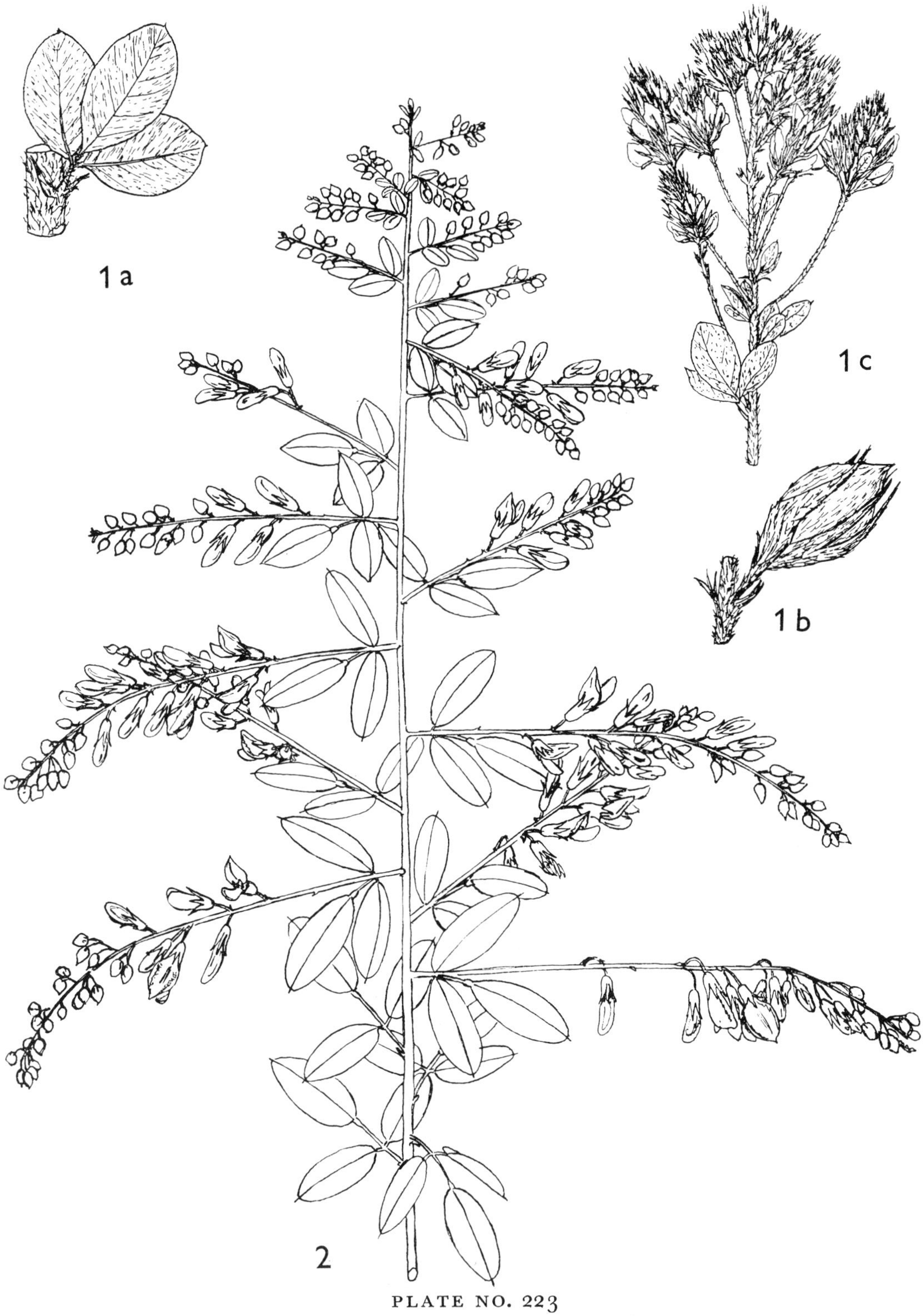


PLATE NO. 223

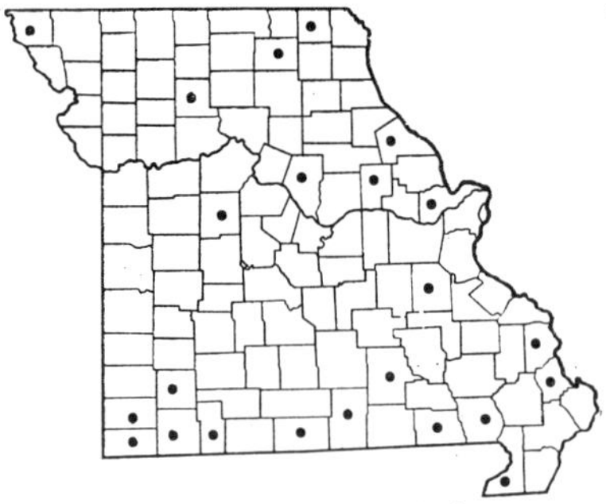

1366 Lespedeza cuneata (Sericea Lespedeza)

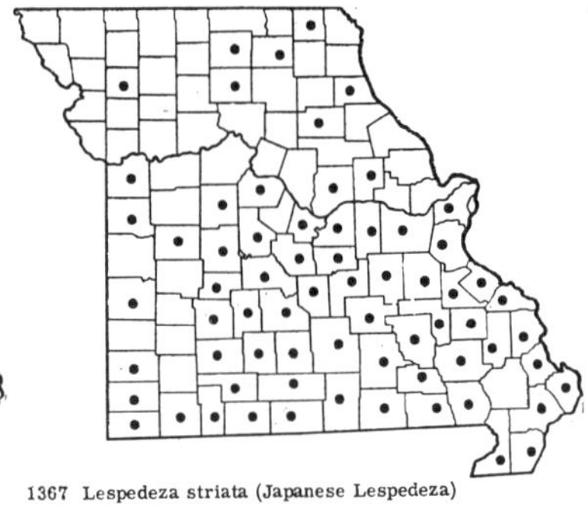

1367 Lespedeza striata (Japanese Lespedeza)

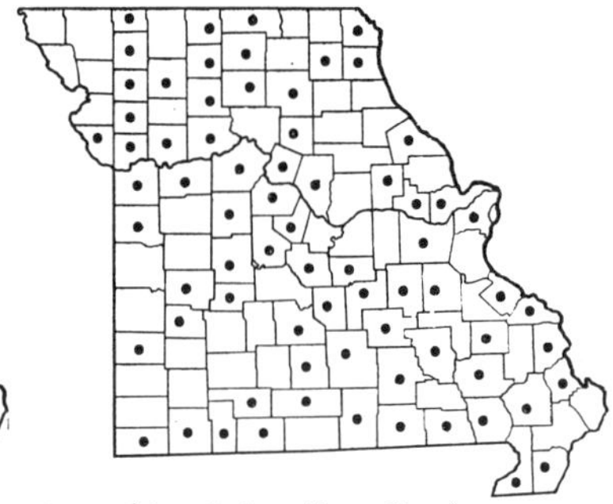

1368 Lespedeza stipulacea (Korean Clover)

102–7) to treat the species as a variable one without recognition of any minor categories. Results from growing plants from seed by Fassett (p. 104) indicated that the offspring may or may not resemble the parent, but the results were admittedly not conclusive. Since additional field, experimental, and growing tests need to be carried out more thoroughly, and since some botanists favor the recognition of several variations, the treatment in this flora has been to present the variations objectively, permitting the collector, if he wishes to distinguish these variations, to be his own judge of the limit of variations indicated. Perhaps future studies may further clarify the real limits of the variations.

The leaves and fruits of *L. capitata* are eaten by wild turkey, and the fruits are eaten by many other birds and wildlife. Where it occurs in prairies and open range, it is considered a very palatable and nutritional forage plant for all classes of livestock.

12. **Lespedeza hirta** (L.) Hornem. var. **hirta** Map 1365

*Lespedeza hirta* (L.) Hornem. [G, BB]

Flowers July–October.

Occurs in acid soils of chert, sandstone, and granite in rocky open woods, glades, and thickets.

Ozark and unglaciated prairie region of southern and central Missouri north to St. Louis, Warren, Gasconade, Cole, Morgan, Benton, St. Clair, and Vernon counties.

Ranges from Maine to Ontario, south to Georgia, Alabama, Arkansas, and Texas.

In Missouri this is an easily recognizable and consistent *Lespedeza*, showing little or none of the wide variability which has led some botanists (Fernald, Rh. 43: 572–87. 1941) to divide some of the local eastern and coastal plain variations into additional recognized varieties. All the Missouri material can be placed without difficulty in typical var. *hirta* with the stems rather softly hairy, the lower surface of the leaflets more or less hairy with spreading to ascending hairs, and with the leaflets of a roundish oblong-ovate to obovate type.

The leaves are eaten by wild turkey.

13. **Lespedeza cuneata** (Dumont) G. Don

Sericea Lespedeza Map 1366

*Lespedeza sericea* (Thunb.) Miq., not Benth.

Flowers August–October.

Planted and now increasingly escaping from plantings and becoming naturalized in woodland, thickets, fields, prairies, open ground, borders of ponds and swamps, meadows, and along roadsides. Throughout Missouri.

Native of eastern Asia; introduced and naturalized in the United States from Maryland, Virginia, Tennessee, Missouri, and Texas, north to Pennsylvania, Ohio, Michigan, Illinois, and Oklahoma.

The first recorded collection in Missouri (Rh. 42: 100. 1940) for this species was made in 1938 (*Steyermark 6635*) from Scott Co. It has been introduced into various areas as a soil cover for erosion control, for soil improvement, as food and cover for bob-white, wild turkey, and other wildlife, and, to a lesser extent, for forage and hay. It shows great resistance to summer drought and ability to form a dense stand on sterile and steep or eroded slopes. Its introduction, however, into natural areas, already furnished with native species of *Lespedeza*, of which there are many in Missouri, appears undesirable and impractical, as unnecessary competition develops with the native vegetation, including the native species of *Lespedeza*.

Plate no. 224. **1.** Lespedeza cuneata; *a.* Leaf and portion of stem, $\times 1^{1}/_{2}$; *b.* Upper portion of plant, $\times 1^{1}/_{2}$. **2.** Lespedeza stipulacea; *a.* Upper part of plant, $\times 1$; *b.* Fruit, $\times 7^{1}/_{2}$. **3.** Lespedeza capitata; *a.* Leaf, $\times {}^{1}/_{2}$; *b.* Flowering top of plant, $\times {}^{3}/_{4}$. **4.** Lespedeza striata; *a.* Upper part of plant, $\times 1$; *b.* Fruit, $\times 7^{1}/_{2}$.

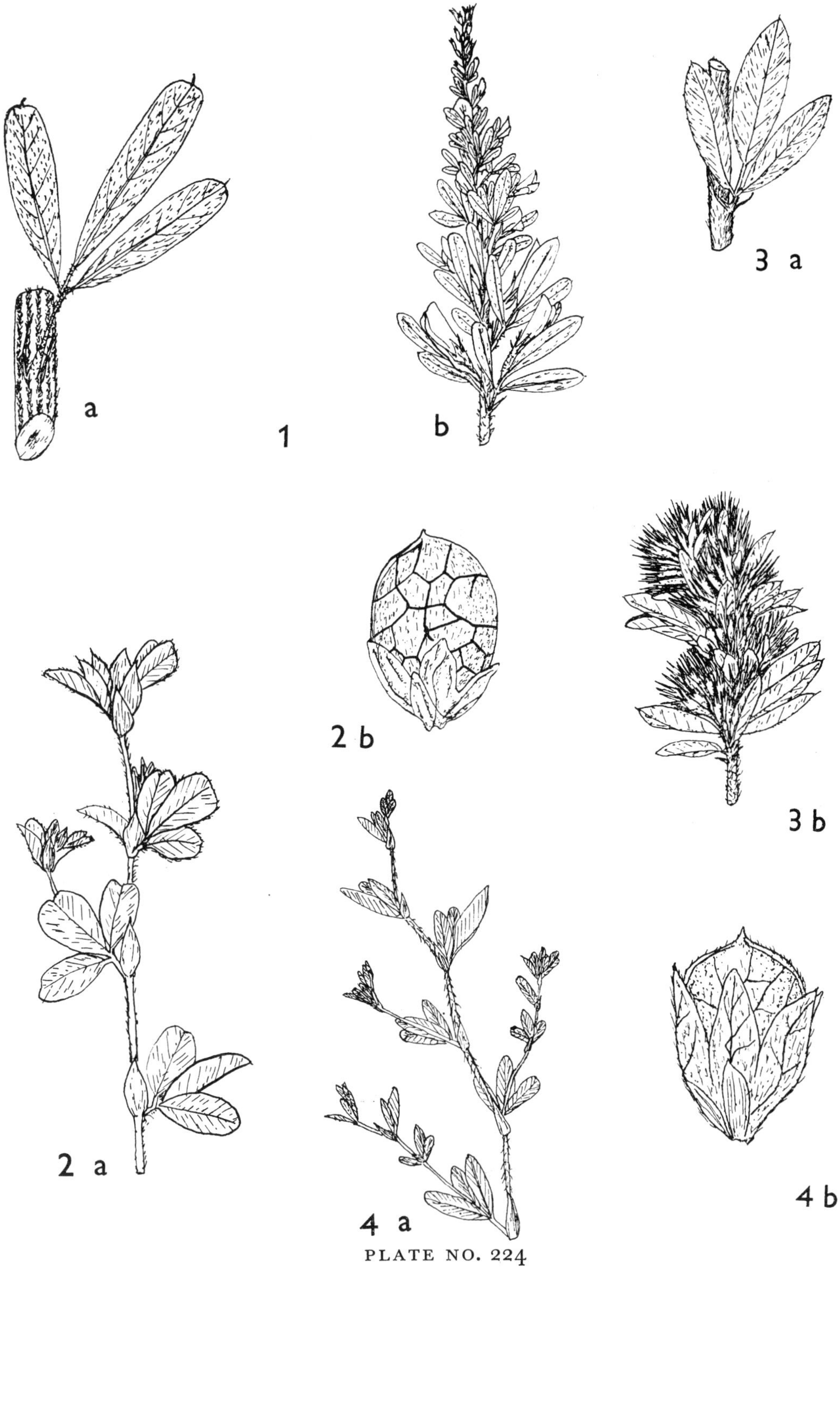


PLATE NO. 224

The pale creamy-yellow flowers are smaller than those of the native species (*L. capitata* and *L. hirta*) with cream-colored or yellowish flowers, and the base of the standard in *L. cuneata* has two broad purplish-rose-colored streaks on the inside of the center portion.

14. **Lespedeza striata** (Thunb.) H. & A.
Japanese Lespedeza Map 1367
Also called Common Lespedeza.
Flowers July–October.

Occurs in dry open woods, rocky open glades, sandy and rocky fields, gravel bars along streams, and along roadsides and waste places.

Common throughout southern and central Missouri north to St. Louis, Montgomery, Callaway, Cooper, Pettis, and Jackson counties; rare or absent in most of northern Missouri, but extending locally north to Scotland and Adair counties.

Native of eastern Asia; introduced and naturalized in the United States from Florida to Texas, north to New Jersey, Pennsylvania, Ohio, Indiana, Illinois, Missouri, and Kansas.

This species was introduced into the United States from Japan in 1846. The flowers are self-fertilized and mature abundant seed. This *Lespedeza* is planted for pasturage, hay, soil improvement, and as a cover crop. It does well in acid soils of low fertility. After Korean Lespedeza (*L. stipulacea*) has finished flowering and stopped growing, *L. striata* continues to grow and provide additional forage, supplementing the value of *L. stipulacea*. When this *Lespedeza* is extensively sown over natural rocky glades, it may take over to the exclusion of other native vegetation, leading to the elimination or depletion of the native flora confined to such areas. This is highly regrettable and undesirable in areas which should be left undisturbed in their natural condition for study and enjoyment.

15. **Lespedeza stipulacea** Maxim.
Korean Clover Map 1368
Flowers July–October.

Occurs along roadsides and railroads, fields, pastures, and open woods. Common throughout Missouri, and in many parts of the state commoner and replacing Japanese Clover (*L. striata*).

Native of eastern Asia; introduced and naturalized from Florida to Texas, north to Pennsylvania, West Virginia, Illinois, and Iowa.

This species was introduced into the United States much later (in 1919) than *L. striata*, Japanese Clover, but has spread quickly over Missouri since about 1930. It replaces *L. striata* as a crop in the northern half of Missouri. Like the Japanese Clover it serves admirably for pasturage, and as a hay and cover crop. It appears to be better adapted to the cooler, more northern sections, starting to grow earlier in the summer and maturing the seeds earlier in the fall. The fruits of both *L. striata* and *L. stipulacea* are eaten extensively by wild turkey and bob-white.

*Excluded Species*

**Lespedeza acuticarpa** Mack. and Bush

Plants identified by this name appear to represent intermediate types belonging to apparent hybrids between *L. violacea* and either *L. virginica*, *L. Stuevei*, or *L. intermedia*, possessing both short and elongated peduncles.

## 31. **Onobrychis** Mill.

**Onobrychis viciaefolia** Scop. Sainfoin Map 1369
Flowers June–July.

Known only from St. Louis County (St. Louis, July 16, 1882, *Schuette*).

Native of Europe, Asia, and North Africa; introduced and naturalized in North America in New Jersey, Missouri, South Dakota, Montana, Wyoming, and Colorado.

This plant has been introduced for its forage value, but is grown only to a rather small extent, since it has not proved as satisfactory as some other legumes. The long-peduncled racemes are much longer than the leaves at their base. The showy flowers have a generally rose-colored standard 6–8 mm. long with dark streaks, and a reddish keel. The fruit is flattened, semi-circular to broadly oval, 5–8 mm. long, toothed on the margins, and strongly nerved or ridged elsewhere.

## 32. **Stylosanthes** Sw. Pencil Flower

**Stylosanthes biflora** (L.) BSP. Pencil Flower
Map 1370

*Stylosanthes biflora* var. *hispidissima* (Michx.) Pollard & Ball [G, BB, P & S, Steyerm.]

1369 Onobrychis viciaefolia (Sainfoin)

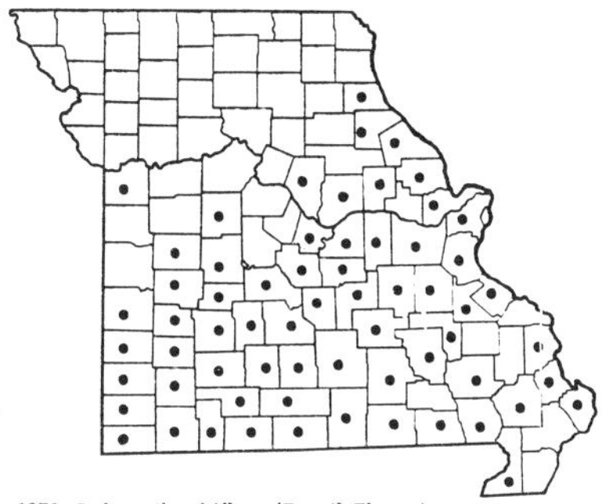

1370 Stylosanthes biflora (Pencil Flower)

1371 Vicia grandiflora

*Stylosanthes riparia* Kearney [G, BB]

Flowers May–September.

Occurs in usually acid soils of chert, cherty limestone, sandstone, or granite in rocky open woods, glades, and prairies. Ozark and unglaciated prairie region of southern and central Missouri north to Marion, Montgomery, Callaway, Boone, Henry, and Jackson counties.

Ranges from Florida to Texas, north to New York, New Jersey, Pennsylvania, West Virginia, Kentucky, Illinois, Missouri, Oklahoma, and Kansas.

Dr. Robert Mohlenbrock has shown (Ann. Mo. Bot. Gard. 44: 327–29. 1957; Bull. Torr. Bot. Club 85: 341–46. 1958) that *S. biflora* var. *hispidissima* cannot be maintained as distinct from *S. biflora* and that *L. riparia*, likewise, must be considered synonymous with *S. biflora*. This latter view had also been advanced previously by Dr. Duane Isely (Iowa State College Jour. Sci. 30: 113. 1955).

## 33. **Arachis** L. Peanut

**Arachis hypogaea** L. Peanut

Flowers July–September.

Occasionally cultivated in Missouri, but rarely escaped in sandy alluvial ground. Noted but not collected as an escape in Madison County, southeastern Missouri.

Native of South America; introduced and naturalized from Florida to Texas, north to Delaware, Missouri, and Kansas.

This well-known edible nut is extensively planted in the southern states. The flower-stalk, after the period of blossoming is completed and fertilization has taken place, curves downward, eventually penetrating below the surface of the soil, where the mature fruit ripens. The oil of the peanuts is used as a salad oil, as an oil for cooking, for illumination, lubrication, oleomargarine, shortenings, packing sardines, adulterant for olive oil, and soapmaking. In the form of oil cakes it is used as a highly nutritious stock food having a higher protein content than similarly prepared types of stock food. In recent years peanut oil in the non-hydrogenated state has been used to treat a condition of excess cholesterol found in persons having a high blood pressure or showing a tendency toward arteriosclerosis.

## 34. **Vicia** L. Vetch

a. Main stalk (peduncle) of inflorescence either lacking and the flowers arising close to the main stem or very short and many times shorter than the leaves . . . . . . . . . . . . . . . . . *b*

b. Flowers 25–35 mm. long, yellow or yellow with lavender; calyx-teeth $\frac{1}{3}$–$\frac{1}{2}$ as long as calyx-tube . . . 1. V. GRANDIFLORA

b. Flowers 10–17 mm. long, predominantly purple; calyx-teeth nearly equaling or longer than the calyx-tube . . . . . . . . . . . . . . . . 2. V. SATIVA var. SEGETALIS and var. NIGRA

a. Main stalk (peduncle) of inflorescence conspicuous, almost as long as or much longer than the leaves . . *c*

c. Flowers 3–8 mm. long; flowers few, usually 2–6 (rarely 1 or 8) in each inflorescence . . . . . . *d*

d. Flowers 3–4 mm. long; all the calyx-teeth as long as or longer than the calyx-tube; fruit hairy, 6–10 mm. long; seeds 2 in each fruit . . . . . . . . . . . . . . 4. V. HIRSUTA

d. Flowers 5–8 mm. long; some or all of the calyx-teeth shorter than the calyx-tube; fruit glabrous (without hairs), 10–30 mm. long; seeds 4–8 in each fruit . . . . . . . . . . . . . . *e*

e. All the calyx-teeth shorter than the calyx-tube and more or less equal, triangular; main stalk (peduncle) of inflorescence shorter than the leaves, 1–2 (rarely 3) cm. long . 6. V. MINUTIFLORA

e. At least 1 of the calyx-teeth as long as the calyx-tube, the teeth unequal, some of them triangular to linear, or subulate; main stalk (peduncle) of inflorescence shorter to longer than the leaves, 1–4 cm. long . . . . . . . . . . . . . . . . . . . . . . *f*

f. Flowers usually 1 or 2 (rarely 3) in each inflorescence, pale lilac-bluish to white; main stalk (peduncle) of inflorescence longer than the leaflets; seeds 4 in each fruit; fruit 10–13 mm. long, rounded to a blunt tip, without a beak . . . . . . 3. V. TETRASPERMA

f. Flowers usually 5–8 (sometimes 2–4) in each inflorescence, of darker lavender-blue to blue-purple; main stalk (peduncle) of inflorescence shorter to longer than the leaflets, seeds usually 4–8 in each fruit, fruit 20–30 mm. long, narrowed to a beak . 5. V. LUDOVICIANA

c. Flowers 8–20 mm. long; flowers usually 8–20 or more, if fewer than 8, then the flowers 15–20 mm. long . . . . . . . . . . . . . . . . . . . . . . . . . . . . *g*

g. Hairs spreading from main stems and axis of inflorescence; lower calyx-teeth hairy . 8. V. VILLOSA

g. Hairs on main stem and axis of inflorescence, if present, appressed (pressed close to or lying parallel to surface), or glabrous (without hairs); lower calyx-teeth practically without hairs or with few hairs . . . . . . . . . . . . . . . . . . . . . . . *h*

h. Stipules (at base of main stalk of leaves) prominently toothed; fully grown inflorescences shorter than the subtending leaves immediately arising at their base, with 3–9 flowers. . 10. V. AMERICANA

h. Stipules (at base of main stalk of leaves) entire (without teeth); fully grown inflorescences as long as or longer than the subtending leaves immediately arising at their base, usually with 8–20 or more flowers, rarely 5–7 . . . . . . . . . . . . . . . . . *i*

i. Calyx-teeth about as broad as long, short-triangular, and all about equal, the mature calyx (including the longest tooth) 3 mm. or less; petals whitish with a blue-tipped keel to pale lilac and whitish; calyx-tube about equal-sided, one side of the tube not enlarged or extended back of the point of insertion of calyx-tube, the flower-stalk (pedicel) appearing to come from the lower end of calyx-tube; fruit 5–7 mm. broad; flowers 7–12 mm. long; native plants of rocky open woods in the Ozark region of southern and central Missouri . . . . . . . . . . . . . . . . . . 7. V. CAROLINIANA

i. At least the lower calyx-teeth much longer than broad and narrowly linear-triangular, unequal, the lower longer than the upper, the mature calyx (including the longest tooth) 4–6 mm. long; petals mostly deep blue-purple or violet; calyx-tube unequal-sided, upper side of calyx-tube enlarged, extended back of the point of insertion of calyx-tube, the flower-stalk (pedicel) appearing to come from one side; fruit 9–11 mm. broad; flowers 12–15 mm. long; introduced along roadsides and railroads . . . 9. V. DASYCARPA

1. **Vicia grandiflora** Scop. Map 1371

Flowers April–June.

Known only from Cape Girardeau County in southeastern Missouri (Cape Girardeau, southwest corner Dunklin and Henderson streets, April 20, 1954, *Shirley G. Hibbs*, in herb. Southeast Mo. State College).

Native of Europe; introduced and naturalized from Delaware to Virginia, and Missouri.

The Missouri collection has 8–10 leaflets, oblong or linear-oblong, truncate at apex with an abrupt mucro, and 4–5 mm. wide. The calyx-tube is 7–8 mm. long, sparsely short-pilose, and somewhat gibbous (unequal-sided); the calyx-teeth are 2.5–3.5 mm. long, about ½ or less the length of the calyx-tube and moderately pilose. The corolla is yellowish with lavender in the lower part near the base, and 2.6–2.8 cm. long. The flowers occur solitary or paired in the upper axils of the leaves.

2. **Vicia sativa** L. Common Vetch Map 1372

Flowers May–August.

Occurs in gardens, grassy open and waste ground, fields, and along railroads.

Missouri material is represented by two variations:

Leaflets of the upper leaves linear-lanceolate to narrowly lanceolate, narrowed or slightly rounded

Plate no. 225. **1.** Stylosanthes biflora, × 2/5; Details from Small, The New York Botanical Garden. **2.** Vicia grandiflora, × 2/5; After Gleason, The New York Botanical Garden. **3.** Vicia sativa var. nigra, × 2/5; Details from Small, The New York Botanical Garden. **4.** Vicia sativa var. segetalis, × 2/5. **5.** Vicia tetrasperma, × 2/5. **6.** Vicia minutiflora, × 2/5. **7.** Vicia caroliniana, × 2/5.

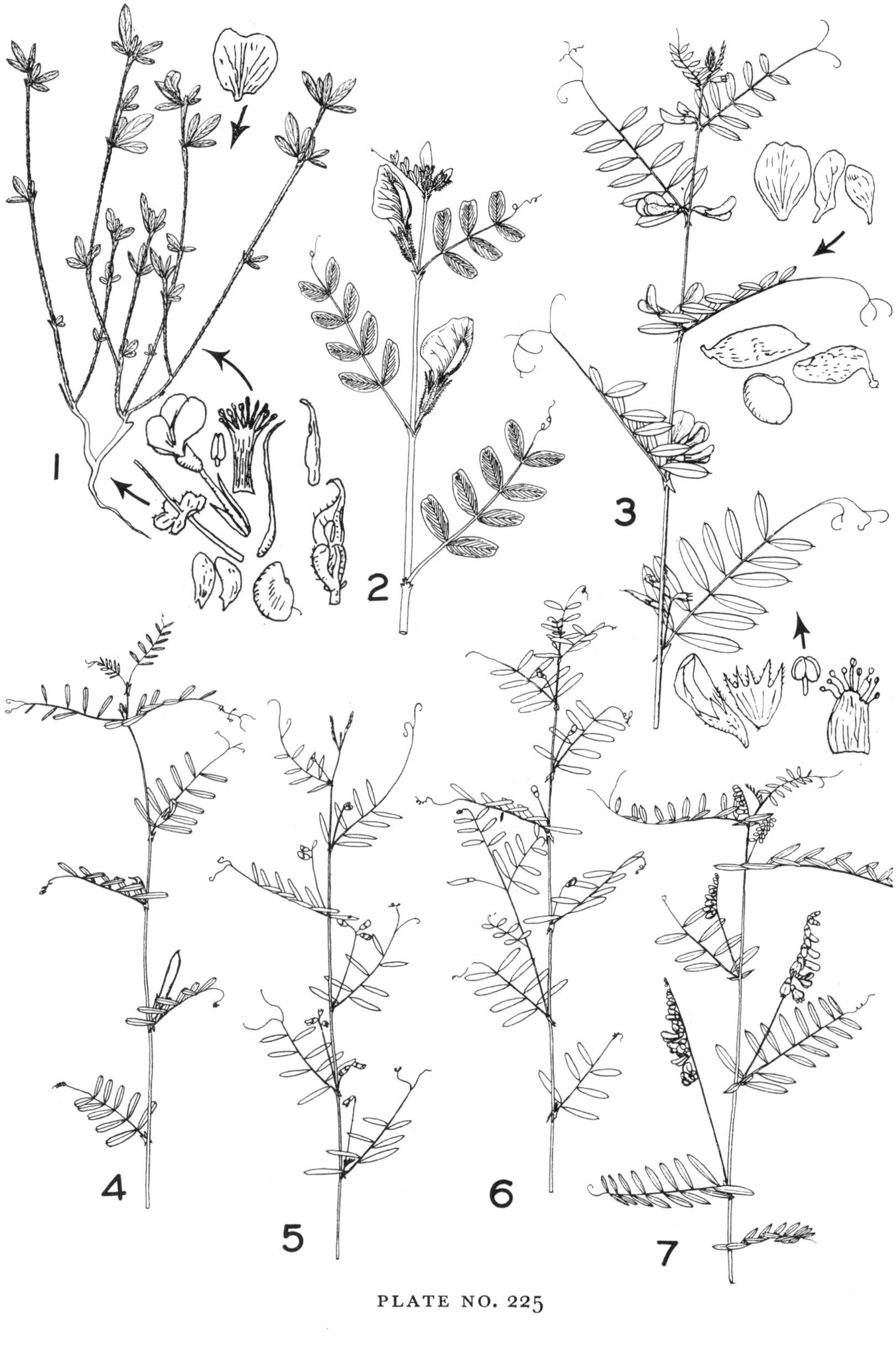


PLATE NO. 225

1372 □ Vicia sativa var. nigra (Common Vetch)
1372 • Vicia sativa var. segetalis

1373 Vicia tetrasperma var. tetrasperma

1374 Vicia hirsuta

at the tip, 1–4 mm. broad . 2a. V. SATIVA var. NIGRA
Leaflets of the upper leaves oblong to oblong-obovate, truncate (as if cut straight across) or emarginate (shallowly notched at the tip) at the summit, 2–9 mm. broad 2b. V. SATIVA var. SEGETALIS

2a. **Vicia sativa** var. **nigra** L. Map 1372

*Vicia sativa* ssp. *angustifolia* (Reichard) Gaudin, emend. Briquet, in part

*Vicia sativa* in part [P & S, Steyerm.], not L.

*Vicia angustifolia* Reichard [G, BB, P & S, Steyerm.]

*Vicia angustifolia* var. *angustifolia* [BB]

This is the commoner of the two variations, but is rare and scattered and occurs in southern and central Missouri.

Native of Europe; introduced and naturalized in North America from Nova Scotia to Michigan and Minnesota, south to Florida, Missouri, Oklahoma, and Texas; Idaho.

2b. **Vicia sativa** var. **segetalis** (Thuillier) Seringe Map 1372

*Vicia sativa* in part [P & S, Steyerm.], not L.

*Vicia angustifolia* var. *segetalis* (Thuill.) W. Koch [G, BB]

*Vicia sativa* ssp. *angustifolia* (Reichard) Gaudin emend. Briquet, in part

Known only from Newton (open rocky ground along creek, Hornet, May 24, 1954, *Palmer 57644*), McDonald (open rocky bank, 4 mi. north of Anderson, July 7, 1955, *Palmer 60459*), Barry (*Palmer 65541*), and Jackson (Courtney, July 17, 1904, *Bush 2124*) counties.

Native of Europe; introduced and established throughout similar range as var. *nigra*.

Typical *V. sativa* var. *sativa* with larger corolla 18–30 mm. long does not occur in Missouri; according to the material examined, the Missouri specimens have corollas 10–17 mm. long and must be placed mainly under *V. angustifolia* var. *segetalis* (= *V. sativa* var. *segetalis*). Since *Vicia angustifolia* differs mainly from *V. sativa* in the dimensions of the flowers and reliance cannot be placed on characters of the fruit, the present treatment follows that of Shinners (*Spring Flora*, p. 224. 1958) and of various European authors who have united the two taxa under *V. sativa* L.

As with *V. grandiflora*, the flowers occur usually in twos in the upper axils of the leaves next to the main stem. The seeds and foliage of this and other species are eaten by domestic animals.

3. **Vicia tetrasperma** (L.) Moench. var. **tetrasperma** Map 1373

*Vicia tetrasperma* (L.) Moench. [G, BB, P & S, Steyerm.]

Flowers May–September.

Occurs in waste ground and along roadsides in Jackson and Lawrence counties.

Native of Europe; introduced and naturalized in North America from Newfoundland to Ontario and Wisconsin, south to Florida, Mississippi, Missouri, and Texas.

4. **Vicia hirsuta** (L.) S. F. Gray Map 1374

Flowers April–September.

Introduced along railroads, where known only from St. Louis County (St. Louis, along Frisco R. R. right-of-way between Macklind Avenue and Evens and Howard Sewer Pipe Company plant, along a siding, May 29, 1954, *Muehlenbach 96*).

Native of Europe; introduced and naturalized from Newfoundland to British Columbia, south to Georgia and California.

A specimen in the Gray Herbarium labeled *V. hirsuta* (June 11, 1891, Courtney, Jackson Co., *Bush*) is misidentified and is referred in the present treatment to *V. micrantha*.

1375 Vicia ludoviciana var. ludoviciana

1376 Vicia minutiflora

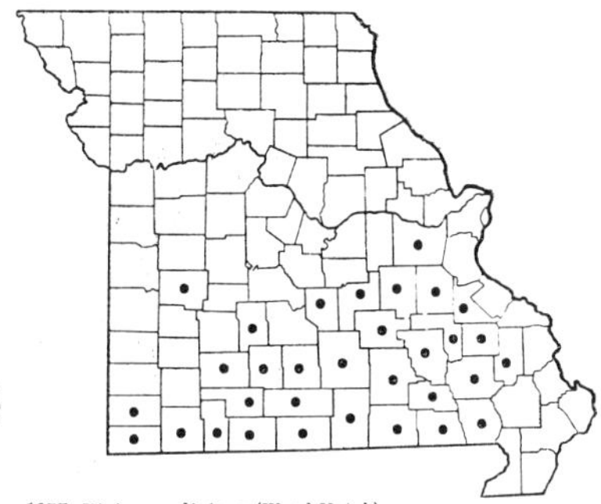
1377 Vicia caroliniana (Wood Vetch)

5. **Vicia ludoviciana** Nutt. var. **ludoviciana** Map 1375

*Vicia ludoviciana* Nutt. [G, BB, P & S, Steyerm.]

Flowers April–May.

Occurs in rich or rocky woods. Southwestern Missouri, known only in Greene (May 19, 1887, *J. W. Blankinship*) and McDonald counties.

Ranges from Florida to Texas, north to Missouri and Oklahoma.

6. **Vicia minutiflora** Dietr. Map 1376

*Vicia micrantha* Nutt. ex T. & G of 1838 [G, BB, P & S, Steyerm.], not *V. micrantha* Hook. & Arn. 1833.

Flowers April–May.

Occurs in rocky open ground, thickets, and along railroads. Western Missouri, south of the Missouri River, in Jackson (adventive from railroad, Courtney, June 11, 1891, *Bush*), Lawrence, and McDonald (gravelly barrens, Noel, May 10, 1915, *Bush 7538*) counties.

Ranges from Florida to Texas, north to Tennessee and Missouri.

The Jackson County collection was misidentified at the Gray Herbarium as *V. hirsuta.* It has 1 or 2 flowers with the short triangular calyx-lobes characteristic of *V. minutiflora.*

7. **Vicia caroliniana** Walt. Wood Vetch Map 1377

Flowers April–June.

Occurs in usually acid soils of sandstone, chert, or granite in rocky open woods, along stream banks, wooded slopes, and upland ridges. Ozark region of southern and east-central Missouri north to Franklin, Crawford, Phelps, Pulaski, Dallas, St. Clair, and Newton counties.

Ranges from New York and Ontario to Wisconsin and Minnesota, south to Florida, Georgia, Alabama, Mississippi, Louisiana, and Texas.

8. **Vicia villosa** Roth f. **villosa** Hairy Vetch Map 1378

Also called Winter Vetch.

*Vicia villosa* Roth [G, BB, P & S, Steyerm.]

Flowers April–October.

Occurs in fields, waste ground, and along roadsides and railroads. Scattered throughout the state, and doubtless to be found in most counties.

Native of Europe; introduced and naturalized in North America from Nova Scotia to British Columbia, south to Georgia, Texas, and California.

Material misidentified as *V. Cracca* L. (*Bush 14777* from Jackson Co., *Kellogg* from Phelps Co., *Daniels* from Boone Co., *Kellogg* from Creve Coeur Lake, St. Louis Co., and *Mrs. Holtzmann* from Mattese, St. Louis Co.) are referred here to *V. villosa.*

This species is sown as a forage crop. Plants often escape in numbers in fallow fields or cover roadside banks, and in flower are quite conspicuous with the dark lavender to purple, many-flowered racemes. The plants form large sprawling mats.

9. **Vicia dasycarpa** Ten. Map 1379

Flowers May–October.

Occurs in open ground, prairie swales, along roadsides and railroads. Scattered in southern and central Missouri north to St. Louis and Clay counties. To be expected in increasing abundance in more counties, where doubtless more common than at present indicated.

Native of Europe; introduced and naturalized in the United States from Maine to Montana, south to Georgia, Missouri, Oklahoma, Texas, and California.

This was first collected in the state in 1939 from Howell County (*Steyermark 27030*), and is probably to be found more commonly throughout the state. The strongly gibbous calyx, glabrous or appressed-pubescent stems and axis of inflorescence, and the nearly glabrous or sparsely hairy lower calyx-teeth easily distinguish this species from the very similar

1378 Vicia villosa f. villosa (Hairy Vetch)

1379 Vicia dasycarpa

1380 • Vicia americana var. americana
1380 □ Vicia americana var. truncata

*V. villosa*, both species resembling one another in the large sprawling clumps with showy many-flowered racemes.

10. **Vicia americana** Muhl. Map 1380

Flowers May–July.

Occurs in open woods and thickets.

Two variations occur in Missouri:

Leaflets of the upper leaves oblong-ovate or elliptic, obtuse to rarely acute . . . . . . . 10a. V. AMERICANA var. AMERICANA

Leaflets of the upper leaves broadly oblong, conspicuously truncate (as if cut straight across) at summit . . . 10b. V. AMERICANA var. TRUNCATA

10a. **Vicia americana** var. **americana** Map 1380

*Vicia americana* Muhl. [G, BB, P & S, Steyerm.]

This is the commoner variation in the state, occurring along the Missouri River counties in the northwestern section from Atchison to Jackson counties and locally east in St. Louis and Jefferson counties.

Ranges from Quebec to Alaska, south to Virginia, Ohio, Indiana, Illinois, Arkansas, Kansas, Oklahoma, New Mexico, and Arizona.

10b. **Vicia americana** var. **truncata** (Nutt.) Brewer Map 1380

Known only from Platte County, west-central Missouri (upper edge of wooded slopes in T54N R33W, sect. 6, just west of Old Union Mill distillery, along Platte River, 1–1¾ mi. northwest and west northwest of Edgerton, May 15, 1949, *Steyermark 67531*).

Ranges from western North America east to Manitoba, Minnesota, and Missouri.

*Excluded Species*

**Vicia Cracca** L. Tufted Vetch, Canada Pea

Specimens from Missouri sometimes identified as this species have proven to be *V. villosa*, as indicated under the discussion of that species. *Vicia Cracca* has a nongibbous calyx, very short and broad upper calyx-teeth, the large expanded blade of the standard of the corolla as long as the claw, and the plant is a perennial, whereas *V. villosa* is an annual or biennial with a gibbous calyx, lanceolate narrow upper calyx-teeth, and the blade of the standard is less than half as long as the claw.

**Vicia sativa** L. var. **sativa**

Specimens from Missouri sometimes identified as typical *V. sativa* have proven to be *V. angustifolia* (= *V. sativa* var. *nigra*) and *V. angustifolia* var. *segetalis* (= *V. sativa* var. *segetalis*), as indicated in the discussion under *V. sativa*.

## 35. **Lathyrus** L. Vetchling, Wild Pea

a. Leaflets of fully grown leaves only 2 . . . . . . . . . . . . . . . . . . . . . . . . . b

b. Stems broadly winged, with the wing 2–5 mm. broad; flowers 13–25 mm. long, 4–10 in each inflorescence . . . . . . . . . . . . . . . . . . . . . 5. L. LATIFOLIUS

---

Plate no. 226. **1.** Vicia villosa, $\times\ ^2/_5$. **2.** Vicia hirsuta, $\times\ ^2/_5$; *a.* Calyx, $\times\ 3^3/_5$; *b.* Fruit, $\times\ 1^1/_5$; After Gleason, The New York Botanical Garden. **3.** Vicia dasycarpa, $\times\ ^2/_5$; *a, b.* Flower, two views, × 2. **4.** Vicia americana, $\times\ ^2/_5$. **5.** Lathyrus venosus var. intonsus, $\times\ ^2/_5$.

PLATE NO. 226

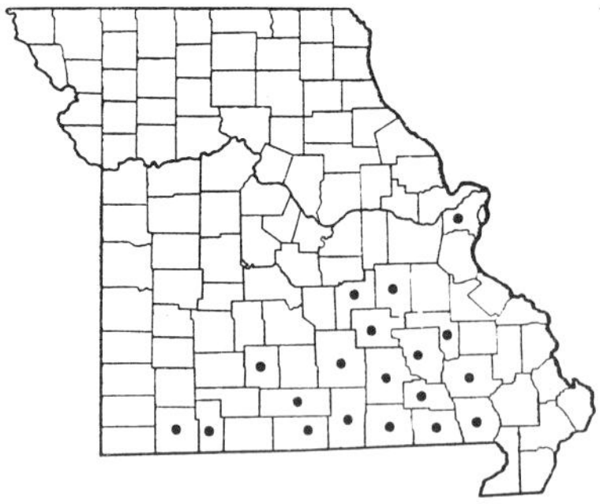

1381 Lathyrus venosus var. intonsus (Bushy Vetch)

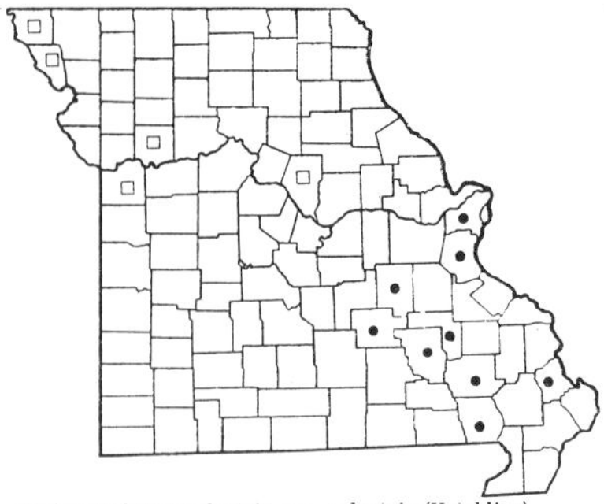

1382 □ Lathyrus palustris var. palustris (Vetchling)
1382 • Lathyrus palustris var. myrtifolius

1383 Lathyrus pusillus

b. Stems narrowly winged with the wing not more than 2 mm. broad; flowers 6–13 mm. long, 1–3 in each inflorescence . . . . . . . . . . . . . . . . . . . . . . . . . . c
c. Calyx-teeth narrowly lanceolate, much longer than the calyx-tube; flowers 5–9 mm. long; fruit smooth or without hairs . . . . . . . . . . . . . . . . . . 3. L. PUSILLUS
c. Calyx-teeth broadly lanceolate or ovate, about equaling the length of calyx-tube; flowers 9–13 mm. long; fruit hairy . . . . . . . . . . . . . . . . . . 4. L. HIRSUTUS
a. Leaflets of fully grown leaves 4–12 . . . . . . . . . . . . . . . . . . . . . . . . . d
d. Hairs present on most of the stem and lower surface of the leaflets . . . 1. L. VENOSUS var. INTONSUS
d. Hairs absent, plant completely glabrous . . . . . . . . . . . . . . . 2. L. PALUSTRIS

1. **Lathyrus venosus** Muhl. var. **intonsus** Butt. & St. John Bushy Vetch Map 1381
*Lathyrus venosus* ssp. *venosus* var. *intonsus* (Butt. & St. John) Hitchc.
Flowers May–June.

Occurs in rocky open woods, usually in acid soils, sometimes along railroad tracks. Ozark region of southern and east-central Missouri, north to St. Louis, Crawford, Phelps, Webster, and Barry counties.

Ranges from Quebec, Ontario to Minnesota and Saskatchewan, south to West Virginia, Indiana, Tennessee, and Missouri.

The rose-purple flowers of this pea are showy (15–20 mm. long) and occur in usually many-flowered (8–19) clusters. Sometimes scattered colonies are found with few or no flower-bearing plants. The plants spread by underground stems.

2. **Lathyrus palustris** L. Vetchling Map 1382
Also known as Marsh Pea.
Flowers June–September.

Occurs in wet meadows and prairies, low open swales, moist banks, and in thickets in valleys and along streams.

The following two variations are encountered in Missouri material:

Stems winged; leaflets usually 6 or 8, usually linear, lanceolate, or narrowly oblong . . . 2a. L. PALUSTRIS var. PALUSTRIS
Stems not winged or very slightly winged, slender; leaflets usually 4, more rarely 6, ovate, elliptic, or broadly lanceolate . . . . 2b. L. PALUSTRIS var. MYRTIFOLIUS

2a. **Lathyrus palustris** var. **palustris** Map 1382
*Lathyrus palustris* L. [G, P & S, Steyerm.]
*Lathyrus palustris* var. *linearifolius* Ser. [G, BB]

This variation occurs mainly in northwestern and central Missouri, in Atchison, Holt, Jackson, Ray, and Boone counties.

Ranges from Lower St. Lawrence River (Anticosti Island) to Alaska, south to New York, Pennsylvania, Ohio, Indiana, Illinois, Missouri, South Dakota, and California; also Eurasia.

2b. **Lathyrus palustris** var. **myrtifolius** (Muhl.) Gray Map 1382

This variation occurs chiefly in eastern Missouri south of the Missouri River, in St. Louis, Jefferson, Crawford, Dent, Iron, Reynolds, Wayne, and Scott counties.

Ranges from New York and Ontario to Wisconsin, south to Indiana, Illinois, and Missouri.

Plate no. 227. **1.** Lathyrus palustris var. palustris, × 2/5; *a.* Lathyrus palustris var. myrtifolius, × 2/5. **2.** Lathyrus pusillus, × 2/5. **3.** Lathyrus hirsutus, × 2/5; *a.* Calyx, × 2; After Gleason, The New York Botanical Garden. **4.** Lathyrus latifolius, × 2/5.

PLATE NO. 227

1384 Lathyrus hirsutus (Caley Pea)

1385 Lathyrus latifolius (Everlasting Pea)

1386 Pisum sativum var. arvense (Field Pea)

Dr. C. L. Hitchcock has recently revised this genus (Publ. U. Wash. Biol. 15: 1–104. 1952). According to his studies, it would appear that the typical variation (var. *palustris*) of this species does not occur in Missouri. However, all the collections cited by him under *L. palustris* var. *palustris* match material of typical *L. palustris* var. *palustris* with 6–8 narrowly oblong or lanceolate leaflets and a conspicuously winged stem. There are numerous intergrading and intermediate specimens which bridge the gap between var. *palustris* and var. *myrtifolius*. *Steyermark 9320* from Jefferson County, for example, has winged stems, but the 6–8 leaflets are oblong-elliptic to elliptic. According to Hitchcock's revision it would probably be placed in var. *palustris*, but following *Gray's Manual*, eighth ed., would be considered intermediate between var. *palustris* and var. *myrtifolius*.

Specimens of *Lathyrus palustris* are sometimes confused with *Vicia americana*, which they resemble. They may be separated on the basis of number of leaflets and character of stipule margin; in *Lathyrus palustris* the leaflets varying generally from 4–8, in *Vicia americana* varying from 8–18 with the stipules in *V. americana* noticeably toothed, whereas in *Lathyrus palustris* they are either not toothed or only faintly so.

3. **Lathyrus pusillus** Ell. Map 1383

Flowers April–May.

Occurs in rocky open woods and chert glades, rarely along railroads. Southwestern Missouri, known only from Jasper (3 mi. northwest of Joplin, May 27, 1922, *Palmer 21490*) and Newton counties.

Ranges from Florida to Texas, north to North Carolina, Missouri, and Kansas; introduced in Oregon.

The corolla is purplish with a paler or lighter color than those of *L. hirsutus*. Some specimens (*Palmer 49123* from Jasper County) have the petioles more winged than usual. As the petioles are slightly winged in some of the material examined from Missouri, this character has not been used in separating *L. pusillus* from *L. latifolius*.

4. **Lathyrus hirsutus** L. Caley Pea Map 1384

Also known as Singletary Pea.

Flowers May–June.

Occurs along railroads and waste ground. Recently introduced in the state, where known only from St. Louis (St. Louis, right-of-way of Frisco R. R. between Macklind Ave. and Evens and Howard Sewer Pipe Company's plant along a siding, May 29, 1954, *Muehlenbach 96*) and Jasper (rich waste ground, along railroad grade, Stockyard Switch, Joplin, June 3, 1949, *Palmer 49123*) counties.

Native of Europe; introduced and naturalized in the United States from Virginia to Alabama, Mississippi, Missouri, Oklahoma, and Texas; introduced also in Oregon.

This annual somewhat resembles the garden sweet pea (*L. odoratus* L.). The corolla varies from a usually dark purple or violet color to pink or white, the upper petal (standard) reddish- or blue-purple or violet with a white spot, the wings similar or slightly paler, and the keel paler or becoming white. In the southern third of the United States, where mild winters prevail, it is grown for pasturage, furnishing forage during the late winter and early spring months. It is also used for hay, as a winter cover, and for soil improvement.

5. **Lathyrus latifolius** L. Everlasting Pea Map 1385

Also called Perennial Pea.

Flowers May–September.

Occurs along roadsides, fence rows, railroads, and prairies, where escaped from cultivation. Rather rare and scattered in the state in Marion, St. Louis, Boone, Howell, Lawrence, and Greene counties.

Native of Europe; introduced and naturalized from New England to Indiana to Missouri, south to Virginia, Oklahoma, and Texas; and in Washington, Oregon, California, Arizona, and Wyoming.

The flowers are showy but not fragrant, and vary from rose-purple and deep rose to pink or white. The species is often planted in old perennial gardens and along fences, forming large masses.

## 36. **Pisum** L. Pea

**Pisum sativum** L. var. **arvense** (L.) Poir.
Field Pea Map 1386

Also called Austrian Winter Pea.

Flowers May–June.

Cultivated and rarely escaped along railroads and waste ground. A single collection from St. Louis County (St. Louis, right-of-way of Terminal R. R. Association, west of Shreve Avenue, May 8, 1955, *Muehlenbach 561*) is the only record for the state. Although the collection is not in flower or fruit, there is no doubt concerning the identity of the plant. Typical *P. sativum* var. *sativum* has white petals and entire leaflets, whereas var. *arvense* has pink or rose-purple and green petals with entire or sparsely toothed leaflets.

Native of Eurasia; introduced and naturalized from New England to North Dakota and the Pacific states, south to Georgia, Missouri, and Texas.

This pea is included in the flora of the state on the basis of the above adventive collection. It is probable that future collecting will reveal a more frequent occurrence of the plant. It is planted for hay, pasturage, silage, green manure, and for the seeds. In the United States the most common variety grown is the Austrian Winter Pea.

## 37. **Clitoria** L. Butterfly Pea

**Clitoria mariana** L. Butterfly Pea Map 1387

Flowers May–September.

Occurs in acid soils of chert, sandstone, or granite in low or upland rocky open woods and border of chert and sandstone glades or openings, along streams, in ravines, and on ridges. Southern and southeastern Ozark region north to Newton, Douglas, Texas, Shannon, Reynolds, Iron, and Ste. Genevieve counties, with a local station northeast in St. Louis County.

Ranges from Florida to Arizona, north to New York, West Virgina, Ohio, Indiana, Illinois, Missouri, and Oklahoma.

The showy flowers of this pea are pale blue and lilac, delicately veined with darker purple markings toward the center of the standard. This plant requires an acid soil and is not usually successfully transplanted because of the long tap root. I have never been able to establish the plant either from seed or through transplanting.

## 38. **Apios** Medic. Groundnut

**Apios americana** Medic. Groundnut Map 1388

Flowers late June–September.

Occurs in wet meadows, low thickets and banks of streams, ponds, and sloughs, and in moist soils of woodland.

Missouri material is represented by the following variations:

a. Stems and leaves more or less hairy . . . b. A. AMERICANA var. AMERICANA f. PILOSA
a. Stems and leaves glabrous (without hairs) or nearly glabrous . . . . . . . . . . *b*
  b. Flower clusters compact, rounded at the summit; common type found . . . . a. A. AMERICANA var. AMERICANA f. AMERICANA
  b. Flower clusters more elongated and pointed or prolonged at the summit, with a lanceolate or prolonged ovoid shape . . . . c. A. AMERICANA var. TURRIGERA

a. **Apios americana** var. **americana** f. **americana** Map 1388

*Apios americana* Medic. [G, BB, P & S, Steyerm.]

Common throughout Missouri, and doubtless in every county. This is the common variation in the state.

Ranges from New Brunswick to Minnesota and Colorado, south to Florida, Louisiana, and Texas.

b. **Apios americana** var. **americana** f. **pilosa** Steyerm. Map 1388

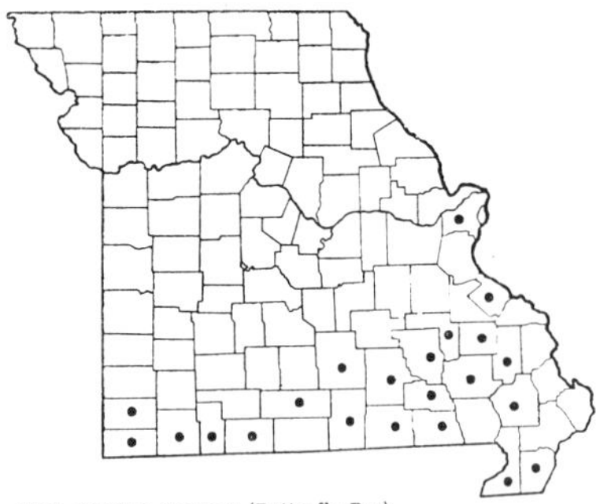

1387 Clitoria mariana (Butterfly Pea)

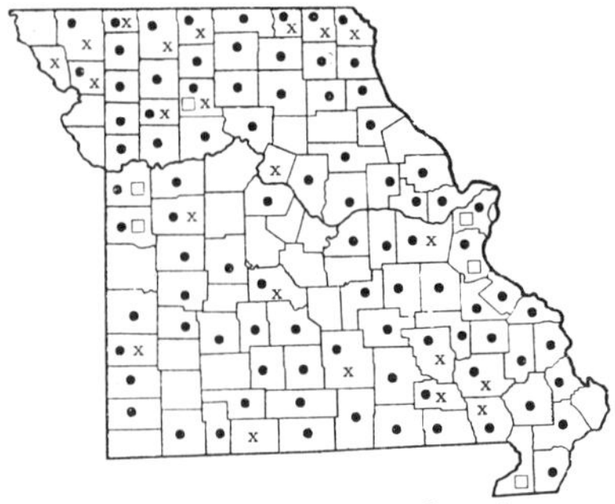

1388 • Apios americana var. americana f. americana (Groundnut)

1388 x Apios americana var. americana f. pilosa

1388 □ Apios americana var. turrigera

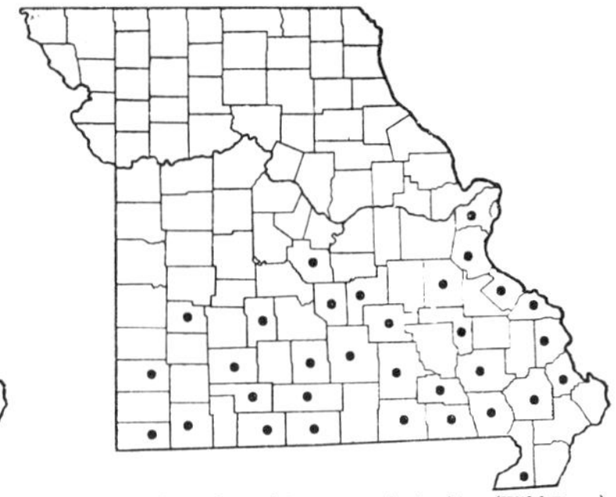

1389 Phaseolus polystachios var. polystachios (Wild Bean)

*Apios americana* f. *pilosa* Steyerm. [G]

Commonly found throughout Missouri and occupying the same general range as f. *americana.*

c. **Apios americana** var. **turrigera** Fern. Map 1388

Rare and scattered in southern and central Missouri in St. Louis (Forest Park, August 21, 1875, *Eggert*), Dunklin, Livingston (*Sparling 486*), Jackson (Sibley, July 19, 1896, *Bush 348*), and Cass (July 1, 1865, *Broadhead*) counties. Probably of more common occurrence.

Ranges from South Carolina to Louisiana and Texas, north to Virginia, Illinois, Missouri, and Kansas.

The cooked starchy tuberous thickenings, sometimes 6 cm. thick, of the slender underground stems make a satisfactory substitute for potatoes, prepared in roasted or fried form. They were eaten by the early New England colonists and held in high esteem by the American Indians. The Indians also used the seeds as a cooked vegetable, similar to other beans.

## 39. **Phaseolus** L. Wild Bean, Kidney Bean

a. Native species of rocky open woods and thickets of the Ozark region; perennial twining, climbing, or trailing plants; main stems more or less hairy; flowers rose or pink; leaflets short and inconspicuously hairy to densely soft hairy on lower surface . . . . . . . . . . . . 1. P. POLYSTACHIOS

a. Cultivated plant, rarely escaping to roadsides, railroad tracks, and waste ground; annual, non-climbing plant of low stature, of the types escaped in Missouri; main stems glabrous to sparsely hairy; flowers white, yellow, green, lavender, or purple; leaflets glabrous (without hairs) or sparsely hairy on lower surface . . . . . . . . . . . . . . . . . . . . . . . . . *b*

  b. Flowers 9–10 mm. long; the 2 bracts at base of calyx inconspicuous, 2 mm. long, 1–1.5 mm. broad, shorter than the calyx-lobes or barely reaching the sinuses of the lobes; calyx-teeth broader than long; fruit 5–8.5 cm. long, 1.8–2 cm. broad, flat; seeds broad and flat . . . . . 3. P. LUNATUS

  b. Flowers 11–20 mm. long; the 2 bracts at base of calyx conspicuous, 5–6 mm. long, 4–5 mm. broad, equaling or longer than calyx-tube and nearly concealing most of calyx; calyx-teeth about as long as broad or longer than broad; fruit 10–20 cm. long, 0.6–1.2 cm. broad, rounded or with curved sides, not flat; seeds not broad and flat . . . . . . . . 2. P. VULGARIS var. HUMILIS

1. **Phaseolus polystachios** (L.) BSP. var. **polystachios** Wild Bean Map 1389

*Phaseolus polystachios* (L.) BSP. [G, BB]

*Phaseolus polystachyus* (L.) BSP. [P & S]

Flowers July–September.

Occurs in low or upland rocky open woods in

Plate no. 228. **1.** Clitoria mariana, × 2/7; Details from Small, The New York Botanical Garden. **2.** Apios americana var. americana, × 2/7; Details from Small, The New York Botanical Garden. **3.** Strophostyles helvola var. missouriensis, × 2/7; Details from Small, The New York Botanical Garden. **4.** Strophostyles leiosperma, × 2/7; *a.* Fruit, × 2/7; After Britton and Brown, details from Small, The New York Botanical Garden. **5.** Strophostyles umbellata, × 2/7; *a.* Fruit, × 2/7; After Britton and Brown, The New York Botanical Garden. **6.** Phaseolus polystachios, × 2/7; Details from Small, The New York Botanical Garden. **7.** Vigna sinensis, × 2/7; Details from Small, The New York Botanical Garden. *a.* Fruit, × 2/7; *b, c.* Seeds, × 2/7. **8.** Glycine Max, × 2/7.

PLATE NO. 228

1390 Phaseolus vulgaris var. humilis (Bush Bean)

1391 Phaseolus lunatus var. lunonanus (Lima Bean)

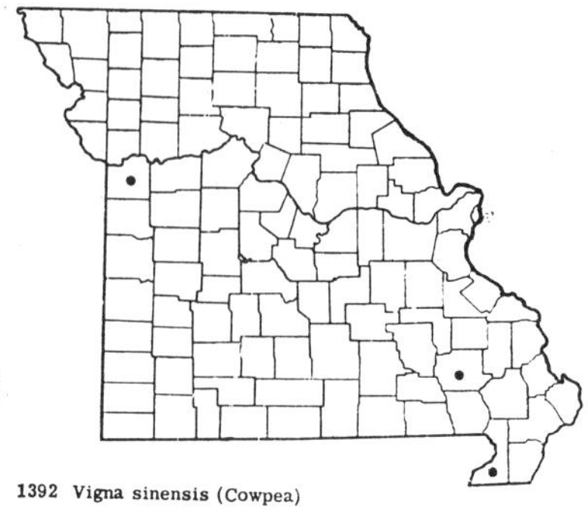

1392 Vigna sinensis (Cowpea)

ravines, along base of bluffs, rocky upper slopes, and thickets. Ozark region of southern and east-central Missouri north to St. Louis, Washington, Phelps, Miller, Dallas, Cedar, and Jasper counties.

Ranges from Florida to Texas, north to New Jersey, West Virginia, Ohio, Indiana, Illinois, and Missouri; reported from Iowa and Nebraska.

This species has been growing in the author's botanical preserve in northern Illinois for the past fifteen years. Some years it does not begin to flower until late September and consequently the fruit does not have a chance to mature before cold weather kills the plant. Other years it flowers earlier and produces fruit. Its clusters of small rose or pink flowers, with corollas 10–12 mm. long, are showy. The leaves have an adherent quality to them, sticking to the clothes or skin. They greatly resemble in shape and size those of *Amphicarpa bracteata*, the Hog Peanut. The seeds may be cooked and used in the manner of regular beans.

2. **Phaseolus vulgaris** L. var. **humilis** Alef.
Bush Bean Map 1390
Flowers May–October.

Commonly planted as a garden vegetable, but rarely escaped or persisting away from cultivation. Known as an adventive in waste ground along railroads in St. Louis County (St. Louis, Baden Freight Yard of M. K. T. R. R. around car-weighing machine in the north part of yard, June 22, 1957, *Muehlenbach 1228*), where a large colony was found.

Native of America, but origin not definitely known; planted in almost all regions of the earth except in the colder areas. Rarely escaping from cultivation, and probably not persisting. The common tall-twining variety of the garden pole bean, *P. vulgaris* var. *vulgaris*, known as Kidney Bean or String Bean and grown like var. *humilis* for its edible green immature pods, has not been found as an escape from cultivation.

3. **Phaseolus lunatus** L. var. **lunonanus** Bailey
Lima Bean, Butter Bean Map 1391
Flowers May–October.

Commonly planted and rarely escaping or persisting along railroads and waste ground. Known as an adventive only in St. Louis County (St. Louis, Montgomery St. at Wabash R. R. crossing, August 11, 1957, *Muehlenbach 1304*).

Native of tropical America. Commonly planted, but rarely escaped from cultivation.

Cases of poisoning have sometimes been reported from this plant which contains a substance, which when acted upon by an enzyme yields hydrocyanic acid. This poisonous property is extracted by prolonged boiling, but this is merely withdrawn and not destroyed, and if the water is absorbed it presents the same dangers as the beans themselves.

The bush or erect lima bean is the kind commonly grown in the United States and is the one escaped from cultivation in Missouri. The climbing or twining type of lima bean is the one grown in tropical America.

## 40. **Vigna** Savi

**Vigna sinensis** Savi Cowpea Map 1392
*Vigna Catjang* Walp.
*Vigna unguiculata* (L.) Walp.
Also known by the seeds as Black-eyed Peas.
Flowers July–September.

Commonly cultivated, and rarely escaped from cultivation in alluvial open ground along streams or in low ground. Southeastern and west-central Missouri in Jackson, Wayne (open sandy ground on bar in zone between edge of St. Francis River and alluvial forest area, T27N, R6E, sect. 10, 3½ mi. northwest of Chaonia, September 1, 1938, *Steyermark 6336*), and Dunklin (common, Malden, August 27, 1894, *Bush*) counties.

Native of Asia; widely cultivated in tropical and warm regions; in the United States introduced and naturalized from Florida to Texas, north to Delaware, Indiana, Illinois, and Missouri.

The pods in Cowpea vary from 17.5–30 cm. long, and are pendent. The stems are erect to trailing or twining. Cowpea is an important forage and food crop in the southeastern United States, where it is planted extensively as a green manure crop. Plants withstand hot wet weather. The leaves and pods furnish excellent forage for all classes of livestock. The seeds are known in the southern states as black-eyed peas, and are there relished as a food in either the immature or dried state. The young pods are often cooked in the manner of snap-beans.

## 41. **Strophostyles** Ell. Wild Bean

a. Calyx-tube densely hairy, 1–1.5 mm. long; the pair of small green bracts (bracteoles) at base of the calyx are densely hairy; flowers 5–7 (–8) mm. long; both sides of leaflets mostly gray with a silky hairiness; seeds shining, smooth; fruit 2–3.5 cm. long, hairy with spreading hairs . . . 3. S. LEIOSPERMA

a. Calyx-tube glabrous (without hairs) or sparsely hairy with a few appressed hairs (pressed close to or lying parallel to the surface), mostly 1.5–2.5 mm. long; the pair of small green bracts (bracteoles) at base of the calyx glabrous or sparsely hairy with a few appressed hairs; flowers mostly 8–13 mm. (7 or 14 also) long; upper side of leaflets usually without hairs (glabrous) or only sparsely hairy; seeds scurfy or with a woolly coating; fruit mostly 4–10 cm. (rarely 3.5) long, mostly sparsely hairy with appressed or strigose hairs (lying flat or parallel to the surface) . . . . . . . . . . . *b*

- b. The pair of small green bracts (bracteoles) at base of calyx are inconspicuous, less than 2.5 mm. long, blunt, ovate or oblong, and reach only half to less than half the length of the calyx-tube; leaflets rarely lobed, never with contracted sides, the larger leaflets up to 2 cm. broad; mature fruits 3.7–5.5 cm. long; seeds 3–5 (–6) mm. long . . . . . . . . . . . 2. S. UMBELLATA
- b. The pair of small green bracts (bracteoles) at base of calyx are conspicuous, 2.8–4.5 mm. long, acute, lanceolate, and reach at least to the base of the calyx-teeth; leaflets lobed or unlobed, usually somewhat contracted on one side, the larger leaflets up to 4 or to 6.5 cm. broad; mature fruits mostly 5–10 (3.5–5) cm. long; seeds 6–12 mm. long . . . . . . . . . . 1. S. HELVOLA

### 1. **Strophostyles helvola** (L.) Ell. Wild Bean Map 1393

Flowers June–October.

Occurs in rich rocky woods and thickets, moist alluvial ground, sand and gravel bars, fallow fields, along roadsides and railroads.

Two variations are found in Missouri:

Common type encountered; main leaflets usually with 1–3 lobes or more or less contracted on one or both sides, often short-pointed, the larger ones 0.8–4 cm. broad; fruits 3.5–8 cm. long; seeds 6–9.5 mm. long . . 1a. S. HELVOLA var. HELVOLA

Rare type encountered; main leaflets unlobed, the sides gradually curved to a blunt or slightly pointed (acutish) tip, the larger ones 3–6.5 cm. broad; fruit 5–10 cm. long; seeds 8–12 mm. long 1b. S. HELVOLA var. MISSOURIENSIS

### 1a. **Strophostyles helvola** var. **helvola** Map 1393

*Strophostyles helvola* (L.) Ell. [G, BB, P & S]

*Phaseolus helvolus* L. [Shinners]

Throughout Missouri, and probably in every county. This is the commoner variation in the state.

Ranges from Florida to Texas, north to Massachusetts, Quebec, Ontario, Michigan, Wisconsin, Minnesota, and South Dakota.

### 1b. **Strophostyles helvola** var. **missouriensis** (S. Wats.) Britt. Map 1393

Scattered throughout Missouri, but less frequent than var. *helvola*; usually in more alluvial soils of woodland and thickets.

The seeds of this species are reported to have nutritive and palatable beanlike quality. Unlike the species of clover (*Trifolium*), alfalfa (*Medicago sativa*), sweet clovers (*Melilotus*), and various other cultivated legumes, this wild bean does well on poor acid soils and is able to enrich the soils upon which it grows with the nitrogen-fixing bacteria contained in the enlarged nodules of its roots. It is, therefore, a good wild plant for fallow and wornout fields.

### 2. **Strophostyles umbellata** (Muhl.) Britt. var. **umbellata** Map 1394

*Phaseolus umbellatus* (Muhl.) Britt. [Shinners]

Flowers July–October.

1393 ● Strophostyles helvola var. helvola (Wild Bean)
1393 □ Strophostyles helvola var. missouriensis

1394 Strophostyles umbellata var. umbellata

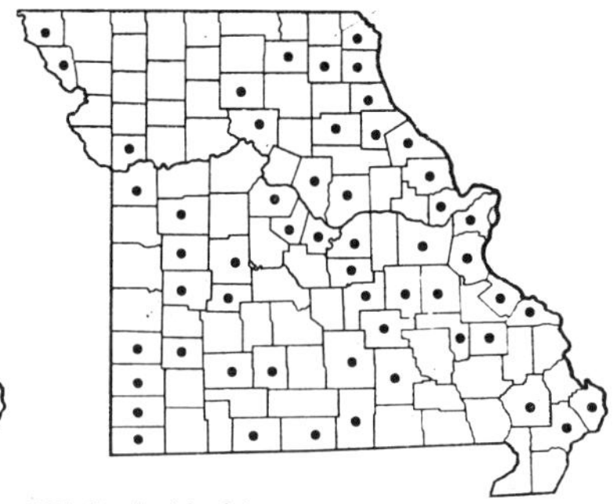

1395 Strophostyles leiosperma

Occurs in low or upland woods and thickets, ravines, along bluffs, streams, wooded slopes and ridges, often in acid soils.

Southern, central, and eastern Missouri, north to Clay, Boone, and Clark counties; absent from most of northern Missouri north of the Missouri River, except in the counties bordering the Mississippi River.

Ranges from Florida to Texas, north to New York, Indiana, Illinois, Missouri, and Oklahoma.

The seeds are eaten by wild turkey.

3. **Strophostyles leiosperma** (T. & G.) Piper Map 1395

*Phaseolus leiospermus* Torr. & Gray [Shinners]

Flowers mid-June–October.

Occurs in prairies, glades, open and upland woods, thickets, fallow fields, along roadsides and railroads, often in acid soils.

Common throughout southern, central, and eastern Missouri, and along the Missouri River northwest to Atchison County, but absent apparently from most of the north-central and northwestern counties.

Ranges from Alabama to Texas and Colorado, north to Ohio, Indiana, Wisconsin, Minnesota, South Dakota, and Nebraska.

A number of specimens from Missouri misidentified as *S. umbellata* are referred in the present flora to *S. leiosperma*. The seeds of *S. leiosperma* average smaller (2–4 mm. long) than those of *S. umbellata* (3.5–6 mm. long), as do also the mature fruits (2–3.5 cm. long in *S. leiosperma*, 3.7–5.5 cm. in *S. umbellata*). The other characters, as presented in the key, should readily distinguish the two species. In general, *S. leiosperma* has leaflets much more pubescent on both sides, but sometimes old specimens occur in which the pubescence is inconspicuous.

## 42. **Glycine** L.

**Glycine Max** (L.) Merr. Soy Bean Map 1396

*Glycine Soja* (L.) Sieb. & Zucc.

*Phaseolus Max* L.

Flowers July–October.

Commonly cultivated, but rarely escaped to gravel bars along streams, waste ground, and along railroads. Known from Callaway (along gravel bed of Stinson Creek, T47N, R9W, sect. 34, 3 mi. southeast of Fulton, September 10, 1937, *Steyermark 26110*), St. Louis (St. Louis, Baden Freight Yard M.K.T. R. R. around the car-weighing machine in north part of yard, June 22, 1957, *Muehlenbach 1228*), and Cape Girardeau (Cape Girardeau, July 15, 1956, *Liz Renner*) counties.

Native of eastern Asia; introduced and naturalized from Florida north to Delaware, Michigan, Illinois, and Missouri.

Easily recognized by the brown hairy stems, leaves, and drooping fruits.

In the midwestern and north-central states, especially in Illinois and Indiana, this is one of the most important field and cash crops. The beans in eastern Asia are the most important legume food. The seed is considered to be the richest natural vegetable food known to man. Since the soy flour is low in carbohydrate and high in protein, it is given to diabetic persons. Soybean milk is also extracted from the seed and used in cooking. Soybean sprouts are used in various chop suey dishes, while soybean sauce, made from the cooked beans, wheat flour, salt, and ferment, is a popular seasoning. In the United States soybean is grown principally for the seeds from which an oil is obtained, used in salads, cooking, margarine and shortenings, in the manufacture of soap, paints,

1396 Glycine Max (Soy Bean)

1397 Canavalia ensiformis (Jack Bean)

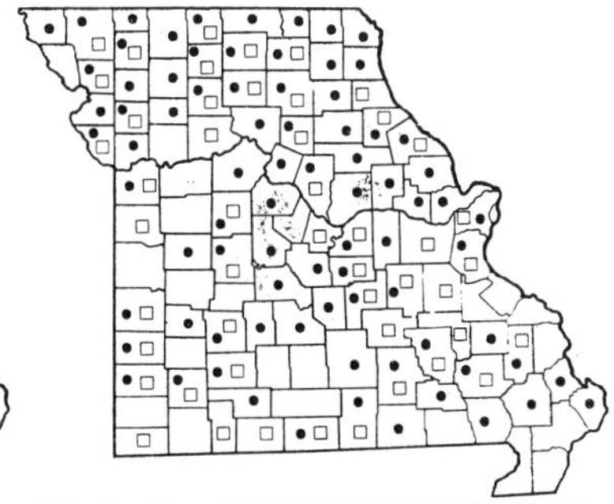

1398 • Amphicarpa bracteata var. bracteata (Hog Peanut)
1398 □ Amphicarpa bracteata var. comosa

varnishes, greases, rubber substitutes, cleaning compounds, disinfectants, and insecticides. The meal in the oil cake contains 40–48 per cent protein and is a valuable livestock food and source of soy flour. It is used also for the making of fertilizers, plastics, adhesives, sizing, foaming solutions for extinguishing oil fires, as a source of synthetic fibers, and has many other uses. The soybean is used in agriculture for green manure, hay, and silage, as well as for the value of the seeds. Hundreds of varieties are in cultivation.

## 43. **Canavalia** Adanson

**Canavalia ensiformis** (L.) DC. Jack Bean
Map 1397

Also known as Wonder Bean.

Known from St. François County, southeastern Missouri (Farmington, April 8, 1939 in U. of Mo. herb.) where established according to information originally received by Dr. Rickett to whom a specimen was sent for determination in 1939.

Probably native of tropical America.

Cultivated in tropical and warm temperate regions as a forage or green manure plant, as well as for its edible seeds and young fruits.

## 44. **Amphicarpa** Ell. Hog Peanut

**Amphicarpa bracteata** (L.) Fern. Hog Peanut
Map 1398

Flowers August–September.

Occurs in dry or moist rich or rocky woods and thickets. Throughout Missouri.

Two variations occur in the state, and may be distinguished as follows:

Middle (terminal) one of the three leaflets 1.3–6 cm. long; leaflets thin, sparsely hairy; hairs of stem pointing away from the tip, mainly appressed (lying flat upon or parallel to the surface) and whitish or pale; stipules midway up the stem 3 mm. long; bracts of the inflorescence 2–2.5 mm. long; sides of mature aerial fruits glabrous (without hairs), with ascending hairs near the base of the margins of the fruit . . . a. A. BRACTEATA var. BRACTEATA

Middle (terminal) one of the three leaflets mostly 5–10 cm. long; leaflets firm, more coarsely and densely hairy; hairs of stem mainly spreading and tawny; stipules midway up the stem 4–5 mm. long; bracts of the inflorescence 2.5–3.5 mm. long; sides of the mature aerial fruits hairy, with backward-pointing (retrorse) hairs near the base of the margins of the fruit . . . b. A. BRACTEATA var. COMOSA

a. **Amphicarpa bracteata** var. **bracteata**
Map 1398

*Amphicarpa bracteata* (L.) Fern. [G, P & S]

Throughout Missouri; probably in every county but not recorded from some western and eastern counties.

Ranges from Quebec to Manitoba and Montana, south to Florida, Louisiana, and Texas.

b. **Amphicarpa bracteata** var. **comosa** (L.) Fern.
Pitcher's Hog Peanut Map 1398

*Amphicarpa comosa* (L.) G. Don [P & S]

Throughout Missouri, but not recorded from nearly half of the counties in the state; undoubtedly in most every county.

This is the coarser and more abundantly pubescent variety. It intergrades into the var. *bracteata*. The varieties were formerly considered as distinct species, but most botanists regard them at present as two varieties of one species connected by intermediate populations of plants. One author has even regarded the var. *comosa* only as 'an extremely vigorous' form of *A. bracteata* (See Schively in Contr. Bot. Lab. Univ. Pa. 1: 356. 1897).

There are two or three types of flowers and fruits present, those of the upper axils and branches with perfect petal-bearing flowers usually producing 3–4-seeded aerial dehiscent fruits, whereas those of the lower axils near the base of the stem and on creeping branches lack petals and produce autumnal dehiscent 1-seeded fruits and fleshy, small 1-seeded fruits usually under the ground which do not open (indehiscent). These subterranean fruits which vary from 6–13 mm. in diameter are quite edible and were commonly used by the American Indians. Cooked and seasoned with salt, pepper, and buttered or creamed, they are said to resemble garden beans. The usual time for gathering these is in late fall and early spring. White-footed and meadow mice (voles) commonly store these underground fruits in their nests and homes. The plants are browsed by white-tailed deer.

Sterile plants of *Amphicarpa* are often difficult to separate from *Strophostyles umbellata* when the two occur together. They may be distinguished by the prolonged midrib at the tip of the leaflet in *Amphicarpa*, not prolonged in *Strophostyles*, by the 10–12-nerved stipules at the base of the petiole in *Amphicarpa*, in *Strophostyles* these are 1–3- or 5–7-nerved, and by differences in angle of arching of the lateral nerves of the leaflets, those in *Strophostyles* arching parallel to the margin of the leaflet, those in *Amphicarpa* not parallel to the margin but directed straight and more divaricate.

## 45. **Galactia** P. Br. Milk Pea

**Galactia volubilis** (L.) Britt. var. **mississippiensis** Vail Milk Pea Map 1399

Flowers July–September.

Occurs on rocky open wooded slopes and glades often on cherty limestone and sandstone strata. Ozark region of southern and east-central Missouri north to St. Charles, Warren, Gasconade, Osage, Cole, Morgan, Benton, St. Clair, and Newton counties.

Ranges from Alabama and Mississippi, north to Kentucky, Illinois, Missouri, and Kansas.

All the Missouri material examined has, for the most part, the upper surface of the leaflets hairy, thus placing it in var. *mississippiensis*. An occasional specimen collected very late in the season, such as *Steyermark 20777* from Stoddard County on November 8, 1936, shows few hairs on the upper surface of some of the leaflets, approaching typical var. *volubilis* thereby, but the other leaflets exhibit more abundant pubescence on the upper surface.

Plants of *Galactia* in leaf only sometimes occur together with sterile plants of *Strophostyles leiosperma* with which they are sometimes confused. The stipule at the base of the petiole in *Galactia volubilis* is setaceous (hairlike), that in *Strophostyles* is spatulate, oblong-spatulate, or oblong-lanceolate. The stipule in *Galactia volubilis* is 1-nerved, in *Strophostyles* is 1–3-, or 5–7-nerved. In *Galactia volubilis* the lowest pair of lateral nerves is inconspicuous at the base of the leaflet, in *Strophostyles* conspicuous; in *Galactia volubilis* the hairs on the stems are ascending, pointing toward the tip of the plant, in *Strophostyles* these are retrorse or pointing backward (See Steyermark, Rh. 42: 213–15. 1940).

The seeds are eaten by wild turkey.

### *Excluded Species*

**Galactia regularis** (L.) BSP.

Specimens once identified as belonging to this species and cited by Palmer and Steyermark in their *Annotated Catalogue* (p. 584) from southern Missouri are misidentified *G. volubilis* var. *mississippiensis* and are now referred to the latter. These specimens (all in the herbarium of the Mo. Bot. Gard.) are as follows: Barry Co. (Eagle Rock, September 19, 1896, *Bush 49*); Shannon Co. (Monteer, October 23, 1911, *Bush 6538*); Dunklin Co. (Crowley Ridge, August 7, 1932, *Kellogg 26021A*); and Butler Co. (July 27, 1892, *Eggert*; Poplar Bluff, September 11, 1919, *Palmer 16348*).

---

Plate no. 229. **1.** Amphicarpa bracteata var. bracteata, × 2/5; Details from Small, The New York Botanical Garden. **2.** Rhynchosia latifolia, × 2/5; *a.* Fruit, × 4/5; Details from Small, The New York Botanical Garden. **3.** Galactia volubilis var. mississippiensis, × 2/5; After Gleason, details from Small, The New York Botanical Garden. **4.** Linum usitatissimum, × 2/5; *a.* Fruiting calyx and capsule, × 4/5; Details from Small, The New York Botanical Garden. **5.** Linum sulcatum, × 2/5; *a.* Fruiting calyx, × 1 1/5. **6.** Pueraria lobata, × 2/5; *a.* Calyx, × 4; After Gleason, details from Small, The New York Botanical Garden. **7.** Linum rigidum, × 2/5; *a.* Calyx with fruit, × 1 1/5.

PLATE NO. 229

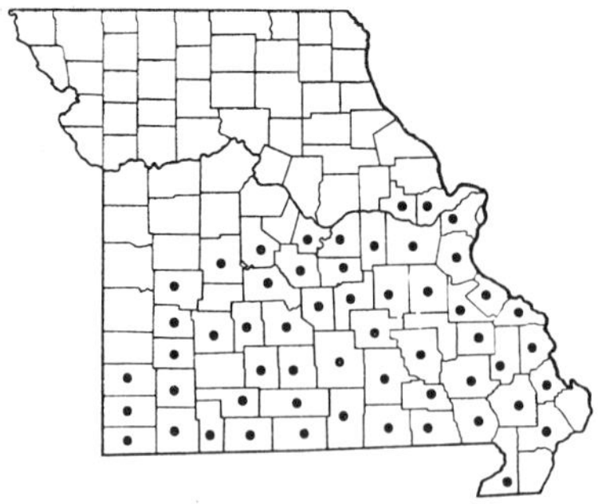

1399 Galactia volubilis var. mississippiensis Vail (Milk Pea)

1400 Pueraria lobata (Kudzu Vine)

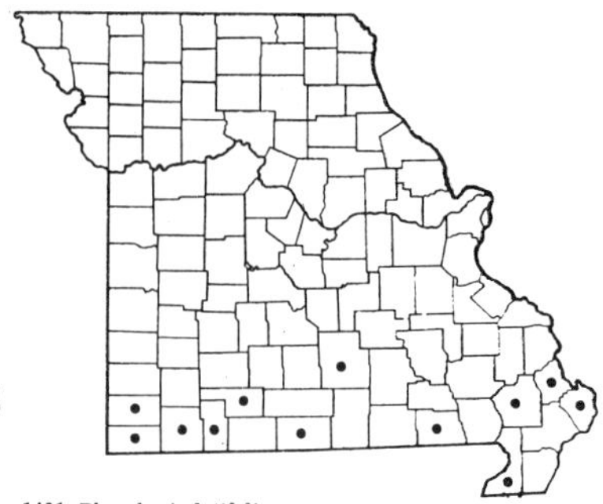

1401 Rhynchosia latifolia

## 46. **Pueraria** DC. Kudzu Vine

**Pueraria lobata** (Willd.) Ohwi Kudzu Vine Map 1400

*Pueraria Thunbergiana* (Sieb. & Zucc.) Benth. 1867 [BB]

*Dolichos lobata* Willd. 1802

*Pachyrrhizus Thunbergianus* Sieb. & Zucc. 1846

Flowers August–September.

Commonly planted and rarely escaping. Known from Ralls (Oakwood, October 17, 1935, *J. E. Burns*), St. Louis (Kirkwood, September 2, 1921, *Drushel 4497*), Lawrence, and Jasper counties.

Native of Japan; introduced and naturalized from southeastern United States, north to Pennsylvania, Tennessee, Illinois, and Missouri.

The Kudzu Vine has been extensively planted along many Missouri highways to cover eroded or open gravelly embankments and freshly exposed road-side grades. It was originally introduced into the United States, however, for its edible starchy root and for its fiber. The starch from the roots is eaten commonly by the Japanese people. The vine produces good hay and forage. The long running stems elongate over the ground for distances of a hundred feet or more. The reddish-purple flowers exude a fragrance resembling that of grapes; the standard is slightly paler purple with a green blotch in the center.

## 47. **Rhynchosia** Lour.

Inflorescence 5–30 cm. long on a long peduncle (main flower-stem) equaling or longer than the subtending leaf which arises immediately at the base of the stalk of the inflorescence; inflorescence loose, interrupted, open; calyx 11–15 mm. long; middle (terminal) one of the three leaflets 3–8 cm. long; leaflets densely and softly hairy on lower surface; more common species . . . . . . . . 1. R. LATIFOLIA

Inflorescence 1–4 cm. long on a short peduncle (main flower-stem) shorter than the subtending leaf immediately arising at the base of the stalk of the inflorescence; inflorescence dense, the flowers close together; calyx 8–11 mm. long; middle (terminal) one of the three leaflets 2–5 cm. long; leaflets sparsely hairy on lower surface; rare species . . . . . . . . . . . . . . . . . 2. R. DIFFORMIS

1. **Rhynchosia latifolia** Nutt. Map 1401

Flowers June–August.

Occurs in sandy prairies, limestone glades, high rocky open woods, sandy alluvial woods, along streams, and thickets. Southern Missouri, north to Scott (prairie northwest of Sikeston), Stoddard, Texas, Christian, and Newton counties.

Ranges from Louisiana and Texas, north to Missouri and Oklahoma.

2. **Rhynchosia difformis** (Ell.) DC. Map 1402

*Rhynchosia tomentosa* [of P & S], not (L.) H. & A.

Flowers June–July.

Occurs in sandy open ground. Known only from Dunklin County, southeastern Missouri.

Ranges from Florida to Texas, north to Virginia and Missouri.

The lowest and earliest leaves of this species have usually only one kidney-shaped leaflet, the later leaves having three rounded to ovate or elliptic leaflets.

1402 Rhynchosia difformis

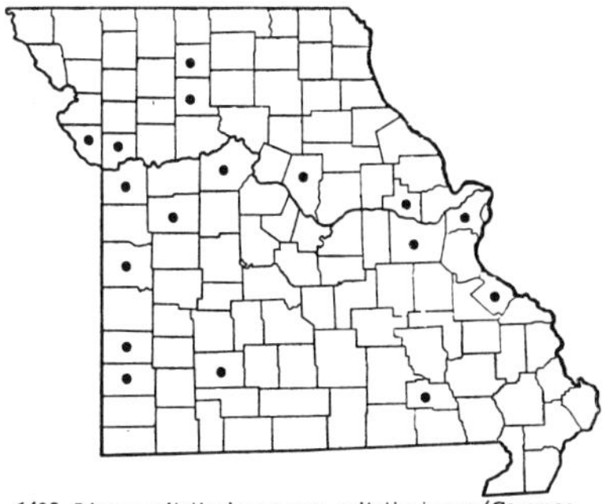

1403 Linum usitatissimum var. usitatissimum (Common Flax)

1404 Linum rigidum var. rigidum

## Order **GERANIALES**

### Fam. **LINACEAE** (Flax Family)

### **Linum** L. Flax

a. Petals blue; fruiting stalks 10–40 mm. long; fruiting capsules usually 7–10 mm. in diameter, 5–10 mm. long; margins of sepals without any glands . . . . . . . . . . . . . 1. L. USITATISSIMUM
a. Petals yellow; fruiting stalks 5–10 mm. long; fruiting capsules 2–3.5 mm. in diameter, 1–5 mm. long; margins of inner or outer sepals, or both, with glands . . . . . . . . . . . . . . . . . *b*
b. Petals 10–15 mm. long; sepals 7–8 mm. long; mature fruiting capsules about 5 mm. long. 2. L. RIGIDUM
b. Petals 5–8 mm. long; sepals 2–6 mm. long; mature fruiting capsules 1–3.5 mm. long . . . . . . *c*
c. Two small round black glands present next to the stem at the base of each leaf; margins of both outer and inner sepals glandular-toothed; outer sepals 4–7 mm. long . . . . . . 3. L. SULCATUM
c. No black glands at the base of each leaf; margins of outer sepals not glandular; outer sepals 1.5–4 mm. long . . . . . . . . . . . . . . . . . . . . . . . . . . . *d*
d. Most of the middle and lower leaves of the stem opposite; larger leaves of stem mainly averaging 5–12 mm. wide; margins of inner sepals mostly entire or with glands or teeth only toward the summit; stem conspicuously angled from the continuing bases on the stem from one leaf to another . . . . . . . . . . . . . . . . . . 5. L. STRIATUM
d. Most of the leaves of stem alternate, but lower ones often opposite and middle ones sometimes opposite; larger leaves of stem mainly averaging 1–5 (–7) mm. broad; margins of inner sepals conspicuously glandular-toothed or ciliate (with short slender hairs) to or below the middle; stem with curved or rounded sides, not conspicuously angled . . . 4. L. MEDIUM var. TEXANUM

1. **Linum usitatissimum** L. var. **usitatissimum**
Common Flax Map 1403
Also known as Linseed.
Flowers May–September.

Occurs along roadsides, railroads, and open waste ground. Scattered throughout Missouri.

Native of Europe; introduced and naturalized throughout most of the United States and Canada.

The seeds of this flax are used for linseed oil and for bird food, while linen cloth, thread, twine, fishing lines, carpet material, canvas, and duck articles are made from the fibers of the stem. A linseed cake is used extensively as an ingredient in stock feeds. Sometimes cases of poisoning occur in cattle and pigs which eat feed containing flax screenings, the immature capsules being considered the most likely to cause poisonous disturbances. The bast fiber of linseed plant is used in paper making primarily for cigarette paper tissue, but also goes into paper money, bond, and fine writing paper. From the meal obtained from the seed a mucilage is extracted which is used in medicinal preparations, water paints, and in the manufacture of certain fibers. The oil is used commonly in Egypt in food. As a drying oil it is used as an ingredient in printer's ink, and in the preparation of linoleum, paints, varnishes, and soaps. Minnesota, Montana, and North and South Dakota lead the United States in the commercial production of this plant.

2. **Linum rigidum** Pursh var. **rigidum** Map 1404
Flowers May–July.

Introduced along railroads and waste ground.

1405 Linum sulcatum

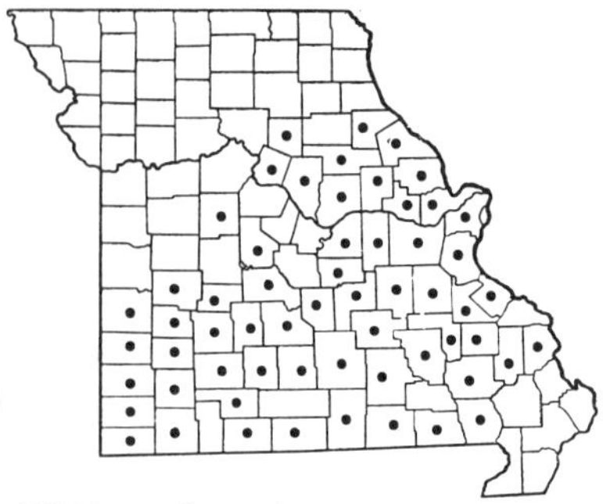

1406 Linum medium var. texanum

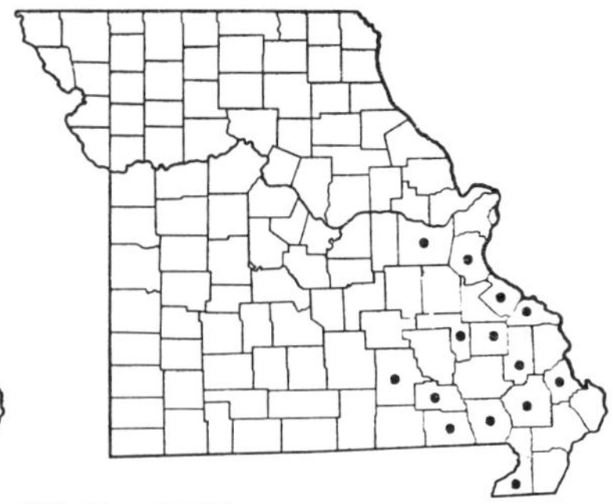

1407 Linum striatum

Known only in west-central Missouri in Jackson County (Kansas City, rare, May 16, 1896, *Mackenzie 102*; locally abundant, Kansas City, May 16, 1896, *Mackenzie 854*).

Ranges from Minnesota to Missouri and Texas, west to Alberta, Colorado, and New Mexico; also in Mexico.

3. **Linum sulcatum** Riddell Map 1405

Flowers May–September.

Occurs on glades, prairies, and rarely in open woods. Throughout most of Missouri, except absent from all the southeastern and some of the northwestern sections of the state.

Ranges from Massachusetts to Michigan, Nebraska, and Manitoba, south to Georgia, Alabama, Arkansas, and Texas.

4. **Linum medium** (Planch.) Britt. var. **texanum** (Planch.) Fern. Map 1406

Flowers May–August.

Occurs in usually acid soils overlying chert, sandstone, or granite substrata in rocky open woods, prairies, and glades. Ozark region of southern and central Missouri north to Ralls, Audrain, Randolph, Howard, Pettis, St. Clair, and Vernon counties.

Ranges from Florida to Texas, north to Maine, Vermont, Pennsylvania, West Virginia, Michigan, Indiana, Illinois, Missouri, and Oklahoma.

5. **Linum striatum** Walt. Map 1407

Flowers late June–August.

Occurs in low wet sandy ground, swampy meadows, wet woods, along streams, and swampy wooded areas. Eastern Missouri south of the Missouri River north to Jefferson and Franklin counties, west to Shannon and Ripley counties.

Ranges from Florida to Texas, north to Massachusetts, New York, Pennsylvania, West Virginia, Michigan, Illinois, Missouri, and Oklahoma.

## Fam. OXALIDACEAE (Wood Sorrel Family)

### Oxalis L. Wood Sorrel, Lady's Sorrel, Sheep Sorrel

a. Petals purple, lavender, rarely white; leaves all arising from the base of the plant; plant arising from underground bulb . . . . . . . . . . . . . . . . . . . . . . 1. O. VIOLACEA

a. Petals yellow; leaves scattered on the stem; no underground bulb present . . . . . . . . . *b*

b. Stems rooting at the nodes, creeping; stipules (at base of leaf-stalk) broad, conspicuous, brown or purplish; plants chiefly of gardens and greenhouses . . . . . . . . . . 2. O. CORNICULATA

b. Stems not rooting, not creeping; stipules narrow, inconspicuous, or lacking; plants of woodland, bluffs, rocky slopes, gravel bars and alluvial ground along streams, as well as fields, waste ground, roadsides, and railroads . . . . . . . . . . . . . . . . . . . . . . . . . . *c*

c. Flowers 1–5 in umbels, all the flower-stalks (pedicels) arising directly from the same point at the top of the main flower-stem (peduncle); fruiting pedicels (stalks) deflexed or downward-spreading but with erect fruits; hairs on fruit pressed close to or parallel to the surface, giving a grayish hairy appearance . . . . . . . . . . . . . . . . . . . 3. O. DILLENII

c. Flowers 1–9 in cymosely unequally branched umbels, some of the flower-stalks arising from branches of the inflorescence, others from the top of the main flower-stem (peduncle); fruiting pedicels (stalks) straight, erect or ascending, not downward-spreading nor deflexed; hairs on fruit spreading out or away from surface, not presenting a grayish hairy appearance . 4. O. STRICTA

1408 • Oxalis violacea var. violacea f. violacea (Violet Wood Sorrel)
1408 x Oxalis violacea var. violacea f. albida
1408 □ Oxalis violacea var. trichophora

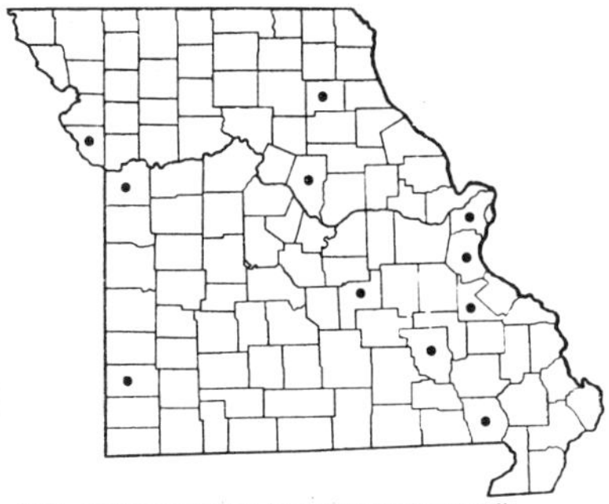

1409 Oxalis corniculata (Creeping Lady's Sorrel)

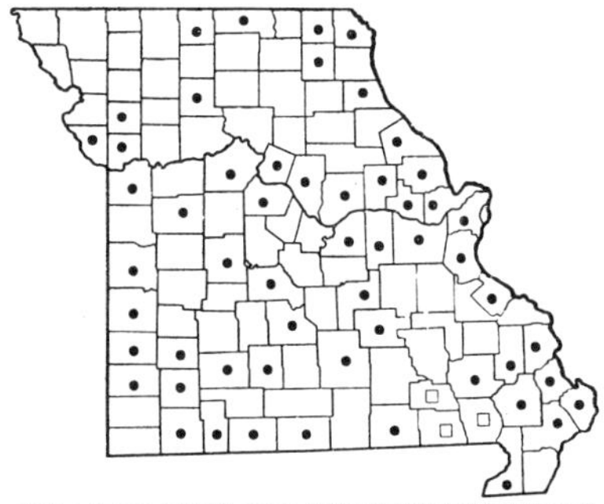

1410 • Oxalis Dillenii subsp. Dillenii (Yellow Wood Sorrel)
1410 □ Oxalis Dillenii subsp. filipes

1. **Oxalis violacea** L. Violet Wood Sorrel
Map 1408

Flowers April–July; September–November.

Occurs in usually acid soils of rocky or open woods, fallow fields, prairies, glades, and along roadsides. Throughout Missouri.

a. Petals white . . . . . . . 1b. O. VIOLACEA var. VIOLACEA f. ALBIDA
a. Petals purple or lavender . . . . . . *b*
  b. Leaf-stalks (petioles) glabrous (without hairs) . . . . . . . . 1a. O. VIOLACEA var. VIOLACEA f. VIOLACEA
  b. Leaf-stalks (petioles) with little knob-tipped (glandular) hairs . 1c. O. VIOLACEA var. TRICHOPHORA

1a. **Oxalis violacea** var. **violacea** f. **violacea**
Map 1408

*Oxalis violacea* L. [G, BB, P & S, Steyerm.]

Throughout the state in nearly every county. This is the commonest variation.

Ranges from Florida to New Mexico, north to Massachusetts, New York, Ohio, Indiana, Michigan, Wisconsin, Minnesota, North Dakota, and Colorado.

1b. **Oxalis violacea** var. **violacea** f. **albida** Fassett
Map 1408

*Oxalis violacea* f. *albida* Fassett [G, Steyerm.]

Scattered in several parts of Missouri, and probably of more frequent occurrence.

Scattered in the range of the typical f. *violacea.*

1c. **Oxalis violacea** var. **trichophora** Fassett
Map 1408

Scattered in southern and east-central Missouri north to Pike and Monroe counties.

Ranges from Arkansas and Missouri to Virginia, Pennsylvania, and Vermont.

Violet Wood Sorrel frequently flowers again during late fall following a renewal of rains after a dry summer. The sour foliage, in small quantities, adds a special tastiness to salads, but the oxalic acid contained in the plant places a restraint on the quantity used.

2. **Oxalis corniculata** L. Creeping Lady's Sorrel
Map 1409

*Oxalis corniculata* var. *viscidula* Wieg. [P & S]
*Oxalis repens* Thunb. [BB]

Flowers April–November.

Occurs around gardens and greenhouses principally, sparingly as a weed in cultivated and waste ground. Scattered in Missouri.

Found throughout the tropical regions of both hemispheres; introduced and naturalized in North America from Florida to Texas, north to Newfoundland, New York, Indiana, Illinois, Missouri, North Dakota, and the Pacific states.

For a discussion of the reason for retaining the name *O. corniculata*, rather than *O. repens*, reference should be made to George Eiten's discussion (Taxon 4: 99–101. 1955).

3. **Oxalis Dillenii** Jacq. Yellow Wood Sorrel
Map 1410

*Oxalis stricta* of Am. auth. [G, BB, P & S, Steyerm.], not L.

Flowers May–November.

Occurs in fields, glades, prairies, gravel bars along streams, roadsides, and along railroads.

Two variations occur in Missouri:

a. Common variation encountered; stipules (at base of leaf-stalk) oblong, evident; fruiting capsules covered with a gray-white close hairiness throughout, usually 15–25 mm. long; sepals 3.5–7 mm. long; leaflets 10–20 mm.

broad; no underground stolons (runners) arising from the old bases of the stem . . . 3a. O. DILLENII subsp. DILLENII

a. Rare variation encountered; stipules none or nearly obsolete; fruiting capsules glabrous (without hairs) or with only a few remote hairs, 8–12 mm. long; sepals 2–4.5 mm. long; leaflets 3–12 mm. broad; underground stolons (runners) present in autumn and arising from the slender old bases of the stem . . . . 3b. O. DILLENII subsp. FILIPES

3a. **Oxalis Dillenii** subsp. **Dillenii** Map 1410

*Oxalis stricta* of Am. auth. [G, BB, P & S, Steyerm.], not L.

*Oxalis stricta* var. *piletocarpa* Wieg. [P & S, Steyerm.]

Throughout Missouri. Common.

Ranges from Prince Edward Island to British Columbia, south to Florida, Texas, and Mexico.

3b. **Oxalis Dillenii** subsp. **filipes** (Small) Eiten Map 1410

*Oxalis filipes* Small [G]

*Oxalis florida* Salisb. var. *strigosifolia* Wieg. [P & S, Wieg.]

*Oxalis stricta* of some auth. in part [BB, in part], not L.

Occurs in low woodland.

Known only from southeastern Missouri in Carter (Grandin, May 6, 1901, *Bush 363*), Ripley (Doniphan, May 16, 1900, *Bush 277*), and Butler (Poplar Bluff, August 14, 1892, *Dewart*) counties.

Ranges from Florida to Louisiana, north to Connecticut, Tennessee, and Missouri.

Dr. Eiten has identified the above-cited specimens. For a discussion of the present name of this species reference should be made to Dr. George Eiten's article in Taxon 4: 99–105. 1955.

4. **Oxalis stricta** L. Yellow Wood Sorrel Map 1411

Also called Lady's Sorrel, Sheep Sorrel.

*Oxalis europaea* Jord. [G, P & S, Steyerm.]

*Oxalis europaea* var. *europaea* f. *europaea* [BB]

*Oxalis europaea* f. *pilosella* Wieg. [G, P & S, Steyerm.]

*Oxalis europaea* var. *europaea* f. *pilosella* [BB]

*Oxalis europaea* f. *cymosa* (Small) Wieg. [G, P & S, Steyerm.]

*Oxalis europaea* var. *europaea* f. *cymosa* [BB]

*Oxalis europaea* f. *villicaulis* Wieg. [G, P & S, Steyerm.]

*Oxalis europaea* var. *europaea* f. *villicaulis* [BB]

*Oxalis europaea* var. *Bushii* (Small) Wieg. [G, P & S, Steyerm.]

*Oxalis europaea* var. *Bushii* f. *Bushii* [BB]

*Oxalis europaea* var. *Bushii* f. *vestita* Wieg. [G, BB, Steyerm.]

*Oxalis europaea* var. *Bushii* f. *subglabrata* Wieg. [G, BB, P & S, Steyerm.]

*Xanthoxalis interior* Small

*Oxalis interior* (Small) Just

Flowers May–October.

Occurs on rocky or open wooded slopes, along bluffs, alluvial banks of streams, thickets, fields, waste ground, along roads and railroads.

Throughout Missouri, doubtless in every county.

Ranges, from Quebec to North Dakota, south to Florida, Tennessee, Arkansas, Oklahoma, and Arizona.

The stems and foliage vary from pale to dark green to copper-purple shades. The latter are quite ornamental and could be introduced as ornamental plants in the garden. The leaves of this and the preceding species may be used fresh in salads, small quantities being suggested because of the oxalic acid content. The young green fruits are often nibbled for their sour taste, and are sometimes called 'pickles' by children.

Dr. Eiten (Taxon 4: 99–105. 1955) has shown that the proper name for this species is *O. stricta* L. The great number of synonyms cited above indicate the extreme variation found in this taxon. Some of the variations are more frequent or abundant in the state than others. For those who are interested in referring their plants to specific forms, the following key, adapted from Dr. Wiegand's original treatment, in which the forms were all placed under *O. europaea*, is presented for convenience:

a. Upper surface of leaflets glabrous (without hairs) . . . . . . . . . . . . . *b*
  b. Hairs of flower- or fruit-stalks appressed, scarcely sticky (viscid) . . . . . . *c*
    c. Stems glabrous (without hairs) or the hairs sparse and strigose (pressed against or parallel to surface) . . O. EUROPAEA var. EUROPAEA f. EUROPAEA
    c. Stems with hairs somewhat spreading . O. EUROPAEA var. EUROPAEA f. PILOSELLA

Plate no. 230. **1.** Linum medium var. texanum, × 2/5; *a.* Fruiting calyx with capsule, × 2. **2.** Linum striatum, × 2/5; *a.* Fruiting calyx with capsule, × 4; After Britton and Brown, details from Small, The New York Botanical Garden. **3.** Oxalis corniculata, × 2/5. **4.** Oxalis violacea, × 2/5; Details from Small, The New York Botanical Garden. **5.** Oxalis stricta, × 2/5. **6.** Oxalis Dillenii, × 2/5; Details from Small, The New York Botanical Garden.

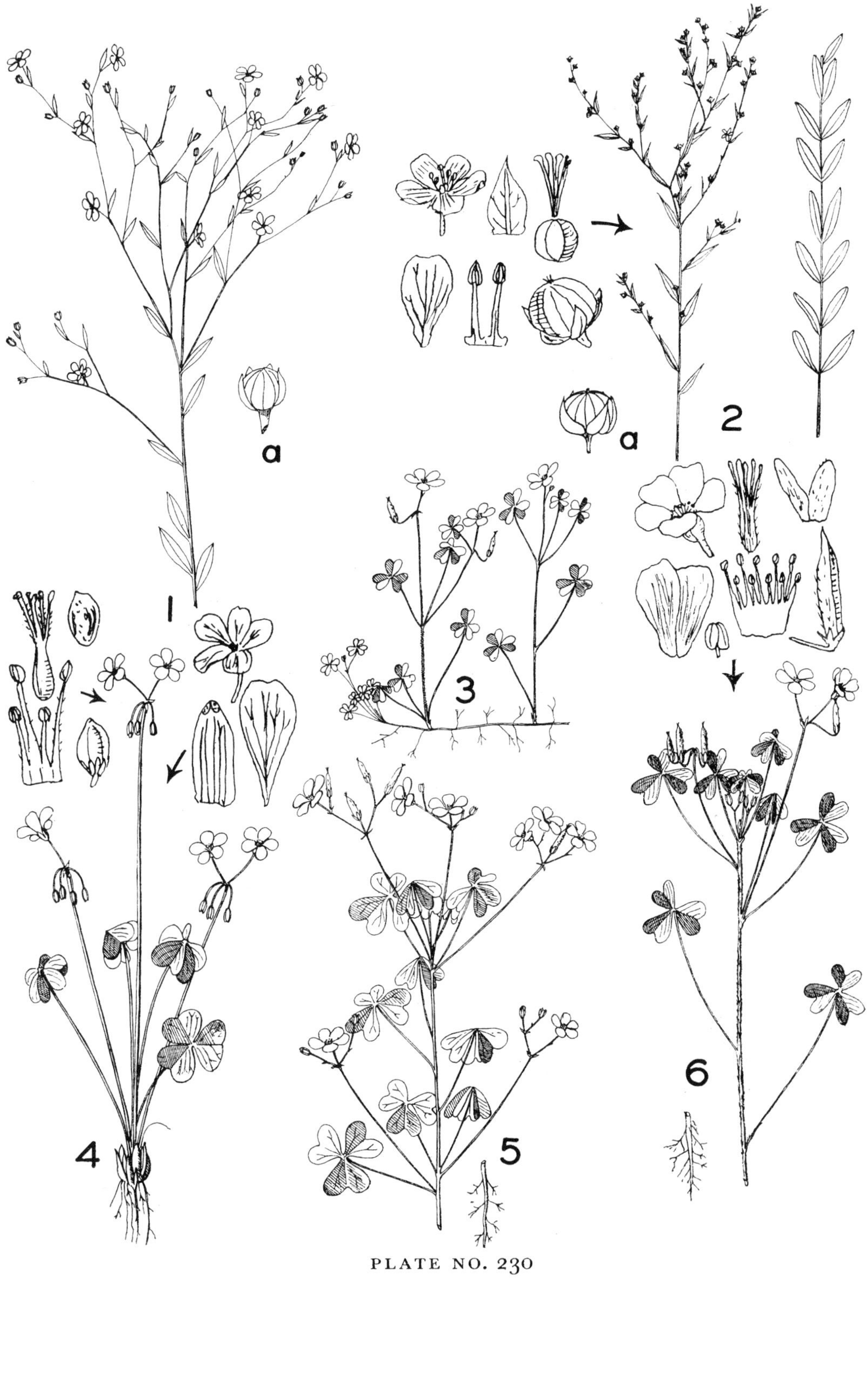


PLATE NO. 230

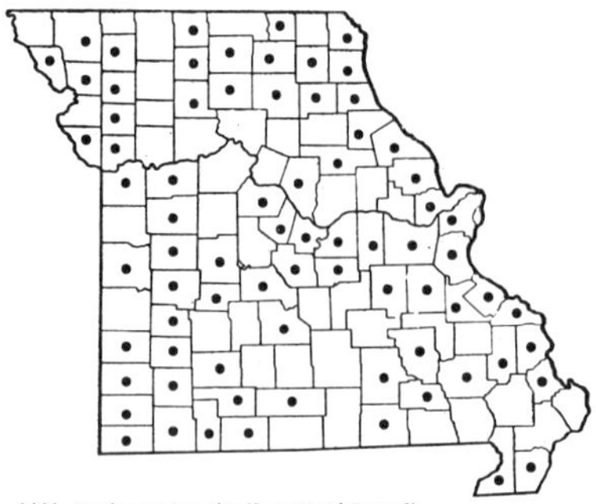

1411 Oxalis stricta (Yellow Wood Sorrel)

1412 Geranium maculatum f. maculatum (Wild Geranium)

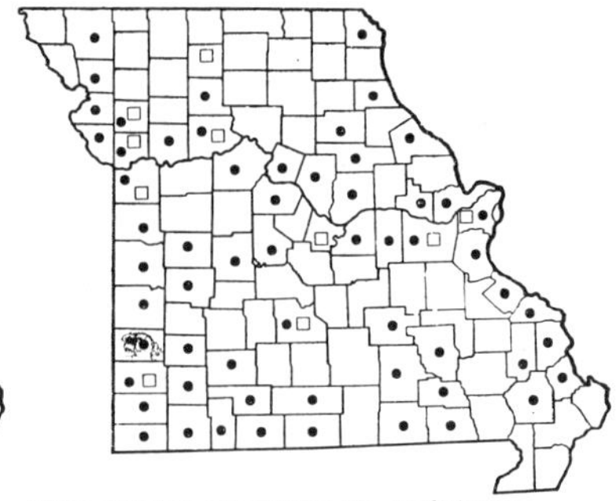

1413 • Geranium carolinianum var. carolinianum
1413 □ Geranium carolinianum var. confertiflorum

b. Hairs of flower- or fruit-stalks spreading, usually sticky (viscid) . . . . . . *d*
  d. Stems without hairs or nearly so . . O. EUROPAEA var. EUROPAEA f. CYMOSA
  d. Stems hairy . O. EUROPAEA var. EUROPAEA f. VILLICAULIS

a. Upper surface of leaflets hairy . . . . . *e*
  e. Hairs of flower- or fruit-stalks appressed (pressed against or lying parallel to surface), scarcely sticky (viscid) . . . . . . *f*
    f. Stems with spreading hairs . O. EUROPAEA var. BUSHII f. BUSHII
    f. Stems without hairs or hairs appressed (pressed against or lying parallel to surface) . . . . . . . O. EUROPAEA var. BUSHII f. SUBGLABRATA
  e. Hairs of flower- or fruit-stalks spreading, usually sticky (viscid); stems with spreading hairs . . O. EUROPAEA var. BUSHII f. VESTITA

*Excluded Species*

**Oxalis recurva** Ell.

This species is recorded from Missouri in *Gray's Manual*, eighth ed. (p. 946). It is based upon a specimen collected from Dunklin County (May 20, 1892, *Bush*) which was labelled by Bush as *O. recurva*. Dr. Eiten has determined this specimen as *O. stricta* L.

**Oxalis albicans** HBK. subsp. **albicans**

*Oxalis Wrightii* Gray (Wieg.) Rh. 27: 119. 1925.
*Oxalis pilosa* Nutt. var. *Wrightii* (Gray) Wieg. [P & S, Wieg.]

A specimen from Jackson County (1888, *Bush*) was cited by Wiegand (Rh. 27: 119. 1925) as of doubtfully determined *O. Wrightii*, a southwestern species. This specimen should be referred to *O. Dillenii* subsp. *Dillenii*.

## Fam. GERANIACEAE (Geranium Family)

Leaves palmately lobed (radiating like the fingers from the palm of a hand); anthers usually 10 (5 in *G. pusillum*) . . . . . . . . . . . . . . . . . . . . . . . . . . . 1. GERANIUM
Leaves pinnately compound (main divisions arranged feather-like one above the other); anthers 5. 2. ERODIUM

### 1. Geranium L. Cranesbill

a. Mostly woodland plants; petals showy, 14–23 mm. long, much longer than the calyx; leaves dark green, those on the stem 5–15 cm. broad, their lobes 3.5–7 cm. long . . . . . . . 1. G. MACULATUM
a. Plants of open ground, fields, and waste places; petals small, 3–7 mm. long, shorter than calyx or about equaling it; leaves pale or dull green, those on the stem mainly 3–6.5 cm. broad, their lobes 1–3.5 cm. long . . . . . . . . . . . . . . . . . . . . . . . . . . . . *b*
  b. Sepals tapering to a slender or awned tip, the tip 0.7–3 mm. long; seeds with honeycombed (reticulate) surfaces . . . . . . . . . . . . . . . . . . . . . . . . . . *c*
    c. Commonly collected species; lobes of the leaves blunt (obtuse); hairs on the main part of the carpels long and ascending, about 1 mm. long; seeds not definitely reticulate (honeycombed) under a 10 × lens, the areolae (honeycombed chambers) irregular, elongate, and thin-walled 2. G. CAROLINIANUM
    c. Rarely collected species; lobes of the leaves acutely pointed; hairs on the main part of the carpels short and horizontally spreading, about 0.5 mm. long; seeds conspicuously reticulate

1414 Geranium dissectum

1415 Geranium pusillum

1416 Geranium molle (Dovesfoot Cranesbill)

(honeycombed), the areolae (honeycombed chambers) rather uniform, square or rounded, and thick-walled . . . . . . . . . . . . . . . . . . . . . 3. G. DISSECTUM

b. Sepals blunt or ending in a minute scarcely noticeable tip; seeds smooth or with a finely granular surface . . . . . . . . . . . . . . . . . . . . . . . . . *d*

d. No beak on fruit or style column; stamens 5; carpels finely hairy, not wrinkled. . . . 4. G. PUSILLUM

d. A beak 1–2 mm. long terminating fruit or style column; stamens 10; carpels without hairs or nearly so, usually cross-wrinkled . . . . . . . . . . . . . . . . . 5. G. MOLLE

1. **Geranium maculatum** L. f. **maculatum**

Wild Geranium Map 1412

Also called Spotted Cranesbill.

*Geranium maculatum* L. [G, BB, P & S, Steyerm.]

Flowers April–June.

Occurs in rich or rocky open woods and thickets. Throughout Missouri; apparently absent only from the lowland counties of southeastern Missouri.

Ranges from Maine to South Dakota and Manitoba, south to Georgia, Alabama, Tennessee, Arkansas, and Oklahoma.

A white-flowered form, f. *albiflorum* (Raf.) House, has not been found as yet in Missouri.

2. **Geranium carolinianum** L. Map 1413

Flowers May–July.

Occurs in meadows, prairies, fallow fields, pastures, glades, waste ground, along roadsides and railroads.

Two variations are encountered in Missouri material:

Hairs on at least the lower part of stem short, 0.5 mm. long, and more or less downwardly pointed (retrorse); flowers in a loosely 4–12-flowered inflorescence . . . 2a. G. CAROLINIANUM var. CAROLINIANUM

Hairs on stem and leaf-stalks spreading, about 1 mm. long; flowers mostly crowded in a 5–25-flowered inflorescence . . . 2b. G. CAROLINIANUM var. CONFERTIFLORUM

2a. **Geranium carolinianum** var. **carolinianum** Map 1413

*Geranium carolinianum* L. [G, P & S, Steyerm.]

This is the commoner variation and is found nearly throughout the state.

Ranges from Florida to Texas and California, north to Massachusetts, New York, West Virginia, Michigan, Illinois, Missouri, Kansas, Wyoming, Idaho, and British Columbia.

2b. **Geranium carolinianum** var. **confertiflorum** Fern. Map 1413

Scattered and less common in the state.

Ranges from Maine to Minnesota, south to North Carolina, Tennessee, and Missouri.

3. **Geranium dissectum** L. Map 1414

Flowers April–August.

Occurs in lawns and grassy open ground. Known only from St. Louis County (lawn at 267 Elm Avenue, Glendale, June 7, 1952, *Steyermark 73314*).

Native of Europe; introduced and naturalized in the United States from Massachusetts to Michigan, North Carolina, and Missouri; also Pacific states.

This was first observed by Mr. Oscar Peterson on his property where it and *G. molle* have been adventive since 1952.

4. **Geranium pusillum** L. Map 1415

Flowers May–October.

Occurs in rocky open and waste ground and along roadsides. Scattered in southern and east-central Missouri north to St. Charles, Crawford, and Miller counties.

Native of Europe; introduced and naturalized in North America from Massachusetts to British Colum-

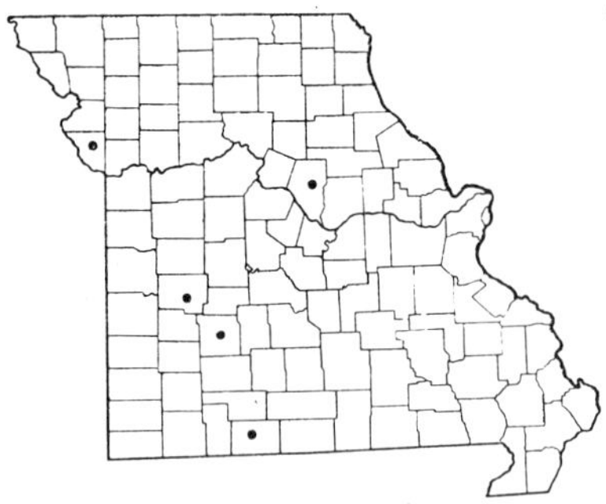

1417 Erodium cicutarium (Pink Needle)

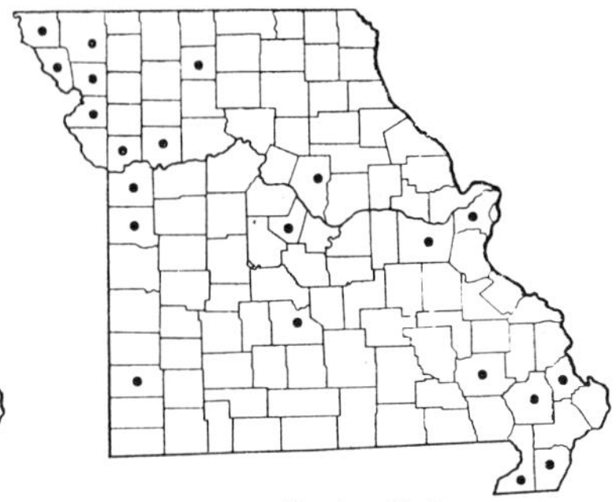

1418 Tribulus terrestris (Puncture Vine)

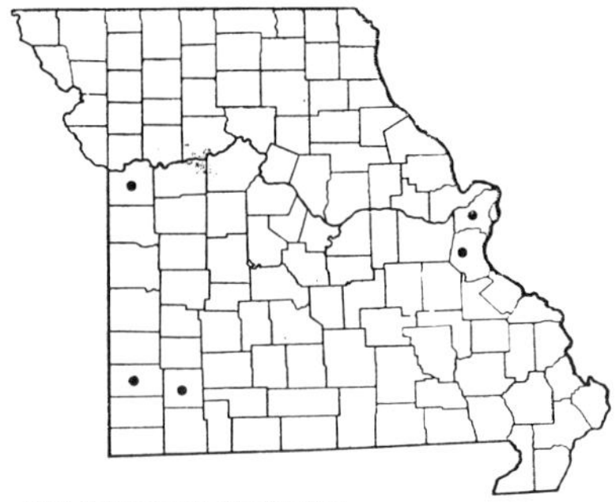

1419 Kallstroemia hirsutissima

bia, south to North Carolina, Indiana, Missouri, Oklahoma, Utah, and Oregon.

The small petals are rose-colored or rose-purplish.

5. **Geranium molle** L. Dovesfoot Cranesbill Map 1416

Flowers May–August.

Occurs on lawns and grassy open ground. Known only in St. Louis County (lawn at 267 Elm Avenue, Glendale, June 7, 1952, *Steyermark 73315*).

Native of Europe; introduced and naturalized in North America from Nova Scotia to British Columbia, southward.

This was first observed in 1952 by Mr. Oscar Peterson who discovered it growing on his property along with *G. dissectum* as an adventive.

## 2. **Erodium** L'Hér Storksbill

**Erodium cicutarium** (L.) L'Hér Pink Needle Map 1417

Also called Pin Clover.

Flowers March–November.

Occurs in grassy open places, waste ground, and along roadsides. Scattered in southwestern and central Missouri in Taney, Polk, St. Clair, Boone, and Platte counties.

Native of Europe; introduced and naturalized in North America from Quebec to Michigan to Illinois and Missouri, south to Virginia, Tennessee, Arkansas, Texas, and California; also in Mexico.

The young leaves are eaten raw or cooked. As an important winter forage plant in the southwestern states it is raised under the name of Alfileria.

## Fam. **ZYGOPHYLLACEAE** (Caltrop Family)

Leaflets 10–16 to a leaf, 5–12 (15) mm. long, 2–5 mm. broad; flowers 8–10 mm. across, the petals 4–6 mm. long; fruit spiny . . . . . . . . . . . . . . . . . . . . 1. TRIBULUS

Leaflets mostly 6–8 (rarely 10) to a leaf, 8–20 mm. long, mostly 5–10 mm. broad; flowers 15–20 mm. across, the petals 5–10 mm. long; fruit not spiny . . . . . . . . . . . . . . 2. KALLSTROEMIA

## 1. **Tribulus** L.

**Tribulus terrestris** L. Puncture Vine Map 1418

Also called Caltrop.

Flowers June–September.

Occurs along railroads and often roadsides or sandy waste ground. Scattered in Missouri, most common in the extreme southeastern and northwestern sectors.

Native of Europe; introduced and naturalized from Florida to Texas and California, north to New York, Ohio, Michigan, Illinois, Iowa, South Dakota, and Utah; also Mexico and the West Indies.

The spiny fruits of this creeping-stemmed plant are

Plate no. 231. **1.** Geranium maculatum, × 2/5. **2.** Geranium carolinianum, × 2/5; Details from Small, The New York Botanical Garden. **3.** Geranium pusillum, × 2/5. **4.** Geranium molle, × 2/5; After Gleason, The New York Botanical Garden. **5.** Erodium cicutarium, × 2/5; Details from Small, The New York Botanical Garden.

PLATE NO. 231

sometimes the cause of mechanical injuries to animals feeding on hay containing the fruit. A photosensitizing agent in the plant causes poisoning in sheep when they feed on flowering plants. The spiny fruits are also a nuisance around playgrounds, lawns, and parking lots, causing punctures in rubber tires of bicycles and automobiles.

## 2. **Kallstroemia** Scop.

**Kallstroemia hirsutissima** Vail Map 1419

*Kallstroemia intermedia* Rydb. [G, BB, P & S]

Flowers June–September.

Occurs along railroads and waste ground. Scattered in southern and central Missouri in St. Louis, Jefferson, Jackson, Lawrence, and Jasper counties.

Ranges from Mexico, New Mexico, and Texas to Illinois, Missouri, Kansas, Oklahoma, and Colorado.

By some authors *K. intermedia* and *K. hirsutissima* are maintained as distinct species as follows: *K. intermedia*: beak of mature fruit 5–6 mm. long, longer than the body of the fruit; petals 7–10 mm. long; *K. hirsutissima*: beak of mature fruit 3–4 mm. long, shorter than the body of the fruit; petals 5–6 mm. long. A specimen from Jackson County (along railroad, Courtney, August 1, 1907, *Bush 767*) has been identified as *K. hirsutissima*, while the other collections in Missouri have been placed under *K. intermedia.* However, it does not seem possible to retain these as distinct species and in the present flora they are merged as one.

## Fam. **RUTACEAE** (Rue Family)

a. Leaves divided into 3 leaflets . . . . . . . . . . . . . . . . . . . . . 2. PTELEA
a. Leaves divided into 5–11 leaflets or several or many narrow lobes . . . . . . . . . . . . *b*
  b. Native shrub to small tree 1.5–8 m. tall with thorns on stems and often on some part of leaf-stalk (petiole); the 5–11 fully grown leaflets of a leaf mostly short-pointed (acute or acutish), 10–30 mm. broad, dark green, with some hairs on lower surface . . . . . . . . . . 1. ZANTHOXYLUM
  b. Low plant up to 1 m. tall without thorns; the several to many leaf lobes or divisions all narrow, mostly broadly rounded to blunt at tip, 2–10 mm. broad, smooth, pale green or bluish, without hairs on either upper or lower surface . . . . . . . . . . . . . . . . . . . 3. RUTA

## 1. **Zanthoxylum** L. Prickly Ash

**Zanthoxylum americanum** Mill. f. **americanum**

Prickly Ash Map 1420

Also called Toothache Tree.

*Zanthoxylum americanum* Mill. [BB, P & S, Steyerm.]

*Xanthoxylum americanum* Mill. [G]

Flowers April–May; fruits June–August.

Occurs in open and rocky woods, along bluffs, and thickets in low moist ground or dry upland, usually in limestone soils. Absent in most of the central and southern Ozark section, otherwise common throughout the remainder of the state.

Ranges from Quebec to Minnesota and North Dakota, south to Georgia, Alabama, Missouri, and Oklahoma.

This shrub usually occurs in small thickets because of its suckering stems which creep with shallow roots just below the surface of the soil. The leaves are bitter-aromatic. The reddish to brown capsules are pitted on the surface and are strongly aromatic, opening eventually to expose the single black, shining seed.

The volatile oil obtained from the bark and fruit is used in medicine as a tonic, stimulant, and diaphoretic, for treating diarrhea and flatulence. The bark and fruit have also been used in some rural sections to relieve toothache.

A spineless form, known as f. *impuniens* Fassett, has not been recorded as yet from Missouri.

---

Plate no. 232. **1.** Tribulus terrestris, × 2/5. **2.** Kallstroemia hirsutissima, × 2/5; After Gleason, The New York Botanical Garden. **3.** Zanthoxylum americanum, × 2/5; *a.* Flowering branch, × 2/5; *b.* Male flower, × 2 4/5; *c.* Female flower, × 2 4/5; *d.* Fruiting branch with leaves, × 2/5. **4.** Ptelea trifoliata var. trifoliata, × 2/5 **5.** Ruta graveolens, × 2/5; *a.* Flower, × 3 3/5. **6.** Ailanthus altissima, × 1; All details from Small, The New York Botanical Garden.

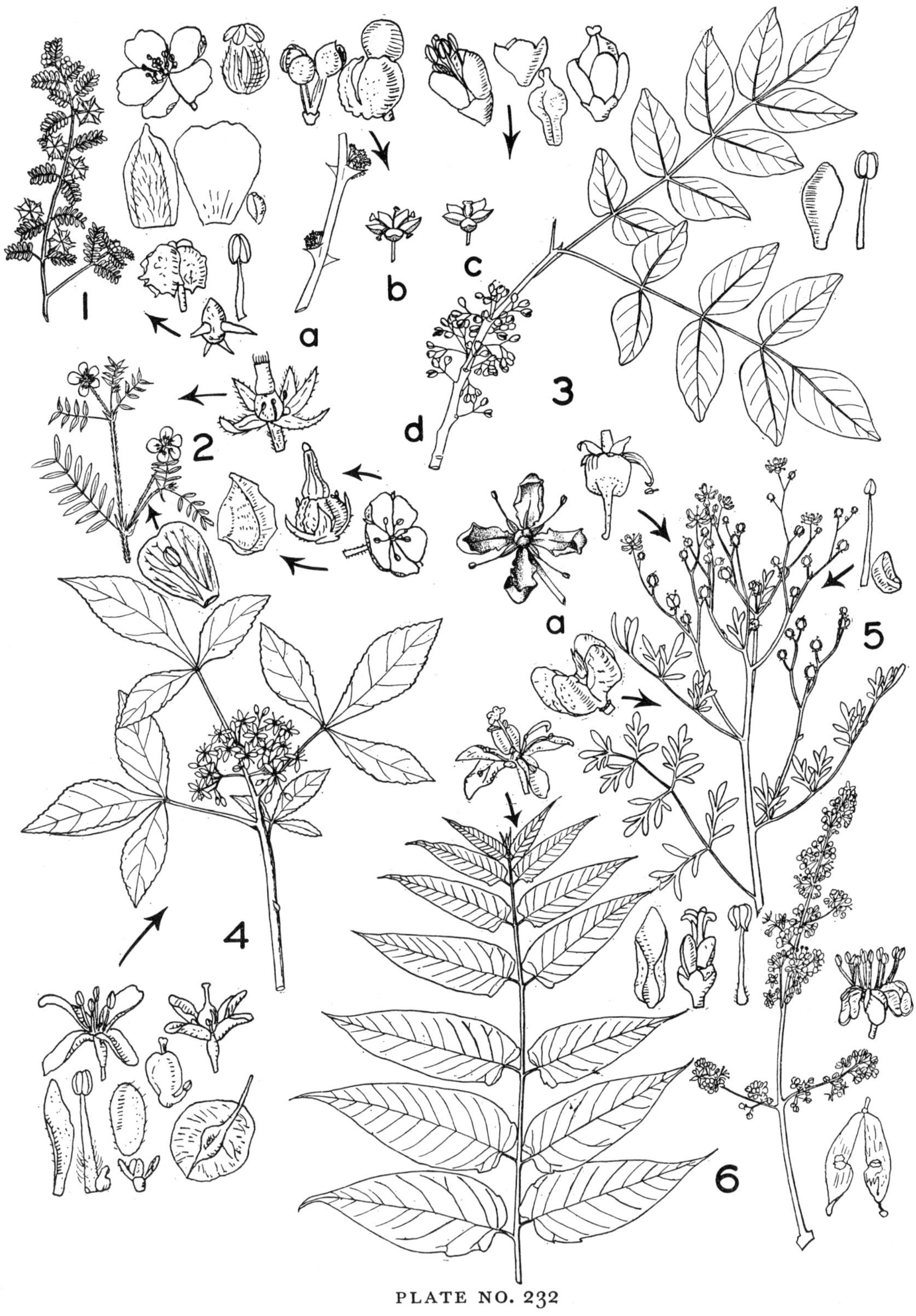


PLATE NO. 232

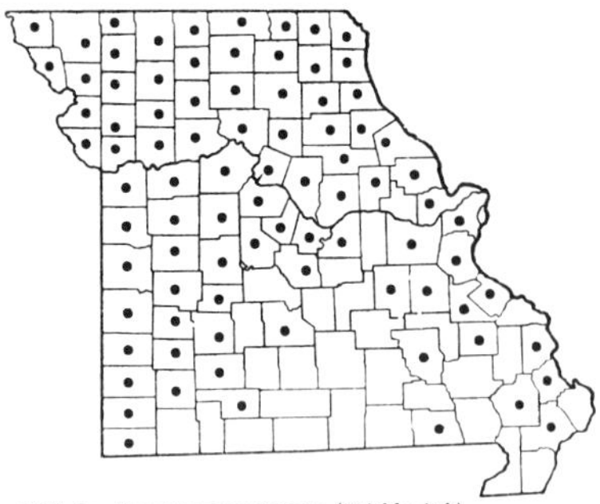

1420 Zanthoxylum americanum (Prickly Ash)

1421 • Ptelea trifoliata var. trifoliata f. trifoliata (Hop Tree)
1421 □ Ptelea trifoliata var. mollis

1422 Ruta graveolens

## 2. **Ptelea** L. Hop Tree

**Ptelea trifoliata** L. Hop Tree Map 1421

Flowers April–June.

Occurs on limestone glades, prairie openings on top of bluffs, rocky open woods, low woods in ravines and valleys, and thickets, usually on limestones soils. Southern, central, and eastern Missouri, east and south of a line drawn from Clark, Knox, Shelby, Randolph, Chariton, Saline, Pettis, Benton, Hickory, and Cedar counties to Vernon County.

Two variations occur in Missouri:

Common type found; branchlets glabrous (without hairs) or nearly so; lower surface of leaflets glabrous (without hairs) or nearly so. . . .
a. P. TRIFOLIATA var. TRIFOLIATA f. TRIFOLIATA

Rare type found; branchlets densely hairy; lower surface of leaflets permanently soft-hairy. . .
b. P. TRIFOLIATA var. MOLLIS

a. **Ptelea trifoliata** var. **trifoliata** f. **trifoliata** Map 1421

*Ptelea trifoliata* L. [G, BB, P & S, Steyerm.]

This is the common type found with the range indicated above.

Ranges from Quebec and New York to Ontario and Nebraska, south to Florida, Alabama, Louisiana, and Texas; also in Mexico.

b. **Ptelea trifoliata** var. **mollis** T. & G. Map 1421

*Ptelea trifoliata* var. *Deamiana* Nieuwl. [Deam, BB as var. *Deamii*]

Known only from Madison County (banks of St. Francois River, May 25, 1918, *Greenman 3879*).

Ranges from North Carolina to Georgia, west to Oklahoma, Arizona, and Mexico; and around Lake Michigan.

The small, greenish flowers and bruised bark of this shrub or small tree have an odor somewhat resembling that of skunk. The foliage turns to yellow-green in autumn. The common name of *Ptelea* refers to its former use as a substitute for hops. The aromatic and bitter roots at one time were used as a substitute for quinine, and because of being employed as a remedy for malaria and fevers, the bark was formerly known as ague bark. It also became employed in medicine as a remedy for indigestion and as a tonic or alterative.

## 3. **Ruta** L. Rue

**Ruta graveolens** L. Common Rue Map 1422

Flowers June–July.

Occasionally planted and rarely escaping from gardens to waste ground. Known only from Franklin and Taney counties.

Native of Europe; introduced and naturalized in the United States from Vermont to Virginia, West Virginia, and Missouri.

The leaves of the plant, when handled, sometimes cause dermatitis and blistering in some sensitive individuals. The plant has a sharp strong odor when bruised and the leaves possess a bitter taste. Formerly, the plant was much used in medicine for its presumed value for its stimulant and irritant properties. A related species, *R. chalepensis* L., is commonly grown in Central America where it is used today by the native inhabitants as a common medicinal plant.

## Fam. **SIMAROUBACEAE** (Quassia Family)

### **Ailanthus** Desf. Ailanthus, Tree of Heaven

**Ailanthus altissima** (Mill.) Swingle
Tree of Heaven Map 1423
Also called Smoke Tree.
Flowers May–June.

Commonly planted as a shade tree, and escaped from cultivation along streams, base of bluffs, thickets, roadsides, waste ground, and along railroads. Scattered in southern and central Missouri north to Pike, Boone, and Platte counties.

Native of Asia; introduced and naturalized in North America from Massachusetts to Ontario and Iowa and Kansas, south to Alabama, Oklahoma, Texas, and California.

The long compound leaves give a tropical appearance to this well-known street tree, which has the ability to withstand drought and high temperatures, as well as smoke, chemical and other gases produced by industry in cities. The name 'Smoke Tree' is sometimes applied because of this successful adaptation. It is also practically free from diseases and insect damage. It is a rapidly growing tree and commonly spreads by root suckers. The greenish-yellow flowers are both perfect as well as staminate and pistillate, and the staminate flowers are the ones which are rather putrid. Nurserymen now usually offer only the pistillate trees, which lack the objectionable odor, and which produce colorful clusters of reddish or brick-colored winged fruits in the f. *erythrocarpa* (Carr.) Rehder. Although the flowers are primarily insect-pollinated, some hay fever is stated to be caused by this plant in China, where it is native, and some cases have been attributed to it in the United States. Contact with the leaves produces dermatitis in some susceptible individuals. The wood is occasionally used together with other hardwoods for pulp in paper-making.

## Fam. **MELIACEAE** (Mahogany Family)

### **Melia** L. China Tree

**Melia Azedarach** L. China Tree, Pride of India
Flowers May–June.

Native of Asia; introduced and naturalized in the United States from Florida to Texas, north to Virginia and Oklahoma.

Although not definitely known to have become escaped and established in Missouri, the tree is commonly planted in the southeastern corner of the state, and is probably to be found in that sector of Missouri naturalized along railroads or in thickets. The pale lilac flowers have the fragrance of lilacs. The leaves are bipinnately compound with toothed leaflets. The rather globe-shaped fruits are yellowish and 1.5–2 cm. in diameter. They, as well as the bark and flowers, contain a narcotic substance; cases of poisoning have been reported from children, poultry, and domestic animals eating the fruit.

## Fam. **POLYGALACEAE** (Milkwort Family)

### **Polygala** L. Milkwort

a. At least the lowest 1–3 sets of leaves of the stem in a whorl (circle) of 3–7 leaves. . . . 4. P. VERTICILLATA
a. All the leaves alternately arranged up and down the stem . . . . . . . . . . . . . . . . . *b*
  b. Stems bare or practically without leaves, which soon fall from the stem; stems bluish- or silvery-green, glaucous (with a 'bloom' which can be rubbed off); leaves linear-subulate, only 0.5–1 mm. broad; corolla 7–10 mm. long . . . . . . . . . . . . . . . . . . . . 2. P. INCARNATA
  b. Stems with many leaves remaining attached to the stem at least during flowering period; stems green, not glaucous; leaves broader, 1–35 mm. broad; corolla 5 mm. or less long . . . . . . . . *c*
    c. Usually only 1 solitary stem to the plant; annual; leaves smooth-edged (not toothed), 1–5 mm. broad, linear; flowers in a rounded or short-cylindrical dense head at most twice as long as broad; inflorescence rose, lavender, greenish-white, or rarely white . . . . . . . 3. P. SANGUINEA
    c. Stems several from the root; perennial; leaves finely toothed (sometimes only seen with a

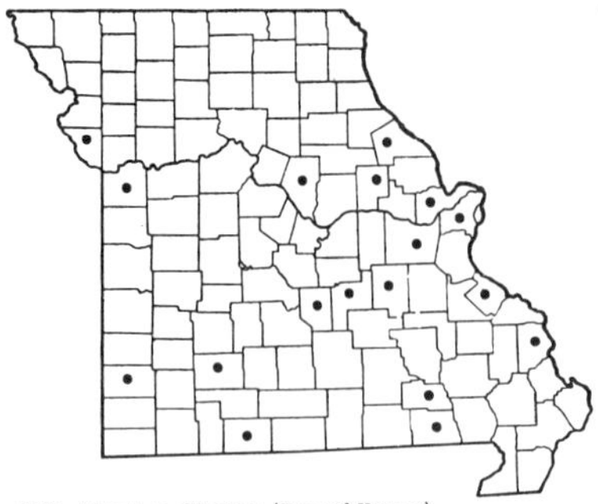

1423 Ailanthus altissima (Tree of Heaven)

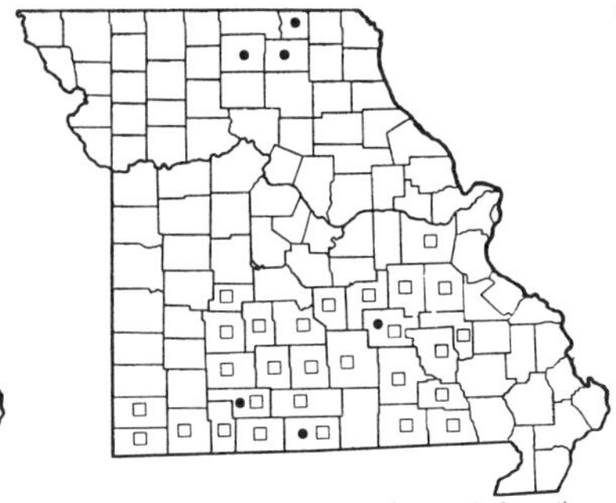

1424 • Polygala Senega var. Senega (Seneca Snakeroot)
1424 □ Polygala Senega var. latifolia

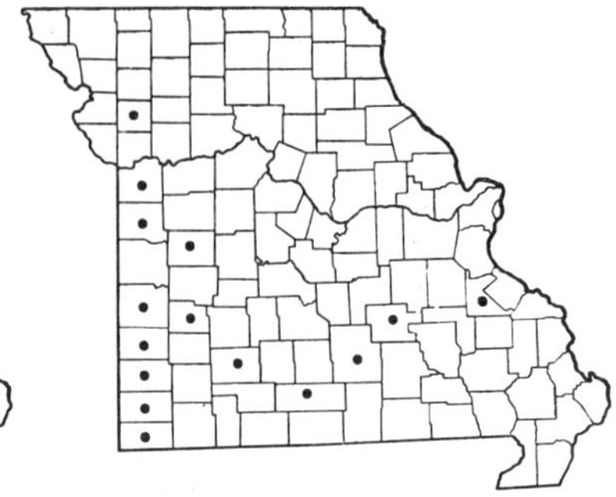

1425 Polygala incarnata

magnifying lens of 10×), mainly 4–35 mm. broad, lanceolate to ovate; flowers in narrow, finger-like inflorescences 3–7 (rarely only $2\frac{1}{2}$) times as long as broad; inflorescence white when in flower . . . . . . . . . . . . . . . . . . . . . . . . 1. P. SENEGA

1. **Polygala Senega** L. Seneca Snakeroot Map 1424

Flowers May–July.

Two varieties are encountered in Missouri:

Upper leaves mostly narrowly lanceolate to lance-elliptical, mostly 3–10 mm. broad, usually pale green on upper surface; wings (2 inner petal-like sepals) 2.2–2.5 mm. broad; fruiting capsules 2.5–3.5 mm. long, 3–3.8 mm. broad . 1a. P. SENEGA var. SENEGA

Upper leaves ovate or lance-ovate, mostly 10–35 mm. broad, dark green on upper surface; wings 2.8–3 mm. broad; fruiting capsules 3.5–4.2 mm. long, 4–4.3 mm. broad. 1b. P. SENEGA var. LATIFOLIA

1a. **Polygala Senega** L. var. **Senega** Map 1424

*Polygala Senega* L. [G, BB, P & S, Steyerm.]

Occurs mostly in open ground, clearings, prairies, thickets, and open woods. Rare and scattered in the state, where known from Schuyler, Adair, Sullivan, Dent, Ozark, and Christian counties.

Ranges from Quebec to Alberta, south to Georgia, Tennessee, Arkansas, and South Dakota.

1b. **Polygala Senega** var. **latifolia** T. & G. Map 1424

Occurs in usually rocky open woods, usually on dry wooded slopes. This is the commoner variation, but is restricted to the Ozark region north to Franklin, Crawford, Phelps, Pulaski, Laclede, Dallas, Hickory, and Newton counties.

Ranges from Pennsylvania and Delaware to Minnesota and South Dakota, south to North Carolina, Tennessee, and Missouri.

Some authors believe that the broader-leaved plants appear to be correlated with a moist or shady habitat, the narrower-leaved ones with a prairie or sandy habitat. In Missouri the two types seem fairly distinct and the broader-leaved var. *latifolia*, while definitely a woodland type, also is geographically restricted in the state to the Ozark region.

2. **Polygala incarnata** L. Map 1425

Flowers late May–November.

Occurs in prairies and on glades. Southern and western Missouri north to St. Francois, Dent, Texas, Henry, and Clinton counties.

Ranges from Florida to Texas and Mexico, north to New York, New Jersey, Pennsylvania, Ontario, Michigan, Wisconsin, Iowa, and Nebraska.

3. **Polygala sanguinea** L. Map 1426

Flowers late May–October.

Occurs in prairies, meadows, fallow fields, and rocky glades, usually in sterile or acid soils. Common throughout Missouri, except not recorded from most of the northwestern counties.

Ranges from Nova Scotia to Ontario and Minnesota, south to South Carolina, Tennessee, Louisiana, Oklahoma, and Texas.

The following variations are encountered in the state:

---

Plate no. 233. **1.** Polygala sanguinea, × $^2/_5$. **2.** Polygala verticillata var. ambigua, × $^2/_5$. **3.** Polygala verticillata var. isocycla, × $^2/_5$. **4.** Polygala Senega var. latifolia, × $^2/_5$. **5.** Polygala incarnata, × $^2/_5$. **6.** Polygala Senega var. Senega, × $^2/_5$. **7.** Melia Azedarach, × $^2/_5$; All details from Small, The New York Botanical Garden.

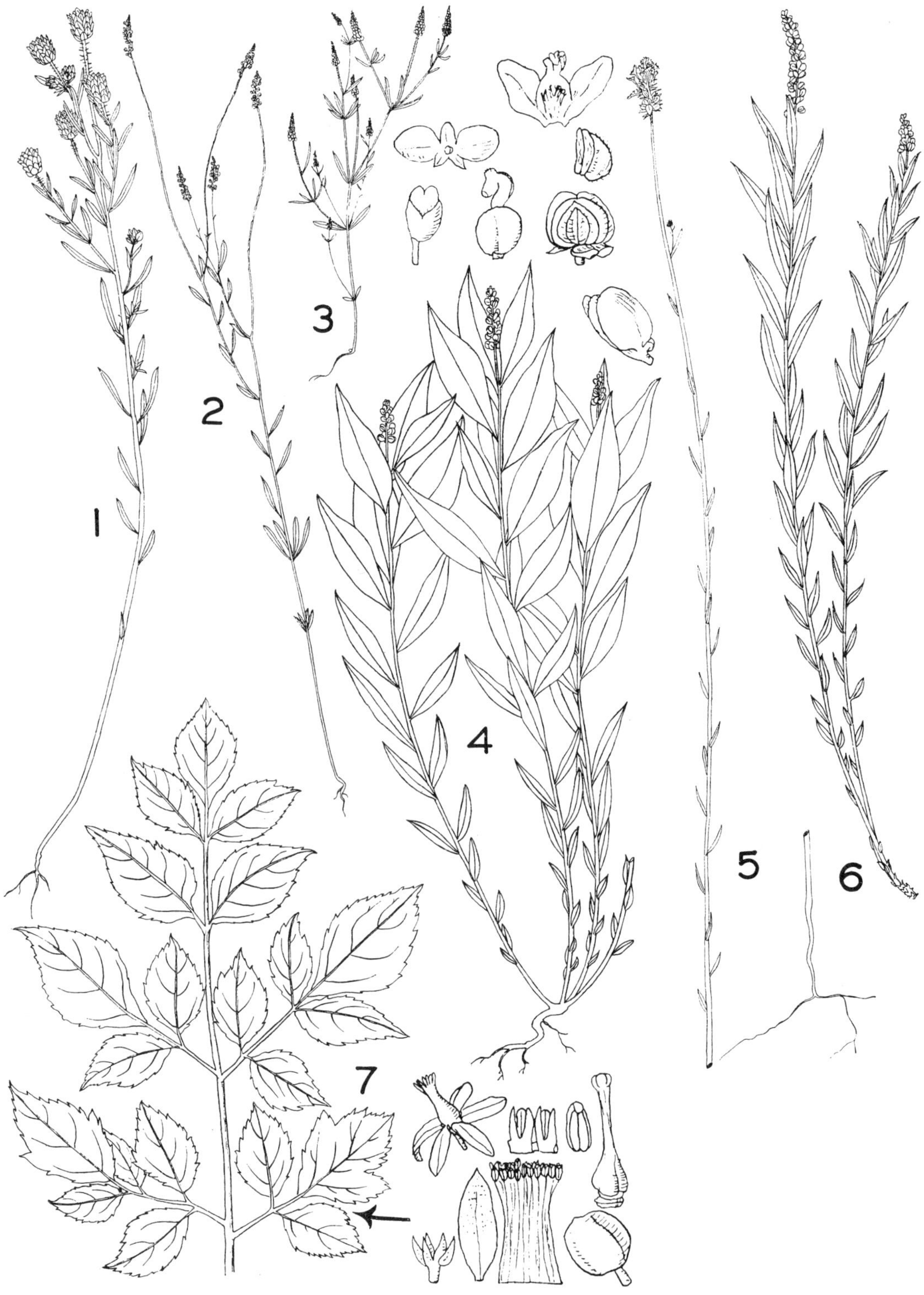


PLATE NO. 233

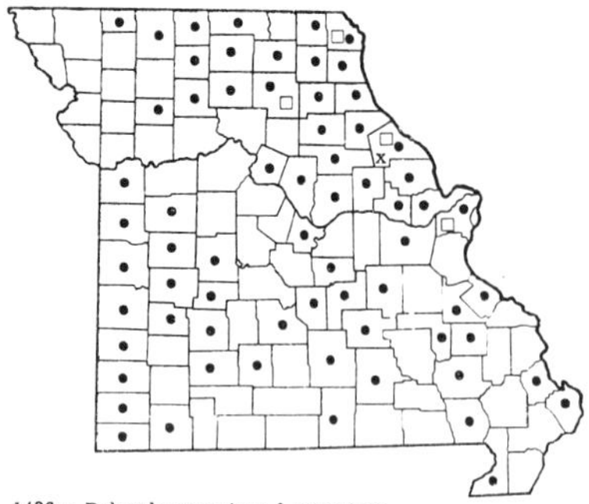

1426 • Polygala sanguinea f. sanguinea
1426 x Polygala sanguinea f. viridescens
1426 □ Polygala sanguinea f. albiflora

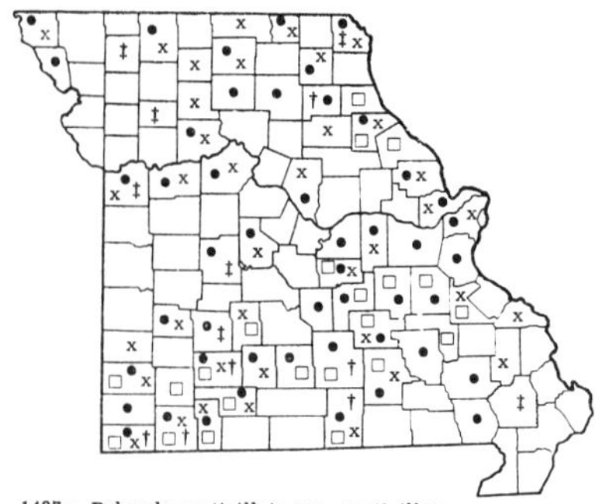

1427 • Polygala verticillata var. verticillata
1427 □ Polygala verticillata var. ambigua
1427 x Polygala verticillata var. isocycla
1427 ‡ Polygala verticillata var. sphenostachya
1427 † Polygala verticillata var. dolichoptera

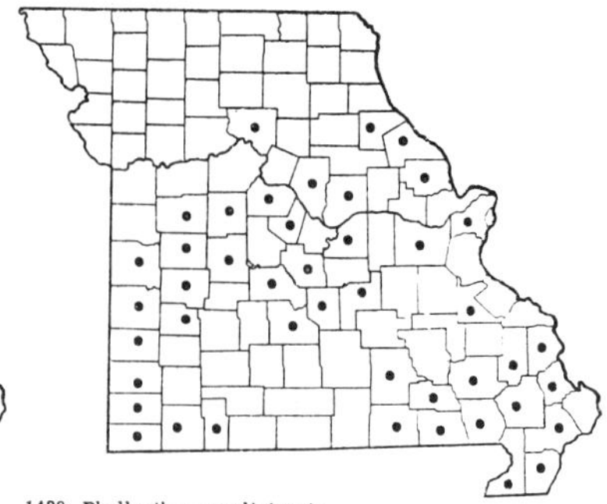

1428 Phyllanthus caroliniensis

a. Flowers pink to rose-purple at the summit . 3a. P. SANGUINEA f. SANGUINEA
a. Flowers white or greenish . . . . . . *b*
  b. Flowers greenish . . . . 3b. P. SANGUINEA f. VIRIDESCENS
  b. Flowers whitish . . . . 3c. P. SANGUINEA f. ALBIFLORA

3a. **Polygala sanguinea** f. **sanguinea** Map 1426

*Polygala sanguinea* L. [G, BB, P & S, Steyerm.]

This is the commonest variation in the state.

3b. **Polygala sanguinea** f. **viridescens** (L.) Farw. Map 1426

Known only from Pike County (4 mi. north of Eolia, July 30, 1949, *Bauer 1017*).

3c. **Polygala sanguinea** f. **albiflora** (Wheelock) Millsp. Map 1426

Known in east-central and northeastern Missouri from Clark (upland prairie, T66N, R8W, sect. 24, along highway C just east of junction with highway 81, 3 mi. northwest of Revere, August 14, 1949, *Steyermark 68976*), Macon, Pike, and St. Louis counties.

4. **Polygala verticillata** L. Map 1427

Flowers late May–October.

Occurs in usually acid or sterile soils of rocky glades of chert, sandstone, or granite, but also leached limestone, prairies, open woodland, fallow fields, eroded open slopes, and upland flats. Scattered throughout Missouri, but apparently absent from the lowland counties of the extreme southeastern section of the state.

The following variations are encountered in the state:

a. Most of leaves of the stem alternate and scattered, only the lowest 1–3 sets of leaves in definite whorls (circles around the stem); inflorescence elongated, loosely-flowered, interrupted with the lower internodes elongated, flower-bearing part 1–5 cm. long; wings (2 inner petal-like sepals) as long as or longer than the fruit . . . . . . . . . *b*
  b. Wings greenish, whitish, or purplish, 1–1.5 mm. long, about as long as the fruit . . 4d. P. VERTICILLATA var. AMBIGUA
  b. Wings milky-white, 2–2.6 mm. long, much longer than the fruit. . 4e. P. VERTICILLATA var. DOLICHOPTERA
a. Most or all of the main sets of leaves in whorls (circles) of 3–7; inflorescence more compact and conical, continuous, the flower-bearing part 0.5–2 cm. long; wings shorter than the fruit . . . . . . . . . . . *c*
  c. Plants with widely spreading branches; sepals greenish-white; inflorescence dense, the flower-bearing part 1–2 cm. long; peduncles (main flower-stem) 5–40 mm. long; seed finely hairy . . . . . . *d*
    d. Plants 5–20 cm. tall; fruits 1–1.6 mm. long; common type encountered . . 4b. P. VERTICILLATA var. ISOCYCLA
    d. Plants 10–30 cm. tall; fruits 1.6–2.3 mm. long . . . . . 4c. P. VERTICILLATA var. SPHENOSTACHYA
  c. Plants with more ascending to erect-ascending branches; sepals often purplish, also white; inflorescence more open, wider and more loosely flowered; peduncles (main flower-stem) 20–70 mm. long; seed hirsute (with rather coarse, stiff hairs) . . 4a. P. VERTICILLATA var. VERTICILLATA

4a. **Polygala verticillata** var. **verticillata** Map 1427

*Polygala verticillata* L. [G, P & S, Steyerm.]

Scattered throughout Missouri, not common.

Ranges from Maine to Ontario and Michigan, south to Virginia, Tennessee, and Louisiana.

4b. **Polygala verticillata** var. **isocycla** Fern. Map 1427

This is the commonest variation in Missouri, throughout the state.

Ranges from Massachusetts to Manitoba, south to Florida, Alabama, Louisiana, Texas, Colorado, and Utah.

4c. **Polygala verticillata** var. **sphenostachya** Pennell Map 1427

Occurs on rocky glades and open ground. Scattered in Missouri.

Ranges from Ohio to North Dakota, south to Missouri, Oklahoma, and Texas.

4d. **Polygala verticillata** var. **ambigua** (Nutt.) Wood Map 1427

*Polygala ambigua* Nutt. [BB, Deam]

Occurs in rocky woods and glades. Ozark region of southern Missouri and northeastward to Marion County.

Ranges from Maine to New York, Indiana, Michigan, Illinois, and Missouri, south to Georgia, Alabama, Louisiana, Oklahoma, and Texas.

4e. **Polygala verticillata** var. **dolichoptera** Fern. Map 1427

*Polygala ambigua* in part [BB], not Nutt.

Scattered in southern Missouri and locally north in Shelby County.

Ranges from Virginia to South Carolina, and from Missouri to Arkansas and Oklahoma.

*Polygala verticillata* is sometimes broken up into two separate species and varieties (by Gleason and by Deam), sometimes kept up as one species with five varieties. The present treatment follows the latter course. These varieties intergrade, with the greatest intergradation occurring between var. *isocycla* and var. *sphenostachya*. Although the differences between them break down with respect to the length of leaves (as presented in *Gray's Manual* by Fernald) and height of stem, they are being retained in their varietal categories pending more intensive field and experimental studies of the varieties of *P. verticillata*. The larger leaves in *P. verticillata* var. *sphenostachya* may attain a length of 3.5 cm. and a width of 5 mm., dimensions slightly larger than those given in some manuals.

*Polygala verticillata* var. *ambigua* and var. *dolichoptera* are more readily separated from the other varieties of *P. verticillata*, especially by the character of the prevailing alternate arrangement of the leaves and elongated inflorescence. While there is some justification for treating them as species, it would seem that they are connected too closely with other variations of *P. verticillata* through *P. verticillata* var. *verticillata* and var. *sphenostachya* to be retained in the specific category.

## Fam. **EUPHORBIACEAE** (Spurge Family)

a. Leaves large and star-shaped or umbrella-shaped, 10–90 cm. wide, 5–11-lobed, peltate (the leaf-stalk joining the leaf-blade on some part of its lower surface, but not at the edge or base of the leaf-blade); tall stout herb, 1–5 m. tall in Missouri . . . . . 7. Ricinus

a. Leaves much smaller, not peltate, the leaf-stalk, if present, attached to the base or edge of the leaf-blade; plant much shorter . . . . . *b*

 b. Low shrub with stems woody or somewhat woody near their base . . . . . 2. Andrachne

 b. Non-woody plants . . . . . *c*

  c. Plants with milky juice (break stem or leaf); flowers in a cup-shaped calyx-like involucre; calyx absent; stamens and pistils (actually staminate flowers and pistillate flowers) all within the same cup-shaped involucre . . . . . 8. Euphorbia

  c. Plants without milky juice (break stem or leaf); flowers not in an involucre; calyx well-developed; stamens and pistils in separate flowers on different parts of the same plant (monoecious) or on separate plants (dioecious) . . . . . *d*

   d. Plant with star-shaped (stellate) or branched hairs or scurfy scales; petals present in either the male or female flowers . . . . . *e*

    e. Use magnifying lens to note that the branches of the star-shaped (stellate) hairs on the lower surface of the leaves are separate all the way to their base; both the upper and lower leaf-surface similarly pubescent with the branches of the star-shaped hairs separate to their base; ovary or fruit (capsule) usually 3- (rarely 2- or 4-) celled; fruit dehiscent (eventually splitting open by valves); flowers at the top of the stem, in dense spike-like inflorescences . . . . . 3. Croton

e. Use magnifying lens to note that the branches of the star-shaped (stellate) hairs on the lower surface of the leaves are united at their base to form a scale; upper and lower leaf-surface unlike in hairiness, the lower with a scaly covering, the upper with a slender hairiness; ovary or fruit (capsule) 1-celled; fruit not dehiscent (not opening); flowers scattered, arising in small inflorescences from the sides or tips of the plant . . 4. CROTONOPSIS

d. Plant without hairs or hairs all simple if present, not branched, star-shaped, or scurfy; petals absent in both male and female flowers . . . . . . . . . . . . . . . . . . . *f*

f. Leaves toothed . . . . . . . . . . . . . . . . . . . . . . . . . *g*

g. Stems and leaves without stinging hairs; stamens 8–16 in the staminate (male) flowers; a conspicuous leafy 5–15-lobed or toothed bract surrounds the pistillate (female) flower; styles of pistillate flower many-cleft; calyx of staminate (male) flower 4-parted . . . . . . . . . . . . . . . . . . 5. ACALYPHA

g. Stems or leaves with stinging hairs; stamens 2 or 3 in the staminate (male) flowers; bract of pistillate (female) flower not lobed nor conspicuous; styles of pistillate flower simple and not cleft; calyx of staminate (male) flower 3–5-parted. . . . . . 6. TRAGIA

f. Leaves without teeth . . . . . . . . . . . . . . . . . . . . . . *h*

h. Plants glabrous (without hairs) . . . . . . . . . . . . . . . 1. PHYLLANTHUS

h. Plants hairy . . . . . . . . . . . . . . . . . . . . . 5. ACALYPHA

## 1. **Phyllanthus** L.

Commonly encountered; annual with 1 main branched stem; leaves broadly obovate or oval, rounded at summit, mainly 2–2½ times as long as broad, the larger leaves mainly 5–10 mm. broad. 1. P. CAROLINIENSIS

Rare, known only from a limestone glade in Stone Co.; perennial with several stems; leaves oblanceolate, slightly pointed at summit, mainly 3–5 times as long as broad, chiefly 2–5 mm. broad . . 2. P. POLYGONOIDES

1. **Phyllanthus caroliniensis** Walt. Map 1428

Flowers June–October.

Occurs in moist alluvial ground and gravel bars along streams, low alluvial thickets, fallow or cultivated fields, and moist depressions in valleys. Southern and central Missouri north to Ralls, Boone, Chariton, Pettis, Johnson, and Bates counties.

Ranges from Florida to Texas, north to Pennsylvania, West Virginia, Ohio, Indiana, Illinois, Missouri, and Kansas; also in Mexico, the West Indies, Central America, and northern South America.

The alternately 2-ranked leaves on slender stems provide ready field identification characters for this species.

2. **Phyllanthus polygonoides** Nutt. Map 1429

Flowers June–October.

Occurs on limestone glades. Known only in southwestern Missouri, Stone County (limestone barren in upland cedar area, T22N, R24W, sect. 18, 4 mi. east of Shell Knob, July 3, 1956, *Steyermark 81927*).

Ranges from Missouri and Kansas to Oklahoma and Texas.

This plant, recently discovered in Missouri, is a southwestern species, which like *Palafoxia callosa*, *Centaurium texense*, *Stenosiphon linifolius*, and *Andrachne phyllanthoides*, occurs from the Edwards Plateau of Texas northeastward, reaching its known northern and eastern limits in the Ozarks of Missouri. The plant grows in bunches with ascending stems and ascending leaves which are gray-green above, silvery beneath. The flowers have a creamy or creamy-greenish color.

## 2. **Andrachne** L.

**Andrachne phyllanthoides** (Nutt.) M. Arg.

Buck Brush Map 1430

*Savia phyllanthoides* (Nutt.) Pax & K. Hoffmann [Shinners]

Flowers May–October.

Occurs on rocky ledges of limestone bluffs, limestone glades, bald limestone knobs, and along dry gravelly washes of rocky stream beds. Restricted to a

Plate no. 234. **1.** Phyllanthus caroliniensis, × 2/5; After Gleason, details from Small, The New York Botanical Garden. **2.** Phyllanthus polygonoides, × 2/5; Details from Small, The New York Botanical Garden. **3.** Andrachne phyllanthoides, × 2/5; *a.* Flower, × 2 4/5. **4.** Croton glandulosus var. septentrionalis, × 2/5; *a.* Male flower, × 2 4/5; After Gleason, The New York Botanical Garden. **5.** Croton capitatus, × 2/5; *a.* Fruit with calyx, × 2. **6.** Crotonopsis elliptica, × 2/5; *a.* Fruit with calyx, × 2 4/5; *b.* Lepidote scales on leaf, highly magnified; Details from Small, The New York Botanical Garden. **7.** Croton monanthogynus, × 2/5; Details from Small, The New York Botanical Garden.

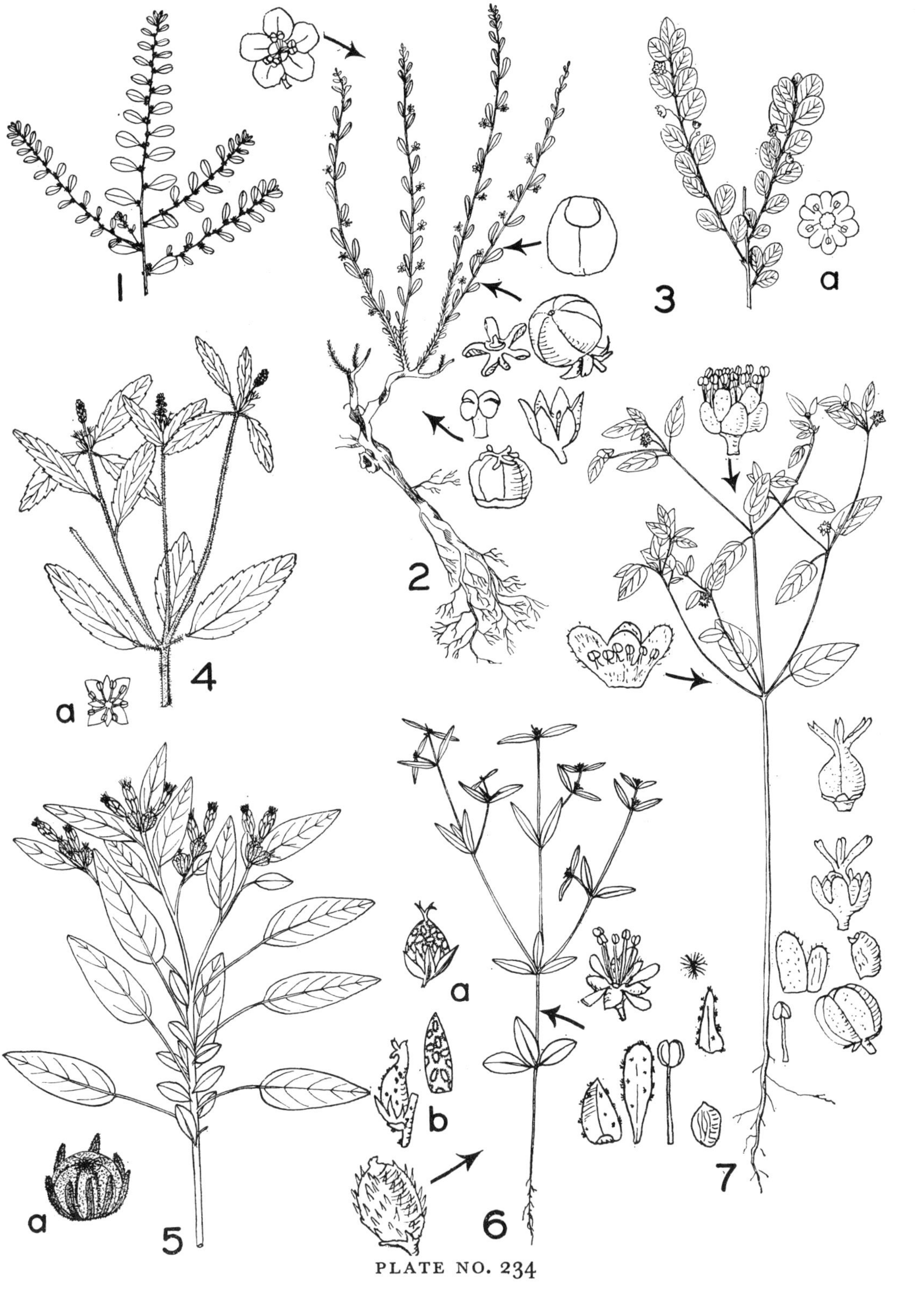


PLATE NO. 234

1429 Phyllanthus polygonoides

1430 Andrachne phyllanthoides (Buck Brush)

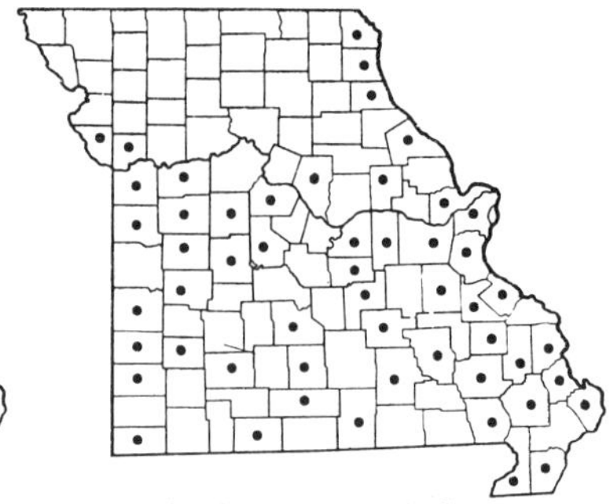

1431 Croton glandulosus var. septentrionalis

small section of the Ozark region, principally in the White River sector locally northeast to Shannon, Texas, and Phelps counties.

Ranges from Missouri and Arkansas to Oklahoma and Texas.

This low shrub is the only woody member of the predominantly tropical Euphorbiaceae to extend so far north in its range as Missouri; the other woody members of the family in the New World all range southward into Mexico, and Central and South America. The presence of this shrub in the Ozarks is indicative of the relic colonies of plants isolated on the bald knobs and limestone glades, whose many generations have survived through adaptation to the peculiar conditions of drought, exposure, and calcareous soils found here, their origin possibly to be traced back to the time preceding the last Tertiary uplift which connected elements of the floras of the Ozarks and the Edwards Plateau of Texas.

### 3. **Croton** L. Croton

a. Leaves toothed . . . . . . . . . . . . . . . . . . 1. C. GLANDULOSUS var. SEPTENTRIONALIS

a. Leaves not toothed (entire) . . . . . . . . . . . . . . . . . . . . . . . *b*

b. Rarely encountered, known only from Jackson County, western Missouri; staminate (male) and pistillate (female) flowers separated on different plants (dioecious); petals absent on both male and female flowers . . . . . . . . . . . . . . . . . . . . 4. C. TEXENSIS

b. Common species, throughout Missouri; staminate (male) and pistillate (female) flowers separated, but on the same plant (monoecious); petals present in the male flowers . . . . . . . . . *c*

c. Leaf-blades mainly 1–4 (rarely 5) cm. long; calyx of female flowers with 5 divisions; female flowers or fruits on recurved peduncle (stalks); styles 2, each split, resulting in 4 stigmas; stamens of male flowers 3–8; fruit 1-seeded . . . . . . . . . . . 3. C. MONANTHOGYNUS

c. Leaf-blades mainly 4–10 cm. long (rarely shorter); calyx of female flowers with 7–12 divisions; female flowers or fruits erect; styles 3, each forked 2–3 times, resulting in 6–12 or more stigmas; stamens of male flowers 7–14; fruit usually 3-seeded . . . . . . . . . . . . . . . *d*

d. Common throughout Missouri; larger leaf-blades averaging mainly 4–7 (sometimes 2) cm. long; hairs of inflorescence brownish or purplish; star-shaped hairs of stem with brown or purplish stalk; seeds flattened on lower ventral surface, with a circular shape . . . . 2a. C. CAPITATUS var. CAPITATUS

d. Rare, known only from southeastern Missouri north to St. Louis and Moniteau counties; larger leaf-blades averaging mainly 6–10 cm. long; hairs of inflorescence white, gray, or yellowish; star-shaped hairs of stem with a white or yellowish stalk; seeds only slightly flattened, longer than broad . . . . . . . . . . . 2b. C. CAPITATUS var. LINDHEIMERI

1. **Croton glandulosus** L. var. **septentrionalis** Muell. Arg. Map 1431

Flowers July–October.

Occurs in prairies, open woods, fallow and cultivated fields, pastures, waste ground, roadsides, and along railroads. Southern, central, and eastern Missouri, northeast to Clark County and north in central Missouri to Montgomery, Boone, Lafayette, and Platte counties.

Ranges from Florida to Texas, north to Delaware, Indiana, Illinois, Iowa, and Kansas; introduced in New Jersey and Pennsylvania.

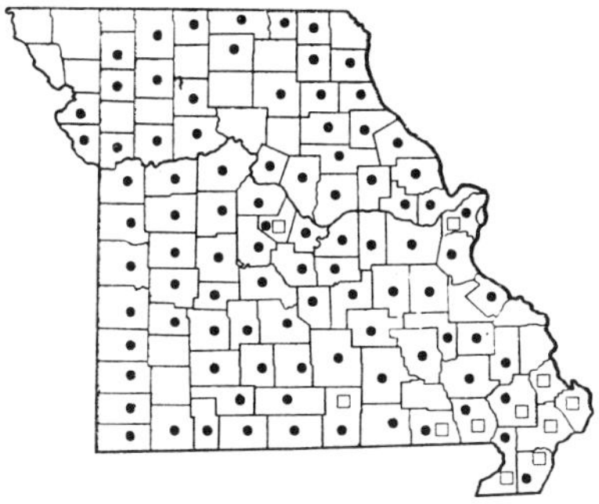

1432 • Croton capitatus var. capitatus (Hogwort)
1432 □ Croton capitatus var. Lindheimeri

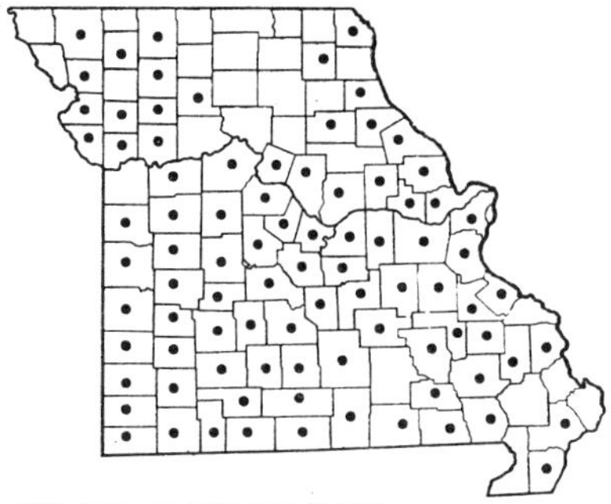

1433 Croton monanthogynus (Croton)

1434 Croton texensis

Typical *C. glandulosus* var. *glandulosus* is native to South and Central America, Mexico, the West Indies, Florida and other states bordering the Gulf of Mexico, and has broader leaf-blades.

2a. **Croton capitatus** Michx. var. **capitatus**
Hogwort Map 1432
*Croton capitatus* Michx. [G, P & S]
Flowers June–October.

Occurs in prairies, glades, fallow fields, pastures, waste ground, and along roadsides and railroads. Throughout Missouri, but not recorded as yet from the extreme northwestern and some of the extreme southeastern counties.

Ranges from Georgia to Texas, north to New York, Ohio, Indiana, Illinois, Iowa, and Kansas.

2b. **Croton capitatus** var. **Lindheimeri** (Engelm. & Gray) Muell. Arg. Map 1432
*Croton Lindheimeri* (Engelm. & Gray) Wood [G]
*Croton Engelmanni* Ferg. [P & S]
Flowers July–October.

Occurs in sandy open ground, fallow fields, along roadsides and railroads. Commonest in southeastern Missouri, where native west to Howell County and north to Stoddard and Scott counties; introduced along railroads in St. Louis and Moniteau counties.

Ranges from Florida to Texas, north to Georgia, Missouri, Kansas, and Oklahoma.

These two varieties are sometimes treated as separate species by some authors, but they intergrade in leaf size and shape and appear to be better regarded as variations belonging to one species.

Cattle are poisoned from eating hay containing plants of *C. capitatus* var. *capitatus* or var. *Lindheimeri*, due to the toxic croton oil contained in the plants. Generally, however, domestic animals avoid eating the plants because of their bitter taste, and it is common to see these plants left untouched in otherwise heavily grazed pastures, fields, and glades. White-tailed deer have been observed browsing on the plants, and wild turkey are known to eat the seeds.

3. **Croton monanthogynus** Michx. Croton Map 1433
Also known as Prairie Tea.
Flowers late May–September.

Occurs in usually acid soils, often overlying sandstone, chert, or granitic substrata, on rocky glades, dry upland prairies, open and waste ground, roadsides, and along railroads, rarely in open woodland. Throughout Missouri, but not recorded as yet from several north-central counties.

Ranges from Florida to Texas, north to Virginia, Ohio, Indiana, Illinois, Iowa, and Kansas; also in Mexico.

Like *C. capitatus*, this plant is poisonous to cattle, if eaten, but is usually avoided because of its disagreeable bitter taste. Grazed pastures, glades, and other areas occupied by livestock usually reveal this species untouched and standing where other vegetation has already been eaten. Wild turkey eat the seeds of this species, and white-tailed deer browse the plant.

4. **Croton texensis** (Klotzsch) Muell. Arg.
Skunk Weed Map 1434
Also known as Texas Croton.
Flowers May–October.

Occurs along streets, railroads, waste ground, and alluvial soils bordering the Missouri River. Known only in west-central Missouri in Jackson County (river bottoms, August 10, 1890, *Bush 3*; waste ground, Kansas City, August 22, 1914, *Bush 7328*; Sheffield, June 29, 1905, *Bush 3049*; along streets, Independence, September 10, 1897, *Bush 240*).

Ranges from Texas to Arizona, north to South Dakota, Colorado, and Wyoming; also in Mexico; introduced eastward in Missouri, Illinois, Alabama,

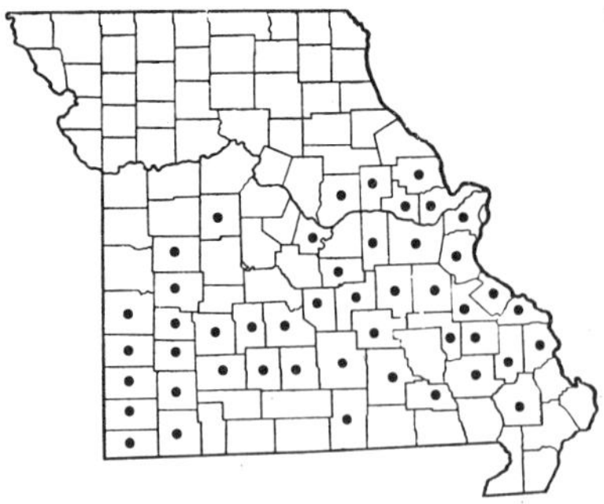

1435 Crotonopsis elliptica

1436 Crotonopsis linearis

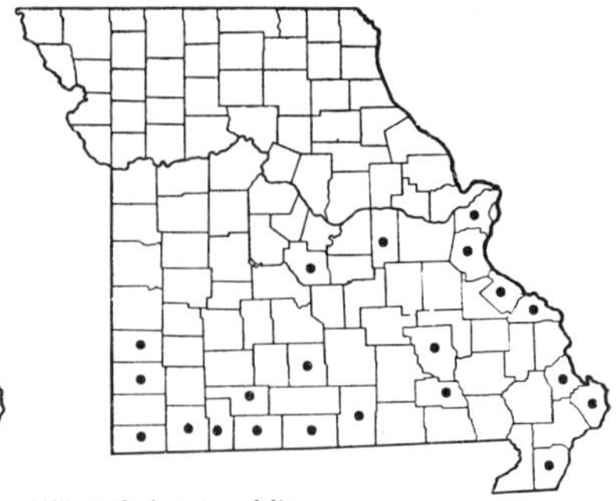

1437 Acalypha ostryaefolia

Delaware, and New England.

Most of the croton poisoning in the western states originates from the eating of this species. Since it is a rare introduced plant in Missouri and known only from a limited area in Jackson County, it is of unimportance to most farmers.

## 4. **Crotonopsis** Michx. Rushfoil

Common throughout southern and central Missouri; fruit smooth or at most covered with a scurfiness; plants rather short, usually 1–2 (up to 4) dm. tall; upper surface of leaf-blade soft to the touch, the star-shaped hairs up to 1 mm. long with overlapping branches; petals and filaments of male flowers scarcely longer than the calyx-divisions . . . . . . . . . . . . . . . . . . . . . . . 1. C. ELLIPTICA

Rarely collected, known only from southeastern Missouri in Stoddard and Dunklin counties; fruits spiny around the summit; plants up to 8 dm. tall; upper surface of leaf-blade not soft to the touch, the star-shaped hairs only 0.3 mm. or less, their branches not overlapping; petals and filaments of male flowers longer than the calyx-divisions . . . . . . . . . . . . . . . . . . . . . . . 2. C. LINEARIS

1. **Crotonopsis elliptica** Willd. Map 1435
Flowers late June–September.

Occurs in acid soils overlying sandstone, chert, or granitic substrata, on rocky glades, sandy fields, and in rocky open woods. Ozark region of southern and central Missouri, north to Lincoln, Montgomery, Callaway, Cole, Pettis, Henry, and Vernon counties.

Ranges from Florida to Texas, north to Connecticut, New Jersey, Pennsylvania, Indiana, Illinois, Missouri, and Kansas.

This species has been merged by Shinners (*Spring Flora of Dallas-Fort Worth Area*, pp. 242–43. 1958) with the next, *C. linearis*. Pending future, more detailed field and experimental studies, these taxa are being retained in the present work in the specific category.

This species is a characteristic plant of dry sterile sandstone or chert glades and outcrops, commonly associated with *Polygonum tenue*, *Oenothera linifolia*, *Hypericum gentianoides*, *Diodia teres*, and other low-growing plants with reduced leaf-surfaces adapted to withstand the arid conditions prevailing on the mainly dry sunny rock exposures.

2. **Crotonopsis linearis** Michx. Map 1436
Flowers July–September.

Occurs in sandy open ground. Known only in southeastern Missouri from Stoddard and Dunklin counties.

Ranges from Florida to Texas, north to Virginia, Illinois, and Missouri.

## 5. **Acalypha** L. Three-seeded Mercury

a. Most or all of leaf-blades broadest at base with a heart-shaped base; each margin of leaf-blade with 18–36 closely placed teeth; surface of fruit covered with little prickles; staminate (male) and pistillate (female) flowers in separate inflorescences, the staminate along the side axils, the pistillate usually at the top of the stem . . . . . . . . . . . . . . . . . . . . 1. A. OSTRYAEFOLIA

a. Leaf-blades not heart-shaped at base, tapering or narrowed to the base; each margin of leaf-blade with 12 or less more scattered teeth; surface of fruit smooth or hairy, but not prickly; staminate (male)

1438 Acalypha rhomboidea var. rhomboidea (Three-seeded Mercury

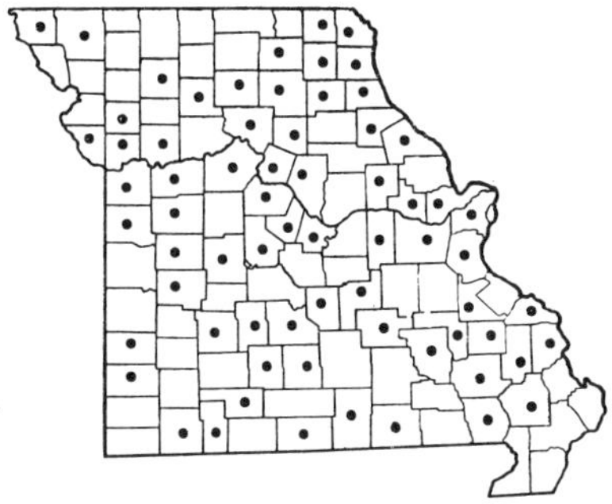

1439 Acalypha virginica

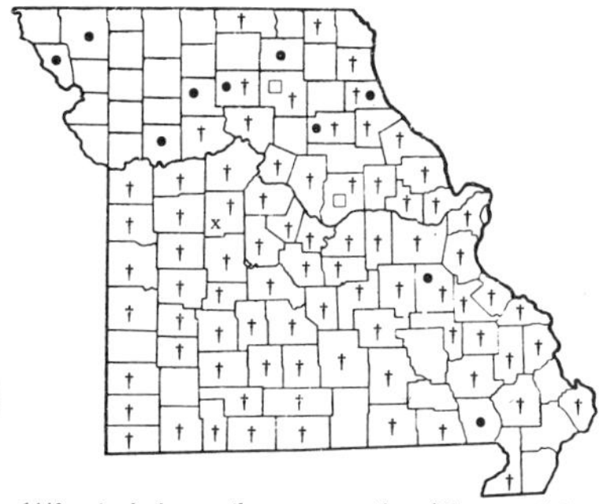

1440 • Acalypha gracilens var. gracilens (Three-seeded Mercury)
1440 □ Acalypha gracilens var. Fraseri
1440 † Acalypha gracilens var. monococca
1440 x Acalypha gracilens X Acalypha virginica

and pistillate (female) flowers all along the sides of the stem and all produced usually in the same inflorescence . . . . . . . . . . . . . . . . . . . . . . . . . . . . . *b*

b. Bracts of the pistillate (female) flowers with usually 5–7 (rarely 9) teeth or lobes; leaf-blades usually not much if any longer than the leaf-stalks (petioles), mostly with only a few scattered hairs or nearly hairless . . . . . . . . . . . . . . . . . . . . 2. A. RHOMBOIDEA

b. Bracts of the pistillate (female) flowers with 9–15 teeth or lobes; leaf-blades usually much longer than the leaf-stalks (petioles), varying from sparsely to densely hairy, at least on lower surface . . . *c*

c. Leaf-stalk (petiole) $\frac{1}{10}$–$\frac{1}{4}$ the length of the leaf-blade, usually shorter than or equal in length to the pistillate bract it accompanies; teeth or lobes of pistillate bract triangular-ovate to broadly oblong, usually with some stalked whitish or non-stalked reddish glands; hairs on stems short, ascending, and curving in toward the stem, rarely spreading; leaf-blades linear to narrowly oblong-lanceolate, 3–15 mm. broad; plants most frequently on rocky open glades, prairies, and open woods . . . . . . . . . . . . . . . . . 4. A. GRACILENS

c. Leaf-stalk (petiole) $\frac{1}{5}$–$\frac{1}{2}$ the length of the leaf-blade, usually longer than the pistillate bract which it accompanies; teeth or lobes of pistillate bract linear or lanceolate, acutely pointed, noticeably hispid (bristly hairy) and without any glands; hairs on stem both spreading and ascending or curved in, rarely with no spreading hairs; leaf-blades narrowly to broadly lanceolate, 8–40 mm. broad; plants of various types of habitats . . . . . . . . 3. A. VIRGINICA

1. **Acalypha ostryaefolia** Riddell Map 1437

Flowers June–October.

Occurs in moist, open, usually alluvial soils in valleys and along streams, and in fallow and cultivated ground. Ozark and southeastern lowland sections of southern and east-central Missouri north to St. Louis, Gasconade, Miller, and Barton counties.

Ranges from Florida to Texas, north to Virginia, Ohio, Indiana, Illinois, Iowa, and Kansas; also in Mexico; introduced in New Jersey.

This is sometimes considered a weed in cultivated ground.

2. **Acalypha rhomboidea** Raf. var. **rhomboidea**

Three-seeded Mercury Map 1438

*Acalypha rhomboidea* Raf. [G, BB]

*Acalypha virginica* of auth. [P & S], not L.

Flowers July–October.

Occurs in open wet or dry woodland, moist alluvial soils and gravel bars along streams, and in fallow and cultivated fields, thickets, prairies, waste ground, roadsides, and along railroads. Throughout Missouri and doubtless in every county.

Ranges from Quebec and Maine to Ontario, Minnesota, and Nebraska, south to Florida, Arkansas, and Oklahoma.

The leaf-stalks (petioles) vary from one-third the length to nearly the length of the leaf-blade. The stems vary from nearly hairless to hairy with short incurved hairs, rarely with spreading hairs. The larger main leaves are mostly hairless, but the nerves are usually finely short-hairy, and the surface often has a few scattered hairs.

This is sometimes a weedy plant in cultivated fields. Plants with drooping leaves and seeds 2.5–3 mm. long are placed in var. *Deamii* Weath., not known from Missouri.

3. **Acalypha virginica** L.

Three-seeded Mercury Map 1439

*Acalypha digyneia* Raf. [P & S]

Flowers July–October.

Occurs in prairies, dry open upland, open valleys, low moist or dry rocky woodland, gravel bars, fallow fields, thickets, roadsides, and along railroads. Throughout Missouri, somewhat less common than *A. rhomboidea.*

Ranges from Georgia to Texas, north to Massachusetts, Indiana, Illinois, Missouri, and Kansas.

This species appears to be spreading northward. Usually it is easily recognized by the combination of spreading and ascending-incurved hairs on the stem, nonglandular pistillate bracts with deeply cut hispid lobes. It is sometimes variable in pubescence and petiole length. The petiole varies from one-fifth as long as to slightly more than one-half as long as the leaf-blades, and the stems, while usually having some spreading hairs, occasionally have all the hairs ascending. Sometimes it apparently hybridizes with the following species, *A. gracilens*, possessing the non-spreading type of pubescence of the stem characteristic of *A. gracilens* but with the deeply cut, hispid, non-glandular, pistillate bracts of *A. virginica.* Such a putative hybrid has been recorded from Pettis County (openings on slopes near top, above north-facing wooded limestone bluffs and slopes along Muddy Creek, T47N, R20W, southwest $\frac{1}{4}$ sect. 25, southeast $\frac{1}{4}$ sect. 26, northeast $\frac{1}{4}$ sect. 35, and northwest $\frac{1}{4}$ sect. 36, $2\frac{1}{2}$–3 mi. southeast of Lookout, 7 mi. [by air] northeast of Beaman, September 26, 1955, *Steyermark 80268*). Specimens are also found which grade into *A. rhomboidea* and are difficult to place in either *A. virginica* or *A. rhomboidea.* Such a specimen as *Steyermark 85577* from Worth County has deeply 12–15-lobed pistillate bracts characteristic of *A. virginica*, but with the shape of the leaf-blade and finely appressed-pubescent stems without spreading hairs characteristic of *A. rhomboidea.*

4. **Acalypha gracilens** Gray

Three-seeded Mercury Map 1440

Flowers last of May–October.

Occurs usually on rocky open glades, either on sandstone, chert, granite, or limestone, rocky prairies, or low meadows, rocky or open woodland, rarely along roadsides and railroads.

The following variations are encountered in Missouri:

a. Leaves oblong to oblong-lanceolate; leaf-stalks (petiole) $\frac{1}{8}$–$\frac{1}{4}$ the length of the blade; seeds 3 in a single fruiting capsule . . . 4a. A. GRACILENS var. GRACILENS

a. Leaves linear to linear-lanceolate; leaf-stalks (petiole) $\frac{1}{10}$–$\frac{1}{8}$ the length of the blade; seed 1 or seeds 3 in a single fruiting capsule . . . *b*

- b. Common type; seed 1 in a fruiting capsule; staminate (male) inflorescence 5–12 mm. long . . 4c. A. GRACILENS var. MONOCOCCA
- b. Rare type; seeds 3 in a fruiting capsule; staminate (male) inflorescence 20–40 mm. long . . . 4b. A. GRACILENS var. FRASERI

4a. **Acalypha gracilens** var. **gracilens** Map 1440

*Acalypha gracilens* Gray [G, BB]

Occurs in northern Missouri in Holt, Nodaway, Ray, Livingston, Linn, Adair, Marion, and Monroe counties, south in eastern Missouri to Butler County.

Ranges from Florida to Texas, north to New Hampshire, Massachusetts, New York, Ohio, Indiana, and Wisconsin.

4b. **Acalypha gracilens** var. **Fraseri** (Muell. Arg.) Weath. Map 1440

Known only from Callaway and Macon (July 30, 1939, *Bauer 349*) counties.

Ranges from Georgia to Texas, north to Illinois, Missouri, and Oklahoma.

4c. **Acalypha gracilens** var. **monococca** Engelm. Map 1440

This is the commonest of the varieties in Missouri, found throughout southern, central, and eastern Missouri west to Putnam, Linn, Carroll, Lafayette, and Jackson counties.

Ranges from Arkansas and Texas, north to Missouri and Kansas.

As noted under *A. virginica*, there appears to be hybridization between *A. gracilens* and *A. virginica. Acalypha gracilens* also grades into *A. virginica.* Some of the specimens of *A. gracilens* var. *gracilens*, for example, from northern Missouri, such as *Steyermark 85627* from Holt County, have the leaf-blades of the upper leaves narrowly elliptic-oblong, 7–8 mm. broad, with a

Plate no. 235. **1.** Acalypha ostryaefolia, × 2/5; *a.* Female flower, × 2 4/5; After Britton and Brown, The New York Botanical Garden. **2.** Acalypha virginica, × 2/5 (Scribner's). **3.** Ricinus communis, leaf, × 2/15; After Gleason, details from Small, The New York Botanical Garden. **4.** Tragia cordata, × 2/5; *a.* Staminate (male) flower, × 2; After Britton and Brown, The New York Botanical Garden. **5.** Euphorbia dentata, × 2/5; After Britton and Brown, details from Small, The New York Botanical Garden. **6.** Tragia urticifolia var. urticifolia, × 2/5; Details from Small, The New York Botanical Garden. **7.** Acalypha rhomboidea, × 2/5; *a.* Bract with flowers, × 1 3/5; *b.* Seed, × 6; After Gleason, details from Small, The New York Botanical Garden. **8.** Acalypha gracilens var. monococca, × 2/5.

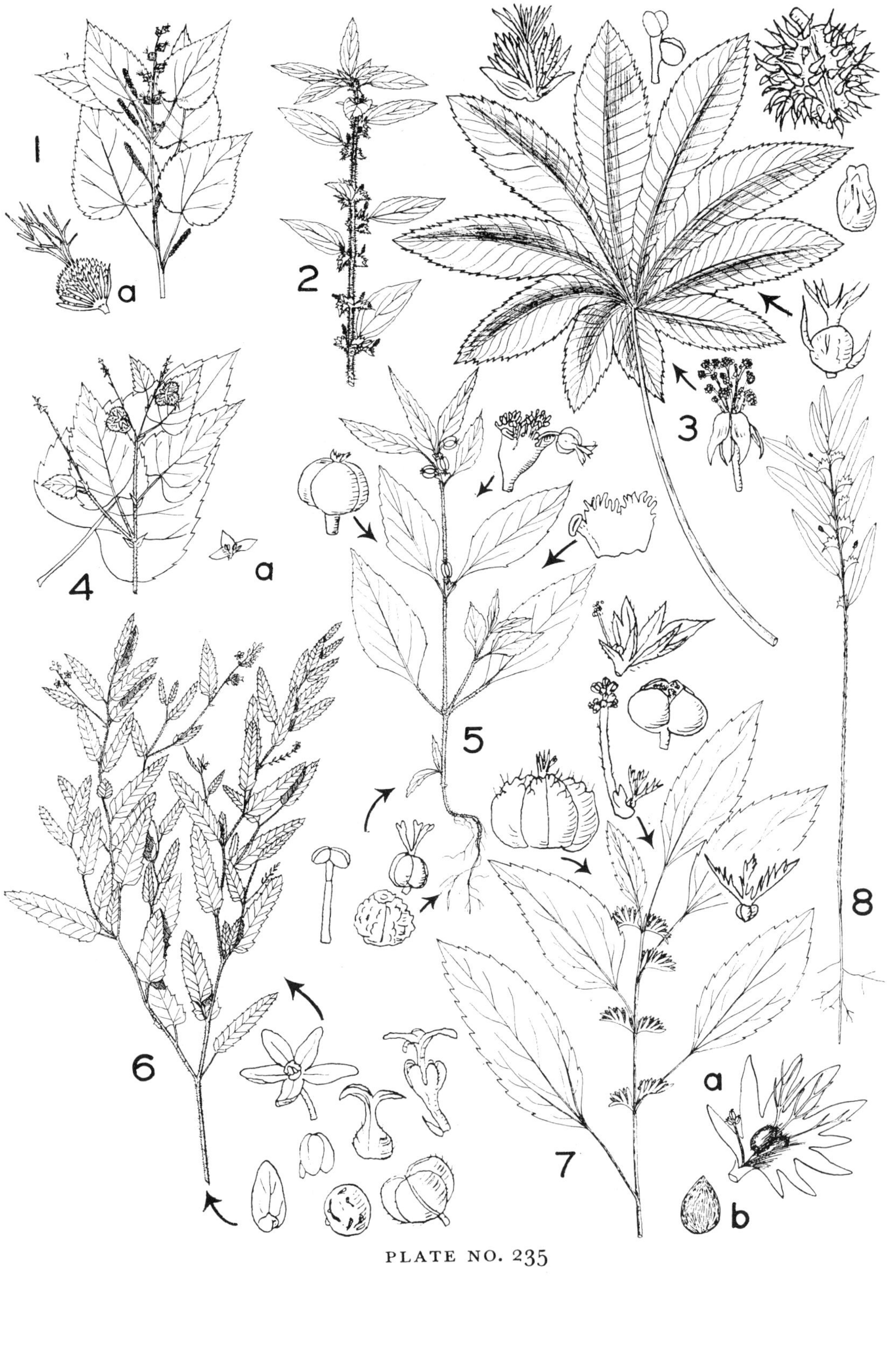


PLATE NO. 235

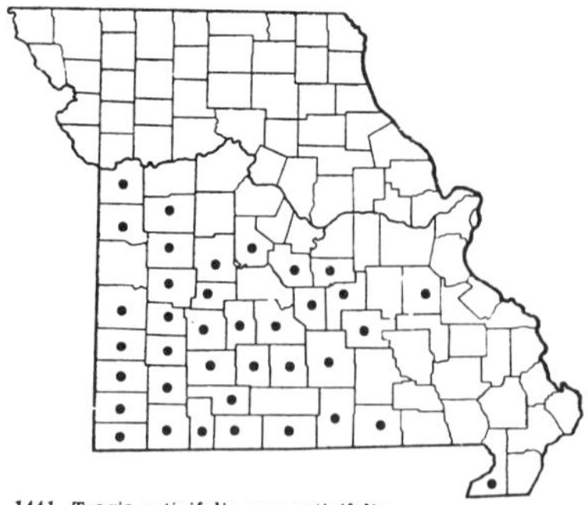

1441 Tragia urticifolia var. urticifolia

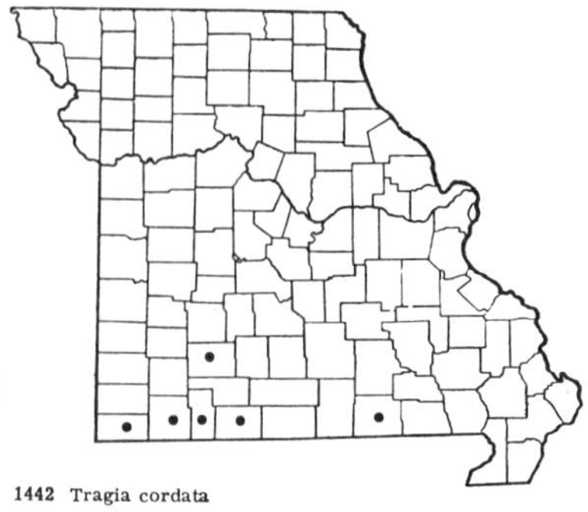

1442 Tragia cordata

1443 Ricinus communis (Castor-oil Plant)

petiole about one-third the length of the blade, the pistillate bracts cut one-third to five-eighths their breadth into 9–11 lobes densely hispid but lacking red and white glands. Such a specimen resembles a narrow-leaved *A. virginica.* The red sessile glands are sometimes absent on the pistillate bracts of specimens of *A. gracilens* var. *gracilens.*

## 6. **Tragia** L.

a. Most of the leaf-blades 2–5 cm. long and less than half as wide as long; larger leaves with petioles (stalks) 0.5–1.5 cm. long; stems erect to reclining to slightly twining; stamens 4–5; mature fruit 5–10 mm. in diameter; seeds 3–4 mm. long . . . . . . . . . . . . . . . 1. T. URTICIFOLIA

a. Most of the leaf-blades 5–13 cm. long and more than half as wide as long; larger leaves with petioles (stalks) 3–7 cm. long; stems twining; stamens 3; mature fruit 12–16 mm. in diameter; seeds about 5 mm. long . . . . . . . . . . . . . . . . . . . . . . . . 2. T. CORDATA

1. **Tragia urticifolia** Michx. var. **urticifolia** Map 1441

*Tragia urticifolia* Michx. [G, BB, P & S, Steyerm.]

Flowers June–September.

Occurs on rocky glades, prairies, sandy open fields, rocky open woods and thickets. Unglaciated prairie and mainly western Ozark sections, north to Jackson, Johnson, Benton, Morgan, Miller, and Maries counties, east in the Ozark section to Washington, Dent, Texas, and Oregon counties, and locally on Crowley Ridge in Dunklin County; absent from most of the eastern Ozark section.

Ranges from Florida to Arizona, north to Missouri, Kansas, and Colorado; also in Mexico. Reported from Virginia.

The stems are usually erect, but vary from a reclining position to slightly twining. They are covered with stinging hairs. Although some manuals state the leaves to be short-petioled to sessile, the petioles in this species may attain lengths 2 cm. or more.

*Tragia urticifolia* has been confused with *T. nepetaefolia* Cav., a species of the southwestern states which does not appear to be in Missouri.

2. **Tragia cordata** Michx. Map 1442

Flowers July–September.

Occurs in rocky open woods, thickets, and glades. Mostly southwestern Missouri in McDonald, Barry, Stone, Taney, and Greene counties and locally eastward in Oregon County.

Ranges from Florida to Texas, north to Georgia, Indiana, Illinois, and Missouri.

With its range eastward, intensive exploration of the southeastern Ozark region should reveal additional stations of this comparatively rare species.

## 7. **Ricinus** L. Castor-oil Plant

**Ricinus communis** L. Castor-oil Plant, Castor Bean Map 1443

Flowers June–October.

Commonly planted, and rarely escaped from cultivation to waste ground, along roadsides, and railroads, where known from Lincoln, St. Louis, Boone, Jackson, and Jasper counties.

Native of tropical Africa and Asia; introduced and

naturalized in parts of the United States.

The color, markings, and shape of the seed show a resemblance to those of the Mediterranean sheep tick, *Ricinus*, accounting for the origin of the generic name of this plant. Castor Bean plants become woody and reach treelike proportions, as much as 12 meters high, in the tropical countries and in Florida and other southern states, where they persist from year to year, unlike their annual habit further north, where they are herbaceous and 1–4.5 meters tall.

All parts of the plant, but especially the seeds, are poisonous to eat, both for human consumption and for domestic animals. Although the best-known use of the plant is as a purgative medicine, obtained from the smaller-seeded varieties with mottled brown seeds, Castor Bean plant has a wide application of uses. The best fiber of the stems is used for paper pulp. In India and China silkworms are fed on the leaves. After the oil has been separated from the seeds, the ground seeds serve as a valuable fertilizer. Likewise, the pomace or oil cake serves for fertilizer. The oil is much used in industry as a lubrication oil for machinery, obtained from the larger black-seeded varieties, for lubricating airplane engines, for dressing tanned hides, in soap manufacture, and as an illuminating oil. Because of its water resistant properties, this oil is used in the dyeing and printing of cotton goods, coating of fabrics, as protective coating for food containers, guns, insulation, and airplanes. The quick-drying properties of the oil lend themselves well to its use in paint and varnish manufacturing. The leaves are sometimes used as an insecticide, and in some regions oil is placed on food exposed for the express purpose of poisoning cockroaches. The oil is used by veterinarians for wounds in livestock and other domestic animals, and also is contained in some hair dressings to provide luster to the hair. Mixed with turpentine the oil is sometimes taken internally to expel tapeworms and the leaves moistened with vinegar are sometimes applied as a dressing to relieve headaches.

A number of horticultural forms are grown, differing chiefly in the color of the stems and leaves.

## 8. **Euphorbia** L. Spurge

Some authors divide this genus into several other genera. For a discussion concerning his reasons for adopting the genus *Chamaesyce* for some of the species included here under *Euphorbia*, see L. H. Shinners (Field and Laboratory 20: 25–26. 1952).

The flower of *Euphorbia* is extraordinarily different from any other in Missouri. Actually what appears to represent a single flower consists of 1 female (pistillate) flower bearing a stalk (gynophore) and ovary, but lacking petals or sepals, and few to many male (staminate) flowers, each male flower consisting of a single stamen with a single bracteole at the base and a filament-like stalk (androphore) tipped by a roundish anther. The entire group of stamens and the single pistil is surrounded by and contained within a small green calyx-like, cup-shaped, and lobed involucre, known as the *cyathium.* The summit of this cyathium has 4 or 5 glands which may or may not be accompanied by colored or often petal-like appendages.

The milky sap of all the species has a blistering irritating effect on many persons, and is poisonous if eaten, especially by cattle.

a. Uppermost leaves of the stem and those at the base of the inflorescences with broad white or petal-like edges; leaf-blades 15–30 mm. broad . . . . . . 9. E. MARGINATA

a. None of the leaves with broad white edges; leaf-blades 0.5–15 mm. broad (dwarfed specimens have narrower leaves 10 mm. or less) . . . . . . *b*

b. Main leaves of the stem chiefly alternate or scattered . . . . . . *c*

c. Upper leaves and those at the base of the inflorescence often red or red-based; upper leaves of stem often lobed; stem not topped by a several-rayed umbel having a whorl (circle) of leaves at its base; involucre with only 1 gland. . . . . . 2. E. HETEROPHYLLA

c. Without the above combination of characters; no red at base of upper leaves or of those at base of the inflorescence; upper leaves of stem never lobed; stem topped by a 3- or more rayed umbel having a whorl (circle) of leaves at its base; involucre with 4 or 5 glands . . . . . . *d*

d. Leaves with tiny teeth along margins (sometimes seen only with magnifying lens). . . . . . *e*

e. Lower side of main leaves of stem with hairs, acutely pointed at tip; topmost umbel supported usually by 5 main branches (rays); outside of involucre usually hairy; rarely found . . . . . . 3. E. PLATYPHYLLA

e. Lower side of main leaves of stem glabrous (without hairs), rounded at tip; topmost umbel supported usually by 5 main branches (rays); outside of involucre glabrous (without hairs); frequently found. . . . . . *f*

f. Seeds smooth, 1.7–2.3 mm. long; styles of very young flowers longer than the ovary; glands 5, usually red; plants of eastern Missouri west to Boone, Morgan, Laclede, Wright, and Howell counties; plants of woodland, sandy and gravel bars along streams . . . . . . 4. E. OBTUSATA

f. Seeds with a delicately honeycombed surface, 1.3–1.5 mm. long; styles of very young flowers shorter than the ovary; glands 4, usually yellow or tawny, one of them replaced by a tuft of hairs or small lobe; plants of western Missouri, east to Boone, Moniteau, Maries, and Howell counties; plants of glades, bluffs, and rocky ledges . 5. E. SPATHULATA

d. Leaves lacking teeth along margins (entire) . . . . . . *g*

g. Leaves of stem all thread-like, narrowly linear or linear-filiform, 0.5–3 mm. wide, very numerous, 2–20 per centimeter of stem . . . . . . 6. E. CYPARISSIAS

g. Leaves of stem lanceolate, oblong, or obovate, or nearly round, mostly 3–20 mm. wide, 0–5 per centimeter of stem . . . . . . *h*

h. Flowers with showy white petal-like parts (glands of involucre with petal-like appendages) . . . . . . 10. E. COROLLATA

h. Flowers without showy white petal-like parts (glands of involucre without any petal-like appendages) . . . . . . *i*

i. Common native plant of the Ozark region; annual; stems mostly 1.5–3 (up to 4) dm. tall; main leaves of stem rounded or blunt at tip, 0.5–3 cm. long; surface of fruit smooth; surface of seed pitted . . . . . . 8. E. COMMUTATA

i. Rare introduced species, known only in Chariton County; perennial; stems mainly 3–9 dm. tall; main leaves of stem acutely pointed, 3–7 cm. long; surface of fruit warty or rough-granular; surface of seed smooth. . . . . . 7. E. ESULA

b. Main leaves of stem chiefly opposite . . . . . . *j*

j. Stems with a rough hairiness usually in the upper part, some of the hairs pointing downward toward the base of the plant; gland of involucre without any petal-like appendage; stipules represented by tiny glands at base of leaf-stalk; seeds 2.2–3 mm. long . . . . . . . . . . . . . 1. E. DENTATA

j. Stem smooth, glabrous (without hairs) or hairs short and appressed (lying parallel to surface) or ascending; glands of involucre with petal-like appendage; stipules definitely developed as scale-like outgrowths at base of leaves or leaf-stalks; seeds 1–2 mm. long . . . . . . . . . . . . . . *k*

k. Stems erect or ascending, never prostrate; leaf-blades either 8–35 mm. long or 2–18 mm. wide or both, either entirely glabrous or with long hairs on lower surface near base; ovary and fruit glabrous; fruiting capsule 1.9–2.5 mm. long . . . . . . . . . . . . . . . . . *l*

l. Stems glabrous (without hairs); leaf-blades without teeth (entire) and completely glabrous; 29–48 staminate flowers inside of each cup-shaped involucre; conspicuous prolonged white or pink petal-like appendages present at summit of cup-shaped involucre . . . . 12. E. MISSURICA

l. At least tip of stem hairy or stem with fine lines of hairs; leaf-blades finely toothed, the lower surface at least near the base with long hairs; 5–11 staminate flowers inside of each cup-shaped involucre; no conspicuous or prolonged petal-like appendages present at summit of involucre 14. E. MACULATA

k. Stems mainly prostrate, lying or creeping along the ground, but if ascending to rarely erect, then either with teeth on some part of leaf or hairs present on stem, leaves, ovary, or fruit, or if plant is completely hairless, then the fruiting capsules are 1.2–2 mm. long and 5–17 staminate flowers are contained within each cup-shaped cyathium, or the involucre lacks conspicuous prolonged appendages; leaf-blades 2–17 mm. long and 2–8 mm. wide . . . . . . . . . . . . . . *m*

m. Hairs present on some part of leaves or stems or both; hairs present on ovary and fruit . . . . *n*

n. Hairs of fruiting capsules spreading; the 3 styles split nearly to their base; seeds with sharp narrow cross ridges whitened on the summit; rarely collected species, known only from Jackson County, west-central Missouri . . . . . . . . . . . . . . 18. E. PROSTRATA

n. Hairs of fruiting capsule finely appressed or strigose (pressed against or lying parallel to surface); the 3 styles either not split at all or split only in their upper $\frac{1}{4}$–$\frac{1}{2}$; seeds either with a smooth, granular, or pitted surface or, if with cross ridges, the ridges low and rounded and not whitened on the summit; includes commonly collected species . . . . . . . . . *o*

o. Rarely collected species, known only from Jackson County, west-central Missouri; the 3 styles only 0.2 mm. long, not at all divided, their tips entire or at most with a shallow notch (emarginate); seeds with little pits or with low ridges . . . 17. E. STICTOSPORA

o. Commonly collected species, throughout Missouri; the 3 styles 0.4–0.7 mm. long, their upper half split from $\frac{1}{4}$–$\frac{1}{2}$; seeds either smooth, granular, or with low rounded cross ridges . . *p*

p. Common in low, moist or alluvial ground along streams and borders of ponds and sloughs; styles 0.7 mm. long, their upper half split $\frac{1}{2}$ to the base; seeds smooth or finely granular, not cross-ridged; leaves remotely and indistinctly toothed, almost smooth-edged, pale green to glaucous on lower surface, the largest leaves averaging 5–8 mm. wide; stems mainly pale green, slightly hairy to nearly hairless; plants averaging larger, coarser, and stiffer than no. 15 . . . . . . . . 16. E. HUMISTRATA

p. Common in dry places, along streets, sidewalks, roads, waste ground, railroads, on rocky glades, prairies, and clearings, more rarely in moist or alluvial ground; styles 0.4 mm. long, their upper half split $\frac{1}{4}$–$\frac{1}{3}$; seeds with low rounded cross ridges; leaves more distinctly toothed, the largest leaves averaging 3–5 mm. wide and averaging smaller than no. 16, usually darker or dull green with a reddish spot, less conspicuously glaucous or pale green beneath; stems usually reddish or brown, usually hairier than no. 16 . . . . . . . . . . . . . . . . . . . . . . . 15. E. SUPINA

m. Leaves and stems without hairs; ovary and fruits without hairs . . . . . . . . . . . *q*

q. Leaves without teeth . . . . . . . . . . . . . . . . . . . . . . . . . *r*

r. Leaves of a roundish to roundish-oblong shape, not much longer than broad, 2–7 mm. long; stipules united into a single scale at base of leaf; fruiting capsules 1–2 mm. long; seeds 1 mm. long, 0.5 mm. thick; common along or near the Missouri, Mississippi, and Osage rivers, in alluvial or moist soils . . . . . . . . . . . . . . 13. E. SERPENS

r. Leaves oblong to elliptic- or ovate-oblong, longer than broad, 4–10 mm. long; stipules usually separate, rarely the ventral ones united; fruiting capsule 2 mm. long; seeds 1.3–1.6 mm. long, 1 mm. thick; known only from Clark County, northeastern Missouri 11. E. GEYERI

q. Leaves with at least a few teeth at their rounded summit or sometimes near their base. . . . *s*

1444 ● Euphorbia dentata f. dentata
1444 □ Euphorbia dentata f. cuphosperma

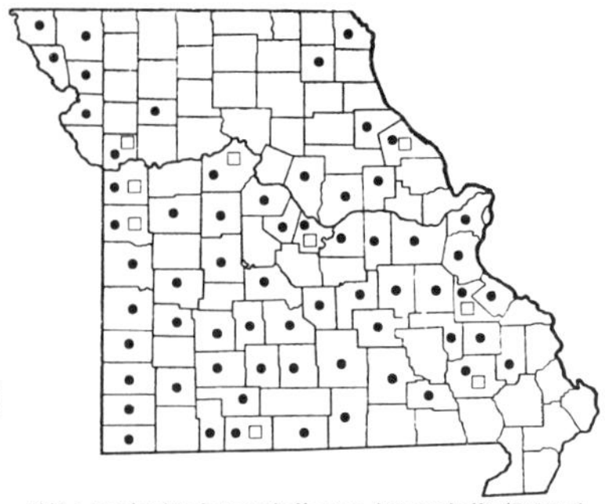

1445 ● Euphorbia heterophylla var. heterophylla (Painted Leaf)
1445 □ Euphorbia heterophylla var. graminifolia

1446 Euphorbia platyphylla

s. Seeds with usually 3–4 (up to 6) conspicuous transverse (cross) ridges extending most of the distance across; leaves oblong to linear; involucre 0.6–0.9 mm. in diameter; 1–5 staminate flowers in each cup-shaped cyathium . . . . . . . . . . . . . . . . . . . . . 20. E. GLYPTOSPERMA

s. Seeds usually with minute pits or smoothish, or more rarely slightly wrinkled with short low ridges not extending across the surface, but not with conspicuous cross ridges; leaves ovate to broadly oblong; involucre 0.8–1.2 mm. in diameter; 5–18 staminate flowers in each cup-shaped cyathium 19. E. SERPYLLIFOLIA

1. **Euphorbia dentata** Michx. Map 1444

Flowers July–October.

Occurs on prairies, glades, waste ground, fallow fields, roadsides, and along railroads. Throughout most of Missouri, except absent from many north-central counties.

Ranges from New York to Minnesota, South Dakota and Wyoming, south to Virginia, Louisiana, Texas, and Mexico.

Two variations occur in Missouri:

Common type found; leaves near the summit of stem and at the main nodes rhombic- to lance-ovate. . . . . . . 1a. E. DENTATA f. DENTATA

Rare type found; leaves near the summit of stem and at the main nodes linear to narrowly lanceolate . . . . 1b. E. DENTATA f. CUPHOSPERMA

1a. **Euphorbia dentata** f. **dentata** Map 1444

*Euphorbia dentata* Michx. [G, BB, P & S]

This is the common form found throughout Missouri.

1b. **Euphorbia dentata** f. **cuphosperma** (Engelm.) Fern. Map 1444

Known only from Jackson (July 29, 1893, *Bush 337*) and Barton (open sandy ground along river, 1 mi. northwest of Lamar, July 30, 1951, *Palmer 52816*) counties, western Missouri.

The plant is poisonous if eaten by livestock, and is sometimes contained in hay. The flowers, fruits, and leaves are eaten by wild turkey.

2. **Euphorbia heterophylla** L. Painted Leaf Map 1445

Also known as Wild Poinsettia, Fire-on-the-Mountain, Painted Spurge.

Flowers July–October.

Occurs in alluvial soils, thickets, open or rocky woods, glades, and along railroads.

Two variations occur in Missouri:

Main leaves of stem and upper leaves linear or narrowly lanceolate, mainly not lobed . . . 2b. E. HETEROPHYLLA var. GRAMINIFOLIA

Main leaves of stem and upper leaves oval to lanceolate, fiddle-shaped or lobed . . . . 2a. E. HETEROPHYLLA var. HETEROPHYLLA

2a. **Euphorbia heterophylla** var. **heterophylla** Map 1445

*Euphorbia heterophylla* L. [G, BB, P & S]

This is the commoner variation and is found commonly in southern, central, eastern, and northwestern Missouri; absent from the lowland counties of southeastern Missouri and from most of the north-central counties.

Ranges from Florida to Texas, north to Virginia, Indiana, Wisconsin, Minnesota, and South Dakota.

Plate no. 236. **1.** Euphorbia spathulata, × 2/5. **2.** Euphorbia Esula, × 2/5; *a.* Flower, × 4; After Gleason, The New York Botanical Garden. **3.** Euphorbia obtusata, × 2/5. **4.** Euphorbia heterophylla, × 2/5; *a, b.* Seeds, × 4/5; *c.* Fruiting cyathium, × 4/5. **5.** Euphorbia Cyparissias, × 2/5; After Britton and Brown, The New York Botanical Garden.

PLATE NO. 236

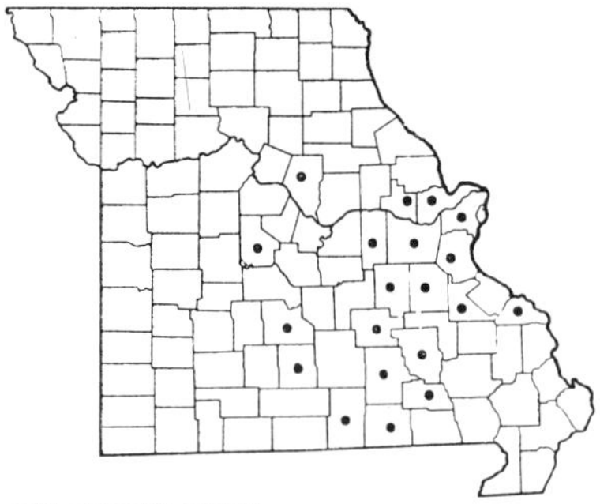
1447 Euphorbia obtusata

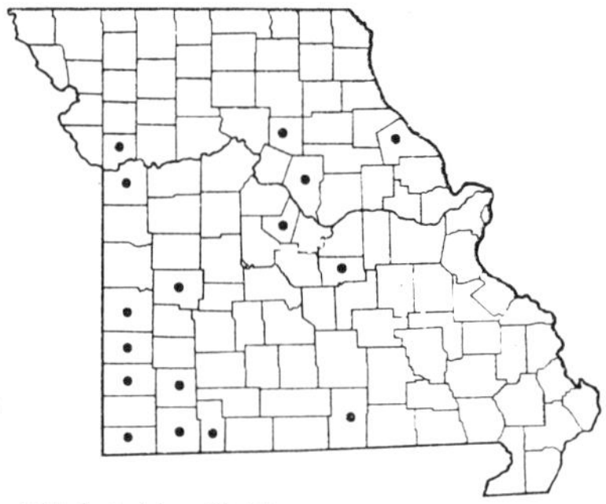
1448 Euphorbia spathulata

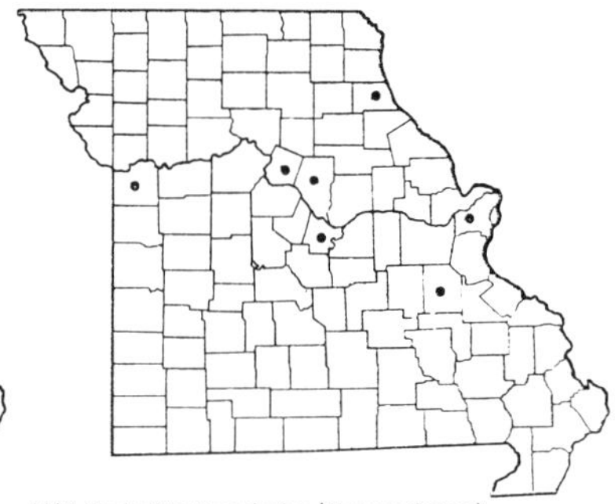
1449 Euphorbia Cyparissias (Cypress Spurge)

2b. **Euphorbia heterophylla** var. **graminifolia** (Michx.) Engelm. Map 1445

Known from southern and central Missouri in St. Francois (northwest of Farmington, August 29, 1898, *Trelease 1187*), Wayne (Mill Spring, August, 1926, *Trelease*), Pike, Cole (Jefferson City, September, 1871, *Dr. P. Krause*), Saline (Blue Lick, September 13, 1934, *Bush 13936*), Clay, Jackson, Cass, and Taney counties.

Ranges from Florida to Texas, north to Indiana, Wisconsin, and Minnesota.

The leaves are quite variable in *E. heterophylla*, differently shaped ones being present on the same plant. Generally, in var. *heterophylla* some of the leaves are broad and fiddle-shaped, whereas in var. *graminifolia*, the more rarely encountered variation, the main and upper leaves are linear and unlobed.

3. **Euphorbia platyphylla** L. Map 1446

Flowers June–August.

Known from St. Louis County in east-central Missouri (St. Louis, July 23, 1910, *Sherff* in Gray Herb.).

Native of Europe; introduced and naturalized in North America from Quebec and Ontario, south to Vermont, New York, Ohio, Michigan, Illinois, and Missouri.

The specimen from St. Louis County was probably from a cultivated or adventive plant, but bears no definite information. Another specimen in the Gray Herbarium labeled *E. platyphylla* (*Bush 202* from Eagle Rock, Barry Co., May 31, 1898) with the lower surface of the leaves glabrous and 3 rays to the terminal umbel is definitely *E. spathulata* Lam. (*E. dictyosperma* Fisch. & Mey.).

4. **Euphorbia obtusata** Pursh Map 1447

Flowers late May–July.

Occurs in low woods, thickets, fallow fields in valleys, sandy ground, low banks, and gravel bars along streams. Eastern Missouri west to Boone, Morgan, Laclede, Wright, and Howell counties, north to St. Charles, Warren, and Boone counties.

Ranges from South Carolina to Texas, north to Pennsylvania, Ohio, Indiana, Illinois, Iowa, and Nebraska.

5. **Euphorbia spathulata** Lam. Map 1448

*Euphorbia dictyosperma* Fisch. & Mey. [G, BB, P & S, Steyerm.]

Flowers May–July.

Occurs on rocky glades, ledges of bluffs, rocky open woods, clearings, and exposed situations. Western Missouri east locally to Pike, Randolph, Boone, Moniteau, Maries, and Howell counties, north to Pike, Randolph, and Clay counties.

Ranges from Minnesota to Washington, south to Alabama, Louisiana, Texas, and Mexico; also in South America.

Usually only a single erect stem is found in this and the preceding species, which closely resemble each other. Although some manuals give the color of the glands of the involucre (red or scarlet in *E. obtusata*, yellow in *E. spathulata*) as one of the differences between the two species, in Missouri specimens, at least, this character cannot be relied upon to separate the two taxa, nor can the presumed difference of floral leaves (cordate-clasping in *E. obtusata*, truncate to subcordate at base in *E. spathulata*) be depended upon.

6. **Euphorbia Cyparissias** L. Cypress Spurge Map 1449

Flowers April–August.

Planted in cemeteries and rock gardens and sometimes escaped from cultivation to fields and roadsides. Scattered in eastern and central Missouri in Marion, St. Louis, Washington, Cole, Boone, Howard, and Jackson counties.

Native of Europe; introduced and naturalized in the United States from Maine to Minnesota and

1450 Euphorbia Esula (Leafy Spurge)

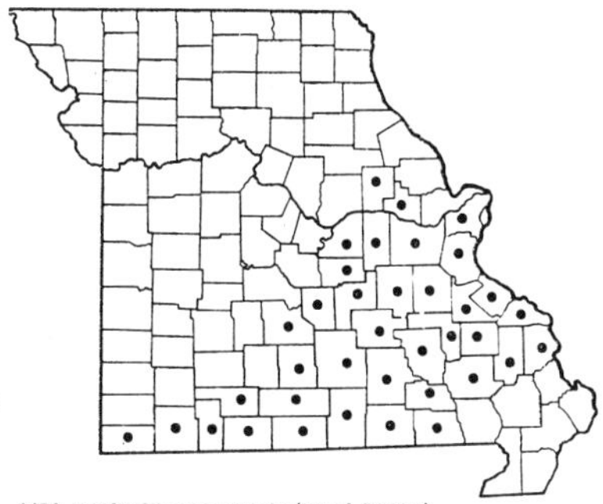

1451 Euphorbia commutata (Wood Spurge)

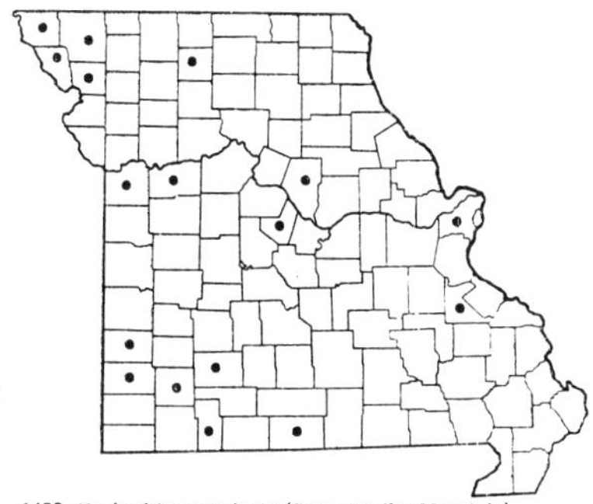

1452 Euphorbia marginata (Snow-on-the-Mountain)

Colorado, south to Virginia, Indiana, Illinois, and Missouri.

The plant spreads rapidly from the underground creeping and forking rootstocks, forming extensive colonies, and, if not checked constantly, can become a weed when planted as a rockery plant. The plant is rarely reported as producing seeds. The dull olive-yellow flower-clusters at the ends of finely leafy stems are attractive.

The roots are sometimes used in medicines as a purgative, but cases of poisoning have been reported where an excess amount was taken.

7. **Euphorbia Esula** L. Leafy Spurge Map 1450
Also called Wolf's Milk.
Flowers May–September.

Occurs in pastures and fields. Known only from central Missouri in Chariton County (upland pasture along west side of route 11, T56N, R20W, sect. 25, 2½ mi. southwest of Rothville, May 21, 1949, *Steyermark 68051*).

Native of Europe; introduced and naturalized in North America from Quebec to Alberta, south to New England and Maryland, Pennsylvania, Indiana, Illinois, Missouri, Nebraska, Kansas, Colorado, and the Pacific states.

Since it spreads rapidly, this is considered a dangerous weed to get started anywhere. Its roots are reported to penetrate to a depth of 5–15 feet. Livestock usually avoid the plant because of its bitter taste, but in fields badly infested with the plant, horses have been reported as having suffered blistering of the skin and hair loss from their ankles where these parts were in contact with the milky sap of the plant.

8. **Euphorbia commutata** Engelm.
Wood Spurge Map 1451
Flowers April–June.

Occurs in rich or rocky woodland, in ravines, valleys, and low ground along streams. Ozark region of southern and east-central Missouri, south and east of a line from St. Louis, Warren, Boone, Osage, Maries, Pulaski, Laclede, Wright, and Christian counties to McDonald County.

Ranges from Pennsylvania to Minnesota, south to Florida, Alabama, Tennessee, Missouri, Oklahoma, and Texas.

This attractive spurge appears in early spring with bright green leaves and compact flower-clusters subtended by symmetrical rounded floral leaves. It is strictly a woodland species. Some red coloring is usually present in the young stems and leaves.

9. **Euphorbia marginata** Pursh
Snow-on-the-Mountain Map 1452
Flowers June–October.

Occurs on loess hills, fields, pastures, roadsides, and along railroads. Native on the loess mounds of northwest Missouri in Atchison, Holt, and Andrew counties, elsewhere in the state escaped from plantings in gardens.

Ranges from Minnesota to Montana, south to Missouri, Oklahoma, Texas, and New Mexico.

The plant is often prized as an ornamental annual for its showy white-margined leaves and floral bracts. It does well in dry, hot clayey soils.

The milky juice of the plant has an irritating effect on the skin of many individuals, causing a dermatitis, so that the plant should be handled with caution. Livestock may be poisoned from eating honey made from the nectar of this species.

10. **Euphorbia corollata** L. Flowering Spurge
Map 1453
Flowers May–October.

Occurs on prairies, rocky glades, ledges of bluffs, open woods, fallow fields, pastures, roadsides, and along railroads. Throughout Missouri, doubtless in every county.

The following variations are encountered in Missouri:

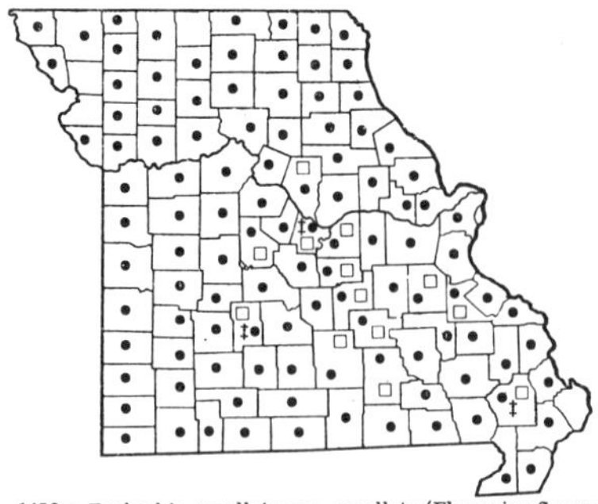

1453 • Euphorbia corollata var. corollata (Flowering Spurge)
1453 □ Euphorbia corollata var. mollis
1453 ‡ Euphorbia corollata var. paniculata

1454 Euphorbia Geyeri

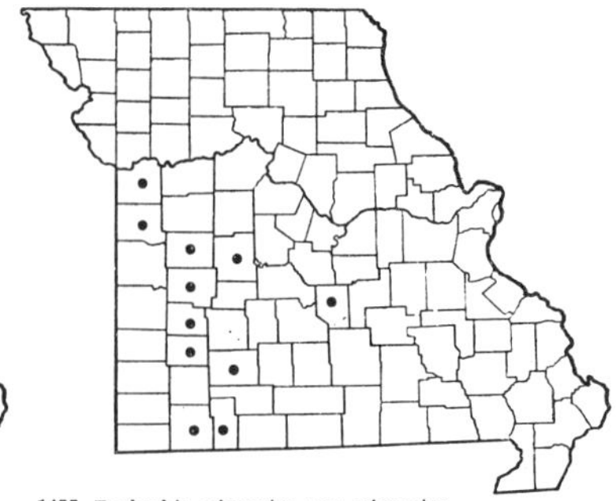

1455 Euphorbia missurica var. missurica

a. Most of the stem, or at least the upper half, and lower surface of leaves, soft-hairy . . . 10b. E. COROLLATA var. MOLLIS

a. Stems and leaves glabrous (without hairs) or essentially so . . . . . . . . . *b*

b. Cup-shaped cyathia, including the conspicuous white appendages, 7–10 mm. broad; cyathia on loosely forked inflorescences . . 10a. E. COROLLATA var. COROLLATA

b. Cup-shaped cyathia, including the conspicuous white appendages, 5–7 mm. broad; cyathia rather crowded . 10c. E. COROLLATA var. PANICULATA

10a. **Euphorbia corollata** var. **corollata** Map 1453

*Euphorbia corollata* L. [G, BB, P & S, Steyerm.]
*Euphorbia corollata* var. *angustifolia* Ell.

This is the commonest variation in Missouri, occurring throughout the state and doubtless in every county.

Ranges from Florida to Texas, north to New York, Ontario, Michigan, Wisconsin, Minnesota, and Nebraska.

Leaves on some plants are only 3–6 mm. broad throughout and could be classified as var. *angustifolia* Ell. However, there is considerable variation in leaf width even on the same plant, so that all gradations can be found.

10b. **Euphorbia corollata** var. **mollis** Millsp. Map 1453

This hairy extreme is most often found on exposed limestone rocky glades and barrens. Apparently restricted to the Ozark section of the state north to St. Francois, Washington, Osage, Boone, and Morgan counties, and west to Dallas County.

Ranges from Alabama to Texas, north to Virginia, North Carolina, Indiana, Missouri, and Oklahoma.

10c. **Euphorbia corollata** var. **paniculata** (Ell.) Boiss. Map 1453

Known from Cole (Jefferson City, August, 1871, *Dr. O. Krause*) and Dallas (near Little Niangua River, $2\frac{1}{2}$ mi. northwest of Tunas, August 5, 1937, *Steyermark 24207*) counties.

Ranges from Georgia and Alabama to Texas, north to Virginia, Indiana, Illinois, Missouri, and Oklahoma.

*Euphorbia corollata* is a showy perennial species with conspicuous white petal-like appendages. It is quite variable in pubescence of stems and leaves, shape and size of leaves, amount of branching, and abundance of flowers. Some authors (Gleason, Deam) regard the variations in pubescence as response to the environment, the hairier plants correlated with drier, more exposed situations, those with loose, irregular inflorescences and few involucres with a shaded habitat. In Missouri the hairier extremes appear to be correlated with the dry open rocky limestone glades and similar habitats, and the relative amount of light does appear to play a definite role in the amount of branching of the inflorescence and abundance of flowers. The true interpretation of these variations must be left for future field and experimental studies.

When included in hay, this species may sometimes cause poisoning when fed to livestock. The flowers, fruits, and leaves are eaten by wild turkey.

11. **Euphorbia Geyeri** Engelm. Map 1454

Flowers July–October.

Occurs in sandy open prairies. Known only from northeastern Missouri in Clark County (sandy prairie

Plate no. 237. **1.** Euphorbia missurica, $\times {}^{2}/_{5}$. **2.** Euphorbia marginata, $\times {}^{2}/_{5}$ (Scribner's); Details from Small, The New York Botanical Garden. **3.** Euphorbia commutata, $\times {}^{2}/_{5}$. **4.** Euphorbia corollata, $\times {}^{2}/_{5}$.

PLATE NO. 237

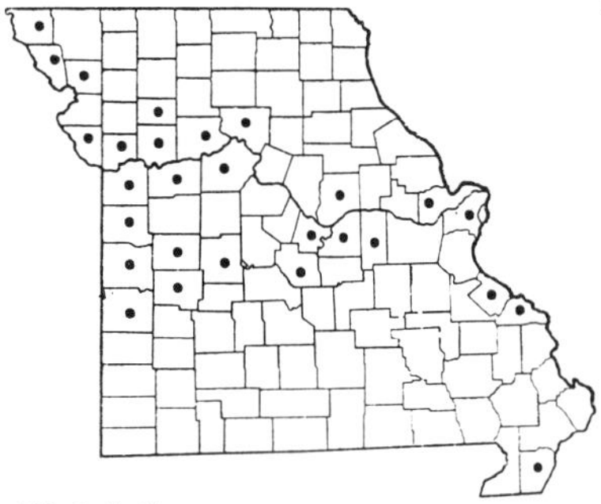

1456 Euphorbia serpens

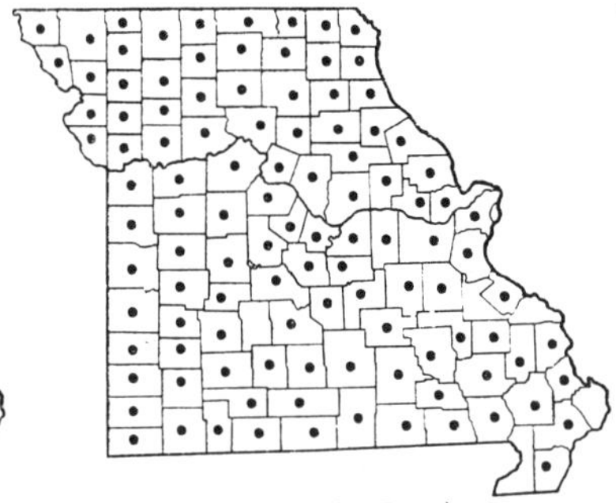

1457 Euphorbia maculata (Nodding Spurge)

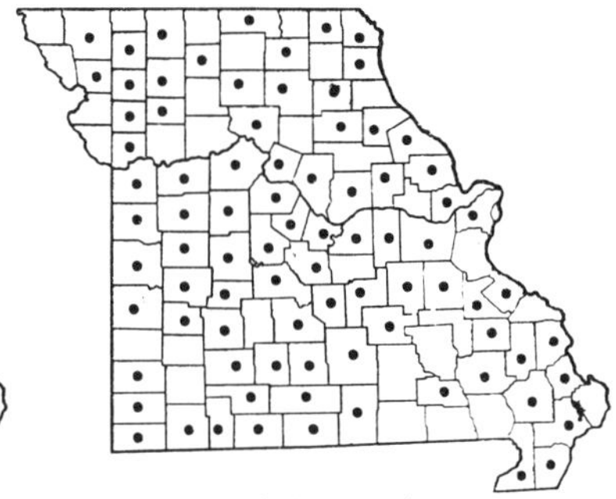

1458 Euphorbia supina (Milk Purslane)

on slopes of old river terrace on beach paralleling Des Moines River, T65N, R6W, sect. 9, 1½ mi. southeast of St. Francisville, August 13, 1949, *Steyermark 68876*).

Ranges from Wisconsin to North Dakota and Colorado, south to Indiana, Illinois, Missouri, Nebraska, Texas, and New Mexico.

12. **Euphorbia missurica** Raf. var. **missurica**

Prairie Spurge Map 1455

*Euphorbia zygophylloides* Boiss. [P & S, Steyerm.]

*Chamaesyce missurica* (Raf.) Shinners [Shinners]

Flowers late May–September.

Occurs on rocky limestone glades, ledges of south- and west-facing limestone bluffs, and rocky open ground. Western Ozark and unglaciated prairie region in western Missouri south of the Missouri River east to Pulaski County, north to Benton, Henry, and Jackson counties.

Ranges from Missouri and Kansas to Arkansas, Oklahoma, and Texas.

This is a very slender-stemmed species with narrow leaves and hairlike peduncles 5–11 mm. long. The flowers are conspicuous because of the showy petal-like, white to pinkish, prolonged appendages of the involucre.

*Euphorbia missurica* var. *intermedia* (Engelm.) L. C. Wheeler has thickish peduncles only 2–4 mm. long, angled seeds, and linear leaves which are truncate and mostly emarginate at the apex instead of obtuse or retuse as in var. *missurica*. Although several collections in Missouri have been identified as this variety, they all appear to be referable to the typical var. *missurica*.

13. **Euphorbia serpens** HBK. Map 1456

Flowers late May–October.

Occurs in low, moist, alluvial soils in open ground bordering sloughs, oxbow lakes of river meanders, spring branches, and the streams of the Missouri, Mississippi, Osage, and tributaries. Western, central, and eastern Missouri, northwest from Atchison County south to Vernon County, east to St. Charles County along the Missouri River, and south along the Mississippi River to Pemiscot County; absent from most of the Ozark region and most of northern Missouri.

Ranges from Ontario to Montana, south to Alabama, Louisiana, Texas, New Mexico, and Mexico; also in Central and South America.

This is a very distinct species, easily recognized in the field by the very slender creeping stems, with mats of gray-green, very small leaves without teeth, rather conspicuous white petal-like appendages of the involucres, and the entirely glabrous plant. Plants often form dense flat spreading masses over the wet soil.

14. **Euphorbia maculata** L. Nodding Spurge Map 1457

*Euphorbia nutans* Lag. [P & S]

*Euphorbia Preslii* Guss. [BB]

*Chamaesyce maculata* (L.) Small [Shinners]

Flowers late May–October.

Occurs in thickets, fallow and cultivated fields, pastures, open woodland, roadsides, and along railroads. Throughout Missouri and doubtless in every county.

Ranges from Florida to Texas and Mexico, north to New Hampshire, New York, Ontario, Michigan, Wisconsin, Minnesota, and North Dakota.

The leaves have a habit of folding late in the day. This species can poison livestock when eaten alone or in hay containing the plants. It more frequently causes cases of poisoning than other species of the genus. The young leaves and buds are eaten by wild turkey.

Plate no. 238. **1.** Euphorbia serpens, × 2/5. **2.** Euphorbia maculata, × 2/5. **3.** Euphorbia supina, × 2/5.

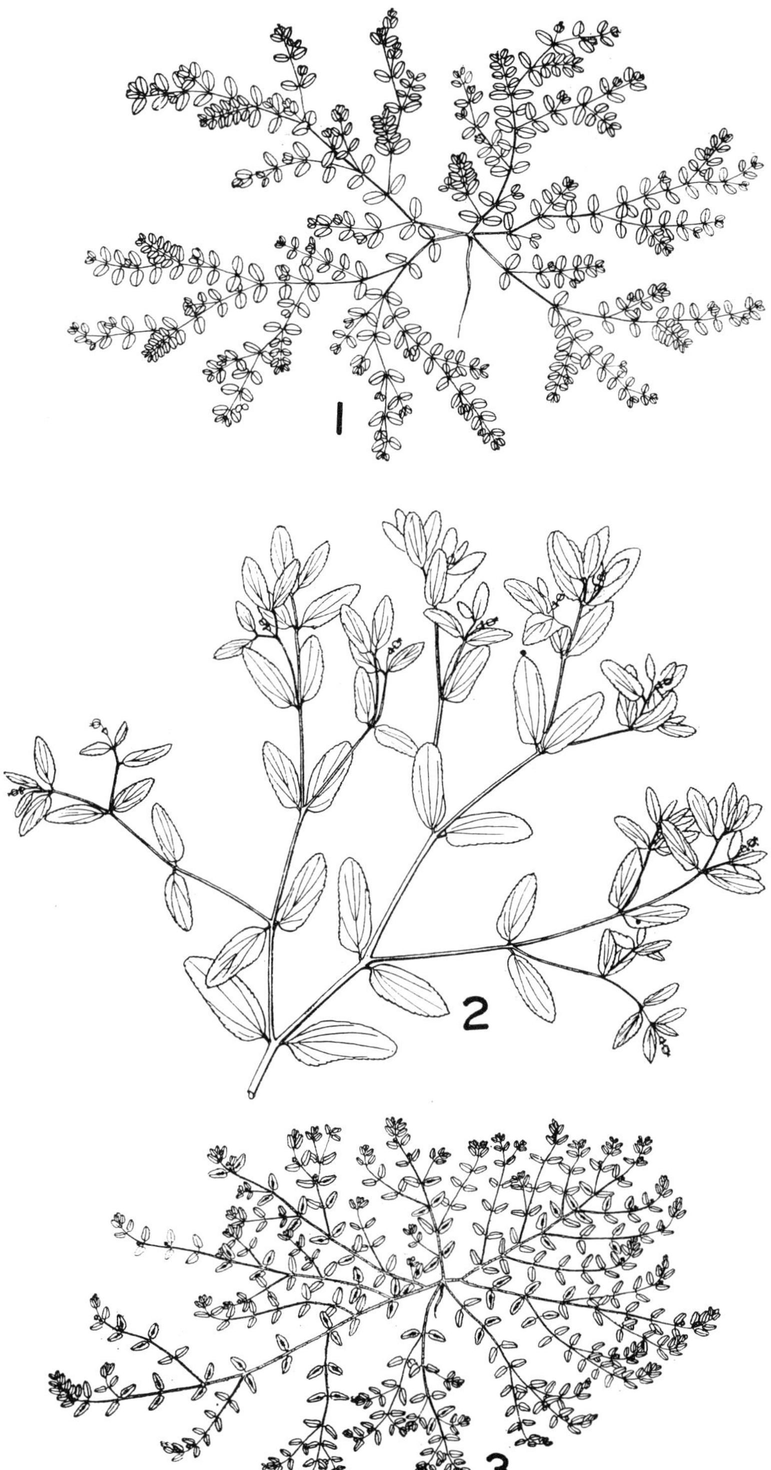

PLATE NO. 238

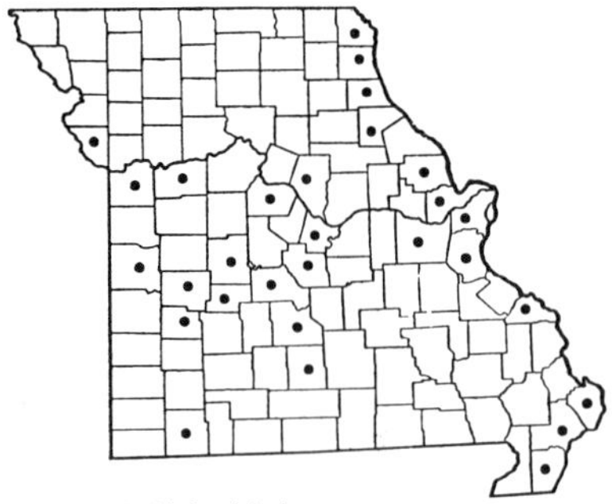

1459 Euphorbia humistrata

1460 Euphorbia stictospora

1461 Euphorbia prostrata

15. **Euphorbia supina** Raf. Milk Purslane Map 1458

*Euphorbia maculata* of auth. [BB, P & S], not L.

*Chamaesyce supina* Raf. [Shinners]

Flowers late May–October.

Occurs in cultivated, waste, or open ground, along sidewalks, streets, roadsides, railroads, rocky glades, most frequently in dry soils, but also in wet ground. Throughout Missouri, doubtless in every county.

Ranges from Quebec and Ontario to North Dakota, south to Florida and Texas; introduced in the western states in Oregon, California, and Arizona.

This is one of the commonest weeds of the eastern half of the United States. Usually a low prostrate plant, but sometimes with ascending stems, it forms large mats up to nearly a meter in diameter. The leaves, which vary from dull green to reddish-purple, often have a reddish spot somewhere near the center. As in other members of the genus, the milky juice is poisonous if eaten, but since the plant grows close to the ground, it is not commonly grazed by cattle.

16. **Euphorbia humistrata** Engelm. Map 1459

Flowers June–October.

Occurs in moist, low, alluvial ground bordering sloughs, ponds, oxbow meanders, and the larger streams of the Missouri, Mississippi, Osage, and certain tributaries. Mainly in eastern and central Missouri along the Missouri and Mississippi rivers, in the Ozark region along the Gasconade and Osage rivers, and in southwestern and unglaciated prairie sections along the White, South Grand, and tributaries.

Ranges from Ohio to Illinois, Missouri, and Kansas, south to Alabama, Louisiana, and Texas; introduced in Virginia.

Although often difficult to distinguish from *E. supina* and related species, this species is usually well-marked in the field by the mainly larger leaves with paler silvery green or glaucous lower surface, the usually paler green stems, the tendency to have the stems and leaves with fewer hairs, and the much less noticeably toothed leaves. The stems are often ascending as well as prostrate.

17. **Euphorbia stictospora** Engelm. Map 1460

Flowers July–October.

Known only from Jackson County, west-central Missouri.

Ranges from Kansas, Nebraska, South Dakota, and Wyoming, south to Texas, Colorado, New Mexico, Arizona, and Mexico; introduced in western Missouri.

18. **Euphorbia prostrata** Ait. Map 1461

*Euphorbia malaca* Small

*Euphorbia Chamaesyce* sensu Wheeler [G, BB], not L.

*Chamaesyce prostrata* (Aiton) Small [Shinners]

Flowers June–October.

Known only from west-central Missouri in Jackson County (Courtney, *Bush 8630* in N. Y. Bot. Gard. Herb.).

Native of tropical America, from South and Central America, to Mexico, and the West Indies; naturalized in the United States from Florida to Texas, north to Virginia, Missouri, and Oklahoma.

For a discussion of the various interpretations of the name of this species, the reader is referred to Rhodora 43: 265. 1941 and Bull. Torr. Bot. Club 72: 312–18. 1945.

19. **Euphorbia serpyllifolia** Pers. Map 1462

Flowers July–October.

Known only from west-central Missouri in Jackson County (dry ground, Courtney, July 10, 1930, *Bush 11824*).

Plate no. 239. **1.** Euphorbia humistrata; *a.* Portion of plant, × 3/4; *b.* Leaf, × 1 1/2; *c.* Fruit and cyathium, × 15. **2.** Euphorbia stictospora; *a.* Fruit and cyathium, × 15; *b.* Seed, × 12; *c.* Leaf, × 3; *d.* Portion of plant, × 1 1/2.

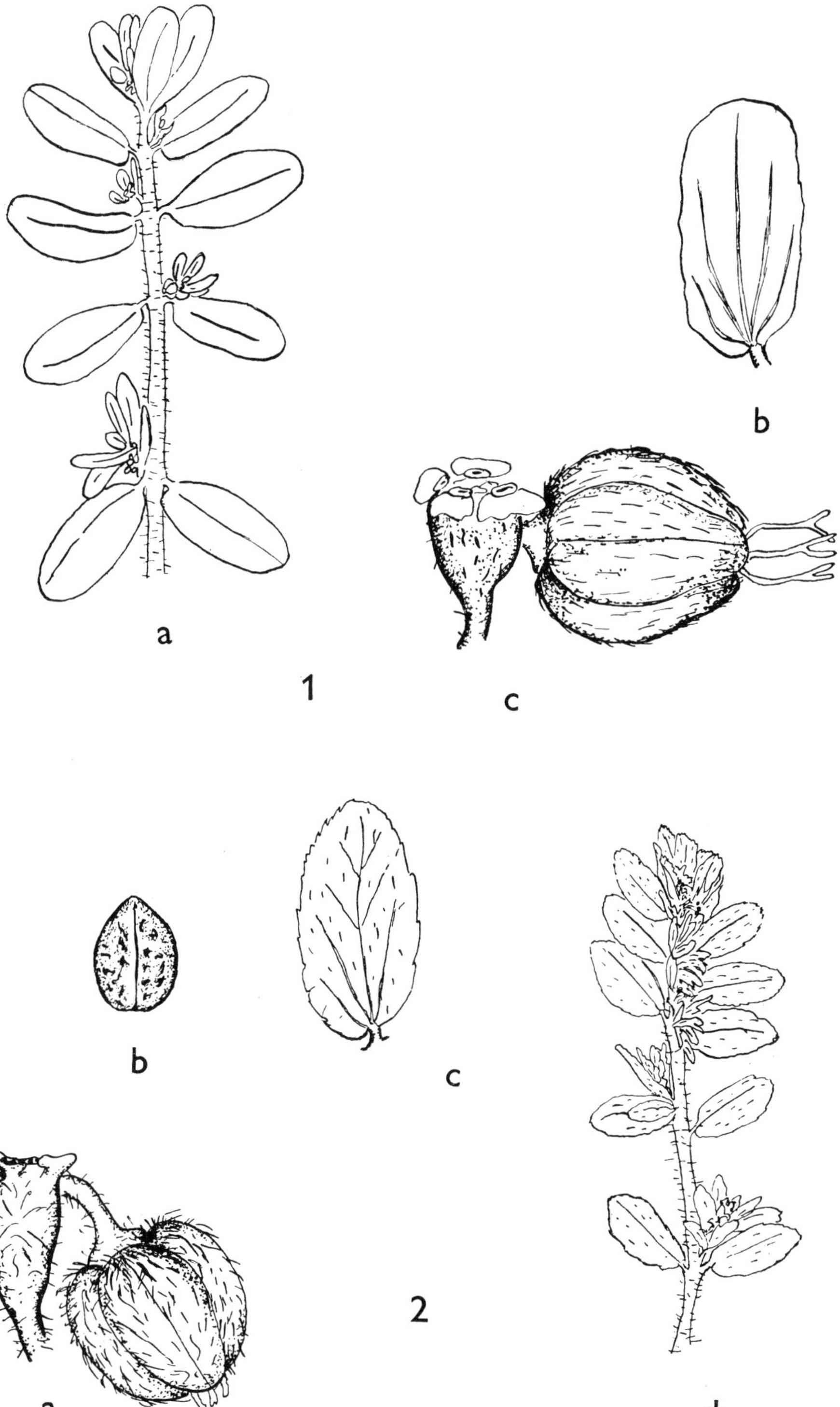


PLATE NO. 239

1462 Euphorbia serpyllifolia

1463 Euphorbia glyptosperma

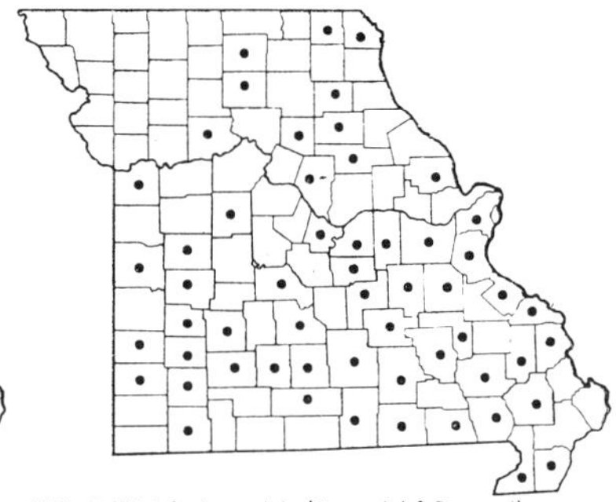

1464 Callitriche terrestris (Terrestrial Starwort)

Ranges from Michigan, Minnesota, and Iowa to British Columbia, south to Missouri, Oklahoma, Texas, New Mexico, California, and Mexico.

Wheeler (Rh. 43: 233. 1941) has the following to remark concerning this species in Missouri: 'Bush's collection of this species from Missouri is not necessarily to be taken as evidence that the species grows there naturally. From this and several other collections of *Euphorbia* made by Bush at Courtney, his home town, I have been led to wonder if he were not careless with the sweepings from his herbarium and consequently found many waifs in this backyard.'

20. **Euphorbia glyptosperma** Engelm. Map 1463
Flowers June–October.

Occurs on sand bars and moist alluvial ground along the Missouri and Mississippi rivers. Northern and central Missouri from Atchison to Jackson counties in northwestern and west-central Missouri, and from Clark to Jefferson counties in northeastern and east-central Missouri; doubtless in all the counties bordering the Missouri River.

Ranges from Quebec and New Brunswick to British Columbia, south to New York, Ontario, Ohio, Indiana, Illinois, Missouri, Oklahoma, Texas, Mexico, and California.

*Excluded Species*

**Mercurialis annua** L. Mercury

This species was included previously for Missouri in Palmer and Steyermark's *Annotated Catalogue* (p. 588), but should be eliminated. The record is based solely upon a cultivated specimen grown at the Missouri Botanical Garden (St. Louis at Missouri Botanical Garden, September 15, 1906, *Kellogg* in Mo. Bot. Gard. Herb.).

## Fam. **CALLITRICHACEAE** (Water Starwort Family)

### **Callitriche** L. Water Starwort

Plants growing on soil, on land out of water; fruit on a short stalk 0.2–0.6 mm. long, broader than high; flowers without bracts; leaves and stems without shield-shaped scales; leaves uniform . . . 1. C. TERRESTRIS

Plants growing in water, entirely submerged or with a rosette of floating leaves, or stranded on mud; fruit sessile, as high as broad or somewhat higher than broad; flowers with 2 bracts at base; leaves and stems with shield-shaped scales; leaves of different shapes on the same plant. . . . . 2. C. HETEROPHYLLA

1. **Callitriche terrestris** Raf. emend. Torr.
Terrestrial Starwort Map 1464
*Callitriche deflexa* A. Br. var. *Austini* (Engelm.) Hegelm. [G, BB, P & S, Steyerm.]
Flowers April–July.

Occurs on damp soil, where somewhat acid, of paths, borders of ponds and swales, shaded ground bordering low woodland, damp open places in upland woods, fallow fields, and waste ground. Common throughout southern, central, and eastern Missouri,

Plate no. 240. **1.** Euphorbia prostrata; *a.* Fruit and cyathium, × 15; *b.* Seed, × 12; *c.* Leaf, × $1^1/_2$; *d.* Portion of plant, × $^3/_4$. **2.** Euphorbia glyptosperma; *a.* Fruit and cyathium, × $11^1/_4$; *b.* Leaf, × $1^1/_2$; *c.* Seed, × 12; *d.* Portion of plant, × $^3/_4$. **3.** Callitriche terrestris; *a.* Portion of plant, × 3; *b.* Habit of plant, × $^3/_4$; Details from Small, The New York Botanical Garden. **4.** Callitriche heterophylla, × $^3/_4$.

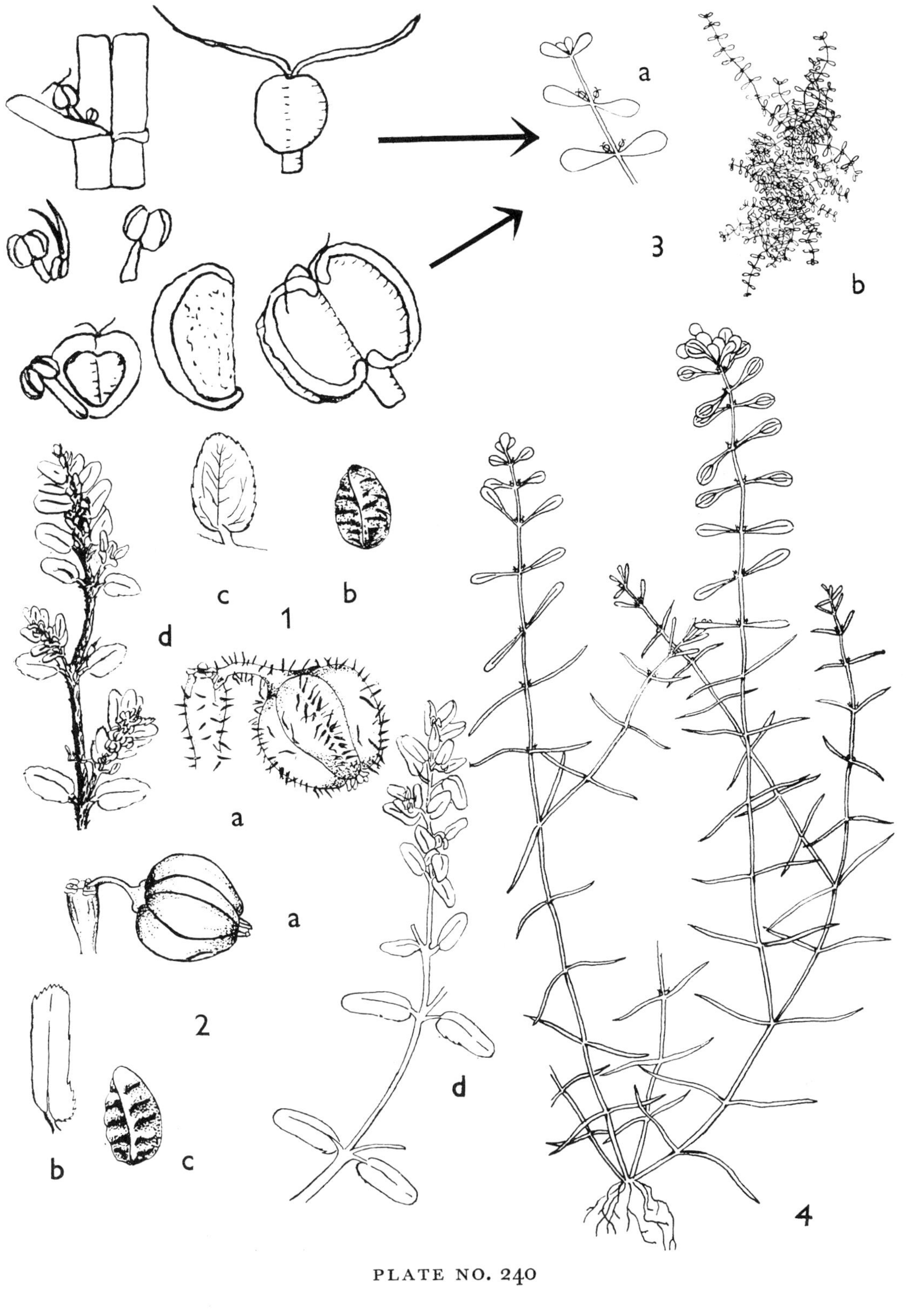


PLATE NO. 240

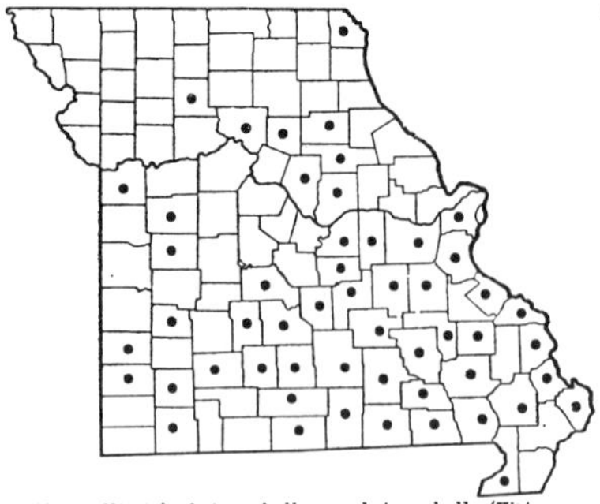

1465 Callitriche heterophylla var. heterophylla (Water Starwort)

1466 Floerkea proserpinacoides (False Mermaid)

1467 Cotinus obovatus (American Smoke Tree)

north and west to Clark, Sullivan, Linn, Carroll, and Jackson counties.

Ranges from Massachusetts, New York, and Pennsylvania, to Ohio, Indiana, Illinois, and Missouri, south to Virginia, West Virginia, Alabama, Louisiana, and Oklahoma.

2. **Callitriche heterophylla** Pursh emend. Darby var. **heterophylla** Water Starwort Map 1465
*Callitriche heterophylla* Pursh [G, BB, P & S, Steyerm.]
Flowers April–December.

Occurs in ponds, streams, springs, and spring branches. Common throughout the Ozark region in southern and part of central Missouri, north locally to Clark, Monroe, Randolph, Livingston, and Jackson counties.

Ranges from Nova Scotia, Maine, New York, Ontario, Ohio, Indiana, Illinois, Wisconsin, Iowa, and Kansas, south to Florida and Texas; in the Pacific states from Montana to Washington, south to Arizona and California; also in Mexico, Central and South America, and the West Indies.

Another variety in the western United States is var. *Bolanderi* (Hegelm.) Fassett.

The submerged leaves of this plant are somewhat needlelike, the terminal floating leaves obovate and rounded at the summit; sometimes on plants stranded on the mud, all the leaves are oblong. The plants are eaten by various ducks.

## Order **SAPINDALES**

### Fam. **LIMNANTHACEAE** (False Mermaid Family)

#### **Floerkea** Willd. False Mermaid

**Floerkea proserpinacoides** Willd.
False Mermaid Map 1466
Flowers April–May.

Occurs in low alluvial woodland in valleys of streams near the Missouri River or tributary streams of that river. Central Missouri, where known from St. Charles (St. Charles, *Egeling*), Howard (Thrall's Plantation near Franklin, May 8, 1820, *James* in N. Y. Bot. Gard. Herb.), and Moniteau (alluvial silver maple-American elm woods on south side of North Moreau Creek, T44N, R15W, eastern part sect. 10, 3¼ mi. south southeast of south boundary of California, May 6, 1957, *Steyermark 84064*) counties.

Ranges from Nova Scotia and Quebec to Vermont, Ontario, Minnesota, and North Dakota, south to Delaware, Virginia, Tennessee, and Missouri.

A report on the rediscovery of this species in Missouri after the original collection by James in 1820 was given in Brittonia 10: 150–52. 1958 by the present author. The plants are pale green, delicate annuals, with weak, sprawling or decumbent stems 1–3 dm. tall. They usually occur in dense colonies and resemble mats of *Galium Aparine* or *Chaerophyllum procumbens*, often occurring with these species. The alternate, pinnately lobed leaves are deeply divided into 3–7 narrow lobes 1–2 cm. long. The peduncles, which arise from the upper axils, at first are about the same length as the petioles, but eventually surpass them in fruit. The minute whitish flowers consist of 3 sepals 3 mm. long, 3 oblong petals which are only 2 mm. long, 6 stamens, 2 or 3 ovaries united only at the base, and 2 or 3 stigmas. The fruits are deeply divided, separating at maturity into 3 fleshy achenes.

Since this plant matures and dries up quickly with the advent of warm weather in late spring, it is likely that it has either been overlooked or collecting has occurred too late in the areas where it occurs. Intensive collecting throughout northern and central Missouri in early spring in alluvial woodland habitats should reveal numerous additional stations for this presently rare species.

## Fam. **ANACARDIACEAE** (Cashew Family)

Leaves undivided, simple . . . . . . . . . . . . . . . . . . . . . 1. Cotinus
Leaves compound, divided into 3 or 5–31 leaflets . . . . . . . . . . . . . . . 2. Rhus

### 1. **Cotinus** Adans. Smoke Tree

**Cotinus obovatus** Raf. American Smoke Tree
Map 1467

*Cotinus americanus* Nutt. [P & S, Steyerm.]

Flowers May.

Occurs on limestone glades and bald knobs, and wooded rocky limestone bluffs. Southwestern Missouri, in the White River and North Fork and tributaries in Stone, Taney, Ozark, and Douglas counties, locally north in Polk County along top of south-facing bluffs bordering the Pomme de Terre River, ½ mi. north of Burns (*Steyermark 13595*, July 17, 1934).

Ranges from Alabama and Tennessee, southwestern Missouri, northwestern Arkansas, eastern Oklahoma, to central Texas.

The foliage of this small tree turns yellow in the autumn. The bluish-gray smokelike appearance of the fruiting sprays, as seen from a distance, account for the common name of Smoke Tree. The wood, which is very durable in contact with the soil, is largely used for fence posts. It also yields an orange-colored dye.

### 2. **Rhus** L. Sumach, Poison Ivy

a. Leaves divided into 3 leaflets . . . . . . . . . . . . . . . . *b*
- b. Non-poisonous plants; leaves fragrant when bruised; middle (terminal) one of the three leaflets without any stalk (petiolule) before joining main leaf-stalk (petiole) or a stalk (petiolule) less than 3 cm. long; flowers yellow, in crowded clusters appearing before or with the leaves; mature fruit red, covered with hairs or glands . . . . . . . . . . . . 4. R. aromatica
- b. Poisonous plants; leaves not fragrant; middle (terminal) one of the three leaflets on a long stalk (petiolule) more than 3 cm. long before joining main leaf-stalk (petiole); flowers greenish, greenish-white, or yellowish-green, in loose open clusters appearing with grown leaves; mature fruit grayish white or yellowish white, greenish or buff, glabrous (without hairs) or inconspicuously short-hairy or papillate-roughened . . . . . . . . . . . . *c*
  - c. Common poisonous plant occurring in every county; middle (terminal) one of the three leaflets with an acute to acuminate (short- to long-pointed) tip; leaflets coarsely or wavy toothed or without any teeth; stems often climbing, mostly more than 6 dm. tall; aerial rootlets present when climbing; leaves alternately scattered along the stem; fruit usually glabrous (without hairs) . . . . . . . . . . . . 5. R. radicans
  - c. Rare poisonous plant known only from extreme southern Missouri, in Scott, Mississippi, and Ozark counties; middle (terminal) one of the three leaflets with a rounded or blunt (obtuse) tip; leaflets with usually 2–4 lobes or large teeth on each side; stems never climbing, only 0.5–6 dm. tall, without aerial roots; leaves mostly clustered near the summit of stem; fruit minutely short hairy and with papillate (warty) projections. . . . . . . 6. R. Toxicodendron

a. Leaflets pinnately divided into 5–31 leaflets . . . . . . . . . . . . *d*
- d. Branches and leaf-stalks (petioles) densely velvety-hairy with spreading and obvious hairs; leaves finely dissected, bipinnatifid or bipinnate; introduced from cultivation. . . . 1. R. typhina f. dissecta
- d. Branches and leaf-stalks (petioles) glabrous (without hairs) or with a minute hairiness; leaves not finely dissected, once-pinnate; native shrubs or trees . . . . . . . . . . . . *e*

1468 Rhus typhina f. dissecta (Dissected Staghorn Sumac)

1469 Rhus glabra var. glabra (Smooth Sumac)

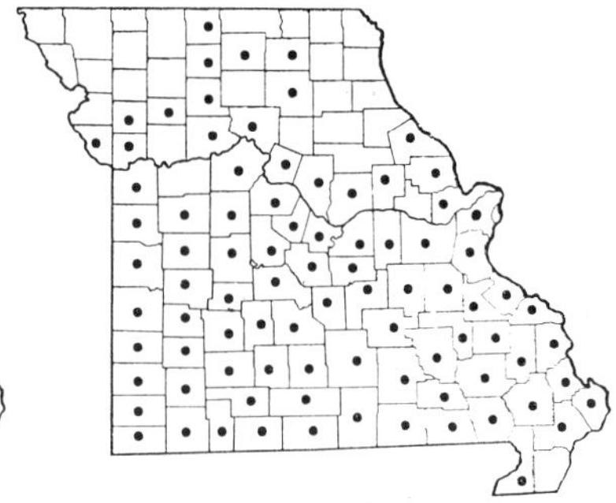

1470 Rhus copallina var. latifolia (Dwarf Sumac)

e. Branches and leaf-stalks glabrous (without hairs); leaflets toothed; main leaf-axis not winged in the spaces between the leaflets . . . . . . . . . . . . . . . . 2. R. GLABRA

e. Branches and leaf-stalks covered with a minute hairiness; leaflets smooth-edged (not toothed); main leaf-axis winged in the spaces between the leaflets . . . . . . . . . . 3. R. COPALLINA

1. **Rhus typhina** L. f. **dissecta** Rehd.
Dissected Staghorn Sumac Map 1468
Flowers June–July.

Commonly planted as an ornamental shrub and rarely escaping from cultivation to railroads. Known only from St. Louis County (St. Louis, Lindenwood [Frisco] Freight Yard on the embankment of northeast corner of Fyler Bridge, June 9, 1956, *Muehlenbach 930*).

Typical *R. typhina* f. *typhina* is native from Nova Scotia and Quebec to Minnesota, south to North Carolina, West Virginia, Kentucky, Illinois, and Iowa. It is not found wild in Missouri, so far as known according to previous exploration of the state, but is often cultivated.

2. **Rhus glabra** L. var. **glabra** Smooth Sumac
Map 1469
Flowers late May–July.

Occurs in upland prairies, thickets, abandoned fields, borders and openings of woods, roadsides, and along railroads. Throughout Missouri in every county.

Ranges from Maine and Quebec to British Columbia, south to Florida, Texas, and Mexico.

The greenish-yellow flowers, closely arranged in a dense pyramid-shaped narrow panicle, are followed by clusters of bright red fruits. In early autumn the upper surface of the leaves assume a bright rose red to vermilion red color contrasting with the silvery-glaucous lower surface. As in the case of the fragrant sumac (*R. aromatica*), pink lemonade can be prepared from the sourish fruits. Formerly, the leaves and berries were used in medicine for various diseases and afflictions. The Indians used this sumac in various ways. They scraped out the pith from the twigs to use the stems as flutes and prepared a red dye from the berries for coloring wool. The leaves were mixed with those of tobacco as a smoke. The leaves and twigs, which contain an abundance of tannin, were used by country people for tanning leather.

The plants are browsed by white-tailed deer, and the fruits are eaten by wild turkey.

The shrub forms dense thickets by virtue of the underground root suckers. Various spellings—Sumach, Shumac, Shumack, Summaque, and Shoemake—and consequent pronunciations are used for this group of plants, said to have had origin in an Arabic name for a Mediterranean species of the genus.

The Missouri plants all have glabrous branches. A more northern variety, var. *borealis* Britt., with short-hairy branches, is not known in Missouri, while plants with bipinnate leaves, also unknown in Missouri at this time, are known as var. *laciniata* Carr.

3. **Rhus copallina** L. var. **latifolia** Engler
Dwarf Sumac Map 1470
Also called Winged Sumac, Shining Sumac.
Flowers late May–November.

Occurs in prairies, thickets, open woods, rocky sandstone, chert, and granite glades, borders of woodland generally in acid soils, fallow and abandoned fields, roadsides and along railroads. Common

Plate no. 241. **1.** Floerkea proserpinacoides, × 2/5; *a.* Flower, × 2 2/5; After Gleason, The New York Botanical Garden. **2.** Cotinus obovatus, × 2/5; *a.* Fruit with fruiting pedicel, × 2/5; *b.* Flowering branch, × 2/5. **3.** Rhus glabra, × 2/5; *a.* Fruit, × 2/5. **4.** Rhus copallina var. latifolia, × 2/5; *a.* Fruiting branch, × 2/5; All details from Small, The New York Botanical Garden.

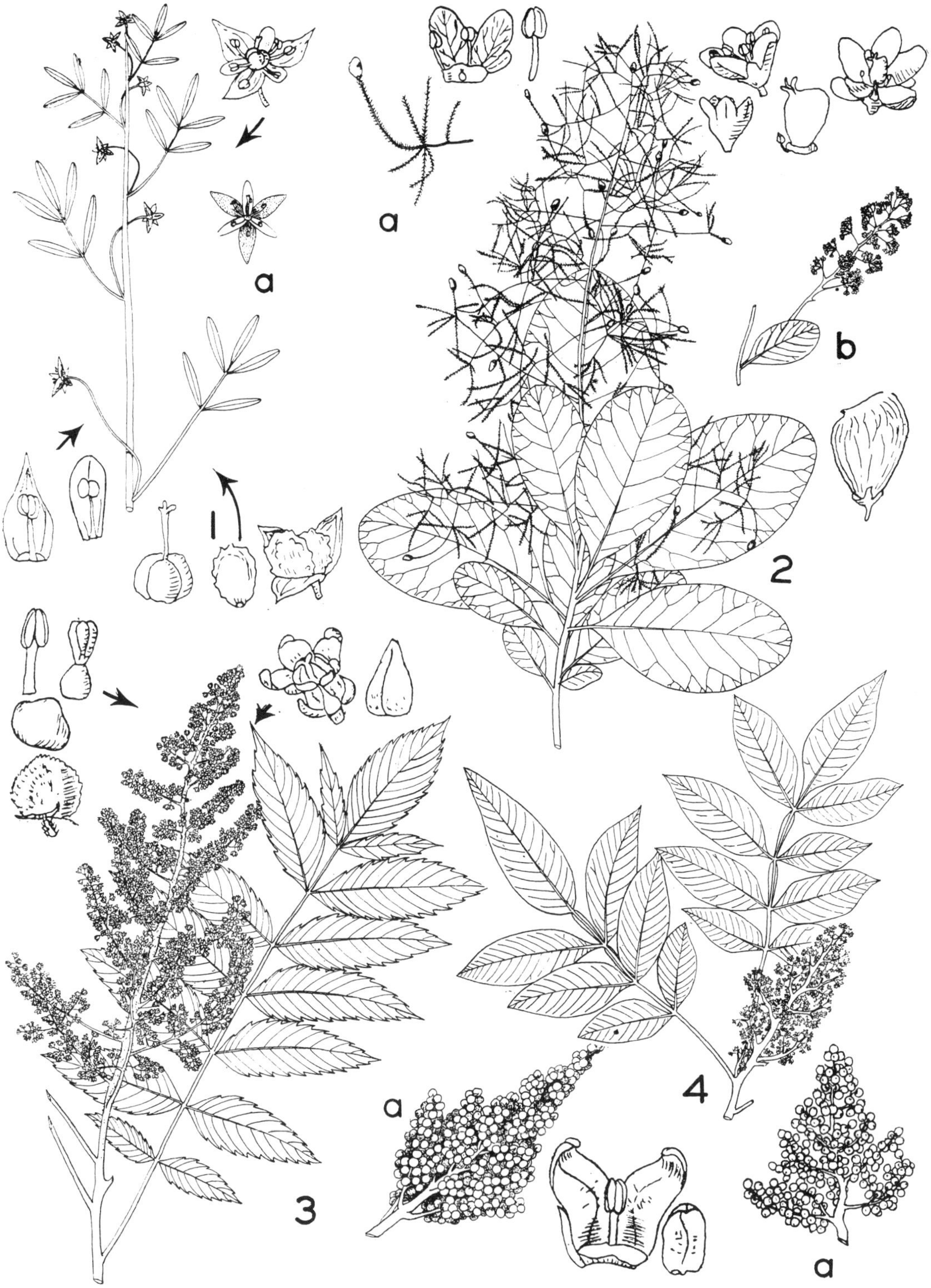


PLATE NO. 241

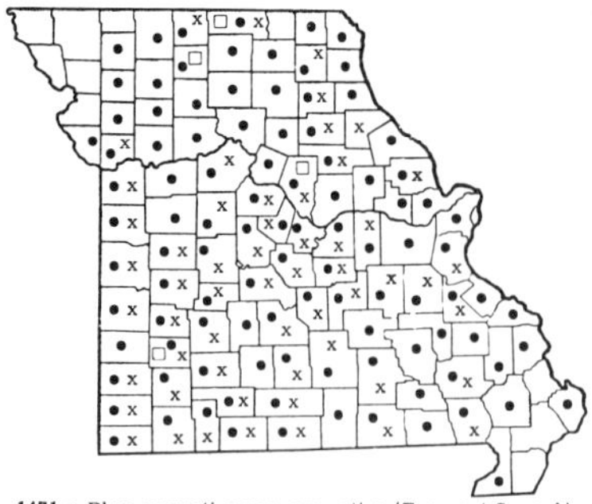

1471 • Rhus aromatica var. aromatica (Fragrant Sumach)
1471 □ Rhus aromatica var. illinoensis
1471 x Rhus aromatica var. serotina

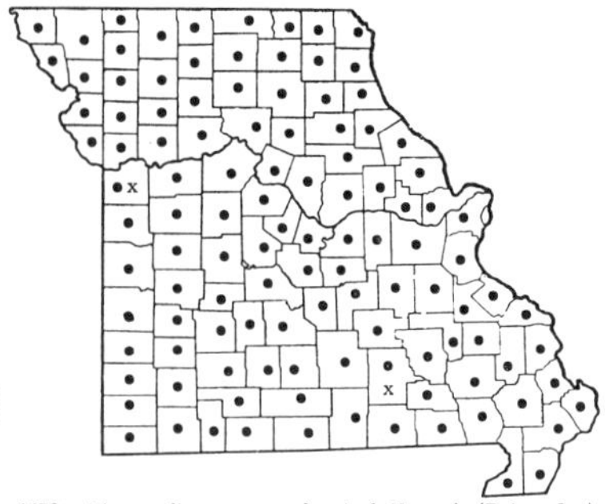

1472 • Rhus radicans var. vulgaris f. Negundo (Poison Ivy)
1472 x Rhus radicans var. radicans f. hypomalaca (Poison Ivy)

1473 Rhus Toxicodendron f. Toxicodendron (Poison Oak)

throughout southern and central Missouri, northward to Pike, Adair, Sullivan, Worth, Caldwell, Clinton, and Platte counties; absent from most of the northeastern and northwestern counties.

Ranges from Maine to Michigan and Illinois, south to Georgia, Alabama, Louisiana, and Texas.

The Missouri material all appears to fall within the limits of var. *latifolia,* distinguished by the leaves with 5–13 broadly oblong to narrowly ovate leaflets 1.5–4 cm. broad, which are strongly rounded at the base on the upper side, as contrasted with typical var. *copallina,* which has 11–23 narrower lance- or linear-oblong leaflets only 1–2 cm. broad and narrowed or attenuate to the base. For discussion, the reader is referred to the article by Fernald and Griscom (Rh. 37: 167–68. 1935).

The foliage of this sumac in autumn becomes a vermilion-red or flame-orange hue. Plants average somewhat shorter than those of *R. glabra,* generally 1.5–2.5 m. (4–7½ feet) tall, those of *R. glabra* usually 1.5–7 m. (4–22 feet) tall. The greenish-yellow to pinkish petals are followed by red sourish fruits, but these are inferior for a pink lemonade preparation compared with those of *R. aromatica* or *R. glabra.*

Like other species of the genus, this shrub makes a desirable ornamental shrub because of its dark green shining summer foliage and brilliant display of autumn color. The stems and foliage are eaten by white-tailed deer.

4. **Rhus aromatica** Ait. Fragrant Sumac Map 1471

Also called Pole-cat Bush, Stinking Hazel.

The following variations are encountered in Missouri, but intergrade with one another:

a. Plant when in flower leafless, the flowers appearing before the leaves; tip of terminal leaflet (middle one of three) acutely pointed . . . *b*
  b. Commonly encountered type; lower surface of leaflets sparsely hairy or nearly hairless (glabrate) . 4a. R. AROMATICA var. AROMATICA
  b. Rarely encountered type; lower surface of leaflets densely hairy or soft-velvety . . 4b. R. AROMATICA var. ILLINOENSIS
a. Plant in flower accompanied by partly grown or more than half grown leaves; tip of terminal leaflet (middle one of three) rounded or blunt 4c. R. AROMATICA var. SEROTINA

4a. **Rhus aromatica** var. **aromatica** Map 1471

*Rhus aromatica* Ait. [G, Steyerm.]

*Rhus canadensis* Marsh., not Mill. [P & S]

Flowers late March–April; fruits May–July.

Occurs in rocky or open woods, thickets, rocky open glades, along railroads, generally in acid soils. Common throughout Missouri; absent from the extreme northwestern and a few of the southeastern lowland counties.

Ranges from Quebec and Vermont to Michigan, Iowa, and Nebraska, south to Florida, Alabama, Mississippi, Louisiana, and Texas.

4b. **Rhus aromatica** var. **illinoensis** (Greene) Rehd. Map 1471

*Rhus canadensis* var. *illinoensis* (Greene) Fern. [P & S]

Flowers late March–April; fruits May–July.

Occurs along rocky bluffs of streams and on dry upland hills. Known only from Putnam (between Livonia and Unionville, *Palmer & Steyermark 41099*), Boone (high bluff above Hinkson Creek, Hermit Hill, south of Columbia, September 15, 1933, *Drouet 115*), and Dade (along rocky bluffs of Sac River, 4½ mi. southeast of Dadeville, May 25, 1953, *Palmer 55674*) counties.

Ranges from Indiana and Illinois to Kansas and Oklahoma.

The Palmer collection from Dade Co. has the leaflets varying from acute- to round-tipped and intergrades with var. *serotina.* Several collections previously identified as *R. aromatica* var. *illinoensis* have the terminal leaflet rounded at the apex and are referred in the present treatment to var. *serotina*; they include the following: *Palmer 39478* from rocky slopes and ledges, terraced dolomite hills, bald knobs, near Eagle Rock, June 3, 1931, from Barry Co.; *Palmer 5874* from dry limestone ledges, Branson, June 5, 1914, from Taney Co.; *Palmer 39496* from high rocky bluffs of Swan Creek, June 4, 1931, from Taney Co.; *Palmer 33093* from Pontiac, Ozark Co. A specimen from Johnson County (along sandstone bluffs near Warrensburg, June 10, 1953, *Palmer 55854*), with the leaflets lightly pubescent on both upper and lower surfaces, is transitional between typical *R. aromatica* var. *aromatica* and var. *illinoensis*, but more nearly approaches var. *aromatica*.

4c. **Rhus aromatica** var. **serotina** (Greene) Rehd. Map 1471

*Schmaltzia serotina* Greene
*Rhus canadensis* var. *serotina* (Greene) Palmer & Steyerm. [P & S]
*Rhus trilobata* var. *serotina* (Greene) Barkley [Steyerm.]

Flowers April–May; fruits June–August.

Occurs along ledges and top of rocky bluff escarpments of usually limestone, limestone glades and bald knobs, and thickets. Common in the Ozark and unglaciated prairie region of southern and central Missouri, also in eastern Missouri north of the Missouri River north to Knox County along rocky bluffs, locally in north-central Missouri in Putnam and Worth counties; absent from the rest of northern Missouri.

Ranges from Illinois and Iowa to Nebraska, south to Arkansas, Oklahoma, and Texas.

Most of the Missouri material referred to this variety has the lower surface of the leaflets sparsely hairy to nearly hairless (glabrate), but some specimens have the lower surface more densely or abundantly hairy and have been confused with var. *illinoensis*. Such specimens are those discussed above under var. *illinoensis* (*Palmer 39478, 39496, 33093, 5874, Steyermark 24365*), but have the terminal leaflet rounded at the tip and are, therefore, placed in var. *serotina*. These densely pubescent-leaved collections could pass as var. *arenaria* (Greene) Fern., but that variety includes low and depressed shrubs with the terminal leaflets only 1.5–3.5 cm. long, whereas in var. *serotina* the plants are upright, up to 2 m. tall, and the terminal leaflets are larger and 2.5–6 cm. long. These Missouri collections have the height and larger terminal leaflets of var. *serotina*, but show the dense leaflet pubescence characteristic of var. *arenaria*. The branchlets of these Missouri specimens which are intermediate between var. *arenaria* and var. *serotina* vary from glabrate to sparsely tomentulose, but the character of tomentulose branches, formerly thought to be characteristic of var. *arenaria*, has been shown by Deam (*Shrubs of Indiana*, second ed., p. 188. 1932) to be unreliable. The co-type specimen of *Schmaltzia serotina* (*Bush 148* from Independence, Jackson Co., May 6, 1894) has the branchlets tomentulose.

*Rhus aromatica* var. *serotina* has been confused by Barkley and some other authors with the more western *Rhus trilobata* Nutt. Fernald has shown, however, (Rh. 43: 599–603. *pls. 686–87.* 1941), that the var. *serotina*, as well as var. *arenaria*, are related more closely to *R. aromatica* than to *R. trilobata*, because they possess the densely long-hairy fruits and relatively plump, somewhat rounded, small (3.8–4.5 mm.) stones of the fruit characteristic of *R. aromatica*; *R. trilobata*, in contrast, has the fruits pruinose (with a waxy surface) either without any hairs or with only a few remote nonglandular hairs, and the stones of the fruit are longer (4.5–6 mm.) and flattened. Furthermore, the flowering bracts of var. *serotina* have a nearly or quite glabrous area within the densely ciliate margin of the bract, as in *R. aromatica*, whereas in *R. trilobata* these bracts are densely hairy all over the back. The pedicels of the flowers of var. *serotina* also show the relatively short length characteristic of *R. aromatica*, whereas in *R. trilobata* they are more elongated. Fernald has also shown that the length of the pedicels in *R. aromatica* var. *aromatica* are as long as in var. *serotina* (p. 602).

The sourish fruit of *R. aromatica* and varieties can be prepared into a pink lemonade drink. Flavored with sugar and served cold, it is quite refreshing. The fruits of fragrant sumac are more acid than those of either the smooth (*R. glabra*) or staghorn sumac (*R. typhina*). The foliage in autumn at first turns to a shade of yellow, then to purplish, and finally to crimson or blood red. The shrub makes a worthy ornamental addition to home plantings.

The fruits are eaten by wild turkey.

5. **Rhus radicans** L. Poison Ivy Map 1472

*Rhus Toxicodendron* of auth. [P & S], not L.
*Toxicodendron radicans* (L.) Kuntze [Steyerm.]

Flowers May–July; fruit August–November.

Occurs in many kinds of habitats, along fence rows, roadsides, railroads, waste ground, low and upland dry or wet woods, alluvial wet soils along streams, in valleys, rocky wooded slopes, ledges of bluffs, and thickets; most commonly in neutral to limestone soils and contrariwise often absent in acid soils of sandstone, chert, or granite glades, frequently invading newly cut-over woodland and disturbed areas. Throughout Missouri in every county.

An exceedingly variable plant, Poison Ivy ranges from low or depressed to bushy, shrubby, or vining types, often climbing to the tops of tall trees. It is recorded as climbing to a height of 150 feet (Ind. Geol. Rept. 7: 523. 1876). The leaflets vary greatly in size, in the amount and size of the teeth, and in hairiness. The following treatment, adapted from that presented by Fernald (Rh. 43: 589–97. *pls. 683–84.* 1941), appears the most satisfactory as based upon presently known data.

Common type encountered; leaflets usually coarsely and noticeably toothed, more rarely nearly smooth-edged (entire), the middle (terminal) leaflet chiefly 11–20 cm. long; leaflets rather thin and membranaceous; leaf-stalks (petioles) mainly 6.5–20 cm. long . . . 5b. R. RADICANS var. VULGARIS f. NEGUNDO

Rarer type; leaflets mostly entire (without teeth), the middle (terminal) leaflet 3.5–10 (up to 14) cm. long; leaflets firm to rather subcoriaceous (thickish); leaf-stalks (petioles) 2–10 (rarely to 18) cm. long . . . . . . 5a. R. RADICANS var. RADICANS f. HYPOMALACA

5a. **Rhus radicans** var. **radicans** f. **hypomalaca** Fern. Map 1472

Known from Jackson (Independence, May 31, 1921, *Bush 9365*) and Shannon (Montier, May 15, 1894, *Bush 144*) counties. Probably widespread in Missouri when more collections have been made.

Ranges from New York to Kentucky and Missouri.

5b. **Rhus radicans** var. **vulgaris** (Michx.) DC. f. **Negundo** (Greene) Fern. Map 1472

This is the common type found throughout the state.

Ranges from Florida to Texas, north to Virginia, Ohio, Indiana, Illinois, and Iowa.

Other varieties and forms are described, but only the above taxa appear to fit Missouri material, all the specimens of which were determined by Fernald at Gray Herbarium. Of the variations which may be expected to occur in Missouri when more intensive collecting of this species has been made, two will probably be found:

**R. radicans** var. **vulgaris** f. **vulgaris** with the petiole and lower surface of the leaflets glabrous, and

**R. radicans** var. **vulgaris** f. **intercursa** Fern., with the petiole glabrous, but the lower surface of the leaflets hairy. In the form commonly found in Missouri, **R. radicans** var. **vulgaris** f. **Negundo,** the petioles and lower surface of the leaflets are hairy.

The plants are browsed by white-tailed deer, and some birds, such as the Northern Flicker, are known to feed on the fruits. Various degrees of immunity are found among individuals who contact the plant. The poisonous property, due to a volatile oil, occurs in the leaves, flowers, fruits, and bark of the stems and roots. Recent experiments show that actual contact with the poisonous juice is necessary to bring on the effects of the poison. The symptoms may start anywhere from usually a few hours to several days afterward. The best preventive appears to be to wash the hands and face thoroughly several times with a strong alkaline soap (yellow laundry or naphtha) immediately or as soon as possible after touching the plant. Another preventive is to lather well in the same kind of soap and allow it to dry on the skin before going out on an excursion. Gelatin capsules, containing the poison in a certain form, are also used to be taken internally, and are claimed to induce a type of immunity. This appears to follow the suggestion claimed by many country people that eating the leaves brings about immunity. Washing the face and hands or affected parts in a solution of 5 per cent ferric chloride in a half-and-half mixture of alcohol and water also serves as a preventive.

For getting rid of the poisoning, once it has started, several remedies are known. One of the best is a phenolated calamine lotion (zinc oxide, glycerin, and phenol or carbolic acid). Others used to treat the poisoning are sugar of lead, iron (ferric) chloride with alcohol, a 5 per cent solution of potassium permanganate, or a strong solution of baking soda or epsom salts (1–2 teaspoonsful to a cup of water). For the relief of itching, nothing is more effective and speedier than holding the affected parts under hot water as hot as can be tolerated. This last remedy has received the approval of many physicians.

The names Poison Ivy and Poison Oak are used indiscriminately for many species. There are a num-

Plate no. 242. **1.** Rhus aromatica var. serotina, × 2/5; *a.* Flowering branch; Details from Small, The New York Botanical Garden. **2.** Rhus aromatica var. aromatica, × 2/5; *a.* Flowering branch, × 2/5. **3.** Rhus radicans var. vulgaris f. Negundo, × 2/5; *a.* Fruiting branch, × 2/5; Details from Small, The New York Botanical Garden. **4.** Rhus Toxicodendron, × 2/5; *a.* Fruiting branch, × 2/5.

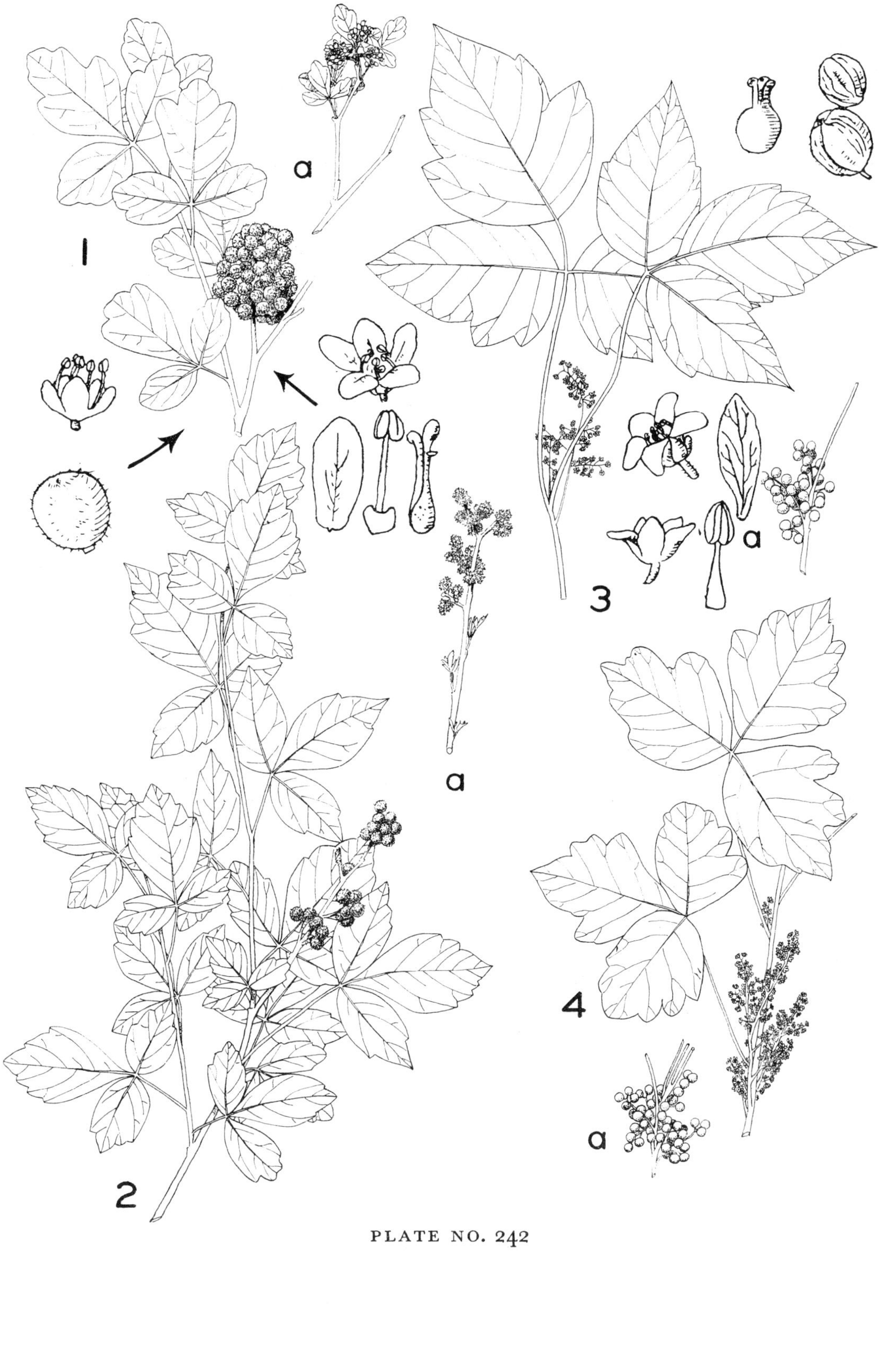


PLATE NO. 242

ber of poison ivies in the United States, but no particular species which may be so designated, and the name Poison Oak is commonly applied to the same species known as Poison Ivy.

6. **Rhus Toxicodendron** L. f. **Toxicodendron**
Poison Oak Map 1473
*Rhus quercifolia* (Michx.) Steud. [P & S]
*Toxicodendron quercifolia* (Michx.) Greene [Steyerm.]
Flowers May–June; fruits August–November.

Occurs on limestone glades, sandy or rocky open woods, and prairies. Known only in southern Missouri, in Scott, Mississippi, and Ozark counties.

Ranges from Florida to Texas, north to New Jersey, Tennessee, Missouri, and Oklahoma.

The typical f. *Toxicodendron* has the fruit with a short papillate-pubescence; in f. *leiocarpa* Fern., not recorded from Missouri, the fruit is glabrous, while in f. *elobata* Fern., the leaves are unlobed.

This Poison Oak grows very low as compared with the common Poison Ivy of Missouri, usually only with a few erect stems 5–60 cm. tall. These do not climb nor produce aerial roots, but spread by underground stems. The leaves appear mostly clustered near the summit of the stem and are densely velvety-hairy on the lower surface.

## Fam. **AQUIFOLIACEAE** (Holly Family)

### **Ilex** L. Holly

a. Leaves thick, leathery, evergreen, usually with a few (0–8) remote, stiff, spiny teeth on each side . 1. I. OPACA
a. Leaves not thick nor leathery, not evergreen, with several to many teeth (usually 8–25) along each margin . . . . . . . . . . . . . . . . . . . . . . . . . . . . . . . . . . *b*
b. Common in southern, central, and eastern Missouri; occurs on limestone glades, wet ground, low woodland in valleys and along streams; teeth of leaves rounded or blunt; tip of leaf rounded or blunt (obtuse); no hairs on margins of calyx-lobes . . . . . . . . . . . . . . 2. I. DECIDUA
b. Rare, known only from the southeastern Ozarks in Ste. Genevieve, St. Francois, Madison, Iron, and Reynolds counties; occurs on sandstone bluffs and along granite 'shut-ins' and granitic rocks; teeth of leaves sharp-pointed; tip of leaf acutely or acuminately rather abruptly pointed; hairs present on margins of calyx-lobes . . . . . . . . . . . . . . . . . 3. I. VERTICILLATA

1. **Ilex opaca** Ait. f. **opaca** American Holly
Map 1474
Also called Christmas Holly.
*Ilex opaca* Ait. [G, BB, P & S, Steyerm.]
Flowers May–June.

Occurs along spring branches in sandy woodland of Crowley Ridge and in low moist ground of woodland at or near the junction of the southeastern lowlands with Crowley Ridge. Southeastern Missouri lowland section, where known definitely only from Stoddard (sandy bogs, Dexter, *Palmer 14272*; sandy low woods along spring-fed creek, near junction of Crowley Ridge and lowland, T25N, R11E, northwest ¼ sect. 6, 3½ mi. southeast of Bloomfield, August 20, 1954, *Steyermark 76767, 76768*; near same locality, *Steyermark 76829*; along creek at Brown's Sand and Gravel Co., T25N, R10E, sect. 1, 3½ mi. northeast of Dexter, June 24, 1955, *Steyermark 78651*; wooded spring branch, T25N, R10E, northeast ¼ sect. 1, 3½–3¾ mi. southeast of Bloomfield, October 18, 1955, *Steyermark 80376*; near same locality *Steyermark 80401, 80419*) and Dunklin (common, Campbell, *Bush 84*; low woods, south of Campbell, *Kellogg 27013*; near creek in open field, T22N, R9E, sect. 4, 1 mi. west of Campbell, June 24, 1955, *Steyermark 78688*) counties.

Recorded by Bush (Rept. Mo. Hort. Soc. [1894] 37: 381. 1895) in his 'List of the trees, shrubs, and vines of Missouri' from Butler, Cape Girardeau, Mississippi, and New Madrid counties, in addition to Stoddard and Dunklin counties, but no preserved herbarium specimens nor presently known wild trees from those counties have been found to substantiate Bush's reports. Until such records can be verified, they are treated in the present flora only as unauthenticated reports.

Ranges from Florida to Texas, north to Massachusetts, New York, Pennsylvania, West Virginia, Ohio, Kentucky, Illinois, Missouri, and Oklahoma.

At the Stoddard County stations, this holly occurred in nearly pure stands along the sandy stream beds of Crowley Ridge, or was associated with beech, sour gum, red maple, sweet gum, and spice bush. The tree in Dunklin County measured 5 ft. 9 in. in circumference, 21.55″ DBH, and had a height of 40 feet.

Various leaf forms have been described. One in which the leaves are without any teeth or have only 1 or 2 small teeth is known as *I. opaca* f. *subintegra* Weath., while trees with yellow fruits instead of the normal red fruits are known as *I. opaca* f. *xanthocarpa* Rehd. Many horticultural forms are in culti-

1474 Ilex opaca f. opaca (American Holly)

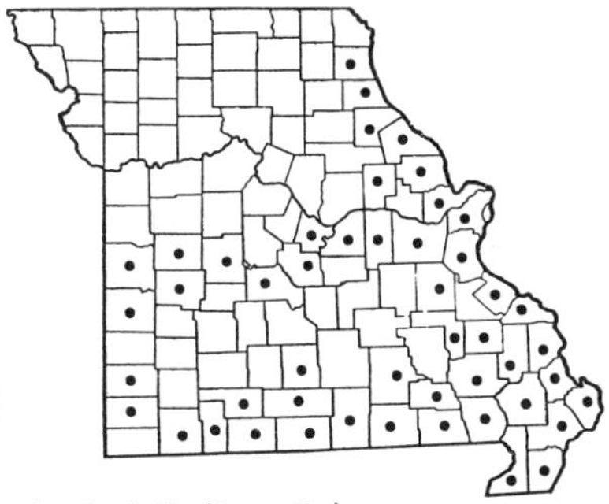

1475 Ilex decidua (Possum Haw)

1476 Ilex verticillata var. padifolia (Winterberry)

vation. The trees do well as far north in the state as St. Louis, Boone, and Jackson counties. Most of the foliage and fruit used as Christmas decorations come from the southern states, but an increasing amount of the English Holly (*I. Aquifolium*) is grown in Washington and Oregon for the Christmas season. As only the pistillate plants produce fruit, it is necessary to plant the pistillate as well as staminate trees. The fruit is reputed to be poisonous. The boiled leaves served as a substitute for tea during the Civil War, but do not possess a caffeine content, such as found in the Paraguay Tea (*Ilex paraguariensis*).

The wood of this holly is very hard, surpassing that of Black Locust (*Robinia Pseudo-Acacia*) but not quite as hard as Flowering Dogwood. The pale, even-grained wood is much used for taking stains and dyes, and for this purpose is employed for furniture and for black and white inlaid lines of musical instruments. It furnishes excellent veneer wood and takes enamel finishes well. When dyed black, it furnishes black keys for pianos. It is also used in art work for its whitish color by engravers. Knife handles and various kinds of turnery result from its hard wood.

2. **Ilex decidua** Walt. Possum Haw Map 1475
Also called Deciduous Holly.
Flowers April–May.

Occurs on upland limestone glades, rocky upland open woods, ledges and top of limestone bluffs, borders of upland and lowland ponds, swamps, sloughs, valleys, and low wet woods along streams. Common in the lowlands of southeastern Missouri, eastern, southern, and northern Ozarks, and north along the counties bordering the Mississippi River to Lewis County; absent from a number of central Ozark counties.

Ranges from Florida to Texas, north to Maryland, Indiana, Illinois, Missouri, and Kansas.

The largest recorded specimen of *Ilex decidua* is at Big Oak Tree State Park in Mississippi County; according to measurements made by Mr. Kendall Laughlin, the circumference measured 1 ft. 1 inch and the height 20 ft.

The foliage turns dull purplish and green in autumn. After the leaves fall, the scarlet fruit on the pistillate trees closely covers the branches, contrasting with the pale gray bark and twigs. They are very colorful in the late fall and early winter landscape, remaining on the tree much of the winter.

It is difficult to get plants started from seeds as germination is very poor. Two shrubs transplanted to the author's botanical preserve in northern Illinois have grown well during the past fifteen years far north of their natural range.

3. **Ilex verticillata** (L.) Gray var. **padifolia** (Willd.) T. & G. Winterberry Map 1476
Also called Black Alder.
Flowers April–May.

Occurs along granite and porphyritic trachyte 'shut-ins,' granite rocky stream beds, and La Motte sandstone bluffs. Known only from the southeastern Ozarks in Ste. Genevieve, Madison, Iron, Reynolds, and St. Francois counties.

Ranges from Nova Scotia to Minnesota, south to Georgia, and Missouri.

Typical var. *verticillata*, which is not known in Missouri, has the lower surface of the leaves with minutely appressed hairs or hairy only along the principal nerves, whereas in var. *padifolia* the lower surface itself as well as the nerves is downy.

The berries are reputed to be poisonous. The boiled leaves are sometimes used as a tea, but are not known to contain caffeine, which is found in the leaves of the southern *Ilex vomitoria*. This shrub, like *I. decidua*, drops the foliage in late autumn, and the bright red berries of the pistillate plants cover the branches with a colorful display which lasts well into winter. This species enters into home plantings in the northern states because of its decorative fruit.

1477 Euonymus atropurpureus (Wahoo)

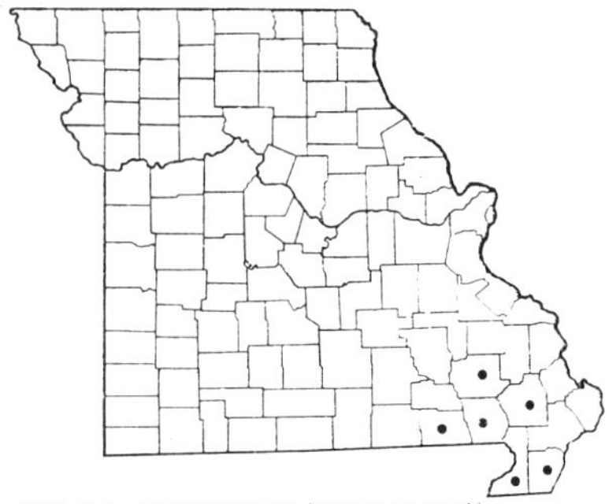

1478 Euonymus americanus (Strawberry Bush)

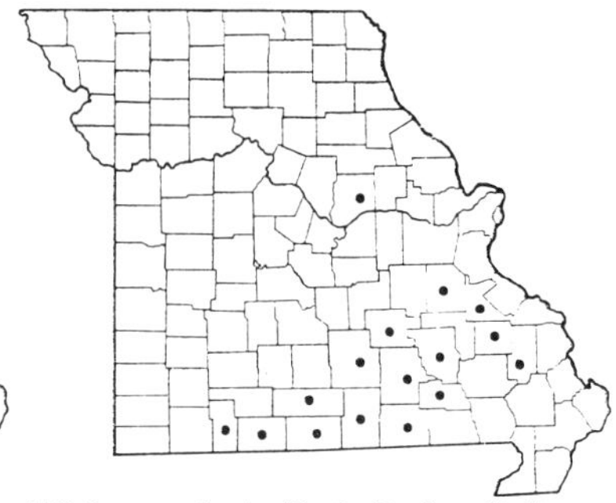

1479 Euonymus obovatus (Running Strawberry Bush)

## Fam. **CELASTRACEAE** (Staff-tree Family)

Leaves alternate; vine, often sprawling on the ground as well as climbing trees . . . . . . 2. CELASTRUS
Leaves opposite; shrubs or small trees or creeping ground plants . . . . . . . . . . . 1. EUONYMUS

### 1. **Euonymus** L. Spindle Tree

a. Common shrub or small tree occurring throughout Missouri; leaf-stalks (petioles) of leaves at tip of stem 5 mm. or more long, usually 5–20 mm. long; outside of fruit smooth . . . 1. E. ATROPURPUREUS
a. Rare shrub or creeping ground plants, known only from southeastern Missouri and the Ozark section of southern and central Missouri; leaf-stalks (petioles) of leaves at tip of twigs 1–4 mm. long; outside of fruit conspicuously warty or with tubercles . . . . . . . . . . . . . . . . *b*
b. Trailing, rooting ground plant of shaded rocky, mainly limestone, north-facing bluffs and slopes of the Ozark region of southern and east central Missouri; leaves thin or membranaceous, those at tip of stems obovate, mostly broadest in upper half, the leaf-stalks mainly 3–4 mm. long . 3. E. OBOVATUS
b. Erect or sprawling shrub or the lower branches sometimes rooting, known only from southeastern Missouri north and west to Wayne and Ripley counties; leaves thick and firm, those at tip of stems lanceolate to ovate, broadest at middle or lower half, the leaf-stalks mainly 0.5–3 mm. (rarely 4 mm.) long . . . . . . . . . . . . . . . . . . . . 2. E. AMERICANUS

1. **Euonymus atropurpureus** Jacq. Wahoo
Map 1477

Also called Burning Bush.

Flowers late April–June.

Occurs on wooded slopes, on bluffs, open woods, alluvial soils along streams, and in thickets. Throughout Missouri, doubtless in every county.

Ranges from New York, Ontario and Minnesota to North Dakota, Nebraska, and Montana, south to Alabama, Tennessee, Arkansas, Oklahoma, and Texas.

The outer surface of the mature fruit is purple becoming reddish, and encloses deep orange-colored arils (seed coverings) of the brown seeds. The autumnal foliage is dull reddish or carmine suffused with green. The leaves and fruits are reported to possess purgative properties, and at one time the bark of the root was used medicinally but is now occasionally found employed as a gastric stimulant and cathartic. The powdered bark was used by the Indians for tobacco. The wood was used for arrows by American Indians.

2. **Euonymus americanus** L. Strawberry Bush
Map 1478

Also known as Brook Euonymus.

Flowers May–June.

Occurs in low sandy woods along spring branches, low rich woods, rich wooded slopes of Crowley Ridge, and moist stream banks. Rare, known only in southeastern Missouri, mainly on Crowley Ridge and low-

---

Plate no. 243. **1.** Ilex opaca, $\times {}^2/_5$; *a.* Fruiting branch, $\times {}^2/_5$; After Britton and Brown, details from Small, The New York Botanical Garden. **2.** Celastrus scandens, $\times {}^2/_5$; After Britton and Brown, details from Small, The New York Botanical Garden. **3.** Ilex dicidua, $\times {}^2/_5$; *a.* Flowering branch, $\times {}^2/_5$. **4.** Euonymus atropurpureus, $\times {}^2/_5$; *a.* Fruiting branch; *b.* Flowering branch. **5.** Euonymus obovatus, $\times {}^2/_5$; Details from Small, The New York Botanical Garden. **6.** Euonymus americanus, $\times {}^2/_5$. **7.** Ilex verticillata var. padifolia, $\times {}^2/_5$; *a.* Fruiting branch; *b.* Flowering branch.

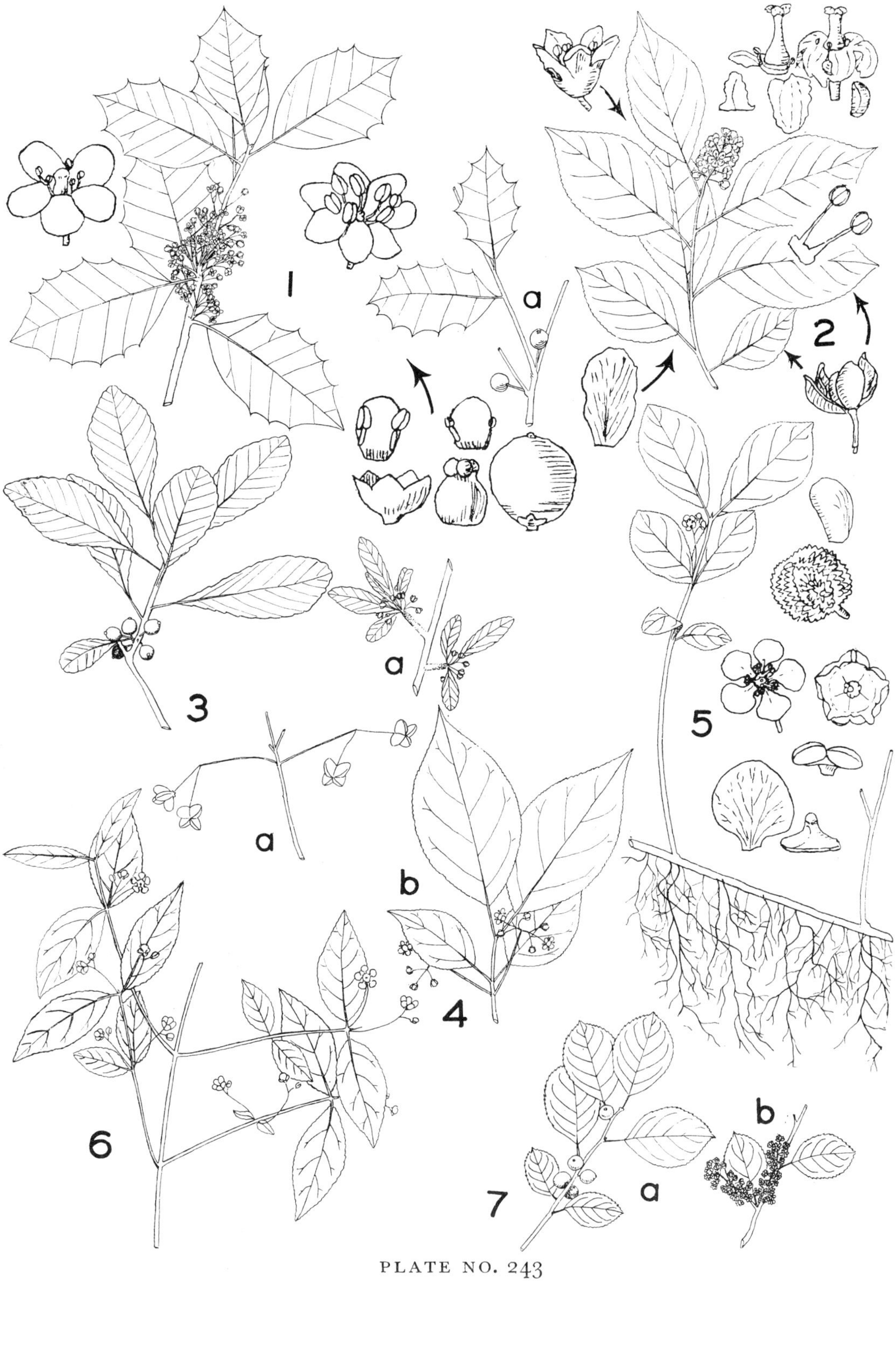


PLATE NO. 243

lands in Pemiscot, Dunklin, Stoddard, Wayne, Butler, and Ripley counties.

Ranges from Florida to Texas, north to New York, Pennsylvania, West Virginia, Indiana, Illinois, Missouri, and Oklahoma.

In the young stages of growth, this plant often has low sprawling or even trailing branches. This habit of growth sometimes causes confusion in identification. A specimen from Wayne County reported in Rh. 42: 101. 1940 as *Euonymus radicans* Sieb. var. *acuta* Rehd. (base of wooded slopes along creek in Possum Hollow, sect. 28, 2½ mi. northwest of Wappapello, September 2, 1938, *Steyermark 6449*, and low swampy woods along road in valley of Lost Creek, T27N, R7E, sect. 4, 1½ mi. south of Shook, September 4, 1938, *Steyermark 6603*) is actually *E. americanus*.

The foliage of this species is dark green, much thicker, and longer persistent on the branches than that of the other native species, remaining on the branches well into late autumn (November–December). It makes an excellent ornamental shrub and has prospered far north of its natural range at the author's botanical preserve in northern Illinois. When mature, the fruit is crimson and the seed-coverings (arils) are scarlet. It grows easily from seed.

3. **Euonymus obovatus** Nutt.
Running Strawberry Bush Map 1479
Flowers late April–June.

Occurs on usually north-facing rich wooded limestone slopes or at the base of limestone bluffs. Southern and eastern Ozark region from Stone County east to Bollinger County and north to St. Francois, Washington, Dent, Texas, and Douglas counties, with a local northern outlier in the Ozark section of Callaway County (north-facing limestone bluffs along Middle River above mouth of Craighead Creek, T46N, R9W, sect. 19, 4 mi. southwest of Ham's Prairie, September 12, 1937, *Steyermark 26209*).

Ranges from New York and Ontario to Michigan and Illinois, south to Tennessee and Missouri.

The plant holds its green color a long time, the foliage persisting and green until December. This species is to be recommended as a ground cover planting for bare shaded slopes, where it eventually forms a dense interlacing mass of stems. On richer soils and on north- or east-facing exposures it can be relied upon as a good ground-cover. The ripe fruits are quite ornamental with an orange-red or scarlet outside portion enclosing the orange-red seed-coverings (arils) of the flesh-colored nutlets.

## 2. **Celastrus** L. Bittersweet

**Celastrus scandens** L. American Bittersweet
Map 1480

Flowers May–June; fruits July–October.

Occurs in woodland, rocky slopes, along bluffs, borders of rocky glades, thickets, and along fence rows. Throughout Missouri, doubtless in every county.

Ranges from Quebec to Ontario and Manitoba and Wyoming, south to Georgia, Alabama, Tennessee, Louisiana, Oklahoma, Texas, and New Mexico.

In autumn the foliage turns to pale greenish-yellow or pale yellow. The leaves are natural food for white-tailed deer, but cases are known where horses have been poisoned by eating the leaves. The bark of this vine was at one time employed medicinally, and from the bark of the root an extract has sometimes been obtained for use as a salve.

When ripe the orange globose, capsular fruit breaks open into 3 valves to expose the fleshy scarlet seed-coverings (arils) surrounding the seeds. These persist all winter and are much sought for indoor decoration. They are eagerly eaten by birds as well. Usually the staminate and pistillate flowers are separated on different plants; occasionally plants are found having both perfect and unisexual flowers on the same vine. The fruit is usually borne on vines having pistillate flowers. The flowers are quite fragrant.

The vine spreads by its orange root suckers, forming large colonies in favorable situations. It may eventually injure trees, if permitted to twine about them. This may take many years to do, however, as I have observed the vine forming tight constrictions around Black Locust (*Robinia Pseudo-Acacia*) for twenty-five years without the trees showing any perceptible effect.

# Fam. **STAPHYLEACEAE** (Bladder-nut Family)

## **Staphylea** L. Bladder-nut

**Staphylea trifolia** L. American Bladder-nut
Map 1481

Flowers April–May.

Occurs in rich woodland, often north- or east-

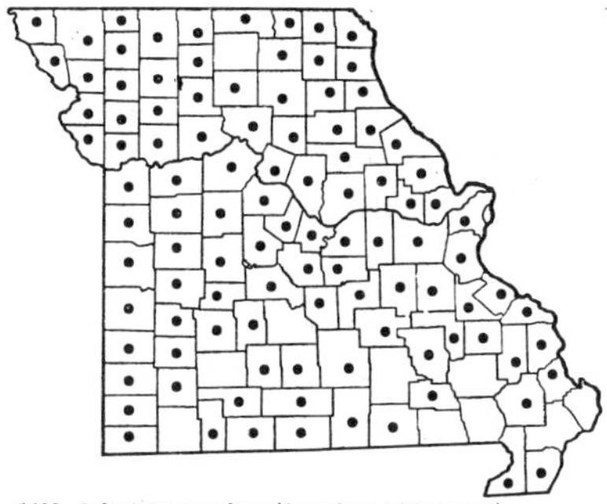

1480 Celastrus scandens (American Bittersweet)

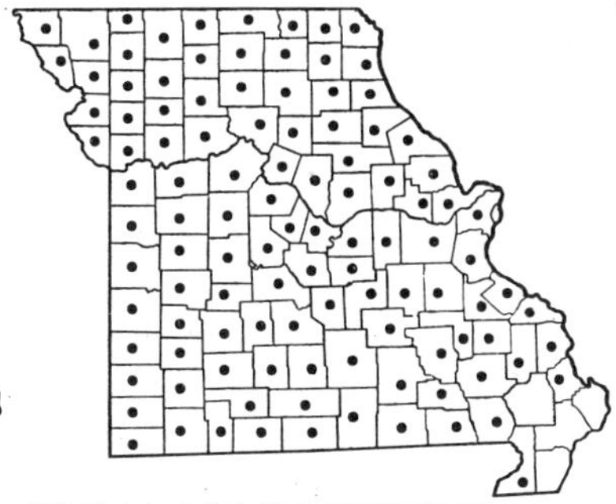

1481 Staphylea trifolia (American Bladder-nut)

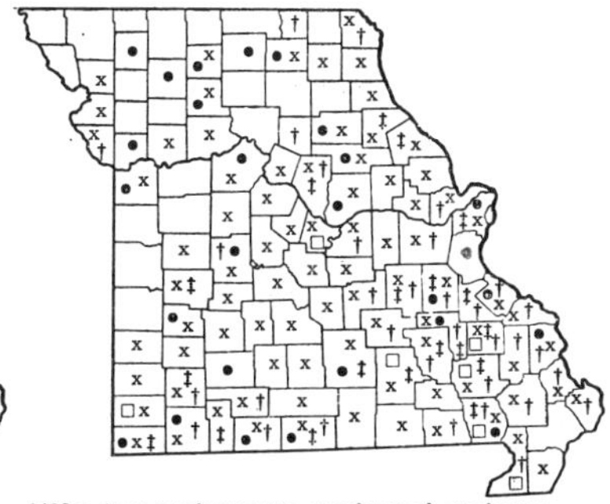

1482 • Acer saccharum var. saccharum f. saccharum (Sugar Maple)
1482 x Acer saccharum var. saccharum f. glaucum
1482 † Acer saccharum var. saccharum f. Schneckii
1482 ‡ Acer saccharum var. saccharum f. Rugelii
1482 □ Acer saccharum var. floridanum (Southern Sugar Maple)

facing wooded bluffs of especially limestone, alluvial ground in valleys, along streams, and in thickets. Throughout Missouri; absent only in a few counties of the lowlands of extreme southeastern Missouri.

Ranges from Quebec to Minnesota, south to Georgia, Alabama, Arkansas, and Oklahoma.

The foliage remains green late into fall, but eventually turns a yellow-green. The shrub is found in dense colonies, due to its natural method of increase by underground root suckers. It makes a good ornamental shrub by virtue of its drooping clusters of white flowers and dark green leaves. The fruit, which becomes inflated and bladder-like at maturity, makes a popping sound when crushed between the fingers. It is reported that the seeds of the Old World species of *Staphylea* resemble pistachio nuts in quality, and the suggestion that the American Bladder-nut has a similar taste has been recorded. The leaves of a Californian species, *S. Baldwinii*, are stated to cause cases of dermatitis when the skin is exposed to strong light, but no cases of poisoning from the leaves of the native eastern *S. trifolia* have come to the author's attention.

## Fam. **ACERACEAE** (Maple Family)

### **Acer** L. Maple

#### I. *Key based mainly on leaves and fruits*

a. Leaves compound, divided into 3–7 leaflets . . . . . . . . . . . . . . . . 5. A. NEGUNDO
a. Leaves simple, merely lobed or cut. . . . . . . . . . . . . . . . . . . . . . *b*
b. Spaces (sinuses) between the main lobes of the leaf rounded, U- or trough-shaped; flowers appearing with or after the leaves develop; fruit ripening in summer and fall (July–September); winter buds usually acutely pointed at tip . . . . . . . . . . . . . . . . . . *c*
c. Lower surface of leaf-blades green or yellowish-green; leaf-blades drooping or turned in at the margins or lower lobes; leaf-blades not deeply lobed; spaces (sinuses) between the main lobes of leaf-blade form angles of more than 90 degrees; leaf-stalks (petioles) of the leaves at end of the twigs abruptly enlarged at base, frequently with small leaf-like outgrowths; axillary buds covered by fully grown stipules . . . . . . . . . . . . . . . . . . . 2. A. NIGRUM
c. Lower surface of leaf-blades gray- or silvery- or blue-green, but not yellow-green; leaf-blades usually with a flat expanded surface, but sometimes also drooping at the margins; leaf-blades more deeply lobed; spaces (sinuses) between the main lobes of leaf-blade form angles of 90 degrees or less; leaf-stalks (petioles) of the leaves at end of the twigs only gradually enlarged at base, without any small leaf-like outgrowths; axillary buds not covered by fully grown stipules . . . . . . . . . . . . . . . . . . . . . . . 1. A. SACCHARUM
b. Spaces (sinuses) between the main lobes of the leaf more or less sharply angled or V-shaped; flowers appearing before the leaves expand; fruit ripening in spring (May–June); winter buds blunt or rounded . . . . . . . . . . . . . . . . . . . . . . . *d*
d. Leaf-blades cut $\frac{1}{2}$ to $\frac{2}{3}$ of the distance to the midrib, the central lobe taking up much more than

half the length of the complete blade; central lobe of leaf conspicuously narrowed at base; fruits densely hairy (when young) to more or less hairy (at maturity), 4–8 cm. long with the main body 1–1.5 cm. long; throughout Missouri in wet soils of bottomlands and along streams . 4. A. SACCHARINUM

d. Leaf-blades usually cut less than ½ of the distance to the midrib, the central lobe taking up less or slightly more than half the length of the complete blade; central lobe of leaf broadest at base; fruits glabrous (without hairs), 1.5–5 cm. long with the main body 0.5–0.9 cm. long; Ozark region of southern and central Missouri in usually dry acid soils of chert, sandstone, or granite, or in swampy soils of lowlands of southeastern Missouri and upland sink-hole ponds 3. A. RUBRUM

II. *Key based mainly on plants with flowers*

a. Flowers appearing long before the leaves appear or immediately with them . . . . . . . . . . *b*

b. Flowers greenish, dioecious, the male (staminate) and female (pistillate) flowers separated on different trees; male flowers in compact close clusters (fascicles), the female in racemes; sepals separated at least to half their length; branchlets olive green or bronze to glaucous or gray-lavender . . . . . . . . . . . . . . . . . . . . . . . . . . 5. A. NEGUNDO

b. Flowers greenish-yellow, reddish-salmon, red or scarlet, the male and female flowers in separate clusters on the same tree (monoecious) or on different trees (dioecious); both male and female flowers in compact close clusters; calyx very shallowly toothed, the sepals nearly completely joined their entire length; branchlets light- to reddish-brown to reddish . . . . . . . . . *c*

c. Branchlets mainly drooping, light- to reddish-brown; flowers mainly greenish-yellow to reddish-salmon; throughout Missouri in wet soils of bottomlands and along streams; petals absent; ovaries and young fruits hairy . . . . . . . . . . . . . . . . . 4. A. SACCHARINUM

c. Branchlets spreading to ascending, at least the youngest are red; flowers red or scarlet; Ozark region of southern and central Missouri in usually dry acid soils of chert, sandstone, or granite, and in swampy soils of lowlands of southeastern Missouri and upland sink-hole ponds; petals present; ovaries and young fruits glabrous (without hairs) . . . . . . . . . . 3. A. RUBRUM

a. Flowers appearing with or after the leaves . . . . . . . . . 1. A. SACCHARUM and 2. A. NIGRUM (these cannot be keyed separately in flower)

1. **Acer saccharum** Marsh Sugar Maple Map 1482

Flowers April–May; fruits July–September.

Occurs in rich or rocky woodland, usually on calcareous or circumneutral soils, along streams, wooded slopes, base of bluffs, borders of limestone upland glades and bluff escarpments. Throughout Missouri.

The following variations, with intergradations, are encountered in the state:

a. Rare type in southern Missouri; mature leaf-blades small, 3–9.5 cm. (averaging 6.5 cm.) long, 3.5–11 cm. (averaging 8 cm.) broad; lobes of the leaf short and blunt; leaf-blades often drooping or turned in at the margins or lower lobes; ovary and young fruit hairy (old fruit losing most of the hairs or becoming glabrous); calyx in flower 1.5–2.5 mm. long with conspicuous long white hairs protruding from its summit; flower-stalks (pedicels) elongating to 3–5 cm. long; bark whitish-gray, smooth, or furrowed in age. . 1e. A. SACCHARUM var. FLORIDANUM

a. Includes common types throughout Missouri; mature leaf-blades, mainly 8–15 cm. (averaging 11.5 cm.) long, 8–20 cm. (averaging 15 cm.) broad; lobes of the leaf long and prominently pointed; leaf-blades mainly with a flat expanded, not drooping, surface; ovary and young fruit glabrous (without hairs); calyx in flower 4–6 mm. long, without any long hairs protruding; flower-stalks (pedicels) elongating to 5–10 cm. long; bark of older trees gray and deeply furrowed . . . . . . . *b*

b. Leaf-blade with 3 main lobes which are long-pointed (acuminate) and usually entire (smooth-edged or without teeth or lobes) or barely lobed . . 1d. A. SACCHARUM var. SACCHARUM f. RUGELII

b. Leaf-blade with usually 5 more or less toothed or angled lobes . . . . . . *c*

c. Veins of lower surface of leaf-blade densely hairy; leaf-stalks (petioles) hairy or glabrous (not hairy) 1c. A. SACCHARUM var. SACCHARUM f. SCHNECKII

c. Lower surface, including veins, of leaf-blade mostly glabrous (without hairs) except for tufts of hairs in the axils of the main veins; leaf-stalks usually glabrous *d*

d. Common type throughout Missouri; lower surface of leaf-blades glaucous

(with a gray- or silvery-green coating)
1b. A. SACCHARUM
var. SACCHARUM f. GLAUCUM

d. Rarer type; lower surface of leaf-blades pale green or green without any glaucous silvery coating . . .
1a. A. SACCHARUM
var. SACCHARUM f. SACCHARUM

1a. **Acer saccharum** var. **saccharum** f. **saccharum** Map 1482

*Acer saccharum* var. *saccharum* [BB]

*Acer saccharum* Marsh. [G, BB, P & S, Steyerm.]

*Acer saccharinum* Wangenh., not L.

This form is scattered throughout Missouri, but much less common than f. *glaucum*.

Ranges from Quebec and Newfoundland to Ontario, Minnesota, Manitoba, and North Dakota, south to Georgia, Alabama, Mississippi, Arkansas, and Texas.

1b. **Acer saccharum** var. **saccharum** f. **glaucum** (Schmidt) Pax Map 1482

*Acer saccharum* f. *glaucum* (Schmidt) Pax [G]

*Acer saccharum* var. *glaucum* (Pax) Sarg. [BB, P & S, Steyerm.]

*Acer nigrum* var. *glaucum* (Schmidt) Fosberg [Fosberg]

*Acer subglaucum* Bush

*Acer saccharinum* var. *glaucum* Schmidt

This is the commonest form in Missouri, found in nearly every county of southern and central Missouri, but absent or not recorded from a number of counties in northwestern and north-central Missouri.

Range similar to typical f. *saccharum*.

1c. **Acer saccharum** var. **saccharum** f. **Schneckii** (Redh.) Deam Map 1482

*Acer saccharum* var. *Schneckii* Rehd. [G, BB, P & S, Steyerm.]

*Acer nigrum* var. *Schneckii* (Rehd.) Fosberg [Fosberg]

*Acer saccharum* subsp. *Schneckii* (Rehd.) Desmarais [Desmarais]

*Acer barbatum Schneckii* (Rehd.) Bush

*Acer saccharophorum* var. *Schneckii* (Rehd.) Rousseau

This is scattered throughout Missouri, but appears to be commonest in the eastern half of the state.

Ranges from Indiana, Illinois, and Wisconsin, south to Tennessee and Missouri.

Although this form appears to have a rather restricted geographical distribution, suggesting varietal status, there is often too great an intergradation in the pubescence of the petioles and lower leaf surface and often lack of correlation in these characters to merit it for varietal recognition. The petioles are often glabrous or glabrescent and specimens are found with densely pubescent petioles and nearly glabrous lower leaf surfaces. Deam also found that seedlings from a tree having originally pubescent petioles did not all come true. His reduction of rank to form appears justified.

1d. **Acer saccharum** var. **saccharum** f. **Rugelii** (Pax) Palmer & Steyermark Map 1482

*Acer saccharum* var. *Rugelii* (Pax) Rehd. [BB, Deam]

*Acer saccharum* subsp. *Schneckii* in part [Desmarais]

*Acer subglaucum Rugelii* (Pax) Bush

*Acer saccharophorum* var. *Rugelii* (Pax) Rousseau

Scattered in southern and central Missouri northeast to Boone and Ralls counties.

Scattered in the range of typical var. *saccharum* f. *saccharum*.

The lower sides of the leaves of this form vary from glaucous (without hairs) to conspicuously hairy.

1e. **Acer saccharum** var. **floridanum** (Chapm.) Small & Heller Southern Sugar Maple Map 1482

*Acer floridanum* (Chapm.) Pax [P & S, Steyerm.]

*Acer barbatum* f. *floridanum* (Chapm.) Fern. [G, BB]

*Acer nigrum* var. *floridanum* (Chapm.) Fosberg [Fosberg]

*Acer saccharum* subsp. *floridanum* (Chapm.) Desmarais [Desmarais]

Occurs in southern and central Missouri, where rare and known from Dunklin (Campbell, October, 1910, *Sargent*), Butler (*Steyermark 69693*), Madison (Fredericktown, *Palmer 30254*; Mine La Motte, *Palmer 39167*), Wayne, Shannon (*Steyermark 69594a*), Cole (Jefferson City, *Palmer 39233*), and Newton (Grand Falls, *Bush 10437*) counties.

Ranges from Florida to Texas, north to Virginia, southern Illinois, Missouri, and southeastern Oklahoma.

From evidence presented by Fernald (Rh. 44: 359–60, 426–28, *pls. 725–27*. 1942 and 47: 156–60. 1945), it would seem that this small-leaved sugar maple with pubescent ovaries and fruits as well as presumed characters of the calyx and pedicels would merit recognition of specific rank as *Acer barbatum* Michx. (as treated by Fernald in the eighth edition of *Gray's Manual* [p. 987]). Other authors who have recently studied this group have accorded this southern sugar maple a rank as subspecies (Desmarais, Ynes. Brittonia 7: 382. 1952) or as variety of *A. nigrum* (Fosberg, F. R. Castanea 19: 27. 1954). Desmarais (p. 383) believes that the original specimen of *Acer barbatum* Michx. was based upon a mixture of several different plants of which only the flowers apply to *Acer saccharum*, and that the bearded calyx, stressed by Fernald as so distinctive of *A. barbatum*, is not a feature restricted solely to *A. barbatum.* As many specimens are collected in mature fruit, lacking hairs, it is difficult, in the absence of flowers, to judge on the basis of leaf size and lobing alone whether to place such fruiting specimens with one of the variations of *A. saccharum* var. *saccharum* or with var. *floridanum.* All the specimens above cited under var. *floridanum* possess hairs on the fruit, and some of these specimens have glabrous petioles and branchlets, which characters Fernald employs to characterize as *A. barbatum* f. *floridanum*, as contrasted with typical *A. barbatum* f. *barbatum* in which the petioles and branchlets are pubescent. More intensive field and experimental studies may eventually shed more light upon the desirability of accepting *A. barbatum* Michx. either as a good species or only as a variety or subspecies of *A. saccharum* (or of *A. nigrum* as held by Fosberg). With present information and data at hand, it would appear that, while this southern sugar maple is rather distinct in a number of respects from the larger-leaved variants of *A. saccharum* and *A. nigrum*, it shows considerable intergradation with varieties of *A. saccharum* and so closely approaches them in mature fruit and leaves as to make positive identification uncertain. By such considerations, the southern sugar maple is treated in the present flora as a variety of *A. saccharum.*

This variety has been observed to be common in southern Illinois, where it has been confused with *A. nigrum* because of a frequent tendency there to show the drooping margins of the leaves, so characteristic of *A. nigrum.* However, the leaves of the trees of southern Illinois are relatively small and are gray- or silvery-green on the lower surface, as is typical of southern sugar maple. This drooping leaf tendency of trees in southern Illinois has not been noted in the field in Missouri, but collectors in the future should record this feature carefully in order to judge its frequency of occurrence in southern and southeastern Missouri trees.

There has been considerable discussion among botanists as to the correct application of the name for the sugar maple, as may be judged by the synonymy. Some authors (Fosberg, Castanea 19: 26–27. 1954) would treat the above taxa as varieties of *A. nigrum*, whereas Desmarais (Brittonia 7: 347–414. 1952) would regard them as subspecies of *A. saccharum.* In the present flora, the name *A. saccharum* is interpreted in the sense employed by Fernald, Gleason, Rehder, Little, and others.

The colorful autumnal foliage of sugar maple is spectacular, varying from brilliant shades of orange and yellow to rose, crimson, and mixtures of these colors, and is one of the reasons for its popularity as a shade tree. It is less susceptible to breakage from ice and wind damage than Norway Maple. As it is affected by city smoke and gases, it prospers best in non-industrial areas. Although the tree has a multitude of uses, it is probably best known for its production of maple sugar. The usual yield from an average-sized tree is 10–15 gallons of sap, equivalent to 3–4 pounds of sugar. From the Indians the early colonists learned the art of preparing sugar from the sap. The sap is also made into a maple beer, for which the botanist Michaux prescribed the following: 'Upon 4 gallons of boiling water, pour 1 quart of maple molasses; add a little yeast or leaven to excite the fermentation, and a spoonful of the essence of spruce: a very pleasant and salutary drink is thus obtained.' In addition to its uses for candies, syrup, flavorings, and cream, it is used for flavoring tobacco. New York, Vermont, and other New England states are the chief producers of maple products. The sugar is considered beneficial for its phosphate content judged as important in bone-building.

The wood is important for flooring, cabinet-making, bedroom furniture, veneering, basketry, violins, gunstocks, and for fuel, in the last case excelled only by hickory. At one time it was widely used for rolling pins. It also enters the papermaking industry as pulpwood. It is also used for shuttles, but not as extensively as the Flowering Dogwood (*Cornus florida*).

2. **Acer nigrum** Michx. f. Black Maple Map 1483
Flowers April–May; fruits July–September.

Occurs in rich woods, on slopes, in ravines, valleys, and near streams. Northern and central Missouri locally south in the Ozark region to Washington, Butler, Dent, Pulaski, Wright, Christian, and Taney counties.

Ranges from Quebec and New Hampshire to Minnesota and South Dakota, south to North Carolina, Kentucky, Tennessee, Missouri, and Kansas.

Missouri material may be divided into the following variations:

Plate no. 244. **1.** Staphylea trifolia, × 2/7; Details from Small, The New York Botanical Garden. **2.** Acer saccharum; *a.* Acer saccharum var. saccharum f. saccharum, flowering branch, × 2/7; *b.* Acer saccharum var. saccharum f. Rugelii, × 2/7; *c.* Acer saccharum var. floridanum, × 2/7; Details from Small, The New York Botanical Garden. **3.** Acer nigrum, × 2/7. **4.** Acer rubrum; *a.* Fruiting cluster, × 2/7; *b.* Female flowering branch, × 2/7; *c.* Male flowering branch, × 2/7.

PLATE NO. 224

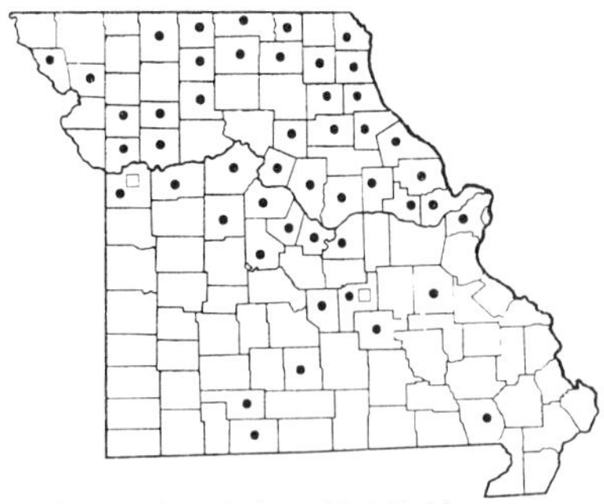

1483 • Acer nigrum f. nigrum (Black Maple)
1483 □ Acer nigrum f. pubescens

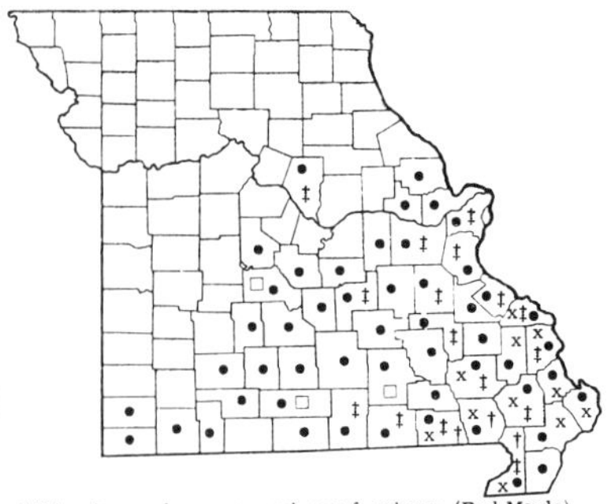

1484 • Acer rubrum var. rubrum f. rubrum (Red Maple)
1484 □ Acer rubrum var. rubrum f. tomentosum
1484 ‡ Acer rubrum var. trilobum
1484 x Acer rubrum var. Drummondii f. Drummondii
1484 † Acer rubrum var. Drummondii f. rotundata

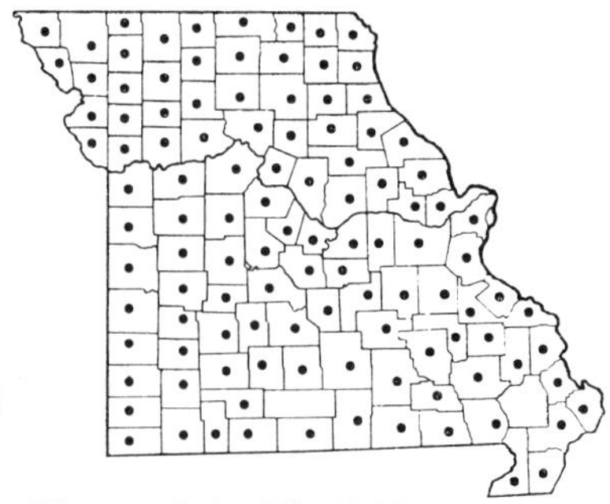

1485 Acer saccharinum (Silver Maple)

Leaf-stalks (petioles) hairy their whole length; uncommon type . . . 2b. A. NIGRUM f. PUBESCENS
Leaf-stalks (petioles) varying from glabrous (without hairs) to partly hairy (usually near base or tip); common type found . . . 2a. A. NIGRUM f. NIGRUM

2a. **Acer nigrum** f. **nigrum** Map 1483

*Acer nigrum* Michx. f. [G, BB, P & S, Steyerm.]
*Acer nigrum* var. *Palmeri* Sarg. [P & S, Deam]
*Acer nigrum* var. *Palmeri* f. *villosum* Deam [Deam]
*Acer saccharum* var. *nigrum* (Michx. f.) Britton
*Acer saccharinum β nigrum* (Michx. f.) T. & G.
*Acer barbatum* Michx. var. *nigrum* (Michx. f.) Sarg.
*Acer saccharophorum* var. *nigrum* (Michx. f.) Rousseau
*Acer saccharum* subsp. *nigrum* (Michx. f.) Desmarais [Desmarais]

This is the common variation of the Black Maple in Missouri. The var. *Palmeri* was originally described to typify a variation of this maple having 3 principal lobes through reduction or suppression of the other 2 lower lobes, the typical *A. nigrum* having 5 evident and developed lobes. However, as noted by Gleason (*New Ill. Fl.* 2: 506. 1952), such an arbitrary division cannot be made to separate a 3-lobed from a 5-lobed type of leaf, all gradations between them being present, sometimes even on the same tree. The var. *Palmeri*, at least in Missouri, does not justify separate recognition. Deam at first (*Trees of Indiana*, first ed. 263. 1932) was unwilling to recognize this variety, but in his *Flora of Indiana* (p. 656–57) accepted it.

Missouri collections originally included under var. *Palmeri* are the following: Clark Co. (Wayland, *Davis 625*; Dumas, *Bush 9487*); Jackson Co. (Atherton, *Bush 1764*); Miller Co. (near Bagnell, *Palmer 39216*); Lewis Co. (*Palmer 35903*); Grundy Co.

2b. **Acer nigrum** f. **pubescens** Deam Map 1483

This form is quite uncommon and is known only from Jackson and Phelps (base of rich wooded limestone slopes with Roubidoux sandstone above, along Dry Fork of Meramec River, T38N, R6W, sect. 33, 4 mi. southeast of St. James, May 5, 1939, *Steyermark 22143*) counties. Scattered through the range of the typical f. *nigrum*.

There has been considerable argument as to whether to accept the Black Maple as a distinct species or as a variation of the Sugar Maple group. Some botanists have considered it a variety of *Acer saccharum*; others, such as Fosberg, consider *A. nigrum* the oldest name for the group, making the common sugar maple a variety [var. *glaucum* (Schmidt) Fosberg] of *A. nigrum*. The author in the present flora has been guided by his personal field observations in retaining *A. nigrum* as a species distinct from *A. saccharum* and varieties. They are readily recognizable in the field, as separated by the characters given in the key, and *Acer nigrum* seems more distinct from all the variations of *A. saccharum* than does *A. saccharum* var. *floridanum* from the other variations of *A. saccharum*.

The autumnal foliage of Black Maple is usually rich bright or pale yellow. It has the same uses as Sugar Maple and is not differentiated by lumbermen from the Sugar Maple.

3. **Acer rubrum** L. Red Maple Map 1484

Flowers March–April; fruits April–June.

The following variations occur in Missouri:

a. Lower surface of fully grown leaves without hairs or with few hairs, the original hairs largely disappearing in age . . . . . *b*
   b. Common type found; central one of the 3 or 5 lobes 4–8 cm. long, the upper lateral lobes 2–5 cm. long; lobes of leaf prominent,

long-pointed and coarsely toothed . . . 3a. A. RUBRUM var. RUBRUM f. RUBRUM

b. Rare type found; central one of the 3 lobes 1–5 cm. long, the upper lateral lobes 0.5–2 (rarely 3.5) cm. long; lobes of leaf short-pointed or sometimes scarcely developed and shallowly toothed . . . 3c. A. RUBRUM var. TRILOBUM

a. Lower surface of fully grown leaves densely or permanently hairy, at least on the main nerves *c*

c. Leaves with 5 lobes (3 larger and 2 smaller); swamps of southeastern lowland counties and sink-hole ponds of southeastern Ozarks *d*

d. Fruits (samaras) 1.5–2.5 (–3) cm. long; trees of the Ozark section, mainly on rocky or upland slopes . 3b. A. RUBRUM var. RUBRUM f. TOMENTOSUM

d. Fruits 3.5–6 (rarely 2.7) cm. long; trees of the swampy country of the southeastern lowlands or of the sink-hole ponds of southeastern Missouri . 3d. A. RUBRUM var. DRUMMONDII f. DRUMMONDII

c. Leaves with 3 principal lobes, the other 2 mostly suppressed; trees of the Ozarks or of the swamps of the southeastern lowland counties . . . . . . . . . . *e*

e. Central lobe 1–5 cm. long, the upper lateral lobes 0.5–2 (rarely 3.5) cm. long; lobes of leaf short-pointed or sometimes scarcely developed and shallowly toothed; fruits (samaras) 1.5–2.5 (–3) cm. long; trees of Ozark hilly country. 3c. A. RUBRUM var. TRILOBUM

e. Central lobe 4–8 cm. long, the upper lateral lobes 2–5 cm. long; lobes of leaf long-pointed and coarsely toothed; fruits mainly 3.5–5 (rarely 2.7) cm. long; trees of the swampy country of the southeastern lowlands . . . . 3e. A. RUBRUM var. DRUMMONDII f. ROTUNDATA

3a. **Acer rubrum** var. **rubrum** f. **rubrum** Map 1484

*Acer rubrum* L. [G, P & S, Steyerm.]

*Acer rubrum* var. *rubrum* [BB]

*Acer stenocarpum* Britton, not Ettinghausen; originally described from St. Louis County, Missouri (dry flint hills, Allenton, 1884, *Letterman*) with fruits 3–6 mm. wide.

Occurs usually in rocky woods and along bluffs, in ravines, upper slopes and narrow upland ridges, in acid soils of chert, sandstone, or granite, rarely in low wet woods. Ozark section of southern and east-central Missouri, north to Lincoln, Warren, Gasconade, Boone, and Moniteau counties, west to Camden, Dallas, Greene, and Newton counties.

Ranges from Newfoundland and Quebec to Ontario, Manitoba, and Minnesota, south to Florida and Texas.

3b. **Acer rubrum** var. **rubrum** f. **tomentosum** (Desf.) Dansereau Map 1484

*Acer rubrum* f. *tomentosum* (Desf.) Dansereau [G]

Known from Shannon (*Bush*), Douglas, and Camden (cherty upper slopes of ravine tributary to Little Niangua River, T38N, R18W, sect. 4, 4½ mi. southeast of Barnumtown, July 6, 1952, *Steyermark 73743*) counties.

Found within the range of typical *A. rubrum* var. *rubrum* f. *rubrum*.

3c. **Acer rubrum** var. **trilobum** K. Koch Map 1484

*Acer rubrum* var. *tridens* Wood [P & S, Steyerm.]

Ozark section of southeastern and east-central Missouri, north to St. Louis and Boone counties, west to Phelps and Howell counties.

Ranges from Florida to Texas, north to Nova Scotia, Quebec, New York, West Virginia, Tennessee, Illinois, and Missouri.

The lower surface of the leaves of this variety may be either glabrous or pubescent.

3d. **Acer rubrum** var. **Drummondii** (H. & A.) Sarg. f. **Drummondii** Map 1484

*Acer rubrum* var. *Drummondii* (H. & A.) Sarg. [G, BB, P & S, Steyerm.]

Occurs in swamps, low wet woods, and margins of upland sink-hole ponds. Lowlands of southeastern Missouri north to Perry County, west to Wayne and Ripley counties.

Ranges from Florida to Texas, north to New Jersey, Indiana, Illinois, and Missouri.

3d. **Acer rubrum** var. **Drummondii** f. **rotundata** Sarg. Bot. Gaz. 67: 237. 1919 Map 1484

Occurs in swampy low woods. Known only from Dunklin, Butler (Poplar Bluff, September 3, 1882, *Letterman*), and Ripley counties.

Found within the range of typical var. *Drummondii*.

The Red Maple is one of the most strikingly beautiful trees in early spring, when the scarlet flowers cover the branches before the leaves appear, and in fall, when the foliage takes on hues of red, crimson, orange-red, or even yellow. The fruits at maturity, the winter buds and twigs, and even the young unfolding leaves are likewise red. This tree, scarlet oak (*Q. coccinea*), and Sour Gum or Tupelo (*Nyssa sylvatica*) provide the three most vivid reds among Missouri's larger trees.

In the eastern and northern states the Red Maple inhabits mainly swamps, river flood plains, and low wet woods. In Missouri it mainly (except for var. *Drummondii*) occupies the drier well-drained slopes of ravines, bluffs, and ridges in acid soils.

The sap of this maple provides a sugar, similar but inferior to that of Sugar Maple. The wood is occasionally mixed with that of other hardwoods for pulpwood in paper making. It is used also for flooring, interior finish, clothes pins, clothes hangers, kitchenware, and veneer. The early colonists used the bark to secure a tannin extract to which was added iron sulphate for making ink. Reddish-brown and black dyes were also prepared from the extract of the bark. The bark was also used by the Indians as a remedy for sore eyes.

A tree of var. *Drummondii*, believed to be the largest of its kind in Missouri, based upon data furnished by Mr. Kendall Laughlin, is found along the St. Francis River west of Kennett in Dunklin County, and measures 6 ft. 5 in. in circumference, 52 feet tall.

Some of the pollen of red maple is wind-borne and may at times be of importance in considering causes of hay fever during the pollinating period.

4. **Acer saccharinum** L. Silver Maple Map 1485

Also called Soft Maple.

Flowers February–April; fruits April–June.

Occurs in low wet woods, in river flood plains and valleys of streams. Throughout Missouri, doubtless in every county.

Ranges from New Brunswick and Quebec to Ontario, Minnesota, and South Dakota, south to Florida, Mississippi, Louisiana, and Oklahoma.

This and the Red Maple have the lower surface of the leaf conspicuously silvery-white. The foliage turns a greenish-pale yellow in general, but sometimes a deep purple-red or purple is prevalent in clusters of leaves among the remainder of the pale yellow foliage.

*Acer saccharinum* is one of the largest forest trees found in Missouri, attaining a trunk diameter (D.B.H.) of 4 feet. A tree measured by Mr. Kendall Laughlin in Big Oak Tree State Park in Mississippi County, considered to be the largest one of its kind in the state, measured 17 feet, 9 inches in circumference. The tree is much planted as a shade tree, but the brittle branches are subject to breakage from wind and ice storms. The sap of the tree yields a sweet pale syrup, but the yield is about half that of the Sugar Maple. A maple beer, similar to that given under *A. saccharum*, can be made from this species. The wood is used for pulpwood in papermaking.

5. **Acer Negundo** L. Box Elder Map 1486

Also called Ash-leaved Maple.

Flowers April–May; fruits May–October.

Occurs usually in alluvial ground along streams, river flood plains, low wet woods, and at the base of bluffs.

The following variations are encountered in Missouri:

a. Twigs from end of branchlets green or olive-green and glabrous (without hairs) . . . 5a. A. NEGUNDO var. NEGUNDO
a. Twigs from end of branchlets either glaucous with a gray-purple coating, which can be rubbed off, or hairy. . . . . . . . *b*
  b. Twigs from end of branchlets glaucous with a gray-purple coating, which can be rubbed off . . 5b. A. NEGUNDO var. VIOLACEUM
  b. Twigs from end of branchlets hairy . . *c*
    c. Fruit minutely hairy; leaf-stalks mainly glabrous (without hairs) . 5c. A. NEGUNDO var. TEXANUM
    c. Fruit usually without hairs; leaf-stalks usually hairy. 5d. A. NEGUNDO var. INTERIUS

5a. **Acer Negundo** var. **Negundo** Map 1486

This is the commonest variety in Missouri, doubtless in every county.

Ranges from Florida to Texas, north to Vermont, New York, Ontario, and Minnesota.

5b. **Acer Negundo** var. **violaceum** (Kirsch.) Jaeg. Map 1486

Northern and central Missouri, south to Franklin, Moniteau, Cooper, Johnson, and Cass counties, and locally south in Dallas and Polk counties.

Ranges from Michigan to Montana, south to Missouri, Kansas, and Colorado; cultivated and established east to New Jersey and New England.

5c. **Acer Negundo** var. **texanum** Pax Map 1486

Western Missouri south of the Missouri River north to Lafayette County, east to Benton, Polk, and Lawrence counties, and locally east to Dunklin County.

Ranges from Texas to North Carolina, north to Oklahoma, Missouri, Indiana, and Ohio.

There is considerable intergradation between this variety and the next, and it is frequently difficult to place material with certainty. Presumed differences in leaf margins (scarcely toothed or slightly or shallowly crenate-dentate in var. *texanum* versus coarsely toothed to lobed leaflets in var. *interius*) do not seem to be constant or reliable criteria for differentiation.

5d. **Acer Negundo** var. **interius** (Britt.) Sarg.
Map 1486

North-central and west-central Missouri, north to Grundy County, east to Chariton and Boone counties, southwest to Cass County.

Ranges from Manitoba to Montana, south to Missouri, Oklahoma, New Mexico, and Arizona.

While most of the other species of maples are pollinated by both wind and insects (amphiphilous), the box elder is wholly wind-pollinated. The pollen of this species, since it is wind-borne, is important from the standpoint of being a causative agent in hay fever cases during the pollinating season.

The leaves in autumn fall when those of most trees are still attached to the trees and do not turn color as do those of other maples. The wood is occasionally mixed with that of other hardwoods as pulpwood in papermaking. Box Elder was formerly much used as a shade tree, but its continued use has been mainly discouraged because of the early leaf-fall and the susceptibility of the trees to various insect pests, such as Box Elder bug and other insects and diseases. The sap, as in most maples, is sweet and is sometimes made into a syrup and sugar. Young saplings are sometimes mistaken for Poison Ivy, because the leaves are similar, but the stems of Box Elder are green and the leaves are opposite and a brighter green. The apetalous flowers are yellow-green and the fruits are greenish.

## Fam. **HIPPOCASTANACEAE** (Horse Chestnut Family)

### **Aesculus** L. Buckeye

I. *Key based on flowers and fruits*

Found throughout Missouri, absent from extreme southeastern Missouri; petals more or less equal, greenish-yellow; calyx yellow or greenish-yellow, campanulate, 3–8 mm. long, rather regular with more or less equal lobes; stamens long-protruding beyond the petals; fruit prickly; leaflets blue-green or grass green on upper surface . . . . . . . . . . . . . . . . . . . . . . . 1. A. GLABRA

Found wild only in southeastern Missouri north to Perry, Cape Girardeau, Bollinger, and Madison counties, west to Wayne, Carter, and Oregon counties; petals unequal in length, width, and shape, red or rarely yellow; calyx dark red, tubular, 8–18 mm. long, irregular and oblique at base, the lobes unequal in length; stamens shorter than or only slightly longer than the upper petals; fruit smooth; leaflets dark green on upper surface . . . . . . . . . . . . . . . . . . . . 2. A. PAVIA

II. *Key based on leaves and fruits*

a. Leaflets mainly 7, rarely 8–11, or 6; fully grown leaflets mainly 1–3 (5) cm. wide. 1. A. GLABRA var. ARGUTA

a. Leaflets mainly 5, rarely 6 or 7; fully grown leaflets mainly 3–8 cm. wide. . . . . . . . . . *b*

b. Fruits prickly; found throughout Missouri, absent from extreme southeastern Missouri . . . 1. A. GLABRA var. GLABRA and other varieties

b. Fruit smooth; found wild only in southeastern Missouri north to Perry, Cape Girardeau, Bollinger, and Madison counties, west to Wayne, Carter, and Oregon counties . . . . . . . . 2. A. PAVIA

1. **Aesculus glabra** Willd. Ohio Buckeye
Map 1487

Flowers April–May.

Occurs in rich or rocky woods of valleys, ravines, gentle or steep slopes, base of bluffs, and thickets. Throughout Missouri, except apparently absent from extreme southeastern Missouri.

The Missouri material may be divided into the following variations:

a. Leaflets mainly 7, rarely 8–11, or 6; leaflets mainly 1–3 (5) cm. wide . . . 1d. A. GLABRA var. ARGUTA

a. Leaflets mainly 5, rarely 6 or 7; leaflets mainly 3–8 cm. wide . . . . . . . . . . *b*

b. Commonest type found; lower surface of mature leaflets densely or lightly hairy, the hairs mostly persisting in age . 1b. A. GLABRA var. GLABRA f. PALLIDA

b. Rare type found; lower surface of mature leaflets glabrous (without hairs) or with only a few hairs on the veins or in their axils . . . . . . . . . . . *c*

c. Lower surface of mature leaflets usually strongly whitened or pale; bark pale or nearly whitish . . . . 1c. A. GLABRA var. LEUCODERMIS

c. Lower surface of mature leaflets green; bark dark gray . . . . 1a. A. GLABRA var. GLABRA f. GLABRA

1486 • Acer Negundo var. Negundo (Box Elder)
1486 x Acer Negundo var. violaceum
1486 □ Acer Negundo var. texanum
1486 ‡ Acer Negundo var. interius

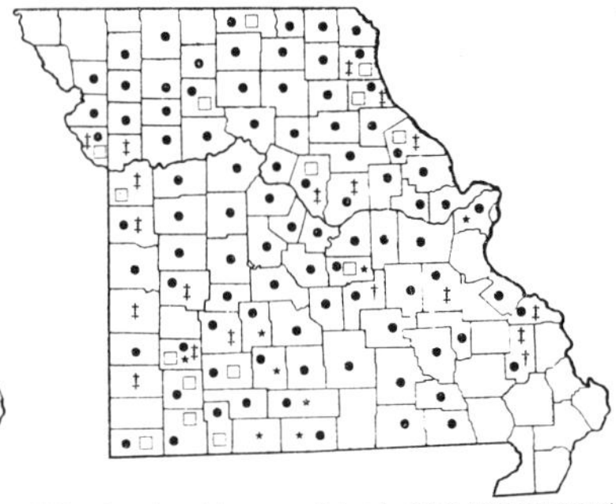

1487 □ Aesculus glabra var. glabra f. glabra (Ohio Buckeye)
1487 • Aesculus glabra var. glabra f. pallida
1487 ⋆ Aesculus glabra var. leucodermis
1487 ‡ Aesculus glabra var. arguta
1487 † Aesculus glabra X Aesculus Pavia

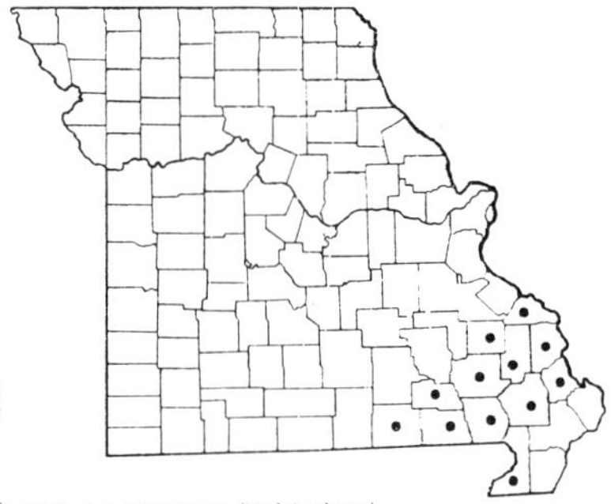

1488 Aesculus Pavia (Red Buckeye)

1a. **Aesculus glabra** var. **glabra** f. **glabra** Map 1487

*Aesculus glabra* Willd. [G, P & S, Steyerm.]

*Aesculus glabra* var. *glabra* [BB, Hardin in part]

This is scattered and rare in Missouri, but occurs in the general range within the state of the common f. *pallida*.

Ranges from Pennsylvania to Michigan, Iowa and Nebraska, south to Alabama, Mississippi, Arkansas, and Oklahoma.

1b. **Aesculus glabra** var. **glabra** f. **pallida** (K. Koch) Fern. Map 1487

*Aesculus glabra* var. *glabra* [Hardin in part]

*Aesculus glabra* f. *pallida* (K. Koch) Fern. [G, Steyerm.]

*Aesculus glabra* var. *pallida* (Willd.) Fern. [P & S]

This is the commonest variation encountered throughout the state, but is apparently absent and has not been recorded from the counties of extreme southeastern and northwestern Missouri.

Ranges from Pennsylvania to Michigan, Iowa, and Nebraska, south to Alabama, Mississippi, Arkansas, and Oklahoma.

Sometimes specimens are collected with some of the leaflets whitened on the lower surface, and others ordinary pale green. Such a collection was found in Pike County, the leaves from the normal end shoots being whitened beneath (*Steyermark 25812*), while those from the lower branches of the same tree were only pale green (*Steyermark 25813*).

1c. **Aesculus glabra** var. **leucodermis** Sarg. Map 1487

*Aesculus glabra* var. *glabra* [Hardin in part]

Rare, and occurs in the Ozark region of southern and east-central Missouri north to Maries and St. Louis counties.

Ranges from Missouri to Arkansas.

1d. **Aesculus glabra** var. **arguta** (Buckl.) Robinson Map 1487

*Aesculus glabra* var. *Sargentii* Rehd. [G, BB, P & S, Steyerm.]

Rare, and scattered in southern, central, and eastern Missouri, north to Platte, Clay, Boone, Callaway, and Lewis counties.

Ranges from Ohio to Iowa, south to Arkansas and Texas.

This genus has recently been revised by Dr. James W. Hardin (Brittonia 9: 173–95. 1957), who does not recognize var. *leucodermis* or var. *glabra* f. *pallida* (f. *pallida*) as meriting separation from typical var. *glabra*, but does separate var. *arguta* (var. *Sargentii*) from the remainder of the *A. glabra* group as a distinct variety.

In the present treatment these variations have been accorded legitimate rank, as they are recognizable as variations even though there is some intergradation. The var. *leucodermis* appears distinct in the field by virtue of its whiter bark and whitened leaflets beneath, and also seems to have a restricted geographic range in the Ozark region. It would seem to merit equal varietal recognition status as that accorded by Hardin to var. *arguta*, and similarly the glabrous and pubescent

Plate no. 245. **1.** Acer saccharinum, $\times {}^2/_7$; *a.* Flowering branch. **2.** Acer Negundo, $\times {}^2/_7$; *a.* Male flowering branch, $\times {}^2/_7$; *b.* Male flower, $\times 1{}^5/_7$; *c.* Female flowering branch, $\times {}^2/_7$; *d.* Female flower, $\times 1{}^1/_7$. **3.** Aesculus glabra var. glabra, $\times {}^2/_7$; *a.* Aesculus glabra var. arguta leaf, $\times {}^2/_7$ **4.** Aesculus Pavia, $\times {}^2/_7$; Details from Small, The New York Botanical Garden.

PLATE NO. 245

forms of var. *glabra* can be distinguished in the field and herbarium as different populations varying toward or away from pubescence.

This buckeye sends out new leaves in earliest spring before any other tree in Missouri and can be easily recognized at that season of the year. It is one of the commonest understory trees of the forest. The large, many-flowered panicles of flowers appear at the ends of the relatively few, nearly or quite simple branches. It also begins to change color in summer and drops its leaves earlier than any other native tree.

The shining, large, dark brown seeds are poisonous when eaten by children, livestock, and domestic animals. Squirrels and birds are not known to eat the seeds. The leaves of early spring, when eaten, are poisonous to cattle, horses, and pigs. The seeds are rendered harmless after boiling or roasting, and were eaten by Indians as a starchy meal after being roasted.

The wood, which is quite brittle and light in weight, is used for violins, artificial limbs, crates, and boxes. It ranks as the softest among all the native woods of the United states, and is sold as tulip tree, the sap wood of which it resembles. The early colonists used hollowed out, short sections of buckeye logs as troughs for catching sugar maple sap. Also cradles of early pioneer families were made from hollowed out logs of this tree, and from the shavings were made summer hats. The wood today is used as pulp for paper making.

Dr. Hardin records hybrids as occurring in Missouri between *A. glabra* and *A. octandra* Marsh., a species which does not occur in Missouri (Rh. 59: 188–92. 1957). *Aesculus octandra* is found from Pennsylvania to Ohio and Indiana, south to Georgia and Alabama, west to southern Illinois. The evidence Dr. Hardin submits for the occurrence of hybrid populations of *A. glabra* and *A. octandra* is based upon the presence of stipitate glands on the flower-stalks (pedicels), considered a characteristic of *A. octandra.* The following Missouri collections are cited by Hardin (p. 190) as hybrids between *A. glabra* and *A. octandra*: Boone Co. (*Drouet 1901*); Marion Co. (*Davis 1462*); St. Louis (*Hardin 670*); Jefferson Co. (April 18, 1869, *Eggert*); Taney Co. (*Bush 13276*); Greene Co. (*Hardin 667*); and Jackson Co. (*Bush 125*). These hybrids are characterized, according to Hardin, by the exserted stamens (characteristic of *A. glabra*), stipitate glands on the pedicel and perianth surfaces (characters of *A. octandra*), spines on the ovary wall (character of *A. glabra*), and much more dissimilar upper and lateral petals than are usually found in *A. glabra.*

2. **Aesculus Pavia** L. Red Buckeye Map 1488
*Aesculus discolor* Pursh [G, P & S, Steyerm.]
*Aesculus discolor* var. *mollis* Sarg. [P & S, Steyerm.]
Flowers April–June.

Occurs in low rich woods in valleys, at the base of bluffs, low slopes, and along streams. Southeastern Missouri north to Perry and Madison counties, west to Wayne, Carter, and Oregon counties.

Ranges from Florida to Texas, north to Virginia, Tennessee, southern Illinois, Missouri, and Oklahoma.

According to Dr. Hardin's recent studies (Brittonia 9: 192–95. 1957), *A. Pavia* cannot be separated from *A. discolor* and var. *mollis*, because of a continuity of intergradation. There is a large amount of variation within this species in shape and size of inflorescences, color of flowers, texture of leaves, form and color of bark, and pubescence of the lower surface of the leaves. Towards the western portion of the range of the species there is an increase in the amount of pubescence found on the lower surface of the leaflets (characteristic of what was called *A. discolor* var. *mollis*) and a tendency for the calyx to become shorter and more campanulate (Hardin, p. 194).

Missouri specimens considered by Dr. Hardin (Rhodora 59: 192–93. 1957) to represent hybrids between *Aesculus glabra* and *A. Pavia* are based upon the following collections: Bollinger Co. (*Steyermark 28404*) and Phelps Co. (Newburg, May 4, 1915, *B. H. Slavin 235*). The Bollinger County collection was originally identified as *A. discolor* with yellow and rose petals and the rose calyx suffused with greenish-yellow. The Phelps County specimen (in Gray Herb.) has a short calyx of *A. glabra*, but has the red color of the calyx characteristic of *A. discolor* = *A. Pavia.* This Phelps County specimen was originally cited by Palmer and Steyermark in their *Annotated Catalogue* as *A. discolor* var. *mollis.* True *A. Pavia* (*A. discolor*) does not occur wild in Phelps County, however, and its nearest station in the state is about 100 miles to the southeast in Madison County. A specimen in the Univ. of Mo. Herb. (*Daniels*, along streams, Columbia, Boone Co., July, 1903) also may be identified as *A. Pavia*, but was probably collected from a cultivated tree; the lower surface of the leaflets are glabrous as are the petioles and petiolules.

Like *A. glabra*, the seeds and young foliage especially are poisonous. The powdered seeds and crushed branches of *A. Pavia*, when placed in ponds or slow water, have been used to catch fish, which become stupefied and float to the water's surface. The roots contain a saponin which has the property of foaming in water, and has been employed for its soap properties.

1489 Cardiospermum Halicababum (Balloon Vine)

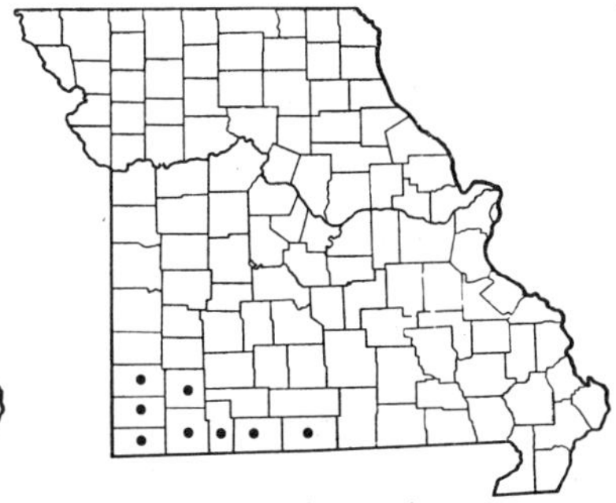

1490 Sapindus Drummondii (Soapberry)

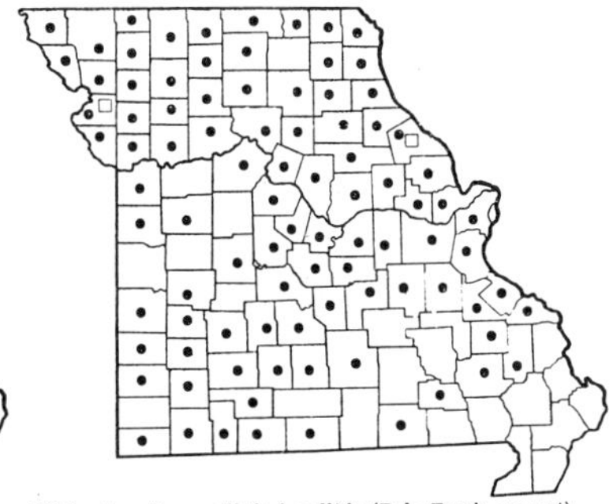

1491 • Impatiens pallida f. pallida (Pale Touch-me-not)
1491 □ Impatiens pallida f. dichroma

The Red Buckeye is of much shorter stature on the average than *A. glabra*, attaining 1–4 meters average height. It is very ornamental when in flower, the dark red flower clusters standing erect and in contrast to the dark green foliage. It is easily raised, as is *A. glabra*, from seed, and flowers within a few years after germination. It prefers a shaded situation in rich deep soil.

*Excluded Species*

**Aesculus Hippocastanum** L. Horse Chestnut

This species is commonly cultivated, but no evidence is at hand to indicate that it has escaped from cultivation or become naturalized. A specimen in the Univ. of Mo. Herb. (*Dunlap*, June 20, 1929, Palmyra, Marion Co. along highway 61, 3 blocks south of the hotel) was taken from a cultivated tree, and cannot be considered as a naturalized specimen.

## Fam. **SAPINDACEAE** (Soapberry Family)

Trees; leaflets smooth-edged (without teeth); native of southwestern Missouri . . . . . . . 2. SAPINDUS

Vine or climbing plant; leaflets toothed; introduced cultivated foreign plant . . . . . 1. CARDIOSPERMUM

### 1. **Cardiospermum** L. Balloon Vine

**Cardiospermum Halicababum** L. Balloon Vine Map 1489

Flowers July–September.

Sometimes planted in gardens, and rarely persisting away from cultivation. Known only from St. Louis and St. Charles (wet meadow near Boschertown, August 13, 1949, *Bauer*) counties.

Native of tropical America; introduced and naturalized in the United States from Florida to Texas, north to New Jersey, Pennsylvania, Ohio, Illinois, Missouri, and Oklahoma; also in the Old World tropics.

This is an ornamental vine with white flowers, compound ternate or biternate leaves, tendrils, and bladdery fruits 2–3.5 cm. long with black seeds. The tender young leaves are cooked in some of the tropical countries as a vegetable.

### 2. **Sapindus** L. Soapberry

**Sapindus Drummondii** Hook. & Arn.
Soapberry Map 1490

Also called Wild China Tree.

Flowers last of May–July.

Occurs at the base of or on the ledges of south- and west-facing limestone bluffs along streams and on limestone glades. Southwestern Missouri, east to Ozark County, north to Lawrence and Jasper counties.

Ranges from Louisiana, Texas, New Mexico, and Arizona, north to southwest Missouri, Kansas, and Oklahoma; also Mexico.

This makes a desirable ornamental shade tree. Two stately specimens are planted on the campus of the University of Missouri. The large panicles of white flowers are showy and in display during June and part of July.

The plant contains saponin, which is the basis for the soapy nature of the fruits. The fruits of a related species, *S. saponaria,* are commonly used by the inhabitants of some of the Latin American countries for washing clothes, because of the abundance of soapy suds produced when the fruits are mashed or cut up in water. This fruit is likewise employed in some parts of Mexico as a fish poison. Some individuals have an allergy to the fruits, and develop a dermatitis when handling them. The globe-shaped large seeds of the related *S. saponaria* are used for making necklaces and rosaries, and the native *S. Drummondii* may be similarly used.

## Fam. **BALSAMINACEAE** (Touch-me-not Family)

### **Impatiens** L. Touch-me-not

a. Leaf-blades sharply toothed, narrowly to broadly lanceolate, gradually narrowed to the base into the leaf-stalk, on short leaf-stalks 0.5–2 cm. long; some or all of leaf-stalks with pin-head-like glands near the base; fruit and spurred sepal hairy; flowers solitary, only 1 from a leaf axil; cultivated plant, rarely found escaped . . . . . . . . . . . . . . . . . . . . 3. I. BALSAMINA

a. Leaf-blades crenately toothed (scallop-edged with the teeth rounded), ovate or elliptic, on long leaf-stalks 0.5–7 cm. or more long, abruptly changing at the base into the leaf-stalk; pin-head-like glands, when present, only at base of leaf-blade, not at base of leaf-stalk; fruit and sepal glabrous (without hairs); flowers few together in racemes from the axils of the leaves, not solitary; wild native plants, commonly found. . . . . . . . . . . . . . . . . . . . . . . . . *b*

b. Flowers pale yellow or yellow and white; spur of sepal bent down (deflexed) at a right angle, $\frac{1}{5}$–$\frac{1}{4}$ the length of the enlarged rear portion of flower (saccate sepal), the latter broader than long; upper surface of leaf-blade grayish- or pale bluish-green, conspicuously glaucous (with a whitish or grayish coating which can be rubbed off) . . . . . . . . . . . . . . . 1. I. PALLIDA

b. Flowers chiefly orange; spur of sepal bent forward close to and parallel with sepal, $\frac{1}{3}$–$\frac{1}{2}$ the length of the enlarged rear portion of flower (saccate sepal), the latter longer than broad; upper surface of leaf-blade less or not at all glaucous, dull or darker green . . . . . . . . . 2. I. CAPENSIS

1. **Impatiens pallida** Nutt. Pale Touch-me-not Map 1491

Also called Jewel Weed.

Flowers late May–October.

Occurs in rich or damp low woodland, base of bluffs, low thickets, banks of streams, and swampy ground. Throughout Missouri, except apparently absent or not recorded from the lowland counties of the extreme southeastern portion.

Ranges from Newfoundland to Saskatchewan, south to Georgia, Tennessee, Missouri, Kansas, and Oklahoma.

Two variations are encountered in Missouri:

Flowers wholly yellow with usually reddish-brown spots; common . . . 1a. I. PALLIDA f. PALLIDA

Flowers yellow in one part (saccate sepal and rear) and white in another part (upper and lateral petals); rare. . . . 1b. I. PALLIDA f. DICHROMA

1a. **Impatiens pallida** f. **pallida** Map 1491

*Impatiens pallida* Nutt. [G, BB, P & S, Steyerm.]

This is the common type found throughout the state.

1b. **Impatiens pallida** f. **dichroma** Steyerm. Map 1491

Known only from Pike (wooded base of bluffs along Mississippi River, T54N, R2W, sect. 11, 1½ mi. northwest of Louisiana, September 6, 1937, *Steyermark 25927*) and Buchanan (base of slopes in alluvial thickets bordering Missouri River bottoms, T56N, R36W, north part sect. 6, 1 mi. southwest of Halls, August 20, 1950, *Steyermark 70063*) counties.

The name 'Touch-me-not' is derived from the fruit, which, when ripe, snaps open suddenly at a mere touch and throws out the seeds. The epithet 'Jewel-weed' is evident when a leaf placed under water shows a shining silvery or quick-silverlike surface. It is reputed that the leaves and stems of this and the next

Plate no. 246. **1.** Cardiospermum Halicababum, × 4/5; *a.* Flower, × 2 2/5; After Gleason, details from Small, The New York Botanical Garden. **2.** Impatiens Balsamina, × 2/5, Details from Small, The New York Botanical Garden. **3.** Sapindus Drummondii, × 2/5; *a.* Flower, × 2 4/5; Details from Small, The New York Botanical Garden. **4.** Impatiens capensis, × 2/5; Details from Small, The New York Botanical Garden. **5.** Impatiens pallida, × 2/5.

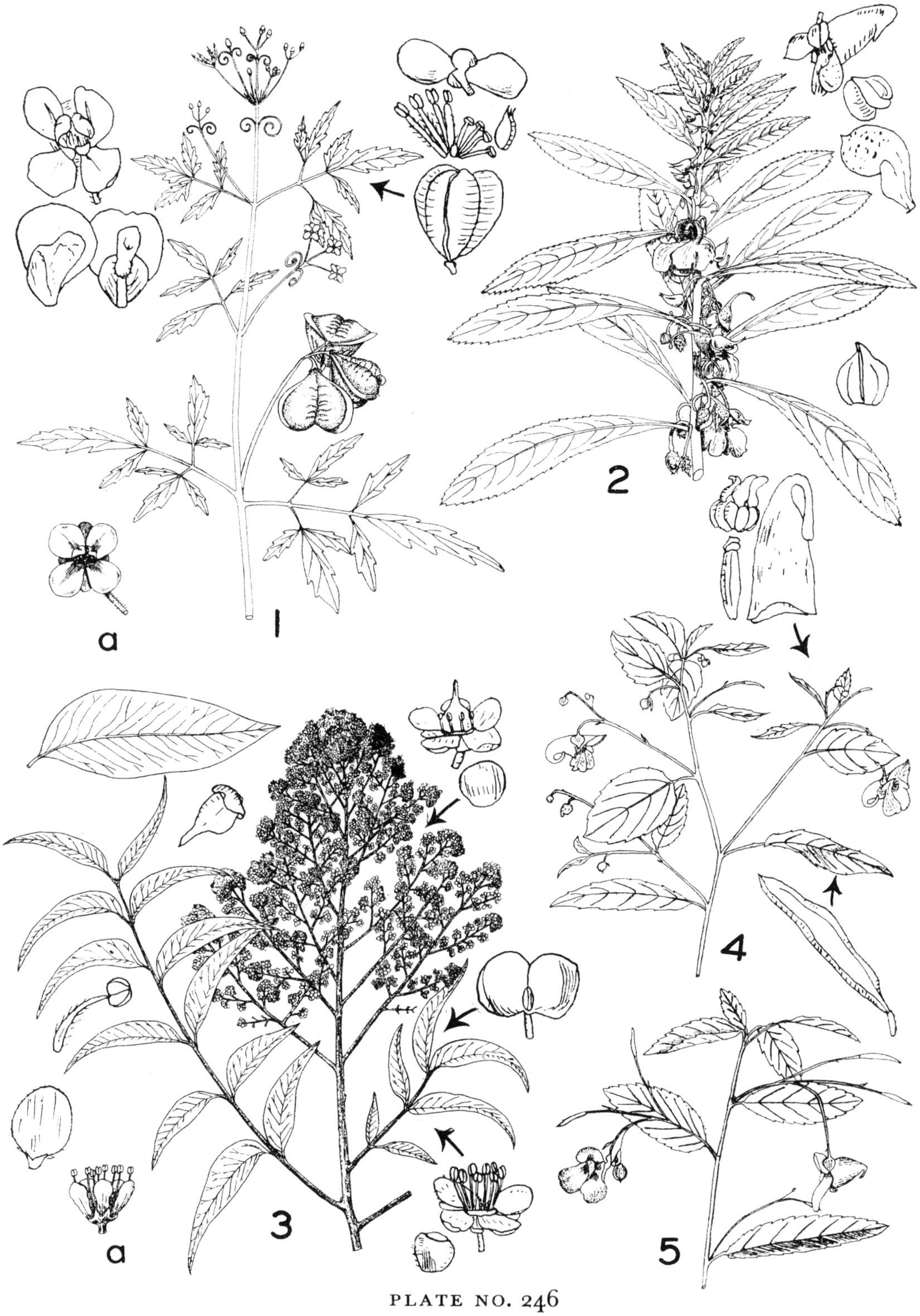


PLATE NO. 246

1492 Impatiens capensis (Spotted Touch-me-not)

1493 Impatiens Balsamina (Balsam)

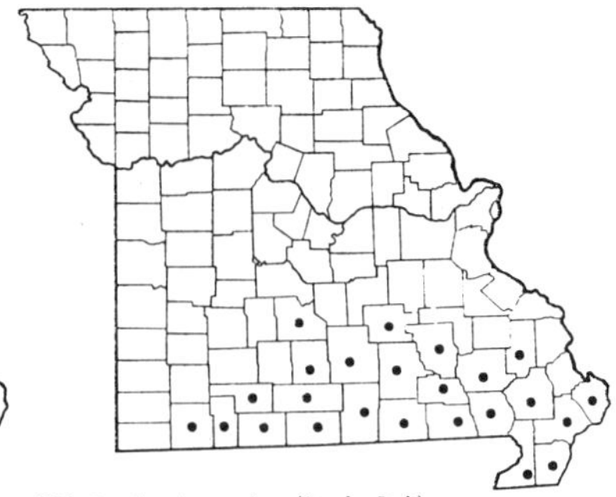

1494 Berchemia scandens (Supple-Jack)

species, rubbed on the skin, will prevent poison ivy from developing after contact with the plant or will even effect a cure. Apparently it has been known to work for some persons. This and spotted Jewel-weed are soft, watery-stemmed annuals. The seeds germinate and usually develop into large colonies.

A form with the flowers creamy-white is known as f. *speciosa* Jennings, but has not been recorded from Missouri.

2. **Impatiens capensis** Meerb.
Spotted Touch-me-not Map 1492
*Impatiens biflora* Walt. [BB, P & S, Steyerm.]
Flowers late May–October.

Occurs in rich or damp low woodland, base of bluffs, low thickets, banks of streams, and swampy ground.

Throughout Missouri, doubtless in every county.

Ranges from Newfoundland to Alaska, south to Florida, Alabama, Arkansas, and Oklahoma.

The plant has emetic properties and is sometimes believed, when eaten, to poison livestock, but white-tailed deer are known to feed on the plants.

A number of color variations of the flowers have been described, but have not been recorded from Missouri. An unspotted variation is f. *immaculata* (Weath.) Fern. & Schub.; one having lemon-yellow flowers and crimson spots is f. *citrina* (Weath.) Fern. & Schub.; a white- or cream-colored variant with pink or brownish-red spots is f. *albiflora* (Rand & Redf.) Fern. & Schub.; a cream-colored form with pink coloring over most of the surface with pink spots is f. *Peasei* (A. H. Moore) Fern. & Schub. The size and color of the leaf varies with the amount of exposure and dryness, those plants exposed to sunlight and in drier habitats often developing smaller and deeper green leaves.

3. **Impatiens Balsamina** L. Balsam Map 1493
Flowers July–September.

Grown as an annual ornamental plant for gardens, and rarely escaped from cultivation. Known only from Dunklin County, southeastern Missouri (July 19, 1900, *R. Dimmitt 286*, in Univ. of Mo. Herb.).

Native of Asia; introduced and naturalized in the United States.

## Order **RHAMNALES**

### Fam. **RHAMNACEAE** (Buckthorn Family)

a. Leaves with 3 prominent nerves arising from the base of the leaf-blade; flowers white; fruit dry, when ripe splitting open into 3 lobes . . . . . . . . . . . . . . . . . 3. CEANOTHUS

a. Leaves feather-nerved (pinnately nerved), all prominent and arising from various parts of the midrib; flowers greenish-yellow or greenish or greenish-white; fruit fleshy . . . . . . . . . *b*

b. Upright shrub or small tree with easily broken twigs; tip of leaf-blade lacking a minute hair-like projection or cusp, but otherwise often pointed or acute at tip; margins of leaves with small but definite rounded or pointed teeth, or without teeth; petals narrowed to a claw, with a notch at tip . . . . . . . . . . . . . . . . . . . . . . . . 2. RHAMNUS

b. Vine or sprawling plant with tough twigs difficult to break; tip of leaf-blade rounded or pointed with a minute hair-like projection (cusp); margins of leaves without teeth or wavy-edged or with blunt rounded teeth; petals not narrowed to a claw, sessile, without any notch at tip. . 1. BERCHEMIA

## 1. **Berchemia** Necker Supple-Jack

**Berchemia scandens** (Hill) K. Koch
Supple-Jack Map 1494

Flowers May–June; fruits August–October.

Occurs in low swampy woods in the southeastern lowland counties, and in the Ozark region occurs on limestone glades, rocky limestone ledges, along bluffs, or in rocky ground along stream beds and small draws of ravines. Southern Ozark section from Barry County north to Laclede, Texas, Dent, and Reynolds counties, and in the southeastern lowland counties north to Wayne, Bollinger, and Mississippi counties.

Ranges from Florida to Texas, north to Virginia, Tennessee, southern Illinois, Missouri, and Oklahoma.

The smooth, oblong leaves are dark green above and pale beneath, and persist well into the winter. On the limestone glades the stems are often sprawling or trail over the ground, but in the wet lowlands of southeastern Missouri, may climb high in the tall trees.

## 2. **Rhamnus** L. Buckthorn

The classical feminine gender is employed to conform to the International Code and to usage.

a. Leaf-blades with mostly 3–4 nerves on each side of the midrib; leaves mainly opposite or nearly opposite; leaf-blades abruptly pointed at tip; twigs often ending in a spiny tip; introduced species . . . 1. R. CATHARTICA

a. Leaf-blades with usually 4–10 nerves on each side of the midrib; leaves mainly alternate; leaf-blades gradually pointed at tip; none of the twigs ending in a spiny tip; native species . . . . . . . . *b*
- b. Margin of leaf-blade with many fine teeth; largest leaf-blades 3.5–9 cm. long; buds for next year covered with scales; fruit 2-seeded; calyx-lobes, petals, and stamens 4 . . . 2. R. LANCEOLATA
- b. Margin of leaf-blade faintly toothed with short not noticeable teeth or nearly without teeth; largest leaf-blades mainly 7.5–15 cm. long; buds for next year naked, densely brown-hairy; fruit 3-seeded; calyx-lobes, petals, and stamens 5 . . . . . . . . . . . . 3. R. CAROLINIANA

### 1. **Rhamnus cathartica** L. Common Buckthorn
Map 1495

*Rhamnus catharticus* L. [BB]

Flowers April–June.

Rarely escaped from cultivation in wooded thickets, along roadsides, and waste ground. Known only from Marion (escaped in Riverview Park and common in wild places in Hannibal, October 7, 1933, *Drouet 1237*), Boone (roadsides near Providence bridge, Columbia, August, 1904, *Daniels*), and Cape Girardeau (Sloan's Creek, Cape Girardeau, April 29, 1954, *Rhodes 301*; same locality, October 11, 1953, *Rhodes 185*) counties.

Native of Europe; introduced and naturalized in North America, from Quebec to Minnesota, south to Virginia, Ohio, Illinois, and Missouri.

This shrub or small tree attains a height of 25 feet. The leaves remain dark green long after other trees and shrubs have changed color, and the leaves persist well into late fall. In the northern states this species was at one time more commonly used as a hedge plant. The bark, leaves, and fruit possess a substance which has purgative properties. The juice of the unripe fruit was at one time used for staining maps, while the ripe fruit was the basis for the color known as sap green used by artists. The plant seeds easily and sometimes becomes too plentiful in some areas.

### 2. **Rhamnus lanceolata** Pursh Buckthorn
Map 1496

Flowers mid-April–June; fruits June–August.

Occurs on open wooded slopes, in thickets, and on glades, usually in and around limestone exposures. Throughout Missouri, but absent from the lowland counties of the extreme southeastern portion.

Two variations occur in Missouri:

Lower surface of leaf-blades minutely hairy . . 2a. R. LANCEOLATA var. LANCEOLATA

Lower surface of leaf-blades glabrous (without hairs) or nearly so . . . . 2b. R. LANCEOLATA var. GLABRATA

### 2a. **Rhamnus lanceolata** var. **lanceolata**
Map 1496

*Rhamnus lanceolatus* var. *lanceolatus* [BB]

*Rhamnus lanceolata* Pursh [G, P & S, Steyerm.]

This is the rarely encountered variation in the state, and is scattered in the eastern section.

Ranges from Alabama to Texas, north to Pennsylvania, West Virginia, Ohio, Indiana, Wisconsin, Iowa, and Nebraska.

### 2b. **Rhamnus lanceolata** var. **glabrata** Gleason
Map 1496

*Rhamnus lanceolatus* var. *glabratus* Gleason [BB]

This is the common variation in Missouri.

1495 Rhamnus cathartica (Common Buckthorn)

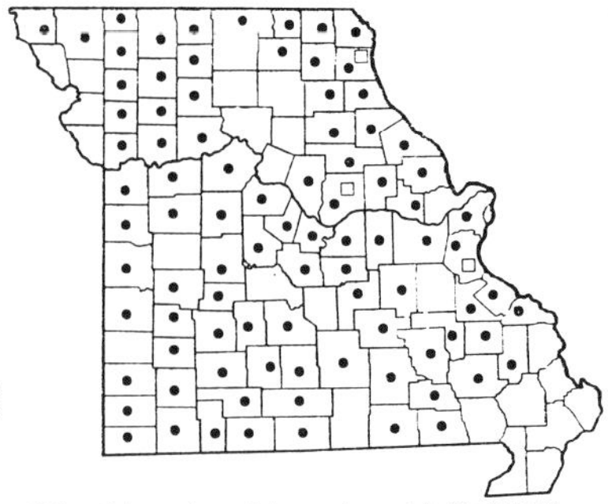

1496 □ Rhamnus lanceolata var. lanceolata (Buckthorn)
1496 ● Rhamnus lanceolata var. glabrata

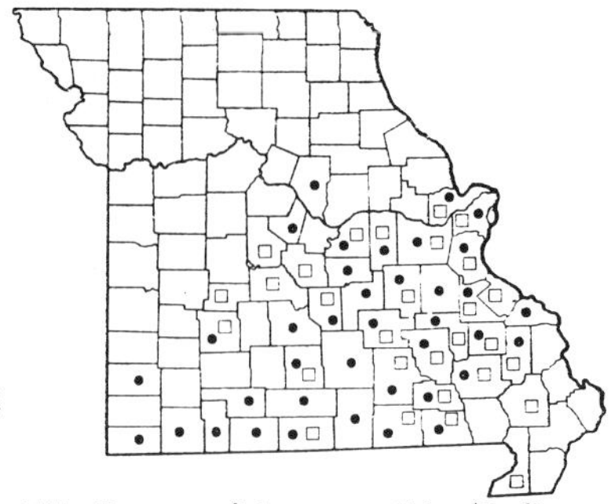

1497 ● Rhamnus caroliniana var. caroliniana (Carolina Buckthorn)
1497 □ Rhamnus caroliniana var. mollis

Ranges from Kentucky to Nebraska, south to Tennessee, Arkansas, and Oklahoma.

3. **Rhamnus caroliniana** Walt.
Carolina Buckthorn Map 1497
Also called Indian Cherry.
Flowers May–June.

Occurs in low woodland in valleys along streams, rocky open wooded slopes, upland ridges, thickets, and glades, chiefly in limestone areas.

Occurs throughout the Ozark region north to St. Charles, Boone, Moniteau, Polk, and Jasper counties; absent from the lowlands of extreme southeastern Missouri, except on Crowley Ridge, where represented.

Two variations are found in Missouri:

Lower surface of fully grown leaves slightly hairy to nearly glabrous (nearly without hairs) . . .
3a. R. CAROLINIANA var. CAROLINIANA
Lower surface of fully grown leaves densely hairy
3b. R. CAROLINIANA var. MOLLIS

3a. **Rhamnus caroliniana** var. **caroliniana**
Map 1497

*Rhamnus caroliniana* Walt. [G, BB, P & S, Steyerm.]

This appears to be the commoner variation in the state, and absent from Crowley Ridge.

Ranges from Florida to Texas, north to Virginia, West Virginia, Ohio, Indiana, Illinois, Missouri, and Nebraska.

3b. **Rhamnus caroliniana** var. **mollis** Fern.
Map 1497

This is less frequent in the southwestern Ozark section, but occurs on Crowley Ridge in Stoddard and Dunklin (low woods, Campbell, *Bush 6376*) counties in southeastern Missouri.

Ranges from Indiana to Tennessee, south to Mississippi, Missouri, and Texas.

As the more densely pubescent-leaved variation seems to be restricted to the more western portion of the range of the species, it is accorded varietal recognition. However, all gradations occur in Missouri between leaves nearly glabrous beneath to those densely and softly hairy, and it is highly questionable whether the variations should be separated with recognized status. They are not clearly cut in Missouri, and in Indiana Deam encountered the same situation, although he recognized both varieties (*Fl. Ind.* 660. 1940; *Shrubs of Ind.*, second ed. 214. 1932).

The dark green shining foliage and red fruits combine to make this a very ornamental shrub to small tree, whose cultivation should be more encouraged in home plantings. The foliage lasts well into mid-winter, after turning orange or orange-yellow and green in late fall. Attempts to establish this attractive plant in northern Illinois at the author's botanical preserve have not succeeded; the winters apparently are too severe.

## 3. **Ceanothus** L. New Jersey Tea

Main flower- or fruit-stalk (peduncle) longer than the leaf arising directly below it; main flower- or fruit-stalk elongated, naked or with a few reduced leafy bracts, arising axillary (from the sides of the

Plate no. 247. **1.** Berchemia scandens, × 2/5; Details from Small, The New York Botanical Garden. **2.** Rhamnus caroliniana, × 2/5; Details from Small, The New York Botanical Garden. **3.** Ceanothis ovatus, × 2/5. **4.** Rhamnus lanceolata, × 2/5. **5.** Ceanothus americanus, × 2/5; *a.* Flower, three views, × 1 1/5; Details from Small, The New York Botanical Garden. **6.** Rhamnus cathartica, × 2/5; *a.* Flowering branch; *b.* Leafy branch; Details from Small, The New York Botanical Garden.

PLATE NO. 247

1498 □ Ceanothus americanus var. intermedius (New Jersey Tea)
1498 • Ceanothus americanus var. Pitcheri

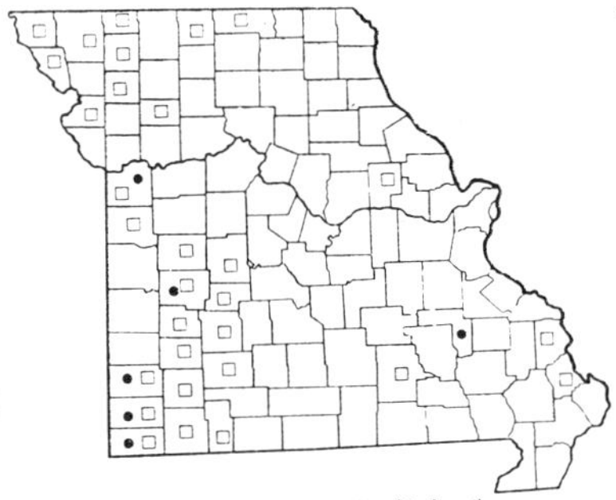

1499 • Ceanothus ovatus var. ovatus (Redroot)
1499 □ Ceanothus ovatus var. pubescens

1500 Ampelopsis cordata (Raccoon Grape)

main stem in the axils of leaf-stalks); leaf-blades broadly ovate or ovate-oblong, mostly short- to long-pointed, rarely blunt . . . . . . . . . . . . . . . . . . . . . . 1. C. AMERICANUS
Main flower- or fruit-stalk (peduncle) usually shorter than the leaf arising directly below it; main flower- or fruit-stalk short, at the tip of the leafy growing shoots; leaf-blades narrowly elliptic or oblong-lanceolate, obtuse (blunt) or short-pointed (acute) . . . . . . . . . . . . . 2. C. OVATUS

1. **Ceanothus americanus** L. New Jersey Tea
Map 1498

Also called Wild Snowball.

Flowers May–November.

Occurs in upland or rocky prairies, glades, open woods, and thickets, sometimes along railroads.

The following varieties occur in Missouri:

Rare type encountered; leaf-blades usually 2–6 cm. long, 1–3.5 cm. broad; upper leaf surface mostly glabrous (without hairs); lower leaf surface more or less hairy . . . . 1a. C. AMERICANUS var. INTERMEDIUS
Common type encountered; leaf-blades usually 5–10 cm. long, 2.5–6 cm. broad; upper leaf surface hairy; lower leaf surface densely and velvety hairy. . . . 1b. C. AMERICANUS var. PITCHERI

1a. **Ceanothus americanus** var. **intermedius** (Pursh) K. Koch Map 1498

This is the rarer variation in Missouri, and is found only in the eastern portion of the state south of the Missouri River.

Ranges from Florida to Louisiana, north to Massachusetts, West Virginia, Illinois, and Missouri.

1b. **Ceanothus americanus** var. **Pitcheri** T. & G.
Map 1498

This is the commoner variation in Missouri, and is found throughout the state, possibly in every county.

Ranges from Georgia to Texas, north to Indiana, Illinois, Iowa, and Kansas.

During the American Revolution the leaves of the plant served as a tea substitute. The leaves should be dried first before using. No caffeine is found in the leaves. The tea has a taste and color slightly similar to that of oriental tea. The root was used at one time in medicine for its expectorant and astringent properties. The bark of the root has been found to contain blood-clotting properties. The leaves are eaten by white-tailed deer, and the fruits by wild turkey.

The clusters of white flowers are very showy and fragrant. The plant is sometimes grown as an ornamental, and its use for such should be more extensively recommended.

2. **Ceanothus ovatus** Desf. Redroot Map 1499

Also called Inland New Jersey Tea.

Flowers late April–June.

Occurs in upland and rocky prairies, loess hills, glades, open and rocky woodland. Mostly in western Missouri, and scattered eastward.

Two variations occur in Missouri:

Lower surface of leaves glabrous (without hairs) or sparsely hairy; uncommon type . 2a. C. OVATUS var. OVATUS
Lower surface of leaves noticeably or densely hairy; common type . 2b. C. OVATUS var. PUBESCENS

2a. **Ceanothus ovatus** var. **ovatus** Map 1499

*Ceanothus ovatus* Desf. [G, BB, P & S, Steyerm.]

*Ceanothus herbaceus* Raf. [Shinners]

Infrequent in Missouri, where known south of the Missouri River in Iron, Jackson, St. Clair, Jasper, Newton, and McDonald counties.

Ranges from Georgia and Texas, north to Maine,

Quebec, New York, Michigan, Minnesota, Manitoba, and South Dakota.

2b. **Ceanothus ovatus** var. **pubescens** T. & G. ex Wats. Map 1499

*Ceanothus ovatus* f. *pubescens* (T. & G. ex Wats.) Soper [G]

*Ceanothus herbaceus* var. *pubescens* (T. & G. ex Wats.) Shinners [Shinners]

The commoner variation, found in western Missouri in the unglaciated and glaciated prairie sections, and east locally in Callaway, Shannon, Cape Girardeau, and Scott counties.

Ranges from Ontario to Wisconsin, Minnesota, Manitoba, South Dakota, and Colorado, south to Missouri, Oklahoma, and Texas.

The more pubescent-leaved variation is apparently more common in the western portion of the range of the species, and because it shows some geographical restriction, is accorded status as variety instead of form, as indicated by Shinners (Field and Labr. 19: 33–34. 1951).

The leaves are thought to possess a similar use as a tea substitute to those of *C. americanus.*

## Fam. **VITACEAE** (Grape Family)

a. All the leaves simple with teeth or lobes, none of them divided into 3 or more leaflets . . . . . . . *b*

b. Pith brown, interrupted at each node (where buds or leaf-stalks arise) by a cross partition (except in *V. rotundifolia*); flowering or fruiting clusters much longer than broad; petals joined at their tips, and falling as one cap-like piece before the flower opens; fruit bluish or blackish, edible; bark of old stems loosening into flaky or ropy strands or shreds (except in *V. rotundifolia*) . . . . . . . . . . . . . . . . . . . . . . . . . . . . . . . . . 4. VITIS

b. Pith white or whitish, continuous, not interrupted at the nodes by a cross partition; flowering or fruiting clusters much broader than long; petals all separate from one another, spreading at flowering time, falling away singly; fruit at first green, orange, rose, or purple, finally a turquoise blue, not edible; bark of old stems tight, not loosening into any shreds or strands. 1. AMPELOPSIS CORDATA

a. Some or all of the leaves divided into 3 or more leaflets, if only 3-parted then leaflets are very fleshy or succulent . . . . . . . . . . . . . . . . . . . . . . . . . . . . . . . . . *c*

c. All the leaves either divided into 3 leaflets or deeply 3-parted; leaves very fleshy or succulent; leaflets and leaf-stalks (petioles) usually falling apart or separating from the main stem when pressed and dried; petals and stamens 4, and disk 4-lobed . . . . . . . . . . 2. CISSUS

c. Most of leaves divided into 5–34 leaflets, rarely only 3; leaves not fleshy or succulent; leaflets and leaf-stalks (petioles) usually remaining attached or staying together when pressed and dried; petals and stamens 5 . . . . . . . . . . . . . . . . . . . . . . . . . . . . *d*

d. All the leaves divided into 3–7 leaflets, palmately arranged (fan-like or like fingers from the hand) . . . . . . . . . . . . . . . . . . . . . . . . . . 3. PARTHENOCISSUS

d. All the leaves once, twice or ternately compound, divided into 11–34 or more leaflets. . . 1. AMPELOPSIS ARBOREA

### 1. **Ampelopsis** Michx.

Leaves all simple with teeth or lobes, none of them divided into leaflets . . . . . . . . 1. A. CORDATA

Leaves compound, divided into leaflets . . . . . . . . . . . . . . . . . 2. A. ARBOREA

1. **Ampelopsis cordata** Michx. Raccoon Grape Map 1500

Also called False Grape, Heartleaf Ampelopsis.

Flowers May–July.

Occurs in rich alluvial soils in valleys, low woods, and slopes near streams, borders of streams, ponds, and oxbow lakes in river flood plains, lowland thickets, borders of low woods, and fence rows. Southern, central, and eastern Missouri, north to Platte, Ray, Chariton, and Lewis counties, and locally northwest along the Missouri River around Garlich Lake, Andrew County.

Ranges from Florida to Texas and Mexico, north to Virginia, Ohio, Indiana, Illinois, Missouri, and Nebraska; naturalized north to Massachusetts.

This vine has grapelike leaves, and is frequently mistaken for that group of plants. In August the fruit, about the size of a pea, begins to ripen, turning from green to changing shades of orange, rose, purple, and finally turquoise blue. It is very ornamental in this multicolored display, and is commonly eaten by songbirds, such as the brown thrasher, wood thrush,

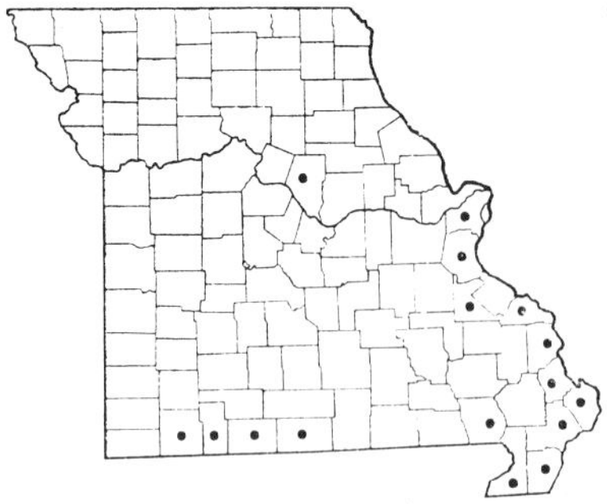

1501 Ampelopsis arborea (Pepper Vine)

1502 Cissus incisa (Marine Vine)

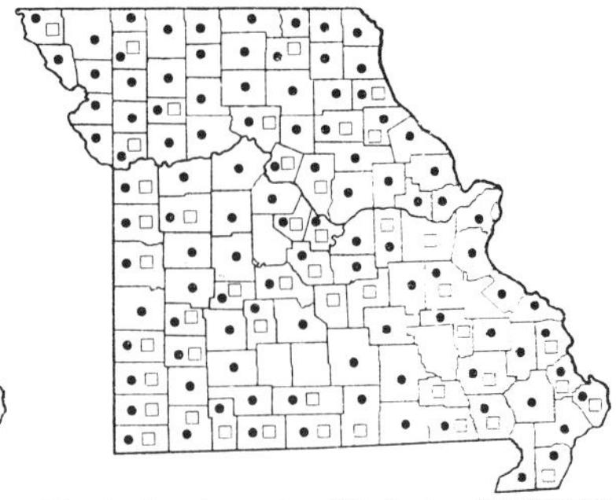

1503 • Parthenocissus quinquefolia f. quinquefolia (Virginia Creeper)

1503 □ Parthenocissus quinquefolia f. hirsuta

and others. In undisturbed forests of tall trees, it may climb to the top of the tallest ones.

2. **Ampelopsis arborea** (L.) Koehne
Pepper Vine — Map 1501
Flowers June–August.

Occurs in low wet thickets and alluvial soils along wooded banks of streams and river forest flood plains and at the base of bluffs along streams. Along the White River and tributaries in Ozark, Taney, Stone, and Barry counties, southwestern Missouri, and in eastern Missouri in the lowlands of southeastern Missouri north along the Mississippi River to the alluvial flood plain of the Meramec River in St. Louis County; introduced and naturalized in Boone County.

Ranges from Florida to Texas and Mexico, north to Maryland, Illinois, Missouri, and Oklahoma.

The lacelike, dark green foliage is very ornamental. The fruit is black. The plant sometimes is found sprawling and trailing along the banks of rivers, and again is found as a high-climbing vine.

## 2. **Cissus** L.

**Cissus incisa** (Nutt.) Des Moulins — Marine Vine
Map 1502

Also called Marine Ivy.

Flowers early June–July.

Occurs along upper portions of south- or west-facing rocky ledges of limestone bluffs and tops of limestone bluff escarpments bordering streams. Known only from southwestern Missouri from McDonald County east to Ozark County.

Ranges from Missouri and Kansas to Texas, Arizona, and Mexico; also in Florida.

This fleshy-leaved vine scrambles over the hot, dry west- and south-facing limestone exposures, where it is often associated with *Juniperus Ashei* and with such species of southwestern affinity as *Acacia angustissima* var. *hirta*, *Sapindus Drummondii*, *Ribes odoratum*, *Notholaena dealbata*, *Mentzelia oligosperma*, and others. In western Texas this same vine is often found along dry rocky canyons. Dr. Shinners records the leaves of this vine as possessing a burnt rubber or sharp nitrogenous odor. Upon drying the leaflets often become separated from one another or from the petiole.

## 3. **Parthenocissus** Planch. Virginia Creeper, Woodbine

Commonly found species; leaves dull or opaque on upper surface, not glossy; tendrils many-branched, ending in enlarged suckers or adhesive disks; inflorescence with 25–200 or more flowers, longer than broad, with a prolonged, well-defined, central axis in each group of flowers, the lower ascending to spreading branch solitary, not dichotomously branched (without regular forking in pairs); fruit 5–7 mm. in diameter . . . . . . . . . . . . . . . . . . . . . 1. P. QUINQUEFOLIA

Plate no. 248. **1.** Ampelopis cordata, × $^2/_7$; *a.* Fruiting branch; *b.* Lobed leaf. **2.** Parthenocissus quinquefolia, × $^2/_7$; Details from Small, The New York Botanical Garden. **3.** Cissus incisa, × $^2/_7$; After Gleason, details from Small, The New York Botanical Garden. **4.** Ampelopsis arborea, × $^2/_7$; *a.* Flower, × $2^6/_7$; After Gleason, details from Small, The New York Botanical Garden. **5.** Vitis labruscana, × $^2/_7$. **6.** Parthenocissus inserta, × $^2/_7$; Details from Small, The New York Botanical Garden.

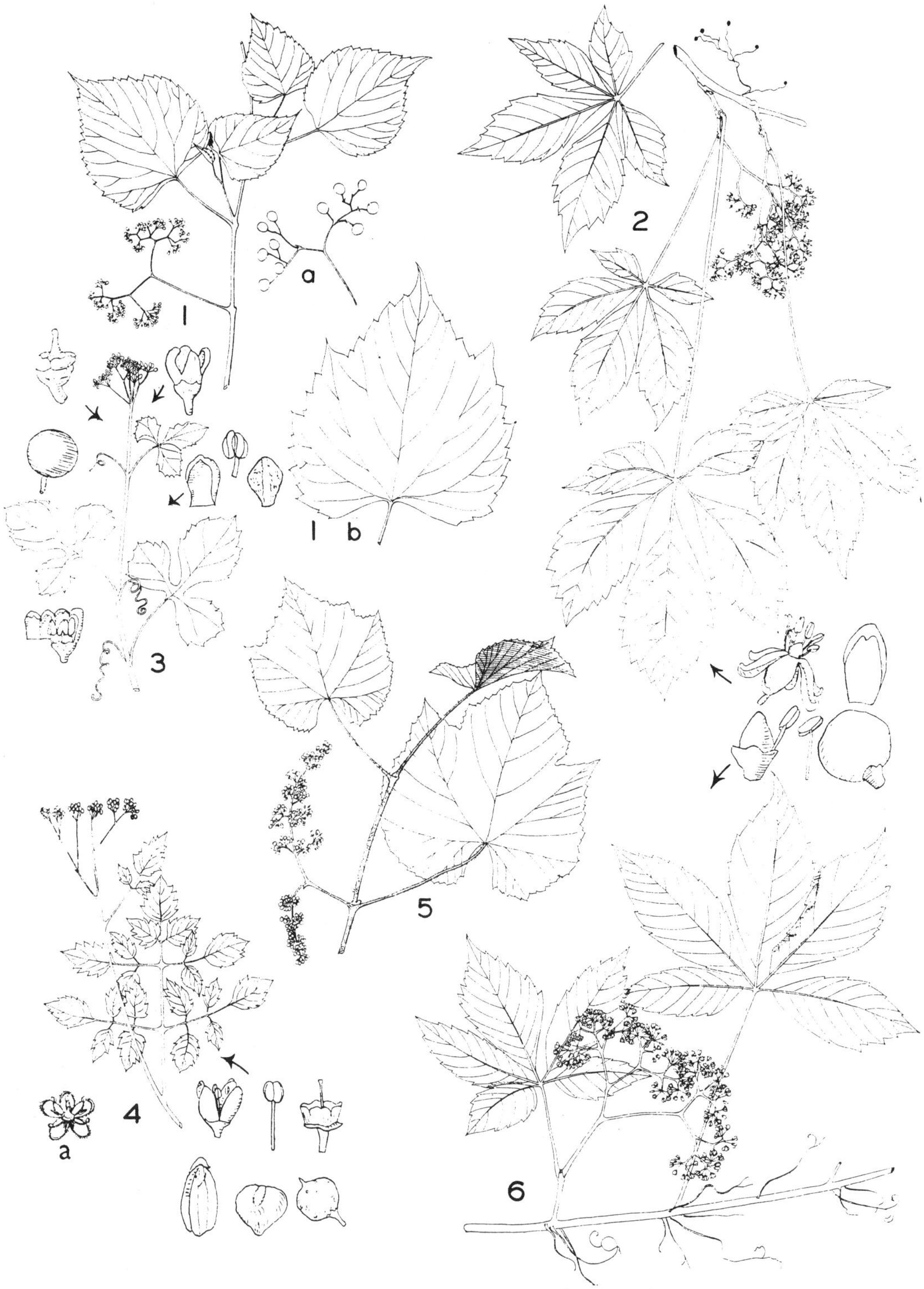


PLATE NO. 248

Rarely found species; leaves glossy on upper surface; tendrils few-branched, slender-tipped, not ending in suckers or adhesive disks or rarely producing them; inflorescence with 10–60 flowers, forming a broad, rounded cluster, the central axis not prolonged, with a pair of nearly equally forking spreading branches, more or less dichotomously branched (with regular forking); fruit 8–10 mm. in diameter
2. P. INSERTA

1. **Parthenocissus quinquefolia** (L.) Planch.
Virginia Creeper, Woodbine, Five-leaved Ivy
Map 1503

Also known as Sour-vine in some sections of the Ozarks.

Flowers late May–August.

Occurs in open and rich woods, valleys, ravines, along rocky bluffs, thickets, and fence rows. Throughout Missouri, doubtless in every county.

Ranges from Florida to Texas, north to Maine, Vermont, Quebec, New York, Ohio, Indiana, Illinois, Wisconsin, and Minnesota; also native in Mexico and Guatemala.

Two variations are found in Missouri:

Lower surface of leaflets glabrous (without hairs)
1a. P. QUINQUEFOLIA f. QUINQUEFOLIA
Lower surface of leaflets more or less hairy . .
1b. P. QUINQUEFOLIA f. HIRSUTA

1a. **Parthenocissus quinquefolia** f. **quinquefolia**
Map 1503

*Parthenocissus quinquefolia* (L.) Planch. [G, P & S, Steyerm.]

This is the commoner variation in Missouri, and is doubtless found in every county.

1b. **Parthenocissus quinquefolia** f. **hirsuta** (Donn) Fern. Map 1503

*Parthenocissus quinquefolia* var. *hirsuta* (Donn) Planch. [P & S, Steyerm.]

*Parthenocissus quinquefolia* var. *Saint-Paulii* [Koehne & Graebn.] Rehder [P & S, Steyerm.]

Found throughout Missouri, but less common than the glabrous-leaved variation.

The autumnal foliage of Virginia Creeper turns to brilliant dull rose and dull purplish to deep red and crimson shades, lighting up the forest with its coloring. It is an excellent vine for covering fences, walls of buildings, trellises, and other objects. The flowers are visited frequently by honey bees, the young tendrils are eaten by wild turkey, and the foliage is browsed by white-tailed deer.

Young growing tips and young leaves of Virginia Creeper with only 3 leaflets may be distinguished from Poison Ivy by cuplike suckers and tendrils on the stems of the Virginia Creeper instead of the aerial, hairlike rootlets which dangle from the woody stems of the Poison Ivy. Moreover, in Poison Ivy the middle (terminal) one of the three leaflets is on a stalk much longer than the other two (lateral) leaflets, whereas in Virginia Creeper, all the leaflets are equally stalked or without stalks (petiolules). At times, Virginia Creeper creeps along the ground, becoming an extensive ground cover, but mostly it climbs to the tops of tall trees.

2. **Parthenocissus inserta** (Kerner) K. Fritsch
Map 1504

*Parthenocissus vitacea* (Knerr) Hitch. [P & S, Steyerm.]

Flowers late May–July.

Occurs along bluffs in rich woods and wooded slopes, thickets, and fence rows. Scattered and infrequent in Missouri; absent from most of the Ozark region.

Ranges from Quebec to Manitoba and Montana, south to New England, Pennsylvania, Ohio, Indiana, Illinois, Missouri, Kansas, New Mexico, Arizona, and California.

Two variations are found in Missouri:

Lower surface of leaflets glabrous (without hairs)
2a. P. INSERTA f. INSERTA
Lower surface of leaflets more or less hairy . .
2b. P. INSERTA f. DUBIA

2a. **Parthenocissus inserta** f. **inserta** Map 1504

*Parthenocissus inserta* (Kerner) K. Fritsch [G, BB]

This is the commoner variation in Missouri.

2b. **Parthenocissus inserta** f. **dubia** Rehd.
Map 1504

Known only from Henry County, western Missouri (covering steep slopes of north-facing limestone bluffs on south and east side of Grand River at McGuiness Bend, T40N, R24W, south half sect. 16, 2½–3 mi. northwest of Finey, July 31, 1949, *Steyermark 68716*).

## 4. **Vitis** L. Grape

a. Lower surface of fully grown leaf-blades either whitened or silvery-bluish-green, or covered with long and cobwebby hairs continuous or in patches lying parallel to the surface or with a close continuous felt of hairs completely hiding the surface . . . . . . . . . . . . . . . . . *b*

b. Lower surface of fully grown leaf-blades with a close continuous felt of rusty, tawny, or whitish hairs, remaining permanently attached and completely hiding the surface; tendrils and/or flowering or fruiting inflorescences arising from each of 3 or more successive nodes on other side of stem opposite the leaf-stalk . . . . . . . . . . . . . . . . . 1. V. LABRUSCANA

b. Lower surface of fully grown leaf-blades with long and cobwebby hairy, either close and forming a continuous covering or often occurring in patches or tufts (in young leaves closer together or hiding the surface), or, if nearly without hairs, then whitened or bluish-silvery-green; tendrils and/or flowering or fruiting inflorescences arising from no more than 2 successive nodes, at least every third node lacking a tendril or inflorescences . . . . . . . . . . . . . . . . *c*

c. Lower surface of fully grown leaf-blades either whitened or silvery-bluish-green with or without whitish or rusty cobwebby hairiness, or green with a rusty or reddish-brown cobwebby hairiness (in young leaves the hairs often closer together and hiding the surface, in old leaves disappearing and exposing much of the surface) . . . . . . . . . . 2. V. AESTIVALIS

c. Lower surface of fully grown leaf-blades dull green with a grayish or ashy more or less uniform, continuous covering of hairs . . . . . . . . . . . . . . . . . . 3. V. CINEREA

a. Lower surface of fully grown leaf-blades green and glabrous (without hairs) or with short straight hairs standing erect from the surface, not cobwebby . . . . . . . . . . . . . . . . . *d*

d. Pith continuous through the nodes, not interrupted by a cross-partition at the nodes (where leaf-stalk, tendril, or inflorescence arises); bark of branches and main stem tight, not shredding off; tendrils simple, not forked; rare species of southeastern Missouri only . . . . . . 8. V. ROTUNDIFOLIA

d. Pith interrupted at the nodes by a cross-partition; bark of older branches and main stem loosening and shredding; tendrils, when present, forked; includes some commonly found species . . . . . *e*

e. Leaf-blades usually broader than long, sometimes shaped like a lima or kidney bean; fully grown leaf-blades mostly 3–10 cm. long, 4–12 cm. broad; tendrils absent or only opposite uppermost leaves or at tips of flowering or fruiting branches; plant bushy, sprawling, or trailing over rocky stream beds or along gravel bars, rarely climbing . . . . . . 7. V. RUPESTRIS

e. Leaf-blades usually longer than broad with a conspicuous triangular tip, generally heart-shaped, broadly ovate or triangular-ovate; fully grown leaf-blades mostly 5–17 cm. long; 6–15 cm. broad; tendrils or flowering or fruiting inflorescences regularly present opposite most leaves, absent opposite each third leaf; plants climbing . . . . . . . . . . . . . . *f*

f. Leaf-blades either merely toothed, or, if lobed, the 2 side (lateral) lobes short, shoulder-like, and usually forming a wide sinus (gap or space) with the middle (terminal) lobe; leaves of flowering or fruiting branches not 3-lobed; fruit black, not glaucous (covered with a grayish 'bloom' which can be rubbed off) . . . . . . . . . . . . . . 6. V. VULPINA

f. Leaf-blades with usually 2 prominent side (lateral) lobes which are generally long-pointed and tapering, 1–4 cm. long and usually form an acute sinus (gap or space) with the middle (terminal) lobe; leaves of flowering or fruiting branches 3-lobed; fruit black and not glaucous or bluish-gray and glaucous (covered with a grayish 'bloom' which can be rubbed off) . . . *g*

g. New branches of the current season bright red or purplish-red; fruit black, without a 'bloom'; diaphragm (cross-partition inside of stem interrupting pith) 4–5 mm. thick; margins of leaf-blades scarcely or not at all ciliolate (with short fringe of hairs on margins); mature leaf-blades thin, those of flowering or fruiting branches 4–9 cm. broad; plants of eastern Missouri from the southeastern lowlands north to Pike County, along the Mississippi River flood plain and west in southeastern Missouri to Ripley County . . . . . . . . . . . . . . . . . . . . . . 4. V. PALMATA

g. New branches of the current season green, gray, or brown, or not red; fruit gray-bluish with a 'bloom'; diaphragm at nodes of stem 0.8–2 mm. thick; margins of leaf-blades ciliolate; mature leaf-blades firmly membranaceous, those of flowering or fruiting branches mainly 7–15 cm. broad; throughout northern and central Missouri south to St. Louis, Franklin, Gasconade, Osage, Cole, Moniteau, Morgan, Benton, Lawrence, and Vernon counties, and locally in Cape Girardeau and Madison counties, but absent from the southeastern lowlands . . . . . . . . . . . . . . . 5. V. RIPARIA

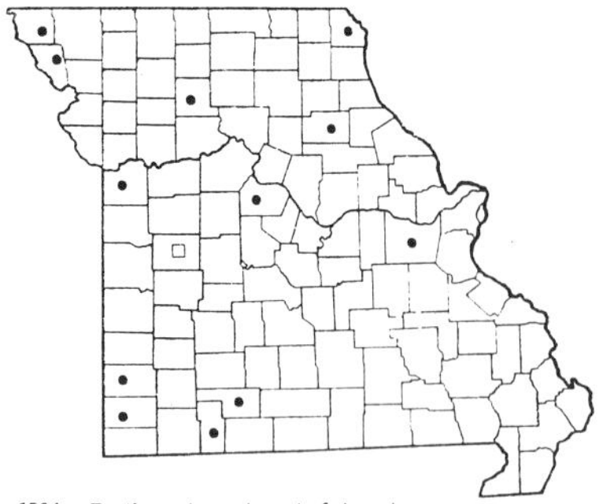

1504 ● Parthenocissus inserta f. inserta
1504 □ Parthenocissus inserta f. dubia

1505 Vitis labruscana (Labruscan Vineyard Grape)

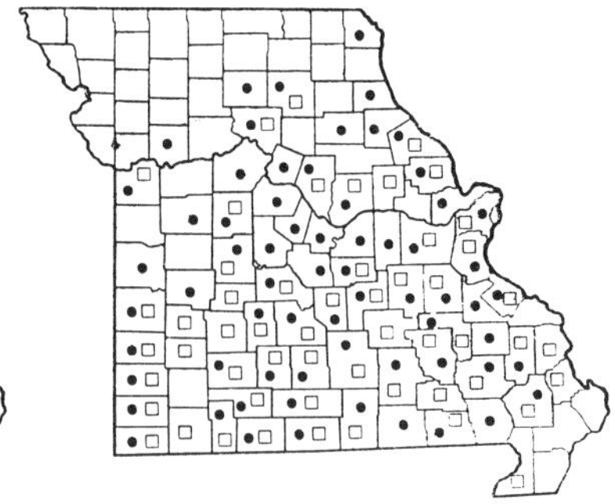

1506 ● Vitis aestivalis var. aestivalis (Summer Grape)
1506 □ Vitis aestivalis var. argentifolia

1. **Vitis labruscana** Bailey
Labruscan Vineyard Grape Map 1505
*Vitis Labrusca* of auth. [P & S, Steyerm.], not L.
Flowers May–June; fruits August–September.

Cultivated and occasionally escaped along railroads, thickets, and waste ground. Known from St. Louis, St. Charles (*Bauer 696*), Boone, and Jasper counties.

This constitutes the group of cultivated strains originating from the Fox Grape, *Vitis Labrusca* L., and includes the varieties known as Concord, Catawba, Niagara, Chautauqua, Worden, and others. The native Fox Grape of the United States, which ranges from Maine to Michigan, south to Georgia, Kentucky, and Tennessee, is not wild in Missouri, and records previously identified as *V. Labrusca* should be referred to the cultivated *V. labruscana.*

2. **Vitis aestivalis** Michx. Summer Grape
Map 1506
Flowers May–July; fruits July–October.

Occurs in dry rocky and upland woods, thickets, glades, along rocky slopes, and top of bluff escarpments.

Missouri material may be divided into the following two varieties:

Leaf-stalk (petiole) and usually young branches with a rusty or reddish-brown hairiness; lower surface of leaf-blade mainly hairy with a loose and rather persistent hairiness . 2a. V. AESTIVALIS var. AESTIVALIS

Leaf-stalk (petiole) and usually young branches glabrous (without hairs); lower surface of leaf-blades whitened, bluish-gray-green, or glaucous, nearly glabrous (without hairs) or with the main nerves or their axils shortly hairy. . . . . 2b. V. AESTIVALIS var. ARGENTIFOLIA

2a. **Vitis aestivalis** var. **aestivalis** Map 1506
*Vitis aestivalis* Michx. [G, P & S, Steyerm.]
*Vitis Lincecumii* Buckl. [G, BB]

Common in southern and central Missouri, and northward to Clark, Macon, Linn, Ray, and Jackson counties.

Ranges from Georgia to Texas, north to Massachusetts, New York, Ohio, Michigan, and Wisconsin.

2b. **Vitis aestivalis** var. **argentifolia** (Munson) Fern. Map 1506
*Vitis Lincecumii* var. *glauca* Munson [G, P & S, Steyerm.]
*Vitis Lecontiana* House
*Vitis bicolor* of auth.
*Vitis aestivalis* var. *bicolor* (LeConte) Britton & Brown [Deam, Shrubs]

Common in southern and central Missouri, and northward to Pike, Macon, and Chariton counties.

Ranges from New Hampshire to Minnesota, south to Virginia, Alabama, Tennessee, Arkansas, Oklahoma, and Texas.

It has not been possible to keep *V. aestivalis* and *V. Lincecumii* as separate species. Fernald likewise showed (Rh. 38: 428. 1936) that it was not possible to maintain *V. argentifolia* specifically apart from *V. aestivalis*, originally separated as distinct by Bailey. There is every gradation in pubescence of the lower leaf surface and in size of fruits and seeds. Dr. U. T. Waterfall of Oklahoma State University has also indicated in a recent communication that there is not justification for separation of the two. Bailey (Gent. Herb. 3: 193. 1934) states that 'relationship of *Vitis Lincecumii* to *V. aestivalis* cannot be made out from existing botanical materials,' but believed *V. Lincecumii* to be distinct from a horticultural standpoint, and was largely influenced by this consideration. Bailey and authors following him (Gleason and Fernald) have attempted to use the character of size of fruit in distinguishing the two as species; fruits mostly larger than 1 cm. in diameter are placed in *V. Lincecumii*,

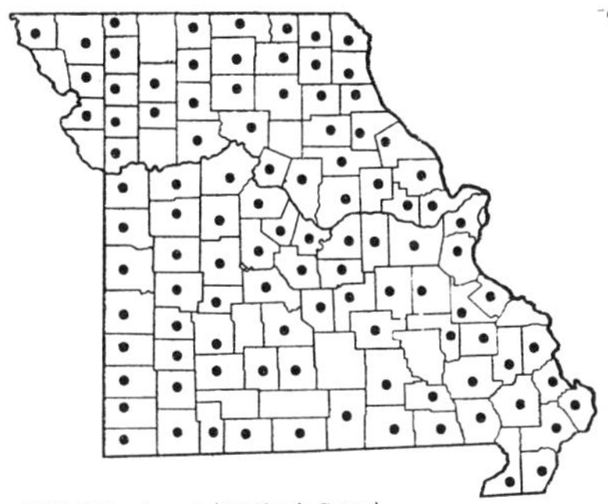

1507 Vitis cinerea (Grayback Grape)

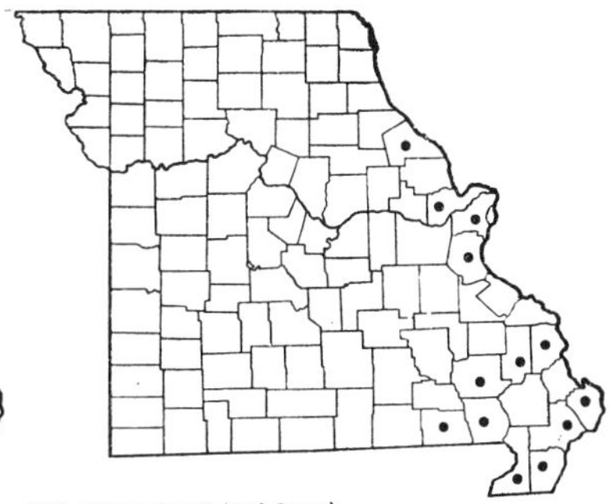

1508 Vitis palmata (Red Grape)

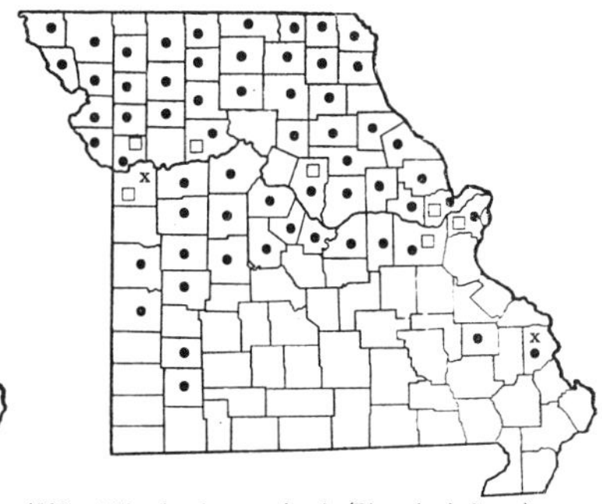

1509 • Vitis riparia var. riparia (River-bank Grape)
1509 □ Vitis riparia var. praecox
1509 x Vitis riparia var. syrticola

those mostly smaller than 1 cm. in diameter are placed in *V. aestivalis.* Fernald characterizes *V. aestivalis* as having fruits 5–12 mm. in diameter and *V. Lincecumii* as having fruits 10–25 mm. in diameter. Fernald (*Gray's Man.*, eighth ed. pp. 996–97. 1950) has attempted to use seed length and breadth also as a character, those of *V. aestivalis* 5–7 mm. long, 4–5 mm. broad, and those of *V. Lincecumii* 7–8 mm. long, 5–6 mm. broad. Bailey, however (p. 193), gives the seeds of *V. Lincecumii* var. *glauca* as only 5–6 mm. long and about 4 mm. broad.

According to the measurements of fruiting material of Missouri specimens studied in the major herbaria, I have found no correlation existing between diameter of fruit and length of seed. Most of the Missouri specimens previously identified as *V. Lincecumii* or var. *glauca* have smaller fruit diameters and seed lengths than those given as distinguishing criteria for *V. Lincecumii* and the majority of specimens so identified would have to be called *V. aestivalis* var. *aestivalis* or var. *argentifolia.* An example of this lack of correlation is to be noted in the specimen of *Bush 3143* from Barry Co. with fruits 12–13 mm. diam. and seeds only 5.5–6 mm. long. The following specimens have the largest fruits seen, but many show the smaller seeds of *V. aestivalis*: *Palmer 4028* from Newton Co. with seeds 7 mm. long; *Eggert*, 1879, from St. Louis Co. with fruits 13–15 mm. in diam. but seeds only 5–5.5 mm. long; *Palmer 21944* from Jasper Co., and *Palmer 5998* from Jasper Co. with fruits 15 mm. or more in diam.; *Bush 5058* from McDonald Co. with seeds 7 mm. long; *Steyermark 65128* from Barry Co. with fruits 15–17 mm. in diam. but seeds only 5–5.5 mm. long.

Similarly, the pubescence on the lower surface of the leaf-blade sometimes used as a criterion for separation of these two taxa cannot be relied upon. Gleason (*New Ill. Fl.* 2: 517. 1952) gives 'Cobwebby pubescence of the lower leaf-surface reddish or rusty' for *V. aestivalis* and 'Cobwebby pubescence of the lower leaf-surface ashy gray' for *V. Lincecumii*; Fernald (pp. 996–97) gives 'pubescence of lower leaf-surface a thin scarcely flocculent web' for *V. Lincecumii* and 'pubescence of lower leaf-surface flocculent or deciduous' for *V. aestivalis,* but this pubescence may be 'cinereous to rufescent' in *V. Lincecumii* (p. 997) so that even the color varies. All gradations of pubescence, from sparse (in age) to heavy (in youth) can be found in both *V. aestivalis* and *V. Lincecumii.* This is stated to be the case by Gleason (p. 518) and Deam (*Fl. Ind.* p. 663) for *V. aestivalis* var. *aestivalis* and var. *argentifolia,* and all degrees of intermediates are likewise found in Missouri material.

At most, therefore, it would appear that Missouri material falls into one species, *V. aestivalis,* variable in size of fruit and pubescence of petioles and lower leaf-surface.

The young tendrils and fruits are eaten by wild turkey and the foliage is eaten by white-tailed deer. The fruits are sweet and edible, and used for preserves and jelly. They do not have the musky taste found in forms of the Fox Grape (*V. Labrusca*).

3. **Vitis cinerea** Engelm. Grayback Grape

Map 1507

Also called Sweet Winter Grape, Pigeon Grape. *Vitis cinerea* var. *canescens* (Engelm.) Bailey [P & S, Steyerm.]

Flowers May–July; fruits September–October.

Occurs in low woods and alluvial soils along streams, in thickets, and fence rows. Throughout Missouri, doubtless in every county.

Ranges from Florida to Texas, north to Virginia, Ohio, Indiana, Illinois, Wisconsin, Iowa, and Nebraska.

The lower surface of the leaves in var. *canescens* is described as being 'more prominently soft-pubescent or canescent even to maturity' (Bailey, p. 216) and with the leaves 'less frequently sharp-lobed and lacking the long deltoid apex and point,' but these characters intergrade freely with typical *cinerea* and cannot be used for separation.

The berries of this grape are numerous but small, only 4–7 mm. in diameter, blackish or purple with only a slight 'bloom,' and occur in long clusters, often 8–12 cm. long. The berries are sweet when ripe.

4. **Vitis palmata** Vahl Red Grape Map 1508

Also called Cat Grape, Catbird Grape, Missouri Grape.

Flowers May–July; fruits September–October.

Occurs in bald cypress swamps, bayous, and low wet woods. Common in the swampy woods of the lowland counties of southeastern Missouri west to Wayne and Ripley counties, north to Cape Girardeau and Bollinger counties, thence north along the alluvial woodlands bordering the Mississippi River to Pike County (bottomland forest in Mississippi River bottoms, vicinity of Bradley Pond and Long Lake, T55N, R2W, sect. 15, $3\frac{1}{2}$–$4\frac{1}{2}$ mi. southeast of Ashburn, October 6, 1956, *Steyermark 82881*).

Ranges from Louisiana and Texas, north to Indiana, Illinois, Missouri, and Oklahoma.

The 1–2 seeds, 4.5–5 mm. long, are very large for the size of the black fruit which is only 5–8 mm. in diameter. Often only one seed is found in a fruit.

5. **Vitis riparia** Michx. River-bank Grape Map 1509

Also called Frost Grape.

*Vitis vulpina* of auth. [P & S, Steyerm.]

Occurs in low woods, thickets, and banks of streams in alluvial soils.

The following variations may be recognized in Missouri:

a. Leaf-stalks (petioles) and lower surface as well as veins of leaf-blades densely hairy . . . 5c. V. RIPARIA var. SYRTICOLA
a. Leaf-stalks (petioles) glabrous (without hairs) or nearly glabrous, and lower surface of leaf-blades mainly glabrous except for hairs on main veins or in the axils of veins . . . . *b*
  b. Common type found; inflorescence in flower 4–12 cm. long; berries 8–12 mm. in diameter; seeds about 5 mm. long . 5a. V. RIPARIA var. RIPARIA
  b. Rare type; inflorescence in flower 4–6 cm. long; berries 6–7 mm. in diameter; seeds about 4 mm. long . 5b. V. RIPARIA var. PRAECOX

5a. **Vitis riparia** var. **riparia** Map 1509

*Vitis riparia* Michx. [G, BB]

*Vitis vulpina* of auth. [P & S, Steyerm.]

Flowers May–June; fruits July–September.

Northern and central Missouri, where common, south to St. Louis, Franklin, Gasconade, Osage, Cole, Moniteau, Morgan, Benton, Lawrence, and Vernon counties, and locally southeast in Cape Girardeau (river banks, Cape Girardeau, *Palmer 17988*) and Madison (open banks along river, near Fredericktown, *Palmer 30257*) counties.

Ranges from Quebec to Manitoba and Montana, south to Virginia, West Virginia, Tennessee, Missouri, Texas, and New Mexico.

5b. **Vitis riparia** var. **praecox** Engelm. Map 1509

*Vitis vulpina* var. *praecox* Bailey [P & S, Steyerm.]

Flowers April; fruits June.

Central Missouri, in the counties bordering the Missouri River from Jackson and Clay counties east to St. Charles and St. Louis counties.

Known only from Illinois and Missouri.

This is an early-flowering and early-fruiting variety, which needs more intensive field and experimental studies.

5c. **Vitis riparia** var. **syrticola** (Fern. & Wieg.) Fern. Map 1509

*Vitis vulpina* var. *syrticola* Fern. & Wieg. [Bailey]

Flowers May–June; fruits August–September.

Known only from Cape Girardeau (thickets at base of east-facing limestone bluffs along tributary to Mississippi River, T32N, R3W, north boundary sect. 14, southwest $\frac{1}{4}$ sect. 11, $\frac{1}{4}$SE$\frac{1}{2}$ sect. 10, $2\frac{3}{4}$–3 mi. southeast of Neelys Landing, 1 mi. northwest of Moccasin Springs, May 9, 1957, *Steyermark 84300*) and Jackson (bottoms, Courtney, August 4, 1933, *Bush 12829*) counties.

Ranges from New York and Ontario to Michigan, Indiana, and Missouri.

Although the axis and branches of the inflorescence are described by Fernald (*Gray's Man.*, eighth ed. p. 998) as glabrous or nearly so, they may be densely pubescent in some specimens of var. *syrticola* (*Steyermark*

---

Plate no. 249. **1.** Vitis aestivalis var. aestivalis, × $^2/_7$; *a.* Vitis aestivalis var. aestivalis, leaf variation. **2.** Vitis cinerea, × $^2/_7$. **3.** Vitis palmata, × $^2/_7$. **4.** Vitis riparia, × $^2/_7$; *a.* Fruiting cluster, **5.** Vitis aestivalis var. argentifolia, × $^2/_7$; *a.* Fruiting cluster. **6.** Vitis rupestris, × $^2/_7$; *a.* Fruiting cluster. **7.** Vitis vulpina, × $^2/_7$. **8.** Vitis rotundifolia; *a.* Fruiting cluster; Details from Small, The New York Botanical Garden.

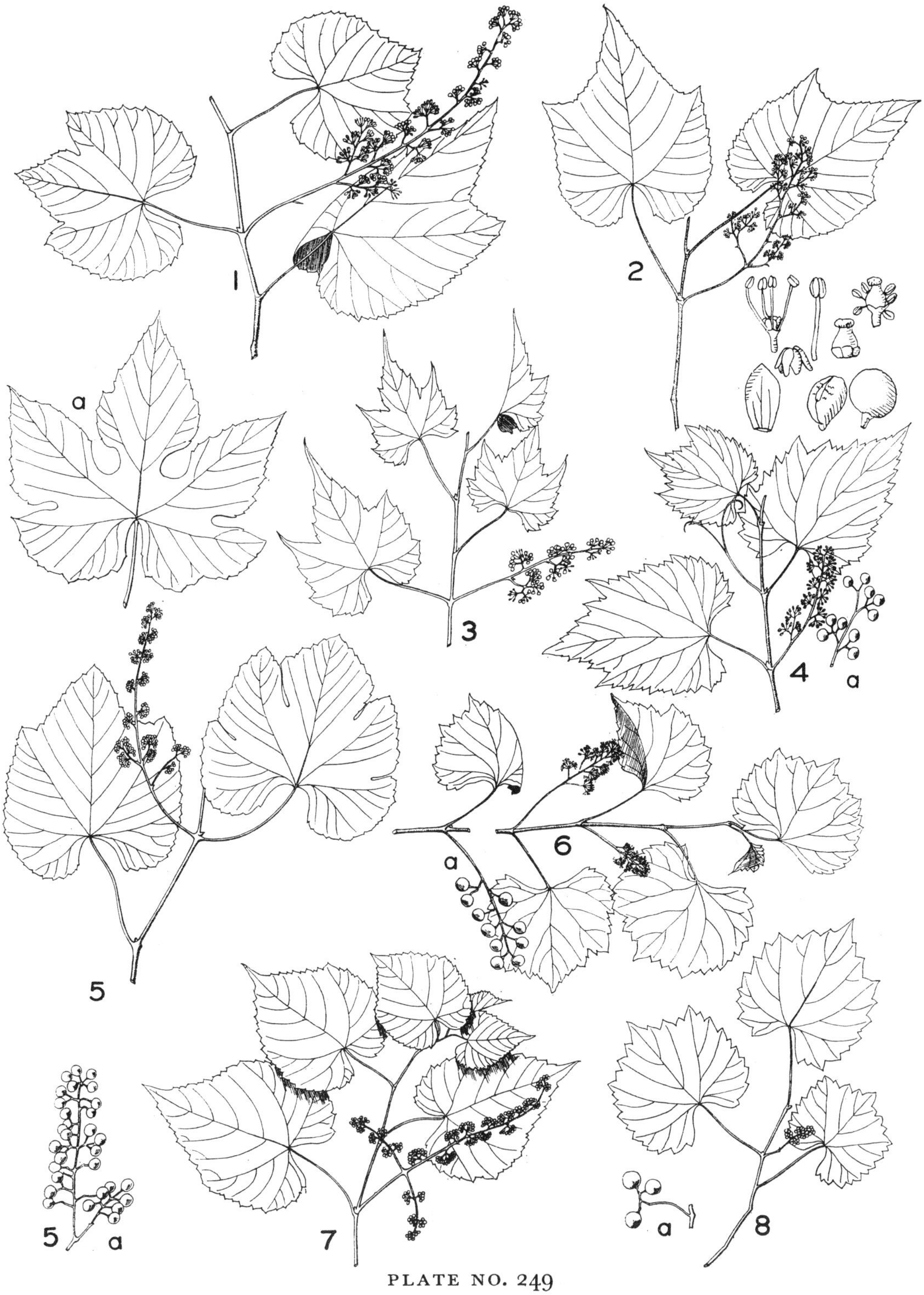


PLATE NO. 249

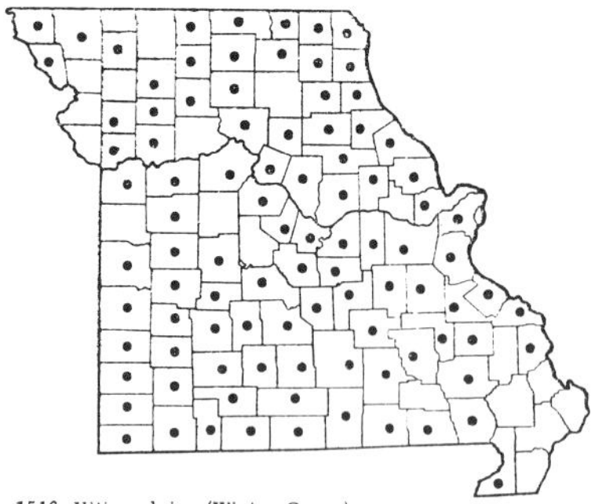

1510 Vitis vulpina (Winter Grape)

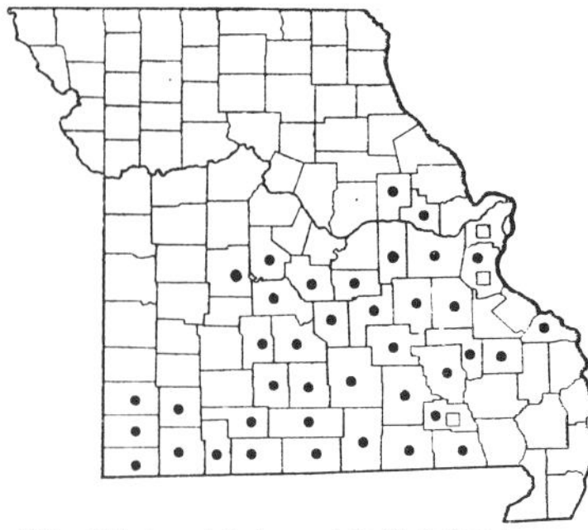

1511 ● Vitis rupestris f. rupestris (Sand Grape)
1511 □ Vitis rupestris f. dissecta

1512 Vitis rotundifolia (Muscadine)

*84300*) as well as in plants of the variety seen from Michigan.

The berries of this grape are fairly sour, but eventually take on a mild flavor. They are used for jellies and preserves as well as for grape juice and other refreshing drinks.

This grape is sometimes confused with *V. vulpina* (*V. cordifolia*), from which it may be distinguished, when flowers and fruits are not available, by the much thinner diaphragm separating the pith, the dull rather than glossy upper surface of the leaves, the much larger, coarser, and sharper pointed teeth, the much broader, more U-shaped instead of generally V-shaped sinus at the base of the leaf-blade, larger stipules (5–8 mm. instead of 3–5 mm. long), and shorter peduncles of the flowering or fruiting clusters. Of course the two lateral lobes of the leaf-blades are much more pronounced in *V. riparia*, whereas in *V. vulpina* (*V. cordifolia*) they are very inconspicuous or none at all.

The autumnal foliage is pale yellow.

6. **Vitis vulpina** L. Winter Grape Map 1510

Also called Frost Grape, Chicken Grape.

*Vitis cordifolia* Michx. [P & S, Steyerm.]

*Vitis cordifolia* var. *foetida* Engelm. [P & S, Steyerm.]

*Vitis Baileyana* Munson [G, BB]

Flowers mid-May–June; fruits September–October.

Occurs in alluvial soils along streams in low wet woods, low wooded slopes, base of bluffs, and thickets. Throughout Missouri, except apparently absent from the lowland counties of southeastern Missouri.

Ranges from Florida to Texas, north to New York, Pennsylvania, West Virginia, Ohio, Indiana, Illinois, Missouri, and Kansas.

*Vitis cordifolia* var. *foetida* was described from specimens collected in St. Louis County. It is described as a vine 'often 4–6 inches in diameter, climbing the highest trees, and bearing fetidly aromatic berries.' It has not been seen nor collected since the original discovery, and the type material appears insufficient, according to Bailey, to render any judgment on the status of the variety. It is, therefore, omitted from the present flora.

*Vitis Baileyana* Munson was separated by Fernald, following Bailey's revision (Gent. Herb. 3: 150–244. 1934), from *V. vulpina* (*V. cordifolia*) as follows: branchlets angled instead of terete; petioles and veins of lower surfaces of leaves permanently pilose instead of glabrous or promptly glabrate; leaf-blades more rounded and less prolonged, with smaller teeth; fruiting pedicels mostly less than 3 mm. long instead of about 5 mm. long. These characters have been found to break down entirely in the Missouri material studied. Complete intergradation in the amount of pubescence of petioles and lower leaf-surface is found. Most of the Missouri specimens of *V. vulpina* exhibit various degrees of pubescence to some extent or another. The striate or angled branchlets, shape of leaves, supposedly smaller teeth, and shorter fruiting pedicels, all fail to correlate satisfactorily. The following specimens in the Gray Herbarium have the pubescence characters associated with *V. Baileyana*: Boone Co.: *Drouet 3053*; St. Louis Co.: September 3, 1878, *Eggert*; May 24, 1878, *Engelmann*; Jackson Co.: rocky barrens, Swope Park, *Bush 10104* and *10215*; Christian Co.: August 1, 1895, *Blankinship*.

Gleason (*New Ill. Fl.* 2: 517, 519. 1952) wrongly described the pubescence on the lower surface of the leaf-blades in *V. Baileyana* as being of 'Cobwebby hairs'; actually, as indicated by Fernald and by Bailey (Gent. Herb. 3: 219. 1934) the hairs are short and upstanding as in *V. vulpina* (*V. cordifolia*), contrasted with *V. cinerea* in which the hairs are cobwebby (araneose).

The leaves and fruits of this grape are eaten by wild turkey. The grapes, which at first are very acid, become sweet after frost.

7. **Vitis rupestris** Scheele Sand Grape Map 1511

Flowers early May–June; fruits July–August.

Occurs along rocky banks and gravelly beds and bars along streams. Ozark region of southern and east-central Missouri north to St. Louis, Franklin, Warren, Montgomery, Maries, Miller, Morgan, Benton, Lawrence, and Jasper counties.

Ranges from Tennessee, Illinois, and Missouri, south to Arkansas, Oklahoma, and Texas; reports from Pennsylvania, D.C., and North Carolina, are believed to be introductions or escaped rather than native plants.

Two variations are found in Missouri:

Commonly found type; leaf-blades usually broader than long, shaped somewhat like a lima bean or kidney; teeth of leaves short-tipped (apiculate), the leaf-blade not irregularly lobed . 7a. V. RUPESTRIS f. RUPESTRIS

Rarely found; leaf-blades somewhat more broadly ovate; leaf-blades with rather long-pointed teeth and irregularly lobed . 7b. V. RUPESTRIS f. DISSECTA

7a. **Vitis rupestris** f. **rupestris** Map 1511

*Vitis rupestris* Scheele [G, BB, P & S, Steyerm.]

This is the commoner variation in Missouri.

7b. **Vitis rupestris** f. **dissecta** (Eggert) Fern. Map 1511

*Vitis rupestris* var. *dissecta* Eggert [P & S, Steyerm.]

Scattered and rare, known from St. Louis, Jefferson, and Carter counties. Originally described by Eggert from along 'brooks in hills near St. Louis.'

Plants are often found only in vegetative condition, with the result that flowers and fruits are frequently difficult to find. The fruit is large and sweet, 6–12 mm. in diameter, black and with or without a 'bloom.' The inflorescence is only 2–5 cm. long. The plant often grows in large colonies on undisturbed creek beds and acts as an efficient cover and binder for loose gravel. It is an attractive plant which should be cultivated more frequently as a ground cover for dry rocky places.

8. **Vitis rotundifolia** Michx. Muscadine Map 1512

Also called Scuppernong.

*Vitis campestris* Bartram (?), See Fernald, Rh. 46: 48–49. 1944, but not subsequently used by Fernald in *Gray's Manual*, eighth edition p. 998.

Flowers May–June; fruits September–October.

Occurs in low wet woods and borders of swamps and bayous. Southeastern Missouri, known only from Madison, Pemiscot, and Dunklin counties.

Ranges from Florida to Texas, north to Delaware, Virginia, West Virginia, Indiana, Missouri, and Oklahoma.

This is the ancestor of the Muscadine grapes, especially the Scuppernong variety with silvery amber-green fruit. The berries, with a thick, tough skin, are purple-black to bronze without a 'bloom,' and are the largest of our native Missouri grapes, attaining 12–25 mm. in diameter. The musky flavor is strong in the wild plants, but excellent flavors are found in cultivated varieties. The Indians formerly gathered quantities of these wild grapes in the southeastern states and dried them to store as future provisions. Their sweet, rich juice is much appreciated.

Fernald (Rh. 46: 48–49. 1944) took up the name *V. campestris* Bartram for the Muscadine, but abandoned it in the present eighth edition of the *Manual*.

This species has a habit, peculiar among grapes, of sending out aerial drooping roots from the branches especially in flooded bottomland habitats.

*Excluded Species*

**Vitis acerifolia** Raf. Bush Grape

*Vitis Longii* Prince

This species is recorded from Missouri by Fernald in the eighth edition of *Gray's Manual* (p. 998). The record is based upon specimens collected from McDonald County in southwestern Missouri (rocky banks, Noel, May 27, 1909, *Bush 5773* and *5773A*). An examination of this material in the Gray Herb. and Mo. Bot. Gard. Herb., originally labeled *V. Longii*, leads me to conclude that the specimen is only a sparsely pubescent variation of *V. rupestris* Scheele, a species common in the Noel region, and should be referred to that taxon. Occasional specimens of *V. rupestris* have the leaves sparsely hairy on the veins beneath (*Steyermark 10664* from Benton County; *Steyermark 41466* from Dent County, and other specimens). The specimen collected by Bush shows only slight hairiness on the lower surface of the leaves and on the lower nerves of the leaves at the tip of the shoots. It is not as strongly pubescent nor shows any of the cobwebby hairiness or close tomentose pubescence nor thick-leaved character found in *V. acerifolia* material seen from Kansas, Oklahoma, Texas, New Mexico, and Colorado.

## Order **MALVALES**

### Fam. **TILIACEAE** (Linden Family)

#### **Tilia** L. Basswood, Linden

This genus is being revised by Dr. G. N. Jones of the University of Illinois. Each of the previous authors who have published before Jones has interpreted the variation within the genus differently, the result being that no two authors have agreed as to the limits of a species. The Missouri material would appear to be separable at most into two species. Many specimens from Missouri have been collected without flower or fruit, thus adding to the difficulty of a satisfactory identification of the specimens. As a more mature evaluation of the genus must await more intensive field and experimental studies, the following treatment is to be considered a provisional one only, based upon present evidence at hand:

a. Lower leaf-surface green, from the beginning glabrous except for tufts of hairs in the axils of the main lateral (side) nerves; peduncles (main stalk of inflorescence), bract of inflorescence, and pedicels (individual flower-stalk) glabrous (without hairs); most commonly found in Missouri . . . . 1a. T. AMERICANA var. AMERICANA

a. Lower leaf-surface green or whitened, from the beginning with scattered simple or star-shaped (stellate) hairs occurring more or less scattered over the surface; peduncles, bract of inflorescence, and pedicels glabrous or more or less hairy; less commonly occurring in Missouri . . . . . . . . *b*

 b. Peduncle, bract of inflorescence, and pedicels glabrous; lower surface of leaves green with scattered star-shaped (stellate) hairs on the surface . . . . . . . 1b. T. AMERICANA var. NEGLECTA

 b. Peduncle, bract of inflorescence, and pedicels usually more or less pubescent (with hairs), sometimes glabrous; lower surface of leaves usually whitened with sparse to dense hairiness of star-shaped (stellate) hairs . . . . . . . . . . . . . . . . . . . . 2. T. HETEROPHYLLA

1. **Tilia americana** L. Basswood, Linden Map 1513

Also called Linn, Bee Tree, Whittle Wood.

Flowers late May–July.

Occurs in rich woods, rocky or nonrocky slopes, base of bluffs, and along streams. Throughout Missouri, doubtless in every county.

1a. **Tilia americana** var. **americana** Map 1513

*Tilia americana* L. [G, BB]

*Tilia glabra* Vent. [P & S, Steyerm.]

This is the common linden found throughout the state, doubtless in every county.

Ranges from Quebec to North Dakota and Manitoba, south to Alabama, Tennessee, Arkansas, and Texas.

1b. **Tilia americana** var. **neglecta** (Spach) Fosberg Map 1513

*Tilia neglecta* Spach [G]

*Tilia glabra* var. *neglecta* [Spach] Bush

*Tilia heterophylla* var. *Michauxii* (Nutt.) Sarg. [P & S, Steyerm.]

Scattered in eastern, western and central Missouri in Marion (north of Hannibal, *Davis 2463*), Jefferson, St. Francois (near Bismarck, *Palmer 19538*), Butler (along Black River, Poplar Bluff, *Kellogg 1524*; *Palmer 14745*), Shannon (*Bauer*), Gasconade (*Steyermark 27840*), Miller (*Steyermark 6848*); Saline (*Steyermark 21573*), Lafayette, Holt (Forest City, *Palmer 25379*), Buchanan (*Palmer 35733*), Jackson (*Palmer 20436*), Lawrence (*Palmer 52395*), and Barry (*Palmer 52793*) counties.

Ranges from Quebec to Minnesota, south to North Carolina, Illinois, Missouri, and Oklahoma.

The lower surface of the leaves are green with some stellate pubescence in the specimens above cited from Marion, St. Francois, and Butler counties. They had been previously identified as *T. heterophylla* var. *Michauxii*. The specimen of *Palmer 52395* has both simple and scattered stellate hairs on the lower leaf-surface.

The flowers of basswood are fragrant and can be used to prepare a flavorful tea. They are much visited by honey bees and a special honey, made exclusively from the basswood flowers, is commonly found for sale. Sugar may be obtained, as in Sugar Maple, from the sap. The ground fruits can be prepared into a

Plate no. 250. **1.** Tilia americana, × 2/7; Details from Small, The New York Botanical Garden. **2.** Althaea rosea, × 2/7; Details from Small, The New York Botanical Garden. **3.** Tilia heterophylla, × 2/7; *a*. Fruiting cluster. **4.** Malva moschata f. laciniata, × 2/7. **5.** Malva sylvestris, × 2/7; Details from Small, The New York Botanical Garden. **6.** Malva verticillata var. crispa, × 2/7; Details from Small, The New York Botanical Garden.

PLATE NO. 250

1513 • Tilia americana var. americana (Basswood)
1513 □ Tilia americana var. neglecta

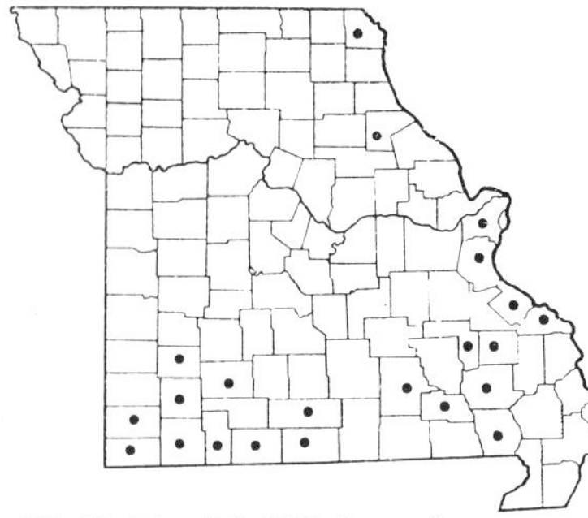

1514 Tilia heterophylla (White Basswood)

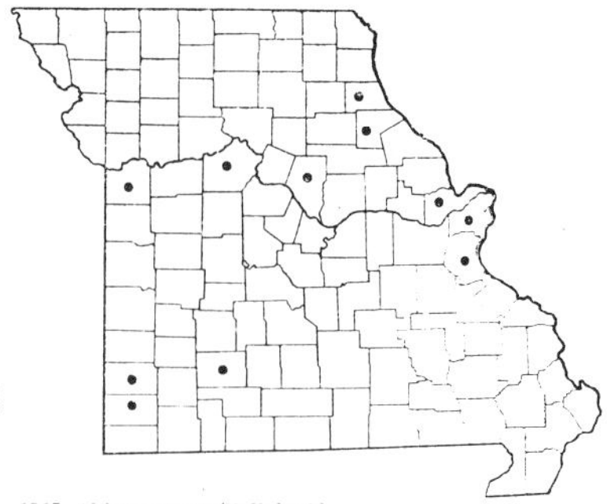

1515 Althaea rosea (Hollyhock)

paste resembling chocolate and used as a substitute; it cannot be kept long as it decomposes after a short time. Some of the American Indians chewed the inner bark and the young buds in spring were used for quenching thirst. They made rope and string out of the tough inner bark. Actually, the word 'basswood' is originally derived from 'bastwood,' referring to the tough fibrous inner bark of the tree.

The wood of basswood is one of the lightest known among North American trees. For this reason it supplies much of the lumber for crates, boxes, toys, excelsior, veneers, dressmaker's yardsticks, frames for comb honey, window sashes, and picture frames. Small amounts of the wood, mixed with other hardwoods are used as pulp in papermaking.

The glabrous, alternately arranged, red winter buds distinguish the tree after the leaves have fallen. Its autumnal colors vary from pale green to pale yellow. With its large, heart-shaped or broadly ovate to nearly rounded, blue-green leaves and fragrant summer flowers, the linden is a desirable ornamental shade tree. However, the native North American species do not adapt well to city conditions and various European species of linden are recommended instead.

A few Missouri towns bear the name of this tree. The leaves and buds are browsed by white-tailed deer.

2. **Tilia heterophylla** Vent. White Basswood
Map 1514

Also called White Linn, Silver-leaved Linn, White Linden.

*Tilia floridana* (V. Engler) Small [P & S, Steyerm.]

*Tilia floridana* var. *hypoleuca* Sarg. [P & S, Steyerm.]

Flowers May–July.

Occurs mainly along rocky woods and bluffs along streams. Southern and east-central Missouri.

Ranges from Florida to Arkansas, north to New York, West Virginia, Ohio, Indiana, Illinois, and Missouri.

A specimen from Jefferson County (Sulphur Springs near St. Louis, June, 1853, *Engelmann*) has the lower leaf-surface densely white-pubescent. Other specimens, such as *Palmer 34795* from Ozark Co., *Palmer 14362, 22815*, and *23878* from Stone Co., placed here under *T. heterophylla*, but originally identified as *T. floridana* var. *hypoleuca*, have white lower leaf-surfaces, but only a sparse pubescence.

*Tilia floridana* has been treated as a species distinct from *T. heterophylla* by Fernald (*Gray's Manual*, eighth ed.) and by Little (Check list of trees, USDA Agri. Handb. 41. pp. 419–20. 1953) and by them credited to Missouri. As indicated above, some Missouri specimens previously identified as *T. floridana* or var. *hypoleuca* are placed in the present treatment under *T. heterophylla*, but most of them do not possess the feltlike pubescence of the lower leaf-surface shown by specimens examined from Kentucky, Florida, and other southeastern states. Other Missouri specimens previously identified as *T. floridana* or var. *hypoleuca* have been identified by Dr. George N. Jones as *T. americana*.

## Fam. **MALVACEAE** (Mallow Family)

a. All the leaves merely toothed or nearly without teeth, either most of them or the lower ones without lobes or divisions . . . . . . . . . . . . . . . . . . . . . . . . . . . . . . *b*

b. All the leaves of a narrow type, linear to narrowly oblong or ovate-lanceolate, 3–25 mm. wide, at most 55 mm. long; no leaves heart-shaped at base. . . . . . . . . . . . . . . . . *c*

c. On the outside and at the base of the 5-lobed calyx are 2–3 narrow bractlets; teeth of leaves mainly 1–7 on each side . . . . . . . . . . . . . . . . . . 4. SPHAERALCEA
c. No bractlets on the outside or at the base of the 5-lobed calyx; teeth of leaves 8–25 on each side . . . . . . . . . . . . . . . . . . . . . . . . . . . 5. SIDA
b. Leaves either of a broad type and rounded or broadly ovate or ovate-triangular, some or all of them at least 40 mm. or more wide and 60 mm. or more long; leaves sometimes heart-shaped at base . . . . . . . . . . . . . . . . . . . . . . . . . . . . *d*
d. Lower leaves different in shape from and noticeably broader than the upper leaves; leaf-blades not soft-hairy or pale on lower surface . . . . . . . . . . . . . . 3. CALLIRHOE
d. Leaves of essentially the same shape throughout; leaf-blades soft-hairy, pale on the lower surface . . . . . . . . . . . . . . . . . . . . . . . . . . . . *e*
e. Plants of the water and of borders of ponds, sloughs, swamps, and wet ditches; on the outside and at the base of the 5-lobed calyx are a circle of narrow bractlets . . . . . . . 8. HIBISCUS
e. Weedy plants of fields, pastures, waste places, along roads and railroads; no bractlets on the outside or at the base of the 5-lobed calyx . . . . . . . . . . . . 7. ABUTILON
a. Some or all of the leaves with shallow or deep lobes or divisions . . . . . . . . . . . . *f*
f. No bractlets on the outside or at the base of the 5-lobed calyx . . . . . . . . . . . . *g*
g. Petals blue-violet to lavender; leaves with only the central lobe prominent or elongated, the other portion of the leaf with short lobes or coarse teeth . . . . . . . . . . 6. ANODA
g. Petals rose-red, red-purple, pinkish, pale lilac, or white; leaves deeply cut or lobed with 3–7 or more prominent or elongated lobes . . . . . . . . . . . . . 3. CALLIRHOE
f. On the outside and at the base of the 5-lobed calyx are narrow or broad bractlets which are separate or united . . . . . . . . . . . . . . . . . . . . . . . . . . *h*
h. 3 bractlets on the outside and at the base of the 5-lobed calyx . . . . . . . . . . 2. MALVA
h. 6 or more bractlets, either separate or united, on the outside and at the base of the 5-lobed calyx . . . . . . . . . . . . . . . . . . . . . . . . . . . . . *i*
i. Bractlets broadly triangular; styles numerous; fruit or ovary consisting of 15–20 carpels arranged in a ring around a central axis and separating when mature . . . . . . 1. ALTHAEA
i. Bractlets narrowly linear or nearly thread-like; styles 5; ovary 5-celled, the fruit becoming a 5-celled capsule of united carpels which do not separate at maturity . . . . . 8. HIBISCUS

## 1. **Althaea** L.

**Althaea rosea** L. Hollyhock Map 1515

Flowers May–September.

Commonly planted throughout Missouri and occasionally escaped along roadsides, railroads, waste ground, and fields. Scattered in the state.

Native of Europe and Asia; introduced and naturalized in various parts of the United States.

## 2. **Malva** L. Mallow

a. At least the upper leaves deeply 5- or more lobed or dissected to below the middle; petals 20–35 mm. long; flowers mainly at the top of the stem in a racemose inflorescence partly solitary in the axils of the upper leaves and longer than the leaves . . . . . . . . . . . . . 1. M. MOSCHATA
a. All the leaves with shallow lobes, not lobed to the middle; petals 6–25 mm. long; flowers crowded in fascicles in the axils of the leaves, the petioles of the latter longer than the flowers . . . . . . . *b*
b. Leaf-blades crisped, puckered, or curled on the margins; most of the flowers and fruits nearly sessile (without stalks) . . . . . . . . . . . . . . . 3. M. VERTICILLATA var. CRISPA
b. Leaf-blades not crisped or puckered on the margins; flowers and fruits mostly on obvious stalks . . *c*
c. Petals 15–25 mm. long, red-purple or rose-purple; bractlets at base of calyx oblong to ovate-lanceolate; leaf-blades with 3–7 conspicuous lobes. . . . . . . . . . . 2. M. SYLVESTRIS
c. Petals 4–12 mm. long, pink, lilac, white, or bluish-white; bractlets at base of calyx linear to linear-lanceolate; leaf-blades shallowly and more inconspicuously lobed . . . . . . . . . *d*
d. Commonly found throughout Missouri; stems lying along the ground, trailing or with only the tips ascending; petals 10–12 mm. long, about twice as long as the calyx; carpels not reticulated, round-margined . . . . . . . . . . . . . . . . . . . . 4. M. NEGLECTA
d. Rarely found species; stems ascending to upright; petals 4–6 mm. long, not quite equaling or barely longer than the calyx; carpels acute-margined, rather rugose-reticulated. . . . . *e*

1515a Malva moschata (Musk Mallow)

1516 Malva sylvestris (High Mallow)

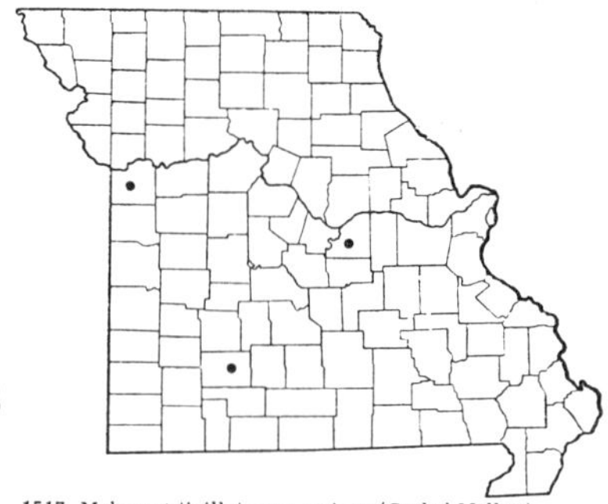

1517 Malva verticillata var. crispa (Curled Mallow)

e. The narrowed base (claw) of petals without hairs; calyx becoming enlarged and with a network of conspicuous veins; carpels without hairs, the angles toothed and thin-margined . . . . . . . . . . . . . . . . . . . . . . 5. M. PARVIFLORA

e. The narrowed base (claw) of petals with hairs; calyx scarcely enlarged nor with a network of conspicuous veins; carpels at first short-hairy, eventually glabrous (without hairs), the angles acute and not winged . . . . . . . . . . . 6. M. ROTUNDIFOLIA

1. **Malva moschata** L. f. **laciniata** (Desr.) Hayek
Musk Mallow Map 1515a
Flowers May–October.

Known only from Boone County (alfalfa field north of Grindstone Creek, Columbia, August, 1904, *F. Daniels*, herb. U. of Mo.), central Missouri.

Native of Europe; introduced and naturalized in North America from Newfoundland to Ontario and British Columbia, south to Virginia, Tennessee, Missouri, and Nebraska.

This species is often cultivated as a garden plant. It occurs in several leaf variations. Typical f. *moschata* has the upper leaves shallowly cut into 5 simply cut, broad rhombic segments; in f. *heterophylla* (Vis.) Hayek only the upper leaves have the blades divided into 5 main portions, these dissected into linear segments, while in f. *laciniata* (Desr.) Hayek all the leaves are finely dissected into linear segments. The latter is the type found in Missouri and commonly found elsewhere.

2. **Malva sylvestris** L. High Mallow Map 1516
Flowers May–July.

Planted in gardens and occasionally escaped along roadsides and waste ground in Warren, Jackson, Greene, and Stone counties.

Native of Europe; introduced and naturalized in North America from Quebec to North Dakota, Colorado, and Oregon, south to Florida, Texas, and California.

Typical *M. sylvestris* var. *sylvestris* has hairy stems and sharply 5–7-lobed leaf-blades with triangular lobes, whereas var. *mauritiana* (L.) Boiss. has glabrous stems and the lobes of the leaf-blades broad and rounded.

3. **Malva verticillata** L. var. **crispa** L.
Curled Mallow Map 1517
*Malva crispa* L. [P & S, Steyerm.]

Occurs in waste ground. Scattered in Osage, Greene, and Jackson counties.

Native of Europe; introduced and naturalized in North America from Quebec to South Dakota and Colorado, south to New Jersey, Illinois, and Missouri.

4. **Malva neglecta** Wallr. Common Mallow
Map 1518
Also called Cheeses.
*Malva rotundifolia* of auth. [P & S], not L.
Flowers April–October.

Occurs in fields, lawns, waste ground, along roads and railroads. Throughout Missouri.

Native of Europe; introduced and naturalized throughout North America.

5. **Malva parviflora** L. Map 1519
Flowers May–October.

Occurs in fields, waste ground, and along roadsides. Scattered in Atchison and Jackson (Sheffield, June 29, 1905, *Bush 3045*) counties.

---

Plate no. 251. **1.** Malva neglecta, $\times {}^{2}/_{7}$. **2.** Malva parviflora, $\times {}^{2}/_{7}$. **3.** Malva rotundifolia, $\times {}^{2}/_{7}$; (Scribner's). **4.** Callirhoe triangulata, $\times {}^{2}/_{7}$; Details from Small, The New York Botanical Garden. **5.** Callirhoe involucrata, $\times {}^{2}/_{7}$. **6.** Callirhoe Papaver var. Papaver, $\times {}^{2}/_{7}$.

PLATE NO. 251

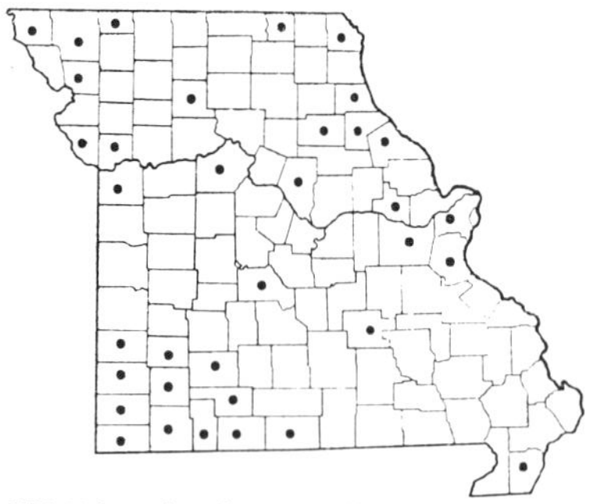

1518 Malva neglecta (Common Mallow)

1519 Malva parviflora

1520 Malva rotundifolia

Native of Europe; introduced and naturalized in North America from Quebec, North Dakota to British Columbia, south to New Jersey, Missouri, Texas, New Mexico, and Mexico.

6. **Malva rotundifolia** L. Map 1520
Flowers May–October.

Occurs in fields, waste ground, and along roadsides and railroads. Scattered in Missouri, known from St. Louis, Clark (near Alexandria, *Palmer & Steyermark 40601*), Clay, Jackson (Sheffield, *Bush 7057*), Nodaway (5 mi. northeast of Clearmont, *Steyermark 5863*), and Dade counties.

Native of Europe; introduced and naturalized from Michigan and Indiana to North Dakota, west to the Pacific states and south to Georgia, Texas, Arizona, and California.

### 3. **Callirhoe** Nutt. Poppy Mallow

a. Most of leaves triangular and merely toothed, most of them without deep lobes or divisions . . . 1. C. TRIANGULATA
a. Most of leaves deeply lobed or parted . . . . . . . . . . . . . . . . . . . . . *b*
  b. Petals either pink, white, or pale lilac, never red, rose-purple, nor wine-red; no bractlets on the outside or at the base of the 5-lobed calyx . . . . . . . . . . . . . . . 4. C. ALCAEOIDES
  b. Petals deep rose, rose-red, purple, or rose-purple; calyx with 3 bractlets on the outside or at the base, or, if lacking bractlets, then most of the stem is glabrous (without hairs) . . . . . . *c*
    c. No bractlets at base of or on outside of the 5-lobed calyx; stem glaucous or gray-green, with a pale coating which can be rubbed off; most of the stem glabrous (without hairs), hairs if present only at very base of stem; stipules (at base of leaf-stalks) falling early and usually absent at flowering time, when present narrowly linear or subulate and less than 5 mm. long 5. C. DIGITATA var. DIGITATA
    c. 3 bractlets present at base of or on outside of the 5-lobed calyx; stem not glaucous as described in preceding alternate section; most of stem hairy, the hairs often more abundant in upper half of stem; stipules (at base of leaf-stalks) prominent, persistent, ovate or rhombic to oblong, mostly 5–20 mm. long . . . . . . . . . . . . . . . . . . . . . . *d*
      d. Stems mostly horizontally spreading (procumbent) or reclining with ascending tips (decumbent); main leaf-divisions deeply cut or toothed; the 3 bractlets remaining close to the 5-lobed calyx and not noticeably separated from it . . . . . . . . 2. C. INVOLUCRATA
      d. Stems mainly ascending or erect; main leaf-divisions entire (without teeth) or with few teeth; at least one of the bractlets slightly separated by 1–3 mm. from the calyx . . . . . . *e*
        e. Hairs on stem pressed closely or lying parallel to surface of stem (appressed) . . . . 3a. C. PAPAVER var. PAPAVER
        e. Hairs on stem spreading away from the stem or pointing toward the base (retrorse) 3b. C. PAPAVER var. BUSHII

1. **Callirhoe triangulata** (Leavenw.) Gray
Clustered Poppy Mallow Map 1521
Flowers July–August.

Occurs in rocky open woods, sandy open ground, sandy prairies, and glades, usually in acid soils. Known only in eastern Missouri south of the Missouri River in St. Louis, Franklin, Scott, and Mississippi counties.

1521 Callirhoe triangulata (Clustered Poppy Mallow)

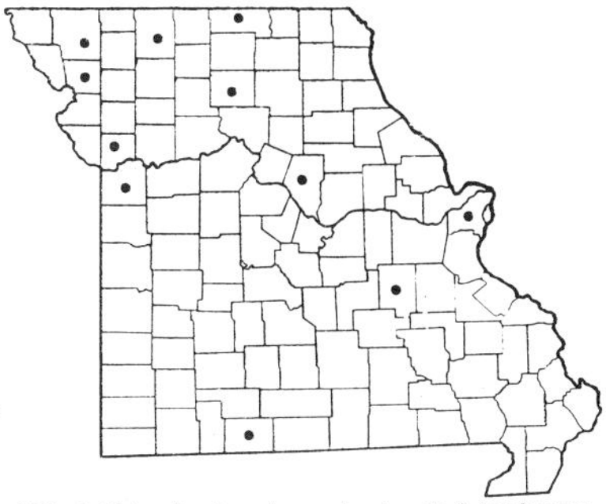

1522 Callirhoe involucrata var. involucrata f. involucrata (Purple Poppy Mallow)

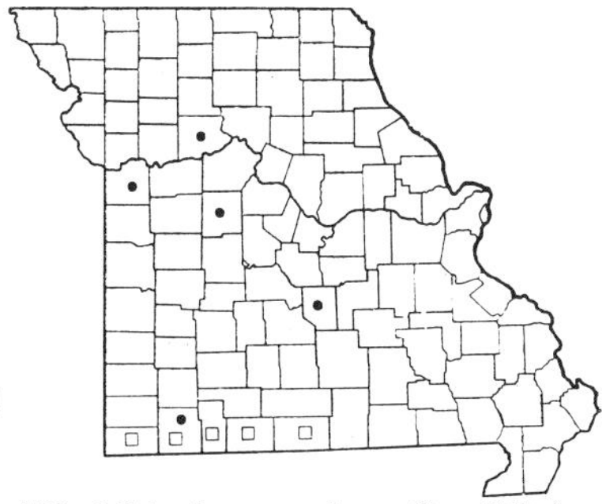

1523 ● Callirhoe Papaver var. Papaver (Poppy Mallow)
1523 □ Callirhoe Papaver var. Bushii

Ranges from Alabama to Missouri, north to North Carolina, Indiana, Illinois, Wisconsin, and Nebraska.

This species has mostly erect stems, which, together with the leaves, flower-stalks and calyx, are densely stellate-hairy. The flowers are quite showy with a deep rose-purple color. The plants do well in open sandy ground, where I have grown them in northern Illinois during the past twelve years.

2. **Callirhoe involucrata** (T. & G.) Gray var. **involucrata** f. **involucrata**
Purple Poppy Mallow — Map 1522
Flowers May–August.

Occurs along roadsides, prairies, fields, and grassy open places. Scattered in the state, mostly in the northern and central sections.

Ranges from North Dakota to Wyoming, south to Missouri, Oklahoma, Texas, and Utah.

Missouri material all falls into typical *C. involucrata* as differentiated from *C. involucrata* var. *involucrata* f. *novomexicana* (Baker) Waterfall and *C. involucrata* var. *lineariloba* (T. & G.) Gray, as treated by Waterfall (Field & Lab. 19: 110–16. 1951).

This is a showy-flowered species which does well in open flower beds, where it forms large spreading patches and increases by seeds. It favors a clayey loam soil.

3a. **Callirhoe Papaver** (Cav.) Gray var. **Papaver**
Poppy Mallow — Map 1523
*Callirhoe Papaver* (Cav.) Gray [G, BB, P & S, Steyerm.]
Flowers May–August.

Occurs along roadsides, low thickets along railroads, fields, prairies, and glades. Rare and scattered in Pulaski, Carroll, Jackson, Pettis, and Barry counties.

Ranges from Florida to Texas, north to Georgia and Missouri.

3b. **Callirhoe Papaver** var. **Bushii** (Fern.) Waterfall — Map 1523
*Callirhoe Bushii* Fern. [P & S, Steyerm.]
*Callirhoe involucrata* var. *Bushii* (Fern.) Martin [G, BB]
Flowers May–August.

Occurs in rocky open woods, wooded valleys, ravine bottoms, and borders of glades. Southwestern Missouri from Ozark County west to McDonald County.

Ranges in Missouri, Arkansas, and Oklahoma.

This variety has been placed in *C. involucrata* by some authors, but its relationship with *C. Papaver* seems more evident as manifested by the separation of the bractlet from the calyx and by the ascending to erect habit of growth. This conclusion has been reached by Dr. Waterfall (Southwest. Nat. 3: 215–16. 1959) with whom the present author is in complete agreement. The shape of the bractlets is stated to be ovate in the eighth edition of *Gray's Manual*, but, as in *C. Papaver* var. *Papaver*, the bractlets are as frequently linear, and in most of the Missouri material examined, they are linear in var. *Bushii* as well as in var. *Papaver*.

Likewise, the width and degree of toothing of the leaf lobes varies considerably in var. *Bushii* and are characters found to be unreliable in separating this variety from var. *Papaver*. Waterfall (pp. 215–16) has also noted that this variety has somewhat larger flower buds just before the opening of the flower than in *C. Papaver* var. *Papaver*.

4. **Callirhoe alcaeoides** (Michx.) Gray
Poppy Mallow — Map 1524
Flowers May–August.

Occurs in prairies, along roadsides, grassy open places, waste ground, and occasionally in open woods. Scattered throughout Missouri.

Ranges from Illinois to Nebraska, south to Ala-

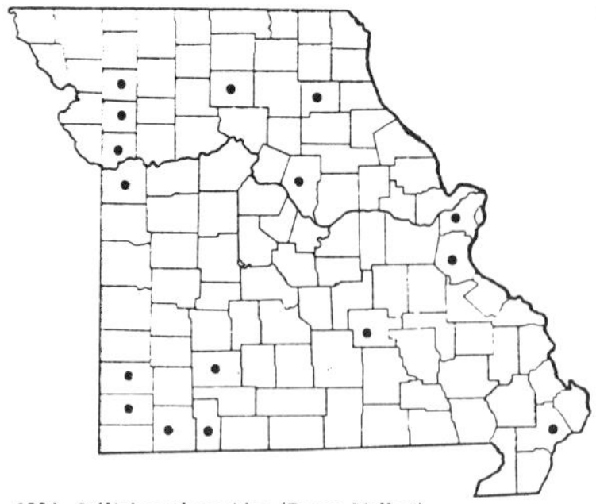

1524 Callirhoe alcaeoides (Poppy Mallow)

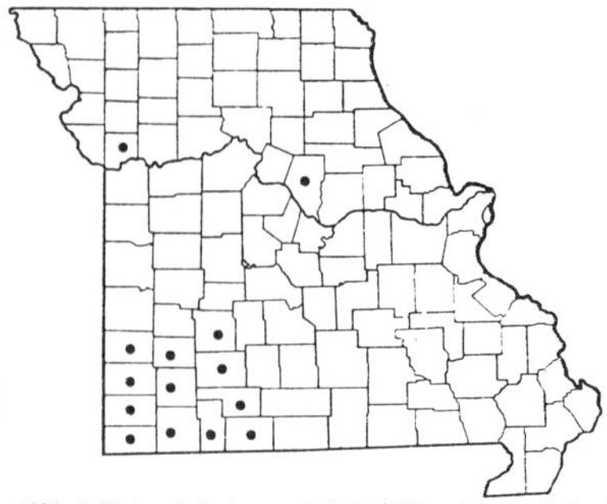

1525 Callirhoe digitata var. digitata (Fringed Poppy Mallow)

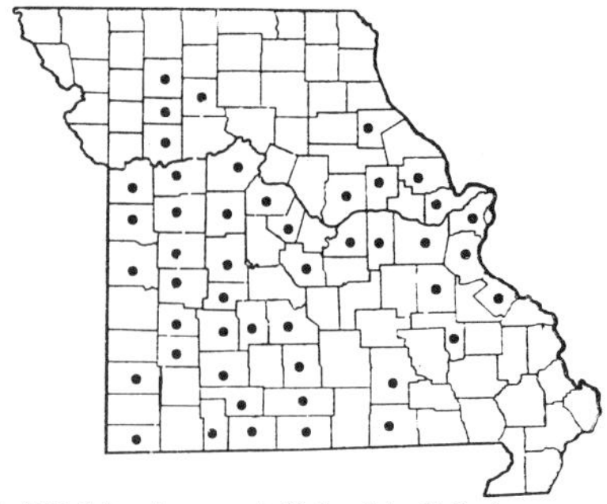

1526 Sphaeralcea angusta (Yellow False Mallow)

bama, Tennessee, Missouri, Oklahoma, and Texas.

This species has several stems growing in a loose clump 1.5–4.5 dm. tall with showy flowers arranged corymbosely. The plant does well in perennial beds exposed to full sun.

5. **Callirhoe digitata** Nutt. var. **digitata**

Fringed Poppy Mallow — Map 1525

*Callirhoe digitata* Nutt. [G, BB, P & S, Steyerm.]

Flowers May–September.

Occurs usually on rocky prairies, glades, and meadows. Southwestern Missouri, where commonest on limestone glades, bald knobs, and barrens or rocky prairies north to Polk and Barton counties and east to Greene, Christian, and Taney counties; introduced in Boone and Clay counties in central Missouri.

Ranges from Missouri and Kansas to Arkansas and Oklahoma.

This is a very showy plant with erect stems usually 7–13 dm. tall. The deep rose- or wine-purple flowers are set on long and very slender peduncles. The plant does well in sunny open beds, and in the author's wildflower preserve in northern Illinois it grows well several hundred miles north of the limits of its natural geographical range.

Another variety, var. *stipulata* Waterfall, occurs in Oklahoma and Texas, distinguished by its large, persistent stipules.

## 4. **Sphaeralcea** St. Hil. False Mallow

[*Malvastrum* of BB in part]

**Sphaeralcea angusta** (Gray) Fern.

Yellow False Mallow — Map 1526

*Malvastrum angustum* Gray [BB, P & S]

Flowers July–September.

Occurs usually on rocky prairies, limestone, sandstone, or cherty limestone glades, edges of bluffs, sometimes open alluvial ground in valleys, and along gravel bars. Southern and central Missouri, mainly in the Ozark and unglaciated prairie sections, locally north to Ralls, Callaway, Saline, Livingston, Caldwell, and Daviess counties; absent from the lowland counties of southeastern Missouri.

Ranges from Illinois to Iowa and Nebraska, south to Alabama, Missouri, and Kansas.

The petals, which are yellow or orange, are barely longer than the inflated calyx, and are sometimes absent.

## 5. **Sida** L.

Main leaf-blades of larger stems ovate-lanceolate or oblong, 8–25 mm. broad; petiole (leaf-stalk) longer than the peduncle (flower-stalk), 6–25 mm. long; carpels 5 in each flower; common throughout Missouri . . . . . . . . . . . . . . . . . . . . . . . . . . 1. S. SPINOSA

---

Plate no. 252. **1.** Callirhoe Papaver var. Bushii, × 2/7. **2.** Callirhoe alcaeoides, × 2/7. **3.** Callirhoe digitata var. digitata, × 2/7. **4.** Sphaeralcea angusta, × 2/7; After Britton and Brown, details from Small, The New York Botanical Garden. **5.** Sida spinosa, × 2/7; (Scribner's). **6.** Anoda cristata var. cristata, × 2/7; *a.* Fruit, × 2; After Gleason, The New York Botanical Garden. **7.** Abutilon Theophrasti, × 2/7; (Scribner's); Details from Small, The New York Botanical Garden.

PLATE NO. 252

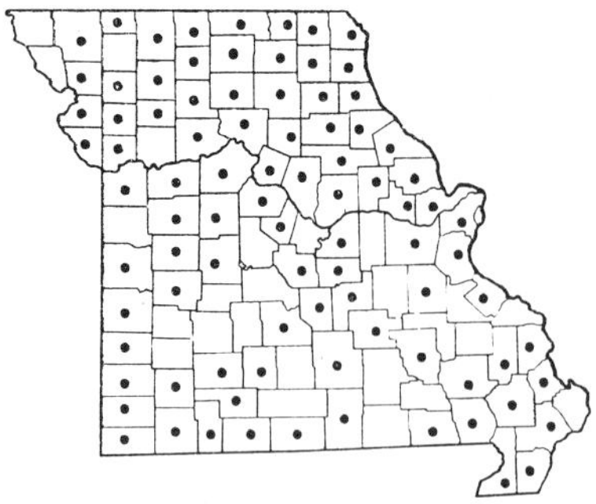
1527 Sida spinosa (Prickly Sida)

1528 Sida Elliottii

1529 Anoda cristata var. cristata

Main leaf-blades of larger stems narrowly linear, 1.5–7 (–11) mm. broad; petiole (leaf-stalk) shorter than the peduncle (flower-stalk), 2–8 mm. long; carpels 8–12, usually 10 in each flower; known only from southeastern Missouri . . . . . . . . . . . . . . . . . . . . . . . . 2. S. Elliottii

1. **Sida spinosa** L. Prickly Sida Map 1527

Also called Prickly Mallow.

Flowers early June–October.

Occurs in fallow or cultivated fields, pastures, along roadsides, railroads, and waste ground, rarely in open woodland. Common throughout Missouri, and doubtless in every county.

Native of the New and Old World tropics; naturalized in the United States north to Massachusetts, Michigan, Nebraska, and Kansas.

A short, slightly prickly projection at the base of the leaves on well-developed plants is responsible for the common name as well as specific latin epithet. However, it cannot be termed a spine in the true sense of the word. The small yellow flowers arise scattered along the leaf axils.

2. **Sida Elliottii** T. & G. Map 1528

Flowers August–October.

Occurs in sandy open ground on Crowley Ridge, sandy prairies, and along roadsides. Southeastern Missouri, where known only from Scott, Stoddard and Dunklin counties.

Ranges from Florida and Alabama, north to South Carolina, Tennessee, and southeastern Missouri.

This plant has a smoother and much more branching stem than *S. spinosa*, often attaining 1 meter in height.

## 6. **Anoda** Cav.

**Anoda cristata** (L.) Schlect. var. **cristata** Map 1529

*Anoda cristata* var. *brachyanthera* of [P & S], not (Reichenb.) Hochr.

Flowers June–September.

Known only from McDonald County, southwestern Missouri (Noel, *Palmer 19069*).

Native of tropical America, Mexico, and southwestern United States; introduced and naturalized in Pennsylvania, and from Iowa and Missouri to Texas.

In typical *A. cristata* var. *cristata* the petals are longer than the calyx, whereas in var. *brachyanthera* (Reichenb.) Hochr., of South American origin, the petals are shorter than or but slightly longer than the calyx.

## 7. **Abutilon** Mill. Indian Mallow

**Abutilon Theophrasti** Medic. Indian Mallow Map 1530

Also called Velvet Leaf, Butter Print.

Flowers late June–October.

Occurs as a weed in cultivated and fallow fields, pastures, waste ground, along roadsides and railroads. Throughout Missouri.

Native of India; introduced and naturalized from Maine to Michigan and South Dakota, south to Florida, Texas, and California.

The plant is a fast-growing, erect annual 6–15 dm. tall with large soft-hairy leaves and scattered flowers with yellow spreading petals. The arrangement of the head of 12–15 carpels resembles the print block used,

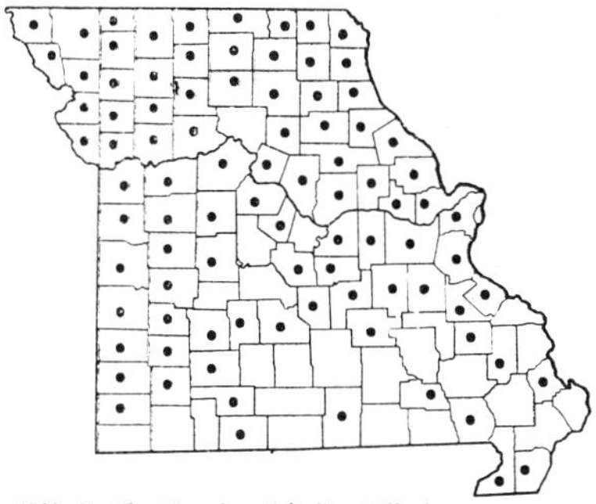

1530 Abutilon Theophrasti (Indian Mallow)

1531 Hibiscus syriacus (Rose of Sharon)

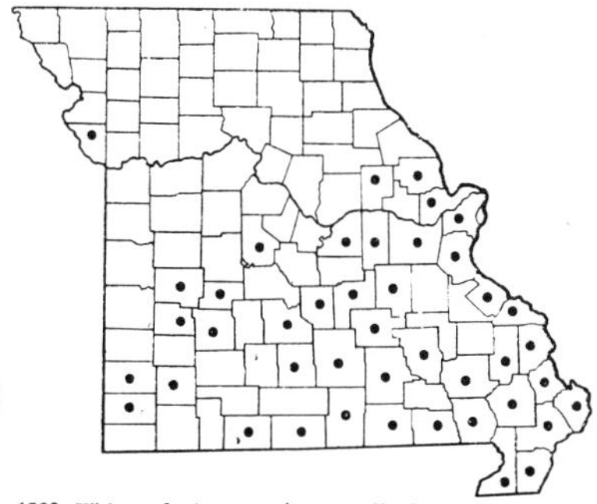

1532 Hibiscus lasiocarpos (Rose Mallow)

on farms for stamping rolls of butter. The seeds are reported to retain their viability for over fifty years and are not killed by siloing. The stems, as in many other members of the same family, possess a strong fiber which has potential use as thread or cord.

## 8. **Hibiscus** L.

a. Low annual herbs 1–5 dm. tall; stems with most of the hairs spreading; leaves deeply 3–5-parted to their base into narrow segments . . . . . . . . . . . . . . . . . . . . 4. H. TRIONUM

a. Tall perennial herbs or woody shrubs 1–6 m. tall; stems glabrous (without hairs) or covered with short stellate (star-shaped) hairs; leaves either not lobed or, if 3–5-lobed most of the leaves with shallow lobes or at least not usually cut to the midrib or base . . . . . . . . . . . . . . . *b*

b. Leaf-blades softly hairy on upper and lower surfaces . . . . . . . . . . . . . 2. H. LASIOCARPOS

b. Leaf-blades glabrous (without hairs) . . . . . . . . . . . . . . . . . . . . . *c*

c. Native herb 1–2 m. tall growing in wet soil or in water; most of leaves widest at the rounded or heart-shaped base or with widely spreading basal lobes; leaves with numerous short, more or less equal teeth . . . . . . . . . . . . . . . . . . . . . . 3. H. MILITARIS

c. Introduced woody shrub up to 6 m. tall growing in dry soils along roadsides, railroads, and thickets; most of leaves wedge-shaped (cuneate) or narrowed at base and broadest around the middle or above the base, the lobes all ascending; leaves with coarse unequal teeth . 1. H. SYRIACUS

1. **Hibiscus syriacus** L. Rose of Sharon

Map 1531

Also called Shrubby Althaea.

Flowers July–September.

Commonly planted as an ornamental shrub, rarely escaping along roadsides, railroads, and thickets. Known only as an escape in St. Louis, Jackson, and Jasper counties.

Native of Asia; introduced and naturalized in the United States from Florida to Texas, north to Connecticut, New York, Ohio, and Missouri.

2. **Hibiscus lasiocarpos** Cav. Rose Mallow

Map 1532

Flowers July–October.

Occurs along borders of sloughs, ponds, bayous, and ditches in wet soil or in water, and in low wet woods. Southern and central Missouri in the southeastern lowland and thoughout the Ozark counties north to Lincoln, Montgomery, Osage, Morgan, Hickory, St. Clair, and Jasper counties, and locally northwest in Platte County.

Ranges from Florida to Texas, north to Kentucky, Indiana, Illinois, Missouri, and Oklahoma.

The hairiness of the leaves varies considerably, on some plants appearing to be very thin, on others quite dense and conspicuous. Without leaves this species may be distinguished in fruit by the long hairs on the capsules, the other tall-growing species found in Missouri having glabrous or minutely hairy capsules. The showy flowers have white or rose-colored petals with a large wine-purple or magenta-crimson base. Seeds germinate readily, but the plants require a fairly moist soil to thrive under cultivation.

This is one of the dominant plants found in and around the borders of upland natural sink-hole ponds of the Ozarks, frequently forming extensive colonies. The seeds of this and the following species are eaten by ducks and bobwhite.

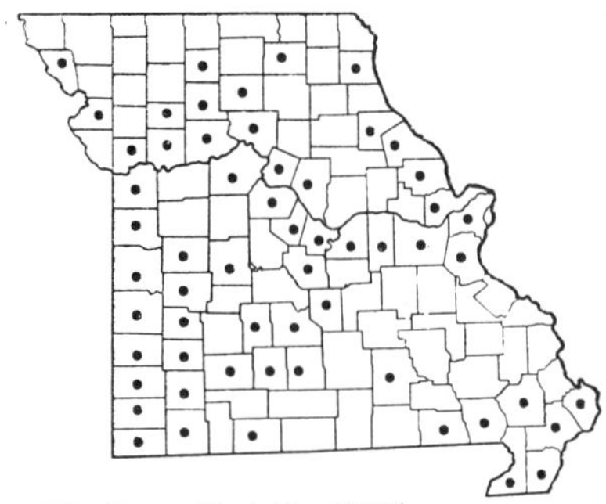

1533 Hibiscus militaris (Rose Mallow)

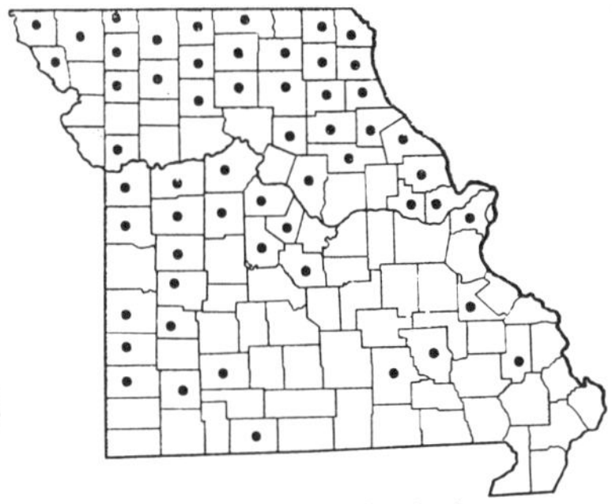

1534 Hibiscus Trionum (Flower-of-an-hour)

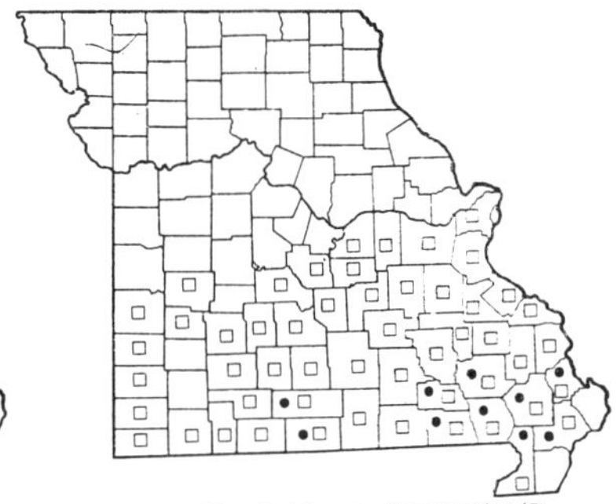

1535 ● Ascyrum Hypericoides var. Hypericoides (St. Andrew's Cross)
1535 □ Ascyrum Hypericoides var. multicaule

3. **Hibiscus militaris** Cav. Rose Mallow
Map 1533

Also called Halberd-leaved Rose Mallow.

Flowers July–October.

Occurs in wet soil or in water along streams, sloughs, ponds, and ditches. Absent from the extreme northern counties and from most of the eastern and central Ozark region, but common in the lowland counties of southeastern Missouri, unglaciated prairie and Ozark border sections of southwestern and central Missouri, in the counties bordering the Missouri and Mississippi rivers and in some of the northern counties, north locally to Lewis, Adair, Grundy, Caldwell, and Holt counties.

Ranges from Florida to Texas, north to Pennsylvania, West Virginia, Ohio, Indiana, Illinois, Minnesota, and Nebraska.

The showy flowers are pink or salmon with a dark wine-purple base. The seeds are eaten by ducks and bobwhite.

4. **Hibiscus Trionum** L. Flower-of-an-hour
Map 1534

Also called Shoo-fly.

Flowers mid-June–September.

Occurs as a weed in cultivated fields, vegetable and flower gardens, waste ground, pastures, and along roadsides and railroads. Throughout Missouri, but recorded most frequently from northern, central, and eastern Missouri, with few records in the Ozark section and none from the lowland counties of southeastern Missouri.

Native of Europe; introduced and naturalized in North America from Nova Scotia to Michigan and Minnesota, south to Florida and Texas.

The showy flowers have pale yellow petals blackish-purple at their base.

*Excluded Species*

**Modiola caroliniana** (L.) G. Don

A specimen belonging to this species was sent from Caruthersville, Pemiscot County, to the University of Missouri for determination, but no data is available concerning the circumstances connected with the finding of the plant.

**Gossypium hirsutum** L. Cotton

Commonly cultivated throughout the lowland counties of southeastern Missouri, and elsewhere in the state locally planted, but no evidence of its becoming established and persisting outside of cultivation is on record.

## Order **PARIETALES**

### Fam. **HYPERICACEAE** (St. John's-wort Family)

(*Guttiferae* [G])

Sepals 4, the outer two much larger; petals 4 . . . . . . . . . . . . . . . . . 1. Ascyrum
Sepals 5; petals 5 . . . . . . . . . . . . . . . . . . . . . . . . . . 2. Hypericum

## 1. **Ascyrum** L.

**Ascyrum Hypericoides** L. St. Andrew's Cross Map 1535

Flowers July–October.

The following variations are found in Missouri:

Uncommon, known only in extreme southeastern Missouri and Douglas and Ozark counties; plants with an erect, usually single main stem 3–8 dm. tall, which is much branched; leaves varying from linear to narrowly oblong or narrowly elliptic . . a. A. HYPERICOIDES var. HYPERICOIDES

Common throughout the Ozark region of southern and east-central Missouri; plants with several spreading or sprawling stems, at least the lower portion of the stems lying upon the ground surface, 1–2 (rarely 3) dm. tall; leaves more or less uniformly oblanceolate. . . . b. A. HYPERICOIDES var. MULTICAULE

a. **Ascyrum Hypericoides** var. **Hypericoides** Map 1535

*Ascyrum Hypericoides* L. (typical) [G]

*Ascyrum Hypericoides* var. *oblongifolium* (Spach) Fern. [G]

Occurs on sandy low wooded knolls, low wet woodland areas, swampy ground, and along streams, in acid soils in southeastern Missouri in the lowland counties north to Scott, Stoddard, Wayne, and Ripley counties; locally west in the extreme southern Ozarks in Carter, Ozark, and Douglas counties, where it occurs in sandy soils along streams or on upper slopes of sandstone bluffs in pine-oak woodland.

Ranges from Florida to Texas, north to Maryland, Virginia, Kentucky, Missouri, and Oklahoma; also in Mexico, the West Indies, and Central America (Guatemala and Honduras); recorded also from New Jersey.

b. **Ascyrum Hypericoides** var. **multicaule** (Michx.) Fern. Map 1535

*Ascyrum Hypericoides* of [P & S] in part, not L.

Occurs in usually dry, rocky open woods, usually on upland slopes and ridges, or in woodland valleys and ravines, in acid soils associated with chert, sandstone, or granite substrata. Southern and east-central Missouri in the Ozark and Crowley Ridge areas, south of the Missouri River north to St. Louis, Franklin, Gasconade, Osage, Miller, Camden, St. Clair, and Vernon counties.

Ranges from Georgia, Alabama, Mississippi, Louisiana, and Texas, north to Massachusetts, New Jersey, Pennsylvania, West Virginia, Ohio, Indiana, Illinois, Missouri, Kansas, and Oklahoma.

Recent work on the genus by Dr. W. P. Adams (Rh. 59: 73–95. 1957) has shown that this species is best divided into two recognizable varieties instead of three (var. *oblongifolium*) as proposed by Fernald (Rh. 38: 433. 1936). The less common variety in Missouri, var. *Hypericoides*, with erect branching stems, has variable leaves, but generally of a narrower type than in var. *multicaule*. I agree with Dr. Adams that the differences between the two varieties must be attributed to other than environmental conditions, as the two types of habit manifested are known to grow within a few feet of one another.

The leaves are browsed by white-tailed deer. They are smooth and evergreen, and the more compact and low-growing var. *multicaule* makes a desirable evergreen shrub for the rock garden. It has not survived the rigors of the winter climate of northern Illinois, however, where efforts to establish it there at the author's wildflower preserve have failed repeatedly. It requires an acid soil and warm climate.

## 2. **Hypericum** L. St. John's-wort

a. All the leaves scale-like or linear-subulate, minute, 1–20 mm. long, not more than 2 mm. broad with only a midrib visible . . . . . . . . . . . . . . . . . . . . . . . . . *b*
  b. Leaves scale-like, 1–3 mm. long; fruit 2–3 times longer than the calyx. . . . 11. H. GENTIANOIDES
  b. Leaves linear-subulate, 6–20 mm. long; fruit slightly longer than or barely equaling the calyx . 10. H. DRUMMONDII
a. Without the above combination of characters; leaves longer and/or broader, usually with either more than 1 main nerve from the base or with side (lateral) nerves; if only the midrib is visible, then leaves are broader than 5 mm. or longer than 20 mm.. . . . . . . . . . . . . . . . *c*
  c. Black dots or spots plainly visible on sepals and/or petals . . . . . . . . . . . . . . *d*
    d. Black dots or spots few on the petals and near or along the edges or at the tip; sepals with few black dots; stems with few black spots, these mainly on the ridges or angles; leafy sterile shoots usually prominently developed at base of plant during time of flowering; stem usually much branched; seeds rough with coarse honeycomb surface . . . . . . 2. H. PERFORATUM

d. Black dots or spots abundant and scattered over the entire length of the petals and sepals; stems with numerous black spots on many parts of the surface; leafy sterile shoots mainly absent at base of plant during time of flowering, but present in late fall, winter, and early spring; stem mainly simple and unbranched below or with simple branches; seeds rather smooth or with only a minutely honeycomb surface . . . . . . . . . . . . . . . . . . . . . *e*

e. Leaves oblong, usually rounded at tip; sepals 2.5–5 mm. long; petals 4–7 mm. long; styles 2–4 mm. long; leaves thin, grass or yellow-green . . . . 3a. H. PUNCTATUM var. PUNCTATUM

e. Uppermost leaves and bracts at base of inflorescence ovate to somewhat broadly triangular, somewhat pointed or acutish; sepals usually 5–7 mm. long; petals 8–12 mm. long; styles 6–10 mm. long; leaves firm, bluish- or glaucous-green . 3c–d. H. PUNCTATUM var. PSEUDOMACULATUM

c. Black dots or spots absent on sepals and petals . . . . . . . . . . . . . . . . *f*

f. Leaves with usually 3–7 nerves arising from the base but if rarely with 1 main nerve and 2 faint lateral nerves at the base, then the leaves only 1–6 mm. broad, thin, and membranaceous, and stamens 5–12 . . . . . . . . . . . . . . . . . . . . . *g*

g. Leaves linear to linear-lanceolate, 1–6 mm. broad, with 3 strong nerves at the base or with only 1 strong nerve and 2 faint lateral nerves at the base; known only from Sullivan County, northern Missouri . . . . . . . . . . . . . . . . . . . 9. H. CANADENSE

g. Leaves ovate-oblong, ovate-triangular or elliptic, the larger leaves of main stem 6–15 mm. broad, with 3–7 nerves arising from the base; plants of southern, central, and much of northern Missouri . . . . . . . . . . . . . . . . . . . . . . . *h*

h. Leaves rounded at tip; stem with usually widely spreading branches; mature fruit 2.5–3.5 (–4) mm. long, shortly ellipsoid, rounded at summit; throughout most of Missouri . . . . . . . . . . . . . . . . . . . . . . . 7. H. MUTILUM

h. Upper and middle leaves pointed or tapering at tip; stem mostly simple and unbranched, or branched only in the upper half or inflorescence; mature fruit 4–5 mm. long, slenderly conical, pointed at summit; plants known only in southern Missouri 8. H. GYMNANTHUM

f. Leaves with 1 midrib with or without lateral nerves pinnately arranged . . . . . . . . *i*

i. Petals pink or salmon; stamens 9; leaves pale green, thin and membranaceous; plants usually of swamps, low wet woods, upland sink-hole ponds, or rarely wet ledges along sandstone bluffs, known only from southeastern Missouri . . . . . . . . . . . . *j*

j. At least the upper leaves, or sometimes all of them rounded to heart-shaped at their base; lower surface of leaves without dark dots or translucent round glands; rare . 12. H. TUBULOSUM

j. All leaves narrowed to the base and tapering into petioles (stalks); lower surface of leaves with dark dots or translucent round glands; more common type found . . 13. H. WALTERI

i. Petals yellow or orange; stamens numerous, 20–100 or more; leaves dark green, thickish; plants of usually dry rocky bluffs, glades, gravel bars of streams, prairies, and dry open, rich, or rocky woodland . . . . . . . . . . . . . . . . . . . . . *k*

k. Petals 2.5–3 cm. long; larger leaves 15–35 mm. wide; styles 5; mature fruit 17–30 mm. long, 5-celled . . . . . . . . . . . . . . . . . 1. H. PYRAMIDATUM

k. Petals 0.5–1 cm. long; leaves 4–19 mm. wide; styles usually 3, sometimes 4–5; mature fruit 3.5–15 mm. long, 1-, 3-, or 4–5-celled . . . . . . . . . . . . . . . *l*

l. Stems mostly herbaceous, only very slightly woody at the base, 3–5 dm. tall; ovary and fruit 1-celled; mature fruit nearly spherical or broadly ovoid; plants with creeping underground rootstock . . . . . . . . . . . . 6. H. SPHAEROCARPUM

l. Stems definitely woody and shrubby for most of its height, 5–25 dm. tall; ovary and fruit 3–5-celled; mature fruit elongated, almost cylindrical to narrowly ovoid or lanceolate; plants without creeping underground rootstock . . . . . . . . . . *m*

m. Common throughout eastern, southern, and central Missouri; sepals 4.5–7.5 mm. long; mature fruit 8–15 mm. long . . . . . . . . . . 4. H. SPATHULATUM

---

Plate no. 253. **1.** Hibiscus syriacus, × 2/7. **2.** Hibiscus lasiocarpos, × 2/7; After Britton and Brown, The New York Botanical Garden. **3.** Hibiscus militaris, × 2/7; After Britton and Brown, The New York Botanical Garden. **4.** Hibiscus Trionum, × 2/7; (Scribner's); Details from Small, The New York Botanical Garden. **5.** Ascyrum Hypericoides var. Hypericoides, × 2/7; Details from Small, The New York Botanical Garden. **6.** Hypericum pyramidatum, × 2/7. **7.** Hypericum punctatum var. punctatum, × 2/7. **8.** Hypericum perforatum, × 2/7. **9.** Hypericum punctatum var. pseudomaculatum, × 2/7; *a.* Flower, × 1 3/7; Details from Small, The New York Botanical Garden. **10.** Hypericum spathulatum, × 2/7; Details from Small, The New York Botanical Garden. **11.** Hypericum sphaerocarpum var. sphaerocarpum, × 2/7; Details from Small, The New York Botanical Garden.

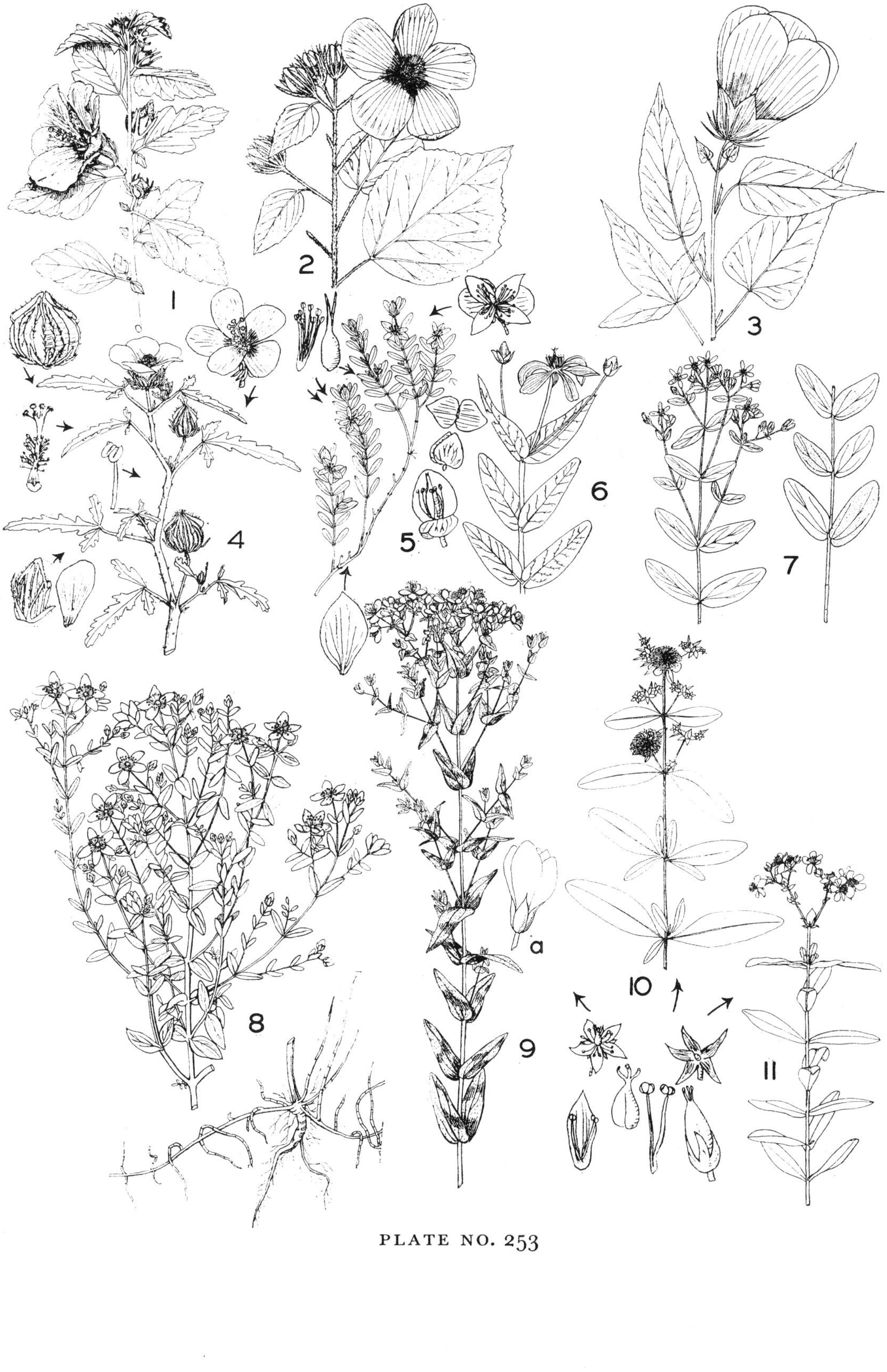


PLATE NO. 253

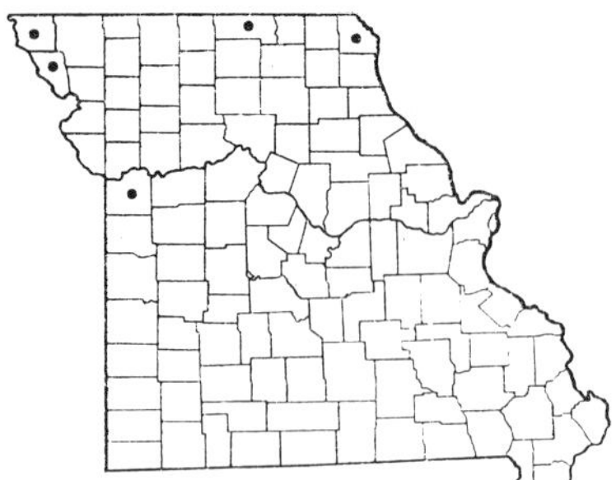

1536 Hypericum pyramidatum (Great St. John's-wort)

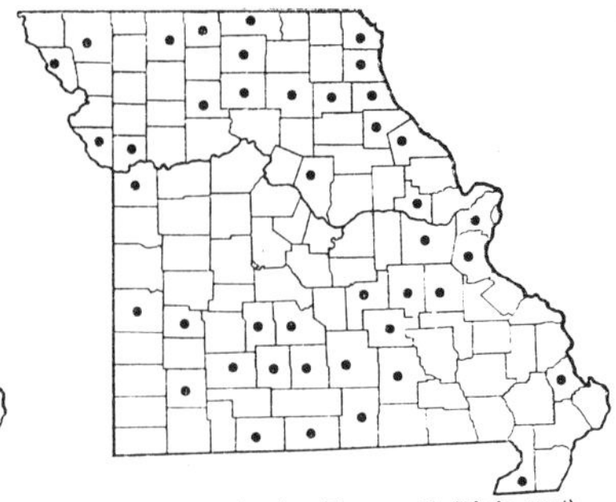

1537 Hypericum perforatum (Common St. John's-wort)

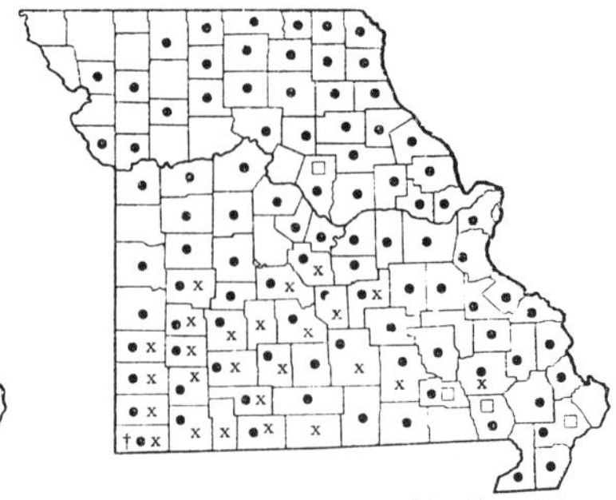

1538 • Hypericum punctatum var. punctatum f. punctatum
1538 □ Hypericum punctatum var. punctatum f. subpetiolatum
1538 x Hypericum punctatum var. pseudomaculatum f. pseudomaculatum
1538 † Hypericum punctatum var. pseudomaculatum f. flavidum

m. Rare known only in southeastern Missouri; sepals 2–5 mm. long; mature fruit 3.5–6.5 mm. long . . . . . . . . . . . . . . . . . . . . *n*
n. Styles 3; fruit with 3 distinct carpels . . . 5a. H. DENSIFLORUM var. DENSIFLORUM
n. Styles 4 or 5; fruit with 4 or 5 distinct carpels. . 5b. H. DENSIFLORUM var. LOBOCARPUM

1. **Hypericum pyramidatum** Ait.
Great St. John's-wort Map 1536
*Hypericum Ascyron* of Am. auth. [P & S], not L.
Flowers late June–September.

Occurs on rich wooded slopes, open woods and thickets. Rare, known only from northern and west-central Missouri in Clark, Putnam, Atchison, Holt, and Jackson counties.

Ranges from Quebec and Maine to Manitoba, south to Vermont, New Jersey, Maryland, Pennsylvania, Ohio, Indiana, Illinois, Missouri, and Kansas.

This is a very showy-flowered species, with flowers larger than in any of the other Missouri species. The nonwoody stems are 6–20 dm. tall.

2. **Hypericum perforatum** L.
Common St. John's-wort Map 1537
Flowers late May–September.

Occurs in pastures, fields, waste ground, along roadsides and railroads. Scattered throughout Missouri.

Native of Europe; introduced and naturalized in North America from Quebec to British Columbia, south to North Carolina, Tennessee, Oklahoma, Colorado, and California.

This plant may be distinguished from *H. sphaerocarpum* when not in flower by the lack of running underground rootstocks found in that species. The plant sometimes poisons grazing animals when eaten in the flowering stage or in strong sunlight, possessing a certain photosensitizing action which causes a dermatitis to the nonpigmented portions of the skin of light-colored horses, sheep, and cattle. If the animal is not exposed to strong sunlight after eating the plant, the symptoms of dermatitis do not appear.

3a–b. **Hypericum punctatum** L. var. **punctatum** Map 1538
*Hypericum punctatum* Lam. [G, BB, P & S]
Flowers June–September.

Occurs in fallow fields, waste ground, along roadsides and railroads, moist open woods, wooded slopes along streams, alluvial thickets, and prairies.

Ranges from Florida to Texas, north to Quebec, Ontario, and Minnesota.

Two variations of *H. punctatum* var. *punctatum* are encountered in Missouri:

Leaves somewhat clasping the stem or rounded at base, sessile, oblong. . . . . . 3a. H. PUNCTATUM var. PUNCTATUM f. PUNCTATUM
Leaves tapering to a narrow base, oblanceolate. 3b. H. PUNCTATUM var. PUNCTATUM f. SUBPETIOLATUM

3a. **Hypericum punctatum** var. **punctatum** f. **punctatum** Map 1538
*Hypericum punctatum* Lam.

This is the common variation found, throughout Missouri.

3b. **Hypericum punctatum** var. **punctatum** f. **subpetiolatum** (Bickn.) Fern. Map 1538
*Hypericum punctatum* f. *subpetiolatum* (Bickn.) Fern. [G]

Scattered in the range of the common variation, and known from New Madrid, Butler (*Dimmitt*),

Carter (*Drouet & Zirkle*), and Boone (*Drouet 77*) counties.

White-tailed deer are known to browse plants of *H. punctatum.*

3c–d. **Hypericum punctatum** var. **pseudomaculatum** (Bush) Fern. Map 1538

*Hypericum pseudomaculatum* Bush [BB, P & S, Steyerm.]

Flowers late May–September.

Occurs on glades, rocky prairies, moist sandy ditches, dry open woods, and along roadsides. Central and western Ozark region, east to Phelps, Shannon, and Wayne counties.

Ranges from Florida to Texas, north to Virginia, Tennessee, Illinois, Missouri, and Oklahoma.

Some authors maintain this as a species separate from *H. punctatum.* In their extremes the plants of *H. pseudomaculatum* with the uppermost pointed leaves somewhat broadly triangular, larger flowers, and longer styles seem well separated from the blunter-leaved, smaller-flowered, shorter-styled *H. punctatum.* However, as Fernald pointed out (Rh. 37: 432. 1935), there is considerable intergradation in these characters and breaking down of presumed distinctions. An example of a transitional or intergrading specimen is *Steyermark 26950* from Texas County, with pointed leaves but sepals only 4–5 mm. long. There also are specimens such as *Steyermark 27259* from Polk County, having the short styles, short sepals and petals of *H. punctatum* but the more pointed leaves of *H. pseudomaculatum.* More intensive field and experimental studies are necessary before the true status of *H. pseudomaculatum* is determined.

Two variations of *H. punctatum* var. *pseudomaculatum* occur in Missouri:

Petals and filaments orange-yellow; common type found . . . . . . . 3c. H. PUNCTATUM var. PSEUDOMACULATUM f. PSEUDOMACULATUM

Petals and filaments pale creamy yellow . . . 3d. H. PUNCTATUM var. PSEUDOMACULATUM f. FLAVIDUM

3c. **Hypericum punctatum** var. **pseudomaculatum** f. **pseudomaculatum** Map 1538

*Hypericum punctatum* var. *pseudomaculatum* (Bush) Fern. [G]

*Hypericum pseudomaculatum* Bush [BB, P & S, Steyerm.]

The common variation in the Ozark region.

3d. **Hypericum punctatum** var. **pseudomaculatum** f. **flavidum** Steyerm. Map 1538

*Hypericum pseudomaculatum* f. *flavidum* Steyerm. Rh. 41: 585. 1939.

Known only from McDonald County, southwestern Missouri (open rocky, cherty barren slopes along Mill Creek, 5 mi. southeast of Pineville, May 31, 1938, *Steyermark 5606*, holotype in Chi Nat. Hist. Mus. Herb.).

Known only from Missouri thus far.

4. **Hypericum spathulatum** (Spach) Steud. Shrubby St. John's-wort Map 1539

*Hypericum prolificum* of Am. auth. [P & S, BB], not L.

Flowers June–September.

Occurs in rocky ground and gravel bars along streams and bluffs, dry wooded slopes, bluff escarpments, fallow fields, and sometimes low moist ground in valleys. Eastern, southern, and central Missouri, north of the Missouri River west to Putnam, Sullivan, Linn, Chariton, and Clay counties.

Ranges from New York and Ontario to Minnesota, south to Georgia, Alabama, Arkansas, and Oklahoma.

When in full flower, this is a handsome plant with numerous, orange-yellow, pomponlike flowers covering the many branches. The plant does well in cultivation and thrives in dry sunny situations in loose, well-drained soil. It should be more commonly planted as a summer-flowering shrub.

5a. **Hypericum densiflorum** Pursh var. **densiflorum** Map 1540

*Hypericum densiflorum* Pursh [G, P & S]

Flowers July–September.

Occurs in sandy open ground in woodland.

Known only from Ripley County, southeastern Missouri.

Ranges from Florida to Texas, north to New York, New Jersey, West Virginia, Indiana, and Missouri.

A specimen from Madison County (*Palmer 31586* from sandstone barrens, Mine La Motte, September 6, 1926), previously recorded as *H. densiflorum* in Palmer & Steyermark's *Annotated Catalogue*, is in a poor condition for certain identification. It has been referred by Dr. W. P. Adams to *H. sphaerocarpum*, with which identification the present author concurs.

5b. **Hypericum densiflorum** var. **lobocarpum** (Gattinger) Svenson Map 1540

*Hypericum lobocarpum* Gattinger [P & S]

*Hypericum oklahomense* Palmer

Flowers June–August.

Known only from Dunklin (low ground, Campbell, April 20, 1912, *Bush 6619*) and Howell (upland meadow adjoining drainage from upland oak-hickory woods, along highway 80, T23N, R9W, sect. 12, 6 mi. southwest of West Plains, June 19, 1941, *Steyermark 40010*) counties.

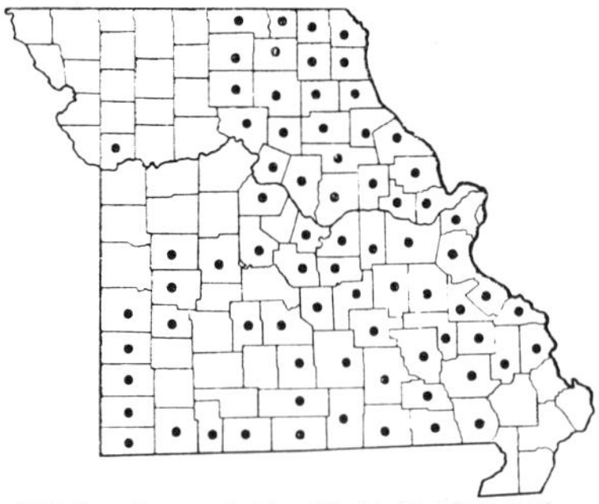

1539 Hypericum spathulatum (Shrubby St. John's wort)

1540 ● Hypericum densiflorum var. densiflorum
1540 □ Hypericum densiflorum var. lobocarpum

1541 ● Hypericum sphaerocarpum var. sphaerocarpum
1541 □ Hypericum sphaerocarpum var. turgidum

Ranges from North Carolina to Missouri, south to Louisiana and Oklahoma.

Bush's specimen is in poor condition, but shows some fruit and crowded, narrow, obtuse leaves. Dr. W. P. Adams has commented on the specimens of *Steyermark 40010* and *Bush 282* that they are 'closest to *H. lobocarpum*' and 'may be a hybrid of this species and *H. prolificum* (*H. spathulatum*).'

6. **Hypericum sphaerocarpum** Michx. Map 1541
*Hypericum cistifolium* of some auth. [P & S, Steyerm.], not Lam.

Flowers late May–September.

Occurs on open rocky glades, crevices of bluffs, prairies, dry open woods, gravel bars, draws, swampy meadows, low thickets, river bottom prairies, and alluvial soils along streams and ponds. Throughout Missouri.

Two variations are found in Missouri:

Leaves narrowly oblong to narrowly elliptic, flat, 5–18 mm. broad, with noticeable spreading side nerves; commoner variation found . . . .
6a. H. SPHAEROCARPUM var. SPHAEROCARPUM

Leaves narrowly linear with leaf margins rather revolute (turned under), 2–7 mm. broad, without lateral nerves; more rarely found variation . .
6b. H. SPHAEROCARPUM var. TURGIDUM

6a. **Hypericum sphaerocarpum** var. **sphaerocarpum** Map 1541
*Hypericum sphaerocarpum* Michx. [G]
*Hypericum cistifolium* of some auth. [P & S, Steyerm.], not Lam.

This is the common variation found throughout Missouri.

Ranges from Alabama and Mississippi to Oklahoma, north to Ohio, Indiana, Wisconsin, Iowa, and Kansas.

6b. **Hypericum sphaerocarpum** var. **turgidum** (Small) Svenson Map 1541

Scattered in eastern and southern Missouri, usually in dry sunny situations.

Ranges from Kentucky and Tennessee to Alabama and Missouri.

This narrow-leaved variation has been confused with and misidentified as *H. dolabriforme*, a species which does not occur in Missouri (see excluded species at end of genus explanation).

*Hypericum sphaerocarpum* has a slightly woody base, producing underground runners or suckers, which distinguishes it from *H. perforatum*, with which it is sometimes confused in sterile condition. It becomes somewhat weedy when transplanted to perennial beds, but its clusters of orange-yellow flowers are attractive. The species varies considerably in width and shape of leaves, abundance of flowers, and height of plant.

7. **Hypericum mutilum** L. Dwarf St. John's-wort Map 1542

Flowers July–October.

Occurs in usually moist ground bordering ponds, ditches, streams, spring branches, wet places in prairies, sandstone glades and ledges, wet meadows, low moist woodland, wooded swamps, occasionally fallow fields, and upland wooded ridges. Throughout most of Missouri, absent apparently from the extreme northwestern and northern counties.

Two variations occur in Missouri:

Leaves ovate to narrowly oblong, the upper ones and those of the branches ovate to somewhat triangular and gradually tapering from their base to their tip . . . 7a. H. MUTILUM var. MUTILUM

Leaves elliptic or oval, mostly broadest in the middle, gradually rounded to the summit . .
7b. H. MUTILUM var. PARVIFLORUM

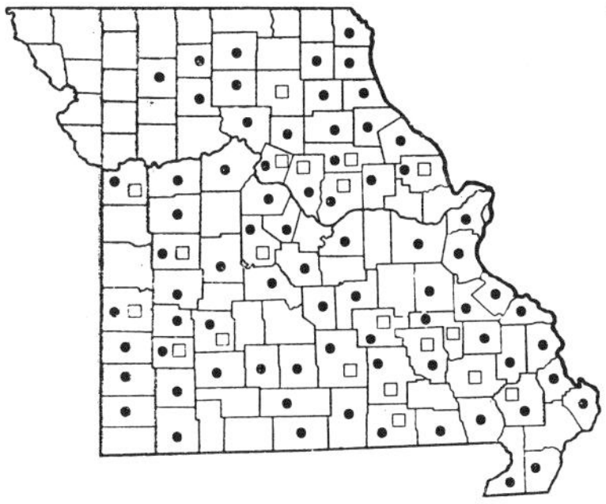

1542 ● Hypericum mutilum var. mutilum
1542 □ Hypericum mutilum var. parviflorum

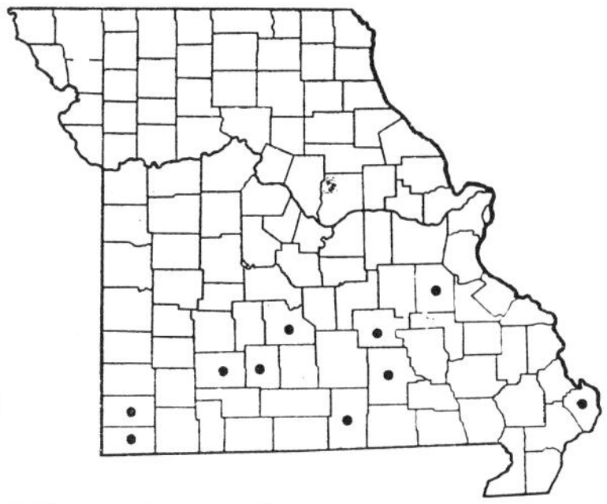

1543 Hypericum gymnanthum

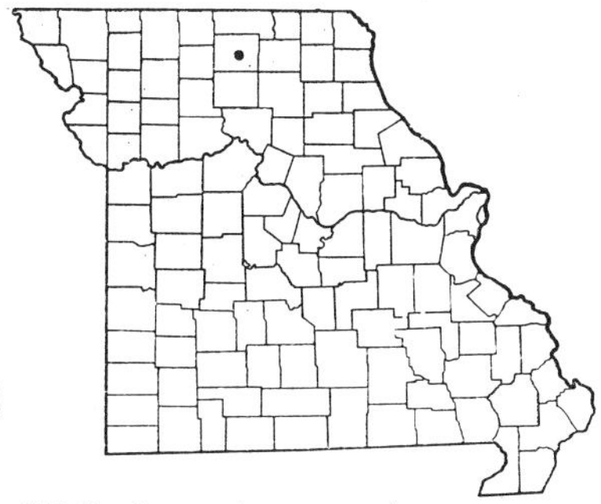

1544 Hypericum canadense var. canadense

7a. **Hypericum mutilum** var. **mutilum** Map 1542

*Hypericum mutilum* L. [G, P & S]

This is the commoner variation found in Missouri.

Ranges from Florida to Texas, north to southern New England, New York, Ohio, Illinois, Missouri, Kansas, and Oklahoma.

7b. **Hypericum mutilum** var. **parviflorum** (Willd.) Fern. Map 1542

This is less common and scattered in eastern, southern, and central Missouri.

Ranges from Florida to Texas, north to Nova Scotia, Newfoundland, Quebec, Ontario, Michigan, Wisconsin, Minnesota, and Manitoba.

In typical *H. mutilum* var. *mutilum* the plants average somewhat taller than in var. *parviflorum*, often up to 7.5–9 dm. tall in var. *mutilum*, whereas in var. *parviflorum* they are more frequently 1–4 dm. tall. However, all intermediate sizes are found in both varieties.

8. **Hypericum gymnanthum** Engelm. & Gray Map 1543

Flowers June–September.

Occurs in moist open ground bordering upland sink-hole ponds, wet upland meadows, sandy low fields, and along moist sandy ditches. Rare and scattered in southern Missouri north to Washington, Dent, Laclede, Greene, and Newton counties.

Ranges from Florida to Texas, north to New York, New Jersey, Pennsylvania, West Virginia, Ohio, Illinois, Missouri, and Kansas.

The leaves sometimes turn a rose-orange or copper color.

Apparent hybrids have been reported between this species and the preceding (*H. mutilum*).

9. **Hypericum canadense** L. var. **canadense** Map 1544

*Hypericum canadense* L. [G]

Flowers July–September.

Known only from Sullivan County, northern Missouri (open slopes along draw, Bookout Branch, tributary to Spring Creek, T64N, R18W, southwest $\frac{1}{4}$ sect. 21 and northeast $\frac{1}{2}$ sect. 28, $4\frac{1}{2}$ mi. northeast of Green City, August 25, 1950, *Steyermark 70133*).

Ranges from Newfoundland to Manitoba, south to Georgia, Alabama, Illinois, and Missouri.

10. **Hypericum Drummondii** (Grev. & Hook.) T. & G. Nits-and-lice Map 1545

Flowers June–September.

Occurs in fallow fields, prairies, rocky sandstone, chert, and granite outcrops and glades, along gravel bars, and dry open woodland, in acid soils. Southern, central, and northeastern Missouri, north to Marion, Knox, Randolph, Pettis, Henry, and Vernon counties.

Ranges from Florida to Texas, north to Maryland, West Virginia, Ohio, Indiana, Illinois, Iowa, Kansas, and Oklahoma.

11. **Hypericum gentianoides** (L.) BSP. Pine-weed Map 1546

Also called Orange Grass.

Flowers July–October.

Occurs on rocky open sandstone, chert, or granite outcrops, bluff escarpments, glades, prairies, and open woodland, in acid soils. Ozark region of southern and east-central Missouri, north to Lincoln, Montgomery, Maries, Pulaski, Laclede, Polk, Henry, and Vernon counties.

Ranges from Florida to Texas, north to Maine, New York, Ontario, Ohio, Indiana, Wisconsin, and Minnesota.

This species is a characteristic plant of open glades of acid rock strata in the Ozarks. Here it is associated with *Polygonum tenue* and *Oenothera linifolia*, annuals having a similar bushy habit of growth with slender,

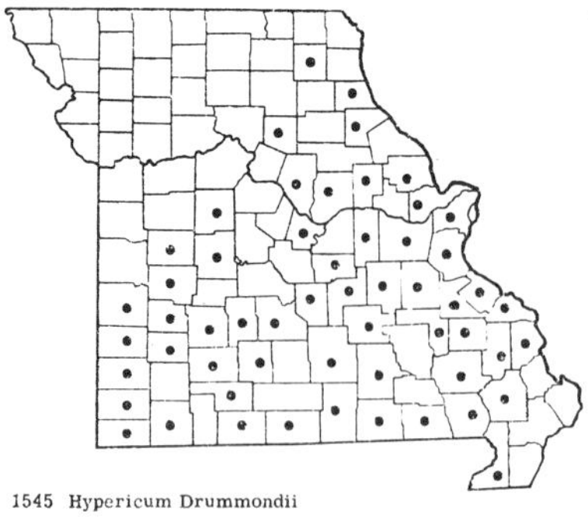

1545 Hypericum Drummondii

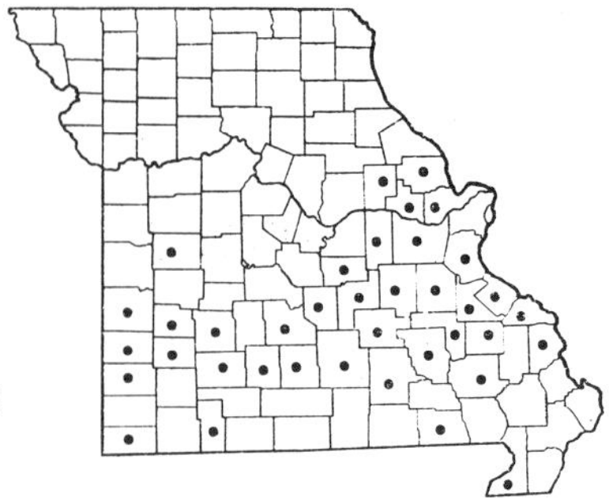

1546 Hypericum gentianoides (Pine Weed)

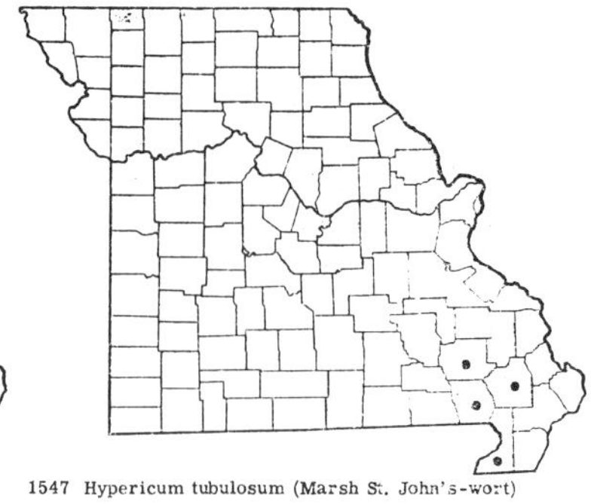

1547 Hypericum tubulosum (Marsh St. John's-wort)

erect branches and similarly reduced, narrow leaves. At maturity the entire plant becomes copper- or brick-colored.

12. **Hypericum tubulosum** Walt.
Marsh St. John's-wort Map 1547
*Triadenum tubulosum* (Walt.) Gl. [BB]
*Hypericum virginicum* of [P & S], not L.
Flowers August–September.

Occurs in low wet woods, bayous, and bald cypress swamps. Known only from Wayne, Butler, Stoddard, and Dunklin counties, in lowlands of southeastern Missouri.

Ranges from Florida to Louisiana, north to Virginia, Ohio, Indiana, Illinois, and Missouri.

Missouri material identified and recorded in Palmer & Steyermark's *Annotated Catalogue* as *Hypericum virginicum* was based upon plants with the leaves clasping or cordate at the base, such as *Bush 6312* from Campbell, Dunklin County, but these specimens are now properly referred to *H. tubulosum.*

13. **Hypericum Walteri** Gmel. Map 1548
*Hypericum tubulosum* var. *Walteri* (Gmel.) Lott. [G]
*Triadenum Walteri* (Gmel.) Gl. [BB]
*Hypericum petiolatum* Walt. [P & S], not L.

Occurs in low wet woods, bald cypress swamps, bayous, sink-hole ponds, and occasionally on wet ledges of sandstone bluffs. Southeastern Missouri in the lowland counties, north on LaMotte sandstone bluffs to Ste. Genevieve County and west in the Ozarks in upland sink-hole ponds to Reynolds, Shannon, and Howell counties.

Ranges from Florida to Texas, north to Maryland, West Virginia, Indiana, Illinois, Missouri, and Oklahoma.

The leaves are usually pale green or silvery or gray-green and rather thin. In the bald cypress swamps this species is frequently found growing as an epiphyte on the decaying bark of large fallen trees. On moist ledges of LaMotte sandstone bluffs in shaded ravines it reaches its northern limits in the state in Ste. Genevieve County, while it extends westward in the drier sections of the Ozarks in upland sink-hole ponds in Reynolds, Ripley, Oregon, Shannon, and Howell counties.

The presence or absence of the dark punctations and translucent glands on the lower surface of the leaves correlated with a rounded or clasping versus petiolate base would appear to justify the treatment of *H. tubulosum* and *H. Walteri* as two distinct species recognized by Gleason (*New Ill. Fl.* 2: 545. 1952) instead of as varieties as evaluated by Fernald in the eighth edition of *Gray's Manual.*

The seeds of this and other swamp species are eaten by ducks.

*Excluded Species*

**Hypericum dolabriforme** Vent

This species is credited to 'se. Mo.' in the eighth edition of *Gray's Manual* (p. 1012). An examination of the specimen at the Gray Herbarium of Harvard University (shelving rocks through the mining region, November, 1845, *Engelmann*) shows it to have small, linear, obtuse leaves, but with a calyx only 3.5–4.5 mm. long and only about 2 mm. broad. The calyx is too short and narrow for *H. dolabriforme.* This specimen is misidentified and is referred in the present flora to *H. sphaerocarpum* var. *turgidum* (Small) Svenson, the

Plate no. 254. **1.** Hypericum mutilum, × 2/5. **2.** Hypericum gymnanthum, × 2/5; After Gleason, The New York Botanical Garden. **3.** Hypriecum Drummondii, × 2/5; After Gleason, The New York Botanical Garden. **4.** Hypericum canadense × 2/5. **5.** Hypericum gentianoides, × 2/5; *a.* Fruit, × 3 1/5; *b.* Flower with leaves, × 8; After Gleason, details from Small, The New York Botanical Garden. **6.** Hypericum tubulosum. **7.** Hypericum Walteri; Details from Small, The New York Botanical Garden.

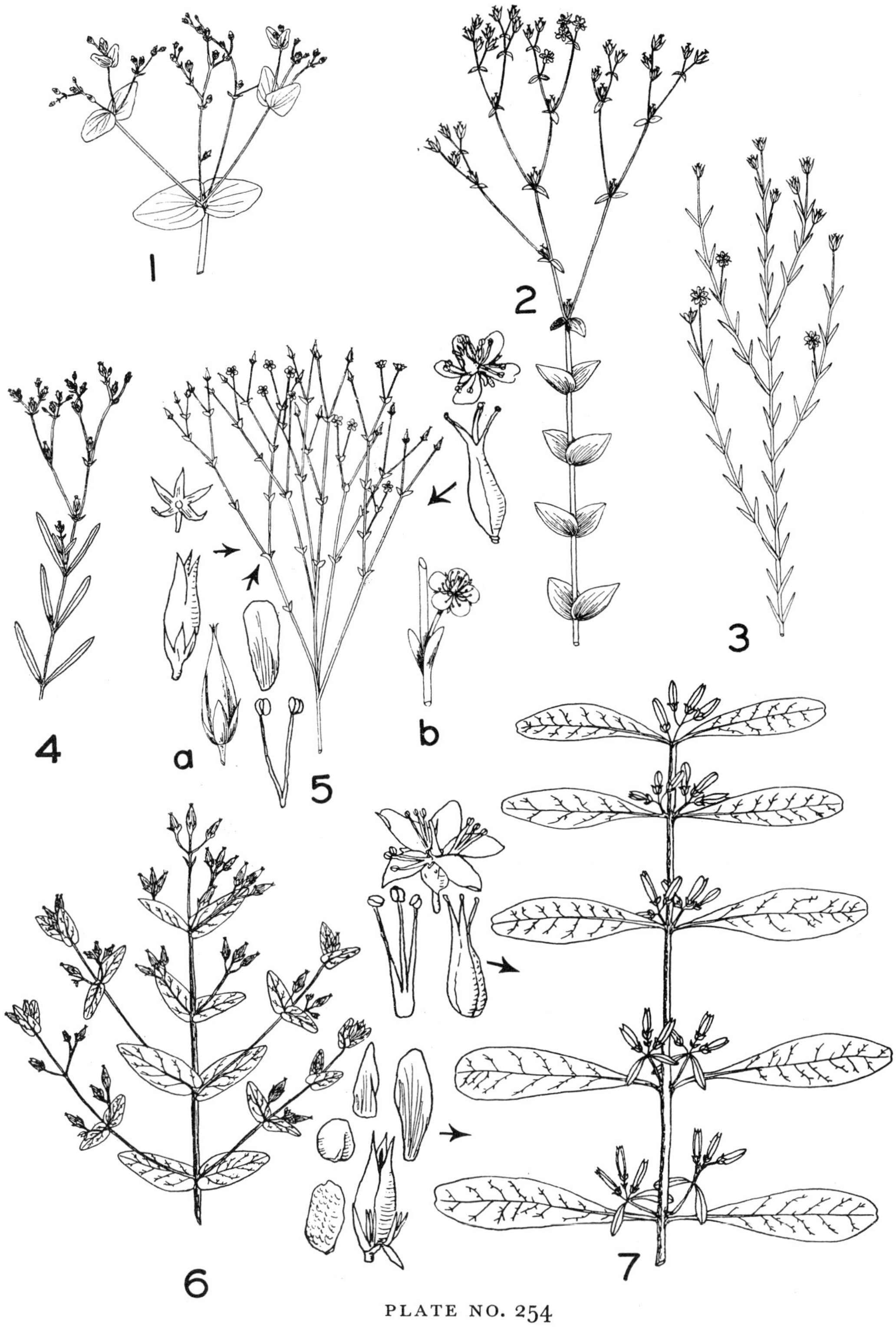


PLATE NO. 254

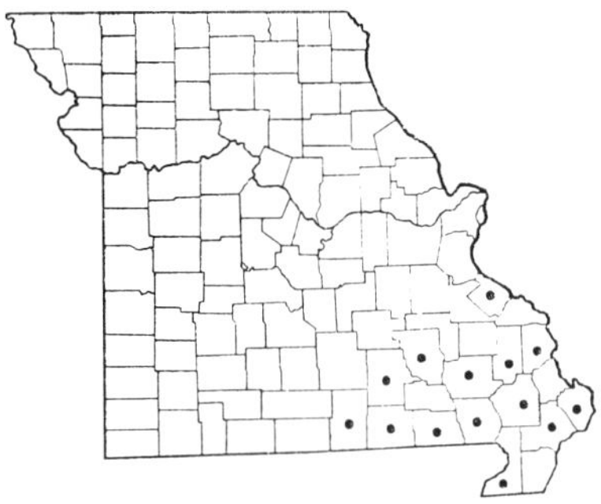

1548 Hypericum Walteri

1549 Elatine triandra var. americana

1550 Bergia texana

narrow-leaved and bushy-branched variation of that species. Similarly, in the herbarium of the Missouri Botanical Garden narrow-leaved specimens labeled *H. dolabriforme*, such as *Davis 1531* and *3608* from Eolia, Pike County, and *Davis 2782* from Oakwood, Ralls County, are misidentified and are referred in the present flora to *H. sphaerocarpum* var. *turgidum*.

**Hypericum virginicum** L.

This species was recorded in Palmer & Steyermark's *Annotated Catalogue* (p. 600) on the basis of specimens from Dunklin (*Bush 6312*) and Butler counties having the leaves clasping or cordate at the base. These have been found instead to belong with the *H. tubulosum*, which possesses this type of leaf base.

## Fam. **ELATINACEAE** (Waterwort Family)

Plants glabrous (without hairs); leaves 3–8 (rarely to 10) mm. long, smooth-edged (entire), rounded at summit . . . . . . . . . . . . . . . . . . . . . . . . . . 1. Elatine

Plants minutely glandular-hairy; leaves 10–45 mm. long, the margins with fine glandular teeth, pointed or tapering at the tip (acute or acutish) . . . . . . . . . . . . . . . . 2. Bergia

### 1. **Elatine** L. Waterwort

**Elatine triandra** Schkuhr var. **americana** (Pursh) Fassett — Map 1549

*Elatine americana* (Pursh) Arn. [G, P & S]

Flowers June–October.

Occurs in upland sink-hole ponds and in muddy swales. Known only from Laclede (upland sink-hole pond along highway 5, 7 mi. north of Lebanon, June 24, 1939, *Steyermark 27219*) and Jackson (common in swales, Atherton, June 27, 1897, *Bush 131*) counties.

Ranges from Quebec and New Brunswick to Quebec to Virginia; Missouri and Oklahoma.

This is a diminutive plant growing in small mats with branches 1–5 cm. long, and somewhat resembling *Callitriche* or *Peplis diandra*. The tiny flowers are sessile and solitary in the leaf-axils, with 3 pinkish petals, 3 sepals, and 3 stamens.

### 2. **Bergia** L.

**Bergia texana** (Hook.) Seubert — Map 1550

Flowers June–October.

Known only from Jackson County, west-central Missouri (uncommon along Missouri River, Courtney, October 24, 1897, *Bush 289*; September 30, 1893, *Bush 20*).

---

Plate no. 255. **1.** Tamarix gallica; *a*. Flowering branch, $\times {}^{2}/_{5}$; *b*. Leafy branch, $\times {}^{2}/_{5}$; *c*. Part of leafy branch, $\times {}^{4}/_{5}$; *d*. Part of flowering branch, $\times {}^{4}/_{5}$; *e*. Flower, $\times 7$; Details from Small, The New York Botanical Garden. **2.** Lechea villosa, $\times {}^{2}/_{5}$; After Britton and Brown, The New York Botanical Garden. **3.** Helianthemum Bicknellii; Details from Small, The New York Botanical Garden. **4.** Bergia texana, $\times {}^{2}/_{5}$; *a*. Flower, $\times 1{}^{3}/_{5}$; *b*. Seed, $\times 8$; After Britton and Brown, The New York Botanical Garden. **5.** Lechea tenuifolia, $\times {}^{2}/_{5}$; *a*. Fruit, $\times 4$; Details from Small, The New York Botanical Garden. **6.** Elatine triandra var. americana, $\times {}^{4}/_{5}$; *a*. Seed, $\times 20$; After Gleason, details from Small, The New York Botanical Garden.

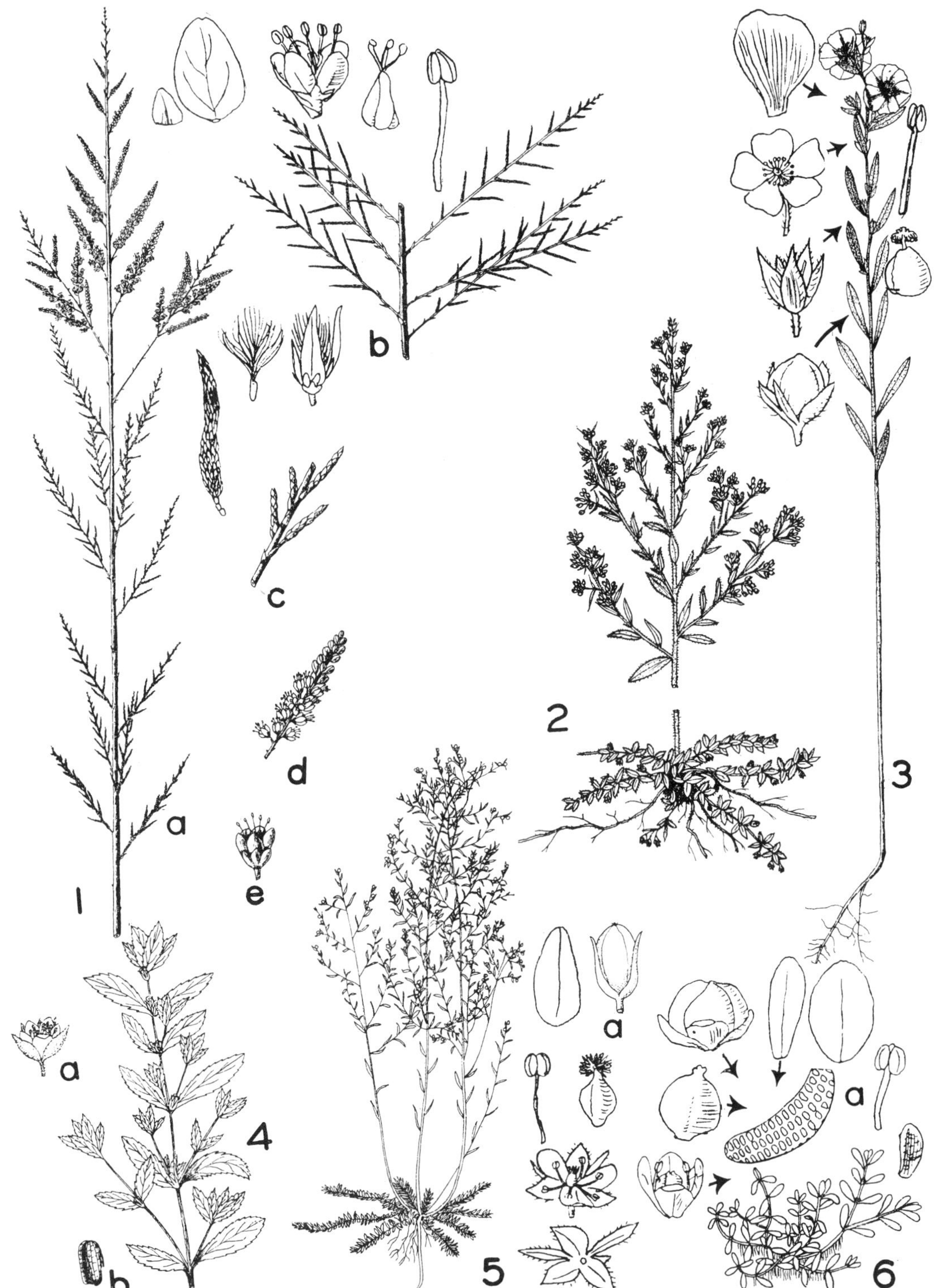


PLATE NO. 255

1551 Tamarix gallica (French Tamarisk)

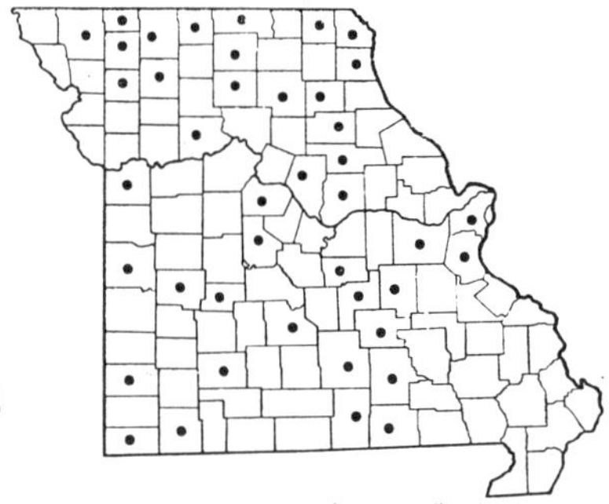

1552 Helianthemum Bicknellii (Frostweed)

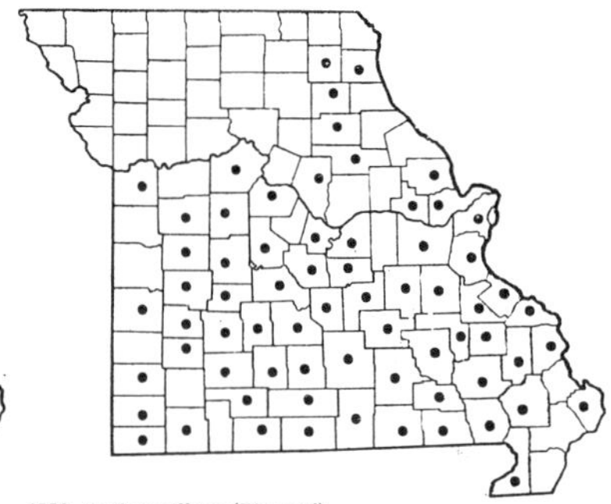

1553 Lechea villosa (Pinweed)

Ranges from Arkansas to Texas and California, north to Illinois, Missouri, South Dakota, and Washington.

The stems are 1–4 dm. long with elliptic-oblong or oblong-spatulate leaves which are narrowed to the base. The flowers occur in the leaf-axils on flower-stalks about 1 mm. long. The 5 sepals are lanceolate, long-pointed and somewhat longer than the 5 oblong petals. The stamens are 5 or 10.

## Fam. **TAMARICACEAE** (Tamarisk Family)

### **Tamarix** L. Tamarisk

**Tamarix gallica** L. French Tamarisk Map 1551

Also called Tamarix, Tamarisk.

Flowers May–September.

Introduced along sand bars and mud flats along the Missouri and Mississippi rivers, and occasionally found along railroads. Known from Atchison, Holt, Jackson, Carroll, Warren, St. Louis, and Pemiscot counties.

Native of southern Europe; introduced and naturalized in the southern states, north to Massachusetts, Indiana, Missouri, Kansas, Oklahoma, and California.

This shrub with bluish- or gray-green cedarlike foliage and small, pale or rose-pink, fragrant flowers is commonly cultivated as an ornamental summer-flowering plant. A number of other species are in cultivation. In the western states this and other species have been planted along roadsides and irrigation ditches.

## Fam. **CISTACEAE** (Rockrose Family)

No leafy rosette or leafy branches present at base of plant; petals yellow, 5 and conspicuous. 1. HELIANTHEMUM

Leafy rosettes or leafy branches present at base of plant; petals dark red, 3, scarcely visible . . 2. LECHEA

### 1. **Helianthemum** Mill. Rockrose, Frostweed

**Helianthemum Bicknellii** Fern. Frostweed Map 1552

Also called Rock Rose.

Flowers April–July.

Occurs in dry open woods, prairies, and eroded open or rocky slopes, in acid soils. Throughout Missouri, except apparently absent from the southeastern lowland counties and the extreme southeastern Ozark section.

Ranges from Maine to Minnesota, South Dakota and Colorado, south to Maryland, North Carolina, Ohio, Indiana, Illinois, Missouri, and Kansas.

Two types of flowers are produced, the earlier ones at the top of the stems with large yellow petals in April and May, followed by smaller inconspicuous later ones, usually without petals and clustered on short branches below the first flowers. The early showy flowers are open for one day only, dropping their petals the next day. The later inconspicuous flowers become fertilized and form seed without the flower opening (cleistogamous condition).

*Excluded Species*

**Helianthemum canadense** (L.) Michx. var. **canadense**

This is recorded from Missouri in the eighth edition of *Gray's Manual* (p. 1017). A specimen labeled *H. canadense* in the Gray Herbarium of Harvard University is without flower or fruit (*Bush 46* from prairie, Exeter, Barry County, September 29, 1896), and I can find nothing to distinguish this sterile specimen from typical *H. Bicknellii*. Similarly, specimens labeled *H. canadense* in the herbarium of the Missouri Botanical Garden (*Bush 51* and *231* from Jackson County, *Bush* from Shannon County, *Bush* from McDonald County, and *Bush 197* from Eagle Rock, Barry County) are referred in the present flora to *H. Bicknellii*. The last cited collection (*Bush 197* from Barry Co.), which is in flower on May 30, 1898, has apparently 2 or more early petaliferous flowers in a terminal corymb rather than a single terminal flower characteristic of true *H. canadense*.

## 2. **Lechea** L. Pinweed

a. At least the upper part of the upright stems with conspicuous spreading or loose or long hairs; leaves of the basal shoots or rosettes 8–15 mm. long, broadly ovate-elliptic; leaves of the stem 5–14 mm. broad . . . . . 1. L. VILLOSA

a. Upright stems with ascending hairs or of hairs pressed to stem (appressed); leaves of the basal shoots or rosettes 4–6 mm. long, elliptic-lanceolate or oblong-ovate to linear; leaves of the stem 1–4 mm. broad . . . . . *b*

  b. Common species nearly throughout Missouri; leaves of the basal shoots or rosettes less than 1 mm. broad, narrowly lanceolate to linear; leaves of the stem very narrowly linear or thread-like, 1–1.5 mm. broad; outer sepals usually noticeably longer than the inner sepals . 2. L. TENUIFOLIA

  b. Rarely found, known only from Marion County, northeastern Missouri; leaves of the basal shoots or rosettes 1.5–2.5 mm. broad, oblong-ovate to elliptic-lanceolate; leaves of the stem elliptic-lanceolate to narrowly oblanceolate, mostly 2–4 (rarely 1.5) mm. broad; outer sepals noticeably shorter than the inner sepals . . . . . 3. L. RACEMULOSA

1. **Lechea villosa** Ell. Pinweed Map 1553

Fruits July–November.

Occurs on rocky open glades of sandstone, chert, or granite, rocky open woods, sandy and fallow fields, and prairies, in acid soils. Eastern, southern, and central Missouri, north of the Missouri River west to Knox, Shelby, Monroe, Audrain, and Boone counties, elsewhere north to Saline and Jackson counties.

Ranges from Florida to Texas, north to New Hampshire, Vermont, New York, Ontario, Michigan, Illinois, Missouri, Kansas, and Oklahoma.

Missouri material is referred to typical var. *villosa*. Other varieties are known from Mexico (var. *Schaffneri* Hodgdon) and Nebraska to Texas (var. *macrotheca* Hodgdon). The flat masses of rosette leafy shoots are quite characteristic of the species and easily distinguish it in the field.

2. **Lechea tenuifolia** Michx. var. **tenuifolia** Pinweed Map 1554

*Lechea tenuifolia* Michx. [G, BB, P & S, Steyerm.]

Fruits June–November.

Occurs in rocky open glades, ledges of bluffs of sandstone, chert, and granite, sandy and fallow fields, open woods, and prairies, in acid soils. Eastern, southern, and central Missouri; absent from the northwestern sector of the state, north of the Missouri River west to Mercer, Livingston, and Chariton counties.

Ranges from South Carolina to Maine; and from Louisiana to Texas, north to Indiana, Wisconsin, and Minnesota. More robust plants with erect instead of openly spreading branches, with the upper surface of the basal and stem-leaves densely hairy, and with larger capsules, seeds, and long outer sepals belong to var. *occidentalis* Hodgdon, but have not been recorded from Missouri, although in adjacent Illinois, Kansas, and Nebraska.

The fernlike basal rosettes with fine, narrow leaves are conspicuous in winter, early spring, and late fall, and easily distinguish the species in the field. The inconspicuous, minute flowers have 5 sepals, of which the outer 2 are narrow and greenish, the inner 3 broad and thin-transparent. The 3 petals are shorter than the sepals, are mostly withered and dried up, rarely seen expanded. The stamens are mostly 5–15, and the fruit, a capsule, opens by 3 valves, maturing 1–6 seeds.

3. **Lechea racemulosa** Michx. Map 1555

Fruits August–November.

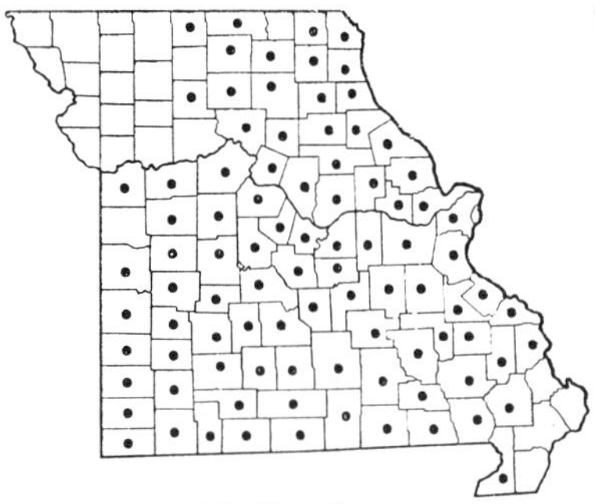

1554 Lechea tenuifolia (Pinweed)

1555 Lechea racemulosa

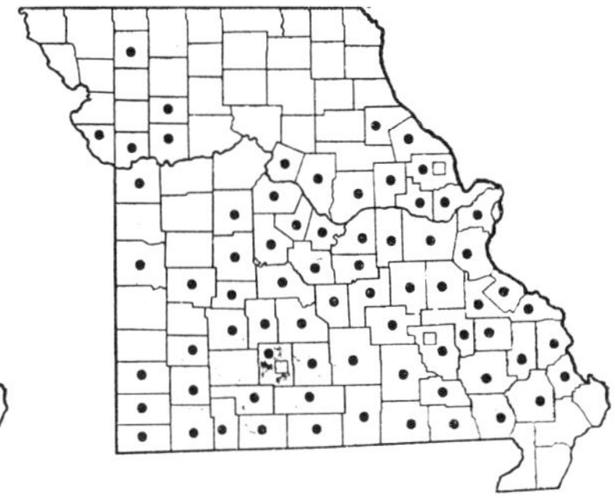

1556 • Hybanthus concolor f. concolor (Green Violet)
1556 □ Hybanthus concolor f. subglaber

Known only from Marion County, northeastern Missouri (River Heights near Hannibal, July 23, 1919, *Davis*).

Ranges from Georgia and Alabama to New York, New Jersey, Pennsylvania, Ohio, Indiana, and Missouri.

The single Missouri collection has the fruits racemosely long-pedicelled, and the outer sepals are shorter than the inner.

*Excluded Species*

**Lechea intermedia** Liggett var. **intermedia**

A specimen so identified by Dr. A. R. Hodgdon, the monographer of the genus, from Adair County, northeastern Missouri (1893, *Link*) is deposited in the herbarium of the University of Missouri. Dr. Hodgdon has written on this sheet 'probably not from Missouri,' and does not cite the specimen nor Missouri in the range of *L. intermedia* in his monograph (Rh. 40: 123–30. 1938). As the species has a more northerly range not approaching Missouri, it is likely that the specimen collected by Link was obtained elsewhere and that the label became mixed in some way. The uncertainty of the specimen precludes its being accepted as a definite member of the flora of the state.

## Fam. **VIOLACEAE** (Violet Family)

Stems 2–6 mm. thick, leafy, 3–10 dm. tall, the leaves conspicuously narrowed at their base and tip and elliptic to ovate-oblong, usually without any teeth at flowering time; flowers 1–3 together on nodding, recurved short stalks (pedicels) 2–15 mm. long at the base of the leaf-stalk (petiole); petals green; none of the petals with a spur (hollow sac) . . . . . . . . . . . . . . . 1. HYBANTHUS

Without the above combination of characters; stems or leaf-stalks 0.5–3 mm. thick; flowers solitary on an ascending to erect usually elongated (often 2.5–30 cm. long) flower-stalks (pedicels or peduncles); petals white, yellow, blue, lilac, or purple, or combinations of these; one of the petals (the lower one) with a spur (hollow sac) . . . . . . . . . . . . . . . . . . . . . . . . 2. VIOLA

### 1. **Hybanthus** Jacq. Green Violet

(*Cubelium* Raf. [BB])

**Hybanthus concolor** (T. F. Forst.) Spreng.

Green Violet Map 1556

*Cubelium concolor* (Forst.) Raf. [BB]

Flowers April–June.

Occurs in rich or rocky woods and along shaded bluffs, usually in limestone soil areas. Throughout the Ozark region and central Missouri north to Ralls, Callaway, Boone, Howard, Caldwell, Clay, and Platte counties, and locally in northwestern Missouri in Gentry County.

Ranges from Georgia to Mississippi, Arkansas, Oklahoma, and Kansas, north to Connecticut, New York, Ontario, Michigan, and Wisconsin.

Two variations are found in Missouri:

Some part of the stem more or less hairy; common type encountered . . a. H. CONCOLOR f. CONCOLOR

Stem glabrous (without hairs) or nearly so; rare type encountered . b. H. CONCOLOR f. SUBGLABER

a. **Hybanthus concolor** f. **concolor** Map 1556

*Hybanthus concolor* (T. F. Forst.) Spreng. [G, P & S, Steyerm.]

This is the common variation found in Missouri.

b. **Hybanthus concolor** f. **subglaber** (Eames) Zenkert Map 1556

Scattered in the range of the typical form. Known from Lincoln (*Steyermark 28477*), Reynolds (*Steyermark 22004*), and Webster (*Steyermark 23912*) counties.

The leaves are ascending, and at flowering time usually entire (without teeth). Later developed leaves or ones near the top of the stem may be toothed. The flowers are on nodding pedicels, but the fruit is erect on ascending stalks. The upper part of the stem usually becomes zigzag in the fruiting stage or late maturity of the plant.

## 2. **Viola** L. Violet

In addition to the usually-encountered larger flowers bearing petals in early spring, there are inconspicuous, self-fertilized, so-called cleistogamous flowers which lack petals. These are produced on usually shorter stalks (peduncles) above or below ground. Of the 5 somewhat unequal petals, the lower one has a spur or backward-projecting protuberance at the base, and the 2 lateral ones often have a tuft of hairs on their inner side.

In using the following key, it should be kept in mind that some of the species which are closely related to one another hybridize in nature, and possess characters intermediate in value, making it difficult to identify such plants. Hybrids are known in Missouri involving species no. 3, 4, 5, 6, 7, and 9, crosses having been recorded between no. 3 and 4, 3 and 5, 3 and 7, 3 and 9, 4 and 5, 5 and 6, 5 and 7, and 6 and 9.

a. All the leaf-stalks (petioles) flower-stalks, and fruit-stalks arising from ground level at the base of the plant or from runners sent out from the base . . . . . . . . . . . . . . . . . . . *b*
  b. None of the leaves with deep cuts, lobes, or divisions . . . . . . . . . . . . . . . . . *c*
    c. Some part of the leaf-blade, leaf-stalk (petiole), or flower-stalk with hairs . . . . . . . . *d*
      d. Some or all of the leaves definitely longer than broad, oblong or arrow-shaped. . 6. V. SAGITTATA
      d. Most or all of the leaves nearly circular, heart-shaped, or broadly ovate . . . 5. V. SORORIA
    c. Leaf-blade, leaf-stalk, and flower-stalk without hairs (glabrous) . . . . . . . . . . . *e*
      e. All the leaves tapering or narrowed to the base of the leaf-blade, lanceolate or oblanceolate, 3–20 times as long as broad . . . . . . . . . . . . . . . . . . . . 11. V. LANCEOLATA
      e. Leaves broadest at the base of the leaf-blade, either oblong, arrowhead-shaped, broadly triangular-heart-shaped, or nearly circular, but not narrowed or tapering to the base, broader than long or at most 2–3 times as long as broad . . . . . . . . . . . . . . *f*
        f. Leaves oblong or arrowhead-shaped, up to 2–3 times as long as broad . . 6. V. SAGITTATA
        f. Leaves broadly triangular-heart-shaped, broadly ovate, or nearly circular. . . . . . *g*
          g. Some or all of the leaves broadly triangular-heart-shaped, and at least some of them somewhat longer than broad and tapering to a narrow triangular summit. 4. V. MISSOURIENSIS
          g. All the leaves broadly ovate or heart-shaped or nearly circular, broader than long to as broad as long or if somewhat longer than broad with a rounded, blunt or short-triangular summit . . . . . . . . . . . . . . . . . . . . . . . . . *h*
            h. Plants sending out thread-like runners (stolons) from the base; petals white; no beard of hairs on inner side of lateral petals; plants of moist sandstone known only from Ste. Genevieve Co., southeastern Missouri . . . . . . . 10. V. PALLENS
            h. Plants without runners (stolons) from the base, the base of plant thickened and fleshy; petals mainly shades of blue and purple; beard of hairs present on inner side of lateral petals; common species throughout all of Missouri or rare species confined to swampy meadows in the southeastern Ozarks . . . . . . . . *i*
              i. Common species throughout Missouri; hairs on inner side of lateral petals mostly slender-tipped, not strongly knobbed; lateral petals about equal in length to spurred petal; flower not darkened toward the center, usually with a pale center; leaves usually taller than the flower-stalks; capsule usually marked with purple, $1\frac{1}{2}$–4 times as long as sepals; auricles (loose flaps or projections) of sepals in fruit broad and rounded, about 1 mm. long; cleistogamous flowers on usually horizontal or prostrate peduncles (stalks). . . 3. V. PAPILIONACEA
              i. Rare species confined to swampy meadows of the southeastern Ozarks; hairs on inner side of lateral petals mostly club-shaped or strongly knobbed or short

and stubby; lateral petals longer than the spurred petal; flower usually with a darker center; leaves usually shorter than the flower-stalks; capsule green, only slightly longer than the sepals; auricles (loose flaps or projections) of sepals in fruit pointing straight back and mostly 2–6 mm. long; cleistogamous flowers on upright peduncles (stalks) . . . . . . . . . . . . 2. V. CUCULLATA

b. Some or all of the leaves with deep cuts, lobes, or divisions . . . . . . . . . . . . . . *j*

j. Most or all of the leaves definitely longer than broad and up to 2–3 times as long as broad, oblong or arrowhead-shaped with coarse teeth or jagged edges only near base of leaf-blade 6. V. SAGITTATA

j. Leaves of a broadly ovate or nearly circular shape, at least one of them with 3 or more prominent lobes or divisions . . . . . . . . . . . . . . . . . . . . . . *k*

k. Leaf-blades, leaf-stalks, and flower-stalks completely glabrous (without hairs) . . . . . . . *l*

l. No hairs on the inner side of any petals; stamens conspicuously protruding from the flowers; all the stalks (peduncle) supporting the fruit erect and more or less equal in length; no cleistogamous flowers (lacking petals) present; includes flower with upper petals dark violet and lower petals lilac or purple as well as flowers of all the same color 1. V. PEDATA

l. Hairs present on the inner side of the lateral petals; stamens not protruding from the flowers; stalks supporting the cleistogamous flowers (lacking petals) often shorter than the stalks supporting the conspicuous flowers; cleistogamous flowers (lacking petals) usually developed; petals all similar in color . . . . . . . . . . . . . . *m*

m. All the leaves of plants in spring growth deeply cut, lobed, or divided into linear or very narrow 2–3-cleft segments, rather uniform in appearance, later summer and fall leaves often less deeply divided, but with usually 5–11 (–15) narrow lobes; leaf-blades more or less orbicular, fan-shaped, or broadly wedge-shaped; lower leaf-segments not toothed on their outer edge; spurred petal with hairs at the base; plants mainly of upland prairies . . . . . . . . . . . . . 9. V. PEDATIFIDA

m. Leaves quite varied on the same plant, the middle lobe usually wider than the others, lobes or segments mostly 3–7, not uniformly narrowly linear; leaf-blades broadly triangular; lower leaf-segments or lower part of leaf toothed on their outer edge; spurred petal lacking hairs at the base; plants of alluvial banks, moist ledges of bluffs, gravel bars along streams, and moist low ground . . . . . . . . . . 8. V. VIARUM

k. Leaf-blades, leaf-stalks, and/or flower-stalks more or less hairy . . . . . . . . . . . . *n*

n. All the leaves usually more or less uniformly deeply cut, lobed, or divided into 5–11 (–15) narrow linear segments, these 2–3-cleft, the leaves of plants in spring all deeply cut, those of late summer and fall less deeply cut, lower leaf-segments not toothed on their outer edge; plants mainly of upland prairies . . . . . 9. V. PEDATIFIDA and hybrids of V. PEDATIFIDA and V. SORORIA

n. Leaves of the same plant quite varied, the middle lobe usually most prominent or widest, the other lobes 3–7 and not uniformly narrowly linear, some of the leaves not lobed but merely more or less heart-shaped, lower leaf-segments or lower part of leaf toothed on their outer edge; plants of either dry or rocky woodland, or of alluvial banks, gravel bars along streams, and moist low ground . . . . . . . . . . . . . *o*

o. Commonly encountered species; leaves definitely heart-shaped (cordate) at base; spurred petal hairy on the inner surface at the base; plants of dry or rocky woodland 7. V. TRILOBA

o. Rarely encountered plant; most of the leaves wedge-shaped or truncate (as if cut straight across) at base, or nearly so, not heart-shaped; spurred petal glabrous (without hairs) at the base; plants of alluvial banks, gravel bars along streams, and moist low ground . . . . . . . . . . . . . . 8. V. VIARUM f. PILIFERA

a. Some of the leaf-stalks, and all the flower- and fruit-stalks arising from the elongated stem . . . . . *p*

p. Stipules (small leaf-like parts at base of leaf-stalk) without teeth or faintly and bluntly toothed toward the tip; flowers yellow . . . . . . . . . . . . . . . . . . . 12. V. PENSYLVANICA

p. Stipules (small leaf-like parts at base of leaf-stalk) prominently fringed, toothed, or deeply cut; flowers white, lavender, bluish, or combining yellow, purple, blue, brown, and white. . . . . . *q*

q. Flowers chiefly white or creamy-white; stipules fringed or with the longer teeth 1.5–5 mm. long. . . . . . . . . . . . . . . . . . . . . . . . . . 13. V. STRIATA

q. Flowers usually lavender, bluish, bluish-white, or combining these colors, or with combinations of yellow, white, or purple; stipules deeply cut into narrow lobes . . . . . . . . . *r*

r. Upper stem-leaves and middle lobe of stipules without teeth (entire) or nearly so; petals lavender or bluish-white to cream-colored, faintly tinged with yellowish, some with purplish veins, 7–10 mm. long, nearly 2 times as long as sepals; commonly encountered wild . . 15. V. KITAIBELIANA var. RAFINESQUII

r. Upper stem-leaves and middle lobe of stipules with rounded teeth; petals with various combinations of yellow, purple, or white, the upper petals of a flower usually darker purple, 10–15 mm. or more long, 2–3 times as long as the sepals; rarely found plant escaped from cultivation . . . . . . . . . . . . . . . . . . . . 14. V. TRICOLOR

1. **Viola pedata** L. Pansy Violet Map 1557

Also called Bird's-foot Violet, Hens and Roosters.

Flowers April–early June; and again frequently in September–December.

Occurs in rocky or dry open woods on upland slopes and ridges, prairies, rocky open glades of usually sandstone, chert, and granite rocks, rarely on limestone, also along chert and gravel slopes along roadsides, most always in acid soils.

The following variations occur in Missouri:

a. Petals all white . . . . . . . 1c. V. PEDATA var. LINEARILOBA f. ALBA

a. Petals blue, lavender, purple and/or violet . *b*

b. Upper petals dark violet, the lower three pale to deep lilac-purple or lavender . . 1a. V. PEDATA var. PEDATA

b. All the petals similarly colored . 1b. V. PEDATA var. LINEARILOBA f. LINEARILOBA

1a. **Viola pedata** var. **pedata** Map 1557

*Viola pedata* L. [G, P & S, Steyerm.]

Eastern, southern, and central Missouri, locally northwest in Harrison and Nodaway counties; absent from the lowland counties of extreme southeastern Missouri and from the west-central section. In some sections this is the commoner variation.

Ranges from South Carolina to Louisiana, north to Connecticut, New York, Pennsylvania, Virginia, Kentucky, Indiana, Illinois, and Missouri.

1b. **Viola pedata** var. **lineariloba** DC. f. **lineariloba** Map 1557

*Viola pedata* var. *lineariloba* DC. [G, BB, P & S, Steyerm.]

Eastern, southern, and central Missouri, locally in Platte County, west-central Missouri; usually less common than var. *pedata.*

Ranges from Florida to Texas, north to New Hampshire, Massachusetts, New York, Ontario, Michigan, Wisconsin, Minnesota, and Kansas.

1c. **Viola pedata** var. **lineariloba** f. **alba** (Thurb.) Britt. Map 1557

Known only from Polk (natural prairie, T34N, R24W, sect. 4, 2.5 mi. northeast of Dunnegan along west side of highway 13, May 3, 1958, *Steyermark 86091*) and Barton (upland prairie, sandstone outcrops, 2 mi. east of Duval, May 1, 1953, *Palmer 55451*) counties, southwestern Missouri, but probably of more widespread distribution.

In the Barton County collection the flowers are pure white; in the collection from Polk County, the petals are mostly white over the surface with some lavender-violet at the base of the lateral petals and with purple streaks down the center of the lower petal.

In the Ozark region these violets are used in a game known as 'Hens and Roosters.' The 'roosters' are the bicolored type having the upper petals dark violet, while the 'hens' are the uniformly colored flowers. One child will hold the flower-stalk with attached 'hen' flower in one hand, another child the stalk with attached 'rooster' flower. The object of the game is to see who can pull away from its stalk the flower of the other person. This is effected by hooking the flower-stalks and attempting to pull the other flower off its stalk by strokes and thrusts of the opponents.

This violet is a handsome gem for rock-gardens and succeeds in any sunny or semishaded, well drained, usually acid soils, where it thrives best on sand, sandstone, chert, granite, and similar acid rocks. It will also grow on limestone. It does best on slopes on soils having good drainage. Along highway cuts through cherty or sandstone substrata, one often sees thousands of these plants growing thickly in rocky or gravelly exposures.

2. **Viola cucullata** Ait. var. **cucullata** Marsh Blue Violet Map 1558

*Viola cucullata* Ait. [G, BB, Steyerm.]

Flowers April–June.

Occurs in swampy, spring-fed, calcareous meadows of the southeastern Ozarks in Iron, Reynolds, Dent, and Shannon counties, and locally in Stoddard County in the Crowley Ridge section.

Ranges from Newfoundland to Ontario and Minnesota, south to Georgia, Tennessee, Arkansas, and Nebraska.

The first authenticated record of this species for Missouri was collected in 1938 from Reynolds County (*Steyermark 5488*). The species was not recorded in

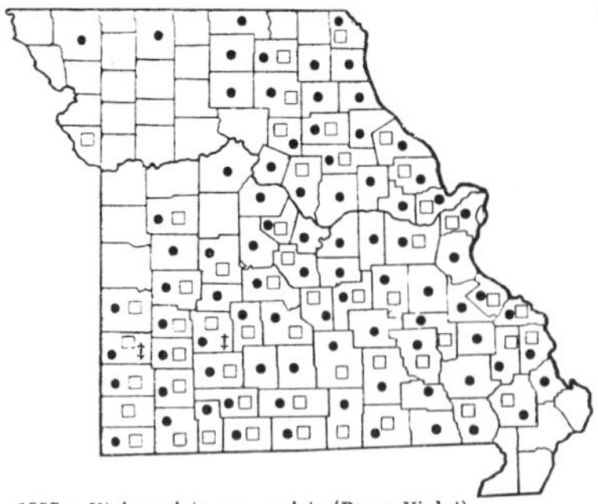

1557 • Viola pedata var. pedata (Pansy Violet)
1557 □ Viola pedata var. lineariloba f. lineariloba
1557 ‡ Viola pedata var. lineariloba f. alba

1558 Viola cucullata var. cucullata (Marsh Blue Violet)

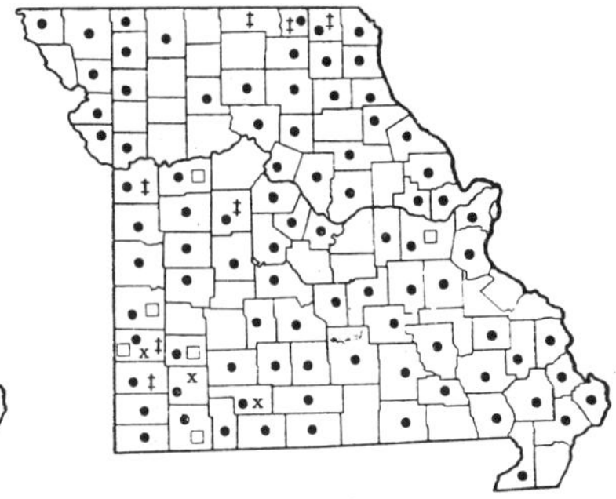

1559 • Viola papilionacea f. papilionacea
1559 □ Viola papilionacea X Viola sororia
1559 x Viola papilionacea X Viola triloba
1559 ‡ Viola X Bernardi

Palmer & Steyermark's *Annotated Catalogue*, since earlier reports had been found to be misidentifications of *V. papilionacea*.

Actually, the occurrence of *V. cucullata* in the swampy meadows of Missouri indicates a northern relict species here near its southwestern limits of geographic dispersal associated with other northern relict species such as *Campanula aparinoides, Liparis Loeselii*, and *Menyanthes trifoliata* var. *minor*. These and others must have reached this section of the Ozarks during one of the southward extensions of the Pleistocene ice lobes, and became stranded in the swampy meadow habitats after the ice sheets retreated.

A variety *microtitis* Brainerd with short auricles of the sepals is known from Nova Scotia to Newfoundland.

The leaves and flowers of this and other stemless blue violets are recommended for their mucilaginous nature for adding to soups, or may be served with dock, lamb's quarters, other greens, and okra.

3. **Viola papilionacea** Pursh f. **papilionacea**
Common Violet — Map 1559
Also called Meadow Violet, Butterfly Violet.
*Viola papilionacea* Pursh [G, P & S, Steyerm.]
*Viola papilionacea* var. *papilionacea* [BB]

Flowers March–June, and frequently October–November.

Occurs along roadsides and railroads, low alluvial ground bordering streams, ponds, wet ditches, meadows, fields, wet ledges of bluffs, thickets, and open ground, sometimes in cultivated and waste ground. Throughout Missouri.

Ranges from Maine and Quebec to North Dakota and Wyoming, south to Georgia and Oklahoma.

The leaves are eaten by white-tailed deer.

Flowers with white flowers are known as f. *alba* (T. & G.) Farw. The so-called Confederate Violet having flowers grayish white with violet or blue veins and more solid patches of these colors on the lower half of the petals forming a broad U-shaped, eyespot is *V. papilionacea* f. *albiflora* Grover (var. *Priceana* [Pollard] Alexander or *V. Priceana* Pollard) and is much cultivated.

This species runs into many variations in size and shape of leaves, coloring and size of petals, and size and robustness of plants. Deam believes (*Fl. Ind.* p. 688) that *V. papilionacea* may best be considered as a variety of the pubescent *V. sororia* Willd.

*Viola papilionacea* hybridizes with several other stemless, blue or purple-flowered violets, of which the following crosses have been recorded for Missouri:

**Viola** × **Bernardi** Greene (*Viola papilionacea* × *pedatifida*) — Map 1559
Occurs in prairies. Scattered in northern and western Missouri in Scotland, Schuyler, Putnam, Pettis, Jackson, Barton, and Jasper counties.

**Viola** × **napae** House (*Viola papilionacea* × *sororia*) — Map 1559
Occurs in woodland. Scattered in Missouri, where known from Franklin, Lafayette, Vernon, Barton, Dade, and Barry counties.

**Viola papilionacea** × **missouriensis**
Known from Dade County, southwestern Missouri (open woods, alluvial soil, along Sons Creek, 4 mi. north of Greenfield, June 10, 1951, *Palmer 52209*).

**Viola papilionacea** × **triloba** — Map 1559
Known from rocky wooded slopes in southwestern

Plate no. 256. **1.** Hybanthus concolor, × $^2/_5$; Details from Small, The New York Botanical Garden. **2.** Viola papilionacea, × $^2/_5$. **3.** Viola pedata, × $^2/_5$; Details from Small, The New York Botanical Garden. **4.** Viola cucullata, × $^2/_5$; *a.* Flower, × $1^4/_5$.

PLATE NO. 256

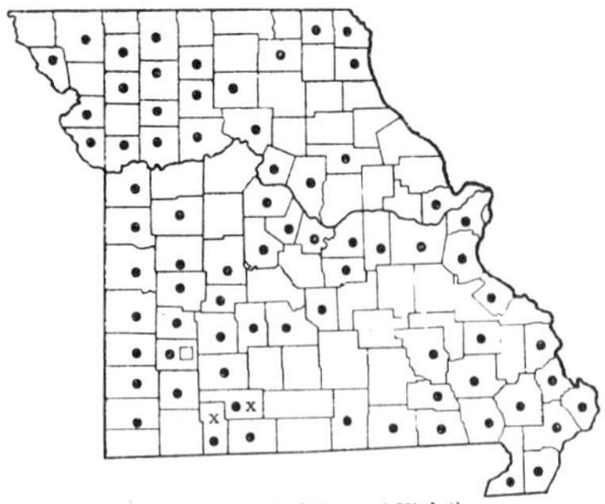

1560 • Viola missouriensis (Missouri Violet)
1560 □ Viola missouriensis X Viola papilionacea
1560 x Viola missouriensis X Viola sororia

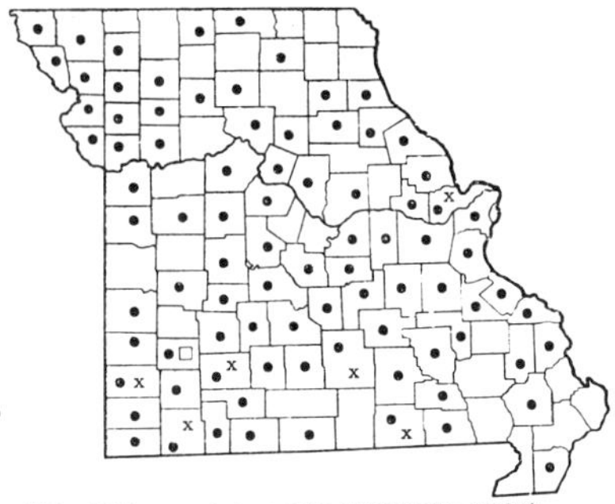

1561 • Viola sororia f. sororia (Woolly Blue Violet)
1561 □ Viola sororia f. Beckwithae
1561 x Viola sororia X Viola triloba

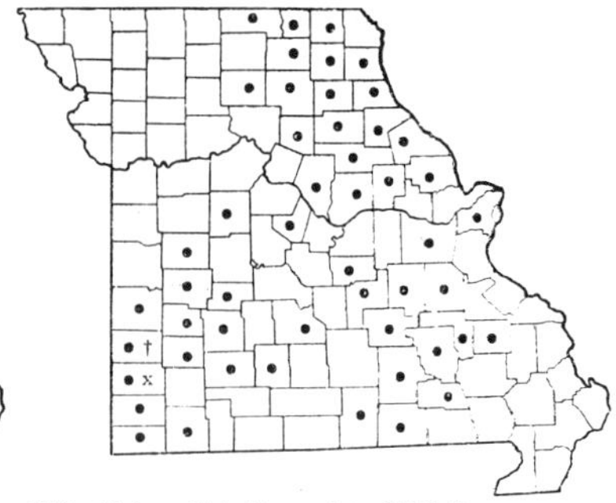

1562 • Viola sagittata (Arrow-leaved Violet)
1562 x Viola sagittata X Viola sororia
1562 † Viola sagittata X Viola papilionacea

Missouri in Christian (*Palmer 61979*), Lawrence (*Palmer 51621*), and Jasper counties.

4. **Viola missouriensis** Greene
Missouri Violet Map 1560
Flowers March–May.

Occurs in low, rich or alluvial soils of woodland in valleys and ravine bottoms, along or near streams, moist ledges of limestone bluffs, and low thickets. Throughout Missouri, but not recorded from some of the Ozark and northeastern counties.

Ranges from Indiana to Iowa and South Dakota, south to Kentucky, Arkansas, and Texas.

Compared with *V. papilionacea*, with which it is sometimes confused, this species has paler flowers of a pale lilac or lavender color with a darker area surrounding a white center and at least some of the leaves taper to a longer, more pronounced apex, suggesting the shape of a Lombardy poplar leaf.

In Missouri the following apparent hybrids are known between *V. missouriensis* and two other species:

**Viola missouriensis × papilionacea**
Already noted above under *V. papilionacea*.

**Viola missouriensis × sororia**
Known from southwestern Missouri in Christian (open alluvial woods along Swan Creek, near Garrison, April 20, 1956, *Palmer 61993*) and Stone (wooded slopes along small stream, May 14, 1951, northeast corner of county, *Palmer 51913*) counties.

5. **Viola sororia** Willd. Woolly Blue Violet
Map 1561

Also called Downy Blue Violet.

Flowers March–June, occasionally October–November.

Occurs in rocky or dry open woods, thickets, and sometimes in sandy alluvial soils along streams. Throughout Missouri.

Ranges from Quebec and Maine to Minnesota and South Dakota, south to North Carolina, Kentucky, Missouri, and Oklahoma.

Two variations occur in Missouri:

Petals violet or lavender . 5a. V. SORORIA f. SORORIA
Petals white . . . 5b. V. SORORIA f. BECKWITHAE

5a. **Viola sororia** f. **sororia** Map 1561
*Viola sororia* Wild. [G, BB, P & S, Steyerm.]
This is the common form encountered throughout the state.

5 b. **Viola sororia** f. **Beckwithae** House Map 1561
Known only from Dade County, southwestern Missouri (moist shaded ledges of sandstone bluff, along Turnback Creek, 5 mi. north of Greenfield, May 6, 1951, *Palmer 51774*).

*Viola sororia* varies in amount of pubescence on the leaf-stalks and leaf-blades. The early spring leaves are quite small and rounded, and frequently purplish on the lower surface. This violet often occurs in the same type of dry woodland habitat as *V. triloba*, frequently growing with that species.

The leaves are eaten by white-tailed deer.

The following apparent hybrids have been recorded for Missouri between *V. sororia* and five other species:

**Viola × napae** House (*Viola papilionacea × sororia*)
Already noted under *V. papilionacea*.

Plate no. 257. **1.** Viola missouriensis, × 2/5. **2.** Viola triloba f. triloba, × 2/5. **3.** Viola sagittata, × 2/5; *a*, *b*. Leaf variations. **4.** Viola sororia, × 2/5. **5.** Viola viarum, × 2/5; *a*. Leaf variation.

PLATE NO. 257

**Viola × populifolia** (*Viola sororia × triloba*)
Known only from St. Charles County, east-central Missouri (dry ridge above north-facing limestone wooded slopes along Callaway Fork, T45N, R2E, sect. 7 and 8, 4–5 mi. southeast of New Melle, 3½–4 mi. northeast of Schluersberg, July 26, 1956, *Steyermark 82050*).

**Viola missouriensis × sororia**
Already noted under *V. missouriensis*.

**Viola sororia × pedatifida**
Known from western Missouri, in Worth, Atchison, Lawrence, and Jasper counties.

**Viola sagittata × sororia**
Known from Jasper County, southwestern Missouri (Webb City, May 16, 1911, *Palmer 3395*). This specimen was labeled as *V. emarginata* by Brainerd with the comment 'appearing crossed with a cordate-leaved species like *V. papilionacea* or *sororia*.'

6. **Viola sagittata** Ait. Arrow-leaved Violet Map 1562

*Viola sagittata* var. *glabra* Pennell [Steyerm.]
*Viola arkansana* Greene
*Viola sagittata* var. *subsagittata* (Greene) Pollard [BB]

Flowers April–June.

Occurs in prairies, swampy meadows, rocky glades, margins of upland ponds, rocky creek beds, open places, and along railroads.

Eastern, southern, and central Missouri, west north of the Missouri River to Putnam, Linn, Randolph, and Boone counties.

Ranges from Massachusetts to Minnesota, south to Georgia, Louisiana, and Texas.

This species varies greatly in shape, size, and pubescence of leaf-blades, which in some Missouri specimens are as much as 4 cm. broad. The commoner variation in Missouri is the one having glabrous or nearly glabrous leaf-blades and leaf-stalks. This is known as *V. sagittata* var. *sagittata* (synonym var. *glabra*), the hairier extreme being treated as var. *subsagittata*. The late fall phase of *V. sagittata* is very deceptive, having ovate-cordate leaves which do not at all resemble the elongated oblong or arrowhead-shaped ones of early spring. The early spring leaves are quite small with short stalks, some being heart-shaped and rounded with a regularly wavy-toothed base. Later leaves, which follow, are larger with a somewhat jagged or coarsely cut base. The variability in width and serration of the leaves of this species has led to misidentification of some Missouri material with *V. emarginata* (referred to under Excluded Species).

*Viola sagittata* produces apparent hybrids with three other species in Missouri:

**Viola sagittata × papilionacea**
Known questionably only from Barton County, southwestern Missouri.

**Viola sagittata × sororia**
Already noted under *V. sororia*.

**Viola pedatifida × sagittata**
Already noted under *V. pedatifida*.

7. **Viola triloba** Schwein. Three-lobed Violet Map 1563

Flowers April–May.

Occurs in dry rocky or open woods on slopes, ridges, above bluffs, along streams, and thickets, frequently in acid soils. Common throughout the Ozark region in southern and east-central Missouri, north to Clay, Saline, Randolph, Monroe, and Marion counties.

The following three variations are known in Missouri:

a. Petals all white . . . 7b. V. TRILOBA f. ALBIDA
a. Petals pale to deep purple . . . . . . *b*
  b. Some of the leaves 3- or 5-lobed with the middle lobe the broadest and most conspicuous . . . 7a. V. TRILOBA f. TRILOBA
  b. Some of the leaves 5–7 or more lobed, the lobes all narrow and deeply cut, the middle lobe only slightly if at all broader than the other lobes . . 7c. V. TRILOBA f. DILATATA

7a. **Viola triloba** f. **triloba** Map 1563

*Viola triloba* Schwein. [G, P & S, Steyerm.]
*Viola triloba* var. *triloba* [BB]

Southern, central, and northeastern Missouri north to Clay, Saline, Howard, Boone, Callaway, and Marion counties; absent from many of the western unglaciated prairie counties south of the Missouri River.

Ranges from Vermont and Massachusetts to Indiana, south to Georgia, Alabama, Missouri, Oklahoma, and Texas.

7b. **Viola triloba** f. **albida** Steyerm. Map 1563

Known only from Stone County, southwestern Missouri (cherty draw in upland along route 13, ½ mi. northeast of Lampe, April 29, 1949, *Steyermark 67440*).

Thus far known only in Missouri. This form has the leaves of typical f. *triloba*.

7c. **Viola triloba** f. **dilatata** (Ell.) Palmer & Steyerm. Map 1563

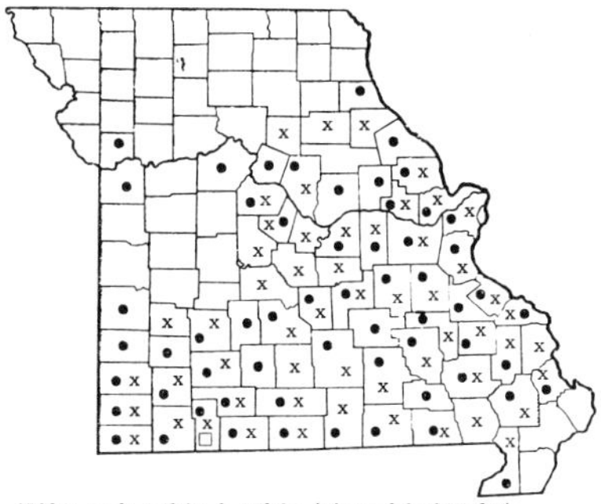

1563 • Viola triloba f. triloba (Three-lobed Violet)
1563 x Viola triloba f. dilatata
1563 □ Viola triloba f. albida

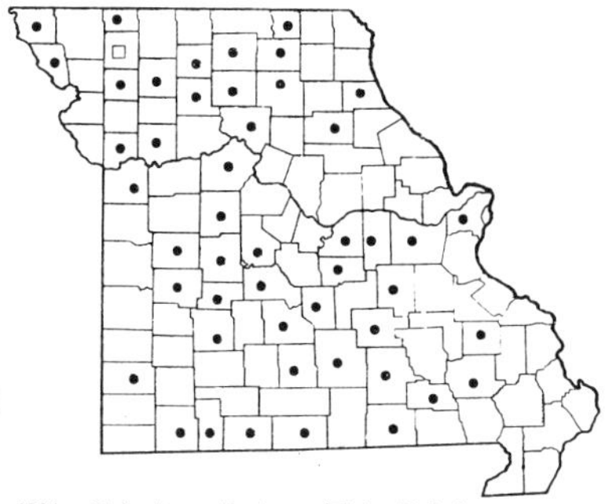

1564 • Viola viarum f. viarum (Plains Violet)
1564 □ Viola viarum f. pilifera

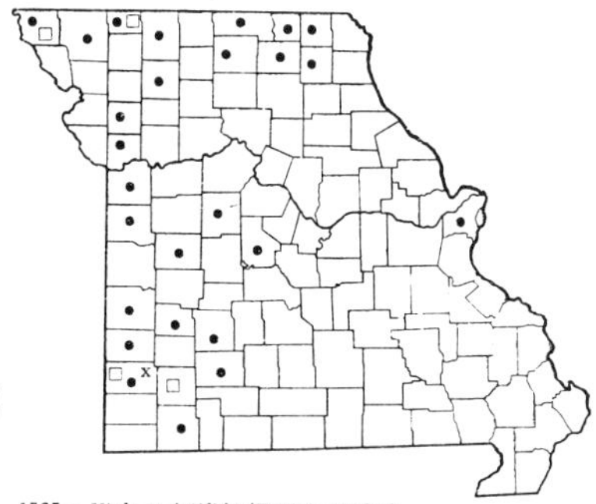

1565 • Viola pedatifida (Prairie Violet)
1565 □ Viola pedatifida X Viola sororia
1565 x Viola pedatifida X Viola sagittata

*Viola triloba* var. *dilatata* (Ell.) Brainerd [G, BB, P & S, Steyerm.]
*Viola falcata* Greene

Ozark region of southern and east-central Missouri, north to Ralls, Monroe, Randolph, Cooper, Cedar, and Jasper counties.

Ranges from Pennsylvania and Indiana to Illinois and Missouri, south to Florida, Louisiana, Oklahoma, and Texas.

There is little difference in the broad geographical ranges occupied by each of these variations. In Missouri both types frequently occur together and have practically the same geographical range. The line of demarcation is often impossible to draw between one and the other variation (Brittonia 10: 115–16. 1958). As the same situation applies in other parts of the range of these variations, they are treated here as forms rather than as varieties.

The more dissected-leaved *V. triloba* f. *dilatata* has been the basis for mistaken identifications of Missouri material reported as *V. palmata* (for which see Excluded Species).

*Viola triloba* apparently hybridizes in Missouri with two other species, recorded as follows:

**Viola papilionacea** × **triloba**
Already noted under *V. papilionacea*.

**Viola** × **populifolia** (*Viola sororia* × *triloba*)
Already noted under *V. sororia*.

8. **Viola viarum** Pollard Plains Violet Map 1564
Flowers April–May.

Occurs along banks of streams, gravel or rocky stream beds, crevices of boulders and bluffs along streams, along railroads and roadsides, and moist low open places or woodland. Throughout Missouri, except absent in the lowland counties of the southeastern section.

Ranges from Illinois, Iowa, to Nebraska, south to Missouri, Kansas, and Oklahoma.

Two variations occur in Missouri:

Common type; plant glabrous (without hairs) . 8a. V. VIARUM f. VIARUM
Rarely encountered; leaf-stalks (petioles) with short spreading hairs . 8b. V. VIARUM f. PILIFERA

8a. **Viola viarum** f. **viarum** Map 1564
*Viola viarum* Pollard [G, BB, P & S, Steyerm.]

This is the common variation found throughout the state.

The species was originally described from a specimen collected at St. Louis (July 15, 1899, *J. B. S. Norton*).

8b. **Viola viarum** f. **pilifera** Palmer & Steyerm. Map 1564

Known only from Gentry County, northern Missouri (along rocky bed of branch tributary to ravines on north and east side of Grand River, T61N, R30W, sect. 13, $1\frac{1}{2}$–2 mi. south of McFall, April 28, 1955, *Steyermark 78500*, holotype in Chi. Nat. Hist. Mus. Herb.).

A specimen from Atchison County, northwestern Missouri (Watson, April 25, 1894, *Bush 20*) referred in the present flora to *V. viarum* f. *viarum* is the basis for the report of *V. triloba* f. *dilatata* from that county in Palmer & Steyermark's *Annotated Catalogue*. Bush originally identified this specimen as *V. palmata*. A specimen collected in Ozark County (*Palmer 34789*) and recorded in Palmer & Steyermark's *Annotated Catalogue* as *Viola Lovelliana* Brainerd is referred in the present flora to *V. viarum*.

This species of violet is quite variable in the leaves, some being more deeply lobed or cut than others, some having narrow lobes, others having broad lobes, or with the middle lobe broader than the other lobes.

However, the species is an easily identifiable and distinct one, after one has become well enough familiar with it. Invariably, it is a frequent habitant of alluvial rocky places bordering streams.

9. **Viola pedatifida** G. Don Prairie Violet
Map 1565

Flowers April–May.

Occurs in prairies and loess hills. Northern, central, and western Missouri, south and east to St. Louis, Morgan, Polk, Greene, and Barry counties; absent from the Ozark and southeastern lowland sections and from most of the east-central and middle counties.

Ranges from Ohio to Alberta, south to Missouri, Oklahoma, Texas, New Mexico, and Arizona.

The petals are usually dark purple or violet. The leaves, as in some other species, are deceptively different in the spring and fall, those of late summer and autumn less deeply cut than the usually deeply cut early spring leaves with narrowly linear lobes. Apparent hybrids between *V. pedatifida* and *V. papilionacea* (*V.* × *Bernardi*) resemble the late summer and autumn, less deeply cut foliage of *V. pedatifida.*

Some plants are more pubescent than others. These pubescent individuals are the basis for some of the reports of *V. palmata* in the state. Such collections have proven to be either pubescent specimens of *V. pedatifida* or hybrids between *V. pedatifida* and *V. sororia* or *V. triloba* f. *dilatata*, or, when the leaves are cordate-based, have been found to be *V. triloba* f. *dilatata.*

*Viola pedatifida* hybridizes with three other species in Missouri:

**Viola** × **Bernardi** (*V. pedatifida* × *papilionacea*)
Already noted under *V. papilionacea.*

**Viola pedatifida** × **sororia**
Already noted under *V. sororia.*

**Viola pedatifida** × **sagittata**
Already noted under *V. pedatifida.*

10. **Viola pallens** (Banks) Brainerd
Smooth White Violet Map 1566
Also called Northern White Violet, Wild White Violet.
*Viola Macloskeyi* Lloyd subsp. *pallens* (Banks) M. S. Baker [Russell]
*Viola pallens* (Banks) Brainerd [G, P & S, Steyerm.]
*Viola pallens* var. *pallens* [BB]

Flowers April–May.

Occurs in moist, shady places in crevices of bluffs and on moist rocky strata of LaMotte sandstone along small wooded streams in narrow ravines and canyons. Known only from Ste. Genevieve County in the southeastern Missouri Ozarks (moist mossy LaMotte sandstone ledge near spring along River aux Vases, at base of bluff, about 5 mi. northeast of Pickle, May 3, 1931, *Steyermark 464*; also collected or known from adjacent areas in the same county from Pickle Creek, Terre Bleue Creek, Hickory Creek, and in many places along River aux Vases).

Ranges from Labrador to Alaska, south to South Carolina and Alabama, Ohio, Indiana, Illinois, Missouri, North Dakota, Montana, and Colorado.

Dr. Norman H. Russell (Am. Midl. Nat. 54: 487–89. 1955) considers *V. pallens* as a variety of the Californian *V. Macloskeyi* Lloyd, and treats it as a subspecies of that, following Baker.

This violet, common in the cooler sections of the northern and eastern United States and Canada, is the rarest one in Missouri. In the LaMotte sandstone ravines of Ste. Genevieve County it has survived as an isolated relict species at one of its southernmost stations in the United States, apparently having reached Missouri during one of the southern glacial movements of Pleistocene times. With the retreat northward of the ice mass, the violet has persisted in the cool, moist habitats of a microclimate provided by the LaMotte sandstone ravines and canyons, habitats more nearly approaching the cooler northern sections to which the species is best adapted. In such a microclimate, surrounded by a generally drier and warmer Ozark upland, the violet, along with other northern and eastern relict species, such as *Lycopodium obscurum* var. *dendroideum*, *L. complanatum* var. *porophilum*, *Bryoxiphium norvegicum*, *Isotria verticillata*, *Goodyera pubescens*, and others, has survived thousands or hundreds of thousands of years in a drier Ozark setting since Pleistocene times.

11. **Viola lanceolata** L. var. **lanceolata**
Lance-leaved Violet Map 1567
*Viola lanceolata* L. [G, BB, P & S, Steyerm.]
*Viola lanceolata* subsp. *lanceolata* (Russell, Am. Midl. Nat. 54: 483. 1955)
Flowers April–June.

Occurs in moist places in sandy prairies, gravelly banks along streams, moist hummocks of low wet

Plate no. 258. **1.** Viola triloba f. dilatata, × 2/5. **2.** Viola pedatifida, × 2/5. **3.** Viola pallens, × 2/5. **4.** Viola lanceolata, × 2/5; *a*. Later developed leaf. **5.** Viola striata, × 2/5. **6.** Viola pensylvanica, × 2/5. **7.** Viola Kitaibeliana var. Rafinesquii, × 2/5.

PLATE NO. 258

1566 Viola pallens (Smooth White Violet)

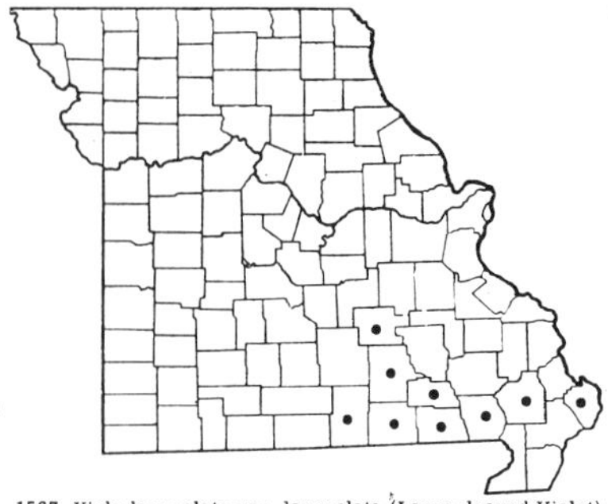

1567 Viola lanceolata var. lanceolata (Lance-leaved Violet)

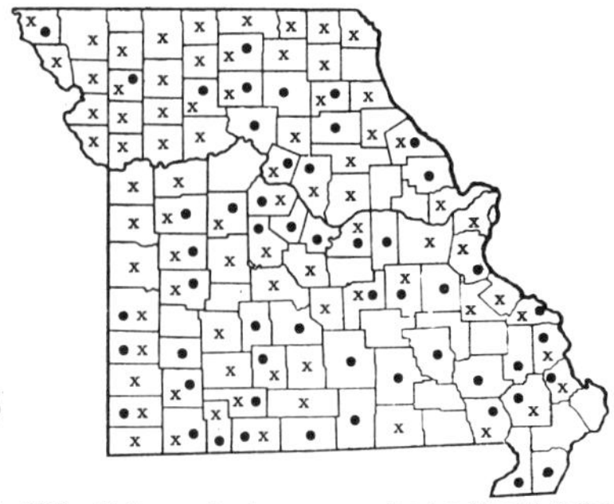

1568 • Viola pensylvanica var. pensylvanica (Smooth Yellow Violet)
1568 x Viola pensylvanica var. leiocarpa

woods, and on tussocks in upland sink-hole ponds. Southeastern Ozark region, where it frequents upland sink-hole ponds in Dent, Shannon, Howell, Oregon, Carter, and Ripley counties, in wet, lowland woods in Butler and Stoddard counties, and sandy prairies in Mississippi County.

Ranges from Florida to Texas, north to Nova Scotia, Quebec, Ontario, Michigan, Wisconsin, Minnesota, and Nebraska.

The petals are all white, except for the three lower ones which are striped with brown-purple veins near the base. The leaves at first in spring are usually not more than 1.5 cm. broad or 15 cm. long, but the later ones of summer and autumn are much larger and up to 3 cm. broad. In summer an abundance of runners (stolons) are produced from the slender rhizomes and root at the nodes.

12. **Viola pensylvanica** Michx.
Smooth Yellow Violet Map 1568
Flowers March–May.

Occurs in rich woods on slopes or in low alluvial ground and thickets. Throughout Missouri.

Two variations are found in the state:

Fruit hairy . . . . . . . 12a. V. PENSYLVANICA var. PENSYLVANICA
Fruit glabrous (without hairs) . 12b. V. PENSYLVANICA var. LEIOCARPA

12a. **Viola pensylvanica** var. **pensylvanica** Map 1568

*Viola pensylvanica* Michx. [G]
*Viola eriocarpa* Schwein. [BB, P & S, Steyerm.]

This variation, although found throughout the state, is less common in Missouri than the next.

Ranges from Connecticut to Minnesota, south to Georgia, Alabama, Arkansas, Oklahoma, and Texas.

12b. **Viola pensylvanica** var. **leiocarpa** (Fern. & Wieg.) Fern. Map 1568

*Viola eriocarpa* var. *leiocarpa* Fern. & Wieg. [BB, P & S, Steyerm.]
*Viola eriocarpa* f. *leiocarpa* (Fern. & Wieg.) Deam [Deam]
*Viola pubescens* of [P & S, Steyerm.], not Ait.

This variation is the more frequent one encountered in the state.

Ranges from Quebec to Manitoba, south to North Carolina, Tennessee, Ohio, Indiana, Illinois, Missouri, and Oklahoma.

A number of Missouri specimens previously identified as *Viola pubescens* Ait. and recorded for the state in Palmer & Steyermark's *Annotated Catalogue* (p. 603) and in Steyermark's *Spring Flora* have been found to be misidentified and are referred in the present flora to *V. pensylvanica* var. *leiocarpa*.

13. **Viola striata** Ait. Pale Violet Map 1569
Also called Cream Violet.
Flowers April–June.

Occurs in low or rich, moist or rocky ground usually in woodland valleys along streams or along banks of streams. Ozark region of southern and east-central Missouri, north to Lincoln, Montgomery, Callaway, Boone, Randolph, and Howard counties, locally northeast in Shelby County, west to Howard, Cooper, Benton, Hickory, Greene, and Barry counties, locally northwest in Clay County (low or shady ground, Greens Farm, Liberty, April 21, 1948, *Shirley Landers 17*).

Ranges from Georgia to Arkansas, north to New York, Ontario, Ohio, Michigan, and Wisconsin.

This species succeeds well in a shaded woodland wildflower garden in neutral to calcareous soils.

14. **Viola tricolor** L. Miniature Pansy Map 1570
Also called Johnny-jump-up, Heart's ease.
*Viola tricolor* var. *hortensis* DC.
Flowers April–May.

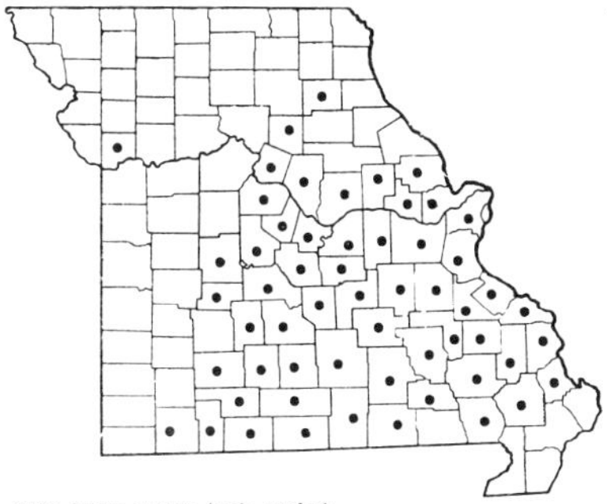

1569 Viola striata (Pale Violet)

1570 Viola tricolor (Miniature Pansy)

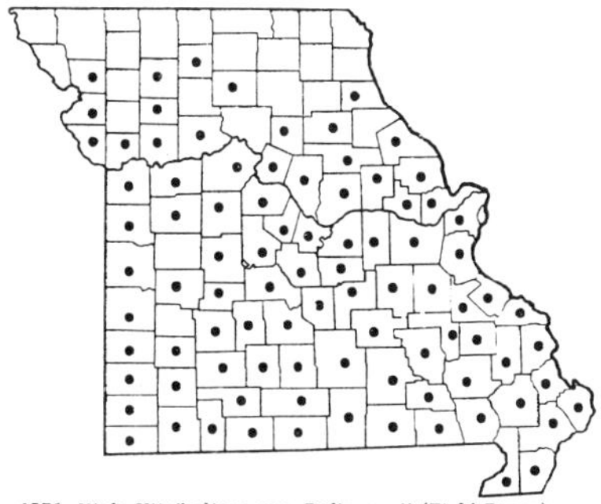

1571 Viola Kitaibeliana var. Rafinesquii (Field Pansy)

Escaping from cultivation in waste ground and along railroads. Known from St. Louis (St. Louis, northeast of Broadway Station of Missouri Pacific Railroad, May 5, 1956, *Muehlenbach 875*) and Clay (escaped, farm north of Liberty, April 20, 1947, *Tom Bray 93*; escaped, Mississippi Street, Liberty, May 3, 1948, *Shirley Landers 72*) counties.

Native of Europe; planted and occasionally escaped from cultivation.

The cultivated large-flowered Pansy is of hybrid origin and is known as *Viola* × *Wittrockiana* Gams. It is believed to have been derived by hybridization of *V. tricolor* with other European (*V. lutea*) and Asian (*V. altaica*) species.

Some Missouri specimens identified as *V. tricolor* have proven to be *V. Kitaibeliana* var. *Rafinesquii* to which they are referred in the present flora.

15. **Viola Kitaibeliana** R. & S. var. **Rafinesquii** (Greene) Fern. Field Pansy Map 1571
Also called Wild Pansy, Johnny-jump-up.
*Viola Rafinesquii* Greene [BB, P & S]
*Viola bicolor* Pursh, not Gilib. [Shinners]
Flowers March–May.

Occurs in fields, meadows, rocky open glades, along roadsides and railroads, and waste ground.

Common throughout southern and central Missouri, and apparently spreading northward, whence known north to Marion, Monroe, Randolph, Linn, Grundy, Daviess, and Andrew counties.

Native of southern Europe or northern Africa (according to Fernald), native of the United States (according to Shinners); introduced and naturalized (according to Fernald) from Georgia to Texas, north to New York, Ohio, Michigan, Illinois, Iowa, and Kansas.

Fernald (Rh. 40: 443–46. 1938) has advanced the theory that this field pansy is an introduction from some part of southern Europe, differing from typical *V. Kitaibeliana* var. *Kitaibeliana* in its usually entire leaf-blades, larger lilac- or lavender-tinged petals, and strongly hairy (ciliate) margins of the sepals. Fernald believes that the plants found in the United States represent a rare, unrecognized extreme variation introduced from southern Europe or northern Africa which has become stabilized in its new environment in the United States. Shinners believes that the North American plant is a native species.

Missouri specimens previously identified as *V. arvensis* and included by Palmer & Steyermark in their *Annotated Catalogue* and by Steyermark in his *Spring Flora* have been found to be referable instead (see Excluded Species) to *V. Kitaibeliana* var. *Rafinesquii*.

*Excluded Species*

**Viola affinis** LeConte

The following specimens, labelled by their collectors as *V. affinis*, are misidentified and referred to *V. papilionacea* in the present flora: Jackson County (Courtney, *Bush 4166, 3859*); Taney County (Swan, *Bush 3457*); Jasper County (Webb City, *Palmer 846*).

**Viola arvensis** Murr. European Field Pansy

This species was recorded for Missouri in Palmer & Steyermark's *Annotated Catalogue* and Steyermark's *Spring Flora* as introduced in Johnson County. A reexamination of the specimen upon which the record was based (*Palmer 36743* from sandy open woods, Pertle Springs, near Warrensburg, June 23, 1930) shows a very mature plant in poor condition without petals and with most of the leaves old and practically past recognition. There are, however, portions of the stipules remaining to show that they are palmately rather than pinnately cut at base, with the middle lobe of the stipule entire or nearly so instead of crenately toothed. The specimen has, therefore, been referred in the present flora to *V. Kitaibeliana* var. *Rafinesquii*.

**Viola canadensis** L. Canada Violet

A specimen of this species exists in the herbarium of William Jewell College from Clinton County (grassy open woods, Cameron, May 11, 1936, *Earl Wilk*). There is reason to doubt that the specimen was obtained from this locality or from a wild plant. The species is of more northern range, the nearest station to Missouri occurring in north-central Iowa along the Des Moines River. Several efforts to relocate the plant at the locality indicated have failed thus far. The species, therefore, is excluded from the present flora until there is more definite documentation of its presence in the state.

**Viola emarginata** (Nutt.) LeConte
Triangle-leaved Violet

This species was credited to Missouri by Palmer & Steyermark in their *Annotated Catalogue*, by Steyermark in his *Spring Flora*, by Fernald in the eighth edition of *Gray's Manual*, and specimens were so identified by Brainerd. A re-study of all material so labeled has shown that the specimens previously interpreted as *V. emarginata* must be transferred mostly to *V. sagittata*, and partly to *V. missouriensis*. Brainerd identified *Palmer 760* from Webb City and *Palmer 3036* from Smithfield, Jasper County, as *V. emarginata*. These specimens have an elongate oblong form of leaf-blade and can be matched by many other collections of *V. sagittata* which at Brainerd's time were not available for study. The specimens of *Palmer 14689* from Oregon County, *Steyermark 4503* from Greene County, *Palmer 3717* from Asbury, Jasper County, and *Bush* from Shannon County, labeled *V. emarginata*, likewise can be matched today by numerous collections of *V. sagittata* showing leaf variations typical of the latter species. The specimen of *Palmer 52021* from Lawrence County, labeled by him as *V. emarginata*, is *V. missouriensis*.

**Viola emarginata × sororia**

Reported by Palmer & Steyermark in their *Annotated Catalogue* from Jasper County. The specimen upon which this was based is referred in the present flora to *V. sagittata*.

**Viola fimbriatula** Sm.

As indicated by Palmer & Steyermark in their *Annotated Catalogue* (p. 602), some Missouri material has been misidentified as this species, the specimens proving to be usually pubescent phases of *V. sagittata* referred sometimes to *V. subsagittata* Greene, considered here synonymous with *V. sagittata*. Brainerd had identified *Palmer 3376* from Carthage, Jasper County as *V. fimbriatula*, noting 'form approaching *V. sagittata* in having incised basal lobes,' but this specimen can be matched by many other collections from Missouri of *V. sagittata* and is a later, more maturely developed stage of the specimen of *Palmer 3343* from Carthage. Likewise, *Palmer 3343* from Carthage and *Palmer 3717* from Asbury, Jasper County, and *Greenman 3867* from Iron County, all previously labeled *V. fimbriatula*, are definitely *V. sagittata*.

**Viola Lovelliana** Brainerd

A specimen collected from Ozark County (*Palmer 34789*) originally identified as this species, proves, upon re-examination to belong to *V. viarum*, to which species it is referred in the present flora.

**Viola palmata** L. Wood Violet

As indicated under *V. pedatifida* and *V. triloba*, a number of Missouri specimens previously identified as *V. palmata* are misidentifications of these other two species. Most of the specimens so misidentified belong usually to *V. triloba* f. *dilatata* (Crystal City, Jefferson Co., *Eggert*; Chain of Rocks, St. Louis Co., *Glatfelter*; Jefferson Co., *Wislizenus*; Chadwick, Christian County, *Bush 4435*; Monteer, Shannon Co., *Bush 4340A*; Swan, Taney Co., *Bush 4208*; Ironton, Iron Co., *Churchill*; Meramec Highland, St. Louis Co., *Churchill*; Prosperity, Jasper Co., *Palmer 1951*; Graniteville, Iron Co., *Steyermark 8698*; Eagle Rock, Barry Co., *Palmer 35695*) or to *V. triloba* (Webb City, Jasper Co., *Palmer 565*; Hannibal, Marion Co., *Davis 1230*).

The collections with deeply divided leaves all have the leaves cuneate or truncate at their base characteristic of *V. viarum* or *V. pedatifida* instead of cordate as in *V. palmata*. In one of the collections cited from Jasper County (upland woods, Prosperity, May 9, 1909, *Palmer 1951*) all the leaves are divided nearly to the base with conspicuously pubescent petioles. However, they are rather truncate instead of definitely cordate, and this specimen is identified as representing a probable cross between *V. pedatifida* and *V. triloba* var. *dilatata* or *V. sororia*.

**Viola pubescens** Ait. Downy Yellow Violet

This species was credited to Missouri by Palmer & Steyermark in their *Annotated Catalogue*, by Steyermark in his *Spring Flora*, and subsequently by Fernald in the eighth edition of *Gray's Manual*, but a re-study of all available material proves that the specimens upon which the report was based are misidentified *V. pensylvanica* var. *leiocarpa* and must be referred to that taxon. These specimens so misidentified include the following: St. Louis County (Creve Coeur, *Wislizenus 464*; *Riehl 207*; *Glatfelter*); Jefferson County (*Kellogg*); Butler

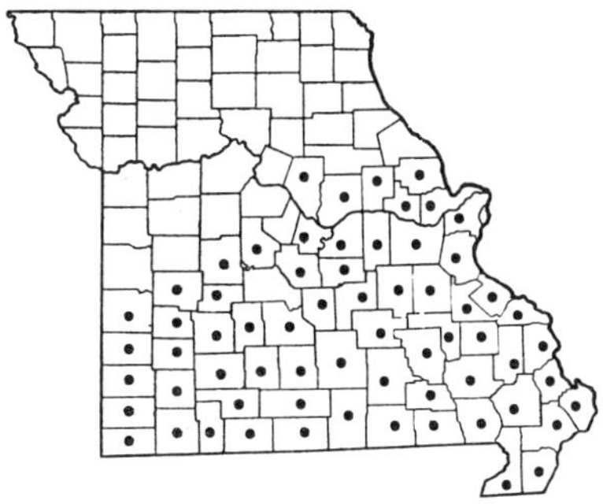

1572 Passiflora lutea var. glabriflora

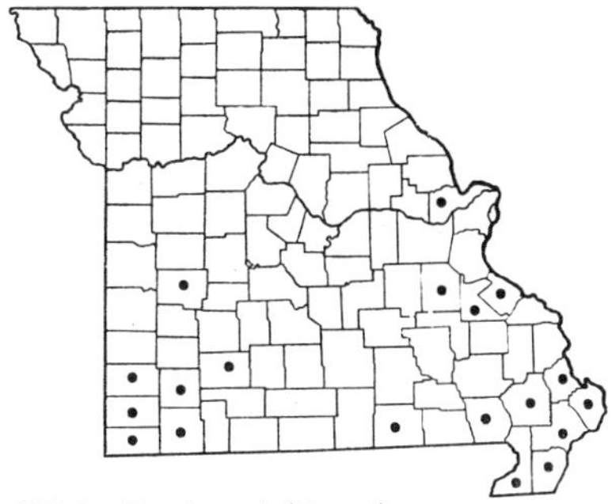

1573 Passiflora incarnata (Maypops)

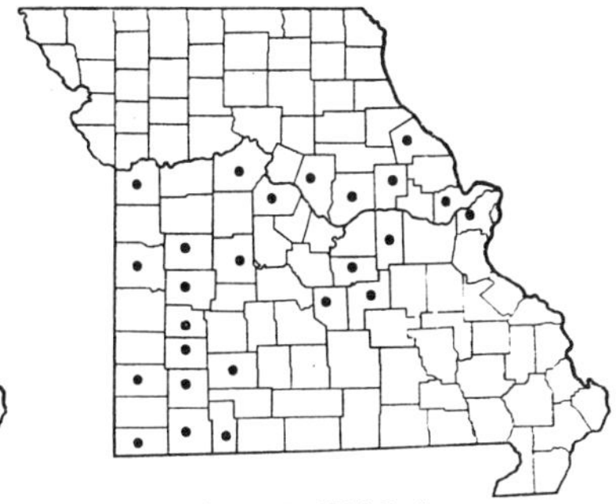

1574 Mentzelia oligosperma (Stick Leaf)

County (*Palmer 14707*); Stoddard County (*Palmer 34875*); Jackson County (Courtney, *Bush 111*); Mercer County (*Palmer & Steyermark 41269*); Harrison County (*Steyermark 10973*); Putnam County (*Palmer & Steyermark 41092*); Grundy County (*Steyermark 11034*). Some of these specimens show pubescence on the uppermost parts of the stem, but most of them have 1–3 basal root-leaves present, stems branched at base, and 25–30 teeth on the margins of the stem-leaves, and all the specimens have glabrous capsules characteristic of *V. pensylvanica* var. *leiocarpa* as well as lanceolate or narrowly ovate stipules rather than broadly ovate or nearly ovate ones as in *V. pubescens*. The Missouri specimens, moreover, do not show the lower leaf surface as densely pubescent as in *V. pubescens*.

## Fam. **PASSIFLORACEAE** (Passion-flower Family)

### **Passiflora** L. Passion-flower

Lobes of leaves blunt or rounded, without teeth (entire); flowers greenish-yellow, small, 1.5–2.5 cm. across, 0.8–1.2 cm. long; fruit 1–1.2 cm. in diameter, 1–1.5 cm. long (slightly longer when pressed) . 1. P. LUTEA var. GLABRIFLORA

Lobes of leaves pointed, with teeth along margins; flowers white and purple or pink, large, 4–8 cm. across, 3–4 cm. long; fruit 3–5 cm. in diameter, 5–8 cm. long . . . . . . . . . . 2. P. INCARNATA

1. **Passiflora lutea** L. var. **glabriflora** Fern. Map 1572

*Passiflora lutea* of auth. in part [P & S], not L.

Flowers early June–September; fruits July–October.

Occurs in low, rich, or open rocky woods, wooded slopes, and thickets, rarely along railroads. Ozark region of southern and east-central Missouri and southeastern lowland section, north to Lincoln, Montgomery, Callaway, Boone, Morgan, Benton, St. Clair, and Vernon counties.

Ranges from West Virginia to Ohio, Illinois, Missouri, and Kansas, south to Florida, Alabama, and Texas.

The young tendrils are eaten by wild turkey.

2. **Passiflora incarnata** L. Maypops Map 1573

Also called Apricot Vine, Passion Flower.

Flowers early June–September; fruits July–October.

Occurs in sandy fields and thickets, fence rows, low alluvial woods, open waste ground, and along roadsides and railroads. Native in the southeastern lowland section and in southwestern Missouri, elsewhere introduced in parts of the southern half of the state north to St. Charles and St. Clair counties.

Ranges from Florida to Texas, north to Maryland and Virginia, Pennsylvania, West Virginia, Ohio, Indiana, Illinois, Missouri, and Oklahoma.

The flowers are very showy with white or nearly white petals and sepals, an outer and inner spreading crown of purple or pink threadlike divisions. The edible fruit, known as 'maypops,' is about the size of a hen's egg and in flavor is slightly acid combined with somewhat of a sweetness. The flowers have a rich fragrance resembling that of carnation.

This is a handsome flowering vine, but does not survive over winter in northern Illinois, where the author has tried it several times without success.

## Fam. **LOASACEAE** (Loasa Family)

### **Mentzelia** L.

**Mentzelia oligosperma** Nutt. Stick Leaf
Map 1574

Flowers late May–August.

Occurs on exposed ledges of south- or west-facing limestone bluffs and escarpments, glades, rocky open places, and along railroads.

Ozark border region and unglaciated prairie sections in central and southwestern Missouri, north to Pike, Montgomery, Callaway, Boone, Saline, and Jackson counties, east to St. Louis, Gasconade, Phelps, Pulaski, Greene, and Stone counties; absent from most of the Ozark section.

Ranges from Illinois and Missouri to South Dakota and Colorado, south to Louisiana, Texas, and Mexico.

The stems are nearly white or pale straw- or buff-colored. The dark green leaves adhere closely to the skin and to clothing. They sometimes affect the grade and market value of wool which contains pieces of leaves or other parts of the plant which were not separated from the woolen hairs.

## Order **CACTALES**

## Fam. **CACTACEAE** (Cactus Family)

### **Opuntia** Mill. Prickly Pear

The fleshy juicy flattened joints are the stems. These stems contain small clusters (areoles) of brown bristles (glochids) scattered over their surface and bear white or brownish spines. The leaves are scale-like and dark green, and soon fall from the stems.

Common species in Missouri; spines spreading, few on the joint, 1 per areole (rarely a second small one present), 1.9–3.1 cm. (or 5.6) long, gray or brownish, 0.5–0.7 (–1) mm. in diameter; joints dark to light green; petal-like parts yellow throughout; margin of seed smooth, regular, 0.5 mm. broad. . 1. O. COMPRESSA

Rarely encountered, known from western Missouri in Caldwell and McDonald counties; spines deflexed (turned down), few to abundant on the joint, 1–6 per areole, 3.8–5.6 cm. long, white to pale gray, rarely brownish or reddish-brown, 0.5 mm. in diameter; joints bluish-green (glaucous); base of yellow to creamy-yellow petal-like parts usually with some red; margin of seed irregular, about 1 mm. broad 2. O. MACRORHIZA

Data for the above differences were kindly supplied the author by Dr. Lyman Benson, cactus authority, of Pomona College, who is currently engaged in a study of this genus of plants.

1. **Opuntia compressa** (Salisb.) Macbr.
Prickly Pear Map 1575
*Opuntia humifusa* Raf. [G]
*Opuntia Rafinesquii* Engelm. [P & S, Steyerm.]

Flowers late May–July.

Occurs in rocky open glades, rocky prairies, sandy and gravelly washes of valleys along streams, fields, open woods, and along railroads. Common throughout southern and central Missouri, north to Ralls, Monroe, Livingston, and Clay counties.

Ranges from Massachusetts to Minnesota, south to South Carolina, Georgia, Alabama, Mississippi, Missouri, Arkansas, and Oklahoma.

The plants in Missouri and the Mississippi Valley have been separated previously as var. *microsperma* (Engelm.) Benson from those of the Atlantic Coast.

---

Plate no. 259. **1.** Viola sororia × triloba, × 2/7. **2.** Viola papilionacea × sororia, × 2/7. **3.** Passiflora lutea, × 2/7; After Britton and Brown, The New York Botanical Garden. **4.** Passiflora incarnata, × 2/7; *a*. Cross-section of fruit, × 2/7; After Britton and Brown, details from Small, The New York Botanical Garden. **5.** Mentzelia oligosperma, × 2/7; Details from Small, The New York Botanical Garden. **6.** Dirca palustris, × 2/7; Details from Small, The New York Botanical Garden. **7.** Opuntia macrorhiza, × 2/7. **8.** Opuntia compressa, × 2/7; Details from Small, The New York Botanical Garden.

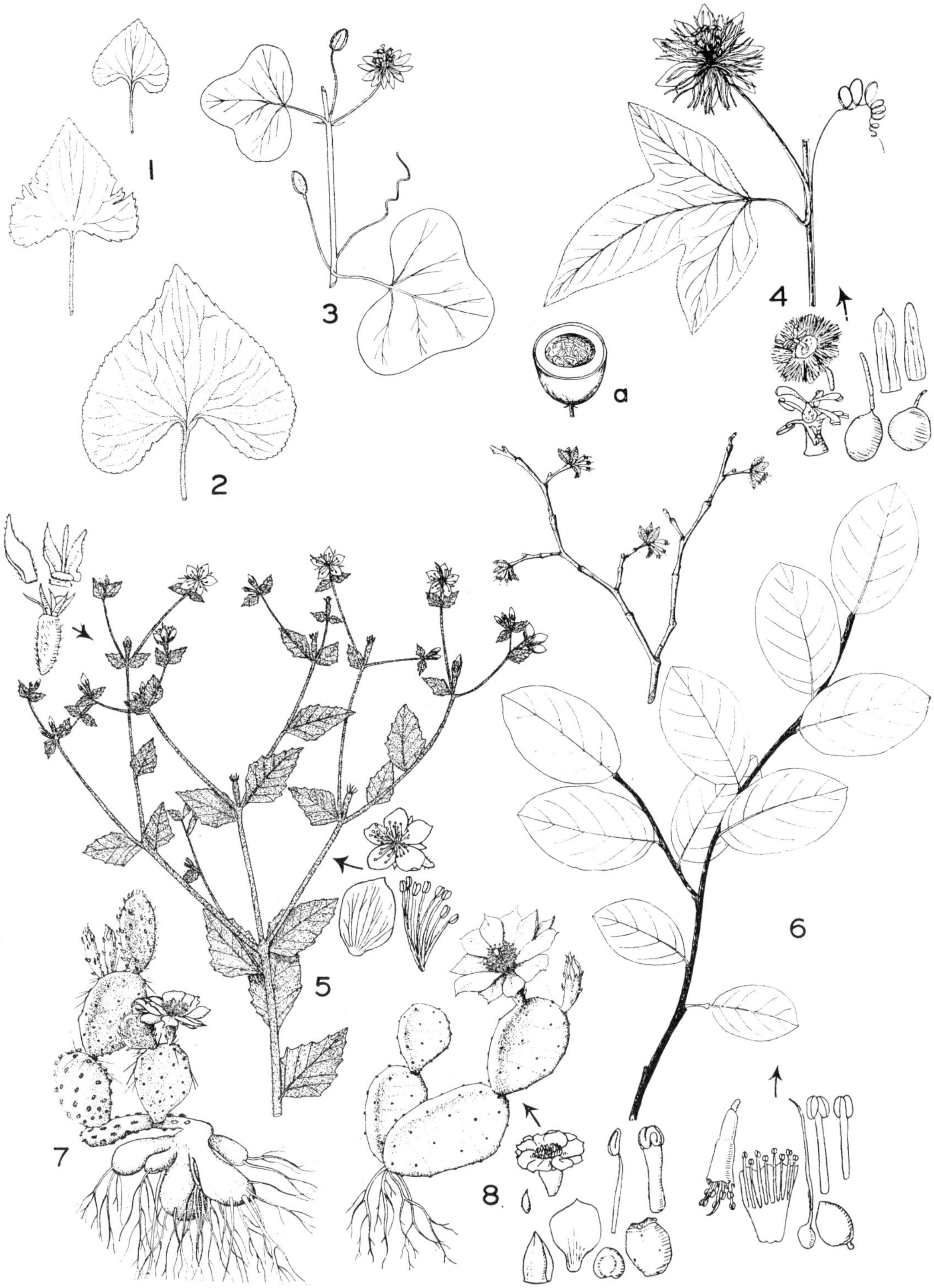


PLATE NO. 259

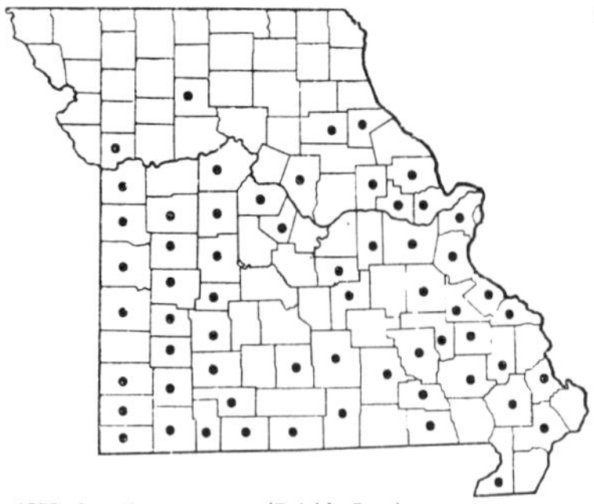

1575 Opuntia compressa (Prickly Pear)

1576 Opuntia macrorhiza

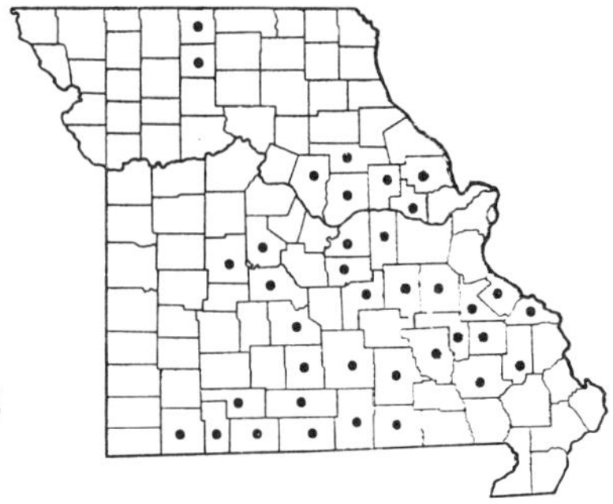

1577 Dirca palustris (Leatherwood)

As this separation may prove not to be justified, the Missouri plants are here being retained as *O. compressa*.

The tiny brown bristles (glochids) penetrate the pores of the skin easily, so that extreme caution must be used if the plant has to be handled. The fruit is red when ripe and may be eaten, but is not as sweet or as good as the larger and less seedy 'tunas' of the southwestern states. In Mexico and some of the southwestern states the seeds are heated, pulverized, and used as a thickening for soups. After the leaflike joints of the stem are roasted, the tough outer skin may be peeled off, leaving the fleshy inner portion to be eaten. This fleshy inner part is sometimes fed to livestock.

2. **Opuntia macrorhiza** Engelm. Map 1576

*Opuntia compressa* var. *macrorhiza* (Engelm.) Benson

Flowers late May–July.

Occurs on rocky glades and ledges, and in rocky open ground. Known only from Caldwell (4 mi. east of Cow Gill, *Steyermark 14892*) and McDonald (Noel, *Bush 5041*; *Palmer*) counties, western Missouri.

Ranges from Missouri and Kansas to Utah, south to Texas and Arizona.

This species is being retained as separate from *O. compressa*, following Dr. Benson's most recent conclusions of his studies of this group of plants. No reliability can be placed on the character of tuberous-thickened roots, used to distinguish *O. macrorhiza* from fibrous, nonthickened roots, supposedly characteristic of *O. compressa*. The characters used to separate these species, if they are distinct, are those supplied by Dr. Benson. Much more detailed field studies must be made before a final delimitation of this taxon can be evaluated.

*Excluded Species*

**Mammillaria missouriensis** Sweet

*Neomammillaria similis* (Sweet) Rydb. [BB]

*Mammillaria missouriensis* var. *caespitosa* (Engelm.) Wats. [P & S]

The basis for the report of this species in Palmer & Steyermark's *Annotated Catalogue* was taken from the identification of a specimen made by Mr. Ladislaus Cutak, cactus authority at the Missouri Botanical Garden. Mr. Cutak received a specimen sent in to him from near Hooker, Pulaski County, with the information that it was found growing in rocky hills of that area. As no preserved specimen is extant to authenticate the material identified, and as repeated attempts have failed to locate any cactus of this type in Missouri, it is best eliminated from the flora of the state.

**Opuntia polyacantha** Haw.

Reported from Missouri in Gleason's *New Ill. Fl.* (2: 570. 1952), but according to Dr. Benson (personal communication of November 4, 1958), there are no specimens to justify its inclusion in the state flora.

## Order **THYMELEALES**

### Fam. **THYMELEACEAE** (Mezereum Family)

#### **Dirca** L. Leatherwood

**Dirca palustris** L. Leatherwood Map 1577

Flowers late March–April.

Occurs in either low bottom woods of alluvial soils along streams or on rich wooded slopes or rocky wooded bluffs, usually on north- or east-facing exposures. Mainly restricted to the Ozark region in southern and east-central Missouri north to Lincoln, Audrain, Boone, Morgan, and Benton counties, west to Barry County, locally in northern Missouri in Grundy and Mercer counties.

Ranges from Quebec to Ontario, and Minnesota, south to Florida, Louisiana, and Oklahoma.

This is one of the earliest shrubs to bloom in Missouri, the pale yellow tubular flowers hanging in clusters on the bare branches, and soon followed by the gray-green leaves. Petals are absent, the tubular-funnelform irregularly 4-toothed calyx resembling a corolla.

The shrub is very slow-growing and usually grows low, from 1–2 (up to 3.3) meters tall. A plant with a trunk 5 cm. (2 inches) in diameter may be over a hundred years old. The twigs are very supple and can be tied into knots without breaking. On account of their toughness, the bark and branches were used by the Indians for making baskets and for tying purposes. The bark, if taken internally, produces vomiting and has purgative properties. Some individuals are sensitive to the bark, their skin becoming irritated or blistered after contact with the fresh bark. The fruit is reputed to have narcotic properties. It matures quickly in spring, and is fully matured by late April. In the northern part of its range the fruit is generally reddish or purplish, while southward it often or always lacks this color and may be greenish or yellowish (for discussion, see Fernald, Rh. 45: 117–19. 1943).

## Fam. **ELAEAGNACEAE** (Oleaster Family)

### **Elaeagnus** L. Oleaster

**Elaeagnus angustifolia** L. Russian Olive
Map 1578

Flowers May–July.

Commonly cultivated and infrequently escaped in thickets, waste ground, and along railroads. Known as an escape in St. Louis and Pike ('Talicon,' near Clarksville, September 21, 1914, *Davis 3230*) counties.

Native of Eurasia; introduced and naturalized in the central and western states, rarely escaping in the eastern states.

**Elaeagnus multiflora** Thunb.

A specimen labeled as this species is in the herbarium of the Missouri Botanical Garden, collected from St. Louis County (escaped on slope near Frisco R. R. tracks, just west of Osage Hills Station, November 19, 1933, *P. M. McGlashon 7733*). It is in a sterile condition, however, and cannot be satisfactorily or definitely determined as this species or *E. umbellata* Thunb., which it also resembles in the olive-green, glabrous upper surface of the leaves.

## Order **MYRTALES**

## Fam. **LYTHRACEAE** (Loosestrife Family)

a. Stem with sticky (viscid) spreading hairs; petals 6, unequal in size and shape; flower irregular and not symmetrical . . . . . . 6. CUPHEA
a. Stem either smooth and glabrous (without hairs) or, if hairy, then without sticky (viscid) spreading hairs and the plants then 6–25 dm. tall; petals none, 4, 5, 6, or 7, equal; flower regular or nearly regular . . . . . . *b*
 b. At least some of the upper leaves of the branches alternate . . . . . . 5. LYTHRUM ALATUM
 b. All the leaves opposite or in whorls of 3's . . . . . . *c*
  c. Water plants, the entire plant usually submerged in water (plants sometimes on dried-up mud where water has disappeared); leaves only 1–3 mm. broad, thin when growing in water, the stem and leaves quickly shrinking and collapsing out of water . . . . . . 1. PEPLIS
  c. Most of plant growing out of water or land plants completely out of water; larger leaves (1.5–) 3–40 mm. broad, firm, the stem and leaves not collapsing when uprooted . . . . . . *d*
   d. Calyx with 5, 6, or 7 teeth or lobes; short fine hairs on some part of stems, leaves, flower-stalks or calyx . . . . . . *e*
    e. Stems woody or half-shrubby, usually spongy at base; leaf-blades narrowed or tapering at the base; calyx-tube globe-shaped or hemispheric, about as broad as long; plants of sink-hole ponds in Reynolds County . . . . . . 4. DECODON

1578 Elaeagnus angustifolia (Russian Olive)

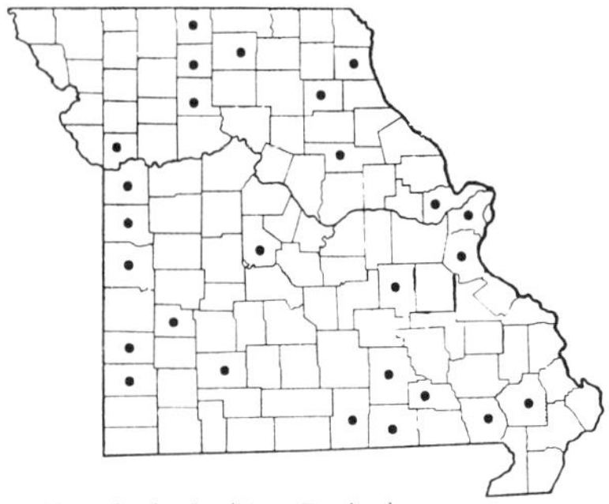

1579 Peplis diandra (Water Purslane)

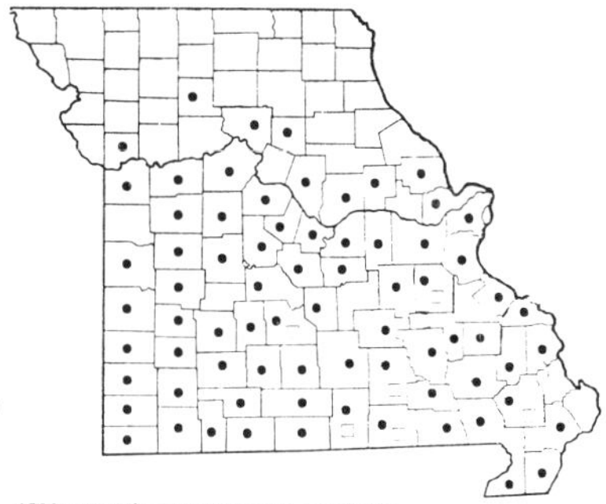

1580 □ Rotala ramosior var. ramosior
1580 • Rotala ramosior var. interior

e. Stems not woody; leaf-blades broadened, rounded, or heart-shaped at base; calyx-tube tubular or cylindrical, much longer than broad; cultivated plant, rarely escaping . . 5. LYTHRUM SALICARIA

d. Calyx with 4 teeth or lobes; plants completely glabrous (without hairs) . . . . . . . . . *f*

f. Leaves more or less broadened and clasping at the base; flowers or fruits usually 3–5 clustered together (rarely 1) at the base of a leaf . . . . . . . . . . 3. AMMANNIA

f. Leaves narrowed or tapering at the base; flowers or fruits usually solitary (rarely up to 3) at the base of a leaf . . . . . . . . . . . . . . . . . . . 2. ROTALA

## 1. **Peplis** L. Water Purslane

**Peplis diandra** Nutt. Water Purslane Map 1579

Also called Water Hedge.

*Didiplis diandra* (Nutt.) Wood [BB]

Flowers May–October.

Occurs in lowland ponds, borders of upland sink-hole ponds, sloughs, and ditches, frequently in oxbow lakes in river flood plains, and in spring-fed basins. Scattered throughout Missouri.

Ranges from Florida to Texas, north to Virginia, Ohio, Indiana, and Wisconsin.

Completely aquatic plants have thin, limp leaves which are not narrowed at the base. Where stranded on mud or growing where the water has disappeared the same plant has leaves somewhat broader, firmer, and somewhat narrowed at each end. The latter form was known as *Didiplis diandra* f. *terrestris* (Koehne) Fassett, the submersed form as f. *aquatica* (Koehne) Fassett. As such variations may be part of the normal life history of the same plant, depending upon the depth of the water, the formal recognition of such variations does not appear justifiable.

The narrow ribbonlike leaves horizontally spreading in regular distichous fashion give an ornamental and symmetrical pattern to the plant, which is sometimes sold under the name of 'Water Hedge' for fresh-water aquaria. The uppermost leaves are often rose-red or purplish-red, especially when exposed to full sun. Ordinarily, the plants are bronze- to dull green.

## 2. **Rotala** L.

**Rotala ramosior** (L.) Koehne Tooth-cup Map 1580

Flowers June–October.

Occurs along margins of streams, spring-branches, sloughs, oxbow lakes of river flood plains, upland sink-hole ponds, ditches, moist ledges, and wet depressions in prairies and open swales.

Two variations are found in Missouri:

Larger leaves 1.5–4 (rarely –5) mm. broad; fruits 2–3.3 mm. broad, 2–4 mm. long; bractlets (at base of flowers) 0.5–1.4 mm. long; less common variety encountered . . . . a. R. RAMOSIOR var. RAMOSIOR

Larger leaves 5–10 mm. broad; fruits 3.8–4.5 (rarely only 3.2) mm. broad, 3.5–5 mm. long; bractlets 1.6–2.5 (up to 4) mm. long; common variety encountered . . . . b. R. RAMOSIOR var. INTERIOR

a. **Rotala ramosior** var. **ramosior** Map 1580

*Rotala ramosior* (L.) Koehne [G]

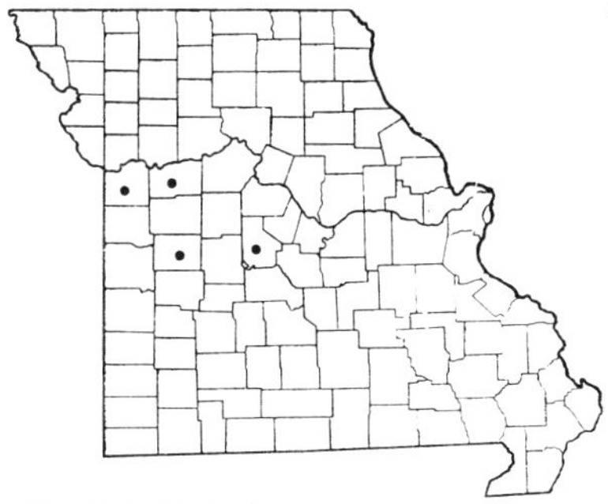

1581 Ammannia auriculata

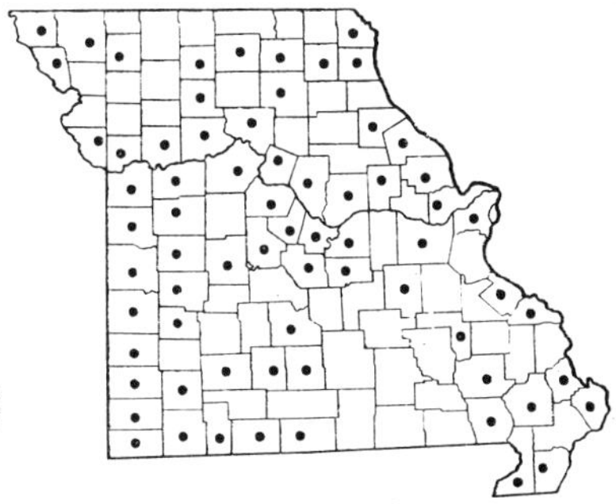

1582 Ammannia coccinea (Tooth-cup)

1583 Decodon verticillatus var. verticillatus (Swamp Loosestrife)

This is more rarely found in Missouri, known in the Ozark and Crowley Ridge sections of the state, in Stoddard, Oregon, Shannon, Howell, Washington, and Laclede counties.

Ranges from Florida to Texas, north to New Hampshire, Massachusetts, New York, and Missouri; Michigan and Indiana to Illinois and Minnesota; Washington and Oregon.

b. **Rotala ramosior** var. **interior** Fern. & Grisc. Map 1580

This is the more commonly encountered variety in Missouri, throughout the southern and central parts north to Lincoln, Montgomery, Callaway, Boone, Randolph, Chariton, Livingston, and Clay counties.

Ranges from Florida to Louisiana, north to New York, West Virginia, Ohio, Illinois, Iowa, and Kansas.

## 3. **Ammannia** L.

Flowers or fruits on stalks (pedicels) mostly 1.5–3 mm. long; peduncles (stalks supporting inflorescences) longer than the pedicels, mostly 3–5 mm. long; fruiting calyx 2–3 mm. in diameter; rare, known only from Jackson, Lafayette, Henry, and Morgan counties, western Missouri . . . 1. A. AURICULATA

Flowers or fruits without stalks or practically so; peduncles, if showing, about as long as the very short pedicels; fruiting calyx 3–5 mm. in diameter; common throughout Missouri . . . . . 2. A. COCCINEA

1. **Ammannia auriculata** Willd. Map 1581

Flowers July–October.

Known only from west-central Missouri, in Morgan (*Steyermark 64925*), Lafayette (*Steyermark 24739*), Henry (*Steyermark 7501*), and Jackson (Grain Valley, September 25, 1898, *Mackenzie 356*) counties.

Ranges from Mississippi to New Mexico and Mexico, north to Indiana, Nebraska, and South Dakota; tropical America, Asia, Africa, and Australia.

The American plants are distinguished as *A. auriculata* var. *arenaria* (HBK.) Koehne, f. *brasiliensis* (St. Hil.) Koehne.

The specimens above cited are the only ones, in all the material examined from Missouri, which approach *A. auriculata*. They have some of the flowers and fruits (especially the lower ones) on peduncles up to 4–5 mm. long, but most of the upper flowers and fruits are sessile or nearly so. The capsules are mostly 3–4 mm. in diameter in these specimens, and are, therefore, broader than typical of *A. auriculata*. Until a more detailed study of the genus has been made, the above-cited specimens are referred to this species.

Other specimens previously assigned to *A. auriculata* in Palmer & Steyermark's *Annotated Catalogue* show sessile or nearly sessile inflorescences and the fruiting calyx 3–5 mm. in diameter. These specimens (*Bush 706* from Swan, Taney Co.; *Palmer 3822* from Carthage, Jasper Co.; *Hoffmann* from Courtney, Jackson Co.; and *Bush 245* from Grain Valley, Jackson Co.) are referred in the present flora to *A. coccinea*.

2. **Ammannia coccinea** Rothb. Tooth-Cup Map 1582

Flowers June–October.

Occurs on muddy margins of ponds, sloughs, ditches, and slow streams, and swales along railroads. Throughout Missouri, but not recorded from some sections of the Ozarks and the northernmost counties of the state.

Ranges from Florida to Texas and Mexico, north to Ohio, Illinois, Minnesota, Nebraska, Montana, and Washington; also in tropical America south to Brazil.

Plants which become submerged in ponds and lake basins, such as *Steyermark 28073* from Gravois Mills Spring, Morgan County, often develop a spongy-thickened epidermis.

One of the characters sometimes used to distinguish this species from *A. auriculata* is the relative length of the capsule and fruiting calyx, the capsule in *A. auriculata* often stated to be longer than the calyx and in *A. coccinea* equaling or shorter than the calyx. This character has been found to break down, as many specimens of *A. coccinea* have fruit longer than the calyx.

## 4. **Decodon** J. F. Gmel. Swamp Loosestrife

**Decodon verticillatus** (L.) Ell. var. **verticillatus**
Swamp Loosestrife Map 1583
Also called Water Willow.
*Decodon verticillatus* (L.) Ell. [G]
Flowers July–September.

Known only from upland sink-hole ponds in Reynolds County, southeastern Ozarks (Bowles Pond, in upland tributary to Harrison Valley, T31N, R1E, sect. 23, 2½ mi. northwest of Redford, May 30, 1947, *Steyermark 64404*; Bowles Pond, September 5, 1949, *Steyermark 69256*; Lily Pond, T31N, R1E, northwest ¼ sect. 23, south of Vinson Hol, 2¾ mi. northwest of Centerville, 7 mi. southeast of Centerville, September 5, 1949, *Steyermark 69241*).

Ranges from Florida to Louisiana, north to Maine, New Hampshire, Massachusetts, New York, Ontario, Illinois, and Missouri.

At the two ponds where the plant occurs, it is associated with *Potamogeton epihydrus* var. *Nuttallii*, which is known nowhere else in the state, and with many species, such as *Hottonia inflata*, *Hypericum Walteri*, *Bidens discoidea*, and *Rosa palustris*, elsewhere known in the lowland swamps of the southeastern part of the state, and with such characteristic sink-hole pond plants as *Glyceria acutiflora* and *Carex decomposita.*

The submerged part of the stem has a spongy-thickened bark. The plant has a habit of rooting at the tips with its arching branches, thus spreading by the rooting of the new stems. The leaves are either in pairs or in whorls of threes. From the axils of the uppermost leaves appear clusters of pink-purple flowers 10–15 mm. long. The seeds of the plant are sometimes eaten by ducks, and the stems sometimes by muskrats.

## 5. **Lythrum** L. Loosestrife

At least some of the upper leaves of the branches alternate; plant glabrous (without hairs); common native species, throughout Missouri . . . . . . . . . . . . . . . . . . . 1. L. ALATUM
All the leaves opposite; hairs present on stems, leaves, and/or calyx; introduced cultivated plant, rarely encountered away from cultivation . . . . . . . . . . . . . . . . . 2. L. SALICARIA

1. **Lythrum alatum** Pursh Winged Loosestrife
Map 1584
*Lythrum dacotanum* Nieuwland [Shinners]
Flowers early June–September.

Occurs in swampy meadows, wet prairies, wet places on rocky open glades, swaley ground along railroads, and margins of streams, ponds, and ditches. Throughout Missouri, apparently absent from most of the lowland counties of southeastern Missouri.

Ranges from Ontario and New York to South Dakota and British Columbia, south to Georgia, Louisiana, and Texas; introduced in New England, New Jersey, and other eastern states.

The leaves may be opposite or in whorls of threes in the lower half of the stem, and alternate on the upper part of the branches. The petals are purple.

Shinners (Field & Lab. 21: 86. 1953) regards true *L. alatum* as a species of the lowland section of Georgia, and the common midwestern prairie species as *L. dacotanum* Nieuwland (Am. Midl. Nat. 3: 266–67. 1914).

Plate no. 260. **1.** Elaeagnus angustifolia, × 2/7; *a.* Scale, highly magnified. **2.** Peplis diandra, × 2/7; Details from Small, The New York Botanical Garden. **3.** Rotala ramosior, × 2/7; Details from Small, The New York Botanical Garden. **4.** Ammannia coccinea, × 2/7. **5.** Ammannia auriculata, × 2/7; Details from Small, The New York Botanical Garden. **6.** Decodon verticillatus var. verticillatus, × 2/7; Details from Small, The New York Botanical Garden. **7.** Lythrum alatum, × 2/7; *a.* Flower, × 1 3/7. **8.** Lythrum Salicaria, × 2/7. **9.** Cuphea petiolata, × 2/7; Details from Small, The New York Botanical Garden. **10.** Rhexia mariana var. mariana, × 2/7; After Britton and Brown, The New York Botanical Garden. **11.** Jussiaea decurrens, × 2/7; After Britton and Brown, The New York Botanical Garden. **12.** Rhexia virginica var. virginica, × 2/7; Details from Small, The New York Botanical Garden.

PLATE NO. 260

1584 Lythrum alatum (Winged Loosestrife)

1585 Lythrum Salicaria var. Salicaria (Purple Loosestrife)

1586 Cuphea petiolata (Clammy Cuphea)

2. **Lythrum Salicaria** L. var. **Salicaria** Purple Loosestrife Map 1585

Also called Spiked Loosestrife.

*Lythrum Salicaria* L. [G, P & S]

Flowers early June–September.

Planted as an ornamental garden species, rarely escaping from cultivation in the state, where known from Franklin and Newton (introduced and spreading along banks of spring-fed pond, near Saginaw, August 6, 1952, *Palmer 54649*) counties.

Native of Europe; introduced and naturalized in North America from Newfoundland and Quebec to Minnesota, south to Virginia, West Virginia, Ohio, Indiana, and Missouri. In many eastern and northern states the plant is abundant in streams, forming dense colonies and often choking out other plants.

Several garden races, some with larger rose-red petals, have been produced from this species and are cultivated.

### 6. **Cuphea** P. Br.

**Cuphea petiolata** (L.) Koehne Clammy Cuphea Map 1586

Also called Blue Waxweed.

Flowers July–October.

Occurs in open woods, thickets, clearings, fallow fields, prairies, pastures, glades, and roadsides. Common in southern and central Missouri north to Marion, Randolph, Saline, Livingston, Ray, and Clay counties, and locally northwest in Nodaway and Atchison counties; absent from the lowland counties of southeastern Missouri.

Ranges from Georgia to Louisiana, north to New Hampshire, New York, Ohio, Indiana, Illinois, Iowa, Kansas, and Oklahoma.

The petals are reddish-purple and the calyx is usually purple-nerved. The stems are usually magenta or purple or have hairs of these colors.

## Fam. **MELASTOMATACEAE** (Melastoma Family)

(*Melastomaceae* [BB, P & S])

### **Rhexia** L. Meadow Beauty

a. Stems without any winged angles or ridges, more or less terete or with two rounded or curved sides broader than and alternating with two flat or concave narrower sides; petals glabrous (without hairs) . . . . . . . . 1. R. MARIANA

a. Stems with winged angles or ridges at least in the upper half; petals bearing few to many gland-tipped hairs . . . . . . . . *b*

  b. Stems with conspicuous wings 0.25–2 mm. broad; roots thickened bearing 1–several tuberous parts (often not collected or seen on dried specimens); leaves frequently broadest near the base; upper leaves usually ovate or broadly lanceolate, usually $1\frac{1}{2}$–$2\frac{1}{2}$ (to $3\frac{1}{2}$) times as long as broad; plants of the Ozark and southeastern section north to Franklin, Laclede, and Jasper counties . 2. R. VIRGINICA

  b. Stems without prominent wings less than 0.25 mm. broad; roots not tuberous; leaves frequently broadest near the middle; upper leaves narrowly elliptic to lanceolate, $2\frac{1}{2}$–$3\frac{1}{2}$ times as long as broad; plants of the unglaciated prairie region of southwestern Missouri . . . . . . 3. R. INTERIOR

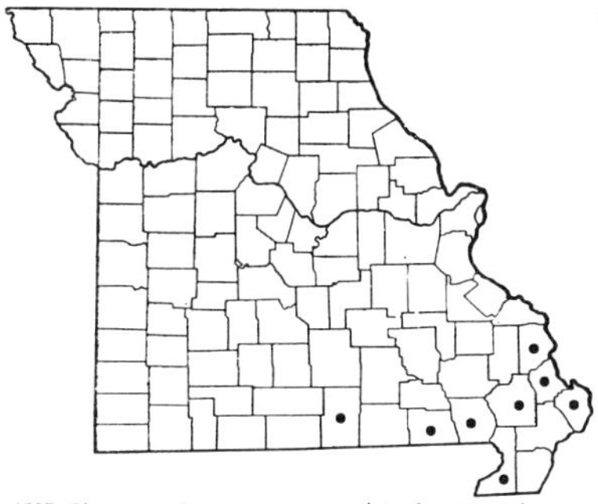

1587 Rhexia mariana var. mariana (Meadow Beauty)

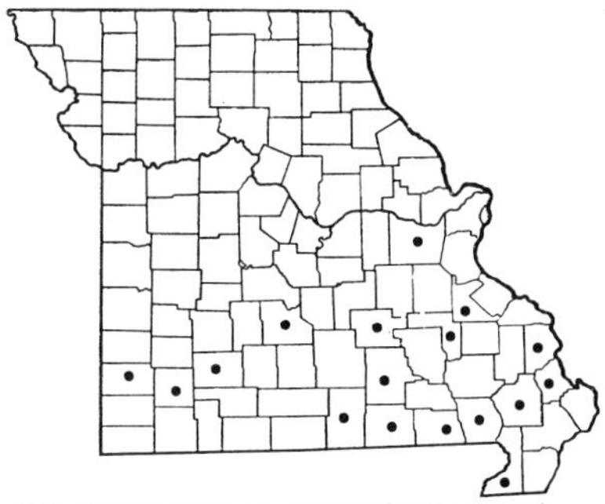

1588 Rhexia virginica var. virginica (Meadow Beauty)

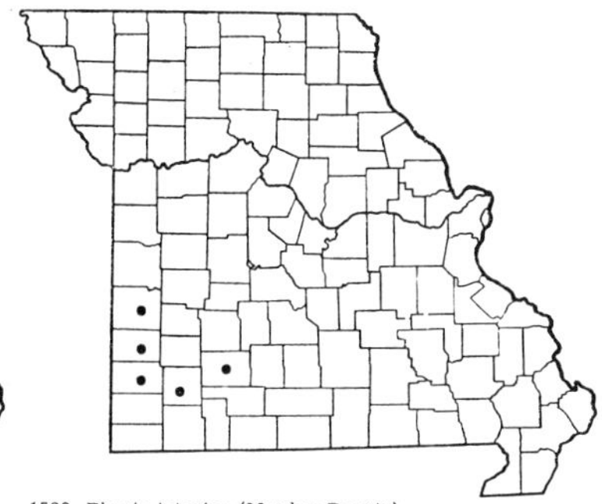

1589 Rhexia interior (Meadow Beauty)

1. **Rhexia mariana** L. var. **mariana** Meadow Beauty Map 1587

*Rhexia mariana* var. *leiosperma* Fern. & Grisc. [G, P & S]

Flowers June–October.

Occurs in open swaley depressions in upland swampy prairie or lowland sandy prairie, low wet woods, along ditches, sandy wet ground along roadsides, and wet depressions along railroads, in acid soils. Southeastern Missouri in the lowlands north to Cape Girardeau, Stoddard, and Butler counties, west to Ripley and Howell counties.

Ranges from Georgia, Alabama, Mississippi, Louisiana, and Texas, north to Massachusetts, New York, New Jersey, Pennsylvania, Maryland, Virginia, Kentucky, Indiana, Illinois, Missouri, and Oklahoma.

2. **Rhexia virginica** L. var. **virginica** Meadow Beauty Map 1588

*Rhexia virginica* L. [G. BB, P & S]

Flowers June–October.

Occurs in sphagnum-covered swampy open meadows in valleys, wet sedge-covered swales, wet meadows, margins of upland sink-hole ponds, along wet ditches and sandy open wet ground along roadsides and railroads, and wet depressions on sandstone in acid soils. Ozark region north to Franklin, Laclede, Greene, and Jasper counties, and Crowley Ridge and lowland section southeast to Scott, Stoddard, and Dunklin counties.

Ranges from Nova Scotia to New York, Ontario, and Wisconsin, south to Georgia, Alabama, Mississippi, Louisiana, and Texas, west to Iowa, Missouri, and Kansas.

The leaves vary from 1–5 cm. broad.

As in the other species of *Rhexia*, the flowers have large showy petals. Those in *R. virginica* are deep rose-red. The plants are always found in wet acid soils and must be grown in a similar type of soil. Plants transplanted to the author's wildflower garden in northern Illinois have succeeded in an open wet prairie situation where the soil has a sand-peat moss mixture.

3. **Rhexia interior** Pennell Meadow Beauty Map 1589

*Rhexia latifolia* Bush, not Aubl.

Flowers June–October.

Occurs in wet places in upland and sandy prairies, moist sandy soil in ditches, and along sandy stream banks, in acid soils. Southwestern Missouri in the unglaciated prairie section in Vernon, Barton, Jasper, Lawrence, and Greene counties.

Ranges from Missouri, Kansas, Arkansas, and Oklahoma, south to Louisiana and Texas, and locally east in Indiana, Kentucky, and Tennessee.

The species was originally described from specimens collected from moist sandy prairies, 4 mi. north of Alba in Jasper County in the same locality where the rare *Geocarpon minimum* Mackenzie was first discovered.

## Fam. **ONAGRACEAE** (Evening Primrose Family)

(In Texas called Texas Buttercup Family)

a. Leaves opposite on the stem . . . . . . . . . . *b*
  b. Leaves without teeth . . . . . . . . . . 2. LUDWIGIA
  b. Leaves with teeth on margins . . . . . . . . . . *c*
    c. Leaf-stalks (petioles) conspicuous, 10–50 mm. long; leaf-blades oblong-ovate, larger ones 30–60 mm. broad; plants of dry or rich woodland. . . . . . . . . . 7. CIRCAEA

c. Leaf-stalks (petioles) very short, not conspicuous, at most 5 mm. long; leaf-blades narrowly lanceolate or oblong-lanceolate, the larger ones 2–25 mm. broad; plants of swampy meadows, bogs, and wet ground along spring branches and margins of ponds . . . . . . . 3. EPILOBIUM

a. Leaves alternate on the stem, or all from the base of the plant . . . . . . . . . . . . . . *d*

d. Calyx-tube not prolonged beyond the ovary . . . . . . . . . . . . . . . . . . . *e*

e. Petals pink to whitish; calyx deciduous (falling) from the fruit; seeds with a silky tuft at one end. . . . . . . . . . . . . . . . . . . . . . 3. EPILOBIUM

e. Petals yellow or greenish; calyx persistent and remaining attached to the fruit; seeds without any silky tuft . . . . . . . . . . . . . . . . . . . . . . . *f*

f. Stamens 4; fruit less than 10 mm. long, mostly as broad as long or slightly longer than broad, rarely cylindrical and elongated . . . . . . . . . . . . . . 2. LUDWIGIA

f. Stamens 8–12; fruit 10–50 mm. long, narrowly cylindrical or obconical, conspicuously longer than broad . . . . . . . . . . . . . . . . . . . . . 1. JUSSIAEA

d. Calyx-tube prolonged beyond the ovary . . . . . . . . . . . . . . . . . . . *g*

g. Petals yellow . . . . . . . . . . . . . . . . . . . . . 4. OENOTHERA

g. Petals white or pink . . . . . . . . . . . . . . . . . . . . . . . *h*

h. Petals more than 1 cm. long, not abruptly narrowed at base into a claw; flowers 1–3, arising at the base of the plant, or a few at the tip of the stem; fruit eventually opening (dehiscing) along definite lines or seams, containing many seeds . . . . . . 4. OENOTHERA

h. Petals less than 1 cm. long, abruptly narrowed at base into a claw; flowers many, in a narrow slender spike or raceme; fruit not opening (indehiscent), containing 1–4 seeds . . . . *i*

i. Stem and leaves glabrous (without hairs); petals remaining white; flowers more or less regular, the petals more or less equal and about equally spreading from all sides of flower; calyx-tube thread-like (filiform), at least twice as long as ovary; fruit with 8 ribs; plants of limestone glades and rocky ledges of southwestern Missouri in Ozark and Taney counties . . . . . . . . . . . . . . . . . . 6. STENOSIPHON

i. Stem and/or leaves usually minutely or conspicuously hairy, rarely without hairs; petals rose or pink or eventually turning pink; flowers irregular, the petals somewhat unequal and turned to one side of the flower and of the stamen cluster; calyx-tube tubular or obconical, about as long as the ovary; fruit with 3–4 ribs or angles; plants in various types of habitats throughout Missouri . . . . . . . . . . . . 5. GAURA

## 1. **Jussiaea** L. Primrose Willow

a. Stems and leaves glabrous (without hairs) . . . . . . . . . . . . . . . . . . *b*

b. Stems usually out of water on land, erect with wings of green tissue running up and down the stem; petals 4; fruit 4-sided . . . . . . . . . . . . . . . . . . 1. J. DECURRENS

b. Stems usually in water, floating or creeping or horizontal, without wings of green tissue; petals 5; fruit cylindric with rounded or curved sides . . . . . . . . . . 2. J. REPENS var. GLABRESCENS

a. Stems and/or leaves hairy . . . . . . . . . . . . . . . . . . . . . . . *c*

c. Petals mostly 15–25 mm. long, much longer than the calyx-lobes; flower-stalks (pedicels) more than half as long as the calyx-tube, eventually 1–2 cm. long . . . . . . . . 3. J. URUGUAYENSIS

c. Petals 5–10 mm. long, about as long as or shorter than the calyx-lobes; flower-stalks less than half as long as the calyx-tube, usually 0.2–0.8 cm. (up to 1.5 cm.) long . . . . 4. J. LEPTOCARPA

1. **Jussiaea decurrens** (Walt.) DC. Primrose Willow Map 1590

Flowers June–October.

Occurs in wet ground of muddy and sandy gravelly soils along sloughs, streams, ponds, ditches, swamps, and bayous.

Southern Missouri, where concentrated in the lowlands of the southeastern section north to Bollinger, Wayne, and Reynolds counties, west to Ripley County, and locally in southwestern Missouri along White River in Stone and Barry counties (where now exterminated by Table Rock Dam), and locally northeast in Franklin County (shade of willows on gravel bar, Gray Summit, Missouri Botanical Garden Arboretum, August 28, 1937, *Edgar Anderson*).

Ranges from Florida to Texas, north to Virginia, West Virginia, Indiana, Illinois, Missouri, and Oklahoma; American tropics south to Argentina.

2. **Jussiaea repens** L. var. **glabrescens** Ktze. Floating Primrose Willow Map 1591

*Jussiaea diffusa* Forsk. [P & S, Steyerm.]

*Ludwigia adscendens* (L.) Hara var. *glabrescens* (Kuntze) Hara [Shinners]

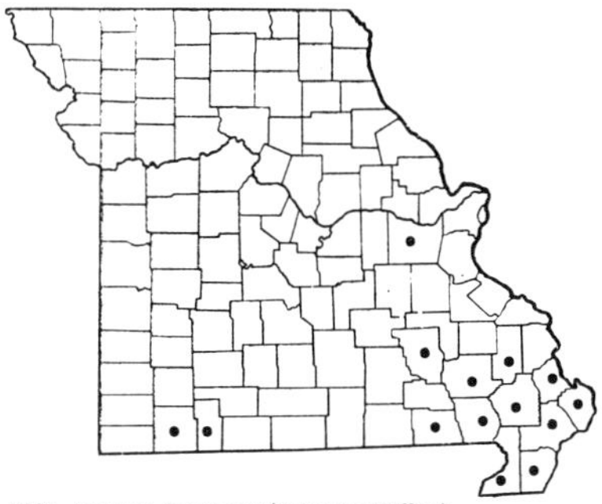

1590 Jussiaea decurrens (Primrose Willow)

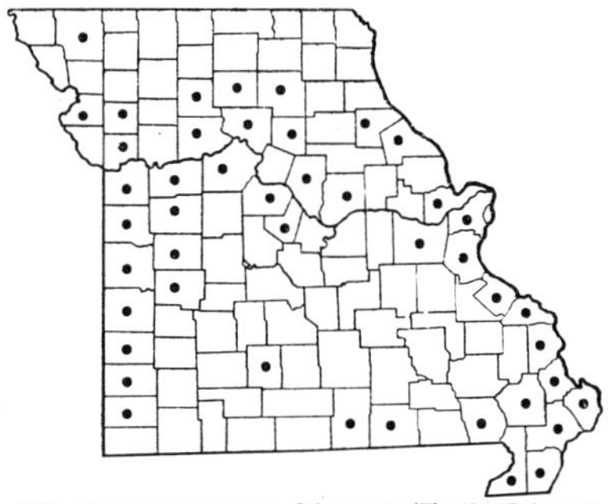

1591 Jussiaea repens var. glabrescens (Floating Primrose Willow)

1592 Jussiaea uruguayensis

Flowers late May–October.

Occurs in shallow water or on muddy margins of sloughs, slow streams, oxbow lakes in river bottoms, artificial and natural ponds, and spring branches.

Scattered over the state mainly in the counties bordering the Missouri and Mississippi rivers, the Grand, Osage, Marmaton, and other large streams in western, central, and eastern Missouri; absent from most of the Ozark section and in the counties of the extreme northern section.

Ranges from Alabama to Texas and Mexico, north to Indiana (introduced), Illinois, Missouri, Kansas, and Oklahoma; introduced in New Jersey, Pennsylvania, and Maryland.

This is a showy plant when in flower. It usually occurs in dense colonies, the soft, spongy stems creeping along the wet mud. It is often seen in artificial ponds. The seeds of this and other aquatic or semi-aquatic species are eaten sometimes by ducks.

3. **Jussiaea uruguayensis** Camb. Map 1592

Flowers August–October.

Occurs in shallow running water and wet margins of small creeks. Known only from Newton County (shallow water of Hickory Creek, Neosho, October 12, 1953, *Palmer 57068*; same locality, October 29, 1953, *Palmer 57156*).

Native of South America; introduced and naturalized from North Carolina to Texas north to New York and Missouri.

4. **Jussiaea leptocarpa** Nutt. var. **leptocarpa** Map 1593

*Jussiaea leptocarpa* Nutt. [G, BB]

Flowers August–October.

Occurs along mudflats and sloughs of the Mississippi River. Known definitely only from Pemiscot (along Mississippi River, Island no. 17, T17N, R14E, sect. 36 and 1, 3–4 mi. southeast of Caruthersville, October 23, 1948, *Steyermark 67032*) and Ste. Genevieve (between Old Ste. Genevieve and Old Kaskaskia, Mississippi River bottoms west of Kaskaskia Island and by slough separating Kaskaskia Island at northwest end from Missouri mainland, T38N, R8W, $3\frac{1}{2}$–$4\frac{1}{2}$ mi. [by air] south southeast of Ste. Genevieve, November 4, 1956, *Steyermark 83628*) counties.

Ranges from South and Central America, the West Indies, and Mexico, northward into the United States from Florida to Texas, north to Georgia, Illinois, Arkansas, Missouri, and Oklahoma.

In his 'A Revision of the New World Species of Jussiaea' Munz (Darwiniana 4: 256. 1942) cited a collection presumably made in Missouri by Frank ('in civitate Missouri, *Frank* in 1837'). The indefiniteness of the locality threw some doubt on the actual occurrence of the plant being in Missouri, but recent discoveries of it in Missouri have confirmed Frank's collection.

## 2. **Ludwigia** L. False Loosestrife

a. Leaves opposite . . . . . . . . . . . . . . . . . . . . . . . . . . . . *b*

b. Common throughout Missouri; calyx-tube 2–4.5 mm. long with 4 green bands extending the length of the tube; bractlets, when present, minute and at the very bases of the calyx-tube; wall of fruit thickened and corky below each calyx-lobe, thin and rather translucent below each sinus (gap between calyx-lobes) . . . . . . . . . . . . . . . . . . 6. L. PALUSTRIS

b. Rarely encountered species, known only from Greene County, southwestern Missouri; calyx-tube

1593 Jussiaea leptocarpa var. leptocarpa

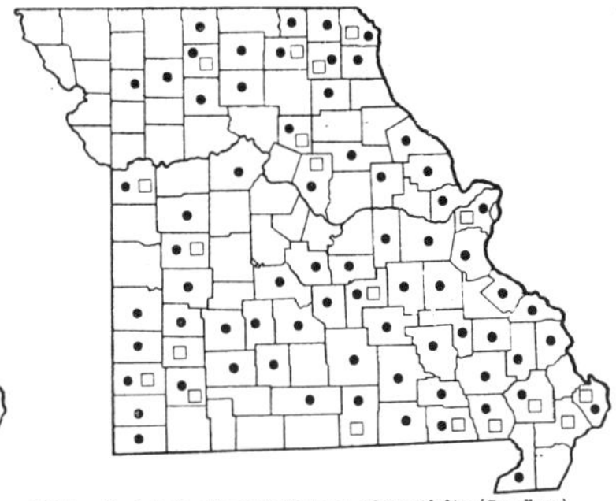

1594 ● Ludwigia alternifolia var. alternifolia (Seedbox)
1594 □ Ludwigia alternifolia var. pubescens

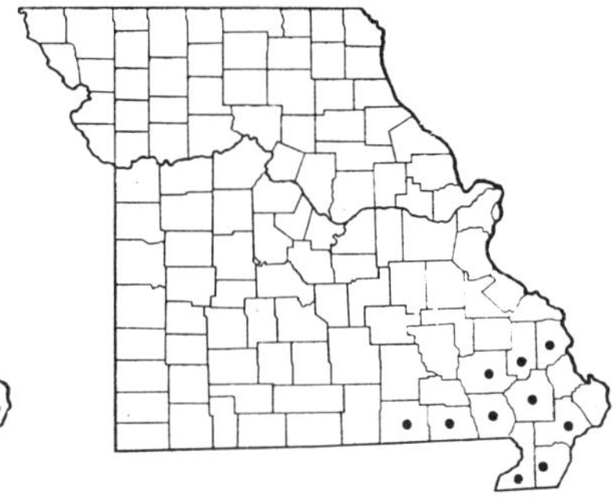

1595 Ludwigia glandulosa

4-6 mm. long without green bands; 1 or more bractlets arising some distance above the base of the calyx-tube; wall of fruit uniformly thickened and corky . . . . . . . . . . 5. L. NATANS

a. Leaves alternate . . . . . . . . . . . . . . . . . . . . . . . . . . . *c*

c. Petals yellow, conspicuous, 7–8 mm. long; mature capsules 4–7 mm. wide (slightly wider when crushed and pressed), 4-sided; flower-stalks noticeable, 3–7 mm. long, longer than the ovary 1. L. ALTERNIFOLIA

c. Petals greenish, minute or absent; mature capsules 1.5–5 mm. wide, rounded or curved on the sides, not 4-angled; flower-stalks absent or 1 mm. or less, very inconspicuous, much shorter than the ovary . . . . . . . . . . . . . . . . . . . . . . . . . *d*

d. Main leaves of the stem 0.5–2.5 cm. long; calyx-tube 1–2 mm. long, shorter than the calyx-lobes . . . . . . . . . . . . . . . . . . . . 3. L. MICROCARPA

d. Main leaves of the stem 3–10 cm. long; calyx-tube in fruit 4–8 mm. long, longer than the calyx-lobes; calyx-tube in flower 3–4 mm. long . . . . . . . . . . . . . *e*

e. Plants of southeastern Missouri swamps and sink-hole ponds west to Oregon County, north to Cape Girardeau and Bollinger counties; stems and branches lacking wings on the angles; fruit about 3–4 times as long as broad, 1.5–2.3 mm. in diameter . . 2. L. GLANDULOSA

e. Plants of the northern half of Missouri, also in western and part of eastern Missouri, but absent from the southeastern lowlands; stems and branches with angles winged; fruit slightly longer than broad, 1½–2 times as long as broad, 3–5 mm. in diameter . 4. L. POLYCARPA

1. **Ludwigia alternifolia** L. Seedbox Map 1594

Flowers June–August.

Occurs along borders of streams and ponds, wet meadows, wet places in prairies and depressions along railroads, low wet woodland, fallow fields, and ditches.

Two variations occur in Missouri:

Plants glabrous (without hairs) or nearly so . . 1a. L. ALTERNIFOLIA var. ALTERNIFOLIA

Summit of stem, leaves and calyx more or less hairy . . . 1b. L. ALTERNIFOLIA var. PUBESCENS

1a. **Ludwigia alternifolia** var. **alternifolia** Map 1594

Throughout Missouri. This is the commoner variation in the state.

Ranges from Florida to Texas, north to Massachusetts, New York, Ontario, Michigan, Illinois, Iowa, and Kansas.

1b. **Ludwigia alternifolia** var. **pubescens** Palmer & Steyerm. Map 1594

Scattered throughout Missouri, but less common than the typical variety. Originally described in Missouri from a collection growing in a sandy prairie in Mississippi County, but since found in many other habitats.

Ranges from Louisiana and Texas, north to Indiana, Illinois, Missouri, Kansas, and Oklahoma.

The erect stems range between 5–10 dm. in height.

2. **Ludwigia glandulosa** Walt. Map 1595

Flowers June–September.

---

Plate no. 261. **1.** Jussiaea repens var. glabrescens, × 2/5; Details from Small, The New York Botanical Garden. **2.** Ludwigia microcarpa, × 2/5; After Britton and Brown, The New York Botanical Garden. **3.** Ludwigia glandulosa, × 2/5; *a.* Fruit, × 2 4/5; After Gleason, details from Small, The New York Botanical Garden. **4.** Ludwigia polycarpa, × 2/5. **5.** Jussiaea uruguayensis, × 2/5; After Gleason, details from Small, The New York Botanical Garden. **6.** Ludwigia alternifolia, × 2/5.

PLATE NO. 261

1596 Ludwigia microcarpa

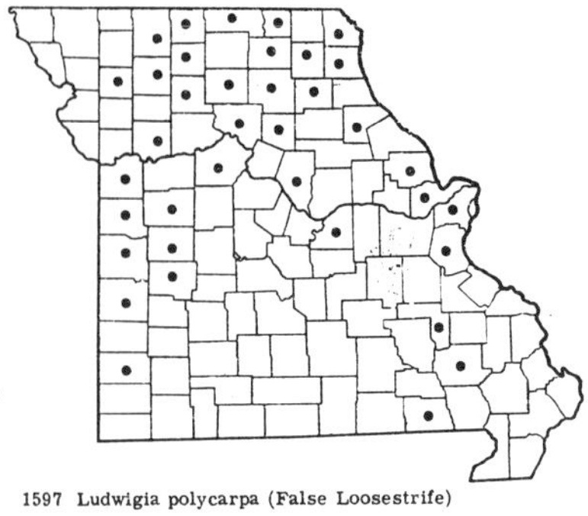
1597 Ludwigia polycarpa (False Loosestrife)

1598 Ludwigia natans

Occurs in low wet woods, bald cypress and tupelo swamps, bayous, wet ditches along roadsides, and margins of upland sink-hole ponds. Southeastern Missouri lowlands north to Cape Girardeau, Bollinger, and Wayne counties, and west in upland sink-hole ponds in the Ozarks to Oregon County.

Ranges from Florida to Texas, north to Virginia, Indiana, Illinois, Missouri, and Kansas.

This species becomes much branched and may attain a height of 1.5 meters. The bases of the stems are often enlarged and have a spongy epidermis when growing in water.

3. **Ludwigia microcarpa** Michx. Map 1596

Flowers July–October.

Occurs in swampy meadows along spring branches. Known only from Oregon County, southeastern Ozarks (swampy meadow bordering 'Hatcher's Spring' along spring branch below Greer Spring, 1 mi. north of Greer, August 26, 1939, *Steyermark 27987*).

Ranges from Florida to Louisiana, north to North Carolina, Tennessee, and Missouri; also in the West Indies.

This southern species was associated with *Panicum agrostoides*, *Eleocharis calva*, *Fuirena simplex*, *Parnassia grandifolia*, *Galium tinctorium*, and *Eupatorium perfoliatum*.

4. **Ludwigia polycarpa** Short & Peter False Loosestrife Map 1597

Flowers June–September.

Occurs in low wet woods, swamps, borders of sloughs, ditches, ponds, and oxbow lakes in river flood plains, wet depressions in prairies and meadows, and along streams in prairies. Common in much of the northern half of the state in the unglaciated prairie section of western Missouri south of the Missouri River, and in part of eastern Missouri; absent from most of the Ozark region, southeastern lowland counties, and extreme northwestern section.

Ranges from Maine to Connecticut; Ontario and Ohio to Minnesota, south to Tennessee, Missouri, and Kansas.

The plant produces leafy offshoots at the base with closely overlapping leaves which remain attached to the plant throughout the late fall and winter and give it a characteristic appearance. The base of the plant is often enlarged and somewhat spongy-thickened.

5. **Ludwigia natans** Ell. Map 1598

Flowers June–September.

Known only from Greene County, southwestern Missouri.

Ranges from Florida to Texas, north to North Carolina, Tennessee, and Missouri.

6. **Ludwigia palustris** (L.) Ell. Water Purslane Map 1599

Flowers late May (May 20)–September.

Occurs in shallow water and muddy margins of natural upland sink-hole and lowland oxbow lakes, artificial ponds, sloughs, slow streams, swamps, ditches, and spring branches. Throughout most of Missouri except absent from northwestern and other northern counties.

Two variations occur in Missouri:

Common variation usually found; fruits 1.8–3.5 mm. broad, 2–4.5 mm. long; calyx-lobes broadly triangular, short-pointed . . . 6a. L. PALUSTRIS var. AMERICANA

Rarely encountered, known only from Stoddard Co., southeastern Missouri; fruits 1.4–2 mm. broad, 2–3 mm. long; calyx-lobes triangular to broadly lanceolate, long-pointed (acuminate) . 6b. L. PALUSTRIS var. NANA

6a. **Ludwigia palustris** var. **americana** (DC.) Fern. & Grisc. Map 1599

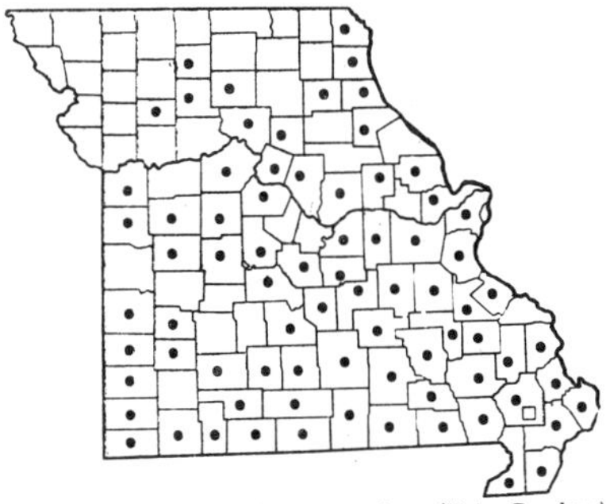

1599 ● Ludwigia palustris var. americana (Water Purslane)
1599 □ Ludwigia palustris var. nana

1600 Epilobium leptophyllum

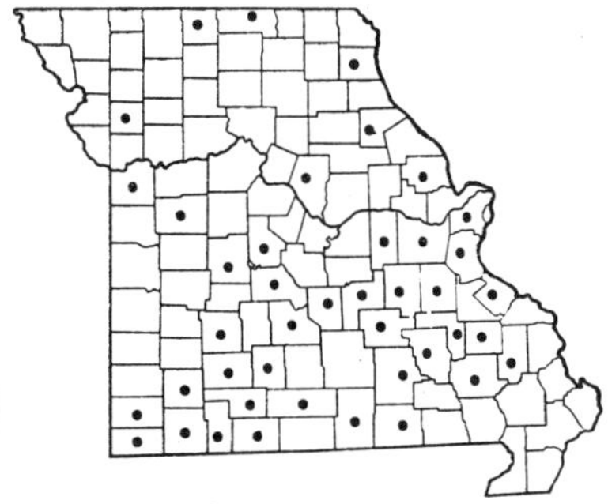

1601 Epilobium coloratum

This is the common variety found with range in Missouri as given above.

Ranges from Nova Scotia to Minnesota and Oregon, south to Florida, Louisiana, Texas, Mexico, and Guatemala; also in Bermuda.

6b. **Ludwigia palustris** var. **nana** Fern. & Grisc. Map 1599

Known only from Stoddard County (open sandy ground in valley of spring branch, Crowley Ridge, near junction with bottomland, T25N, R10E, SE¼ sect. 11, 4–4½ mi. [by air] south of Bloomfield, 2 mi. northeast of Dexter, October 18, 1955, *Steyermark 80436*).

Water Purslane grows either submerged in running water or may withstand drying out in exsiccated ponds and stream borders. The foliage varies from grass- or olive-green to tinged with red or rose-purple, and in size varies considerably, becoming quite small when growing in dry situations. The petals vary from greenish to purplish or reddish. This species is often grown as an aquarium plant. The base of the stems is often eaten by muskrats.

## 3. **Epilobium** L. Willow Herb

Leaves 1–4 mm. wide, linear or linear-lanceolate, mostly entire (without teeth), minutely hairy on upper surface; stem lacking decurrent lines running from the base of each leaf, terete (round in cross-section): rarely encountered, known only in Jackson County, west-central Missouri . . 1. E. LEPTOPHYLLUM

Larger leaves 5–25 mm. wide, lanceolate, closely and rather regularly toothed, glabrous (without hairs) on upper surface; stem with decurrent lines starting from the base of each leaf, faintly or obscurely 4-angled; commonly collected, ranging throughout Missouri . . . . . . . . . . 2. E. COLORATUM

1. **Epilobium leptophyllum** Raf. Map 1600

*Epilobium densum* of auth. [P & S], not Raf.

Flowers July–September.

Occurs in wet ground along streams and in bogs.

Known only from Jackson County, west-central Missouri.

Ranges from Quebec to Alberta, south to Virginia, West Virginia, Ohio, Indiana, Illinois, Missouri, Kansas, and Colorado.

2. **Epilobium coloratum** Biehler Map 1601

Flowers July–October.

Occurs in swampy meadows, calcareous spring branches, and wet ground along streams and ditches. Commonest in the Ozarks of southern and east-central Missouri, north locally to Lewis, Putnam, Mercer, and Clinton counties.

Ranges from Quebec to Minnesota and South Dakota, south to Georgia, Alabama, Tennessee, Arkansas, Kansas, and Oklahoma.

The petals, usually pink, are 3–5 mm. long.

### 4. **Oenothera** L. Evening Primrose

In Texas known as Texas Buttercup

a. Leaves and flower-stalks all arising at the base of the plant, the plant stemless . . . . 14. O. TRILOBA
a. Leaves and flowers all arising from the main stem or its branches, the plant with a stem . . . . . *b*
  b. Leaves all thread-like (filiform), 0.5–1 mm. broad . . . . . . . . . . . . 6. O. LINIFOLIA
  b. Leaves all broader . . . . . . . . . . . . . . . . . . . . . . . *c*
    c. Petals white to rose-colored or pink . . . . . . . . . . . . . . . . . *d*
      d. Flower-buds drooping or nodding; petals 2–4 cm. long . . . . . . . . . 11. O. SPECIOSA
      d. Flower-buds erect or upright; petals 1–2 cm. long . . . . . . . . . . 12. O. KUNTHIANA
    c. Petals yellow . . . . . . . . . . . . . . . . . . . . . . . *e*
      e. Stems sprawling or spreading for most of their length; flowers very large, the petals 3–7 cm. long; fruit 5–8 cm. long with 4 broad wings 1–2.5 cm. wide; plants of limestone glades of the Ozark region . . . . . . . . . . . . . . . . 13. O. MISSOURIENSIS
      e. Stem erect or spreading; flowers smaller, the petals 0.5–3.5 cm. long; fruit rounded on the sides, ridged, or angled, but not winged, 0.4–3.5 cm. long . . . . . . . . . . *f*
        f. At least some of the leaves sinuate-pinnatifid, coarsely toothed or deeply cut toward the midrib, rarely nearly smooth-edged (subentire). . . . . . . . . . 4. O. LACINIATA
        f. Leaves either smooth-edged (entire) or with few or many fine or short teeth, but none of the leaves sinuate-pinnatifid, coarsely toothed or deeply cut toward the midrib . . . . *g*
          g. Ovary terete (with rounded or curved sides or rounded in cross-section) or nearly so; fruit cylindrical, mostly terete, if 4-sided the angles rounded . . . . . . . . . . *h*
            h. All the leaves narrowed at the base, and the uppermost narrowest at the base; leaves 1.5–5 mm. broad, with usually 4–6 conspicuous or inconspicuous slender, long, or short teeth per cm., rarely without teeth . . . . . . 5. O. SERRULATA
            h. Uppermost leaves broadest at the base and/or clasping or rounded at base; main leaves from middle and lower part of stem 5–40 mm. broad, with usually entire (without teeth) or with 1–4 inconspicuous short broad or triangular teeth per cm. . . *i*
              i. Petals 4–6 cm. long; calyx-lobes 2.2–5 cm. long . . . . . . 2. O. GRANDIFLORA
              i. Petals 1.5–2.5 cm. long; calyx-lobes 1–2.5 cm. long . . . . . . . . . . *j*
                j. Main leaves from middle part of stem linear or linear-lanceolate, not much if any broader in the middle than at the base, mainly 5–10 mm. broad; flowering spike dense, narrow and elongated, with 15–50 or more flowers crowded between the bracts; bracts of flowers and fruits rather uniform, closely crowded, overlapping, with 6–10 or more flowers or 3–4 fruits per 2.5 cm. of axis (rachis) of inflorescence; fruit linear, uniformly narrow its entire length, the mature fruits 2–3 mm. broad; seeds not angled . 3. O. RHOMBIPETALA
                j. Main leaves from middle part of stem elliptic or oblong-lanceolate, mostly broadest around the middle and narrowed or tapering at base, mainly 10–40 mm. broad; flowering spike less closely flowered, with mainly 3–20 flowers scattered between the bracts; bracts of flowers and fruits more widely separated, with 1–4 flowers or 1–3 fruits per 2.5 cm. of axis (rachis) of inflorescence; fruit thickest usually near the base, tapering to the tip, mature fruits 4–6 mm. broad; seeds sharply angled . . . . . 1. O. BIENNIS
          g. Ovary 4-angled; fruit sharply 4-angled . . . . . . . . . . . . . . . *k*
            k. Petals 5–9 mm. long; buds nodding or drooping; inflorescence nodding; anthers 1.5–2.6 mm. long; calyx-tube 4–8 mm. long, the lobes 5–8 mm. long . . 10. O. PERENNIS
            k. Petals 10–30 mm. long; buds erect; inflorescence erect; anthers 4–8 mm. long; calyx-tube 5–25 mm. long, the lobes 5–20 mm. long . . . . . . . . . . . *l*
              l. Stems with noticeable long spreading hairs; fruit without knob-tipped (glandular) hairs . . . . . . . . . . . . . . . . . 7. O. PILOSELLA
              l. Stems either with short closely appressed or short spreading hairs; fruit with or without knob-tipped (glandular) hairs . . . . . . . . . . . . . *m*
                m. Stems with short closely appressed hairs; fruit without knob-tipped hairs 8. O. FRUTICOSA
                m. Stems with short spreading hairs; fruit with knob-tipped hairs . 9. O. TETRAGONA

1. **Oenothera biennis** L. Evening Primrose
Map 1602

Flowers June 1–October.

Occurs in fields, prairies, glades, thickets, waste

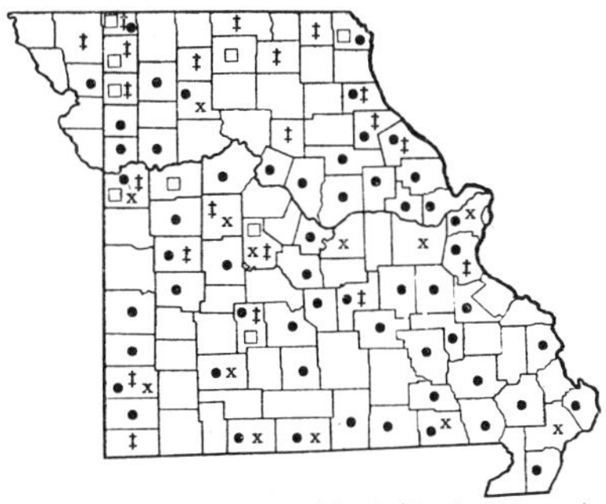

1602 • Oenothera biennis var. biennis (Evening Primrose)
1602 x Oenothera biennis var. pycnocarpa
1602 □ Oenothera biennis var. canescens
1602 ‡ Oenothera biennis var. hirsutissima

1603 Oenothera grandiflora

1604 Oenothera rhombipetala

and cultivated ground, and along roadsides and railroads.

The following variations are found in Missouri:

a. Hairs few and scattered or absent on surface of calyx-lobes, ovaries, and fruits; leaves membranaceous, soft-hairy on lower surface, spreading or loosely ascending . . . . . *b*
  b. Bracts of the inflorescence falling by fruiting time, leaving the fruiting spike without bracts . . . 1a. O. BIENNIS var. BIENNIS
  b. Bracts of the inflorescence remaining attached and persistent in fruit . . . . 1b. O. BIENNIS var. PYCNOCARPA
a. Hairs abundant and conspicuous on surface of calyx-lobes, ovaries, and fruits, producing a grayish or whitish covering; leaves firm, strongly ascending . . . . . . . . . *c*
  c. Hairs of calyx and of fruit short and closely appressed (lying flat against surface) . . 1c. O. BIENNIS var. CANESCENS
  c. Hairs of calyx and of fruit long and loosely ascending to spreading . . . 1d. O. BIENNIS var. HIRSUTISSIMA

1a. **Oenothera biennis** var. **biennis** Map 1602

Occurs throughout Missouri.

Ranges from Quebec to Alberta, south to Florida, Tennessee, Arkansas, North Dakota, and Idaho.

1b. **Oenothera biennis** var. **pycnocarpa** (Atkinson & Bartlett) Wieg. Map 1602

*Oenothera pycnocarpa* Atkinson & Bartlett [Gates]

Scattered throughout Missouri.

Ranges from New England to Minnesota, south to Florida, Indiana, Arkansas, and Oklahoma.

1c. **Oenothera biennis** var. **canescens** T. & G. Map 1602

Scattered in the northern and western half of Missouri.

Ranges from Quebec to Alberta, south to New York, Ontario, Ohio, Illinois, Missouri, and Oklahoma.

1d. **Oenothera biennis** var. **hirsutissima** Gray Map 1602

*Oenothera strigosa* (Rydb.) Mackenz. & Bush [P & S, BB]

Occurs mainly in the northern and western half of Missouri, south and east to Jefferson, Phelps, and Dallas counties; absent from most of the Ozark region.

Ranges from Quebec to British Columbia, south to New Jersey, Pennsylvania, Michigan, Illinois, Kansas, Texas, New Mexico, Arizona, and Mexico.

The specimens previously classified as *O. parviflora* L. in Palmer & Steyermark's *Annotated Catalogue* are referred in the present flora to the varieties of *O. biennis*, true *O. parviflora* apparently not occurring in Missouri.

The common evening primrose is included here among the different varieties enumerated. Experiments carried on with members of this group have shown the existence of a large number of races, which for all practical purposes are best discussed outside the scope of this work. They serve to illustrate how new races of these plants come into existence through hybridization. For a discussion of this group and its complexity the reader is referred to the work of Cleland (Am. Nat. 48: 5–28. 1944), Fernald (Rh. 51: 61–67. 1949), and R. R. Gates (Rh. 59: 9–17. 1957; Phil. Trans. Roy. Soc. London, Ser. B. 226: 239–355. 1950).

The fragrant flowers of this biennial plant open mainly late in the afternoon, at night, and early in the morning, or on cloudy days may remain open during the day, being pollinated mainly by night-flying sphynx moths. The newly formed roots of the first year are sometimes cooked as a vegetable and somewhat

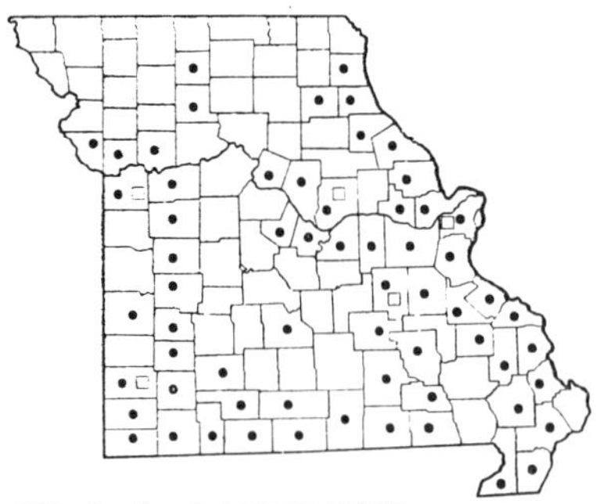

1605 • Oenothera laciniata var. laciniata
1605 □ Oenothera laciniata var. grandiflora

1607 Oenothera serrulata
. . .
1606 Excluded species

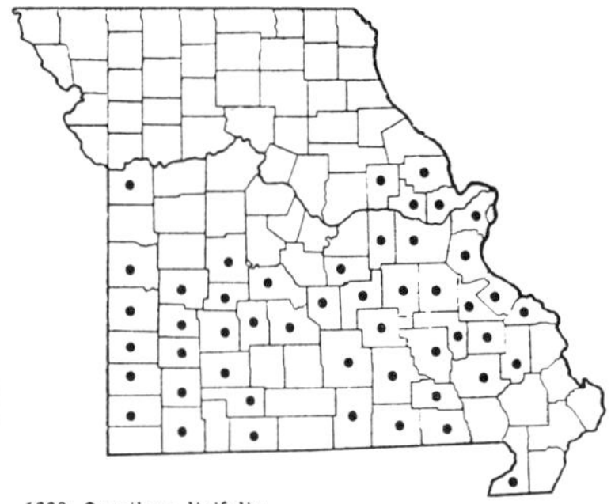

1608 Oenothera linifolia

resemble in taste the salsify or even parsnip, if gathered in the proper stage of development.

2. **Oenothera grandiflora** Ait. Map 1603

*Oenothera biennis* var. *grandiflora* Lindl.

Flowers July–November.

Cultivated, and rarely escaped. Known only from Greene County (Willard).

Introduced via Europe, and naturalized from Quebec south and westward.

3. **Oenothera rhombipetala** Nutt. Map 1604

Flowers June–October.

Occurs in sandy open ground of prairies and fields, and along railroads. Rare and scattered in Clark, Ralls, St. Louis, Jackson, and Phelps counties.

Ranges from Florida to Texas, north to Georgia, Indiana, Wisconsin, Minnesota, and North Dakota.

This is a biennial plant with erect stems up to 1 meter high. As in *O. biennis* the flowers are in dense spikes.

4. **Oenothera laciniata** Hill Map 1605

Flowers May–October.

Occurs in fallow fields, sand flats, alluvial soils, prairies, roadsides, along railroads, and waste ground.

Two variations occur in Missouri:

Calyx-lobes 6–12 mm. long with free tips 1–2 mm. long; petals 5–18 mm. long . . . 3a. O. LACINIATA var. LACINIATA

Calyx-lobes 20–30 mm. long with free tips 2–5 mm. long; petals 20–35 mm. long . . . . 3b. O. LACINIATA var. GRANDIFLORA

4a. **Oenothera laciniata** var. **laciniata** Map 1605

*Oenothera laciniata* Hill [G, P & S, Steyerm.]

Common throughout southern and central Missouri, north to Lewis, Shelby, Livingston, Grundy, Ray, Clay, and Platte counties.

Ranges from New Jersey to North Dakota, south to Florida, Louisiana, and Texas; introduced northeast to New York and Massachusetts.

4b. **Oenothera laciniata** var. **grandiflora** (S. Wats.) Robins. Map 1605

*Oenothera grandis* (Britton) Rydb. [Rydb.]

Scattered in Missouri, in St. Louis, Crawford, Boone, Jackson, and Jasper counties.

Ranges from Texas, New Mexico, and Mexico, northeast to Missouri, Kansas, and Oklahoma.

This evening primrose has pale yellow petals in contrast to the bright yellow of *O. biennis*.

5. **Oenothera serrulata** Nutt. Map 1607

Flowers late May–September.

Occurs on open loess hills, fields, and prairies. Rare and scattered in northern and central Missouri, south to St. Louis, Boone, and Jackson counties, locally adventive along railroads south to Iron County (*H. H. Smith 460*), native on the loess hills of the northwestern part of the state.

Ranges from Saskatchewan to Montana, south to Wisconsin, Missouri, Texas, New Mexico, and Arizona; introduced farther east into New England.

The narrow linear to linear-oblanceolate leaves vary from nearly entire to shallowly and finely toothed.

---

Plate no. 262. **1.** Ludwigia palustris var. americana, × $^{2}/_{7}$; *a.* Flower, × $1^{3}/_{7}$; *b*, *c*. Two leaf and habit variations; Details from Small, The New York Botanical Garden. **2.** Epilobium coloratum, × $^{2}/_{7}$; Details from Small, The New York Botanical Garden. **3.** Oenothera biennis var. biennis, × $^{2}/_{7}$ (Scribner's); Details from Small, The New York Botanical Garden. **4.** Oenothera pilosella, × $^{2}/_{7}$. **5.** Oenothera serrulata, × $^{2}/_{7}$. **6.** Oenothera tetragona var. hybrida, × $^{2}/_{7}$. **7.** Oenothera fruticosa var. linearis, × $^{2}/_{7}$. **8.** Oenothera speciosa, × $^{2}/_{7}$; Details from Small, The New York Botanical Garden. **9.** Oenothera linifolia, × $^{2}/_{7}$; Details from Small, The New York Botanical Garden. **10.** Oenothera triloba, × $^{2}/_{7}$; Details from Small, The New York Botanical Garden.

PLATE NO. 262

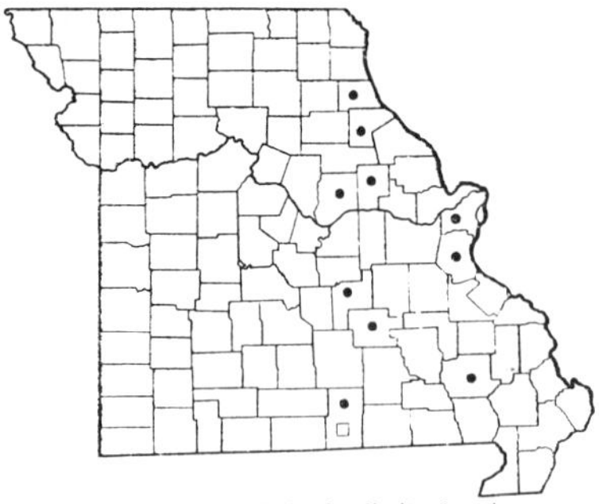

1609 • Oenothera pilosella f. pilosella (Sundrops)
1609 □ Oenothera pilosella f. laevigata

1610 Oenothera fruticosa var. linearis

1611 Oenothera tetragona var. hybrida

6. **Oenothera linifolia** Nutt. Sundrops Map 1608

Flowers May–July.

Occurs on sandstone, chert, or granite rocky glades, prairies, and fallow fields, in acid soils.

Occurs in southern and east-central Missouri, restricted to the Ozark and unglaciated prairie sections, north to Lincoln, Montgomery, Maries, Camden, Benton, Bates, and Jackson counties.

Ranges from Georgia to Texas, north to Illinois, Missouri, and Kansas.

This thread-leaved annual is characteristic of acid soil rocky glades, where it is associated with such other narrow-leaved annuals of similar habit and ecological adaptation as *Polygonum tenue*, *Hypericum gentianoides*, and *Bulbostylis capillaris*.

7. **Oenothera pilosella** Raf. Sundrops Map 1609

Flowers May–July.

Occurs in prairies, fallow fields, wet upland, open woodland, open wet grassy places, and along railroads. Eastern Missouri.

Ranges from Ontario to Michigan and Iowa, south to Pennsylvania, West Virginia, Ohio, Indiana, Illinois, and Arkansas; introduced eastward to New England and Virginia.

Two variations occur in Missouri:

Common type encountered; fruit with few to many spreading hairs . 7a. O. PILOSELLA f. PILOSELLA

Rarely encountered; fruit entirely glabrous (without hairs) or nearly so . . . . 7b. O. PILOSELLA f. LAEVIGATA

7a. **Oenothera pilosella** f. **pilosella** Map 1609

*Oenothera pilosella* Raf. [G, BB, Steyerm.]

*Oenothera pratensis* (Small) Robinson [P & S]

This is the commoner variation, scattered in eastern Missouri north to Marion County, and west to Callaway, Phelps, Dent, and Howell counties.

7b. **Oenothera pilosella** f. **laevigata** Palmer & Steyerm. Map 1609

Known only from Howell County, southern Missouri (swampy sink depression and tributary drainage in wooded upland, T23N, R8W, east part sect. 16, 4 mi. south of West Plains, June 25, 1955, *Steyermark 78703*, holotype at Chi. Nat. Hist. Mus. herb.).

The mature fruiting capsules of *O. pilosella* vary from 8–20 mm. long and 2–5 mm. thick; they are usually thickly club-shaped, changing somewhat in shape as they mature from slenderly club-shaped, and much longer in proportion to their thickness in the ovaries and young or immature fruit in the upper axils of the plant, to thickly club-shaped or even quadrate-oblong in more mature fruit found in the lower axils.

This showy-flowered perennial does well in cultivation, forming loose beds. It prefers sunny situations and slightly acid clayey soils.

8. **Oenothera fruticosa** L. var. **linearis** (Michx.) S. Wats. Map 1610

*Oenothera fruticosa* var. *vera* Hook. f. *angustifolia* Léveillé [Steyerm.]

*Oenothera fruticosa* of auth. in part [P & S], not L.

*Oenothera fruticosa* var. *fruticosa* f. *angustifolia* Léveillé [BB]

Flowers latter part of May–August.

Occurs in rocky or sandy open woods in usually upland level areas on ridges or flats. Ozark region of southern and east-central Missouri northeast to Jefferson County.

Ranges from Florida to Oklahoma, north to southern New England, New York, New Jersey, Maryland, Michigan, and Missouri.

9. **Oenothera tetragona** Roth var. **hybrida** (Michx.) Fern. Map 1611

*Oenothera hybrida* Michx. [P & S]

*Oenothera tetragona* var. *Fraseri* (Pursh) Munz f. *hybrida* (Michx.) Munz [Steyerm.]

1612 Oenothera perennis var. perennis

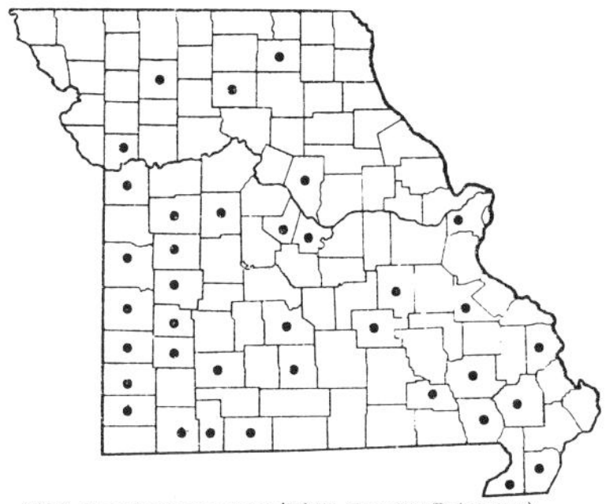

1613 Oenothera speciosa (White Evening Primrose)

1613A Oenothera Kunthiana

*Oenothera tetragona* var. *Fraseri* (Pursh) Munz [BB]

Flowers June–August.

Known only from Henry County, unglaciated prairie region of west-central Missouri.

Ranges from North Carolina and Tennessee to Virginia, west to Missouri.

10. **Oenothera perennis** L. var. **perennis** Map 1612

*Oenothera perennis* L. [G, BB]

*Oenothera perennis* var. *typica* Munz [Munz]

Flowers June–August.

Occurs in wet upland flat woods near sink-hole ponds. Known only in a wild state from Dent County, eastern Ozark region (T32N, R4W, sect. 30, $\frac{3}{4}$ mi. south of Turtle P.O., June 25, 1951, *Steyermark 71808*).

Ranges from Newfoundland and Quebec to Minnesota and Manitoba, south to North Carolina, Georgia, West Virginia, Ohio, Indiana, and Missouri.

Two other collections of this species from Missouri should be noted here: one from St. Louis County (St. Louis, July 2, 1910, *Sherff*) was taken from a cultivated specimen (personal communication from Dr. Sherff; see Rh. 54: 256. 1952) deposited in Gray Herbarium and Missouri Botanical Garden Herbarium; the other doubtfully labeled as from Adair County (July 23, 1888 ?, *E. S. Link*) is deposited in the herbarium of the University of Missouri.

11. **Oenothera speciosa** Nutt. White Evening Primrose Map 1613

Flowers May–July.

Occurs in prairies, fields, glades, open waste ground, and along roadsides and railroads. Commonest in southern and central Missouri and rare in northern Missouri in Adair, Linn, and Daviess counties.

Ranges from Missouri and Kansas to Texas and Mexico; naturalized from Illinois to Louisiana, eastward to Indiana, Pennsylvania, Virginia, North and South Carolina, Georgia, and Florida.

This species does well in cultivation if not crowded by other plants, and eventually forms beds.

12. **Oenothera Kunthiana** (Spach) Munz Map 1613A

*Oenothera tetraptera* Cav. var. *Childsii* Bailey

Flowers May–June.

Occurs along railroad tracks. Known only from St. Louis County (St. Louis, Wabash R.R., north of Palm Street, May 24, 1959, *Muehlenbach 1505*).

Ranges from Texas to Mexico; introduced in Missouri and naturalized in Africa.

This is a low-growing plant 1–2 dm. tall, with attractive flowers resembling those of *O. speciosa* but smaller, and with erect flower-buds. According to the studies of Dr. Munz (Bull. Torr. Bot. Club 19: 759–60 1932), *O. Kunthiana* intergrades with the closely related *O. tetraptera* Cav., from which it differs in the smaller, pinker petals (1–2 cm. instead of 2–3.5 cm long), more prostrate habit, longer calyx-tube (1–2 cm. instead of 1 cm. long), more slender capsule 4–5 mm. instead of 6–8 mm. thick, with wings 1 mm. instead of 2–3 mm. wide, and the upper parts of the plant with an appressed instead of spreading type of hairiness. From *O. rosea* Cav., *O. Kunthiana* differs in having longer petals and longer calyx-tube.

13. **Oenothera missouriensis** Sims var. **missouriensis** Missouri Primrose Map 1614

Also called Glade Lily.

*Oenothera missouriensis* Sims [G, BB, P & S, Steyerm.]

Flowers May–August.

Occurs on rocky open limestone glades and bald knobs, edges of limestone bluff escarpments, rocky prairies, and rocky mounds. Ozark and unglaciated prairie regions of southern and east-central Missouri, north to St. Louis, Franklin, Gasconade, Callaway, Moniteau, Morgan, Benton, and Bates counties.

Ranges from Missouri and Kansas to Texas.

Other varieties are known to the west of Missouri.

1614 Oenothera missouriensis var. missouriensis (Missouri Primrose)

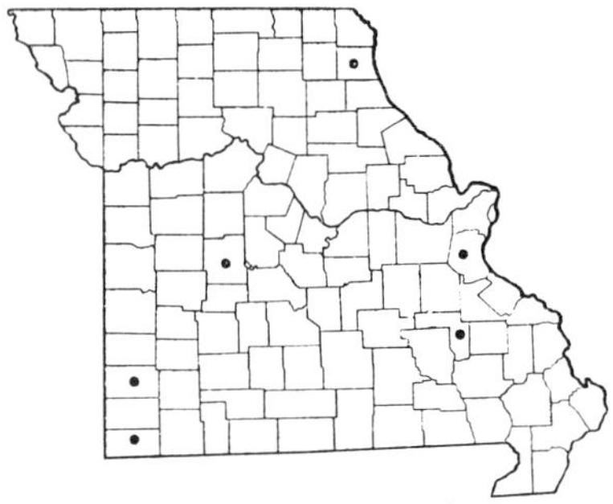

1615 Oenothera triloba (Evening Primrose)

1616 • Gaura biennis var. biennis (Biennial Gaura)
1616 □ Gaura biennis var. Pitcheri

The Missouri Primrose is one of the showiest wild flowers of the state. Each flower, produced in successive axils of the petioles, lasts but one day, usually opening in late afternoon or early evening and remaining open until the following morning.

The plant is easily raised from seed, but, because of the deep root, is difficult to transplant. It succeeds well in sunny, well-drained, limey soils and seeds itself. In the author's wildflower garden in northern Illinois, it has prospered during twenty years.

14. **Oenothera triloba** Nutt. Evening Primrose Map 1615

Flowers April–May.

Occurs on limestone glades, prairies, and open grassy or rocky ground along roadsides. Rare and scattered in the Ozark region northeast to Lewis County.

Ranges from Virginia to Indiana and Kansas, south to Alabama, Tennessee, Oklahoma, and Texas.

This is mainly a biennial plant, which forms a rosette of oblanceolate, deeply cut, runcinate-pinnatifid leaves up to 3 dm. in length and showy pale yellow to whitish or rose-colored flowers. The plant makes a fine addition to a rock garden or perennial border.

## 5. **Gaura** L.

a. Larger leaves 2–8 mm. broad, the leaves of the stem linear to narrowly lanceolate; hairs of stem, if present, ascending or pressed against or lying upon surface of stem; plant usually branched from the base; hairs on fruit pressed against or lying parallel to surface; fruit terete (with curved or rounded sides) at base, 4-angled above the base . . . . . . . . . . . . . . 3. G. COCCINEA

a. Larger leaves 10–50 mm. broad, the leaves of the stem oblong- to ovate-lanceolate; hairs of stem spreading or loose; plant usually branched above the middle portion; hairs on fruit, if present, spreading, or the fruit glabrous (without hairs); fruit 4-angled from top to base. . . . . . . . *b*

b. Commonly occurring and encountered throughout Missouri; petals 5–8 mm. long; calyx-tube 5–7 mm. long; calyx-lobes 6–8 mm. long; anthers 2.5–3 mm. long . . . . . . . . 1. G. BIENNIS

b. Less commonly collected, mostly in the northern and western half of the state; petals 1.5–2 mm. long; calyx-tube 1.5–3 mm. long; calyx-lobes 1.5–3 mm. long; anthers about 1 mm. long . . 2. G. PARVIFLORA

1. **Gaura biennis** L. Biennial Gaura Map 1616

Flowers June–October.

Occurs in prairies, glades, thickets, fallow fields, openings in dry woodland, along gravel and alluvial washes in valleys, and along roadsides and railroads.

Two variations occur in Missouri:

Hairs on stem spreading, conspicuous, downy; leaves with few hairs or nearly glabrous; gland-tipped hairs occurring on inflorescence . . . 1a. G. BIENNIS var. BIENNIS

Hairs on stem short; leaves with a close short pubescence; no glands present on inflorescence . 1b. G. BIENNIS var. PITCHERI

1a. **Gaura biennis** var. **biennis** Map 1616

This is the common variation and is found throughout Missouri, probably in every county, at present unknown from most of the lowland counties of southeastern Missouri.

Ranges from Quebec to Minnesota, south to Virginia, North Carolina, Tennessee, and Missouri.

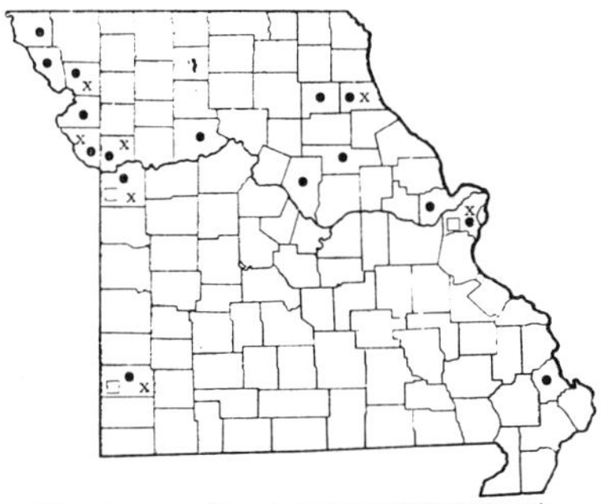

1617 • Gaura parviflora f. parviflora (Velvety Gaura)
1617 x Gaura parviflora f. glabra
1617 □ Gaura parviflora f. lachnocarpa

1618 Gaura coccinea var. coccinea (Scarlet Gaura)

1619 Stenosiphon linifolius

1b. **Gaura biennis** var. **Pitcheri** Pickering Map 1616

Less commonly encountered and scattered throughout the state.

Ranges from Illinois to Nebraska, south to Arkansas, Oklahoma, and Texas.

As the common name implies, this is a biennial plant, producing a rosette of leaves which persist through the winter and from the deep tap root sending up the following year a long branched stem up to 3.5 meters tall.

2. **Gaura parviflora** Doug. Velvety Gaura Map 1617

Flowers June–October.

Occurs in prairies, waste ground, and along roadsides and railroads. Mainly in northern, western, and central Missouri along the Missouri and Mississippi rivers.

Ranges from Texas and Mexico, north to Indiana, Illinois, Iowa, South Dakota, Wyoming, Idaho, and Washington; introduced eastward to New England.

The following variations are represented by Missouri material:

a. Ovary and fruit hairy . . 2c. G. PARVIFLORA f. LACHNOCARPA
a. Ovary and fruit glabrous (without hairs) . . *b*
  b. Calyx-tube (hypanthium) minutely short-hairy . . 2a. G. PARVIFLORA f. PARVIFLORA
  b. Calyx-tube (hypanthium) glabrous (without hairs) . 2b. G. PARVIFLORA f. GLABRA

2a. **Gaura parviflora** f. **parviflora** Map 1617

Scattered mainly in the northern, central and western parts of the state, chiefly along the Missouri and Mississippi river counties.

2b. **Gaura parviflora** f. **glabra** Munz Map 1617

Scattered in the range of *G. parviflora*.

2c. **Gaura parviflora** f. **lachnocarpa** Weath. Map 1617

Known from St. Louis, Jackson, and Jasper counties. This is also a biennial species.

3. **Gaura coccinea** Pursh var. **coccinea** Scarlet Gaura Map 1618

Flowers May–August.

Occurs in loess hills, and along railroads and waste ground. Known from northwestern Missouri where native in Atchison and Holt counties, and introduced along railroads in Clay, Jackson, and St. Louis (Creve Coeur Lake, May 15, 1925, *Woodson 215*; Terminal R. R. Association, east of West Florissant Ave., St. Louis, June 18, 1955, *Muehlenbach 632*) counties.

Ranges from Alberta to California, east to Manitoba, Minnesota, Missouri, Oklahoma, and Texas; introduced eastward to New York.

This is a perennial species, producing several stems usually 1–3 dm. tall. The pink, reddish, or white petals are 3–6 mm. long.

## 6. **Stenosiphon** Spach

**Stenosiphon linifolius** (Nutt.) Britt. Map 1619

Flowers July–October.

Occurs on rocky limestone glades and bald knobs. Known only from Ozark and Taney counties, southern Missouri.

Ranges from Missouri to Nebraska and Colorado,

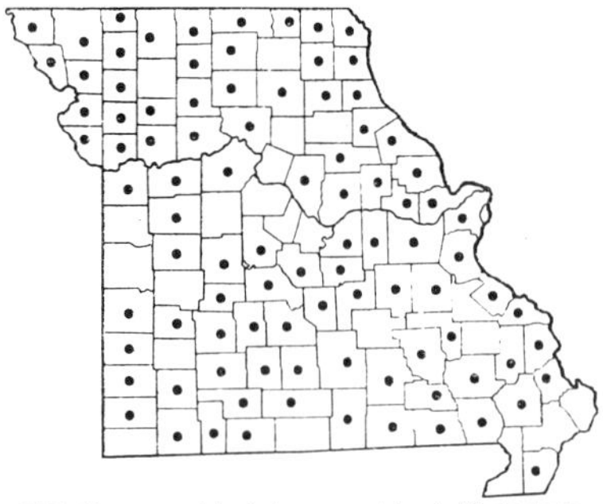

1620 Circaea quadrisulcata var. canadensis (Enchanter's Nightshade)

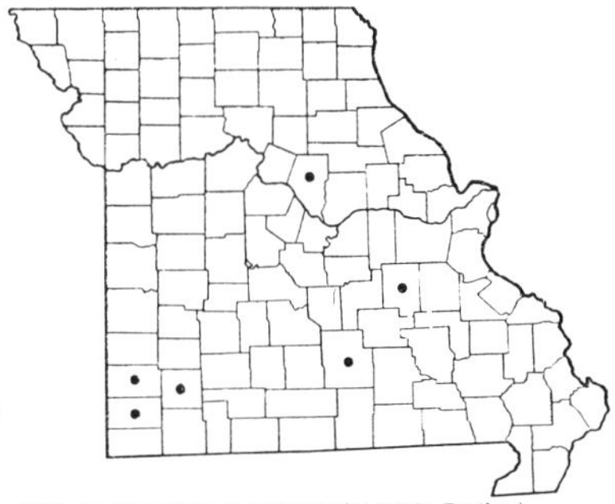

1621 Myriophyllum brasiliense (Parrot's Feather)

1622 Myriophyllum heterophyllum (Water Milfoil)

south to Arkansas, Oklahoma, Texas, and Mexico.

This is a glabrous biennial plant with smooth pale or gray-green leaves. The first year there is produced a small rosette of leaves, which overwinter. From these, the following year, is sent up a single, often unbranched stem as much as 1.5 meters tall. The flowers are arranged in a slender showy elongated spike, producing a conspicuous ornamental plant in a wild or perennial garden. Rosette first-year plants have overwintered in the author's wildflower garden in northern Illinois and bloomed the following autumn, but they have failed to persist over the years through self-sowing.

### 7. **Circaea** L. Enchanter's Nightshade

**Circaea quadrisulcata** (Maxim.) Franch. & Sav. var. **canadensis** (L) Hara Enchanter's Nightshade Map 1620

*Circaea latifolia* Hill [P & S, Steyerm.]

Flowers June–August.

Occurs in rich woods and thickets. Common throughout Missouri.

Ranges from Nova Scotia to Ontario, Minnesota, and North Dakota, south to Georgia, Tennessee, Missouri, and Oklahoma.

Typical *C. quadrisulcata* var. *quadrisulcata* is native of Asia.

## Fam. **HALORAGIDACEAE** (Water Milfoil Family)

All the leaves alternate on the stem; some or all of the main leaves of the stem only shallowly toothed; flower with parts arranged in 3's; stamens 3; carpels 3; fruit 3-angled . . . . . . . 2. PROSERPINACA

Leaves either in distinct whorls (circles) of 3's, 4's, 5's, or 6's on the stem, or scattered and alternate; all the main leaves of the stem deeply dissected to the midrib into feather- or comb-like divisions (only the uppermost ones subtending the flowers and fruits in *M. heterophyllum* toothed); flower parts arranged in 4's; stamens 4 or 8; carpels 4; fruit separating into 4 parts (nutlets) . . . 1. MYRIOPHYLLUM

### 1. **Myriophyllum** L. Water Milfoil

a. At least some of the leaves of the stem alternately scattered. . . . . . . . . . 3. M. PINNATUM

a. All the leaves in distinct whorls (circles) of 4's, 5's, or 6's . . . . . . . . . . . . . *b*

b. Leaves bluish- or grayish-green, all of them similar in appearance; each leaf with 10–18 closely spaced short divisions 3–5 mm. long on each side; flowers produced from the axils of submerged foliage leaves; rare introduced plant . . . . . . . . . . . . . . . . 1. M. BRASILIENSE

---

Plate no. 263. **1.** Oenothera laciniata var. laciniata; *a.* Oenothera laciniata var. grandiflora, × 2/7. **2.** Gaura biennis, × 2/7; Details from Small, The New York Botanical Garden. **3.** Oenothera missouriensis, × 2/7; *a.* Fruit, × 2/7. **4.** Myriophyllum heterophyllum, × 2/7; *a.* Upper emerged leaves; *b.* Submerged leaves. **5.** Gaura coccinea, × 2/7; Details from Small, The New York Botanical Garden. **6.** Gaura parviflora, × 2/7; After Britton and Brown, details from Small, The New York Botanical Garden. **7.** Stenosiphon linifolius, × 2/7; **8.** Myriophyllum brasiliense; *a.* Leafy portion of plant, × 2/7; *b.* Leaf, × 2 6/7; *c.* Fruit, × 2 6/7; After Gleason, details from Small, The New York Botanical Garden. **9.** Myriophyllum pinnatum; *a.* Habit, × 2/7; *b.* Portion of fruiting plant, × 4/7; Details from Small, The New York Botanical Garden. **10.** Circaea quadrisulcata var. canadensis, × 2/7; Details from Small, The New York Botanical Garden.

PLATE NO. 263

b. Leaves dark or grass-green, bronze-green or brownish-green, the upper emersed (out of water) leaves greatly reduced in size and toothed only; each leaf with 3–11 mostly widely separated elongated divisions mainly 5–15 mm. long on each side; flowers produced from the axils of the reduced bracts above water; common native species of springs, spring branches, ponds, and sloughs . . . . . . . . . . . . . . . . . . . . . . 2. M. HETEROPHYLLUM

1. **Myriophyllum brasiliense** Camb. Parrot's Feather Map 1621

Also called Water Feather.

*Myriophyllum proserpinacoides* Gill. [P & S]

Flowers June–September.

Planted as an ornamental aquatic plant in fish-pools, waterlily ponds, and around spring branches, sometimes escaping and becoming established in shallow water and along muddy margins of ponds, small streams, and spring branches. Scattered in southern and central Missouri north to Boone County.

Native of South America; introduced and naturalized in the southern states and north to New York, New Jersey, Virginia, Missouri, and Oklahoma.

This species is often seen in outdoor fish tanks and lily pools, and is sometimes used indoors in large tanks and pools in conservatories and display houses.

2. **Myriophyllum heterophyllum** Michx. Water Milfoil Map 1622

Also called Coontail.

Flowers May–September.

Occurs in spring branches, spring-fed ponds, sloughs, oxbow ponds of river bottoms, ditches, and streams. Ozark region of southern and east-central Missouri, north to St. Charles, Maries, Camden, and Henry counties, and north locally in northern Missouri to Grundy County (Gooseneck Lake, in bottoms of Thompson River, T60N, R24W, northwest sect. 34, 6½ mi. south of Trenton, August 10, 1952, *Steyermark 74222*).

Ranges from Florida to Texas and New Mexico, north to Quebec, Ontario, New York, Minnesota, and South Dakota.

Sometimes sterile specimens of *M. heterophyllum* are confused with sterile, luxuriant, completely submerged plants of *M. pinnatum.* One means of distinguishing them, aside from the whorled arrangement of all the leaves in *M. heterophyllum* and its habitat usually in cool spring-fed waters, is by the shrunken edges and darker lines extending along the stem in the dried plants of *M. pinnatum*, these lines not being evident in *M. heterophyllum.*

This species generally grows in colonies or beds of numerous plants. It is one of the commonest species encountered in the cool waters of natural springs and spring branches in the Ozarks, having been found at 53 different springs and ranking third in frequency of occurrence of species growing in such habitats (Steyermark, Phanerogamic Flora of the Fresh-water Springs in the Ozarks of Missouri, Bot. Ser. Field Mus. Nat. Hist. 9: 493. 1941). In such waters it is most commonly found in the deeper and more quiet sections, and vegetative or sterile leafy shoots are frequently the only kinds produced. The plants remain green and alive throughout the year in the springs, forming vegetative strands several feet long (Steyermark, pp. 496–501). The leaves are usually dark green, but vary greatly according to the season of the year and depth of the water. Sometimes the color is bronze-red or purple-red. In winter, plants are often brown-red or brown. A bronze tinge is characteristic of many plants of this species at certain times of the year. In spring, dark green is the prevailing shade. Sometimes the stems are entirely purple-red.

At the natural pond in Grundy County, northern Missouri, where the species occurs at its northern limit in the state, it is associated with *Sparganium eurycarpum*, *S. androcladum*, *Spirodela polyrhiza*, *Anacharis Nuttallii*, *Pontederia cordata*, and *Bacopa rotundifolia.*

Mallard ducks are reported to eat the foliage and fruits of this species.

This and related species are grown as aquarium plants for tropical and goldfish varieties, but in very warm waters tend to drop the leaves.

3. **Myriophyllum pinnatum** (Walt.) BSP. Map 1623

*Myriophyllum scabratum* Michx. [P & S]

Flowers June–October.

Occurs along muddy margins of natural and artificial upland ponds and oxbow lakes of river flood plains, sloughs, slow streams, ditches, and spring branches. Scattered over the state, but absent from northwestern Missouri and most of central Missouri and western Ozark region.

Ranges from Florida to Texas, north to Massachusetts, West Virginia, Kentucky, Illinois, Iowa, and Kansas; also in the West Indies.

This species has a prostrate or procumbent habit of growth, usually creeping along wet, muddy places at the edge of the water. Unlike *M. heterophyllum*, which most frequently inhabits cool spring water having a temperature of 54°–60° F., *M. pinnatum* is much commoner around the margins of sloughs, ponds, and streams where the temperatures exceed 70° F. *Myrio-*

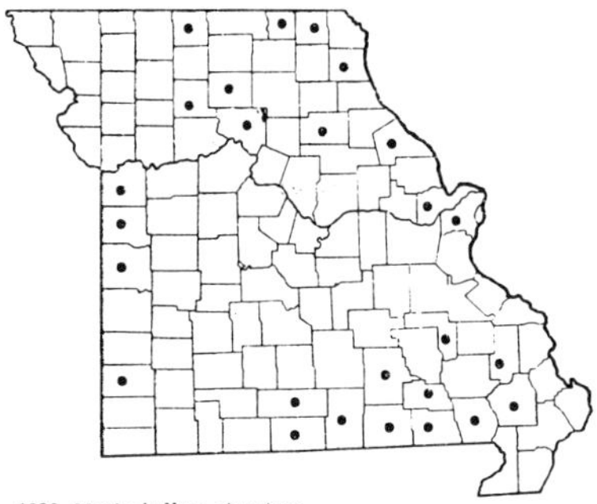

1623 Myriophyllum pinnatum

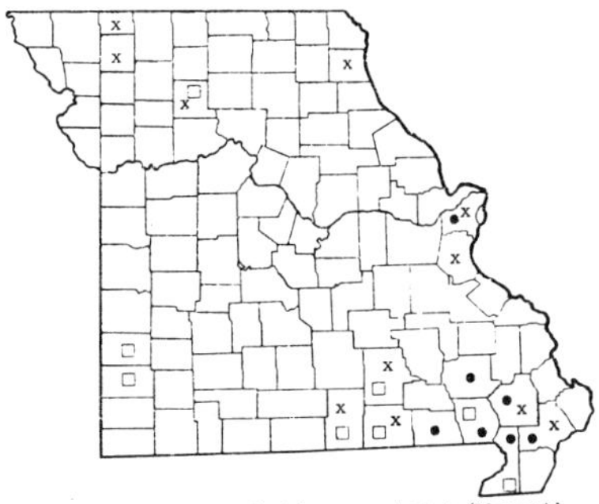

1624 • Proserpinaca palustris var. palustris (Mermaid Weed)
1624 x Proserpinaca palustris var. crebra
1624 □ Proserpinaca palustris var. amblyogona

1625 Aralia spinosa (Hercules' Club)

*phyllum pinnatum*, although occurring in colonies, does not form the dense mats with stems as close together as is the habit with *M. heterophyllum*, and the stems are usually more slender than in *M. heterophyllum* with the flowering stems only 1–3 dm. tall. The leaf-divisions are usually fewer than in *M. heterophyllum*, with mostly 2–6 on each side. The bracts subtending the flowers and fruits in *M. pinnatum* are usually narrower and more deeply cut with fewer, longer, and more slender teeth. Sterile specimens of *M. pinnatum* are sometimes confused with aquatic stages of *Proserpinaca palustris* having all the leaves dissected. In such cases the *Proserpinaca* shows all the leaves alternate, whereas in *M. pinnatum*, some of them are whorled.

The fruit and foliage of this and other species are eaten by wildfowl, and the foliage by muskrats.

## 2. **Proserpinaca** L. Mermaid Weed

**Proserpinaca palustris** L. Mermaid Weed
Map 1624

Flowers June–October.

Occurs in bald cypress swamps, sloughs, margins of upland sink-hole ponds and of oxbow lakes and natural ponds in the meander of streams and river flood plains.

Three varieties are known in Missouri:

a. Sides of fruit rounded with rounded angles . c. P. PALUSTRIS var. AMBLYOGONA
a. Sides of fruit flat or concave (sunken) with acute or wing-like angles . . . . . . . *b*
  b. Fruits 3–4.5 mm. wide, 1¼–2 times as broad as long, narrowed to a short neck at the summit . . . a. P. PALUSTRIS var. PALUSTRIS
  b. Fruits 2–3.5 mm. wide, as broad as or at most 1½ times as broad as long, tapering only or rounded at the summit . . . b. P. PALUSTRIS var. CREBRA

a. **Proserpinaca palustris** var. **palustris**
Map 1624

Occurs in the lowland counties of southeastern Missouri, locally northward in St. Louis County.

Ranges from Florida to Texas, north to Massachusetts, New York, and Missouri; also in the West Indies and Colombia.

b. **Proserpinaca palustris** var. **crebra** Fern. & Grisc.
Map 1624

Occurs in scattered sections throughout Missouri, but uncommon.

Ranges from Nova Scotia and Ontario to Wisconsin, south to Florida, Alabama, Louisiana, and Texas; also in Mexico and Central America.

c. **Proserpinaca palustris** var. **amblyogona** Fern.
Map 1624

Occurs mainly in southern Missouri, and locally northward in Livingston County (*Palmer & Steyermark 41251*), northern Missouri.

Ranges from Georgia to Texas, north to Ontario, Michigan, Indiana, Illinois, and Missouri.

The lower leaves are finely dissected, the upper ones and those out of the water toothed.

The geographical ranges cited are in accord with Dr. Fassett's most recent studies of the genus (Communicaciones del Inst. Trop. Invest. Cient. de Univ. El Salvador 2: 139–62. 1953).

Canvasback and other ducks are reported to eat the seeds of this species.

## Order **UMBELLALES**

### Fam. **ARALIACEAE** (Ginseng Family)

a. Leaves simple, roundish, angled or slightly lobed or entire (without teeth or lobes); introduced plant . . . 3. HEDERA
a. Leaves compound, divided into several to many divisions, the margins toothed; native species . . . . *b*
  b. Each leaf palmately compound, all the 5 leaflets from one leaf arising from the same place at the very tip of the leaf-stalk (petiole); inflorescence of only 1 simple umbel . . . . . . 2. PANAX
  b. Each leaf pinnately compound, some of the leaflets arising at a lower level from that of the others; inflorescence compound, of 2 or more umbels . . . . . . . . . . . . . 1. ARALIA

### 1. **Aralia** L.

a. Shrub or small tree with woody stem; stems, branches, and main stalks (petioles) of the leaves armed with prickles . . . . . . . . . . . . . . . . . . . . . . . . 1. A. SPINOSA
a. Soft-stemmed herbaceous plants; no prickles on plant . . . . . . . . . . . . . . . . *b*
  b. Leaflets 3–7 on each of the 3 leaves; main peduncle (stem supporting inflorescence) arising directly from the ground, bearing usually 3 (from 2–7) umbels . . . . . . . . 3. A. NUDICAULIS
  b. Leaflets 9–21 on each leaf; peduncle arising from the main stem, bearing numerous umbels . 2. A. RACEMOSA

1. **Aralia spinosa** L. Hercules' Club Map 1625
Also called Angelica Tree, Devil's Walking Stick, Tear Blanket.

Flowers July–September.

Occurs in low upland woods, thickets, wooded slopes, bluffs, and ravines. Southeastern Missouri on Crowley Ridge in Scott, Stoddard, New Madrid, and Dunklin counties, and in other parts of the southeastern lowlands north to Cape Girardeau, Bollinger, and Madison counties, west to Wayne and Ripley counties, locally north in St. Louis County (Allenton, June 28, 1885, *Kellogg*; St. Louis, *Sherff 1225*; the Kellogg specimen doubtfully native, the Sherff specimen from cultivated plants).

Ranges from Florida to Texas, north to New Jersey, Pennsylvania, New York, Ohio, Indiana, Illinois, and Missouri; reported from Iowa and Oklahoma.

This species attains a height of 10–12 meters and often forms small colonies. It has soft and brittle wood. The foliage and large white inflorescences make it a desirable shrub for summer flowering. The black fleshy fruits, attractive and eaten by birds, at one time were used for dyeing hair black. The foliage turns a dull brown-purple in autumn. Some individuals are allergic to the bark and roots, developing a dermatitis after handling these parts. It was reported that at one time the inner bark was used for curing toothaches. The wood is said to have been used for small articles such as pen racks, button boxes, photograph frames, stools, and rocking chair arms, but today the wood appears not to be used in any way. In Japan the young leaves of a related species are cooked as a vegetable.

The plant is easily grown from seed or by root cuttings. Plants have thrived in the author's botanical preserve in northern Illinois for the past fifteen years, although in some winters part of the upper stem may die back.

2. **Aralia racemosa** L. Spikenard Map 1626
Also called American Spikenard.

Flowers June–August.

Occurs on rich wooded, often north-facing slopes, ravines, and shaded usually moist ledges and bluffs. Northern and eastern Missouri and Ozark section, west, south of the Missouri River to Moniteau, Morgan, Benton, Dallas, Wright, Christian, and Ozark counties; absent from the unglaciated prairie region of southwestern Missouri and from west-central Missouri.

Ranges from Quebec to Minnesota and South Dakota to Manitoba, south to Georgia, Alabama, Mississippi, Missouri, and Kansas.

The large roots are aromatic and spicy. They are used as one of the ingredients of root beer. The fruit, not considered edible, is juicy, at first reddish-purple, turning blackish-purple. The roots were used formerly

---

Plate no. 264. **1.** Aralia racemosa, × 2/7, showing part of leaf and inflorescence; *a.* Portion of inflorescence; *b.* Portion of fruiting spray; *c.* Portion of leaf. **2.** Hedera Helix, × 2/7. Details from Small, The New York Botanical Garden. **3.** Panax quinquefolius; Details from Small, The New York Botanical Garden. **4.** Proserpinaca palustris, × 2/7; *a.* Fruit, × 2 6/7; Details from Small, The New York Botanical Garden. **5.** Aralia nudicaulis, × 2/7; *a.* Flower, × 1 5/7. **6.** Aralia spinosa; *a.* Portion of inflorescence, × 2/7; *b.* Leaf, × 1/14; Details from Small, The New York Botanical Garden.

PLATE NO. 264

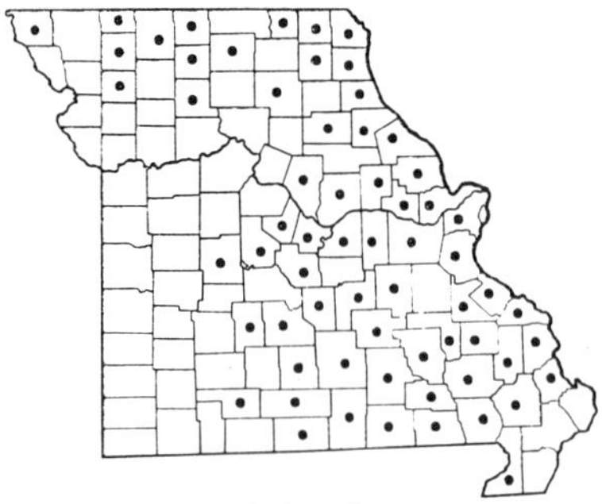

1626 Aralia racemosa (Spikenard)

1627 Aralia nudicaulis (Wild Sarsaparilla)

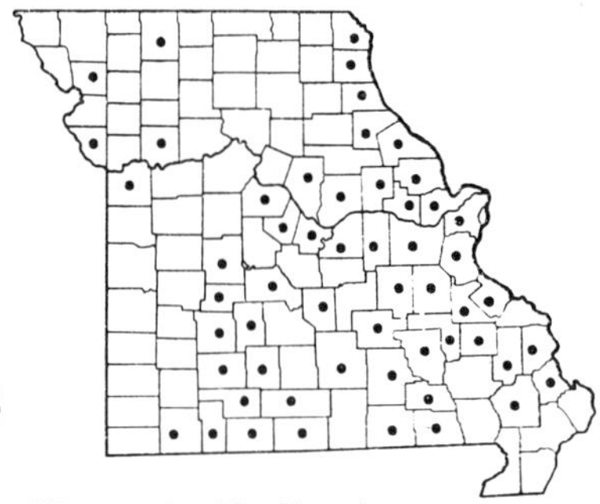

1628 Panax quinquefolius (Ginseng)

as a remedy for respiratory ailments for man and domesticated animals.

3. **Aralia nudicaulis** L. Wild Sarsaparilla Map 1627

Flowers May–June.

Occurs on rich wooded slopes of limestone soils. Known only from northeastern Missouri in Ralls (talus of wooded limestone bluffs along Mississippi River, T56N, R3W, sect. 12, 1 mi. southeast of Ilasco, May 12, 1939, *Steyermark 22343*) and Pike ($\frac{3}{4}$ way up east-facing limestone bluff slopes along Mississippi River, T55N, R3W, sect. 18 and 20, $\frac{3}{4}$–1 mi. southeast of Ashburn, May 13, 1939, *Steyermark 22371*) counties.

Ranges from Newfoundland and Labrador to British Columbia, south to Georgia, Tennessee, Illinois, Missouri, Nebraska, Colorado, and Idaho.

The plant usually forms colonies from the long horizontal roots, which are used as one of the ingredients of root beer.

## 2. **Panax** L.

**Panax quinquefolius** L. Ginseng Map 1628

*Panax quinquefolium* L. [P & S, Steyermark, BB]

Flowers June 1–July.

Occurs on usually north-facing rich, wooded slopes, often of steep limestone bluffs and outcrops. Frequent throughout the Ozark section and in extreme eastern Missouri, rare elsewhere in the state in Harrison, Andrew, Platte, Ray, and Jackson counties.

Ranges from Quebec to Manitoba, south to Florida, Alabama, Louisiana, and Oklahoma.

The leaves are sometimes used for tea. The bright red shining fruit is conspicuous in late summer and fall, and has a color similar to the fruit of *Arisaema atrorubens*. The principal leaves are usually 3, occasionally 4, to a plant. Each has 5 sharply toothed, obovate-oblong leaflets on long stalks (petiolules).

The demand for the root of this plant by the Chinese for its supposed medicinal value as an aphrodisiac, as heart stimulant, and as a cure-all for most diseases, has threatened it with extinction in many parts of Missouri and elsewhere in its range. There has been no scientific basis, however, to support any belief that the plant possesses the properties ascribed to it by the Chinese. In some parts of the Ozarks, as near Licking, in Texas County, it is grown commercially. It does well in a rich, shaded soil. Usually at any given locality in Missouri, the plants are few to several in number, and occur as scattered individuals, well separated from one another.

## 3. **Hedera** L. Ivy

**Hedera Helix** L. English Ivy Map 1629

Commonly planted and rarely escaping from cultivation. Known in Missouri only from St. Louis County (St. Louis, Lindenwood freight yard of Frisco Railroad, on northeastern embankment of the Fyler Bridge, September 10, 1955, *Muehlenbach 764*).

Native of Europe; introduced and naturalized in Virginia, Missouri, and elsewhere in the southern states.

The leaves, berries, and other parts of the plant are poisonous if eaten. Some persons develop a dermatitis after contact with the leaves.

## Fam. **UMBELLIFERAE** (Parsley Family)

In this family the flowers are typically arranged in *umbels* or umbrella-like clusters in which all or most of the flower-stalks (pedicels) arise from the same place, as spokes in an umbrella or rays of a wheel. Each group of flower-stalks may arise from larger ones attached at some central point; rarely all the flowers are without stalks and in dense heads, as in *Eryngium*. The umbel is often subtended at its base by a circle or whorl of bracts, the *involucre*; the smallest clusters (umbellules or umbellets) may also be subtended by bractlets, the *involucel*. The fruit, which is important in classification, is a dry structure, consisting of 2 equal parts (*1-seeded mericarps*) facing each other, and eventually separating from the base upward and at the summit connected to a slender *carpophore*. The inner face of each mericarp, along which the pair are connected, is known as the *commisure*. The fruit may be flattened laterally, perpendicular to the rather narrow commissure, or flattened dorsally parallel to the rather broad commissure, or the fruit may be nearly terete (circular in cross-section) with each mericarp nearly semicircular in cross-section. Usually at least 5 main (primary) *ribs* run longitudinally over each mericarp, of which the median is known as the *dorsal* rib, the two nearest the commissure known as the *lateral*, and the others known as the *intermediate* ribs.

The stems are usually hollow. The oil secreted by the members of this family from the fruit particularly is responsible for the characteristic odor or taste.

a. None of the leaves divided or dissected, all simple and in one piece . . . . . . . . . . . . *b*
  b. Upper leaves of the stem perfoliate, their lower portion clasping and surrounding all sides of the stem; all of the leaves lacking teeth. . . . . . . . . . . . . . . . . . 13. Bupleurum
  b. No leaves perfoliate; some or all of the leaves possessing teeth or lobes. . . . . . . . . . . *c*
    c. Leaf-blades round, umbrella-shaped, or broadly lima-bean-shaped, peltate (leaf-stalk joining the leaf-blade near its center) . . . . . . . . . . . . . . . . . . 1. Hydrocotyle
    c. Leaf-blades either long, narrow, strap-shaped, or oblong to ovate, not peltate, the leaf-stalk, if present, joining the leaf-blade at its lower end . . . . . . . . . . . . 3. Eryngium
a. Some or all of the leaves compound, divided to their base or midrib into separate divisions (leaflets), or these divisions also subdivided . . . . . . . . . . . . . . . . . . . . . . . . *d*
  d. Leaves only once compound, either divided to their base (palmately) or to their midrib (pinnately), the main leaf-divisions (leaflets) remaining simple and undivided, and either with toothed or non-toothed (entire) margins . . . . . . . . . . . . . . . . . . . . . *e*
    e. The main leaf-divisions (leaflets) entire (smooth-edged, non-toothed) . . . . . . . . . *f*
      f. All the leaf-divisions thread- or hair-like, less than 1 mm. broad . . . . . . . 24. Ptilimnium
      f. Leaf-divisions 2–40 mm. broad, not thread- or hair-like . . . . . . . . . . . . . *g*
        g. Leaflets 2–7 mm. broad; main leaf-stalk (petiole) of leaves of the stem 5–40 mm. long; fruit without any wings; bractlets at base of main (primary) rays several . 25. Cynosciadium
        g. Leaflets 2–45 mm. broad; main leaf-stalk (petiole) of leaves of the stem 40–150 mm. long; fruit with wings on the lateral ribs; bractlets absent at base of main (primary) rays or 1 or 2 . . . . . . . . . . . . . . . . . . . . 31. Oxypolis
    e. The main leaf-divisions (leaflets) with teeth . . . . . . . . . . . . . . . . . *h*
      h. Stems conspicuously hairy . . . . . . . . . . . . . . . . . . . . . . . . *i*
        i. Leaf-divisions less than 1 cm. broad; fruit or ovary covered with few or many spreading hooked or barbed bristles or prickles . . . . . . . . . . . . . . . . 7. Torilis
        i. Leaf-divisions 3 cm. or more broad; no bristles or prickles on fruit or ovary. . 36. Heracleum
      h. Plant completely glabrous (without hairs) . . . . . . . . . . . . . . . . . . *j*
        j. Leaves of stem divided into 3 leaflets. . . . . . . . . . . . . . . . . . . *k*
          k. All the leaf-divisions more or less equally attached to the main leaf-stalk (petiole) with more or less equal stalks (petiolules) or without any stalks . . . . . . . . . *l*
            l. Flowers greenish or greenish-yellow, rarely greenish-white; ovary or fruits covered with hooked prickles . . . . . . . . . . . . . . . . . 2. Sanicula
            l. Flowers white; ovary or fruits smooth or without prickles . . . . . . . . . . *m*
              m. Leaflets not simply toothed, but each principal tooth split or broken (doubly serrate) along its margin . . . . . . . . . . . . . . . 16. Cryptotaenia
              m. Leaflets mostly simply toothed, each tooth remaining uncut or unbroken between its connections with adjacent teeth . . . . . . . . 17. Falcaria
          k. Middle leaf-division of at least some of the leaves on a longer stalk (petiolule) than the other two (lateral) leaflets, or the middle leaflet stalked and the lateral leaflets sessile, (without stalks), the leaf-divisions not equally attached to the main leaf-stalk (petiole) . . *n*

n. Flowers purple. . . . . . . . . . . . . . . 29. THASPIUM TRIFOLIATUM var. TRIFOLIATUM

n. Flowers yellow, white, or greenish-white . . . . . . . . . . . . . . . . . . *o*

o. Main veins of all leaflets palmate (arising at the base from same point); leaflets lobed or coarsely incised; introduced plant . . . . . . . . . . . . . . . 12. APIUM

o. Main veins of some of leaflets pinnate (arising one above the other from base to tip on each side of midrib); leaflets more or less evenly toothed . . . . . . . . . . . . *p*

p. Center flower and fruit of each ultimate umbel (umbellet) sessile (without a stalk); fruit not winged . . . . . . . . . . . . . . . . . . . . . 14. ZIZIA

p. Center flower and fruit of each ultimate umbel (umbellet) with a stalk (the stalk may be very short or inconspicuous); fruit with wings on the dorsal and lateral ribs. 29. THASPIUM

j. Leaves of stem divided into 5 or more leaflets . . . . . . . . . . . . . . . . . *q*

q. Leaf-divisions palmately arranged, all arising from the same point at the tip of the leaf-stalk (petiole) . . . . . . . . . . . . . . . . . . . . . . . . . . . *r*

r. Flowers greenish, greenish-yellow or rarely greenish-white; ovary or fruits covered with hooked prickles; leaflets coarsely toothed, the teeth not simple but each principal tooth split or broken (doubly serrate) along its margin . . . . . . . . . . 2. SANICULA

r. Flowers white; ovary or fruits smooth or without prickles; leaflets shallowly and more finely toothed, the teeth simple, remaining uncut or unbroken between connections with adjacent teeth . . . . . . . . . . . . . . . . . . . . . 17. FALCARIA

q. At least some of the leaf-divisions pinnately arranged, arising one above the other on each side of midrib . . . . . . . . . . . . . . . . . . . . . . . . . *s*

s. Main veins of all leaflets palmate (arising at the base from same point) . . . . 12. APIUM

s. Main veins of some of leaflets pinnate (arising one above the other from base to tip on each side of midrib) . . . . . . . . . . . . . . . . . . . . . *t*

t. Leaflets with no teeth or with at most 8 teeth along each margin . . . . 31. OXYPOLIS

t. All the leaflets with numerous teeth along each margin . . . . . . . . . . . *u*

u. Flowers yellow; at least some of leaflets lobed or coarsely or irregularly toothed; fruit conspicuously winged; no bracts or bractlets present at base of umbels or umbellets . . . . . . . . . . . . . . . . . . . . . 35. PASTINACA

u. Flowers white; none of the leaflets lobed or coarsely toothed, all with fine, regular teeth; fruit not winged; usually 5 or 8 bracts and 5 or 8 bractlets present at base of umbels and umbellets . . . . . . . . . . . . . . . . . 23. SIUM

d. Leaves more than once compound, the main leaf-divisions subdivided into additional leaf-divisions . . *v*

v. All the leaf-divisions (leaflets) entire (smooth-edged) . . . . . . . . . . . . 20. TAENIDIA

v. At least some of the leaf-divisions (leaflets) finely or coarsely toothed, lobed, or divided . . . . . *w*

w. Hairs present around the upper nodes (joints) of the stem at base of leaf-stalk (petiole) . . 29. THASPIUM BARBINODE

w. Plants otherwise, either completely glabrous (without hairs) or hairs present on parts of plant other than the nodes . . . . . . . . . . . . . . . . . . . . . . *x*

x. Some or all parts of stems, leaves, or flower-bearing branches hairy . . . . . . . . . *y*

y. Leaves arising only from the base of the plant . . . . . . . . . 34. LOMATIUM

y. Leaves present on stem as well as at the base . . . . . . . . . . . . . . *z*

z. Flower-cluster encircled at the base by a crown (involucre) of conspicuous, pinnatifid (deeply cut toward the center, feather-like) bracts . . . . . . . . . . 37. DAUCUS

z. Bracts of involucre at base of flower-clusters either absent or not pinnatifid . . . . . *1*

1. Most of the ultimate (smallest) leaf-divisions of all leaves 10 mm. or more broad. . . *2*

2. Main (primary) rays of the umbel (flower or fruit cluster) 12–35; fruit winged, heart-shaped or rounded at base . . . . . . . . . . . . . 30. ANGELICA

2. Main (primary) rays of the umbel (flower or fruit cluster) 1–6; fruit not winged, long-tapering and narrowed at base . . . . . . . . . 6. OSMORHIZA

1. Ultimate (smallest) leaf-divisions of at least the middle and upper leaves 0.5–10 mm. broad . . . . . . . . . . . . . . . . . . . . . . . . . *3*

3. Flowers yellow; fruit with thick corky wings; main (primary) rays of the umbel (flower or fruit cluster) mostly 8–21 . . . . . . . . . 33. POLYTAENIA

3. Flowers white; fruit without wings, but either ribbed, or rough with short hard points or covered with hooked or barbed bristles or prickles; main (primary) rays of the umbel (flower or fruit cluster) 1–10 . . . . . . . . *4*

4. Fruit or ovary (hypanthium) glabrous (without hairs), or rarely hairy, but not covered by hooked or barbed bristles or prickles nor with short hard

points; bractlets (tiny green leaf-like parts) at base of ultimate (smallest) umbellet (flower or fruit-cluster) rounded or obtuse at tip . . . . . . . . . . . . . . . . 4. CHAEROPHYLLUM

4. Fruit or ovary (hypanthium) covered partly or wholly by hooked or barbed bristles or prickles or by short hard points; bractlets acutely to long-pointed at tip . . . . . . . . . . . . *5*

5. At least the upper leaves with hairy-margined leaf-sheath; rays of the flower or fruit cluster glabrous (without hair) . . . . . . . . . . . . . . . . . . . . . . 5. ANTHRISCUS

5. Upper leaves lacking a hairy-margined leaf-sheath, the hairs covering the sides but not fringing the margins; rays of the flower or fruit cluster hairy . . . . . . . . . . . . 7. TORILIS

x. All parts of plant glabrous (without hairs) . . . . . . . . . . . . . . . . . . . . *6*

6. No leaves present on flowering or fruiting stem, all leaves at base of plant or giving that appearance . . . . . . . . . . . . . . . . . . . . . . . . . . . . *7*

7. Woodland plant; main (primary) rays of flower or fruit cluster 1–5; flower or fruit cluster accompanied at base by a compound leaf; flowers white with purplish-black; fruit as broad as or broader than long, 2–3 mm. long . . . . . . . . . . . . . . . . . . 9. ERIGENIA

7. Plants of open rocky ground and prairies; main (primary) rays of flower or fruit cluster 8–30; flower or fruit cluster not accompanied by any leaf; flowers yellow; fruit longer than broad, 5–8 mm. long . . . . . . . . . . . . . . . . . . . . . . . 34. LOMATIUM

6. Leaves present on flowering or fruiting stem as well as at base of plant . . . . . . . . . *8*

8. Ultimate (smallest) leaf-divisions of at least the upper leaves of the stems thread- or hair-like (filiform), or linear-oblong, elliptic, or the divisions only 0.5–3 (up to 4.5) mm. broad. . . . . *9*

9. Key for plants with flowers . . . . . . . . . . . . . . . . . . . . . . *10*

10. Flowers yellow. . . . . . . . . . . . . . . . . . . . . . . . . . *11*

11. Leaf-stalk (petiole) of larger leaves with a sheathing portion 1–3 cm. long; plants annual or biennial . . . . . . . . . . . . . . . . . . . . 27. ANETHUM

11. Leaf-stalk (petiole) of larger leaves with a sheathing portion 3–10 cm. long; plant perennial . . . . . . . . . . . . . . . . . . . . . . 26. FOENICULUM

10. Flowers white . . . . . . . . . . . . . . . . . . . . . . . . . . *12*

12. Plants of rich woodland; stems 1–2.3 dm. tall; leaves chiefly at the base of the stem . . . . . . . . . . . . . . . . . . . . . . . . . . 9. ERIGENIA

12. Without the above combination of characters; leaves usually scattered along the length of the stem . . . . . . . . . . . . . . . . . . . . . . . . *13*

13. Base of main rays of inflorescence subtended by many green thread-like long bracts, which may be pinnately cleft or divided . . . . . . . . . . . . *14*

14. Ovary (hypanthium) and young fruit with barbed prickles or bristles or hairs; one central flower of the main flower-cluster dark purple; some of the leaf-divisions broader than others. . . . . . . . . . . . . 37. DAUCUS

14. Ovary (hypanthium) and young fruit glabrous (without hairs) and without any prickles or bristles; no dark purple flower in center of main flower-cluster; leaf-divisions uniformly narrow . . . . . . . . . . . . . *15*

15. At least some of the leaf-divisions not thread- or hair-like and 1–10 mm. broad; leaves mostly divided into 3 main stalked leaf-divisions, the 3 principal stalked branches arising together at the summit of the main petiole and dividing into usually 7–15 smaller leaf-divisions. 18. AMMI

15. Some or all of the leaf-divisions thread- or hair-like, 0.1–1.5 mm. broad; leaves pinnately compound, the leaf-divisions arranged feather-like, one above the other on a common axis, the leaf-divisions very numerous, 20 or more . . . . . . . . . . . . . . . . *16*

16. Sepals absent; main leaves of the stem with 2 finely divided leaf-like stipules below the sheathing part of the leaf-stalk (petiole); introduced biennial plants of roadsides and railroads. . . . 19. CARUM

16. Sepals lanceolate to triangular, often minute, but present; stipules not as above; native annual plants of swamps, wet meadows and prairies, and wet places in glades, sometimes in fallow fields 24. PTILIMNIUM

13. Base of main rays of inflorescence subtended by no bracts or by only 1, 2, or few short bracts, which are not cut or divided . . . . . . . . . . . *17*

17. Flowers not all alike or irregular, those on outside with 2 or more petals larger than the others . . . . . . . . . . . . . . . 8. CORIANDRUM

17. Flowers all alike, regular with none of the petals larger than the others . . *18*

18. Ovary (hypanthium) or young fruit roughened by tubercles (short bumps or projections), sharp points, or bristles . . . . . . . . . . . . . . . . . . . . . . . . . . . . . *19*
    19. Ultimate (smallest) leaf-divisions only 1–3 mm. long, linear-oblong, at most only twice as long as broad, pinnatifid (cut nearly to their midrib); individual flower or fruiting stalks (pedicels) mostly equal. . . . . . . . . . . . . . . . . . . . . . . 5. ANTHRISCUS
    19. Ultimate (smallest) leaf-divisions of the main middle and upper leaves 5–30 mm. long, filiform or narrowly linear, 5-many times as long as broad, finely pinnately divided to the midrib; individual flower or fruiting stalks (pedicels) mostly unequal. . . . . 11. SPERMOLEPIS
18. Ovary (hypanthium) or young fruit glabrous (without hair) or without tubercles or bris tles . . *20*
    20. Main (primary) rays of the flower- or fruit-cluster 1–4; bractlets at the base of the ultimate flower- or fruit-cluster (umbellet) conspicuous, oblong, elliptic, or obovate . . 4. CHAEROPHYLLUM
    20. Main (primary) rays of the flower- or fruit-cluster 7–20 or more; no bractlets at the base of the ultimate flower- or fruit-cluster (umbellet), or, if present, not conspicuous and either narrow or hair-like . . . . . . . . . . . . . . . . . . . . . . . . . . . . *21*
        21. Upper leaves of stem ternate (divided into 3 main branches); stems often unbranched, if branched, branched above the middle; ultimate leaf-divisions of middle and upper leaves mostly 20–50 mm. long and 1.5–3 mm. broad . . . . . . . . 22. PERIDERIDIA
        21. All the leaves pinnately divided, feather-like in arrangement; stems usually branched below the middle; ultimate leaf-divisions of middle and upper leaves mostly 5–15 mm. long and 0.5–1 mm. broad . . . . . . . . . . . . . . . . . . . 19. CARUM

9. Key for plants with fruits . . . . . . . . . . . . . . . . . . . . . . . . . . . . *22*
    22. Base of main rays of inflorescence subtended by many green thread-like bracts, which may be pinnately cleft or divided . . . . . . . . . . . . . . . . . . . . . . . . . . *23*
        23. Fruit with barbed prickles, bristles, or hairs . . . . . . . . . . . . . . . 37. DAUCUS
        23. Fruit ribbed, glabrous (without hairs) . . . . . . . . . . . . . . . . . . . *24*
            24. At least some of the leaf-divisions not thread- or hair-like and 1–10 mm. broad; leaves mostly divided into 3 main stalked leaf-divisions, the 3 principal stalked branches arising together at the summit of the main petiole and dividing into usually 7–15 smaller leaf-divisions . . . . . . . . . . . . . . . . . . . . . . 18. AMMI
            24. Some or all of the leaf-divisions thread- or hair-like, 0.1–1.5 mm. broad; leaves pinnately compound, the leaf-divisions arranged feather-like, one above the other on a common axis, the leaf-divisions very numerous, 20 or more . . . . . . . . . . . . . *25*
                25. Sepals absent; main leaves of the stem with 2 finely divided leaf-like stipules below the sheathing part of the leaf-stalk (petiole); introduced biennial plants of roadsides and railroads . . . . . . . . . . . . . . . . . . . . . 19. CARUM
                25. Sepals lanceolate to triangular, often minute, but present; stipules not as above; native annual plants of swamps, wet meadows and prairies, and wet places in glades, sometimes in fallow fields . . . . . . . . . . . . . . 24. PTILIMNIUM
    22. Base of main rays of inflorescence subtended by no bracts or by only 1 or 2 bracts which are not further divided . . . . . . . . . . . . . . . . . . . . . . . . . . . . . *26*
        26. Fruit with tubercles (short bumps or projections), sharp points, or bristles . . . . . . . *27*
            27. Rays of umbels usually unequal; fruit ovoid, without a noticeable beak; fruit-stalks (pedicels) usually unequal; ultimate (smallest) leaf-divisions of the main middle and upper leaves 5–30 mm. long, filiform or narrowly linear, 5-many times as long as broad, finely pinnately divided to the midrib . . . . . . . . . . . . 11. SPERMOLEPIS
            27. Rays of umbel more or less equal; fruit ovoid-lanceolate, with a short but noticeable beak; fruit-stalks (pedicels) more or less equal; ultimate (smallest) leaf-divisions only 1–3 mm. long, linear-oblong, at most only twice as long as broad, pinnatifid (cut nearly to their midrib) . . . . . . . . . . . . . . . . . . . . . . 5. ANTHRISCUS
        26. Fruit glabrous (without hairs), without prickles, bristles, barbs or tubercles . . . . . . . *28*
            28. Fruit with thin wings on the lateral ribs . . . . . . . . . . . . . 27. ANETHUM
            28. Fruit not winged . . . . . . . . . . . . . . . . . . . . . . . . . . . *29*
                29. Bractlets at the base of the ultimate fruiting cluster (umbellet) conspicuous, oblong, elliptic, or obovate . . . . . . . . . . . . . . . . 4. CHAEROPHYLLUM
                29. No bractlets at the base of the ultimate fruiting cluster (umbellet), or, if present, not conspicuous and either narrow or hair-like . . . . . . . . . . . . *30*
                    30. Divisions of lower and basal leaves obovate and much broader than those of main leaves of stem; fruit nearly globose, the two halves (mericarps) difficult to separate . . . . . . . . . . . . . . . . . . . 8. CORIANDRUM

30. Divisions of all the leaves more or less equal and narrowly linear or filiform (thread- or hair-like); fruit of other shapes, but, if somewhat globose, at least the two halves (mericarps) easily separated . . . . . . 31

31. Styles of fruit short, erect; leaf-stalk (petiole) of larger leaves with a sheathing portion 3–10 cm. long . . . . . . 26. FOENICULUM

31. Styles of fruit widely spreading or recurved; leaf-stalk (petiole) of larger leaves with a sheathing portion 1.5–4 cm. long . . . . . . 32

32. Upper leaves of stem ternate (divided into 3 main branches); stems often unbranched, if branched, branched above the middle; ultimate leaf-divisions of middle and upper leaves mostly 20–50 mm. long and 1.5–3 mm. broad . . . . . . 22. PERIDERIDIA

32. All the leaves pinnately divided, feather-like in arrangement; stems usually branched below the middle; ultimate leaf-divisions of middle and upper leaves mostly 5–15 mm. long and 0.5–1 mm. broad . . . . . . 19. CARUM

8. Ultimate (smallest) leaf-divisions of the main leaves of the stems broader, mainly 3–80 mm. broad . 33

33. Key for plants with flowers. . . . . . 34

34. Flowers purple, yellow, or greenish-yellow . . . . . . 35

35. Base of main rays of inflorescence subtended by small leaf-like bracts . . . 32. LEVISTICUM

35. Base of main rays of inflorescence not subtended by any bracts . . . . . . 36

36. Divisions of most of stem-leaves (but not root-leaves) mainly narrowed or wedge-shaped at base, irregularly and unequally toothed, 3–10 mm. broad. . . 33. POLYTAENIA

36. Divisions of at least some of stem-leaves rounded or curved at base, rather regularly and equally toothed, mostly more than 10 mm. broad . . . . . . 37

37. Center flower and fruit of each ultimate umbel (umbellet) sessile (without a stalk); fruit not winged . . . . . . 14. ZIZIA

37. Center flower and fruit of each ultimate umbel (umbellet) with a stalk (the stalk may be very short or inconspicuous); fruit with wings on the dorsal and lateral ribs . . . . . . 29. THASPIUM

34. Flowers white . . . . . . 38

38. Stems 1–2.3 dm. tall; leaves chiefly at base of stem . . . . . . 9. ERIGENIA

38. Stems 3–12 dm. tall; leaves usually scattered on the stem as well as at the base. . . . 39

39. Main (primary) rays of the flower- or fruit-cluster 1–6 . . . . . . 6. OSMORHIZA

39. Main (primary) rays of the flower- or fruit-cluster 10–25. . . . . . 40

40. Margin of leaf-divisions with a narrow cartilaginous border; main axis (rachis) of the leaf with a leafy wing-like part on each side below each group of 3 leaflets . . . . . . 17. FALCARIA

40. Margin of leaf-divisions without a cartilaginous border; no wing-like development along rachis of leaf . . . . . . 41

41. Ultimate (smallest) leaf-divisions of the main leaves of the stems 3–10 mm. broad . . . . . . 42

42. Base of main rays of inflorescence subtended by several narrow simple (undivided) bracts which eventualy fall off; stem purple- or dark-spotted . . . . . . 10. CONIUM

42. Base of main rays of inflorescence subtended by divided bracts which persist in fruit; stem not purple- nor dark-spotted . . . . . . 18. AMMI

41. Ultimate (smallest) leaf-divisions of the main leaves of the stems mostly 10–80 (rarely 5–10) mm. broad . . . . . . 43

43. No bractlets present at base of ultimate flower- or fruit-clusters (umbellets) . . . . . . 21. AEGOPODIUM

43. Several narrow bractlets present at base of ultimate flower- or fruit-clusters (umbellets) . . . . . . 44

44. Ultimate (smallest) divisions of middle stem-leaves narrowly lanceolate to lance-oblong, 5–15 mm. broad, of the lower ones 6–40 mm. broad; styles much longer than the stylopodium (disk-like swelling at base of style); lateral veins of the leaf-division running from the midnerve to the depression (sinus) between the teeth; plants of wet ground . . . . . . 15. CICUTA

44. Ultimate (smallest) divisions of middle stem-leaves mostly ovate or ovate-oblong, mainly 15–80 mm. broad; styles shorter than to slightly longer than the stylopodium (disk-like swelling at base of

style); lateral veins of the leaf-divisions running directly from the midnerve to the tip of the teeth; plants of usually dry ground or rocky dry woodland . . . . . . . . . 27. LIGUSTICUM

33. Key for the plants with fruits . . . . . . . . . . . . . . . . . . . . . . . . . *45*

45. Fruit narrowly long tapering to base with appressed bristles (lying pressed against or parallel to surface) . . . . . . . . . . . . . . . . . . . . . . . . . 6. OSMORHIZA

45. Fruit rounded or curved at base, or if slightly narrowed, glabrous (without hairs) and not bristly . . . . . . . . . . . . . . . . . . . . . . . . . . *46*

46. No bracts subtending base of main rays of inflorescence and no bractlets subtending base of ultimate fruiting clusters (umbellets) . . . . . . . . . . . . . 20. AEGOPODIUM

46. Either bracts or bractlets or both present in some part of inflorescence or ultimate fruiting clusters . . . . . . . . . . . . . . . . . . . . . . . . . *47*

47. Margin of leaf-divisions with a narrow cartilaginous border; main axis (rachis) of the leaf with a leafy wing-like part on each side below each group of 3 leaflets . 17. FALCARIA

47. Margin of leaf-divisions without a cartilaginous border; no wing-like development along rachis of leaf . . . . . . . . . . . . . . . . . . . . . . *48*

48. Stems 1–2.3 dm. tall; leaves chiefly at base of stem . . . . . . . . . 9. ERIGENIA

48. Stems 3–12 dm. tall; leaves scattered along the stem as well as at the base. . . . *49*

49. Center fruit of each ultimate umbel (umbellet) sessile (without a stalk) . 14. ZIZIA

49. Center fruit of each ultimate umbel (umbellet) with a stalk . . . . . . *50*

50. Ultimate (smallest) leaf-divisions of the main leaves of the stem 3–10 mm. broad . . . . . . . . . . . . . . . . . . . . . . *51*

51. Bracts present at base of main rays of inflorescence . . . . . . *52*

52. Bracts at base of main rays of inflorescence narrow and simple (undivided), eventually falling off; stem purple- or dark-spotted; fruit with 5 prominent ribs . . . . . . . . . . . 10. CONIUM

52. Bracts at base of main rays of inflorescence divided and persisting in fruit; stem not purple- nor dark-spotted; fruit lacking prominent ribs . . . . . . . . . . . . . . . . . . 18. AMMI

51. Bracts absent at base of main rays of inflorescence. . . . . . . *53*

53. Stem purple- or dark-spotted; base of main rays of inflorescence subtended by several narrow bracts which eventually fall off; plants of waste ground, roadsides, and railroads; fruit with 5 prominent ribs . . . . . . . . . . . . . 10. CONIUM

53. Stem not spotted; no bracts at base of main rays of inflorescence; plants of prairies and glades; fruit with a thick corky wing on each side . . . . . . . . . . . . . 33. POLYTAENIA

50. Ultimate leaf-divisions of the main leaves of stem mainly 10–80 mm. broad (rarely 5–10 mm.) . . . . . . . . . . . . . . . . *54*

54. Base of main rays of inflorescence subtended by small leaf-like bracts 32. LEVISTICUM

54. Base of main rays of inflorescence not subtended by any bracts . . . *55*

55. Fruit with conspicuous wings developed on the ribs; main rays of the fruiting cluster 4–11; leaves of stem with 3–7 leaflets. 28. THASPIUM

55. Fruit without wings; main rays of the fruiting cluster 9–20; leaves of the stem with 9 or more leaflets . . . . . . . . . . *56*

56. Ultimate (smallest) divisions of middle stem-leaves narrowly lanceolate to lance-oblong, 5–15 mm. broad, of the lower ones 6–40 mm. broad; styles much longer than the stylopodium (disk-like swelling at base of style); lateral veins of the leaf-division running from the midnerve to the depression (sinus) between the teeth and then branching to the end of the tooth; plants of wet ground . . . . . . . . 15. CICUTA

56. Ultimate (smallest) divisions of middle stem-leaves mostly ovate or ovate-oblong, mainly 15–80 mm. broad; styles shorter than to slightly longer than the stylopodium (disk-like swelling at base of style); lateral veins of the leaf-divisions running directly from the midnerve to the teeth; plants of usually dry ground or rocky dry woodland . . . 28. LIGUSTICUM

1629 Hedera Helix (English Ivy)

1630 Hydrocotyle verticillata var. verticillata (Water Pennywort)

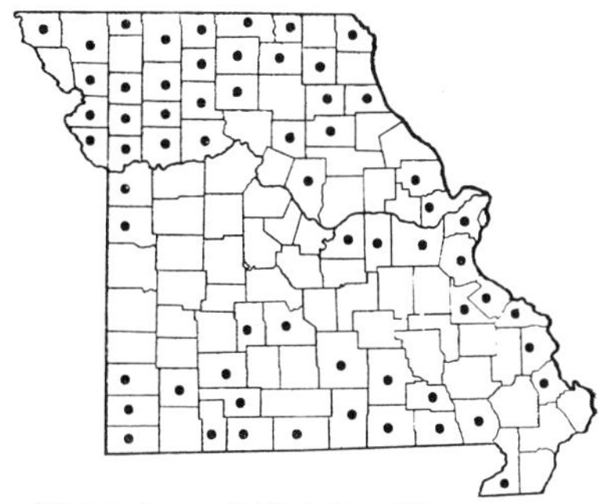

1631 Sanicula gregaria (Black Snakeroot)

## 1. **Hydrocotyle** L. Water Pennywort

**Hydrocotyle verticillata** Thunb. var. **verticillata**
Water Pennywort Map 1630
*Hydrocotyle verticillata* Thunb. [G]
Flowers May–August; fruits July–November.

Occurs along moist banks of spring-fed streams. Known only from Ozark County, southern Missouri (along North Fork of White River, north of Rainbow [Double] Spring, 4 mi. southeast of Dora, July 18, 1936, *Steyermark 11982*; same locality, May 16, 1936, *Steyermark 10400*).

Ranges from Florida to Texas and Mexico, north to Massachusetts, Missouri, Oklahoma, Utah, and Oregon.

The seeds and leaves of this and other species are eaten by wildfowl.

## 2. **Sanicula** L. Black Snakeroot

a. Styles much longer than the bristles of the fruit, and recurved-spreading . . . . . . . 1. S. GREGARIA
a. Styles shorter than the bristles of the fruit, concealed by the bristles, and erect . . . . . . . . . *b*
  b. Common species, throughout Missouri; roots slender, fibrous, cord-like; fruits 3–5 mm. long, on stalks 1–1.5 mm. long; styles not longer than the calyx . . . . . . . . . . 2. S. CANADENSIS
  b. Rarely found, known only from Ripley County, southeastern Missouri; roots tuberous-thickened; fruits 5–6 mm. long, without any stalks (sessile); styles longer than the calyx . . . . 3. S. SMALLII

1. **Sanicula gregaria** Bicknell Black Snakeroot
Map 1631
Also called Sanicle.
Flowers April–June.

Occurs in rich or rocky open woods and thickets. Throughout Missouri, but not recorded from a large sector of the unglaciated prairie portion of southwestern Missouri, and from a number of the Ozark counties.

Ranges from Quebec to Minnesota and South Dakota, south to Florida, Alabama, Missouri, Kansas, and Texas.

2. **Sanicula canadensis** L. Black Snakeroot
Map 1632
Also called Sanicle.
Flowers May–July.

Occurs in rich or rocky open woods and thickets. Two variations are encountered in the state:

Common type found; larger leaflets of the petioled leaves 3.5–8 cm. long, 1.5–4 (up to 5) cm. broad, those of the lower nearly sessile leaves 3–7 cm. long; cluster of fruits 7–9 mm. broad . . . .
2a. S. CANADENSIS var. CANADENSIS

More rarely found; larger leaflets of the petioled leaves 5.5–13 cm. long, 2.5–8 cm. broad, those of the lower nearly sessile leaves 4.5–12 cm. long; cluster of fruits 10–15 mm. broad . . . .
2b. S. CANADENSIS var. GRANDIS

2a. **Sanicula canadensis** var. **canadensis**
Map 1632
*Sanicula canadensis* L. [G, P & S, Steyerm.]

This is the commoner variation and occurs throughout Missouri, probably to be found in every county.

Ranges from Florida to Texas, north to Massachusetts, New Hampshire, Vermont, Pennsylvania, Ohio, Kentucky, Missouri, and Oklahoma.

1632 • Sanicula canadensis var. canadensis (Black Snakeroot)
1632 ‡ Sanicula canadensis var. grandis

1633 Sanicula Smallii

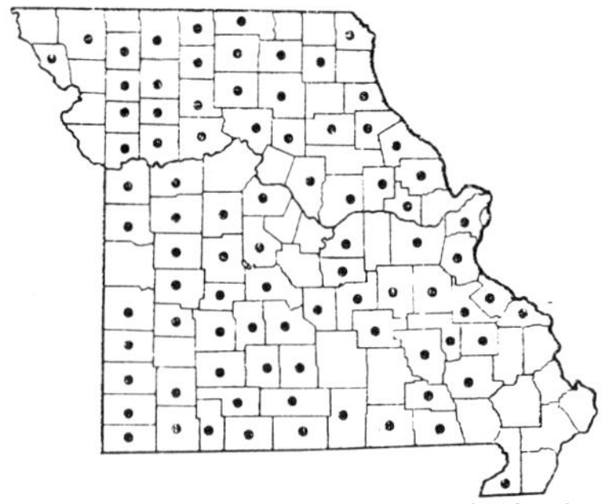

1634 Eryngium yuccifolium var. yuccifolium (Rattlesnake Master)

2b. **Sanicula canadensis** var. **grandis** Fern. Map 1632

This more robust and larger-leaved variety is scattered in several parts of southern and central Missouri, where known from Lincoln (*Steyermark 24556*), Johnson, Clay (*Bush 12751A*), and Jasper (*Palmer 2133*) counties.

More detailed field and experimental studies need to be carried out to determine the status of var. *grandis*.

The two common Sanicles of Missouri may sometimes become too numerous in woodland wildflower gardens and have to be weeded out. The hooked prickles of the fruits attach themselves to clothing in late summer and fall.

3. **Sanicula Smallii** Bickn. Map 1633

Flowers May–June.

Occurs in rich woods. Known only from Ripley County, southeastern Missouri (Pleasant Grove, May 21, 1900, *Bush 358*).

Ranges from Florida to Texas, north to Virginia, Tennessee, and Missouri.

The specimen cited has been verified by Drs. Constance and Shan.

## 3. **Eryngium** L. Eryngo

Stem erect; leaves linear, long and narrow, parallel-veined, with usually spiny projections along margins; heads of whitish flowers . . . . . . . . . . . . . . . . . . . . . 1. E. YUCCIFOLIUM
Stem lying on the ground, creeping; leaves ovate to lanceolate, not parallel-veined, without spiny projections along margins; heads of blue flowers. . . . . . . . . . . . . . . 2. E. PROSTRATUM

1. **Eryngium yuccifolium** Michx. var. **yuccifolium** Rattlesnake Master Map 1634

Also called Button Snakeroot.

*Eryngium yuccifolium* Michx. [G, BB, P & S, Steyerm.]

Flowers July–August; fruits September–November.

Occurs in prairies, glades, and rocky open woodland. Throughout Missouri, except absent from the lowlands of extreme southeastern Missouri.

Ranges from Florida to Texas, north to New Jersey, Connecticut, Ohio, Michigan, Wisconsin, Minnesota, and Kansas.

The yuccalike leaves, clustered at the base of the plant, and encountered for the first time, are unlike most other members of the family found in the eastern and central United States. They are silvery-green and of firm texture. An infusion of the leaves and root of the plant was at one time believed to be an effective remedy in treating bites of rattlesnakes.

A variety *synchaetum* Gray occurs in the southwestern portion of the range of the species.

2. **Eryngium prostratum** Nutt. var. **prostratum** Map 1635

---

Plate no. 265. **1.** Hydrocotyle verticillata, × 2/7; Details from Small, The New York Botanical Garden. **2.** Eryngium prostratum, × 2/7; Details from Small, The New York Botanical Garden. **3.** Eryngium yuccifolium, × 2/7. **4.** Chaerophyllum Tainturieri, × 2/7; *a.* Fruit, × 3 3/7. **5.** Chaerophyllum texanum, × 2/7; *a.* Fruit, × 5 5/7. **6.** Chaerophyllum procumbens; *a.* Chaerophyllum procumbens var. procumbens, × 2/7; *b.* Fruit, × 5 5/7; *c.* Chaerophyllum procumbens var. Shortii, × 5 5/7; Details from Small, The New York Botanical Garden. **7.** Anthriscus scandicina, × 2/7; *a.* Fruit, × 4 4/7. **8.** Sanicula gregaria, × 2/7; *a.* Fruit, × 2 6/7; Details from Small, The New York Botanical Garden.

PLATE NO. 265

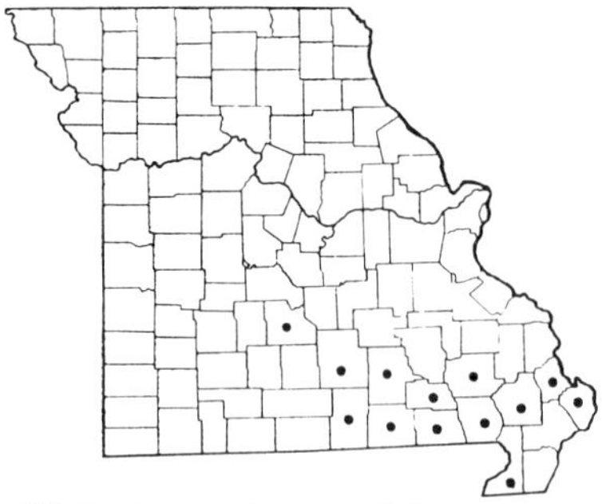

1635 Eryngium prostratum var. prostratum

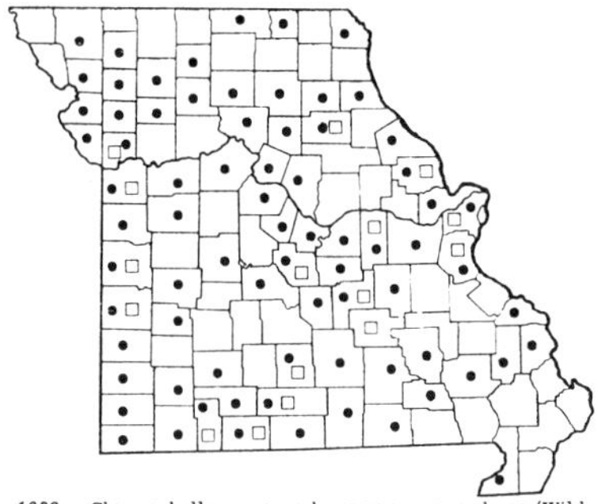

1636 ● Chaerophyllum procumbens var. procumbens (Wild Chervil)
1636 □ Chaerophyllum procumbens var. Shortii

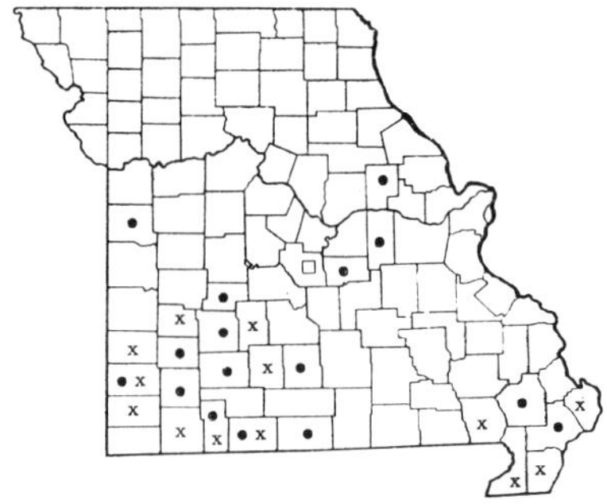

1637 ● Chaerophyllum Tainturieri var. Tainturieri
1637 x Chaerophyllum Tainturieri var. floridanum
1637 □ Chaerophyllum Tainturieri var. dasycarpum

*Eryngium prostratum* Nutt. [G, P & S, Steyerm.]
Flowers May–November.

Occurs in low wet woods, borders of swamps and ponds, and moist sandy prairies. Lowlands of southeastern Missouri, and westward in the Ozarks of southern Missouri, where isolated around upland sink-hole ponds in Carter, Ripley, Shannon, Oregon, Texas, Howell, and Laclede counties. The species reaches its northwestern limit in a sink-hole pond in Laclede County.

Ranges from Florida to Texas, north to South Carolina, Georgia, Kentucky, Missouri, and Oklahoma.

*Excluded Species*

**Eryngium Leavenworthii** T. & G.

A specimen lacking clear locality data was collected by Broadhead in 1879. The writing on the label suggests possibly 'Anderson Co.' with an indication possibly of 'Mo.' appearing below the name of the county. There is no Anderson Co. in Missouri, and the species is eliminated from the flora of the state.

## 4. **Chaerophyllum** L.

a. Very rarely found; fruit and ovary (hypanthium) covered with numerous, short hairs; ribs of fruit broader than spaces between the ribs . . . . . . . . . . . . 2. C. TAINTURIERI var. DASYCARPUM

a. Commonly collected; fruit and ovary (hypanthium) glabrous (without hairs), or if rarely sparsely or minutely hairy, the ribs of fruit are narrower than the spaces between the ribs . . . . . . . *b*

b. Ultimate (smallest) divisions of the leaflets very narrow, linear, of about the same width from one end to the other, most of those of the middle and upper leaves entire (not toothed or further cut), 0.5–1 mm. broad; stylopodia (enlarged base of styles) joined along their inner length or converging; bractlets at base of inflorescences turned down (reflexed) in fruit . . . . . 3. C. TEXANUM

b. Ultimate (smallest) divisions of the leaflets oblong, lanceolate, elliptical, or oblanceolate, broader in one part than another, not uniformly of the same width throughout, some of those of the middle and upper leaves with 1 or more teeth, up to 3 mm. broad; stylopodia (enlarged base of styles) separated from one another, erect or slightly diverging; bractlets at base of inflorescences spreading in fruit . . . . . . . . . . . . . . . . . . . . . . . . . . *c*

c. Pedicels (stalks) of fruits of uniform thickness throughout their length; fruit broadest at or near the middle; ribs on fruit narrower than the spaces between the ribs; common species, throughout Missouri . . . . . . . . . . . . . . . . . . . . . 1. C. PROCUMBENS

c. Pedicels (stalks) of fruits thicker toward the top; fruit broadest below the middle; ribs on fruit broader than the spaces between the ribs, or if narrower than the spaces, then the stems and lower surface of leaf-segments rather hairy; plants of southern and central Missouri . 2. C. TAINTURIERI

1. **Chaerophyllum procumbens** (L.) Crantz
Wild Chervil Map 1636
Flowers early April–June.

Occurs in rich or rocky open woods, alluvial soils along streams and in valleys, thickets, glades, roadsides, and along railroads.

Two variations are found in the state:

Fruits narrowly oblong, 6–10 mm. long, 1.5–2 mm. broad, contracted into a thick neck near the summit, glabrous . . . . 1a. C. PROCUMBENS var. PROCUMBENS

Fruits broadly oblong to narrowly ovate, 4.5–6.5

mm. long, 2–2.5 mm. broad, not contracted at the summit, glabrous or minutely hairy . . . . 1b. C. PROCUMBENS var. SHORTII

1a. **Chaerophyllum procumbens** var. **procumbens** Map 1636

*Chaerophyllum procumbens* (L.) Crantz [G, P & S, Steyerm.]

This is the commoner variation and occurs throughout the state, although not recorded from most of the lowland counties of southeastern Missouri.

Ranges from New York and Ontario to Michigan and Iowa, south to Georgia, Alabama, Mississippi, Arkansas, Kansas, and Oklahoma.

1b. **Chaerophyllum procumbens** var. **Shortii** T. & G. Map 1636

Scattered in southern and central Missouri north to Lincoln, Monroe, and Clay counties.

Ranges from Pennsylvania to Indiana, south to West Virginia, Tennessee, and Louisiana.

The two varieties are not distinguishable unless mature fruit is at hand. Deam (*Fl. Ind.* p. 718. 1940) finds that the var. *Shortii* flowers about 10 days earlier than var. *procumbens* in Indiana, but no similar phenological data is at hand in Missouri to support this observation. There is considerable variation of size of plants and pubescence of leaves and stems in both varieties. Generally, the stem is glabrous, but hairs are often present at the base of the plant and around the nodes of the stem. Likewise, while the leaf-surfaces are generally glabrous, they are often hairy below on the nerves.

2. **Chaerophyllum Tainturieri** Hook. Map 1637

Flowers March–May.

Occurs in rocky open glades, fallow fields, waste ground, and along roadsides.

a. Very rarely found; fruit and ovary (hypanthium) covered with numerous, short hairs . 2c. C. TAINTURIERI var. DASYCARPUM
a. Common types encountered; fruit and ovary (hypanthium) glabrous (without hairs) . . *b*
  b. Ribs of fruit broader than the spaces between the ribs . . . 2a. C. TAINTURIERI var. TAINTURIERI
  b. Ribs of fruit narrower than the spaces between the ribs . . . 2b. C. TAINTURIERI var. FLORIDANUM

2a. **Chaerophyllum Tainturieri** var. **Tainturieri** Map 1637

*Chaerophyllum Tainturieri* Hook. [G, BB]

Scattered in southern and central Missouri north to Montgomery and Cass counties, commonest in the southwestern quarter of the state.

Ranges from Florida to Texas, north to Virginia, Ohio, Indiana, Missouri, and Kansas.

2b. **Chaerophyllum Tainturieri** var. **floridanum** C. & R. Map 1637

Scattered in southern Missouri north to Mississippi, Butler, Dallas, Cedar, and Barton counties; absent from most of the Ozark area.

Ranges from Florida to Virginia and Missouri.

2c. **Chaerophyllum Tainturieri** var. **dasycarpum** Wats. Map 1637

Known only from Miller County, northern Ozark region of middle Missouri (along highway 52, 2½ mi. northwest of Tuscumbia, May 6, 1957, *Steyermark 83986*).

Ranges from Alabama and Mississippi to Louisiana and Texas, north to Missouri and Oklahoma.

The above cited specimen has been verified by Drs. Mildred Mathias and Lincoln Constance.

*Chaerophyllum Tainturieri* often occurs in dense masses with erect stems. The stems and leaves are generally more conspicuously hairy, and the stems average usually taller, stiffer, and more erect than in *C. procumbens*, but are about as tall as in *C. texanum*.

3. **Chaerophyllum texanum** Coult. & Rose Map 1638

Flowers April–June.

Occurs in rocky prairies and glades, also introduced along roadsides and railroads. Mainly in the southwestern third of the state north to Caldwell County, introduced eastward to Bollinger, St. Louis, and Boone counties.

Ranges from Missouri and Kansas to Louisiana and Texas.

This species has more numerous pedicels in a single inflorescence than in either of the other two Missouri species, generally 6–15 pedicels being present, whereas in *C. Tainturieri* 3–10 pedicels are developed, and in *C. procumbens* only 1–6.

## 5. **Anthriscus** Bernh. Chervil

**Anthriscus scandicina** (Weber) Mansf. Chervil Map 1639

Flowers May–June; fruits June–July.

Occurs along railroads. Known only from St. Louis

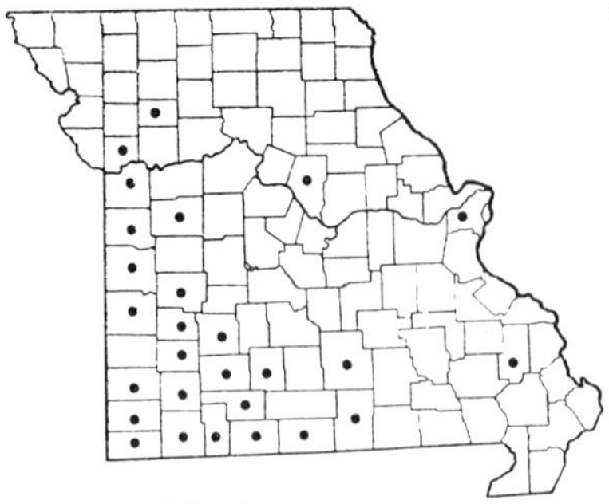
1638 Chaerophyllum texanum

1639 Anthriscus scandicina (Chervil)

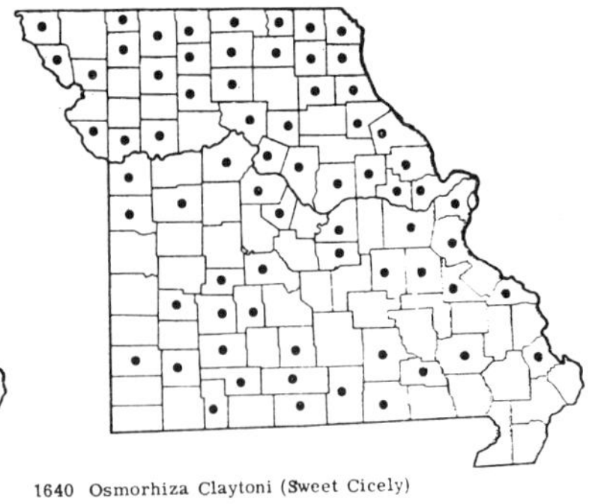
1640 Osmorhiza Claytoni (Sweet Cicely)

County, east-central Missouri (St. Louis, between track on First Street and Plum Street, May 25, 1957, *Muehlenbach 1150*; same locality, May 13, 1956, *Muehlenbach 887*).

Native of Europe; introduced and naturalized in North America in Nova Scotia, Virginia, Missouri, and in the Pacific states.

### 6. **Osmorhiza** Raf. Sweet Cicely

Styles in flower and fruit not more than 1.5 mm. long, not longer than the petals, recurved in flower, erect and straight in fruit . . . . . . . . . . . . . . . . . . . . . . 1. O. CLAYTONI
Styles in flower at least 2 mm. long, in fruit 2–4 mm. long, longer than the petals and spreading in fruit 2. O. LONGISTYLIS

1. **Osmorhiza Claytoni** (Michx.) Clarke
Sweet Cicely Map 1640

Also known as Woolly Sweet Cicely.

Flowers April–June; fruits June–August.

Occurs in rich woods on slopes of hills, ravines, in valleys, and in thickets. Throughout Missouri, except absent from most of the lowland counties of southeastern Missouri.

Ranges from Quebec and Nova Scotia to Manitoba and Saskatchewan, south to North Carolina, Alabama, Missouri, Arkansas, and Kansas.

The roots of this species are only slightly aromatic as compared with the next species.

2. **Osmorhiza longistylis** (Torr.) DC. Anise Root
Map 1641

Also called Sweet Anise, Smooth Sweet Cicely.

Flowers April–June; fruits June–August.

Occurs in rich woods and thickets, often on north-facing wooded slopes, ravines, and valley woodland.

a. Stems glabrous (without hairs) or nearly so . 2a. O. LONGISTYLIS var. LONGISTYLIS
a. Stems, leaf-stalks (petioles), and base of the branches more or less hairy . . . . . *b*
b. Rarely found variation; hairs very short, 0.5 mm. long. . . . 2b. O. LONGISTYLIS var. BRACHYCOMA
b. More commonly found; hairs more conspicuous, 0.5–2 mm. long . 2c. O. LONGISTYLIS var. VILLICAULIS

2a. **Osmorhiza longistylis** var. **longistylis**
Map 1641

*Osmorhiza longistylis* (Michx.) Clarke [G, BB, P & S, Steyerm.]

Scattered throughout Missouri.

Ranges from Quebec to Alberta, south to Virginia, Kentucky, Missouri, Oklahoma, and New Mexico.

2b. **Osmorhiza longistylis** var. **brachycoma** Blake Map 1641

Rare and scattered, in Putnam, Lincoln (*Davis 2388*), Dent (*Lodewkys 283*), and Phelps (*Steyermark 22179*) counties.

Ranges from Connecticut to Ontario and Michigan, south to Virginia, West Virginia, Ohio, Indiana, and Missouri.

Plate no. 266. **1.** Osmorhiza Claytoni, $\times$ 2/7; *a.* Fruit, $\times$ 4; Details from Small, The New York Botanical Garden. **2.** Osmorhiza longistylis, $\times$ 2/5; *a.* Fruit, $\times$ 4. **3.** Torilis nodosa, $\times$ 2/7; After Gleason, details from Small. The New York Botanical Garden.

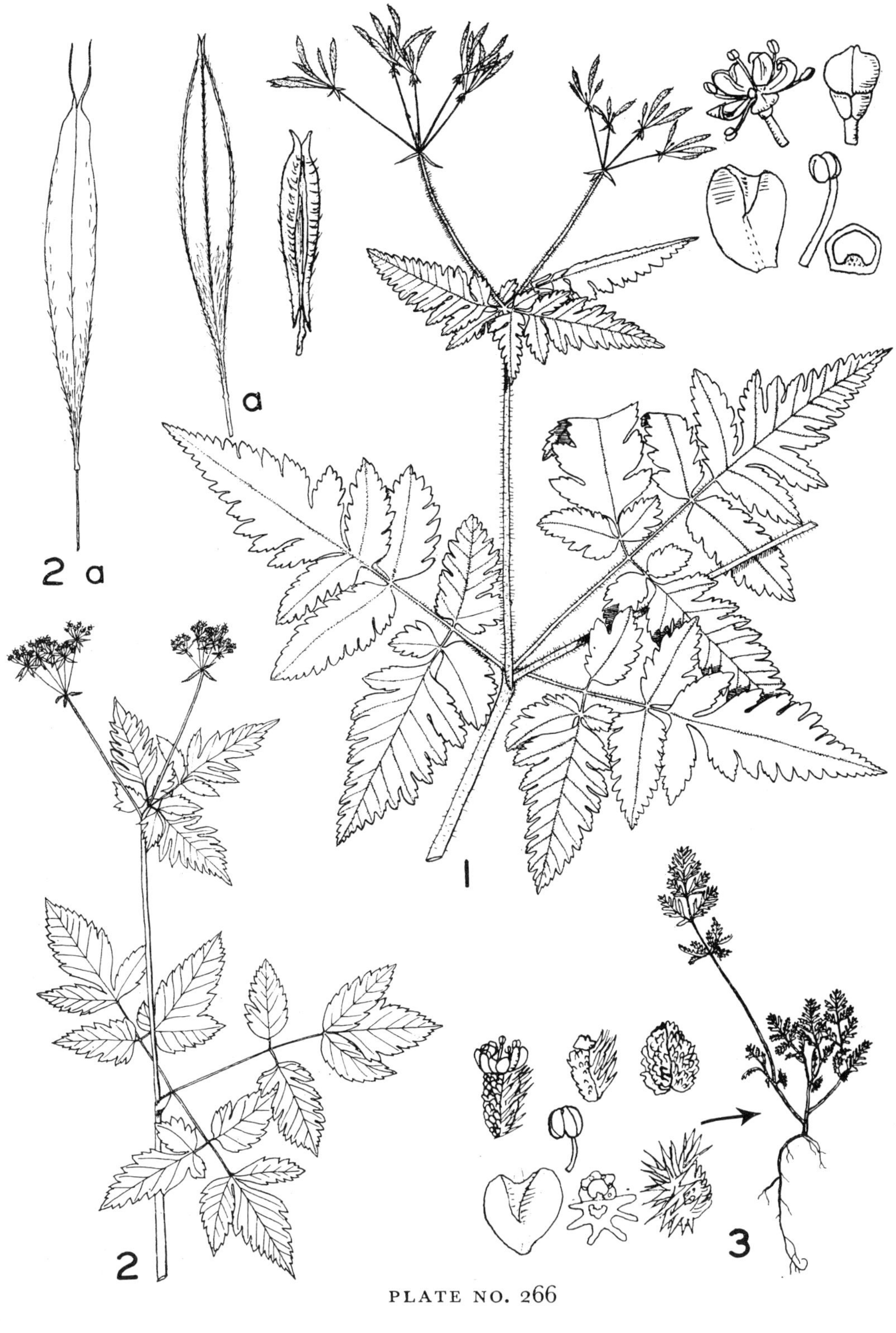


PLATE NO. 266

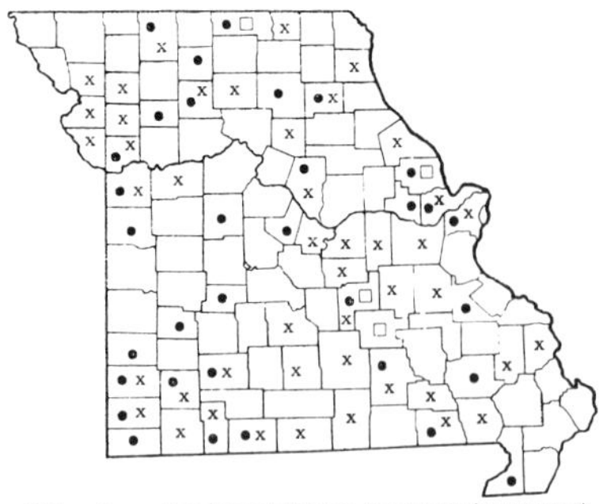

1641 • Osmorhiza longistylis var. longistylis (Anise Root)
1641 □ Osmorhiza longistylis var. brachycoma
1641 x Osmorhiza longistylis var. villicaulis

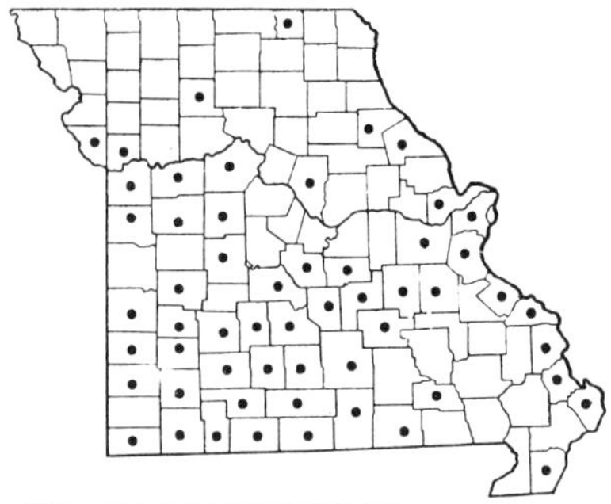

1642 Torilis japonica (Hedge Parsley)

1643 Torilis nodosa

2c. **Osmorhiza longistylis** var. **villicaulis** Fern. Map 1641

Scattered throughout Missouri, and sometimes commoner in some localities than var. *longistylis.*

Ranges from Alabama to Texas, north to Connecticut, New York, Ontario, Michigan, Illinois, and Missouri.

*Osmorhiza longistylis* has pleasantly sweet aromatic roots, which are sometimes used as a flavoring. The roots yield an oil similar to that of anise (*Pimpinella Anisum*).

## 7. **Torilis** Adans. Hedge Parsley

Inflorescences on long peduncles (main flower-stem) raised above the leaves and arising at the tip of the stem as well as from the sides; inflorescence loose and open, 15–50 mm. broad; main (primary) rays of inflorescence 1–3 cm. long; common throughout the southern half of the state . . . . . 1. T. JAPONICA

Inflorescence nearly sessile (stalkless) or on a short peduncle, hidden among and shorter than the leaves, arising opposite the leaves along the sides of the stem; inflorescence congested, head-like, 5–10 mm. broad; main (primary) rays of inflorescence 0.1–0.3 cm. long; rare, known only from St. Louis County 2. T. NODOSA

1. **Torilis japonica** (Houtt.) DC. Hedge Parsley Map 1642

Also called Hemlock Chervil.

*Torilis Anthriscus* of auth. [P & S, Steyerm.], not (L.) Bernh.

Flowers early June–August.

Occurs in open and waste ground, roadsides, along railroads, thickets, alluvial ground, gravel bars, and recently cleared woodland. Common throughout the southern and central sections, north to Ralls, Boone, Livingston, and Platte counties, and locally north in Schuyler County.

Native of Europe and Asia; introduced and naturalized in the United States from New York to Ohio, Indiana, Illinois, and Iowa, south to Florida and Texas; also in the Pacific states and West Indies.

This species has been introduced in Missouri relatively recently, since 1900. A specimen (*Palmer 2221*) collected in 1909 appears to be the earliest record for this species in Missouri. The fruit, which is densely covered with rough hooked bristles, clings easily to clothes and hairy skin.

2. **Torilis nodosa** (L.) Gaertn. Map 1643

Flowers May–June; fruits May–July.

Occurs along railroads. Known only from St. Louis County, east-central Missouri (right-of-way of Wabash Railroad, north of Palm Street, St. Louis, May 30, 1958, *Muehlenbach 1403*).

Native of Europe; introduced and naturalized in the southern United States, California, and Oregon, north to Missouri.

This species grows much lower than *T. japonica*, mainly 1–2 dm. tall, whereas *T. japonica* has stems usually 3–8 dm. tall. Some of the fruits of *T. nodosa* are merely covered with tubercles, others are bristly; in the Muehlenbach specimen cited, some of the central fruits are mainly tuberculate with short hispid hairs, while the other fruits have long bristly hairs on one side (or half) and tubercles only on the other side.

1643-A Spermolepis divaricata

1644 Coriandrum sativum (Coriander)

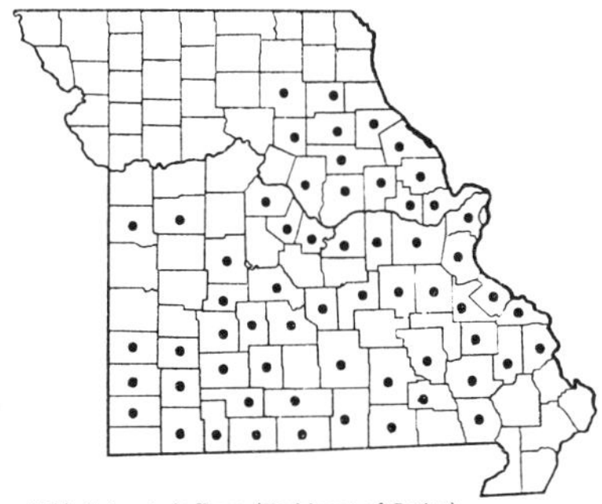

1645 Erigenia bulbosa (Harbinger of Spring)

## 8. **Coriandrum** L. Coriander

**Coriandrum sativum** L. Coriander Map 1644

Flowers May–June.

Occurs in waste ground, along roadsides and railroads. Known only from central Missouri in St. Louis, Boone, and Jackson counties.

Native of Europe and Asia; cultivated as a spice and occasionally escaping from cultivation and naturalized in many parts of the United States.

The dried fruits are pleasantly scented and are used either dried or as an extract of oil for flavoring dishes, especially in Europe and Asia. During World War II it was used by the Jugoslavians as an acceptable substitute for pepper. An essential oil produced from the fruit is used in medicine and in flavoring gin, whiskey, and various beverages.

## 9. **Erigenia** Nutt. Harbinger of Spring

**Erigenia bulbosa** (Michx.) Nutt. Harbinger of Spring Map 1645

Also called Pepper and Salt.

Flowers January–April.

Occurs in rich woods usually at or near the base of slopes or in alluvial woodland soils along streams and in valleys, alluvial thickets, and at the base of rocky bluffs. Common throughout the Ozark region of southern and east-central Missouri, north in the eastern half of the state only to Ralls, Shelby, and Macon counties, in western Missouri south of the Missouri River only to Benton, Johnson, and Cass counties.

Ranges from Ontario and New York to Michigan, Wisconsin, and Minnesota, south to Alabama, Mississippi, Arkansas, and Missouri.

This species and Snow Trillium (*Trillium nivale*) are the two earliest of the herbaceous native plants to bloom in Missouri. Usually, however, Harbinger of Spring precedes the Snow Trillium by as much as three weeks. Because of its very early bloom and low growth, the tiny white and purplish-black flower-clusters barely protruding a few inches above the ground, it is easily overlooked and infrequently seen or collected in the flowering stage. The small globose, fleshy tuberous root is stated to be edible.

## 10. **Conium** L. Poison Hemlock

**Conium maculatum** L. Poison Hemlock Map 1646

Flowers May–August.

Occurs in waste ground, fields, pastures, fence rows, thickets, low ground, and along roadsides and railroads. Scattered throughout Missouri.

Native of Europe; introduced and naturalized in North America, where ranging from Quebec to Florida west to the Pacific states.

All parts of this plant are reputed to be poisonous if eaten, especially the flowers, fruits, and leaves. The root is reported to be nearly harmless in early spring, but poisonous later. Stock are poisoned especially in spring from eating the fresh young leaves and stems. If a person has eaten some part of the plant, a strong emetic to induce vomiting should be given to rid the system of the poison.

Specimens sometimes misidentified as *Conioselinum chinense* have been found referable to *Conium maculatum*.

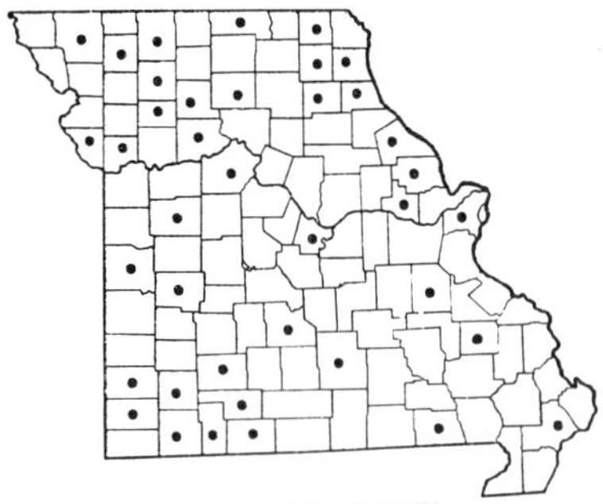

1646 Conium maculatum (Poison Hemlock)

1647 Spermolepis divaricata

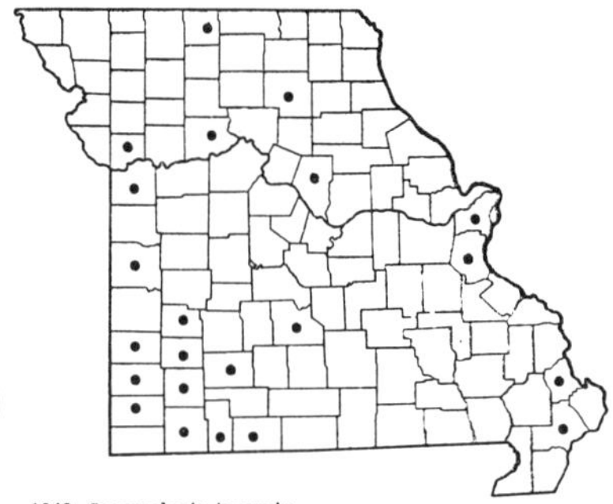

1648 Spermolepis inermis

## 11. **Spermolepis** Raf.

a. Ovary (hypanthium) and fruit covered with hooked short bristles . . . . . . . . . 3. S. ECHINATA
a. Ovary (hypanthium) and fruit smooth or roughened with tubercles (short wart-like projections) . . . *b*
 b. Main (primary) rays of the inflorescence 2–4, more or less equal, spreading and distant from one another, 5–20 mm. in flower, up to 30 mm. in fruit . . . . . . . . . . . 1. S. DIVARICATA
 b. Main (primary) rays of the inflorescence 2–9, very unequal, erect to strongly ascending, close to one another, 2–7 mm. long in flower, up to 20 mm. long in fruit . . . . . . . . 2. S. INERMIS

1. **Spermolepis divaricata** (Walt.) Britt. Maps 1643A and 1647

Flowers April–June.

Occurs in sandy and fallow fields and along railroads. Known only from Dunklin County, southeastern Missouri (open sandy ground, 2 mi. south of Malden, May 26, 1939, *Steyermark 26555*; *Steyermark 71222*; *Steyermark 26568*; Malden, August 21, 1894, *Bush 325*).

Ranges from Florida to Texas, north to Virginia, Missouri, Kansas, and Oklahoma.

2. **Spermolepis inermis** (Nutt.) Math. & Const. Map 1648

*Spermolepis patens* (Nutt.) Robins. [P & S, Steyerm.]

Flowers May–June.

Occurs in rocky prairies, glades, sandy and fallow fields, and rarely along railroads. Scattered mostly in southern and central Missouri, and north locally in Macon and Mercer counties.

Ranges from Illinois to Nebraska, southwest to Oklahoma, Texas, and Mexico; occasionally introduced eastward.

3. **Spermolepis echinata** (Nutt.) Heller Map 1649

Flowers April–June.

Occurs in rocky prairies, glades, sandy or gravelly ground along streams, rarely along railroads. Mainly southern Missouri to Ste. Genevieve, Iron, and Polk counties, locally north to Boone County (railroad tracks, Rocheport, May 18, 1935, *Rickett 731*, U. of Mo. Herb.).

Ranges from Mississippi to Arizona and California, north to Missouri and Oklahoma; also in Mexico.

All three species of *Spermolepis* have slender erect stems often only 1–3 dm. tall, but attaining a height of 5–6 dm. The leaves in all species are similarly finely dissected.

## 12. **Apium** L.

**Apium graveolens** L. Celery Map 1650

Flowers May–July.

Commonly cultivated and rarely escaping to railroads or waste ground. Known as an escape only from St. Louis County, east-central Missouri (St. Louis, reserve tracks by Federal Cold Storage Division of

Plate no. 267. **1.** Torilis japonica, × 2/5; *a.* Fruit, × 10; *b.* Basal leaf, × 2/5. **2.** Erigenia bulbosa, × 2/5; Details from Small, The New York Botanical Garden. **3.** Spermolepis echinata, × 2/5; *a.* Fruit, × 8; Details after Small, The New York Botanical Garden. **4.** Spermolepis inermis, × 2/5; *a.* Fruit, × 8; Details from Small, The New York Botanical Garden. **5.** Coriandrum sativum, × 2/5; *a.* Fruit, × 2 4/5; Details from Small, The New York Botanical Garden.

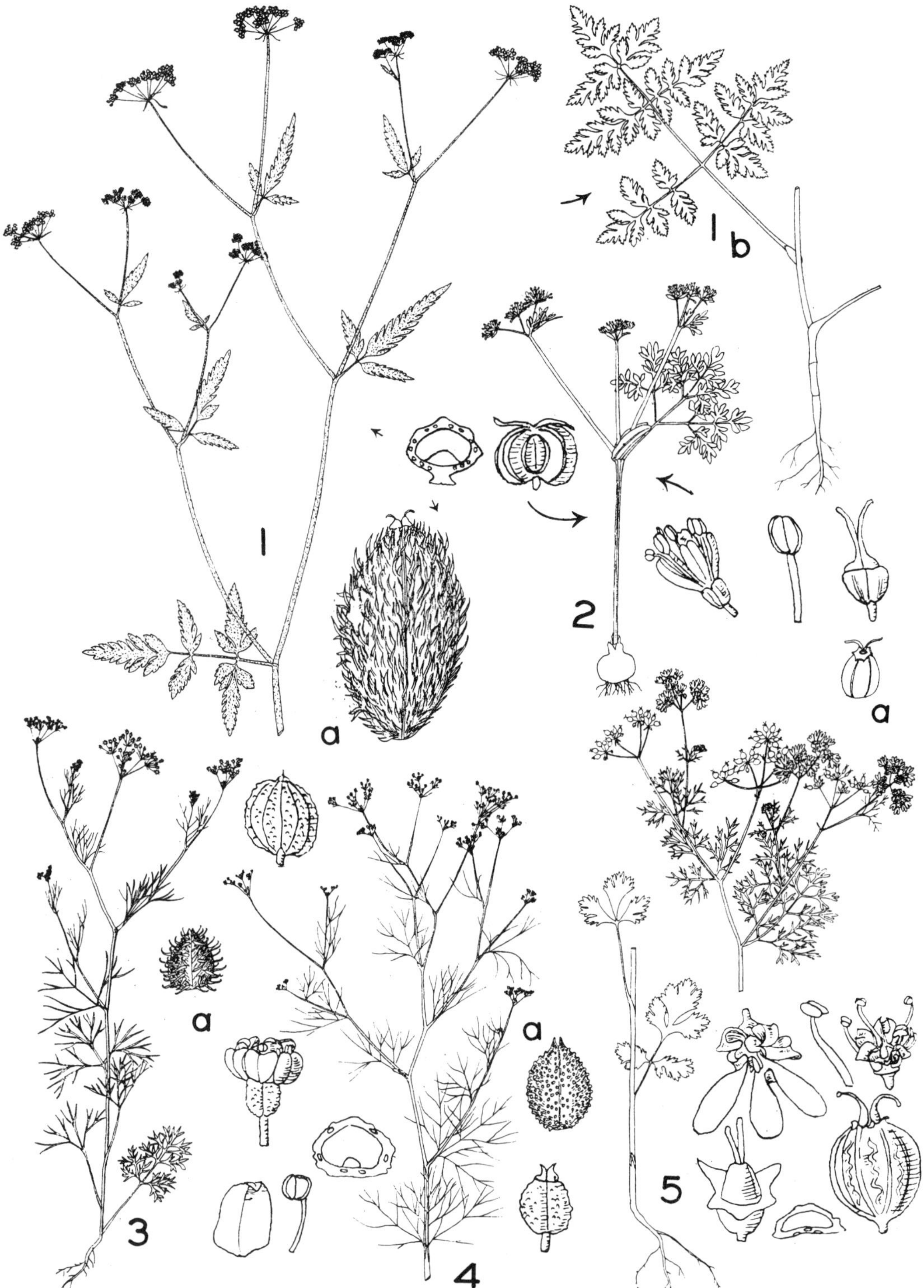


PLATE NO. 267

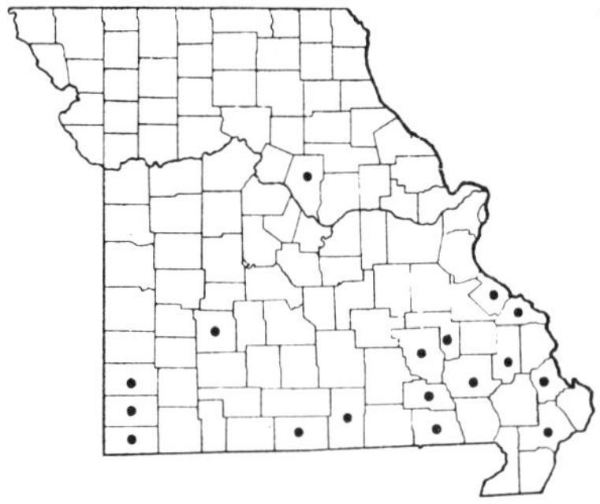

1649 Spermolepis echinata

1650 Apium graveolens (Celery)

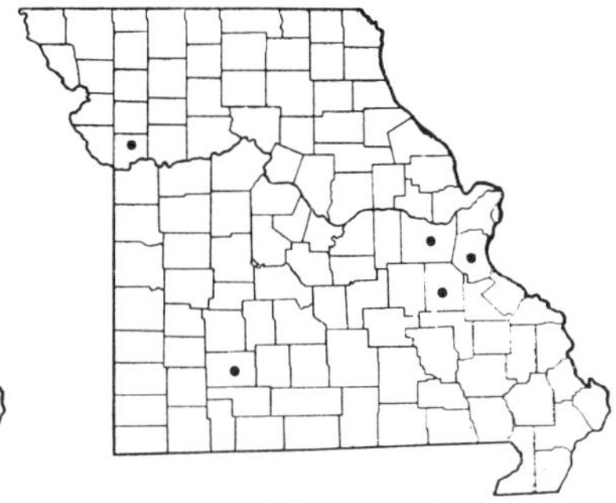

1651 Bupleurum rotundifolium (Thoroughwax)

City Products Corporation Building, September 7, 1957, *Muehlenbach 1349*). Two sterile specimens were found at this locality.

Native of Europe and Asia; well established in the southwestern states, but only occasionally naturalized in the eastern states.

## 13. **Bupleurum** L. Thoroughwax

**Bupleurum rotundifolium** L. Thoroughwax Map 1651

Also called Hare's Ear.

Flowers May–July.

Occurs in fields, waste ground, and along roadsides. Rare and scattered in southern and central Missouri in Jefferson, Franklin, Washington, Greene, and Clay counties.

Native of Europe and Asia, introduced and naturalized in the United States from Alabama to Tennessee and Arkansas, north to New England, New York, West Virginia, Indiana, Missouri, and South Dakota.

Like *Eryngium yuccifolium*, this species has an aspect wholly unlike the usual appearance believed to be characteristic of a member of the Carrot family. The mustard yellow umbels of nearly sessile flowers surrounded by the large, broadly ovate bractlets resemble some of the species of the genus *Euphorbia* having a similar habit. The perfoliate upper leaves are also quite unusual for a member of this family.

## 14. **Zizia** Koch Golden Alexanders

Leaves at the base of plant not divided, broadly ovate to nearly round, heart-shaped at base; leaflets of the stem-leaves not more than 3, rather blunt or somewhat acutely pointed . . . . . . 1. Z. APTERA

Leaves at the base of plant all or nearly all divided into 3–11 leaflets; leaflets of the stem-leaves 3–11, noticeably acute (pointed) at tip . . . . . . . . . . . . . . . . . . . . 2. Z. AUREA

1. **Zizia aptera** (Gray) Fern. Golden Alexanders Map 1652

*Zizia cordata* of auth. [P & S, Steyerm.], not *Smyrnium cordatum* Walt., the basis for the name.

Flowers April–June; fruits July–September.

Occurs in prairies, glades, rich or rocky woods, along bluff escarpments, and thickets. Southern and central Missouri, north to Ralls, Callaway, Boone, and Jackson counties; absent from the lowland counties of southeastern Missouri.

Ranges from New York to Manitoba and British Columbia, south to Georgia, Tennessee, Missouri, Oklahoma, and Texas.

This species is often confused with *Thaspium trifoliatum* var. *flavum*, both having thickened, white, glabrous leaf margins, but the central flower of each umbellet in *Zizia* is sessile, stalked in *Thaspium*.

2. **Zizia aurea** (L.) Koch f. **aurea** Golden Alexanders Map 1653

Plate no. 268. **1.** Conium maculatum, × 2/7; Details from Small, The New York Botanical Garden. **2.** Zizia aurea, × 2/7; *a.* Fruiting cluster, × 6/7; *b.* Fruit, × 4 2/7; Details from Small, The New York Botanical Garden. **3.** Bupleurum rotundifolium, × 2/7; After Gleason, details from Small, The New York Botanical Garden. **4.** Cicuta maculata, × 2/7; Details from Small, The New York Botanical Garden. **5.** Zizia aptera, × 2/7; *a.* Fruit, × 4 2/7; *b.* Fruiting cluster, × 6/7; *c.* Basal leaves, × 2/7. **6.** Apium graveolens, × 2/7; After Gleason, The New York Botanical Garden.

PLATE NO. 268

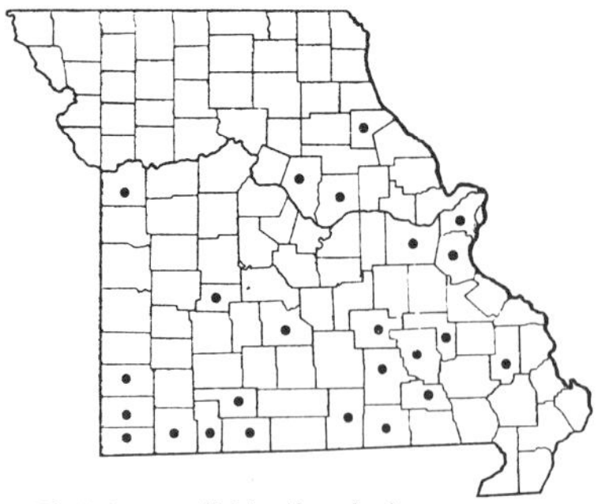

1652 Zizia aptera (Golden Alexanders)

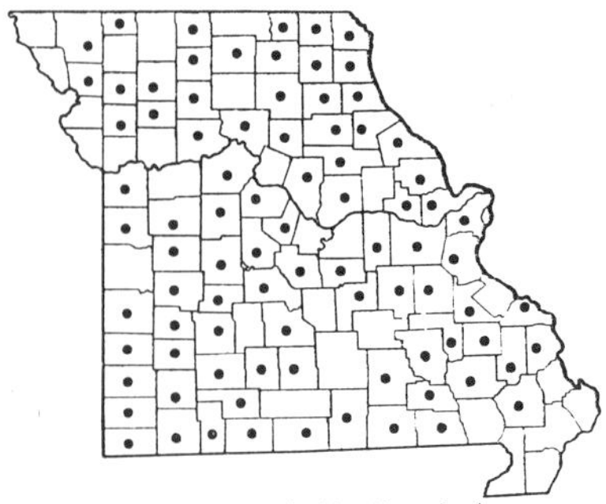

1653 Zizia aurea f. aurea (Golden Alexanders)

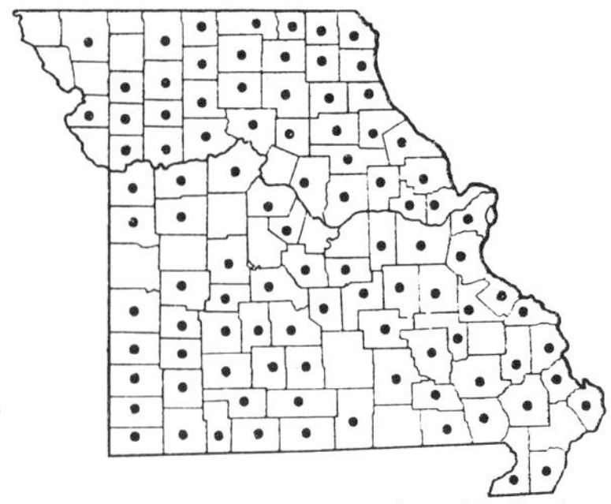

1654 Cicuta maculata var. maculata (Spotted Cowbane)

Also called Meadow Parsnip.

*Zizia aurea* (L.) Koch [G, BB, P & S, Steyerm.]

Flowers April–June; fruits July–September.

Occurs in prairies, glades, moist meadows, wooded thickets, fields, rich woods at base of bluffs, alluvial soils in woodland, and open woodland. Throughout Missouri, except not recorded from most of the southeastern lowland counties.

Ranges from Quebec and Maine to Saskatchewan, south to Florida, Tennessee, Missouri, Arkansas, Oklahoma, and Texas.

This species generally occurs in small colonies, where its yellow flower-clusters are conspicuous in prairies or meadows in early spring, standing out well above the generally lower growing flowering plants.

The leaflets are usually ovate to lanceolate with acute or sharp-pointed summits, but in f. *obtusifolia* (Bissell) Fern. they are rounded-elliptical to obovate or suborbicular and blunt at the tip. This form has not been found in Missouri thus far.

## 15. **Cicuta** L. Water Hemlock

**Cicuta maculata** L. var. **maculata** Spotted Cowbane Map 1654

Also called Water Hemlock.

*Cicuta maculata* L. [G, P & S, Steyerm.

Flowers late May–September.

Occurs along borders of ponds, sloughs, ditches, spring branches, and streams, and in wet depressions of prairies and low and upland meadows, rarely along railroads. Throughout Missouri, doubtless in every county.

Ranges from Quebec to Minnesota and Manitoba, south to North Carolina, Tennessee, Missouri, Oklahoma, and Texas.

The stems especially in the lower part, are often mottled purple as in *Conium maculatum.* All parts of the plant, but especially the roots and rootstocks, are poisonous. In fact, the plant is considered to contain one of the most violent poisons of any native North American species. However, the leaves and fruits when still green and in hay may be eaten by stock without causing poisoning. It is stated that a walnut-sized portion of the root, if eaten, is sufficient to kill a cow. The fleshy roots, with the fragrance of parsnips and resembling small sweet potatoes, sometimes mistaken by children and adults for parsnips, sweet anise (*Osmorhiza longistylis*), Jerusalem artichokes (*Helianthus tuberosus*), and others, have led to cases of fatal poisoning. If a strong emetic followed by a purgative medicine is given to anyone known to have eaten any of the plant, recovery is likely to occur. The seeds of this and other species are eaten by wildfowl.

The var. *Curtissii* (C. & R.) Fern. is distinguished from typical var. *maculata* by having the fruits with a dark furrow separating the marginal or lateral ribs, larger leaflets of the lower leaves 3–5.5 cm. broad instead of 0.6–3 (–4) cm. broad, and with the margins of the leaflets coarsely crenate or dentate with broad-based teeth instead of sharply serrate with long-pointed lanceolate teeth. It has not been found, thus far, in Missouri.

## 16. **Cryptotaenia** DC. Honewort

**Cryptotaenia canadensis** (L.) DC. Honewort Map 1655

Also called Wild Chervil.

Flowers May–August.

1655 Cryptotaenia canadensis (Honewort)

1656 Falcaria sioides (Sickleweed)

1656A Ammi majus

Occurs in rich or rocky woods, often in ravines, low ground in wooded valleys and along streams, ledges along bluffs, and rarely along railroads. Throughout Missouri.

Ranges from Quebec and New Brunswick to Manitoba, south to Georgia, Alabama, Arkansas, and Texas.

This plant can become aggressive and weedy in woodland wildflower gardens, and, like *Sanicula gregaria* and *canadensis*, often has to be eradicated each year to maintain upkeep of the more prized wildflower specimens.

In Japan a closely related or nearly identical species is used as an important vegetable, the tops serving as greens and for flavoring soups, while the blanched stems are either boiled as a vegetable or eaten raw in salads. The root, boiled and served with oil, is also eaten.

## 17. **Falcaria** Bernh. Sickleweed

**Falcaria sioides** (Wibel) Aschers. Sickleweed Map 1656

*Falcaria vulgaris* Bernh. [G]

Flowers July–September; fruits August–October.

Occurs in waste places and roadsides. Known only from Boone County, central Missouri (along a fence row in bottomland soil on the roadside between Hartsburg and Welton, July 20, 1940, *Mrs. Grace Ernst Miller*, Univ. of Mo. Herb.).

Native of Europe and Asia; introduced and naturalized in the United States from New York to Iowa and Nebraska, south to Pennsylvania, Missouri, and Kansas.

## 18. **Ammi** L.

**Ammi majus** L. Map 1656A

Flowers May–July.

Occurs along railroads. Known only from St. Louis County, east-central Missouri (St. Louis, Wabash R.R. along the most western siding north of Palm Street, July 4, 1959, *Muehlenbach 1539*).

Native of Europe; introduced into the United States from South Dakota to Texas, east to Missouri.

## 19. **Carum** L. Caraway

**Carum Carvi** L. f. **Carvi** Caraway Map 1657

*Carum Carvi* L. [G, BB, P & S, Steyerm.]

Flowers May–July.

Sometimes cultivated and in Missouri rarely escaped from gardens to roadsides and waste ground. Known only from central Missouri in Boone and Jackson counties.

Native of Europe and Asia; introduced and naturalized in North America from Newfoundland to British Columbia, south to Virginia, Pennsylvania, Indiana, Illinois, and Missouri.

The fruits are commonly used for seasoning bread, cakes, and cheese. They are used in the alcoholic liqueur, kümmel, and in other beverages, as well as in medicine and various perfumes. In Europe the young leaves are eaten in salad, while the older leaves are cooked as a green vegetable. The thickish biennial root is sometimes eaten, after being cooked, and is said to be superior to parsnips.

1657 Carum Carvi f. Carvi (Caraway)

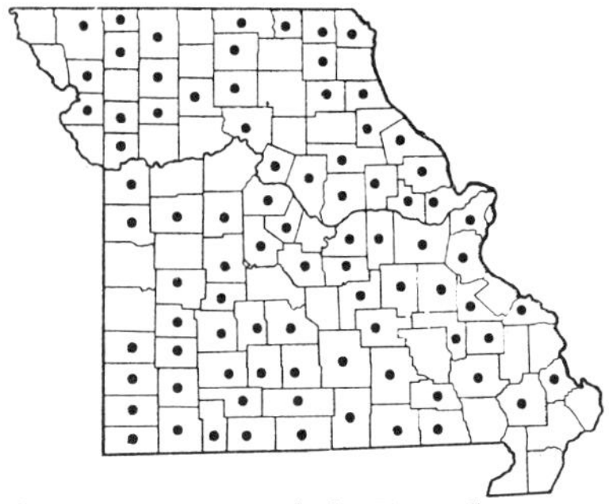

1658 Taenidia integerrima (Yellow Pimpernel)

1659 Aegopodium Podagraria (Goutweed)

### 20. **Taenidia** Drude Yellow Pimpernel

**Taenidia integerrima** (L.) Drude Yellow Pimpernel Map 1658

Flowers May–July.

Occurs in rocky or dry open or upland woods, prairies, and prairie openings, and ledges along bluffs. Throughout Missouri; commonest in the Ozark region, absent from most of the lowland counties of extreme southeastern Missouri.

Ranges from Quebec to Minnesota, south to Georgia, Alabama, Mississippi, Louisiana, and Texas.

### 21. **Aegopodium** L. Goutweed

**Aegopodium Podagraria** L. Goutweed Map 1659

Flowers May–August.

Planted in gardens and rarely escaping to waste ground. Known only from Jefferson County, east-central Missouri (Festus, May 20, 1940, *Bauer*).

Native of Europe and Asia; introduced into gardens and naturalized in North America from Newfoundland to Michigan, south to North Carolina and Missouri.

This species spreads in a short time and covers the ground, especially where there is shade and moisture. The var. *variegatum* Bailey, with white-margined leaves, is the most common horticultural variety, and is much used as a ground cover for shaded areas.

### 22. **Perideridia** Reichenb.

**Perideridia americana** (Nutt.) Reichenb. Map 1660

*Eulophus americanus* Nutt. [P & S, Steyerm.]

Flowers late April–July.

Occurs in rocky open woods on limestone and cherty limestone substrata, borders of open woods, along rocky bluffs and glades, usually of limestone. Ozark region of southern and east-central Missouri north to Lincoln, Montgomery, Phelps, Laclede, Dallas, Polk, and Vernon counties, with an isolated station in extreme northeastern Missouri in Scotland County.

Ranges from Ohio to Michigan and Illinois, south to Tennessee, Arkansas, Kansas, and Oklahoma.

### 23. **Sium** L. Water Parsnip

**Sium suave** Walt. f. **suave** Water Parsley Map 1661

Also called Water Parsnip.

*Sium suave* Walt. [G, BB, P & S, Steyerm.]

---

Plate no. 269. **1.** Carum Carvi, $\times {}^{2}/_{7}$. **2.** Cryptotaenia canadensis, $\times {}^{2}/_{7}$; *a.* Fruit, $\times 3^{3}/_{7}$; Details from Small, The New York Botanical Garden. **3.** Falcaria sioides, $\times {}^{2}/_{7}$; *a.* Fruit, $\times 2^{6}/_{7}$; After Gleason, The New York Botanical Garden. **4.** Aegopodium Podagraria, $\times {}^{2}/_{7}$; Details from Small, The New York Botanical Garden. **5.** Perideridia americana, $\times {}^{2}/_{7}$; Details from Small, The New York Botanical Garden. **6.** Taenidia integerrima, $\times {}^{2}/_{7}$; *a.* Fruit, $\times 4^{4}/_{7}$; Details from Small, The New York Botanical Garden. **7.** Sium suave, $\times {}^{2}/_{7}$; Details from Small, The New York Botanical Garden.

PLATE NO. 269

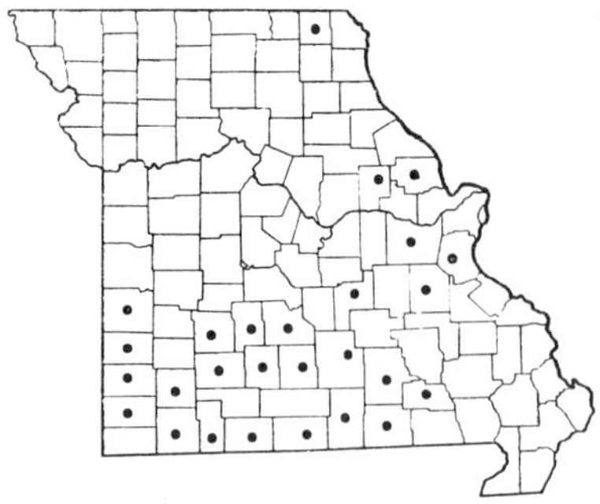

1660 Perideridia americana

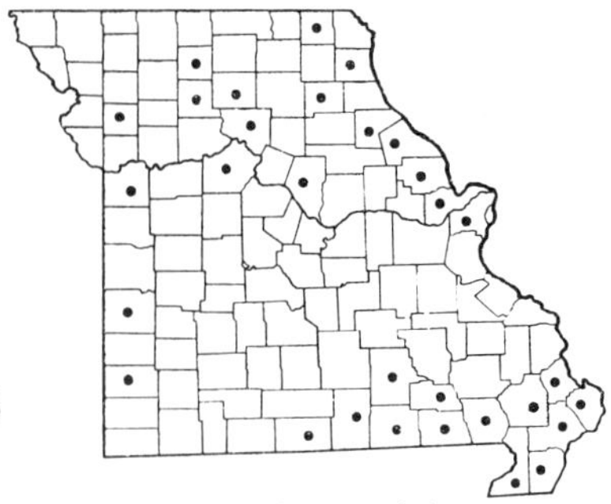

1661 Sium suave f. suave (Water Parsley)

1662 Ptilimnium capillaceum (Mock Bishop's Weed)

Flowers July–September.

Occurs in wet prairies, borders of swamps, ponds, sloughs, wet woodland, and spring-fed streams. Scattered in Missouri, chiefly in the northern, central, and southeastern sectors, absent from most of the Ozark and unglaciated prairie region.

Ranges from Newfoundland to British Columbia, south to Florida, Ohio, Indiana, Illinois, Missouri, Kansas, Colorado, Utah, Nevada, and California.

An aquatic form, f. *Carsonii* (Durand) Fassett, not known in Missouri, is distinguished by having fewer (3–9) oblong to ovate shorter leaflets 1–4 cm. long.

The lowermost and basal leaves, when growing or submerged in water, are usually finely and deeply dissected into filiform or threadlike divisions. The stems of *S. suave* are corrugated and not purple-mottled as in *Cicuta maculata* or *Conium maculatum*. Although *S. suave* has been suspected of causing cases of cattle poisoning, no real evidence is at hand to substantiate the reports. Some Indian tribes of Canada have been reported to have eaten the roots of this species, which is related to the edible Old World Skirret, *Sium Sisarum*

## 24. **Ptilimnium** Raf. Mock Bishop's Weed

a. Styles shorter than the stylopodia (enlarged base of style); calyx-teeth broadly triangular; bracts at base of main (primary) rays mainly 3-cleft . . . . . . . . . . . . . . 1. P. CAPILLACEUM
a. Styles longer than the stylopodia (enlarged base of style); calyx-teeth narrowly triangular or lanceolate; bracts at base of main (primary) rays mostly remaining undivided or rarely cleft . . . . . . *b*
  b. Main (primary) leaf-divisions arising either alternately or opposite on the axis (rachis) of the leaf 2. P. NUTTALLII
  b. Main (primary) leaf-divisions arising in whorls (verticillate) on the axis (rachis) of the leaf. 3. P. COSTATUM

1. **Ptilimnium capillaceum** (Michx.) Raf. Mock Bishop's Weed Map 1662

Flowers June–August; fruits July–October.

Occurs in wet ground. Known only from Iron County, southeastern Ozark region (Pilot Knob, June 17, 1888, *L. A. Pammel*, in Mo. Bot. Gard. Herb.).

Ranges from Florida to Texas, north to Massachusetts, inland north to Missouri, Kansas, and Oklahoma.

The specimen cited is the only one referred to this species by Dr. William Easterly who has recently revised the genus. Other specimens previously referred to this species in Palmer and Steyermark's *Annotated Catalogue* are now assigned to *P. Nuttallii.*

2. **Ptilimnium Nuttallii** (DC.) Britt. Mock Bishop's Weed Map 1663

Flowers mid-June–August.

Occurs in wet places in prairies, swampy meadows, wet depressions in rocky glades, and swampy ground, and rarely along railroads. Southern and central Missouri north to St. Louis, Franklin, Osage, and Clay counties, but mainly in the southern portion of the state.

---

Plate no. 270. **1.** Ptilimnium capillaceum, × 2/5; *a.* Fruit, × 4; After Gleason, details from Small, The New York Botanical Garden. **2.** Ptilimnium Nuttallii, × 2/5; *a.* Fruit, × 4; After Gleason, details from Small, The New York Botanical Garden. **3.** Ptilimnium costatum, × 2/5; *a.* Fruit, × 4; After Gleason, details from Small, The New York Botanical Garden. **4.** Cynosciadium pinnatum, × 2/5. **5.** Cynosciadium digitatum, × 2/5; Details from Small, The New York Botanical Garden.

PLATE NO. 270

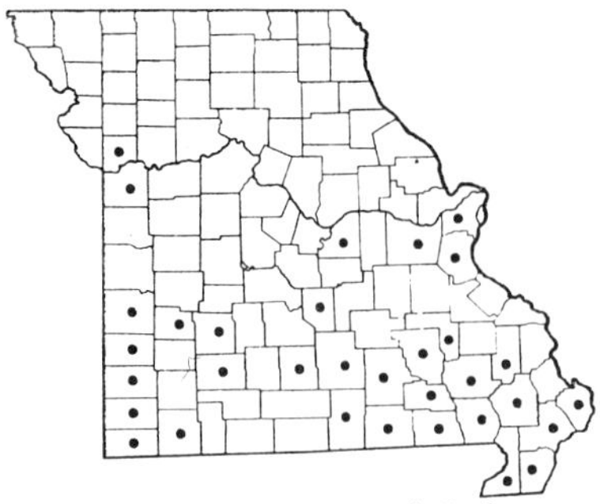

1663 Ptilimnium Nuttallii (Mock Bishop's Weed)

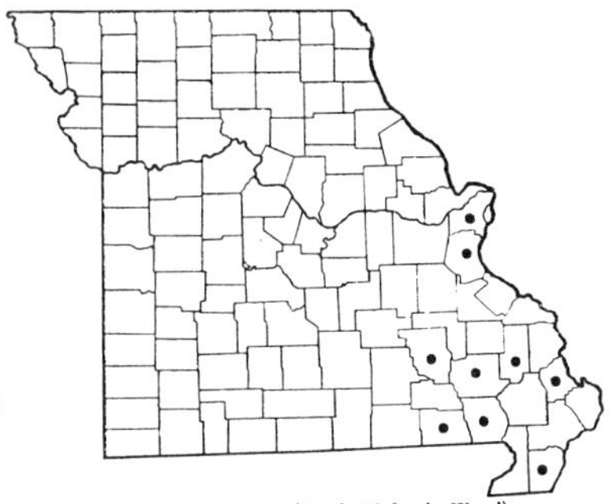

1664 Ptilimnium costatum (Mock Bishop's Weed)

1665 Cynosciadium pinnatum

Ranges from Alabama to Texas, north to Kentucky, Illinois, Missouri, and Kansas.

All the specimens previously identified as *P. capillaceum* and so included in Palmer and Steyermark's *Annotated Catalogue* are treated in the present flora under *P. Nuttallii*, following the recent study of the genus by Dr. William Easterly.

3. **Ptilimnium costatum** (Ell.) Raf. Mock Bishop's Weed Map 1664

Flowers July–September.

Occurs in swampy meadows, wooded swamps, and low wet woodland bordering sloughs, bayous, and streams. Eastern Missouri south of the Missouri River north to St. Louis County, west to Reynolds and Ripley counties.

Ranges from Georgia to Louisiana and Texas, north to North Carolina, Illinois, Missouri, and Oklahoma.

## 25. **Cynosciadium** DC.

This genus has been divided into *Limnosciadium* and *Cynosciadium* by Drs. Mathias and Constance. The differences used by these authors to separate the species involved, do not appear to justify, in the present author's opinion, their segregation.

Main leaves of stem pinnately compound, their 3–11 divisions arising from different levels along the leaf-axis (rachis); fruit without a noticeable beak . . . . . . . . . . . . . . . 1. C. PINNATUM
Main leaves of stem palmately compound, their 3–5 divisions all arising at the same point or level at tip of the leaf-axis (rachis); fruit with a noticeable beak because of the constriction below the stylopodia 2. C. DIGITATUM

1. **Cynosciadium pinnatum** DC. Map 1665
*Limnosciadium pinnatum* (DC.) Math. & Const. [BB]
Flowers May–August.

Occurs in wet depressions and borders of shallow pools in rocky glades, rarely introduced along railroad tracks. Southwestern Missouri in Jasper, Newton, and McDonald counties, and locally introduced along railroads in St. Louis County, east-central Missouri (St. Louis, between Biddle and Ashley streets, west of Second Street, June 16, 1956, *Muehlenbach 938*).

Ranges from Louisiana and Texas, north to Missouri, Iowa, Kansas, and Oklahoma.

2. **Cynosciadium digitatum** DC. Map 1666
Flowers May–June.

Occurs in swamps and low wet woods bordering bayous, sloughs, and slow streams. Southeastern Missouri lowlands in Pemiscot, Dunklin, Butler, and Ripley counties.

Ranges from Mississippi to Texas, north to Missouri and Oklahoma.

Some authors have attributed to *C. pinnatum* the character of conspicuous septations (transverse lines) on the leaflets as contrasted with inconspicuous or faint septations or the lack of them in *C. digitatum*. Actually, both species may show conspicuous septations.

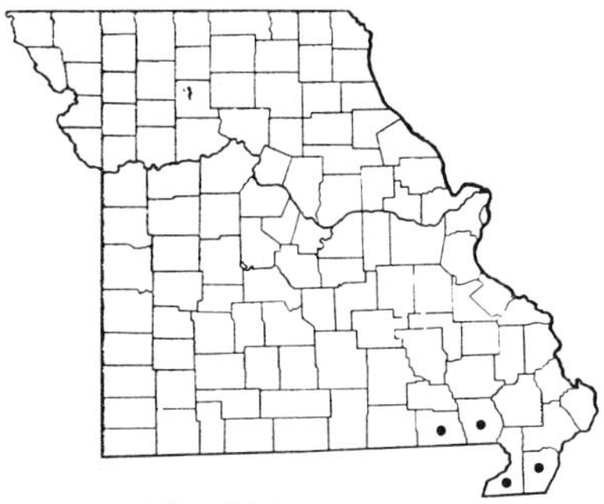
1666 Cynosciadium digitatum

1667 Foeniculum vulgare (Fennel)

1668 Anethum graveolens (Dill)

## 26. **Foeniculum** Mill. Fennel

**Foeniculum vulgare** Mill. Fennel Map 1667

Flowers late May–September.

Sometimes cultivated and escaped along roadsides and waste ground. Known only from St. Louis, Boone, and Jackson counties in central Missouri.

Native of Europe; introduced and naturalized in the United States from Connecticut to Michigan and Nebraska, south to Florida and Texas.

All parts of the plant are strongly aromatic. The young, tender leafy shoots are eaten like celery or used as a salad. The leafy shoots themselves as well as the aromatic fruits are used to season various dishes, and the roots and leaves are cooked as a vegetable and are especially popular among French and Italian cooks. The fruits are used also to flavor candy and various liqueurs. An oil from the fruits is used in medicine, and in soaps and perfumery.

The enlarged, fleshy, bleached leaf-stalks of one variety of fennel, known as finocchio or Florence Fennel (*F. vulgare* var. *dulce* [Mill.] Fiori) are eaten as a vegetable.

## 27. **Anethum** L. Dill

**Anethum graveolens** L. Dill Map 1668

Flowers June–August.

Cultivated and rarely escaping to roadsides, railroad tracks, and waste ground. Known from scattered areas in northern and central Missouri in St. Louis (*Muehlenbach 939*), Grundy, Nodaway (*Singleton 152*), and Jackson (*Bush*, July 15, 1892) counties.

Native of Europe and Asia; introduced and naturalized in the United States from Connecticut and Pennsylvania to Minnesota and North Dakota, south to Florida, Missouri, and Texas.

In the United States the flower- and fruit-stalks and flower-clusters of this aromatic plant are used for flavoring pickles. In some countries of Europe and Asia dill is employed in sauces, stews, and soups. An oil expressed from the plant and the fruit is used in medicine.

## 28. **Ligusticum** L. Lovage

**Ligusticum canadense** (L.) Britt. Angelico Map 1669

Flowers mid-May–July.

Occurs in rocky woodland in moist or dry, low or upland, well-drained soils, in ravines, along creeks, wooded slopes and bluffs. Ozark region of southern Missouri north to Ste. Genevieve, Iron, Dent, Texas Wright, Webster, Christian, Stone, and McDonald counties.

Ranges from Pennsylvania and West Virginia to Kentucky, south to Georgia and Alabama; and in Missouri and Arkansas.

## 29. **Thaspium** Nutt. Meadow Parsnip

a. Stem with a beard of short hairs at least at the upper nodes where the stem joins the leaf; margins of leaves and leaflets green and fringed with short hairs; leaves at base of plant divided into 3 main

1669 Ligusticum canadense (Angelico)

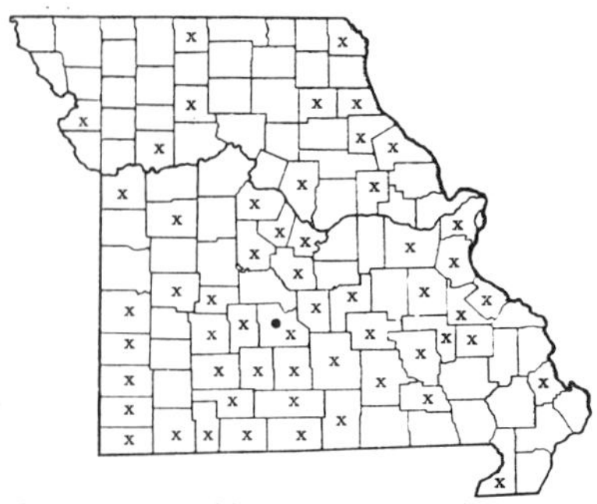

1670 • Thaspium trifoliatum var. trifoliatum (Meadow Parsnip)

1670 x Thaspium trifoliatum var. flavum (Meadow Parsnip)

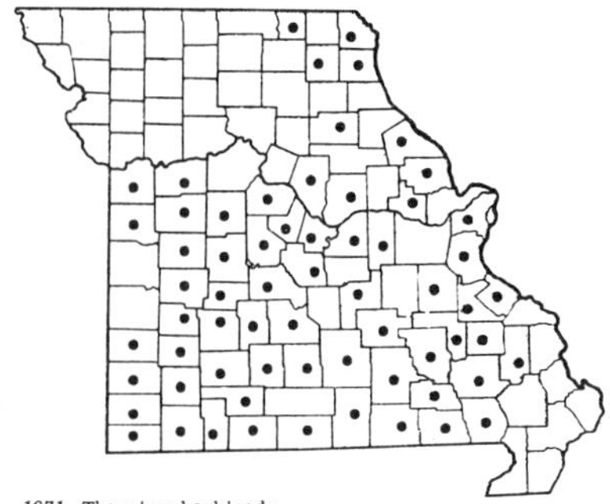

1671 Thaspium barbinode

divisions which are again divided either 2 or 3 times (twice or thrice ternate); flowers creamy- or pale yellow . . . . . . . . . . . . . . . . . . . . . . . . 2. T. BARBINODE

a. Plants completely glabrous (without hairs); margins of leaves and leaflets white and glabrous; leaves at base of plant simple and heart-shaped or only divided usually once into 3 leaflets (ternate); flowers butter yellow or rarely purple or brownish-purple . . . . . . . . . . . . . . *b*

b. Flowers purple or brownish-purple; very rarely found . . . 1a. T. TRIFOLIATUM var. TRIFOLIATUM

b. Flowers yellow; commonly found . . . . . . . . . . . . . 1b. T. TRIFOLIATUM var. FLAVUM

1a. **Thaspium trifoliatum** (L.) Gray var. **trifoliatum** Meadow Parsnip Map 1670

*Thaspium trifoliatum* (L.) Gray [G, BB, Steyerm.]

Flowers April–June; fruits August–October.

Occurs in rocky open woods. Known only from Laclede County, western Ozark region (1936, *George Moore*).

Ranges from Florida to Louisiana, north to New Jersey and Pennsylvania, Kentucky, and Missouri.

The vegetative as well as the floral parts of the Missouri specimen were purplish. No other specimen than that discovered by Mr. Moore in Laclede County has been found.

1b. **Thaspium trifoliatum** var. **flavum** Blake Meadow Parsnip Map 1670

*Thaspium trifoliatum* of [BB], not (L.) Gray

Flowers May–June; fruits July–September.

Occurs in prairies, rocky open woods, ledges along bluffs, and thickets. Commonest in southern, central, and eastern Missouri, and northwest locally in Mercer, Livingston, Ray, and Buchanan counties.

Ranges from New York to Minnesota and South Dakota, south to Georgia, Alabama, Arkansas, and Oklahoma.

This is frequently confused with both *Zizia aurea* and *Z. aptera*; it is distinguished from both by the stalked central flower or fruit of each umbellet and by the broadly winged fruits.

2. **Thaspium barbinode** (Michx.) Nutt. var. **barbinode** Map 1671

Flowers April–June; fruits July–October.

Occurs in rich or rocky open woods, generally on steep slopes and ledges of wooded bluffs and along streams.

Eastern, southern, and central Missouri, north to Schuyler, Boone Cooper, Lafayette, and Jackson counties.

Ranges from New York to Ontario and Minnesota, south to Florida, Georgia, Alabama, Mississippi, Arkansas, and Oklahoma.

Another variation, var. *angustifolium* C. & R., not known from Missouri, has 15–75 leaflets at the first fork of the stem instead of the usual 9–15, and these leaflets are 2–10 mm. instead of 0.5–5 mm. broad; also the leaflets of the basal leaves are narrower and cut into linear or oblanceolate portions.

Occasionally plants are found in which most of the nodes are glabrous, but generally they are short-hairy. In addition, the peduncles (main flower-stems), rays of the inflorescence, and furrows of the fruits are short-hairy, as well as the veins on the lower leaf-surface.

---

Plate no. 271. **1.** Foeniculum vulgare, $\times$ 2/7; Details from Small, The New York Botanical Garden. **2.** Ligusticum canadense, $\times$ 2/7; *a.* Fruit, $\times$ 2 6/7; Details from Small, The New York Botanical Garden. **3.** Thaspium trifoliatum var. flavum, $\times$ 2/7. **4.** Thaspium barbinode, $\times$ 2/7; *a.* Fruit, $\times$ 5 1/7; Details from Small, The New York Botanical Garden.

PLATE NO. 271

1672 Angelica venenosa (Wood Angelica)

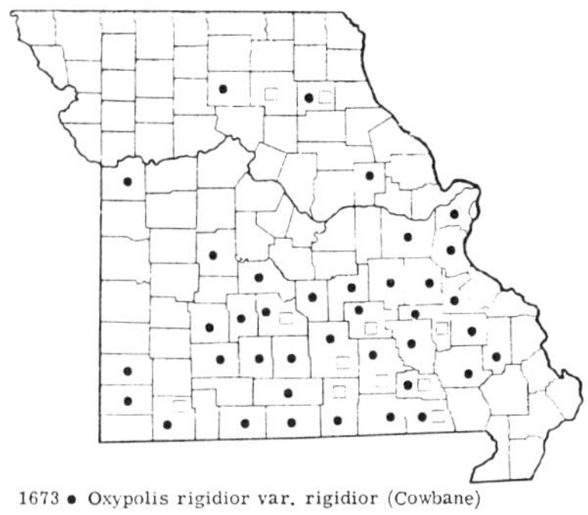

1673 • Oxypolis rigidior var. rigidior (Cowbane)
1673 □ Oxypolis rigidior var. ambigua

1674 Levisticum paludapifolium (Lovage)

## 30. **Angelica** L. Angelica

**Angelica venenosa** (Greenway) Fern. Wood Angelica — Map 1672

Also called Hairy Angelica.

*Angelica villosa* (Walt.) BSP. [P & S, Steyerm.]

Flowers late May–July; fruits August–October.

Occurs in dry rocky woodland, low rich woods along streams, cherty and sandstone wooded slopes and ridges, and prairie openings. Ozark region east to Washington, Iron, and Carter counties, north to Crawford, Phelps, Pulaski, Laclede, Greene, Lawrence, and Newton counties.

Ranges from Massachusetts and Connecticut to Michigan, Illinois, and Minnesota, south to Florida, Alabama, Mississippi, Missouri, and Arkansas.

This species is stated to be quite poisonous when eaten, and a detailed account of the effects of the poison is given by Dr. James Greenway (Trans. Am. Phil. Soc. 3: 235. 1793).

## 31. **Oxypolis** Raf. Hog Fennel

**Oxypolis rigidior** (L.) C. & R. Cowbane — Map 1673

Flowers July–September; fruits September–November.

Occurs in wet soils along streams and spring branches, swampy meadows, wet places in prairies, moist shaded bluffs, and swampy places.

Two variations occur in Missouri:

Leaflets 4–40 mm. broad, oblong, lanceolate, or narrowly obovate . a. O. RIGIDIOR var. RIGIDIOR
Leaflets 2–4 mm. broad, narrowly linear . . b. O. RIGIDIOR var. AMBIGUA

a. **Oxypolis rigidior** var. **rigidior** — Map 1673

*Oxypolis rigidior* (L.) Raf. [G, BB, P & S]

Mainly in the Ozark region of southern and east-central Missouri, north locally in Montgomery, Shelby, Linn, and Jackson counties. Regarding the occurrence of the species in Jackson County, Bush (letter of September 15, 1933, to the author) writes '*Oxypolis* is Ozarkian, and does not occur in Jackson County.'

Ranges from Florida to Louisiana, north to New York, Ontario, Ohio, Michigan, Wisconsin, and Minnesota.

b. **Oxypolis rigidior** var. **ambigua** (Nutt.) Robins. — Map 1673

Mainly in the southern Ozark region, and locally in northern Missouri in Shelby and Macon counties.

Ranges from New Jersey and Delaware to Missouri and Louisiana.

Deam (*Fl. Ind.* p. 727. 1940) did not recognize var. *ambigua*, basing his conclusions on a study of variations which to him indicated that the smaller plants had fewer or no teeth on the leaflets, and that leaflets of the nontoothed plants varied from 2.5–15 mm. broad. More detailed field and experimental studies need to be carried on before a final evaluation of this narrow-leaved variation can be determined.

The roots and leaves of *Oxypolis rigidior* are poisonous, and are reported to poison cattle. Cases of dermatitis have been reported from persons whose skin is allergic to the plant.

---

Plate no. 272. **1.** Angelica venenosa, $\times {}^2/_7$; *a.* Fruit, $\times 7^1/_7$; Details from Small, The New York Botanical Garden. **2.** Lomatium foeniculaceum, $\times {}^2/_7$; *a.* Fruit, $\times 3^5/_7$. **3.** Polytaenia Nuttallii, $\times {}^2/_7$; Details from Small, The New York Botanical Garden. **4.** Oxypolis rigidior, $\times {}^2/_7$; Details from Small, The New York Botanical Garden.

PLATE NO. 272

1675 Polytaenia Nuttallii (Prairie Parsley)

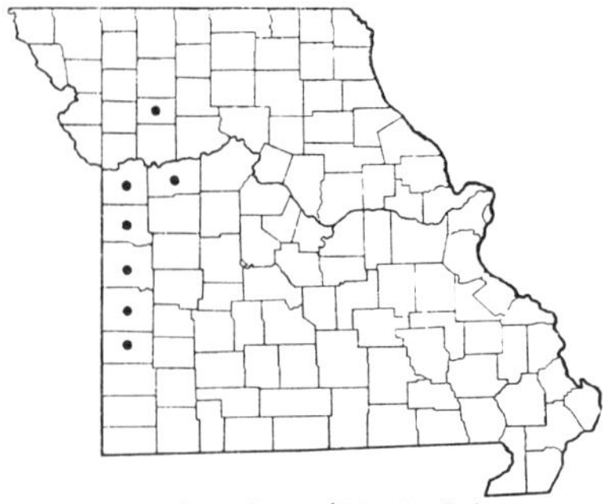

1676 Lomatium foeniculaceum (Hairy Parsley)

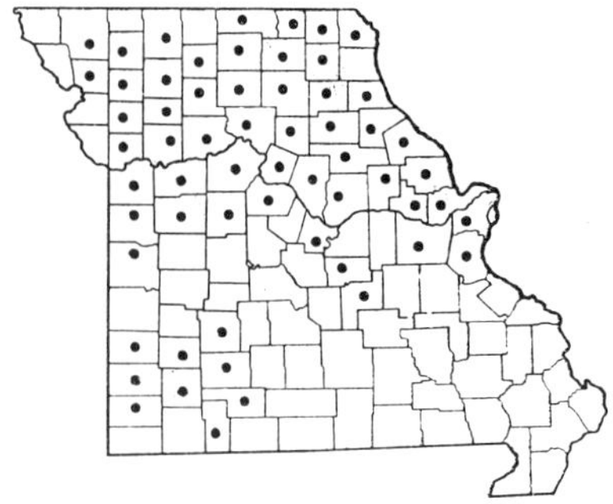

1677 Pastinaca sativa (Parsnip)

## 32. **Levisticum** Hill. Lovage

**Levisticum paludapifolium** (Lam.) Asch. Lovage Map 1674

*Levisticum officinale* Koch

Flowers May–July.

Escaped from cultivation in waste ground and along roadsides. Known only from Washington County, eastern Ozark region (Old Mines, 1932, *Kellogg*).

Native of Europe; introduced and naturalized in the United States from Vermont to Pennsylvania and New Jersey, south to Indiana, Missouri, and New Mexico.

The fruits contain an aromatic oil which is used for flavoring and in medicine.

## 33. **Polytaenia** DC.

**Polytaenia Nuttallii** DC. Prairie Parsley Map 1675

Flowers April–June; fruits June–August.

Occurs in prairies and rocky open glades and prairie openings. Chiefly southern and central Missouri in the Ozark and unglaciated prairie sections, and locally in northern Missouri in Ralls, Macon, Sullivan, De Kalb, and Nodaway counties; absent from the lowlands of southeastern Missouri.

Ranges from Indiana, Michigan, and Wisconsin to North Dakota and Nebraska, south to Alabama, Mississippi, Louisiana, Texas, and New Mexico.

## 34. **Lomatium** Raf.

**Lomatium foeniculaceum** (Nutt.) C. & R. Hairy Parsley Map 1676

*Lomatium daucifolium* (Nutt.) C. & R. [Steyerm.]

*Cogswellia daucifolia* (Nutt.) M. E. Jones [P & S]

Flowers April–June.

Occurs in prairies, limestone glades, and rocky open limestone exposures along roads. Western Missouri from Barton County north along the western border of the state to Lafayette and Caldwell counties.

Ranges from North Dakota to Manitoba, south to Missouri, Oklahoma, and Texas.

The young fruits and ovaries of most of the Missouri material are glabrous or nearly so, a character which would place such specimens with *L. daucifolium*, according to the work of Mathias and Constance (N. Am. Fl. 28B: 241. 1945). In the present flora, this species is combined with *L. foeniculaceum.*

## 35. **Pastinaca** L. Parsnip

**Pastinaca sativa** L. Parsnip Map 1677

Flowers May–October.

Occurs in fields, waste ground, and along roadsides and railroads. Throughout northern, central, and western Missouri, but not recorded from most of the Ozark counties nor from any of the counties of the lowlands of southeastern Missouri.

Native of Europe and Asia; introduced and naturalized throughout North America.

Some persons are poisoned if their skin comes into

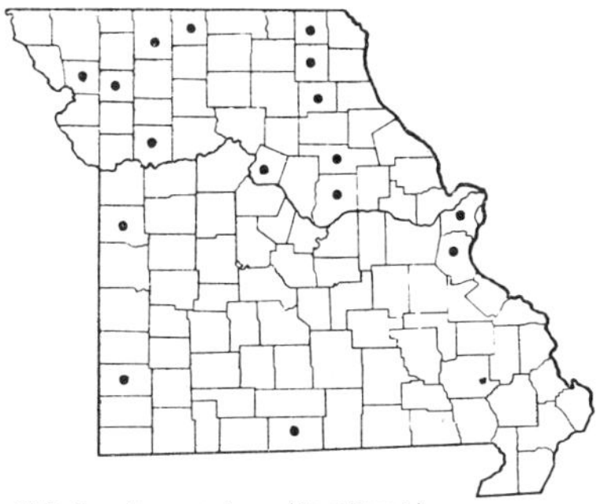

1678 Heracleum maximum (Cow Parsnip)

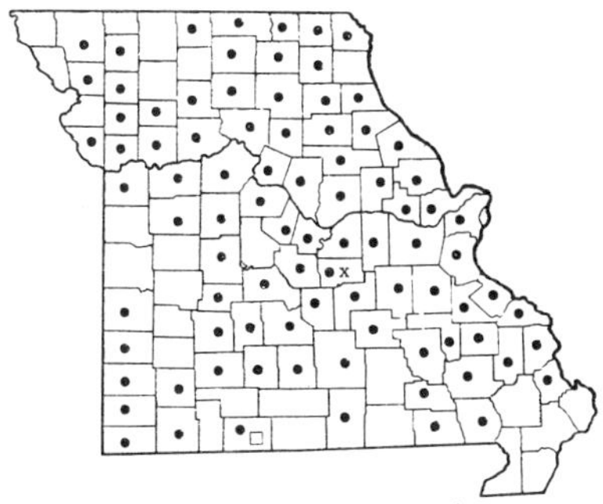

1679 • Daucus Carota f. Carota (Wild Carrot)
1679 x Daucus Carota f. roseus
1679 □ Daucus Carota f. epurpuratus

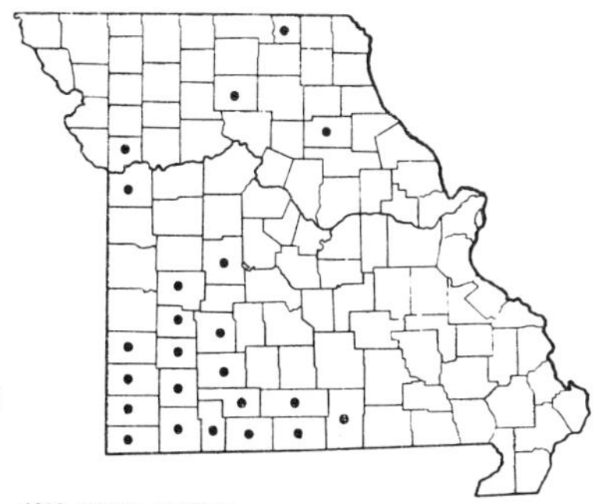

1680 Daucus pusillus

contact with the plant, and develop a dermatitis resembling poison ivy.

Apparently two kinds of parsnip may be distinguished, an edible kind grown for its thick roots, the well-known garden parsnip, recognized by some botanists as *P. sativa* var. *hortensis* Ehrh., and a wild race with more slender roots, known as var. *pratensis* Pers. which is sometimes reported to be poisonous when eaten. However, Dr. Pammel (Man. Poisonous Pl. part 2. pp. 662–63. 1911) gives evidence to indicate that the wild form can be eaten without any ill effects, and those cases of poisoning which have been reported are believed to be mistaken identification for such plants as Water Hemlock (*Cicuta maculata*).

## 36. **Heracleum** L. Cow Parsnip

**Heracleum maximum** Bartr. Cow Parsnip Map 1678

Also called Masterwort.

*Heracleum lanatum* Michx. [BB, P & S, Steyerm.]

Flowers May–July.

Occurs in moist rich woods in valleys near streams and in ravines and thickets. Chiefly in northern and central Missouri, and locally south in Ozark and Jasper counties in southern Missouri.

Ranges from Labrador to Alaska, south to Georgia, Ohio, Indiana, Illinois, Missouri, Kansas, New Mexico, Arizona, and California.

The dried plant possesses an odor resembling slippery elm or parsnip.

Cases of dermatitis have been reported by some persons who have come into contact with the foliage. Also there are reports of cattle having been poisoned from eating the leaves of the plant.

The plant, however, can be eaten without any ill effects. The cooked root is eaten by some Indian tribes, and is stated to resemble the rutabaga somewhat in taste. Some of the Indians of northern California use the hollow basal part of the stem as a substitute for salt either drying it first before eating or placing it directly with other food to be cooked. The young leafstalks and young stems, when cooked, are said to be a tasty vegetable with a resemblance to stewed celery. Indians of the northern states and Canada eat the peeled stalks of the plant either raw or cooked.

## 37. **Daucus** L. Carrot

Complete spread of flower- or fruit-clusters 5–12 cm. broad, with 7–30 or more main (primary) rays; main bracts of involucre at base of long rays once pinnately divided, the ultimate divisions 5–20 mm. long; ultimate divisions of leaves lanceolate, oblanceolate, or oblong, or broader in one part than in another . . . . . . . . . . . . . . . . . . . . . . . . . . . . 1. D. Carota

Complete spread of flower- or fruit-clusters 1–6 cm. broad, with 3–15 main (primary) rays; main bracts of involucre at base of long rays bipinnatifid (the main divisions cleft again), the ultimate divisions 2–8 mm. long; ultimate divisions of leaves more uniformly filiform or thread-like . . . 2. D. pusillus

1. **Daucus Carota** L. Wild Carrot Map 1679

Also called Queen Anne's Lace.

Flowers May–October.

Occurs in fields, waste ground, roadsides, along railroads, thickets, gravel bars, and in woodland.

Three variations are found in Missouri:

a. All the flowers pink, rose-colored or purplish . 1b. D. CAROTA f. ROSEUS
a. All the flowers white, except for the centermost one . . . . . . . . . . . . . . *b*
  b. Common type found; central-most flower of the inflorescence dark purple, the other flowers white or pinkish. . . 1a. D. CAROTA f. CAROTA
  b. Rarely found; central-most flower and all the other flowers of the inflorescence white or pinkish . 1c. D. CAROTA f. EPURPURATUS

1a. **Daucus Carota** f. **Carota** Map 1679

*Daucus Carota* L. [G. BB, P & S, Steyerm.]

Throughout Missouri, probably in every county, but not recorded from most of the lowland counties of the extreme southeastern section.

Native of Europe and Asia; introduced and naturalized in North America from southern Canada south throughout the United States.

1b. **Daucus Carota** f. **roseus** Millsp. Map 1679

Known only from Maries County, northern Ozark section (alluvial thickets along banks of Little Tavern Creek, T40N, R11W, sect. 5 and 8, 1½ mi. west of Van Cleve, June 29, 1939, *Steyermark 27578*).

1c. **Daucus Carota** f. **epurpuratus** Farw. Map 1679

Known only from Taney County, southwestern Missouri.

For a discussion of color variation in this species, reference should be made to the article by Grier (Torreya 22: 64–66. 1922).

*Daucus Carota* is one of the three commercial sources for the extraction of *carotene*, a synthetic vitamin A product. Some persons develop a dermatitis when their skin comes into contact with the leaves of the wild carrot. If cows eat large amounts of the plant, their milk acquires a bitter flavor.

The cultivated carrot is a race of the wild species. A belief is sometimes held that the wild form of the species is not edible and even poisonous, but various individuals who have eaten roots raised from wild seed have reported them as sweet and tender.

2. **Daucus pusillus** Michx. Map 1680

Flowers April–June.

Occurs in rocky prairies, openings in woodland, and rocky open, frequently limestone, glades, edges of bluff escarpments, sometimes along roadsides and railroads. Mainly in southwestern Missouri, and locally northeastward in Monroe, Schuyler, and Linn counties.

Ranges from Florida to Texas and California, north to South Carolina, Missouri, Kansas, Oklahoma, and British Columbia; also in Mexico.

This species averages shorter in stature than *D. Carota*, the stems usually 1–7 dm. tall instead of 4.5–12 dm. tall (rarely as low as 3 or as high as 16 dm.) in *D. Carota*. The stems in *D. pusillus* are short-hairy and in *D. Carota* vary from glabrous to long-hairy or bristly.

*Excluded Species*

**Ammoselinum Butleri** (Engelm.) C. & R.

Reported for Missouri, but no specimens are extant.

**Ammoselinum Popei** T. & G.

Reported from Missouri by Rydberg (*Fl. Prairies and Plains*, p. 599) and others, but no specimen appears to be extant.

**Bifora americana** (DC.) Benth. & Hook.

Reported at various times for Missouri, but no specimen is extant. Bush (in letter of November 27, 1935, to the present author) states 'As to the name *Bifora* and *Ammoselinum*..., he [Britton] gave me these names, and the specimens should be in the New York Botanical Garden Herbarium.' Search in the collections at the latter institution have not revealed any specimens from Missouri.

**Conioselinum chinense** (L.) BSP.

Reported by Palmer and Steyermark from Christian, Lawrence, and Jasper counties in their *Annotated Catalogue* (p. 612), and the report entered in *Gray's Manual* and Gleason's *New Illustrated Flora*. A re-examination of the material by the present author shows that the specimens are misidentified *Conium maculatum*. Likewise, specimens from Stone County (*Bush 14985*) and Barry County (*Bush 14975*), labeled as *Conioselinum chinense* by Bush are also *Conium maculatum*.

---

Plate no. 273. **1.** Daucus Carota, × 2/7; Details from Small, The New York Botanical Garden. **2.** Pastinaca sativa, × 2/7; *a.* Fruit, × 5 5/7; Details from Small, The New York Botanical Garden. **3.** Heracleum maximum, × 2/7; *a.* Fruit, × 2 6/7; Details from Small, The New York Botanical Garden. **4.** Daucus pusillus, × 1; *a.* Fruit, × 7 1/7. **5.** Ammi majus; Details from Small, The New York Botanical Garden. **6.** Anethum graveolens; Details from Small, The New York Botanical Garden.

PLATE NO. 273

**Hydrocotyle umbellata** L.

At one time reported by Bush and others from Missouri, but no specimen has ever been found to confirm the report. Bush wrote the present author years ago that he had found this species at 'Kennett, Missouri, at Varner cut-off.' This is a logical locality for the species in Dunklin County, in the lowlands of southeastern Missouri, but, lacking any documentary evidence, the species cannot be accepted as part of the state flora.

**Sium floridanum** Small

Some Missouri material from Ripley and Ozark counties identified as this species has been found to be *S. suave*. The species is known only from Virginia to Florida.

## Fam. **CORNACEAE** (Dogwood Family)

(including *Nyssaceae* (Sour Gum Family) [G]

Leaves opposite on the branch, except alternate in one species; leaf-blades always smooth-edged, without any teeth on margins; petals and stamens 4; side nerves of leaf all curving upward toward tip of leaf at an angle of about 60°, nearly parallel with the margins, the 2 uppermost side nerves ending in the tip of the leaf; shrubs or small trees . . . . . . . . . . . . . . . . . . 1. CORNUS

Leaves alternate on the branch; leaf-blades smooth-edged or with 1 or more teeth; petals and stamens 5; side nerves of leaf spreading-ascending more or less directly to the margin at an angle of about 45° or less, sometimes nearly horizontal, not paralleling the margin, all the side nerves ending at or near the margins, none ending in the tip of the leaf; large trees . . . . . . . . . . . . . 2. NYSSA

### 1. **Cornus** L. Dogwood

a. Leaves alternate, but crowded near the end of the branches and giving the appearance of arising in whorls or even nearly opposite; most of leaf-stalks 2–6 cm. long (some of them sometimes only 1 cm. long) . . . . . . . . . . . . . . . . . . . . . . . . . 6. C. ALTERNIFOLIA

a. Leaves opposite; leaf-stalks mostly 0.5–1.5 cm. (up to 2.5) long . . . . . . . . . . . . *b*

b. True flowers greenish-yellow in a dense head-like cluster surrounded by 4 large white or pink petal-like bracts; fruit bright red; trees up to 40 ft. (12 meters) tall. . . . . . . . 1. C. FLORIDA

b. Flowers white or creamy-white in an open broad cymose inflorescence, not surrounded by bracts; fruit blue or white; shrubs 3–20 ft. (0.9–6 meters) tall . . . . . . . . . . . . . . . *c*

c. Lower surface of leaves woolly with loose curled or curving hairs over 0.5 mm. long; upper surface of leaves rather rough-hairy (rarely smooth) . . . . . . . . . . 3. C. DRUMMONDI

c. Lower surface of leaves either lacking hairs or inconspicuously hairy with minute colorless appressed hairs (pressed against or lying parallel to surface); upper surface of leaves smooth. . . . *d*

d. Lower surface of leaves green and without any hairs; pith of branchlets of the present year and of one or two year old branches white and less than $^1/_3$ the diameter of the branch or twig; shrubs of swamps of southeastern Missouri lowlands . . . . . . . . 2. C. FOEMINA

d. Lower surface of leaves whitish, grayish, or gray-silvery green with minute colorless appressed hairs (pressed against or lying parallel to surface); pith of branchlets of the present year and of one- or two-year-old branches pale brown or tawny, or, if white, the pith at least half the diameter of the branch or twig; widespread shrubs in various parts of Missouri . . . . *e*

e. Youngest twigs or branchlets densely hairy; older branches reddish-brown; fruit bluish; inflorescence flat or convexly curved at the summit, broader than high; branches of inflorescence yellow-brown; calyx-lobes 0.75–1 mm. or more long . . . . 4. C. OBLIQUA

e. Youngest twigs or branchlets glabrous (without hairs) or very sparsely hairy; older branches gray; fruit white on red stalks; inflorescence dome-shaped or elongated so that it is usually nearly or quite as high as broad; branches of inflorescence red; calyx-lobes not over 0.5 mm. long . . . . . . . . . . . . . . . . . . . 5. C. RACEMOSA

1. **Cornus florida** L. Flowering Dogwood

Map 1681

Flowers mid-April–mid-May; fruits August–November.

Occurs in usually acid soils overlying chert, gravel, sandstone, or granitic substrata, but also on limestone, generally on wooded slopes, ravines, along bluffs, upland ridges and in thickets, less commonly in valleys and low ground; usually in well-drained soils.

Ranges from Florida to Texas and Mexico, north

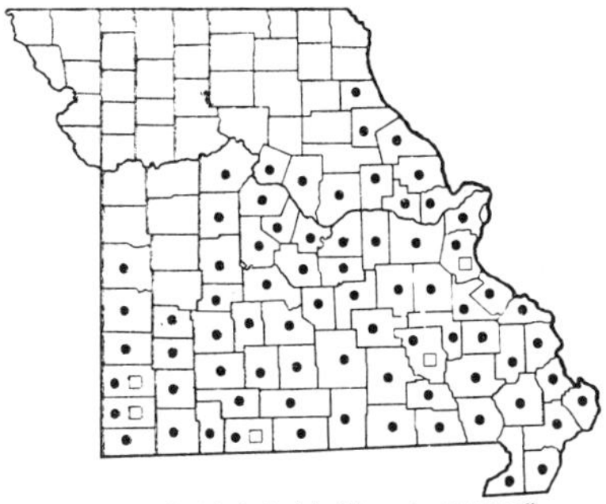

1681 • Cornus florida f. florida (Flowering Dogwood)
1681 □ Cornus florida f. rubra (Pink Dogwood)

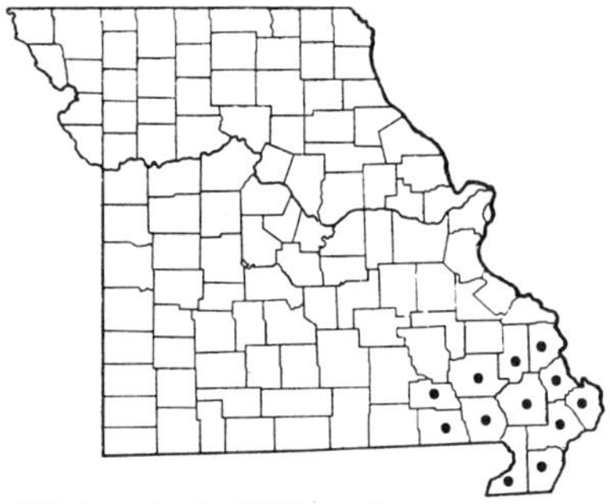

1682 Cornus foemina (Stiff Dogwood)

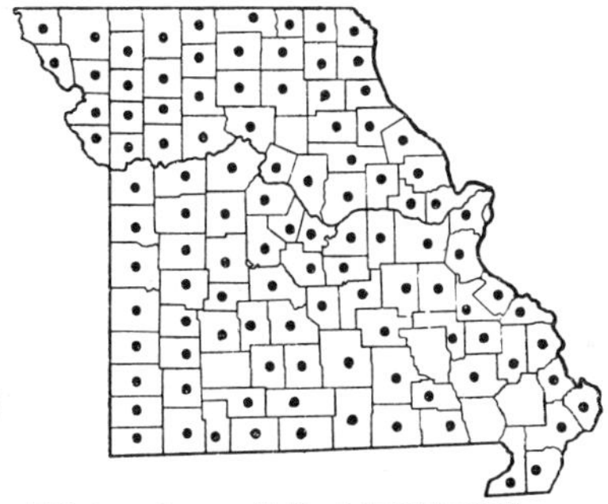

1683 Cornus Drummondi (Rough-leaved Dogwood)

to Maine, New Hampshire, Vermont, New York, Ontario, Ohio, Indiana, Illinois, Missouri, Kansas, and Oklahoma.

Two variations occur in Missouri:

Large petal-like bracts white or creamy with greenish or tinge of pink at tip . 1a. C. FLORIDA f. FLORIDA

Large petal-like bracts with pink or rose-color predominant . . . . . 1b. C. FLORIDA f. RUBRA

1a. **Cornus florida** f. **florida** Map 1681

*Cornus florida* L. [G, BB, P & S, Steyerm.]

This is the common variation in Missouri, ranging throughout the Ozark region of southern and central Missouri and the Ozark border counties south of a line drawn from Marion, Ralls, Pike, Montgomery, Callaway, Boone, Howard, Saline, Pettis, and Cedar counties to Bates County.

1b. **Cornus florida** f. **rubra** (Weston) Palmer & Steyerm. Pink Dogwood Map 1681

*Cornus florida* var. *rubra* Weston

Rare and scattered in the Ozark region in Jefferson, Reynolds, Taney (shallow draw in limestone upland, 0.2 mi. south of Cedar Creek P.O., T22N, R19W, sect. 24, west of road, April 26, 1949, *Steyermark 67343*), Jasper, and Newton (rocky banks of ravine near Indian Creek, 4 mi. southeast of Christopher, April 29, 1954, *Palmer 57380*) counties.

This pink-flowered variation is rarely encountered in the wild state, but has become very popular in cultivation and is now commonly seen in many gardens and around homes especially in the southern states north to the vicinity of Boston, New York, Philadelphia, Cleveland, Springfield, Illinois, St. Louis and Kansas City. The cultivated form generally has deeper rose or rose-red coloring of the bracts than the wild form.

Flowering dogwood is the state tree of Virginia. It starts flowering in the Ozarks somewhat later than redbud, but the time of flowering usually overlaps so that in mid to late April the two are usually seen together in flower, producing the greatest mass display of flower color to be found at any one time in the Ozarks. Generally, the pale yellow flowers of sassafras, and the white masses of flowers of wild plums are conspicuous at the same time as redbud and dogwood, all of them together creating a truly dramatic color scene.

In autumn the foliage turns from a dull yellow and dull rose mixed with pale or dull green to finally dull or deep rose-red throughout. Together with the bright red fruits the autumnal coloring is strikingly beautiful. Because of the showy flowers, good autumnal coloration, freedom from most diseases, and decorative winter branching effect, the flowering dogwood is considered the best ornamental of all native North American trees.

The buds and twigs are eaten by the white-tailed deer and the fruits by the wild turkey in the Ozarks.

The tree has many uses. Formerly, a scarlet dye was prepared from the roots by North American Indians for coloring their quills and feathers. The aromatic bitter inner bark was used as a remedy for malaria by the Indians and by the early colonists for chills and fevers, at one time was even employed as a quinine substitute.

The wood is especially strong, hard, and resistant to shock. It has the ability to remain smooth under constant friction and is, therefore, admirably suited for the making of shuttles; it supplies 90 per cent of the wood for that industry. Because of similar properties of resilience, resistance, and smooth wear, it is employed in the making of roller skates, heads for golf sticks and mallets, chisel handles, mauls, wedges, spindles, small pulleys, knitting needles, runners for sledges, engraver's blocks, and brush blocks. Jewelers also favor sticks made from dogwood for cleaning long lenses.

2. **Cornus foemina** Mill. Stiff Dogwood Map 1682

*Cornus femina* Mill. [P & S]

*Cornus stricta* Lam. [BB, Steyerm.]

Flowers May–June; fruits August–October.

Occurs in swampy and low wet woodland and wet open ground. Occurs in the lowland counties of southeastern Missouri north to Cape Girardeau, Bollinger, Wayne, and Carter counties.

Ranges from Florida to Texas, north to Virginia and Delaware, Indiana, Illinois, and Missouri.

The foliage becomes a russet bronze-green in late fall. The main peduncles are rose-purple. This is the species commonly found in swamps where bald cypress, swamp tupelo, swamp red maple, and pin and basket oak abound, and is associated with such species as *Planera aquatica, Styrax americana, Brunnichia cirrhosa, Wisteria frutescens* var. *macrostachya, Itea virginica, Rosa palustris, Trachelospermum difforme*, and other denizens of the southeastern lowlands.

The white pith of the young branchlets and older branches is less than $\frac{1}{3}$ the diameter of the branch, whereas in *C. racemosa*, which may have white as well as brown pith, the pith is about $\frac{1}{2}$ the diameter.

3. **Cornus Drummondi** Meyer Rough-leaved Dogwood Map 1683

*Cornus asperifolia* of auth. [P & S, Steyerm.], not Michx.

Flowers May–June; fruits August–October.

Occurs in dry or rocky woods, thickets, limestone glades and borders of rocky open glades, prairies, bluff escarpments, occasionally low wet ground along ponds, streams, and at the base of bluffs. Throughout Missouri, doubtless in every county.

Ranges from Mississippi to Texas, north to Ontario, Ohio, Indiana, Illinois, Iowa, and Nebraska.

True *C. asperifolia* Michx. has been shown (Rickett, Am. Midl. Nat. 27: 259–61. 1942) to be a shrubby species of the coastal plain from South Carolina to Florida and Alabama.

This dogwood is able to survive drier and more exposed conditions than any other species in Missouri, and is often found around limestone glades and rocky exposed outcrops and sites. The pith is brown on two-year-old wood, and usually brown on one-year-old branches, but sometimes white. It is brown or white on the young branchlets.

4. **Cornus obliqua** Raf. Swamp Dogwood Map 1684

Also called Pale Dogwood, Silky Dogwood, Kinnikinnik.

*Cornus Purpusi* Koehne [BB]

Flowers May–July; fruits June–October.

Occurs generally along rocky beds and among boulders of small streams, spring branches, wet places in prairies, swampy meadows, wet thickets, swamps and low woodland. Throughout Missouri, but apparently absent from the lowlands of southeastern Missouri.

Ranges from New Brunswick to North Dakota, south to New Jersey, Pennsylvania, West Virginia, Kentucky, Arkansas, and Oklahoma.

This species resembles *C. racemosa*, but the leaf-blades are generally narrower and more lanceolate to narrowly ovate than those of *C. racemosa* or of any other species in Missouri, and conspicuously narrowed (cuneate) at the base. The reddish-brown or dark brown young branchlets and blue fruit, conspicuous in late fall and winter, make this shrub a meritorious subject for ornamental plantings.

The Indian name for tobacco, Kinnikinnik, is applied to this and several other species, from which the Indians used the bark for tobacco.

5. **Cornus racemosa** Lam. Gray Dogwood Map 1685

Flowers May–July; fruits July–October.

Occurs in moist or rocky ground along streams and ponds, wet meadows and borders of prairies, thickets along fence rows and roadsides, upland rocky thickets bordering glades, and along bluffs.

Ranges from Maine to Ontario and Manitoba, south to Florida, Alabama, Kentucky, Illinois, Missouri, Arkansas, and Oklahoma.

A hybrid between *Cornus racemosa* and *C. obliqua*, known as *Cornus* X *arnoldiana* Rehd. is reported from Missouri. It has slightly bluish or white fruits, purplish color of the one-year-old branchlets, pale and sparsely appressed-hairy on the lower leaf-surface, grayish color of older branches, and inflorescence not elongated.

The foliage turns a purplish-red to dull rose-red or purplish-brown in autumn. This species forms thickets by virtue of the suckering underground stems. The clusters of white fruits on red stalks and supporting branches of the peduncle, together with the showy

Plate no. 274. **1.** Cornus florida, × 2/5; Details from Small, The New York Botanical Garden. **2.** Cornus foemina, × 2/5. **3.** Cornus obliqua, × 2/5. **4.** Cornus Drummondi, × 2/5. **5.** Cornus racemosa, × 2/5. **6.** Cornus alternifolia, × 2/5; Details from Small, The New York Botanical Garden.

PLATE NO. 274

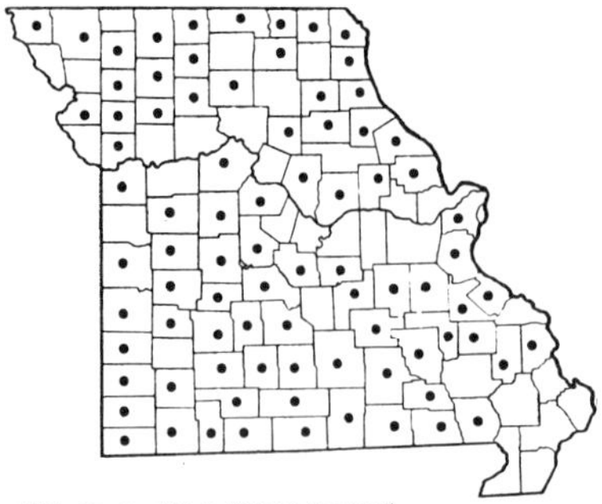

1684 Cornus obliqua (Swamp Dogwood)

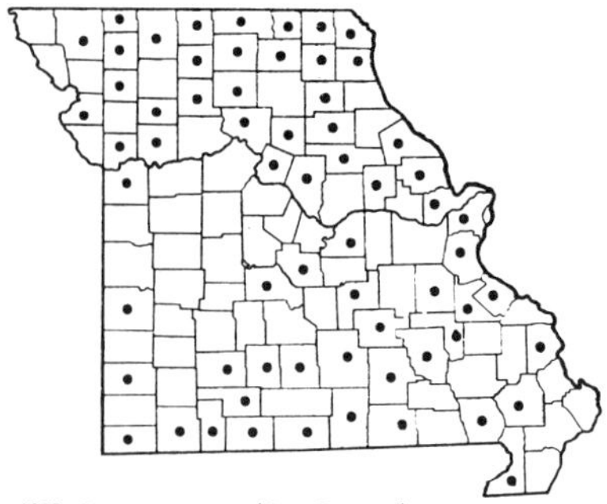

1685 Cornus racemosa (Gray Dogwood)

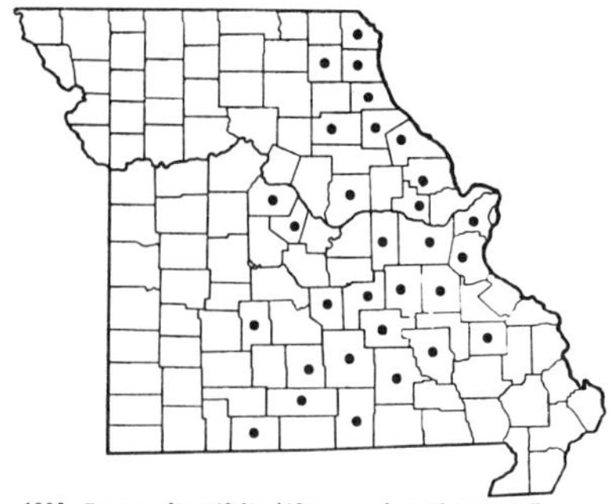

1686 Cornus alternifolia (Alternate-leaved Dogwood)

flower-clusters, give this species an ornamental value.

This dogwood usually comes into flower at an earlier and smaller stage of growth than any of the other Missouri species, plants often flowering when only 1 meter or less tall. The leaf-blades average broader and often more ovate than in *C. obliqua*, with which it is occasionally confused.

6. **Cornus alternifolia** L. f. Alternate-leaved Dogwood Map 1686

Also known as Pagoda Dogwood, Green Osier, Pigeon Berry.

Flowers May–July; fruits July–September.

Occurs usually on rich steep north-facing, wooded slopes and bluffs, and wooded banks of streams. Eastern Missouri west to Knox, Monroe, Callaway, Cooper, Moniteau, Pulaski, Dallas, Wright, Douglas, and Taney counties.

Ranges from Newfoundland and Nova Scotia to Ontario and Minnesota, south to Florida, Georgia, Alabama, and Arkansas.

The leaves are pale or whitened on the under surface. In fall they turn dull purple color suffused with yellow or green to a rose-red and green. The fruits are blue-black on reddish stalks. The graceful horizontally spreading greenish branches arranged in whorls and arching at the tips produce a somewhat flat-topped pagodalike tree with an effect similar to that of the flowering dogwood, for which it is sometimes recommended as a substitute in cold northern areas where *C. florida* cannot survive the winters. The fruits are eaten by warbling vireos and other birds.

## 2. **Nyssa** L. Tupelo, Sour Gum

Leaves usually with 1 or more teeth on the margins, sometimes entire (without teeth); leaf-stalks (petioles) mostly 3–5 cm. long; leaves of the pistillate flowering branches 10–30 cm. long; leaf long-pointed with an elongated projection (mucro) at tip; pistillate (female) flowers solitary; fruit 2–3 cm. long; trees of the swamps of southeastern Missouri and upland sink-hole ponds of Oregon and Ripley counties . . . . . . . . . . . . . . . . . . . . . . . . . . . . . . . . 1. N. AQUATICA

Leaves usually without teeth (entire) on margins, rarely with a few teeth; leaf-stalks (petioles) mostly 0.5–2.5 cm. long; leaves of the pistillate flowering branches 2.5–15 cm. long; leaf blunt or acutely pointed but without an elongated projection (mucro) at tip; pistillate (female) flowers 2 or more on each peduncle (main flower-stem); fruits 1–1.5 cm. long; trees of the Ozark of southern and east-central Missouri and in the swampy woods of the southeastern lowlands . . . . . . . . 2. N. SYLVATICA

1. **Nyssa aquatica** L. Swamp Tupelo Map 1687

Also known as Cotton Gum, Tupelo, Tupelo Gum, and Water Tupelo.

*Nyssa uniflora* Wang. [BB]

Flowers April–May.

Occurs in bald cypress swamps, bayous, and low wet woods of the southeastern Missouri lowlands north to Cape Girardeau, Bollinger, and Wayne counties, locally west in upland sink-hole ponds in the southeastern Ozarks to Ripley and Oregon counties.

Ranges from Florida to Texas, north to Virginia, Tennessee, Kentucky, Illinois, and Missouri.

The name *N. aquatica* L. has been rejected by Rickett (N. Am. Fl. 28B: 313, 315. 1945) as a *nomen confusum* based upon two species.

The base of the trunk of this species is usually enlarged or buttressed as in bald cypress. As in Sour Gum or Black Tupelo (*Nyssa sylvatica*) the wood is full of

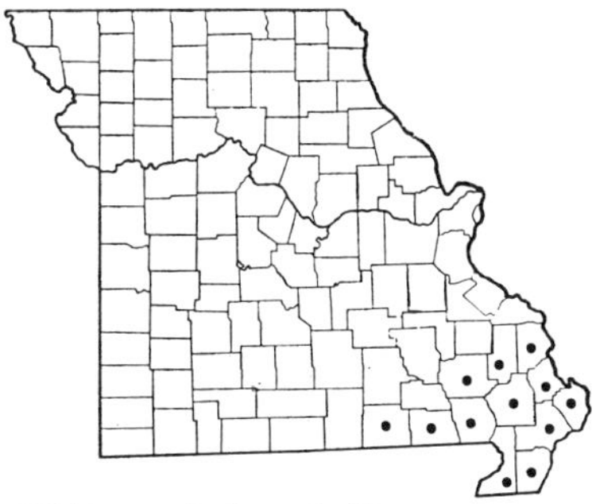

1687 Nyssa aquatica (Swamp Tupelo)

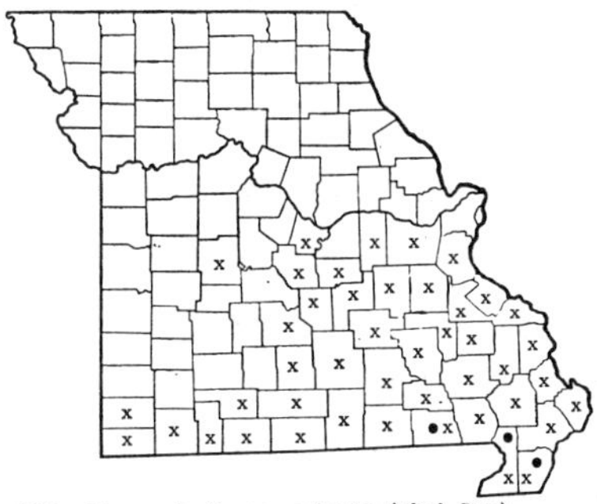

1688 • Nyssa sylvatica var. sylvatica (Black Gum)
1688 x Nyssa sylvatica var. caroliniana (Black Gum)

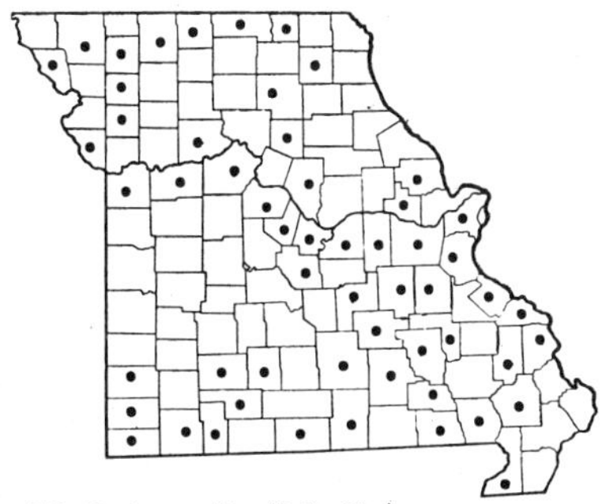

1689 Monotropa uniflora (Indian Pipe)

interlocking tough fibers which makes it very difficult to split by ax. In the southern states this species is used for its pulp in papermaking. The fruit is very bitter.

2. **Nyssa sylvatica** Marsh. Black Gum Map 1688

Also known as Sour Gum, Black Tupelo, and Tupelo.

Flowers April–June.

Occurs in acid soils overlying chert, sandstone, or granitic substrata of dry, rocky wooded slopes, ravines, upland ridges, and borders of sink-hole ponds in the Ozark area, and in low wet woods in the southeastern lowlands. Ozark and southeastern lowland regions north to Jefferson, Franklin, Gasconade, Maries, Cole, Miller, Benton, Laclede, Wright, Christian, and Newton counties.

Two variations occur in Missouri:

Common type encountered; lower surface of leaves (seen best when dried) papillose (covered with tiny round bumps and best seen with magnifying lens); green branchlets usually breaking when turned to a right angle; leaves usually long-pointed (acuminate) at the tip . . . . .
2b. N. SYLVATICA var. CAROLINIANA

Rare type, known only from Pemiscot, Dunklin, and Ripley counties, southeastern Missouri; lower surface of leaves (best seen when dried) not papillose; green branchlets usually bending and not breaking when turned to a right angle; leaves usually short-tipped (acute) or blunt . . .
2a. N. SYLVATICA var. SYLVATICA

2a. **Nyssa sylvatica** var. **sylvatica** Map 1688

*Nyssa sylvatica* Marsh [G, BB, P & S in part]

This is the rarely found variation in Missouri, known only in the southeastern portion from Pemiscot, Dunklin, and Ripley counties.

Ranges from Maine to Ontario, Michigan and Wisconsin, to Missouri, south to Florida and Texas.

Fernald (Rh. 37: 434–35, map 1. 1935) shows a dot for northern Missouri in the range of this variety. This dot is undoubtedly based upon a mislabeled specimen or a cultivated one or probably due to an error in locating an obscure local name on a collector's label, as the genus does not occur north of the Ozark area in Missouri.

2b. **Nyssa sylvatica** var. **caroliniana** (Poir.) Fern. Map 1688

*Nyssa sylvatica* of auth. in part [P & S], not Marsh.

This is the common variation in Missouri and has the range as indicated in the Ozarks. A few trees originally planted in Pershing State Park, Linn County, northern Missouri, during 1936, appear along the margin of woodland bordering a major road into the park, and should be discounted as not to be included within the natural range of the variety in Missouri.

Ranges from Pennsylvania to Ontario, south to Virginia, North Carolina, Tennessee, Mississippi, and Texas.

The Sour Gum is one of the most strikingly beautiful trees in Missouri by virtue of its shining green leaves, pyramidal habit, and brilliant deep red or scarlet late summer to early fall foliage. It is one of the first trees to turn color in late summer as well as one of the first to lose the leaves. After the leaves fall, the bare more or less horizontal branches stand out more prominently. The blue-black or dark blue fruits are sour, but eaten by robins, wild turkey, and other birds. White-tailed deer also browse young leafy branches. The male and female flowers are produced separately on different trees.

Some woodsmen, according to Deam (*Fl. Ind.* p. 729. 1940), recognize two varieties of Sour Gum, the 'yellow gum' which splits easily and considered to be the dry upland variety *caroliniana* found in the Ozarks, and the 'black gum' which is exceedingly difficult to split and found in low wet ground. This distinction has not been tested in Missouri, so far as known, nor have botanists correlated these differences between the splitting of the wood and the papillose vs. nonpapillose lower leaf-surfaces.

In the southern states Sour Gum is one of the principal hardwoods used for pulpwood in paper-making. Because the tough fibers of the wood of the wet low woodland variety of 'black gum' are interlaced, the wood is exceedingly difficult to split. For this reason it is used for a number of purposes out of the ordinary, such as mine rollers, scaffolding, hatter's blocks, water pipes, chopping bowls and other wooden ware, as well as for gunstocks, musical instruments, berry crates, and floors of factories receiving rough wear.

The so-called bee-gums for beehives owe their name to the original use of the sour gum tree for this purpose. The hives were prepared by taking the decayed hollow trunks, cutting them into sections placed on end on boxes, and covered over by a board.

Although the tree is one of the most ornamental and beautiful of all native species, it is one of the most difficult to transplant and repeated efforts and perseverance are often necessary to establish the Sour Gum.

A specimen collected in Christian County (upland woods along Pine Ridge Road, 5 mi. southwest of Chadwick, July 17, 1954, *Palmer 58258*) has been questionably referred to *N. sylvatica* var. *dilatata* Fern. This variety differs from var. *caroliniana* in the non-papillose lower leaf-surface and the broadly oval to round-oblong leaf-blades of the pistillate trees. Since the specimen as well as the variety are doubtfully distinguishable, the variety is not accepted as part of the state flora on the basis of this particular specimen.

## SYMPETALAE

(Gamopetalae, Metachlamydeae)

### Order **ERICALES**

### Fam. **PYROLACEAE** (Wintergreen Family)

#### **Monotropa** L. Indian Pipe

Stem with 1 flower; all parts of living plant pure white, rarely flesh-pink or reddish, glabrous (without hairs) . . . . . . . . . . . . . . . . . . . . . . . . . . . . 1. M. UNIFLORA

Stem with mainly 3 or more flowers; living plant never white, but rather yellowish, buff, tawny, or salmon, more or less hairy . . . . . . . . . . . . . . . . . . . . . . 2. M. HYPOPITHYS

1. **Monotropa uniflora** L. Indian Pipe Map 1689

Also called Ghost Flower.

Flowers August–October.

Occurs in dry upland woods and thickets, ravines, on slopes or ridges, usually in dense leaf-mulch and leaf-mold of oak or sugar maple woodland.

Scattered throughout Missouri, but apparently absent or unrecorded from most of the counties of the unglaciated prairie region of southwestern Missouri; nowhere common where seen.

Ranges from Newfoundland to Alaska, south to Florida, California, and Mexico, and Central America. Also in Asia.

This saprophytic plant quickly blackens upon drying. It is usually found in oak woods. It is stated that the plant can be eaten after being boiled or roasted. The stems, at first pure white, eventually turn pink and lilac, and finally coal-black.

2. **Monotropa Hypopithys** L. Pinesap Map 1690

Also called False Beech Drops.

Flowers late June–October.

Occurs in rich or dry rocky woods, along streams, ravines, uplands, ledges of sandstone, chert, or limestone bluffs under red cedar, pine, oak, hickory, chestnut, on gravelly banks, and alluvial valley bottoms. Rare and scattered in portions of the Ozark region and north in Linn, Adair, and Schuyler counties, northern Missouri.

Ranges from Newfoundland to British Columbia, south to Florida and Mexico; also in Eurasia.

This is also a saprophytic plant. It occurs in a

Plate no. 275. **1.** Nyssa sylvatica var. caroliniana, × 2/5; *a.* Male flowering branch; *b.* Female flowering branch. **2.** Nyssa aquatica, × 2/5; *a.* Male flowering branch; *b.* Fruiting branch; *c.* Female flowering branch; Details from Small, The New York Botanical Garden. **3.** Rhododendron roseum, × 2/5; Details from Small, The New York Botanical Garden. **4.** Monotropa uniflora, × 2/5; After Gleason, details from Small, The New York Botanical Garden. **5.** Monotropa Hypopithys, × 2/5; After Gleason, details from Small, The New York Botanical Garden.

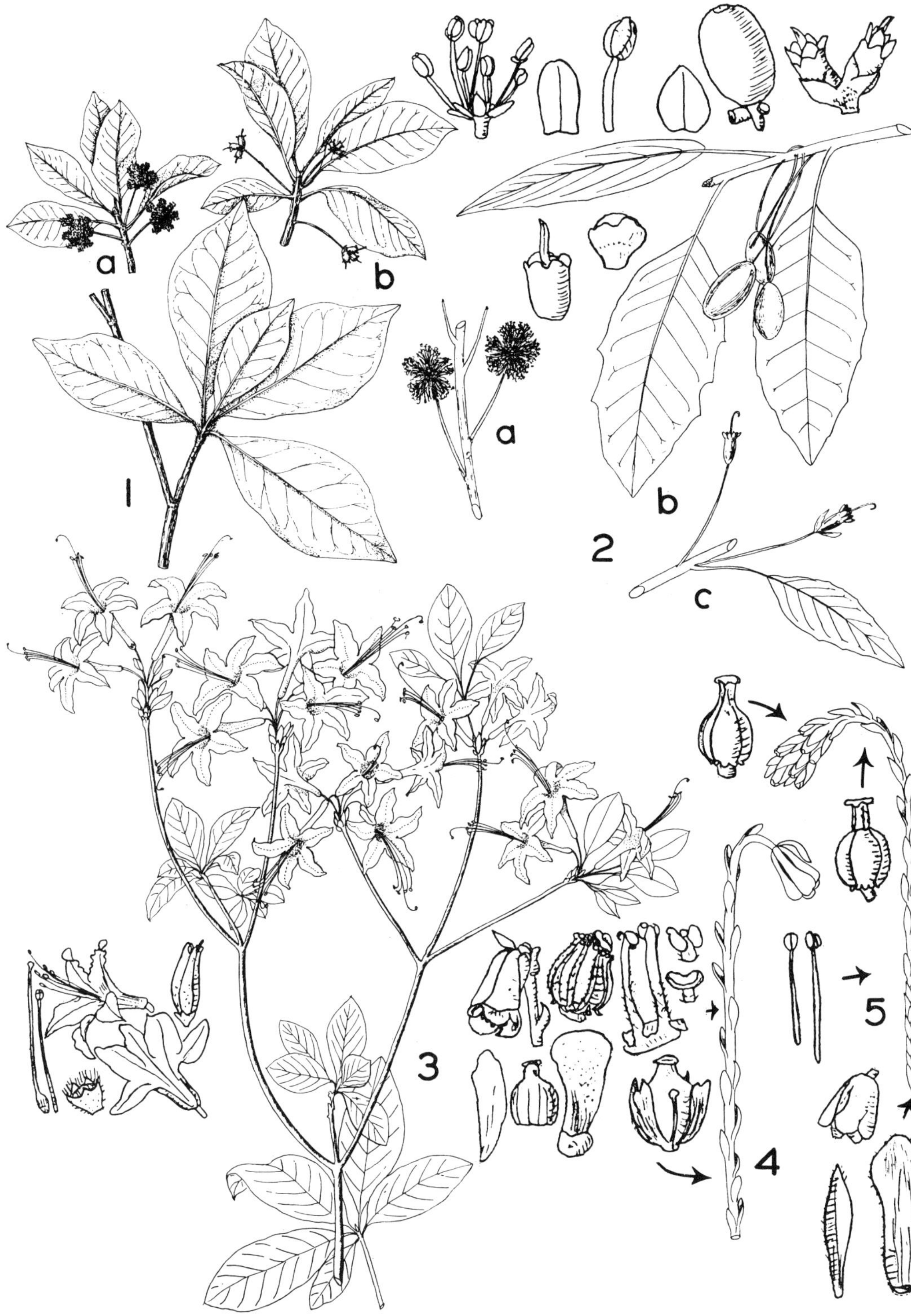


PLATE NO. 275

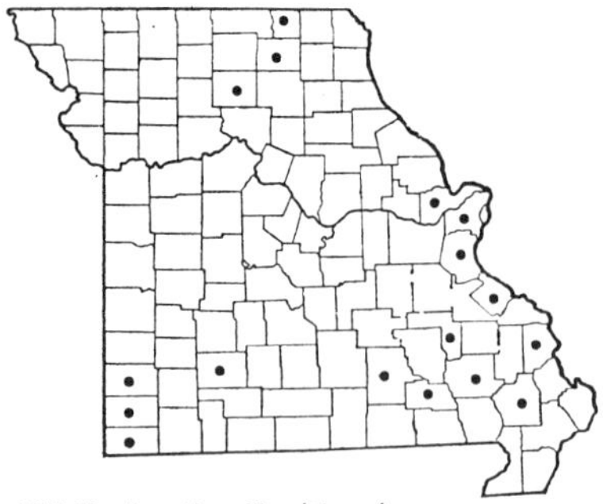

1690 Monotropa Hypopithys (Pinesap)

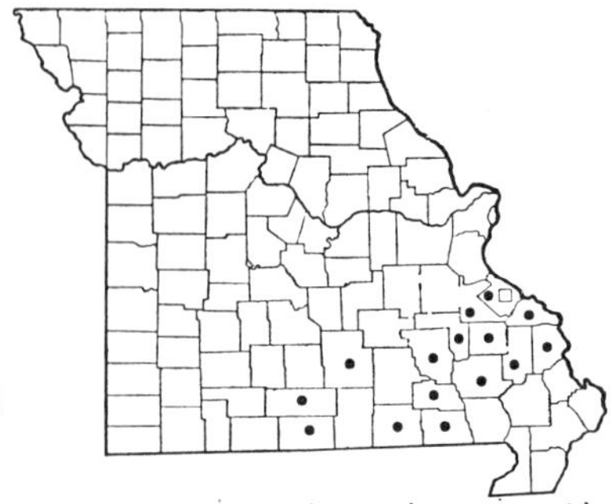

1691 ● Rhododendron roseum f. roseum (Wild Honeysuckle)
1691 □ Rhododendron roseum f. albidum

1692 Lyonia mariana (Stagger Bush)

variety of habitats, ranging from mixed beech-*Ilex opaca* low woodlands of Crowley Ridge to southern yellow pine, limestone bluffs under red cedar, to oak forest in northern Missouri.

The plants vary somewhat in color. Some Missouri material is recorded as having coral rose–red stems, buff-stramineous bracts, and flowers coral rose-red in the lower portion, buff-stramineous at tip.

## Fam. **ERICACEAE** (Heath Family)

(includes Vacciniaceae)

a. Flowers large and showy, at least 30 mm. long, 40–50 mm. broad; leaves usually clustered at or near the tips of the twigs . . . . . . . . . . . . . . . . . . . . . . . 1. RHODODENDRON
a. Flowers smaller, 4–13 mm. long, 3–7 mm. broad; leaves scattered all along the length of the twigs, not clustered at the tip . . . . . . . . . . . . . . . . . . . . . . . . . *b*
  b. At least the lower leaf-surfaces marked with resin dots or scales. . . . . . . 3. GAYLUSSACIA
  b. Lower and upper leaf-surface without any resin dots or scales . . . . . . . . . . . . *c*
    c. Corolla 8–13 mm. long; fruit a dry capsule; ovary superior, the calyx at the base of the ovary and remaining free from it; rare plant known only from Dent County . . . . . . 2. LYONIA
    c. Corolla 4–6 mm. long; fruit a dry or fleshy berry; ovary inferior, the calyx-tube united its entire length with the ovary; common plants of the Ozark region . . . . . . 4. VACCINIUM

### 1. **Rhododendron** L. Rhododendron, Azalea

**Rhododendron roseum** (Loisel.) Rehd. Wild Honeysuckle Map 1691

Also called Azalea, Mountain Azalea, Election Pink.

Flowers late April (April 20)–late May.

Occurs in acid soils overlying sandstone, chert, or granitic substrata, north-facing, steep wooded slopes of bluffs, ravines, slopes along streams, upland ridges, and shallow draws. Southeastern and southern Ozark region, west to Texas, Douglas, and Ozark counties, north to Ste. Genevieve, St. Francois, Iron, and Reynolds counties; commonest in the southeastern Ozark section.

Ranges from Maine to Quebec, south to Virginia, Tennessee, and Kentucky; also in southern Illinois and Missouri.

Two variations occur in Missouri:

Corolla pink, rose, or pink with white . . . a. R. ROSEUM f. ROSEUM
Corolla white throughout . b. R. ROSEUM f. ALBIDUM

a. **Rhododendron roseum** f. **roseum** Map 1691

*Rhododendron roseum* (Loisel.) Rehd. [G, BB, P & S, Steyerm.]

This is the common form encountered.

b. **Rhododendron roseum** f. **albidum** Steyermark Map 1691

Known only from Ste. Genevieve County (sandy pine-oak woods east of Chimney Rocks, along River aux Vases, 5 mi. east of Pickle, May 24, 1933, *Steyermark 8522*).

The flowers are very fragrant. This shrub is one of the most ornamental and showy-flowered in Missouri.

Provided it is given acid, well-drained soils, it can be grown in either shade or full sun. It should be transplanted with the original soil in as undisturbed condition as possible. Plants moved from Missouri to the author's wildflower preserve have survived northern Illinois winters for the past ten years.

The usual habitat in Missouri for *R. roseum* is along wooded ravines and north-facing wooded bluffs, but it adapts itself well to fully exposed and sunlit conditions, as well exemplified by the hundreds of profusely flowering bushes seen in an area along the main highway between Fredericktown and the Castor River in Madison County near the Bollinger County line. Here the owner has removed the timber for light grazing, but the azaleas have been left to grow in full sunlight, where they adorn the open slopes with gorgeous masses of pink flowers in May. The author has watched this area with intense interest for the past fifteen years to note the success these particular azaleas have had in the sunny environment.

## 2. **Lyonia** Nutt.

**Lyonia mariana** (L.) DC. Stagger Bush
Map 1692

Flowers May–June.

Occurs in sandy pine-oak woodland. Known only from the eastern section of the Ozarks in Dent County (sandy pine-oak woods, T35N, R7W, sect. 15 and 16, 2 mi. south of Lake Spring, June 11, 1953, *C. L. Kucera*).

Ranges from Florida to Texas, north to Rhode Island, Connecticut, New York, New Jersey, Pennsylvania, Tennessee, Arkansas, Missouri, and Oklahoma.

The foliage has been reported to poison sheep when browsed by them. However, the plant is such a great rarity in Missouri that it should be protected as a relict plant which is surviving in Missouri at one of its westernmost outposts of geographical range.

## 3. **Gaylussacia** HBK. Huckleberry

**Gaylussacia baccata** (Wang.) K. Koch Black Huckleberry
Map 1693

Flowers April–May; fruits June–September.

Occurs on rocky wooded ridges above bluffs, usually in cherty soils. Known only from northeastern and eastern Ozark region in Montgomery (cherty upper slopes above north-facing limestone bluffs along West Fork of Cuivre River, T50N, R3W, sect. 32, 8½ mi. northeast of Bellflower, September 16, 1954, *Steyermark 77419*) and Perry (cherty ridge top along Mississippi River, T33N, R14E, near Gerler Station, 5 mi. southeast of Altenburg, April 22, 1938, *Steyermark 4948*) counties.

Ranges from Newfoundland and Quebec to Ontario, Manitoba, and Saskatchewan, south to Georgia, Alabama, Kentucky, Illinois, Missouri, and Louisiana.

The Perry County specimen shows the resin dots mostly confined to the lower surface of the leaf, but there are few and occasional dots on the upper surface, thereby suggesting an affinity with *G. frondosa.* However, the corolla in the specimen is cylindric, not campanulate, and the pedicels measure only 3–4 mm. long, characteristic of *G. baccata.*

This plant is easily mistaken superficially for *Vaccinium vacillans*, the common Low-bush Blueberry of the Ozark region, but the occurrence of resin-dots on the leaves quickly serve to distinguish the *Gaylussacia.*

In Canada and the northeastern and eastern United States, where this plant is abundant, the fruit is eaten for its sweet, spicy flavor. The seeds are harder than in the edible species of *Vaccinium* found in Missouri.

In the eastern states *Gaylussacia* is known as Huckleberry and many of the species of *Vaccinium* as Blueberry. In Missouri, most of the species of *Vaccinium* are referred to by country folk as Huckleberry.

## 4. **Vaccinium** L. Blueberry

### I. *Key based mainly on foliage and flowers*

a. Leaves mostly rounded or blunt at tip, or with a short abrupt point (mucro) at tip, coriaceous, thick and firm, dark green on upper side, sometimes evergreen; tall shrubs or small trees up to 9 meters (20 ft.) tall . . . . . . . . . . . . . . . . . . . . . . . 1. V. ARBOREUM

1693 Gaylussacia baccata (Black Huckleberry)

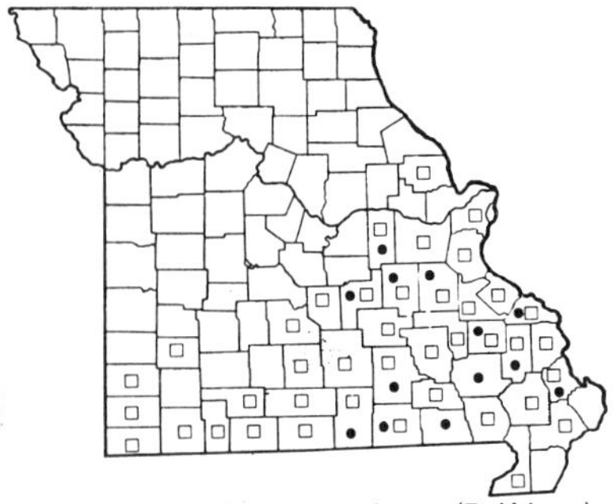

1694 ● Vaccinium arboreum var. arboreum (Farkleberry)
1694 □ Vaccinium arboreum var. glaucescens

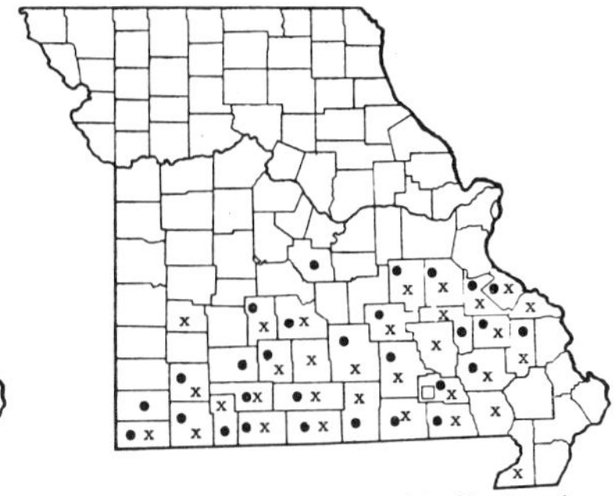

1695 ● Vaccinium stamineum var. interius (Deerberry)
1695 x Vaccinium stamineum var. neglectum
1695 □ Vaccinium stamineum var. melanocarpum (Southern Gooseberry)

a. Leaves mostly short- to long-pointed at tip, firmly membranous, pale green or pale grass green on upper side, deciduous; small shrubs 0.3–3 meters (1–10 feet) tall . . . . . . . . . . . . . . *b*
  b. Small leaves (bracts) present at the base of each flower- or fruit-stalk (pedicel); flowers broadly bell-shaped, wider than long, with the long anthers projecting beyond the corolla . 2. V. STAMINEUM
  b. No small leaves (bracts) present at the base of any flower- or fruit-stalk; flower barrel-shaped or cylindrical, longer than broad, the anthers remaining within the corolla tube . . . . 3. V. VACILLANS

II. *Key based mainly on flowers and fruits*

a. No small leaves (bracts) present at the base of any flower-stalk; anthers without 2 awns on the back; corolla longer than broad, cylindrical or barrel-shaped; flower- and fruit-stalks (pedicels) 2–7 mm. long; berries blue or black, fleshy, sweet-sour, edible; shrubs 0.3–0.9 m. tall . . . . . 3. V. VACILLANS
a. Small leaves (bracts) present at the base of each flower- or fruit-stalk (pedicel); anthers with 2 awns on the back besides the 2 slender tubular tips; corolla as broad as or broader than long, open bell-shaped; flower- and fruit-stalks 5–15 mm. long; berries black, purple, green, yellowish, or blue, dry or fleshy, rather tasteless and not edible; shrub to small tree, mostly 1–9 m. tall . . . . . . . . . *b*
  b. Anthers projecting beyond the corolla lobes, the 2 awns much shorter than the slender tubular tips; flowers and fruits not on jointed stalks (pedicels); fruit juicy or fleshy with soft seeds; lobes of corolla prominent, spreading . . . . . . . . . . . . . . . . . . 2. V. STAMINEUM
  b. Anthers remaining within the corolla tube, the 2 awns more than half the length of the slender tubular tips; flowers and fruits on jointed stalks (pedicels); fruit dry with hard seeds; lobes of corolla short, reflexed. . . . . . . . . . . . . . . . . . . . 1. V. ARBOREUM

1. **Vaccinium arboreum** Marsh. Farkleberry
Map 1694

Flowers May–June; fruits July–October.

Occurs in acid soils overlying sandstone, chert, or granitic substrata, rocky open woods of pine, pine-oak, or oak-hickory, upland slopes and ridges, along bluffs and glades, occasionally in low woods along creeks and near swamps. Ozark region south and east of a line drawn from Lincoln, Gasconade, Phelps, Pulaski, Laclede, and Dade counties to Jasper County.

Two variations occur in Missouri:

Bracts at base of flower-stalks (pedicels) smaller and often different in shape from the true leaves of the branches . . 1a. V. ARBOREUM var. ARBOREUM
Bracts at base of flower-stalks (pedicels) similar and rather equal in size and shape to those of the true leaves of the branches . . 1b. V. ARBOREUM var. GLAUCESCENS

1a. **Vaccinium arboreum** var. **arboreum**
Map 1694

*Vaccinium arboreum* Marsh [G, BB, P & S, Steyerm.]

This is scattered in the Ozark region and apparent-

Plate no. 276. **1.** Lyonia mariana, × $^2/_5$; *a.* Flower, × 2; After Gleason, details from Small, The New York Botanical Garden. **2.** Gaylussacia baccata, × $^2/_5$; *a.* Flowering branch, × $^4/_5$; After Gleason, The New York Botanical Garden. **3.** Vaccinium arboreum, × $^2/_5$; Details from Small, The New York Botanical Garden. **4.** Vaccinium stamineum var. neglectum, × $^2/_5$. **5.** Vaccinium stamineum var. melanocarpum, × $^2/_5$; *a.* Leafy branch; Details from Small, The New York Botanical Garden. **6.** Vaccinium vacillans var. vacillans, × $^2/_5$, fruiting branch; *a.* Flowering branch, × $^2/_5$. **7.** Androsace occidentalis, × $^2/_5$.

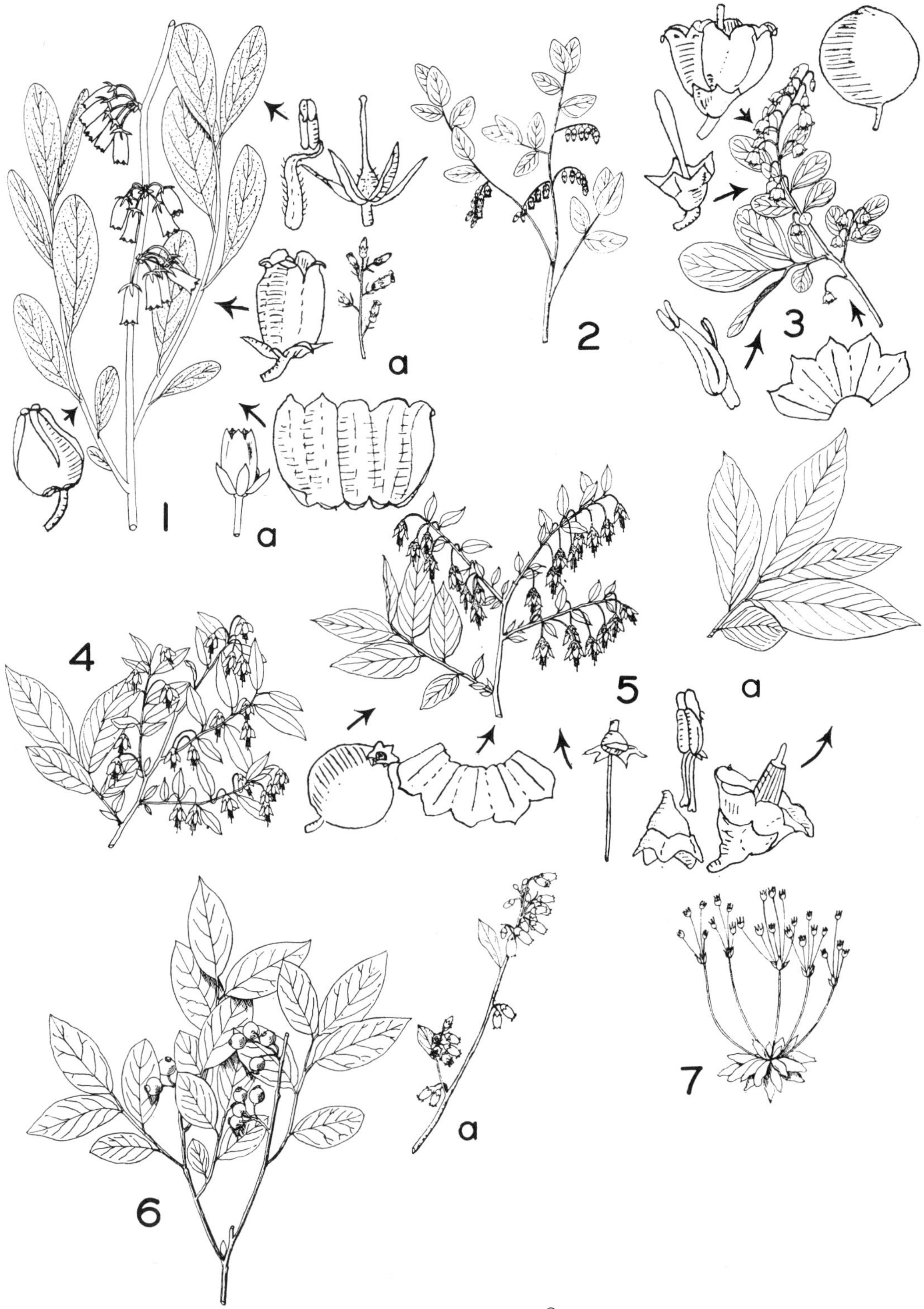


PLATE NO. 276

ly is less common than the var. *glaucescens*.

Ranges from Florida to Texas, north to Virginia, Indiana, Illinois, Missouri, and Oklahoma.

1b. **Vaccinium arboreum** var. **glaucescens** (Greene) Sarg. Map 1694

Apparently the commoner variety in Missouri.

Ranges from Alabama to Texas, north to Illinois, Missouri, and Oklahoma.

It is questionable whether there are actually two distinguishable varieties of this species. Specimens cannot be separated into either variety if collected in the sterile leafy stage or in the late fruiting stage when most of the bracts have disappeared. The two varieties apparently merge and intergrade, and are retained in the present flora until more detailed field and experimental studies are at hand. There is some variation in flower shape, corollas having a more urceolate-campanulate shape corresponding to the description of *Batodendron andrachneforme* Small, considered here synonymous with *Vaccinium arboreum*.

The thick dark green foliage which is long-lasting and partially evergreen, combined with the numerous sprays of white flowers, give this shrub a highly ornamental appearance. Unfortunately, it is difficult to transplant and, like other members of the blueberry family, requires a highly acid soil. It does well in full sun.

Occasional specimens are found (*Steyermark 63856*) with stems only $2\frac{1}{2}$–$3\frac{1}{2}$ feet tall.

2. **Vaccinium stamineum** L. Deerberry Map 1695

Also called Highbush Huckleberry, Squaw Huckleberry.

Flowers April–June, rarely October–November; fruits July–September.

Occurs in acid soils of sandstone, chert, and granitic substrata, in rocky open dry woods, ridges, upland slopes, and glades, usually in pine-oak or oak-hickory woodland.

The following variations occur in Missouri:

a. Calyx (hypanthium) hairy; fruit black-purple
2c. V. STAMINEUM var. MELANOCARPUM

a. Calyx (hypanthium) glabrous (without hairs); fruit variously colored, from green, yellow, reddish-tinged to dark purple . . . . . *b*

b. Branchlets, leaf-stalks, and lower surface of leaves glabrous (without hairs) . . . .
2b. V. STAMINEUM var. NEGLECTUM

b. Branchlets more or less hairy; lower surface of leaves glabrous or with only the midnerve more or less hairy . . . 2a. V. STAMINEUM var. INTERIUS

2a. **Vaccinium stamineum** var. **interius** (Ashe) Palmer & Steyermark Map 1695

*Vaccinium stamineum* of [P & S, Steyerm.], not L.

Ozark region north to Ste. Genevieve, St. Francois, Washington, Crawford, Dent, Laclede, Dallas, Greene, Lawrence, and Newton counties.

Ranges from Maryland to Kansas, south to Virginia, Arkansas, and Oklahoma.

Much of the material previously identified as *V. stamineum* var. *stamineum* in Palmer and Steyermark's *Annotated Catalogue* belongs here. No true *V. stamineum* var. *stamineum* occurs in Missouri if that variety is reserved for plants having the pubescence on the lower surface of the leaves persistent. In all the Missouri material, the lower surface is either glabrous or only the midnerve on the lower side of the leaf is pilose.

2b. **Vaccinium stamineum** var. **neglectum** (Small) Deam Map 1695

Common in the Ozark region north to Ste. Genevieve, St. Francois, Washington, Crawford, Texas, Laclede, Dallas, and Cedar counties.

Ranges from Georgia, Alabama, Tennessee, Arkansas, and Oklahoma, north to New Jersey, Pennsylvania, West Virginia, Ohio, Indiana, Missouri, and Kansas.

2c. **Vaccinium stamineum** var. **melanocarpum** Mohr Southern Gooseberry Map 1695

*Vaccinium melanocarpum* Mohr [P & S, Steyerm., BB]

Known only from Carter County, southeastern Ozarks (Grandin, May 7, 1905, *Bush 2728*).

Ranges from Virginia, Kentucky, and Missouri, southward.

Material previously referred to this variety by Palmer and Steyermark in their *Annotated Catalogue* from Wayne, Barry, and McDonald counties has been reexamined and is placed in the present treatment in either var. *interius* (*Bush 3175* from Barry Co.) or in var. *neglectum* (*Bush 6639* from Wayne Co.). Other collections labeled *V. melanocarpum* (*Palmer 19181* from Stone Co., *Palmer 3995* from Newton Co., and *Bush 2786* from Shannon Co.) are here referred to *V. stamineum* var. *interius* or var. *neglectum*. These collections have either a glabrous calyx or pale fruits, or are sterile and not identifiable.

Much more intensive field and experimental work needs to be carried out on the variations of *V. stamineum*. Some authors, such as Gleason (see *New Ill. Fl.* 3: 25. 1952) treats *V. stamineum*, *V. neglectum*, and *V. melanocarpum* as separate and distinct species, others, such as Fernald relegate these to one variable species. Of the various taxa, *V. melanocarpum* appears more readily separable as a species from *V. stamineum* on the basis of its densely pubescent calyx. It is highly questionable whether var. *interius* can be kept separate from var. *stamineum*, and probably the two should be combined as var. *stamineum.*

The buds and twigs of *V. stamineum* and varieties are eaten by white-tailed deer. The fruit, served cold, after being cooked, has a flavor suggestive of cranberry and gooseberry sauce combined with grapefruit marmalade, but freshly stewed berries, eaten warm, are said to have a poor taste. A jelly or marmalade, stated to have an agreeable flavor, can be prepared from the fruit by cooking the pulp or juice with an equivalent measure in weight of sugar.

3. **Vaccinium vacillans** Torr. Lowbush Blueberry
Map 1696

Flowers April–May; fruits late June–August.

Occurs in dry rocky open woods, ledges of bluffs, glades, and upland level flats, ridges, or slopes, in acid soils overlying sandstone, chert, or granitic substrata. Ozark region north to Lincoln, Montgomery, Callaway, Boone, Morgan, Benton, Henry, and Vernon counties; on Crowley Ridge but elsewhere absent from the lowlands of southeastern Missouri.

The following variations occur in Missouri:

a. Fruit black without a bloom (grayish or silvery coating which can be rubbed off) . . . . 3c. V. VACILLANS var. MISSOURIENSE
a. Fruit blue with more or less of a bloom . . *b*
  b. Young branchlets and lower surface of leaves glabrous (without hairs) . 3a. V. VACILLANS var. VACILLANS
  b. Young branchlets and lower surface of leaves more or less hairy . . . 3b. V. VACILLANS var. CRINITUM

3a. **Vaccinium vacillans** var. **vacillans**
Map 1696

*Vaccinium vacillans* Torr. [G, BB, P & S, Steyerm.]
*Vaccinium alto-montanum* Ashe [BB]

This glabrous variety is less common in the Ozarks than the pubescent-stemmed and leaved variety *crinitum.*

Ranges from Georgia to Missouri, north to Nova Scotia, Maine, New Hampshire, Vermont, New York, Ontario, Ohio, Michigan, Illinois, and Iowa.

The leaves vary from entire (smooth-edged) to finely toothed.

3b. **Vaccinium vacillans** var. **crinitum** Fern.
Map 1696

This pubescent variety is the commonest variation in Missouri, possessing both entire or finely toothed leaves.

Ranges from Georgia to Missouri and Oklahoma, north to Nova Scotia, Maine, New Hampshire, Vermont, New York, Ontario, Michigan, Illinois, and Iowa.

A specimen collected from St. Louis County (Eureka, 1900, *G. Pauls*) has indicated on the label that it was taken from a plant with white fruit. Its pubescent branchlets would place it with var. *crinitum*, of which it represents an undescribed form. A description of such a white-fruited form must await further and additional discovery and collections for better data.

3c. **Vaccinium vacillans** var. **missouriense** Ashe
Map 1696

*Vaccinium missouriense* Ashe

The black-fruited variety is rare and is known only from the southern section of Missouri in Shannon (slopes of mountain opposite Bee Bluff, along Current River, north of Eminence, July 4, 1935, *Steyermark 19300, 19301*), Ozark (rocky glades near Tecumseh, June 28, 1928, *Palmer 34804*), Barton (upland woods along top of sandstone ledge, 3½ mi. northeast of Milford, July 9, 1951, *Palmer 52621*), and Dade (along partially shaded sandstone ledges near Turnback Creek, 4 mi. east of Greenfield, May 6, 1951, *Palmer 51763*) counties. These collections have the fruit black or purple-black and without a bloom, and are the only ones which can be identified to var. *missouriense* on the basis of the fruit character. All others so labeled as *V. missouriense* by Dr. Camp cannot be differentiated from *V. vacillans* var. *crinitum.*

*Vaccinium vacillans* is interpreted in the present treatment as a variable species in the pubescence of branchlets and lower surface of leaves, size, shape and toothing of leaves, and size of plant, varying from 3–9 dm. tall. Dr. Camp has labeled most of the material, referred here to *V. vacillans* var. *crinitum*, as *V. missouriense*, but I have been unable to follow this treatment or to keep the specimens so annotated separate from other variations of *V. vacillans* on the basis of flowering or vegetative material examined.

Specimens from Missouri, previously identified as *V. virgatum* Ait. var. *tenellum* (Ait.) Gray or as *V. tenellum* Ait., are here referred to *V. vacillans* var. *crinitum*, as are also specimens labeled by Camp as *V. viride* Ashe.

The country folk in Missouri usually refer to this plant as Low-bush Huckleberry, whereas in the eastern and northern states it is referred to as Blueberry. The plants in Missouri often produce few ripe fruits at one

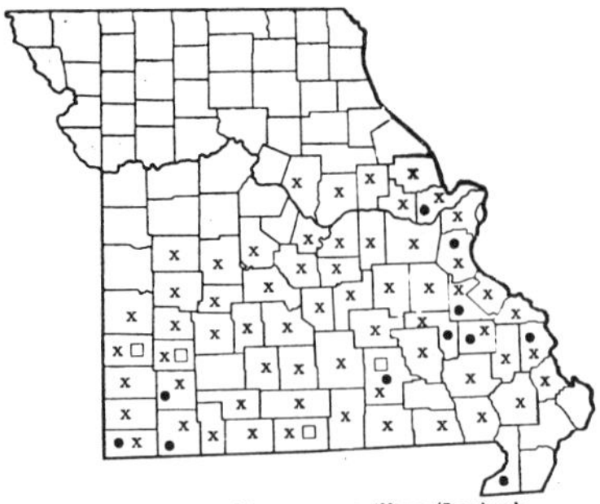

1696 • Vaccinium vacillans var. vacillans (Lowbush Blueberry)
1696 x Vaccinium vacillans var. crinitum
1696 □ Vaccinium vacillans var. missouriense

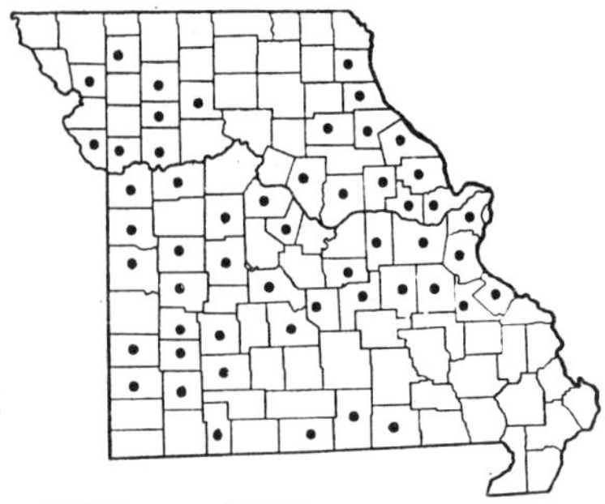

1697 Androsace occidentalis

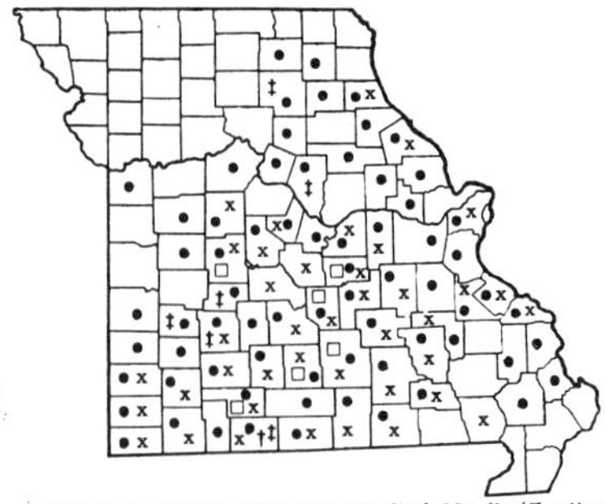

1698 • Dodecatheon Meadia var. Meadia f. Meadia (Shooting Star)
1698 † Dodecatheon Meadia var. Meadia f. sedens
1698 ‡ Dodecatheon Meadia var. Meadia f. album
1698 x Dodecatheon Meadia var. brachycarpum f. brachycarpum
1698 □ Dodecatheon Meadia var. brachycarpum f. pallidum

time as compared with the abundance of production of other species, such as *V. corymbosum* and *V. angustifolium*, in the northern and eastern states. However, the fruits, though often sparse, are sweet and delicious, and are eagerly eaten when available from June on. The uses of blueberry are familiar in fresh or cooked form for desserts, pies, cakes, muffins, juice, jelly, preserves, and jams. The fruits and leaves of the Lowbush Blueberry are eaten by wild turkey, and the buds and twigs are eaten by white-tailed deer.

*Excluded Species*

**Vaccinium pallidum** Ait.

Reported for Missouri by both *Gray's Manual* (p. 1134) and Gleason's *New Illustrated Flora* (3: 29. 1952).

No specimens have been seen from Missouri which agree with material identified as this species. Gleason differentiates *V. pallidum* from *V. vacillans* on the basis of having well-developed serrulate leaves 3–5 cm. instead of 1–3 cm. long, but such larger leaves would be included categorically within the measurements (1.5–5 cm. long) given by Fernald for *V. vacillans*. It is believed that such specimens as have been identified as *V. pallidum* should be referred to *V. vacillans*, and are so interpreted in the present flora.

**Vaccinium tenellum** Ait.

*Vaccinium virgatum* Ait. var. *tenellum* (Ait.) Gray [P & S, Steyerm.]

Specimens from southeastern Ozark counties (Ste. Genevieve, Bollinger, Madison, and Butler) have been identified as this species at one time or another, but a careful study of them reveals that they are more properly placed in *V. vacillans* var. *crinitum*. Such specimens as *Engelmann*, from granite quarry near Pilot Knob, September 9, 1859, have small leaves which approach the leaves of *V. tenellum*, but are slightly broader and have a different shape than are characteristic of *V. tenellum*. The specimen, moreover, is without flower or fruit, so that it cannot be placed with certainty under *V. tenellum*. Other specimens are similarly regarded as showing greater affinity to *V. vacillans* than to *V. tenellum*, so that the latter species is best regarded as an excluded one for Missouri.

## Fam. **PRIMULACEAE** (Primrose Family)

a. Leaves deeply cut (pinnately dissected) into narrow segments; plants usually floating in the water of swamps and sink-hole ponds; stems hollow and inflated . . . . . . . . . . . 7. HOTTONIA
a. Leaves entire (without teeth) or slightly toothed; plants growing on land, or if in wet places not floating; stems solid, not enlarged . . . . . . . . . . . . . . . . . . . . . *b*
  b. Leaves all arising at the base of the plant only . . . . . . . . . . . . . . . *c*
    c. Complete plant less than 10 cm. (4″) tall; leaves tiny, at most 2 cm. long . . . . 1. ANDROSACE
    c. Complete plant usually 15–60 cm. (6–24″) tall; leaves conspicuous, 6–30 cm. long. 2. DODECATHEON
  b. Leaves on the main stem as well as at the base of the plant . . . . . . . . . . . . . *d*
    d. Leaves alternate on the stem. . . . . . . . . . . . . . . . . . . . . . *e*
      e. Complete plant less than 10 cm. (4″) tall; leaves 4–8 mm. long; flowers and fruits sessile

(without stalks) in the axils of the leaves; mature fruit circumscissile (splitting into an upper and lower half) . . . . . . . . . . . . . . . . . . . . 5. CENTUNCULUS

e. Complete plant usually 10–30 (rarely 5) cm. or more tall; leaves mainly 20–80 (up to 100) mm. long; flowers and fruit on conspicuous stalks 10–20 mm. long; mature fruit splitting into 5 valves . . . . . . . . . . . . . . . . . . . . . . . . . 6. SAMOLUS

d. Leaves opposite or in whorls of 3–4 on the stem . . . . . . . . . . . . . . . . . *f*

f. Flowers red or blue, rarely white; leaves 1–2 cm. long; plants annual; mature fruit circumscissile (splitting into an upper and lower half) . . . . . . . . . . 4. ANAGALLIS

f. Flowers yellow, rarely white; leaves usually 3–15 cm. long (except in *Lysimachia Nummularia*, a creeping plant with somewhat round leaves); plants perennial; mature fruit splitting vertically by valves . . . . . . . . . . . . . . . . . . . . 3. LYSIMACHIA

## 1. **Androsace** L.

**Androsace occidentalis** Pursh. Map 1697

Flowers March–May.

Occurs in fallow fields, rocky open glades, bluff escarpments, rocky open woods, and prairies, mostly on limestone soils. Mainly in the northern Ozark, Ozark border, and unglaciated prairie sections in central and western Missouri, north locally to Lewis, Monroe, Livingston, Daviess, Gentry, and Andrew counties; absent or at least not recorded from most of the southern Ozarks or from the lowland counties of southeastern Missouri.

Ranges from Ontario to Manitoba and British Columbia, south to Indiana, Illinois, Arkansas, Texas, New Mexico, and Arizona.

This diminutive plant is the only representative in Missouri of this arctic-alpine group and is appropriate for a rock garden, especially favoring a gritty limestone open exposure. It can be easily raised from seed.

## 2. **Dodecatheon** L. Shooting Star

Base of leaves along the petiole and base of midrib more or less tinged with red or brick color (lacking in var. *Meadia* f. *sedens* and var. *brachycarpum* f. *pallidum*); mature fruit firm-walled, ovoid or conical, mostly 2–3 times as long as thick (broad), dark reddish-brown; commonly encountered and abundant 1. D. MEADIA

Base of leaves pale green or greenish-white along the petiole and base of midrib, not red- or brick-tinged; mature fruit thin-walled, narrowly ellipsoid or cylindric, usually more than 3 times as long as thick (broad), pale brown to yellow; rare species uncommonly encountered . . . . 2. D. AMETHYSTINUM

1. **Dodecatheon Meadia** L. Shooting Star Map 1698

Also called Bird's Bill, American Cowslip.

Flowers April–June.

Occurs in prairies, rocky open glades, ledges of bluffs, and high wooded banks and slopes.

Eastern, southern, and central Missouri.

The following variations are known from Missouri:

a. Anthers 6.5–10 mm. (mainly 7–8 mm.) long; calyx-lobes 3–9 mm. long; fruit 10.5–18 mm. long; 4–50 or more flowers arising from the top of the flower-stem . . . . . . . . *b*

b. Base of leaves without red or brick color . 1b. D. MEADIA var. MEADIA f. SEDENS

b. Base of leaves more or less tinged with red or brick color . . . . . . . . . . *c*

c. Corolla chiefly white. . . 1c. D. MEADIA var. MEADIA f. ALBUM

c. Corolla pink, rose-colored, lavender, purplish, or of various combinations of these colors . 1a. D. MEADIA var. MEADIA f. MEADIA

a. Anthers 4–7 mm. (mainly 5.5–6.5 mm.) long; calyx-lobes 2.5–5 mm. long; fruit 7.5–10 mm. long; usually only 1–14 flowers arising from the top of the flower-stem . . . . . . *d*

d. Base of leaves with red or brick color . . 1d. D. MEADIA var. BRACHYCARPUM f. BRACHYCARPUM

d. Base of leaves without red or brick color . 1e. D. MEADIA var. BRACHYCARPUM f. PALLIDUM

1a. **Dodecatheon Meadia** var. **Meadia** f. **Meadia** Map 1698

*Dodecatheon Meadia* L. [G, BB, P & S, Steyerm.]
*Dodecatheon Meadia* subsp. *Meadia*
*Dodecatheon Meadia* subsp. *eumeadia* R. Knuth
*Dodecatheon Meadia* var. *genuinum* Fassett

Eastern, southern, and central Missouri, east and

south of a line drawn from Adair, Macon, Randolph, Howard, and Saline counties to Jackson County.

Ranges from Georgia, Alabama, Arkansas, Louisiana, and Texas, north to D.C., Pennsylvania, Indiana, Illinois, and Wisconsin.

1b. **Dodecatheon Meadia** var. **Meadia** f. **sedens** Fassett Map 1698

Scattered in the Ozark area, known from Taney County (Forsyth, April 19, 1938, *Edith Seymour Jones*, holotype in U. Wis. Herb.), and undoubtedly occurring elsewhere in the state.

1c. **Dodecatheon Meadia** var. **Meadia** f. **album** Macbr. Map 1698

*Dodecatheon Meadia* f. *album* Macbr. [G, Steyerm.]

Scattered throughout the range of var. *Meadia* in Missouri.

1d. **Dodecatheon Meadia** var. **brachycarpum** (Small) Fassett f. **brachycarpum** Map 1698

*Dodecatheon Meadia* var. *brachycarpum* (Small) Fassett [G, BB]

*Dodecatheon brachycarpum* Small [Steyerm.]

*Dodecatheon Meadia* subsp. *brachycarpum* (Small) Knuth

Common throughout the Ozark region and north to Marion, Gasconade, Osage, Moniteau, Pettis, and Jasper counties.

Ranges from Alabama, Arkansas, and Texas, north to Virginia, Missouri, and Oklahoma.

1e. **Dodecatheon Meadia** var. **brachycarpum** f. **pallidum** Fassett Map 1698

Scattered in the south-central and southwestern Ozark region, where known east to Maries, Pulaski, and Texas counties, northwest to Benton County, southwest to Christian County, but to be expected over a wider range when more intensively collected.

The above treatment follows the work of Fassett (Am. Midl. Nat. 31: 455–86. 1944). The flowers vary considerably in range of color and the leaves vary from entire to toothed. Plants growing in moist crevices of bluffs sometimes exhibit a toothed leaf-blade which tends to be more abruptly contracted into a petiole, suggesting var. *Frenchii* Vasey. The latter is considered by Dr. Voigt to be a separate species, *D. Frenchii* (Vasey) Rydb. (Rh. 57: 325–32. 1955), based upon the distinctive broadly oval or ovate subcordate leaf-blades abruptly contracted into a petiole combined with an unchanged morphological appearance when transplanted.

Flowers of Shooting Star have a fragrance reminiscent of the odor of grape juice, and the roots have an odor suggestive of that of canned corned beef.

2. **Dodecatheon amethystinum** Fassett Map 1699

Flowers May–June.

Occurs in moist crevices and on steep slopes of north-facing wooded limestone bluffs. Found in Marion County, northeastern Missouri, and in the northern and western Ozark border counties.

Ranges in eastern Pennsylvania, West Virginia, and Kentucky; and the 'driftless area' of southwestern Wisconsin, northwestern Illinois, southeastern Minnesota, northeastern Iowa, and Missouri.

Two variations occur in Missouri:

Corolla lilac, rose, orchid-pink, or red-purple
 2a. D. AMETHYSTINUM f. AMETHYSTINUM
Corolla chiefly white (some lavender may occur at very base) . . . . . 2b. D. AMETHYSTINUM f. MARGARITACEUM

2a. **Dodecatheon amethystinum** f. **amethystinum** Map 1699

*Dodecatheon amethystinum* Fassett [G, Steyerm.]

*Dodecatheon radicatum* Greene subsp. *radicatum* [Thompson]

Known from Marion (between Lover's Leap and Riverside Cemetery, T56N, R4W, sect. 27, ½ mi. southeast of Hannibal, May 15, 1938, (*Steyermark 5318*; *Davis*), Osage (north-facing slopes bordering Missouri River, T45N, R8W, sect. 10, 1½ mi. west of Chamois, May 20, 1950, *Steyermark 69756*), Cole, and Dallas (base of moist north-facing limestone bluffs along Niangua River, T34N, R18W, sect. 8, 1½ mi. south of Windyville, June 2, 1951, *Steyermark 71468*) counties.

2b. **Dodecatheon amethystinum** f. **margaritaceum** Fassett Map 1699

Known only from Cole County, northern Ozarks (moist crevices of bluffs, on steep shaded slopes above bluffs, and at base of north-facing steep wooded bluffs along Missouri River, T44N, R10W, sect. 16, just

Plate no. 277. **1.** Dodecatheon Meadia var. Meadia, × 2/7; Details from Small, The New York Botanical Garden. **2.** Lysimachia Nummularia, × 2/7; Details from Small, The New York Botanical Garden. **3.** Lysimachia ciliata, × 2/7. **4.** Dodecatheon amethystinum, × 2/7. **5.** Lysimachia thyrsiflora, × 2/7. **6.** Lysimachia lanceolata, × 2/7; Details from Small, The New York Botanical Garden. **7.** Lysimachia hybrida, × 2/7.

PLATE NO. 277

1699 • Dodecatheon amethystinum f. amethystinum
1699 □ Dodecatheon amethystinum f. margaritaceum

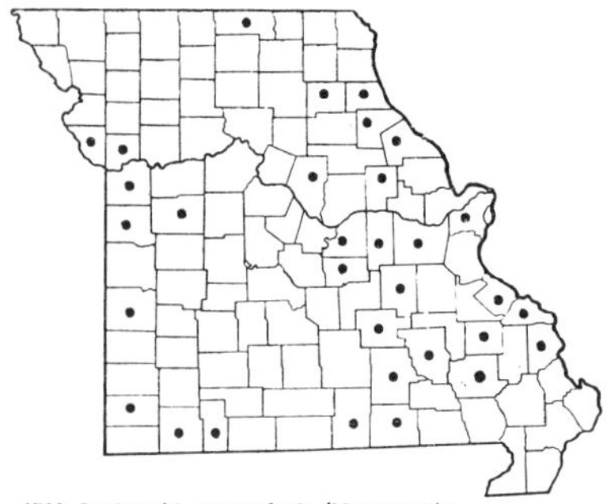

1700 Lysimachia nummularia (Moneywort)

1701 Lysimachia thyrsiflora (Tufted Loosestrife)

north and northwest of Osage City, May 20, 1950, *Steyermark 69732*).

This albino form is quite rare within the range of *D. amethystinum* as noted by Fassett (Am. Midl. Nat. 31: 475. 1944).

H. J. Thompson has annotated material of *D. amethystinum* as *D. radicatum* subsp. *radicatum*, considering the two taxa as the same. This view is not followed in the present flora, as *D. radicatum*, a western species, appears sufficiently distinct from *D. amethystinum* to be treated as a separate species. I have plants of *D. radicatum* from the Black Hills of South Dakota growing and under observation together with plants of *D. amethystinum*, the latter now fifteen years old, originally collected in southwestern Wisconsin by Dr. Fassett. The two species appear distinct in a number of ways, *D. radicatum* from the Black Hills and other western states having much shorter leaves of a different shape than those of *D. amethystinum*, much shorter scapes (flower-stems), and smaller and fewer flowers, with much shorter corolla-lobes, shorter calyx-lobes, and shorter anthers. The base of the leaf-blades in *D. radicatum* is also usually marked with red.

## 3. **Lysimachia** L. Loosestrife

a. Leaves nearly round or roundish, heart-shaped at base, rounded at tip, 1.5–3 cm. long; stems creeping and trailing . . . . . . . . . . . . . . . . . . . . . . . 1. L. NUMMULARIA
a. Leaves of other shapes, linear, lanceolate, elliptic, or ovate, mostly 3–15 cm. long; stems upright, spreading, or reclining, or if rooting and creeping at the nodes, then the leaves ovate or lanceolate and pointed at tip and 3–9 cm. long . . . . . . . . . . . . . . . . . . . . . . . *b*
  b. Leaves from the middle and upper part of stem narrowly linear, 2–7 mm. broad, with only 1 main nerve, side nerves either not showing or very faint . . . . . . . . . . 7. L. QUADRIFLORA
  b. Leaves from the middle part of the stem lanceolate, elliptic, or ovate, usually 9–55 mm. broad, with the side nerves plainly visible . . . . . . . . . . . . . . . . . . . . . . *c*
    c. Leaves dotted on upper surface, not ciliate (hairs along margins) at base; flowers crowded into head- or club-shaped clusters; individual flowers small with narrowly linear smooth-edged divisions 3–5 mm. long and mostly conspicuously dotted with black-purple. . . . 2. L. THYRSIFLORA
    c. Leaves not dotted on upper surface, more or less ciliate at base; flowers solitary or few together, not crowded together in dense clusters; individual flowers showy with toothed or uneven-edged broadly ovate or elliptical divisions mostly 6–13 mm. long, or 3–5 mm. long in *L. radicans*, not dotted or not conspicuously dotted . . . . . . . . . . . . . . . . . . . . . *d*
      d. Divisions of corolla 3–5 mm. long; divisions of calyx 3–5 mm. long; stems reclining, rooting at the nodes and eventually creeping; plants known only from the swamps of southeastern Missouri and the sink-hole ponds of the southeastern Ozarks west to Howell County. 3. L. RADICANS
      d. Divisions of corolla 7–13 mm. long; divisions of calyx 3.5–10 mm. long; stems mainly erect or ascending, sometimes reclining but not rooting; plants of various parts of Missouri throughout the state . . . . . . . . . . . . . . . . . . . . . . . . *e*
        e. Main leaf-blades from the middle part of the stem 2–6 cm. broad, ovate, rounded or somewhat heart-shaped at base, abruptly contracted at base into a distinct leaf-stalk (petiole); leaf-stalk fringed with hairs throughout its length . . . . . . . 4. L. CILIATA

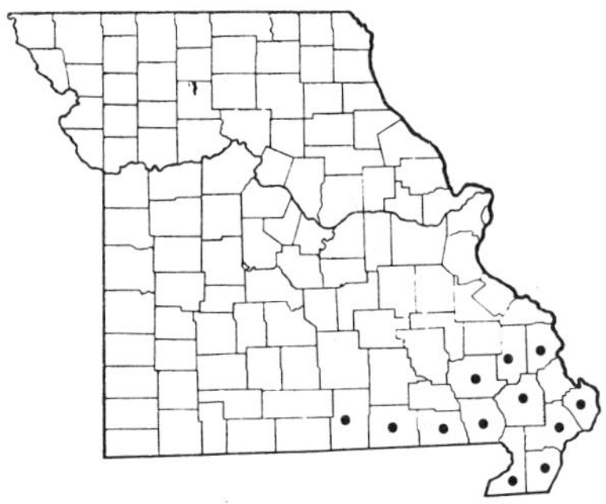

1702 Lysimachia radicans

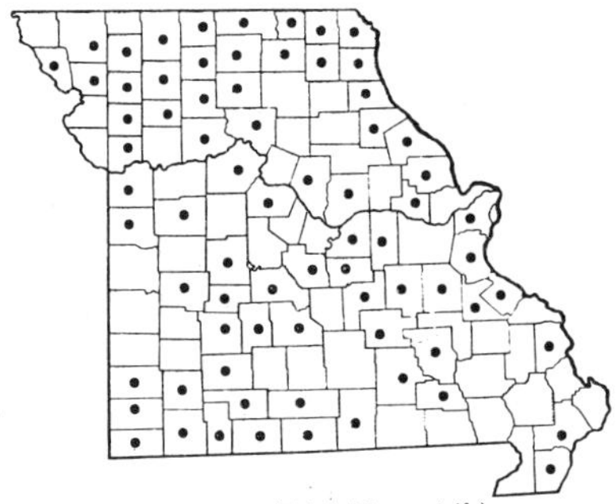

1703 Lysimachia ciliata (Fringed Loosestrife)

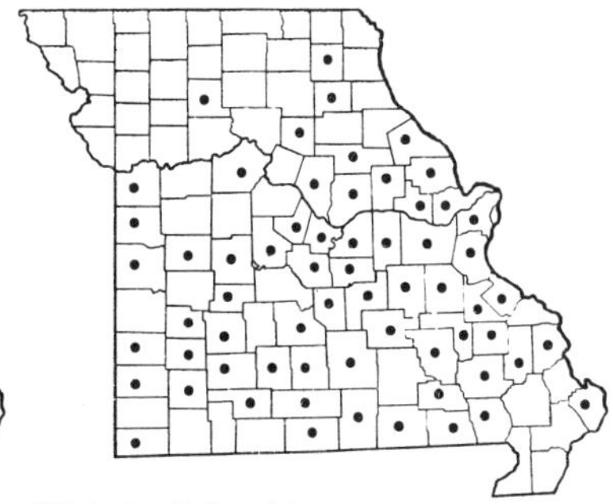

1704 Lysimachia lanceolata

e. Main leaf-blades from the middle part of the stem 0.5– 2 cm. (up to 3) broad, narrowly oblong to linear-lanceolate, gradually narrowed and tapering to the base, not contracted into a definite or distinct leaf-stalk (petiole); fringe of hairs, when present, only at the base of the leaf-stalk or leaf . . . . . . . . . . . . . . . . . . . . . . . *f*

f. Divisions of calyx thin with 3 nerves (1 main nerve down the center and 2 side nerves) plainly visible (use magnifying lens of 8 ×); plant without runners (stolons) at the base; stem usually more than 4 mm. in diam.; lower leaves not persisting; middle and upper leaves green on lower surface, gradually tapering into a somewhat winged leaf-stalk (petiole), slightly fringed at the base; smallest leaf-blades in the upper third of plant . . . . . . . . . . . . . . . . . . . . . 6. L. HYBRIDA

f. Divisions of calyx thicker with nerves either not showing or only faint (use magnifying lens of 8 ×); plant with slender elongate runners (stolons) at base; stem usually less than 4 mm. in diam.; lower leaves persisting; middle and upper leaves pale or gray-green on lower surface, gradually tapering to the sessile (stalkless) or nearly sessile base, conspicuously fringed at the base; smallest leaf-blades at the base of the stem . . . . . . . . . . . . . . . . . . . . . 5. L. LANCEOLATA

1. **Lysimachia Nummularia** L. Moneywort Map 1700

Flowers May–August.

Escaped from gardens and lawns to low wet ground in valley woodland, banks of streams, along ditches and roadsides. Scattered throughout the state, apparently commonest in the southern half.

Native of Europe; introduced and naturalized in North America from Newfoundland to Ontario, south to Georgia, Illinois, Missouri, and Kansas; also in the Pacific states.

Plants seldom flower, spreading by means of the creeping stem which roots in the damp soil. The plant is a good ground cover for damp shaded ground.

2. **Lysimachia thyrsiflora** L. Tufted Loosestrife Map 1701

Also called Water Loosestrife.

*Naumburgia thyrsiflora* (L.) Duby [BB]

Flowers late May–July.

Occurs in swamps, boggy ground, and wet river bottom prairies and swales. Northwestern and west-central Missouri in Jackson (swamps, Courtney, *Bush 26*; bogs, Sibley, *Mackenzie 72*) and Atchison (swales and marshy ground in valley of Big Tarkio River, along east side of route 59 and 275, T64N, R40W, north part of sect. 10, 3 mi. north of Fairfax, July 21, 1952, *Steyermark 73883*) counties.

Ranges from Quebec to Alaska, south to New Jersey, Pennsylvania, Ohio, Indiana, Illinois, Missouri, Colorado, and California.

3. **Lysimachia radicans** Hook. Map 1702

*Steironema radicans* (Hook.) Gray [P & S, BB]

Flowers June–August.

Occurs in low wet woods, bald cypress swamps and bayous, and borders of upland sink-hole ponds. Southeastern lowland counties north to Cape Girardeau, Bollinger, and Wayne counties, west to the Ozarks in sink-hole ponds in Ripley, Oregon, and Howell counties.

Ranges from Mississippi to Texas, north to Tennessee, Illinois, and Missouri.

In his revision of *Lysimachia* (Ill. Biol. Monog. 24: 1–160. 1956) Dr. Ray cites specimens of this species from two western Missouri counties. One of them (Blue Spring, August 18, 1892, *Eggert*) referred by Ray to Jackson County is not Blue Spring, Jackson County, since Eggert never collected in any part of western Missouri, but refers to Blue Spring in either Carter,

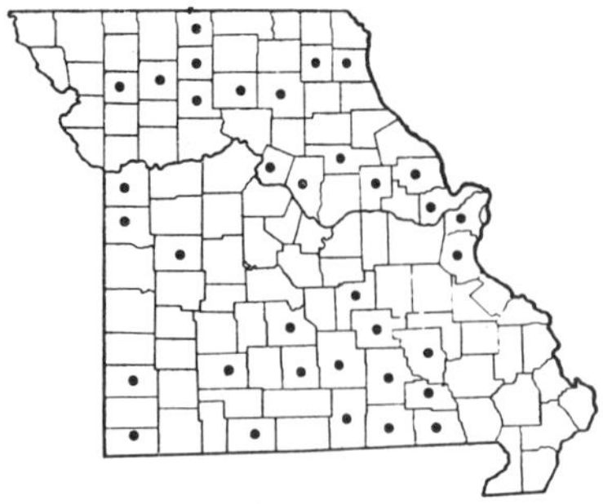

1705 Lysimachia hybrida

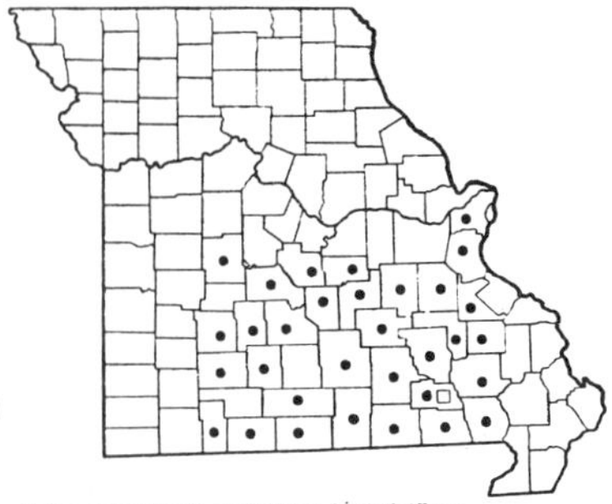

1706 • Lysimachia quadriflora f. quadriflora
1706 □ Lysimachia quadriflora f. albescens

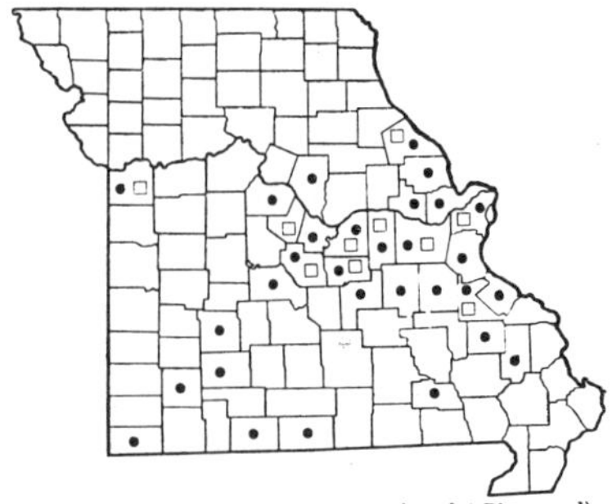

1707 • Anagallis arvensis f. arvensis (Scarlet Pimpernel)
1707 □ Anagallis arvensis f. caerulea (Poor Man's Weatherglass)

Shannon, Wayne, Oregon, or Ripley counties, southeastern Missouri Ozarks, a section of the state Eggert visited on a number of occasions. The other specimen cited by Ray from Cass County (bottoms, *Broadhead*) proves on re-examination to be *L. hybrida.*

4. **Lysimachia ciliata** L. Fringed Loosestrife Map 1703

*Steironema ciliatum* (L.) Raf. [P & S, BB]

Flowers May–July.

Occurs in moist low woodland, wet places in prairies, and borders of streams and ponds. Throughout Missouri.

Ranges from Quebec to British Columbia, south to Florida, Texas, New Mexico, and Arizona.

5. **Lysimachia lanceolata** Walt. Map 1704

*Steironema lanceolatum* (Walt.) Gray [P & S, BB]

*Steironema lanceolatum* var. *angustifolium* (Lam.) Gray [P & S]

*Lysimachia lanceolata* subsp. *lanceolata* [Ray]

Flowers May–August.

Occurs in dry or wet rocky woods, slopes, and ridges, or in valley bottoms, thickets, wet places in prairies, borders of streams, ponds, and swamps, rarely along railroads. Common in southern and central Missouri north to Pike, Shelby, Knox, Randolph, Livingston, and Jackson counties.

Ranges from Florida to Louisiana and Oklahoma, north to Pennsylvania, Ohio, Michigan, and Wisconsin.

6. **Lysimachia hybrida** Michx. Map 1705

*Lysimachia lanceolata* subsp. *hybrida* (Michx.) Ray [Ray]

*Steironema lanceolatum* var. *hybridum* (Michx.) Gray [P & S]

*Steironema hybridum* (Michx.) Raf. [BB]

Flowers June (rarely in late May)–September.

Occurs in wet or swampy meadows, river bottom prairies, margins of sloughs, pools, ponds, spring branches, occasionally low wet woods. Scattered throughout Missouri.

Ranges from Quebec to Ontario and North Dakota, south to Florida, Mississippi, and Texas.

A specimen collected by Broadhead from 'bottoms, Cass County' and identified by Ray as *L. radicans* is a portion or fragment of a plant with calyx-lobes 5 mm. long, and is referred in the present flora to *L. hybrida.*

As there appear to be sufficient differences for separating this taxon from *L. lanceolata,* I am treating it in the present flora as a separate species rather than as a subspecies of *L. lanceolata* as was done by Dr. Ray.

7. **Lysimachia quadriflora** Sims Map 1706

*Steironema quadriflorum* (Sims) Hitchc. [BB, P & S]

*Lysimachia longifolia* Pursh

Flowers June–August.

Occurs in swampy calcareous meadows fed by springs, along spring branches, rocky banks and beds of small streams, and also wet thickets around springs. Ozark region of southern and east-central Missouri north to St. Louis, Washington, Crawford, Maries, Miller, Camden, and Benton counties, west to Benton, Polk, Greene, and Stone counties.

Ranges from New York and Ontario to Manitoba, south to Virginia, Kentucky, Illinois, and Missouri.

Two variations occur in Missouri:

Corolla yellow . 7a. L. QUADRIFLORA f. QUADRIFLORA
Corolla whitish or cream-colored . . . . 7b. L. QUADRIFLORA f. ALBESCENS

7a. **Lysimachia quadriflora** f. **quadriflora** Map 1706

*Lysimachia quadriflora* Sims [G]

*Steironema quadriflorum* (Sims) Hitchc. [BB, P & S]

This is the common form encountered in Missouri and elsewhere throughout its range.

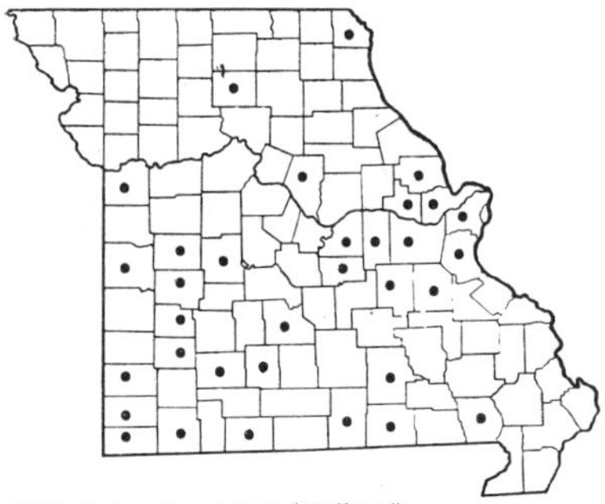

1708 Centunculus minimus (Chaffweed)

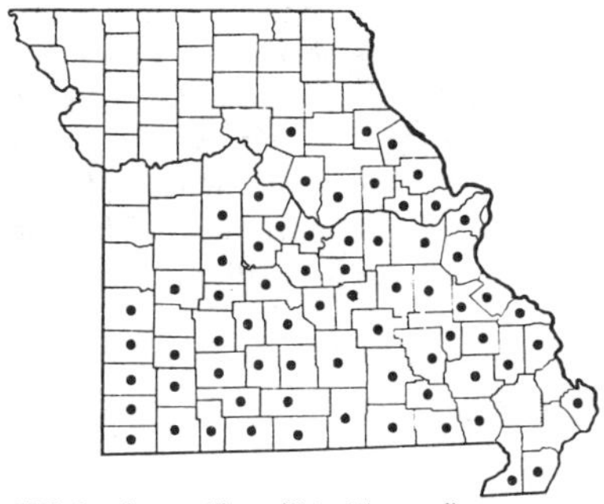

1709 Samolus parviflorus (Water Pimpernel)

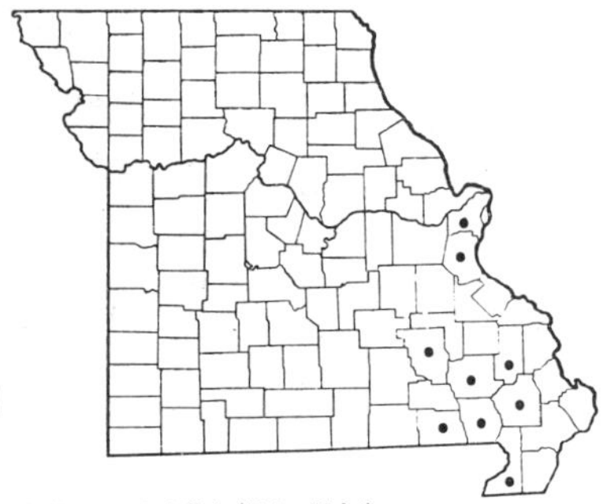

1710 Hottonia inflata (Water Violet)

7b. **Lysimachia quadriflora** f. **albescens** Steyerm. Map 1706

Known only from Carter County, southeastern Ozarks (swampy meadow along route 60, spring branch tributary to Right Fork of Carter Creek, T27N, R1E, western part sect. 2, 6.7 mi. northeast of Van Buren, July 4, 1949, *Steyermark 68397*).

*Lysimachia quadriflora* is a characteristic species of wet limey places, usually following spring branches or rocky limestone beds of small streams, or in calcareous spring-fed swampy meadows.

## 4. **Anagallis** L. Pimpernel, Poor Man's Weatherglass

**Anagallis arvensis** L. Map 1707

Flowers late May–September.

Occurs in fields, pastures, rocky open glades, edges of bluffs, waste ground, along roadsides and railroads, gravel bars and sandy alluvium along streams, and around gardens.

Two variations occur in Missouri:

Flowers red or scarlet, rarely white . . . . a. A. ARVENSIS f. ARVENSIS
Flowers blue . . . b. A. ARVENSIS f. CAERULEA

a. **Anagallis arvensis** f. **arvensis** Scarlet Pimpernel Map 1707

*Anagallis arvensis* L. [G, P & S, Steyerm.]

*Anagallis arvensis* var. *arvensis* [BB]

Southern and central Missouri north to Pike, Boone, and Jackson counties.

Native of Europe; introduced and naturalized in North America from Newfoundland to Minnesota and British Columbia, south to Florida, Texas, and California; also in Mexico, Central, and South America.

b. **Anagallis arvensis** f. **caerulea** (Schreb.) Baumg. Poor Man's Weatherglass Map 1707

Also called Pimpernel.

*Anagallis arvensis* var. *caerulea* (Schreb.) Gren. & Godr. [BB, P & S, Steyerm.]

Less common in Missouri than the red-flowered form, and known chiefly from central Missouri, north to Pike, Moniteau, and Jackson counties, south to St. Francois, Franklin, Gasconade, Maries, and Miller counties.

Native of Europe; introduced and naturalized in the United States from Vermont to Ohio and Missouri, south to Virginia and Texas to California.

The name Weatherglass refers to the closing of the flowers with the approach of bad weather.

## 5. **Centunculus** L. Chaffweed

**Centunculus minimus** L. Chaffweed Map 1708

Flowers May–August.

Occurs in damp places in prairies, meadows, glades, along paths in woods, clearings, and bare ground in pastured woods, usually in acid soils near sandstone, chert, or granite. Chiefly in southern and central Missouri, north locally in Clark and Linn counties.

Ranges from Florida to Texas, Mexico, and California, north to Georgia, Delaware, Nova Scotia, Ohio, Illinois, Minnesota, North Dakota, Saskatchewan, and British Columbia; also in Central America; widely distributed in the Old World.

### 6. **Samolus** L. Water Pimpernel

**Samolus parviflorus** Raf. Water Pimpernel Map 1709

*Samolus floribundus* HBK. [BB]

Flowers late April–June.

Occurs on wet ledges and crevices of moist shaded rocky bluffs, muddy and sandy banks of streams, margins of ponds, sloughs, and swamps, and low wet woodland. Chiefly in the Ozark region of southern and central Missouri, north to Ralls, Randolph, Cooper, Pettis, St. Clair, and Vernon counties.

Ranges from Florida to Texas, California, Mexico, north to New Brunswick, Quebec, Ontario, Michigan, Illinois, Missouri, Kansas, and British Columbia; also in the West Indies, Central and South America.

The leaves are recorded as one of the food plants of white-tailed deer.

### 7. **Hottonia** L. Water Violet

**Hottonia inflata** Ell. Water Violet Map 1710

Also called Featherfoil.

Flowers early April–mid-May.

Occurs in bald cypress and swamp tupelo swamps, bayous, sloughs, and upland sink-hole ponds. Southeastern Missouri lowlands north locally to St. Louis County and locally west in the Ozarks in sink-hole ponds in Reynolds and Ripley counties.

Ranges from Florida to Texas, north to Maine, New Hampshire, New York, Ohio, Indiana, Illinois, Missouri, and Oklahoma.

The white or greenish-white flowers are arranged in whorls (circles) at the joints in an interrupted inflorescence.

This species usually is found in swamps in low ground, but is remarkably isolated on the dry Ozark upland in sink-hole ponds in Reynolds (Lily and Vinson Ponds) and Ripley (Cupola Pond) counties, where it is associated with other species characteristic of bald cypress and tupelo swamps of southeastern Missouri, the flora a part of that usually found in the Mississippi Embayment of the Gulf and Atlantic Coastal Plain.

## Order **EBENALES**

## Fam. **SAPOTACEAE** (Sapodilla Family)

### **Bumelia** Sw. Southern Buckthorn, Chittim-wood

Lower surface of leaves hairy; twigs and young branchlets more or less hairy . . . . . . . . 1. B. LANUGINOSA var. ALBICANS

Lower surface of leaves glabrous (without hairs) or nearly glabrous; twigs and young branchlets glabrous or nearly so . . . . . . . . . . . . . . . . . . . . . . . . 2. B. LYCIOIDES

1. **Bumelia lanuginosa** (Michx.) Pers. var. **albicans** Sarg. Chittim-wood Map 1711

Also called False Buckthorn, Gum-elastic.

*Bumelia lanuginosa* var. *oblongifolia* (Nutt.) Clark [G, BB]

*Bumelia lanuginosa* in part of auth. [P & S], not (Michx.) Pers.

*Bumelia lanuginosa* subsp. *oblongifolia* (Nutt.) Cronq.

Flowers June 1–July.

Occurs in dry or open rocky woodland, rocky open glades, and bluff escarpments, crevices of bluffs, usually on upland ridges and slopes, rarely in valleys and ravine bottoms. Ozark region of southern and central Missouri north to Lincoln, Montgomery, Callaway, Boone, Cooper, Pettis, Henry, and Barton counties.

Ranges from Louisiana and Texas, north to Illinois, Missouri, Kansas, and Oklahoma; also in Mexico.

In order of flowering, this is next to the last of the native shrubs or small trees to come into bloom in Missouri; buds are in evidence in late May and first

---

Plate no. 278. **1.** Lysimachia quadriflora, × 2/5. **2.** Anagallis arvensis, × 2/5; Details from Small, The New York Botanical Garden. **3.** Hottonia inflata, × 2/5; Details from Small, The New York Botanical Garden. **4.** Bumelia lycioides, × 2/5; After Gleason, details from Small, The New York Botanical Garden. **5.** Samolus parviflorus, × 2/5; Details from Small, The New York Botanical Garden. **6.** Centunculus minimus, × 2/5; *a.* Fruit with calyx, × 6. **7.** Bumelia lanuginosa var. albicans, × 2/5; After Gleason, details from Small, The New York Botanical Garden.

PLATE NO. 278

1711 Bumelia lanuginosa var. albicans (Chittim-wood)

1712 Bumelia lycioides (Southern Buckthorn)

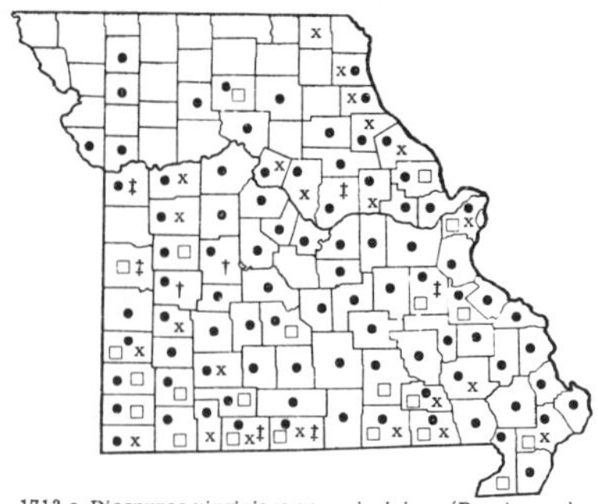

1713 • Diospyros virginiana var. virginiana (Persimmon)
1713 x Diospyros virginiana var. pubescens f. pubescens
1713 ‡ Diospyros virginiana var. pubescens f. pumila
1713 □ Diospyros virginiana var. platycarpa f. platycarpa
1713 † Diospyros virginiana var. platycarpa f. atra

flowers have been noted by the first of June. Species of Basswood (*Tilia*) are almost as late to start flowering, but are seen in the flowering stage from late May on. The Vernal Witch Hazel (*Hamamelis vernalis*) is the earliest woody plant to flower in Missouri (late December to early April), while the Eastern Witch Hazel (*Hamamelis virginiana*) is the last to flower (September–December).

The flowers of *Bumelia lanuginosa* var. *albicans* occur along the sides of the branchlets in the axils of the leaves, and are grouped into clusters of usually 12–30. They are whitish or cream-colored with hairy pedicels (stalks). The small globular fruit is black. The foliage is long-persisting and in late autumn turns a yellowish-green.

Typical *B. lanuginosa* var. *lanuginosa* occurs in the southern states from Georgia and Florida to Louisiana.

2. **Bumelia lycioides** (L.) Pers. Southern Buckthorn Map 1712

Also called Carolina Buckthorn, Smooth Bumelia, Ironwood, and Buckthorn Bumelia.

*Bumelia Smallii* Clark [G]

Flowers June–July.

Occurs in low alluvial woods of flood plains and river bottom land, and thickets along streams. Known only from the lowlands of southeastern Missouri in New Madrid, Pemiscot, and Butler counties.

Ranges from Florida to Texas, north to Virginia, Kentucky, Indiana, Illinois, and Missouri.

This shrub or small tree, like *B. lanuginosa* var. *albicans*, has spiny branches. It is known in the southeastern lowland of the state. In New Madrid County it is of rare occurrence along the sandy alluvium in wooded thickets of the flood plain forest bordering the Mississippi River.

Clark (Ann. Mo. Bot. Gard. 29: 155–82. 1942) determined specimens from Missouri with small leaves (2–6 cm. long) as *B. Smallii*, and Fernald has attempted to differentiate these smaller-leaved plants as specifically distinct from those of *B. lycioides* having leaves 6–15 cm. long. There seems, however, to be no basis for the separation on foliage differences, nor on those of presumed differences in the shape of the fruit.

## Fam. **EBENACEAE** (Ebony Family)

### **Diospyros** L. Persimmon

**Diospyros virginiana** L. Persimmon Map 1713

Flowers late May–June.

Occurs in rocky or dry open woods, borders of woods, rocky glades, limestone bald knobs, prairies, fallow and abandoned fields, thickets, alluvial woods, and valleys along streams. Common throughout southern and central Missouri, north to Lewis, Macon, Linn, Livingston, Gentry, and Platte counties.

The following variations occur in Missouri:

a. Lower surface of mature leaves glabrous (without hairs) or nearly so . . . a. D. VIRGINIANA var. VIRGINIANA

a. Lower surface of mature leaves more or less hairy . . . . . . . . . . . . . *b*

b. Mature fruit dark purple, blue-purple, or plum blue . . . . . . e. D. VIRGINIANA var. PLATYCARPA f. ATRA

b. Mature fruit orange-yellow to salmon-yellow, with a gray-silvery 'bloom' . . . *c*

c. Base of leaf-blades broadly rounded to somewhat heart-shaped; fruit depressed-globose or flattened, up to 7.5 cm. across, early-ripening from July to September . . . . . . . d. D. VIRGINIANA var. PLATYCARPA f. PLATYCARPA
c. Base of leaf-blade more or less narrowed, acute, obtuse, or slightly rounded; fruit oblong or globe-shaped, mostly 2–4 cm. across, usually ripening after frost . . *d*
d. Mature fruiting shrub usually 1–4 meters (3 feet 4 inches to 13½ feet) tall, rarely a tree up to 7–8 m. tall; leaves 3–8 cm. long, 2–4 cm. broad . c. D. VIRGINIANA var. PUBESCENS f. PUMILA
d. Mature tree usually over 4.5 meters (15 feet) tall; leaves 8–11 cm. long, mainly 5–7 cm. broad. b. D. VIRGINIANA var. PUBESCENS f. PUBESCENS

a. **Diospyros virginiana** var. **virginiana** Map 1713
*Diospyros virginiana* L. [G, P & S, Steyerm.]

This is found throughout the range and is the commonest variation in the state.

Ranges from Florida to Texas, north to New York, Pennsylvania, West Virginia, Ohio, Indiana, Illinois, Iowa, Kansas, and Oklahoma.

b. **Diospyros virginiana** var. **pubescens** (Pursh) Dippel f. **pubescens** Map 1713
*Diospyros virginiana* var. *pubescens* (Pursh) Dippel [G, BB, P & S, Steyerm.]

Scattered throughout the range of var. *virginiana*.

Ranges from Florida to Arkansas, north to Virginia, Kentucky, Illinois, Missouri, Iowa, Kansas, and Oklahoma.

c. **Diospyros virginiana** var. **pubescens** f. **pumila** (Palmer & Steyermark) Steyerm. Map 1713
*Diospyros virginiana* f. *pumila* Palmer & Steyerm. [P & S, Steyerm.]

Scattered in the range of var. *pubescens*, where known in southern and central Missouri in Callaway (3 mi. east of Ham's Prairie, September 11, 1937, *Steyermark 26151*), Washington, Ozark (*Steyermark 22913*), Taney, Jackson (Dodson, June 6, 1913, *Bush 7041*, holotype in herb. Arn. Arb., Harvard Univ.; Westport, *Bush 787*, Greenwood, *Bush 10092*, Greenwood, *Palmer 26030*, paratypes in herb. Arn. Arb.) and Bates (Monteith Junction, *Palmer 26080*, paratype in herb. Arn. Arb.) counties.

Known only from Missouri and Arkansas.

This form is often found on limestone bald knobs and glades.

d. **Diospyros virginiana** var. **platycarpa** f. **platycarpa** Sarg. Map 1713
*Diospyros virginiana* var. *platycarpa* Sarg. [G, P & S, Steyerm.]

Scattered in the range in Missouri of the var. *virginiana* and var. *pubescens*, north to Linn County.

Ranges from Missouri to Kansas, south to Arkansas and Oklahoma.

In his original description of var. *platycarpa*, Sargent was influenced by E. J. Palmer's observations. The fruit was described as having a yellow skin, and a flesh which was edible and soft without the action of frost. The fruit was described as being larger, more depressed-globose, and attaining a breadth of 4.5–7.5 cm. and a height of 2.5 cm. Mr. Palmer noted that, whereas the typical persimmon has a tougher skin, more fibrous flesh, and often candies on the tree, sometimes remaining attached all winter, the fruit of var. *platycarpa* is often so soft that it becomes crushed or completely squashed in falling to the ground when fully ripe. Also Palmer noted that var. *platycarpa* is usually a small tree seldom over 4–8 m. tall, never attaining the size of the trunk nor the height of the smaller-fruited typical *D. virginiana*.

e. **Diospyros virginiana** var. **platycarpa** f. **atra** Sarg. Map 1713

Known only from Benton (west side of Tebo Branch of Grand River, T41N, R23W, northwest ¼ sect. 28, 1½ mi. northeast of Racket, July 31, 1949, *Steyermark 68653*) and St. Clair (upland woods near clubhouse, on top of wooded limestone slopes along Osage River, T37N, R28W, sect. 6, 2½ mi. west of Taberville, September, 27, 1938, *Steyermark 9614*) counties.

Known only from specimens from Missouri and Oklahoma, and reported from Indiana and Kentucky. The fruit of this form was tested by the present author and found to have poor-tasting flesh, but recently collected specimens from Kentucky are said to have a superior taste.

The persimmon is quite variable with respect to size of plant, amount of hair present on the surface of the leaves, curvature of base and tip of leaves, and shape and ripening of fruits. Much more detailed field and experimental studies need to be carried out before any final judgment can be passed on the relative merits of the variations given above. There may be some correlation between the depressed-globose type of fruit of var. *platycarpa* with a rounded or heart-shaped base of leaf as was originally supposed, but it is highly questionable whether the more pubescent-leaved *D. virginiana* var. *pubescens* can really be set apart from typical var. *virginiana* having mainly glabrous, lower leaf surfaces. While most specimens of var. *platycarpa,* with the base of the leaf-blade rounded to subcordate, have the lower surface and midrib more or less densely pubescent, trees are found with the leaves nearly or quite glabrous or sparsely pubescent beneath (*Steyermark 22922* from Taney County, for example). Furthermore, while dwarf mature fruiting plants, conforming to the description of var. *pubescens* f. *pumila,* appear distinctive, intergradations may be found so frequently as to make the separation of the dwarf form appear inadvisable. Until a more critical study of the persimmons has been made, it seems best to consider the above treatment as tentative only. Furthermore, it would appear unscientific at the present time, merely because of apparent intergradations and noncorrelations between the variations, to combine all variations into one highly variable taxon.

Statements are sometimes made that the fruit of the persimmon is not edible or ripe until after frosts have occurred. This is erroneous, of course, as any one who lives in persimmon country knows, there being varieties, such as described under var. *platycarpa,* which ordinarily ripen in July, August, and September, as well as early-ripening individuals of typical *D. virginiana* var. *virginiana* and var. *pubescens* with oblong or subglobose fruits. Deam also notes in his *Trees of Indiana* (second ed., p. 284. 1931) that the 'best and largest' persimmon he had eaten 'ripened without a frost.'

Persimmons, like Sassafras, spread vegetatively by the formation of suckering surface roots, eventually forming thickets. These colonies or clones show similarities of branching and a general uniformity of habit and growth. The cream-colored or pale yellow flowers are either all staminate (male) or all pistillate (female), occurring on separate trees, but may also be perfect, both perfect and unisexual flowers appearing together on the same tree. The pistillate flowers are solitary and 1–1.5 cm. long, while the staminate flowers are clustered usually in 2's and smaller.

The leaves turn pale yellow in autumn. They are eaten by wild turkey, but are avoided by cattle. The fruits are eaten by hogs and by many native fur-bearing animals, such as fox, raccoon, skunk, possum, woodchuck, squirrels, wild mice, and white-tailed deer. Bobwhite are also known to be fond of the fruit.

The ripe fruit has a pleasant sweet flavor, unlike any other fruit. Unfortunately, it does not ship without bruising and spoils easily. Many people are acquainted only with the cultivated Japanese persimmon, which is of inferior flavor and lacks the more pleasant sweeter quality of the native fruit. Puddings, syrups, jellies, pies, and ice cream desserts can be prepared from the fruits, recipes for which are given in Fernald and Kinsey's *Edible Wild Plants of North America* (pp. 320–21). It is stated that a vinegar may be prepared from the ripe fruits to which whiskey and water have been added. The dried leaves are used for tea with a flavor resembling that of sassafras, and the roasted seeds have been used as a substitute for coffee. Some of the Indian tribes of the southern United States ground the dried fruits into a meal for making into bread.

The well-seasoned (not green) wood of the persimmon is one of the hardest known, exceeded in hardness only by Black Locust (*Robinia*), Honey Locust (*Gleditsia*), Dogwood, and Ironwood. The heartwood is dark blackish-brown on very old trees. Together with *Cornus florida* it is one of the main woods used for making shuttles and for the heads of golf clubs. It is also used for billiard cues and parquet flooring.

## Fam. **STYRACACEAE** (Storax Family)

**Styrax americana** Lam. var. **americana**

Snow-bell Map 1714

Also called Mock Orange.

*Styrax americana* Lam. [G, BB, P & S, Steyerm.]

Flowers early May–late May.

Occurs in bald cypress and tupelo swamps and low wet woods. Southeastern Missouri in the lowland counties north to Cape Girardeau, Bollinger, and Wayne counties, west to Ripley County.

Ranges from Florida to Louisiana, north to Virginia, Ohio, Indiana, Illinois, Missouri, and Oklahoma.

A variety *pulverulenta* (Michx.) Perkins, with the leaf-blades sparsely short-hairy above and scurfy beneath, has not been collected in Missouri, all the plants thus far found in the state having the leaves glabrous or only barely short-hairy beneath, characteristic of var. *americana.*

This is a very showy-flowered shrub when in full

---

Plate no. 279. **1.** Diospyros virginiana, × $^2/_5$; *a.* Fruiting branch; *b.* Flowering branch; Details from Small, The New York Botanical Garden. **2.** Fraxinus tomentosa, × $^2/_5$, *a.* Fruit, × $^2/_5$. **3.** Fraxinus pennsylvanica var. subintegerrima, × $^2/_5$; *a.* Female flowering branch; *b.* Male flowering branch; *c.* Fruiting branch. **4.** Styrax americanus var. americanus, × $^2/_5$; Details from Small, The New York Botanical Garden.

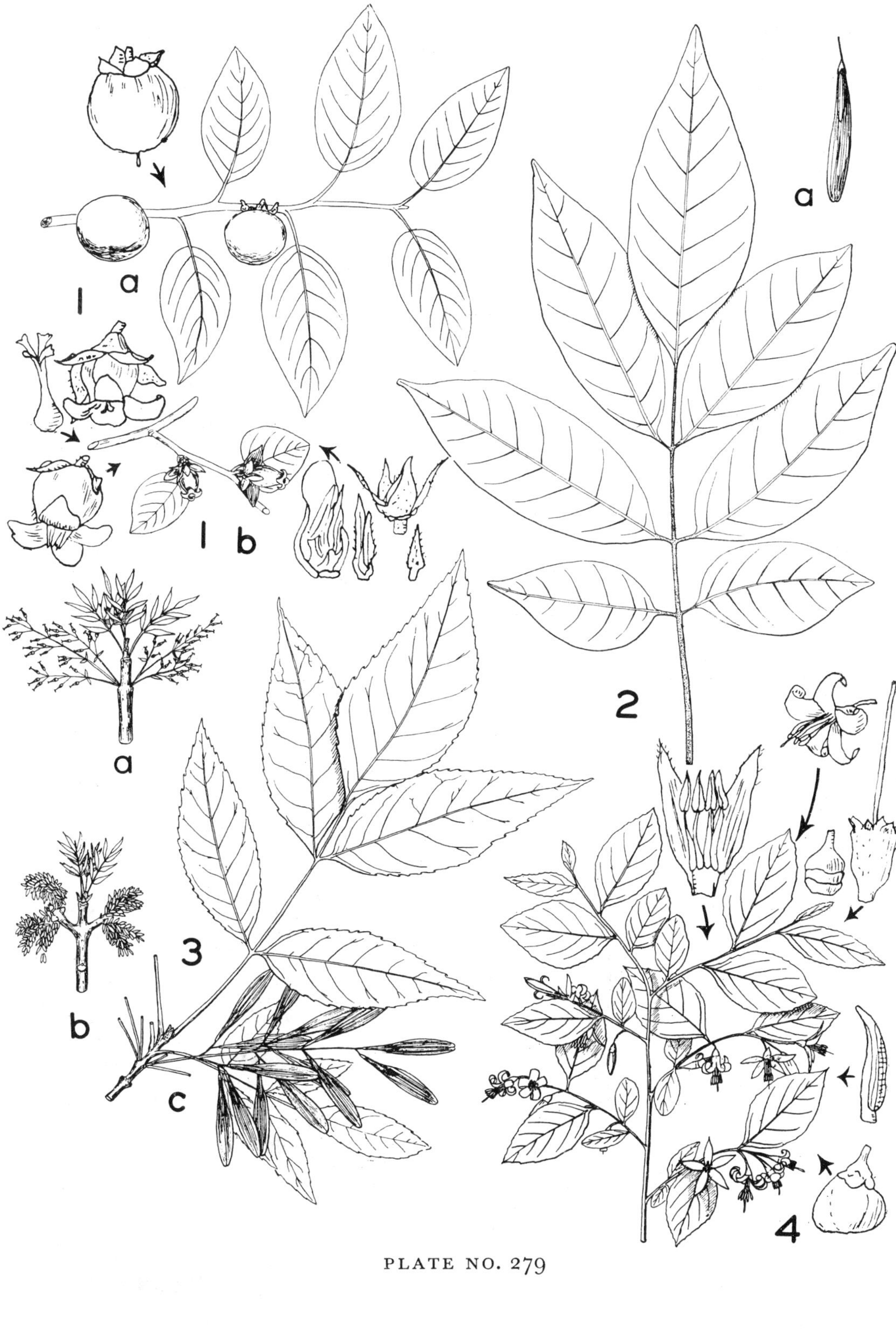


PLATE NO. 279

bloom, bearing white, drooping flowers on the leafy branches, and should be more commonly grown as an ornamental plant.

*Excluded Species*

**Halesia carolina** L. Silverbell Tree, Snowdrop Tree

Reported from southeastern Missouri by Fernald in the eighth edition of *Gray's Manual* (p. 1146), but no specimens have been found to authenticate this report, and the species should be eliminated from any list pertaining to the flora of the state.

## Order **GENTIANALES**

### Fam. **OLEACEAE** (Olive Family)

a. Leaves pinnately compound, divided into 5–11 leaflets with the odd one at the top of the leaf-stem; fruit winged (a samara) . . . . . 1. FRAXINUS
a. Leaves simple, not divided into leaflets; fruit a dry 2-celled capsule or a fleshy drupe. . . . . *b*
  b. Some part of the leaf margins with fine or short rounded teeth; corolla absent; calyx minute or scarcely evident . . . . . 3. FORESTIERA
  b. No teeth on any part of leaf margins; corolla present; calyx evident and well developed. . . . . *c*
    c. Lower surface of leaves hairy; flowers in drooping sprays (panicles), they and the fruits arising from the sides of the branches; lobes of corolla linear, very narrow, much longer than the short tube; native shrubs or small trees mainly of bluffs and glades of the White River country of southwestern Missouri, and locally in southeastern Missouri . . . . . 4. CHIONANTHUS
    c. Lower surface of leaves glabrous (without hairs); flowers in mainly erect dense panicles, they and the fruits arising from the tip of the branches; lobes of corolla broad, equaling or shorter than the tube; garden shrubs escaped from cultivation, persisting in abandoned fields, farmyards, or along roadsides . . . . . *d*
      d. Flowers white; fruit a fleshy 2-celled berry; leaves 3–6 cm. long; leaf-stalk (petiole) 3–10 mm. long . . . . . 5. LIGUSTRUM
      d. Flowers usually lilac, blue-lavender, sometimes white; fruit a dry capsule; leaves 5–12 cm. long; leaf-stalk (petiole) 15–30 mm. long . . . . . 2. SYRINGA

#### 1. **Fraxinus** L. Ash

a. Some part of the twig 4-angled. . . . . 4. F. QUADRANGULATA
a. No part of any twig 4-angled . . . . . *b*
  b. Branchlets, leaf-stalks (petioles), and main axis (rachis) of leaf glabrous (without hairs). . . . . *c*
    c. Lower surface of fully grown leaflets conspicuously whitened or gray-green; base of lateral leaflets curved or rounded, forming an angle with the midrib of more than 45 degrees; stalks (petiolules) of the lower and middle fully grown leaflets wingless mostly their whole length, 3–20 mm. long; terminal buds broadly triangular (deltoid); main body of fruit (containing seed) winged no more than 1/3 its length, with more than 4 usually indistinct ridges; leaf scars rarely more than 1/3 as broad as long, V-shaped . . . . . 1a. F. AMERICANA var. AMERICANA
    c. Lower surface of fully grown leaflets dull or grass green; base of lateral leaflets more tapering, straighter, and less curved, forming an angle with the midrib of less than 45 degrees; stalks (petiolules) of the lower and middle fully grown leaflets winged with green leafy tissue mostly their whole length, 1–5 (–9) mm. long; terminal buds longer than broad; main body of fruit (containing seed) mostly winged more than 1/3 its length, with 2–4 distinct ridges on each side; leaf scars usually more than ½ as broad as long, nearly straight across at top . . . . . 3b. F. PENNSYLVANICA var. SUBINTEGERRIMA
  b. Branchlets, leaf-stalks (petioles), and main axis (rachis) of leaf soft- or short-hairy . . . . . *d*
    d. Stalks (petiolules) of the lower and middle fully grown leaflets winged with green leafy tissue mostly their whole length, mostly 1–5 (–9) mm. long . . 3a. F. PENNSYLVANICA var. PENNSYLVANICA
    d. Stalks (petiolules) of the lower and middle fully grown leaflets wingless mostly their whole length, 3–20 mm. long . . . . . *e*
      e. Lower surface of fully grown leaflets conspicuously whitened or gray-green; main body of fruit (containing seed) winged no more than 1/3 its length; wing of fruit 2–7 (–10) mm. broad; fully grown fruit 2.5–5 cm. long, the body 0.3–1.6 cm. long . 1b. F. AMERICANA var. BILTMOREANA

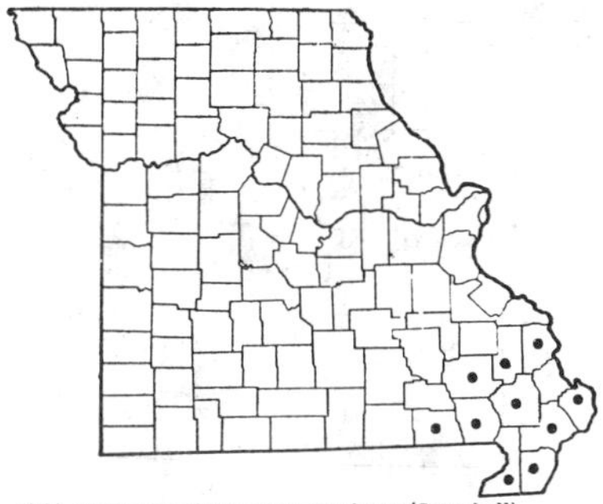

1714 Styrax americana var. americana (Snow-bell)

1715 • Fraxinus americana var. americana (White Ash)
1715 □ Fraxinus americana var. biltmoreana (Biltmore Ash)

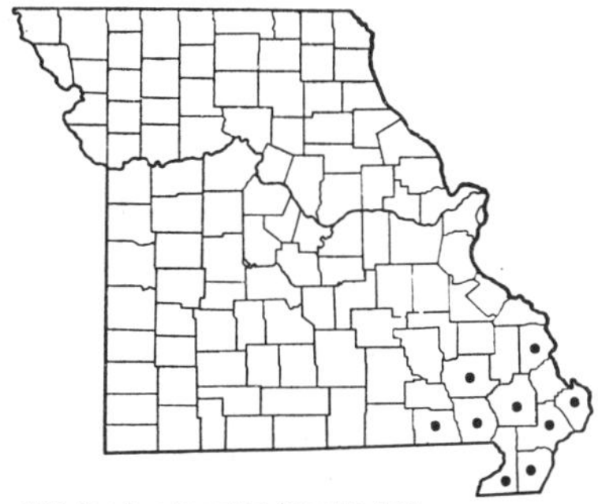

1716 Fraxinus tomentosa (Pumpkin Ash)

e. Lower surface of fully grown leaflets yellowish-brown or brownish-green; main body of fruit (containing seed) winged to about ½ its length; wing of fruit 6–12 mm. broad; fully grown fruit 4–7.5 cm. long, the body 1–3 cm. long . . . . . . . . . . . . . 2. F. TOMENTOSA

1a. **Fraxinus americana** L. var. **americana**
White Ash Map 1715
*Fraxinus americana* L. [G, P & S, Steyerm.]
*Fraxinus americana* var. *subcoriacea* Sarg. [P & S, Steyerm.]
Flowers April–May.

Occurs in rich low woods along streams, slopes, base of bluffs, upland and rocky woods, and glades. Throughout Missouri and doubtless in every county.

Ranges from Nova Scotia and Quebec to Ontario and Minnesota, south to Florida and Texas.

1b. **Fraxinus americana** var. **biltmoreana** (Beadle) J. Wright Biltmore Ash Map 1715
*Fraxinus biltmoreana* Beadle [P & S, Steyerm.]

Known only from Crawford, Carter, Dunklin (wooded northwest- and southwest-facing clay slopes of Crowley Ridge along St. Francis River, sect. 1, 4 mi. west of Campbell, September 18, 1938, *Steyermark 6664*) and Wayne (valley, Blue Spring Hollow, T27N, R7E, sect. 20, 2 mi. east of Chaonia, September 2, 1938, *Steyermark 6378*) counties, southeastern Missouri.

Ranges from Georgia and Alabama, north to New Jersey, Pennsylvania, Indiana, Illinois, and Missouri.

The autumnal foliage of White Ash is quite beautiful and varied, ranging from purple with yellow and pale green to wine-red and purple, or simply rich yellow throughout. This ash, together with the other species in Missouri, is wind-pollinated, shedding large quantities of pollen which becomes a contributing cause of hay fever during the blossoming period of the tree. The flowers lack petals and are arranged in crowded greenish or purplish clusters, the color coming from the more showy male flowers. The sexes of the flowers are generally separated, the male (staminate) appearing on one tree, the female (pistillate) on another, but the sexes also appear either on different parts of branches of the same tree or with the sexes together in perfect flowers.

The White Ash is one of the most valuable commercial woods, especially prized for use in games and sports, for it is the principal wood used for superior baseball bats, tennis racquet frames, hockey sticks, polo mallets, and equipment for playgrounds. It is widely used for handles of many garden tools and agricultural implements, such as spades and shovels. Other uses are for butter tubs, oars for small boats, automobile and airplane construction, light-weight furniture for gardens, porches, and the like. Small quantities are also used for pulp in papermaking mixed with other hardwoods.

This and other species of ash are often attacked by the oyster-shell scale.

2. **Fraxinus tomentosa** Michx. f. Pumpkin Ash
Map 1716
Also called Red Ash (but not to be confused with *Fraxinus pennsylvanica*).
*Fraxinus profunda* (Bush) Bush [P & S, Steyerm., Little]
*Fraxinus profunda* var. *Ashei* Palmer [P & S, Steyerm.]
Flowers April–May.

Occurs in swamps and low, wet woods. Lowlands of southeastern Missouri and border zone of the Ozark escarpment north to Cape Girardeau and Wayne counties, west to Ripley County.

Ranges from Florida to Louisiana, north to New York, Ohio, Indiana, Illinois, and Missouri.

3a. **Fraxinus pennsylvanica** Marsh. var. **pennsylvanica** Red Ash Map 1717

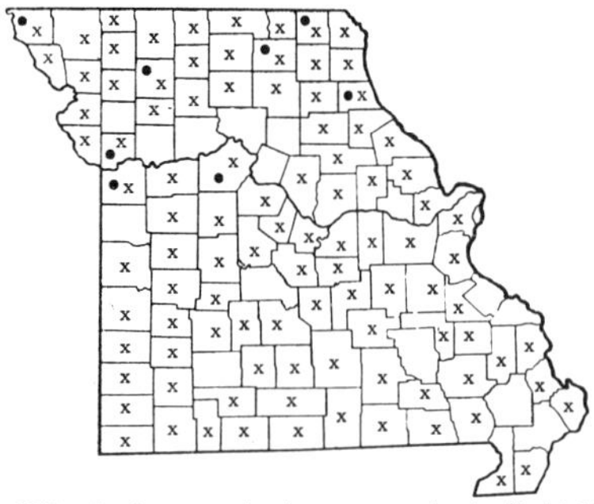

1717 • Fraxinus pennsylvanica var. pennsylvanica (Red Ash)
1717 x Fraxinus pennsylvanica var. subintegerrima (Green Ash)

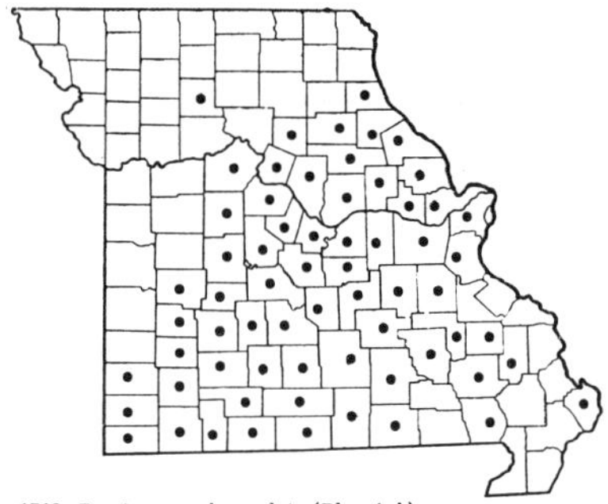

1718 Fraxinus quadrangulata (Blue Ash)

1719 Syringa vulgaris (Common Lilac)

*Fraxinus pennsylvanica* Marsh. [G, P & S, Steyerm.]
Flowers April–May.

Occurs in low woods and wet ground along streams, and borders of ponds. Rare and scattered in northern and central Missouri south to St. Louis, Saline, and Jackson counties.

Ranges from Quebec and New York to Ontario, Michigan, Wisconsin, and Minnesota, south to Georgia, Alabama, Arkansas, and Oklahoma.

3b. **Fraxinus pennsylvanica** var. **subintegerrima** (Vahl.) Fern. Green Ash Map 1717
*Fraxinus pennsylvanica* var. *lanceolata* (Borkh.) Sarg. [P & S, Steyerm.]

Occurs in alluvial soils of flood plain forests, commonly along streams, borders of sloughs, swamps, and ponds. Throughout Missouri and doubtless in every county.

Ranges from Quebec and Maine to Ontario, Manitoba, Saskatchewan, and Montana, south to Georgia, Alabama, Louisiana, and Texas.

The above varieties of Ash are sometimes combined as one species, *F. pennsylvanica*. They are retained as separate varieties in the present flora, since they may be distinguished on the basis of their pubescence or lack of pubescence and on the basis of the geographical range, the pubescent variation, so far as records indicate, being known from only the northern half of Missouri. Recent progeny tests by Jonathan W. Wright (Jour. For. 42: 591–97. 1944) indicate that the character of pubescence of leaves and twigs, used to distinguish *F. pennsylvanica* var. *pennsylvanica*, may occur in seedlings of both var. *pennsylvanica* and var. *subintegerrima*.

The foliage in autumn often turns a brilliant yellow.

The Red Ash owes its name to the red or cinnamon-colored fresh inner surface of the bark. In addition to the hairy twigs, petioles, and leaf-rachises, the Red Ash may usually be distinguished from the Green Ash by the lack of teeth on the leaf margins or these may be only slightly undulately toothed; in Green Ash the margins are usually toothed.

Green Ash is often a weed tree, and along river bars and mudflats, seedlings from parent trees sprout by the hundreds, soon becoming one of the pioneer tree species, along with species of willow, silver maple, cottonwood, and river birch, in the natural succession of a flood plain forest gradually built up from a sandy muddy alluvium.

The Green Ash supplements the White Ash in the lumber market, providing most of the paddles for canoes and oars for small boats. Like other species of ash, Green Ash is an excellent fuel for camp-fires and outdoor cookery, burning with little or no smoke. The Red Ash is often more commonly used for flooring, boxes, interior finish, and butter tubs.

4. **Fraxinus quadrangulata** Michx. Blue Ash Map 1718

Flowers March–April.

Occurs on ledges and escarpments of limestone bluffs, limestone glades, dry rocky woodland, and more rarely in alluvial soils along streams and at the

---

Plate no. 280. **1.** Fraxinus americana, × $^2/_7$; *a.* Fraxinus americana var. biltmoreana; *b.* Fraxinus americana var. americana, **2.** Ligustrum vulgare, × $^2/_7$; *a.* Flower, × $2^2/_7$; After Gleason, details from Small, The New York Botanical Garden. **3.** Fraxinus quadrangulata, × $^2/_7$; *a.* Fruiting branch; *b.* Female flowering branch; Details from Small, The New York Botanical Garden. **4.** Forestiera acuminata; *a.* Female flowering branch, × $^2/_7$; *b.* Group of female flowers, × $^6/_7$; *c.* Male flowering branch, × $^2/_7$; *d.* Male flower, × $^6/_7$; *e.* Leafy branch, × $^2/_7$; Details from Small, The New York Botanical Garden. **5.** Chionanthus virginica, × $^2/_7$. **6.** Syringa vulgaris, floral details, highly magnified (Details from Small, The New York Botanical Garden).

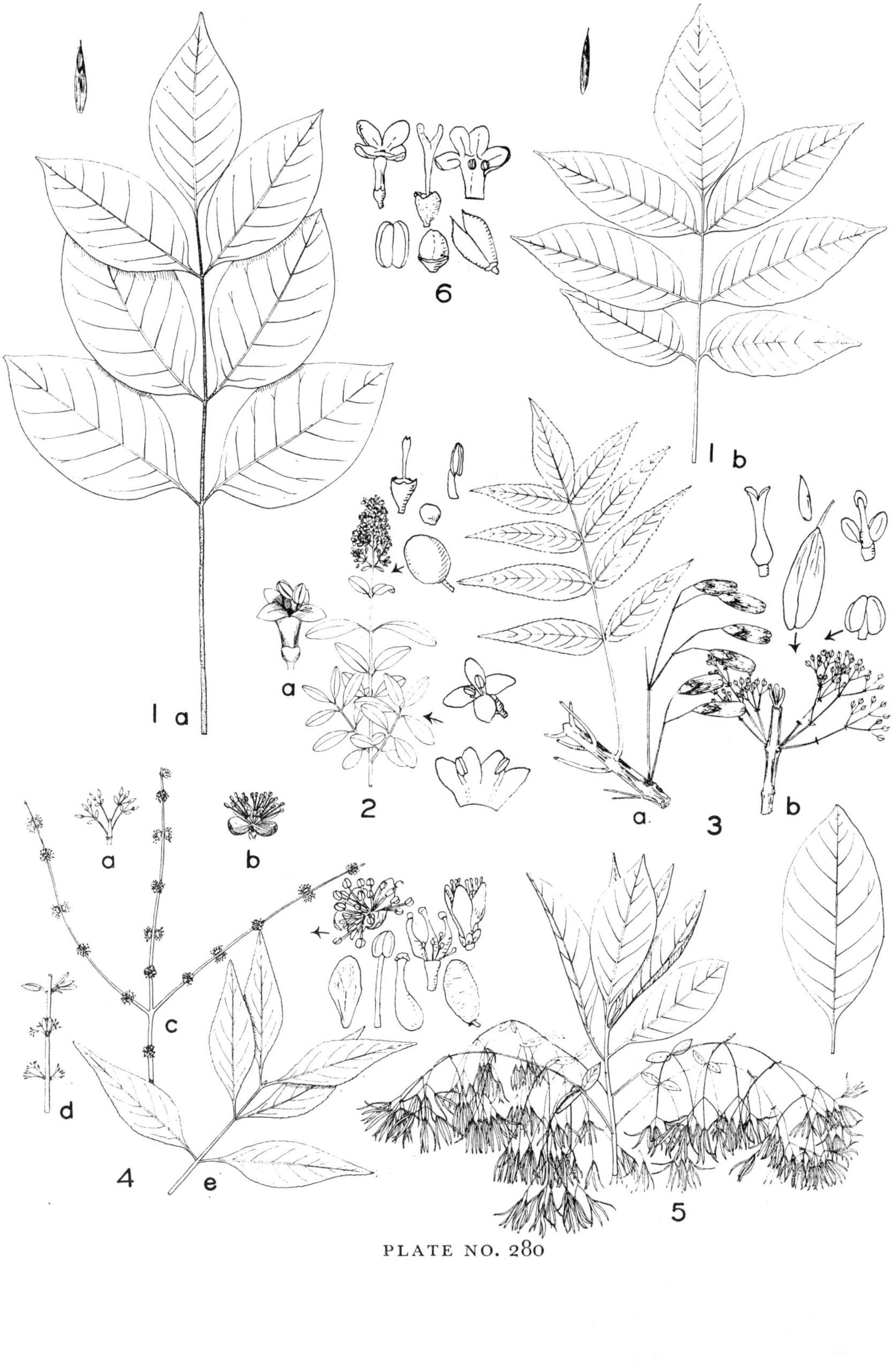


PLATE NO. 280

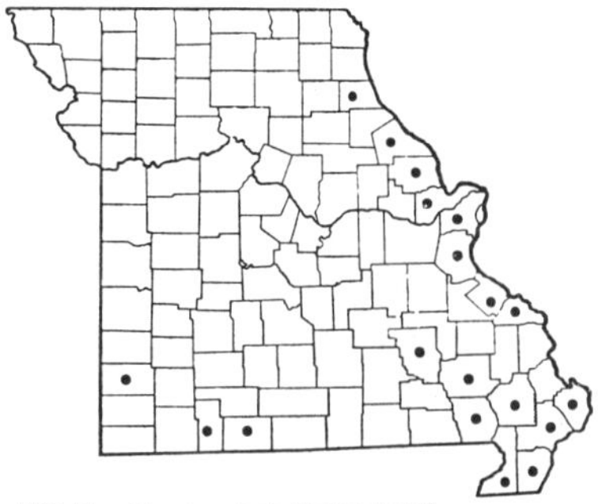

1720 Forestiera acuminata (Swamp Privet)

1721 Chionanthus virginica (Fringe Tree)

1722 Ligustrum ovalifolium (California Privet)

base of bluffs. Common throughout the Ozark region of southern and central Missouri northeast to Marion, Monroe, and Randolph counties, and northwest locally to Livingston County; absent from most of the lowlands of southeastern Missouri and from most of the unglaciated prairie section.

Ranges from Ontario, Michigan, and Wisconsin, south to Alabama, Arkansas, and Oklahoma.

The foliage turns a pale green or yellow in autumn.

When the inner bark or the branches are cut and moved around in the water, a blue color is reflected to the water. A blue dye from the bark was used by some of the pioneers in this country.

The wood usually passes in the lumber market as white ash and is used for most of the same purposes given for that species, especially for handles of garden and agricultural implements.

## 2. **Syringa** L. Lilac

**Syringa vulgaris** L. Common Lilac Map 1719

Flowers April–June.

Commonly planted, but rarely found as an escape or persisting in abandoned farmyards. Known only from Boone County, central Missouri (abandoned farmyards, federal wildlife area east of Ashland, May 19, 1936, *Drouet 2960*). According to the collector, the plants at this locality had persisted and were spreading.

Native of Europe; introduced and naturalized in the United States, escaped from cultivation from New York to North Dakota, south to Georgia and Kansas.

## 3. **Forestiera** Poir.

**Forestiera acuminata** (Michx.) Poir. Swamp Privet Map 1720

Flowers late March–April.

Occurs in swamps, low wet woods, alluvial and rocky borders of streams, ponds, sloughs, and bayous, mainly in alluvial soils in flood plain forest along the Mississippi River or at the base of rocky slopes or bluffs and rocky banks in the Ozarks, rarely found along railroad embankments. Mainly eastern Missouri along the Mississippi and larger streams of the southeastern Ozarks, locally in southwestern Missouri along White River in Taney and Stone counties (where mainly destroyed by Table Rock and Bull Shoals dams), and locally in Jasper County.

Ranges from Florida to Texas, north to South Carolina, Indiana, Illinois, Missouri, Kansas, and Oklahoma.

The swamp privet often occurs in small stands. It is quite conspicuous in early spring with the clusters of yellow flowers appearing along the gray branches before the leaves come out. In this respect, it resembles and much appears like Spice Bush (*Lindera Benzoin*). It should be planted more extensively for its early flowering.

## 4. **Chionanthus** L. Fringe Tree

**Chionanthus virginica** L. Fringe Tree Map 1721

Also called Old Man's Beard.

Flowers April–May.

Occurs along rocky limestone wooded ledges and

bluff escarpments, borders of limestone glades and bald knobs in southwestern Missouri and along wooded slopes of small creeks and wet woods in southeastern Missouri. Frequent in the southwestern Ozarks of the White River country and tributaries of White River in Ozark, Douglas, and Taney counties, and locally in southeastern Missouri in Mississippi and Cape Girardeau (Sloan's Creek, Cape Girardeau, April 29, 1954, *Donald G. Rhodes 302*) counties.

Ranges from Florida to Texas, north to New Jersey, Pennsylvania, West Virginia, Ohio, Missouri, and Oklahoma.

This is one of the showiest and most handsome shrubs in Missouri and deserves to be even more widely cultivated than it is. Although the autumnal foliage is merely a pale to rich yellow color, the drooping panicles of deeply 4-parted white flowers, producing an effect of fringe, are quite striking.

### 5. **Ligustrum** L. Privet

Branchlets minutely hairy or glabrous; corolla-tube about equal in length to or somewhat shorter than the corolla-lobes; anthers shorter than the corolla-lobes . . . . . . . . 1. L. VULGARE (excluded species)
Branchlets glabrous; corolla-tube about 3 times the length of the corolla-lobes; anthers equal in length to the corolla-lobes . . . . . . . . . . . . . . . . . . . . . . 2. L. OVALIFOLIUM

1. **Ligustrum vulgare** L. Common Privet

Flowers June–July.

Commonly planted as a hedge plant. Commonly established as an escape from cultivation in New England and Pennsylvania, and west to Illinois and Michigan, but not recorded from Missouri, although to be expected.

The plant has been recorded as having poisoned horses which have browsed the leaves, and children in Europe have died as the result of eating the fruits.

2. **Ligustrum ovalifolium** Hassk. California Privet Map 1722

Flowers May–July.

Planted and rarely escaped from cultivation in Missouri, where known only from Texas County, central Ozarks (among shrubs along ravine of Roubidoux sandstone tributary to Big Piney River, T31N, R9W, sect. 31, 1 mi. north of Houston, May 8, 1957, *Steyermark 84193*).

Native of Japan; introduced and naturalized in the United States but rarely escaped from cultivation southward.

In regions where this and other species of *Ligustrum* are planted in abundance, the pollen may become important in causing hay fever.

## Fam. **LOGANIACEAE** (Logania Family)

a. Leaves narrowly linear, thread-like, 0.5–2 mm. broad . . . . . . . . . . 3. POLYPREMUM
a. Leaves lanceolate, elliptic, or ovate, 9–55 mm. broad . . . . . . . . . . *b*
  b. Flowers showy, red with yellow, 3–5 cm. long; leaves broadest and rounded at base or below middle, all sessile (without stalks) . . . . . . . . . . 1. SPIGELIA
  b. Flowers tiny, white, about 0.3 cm. long; leaves narrowed and tapering at base, broadest at the middle, the lower ones petioled (on stalks). . . . . . . . . . 2. CYNOCTONUM

### 1. **Spigelia** L. Pink-root

**Spigelia marilandica** L. Pink-root Map 1723

Flowers May–August.

Occurs in low rich or moist woods, ravine bottoms, moist thickets, and wooded banks along streams. Southeastern Missouri north to Cape Girardeau and St. Francois counties, west to Carter and Oregon counties.

Ranges from Florida to Texas, north to Maryland, Ohio, Indiana, Illinois, Missouri, and Oklahoma.

This is one of the most vividly colored wild flowers, the erect, tubular deep red corolla-tube ending in 5 yellow lobes. Plants are usually found in several-stemmed clumps. They thrive in wildflower gardens if given rich, damp soil and a shaded location.

The root, which contains an alkaloid, is used in medicine to get rid of intestinal parasites. Some cases of poisoning have resulted from using an excess amount of the medicine for this purpose.

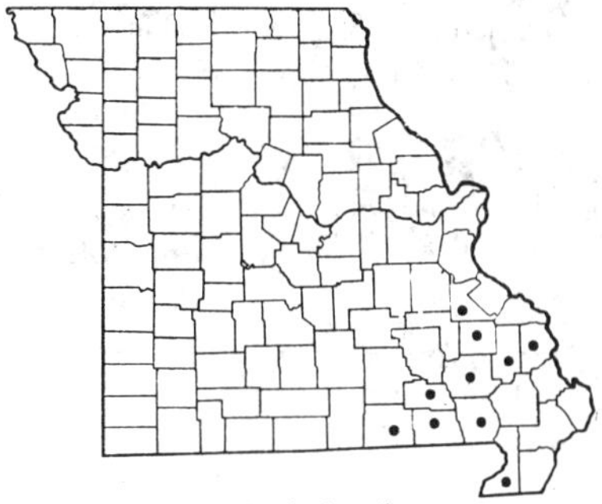

1723 Spigelia marilandica (Pink-root)

1724 Cynoctonum Mitreola (Miterwort)

1725 Polypremum procumbens

### 2. **Cynoctonum** Gmel. Miterwort

**Cynoctonum Mitreola** (L.) Britt. Miterwort Map 1724

Flowers June–November.

Occurs in swamps and wet shores of ponds, streams, and ditches. Known only from the southeastern lowlands in Butler (swamps, October 19, 1905, *Bush 3765*; St. Francis River, October 21, 1899, *Trelease*) and Dunklin (E. Bertig, October 28, 1897, *Trelease 456*) counties.

Ranges from Florida to Texas, north to Virginia, Tennessee, Missouri, and Oklahoma; also in Mexico, the West Indies, and Central America (Guatemala and Honduras).

### 3. **Polypremum** L.

**Polypremum procumbens** L. Map 1725

Flowers June–October.

Occurs in sandy open ground. Southeastern Missouri, north to Scott and Stoddard counties, west to Dunklin County, locally in Clark County, northeastern Missouri (Calma, July 21, 1921, *John Davis*).

Ranges from Florida to Texas, north to New York, Pennsylvania, and Missouri; also in tropical America.

## Fam. **GENTIANACEAE** (Gentian Family)

a. Leaves divided into 3 leaflets (compound) . . . . . . . . . . . . . . . . 7. MENYANTHES
a. Leaves simple and undivided . . . . . . . . . . . . . . . . . . . . . . . *b*
  b. Water plants with floating leaves; leaf-blades round, their petioles (leaf-stalk) alternate . 8. NYMPHOIDES
  b. Land plants, or if growing in moist ground, the leaves not floating; leaves opposite or in whorls (circles) of usually 4's or, if alternate, the leaves scale-like and only 1–2.5 mm. long . . . . . . *c*
    c. Leaves in whorls (circles) of mostly 4's on the stem; flowering and fruiting stems 1–2.5 m.tall 4. SWERTIA
    c. Leaves opposite, in pairs or alternate; flowering and fruiting stems 0.3–8 dm. tall. . . . . . . *d*
      d. Stems thread-like; leaves of the stem mostly alternate, scale-like, 1–2.5 mm. long; corolla 2–7 mm. long . . . . . . . . . . . . . . . . . . . . . . . 5. BARTONIA
      d. Stems not thread-like; leaves of the stem always opposite, in pairs, 8–150 mm. long; corolla 7–50 mm. long. . . . . . . . . . . . . . . . . . . . . . . *e*
        e. Flowering in early spring, April–May; leaves narrowed at base, broadest in upper half; calyx-lobes 2, leaf-like . . . . . . . . . . . . . . . . . . 6. OBOLARIA
        e. Flowering in summer and fall, July–November; leaves either broadest at the base or

---

Plate no. 281. **1.** Spigelia marilandica, × 2/5; Details from Small, The New York Botanical Garden **2.** Polypremum procumbens, × 1 3/5; *a*. Flower, × 4; After Gleason, details from Small, The New York Botanical Garden. **3.** Cynoctonum Mitreola, × 2/5; *a*. Flower, × 5 3/5; After Gleason, details from Small, The New York Botanical Garden. **4.** Sabatia campestris, × 2/5; After Gleason, The New York Botanical Garden.

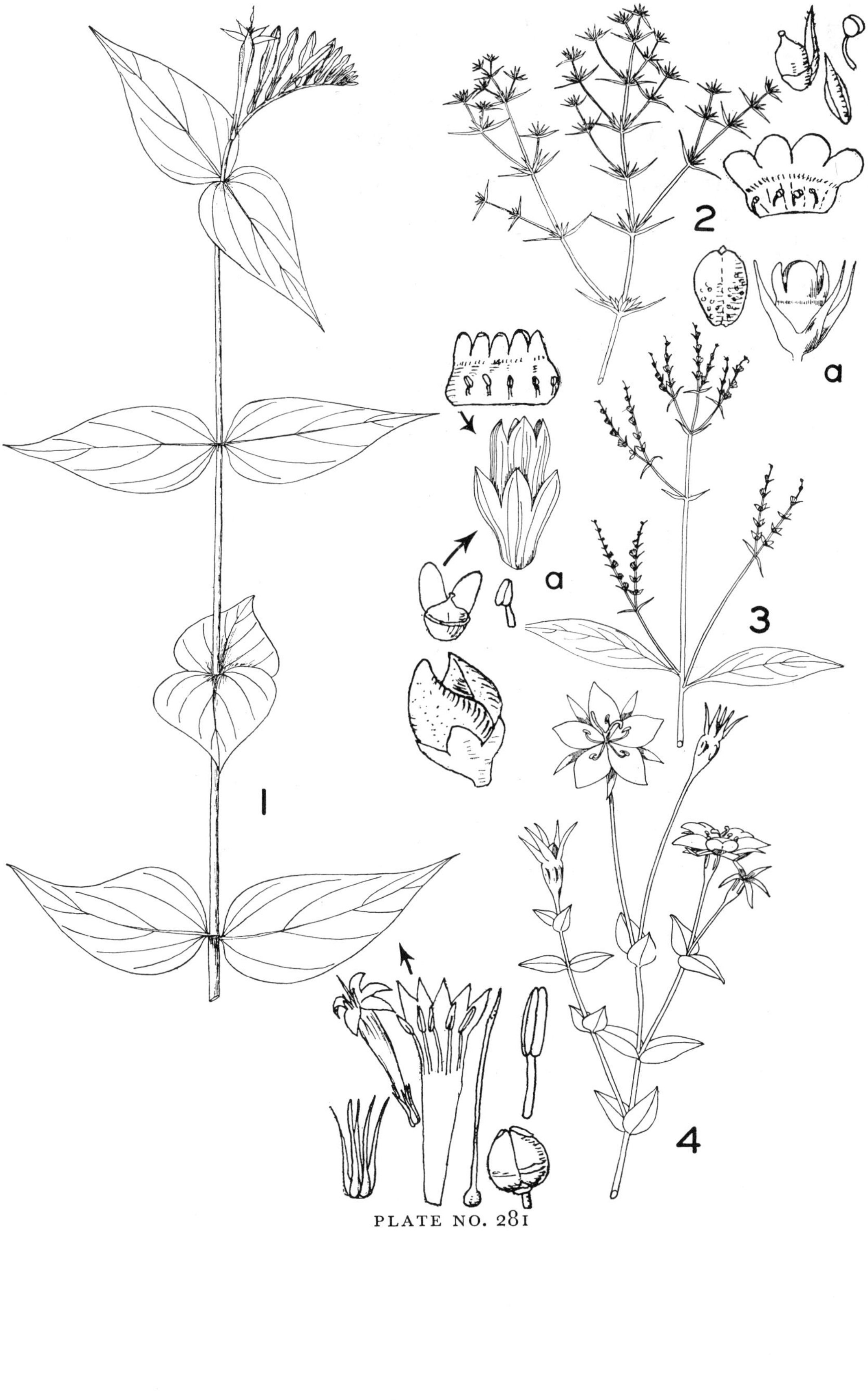


PLATE NO. 281

1726 Sabatia brachiata

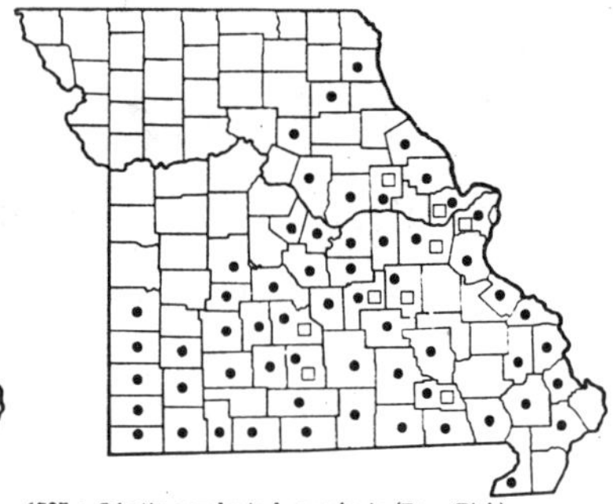

1727 • Sabatia angularis f. angularis (Rose Pink)
1727 □ Sabatia angularis f. albiflora

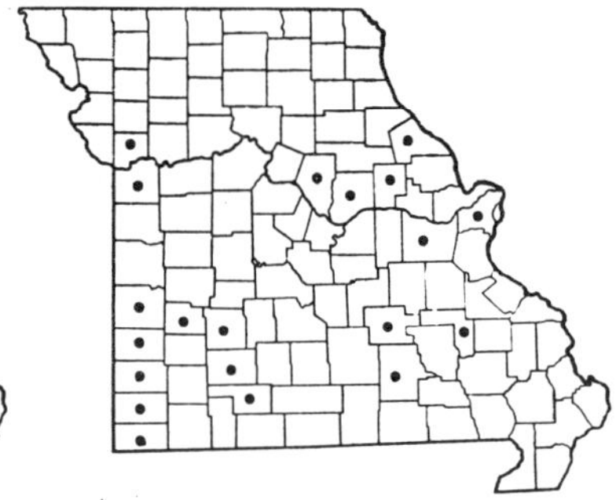

1729 Sabatia campestris
. . .
1728 Excluded species

middle and narrowest in upper half or the same width throughout, but not broadest in upper half; calyx-lobes 4–6. . . . . . . . . . . . . . . . . . . . . . . . *f*

f. Corolla-lobes much longer than the short corolla-tube; at least the upper part of the stem and branches conspicuously 4-angled . . . . . . . . . . . 1. SABATIA

f. Corolla-lobes shorter than or equaling the long or conspicuous corolla-tube; all parts of stems and branches terete (round in cross-section), with rounded sides. . . . *g*

g. Corolla rose-red or pink; corolla-tube very narrow and slender, 2 mm. or less in diameter; calyx-lobes with a ridge (keel) running down the center; leaves chiefly 1–3 cm. long, at most 8 mm. broad . . . . . . . . . . . . 2. CENTAURIUM

g. Corolla blue, purple, or white; corolla-tube much broader, funnel-shaped, tubular-bell-shaped, or ovate, 5 mm. or more in diameter; calyx-lobes without a ridge (keel) running down the center; leaves chiefly 3–12 cm. long, 8–50 mm. broad 3. GENTIANA

## 1. **Sabatia** Adans. Marsh Pink

a. Branches of the inflorescence alternate; calyx-tube prominently 5-ribbed or winged, 4–8 mm. long 3. S. CAMPESTRIS

a. Branches of the inflorescence opposite; calyx-tube smooth or at least not 5-ribbed or winged, 1–3.5 mm. long . . . . . . . . . . . . . . . . . . . . . . . . . . . . . *b*

b. Commonly encountered throughout the Ozark, east-central, and eastern portions of the state; lower part of stem conspicuously 4-angled; main stem-leaves broadest near the rounded or somewhat heart-shaped and clasping base, 3–7-nerved, 1½–3 times as long as broad, ovate to ovate-lanceolate or almost round; angles of stem with conspicuous wings (expanded leafy tissue). 2. S. ANGULARIS

b. Rarely encountered, known only from Butler County, southeastern Missouri; lower part of stem terete (rounded in cross-section) with curved sides; main stem-leaves broadest at or near the middle, not clasping, 1–3-nerved, usually 3–4 times as long as broad, lanceolate to linear; angles of stem without wings . . . . . . . . . . . . . . . . . . 1. S. BRACHIATA

1. **Sabatia brachiata** Ell. Map 1726

Flowers June–August.

Known only from Butler County, southeastern Missouri (upland oak-hickory woods, 12 mi. north of Poplar Bluff, north of Rombauer, T26N, R7E, sect. 20, near head of Mud Creek, July 10, 1936, *Steyermark 11593*).

Ranges from Georgia to Louisiana, north to Virginia, Tennessee, and Missouri.

2. **Sabatia angularis** (L.) Pursh Rose Pink Map 1727

Also called Rose Gentian.

Flowers late June–September.

Occurs in rocky open woods, glades, thickets, fallow fields, and upland ridges, usually in acid soils.

Ranges from Florida to Texas, north to Massachusetts, New York, Pennsylvania, Ohio, Ontario, Michigan, Indiana, Illinois, Missouri, and Kansas.

Two variations are encountered in Missouri:

Corolla pink or rose-colored . . 2a. S. ANGULARIS f. ANGULARIS

Corolla white . . . 2b. S. ANGULARIS f. ALBIFLORA

2a. **Sabatia angularis** f. **angularis** Map 1727

*Sabatia angularis* (L.) Pursh [G, BB, P & S]

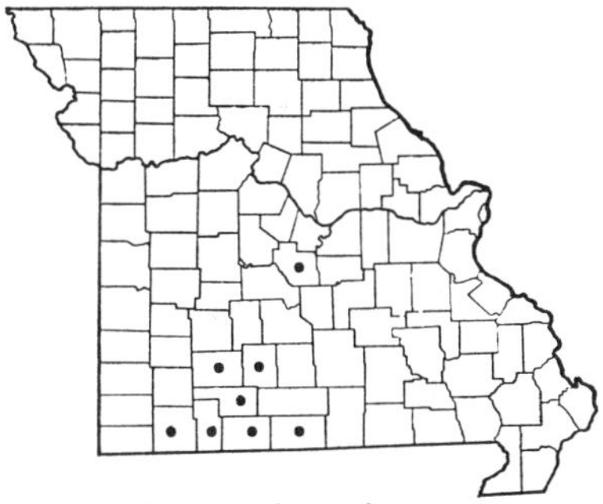

1730 Centaurium texense (Centaury)

1731 Centaurium calycosum

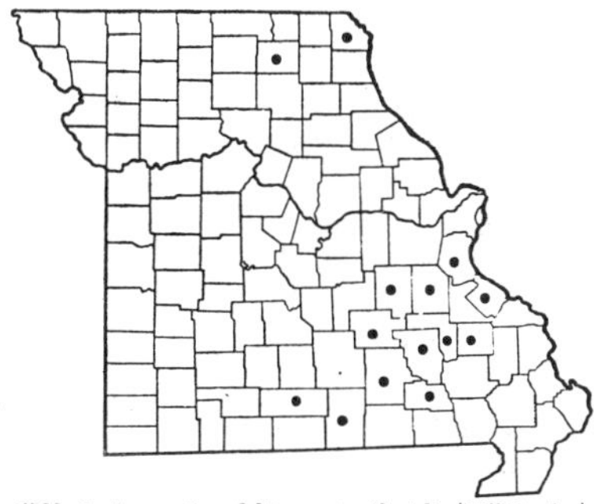

1732 Gentiana quinquefolia var. occidentalis (Stiff Gentian)

This is the commoner variation in Missouri, found throughout southern, east-central, and northeastern Missouri, north to Lewis, Shelby, Randolph, Boone, Moniteau, Benton, and Vernon counties.

2b. **Sabatia angularis** f. **albiflora** (Raf.) House Map 1727

This is less common, and scattered throughout the Ozark section of the state north to St. Charles and Montgomery counties.

Although sometimes stated to be an annual (Wilbur in Rh. 57: 19. 1955), this is definitely a biennial plant, the basal rosette of leaves spreading flat upon the surface of the ground the first year and not flowering, the second year sending up a flowering stem. I have observed this biennial habit in my wildflower preserve for fifteen years.

The flowers are delicately fragrant. Once the plants become established in a wildflower bed by seeding themselves, they may be regarded as rather permanent. Seeds are sown in late fall and spread over a slight depression covered with fine sand. They require a fairly acid soil in an open situation. By planting of seed in successive years, flowering will ensue each year thereafter.

3. **Sabatia campestris** Nutt. Map 1729

Flowers July–September.

Occurs in prairies, fields, glades, and along roadsides. Southern and central Missouri, commonest in the unglaciated prairie section of southwestern Missouri, north to Pike, Montgomery, Callaway, Boone, and Clay counties.

Ranges from Mississippi to Texas, north to Illinois, Iowa, Missouri, Kansas, and Oklahoma; rarely adventive eastward.

This species has longer calyx-lobes than *S. angularis*, and often occurs in stands of more numerous individuals.

## 2. **Centaurium** Hill Centaury

Corolla-lobes 4–6 mm. long; leaves of stem linear or narrowly lanceolate, 1–8 mm. broad; limestone glades of southwestern Missouri northeast to Miller County . . . . . . . . . . . . 1. C. TEXENSE

Corolla-lobes 7–10 mm. long; leaves of stem narrowly oblong to lanceolate, 4–15 mm. broad; introduced along railroads in Jackson County, west-central Missouri . . . . . . . . . 2. C. CALYCOSUM

1. **Centaurium texense** (Griseb.) Fern. Centaury Map 1730

Flowers late June (June 20)–early September.

Occurs on limestone glades and bald knobs. Southwestern Missouri from Ozark to Barry County, north to Miller, Webster, and Greene counties.

Ranges from Missouri and Oklahoma to Texas.

This dwarf annual plant, generally 1–3 dm. tall, has pale green leaves and attractive star-shaped rosy-pink corollas. It starts blooming about July 4 on the hot dry limestone barrens, associated with such other summer-flowering glade annuals as *Heliotropium tenellum*, *Isanthus brachiatus*, *Croton capitatus*, and *Linum sulcatum*.

2. **Centaurium calycosum** (Buckl.) Fern. Map 1731

Flowers July–August.

Introduced along railroads. Known only from Jackson County, west-central Missouri (Sheffield, July 21, 1892, *Bush*).

Ranges from Texas to Utah; adventive in Missouri.

## 3. **Gentiana** L. Gentian

a. Corolla 1.5–2.5 cm. long; calyx 3–9 mm. long; no fringed or toothed folds or extra petal-like parts in the sinuses (space or interval) between the corolla-lobes; annuals or biennials . . . . . . 1. G. QUINQUEFOLIA var. OCCIDENTALIS

a. Corolla 3–5 cm. long; calyx 12–30 mm. long; fringed or toothed folds or extra petal-like parts in the sinuses (space or interval) between the corolla-lobes; perennials . . . . . . . . . . . . . . *b*

b. Lobes of corolla conspicuous, spreading, mainly 6–10 mm. long; corolla funnel-shaped, open in the center and at the top; flowers chiefly at the top of the stem; calyx-lobes linear-lanceolate or linear, 1–1.5 mm. broad; leaves 5–15 mm. broad, mainly lanceolate to linear-lanceolate, 3–7 times as long as broad, with usually 1 main nerve, occasionally with 1 faint lateral nerve on each side; anthers not connected with one another; stems with lines of short hairs . . . . 2. G. PUBERULA

b. Lobes of corolla inconspicuous, erect or curved inward, 2–5 mm. long; corolla barrel-shaped or ovoid-cylindrical, closed or narrowed or scarcely open at summit; flowers often in the upper axils along the sides of the stem as well as at the top; calyx-lobes ovate-lanceolate, lanceolate, ovate, or obovate, 3–5 mm. broad; leaves mainly 15–50 mm. broad, ovate-lanceolate to lanceolate, usually 2½–4 (rarely 5) times as long as broad, with usually 3–5 main nerves; anthers more or less connected with one another; stem completely smooth, without lines of short hair. . . . . . . . *c*

c. Corolla creamy white or white with greenish; margins of leaves and of calyx-lobes smooth, without a fringe of minute hairs (use magnifying lens of at least 8×) . . . . . . 5. G. FLAVIDA

c. Corolla mostly blue-violet or blue; margins of leaves and of calyx-lobes with a fringe of minute hairs (use magnifying lens of at least 8×) . . . . . . . . . . . . . . . . . . . . . *d*

d. Corolla-lobes rounded at top, as broad as or broader than the 2–3-cleft petal-like appendages between the lobes . . . . . . . . . . . . . . . . . . . . 4. G. CLAUSA

d. Corolla-lobes narrowed at top, narrower and shorter than the broad irregularly toothed and fringed petal-like border (plait) between the lobes . . . . . . . . 3. G. ANDREWSII

1. **Gentiana quinquefolia** L. var. **occidentalis** (Gray) Hitchc. Stiff Gentian Map 1732

Also called Ague Weed.

Flowers August–November.

Occurs along moist rocky wooded banks, moist rocky ledges and low thickets along small streams, and low, moist woodland. Mainly in the eastern Ozark section west to Douglas, Dent, and Crawford counties, north locally in Clark and Adair counties.

Ranges from New York and Ontario to Michigan and Minnesota, south to North Carolina, Tennessee, Louisiana, and Missouri.

The flowers are mainly lilac to pale lavender-blue.

2. **Gentiana puberula** Michx. Downy Gentian Map 1733

Flowers September–November.

Occurs in usually dry upland prairies, prairie remnants along railroads, and on limestone or cherty limestone rocky glades bordering bluff escarpments or on rocky open slopes. Throughout Missouri, but absent from the lowland section of southeastern Missouri.

Ranges from New York and Ontario to Michigan, Wisconsin, Minnesota, Manitoba, and North Dakota, south to Georgia, Ohio, Kentucky, Illinois, Missouri, and Kansas.

The roots have sometimes been used in medicine as a bitter tonic.

This gentian has the deepest dark blue-purple color of any of the native species of Missouri. It often occurs in great numbers in undisturbed and unburnt prairie along railroads in northern and central Missouri. Although thousands of seeds are produced by one plant, the ripening capsule is often attacked by certain beetles which pierce holes through the ovary wall and destroy the seeds. Many seeds are also apparently not fertile, so that it is difficult to raise this gentian from seed. They may be transplanted successfully if the roots are not too much disturbed, and plants usually succeed if provided with a well-drained soil in an open sunny situation. A spectacular, luxuriant hybrid plant between this species and the next (*G. Andrewsii*) has developed at the author's wildflower preserve, and has maintained itself for the past fifteen years. A recent investigation by Dr. Chas. T. Mason (Brittonia 11:

---

Plate no. 282. **1.** Sabatia angularis, × 2/5; Details from Small, The New York Botanical Garden. **2.** Centaurium texense, × 2/5; After Gleason, details from Small, The New York Botanical Garden. **3.** Gentiana quinquefolia, × 2/5; After Gleason, The New York Botanical Garden. **4.** Centaurium calycosum, × 2/5; After Gleason, The New York Botanical Garden. **5.** Gentiana flavida, × 2/5; After Gleason, The New York Botanical Garden. **6.** Gentiana Andrewsii, × 2/5; After Gleason, The New York Botanical Garden. **7.** Gentiana puberula, × 2/5; Details from Small, The New York Botanical Garden.

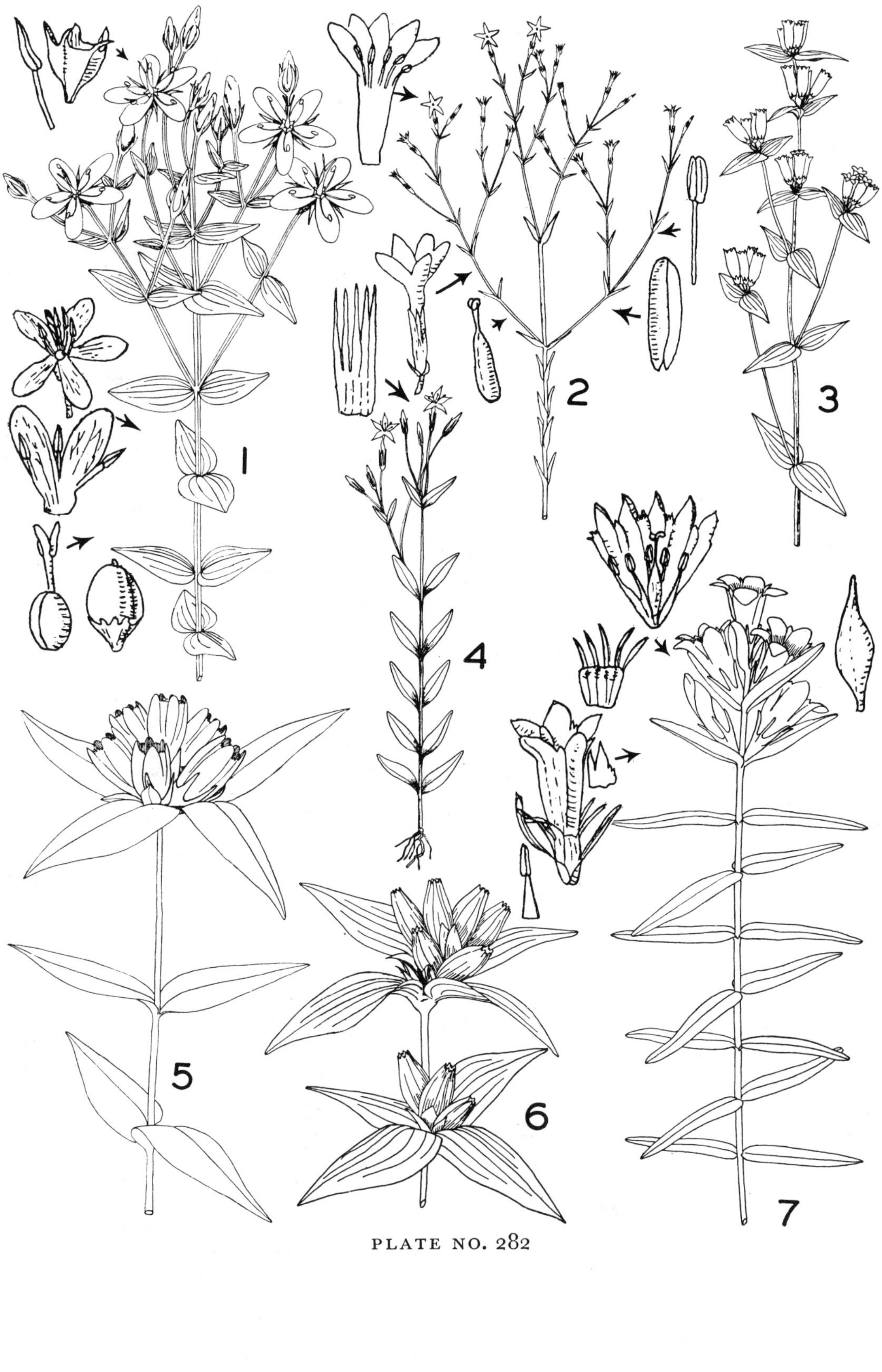


PLATE NO. 282

1733 Gentiana puberula (Downy Gentian)

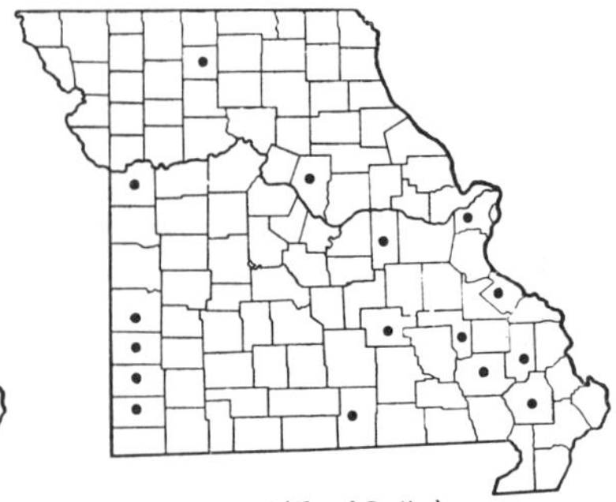

1734 Gentiana Andrewsii (Closed Gentian)

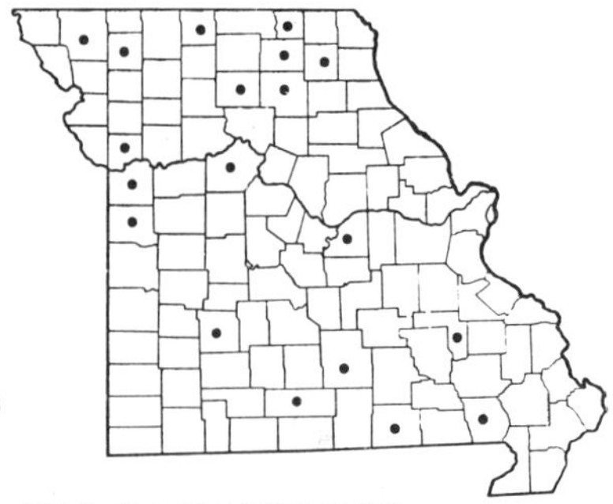

1735 Gentiana clausa (Closed Gentian)

40–43. 1959) brings forth some evidence to indicate that the gentian known as *G. saponaria*, as it occurs in Wisconsin, appears to be produced by hybridization between *G. puberula* and *G. Andrewsii.*

What appears to be a hybrid between *G. puberula* and *G. clausa* or *G. Andrewsii* was found by the author in Ozark County (edge of limestone glades above bluffs of Little North Fork of White River, vicinity of Sand Rock Bluff, T23N, R15W, sect. 20, 1–2 mi. southeast of Hammond, September 27, 1949, *Steyermark 69413*). The leaves were thinner than in *G. puberula*, with the texture of *G. clausa* or *G. Andrewsii*, and the flowers were lavender instead of blue-violet. The flowers had the characteristic appearance of *G. puberula* and were bunched at the summit of the stem as in *G. clausa* or *G. Andrewsii.* The area is probably now destroyed by the Bull Shoals Dam.

3. **Gentiana Andrewsii** Griseb. Closed Gentian
Map 1734

Flowers August–October.

Occurs in low woods or wooded slopes along ravines and bluffs bordering or near streams, and borders of wooded ponds. Scattered and rare in southern and central Missouri north locally in Grundy County.

Ranges from Georgia to Arkansas, north to Massachusetts, Vermont, Quebec, Ontario, Manitoba, and Saskatchewan.

This and the next species (*G. clausa*) are often confused and greatly resemble one another in general aspect. In the author's wildflower garden, plants of this species have increased over the past fifteen years by seed reproduction.

The roots have sometimes been used in medicine as bitter tonics.

4. **Gentiana clausa** Raf. Closed Gentian
Map 1735

Also called Bottle Gentian.

Flowers August–October.

Occurs in river bottom prairies, prairie swales along railroads, moist banks and wooded slopes of ravines, bluffs, and ledges near or bordering streams, and low woodland. Throughout Missouri and commoner than *G. Andrewsii* in the state.

Ranges from New England and Quebec to Minnesota and Manitoba, south to North Carolina, Tennessee, and Missouri.

This species and *G. Andrewsii* succeed in a wildflower bed if given an open sunny situation in a well-drained, slightly damp soil. The plants should be transplanted with as little disturbance to the long, cordlike roots as possible. They can be grown successfully from seed also, but the proportion of viable seed is apparently very small. Specimens are occasionally found which appear to be hybrids between *G. clausa* and *G. flavida.* A specimen from Gentry County, for example (*Steyermark 72915*), has lilac-white on the corolla lobes and pale lavender in the sinuses, while in Mercer County plants of both species (*Steyermark 70328* and *70729*) were found growing only a few feet apart.

5. **Gentiana flavida** Gray Pale Gentian
Map 1736

Also called Yellowish Gentian.

Flowers August–October.

Occurs in rocky prairies, rich wooded slopes, open or rocky wooded slopes, ledges and open escarpments along bluffs, rocky open limestone and cherty limestone glades, and openings of woods. Throughout Missouri, apparently absent from the lowlands of southeastern Missouri.

Ranges from Pennsylvania and Ontario to Michigan, Wisconsin, Minnesota, and Manitoba, south to North Carolina, Kentucky, and Arkansas.

This species succeeds well in a wildflower garden, producing larger clumps with more numerous flowers when grown in full sun. The flowers also have a clearer

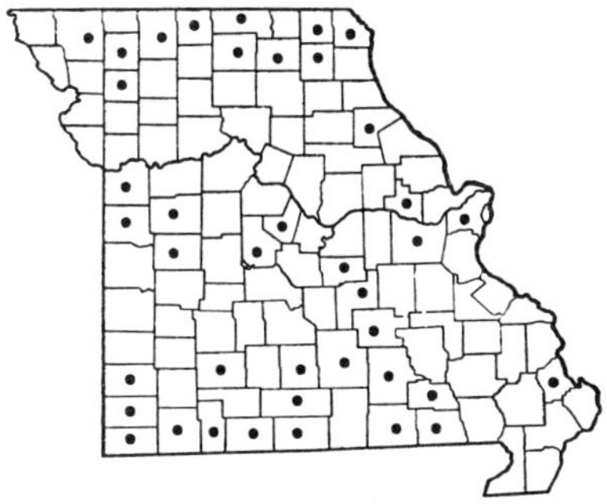

1736 Gentiana flavida (Pale Gentian)

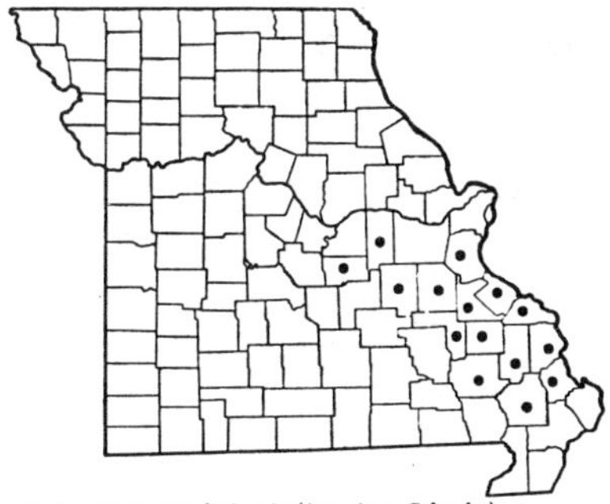

1737 Swertia caroliniensis (American Columbo)

1738 Bartonia paniculata var. paniculata

white color in full sun. In Johnson County a color variant was noted (*Steyermark 72749*) in which some of the flowers have corollas which are whitish in their lower half, but lilac in the upper half, especially on the inside.

## 4. **Swertia** L. Columbo

**Swertia caroliniensis** (Walt.) Kuntze American Columbo Map 1737

Also called Green Gentian.

*Frasera caroliniensis* Walt. [P & S, Steyerm.]

Flowers May–June.

Occurs in rich, low, or rocky open woods and thickets, generally in ravine bottoms and wooded valleys along streams, in acid soils. Eastern Ozark region west to Maries, Crawford, Iron, Wayne, and Stoddard counties, north to Jefferson and Gasconade counties.

Ranges from Georgia to Louisiana and Oklahoma, north to New York, Ontario, Michigan, and Wisconsin.

This species produces a large cluster of oblanceolate, pale green rosette leaves up to 4 dm. long. Colonies of individuals are frequently found with only these sterile rosette leaves present. The flowering stem is brown-purple and may reach a height of 2.5 meters (8 feet). The pale greenish-yellow divisions of the corolla, marked with brown-purple dots, bear below the middle a large round gland surrounded by long fringe. After flowering, the plant dies, and new plants, generally in colonies, appear the following year. Despite statements sometimes made in certain manuals that the plant is a biennial or triennial, it is definitely a perennial. The plant may live for many years without flowering. A plant, moved fifteen years ago to the author's wildflower garden, has produced only a basal rosette of leaves each year, and has yet to flower.

The root has been used in medicine as a tonic, and as an emetic and cathartic.

## 5. **Bartonia** Muhl.

**Bartonia paniculata** (Michx.) Muhl. var. **paniculata** Map 1738

*Bartonia paniculata* (Michx.) Muhl. [G, BB]

*Bartonia paniculata* subsp. *paniculata* [Gillett]

Flowers August–October.

Occurs on wet mossy hummocks usually of sphagnum moss around springs and spring branches in acid soils in low woods of *Ilex opaca* (American Holly) at the base of Crowley Ridge bordering the lowlands. Known only from Stoddard County, southeastern Missouri (at head of spring branch, at base of sandy ravine near junction of Crowley Ridge and lowland, T25N, R11E, northwest ¼ sect. 6, 3½ mi. southeast of Bloomfield, August 20, 1954, *Steyermark 76784*; wooded spring branch between property of Mr. Adam and Mr. Harvey, T25N, R10E, sect. 1, 3¾ mi. (by air) south southeast of Bloomfield, October 18, 1955, *Steyermark 80402*; spring branch in valley of Crowley Ridge, T25N, R10E, southeast ¼ sect. 11, 4–4¼ mi. (by air) south of Bloomfield, 2 mi. northeast of Dexter, October 18, 1955, *Steyermark 80432*).

Ranges from Florida to Louisiana, north to Massachusetts, New York, Kentucky, Missouri, and Oklahoma.

This is an annual plant with saprophytic habit. The stems are green or purple with small flowers. The

1739 Obolaria virginica (Pennywort)

1740 Menyanthes trifoliata var. minor (Buckbean)

1741 Nymphoides peltatum (Floating Heart)

corolla lobes are creamy-white above, about 1 mm. wide. The filaments of the stamens are white below and lavender above, and the anthers are yellow.

This genus is thus far known to occur in only a very small sector of Missouri, where it appears to be confined to a few valleys at the base of Crowley Ridge, often growing with the few remaining stands of American Holly (*Ilex opaca*) in the state. Here, on hummocks of moist sphagnum moss, it is associated with *Habenaria clavellata, Oldenlandia uniflora*, and *Polygonum arifolium.* At one of its localities, it frequents the head of the same spring branch along which have been recently added to the flora of the state such rarities as *Pyrus melanocarpa* and *Trisetum pensylvanicum.*

For a more complete discussion of the genus, the reader is referred to a recent revision by Dr. John M. Gillett (Rh. 61: 43–57. 1959).

## 6. **Obolaria** L. Pennywort

**Obolaria virginica** L. Pennywort Map 1739

Flowers April–May.

Occurs in rich low ravines of beech-sugar maple on Crowley Ridge. Known only from Stoddard County, southeastern Missouri (head of ravine on southeast slopes of Ringer Hill, Crowley Ridge, T26N, R11E, southwest $\frac{1}{4}$ sect. 23, $\frac{1}{2}$ mi. west of Messler Baptist Church, $\frac{1}{2}$ mi. west of Messler, 3 miles north northeast of Bell City, April 19, 1959, *Steyermark 86154*).

Ranges from Florida to Texas, north to New Jersey, Pennsylvania, Ohio, Indiana, Illinois, and Missouri.

This is a smooth perennial herb, only 4–17 cm. tall. The plants, growing in leaf mold, are difficult to find, as the stems barely protrude above the carpet of leaves. The stems arise from deep-seated, fleshy, coral-like roots which depend upon a mycorrhizal fungus for their welfare. The lower part of the stems are pale greenish-white. The leaves are pale gray green suffused with a slight lavender color on the lower surface, and are spreading to ascending. They are crowded in the upper portion of the plant, nearly concealing the small clusters of pale lilac and white flowers. The calyx has the same color as the leaves. The corolla-tube is pale greenish in the lower half, pale lilac-whitish in the upper half and on the 4 oblong-obovate lobes.

It is probable that with more intensive exploration this species will eventually be found to be more common throughout the Crowley Ridge section of the state in rich beech-sugar maple forest. For a more complete discussion of this plant, the reader is referred to a recent revision of the genus by Dr. John M. Gillett (Rh. 61:59–62 1959).

## 7. **Menyanthes** L. Buckbean

**Menyanthes trifoliata** L. var. **minor** Raf. Buckbean Map 1740

Also called Bogbean.

*Menyanthes trifoliata* [of BB], not L.

Flowers April–May.

Occurs in calcareous bogs. Known only from Reyn-

---

Plate no. 283. **1.** Swertia caroliniensis, × $^2/_5$; Details from Small, The New York Botanical Garden. **2.** Bartonia paniculata, × $^2/_5$; *a.* Flower, × 6; After Gleason, details from Small, The New York Botanical Garden. **3.** Obolaria virginica, × $^2/_5$; Details from Small, The New York Botanical Garden. **4.** Menyanthes trifoliata var. minor, × $^2/_5$.

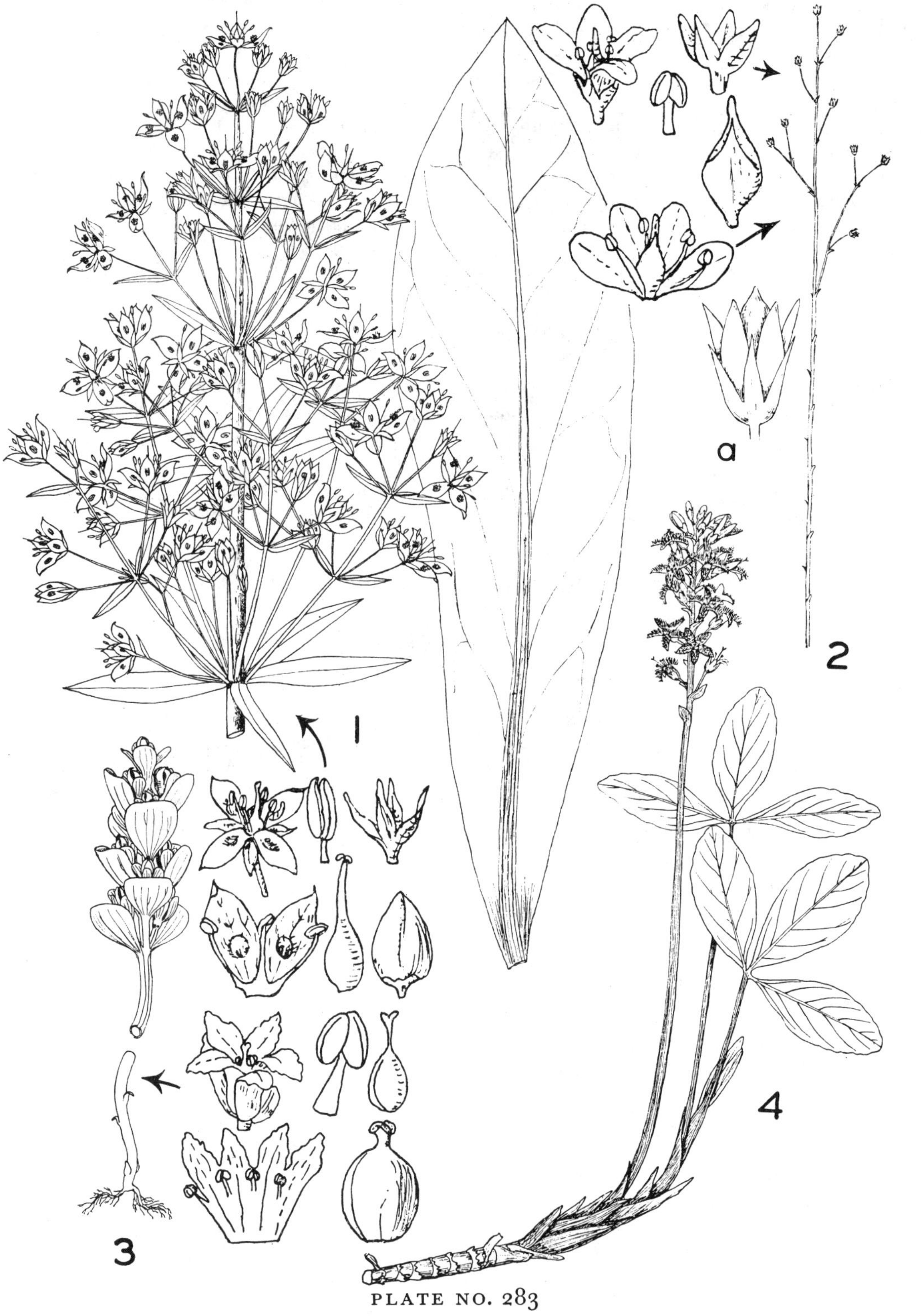


PLATE NO. 283

olds County, southeastern Ozarks (along north prong of Bee Fork, 5 mi. east of Bunker, T32N, R2W, sect. 23, August 5, 1936, *Steyermark 12564*; same locality, May 26, 1938, *Steyermark 5480*; same locality, April 26, 1952, *Steyermark 73111*).

Ranges from Labrador to Alaska, south to Virginia, Maryland, Ohio, Indiana, Illinois, Missouri, Nebraska, and Wyoming.

The flowers are white or purple-tinged and the corolla-lobes are conspicuously bearded on the upper or inner surface. The only flowering collection in Missouri is the one cited above for April 26, 1952. This is another one of the several relict species of a boreal flora isolated in one of the valleys of the southeastern Ozarks. This species is surviving at one of its southernmost outposts, where it has been stranded south of one of the advances of Pleistocene ice sheets. Following the retreat of the glaciers northward north of the Missouri River, this species, together with *Campanula aparinoides*, *Liparis Loeselii*, *Filipendula rubra*, and others of similar northern and eastern geographical range, have been left behind in various calcareous bogs and swampy meadows in sections of the southeastern Ozark region.

The plant is bitter and considered poisonous.

## 8. **Nymphoides** Hill Floating Heart

**Nymphoides peltatum** (Gmel.) Britten & Rendle
Floating Heart Map 1741
*Nymphoides peltata* (Gmel.) Ktze. [G]
Flowers June–September.

Sometimes planted and introduced in artificial ponds, where naturalized in St. Louis, Iron, and Newton counties in southern and central Missouri.

Native of Europe; introduced and naturalized in the United States from New York and D.C. to Missouri.

The plants are sometimes eaten by wildfowl. Used as an ornamental plant for pools and outdoor aquaria.

## Fam. **APOCYNACEAE** (Dogbane Family)

a. Leaves alternate . . . . . . . . . . 1. Amsonia
a. Leaves opposite . . . . . . . . . . *b*
  b. Stems erect . . . . . . . . . . 4. Apocynum
  b. Stems trailing, creeping, or climbing . . . . . . . . . . *c*
    c. Leaves mostly long-pointed (acuminate) at tip, not evergreen; flowers pale yellow or cream-colored, more than 1 in the inflorescence; woody climbing plant; seeds with hairy tufts at one end; native plants of swamps and low wet ground in southeastern Missouri west to Taney County . . . . . . . . . . 3. Trachelospermum
    c. Leaves short-pointed (acute) or blunt at tip, evergreen; flowers blue or blue-violet, only 1 in each leaf axil; non-woody creeping or trailing plant; seeds without any hairy tufts; introduced ornamental plant, found throughout Missouri . . . . . . . . . . 2. Vinca

## 1. **Amsonia** Walt. Blue Star

a. Leaves thread-like or narrowly linear, 0.5–5 mm. broad . . . . . . . . 3. A. ciliata var. filifolia
a. Leaves not thread-like, 10–60 mm. broad . . . . . . . . . . *b*
  b. Leaves shining on upper surface; calyx sparsely hairy; mature fruit drooping; plants of gravel bars . . . . . . . . . . 2. A. illustris
  b. Leaves dull on upper surface; calyx glabrous (not hairy); mature fruit erect; plants of woodland, rocky open slopes, river-banks, along ditches, and moist thickets . . . . 1. A. Tabernaemontana

1. **Amsonia Tabernaemontana** Walt. Map 1742
Flowers April–May.
The following varieties occur in Missouri:

a. Leaves elliptic-oblong to ovate, 3–6 cm. broad. . . . . . 1a. A. Tabernaemontana var. Tabernaemontana
a. Leaves lanceolate, 1–3 cm. broad . . . . *b*
  b. Lower surface of leaves glaucous (gray-green with a pale coating which can be rubbed off); inflorescence loose, few-flowered . . . 1b. A. Tabernaemontana var. salicifolia
  b. Lower surface of leaves green; inflorescence dense, many-flowered . . . . . . 1c. A. Tabernaemontana var. Gattingeri

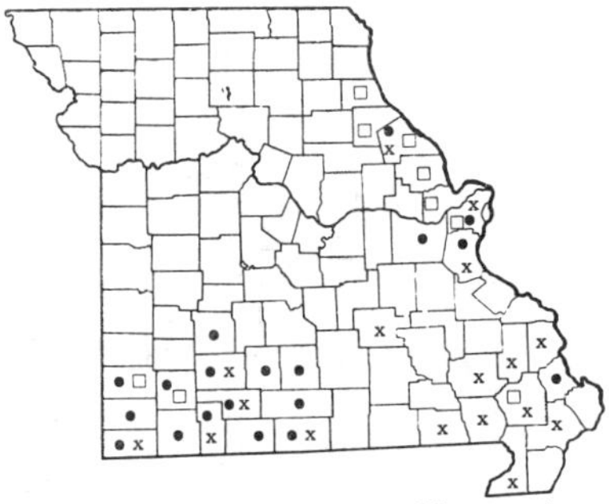

1742 • Amsonia Tabernaemontana var. Tabernaemontana
1742 x Amsonia Tabernaemontana var. salicifolia
1742 □ Amsonia Tabernaemontana var. Gattingeri

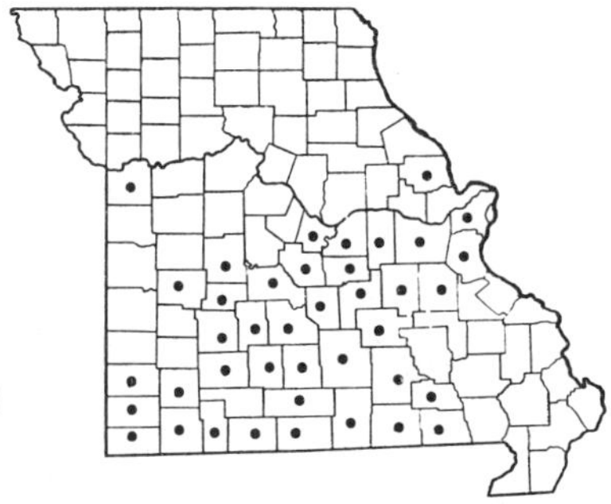

1743 Amsonia illustris

1744 Amsonia ciliata var. filifolia

1a. **Amsonia Tabernaemontana** var. **Tabernaemontana** Map 1742

Occurs in rich or rocky woods and thickets, on slopes and in ravines.

Mainly in southwestern Ozark region and locally in eastern Missouri north to Pike County.

Ranges from Georgia to Louisiana and Oklahoma, north to Virginia, Illinois, Missouri, and Kansas.

1b. **Amsonia Tabernaemontana** var. **salicifolia** (Pursh) Woodson Map 1742

Occurs in alluvial thickets and rich ground along bluffs, and railroads.

Ozark region northeast to Pike County.

Ranges from Georgia to Texas, north to Virginia, Indiana, Illinois, and Missouri.

1c. **Amsonia Tabernaemontana** var. **Gattingeri** Woodson Map 1742

Occurs in rich open woods and along bluffs bordering streams.

Mainly in eastern Missouri from Stoddard County north to Marion County, and locally in southwestern Missouri in Jasper and Lawrence counties.

Ranges from Georgia to Texas, north to Tennessee, Illinois, Missouri, and Kansas.

*Amsonia Tabernaemontana* and varieties is a showy pale blue-flowered plant which responds well to cultivation. Plants are easily grown from seed. *Amsonia Tabernaemontana* var. *Tabernaemontana* does well when treated as a woodland plant grown in rich soil, whereas var. *salicifolia* and var. *Gattingeri* thrive better in open situations. The leaves turn a pale yellow in autumn.

2. **Amsonia illustris** Woodson Map 1743

Flowers April–May.

Occurs on gravel bars, crevices of bluffs, and rocky open places along streams. Throughout the Ozark region of southern and central Missouri north to Lincoln, Gasconade, Osage, Cole, Camden, Benton, St. Clair, and Jasper counties, and locally in Jackson County.

Ranges from Missouri and Kansas to Oklahoma and Texas.

This species succeeds well when grown in open ground in a wildflower or perennial garden. The shining leaves are quite handsome.

3. **Amsonia ciliata** Walt. var. **filifolia** Wood Map 1744

*Amsonia ciliata* var. *tenuifolia* (Raf.) Woodson [P & S, Steyerm.]

Flowers April–May.

Occurs on limestone glades and bald knobs and edges of limestone escarpments of bluffs along streams. Known only from North Fork of White River and Bryant Creek in Ozark and Douglas counties, southwestern Ozarks.

Ranges from Georgia to Texas, north to North Carolina and Missouri; also in Mexico.

The stems of this species average shorter than in the other Amsonias of Missouri, usually growing only 2–4 dm. tall. The plants are quite handsome, producing dense clusters of pale blue flowers held high above the numerous, narrowly linear leaves. I have grown the plant from seed in my northern Illinois wildflower garden, where it has prospered for the past ten years. It makes an appropriate subject for a limestone rock garden.

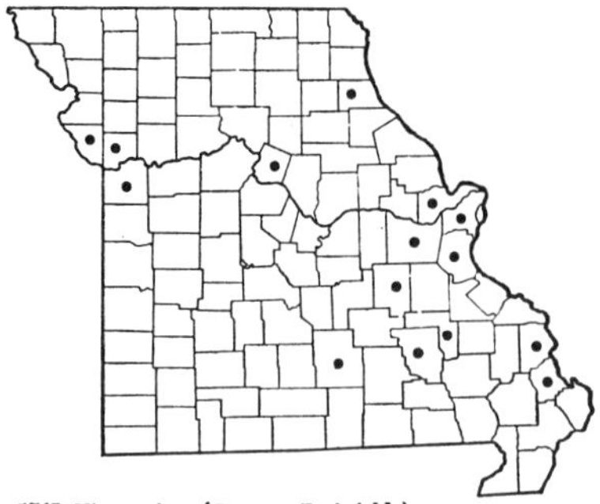

1745 Vinca minor (Common Periwinkle)

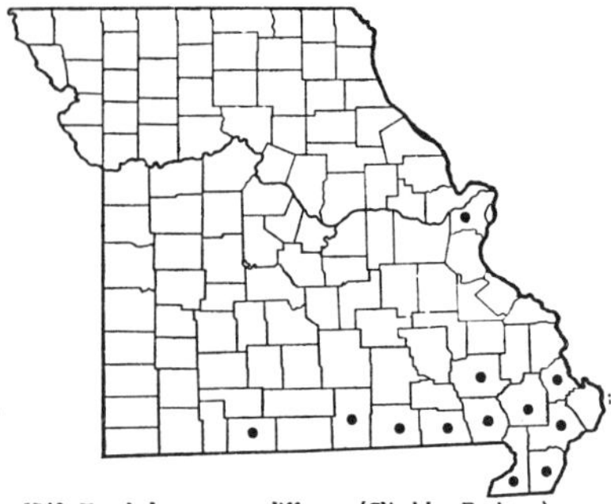

1746 Trachelospermum difforme (Climbing Dogbane)

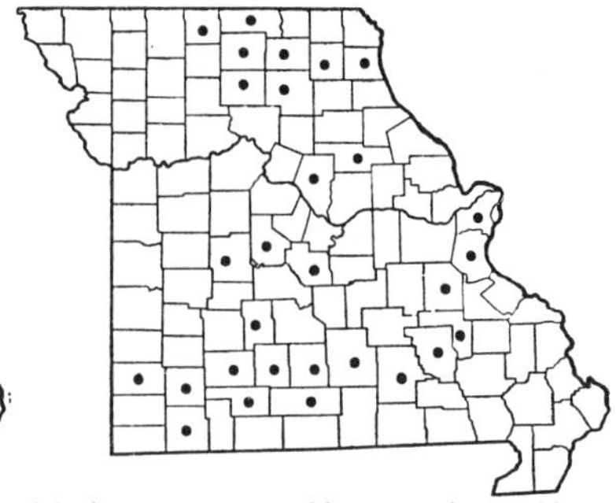

1747 Apocynum androsaemifolium var. androsaemifolium (Spreading Dogbane)

## 2. **Vinca** L. Periwinkle

**Vinca minor** L. Common Periwinkle Map 1745

Also called Myrtle.

Flowers April–May.

Commonly cultivated and sometimes escaped into woods, ledges of bluffs, rocky banks, and waste ground. Scattered in parts of central and eastern Missouri.

Native of Europe; introduced and naturalized in North America from Nova Scotia to Minnesota, south to Georgia, Arkansas, and Kansas.

This species is commonly planted as a ground cover for cemetery beds and shaded places.

## 3. **Trachelospermum** Lemaire Climbing Dogbane

**Trachelospermum difforme** (Walt.) Gray

Climbing Dogbane Map 1746

Flowers last of May–July.

Occurs in swamps, borders of slow streams and bayous, low wet woods, and thickets, in the lowland section, and along moist rocky stream banks in sections of the Ozarks. Southeastern Missouri lowlands west in the Ozarks to Howell and Taney (*Steyermark 80742*) counties, and locally north in St. Louis County.

Ranges from Florida to Texas, north to Delaware, D.C., Indiana, Illinois, Missouri, and Oklahoma.

The isolated westernmost Missouri station in Taney County was an unique swamp along an old creek meander of a valley on the south side of Long Creek, T21N, R22W, sect. 3, $1\frac{1}{2}$ mi. (by air) north northwest of Oasis (Cedar Valley). With the impoundment of White River and tributaries by the Table Rock Dam, this station is now destroyed and under one hundred feet of water.

The flowers are very fragrant.

## 4. **Apocynum** L. Dogbane, Indian Hemp

a. Corolla pink, pink-striped, or white, campanulate (bell-shaped), 4–9 mm. long, at least twice as long as the calyx-lobes; leaves drooping or wide-spreading; seeds 2.5–4 mm. long; topmost inflorescence usually projecting above the foliage or leafy branches . . . . . . . . . . . . . . . . . . *b*

b. Corolla pink or white striped pink, mostly 6–9 mm. long, 5–10 mm. broad, with recurving lobes; flowers mostly nodding; inflorescences at the tip and along the sides of the branches; leaves of the main stem usually drooping; seeds 2.5–3 mm. long . . . . . . . . . 1. A. ANDROSAEMIFOLIUM

b. Corolla white or pink-tinged, 4–7 mm. long, 3.5–6 mm. broad, with spreading but not recurving lobes; flowers ascending to spreading or slightly drooping; inflorescences at the tip of the stem

---

Plate no. 284. **1.** Nymphoides peltatum, × $^{4}/_{7}$; After Gleason, details from Small, The New York Botanical Garden. **2.** Amsonia Tabernaemontana var. Tabernaemontana, × $^{2}/_{7}$; Details from Small, The New York Botanical Garden. **3.** Amsonia Tabernaemontana var. Gattingeri, × $^{2}/_{7}$. **4.** Amsonia Tabernaemontana var. salicifolia, × 1. **5.** Amsonia illustris, × $^{2}/_{7}$. **6.** Amsonia ciliata var. filifolia, × $^{2}/_{7}$. **7.** Apocynum androsaemifolium, × $^{2}/_{7}$; Details from Small, The New York Botanical Garden. **8.** Vinca minor, × $^{2}/_{7}$; Details from Small, The New York Botanical Garden. **9.** Trachelospermum difforme, × $^{2}/_{7}$; Details from Small, The New York Botanical Garden.

PLATE NO. 284

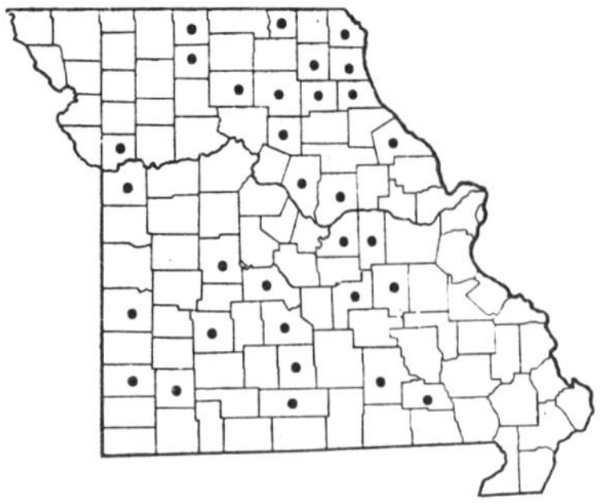

1748 Apocynum medium

1749 • Apocynum cannabinum var. cannabinum (Indian Hemp)
1749 □ Apocynum cannabinum var. pubescens

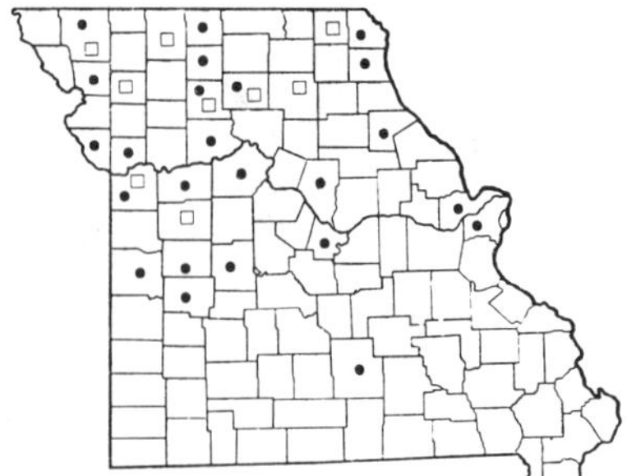

1750 • Apocynum sibiricum var. sibiricum (Indian Hemp)
1750 □ Apocynum sibiricum var. cordigerum

and at the ends of leafy branches; leaves of the main stem usually wide-spreading; seeds 3–4 mm. long . . . . . . . . . . . . . . . . . . . . . . . . . 2. A. MEDIUM

a. Corolla white to greenish- or creamy-white, tubular-cylindric to ovoid, 2–4 mm. long, equaling or at most only slightly longer than the calyx-lobes; leaves ascending, erect, or wide-spreading; seeds 4–6 mm. long; topmost inflorescences prominently surpassed or concealed by the longer leafy branches . . . . . . . . . . . . . . . . . . . . . . . . . . *c*

c. Leaves of the main stem with a stalk (petiole) 4–10 mm. long; hair protruding at one end of seed 2–3 cm. long; mature fruits (follicles) 12–20 cm. long . . . . . . . . . 3. A. CANNABINUM

c. Leaves of the main stem without a stalk (sessile) or with an inconspicuous stalk rarely up to 4 mm. long; hair protruding at one end of seed 0.8–2 cm. long; mature fruits (follicles) 4–10 cm. long 4. A. SIBIRICUM

1. **Apocynum androsaemifolium** L. var. **androsaemifolium** Spreading Dogbane Map 1747

Also called Pink-flowered Dogbane.

*Apocynum androsaemifolium* L. [G, BB, P & S, Steyerm.]

Flowers May–July.

Occurs in dry rocky or open woods, prairie openings, and thickets. Scattered throughout Missouri, but absent from the extreme southeastern quarter and in most of extreme western and northwestern Missouri.

Ranges from Newfoundland to Alaska, south to North Carolina, West Virginia, Ohio, Indiana, Illinois, Arkansas, Nebraska, New Mexico, Arizona, and Mexico.

A western United States variety with the leaves glabrous on the lower surface is known as var. *glabrum* Macoun. The flowers of *A. androsaemifolium* are quite fragrant. Their fragrance resembles that of the Lily-of-the-Valley. The plant has poisonous effects similar to those produced by *Apocynum cannabinum*, described below.

2. **Apocynum medium** Greene Map 1748

*Apocynum medium* var. *leuconeuron* (Greene) Woodson [P & S, Steyerm.]

Flowers May–July.

Occurs in rocky open woods, prairie openings, and thickets. Scattered throughout Missouri south to Virginia, Tennessee, Missouri, Texas, and New Mexico.

This species is considered to be a fertile hybrid between *A. androsaemifolium* and *A. cannabinum* (Anderson, E., Ann. Mo. Bot. Gard. 23: 159–68. 1936). Gleason treats *A. medium* as representing hybrids produced between either *A. androsaemifolium* and *A. cannabinum* or between *A. androsaemifolium* and *A. sibiricum*. In the field it is a recognizable taxon distinct from either *A. androsaemifolium* or *A. cannabinum*.

3. **Apocynum cannabinum** L. Indian Hemp Map 1749

Also called Dogbane.

Flowers early May–August.

Occurs in prairies, glades, rocky open woods, thickets, waste ground, and along railroads.

Two variations occur in Missouri:

Mostly glabrous (hairless) plant; stems glabrous; lower surface of leaf-blades glabrous or sparsely hairy; branches of inflorescence and calyx glabrous . . . 3a. A. CANNABINUM var. CANNABINUM

Plate no. 285. **1.** Apocynum medium, × $^2/_7$; *a.* Flower, × $1^1/_7$. **2.** Asclepias tuberosa, × $^2/_7$; Details from Small, The New York Botanical Garden. **3.** Asclepias viridis, × 1; *a.* Flower, × 4; Details from Small, The New York Botanical Garden. **4.** Apocynum sibiricum var. sibiricum, × $^2/_7$. **5.** Apocynum cannabinum var. cannabinum, × $^2/_7$.

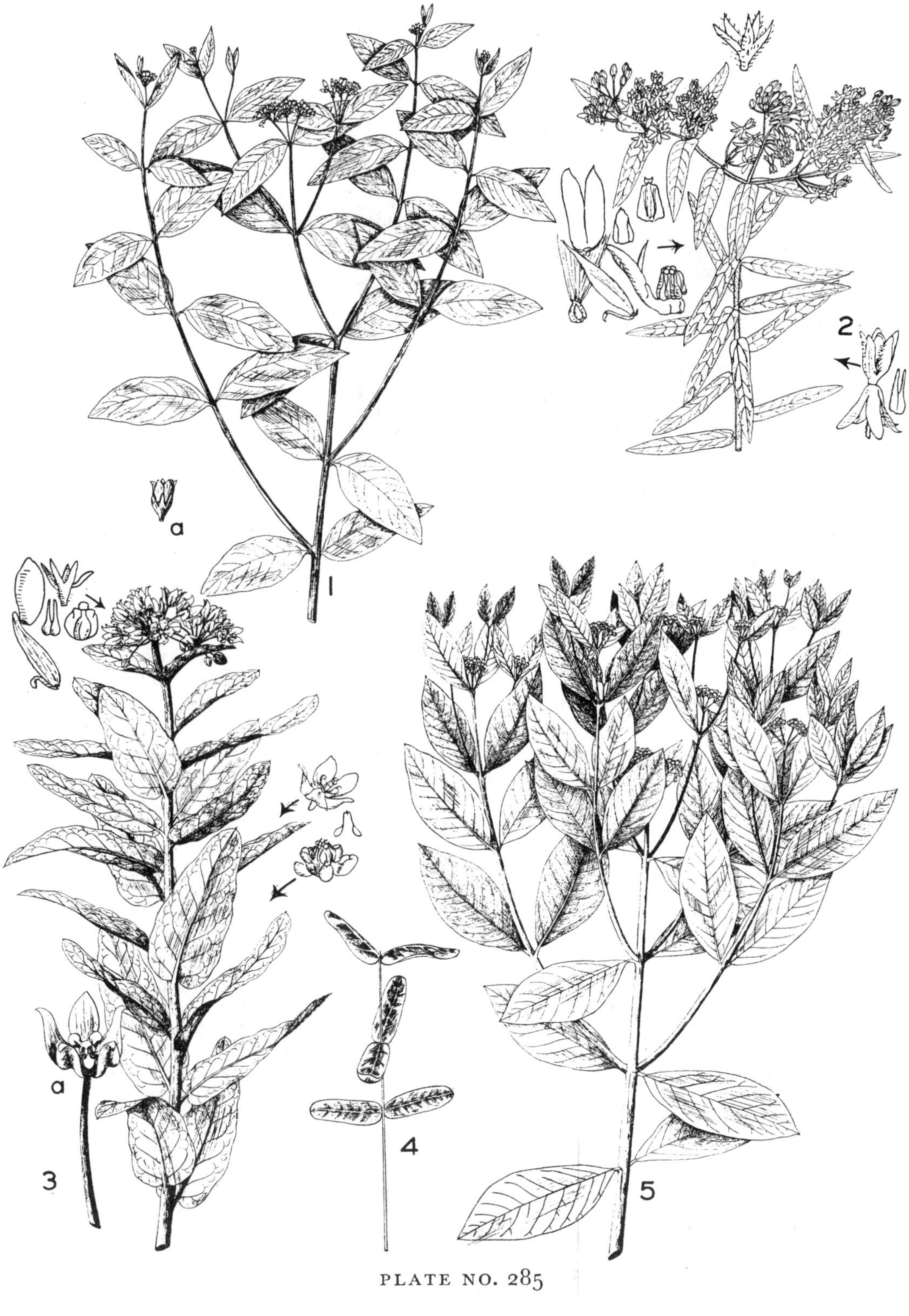


PLATE NO. 285

Upper part of stems hairy; lower surface of leaf-blades hairy; branches of inflorescence and calyx hairy . . . 3b. A. CANNABINUM var. PUBESCENS

3a. **Apocynum cannabinum** var. **cannabinum** Map 1749

*Apocynum cannabinum* L. [G, P & S, Steyerm.]

*Apocynum cannabinum* var. *glaberrimum* A. DC. [P & S, Steyerm., BB, Woodson]

Throughout Missouri.

Ranges from Quebec to Alberta and Washington, south to Florida, Alabama, Mississippi, Louisiana, Texas, New Mexico, Arizona, and California.

3b. **Apocynum cannabinum** var. **pubescens** (Mitchell) A. DC. Map 1749

Scattered throughout Missouri.

Ranges from Massachusetts to Ontario and North Dakota, south to Georgia, Alabama, Mississippi, and Texas.

Horses, cattle, and sheep are sometimes poisoned by eating green or dried leaves of this species.

Both *A. cannabinum* and *A. androsaemifolium* have been used in medicine for their emetic, cathartic, and diuretic properties. These and other species of the genus contain a strong bast fiber which may be used for tying articles.

4. **Apocynum sibiricum** Jacq. Indian Hemp Map 1750

Flowers June–August.

Occurs in river bottom prairies, wet meadows, prairie swales and depressions, alluvial soils and banks along streams, ponds, sloughs, and ditches.

Two variations are found in Missouri:

Main leaves of the stem 1–4 cm. broad, rounded or heart-shaped at base, oblong to oblong-lanceolate or narrowly ovate . . . . 4a. A. SIBIRICUM var. SIBIRICUM

Main leaves of the stem 2.5–5.5 cm. broad, deeply heart-shaped and clasping at the base, ovate to oval-oblong . . . 4b. A. SIBIRICUM var. CORDIGERUM

4a. **Apocynum sibiricum** var. **sibiricum** Map 1750

*Apocynum sibiricum* Jacq. [G]

*Apocynum hypericifolium* Ait. [P & S]

Northern and central Missouri, south to St. Louis, Cole, Benton, St. Clair, and Bates counties.

Ranges from Newfoundland to British Columbia, south to Virginia, West Virginia, Ohio, Indiana, Illinois, Missouri, and Texas.

4b. **Apocynum sibiricum** var. **cordigerum** (Greene) Fern. Map 1750

*Apocynum hypericifolium* var. *cordigerum* (Greene) Beg. & Bel. [P &S]

Northern and central Missouri, south to Macon, Livingston, Johnson, and Jackson counties.

Ranges from New England to Saskatchewan, south to Ohio, Indiana, Illinois, Missouri, and Kansas.

This species prefers to grow in moister situations than the other species of *Apocynum.*

## Fam. **ASCLEPIADACEAE** (Milkweed Family)

The following classification follows that of Woodson's recent works (Ann. Mo. Bot. Gard. 28: 193–244. 1941; 41: 47–201. 1954).

a. Stems erect or spreading from the base, but never climbing or twining . . . . . . . 1. ASCLEPIAS
a. Stems climbing or twining . . . . . . . . . . . . . . . . . . . . . . . . . . *b*
  b. Stem glabrous (without hairs); corolla 4–6 mm. long; lobes of corolla erect during flowering; lobes of crown (raised disk in center of flower) erect, nearly as long as the corolla-lobes and much longer than the other portion of the crown (where anthers surround stigma). . . . . 2. CYNANCHUM
  b. Stem more or less hairy; corolla 7–25 mm. long; lobes of corolla mostly spreading or ascending, but not erect; lobes of crown (raised disk in center of flower) short or scarcely developed, much shorter than corolla-lobes and shorter than or equaling the other portion of the crown (where anthers surround stigma) . . . . . . . . . . . . . . . . . . 3. MATELEA

## 1. **Asclepias** L. Milkweed

(*Asclepiodora* Gray [G, BB, P & S, Steyerm.])

(*Acerates* Ell. [BB, P & S, Steyerm.])

The complicated flower consists of an outermost calyx with 5 small, reflexed lobes. The corolla is deeply parted into 5 lobes turned down or back, exposing to view the most conspicuous portion of the flower, the 5 hoods, together comprising the crown or corona. Each hood contains a cavity from which a needle-like horn may protude, or the horn may be absent in certain species. The 5 stamens, situated on the bottom of the corolla, consist of 5 filaments joined into a tube, the latter surrounding the pistil with 2 ovaries. The anthers touch the stigma, each of the anthers having 2 vertical chambers tipped with a thin appendage and containing the pollen masses. The pollen is collected in masses of pollinia which are united in pairs suspended by means of a slender prolongation (translator) of their summits from adjacent halves of the margins of the anthers.

Woodson has united *Asclepiodora*, *Acerates*, and *Asclepias* into one genus, *Asclepias*, a procedure followed in the present treatment.

a. Some or all of the leaves alternate . . . . . *b*
  b. Leaves narrowly linear, 0.1–1.5 cm. broad . . . . . *c*
    c. Leaves 1.5–5 mm. broad; lower surface of leaves with no lateral nerves or veins showing or these very faint at most; flower-stalks 9–13 mm. long; hoods containing a very small horn; crown without a stalk (sessile) at its base; hairs on fruit appressed (pressed against or lying parallel to surface), minute, inconspicuous . . . . . 14. A. STENOPHYLLA
    c. Main leaves 4–20 mm. broad; lower surface of leaves with visible and conspicuous lateral (side) nerves and veins showing; flower-stalks during flowering time 13–20 mm. long; hoods lacking a horn; crown with a short stalk at its base; hairs of fruit spreading . . . 15. A. HIRTELLA
  b. Leaves linear to broadly oblong or oblong-ovate, 2–13 cm. broad . . . . . *d*
    d. Plant glabrous (without hairs) or nearly so; flowers green with purplish, the corolla-lobes spreading to ascending; larger leaves 2–13 cm. broad, broadly oblong . . . . . 1. A. VIRIDIS
    d. Plant hairy; flowers orange, orange-red, or yellow, the corolla-lobes reflexed or turned down; leaves mainly 0.5–2 (rarely 2.5–3) cm. broad, linear to broadly oblanceolate . . . 3. A. TUBEROSA
a. All the leaves opposite or in whorls (arranged in circles) of 3–6 on the stem . . . . . *e*
  e. All the leaves thread-like, narrowly linear, at most 3 mm. broad, and in whorls (circles) of 3–6 13. A. VERTICILLATA
  e. Leaves not thread-like or narrowly linear, at least 7 mm. broad, or, if narrower, not in whorls (circles of 3 or more) . . . . . *f*
    f. Some of the leaves in whorls (circles) of 3, 4, or 6 . . . . . *g*
      g. Most of the lower surface of the leaf, including the nerves, minutely hairy; corolla and hoods red-purple; hoods 4.5–7 mm. high . . . . . 4. A. PURPURASCENS
      g. All or most of the lower surface of the leaf glabrous (without hairs), hairs, if present, only on main nerves; corolla and hoods pink and white or white with a purplish center; hoods 2–4.5 mm. high . . . . . *h*
        h. Leaves blunt, rounded, or short-pointed at tip . . . . . 8. A. VARIEGATA
        h. Leaves long-pointed or long-tapering at tip . . . . . 7. A. QUADRIFOLIA
    f. All or most of the leaves normally opposite in pairs . . . . . *i*
      i. No slender horn arising from the cavity of any hoods of the flower . . . . 2. A. VIRIDIFLORA
      i. A slender horn arising from the cavity of each hood of the flower . . . . . *j*
        j. Two to six pairs of leaves (4–12 leaves) on the main stem . . . . . *k*
          k. Leaves without stalks (sessile), their base broad, rounded, heart-shaped, or clasping . . *l*
            l. Leaves blunt or rounded at tip, about the same width for most of their length; leaves mainly 7–15 cm. long; flowers 13–18 mm. long . . . . 11. A. AMPLEXICAULIS
            l. Leaves more or less tapering or narrowed at the tip, broadest at their base; leaves 3–8.5 cm. long; flowers 12–14 mm. long . . . . . 12. A. MEADII
          k. Leaves with definite stalks (petioles), narrowed or tapering to the base . . . . . *m*
            m. Corolla and hoods red-purple; hoods 4.5–7 mm. high; lower surface of leaves, including nerves, minutely hairy . . . . . 4. A. PURPURASCENS
            m. Corolla and hoods pink and white, white, or white with a purplish center; hoods

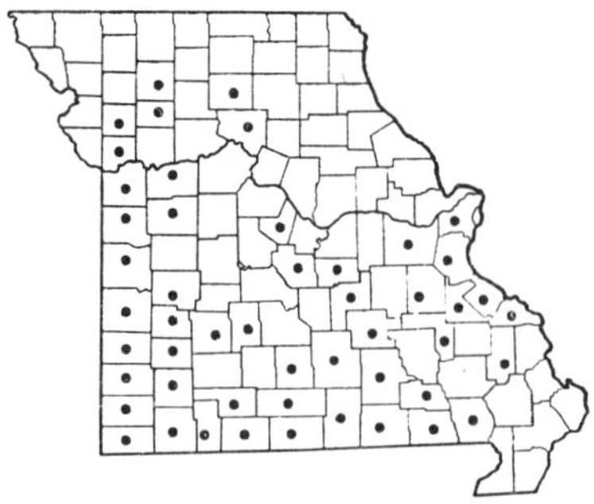

1751 Asclepias viridis (Green-flowered Milkweed)

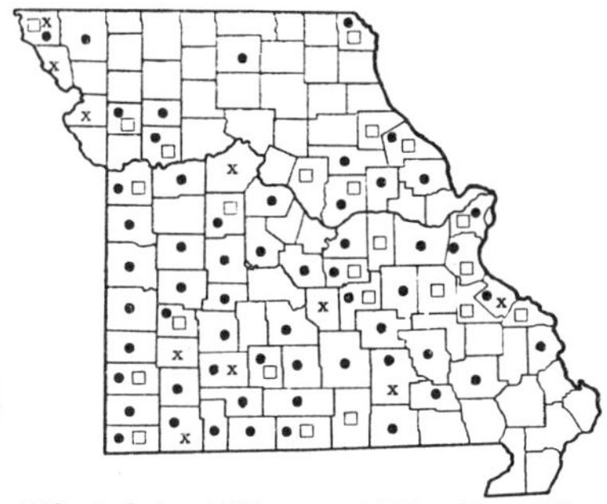

1752 • Asclepias viridiflora var. viridiflora (Green Milkweed)
1752 □ Asclepias viridiflora var. lanceolata
1752 x Asclepias viridiflora var. linearis

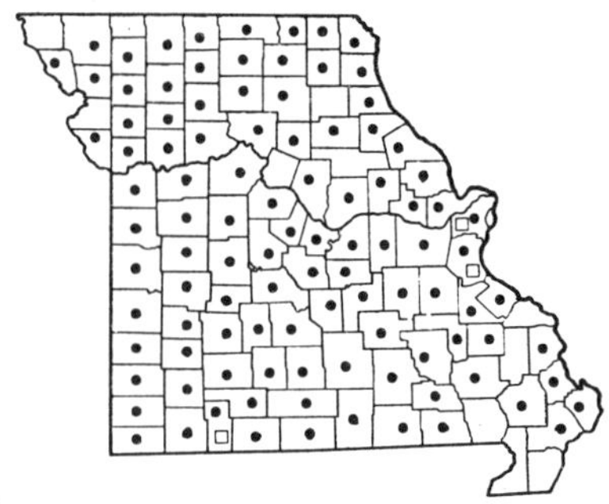

1753 • Asclepias tuberosa var. interior f. interior
1753 □ Asclepias tuberosa var. interior f. lutea

2–4.5 mm. high; lower surface of leaves mainly glabrous (without hairs), the only hairs, if present, on main nerves . . . . . . . . . . . . . . . . . . *n*

n. Leaves blunt, rounded, or short-pointed at tip . . . . . . . 8. A. VARIEGATA

n. Leaves long-pointed or long-tapering at tip . . . . . . . . . . . . . . *o*

o. Only 3–5 pairs of leaves present; stem solitary, usually one to a plant and leafless below; plants of dry or rocky woods in eastern and southern Missouri 7. A. QUADRIFOLIA

o. At least 6 pairs of leaves present, stems usually 2 or more to a plant, leafy below; plants of swamps and low wet woods of southeastern Missouri. 6. A. PERENNIS

j. Seven to twenty-one or more pairs of leaves normally present on main stem. . . . . . *p*

p. Main side nerves of leaves ascending, their ends directed toward the tip of the leaf. . . *q*

q. Flowers white, 3–6 mm. long, the reflexed corolla-lobes 2.5–4 mm. long; stems 3–8 dm. tall; outside of fruits glabrous (without hairs); seeds without any tufts of hairs; plants of southeastern Missouri swamps . . . . . . . 6. A. PERENNIS

q. Flowers pink to rose-red, rarely white, 8–9 mm. long, the reflexed corolla-lobes 4–5 mm. long; stems up to 15 dm. tall; outside of fruits minutely hairy; seeds with a conspicuous tuft of hairs (coma); throughout Missouri . . . . . . 5. A. INCARNATA

p. Main side nerves of leaves horizontally or transversely spreading toward the margin of the leaf . . . . . . . . . . . . . . . . . . . . . . . *r*

r. Lower surface of leaves glabrous (without hairs); base of leaf somewhat heart-shaped, the leaf nearly sessile (without a stalk) but with a very short stalk; flower-stalks glabrous (without hairs); outside of fruit glabrous or nearly so. 9. A. SULLIVANTII

r. Lower surface of leaves minutely hairy; base of leaf rounded or tapering to a definite stalk (petiole) 5–15 mm. long; flower-stalks hairy; outside of fruit minutely hairy . . . . . . . . . . . . . . . . . . . . . . . *s*

s. Flowers dark purplish-red, in 1–3 clusters at or near top of the stem; upper surface of leaves dark green; fruits not covered with slender appendages or projections . . . . . . . . . . . . . . . . 4. A. PURPURASCENS

s. Flowers lavender and green mixed with rose-color and white, in usually 2–6 clusters scattered along the stem; upper surface of leaves olive or dull pale green; fruits covered with slender appendages or projections . 10. A. SYRIACA var. KANSANA

1. **Asclepias viridis** Walt. Green-flowered Milkweed Map 1751

Also called Spider Milkweed, Antelope Horn.

*Asclepiodora viridis* (Walt.) Gray [G, BB, P & S, Steyerm.]

Flowers May–June.

Occurs on rocky prairies and glades, usually on limestone strata. Mainly southern and central Missouri in the Ozark and unglaciated prairie regions, north to St. Louis, Franklin, Maries, Moniteau, Chariton, Linn, Daviess, Clinton, and Jackson counties.

Ranges from Florida to New Mexico, north to South Carolina, Ohio, Illinois, Missouri, and Nebraska.

The flowers have the lower half (corolla-lobes) green and the upper half (crown) purplish.

2. **Asclepias viridiflora** Raf. Green Milkweed Map 1752

*Acerates viridiflora* (Raf.) Eaton [BB, P & S, Steyerm.]

Flowers late May–August.

Occurs on rocky prairies and glades. Throughout Missouri, commonest in the southern and central sections.

Three variations are encountered in the state:

a. Leaves linear or linear-lanceolate, 2–10 mm. broad . . . 2c. A. VIRIDIFLORA var. LINEARIS
a. Leaves oval, oblong, narrowly ovate, or lanceolate, 12–50 mm. broad . . . . . . . *b*
  b. Leaves lanceolate to narrowly ovate, tapering to a short- (acute) or long- (acuminate) pointed tip . . . . 2b. A. VIRIDIFLORA var. LANCEOLATA
  b. Leaves oval, obovate, or oblong, blunt (obtuse) or slightly tipped with a short point . 2a. A. VIRIDIFLORA var. VIRIDIFLORA

2a. **Asclepias viridiflora** var. **viridiflora** Map 1752

*Acerates viridiflora* (Raf.) Eat. [BB, P & S, Steyerm.]

This is the commonest variation in Missouri, throughout the state, most frequent in the southern and central sections, rare in the northern part of the state, and apparently absent from the southeastern lowlands.

Ranges from Florida to Texas and New Mexico, north to Massachusetts, New York, Pennsylvania, Kentucky, and Kansas.

2b. **Asclepias viridiflora** var. **lanceolata** (Ives) Torr. Map 1752

*Acerates viridiflora* var. *lanceolata* (Ives) Gray [P & S, Steyerm.]

Scattered throughout the state, rare in northern Missouri, and apparently absent in the southeastern lowlands. Ranges from Louisiana to Texas and New Mexico, north to Virginia, Pennsylvania, Kentucky, Ontario, Manitoba, and Wyoming.

2c. **Asclepias viridiflora** var. **linearis** (Gray) Fern. Map 1752

*Acerates viridiflora* var. *linearis* Gray [P & S, Steyerm.]

Scattered in western and southern portions of the state.

Ranges from Louisiana to New Mexico, north to Ontario and Manitoba.

The above varieties intergrade into each other, and it is often difficult to place specimens certainly into one or another of these variations, which, however, are distinct in their extremes. Gleason (*New Britton & Brown Ill. Fl.* 3: 81. 1952) believes the variations may be correlated with relative dryness of habitat, and Woodson does not give them taxonomic rank. More intensive field and experimental studies need to be carried on, however, before the final status of these variations can be evaluated. As the var. *linearis*, especially, appears to be geographically orientated more inland and westward within the range of *A. viridiflora*, there would seem to be some geographical isolation involved. Pending more detailed studies, the present treatment has retained these three variations.

3. **Asclepias tuberosa** L. Butterfly Weed Map 1753

Also called Pleurisy Root, Chigger Flower.

Flowers late May–September.

Occurs in rocky or dry open woods, prairies, glades, rocky open places, fields, and along roadsides. Throughout Missouri.

Two variations are encountered in the state:

Flowers orange or reddish-orange . . . . 3a. A. TUBEROSA var. INTERIOR f. INTERIOR
Flowers yellow . . . . . . 3b. A. TUBEROSA var. INTERIOR f. LUTEA

3a. **Asclepias tuberosa** var. **interior** (Woodson) Shinners f. **interior** Map 1753

*Asclepias tuberosa* subsp. *interior* Woodson [Woodson, BB]

*Asclepias tuberosa* [of G, P & S, Steyerm.], not L.

Throughout Missouri, and doubtless in every county.

Ranges from Ontario to Minnesota and Nebraska, south to Kentucky, Mississippi, Oklahoma, and Texas.

3b. **Asclepias tuberosa** var. **interior** f. **lutea** (Clute) Steyerm. Map 1753

*Asclepias tuberosa* f. *lutea* Clute, Am. Bot. 18: 73 1912.

*Asclepias tuberosa* subsp. *interior* Woodson [Woodson]

Rare and scattered in the southern and central sectors in St. Louis, Jefferson (*Steyermark 1039*), and Stone (*Steyermark 22500*) counties.

Woodson includes this yellow-flowered form as a synonym under *A. tuberosa* subsp. *interior*. I have grown this form in my garden for ten years, and as the same plant has constantly shown the yellow color, it would appear to merit formal recognition, which status it is accorded in the present flora.

Woodson has separated *A. tuberosa* into four subspecies, subsp. *tuberosa*, *terminalis*, *rolfsii*, and *interior*, of which only the subsp. *interior* occurs in Missouri (in the present flora treated as a variety). For his discussion of these subspecies, the reader is referred to Woodson's publications (Ann. Mo. Bot. Gard. 31: 366–69. 1944; 34: 353–432. 1947; 41: 74–75. 1954).

1754 Asclepias purpurascens (Purple Milkweed)

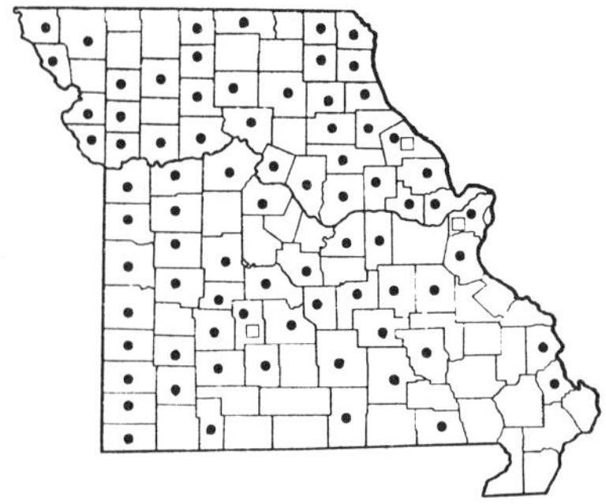

1755 • Asclepias incarnata var. incarnata f. incarnata (Swamp Milkweed)

1755 □ Asclepias incarnata var. incarnata f. albiflora

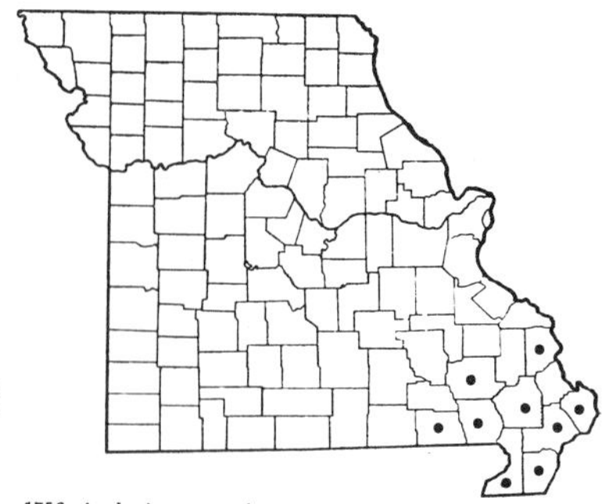

1756 Asclepias perennis

Butterfly Weed is one of the showiest wildflowers of the native flora. It responds well to cultivation if given an exposed sunny situation. The deep fleshy roots are easily broken when the plant is dug, but if sufficient care is used to remove as large a quantity as possible of the original soil with the roots, success in transplanting frequently occurs. However, it is easier to grow the plants from seeds, collecting these in late fall and leaving them in the ground over winter to germinate the following spring. Such plants may take two to three years to produce flowers.

The root has been used in medicine as an emetic and diuretic. The leaves and stems are reported to be poisonous if eaten by grazing animals.

4. **Asclepias purpurascens** L. Purple Milkweed
Map 1754

Flowers late May–July.

Occurs in rocky open woods, glades, prairie openings, stream banks, and wet meadows in valleys, upland dry ridge tops, thickets. Throughout Missouri, but apparently absent from the lowlands of southeastern Missouri.

Ranges from New Hampshire to Ontario, Minnesota, and South Dakota, south to North Carolina, Tennessee, Mississippi, Arkansas, and Oklahoma.

The unopen and open inflorescences of this and other species of this genus may be boiled and used as a substitute for broccoli.

This species is sometimes confused with *A. syriaca.* In addition to the characters given in the key, it should be noted that the hoods of *A. syriaca* have a tooth projecting from about the middle of each margin; these lateral teeth are absent from the hoods of *A. purpurascens.* The leaves normally are all in pairs opposite each other. Rarely there is encountered a specimen having some of the leaves whorled, as represented by *Steyermark 73436* from Livingston County.

5. **Asclepias incarnata** L. Swamp Milkweed
Map 1755

Flowers June–August.

Occurs in wet meadows, river bottom prairies, borders of ponds, sloughs, and streams, rarely in low wet woods, and along railroads.

Two variations occur in Missouri:

Flowers pink to rose-purple . 5a. A. INCARNATA var. INCARNATA f. INCARNATA

Flowers whitish . . . . . 5b. A. INCARNATA var. INCARNATA f. ALBIFLORA

5a. **Asclepias incarnata** var. **incarnata** f. **incarnata** Map 1755

*Asclepias incarnata* var. *incarnata* [BB]

*Asclepias incarnata* subsp. *incarnata* [Woodson]

*Asclepias incarnata* L. [G, P & S]

This is the common variation throughout Missouri.

Ranges from Maine and Quebec to Manitoba, Wyoming, and Utah, south to Florida, Louisiana, Texas, and New Mexico.

5b. **Asclepias incarnata** var. **incarnata** f. **albiflora** Heller Map 1755

*Asclepias incarnata* f. *albiflora* Heller [G]

*Asclepias incarnata* subsp. *incarnata* [Woodson]

Rarely encountered in the state, where known only from Pike, St. Louis, and Dallas counties.

The silky floss of the fruit has been used as a substitute for kapok in life preservers and other articles

Plate no. 286. **1.** Asclepias viridiflora, × $^2/_7$; *a.* Flower, × 10; Details from Small, The New York Botanical Garden. **2.** Asclepias incarnata, × $^2/_7$. **3.** Asclepias quadrifolia, × $^2/_7$. **4.** Asclepias purpurascens, × $^2/_7$; *a.* Flower, × $^6/_7$. **5.** Asclepias perennis, × $^2/_7$; *a.* Flower, × $1^1/_7$. **6.** Asclepias variegata, × $^2/_7$.

PLATE NO. 286

1757 Asclepias quadrifolia

1758 Asclepias variegata

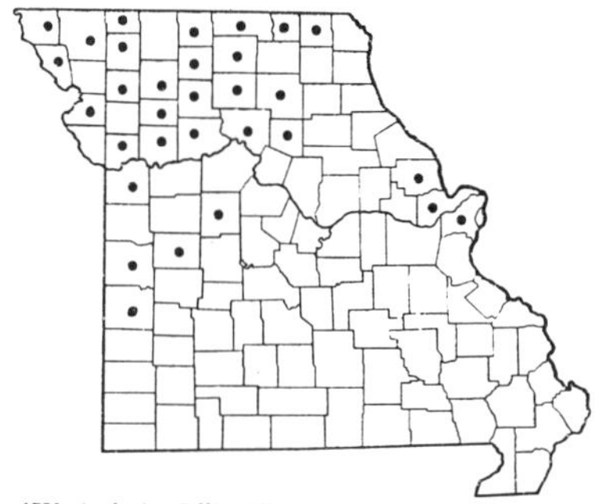

1759 Asclepias Sullivantii

and has possibilities as a waste material for use in the paper industry. As in the case of other species of milkweed, the young shoots, when only a few inches high, may be gathered and cooked as an asparagus substitute. The inflorescences and young leaves, likewise, while still tender, may be cooked and used as a vegetable, the original water being replaced to get rid of any bitter quality present. Muskrats eat the roots of this species.

6. **Asclepias perennis** Walt. Milkweed Map 1756
Flowers last of May–September.

Occurs in low wet woods, bald cypress swamps, and borders of bayous, lakes, and slow streams. Lowlands of southeastern Missouri, north to Cape Girardeau and Wayne counties, west to Ripley County.

Ranges from Florida to Texas, north to South Carolina, Indiana, Illinois, and Missouri.

7. **Asclepias quadrifolia** Jacq. Milkweed Map 1757
Flowers May–July.

Occurs in rich or dry or rocky open woods, generally on upland slopes and ridges. Throughout the eastern half of the state and the Ozark region, absent from the lowlands of southeastern Missouri and from the northwestern quarter.

Ranges from New Hampshire to Ontario and Minnesota, south to North Carolina, Alabama, Arkansas, and Oklahoma.

The flowers of this milkweed are delicately fragrant. The hoods are white and the corolla-lobes are pink to pinkish-white. Usually one or two whorls of the four leaves occur on the stem, with an additional one or two pairs of leaves present. Rarely all the leaves are in pairs and none in whorls.

8. **Asclepias variegata** L. Milkweed Map 1758
Flowers late May–July.

Occurs in dry or rocky woods and sandy open ground, ravine bottoms, wet depressions in lowland forest, upland slopes and ridges, and along roadsides. Southeastern Missouri, mainly on Crowley Ridge, north to Cape Girardeau and Wayne counties, west to Ripley County.

Ranges from Florida to Texas, north to Connecticut, New York, New Jersey, Pennsylvania, West Virginia, Kentucky, Illinois, Missouri, and Oklahoma.

The usually solitary stems are slightly hairy in the upper half or have lines of pubescence. Plants occasionally have some of the leaves in whorls, but normally they are opposite in pairs.

9. **Asclepias Sullivantii** Engelm. Milkweed Map 1759
Flowers June–July.

Occurs in wet meadows, upland, river bottom and low prairies, and alluvial thickets. Northern and central Missouri in the glaciated and unglaciated prairie sections south to St. Louis, Pettis, Henry, and Vernon counties.

Ranges from Ontario to Minnesota, south to Ohio, Indiana, Illinois, Missouri, Kansas, and Oklahoma.

This milkweed generally resembles *A. syriaca* in habit, but is glabrous on all parts. The corolla is purplish-rose with pale rose or pink hoods.

10. **Asclepias syriaca** L. var. **kansana** (Vail) Palmer & Steyerm. Common Milkweed Map 1760
Flowers late May–August.

Occurs in fields, open woods, waste ground, along roadsides, and railroads. Throughout Missouri, and probably in every county, but not recorded from a large number of counties of the western and southern Ozarks.

Ranges from Indiana and Illinois to Iowa, Missouri, Nebraska, Kansas, and Oklahoma; adventive in New York.

Typical var. *syriaca* f. *syriaca* has the surface of the

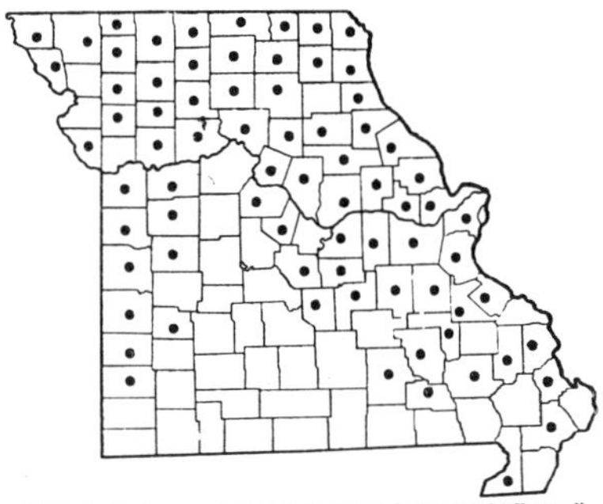

1760 Asclepias syriaca var. kansana (Common Milkweed)

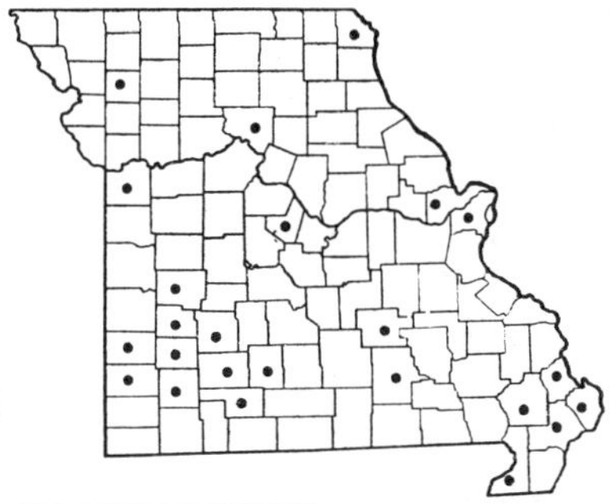

1761 Asclepias amplexicaulis

1762 Asclepias Meadii

fruits covered with short conical appendages 1–3 mm. long, and in var. *syriaca* f. *inermis* the surface is smooth. In var. *kansana* these appendages are long and narrow (subulate-filiform), varying from 3–10 mm. long. Typical var. *syriaca* is more easterly in its distribution, ranging from New Brunswick south to Georgia and Tennessee and west to Saskatchewan and Iowa. Woodson considers these variations as one variable species without according varietal rank to the variations (Ann. Mo. Bot. Gard. 41: 105–8. 1954). For a discussion of the present author's interpretation, the reader should refer to Rh. 58: 197–98. 1956.

Cases of poisoning have been recorded from livestock feeding on the leaves and stems of this milkweed. However, the plants are usually left untouched, due to the bitter quality of the juice in the leaves and stems. An alkaloid, similar to that of nicotine, has been found in the plant. The silky floss of the seed, as in some of the other species of milkweed, has been used as a substitute for kapok to serve in life jackets, life preservers, and other articles where buoyancy is needed. As a waste material this floss also has possibilities for use in the paper industry.

As with *A. incarnata* and other species of the genus, the young stems of *A. syriaca*, only a few inches high, may be gathered and cooked as a substitute for asparagus, while the tender young leaves may be boiled and eaten, after throwing away the original water, as a cooked vegetable. It is recorded that the Indians of Nebraska and the Dakotas cooked the undeveloped flower-clusters and immature fruits as greens. The latter are stated to resemble okra in flavor. These young fruits, after being boiled in salt water, may also be canned. It has also been recorded that the milky sap of the leaves, after hardening, is sometimes used as a chewing gum.

The flowers are quite fragrant. Their nectar attracts honeybees and other insects. In some instances, these flowers become insect traps by attracting the flies and bees to their nectar which is lodged in the hollow hoods connected with the anthers. As the claws of the insect enter the slit or cleft leading to the nectar, they pick up the two masses of pollen, quite accidentally, and, upon leaving, lift these pollen masses out of the pouches of the anther. In most cases the insects procure their nectar and fly away, but in rare instances their feet get entangled in the process and become trapped, the insects eventually remaining attached and dying on the plant. I have witnessed this phenomenon only a few times. The reader is referred to an article published by the author ('Milkweed Death Trap' in Bull. Chi. Nat. Hist. Mus. 27, no. 5: 8. May, 1956).

11. **Asclepias amplexicaulis** Sm. Milkweed Map 1761

*Asclepias humistriata* Walt. [P & S]

Flowers late April–July.

Occurs in prairies, glades, rocky open woods, roadsides, and prairie remnants along railroads. Scattered throughout Missouri, but mainly in the southern third of the state.

Ranges from Florida to Texas, north to New Hampshire, Massachusetts, New York, West Virginia, Ohio, Michigan, Wisconsin, Minnesota, and Nebraska.

The flowers occur in usually one cluster terminating the stem. They have greenish-purple corolla-lobes and pinkish or flesh-colored hoods.

12. **Asclepias Meadii** Torr. Milkweed Map 1762

Flowers May–June.

Occurs in prairies and glades. Rare and local, known from the southern half of the state north to St. Louis, Pettis, Johnson, and Cass counties.

Ranges from Indiana, Illinois, Wisconsin, and Iowa, south to Missouri and Kansas.

The flowers are quite fragrant and have greenish corolla-lobes with purplish hoods. Like *A. amplexicaulis*, they are arranged in a solitary cluster at the tip of the stem.

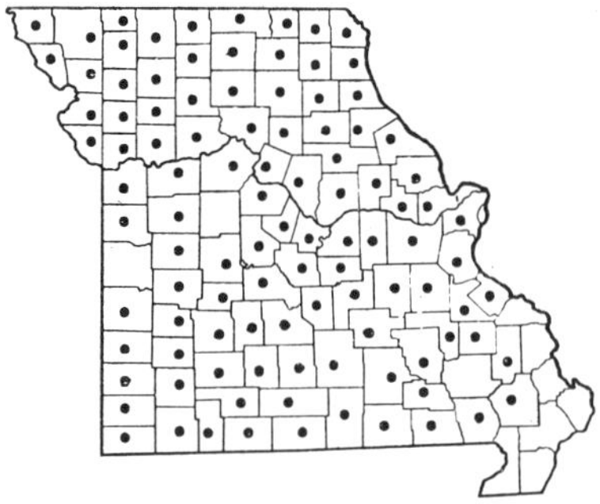

1763 Asclepias verticillata (Whorled Milkweed)

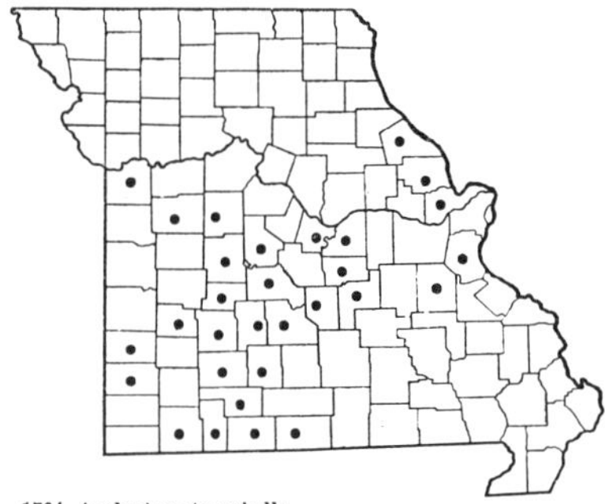

1764 Asclepias stenophylla

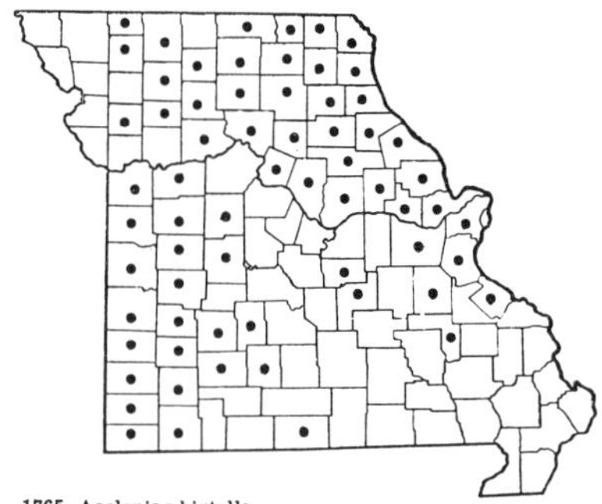

1765 Asclepias hirtella

13. **Asclepias verticillata** L. Whorled Milkweed Map 1763

Also called Horsetail Milkweed.

Flowers late May–September.

Occurs in dry prairies, glades, dry open woods, rocky escarpments of bluffs, roadsides, fields, and along railroads. Throughout Missouri.

Ranges from Florida to Texas, north to Massachusetts, New York, Ontario, Michigan, Wisconsin, Manitoba, and Saskatchewan.

The flowers have greenish-white corolla-lobes and white hoods and occur in several separate clusters scattered along the upper part and at the tip of the stem.

14. **Asclepias stenophylla** Gray Map 1764

*Acerates angustifolia* (Nutt.) Decne. [BB]

Flowers June–July.

Occurs in dry prairies, rocky glades, ledges and escarpments of rocky bluffs, usually of limestone strata. Mostly in the western and northern Ozark and unglaciated prairie regions, locally northeast to Pike County; absent from most of the eastern and central Ozark sections.

Ranges from South Dakota to Missouri and Oklahoma, west to Colorado, New Mexico, and Arizona.

The flowers have greenish corolla-lobes and whitish hoods. The plants somewhat resemble *A. verticillata*, but are usually taller, and the leaves, instead of being arranged in whorls of 3–6 as in *A. verticillata*, have the lower usually opposite and the upper usually alternate. In the somewhat smaller flowers of *A. verticillata* the horn projects prominently from the hood, whereas in *A. stenophylla* it is inconspicuous and does not project out.

The saclike hoods have a toothlike horn, which character would justify the placing of the species with other members of *Asclepias*. The species is considered by Woodson to be a connecting link between the species placed in *Asclepias* on the one hand and those previously placed in *Acerates* on the other.

15. **Asclepias hirtella** (Pennell) Woodson Map 1765

*Acerates hirtella* Pennell [BB, P & S]

Flowers late May–August.

Occurs in prairies and glades. Mostly in northern, central, and western Missouri; absent from much of the Ozark and all of the southeastern lowland sections.

Ranges from West Virginia to Wisconsin, south to Alabama, Arkansas, Kansas, and Oklahoma.

The flowers vary from 6.5–8.5 mm. long and have greenish corolla-lobes with white margins. The peduncles (main flower-stems), pedicels (individual flower-stalks), and fruits have a conspicuous spreading pubescence, and the plant is more conspicuously pubescent than either *A. verticillata* or *A. stenophylla*. The stems are also stouter and the leaves broader and more of them arranged alternately than in either of the other species.

*Excluded Species*

**Asclepias latifolia** (Torr.) Raf.

This species was credited to the state by Palmer and Steyermark in their *Annotated Catalogue* (p. 622) on the basis of a collection from Jackson County, west-central Missouri. This collection has been confused with *A. amplexicaulis* which it resembles.

Woodson does not cite any material from the state

Plate no. 287. **1.** Asclepias syriaca var. kansana, $\times\ ^2/_7$. **2.** Asclepias amplexicaulis, $\times\ ^2/_7$. **3.** Asclepias Meadii, $\times\ ^2/_7$; *a.* Lower half of stem. **4.** Asclepias verticillata, $\times\ ^2/_7$. **5.** Asclepias stenophylla, $\times\ ^2/_7$; *a.* Flower, $\times\ 1^5/_7$; After Gleason, The New York Botanical Garden. **6.** Cynanchum laeve, $\times\ ^2/_7$; After Gleason, details from Small, The New York Botanical Garden. **7.** Asclepias hirtella, $\times\ ^2/_7$; *a.* Flower, $\times$ 10.

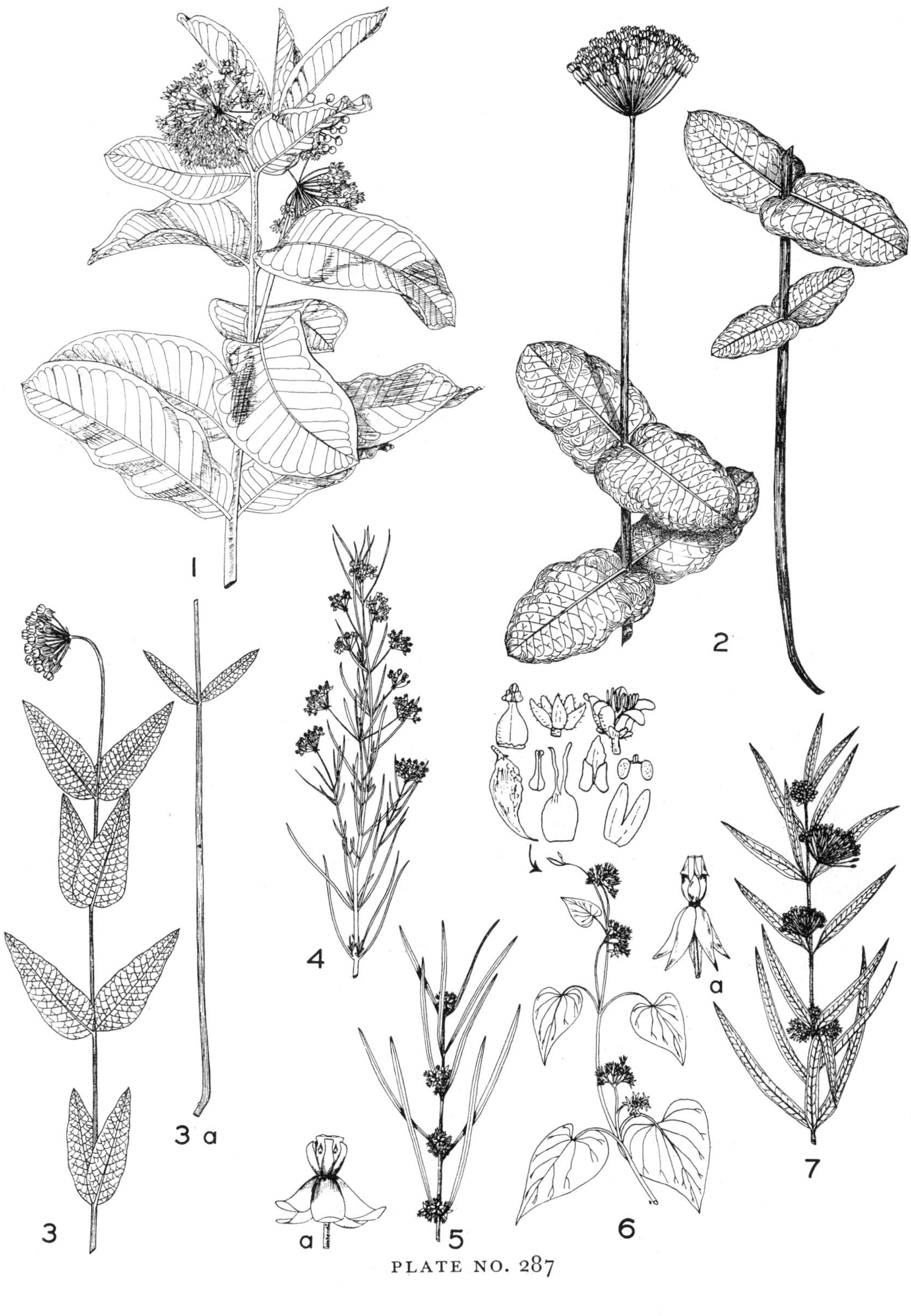


PLATE NO. 287

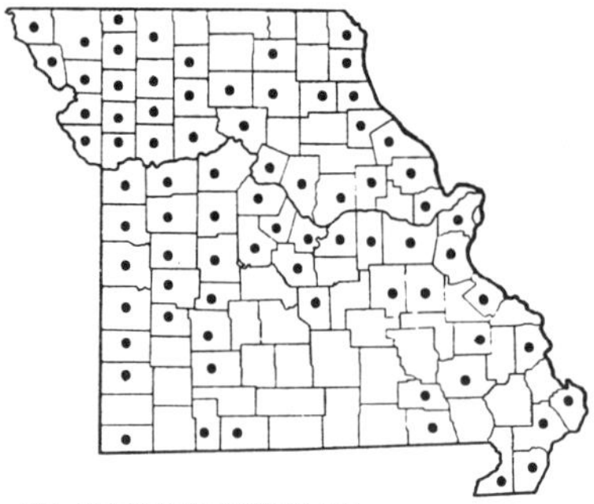

1766 Cynanchum laeve (Angle-pod)

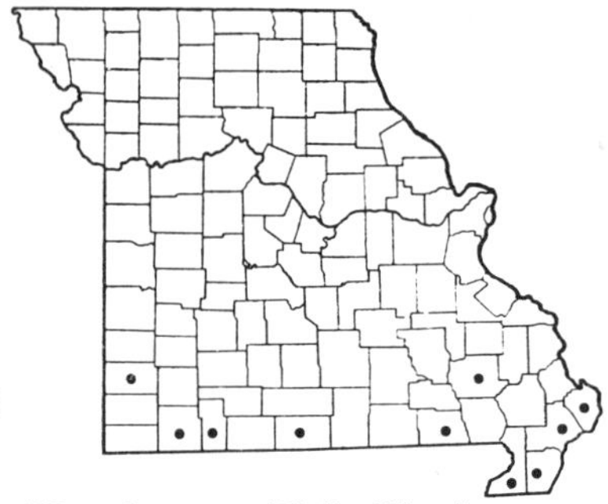

1767 Matelea gonocarpa (Climbing Milkweed)

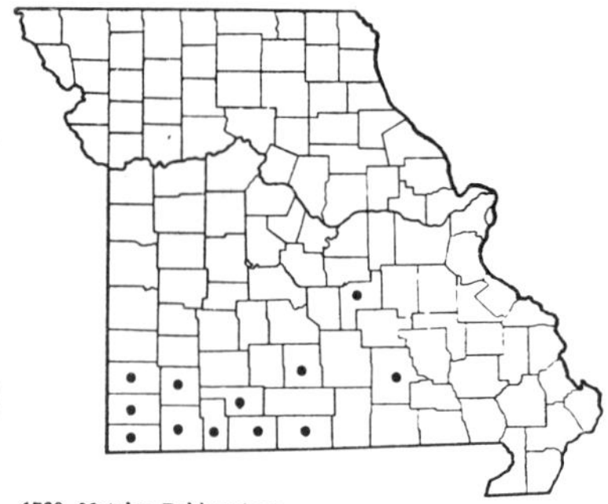

1768 Matelea Baldwyniana

in his recent monograph, and the species should be eliminated from the Missouri flora.

**Asclepias speciosa** Torr.

This species was credited to the state by Palmer and Steyermark in their *Annotated Catalogue* (p. 621) on the basis of a specimen from St. Louis County. This collection should be referred instead to *A. syriaca* var. *kansana*. Woodson does not cite any material from Missouri in his recent monograph of the genus.

## 2. **Cynanchum** L.

(*Ampelamus* Raf. [G, BB])

**Cynanchum laeve** (Michx.) Pers. Angle-pod
Map 1766

Also called Sand Vine, Climbing Milkweed, Blue Vine, Honey Vine.

*Ampelamus albidus* (Nutt.) Britt. [G, BB]

*Gonolobus laevis* Michx. [P & S]

Flowers July–September.

Occurs in moist alluvial woodland, thickets, cultivated fields, fence rows, roadsides, and along railroads. Throughout Missouri, but not recorded from most of the southern and central Ozark counties.

Ranges from Georgia and Alabama to Texas, north to Pennsylvania, West Virginia, Ohio, Indiana, Illinois, Iowa, and Nebraska.

Woodson (Ann. Mo. Bot. Gard. 28: 208–13. 1941) has given logical reasons for placing this species under *Cynanchum*, and the present treatment is in accord with his studies.

This vine has dark green leaves and fragrant small white flowers. The plant has been recommended by beekeepers as an excellent honey plant. It is likely to become well established and later behave as a weed, making it difficult to eradicate. In areas where it infests fields, it interferes with crop yields and is considered a more vicious pest than bindweed, and more difficult to get rid of.

## 3. **Matelea** Aubl.

(*Gonolobus* [of G, BB, Steyerm.])

(*Vincetoxicum* [of P & S])

The treatment here presented follows for the most part the interpretation of Woodson (Ann. Mo. Bot. Gard. 28: 217–37. 1941), and more particularly that of Shinners (Field & Lab. 18: 73–76. 1950).

a. Calyx glabrous (without hairs) or only slightly hairy near tip of lobes; flower-stalks (pedicels) glabrous (without hairs); surface of fruit smooth, angled and ribbed, not covered with short hard points; crown shorter than anther-column; leaves oblong-ovate or broadly oblong . . 1. M. GONOCARPA
a. Calyx hairy throughout; flower-stalks (pedicels) hairy; surface of fruit covered with short hard points, not angled and ribbed; crown as long as or longer than anther-column; leaves round-ovate to ovate . . . . . . . . . . . . . . . . . . . . . . . . . . *b*
 b. Corolla-lobes whitish . . . . . . . . . . . . . . . . . . . 2. M. BALDWYNIANA
 b. Corolla-lobes brownish-purple or purple with greenish-brown . . . . . . . . . . . . . . *c*

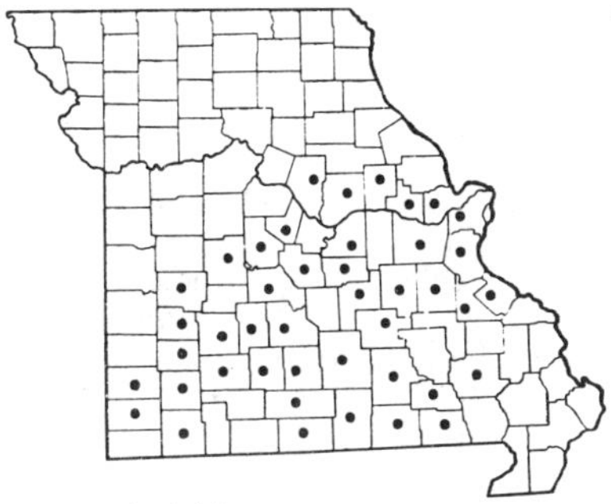
1769 Matelea decipiens

1770 Matelea obliqua

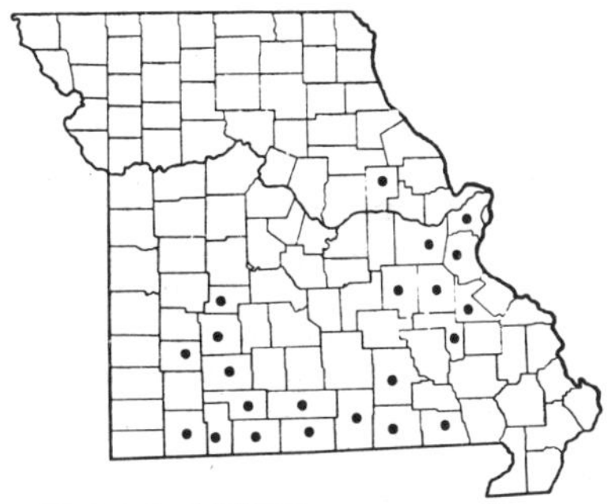
1771 Evolvulus Nuttallianus

c. Corolla-lobes 1.5–2 mm. broad; all the lobes of the crown truncate (as if cut straight across) or slightly wavy-edged at summit, the crown only as long as or slightly longer than the anther-column . . . . . . . . . . . . . . . . . . . . . . 4. M. OBLIQUA

c. Corolla-lobes 3–6 mm. broad; every other lobe of the crown prolonged into 2 short-pointed (acute) or long-pointed (acuminate) tips, the long 2-toothed lobes longer than the anther-column . . . . . . . . . . . . . . . . . . . . . . 3. M. DECIPIENS

1. **Matelea gonocarpa** (Walt.) Shinners
Climbing Milkweed Map 1767
Also called Angle-pod.
*Gonolobus gonocarpos* (Walt.) Perry [G, BB, Woodson]
*Vincetoxicum gonocarpos* Walt. [P & S]
Flowers July–August.

Occurs in rocky woods and thickets along streams. Southern Missouri north to Mississippi, Wayne, and Jasper counties.

Ranges from Georgia to Texas, north to Virginia, Indiana, Illinois, Missouri, and Oklahoma.

2. **Matelea Baldwyniana** (Sweet) Woodson
Climbing Milkweed Map 1768
Also called Angle-pod.
*Gonolobus Baldwynianus* Sweet [G, BB, Steyerm.]
*Vincetoxicum Baldwinianus* (Sweet) Britton [P & S]
Flowers late May–June.

Occurs in open rocky woods and thickets along streams. Ozark region, commonest in the southwestern portion from Jasper and Wright counties east to Phelps and Shannon counties.

Ranges from Georgia and Alabama to Missouri and Oklahoma.

3. **Matelea decipiens** (Alex.) Woodson
Climbing Milkweed Map 1769
Also called Angle-pod.
*Gonolobus decipiens* (Alex.) Perry [G, BB]
*Gonolobus carolinensis* [of Steyerm.], not (Jacq.) Schultes
*Vincetoxicum carolinense* [of P & S], not (Jacq.) Britton
Flowers late May–June.

Occurs in rocky open woods, base of and ledges of bluffs, low or upland thickets, and glades, often in valleys and thickets along streams. Common throughout the Ozark region of southern and central Missouri, north to St. Charles, Warren, Montgomery, Callaway, Boone, Moniteau, Morgan, Benton, St. Clair, and Jasper counties.

Ranges from Louisiana to Oklahoma, north to South Carolina, Maryland, and Missouri.

This species has been separated from *Matelea carolinensis* (Jacq.) Woodson on the basis of the oblong-conical instead of bluntly ovoid flower-buds, by the ascending instead of rotate limb of the corolla, and by the longer corolla-lobes.

4. **Matelea obliqua** (Jacq.) Woodson
Climbing Milkweed Map 1770
Also called Angle-pod.
*Gonolobus obliquus* (Jacq.) Schultes [G, BB, Steyerm.]
*Vincetoxicum obliquum* (Jacq.) Britton [P & S]
Flowers May–June.

Occurs in rocky woods and thickets. Rare and scattered in the eastern Ozark region from St. Louis County southwest to Shannon and Howell counties.

Ranges from Georgia and Tennessee to Illinois and Missouri, north to Maryland, Virginia, Pennsylvania, Ohio, and Indiana.

## Order **POLEMONIALES**

### Fam. **CONVOLVULACEAE** (Morning Glory Family)

a. Parasitic twining plants with string-like yellow, orange, or brown stems; leaves not present or reduced to minute scales . . . . . . . . . . . . . . . . . . . . . . . . 4. CUSCUTA
a. Plants with well-developed leaves, not parasitic, with green stem, leaves, and calyx, except when covered by dense gray hairiness . . . . . . . . . . . . . . . . . . . . . . . . *b*
  b. Main leaves of the stem small, 1–2 cm. long; leaves neither heart-shaped at base nor lobed; each branch of style split, appearing like 4 stigmas. . . . . . . . . . . . . 1. EVOLVULUS
  b. Main leaves of the stem larger, 2.5–10 cm. or more long; leaves either heart-shaped at base or lobed; style not split or only slightly split . . . . . . . . . . . . . . . . . . *c*
    c. No bracts present at the very base of or surrounding the calyx . . . . . . . . . . . *d*
      d. Calyx 3–5 mm. long, the sepals obtuse (blunt) or rounded at tip; style slightly split at tip into 2 thread-like stigmas . . . . . . . . . . . . . . . . . 3. CONVOLVULUS ARVENSIS
      d. Calyx 9–25 mm. long, the sepals rounded to long-pointed (acuminate) at tip; style enlarged or head-like at tip into an undivided or 2- or 3-lobed stigma. . . . . . . . . 2. IPOMOEA
    c. Two small leaf-like bracts present at the very base of or surrounding the calyx. . 3. CONVOLVULUS

### 1. **Evolvulus** L.

Native plant of dry rocky limestone glades of the Ozark region north to Montgomery County; stems erect or stiffly ascending; stalks (pedicels) of flowers or fruits scarcely showing or much shorter than the leaves accompanying them. . . . . . . . . . . . . . . . . . . . . 1. E. NUTTALLIANUS
Introduced plant, known only from St. Louis County; stems lying on the ground (prostrate) or loosely ascending; stalks (pedicels) of flowers or fruits conspicuous and longer than the leaves accompanying them . . . . . . . . . . . . . . . . . . . . . . . . . . 2. E. ALSINOIDES

1. **Evolvulus Nuttallianus** R. & S. Map 1771
*Evolvulus pilosus* Nutt. [BB]
*Evolvulus argenteus* Pursh, not R. Br. [P & S]
Flowers late April–July.

Occurs on dry rocky limestone glades and bald knobs. Ozark region north to St. Louis, Montgomery, Hickory, and Dade counties; not recorded from a large sector of the central Ozarks.

Ranges from North Dakota to Montana and Colorado, south to Missouri, Kansas, Texas, and Arizona; recorded from Tennessee by Gleason.

The entire plant is covered with a dense appressed silky hairiness, giving it a dusty appearance. The wide funnel-shaped flowers are dark lavender and about 1 cm. across.

The thick deep root makes the plant extremely difficult to transplant. It may be raised from seed, however, and when established, is an ideal plant for the limestone rock garden.

2. **Evolvulus alsinoides** L. Map 1772
Flowers June–August.

Known only from St. Louis County, east-central Missouri (July 7, 1883, *Letterman*, in U. of Mo. Herb.).

Native to the warmer sections of the Old and New World and in the southern United States; accidentally introduced in Missouri.

No recent collection of this species has come to light in Missouri since Letterman's original discovery. The specimen collected by Letterman was originally mislabeled 'Breweria,' and undoubtedly is the basis for the report of that genus for Missouri.

### 2. **Ipomoea** L. Morning Glory

a. Corolla deep red or scarlet, the tube slender and abruptly widened at the summit; anthers and style projecting from the corolla-tube . . . . . . . . . . . . . . . . . . . . *b*
  b. Leaves divided into narrowly linear parallel lobes . . . . . . . . . . . 1. I. QUAMOCLIT
  b. Leaves heart-shaped, either entire (without teeth or lobes) or slightly lobed . . . . 2. I. COCCINEA

---

Plate no. 288. **1.** Matelea gonocarpa, × 2/7; *a.* Flower, × 4/7; *b.* Bud, × 4/7; After Gleason, details from Small, The New York Botanical Garden. **2.** Matelea decipiens, × 2/7; *a.* Flower, × 4/7; *b.* Bud, × 4/7; After Gleason, details from Small, The New York Botanical Garden. **3.** Matelea Baldwyniana, × 2/7. **4.** Matelea obliqua, × 2/7.

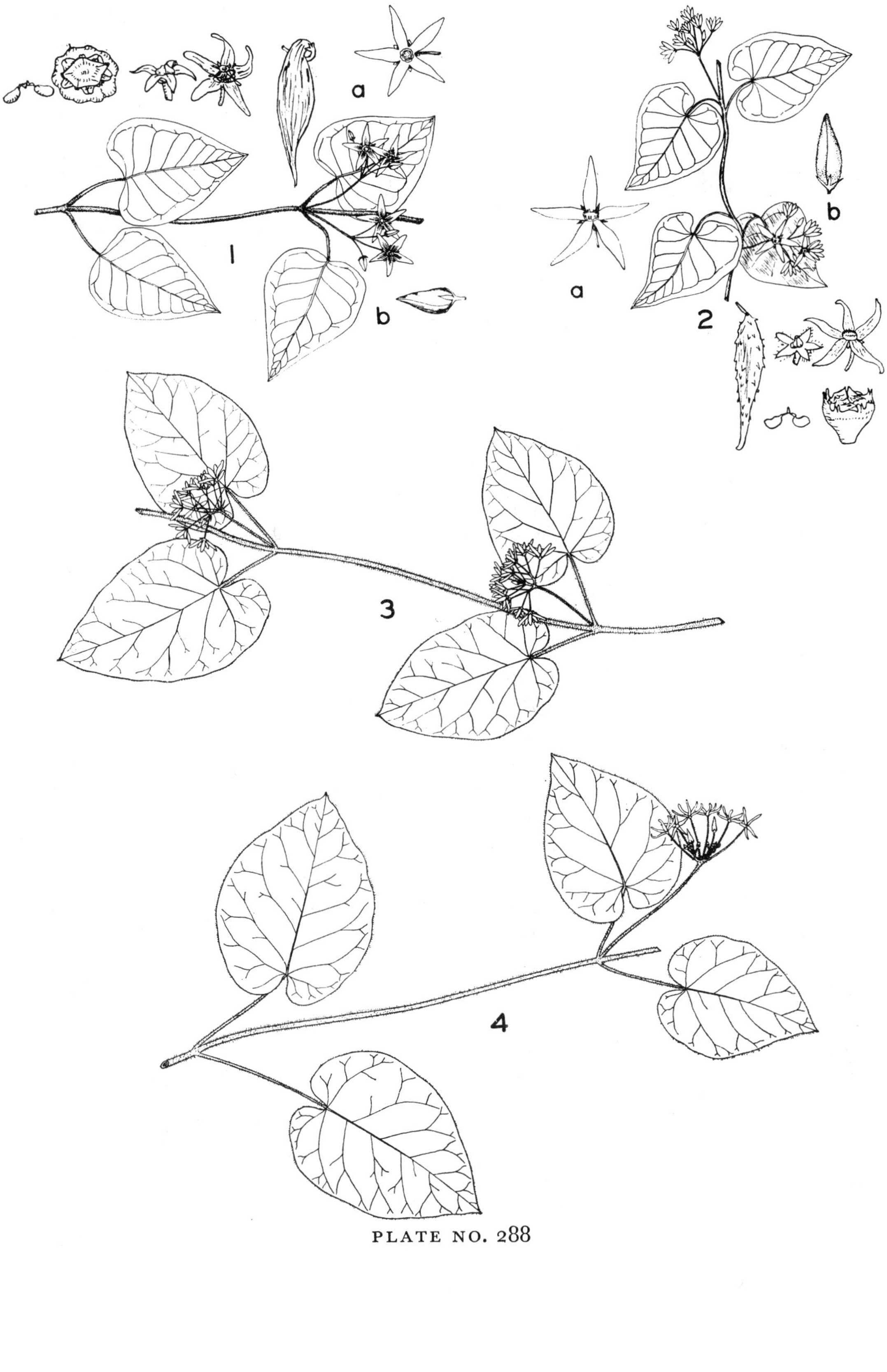


PLATE NO. 288

1772 Evolvolus alsinoides

1773 Ipomoea Quamoclit (Cypress Vine)

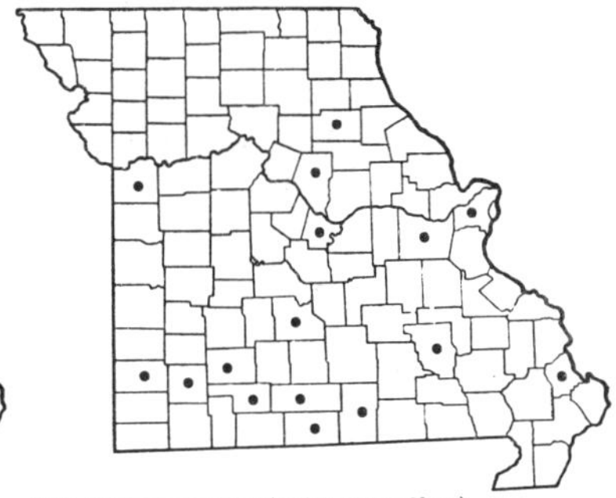

1774 Ipomoea coccinea (Red Morning Glory)

a. Corolla rose-colored, purple, blue, white, or white with a dark purple center, the tube funnel- or somewhat bell-shaped, gradually widened toward the summit; anthers and style remaining within and not projecting from the corolla-tube . . . . . . . . . . . . . . . . . . . . *c*
c. Sepals glabrous (without hairs) or nearly so, obtuse (blunt) or rounded at tip . . . 5. I. PANDURATA
c. Sepals hairy, with short- to long-pointed tip . . . . . . . . . . . . . . . . . *d*
d. Corolla small, 1.5–2 cm. long; fruit (capsule) hairy; seed glabrous (without hair); leaf-stalks (petioles) and flower-stems (peduncles) glabrous or sparsely hairy . . . . . . 6. I. LACUNOSA
d. Corolla showy, 3–7 cm. long; fruit (capsule) glabrous; seed more or less hairy; leaf-stalks (petioles) and flower-stems (peduncles) with backward-pointing hairs . . . . . . . . . . *e*
e. Sepals with long slender linear tips; leaves usually with 3 lobes, less frequently without any lobes; corolla 3–4.5 cm. long, blue with a white tube, changing to rose-purple; sepals 15–25 mm. long . . . . . . . . . . . . . . . . . . . . 3. I. HEDERACEA
e. Sepals with short-pointed (acute) or rather short-pointed (acutish) tips; leaves not lobed; corolla 4.5–7 cm. long, purple, bluish, rose-red, white, or variegated; sepals 10–15 mm. long . . . . . . . . . . . . . . . . . . . . . . 4. I. PURPUREA

1. **Ipomoea Quamoclit** L. Cypress Vine
Map 1773

*Quamoclit vulgaris* Choisy [BB]
*Quamoclit pennata* (Desr.) Bojer [Bailey]
Flowers August–October.

Rarely escaped from cultivation to waste ground, fence rows, or roadsides. Known only from Grundy and Greene counties.

Native of tropical America; planted as an ornamental vine and naturalized in the southern United States north to Virginia and Missouri.

2. **Ipomoea coccinea** L. Red Morning Glory
Map 1774

Also called Scarlet Starglory.
*Quamoclit coccinea* (L.) Moench. [BB]
Flowers July–October.

Occurs in alluvial low ground along muddy banks and gravel bars of streams, low thickets, fields, waste ground, and along railroads. Scattered in southern and central Missouri north to St. Louis, Monroe, Boone, and Jackson counties.

Native of tropical America; introduced and naturalized in the United States from Florida to Arizona, north to Massachusetts, New York, West Virginia, Ohio, Michigan, Illinois, Missouri, and Kansas.

3. **Ipomoea hederacea** (L.) Jacq. Blue Morning Glory
Map 1775

Also called Ivy-leaf Morning Glory.
Flowers July–October.

Occurs in fallow and cultivated fields, waste ground, and along railroads.

Two variations occur in Missouri:

Leaves with 3 lobes . . . . 3a. I. HEDERACEA var. HEDERACEA
Leaves not lobed . . . . . 3b. I. HEDERACEA var. INTEGRIUSCULA

3a. **Ipomoea hederacea** var. **hederacea**
Map 1775

*Ipomoea hederacea* (L.) Jacq. [G, BB, P & S]

Plate no. 289. **1.** Evolvulus Nuttallianus, × 2/5; Details from Small, The New York Botanical Garden. **2.** Ipomoea pandurata, × 2/5; *a, b.* Leaf variations. **3.** Evolvulus alsinoides, × 2/5; After Gleason, The New York Botanical Garden. **4.** Ipomoea hederacea, × 2/5 (Scribner's). **5.** Ipomoea lacunosa, × 2/5; After Gleason, details from Small, The New York Botanical Garden. **6.** Ipomoea purpurea, × 2/5; After Gleason, details from Small, The New York Botanical Garden.

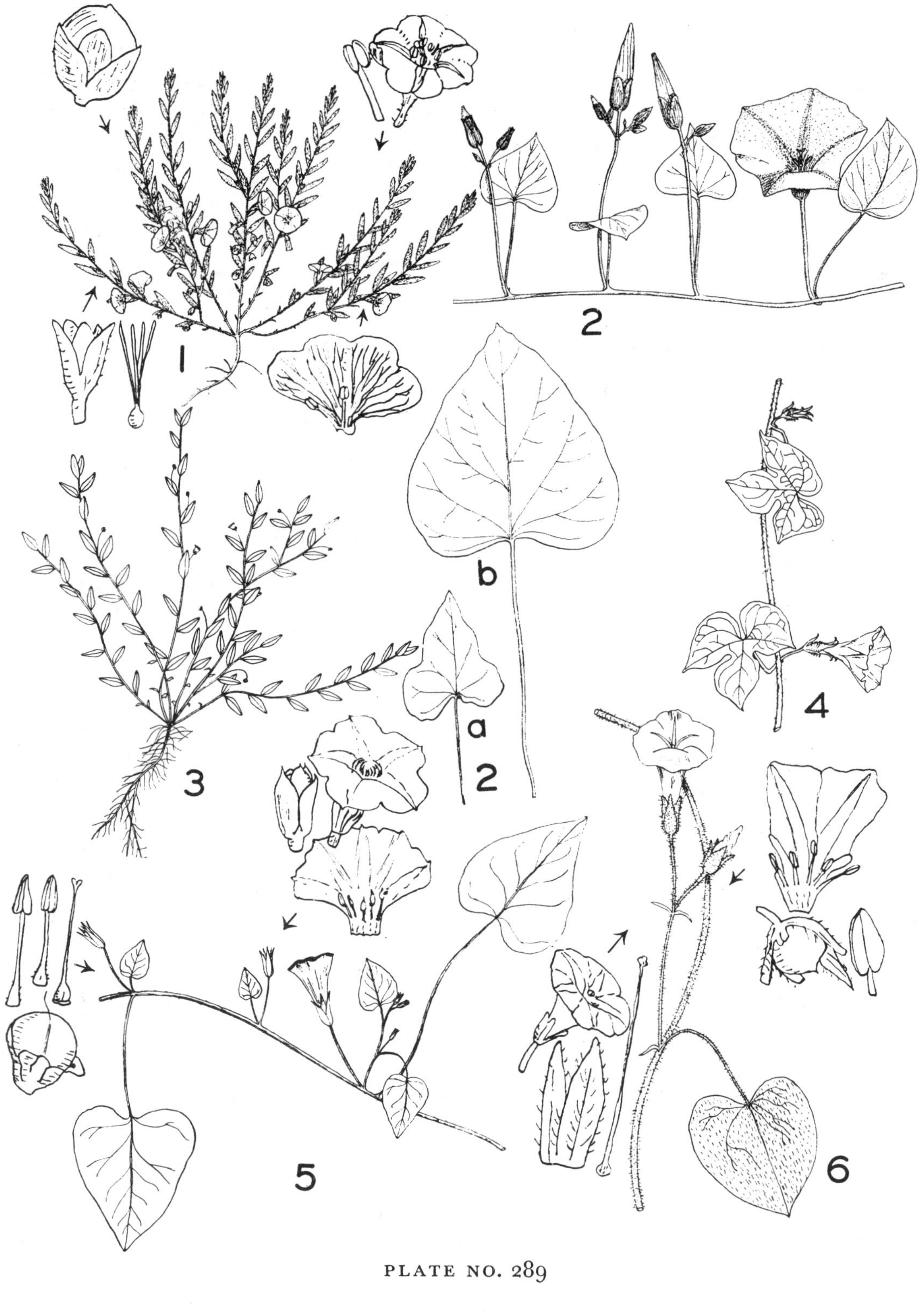


PLATE NO. 289

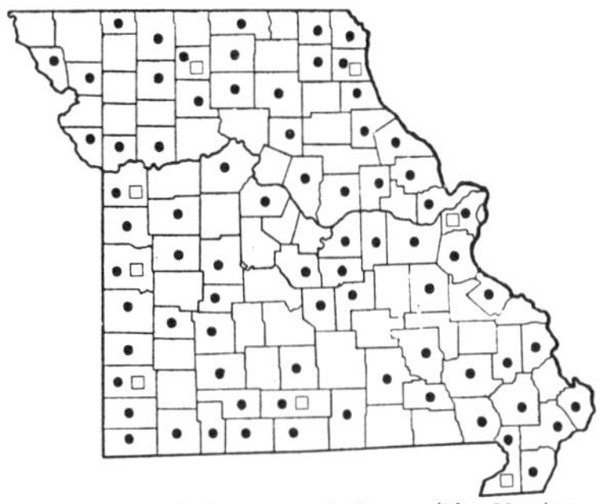

1775 ● Ipomoea hederacea var. hederacea (Blue Morning Glory)
1775 □ Ipomoea hederacea var. integriuscula

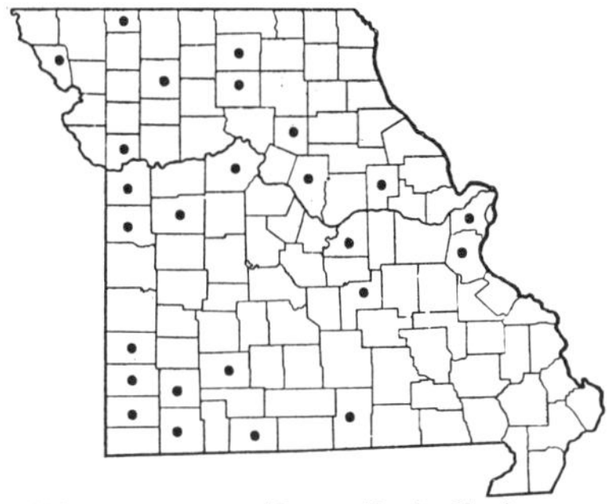

1776 Ipomoea purpurea (Common Morning Glory)

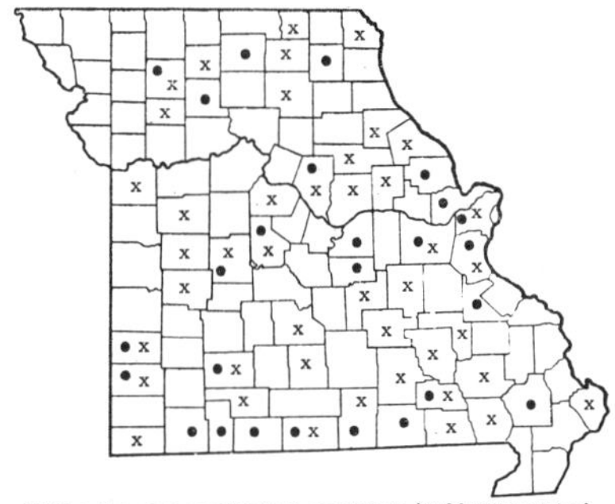

1777 ● Ipomoea pandurata f. pandurata (Wild Potato Vine)
1777 x Ipomoea pandurata f. leviuscula

This is the commoner variation and is found throughout Missouri, doubtless in every county.

Native of tropical America; naturalized in North America from Florida to Arizona, north to New England, New York, Ontario, Ohio, Indiana, Illinois, Minnesota, and North Dakota.

3b. **Ipomoea hederacea** var. **integriuscula** Gray Map 1775

Scattered throughout the state, and less common.

Ranges in the United States north to Maryland and Kansas.

4. **Ipomoea purpurea** (L.) Roth Common Morning Glory Map 1776

Flowers July–October.

Occurs in fallow and cultivated fields, waste ground, and along railroads. Scattered throughout Missouri, and much less frequently encountered than *I. hederacea*.

Native of tropical America; introduced and naturalized in North America from Nova Scotia to New York, Wisconsin, and Nebraska, south to Florida and Texas.

5. **Ipomoea pandurata** (L.) G. F. W. Mey. Wild Potato Vine Map 1777

Also called Man-of-the-earth.

Flowers late May–September.

Occurs in fallow and cultivated fields, thickets, alluvial or low ground along streams and borders of lakes, along roadsides, and railroads, Throughout Missouri.

Ranges from Florida to Texas, north to Connecticut, New York, Ontario, Ohio, Michigan, Illinois, Iowa, and Kansas.

Two variations are found in Missouri:

Leaves with short hairs on lower and/or upper surface . . . 5a. I. PANDURATA f. PANDURATA

Leaves glabrous (without hairs) on both lower and upper surfaces . . . . . 5b. I. PANDURATA f. LEVIUSCULA

5a. **Ipomoea pandurata** f. **pandurata** Map 1777

*Ipomoea pandurata* (L.) G. F. W. Mey. [G, BB, P & S, Steyerm.]

Scattered throughout Missouri, but less common than the next form.

5b. **Ipomoea pandurata** f. **leviuscula** Fern. Map 1777

*Ipomoea pandurata* var. *rubescens* Chois. [BB, P & S, Steyerm.]

This is the commoner variation, found throughout Missouri, but not recorded thus far from any of the counties in the extreme northwestern section of the state.

The large vertical tuberlike root of this morning glory may attain two feet in length and weighs 4–11 kilograms.

Although the fresh root is said to be somewhat poisonous because of its purgative effects, it is recorded as having been used as a cooked vegetable by some North American Indians. The long trailing stems may attain 15 feet or more in length or may climb bushes 10 feet or so tall.

This may become a bad weed in cultivated fields, and is difficult to eliminate because of the large fleshy roots.

6. **Ipomoea lacunosa** L. Small White Morning Glory Map 1778

Flowers July–October.

Occurs in moist alluvial soils and gravel bars along streams, thickets, moist places in natural prairie, wet meadows, waste ground, roadsides, and along railroads. Common throughout southern and central Missouri, north locally in northern Missouri to Clark,

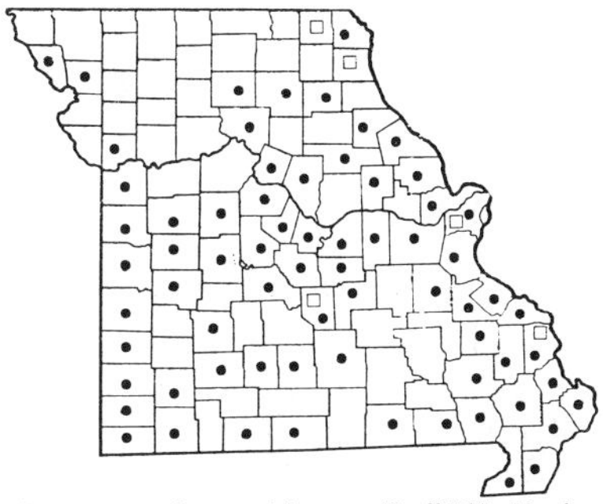

1778 • Ipomoea lacunosa f. lacunosa (Small White Morning Glory)
1778 □ Ipomoea lacunosa f. purpurata

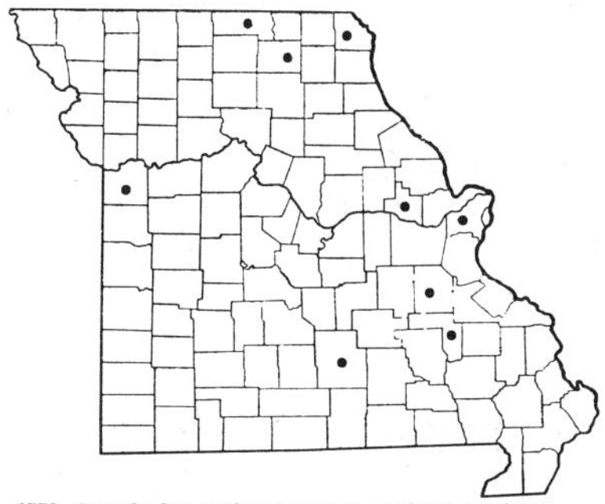

1779 Convolvulus spithamaeus var. spithamaeus (Low Bindweed)

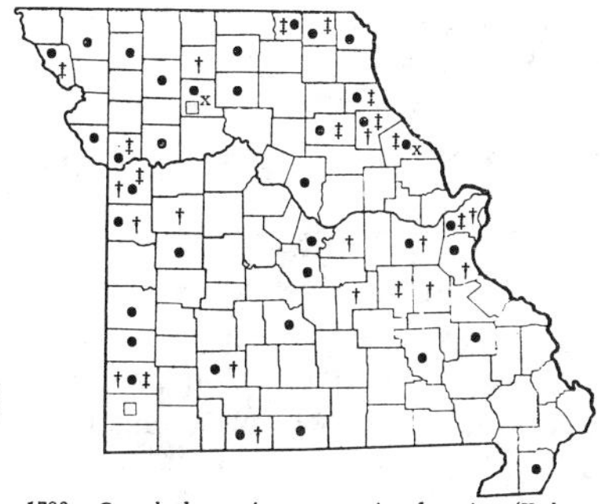

1780 • Convolvulus sepium var. sepium f. sepium (Hedge Bindweed)
1780 □ Convolvulus sepium var. sepium f. coloratus
1780 x Convolvulus sepium var. sepium f. malacophyllus
1780 † Convolvulus sepium var. fraterniflorus
1780 ‡ Convolvulus sepium var. repens

Shelby, Macon, Linn, Andrew, and Holt counties.

Ranges from Florida to Texas, north to New Jersey, Pennsylvania, Ohio, Indiana, Illinois, Missouri, and Kansas.

Two variations are encountered in Missouri:

Corolla white . . . 6a. I. LACUNOSA f. LACUNOSA
Corolla lavender or purplish . . 6b. I. LACUNOSA f. PURPURATA

6a. **Ipomoea lacunosa** f. **lacunosa** Map 1778

*Ipomoea lacunosa* L. [G, BB, P & S]

This is the commoner variation in Missouri.

6b. **Ipomoea lacunosa** f. **purpurata** Fern. Map 1778

Infrequently encountered within the range of the common variation, and thus far known from Clark, Lewis, St. Louis, Cape Girardeau, and Pulaski counties.

### 3. **Convolvulus** L. Bindweed

a. No leaf-like bracts at the very base of or surrounding the calyx; sepals 3–5 mm. long . . 4. C. ARVENSIS
a. Two leaf-like bracts at the very base of or surrounding the calyx; sepals or bracts 8–35 mm. long. . . *b*
  b. Stem mainly erect or ascending, at least in the lower half, not twining or trailing, usually 1.5–4 dm. (varying from 0.7 to 5) tall; leaf-stalks (petioles) very short and usually less than $\frac{1}{4}$ as long as the leaf-blade . . . . . . . . . . . . . . . . . . . . 1. C. SPITHAMAEUS
  b. Stem twining, climbing, or trailing, elongated, usually 6–30 dm. long; leaf-stalks (petioles) from $\frac{1}{4}$ to as long as length of leaf-blade . . . . . . . . . . . . . . . . . . *c*
    c. Flowers double (all parts petal-like), corolla 2.5–4.5 cm. long; leaf-stalks (petioles) $\frac{1}{4}$–$\frac{1}{3}$ as long as leaf-blade; stems, leaves, and bracts more or less short-hairy . . 3. C. PELLITUS f. ANESTIUS
    c. Flowers single (with corolla, stamens, and pistil), corolla 4–8 cm. long; leaf-stalks (petioles) from more than $\frac{1}{2}$ as long as to equaling length of leaf-blade; stems, leaves, and bracts glabrous (without hairs) or sparsely hairy. . . . . . . . . . . . . . . . 2. C. SEPIUM

1. **Convolvulus spithamaeus** L. var. **spithamaeus**
Low Bindweed Map 1779

Sometimes called Dwarf Morning Glory, a name also applied to *Convolvulus tricolor* L.

*Convolvulus spithamaeus* L. [G, BB, P & S, Steyerm.]

Flowers May–July.

Occurs in rocky or dry open woodland, on upland slopes and ridges, prairie openings in woodland, and banks of ravines. Rare and scattered in the state, where known from Clark and Putnam counties south to Iron and Texas counties west to Jackson County.

Ranges from Quebec and Maine to Ontario and Minnesota, south to Virginia, D.C., Tennessee, Kentucky, and Missouri.

Another variety, var. *pubescens* (Gray) Fern., occurring in the southeastern states, has longer-petioled, hairier leaves on a more elongated stem.

Missouri specimens vary from stems and leaves nearly glabrous to more or less hairy. This little Convolvulus is quite showy when in flower, with the large usually white corollas scattered along the stem. However, if introduced into a rock garden, the plant may become weedlike, as is characteristic of other members of the genus.

2. **Convolvulus sepium** L. Hedge Bindweed Map 1780

Also called Wild Morning Glory, Rutland Beauty. Flowers May–September.

Occurs in alluvial moist ground of valleys, along streams, ponds, and sloughs, fields, meadows, waste ground, roadsides, and along railroads. Throughout Missouri.

Fernald's interpretation of the variations of this species, as given in the eighth edition of *Gray's Manual* (p. 1181), has been followed in the present treatment.

The following variations occur in Missouri:

a. Main portion of the leaf-blade of the upper and middle leaves 1/5–rarely 1/2 as broad as long, 1–3 cm. broad, lanceolate to narrowly ovate . . . . . 2e. C. SEPIUM var. REPENS

a. Main portion of the leaf-blade of the upper and middle leaves 2/3 as broad as or broader than long, 3–10 cm. broad, broadly ovate to ovate-triangular to broadly oval . . . . *b*

b. The space (sinus) between the two lobes or basal extensions of the leaf squarish (quadrate) with nearly parallel sides . . . . 2d. C. SEPIUM var. FRATERNIFLORUS

b. The space (sinus) between the two lobes or basal extensions of the leaf U- or V-shaped, with sides sloping or spreading outwards from the base of the leaf-blade . . . . *c*

c. Stems, leaf-stalks (petioles), and surfaces of the leaf-blades soft-hairy . 2c. C. SEPIUM var. SEPIUM f. MALACOPHYLLUS

c. Stems, leaf-stalks (petioles), and surfaces of the leaf-blades glabrous (without hairs) or nearly so . . . . . . . *d*

d. Corolla white . . . . 2a. C. SEPIUM var. SEPIUM f. SEPIUM

d. Corolla rose-colored . . 2b. C. SEPIUM var. SEPIUM f. COLORATUS

2a. **Convolvulus sepium** var. **sepium** f. **sepium** Map 1780

*Convolvulus sepium* L. [G, P & S, Steyerm.]

*Convolvulus sepium* var. *americanus* Sims [BB, Steyerm., Tryon]

*Convolvulus sepium* var. *communis* Tryon [BB, Tryon] in part

Scattered throughout Missouri, probably to be found in every county when intensively collected.

Ranges from Newfoundland to British Columbia, south to Florida, Alabama, Missouri, Oklahoma, New Mexico, and Oregon.

2b. **Convolvulus sepium** var. **sepium** f. **coloratus** Lange Map 1780

*Convolvulus sepium* f. *coloratus* Lange [G]

*Convolvulus sepium* var. *communis* Tryon, in part [Tryon]

Known from Livingston (*Sparling 1314*) and Newton counties.

2c. **Convolvulus sepium** var. **sepium** f. **malacophyllus** Fern. Map 1780

*Convolvulus sepium* f. *malacophyllus* Fern. [G]

Known from Pike (*Etter 243*) and Livingston (*Sparling 916*) counties.

2d. **Convolvulus sepium** var. **fraterniflorus** Mackenz. & Bush Map 1780

Scattered in Missouri.

Ranges from Pennsylvania to North Dakota, south to Virgina, Kentucky, and Arkansas.

2e. **Convolvulus sepium** var. **repens** (L.) Gray Map 1780

*Convolvulus sepium* var. *pubescens* (Gray) Fern. [P & S, Steyerm.]

Scattered in Missouri.

Ranges from Florida to Texas, north to New Brunswick, Ohio, Indiana, and Wisconsin.

The treatment of this species by Gleason in the *New Ill. Fl.* (3: 90. 1952) follows the work of Tryon (Rh. 41: 418–22. 1939). These treatments are at variance with the present one, which follows the studies by Fernald. Specimens, which in the present treatment key out to *C. sepium* var. *sepium* f. *malacophyllus*, would key in part to *C. sepium* var. *fraterniflorus* in Gleason's and Tryon's works, and specimens which in the present treatment key out to *C. sepium* var. *fraterniflorus*, would key in part to *C. sepium* var. *americanus* in Gleason's and Tryon's works.

The roots of this bindweed are reputed to be poisonous to swine. As in the case of *C. arvensis*, this species can become an objectionable weed in many areas where it invades cultivated fields.

3. **Convolvulus pellitus** Ledeb. f. **anestius** Fern. Map 1781

Plate no. 290. **1.** Convolvulus spithamaeus, × 2/5. **2.** Convolvulus sepium, × 2/5; *a.* Leaf variation; Details from Small, The New York Botanical Garden. **3.** Convolvulus pellitus, × 2/5. **4.** Convolvulus arvensis, × 2/5; Details from Small, The New York Botanical Garden.

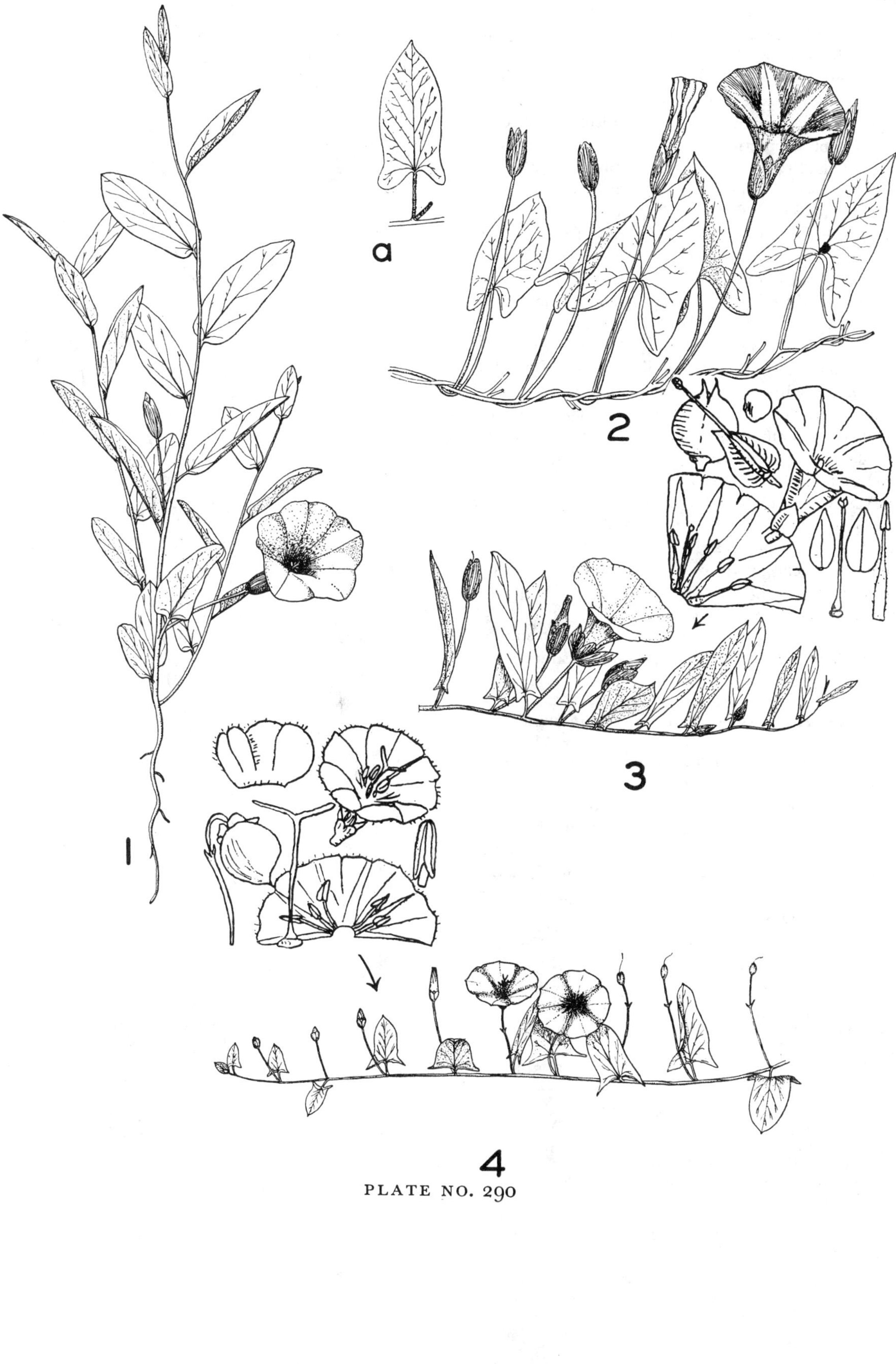


PLATE NO. 290

*Convolvulus japonicus* of many authors, not *C. japonicus* Thunb. [BB, P & S, Steyerm.]

Flowers May–September.

Occurs in waste and cultivated ground, usually near dwellings, sometimes along roadsides. Scattered in parts of northern and central Missouri, where known from St. Louis, Gasconade, Shelby, Caldwell, Johnson, Jackson, and Platte counties.

Introduced from China to England, into the United States, where naturalized as an escape from cultivation from Maine to Michigan, south to Virgina, Tennessee, and Missouri.

This double form of morning glory has pink flowers. As Fernald pointed out (Rh. 51 : 73–75. 1949), this has been mistaken for the quite different *Convolvulus japonicus* Thunb., which is also a double form, but is glabrous with longer leaves with 10–12 pairs of conspicuous side nerves, longer, more cordate-based bracts up to 3 cm. long surrounding the calyx, and a corolla 4–8 cm. long. In *C. pellitus* f. *anestius* the leaves are much smaller, more or less truncate at the base, with 3–6 pairs of side nerves, bracts 1–2.5 cm. long only gradually rounded at base, and a corolla 4–4.5 cm. long. Both species are quite showy and deserve to be more widely grown for the showy double flowers. *Convolvulus japonicus* is usually known as Japanese Morning Glory or German Rose.

The single form of *C. pellitus* f. *pellitus*, with stamens and style present, has not been found thus far in Missouri.

4. **Convolvulus arvensis** L. Field Bindweed
Map 1782

Also known as Small Bindweed.

Flowers May–September.

Occurs along open banks, waste ground, roadsides, and along railroads. Throughout Missouri.

Native of Europe and Asia; introduced and naturalized in North America from Quebec to Nova Scotia and Ontario to British Columbia, south to Florida and California.

The following variations occur in Missouri:

Lobes at base of leaf-blades acutely pointed; leaf-blade arrowhead-shaped, triangular or broadly ovate-triangular . . 4a. C. ARVENSIS f. ARVENSIS
Lobes at base of leaf-blades rounded; leaf-blade broad and heart-shaped . . . 4b. C. ARVENSIS f. CORDIFOLIUS

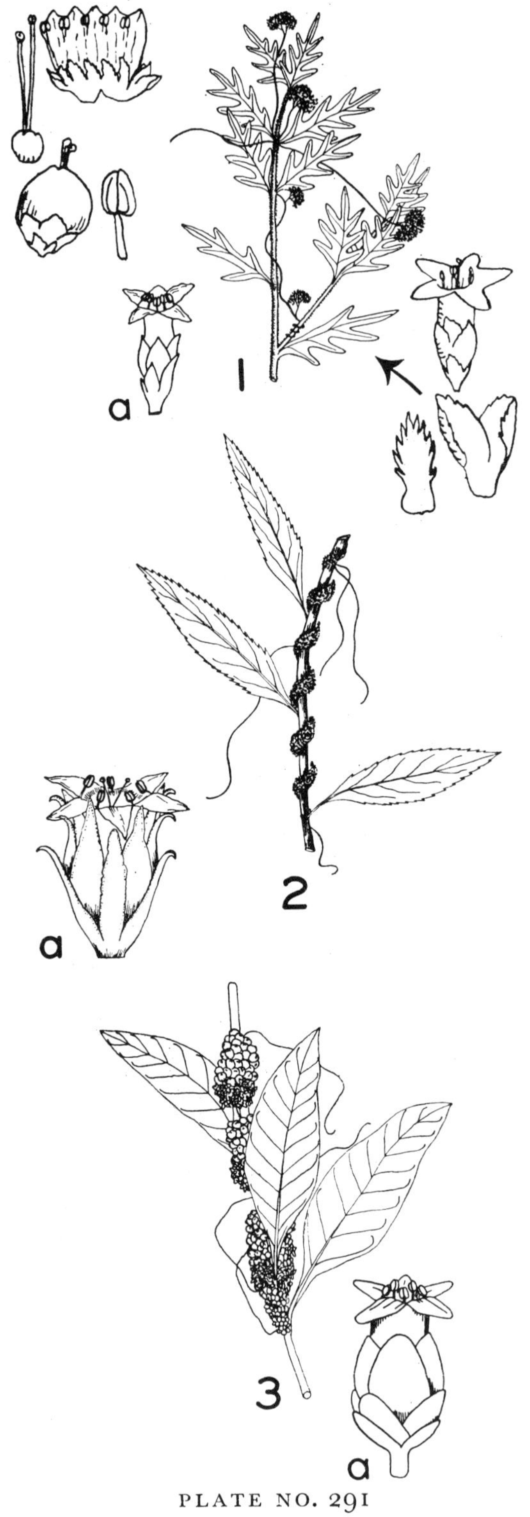


PLATE NO. 291

Plate no. 291. **1.** Cuscuta cuspidata, × 2/5, habit; *a.* Flower, × 3 1/5; Details from Small, The New York Botanical Garden. **2.** Cuscuta glomerata, × 2/5, habit; *a.* Flower, × 3 1/5. **3.** Cuscuta compacta, × 2/5, habit; *a.* Flower, × 3 1/5. All after Gleason, The New York Botanical Garden.

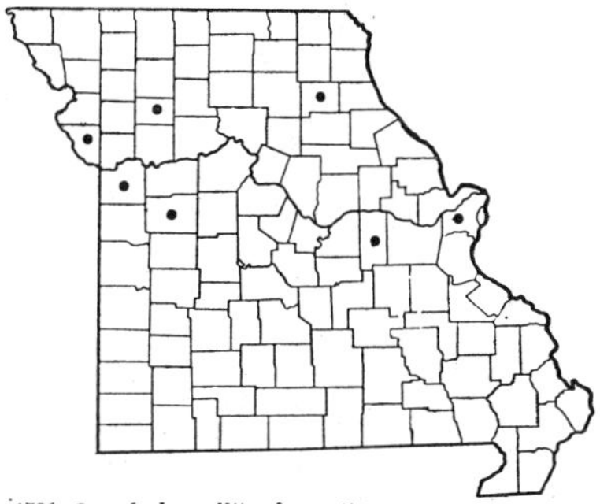

1781 Convolvulus pellitus f. anestius

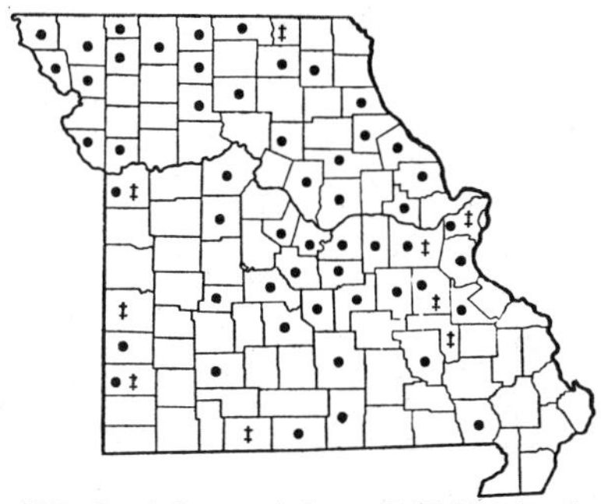

1782 • Convolvulus arvensis f. arvensis (Field Bindweed)
1782 ‡ Convolvulus arvensis f. cordifolius

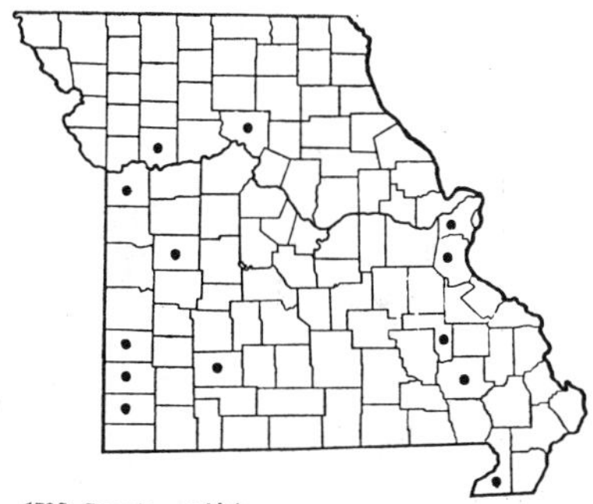

1783 Cuscuta cuspidata

4a. **Convolvulus arvensis** f. **arvensis** Map 1782

*Convolvulus arvensis* L. [G, BB, P & S, Steyerm.]

This is the commoner variation in Missouri.

4b. **Convolvulus arvensis** f. **cordifolius** Map 1782

*Convolvulus arvensis* var. *obtusifolius* Chois. [P & S, Steyerm.]

Scattered within the range of typical f. *arvensis*.

Another variation, f. *auriculatus* Desr., distinguished by the linear-oblong or lanceolate leaf-blades with acute lobes, has not been recorded from Missouri thus far.

The roots and rootstocks of this species are reported to be poisonous to swine. They have purgative effects when eaten. In some states this is considered the worst weed of all, infesting millions of acres of cultivated fields throughout the country. Flowers vary from white to pink.

## 4. **Cuscuta** L. Dodder

Identification of plants of this family should be attempted only with the aid of a magnifying lens of at least 9 ×. While the different species are parasitic on certain kinds and species of plants, with the host-parasite relationship a constant and readily identifiable one, use must be made of floral and fruit characters for certain identification.

a. Sepals distinct to their base; 1, 2, or more bracts at the base of calyx . . . . . . . . . . *b*
  b. Flowers on stalks (pedicels) in loose inflorescences; seeds 1.4 mm. long. . . . . . . 1. C. CUSPIDATA
  b. Flowers sessile (without stalks) in thick, rope-like masses or densely compact clusters; seeds 1.7–2.6 mm. long . . . . . . . . . . . . . . . . . . . . . . . . . . *c*
    c. Bracts at base of sepals and sepals rounded at tip, oval or nearly round, appressed (more or less flattened or erect); ovary and fruiting capsule globose-ovoid; seeds 2–2.6 mm. long . 3. C. COMPACTA
    c. Bracts at base of sepals and sepals slender-pointed, narrowly oblong or lanceolate, recurving-spreading at the tips; ovary and fruiting capsule flask-shaped; seeds 1.6–1.8 mm. long . . 2. C. GLOMERATA
a. Calyx with the sepals united, forming a calyx-tube and calyx-lobes; no bracts present at base of the calyx . . . . . . . . . . . . . . . . . . . . . . . . . . . . *d*
  d. Calyx and corolla with usually 5 lobes . . . . . . . . . . . . . . . . . . . . . *e*
    e. Flowers fleshy, the outer surface with tiny, wart-like or bump-like projections (papillate or granular); seeds 1.7 mm. long; flowers up to 5 mm. long. . . . 6. C. INDECORA var. NEUROPETALA
    e. Flowers with a smooth outer surface, or at least not with bumpy or wart-like projections; seeds 1–1.5 mm. long; flowers 1.5–4 mm. long . . . . . . . . . . . . . . . . . . *f*
      f. Lobes of corolla obtuse (blunt) at tip; ovary and capsule (fruit) with a thickened summit, the fruit rather pointed at summit; the dried or withered corolla forms a cap over the fruit; flowers in loose clusters. . . . . . . . . . . . . . . . . . 7. C. GRONOVII
      f. Lobes of corolla short- to long-pointed (acute to acuminate); ovary and capsule (fruit) with a thin, not thickened, summit, the fruit depressed or rounded, not pointed, at summit; the dried or withered corolla remaining at base of capsule; flowers in rather dense clusters . . *g*
        g. Lobes of calyx rather rounded or obtuse at tip, ovate or somewhat round, overlapping

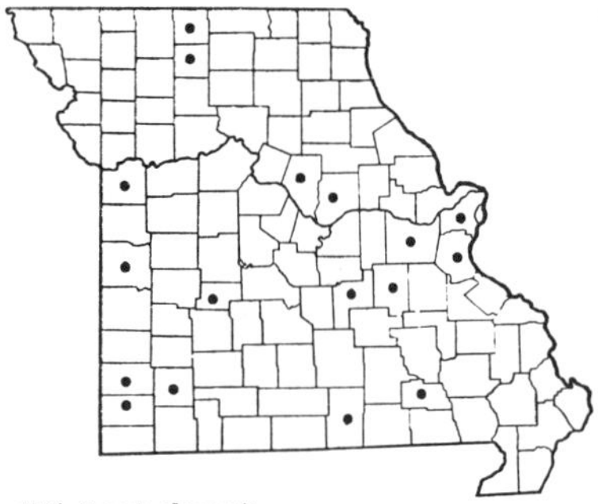

1784 Cuscuta glomerata

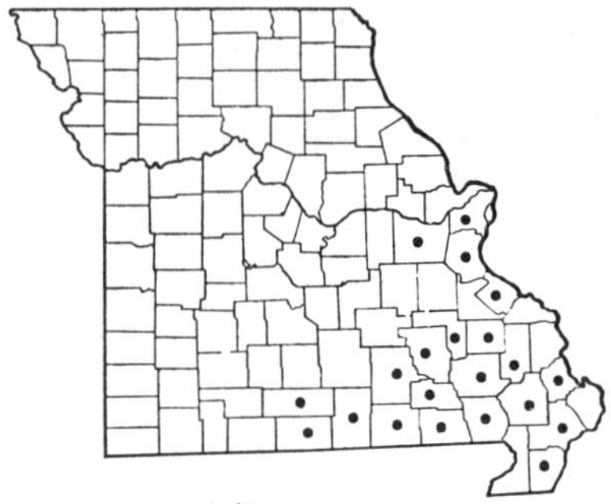

1785 Cuscuta compacta

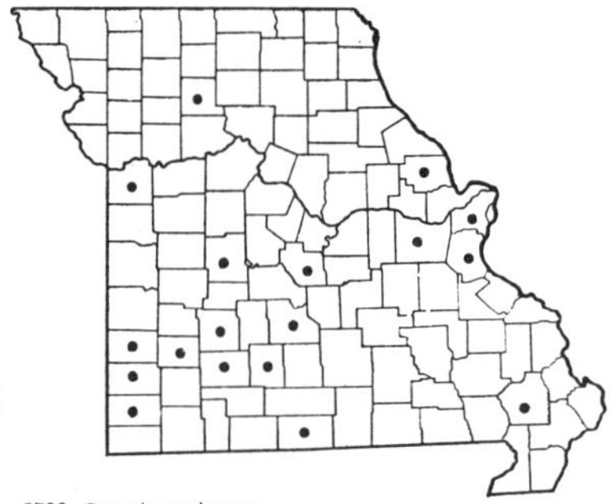

1786 Cuscuta pentagona

at their sinuses; lobes of corolla with slender, lanceolate, long-pointed (acuminate) tip; scales inside corolla-tube reaching somewhat above the middle of the tube; seeds about 1 mm. long . . . . . . . . . . . . . . . . . . . . 4. C. PENTAGONA

g. Lobes of calyx acute or nearly so (short-pointed), triangular-ovate, not overlapping at their sinuses; lobes of corolla triangular-ovate, with short acute tip; scales inside corolla-tube reaching the top of the corolla-tube; seeds 1.5 mm. long. . . . . . 5. C. CAMPESTRIS

d. Calyx and corolla with usually 4 lobes . . . . . . . . . . . . . . . . . . . . . . *h*

h. Lobes of corolla obtuse (blunt) or rounded; lobes of calyx much shorter than the corolla-tube; flowers about 3 mm. long. . . . . . . . . . . . . . . . . . . 8. C. CEPHALANTHI

h. Lobes of corolla acute (short-pointed) or essentially so; lobes of calyx equaling or longer than the corolla-tube; flowers 2–2.5 mm. long . . . . . . . . . . . . . . . . . . *i*

i. Flowers fleshy, the outer surface with tiny, wart-like or bump-like projections (papillate or granular), about 2 mm. long; lobes of calyx acute (short-pointed) or essentially so; tips of corolla-lobes turned in toward center of flower (inflexed) . . . . . . . 9. C. CORYLI

i. Flowers with a smooth outer surface, not bumpy or wart-like, 2–2.5 mm. long; lobes of calyx obtuse (blunt); tips of corolla-lobes erect to ascending, not turned inward. 10. C. POLYGONORUM

1. **Cuscuta cuspidata** Engelm. Map 1783

Flowers July–October.

Occurs on *Ambrosia*, *Iva ciliata*, and various compositae in river bottom prairies, low open woods, and borders of streams. Scattered in southern and central Missouri north to St. Louis, Chariton, Ray, and Jackson counties.

Ranges from Indiana and Wisconsin to North Dakota, Colorado, and Utah, south to Louisiana and Texas.

The flowers occur in loose or dense clusters.

2. **Cuscuta glomerata** Choisy Map 1784

Flowers July–September.

Occurs on many kinds of compositae, including *Vernonia*, *Solidago*, *Helianthus*, and other tall herbaceous plants in upland prairies, prairie swales, river bottom meadows, along streams, and in wet ground. Scattered throughout Missouri.

Ranges from Indiana and Michigan to Wisconsin to South Dakota, south to Mississippi, Missouri, Kansas, and Texas.

This species has globose, dense clusters of fragrant flowers.

3. **Cuscata compacta** Juss. Map 1785

Flowers July–October.

Occurs on various shrubs (*Hydrangea*, *Rosa*, *Lindera*, *Alnus*, *Decodon*, *Cephalanthus*) and herbs in low wet ground of woodland and thickets. Eastern Ozark and southeastern lowland region north to St. Louis and Franklin counties, west to Iron, Shannon, Douglas, and Ozark counties.

Ranges from Florida to Texas, north to Massachusetts, New Hampshire, New York, Pennsylvania, Virginia, Indiana, Illinois, Missouri, and Oklahoma.

This species has dense clusters of flowers which may run together in ropelike masses.

4. **Cuscuta pentagona** Engelm. Map 1786

*Cuscuta pentagona* var. *calycina* Engelm. [P & S]

*Cuscuta arvensis* Beyrich

Flowers June–October.

Occurs on various herbaceous plants, such as *Solidago*, *Erigeron*, *Petalostemon*, *Ipomoea*, *Verbena stricta*, *Penstemon*, and other plants. Scattered mainly in the southern and central portions of the state north to Lincoln, Livingston, and Jackson counties.

Ranges from Florida to California, north to Massachusetts, New York, Pennsylvania, Ontario, Indiana, Wisconsin, Minnesota, South Dakota, and Montana.

The flowers occur in rather globe-shaped clusters.

5. **Cuscuta campestris** Yuncker Map 1787

Flowers June–October.

Occurs on various herbaceous plants, including compositae and leguminosae, and such genera as *Oenothera*, *Euphorbia*, *Asclepias*, *Perilla*, and *Saururus*. Common throughout Missouri.

Ranges from South Carolina to Texas and California, north to Indiana, Illinois, Minnesota, Manitoba, Saskatchewan, Colorado, and Washington; also in Mexico and tropical America; adventive eastward to Massachusetts and Quebec.

Many species previously identified as *C. pentagona* or *C. arvensis* are here referred to *C. campestris*, which is much more common in Missouri. The flowers occur in somewhat larger clusters than in *C. pentagona*.

6. **Cuscuta indecora** Choisy var. **neuropetala** (Engelm.) Hitchc. Map 1788

Flowers June–August.

Occurs on various herbaceous plants, such as *Erysimum repandum* and *Convolvulus arvensis*. Introduced along railroads, where known only from St. Louis County, east-central Missouri (St. Louis, along Burlington tracks south of E. Grand Ave., June 30, 1956, *Muehlenbach 973*; Burlington R.R., between Ferry St. and E. Grand Ave., opposite the northern Carrier, August 9, 1958, *Muehlenbach 1462*).

Ranges from Florida to Texas and Mexico, north to Illinois, Minnesota, South Dakota, Wyoming, Idaho, and California; also in the West Indies.

The flowers occur in rather open panicles.

In var. *indecora* the calyx is much shorter than the corolla-tube, whereas in var. *neuropetala* it nearly or quite equals the corolla-tube; also the flowers in var. *indecora* are only 2–2.5 mm. long, whereas in var. *neuropetala* they are 2–5 mm. long. This dodder is easily distinguished by the relatively large flowers with a conspicuously papillate surface, and inflexed, acute corolla-lobes.

7. **Cuscuta Gronovii** Willd. Love Vine, Dodder Map 1189

*Cuscuta Gronovii* var. *vulgivaga* Engelm. [P & S]

Flowers July–October.

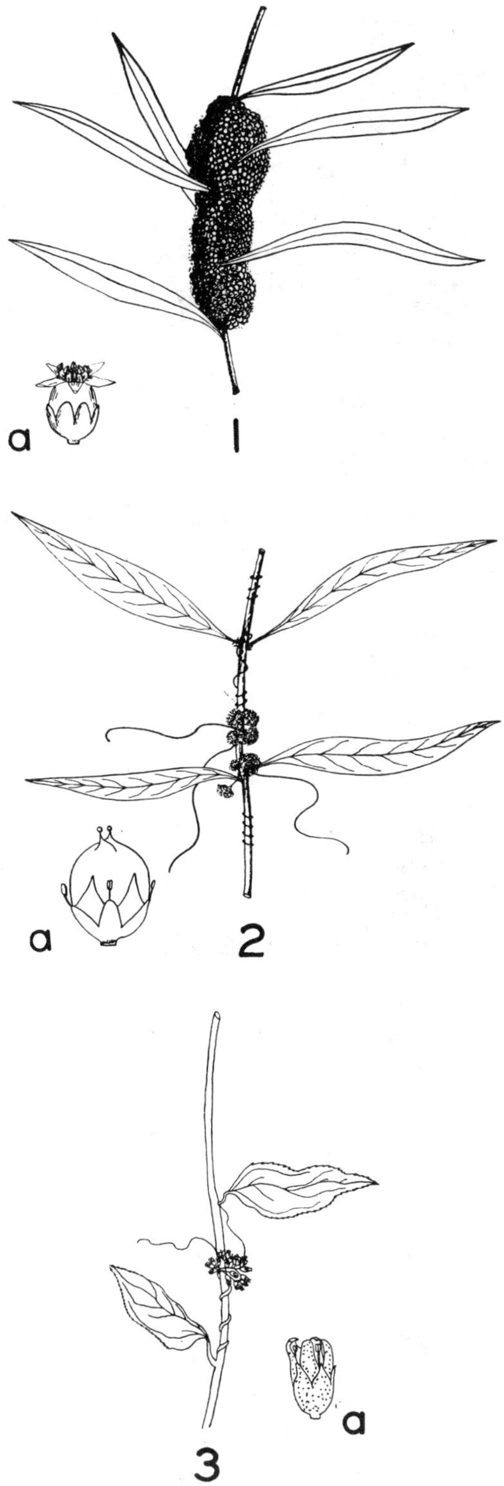


PLATE NO. 292

Plate no. 292. **1.** Cuscuta Gronovii, × 2/5, habit; *a.* Flower, × 3 1/5. **2.** Cuscuta polygonorum, × 2/5, habit; *a.* Fruit, with withered corolla, × 3 1/5. **3.** Cuscuta Coryli, × 2/5, habit; *a.* Flower, × 3 1/5. All after Gleason, The New York Botanical Garden.

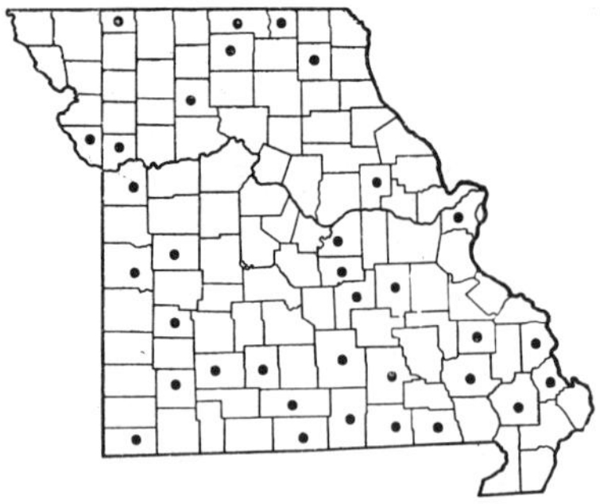

1787 Cuscuta campestris

1788 Cuscuta indecora var. neuropetala

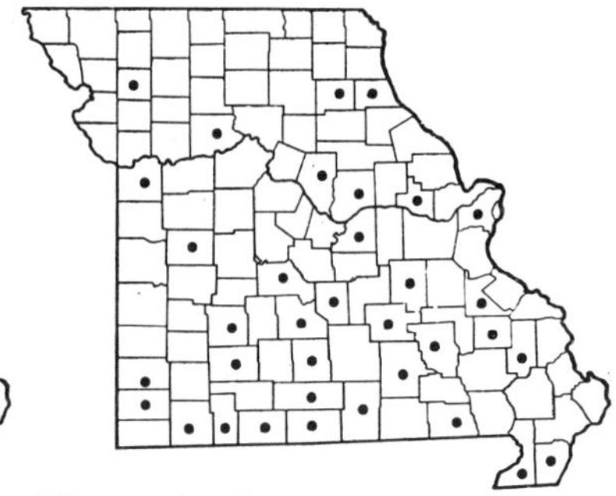

1789 Cuscuta Gronovii

Occurs on various herbaceous (*Agrimonia, Polygonum, Pilea, Amphicarpa, Perilla, Actinomeris, Aster, Eupatorium*) and woody plants, in moist ground, thickets, and prairies.

Scattered throughout Missouri, commonest apparently in the southern and central portions of the state.

Ranges from Quebec to Manitoba, south to Florida, Texas, New Mexico, and Arizona.

The flowers occur in loosely to densely flowered inflorescences.

8. **Cuscuta Cephalanthi** Engelm. Buttonbush Dodder — Map 1790

Flowers July–September.

Occurs on various herbaceous (commonly on *Justicia americana*) and woody plants, including *Cephalanthus*, in moist ground, low woods, and thickets. Scattered throughout Missouri, but infrequently encountered.

Ranges from Nova Scotia to British Columbia, south to Virginia, Tennessee, Missouri, Texas, and New Mexico.

This species forms masses of flowers even more crowded and bunched together than in *C. Polygonorum*.

9. **Cuscuta Coryli** Engelm. Hazel Dodder — Map 1790A

Flowers July–September.

Occurs on various herbaceous and woody plants, including *Corylus*, in thickets and open ground. Scattered in eastern, southern, and central Missouri.

Ranges from southern New England to Montana, south to North Carolina, Tennessee, Missouri, Texas, New Mexico, and Arizona.

The flowers occur in rather open panicles.

10. **Cuscuta Polygonorum** Engelm. Smartweed Dodder — Map 1791

Flowers July–September.

Occurs on various herbaceous species, principally belonging to *Polygonum*, *Justicia americana*, and compositae, in moist ground along streams, ponds, sloughs, in prairies, and thickets. Throughout Missouri.

Ranges from Quebec to Minnesota and Nebraska, south to Maryland, Tennessee, Missouri, Kansas, and Oklahoma.

The flowers of this species occur in rather closely crowded clusters.

## Fam. **POLEMONIACEAE** (Phlox Family)

a. Leaves dissected, lobed, or divided into several or many separate leaflets (compound) . . . . . . . *b*
- b. Leaf-divisions thread-like (filiform); corolla usually red or pink; calyx with narrowly triangular lobes . . . . . . . . . . . . . . . . . . . . . . . . . . . . . 2. Gilia
- b. Leaf-divisions oval or lanceolate; corolla blue or blue-lavender; calyx with broader lanceolate lobes . . . . . . . . . . . . . . . . . . . . . . . . . . . . 1. Polemonium

a. Leaves not divided, not lobed, nor dissected . . . . . . . . . . . . . . . . . . . *c*
- c. All the main leaves of the stem alternate . . . . . . . . . . . . . . . . . 3. Collomia
- c. At least the lower leaves of the stem opposite, the upper ones alternate in one cultivated species, otherwise all of them opposite in all the native species . . . . . . . . . . . . . 4. Phlox

### 1. **Polemonium** L. Greek Valerian, Jacob's Ladder

**Polemonium reptans** L. Jacob's Ladder, Greek Valerian — Map 1792

Flowers April–June.

Occurs in rich, low or rocky woods and thickets, at

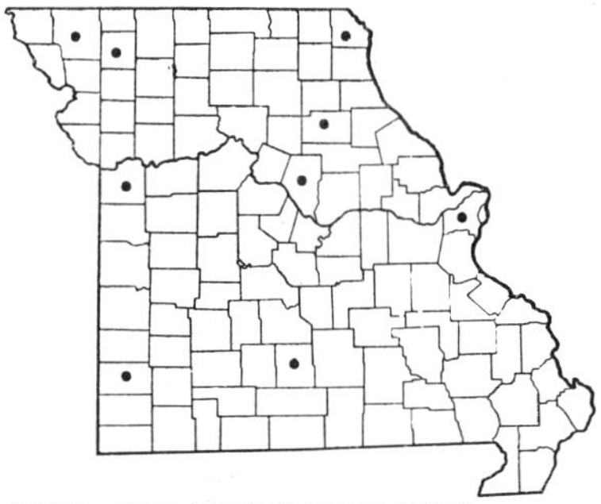

1790 Cuscuta Cephalanthi (Buttonbush Dodder)

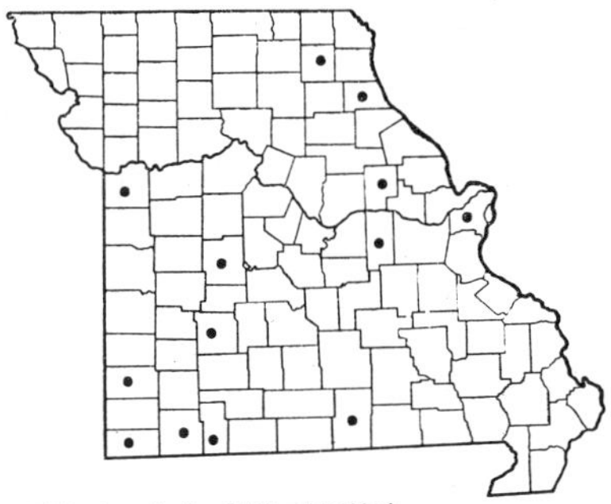

1790A Cuscuta Coryli (Hazel Dodder)

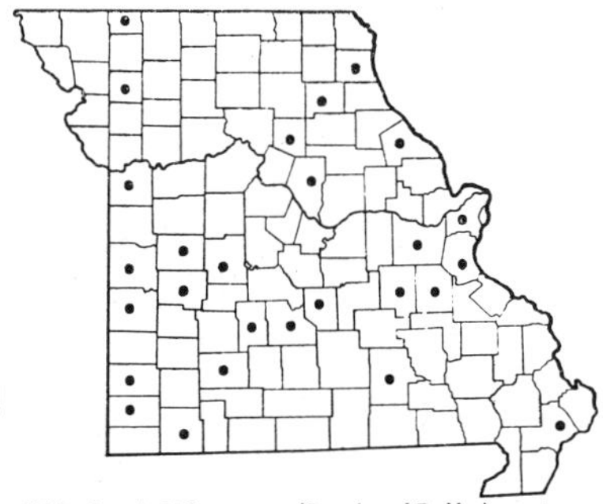

1791 Cuscuta Polygonorum (Smartweed Dodder)

the base of bluffs, alluvial ground near streams, and upland woods.

Throughout Missouri, except absent from the extreme northwestern counties.

Ranges from New York to Minnesota, south to Georgia, Alabama, Mississippi, Arkansas, and Oklahoma.

Plants vary from glabrous to sparsely hairy. The root has been used as a remedy for kidney troubles and as a diuretic.

## 2. **Gilia** R. & P.

**Gilia rubra** (L.) Heller Standing Cypress Map 1793

Flowers late May–August.

Occurs in waste, cultivated, and open ground, as an escape from cultivation. Rare, scattered in the state in Clark, Greene, and Holt counties.

Native of the southern United States from Florida to Texas, north to North Carolina and Oklahoma; introduced and naturalized northward to Massachusetts, New York, Ohio, Michigan, Illinois, and Missouri.

This is a tall biennial plant attaining a meter or so in height. The numerous tubular red or pink flowers arranged in a long showy inflorescence are very attractive among the lace-cut leaves.

## 3. **Collomia** Nutt.

**Collomia linearis** Nutt. Map 1794

Flowers late May–August.

Occurs in open grassy places, meadows, and rocky places along bluffs. Known only from Marion (along bluffs, Palmyra Junction, June 1, 1918, *Davis 8924*; same locality and date, *Davis 3617*) and Christian counties.

Ranges from Quebec to New Brunswick; Ontario to British Columbia, south to Wisconsin, Minnesota, Nebraska, New Mexico, Arizona, and California; introduced eastward from Nova Scotia to Iowa, south to New Jersey, Pennsylvania, and Missouri.

The flowers vary from lavender-purple to nearly white, are very slender, and 10–12 mm. long.

## 4. **Phlox** L. Phlox

For a complete elaboration of this genus, the reader is referred to the most recent comprehensive work by Dr. Edgar T. Wherry (The genus Phlox. Morris Arboretum Monographs III. 174 pp. Philadelphia. 1955).

a. Stems sprawling, spreading, or trailing, in mats; leaves linear to very narrow, 1.5–3 mm. broad; each of the 5 lobes of corolla deeply split or notched nearly to or below the middle, appearing almost 10-lobed . . . . . . . . . . . . . . . . . . . . . . . . . 1. P. BIFIDA

a. At least the flowering stems upright or ascending for the most part; leaves linear to lanceolate, oblong, or ovate, mainly 3–50 mm. broad; lobes of corolla not split . . . . . . . . . . . *b*

  b. Upper leaves of stem mostly alternate; annual with slender root . . . . . . . 9. P. DRUMMONDII

  b. All the stem-leaves opposite; perennial plants . . . . . . . . . . . . . . . . . *c*

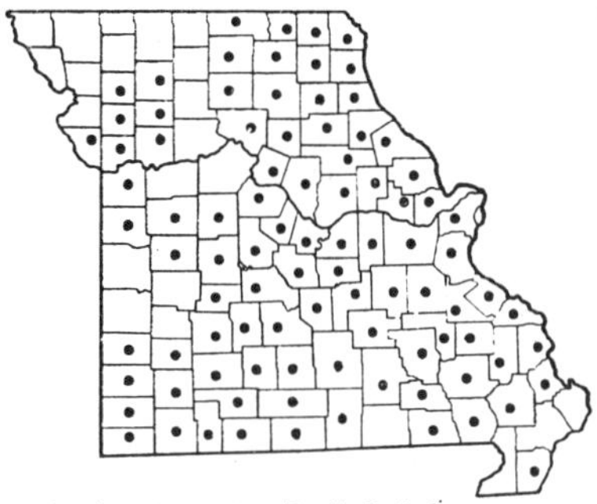

1792 Polemonium reptans (Jacob's Ladder)

1793 Gilia rubra (Standing Cypress)

1794 Collomia linearis

c. Dark green leafy sterile (non-flowering) shoots sprawling or spreading from the base of the plant, often rooting at the joints . . . . . . . . . . . . . . . . . . 2. P. DIVARICATA

c. All shoots upright, not sprawling nor rooting at joints . . . . . . . . . . . . . . *d*

d. Leaves with plainly visible lateral (side) nerves, which are joined together toward the leaf-margin; leaves 10–60 mm. broad . . . . . . . . . . . . . . . . . . . *e*

e. Common species; usually 14–25 or more nodes (joints where leaves originate) present on the stem below the flowering or fruiting portion; hairs of the branches and stalks of the inflorescence simple, not gland-tipped or very rarely so; corolla-tube more or less hairy; calyx-lobes often glabrous (without hairs); upper surface of leaves usually glabrous or not rough-hairy . . . . . . . . . . . . . . . . . . . . 7. P. PANICULATA

e. Uncommon species; usually 7–15 nodes present on the stem below the flowering or fruiting portion; hairs of the branches and stalks of the inflorescence gland-tipped; corolla-tube glabrous (without hairs); calyx-lobes glandular-hairy; upper surface of leaves usually bristly- or rough-hairy. . . . . . . . . . . . . 8. P. AMPLIFOLIA

d. Leaves with either no plainly visible lateral (side) nerves or these very indistinct and not connecting with one another; leaves chiefly 3–15 (–25) mm. broad . . . . . . . . *f*

f. Flowers arranged in a somewhat broadly cylindric inflorescence which is longer (twice or more) than thick, the lower branches of the inflorescence short and not reaching the topmost (terminal) portion of inflorescence; stems with usually conspicuous purplish-red or dark wine-colored spots; calyx-lobes during flowering period much shorter than the calyx-tube . . . . . . . . . . . . . . . . . . . . . . 6. P. MACULATA

f. Flowers arranged in a flattened or rounded inflorescence which is as broad as long or broader than long, the lower peduncles (stalks supporting groups of flowers) longer than the upper peduncles; stems green, not spotted with purple or reddish or with these colors only slightly showing among the green; calyx-lobes during the flowering period either nearly equaling or longer than the calyx-tube . . . . . . . . . . . . . . . *g*

g. Stems, leaves, and/or calyx more or less hairy with visible or spreading hairs; corolla-tube more or less hairy on the outer surface; style and stigmas very short, together shorter than half the length of the corolla-tube; calyx-lobes during flowering period longer than calyx-tube . . . . . . . . . . . . . . . . . . . 3. P. PILOSA

g. Stems glabrous (without hairs) or somewhat hairy; leaves and calyx usually glabrous; style elongated, nearly equaling the length of the corolla-tube; calyx-lobes during flowering period about equaling or even shorter than calyx-tube. . . . . . . . *h*

h. Calyx nearly cylindrical, the sepals rather broad with thinnish membranes and a moderately developed costa (midrib), becoming conspicuously folded into plaits lengthwise (plicate) in the early stages; calyx 6–8 mm. long; usually 10–15 leaf-nodes present on stem . . . . . . . . . . . . . . . . 4. P. GLABERRIMA

h. Calyx nearly campanulate, the sepals narrow with firm membranes and a strongly developed costa (midrib), flat to somewhat keeled and folded into plaits lengthwise; calyx 7–9 (–11) mm. long; usually 6–12 (–15) leaf-nodes present on stem 5. P. CAROLINA

---

Plate no. 293. **1.** Phlox bifida var. bifida, × 2/5. **2.** Polemonium reptans, × 2/5; Details from Small, The New York Botanical Garden. **3.** Collomia linearis, × 2/5. **4.** Gilia rubra, × 2/5; Details from Small, The New York Botanical Garden.

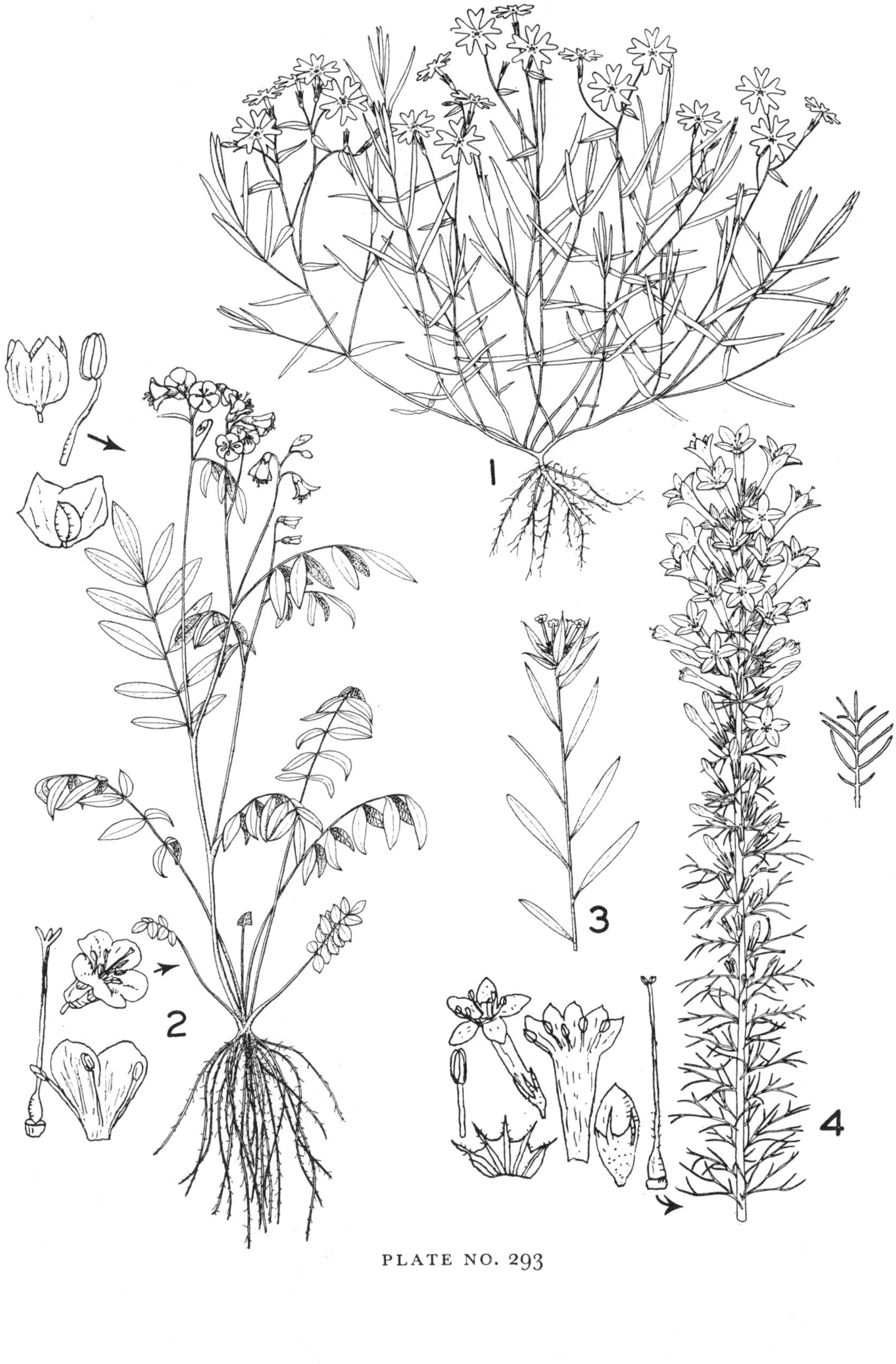


PLATE NO. 293

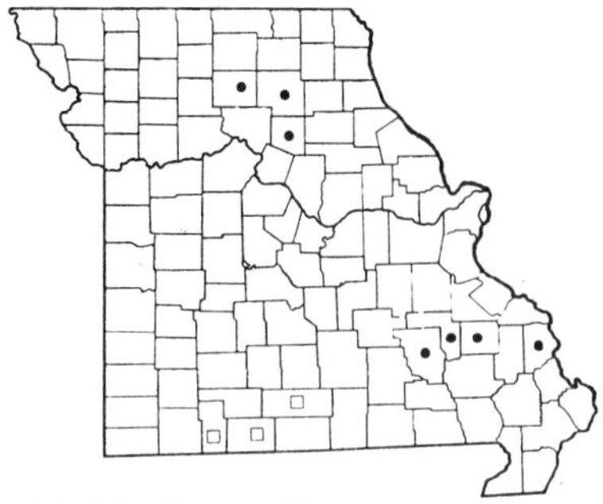

1795 • Phlox bifida var. bifida
1795 □ Phlox bifida var. stellaria

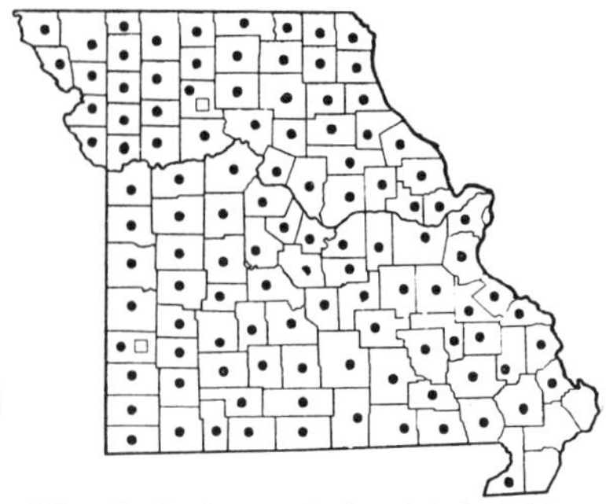

1796 • Phlox divaricata var. Laphamii f. Laphamii (Blue Phlox)
1796 □ Phlox divaricata var. Laphamii f. candida

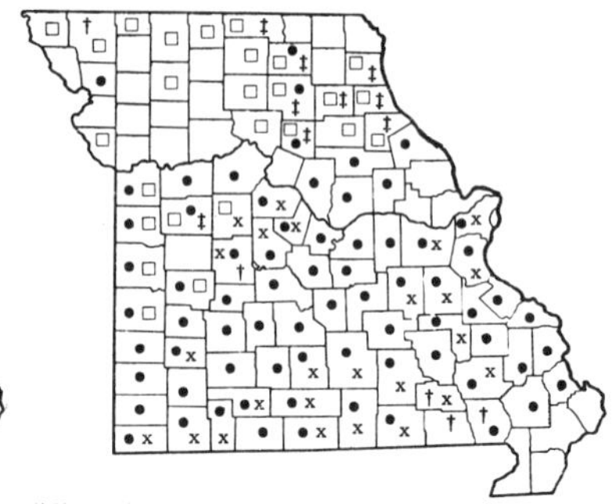

1797 • Phlox pilosa var. pilosa
1797 x Phlox pilosa var. ozarkana
1797 □ Phlox pilosa var. fulgida f. fulgida
1797 ‡ Phlox pilosa var. fulgida f. albiflora
1797 † Phlox pilosa var. amplexicaulis

1. **Phlox bifida** Beck Sand Phlox Map 1795
Flowers late March–May.

Occurs in rocky or dry upland woods, slopes of ravines, ledges of bluffs, along boulders, outcrops and rocky wooded banks and bluffs along streams, on limestone, chert, or granite.

Two variations occur in Missouri:

Stalks of flowers and fruits, calyx, and/or stems and branches bearing gland-tipped hairs . . 1a. P. BIFIDA var. BIFIDA
No gland-tipped hairs present . . 1b. P. BIFIDA var. STELLARIA

1a. **Phlox bifida** var. **bifida** Map 1795
*Phlox bifida* subsp. *bifida* [Wherry]
*Phlox bifida* Beck [G, P & S, Steyerm.]
*Phlox bifida* var. *glandifera* Wherry

Occurs in two separate sectors of the state: (1) unglaciated southeastern Ozark region in Cape Girardeau, Madison, Iron, and Reynolds counties, and (2) glaciated prairie region of north-central Missouri in Randolph, Macon (*Steyermark 40512*), and Linn (*Steyermark 40502*) counties.

Ranges from Michigan to Iowa, south to Kentucky, Tennessee, Arkansas, Kansas, and Oklahoma.

1b. **Phlox bifida** var. **stellaria** (Gray) Wherry Map 1795
*Phlox bifida* subsp. *stellaria* (Gray) Wherry [Wherry]
*Phlox bifida* var. *cedaria* (Brand) Fern. [G]

Southwestern Ozark region in Douglas (upper cherty slopes along Spring Creek, near Roosevelt, T25N, R11W, sect. 23, July 17, 1937, *Steyermark 23307*), Taney (very common, all over rocky woods in shade from base to $\frac{1}{2}$ way up on northwest-facing steep, wooded, limestone slopes along Swan Creek, $\frac{1}{2}$ mi. northeast of Forsyth, April 27, 1949, *Steyermark 67358*), and Stone (cherty draw in upland along highways 13 and 86, 4 mi. southeast of Lampe, April 29, 1949, *Steyermark 67443*) counties.

Ranges from Indiana to Tennessee, Missouri, and Arkansas.

The corolla varies from pale lavender to rose-lavender to white with a purplish tube, or rarely all white. Plants from the granitic areas of southeastern Ozarks usually have the deeper lilac to rose-lavender shades. Plants from sandy prairies or other prairie areas in northern Illinois often show predominantly white flowers.

This is a fine plant for rock gardens, soon producing large creeping mats which eventually cover a large area. The plant does best in slightly shaded situations and on slopes with some drainage. Some strains of the species from southern Illinois appear to grow more rampant and spread more than other strains.

2. **Phlox divaricata** L. var. **Laphamii** Wood
Blue Phlox Map 1796
Also called Wild Sweet William.
Flowers April–June.

Occurs in rich or rocky woods and thickets, frequently in low wet ground and alluvial soils along streams. Throughout Missouri, not recorded from the extreme southeastern counties of Mississippi, New Madrid, and Pemiscot.

Two variations are found in Missouri:

Corolla blue, purple, lilac, or rose-lavender, sometimes greenish . . . . 2a. P. DIVARICATA var. LAPHAMII f. LAPHAMII
Corolla white throughout or nearly so . . . 2b. P. DIVARICATA var. LAPHAMII f. CANDIDA

2a. **Phlox divaricata** var. **Laphamii** Wood f. **Laphamii** Map 1796

*Phlox divaricata* subsp. *Laphamii* (Wood) Wherry [Wherry]

*Phlox divaricata* var. *Laphamii* Wood [G, BB, P & S, Steyerm.]

This is the common type found in Missouri.

Ranges from Florida to Texas, north to Virginia, Alabama, Wisconsin, Minnesota, and Nebraska.

2b. **Phlox divaricata** var. **Laphamii** f. **candida** Palmer & Steyerm. Map 1796

*Phlox divaricata* subsp. *Laphamii* f. *albiflora* (D. M. Moore) McGregor

*Phlox divaricata* f. *albiflora* D. M. Moore, not f. *albiflora* House

Known only from Livingston (terrace along alluvial bottom of Grand River along route 36, 4 mi. southwest of Chillicothe, May 1, 1950, *Steyermark 69584*, holotype in Chi. Nat. Hist. Mus. Herb.) and Barton (low, rich, alluvial woods, 1 mi. west of Nashville, April 30, 1953, *Palmer 55411*) counties. Undoubtedly of wider distribution in the state.

Typical *P. divaricata* var. *divaricata* f. *divaricata* has the corolla-lobes notched or emarginate at the tip and occurs east of Missouri; the variety found in Missouri, var. *Laphamii*, has the corolla-lobes without any notch at the tip.

This phlox does well in wildflower gardens and is easily grown. It thrives in shaded or partly shaded situations in rich or neutral soils.

3. **Phlox pilosa** L. Map 1797

Flowers April–July, sometimes again October–November.

Occurs in rocky or dry open woods, sandy alluvial soils of valleys, thickets, meadows, prairies, and glades.

The following variations are encountered in Missouri:

a. Hairs of the calyx and branches of the inflorescence with simple hairs, not gland-tipped . *b*
  b. Hairs of the inflorescence and calyx abundant, fine, about 0.5 mm. long and producing a whitish appearance; common in prairies of northern, central, and western Missouri . . . . . . . . . . . *c*
    c. Corolla rose, pink, or shades of purple. 3c. P. PILOSA var. FULGIDA f. FULGIDA
    c. Corolla white throughout or nearly so. 3d. P. PILOSA var. FULGIDA f. ALBIFLORA
  b. Hairs of the inflorescence and calyx sparse or nearly wanting, coarse, 1–1.5 mm. long. 3e. P. PILOSA var. AMPLEXICAULIS
a. Hairs of the calyx and branches of the inflorescence gland-tipped . . . . . . . . *d*
  d. Upper leaves of stem ovate or ovate-lanceolate, broadest and somewhat heart-shaped at base; at least the upper leaves with gland-tipped hairs; stems with gland-tipped hairs, sometimes branched . . 3b. P. PILOSA var. OZARKANA
  d. Upper leaves of stem mostly linear to narrowly oblanceolate or linear-lanceolate, rounded or truncate but not noticeably broader at base; stems with usually simple, not gland-tipped, hairs, usually unbranched 3a. P. PILOSA var. PILOSA

3a. **Phlox pilosa** var. **pilosa** Map 1797

*Phlox pilosa* L. [G. Steyerm.]

*Phlox pilosa* var. *virens* (Michx.) Wherry [BB, P & S, Steyerm.]

*Phlox argillacea* Clute & Ferriss

Throughout southern and central Missouri north to Pike, Adair, Saline, Lafayette, Jackson, and Andrew (*Steyermark 84550*) counties.

Ranges from Florida to Texas, north to Connecticut, New York, Ontario, Michigan, Iowa, Missouri, and Kansas.

The corollas in var. *pilosa* vary from shades of rose, red-purple, violet, and white throughout, or white with a purple center. The pure white-flowered plants were referred by Steyermark (*Spring Flora*, p. 435. 1940) to typical *P. pilosa* var. *pilosa*, and have been recorded from Pike, St. Louis, St. Francois, Crawford, Pulaski, Ozark, Cedar, and Jackson counties; plants with other than white corollas were treated as var. *virens*. In the present treatment both are combined under var. *pilosa*.

3b. **Phlox pilosa** var. **ozarkana** Wherry Map 1797

*Phlox pilosa* subsp. *ozarkana* (Wherry) Wherry [Wherry]

Occurs throughout the Ozark region north to St. Louis, Franklin, Moniteau, Cooper, Pettis, and Dade counties.

Ranges from Missouri, Arkansas, and Oklahoma to Louisiana; also in Alabama and Georgia.

Although Wherry states in his key to the varieties of *P. pilosa* (Morris Arb. Monog. III: 38. 1955) that the stems of this variety tend to 'be branched,' actually most plants I have seen in the field have simple, unbranched stems, and the numerous plants of this variety growing in my wildflower garden all have simple stems.

The corolla-lobes of this variety are usually a deep rose or rose-lavender color.

3c. **Phlox pilosa** var. **fulgida** Wherry f. **fulgida** Map 1797

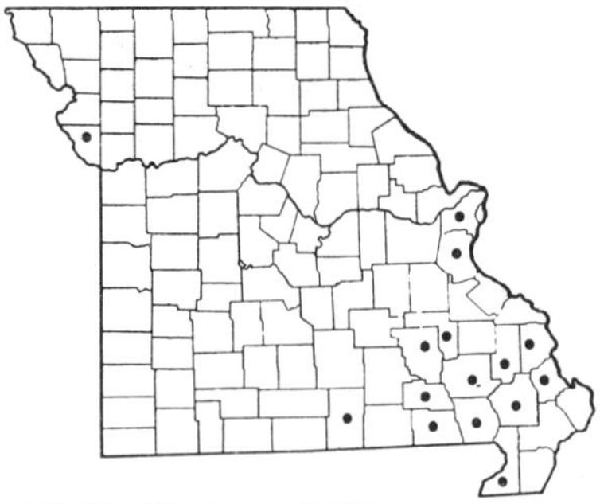

1798 Phlox glaberrima var. interior

1799 • Phlox carolina var. carolina
1799 □ Phlox carolina var. angusta

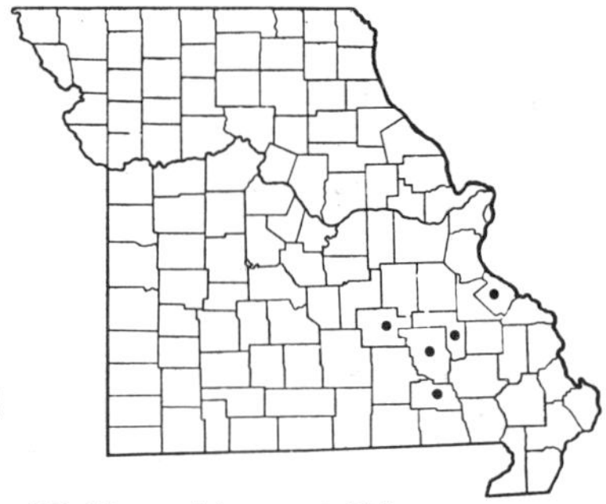

1800 Phlox maculata var. pyramidalis

*Phlox pilosa* var. *fulgida* Wherry [G, BB, P &S, Steyerm.]

*Phlox pilosa* subsp. *fulgida* (Wherry) Wherry [Wherry]

Common in the prairies and prairie remnants along railroads in northern, central, and western Missouri, north and west of a line extending from Ralls, Monroe, Randolph, Pettis, and St. Clair counties to Vernon County.

Ranges from Wisconsin to Saskatchewan, south to Indiana, Illinois, Missouri, and Kansas.

3d. **Phlox pilosa** var. **fulgida** f. **albiflora** (MacM.) Standl. Map 1797

Within the range of the more common f. *fulgida*, mainly in northeastern Missouri, locally west in Johnson County.

3e. **Phlox pilosa** var. **amplexicaulis** (Raf.) Wherry Map 1797

*Phlox pilosa* subsp. *pulcherrima* Lundell [Wherry]

*Phlox pulcherrima* Lundell

Rare and scattered in the state in Nodaway, Benton (*Steyermark 10719*), St. Clair (*Steyermark 71421*), and Butler counties.

Ranges from Indiana and Tennessee to Missouri, south to Louisiana and Texas.

This variety has nearly glabrous stems and leaves. The flowers vary from a brilliant carmine-rose to pink, purple, lavender, or white. It is a very showy-flowered variety.

*Phlox pilosa* shows much variation in time of flowering, height of plant, shape, length and breadth of leaves, amount and type of pubescence on stems, leaves, and inflorescence, and color of flowers. The var. *ozarkana* blooms earliest, on the average, of any of the varieties, and is usually the most fragrant, resembling the delicate sweetness of *Saponaria officinalis* (Bouncing Bet) or *Hesperis matronalis* (Dame's Rocket). The later-flowering prairie variety, var. *fulgida*, lacks the fragrance characteristic of var. *ozarkana*. Typical var. *pilosa* is fragrant and has late-flowering types that bloom late into June and even July. All the varieties do well in cultivation and in wildflower gardens in open to partly shaded dry, well-drained soils, of clayey or rocky types.

4. **Phlox glaberrima** L. var. **interior** Wherry Map 1798

*Phlox glaberrima* subsp. *interior* (Wherry) Wherry [Wherry]

Flowers May–June.

Occurs in wet depressions of low or swampy woods, wet meadows, and low prairies. Occurs in eastern Missouri from St. Louis County south to Dunklin, Butler, Ripley, and Howell counties, locally northwest in Platte County, west-central Missouri (wet depression in light woods, Lakeside Country Club, Parkville, May 12, 1956, *Howard Hoffmaster 46*).

Ranges from Ohio to Illinois and Wisconsin, south to Tennessee, Alabama, Arkansas, and Missouri.

This has been separated from typical var. *glaberrima* (var. *melampyrifolia* [Salisb.] Wherry) on the basis of a shorter calyx (6–7.5 mm. long as contrasted with 7.5–10 mm. long for var. *glaberrima*) with shorter and broader calyx-lobes. The slight differences do not appear to warrant their separation, in the opinion of Gleason and also the present author.

---

Plate no. 294. **1.** Phlox divaricata, × 2/7; Details from Small, The New York Botanical Garden. **2.** Phlox glaberrima var. interior, × 2/7. **3.** Phlox maculata var. pyramidalis, × 2/7. **4a.** Phlox pilosa var. virens, × 2/7. **4b.** Phlox pilosa var. ozarkana, × 2/7. **5.** Phlox paniculata, × 2/7; After Gleason, The New York Botanical Garden. **6.** Phlox amplifolia, × 2/7. **7.** Hydrophyllum virginianum, × 2/7.

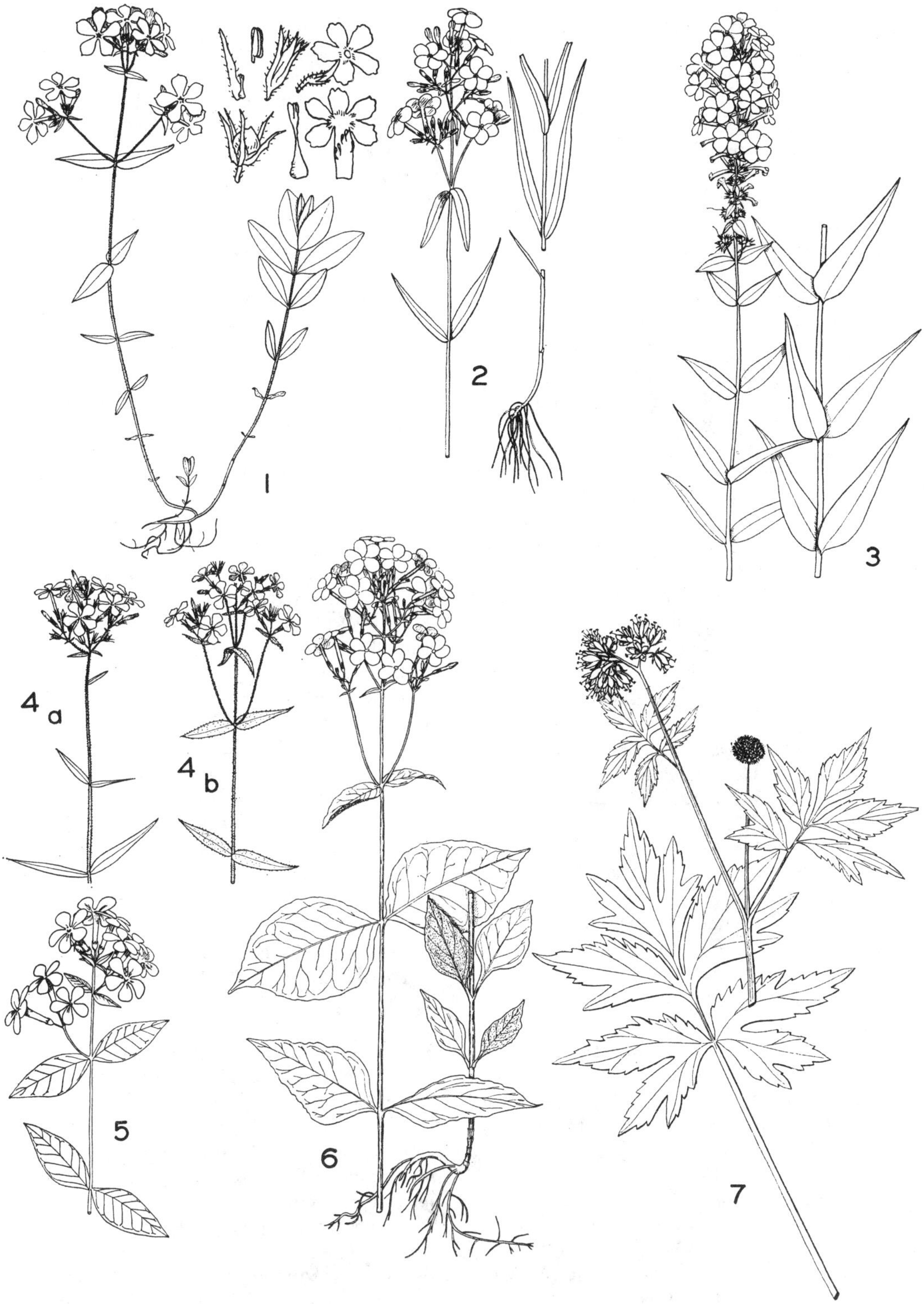


PLATE NO. 294

This is an attractive species with large clusters of usually deep rose-red to reddish-purple flowers. White- and violet-colored plants have also been recorded.

As this species prefers wet soils, it does not succeed if transplanted to dry situations.

5. **Phlox carolina** L. Map 1799

Flowers May–July.

Occurs in swampy ground, bogs, and low wet woods along streams.

Known only from southeastern Missouri.

Two variations occur in Missouri:

Largest leaves in upper half of stem, lanceolate to ovate-oblong . . 5a. P. CAROLINA var. CAROLINA
Largest leaves in lower half of stem, linear to linear-lanceolate . 5b. P. CAROLINA var. ANGUSTA

5a. **Phlox carolina** var. **carolina** Map 1799

*Phlox carolina* subsp. *typica* Wherry

Known only from Carter County (Big Barren Creek, June 5, 1948, *Bill Bauer 12834*; low woods along Big Barren Creek, T25N, R1W, sect. 4, 10 mi. northwest of Bennett, May 22, 1938, *Steyermark 5339*).

Ranges from Florida to Mississippi, north to North Carolina, Tennessee, and Missouri.

5b. **Phlox carolina** var. **angusta** (Wherry) Steyerm. Map 1799

*Phlox carolina* subsp. *angusta* Wherry [Wherry]

Known only from Ripley County (bogs, Bay Mills, locally common, July 24, 1897, *Mackenzie 367* in Acad. Nat. Sci. Phila. Herb.).

Ranges from Florida to Texas, north to South Carolina, Georgia, Alabama, Kentucky, Illinois, Missouri, and Arkansas.

The Mackenzie specimen had been previously identified as *P. glaberrima*.

This species has been generally confused previously with *P. glaberrima*, which it resembles.

6. **Phlox maculata** L. var. **pyramidalis** (Smith) Wherry Wild Sweet William Map 1800

Also called Sweet William Phlox, Meadow Phlox.

*Phlox maculata* subsp. *pyramidalis* (Smith) Wherry [Wherry]

*Phlox maculata* L. [G, BB]

Flowers late May–October.

Occurs in swampy meadows and wet open ground, usually in calcareous spring-fed soils. Southeastern Ozark region, where known only from Ste. Genevieve, Iron, Dent, Reynolds, and Carter counties.

Ranges from Virginia and North Carolina, west to Tennessee, Indiana, Illinois, and Missouri.

Wherry divides *P. maculata* into two subspecies, distinguishing them on the basis of 7–15 spaced nodes and oblong to ovate, cordate upper leaves of early summer-blooming subsp. *maculata* versus 16–35 crowded nodes and lanceolate to oblong-ovate, truncate to subcordate leaves of a later-blooming subsp. *pyramidalis*. The present author does not believe these differences hold.

Wherry has annotated a specimen (*Steyermark 5490*) from Reynolds County, as *P. maculata* subsp. *maculata*. It was blooming on May 26.

This is a mostly late-blooming species (July–October). The tall, erect, usually simple stems attain a height of nearly 1 meter. The long, closely-flowered inflorescence is quite showy with the deep rose-red to reddish-purple colored corollas.

7. **Phlox paniculata** L. Perennial Phlox Map 1801

Also called Garden Phlox, Summer Phlox.

Flowers July–October.

Occurs in moist or rich low woods and thickets, alluvial banks and gravel bars along streams, and at the base of bluffs. Ozark region of southern and central Missouri northeast to Marion County, westward north to Callaway, Cole, Miller, Camden, Pettis, and Jackson counties, locally northwest in Holt County, where escaped from cultivation.

Ranges from New York to Iowa, south to North Carolina, Georgia, Mississippi, and Arkansas; elsewhere naturalized as an escape from cultivation.

Two variations occur in Missouri:

Corolla rose-purple or deep pink . 7a. P. PANICULATA f. PANICULATA
Corolla white . . . 7b. P. PANICULATA f. ALBA

7a. **Phlox paniculata** f. **paniculata** Map 1801

*Phlox paniculata* L. [G, BB, P & S, Steyerm.]

This is the common form in Missouri.

7b. **Phlox paniculata** f. **alba** Don Map 1801

Rare and scattered, known only from St. Charles, Ripley, and Barton counties.

Numerous horticultural varieties have been selected from this species and are grown for perennial beds for late summer and fall flowering. These are all adapted for full sun; however the wild species is primarily a woodland plant and shade dweller.

8. **Phlox amplifolia** Britt. Broadleaf Phlox Map 1802

Flowers last of May–August.

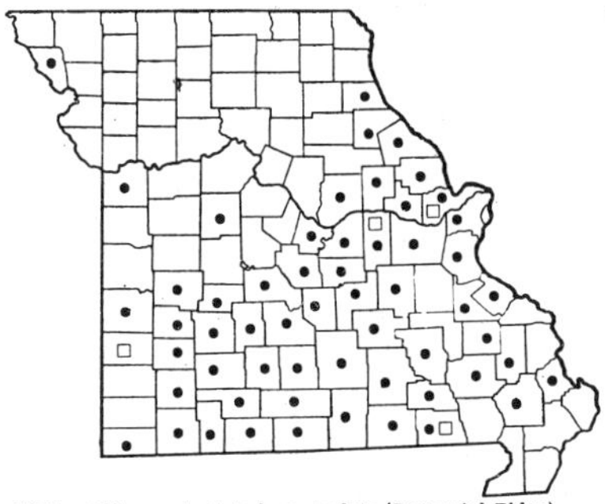

1801 • Phlox paniculata f. paniculata (Perennial Phlox)
1801 □ Phlox paniculata f. alba

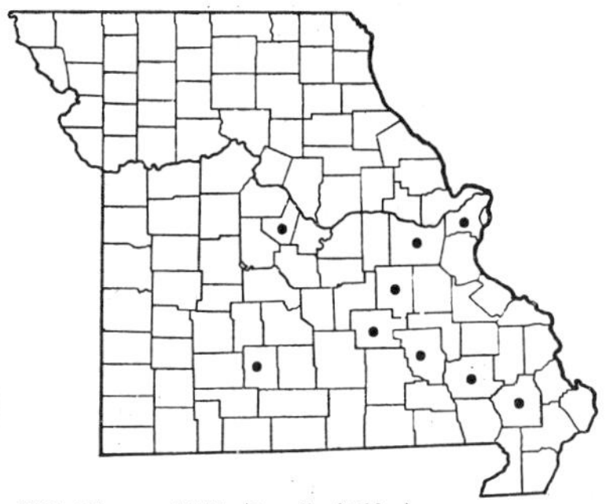

1802 Phlox amplifolia (Broadleaf Phlox)

1803 Phlox Drummondii (Annual Phlox)

Occurs in rich, rocky, or alluvial woods, generally on slopes along streams. Ozark region mainly of eastern Missouri northeast to St. Louis County, locally west to Moniteau and Webster counties.

Ranges from Virginia to Indiana, west to Missouri, south to North Carolina, Tennessee, and Alabama.

The stems in this species and in *P. paniculata* vary from glabrous to pubescent, but in *P. paniculata* the stems are more frequently glabrous, whereas in *P. amplifolia* they are more frequently hairy with often conspicuous spreading hairs. Moreover, in addition to the characters mentioned in the key, the anthers of *P. amplifolia* are all included within the corolla-tube, whereas in *P. paniculata* one or more of the anthers protrude (exsert) from the corolla-tube.

9. **Phlox Drummondii** Hook. Annual Phlox
Map 1803

Flowers June–September.

Introduced along railroads in Dunklin County, southeastern Missouri (Kennett, along railroad, July 26, 1900, *R. Dimmitt 438*, in U. Mo. Herb.).

Native of Texas and adjacent states; planted as an ornamental garden annual and occasionally escaped.

The flowers are typically rose-red and 2–2.5 cm. across. Cultivated strains are more dwarfed and compact growing, come in 'double' flowers and in shades of red, purple, pink, buff, and white.

*Excluded Species*

**Phlox subulata** L. Moss Pink, Rock Pink

This is a commonly cultivated, well-known, spring-flowering species grown in gardens throughout Missouri. There is no indication that the two specimens from Missouri brought in by students for their herbarium collections represent escaped or naturalized plants, but rather they appear to be cultivated plants taken from a garden source. For this reason they are excluded from the present flora and not considered adventive. The two specimens concerned are from Platte County (1½ mi. west on Hoy Road ¼ mi. north of Parkville, May 1, 1957, *Kent Drew 61*, Park College Herb.) and Clay County (escaped and cultivated, Melrose Hall, William Jewell College, Liberty, April 22, 1947, *Tom Bray 109*, Wm. Jewell Coll. Herb.).

## Fam. **HYDROPHYLLACEAE** (Waterleaf Family)

a. Leaf-margin entire (without teeth); spines usually occurring on the stem at the base of the leaves . . . 4. HYDROLEA
a. Leaf-margin toothed, lobed, or deeply divided; no spines present on the stems . . . *b*
- b. At least the lower leaves of the stem opposite; each flower arising singly opposite a leaf, the flowers not clustered . . . 2. ELLISIA
- b. All the leaves of the stem alternate; flowers arranged in clusters of 2–6 or more . . . *c*
  - c. Middle and lower stem-leaves relatively small, mostly 1–3.5 cm. broad and less than 7 cm. long (except *Phacelia bipinnatifida*) . . . 3. PHACELIA
  - c. Middle and lower stem-leaves, including the divisions, relatively large, mostly 4–20 cm. broad and usually more than 7 cm. long . . . *d*
    - d. Most of the stem-leaves with 5–9 main points or lobes, usually not divided to the central midrib . . . 1. HYDROPHYLLUM
    - d. Most of the stem-leaves divided to the center or midrib into 3–7 divisions . . . *e*

1804 Hydrophyllum virginianum var. virginianum (Waterleaf)

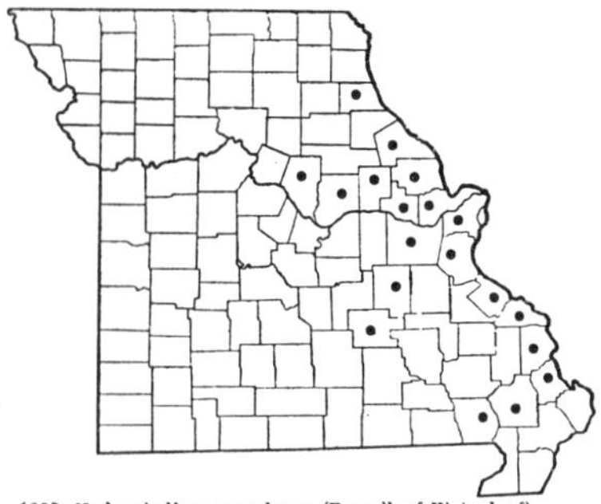

1805 Hydrophyllum canadense (Broadleaf Waterleaf)

1806 • Hydrophyllum appendiculatum f. appendiculatum (Woollen Breeches)

1806 □ Hydrophyllum appendiculatum f. album

e. Stem hairy, at least the upper portion with visible spreading hairs; gland-tipped hairs covering calyx-lobes, flower-stalks (pedicels), and branches of inflorescence; axis of the inflorescence unbranched, in age elongating and much longer than broad . . . . 3. PHACELIA BIPINNATIFIDA

e. Stem mainly glabrous (without hairs) or if hairy, the hairs inconspicuous or scattered; glandless hairs present on calyx-lobes, flower-stalks (pedicels), and branches of inflorescence; axis of inflorescence forked and branched into more forks, the inflorescence as broad as or broader than long . . . . . . . . . . 1. HYDROPHYLLUM VIRGINIANUM

## 1. **Hydrophyllum** L. Waterleaf

a. All the leaves of the stem deeply parted or divided into 3–7 divisions, the lowest divisions separated from those above . . . . . . . . . . . . . . . . . . . . . . . . . . 1. H. VIRGINIANUM

a. Leaves of the stem merely with 5–9 main points or lobes . . . . . . . . . . . . . . . *b*

b. Found throughout most of Missouri; flower-stalks (pedicels), calyx-lobes, and branches of inflorescence bristly-hairy, some of the hairs 1–3 mm. long; stamens slightly protruding (exserted) beyond the corolla; biennial. . . . . . . . . . . . . . . . . . . 3. H. APPENDICULATUM

b. Eastern Missouri only, west to Boone, Phelps, and Butler counties; flower-stalks (pedicels), calyx-lobes and branches of inflorescence mostly glabrous (without hairs) or sparsely hairy with short inconspicuous hairs mostly less than 1 mm. long; stamens long protruding (exserted) beyond the corolla; perennial with a creeping rhizome . . . . . . . . . . . . . . 2. H. CANADENSE

1. **Hydrophyllum virginianum** L. var. **virginianum** Waterleaf Map 1804

Sometimes called John's Cabbage, Virginia Waterleaf.

*Hydrophyllum virginianum* L. [G, BB, P & S, Steyerm.]

Flowers late April–July.

Occurs in moist, rich, or low woods and thickets in ravine bottoms, base of bluffs, valleys along streams, and river flood plain. Throughout northern, central, and western Missouri, but not recorded from the southeastern lowlands or the eastern Ozarks.

Ranges from Quebec to Manitoba, south to Virginia, Tennessee, Arkansas, Kansas, and Oklahoma.

The creeping scaly-toothed rhizomes form dense colonies. The lowest and basal leaves as well as those of early spring and late fall are usually spotted or blotched with gray, which usually disappears in age. The young leaves and tender tips of the stems can be cooked and eaten as a green vegetable.

2. **Hydrophyllum canadense** L. Broadleaf Waterleaf Map 1805

Flowers late May–July.

Occurs in rich woods, moist wooded slopes and ledges, and base of bluffs in ravines and along streams. Eastern Missouri north to Marion County, west to Boone, Phelps, and Butler counties.

Ranges from Vermont to Michigan and Ontario, Illinois and Missouri, south to Georgia and Alabama.

---

Plate no. 295. **1.** Hydrophyllum canadense, × $^2/_7$. **2.** Hydrophyllum appendiculatum, × $^2/_7$; Details from Small, The New York Botanical Garden. **3.** Ellisia Nyctelea, × $^2/_7$; Details from Small, The New York Botanical Garden. **4.** Phacelia bipinnatifida, × $^2/_7$.

PLATE NO. 295

The dark green, maple-shaped leaves form a good ground cover for woodland wildflower gardens.

Like *H. virginianum*, the young leaves and tender tips of the stems can be cooked and eaten as a green vegetable.

3. **Hydrophyllum appendiculatum** Michx.
Woollen Breeches Map 1806
Flowers April–July.

Occurs in rich woods, on steep slopes and at the base of bluffs, thickets, and wooded valleys. Throughout Missouri, but not recorded from a number of western counties.

Ranges from Ontario and Minnesota, south to Pennsylvania, Tennessee, Missouri, and Kansas.

Two variations occur in the state:

Corolla lavender, pink-purple, or dark purplish . . . 3a. H. APPENDICULATUM f. APPENDICULATUM
Corolla white . 3b. H. APPENDICULATUM f. ALBUM

3a. **Hydrophyllum appendiculatum** f. **appendiculatum** Map 1806
*Hydrophyllum appendiculatum* Michx. [G, BB, P & S, Steyerm.]
This is the common type found in Missouri.

3b. **Hydrophyllum appendiculatum** f. **album** Steyermark Map 1806

Known only from Johnson County, west-central Missouri (slopes bordering wooded valleys, Knobnoster State Park, 3–5 mi. southwest of Knobnoster, May 19, 1949, *Steyermark 67915*, holotype in Chi. Nat. Hist. Mus. Herb.).

The upper surface of the basal and young leaves are usually mottled gray and pale green. The species is often dominant as a ground cover on woodland slopes and at base of bluffs.

## 2. **Ellisia** L.

**Ellisia Nyctelea** L. Aunt Lucy Map 1807
Also called Nyctelea.
Flowers April–June.

Occurs in moist woods, thickets, alluvial soils of river flood plains and stream banks, waste and cultivated ground, and along railroads. Throughout Missouri, except in some of the southeastern Ozark and southeastern lowland counties.

Ranges from Indiana, Illinois, Michigan, and Manitoba to Alberta, south to Arkansas, Oklahoma, and Colorado; also from New York, New Jersey, and Pennsylvania to North Carolina, and locally introduced in New England.

The tiny flowers are whitish to pale bluish.

## 3. **Phacelia** Phacelia

The following treatment is in accord with the most recent revision of a section of the genus by Dr. Lincoln Constance (Contr. Gray Herb. 168: 1–46. 1949; Proc. Am. Acad. Arts & Sci. 78: 135–47. 1950).

a. Divisions of leaves mostly 10–40 mm. broad; complete leaf, including the divisions, mostly 40–130 mm. broad and 70 mm. or more long (including petiole); gland-tipped hairs covering calyx-lobes, flower-stalks (pedicels), and branches of inflorescence . . . . . . . . . . 5. P. BIPINNATIFIDA
a. Divisions of leaves 1–15 mm. broad; complete leaf, including the divisions, mostly 5–35 mm. broad and 10–60 mm. long; only glandless hairs usually present on calyx, flower-stalks (pedicels) and branches of inflorescence, or, if gland-tipped hairs occur, then very sparsely so . . . . . . . . *b*
  b. Flowers 2–6 in each inflorescence; all the leaves with petioles (leaf-stalks), the petioles of the upper leaves nearly as long as the leaf-blade; corolla with a short tube, tubular-bell-shaped, 2–4 mm. broad, 3–5 mm. long . . . . . . . . . . . . . . . . . . . . 4. P. RANUNCULACEA
  b. Flowers 8–30 in each inflorescence; upper leaves without stalks (sessile) or with very short stalks (petioles) much shorter than length of leaf-blade; corolla without a definite tube, more or less rotate (flat and wheel-shaped), 5–13 mm. broad, 6–10 mm. long . . . . . . . . . . . . *c*
    c. Outer surface of corolla-lobes glabrous (without hairs) or nearly glabrous; ovules or seeds 2 on each placenta (4 ovules or 4 seeds altogether in ovary or capsule) . . . . . . 3. P. PURSHII
    c. Outer surface of corolla-lobes hairy; ovules usually 4 to each placenta; seeds 6–8 altogether in the capsule . . . . . . . . . . . . . . . . . . . . . . . . . . *d*
      d. Hairs of uppermost part of stem and of flower-stalks (pedicels) mostly spreading; corolla-

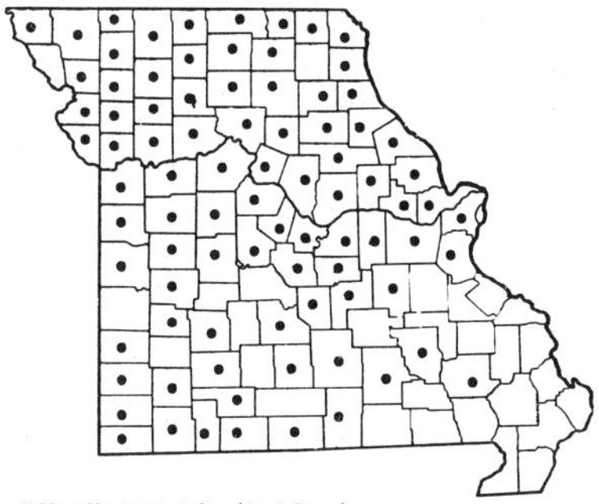

1807 Ellisia Nyctelea (Aunt Lucy)

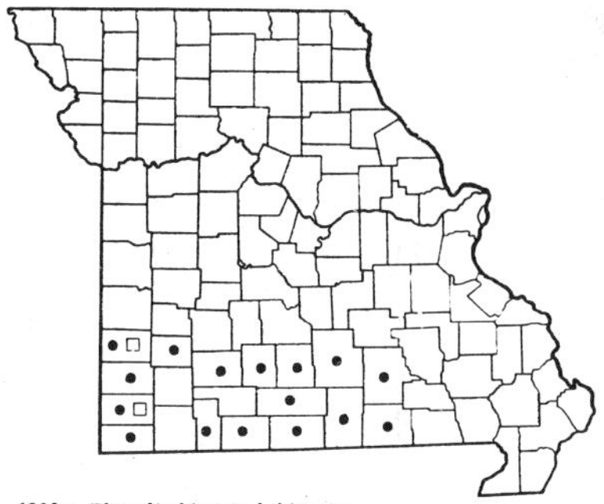

1808 • Phacelia hirsuta f. hirsuta
1808 □ Phacelia hirsuta f. albiflora

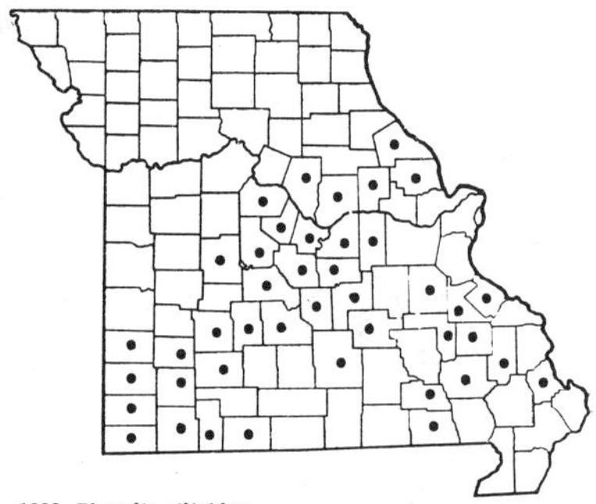

1809 Phacelia gilioides

lobes with entire (untoothed) margins; lobes of the main stem-leaves mostly obtuse (blunt) 1. P. HIRSUTA

d. Hairs of uppermost part of stem and of flower-stalks (pedicels) strigose with hairs appressed (lying parallel to or pressed against surface); corolla-lobes with toothed or somewhat irregularly fringed margins; lobes of the main stem-leaves usually acute (short-pointed) . 2. P. GILIOIDES

1. **Phacelia hirsuta** Nutt. Map 1808

Flowers late April–June.

Occurs in rocky fallow fields and prairies, sandy and rocky alluvial soils in valleys, glades, ledges of bluffs, rocky or sandy open woods, and roadside banks. Southern Missouri east to Shannon and Oregon counties, north to Shannon, Texas, Wright, Webster, Greene, Dade, and Barton counties.

Ranges from Missouri and Kansas, south to Louisiana and Texas.

Two variations occur in Missouri:

Corolla bluish-lavender with a whitish center 1a. P. HIRSUTA f. HIRSUTA
Corolla white . . . 1b. P. HIRSUTA f. ALBIFLORA

1a. **Phacelia hirsuta** f. **hirsuta** Map 1808

*Phacelia hirsuta* Nutt. [G, BB, P & S, Steyerm.]

This is the common type found.

1b. **Phacelia hirsuta** f. **albiflora** Palmer & Steyerm. Map 1808

Known only from Barton (sandy open woods, 2½ miles southeast of Verdella, May 5, 1955, *Palmer 59803*) and Newton (open upland bank along roadside, 1 mile south of Granby, May 15, 1957, *Palmer 65070*, lectotype in Chi. Nat. Hist. Mus. Herb.) counties, southwestern Missouri.

Many of the specimens previously included under this species by Palmer & Steyermark in their *Annotated Catalogue* are now referred to *P. gilioides*.

This and the following two species do well in rock or woodland wildflower gardens. Although they are winter annuals, they may be grown from collected seed or transplanted individuals allowed to set seed. Once established, they make a desirable addition to the spring flower display. They favor open situations in loose or well-drained soils.

2. **Phacelia gilioides** Brand Map 1809

*Phacelia dubia* (L.) Small var. *gilioides* (Brand) Gl. [BB]

Flowers late April–June.

Occurs in openings in woodland, ledges of bluffs, rocky open glades, usually of limestone, also sandstone, low rich woodland, wet meadows, alluvial soils and gravelly washes along streams and in valley bottoms, along railroads and roadsides. Throughout most of the Ozark region north to Pike, Montgomery, Callaway, Boone, Cooper, Benton, and Barton counties; absent from a few southern counties.

Ranges from Missouri and Kansas to Oklahoma.

This species has previously been confused with both *P. hirsuta* and *P. Purshii*, and some of the Missouri records, previously assigned to *P. hirsuta* and *P. Purshii* by Palmer and Steyermark in their *Annotated Catalogue* and by Steyermark in his *Spring Flora* are here referred to *P. gilioides*.

Constance (Contr. Gray Herb. 168: 31–32. 1949, and Proc. Am. Acad. Arts & Sci. 78: 144. 1950) has suggested that this species probably originated as a result of hybridization between *P. hirsuta* and *P. Purshii* because of its intermediate position geographically and morphologically between the other species.

3. **Phacelia Purshii** Buckl. Miami Mist Map 1810

Flowers April–June.

Occurs in low, rich or alluvial sandy or gravelly soils bordering streams, alluvial thickets, open wood-

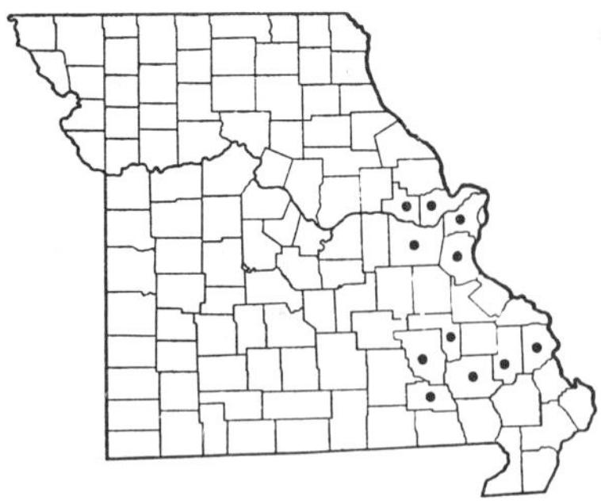

1810 Phacelia Purshii

1811 Phacelia ranunculacea

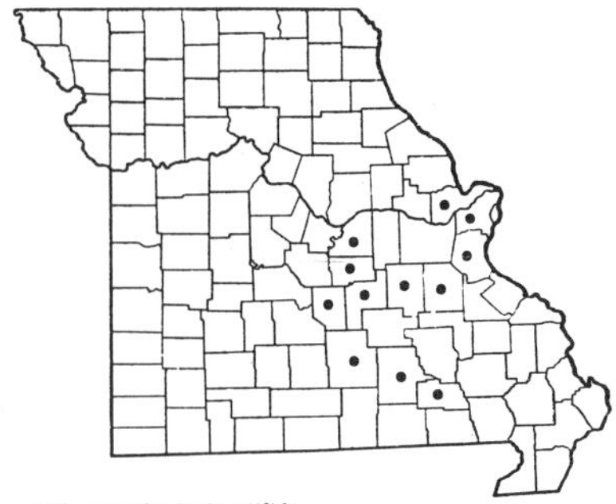

1812 Phacelia bipinnatifida

land, ledges and slopes along bluffs. Eastern Missouri north to St. Charles and Warren counties, south to Cape Girardeau, Bollinger, Wayne, and Carter counties, and west to Franklin, Iron, and Reynolds counties.

Ranges from Georgia and Alabama, north to Pennsylvania, Ohio, Indiana, Illinois, and Missouri.

Most of the records previously cited for this species by Palmer and Steyermark in their *Annotated Catalogue* are now referred to *P. gilioides*. Both *P. gilioides* and *P. Purshii* have the margins of the corolla-lobes toothed, but in *P. Purshii* the margins are more deeply fringed (fimbriate) with longer, more slender, projections, whereas in *P. gilioides* the teeth are usually shorter and less prominent.

4. **Phacelia ranunculacea** (Nutt.) Const. Map 1811

Flowers April–May.

Occurs in rich low woodlands in valleys bordering streams, at the base of bluffs and slopes, and moist low woods. Southeastern Missouri in the lowland section of Cape Girardeau, Scott, Stoddard, Pemiscot, Dunklin, and Butler conties, and locally in the Ozark region in Texas County (along Roubidoux Creek, T33N, R12W, sect. 10, 6–6½ mi. [by air] southwest of Evening Shade, April 28, 1952, *Steyermark 73166*).

Ranges from Maryland, D.C., to Virginia; and from Indiana, Illinois, and Missouri, south to Tennessee and Arkansas.

At the Texas County locality the plants grew at the base of rich, wooded, north-facing slopes. As is characteristic of this annual, the plants occurred in dense stands, thousands of individuals being found in a small area. The stems are weak and with erect to spreading or even prostrate branches bearing small, dull green leaves mottled with gray on the upper surface. The small corollas are lavender, pale purple, or bluish-lilac.

The Missouri records of this species have been previously misidentified and erroneously recorded as *Nemophila microcalyx* (Nutt.) Fisch. & Mey. by Palmer and Steyermark in their *Annotated Catalogue* and by Steyermark in his *Spring Flora of Missouri*, as pointed out by Constance (Contr. Gray Herb. 168: 43. 1949).

5. **Phacelia bipinnatifida** Michx. Map 1812

Flowers April–June.

Occurs in rich rocky woods on slopes and ledges of bluffs, and along rocky banks of streams. Eastern Missouri north to St. Charles and Osage counties, west to Maries, Pulaski, and Texas counties, south to Carter County.

Ranges from South Carolina, Georgia, Alabama, and Arkansas, north to Virginia, West Virginia, Ohio, Indiana, Illinois, and Missouri.

This is a biennial plant with bluish-lavender or bluish-purple corollas. The early spring, basal, and seedling leaves are blotched above with gray. The author has grown this species for the past twelve years in his northern Illinois wildflower preserve, where it has prospered at a northern latitude, new flowering plants forming each year from natural annual reproduction.

Plate no. 296. **1.** Phacelia Purshii; *a.* Flower, × $1\frac{1}{5}$; *b.* Portion of stem, × $1\frac{1}{5}$. Details from Constance. **2.** Phacelia hirsuta, × $\frac{2}{5}$; *a.* Flower, × $1\frac{1}{5}$; *b.* Portion of stem, × $1\frac{1}{5}$. Details from Constance. **3.** Phacelia gilioides, × $\frac{2}{5}$; *a.* Flower, × $1\frac{1}{5}$; *b.* Portion of stem, × $1\frac{1}{5}$. Details from Constance. **4.** Phacelia ranunculacea, × $\frac{2}{5}$. Details from Small, The New York Botanical Garden.

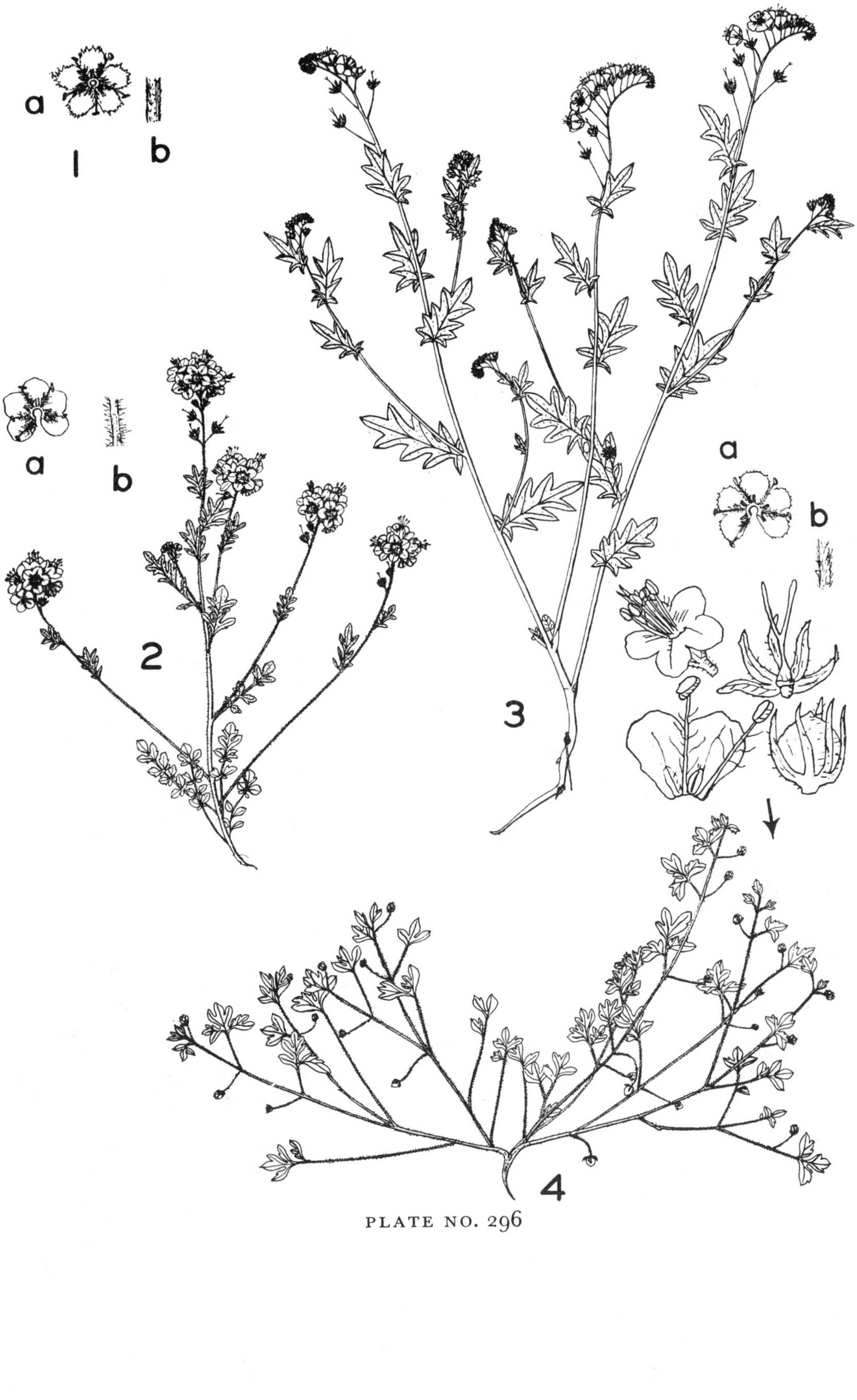


PLATE NO. 296

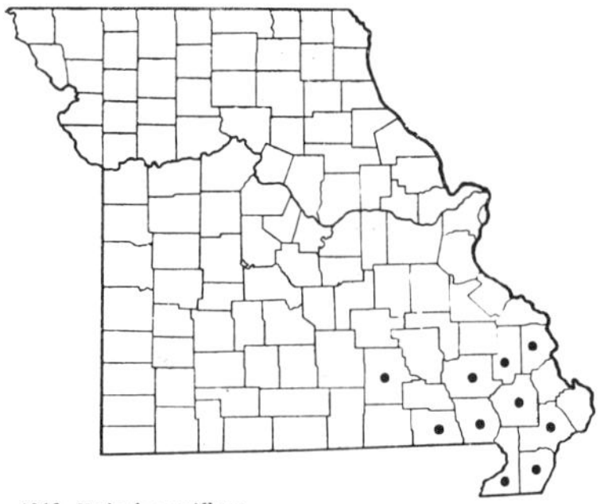

1813 Hydrolea uniflora

1814 Hydrolea ovata

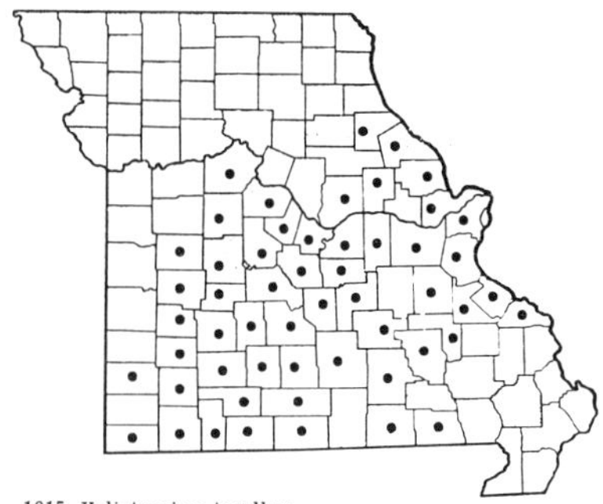

1815 Heliotropium tenellum

### 4. **Hydrolea** L.

Stems and leaves glabrous or nearly so; leaves lanceolate or narrowly elliptic; flowers along the sides of the stems in the axils (at the base) of the leaves; calyx glabrous (without hairs) or with short and inconspicuous hairs, the lobes of the calyx equaling the corolla . . . . . . . . . . . 1. H. UNIFLORA

At least the upper half of stem and midrib of lower surface of leaves densely hairy; leaves ovate; flowers at the top of the leafy stem; calyx conspicuously hairy with spreading hairs, the lobes of the calyx shorter than the corolla . . . . . . . . . . . . . . . . . . . . . . 2. H. OVATA

1. **Hydrolea uniflora** Raf. Map 1813

*Hydrolea affinis* Gray [P & S]

Flowers June–September.

Occurs in swampy woods, sloughs, along bayous and ditches. Southeastern Missouri lowlands north to Cape Girardeau, Bollinger, and Wayne counties, west to Ripley County, and locally in the Ozark area in upland sink-hole pond in Shannon County (around marshy edge of McCormack Pond, T27N, R5-4W, sect. 7, 3 mi. northwest of Bartlett, July 23, 1936, *Steyermark 12168*).

Ranges from Mississippi to Texas, north to Indiana, Illinois, Missouri, and Oklahoma.

In the southern states the bitter leaves are sometimes reduced to a pulpy mass applied as a poultice. But in the East Indies, a species, related to this and the following species, is used as a cooked vegetable, the young tips of the plants being employed for this purpose.

2. **Hydrolea ovata** Nutt. Map 1814

Flowers June–September.

Occurs in swamps, sloughs, and margins of upland ponds. Known only from the southeastern lowlands in Dunklin (Malden, July 20, 1893, *Bush 366*; swampy places, August 8, 1893, *Eggert*) and Butler (swamps near Neeleyville, August 8, 1893, *Eggert*) counties, and locally west in Howell (margin of sink-hole pond of Tingle Lake, T23N, R8W, south part sect. 32, 8 mi. south of West Plains, September 4, 1949, *Steyermark 69106*; small upland pond along south side of highway 80, 7½ mi. southwest of West Plains, July 24, 1949, *Steyermark 68590*) County. Two sterile collections from Howell County (*Steyermark 27048* and *Steyermark 40007*), previously identified as *H. uniflora*, are here referred to *H. ovata*. Both collections show a slight pubescence on the upper part of the stems and no. *40007* has the midrib of the lower leaf surface somewhat pubescent.

Ranges from Georgia to Texas, north to Missouri and Oklahoma.

The flowers of both species are showy with the blue broadly bell-shaped, open corollas. Spines are usually present in the axils of the leaves.

*Excluded Species*

**Nemophila microcalyx** (Nutt.) Fisch. & Mey.

As indicated under *Phacelia ranunculacea*, the Missouri records were misidentified and erroneously recorded by Palmer and Steyermark in their *Annotated Catalogue* and by Steyermark in his *Spring Flora of Missouri*, and are now referred to *Phacelia ranunculacea*.

---

Plate no. 297. **1.** Hydrolea uniflora, × 2/5; After Gleason, details from Small, The New York Botanical Garden. **2.** Hydrolea ovata, × 2/5; After Gleason, details from Small, The New York Botanical Garden. **3.** Heliotropium tenellum, × 2/5; Details from Small, The New York Botanical Garden **4.** Heliotropium indicum, × 2/5; *a.* Fruit, × 1 1/5. **5.** Symphytum officinale, × 2/5; *a.* Nutlet, × 2 2/5; Details from Small, The New York Botanical Garden. **6.** Coldenia Nuttallii, × 2/5.

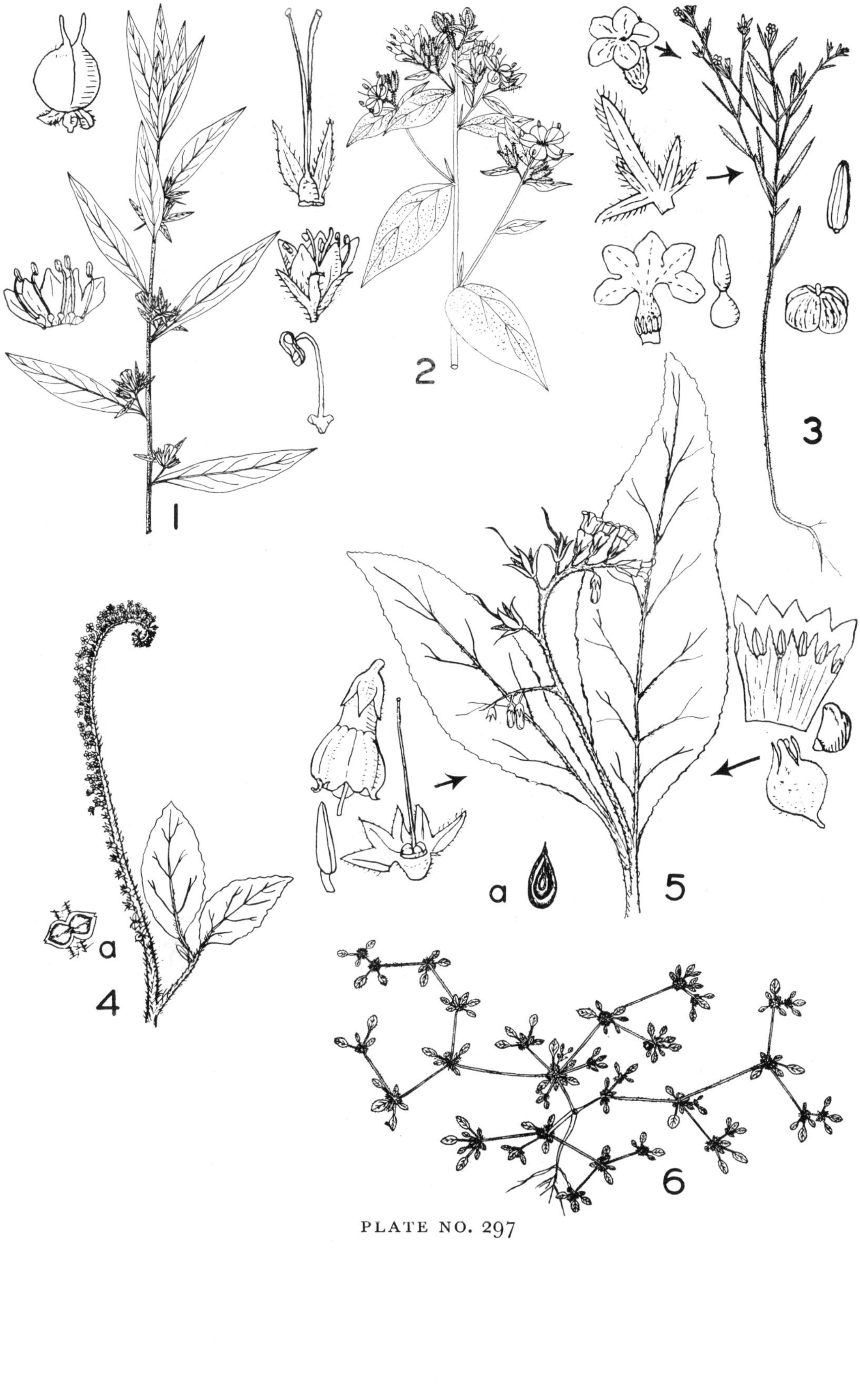


PLATE NO. 297

## Fam. **BORAGINACEAE** (Borage Family)

a. Stems and leaves completely smooth and glabrous (without hairs) . . . . . . . . . 10. MERTENSIA
a. Stems or leaves or both more or less hairy . . . . . . . . . . . . . . . . . . *b*
b. Leaves of stem very small, 3–7 mm. long, most of them on distinct petioles (stalks) . . 2. COLDENIA
b. Leaves much larger, 25 or more mm. long, most of them sessile (without petioles), or, if petioled (stalked), then the leaf blades 25 mm. or more broad . . . . . . . . . . . . . . . . *c*
c. White-hairy plants with white flowers; leaves very narrow, 1–4 mm. broad; plants of limestone glades only . . . . . . . . . . . . . . . . . . . . 1. HELIOTROPIUM TENELLUM
c. Without the above combination of characters . . . . . . . . . . . . . . . . *d*
d. Upper and middle stem-leaves clasping the stem at their heart-shaped (cordate) base; flowers pale blue . . . . . . . . . . . . . . . . . 7. CYNOGLOSSUM VIRGINIANUM
d. Stem-leaves not clasping the stem with a heart-shaped base, or, if rarely clasping the flowers yellow or orange . . . . . . . . . . . . . . . . . . . . . . . . . *e*
e. All the leaves on definite petioles (stalks) . . . . . . . . . 1. HELIOTROPIUM INDICUM
e. Main stem-leaves sessile (without stalks) or only the lowest leaves with stalks . . . . . *f*
f. Leaf-tissue of base of leaves continuing down (decurrent) the stem . . . 3. SYMPHYTUM
f. Stem lacking any leaf-tissue from the base of the leaves . . . . . . . . . . *g*
g. None of the flowers or fruits or their stalks (pedicels) with a green leaf-like bract at their base . . . . . . . . . . . . . . . . . . . . . . . . . *h*
h. Corolla blue with a yellow eye; fruits smooth and shining; plants usually of wet ground, in or bordering water . . . . . . . . . 8. MYOSOTIS SCORPIOIDES
h. Corolla reddish-purple or dull red, rarely white; fruits covered with hooked prickles; plants of usually dry rocky fields, waste ground, open places. . . 7. CYNOGLOSSUM OFFICINALE
g. At least the lower flowers, fruits, or their stalks (pedicels) accompanied by a green leaf-like bract at their base . . . . . . . . . . . . . . . . . . . . *i*
i. Corolla brilliant blue, rarely pink or white, irregular, longer on the upper side, the lobes of unequal length; stamens of unequal length, all protruding (exserted) beyond the corolla . . . . . . . . . . . . . . . . . . . 4. ECHIUM
i. Without the above combination of characters; corolla blue, white, greenish, cream-colored, orange, or yellow, regular, with the lobes of more or less equal length; stamens of equal length, included within and not protruding beyond the corolla . . . . . . . . . . . . . . . . . . . . . . . *j*
j. Lobes of corolla acutely pointed at tip, erect; corolla greenish-white or greenish; styles extended (exserted) beyond the corolla and persisting on the young fruits . . . . . . . . . . . . . . . . . 5. ONOSMODIUM
j. Lobes of corolla rounded or obtuse (blunt), spreading; corolla white, yellowish-white, greenish-white, blue, orange, or pale to bright yellow; styles not extended beyond the corolla or persisting on the young fruit. . . . *k*
k. Main leaves of stem with prominent side (lateral) nerves, ovate-oblong or ovate-lanceolate, mainly 15–70 mm. broad, broadest near the middle . . *l*
l. Corolla yellow, the tube longer than the calyx; stems and leaves conspicuously rough-hairy . . . . . . . . . . . 9. AMSINCKIA
l. Corolla yellowish-white, creamy, pale yellow, greenish-white, white or pale blue, the tube shorter than or only equaling the calyx; stems and leaves not conspicuously hairy or rough-hairy . . . . . . . *m*
m. Corolla white or pale blue; stigma 1, solitary, not split; fruiting stalks recurved or nodding; fruits bearing prickles . . . 12. HACKELIA
m. Corolla yellowish-white, creamy, pale yellow, or greenish-white; stigmas 2 or style split; fruiting stalks erect; fruits smooth. . . 6. LITHOSPERMUM LATIFOLIUM
k. Main leaves of stem with 1 main midrib, but side (lateral) nerves absent or only faint, if present, mostly linear, linear-oblong or lanceolate, mainly 3–15 (rarely to 18) mm. broad, about the same width throughout or broadest at the base . . . . . . . . . . . . . . . . . . . *n*
n. Corolla blue or sometimes white; fruits bearing prickles . . 11. LAPPULA
n. Corolla white, pale bluish-white, yellow or orange; fruits smooth, wrinkled, or rough, but not bearing prickles . . . . . . . . . *o*

o. Each flower, flower-stalk, or fruit-stalk with a leaf-like bract at its base; stigmas 2 or style split into 2 . . . 6. LITHOSPERMUM

o. A leaf-like bract absent from the base of some or most of the flowers, flower-stalks, or fruit-stalks; stigma 1, solitary, not split . . . *p*

p. Corolla yellow; corolla-tube much longer than calyx; throat of corolla open; calyx equally 5-lobed; none of the hairs on the calyx hooked or gland-tipped; stems and leaves conspicuously rough-hairy with long hairs . . . 9. AMSINCKIA

p. Corolla white to pale blue; corolla-tube about equaling or only slightly longer than calyx; throat of corolla partly closed by 5 obtuse appendages; calyx unequally lobed, 3 of the lobes shorter than the other 2; at least some of the hairs of the calyx distinctly hooked or gland-tipped; stems and leaves short-hairy . . . 8. MYOSOTIS

## 1. **Heliotropium** L. Heliotrope

Leaves linear, 1–4 mm. broad, sessile; plants grayish, covered with white hairs; corolla white; plants of limestone glades . . . 1. H. TENELLUM

Leaves ovate or oval, 20–80 mm. broad, petioled (with stalks); plants not presenting a gray or white-hairy appearance; corolla bluish; plants of moist, low, or alluvial soils, along streams and in valley bottoms . . . 2. H. INDICUM

### 1. **Heliotropium tenellum** (Nutt.) Torr. Map 1815

Flowers June 1–August.

Occurs on limestone glades and rocky prairies of calcareous substrata. Ozark region of southern and central Missouri north to Ralls, Montgomery, Callaway, Cole, Moniteau, Cooper, Saline, Henry, and Jasper counties.

Ranges from Alabama to Texas, north to Kentucky, Missouri, Kansas, and Oklahoma.

This annual is a characteristic plant of the limestone barrens of the unglaciated Ozark region, where it endures the hot dry summers and parched rocky open exposures which it favors. It is well-adapted to such arid conditions by virtue of the annual habit, slender wiry stems, linear leaves, and dense, protective covering of hairs over the entire plant. During extremes of drouth the margins of the leaves become revolute to lessen the area of leaf surface exposed and consequent reduced transpiration.

### 2. **Heliotropium indicum** L. Indian Heliotrope Map 1816

Also called Turnsole.

Flowers July–November.

Occurs in low, moist alluvial woods, along muddy banks, gravel bars, and waste ground.

Southern and central Missouri north to Lincoln, Gasconade, Osage, Cole, Saline, and Jackson counties.

Naturalized from Asia, but supposedly native to Brazil; introduced into the United States from Florida to Texas, north to Virginia, West Virginia, Kentucky, Indiana, Illinois, Missouri, and Oklahoma; found as a weed in all tropical and subtropical countries of both hemispheres.

The leaves are rugose-wrinkly on the upper surface and dark green.

## 2. **Coldenia** L.

### **Coldenia Nuttallii** Hook. Map 1817

Flowers May–July.

Known only from Crawford County, eastern Ozarks (base of open rocky, south-facing slopes of hill in former Steelville High School grounds, above Frisco R.R., Steelville, May, 1935, *Cora Shoop*, in Chi. Nat. Hist. Mus. Herb.).

Introduced in Missouri; native of the western United States, from New Mexico to California, north to Wyoming, Idaho, and Washington.

This is a sprawling finely gray-hairy annual plant with forking, dichotomous branches bearing small, petioled conspicuously impressed-veined leaves, which give the appearance of growing nearly in whorls (falsely verticillate). The small lavender-blue, pink, or white corollas are 3–4 mm. long.

Development by the Works Progress Administration (WPA) the following year after the plant was originally collected destroyed the only known station. It is possible that the species may eventually be found as an introduced plant along other railroads in Missouri.

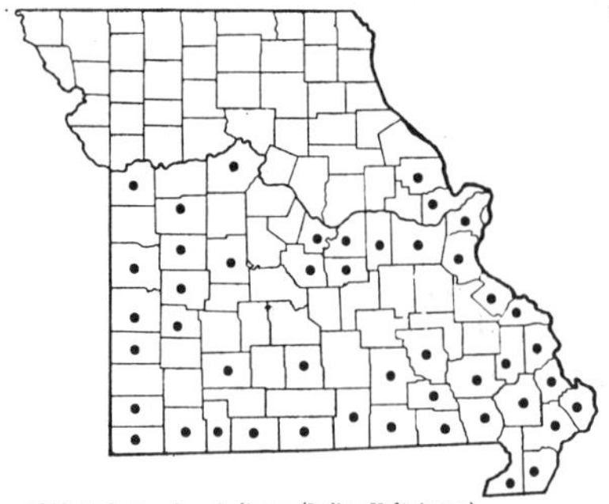

1816 Heliotropium indicum (Indian Heliotrope)

1817 Coldenia Nuttallii

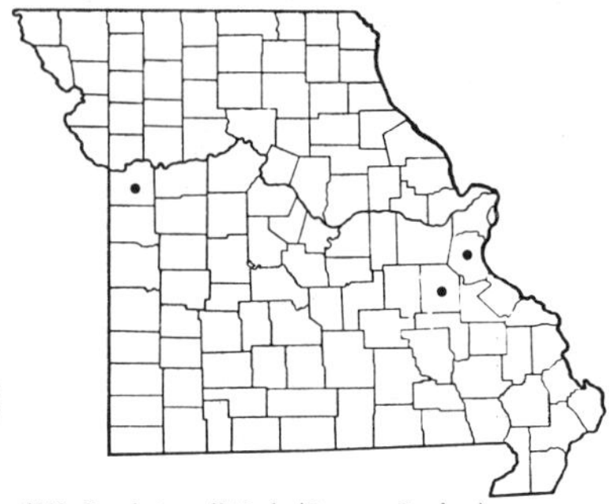

1818 Symphytum officinale (Common Comfrey)

## 3. **Symphytum** L. Comfrey

**Symphytum officinale** L. Common Comfrey Map 1818

Flowers June–August.

Occasionally escaped from cultivation along roadsides and waste ground. Known from Jefferson (roadside below Kimmswick, 1894, no collector, *N. M. Glatfelter* herb.), Washington, and Jackson (July 10, 1893, *Bush*) counties.

Native of Europe; introduced and naturalized in North America from Newfoundland to Ontario, south to Georgia, Tennessee, and Louisiana.

The young leaves and stems are sometimes cooked as a vegetable. The plant has been used for forage in Europe.

## 4. **Echium** L. Viper's Bugloss

**Echium vulgare** L. var. **vulgare** Blue-weed Map 1819

*Echium vulgare* L. [G, P & S, Steyerm.]

Also called Blue Devil, Blue Thistle.

Flowers mid-May–September.

Occurs in open waste ground, along roadsides, gravel bars, and gravelly sandy washes in valleys. Scattered in northern, central, and eastern Missouri in a few sections, most common in the eastern Ozarks.

Native of Europe; introduced and naturalized in North America from Quebec to Ontario and Minnesota, south to Georgia and Texas.

The flowers of this biennial plant are quite handsome, a deep blue when fully open, pink in bud. Entirely pink (*E. vulgare* var. *vulgare* f. *roseum* F. Zimm.) and white (*E. vulgare* var. *vulgare* f. *albiflorum* R. Hoffm.) forms are known, but have not been found thus far in Missouri.

Typical *E. vulgare* var. *vulgare* has the hairs of the leaves and inflorescence slender and without swelling at the base, and the inflorescence is short and dense, 3–7 cm. thick, and rarely more than 5 cm. long. In var. *pustulatum* (Sibth. & Sm.) Coincy the hairs are swollen at the base (pustular) and the looser, broader inflorescence is 7–30 cm. thick, and up to 15 cm. long.

The stiff, bristly hairs, which cover the entire plant, may be irritating to some persons, causing upon contact a dermatitis with itching and inflammation.

## 5. **Onosmodium** Michx. False Gromwell

a. Lower part of stem without hairs or with only a few scattered hairs . . . . . . . 3. O. SUBSETOSUM
a. Stem densely hairy from base to summit . . . . . . . . . . . . . . . . . . . . . . *b*
  b. Hairs on both sides of leaves closely appressed (lying parallel to or closely pressed against surface); fruits rounded, not constricted at the base, usually without any pits on the surface . 1. O. OCCIDENTALE
  b. Hairs on both sides of leaves spreading; fruits constricted to a very short neck at the very base, usually with tiny pits appearing on the surface . . . . . . . . . . . 2. O. HISPIDISSIMUM

---

Plate no. 298. **1.** Onosmodium occidentale, × 2/5; Details from Small, The New York Botanical Garden. **2.** Onosmodium hispidissimum, × 2/5. **3.** Echium vulgare, × 2/7; Details from Small, The New York Botanical Garden. **4.** Onosmodium subsetosum, × 2/5.

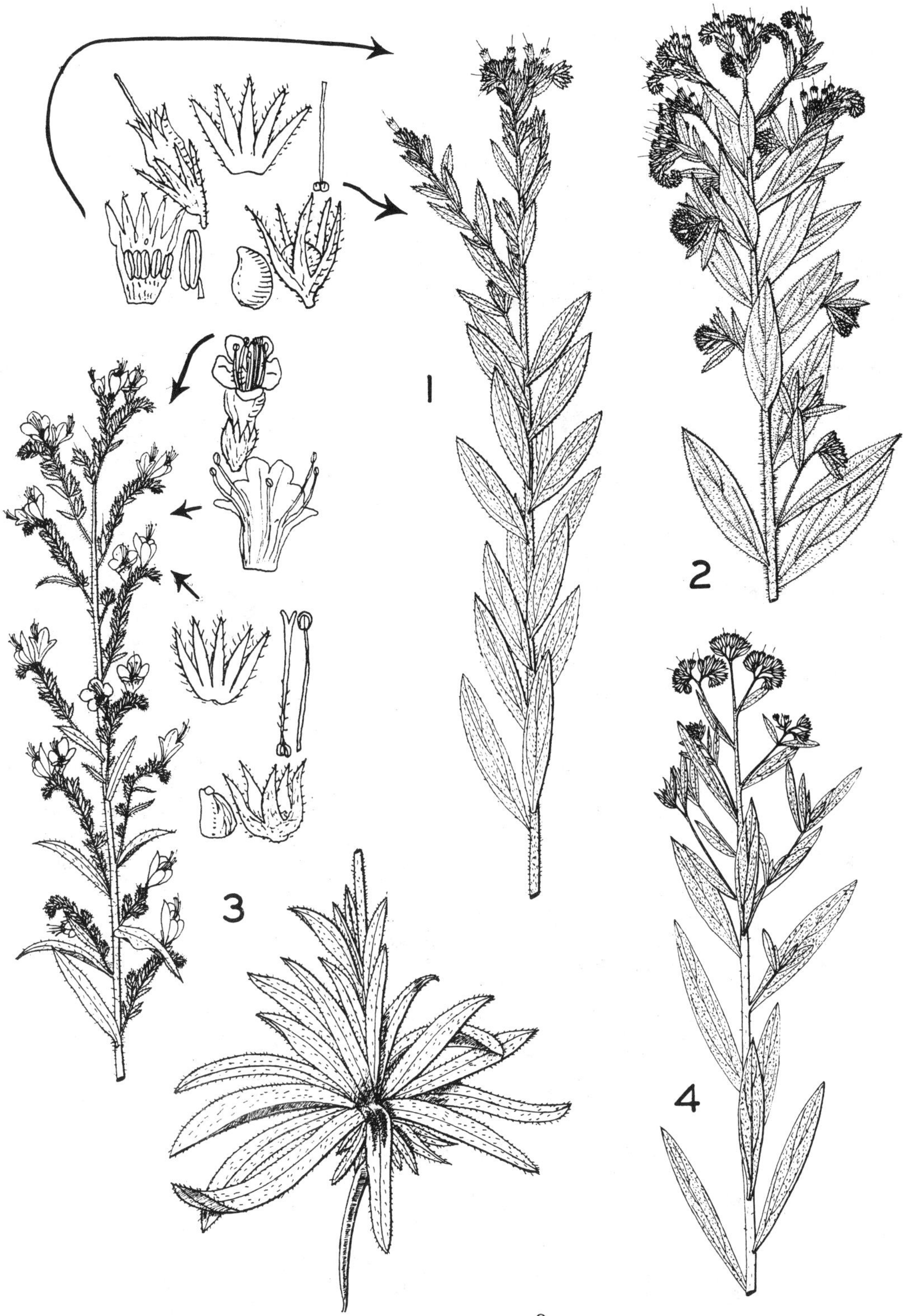


PLATE NO. 298

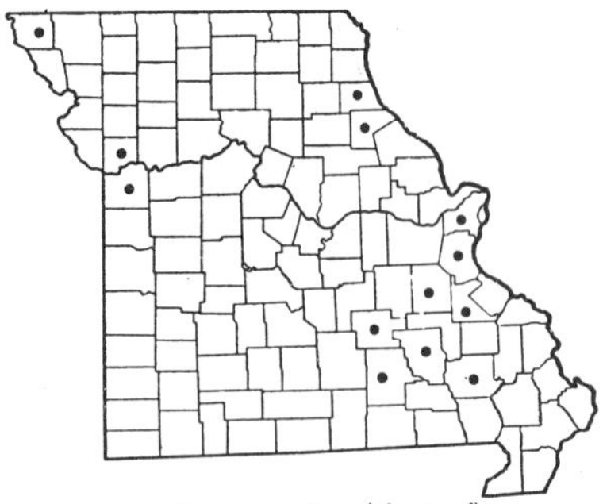

1819 Echium vulgare var. vulgare (Blue-weed)

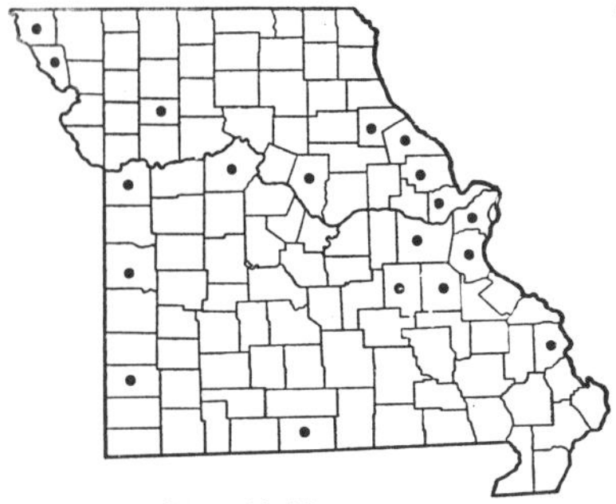

1820 Onosmodium occidentale

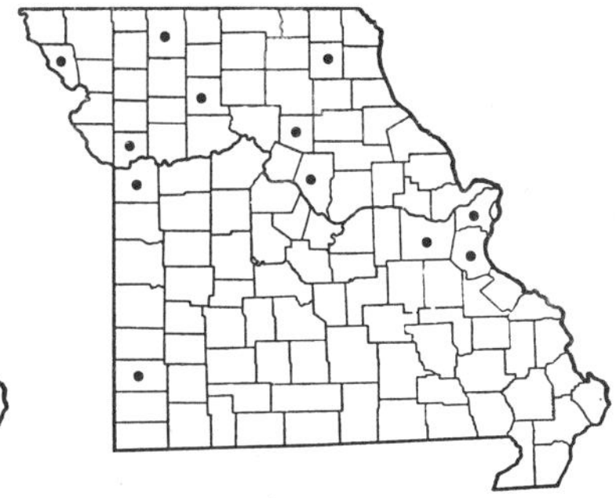

1821 Onosmodium hispidissimum

1. **Onosmodium occidentale** Mackenz.
Map 1820

*Onosmodium occidentale* var. *sylvestre* Mackenz. [P & S, Steyerm.]

Flowers May–July.

Occurs in rocky prairies, loess hills, glades, open rocky woods, thickets, and open ground. Scattered in Missouri, not recorded from most of the Ozark region.

Ranges from Minnesota to Saskatchewan, south to Illinois, Missouri, Oklahoma, Texas, and New Mexico.

There appears to be no justification for retaining var. *sylvestre* as a recognizable variation as there is no constancy in the character of more shaggy hairiness previously considered for this taxon.

2. **Onosmodium hispidissimum** Mackenz.
Map 1821

*Onosmodium hispidissimum* var. *macrospermum* Mackenz. & Bush [G, P &S, Steyerm.]

Flowers May–July.

Occurs on dry rocky prairies, glades, rocky outcrops, thickets, and sometimes open woodland. Scattered in northern, central, and western Missouri, apparently absent from most of the Ozark section.

Ranges from New York and Ontario to Minnesota and Nebraska, south to North Carolina, Tennessee, Louisiana, and Texas.

There appears to be no justification for retaining var. *macrospermum* as a recognizable variation characterized by white fruits 4 mm. long. Plants, having the nutlets 3–4 mm. long and varying from brownish-tinged to white, completely bridge the gap in variability.

3. **Onosmodium subsetosum** Mackenz. & Bush
Map 1822

Flowers May–June.

Occurs on limestone glades, barrens, and rocky prairies overlying calcareous substrata. Ozark region north to St. Louis, Montgomery, Boone, Cooper, Pettis, St. Clair, and Jasper counties.

Ranges from Missouri and Arkansas to Oklahoma.

This species presents an almost glabrous appearance in the yellowish-green or greenish-white stems and in the leaves covered with a fine, inconspicuous, appressed hairiness. The leaves vary from 6–10 cm. long and 1–3 cm. broad, the upper leaves generally having the broader dimensions. The fruits vary from 2–3.5 mm. long.

Like *Scutellaria Bushii*, *Penstemon Cobaea* var. *purpureus*, *Delphinium Treleasei*, and several others, this species is one of the endemics which has apparently originated in the Ozark region itself, and, therefore, has a long antecedent history.

## 6. **Lithospermum** L. Puccoon, Gromwell

a. Main leaves of the stem with prominent lateral (side) nerves, 15–70 mm. broad, mostly broadest near the middle, narrowly ovate or broadly elliptic and tapering at each end . . . . . . 2. L. LATIFOLIUM

a. Main leaves of the stem with 1 main midrib, the lateral (side) nerves absent or faint at most, mainly 3–15 (rarely to 18) mm. broad, about the same width throughout or broadest at the base, mostly linear, linear-oblong, or lanceolate . . . . . . . . . . . . . . . . . . . . b

b. Corolla white, very small, 5–8 mm. long, scarcely longer than the calyx; annual plant with slender root easily removed from the ground; fruits gray or grayish-brown, deeply wrinkled and roughened
1. L. ARVENSE

b. Corolla pale to bright yellow or orange, showy, 15–35 mm. long, the tube 7–30 mm. long, much

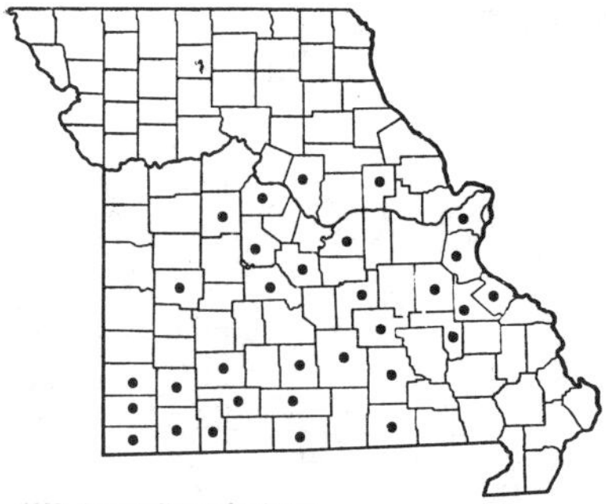

1822 Onosmodium subsetosum

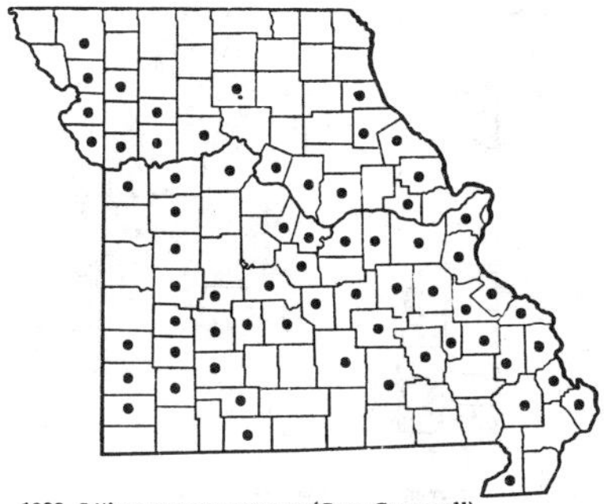

1823 Lithospermum arvense (Corn Gromwell)

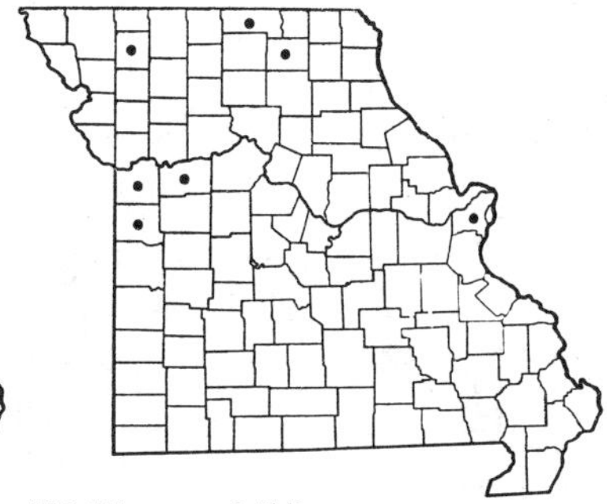

1824 Lithospermum latifolium

longer than the calyx; perennial plants with thick tough roots difficult to remove from the ground; fruits white, smooth . . . . . . . . . . . . . . . . . . . . . . . . . . *c*

c. Corolla with a tube 13–30 mm. long, 2–4 times the length of the calyx; corolla-lobes finely fringed or toothed on the margins; leaves narrowly linear, acutely pointed, 3–5 (–8) mm. broad 3. L. INCISUM

c. Corolla with a tube 7–14 mm. long, one-half to twice as long as the calyx; corolla-lobes entire (without teeth or irregularities on margins); leaves oblong, linear-oblong, or lanceolate, mostly obtuse (blunt) at tip, 7–15 mm. broad . . . . . . . . . . . . . . . . . *d*

d. Common species throughout Missouri; calyx-lobes 3–6 mm. long in flower, 6–8 mm. long in fruit, the hairs soft and ascending; hairs of stems and leaves soft and mainly appressed (lying parallel to or pressed against surface), without swollen bases; corolla-limb (expanded portion) 10–15 mm. broad, the lobes 5–6 mm. long; inside of corolla-tube glabrous (without hairs) at base . . . . . . . . . . . . . . . . . . . . 5. L. CANESCENS

d. Rare species of sandy soils of eastern and northern Missouri; calyx-lobes 8–11 mm. long in flower, 10–15 mm. long in fruit, the hairs stiff or spreading; hairs of stems and leaves rough-hairy, stiff, or spreading, with or without swollen bases; corolla-limb (expanded portion) 15–25 mm. broad, the lobes 7–9 mm. long; inside of corolla-tube hairy at base 4. L. CAROLINIENSE

1. **Lithospermum arvense** L. Corn Gromwell Map 1823

Also called Bastard Alkanet.

Flowers April–June.

Occurs in fallow and cultivated fields, waste ground, along roadsides and railroads. Throughout Missouri, but not recorded from a large section of northern Missouri.

Native of Europe; introduced and naturalized in North America from Nova Scotia and Maine to British Columbia, south to Florida, Louisiana, and California.

2. **Lithospermum latifolium** Michx. Map 1824

Flowers May–June.

Occurs in rich woods, on slopes, level valleys of alluvial soils near streams, and in thickets. Northern and central Missouri, south to St. Louis, Lafayette, and Cass counties. Rare and scattered in this range.

Ranges from New York to Minnesota, south to Tennessee, Arkansas, and Kansas.

The leaves, which vary in Missouri specimens from 2–6 cm. broad, are dark green above and pale gray beneath. The shining white fruits, which are 2.8–5 mm. long, and the general aspect of the plant resemble an *Onosmodium*, with which genus it has occasionally been confused, one of the specimens from Jackson County collected by Bush actually being referable to *Onosmodium occidentale*.

3. **Lithospermum incisum** Lehm. Yellow Puccoon Map 1825

*Lithospermum angustifolium* Michx. [P & S, Steyerm.]

Flowers April–June.

Occurs in prairies, sandy open ground, rocky open glades and barrens, dry open woods, and roadsides. Scattered in the state, but mainly in the western and southern half.

Ranges from Ontario to Manitoba, Wisconsin, and British Columbia, south to Indiana, Illinois, Missouri, Texas, New Mexico, and Utah.

This is a showy, bright yellow-flowered species. It is difficult to transplant. Plants which have been transferred to the author's wildflower garden in northern Illinois have not persisted after two or three years.

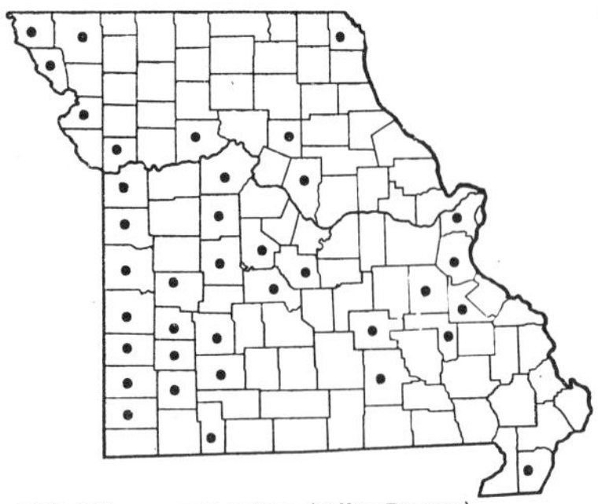

1825 Lithospermum incisum (Yellow Puccoon)

1826 Lithospermum caroliniense (Puccoon)

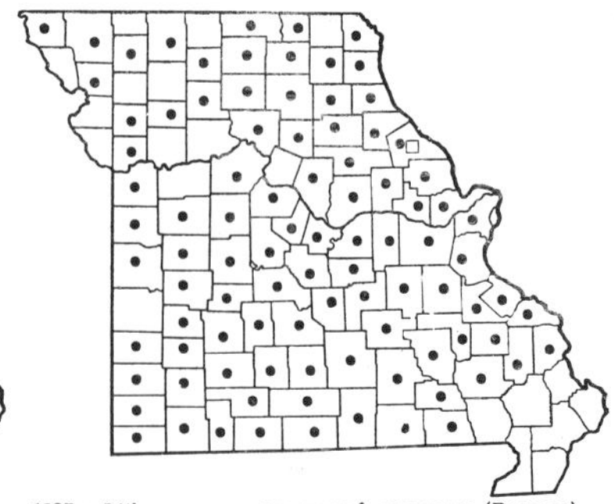

1827 ● Lithospermum canescens f. canescens (Puccoon)
1827 □ Lithospermum canescens f. pallidum

4. **Lithospermum caroliniense** (Walt.) MacM.
Puccoon Map 1826
*Lithospermum croceum* Fern. [G, P & S, Steyerm.]
*Lithospermum Gmelini* (Michx.) Hitchc.
Flowers April–June.

Occurs in sandy open ground, sandy prairies, sandy banks along roadsides and railroads. Mainly eastern Missouri from Clark to Stoddard and Dunklin counties, and locally in Nodaway County, northwestern Missouri.

Ranges from New York, Ontario, Ohio to Minnesota and Montana, south to Florida, Texas, and Nevada; also in Mexico.

This species has showy orange flowers, larger than in the following species. On repeated occasions the author has attempted to grow this from seed as well as from transplanted individuals, especial care having been taken not to disturb the long thick root and to bring back intact a large clump of the original sand. Notwithstanding these precautions, the species has never successfully withstood any transplantings, and has defied every effort to become established. On the other hand, the following common species of puccoon, *L. canescens*, transplants easily and becomes rapidly established in a wildflower garden.

The differences attributed to *L. croceum* do not hold true when contrasted with *L. caroliniense*, and I agree with the results of Dr. I. M. Johnston's recent studies (Journ. Arn. Arb. 33: 339–40. 1952), combining the two species under *L. caroliniense*.

5. **Lithospermum canescens** (Michx.) Lehm.
Puccoon Map 1827
Also called Orange Puccoon.
Flowers March–June.

Occurs in prairies, glades, dry open or rocky woods along roadsides and railroads. Throughout Missouri, except absent from the extreme southeastern lowland counties.

Ranges from Pennsylvania and Ontario to Manitoba and Saskatchewan, south to Georgia, Alabama, Mississippi, and Texas.

Two variations occur in Missouri:

Corolla orange or orange-yellow . 5a. L. CANESCENS f. CANESCENS
Corolla pale yellow or creamy-yellow . . . 5b. L. CANESCENS f. PALLIDUM

5a. **Lithospermum canescens** f. **canescens**
Map 1827

This is the common form found throughout the state.

5b. **Lithospermum canescens** f. **pallidum**
Steyerm. Map 1827

Known only from Pike County, northeastern Missouri (along southeast side of highway 54, ½ mi. southwest of junction with highway 154, T53N, R4W, west part sect. 25, 4 mi. west southwest of Curryville, April 30, 1952, *Steyermark 73223*).

Specimens transplanted in 1952 to the author's wildflower preserve in northern Illinois have retained the characteristic pale color of the corollas up to the present time.

A specimen collected in Polk County (natural prairie, T34N, R24W, sect. 4, 2.5 mi. northeast of Dunnegan along west side of highway 13, May 3, 1958, *Steyermark 86092*) approaches f. *pallidum*, but has the corolla deep yellow. It is somewhat intermediate in color intensity between the pale or creamy-yellow

Plate no. 299. **1.** Lithospermum arvense, × 2/7; Details from Small, The New York Botanical Garden. **2.** Lithospermum latifolium, × 2/7; Details from Small, The New York Botanical Garden. **3.** Lithospermum incisum, × 2/7. **4.** Lithospermum caroliniense, × 2/7. **5.** Lithospermum canescens, × 2/7. **6.** Cynoglossum officinale, × 2/7.

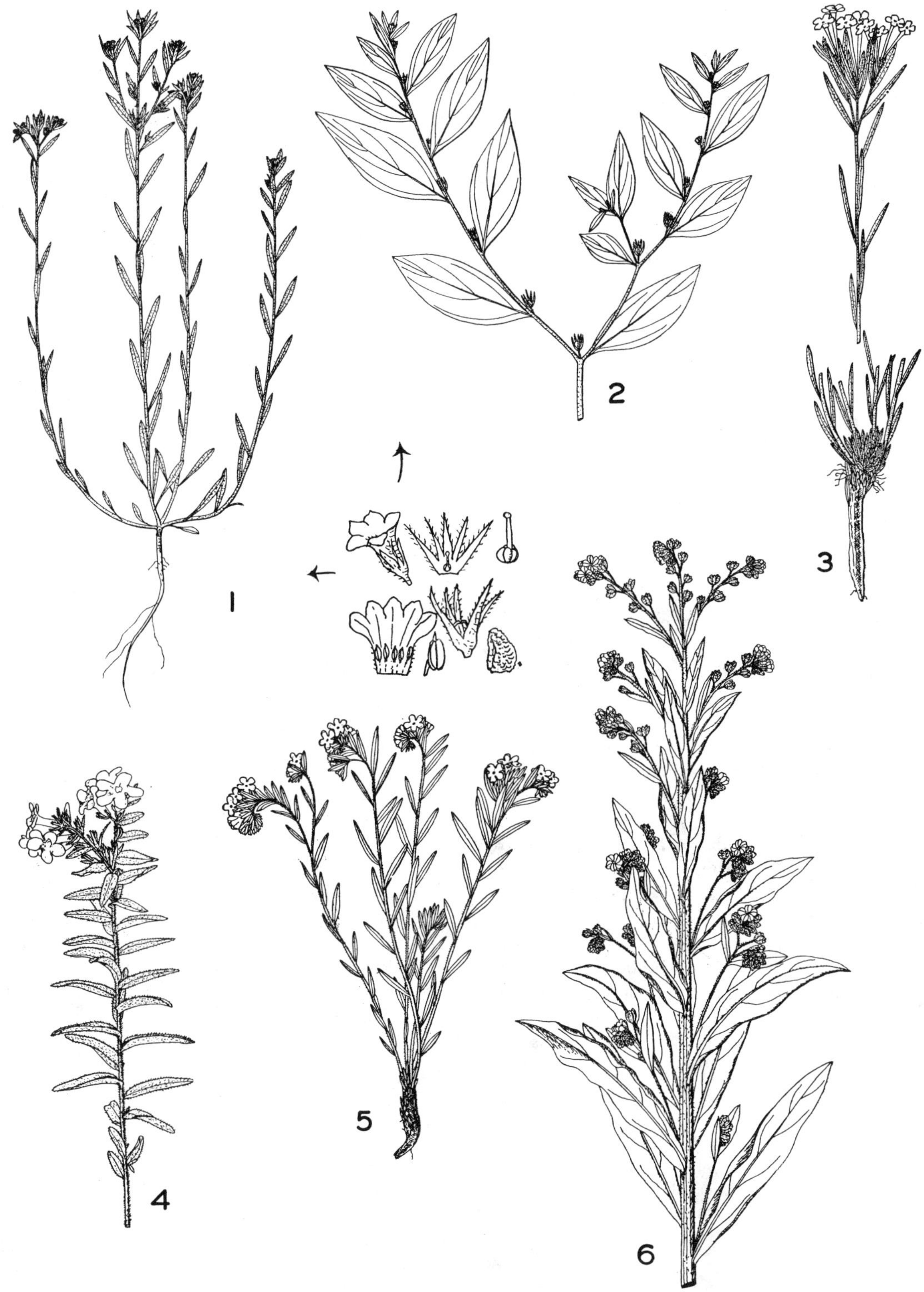


PLATE NO. 299

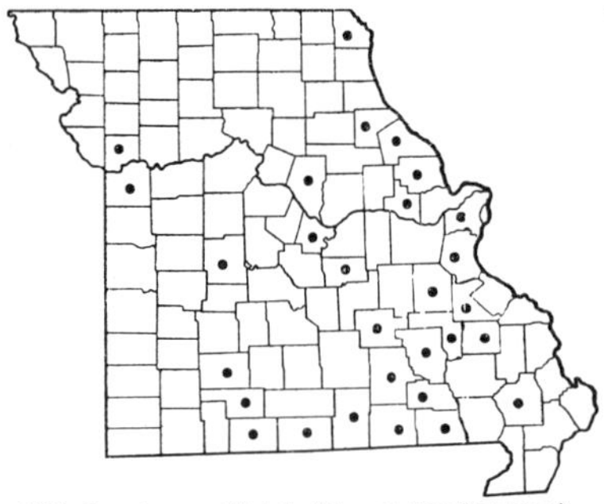

1828 Cynoglossum officinale (Common Hound's Tongue)

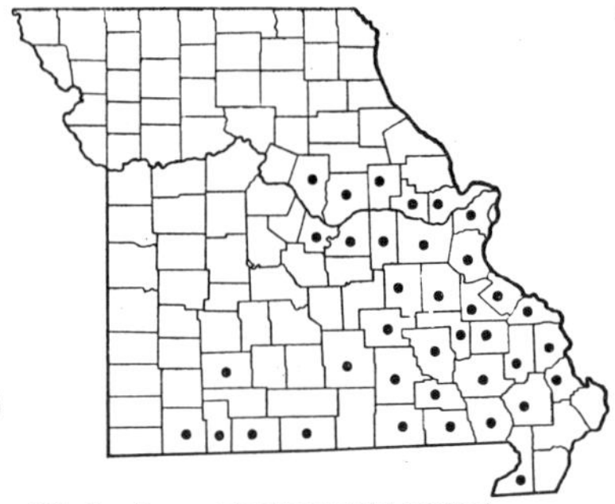

1829 Cynoglossum virginianum (Wild Comfrey)

1830 Myosotis scorpioides (Forget-me-not)

color of f. *pallidum* and the orange color of typical *L. canescens* f. *canescens*.

*Lithospermum canescens* transplants easily and makes a desirable addition to a rock garden or open sunny sections of a wildflower bed. The red juice of the roots of this species and *L. incisum* was at one time employed by some Ozark folk as a dye for coloring cloth.

## 7. **Cynoglossum** L. Hound's Tongue

15 to 40 leaves on the stem, those on the stem crowded; corolla reddish-purple or dull red, rarely white; fruits (nutlets) flattened on the upper (outer) face; plants of fields and open waste ground . 1. C. OFFICINALE
Only 2–10 leaves on the stem, those on stem rather widely separated; corolla pale blue or bluish-white; fruits (nutlets) convex on the upper (outer) face; plants of woodland . . . . . . . . 2. C. VIRGINIANUM

1. **Cynoglossum officinale** L. Common Hound's Tongue Map 1828

Flowers May–July.

Occurs in fields, rocky waste ground, eroded slopes, and along roadsides. Scattered in eastern, southern, and central Missouri, north to Clark, Boone, and Platte counties.

Native of Europe; introduced and naturalized in North America, from Quebec, Ontario, Manitoba, and British Columbia, south to Georgia, Alabama, Arkansas, and New Mexico.

The prickle-covered fruits become attached to the fur of animals and to clothing. The plant has a mouse-like, fetid odor. It contains an alkaloid stated to resemble *curare* in its action and occasionally is reported to cause dermatitis upon contact with sensitive individuals.

2. **Cynoglossum virginianum** L. Wild Comfrey Map 1829

Also known as Giant Forget-me-not.

Flowers April–June.

Occurs in rich or rocky woods, on slopes, ravine bottoms, and low bottom thickets. Ozark region north to St. Charles, Warren, Montgomery, Callaway, and Boone counties, west to Cole, Crawford, Dent, Texas, Greene, and Barry counties.

Ranges from Florida to Texas, north to Connecticut, Pennsylvania, West Virginia, Ohio, Indiana, Illinois, Missouri, and Oklahoma.

This is a showy species, when in flower. It does not transplant readily and is difficult to become established in a woodland wildflower garden.

## 8. **Myosotis** L. Forget-me-not

Corolla showy, blue with a yellow center, 6–9 mm. broad; stems smoothish or sparsely hairy; hairs on calyx short, straight, appressed (lying parallel to or closely pressed to surface); calyx equally cut, the 5 lobes equal in length; plants of wet or moist soil, growing in or bordering water . . . 1. M. SCORPIOIDES

---

Plate no. 300. **1.** Myosotis virginica var. virginica, × 2/5; *a*. Flower, × 2/5; Details from Small, The New York Botanical Garden. **2.** Myosotis scorpioides, × 2/5. **3.** Cynoglossum virginianum, × 2/5; Details from Small, The New York Botanical Garden.

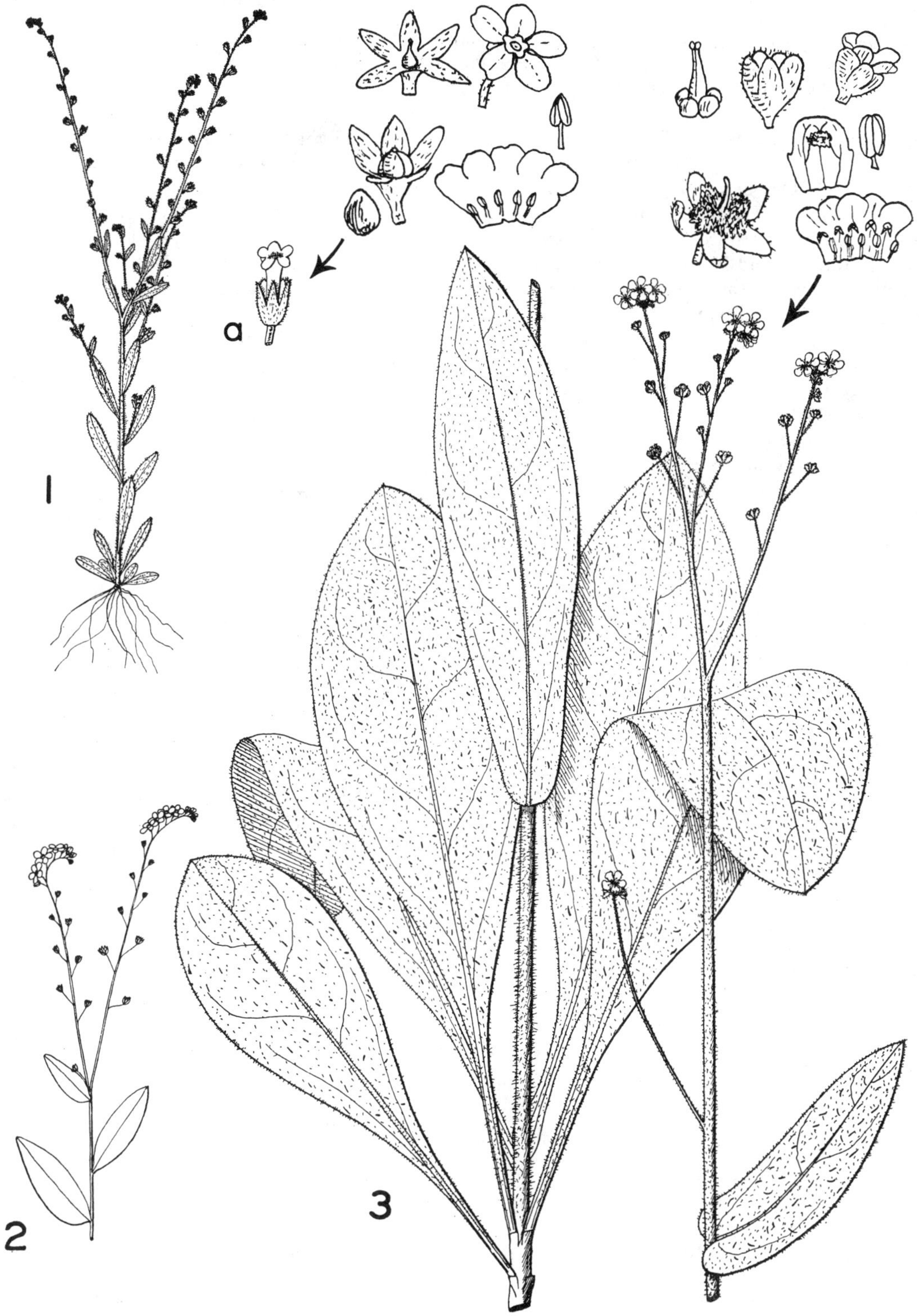


PLATE NO. 300

Corolla not conspicuous, white or tinged with pale blue, 1–2 mm. broad; stems noticeably hairy; at least the lowest hairs of the calyx with hooked tips, the hairs of the calyx spreading; calyx unequally cut, 3 of the lobes shorter than the other 2; plants of dry soils . . . . . . . . . . . 2. M. VIRGINICA

1. **Myosotis scorpioides** L. Forget-me-not Map 1830

Flowers April–October.

Occurs along wet margins of streams and spring-branches. Rarely escaped from cultivation, known in St. Louis, Franklin, and Newton (wet gravelly margins of Hickory Creek, near Neosho, October 29, 1953, *Palmer 57157*; same locality, October 12, 1953, *Palmer 57017*; April 19, 1954, *Palmer 57396*) counties.

Native of Europe; introduced and naturalized in North America from Newfoundland to Ontario, south to Georgia, Tennessee, and Louisiana; and from British Columbia to California.

This well-known cultivated plant, often planted around pools, along streams and wet places, rarely escapes to become naturalized.

2. **Myosotis virginica** (L.) BSP. Scorpion Grass Map 1831

Flowers April–May.

Occurs in fallow and cultivated fields, moist or dry woodlands, slopes above bluff escarpments, in upland or lowland, along roadsides, and railroads.

Two variations occur in Missouri:

Fruiting pedicels erect and nearly parallel the axis of the inflorescence; fruiting calyx 4–6 mm. long, persisting on the pedicel (stalk); fruits (nutlets) 1–1.5 mm. broad; main leaves of stem 2–10 mm. broad; lowest fruiting pedicels 5–20 mm. apart . . . . . 2a. M. VIRGINICA var. VIRGINICA

Fruiting pedicels spreading away from the axis of the inflorescence; fruiting calyx 5.5–9 mm. long, quickly falling; fruits (nutlets) 1.4–2.2 mm. broad; main leaves of stem 5–17 mm. broad; lowest fruiting pedicels 20–50 mm. apart . . 2b. M. VIRGINICA var. MACROSPERMA

2a. **Myosotis virginica** var. **virginica** Map 1831

*Myosotis verna* Nutt. [G]

*Myosotis virginica* (L.) BSP. [BB, P & S, Steyerm.]

Throughout Missouri, and probably in every county.

Ranges from Maine to Michigan and Minnesota, south to Florida, Tennessee, Oklahoma, and Texas; also Montana and Idaho to British Columbia, south to Wyoming and California.

2b. **Myosotis virginica** var. **macrosperma** (Engelm.) Fern. Map 1831

*Myosotis macrosperma* Engelm. [G, BB]

Scattered in southern and central Missouri in Dunklin, Cape Girardeau, St. Louis, Shannon, and Jasper counties.

Ranges from Florida to Texas, north to Maryland, D.C., Kentucky, Indiana, Illinois, and Missouri.

The status of *M. virginica* var. *macrosperma* has been shifted from species to variety and back to species (Fernald, Rh. 43: 636–37. 1941). A number of characters stressed by Fernald (p. 637. and *Gray's Man.*, eighth ed. p. 1204. 1950) overlap when applied to individual plants, and I do not believe that the specific segregation of *M. virginica* and *M. macrosperma* is justified. The measurements given for the length of the nutlets are variously interpreted by Fernald, Gleason, Deam, and other authors, as are also the dimensions of the length of the fruiting calyx (Deam, *Fl. Ind.*, p. 790. 1940; Gleason, *New Ill. Fl.* 3: 118. 1952; Fernald, *Gray's Manual*, eighth edition, p. 1204. 1950; and other authors). So far as the width of the leaves and length of the central fruiting racemes are concerned, there is definite overlapping in these characters.

The corolla is described in current manuals as white, but varies to pale bluish in some specimens, as noted in the fresh condition (*Steyermark 86066* from Polk County, May 3, 1958).

Birds eat the seeds of this and other species of the genus.

## 9. **Amsinckia** Lehm. Fiddleneck, Tarweed

This genus is greatly in need of revision and the following treatment is based upon identification of specimens by Dr. I. M. Johnston. I am greatly indebted to him for his suggestions which I have followed in the present treatment.

The key to the species is taken from Abrams, *Illustrated Flora of the Pacific Coast* 3: 604. 1951.

a. Corolla-tube 20-nerved below the insertion of stamens; calyx-lobes unequal in width and reduced in numbers (2, 3, or 4) by fusion . . . . . . . . . . . . . . . . . . 4. A. TESSELLATA

a. Corolla-tube 10-nerved below insertion of stamens; calyx-lobes 5, distinct. . . . . . . . . . *b*

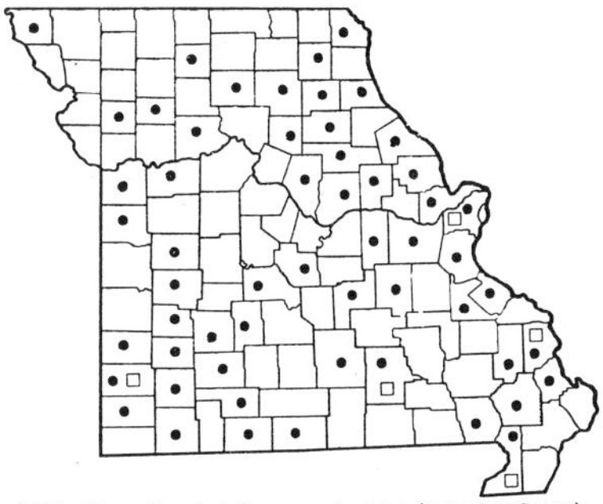

1831 • Myosotis virginica var. virginica (Scorpion Grass)
1831 □ Myosotis virginica var. macrosperma

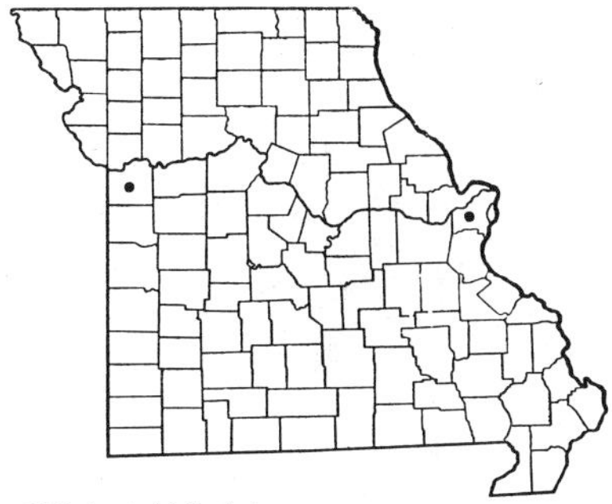

1832 Amsinckia barbata

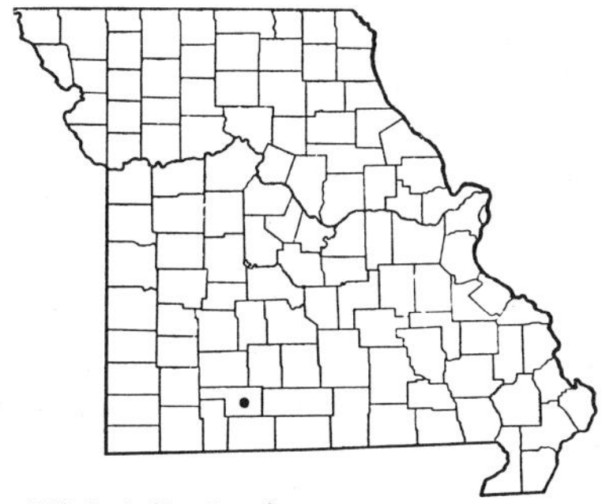

1833 Amsinckia micrantha

b. Corolla with a slender tube abruptly expanded into the lobes (salverform), the throat constricted and closed or nearly closed by hairy sac-shaped processes; stamens inserted evenly at one level on the corolla-tube well below its constriction . . . . . . . . . . . . . 1. A. LYCOPSOIDES

b. Corolla gradually expanded above (funnelform), the throat open and glabrous (without hairs); stamens inserted more or less irregularly at different levels on the corolla-throat . . . . . . . *c*

c. Corolla-limb 2–3 mm. broad; leaves with the hairs ascending or appressed (pressed against or lying parallel to surface); nutlets usually with roughened projections running across the surface between the keel and the lateral angles . . . . . . . . . . . . 3. A. RETRORSA

c. Corolla-limb about 1 mm. broad; leaves sparsely hairy with long, spreading, bristly hairs; nutlets uniformly roughened with short hard points extending between the keel and the lateral angles . . . . . . . . . . . . . . . . . . . . . 2. A. MICRANTHA

1. **Amsinckia lycopsoides** Lehm. Map 1832

*Amsinckia barbata* Greene [G]

Flowers May–July.

Occurs along railroads, where introduced from the western United States.

Known from St. Louis (St. Louis, along tracks of Mo.-Pac. R.R., opposite Plum Street freight sheds, June 17, 1956, *Muehlenbach 959*; Carrie Avenue freight yard of Terminal Railroad Association, north of station buildings, May 30, 1956, *Muehlenbach 893*) and Jackson (Courtney, June 2, 1914, *Bush 7123*; May 30, 1918, *Bush 8316*; June 20, 1918, *Bush 8421, 8326*; June 7, 1918, *Bush 8367*; June 7, 1918, *Bush 8366*; and May 18, 1918, *Bush 8278*) counties.

Ranges from Washington and Oregon, south to California and Nevada; introduced in Missouri and elsewhere eastward.

The identity of the Jackson County specimens is uncertain, having been interpreted variously at different times by different specialists as *A. modesta* Suksdorf, *A. menziesii* (Lehm.) Nels. & Macbride, and *A. arenaria* Suksdorf.

2. **Amsinckia micrantha** Suksd. Map 1833

Flowers May–July.

Occurs along railroads, where introduced in Christian County, southwestern Missouri (open waste ground along railway siding near station, Billings, June 24, 1954, *Palmer 57908*).

Ranges from British Columbia and Idaho south to Oregon; introduced eastward in Missouri and elsewhere.

3. **Amsinckia retrorsa** Suksd. Map 1834

Flowers May–July.

Occurs along railroads, where introduced in Christian County, southwestern Missouri (open waste ground along railway siding, near station, Billings, May 17, 1955, *Palmer 59940*).

Ranges from Washington south to California, east to Idaho and Nevada; introduced eastward in Missouri and elsewhere.

4. **Amsinckia tessellata** Gray Map 1835

Flowers May–July.

Occurs along railroads, where introduced in St. Louis County (St. Louis, Manufacturers Railway, DeKalb Street, May 23, 1954, *Muehlenbach 44*).

Ranges from California and Washington, east to Idaho, Arizona, and Utah; introduced eastward in Missouri and elsewhere.

According to Dr. I. M. Johnston, some of these introduced species may come from the central valley of California, but most of them probably come from eastern Washington, where the genus is a serious weed in the wheat fields, the hairs from the plants causing

1834 Amsinckia retrorsa

1835 Amsinckia tessellata

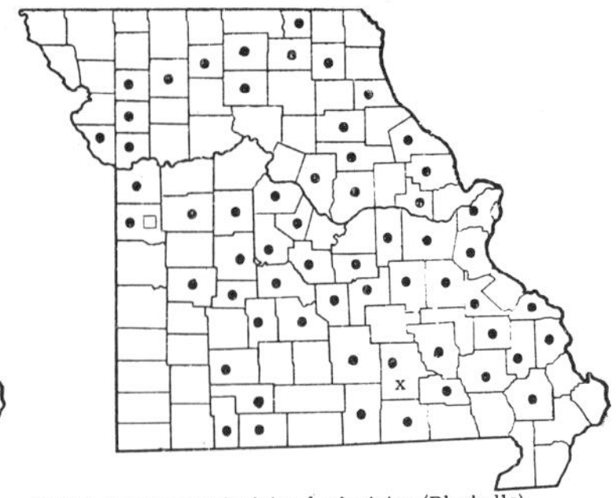

1836 • Mertensia virginica f. virginica (Bluebells)
1836 □ Mertensia virginica f. rosea
1836 x Mertensia virginica f. Berdi

silicosis to the thrashers. The screenings from such wheat, when used for feed for horses, cattle, and swine, can cause poisoning to the animals. The introduced species, which reach Missouri, enter the state as impurities in chicken-feed, hay, or straw, and although they are recent introductions and may not become permanent members of the Missouri flora, nevertheless, the fact that they do come into an area from time to time, and are picked up by collectors, who have been previously unacquainted with them, would indicate the necessity of recording their identification and region of introduction.

## 10. **Mertensia** Roth Bluebell

**Mertensia virginica** (L.) Pers. Bluebells
Map 1836

Also known as Virginia Cowslip.

Flowers late March–early June.

Occurs in moist rich woods and low slopes, usually in rich bottom soils along streams and in valleys, sometimes on adjacent low or rocky slopes or bluffs. Scattered throughout Missouri, but not recorded from many counties.

Ranges from New York and Ontario, Michigan to Minnesota and Iowa, south to Alabama, Tennessee, Arkansas, and Kansas.

The following variations are known in Missouri:

a. Corolla pure white throughout . . . . c. M. VIRGINICA f. BERDI
a. Corolla pink, rose-colored, or pale to deep blue . . . . . . . . . . . . . *b*
  b. Corolla remaining pink to the end of flowering . . . . . b. M. VIRGINICA f. ROSEA
  b. Corolla pink in bud, but eventually turning light to deep blue . . . a. M. VIRGINICA f. VIRGINICA

a. **Mertensia virginica** f. **virginica** Map 1836

*Mertensia virginica* (L.) Pers. [G, BB, P & S, Steyerm.]

This is the common form occurring in the state.

b. **Mertensia virginica** f. **rosea** Steyerm.
Map 1836

Known only from Cass County, west-central Missouri (rich alluvial woods along north side of Big Creek, T46N, R30W, sect. 35, 4 mi. southeast of Pleasant Hill, April 17, 1949, *Steyermark 67235*, holotype in Chi. Nat. Hist. Mus. Herb.).

This rare form has been found at only one locality in Missouri. At this station several plants of the pink-flowered form occurred within the midst of thousands of individuals of the normal blue-flowered type. Plants brought back from this station to the author's wild-flower preserve in northern Illinois have maintained their pink color throughout from 1949 to the present time.

c. **Mertensia virginica** f. **Berdi** Moldenke, Phytologia 3: 420. June, 1951. Map 1836

*Mertensia virginica* f. *alba* Allard, Castanea 3: 112. September, 1951.

---

Plate no. 301. **1.** Amsinckia barbata, × 1; Details from Small, The New York Botanical Garden. **2.** Mertensia virginica, × 2/5; Details from Small, The New York Botanical Garden. **3.** Lappula Redowskii var. occidentalis, × 2/5; Details from Small, The New York Botanical Garden. **4.** Lappula echinata, × 2/5; Details from Small, The New York Botanical Garden. **5.** Hackelia virginiana, × 2/5; *a.* Nutlet, × 4/5.

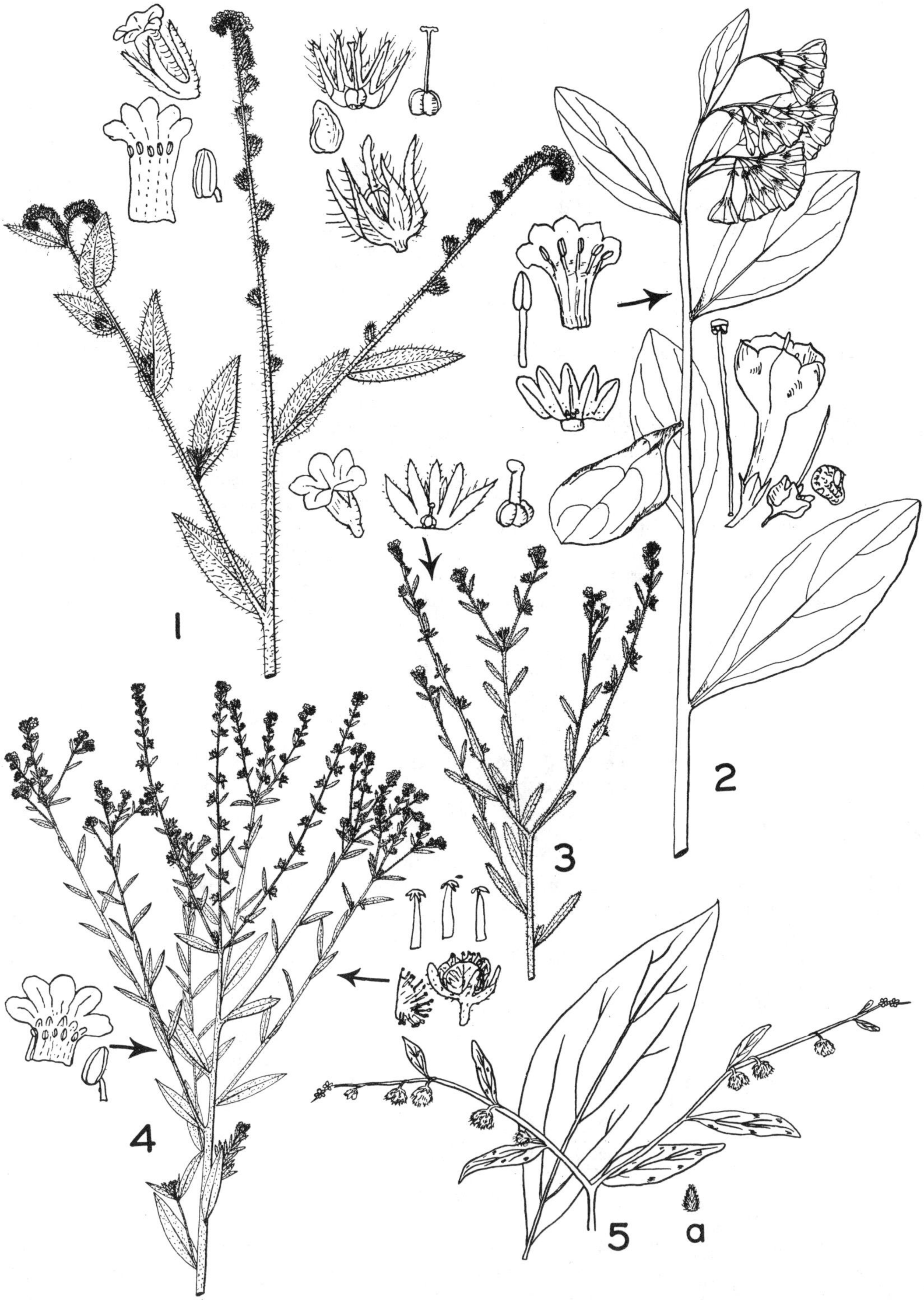


PLATE NO. 301

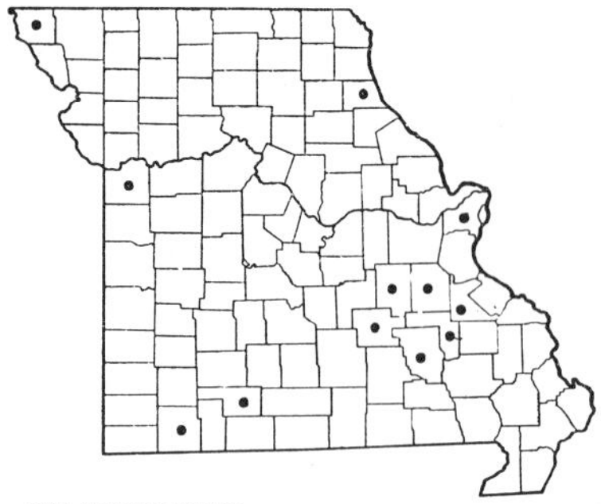

1837 Lappula echinata

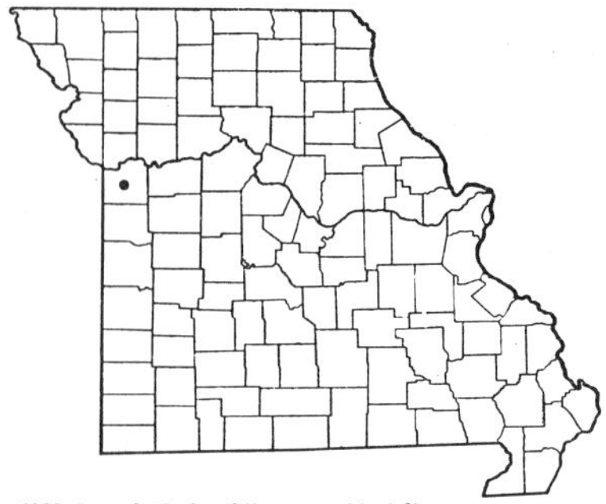

1838 Lappula Redowskii var. occidentalis

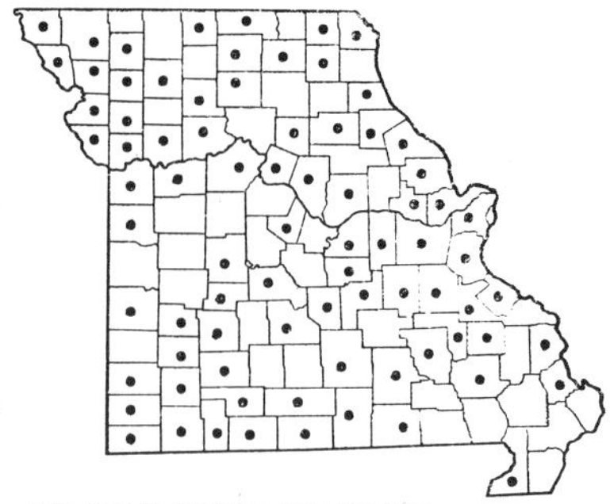

1839 Hackelia virginiana (Beggar's Lice)

The white-flowered form is not represented by any herbarium specimen, but was seen by Mrs. C. L. Kucera in woods along the Current River near Round Spring State Park, Shannon County, April 19, 1959. As only one plant was seen, Mrs. Kucera did not take a specimen for preservation. It is probable that the white-flowered form will be found in several other localities with more intensive early spring exploration in the state.

The white-flowered form maintains itself true to color over the years, as testified by plants grown in the author's wildflower garden in northern Illinois for the past fifteen years and by the experiences of other wildflower gardeners.

*Mertensia virginica* is one of the more easily grown wild flowers and succeeds in any good garden loam. It prefers shaded to semi-shaded situations, eventually increasing over the years through self-seeding. The numbers of stems to a clump increase each year. In the rich bottoms bordering Missouri streams and in valleys this species generally grows in large masses, often covering acres in favorable undisturbed sites.

## 11. **Lappula** Moench Stickseed, Beggar's Lice

A double row of prickles present on each of the 8 edges of the fruit (nutlet); fruits 3–4 mm. long; hairs of stem tending to be appressed; leaves rough-hairy. . . . . . . . . . . . . . . 1. L. ECHINATA

A single row of prickles present on each of the 8 edges of the fruit (nutlet); fruits 2–3 mm. long; hairs of stems tending to be spreading; leaves more softly hairy . . . . . 2. L. REDOWSKII var. OCCIDENTALIS

1. **Lappula echinata** Gilib. Map 1837

Flowers May–September.

Occurs in fields, waste and cultivated ground, along roadsides and railroads. Scattered over the state, but uncommonly found.

Native of Europe; introduced and naturalized in North America from Newfoundland to Alaska, south to New Jersey, Pennsylvania, West Virginia, Kentucky, Illinois, Missouri, Kansas, and Texas.

2. **Lappula Redowskii** (Hornem.) Greene var. **occidentalis** (Wats.) Rydb. Map 1838

Flowers May–September.

Occurs along railroads where introduced.

Known only from Jackson County, west-central Missouri.

Native of the western United States and Canada; introduced eastward to Michigan, Iowa, Missouri, and New England.

The fruits of this species and *L. echinata* may cause inflammation when eaten by animals, and are frequently found in wool.

## 12. **Hackelia** Opiz Stickseed, Beggar's Lice

(*Lappula* in part [P & S])

**Hackelia virginiana** (L.) I. M. Johnston

Beggar's Lice Map 1839

*Lappula virginiana* (L.) Greene [P & S]

Flowers June–September.

Occurs in open dry or moist woodland, alluvial soils, wooded thickets, and sometimes waste places.

Throughout Missouri, probably in every county.

Ranges from Quebec and Maine to Minnesota and

South Dakota, south to Georgia, Alabama, Louisiana, and Oklahoma.

The fruits of this species are a nuisance in late fall, clinging closely to skin, hair, and clothing. As in *Lappula*, animals eating the prickly fruits may suffer inflammation.

## Fam. **VERBENACEAE** (Vervain Family)

a. Shrub; fruit fleshy, purple, blue, or rose-pink; flowers and fruits in the axils of the leaves along the stem; stamens protruding (exserted) from the corolla . . . . . . . . . . . . 3. CALLICARPA
a. Herbaceous plants; fruit dry, brown; flowers and fruits arranged in spikes or heads at the tips of the stems and branches; stamens included, not protruding from the corolla . . . . . . . . . . *b*
  b. Flowers in dense knob-shaped heads 3–20 mm. long borne on special leafless flower-stems (peduncles); corolla 4-lobed, 2-lipped; calyx with 2 or 4 teeth . . . . . . . . . 2. LIPPIA
  b. Flowers in long or short rather loose, pencil- or tail-shaped inflorescences or in broad, loose, flat-topped clusters; corolla 5-lobed, only slightly irregular; calyx with 5 teeth . . . . . . 1. VERBENA

### 1. **Verbena** L. Vervain

A number of hybrids are known in Missouri, and due allowance should be made for them in the following key.

a. Individual flowers 12–25 mm. long, the expanded limb 8–25 mm. broad; calyx 8–13 mm. long; inflorescence a broad, flat-topped cluster (corymbose) . . . . . . . . . . . . . . . . *b*
  b. Common throughout southern and central Missouri; calyx bearing gland-tipped hairs; bracts mainly shorter than the calyx; corolla-tube about 2 times as long as the calyx; expanded limb of corolla 10–25 mm. broad; leaves usually not cut as far as the midrib, the divisions of unequal width . . . . . . . . . . . . . . . . . . . . . . . . 6. V. CANADENSIS
  b. Rare and known only from Jackson County, west-central Missouri; calyx without gland-tipped hairs; bracts mainly longer than the calyx; corolla-tube $1\frac{1}{3}$–$1\frac{1}{2}$ times as long as the calyx; expanded limb of corolla 8–11 mm. broad; leaves parted to or nearly to the midrib into very narrow divisions of uniform width . . . . . . . . . . . . . . . . . 7. V. BIPINNATIFIDA
a. Individual flowers 4–10 mm. long, the expanded limb 2–9 mm. broad; calyx 1.7–5 mm. long; inflorescence in long or short narrow clusters . . . . . . . . . . . . . . . . . . *c*
  c. Stems sprawling or spreading; dwarf plants less than 2 dm. tall; bracts longer than the flowers; leaves deeply lobed or deeply cut with usually 2 deep cuts near their base . . . 5. V. BRACTEATA
  c. Stems erect; plants 3–15 dm. tall; bracts shorter than the flowers; leaves shallowly or coarsely toothed, usually without deep lobes or cuts near the base, but, if lobed, then the stems erect and the bracts shorter than the flowers . . . . . . . . . . . . . . . . . . . . . *d*
    d. Main leaves of stem 3–10 (up to 15) mm. broad, narrowly lanceolate to oblanceolate. 4. V. SIMPLEX
    d. Main leaves of stem mostly 15–60 mm. broad, ovate, broadly elliptic, oblong or lanceolate. . . *e*
      e. Main leaves of the stem sessile (without stalks or on petioles up to 5 mm. long); leaves densely hairy, the hairs close together and producing a gray effect; flowering or fruiting spikes 1 or in 3's at tips of the stem or branches, the flowering spikes 7–10 mm. thick; fruiting calyx 2.5–5 mm. long; expanded limb of corolla 8–9 mm. broad . . . 3. V. STRICTA
      e. At least some of the main leaves of the stem on petioles (stalks) 1–2 cm. long; leaves sparsely hairy, the hairs scattered over the surface, not close enough to give a gray effect, or without hairs; flowering or fruiting spikes numerous, in panicled arrangement at the top of the stems or branches, the flowering spikes 3–6 mm. thick; fruiting calyx 2–3 mm. long; expanded limb of corolla about 2 mm. broad . . . . . . . . . . . . . . . *f*
        f. Corolla blue; flowering or fruiting spikes dense, most of the flowers and fruits overlapping one another (imbricate); leaves mainly 2–3 (up to 4) cm. broad, lanceolate, narrowly oblong to narrowly ovate; expanded limb of corolla 3–4.5 mm. broad; calyx 2.5–3 mm. long . . . . . . . . . . . . . . . . . . . . 2. V. HASTATA
        f. Corolla mainly white; flowering or fruiting spikes loose, most of the flowers and fruits scattered, not overlapping; leaves mainly 4–7 (2) cm. broad, ovate to lanceolate; expanded limb of corolla 2 mm. broad; calyx 2–2.3 mm. long . . . . . . 1. V. URTICIFOLIA

1. **Verbena urticifolia** L. var. **urticifolia**
White Vervain — Map 1840

*Verbena urticifolia* L. [G, P & S]
Flowers early June–October.

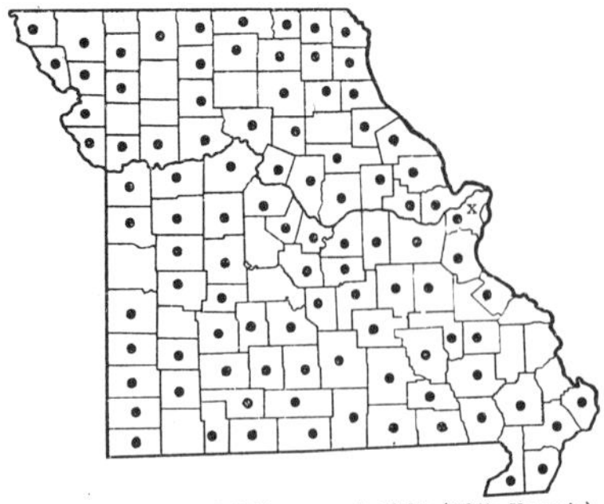

1840 • Verbena urticifolia var. urticifolia (White Vervain)
1840 x Verbena X Engelmannii (V. urticifolia X hastata)

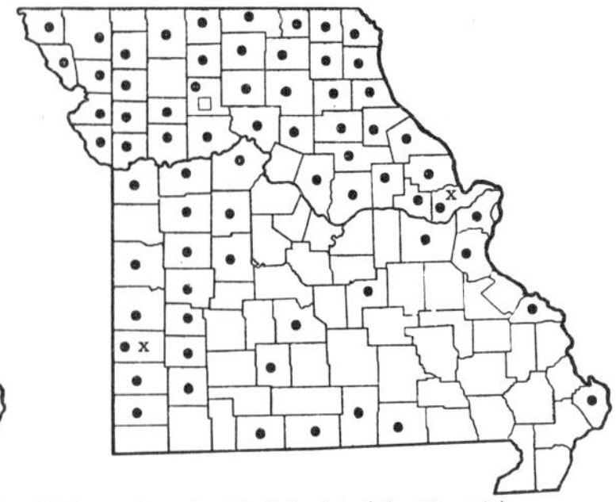

1841 • Verbena hastata f. hastata (Blue Vervain)
1841 □ Verbena hastata f. rosea
1841 x Verbena hastata f. albiflora

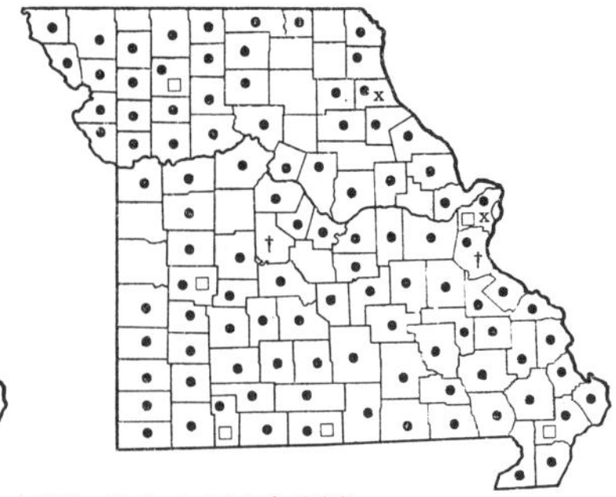

1842 • Verbena stricta f. stricta
1842 □ Verbena stricta f. albiflora
1842 x Verbena X Rydbergii (V. stricta X hastata)
1842 † Verbena X illicita (V. stricta X urticifolia)

Occurs in fallow or cultivated fields, alluvial thickets along streams, waste ground, along roadsides and railroads, and pastured and open woodland. Throughout Missouri, doubtless in every county.

Ranges from Quebec and Maine to Ontario and South Dakota, south to Florida, Alabama, Louisiana, and Texas.

In var. *leiocarpa* Perry and Fern., not recorded from Missouri, the lower surface of the leaves is densely but short-hairy with hairs up to 0.3 mm. long instead of with stiff long hairs up to 1.3 mm. as in var. *urticifolia*, the mature calyx is shorter (1.7–2 mm. long instead of 2–2.3 mm. long as in var. *urticifolia*), and the small nutlets are 1.5 mm long and smooth, whereas in var. *urticifolia* they are 2 mm. long and ribbed.

This species hybridizes with *V. stricta* and *V. hastata*, and may be recorded as follows:

**Verbena X illicita** Moldenke (*V. urticifolia* X *V. stricta*) is known from Morgan (pastures, August 28, 1933, *Bush 12927*) and Jefferson (Pacific, July 4, 1896, *Eggert*) counties. This resembles *V. urticifolia* in habit, but has denser spikes with the mature calyces mostly touching, larger flowers, broader leaves, and densely long-hairy stems, branches, and inflorescences.

**Verbena X Engelmanni** Moldenke (*V. urticifolia* X *V. hastata*) is known from St. Louis County (St. Louis, August 4, 1875, *Eggert*). It resembles *V. urticifolia* in habit, but has denser spikes with the mature calyces mostly touching, small purplish corollas, coarse teeth of the leaves, and nearly glabrous or sparsely hairy stems, branches, leaves, and inflorescences.

2. **Verbena hastata** L. Blue Vervain — Map 1841
Flowers June–October.

Occurs in wet meadows, wet or river bottom prairies, low open wet woodland, borders of streams, lakes, sloughs, wet ledges of bluffs, fields, waste ground, and along roadsides and railroads. Absent from most of the Ozark region, otherwise common throughout northern, central, and western Missouri, especially in the glaciated and unglaciated prairie sections.

Ranges from Nova Scotia to British Columbia, south to Florida, Nebraska, Texas, Arizona, and California.

Three variations occur in Missouri:

a. Corolla white . . 2c. V. HASTATA f. ALBIFLORA
a. Corolla purple, blue, rose-colored, or pink . *b*
  b. Corolla purple or blue . . . 2a. V. HASTATA f. HASTATA
  b. Corolla pink or rose-colored . 2b. V. HASTATA f. ROSEA

2a. **Verbena hastata** f. **hastata** — Map 1841
*Verbena hastata* L. [G, BB, P & S]
This is the common variation in Missouri.

2b. **Verbena hastata** f. **rosea** Cheney — Map 1841
Known only from Livingston County, northern Missouri.

2c. **Verbena hastata** f. **albiflora** Moldenke — Map 1841

Known from Barton (prairies and roadsides, near Nashville, August 6, 1933, *Palmer & Steyermark 42093*) and St. Charles (Belleau swamps, St. Peters, September 23, 1919, *Jensen*) counties.

A hybrid between this species and *V. stricta* has been recorded for Missouri. It is:

**Verbena X Rydbergii** Moldenke (*V. hastata* X *V. stricta*) known from Marion and St. Louis (St

Plate no. 302. **1.** Verbena urticifolia, × 2/5 (Scribner's). **2.** Verbena simplex, × 2/5. **3.** Verbena bracteata, × 2/5. **4.** Verbena stricta, × 2/5. **5.** Verbena hastata, × 2/5. All details from Small, The New York Botanical Garden.

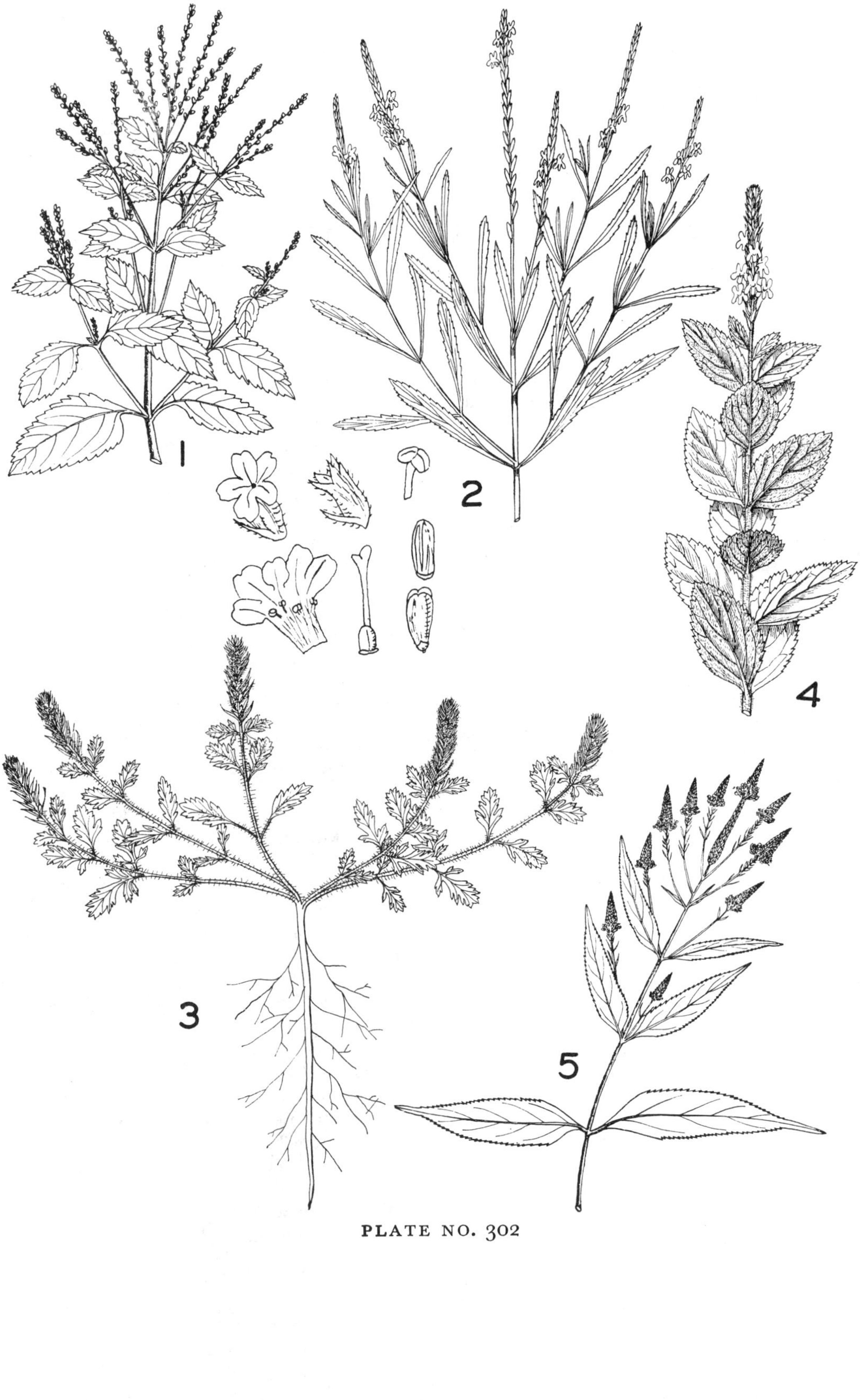


PLATE NO. 302

1843 • Verbena simplex
1843 x Verbena X moechina (V. simplex X stricta)

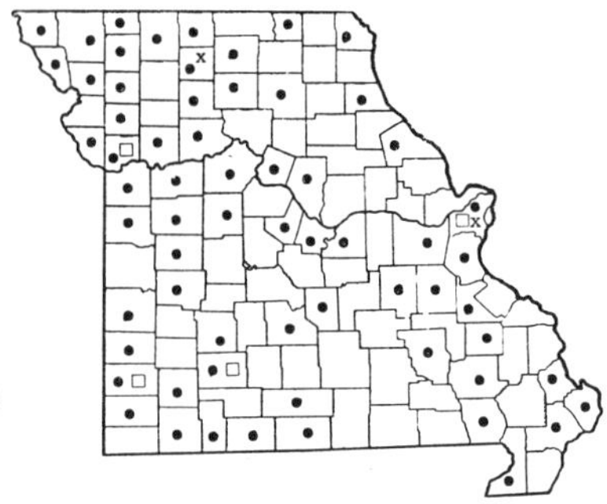

1844 • Verbena bracteata
1844 □ Verbena X Deamii (V. bracteata X stricta)
1844 x Verbena X Perriana (V. bracteata X urticifolia)

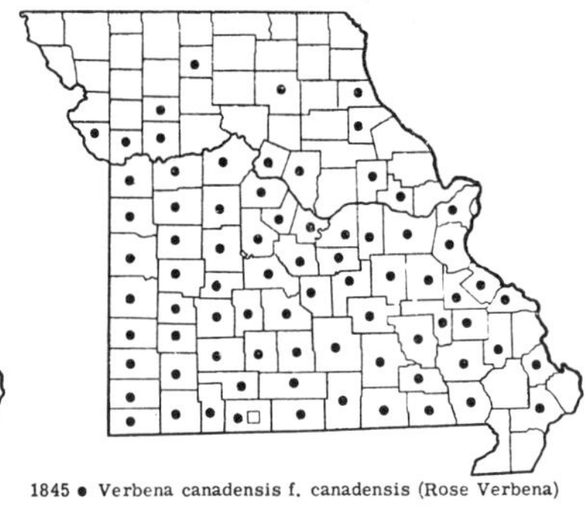

1845 • Verbena canadensis f. canadensis (Rose Verbena)
1845 □ Verbena canadensis f. candidissima

Louis, August 4, 1875, *Eggert*) counties. It resembles *V. hastata* in habit, but has more reticulate-veined leaves, more densely soft hairy on the lower surface, larger corollas, calyx 4–5 mm. long, and non-keeled bracts about 4 mm. long.

*Verbena hastata* is a showy-flowered species and attains a height of 1.5 meters. It is recorded that the seeds are roasted and used as a breadstuff by some of the Californian Indians.

3. **Verbena stricta** Vent. Map 1842

Flowers late May (May 20)–September.

Occurs in prairies, glades, thickets, waste ground, pastures, fallow fields, along roadsides and railroads. Throughout Missouri, doubtless in every county.

Ranges from Ontario to Minnesota and Montana, south to Ohio, Tennessee, Arkansas, Oklahoma, Texas, and New Mexico; introduced from Massachusetts to Delaware and West Virginia, and in Washington.

The following variations occur in Missouri:

a. Corolla white . . . 3c. V. STRICTA f. ALBIFLORA
a. Corolla deep blue, purple, rose-colored, or pink . . . . . . . . . . . . . . *b*
  b. Corolla deep blue or purple . 3a. V. STRICTA f. STRICTA
  b. Corolla rose-colored or pink . 3b. V. STRICTA f. ROSEIFLORA

3a. **Verbena stricta** f. **stricta** Map 1842

*Verbena stricta* Vent [G, BB, P & S, Steyerm.]

This is the common form throughout the state.

3b. **Verbena stricta** f. **roseiflora** Benke

Known only from 'southwestern Missouri' ('S.W. Mo., August, 1889, *Spillman*' in Herb. State Coll. Wash., Pullman). The label gives no indication as to county collected.

3c. **Verbena stricta** f. **albiflora** Wadmond Map 1842

Scattered within the range of typical f. *stricta*, known from St. Louis, New Madrid, Daviess, St. Clair, Ozark, and Stone counties.

*Verbena stricta* hybridizes in Missouri with *V. bracteata*, *V. urticifolia*, and *V. simplex*. The hybrid with *V. urticifolia* has already been noted under *V. urticifolia*. The other hybrids will be discussed under the other species.

The roots of this plant go down 4 or 5 feet deep, making the plant very drouth-resistant. It is very bitter and therefore not eaten by cattle.

4. **Verbena simplex** Lehm. Map 1843

Flowers May–September.

Occurs on glades, prairies, fields, open waste ground, along roadsides and railroads. Common in southern and central Missouri, except absent in the lowland counties of southeastern Missouri, north to Marion, Monroe, Boone, Saline, Grundy, Daviess, Nodaway, and Atchison counties.

Ranges from Quebec and New Hampshire to Ontario, Minnesota, and Nebraska, south to Florida, Alabama, Mississippi, Louisiana, Kansas, and Oklahoma.

This species hybridizes with *V. stricta* and is known as **Verbena X moechina** Moldenke (*V. simplex* X *V. stricta*). It occurs in the Ozark region north to St. Louis, Franklin, Phelps, Greene, and Jasper counties. It resembles *V. simplex* in habit, but has broader, more often elliptic or elliptic-ovate leaves, which are short-hairy on the lower surface, densely hairy stems and branches; from *V. stricta* it differs in the more slender inflorescences, smaller flowers, and narrower leaves.

5. **Verbena bracteata** Lag. & Rodr. Map 1844

Flowers April–October.

Occurs in waste ground along roadsides, railroads, fields, pastures, open slopes, crevices of sidewalks, and gravelly or sandy open ground. Throughout Missouri and probably in every county.

Ranges from Florida to Texas and Mexico, north to Virginia, Ontario, Michigan, Wisconsin, Minnesota, North Dakota, Alberta, and British Columbia; introduced along the coast in New England.

This species hybridizes with *V. stricta* and *V. urticifolia* in Missouri. They may be recorded as follows:

**Verbena X Deamii** Moldenke (*V. bracteata* X *V. stricta*). Known from St. Louis (*Eggert 5321*; August, 1886, *Eggert*; St. Louis, August, 1843, *Engelmann*; 1861, *Engelmann*; *M. Martens*), Greene (Brookline, July 13, 1879, *Eggert*), Clay, and Jasper counties. This resembles *V. bracteata* in habit, but has broad, coarsely cut, strongly nerved leaves, larger flowers, smaller bracts about 5 mm. long, and inflorescences narrower, 6–10 mm. wide.

**Verbena X Perriana** Moldenke (*V. bracteata* X *V. urticifolia*). Known from St. Louis County (Glencoe, July 28, 1879, *Eggert*). This resembles *V. bracteata* in habit, but is a more erect, stouter plant with broad coarsely cut leaves, long, slender inflorescences, and small bracts only 3–4 mm. long which are only slightly longer than the calyx.

A specimen from Grundy County (pastures, July 28, 1937, *D. R. Crookshanks 142*, in U. Mo. Herb.), appears to be a hybrid between *V. bracteata* and *V. hastata*.

6. **Verbena canadensis** (L.) Britt. Rose Verbena Map 1845

Also called Eastern Verbena.

Flowers March–November.

Occurs on rocky glades (especially limestone), open and rocky or sandy wooded slopes, ledges of bluffs, fields, pastures, prairies, gravel bars along streams, and along roadsides and railroads.

Ranges from Florida to Texas, north to Virginia, Tennessee, Illinois, Iowa, Missouri, Kansas, and Colorado; introduced in Pennsylvania, Indiana, Michigan, and Minnesota.

Two variations occur in Missouri:

Corolla purple, blue, lavender, lilac, or rose-red . . . . 6a. V. CANADENSIS f. CANADENSIS
Corolla white . 6b. V. CANADENSIS f. CANDIDISSIMA

6a. **Verbena canadensis** f. **canadensis** Map 1845

*Verbena canadensis* (L.) Britt. [G, BB, P & S, Steyerm.]

This is the common form in the state.

6b. **Verbena canadensis** f. **candidissima** (Haage & Schmidt) Palmer & Steyerm. Map 1845

Known only from Taney County, southwestern Missouri (exposed limestone bluff along road along Long Creek, just north of Oasis, T22N, R22W, sect. 10, 11, and 3, April 28, 1949, *Cora Steyermark*).

This locality has now been exterminated by the impoundment of the Table Rock Dam.

This species does well under cultivation and is a good subject for partially shaded to sunny rock gardens. It spreads rapidly. Some plants exhibit more intensely colored flowers than others, although the deep lavender to blue-purple colors predominate.

7. **Verbena bipinnatifida** Nutt. Map 1846

Flowers last of May–September.

Introduced along railroads and open ground in Jackson County, west-central Missouri (Sheffield, September 5, 1895, *Bush 440*; Courtney, June 6, 1894, *Bush 351*; Courtney, July 13, 1906, *Bush 4029*).

Ranges from Georgia and Alabama to Arizona and Mexico, north to Missouri, South Dakota, and Colorado.

## 2. **Lippia** L. Fog Fruit

(*Phyla* Lour. [BB])

Common species encountered; leaves conspicuously pointed (acuminate) at tip, broadest at or below the middle, with 5–11 teeth extending on each side from the tip to below the middle . . . 1. L. LANCEOLATA
Rare species, known only from Dunklin County, southeastern Missouri; leaves rounded or obtuse (blunt) at summit, broadest above the middle, with 3–7 teeth on each margin . . . . . . . 2. L. NODIFLORA

1. **Lippia lanceolata** Michx. Fog Fruit Map 1847

*Lippia lanceolata* var. *recognita* Fern. & Grisc. [G, P & S, Steyerm.]

*Phyla lanceolata* (Michx.) Greene [BB]

Flowers late May–September.

Occurs along muddy or gravelly margins of streams, ponds, sloughs, bayous, ditches, prairie swales, and along railroads. Throughout Missouri, and probably occurring in every county.

Ranges from Florida to California and Mexico,

1846 Verbena bipinnatifida

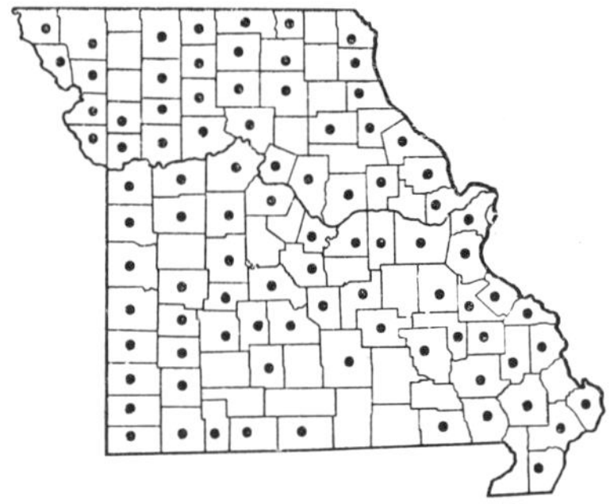

1847 Lippia lanceolata (Fog Fruit)

1848 Lippia nodiflora

north to New Jersey, Pennsylvania, Ontario, Michigan, Wisconsin, Minnesota, and Nebraska.

There seems to be no justification in maintaining var. *recognita* as a separate variety, as leaves in Missouri plants vary in the same colony from narrowly lanceolate of typical var. *lanceolata* to broader ones characteristic of var. *recognita*. The seeds of the plant are eaten by ducks.

2. **Lippia nodiflora** (L.) Michx. Map 1848
*Phyla nodiflora* (L.) Greene [BB]
Flowers late May–October.

Known only from Dunklin County, southeastern Missouri (sandy ground, October 26, 1892, *Bush*; Campbell, August 13, 1895, *Bush 471*).

Ranges from Florida to Texas and Mexico, north to Virginia, Missouri, Oklahoma, and California; also in tropical and subtropical America and Old World; introduced in Pennsylvania.

Bush notes on his label 'common throughout southeast Missouri'. There appears to be no basis for this statement, as no other material, except Bush's collection, is available in herbaria.

*Excluded Species*

**Lippia cuneifolia** (Torr.) Steud.

Known from a specimen from St. Louis County (St. Louis, June 22, 1876, *Eggert* in Herb. Cath. Univ. Am.). The specimen has been determined by Dr. Harold N. Moldenke, but he informs me that the material is represented by a fragment of a specimen mixed with *Lippia lanceolata* and mounted on the same sheet. It would appear that this fragment may have been taken originally from another source, dropped out from the original folder and eventually mounted with the specimen of *L. lanceolata*, thereby creating confusion of data.

### 3. **Callicarpa** L. Beauty Berry

**Callicarpa americana** L. French Mulberry, Beauty Berry Map 1849
Flowers June–August.

Occurs on wooded limestone slopes bordering White River. Known only from Taney County, southwestern Missouri (rather frequent on south-facing slopes from base to lowest line of bluffs about 60 feet above water level, along north and west side of White River, north and east of Brown Ferry, downstream to Cedar Hol, T21N, R16W, sect. 13, $5\frac{1}{2}$–6 mi. southeast of Protem, September 29, 1949, *Steyermark 69453*).

Ranges from Florida to Texas, north to Maryland, North Carolina, Tennessee, Missouri, Arkansas, and Oklahoma, and in northern Mexico.

The only station known in Missouri and reported by the author (Rh. 55: 238–41. 1953) has now been exterminated by the impounded waters of Table Rock Dam.

This is undoubtedly the shrub having the most unusual colored fruit of any of the native species, the fleshy fruits varying from violet to ultramarine blue-purple or pink. They are juicy, somewhat sweet in the beginning but become rather pungent later. They are now and then eaten more as a curiosity than for any satisfying effect. The shrub is one which merits cultiva-

Plate no. 303. **1.** Verbena canadensis, × $^2/_5$. **2.** Lippia lanceolata, × $^2/_5$; Details from Small, The New York Botanical Garden. **3.** Verbena bipinnatifida, × $^2/_5$. **4.** Callicarpa americana, × $^2/_5$; After Gleason, details from Small, The New York Botanical Garden. **5.** Lippia nodiflora, × $^2/_5$.

PLATE NO. 303

tion. It is a tragedy that the only station known in Missouri has been exterminated.

In the University of Illinois Herbarium appears a specimen of this labeled '*Oxybaphus albidus*, Jackson County, May 29, 1884,' collected by Bush. The label is from a Mo. State University sheet (no. 17619) and obviously is a mixed label. The present author has already called attention (Rh. 55: 238–39. 1953) to another error on a label presumed to have come from Missouri, but in reality from Arkansas.

## Fam. **LABIATAE** (Mint Family)

a. One side (the upper) of the calyx with a helmet- or cap-like projection, sometimes shaped like the seat of a tractor . . . . . . . . . . . . . . . . . . . . . . . . . . . *b*

b. Calyx with only 2 lobes, both rounded at tip; a helmet- or tractor seat-like projection extending from the upper side of calyx; common native plants . . . . . . . . . . . . 4. SCUTELLARIA

b. Calyx with 5 lobes, some of them acutely pointed at tip; one (upper) of the calyx-lobes extending over the others and forming a cap; rare introduced plants . . . . . . . . . . . 29. OCIMUM

a. Calyx without any projection or extension on any side . . . . . . . . . . . . . . . . *c*

c. All the flowers and fruits appearing along or next to the main stem in the axils or bases of the stem-leaves, not in inflorescences at the tip of the stem or at the tips of branches . . . . . . . *d*

d. Flowers and fruits sessile (without stalks or pedicels) . . . . . . . . . . . . . . . *e*

e. Calyx with 10 teeth, the teeth hooked or recurved at their tip; stems and tips of shoots covered with a dense white woolliness . . . . . . . . . . . . . . . 5. MARRUBIUM

e. Calyx with 5 teeth, the teeth not hooked or recurved at their tip; stems and tips of shoots without white woolly hairs . . . . . . . . . . . . . . . . . . . . . . *f*

f. Corolla nearly regular with all the lobes similar; corolla entirely white, very small and inconspicuous, 1–3 mm. long . . . . . . . . . . . . . . . . . . 24. LYCOPUS

f. Corolla irregular with an upper and lower lip, 1 or 2 of the lobes quite different in shape from the others; corolla blue, lavender, red, pink, yellow, or white spotted with purple or yellow, usually not entirely white, 11–45 mm. long . . . . . . . . . . . . *g*

g. Anther-bearing stamens 2; clusters (verticils) of flowers or fruits 20–80 mm. or more thick . . . . . . . . . . . . . . . . . . . . . . . . . 17. MONARDA

g. Anther-bearing stamens 4; clusters (verticils) of flowers or fruits 10–20 mm. thick . . . *h*

h. At least some of the leaves, especially the middle and lower ones, with 3–5 conspicuous lobes, maple-leaf shaped or the leaves all 3-parted with cleft divisions . 12. LEONURUS

h. None of the leaves lobed . . . . . . . . . . . . . . . . . . . *i*

i. Corolla 10–18 mm. long, purple, rose-colored, or reddish-purple; calyx-lobes narrowly pointed (acuminate) but not prolonged into short stiff spines; stems glabrous (without hairs) or only sparsely hairy . . . . . . . . . 14. LAMIUM

i. Corolla about 20 mm. long, white or pink marked with purple and yellow; calyx-lobes prolonged at tip into short stiff spines; stems bristly-hairy . 13. GALEOPSIS

d. Flowers and fruits on stalks (pedicels), the stalks very short to conspicuous, 1 mm. or more long . . *j*

j. Stems creeping or trailing, rooting at the nodes; leaves round or lima-bean shaped (reniform) . . . . . . . . . . . . . . . . . . . . . . . 8. GLECHOMA

j. Stems erect, not rooting at the nodes; leaves not rounded or lima-bean shaped . . . . . *k*

k. All teeth of calyx more or less equal and similar, the calyx regular, not 2-lipped . . . . *l*

l. Anther-bearing stamens 2, conspicuously long-protruding beyond the corolla; corolla 6–8 mm. long . . . . . . . . . . . . . . . . . . . 23. CUNILA

l. Anther-bearing stamens 4; corolla 3–6 mm. long, or if 10 mm. long, then the leaves are linear-lanceolate and 1–5 mm. broad . . . . . . . . . . . . . . . . *m*

m. Corolla irregular, with an upper and lower lip, 1 or 2 of the lobes quite different in shape from the others . . . . . . . . . . . . . . 21. SATUREJA

m. Corolla nearly regular with the lobes similar . . . . . . . . . . . . . *n*

n. Leaves nearly or quite entire (without teeth); leaves 3–16 mm. broad . . 1. ISANTHUS

n. Leaves with teeth scattered along margins; leaves 10–40 mm. broad . 25. MENTHA

k. Calyx-teeth not all equal, some longer or of a different shape than the others, or the calyx 2-lipped and irregular . . . . . . . . . . . . . . . . . . . *o*

o. Main leaves of stem 20–50 mm. or more broad . . . . . . . . . . 20. MELISSA

o. Main leaves of stem 1–13 mm. broad . . . . . . . . . . . . . . . *p*

p. Anther-bearing stamens 2 . . . . . . . . . . . . . . 19. HEDEOMA

p. Anther-bearing stamens 4 . . . . . . . . . . . . . . 21. SATUREJA

c. Flowers and fruits mainly in loose or dense inflorescences at the tip of the stem or branches of the stem, sometimes also with inflorescences occurring in the axils or at the base of the stem-leaves. . . *q*

q. Corolla nearly regular with all its lobes similar or nearly equal . . . . . . . . . . *r*

r. Leaves entire (without teeth) or nearly so . . . . . . . . . . . . . . . . *s*

s. All teeth of calyx more or less equal and similar; stamens slightly protruding (exserted) from the corolla-tube or included in it; ovary 5-lobed . . . . . . . . . . 1. ISANTHUS

s. Calyx-teeth not all equal, some longer or of a different shape than the others; stamens conspicuously protruding (exserted) beyond the corolla-tube; ovary 4-lobed . . . 2. TRICHOSTEMA

r. Leaves shallowly or coarsely toothed or deeply cut . . . . . . . . . . . . . . . *t*

t. Corolla 5-lobed, one of the lobes (the lower one) fringed . . . . . . . . . . . 27. PERILLA

t. Corolla 4-lobed, none of the lobes fringed . . . . . . . . . . . . . . . 25. MENTHA

q. Corolla irregular with an upper and lower lip, 1 or 2 of the lobes quite different in shape from the others . . . . . . . . . . . . . . . . . . . . . . . . . . . . . *u*

u. At least the lower and middle leaves of the stem maple-leaf shaped with 3–5 conspicuous sharp-pointed lobes or the leaves all 3-parted with cleft divisions . . . . . . . . . . 12. LEONURUS

u. Leaves shallowly or coarsely toothed or entire (without teeth), but if lobed, the lobes rounded and pinnately arranged . . . . . . . . . . . . . . . . . . . . . . . . . *v*

v. Flowers and fruits on short or long stalks (pedicels) 1 mm. or more long . . . . . . . . . *w*

w. Lower lip or lobe of the corolla fringed; corolla yellow with purple and brown spots; calyx and corolla lemon-scented; all the lobes of calyx nearly equal in size and shape, the calyx more or less regular . . . . . . . . . . . . . . . . . . 26. COLLINSONIA

w. No lobe of the corolla fringed; corolla purple, blue, lavender, rose, or pink; calyx and corolla not lemon-scented; some of the lobes or teeth of calyx longer or of a different shape than the others or the calyx split into an upper and lower lip and therefore irregular. . . . . *x*

x. Corolla 4–7 mm. long; all 4 stamens conspicuously long-protruding (exserted) beyond the corolla . . . . . . . . . . . . . . . . . . 2. TRICHOSTEMA

x. Corolla 8–30 mm. long; stamens 2 or 4, only slightly if at all longer than the corolla, but not conspicuously long-protruding beyond it . . . . . . . . . . . . . *y*

y. Stamens 4; upper lip of corolla scarcely noticeable, very short; calyx 10-ribbed 3. TEUCRIUM

y. Stamens 2; upper lip of corolla prominent and at least noticeable; calyx 12–13-ribbed . . . . . . . . . . . . . . . . . . . . 16. SALVIA

v. Flowers and fruits sessile (without stalks or pedicels) . . . . . . . . . . . . . *z*

z. All teeth of calyx more or less equal and similar, the calyx regular, not 2-lipped . . . . . *1*

1. Stamens 2; flowers in dense round or head-like clusters 2.5–8 cm. broad . . 17. MONARDA

1. Stamens 4; flowers in either long, narrow, dense spikes, or looser interrupted elongated inflorescences, or short, condensed inflorescences, or round head-like clusters 0.3–3.5 cm. broad . . . . . . . . . . . . . . . . . . . . . . . . *2*

2. Flowers and fruits in round head-like masses as broad as long or broader than long, forming crowded or flattened inflorescences at the top of the stem, the individual head-like group of flowers 3–35 mm. broad. . . . . . . . . . . 22. PYCNANTHEMUM

2. Flowers and fruits not as above . . . . . . . . . . . . . . . . . . *3*

3. Stems with a close mat of short, soft, white or gray hairs; lower surface of leaves with short white hairiness . . . . . . . . . . . . . . . . 7. NEPETA

3. Stems glabrous, or if hairy, the hairs either longer, or reflexed or stiffer; lower surface of leaves without a short white hairiness . . . . . . . . . . . . *4*

4. Leaf-stalks (petiole) 1–6 cm. long; flowers and fruits in a long, slender, densely crowded spike without gaps or spaces between the flowers; corolla greenish-yellow to lavender, 6–10 mm. long; calyx 15-nerved; all the stamens long-protruding (exserted) beyond the corolla . . . . . . . . . 6. AGASTACHE

4. Leaf-stalks (petiole) none or at most 2.5 (rarely 3.2) cm. long; flowers and fruits in an interrupted or loose inflorescence, or at least the lower groups of flowers and fruits with gaps or spaces between them; corolla rose, lavender, pink, purplish, or rarely white, 10–30 mm. long; calyx 5–10-nerved; stamens not protruding (exserted) beyond the corolla . . . . . . . . . . . . . . *5*

5. Leaves glabrous (without hairs) and sessile (without stalk), linear-oblong, lanceolate or oblanceolate, usually narrowed at the base, 4–10 times as long as broad; stems glabrous; flowers and fruits arising singly in the axils of small bracts; corolla much longer than bracts. . . . . . . 11. PHYSOSTEGIA

5. Without the above combination of characters; leaves hairy or stems hairy, or if leaves and stems are glabrous, then on distinct leaf-stalks up to 7 cm. long, the leaf-blades narrowly oblong, oblanceolate, elliptic, or ovate, $2\frac{1}{2}$–4 times as long as broad; stems glabrous or hairy; flowers and fruits arising 2 or more together in the axils of leaf-like bracts; corolla equaling or usually shorter than the leaf-like bracts . . . . . . . . . 15. STACHYS

1849 Callicarpa americana (French Mulberry)

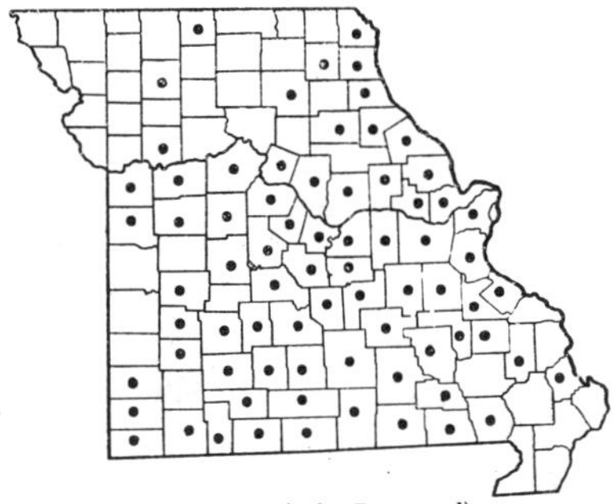

1850 Isanthus brachiatus (False Pennyroyal)

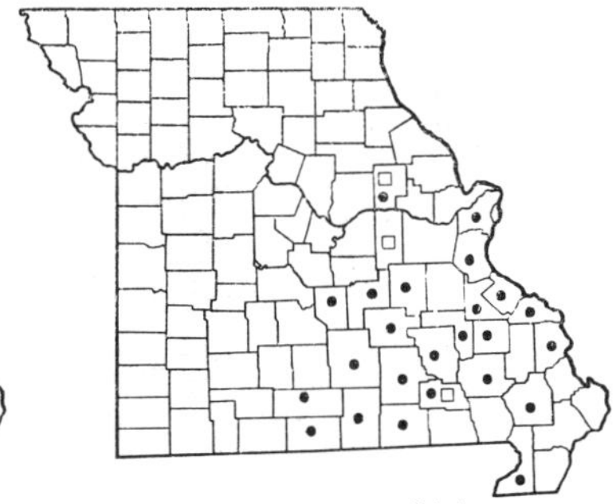

1851 • Trichostema dichotomum var. dichotomum
1851 □ Trichostema dichotomum var. puberulum

z. Calyx-teeth or lobes not all equal, some longer or of a different shape than the others, the calyx 2-lipped or irregular . . . . . . . . . . . . . . . . . . . . . . . . . . . . . . 6
6. Flowers or fruits in dense, compact head-like or short-cylindric clusters, no gaps or spaces occurring between the flowers . . . . . . . . . . . . . . . . . . . . . . . 7
7. Bracts of inflorescence spiny-tipped, and spiny-margined, with several prominent teeth or spines on each bract; 1 of the calyx-teeth (the upper middle one) longer and broader than the other 4 . . . . . . . . . . . . . . . . . . . . . . 9. DRACOCEPHALUM
7. Bracts of inflorescence not spiny-tipped, or with only 1 spine or point at the very tip, the margin otherwise without teeth or spines; 3 of the calyx-teeth differing from the other 2 . . . . 8
8. Anther-bearing stamens 2; flower-heads in dense clusters of 3–5 arranged one above the other at the top of the stem or branches, crowded in the axils of the upper leaves. 18. BLEPHILIA
8. Anther-bearing stamens 4; flower-heads not as described above, but in other types of arrangements . . . . . . . . . . . . . . . . . . . . . . . . . . . 9
9. Corolla usually lavender, rarely white, 10–16 mm. long; stamens not protruding (exserted) beyond the corolla; heads of flowers and fruits usually longer than thick (broad) . . . . . . . . . . . . . . . . . . . . . . . 10. PRUNELLA
9. Corolla mainly white or white with lilac or lavender dots, 3–8 mm. long; at least some of the stamens conspicuously protruding (exserted) beyond the corolla; heads of flowers and fruits about as broad as or broader than long . . . . . . . . . 22. PYCNANTHEMUM
6. Fruits or flowers in an elongated, interrupted or continuous, more or less loosely-flowered inflorescence, with gaps or open spaces occurring between clusters of flowers . . . . . . . 16. SALVIA

## 1. **Isanthus** Michx. False Pennyroyal

**Isanthus brachiatus** (L.) BSP. var. **brachiatus**
False Pennyroyal Map 1850
*Isanthus brachiatus* (L.) BSP. [G, P & S]
Flowers July–October.

Occurs along gravel bars of streams, on rocky open glades, usually of limestone, prairies, and rocky banks. Chiefly in the southern and eastern halves of the state and northwest locally in Mercer and Daviess counties.

A variety *linearis* Fassett with linear leaves 2–4 mm. broad, occurs in Ontario.

The corolla is blue and about 5 mm. long, only slightly longer than the calyx. The genus is sometimes placed under the next one, *Trichostema.*

## 2. **Trichostema** L. Blue Curls

Leaf-blades oblong to elliptic or ovate, usually 5–20 mm. wide, lateral (side) nerves evident . 1. T. DICHOTOMUM
Leaf-blades linear or linear-lanceolate, 1–6 mm. wide with 1 main mid-nerve, lateral nerves absent or not evident . . . . . . . . . . . . . . . . . . . . . . . . . . . 2. T. SETACEUM

1. **Trichostema dichotomum** L. Blue Curls
Map 1851

Also called Bastard Pennyroyal.

Flowers August–October.

Occurs in acid soils of rocky open woods and glades, sandy gravel and sandy alluvial soils, and particularly

1852 Trichostema setaceum

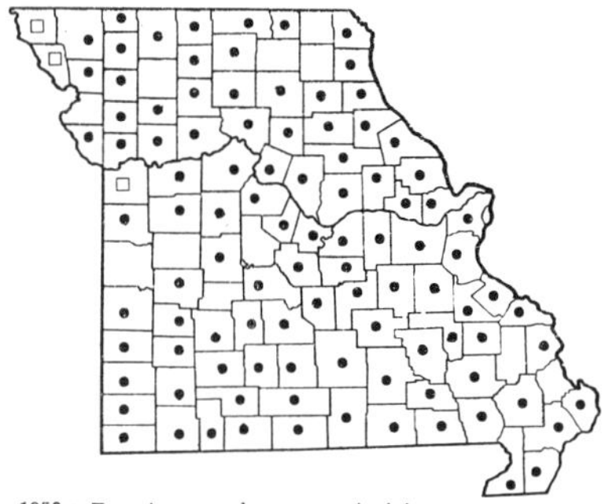

1853 • Teucrium canadense var. virginicum
1853 □ Teucrium canadense var. occidentale

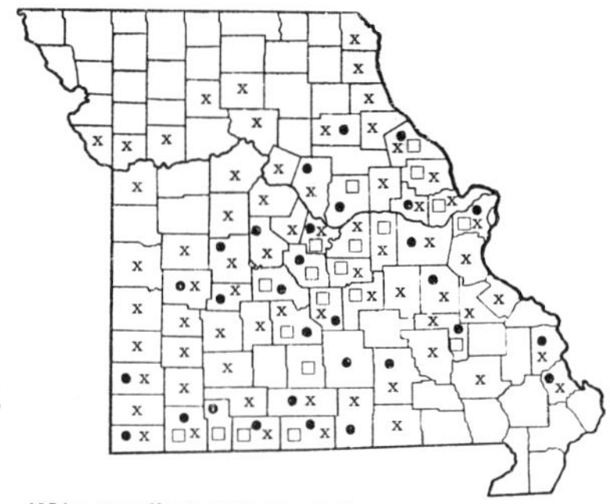

1854 • Scutellaria ovata var. ovata
1854 x Scutellaria ovata var. versicolor
1854 □ Scutellaria ovata var. rugosa

on sandstone, chert, and granite substrata. Confined to the Ozark section and Crowley Ridge.

Two varieties are encountered:

Stems with noticeable spreading hairs and many long-stalked glands; common type encountered . 1a. T. DICHOTOMUM var. DICHOTOMUM

Stems with minute short hairs and/or minute glands; more rarely encountered . . . . . 1b. T. DICHOTOMUM var. PUBERULUM

1a. **Trichostema dichotomum** var. **dichotomum** Map 1851

*Trichostema dichotomum* L. [G, BB, P & S]

Ozark region of southern and east-central Missouri north to St. Louis and Montgomery counties, west to Pulaski, Texas, Douglas, and Ozark counties, and on Crowley Ridge in Stoddard and Dunklin counties.

Ranges from Maine to Michigan, south to North Carolina, Kentucky, and Missouri.

1b. **Trichostema dichotomum** var. **puberulum** Fern. & Grisc. Map 1851

Known from Ste. Genevieve, Cape Girardeau, Montgomery, Gasconade, and Carter counties.

Ranges from Florida to Texas, north to Virginia, Tennessee, Missouri, and Oklahoma.

The lavender-blue corollas with the long curving slender filaments of the stamens produce a rather striking little flower.

2. **Trichostema setaceum** Houtt. Map 1852

*Trichostema dichotomum* var. *lineare* (Walt.) Gl. [BB]

Flowers September–October.

Occurs on sandy open slopes of Crowley Ridge, where known only from Stoddard County, southeastern Missouri (sandy openings on slopes of blowout of sandy prairie and slopes on east side of road and eroded gully, Crowley Ridge, T25N, R11E, sect. 31, 3½ mi. southeast of Bloomfield, October 17, 1955, *Steyermark 80349*).

Ranges from Florida to Louisiana and Texas, north to Connecticut, Pennsylvania, Ohio, and Missouri.

This rare species is another one of the relicts of a more southern and eastern flora which has survived on this restricted sandy habitat of Crowley Ridge, together with other similar rarities, such as *Aristida lanosa*, *Trisetum pensylvanicum*, *Pyrus melanocarpa*, *Bartonia paniculata*, and *Obolaria virginica*.

## 3. **Teucrium** L. Germander

**Teucrium canadense** L. Wood Sage Map 1853

Also called American Germander.

Flowers mid-June–September.

Occurs in prairies, wet meadows, low woodland, thickets, fields, along streams, and railroads.

Two varieties are found in Missouri:

Common type encountered; no glands at the tip of the hairs which cover the calyx, bracts, and upper part of the stem; hairs of the upper part of stem, calyx, and bracts short (0.5–0.75 mm. long)

---

Plate no. 304. **1.** Trichostema dichotomum, × 2/5; *a.* Flower, × 3 1/5. **2.** Isanthus brachiatus, × 2/5; Details from Small, The New York Botanical Garden. **3.** Scutellaria elliptica, × 2/5. **4.** Scutellaria ovata var. rugosa, × 1. **5.** Scutellaria ovata var. versicolor, × 2/5. **6.** Teucrium canadense var. virginicum, × 2/5; *a.* Flower, × 4/5; Details from Small, The New York Botanical Garden.

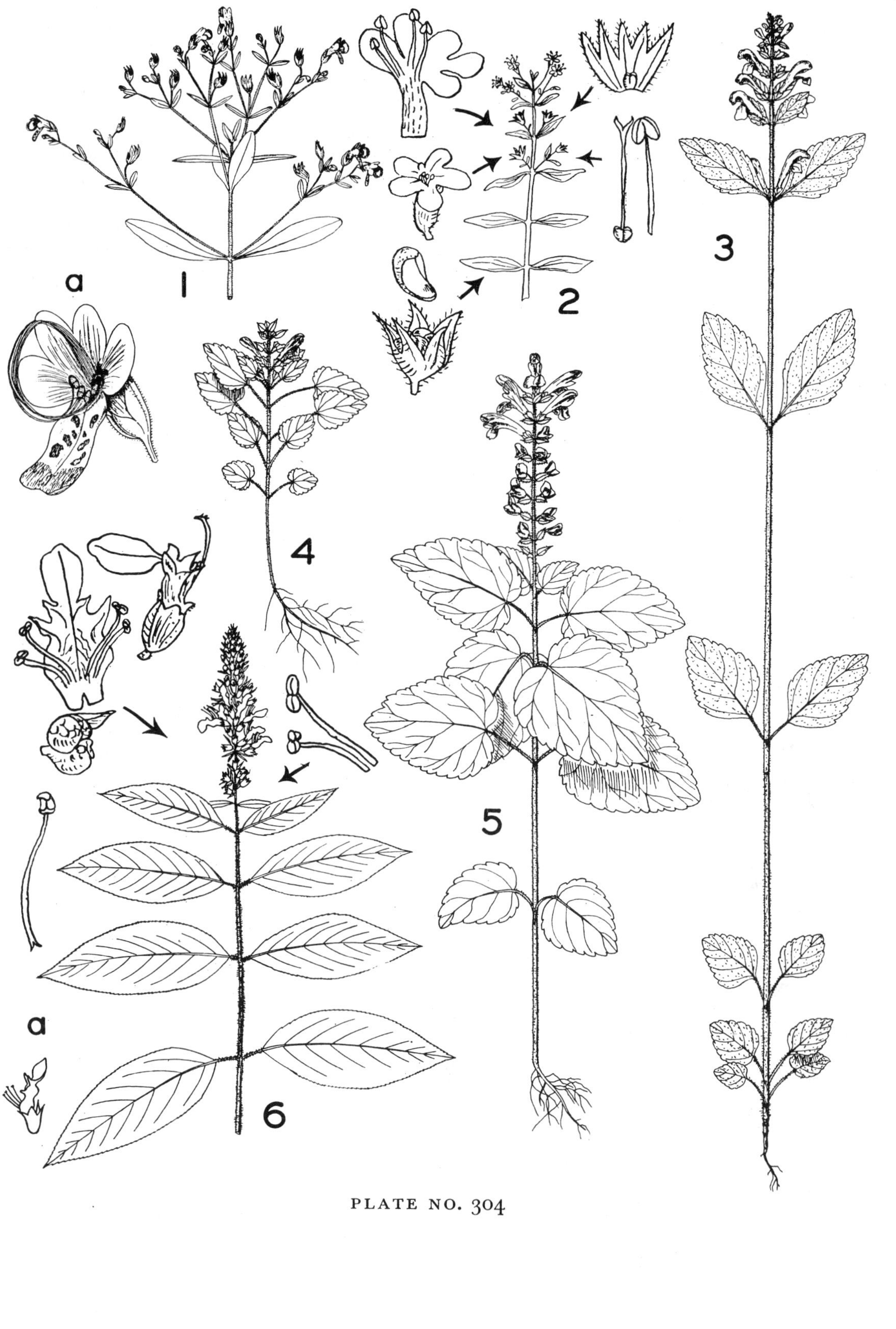


PLATE NO. 304

and appressed (pressed close to or lying parallel to surface); lower surface of leaves green, sparsely hairy with appressed hairs. . . . a. T. CANADENSE var. VIRGINICUM

Rare type, found in northwestern and west-central Missouri; gland-tipped hairs present on calyx, bracts, and upper part of stem; hairs of the upper part of stem, calyx, and bracts spreading and rather conspicuous, 1 mm. or more long; lower surface of leaves grayish with loose spreading hairs . . . . b. T. CANADENSE var. OCCIDENTALE

a. **Teucrium canadense** var. **virginicum** (L.) Eat. Map 1853

Throughout Missouri, and doubtless in every county.

Ranges from Quebec and Vermont to Minnesota and Nebraska, south to Georgia, Alabama, Mississippi, and Texas.

The usual color of the corolla is purplish, pink-lavender, or pink and cream-colored. A white-flowered variant, *T. canadense* var. *virginicum* f. *albiflorum* House, has not been recorded as yet from Missouri.

The plant has aromatic properties and was formerly used as a stimulant in medicine.

b. **Teucrium canadense** var. **occidentale** (Gray) McClintock & Epling Map 1853

*Teucrium occidentale* Gray [G, P & S]

Known only from northwestern and west-central Missouri in Atchison (Watson, July 11, 1894, *Bush 483*), Holt (Corning, July 13, 1894, *Bush 842*), and Jackson (low grounds, Atherton, July 18, 1896, *Mackenzie 406*; prairie, Atherton, August 1, 1902, *Bush 1753*; low ground, Atherton, July 18, 1896, *Bush 754*) counties.

Ranges from Maine to British Columbia, south to Pennsylvania, Ohio, Indiana, Illinois, Missouri, Kansas, New Mexico, and California.

I agree with the results of studies made by Drs. McClintock, Epling, and Gleason that the var. *occidentale* constitutes a northern and western variety of *T. canadense,* rather than a separate species, with sufficient intergradation occurring between the *T. canadense* var. *virginicum* and *T. occidentale* to warrant this point of view.

## 4. **Scutellaria** L. Skullcap

a. Leaf-margin of all the leaves entire, not broken by any teeth or irregularities. . . . . . . . . *b*
  b. Leaves gradually narrowed to the base which is narrower than upper half of leaf-blade, 2–4 cm. long, 5–8 times as long as broad . . . . . . . . . . . . . . . . . . . . 7. S. BUSHII
  b. Leaves rounded or broadened at the base which is broader than upper half of leaf-blade, 0.5–2 cm. long, as broad as long or at most 3½ times as long as broad. . . . . . . . 9. S. PARVULA
a. Leaf-margins of at least the main stem-leaves shallowly or coarsely toothed with 1 or more rounded or pointed teeth . . . . . . . . . . . . . . . . . . . . . . . . . . . *c*
  c. Only 1 flower or fruit in the axil or base of either an ordinary stem-leaf or a reduced upper leaf. . . *d*
    d. Leaf-stalks (petioles) conspicuous, 5–30 mm. long . . . . . . . . . . . 5. S. LATERIFLORA
    d. Leaf-stalks none or 1–4 mm. long . . . . . . . . . . . . . . . . . . . . *e*
      e. Corolla 15–25 mm. long; leaves 2–4½ times as long as broad; hairs on stem recurved, pointing down; plants of swamps; rare species, known only from Jackson County, west-central Missouri . . . . . . . . . . . . . . . . . . . . . 6. S. EPILOBIIFOLIA
      e. Corolla 5–11 mm. long; leaves as broad as long or at most 2½ times as long as broad; at least some of the hairs, if present, on stem spreading or ascending; plants not of swamps; occurring in various parts of Missouri . . . . . . . . . . . . . . . . . . *f*
        f. Stem glabrous (without hairs) or nearly so or with only a few hairs; main stem-leaves 2–5 cm. long; corolla pale blue to whitish; plants of low or moist woodland. . 8. S. NERVOSA
        f. Stem more or less hairy; main stem-leaves 0.5–2 cm. long; corolla dark blue-lavender or purple; plants of dry or rocky glades, prairies, and upland open wooded ridges. 9. S. PARVULA
  c. Flowers and fruits mostly 2 to several or many either in an inflorescence at the top or from the sides of the stem . . . . . . . . . . . . . . . . . . . . . . . . . *g*
    g. Flowers and fruits arising mainly or entirely from the sides of the stem at the base of the leaves . . . . . . . . . . . . . . . . . . . . . . . 5. S. LATERIFLORA
    g. Flowers and fruits arising mainly or entirely at the top of the stem . . . . . . . . . *h*
      h. Main stem-leaves with a heart-shaped (cordate) or rounded base, the leaf-tissue not extending down upon the petiole (leaf-stalk) . . . . . . . . . . . . . . 1. S. OVATA
      h. Main stem-leaves tapering or narrowed at the base, or if rounded or slightly heart-shaped

at base, always more or less V-shaped at the tip of the leaf-stalk or the leaf-tissue extending down the petiole (leaf-stalk) . . . . . . . . . . . . . . . . . . . . . . . . *i*

i. Spreading gland-tipped hairs occur on calyx; 2–5 pairs of stem-leaves occur below the flowering or fruiting portion of plant; main stem-leaves 3–6 cm. long . . . 2. S. ELLIPTICA

i. No gland-tipped hairs on calyx, the hairs all appressed (lying parallel to or pressed against surface); 4–15 or more pairs of stem-leaves occur below the flowering or fruiting portion of plant; main stem-leaves 4–14 cm. long . . . . . . . . . . . . . *j*

j. Commonly encountered throughout the Ozark region; stems and sometimes leaves finely hairy with appressed hairs (lying parallel to or pressed closely to surface) which are not gland-tipped; lowest pair of flowers accompanied by reduced bract-like leaves; usually 6–15 or more firm leaves occur on stem below the flowering or fruiting portion of plant; leaf-blades from the middle of the stem the largest . . . . 4. S. INCANA

j. Rarely found species of the Ozark region; stems and leaves with gland-tipped hairs; lowest pair of flowers accompanied by normal foliage leaves; usually 3–6 pairs of thin leaves occur on stem below the flowering or fruiting portion of plant; leaf-blades from the uppermost part of the stem the largest . . . . . . 3. S. SERRATA var. MONTANA

1. **Scutellaria ovata** Hill Map 1854

Flowers late May–October.

Occurs in rocky open woods, glades, rocky ledges, and bluff escarpments.

The following varieties occur in Missouri:

a. Largest leaf-blades 1–4 cm. long, 0.5–2.5 (–3) cm. broad; dwarfed plants with mostly sprawling or depressed stems 0.5–2.5 (–3) dm. tall . 1c. S. OVATA var. RUGOSA

a. Largest leaf-blades 3.5–13 cm. long, 2.5–11 cm. broad; plants with erect stems mainly 2.5–8.5 m. tall (rarely 1.5 dm.) . . . . *b*

b. Bract (leaf-like part) at base of flower prominent, the upper ones longer than the calyx, 8–20 mm. long . 1a. S. OVATA var. OVATA

b. Bract (leaf-like part) at base of flower not prominent, all but the lowest ones shorter than the calyx, 3–12 mm. long . . . 1b. S. OVATA var. VERSICOLOR

1a. **Scutellaria ovata** var. **ovata** Map 1854

*Scutellaria ovata* Hill [G]

*Scutellaria ovata* var. *bracteata* (Benth.) Blake [P & S, Steyerm.]

Ozark region of southern and east-central Missouri north to Pike, Monroe, Morgan, Benton, St. Clair, and Jasper counties.

Ranges from Louisiana to Texas and Mexico, north to Tennessee, Illinois, Missouri, and Kansas.

The leaf-blades of var. *ovata* are stated by various botanists as averaging smaller than those of the next variety, the larger ones 4–7.5 cm. long and 2.5–6 cm. broad with petioles 1–5 cm. long, but there is too much intergradation in these measurements and those of the next variety to be of significance.

1b. **Scutellaria ovata** var. **versicolor** (Nutt.) Fern. Map 1854

*Scutellaria ovata* of [P & S, Steyerm., and other authors], not Hill

This is the commonest variety in Missouri, occurring throughout southern and central Missouri north to Clark, Linn, Livingston, Ray, Clay, and Platte counties.

Ranges from South Carolina to Louisiana, north to Maryland, West Virginia, Ohio, Indiana, Wisconsin, and Minnesota.

The leaf-blades of var. *versicolor* are stated to average larger than those of the preceding variety, the larger ones being 6–13 cm. long and 5–11 cm. broad, with the petioles 2.5–8 cm. long, but intermediate specimens are frequently found which are not possible to identify on the basis of leaf size.

1c. **Scutellaria ovata** var. **rugosa** (Wood) Fern. Map 1854

*Scutellaria ovata* var. *pilosior* (Benth.) Leonard [P & S, Steyerm.]

*Scutellaria ovata* subsp. *rupestris* Epl. [Epling]

*Scutellaria ovata* subsp. *rugosa* (Wood) Epl. [Epling]

Occurs on rocky glades, exposed ledges, and bluff escarpments in the Ozark region north to Pike, Callaway, Cole, Miller, and Camden counties.

Ranges from West Virginia and Virginia west to Missouri and Arkansas.

The variation is considerable in *S. ovata* in the height of the stem, shape and size of leaves, length of bracts at base of flowers, and length of corolla. Various botanists have studied the variations of this species complex with entirely different results. The var. *rugosa*, in the present treatment maintained for dwarf plants with small leaves, is not recognized by Gleason, who treats such dwarfed specimens under both var. *ovata* and var. *versicolor*. This may be the best solution, but much more field and experimental studies need to be carried out to show the validity of

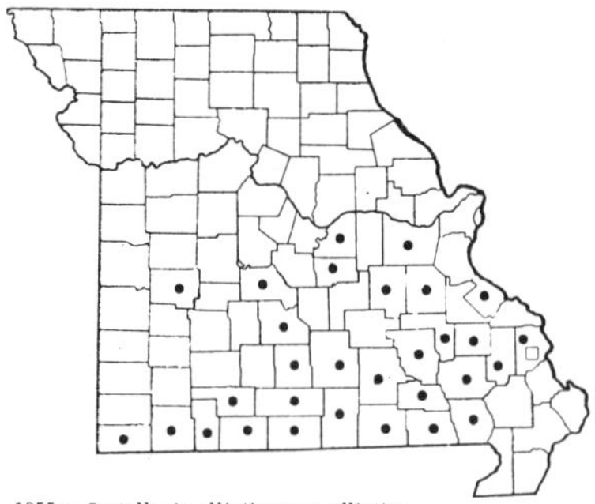

1855 • Scutellaria elliptica var. elliptica
1855 □ Scutellaria elliptica var. hirsuta

1856 Scutellaria serrata var. montana

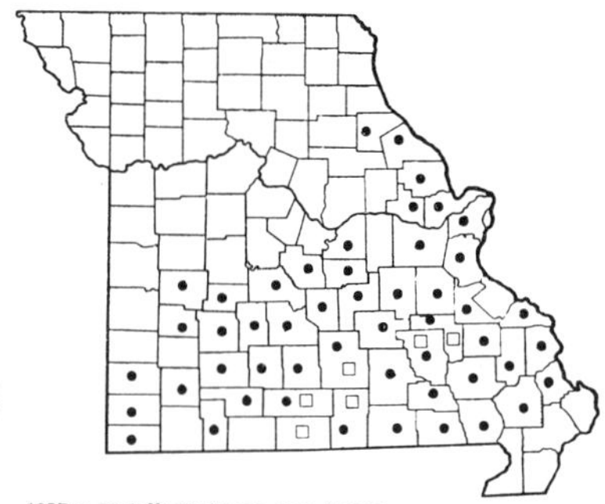

1857 • Scutellaria incana var. incana
1857 □ Scutellaria incana var. punctata

this judgment. Fernald (*Gray's Manual*, p. 1219) uses the character of a short corolla (1 cm. long) as an additional diagnostic feature of var. *rugosa*, but many specimens from Missouri have the small leaves of var. *rugosa* but the corollas 1.5 cm. long, which would place them in the size category used by Fernald for vars. *ovata* and var. *versicolor* having corollas 1–2.3 cm. long.

Likewise, there is poor correlation in the presumed smaller size of the leaves with longer bracts of *S. ovata* var. *ovata* as contrasted with somewhat larger leaves with shorter bracts of *S. ovata* var. *versicolor*, the two overlapping in these characters to such an extent that they are scarcely recognizable.

2. **Scutellaria elliptica** Muhl. Map 1855
Flowers late May–July.

Occurs in usually dry and rocky woods, along bluffs, and wooded slopes along streams.

Two varieties occur in Missouri:

Hairs of stem short, curving . . 2a. S. ELLIPTICA var. ELLIPTICA
Hairs of stem straight, spreading, 1–2 mm. long . 2b. S. ELLIPTICA var. HIRSUTA

2a. **Scutellaria elliptica** var. **elliptica** Map 1855

Ozark region north to Franklin, Osage, Camden, and St. Clair counties.

Ranges from Florida to Texas, north to New York, Pennsylvania, West Virginia, Ohio, Missouri, and Oklahoma.

2b. **Scutellaria elliptica** var. **hirsuta** (Short) Fern. Map 1855

Known only from Cape Girardeau County, southeastern Missouri (steep, wooded, mostly acid-soil ravines tributary to Dog Hollow, T11N, R14E, sect. 29, 2 mi. northwest of Neelys Landing, August 29, 195[illegible], *Steyermark 76984*).

Ranges from Georgia to Texas, north to Pennsylvania, Michigan, Illinois, and Missouri.

3. **Scutellaria serrata** Andr. var. **montana** (Chapm.) Penland Map 1856
*Scutellaria montana* Chapm. [P & S]
Flowers May–June.

Occurs in rocky open woods in the eastern Ozark section, in Iron County (*Pammel*).

Ranges in Georgia, Tennessee, Alabama, and Missouri.

Typical *S. serrata* var. *serrata*, which is not known from Missouri, has the stem glabrous or sparsely hairy toward the summit, and the leaf-blades are glabrous or very sparsely hairy. In the var. *montana* the stem and leaves are glandular-hairy.

4. **Scutellaria incana** Biehler Map 1857
Flowers June–September.

Occurs in rocky open woods, ledges along bluffs, ravines, wooded slopes along streams, and thickets.

Two intergrading variations occur in the state:

Stem with gray appressed hairs (lying parallel to surface or pressed closely against it); lower surface of leaf-blades with a fine white, short hairiness . 4a. S. INCANA var. INCANA
Stem less hairy; lower surface of leaf-blades glabrous (without hairs) except somewhat hairy along the nerves . . 4b. S. INCANA var. PUNCTATA

4a. **Scutellaria incana** var. **incana** Map 1857

Ozark region of southern, central, and east-central Missouri north to Ralls, Warren, Osage, Miller, Camden, Hickory, St. Clair, and Jasper counties. This is the commoner variation in Missouri.

Ranges from Alabama, Tennessee, Arkansas, and Oklahoma, north to New Jersey, New York, Indiana, Illinois, Iowa, and Kansas.

1858 • Scutellaria lateriflora f. lateriflora (Mad-dog Skullcap)
1858 □ Scutellaria lateriflora f. albiflora

1859 Scutellaria epilobiifolia

1860 Scutellaria Bushii

4b. **Scutellaria incana** var. **punctata** (Chapm.) Mohr — Map 1857

*Scutellaria punctata* (Chapm.) Leonard [P & S]

Southern Ozark region east and north to Iron, Reynolds, and Texas counties, west to Douglas and Ozark counties.

Ranges from Florida to Alabama, north to Virginia, West Virginia, Kentucky, and Missouri.

5. **Scutellaria lateriflora** L. Mad-dog Skullcap — Map 1858

Flowers June–October.

Occurs in low wet woods, swampy meadows, alluvial thickets, gravel bars, river flood plain forests, borders of sloughs, spring branches, streams, swamps, lowland and sink-hole ponds, and moist ledges of limestone and sandstone bluffs. Throughout Missouri, probably in every county.

Ranges from Quebec and Newfoundland to Ontario, Manitoba, Saskatchewan, Alberta, and British Columbia, south to Florida, Louisiana, Arizona, and California.

Two variations occur in Missouri:

Corolla blue-purple . . . 5a. S. LATERIFLORA f. LATERIFLORA
Corolla white . 5b. S. LATERIFLORA f. ALBIFLORA

5a. **Scutellaria lateriflora** f. **lateriflora** — Map 1858

*Scutellaria lateriflora* L. [G, BB, P & S]

The common variation throughout Missouri.

5b. **Scutellaria lateriflora** f. **albiflora** (Farw.) Fern. — Map 1858

Known only from St. Charles County, east-central Missouri (highway 61 bridge over Missouri River, near St. Charles, August 5, 1939, *Bauer 512* in Chi. Nat. Hist. Mus. Herb.), but probably more frequent.

*Scutellaria lateriflora* has been used in medicine as a nerve remedy and tonic; the entire plant is employed for this purpose. A pink-flowered form, f. *rhodantha* Fern., occasionally occurs within the range of the species, but has not yet been recorded for Missouri. Seeds and other parts of the plants are sometimes eaten by pheasants.

6. **Scutellaria epilobiifolia** A. Hamilton — Map 1859

*Scutellaria galericulata* of Am. auth., not L. [BB]

Flowers June–September.

Known only from Jackson County, west-central Missouri (July 22, 1885, *Bush*, in U. of Mo. Herb.).

Ranges from Newfoundland and Labrador to James Bay, and Alaska, south to Delaware, New York, Pennsylvania, West Virginia, Ohio, Indiana, Illinois, Missouri, Kansas, New Mexico, Arizona, and California.

The Missouri sheet cited is one of the collection of World's Fair specimens. I was assured by Mr. B. F. Bush that all the specimens exhibited were from Missouri collections. Except for this collection, the species has not been re-collected in the state, and some doubt may be expressed as to the actual source of this and other World's Fair material which has not been found in Missouri since that exhibition.

7. **Scutellaria Bushii** Britt. — Map 1860

Flowers May–June.

Occurs on limestone glades and bald knobs. Known from the southern Ozark section east to Carter County, north to Dent and Pulaski counties, west to Christian and Taney counties.

Ranges in Missouri and Arkansas.

This is one of the species of plants endemic to the Ozark section of Missouri and Arkansas. Others of similar endemism, which occur on limestone glades with *Scutellaria Bushii*, are *Penstemon Cobaea* var. *purpureus*, *Delphinium Treleasei*, *Onosmodium subsetosum*, *Valerianella Bushii*, and *V. ozarkana*.

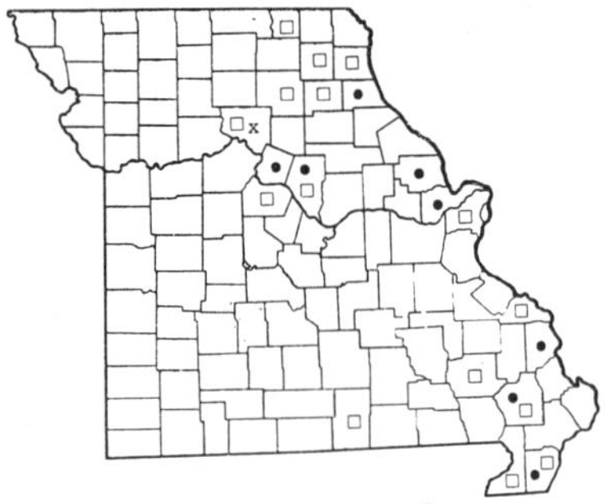

1861 • Scutellaria nervosa var. nervosa f. nervosa
1861 x Scutellaria nervosa var. nervosa f. alba
1861 □ Scutellaria nervosa var. calvifolia

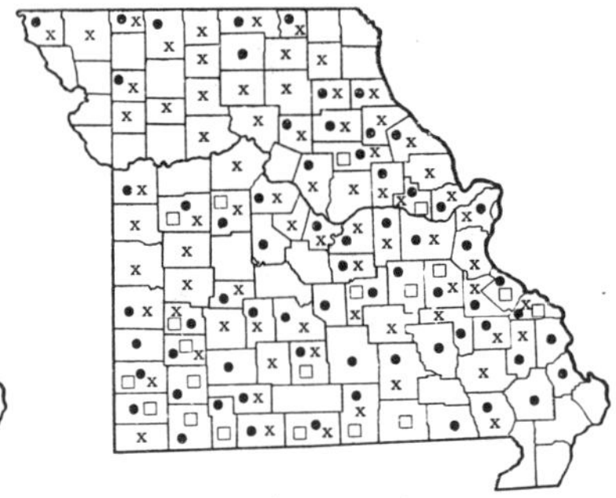

1862 • Scutellaria parvula var. parvula
1862 □ Scutellaria parvula var. australis
1862 x Scutellaria parvula var. Leonardi

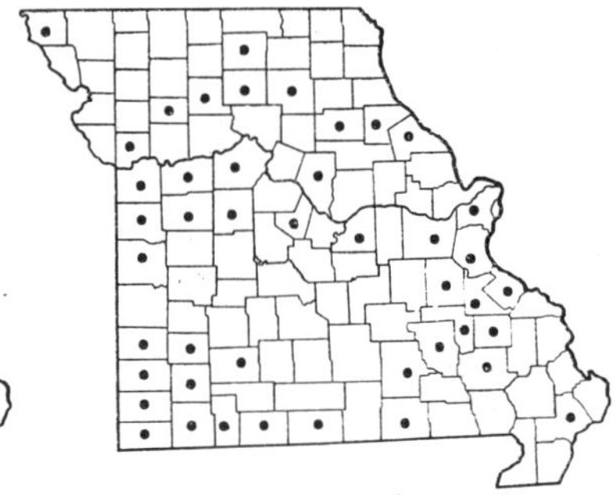

1863 Marrubium vulgare (Horehound)

8. **Scutellaria nervosa** Pursh Map 1861

Flowers late April–July.

Occurs in rich, moist or low woodland bordering swamps and streams, base of rich wooded bluffs, and thickets.

The following variations occur in Missouri:

a. Upper surface of leaves glabrous (without hairs) . . . . 8c. S. NERVOSA var. CALVIFOLIA
a. Upper surface of leaves more or less hairy with hairs parallel to or lying upon the surface . *b*
 b. Corolla pale bluish . . . 8a. S. NERVOSA var. NERVOSA f. NERVOSA
 b. Corolla white . 8b. S. NERVOSA var. NERVOSA f. ALBA

8a. **Scutellaria nervosa** var. **nervosa** f. **nervosa** Map 1861

*Scutellaria nervosa* Pursh [G, P & S, Steyerm.]

*Scutellaria nervosa* var. *nervosa* [BB]

This is the rarer variation of the two main varieties in Missouri, known only from Marion, Lincoln, St. Charles, Cape Girardeau, Stoddard, Pemiscot, Boone, and Howard counties in the eastern half of the state.

Ranges from Pennsylvania to Ohio, Indiana, Illinois, and Missouri, south to North Carolina, Tennessee, and Louisiana.

8b. **Scutellaria nervosa** var. **nervosa** f. **alba** Steyerm. Map 1861

*Scutellaria nervosa* f. *alba* Steyerm.

Known only from Chariton County, central Missouri (bottom woods along Yellow Creek, T56N, R19W, sect. 20 and 17, at and just north of Rothville, May 21, 1949, *Steyermark 68024*, holotype in Chi. Nat. Hist. Mus. Herb.).

8c. **Scutellaria nervosa** var. **calvifolia** Fern. Map 1861

This is the commoner variation in Missouri, known in the eastern half of the state west to Chariton, Cooper, and Howell counties.

Ranges from New Jersey to Ontario and Iowa, south to Virginia, Ohio, Indiana, Tennessee, and Missouri.

9. **Scutellaria parvula** Michx. Map 1862

Flowers May–July.

Occurs on rocky open glades, prairies, meadows, fallow fields, bluffs, openings on wooded ridges and uplands, and low woods at base of bluffs.

The following variations occur in Missouri:

a. Hairs on stem not present or if present, then curved-ascending, and not gland-tipped; upper leaf-surface glabrous (without hairs) or minutely and sparsely hairy; the upper and middle leaves 2–10 mm. broad with conspicuously revolute (turned under) margins; usually 2 (or 3) lateral nerves on each side of the midnerve; no glands on calyx . 9c. S. PARVULA var. LEONARDI
a. Hairs on stem abundant, some of them spreading and gland-tipped; upper leaf-surface more or less noticeably hairy; the upper and middle leaves 3–14 mm. broad, the margins not or only slightly revolute; usually 3–5 lateral nerves on each side of the midnerve; glandular hairs present on calyx . . . . . . . *b*
 b. Sessile (stalkless) resinous dots or glands present and numerous on lower leaf-surface mixed with longer hairs; lateral nerves of leaf not or inconspicuously connecting along the margin of leaf . . . 9a. S. PARVULA var. PARVULA

Plate no. 305. **1.** Scutellaria incana, × 2/5. **2.** Scutellaria Bushii, × 2/5. **3.** Scutellaria epilobiifolia, × 2/5. **4.** Scutellaria lateriflora, × 2/5. **5.** Scutellaria parvula, × 2/5. **6.** Scutellaria nervosa, × 2/5. All details from Small, The New York Botanical Garden.

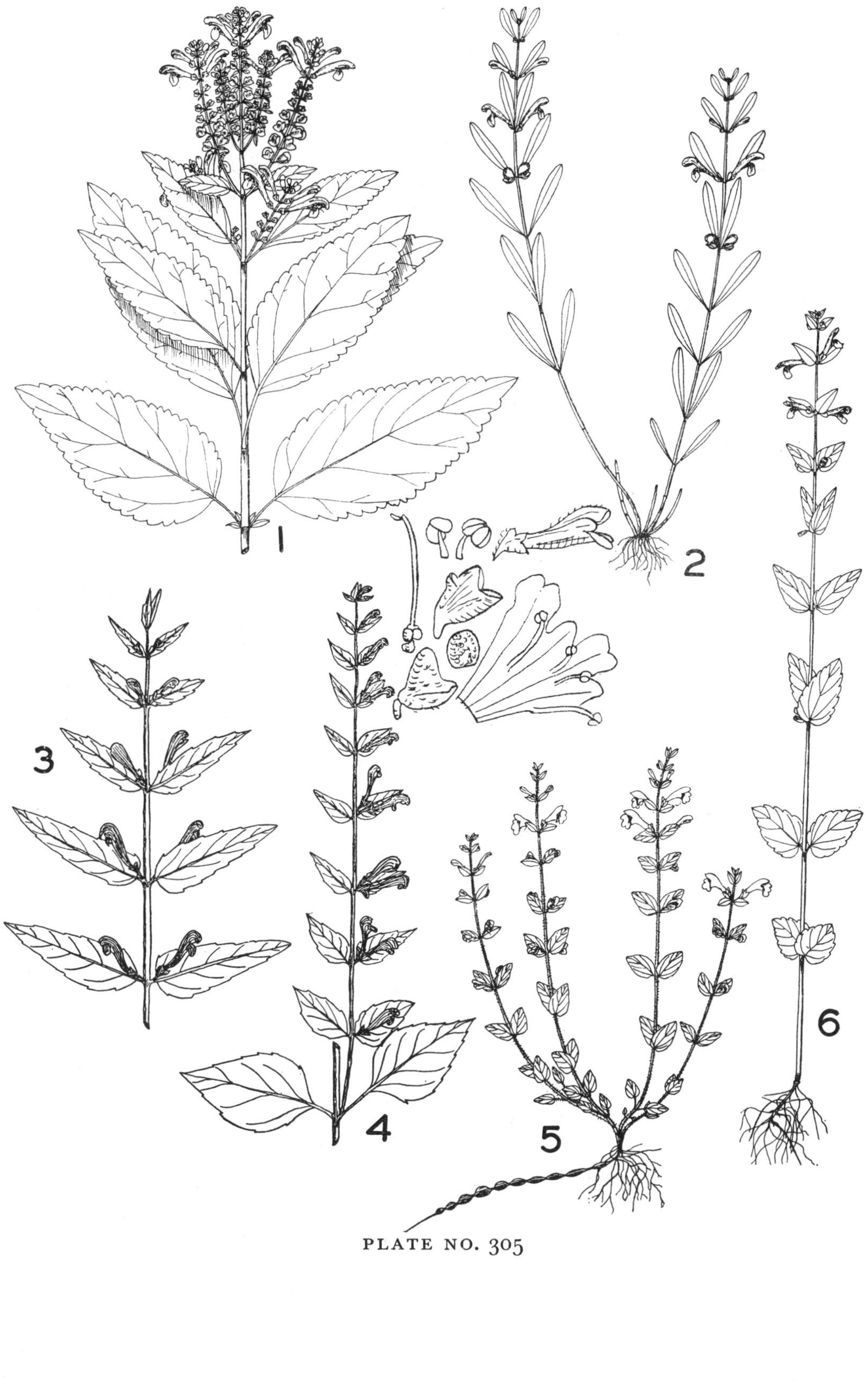


PLATE NO. 305

b. Mainly long hairs present on nerves of lower leaf-surface, no sessile glands or essentially none present; lateral nerves of leaf connecting to form a more or less continuous nerve along the margin of leaf . . 9b. S. PARVULA var. AUSTRALIS

9a. **Scutellaria parvula** var. **parvula** Map 1862

*Scutellaria parvula* Michx. [G, BB, Steyerm., P & S in part]

Scattered over the state, commonest in southern and central Missouri.

Ranges from Quebec and Ontario to Wisconsin and Iowa, south to Georgia, Alabama, Mississippi, Arkansas, and Texas.

9b. **Scutellaria parvula** var. **australis** Fassett Map 1862

*Scutellaria australis* (Fassett) Epl. [BB]

Ozark region of southern Missouri north to Ste. Genevieve, Wright, and Cedar counties.

Ranges from Florida to Texas, north to Connecticut, New York, West Virginia, Virginia, Indiana, Illinois, Missouri, Kansas, and Oklahoma.

9c. **Scutellaria parvula** var. **Leonardi** (Epling) Fern. Map 1862

*Scutellaria parvula* var. *ambigua* of some auth. [Steyerm., P & S], not *S. ambigua* Nutt.

*Scutellaria Leonardi* Epl. [BB]

Common throughout Missouri, except absent from the lowlands of extreme southeastern Missouri.

Ranges from Maine to Michigan and North Dakota, south to Virginia, Tennessee, Louisiana, and Texas.

The var. *Leonardi* is the most easily recognizable of the three varieties of *S. parvula*, and, as has been treated by some botanists, might be considered a distinct species. The differences separating var. *Leonardi* from the other two varieties are well-marked, but occasional intermediate specimens occur which combine some of the characters of the taxa. The varieties *parvula* and *australis* appear to be less distinct from each other than from var. *Leonardi*, and intergradations exist, making their specific status, as given to them by Gleason, Leonard, and Epling, appear doubtful and unjustified.

This species complex produces from the slender underground stems a series of tubers resembling a string of sausages (moniliform) or beads. These are most evident in early spring and late fall, but are often not collected with the flowering or fruiting plant.

## 5. **Marrubium** L. Horehound

**Marrubium vulgare** L. Horehound Map 1863

Flowers May–September.

Occurs in fields, waste ground, pastures, along roadsides, and along exposed ledges of bluffs and escarpments, often persisting around old dwellings. Throughout Missouri.

Native of Europe; introduced and naturalized in North America throughout the United States north in Canada to Ontario and British Columbia.

The flowering part and dried leaves of this plant enter into medicinal use as a stimulant, tonic, and laxative, as a remedy for colds, rheumatism, and dyspepsia. Horehound candy or coughdrop lozenges are often used for coughs and colds.

## 6. **Agastache** Clayton. Giant Hyssop

Corolla greenish-yellow or yellowish; teeth of calyx ovate, obtuse (blunt) or slightly acute (pointed), 1–2 mm. long; inflorescence 1–1.5 cm. thick; bracts and calyx-teeth purplish; common throughout Missouri . . . . . . . . . . . . . . . . . . . . . . . . 1. A. NEPETOIDES

Corolla pink-purplish or blue-lavender; teeth of calyx triangular-lanceolate, acute to acuminate (long-pointed), 2–3 mm. long; inflorescence 1.5–2 cm. thick; bracts and calyx-teeth green or buff; rare, known only from northern and west-central Missouri . . . . . . . . . 2. A. SCROPHULARIAEFOLIA

1. **Agastache nepetoides** (L.) Ktze. Yellow Giant Hyssop Map 1864

Flowers July–September.

Occurs in open woods, low moist or rich woods, base of bluffs, borders of woods, and thickets. Throughout Missouri.

---

Plate no. 306. **1.** Marrubium vulgare, × 2/5. **2.** Agastache scrophulariaefolia, × 2/5. **3.** Dracocephalum parviflorum, × 1. **4.** Glechoma hederacea var. hederacea, × 2/5. **5.** Nepeta Cataria, × 2/5. All details from Small, The New York Botanical Garden.

PLATE NO. 306

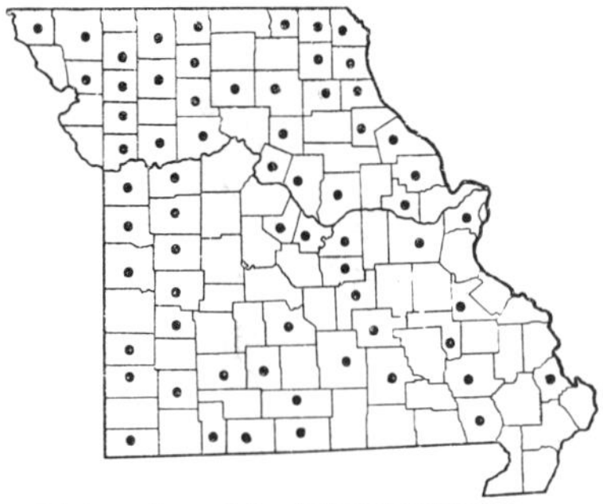

1864 Agastache nepetoides (Yellow Giant Hyssop)

1865 • Agastache scrophulariaefolia f. scrophulariaefolia (Purple Giant Hyssop)
1865 □ Agastache scrophulariaefolia f. mollis

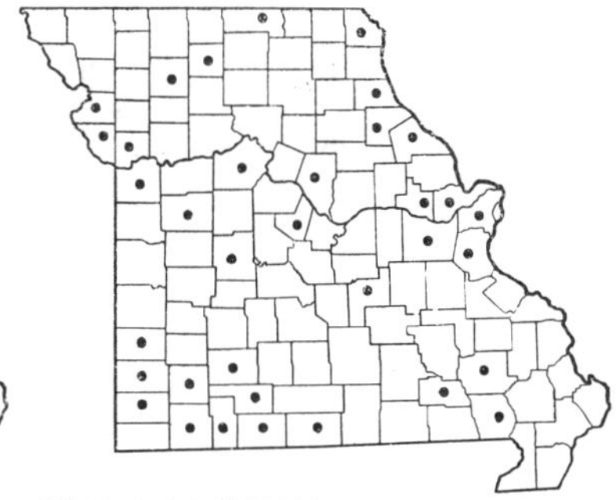

1866 Nepeta Cataria (Catnip)

Ranges from Quebec to Ontario and South Dakota and Nebraska, south to Georgia, Kentucky, Arkansas, Kansas, and Oklahoma.

2. **Agastache scrophulariaefolia** (Willd.) Ktze. Purple Giant Hyssop Map 1865

Also called Figwort Giant Hyssop.

Flowers July–September.

Occurs in open and cut-over woods, low alluvial ground near streams, and thickets. Northern and west-central Missouri.

Two variations are encountered in Missouri:

Stem and lower surface of leaves slightly hairy .
2a. A. SCROPHULARIAEFOLIA f. SCROPHULARIAEFOLIA
Stem and lower surface of leaves densely hairy .
2b. A. SCROPHULARIAEFOLIA f. MOLLIS

2a. **Agastache scrophulariaefolia** f. **scrophulariaefolia** Map 1865

*Agastache scrophulariaefolia* (Willd.) Ktze. [G, BB, P & S]

Known from Scotland, Mercer, Atchison, Jackson and Cass counties.

Ranges from Vermont and Massachusetts to Ontario, south to North Carolina, Kentucky, Missouri, and Kansas.

2b. **Agastache scrophulariaefolia** f. **mollis** (Fern.) Steyerm. Map 1865

*Agastache scrophulariaefolia* var. *mollis* (Fern.) Heller [G, P & S]

Occurring with the typical form.

Ranges from Vermont to Minnesota, south to Connecticut, New York, Ohio, and Missouri.

## 7. **Nepeta** L. Catnip

**Nepeta Cataria** L. Catnip Map 1866

Also called Catmint.

Flowers early June–September.

Occurs in fields, waste ground about old dwellings, open woodland, along roadsides, streams, and railroads. Throughout Missouri.

Native of Europe; introduced and naturalized throughout most of the United States and southern Canada from Newfoundland and Quebec to British Columbia.

The leaves, steeped in water, make a pleasant-tasting tea. The plant was at one time cultivated for its supposed medicinal properties, especially as a tonic or for colds. It is well-known that cats have a peculiar liking for the plant.

## 8. **Glechoma** L. Ground Ivy

(*Nepeta* in part [P & S, Steyerm.])

**Glechoma hederacea** L. Ground Ivy Map 1867

Also called Gill-over-the-ground.

Flowers April–July.

Occurs in low woodland, in valleys and banks along streams, waste ground, about dwellings, garden plots, fields, along roadsides and railroads. Throughout Missouri:

Two varieties are found in Missouri:

More rarely found; corolla 16–22 mm. long. .
a. G. HEDERACEA var. HEDERACEA

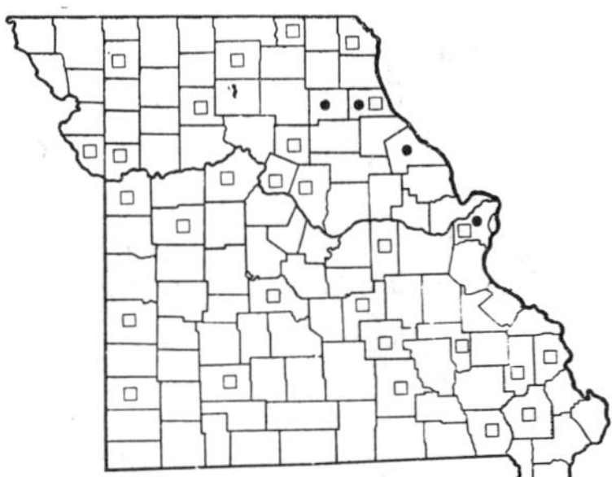

1867 • Glechoma hederacea var. hederacea (Ground Ivy)
1867 □ Glechoma hederacea var. micrantha

1868 Dracocephalum parviflorum (Dragonhead)

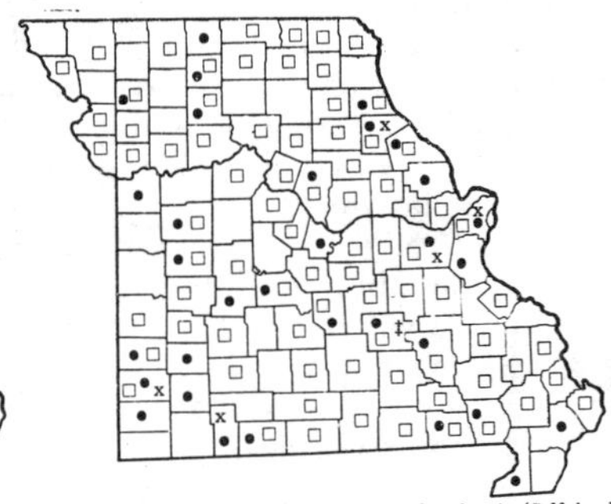

1869 • Prunella vulgaris var. vulgaris f. vulgaris (Self-heal)
1869 □ Prunella vulgaris var. lanceolata f. lanceolata
1869 ‡ Prunella vulgaris var. lanceolata f. candida
1869 x Prunella vulgaris var. lanceolata f. iodocalyx

More frequent in occurrence; corolla 10–15 mm. long . . . . b. G. HEDERACEA var. MICRANTHA

a. **Glechoma hederacea** var. **hederacea** Map 1867

*Glechoma hederacea* L. [G, BB]

*Nepeta hederacea* (L.) Trevisan [P & S, Steyerm.]

Known only from Marion, Shelby, Pike, and St. Louis counties, eastern Missouri.

Native of Europe; introduced and naturalized in North America from Newfoundland to Ontario, south to Virginia, Tennessee, and Missouri.

b. **Glechoma hederacea** var. **micrantha** Moricand Map 1867

*Glechoma hederacea* var. *parviflora* (Benth.) Druce

*Nepeta hederacea* var. *parviflora* Benth. [P & S, Steyerm.]

*Glechoma heterophylla* Waldst. & Kit. [Jones *et al.*]

The common variation, throughout Missouri.

Native of Europe; introduced and naturalized in North America from Newfoundland to Ontario and North Dakota, south to Georgia, Alabama, Missouri, Kansas, Colorado; also from Alaska to California.

The round scalloped leaves and blue-lavender flowers make this an attractive ground-cover plant. However, it has the propensities of being weedy, and, once established, it increases so rapidly into adjacent ground that it becomes undesirably abundant. It is practically impossible to eradicate when once established as each segment of the creeping stem may take root and develop.

Fresh plants or hay containing the dried plants of *Glechoma hederacea*, if eaten in large quantity, are reported to be poisonous to horses.

## 9. **Dracocephalum** L. Dragonhead

(*Moldavica* Adans. [BB])

**Dracocephalum parviflorum** Nutt. Dragonhead Map 1868

*Moldavica parviflora* (Nutt.) Britt. [BB]

Flowers late May–August.

Occurs in waste ground and along railroads. Known only from St. Louis (St. Louis, Baden Freight Yard of Mo.-Kan.-Tex. R.R., July 5, 1958, *Muehlenbach 1443*) and Jackson counties.

Ranges from Quebec to Yukon and Alaska, south to New York, Michigan, Illinois, Wisconsin, Iowa, Nebraska, New Mexico, and Arizona; introduced in Missouri and New England.

## 10. **Prunella** L. Self-heal, Heal-all

**Prunella vulgaris** L. Self-heal Map 1869

Also called Heal-all.

Flowers May–September.

Occurs in low woodland, borders, banks and gravel bars of streams, ponds, and ditches, pastures, prairies, fields, waste ground, along roadsides, and railroads. Throughout Missouri.

The following variations are encountered in Missouri:

a. Main leaves or those from half-way up stem

rounded at their base, 2/5–2/3 as broad as long
a. P. VULGARIS var. VULGARIS f. VULGARIS

a. Main leaves or those from half-way up stem gradually tapering or wedge-shaped at their base, 1/5–1/2 (average 1/3) as broad as long . *b*
 b. Corolla white . . . . c. P. VULGARIS var. LANCEOLATA f. CANDIDA
 b. Corolla bluish, lavender, or violet . . . *c*
  c. Calyx green or at most with purple-tinged edges . . . . b. P. VULGARIS var. LANCEOLATA f. LANCEOLATA
  c. Calyx purple . . . . d. P. VULGARIS var. LANCEOLATA f. IODOCALYX

1a. **Prunella vulgaris** var. **vulgaris** f. **vulgaris** Map 1869

*Prunella vulgaris* L. [G, P & S, Steyerm.]

Less common than var. *lanceolata*, but scattered throughout the state.

Native of Europe; naturalized in North America from Newfoundland and Quebec to Minnesota and British Columbia, south to North Carolina and Mexico.

b. **Prunella vulgaris** var. **lanceolata** (Bart.) Fern. f. **lanceolata** Map 1869

*Prunella vulgaris* var. *lanceolata* (Bart.) Fern. [G, BB, P & S, Steyerm.]

*Prunella vulgaris* ssp. *lanceolata* (Bart.) Hultén

The common variation in Missouri, where common throughout the state.

Ranges from Newfoundland to Alaska, south to Florida, Louisiana, New Mexico, Arizona, and California; also in Asia.

c. **Prunella vulgaris** var. **lanceolata** f. **candida** Fern. Map 1869

Known only from Dent County (upland north of Elk Hollow, T34N, R7W, west part sect. 1, 4½ mi. southeast of Anutt, June 25, 1951, *Steyermark 71782*).

d. **Prunella vulgaris** var. **lanceolata** f. **iodocalyx** Fern. Map 1869

Scattered in Ralls, St. Louis, Franklin, Stone, and Jasper counties.

The common name Heal-all refers to the idea, held by some Indians, that the plant could be used to cure most all ailments and was therefore used as a general remedy.

## 11. **Physostegia** Benth. False Dragonhead

(*Dracocephalum* L. [BB, McClintock])

Dr. Elizabeth McClintock has recently studied this genus and accepts *Dracocephalum* as the proper generic name instead of *Physostegia*. The present author agrees with the views expressed by Shinners (Rh. 51: 120–22. 1949) to retain the name *Physostegia*. In my opinion, the species as accepted thus far appear excessive, and are not satisfactorily separated in any of the present manuals. The present elaboration of the species is at best only a tentative one, with much more field and experimental work needed to be carried out and correlated with the correct names of the various taxa.

a. Teeth of the main leaves of stem almost absent or inconspicuous or 3–11 (up to 14) low rounded or wavy projections on each side; plants of swamps and water of the lowlands of southeastern Missouri 4. P. INTERMEDIA
a. Teeth of the main leaves of stem conspicuous, sharp-tipped or acutely pointed, 10–30 on each side; plants found throughout Missouri . . . . . . . . . . . . . . . . . . . . . . . . *b*
 b. Corolla 8–15 mm. long; larger upper leaves not much reduced in size below the flowering or fruiting portion of plant; inflorescence appearing sessile or nearly so above the leaves; upper internodes (intervening space just below flowering or fruiting portion) not longer than the ones at the middle of the stem, often shorter . . . . . . . . . . . . 3. P. FORMOSIOR
 b. Corolla 20–30 mm. long; larger upper leaves much reduced in size below the flowering or fruiting portion of plant; inflorescence appearing elevated or raised on a long flowering-stem; upper internodes (intervening space just below flowering or fruiting portion) slightly shorter than to much longer than the ones at the middle of the stem . . . . . . . . . . . . *c*
  c. Main leaves of stem 10–40 mm. broad, rather membranaceous to firm, lanceolate to oblanceolate or oblong; inflorescence rather closely flowered, the pairs of calyces of the inflorescence near one another or overlapping . . . . . . . . . . . . . 1. P. VIRGINIANA
  c. Main leaves of stem 3–12 mm. broad, thick, and rather stiff, linear to narrowly lanceolate; inflorescence remotely flowered, the lower calyces or empty bracts of the inflorescence separated by a distance of 1–5 cm. . . . . . . . . . . . . . . . . . . 2. P. ANGUSTIFOLIA

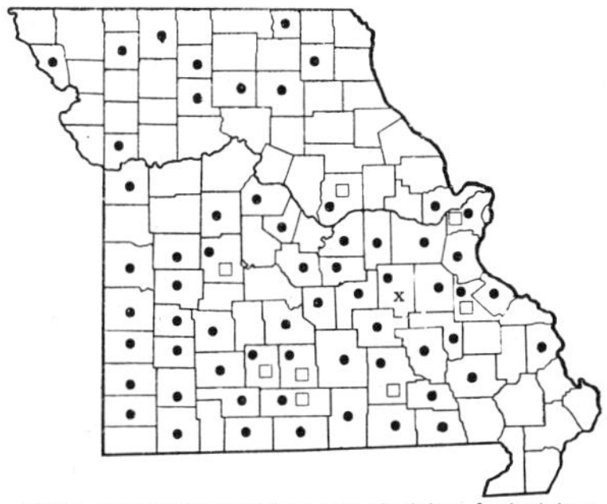

1870 • Physostegia virginiana var. virginiana f. virginiana (False Dragonhead)
1870 x Physostegia virginiana var. virginiana f. candida
1870 □ Physostegia virginiana var. speciosa

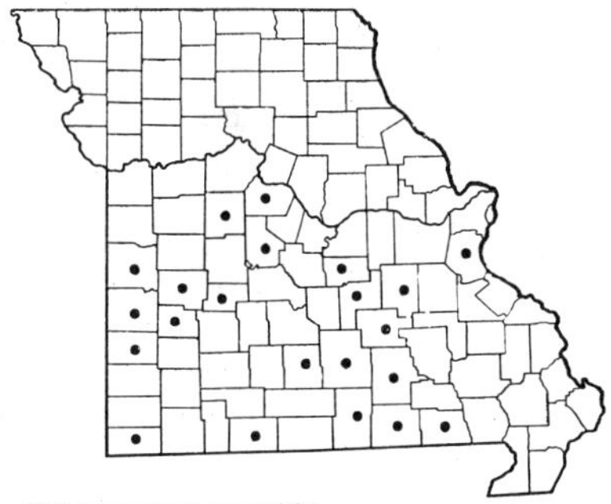

1871 Physostegia angustifolia

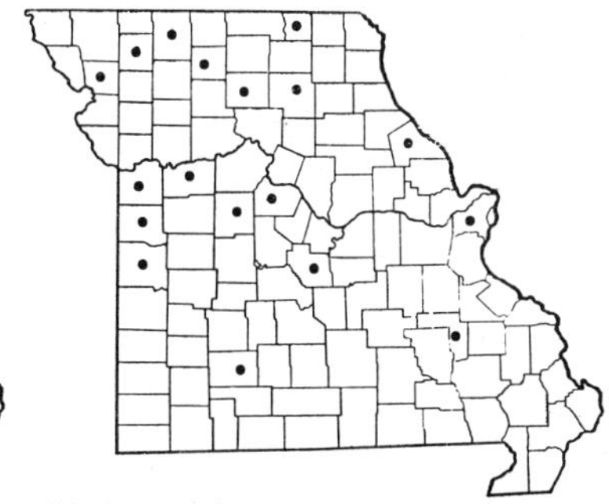

1872 Physostegia formosior

1. **Physostegia virginiana** (L.) Benth. False Dragonhead Map 1870

Also called Obedient Plant.

Flowers May–September.

Occurs in moist ground of prairies, along prairie streams and spring branches, glades, base of wooded bluffs, gravel bars, rocky banks of streams, and along railroads. Common throughout Missouri.

The following variations are encountered in the state:

a. Leaves 20–40 mm. broad, broadly lanceolate to oblanceolate or oblong; calyx with few or no glands; upper leaves gradually or scarcely reduced in size as compared with the ones at the middle of stem . . . 1c. P. VIRGINIANA var. SPECIOSA

a. Leaves 10–25 mm. broad, lanceolate; calyx usually with abundant glands; upper leaves much reduced in size as compared with the ones below . . . . . . . . . . b

b. Corolla rose-purple or pinkish flushed with lilac . 1a P. VIRGINIANA var. VIRGINIANA f. VIRGINIANA

b. Corolla whitish . . . 1b. P. VIRGINIANA var. VIRGINIANA f. CANDIDA

1a. **Physostegia virginiana** var. **virginiana** f. **virginiana** Map 1870

*Physostegia virginiana* (L.) Benth. [G, P & S, Steyerm.]

*Dracocephalum virginianum* L. [BB]

Common throughout Missouri.

Ranges from Quebec to Minnesota, south to Maryland, North Carolina, Tennessee, Missouri, and Oklahoma.

1b. **Physostegia virginiana** var. **virginiana** f. **candida** Benke Map 1870

*Physostegia virginiana* f. *candida* Benke [G]

Known only from Washington (near Old Mines, August 30, 1948, *Bauer*, where originally collected and later grown in garden) and Cedar (limestone glade along slopes of Brush Creek, T35N, R25–26W, sect. 7, 4½ mi. northwest of Arnica, June 25, 1939, *Steyermark 27326*) counties.

1c. **Physostegia virginiana** var. **speciosa** (Sweet) Gray Map 1870

Scattered in southern and east-central Missouri in St. Louis, St. Francois, Shannon, Callaway, Wright, Webster, Benton, and Douglas counties.

Ranges from Ohio to North Dakota, south to North Carolina, Tennessee, and Louisiana; spread from cultivation to New England and Pennsylvania.

According to Gleason's key (in *New Ill. Fl.* 3: 157. 1952) specimens with the upper leaves not reduced in size would key to *Dracocephalum formosius*, whereas in Fernald's key (in *Gray's Man.* p. 1226. 1950) they could be placed as well in *Physostegia virginiana* var. *speciosa*.

This species and its varieties are commonly cultivated, eventually forming dense stands by virtue of the underground creeping stems or rhizomes. Obedient plant owes its name to the ease with which a corolla and calyx can be turned or directed to a new position, in which new position the flower remains.

2. **Physostegia angustifolia** Fernald Map 1871

*Dracocephalum purpureum* [BB], as to Missouri reference, not (Walt.) McClintock

Flowers June–September.

Occurs on rocky open glades, prairies, moist rocky ledges of bluffs, gravel bars along streams, and rocky open ground. Ozark and unglaciated prairie regions of southern and central Missouri north to Jefferson, Crawford, Maries, Cooper, Hickory, St. Clair, and Vernon counties.

Ranges from Illinois, Missouri, and Oklahoma, south to Tennessee, Mississippi, Louisiana, and Texas.

The status of this taxon as a species distinct from *P. virginiana* is quite doubtful in the present author's opinion. Although Shinners considers it distinct as an annual plant without rhizomatous underground extensions of the stem, as contrasted with the perennial rhizomatous habit of *P. virginiana*, field observations and collections of *P. angustifolia* from Missouri are insufficient at present to warrant this conclusion. Actually, *P. angustifolia* suggests only a more rigid, narrower-leaved variation of *P. virginiana* which may occur in drier and more exposed situations than *P. virginiana*. The two have nearly the same corolla length and intergrade in length of inflorescence. Whether the more remotely flowered inflorescence attributed to *P. angustifolia* is actually a real difference remains to be further studied.

A specimen in the New York Botanical Garden from Vernon County (Sheldon, June 27, 1927, *Hugo Osvald*) has been questionably referred by Dr. McClintock to *Dracocephalum purpureum*, and is the probable basis for the report of this species from Missouri by Gleason (*New Ill. Fl.* 3: 157. 1952). This specimen, however, has linear to narrowly lanceolate leaves which are acute at the apex and the teeth of the leaves are short and acute. Gleason's description of *D. purpureum* calls for leaves which are obtuse or rounded at the apex with few or nearly obsolete, obtuse teeth. The Missouri specimen I have referred, because of its narrow, acute leaves and acute teeth to *P. angustifolia*, and other specimens from Missouri questionably referred to *Dracocephalum purpureum* by Dr. McClintock are referred in the present flora to *P. angustifolia*.

3. **Physostegia formosior** Lunell Map 1872

*Dracocephalum formosius* (Lunell) Rydb. [BB, McClintock]

*Physostegia parviflora* [G], as to Missouri reference, not Nutt.

Flowers July–October.

Occurs in prairies, glades, rocky or open ground, moist ground bordering lakes and spring branches, alluvial meadows, open low marshy ground, and gravel bars.

Scattered in Missouri, mostly in the northern and central portions south to Iron, Miller, and Bates counties.

Ranges from Maine to Manitoba and Alberta, south to New York, Ohio, Missouri, and Nebraska.

Like *P. virginiana*, this species is often cultivated in perennial gardens.

At the Gray Herbarium Fernald has labeled as *Physostegia parviflora* the following Missouri collections: *Demetrio 118* from Emma, Lafayette Co.; St. Louis, St. Louis County, collector uncertain. These were determined by Dr. McClintock as *Dracocephalum formosius*, and I agree with her determination. Likewise, Dr. McClintock identified a number of Missouri specimens in the Missouri Botanical Garden Herbarium as *D. formosius*, shown on the distribution map. Two specimens at the Missouri Botanical Garden Herbarium (*Steyermark 6804* from Miller Co., and *Mathias* from St. Louis Co.) and two at the Chicago Natural History Museum Herbarium (*Steyermark 70308* from Schuyler County and *Steyermark 70052* from Andrew County) were identified as *Physostegia parviflora* and are here referred to *P. formosior*.

This species is similar to *P. virginiana*, but has smaller corollas and the upper leaves are not or scarcely reduced below the inflorescence. However, considerable doubt remains in this author's mind as to the justification of maintaining *P. formosior* as anything more than a variety of *P. virginiana*.

4. **Physostegia intermedia** (Nutt.) Engelm. & Gray Map 1873

*Dracocephalum intermedium* Nutt. [BB, McClintock]

Flowers late May–July.

Occurs in low wet woods, swamps, and along wet ditches. Lowlands of southeastern Missouri, north to Cape Girardeau and Stoddard counties, west to Ripley County.

Ranges from Alabama to Texas, north to Kentucky, Illinois, Missouri, and Oklahoma.

The corolla is small in this species, only 10–15 mm long.

According to Shinners (Rh. 51: 121–22. 1949; Field and Lab. 19: 167. 1951) this is an annual species lacking the rhizomatous underground extensions of

---

Plate no. 307. **1.** Prunella vulgaris var. lanceolata, × 2/7; Details from Small, The New York Botanical Garden. **2.** Physostegia virginiana var. virginiana, × 2/7. **3.** Physostegia intermedia, × 2/7. **4.** Physostegia formosior, × 2/7. **5.** Leonurus Cardiaca, × 2/7; Details from Small, The New York Botanical Garden. **6.** Stachys palustris var. phaneropoda, × 2/7; *a.* Stem showing pubescence, × 1 5/7; After Gleason, The New York Botanical Garden. **7.** Stachys tenuifolia, × 2/7; Details from Small, The New York Botanical Garden. **8.** Lamium purpureum, × 2/7. **9.** Leonurus sibiricus, × 2/7; *a.* Calyx, × 2 6/7; After Gleason, details from Small, The New York Botanical Garden. **10.** Galeopsis Tetrahit var. Tetrahit, × 2/7; *a.* Flower, × 1 1/7; Details from Small, The New York Botanical Garden. **11.** Lamium amplexicaule, × 2/7; Details from Small, The New York Botanical Garden.

PLATE NO. 307

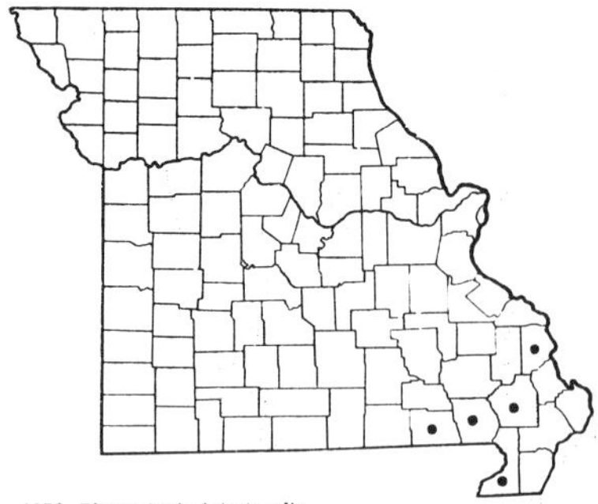

1873 Physostegia intermedia

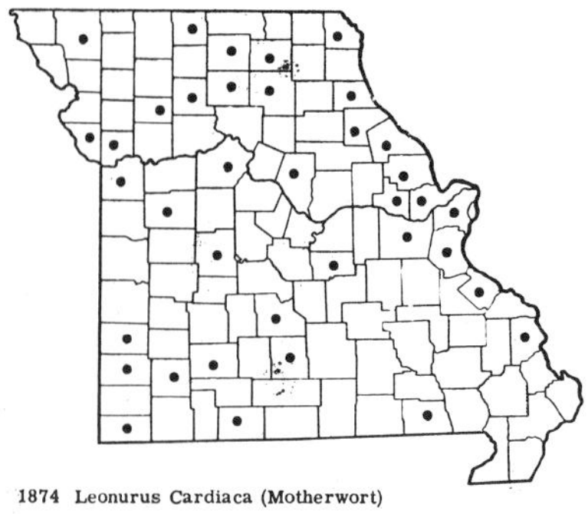

1874 Leonurus Cardiaca (Motherwort)

1875 Leonurus sibiricus

*P. virginiana.* This point remains to be verified in Missouri material.

*Excluded Species*

**Physostegia parviflora** Nutt.

As indicated under *P. formosior* Lunell, specimens at Gray Herbarium identified by Fernald as *P. parviflora* were determined by Dr. McClintock as *Dracocephalum formosior*, with which species (as *Physostegia*) they are referred in the present flora. Gleason places *Physostegia parviflora* as a synonym under *Dracocephalum Nuttallii* Benth., but excludes Missouri from the range. Fernald considers *D. Nuttallii* as synonymous with *Physostegia parviflora.*

**Dracocephalum purpureum** (Walt.) McClintock

Reported for Missouri in Gleason's *New Ill. Fl.* (3: 157. 1952), but as shown under *Physostegia angustifolia* a specimen in the New York Botanical Garden Herbarium questionably identified by Dr. McClintock as *D. purpureum* is referred in the present flora to *Physostegia angustifolia. Physostegia denticulata* (Ait.) Britt., maintained by Fernald as a distinct species of Virginia to Florida, is placed in synonymy by Gleason under *Dracocephalum purpureum* (Walt.) McClintock.

**Physostegia obovata** (Ell.) R. K. Godfrey

Reported by Fernald from Missouri, but no specimens at Gray Herbarium so labeled have been seen. It is believed that specimens so identified by Fernald should be referred to the similar *P. intermedia.*

## 12. **Leonurus** L. Motherwort

Calyx with 5 ribs and 5 angles; lower and middle leaves of stem maple-leaf shaped with 3–5 conspicuous large lobes; upper lip of corolla densely long-hairy; common species throughout Missouri . . 1. L. CARDIACA

Calyx with 10 ribs and scarcely any angles; leaves of stem deeply 3-parted, each division 2–5-cleft, or deeply 3–7 cleft and incised, the divisions narrow; upper lip of corolla short-hairy; rare, known only from Stone County . . . . . . . . . . . . . . . . . . . . . . . . 2. L. SIBIRICUS

1. **Leonurus Cardiaca** L. Motherwort Map 1874

Flowers late May–August.

Occurs in pastures, waste ground, fields, around old dwellings, along roadsides and railroads. Throughout Missouri.

Native of Europe and Asia; introduced and naturalized in North America from Quebec to Montana, south to North Carolina, Tennessee, Texas, and Utah.

Young plants, when boiled, make a tender vegetable, and in the raw state are sometimes nibbled for their flavor.

The dried leaves and flowering tops of this species are used as a stimulant bitter tonic and in the treatment of amenorrhoea or absence of menses. Some individuals are susceptible to the plant, contact causing a dermatitis. Mechanical injuries are sometimes caused to grazing animals by the sharp teeth of the calyx entering the membranes of the mouth.

2. **Leonurus sibiricus** L. Map 1875

Flowers June–August.

Known only from Stone County (along roadside bank near White River, just above the old highway 13 bridge at station 8, July 18, 1953, *Moore & Iltis 261*, in U. of Ark. Herb.).

Native of Eurasia; introduced and naturalized in North America from Florida to Louisiana, north locally to Quebec, Pennsylvania, Delaware, Kentucky, Missouri, and Iowa.

The Missouri station has been exterminated by the

1876 Galeopsis Tetrahit var. Tetrahit (Hemp Nettle)

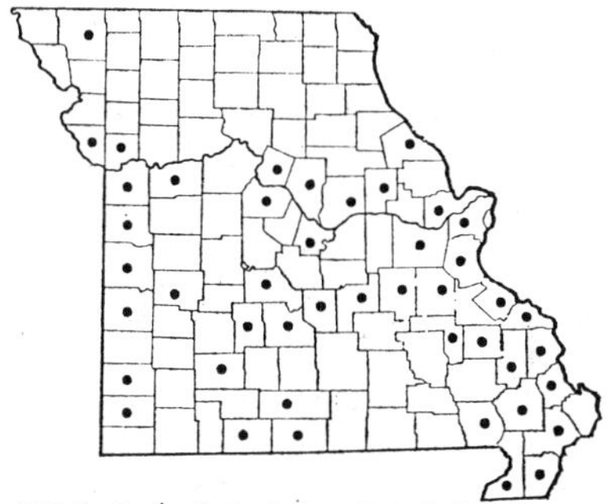

1877 Lamium amplexicaule f. amplexicaule (Henbit)

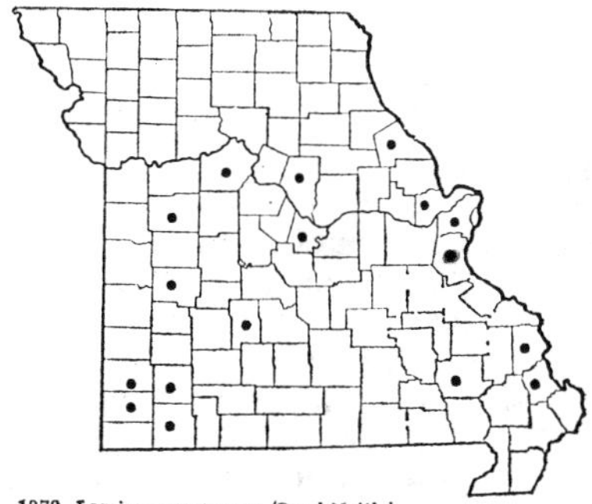

1878 Lamium purpureum (Dead Nettle)

impounding of waters from the Table Rock Dam, but it is probable that it will be found eventually somewhere else in the state.

The plant is sometimes used to relieve pains and difficulties in menstruation.

## 13. **Galeopsis** L. Hemp Nettle

**Galeopsis Tetrahit** L. var. **Tetrahit**
Hemp Nettle Map 1876

*Galeopsis Tetrahit* L. [G]

Flowers June–September.

Occurs along railroads. Known only from St. Louis County (St. Louis, Luther St. freight yard of Wabash R.R., south of Humboldt Ave., June 30, 1956, *Muehlenbach 978*; Wabash Fruit Terminal, south of Carr St., St. Louis, June 16, 1956, *Muehlenbach 934*; about 1 mile from tracks of Terminal R.R. Assoc. east of Shreve Ave., St. Louis, July 1, 1956, *Muehlenbach 992*).

Native of Eurasia; introduced and naturalized in North America from Newfoundland to Alberta, south to New York, Michigan, Wisconsin, Iowa, and Missouri.

## 14. **Lamium** L. Dead Nettle

Upper leaves without stalks (sessile) . . . . . . . . . . . . . . . . . 1. L. AMPLEXICAULE
Upper leaves with stalks (petioles). . . . . . . . . . . . . . . . . . 2. L. PURPUREUM

1. **Lamium amplexicaule** L. f. **amplexicaule**
Henbit Map 1877

*Lamium amplexicaule* L. [G, BB, P & S, Steyerm.]

Flowers February–November.

Occurs in lawns, meadows, pastures, cultivated and fallow fields, waste ground, along roadsides and railroads. Recorded only from southern, central, and extreme northwestern Missouri, but probably in all the counties of northern Missouri.

Native of Eurasia and Africa; introduced and naturalized in North America from Florida to California north to Newfoundland, Labrador, New England, New York, Ontario, Michigan, and British Columbia.

The normal color of the petals is purplish-pink or purplish tinged with rose. In f. *albiflorum* D. M. Moore, not yet recorded from Missouri, the corolla is white. Still another form, f. *clandestinum* (Reichenb.) G. Beck, has small, crowded flowers which do not open (cleistogamous) of a tubular, round-tipped type.

The plant is sometimes regarded as poisonous to cattle.

2. **Lamium purpureum** L. Dead Nettle
Map 1878

Flowers April–October.

Occurs in waste and open alluvial ground, and along roadsides and railroads. Scattered in southern and central Missouri.

Native of Eurasia; introduced and naturalized from Newfoundland and Nova Scotia to New England and Michigan, south to South Carolina, West Virginia, Ohio, Indiana, Illinois, Missouri, and Oklahoma.

## 15. **Stachys** L. Hedge Nettle

The following treatment is in accordance with the studies of Fernald (Rh. 45: 465–75. 1943) and *Gray's Manual*, eighth edition.

a. Main leaves of the stem on petioles (stalks) 10–70 mm. long . . . . . . . . . . . . . . . *b*
  b. Sides of middle and upper parts of stem glabrous (without hairs) or with few and distantly spaced hairs; lower surface of leaves glabrous; surface of calyx glabrous; petioles of middle leaves 10–70 mm. long . . . . . . . . . . . . . 2a. S. TENUIFOLIA var. TENUIFOLIA
  b. Sides of middle and upper parts of stem more or less hairy; lower surface of leaves more or less hairy; surface of calyx hairy. . . . . . . . . . . . 1c. S. PALUSTRIS var. PHANEROPODA
a. Main leaves of the stem sessile (without petioles) or the petioles at most 8 mm. long. . . . . . . *c*
  c. Sides of middle and upper lengths of the stem more or less covered with hairs; lower surface of leaves, including the nerves, more or less hairy . . . . . . . . . . . . . . 1. S. PALUSTRIS
  c. Sides of middle and upper lengths of the stem glabrous or rarely with a few widely separated hairs; lower surface of leaves glabrous, if hairs are present they are confined to the main nerves. . . . *d*
    d. Largest leaves 5–15 mm. broad, blunt (obtuse) or only somewhat pointed (acutish) at tip; leaves faintly veined only; teeth of leaves low and depressed. . . . 3. S. HYSSOPIFOLIA var. AMBIGUA
    d. Largest leaves 8–24 mm. broad, long-pointed (acuminate); leaves conspicuously veined; teeth of leaves conspicuous, more or less pointed. . . . . . . . . 2b. S. TENUIFOLIA var. HISPIDA

1. **Stachys palustris** L. Woundwort Map 1879

Flowers July–September.

Occurs in wet meadows, river bottom prairies, swampy open ground along sloughs, prairie streams, and margins of oxbow lakes in river flood plains.

The following varieties are found in Missouri:

a. Main leaves of the stem on petioles (stalks) 10–14 mm. long. . . . . 1c. S. PALUSTRIS var. PHANEROPODA
a. Main leaves of the stem sessile (without petioles) or the petioles at most 8 mm. long . . *b*
  b. Main leaves of the stem obtuse (blunt) or only slightly pointed (subacute), oblong to oblong-ovate, or oval, 1.5–5 cm. broad, 3/10–1/2 as broad as long . 1a. S. PALUSTRIS var. PILOSA
  b. Main leaves of the stem long-pointed (acuminate), narrowly to broadly lanceolate or narrowly oblong, 0.4–3 cm. broad, 1/7–1/3 as broad as long . . . . 1b. S. PALUSTRIS var. NIPIGONENSIS

1a. **Stachys palustris** var. **pilosa** (Nutt.) Fern. Map 1879

*Stachys palustris* subsp. *pilosa* (Nutt.) Epling [Epling]

Scattered in northern and central Missouri.

Ranges from British Columbia to Arizona, east to James Bay, Quebec, Wisconsin, Illinois, Missouri, and Nebraska.

1b. **Stachys palustris** var. **nipigonensis** Jennings Map 1879

*Stachys palustris* [of P & S], not L.

*Stachys ambigua* [of P & S in part], not Sm.

This is the commonest variation in the state, known from northern, central, and eastern Missouri, south to Jackson, Ray, St. Charles, New Madrid, and Pemiscot counties.

Ranges from Quebec to Alaska, south to New England, New York, Ohio, Michigan, Illinois, Iowa, Missouri, New Mexico, and Arizona.

The Daviess County record cited under *Stachys palustris* and some of the records cited under *S. ambigua* in Palmer and Steyermark's *Annotated Catalogue* are referred here to this variety.

1c. **Stachys palustris** var. **phaneropoda** Weath. Map 1879

*Stachys ambigua* [of P & S in part], not Sm.

Northwestern, west-central, and eastern Missouri, south to Jackson, St. Louis, and Perry counties. Probably more common in the northern counties than now indicated.

Ranges from Ohio to Wisconsin, Iowa, and Missouri.

The crisp tubers and rootstocks of this species and its varieties, developed by autumn, may be used fresh in salads or boiled as a cooked vegetable. A related Asiatic species, *Stachys Sieboldii* Miq., is cultivated and eaten in China and Japan especially for the fleshy tubers, known by the name of Chinese Artichoke or Japanese Potato.

2. **Stachys tenuifolia** Willd. Map 1880

Flowers June–September.

Occurs in moist or low woodland, in wooded valleys and ravines, banks of streams and margins of ponds and sloughs, bald cypress swamps, thickets, alluvial meadows, rarely along roadsides.

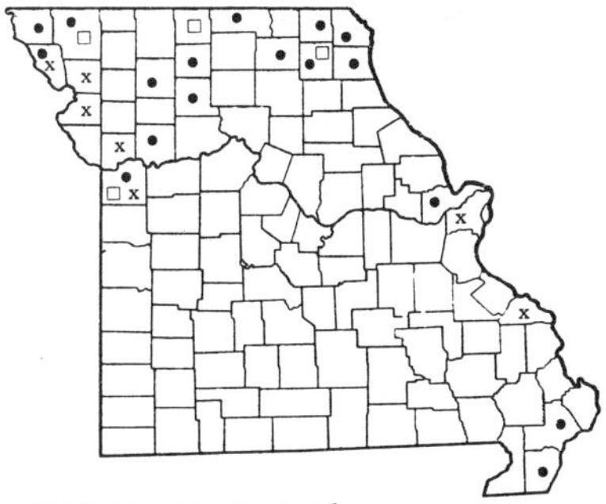

1879 □ Stachys palustris var. pilosa
1879 • Stachys palustris var. nipigonensis
1879 x Stachys palustris var. phaneropoda

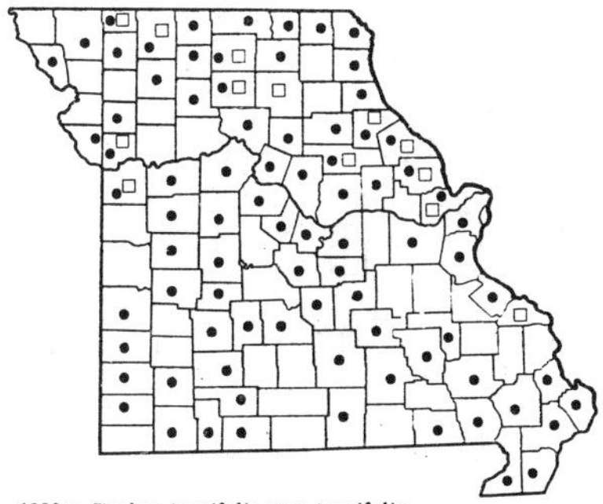

1880 • Stachys tenuifolia var. tenuifolia
1880 □ Stachys tenuifolia var. hispida

1881 Stachys hyssopifolia var. ambigua

Two variations occur in Missouri:

Petioles (leaf-stalks) of the main leaves 5–30 mm. long; leaves glabrous (without hairs) on both surfaces; calyx glabrous throughout; margins of the bracts at the base of the clusters of flowers not fringed with hairs . . . . 2a. S. TENUIFOLIA var. TENUIFOLIA

Petioles (leaf-stalks) of the main leaves not more than 8 mm. long; veins of lower surface of leaf often hairy, upper side of leaf often hairy, the hairs straight, stiff, and appressed (lying parallel or along surface); calyx often bristly-hairy on the angles; margins of the bracts at the base of the clusters of flowers fringed with hairy bristles . 2b. S. TENUIFOLIA var. HISPIDA

2a. **Stachys tenuifolia** var. **tenuifolia** Map 1880

*Stachys tenuifolia* Willd. [G, BB, P & S]

Throughout Missouri and probably in every county.

Ranges from New York to Minnesota, south to South Carolina, Tennessee, Louisiana, and Texas.

2b. **Stachys tenuifolia** var. **hispida** (Pursh) Fern. Map 1808

*Stachys hispida* Pursh [BB, P & S]

*Stachys tenuifolia* var. *aspera* of auth., not *S. aspera* Michx.

Northern, central, and eastern Missouri, south to Jackson, St. Charles, and Perry counties.

Ranges from New Hampshire south to South Carolina; Michigan to Minnesota, south to Indiana, Illinois, and Missouri.

This is treated by Gleason, following Epling, as a distinct species.

3. **Stachys hyssopifolia** Michx. var. **ambigua** Gray Map 1881

*Stachys aspera* Michx. [BB, P & S]

*Stachys ambigua* (Gray) Britt., not Sm.

Flowers June–August.

Occurs in low moist ground and open grassy places. Known only from St. Louis County, east-central Missouri (near St. Louis, Chain-of-Rocks, August, 1915, *Florence Beckwith*; grassy places, August, 1876, *Eggert*; Forest Park, St. Louis, August 16, 1891, *Glatfelter herb.*, no collector indicated; St. Charles Rock Road, August 25, 1875, *Eggert*; near Shaw's Garden, 1833, *Engelmann*).

Ranges from Pennsylvania to Iowa, south to Florida, Georgia, Kentucky, Tennessee, Illinois, and Missouri.

This species has not been collected in Missouri in recent years, and all the collections have come from St. Louis County.

As in *S. palustris* and varieties the white thickened tubers are crisp with a nutty flavor and may be eaten raw in salad or cooked as a vegetable.

This is treated as a distinct species by Gleason, following Epling's studies in Fedde Rep. Sp. Nov. Beih. 80: 63–71. 1934.

## 16. **Salvia**

a. Leaves chiefly at base of plant; leaves at base of plant irregularly long-lobed or wavy-edged; 2–6 leaves present on the flowering stem . . . . . . . . . . . . . . . . . . 1. S. LYRATA

a. Leaves well-developed on stem; leaves not irregularly long-lobed, with or without short or coarse teeth; 8–20 or more leaves present on the flowering stem . . . . . . . . . . . . . . *b*

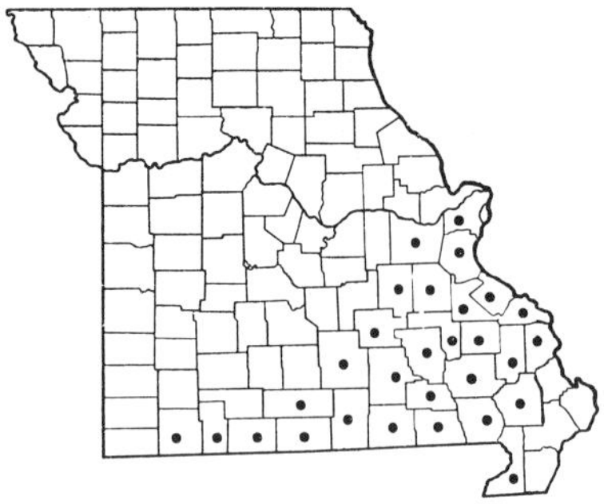

1882 Salvia lyrata (Cancer Weed)

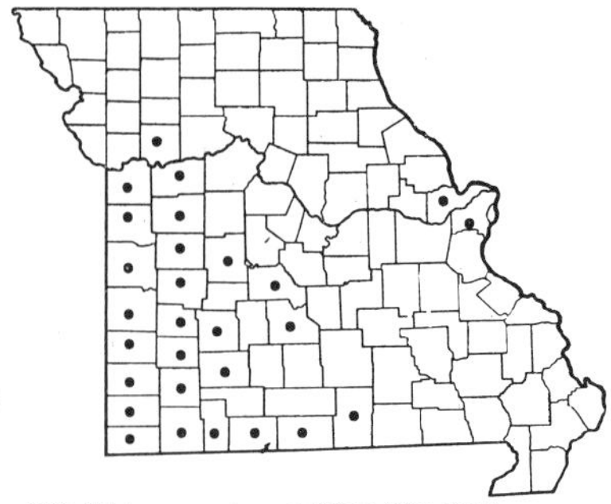

1883 Salvia azurea var. grandiflora (Blue Sage)

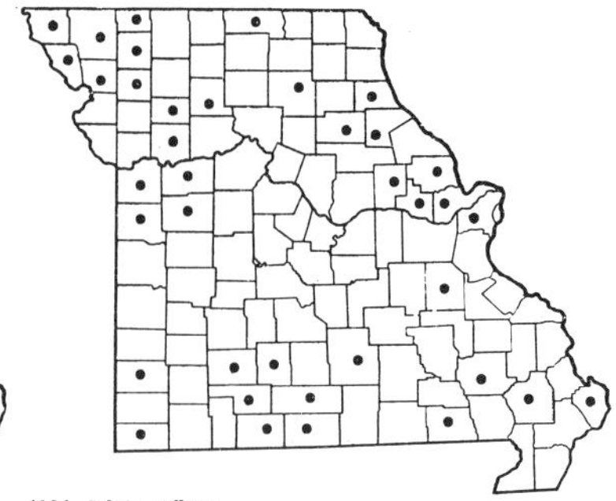

1884 Salvia reflexa

b. Leaves ovate or ovate-oblong, 2–6 cm. broad; upper lip of calyx with 3 sharp-tipped lobes 4. S. SCLAREA
b. Leaves linear-lanceolate, lanceolate, or narrowly oblong, less than 2 cm. broad . . . . . . . *c*
  c. Main leaves of stem stalkless (sessile) or tapering to short petioles (stalks) 1–5 mm. long; corolla 15–25 mm. long; flowers usually 6–12 at each node of the inflorescence; perennial plant . . . . . . . . . . . . . . . . . . . 2. S. AZUREA var. GRANDIFLORA
  c. Main leaves of stem on long petioles (stalks) 5–20 mm. long; corolla about 10 mm. long; flowers usually 2 (rarely 3–4) at each node of the inflorescence; annual plant . . . 3. S. REFLEXA

1. **Salvia lyrata** L. Cancer Weed Map 1882
Also called Lyre-leaved Sage.
Flowers April–June.

Occurs in rich or rocky, usually open woods, sandy and gravelly alluvial soils in valleys and bars along streams, wet grassy swampy ground, pastures, thickets, moist open ground, and along railroads. Ozark region of southern and east-central Missouri, south and east of a line drawn from St. Louis, Franklin, Crawford, Dent, Texas, Douglas, and Taney counties to Barry County.

Ranges from Florida to Texas, north to Connecticut, New York, Pennsylvania, West Virginia, Ohio, Indiana, Illinois, Missouri, and Oklahoma.

2. **Salvia azurea** Lam. var. **grandiflora** Benth.
Blue Sage Map 1883
*Salvia Pitcheri* Torr. [BB]
Flowers July–September.

Occurs chiefly on limestone glades, rocky prairies, prairie openings, bluff escarpments, and occasionally escaped to roadsides and open waste ground. Native to the unglaciated prairie and western Ozark regions north to Jackson, Ray, and Lafayette counties, east to Johnson, Benton, Camden, Laclede, and Howell counties; introduced and escaped from cultivation in east-central Missouri in St. Louis (colony at Gravois and Laclede Station road, summer, 1953, *Bauer*) and St. Charles (August Busch Wildlife area, Weldon Springs, 1957, *Bruce Dowling*, U. of Mo. Herb.) counties.

Ranges from Minnesota to Nebraska, south to Kentucky, Arkansas, and Texas; escaped from cultivation east to New England.

The stems and axis of the inflorescence are covered with a close gray-white hairiness. The flowers are showy with a deep blue, rarely white, corolla. The plant is a striking one for the summer and fall garden, doing well in dry, sunny open situations. The plant is nutritious and palatable, being eaten by all classes of livestock, especially in the early spring.

3. **Salvia reflexa** Hornem. Map 1884
*Salvia lanceaefolia* [of P & S], not Poir.
Flowers late May–October.

Occurs in dry pastures, fields, alluvial ground, dry rocky open ground, or open woods, along roadsides, and railroads. Scattered over the state, commonest in western and southern Missouri.

Ranges from Wisconsin to Montana, south to Arkansas, Texas, Utah, and Mexico; introduced east to Michigan, Ohio, West Virginia, and New Jersey.

4. **Salvia Sclarea** L. Clary Map 1885
Also called Clear-eye.
Flowers June–August.

Known from St. Louis County, east-central Missouri (St. Louis, July 19, 1910, *Sherff 570*, in Gray Herb.).

---

Plate no. 308. **1.** Salvia azurea var. grandiflora, × 2/5; *a.* Corolla, × 3; *b.* Calyx, × 1 1/5; After Gleason, The New York Botanical Garden. **2.** Salvia reflexa, × 2/5. **3.** Salvia lyrata, × 2/5. **4.** Monarda Russeliana, × 2/5. All details from Small, The New York Botanical Garden.

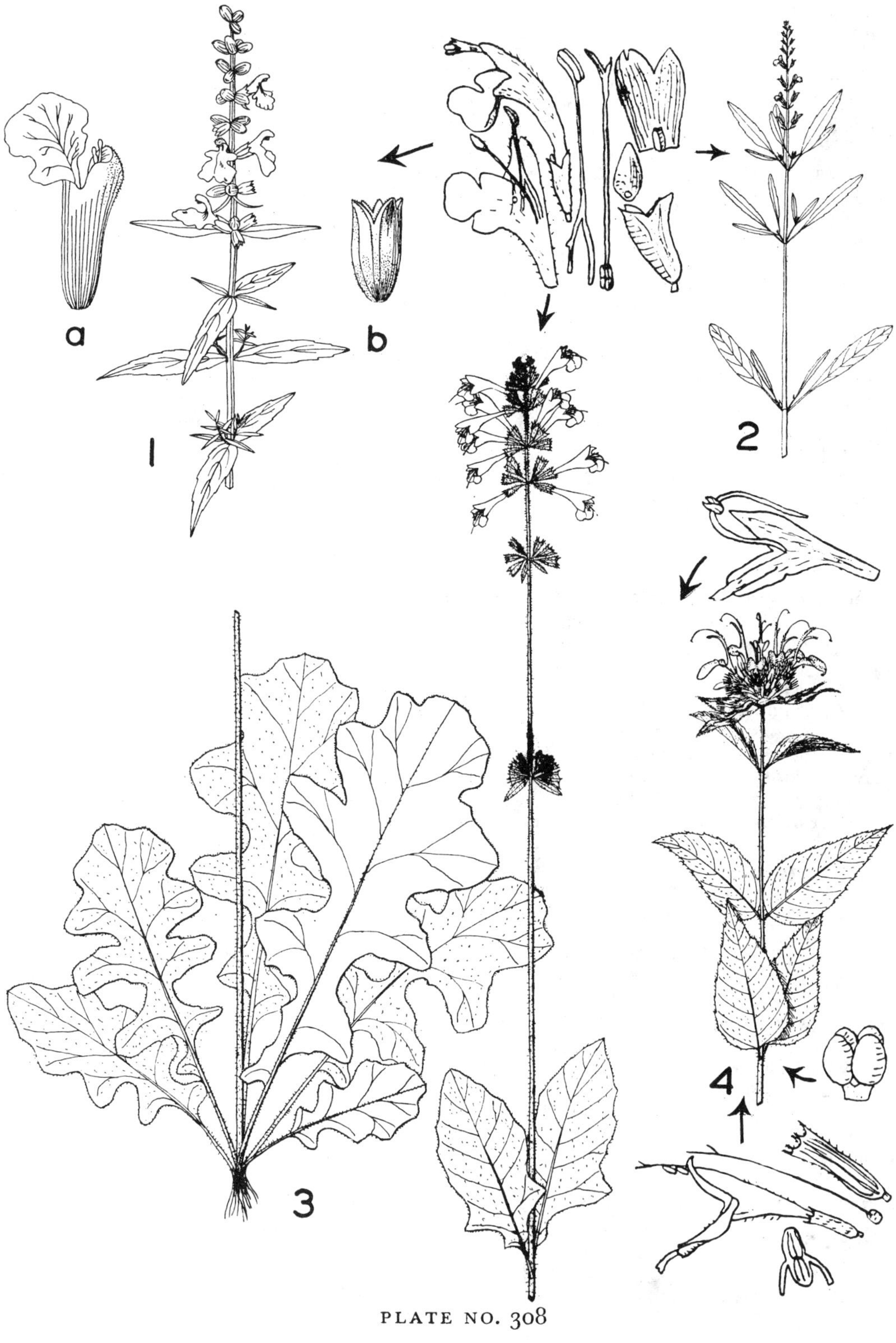


PLATE NO. 308

1885 Salvia Sclarea (Clary)

1886 Monarda Russeliana

1887 Monarda didyma (Bee Balm)

Native of Europe; grown as an ornamental plant and rarely escaped from cultivation.

The above-cited collection was probably collected from a garden specimen, and is here included on the assumption that it may have been from an escaped plant or that other collections of the species will eventually be found. The corolla is bluish-white, varying to rose or purple, often marked with yellow, and is 20–25 mm. long. The hairy leaves are 10–20 cm. long with coarse and irregular toothing.

The leaves of this species are sometimes used for flavoring purposes.

## 17. **Monarda** L. Horsemint, Wild Bergamot

a. Usually only 1 flower-cluster at the top of the stem or branch; stamens conspicuously protruding from and longer than the straight upper lip of the corolla . . . . . . . . . . . . . . . . *b*

b. Leaves without petioles (stalks) or nearly so, the petioles at most 5 mm. long; calyx-teeth 1.5–4 mm. long . . . . . . . . . . . . . . . . . . . . . . . . . 1. M. RUSSELIANA

b. Leaves with definite petioles (stalks) usually 10–40 mm.; calyx-teeth 1–2 mm. long . . . . . . *c*

c. Leaves gray-green, usually firm (except in deep shade); corolla covered with a fine, close hairiness; bracts at base of flower-cluster pale green to grayish or lilac-tinged; corolla lilac, lavender, or rose-tinged; commonly encountered species . . . . . . . . . . 3. M. FISTULOSA

c. Leaves dark or dull green, thin and membranous; corolla glabrous (without hairs) or nearly so; bracts at base of flower-cluster red-tinged, white or white-tinged; corolla vermilion to scarlet, or dull white to flesh-pink with dark spots; rarely encountered species. . . . . . . *d*

d. Corolla vermilion to scarlet, 3–4.5 cm. long; bracts at base of flower-cluster red-tinged; calyx 9–14 mm. long without a beard of hairs in the throat . . . . . . . . 2. M. DIDYMA

d. Corolla dull white to flesh-pink with dark spots, 1.5–3 cm. long; bracts at base of flower-cluster white or white-tinged; calyx 6–10 mm. long with a beard of hairs in the throat. . 4. M. CLINOPODIA

a. Two or more flower-clusters on the same stem or branch, situated one above the other (terminal and axillary); stamens not protruding from or longer than the strongly curved upper lip of the corolla. . . . . . . . . . . . . . . . . . . . . . . . . . . . . . *e*

e. Corolla yellowish, strongly marked with brown-purple dots; calyx-teeth triangular, short-pointed, 1–1.5 mm. long; bracts and leaves at base of flower-clusters not ending in long awns; perennial plants with strong forking crowns . . . . . . . . . . . . . . . . . 5. M. PUNCTATA

e. Corolla whitish to pinkish, slightly or not at all spotted; calyx-teeth with bristle-like tips, 2–8 mm. long; bracts and leaves at base of flower-clusters ending in long awns; annual plants without strong crowns . . . . . . . . . . . . . . . . . . . . . . . . . . . . *f*

f. Leafy bracts at base of flower-clusters densely hairy, the upper surface white to purple; more frequently encountered . . . . . . . . . . . . . . . . . . . 6. M. CITRIODORA

f. Leafy bracts at base of flower-clusters glabrous (without hairs) or nearly so, the upper surface green or greenish; known only from Jackson County, west-central Missouri . . . 7. M. PECTINATA

1. **Monarda Russeliana** Nutt. Map 1886
*Monarda Bradburiana* Beck [P & S, Steyerm.]
Flowers late April–June.

Occurs in usually acid soils of rocky or dry open woods and borders of glades, usually in sandstone, chert, or granitic substrata, rarely along railroads.

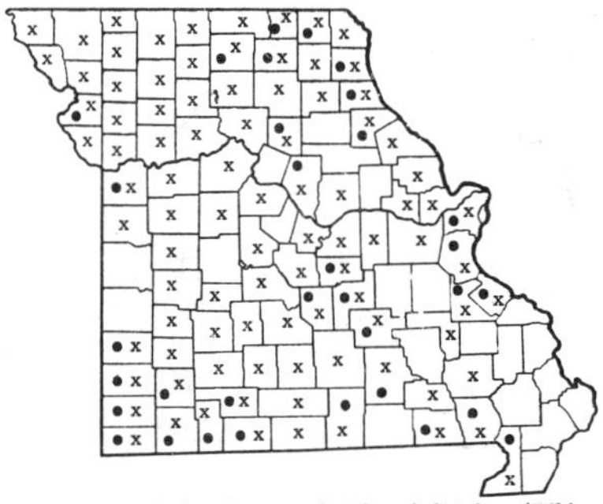

1888 • Monarda fistulosa var. fistulosa f. fistulosa (Wild Bergamot)
1888 x Monarda fistulosa var. mollis

1889 Monarda clinopodia

1890 • Monarda punctata var. villicaulis
1890 □ Monarda punctata var. occidentalis

Ozark region east and south of a line drawn from Marion, Montgomery, Callaway, Boone, Moniteau, Pettis, Benton, Hickory, and Cedar counties to Jasper County.

Ranges from Alabama to Texas, north to Kentucky, Indiana, Illinois, Iowa, and Kansas.

This is the earliest-flowering of the genus in Missouri. The corolla is pale lavender or white with purple spots. A tea may be prepared from the leaves of the plant.

2. **Monarda didyma** L. Bee Balm Map 1887

Also called Oswego Tea.

Flowers late June–September.

Planted as a garden ornamental and occasionally escaped from cultivation in moist soil along banks of streams and woodland. Known from Ralls, Pike, and Clay (creek bank, Holt, July 13, 1929, no collector indicated, Wm. Jewell College Herb.) counties.

Native in rich woods from New York to Michigan, south to Georgia, Tennessee, and Ohio; escaped from cultivation east to Quebec, Maine, New York, and New Jersey.

The leaves vary from ovate to nearly lanceolate and the stems from nearly glabrous to conspicuously hairy. This perennial plant spreads by runners from the base of the stem, eventually forming large stands in rich, loose soil and shaded situations which it favors. The plant is an excellent one to lend vivid coloring to the woodland wildflower garden in summer, when the scarlet, showy flowers are in bloom. When combined with white flowers of *Cimicifuga racemosa* and *Hydrangea arborescens* and blue flowers of *Campanula americana*, which bloom at about the same period, a combination of red, white, and blue can be obtained.

The leaves are pleasantly aromatic, and a fragrant, delicious tea is obtained after the leaves are placed in boiling water.

White-, rose-, and purplish-flowered variants are known in cultivation.

3. **Monarda fistulosa** L. Wild Bergamot Map 1888

Also called Horsemint.

Flowers late May–August.

Occurs in prairies, fallow fields, open and dry rocky woods, borders of rocky glades, alluvial thickets, and along roadsides and railroads. Throughout Missouri, except apparently absent from the lowlands of the extreme southeastern section.

Two varieties are encountered in the state:

Lower surface of leaves with spreading hairs and the nerves beneath with long hairs; upper lengths of stem with loose or spreading hairs; less common type encountered . . . 3a. M. FISTULOSA var. FISTULOSA

Lower surface of leaves minutely hairy with short hairs or practically hairless; stem covered with close incurved hairs pressed close to the surface; common type encountered . . 3b. M. FISTULOSA var. MOLLIS

3a. **Monarda fistulosa** var. **fistulosa** f. **fistulosa** Map 1888

*Monarda fistulosa* L. [G, BB, P & S]

Scattered over the state.

Ranges from Quebec to Minnesota, south to Georgia, Alabama, Louisiana, and Texas.

The normal color of the corolla is lilac to deep purple or rose-lavender. A form with whitish corollas, *M. fistulosa* var. *fistulosa* f. *albescens* Farw., has not been recorded thus far from Missouri.

3b. **Monarda fistulosa** var. **mollis** (L.) Benth. Map 1888

The common variety found throughout Missouri.

Ranges from Maine to Saskatchewan, south to

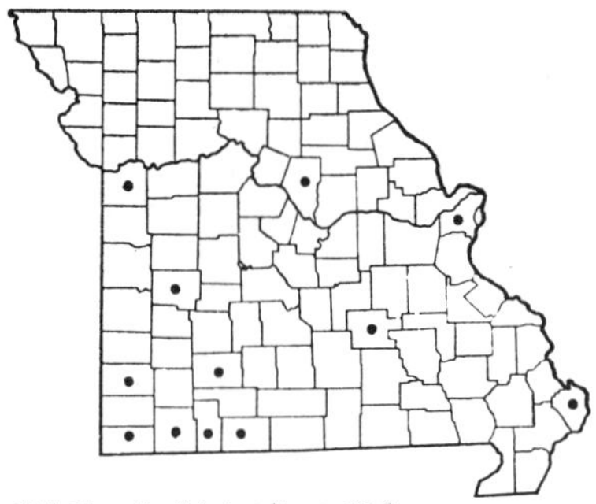

1891 Monarda citriodora (Lemon Mint)

1892 Monarda pectinata

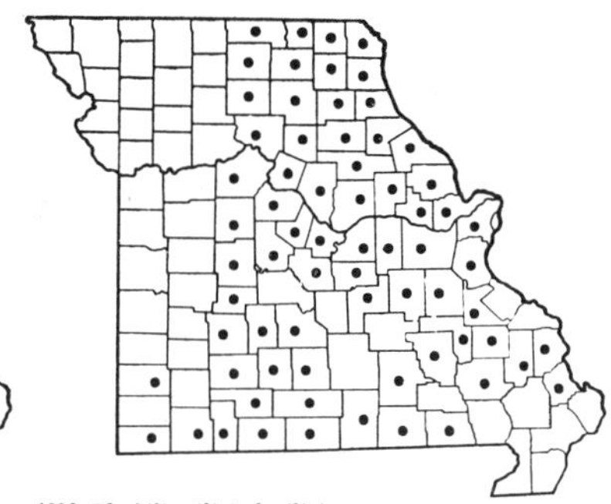

1893 Blephilia ciliata f. ciliata

Georgia, Alabama, Mississippi, Louisiana, Texas, and Mexico.

The leaves vary from ovate with rounded bases to lanceolate with rounded truncate, or wedge-shaped bases. As noted by Deam for Indiana specimens (*Fl. Ind.* p. 815. 1940), plants in Missouri found in dry open habitats possess narrower leaves on the average, and those of more dense woodland or shaded situations possess broader and often thinner leaves. There is intergradation between the two varieties in Missouri in pubescence of leaves and stems.

A tea may be prepared from the dried or fresh leaves. The species has been used in medicine as a stimulant to remove pain in colic.

4. **Monarda clinopodia** L. Map 1889

Flowers late June–July.

Occurs in low rich woods along streams. Known only from Butler County, southeastern Missouri (rich wooded valley just below junction of the two forks of West Prong and East Prong of Indian Creek, T25N, R6E, northeast ¼ sect. 1 and northwest ¼ sect. 6, 6½ mi. north northeast of Poplar Bluff, October 29, 1956, *Steyermark 83306*; same locality, July 19, 1957, *Steyermark 85378*).

Ranges from New York to Ohio, Indiana, Illinois, and Missouri, south to Alabama, and Kentucky.

5. **Monarda punctata** L. Dotted Monarda Map 1890

Flowers late June–October.

Occurs in prairies, and sandy and open ground.

Two variations occur in Missouri:

Hairs on stem spreading or recurving and rather long; lower surface of leaves densely hairy and concealing the glands . . . . 5a. M. PUNCTATA var. VILLICAULIS

Hairs on stem closely pressed to stem and minute; lower surface of leaves minutely hairy and not concealing the glands . . . . 5b. M. PUNCTATA var. OCCIDENTALIS

5a. **Monarda punctata** var. **villicaulis** (Pennell) Shinners Map 1890

*Monarda punctata* subsp. *villicaulis* Pennell [BB]

Eastern Missouri, where known from Clark, Pike, Scott, and Dunklin counties.

Ranges from Vermont to Minnesota, south to New Jersey, Tennessee, Missouri, and Oklahoma.

5b. **Monarda punctata** var. **occidentalis** (Epling) Palmer & Steyerm. Map 1890

*Monarda punctata* subsp. *occidentalis* Epling [BB]

*Monarda punctata* var. *lasiodonta* [of G], not Gray

Northern half of Missouri, where known only from Clark (natural sandy prairie paralleling Des Moines River, T65N, R6W, sect. 9, 1½ mi. southeast of St. Francisville, August 13, 1949, *Steyermark 68882*) and Jackson counties.

Ranges from Illinois to Kansas, south to Texas and New Mexico.

The varieties of this species contain aromatic oils sometimes used in stimulant medicines.

6. **Monarda citriodora** Cerv. Lemon Mint Map 1891

*Monarda dispersa* Small

Flowers late May–August.

Occurs on limestone glades, dry limestone ledges, bald knobs, and rocky prairies, southern and central

Plate no. 309. **1.** Monarda fistulosa var. mollis, × 2/5. **2.** Monarda citriodora, × 2/5; Details from Small, The New York Botanical Garden. **3.** Monarda punctata, × 2/5; *a.* Flower, × 1 1/5; *b.* Calyx, × 1 3/5; After Gleason, The New York Botanical Garden. **4.** Monarda pectinata, × 2/5; After Gleason, The New York Botanical Garden. **5.** Blephilia ciliata × 2/5; Details from Small, The New York Botanical Garden.

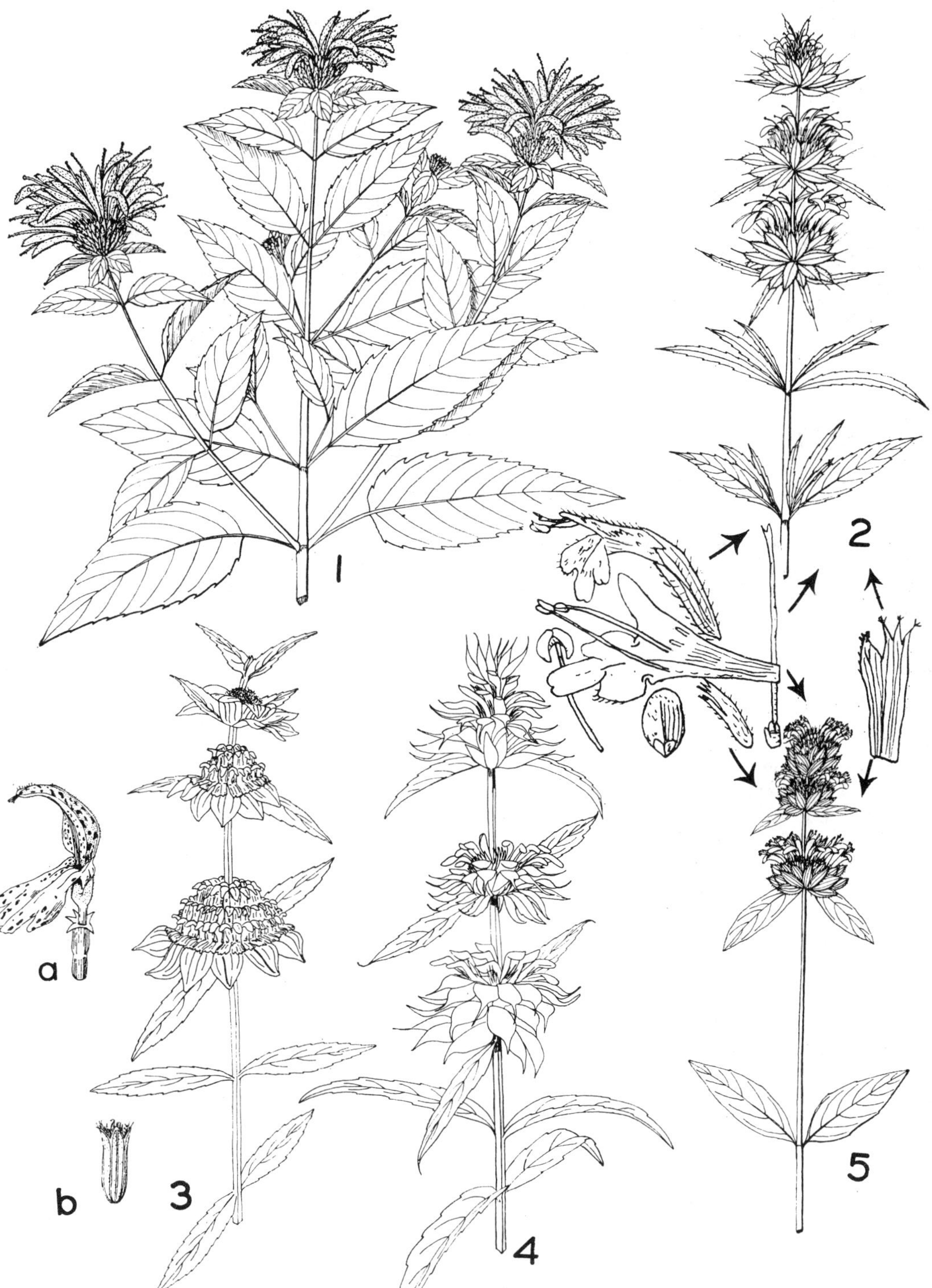


PLATE NO. 309

Missouri north to St. Louis, Boone, and Jackson counties.

Ranges from Texas and Mexico north to Missouri and Kansas; introduced east to Michigan, Tennessee, and Georgia.

Specimens from Mississippi, St. Louis, Dent, Boone, and Jackson counties have been introduced in those counties along railroads and waste ground, but plants from the southwestern part of the state are native there on the glades and bald knobs.

7. **Monarda pectinata** Nutt. Map 1892

Flowers June–July.

Known only from Jackson County, west-central Missouri (waste ground, Courtney, June 24, 1927, *Bush 11341*).

Ranges from Nebraska to Colorado and Utah, south to Texas, New Mexico, and Arizona; introduced in Missouri.

The Missouri collection, labeled by Bush as *M. pectinata*, has been questionably referred to this species by Dr. Epling, who notes that the collection has 'wide bracts'.

## 18. **Blephilia** Raf.

Upper leaves usually lanceolate or oblong, mostly narrowed to base, sessile (without stalks) or on petioles (stalks) up to 10 mm. (rarely more) long; leaves only slightly toothed; outer bracts surrounding base of flower-clusters ovate, short- to long-pointed (acute to acuminate) and as long as the calyx; stem usually simple and unbranched; the 2 lower calyx-teeth reaching beyond the sinuses of the 3 upper calyx-teeth . . . . . . . . . . . . . . . . . . . . . . . . . . 1. B. CILIATA

Upper leaves ovate to ovate-lanceolate, usually broadest at or near the base, on petioles usually 10–30 mm. long; leaves usually conspicuously or sharply toothed; outer bracts surrounding base of flower-clusters mainly linear with bristle-like tips shorter than the calyx; stem usually branched with 2 or more branches; the 2 lower calyx-teeth not reaching to the sinuses of the upper calyx-teeth . . . 2. B. HIRSUTA

1. **Blephilia ciliata** (L.) Benth. f. **ciliata**

Ohio Horse Mint Map 1893

*Blephilia ciliata* (L.) Benth. [G, BB, P & S, Steyerm.]

Flowers May–August.

Occurs in open woods, limestone glades, ledges of bluffs, thickets, fields, clearings, and along roadsides. Southern and eastern Missouri west to Putnam, Sullivan, Linn, Chariton, Saline, Pettis, Benton, Hickory, Polk, Jasper, and McDonald counties.

Ranges from Vermont to Wisconsin and Iowa, south to Georgia, Mississippi, Arkansas, Oklahoma, and Texas.

The corollas are usually pale blue-lavender with purple spots. In f. *albiflora* House, not yet recorded from Missouri, they are white. The basal shoots and root-leaves remain green over winter and are sometimes gathered as 'spring sprouts' or greens in the spring by some of the Ozark inhabitants. Some specimens with elongate petioles up to 3.5 cm. long (*Steyermark 73573* from Putnam County) appear intermediate between this species and *B. hirsuta* and may be of hybrid origin.

This species has a relatively long flowering period, and makes a desirable ornamental flowering plant.

2. **Blephilia hirsuta** (Pursh) Benth. var. **hirsuta**

Wood Mint Map 1894

*Blephilia hirsuta* (Pursh) Benth. [G, BB, P & S, Steyerm.]

Flowers late May–September.

Occurs in rich woods in ravines, valley bottoms, north- and east-facing wooded slopes and at the base of bluffs. Scattered throughout Missouri, but commonest in the Ozark region and along rich wooded slopes in counties bordering the Missouri and Mississippi rivers, north locally to Sullivan and Putnam counties; mostly absent from the unglaciated prairie region.

Ranges from Quebec and Vermont to Minnesota, south to Georgia, Tennessee, Missouri, and Texas.

In typical var. *hirsuta* the stems and leaves are hairy; in var. *glabrata* Fern., not known in Missouri, the stems and leaves are nearly or quite glabrous.

As noted by Deam (*Fl. Ind.* p. 817. 1940) the leaves of this species are generally eaten by insects, making it difficult to find a specimen with leaves not damaged.

---

Plate no. 310. **1.** Blephilia hirsuta, × 2/5. **2.** Hedeoma pulegioides, × 4/5; *a.* Flower, × 1 3/5; Details from Small, The New York Botanical Garden. **3.** Hedeoma hispida, × 2/5. **4.** Melissa officinalis, × 2/5; *a.* Flower, × 1 1/5; After Gleason, details from Small, The New York Botanical Garden. **5.** Satureja Acinos, × 4/5; After Gleason, The New York Botanical Garden. **6.** Satureja arkansana, × 2/5.

PLATE NO. 310

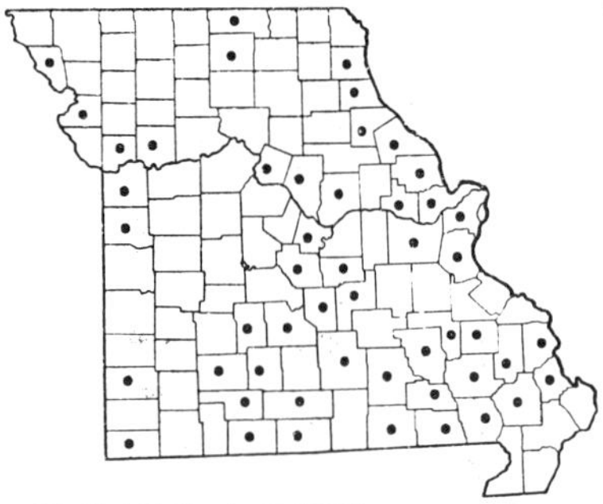

1894 Blephilia hirsuta var. hirsuta

1895 Hedeoma pulegioides (Pennyroyal)

1896 Hedeoma hispida

## 19. **Hedeoma** Pers. Mock Pennyroyal

Plant strongly aromatic; leaves 3–13 mm. or more broad, oblong-ovate, elliptic, or obovate to lanceolate, the main leaves on distinct petioles 3–9 mm. long, the broader leaves usually toothed; bracts at base of flower-stalks (pedicels) elliptical or oblong-elliptic, obtuse (blunt) or subacute; flowering July–September . . . . . . . . . . . . . . . . . . . . . . . . . . 1. H. PULEGIOIDES

Plant with only a slight odor; leaves mainly 1.5–2.5 mm. (rarely 4) broad, linear, sessile (without stalks), entire (without teeth); bracts at base of flower-stalks (pedicels) subulate (very narrow and long-tapering); flowering May–July . . . . . . . . . . . . . . . . . . . . 2. H. HISPIDA

1. **Hedeoma pulegioides** (L.) Pers. Pennyroyal Map 1895

Flowers July–September.

Occurs in dry open or rocky woods, often on ridge tops, upland slopes, clearings, pastured thin soils, glades, roadsides, and thickets. Throughout Missouri, doubtless in every county.

Ranges from Quebec to Minnesota and South Dakota, south to Florida, Alabama, Arkansas, and Oklahoma.

A pleasantly aromatic tea may be prepared from the fresh or dried plants. An oil obtained from the dried leaves and upper part of the plant is an ingredient in certain insect repellents, and some persons find rubbing the fresh plant on their skin an effective preventive against mosquitoes and gnats. The oil is also used sometimes as an ingredient in liniments and for medicines to be taken internally.

Although termed Pennyroyal in the United States, the Pennyroyal of Europe, which first received the common name, is *Mentha Pulegium* L., with reference to its supposed efficacy in driving away fleas.

2. **Hedeoma hispida** Pursh Map 1896

Flowers late May–July.

Occurs in prairies, rocky open glades, rocky leached limestone or sandstone, chert, or granite substrata, also on limestone glades, fallow fields, sandy open ground, and along roadsides and railroads. Throughout Missouri, except apparently absent from the lowland counties of the extreme southeastern section.

Ranges from New York and Ontario to Saskatchewan and Alberta, south to Tennessee, Mississippi, Louisiana, and Texas; introduced in Vermont and Connecticut.

This species has only a slight minty odor and begins to flower during the latter part of May.

*Excluded Species*

**Hedeoma acinoides** Scheele

This species was recorded in Palmer and Steyermark's *Annotated Catalogue* on the basis of a specimen from Jasper County (Webb City, May 14, 1909, *Palmer 2007*). Critical restudy of this specimen reveals its identity as *Satureja Acinos* (L.) Scheele.

## 20. **Melissa** L. Balm

**Melissa officinalis** L. Lemon Balm Map 1897

Also called Bee Balm.

Flowers June–September.

Occurs in waste ground, thickets, abandoned fence rows, near old dwellings and homesites, wooded slopes along streams, along roadsides and railroads. Scattered in southern and central Missouri, north to St. Louis, Boone, and Jackson counties.

Native of Europe; introduced in North America where naturalized from Maine to New York, Ohio,

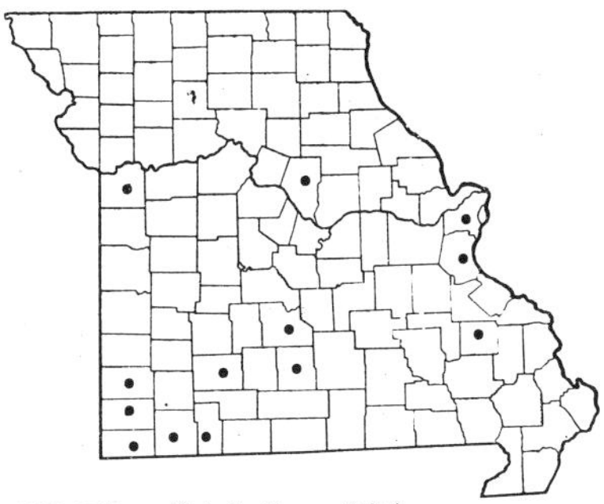

1897 Melissa officinalis (Lemon Balm)

1898 Satureja Acinos (Mother-of-Thyme)

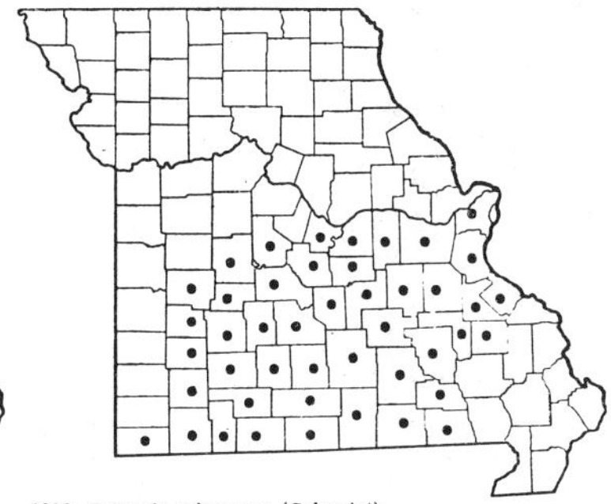

1899 Satureja arkansana (Calamint)

Indiana, Illinois, and Missouri, south to Florida, Arkansas, and Oklahoma; also in California, Oregon, and British Columbia.

This plant has a pronounced lemonlike fragrance, and is grown as a pot-herb around home gardens for use in seasoning, flavoring liqueurs, and domestic medicine. The plant is also a source of honey.

## 21. **Satureja** L. Calamint, Savory

Main stem and branches and leaves more or less hairy; leaves elliptic-ovate or ovate-oblong, 2–2½ times as long as broad; rare introduced species . . . . . . . . . . . . . . . 1. S. ACINOS

Most of stem and branches and leaves nearly or quite glabrous (without hairs); leaves on the stem linear, usually 4 or more times as long as broad . . . . . . . . . . . . . . . 2. S. ARKANSANA

1. **Satureja Acinos** (L.) Scheele
Mother-of-Thyme Map 1898
Flowers May–September.

Occurs in waste ground, where known only from Jasper County, southwestern Missouri (uncommon, Webb City, May 14, 1909, *Palmer 2007*, in Mo. Bot. Gard. Herb.).

Native of Europe; introduced and naturalized in North America from Quebec and Ontario south to New England, New Jersey, Pennsylvania, Michigan, and Missouri.

This plant was previously erroneously identified as *Hedeoma acinoides* Scheele in Palmer and Steyermark's *Annotated Catalogue*. The Missouri specimen shows the characteristic 4 stamens of *Satureja*. The corolla is shorter than in *Hedeoma acinoides* and without the long tube of that species.

2. **Satureja arkansana** (Nutt.) Briq. Calamint
Map 1899
*Satureja glabra* (Nutt.) Fern. [P & S, Steyerm.]
*Satureja glabella* (Michx.) Briq. var. *angustifolia* (Torr.) Svenson [BB]
Flowers late May–October.

Occurs on limestone glades and bald knobs, rocky open and moist ledges of limestone bluffs and escarpments, wet meadows along spring branches, gravel bars along streams. Ozark region of southern and central Missouri north to St. Louis, Franklin, Gasconade, Osage, Cole, Morgan, Benton, St. Clair, and McDonald counties.

Ranges from Ontario to Minnesota, south to New York, Ohio, Indiana, Illinois, Arkansas, and Texas.

Two variations occur in Missouri:

Corolla lavender or purplish . . 2a. S. ARKANSANA f. ARKANSANA
Corolla white . . . 2b. S. ARKANSANA f. ALBA

2a. **Satureja arkansana** f. **arkansana** Map 1899
*Satureja arkansana* (Nutt.) Briq. [G]

This is the common form encountered throughout the Ozarks.

2b. **Satureja arkansana** f. **alba** Steyermark
No Map

Known only from Washington County, east-central Missouri (creek banks along Courtois Creek, T35N, R1W, sect. 32 and 5, ½–1½ mi. south of Ishmael, June 13, 1941, *Steyermark 41237*, holotype in Chi. Nat. Hist. Mus. Herb.).

This is a perennial species which sends out from the base of the plant little creeping runners (stolons) with short ovate to elliptic leaves, usually rose-purple on the lower surface, in contrast to the elongated linear or linear-oblanceolate leaves of the main stem and branches.

1900 Pycnanthemum tenuifolium (Slender Mountain Mint)

1901 Pycnanthemum Torrei var. Torrei

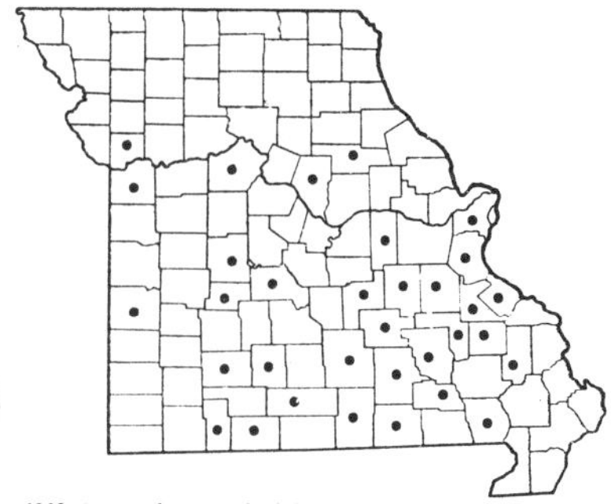

1902 Pycnanthemum virginianum

The plant is highly aromatic with a strong odor of menthol. The fresh plant, rubbed upon the skin, is sometimes used by country people as a repellent against mosquitoes and other insects. A pleasantly aromatic tea may be prepared from the fresh or dried plant.

*Excluded Species*

**Satureja glabella** (Michx.) Briquet
*Satureja glabella* var. *glabella* [BB]

This has been reported for Missouri by both Gleason and Fernald in the current manuals.

I have seen no material from Missouri of this species which can be positively identified as *S. glabella.* Certain specimens from Lawrence (*Steyermark 22255*) and Stone (*Steyermark 22583*) counties show a tendency toward a broader leaf (sometimes 6–7 mm. broad), but have mostly beardless nodes of the stem, characteristic small corollas, long slender filaments, elongate style branches, and shorter and narrower leaves characteristic of *S. arkansana*, or represent young luxuriant stages of growth. On moist shaded limestone ledges, plants of *S. arkansana* often develop a more lax habit with slightly broader and more elongated leaves, and it is believed that such plants may be the basis for reports of *S. glabella* from Missouri.

Whether or not the two taxa should be maintained as separate species or treated as varieties remains for future studies to bring out.

## 22. **Pycnanthemum** Michx. Mountain Mint

a. Upper floral leaves or bracts below the flower-clusters and at least the upper leaves of the stem and branches strongly whitened, at least on upper surface; calyx distinctly 2-lipped with the upper and lower lips quite unequal with some of the teeth longer than the others; inflorescence rather open and branching, with loose, elongate branches, the flower-clusters somewhat separated from one another . . . 6. P. ALBESCENS

a. Without the above combination of characters; no leaves or bracts whitened on upper surface, or if rarely whitened, then the calyx more or less regular and not 2-lipped, and with all the teeth of about the same length; inflorescence dense with a flat congested appearance, with short branches, the flower-clusters close together . . . *b*

b. Upper surface of leafy bracts at base of flower-clusters whitened; leaf-blades ovate or broadly ovate-lanceolate, rounded or heart-shaped at base, the main ones 15–40 mm. broad, distinctly toothed with 5–10 teeth on each margin . . . 5. P. MUTICUM

b. Upper surface of leafy bracts at base of flower-clusters not whitened; leaf-blades linear to lanceolate, the largest leaves 1.5–15 mm. (rarely 20) broad, entire (without teeth) or with few remote teeth . . . *c*

c. Sides and angles of at least the upper part of stem hairy . . . *d*

d. Outermost leaf-like bracts at base of flower-clusters glabrous on upper surface; calyx-teeth 1.5–2 mm. long (1 mm. in young flowers); lower surface of leaves glabrous or slightly hairy on midrib; stems minutely hairy; rare species of southeastern lowland counties . 2. P. TORREI

d. Outermost leaf-like bracts at base of flower-clusters densely hairy on upper surface; calyx-teeth 1 mm. long; lower surface of leaves usually densely hairy or with noticeable hairiness; stems usually noticeably hairy; common species throughout Missouri . . . . 4. P. PILOSUM

c. Sides of stem glabrous (without hairs), angles glabrous or hairy . . . *e*

e. Larger leaves 1.5–5.5 mm. broad; angles of stem without hairs or minutely hairy; calyx-

1903 Pycnanthemum pilosum (Hairy Mountain Mint)

1904 Pycnanthemum muticum

1905 Pycnanthemum albescens

teeth glabrous or nearly so, 1–2 mm. long, long and slender-tipped; plants mostly of dry or rocky open woods, fields, prairies, and dry open ground, sometimes in wetter habitats . 1. P. TENUIFOLIUM

e. Larger leaves 6–11 mm. broad; angles of stem noticeably hairy; calyx-teeth hairy at tip, less than 1 mm. long, triangular with relatively short tips; plants generally of wet meadows, swamps, moist ground along streams and spring branches, and moist ledges of bluffs . . 3. P. VIRGINIANUM

1. **Pycnanthemum tenuifolium** Schrad. Slender Mountain Mint Map 1900

*Pycnanthemum flexuosum* [of P & S, BB], not (Walt.) BSP.

Flowers June–September.

Occurs in dry or rocky open woods, dry fields and prairies, open wet thickets, meadows, alluvial ground and gravel bars along streams, along roadsides and railroads. Throughout Missouri, doubtless, in every county.

Ranges from Georgia to Texas, north to New England, New York, Ontario, Ohio, Michigan, Wisconsin, and Minnesota.

2. **Pycnanthemum Torrei** Benth. var. **Torrei** Map 1901

*Pycnanthemum Torrei* Benth. [G, BB, P & S]

Flowers late June–October.

Occurs in dry open woods. Known only from Dunklin County, southeastern Missouri (October 21, 1892, *Bush*).

Ranges from Connecticut and New York south to Virginia and Georgia; Illinois and Missouri to Kansas and Arkansas.

The specimen cited above has finely pubescent stems and the lower leaf-surface is glabrous, characteristic of var. *Torrei.* The var. *leptodon* (Gray) Boomhour, reported for Missouri by Fernald in *Gray's Manual,* is based upon a specimen in the Gray Herbarium ('W. Missouri, December, 1870, *Geyer,* ex. Engelmann'), but the origin of the specimen is extremely doubtful and I am excluding it as not having been actually collected in the state.

3. **Pycnanthemum virginianum** (L.) Durand & Jackson Map 1902

Flowers July–September.

Occurs in wet meadows, swamps, moist ground along streams and spring branches, and along moist ledges of bluffs, generally of limestone and other calcareous strata. Southern and central Missouri, north to St. Louis, Audrain, Boone, Saline, and Clay counties.

Ranges from Maine to North Dakota, south to Virginia, North Carolina, Tennessee, Ohio, Indiana, Illinois, Missouri, and Kansas.

4. **Pycnanthemum pilosum** Nutt. Hairy Mountain Mint Map 1903

Flowers July–September.

Occurs in prairies, open dry and upland woodland, thickets, and along railroads. Throughout Missouri, except absent from the lowland counties of the extreme southeastern section.

Ranges from Ontario and Michigan to Iowa, south to Tennessee, Arkansas, and Oklahoma; introduced from Massachusetts to Pennsylvania.

This is a favorite bee plant.

5. **Pycnanthemum muticum** (Michx.) Pers. Map 1904

Flowers July–September.

Occurs in grassy open places and low or dry woodland. Known only from southeastern Missouri in Dunklin (Malden, July 20, 1895, *Bush 472*; low ground, Campbell, October 7, 1910, *Bush 6370A*) and Butler (grassy places, August 2, 1892, *Eggert*) counties.

Ranges from Florida to Texas, north to Maine, Michigan, Illinois, and Missouri.

6. **Pycnanthemum albescens** T. & G. Map 1905
*Pycnanthemum incanum* [of P & S], not (L.) Michx.
Flowers July–September.

Occurs in rocky open woods, grassy slopes, and clearings in acid soils overlying chert, sandstone, and granite substrata. Ozark region of southern Missouri and on Crowley Ridge in Dunklin and Stoddard counties, north to Ste. Genevieve, Washington, Dent, Phelps, Pulaski, Wright, Christian, and Newton counties.

Ranges from Florida to Texas, north to Missouri and Oklahoma.

This species has been confused with *P. incanum*, a species of more easterly range, which has not been found in Missouri. Some specimens from Dunklin, Butler, Ripley, Shannon, and Barry counties, previously recorded as *P. incanum* in Palmer and Steyermark's *Annotated Catalogue*, are here included under *P. albescens*. A specimen from Dunklin County (sands, Campbell, September 7, 1910, *Bush 6256*) questionably labeled as a hybrid between *P. albescens* and *P. muticum* appears to be best referred to *P. albescens*. It has the stems with spreading hairs and the lower surface of the leaves pubescent with short spreading hairs.

This is a conspicuous plant of dry, rocky acid woodland in the Ozarks, easily recognized by the strongly whitened upper surface of the upper leaves and floral bracts, and the generally white-hairy upper part of stems, leaves, bracts, and calyces.

## 23. **Cunila** L. Dittany

**Cunila origanoides** (L.) Britt. Dittany Map 1906
Flowers July–November.

Occurs in acid soils of sandstone, chert, or granite usually in dry, rocky or open woods on upland slopes and ridges, sandy open ground, and prairies. Throughout the Ozark region and on Crowley Ridge in Scott, Stoddard, and Dunklin counties, north to Pike, Montgomery, Callaway, Boone, Morgan, Benton, Dade, and Jasper counties.

Ranges from Florida to Texas, north to New York, Pennsylvania, West Virginia, Ohio, Indiana, Illinois, Missouri, and Oklahoma.

The flowers are usually purple or lavender, rarely white, and occur in clusters in the axils of the leaves. The leaves have a characteristic mintlike aroma, and may be used, either fresh or dried, for tea. At one time the plant was used as a remedy for fever, and headaches, and as an application in snake bites.

The plant is attractive when in flower, and makes a desirable addition to the rock garden or wildflower bed. During the first frosty days of fall this species is one which produces the so-called 'frost flowers,' those white, ribbonlike, fluted, and sometimes twisted ice formations about 2.5–5 cm. broad sent up from the base of the plant. These formations of ice result from the rise of the cell sap and moisture from the still active root into the dead and dry stem. The sap which oozes above the ground from the cracks of the stem becomes solidified, forming the ice and continually freezing on the inner edge and forcing that already frozen toward the anterior edge of the ribbon. This phenomenon occurs also in *Helianthemum canadense*, *Pluchea camphorata*, *Verbesina virginica*, and a few other species.

## 24. **Lycopus** L. Bugle Weed, Water Horehound

a. At least the lower leaves more or less deeply and irregularly or sharply cut (incised) or pinnately cleft . . . . . . . . . . . . . . . . . . . . . . . . . . 2. L. AMERICANUS
a. Leaves more or less toothed, but none deeply or irregularly incised or pinnately cleft . . . . . . *b*
  b. Calyx-teeth blunt (obtuse) or slightly pointed, broadly triangular, less than 1 mm. long, without a conspicuous midnerve, shorter than the nutlets at maturity . . . . . . . . 1. L. VIRGINICUS
  b. Calyx-teeth long-pointed (distinctly acute to acuminate), narrowly triangular, 1–2 mm. long,

---

Plate no. 311. **1.** Pycnanthemum pilosum, × 2/5; *a.* Bract, × 1 1/5; After Gleason, The New York Botanical Garden. **2.** Pycnanthemum muticum, × 2/5; *a.* Leaf, × 2/5; *b.* Calyx, × 2; After Gleason, The New York Botanical Garden. **3.** Pycnanthemum tenuifolium, × 2/5; *a.* Flower-cluster, × 1 1/5; Details from Small, The New York Botanical Garden. **4.** Lycopus virginicus, × 2/5; *a.* Fruit with calyx, × 1 1/5; *b.* Flower, × 2; *c.* Fruiting cluster, × 1 1/5; After Gleason, The New York Botanical Garden. **5.** Pycnanthemum albescens, × 2/5; *a.* Calyx, × 2 4/5; After Gleason, The New York Botanical Garden. **6.** Cunila origanoides, × 2/5; After Gleason, details from Small, The New York Botanical Garden.

PLATE NO. 311

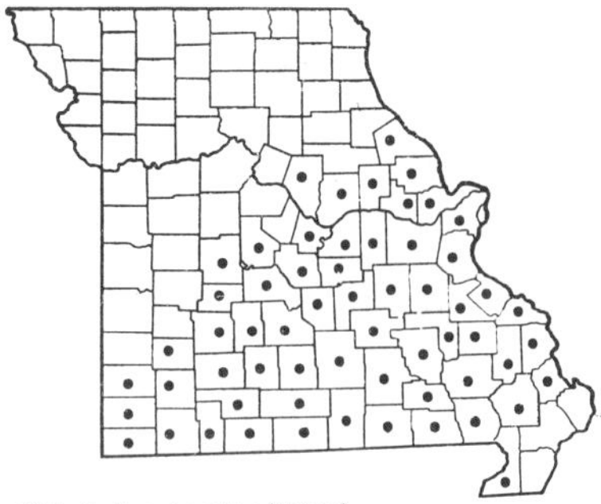
1906 Cunila origanoides (Dittany)

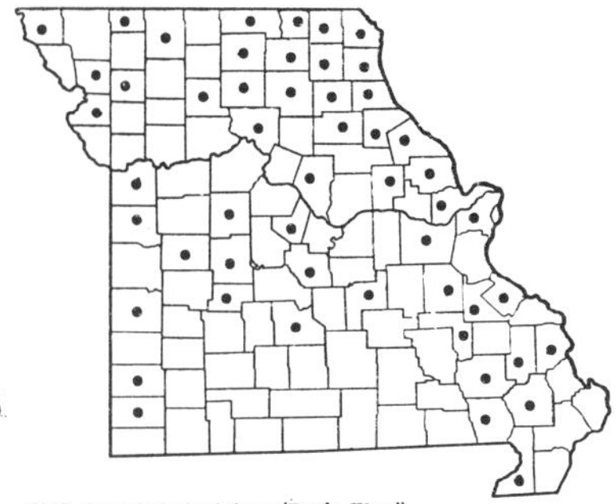
1907 Lycopus virginicus (Bugle Weed)

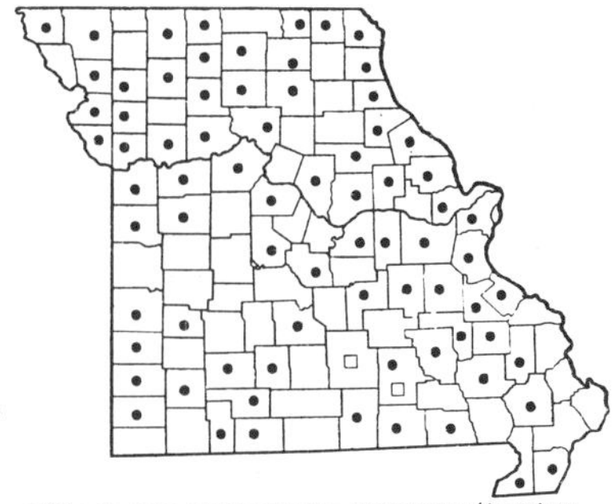
1908 ● Lycopus americanus var. americanus (American Bugle Weed)
1908 □ Lycopus americanus var. Longii

with a conspicuous midnerve often protruding beyond the tip, longer than the nutlets at maturity . . *c*
c. Leaves with definite petioles (stalks) . . . . . . . . . . . . . . . 3. L. RUBELLUS
c. At least the lower and middle leaves of the stem sessile (without stalks) . . . . . . . 4. L. ASPER

1. **Lycopus virginicus** L. Bugle Weed Map 1907
Flowers July–October.

Occurs in wet ground in wet meadows, low woods and alluvial thickets along streams, sloughs, oxbow lakes of river flood plains, along ditches, and sometimes moist ledges of bluffs. Throughout most of Missouri, but apparently absent from much of the southern Ozark section.

Ranges from Maine and Quebec to Minnesota and Nebraska, south to Georgia, Alabama, Arkansas, Oklahoma, and Texas.

2. **Lycopus americanus** Muhl. American Bugle Weed Map 1908
Flowers June–October.

Occurs in low wet woods, alluvial meadows and thickets along streams and ditches, borders of sloughs and oxbow lakes of river flood plains, upland ponds, and along railroads.

The following varieties occur in Missouri:

Stems glabrous (without hairs) or sparsely covered with appressed hairs (lying parallel to or pressed close to surface); commonly encountered . 2a. L. AMERICANUS var. AMERICANUS

Upper lengths of the stem with long spreading hairs; rarely encountered, around sink-hole ponds of the southeastern Ozarks . . . . 2b. L. AMERICANUS var. LONGII

2a. **Lycopus americanus** var. **americanus** Map 1908

*Lycopus americanus* Muhl. [G, BB, P & S]

The common variation found throughout Missouri.

Ranges from Newfoundland to British Columbia, south to Florida, Mississippi, Texas, New Mexico, and California.

2b. **Lycopus americanus** var. **Longii** Benner Map 1908

Known only from sink-hole ponds of the southeastern Ozarks in Shannon (Gilmore Pond [Grassy Pond], T27N, R6W, sect. 32, between Jack's Fork of Current River and Flat Rock Hollow, 6½ mi. northwest of Montier, July 8, 1951, *Steyermark 72116*) and Texas (natural upland pond on west side of highway 32, T32N, R10W, sect. 20, 2½ mi. northeast of Success, November 3, 1956, *Steyermark 83586*) counties.

Ranges from New York to Virginia; and Missouri.

*Lycopus americanus* var. *scabrifolius* Fern., not recorded from Missouri thus far, is similar to var. *Longii*, but has the upper surface of the leaves scabrous (rough), whereas in var. *Longii* the upper surface is glabrous or smooth. A faint indication of some scabridity is found on the upper leaf surface of a collection from Washington County (*Kellogg 1965* from Old Mines, August 19, 1928), but it is only sparingly so and not conspicuously uniform; moreover, since many of the upper leaf surfaces of the other leaves of this specimen are smooth, it is preferable to exclude the variety from the Missouri flora.

3. **Lycopus rubellus** Moench Map 1909
Flowers July–October.

Occurs in low or swampy woods, bald cypress swamps, border of lowland or upland sink-hole ponds, streams, sloughs, spring branches, and wet meadows.

The following 3 varieties occur in Missouri:

a. Main leaves of stem and branches 5–10 mm. (rarely 20) broad; stem and lower surface of leaves glabrous . . . . . 3c. L. RUBELLUS var. LANCEOLATUS

a. Main leaves of stem and branches 10–40 mm. broad; stem and lower surface of leaves gla-

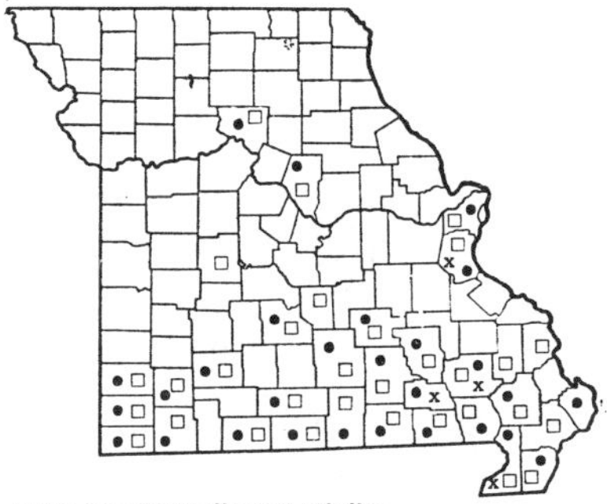

1909 • Lycopus rubellus var. rubellus
1909 □ Lycopus rubellus var. arkansanus
1909 x Lycopus rubellus var. lanceolatus

1910 Lycopus asper

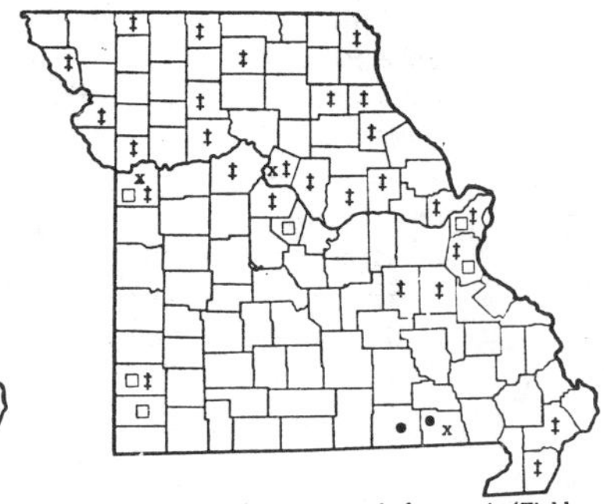

1911 • Mentha arvensis var. arvensis f. arvensis (Field Mint)
1911 x Mentha arvensis var. arvensis f. glabra
1911 □ Mentha arvensis var. villosa f. villosa
1911 ‡ Mentha arvensis var. villosa f. glabrata

brous or hairy . . . . . . . . . . *b*

b. Stem glabrous or nearly so; lower surface of leaves glabrous; main leaves of stem with conspicuous sharp teeth . . 3a. L. RUBELLUS var. RUBELLUS

b. Upper portions of stem hairy; lower surface of leaves hairy; main leaves of stem with short, low, depressed teeth . 3b. L. RUBELLUS var. ARKANSANUS

3a. **Lycopus rubellus** var. **rubellus** Map 1909

*Lycopus rubellus* Moench [G, BB, P & S]

Known from southern and east-central Missouri, northeast to St. Louis County.

Ranges from Florida to Texas, north to New England, New York, Ohio, Michigan, Illinois, and Missouri.

3b. **Lycopus rubellus** var. **arkansanus** (Fresn.) Benner Map 1909

The commonest variation in the state, in southern and central Missouri north to St. Louis, Boone, and Chariton counties.

Ranges from Arkansas to Texas, north to Michigan and Missouri.

3c. **Lycopus rubellus** var. **lanceolatus** Benner Map 1909

Known from eastern Missouri in Jefferson (August 30, 1887, *Eggert*), Wayne (1½ mi. southwest of Wappapello, August 29, 1938, *Steyermark 6242*), Dunklin (Campbell, September 11, 1893, *Bush*), and Carter (Big Spring, Current River, September 10, 1897, *Trelease 741*) counties.

Ranges from Florida to Louisiana, north to Virginia, Tennessee, and Missouri.

4. **Lycopus asper** Greene Map 1910

*Lycopus lucidus* Turcz. var. *americanus* Gray [P & S]

Flowers July–August.

Occurs along alluvial soils in flood plain of the Missouri River near Courtney, July 7, 1933, *Bush and Steyermark 12672*).

Ranges from Alaska to California, east to Manitoba, Michigan, Illinois, and Missouri.

The tuberous roots of this and other species of the genus are eaten by muskrats.

## 25. **Mentha** L. Mint

a. Clusters of flowers and fruits arranged along the sides of the stem at the base (axil) of the leaves . . . *b*

b. Calyx-tube glabrous (without hairs), calyx-lobes hairy; upper leaves accompanying flower-clusters conspicuously reduced to only 2 or 3 times the length of the flower-clusters . 3. M. CARDIACA

b. Calyx-tube and lobes hairy; upper leaves not or scarcely reduced as compared with the lower ones . . . . . . . . . . . . . . . . . . . . . . . . . 1. M. ARVENSIS

a. Clusters of flowers and fruits in elongated continuous or interrupted inflorescences at the tops of the stems and branches, or some in the axils of the upper leaves . . . . . . . . . . . . . *c*

c. Lower surface of leaves hairy; stems, bracts of inflorescence, and calyx more or less conspicuously gray- or white-hairy . . . . . . . . . . . . . . . . 6. M. LONGIFOLIA

c. Leaves glabrous; stem glabrous (without hairs) or nearly so; bracts of inflorescence and calyx green and glabrous, or, if hairy, not with a gray or white hairy covering . . . . . . . . . . d
d. Calyx-tube and lobes glabrous; inflorescence composed of 1–3 clusters of flowers or fruits. . 2. M. CITRATA
d. Calyx-tube or calyx-lobes or both hairy; inflorescence composed of several to many (4 or more) clusters of flowers or fruits . . . . . . . . . . . . . . . . . . . . . . . . e
e. Bracts or leaf-like bracts accompanying the flower-clusters resembling the leaves lower on the stem, 2–3 times the length of the flower clusters . . . . . . . . 3. M. CARDIACA
e. Bracts accompanying the flower-clusters linear to linear-lanceolate, scarcely or little longer than the length of the flower-clusters . . . . . . . . . . . . . . . . . f
f. Calyx 1.5–2 mm. long; leaves nearly or quite sessile (without stalks), the petioles (stalks) at most 3 mm. long. . . . . . . . . . . . . . . . . . . . . 4. M. SPICATA
f. Calyx 2.5–4 mm. long; leaves on definite petioles (stalks) 4–15 mm. long . . 5. M. PIPERITA

1. **Mentha arvensis** L. Field Mint Map 1911

Flowers July–September.

Occurs in moist low ground along streams, ponds, spring branches, limestone outcrops and ledges, alluvial woods, and along railroads.

Scattered throughout Missouri.

The treatment of *Mentha arvensis* here is based upon the work of Sara R. Stewart (Mrs. George Metcalf Hinkley) in Rh. 46: 331–35. 1944. The following variations are known from Missouri:

a. The ordinary leaves and those at base of lower flower-clusters broadest near or somewhat above middle, 1/4–1/2 as broad as long, narrowed or wedge-shaped to a petiole(leaf-stalk); petiole usually longer than the flower- or fruit-clusters . . . . . . . . . . . b
b. Leaves more or less hairy; both sides and angles of the stem near the lowest flower-clusters hairy. 1c. M. ARVENSIS var. VILLOSA f. VILLOSA
b. Leaves glabrous (without hairs) or nearly so; sides of stem near the lowest flower-clusters glabrous, the angles minutely hairy . . . 1d. M. ARVENSIS var. VILLOSA f. GLABRATA
a. The ordinary leaves and those at the base of the lower flower-clusters broadest mostly below the middle, 3/7–4/5 as broad as long, rounded to a short petiole; petiole shorter than the flower- or fruit-clusters . . . . . . . c
c. Leaves more or less hairy; both the sides and angles of the stem near the lowest flower-clusters hairy . . . 1a. M. ARVENSIS var. ARVENSIS f. ARVENSIS
c. Leaves glabrous (without hairs) or nearly so; sides of stem near the lowest flower-clusters glabrous, the angles minutely hairy. 1b. M. ARVENSIS var. ARVENSIS f. GLABRA

1a. **Mentha arvensis** var. **arvensis** f. **arvensis** Map 1911

*Mentha arvensis* var. *arvensis* [BB]

*Mentha arvensis* L. [G, P & S]

Known only from Ripley and Oregon counties, southeastern Missouri.

Ranges from Labrador to Washington, south to Virginia, Kentucky, Missouri, Nebraska, New Mexico, and California.

1b. **Mentha arvensis** var. **arvensis** f. **glabra** (Benth.) Stewart Map 1911

*Mentha arvensis f. glabra* (Benth.) Stewart [G]

Known from Ripley (low woods along Current River, 1½ mi. north of mouth of Buffalo Creek, 5 mi. east of Bennett, August 6, 1934, *Steyermark 14272*), Howard (*Steyermark 26281*), and Jackson (Courtney, September 5, 1917, *Bush 8106*) counties.

Ranges from Newfoundland and Quebec to Washington, south to New England, New York, Missouri, and Kansas.

1c. **Mentha arvensis** var. **villosa** (Benth.) Stewart f. **villosa** Map 1911

*Mentha arvensis* var. *villosa* (Benth.) Stewart [G, BB]

*Mentha arvensis* var. *canadensis* (L.) Briq. [P & S]

Scattered in southern and central Missouri.

Ranges from Newfoundland to Alaska, south to Virginia, Ohio, Indiana, Illinois, Missouri, Nebraska, New Mexico, and California.

1d. **Mentha arvensis** var. **villosa** f. **glabrata** (Benth.) Stewart Map 1911

*Mentha arvensis* var. *glabrata* (Benth.) Fern. [P & S]

This is the commonest variation in Missouri, scat-

Plate no. 312. **1.** Lycopus americanus, × 2/5; *a.* Corolla, × 2; *b.* Fruiting cluster, × 2; After Gleason, details from Small, The New York Botanical Garden. **2.** Mentha cardiaca, × 2/5; *a.* Calyx, × 4; After Gleason, The New York Botanical Garden. **3.** Lycopus asper, × 2. **4.** Mentha arvensis var. villosa f. glabrata, × 1; After Gleason, The New York Botanical Garden. **5.** Mentha citrata, × 2/5; *a.* Calyx, × 4 4/5. **6.** Mentha piperita, × 2/5.

PLATE NO. 312

1911-A Mentha citrata (Bergamot Mint)

1911-B Mentha cardiaca

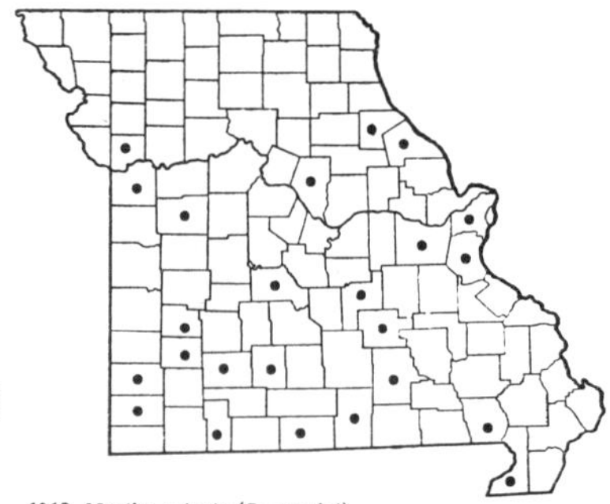

1912 Mentha spicata (Spearmint)

tered throughout the state, but not recorded from most of the Ozark region.

Ranges from Labrador to British Columbia, south to Maryland, West Virginia, Ohio, Illinois, Missouri, New Mexico, Arizona, and California.

The fresh or dried leaves of *M. arvensis* may be used as a tea.

2. **Mentha citrata** Ehrh. Bergamot Mint Map 1911A
Flowers July–October.

Occurs in wet meadows and open wet places. Known only from southern Missouri in Howell County (swampy meadow along north side of highway 80, 4.9 mi. southwest of West Plains, July 24, 1949, *Steyermark 68613*).

Introduced and naturalized from Europe to North America from Nova Scotia to Michigan, south to Virginia and Missouri.

This is sometimes treated as a variety of *M. piperita*, and is also considered to be of hybrid origin between *M. aquatica* and *M. spicata*. A collection from Jasper County (wet ground along railroad, Sarcoxie, September 18, 1910, *Palmer 3185*) previously recorded as *M. citrata* in Palmer and Steyermark's *Annotated Catalogue* is referred in the present flora to *M. piperita* because of the hirsute margins of the calyx-teeth.

3. **Mentha cardiaca** Gerarde Map 1911B
Flowers July–October.

Occurs along margins of streams. Known only from Howard (along gravel bed of Moniteau Creek near 'Blue Bluff,' T51N, R14W, sect. 10, 8 mi. east of Burton, September 14, 1937, *Steyermark 26281*) and Jasper (waste ground along Joplin Creek, Joplin, August 17, 1920, *Palmer 18746*; margins of perennial brook, Webb City, September 20, 1923, *Palmer 23787*) counties.

Introduced from Europe and naturalized in North America from Newfoundland to Michigan, south to Virginia, Ohio, Indiana, Illinois, and Missouri.

This is believed to be a hybrid between *M. arvensis* and *M. spicata*.

4. **Mentha spicata** L. Spearmint Map 1912
Flowers late June–October.

Occurs in wet ground bordering streams, spring branches, ponds, ditches, meadows in low valleys, along roadsides and railroads. Scattered throughout southern and central Missouri, north to Ralls, Boone, and Clay counties; probably found throughout the state but not recorded from most of northern Missouri.

Native of Europe; introduced and naturalized in North America from Nova Scotia to Washington, south to Florida and California.

The fresh, fragrant leaves are used in flavoring iced tea and other drinks, such as juleps. The fresh or dried leaves may also be used for making into hot tea. Mint sauces and jellies, stews, and soups are often seasoned with the fresh or dried leaves of this plant. The oil from the plant is used in flavoring various candies, medicine, chewing gum, toothpastes, and powders.

5. **Mentha piperita** L. Peppermint Map 1913
Flowers June–October.

Occurs in wet ground in meadows, along spring branches, streams, borders of ponds, sloughs, and ditches, roadsides and railroads, occasionally found submerged in spring water. Scattered over southern and central Missouri north to Ralls, Boone, Livingston, and Clay counties.

Introduced and naturalized from Europe in North America from Nova Scotia to Minnesota, south to Florida, Arkansas, Oklahoma, and California.

Considered to be of hybrid origin between *M. spicata* L. and *M. aquatica* L.

As with *M. spicata*, the leaves are used fresh in flavoring iced tea and other drinks, including juleps. The fresh or dried leaves, likewise, may be prepared for a hot tea. The plant is cultivated for its oil, used as a flavoring for candies, chewing gum, toothpastes and

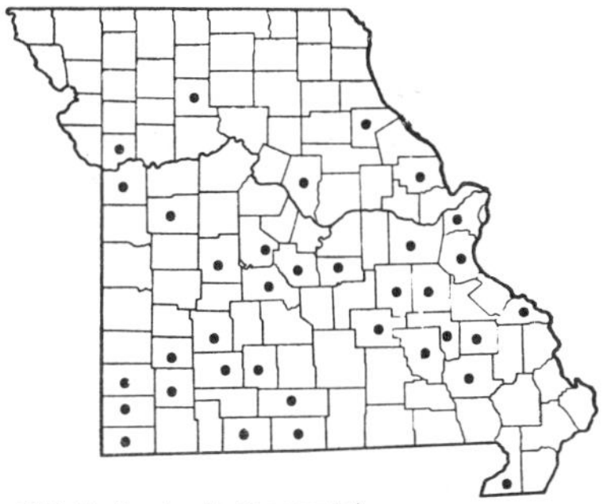

1913 Mentha piperita (Peppermint)

1914 • Mentha longifolia var. mollissima (Horsemint)
1914 □ Mentha longifolia var. undulata

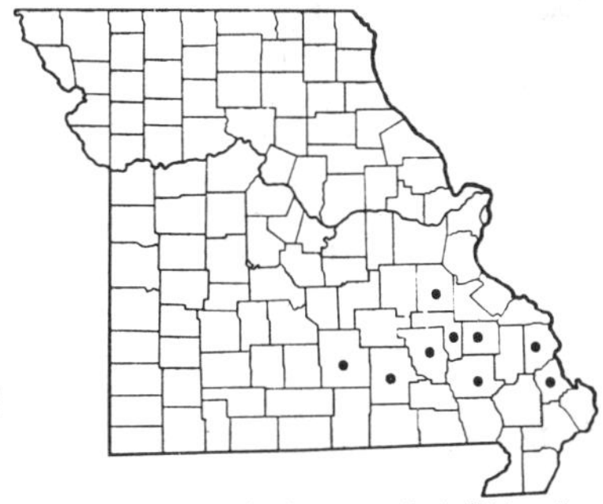

1915 Collinsonia canadensis var. canadensis (Richweed)

powders, pharmaceutical preparations, medicines, perfumery, and soaps. Menthol, derived from the oil, is used as an antiseptic and in the treatment of colds.

6. **Mentha longifolia** (L.) Huds. Horsemint Map 1914

Flowers July–September.

Occurs in wet places along roadsides, thickets, and waste ground.

The following variations occur in Missouri:

Leaves sharply toothed, ovate-oblong to oblong-lanceolate . . 6a. M. LONGIFOLIA var. MOLLISSIMA
Leaves with coarse, jagged toothing, round-ovate and somewhat heart-shaped at base . . . . 6b. M. LONGIFOLIA var. UNDULATA

6a. **Mentha longifolia** var. **mollissima** (Borkh.) Rouy Map 1914

Known only from western Missouri, in Clay, Jackson, and Jasper (Carthage, July 24, 1910, *Palmer 6068, 6068A*; established along roadside, 2 mi. southeast of Carthage, *Bush & Palmer 3042*; along roadside, Carthage, May 12, 1909, *Palmer 1987*) counties.

Native of Eurasia; introduced and naturalized from Massachusetts to Indiana, south to Pennsylvania and Missouri.

The *Palmer 6068* and *6068A* collections were originally labeled *M. alopecuroides*. The leaves on the specimens are oblong and small-toothed. Other collections cited were labeled originally as *M. rotundifolia*, but do not possess the rugose, round-ovate to broadly elliptical, crenately toothed leaves of the latter species.

6b. **Mentha longifolia** var. **undulata** (Willd.) Fiori & Paoletti Map 1914

*Mentha alopecuroides* [of P & S], not Hull

Known only from Jackson County, west-central Missouri (Kansas City, August 7, 1895, *Bush 313*).

Introduced from Europe into the United States.

The Missouri collection was originally labeled *M. alopecuroides* Hull and so recorded in Palmer and Steyermark's *Annotated Catalogue*. The specimen agrees with *M. alopecuroides*, if that name is accepted, in having prominent veins on the lower surface of the leaf, and in the open-dentate leaves rounded at the tip. The jagged toothing of the specimen, however, would place it with *M. longifolia* var. *undulata*. *Mentha alopecuroides* is sometimes considered a hybrid between *M. longifolia* and *M. rotundifolia*, and the Bush specimen exhibits characters of the two species.

*Excluded Species*

**Mentha gentilis** L.

Recorded from Jackson County in Palmer and Steyermark's *Annotated Catalogue* on the basis of the collection of *Bush 12287* (waste ground, Courtney, August 10, 1931). The stems on this specimen show retrorse hairs, by virtue of which the specimen is referred in the present flora to *M. arvensis* var. *villosa glabrata*.

This mint is considered to be of hybrid origin between *M. spicata* and *M. arvensis*.

## 26. **Collinsonia** L. Horse Balm

**Collinsonia canadensis** L. var. **canadensis** Richweed Map 1915

Also called Citreonlla Horse Balm.

*Collinsonia canadensis* L. [G, BB, P & S]

Flowers July–September.

Occurs in rich or rocky woods of usually limestone

soils, in ravines and slopes. Eastern Ozark region north to Cape Girardeau, Madison, and Washington counties, west to Texas and Shannon counties.

Ranges from Florida to Arkansas, north to Quebec, Massachusetts, Vermont, New York, Ontario, and Wisconsin.

The flowers possess a pleasant lemon fragrance. The greater part of the corolla is dull brownish-yellow with dull purple lines on the lower part of the tube and a fringed white lower lip. The thickened rootstock has been used in medicine as a remedy for kidney and urinary troubles.

## 27. **Perilla** L.

**Perilla frutescens** (L.) Britt. Beef-steak Plant Map 1916

Flowers August–October.

Occurs in rich or alluvial moist or dry ground along streams, spring branches, often in pastured valleys and meadows, along sand and gravel bars, in dry woodland, rarely along railroads. Throughout southern and central Missouri north to Marion, Howard, Saline, Lafayette, and Jackson counties.

Two variations are encountered in the state:

Common type encountered; margins of leaves regularly but bluntly toothed . . a. P. FRUTESCENS var. FRUTESCENS

Rarely encountered; margins of leaves with coarse, irregular, and larger teeth or lobes . . b. P. FRUTESCENS var. CRISPA

a. **Perilla frutescens** var. **frutescens** Map 1916

*Perilla frutescens* (L.) Britt. [G, BB, P & S]

Throughout the range indicated in Missouri.

Native of India; introduced and naturalized from Florida to Texas, north to Massachusetts, New York, Ohio, Indiana, Illinois, Iowa, and Kansas.

b. **Perilla frutescens** var. **crispa** (Benth.) Deane Map 1916

Known only from southwestern Missouri in Greene (Willard, September 4, 1889, *Blankinship*) and Barry (gravelly streams, Eagle Rock, August 14, 1905, *Bush 3242*) counties.

Native of India; introduced and naturalized in the United States north to Massachusetts, New York, West Virginia, Illinois, and Missouri.

The plant is an annual which is sometimes grown as a border ornamental for the leaves which are purple or purple-brown with a bronze luster, variegated or suffused, when young, with green, rose or pink colors. The bruised plant has a strong somewhat sweetly fetid odor. It is not browsed by hogs or cattle. The flowers have a white or purplish-red corolla and a white-hairy calyx.

The seeds and leaves of the cultivated strain of this species are eaten by the Japanese people in salads or in a pickled form.

Perilla oil, expressed from the crushed and roasted seeds, is not only used for food purposes, but is important in being used in the manufacture of cheap lacquer, artificial leather, paper umbrellas, Japanese oil papers, printer's ink, and waterproof clothes. A large amount is imported into the United States as a substitute for linseed oil in the varnish and paint industries.

## 28. **Ocimum** L.

**Ocimum basilicum** L. Basil Map 1917

Flowers August–September.

Occurs in waste ground and along railroads. Known only from St. Louis County, east-central Missouri (along Gimillin road, east of Water Works Conduit track, St. Louis, September 1, 1957, *Muehlenbach 1333*).

Native of tropical Asia and Africa; introduced and rarely escaped from cultivation in the United States.

The plant is often cultivated in herb gardens for the sweetly fragrant leaves, which are used as an ingredient in various kinds of dressings, stews, soups, and sausage. An oil, derived from the plant, finds a place in perfumery and beverages.

---

Plate no. 313. **1.** Mentha spicata, × 2/7; *a.* Corolla, × 2 2/7; Details from Small, The New York Botanical Garden. **2.** Mentha longifolia var. mollissima, × 2/7; *a.* Corolla, × 2 2/7. **3.** Perilla frutescens, × 2/7; *a.* Portion of flowering branch, × 1 1/7; Details from Small, The New York Botanical Garden. **4.** Solanum Dulcamara, × 2/7. **5.** Solanum tuberosum, × 2/7. **6.** Ocimum basilicum, × 2/7; *a.* Calyx, × 1 1/7; *b.* Flower, × 4/7; After Gleason, details from Small, The New York Botanical Garden. **7.** Solanum triflorum, × 1; After Gleason, The New York Botanical Garden. **8.** Collinsonia canadensis, × 2/7; Details from Small, The New York Botanical Garden.

PLATE NO. 313

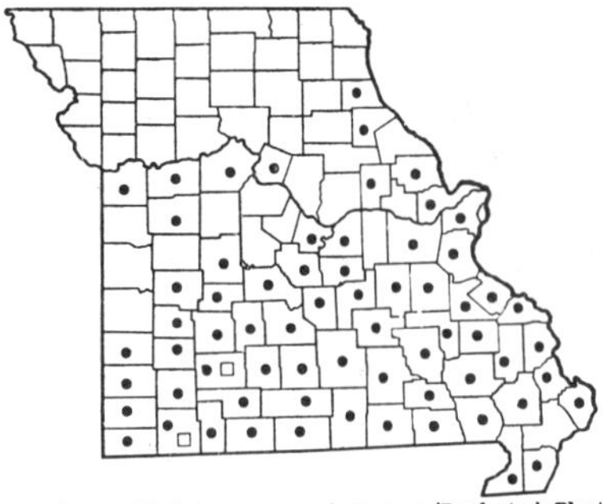

1916 • Perilla frutescens var. frutescens (Beef-steak Plant)
1916 □ Perilla frutescens var. crispa

1917 Ocimum basilicum (Basil)

1918 Solanum Dulcamara var. Dulcamara f. Dulcamara (Climbing Nightshade)

## Fam. **SOLANACEAE** (Nightshade Family)

a. Most or some of the stem woody . . . . . *b*
- b. No lobes or other indentations on any leaves; leaves usually narrowed or wedge-shaped at base; spines sometimes present on stems . . . . . 5. LYCIUM
- b. Some of leaves usually with 1–4 lobes in the lower half; leaves rounded or heart-shaped at base; no spines present on stems . . . . . 1. SOLANUM DULCAMARA

a. Stem not woody . . . . . *c*
- c. Corolla-tube either much shorter than the spreading or expanded corolla-lobes or nearly absent, the corolla less than 3.5 cm. long; fruit a fleshy or dry berry, sometimes enclosed within the calyx . . *d*
  - d. Plants with spines or prickles . . . . . 1. SOLANUM
  - d. Plants without spines or prickles. . . . . . *e*
    - e. Leaves compound, divided all the way to the midrib into 5–9 main, separate divisions (leaflets) with smaller divisions sometimes present in between . . . . . *f*
      - f. Stems and branches with long, conspicuous, spreading hairs; leaflets irregularly toothed; corolla yellow; anthers opening along their length (longitudinally) . . . 2. LYCOPERSICON
      - f. Stems and branches mostly with short and appressed (lying parallel to or pressed against surface) hairs or glabrate (nearly glabrous); leaflets entire (with teeth); corolla white or pale purple or blue; anthers opening by pores at theirs tips . . . 1. SOLANUM TUBEROSUM
    - e. Leaves not compound, not divided to their midrib into separate divisions (leaflets), but sometimes deeply cut or lobed, but then these lobes connected at their bases by the main leaf tissue . . . . . *g*
      - g. Anthers conspicuously protruding from the corolla, opening by pores at their tips; fruiting calyx not 5-angled nor 10-ribbed, not bladder-like in fruit with the berry inside 1. SOLANUM
      - g. Anthers included within the corolla-tube or just at the opening, but not conspicuously protruding, opening along their length (longitudinally); fruiting calyx 5-angled or 10-ribbed, becoming bladder-like in fruit with the berry inside . . . . . *h*
        - h. Calyx deeply 5-parted nearly to the base into almost distinct divisions, the calyx-tube almost not evident; corolla blue, 20–25 mm. long; berry dry; rarely encountered genus . . . . . 4. NICANDRA
        - h. Calyx with the 5 lobes obviously connected to a calyx-tube; corolla yellow or yellow with a brown-purple center, 5–20 mm. long; berry fleshy, juicy; commonly encountered genus. . . . . . 3. PHYSALIS
- c. Corolla-tube much longer than the spreading or expanded corolla-lobes, the corolla 1.5–20 cm. long; fruit a dry capsule . . . . . *i*
  - i. Calyx 4–10 cm. long with a long tube; fruiting capsule covered with spines or prickles . 6. DATURA
  - i. Calyx 0.5–2 cm. long with a relatively short tube; fruiting capsule not covered with spines or prickles. . . . . . *j*
    - j. Flowers or fruits occurring solitary, not arranged in 2-several-flowered inflorescences; 1 of the stamens smaller than the others . . . . . 8. PETUNIA
    - j. Flowers or fruits occurring in 2-several-flowered inflorescences, not solitary; all the stamens alike. . . . . . 7. NICOTIANA

### 1. **Solanum** L. Nightshade

a. Plants with spines or prickles . . . . . . . . . . . . . . . . . . . . . . . . . . *b*
  b. Corolla yellow; calyx almost concealed by spine-like prickles; leaves deeply pinnately lobed nearly to the midrib, or the main segments further lobed . . . . . . . . . 10. S. ROSTRATUM
  b. Corolla violet, purple, or white; calyx mostly covered by scurfy scales, not concealed by prickles; leaves shallowly lobed or with a few large teeth or entire (without teeth) . . . . . . . . . *c*
    c. Stem with spreading hairs; hairs of the lower leaf-surface sessile (without a stalk), with 4–8 branches; common species in every county of the state . . . . . . . . . 7. S. CAROLINENSE
    c. Stem covered with matted, short hairs; hairs of the lower leaf-surface on distinct stalks with 9–13 or more branches; rarely found species . . . . . . . . . . . . . . . . . *d*
      d. Coating of hairs producing a gray-silvery appearance to stems and leaves; leaves linear-oblong or oblong-lanceolate, $\frac{1}{7}$–$\frac{1}{3}$ as wide as long, wavy-margined or entire (without teeth), gradually narrowed to the base . . . . . . . . . . . . . . 9. S. ELAEAGNIFOLIUM
      d. Coating of hairs not producing a gray-silvery appearance to stems and leaves; leaves ovate, about $\frac{1}{2}$–$\frac{2}{3}$ as wide as long, with a few large teeth or lobes, rounded to heart-shaped at base 8. S. TORREYI

a. Plants without spines or prickles . . . . . . . . . . . . . . . . . . . . . . . . . . *e*
  e. Leaves compound, divided all the way to the midrib into 7–9 main, separate divisions (leaflets) with smaller divisions sometimes present in between . . . . . . . . . . 2. S. TUBEROSUM
  e. Leaves not compound, not divided to their midrib into separate divisions (leaflets), but sometimes deeply cut or lobed, but then these lobes connected at their bases by the main leaf tissue. . . . . *f*
    f. Stems climbing, twining, or long trailing; lobes, when present, occurring at the base of or below the middle of the leaf, 1–4 . . . . . . . . . . . . . . . . 1. S. DULCAMARA
    f. Stem not climbing, twining, nor long trailing; lobes, when present, occurring all along the leaf from base to tip . . . . . . . . . . . . . . . . . . . . . . . . . *g*
      g. Leaves deeply regularly lobed with 2–6 narrow lobes on each side . . . . 3. S. TRIFLORUM
      g. Leaves entire (without teeth), or with teeth or wavy margins, but not deeply or regularly lobed . . . . . . . . . . . . . . . . . . . . . . . . . . . *h*
        h. Stems glabrous (without hairs), or with few and distant hairs; leaves thin and membranaceous, the lower surface sparsely hairy or nearly glabrous . . . . 4. S. AMERICANUM
        h. Stems conspicuously and abundantly hairy or scurfy; leaves rather thick and opaque, at least the lower surface densely hairy . . . . . . . . . . . . . . . . . *i*
          i. Corolla whitish, small, 3–5 mm. long; stem and leaves covered with spreading or appressed hairs, without a gray-silvery appearance; calyx in flower 2–3 mm. long, in fruit up to 6 mm. long . . . . . . . . . . . . . . . . . . . . . *j*
            j. Hairs of stems and leaves dense, some of them more or less appressed (pressed against or lying parallel to surface), others spreading, the stems and leaves appearing grayish to the naked eye; pedicel in fruit gradually thickened upward (clavate); mature fruit yellow to red, protruding from and not enclosed by the calyx which is not enlarged in fruit . . . . . . . . . . . . 5. S. VILLOSUM
            j. Hairs of stems and leaves spreading, more scattered and not so dense, the stems and leaves appearing green to the naked eye; pedicel in fruit enlarged upward into the base of the calyx; mature fruit green, mostly enclosed by the calyx which is enlarged in fruit . . . . . . . . . . . . . . . 6. S. SARACHOIDES
          i. Corolla usually violet or bluish, rarely white, showy, 10–20 mm. long; stem and leaves covered with matted short hairs producing a gray-silvery appearance; calyx 7–8 mm. long . . . . . . . . . . . . . . . . . 9. S. ELAEAGNIFOLIUM

1. **Solanum Dulcamara** L. var. **Dulcamara** f. **Dulcamara** Climbing Nightshade Map 1918
Also called Bittersweet, Bitter Nightshade.
*Solanum Dulcamara* L. [G, BB, P & S, Steyerm.]
Flowers May–November.

Occurs in low woods, thickets, roadsides, and fence rows. Scattered throughout Missouri, but recorded from only several counties; doubtless to be found in most of the counties.

Native of Europe and Asia; introduced and naturalized in North America from Nova Scotia to Ontario and Minnesota, south to Georgia, Missouri, Kansas; also Idaho to Washington and California.

The ordinary color of the corolla is violet or purple; in var. *Dulcamara* f. *albiflorum* House, not recorded thus far from Missouri, the corolla is white. Typical var. *Dulcamara* has mostly glabrous branches and leaves. In var. *villosissimum* Desv. the branches and

1919 Solanum tuberosum (Potato)

1920 Solanum triflorum

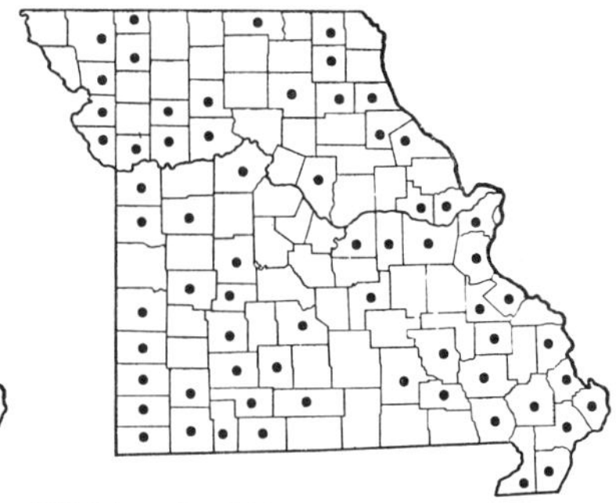

1921 Solanum americanum

leaves are quite hairy, but this variation has not been recorded from the state.

Eating of the new shoots and leaves is stated to cause occasional poisoning to cattle, horses, and sheep, but the stems serve as food for muskrat and the berries as winter food for pheasants. The berries have been used to poison rabbits and dogs. For some skin eruptions, muscular rheumatism, and bronchial and pulmonary affections the plant has been used with some beneficial results.

This is not to be confused with *Celastrus scandens*, likewise called Bittersweet. *Solanum Dulcamara* is the Bittersweet of medicine.

2. **Solanum tuberosum** L. Potato Map 1919

Flowers June–August.

Commonly cultivated, but rarely escaped from cultivation along railroad tracks, where known only from St. Louis County, east-central Missouri (right-of-way of the Terminal Railroad just opposite the Continental Grain Co., Brooklyn St. elevator, St. Louis, May 15, 1955, *Muehlenbach 584*; Fyler Ave., St. Louis, July 20, 1939, *Bauer 162*).

Native of South America.

Cases of fatal poisoning among humans have resulted from eating the green tubers of potato which grew at the surface of the soil exposed to sunlight.

3. **Solanum triflorum** Nutt. Map 1920

Flowers June–August.

Occurs in waste ground and along railroads. Known only from Laclede and Jackson (Courtney, July 21, 1910, *Bush 6007*; Sheffield, June 14, 1905, *Bush 3007*; August 20, 1913, *Bush 7096A*) counties.

Native of the western U.S., from North Dakota to British Columbia, south to Kansas, New Mexico, and California; introduced in Missouri and east to Connecticut.

Under certain conditions, cases of poisoning occur among animals feeding on the berries, stems, and leaves of this species.

4. **Solanum americanum** Mill. Black Nightshade Map 1921

*Solanum nigrum* of many auth. [P & S, Steyerm.], not L.

*Solanum nigrum* var. *vulgare* L. [BB]

Flowers May–November.

Occurs in open woodland, at base of and ledges of bluffs, along streams, about dwellings, fallow fields, pastures, along roadsides and railroads. Throughout Missouri.

Ranges from Maine to North Dakota, south to Florida, Louisiana, and Texas.

The ripe, black berries may be eaten raw or cooked for preserves and pies. The cooked young leaves and stems may be eaten as a tender vegetable. The green and unripe berries and fresh leaves are poisonous to sheep and other grazing animals. Usually, however, the plants are not grazed. Young children have been poisoned by eating the unripe berries.

As here treated, following the studies of G. L. Stebbins and E. F. Paddock (Madroño 10: 70–80. 1949), true *S. nigrum* L. is a European species very rarely encountered in North America, differing from the native North American *S. americanum* in having larger fruits and seeds, thicker opaque leaves, and longer anthers.

5. **Solanum villosum** Mill. Map 1922

Flowers June–September.

Occurs in waste ground along railroads. Known only from St. Louis County (St. Louis, right-of-way of Burlington R.R., south of grain elevator, August 20, 1955, *Muehlenbach 729*; Bremen Ave. freight yard of Terminal R.R. Assoc., July 4, 1958, *Muehlenbach 1436*).

Native of Eurasia; introduced and naturalized in the United States from Massachusetts southward.

6. **Solanum sarachoides** Sendtner Map 1923

Flowers July–October.

Occurs in waste ground and along railroads, where

1922 Solanum villosum

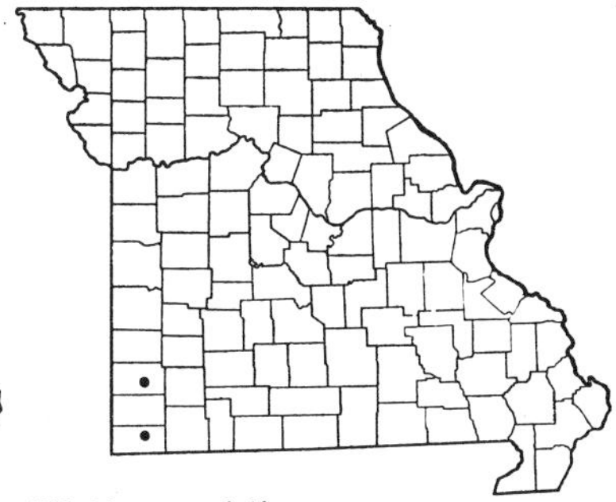

1923 Solanum sarachoides

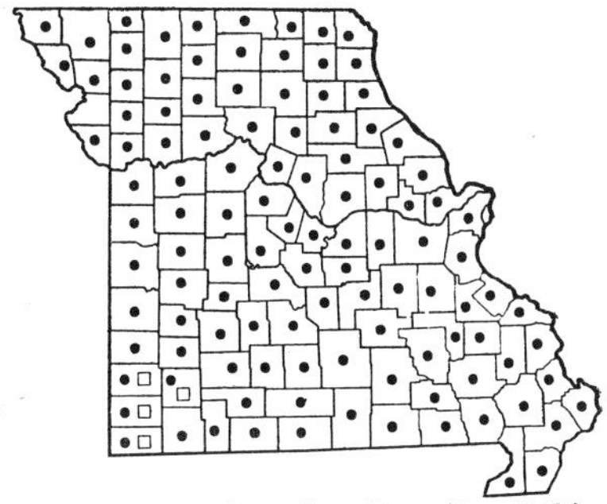

1924 ● Solanum carolinense f. carolinense (Horse Nettle)
1924 □ Solanum carolinense f. albiflorum

known only from Jasper (waste ground along railway, stockyard switch, Joplin, August 12, 1951, *Palmer 52976*) and McDonald (rocky waste ground, McNatt, October 1, 1955, *Palmer 61489*; same locality, August 15, 1955, *Palmer 60872*) counties.

Native of South America; introduced and naturalized in the United States from New England to Minnesota, southward to Missouri and Oklahoma; also in western North America.

7. **Solanum carolinense** L. Horse Nettle Map 1924

Also called Bull Nettle.

Flowers May–October.

Occurs in waste ground, fallow and cultivated fields, prairies, roadsides, and along railroads.

Throughout Missouri, in every county.

Ranges from Florida to Texas, north to Vermont, New York, Ontario, Ohio, Indiana, Illinois, Iowa, Nebraska, and Minnesota, west to Idaho, and Washington.

Two variations occur in Missouri:

Corolla violet or purplish-blue . 7a. S. CAROLINENSE f. CAROLINENSE
Corolla white . . . . . . 7b. S. CAROLINENSE f. ALBIFLORUM

7a. **Solanum carolinense** f. **carolinense** Map 1924

*Solanum carolinense* L. [G, BB, P & S, Steyerm.]

This is the common variation found throughout the state.

7b. **Solanum carolinense** f. **albiflorum** Benke Map 1924

Known only from Lawrence, Jasper, Newton, and McDonald counties, but of expected occurrence in other sections of the state.

Under certain conditions, animals eating the fruits, stems, and leaves of *Solanum carolinense* are reported to be poisoned.

8. **Solanum Torreyi** Gray f. **Torreyi** Map 1925

*Solanum Torreyi* Gray [G, BB, P & S]

Flowers last of May–September.

Occurs in rocky or sandy open ground and along railroads. Scattered and rare in southern and central Missouri north to St. Louis and Jackson counties.

Ranges from Kansas and Arkansas, south to Oklahoma and Texas; introduced in Missouri.

The ordinary color of the corolla is violet or bluish. A white-flowered form, f. *album* Waterfall, has not been recorded thus far in Missouri.

9. **Solanum elaeagnifolium** Cav. f. **elaeagnifolium** White Horse Nettle Map 1926

Also known as Silverleaf Nettle.

*Solanum elaeagnifolium* Cav. [G, BB, P & S]

Flowers July–September.

Occurs in rocky prairies and pastures, eroded slopes and rocky escarpments, waste ground, fields, roadsides, and railroads. Southern and central Missouri north to St. Louis, Boone, Howard, Saline, and Clay counties.

Ranges from Mexico to Arizona and Texas, north to Kansas and Missouri; introduced east to Florida, Indiana, and Ohio.

Under certain conditions, animals eating the fruits, stems, and leaves of this species are reported to be poisoned.

The ordinary color of the corolla is violet or bluish. A white-flowered form, f. *albiflorum* Cockerell, has not been recorded yet from Missouri.

10. **Solanum rostratum** Dunal Buffalo Bur Map 1927

Also called Kansas Thistle.

Flowers May–October.

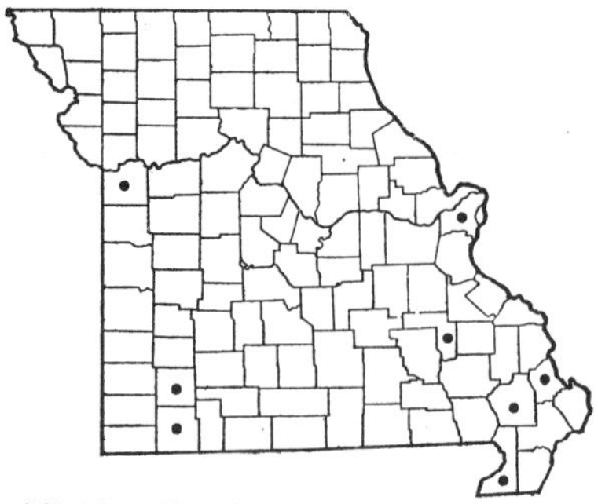

1925 Solanum Torreyi

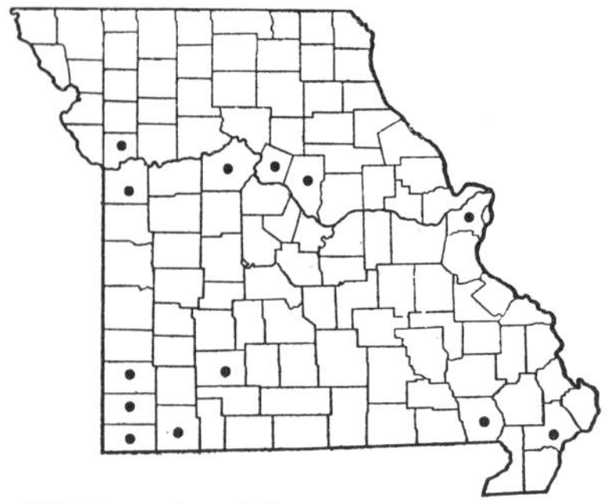

1926 Solanum elaeagnifolium

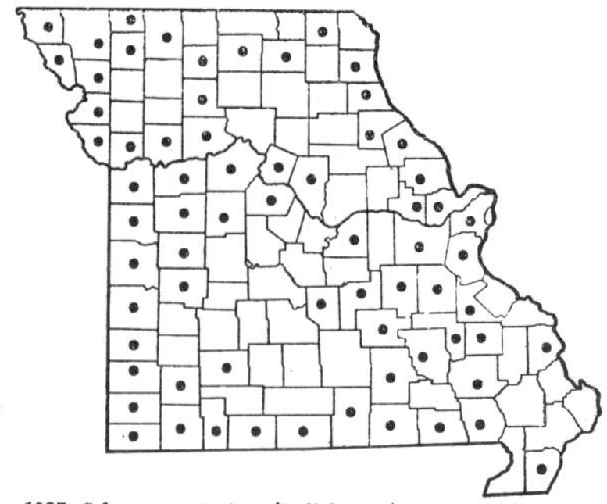

1927 Solanum rostratum (Buffalo Bur)

Occurs in pastures, waste ground, rocky open exposures, alluvial soils, and along roadsides and railroads. Throughout Missouri.

Native from North Dakota to Wyoming and Utah, south to Mexico; introduced in the Middle West and occasionally east to New Hampshire and Florida.

As with some of the other species of the genus, under certain conditions poisoning has been reported from animals feeding on the berries, stems, and leaves of *S. rostratum.* The prickles of the plant may cause mechanical injuries to grazing animals, while the burs, which often cling to the fleece of sheep, cause damage and loss in the value of the wool.

## 2. **Lycopersicon** Mill. Tomato

**Lycopersicon esculentum** Mill. Tomato

Map 1928

Flowers May–September.

Occurs in waste ground, along gravel bars, and along railroads. Occasionally found as a waif scattered in southern and central Missouri, north to St. Louis, Boone, and Jackson counties.

Native of western South America.

Although rarely persisting in the wild state, the tomato is occasionally encountered as an escape into wild habitats, and is therefore included in this flora.

Escaped from cultivation in the United States from Florida to California, north to New York, Illinois, Missouri, and Colorado.

## 3. **Physalis** L. Ground Cherry, Husk Tomato

The most recent studies on this genus have been published by Dr. U. T. Waterfall (Rh. 60: 107–14; 128–42; 152–73. 1958). His work has been partly followed in the present treatment, modified by observations and studies made by the present author.

a. Stems and leaves mainly glabrous (without hairs) or nearly so, the hairs, if present, on the young growth, upper part of stem, and flower- or fruit-stalks (peduncles) . . . . . . . . . . . . . *b*

b. Flower- or fruit-stalk (peduncle) glabrous; corolla without a dark center or dark spots towards base or center . . . . . . . . . . . . . . . . . . . . . . . . 7. P. ANGULATA

b. Flower- or fruit-stalk more or less hairy; corolla with a dark center or dark-spotted around center or base . . . . . . . . . . . . . . . . . . . . . . . . . . *c*

c. Hairs on flower- or fruit-stalk spreading or upwardly curved, but not appressed (pressed closely against or lying parallel to surface) . . . . . . . . . . . . . . . . . . *d*

d. Anthers yellow, 3–5 mm. (rarely 2) long; leaves lanceolate or ovate-lanceolate, thick and firm; perennial plants (when plant is pulled, the root remains in ground) . . . . . 5b. P. LONGIFOLIA var. HISPIDA

Plate no. 314. **1.** Solanum americanum, × 2/7. **2.** Lycopersicon esculentum, × 2/7; Details from Small, The New York Botanical Garden. **3.** Solanum carolinense, × 2/7; Details from Small, The New York Botanical Garden. **4.** Solanum rostratum, × 2/7. **5.** Solanum elaeagnifolium, × 2/7; After Gleason, The New York Botanical Garden. **6.** Physalis pumila, × 2/7. **7.** Physalis heterophylla var. heterophylla, × 2/7.

PLATE NO. 314

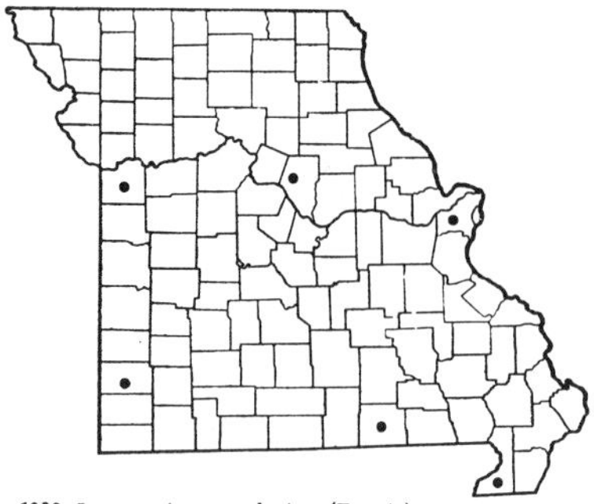

1928 Lycopersicon esculentum (Tomato)

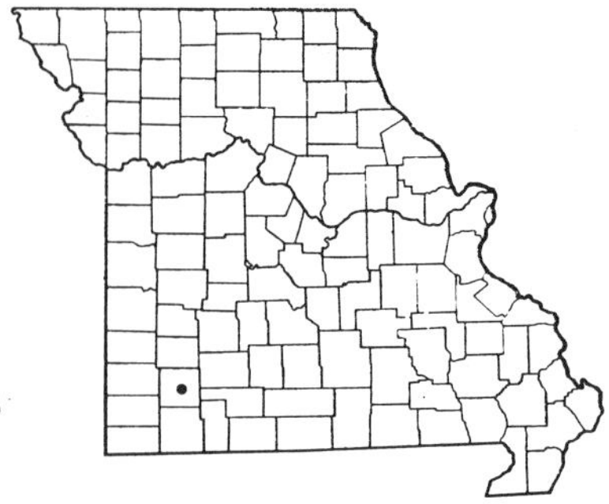

1929 Physalis viscosa subsp. mollis var. cinerascens

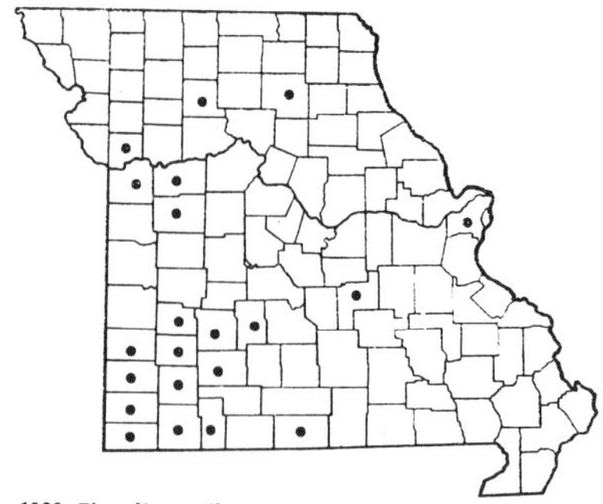

1930 Physalis pumila

d. Anthers bluish, 1.8–2.4 mm. long; leaves ovate or nearly round (suborbicular); annual plants (when plant is pulled from ground, roots are usually pulled out) . . . . . . 8. P. PUBESCENS var. GLABRA

c. Hairs on flower- or fruit-stalk appressed . . . . . . . . . . . . . . . . . . . . *e*

e. Flowering peduncles (stalk) 3–5 mm. long, fruiting peduncles 3–8 mm. long; corolla 7–15 mm. long; annuals (when plant is pulled from ground, roots are usually pulled out); fruiting calyx weakly or obscurely angled; hairs of peduncle usually downwardly appressed or spreading . . . . . . . . . . . . . . . . . . . . . . . 6. P. IXOCARPA

e. Flowering peduncles (stalk) 10–15 mm. long, fruiting peduncles 10–25 mm. long; corolla 15–25 mm. long; perennials (when plant is pulled, the root remains in ground); fruiting calyx more noticeably angled; hairs of peduncle usually upwardly appressed . 5. P. LONGIFOLIA

a. Stems or leaves or both long- or short-hairy . . . . . . . . . . . . . . . . . . . *f*

f. Hairs on stems and leaves forked, branched, or star-shaped (stellate) . . . . . . . . . . *g*

g. Hairs forked or irregularly branched, spreading and mostly 1–2 mm. long, producing a bristly-hairy effect; leaves usually entire (without teeth), rarely slightly and irregularly wavy-margined 2. P. PUMILA

g. Hairs star-shaped (stellate), with a dense soft mat at least on lower surface of leaves, not spreading nor producing a bristly-hairy effect; leaves conspicuously toothed (dentate) . . 1. P. VISCOSA subsp. MOLLIS var. MOLLIS

f. Hairs on stems and leaves mainly simple and unbranched, the branched hairs, if present, very inconspicuous . . . . . . . . . . . . . . . . . . . . . . . . . . *h*

h. Annuals, the slender roots easily pulled from the ground; anthers 1.5–2 (–2.4) mm. long; corolla 6–10 mm. long; flowering stalks (peduncles) 2–7 mm. long . . . . . . . . . . *i*

i. Flower completely yellow without a dark center or without dark spots near base or center; anthers 1–1.2 mm. long; flowering calyx 3–4 mm. long, the calyx-lobes 1–2 mm. long; fruiting calyx 1.5–2.5 cm. long . . . . . . . . . . . . . . . 9. P. MISSOURIENSIS

i. Flower with dark spots or dark center or base; anthers 1.5–2 mm. long; flowering calyx 4–7 mm. long, the calyx-lobes 2–4 mm. long; fruiting calyx 2–4 cm. long . . . 8. P. PUBESCENS

h. Perennials, the stem breaking when pulled from the ground, the thick roots remaining in the ground; anthers 2–4.5 mm. long; corolla 10–15 mm. long; flowering stalks (peduncles) 10–15 mm. long . . . . . . . . . . . . . . . . . . . . . . . . . *j*

j. Hairs on stem usually widely spreading, some of them gland-tipped and sticky, and some of them long-jointed; leaves mostly rounded, somewhat heart-shaped, or broadened at base, if narrowed or tapering usually abruptly so; anthers averages 3.5–4.5 mm. long (rarely 3 mm.); filaments usually broadened gradually from base to summit (clavate) . . 3. P. HETEROPHYLLA

j. Hairs on stem usually reflexed (turned downward), none of them gland-tipped or sticky; leaves mostly narrowed or tapering to the base; anthers averaging 2.5 mm. (rarely 2 or 3) long; filaments not gradually broadened from the base to summit . . . . 4. P. VIRGINIANA

1. **Physalis viscosa** L. subsp. **mollis** (Nutt.) Waterfall var. **cinerascens** (Dunal) Waterfall Map 1929

*Physalis mollis* Nutt. var. *cinerascens* (Dunal) Gray

Flowers May–October.

Occurs in dry open ground along railroads. Known only from Lawrence County (1 mi. south of Verona, October 6, 1953, *Palmer 56949*).

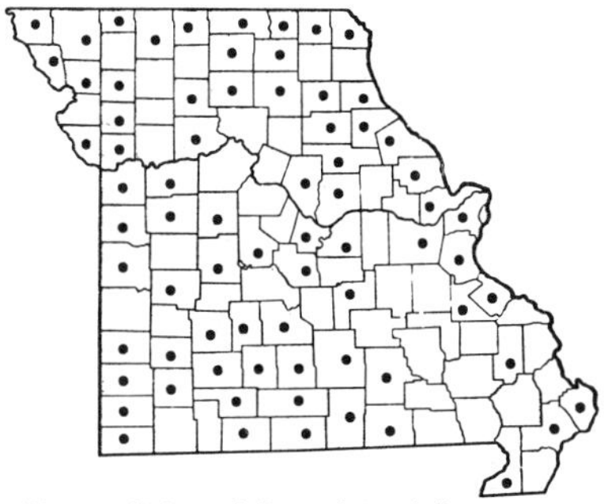

1931 Physalis heterophylla var. heterophylla

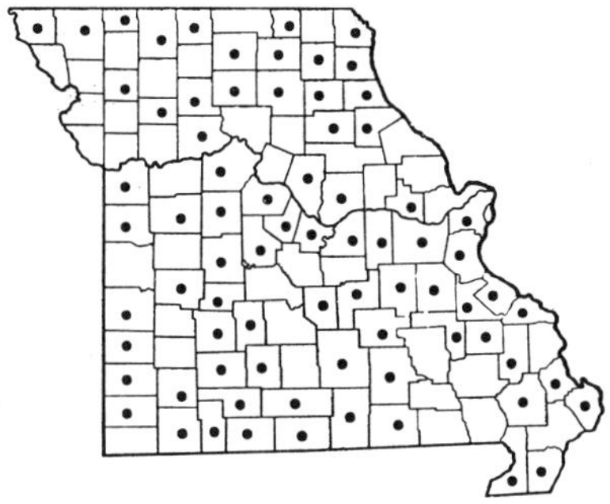

1932 Physalis virginiana

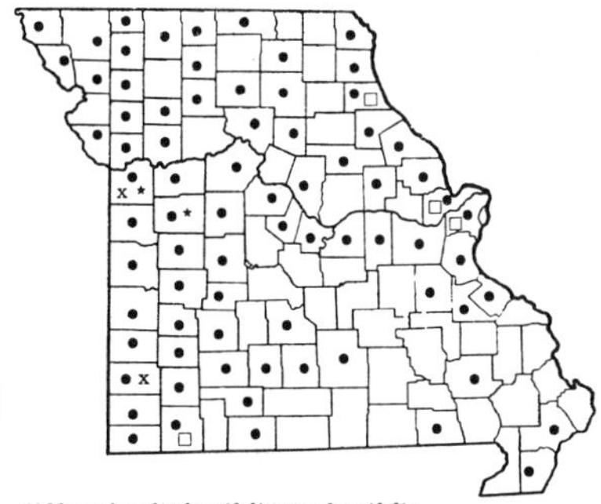

1933 □ Physalis longifolia var. longifolia
1933 • Physalis longifolia var. subglabrata f. subglabrata
1933 x Physalis longifolia var. subglabrata f. macrophysa
1933 ⋆ Physalis longifolia var. hispida

Ranges from Mexico, Texas, Oklahoma, and Arkansas, north to Kansas and Missouri; introduced in Missouri.

This was recently reported by Palmer and Steyermark (Brittonia 10: 118. 1958) as *P. mollis* var. *cinerascens*.

2. **Physalis pumila** Nutt. Map 1930

Flowers late May–August.

Occurs on limestone glades, dry upland prairies, dry or rocky open ground, and along railroads. Mostly in unglaciated prairie region of the southwestern part of the state, north to Macon, Livingston, and Clay counties in northern and central Missouri, and east to St. Louis and Phelps counties.

Ranges from Texas, Oklahoma, and Arkansas, north to Missouri, and Kansas; introduced in Illinois; reported from Iowa and Colorado.

The ripe fruit is edible raw or cooked into a preserve.

3. **Physalis heterophylla** Nees var. **heterophylla** Map 1931

*Physalis heterophylla* Nees [G, P & S, Steyerm.]

*Physalis heterophylla* var. *ambigua* (Gray) Rydb. [G, BB, P & S, Steyerm.]

*Physalis heterophylla* var. *nyctaginea* (Dunal) Rydb. [G, P & S, Steyerm.]

Flowers May–August.

Occurs in prairies, open or rocky wooded slopes, fields, roadsides, waste ground, thickets, open places in upland woods and ridges above rocky slopes, and alluvial sandy soils and gravel bars along streams. Throughout Missouri.

Ranges from Quebec and New Hampshire to Saskatchewan, Idaho, and Utah, south to Georgia and Texas.

The species is a variable one, principally in the amount and length of pubescence, thickness and opaqueness of leaves, and degree and extent of toothing on the leaf-margins. The anthers vary from yellow or green to brown-purple. Following Dr. Waterfall's studies of this species, three variations previously considered as varieties have been combined as var. *heterophylla*. Two specimens in the Missouri Botanical Garden Herbarium from St. Louis County (Allenton, June, 1886, *Kellogg*) and Jefferson County (canyon north of spring, east of Sulphur Spring, October, 1912, *Craig*) resemble *P. viscosa* subsp. *mollis*, but close study of the material proves them to be referable to *P. heterophylla* var. *heterophylla*.

Grazing animals usually avoid this species, but under conditions where forage is scarce, cases of poisoning have been reported from eating large quantities of unripe berries and tops of the plant. The ripe fruit is edible either raw or cooked into a tasty preserve.

4. **Physalis virginiana** Mill. Map 1932

*Physalis virginiana* var. *virginiana* [Waterfall]

Flowers April–July.

Occurs in rocky or dry open woods, dry prairies, glades, thickets, waste ground, and along railroads. Throughout Missouri.

Ranges from Florida to Texas, north to Maine, Quebec, Ontario, Michigan, Wisconsin, Minnesota, and Manitoba, west in the United States to Colorado and Utah.

This is a well-marked and distinct species, in Missouri at least not showing intergradation with other species. Dr. Waterfall, however, has placed, as varieties under *P. virginiana*, the wholly different *P. subglabrata*, *P. longifolia*, and *P. lanceolata* which have quite a uniformly different type of pubescence than is found in typical *P. virginiana*. As noted in the key, *P. virginiana* has the hairs of the stem reflexed or retrorse, whereas the group of plants referred to *P. longifolia*, including what have been known as *P. subglabrata* (and Dr. Waterfall's *P. virginiana* var. *sonorae*) and *P. lanceolata* (Dr. Waterfall's *P. virginiana* var. *hispida*) have upwardly appressed hairs on the upper part of stem, flowering and fruiting peduncles.

I am retaining in the present flora, therefore, *P. virginiana* as a species easily recognized and sufficiently distinct from *P. longifolia*. Further discussion will be found under the next species, *P. longifolia*.

The fruit of this species is edible either fresh or cooked into a tasty preserve.

5. **Physalis longifolia** Nutt. Map 1933
Flowers May–September.

Occurs in rich or low bottom woods, in ravines, base of slopes, alluvial soils along streams, ponds, sloughs, and ditches, thickets, meadows, fallow and cultivated fields, pastures, roadsides, and along railroads. Throughout Missouri.

The following variations occur in Missouri:

a. Anthers yellow . . . . . . . . . . *b*
 b. At least some of the hairs stiff, spreading, and about 1 mm. long on the calyx, flower-buds, margins, or some part of leaf-blade, or stems; leaves ovate-lanceolate or lanceolate or spatulate-lanceolate, more or less hairy . . . 5b. P. LONGIFOLIA var. HISPIDA
 b. Hairs all appressed (pressed against or lying parallel to surface), and short, mostly 0.5 mm. or less long; leaves usually lanceolate to lanceolate-linear, glabrous or nearly so . 5a. P. LONGIFOLIA var. LONGIFOLIA
a. Anthers gray-lilac, blue, or purple . . . *c*
 c. Fruiting calyx 2.5–3.5 cm. long, 2–3 cm. broad . . 5c. P. LONGIFOLIA var. SUBGLABRATA f. SUBGLABRATA
 c. Fruiting calyx 4–5 cm. long, 3–4 cm. broad . . 5d. P. LONGIFOLIA var. SUBGLABRATA f. MACROPHYSA

5a. **Physalis longifolia** var. **longifolia** Map 1933
*Physalis longifolia* Nutt. [G, BB, P & S]
*Physalis virginiana* Mill. var. *sonorae* (Torr.) Waterfall [Waterfall]

Known from Marion (old gardens, 'Douglastown,' Hannibal, September 14, 1916, *Davis 1874*; Mississippi bluffs near Hannibal, September 14, 1916, *Davis 1874*; dry bluffs, suburbs of Hannibal, September 14, 1916, *Davis 4781*), St. Charles (Matoon, October 21, 1897, *Trelease 505*), St. Louis, and Barry (fields, Eagle Rock, June 27, 1897, *Bush 165*) counties.

Ranges from Oregon to California, Wyoming and Utah to Arizona and Mexico, east to Iowa, Missouri, and Arkansas; also in Pennsylvania, Virginia, West Virginia, North Carolina, Tennessee, Indiana, Illinois, and Wisconsin.

There is little if any difference, except in anther color, between var. *longifolia* and var. *subglabrata*, both having glabrous or glabrate leaf surfaces and similar pubescence when hairs are present. The leaves of var. *longifolia* average somewhat more elongated and narrower than those of var. *subglabrata*

5b. **Physalis longifolia** var. **hispida** (Waterfall) Steyermark Map 1933
*Physalis virginiana* Mill. var. *hispida* Waterfall, Rh. 60: 154–56. 1958.
*Physalis lanceolata* of auth., not Michx. [G, BB, P & S, Steyerm.]

Known from west-central Missouri in Jackson (Sheffield, June 3, 1896, *Bush 291*; August 29, 1906, *Bush 4094*; Atherton, June 17, 1908, *Bush 4970*; Courtney, August, 1900, *Bush 829*; Sheffield, May 23, 1896, *Mackenzie*) and Johnson (*Stevens 4166*) counties.

Ranges from Minnesota and Nebraska to Colorado and Utah, south to Oklahoma, Texas, and New Mexico; introduced in Indiana and Missouri.

In its dwarf habit and leaf-shape this variety closely resembles *P. pumila*, but the latter species has forked or irregularly branched pubescence instead of the simple hairs found in var. *hispida*. The var. *hispida* suggests a possible hybrid or a hybrid having originated between *P. longifolia* and *P. pumila*.

5c. **Physalis longifolia** var. **subglabrata** (Mackenz. & Bush) Steyermark f. **subglabrata** Map 1933
*Physalis subglabrata* Mackenz. & Bush, Trans. Acad. Sci. St. Louis 12: 86–87. 1902 [G, P & S, Steyerm.]
*Physalis longifolia* in part [of BB]
*Physalis virginiana* var. *subglabrata* (Mackenz. & Bush) Waterfall f. *subglabrata*, Rh. 60: 152. 1958 (as *P. virginiana* var. *subglabrata*)

This is the common variety found throughout Missouri.

Ranges from Georgia to Texas and New Mexico, north to Vermont, New York, Ontario, Michigan, Illinois, Wisconsin, Iowa, and Nebraska; also in Colorado, Utah, Idaho, and Oregon.

---

Plate no. 315. **1.** Physalis longifolia var. subglabrata, × 2/5. **2.** Physalis virginiana, × 2/5. **3.** Physalis pubescens, × 2/5. **4.** Physalis ixocarpa, × 2/5. **5.** Physalis virginiana var. hispida, × 2/5. **6.** Physalis heterophylla, × 2/5; *a.* Fruiting calyx, × 2/5. All details from Small, The New York Botanical Garden.

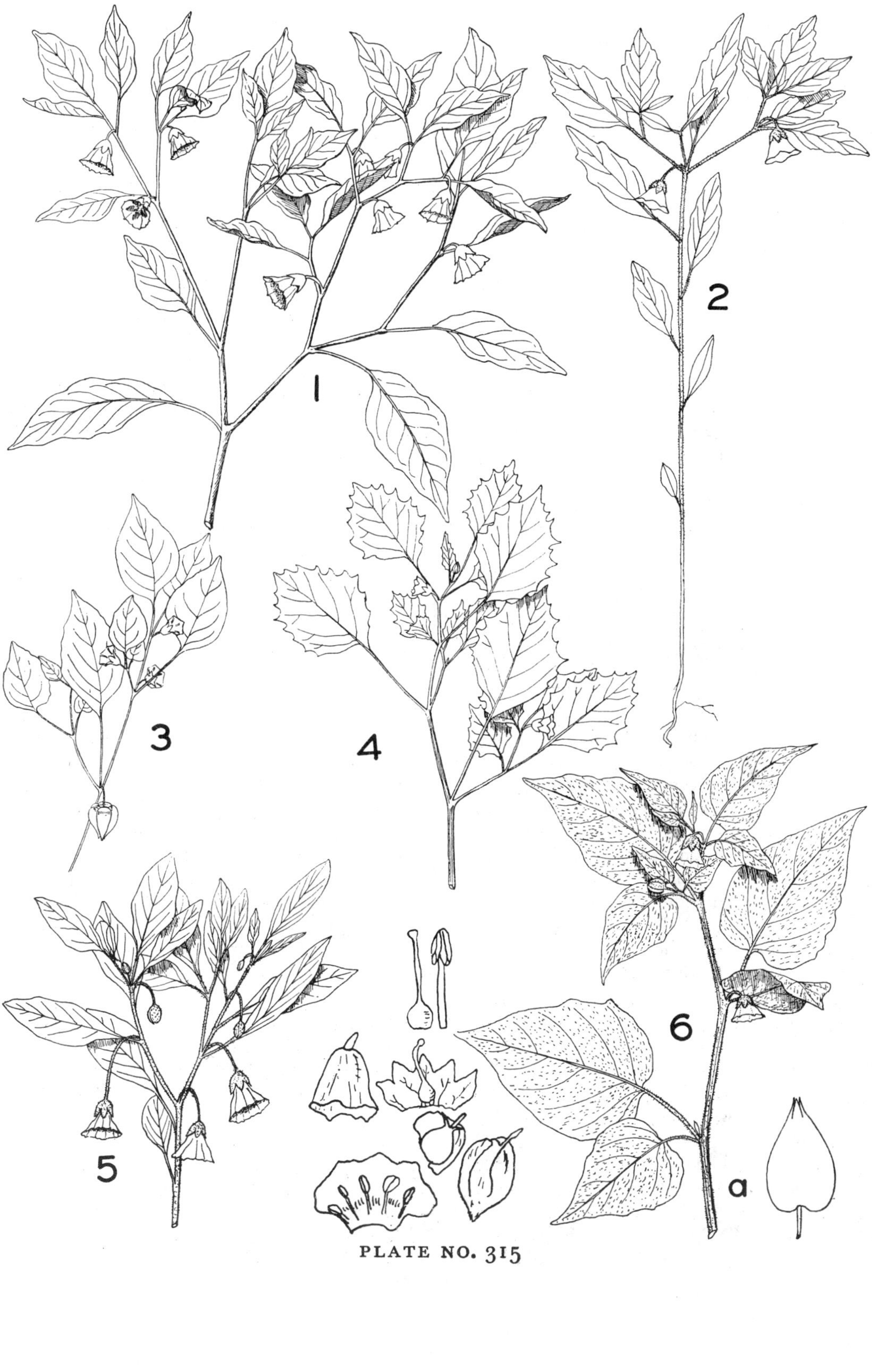


PLATE NO. 315

1934 Physalis ixocarpa (Tomatillo)

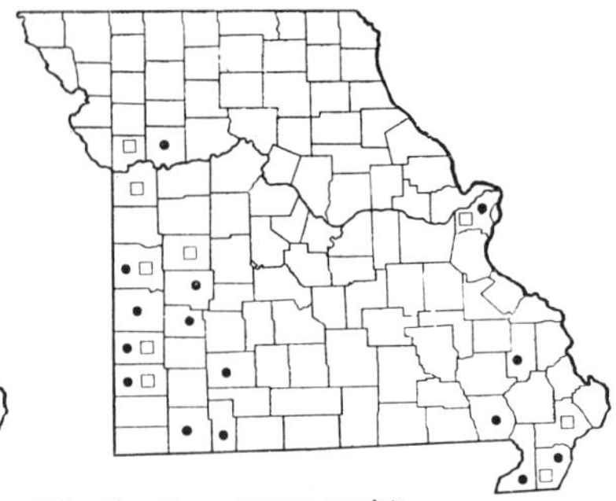

1935 ● Physalis angulata var. angulata
1935 □ Physalis angulata var. pendula

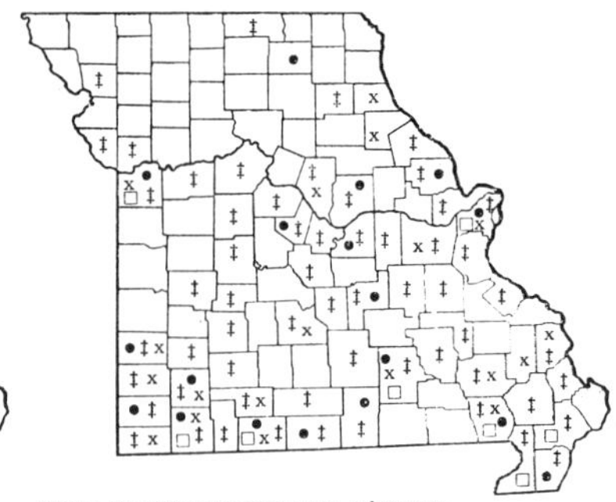

1936 ● Physalis pubescens var. pubescens
1936 □ Physalis pubescens var. glabra
1936 ‡ Physalis pubescens var. integrifolia
1936 x Physalis pubescens var. grisea

5d. **Physalis longifolia** var. **subglabrata** f. **macrophysa** (Rydb.) Steyermark

Map 1933

*Physalis macrophysa* Rydb., Bull. Torr. Bot. Club 22: 308. 1895 [G, P & S]

*Physalis virginiana* var. *subglabrata* f. *macrophysa* (Rydb.) Waterfall, Rh. 60: 153. 1958 (as *P. virginiana* f. *macrophysa*)

Known only from western Missouri in Jackson (*Bush 12483* and *12483A*) and Jasper (low open ground along Spring River, near Neck City, July 1, 1920, *Palmer 18171*) counties.

Ranges from Arkansas to Texas, north to Nebraska, Indiana, and Iowa; also in Illinois, Michigan, and New Jersey.

In addition to the larger fruiting calyx, the leaves of f. *macrophysa* are often more menbranaceous and more coarsely toothed than those of f. *subglabrata*.

In Dr. Waterfall's treatment the *P. longifolia* group was combined with *P. virginiana*. In the present flora these taxa are retained as distinct species, *P. virginiana*, as represented by Missouri material, having the hairs on the stem, petioles, and peduncles reflexed or retrorse, and *P. longifolia* (including what has been known as *P. subglabrata* and *P. lanceolata*) having the hairs on stem, petioles, and peduncles upwardly appressed or ascending. Both Dr. Waterfall's *P. virginiana* var. *hispida* and var. *sonorae* have the same type of upwardly appressed or ascending hairs on the peduncles, petioles, and upper part of the stems as are shown by *P. subglabrata* and *P. macrophysa*, so that all seem properly referable to the same complex. It has, therefore, seemed in keeping with a more natural classification of the species of *Physalis*, as at least encountered in Missouri, and apparently elsewhere in the range of variation, to retain *P. virginiana* as a separate species from *P. longifolia* and its varieties, including var. *subglabrata* and var. *hispida*. In Missouri both *P. virginiana* and *P. longifolia* stand out, both in field and herbarium and under growing conditions, as distinct and readily recognizable, and are, therefore, treated accordingly in the present flora.

Some cases of poisoning have been reported for *P. longifolia* var. *subglabrata* f. *subglabrata* from grazing animals which have consumed large quantities of the unripe berries and upper parts of the stems. The ripe berries may be cooked into a tasty preserve in this species and in others of the genus.

6. **Physalis ixocarpa** Brotero Tomatillo

Map 1934

Flowers July–September.

Occurs in waste ground. Known only from Jackson County in west-central Missouri (Courtney, September 30, 1921, *Bush 9698, 9698A*; July 25, 1926, *Bush 10539, 10540*; September 8, 1927, *Bush 11536*).

Native of Mexico; cultivated in the southwest United States, escaped from cultivation and naturalized from California, Arizona, New Mexico, and Texas to Virginia, north to Massachusetts, Quebec, Pennsylvania, Minnesota, Oregon, and Washington.

The plant is sometimes cultivated for the large, sticky, purplish berry, which completely fills the enlarged fruiting calyx.

7. **Physalis angulata** L. Map 1935

Flowers May–September.

Occurs in alluvial soils along streams and in valleys, waste and cultivated ground, and along railroads.

Two variations occur in Missouri:

Flower-stalks (peduncles) mainly 5–15 mm. long, up to 3 times as long as length of calyx; fruit-stalks (peduncles) mainly 20–30 mm. long, shorter than to equaling the fruiting calyx; fruiting calyx 25–35 mm. long; flowering calyx mostly

4–5 mm. long, the teeth 2–2.5 mm. long . . . 7a. P. ANGULATA var. ANGULATA

Flower-stalks (peduncles) mainly 15–40 mm. long, 3–13 times as long as length of calyx; fruit-stalks (peduncles) 20–45 mm. long, equaling to 3 times as long as fruiting calyx; fruiting calyx 20–25 mm. long; flowering calyx mainly 3 mm. (sometimes 4) long, the teeth about 1 mm. long . 7b. P. ANGULATA var. PENDULA

7a. **Physalis angulata** var. **angulata** Map 1935

*Physalis angulata* L. [G, BB, P & S]

Scattered in southern and central Missouri north to St. Louis and Ray counties.

Ranges from Florida to Texas, north to Virginia and Oklahoma; introduced north to Connecticut, Illinois, and Missouri; also in tropical America.

7b. **Physalis angulata** var. **pendula** (Rydb.) Waterfall Map 1935

*Physalis pendula* Rydb. [G, BB, P & S]

Scattered in southern and central Missouri north to St. Louis and Clay counties.

Ranges from Louisiana and Texas, north to Illinois, Missouri, Kansas, and Oklahoma.

The ripe fruit is edible, raw or cooked.

8. **Physalis pubescens** L. Map 1936

Flowers May–November.

Occurs on sand and gravel bars and alluvial soils bordering streams, moist ledges along bluffs, rocky woods, open fallow and cultivated fields, waste ground, and along railroads.

The following variations are encountered in the state, the key and classification following that of Dr. Waterfall's studies (Rh. 60: 128, 132–33; 164–68. 1958).

a. Plants glabrous (without hairs) or sparsely short-hairy . . . 8b. P. PUBESCENS var. GLABRA

a. Plants with long and conspicuous hairs, sometimes with shorter hairs or glands intermixed . *b*

b. Leaves gray-green, with usually sessile (stalkless) glands on surface; leaf-blades toothed nearly to the base . 8d. P. PUBESCENS var. GRISEA

b. Leaves grass-green or dull green, not gray-green, without any sessile glands on surface; leaf-blades toothed or entire (without teeth) . . . . . . . . . . . *c*

c. Leaves toothed nearly to the base with 5–8 teeth on each side; leaf-blade seldom translucent . . . 8a. P. PUBESCENS var. PUBESCENS

c. Leaves with 3–4 teeth on each side or entire (without teeth); leaf-blades thin and translucent . . . 8c. P. PUBESCENS var. INTEGRIFOLIA

8a. **Physalis pubescens** var. **pubescens** Map 1936

*Physalis barbadensis* Jacq. [G, BB, P & S]

Scattered in southern and central Missouri, north to Lincoln, Callaway, and Jackson counties, and locally north in Adair County.

Native of tropical America; introduced and naturalized from Florida to Texas and Mexico, north to Pennsylvania, Illinois, and Iowa.

8b. **Physalis pubescens** var. **glabra** (Michx.) Waterfall Map 1936

*Physalis barbadensis* var. *obscura* (Michx.) Rydb. [P & S]

*Physalis barbadensis* var. *glabra* (Michx.) Fern. [G]

Scattered in southern and central Missouri, north to St. Louis and Jackson counties.

Ranges from Florida to Texas, north to North Carolina and Missouri; introduced in Pennsylvania and California.

8c. **Physalis pubescens** var. **integrifolia** (Dunal) Waterfall Map 1936

*Physalis pubescens* of most auth., not L. [G, BB, P & S]

This is the commonest variety in Missouri, found throughout the southern and central portions of the state, north to Pike, Shelby, Putnam, and Andrew counties.

Ranges from Florida to Texas, north to Massachusetts, Pennsylvania, Ohio, Indiana, Illinois, Wisconsin, Iowa, and Kansas; also in California.

8d. **Physalis pubescens** var. **grisea** Waterfall Map 1936

*Physalis pruinosa* of auth. [G, BB, P & S, Steyerm.], probably not L.

Southern and central Missouri, north to Marion, Boone, and Jackson counties.

Ranges from Florida to Texas, north to Maine, Vermont, Ontario, New York, Michigan, Wisconsin, Iowa, and Kansas; also introduced from California to Washington.

The ripe fruit is edible raw, or may be cooked into a tasty preserve.

9. **Physalis missouriensis** Mackenz. & Bush Map 1937

Flowers June–October.

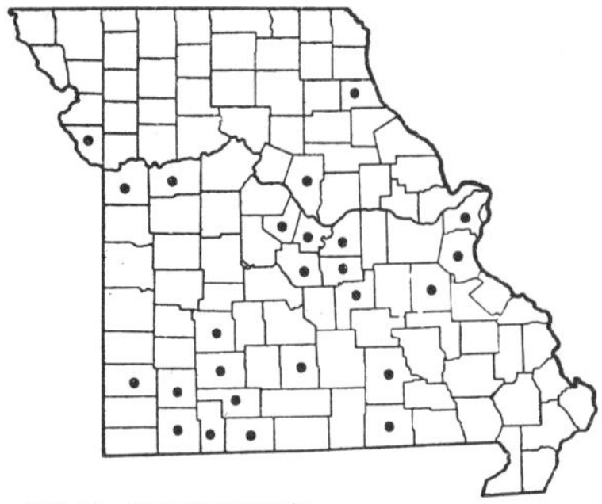

1937 Physalis missouriensis

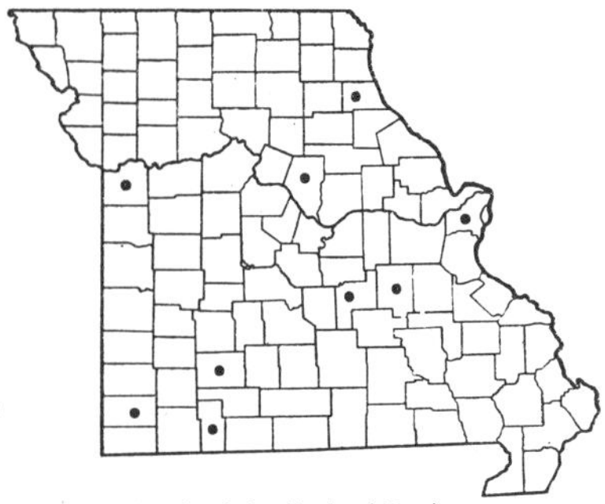

1938 Nicandra physalodes (Apple-of-Peru)

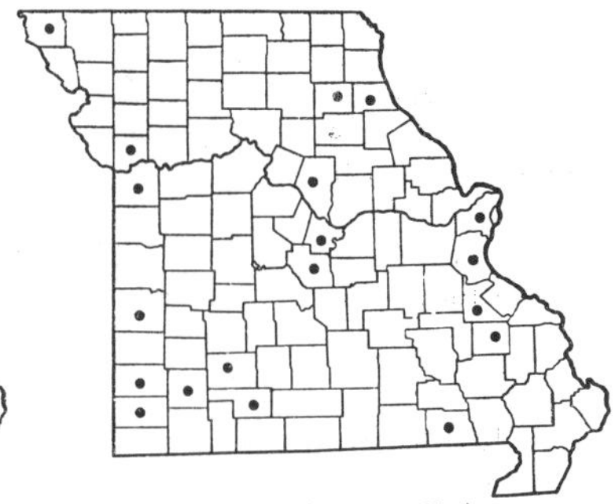

1939 Lycium halimifolium (Matrimony Vine)

Occurs in rocky or open woods, along wooded bluffs, talus slopes, and glades, usually on limestone strata. Southern and central Missouri, north to St. Louis, Boone, and Platte counties, locally northeast in Marion County; holotype from Jackson County (rocky soil, Red Bridge, *Mackenzie 485*).

Ranges from Arkansas and Texas, north to Missouri and Kansas; introduced in Massachusetts.

*Excluded Species*

**Physalis lobata** Torr.

A specimen belonging to this species supposedly collected from St. Louis County in the Missouri Botanical Garden Herbarium bears the data 'Allenton, July 28, 1885, *J. H. Kellogg*,' but the locality has been questioned on the label. Since there is doubt as to the authenticity of the locality in question, the species is excluded. The species is found in adjacent Kansas and Oklahoma.

**Physalis peruviana** L. Cape Gooseberry

This species was included in Palmer & Steyermark's *Annotated Catalogue* and in Steyermark's *Spring Flora* on the basis of two specimens from St. Louis County (Allenton, July 1, 1894, *Letterman*) in the Missouri Botanical Garden Herbarium. Both have large, nearly entire leaves, but the specimens can be matched by material of *P. heterophylla* and are here referred to that species.

## 4. **Nicandra** Adans. Apple-of-Peru

**Nicandra physalodes** (L.) Pers. Apple-of-Peru
Map 1938

Also called Shoo-Fly Plant.

Flowers July–September.

Occurs in cultivated and fallow fields and waste ground. Scattered in southern and central Missouri to Marion, Boone, and Jackson counties.

Native of Peru; escaped from cultivation and naturalized in North America from Nova Scotia and Ontario south to Florida, Indiana, and Missouri.

The large blue flowers are 2.5–3.5 cm. long. This plant at one time was grown exclusively for eradicating house flies, and was known as Fly-Poison Plant. The green stems of the plant were beaten into a pulp and moistened with a little milk. The flies attracted to this mixture were reported to die within half an hour after sampling this food.

## 5. **Lycium** L. Matrimony Vine

Leaves lanceolate or oblanceolate; flower-stalks (pedicels) 10–23 mm. long; calyx-lobes obtuse (blunt); corolla-lobes shorter than corolla-tube . . . . . . . . . . . . . . . . . . 1. L. HALIMIFOLIUM

Leaves elliptic, ovate, or rhombic-ovate; flower-stalks (pedicels) 3–12 mm. long; calyx-lobes acute; corolla-lobes longer than the corolla-tube. . . . . . . . . . . . . . . . . . 2. L. CHINENSE

Plate no. 316. **1.** Physalis pubescens var. grisea, $\times {}^{2}/_{5}$. **2.** Nicandra physalodes, $\times {}^{2}/_{5}$; After Gleason, details from Small, The New York Botanical Garden. **3.** Datura Stramonium var. Tatula, $\times {}^{2}/_{5}$; Details from Small, The New York Botanical Garden. **4.** Lycium halimifolium, $\times {}^{2}/_{5}$; *a.* Flowering branch; *b.* Fruiting branch; Details from Small, The New York Botanical Garden.

PLATE NO. 316

1940 Lycium chinense (Chinense Matrimony Vine)

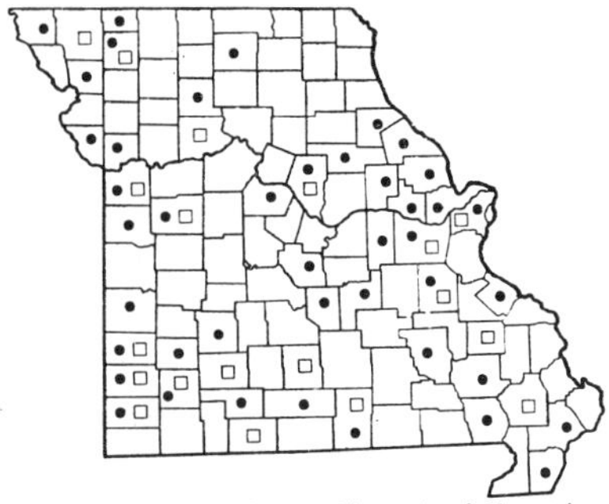

1941 ● Datura Stramonium var. Stramonium f. Stramonium (Jimson Weed)

1941 □ Datura Stramonium var. Tatula

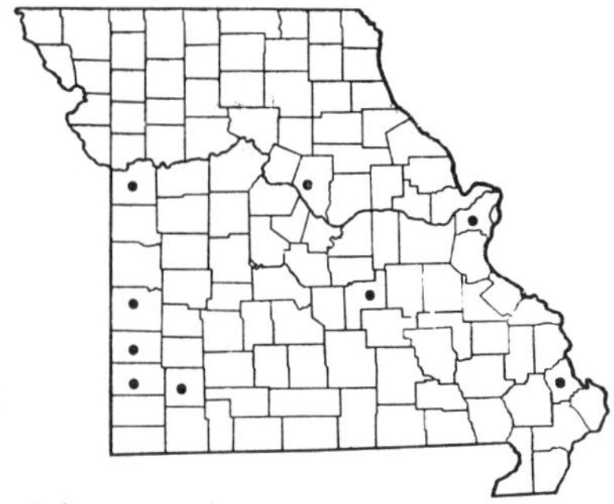

1942 Datura innoxia

1. **Lycium halimifolium** Mill. Matrimony Vine Map 1939

Flowers May–September.

Occurs about old dwellings, old fields, waste ground, roadsides and along railroads. Scattered throughout Missouri.

Native of Europe; introduced and naturalized in North America from Ontario to Minnesota, south to Virginia, Missouri, Kansas, and Oklahoma.

It is reported that the young shoots and leaves of this species, when eaten, may poison sheep and cattle. In Europe and Asia the young leaves are occasionally eaten as a cooked vegetable.

2. **Lycium chinense** Mill. Chinese Matrimony Vine Map 1940

Flowers June–October.

Occurs in waste ground. Known only from Lawrence County, southwestern Missouri (waste open ground along Clear Creek, Pierce City, August 22, 1950, *Palmer 50766*).

Native of China; introduced and naturalized from Massachusetts to Michigan, south to Virginia, Louisiana, and Oklahoma.

This is sometimes planted for its ornamental fruit. In the specimen cited above, some of the calyx-lobes are acute, and others vary from obtuse to rounded.

## 6. **Datura** L. Jimson Weed, Thorn Apple

Stem green or purple, glabrous (without hairs); leaves dark green, glabrous, coarsely toothed; corolla 7–10 cm. long; fruit erect . . . . . . . . . . . . . . . . . . . . 1. D. STRAMONIUM

Stem grayish, finely short-hairy; leaves gray-green, finely short-hairy, entire (without teeth) or with slightly wavy or inconspicuously low-toothed margins; corolla 15–20 cm. long; fruit nodding . 2. D. INNOXIA

1. **Datura Stramonium** L. Jimson Weed Map 1941

Flowers late May–October.

Occurs in pastures, barnyards, fields, waste and cultivated ground, along roadsides, railroads, and rocky open places.

Two variations are found in Missouri:

Stem green; corolla white; lower prickles of the fruit mostly shorter than upper ones . . . . 1a. D. STRAMONIUM var. STRAMONIUM f. STRAMONIUM

Stem purple or purple-tinged; corolla lavender or pale violet; prickles of fruit often nearly equal . 1b. D. STRAMONIUM var. TATULA

1a. **Datura Stramonium** var. **Stramonium** f. **Stramonium** Map 1941

*Datura Stramonium* L. [G, BB, P & S]

Throughout Missouri.

Native probably of tropical America; introduced and naturalized in North America from Nova Scotia to Minnesota, south to Florida and Texas; Colorado to California; introduced in Eurasia and Africa.

A form without prickles on the fruit is known as *D. Stramonium* var. *Stramonium* f. *inermis* (Juss.) Hupka, but has not been recorded from Missouri.

1b. **Datura Stramonium** var. **Tatula** (L.) Torr. Map 1941

*Datura Tatula* L. [P & S]

Scattered throughout Missouri.

Ranges from tropical America, where native, to the southern United States from Florida to California north in the northern United States.

Children have been poisoned by eating the unripe seeds, which are especially poisonous, and sheep,

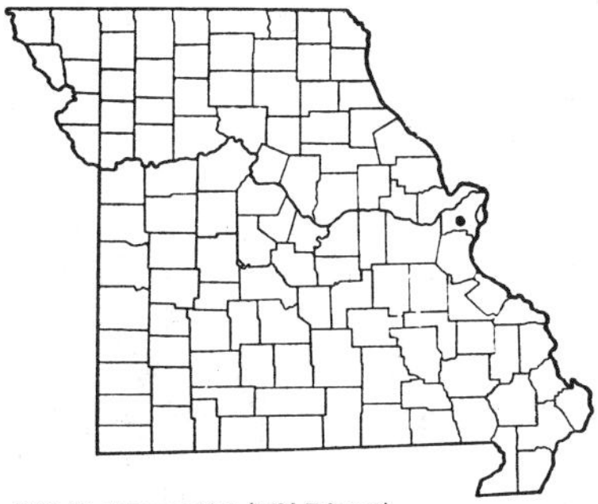

1943 Nicotiana rustica (Wild Tobacco)

1944 Nicotiana longiflora

1945 Petunia axillaris (Petunia)

horses, and cattle are poisoned from eating the plant. Some persons are sensitive to the plant, acquiring a dermatitis upon contact with the leaves or flowers.

One of the constituents found in belladonna (*Atropa Belladonna*) occurs also in *Datura Stramonium* and *D. innoxia*. *Datura Stramonium* has been used since early times as an internal medicine, principally as a belladonna substitute in treatment of asthma, and also for its narcotic properties.

2. **Datura innoxia** Mill. Map 1942

*Datura Metel* [of BB, P & S], not L.

Flowers August–October.

Occurs in waste ground, along roadsides and railroads. Scattered in southern and central Missouri north to St. Louis, Boone, and Jackson counties.

Native of tropical America; introduced and naturalized in the United States from Florida north to New York, Illinois, and Minnesota, west to Kansas.

This species has narcotic properties similar to those of *D. Stramonium* and has been used in medicine by the Aztec Indians as well as by the Indians of Mexico.

## 7. **Nicotiana** L. Tobacco

Corolla 1.5–2 cm. long, yellow or yellow-green, cylindrical, $\frac{2}{3}$ longer than the calyx; leaves ovate, all on petioles (stalks) . . . . . . . . . . . . . . . . . . . . . . . . . . 1. N. RUSTICA

Corolla 7–11 cm. long, white or greenish-white turning purplish, funnel-form, many times longer than the calyx; leaves sessile (without stalk), the middle and lower leaves narrowed below the middle to the base . . . . . . . . . . . . . . . . . . . . . . . . . . 2. N. LONGIFLORA

1. **Nicotiana rustica** L. Wild Tobacco Map 1943

Flowers July–October.

Occurs in waste and cultivated ground. Known only from St. Louis County (St. Louis, July 8, 1910, *Sherff 335*).

Native of Peru; occasionally naturalized in North America from Florida to New Mexico, north to Massachusetts, New York, Ontario, and Minnesota.

This plant possesses alkaloids and poisonous properties similar to those of *N. Tabacum* L., the cultivated tobacco plant. Before the settlement of the United States by white man, this species was smoked by Indians of the eastern United States. It is still smoked locally, but is used chiefly as a source of nicotine for insecticides.

2. **Nicotiana longiflora** Cav. Map 1944

Flowers June–September.

Occurs in waste and cultivated ground, sometimes on dump heaps. Known from St. Louis (ballast ground near Mississippi River, about 7500 south, St. Louis, August 25, 1933, *Steyermark 8966*) and Jackson counties.

Native of tropical America. Occasionally escaped from cultivation and naturalized in parts of the eastern United States.

## 8. **Petunia** Juss. Petunia

Corolla white, 5–6.5 cm. long, the expanded portion 4–6 cm. broad; corolla-tube narrow, abruptly widened at summit; calyx 1.2–2 cm. long. . . . . . . . . . . . . . . . . . . . 1. P. AXILLARIS

Corolla purple, 2.5–4 cm. long, the expanded portion 3–4 cm. broad; corolla-tube funnel-form, gradually widened to summit; calyx 0.7–1.5 cm. long . . . . . . . . . . . . . . 2. P. VIOLACEA

1. **Petunia axillaris** (Lam.) BSP. Map 1945

Flowers July–September.

Occurs in waste ground, dump heaps, and along railroads. Known from Marion (railroad banks, Collier St., Hannibal, *Davis 7517*) and St. Louis (dumping ground between right-of-way of Wabash R.R. and the Waterworks conduit track, north of Adelaide Ave., St. Louis, August 6, 1955, *Muehlenbach 707*) counties.

Native of Argentina; escaped from cultivation in old gardens and occasionally naturalized in the United States from New York and Pennsylvania to Illinois and Missouri.

This is considered to be one of the parent species of the common cultivated garden petunia (*Petunia* X *atkinsiana* D. Don [*P. hybrida* Vilm.]).

2. **Petunia violacea** Lindl. Map 1946

Flowers July–September.

Occurs in waste ground and along railroads. Known from St. Louis (*Muehlenbach 338*), Boone (waste places, escaped, Columbia, August, 1904, *Daniels*), and Jasper (*Palmer 56463*) counties.

Native of Argentina; escaped from cultivation and occasionally naturalized in the United States from Pennsylvania, Illinois, and Missouri southward.

This is considered to be one of the parent species of the common cultivated garden petunia (*Petunia* X *atkinsiana* D. Don [*P. hybrida* Vilm.]).

Plate no. 317. **1.** Datura innoxia, × 3/5; After Gleason, details from Small, The New York Botanical Garden. **2.** Petunia axillaris, × 4/5; After Gleason, The New York Botanical Garden. **3.** Nicotiana rustica, × 2/5; After Britton and Brown, details from Small, The New York Botanical Garden. **4.** Nicotiana longiflora, × 2/5; Details from Small, The New York Botanical Garden.

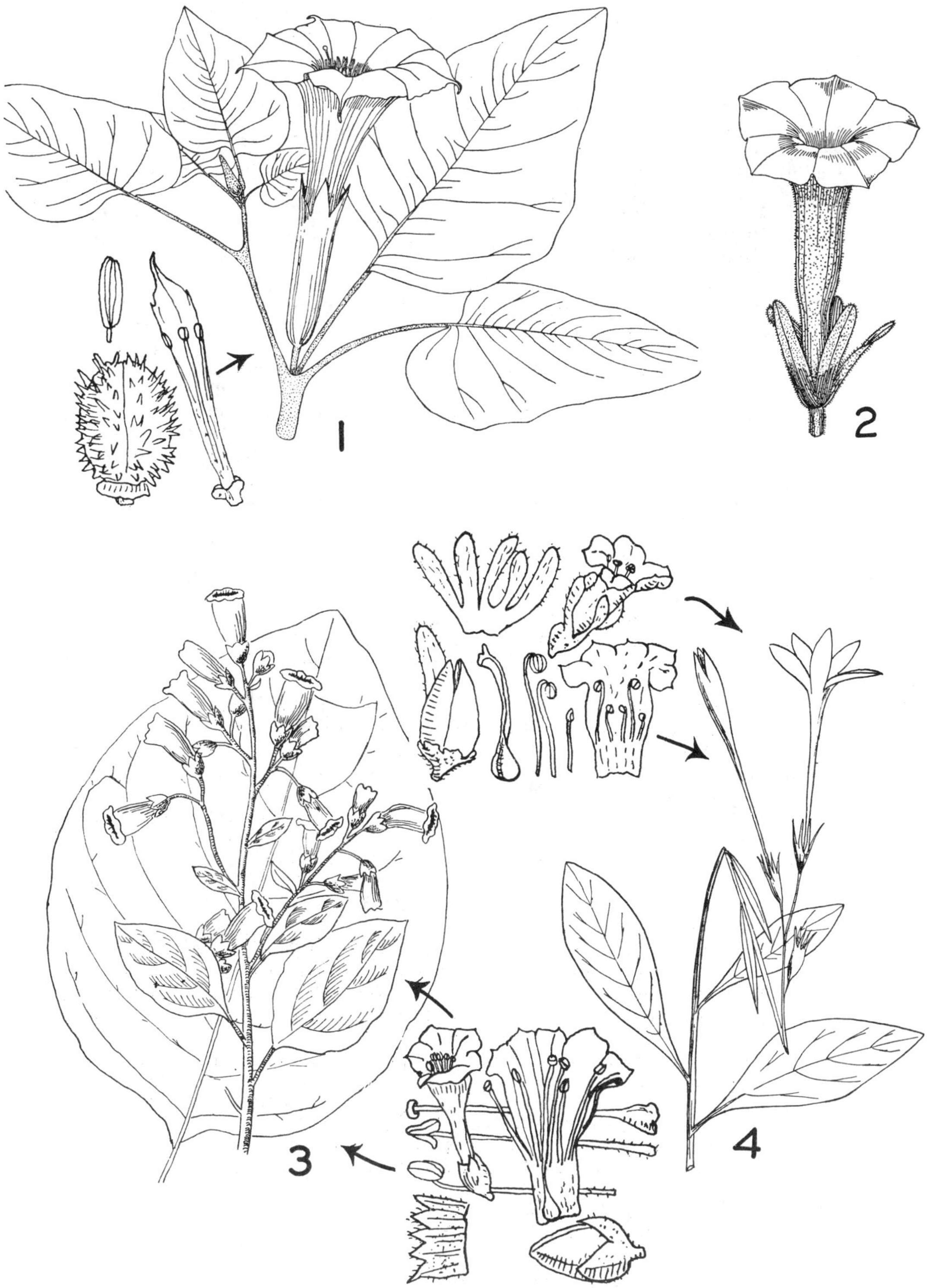


PLATE NO. 317

## Fam. **SCROPHULARIACEAE** (Figwort Family)

The sequence of genera follows the treatment of Pennell (Monog. no. 1. Acad. Nat. Sci. Phila. 650 pp. 1935).

a. Trees with large broadly ovate leaves 15–25 cm. long and 15–40 cm. broad, entire (without teeth) or slightly 3–5-angled; corolla 5–6 cm. long, violet with yellow stripes . . . . . . . 1. PAULOWNIA
a. Herbs or non-woody plants . . . . . . . . . . . . . . . . . . . . . . . . *b*
b. Leaves in a tuft all arising at the base of the plant; a solitary flower at the tip of a leafless peduncle (stalk) less than 2 cm. long and much shorter than the leaves . . . . . . . . . 8. LIMOSELLA
b. Some or all of the leaves occurring on the main stem and branches; flower not as above. . . . . *c*
c. At least some (upper) of the main leaves of the stem alternate (do not consider those at the base of or accompanying flowers) . . . . . . . . . . . . . . . . . . . . *d*
d. Stem completely glabrous or the main part of it glabrous (without hairs) . . . . . . . . *e*
e. Anther-bearing stamens 5; corolla nearly symmetrical, with a short tube, about as broad as or broader than long, the 3 lower lobes only slightly larger than the 2 upper ones . 9. VERBASCUM BLATTARIA
e. Anther-bearing stamens 4; corolla quite irregular, with an elongated tube, much longer than broad, divided into a lower and an upper lip . . . . . . . . . . . . . . *f*
f. Leaves roundish, with 5–9 lobes or rounded teeth, main nerves all arising from the base of the blade; plant trailing or twining . . . . . . . . . . 16. CYMBALARIA
f. Leaves linear or lanceolate, entire (without teeth), some of the nerves arising from various levels of the midrib; stems erect or ascending . . . . . . . . . . . *g*
g. Leaves linear, at most 7 mm. broad; corolla possessing a tail-like spur at the base 14. LINARIA
g. Larger leaves lanceolate, 7–13 mm. broad; corolla without any spur at the base 17. ANTIRRHINUM
d. Stem more or less hairy . . . . . . . . . . . . . . . . . . . . . . *h*
h. Flowers solitary, along the sides or near the top of the stem, 1 or 2 in the axil of a leaf or leaf-like bract; corolla possessing a tail-like spur at the base . . . . . . . . . . *i*
i. Leaves mostly linear or narrowly linear-lanceolate, 5–10 times as long as broad; stem erect; fruit hairy . . . . . . . . . . . . . . . . . . 18. CHAENORRHINUM
i. Leaves ovate, oblong, or roundish, as broad as long or not more than twice as long as broad; stems sprawling, widely spreading, or lying along the ground; fruit glabrous (without hairs) . . . . . . . . . . . . . . . . . . . . . . 15. KICKXIA
h. Flowers several to many clustered in an inflorescence at the top of the stem or branches; corolla without any spur at the base . . . . . . . . . . . . . . . . *j*
j. Anther-bearing stamens 5; corolla nearly symmetrical, with a short tube, about as broad as or broader than long, the 3 lower lobes only slightly longer than the 2 upper ones; stems often 1–2 m. tall, but not usually less than 4 dm. tall . . . . 9. VERBASCUM
j. Anther-bearing stamens 4; corolla quite irregular or with an elongated tube, much longer than broad, divided into a lower and an upper lip; stems usually 1.5–4 dm. (up to 6) tall . . . . . . . . . . . . . . . . . . . . . . *k*
k. Main lobes of leaf rounded at their tips, their margins finely toothed, the leaves pinnately parted into 14–32 lobes; lower and basal leaves 4–10 cm. long, or the basal elongating to 30 cm. long after flowering; all but the uppermost leaves on petioles (stalks); bracts at base of flowers green, not concealing the flowers . 24. PEDICULARIS
k. Main lobes of leaves long and sharp-pointed at their tips, their margins entire (without teeth), the leaves 3–7 cleft or parted; leaves 2–7 cm. long; principal leaves on stem sessile (without stalks); bracts at base of flowers mostly concealing the flowers, green, yellow, orange, red, purple, or white . . . . . . 25. CASTILLEJA
c. All the leaves on the stem opposite in pairs or whorled (arranged in circles of 3 or more) . . . . *l*
l. Leaves chiefly arranged in whorls (circles) of 3–7, rarely opposite in pairs . 19. VERONICASTRUM
l. Leaves opposite in pairs . . . . . . . . . . . . . . . . . . . . . *m*
m. Some or most of the leaves divided, deeply parted, or cut at least halfway to the midrib . . *n*
n. Low plant 1–2 dm. tall; leaves of stem 1–3 cm. long; corolla pale lavender, 4–5 mm. long . . . . . . . . . . . . . . . . . . . . . . . 4. CONOBEA
n. Tall plants 3–20 dm. tall; leaves of stem 3–20 cm. long; corolla yellow, 15–50 mm. long . . . . . . . . . . . . . . . . . . . . . . . . *o*

o. Corolla less than 20 mm. long, the tube shorter than the expanded limb; flowers sessile (without stalks) or on short stalks up to 2 mm. long; anthers glabrous (not hairy); stamens nearly equal, protruding from the throat of the corolla . . . . . . . . . . . . . . . . 21. SEYMERIA

o. Corolla mostly 30–50 mm. long, the tube longer than the expanded limb; flowers and fruits on pedicels (stalks) 4–20 mm. long; anthers hairy; stamens in 2 unequal pairs, included within the corolla . . . . . . . . . . . . . . . . . . . . . . . . 22. GERARDIA

m. Leaves entire (without teeth) or toothed, if lobed or coarsely cut, then cut less than halfway to the midrib. . . . . . . . . . . . . . . . . . . . . . . . . . . . . *p*

p. Calyx 4-lobed or 4-parted; corolla 4-lobed; fruit flattened, somewhat notched or heart-shaped at tip . . . . . . . . . . . . . . . . . . . . . . . . 20. VERONICA

p. Calyx 5-lobed, 5-parted, or 5-toothed; corolla with 5 plain or obscure lobes, the lobes either nearly equal or part of a 2-lipped corolla; fruit not flattened, more or less inflated, usually tapering or narrowed, but also rounded at summit . . . . . . . . . . . . . . . . . . . . *q*

q. Flowers solitary, along the sides or near the top of the stem, 1 or 2 in the axil of a leaf or leaf-like bract, not arranged in inflorescences of 2 to many . . . . . . . . . . . . . . *r*

r. Calyx parted to the base into separate and distinct sepals or nearly to the base, the calyx-tube not evident or lacking . . . . . . . . . . . . . . . . . . . . . *s*

s. Water plants with nearly round to round-obovate, entire (toothless) leaves, the main nerves all arising at the base of the blade . . . . . . . . . 2. BACOPA ROTUNDIFOLIA

s. Without the above combination of characters; nerves of leaves pinnately arranged, the side nerves arising from various levels of the midrib . . . . . . . . . . . . *t*

t. Anther-bearing stamens 4 . . . . . . . . . . . . . . 2. BACOPA ACUMINATA

t. Anther-bearing stamens 2, sometimes 2 other filaments are present but these are without anthers . . . . . . . . . . . . . . . . . . . . . . . *u*

u. Pedicels (flower- or fruit-stalks) with 2 bractlets immediately below the calyx; stems and leaves glabrous (without hairs); in addition to the 2 filaments bearing anthers, there are 2 other filaments without anthers . . . . . . . . 7. LINDERNIA

u. No bractlets at summit of pedicels; stems and/or leaves glabrous or hairy; only 2 filaments present, any additional ones either not present or not evident. . 3. GRATIOLA

r. Calyx with a definite and evident well-developed tube . . . . . . . . . . . . *v*

v. Corolla very irregular, 2-lipped into an upper and lower lip which are close together, the enlarged elevation from the lower lip nearly closing the opening to the corolla; flowers mostly all axillary, arising at the sides of the stem at the base of the leaves; calyx with 5 prominent angles . . . . . . . . . . . . . . . . . . . . . 5. MIMULUS

v. Corolla slightly irregular, not 2-lipped, the 5 lobes slightly unequal, the interior of the corolla visible and not closed by any enlarged projection from the lower lip; flowers not all axillary, some of them also at the top of the stem or branches; calyx with rounded or curved sides, not angled . . . . . . . . . . . . . . . . 22. GERARDIA

q. Some or all of the flowers arranged in an inflorescence of 2–many at the top of the stem or branches . . . . . . . . . . . . . . . . . . . . . . . . . . . *w*

w. Flowers and fruits sessile (without pedicels or stalks) . . . . . . . . . . . . . *x*

x. Stems glabrous; corolla 2.3–3.5 cm. long, very irregular, definitely 2-lipped with an upper and lower lip . . . . . . . . . . . . . . . . . . . 10. CHELONE

x. Stems hairy; corolla about 2 cm. long, only slightly irregular, not 2-lipped, the 5 lobes only slightly unequal . . . . . . . . . . . . . . . . . . . . . . *y*

y. Corolla deep purple, the tube narrow and slender, abruptly enlarged at summit; leaves with few coarse teeth; calyx-tube much longer than the calyx-lobes . 23. BUCHNERA

y. Corolla rose-colored or lavender, the tube bell-shaped and gradually enlarged upward; leaves either entire (without teeth) or with 1 or 2 small lobes near the base; calyx-tube much shorter than the calyx-lobes . . . . . . 22. GERARDIA AURICULATA

w. Flowers and fruits either on conspicuous pedicels (stalks) or the pedicels as short as 1 mm. long and scarcely evident, but not absolutely sessile (stalkless) . . . . . . . . . . *z*

z. Some or all of the leaves deeply lobed or deeply toothed . . . . . . . . . . . *1*

1. Flowers and fruits crowded, closely clustered in a spike; corolla greenish-white or pale yellow, very irregular, definitely 2-lipped, with an upper and lower lip, 15–25 mm. long; plants of calcareous swamps and spring-fed swampy meadows . . . 24. PEDICULARIS LANCEOLATA

1. Flowers and fruits loosely arranged and separated by intervals throughout the inflorescence; corolla yellow, only somewhat irregular, not 2-lipped, 30–55 mm.

1946 Petunia violacea (Petunia)

1947 Paulownia tomentosa (Empress Tree)

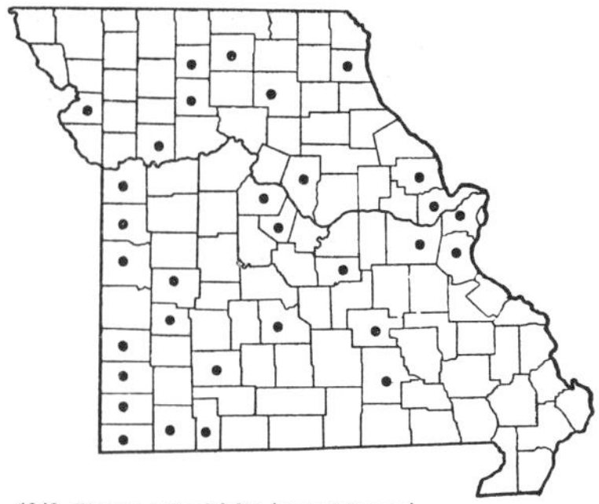

1948 Bacopa rotundifolia (Water Hyssop)

long, the 5 lobes only slightly unequal; plants of dry, rocky, or rich woodland and openings in woods . . . . . . . . . . . . . . . . . . . . . . . . . 22. GERARDIA

1. Leaves either entire (without teeth) or toothed, but not lobed or deeply cut . . . . . . . . . *2*
   2. All the leaves on well-developed petioles (leaf-stalks) 15–60 mm. long; all the leaves sharply or coarsely toothed from base to tip with numerous teeth up to 10 mm. long, the leaf-blades 8–25 cm. long; stems up to 2–3 m. tall . . . . . . . . . . . . . . . . 12. SCROPHULARIA
   2. Leaves either sessile (without petioles) or on petioles at most 25 mm. long; leaves entire (without teeth) or teeth either inconspicuous or few; stems 5 cm. to 1.5 m. tall. . . . . . . . . *3*
      3. Stem creeping; flowers alternately arranged in a simple, unbranched raceme, blue, 7–10 mm. long; rare introduced plant of lawns and grassy open ground . . . . . . . . 6. MAZUS
      3. Stem erect; flowers in a branched or compound inflorescence or otherwise, but not alternately arranged in a simple unbranched raceme, white, rose, pink, purple, lavender, blue, or combinations of these; native species . . . . . . . . . . . . . . . . . . *4*
         4. Principal leaves narrowly linear or thread-like, 1–8 mm. broad; corolla usually rose, pink, lavender, or flesh-color . . . . . . . . . . . . . . . . . 22. GERARDIA
         4. Principal leaves oblong, elliptic, triangular-ovate, or round-ovate, 5–60 mm. broad; corolla white, purple, blue, purple and white, or blue and white . . . . . . . . . *5*
            5. Upper and lower halves of corolla of different colors, the upper half usually white, the lower half blue or purple; corolla 10–15 mm. long, split nearly to the base, the short tube 4–7 mm. long and much shorter than the lobes; low annual plants mostly 0.5–4 dm. (rarely 6) tall . . . . . . . . . . . . . . . . . 13. COLLINSIA
            5. Lower and upper halves of corolla of the same general color, either white, lavender, purple, blue, or combinations of these; corolla 15–50 mm. long, lobed only above the middle, the long tube at least 7 mm. long and much longer than the lobes; perennial plants 3–15 dm. tall . . . . . . . . . . . . . . . . . 11. PENSTEMON

## 1. **Paulownia** Sieb. & Zucc. Empress Tree

**Paulownia tomentosa** (Thunb.) Steud. Empress Tree Map 1947

Also called Princess Tree.

Flowers April–May.

Occasionally escaped from cultivation in thickets, fencerows, and waste places. Known from southern and central Missouri in Perry, Dunklin, Ripley, Pettis, and McDonald counties, and north locally in Marion County.

Native of China; introduced and naturalized in the United States from Florida to Oklahoma, north to Connecticut, New York, Pennsylvania, Ohio, Indiana, Illinois, and Missouri.

This is a strikingly handsome tree in flower, the large violet, yellow-striped corollas appearing in erect, pyramidal panicles at the tips of the branches. The large leaves are catalpa-like. The sprays of the dry fruit are conspicuous on the tree throughout the winter, as are the clusters of flower-buds which form late in autumn and expand in spring before the leaves are

---

Plate no. 318. **1.** Paulownia tomentosa, × 2/5; Details from Small, The New York Botanical Garden. **2.** Bacopa rotundifolia, × 2/5; Details from Small, The New York Botanical Garden. **3.** Bacopa acuminata, × 2/5; *a.* Flower, × 2; After Gleason, details from Small, The New York Botanical Garden. **4.** Gratiola viscidula, × 2/5; *a.* Flower, × 1 3/5; *b.* Leaf, × 1 3/5; After Gleason, details from Small, The New York Botanical Garden. **5.** Gratiola neglecta, × 2/5.

PLATE NO. 318

1949 Bacopa acuminata

1950 Gratiola viscidula

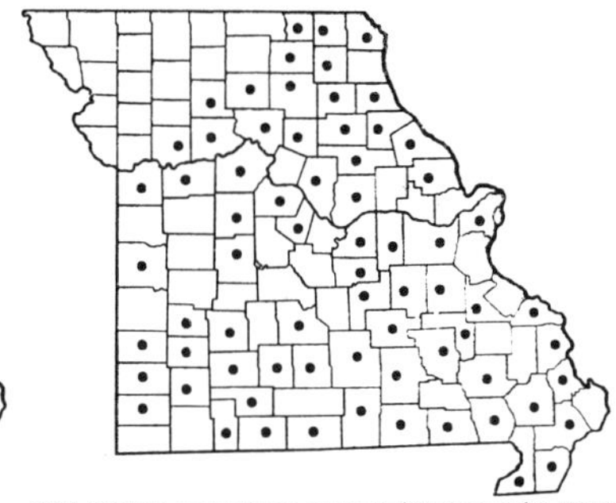

1951 Gratiola neglecta var. neglecta (Clammy Hedge Hyssop)

fully formed. The tree grows fast and has very light but strong wood which is used for crates and boxes for lightweight shipments carried by air express. Some botanists place this genus in the Bignoniaceae (Campbell, D. H. Bull. Torr. Bot. Club 57: 17–50. 1930; Jones *et al., Vascular Plants of Illinois*, p. 441. 1955), which may eventually prove to be correct, although the presence of endosperm in the embryo and the structure of the placenta and capsule indicate a sound relation in the Scrophulariaceae, as that family is at present understood.

## 2. **Bacopa** Aubl. Water Hyssop

Plants growing in the water; leaves without teeth, round to round-obovate, all the main nerves arising from the base of the leaf-blade; no bractlets at base of pedicel (flower-stalk) . . . . . . 1. B. ROTUNDIFOLIA

Plants usually growing out of water; leaves toothed above the middle, oblanceolate, wedge-shaped at base, the nerves of leaves pinnately arranged, the side nerves arising from various levels of the midrib; 2 bractlets occurring at the base of pedicel . . . . . . . . . . . . . . . 2. B. ACUMINATA

1. **Bacopa rotundifolia** (Michx.) Wettst. Map 1948

*Macuillamia rotundifolia* (Michx.) Raf. [P & S, Steyerm.]

Flowers May–September.

Occurs in borders of sloughs, slow streams, natural and artificial ponds and ditches, also in swamps and low wet woods. Scattered throughout Missouri, although not recorded from the extreme southeastern and northwestern sectors of the state.

Ranges from Mississippi to Texas, north to Indiana, Illinois, Minnesota, North Dakota, and Montana.

This plant is sometimes used in tropical aquaria and outdoor pools. When growing under water the stems are usually elongated with widely separated pairs of leaves, the lower submerged ones generally much smaller than the uppermost floating leaves. Ducks eat the seeds and leaves of this species.

2. **Bacopa acuminata** (Walt.) Robins. Map 1949

*Mecardonia acuminata* (Walt.) Small [BB, P & S]

*Pagesia acuminata* (Walt.) Pennell [Pennell]

Flowers July–September.

Occurs in swamps, sloughs, low wet woods, and sandy open knolls. Southern Missouri, in the southeastern lowlands in Mississippi, Dunklin, and Ripley (*Steyermark 66941*) counties, and in southwestern Missouri in the unglaciated prairie section in Jasper and Newton counties.

Ranges from Florida to Texas, north to Maryland, Tennessee, Missouri, and Oklahoma.

The plant turns black upon drying.

## 3. **Gratiola** L. Hedge Hyssop

a. Main leaves of the stem broadest at or near the rounded or clasping base; fruit 1–3 mm. long, much shorter than the calyx-divisions; rare species of a sink-hole pond in Shannon County; perennial plants with creeping, rooting, underground stems . . . . . . . . . . . . . . 1. G. VISCIDULA

a. Main leaves of the stem mostly broadest near or above the middle, tapering or narrowed to the base; fruit 3–7 mm. long, nearly equaling to slightly longer than the calyx-divisions; common in many parts of the state; annual or biennial plants without creeping stems . . . . . . . . . . . *b*

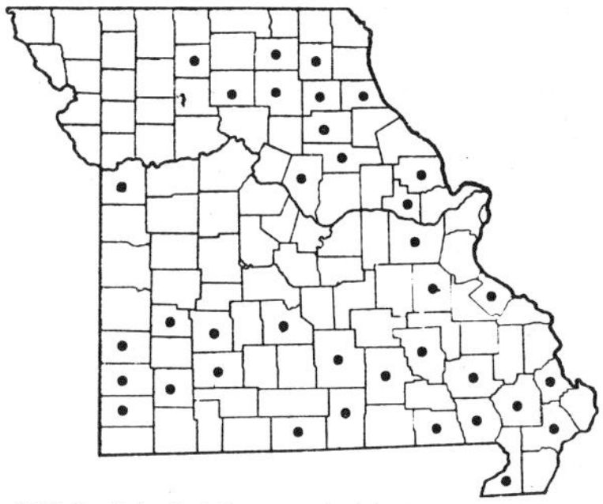

1952 Gratiola virginiana var. virginiana

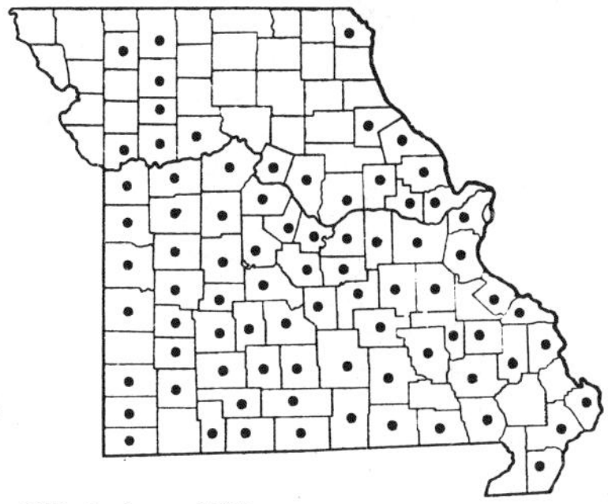

1953 Conobea multifida

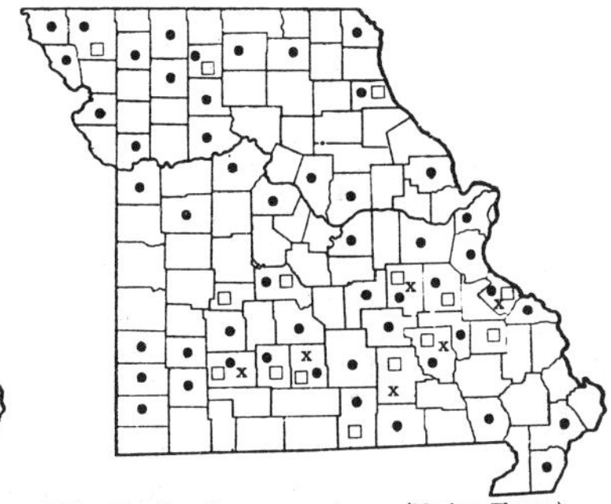

1954 • Mimulus ringens var. ringens (Monkey Flower)
1954 □ Mimulus ringens var. minthodes
1954 x Mimulus ringens X alatus

b. Flower- and fruit-stalks (pedicels) thread-like, 10–25 mm. long; upper part of stem more or less hairy; stem slender; usually growing near, but not in the water, except during floods . . 2. G. NEGLECTA

b. Flower- and fruit-stalks (pedicels) stouter, 1–4 mm. in flower, up to 10 (rarely 14) mm. long in fruit; stem mostly glabrous (without hairs) throughout its length; stem thick and fleshy; usually growing in the water . . . . . . . . . . . . . . . . . . . . . . 3. G. VIRGINIANA

1. **Gratiola viscidula** Pennell Map 1950

Flowers June–September.

Occurs in an upland sink-hole pond, where known only from the southeastern Ozarks in Shannon County (Gilmore Pond [Grassy Pond], T27N, R6W, sect. 34, between Jack's Fork of Current River and Flat Rock Hollow, 6½ mi. northwest of Monteer, July 8, 1951, *Steyermark 72109*).

Ranges from Georgia to Missouri, north to Delaware, Maryland, Ohio, Tennessee, and Kentucky.

The plants grow along the border of this pond with their bases partly submerged. They occur in dense masses. This is another one of the relict species surviving in remote sink-hole ponds of the Ozarks (See Rh. 54: 258. 1952), isolated by several hundred miles from other stations of this species found in the Coastal Plain, Piedmont, southern Appalachians, and Interior Low Plateau sections to the east and south.

2. **Gratiola neglecta** Torr. var. **neglecta**

Clammy Hedge Hyssop Map 1951

*Gratiola lutea* Raf. [P & S]

*Gratiola neglecta* Torr. [BB, G, Steyerm.]

Flowers May–October.

Occurs on muddy borders of upland and lowland natural or artificial ponds, sloughs, streams, ditches, low moist woodland, wet meadows, wet depressions along paths and roads, around hog wallows, and moist open ground. Eastern, southern, and central Missouri, south and east of a line drawn from Schuyler, Adair, Linn, Livingston, and Ray counties to Jackson County.

Ranges from Quebec and Maine to British Columbia, south to Georgia, Texas, and Arizona.

3. **Gratiola virginiana** L. var. **virginiana** Map 1952

*Gratiola virginiana* L. [G, BB, P & S, Steyerm.]

Flowers last of April–October.

Occurs in ponds, sloughs, ditches. Scattered in eastern, southern, and central Missouri, north to Knox, Adair, Linn, Grundy, and Jackson counties.

Ranges from Florida to Texas, north to New Jersey, West Virginia, Ohio, Indiana, Illinois, Iowa, Kansas, and Oklahoma.

This species is more aquatic than *G. neglecta*.

## 4. **Conobea** Aubl.

**Conobea multifida** (Michx.) Benth. Map 1953

*Leucospora multifida* (Michx.) Nutt. [BB, P & S, Steyerm.]

Flowers last of May–October.

Occurs along sand and gravel bars and mud flats along streams and spring branches, open fields, moist prairies, ledges of bluffs, low open woods, and roadsides. Throughout southern and central Missouri and locally north in Clark, Harrison, Daviess, and Gentry counties.

Ranges from Ontario to Iowa and Kansas, south to Georgia, Alabama, Louisiana, and Texas.

## 5. **Mimulus** L. Monkey Flower

a. Corolla yellow, 8–12 mm. long; lower part of stem spreading or creeping along the ground; leaves nearly round or oval, nearly as broad as long or at most only slightly longer than broad; rarely encountered species . . . . . . . . . . . . . . . . . . 3. M. GLABRATUS var. FREMONTII

a. Corolla usually blue or blue-purple, rarely white or pinkish, 20–40 mm. long; stem erect from the base; leaves oblong, elliptical, ovate, or obovate, 3–5 times as long as broad; commonly encountered species . . . . . . . . . . . . . . . . . . . . . . . . b

b. Leaves without a petiole (stalk) with the base either rounded and clasping the stem or simply narrowed and tapering; angles of stem without wings (strips of green tissue); fruiting pedicels (stalks of fruit) 2–6 cm. long; fully grown teeth of calyx 1.5–8 mm. long . . . . . . 1. M. RINGENS

b. Leaves definitely on petioles (stalks); angles of stem more or less winged (with strips of green tissue); fruiting pedicels (stalks of fruit) 0.2–1.7 cm. long; fully grown teeth of calyx 1–1.5 mm. long . . . . . . . . . . . . . . . . . . . . . 2. M. ALATUS

### 1. **Mimulus ringens** L. Monkey Flower
Map 1954

Flowers late June–September.

Occurs in wet soils along borders of streams, sloughs, ponds, ditches, swamps, swampy meadows and low woodland.

The following variations are found in Missouri:

Most of the main leaves of stem rounded or clasping the stem at their base . . . 1a. M. RINGENS var. RINGENS

Most of the main leaves of stem narrowed and tapering to the base . . . . 1b. M. RINGENS var. MINTHODES

### 1a. **Mimulus ringens** var. **ringens** Map 1954

*Mimulus ringens* L. [G, BB, P & S]

Throughout Missouri, but less common than *M. alatus* and not recorded from a number of counties.

Ranges from Quebec and Nova Scotia to Saskatchewan, south to Georgia, Louisiana, Texas, and Colorado.

### 1b. **Mimulus ringens** var. **minthodes** (Greene) Grant
Map 1954

Scattered throughout Missouri and less common than var. *ringens*.

Ranges from Massachusetts to Georgia, Alabama, and Kansas.

The calyx-teeth in var. *minthodes* are stated by some authors to be longer (4–8 mm. long) than those of var. *ringens* (1.5–5 mm. long), but there is intergradation in this character, some Missouri specimens of var. *minthodes* (*Steyermark 85519* from Madison County, for example) having calyx-teeth only 3–4 mm. long. The teeth of var. *minthodes* do have an average greater length on the whole, however.

### 2. **Mimulus alatus** Ait. Monkey Flower
Map 1955

Flowers June–September.

Occurs along borders of streams, spring branches, sloughs, ponds, ditches, swamps, swampy meadows, and in low moist woodland. Throughout Missouri, except absent apparently in the extreme northwestern sector.

Ranges from Florida to Texas, north to Connecticut, New York, Ontario, Michigan, Illinois, Iowa, and Nebraska.

The following 2 variations occur in Missouri:

Corolla blue-lavender or pink. . . 2a. M. ALATUS f. ALATUS

Corolla white . . . 2b. M. ALATUS f. ALBIFLORUS

### 2a. **Mimulus alatus** f. **alatus** Map 1955

*Mimulus alatus* Ait. [G, BB, P & S]

This is the common variation found throughout Missouri.

### 2b. **Mimulus alatus** f. **albiflorus** House
Map 1955

Known only from Maries (alluvial woods along the Gasconade River, T40N, R9W, sect. 15, 2½ mi. northeast of Vienna, August 29, 1937, *Steyermark 25590*) and Lawrence (Red Mill, near Verona, August 22, 1950, *Palmer 50783*) counties.

Specimens which combine characters of both species and appear to represent hybrids are known from

---

Plate no. 319. **1.** Gratiola virginiana, × 2/5. **2.** Conobea multifida, × 2/5; Details from Small, The New York Botanical Garden. **3.** Mimulus ringens, × 2/5. **4.** Mimulus alatus, × 2/5; Details from Small, The New York Botanical Garden. **5.** Mimulus glabratus var. Fremontii, × 2/5; Details from Small, The New York Botanical Garden. **6.** Mazus japonicus, × 2/5; *a.* Flower, × 1; After Gleason, The New York Botanical Garden. **7.** Limosella aquatica, × 1 1/5; *a.* Flower, × 2 2/5; After Gleason, The New York Botanical Garden.

PLATE NO. 319

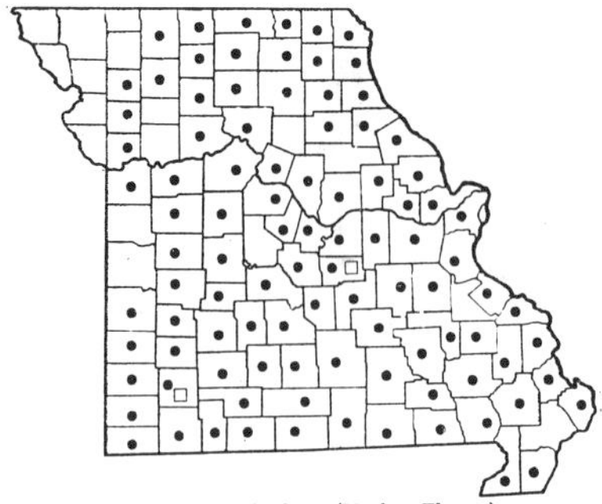

1955 • Mimulus alatus f. alatus (Monkey Flower)
1955 □ Mimulus alatus f. albiflorus

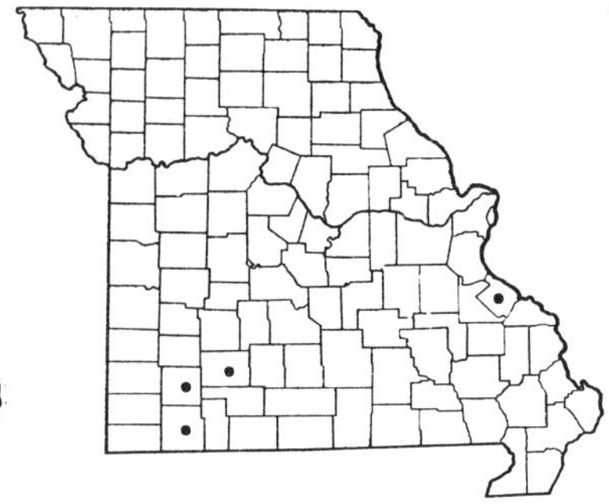

1955-A Mimulus glabratus var. Fremontii

1956 Mazus japonicus

southern Missouri in Ste. Genevieve, Crawford, Reynolds, Shannon, Wright, and Greene counties.

3. **Mimulus glabratus** HBK. var. **Fremontii** (Benth.) Grant — Map 1955A

Flowers May–October.

Occurs around springs, spring branches, and we ledges along bluffs. Rare, southern Missouri in Ste. Genevieve, Greene, Lawrence, and Barry counties.

Ranges from Mexico north to Ontario, Michigan, Wisconsin, Minnesota, Manitoba, and Montana.

## 6. **Mazus** Lour.

**Mazus japonicus** (Thunb.) Ktze. — Map 1956

Flowers July–October.

Occurs in grassy open places on lawns. Known only from St. Louis County.

Native of eastern Asia; introduced and escaped from cultivation in the United States from Pennsylvania to D.C., Illinois, Missouri, and Louisiana; also in Washington and Oregon.

This species is sometimes grown as a ground-cover or rock garden ornamental plant. The flowers, 8–14 mm. long, are blue.

## 7. **Lindernia** All. False Pimpernel

(*Ilysanthes* Raf. [P & S])

Both the lower and upper flower- and fruit-stalks (pedicels) longer than the leaves accompanying them; lowest leaves rounded or somewhat heart-shaped at the broader base; calyx-lobes usually shorter than the capsules . . . . . . . . . . . . . . . . . . . . . . . . . . 1. L. ANAGALLIDEA

Lower flower- and fruit-stalks (pedicels) shorter than the leaves accompanying them; lowest leaves narrowed to the base which is narrower than upper and middle part; calyx-lobes mostly equaling or longer than the capsules . . . . . . . . . . . . . . . . . . . . . 2. L. DUBIA

1. **Lindernia anagallidea** (Michx.) Pennell — Map 1957

*Ilysanthes inaequalis* (Walt.) Pennell [P & S]

Flowers June–October.

Occurs in wet muddy, gravelly, or sandy soils on flats, bars, banks along streams, ponds, sloughs, and ditches, moist places in woodland, and occasionally along railroads.

Scattered throughout eastern, southern, and central Missouri, apparently absent from most of northwestern Missouri; known in northern Missouri from Lewis, Clark, Scotland, Ralls, Monroe, Macon, Grundy, and Buchanan counties.

Ranges from Florida to Texas and Mexico, north to New Hampshire, New York, Ohio, Indiana, Wisconsin, Minnesota, North Dakota, Colorado, and Washington.

---

Plate no. 320. **1.** Verbascum Thapsus, × $^{2}/_{5}$; Details from Small, The New York Botanical Garden. **2.** Lindernia anagallidea, × $^{2}/_{5}$; *a*. Flower, × $1^{3}/_{5}$; After Gleason, details from Small, The New York Botanical Garden. **3.** Lindernia dubia, × $^{2}/_{5}$; *a*. Flower, × $1^{3}/_{5}$; After Gleason, The New York Botanical Garden.

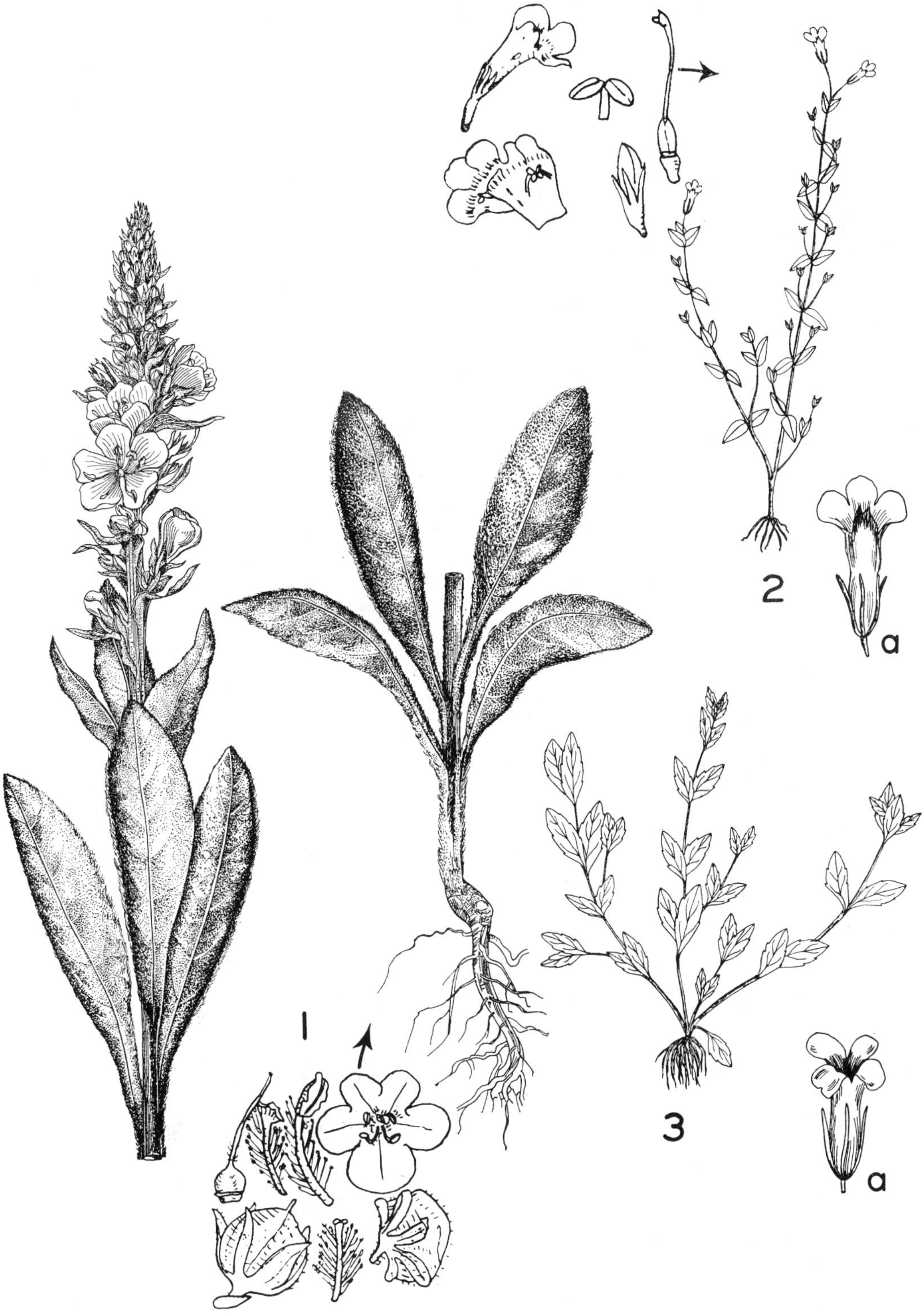


PLATE NO. 320

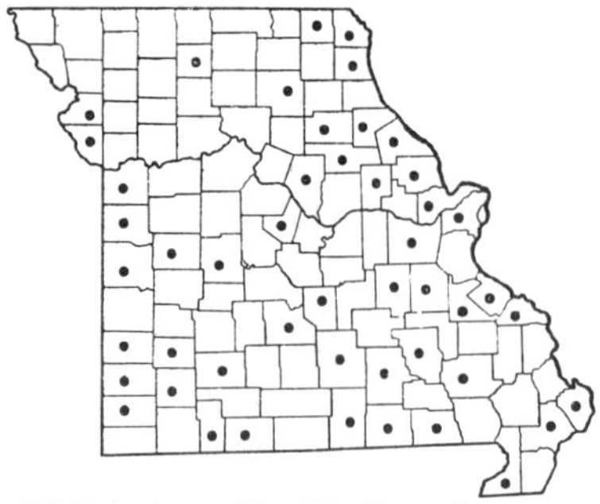
1957 Lindernia anagallidea (False Pimpernel)

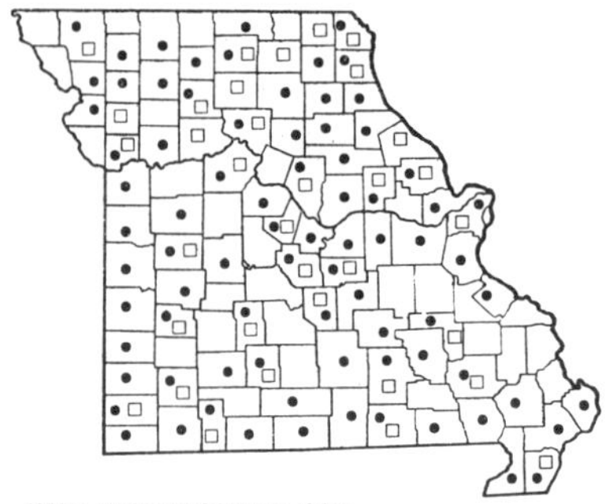
1958 ● Lindernia dubia var. dubia
1958 □ Lindernia dubia var. riparia

1959 Limosella aquatica (Mudwort)

2. **Lindernia dubia** (L.) Pennell Map 1958

Flowers June–October.

Occurs in wet muddy or sandy soils on flats, bars, banks along streams, ponds, sloughs and ditches, moist places in fallow fields and woodland, and occasionally along railroads; more frequent in muddy soils than *L. anagallidea*. Throughout Missouri.

The following intergrading variations occur in Missouri:

Flower- and fruit-stalks (pedicels) shorter than the leaves accompanying them . . 2a. L. DUBIA var. DUBIA

Upper flower- and fruit-stalks (pedicels) longer than the leaflike bracts accompanying them . . 2b. L. DUBIA var. RIPARIA

2a. **Lindernia dubia** var. **dubia** Map 1958

*Lindernia dubia* (L.) Pennell [G]

*Ilysanthes dubia* (L.) Barnhart [P & S in part]

*Lindernia dubia* subsp. *major* Pennell [Pennell]

This is the more common variation throughout the state.

Ranges from Quebec to Ontario and Minnesota, south to Georgia, Texas, and Mexico; also Washington to California.

2b. **Lindernia dubia** var. **riparia** (Raf.) Fern. Map 1958

*Ilysanthes dubia* [of P & S in part], not (L.) Barnhart

*Lindernia dubia* subsp. *typica* Pennell [Pennell]

Scattered throughout Missouri but less common than var. *dubia*.

Ranges from Maine to Minnesota, North Dakota, and Nebraska, south to Florida and Texas.

These two varieties intergrade in various other characters used to separate them. In *L. dubia* var. *dubia* the leaves of the main stem and branches are about the same size as the leafy bracts accompanying the flower- and fruit-stalks, whereas in var. *riparia* the foliage-leaves are usually larger than the leaflike bracts above, which become reduced in size, usually 1–6 mm. broad as contrasted with 5–10 mm. broad for var. *dubia*. However, these characters fluctuate in their degree of constancy, and are not used, therefore, in the above key differences.

## 8. **Limosella** L. Mudwort

**Limosella aquatica** L. Mudwort Map 1959

Flowers July–September.

Occurs on mud or sand flats along the Missouri River, where known only from west-central Missouri in Jackson County.

Ranges in Newfoundland and Labrador; Quebec; Manitoba to British Columbia, south locally to Minnesota, Missouri, Nebraska, New Mexico, Arizona, and California; also in Alaska and Eurasia.

This dwarf plant has lanceolate to elliptic leaf-blades 2.5–6 mm. broad on long petioles. The leaves occur in little tufts. The corolla, about 3 mm. long, is pinkish or white tinged with purplish.

The plant has not been collected in the state since the original collection by Bush.

1960 Verbascum Thapsus f. Thapsus (Mullein)

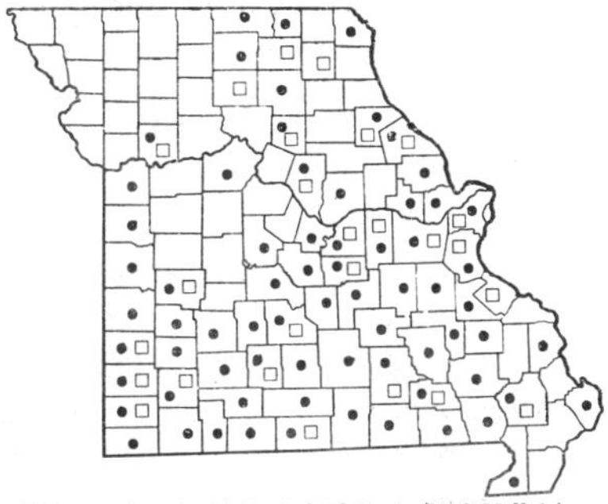

1961 ● Verbascum Blattaria f. Blattaria (Moth Mullein)
1961 □ Verbascum Blattaria f. erubescens

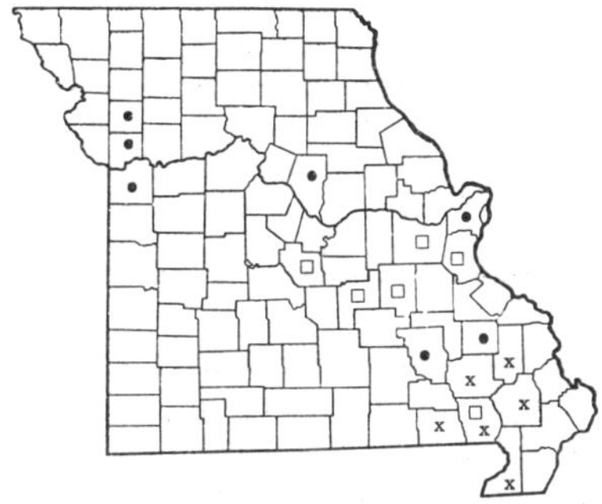

1962 ● Chelone glabra var. glabra f. glabra (White Turtlehead)
1962 □ Chelone glabra var. glabra f. tomentosa
1962 x Chelone glabra var. linifolia f. linifolia

## 9. **Verbascum** L. Mullein

Flowers closely crowded into a long, narrow, inflorescence with no gaps or spaces between the flowers; leaves pale gray green, soft and like flannel to the touch; hairs covering calyx and other parts of plants dense, woolly, branched and not gland-tipped; stem winged by the prolonged bases of the leaves which extend down the surface of the stem . . . . . . . . . . . . . . . . . . . . . 1. V. THAPSUS
Flowers loosely arranged with wide gaps or spaces showing between the flowers; leaves dark green, mainly glabrous (without hairs); hairs on calyx and often on upper part of stem simple, unbranched, and gland-tipped; stem without wings, the leaf-bases not prolonged . . . . . . . . 2. V. BLATTARIA

1. **Verbascum Thapsus** L. f. **Thapsus** Mullein Map 1960

Also called Flannel Plant.

*Verbascum Thapsus* L. [G, BB, P & S, Steyerm.]

Flowers late May–September.

Occurs in dry fields, pastures, waste ground, along roadsides and railroads. Throughout Missouri, doubtless in every county.

Native of Europe; introduced and naturalized in North America from Nova Scotia and Quebec to British Columbia, south to Florida, Texas, and California.

This species is not eaten by livestock. It has been used medicinally as a stimulant because of its mucilaginous properties. A form with white corollas, f. *candicans* House, has not been recorded thus far from Missouri.

2. **Verbascum Blattaria** L. Moth Mullein Map 1961

Flowers May–September.

Occurs in pastures, fallow and cultivated ground, rocky open places, waste ground, gravel bars along streams, roadsides, and along railroads. Common throughout eastern, southern, and central Missouri, but not recorded from the northwestern section of the state.

Ranges from New England and Ontario to British Columbia, south to Florida, Texas, and California.

Two variations occur in Missouri:

Corolla yellow . . 2a. V. BLATTARIA f. BLATTARIA
Corolla white with purple at the base . . . 2b. V. BLATTARIA f. ERUBESCENS

2a. **Verbascum Blattaria** f. **Blattaria** Map 1961

*Verbascum Blattaria* L. [G, P & S, Steyerm.]

This is the more common variation in Missouri, and is known in northern Missouri from Clark, Schuyler, Putnam, Sullivan, and Ralls counties.

2b. **Verbascum Blattaria** f. **erubescens** Brugger Map 1961

*Verbascum Blattaria* f. *albiflora* (G. Don) House [G, P & S, Steyerm.]

This is less common than f. *Blattaria*, and is scattered in southern and central Missouri, and in the northern part of the state occurs in Knox, Adair, and Linn counties.

## 10. **Chelone** L. Turtlehead

Corolla white or cream-colored; leaf-stalks (petioles), if present, mostly very short or scarcely noticeable, up to 5 mm. long; leaves narrowly or linear-lanceolate to elliptic-lanceolate, the largest

8–30 mm. broad (in Missouri specimens); upper side of leaf-stalk and upper midrib green; sepals barely or faintly fringed with short hairs on margins; more commonly encountered . . . . . . . 1. C. GLABRA
Corolla deep rose, lavender-rose, or deep orchid-colored; leaf-stalks (petioles) evident, 5–15 mm. long; leaves broadly lanceolate to oblong-elliptical or ovate-oblong, the largest mostly 30–60 mm. broad; upper side of leaf-stalk and upper midrib often wine-purple; sepals fringed with short hairs on margins; very rarely encountered. . . . . . . . . . . . . . . . . . . . . 2. C. OBLIQUA var. SPECIOSA

1. **Chelone glabra** L. White Turtlehead Map 1962

Flowers July–October.

Occurs in wet boggy ground along small streams and spring branches, swampy woods, swampy meadows, moist ledges of bluffs, and wet alluvial thickets.

The following variations occur in Missouri:

a. Leaves narrowly or linear-lanceolate, the larger ones 8–23 mm. broad; corolla greenish-yellow at summit or externally, and white on inside. . . . 1c. C. GLABRA var. LINIFOLIA
a. Leaves lanceolate to elliptic-lanceolate, the larger ones 15–30 mm. broad; corolla creamy-white to pinkish at summit or pinkish on the inside of the lip . . . . . . . . . *b*
  b. Lower surface of leaves glabrous (without hairs) . 1a. C. GLABRA var. GLABRA f. GLABRA
  b. Lower surface of leaves minutely hairy . 1b. C. GLABRA var. GLABRA f. TOMENTOSA

1a. **Chelone glabra** var. **glabra** f. **glabra** Map 1962

*Chelone glabra* L. [G, P & S]
*Chelone glabra* var. *glabra* [BB]

Known from Jefferson County.

Ranges from Newfoundland to Ontario and Minnesota, south to Georgia, Alabama, and Missouri.

1b. **Chelone glabra** var. **glabra** f. **tomentosa** (Raf.) Pennell Map 1962

*Chelone glabra* f. *tomentosa* (Raf.) Pennell [G]

Scattered in eastern Missouri south of the Missouri River in Jefferson, Franklin, Crawford, Butler, Phelps, and Miller counties.

Range as in f. *glabra*.

1c. **Chelone glabra** var. **linifolia** Coleman f. **linifolia** Map 1962

*Chelone glabra* var. *linifolia* Coleman [G]

Southeastern Missouri in Wayne, Bollinger, Stoddard, Butler, and Ripley counties.

Ranges from Ontario and Manitoba, south to Ohio, Indiana, Illinois, and Missouri.

A form of var. *linifolia* with the lower surface of the leaves minutely hairy is known as f. *velutina* Pennell & Wherry, but has thus far not been collected in Missouri.

There is considerable variation in width of leaves, and color of corolla. Much of the Missouri material has the corollas greenish-yellow or creamy at the summit and white within. More intensive field and experimental studies need to be carried out to evaluate the various forms described.

This species does well in bog gardens or where the soil is sufficiently damp most of the year. It is sometimes difficult to transplant.

2. **Chelone obliqua** L. var. **speciosa** Pennell & Wherry Rose Turtlehead Map 1963

*Chelone obliqua* L. [BB]

Flowers August–October.

Occurs in swampy meadows, margins of springs in calcareous open meadows, and low or swampy woods in valleys along or near streams. Known only from Lewis (La Grange, *Davis 1055*), Macon (woods along East Fork of Chariton River, T56N, R15W, east part sect. 13 and 18, 1¾–2 mi. northeast of Ardmore, September 15, 1954, *Steyermark 77344*), Chariton (Chariton River, September, 1869, *E. Hall*), and Wayne (swampy meadow in valley of Stanley Creek, T27N, R7E, sect. 18, 6–7 mi. northeast of Wappapello, September 11, 1949, *Steyermark 69263*; same locality, October 22, 1948, *Steyermark 66980*; low thickets in valley of Mill Creek, T26N, R7E, sect. 4, 1½ mi. southwest of Wappapello, August 29, 1938, *Steyermark 6243*) counties in eastern Missouri.

Ranges from Indiana to Minnesota, south to Iowa, Missouri, and Arkansas.

This is a handsome-flowered species. Although native to swamps, it succeeds in any rich soil in shade or sun. It is a welcome addition to the woodland wildflower garden where it forms larger clumps each year. It is easily grown from seed.

---

Plate no. 321. **1.** Verbascum Blattaria, × 2/5. **2.** Chelone glabra, × 2/5. **3.** Chelone obliqua var. speciosa, × 2/5; Details from Small, The New York Botanical Garden. **4.** Penstemon grandiflorus, × 2/5; Details from Small, The New York Botanical Garden.

PLATE NO. 321

## 11. **Penstemon** Mitchell Beard-tongue

a. Corolla with pale lilac or purple as the principal color, 3.5–5 cm. long; calyx during flowering period 7.5–13 mm. long . . . . . . . . . . . . . . . . . . . . . . . . . . *b*
  b. Stems completely glabrous; calyx and pedicels (flower- and fruit-stalks) glabrous; leaves entire (without teeth); native to northwestern Missouri . . . . . . . . . . . 1. P. GRANDIFLORUS
  b. Stem finely and minutely hairy; calyx and pedicels (flower- and fruit-stalks) hairy; leaves with short teeth; native to southwestern Missouri north to Bates County . . . . . . 2. P. COBAEA
a. Corolla with white as principal color, 1.5–3 cm. long, except in *P. Cobaea* var. *Cobaea* which is 3.5–5 cm. long; calyx during flowering period mainly 2–6 mm. long, except in *P. Cobaea* var. *Cobaea* which is 8.5–13 mm. long . . . . . . . . . . . . . . . . . . . . . . . . . *c*
  c. Corolla 3.5–5 cm. long; calyx 9–11 mm. long . . . . . . . . . . 2a. P. COBAEA var. COBEAEA
  c. Corolla 1.5–3 cm. long; calyx 2–8 mm. long . . . . . . . . . . . . . . . . . . *d*
    d. At least the lower half of the stem glabrous (without hairs) . . . . . . . . . . . . *e*
      e. Corolla 2.3–3 cm. long; axis and branches of inflorescence conspicuously covered with gland-tipped hairs during the flowering period; sepals 5–8 mm. long during the flowering period . . . . . . . . . . . . . . . . . . . . . . . . . 4. P. DIGITALIS
      e. Corolla 1.5–2.3 cm. long; axis and branches of inflorescence either glabrous (without hairs or glands) or sometimes covered with gland-tipped hairs during the flowering period; sepals 3–5 mm. long during the flowering period . . . . . . . . . . . . . . . *f*
        f. Plants usually of dry upland, prairies, or prairie remnants along railroads in many parts of Missouri, principally in the southwestern and Ozark sections; leaves usually entire (without teeth); uppermost main leaf of stem noticeably separated and distant from the lowest branch of the inflorescence; inflorescence narrow, the branches and flower-stalks (pedicels) erect and remaining parallel or close to the central axis of the inflorescence and encircling it; corolla trumpet-shaped, gradually enlarged at the throat, inside of throat minutely and microscopically covered with gland-tipped hairs . . 3. P. TUBAEFLORUS
        f. Plants of low moist woodland, swamps and wet ground in southeastern Missouri only; leaves usually toothed; uppermost leaves and bracts of stem close to the base of the lowest branch of the inflorescence; inflorescence usually much branched and uninterrupted, the branches ascending at an angle directed away from the central axis of the inflorescence; corolla conspicuously enlarged at the throat, inside of throat and corolla surface lacking glands . . . . . . . . . . . . . . . 5. P. ALLUVIORUM
    d. Lower half of stems minutely or more noticeably hairy . . . . . . . . . . . . . . *g*
      g. Corolla 1.5–1.8 cm. long; stem minutely hairy or nearly glabrous throughout; leaves glabrous (without hairs) or nearly glabrous or minutely hairy (puberulent); plants native to southwestern Missouri . . . . . . . . . . . . . . . . 7. P. ARKANSANUS
      g. Corolla 1.7–2.3 cm. long; leaves more or less hairy; plants native either to southeastern Missouri or in many portions of the state . . . . . . . . . . . . . . . . . *h*
        h. Plants of dry or open woods, bluffs, and prairies throughout much of Missouri; stem usually densely and noticeably hairy; axis and branches of inflorescence densely covered with hairs; leaves conspicuously soft-hairy on both upper and lower surface; corolla gradually and only slightly enlarged from base to summit, the lower lobes extending beyond the upper lobes, its throat much longer than the tube . . . . . . 6. P. PALLIDUS
        h. Plants of low moist woodland, swamps and wet ground in southeastern Missouri only; stem glabrous or inconspicuously short-hairy; axis and branches of inflorescence glabrous or with a few gland-tipped hairs during the flowering period; leaves usually completely glabrous, sometimes hairy; corolla conspicuously enlarged at the throat, the lower and upper lobes about equal . . . . . . . . . . . . . . 5. P. ALLUVIORUM

1. **Penstemon grandiflorus** Nutt. Map 1964
Flowers May–June.

Occurs on grassy open places of loess hills, where known only from Atchison County, northwestern Missouri (Watson, June 7, 1894, *Bush 476*; August 5, 1893, *Bush*; prairie bluffs, June 3, 1872, *Broadhead*; loess hills, 6 mi. northeast of Watson, September 1, 1932, *Steyermark 4449*).

Ranges from Wyoming to Texas, east to Wisconsin, Illinois, Missouri, and Oklahoma.

The glaucous blue-green smooth leaves and large blue-lavender corollas combine to make this a strikingly beautiful species. It does well when planted in open dry situations with a calcareous or alkaline soil. It is easily grown from seed.

1963 Chelone obliqua var. speciosa (Rose Turtlehead)

1964 Penstemon grandiflorus

1965 □ Penstemon Cobaea var. Cobaea
1965 • Penstemon Cobaea var. purpureus

2. **Penstemon Cobaea** Nutt. Map 1965

Flowers late April–June.

Two variations occur in Missouri:

Corolla white to pale lilac, irregularly lined with purple; known only from Bates County . . . 2a. P. COBAEA var. COBAEA

Corolla rich violet or rose-purple throughout; known from limestone glades of the White River area in southwestern Missouri . . 2a. P. COBAEA var. PURPUREUS

2a. **Penstemon Cobaea** var. **Cobaea** Map 1965

*Penstemon Cobaea* Nutt. [G, BB, P & S, Steyerm.]

*Penstemon Cobaea* subsp. *typicus* [Pennell]

Known only from open shaly limestone mounds of western Missouri in Bates County (on top of open shaly limestone outcrops on mound along Miami Creek, 2 mi. southeast of Merwin, T41N, R33W, sect. 13, June 2, 1938, *Steyermark 5715*).

Ranges from Texas, Oklahoma, Kansas, and Nebraska to Missouri; introduced in Ohio and Arizona from cultivated specimens.

2b. **Penstemon Cobaea** var. **purpureus** Pennell Map 1965

*Penstemon Cobaea* subsp. *purpureus* Pennell [Pennell]

Known only from limestone glades and bald knobs of the White River and tributaries of southwestern Missouri in Ozark, Taney, Christian, Stone, and Barry counties.

Known only from the White River section of Missouri and adjacent Arkansas.

The violet or rose-purple var. *purpureus* is a strikingly beautiful plant. It is a conspicuous member of the flora of the limestone glades of southwestern Missouri, where it is associated with such other showy species as *Callirhoe digitata* var. *digitata*, *Oenothera missouriensis*, *Baptisia australis* var. *minor*, *Stenosiphon linifolius*, and *Echinacea paradoxa* and *E. pallida*. In the states west of Missouri, *P. Cobaea* var. *Cobaea* is eaten by livestock and is nutritious.

Some current manuals combine var. *purpureus* and var. *Cobaea* into one taxon. I have grown both variants and studied them in the field for the past twenty years and they have maintained their characteristic colors. Since the purple-flowered variation is locally restricted to the glades of southwestern Missouri and adjacent Arkansas, and because the color is maintained within a distinct geographical area, it should merit varietal recognition. It is a handsome addition to the limestone rock garden or perennial bed and prospers in a sunny open situation in dry calcareous soil. It can be grown easily from seed and new rosettes are produced yearly. Care must be taken that these are not overcrowded by other plants, as severe competition tends to weaken and kill out the plants.

3. **Penstemon tubaeflorus** Nutt. Map 1966

Flowers May–June.

Occurs in usually dry soils of prairies, prairie remnants along railroads, rocky glades of limestone and cherty limestone, openings in woods, and open woods. Common in the western Ozarks and unglaciated prairie region of the southwestern portion of the state, locally east and north to Dunklin, Cape Girardeau, Franklin, Boone, Monroe, Marion, Shelby, and Putnam counties.

Ranges from Nebraska to Texas, east to Mississippi, Tennessee, Indiana, and Wisconsin; introduced from Maine to Pennsylvania.

The flower-stalks vary from glabrous to densely glandular.

4. **Penstemon Digitalis** Nutt. Map 1967

Flowers May–July.

Occurs in rich or low alluvial woodland, borders of woods, thickets, meadows, prairies, fallow fields, and

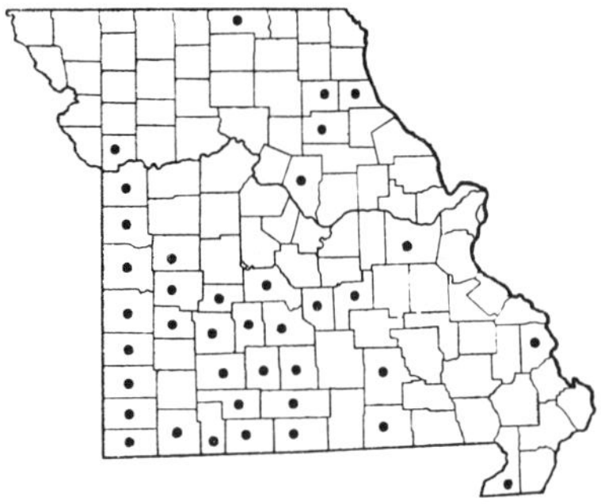

1966 Penstemon tubaeflorus

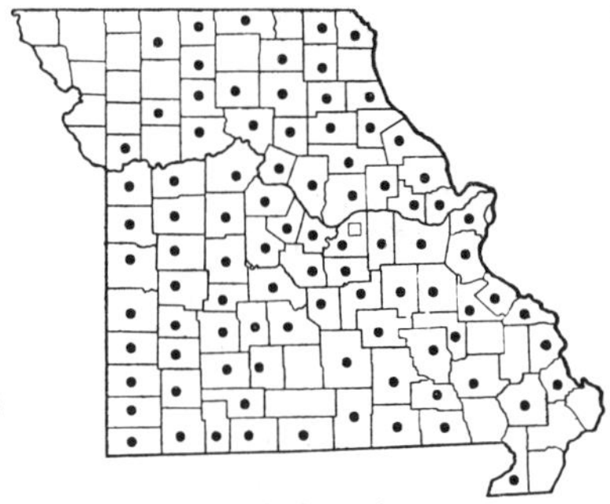

1967 ● Penstemon Digitalis f. Digitalis
1967 □ Penstemon Digitalis f. Baueri

1968 Penstemon alluviorum

along railroads. Throughout Missouri, except not recorded and apparently absent from the extreme northwestern sector.

Ranges from Maine and Quebec to South Dakota, south to Alabama, Louisiana, and Texas.

Two variations occur in Missouri:

Leaves occurring opposite in pairs . . . . 4a. P. DIGITALIS f. DIGITALIS
Leaves whorled, in 3's . 4b. P. DIGITALIS f. BAUERI

4a. **Penstemon Digitalis** f. **Digitalis** Map 1967
*Penstemon Digitalis* Nutt. [G, BB, P & S, Steyerm.]
This is the common variation in the state.

4b. **Penstemon Digitalis** f. **Baueri** Steyerm. Map 1967

Known only from Osage County (wooded southwest-facing limestone bluffs along Maries River, T43N, R10W, sect. 18, 3 mi. northeast of Westphalia, July 1, 1939, *Steyermark 27665*).

This species often forms large colonies, especially in low ground, creating a conspicuous effect with the large, white-flowered panicles.

5. **Penstemon alluviorum** Pennell Map 1968
Flowers May–June.

Occurs in wet soil in low moist woodland or open swampy places. Lowlands of southeastern Missouri in Cape Girardeau, Stoddard, Wayne, Butler, and Dunklin counties.

Ranges from Mississippi and Arkansas, north to Indiana, Illinois, and Missouri.

This is separated from *P. Digitalis* mainly by the size of the corolla. Future studies may show that this is best regarded as a small-flowered variation of *P. Digitalis*.

6. **Penstemon pallidus** Small Map 1969
*Penstemon arkansanus* Pennell var. *pubescens* Pennell [Pennell]
Flowers April–July.

Occurs in dry or rocky open woods or glades, usually in acid soils of sandstone, chert, or granite, but also on limestone, along bluff ledges and escarpments, prairies, openings in woods, and along railroads.

Common throughout eastern, southern, and central Missouri, west to Harrison, Grundy, Livingston, Carroll, Saline, Johnson, and Cass counties.

Ranges from Maine to New York, Michigan, and Iowa, south to Georgia, Tennessee, Arkansas, and Kansas.

The pubescence of the stems is usually longer (with hairs 1 mm. or more long) and more conspicuous in this species than in any of the other species in Missouri. However, it varies considerably in the length and density of the hairs. The pubescence of the leaves also varies considerably, being usually denser on the lower surface. The corolla varies from white to tinged with lavender.

7. **Penstemon arkansanus** Pennell Map 1970
Flowers April–June.

Occurs in rocky open woods, bluff ledges and escarpments, and glades. Southwestern Missouri, east to Ozark County and north to Christian and Newton counties.

Ranges from Missouri and Arkansas to Oklahoma and Texas.

Gleason (*New Ill. Fl.* 3: 224. 1952) reduces this to

Plate no. 322. **1.** Penstemon Cobaea var. purpureus, × 2/7; *a, b.* Upper and lower portions of plant. **2.** Penstemon Digitalis, × 2/7; *a, b.* Upper and lower portions of plant. **3.** Penstemon tubaeflorus, × 2/7; *a, b.* Upper and lower portions of plant. **4.** Penstemon alluviorum, × 2/7; *a, b.* Upper and lower portions of plant.

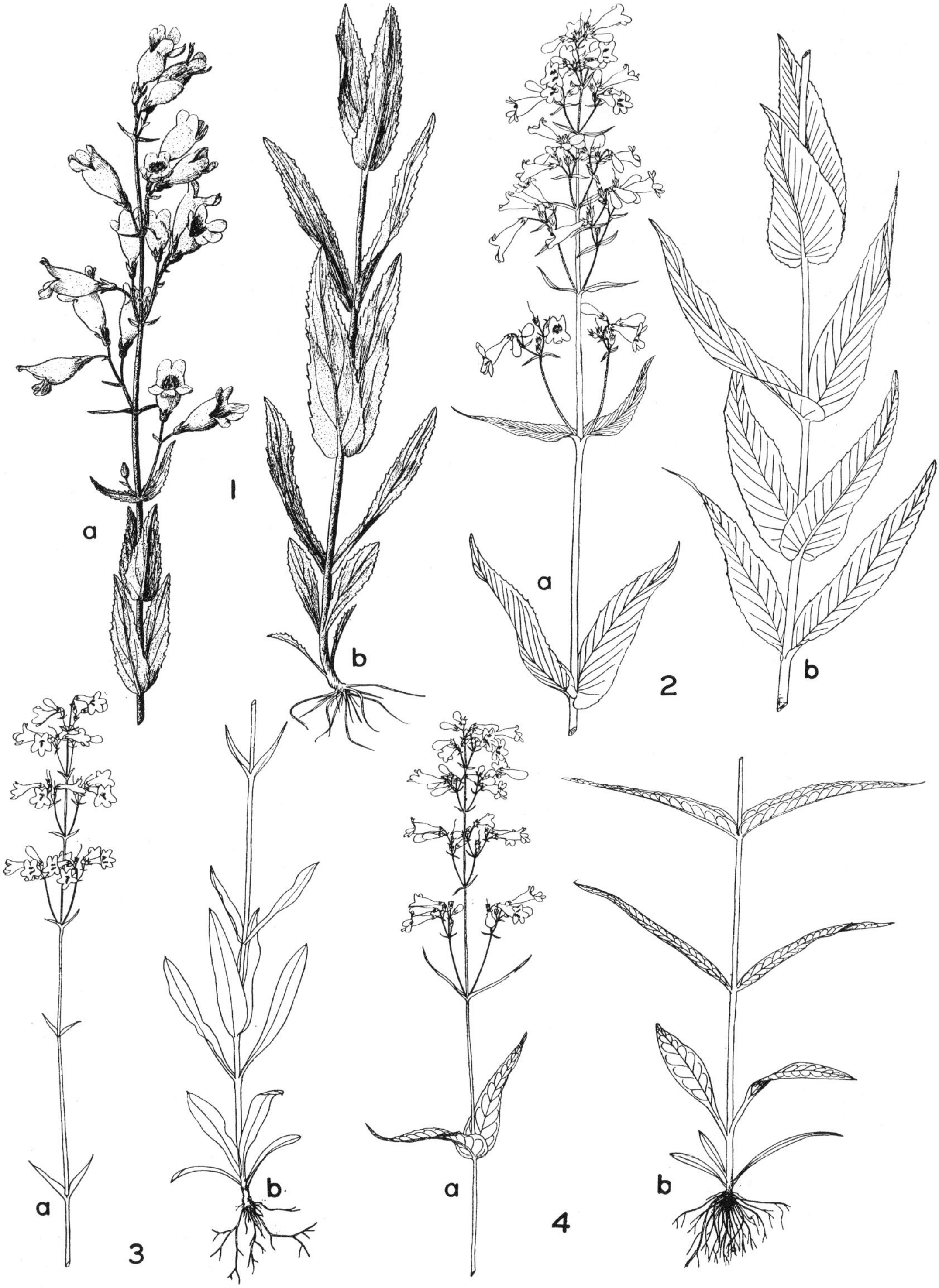


PLATE NO. 322

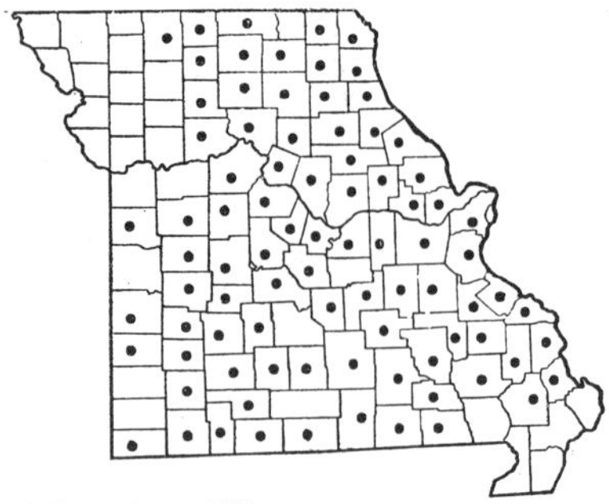

1969 Penstemon pallidus

1970 Penstemon arkansanus

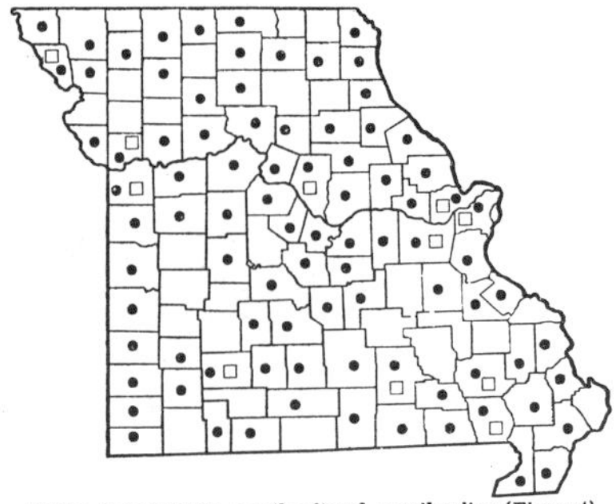

1971 • Scrophularia marilandica f. marilandica (Figwort)
1971 □ Scrophularia marilandica f. neglecta

synonymy under *P. pallidus*. Until more detailed field and experimental studies have been carried out, I am retaining it in specific rank. In addition to the characters given in the key, it usually has reddish-purple instead of green stems, and the stems are often clustered instead of being solitary. The leaves are usually thinner than those of *P. pallidus* and darker olive-green instead of pale green. The leaves and stems vary from glabrous to minutely hairy.

*Excluded Species*

**Penstemon calycosus** Small

This species was reported by Palmer and Steyermark in their *Annotated Catalogue* from St. Louis and Texas counties. Specimens so identified have now been referred to *P. Digitalis*. The species is distinguished from *P. Digitalis* by the glabrous instead of bearded anthers, and the purplish instead of predominantly white corolla.

## 12. **Scrophularia** L. Figwort

Commonly encountered, found throughout Missouri; plant flowering from July–October; corolla 5–8 mm. long; petioles (leaf-stalk) mainly 3–8 cm. long, without wing-like margins; inflorescence very broad, 5–18 cm. broad, the branches usually spreading; the filament of the stamen which does not bear an anther is dark purple or brown . . . . . . . . . . . . . . . . . . 1. S. MARILANDICA

Known only from Jackson County, west-central Missouri; plant flowering May–June; corolla 7–11 mm. long; petioles (leaf-stalk) mainly 1–3 cm. long and with conspicuous wing-like margins; inflorescence more narrowly elongated, 4–8 cm. broad, the branches strongly ascending; the filament of the stamen which does not bear an anther is greenish-yellow . . . . . . . . . . . . . 2. S. LANCEOLATA

1. **Scrophularia marilandica** L. Figwort Map 1971

Also called Carpenter's Square.

Flowers July–October.

Occurs in rich woodland, often in low or alluvial ground, ravines, thickets, and borders of woods. Throughout Missouri, doubtless in every county.

Ranges from Maine and Quebec to Minnesota, south to Georgia, Alabama, Louisiana, and Oklahoma.

Two variations occur in Missouri:

Lower surface of leaf-blade glabrous (without hairs) or nearly so . . . 1a. S. MARILANDICA f. MARILANDICA

Lower surface of leaf-blade more or less densely hairy. . . . 1b. S. MARILANDICA f. NEGLECTA

1a. **Scrophularia marilandica** f. **marilandica** Map 1971

*Scrophularia marilandica* L. [G, BB, P & S]

This is the form commonly encountered throughout the state.

1b. **Scrophularia marilandica** f. **neglecta** (Rydb.) Pennell Map 1971

Scattered throughout the range in Missouri.

The status of this hairy-leaved extreme is uncertain, and I am retaining it subject to future work which may furnish additional information as to its relative merits.

2. **Scrophularia lanceolata** Pursh Map 1972

Flowers May–June.

Known only from west-central Missouri in Jackson County (Sheffield, *Bush 812*).

1972 Scrophularia lanceolata

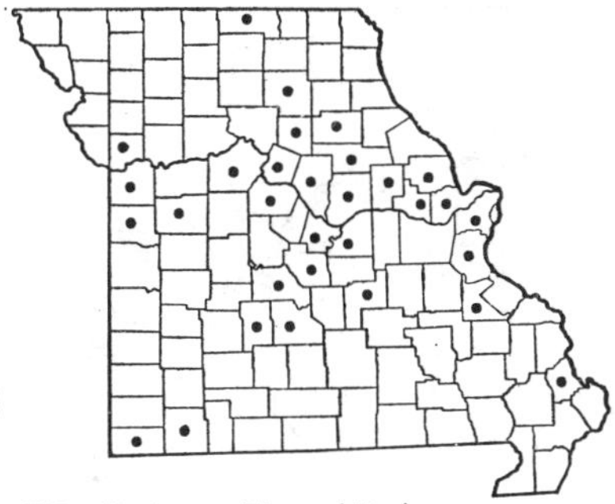

1973 Collinsia verna (Blue-eyed Mary)

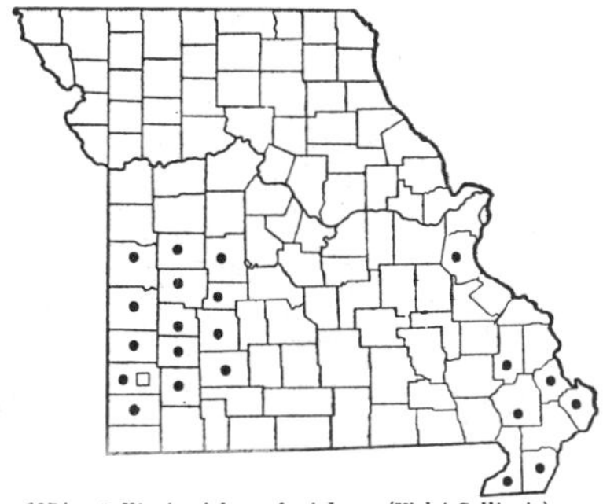

1974 • Collinsia violacea f. violacea (Violet Collinsia)
1974 □ Collinsia violacea f. pallida

Ranges from Quebec to British Columbia, south to Virginia, Indiana, Illinois, Missouri, Oklahoma, New Mexico, Utah, and California.

The only certain character for distinguishing this species from the preceding is the color of the filament of the sterile stamen. However, the additional characters included in the key should serve to distinguish this species, when encountered.

## 13. **Collinsia** Nutt.

Plants usually of low, rich, or moist woodland, usually in valleys; lower lip of corolla bright blue; lobes of corolla with an indentation or notch 0.5–1 mm. deep; main stems often with minute hairs occurring in lines, sometimes hairs scattered over most of surface; main leaves of stem usually broadest just above the base, ovate or ovate-lanceolate, partly clasping by a heart-shaped base; seeds 2–4 in the capsule . 1. C. VERNA

Plants of open sandstone or chert glades and barrens, sandy soils, along roadsides, and in open dry upland woods and ridges; lower lip of corolla violet or purple or rarely the corolla all white; lobes of corolla with an indentation or notch 2–3 mm. deep; stems with hairs more uniformly occurring over the surface, not in lines; main leaves of stem oblong, lanceolate, or elliptic, more uniform in width, often broader or broadest near the middle or well above the base, but also sometimes broadest at base; seeds 6–12 in the capsule . . . . . . . . . . . . . . . . . . . . . . 2. C. VIOLACEA

1. **Collinsia verna** Nutt. Blue-eyed Mary
Map 1973

Flowers April–June.

Occurs in moist, low or rich woodland, generally in valleys and ravine bottoms near or adjacent to streams. Commonest in middle and central Missouri, absent from most of the Ozark region, locally southeast in Scott County, locally southwest in Barry and McDonald counties, and north locally in Macon and Putnam counties.

Ranges from New York to Wisconsin and Iowa, south to Virginia, Kentucky, Arkansas, and Kansas.

This is a winter annual, the seeds germinating in late fall, overwintering as young plants which commence to grow in early spring. In favorable situations in low rich woods acres are covered by dense stands of this attractive plant. It can be easily grown from seed, but is killed out unless competing plants, which crowd the ground necessary for seedling development by the *Collinsia*, are weeded out.

2. **Collinsia violacea** Nutt. Violet Collinsia
Map 1974

Flowers April–June.

Occurs in usually acid soils of sandstone or chert glades, barrens, and outcrops, sandy soils along roadsides, in open dry upland woods and ridges, and openings of woodland. Mainly in the unglaciated prairie region of southwestern Missouri north to Benton, Henry, and Bates counties, and in sandy areas of southeastern Missouri north to Scott and Bollinger counties, locally northeast in Jefferson County.

Ranges from Texas to Kansas, east to Missouri and Illinois.

Two variations occur in Missouri:

Corolla completely white . . 2b. C. VIOLACEA f. PALLIDA

Lower lip of corolla violet or purple, upper lip white, rarely pale lilac-tinged . . 2a. C. VIOLACEA f. VIOLACEA

2a. **Collinsia violacea** f. **violacea** Map 1974

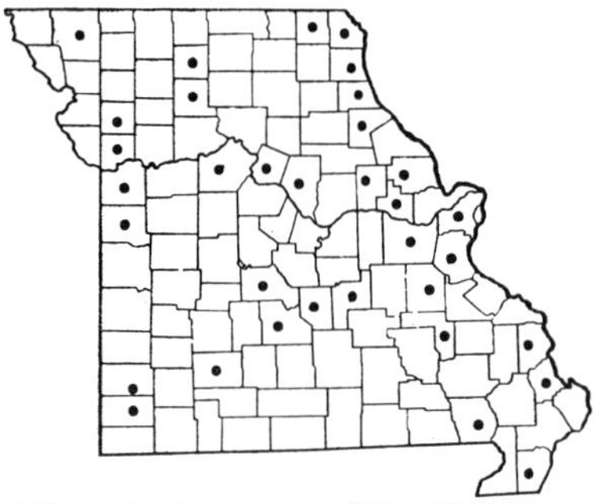

1975 Linaria vulgaris f. vulgaris (Butter-and-Eggs)

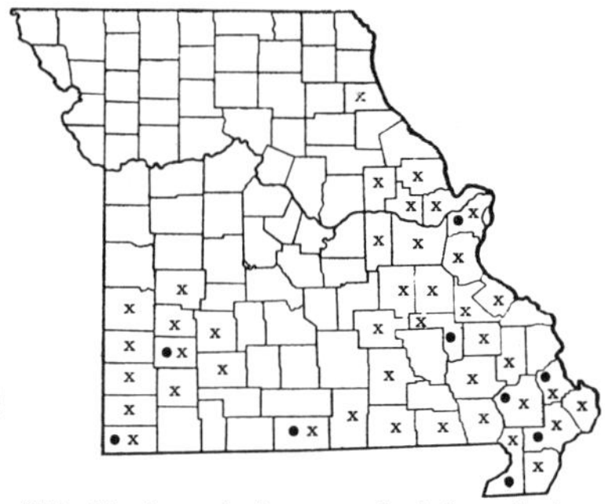

1976 • Linaria canadensis var. canadensis f. canadensis (Blue Toadflax)

1976 x Linaria canadensis var. texana

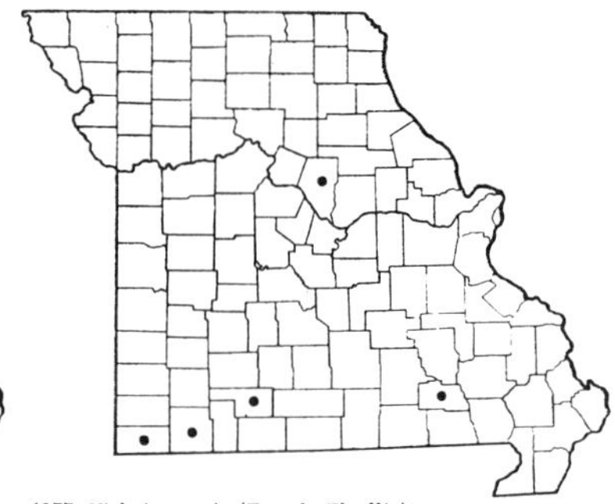

1977 Kickxia spuria (Female Fluellin)

*Collinsia violacea* Nutt. [G, BB, P & S, Steyerm.]
This is the common variation in Missouri.

2b. **Collinsia violacea** f. **pallida** Palmer Map 1974

Known only from Jasper County, southwestern Missouri (rocky open upland woods and clearings, 3 mi. southwest of Webb City, April 28, 1949, *Palmer 45867*, holotype in Palmer Herb., isotype at Chi. Nat. Hist. Mus.).

Like *C. verna*, this is also a winter annual. The flowers are slightly smaller than in *C. verna*. The upper lip is usually white or tinged with lilac, while the lower lip is violet or purple. This species often occurs in thick, dense stands along roadsides in the southwestern portion of the state, creating a spectacular color effect.

## 14. **Linaria** Mill. Toadflax

Corolla yellow and orange, 2–3 cm. long (including spur); capsule 7–12 mm. long; leaves 2–7 cm. long, 1.5–5 mm. broad . . . . . . . . . . . . . . . . . . . . . . . . . 1. L. VULGARIS
Corolla bluish or purple 0.8–2 cm. long (including spur); capsule 2–5 mm. long; leaves 1–3 cm. long, 0.5–2 mm. broad . . . . . . . . . . . . . . . . . . . . . . . . 2. L. CANADENSIS

1. **Linaria vulgaris** Hill f. **vulgaris**
Butter-and-Eggs Map 1975

*Linaria vulgaris* Hill [G, BB, P & S, Steyerm.]

Flowers May–November.

Occurs in fields, pastures, waste ground, along road sides, and railroads.

Native of Europe; introduced and naturalized from Newfoundland to British Columbia, south to Florida, Texas, and California.

Although an attractive plant, it spreads rapidly in a garden, once it becomes established, and is then difficult to get rid of.

A form with whitish corolla, f. *leucantha* Fern., and another with a 3- or 5-spurred more or less symmetrical corolla, f. *Peloria* (L.) Rouleau, sometimes occur, but have not yet been recorded from Missouri.

2. **Linaria canadensis** (L.) Dumort. Blue Toadflax Map 1976

Also called Old-field Toadflax.

Flowers April–September.

Occurs in sandy open ground, sandstone, chert, and granite glades, outcrops, bluff escarpments, rocky prairies, along roadsides, and railroads.

Two variations occur in Missouri:

Corolla 5–10 mm. long, the spur 2–6 mm. long; less commonly encountered . 2a. L. CANADENSIS var. CANADENSIS f. CANADENSIS
Corolla 10–12 mm. long, the spur 5–9 mm. long; commonly encountered . . 2b. L. CANADENSIS var. TEXANA

2a. **Linaria canadensis** var. **canadensis** f. **canadensis** Map 1976

---

Plate no. 323. **1.** Penstemon pallidus, × 2/7; *a.* Flower, × 1 1/7. **2.** Scrophularia marilandica, × 2/7; *a.* Flower, × 4/7; Details from Small, The New York Botanical Garden. **3.** Collinsia violacea, × 2/7. **4.** Collinsia verna, × 2/7; Details from Small, The New York Botanical Garden. **5.** Linaria vulgaris, × 2/7; Details from Small, The New York Botanical Garden. **6.** Antirrhinum majus, × 2/7; *a.* Seed, × 7 1/7; After Gleason, The New York Botanical Garden. **7.** Chaenorrhinum minus, × 2/7; *a.* Fruit, × 1 3/7; *b.* Flower, × 6/7. **8.** Linaria canadensis, × 2/7; *a.* Linaria canadensis var. canadensis; *b.* Linaria canadensis var. texana; Details from Small, The New York Botanical Garden. **9.** Penstemon arkansanus, × 2/7; *a, b.* Upper and lower portions of plant. **10.** Scrophularia lanceolata, × 2/7.

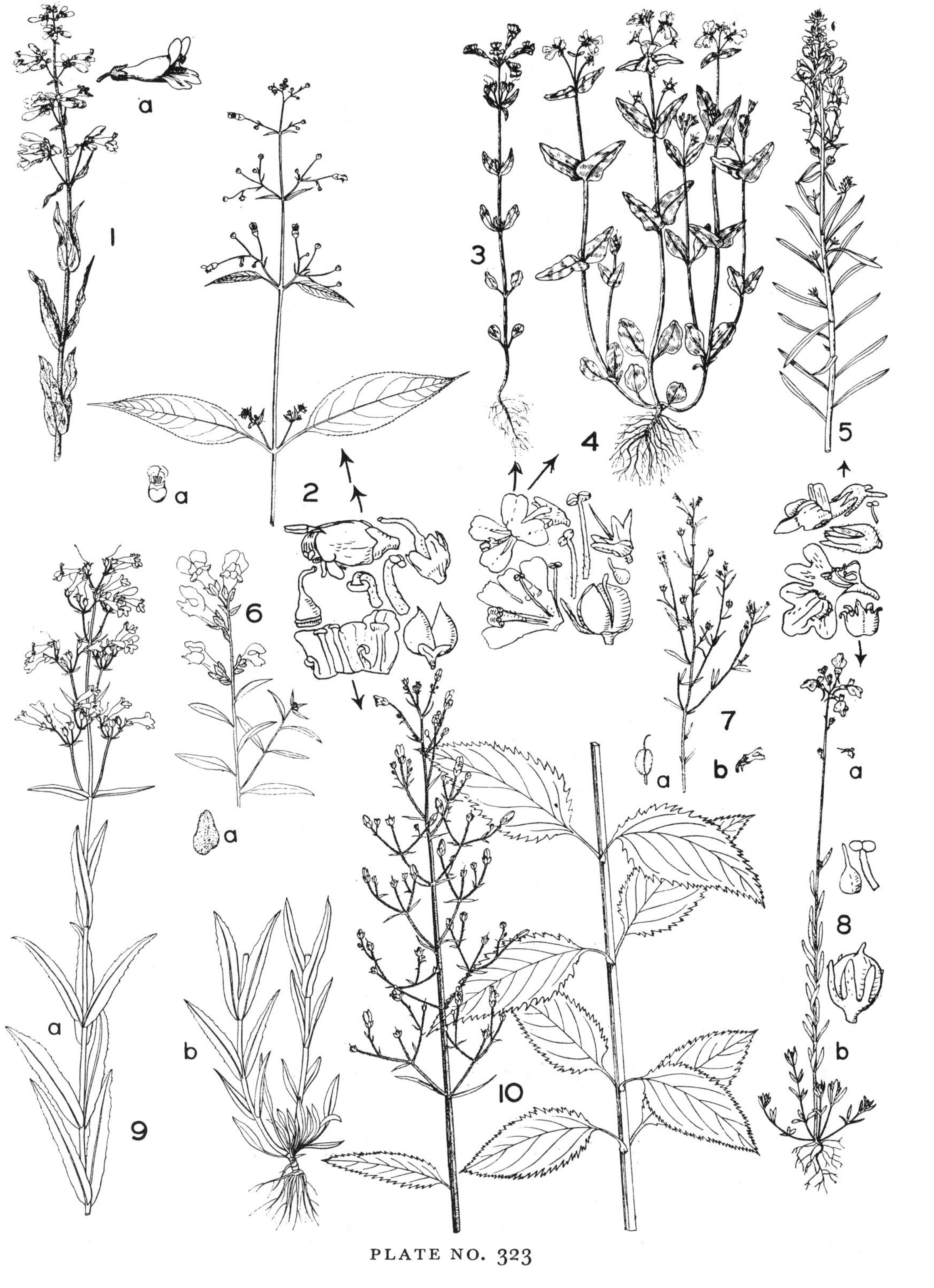


PLATE NO. 323

*Linaria canadensis* (L.) Dumort. [G, P & S, Steyerm.]

*Linaria canadensis* var. *canadensis* [BB]

Scattered in southern and central Missouri northeast to St. Louis County.

Ranges from Florida to Texas, north to Nova Scotia, Quebec, Ontario, Michigan, Wisconsin, Minnesota, and South Dakota.

A form of var. *canadensis* with white corollas, f. *albina*, has not been recorded yet from Missouri. Another form has been described, f. *cleistogama* Fern., in which the corolla is reduced to a closed small stub, but this appears to be part of a normal stage found in the plant, so that I believe that it should not be recognized for more than this.

2b. **Linaria canadensis** var. **texana** (Scheele) Pennell Map 1976

Southern and east-central Missouri, north to Lincoln, Montgomery, Gasconade, St. Clair, and Vernon counties, and locally north in Marion County; absent apparently from the central Ozark section.

Ranges from Florida to Mexico, north to Virginia, Missouri, South Dakota, and British Columbia.

This is an attractive plant when showing the slender racemes of bluish-lavender flowers. Slender leafy gray- or silvery-green runners forming rosettes are produced from the base of the plant and overwinter.

## 15. **Kickxia** Dum. Cancerwort

Leaves rounded or somewhat heart-shaped at base, without lobes projecting from the base and without any teeth; flower- and fruit-stalks hairy their entire length; calyx-lobes lanceolate . . . . . . 1. K. SPURIA

Upper and middle leaves arrowhead-shaped or with lobes projecting from the base, lowest leaves more or less few-toothed; flower- and fruit-stalks glabrous (without hairs) for most of their length or hairy; calyx-lobes becoming ovate and somewhat heart-shaped . . . . . . . . . . . . 2. K. ELATINE

1. **Kickxia spuria** (L.) Dumort. Female Fluellin Map 1977

Flowers May–September.

Occurs in rocky waste ground, and along sand and gravel bars of streams. Scattered and rare, known only from Carter, Christian, Barry, McDonald, and Boone counties.

Native of Europe; introduced and naturalized in the United States from Florida and Alabama to Rhode Island; also Missouri.

This species is doubtfully distinct from the following and is being retained for the present, subject to future studies in field and herbarium. Specimens are found which combine characters of this and the following species. The pubescence of the pedicels is a character of questionable value. Specimens which can be referred to either species often occur together. *Steyermark 85859* from Christian County, identified as *K. Elatine*, has hastate leaves and the middle portion of some pedicels glabrous, with the base and summit of the same pedicels pubescent, while other pedicels on the same specimen are entirely pubescent as in *K. spuria*. In *Steyermark 85864* from the same creek bed and occurring near no. *85859*, the plant is more luxuriant with none of the leaves hastate and the pedicels are pubescent throughout, although in a younger stage. This latter collection has been referred doubtfully to *K. spuria*, although it may well be only a variation of *K. Elatine*.

2. **Kickxia Elatine** (L.) Dumort. Cancer Root Map 1978

Flowers late May–October.

Occurs along sand and gravel beds of streams, waste ground, and along railroads. Common in the Ozark region along most of the streams, north in central Missouri to St. Louis, Franklin, Gasconade, Cooper, Jackson, and Clay counties.

Native of Europe; introduced and naturalized in the United States from Georgia to Louisiana, north to Massachusetts, Indiana, Illinois, Missouri, and Kansas.

See discussion under previous species, *K. spuria*, concerning intergradation of the two species.

## 16. **Cymbalaria** Hill Kenilworth Ivy

**Cymbalaria muralis** Gaertn. Kenilworth Ivy Map 1979

*Linaria Cymbalaria* (L.) Mill. [P & S]

Flowers May–October.

Occurs in waste ground and along roadsides. Rare, known only from Marion (Mississippi banks, Hanni-

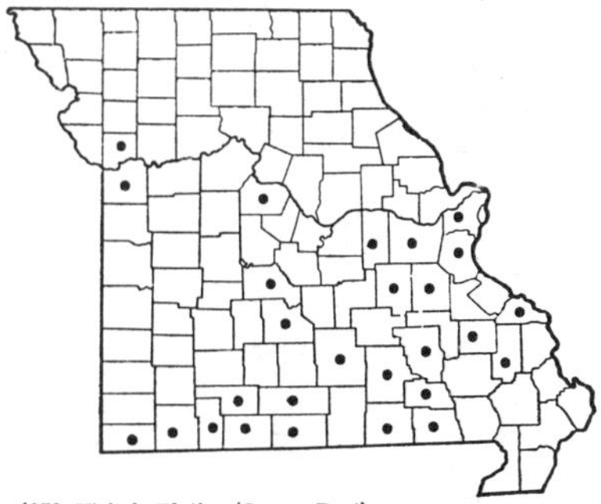

1978 Kickxia Elatine (Cancer Root)

1979 Cymbalaria muralis (Kenilworth Ivy)

1980 Antirrhinum majus (Snapdragon)

bal, June 11, 1918, *Davis 8762*; Johnson Terrace, south of Hannibal, June 12, 1912, *Davis 8962*; south Hannibal, May 16, 1921, *Davis 4434*) and St. Louis (Kirkwood, August, 1913, *J. M. & Milton Greenman 3697*) counties.

Native of Europe; introduced from cultivation and escaped and naturalized from Massachusetts to Ontario and South Dakota, south to South Carolina, Illinois, and Missouri; also British Columbia to California.

This is an attractive trailing perennial plant with long-stalked flowers. The corolla is blue-lavender with a yellow palate protruding from the lower lip. The plant does well in shade and in cool, moist situations, seeding readily. It is a good subject for hanging baskets for indoors.

## 17. **Antirrhinum** L. Snapdragon

**Antirrhinum majus** L. Snapdragon Map 1980

Flowers June–November.

Escaped from cultivation and rarely naturalized along railroads and waste ground. Known as such only from St. Louis County (St. Louis, Carrie Ave. freight yard of Rock Island R.R., August 25, 1956, *Muehlenbach 1050*).

Native of Europe; naturalized from Nova Scotia to Ontario, south to Virginia, Illinois, Missouri, and Kansas.

This commonly grown ornamental perennial is rarely found escaped from gardens.

## 18. **Chaenorrhinum** Reichenb.

**Chaenorrhinum minus** (L.) Lange Dwarf Snapdragon Map 1981

*Linaria minor* (L.) Desf.

Flowers June–September.

Occurs along railroads. Northern and central Missouri, south to St. Louis, Montgomery, Livingston, and Clay counties.

Native of Europe; introduced and naturalized in North America from Quebec to Ontario, south to Virginia, Ohio, Indiana, Illinois, and Missouri.

This was first reported as introduced in the state in 1946 in Pike and Audrain counties by Mr. Bill Bauer, and in 1947 was found in Putnam and Clay counties (Rh. 51: 118. 1949).

The plant is an annual with glandular-hairy stems and branches, 0.5–3 dm. tall, with linear leaves 1–2 cm. long, flower- and fruit-stalks 10–15 mm. long borne from most of the leaf-axils, and small blue-lavender corollas 5–8 mm. long with a yellow throat.

## 19. **Veronicastrum** Fabr. Culver's Root

**Veronicastrum virginicum** (L.) Farw. Culver's Root Map 1982

Flowers June–September.

Occurs in open woods, thickets, prairies, swampy meadows, and sometimes along railroads. Throughout Missouri.

Ranges from Vermont to Ontario and Manitoba, south to Florida, Louisiana, and Texas.

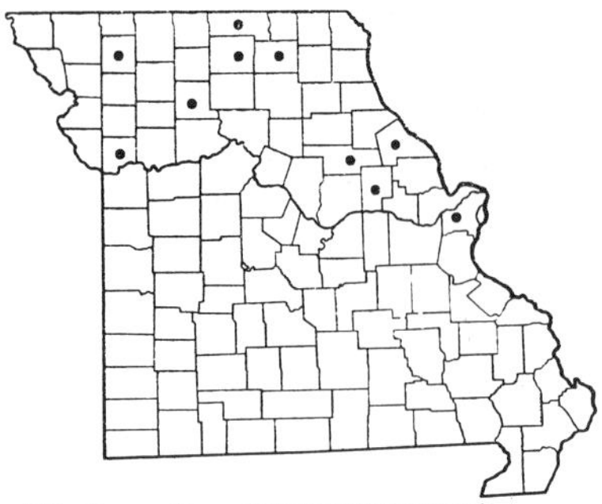

1981 Chaenorrhinum minus (Dwarf Snapdragon)

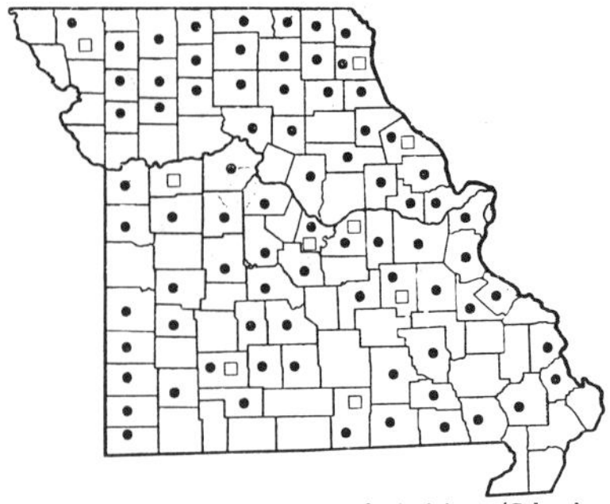

1982 ● Veronicastrum virginicum f. virginicum (Culver's Root)

1982 □ Veronicastrum virginicum f. villosum

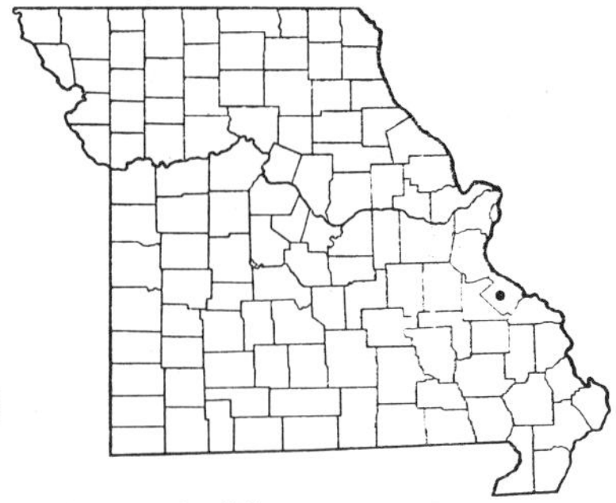

1983 Veronica longifolia

Two intergrading variations occur in the state:

Lower surface of leaves densely hairy with relatively long or noticeable hairs; stems usually more or less hairy but also glabrous . . . . b. V. VIRGINICUM f. VILLOSUM

Lower surface of leaves glabrous (without hairs) or with short and inconspicuous hairiness; stems usually glabrous (without hairs) or with few hairs . . . . . a. V. VIRGINICUM f. VIRGINICUM

a. **Veronicastrum virginicum** (L.) Farw. f. **virginicum** Map 1982

*Veronicastrum virginicum* (L.) Farw. [G, BB, P & S]

This is the more common variation and is found throughout the state.

b. **Veronicastrum virginicum** f. **villosum** (Raf.) Pennell Map 1982

Rare and scattered over the state.

The root of this species contains a substance, *leptandrin*, having powerful cathartic and emetic properties, which have been employed in medicine for digestive disorders, although the use of the drug has been suspected of being unsafe.

### 20. **Veronica** L. Speedwell

a. Flowers closely crowded in a long upright, spike-like inflorescence 7–20 cm. long; leaves long-pointed and sharply toothed with numerous teeth; plant erect, 3–15 dm. tall . . . . . . 1. V. LONGIFOLIA
a. Without the above combination of characters . . . . . . . . . . . . . . . . . *b*
  b. Most of the main foliage leaves of the stem and branches less than 1.5 cm. (sometimes up to 2 cm.) long . . . . . . . . . . . . . . . . . . . . . . . . . *c*
    c. Fruiting pedicels (stalks) 5–12 mm. long; flower-pedicels (stalks) longer than the calyx; corolla about 5–6 mm. broad . . . . . . . . . . . . . . . . . . 7. V. POLITA
    c. Flowers and fruits nearly sessile (without stalks) or the pedicels (stalks) up to 4 mm. long; corolla 2–4.5 mm. broad . . . . . . . . . . . . . . . . . . . . . . . *d*
      d. Stems noticeably hairy with spreading hairs; leaves mainly hairy or nearly so; flowers and fruits nearly sessile (without stalks) or the pedicels (stalks) up to 1.5 mm. long; corolla blue, about 2 mm. broad . . . . . . . . . . . . . . . . . . . 6. V. ARVENSIS
      d. Stem nearly smooth or with very short upwardly curved hairs 0.2–0.4 mm. long, visible with a magnifying lens; leaves mainly glabrous; flowers and fruits on pedicels (stalks) 2–4 mm. long; corolla white or pale blue, 2–4 mm. broad . . . . . 2. V. SERPYLLIFOLIA
  b. Most of the main foliage leaves of the stem and branches more than 1.5 cm. long . . . . . . *e*
    e. Flowers occurring solitarily or singly in the axils of the upper leaves, which are alternate; main leaves of stem 0.8–3 cm. long; weak-rooted annuals . . . . . . . . . . . . . *f*
      f. Pedicels of flowers or fruits absent or up to 1 mm. long, shorter than the calyx; stem erect or mainly so; corolla white, scarcely longer than the calyx; seeds smooth, less than 1 mm. long . . . . . . . . . . . . . . . . . . . . . . . . 5. V. PEREGRINA

Plate no. 324. **1.** Kickxia spuria, × 2/7; Details from Small, The New York Botanical Garden. **2.** Kickxia Elatine, × 2/7; *a.* Flower, × 1 1/7. **3.** Veronicastrum virginicum, 2/7; Details from Small, The New York Botanical Garden. **4.** Cymbalaria muralis, × 2/7. **5.** Veronica americana, × 2/7. **6.** Veronica serpyllifolia, × 1; Details from Small, The New York Botanical Garden. **7.** Veronica comosa, × 2/7; Details from Small, The New York Botanical Garden.

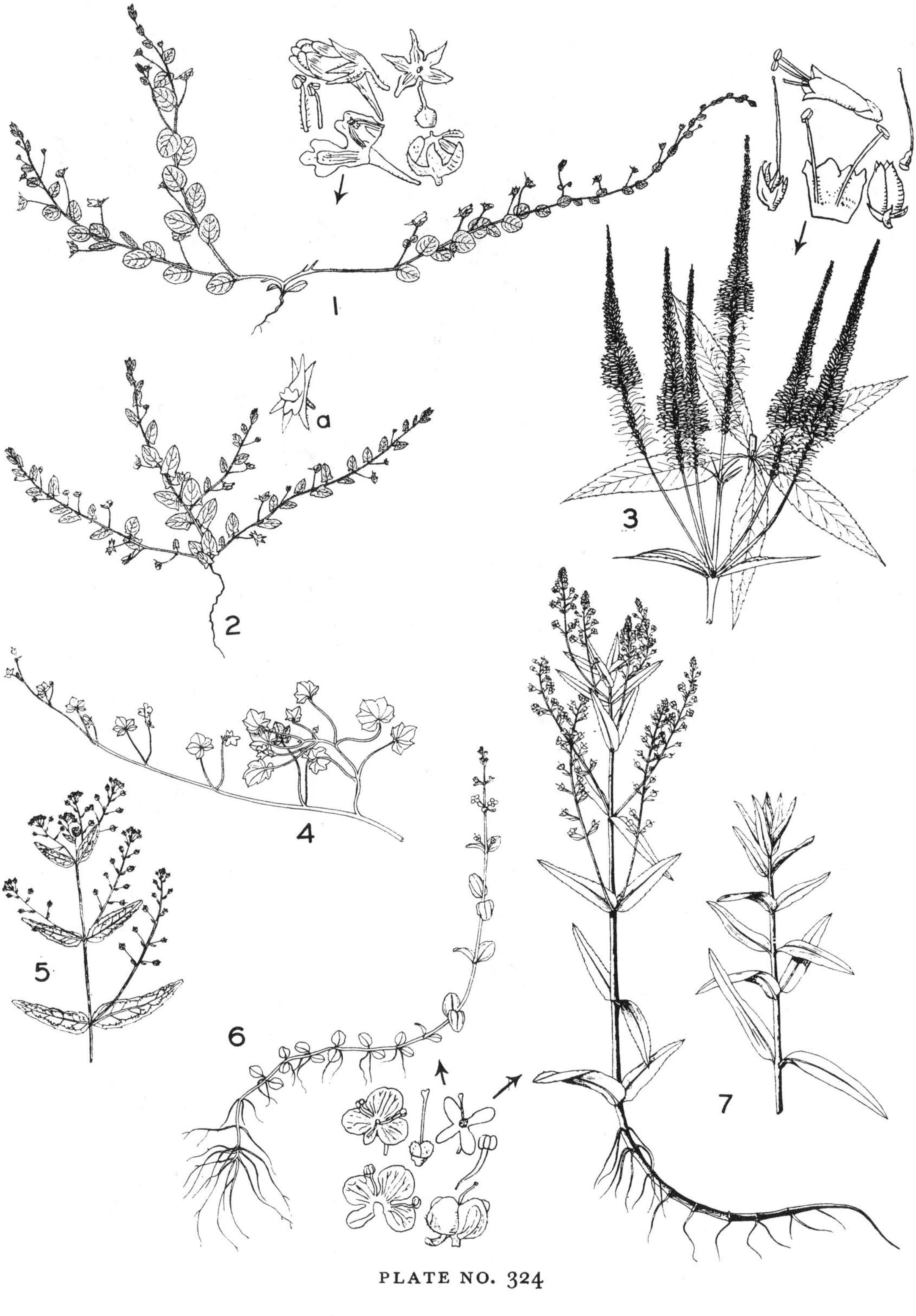


PLATE NO. 324

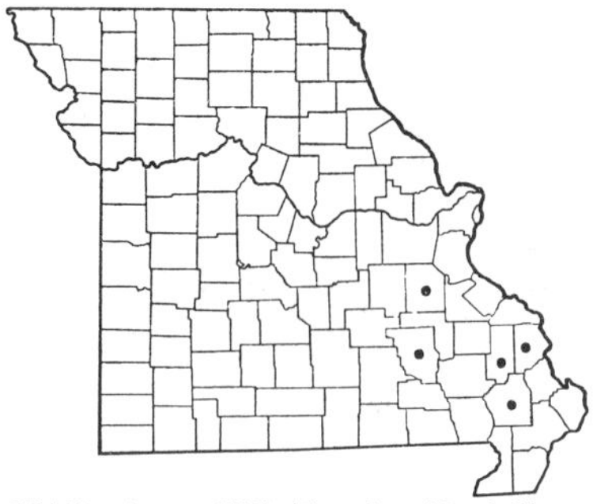

1984 Veronica serpyllifolia (Thyme-leaved Speedwell)

1985 Veronica americana (American Brooklime)

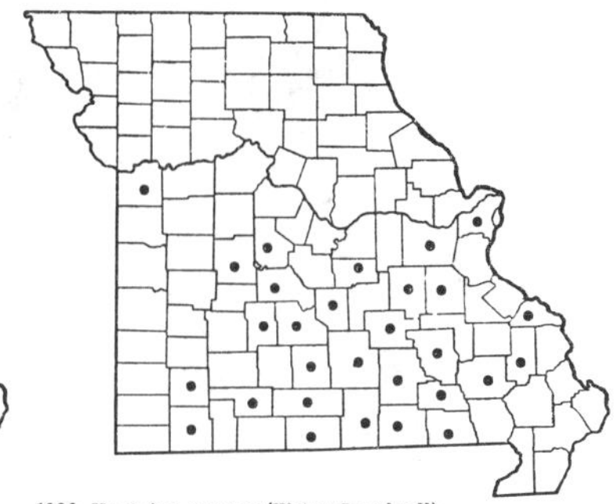

1986 Veronica comosa (Water Speedwell)

f. Pedicels of flowers or fruits 4–12 mm. long, longer than the calyx; stem lying on the ground or reclining; corolla blue, longer than calyx; seeds roughened, 1.3–1.5 mm. long . 7. V. POLITA

e. Flowers occurring in definite inflorescences (racemes) on special branches (axes) produced from the sides of the stems and from the axils of opposite leaves; main leaves of stem usually 3–8 cm. long (sometimes shorter); perennial . . . . . . . . . . . . . . . . . . . *g*

g. Most of the leaves on petioles (stalks) 3–10 mm. long; flowers and fruits on pedicels (stalks) 6–13 mm. long . . . . . . . . . . . . . . . . . . . . . 3. V. AMERICANA

g. At least the upper leaves and those subtending the flowering stems sessile (without stalks); flowers and fruits on pedicels (stalks) 3–6 mm. long . . . . . . . . . . 4. V. COMOSA

1. **Veronica longifolia** L. Map 1983

Flowers April–August.

Escaped from cultivation along roadsides. Known only from Ste. Genevieve County, southeastern Ozark region (along road to Chimney Rocks, April 30, 1955, *Robert H. Mohlenbrock 5926*).

Native of Europe; introduced and naturalized in North America from Newfoundland to Quebec and North Dakota, south to Maryland, Ohio, and Missouri.

This is often cultivated as a border plant for perennial beds. The flowers are usually deep blue.

2. **Veronica serpyllifolia** L. Thyme-leaved Speedwell Map 1984

*Veronica serpyllifolia* var. *serpyllifolia* [BB]

Flowers April–July.

Occurs in low wet woods, along spring branches, sandy wet shaded slopes, grassy places, and waste ground. Rare, and known only from southeastern Missouri in Washington, Cape Girardeau, Bollinger, Reynolds, and Stoddard counties.

Native of Europe; introduced and naturalized in North America from Newfoundland to Ontario, south to Georgia and Missouri; also from British Columbia to California.

This plant has dark green, smoothish leaves. The creeping stems may form loose mats.

3. **Veronica americana** (Raf.) Schwein. American Brooklime Map 1985

*Veronica Beccabunga* L. var. *americana* Raf.

Flowers June–August.

Occurs in wet ground. Known only from Boone County, central Missouri (limose places, Columbia, July, 1903, *Francis Daniels*, in Univ. of Mo. Herb.).

Ranges from Newfoundland to Alaska, south to North Carolina, Tennessee, Illinois, Missouri, Nebraska, Mexico, and California.

The plant has not been found since its original discovery in the state. The closely related *V. Beccabunga* L. of Europe is used for salads, somewhat like watercress, and as a cooked vegetable. It is reputed to have properties for warding off scurvy and other illnesses due to malnutrition.

4. **Veronica comosa** Richter Water Speedwell Map 1986

*Veronica salina* Schur. [BB]

*Veronica salina* f. *laevipes* (G. Beck) Fern.

*Veronica connata* Raf. [P & S, Steyerm.]

*Veronica connata* subsp. *glaberrima* Pennell [Steyerm.]

*Veronica catenata* Pennell

*Veronica catenata glandulosa* (Farw.) Pennell

Flowers May–October.

Occurs in springs, spring branches, and spring-fed streams. Ozark region north to St. Louis, Franklin,

Plate no. 325. **1.** Veronica peregrina, × 3/4; *a.* Plant, × 3/4; *b.* Fruit and calyx, × 5 1/4; *c.* Corolla, × 5 1/4. **2.** Veronica polita, × 3/4. **3.** Veronica arvensis; *a.* Habit, × 3/4; *b.* Flower, × 3 3/4; *c.* Fruit, × 3 3/4.

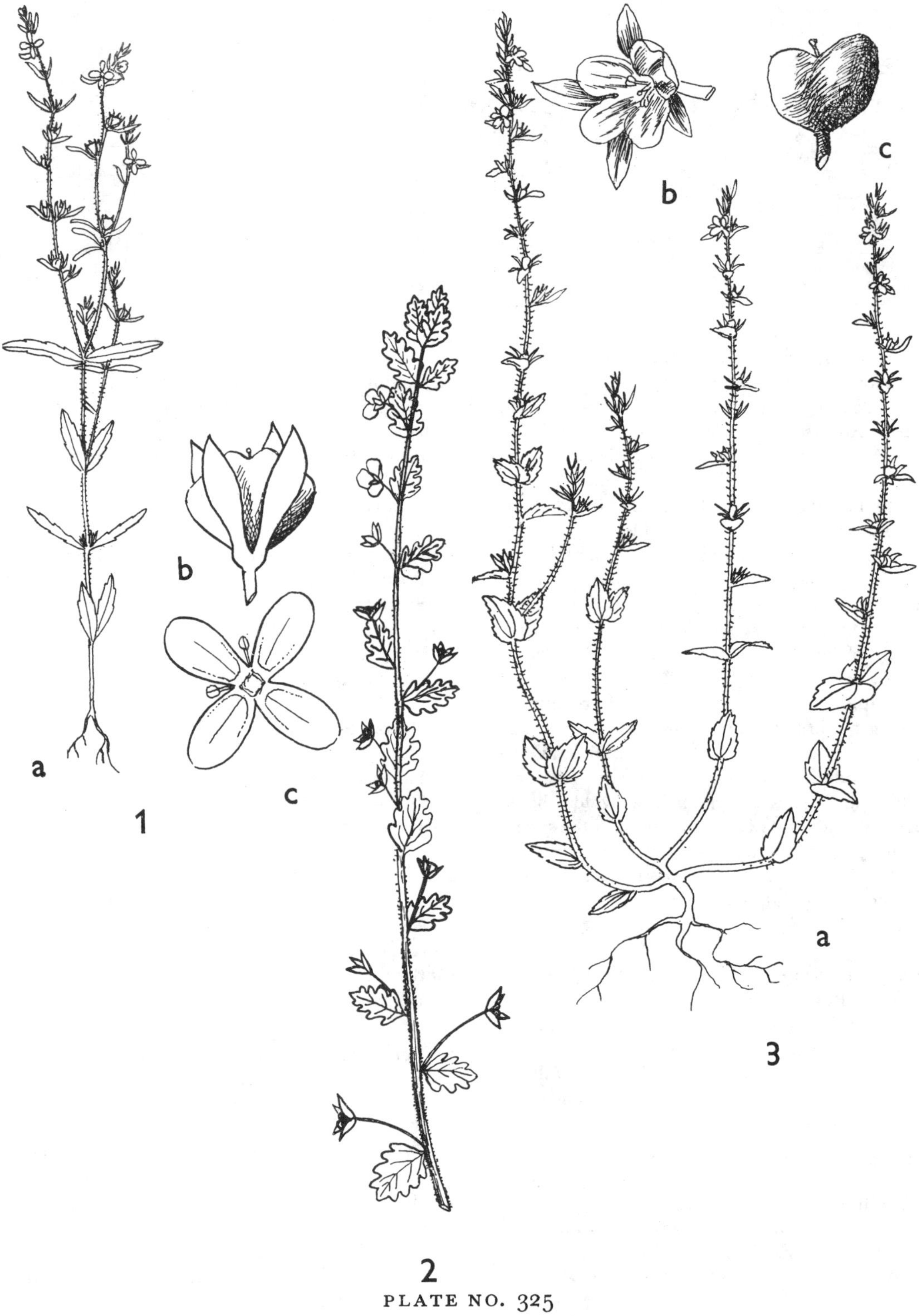


PLATE NO. 325

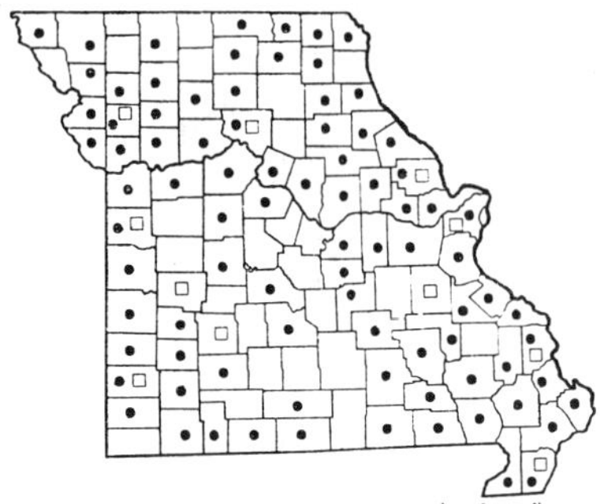

1987 • Veronica peregrina var. peregrina (Neckweed)
1987 □ Veronica peregrina var. xalapensis

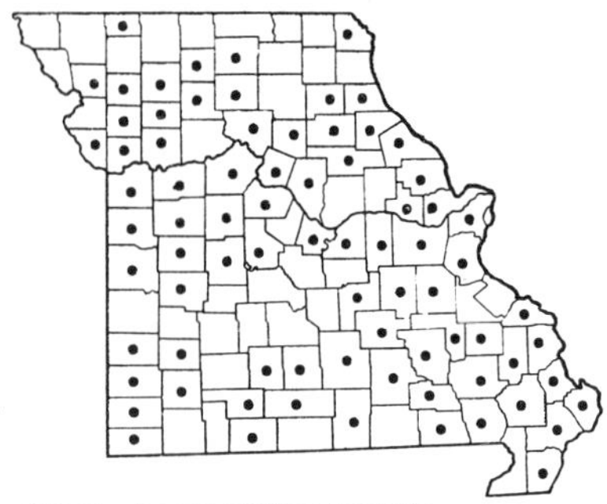

1988 Veronica arvensis (Corn Speedwell)

1989 Veronica polita

Maries, Morgan, Benton, and Lawrence counties, locally northwest in Jackson County.

Ranges from Quebec to Ontario, Saskatchewan, and Washington, south to Pennsylvania, Tennessee, Missouri, Oklahoma, New Mexico, Arizona, and California.

The fresh leaves are reported to be used in salads or the plant may be cooked as a vegetable.

This is one of the characteristic plants found in the fresh, cool spring water of the Ozarks, ranking sixth in frequency of occurrence among the flowering plants present in Ozark springs (see Steyermark, Phanerogamic Flora of the Fresh-water Springs in the Ozarks of Missouri, Bot. Ser. Field Mus. Nat. Hist. 9: 493. 1941). In rapid-running water the plants are usually sterile and grow in larger and more compact clusters with generally smaller leaves. They remain alive in vegetative condition during the winter months. Generally, this species favors shallow water, being found most abundantly near the margins of the spring branch, either in mud or on a sand-gravel substratum. The usual color of the leaves is olive-green, but varies to light green, or, in late fall and winter, to an olive-green tinged with brick.

5. **Veronica peregrina** L. Neckweed Map 1987

Also called Purslane Speedwell.

Flowers April–August.

Occurs in fallow and cultivated ground, garden plots, low moist woodlands, alluvial soils of valleys, waste ground, and along roadsides and railroads.

Two variations occur in Missouri:

Stem glabrous (without hairs); fruit glabrous . 5a. V. PEREGRINA var. PEREGRINA

Stem glandular-hairy; fruit glandular-hairy . 5b. V. PEREGRINA var. XALAPENSIS

5a. **Veronica peregrina** var. **peregrina** Map 1987

*Veronica peregrina* L. [G, P & S, Steyerm.]

Common throughout Missouri, and probably in every county. This is the commoner variation in the state.

Ranges from Quebec to Minnesota, south to Florida and Texas; Alaska to Oregon. Also in the West Indies and Europe.

5b. **Veronica peregrina** var. **xalapensis** (HBK.) Pennell Map 1987

Scattered through southern and central Missouri, north to Lincoln, Chariton, and Clinton counties.

Ranges from Quebec to Alaska, south to Georgia, Texas, and Mexico.

6. **Veronica arvensis** L. Corn Speedwell Map 1988

Flowers March–August.

Occurs in open or rocky woods, on rocky glades and barrens, ledges of bluffs, waste ground, fallow and cultivated fields, pastures, lawns, and along roadsides and railroads. Throughout Missouri, except not recorded from most of the northern tier of counties.

Native of Europe; introduced and naturalized in North America from Newfoundland to Minnesota, south to Alabama and Texas; Colorado, Utah, and from British Columbia to California.

7. **Veronica polita** Fries Map 1989

*Veronica didyma* Ten. (1830), not Ten. (1811).

*Veronica agrestis* L. [BB]

Flowers March–May.

Occurs in lawns, border of woodland, moist low woodland, and waste places. Scattered in the state in Adair (Kirkwood, April 25, 1906, *G. Pauls*), St. Louis, Jefferson, Cape Girardeau (woodland ½ mile north of campus, Cape Girardeau, March 12, 1954, *Donald Rhodes 153*), and Taney (moist soil of bottomland woods, Drury Wildlife Refuge, March 27, 1941,

*Spencer*) counties. This species was first noted in Missouri in 1952 in Jefferson County by Harry Ahles, as reported in Rhodora 57: 317. 1955.

Native of Europe; introduced and naturalized in the United States from New York to Michigan, south to Florida and Texas.

This is an attractive little plant with bright blue corolla striped with a deeper blue color.

## 21. **Seymeria** Pursh

**Seymeria macrophylla** Nutt. Mullein Foxglove
Map 1990

Also called False Foxglove.

*Dasistoma macrophylla* (Nutt.) Raf. [BB, P & S]

Flowers June–September.

Occurs in rich or low woods, in valleys, ravines, along rocky slopes, and bluffs, and in thickets. Throughout Missouri, except not recorded from many of the northeastern counties.

Ranges from West Virginia to Wisconsin, Iowa, and Nebraska, south to Georgia, Alabama, Mississippi, Louisiana, and Texas.

## 22. **Gerardia** L. Gerardia

(including *Aureolaria* Raf., *Tomanthera* Raf., and *Agalinis* Raf.)

There are arguments for and against splitting this genus into *Aureolaria* for the large yellow-flowered species, and into *Gerardia* (Agalinis) for the pink- or purple-flowered species. There has been considerable shifting around of the species from one genus to another. The present treatment is in accordance with that followed by Fernald in the eighth edition of *Gray's Manual.*

a. Corolla yellow; some or all of the leaves deeply lobed or deeply cut into various segments; capsules acutely pointed at tip, 12–23 mm. long . . . . . . . . . . . . . . . . . . . . . *b*
  b. Stem glabrous (without hairs) . . . . . . . . . . . . . . . . . 11. G. FLAVA
  b. Stem hairy or glandular-hairy . . . . . . . . . . . . . . . . . . . . . . *c*
    c. Most or all of the leaves very finely cut into narrow segments and bipinnatifid (the main lobes once again deeply and finely pinnately cleft into smaller lobes); hairs at least on calyx-tube, flower- and fruit-stalks, and leaves gland-tipped; annuals . . 12. G. PEDICULARIA var. PECTINATA
    c. Leaves at most only simply and deeply cleft into coarse lobes and pinnatifid (pinnately cleft into primary lobes); hairs simple, not gland-tipped; perennials . . . . . 10. G. GRANDIFLORA
a. Corolla purple, pink, rose-colored, or rarely white; all the leaves entire (without teeth) or the upper sometimes with basal projecting lobes in *G. auriculata*, or the lower leaves sometimes 3-cleft in *G. heterophylla*; capsules rounded at summit with sometimes a point projecting, 3–13 mm. long. . . . *d*
  d. Leaves lanceolate to lance-ovate, 8–20 mm. broad, the uppermost leaves auricled at base (with 1 or 2 small spreading lobes); hairs on stem downward-pointing (retrorse). . . . . 9. G. AURICULATA
  d. Leaves narrowly linear, lanceolate, or thread-like, 1–6 mm. (rarely to 8) broad, none of the leaves with any lobes at base, or the lower leaves often 3-cleft in *G. heterophylla*; stem either smooth or with minute ascending hairs. . . . . . . . . . . . . . . . . . . . . *e*
    e. Stalks (pedicels) of flowers and fruits usually shorter than or equaling the calyx, 1–5 mm. long . . *f*
      f. Calyx-lobes longer than the calyx-tube, 3.5–6.5 mm. long; lower leaves often 3-cleft. . 1. G. HETEROPHYLLA
      f. Calyx-lobes shorter than the calyx-tube, 0.5–2 mm. long; all the leaves entire, never cleft . . *g*
        g. Stem with short rough hairs or projections directed upward and outward. 4. G. FASCICULATA
        g. Stem glabrous (without hairs or projections), smooth to the touch . . . . 3. G. PURPUREA
    e. Stalks (pedicels) of flowers and fruits noticeably longer than the calyx, mostly 5–15 mm. or more long during flowering period . . . . . . . . . . . . . . . . . . . *h*
      h. Calyx-tube either without a network of veins or with only faint longitudinal nerves showing; plant usually drying blackish or showing black on some part, the fresh plant deep to dark green tinged purplish; corolla usually rose-colored or rose-purple . . . . . . . . . *i*
        i. Stalks (pedicels) of flowers and fruits rather thick, nearly erect or strongly upward spreading, often roughened with minute projections; calyx-tube 4–6 mm. long; corolla 18–27 mm. long; at least the upper part of stem roughened with minute hairs; fruit noticeably longer than thick, ellipsoid or shortly cylindrical, 7–11 mm. long. . 2. G. ASPERA

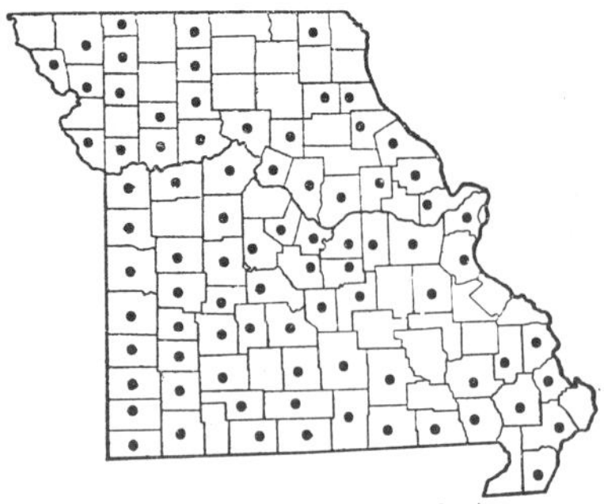

1990 Seymeria macrophylla (Mullein Foxglove)

1991 Gerardia heterophylla

1992 Gerardia aspera

i. Stalks (pedicels) of flowers and fruits slender and hair-like, almost horizontally spreading to somewhat upward-curving, smooth and glabrous (without projections or hairs); calyx-tube 2.5–3 mm. long; corolla 10–15 mm. long; stem smooth and glabrous (without hairs); fruit globose, about as long as thick, 3–7 mm. long . . . . . . . . 5. G. TENUIFOLIA

h. Calyx-tube with a noticeable network of veins showing; plant usually not drying blackish, the fresh plant pale green or yellow green; corolla pink or salmon . . . . . . . . . *j*

j. Branches or most of them usually ending in a solitary flower or fruit; plant abundantly branched with long widely spreading branches . . . . . . . . . . 7. G. GATTINGERI

j. Branches with 2 or more flowers or fruits, the main stem ending in a 2–several-flowered inflorescence (raceme); plant more sparsely branched . . . . . . . . . . . . . *k*

k. Stalks (pedicels) of flowers or fruits longer than their accompanying leaves; angles of stems with minute rough projections; calyx-lobes 0.3–0.8 mm. long, calyx-tube 1.5–3.5 mm. long; corolla 12–15 mm. long; fruit globose, about as thick as long, 4–5 mm. long . . . . . . . . . . . . . . . . . . . 6. G. SKINNERIANA

k. Stalks (pedicels) of flowers or fruits mostly shorter than or equaled by their accompanying leaves; angles of stems usually smooth; calyx-lobes 1.5–2 mm. long, calyx-tube 3.5–4 mm. long; corolla 8–12 mm. long; fruit obovoid, slightly longer than thick, 5–7 mm. long . . . . . . . . . . . . . . . . . . 8. G. VIRIDIS

1. **Gerardia heterophylla** Nutt. Map 1991

*Agalinis heterophylla* (Nutt.) Small [P & S]

Flowers August–September.

Occurs in prairies and moist swales. Known only from southeastern Missouri in Dunklin (Campbell, September 9, 1910, *Bush 6280, 6280A*) and Lawrence (moist swales, thickets and borders of woods, ½ mi. east of Aurora, September 9, 1952, *Palmer 55021*) counties.

Ranges from Louisiana and Texas, north to Missouri and Oklahoma.

This is a very distinctive species, especially if the lower leaves, which are often 3-cleft, are present. A specimen in the Mo. Bot. Gard. Herbarium (Allenton, St. Louis County, August 15, 1886, *Kellogg*) is not included in the cited specimens for Missouri, as the locality data on the label is questionable; Pennell, the monographer of the genus, wrote 'surely wrong locality' for this label.

2. **Gerardia aspera** Dougl. Map 1992

*Agalinis aspera* (Dougl.) Britton [P & S]

Flowers August–October.

Occurs on rocky limestone glades, bluff escarpments, prairie swales, and dry prairies. Mainly in western Missouri, east in the Ozarks to Pulaski and Crawford (limestone glade along bluffs along Meramec River, 7 mi. southeast of Bourbon, September 21, 1934, *Steyermark 15348*) counties.

Ranges from North Dakota to Oklahoma, east to Manitoba, Minnesota, Wisconsin, and Illinois.

This species is sometimes confused with *G. fasciculata*, from which it is readily distinguished by the longer pedicels and calyx-tube, stems and pedicels with much fewer hairs and less rough pubescence, longer capsule, and a shorter corolla.

3. **Gerardia purpurea** L. f. **purpurea** Map 1993

*Agalinis purpurea* (L.) Pennell [P & S]

Plate no. 326. **1.** Gerardia heterophylla, calyx, × 1 1/8; *a.* Calyx, × 5 1/4; *b.* Calyx, with fruit, × 5 1/4. **2.** Gerardia Skinneriana, *a.* Calyx, × 7 1/2; *b.* Fruit with calyx, × 7 1/2. **3.** Gerardia aspera; *a.* Fruit with calyx, × 7 1/2; *b.* Calyx, × 4 1/2. **4.** Gerardia Gattingeri, calyx, × 4 1/2. **5.** Gerardia fasciculata, × 3 3/4. **6.** Gerardia viridis; *a.* Fruit with calyx, × 4 1/2; *b.* Calyx, × 4 1/2.

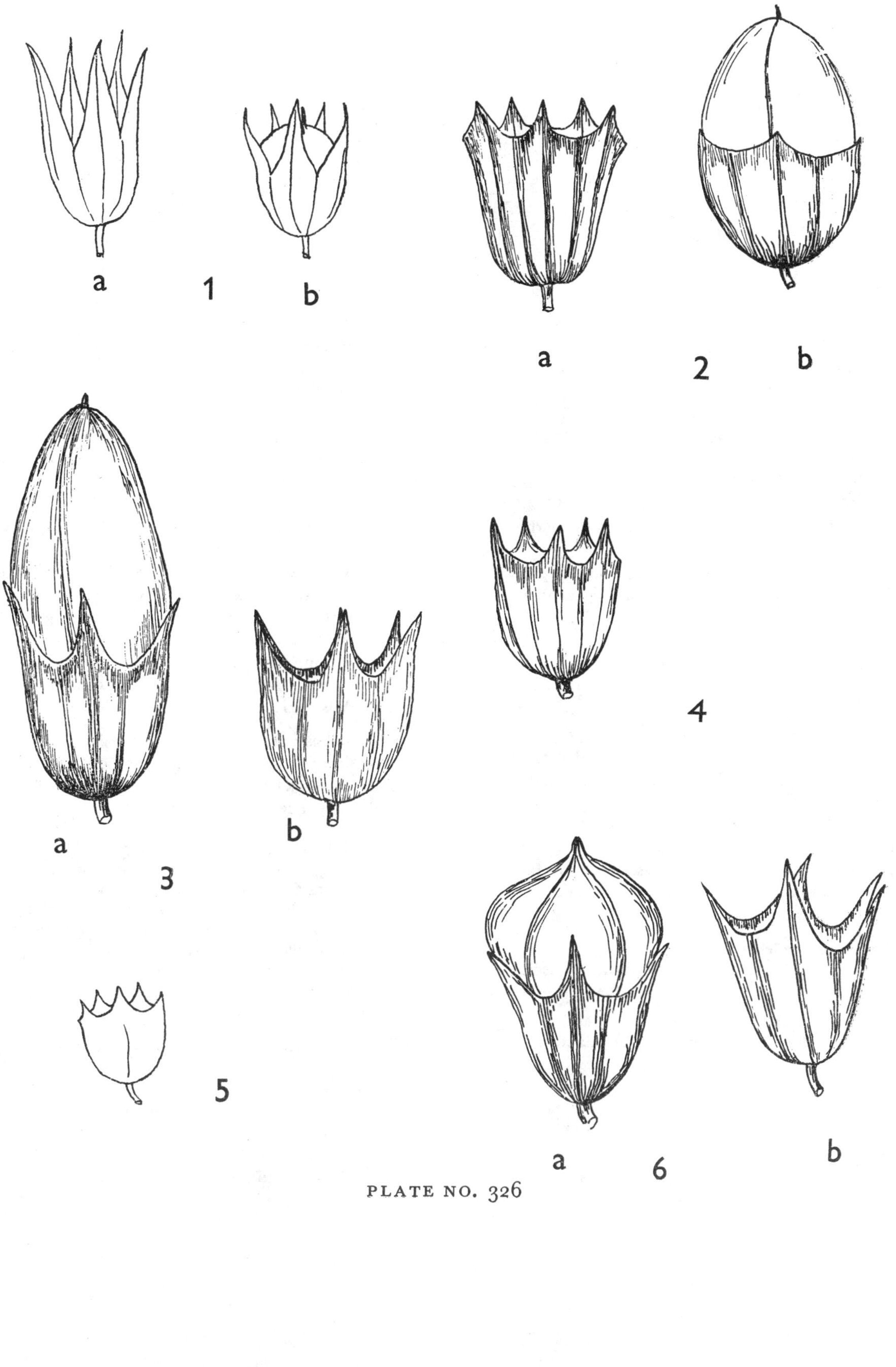


PLATE NO. 326

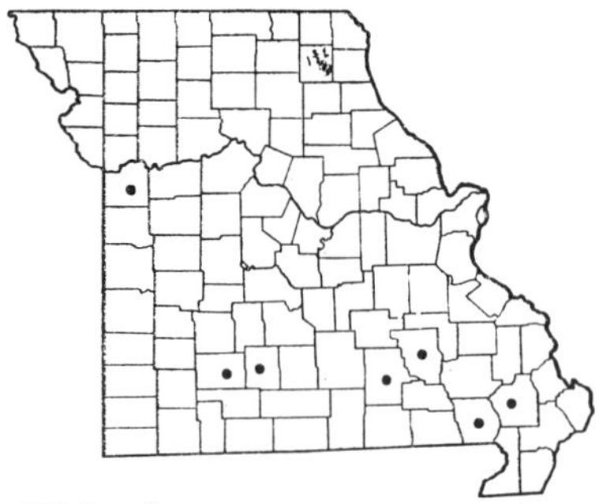

1993 Gerardia purpurea

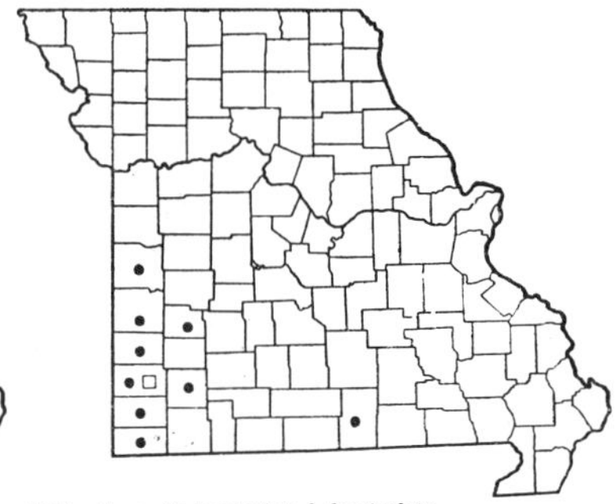

1994 • Gerardia fasciculata f. fasciculata
1994 □ Gerardia fasciculata f. albiflora

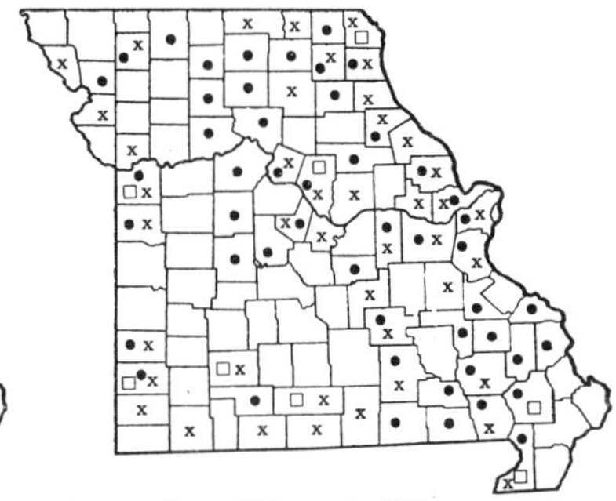

1995 • Gerardia tenuifolia var. tenuifolia
1995 x Gerardia tenuifolia var. macrophylla
1995 □ Gerardia tenuifolia var. parviflora

*Gerardia purpurea* L. [G]
*Gerardia purpurea* var. *purpurea* [BB]
Flowers July–September.

Occurs in swampy meadows, wet prairies, and springy ground. Southern Missouri and locally northwest in central Missouri in Jackson County.

Ranges from Florida to Texas, north to Massachusetts, New York, Pennsylvania, Ohio, Michigan, Wisconsin, Minnesota, and Nebraska; also in the West Indies.

A form with white-flowered corollas, *G. purpurea* f. *albiflora* Britt., has not been recorded yet from Missouri.

4. **Gerardia fasciculata** Ell. Map 1994

*Agalinis fasciculata* (Ell.) Raf. [P & S]
Flowers August–October.

Occurs in prairies, prairie swales, sandy open ground, and borders of woods and thickets. Southwestern Missouri north to Bates County and locally east to Howell County (meadow along north side of highway 80, 4.9 mi. southwest of West Plains, September 25, 1949, *Steyermark 69335*).

Ranges from Florida to Texas, north to Maryland, and inland north to Arkansas and Missouri.

Two variations occur in Missouri:

Corolla rose-colored or rose-purple . . . . 4a. G. FASCICULATA f. FASCICULATA
Corolla white . 4b. G. FASCICULATA f. ALBIFLORA

4a. **Gerardia fasciculata** f. **fasciculata** Map 1994

*Gerardia fasciculata* Ell. [G]
*Agalinis fasciculata* (Ell.) Raf. [P & S]
The common variation in Missouri.

4b. **Gerardia fasciculata** f. **albiflora** Palmer Map 1994

Known only from Jasper County (swales of upland prairie, ½ mi. north of Webb City, August 30, 1950, *Palmer 50841*), holotype in Palmer herb., isotype in Chi. Nat. Hist. Mus. Herb.).

The seeds are stated in some manuals as being 0.5–0.8 mm. long, but may be as much as 1 mm. long in some Missouri specimens.

5. **Gerardia tenuifolia** Vahl Map 1995

Flowers August–October.

Occurs in dry, moist, or open woodland, ledges of bluffs, prairies, wet depressions, swampy meadows, borders of streams, ponds, and ditches, open ground, and wet thickets.

Throughout Missouri.

The following three much intergrading varieties occur in Missouri:

a. Calyx-lobes only 0.2–1 mm. long; fruit 3–5 mm. long, mostly 3–4 mm. in diameter; clusters (fascicles) of smaller leaves in the axils of the larger leaves rare or absent . . . . 5a. G. TENUIFOLIA var. TENUIFOLIA
a. Calyx-lobes 1–2 mm. long; fruit mostly 5–7 mm. long, commonly 5–7 mm. in diameter; clusters (fascicles) of smaller leaves in the axils of the larger leaves sometimes or often present . *b*
    b. Anthers densely hairy; main leaves of the central stem often 4–6 mm. broad, sometimes with clusters (fascicles) of smaller leaves in their axils; leaves and branches usually spreading . . 5b. G. TENUIFOLIA var. MACROPHYLLA
    b. Anthers sparingly hairy; main leaves of the central stem 1–4 mm. broad, often with clusters (fascicles) of smaller leaves in their axils; leaves and branches usually more ascending . 5c. G. TENUIFOLIA var. PARVIFLORA

5a. **Gerardia tenuifolia** var. **tenuifolia** Map 1995

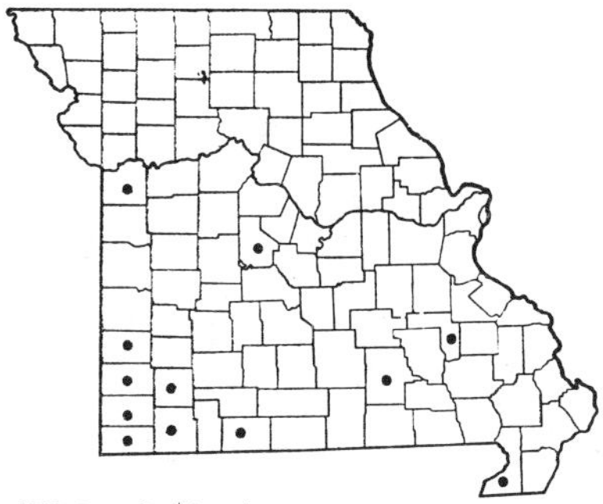

1996 Gerardia Skinneriana

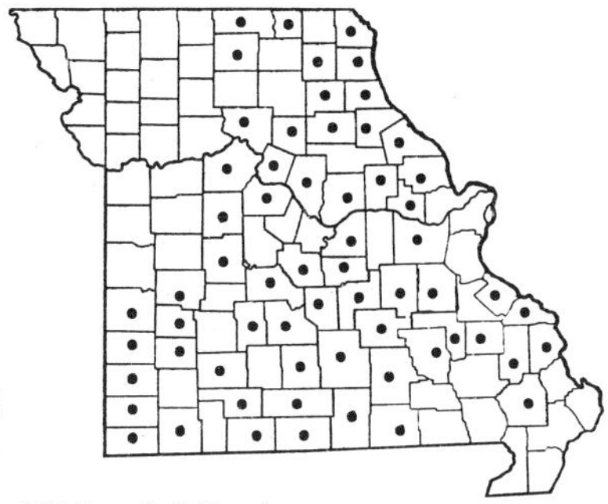

1997 Gerardia Gattingeri

1998 Gerardia viridis

*Agalinis tenuifolia* (Vahl) Raf. [P & S]
*Gerardia tenuifolia* Vahl [G]
Throughout Missouri.

Ranges from Maine to Michigan, south to Georgia, Mississippi, and Louisiana.

5b. **Gerardia tenuifolia** var. **macrophylla** Benth. Map 1995
*Agalinis tenuifolia* var. *macrophylla* (Benth.) Blake [P & S]
Occurs throughout Missouri.

Ranges from Connecticut to Minnesota and Nebraska, south to Alabama, Mississippi, Arkansas, and Texas.

5c. **Gerardia tenuifolia** var. **parviflora** Nutt. Map 1995
*Agalinis tenuifolia* var. *parviflora* (Nutt.) Pennell [P & S]

Scattered in southern and central Missouri and northeast locally in Clark County.

Ranges from Quebec and Vermont to Manitoba and Wyoming, south to New York, Ohio, Indiana, Illinois, Missouri, Oklahoma, and Colorado.

The varieties presented here are not very distinct, except in their extremes, often showing overlapping characters, thus making their determination doubtful or unsatisfactory. The eventual status of these varieties must await more detailed field and experimental studies.

The plants of this species are reported to be poisonous to sheep.

6. **Gerardia Skinneriana** Wood Map 1996
*Agalinis Skinneriana* (Wood) Britton [P & S]
Flowers July–September.

Occurs in dry prairies, rocky open glades, dry open woods and thickets. Southern and central Missouri north to Iron, Morgan, and Jackson counties.

Ranges from Ontario to Wisconsin, south to Ohio, Indiana, Illinois, Arkansas, and Oklahoma.

7. **Gerardia Gattingeri** Small Map 1997
*Agalinis Gattingeri* (Small) Small
Flowers August–October.

Occurs in dry acid soils of open woodland, on the upper leached slopes or upland ridges and crests of hills, frequently overlying sandstone, chert, or granite exposures, sometimes bordering rocky glades of these substrata.

Throughout the eastern half and Ozark sections of Missouri, east and south of a line drawn from Putnam, Sullivan, Chariton, Saline, Pettis, Benton, St. Clair, and Vernon counties.

Ranges from Ontario and Michigan to Minnesota and Nebraska, south to Alabama, Louisiana, and Texas.

This is a characteristic fall-flowering inhabitant of dry sterile acid woodland, where it is frequently associated with *Tephrosia virginiana*, *Danthonia spicata*, and other species of acid soils.

The leaves on some Missouri specimens are up to 4 cm. long.

8. **Gerardia viridis** Small Map 1998
*Agalinis viridis* (Small) Pennell [P & S]
Flowers September–October.

Occurs in prairies, on open banks, borders of woods, and thickets. Southern Missouri, in Stoddard, Dunklin, Shannon, and Jasper counties.

Ranges from Louisiana and Texas, north to Arkansas and Missouri.

The leaves in some Missouri specimens are 2.5 cm. long.

9. **Gerardia auriculata** Michx. Map 1999
*Otophylla auriculata* (Michx.) Small [P & S]
*Tomanthera auriculata* (Michx.) Raf. [BB]
Flowers August–September.

Occurs in dry prairies, fallow fields, borders of up land sterile woods and thickets.

Rare and scattered in Missouri.

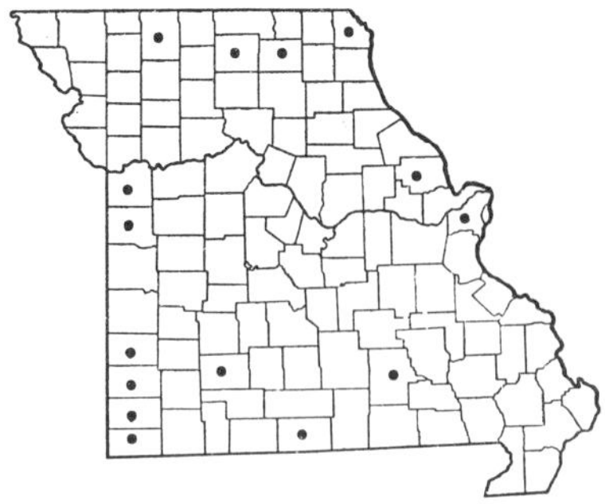

1999 Gerardia auriculata

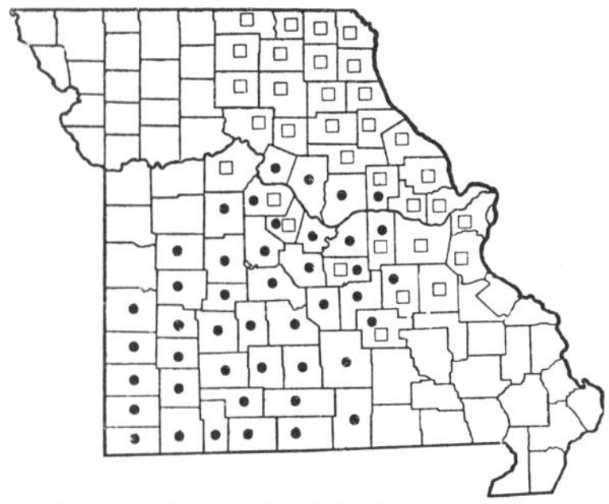

2000 • Gerardia grandiflora var. cinerea
2000 □ Gerardia grandiflora var. pulchra

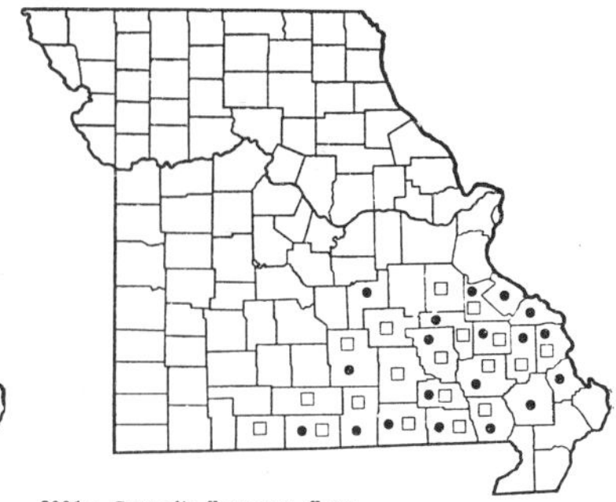

2001 • Gerardia flava var. flava
2001 □ Gerardia flava var. calycosa

Ranges from New Jersey to Minnesota, south to Virginia, Alabama, Tennessee, Arkansas, and Texas.

This species has attractive rose-purple or pink corollas clustered along the axils of the dull or dark green leaves.

10 **Gerardia grandiflora** Benth. Map 2000
Flowers July–September.

Occurs in usually dry or rocky oak-hickory woodland, on upper slopes and ridges in acid soils, usually overlying sandstone, chert, or granite substrata, also on bluff escarpments and on limestone and borders of glades.

Two varieties occur in Missouri:

Upper leaf-blades and bracts accompanying flowers deeply cut (laciniate) or lobed at the base; leaf-blades deeply cut, the lower ones cut usually nearly to the midrib; fruit 15–23 mm. long; northeastern and east-central Missouri south and west to Washington, Dent, Maries, Moniteau, Cooper, Saline, Chariton, Linn, Sullivan, and Putnam counties . . . 10a. G. GRANDIFLORA var. PULCHRA

Upper leaf-blades and bracts accompanying flowers entire (without teeth), shallowly toothed or only slightly cut at base; leaf-blades less deeply lobed, the lower ones only occasionally cut nearly to the midrib; fruit 10–17 mm. long; western and central Ozark and unglaciated prairie sections of the state north and east to Howell, Texas, Dent, Crawford, Montgomery, Callaway, Boone, Howard, Pettis, Henry, and Vernon counties . . 10b. G. GRANDIFLORA var. CINEREA

10a. **Gerardia grandiflora** var. **pulchra** (Pennell) Fern. Map 2000

*Aureolaria grandiflora* (Benth.) Pennell var. *pulchra* Pennell [BB, P & S]

Northeastern and east-central Missouri south to Washington, Dent, Maries, Moniteau, and Saline counties.

Ranges from Wisconsin and Minnesota, south to Indiana, Illinois, and Missouri.

10b. **Gerardia grandiflora** var. **cinerea** (Pennell) Cory Map 2000

*Aureolaria grandiflora* var. *cinerea* Pennell [BB, P & S]

Western and central Ozark region and unglaciated prairie section north to Montgomery, Callaway, Boone, Howard, Pettis, Henry, and Vernon counties, east to Gasconade, Crawford, Dent, Texas, and Howell counties.

Ranges from Missouri and Arkansas to Oklahoma and Texas.

The two varieties, presented above, have large yellow corollas varying from 40–55 mm. long. Those of var. *cinerea* average slightly smaller than in var. *pulchra*, but overlapping in this respect prevents ready separation of the two on this character. Intermediate specimens are found along the margins of the range of the two varieties in Missouri and are difficult to assign to one or the other variety.

I have transplanted this foxglove with the original soil and it lived and flowered for three successive years. It is considered to be somewhat parasitic on the roots of *Quercus*, but in this instance continued to live after transplanting.

11. **Gerardia flava** L. Map 2001
Flowers late June–September.

Plate no. 327. **1.** Seymeria macrophylla, × 2/5; *a.* Lower leaf; *b.* Middle leaves; *c.* Upper part of inflorescence; *d.* Fruit, × 1 3/5; Details from Small, The New York Botanical Garden. **2.** Gerardia purpurea, × 2/5; *a.* Flower, × 3/5; After Gleason, The New York Botanical Garden. **3.** Gerardia tenuifolia, × 2/5; Details from Small, The New York Botanical Garden. **4.** Gerardia auriculata, × 2/5; *a.* Calyx, × 2; After Gleason, The New York Botanical Garden. **5.** Gerardia grandiflora var. pulchra, × 2/5; *a.* Leaves from middle of stem, × 2/5.

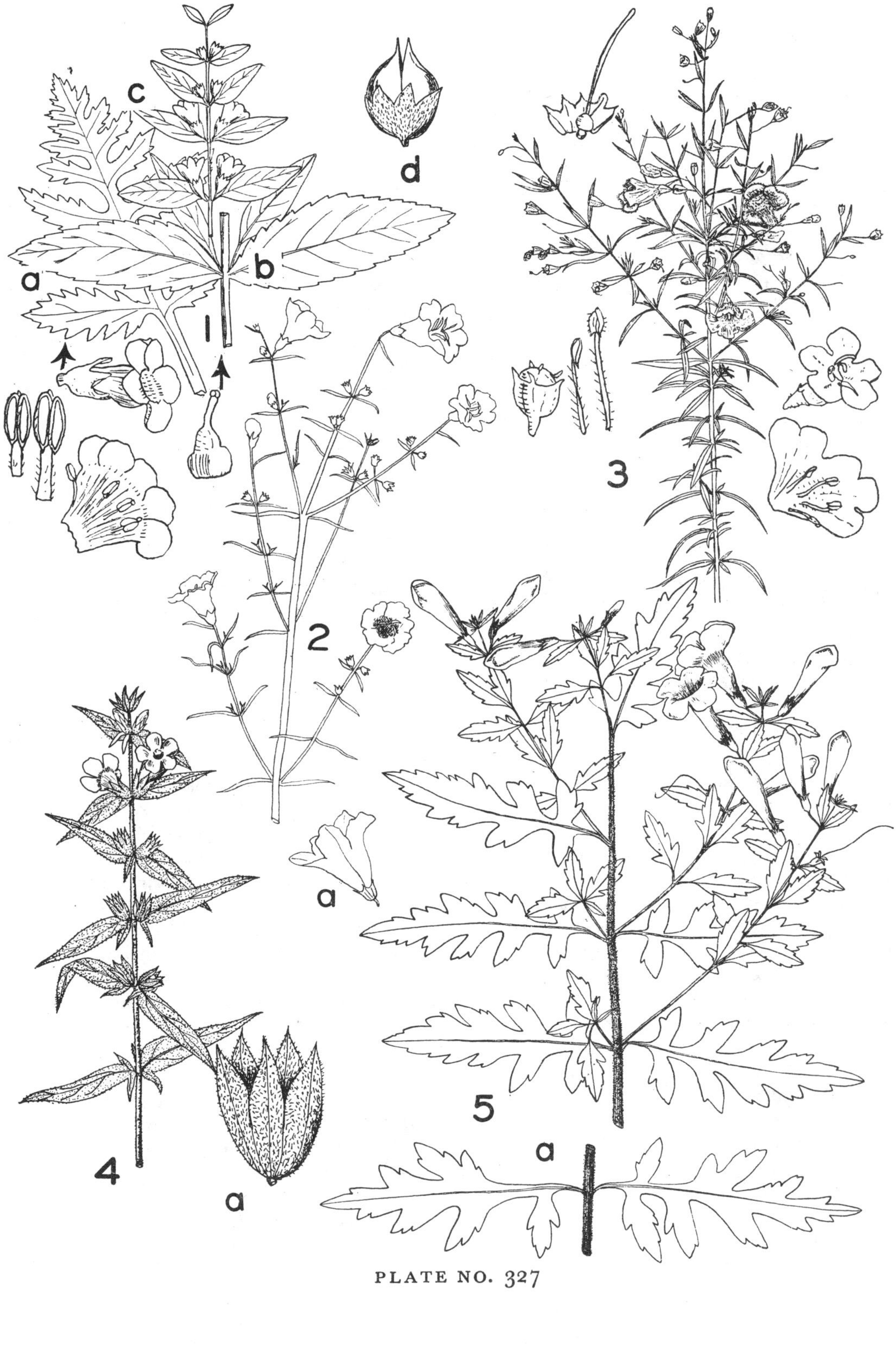


PLATE NO. 327

Occurs in rocky open woods and borders of glades.

Two variations occur in Missouri:

Middle and upper leaves of the stem shallowly to deeply cut into lanceolate, oblong or ovate segments mainly 5–15 mm. broad; some of the principal lower and middle leaves of the stem only shallowly cut . . . 11a. G. FLAVA var. FLAVA

At least the lower half of the middle and upper leaves of the stem usually deeply cut or parted nearly to the midrib into mostly 6–12 linear or linear-lanceolate segments 1–5 mm. broad; most of the lower and middle leaves of the stem deeply cut or parted for most of their length . . . 11b. G. FLAVA var. CALYCOSA

11a. **Gerardia flava** var. **flava** Map 2001

*Gerardia flava* L. [G]

*Gerardia flava* var. *macrantha* (Pennell) Fern. [G]

*Aureolaria flava* var. *macrantha* Pennell [P & S]

Ozark region north and west to Ste. Genevieve, St. Francois, Iron, Phelps, Texas, and Ozark counties.

Ranges from Maine to Ontario, Michigan, Wisconsin, and Minnesota, south to Florida and Louisiana.

11b. **Gerardia flava** var. **calycosa** (Mackenz. & Bush) Steyerm. Map 2001

*Dasystoma calycosa* Mackenz. & Bush, Rep. Mo. Bot. Gard. 16: 105. 1905 (holotype in Mo. Bot. Gard. Herb. from Monteer, Shannon Co., July 31, 1899, *Bush 219*).

*Aureolaria calycosa* (Mackenz. & Bush) Pennell [P & S]

*Gerardia calycosa* (Mackenz. & Bush) Fern. [G]

Ozark region north and west to Cape Girardeau, St. Francois, Washington, Dent, Texas, Douglas, and Taney counties.

Ranges from Missouri to Arkansas.

It has not been possible to maintain as separate species the above variations, and I agree with Gleason's interpretation in general (New Ill. Fl. 3: 241. 1952) that this series of taxa should be united under the same species. Fernald (eighth ed. *Gray's Manual*, pp. 1287–88, 1292) attempted to separate *G. flava* from *G. calycosa* on the basis of presumed differences in length of anther awns, seed length, and absence or presence of hairs on the inside of the corolla above the base of the filaments. All these characters have been found to overlap and do not correlate with the relative breadth of the leaf-segments. Some specimens, for example, have the broader leaf-segments of *G. flava*, but the corolla is glabrous within above the base of the filaments as given for *G. calycosa*. Nor are the differences in the seed length additional ones sufficient to separate *G. flava* from *G. calycosa*, as specimens with deeply cut narrow segments have seeds 2 mm. long, bridging the gap between the two taxa.

The supposed differences also between *G. flava* var. *flava* and var. *macrantha* are even less pronounced than those supposedly separating *G. flava* and *G. calycosa*. The relative differences in length of calyx-lobes, corolla, and capsule given for *G. flava* and var. *macrantha* overlap and, as Gleason indicates (New Ill. Fl. 3: 241. 1952), great variation in length of calyx-lobes often is found on the same plant.

12. **Gerardia pedicularia** var. **pectinata** Nutt. Map 2002

*Gerardia pectinata* (Nutt.) Benth. [G]

*Aureolaria pectinata* (Nutt.) Pennell [P & S]

*Aureolaria pedicularia* (L.) Raf. [BB in part]

*Aureolaria pectinata* var. *ozarkensis* Pennell (holotype from Mountain View, Howell Co., August 31, 1913, *Pennell 5337*, in U. of Pa. Herb.).

Flowers August–September.

Occurs in dry or open rocky woodland and borders of glades, in acid soils overlying chert, sandstone, or granite substrata. Ozark region north and west to St. Louis, Franklin, Gasconade, Maries, Pulaski, Laclede, Greene, Barry, and McDonald counties.

Ranges from Florida and Georgia to Louisiana, north to North Carolina, Kentucky, and Missouri.

The differences presumably separating *G. pedicularia* from *G. pectinata* seem best relegated to one variable species, a view which Gleason (New Ill. Fl. 3: 240. 1952) apparently favors. The calyx-tube in *G. pedicularia* var. *pectinata* is shorter and more hemispherical than in var. *pedicularia*, with a denser and longer grandular pubescence, and the more ovoid capsule of var. *pectinata* is enclosed only at its base by the calyx-tube, whereas in var. *pedicularia* the capsule is ellipsoid and its inner half is enclosed by the calyx-tube. The lobes of the leaves in var. *pectinata* are narrower and more sharply and finely toothed than in var. *pedicularia*, and the stems in var. *pectinata* more glandular in the upper portion with more abundant hairs than in var. *pedicularia*. These differences are not absolute and tend to intergrade with other varieties of *G. pedicularia*.

---

Plate no. 328. **1.** Gerardia flava var. flava, × 2/5; Details from Small, The New York Botanical Garden. **2.** Gerardia flava var. calycosa, × 2/5; Details from Small, The New York Botanical Garden. **3.** Gerardia pedicularia var. pectinata, × 2/5. **4.** Buchnera americana, × 1/5; *a.* Flower, × 2 2/5; After Gleason, details from Small, The New York Botanical Garden.

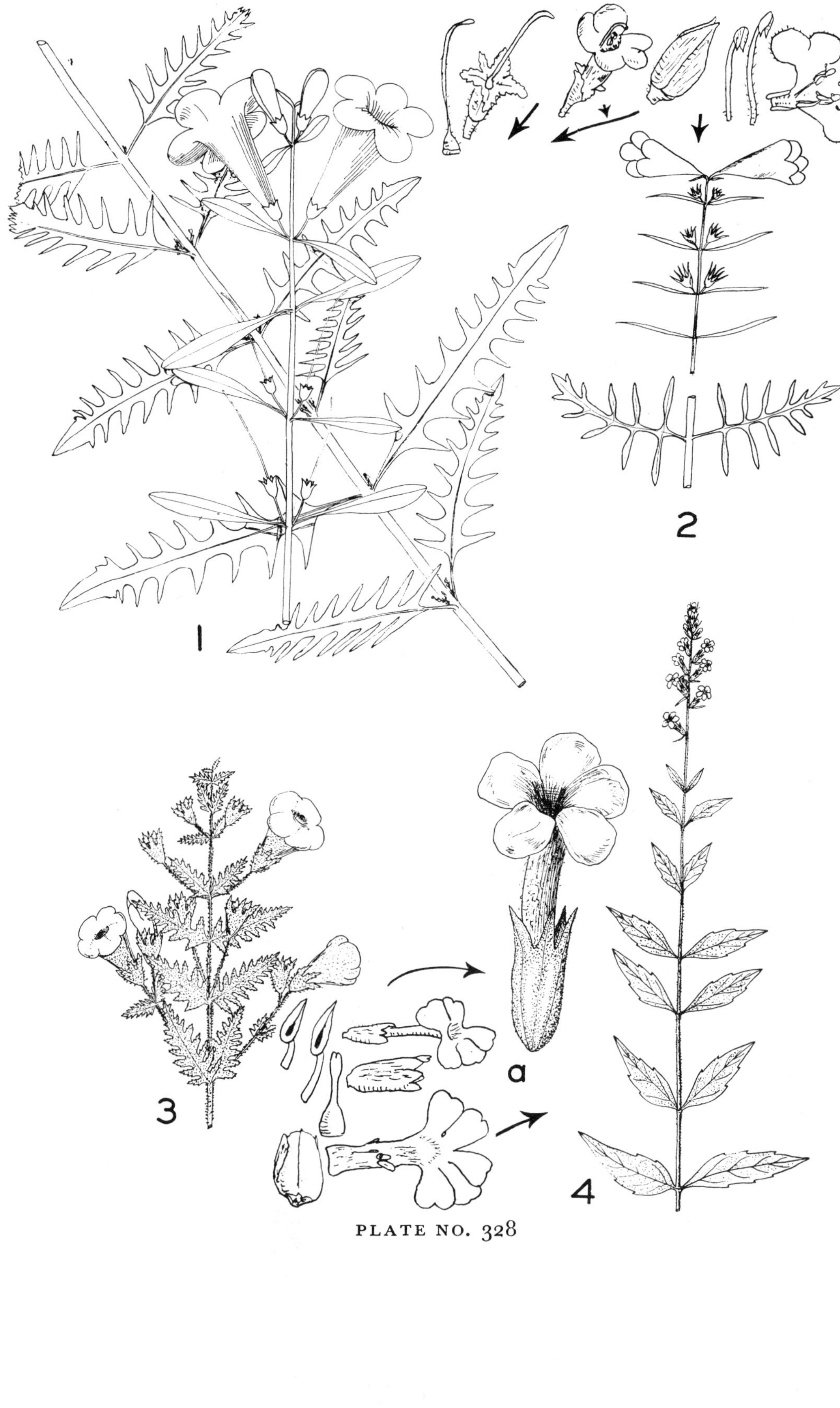


PLATE NO. 328

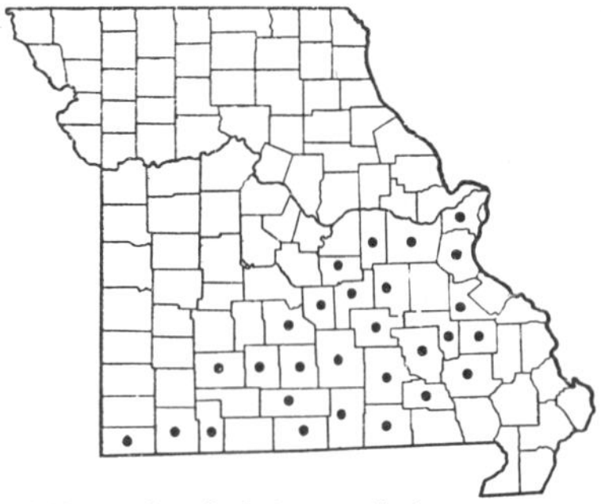

2002 Gerardia pedicularia var. pectinata

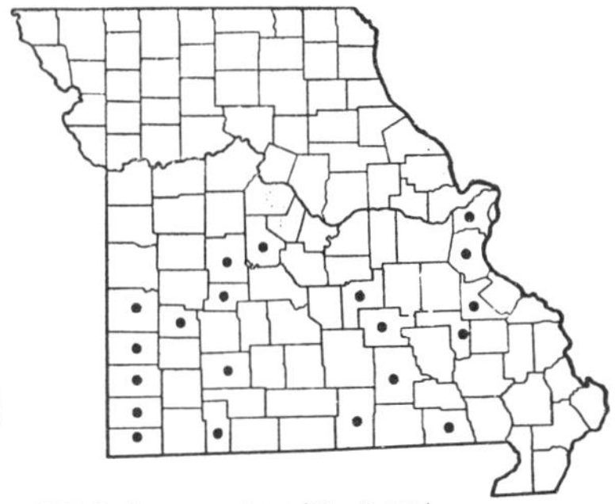

2003 Buchnera americana (Blue Hearts)

2004 ● Pedicularis canadensis f. canadensis (Wood Betony)
2004 □ Pedicularis canadensis f. albescens

### 23. **Buchnera** L. Blue Hearts

**Buchnera americana** L. Blue Hearts Map 2003

Flowers mid-June–September.

Occurs in upland prairies and limestone glades and barrens. Ozark and unglaciated prairie region of southern and east-central Missouri north to St. Louis, Phelps, Morgan, Benton, Cedar, and Vernon counties.

Ranges from Florida to Texas, north to New Jersey, New York, Ontario, Michigan, Illinois, Missouri, and Kansas.

This is a very attractive species when in flower with blue-purple corollas about 2 cm. long. The foliage and flowers turn black upon drying.

### 24. **Pedicularis** L. Wood Betony, Lousewort

Flowering in spring from April to June; plants of dry woodland and prairies, in acid soils; corolla mainly yellow or tinged pink, reddish, or purplish, rarely all white; leaves of main stem alternate, the lower with long stalks (petioles); at least the upper part of the stem more or less hairy . . 1. P. CANADENSIS

Flowering in summer and autumn from August to October; plants of swampy open meadows, in limey soils; corolla cream-colored or yellowish white; leaves of main stem opposite or nearly opposite, sessile (without stalks); upper part of stem mainly glabrous (without hairs) or nearly so . . . . 2. P. LANCEOLATA

1. **Pedicularis canadensis** L. Wood Betony Map 2004

Flowers April–May.

Occurs in dry open woods usually on upland slopes, ridges, and above bluff escarpments, also along mossy slopes bordering creeks, prairies, and thickets, usually in leached or sterile acid soils, also in low ground in valleys. Throughout Missouri, except absent from some of the lowland counties of extreme southeastern Missouri.

Ranges from Maine and Quebec to Manitoba, south to Florida, Texas, and Mexico.

The following two variations occur in Missouri:

Corolla mostly all white . . . 1b. P. CANADENSIS f. ALBESCENS

Corolla yellow or yellowish, or tinged with red, pink, or purple . . . . . . 1a. P. CANADENSIS f. CANADENSIS

1a. **Pedicularis canadensis** f. **canadensis** Map 2004

*Pedicularis canadensis* L. [G, BB, P & S, Steyerm.]

*Pedicularis canadensis* var. *Dobbsii* Fern. [G]

This is the common variation found throughout the state.

1b. **Pedicularis canadensis** f. **albescens** Steyerm. Map 2004

Known only from southwestern Missouri in Wright County (low mossy slopes along creek, cherty shallow ravine along route 5, 2½ mi. south of Mansfield, May 1, 1949, *Steyermark 67499*, holotype in Chi. Nat. Hist. Mus. Herb.).

Other color variations, not recorded from Missouri, are f. *bicolor* Farw., with the corolla crimson on the back, and f. *praeclara* A. H. Moore, with the corolla crimson throughout. A variation described as var. *Dobbsii* Fern. is based upon plants having prolonged, often creeping offsets arising from the base of the plant, the stems therefore scarcely clustered, as contrasted with closely clustered stems of typical *P. canadensis*. Although plants agreeing with var. *Dobbsii* are found

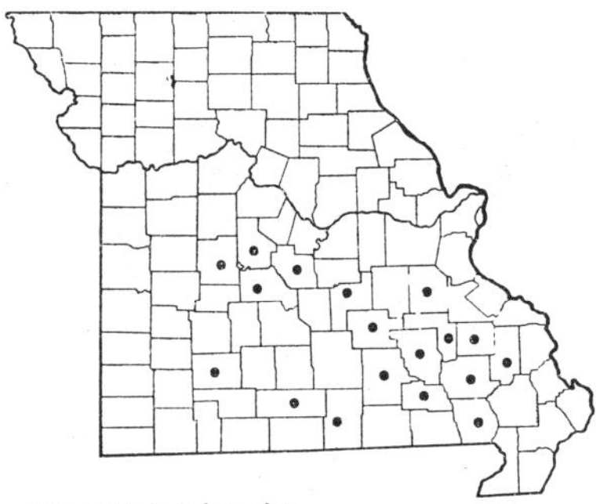

2005 Pedicularis lanceolata

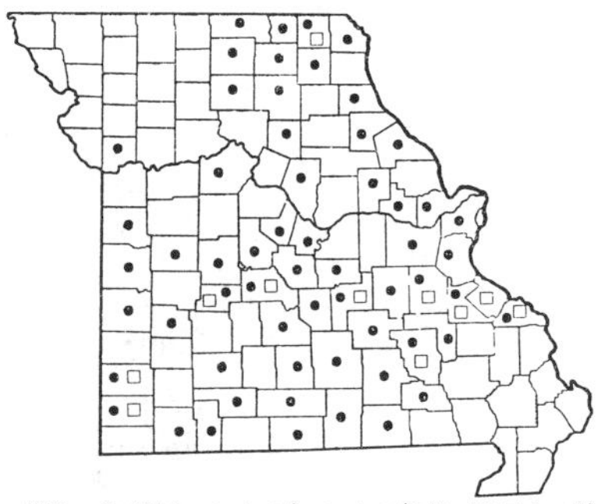

2006 • Castilleja coccinea f. coccinea (Indian Paint-brush)
2006 □ Castilleja coccinea f. lutescens

2007 Castilleja sessiliflora (Downy Painted Cup)

in Missouri (collections are known from Nodaway, Putnam, Randolph, Warren, Ste. Genevieve, and Ozark counties), the production of prolonged basal offsets appears to be merely a vegetative growth form or phase in the life history of the plant, which may or may not occur, depending upon local conditions. It does not seem to this author to merit recognition and is being treated as synonymous under typical *P. canadensis*. Perhaps future experimental studies may throw more light on this growth form.

This species is reported to be poisonous to sheep.

2. **Pedicularis lanceolata** Michx. Swamp Wood Betony Map 2005

Flowers August–October.

Occurs in calcareous and spring-fed swampy meadows and swales in valleys and lowland, occasionally on moist limestone ledges. Ozark region, especially in the southeastern sector, north and west to Bollinger, Madison, Washington, Miller, Morgan, Benton, and Greene counties.

Ranges from Massachusetts to Ontario and Manitoba, south to Virginia, North Carolina, Tennessee, Missouri, and Nebraska.

## 25. **Castilleja** Mutis Painted Cup, Paint Brush

a. Green bracts without other color surrounding the flowers; corolla 4–5.5 cm. long, conspicuously protruding beyond the tip of the bract . . . . . . . . . . . . . . . . 2. C. SESSILIFLORA
a. Red, orange, yellow, white, or purple bracts surrounding the flowers; corolla 2.5–3.5 cm. long, shorter than the bracts or only slightly protruding beyond the tip of the bract . . . . . . . . *b*
  b. Common in many parts of Missouri; red, orange, yellow, or white bracts surrounding the flowers; flowering stem usually solitary from the base; biennial plant with weak roots; calyx split into 2 broad, rounded lobes . . . . . . . . . . . . . . . . . . . . 1. C. COCCINEA
  b. Rare, known only from Greene County, southwestern Missouri; purple or violet-colored bracts surrounding the flowers; flowering stems clustered, usually 2 or more from the base; perennial plant, the root remaining in the ground when the stem is pulled; calyx split into 2 narrow, acute lobes . . . . . . . . . . . . . . . . . . . . . 3. C. PURPUREA

1. **Castilleja coccinea** (L.) Spreng. Indian Paint-brush Map 2006

Also known as Painted Cup, Indian Blanket.

Flowers April–July.

Occurs in prairies, swampy meadows, rocky open glades of usually limestone, but also of sandstone, open and moist woodland, thickets, and wet ground along streams. Southern, central, and eastern Missouri east and south of a line drawn from Putnam, Sullivan, Linn, and Saline counties to Clay County.

Ranges from New Hampshire to Manitoba, south to Florida, Louisiana, and Oklahoma.

Two variations occur in Missouri:

Bracts surrounding flowers red or orange-red or orange . . . . . 1a. C. COCCINEA f. COCCINEA
Bracts surrounding flowers yellow to cream-yellow . . . . . 1b. C. COCCINEA f. LUTESCENS

1a. **Castilleja coccinea** f. **coccinea** Map 2006

*Castilleja coccinea* (L.) Spreng. [G, BB, P & S, Steyerm.]

This is the common variation found.

1b. **Castilleja coccinea** f. **lutescens** Farw. Map 2006

2008 Castilleja purpurea

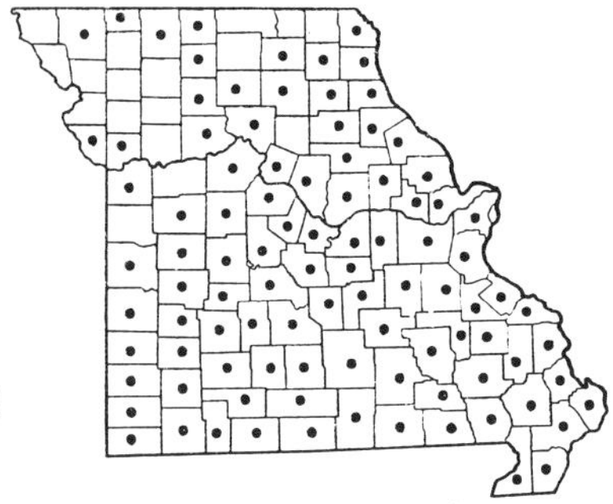

2009 Campsis radicans (Trumpet Creeper)

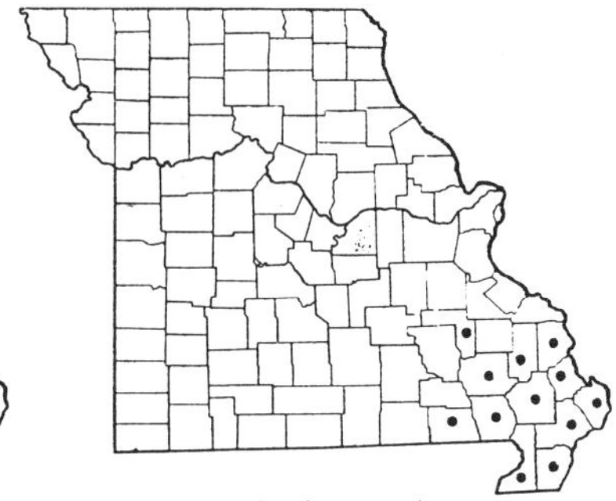

2010 Bignonia capreolata (Cross-vine)

Scattered in the range in southern Missouri mainly, and locally north in Scotland County.

This is the most frequently encountered species of the genus in Missouri. In some favorable situations in undisturbed prairies and ungrazed glades, it may occur in dense stands by the hundreds or even thousands. At virgin prairies near Elk Prairie in Phelps County and near Cross Timbers in Hickory County thousands of plants, flashing their brilliant red bracts, have been observed for a number of years. Populations of the plants shift sometimes, leaving one site for another.

The plant is sometimes stated to be an annual, but from experience in raising plants from seed and observing them over a period of fifteen years, they are, in my judgment, always or mainly a biennial, producing a basal rosette of leaves the first year and sending up a flowering stem the second year. The plant dies after the second year, and the seeds germinate the same year in the fall.

2. **Castilleja sessiliflora** Pursh Downy Painted Cup Map 2007
Flowers April–July.

Occurs on exposed dry soils of loess hills. Known only from Atchison and Holt counties, northwestern Missouri.

Ranges from Wisconsin and Illinois to Manitoba and Montana, south to Missouri, Kansas, Oklahoma, Texas, New Mexico, and Arizona.

The flowers have a sweet fragrance reminiscent of Rose Gentian (*Sabatia angularis*). The corolla is creamy yellow with olive green along the dorsal lobes and rose tint on the ventral lobe.

3. **Castilleja purpurea** (Nutt.) G. Don Map 2008
Flowers April–May.

Occurs in rocky open ground. Known only from Greene County, southwestern Missouri (Percy Cave, *Standley*, in U.S. Nat. Herb.).

Ranges from Texas and Oklahoma to Missouri.

## Fam. **BIGNONIACEAE** (Trumpet Creeper Family)

a. Trees, but sometimes flowering when only a few feet tall; leaves simple and undivided; corolla white, spotted purple, and yellow; calyx splitting into 2 unequal lobes . . . . . . . . 3. CATALPA

a. Climbing shrubs or vines; leaves compound, divided into 2 or into 7–11 leaflets; corolla scarlet or orange-yellow tinged reddish; calyx with either 5 lobes or the lobes scarcely showing . . . . . . *b*

  b. Leaves divided into 2 entire (without teeth) leaflets, or rarely 1; leaves bearing tendrils (slender twining outgrowth); corolla orange-yellow tinged with dull red . . . . . . . . . 2. BIGNONIA

  b. Leaves divided into 7–11 sharply toothed leaflets; leaves not bearing tendrils; corolla scarlet . 1. CAMPSIS

### 1. **Campsis** Lour. Trumpet Creeper

**Campsis radicans** (L.) Seem. Trumpet Creeper Map 2009

Also called Trumpet Vine, Devil's Shoe Laces, Shoe Strings, and Hell Vine.

---

Plate no. 329. **1.** Pedicularis canadensis, × 2/7; Details from Small, The New York Botanical Garden. **2.** Pedicularis lanceolata, × 2/7; After Gleason, details from Small, The New York Botanical Garden. **3.** Castilleja coccinea, × 2/7; Details from Small, The New York Botanical Garden. **4.** Castilleja sessiliflora, × 2/7; Details from Small, The New York Botanical Garden. **5.** Castilleja purpurea, × 2/7; Details from Small, The New York Botanical Garden. **6.** Campsis radicans, × 2/7. **7.** Catalpa bignonioides, × 2/7; Details from Small, The New York Botanical Garden. **8.** Bignonia capreolata, × 2/7; Details from Small, The New York Botanical Garden.

PLATE NO. 329

Flowers May–August.

Occurs in open woods, thickets, stream banks, fallow and old fields, and along roadsides and railroads. Throughout Missouri except unrecorded from most of the northwestern counties.

Ranges from Florida to Texas, north to New Jersey, Pennsylvania, West Virginia, Kentucky, Indiana, Illinois, and Iowa; naturalized north to Connecticut and Michigan.

This vine is native in the Ozarks, and is planted and escaped from cultivation northward. Its bright red flowers are much visited by hummingbirds. It can be transplanted easily, but may not flower for years after transplanting. It is a good plant for fencerows, and for covering bare trunks and poles. However, it may become aggressive and take over old fields and pastures, especially in bottom lands, where its interlacing, sprawling, and trailing stems make walking difficult.

Some cases of dermatitis are known where susceptible individuals have handled the leaves and flowers. The vine contains narcotic properties.

## 2. **Bignonia** L. Cross-vine

(*Anisostichus* Bureau [BB])

**Bignonia capreolata** L. Cross-vine Map 2010

Sometimes called Quarter Vine.

*Anisostichus capreolata* (L.) Bureau [BB]

Flowers April–June.

Occurs in low bottom woods, swampy ground, thickets along streams, low fields, and fence rows. Southeastern Missouri lowlands and Crowley Ridge north to Cape Girardeau, Bollinger, Wayne, and Ripley counties, and locally in the southeastern Ozarks in Iron County.

Ranges from Florida to Louisiana, north to Maryland, West Virginia, Ohio, Indiana, Illinois, Missouri and Oklahoma.

The large corollas have a slightly bent tube. The flowers have a characteristic sweet odor resembling that of maple sugar or slippery elm. The leaves are evergreen or nearly so. The vine may be a high-climber or sprawls along the ground. The author has succeeded in getting this species to grow in northern Illinois, where it has maintained itself for fifteen years, but shows no tendency to flower. The name Cross-vine is derived from a form resembling a Greek cross which is visible in the wood when the stem is cut crosswise.

## 3. **Catalpa** Scop. Catalpa, Lady Cigar, Indian Bean

a. Leaves glabrous (without hairs) or the lower surface soon becoming glabrous, often sharply lobed or angled; corollas 1–2 cm. broad, yellow with orange and spotted violet; capsules 5–8 mm. in diameter; seeds 2–4 mm. broad; small shrubs or trees, sometimes flowering when only 2½ feet tall . . . 3. C. OVATA

a. Leaves permanently soft-hairy on lower surface, mostly without lobes or angles; corollas 2–4 cm. broad, white, marked with yellow and purple-brown; capsules 8–15 mm. in diameter; seeds 2.5–5 mm. broad; large trees . . . *b*

b. Odor of bruised leaves unpleasant, somewhat musky-scented; bark of old trees thin and scaly; corolla conspicuously spotted on the inside with purple, 2–3 cm. across the upper part of tube (not including lobes); lower lobe of corolla without any notch in the center; lobes of corolla with creases and wrinkles; valves of the fruit flat and thin after the fruit opens; seeds 2.5–4.5 mm. broad, the tufts of hairs at the ends coming together to form a point; fruit thin-walled, 5–8 mm. in diameter; inflorescence often 20–25 cm. long . . . 1. C. BIGNONIOIDES

b. Odor of bruised leaves not unpleasant; bark of old trees fissured and prominently ridged; corolla lined with purple, only slightly spotted, nearly white, about 4 cm. across at upper part of tube (not including lobes); lower lobe of corolla with a notch in the center; lobes of corolla not wrinkled; valves of the fruit remaining semiterete (somewhat cylindrical or rounded in cross-section) after the fruit opens; seeds 4–5 mm. broad, the tufts of hairs at the ends remaining separated and not coming together or uniting; fruit thicker-walled, 12–15 mm. in diameter; inflorescence mainly 15 cm. or less long . . . 2. C. SPECIOSA

1. **Catalpa bignonioides** Walt. Catalpa Map 2011

Flowers May–June.

Often planted as a street tree, and sometimes escaped along roadsides, railroads, waste ground, alluvial banks and gravel bars along streams. Southern and central Missouri north to St. Louis, Callaway,

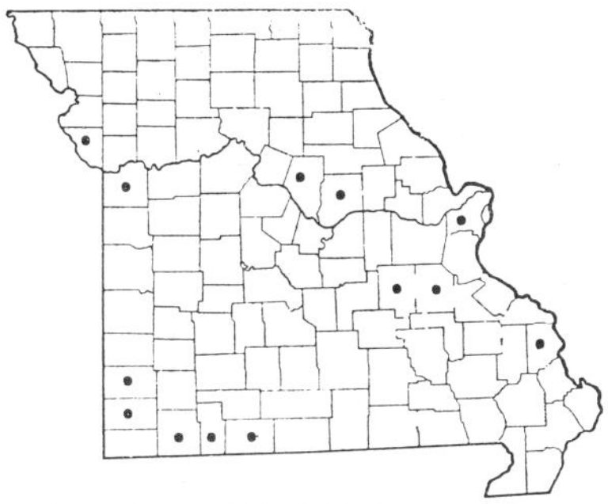

2011 Catalpa bignonioides (Catalpa)

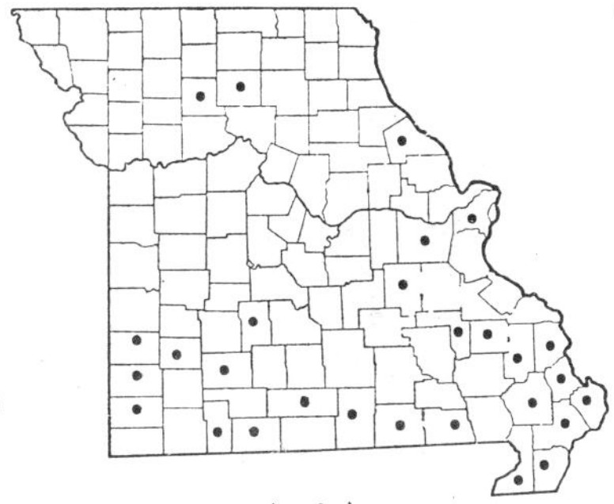

2012 Catalpa speciosa (Catalpa)

2012 A Catalpa ovata (Chinese Catalpa)

Boone, Jackson, and Platte counties.

Native in Florida, Georgia, Alabama, Mississippi, and Louisiana; introduced and naturalized elsewhere in the United States north to southern New England, New York, Pennsylvania, Ohio, Michigan, Missouri, and Oklahoma.

2. **Catalpa speciosa** Warder Catalpa Map 2012

Called Catawba Tree, Cigar Tree, Hardy Catalpa, Indian Bean.

Flowers May–June.

Occurs in low or alluvial woods generally along streams and base of bluffs, also in upland woods, and introduced in waste ground and streams northward in the state. Native in southeastern Missouri in the lowland counties; elsewhere in the state introduced northward to Pike, Linn, and Livingston counties.

Ranges from Tennessee to Arkansas and Texas, north to Indiana, Illinois, and Missouri; elsewhere introduced from cultivation and naturalized north to Virginia, West Virginia, Ohio, Iowa, and Kansas.

The two species of Catalpa treated here are very difficult to distinguish and all the characters given in the key must be considered for final determination. The leaves in autumn turn yellowish-green to brownish-yellow. The wood is durable and is used for railroad ties and fence posts. Small amounts of this and the preceding species are mixed with other hardwoods as pulp for papermaking.

Some individuals develop a dermatitis after handling the flowers, and even inhalation of the odor of the flowers is reported to be poisonous to some people. Nectar glands at the base of the leaf-blades secrete nectar as do also the flowers. The larvae of the catalpa sphinx moth prey upon the leaves of this tree and often strip it twice a year, so that trees sometimes seldom produce additional leafy growth as a consequence. The trees are often recommended for planting in groves for shade for hog lots.

3. **Catalpa ovata** G. Don Chinese Catalpa Map 2012A

*Catalpa Kaempferi* Sieb. & Zucc.

Flowers June–August.

Escaped from cultivation along roadsides. Known only from Crawford County (roadside, highway 49, $\frac{1}{2}$ mi. east of Dillard, July 6, 1951, *Steyermark 71924A*).

Native of Asia; introduced and naturalized in the United States from Connecticut to Ontario, south to Maryland, Ohio, and Missouri.

This species was found flowering when only $2\frac{1}{2}$ feet tall. The leaves are fetid as in *C. bignonioides*, for which the plant was at first identified.

## Fam. **PEDALIACEAE** (Pedalium Family)

### **Sesamum** L. Sesame, Bené

**Sesamum indicum** L. Map 2013

*Sesamum orientale* L. [P & S]

Flowers July–September.

Occurs in sandy waste ground, where introduced. Known only from Jackson County, west-central Missouri (sands, Kansas City, September 12, 1933, *Bush 13003A*).

Native probably of India, but not known in a wild state; introduced and naturalized in New York and Missouri, and recorded from Texas.

The 2-lipped, 5-parted corolla varies in color from pink to yellowish or white and is 2–2.5 cm. long. The oblong fruit is somewhat 4-sided.

In many areas the seed is used as an edible oil, and

2013 Sesamum indicum (Sesame)

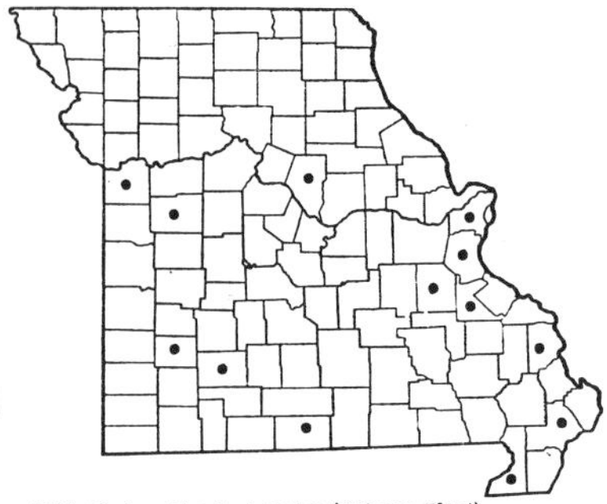

2014 Proboscidea louisianica (Unicorn Plant)

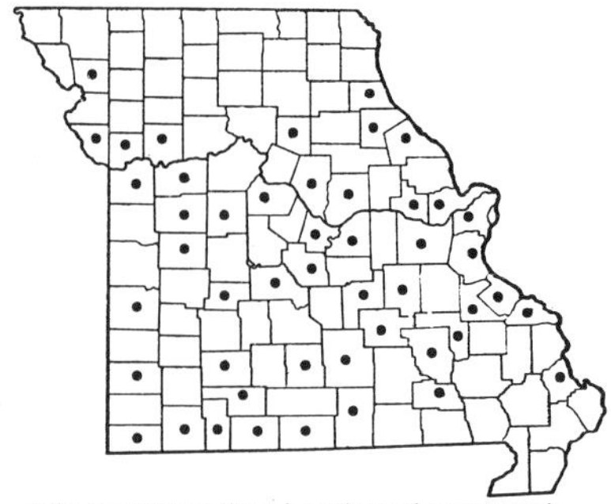

2015 Orobanche uniflora (One-flowered Cancer-root)

especially in Africa and in the Orient the whole seed is eaten in a wide range of foods. In the United States the whole seed is used chiefly in the bakery industry as topping for breads and rolls.

The sesame oil, obtained from the seed, is used as a salad or cooking oil, in margarine, shortening, in the manufacture of soap, as a fixative in the perfume industry, and as a carrier for fat-soluble substances in various pharmaceuticals. One of the lesser constituents in sesame oil, *sesamin*, is added to insecticides containing pyrethrum for its marked increase in effectiveness for fly sprays.

## Fam. MARTYNIACEAE (Unicorn Plant Family)

### **Proboscidea** Schmidel Unicorn Plant

**Proboscidea louisianica** (Mill.) Thell. Unicorn Plant Map 2014

Also called Ram's Horn, Proboscis Flower, Devil's Claws.

*Martynia louisiana* Mill. [P & S]

*Martynia louisianica* Mill. See Rh. 34: 215. 1915.

*Proboscidea Jussieui* Keller [Bailey]

Flowers June–September.

Occurs in cultivated and waste ground, pastures, sandy ground and gravel bars along streams, and along railroads. Scattered in southern and central Missouri, north to St. Louis, Boone, and Jackson counties.

Ranges from Georgia to Mexico, north to West Virginia, Ohio, Indiana, Illinois, and Minnesota.

This species is sometimes cultivated for the fruits which are pickled for consumption. The curiously shaped fruit is unlike any other found in the flora of the state. The flowers are very showy with a whitish or lilac-tinged corolla spotted and mottled with yellow and purple. The plant is covered with sticky hairs and has a rank odor. The leaves are thin.

## Fam. OROBANCHACEAE (Broom-rape Family)

Found only under beech (Fagus) trees in southeastern Missouri; corolla 4-toothed, 1–1.3 cm. long; calyx 2.5–3 mm. long; stem branched, the numerous flowers or fruits scattered along the elongated branches . . . . . . . . . . . . . . . . . . . . . . . . . . . . . . 2. EPIFAGUS

Found in southern and central Missouri in oak-hickory and other woodland types and on rocky soils; corolla 5-lobed, mainly 1.4–2.5 cm. long; calyx 4–11 mm. long; stem usually underground and unbranched, the simple stalks (pedicels) above ground bearing only 1 flower or fruit, rarely (in one species) with a forking thickened stem above ground . . . . . . . . . . . . . 1. OROBANCHE

Plate no. 330. **1.** Proboscidea louisianica, × 2/5; *a.* Fruit, × 2/5; After Gleason, details from Small, The New York Botanical Garden. **2.** Sesamum indicum, × 2/5; After Gleason, details from Small, The New York Botanical Garden. **3.** Utricularia gibba, × 2/5. **4.** Orobanche ludoviciana, × 2/5; *a.* Calyx, × 2 4/5; After Gleason, details from Small, The New York Botanical Garden. **5.** Epifagus virginiana, × 2/5; *a.* Flower, × 2; After Gleason, details from Small, The New York Botanical Garden. **6.** Orobanche uniflora, × 2/5. **7.** Utricularia vulgaris, × 2/5.

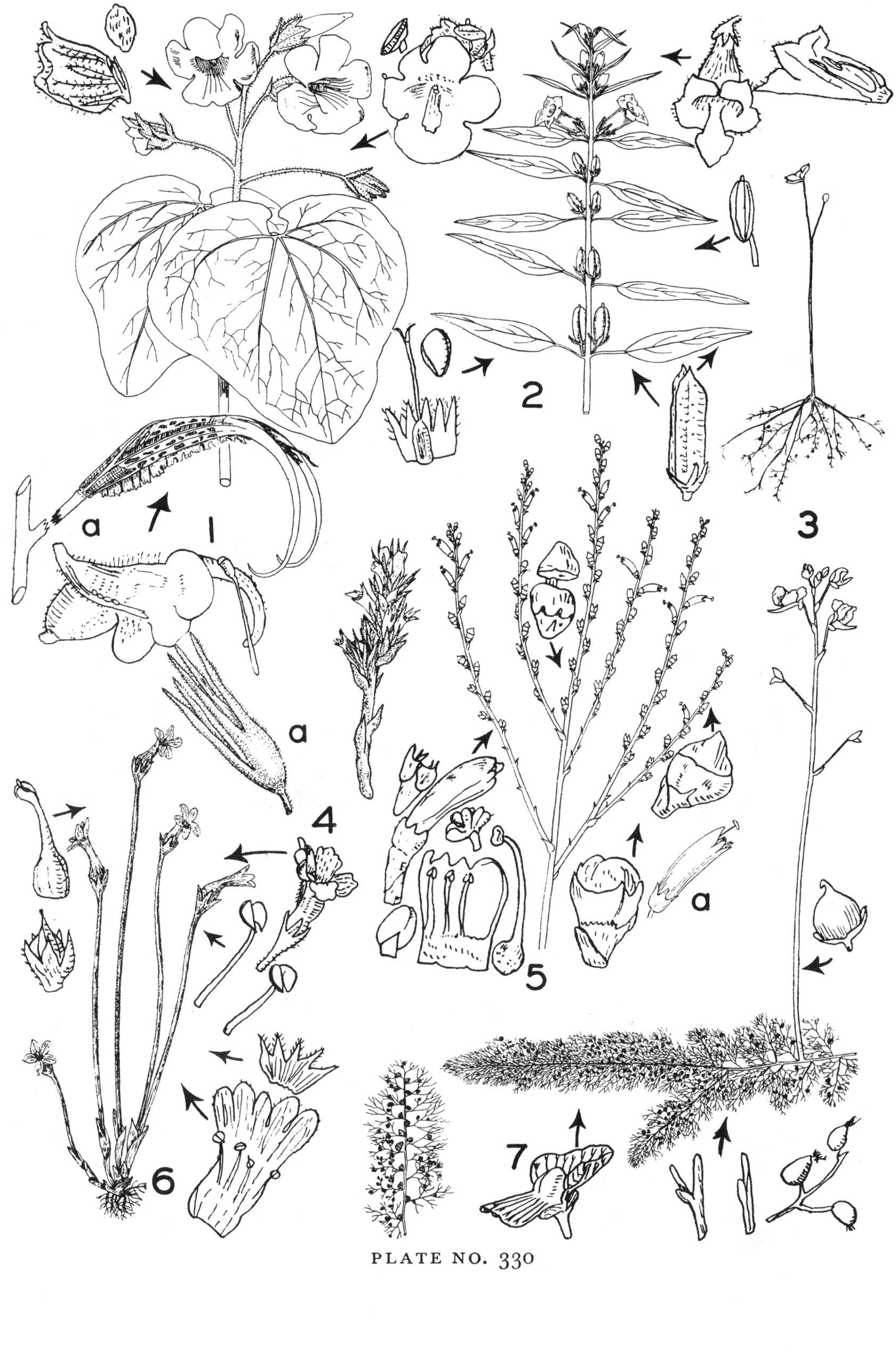


PLATE NO. 330

2015A Orobanche ludoviciana

2016 Epifagus virginiana (Beech-drops)

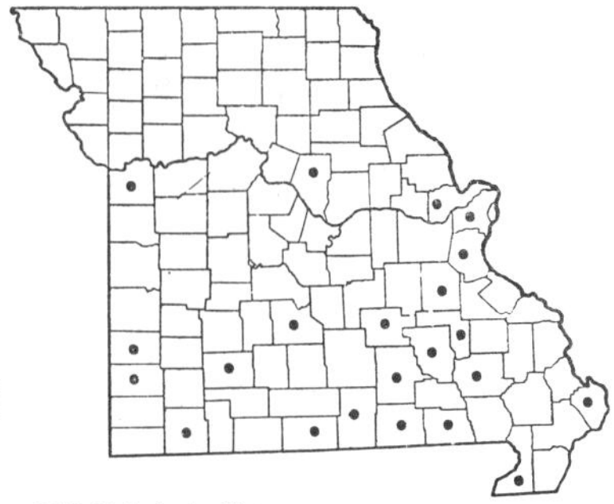

2017 Utricularia gibba

## 1. **Orobanche** L. Broom-rape

Commonly encountered species; only 1 flower or fruit on the unbranched, simple, slender stalk (pedicel); flowers on elongated stalks (pedicels) 5–25 cm. long; no bracts at base of calyx; lobes of the corolla more or less equal and rounded . . . . . . . . . . . . . . . . . . . . 1. O. UNIFLORA

Rarely encountered, known only from St. Louis County; 3 or more flowers arising from the forking, thickened, above-ground stem; flowers on short stalks (pedicels) up to 1 cm. long; 1 or 2 bracts at base of each calyx; lobes of the corolla unequal and acutely pointed . . . . . . . . . 2. O. LUDOVICIANA

1. **Orobanche uniflora** L. One-flowered Cancer-root Map 2015

Also called Small Cancer-root, One-flower Broom-rape.

Flowers April–June.

Occurs in rocky or dry open woods, upland slopes, ridges, above bluff escarpments, low woods at base of bluffs, and in rocky glades, usually in acid soils. Chiefly in southern and central Missouri, locally north to Marion, Randolph, Ray, Clay, Jackson, and Andrew counties.

Ranges from Newfoundland and Quebec to British Columbia, south to Florida, Texas, and California.

The roots of this species are parasitic in Missouri on such plants as *Quercus*, *Aster*, *Solidago*, and *Potentilla simplex*.

2. **Orobanche ludoviciana** Nutt. Map 2015A

*Myzorrhiza ludoviciana* (Nutt.) Rydb. [Rydberg]

Flowers July–September.

Known only from St. Louis County, east-central Missouri (near Crescent, August 20, 1893, and August, 1890, *Letterman*, in Mo. Bot. Gard. Herb.).

Ranges from Minnesota and Manitoba to Alberta and British Columbia, south to Ohio, Indiana, Illinois, Missouri, Kansas, Texas, and California; also in Mexico.

In Indiana this is usually a parasite on *Ambrosia trifida*, the Giant Ragweed. It has also been found in neighboring states as a parasite on *Grindelia squarrosa*, *Xanthium pensylvanicum*, *Artemisia*, and other Compositae. No data are known concerning the host in Missouri.

## 2. **Epifagus** Nutt. Beech-drops

**Epifagus virginiana** (L.) Bart. Beech-drops Map 2016

Flowers August–October.

Occurs in beech woods on Crowley Ridge in southeastern Missouri, in Cape Girardeau, Scott, and Stoddard counties.

Ranges from Quebec and Nova Scotia to Ontario and Wisconsin, south to Florida, Alabama, and Louisiana.

The species is found only in beech woods, where it was first discovered in Missouri in 1933 in Scott County by John H. Kellogg. It probably occurs in all the counties where beech is native in the southeastern portion of the state.

Two kinds of flowers are present, the upper and larger, perfect but sterile flowers with white corolla striped brown-purple, and the lower and smaller, fertile pistillate (female) ones which do not open (cleistogamous) and lack stamens. In forma *pallida* Weath., not yet recorded from Missouri, the plant lacks the brown-purple color of the stems and corollas and is pale brown except for the white to dull pink corollas.

## Fam. **LENTIBULARIACEAE** (Bladderwort Family)

### **Utricularia** L. Bladderwort

Flowers and fruits only 1–3 on a stem; main stem and peduncle (flower-branch) delicate and thread-like, less than 0.5 mm. thick; complete leaf 0.5–2 cm. across, the leaves 0.2–2 cm. long with usually only 2 divisions; corolla 6–12 mm. long . . . . . . . . . . 1. U. GIBBA

Flowers and fruits 6–20 or more on a stem; main stem and peduncle (flower-branch) thicker, 0.5–3 mm. thick; complete leaf 3–12 cm. across, the leaves 1.5–7 cm. long with numerous divisions; corolla 14–25 mm. long . . . . . . . . . . 2. U. VULGARIS

1. **Utricularia gibba** L. Map 2017

Flowers May–September.

Occurs in stagnant water of ponds, lakes, ditches, and sloughs. Southern and central Missouri, north to St. Charles, Boone, and Jackson counties.

Ranges from Florida to Texas and Mexico, north to Quebec, New York, Ontario, Michigan, Wisconsin, Minnesota, Oklahoma, and California; also West Indies and Central America.

This species is often found floating in masses on mucky debris and organic detritus on the surface of upland sink-hole ponds in the Ozarks. In such masses, the small yellow corollas are showy.

2. **Utricularia vulgaris** L. Map 2018

*Utricularia vulgaris* var. *americana* Gray [P & S, Steyerm.]

Flowers late May–September.

Occurs in slow streams, sloughs, oxbow lakes of river meanders, ponds, and ditches in alluvial wooded bottoms. Scattered throughout Missouri, but commonest in the southern and central sections in northern Missouri known from Putnam, Grundy, and Buchanan counties.

Ranges from Labrador to Alaska, south to Virginia, Ohio, Indiana, Missouri, Texas, Colorado, California, and Mexico.

## Fam. **ACANTHACEAE** (Acanthus Family)

a. Calyx-divisions 10–30 mm. long; corolla with 5 more or less equal lobes, mostly 3–8 cm. long; stamens 4 . . . . . . . . . . 1. RUELLIA

a. Calyx-divisions 3–9 mm. long; corolla with a 2-lipped tube, the lips unlike, the whole corolla only 12–18 mm. long; stamens 2 . . . . . . . . . . *b*

b. Leaves long-petioled (stalked), the petioles 10–65 mm. long; special branch (peduncle) bearing flowers or fruits shorter than the accompanying leaf-stalks (petioles); land plants not growing in or immediately alongside water . . . . . . . . . . 3. DICLIPTERA

b. Leaves without petioles (stalks) or the petiole scarcely developed to more than 5 (rarely 10–17) mm. long; special branch (peduncle) bearing flowers or fruits elongated and very noticeable, always longer than the leaf-stalks and from $\frac{1}{2}$ as long as to longer than complete accompanying leaf; plants usually growing in or immediately alongside water . . . . . . . . . . 2. JUSTICIA

### 1. **Ruellia** L. Ruellia

a. Stems either glabrous (without hairs) or hairs appearing in a narrow strip on two opposite sides; leaves dark green, glabrous (without hairs) or sparsely and remotely hairy; calyx-divisions 2–4 mm. broad, lanceolate or lance-linear . . . . . . . . . . 1. R. STREPENS

a. Stems short- to long-hairy, equally hairy on all sides; leaves olive- or grass-green, more or less densely hairy with short or long hairs; calyx-divisions 1.5 mm. or less broad, linear, needle-like, or bristle-like . . . . . . . . . . *b*

b. Leaves sessile (without stalks) or the stalks (petioles) barely developed and only 1–3 mm. long; flowers and fruits sessile (without stalks) or nearly so, in clusters from the base of the leaves; fruit glabrous . . . . . . . . . . 3. R. HUMILIS

b. Leaf-blades narrowed to a petiole (stalk) 3–15 mm. long; flowers and fruits at the ends of special slender stalks (peduncles) arising from the nodes of the stems; fruit hairy . . . 2. R. PEDUNCULATA

1. **Ruellia strepens** L. Wild Petunia Map 2019

Flowers May–October.

Occurs in rich, usually lowland and alluvial forest at the base of bluffs in ravines and valleys, moist forest,

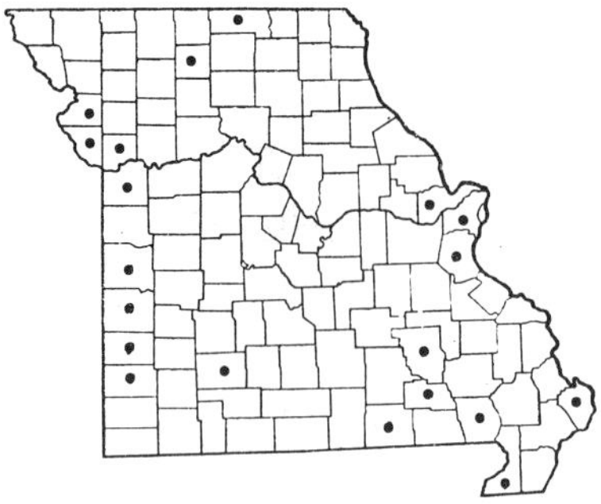

2018 Utricularia vulgaris

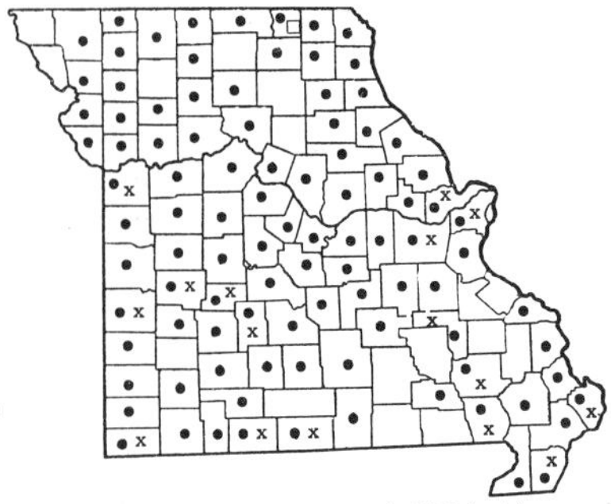

2019 • Ruellia strepens f. strepens (Wild Petunia)
2019 □ Ruellia strepens f. alba
2019 x Ruellia strepens f. cleistantha

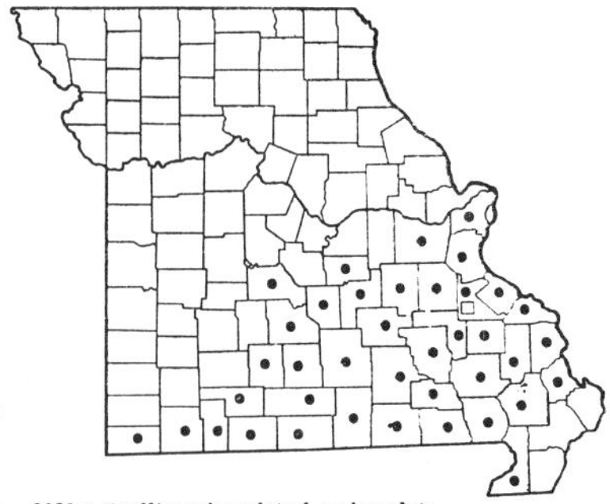

2020 • Ruellia pedunculata f. pedunculata
2020 □ Ruellia pedunculata f. Baueri

borders of streams and ponds, open upland woods, and thickets. Throughout Missouri and doubtless in every county.

Ranges from South Carolina and Alabama to Texas, north to New Jersey, Pennsylvania, Ohio, Indiana, Illinois, Iowa, and Kansas.

The following variations occur in the state:

a. Corolla white . . . 1b. R. STREPENS f. ALBA
a. Corolla lavender to blue-purple . . . . *b*
  b. Common type encountered; corolla 3.5–6.5 cm. long, open with spreading lobes and funnelform tube . 1a. R. STREPENS f. STREPENS
  b. Rarely encountered type; corolla very small, 1.2–2.5 cm. long, remaining more or less closed (cleistogamous), small, club-shaped, and tubular . . . 1c. R. STREPENS f. CLEISTANTHA

1a. **Ruellia strepens** f. **strepens** Map 2019

*Ruellia strepens* L. [G, BB, P & S, Steyerm.]

This is the common variation throughout the state.

1b. **Ruellia strepens** f. **alba** Steyerm. Map 2019

Known only from Schuyler County, northeastern Missouri (base of rich, wooded ravines tributary to Chariton River, T66N, R16W, south part sect. 22, northeast of Livonia, 6 mi. southwest of Glenwood, August 26, 1950, *Steyermark 70292*, holotype in Chi. Nat. Hist. Mus. Herb.).

1c. **Ruellia strepens** f. **cleistantha** (Gray) McCoy Map 2019

Scattered throughout southern and central Missouri north to St. Louis, St. Charles, and Jackson counties.

This cleistogamous form is believed to be only an autumnal phase of the species, according to studies of McCoy (Am. Bot. 43: 22–24. 1937).

2. **Ruellia pedunculata** Torr. Wild Petunia Map 2020

Flowers May–September.

Occurs in dry or rocky open woods, lowland, ravines, upland slopes, and rocky glades. Ozark region south and east of a line drawn from St. Louis, Franklin, Maries, Pulaski, Camden, Laclede, Webster, Christian, and Barry counties to McDonald County.

Ranges from Louisiana and Texas, north to Illinois, Missouri, and Oklahoma.

Two variations occur in Missouri:

Corolla lavender or purple . 2a. R. PEDUNCULATA f. PEDUNCULATA
Corolla white . . 2b. R. PEDUNCULATA f. BAUERI

2a. **Ruellia pedunculata** f. **pedunculata** Map 2020

*Ruellia pedunculata* Torr. [G, BB, P & S, Steyerm.]

This is the common form encountered throughout the Ozarks.

2b. **Ruellia pedunculata** f. **Baueri** Steyerm. Map 2020

Known only from St. Francois County, southeastern Ozark region (in colony, Koester, June 20, 1948, *Bill Bauer*, holotype in Chi. Nat. Hist. Mus. Herb.).

3. **Ruellia humilis** Nutt. Wild Petunia Map 2021

*Ruellia caroliniensis* of some auth. [P & S, Steyerm.], not (Walt.) Steud.

Flowers May–October.

Occurs in prairies, glades, dry or rocky open woods, and escarpments of bluffs.

Generally throughout Missouri, except absent from the lowlands of extreme southeastern Missouri.

The following variations occur in Missouri:

a. Hairs of stem all short or lying parallel to or

close upon the surface of the stem, long hairs absent or few . . 3b. R. HUMILIS var. HUMILIS f. GRISEA

a. Hairs of stem long, spreading, conspicuous . *b*

b. Corolla mostly 3–4.5 cm. (rarely 5) long, the tube 1.2–2.5 cm. long . . . . . *c*

c. Larger leaves of the main stem or axis 2–4 cm. broad, usually rounded at tip, ovate or broadly elliptic . . 3c. R. HUMILIS var. FRONDOSA

c. Larger leaves of the main stem or axis 1–2.5 cm. broad, blunt (obtuse) to somewhat acutely pointed at tip, elliptic-oblong to oblong-lanceolate . . . . 3a. R. HUMILIS var. HUMILIS f. HUMILIS

b. Corolla 5–8 cm. long, the tube 3–6 cm. long *d*

d. Larger leaves of main stem or axis 2.5–4 cm. broad, rounded at tip, ovate-oblong to broadly oval . . . . 3f. R. HUMILIS var. EXPANSA

d. Larger leaves of main stem or axis 1–2.5 cm. broad, blunt (obtuse) to somewhat acutely pointed at tip, elliptic-oblong to oblong-lanceolate . . . . . . *e*

e. Corolla lavender or bluish-purple . 3d. R. HUMILIS var. LONGIFLORA f. LONGIFLORA

e. Corolla white. . . . 3e. R. HUMILIS var. LONGIFLORA f. ALBA

3a. **Ruellia humilis** var. **humilis** f. **humilis** Map 2021

*Ruellia humilis* Nutt. [G, BB]

*Ruellia caroliniensis* of some auth. [P & S, Steyerm. in part], not (Walt.) Steud.

Scattered throughout Missouri.

Ranges from Iowa to Texas, east locally to Pennsylvania, Maryland, Virginia, Tennessee, and Alabama.

3b. **Ruellia humilis** var. **humilis** f. **grisea** Fern. Map 2021

*Ruellia humilis* f. *grisea* Fern. [G]

Known only from Jackson County (dry open bank, Kansas City, September 15, 1916 (*R. Hoffmann*). A specimen collected in Phelps County (*Kellogg 157*) determined by Fernald as f. *grisea* has numerous spreading hairs and is here referred to var. *humilis f. humilis*.

Scattered in the range of var. *humilis* f. *humilis*.

3c. **Ruellia humilis** var. **frondosa** Fern. Map 2021

Scattered throughout Missouri.

Ranges from Nebraska to Texas and Louisiana, east to Pennsylvania and Virginia.

3d. **Ruellia humilis** var. **longiflora** (Gray) Fern. f. **longiflora** Map 2021

*Ruellia humilis* var. *longiflora* (Gray) Fern. [G]

Scattered throughout Missouri.

Ranges from Illinois, Missouri, and Kansas, to Louisiana and Texas.

3e. **Ruellia humilis** var. **longiflora** f. **alba** (Steyerm.) Steyerm. Map 2021

*Ruellia humilis* f. *alba* (Steyerm.) Fern. Rh. 47: 54. 1945 [G]

*Ruellia caroliniensis* (Walt.) Steud. f. *alba* Steyerm., Rh. 41: 585. 1939.

Known only from Caldwell County, northwestern Missouri (prairie slopes above limestone bluffs along Long Creek, 1½ mi. south of Kingston, June 23, 1938, *Steyermark 6058*, holotype in Chi. Nat. Hist. Mus. Herb.). The corolla in the holotype measures 5.1 cm. long with the tube 2.8 cm. long. Some of the larger leaves of the main stem are 2.5–2.7 cm. broad, others are only 2–2.2 cm. broad, making it somewhat intermediate in leaf-breadth between var. *longiflora* and var. *expansa*.

3f. **Ruellia humilis** var. **expansa** Fern. Map 2021

Scattered throughout Missouri.

Ranges from Iowa and Nebraska to Texas, east to Michigan, Indiana, Illinois, Missouri, and locally to Florida.

There is great variation in this species, and intergrading occurs between some of the varieties, making the determination of specimens uncertain. Intergradation between *R. humilis* var. *humilis* and var. *longiflora* occurs, plants being found with short corolla-tubes of *R. humilis* var. *humilis* but with over-all corolla length of var. *longiflora*. Also between var. *longiflora* and var. *expansa* there are intermediate specimens difficult to place in one or the other variety.

The plant is palatable and nutritious, and is eaten by all classes of livestock.

## 2. **Justicia** L. Water Willow

(*Dianthera* L. [P & S, Steyerm., Shinners])

Flowers rather loosely scattered along the axis of the flower-stem (peduncle); larger leaves elliptic-ovate to rhombic-oblong, usually broadest around the middle and tapering to base and tip, 2½–6 times as long as broad; known only from swamps and wet lands of southeastern Missouri lowlands . . . . 2. J. OVATA

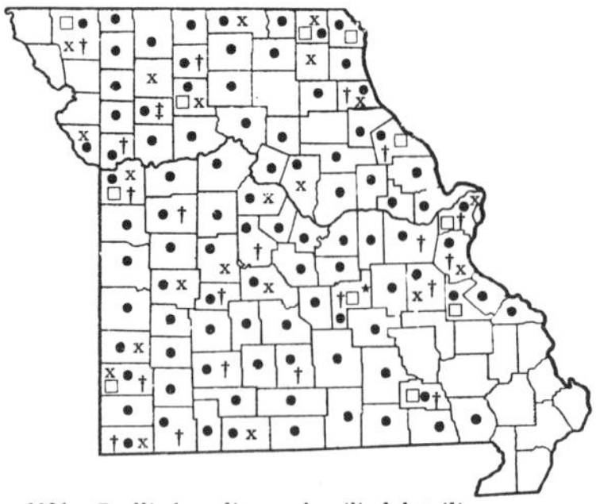

2021 ● Ruellia humulis var. humilis f. humilis
2021 ⁎ Ruellia humilis var. humilis f. grisea
2021 □ Ruellia humilis var. frondosa
2021 x Ruellia humilis var. longiflora f. longiflora
2021 ‡ Ruellia humilis var. longiflora f. alba
2021 † Ruellia humilis var. expansa

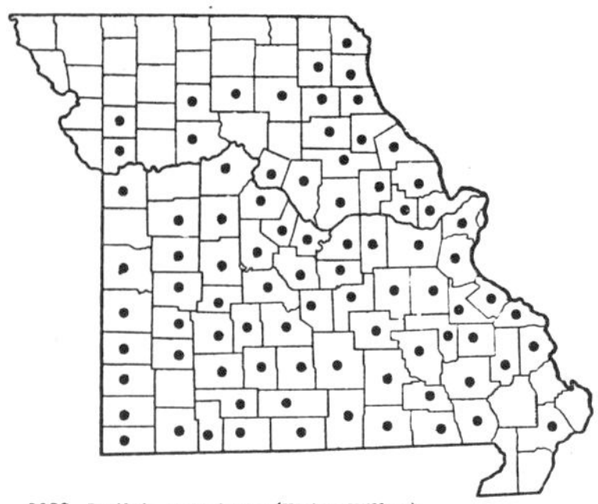

2022 Justicia americana (Water Willow)

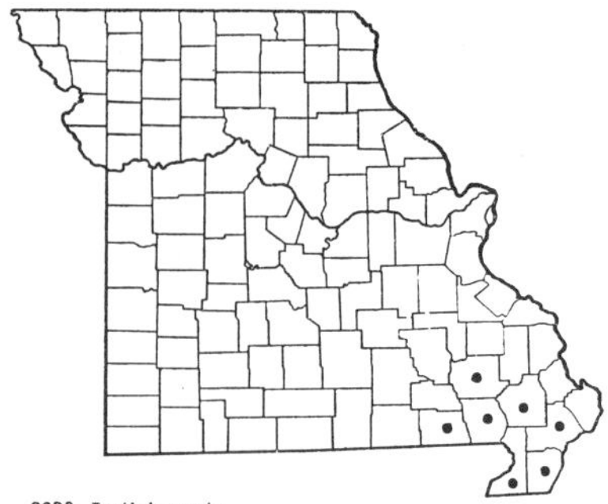

2023 Justicia ovata

Flowers densely clustered in a head-like spike at the tip of the flower-stem (peduncle); larger leaves lanceolate, lance-linear, or oblanceolate, not much, if any, broader at the middle, 5–12 times as long as broad; throughout much of Missouri, but mostly absent from the southeastern Missouri lowlands.
1. J. AMERICANA

1. **Justicia americana** (L.) Vahl    Water Willow
Map 2022

*Dianthera americana* L. [P & S, Steyerm.]

Flowers late May–October.

Occurs along gravelly and muddy margins of streams, sloughs, ponds, and ditches, generally in running water. Common throughout southern, central, and northeastern Missouri, north to Clark, Knox, Macon, Linn, Livingston, and Clinton counties.

Ranges from Georgia to Texas, north to Quebec, Vermont, New York, Ontario, Wisconsin, Missouri, and Kansas.

This is a characteristic plant of the margins of Ozark streams, where it forms large colonies rooting in the gravel, sand, or mud along the shores.

It is especially common around shallow running water, in the vicinity of riffles, rapids, and fords, frequently remaining throughout much of the year with the lower part of the stems submerged. In periods of high water, the plants may be either completely submerged or only their flowering peduncles may be seen above the water for a period of a day or two.

Fernald (Rh. 43: 641. 1941) has described a var. *subcoriacea* presumed to differ from typical *J. americana* in having the leaves firmer, more crowded, broader, shorter, and blunter (1–3 cm. broad instead of 0.5–2.5 cm.), and with the peduncles arising mostly from the upper axils elevating the flowering heads well above the upper leaves instead of being overtopped by the developing leafy tips. He gives the range of this variety from Texas and Oklahoma, north to Kansas and Missouri, citing a Missouri specimen (Meramec River, *N. M. Glatfelter*) as pertaining to the var. *subcoriacea*. Both types of variations occur in Missouri, but I am unable to place any merit on the characters stressed by Fernald to justify separating this species into two varieties. The variations noted by Fernald appear to be merely normal ecological variation encountered in the habitat. The fluctuating water level of the environment in which the plant grows, with relative variation in its exposure to sun, wind, and atmospheric currents, affects the growing habit; prolonged dry sunny spells out of the water produce firmer, shorter leaves with peduncles more conspicuously elongated above the upper leaves as stated for var. *subcoriacea*, while plants growing under shadier or more submerged conditions develop thinner, narrower leaves with leafy tips overtopping the inflorescences. Also, it should be noted that late in the season the leafy tips develop and elongate, normally surpassing the flowering and fruiting peduncles, and more or less concealing them.

Shinners (Field & Lab. 25: 79. 1957) places this and the following species in *Dianthera*, following Bremekamp, who considers *Dianthera* to be an exclusively American group of swamp plants, while *Justicia* is

Plate no. 331. **1.** Ruellia strepens, × 2/7. **2.** Ruellia pedunculata, × 2/7; Details from Small, The New York Botanical Garden. **3.** Ruellia humilis, × 2/7. **4.** Justicia americana, × 2/7; *a.* Flower, × 2/7; Details from Small, The New York Botanical Garden. **5.** Justicia ovata, × 2/7; *a.* Flower, front view, × 1 5/7; *b.* Flower, side view, × 1 5/7; Details from Small, The New York Botanical Garden.

PLATE NO. 331

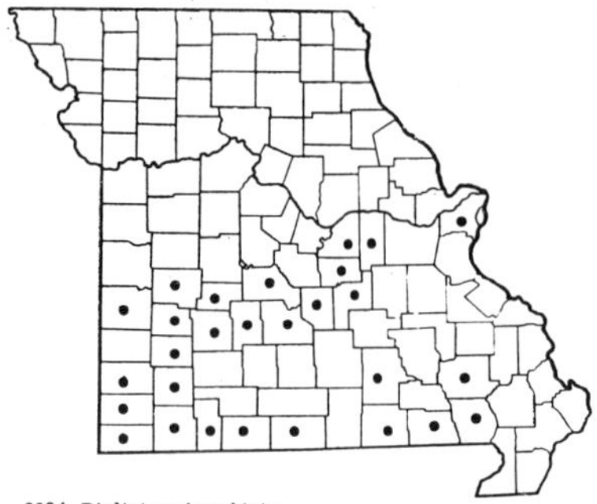

2024 Dicliptera brachiata

2025 Phryma Leptostachya var. Leptostachya (Lopseed)

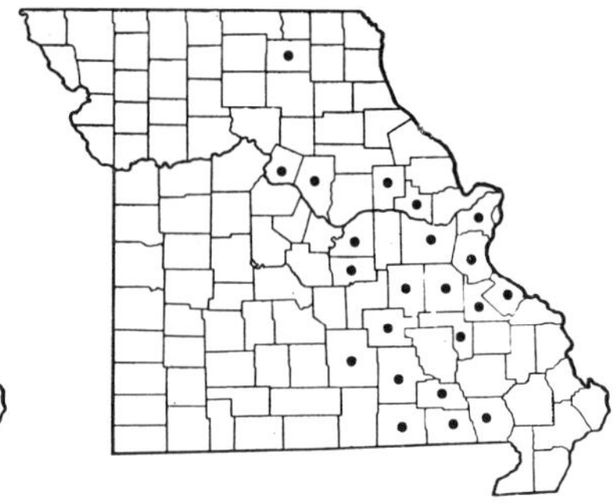

2026 Plantago cordata (Heartleaf Plantain)

considered to include *J. hyssopifolia* and a few species occurring in Africa.

2. **Justicia ovata** (Walt.) Lindau Map 2023
*Dianthera ovata* Walt. [P & S, Steyerm.]
*Justicia humilis* Michx. var. *lanceolata* (Chapm.) Gl. [BB]

Flowers May–June.

Occurs in swamps and low wet woods. Lowlands of southeastern Missouri in Pemiscot, Dunklin, Scott, Stoddard, Wayne, Butler, and Ripley counties.

The flowers are pale purple with darker violet spots on the lower lip.

### 3. **Dicliptera** Juss.

**Dicliptera brachiata** (Pursh) Spreng. Map 2024

Flowers August–October.

Occurs in low muddy or gravelly alluvial ground of wooded valleys and at the base of bluffs along streams. Ozark region north to St. Louis, Gasconade, Osage, Camden, Hickory, St. Clair, and Vernon counties; not recorded from many of the eastern and central Ozark counties.

Ranges from Florida to Louisiana and Oklahoma, north to Virginia, Indiana, Missouri, and Kansas.

The flowers are often cleistogamous and do not open, but the well developed ones are 12–18 mm. long and have showy rose-pink or rose-purplish corollas. Some related species of this genus from the Dutch East Indies are eaten as cooked vegetables or as salad greens, the young leafy tips being used for this purpose.

## Fam. **PHRYMACEAE** (Lopseed Family)

### **Phryma** L. Lopseed

**Phryma Leptostachya** L. var. **Leptostachya**
Lopseed Map 2025
*Phryma Leptostachya* L. [G, BB, P & S]

Flowers late June–September.

Occurs in rich woodland, usually on slopes or in valleys, or in alluvial soils of woodland, sometimes in rocky soils and thickets. Throughout Missouri, probably in every county.

Ranges from Quebec to Manitoba, south to Florida, Alabama, Louisiana, and Texas; also in Asia.

A var. *confertifolia* Fern., ranging from Virginia to Delaware, has hairier stems and leaves, with the teeth simple, crenate, and short.

## Fam. **PLANTAGINACEAE** (Plantain Family)

### **Plantago** L. Plantain

a. Leaves narrowly linear, about the same width from one end to the other, 0.7–9 mm. broad . . . . . *b*
b. Sepals and/or bracts which accompany each flower more or less hairy; corolla-lobes spreading or reflexed in fruit, not closed over the top of the fruit; stamens 4 . . . . . . . . . . . . *c*

c. Bracts slightly or not longer than flower, not conspicuous; bracts usually hidden by the densely silky hairy inflorescence . . . . . . . . . . . . . . . . . . . . 7. P. PURSHII

c. Bracts conspicuous, much longer than the flowers; bracts not hidden by the woolly inflorescence . . *d*

d. Commonly encountered species; bracts glabrous (without hairs) to short-hairy, tipped with awns 1–5 cm. long; longer bracts several times longer than the flowers; corolla-lobes about 2 mm. broad . . . . . . . . . . . . . . . . . 5. P. ARISTATA

d. Rarely encountered, known only from Jackson Co., west-central Missouri; bracts long-hairy, tipped with awns 0.5–1 cm. long; longer bracts 2–4 times longer than the flowers; corolla-lobes about 1 mm. broad . . . . . . . . . . . . . . . . 6. P. SPINULOSA

b. Sepals and bracts which accompany each flower glabrous (without hairs) or slightly ciliate (with fringe of hairs on margins); corolla-lobes erect and closed over the top of the fruit forming a beak; stamens 2 . . . . . . . . . . . . . . . . . . . . . . . *e*

e. Commonly encountered species; seeds 4, symmetrical, 1.2–1.8 mm. long; bracts usually shorter than to scarcely longer than the calyx; sepals obovate; corolla-lobes 0.5 mm. long . 10. P. PUSILLA

e. Rarely collected, known only from Scott Co., southeastern Missouri; seeds 10–30, unsymmetrical, about 0.75 mm. long; bracts usually well surpassing the calyx; sepals ovate, corolla-lobes 0.25 mm. long . . . . . . . . . . . . . . . . . . 1. P. HETEROPHYLLA

a. Leaves lanceolate, elliptic, to ovate, oval, or ovate with a heart-shaped base, 5–170 mm. broad. . . . *f*

f. Main side (lateral) nerves of leaves arising from the midrib within the blade of the leaf; flowering stems (scapes) hollow; roots fleshy, 0.5–1 cm. thick . . . . . . . . . . . 1. P. CORDATA

f. Main side (lateral) nerves of leaves arising at the base of the midrib within the petiole (leaf-stalk) or sometimes only 1 main nerve present; flowering stems (scapes) solid; roots slender and fibrous . . *g*

g. Leaves lanceolate, oblanceolate, to elliptical-oblong, 3–50 mm. broad; seeds 1–2 in each fruit . . *h*

h. Bracts accompanying flowers narrowed to a long transparent and dry or thin (scarious) whitish tip, only the base and center of the bract not thin and transparent; sepals and bracts accompanying flowers glabrous (without hairs) or ciliate (with fringe of hairs on margins); corolla-lobes spreading or reflexed in flower or fruit, not closed over the top of the fruit . . . . . . . . . . . . . . . . . . . . . 4. P. LANCEOLATA

h. Bracts accompanying flowers without slender transparent and thin (scarious) tips, only the margins slightly if at all scarious; sepals and/or bracts accompanying flowers more or less noticeably hairy; corolla-lobes erect and closed over the top of the fruit forming a beak . . . *i*

i. Commonly encountered species; sepals obtuse or rounded at tip or with the midrib scarcely protruding from tip, 2–2.5 mm. long; bracts accompanying flowers 1–2.5 mm. long; mature seeds usually pale brown or yellow-brown, convex on one side, concave on the other, 1.5–2 mm. long; leaves entire (without teeth) or slightly wavy or shallowly toothed . . . . . . . . . . . . . . . . . . . . . . 8. P. VIRGINICA

i. Rarely collected, known only from Jackson County, west-central Missouri; sepals tapering at the tip, the midrib prolonged conspicuously from tip, 2.5–3 mm. long; bracts accompanying flowers 3–4.5 mm. long; mature seeds usually red to reddish, nearly flat on both sides, 2.5–3 mm. long; leaves mostly toothed . . . 9. P. RHODOSPERMA

g. Leaves broadly elliptic, oval, or broadly ovate with or without a heart-shaped base, 25–170 mm. broad; seeds 4–15 in each fruit . . . . . . . . . . . . . . . . . . *j*

j. Common native species found throughout Missouri; base of petioles (leaf-stalks) dark red-purple; leaf-blades usually glabrous (without hairs) or rarely sparsely hairy; sepals and bracts accompanying flowers narrowed and somewhat pointed (subacute to acute) at tip, the bracts triangular-lanceolate; mature fruit opening (dehiscing transversely) between the base and the middle or in the lower half, with 4–9 seeds . . . . . . . . . . 3. P. RUGELII

j. Rarely collected species, usually found around towns and streets; base of petioles (leaf-stalks) usually green; leaf-blades usually minutely hairy on one or both surfaces; sepals and bracts accompanying flowers rounded or obtuse (blunt) at tip, the bracts broadly ovate; mature fruit opening (dehiscing transversely) near the middle, with 6–15 seeds . . 2. P. MAJOR

1. **Plantago cordata** Lam. Heartleaf Plantain
Map 2026

Flowers May–July.

Occurs in wet woods, along sloughs, rocky stream beds, and spring branches. Eastern Missouri, commonest in the eastern and northern Ozark sections, west to Oregon, Texas, Crawford, Maries, Osage, Boone, and Howard counties, and locally in Adair County, northeastern Missouri.

Ranges from Alabama to Louisiana, north to Ohio,

2027 Plantago major var. major f. major (Common Plantain)

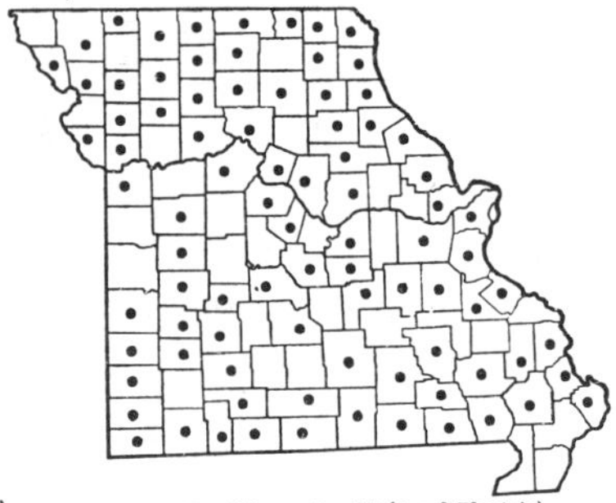

2028 Plantago Rugelii var. Rugelii (Rugel Plantain)

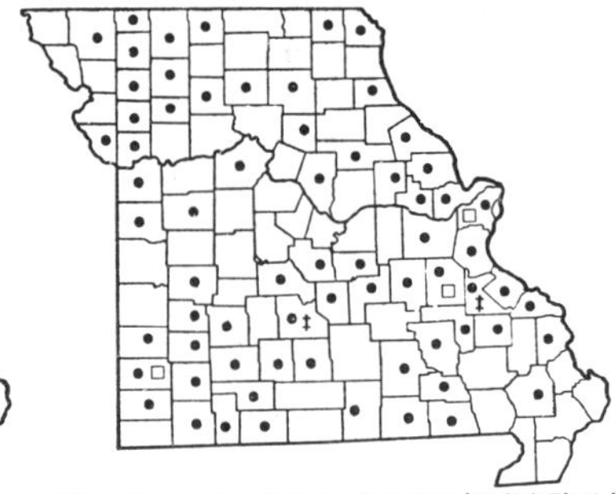

2029 ● Plantago lanceolata var. lanceolata (English Plantain)
2029 □ Plantago lanceolata var. sphaerostachya f. sphaerostachya
2029 ‡ Plantago lanceolata var. sphaerostachya f. eriophora

Ontario, Michigan, Wisconsin, and Minnesota, west to Missouri; locally along the Atlantic Coast from Virginia and D.C. to New York.

Of all the plantains found in the state, this is the most ornamental, from the standpoint of the dark green, large, heart-shaped leaves which form an attractive rosette. In favored sites along stream beds, this species often occurs in large numbers. The young fleshy leaves and tender petioles may be cooked as a vegetable, and of all the native plantains is the most tender.

2. **Plantago major** L. var. **major** f. **major** Common Plantain Map 2027

*Plantago major* L. [G, P & S, Steyerm.]

*Plantago major* subsp. *pleiosperma* Pilger var. *paludosa* Béguinot [BB]

Flowers May–October.

Occurs in waste ground, along streets, door-yards, rubbish-heaps, and railroads. Rare and scattered in southern and central Missouri, where known only from St. Louis, Phelps, Johnson, Jackson, Jasper, and Newton counties.

Native of Europe; introduced and naturalized throughout the United States and southern Canada.

The species has been divided recently into two subspecies, twelve varieties, ten subvarieties, and eleven forms. The Missouri plants and those most commonly found in this country would fall, following the most recent classification, into *P. major* subsp. *pleiosperma* Pilger var. *paludosa* Béguinot.

The pollen of this species has potentialities of causing hay fever, but so little is produced and the plant is so relatively rare in Missouri that it may be regarded as unimportant. The seeds are sometimes used for bird seed.

3. **Plantago Rugelii** Dcne. var. **Rugelii** Rugel Plantain Map 2028

*Plantago Rugelii* Dcne.

Flowers late May–October.

Occurs in fields, waste ground, lawns, sidewalks, gardens, low or moist woods, alluvial ground along streams, roadsides and railroads. Throughout Missouri, and probably in every county.

Ranges from Quebec to North Dakota, south to Florida and Texas.

This is the common broad-leaved plantain in Missouri. A var. *asperula* Farw., with the leaf-blades and/or part of the scapes (flowering stems) more or less hairy, may occur with the typical glabrous (hairless) type, but is quite rare, and is not recorded for Missouri.

The pollen possesses potentialities for causing hay fever irritation, but is produced in such small quantity that it is not considered of importance in this respect.

Young tender leaf-blades and petioles can be cooked as a vegetable, but are not as tender as those of *P. cordata*. The seeds are sometimes one of the ingredients of wild bird seed.

4. **Plantago lanceolata** L. English Plantain Map 2029

Also called Buckhorn, Rib Grass, Ripple Grass.

Flowers late April–October.

Occurs in usually dry soils in pastures, lawns, cultivated and fallow fields, gardens, waste ground,

Plate no. 332. **1.** Dicliptera brachiata, × 2/5; Details from Small, The New York Botanical Garden. **2.** Plantago heterophylla, × 2/5; *a.* Variation in size and leaves. **3.** Phryma Leptostachya, × 2/5; *a.* Flower, × 4; *b.* Fruit, × 4; After Gleason, details from Small, The New York Botanical Garden. **4.** Plantago Purshii, × 2/5. **5.** Plantago pusilla, × 2/5. **6.** Plantago virginica, × 2/5. **7.** Plantago spinulosa, × 2/5. **8.** Plantago rhodosperma, × 2/5.

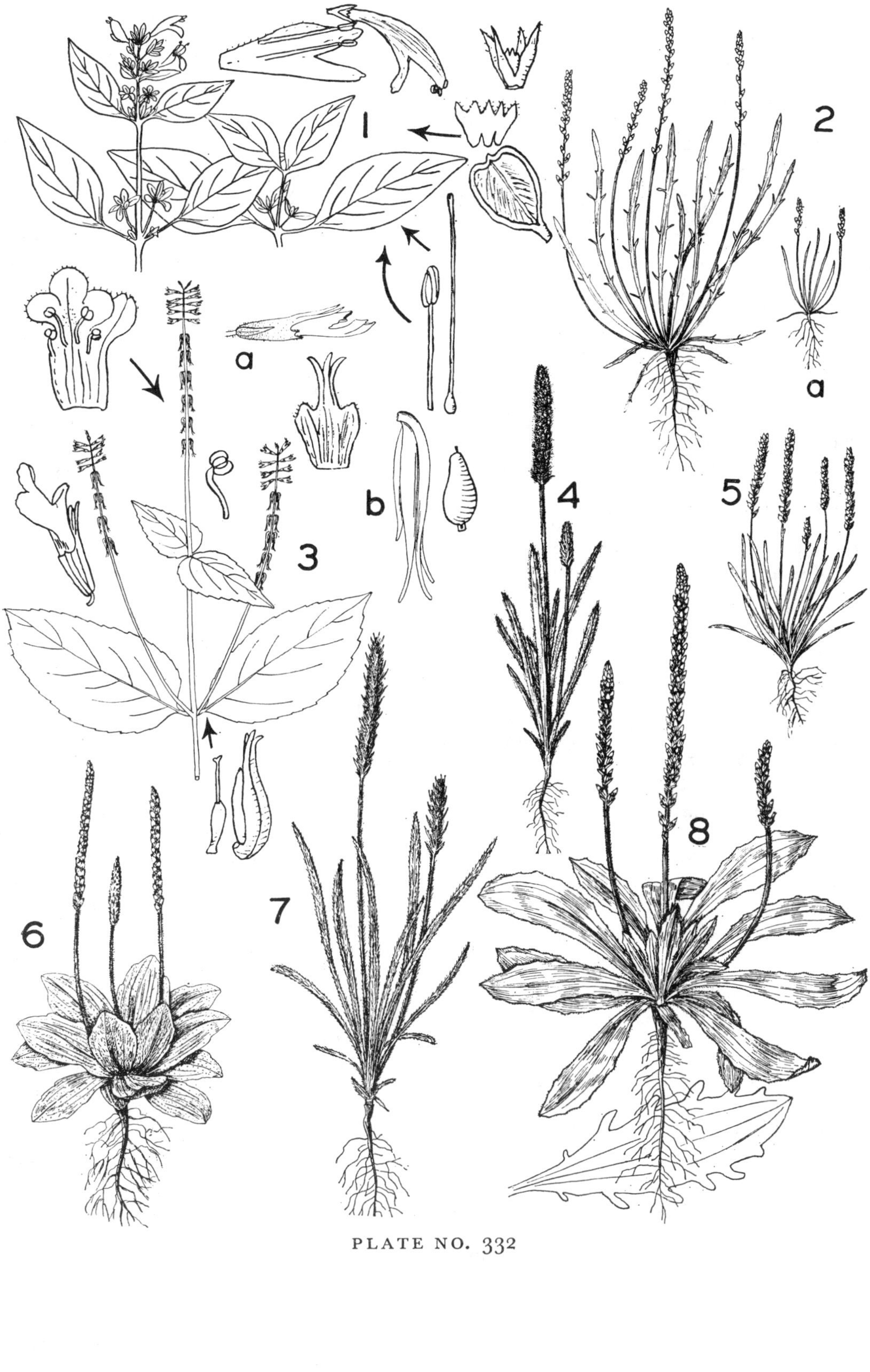


PLATE NO. 332

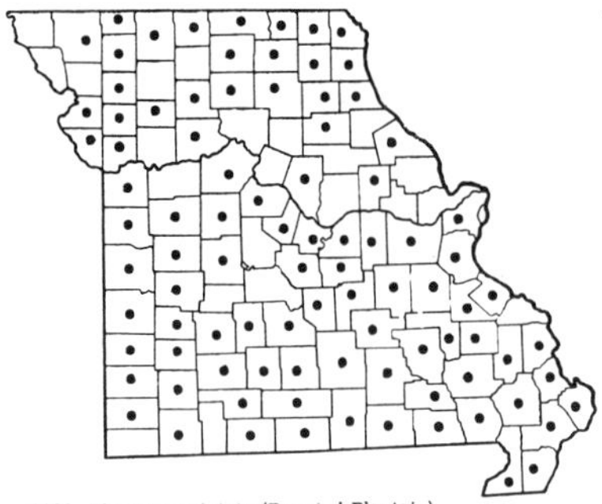
2030 Plantago aristata (Bracted Plantain)

2031 Plantago spinulosa

2032 Plantago Purshii (Salt-and-Pepper Plant)

streets, roadsides, and railroads. Throughout Missouri.

Native of Europe.

The following variations occur in Missouri:

a. Commonly encountered type; flower-spike tapering at summit during period of flowering, ovoid-conic, becoming obtuse (blunt) and cylindric in fruit, 1.5–8 cm. long . . . . 4a. P. LANCEOLATA var. LANCEOLATA

a. More rarely encountered; flower-spike rounded at summit during period of flowering, round-tipped and nearly globose to cylindric in fruit, 0.5–2.3 cm. long . . . . . . *b*

b. Upper surface of leaf-blades green and glabrous (without hairs) or only sparsely hairy . . . . . . 4b. P. LANCEOLATA var. SPHAEROSTACHYA f. SPHAEROSTACHYA

b. Upper surface of leaf-blades gray with abundant hairiness . . 4c. P. LANCEOLATA var. SPHAEROSTACHYA f. ERIOPHORA

4a. **Plantago lanceolata** var. **lanceolata** Map 2029

*Plantago lanceolata* L. [G, P & S, Steyerm.]

This is the common variation throughout the state.

Introduced and naturalized throughout the United States and southern Canada.

4b. **Plantago lanceolata** var. **sphaerostachya** Mert. & Koch f. **sphaerostachya** Map 2029

*Plantago lanceolata* var. *sphaerostachya* Mert. & Koch [G]

*Plantago lanceolata* var. *communis* Schlecht. subvar. *sphaerostachya* (Mert. & Koch) Pilger [Pilger]

Known only from St. Louis (St. Louis, May 3, 1892, *Glatfelter*; St. Louis, April, 1887, *Pammel*), Washington (Potosi, July 24, 1885, *Wislizenus 249*), and Jasper (Webb City, May 3, 1909, *Palmer 1854*) counties.

Introduced and naturalized in North America from Newfoundland to Michigan, south to New England, New York, and Missouri.

4c. **Plantago lanceolata** var. **sphaerostachya** f. **eriophora** (Hoffmgg. & Link) G. Beck Map 2029

*Plantago lanceolata* var. *dubia* (L.) Wahlenb. subvar. *eriophora* (Hoffmgg. & Link) Pilger [Pilger]

*Plantago lanceolata* var. *angustifolia* Poir. [BB]

Known only from St. Francois (Flat River, October 13, 1897, *Trelease 489*) and Laclede (dry grazed fields near Gasconade River, between Falcon and Nebo, July 22, 1934, *Steyermark 13907*) counties.

Introduced and naturalized in North America from Newfoundland to Michigan, south to New England, New York, Indiana, and Missouri.

There is some evidence that the var. *sphaerostachya* represents either young plants or those found on dry sterile soils. Until more definite results from experimental studies are at hand, the variations are here retained in their accustomed categories.

This species sheds sufficient pollen to be considered an important cause of hay fever, especially during the months of May and June, when flowering of the species is at its height. It is not considered to be as important a cause of hay fever, however, as compared with the grasses with which it may be associated.

5. **Plantago aristata** Michx. Bracted Plantain Map 2030

Flowers May–November.

Occurs in fallow pastures, glades, barrens, open upland wooded ridges, waste ground, roadsides and along railroads, usually in acid soils. Throughout Missouri, doubtless in every county.

Ranges from Maine to Michigan, North Dakota, Montana, and Oregon, south to Florida, Texas, and Mexico.

The species is quite variable in size of plant, leaf shape, and amount of hairiness. Very small individuals

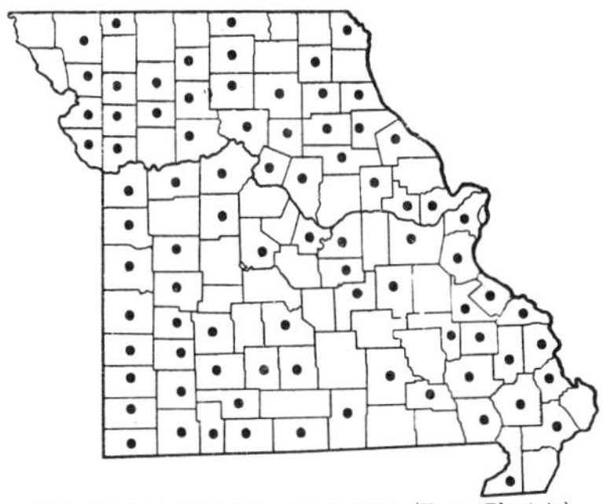

2033 Plantago virginica var. virginica (Hoary Plantain)

2034 Plantago rhodosperma

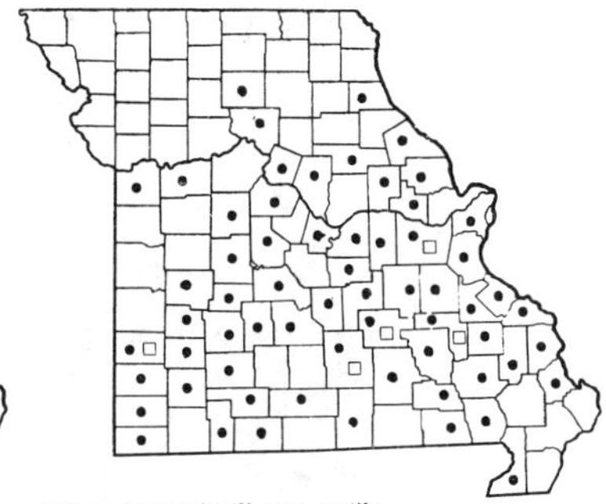

2035 • Plantago pusilla var. pusilla
2035 □ Plantago pusilla var. major

are often the result of delayed germination or crowding, and are characterized by nearly hairlike leaves and few-flowered spikes.

6. **Plantago spinulosa** Dcne. Map 2031
*Plantago Purshii* var. *spinulosa* (Dcne.) Shinners [Shinners]
Flowers May–July.
Occurs in waste ground. Known only from Jackson County, west-central Missouri.
Ranges from Minnesota to Alberta, south to Missouri, Oklahoma, and Texas.

7. **Plantago Purshii** R. & S. Salt-and-pepper Plant Map 2032
Also called Pursh Plantain.
Flowers May–August.
Occurs on loess hills in northwestern Missouri, on glades, dry prairies, and in waste ground. Mainly in western Missouri in Atchison, Holt, Clay, Jackson, St. Clair, and Barton counties, introduced in St. Louis County in eastern Missouri.
Ranges from Ontario to British Columbia, south to Ohio, Indiana, Illinois, Missouri, Oklahoma, Texas, and Arizona; introduced east to New England and New Jersey.

8. **Plantago virginica** L. var. **virginica**
Hoary Plantain Map 2033
*Plantago virginica* L. [G, BB, P & S, Steyerm.]
Flowers April–June.
Occurs in fallow fields, pastures, glades, barrens, rocky waste ground, sterile open slopes, and along roadsides and railroads. Throughout Missouri.
Ranges from Florida to Texas and California, north to Maine, New York, Ohio, Michigan, Illinois, Missouri, Kansas, and Oregon.
This species varies greatly in size of plant and leaf-shape, amount of toothing of leaf, and degree of hairiness. Fernald has described a var. *viridescens*, ranging from Florida to Texas north to New Jersey, characterized by less pubescent or nearly glabrous leaves, more shortly pubescent scapes, and smaller seeds.

9. **Plantago rhodosperma** Dcne. Map 2034
*Plantago virginica* var. *longifolia* Gray
Flowers May–June.
Occurs in waste ground. Known only from Jackson County, west-central Missouri (introduced, Courtney, June 9, 1905, *Bush 2986*).
Ranges from Arizona and Texas, north to Oklahoma and Kansas; introduced in Missouri and Tennessee.
This species resembles *P. virginica*, but is generally more robust and luxuriant with denser hairiness and longer stiffer hairs.

10. **Plantago pusilla** Nutt. Map 2035
*Plantago elongata* [of P & S, Steyerm.], not Pursh
Flowers April–June.
Occurs in fallow fields, sandy alluvial soils in valleys, pastures, prairies, rock outcrops, glades, upland slopes and ridge tops, usually in dry acid soils, but sometimes in moist soils around upland sink-hole ponds.
Two variations are found in Missouri:

Leaves usually entire (without teeth), 1–5 cm. long; plant mainly less than 15 cm. high; spikes 1–6 cm. long, the flowers at first closely overlapping, later becoming scattered; seeds 1.2 mm. long . . . . . . . 10a. P. PUSILLA var. PUSILLA
Leaves often slightly toothed, 5–9 cm. long; plant 12–25 cm. high; spikes 6–12 cm. long, the flowers scattered from the beginning; seeds 1.7–1.8 mm. long . . . . . . 10b. P. PUSILLA var. MAJOR

10a. **Plantago pusilla** var. **pusilla** Map 2035
*Plantago pusilla* Nutt. [G, BB]
*Plantago elongata* [of P & S, Steyerm.], not Pursh

This is the commoner variation in the state, found in southern and central Missouri locally north to Marion, Audrain, Linn, Lafayette, and Jackson counties.

Ranges from Georgia to Texas, north to Massachusetts, New York, Indiana, Illinois, Minnesota, Nebraska, and Oregon.

10b. **Plantago pusilla** var. **major** Engelm. Map 2035

Often occurring in moist areas bordering sink-hole ponds. Known in southern and central Missouri from Franklin (Gray Summit, May 3, 1927, *Kellogg 1186*), Dent (1 mi. north of Shannon Co. line bordering sink-hole pond, May 11, 1935, *Steyermark 18933* in part), Texas (base of dead stump in sink-hole pond, between Eunice and Yukon, May 25, 1935, *Steyermark 19097*), Iron (Ironton, May 5, 1925, *Epling 5182*), and Barton (upland prairie with sandstone outcrops, 3 mi. northeast of Iantha, May 3, 1955, *Palmer 59728*) counties.

Ranges from Louisiana and Oklahoma to Missouri and Indiana.

11. **Plantago heterophylla** Nutt. Map 2036

Flowers April–May.

Occurs in sandy open ground. Known only from Scott (sandy ground, May 20, 1894, *Eggert*) and St. Louis (St. Louis, 1842, *Engelmann*) counties, eastern Missouri south of the Missouri River.

Ranges from Florida to Lower California, north to Virginia, Kentucky, Illinois, Missouri, and California; introduced north to New Jersey and Pennsylvania.

The leaves of this species often have a few teeth.

## Order **RUBIALES**

### Fam. **RUBIACEAE** (Madder Family)

a. Shrubs 1–3 m. ($3\frac{1}{3}$–10 feet) tall, with woody stems . . . . . . . . . . . 6. CEPHALANTHUS
a. Herbaceous non-woody plants . . . . . . . . . . . . . . . . . . . . . . . *b*
  b. Leaves in whorls (circles) of 4–8 . . . . . . . . . . . . . . . . . . . . . *c*
    c. Corolla funnelform with a conspicuous long tube, 4–5-lobed, blue or pink; calyx-lobes evident, lanceolate; flowers surrounded by an involucre of narrow, green, leaf-like divisions; rarely encountered . . . . . . . . . . . . . . . . . . . . . 1. SHERARDIA
    c. Corolla with scarcely any tube or the tube short and inconspicuous, 3–4-lobed, white; calyx with teeth or lobes not evident; flowers not surrounded by any involucre of green leaf-like divisions; commonly encountered . . . . . . . . . . . . . . . . . . . 2. GALIUM
  b. Leaves opposite in pairs . . . . . . . . . . . . . . . . . . . . . . . . *d*
    d. Bases of leaves or petioles (leaf-stalks) connected by a membrane (stipule) bearing several to many bristles on each side of stem . . . . . . . . . . . . . . . . . . *e*
      e. Flowers and fruits only 1–3 in one leaf axil, 2–6 at a node; some part of the leaves, stems, or fruits with hairs; corolla 4–10 mm. long, the conspicuous tube elongated and much longer than the calyx-lobes; fruit separating into 2 or 3 carpels which do not split open. 4. DIODIA
      e. Flowers and fruits 10–40 at a node; all parts of plant glabrous (without hairs); corolla 2–3 mm. long, scarcely longer than the calyx; fruit a capsule of 2 carpels, one of them opening, the other remaining closed . . . . . . . . . . . . . . . 3. SPERMACOCE
    d. Bases of leaves or petioles (leaf-stalks) connected by a membranous stipule, but this part either lacking bristles or with only 1–2 bristles at tip on each side of stem . . . . . . . . *f*
      f. Flowers or fruits sessile (without stalks) or nearly so, the stalks (pedicels) if present at most 1.5 mm. long; corolla 1–1.5 mm. long, with a short scarcely apparent tube, shorter than the calyx . . . . . . . . . . . . . . . . . . . . . 8. OLDENLANDIA
      f. Flowers or fruits on definite stalks (pedicels), the stalks 1–20 or more mm. long; corolla 5–15 mm. long, with usually a long noticeable tube longer than the calyx . . . . . . . . *g*
        g. Plant trailing or creeping with evergreen round-ovate leaves; flowers in pairs, their ovaries united; fruit usually scarlet and fleshy containing 8 seeds; ovule 1 in each of the 4 cells of the ovary . . . . . . . . . . . . . . . . . . . 5. MITCHELLA
        g. Plant erect or ascending, the leaves not evergreen and usually of other shapes; flowers 1 or several to many, their ovaries not united; fruit brown, dry, containing 8–40 seeds; ovules numerous in each of the 2 cells of the ovary. . . . . . . . . . . 7. HOUSTONIA

Plate no. 333. **1.** Plantago cordata, × 2/7; *a.* Flower, × 1 3/7. **2.** Plantago major, × 2/7. **3.** Plantago lanceolata, × 2/7; *a.* Flower, × 1 3/7. **4.** Plantago Rugelii, × 2/7. **5.** Plantago aristata, × 2/7. All details from Small, The New York Botanical Garden.

PLATE NO. 333

2036 Plantago heterophylla

2037 Sherardia arvensis (Field Madder)

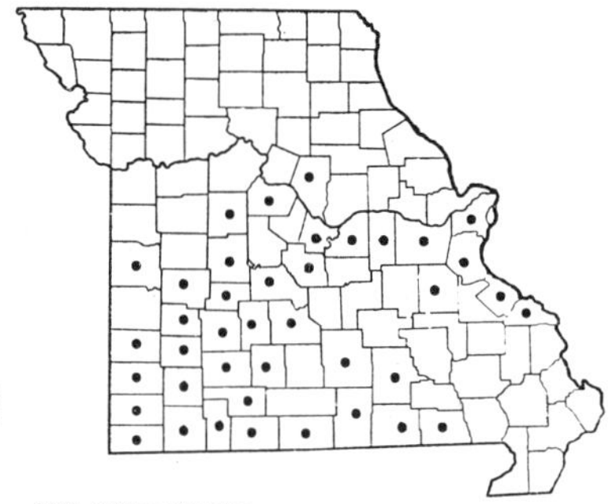

2038 Galium virgatum

## 1. **Sherardia** L. Field Madder

**Sherardia arvensis** L. Field Madder Map 2037
Flowers April–May.

Occurs in openings in woods along creeks, along roadsides, and clearings. Known only from southwestern Missouri in Barry (roadside bank, Eagle Rock, April 28, 1936, *Kellogg 27059*) and Newton (openings in woods, along small draw, along south side of Shoal Creek, ½ mi. east of Mo.-Kan. state line, May 20, 1953, *Palmer 55578*) counties.

Native of Europe; introduced and naturalized in North America from Nova Scotia and Quebec to Ohio, south to North Carolina, Tennessee, Missouri, and Oklahoma.

## 2. **Galium** L. Bedstraw, Cleavers

a. Flowers and fruits solitary, without stalks (pedicels), arising directly from the stem at the base of the foliage leaves (axillary) . . . . . 1. G. VIRGATUM
a. Flowers and fruits either at the end of long or slender definite or special stalks (pedicels), or 2 or more together along the sides or at the ends of an inflorescence . . . . . *b*
  b. Leaves 3-nerved . . . . . *c*
    c. Ovary and fruit smooth, glabrous (without hairs), or merely roughened . . . . . *d*
      d. Flowers white, numerous, in compact upright inflorescences; lower surface of leaves obscurely or not at all marked with narrow or elongated pits or depressions, the lateral nerves usually conspicuous, or well-developed; largest leaves 2–5 mm. broad; plants of moist north-facing limestone bluffs and chert in Shannon and Texas counties . . 7. G. BOREALE
      d. Flowers dull purple, separated in loosely branched inflorescences; lower surface of leaves conspicuously marked with narrow or elongated depressions or pits, the lateral nerves usually weak and poorly developed; largest leaves 4–10 mm. broad; plants of dry acid soils of chert, sandstone, or granite in woods of the southern Ozarks . . . . . 6. G. ARKANSANUM
    c. Ovary and fruit bristly . . . . . *e*
      e. Flowers all on stalks (pedicels) at the ends of the branches of the inflorescence; leaves oval, 1.5–2.5 cm. long, 6–12 mm. broad . . . . . 4. G. PILOSUM
      e. At least some of the flowers sessile (without stalks) along the side of the branches of the inflorescence; leaves ovate or ovate-lanceolate, mainly 2–5 cm. (1.5) long, 10–25 (7) mm. broad . . . . . 5. G. CIRCAEZANS
  b. Leaves 1-nerved . . . . . *f*
    f. Leaves in whorls (circles) of 4 . . . . . *g*
      g. Corolla dull purple; leaves acutely pointed, 2–5 cm. long . . . . . 6. G. ARKANSANUM
      g. Corolla white; leaves obtuse (blunt) at tip, 0.8–3 cm. long . . . . . 10. G. OBTUSUM
    f. Leaves in whorls (circles) mostly of 5, 6, or 8, sometimes with whorls of 4, 5, and 6, or of 4 and 5 on the same plant . . . . . *h*
      h. Ovary or fruit bristly or rough-hairy . . . . . *i*
        i. Leaves mainly in whorls of 8's; leaves broadest near the tip or above the middle. 2. G. APARINE
        i. Leaves mainly in whorls of 6's; leaves broadest near middle or base . . . 3. G. TRIFLORUM
      h. Ovary or fruit smooth or merely roughened . . . . . *j*

2039 Galium Aparine (Cleavers)

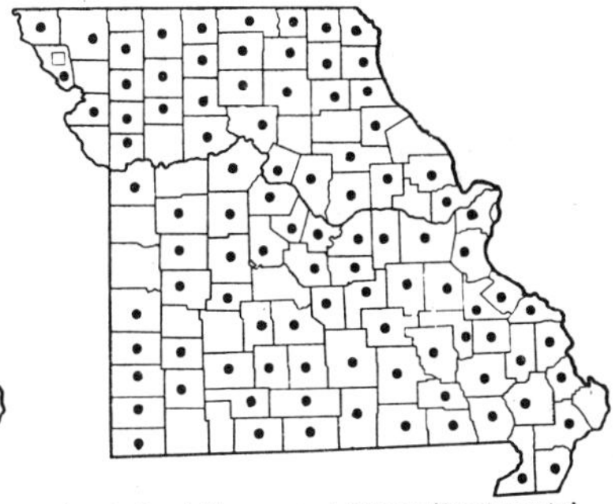

2040 • Galium triflorum var. triflorum (Sweet-scented Bedstraw)
2040 □ Galium triflorum X circaezans var. hypomalacum

2041 • Galium pilosum var. pilosum
2041 □ Galium pilosum var. puncticulosum

j. Flowers yellow; stem, at least in the inflorescence, more or less hairy; leaves linear to needle-like . . . . . 8. G. VERUM

j. Flowers white or greenish-white; stem smooth or rough on the angles; leaves linear to oblanceolate. . . . . . *k*

k. Leaves sharply pointed or bristle-tipped; leaves always in 6's . . . . 11. G. CONCINNUM

k. Leaves rounded or blunt at tip, without bristle tips; leaves in whorls of 4, 5, or 6, or combinations of these . . . . . *l*

l. Corollas 2–2.5 mm. broad, usually with 4 acute or somewhat acute lobes 1 mm. or more long; stems smooth, stiff and erect; leaves mainly in 4's, rarely 5's or 6's. 10. G. OBTUSUM

l. Corollas 1.5 mm. or less broad, with usually 3, sometimes 4, obtuse (blunt) lobes 0.5–1 mm. long; stems, especially the younger portions, rough with backward pointing hairs, reclining or ascending; leaves of main stem mostly in 5's or 6's, sometimes 4's. . . . . . 9. G. TINCTORIUM

1. **Galium virgatum** Nutt. Map 2038

Flowers April–June.

Occurs on limestone glades, rocky ledges, edges of limestone bluff escarpments, and prairies. Ozark and unglaciated prairie sections north to St. Louis, Franklin, Gasconade, Osage, Cole, Boone, Cooper, Pettis, and Bates counties.

Ranges from Louisiana and Texas, north to Tennessee, Missouri, and Oklahoma.

After flowering and fruiting, this annual bedstraw dries up in early summer.

2. **Galium Aparine** L. Cleavers Map 2039

Also called Goose Grass.

*Galium Aparine* var. *Vaillantii* (DC.) Koch [P & S, Steyerm.]

Flowers May–July.

Occurs in alluvial soils in valleys, moist or rich woodland and thickets, in waste ground, and along railroads. Throughout Missouri, doubtless in every county.

Ranges from Newfoundland to Alaska, south to Florida and Texas; also in Eurasia.

A variation, var. *Vaillantii*, with small leaves less than 25 mm. long, occurs on plants growing in a drier situation than usual, such as dry open sandy ridges or upland wooded slopes, or may also be found on stunted or weak individuals. As the variation appears to be correlated with poorer growing conditions, it does not seem to merit varietal recognition.

The plant forms dense tangles in early spring, especially in moist low woodland. In late spring and early summer the ripened bristly fruits adhere to clothing and skin, forerunners of a long list of various species whose fruits have developed hooks, burs, hairs, bristles, spines, and other means of attachment for securing wider geographical distribution and spread of the plant. After maturing the fruit in late spring and early summer, the entire plant, an annual, dries up and is no longer in evidence in the woodland until the following spring.

If the mature fruits are dried and slightly roasted, they yield a coffeelike drink, stated to be the best substitute for coffee in North America. An easy way to secure the fruits is to allow them to adhere to the clothing and then scrape them off. Quantities may also be collected by allowing the fruits to drop into a receptable as one, with light strokes, pulls the fruit away from the stems.

3. **Galium triflorum** Michx. var. **triflorum**

Sweet-scented Bedstraw Map 2040

Also called Fragrant Bedstraw.

*Galium triflorum* Michx. [G, BB, P & S, Steyerm.]

Flowers last of May–September.

Occurs in rich or moist woodland, low open ground, along streams and spring branches, moist rocky ledges of bluffs, rocky talus slopes at the base of bluffs, rarely in wet meadows. Throughout Missouri, doubtless in every county.

Ranges from Newfoundland to Alaska, south to Virginia, Tennessee, Louisiana, Texas, and Mexico.

A variety *asprelliforme* Fern. with leaves 3–10 mm. long and with elongated, many-flowered, paniculately branched inflorescences with 3–8 nodes, occurs from Quebec and New York to Florida and Tennessee.

*Galium triflorum* is used in making an aromatic drink, and, like *Asperula odorata*, the Sweet Woodruff, may be added to wine to impart a flavor.

The dark green smooth leaves often form a dense carpet over the moist soil and are attractive as a ground cover. The plant is perennial.

4. **Galium pilosum** Ait. Hairy Bedstraw Map 2041

Flowers June–August.

Occurs in dry or rocky usually upland woods, borders of glades, and thickets, usually in acid soils overlying sandstone, chert, or granite substrata.

Two variations occur in Missouri:

Hairs on at least lower part of stem spreading and more or less straight; common type encountered . 4a. G. PILOSUM var. PILOSUM

Hairs on lower part of stem upwardly incurved; rare type . . 4b. G. PILOSUM var. PUNCTICULOSUM

4a. **Galium pilosum** var. **pilosum** Map 2041

*Galium pilosum* Ait. [G, P & S, Steyerm.]

Throughout the Ozark and unglaciated prairie regions of southern and central Missouri north to Marion, Montgomery, Callaway, Boone, Saline, and Jackson counties.

Ranges from New Hampshire to Ontario and Michigan, south to Virginia, North Carolina, Tennessee, Missouri, Oklahoma, and Texas.

4b. **Galium pilosum** var. **puncticulosum** (Michx.) T. & G. Map 2041

Known only from Jefferson (sandy woods opposite Pacific, July 4, 1896, *Eggert*), St. Francois (stony hills near Big River, August 31, 1891, *Eggert*), and Jasper (rocky open ground, 3 mi. northwest of Joplin, Sept. 12, 1920, *Palmer 19094*) counties.

Ranges from Florida to Mississippi, north to New Jersey, Missouri, and Oklahoma.

The dark green leaves of this perennial persist until late in fall. The roots of this and other species of the genus contain a red coloring matter, which formerly was used for dyeing clothes.

5. **Galium circaezans** Michx. Wild Licorice Map 2042

Flowers May–July.

Occurs in rich or rocky wooded slopes and bluffs and thickets. Throughout Missouri.

Two variations are found in the state:

Rarely encountered; stems glabrous (without hairs) or sparsely hairy on the angles; larger leaves 1.5–2.5 cm. (rarely to 4 cm.) long, 0.7–1.4 cm. (rarely to 1.8 cm.) broad, the lower surface glabrous or sparsely short-hairy on the main nerves . . . 5a. G. CIRCAEZANS var. CIRCAEZANS

Common type throughout the state; stems usually hairy on the angles; larger leaves 2–5 cm. long, 1–2.5 cm. broad, the lower surface with appressed (lying parallel to surface) hairs with the nerves long-hairy . 5b. G. CIRCAEZANS var. HYPOMALACUM

5a. **Galium circaezans** var. **circaezans** Map 2042

Rare and scattered, known only from Pemiscot, Barton (rocky open woods, 2½ mi. southwest of Lamar, June 13, 1950, *Palmer 50109*), Clay, and Atchison (*Steyermark 85694*) counties.

Ranges from Florida to Texas, north to Virginia, New England, New York, Kentucky, Michigan, and Missouri.

5b. **Galium circaezans** var. **hypomalacum** Fern. Map 2042

Common throughout Missouri, absent from some of the lowland counties of the extreme southeastern section, but probably to be found there and in other counties where absent.

Ranges from Quebec to Minnesota and Nebraska, south to Virginia, North Carolina, Kentucky, Missouri, Oklahoma, and Texas.

The glabrate extreme with smaller leaves is rare in Missouri, but intergrading individual plants may be found. *Steyermark 85694* from Atchison County has the glabrate or sparsely pilose stems of *G. circaezans* var. *circaezans*, with the lower leaf surface having the mostly long-hirsute nerves of *G. circaezans* var. *hypomalacum*,

Plate no. 334. **1.** Sherardia arvensis, × 2/5; *a.* Flower, × 2 2/5. **2.** Galium virgatum, × 2/5. **3.** Galium triflorum, × 2/5. **4.** Galium Aparine, × 2/5. **5.** Galium arkansanum, × 2/5. **6.** Galium concinnum, × 2/5.

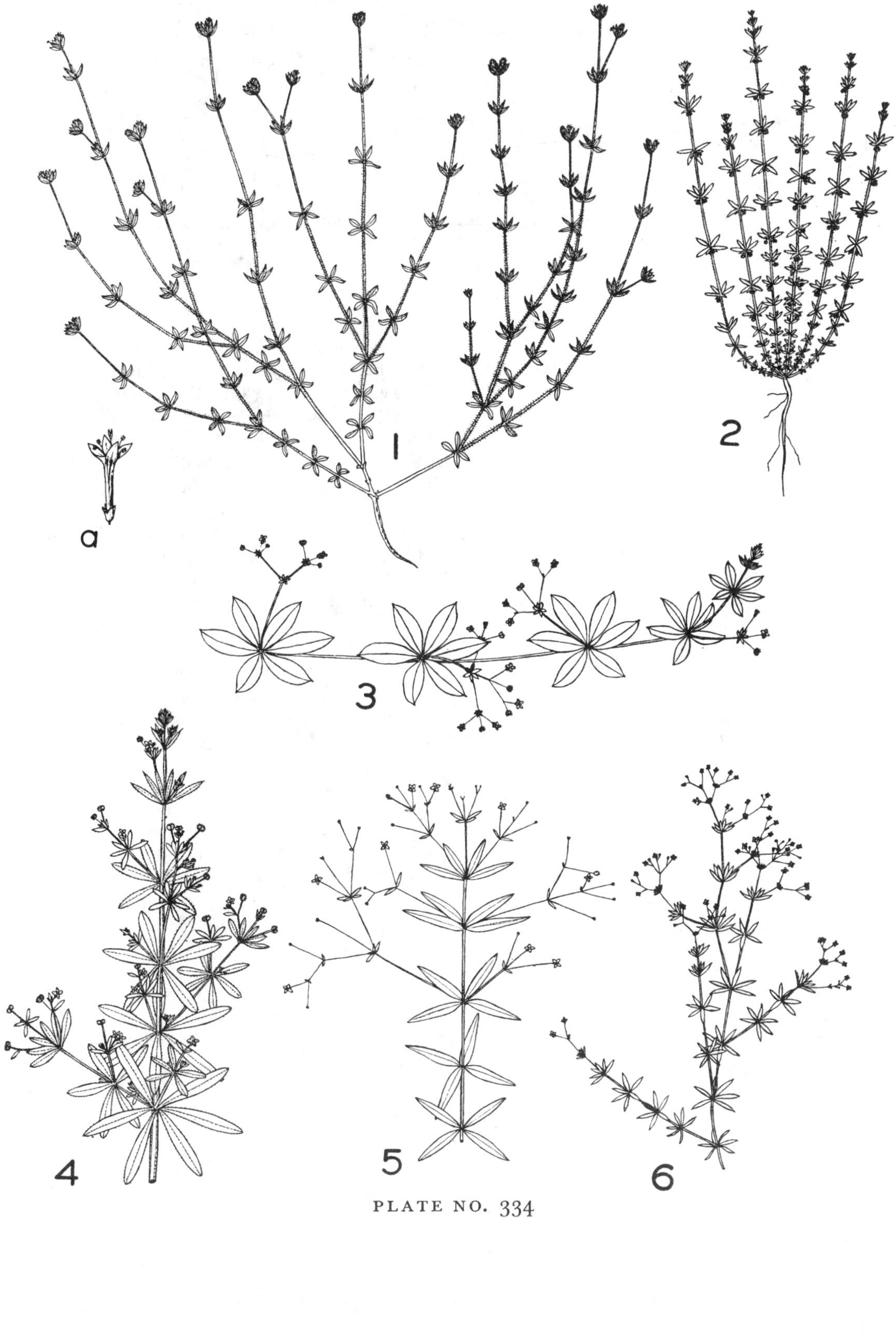


PLATE NO. 334

2042 □ Galium circaezans var. circaezans (Wild Licorice)
2042 ● Galium circaezans var. hypomalacum

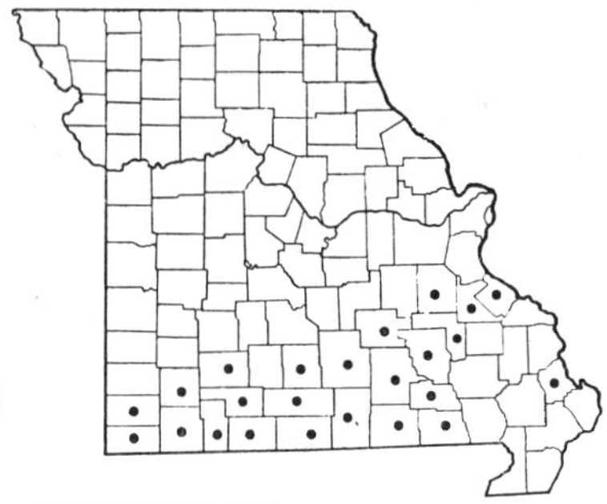

2043 Galium arkansanum

2044 Galium boreale var. hyssopifolium (Northern Bedstraw)

but the lower surface itself is not as densely appressed-pubescent as in characteristic var. *hypomalacum.* Gleason (*New Ill. Fl.* 3: 283. 1952) treats the var. *hypomalacum* as an intergrading variation probably not meriting varietal recognition. However, in Missouri, the extremely pubescent, larger-leaved variation is the usual type encountered and is recognized in the present flora.

An apparent hybrid between this *G. circaezans* var. *hypomalacum* and *G. triflorum* Michx. has been collected in Holt County, northwestern Missouri (wooded ravines tributary to Thorp Branch, T59N, R8W, northeast sect. 12, 4$\frac{1}{2}$ mi. southeast of Oregon, July 20, 1952, *Steyermark 73804*). This plant grew associated with *G. circaezans* var. *hypomalacum* (*Steyermark 73804A*) and *G. triflorum.* The leaves occurred in whorls of 6 and were elliptic-lanceolate as in *G. triflorum* but the stems were pubescent and the lower leaf surface pilose as in *G. circaezans* var. *hypomalacum.*

6. **Galium arkansanum** Gray — Map 2043

Flowers last of May–June.

Occurs in rocky open woods, openings in woods, and borders of rocky open glades, in acid soils overlying sandstone, chert, or granite substrata. Ozark region of southern Missouri north to Ste. Genevieve, St. Francois, Washington, Dent, Texas, Wright, Greene, Lawrence, and Newton counties.

Ranges from Missouri to Arkansas and Oklahoma.

This is one of the endemic species confined to the Ozark, Boston, and Ouachita mountain area. It is related to and probably evolved from *G. latifolium* Michx., a species of the Appalachian mountains.

7. **Galium boreale** L. var. **hyssopifolium** (Hoffm.) DC. Northern Bedstraw — Map 2044

Flowers late May–July.

Occurs on north-facing ledges and crevices of limestone bluffs along Jack's Fork of Current River, in the southeastern Ozarks, where known only from Shannon (on top of bluff in shaded or exposed ledges of Roubidoux sandstone and boulders, also on upper and lower part of limestone part of bluff, Jam-up Bluff, 6 mi. northwest of Montier, June 8, 1941, *Steyermark 28836*; same locality, June 4, 1939, *Steyermark 26887*; dry ledges of limestone bluffs at Jam-up Bluff, 6 mi. north of Montier, July 17, 1932, *Steyermark 7113*; high bluffs, Montier, May 27, 1907, *Bush 4745*) and Texas (near base of moist north-facing limestone bluffs along Jack's Fork of Current River, T27N, R7W, sect. 36, near Shannon Co. line, 6 mi. southeast of Arroll, June 10, 1950, *Steyermark 69938*) counties.

Ranges from Quebec to North Dakota and Colorado, south to New Jersey, New York, Ohio, Indiana, Illinois, and Missouri; also on the Pacific Coast and in Eurasia.

The *Steyermark 7113* collection shows some immature fruits glabrous, as characteristic of var. *hyssopifolium,* others partly appressed-pubescent, as characteristic of var. *intermedium* DC. Since the other collections cited from Jam-up Bluff have mainly glabrous fruits, the *Steyermark 7113* is believed to be part of the same population here judged to be better placed with var. *hyssopifolium.*

This is one of the rare survivors or relicts of a more northern flora, which during relatively cooler periods of the Pleistocene period entered the Ozark region, and, then, subsequent to the retreat of the glaciers northward, has persisted since that time in remote, favorable, cool situations provided by north-facing bluff exposures.

---

Plate no. 335. **1.** Galium verum, × $^2/_5$. **2.** Galium boreale var. hyssopifolium, × $^2/_5$. **3.** Galium obtusum, × $^2/_5$. **4.** Galium tinctorium, × $^2/_5$. All details from Small, The New York Botanical Garden.

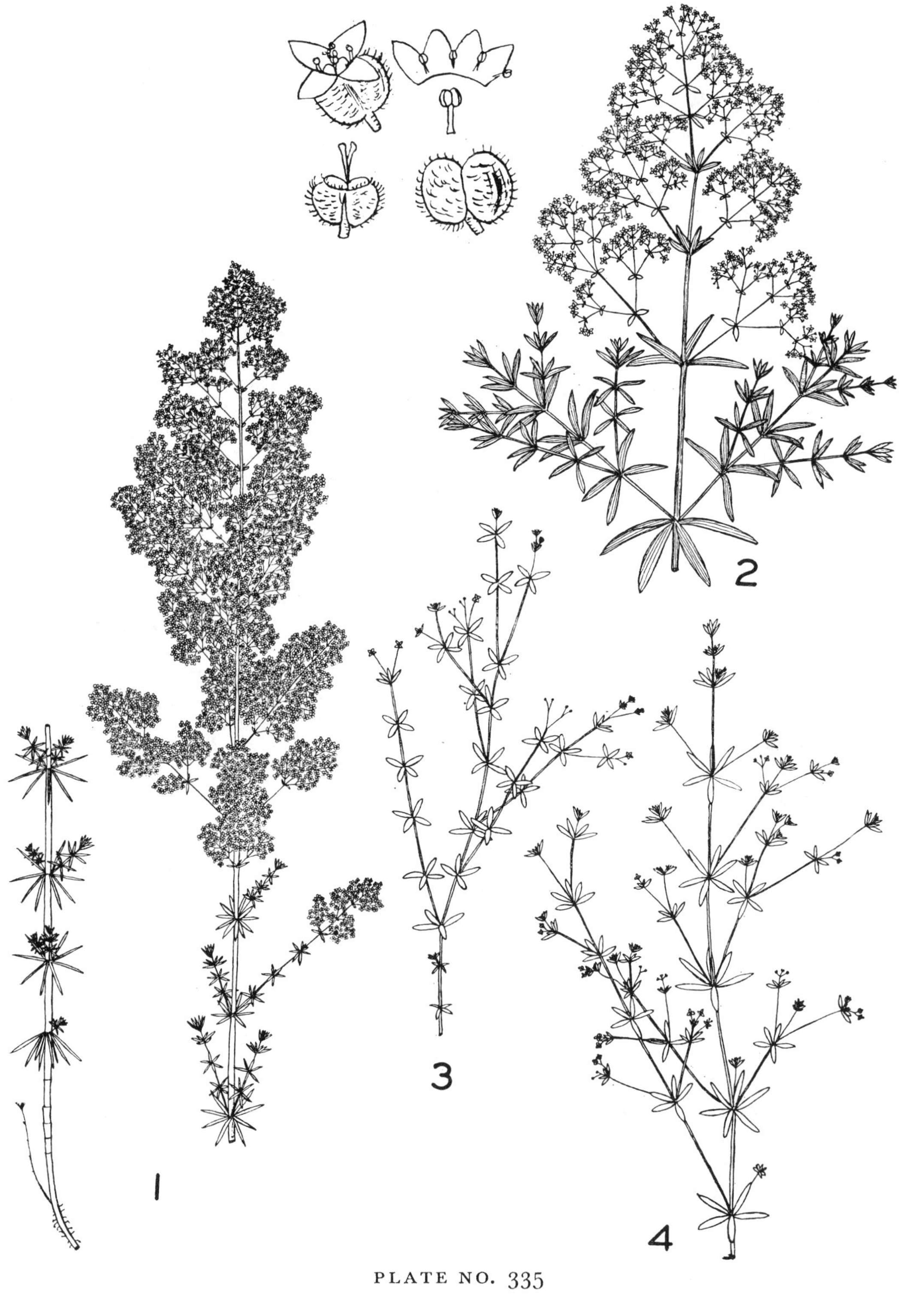


PLATE NO. 335

2045 Galium verum f. verum (Yellow Bedstraw)

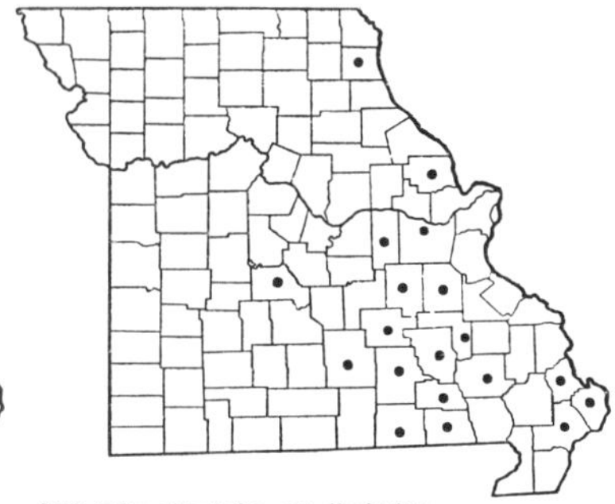
2046 Galium tinctorium var. tinctorium

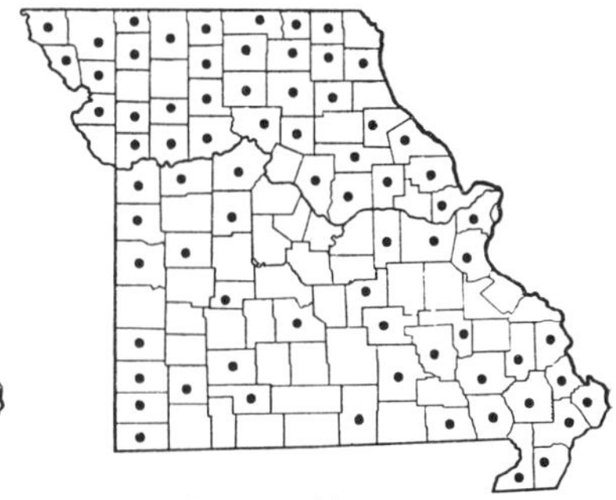
2047 Galium obtusum var. obtusum

8. **Galium verum** L. f. **verum** Yellow Bedstraw Map 2045

Also called Our Lady's Bedstraw.

*Galium verum* L. [G, BB, Steyerm.]

Flowers late May–July.

Known only from Adair County, northern Missouri (Kirksville, June 6, 1939, collector not known, in U. of Mo. Herb.).

Native of Eurasia and Africa; introduced and naturalized from Newfoundland to Ontario and North Dakota south to Virginia, West Virginia, Ohio, Indiana, Illinois, Missouri, and Kansas.

The panicles are densely flowered and the lower branches of the panicle which rise from below the internodes, often surpass them. A form with white corollas, f. *albidum* (Hartm.) Lindm., has not been recorded from Missouri.

This is the species of bedstraw which is claimed to have been the one which filled the Christ Child's manger at Bethlehem, and later on was used for stuffing mattresses. The distilled flowering tops are made into a beverage, and a type of cheese, made from the milk of sheep and goats, is prepared from the plant when mixed with rennet from calves.

9. **Galium tinctorium** L. var. **tinctorium** Map 2046

*Galium tinctorium* L. [G, BB, Steyerm.]

*Galium Claytoni* Michx. [P & S]

Flowers May–September.

Occurs in swampy ground, among hummocks and wet woods in and bordering upland sink-hole ponds, around swampy spring-fed meadows of low valleys, and in swamps and ditches. Eastern Missouri, mostly in the southeastern Ozark region around upland sink-hole ponds, west to Howell, Texas, Camden, and Gasconade counties, and locally in northeastern Missouri to Lincoln and Lewis counties.

Ranges from Newfoundland to Ontario and Nebraska, south to South Carolina, Kentucky, Missouri, and Texas.

This perennial plant at first is more or less upright with ascending or curving stems, but eventually becomes much branched, sprawling and spreading over adjacent plants. In sink-hole ponds it is often found on mossy hummocks at the base of bushes of *Cephalanthus occidentalis* or *Hibiscus lasiocarpos*, where it is associated with *Viola lanceolata*, *Carex decomposita*, *Bidens discoidea*, and other characteristic species.

Upon drying the plant becomes blackened. Some collections are intermediate between this species and the following, *G. obtusum*. One of these, *Steyermark 85523* from a sink-hole pond (Spaugh Pond) in Wayne County has flowers about 1.5 mm. across with the corollas 3- or 4-lobed, but the lobes are acutish as in *G. obtusum*, and the leaves vary from 4 to 5 in a whorl.

10. **Galium obtusum** Bigel. var. **obtusum** Map 2047

*Galium obtusum* Bigel. [G, Steyerm.]

*Galium tinctorium* [of P & S], not L.

*Galium tinctorium* var. *ramosum* Gl. [BB]

Flowers May–July.

Occurs in swamps and swampy ground, wet places in prairies, low swampy or wet woods, alluvial thickets, and wet places along ditches and railroads. Throughout Missouri, commonest in the northern half of the state, rare and scattered in most of the Ozark section.

Ranges from Florida to Arizona, north to Nova Scotia, Ontario, Michigan, Wisconsin, Minnesota, and Nebraska.

Gleason includes for Missouri a var. *ramosum* Gl., differing from typical var. *obtusum* in the leaves averaging 4 instead of 7 times as long as broad, and with the leaves thinner and with hairs 0.2 instead of 0.1 mm. long. This range of variation in leaf-size, relative thickness, and extent of pubescence can be found within a given colony, fluctuating from relatively broader and thinner leaves in shaded wooded situations to

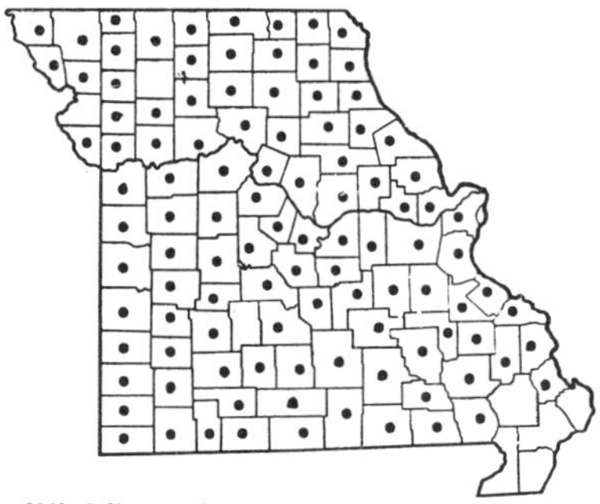

2048 Galium concinnum

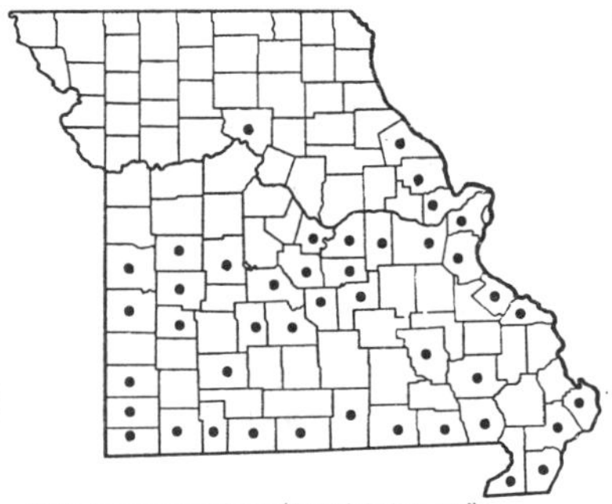

2049 Spermacoce glabra (Smooth Buttonweed)

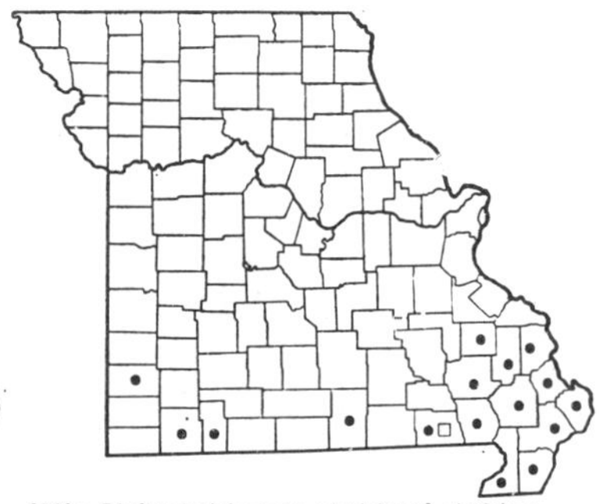

2050 ● Diodia virginiana var. virginiana f. virginiana
2050 □ Diodia virginiana var. virginiana f. hirsuta

narrower and somewhat thicker leaves in more exposed prairies and meadow habitats. Furthermore, there appears to be no correlation in Missouri material between amount of pubescence and leaf proportions, so that no recognition is given in the present treatment to this variety. Most of the Missouri material has leaves 7–10 times as long as broad, with a smaller proportion having leaves 4–6 times as long as broad.

*Galium obtusum* usually grows in dense masses in moist prairies. Upon drying, the plants turn blackish.

11. **Galium concinnum** T. & G. Map 2048

Flowers late April–July.

Occurs in rich, dry or moist woodland, moist or rocky ledges of bluffs, and wet places in prairies, low thickets, and along railroads. Throughout Missouri, except absent from the extreme southeastern lowland section.

Ranges from New Jersey to Minnesota, south to Virginia, Kentucky, Arkansas, Kansas, and Oklahoma.

The delicate, airy, loosely branched sprays of small white flowers lend a graceful and ornamental feature to the woodland flora, and such sprays add much interest as interior adornment. It is reported that tea prepared from the plant is a remedy for dropsy and kidney disorders.

## 3. **Spermacoce** L. Buttonweed

**Spermacoce glabra** Michx. Smooth Buttonweed Map 2049

Flowers June–October.

Occurs in low wet woods, alluvial fields and bottoms, and borders of ponds, sloughs, and streams. Southern and central Missouri north to Marion, Franklin, Gasconade, Osage, Cole, Camden, Benton, Henry, and Bates counties, and locally north in Chariton County.

Ranges from Florida to Texas, north to Ohio, Indiana, Illinois, Missouri, and Kansas.

## 4. **Diodia** L. Buttonweed

Plants of wet soils of swamps, margins of ponds and sloughs, and swampy meadows, southeastern Missouri lowlands and extreme southern Missouri elsewhere; leaves 6–20 mm. broad; corolla about 10 mm. long; calyx-teeth 2; fruits 7–10 mm. long, glabrous (without hairs) to hairy. . . . 1. D. VIRGINIANA

Plants of dry soils, glades, fields, prairies, and sandy or rocky ground, throughout southern, central, and northeastern Missouri; leaves mostly 1.5–6 mm. (up to 9) broad; corolla 4–6 mm. long; calyx-teeth 4; fruits 2.5–4 mm. long, more or less densely hairy . . . . . . . . . . . . . . . . 2. D. TERES

1. **Diodia virginiana** L. Map 2050

Flowers June–September.

Occurs in wet soils of swamps, margins of lowland and upland sink-hole ponds, streams, and sloughs, low fields and swales, and swampy meadows. Lowlands of southeastern Missouri and Ozark border north to Cape Girardeau, Bollinger, and Madison counties, west to Wayne and Ripley counties, locally west to swampy meadows and upland sink-hole ponds in Howell County, along White River in Stone and Barry counties,

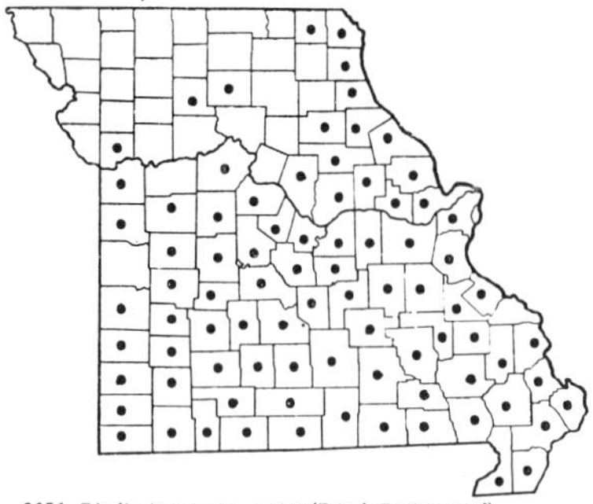

2051 Diodia teres var. teres (Rough Buttonweed)

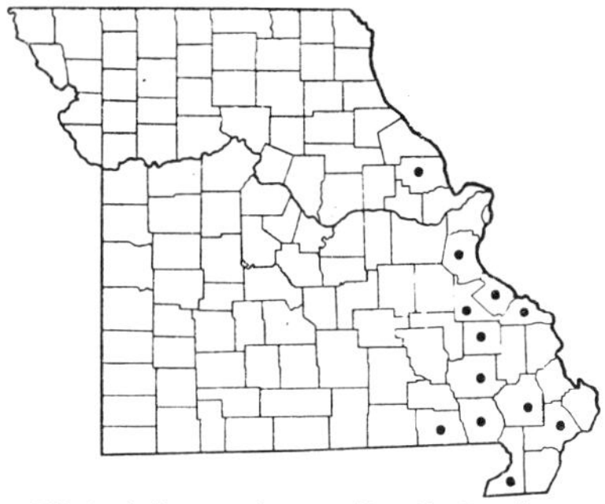

2052 Mitchella repens f. repens (Partridge Berry)

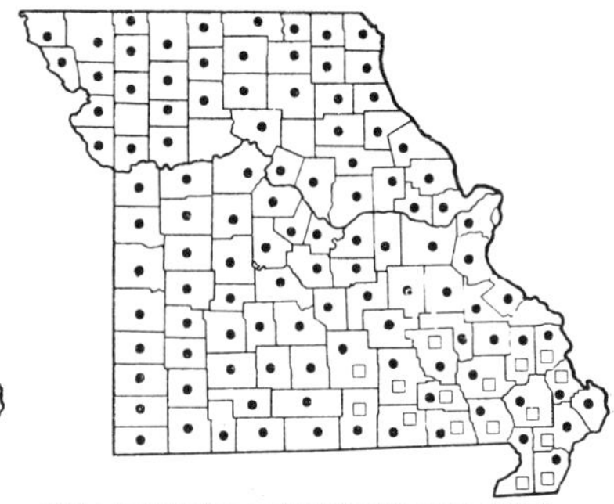

2053 • Cephalanthus occidentalis var. occidentalis f. occidentalis (Buttonbush)

2053 □ Cephalanthus occidentalis var. pubescens

and introduced locally in Jasper County (rich waste ground, along railway grade, Stockyard Switch, Joplin, August 10, 1949, *Palmer 59759*), southwestern Missouri. The Jasper County collection cited was introduced in stock car cleanings.

Two variations occur in Missouri:

Stems glabrous (without hairs) or sparsely hairy; leaves and fruits glabrous or sparsely hairy . . .
1a. D. VIRGINIANA var. VIRGINIANA f. VIRGINIANA

Stems, leaves, and fruits noticeably hairy . . .
1b. D. VIRGINIANA var. VIRGINIANA f. HIRSUTA

1a. **Diodia virginiana** var. **virginiana** f. **virginiana** Map 2050

*Diodia virginiana* L. [G, BB, P & S]

This is the usual form found in Missouri.

1b. **Diodia virginiana** var. **virginiana** f. **hirsuta** (Pursh) Fern. Map 2050

Known only from Ripley County, southeastern Missouri (creek bed of valley of 'Goose Lake' meander of Current River, T22N, R2E, sect. 25 and 36, 2¼ mi. northwest of Current View, 7½ mi. south of Doniphan, August 18, 1955, *Steyermark 73378*).

The above forms of var. *virginiana* range from Florida to Texas, north to New Jersey, Illinois, Missouri, and Oklahoma. Another variety, *attenuata* Fern., with nearly cylindrical, narrower fruits, is found in Virginia.

2. **Diodia teres** Walt. var. **teres** Rough Buttonweed Map 2051

*Diodia teres* Walt. [G, P & S]

*Diodia teres* var. *setifera* Fern. & Grisc. [G]

Flowers June–October.

Occurs in dry rocky and sterile soils of gravelly and sandy washes and alluvial flats along streams and creek beds, sandy open ground, glades, fallow fields, pastures, prairies, waste ground, and along railroads. Common throughout southern, central, and eastern Missouri, south and east of a line drawn from Scotland, Monroe, Audrain, Boone, and Saline counties to Clay County, and locally north in Linn and Livingston counties.

Ranges from Florida to Texas, north to Connecticut, New York, Pennsylvania, Ohio, Michigan, Wisconsin, Illinois, Missouri, and Oklahoma.

Most of the Missouri material falls under *D. teres* var. *setifera* Fern. & Grisc., characterized by having slender bristly tips to the young leaves. However, this character seems scarcely to have value, intergrading specimens of typical var. *teres* (without long bristle-tips) and var. *setifera* being frequent.

## 5. **Mitchella** L. Partridge Berry

**Mitchella repens** L. f. **repens** Partridge Berry, sometimes called Twin Berry Map 2052

*Mitchella repens* L. [G, BB, P & S, Steyerm.]

Flowers latter part of May–July.

Occurs in acid soils along moist ledges of bluffs of St. Peter and La Motte sandstones, sandy banks of streams, among sphagnum moss, sandy bogs, and low moist woodland. Eastern Missouri from the south-

Plate no. 336. **1.** Spermacoce glabra, × 2/5; *a.* Fruit, × 4; After Gleason, details from Small, The New York Botanical Garden. **2.** Diodia virginiana, × 2/5; *a.* Fruit, × 1 1/5; *b.* Flower, × 2; After Gleason, details from Small, The New York Botanical Garden. **3.** Mitchella repens, × 2/5; Details from Small, The New York Botanical Garden. **4.** Diodia teres, × 2/5 (Scribner's). **5.** Cephalanthus occidentalis, × 2/5; *a.* Fruiting cluster, × 2/5; *b.* Flower, × 2 4/5; After Gleason, details from Small, The New York Botanical Garden.

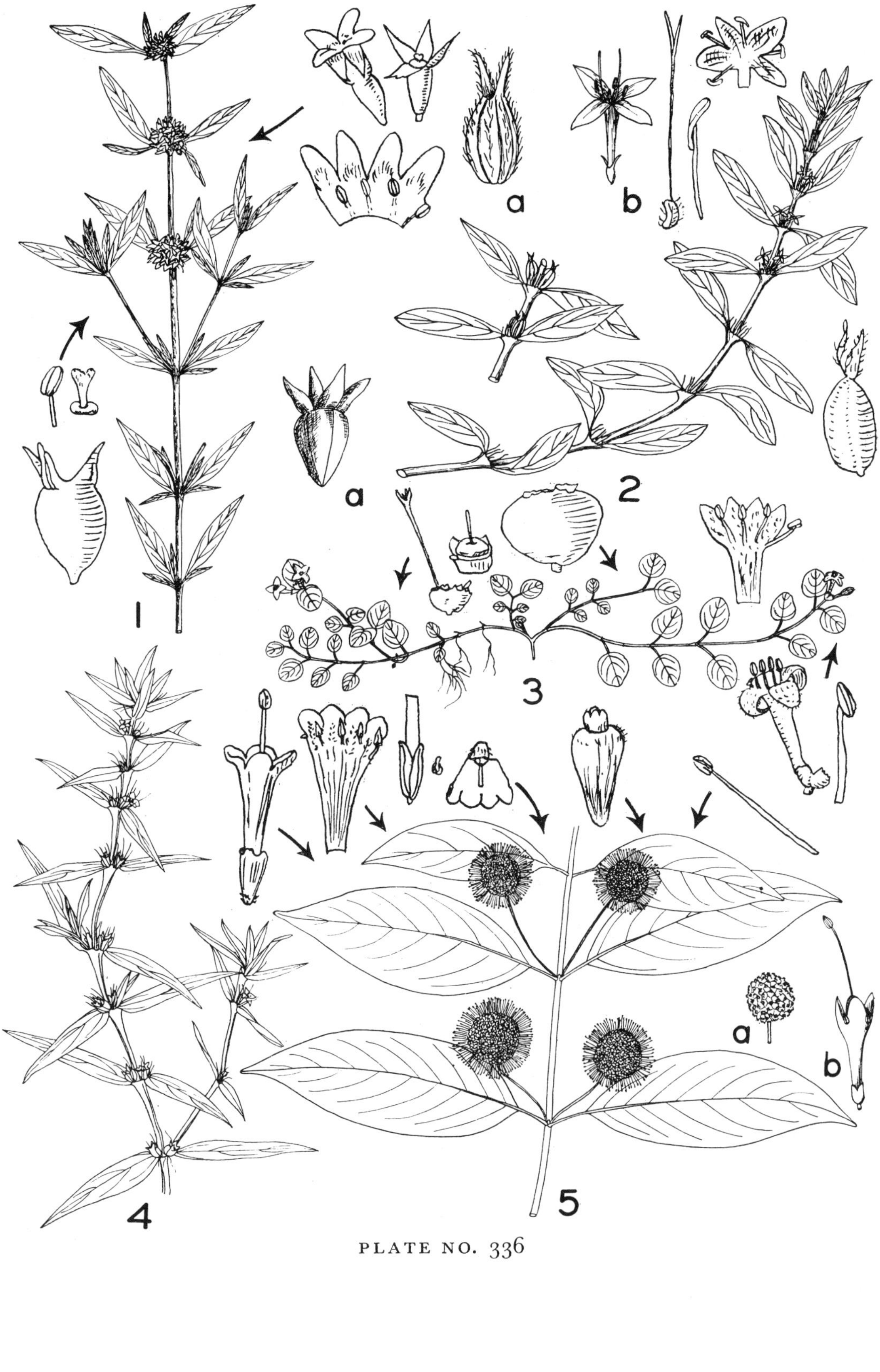


PLATE NO. 336

eastern lowlands north in the Ozarks locally to Lincoln County; reported from the central and west-central sections, but no specimens seen.

Ranges from Florida to Texas, north to Newfoundland, Quebec, Ontario, Minnesota, west to Missouri and Oklahoma; also in Mexico and Guatemala.

A form with white instead of red berries, f. *leucocarpa* Bissell, has not yet been recorded from Missouri. The berries are edible.

This attractive, evergreen-leaved species with red berries makes a handsome subject for the Wardian Case or winter terrarium. An addition of small pieces of charcoal in these glass containers prevents formation of mildew and mold.

## 6. **Cephalanthus** L. Buttonbush

**Cephalanthus occidentalis** L. Buttonbush — Map 2053

Also called Globe Flower, Honeyball, Swamp Sycamore, and Pond Dogwood.

Flowers early June–August.

Occurs in low swampy woods and thickets, borders of streams and sloughs, and in upland sink-hole ponds or river bottom oxbow lakes and ponds.

Two variations are encountered in the state:

Lower surface of leaves glabrous (without hairs); branches and twigs glabrous; commonly encountered throughout Missouri . . a. C. OCCIDENTALIS var. OCCIDENTALIS f. OCCIDENTALIS

Lower surface of leaves more or less soft hairy; young branches more or less hairy; southeastern Missouri lowlands west to sink-hole ponds of the southeastern Ozarks . . . b. C. OCCIDENTALIS var. PUBESCENS

a. **Cephalanthus occidentalis** var. **occidentalis** f. **occidentalis** — Map 2053

*Cephalanthus occidentalis* L. [G, BB, P & S]

Commonly occurring throughout Missouri, doubtless in every county.

Ranges from Florida to Mexico, north to Nova Scotia, Quebec, Maine, New York, Ontario, and California; also West Indies and Central America.

Typical var. *occidentalis* f. *occidentalis* has oblong-ovate leaves 2.3–15 cm. broad, which are abruptly short-acuminate at the tip. In var. *occidentalis* f. *lanceolatus* Fern., not recorded from Missouri, the leaves are lanceolate, only 1–3 cm. broad, and taper at each end.

b. **Cephalanthus occidentalis** var. **pubescens** Raf. — Map 2053

Lowlands of southeastern Missouri north to Cape Girardeau, Bollinger, and Wayne counties, thence west in the Ozarks in sink-hole ponds to Texas and Howell counties.

Ranges from Georgia to Texas, north to Virginia, Ohio, Indiana, Illinois, Missouri, and Oklahoma.

While some variation in pubescence occurs, the species manifests definite geographical segregation, the plants throughout the state being generally completely glabrous, whereas the var. *pubescens* is confined to the southeastern sector.

This shrub is a characteristic one of the upland sink-hole ponds of the Ozarks and also in bald cypress swamps and other kinds of undrained wet situations. The fragrant, round clusters of white flowers are much visited by honey bees, and it is one of the favorite bee plants.

The leaves contain a bitter substance used in medicine. A tea made from the root of the bark has sometimes been employed in treatment of diabetes. Animals may be poisoned by feeding on the leaves.

## 7. **Houstonia** L. Houstonia, Bluets

Some botanists, notably Fosberg (Castanea 19: 29–37. 1954; Va. Jour. Sci. 2: 110–11. 1941), Shinners (Field & Lab. 17: 166–69. 1949), and Lewis (Rh. 63:216–23. 1961) have favored the transfer of *Houstonia* to *Hedyotis*, while another recent student of the genus, Dr. Terrell (Rh. 61: 157–80; 188–207. 1959), retains the genus *Houstonia* in its usual traditional sense apart from *Hedyotis*. There is considerable difference in the various interpretations of the limits of each species within the genus.

a. Only 1 solitary flower at the end of a peduncle (flower-bearing branch); stems delicate, thread-like, mainly 2–10 cm. (up to 15) tall; corolla with a narrow tube abruptly expanded at the summit (salverform); stamens included within the corolla-tube, not protruding . . . . . . . . . . *b*

b. Corolla-tube 5–10 mm. long, corolla-lobes 2.5–5 mm. broad; corolla pale blue to white with yellow center . . . . . . . . . . . . . . . . . . . . . . . . 1. H. CAERULEA

b. Corolla-tube 2–5 mm. long, corolla-lobes 1–2.5 mm. broad; corolla mainly pale to deep purple with dark center . . . . . *c*

c. Calyx-lobes broadly lanceolate, 2–4 mm. long, nearly as long as or scarcely shorter than corolla-tube . . . . . 3. H. MINIMA

c. Calyx-lobes narrowly triangular or lance-linear, 1–2 mm. long, $\frac{1}{2}$–$\frac{2}{3}$ as long as corolla-tube . . . . . 2. H. PUSILLA

a. Several or many flowers clustered at the tip of a flower-bearing branch; stems stouter and firmer, mainly 15–75 cm. tall (rarely 10); corolla with a funnelform tube which is gradually enlarged toward the summit; stamens protruding from mouth of corolla . . . . . *d*

d. Connecting membranes (stipules) of upper and middle leaves tapering to 1 or 2 slender or long tips; calyx-lobes much shorter than calyx-tube in fruit; extra leaves usually present at the base of most of the main stem-leaves, the leaves narrowly linear or linear-lanceolate; root hard or woody; some of the flowers sessile (without stalks); fruit included two-thirds its length within the surrounding calyx-tube . . . . . 6. H. NIGRICANS

d. Connecting membranes (stipules) of upper and middle leaves rounded or triangular, at most only short-pointed; calyx-lobes longer than calyx-tube in fruit; most of stem-leaves lacking additional leaves at their bases, the leaves varying from linear or lanceolate to ovate; root system fine and thread- or hair-like, not hard or woody; flowers all on stalks (pedicels); fruit included at most one-half its length within the surrounding calyx-tube . . . . . *e*

e. Leaves of the stem with 3–5 main nerves; leaves 5–30 mm. broad, rounded to heart-shaped at base; calyx-lobes 1.7–6.5 mm. long during flowering period . . . . . 4. H. PURPUREA

e. Leaves of the stem with only 1 main nerve; leaves mainly 1–7 mm. broad, narrowed or tapering at the base; calyx-lobes 1–2.3 mm. long during flowering period . . . . . *f*

f. Commonly encountered throughout the Ozark region; mature fruit longer than broad or as long as broad; flower- and fruit-stalks (pedicels) 1.5–4 mm. long; stem-leaves 2–5 mm. broad, narrowly oblong to broadly linear . . . . . 5a. H. LONGIFOLIA var. LONGIFOLIA

f. Rarely encountered, known only from Ste. Genevieve and Madison counties, southeastern Ozarks; fruit broader than long; flower- and fruit-stalks (pedicels) 5–15 mm. long; stem leaves 1–3 mm. broad, narrowly linear . . . . . 5b. H. LONGIFOLIA var. TENUIFOLIA

1. **Houstonia caerulea** L. var. **caerulea** f. **caerulea**
Bluets Map 2054

Also called Innocence, Quaker Ladies.
*Houstonia caerulea* L. [G, BB, P & S, Steyerm.]
*Hedyotis caerulea* (L.) Hook. var. *caerulea* [Fosberg]
Flowers April–May.

Occurs in acid soils on moist ledges of sandstone and granite, swampy meadows, low prairies, sandy banks, and sandy open woods and glades. Ozark region north to Ste. Genevieve, St. Francois, Iron, Dent, and Phelps counties, west to Texas, Douglas, and Howell counties.

Ranges from Nova Scotia and Quebec to Ontario and Wisconsin, south to Georgia, Alabama, and Arkansas.

The form of var. *caerulea*, f. *albiflora* Millsp., has not been found as yet in Missouri. Another variety, var. *Faxonorum* Pease & Moore, with white corolla and shorter corolla-tube and of stiffer habit, occurs in the White Mountains of New Hampshire and on St. Pierre and Miquelon.

This species is quite distinct from either of the following two species of *Houstonia*, at least in Missouri, and shows no intergrading forms as reported by Dr. Fosberg (Castanea 19: 31. 1954).

2. **Houstonia pusilla** Schoepf Star Violet
Map 2055

Also called Small Bluets.
*Houstonia patens* Ell. [G, P & S]
*Hedyotis caerulea* var. *minor* (Michx.) T. & G. [Fosberg]
*Hedyotis crassifolia* Raf. [Shinners]
Flowers March–April.

Occurs in acid soils of sandstone, chert, or granite glades and rocky ledges and outcrops, fallow fields in valleys or on upland, and dry open places. Ozark region south and east of a line drawn from Lincoln, Warren, Gasconade, Phelps, Pulaski, Laclede, Dallas, Wright, and Douglas counties to Ozark County.

Ranges from Florida to Texas, north to Virginia, Illinois, Missouri, and Oklahoma.

The following variations occur in Missouri:

a. Corolla deep lavender or purple with a dark center; common type found . . 2a. H. PUSILLA f. PUSILLA

a. Corolla white, rose, or carmine . . . . *b*

b. Corolla white. . 2b. H. PUSILLA f. ALBIFLORA

b. Corolla rose or carmine. . . 2c. H. PUSILLA f. ROSEA

2a. **Houstonia pusilla** f. **pusilla** Map 2055
*Houstonia pusilla* Schoepf [BB, Steyerm.]

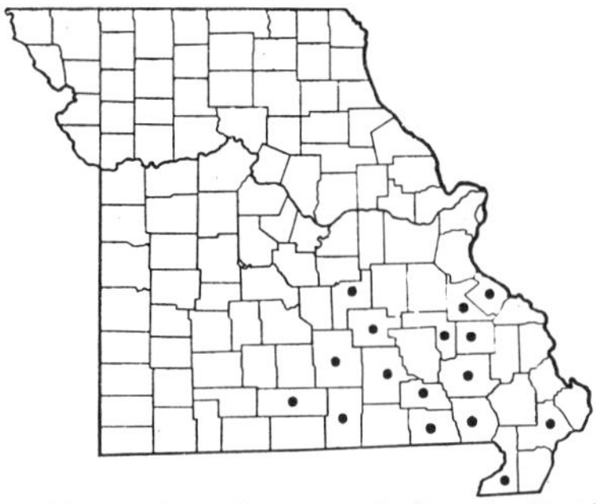

2054 Houstonia caerulea var. caerulea f. caerulea (Bluets)

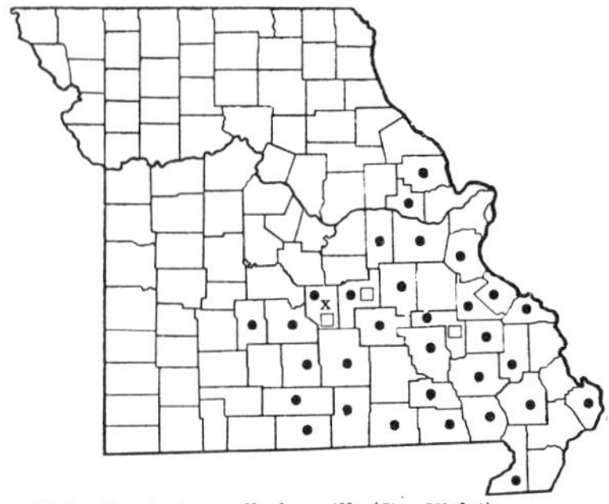

2055 • Houstonia pusilla f. pusilla (Star Violet)
2055 □ Houstonia pusilla f. albiflora
2055 x Houstonia pusilla f. rosea

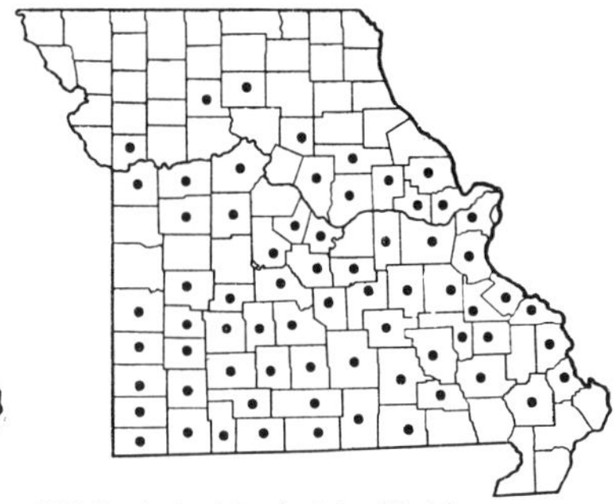

2056 Houstonia minima f. minima (Bluets)

*Houstonia patens* Ell. [G, P & S]
*Hedyotis caerulea* var. *minor* (Michx.) T. & G. [Fosberg]
*Hedyotis crassifolia* Raf. [Shinners]

This is the commonest of the variations in Missouri.

2b. **Houstonia pusilla** f. **albiflora** Standl. Map 2055

*Hedyotis caerulea* var. *minor* f. *Benkei* Fosberg [Fosberg]

Known from Iron (*Steyermark 73101*), Phelps (*Steyermark 4592*), and Pulaski (*Steyermark 4599*) counties.

2c. **Houstonia pusilla** f. **rosea** Steyerm. Map 2055

Known only from Pulaski County (along road D, in sect. 8, 4 mi. west of Jerome, April 17, 1937, *Steyermark 4599A*, holotype in Chi. Nat. Hist. Mus. Herb.).

In Missouri this species shows some intergradation with the following, *H. minima*, but none at all with *H. caerulea*. Dr. Fosberg (Castanea 19: 31–32. 1954) treats *Houstonia minima* and *H. pusilla* as varieties of *H. caerulea*, a view which cannot be accepted in the present flora.

3. **Houstonia minima** Beck f. **minima** Bluets Map 2056

*Hedyotis minima* (Beck) T. & G. [Shinners]
*Hedyotis caerulea* var. *minima* (Beck) Fosberg [Fosberg]
*Houstonia minima* Beck [G, BB, P & S, Steyerm.]

Flowers January–April.

Occurs in prairies, pastures, alluvial soils in river flood plains, fallow fields, rocky ledges, bluff escarpments, and glades usually of sandstone, chert, or granite. Common throughout southern and central Missouri, north to Lincoln, Montgomery, Audrain, Randolph, Linn, Livingston, and Clay counties.

Ranges from Arkansas to Texas, north to Illinois, Iowa, and Kansas.

A form with white corollas, f. *albiflora* Lathrop, has not been found as yet in Missouri.

This is the commonest of the small bluets found in the spring, occurring in abundance and often making carpets of purple-blue in grassy open places. Like *H. pusilla*, it is a winter annual, maturing its seeds in early spring, germinating in the fall of the same year, then remaining as green rosettes during winter to become flowering plants the following spring.

Dr. Fosberg treats this and the preceding species, *H. pusilla*, as varieties of *H. caerulea*, a course which is not accepted in the present flora. While *H. pusilla* and *H. minima* are closely related and sometimes difficult to separate, neither species is readily confused with nor intergrades with *H. caerulea*, at least so far as observations with Missouri material are concerned.

I have tried on numerous occasions to acclimate this species in my wildflower preserve in northern Illinois by transporting the original sods in which the plants occur, but have had no success in getting them to overwinter at that latitude.

4. **Houstonia purpurea** L. Map 2057

Sometimes called Mountain Houstonia.

*Hedyotis purpurea* (L.) T. & G. [Shinners, Fosberg]

Flowers May–June.

Occurs in moist or rocky open woods, ledges along bluffs, and rocky slopes and banks along streams.

Two varieties occur in Missouri:

Plate no. 337. **1.** Houstonia caerulea, × 3/4. **2.** Houstonia pusilla, × 3/4. **3.** Houstonia minima, × 3/4. **4.** Houstonia longifolia var. longifolia, × 3/4.

PLATE NO. 337

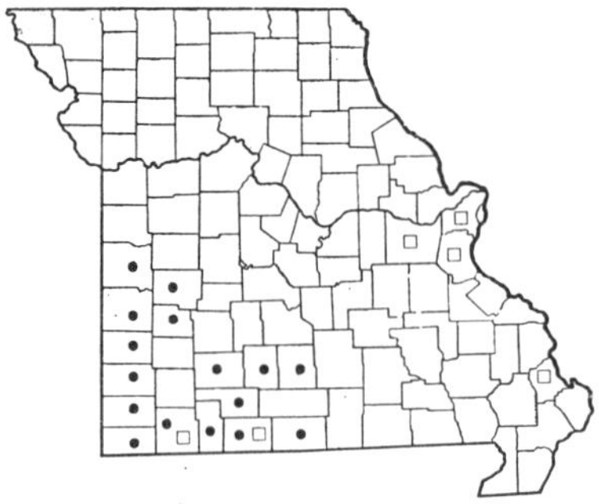

2057 • Houstonia purpurea var. purpurea
2057 □ Houstonia purpurea var. calycosa

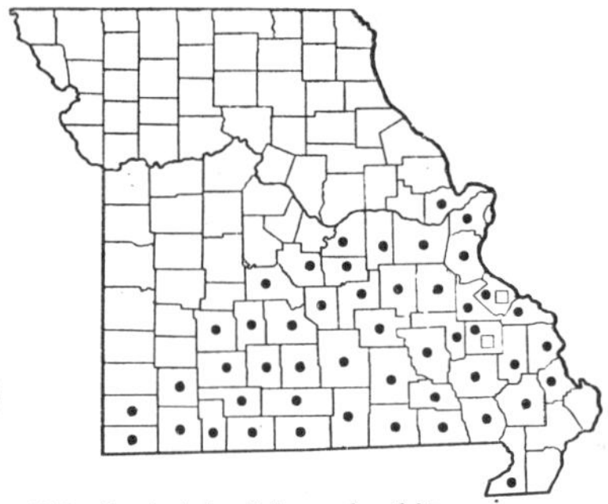

2058 • Houstonia longifolia var. longifolia
2058 □ Houstonia longifolia var. tenuifolia

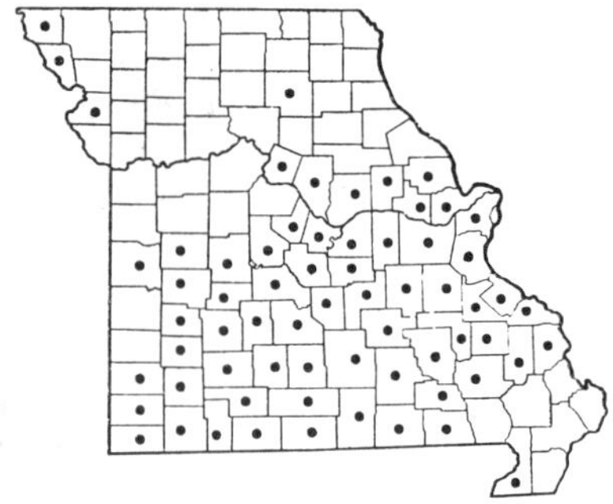

2060 Houstonia nigricans
2059 Excluded species

Main leaves of stem chiefly 10–35 mm. broad, broadly ovate to ovate-lanceolate, $1\frac{1}{2}$–3 times as long as broad; calyx-lobes 1.7–4.5 mm. long during flowering period . . 4a. H. PURPUREA var. PURPUREA

Main leaves of stem chiefly 4–12 mm. broad, oblong-lanceolate or broadly lanceolate, $2\frac{3}{4}$–6 times as long as broad; calyx-lobes 4–6.5 mm. long during flowering period . 4b. H. PURPUREA var. CALYCOSA

4a. **Houstonia purpurea** var. **purpurea** Map 2057

*Houstonia purpurea* L. [G, P & S, Steyerm.]
*Hedyotis purpurea* (L.) T. & G. [Shinners]
*Hedyotis purpurea* var. *purpurea* [Fosberg]

Southwestern Missouri east to Wright and Ozark counties, north to St. Clair and Bates counties.

Ranges from Georgia, Alabama, Louisiana, and Oklahoma north to Delaware, Pennsylvania, Ohio, Indiana, and Missouri.

4b. **Houstonia purpurea** var. **calycosa** Gray Map 2057

*Houstonia lanceolata* (Poir.) Britt. [G, P & S, Steyerm.]
*Hedyotis purpurea* var. *calycosa* (Gray) Fosberg [Fosberg]

Scattered in the Ozark region, in St. Louis, Jefferson, Franklin, Scott, Taney, and Barry counties.

Ranges from Alabama to Oklahoma, north to Maine, New York, West Virginia, Ohio, Kentucky, Illinois, and Missouri.

5a. **Houstonia longifolia** Gaertn. var. **longifolia** Map 2058

*Houstonia longifolia* Gaertn. [G, BB, P & S, Steyerm.]
*Hedyotis canadensis* (Willd.) Fosberg [Fosberg]
*Hedyotis purpurea* var. *longifolia* (Gaertn.) Fosberg [Fosberg]
*Houstonia purpurea* var. *longifolia* Gray

Flowers April–July; occasionally blooming in October.

Occurs in rocky open woods, prairies, glades, fallow fields, and ledges of bluffs, usually in acid soils overlying chert, sandstone, or granite substrata. Ozark region south and east of a line drawn from St. Charles, Franklin, Gasconade, Osage, Miller, Camden, Dallas, Polk, and Lawrence counties to Newton County.

Ranges from Maine to Saskatchewan, south to Georgia, Alabama, Mississippi, Arkansas, and Oklahoma.

Recent evaluation of the status of this Houstonia has had various interpretations. Dr. Fosberg (Castanea 19: 34. 1954) treats this and the following taxon (*H. longifolia* var. *tenuifolia*), as varieties of *Hedyotis purpurea*. Since *Houstonia longifolia* and *H. purpurea* are readily distinguishable and show no intergradation throughout the Ozarks and in other areas where the present writer has observed them, Dr. Fosberg's reduction of the former to varietal rank obscures the distinctness of *H. longifolia* as a separate species. Dr. Terrell would place into *H. tenuifolia* all the Missouri material, previously referred to *Houstonia longifolia*, keeping *H. longifolia* as a distinct species (Rh. 61: 195–207. 1959) restricted to the northern and eastern portions of the United States, but not found in Missouri. Dr. Terrell has also named a number of slightly

Plate no. 338. **1.** Houstonia purpurea var. purpurea, × 2/5. **2.** Houstonia purpurea var. calycosa, × 2/5. **3.** Houstonia nigricans, × 2/5. **4.** Oldenlandia Boscii, × 2/5; *a.* Fruit, × 3 3/5; After Gleason, The New York Botanical Garden. **5.** Houstonia longifolia var. tenuifolia, × 2/5. **6.** Oldenlandia uniflora, × 2/5. All details from Small, The New York Botanical Garden.

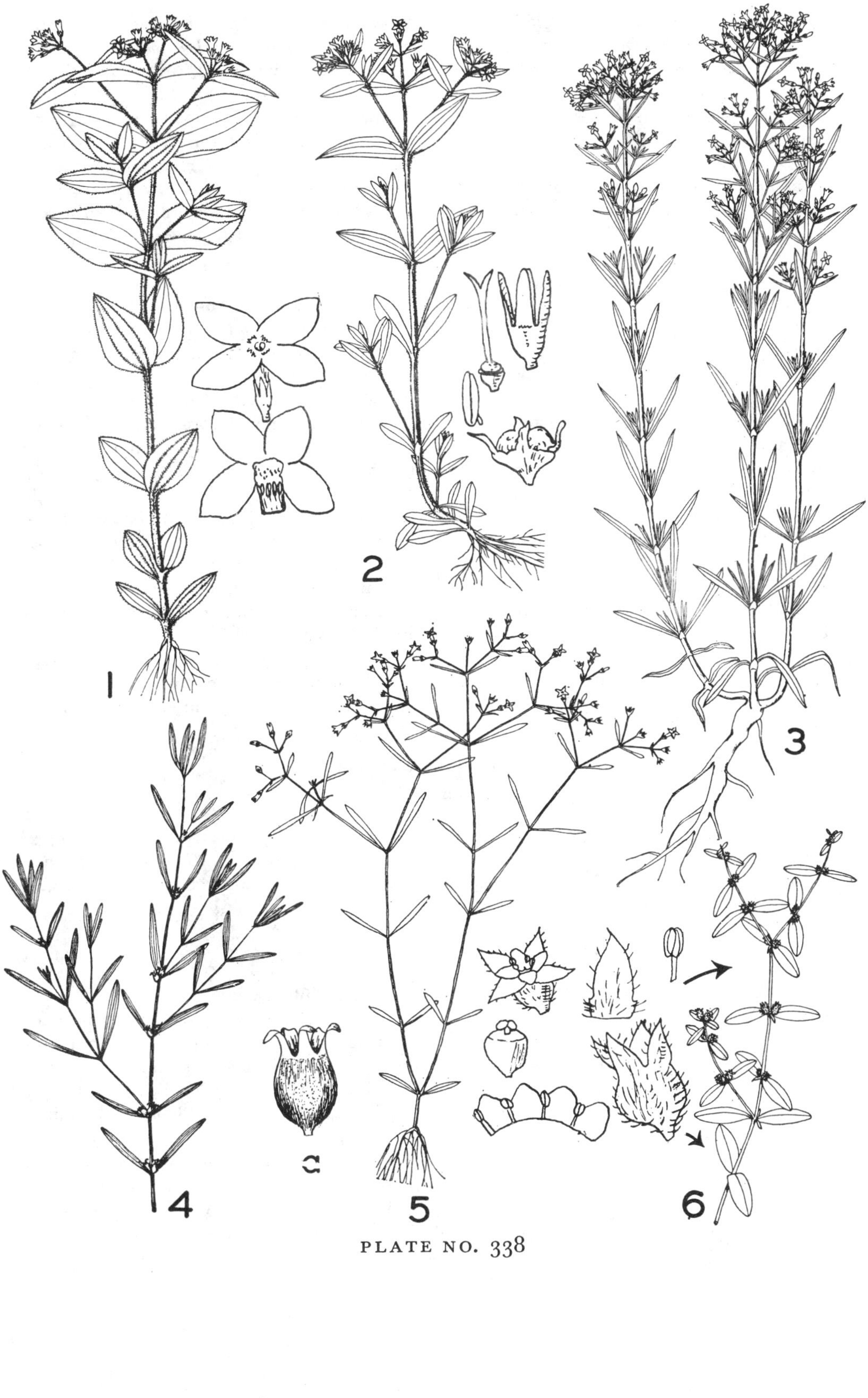


PLATE NO. 338

broader-leaved specimens of *Houstonia longifolia* var. *longifolia* as putative hybrids between *H. purpurea* and *H. tenuifolia* (= *H. longifolia*), but I am unable to consider them as anything other than the normal leaf variation of *H. longifolia* var. *longifolia* with somewhat broader-leaved plants commonly found in populations of the taxon.

Missouri material has stems varying from glabrous to puberulous, and leaves varying from non-ciliate to puberulous.

5b. **Houstonia longifolia** var. **tenuifolia** (Nutt.) Wood Map 2058

*Houstonia tenuifolia* Nutt. [G, BB, P & S, Steyerm.]
*Houstonia Nuttalliana* Fosberg [Fosberg, Shinners]
*Hedyotis purpurea* var. *tenuifolia* (Nutt.) Fosberg [Fosberg]

Flowers April–July, occasionally into October.

Occurs on sandstone ledges of La Motte sandstone in woods. Known only from Ste. Genevieve (wooded sandstone slopes of La Motte sandstone bluffs along Terre Bleue Creek, T37N, R6E, south part sect. 20 and north part sect. 29, 2–2½ mi. south of Thurman, 5–6 mi. northwest of Sprott, October 3, 1950, *Steyermark 71037*) and Madison counties.

Ranges from Georgia to Texas and Mexico, north to Pennsylvania, West Virginia, Kentucky, and Missouri.

Although Dr. Terrell refers all Missouri material, previously identified as *H. longifolia*, to this species, I am restricting the use of the name to the above-cited specimens, retaining *H. longifolia* and *H. tenuifolia* in the sense used by Fernald (*Gray's Manual*, eighth ed.), by Gleason (*New Ill. Fl.*) and by Standley (N. Am. Fl. 32, part 1: 25, 34, 37. 1918). Dr. Fosberg reduced *Houstonia tenuifolia* to a variety under *Hedyotis purpurea*, a procedure which, as similarly indicated under the discussion of *H. longifolia* var. *longifolia*, obscures, in my opinion, rather than clarifies, the distinctness and true status between *H. purpurea* and *H. longifolia* and variety *tenuifolia*.

6. **Houstonia nigricans** (Lam.) Fern. Map 2060

*Houstonia angustifolia* Michx. [P & S, Steyerm.]
*Hedyotis nigricans* (Lam.) Fosberg [Fosberg, Shinners]

Flowers last of May–October.

Occurs on exposed grassy loess hills, rocky stream banks, rocky ledges and escarpments of limestone bluffs, and limestone glades. Mostly in the Ozark region in southern and central Missouri north to Lincoln, Montgomery, Callaway, Boone, Howard, and Bates counties, locally north in Macon, Buchanan, Holt, and Atchison counties.

Ranges from Florida to Texas and Mexico, north to Georgia, Ohio, Indiana, Michigan, Illinois, Iowa, and Nebraska.

This is an attractive species for the rock garden, thriving in full sun in a gritty limestone bed. The pale lilac to white flowers are produced in showy clusters.

## 8. **Oldenlandia** L.

(*Hedyotis* L. [G])

This genus has been combined, together with *Houstonia*, by Fosberg (Castanea 19: 37. 1954), by Shinners (Field & Lab. 17: 167, 169. 1949), and by Lewis (Rh. 63:216–23. 1961) with *Hedyotis* L. In the present treatment *Oldenlandia* is maintained in its traditional sense, pending monographic treatment of the group.

Plant glabrous (without hairs); leaves linear, 1–2.5 mm. broad; fruit smooth or slightly roughened . . . 1. O. BOSCII
Plant hairy; leaves ovate to oblong or lanceolate, 5–10 mm. broad; fruit hairy . . . . . 2. O. UNIFLORA

1. **Oldenlandia Boscii** (DC.) Chapm. Map 2061

*Hedyotis Boscii* DC. [G]

Flowers July–October.

Occurs in swampy ground. Known only from Ripley County, southeastern Missouri (exsiccated depression, T22N, R4E, sect. 36, 4 mi. south of Naylor, October 20, 1948, *Steyermark 66958*).

Ranges from Florida to Texas, north to Virginia, Tennessee, and Missouri.

2. **Oldenlandia uniflora** (L.) Lam. Map 2062

*Hedyotis uniflora* (L.) Lam. [G]

Flowers July–October.

Occurs in swampy ground and along sandy spring branches on Crowley Ridge. Known only from Stoddard (swampy seepage of spring branch in valley of Crowley Ridge, T25N, R10E, southeast ¼ sect. 11, 4–4½ mi. [by air] south of Bloomfield, 2 mi. northeast of Dexter, October 18, 1955, *Steyermark 80424*) and Dunklin (common, Kennett, August 22, 1894, *Bush 366*) counties.

Ranges from Florida to Texas, north to New York and Missouri.

2061 Oldenlandia Boscii

2062 Oldenlandia uniflora

2063 Lonicera Morrowi

## Fam. **CAPRIFOLIACEAE** (Honeysuckle Family)

a. Leaves compound, divided into several leaflets . . . . . . . . . . . . . . . 5. SAMBUCUS
a. Leaves simple, not divided . . . . . . . . . . . . . . . . . . . . . . . *b*
 b. Herbaceous, non-woody plants with erect stems . . . . . . . . . . . . . . 3. TRIOSTEUM
 b. Woody shrubs, vines, or trees . . . . . . . . . . . . . . . . . . . . . *c*
  c. All or most of the leaves normally toothed (young or tip leaves sometimes show no teeth) . . 4. VIBURNUM
  c. All or most of the leaves normally without teeth (young or juvenile leaves sometimes show teeth, lobing, or wavy edges) . . . . . . . . . . . . . . . . . . . . . . *d*
   d. Usually vining or sprawling plants, but if upright shrubs or bushes, then without the characters of the other *d*; corolla showy, 10 mm. or more long, funnelform to tubular and irregular (corolla lobes unequal); ovary 2–3-celled, each cell with several ovules; fruit a several-seeded berry . . . . . . . . . . . . . . . . . . . . 1. LONICERA
   d. Shrubs or bushes with erect or spreading branches; corolla small, not showy, usually 4–7 mm. long, bell-shaped and without a well-developed tube, more or less regular (symmetrical and with corolla lobes more or less equal); ovary 4-celled, 2 of the cells with 1 fertile ovule and 2 with several nonfunctioning ovules; fruit with 2 seeds maturing . . . 2. SYMPHORICARPOS

### 1. **Lonicera** L. Honeysuckle

a. None of the leaves joined at their bases . . . . . . . . . . . . . . . . . . *b*
 b. Stems with noticeable spreading hairs; twining, trailing, or sprawling plants; corolla 3–4 cm. long; bracts of flowers similar to, but smaller than, the leaves; fruit black. . . . . . 2. L. JAPONICA
 b. Stems with short hairs; erect or ascending bushy shrubs; corolla 1–2 cm. long; bracts of flowers much smaller than the normal leaves, linear to lance-oblong; fruit red or yellow . . . 1. L. MORROWI
a. Uppermost leaves joined at their bases (connate) . . . . . . . . . . . . . . *c*
 c. Corolla deep red on outside, 3.5–4 cm. long, the 5 lobes more or less equal; stamens and style included within the tube or barely protruding from the summit; cultivated and escaped species 3. L. SEMPERVIRENS
 c. Corolla orange, yellow, cream-color, white, brick-red or purplish, or yellow with brick-red or purplish, 1.5–3 cm. long, 2-lipped, the upper broad and 4-lobed, the lower narrow; stamens and style conspicuously protruding at the summit; native species of woods, bluffs, and along streams . . *d*
  d. A white coating occurs on the upper surface of the uppermost leaves which are joined at their base (connate); the uppermost joined leaves forming a more or less circular or nearly circular form, broader than long; flowering and fruiting spikes arranged in 2–6 whorls (in circles around the axis), which are usually separated . . . . . . . . . . . . . . . 6. L. PROLIFERA
  d. Upper surface of the uppermost joined leaves green or barely whitened; the uppermost joined leaves forming a longer than broad oblong, elliptic, or diamond-shaped (rhombic) disk; flower- and fruit-spikes in 1–3 crowded whorls (in circles around the axis) . . . . . . . . . *e*
   e. Lower surface of leaves conspicuously whitened or silvery-blue; corolla-tube slightly enlarged on one side at the base; corolla yellow or greenish-yellow tinged with purple, rose, or brick color, 1.5–2.3 cm. long . . . . . . . . . . . . . . . . 4. L. DIOICA

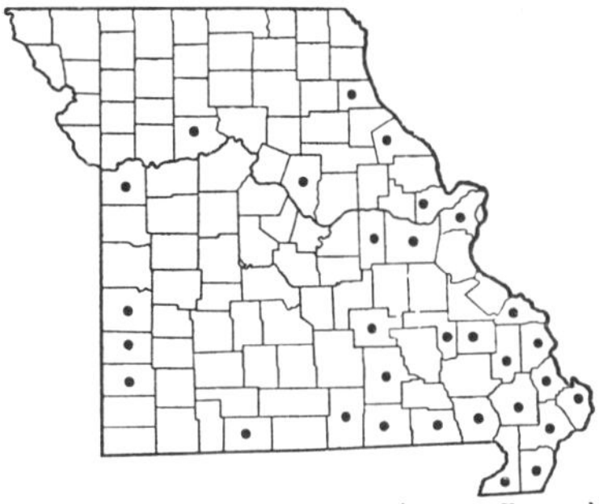

2064 Lonicera japonica var. japonica (Japanese Honeysuckle)

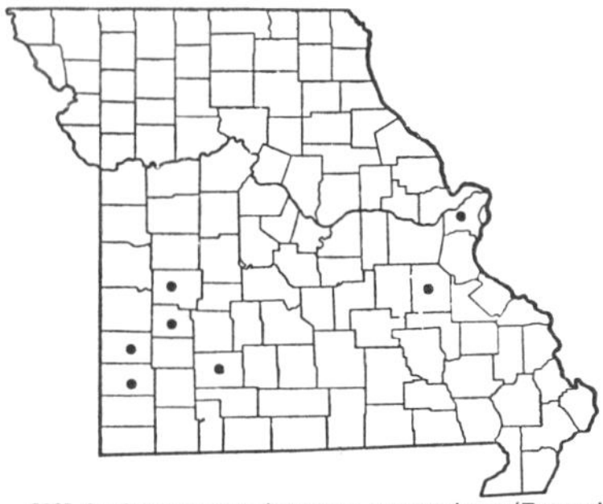

2065 Lonicera sempervirens var. sempervirens (Trumpet Honeysuckle)

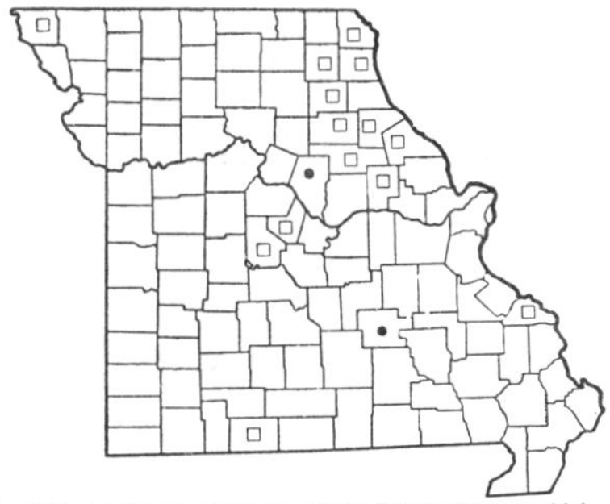

2066 • Lonicera dioica var. dioica (Limber Honeysuckle)
2066 □ Lonicera dioica var. glaucescens f. glaucescens

e. Lower surface of leaves slightly gray green or pale but not noticeably whitened; corolla-tube slender without any enlargement on one side at the base; corolla orange, orange-yellow, or creamy-yellow, but lacking purple, rose, or brick color, 2–3 cm. long . . 5. L. FLAVA

1. **Lonicera Morrowi** Gray Map 2063
Flowers May–June.

Planted and occasionally escaped from cultivation along roadsides and wooded sections. Known from northern and central Missouri in Livingston (roadside, T58N, R24W, sect. 21, Chillicothe, May 13, 1951, *Sparling 782* and *1104*), Marion (abundantly escaped in woods, Riverview Park, Hannibal, October 18, 1933, *Drouet 1286*), and St. Louis (30 mi. west of St. Louis along highway 40, May 5, 1957, *Donald Hoss*) counties.

Native of Japan; introduced and naturalized in the United States from Maine to Michigan, south to New York, Pennsylvania, and Missouri.

The corolla is at first white, eventually turning yellow.

2. **Lonicera japonica** Thunb. var. **japonica**
Japanese Honeysuckle Map 2064
*Lonicera japonica* Thunb.
*Lonicera japonica* var. *repens* (Lav.) Rehder [BB]
Flowers May–June.

Escaped from cultivation into thickets, fence rows, open woods, rocky slopes, ditches, roadsides, and along railroads. Commonest in southeastern Missouri, scattered elsewhere, north to Marion, Boone, Carroll, and Jackson counties.

Native of Asia; introduced and naturalized in the United States from Florida to Texas, north to Massachusetts, New York, Ohio, Indiana, Missouri, and Kansas.

The young or early leaves are often lobed or irregularly cut. A var. *chinensis* (P. W. Wats.) Baker with glabrous purple leaves and branchlets and with the corolla rosy-purple on the outside, has not been recorded as yet from Missouri.

In many parts of the southern and eastern states this vine has penetrated woodlands, smothering and choking out native shrubs and small trees. It should not be recommended for extensive roadside plantings in Missouri, as it quickly becomes established and gets out of control. The flowers, which are very fragrant, vary from white or creamy to yellow.

3. **Lonicera sempervirens** L. var. **sempervirens**
Trumpet Honeysuckle Map 2065
Also called Coral Honeysuckle.
Flowers April–July.

Escaped from cultivation along roadsides, sandy or rocky stream banks, and in thickets in parts of southern and central Missouri north to St. Louis and St. Clair counties.

Ranges from Florida to Texas, north to Maine, Massachusetts, New York, Ohio, Iowa, and Nebraska.

The leaves are evergreen in the southern part of the range. The species is generally encountered in the state as an escape from cultivation, often being found near dwellings. So far as known, no truly wild plants have been recorded in the state.

4. **Lonicera dioica** L. Limber Honeysuckle
Map 2066
Flowers April–June.
Two varieties are encountered in the state:

Lower surface of leaves glabrous (without hairs); outside of corolla-tube glabrous or nearly so; rarely encountered . . . 4a. L. DIOICA var. DIOICA

Lower surface of leaves more or less hairy; outside of corolla-tube hairy and glandular . . . . 4b. L. DIOICA var. GLAUCESCENS

4a. **Lonicera dioica** var. **dioica** Map 2066
*Lonicera dioica* L. [G, Steyerm.]

Occurs along north-facing ledges of bluffs and rocky banks of streams. Southern and central Missouri, known only from Boone (rocky banks, Columbia, May, 1902, *Daniels*) and Dent (hanging over and near top of north-facing limestone bluffs along spring branch from Montauk Spring, Montauk State Park, May 5, 1939, *Steyermark 22126*; on north-facing bluffs along creek tributary to Crooked Creek, Montauk State Park, May 5, 1939, *Steyermark 22132*) counties.

Ranges from Maine and Quebec to Manitoba, south to Georgia and Missouri.

The duration of flowering of this variety is only about one week so that it is often difficult to see the flowers in good flowering condition (see Rh. 42: 103–4. 1940).

4b. **Lonicera dioica** var. **glaucescens** (Rydb.) Butters f. **glaucescens** Map 2066

*Lonicera dioica* var. *glaucescens* (Rydb.) Butters [G, BB, P & S, Steyerm.]

Occurs along wooded bluffs and ledges, alluvial woods, rocky banks of streams, and thickets. Mainly in northeastern Missouri, and elsewhere in northern Missouri in Atchison County, locally southward in central Missouri in Moniteau and Morgan counties, and in southern Missouri in Taney County (*Steyermark 67494*).

Ranges from Quebec to British Columbia, south to North Carolina, Kentucky, Missouri, and Kansas.

Sterile or fruiting specimens of *L. flava*, *L. prolifera*, and *L. dioica* var. *glaucescens* with pubescence on the lower surface of the leaf-blades are frequently confused. In *L. dioica* var. *glaucescens* the lower surface is more uniformly pubescent with long hairs than in *L. prolifera*, which has short flat white hairs on the surface with or without some spreading hairs on the midrib and main lateral nerves. In *L. flava* the lower surface of the leaves is either glabrous or has spreading hairs on the midrib. The collection from Taney County, cited above, is placed with var. *glaucescens*, since the upper connate leaves are not glaucous above, are pointed at the ends, and the flowers are in closely approximate whorls.

5. **Lonicera flava** Sims Yellow Honeysuckle Map 2067

Flowers April–May.

Occurs in rocky woods, on bluff ledges and escarpments, and rocky ground along streams.

Two intergrading variations occur in Missouri:

Commonly encountered; corolla orange or orange-yellow, usually 2.8–3 cm. long . . . 5a. L. FLAVA var. FLAVA

Rarely encountered; corolla pale yellow or cream-colored, usually 2–2.5 cm. long . . . 5b. L. FLAVA var. FLAVESCENS

5a. **Lonicera flava** var. **flava** Map 2067

*Lonicera flava* Sims [G, BB, P & S, Steyerm.]

Ozark region north to Jefferson, Gasconade, Camden, Hickory, and Dade counties.

Ranges from Georgia, Alabama, Arkansas, and Oklahoma, north to North Carolina, Kentucky, and Missouri.

5b. **Lonicera flava** var. **flavescens** (Small) Gl. Map 2067

*Lonicera flavescens* Small, not Dippel

*Lonicera flavida* Cockerell [G, P & S, Steyerm.]

Southwestern Missouri in Greene, Christian, Taney, Stone, Barry, and McDonald counties.

Ranges from South Carolina and Kentucky to Tennessee and Missouri.

*Lonicera flava* and *L. flavida* cannot be distinguished except by color and length of corolla, averaging longer in *L. flava*. Gleason's procedure in transferring *L. flavida* to varietal rank, therefore, is in accord with the intergradation found in Missouri material, in which the corolla-tube is densely hairy within in both the orange-flowered *L. flava* and the pale yellow or cream-flowered *L. flavida*. None of the Missouri specimens with orange corollas have the corolla-tube glabrous within as sometimes occurs in the range of *L. flava* east of Missouri. It is recorded also (*Gray's Man.*, eighth ed. p. 1335) that in *L. flava* the leaves immediately below the disk have rounded to sessile bases, while in *L. flavida* they taper to short petioles, but this difference is not real and does not hold true.

A specimen with pale yellow corollas but growing with typical orange-flowered *L. flava* var. *flava* and having the corolla length of that variation has been collected in Texas County (½ way up on north-facing limestone bluffs along Big Piney River, T31N, R9–10W, sect. 30 and 25, 3 mi. north-northwest of Houston, May 7, 1957, *Steyermark 84182*).

Specimens found in leaf or fruiting condition are often confused as *L. prolifera*, *L. dioica* var. *glaucescens*, and *L. flava*. In general, the fruiting spikes of *L. prolifera* have 2–6 whorls of fruit and the leaves beneath have a shorter hairiness, when this is present, than either *L. dioica* var. *glaucescens* or *L. flava*. In *L. dioica* var. *glaucescens* the hairs on the lower surface of the leaf are more numerous, longer, and more uniformly distributed than in *L. flava*, whereas in *L. flava* they occur mainly on the midrib or the leaf is glabrous beneath.

2067 • Lonicera flava var. flava (Yellow Honeysuckle)
2067 □ Lonicera flava var. flavescens

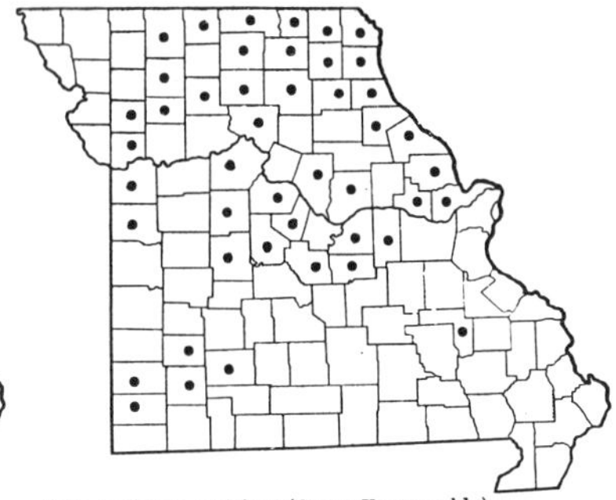

2068 Lonicera prolifera (Grape Honeysuckle)

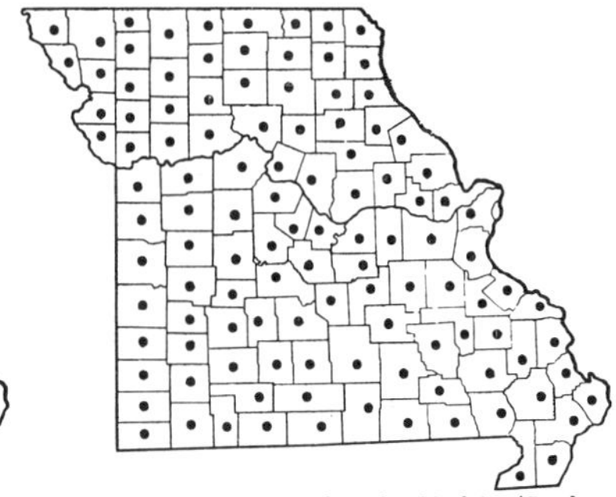

2069 Symphoricarpos orbiculatus f. orbiculatus (Coral Berry)

This is a very showy-flowered species and has very fragrant flowers. It does well in cultivation in either alkaline or acid soils, although it is found most frequently in acid soils associated with chert, sandstone, or granite. It occurs both in exposed sunny situations as well as in woodland shade.

6. **Lonicera prolifera** (Kirchn.) Rehder Grape Honeysuckle Map 2068
*Lonicera prolifera* var. *prolifera* [BB]
*Lonicera prolifera* var. *glabra* Gl. [BB], holotype described from Clark County, Missouri (*Bush 10135*)
Flowers April–June.

Occurs in open woods, rich wooded slopes, ledges of bluffs and escarpments, and wooded thickets. Mainly in northern and central Missouri, locally south to Iron (*Epling 6196*), Greene, Dade, Lawrence, Jasper, and Newton counties.

Ranges from New York and Ontario to Wisconsin and Manitoba, south to Tennessee, Arkansas, and Kansas.

Gleason described a glabrous-leaved variation, var. *glabra*, based upon a Missouri specimen from Clark County (open woods, Dumas, *Bush 10135*, in N. Y. Bot. Gard. Herb.), but many specimens have the lower surface of the leaves slightly pubescent in early spring and glabrous in later and more mature stages. It is believed that the glabrous-leaved types are occasional late-season plants which have lost their hairs and do not merit recognition and are, therefore, not accorded varietal status in the present flora. Much Missouri material, previously identified as *L. dioica*, is now referred to *L. prolifera* in the present flora.

*Excluded Species*

**Lonicera Caprifolium** L.

A specimen from St. Louis County (vine climbing along fence, St. Louis, summer, 1931, collected by *Mrs. F. Springmeyer*, Moldenke Herb. *7260* in N. Y. Bot. Gard. Herb.) is a fragmentary one with only two upper pairs of connate leaves present with the remains of a fruiting rachis 1.1 cm. long bearing two whorls where fruit occurred, separated by an interval of 5 mm. One undeveloped carpel remains at the top of the uppermost whorl. The fragmentary condition of the specimen precludes its positive identification, and, therefore, exclusion from the present flora.

**Lonicera tatarica** L. f. **albiflora** (DC.) House Tartarian Honeysuckle

A collection in Park College herbarium from Platte County (1 mi. north of Parkville on highway 9, May 1, 1957, *Kent Drew 56*) was obtained probably from a cultivated plant. Lacking any further definite evidence that the plant was established and naturalized at this locality, it is excluded from the flora. Both pink- and white-flowered forms are commonly cultivated in Missouri.

Plate no. 339. **1.** Lonicera japonica, × 2/5. **2.** Lonicera sempervirens, × 2/5. **3.** Lonicera dioica var. dioica, × 2/5. **4.** Lonicera flava var. flava, × 2/5. **5.** Lonicera flava var. flavescens, × 2/5. **6.** Lonicera prolifera, × 2/5. **7.** Symphoricarpos occidentalis, × 2/5; After Gleason, The New York Botanical Garden. **8.** Symphoricarpos orbiculatus, × 2/5; *a.* Flower, × 2 2/5. After Gleason, details from Small, The New York Botanical Garden.

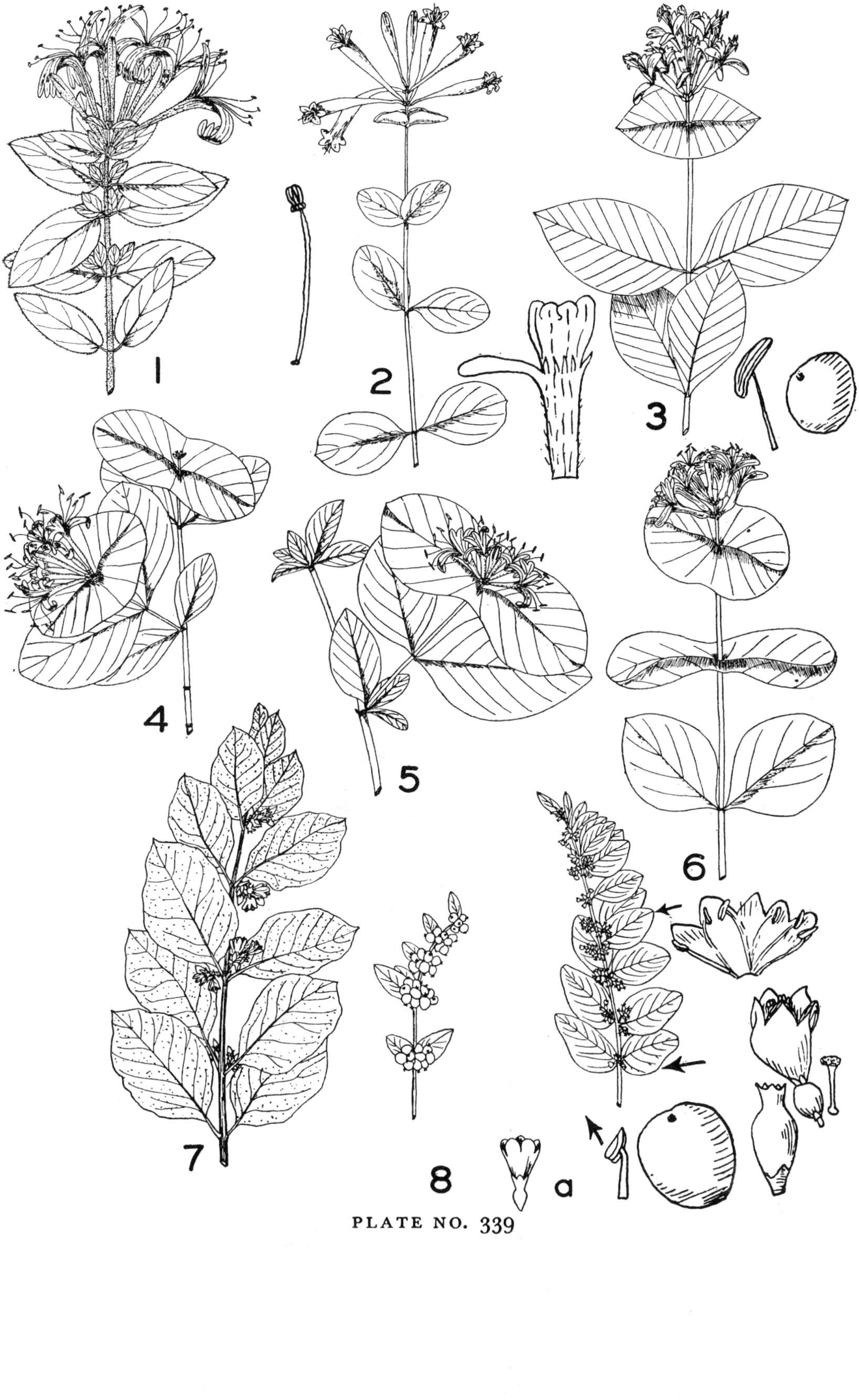


PLATE NO. 339

## 2. **Symphoricarpos** Duham. Snowberry

Common species throughout Missouri; corolla 3–4 mm. long, greenish and purplish; fruit usually coral-pink, purplish-red, or red . . . . . . . . . . . . . . . . . . . . 1. S. ORBICULATUS
Rare, known only in northern Missouri; corolla 6–9 mm. long, pale pink; fruit white or greenish-white 2. S. OCCIDENTALIS

1. **Symphoricarpos orbiculatus** Moench f. **orbiculatus** Coral Berry Map 2069
Also called Indian Currant, Buck Brush.
*Symphoricarpos orbiculatus* Moench [G, BB, P & S]
Flowers July–August.

Occurs in pastured and 2nd-growth dry or rocky woodland, old fields, pastures, thickets, rocky bluffs, and along railroads. Throughout Missouri, known from every county.

Ranges from Florida to Texas and Mexico, north to Pennsylvania, Ohio, Indiana, Illinois, Minnesota, South Dakota, and Colorado; escaped from cultivation north to Connecticut and New York.

This shrub often becomes thick in woodland that has been pastured and cut over as the dominant understory shrub, excluding the growth of other understory plants.

A form with white fruits, f. *leucocarpus* (D. M. Andrews) Rehder, is not recorded yet from Missouri.

The reddish or purple-red fruit remains on the twigs most of the winter. Birds do not eat the fruit. In some sections of the South, baskets are made from the suckers of the stems.

2. **Symphoricarpos occidentalis** Hook.
Wolfberry Map 2070
Flowers June–August.

Occurs on loess hills, open banks, and thickets. Northern Missouri, in Adair, Nodaway, and Atchison counties.

Ranges from Ontario to British Columbia, south to Michigan, Illinois, Missouri, Kansas, and New Mexico; introduced east to New England and Pennsylvania.

A cultivated plant of this species was collected near Kansas City, Jackson County, and is in the herbarium of Wm. Jewell College. The white fruits resemble those of the commonly cultivated Snowberry, *S. albus* (L.) Blake var. *laevigatus* (Fern.) Blake, but in the latter are much longer (6–20 mm. long) and the flowers are short-pedicelled instead of sessile.

## 3. **Triosteum** L. Horse Gentian

a. Stem conspicuously bristly-hairy, the stiff hairs nearly all long and 1.5–3 mm. long; margins of calyx-lobes bristly-hairy, rest of calyx-lobes glabrous (without hairs) or nearly so; upper surface of leaves with prominent hairs 0.8–1.8 mm. long; larger fully grown leaf-blades 2–6 cm. broad; usually only 2 flowers present at the base of a pair of leaves; corolla yellowish, creamy, or greenish, with loose hairs . . . . . . . . . . . . . . . . . . . . 3. T. ANGUSTIFOLIUM
a. Stem glabrous (without hairs), short-hairy, or with soft hairs up to 2.5 mm. long; calyx-lobes uniformly short-hairy and also with the margins bristly-hairy; upper surface of leaves with minute hairs less than 1 mm. long or glabrous (without hairs); larger fully-grown leaf-blades 5–15 cm. broad; usually 2–8 flowers present at the base of a pair of leaves; corolla purple-red, yellowish, greenish, or dull purple, covered with very short hairs . . . . . . . . . . . . . . *b*
b. Usually 3–5 pairs of the main or middle leaves of the stem united at their bases (connate), the stem appearing to pierce the leaf, 3–9 cm. broad where united; lower surface of leaf velvety-hairy; corolla yellowish, green, or dull purple; fruit nearly globose, orange-yellow usually 6 maturing at some of the leaf-bases; stem densely short-hairy, at least in the upper half, with a mixture of many gland-tipped shorter hairs and longer, non-glandular hairs . . . . . . . 1. T. PERFOLIATUM
b. Usually all the leaves narrowed to their bases, rarely with 1–3 pairs of upper leaves united at their bases and there 1–2 cm. broad; lower surface of leaf less hairy; corolla purple-red; fruit ellipsoid-ovoid, orange or crimson-red, usually less than 6 maturing at the leaf-bases; stem with mostly all long non-glandular hairs 1.5–2.5 mm. long, the short hairs few or none . . . . . . 2. T. AURANTIACUM var. ILLINOENSE

1. **Triosteum perfoliatum** L. Common Horse Gentian Map 2071
Also called Wild Coffee, Tinker's Weed.
Flowers May–July; fruits August–October.

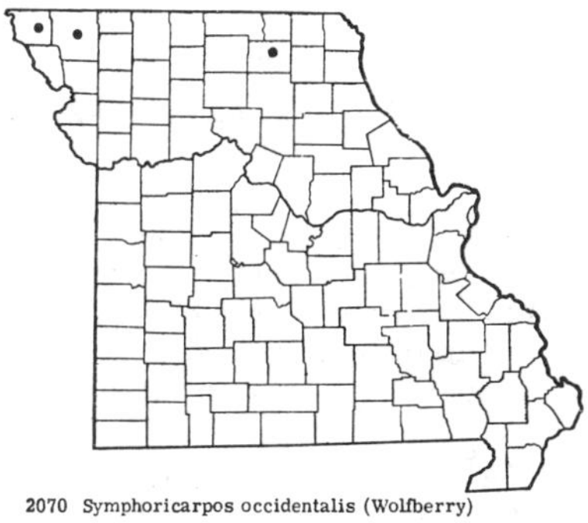

2070 Symphoricarpos occidentalis (Wolfberry)

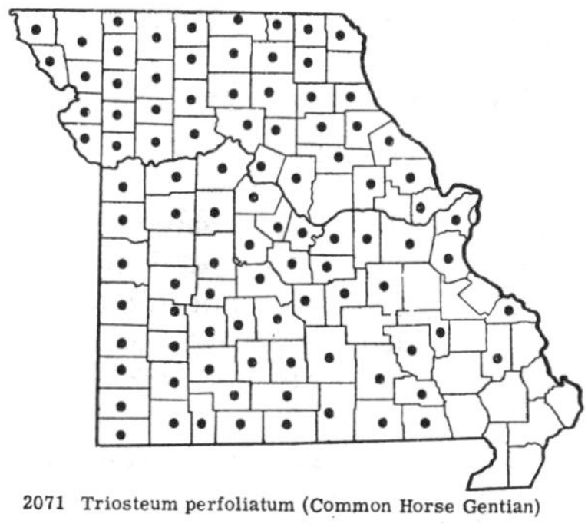

2071 Triosteum perfoliatum (Common Horse Gentian)

2072 Triosteum aurantiacum var. illinoense

Occurs in usually dry or open woods and thickets.

Throughout Missouri, except absent from the lowlands of southeastern Missouri.

Ranges from Georgia to Kansas, north to Massachusetts, New York, West Virginia, Michigan, Wisconsin, Minnesota, and Nebraska.

The dried and roasted fruits have been used by some of the early Pennsylvania German settlers in the United States as a substitute for coffee. The bitter roots have been used as a cathartic and emetic. The Indians used this and other species as a cure for fevers.

2. **Triosteum aurantiacum** Bickn. var. **illinoense** (Wieg.) Palmer & Steyerm. Map 2072

*Triosteum illinoense* (Wieg.) Rydb. [Lane]

*Triosteum perfoliatum* var. *illinoense* Wieg. [BB]

Flowers May–July; fruits August–October.

Occurs in rich woods, often on north-facing slopes, rocky wooded slopes, and bluffs. Found mainly in northern, central, and eastern Missouri and from most of the northern and central Ozarks, but absent from southeastern, southwestern, and parts of northwestern Missouri.

Ranges from Ohio to Illinois, Missouri, and Kansas.

Dr. Franklin C. Lane, who has recently monographed the genus, considers this a species distinct from *T. aurantiacum*, whereas Gleason (*New Ill. Fl.* 3: 303. 1952) places the *T. aurantiacum* group in varietal rank under *T. perfoliatum* because of connection by intermediate forms. However, Deam (*Fl. Ind.* p. 885. 1940), on the basis of field, herbarium, and garden observations, maintains them as distinct species in Indiana, producing evidence in differences of fruit maturity and period of flowering between the two. While there are occasional intergradations in Missouri between *T. perfoliatum* and *T. aurantiacum* var. *illinoense*, in the main they can be recognized as distinct, non-intergrading taxa and are treated here as species.

3. **Triosteum angustifolium** L. Yellow-flowered Horse Gentian Map 2073

Flowers April–May; fruits July–September.

Occurs in rocky or dry open woods, at the base of and on ledges of bluffs, low ground along streams and in ravine bottoms. Ozark region north to St. Louis, Franklin, Cole, Morgan, and Benton counties.

Ranges from Alabama to Louisiana, north to Connecticut, Pennsylvania, West Virginia, Indiana, Illinois, and Missouri.

Two variations occur in Missouri:

Lower surface of leaves glabrous (without hairs) or merely hairy on the nerves . 3a. T. ANGUSTIFOLIUM var. ANGUSTIFOLIUM

Lower surface and nerves of leaves more or less short-hairy . . 3b. T. ANGUSTIFOLIUM var. EAMESII

3a. **Triosteum angustifolium** var. **angustifolium** Map 2073

*Triosteum angustifolium* L. [G, BB, P & S, Steyerm.]

This is the common variation in the state and is found throughout the range in Missouri.

3b. **Triosteum angustifolium** var. **Eamesii** Wieg. Map 2073

Known from Maries (½ way up wooded west- and south-facing limestone slopes along Gasconade River in region of Thox Rock, T38N, R9W, sect. 5 and 6, 2½ mi. southeast of Hayden, June 29, 1939, *Steyermark 27543, 27553*) and Benton (wooded slopes along north- and east-facing limestone bluffs along Cole Camp Creek, T41N, R21W, sect. 16 and part of 15, 1½ mi. northwest of Edmonson, September 24, 1938, *Steyermark 7287*) counties.

## 4. **Viburnum** L. Viburnum, Black Haw, Arrow-wood

The most recent study of species of this genus in temperate North America is that of W. L. McAtee (*A Review of the Nearctic Viburnum.* 125 pp. Chapel Hill, N.C. 1956).

a. Margins of leaves normally with many fine, narrow, closely crowded teeth, 6–12 per cm.; lateral nerves (those extending from midrib) curved, branched and connecting before reaching the margin; leaves often glabrous (without hairs) on both surfaces, if not, then with short scurfy rusty or chestnut-brown hairs; petioles (leaf-stalks) flat, more or less narrowly bordered by a strip of tissue extending from the leaf-blade; no stipules at base of leaf-stalks; stone of fruit flat, not grooved or ridged on the sides; winter-buds with 1 pair of outer scales . . . . . . . . . . . . . . . . . . . *b*

b. Rusty or chestnut-brown short scurfy hairs appearing especially in younger growth on some part of petiole (leaf-stalk), midrib of lower surface and/or lower surface itself of leaf-blade, rays of inflorescence, and on winter-buds; leaf-blades rather thick and somewhat leathery (subcoriaceous), glossy on upper surface . . . . . . . . . . . . . . . . . . 3. V. RUFIDULUM

b. Few or no rusty or chestnut-brown short scurfy hairs appearing on midrib of lower surface or lower surface itself of leaf-blade, nor on most of the petioles (but some petioles below the inflorescences may have rusty scurfiness); leaf-blades rather thin and more membranaceous, dull on upper surface . . . . . . . . . . . . . . . . . . . . . . . . *c*

c. Rarely encountered, known only from Schuyler County, northeastern Missouri; all or most of leaf-blades abruptly slender- and long-pointed (caudate-acuminate) at tip; teeth of leaf-blades conspicuously pointing outward, the free part projecting 0.5–1 mm. long; margins of petioles (leaf-stalks) undulate (wavy) or wrinkled and revolute (edges turned under); those immediately below the inflorescences usually 1.5–3.5 cm. long . . . . . . . . 1. V. LENTAGO

c. Commonly encountered throughout Missouri, in nearly every county; leaf-blades blunt (obtuse) or short-pointed (acute), rarely long-pointed (acuminate) at tip; teeth of leaf-blades curving inward or upward toward the tip of the leaf, shorter and less conspicuous, the free part projecting 0.1–0.2 mm. long; margins of petioles (leaf-stalks) not at all or only slightly undulate, the margins not revolute, those immediately below the inflorescences usually 0.5–2 cm. long, sometimes longer . . . . . . . . . . . . . . . . . 2. V. PRUNIFOLIUM

a. Margins of leaves normally with coarse, broad, or triangular teeth, only 1–3 per cm.; lateral nerves (those extending from midrib) more or less straight and each ending in a tooth; leaves more or less hairy on one or both surfaces, the hairs never rusty or chestnut-brown, but colorless and scattered or in fascicles (clusters); petioles (leaf-stalks) rounded; stipules present or absent at base of leaf-stalk; stone of fruit more or less grooved or ridged on the sides; winter-buds with 2 pairs of outer scales . . . *d*

d. Usually 4–10 (rarely 12) teeth on each half of the margin of the leaf-blade; a pair of stipules (narrow green appendages) present at base of leaves; petioles (leaf-stalks) 3–10 (up to 15) mm. long, those on the leaves immediately below peduncles (stalk supporting inflorescence) 0–7 mm. long; peduncles 0.5–2 cm. long; inflorescence 1.5–5 (up to 7) cm. broad . . . 4. V. RAFINESQUIANUM

d. Usually 10–30 (sometimes as few as 7) teeth on each half of the margin of the leaf-blade; stipules present or absent; petioles (leaf-stalks) 5–50 mm. long, those on the leaves immediately below peduncles 10–50 mm. long; peduncles mostly 2–10 cm. (rarely 1) long; inflorescence mostly 3.5–10 cm. (rarely less) broad . . . . . . . . . . . . . . . . . . . . . . *e*

e. Leaves usually deeply heart-shaped (cordate) at the base; usually 16–30 (rarely 14) teeth on each half margin of the leaf-blade; stipules usually present during flowering period; outer bark loose and peeling off; fruit 8–12 mm. long . . . . . . . . . . . . . . 5. V. MOLLE

e. Leaves usually only rounded or narrowed at base, or somewhat but not deeply heart-shaped; mostly 10–15 (rarely as few as 7 or as many as 23) teeth on each half margin of the leaf-blade; stipules absent or present; outer bark close, not peeling off; fruit 4–7 mm. long . . . . . . *f*

f. Known only from Oregon County, southern Missouri; stipules usually absent; lower surface of leaf-blades glabrous or hairy on nerves; peduncle and branches of inflorescence glabrous. . . . . . . . . . . . . . . . . . . . . . . . 7. V. RECOGNITUM

f. Known only from Shelby County, northeastern Missouri; stipules frequently present; at least the lower surface of leaf-blades more or less hairy; peduncle and branches of inflorescence more or less hairy . . . . . . . . . . . . . . 6. V. PUBESCENS var. DEAMII

---

Plate no. 340. **1.** Triosteum perfoliatum, × 2/7; Details from Small, The New York Botanical Garden. **2.** Triosteum aurantiacum var. illinoense, × 2/7. **3.** Viburnum Lentago, × 2/7. **4.** Viburnum prunifolium, × 2/7; *a*, *b*. Leaf variations; Details from Small, The New York Botanical Garden. **5.** Triosteum angustifolium, × 2/7.

PLATE NO. 340

2073 • Triosteum angustifolium var. angustifolium (Yellow-flowered Horse Gentian)

2073 □ Triosteum angustifolium var. Eamesii

2074 Viburnum Lentago (Nannyberry)

2075 Viburnum prunifolium (Black Haw)

1. **Viburnum Lentago** L. Nannyberry Map 2074

Also called Sheepberry, Wild Raisin.

Flowers May–June.

Occurs in low woods, wooded slopes, and rich valleys bordering or near streams. Known only in a wild state from Schuyler County, northeastern Missouri (wooded slopes along east side of Chariton River, east of Livonia, May 27, 1934, *Steyermark 7802*; open low ground in valley of Brush Creek, just north of road E, T65N, R14W, sect. 35, ½ mi. south of Clifton Hill, May 23, 1950, *Steyermark 69919*).

Ranges from Quebec to Manitoba, south to New Jersey, Georgia, Ohio, Indiana, Illinois, Missouri, South Dakota, and Colorado.

A specimen, presumably originating from cultivated nursery stock, was collected from Franklin County (Gray Summit, Aug. 1, 1928, *Kellogg 2020*). As Kellogg doubtless collected the specimen at the Missouri Botanical Garden's Arboretum at Gray Summit, the record is excluded in the present flora.

A specimen of this species in Arnold Arboretum Herbarium from St. Louis County (Allenton, October, 1882, *Letterman*) probably represents a mixed label resulting from the distribution of duplicate material. Another, in the Missouri Botanical Garden Herbarium from St. Louis County, distributed as *V. Lentago* (Allenton, May 8, 1898, *Letterman*) has the shorter appressed-serrulate margins of leaves with nonundulate petioles characteristic of *V. prunifolium*, to which species it is referred in the present flora.

The fall coloring of this species is very pronounced, taking on a dull or deep purple-red with dull green to dull purple-rose. The edible fruit is blue-black with a 'bloom' and has a slightly sweet but dryish pulp. The shrub sends out root shoots, eventually forming thickets. It is sometimes attacked by the woolly aphis, scale insects, and buffalo hoppers. Its ornamental foliage, flowers, and fruits, together with good autumnal coloring, recommend it for more extensive planting.

2. **Viburnum prunifolium** L. Black Haw Map 2075

*Viburnum prunifolium* var. *Bushii* (Ashe) Palmer & Steyerm. [G, P & S, Steyerm.]

*Viburnum Bushii* Ashe

Flowers April–May.

Occurs in low woods along streams, at the base of and edge of bluffs, dry upper slopes of ravines, and thickets. Throughout Missouri and doubtless in every county.

Ranges from Florida to Texas, north to Connecticut, New York, Ohio, Michigan, Illinois, Iowa, and Kansas.

This species varies in shape and relative pointedness of the leaf-tip, variations of leaves which are oblong-lanceolate, thin, and membranaceous at maturity, and with acute to acuminate tips having been described as *V. Bushii* and *V. prunifolium* var. *Bushii*. This variation has been the basis for much confusion and misidentification of *V. prunifolium* with *V. Lentago*. In addition to the characters of more abruptly acuminate leaf-blades with longer and more conspicuously spreading teeth and broader undulate-margined petioles of *V. Lentago*, the stamens of *V. Lentago* protrude half their length or more from the flower, whereas in *V. prunifolium* they protrude about one-fourth their length. The variations in leaf-shape and extent of pointedness of leaf-tip fluctuate so greatly in Missouri among shrubs and even on the same plant that no reliance can be placed in the extent of acuteness or acumination of the leaf-tip of *V. prunifolium* var. *Bushii*, and this variation is, therefore, not recognized in the present flora. Specimens of *V. prunifolium* with young or immature foliage sometimes approach *V. rufidulum* in having rusty tomentum on the petioles and midrib of the lower leaf-surface, such as *Sparling 789* from Livingston County.

The species has beautiful fall color, the foliage turning from deep lavender or maroon-purple to

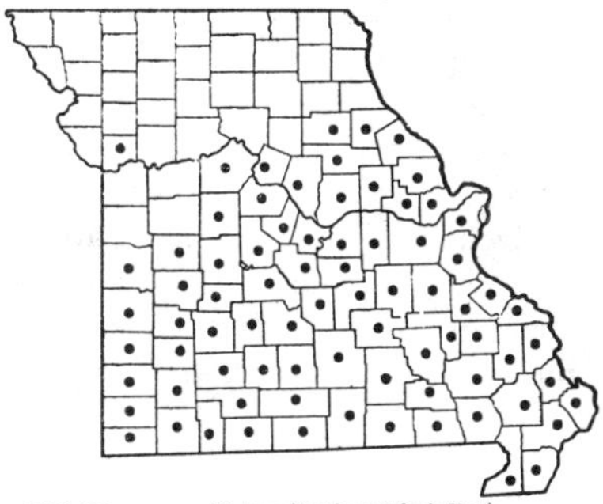

2076 Viburnum rufidulum (Southern Black Haw)

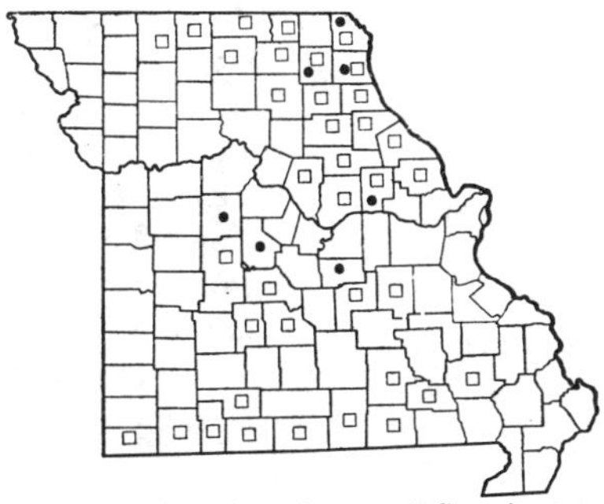

2077 • Viburnum Rafinesquianum var. Rafinesquianum (Downy Arrow-wood)
2077 □ Viburnum Rafinesquianum var. affine

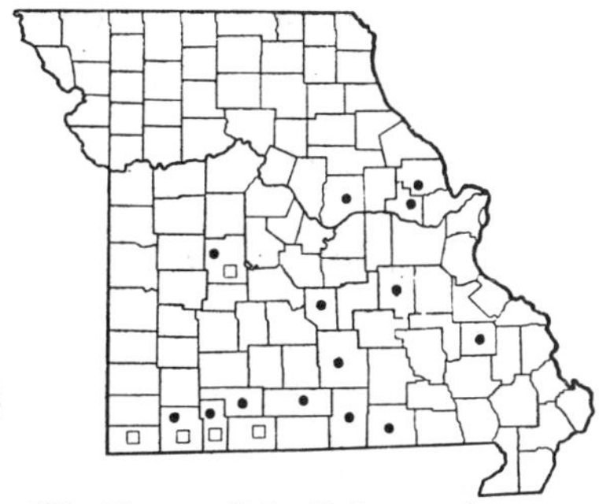

2078 • Viburnum molle f. molle (Arrow-wood)
2078 □ Viburnum molle f. leiophyllum

finally deep rose-red. The blackish-blue fruits are edible and have a similar taste to those of *V. Lentago.* The species can easily be grown from seed and becomes a flowering shrub in several years.

The bark of *V. prunifolium* is used in medicine as an astringent, nerve tonic, and antispasmodic. It is also employed as a uterine sedative.

3. **Viburnum rufidulum** Raf. Southern Black Haw Map 2076

Also called Wild Raisin.

Flowers April–May.

Occurs in rocky or dry woods, bluff escarpments, rich valleys and alluvial ground along streams, rocky glades, and thickets. Ozark region of southern and central Missouri, north to Ralls, Monroe, Boone, Howard, Saline, Henry, and Bates counties, locally introduced in Clay County.

Ranges from Florida to Texas, north to Virginia, Ohio, Indiana, Illinois, and Missouri.

The edible black-blue fruits of this species have a sweet dry pulp similar to those of the two preceding species. *Viburnum rufidulum* usually grows to the size of a small tree in Missouri. From data furnished the author by Mr. Kendall Laughlin, a specimen measuring 2 feet in circumference and 23 feet high in the Big Oak Tree State Park in Mississippi County is considered to be the largest of this species known in the United States. The dark green, lustrous leaves, showy clusters of flowers, and rich fall coloring highly commend this species in ornamental plantings. The foliage turns a deep rose-purple to rose-red or bright red.

In northern Illinois, where the author has his botanical garden, this *Viburnum* is the only one of the Missouri species which has failed to prove hardy, and he has not been able to establish it at this latitude either by transplanting or from seed.

4. **Viburnum Rafinesquianum** Schultes Downy Arrow-wood Map 2077

Also called Missouri Viburnum.

Flowers May–June.

Occurs on steep wooded bluffs and escarpments and rocky open woods generally bordering streams, sometimes in thickets along rocky stream beds.

Two variations occur in Missouri:

Lower surface of leaf-blades hairy over most of the surface . . . . 4a. V. RAFINESQUIANUM var. RAFINESQUIANUM

Lower surface of leaf-blades glabrous (without hairs) or only the nerves hairy . 4b. V. RAFINESQUIANUM var. AFFINE

4a. **Viburnum Rafinesquianum** var. **Rafinesquianum** Map 2077

*Viburnum affine* Bush var. *hypomalacum* Blake [P & S, Steyerm., McAtee]

This is the less common variety in the state, scattered in northeastern and middle Missouri south to Maries, Morgan, and Pettis counties.

Ranges from Quebec to Manitoba, south to Georgia, Kentucky, and Missouri.

4b. **Viburnum Rafinesquianum** var. **affine** (Bush) House Map 2077

*Viburnum affine* Bush [P & S, Steyerm., McAtee]

This is the commoner variation in Missouri, being found in many parts of the Ozark section and in northern Missouri; it is absent from the southeastern lowlands and much of the western sections of the state.

The autumnal coloring varies from dull purplish-red to dull wine purple with dull red. It is an attractive shrub and recommended for small shrub plantings. On north-facing bluffs it often is one of the dominant shrubs in an understory consisting of *Carpinus caroli-*

*niana*, *Ostrya virginiana*, *Staphylea trifolia*, and *Cornus alternifolia*.

5. **Viburnum molle** Michx. Arrow-wood Map 2078

Also called Kentucky Viburnum.

Flowers May–June.

Occurs at the base of and on steep wooded slopes of bluffs, escarpments, talus slopes, and rocky woods along streams.

Two variations occur in the state:

Lower surface of leaves hairy . 5a. V. MOLLE f. MOLLE
Lower surface of leaves glabrous, only the main nerves hairy . . . 5b. V. MOLLE f. LEIOPHYLLUM

5a. **Viburnum molle** f. **molle** Map 2078

*Viburnum molle* Michx. [G, BB, P & S, Steyerm.]

*Viburnum ozarkense* Ashe [McAtee]

*Viburnum demetrionis* Deane & Robinson (holotype from near 'Big Cave,' bluffs of Cole Camp Creek, May 29, 1896, and July, 1894, Benton Co., Missouri, *C. H. Demetrio* in Gray Herb.).

This is the more commonly encountered variation, scattered throughout the Ozarks north to Lincoln, Callaway, and Benton counties.

Ranges from Indiana to Iowa, south to Kentucky, Arkansas, and Oklahoma.

5b. **Viburnum molle** f. **leiophyllum** Rehder Map 2078

Rare and known only from the western Ozarks in Taney, Stone, Barry, and McDonald counties, locally north in Benton County.

Scattered in the range of f. *molle*.

McAtee (Revision, p. 63) attempts to maintain *V. ozarkense* Ashe as distinct from *V. molle*, but I have not been able to keep it separate from the latter species. It has the same type of winter-buds, fruits of the same length, large, cordate leaves of the same shape, size, and pubescence, and similar elongated peduncles of *V. molle*. McAtee differentiates it from *V. molle* on the basis of having the leaves near the inflorescences with short petioles, with some of these leaves almost clasping. This character does not hold true, however, there being variation in length of petioles, varying from 5–8 (up to 15) mm. long for the *V. ozarkense* type of leaf subtending the inflorescence to 10–45 mm. long for the *V. molle* type.

The bark of the shrubs loosens and peels off like that of birch, extending up to the stem to two-year-old branches. The smooth bark, which remains, is reddish-brown with the branches becoming grayish. This is an attractive shrub with its large, heart-shaped leaves, showy flowers, blue-black fruits, and vase-like habit of growth. It has purple-red and yellow-orange autumn coloring, and succeeds in usually dry, shaded situations. It usually comes into flower later than *V. Rafinesquianum* var. *affine*.

6. **Viburnum dentatum** L. var. **Deamii** (Rehd.) Fern. Southern Arrow-wood Map 2079

*Viburnum pubescens* (Ait.) Pursh var. *Deamii* Rehder [P & S, Steyerm.]

*Viburnum carolinianum* Ashe var. *Deamii* Rehder [McAtee], but requiring new combination

Flowers May–June.

Occurs in low alluvial woods and wooded slopes along Salt River. Known only from Shelby County, northeastern Missouri (low open woods along Salt River, 3 mi. north of Shelbina, June 28, 1933, *Palmer & Steyermark 40845*; alluvial woods along south side of North River, west of highway 15, T57N, R10W, sect. 17, 2¾ mi. north of Shelbina, September 25, 1948, *Steyermark 66566*; sect. 9, 3½ mi. north of Shelbina, September 25, 1948, *Steyermark 66579*).

Ranges from Pennsylvania to Missouri and Tennessee.

This is a handsome shrub, the showy flower-clusters standing on elongated peduncles. I have shrubs growing in my wildflower preserve in northern Illinois which were originally given me by Dr. Deam from his arboretum, and these have done well during the past fifteen years.

*Viburnum dentatum* L. is a variable species. McAtee (Revision, pp. 11–12, 89–107. 1956) attempts to keep *V. dentatum*, *V. carolinianum*, and *V. pubescens* separated on characters of pubescence of leaves and immature fruit. These characters have been found unstable and intergrading in all respects.

7. **Viburnum recognitum** Fern. Arrow-wood Map 2080

*Viburnum recognitum* var. *recognitum* [McAtee]

*Viburnum dentatum* var. *lucidum* Ait. [BB]

Flowers May–June.

Occurs along margins of gravel bars of small streams. Known only from southern Missouri, Oregon

Plate no. 341. **1.** Viburnum rufidulum, × 2/7. **2.** Viburnum Rafinesquianum var. affine, × 2/7. **3.** Viburnum molle, × 2/7. **4.** Sambucus canadensis, × 2/7; Details from Small, The New York Botanical Garden. **5.** Sambucus pubens, × 2/7; Details from Small, The New York Botanical Garden. **6.** Viburnum dentatum var. Deamii, × 2/7.

PLATE NO. 341

2079 Viburnum dentatum var. Deamii (Southern Arrow-wood)

2080 Viburnum recognitum (Arrow-wood)

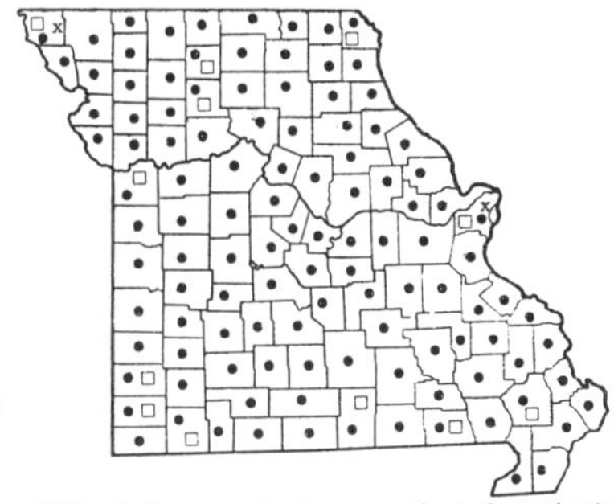

2081 • Sambucus canadensis var. canadensis f. canadensis (Common Elderberry)
2081 x Sambucus canadensis var. canadensis f. rubra
2081 □ Sambucus canadensis var. submollis

County (margin of dry gravel bar of creek opposite south-facing bluffs, along Little Hurricane Creek, vicinity of 'Blue Hole,' T24N, R3W, sect. 5, 2½–3 mi. [by air] east of Greer, July 23, 1955, *Steyermark 78893*).

Ranges from Maine and New Brunswick to Ontario, south to South Carolina, Ohio, Illinois, Michigan, and Missouri.

### 5. **Sambucus** L. Elderberry

Common species, found throughout the state; flowering and fruiting spray flat-topped, umbelliform; fruits usually purple-black, rarely red; soft inside (pith) of stem white . . . . . . . 1. S. CANADENSIS
Rarely found, known only from Marion County, northeastern Missouri; flowering and fruiting spray cone- or pyramid-shaped, paniculiform; fruits usually red; soft inside (pith) of stem brown . . . 2. S. PUBENS

1. **Sambucus canadensis** L. Common Elderberry Map 2081

Flowers late May–July; fruits August–October.

Occurs in open woods, thickets, along streams, fence rows, roadsides, and railroads.

Throughout Missouri in every county.

The following three variations occur in the state:

a. Hairs on lower surface of leaf present on both the veins as well as the actual surface . . 1c. S. CANADENSIS var. SUBMOLLIS
a. Hairs when present on lower surface of leaf mainly confined to the veins, or hairs absent. *b*
  b. Common type; fruit purple-black . . . 1a. S. CANADENSIS var. CANADENSIS f. CANADENSIS
  b. Rare type; fruit red . . 1b. S. CANADENSIS var. CANADENSIS f. RUBRA

1a. **Sambucus canadensis** var. **canadensis** f. **canadensis** Map 2081

*Sambucus canadensis* L. [G, BB, P & S, Steyerm.]

This is the common type in Missouri, occurring in every county.

Ranges from Nova Scotia and Quebec to Manitoba and South Dakota, south to Florida, Louisiana, and Oklahoma.

1b. **Sambucus canadensis** var. **canadensis** f. **rubra** Palmer & Steyerm. Map 2081

*Sambucus canadensis* f. *rubra* Palmer & Steyerm. [G, Steyerm.]

Known only from St. Louis (from cultivated plants in grounds of Normandy High School, originally collected in northern St. Louis County, September 14, 1936, *Steyermark 20068*) and Atchison (Watson, September 3, 1920, *Palmer 18928*, holotype in Arn. Arb. Herb.) counties.

1c. **Sambucus canadensis** var. **submollis** Rehder Map 2081

Scattered but rare, throughout Missouri.

Ranges from Illinois to Iowa, south to Oklahoma and Texas.

Deam (*Fl. Ind.* p. 880. 1940) does not recognize the var. *submollis* because his observations show that the pubescence varies greatly even on the same plant. More detailed field observations and experimental studies are needed in Missouri and other parts of the western sector of the range of var. *submollis* to substantiate the relative merits of this variety.

Elderberry plants persist for only three to five years, then die. New plants come up from the root sprouts, however, perpetuating a plant in a given locality. The flowers are highly fragrant. The fruit is eaten by various birds.

Elderberry can be used to advantage in several ways. Formerly, the pith was widely employed for sectioning parts of plants in histological and cytological studies. The flowers and fruits especially are made into a number of foods and beverages. The flower-clusters, while still fresh, may be dipped in egg batter for fritters or into pancake or muffin mix to lend a sweet, pleasant flavor to the cakes. The unopened flowers and young green fruits are sometimes pickled and used like capers for seasoning. A tea is sometimes prepared from the flowers, with mint often added, for use in dyspepsia. An elderberry 'white' wine is prepared from the flowers. The ripe fruits, in a fermented state, are prepared into wine, imparting a pleasant flavor. When boiled with some sugar and the rind of lemon, the fruits may be canned and used in pies. Mixed with grape or other kinds of fruit, the elderberry fruits make a tasty jelly. The author has eaten a delicious pudding of elderberry which was cooked and thickened with cornstarch to which lemon juice was added. Elderberry juice cooked with sugar makes a tasty beverage; this can be kept over winter and served with lemon and water. In Europe the fruit of a related species of elderberry is dried and used for soup, thickened with flour. A type of chutney and a spice dish, known as elderberry rob, are prepared also from the fruits.

The plant has been used medicinally. The inner bark, blossoms, and bruised leaves served for various sicknesses, the flowers being used as a sudorific. Small amounts of prussic acid can be produced by the plant, causing poisoning under certain conditions. Cattle and sheep have been fatally poisoned by eating the leaves, young shoots, and opening buds of the plant. Children are believed to be poisoned at times by chewing on the bark of elderberry. The green leaves are reported to keep mice away, and when the leaves are placed around plants of cabbages, cucumbers, and melons, they are reported to keep various destructive insects and their larvae away. The flowers are reported to be poisonous to turkeys and the fresh berries poisonous to poultry. However, it is known that wild turkeys eat the fruits. Cases of poisoning among persons have been reported from eating fresh leaves and flowers. The green bark and root are considered to be poisonous and give cathartic and hydragogue reactions.

2. **Sambucus pubens** Michx. f. **pubens**
Red-berried Elder Map 2082
*Sambucus pubens* Michx. [G, BB, P & S, Steyerm.]
Flowers April–May; fruits June–August.

Occurs on shaded, north-facing, wooded limestone bluffs and ledges. Known only from northeastern Missouri, Marion County (shaded upper banks of limestone bluffs, below uppermost ledges adjoining and just southeast of Riverside Cemetery, vicinity of Hannibal, T56N, R4W, sect. 34, April 15, 1938, *Steyermark 5317*; same locality, rocky banks, Hannibal, July 19, 1930, *Kellogg 15305*, in sterile condition).

Ranges from Newfoundland to Alaska, south to Georgia, Tennessee, West Virginia, Ohio, Indiana, Illinois, Missouri, South Dakota, Colorado, and Oregon.

This rare species is another one of the representatives of a northern flora which have migrated into Missouri at an earlier time, probably during one of the advances of the Pleistocene Ice, and have persisted since in scattered localities. The station where this species of elder occurs in Missouri is near the one where *Dodecatheon amethystinum*, a rare species of the 'Driftless Area,' has likewise persisted.

The red-berried elder flowers in early spring in April a couple of months ahead of *S. canadensis*. The red berries mature before *S. canadensis* flowers. These berries are reported as being poisonous, but definite evidence to substantiate this is lacking.

## Fam. **VALERIANACEAE** (Valerian Family)

### **Valerianella** Mill. Corn Salad

The species are separated chiefly by characters of the fruit, so that it is necessary to have fruiting specimens available. The following treatment follows the most recent work by Sarah C. Dyal (Rhodora 40: 186–212. 1938). The fruit is 3-celled, 2 of the cells (the ventral) being empty and sometimes confluent into 1 cell, the third cell (the dorsal) being fertile and 1-seeded.

a. Corolla lilac or rose-purplish, 10–12 mm. long, the long slender tube 7–10 mm. long and 3–4 times as long as the expanded upper portion; bracts surrounding flower-clusters with gland-tipped tiny teeth . . *b*
b. Fruits 1.7–2.5 mm. broad, ovoid-ellipsoid, laterally compressed, each cell keeled; fruit with

2082 Sambucus pubens f. pubens (Red-berried Elder)

2083 Valerianella olitoria (Corn Salad)

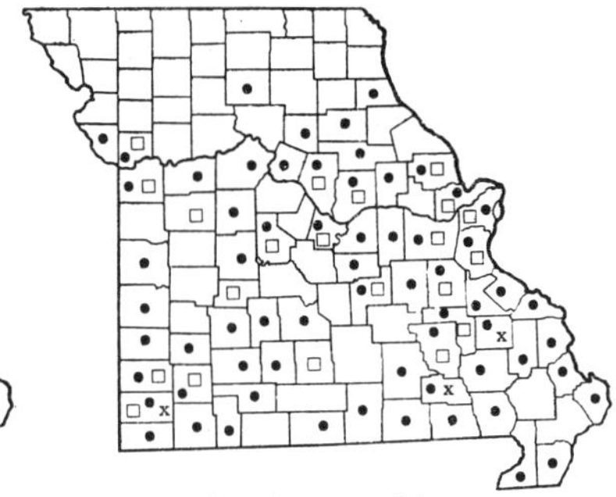

2084 • Valerianella radiata var. radiata
2084 □ Valerianella radiata var. missouriensis
2084 x Valerianella radiata var. Fernaldii

3 lines of long hairs, 1 down the dorsal side of the fertile cell and 1 down each of the sterile cells; combined width of the sterile cells greater than the width of the fertile cell . . . 4. V. OZARKANA

b. Fruits 0.7–1 mm. broad, narrowly oblong or oblong, dorsiventrally compressed, the cells not keeled; fruit glabrous or covered with short fine hairs; combined width of the sterile cells narrower than the width of the fertile cell . . . . . . . . . . . . . . . . . . . 5. V. BUSHII

a. Corolla white or tinged blue or pink, 1.5–5 mm. long, the tube shorter than or about as long as the expanded upper portion; bracts surrounding flower-clusters either not fringed with hairs on their margins or fringed with glandless spiny hairs. . . . . . . . . . . . . . . . . . . . *c*

c. Bracts and bractlets surrounding flower-clusters not fringed with hairs on their margins. . . 3. V. STENOCARPA var. PARVIFLORA

c. Some or all of the bracts and bractlets surrounding flower-clusters more or less fringed on their margins . . . . . . . . . . . . . . . . . . . . . . . . . . . . . *d*

d. Corolla-lobes bluish; stamens included within the corolla or barely protruding; all the bracts surrounding flower-clusters fringed on their margins, the outer ones spatulate and rounded at tip; fertile cell of fruit with an enlarged corky mass on the back. . . . . . . . 1. V. OLITORIA

d. Corolla-lobes white or tinged pink; stamens noticeably protruding from the corolla; outer bracts surrounding flower-clusters fringed on their margins, but inner ones usually not fringed, lanceolate or oblanceolate and acutely pointed at tip; fertile cell of fruit without any enlarged corky mass on the back . . . . . . . . . . . . . . . . . . . . . . . . . *e*

e. Rarely encountered species; fruits ellipsoid-oblong, $2\frac{1}{2}$–3 times as long as broad, 2 mm. long, 0.7 mm. broad; ventral groove of fruit extending to the base, dorsal side of the fertile cell of fruit rounded . . . . . . . . . . . . . . 3. V. STENOCARPA var. PARVIFLORA

e. Commonly encountered species throughout southern and central Missouri; fruit ovoid, $1\frac{1}{2}$–$2\frac{1}{4}$ times as long as broad, 2–2.5 mm. long, 1–1.5 mm. broad; ventral groove of fruit extending only part way to the base, dorsal side of the fertile cell of fruit flattened . 2. V. RADIATA

1. **Valerianella olitoria** (L.) Poll. Corn Salad Map 2083

Also called Lamb's Lettuce.

Flowers April–May.

Occurs in waste ground, borders of fields, and along roadsides. Known only from southwestern Missouri, Lawrence County (waste ground along border of field and small stream near Verona, May 10, 1951, *Palmer 51807*; open ground along roadsides and ditches, $1\frac{1}{2}$ mi. northwest of Verona, May 8, 1953, *Palmer 55454*).

Native of Europe; cultivated and naturalized in the United States from New England, New York, and Pennsylvania to Indiana and Missouri, south to North Carolina and Tennessee; Utah, Idaho, Oregon, and California.

The Corn Salad is commonly grown in Europe as a salad green, like lettuce, but is relatively unknown in the United States, although seeds are offered for sale by leading seed supply houses. The author has been acquainted from childhood with this plant and has grown it in his vegetable garden each year. The basal rosettes of pale green, thin leaves make an excellent salad, especially when prepared with olive oil and vinegar. Some of the native species in the Missouri flora may be similarly used in a young state, either fresh as a salad or cooked as a vegetable.

2. **Valerianella radiata** (L.) Dufr. Map 2084

Flowers April–May.

Occurs in fallow fields, prairies, wet meadows, alluvial open soils in valleys, creek bottoms, rocky glades, rocky or open ground, roadsides and railroads.

2085 Valerianella stenocarpa var. parviflora

2086 Valerianella ozarkana

2087 Valerianella Bushii

The following varieties, found in Missouri, can be distinguished only in fruit:

a. The 2 empty sterile (ventral) cells of the fruit separated by only a slight groove between them, their combined width about half the width of the fertile 3rd (dorsal) cell containing the seed . . . 2c. V. RADIATA var. FERNALDII

a. The 2 empty sterile (ventral) cells of the fruit separated by a prominent wide groove between them, their combined width slightly less than or broader than the fertile 3rd (dorsal) cell containing the seed . . . . . *b*

b. Fertile (dorsal) cell which contains the seed as broad as or broader than the combined width of the two empty sterile (ventral) cells . . . 2a. V. RADIATA var. RADIATA

b. Fertile (dorsal) cell which contains the seed slightly less than the combined width of the two empty sterile (ventral) cells. . . . 2b. V. RADIATA var. MISSOURIENSIS

**2a. Valerianella radiata var. radiata** Map 2084

*Valerianella radiata* (L.) Dufr. [G, P & S, Steyerm.]

*Valerianella radiata* var. *leiocarpa* (T. & G.) Krok [P & S]

Common throughout southern and central Missouri north to Marion, Monroe, Randolph, Linn, Lafayette, Clay, and Platte counties.

Ranges from Florida to Texas, north to New Jersey, Pennsylvania, Ohio, Illinois, Missouri, and Kansas.

**2b. Valerianella radiata var. missouriensis** Dyal Map 2084

Frequent in southern and central Missouri.

Ranges from Missouri to Arkansas.

**2c. Valerianella radiata var. Fernaldii** Dyal Map 2084

Known in southern Missouri in Madison, Carter, and Newton counties.

Ranges from Florida to Connecticut, and Missouri and Kansas.

The species is variable in height of plant, size and shape of leaves and amount of toothing at the base, size of corolla, pubescence of fruits, and the relative depth and width of the empty sterile cells of the fruits. Individual plants with glabrous or hairy fruits occur among the three varieties of this species.

**3. Valerianella stenocarpa** (Engelm.) Krok var. **parviflora** Dyal Map 2085

*Valerianella stenocarpa* [of P & S], not (Engelm.) Krok

Flowers April–May.

Occurs in rocky prairies and glades. Known only from western Missouri in Jackson (Dodson, June 10, 1896, *Bush 793*; same locality, May 11, 1895, *Mackenzie 293*) and Dade (sandstone glades, 1½ mi. west of Bona, May 11, 1953, *Palmer 55564*) counties.

Ranges from Missouri and Oklahoma to Texas.

Typical *V. stenocarpa* var. *stenocarpa* occurs in Texas and has longer corollas 2–3 mm. long.

**4. Valerianella ozarkana** Dyal Map 2086

*Valerianella longiflora* in part [of P & S], not (T. & G.) Walp.

Flowers April–May.

Occurs in rocky glades and rocky open woods. Known only from southwestern Missouri in Barry (barrens, Eagle Rock, April, 1898, *Bush 232* in part; rocky cedar woods near stream, Eagle Rock, May 9, 1936, *Dyal, Fisher, & Hazard 225*, holotype in Cornell Univ. Herb.; rocky open woods near Seligman, May, 1926, *Palmer 29780*) and McDonald (barrens, Noel, May, 1915, *Bush 7524*) counties.

Ranges from Missouri to Arkansas.

**5. Valerianella Bushii** Dyal Map 2087

*Valerianella longiflora* in part [of P & S], not (T. & G.) Walp.

Flowers April–May.

Occurs in rocky glades and rocky open woods. Known in southwestern Missouri in Taney, Stone, Barry (rocky cedar woods near stream, Eagle Rock, May 9, 1936, *Dyal, Fisher, & Hazard 218*, holotype in Cornell Univ. Herb.), and McDonald counties, and locally east in Madison County (head of St. Francis River 'shutins,' 14 mi. south of Fredericktown, April 27, 1930, *Steyermark 2086*).

Ranges from Missouri to Arkansas, and Oklahoma.

The specimen cited above from Madison County has characteristic narrowly oblong fruits of *V. Bushii*, but are 1 mm. broad and covered with a short fine pubescence.

It is highly questionable whether *V. ozarkana* and *V. Bushii* should be maintained as distinct species although their fruits are definitely different. Both are often found together in the same colony, but are judged to be distinct species by Dr. Dyal on the basis of their fruit characters and their breeding true from seed. However, there is still some doubt as to the final acceptance of these two very similar taxa, and they may eventually be considered as one species.

Although the corolla of *V. ozarkana* and *V. Bushii* has been described as having a white limb with a reddish corolla-tube, the present author's field notes indicate that the color is lilac or rose-purplish more or less throughout.

*Excluded Species*

**Valeriana pauciflora** Michx. Large-flower Valerian

A specimen obtained from Mr. Bill Bauer in 1946 from his wild flower garden was reported by him to have come from Lincoln County (alluvial woods along Big Creek, T48N, R1W, sect. 25, 6½ mi. south of Troy, April 21, 1946, *Bauer 11279*) and reported by the author in Rh. 51: 118. 1949. As repeated trips to this locality have failed to substantiate the occurrence of the species at this locality, it should be excluded from the flora.

Asa Gray (*Syn. Fl. N. Am. 1²*: 44. 1884) cited this species from Missouri, and Tracy (*Cat. Phaen. and Vasc. Crypt. Pl. Mo.* p. 39. 1886) included it, apparently on the basis of Gray's reference. It is common on the east side of the Mississippi River in southern Illinois and records may eventually be found in eastern Missouri to substantiate its occurrence in Missouri.

## Fam. **DIPSACACEAE** (Teasel Family)

### **Dipsacus** L. Teasel

**Dipsacus sylvestris** Huds. f. **sylvestris** Teasel Map 2088

*Dipsacus sylvestris* Huds. [G, BB, P & S]

Flowers June–October.

Occurs in fields, thickets, pastures, dry and open waste ground, open woods, roadsides, railroads, and gravel bars along streams. Southern and central Missouri, north to Ralls, Boone, and Clay counties.

Native of Europe; introduced and naturalized in North America from Quebec and Maine to Ontario and Michigan, south to North Carolina, Tennessee, and Missouri.

The usual color of the flowers is lavender. A form with white corollas, f. *albidus* Steyerm., has not yet been recorded from Missouri.

A related species, *D. fullonum* L., the Fuller's Teasel, is cultivated for the fruiting heads with their spine-tipped bracts, used for raising the nap (fulling) on woollen cloth in textile manufacturing.

## Fam. **CUCURBITACEAE** (Gourd Family)

Plants of this family have the sexes of the flowers separated, so that either the male and female flowers occur on separate plants (dioecious) or on different parts of the same plant (monoecious).

---

Plate no. 342. **1.** Valerianella Bushii, × 2/5; *a, b, c.* Fruit, various views, × 8; *d.* Cross-section of fruit, × 8. **2.** Valerianella ozarkana; *a.* Fruit, × 8; *b.* Fruit, cross-section, × 2/5. **3.** Valerianella stenocarpa var. parviflora, × 2/5; *a, b, c.* Fruit, various views, × 8; *d.* Cross-section of fruit, × 8. **4.** Valerianella olitoria, × 2/5; After Gleason, The New York Botanical Garden. **5.** Valerianella radiata var. radiata × 2/5 *a, b.* Fruit, two views, × 6; *c.* Cross-section of fruit, × 6. **6.** Valerianella radiata var. missouriensis; *a, b, c.* Fruit, various views, × 6; *d.* Cross-section of fruit, × 6. **7.** Valerianella radiata var. Fernaldii; *a, b.* Fruit, two views, × 6; *c.* Cross-section of fruit, × 6. All details from Small, The New York Botanical Garden.

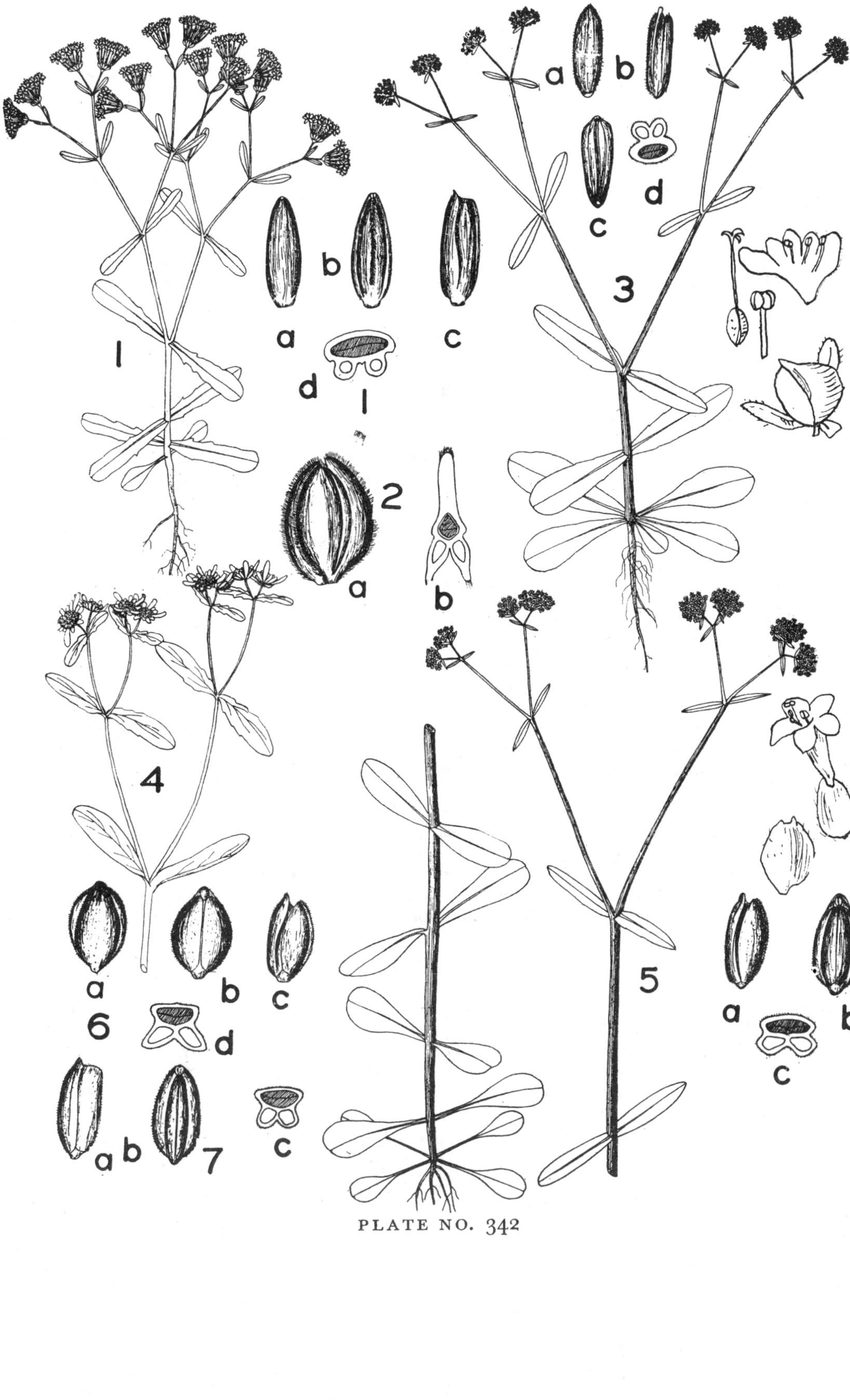


PLATE NO. 342

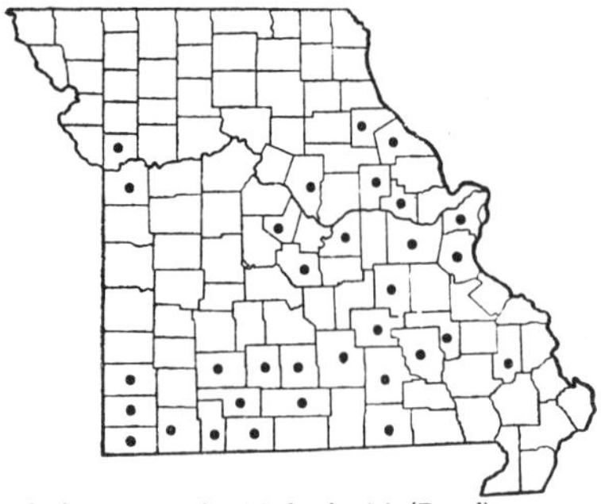

2088 Dipsacus sylvestris f. sylvestris (Teasel)

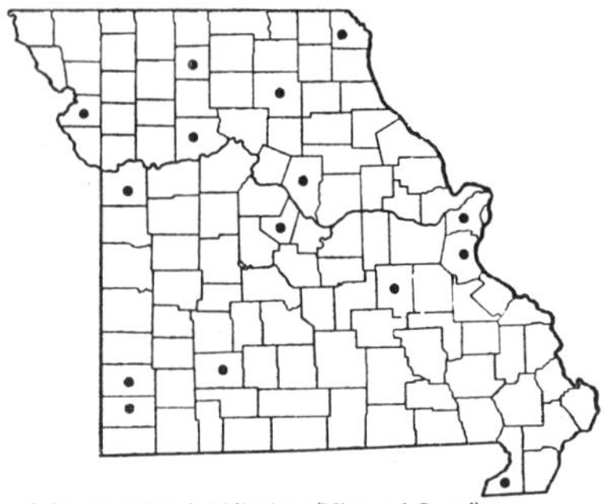

2089 Cucurbita foetidissima (Missouri Gourd)

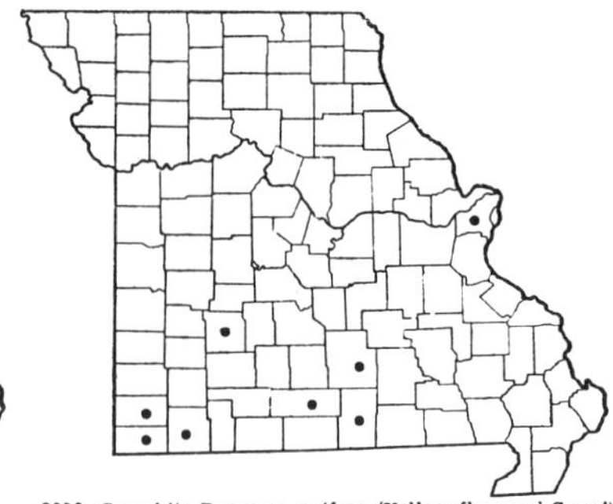

2090 Cucurbita Pepo var. ovifera (Yellow-flowered Gourd)

a. Leaves deeply cut and pinnatifid into 3–4 pairs of lobes which are again lobed, the narrowest lobes 0.3–1.5 cm. broad . . . . . . . . . . . . . . . . . . . . . 3. CITRULLUS

a. Leaves either only toothed, or, if angled or lobed, the lobes mainly 1.5–6 or more cm. broad . . . . *b*

- b. Stems, leaves, flower-stalk, and calyx covered with a soft-hairiness of dense sticky (viscid) hairs; a disk-shaped gland occurs on each side of the base of the leaf-blade above its junction with the petiole (leaf-stalk) . . . . . . . . . . . . . . . . . . . . . 2. LAGENARIA
- b. Stems, leaves, flower-stalk, and calyx either glabrous (without hairs) or hairy, but not covered with a soft dense hairiness; no disk-shaped gland occurring at the base of leaf-blade . . . . . . *c*
  - c. Ovaries and fruits with prickles or spines; filaments of the staminate (male) flowers united into a column . . . . . . . . . . . . . . . . . . . . . . *d*
    - d. Stems and leaves glabrous (without hairs) or nearly so; leaves deeply 5-lobed; fruit solitary, inflated, 4-seeded . . . . . . . . . . . . . . . . . . . . 7. ECHINOCYSTIS
    - d. Stems and leaves noticeably hairy; leaves shallowly 3–5-angled or lobed; fruit in clusters of 2 or more, firm, not inflated, 1-seeded . . . . . . . . . . . . . . 8. SICYOS
  - c. Ovaries and fruits glabrous (without hairs), or hairy, not prickly; filaments of the staminate (male) flowers free or distinct at least in the upper portion . . . . . . . . . . . *e*
    - e. Corolla 5–12 cm. long, 5–10 cm. broad; fruit 5–12 cm. long . . . . . . . . 1. CUCURBITA
    - e. Corolla 0.3–2.5 cm. long, 0.3–2.5 cm. broad; fruit 1–2 cm. long (except in *Cucumis Melo* which is 15 or more cm. long) . . . . . . . . . . . . . . . . . . . *f*
      - f. Corolla yellow, 2.5 cm. long, 2.5–3.5 cm. broad; leaves not deeply lobed, the margins only wavy and toothed . . . . . . . . . . . . . . . 4. CUCUMIS MELO
      - f. Corolla white, greenish-white, or greenish 0.3–1 cm. long, 0.3–1 cm. broad; leaves shallowly or deeply 3–5-lobed. . . . . . . . . . . . . . . . . . . *g*
        - g. Tendrils simple and unbranched; leaves (including lobes) 3–7 cm. broad; fruit and pistillate flowers on elongated stalks (peduncles) 2.5–4 cm. long; staminate flowers with anther-cells straight or slightly curved . . . . . . . . . . . 5. MELOTHRIA
        - g. Tendrils forked or branched; leaves (including lobes) 10–20 cm. broad; fruit and pistillate flowers on very short stalks up to 2 mm. long or nearly stalkless (sessile); staminate flowers with anther-cells strongly twisted or contorted . . . . 6. CAYAPONIA

### 1. **Cucurbita** L. Gourd, Squash

Wild species found along railroads; leaves longer than broad, thick, stiff, usually lacking any prominent lobes, triangular-ovate . . . . . . . . . . . . . . . . . . . . . . 1. C. FOETIDISSIMA

Cultivated species found usually in woods in valleys and gravel bars near streams, rarely along railroads; leaves either as broad as long or only slightly longer than broad, rather thin and flexible, prominently 3–5-lobed, broadly ovate or orbicular-ovate . . . . . . . . . . . . . 2. C. PEPO var. OVIFERA

---

Plate no. 343. **1.** Dipsacus sylvestris, × 1/7; *a*. Flower, × 6/7; *b*. Inflorescence, × 2/7; After Gleason, The New York Botanical Garden. **2.** Cucurbita foetidissima, × 1/7; After Gleason, The New York Botanical Garden. **3.** Cucurbita Pepo var. ovifera, × 1/7; Details from Small, The New York Botanical Garden. **4.** Citrullus vulgaris, × 1. **5.** Lagenaria Siceraria, × 1/7. **6.** Cucumis Melo var. Melo, × 1/7, with female flower; *a*. Male flower, × 1/7; Details from Small, The New York Botanical Garden. **7.** Melothria pendula, × 1/7; After Gleason, details from Small, The New York Botanical Garden.

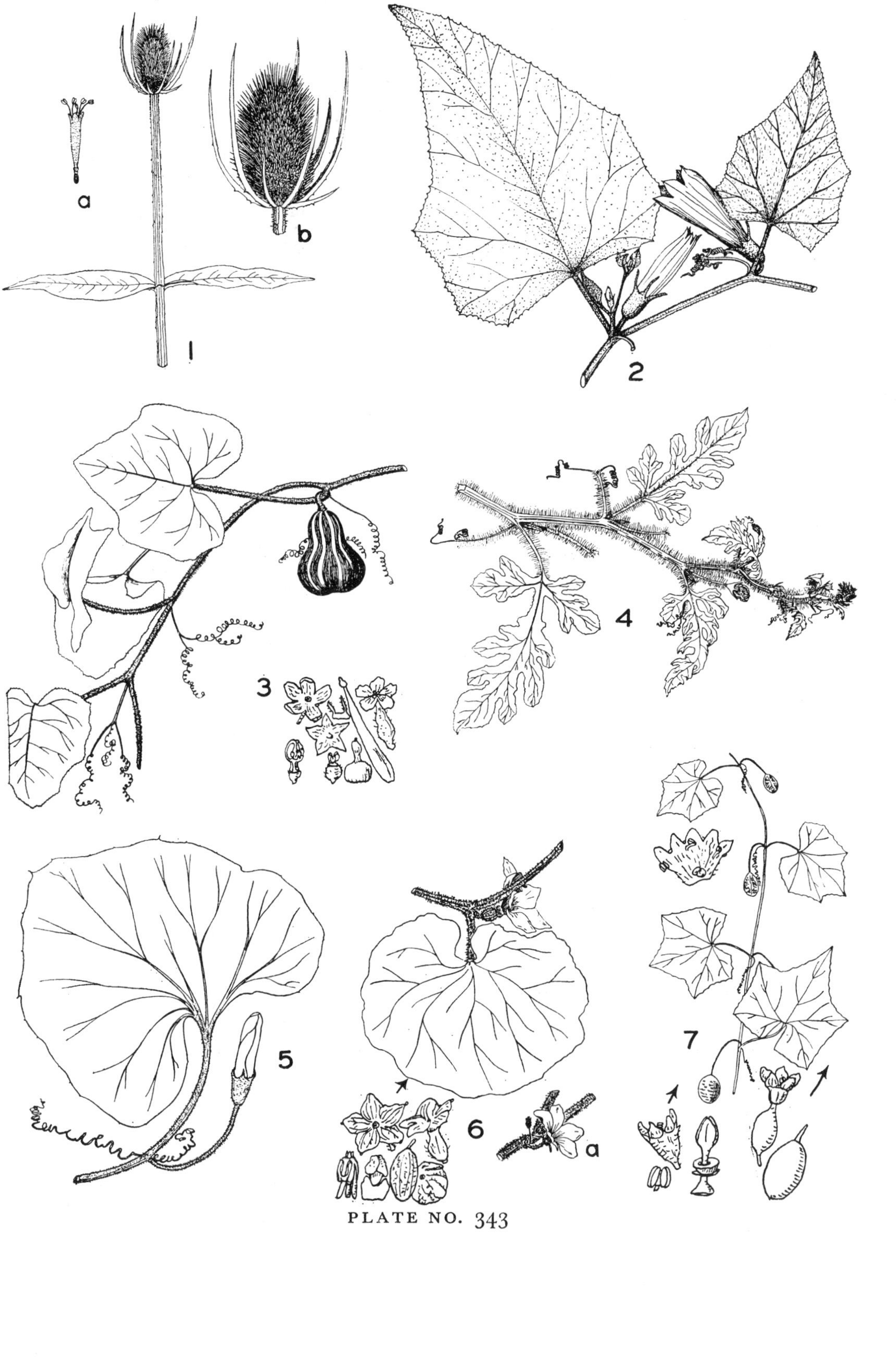


PLATE NO. 343

2091 Lagenaria Siceraria (Bottle Gourd)

2092 Citrullus vulgaris var. vulgaris (Watermelon)

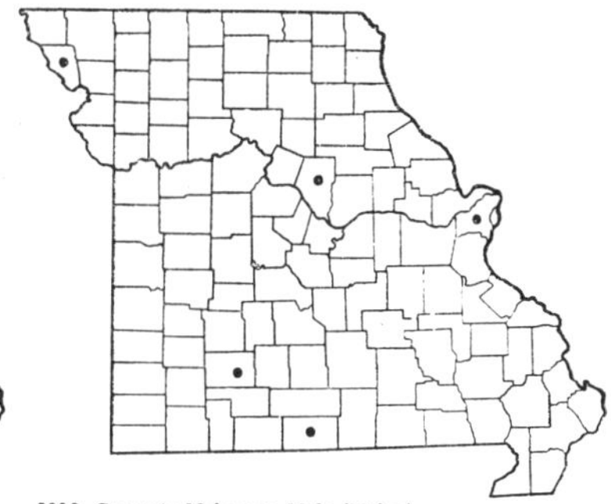

2093 Cucumis Melo var. Melo (Melon)

1. **Cucurbita foetidissima** HBK. Missouri Gourd Map 2089

Also called Wild Pumpkin.

Flowers mid-June–July.

Occurs usually along railroad embankments and roadsides, also in fields and waste ground. Scattered throughout Missouri.

Ranges from Missouri and Nebraska to Texas, Mexico, and California; introduced in Illinois and Indiana.

The fruit is nearly globose and greenish-orange when ripe. The large root sometimes attains a foot in diameter. The fruit contains a *saponin* substance which has the property of foaming in water.

2. **Cucurbita Pepo** L. var. **ovifera** (L.) Alef.
Yellow-flowered Gourd Map 2090

Also known as Pear Gourd.

Flowers August–October.

Occurs in rich woods of valleys and moist low ground along streams, gravel bars, and low woodland, sometimes along railroads. Scattered in southern and central Missouri, where known from St. Louis, Texas, Howell, Douglas, Polk, Barry, Newton, and McDonald counties.

Native of tropical America, but not known in the wild state.

Typical *C. Pepo* var. *Pepo,* the Pumpkin, has larger leaves (15–30 cm. long), and much larger fruits.

Other species of *Cucurbita,* cultivated in Missouri, *C. maxima* Duchesne (Squash) and *C. moschata* Duchesne (Crookneck Squash), have not been found naturalized or established as escapes from cultivation.

## 2. **Lagenaria** Ser. Bottle Gourd

**Lagenaria Siceraria** (Molina) Standl. Bottle Gourd Map 2091

*Lagenaria vulgaris* Ser. [G]

*Lagenaria leucantha* (Duchesne) Rusby

Flowers August–September.

Occurs along railroad tracks. Known only from St. Louis County (St. Louis, on the Mo. Pac. R.R. embankment between Hurck and Marcean streets, September 2, 1956, *Muehlenbach 1055*; St. Louis, along the overpass of Kingshighway, south of the Frisco R.R. tracks, August 31, 1958, *Muehlenbach 1476*).

Native of tropical Asia and Africa; introduced and naturalized in the New World, in the United States from Florida to Texas north to Illinois and Missouri.

The large leaves are somewhat orbicular kidney-shaped and nearly entire with only minute teeth on the margins. The flowers are solitary and showy with white corollas, with a long hairy ovary on the pistillate flower. The tendrils are branched. The fruit varies from 7.5 cm. to 90 cm. long and occurs in many sizes and shapes. These are used for drinking receptacles, utensils, decoration, and bird-houses. When small, the young fruits may be cooked and eaten. The seeds contain a *saponin* substance.

## 3. **Citrullus** Neck

**Citrullus vulgaris** Schrad. var. **vulgaris**
Watermelon Map 2092

*Citrullus vulgaris* Schrad. [Bailey, G]

Flowers May–October.

Occurs in waste ground, along sand and gravel bars and alluvial soils in valleys along streams, road-

2094 Melothria pendula (Creeping Cucumber)

2095 Cayaponia grandifolia

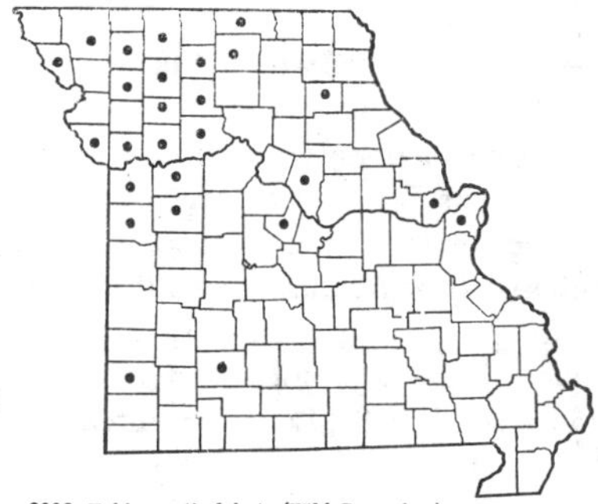
2096 Echinocystis lobata (Wild Cucumber)

sides, and along railroads. Scattered in southern, central, and northwestern Missouri northwest to Holt County.

Native of Africa; commonly planted, and naturalized in the United States.

The citron or Preserving Melon, var. *citroides* Bailey, is a variety of *C. vulgaris.*

## 4. **Cucumis** L. Melon

**Cucumis Melo** L. var. **Melo** Melon, Muskmelon
Map 2093

*Cucumis Melo* L. [Bailey, B]

Flowers May–October.

Occurs in alluvial soils along streams and lakes, waste ground, roadsides, and along railroads. Scattered in the state, where known from St. Louis, Boone, Ozark, Greene, and Holt counties.

Native probably of Asia; commonly planted and naturalized in the United States.

The cantaloupe, var. *cantalupensis* Naud., is a variety of *C. Melo.* Other species of the genus, such as *C. sativus* L. (Cucumber), are commonly cultivated and are occasionally found naturalized, but have not yet been recorded from Missouri as escaped from cultivation.

## 5. **Melothria** L.

**Melothria pendula** L. Creeping Cucumber
Map 2094

Flowers July–September.

Occurs in rich or rocky low woods, at the base of limestone bluffs or wooded slopes, alluvial woodland, and along streams. Southern Missouri, chiefly in southeastern Missouri lowlands north to Scott and Stoddard counties, locally west in the Ozarks along Eleven Point River in Oregon County, thence reappearing in southwestern Missouri in Barry and McDonald counties.

Ranges from Florida to Texas and Mexico, north to Virginia, Indiana, Illinois, Missouri, and Oklahoma.

The seeds are reported to have purgative properties, only half a seed being sufficient to affect an adult per son.

## 6. **Cayaponia** Silva Manso

**Cayaponia grandifolia** (T. & G.) Small
Map 2095

Flowers June–August.

Occurs in rich low alluvial woodland, along bayous and wet depressions. Lowlands of southeastern Missouri in Pemiscot (alluvial woods around Robinson Lake, east of Big Lake, T19N, R13E, $3\frac{1}{2}$ mi. east of Concord, September 18, 1938, *Steyermark 6707*), New Madrid, and Mississippi counties.

Ranges from Florida and Georgia to Mississippi and Louisiana, north to Arkansas and Missouri.

## 7. **Echinocystis** T. & G. Wild Balsam Apple

**Echinocystis lobata** (Michx.) T. & G.
Wild Cucumber Vine Map 2096
Also called Balsam Apple.
Flowers June–October.

Occurs in rich low woods, base of wooded slopes, thickets in alluvial soils of valleys and along streams, sometimes along roadsides, fence rows, and in waste places. Mostly in northwestern and west-central Missouri, south to St. Louis, Moniteau, Johnson, and Cass counties, with local stations in southwestern Missouri in Greene and Jasper counties.

Ranges from New Brunswick to Saskatchewan, south to Florida and Texas.

This vine is an annual and is sometimes grown for arbors and verandas as an ornamental. The white to greenish-white staminate flowers are in long showy racemes.

## 8. **Sicyos** L.

**Sicyos angulatus** L. Bur Cucumber Map 2097
Flowers July–October.

Occurs in low alluvial woodland, river bottom forest, rich ground at the base of bluffs, fields, thickets, and along railroads. Common in southern and central Missouri and along the Mississippi, Missouri, and Grand rivers in eastern and northwestern Missouri; absent from the north-central section.

Ranges from Maine and Quebec to Minnesota, south to Florida, Louisiana, and Texas.

Some persons are reported receiving a dermatitis from handling the fruits of this species. The spines of the fruit sometimes puncture the hands of workmen who are engaged in husking corn in the bottoms where *Sicyos* grows. The vine is occasionally grown as a climber. It forms dense mats. Like *Echinocystis*, it is an annual. It often covers extensive areas in alluvial bottoms of the Mississippi and Missouri rivers, especially in cut-over land and thickets.

# Order **CAMPANULALES**

## Fam. **CAMPANULACEAE** (Bellflower Family)

(*Lobeliaceae* [BB, P & S, Steyerm.])

As here interpreted, this family includes the Lobeliaceae, sometimes treated by authors as a separate family. Other authors divide the family into two subfamilies, *Campanuloideae* and *Lobelioideae*, characterized respectively by (1) regular (actinomorphic) flowers, distinct filaments, and determinate or mixed inflorescences, and (2) irregular (zygomorphic) flowers, united filaments, and indeterminate inflorescences.

a. Corolla strongly irregular, 2-lipped, the tube split nearly to its base on one side; stamens united into a tube surrounding the style; fruit opening by valves at the summit . . . . . . . . . 3. Lobelia
a. Corolla regular with 5 equal lobes; stamens distinct, not united; fruit opening on the sides by outwardly curled elastic valves or by longitudinal slits . . . . . . . . . . . . . . . . . . . . *b*
  b. Corolla 15 mm. or less long, 10–15 mm. broad, wheel-shaped (rotate), rather flat with widely spreading lobes, which are much longer than the short scarcely evident corolla-tube; flowers sessile (without stalks) or nearly sessile in the axils of the leaves; at least the lower flowers cleistogamous (not opening), with only imperfectly developed corollas and stamens; fruit slenderly cylindric or prism-shaped . . . . . . . . . . . . . . . . . . . . 2. Specularia
  b. Corolla 5–30 mm. long, 10–25 mm. broad, usually bell-shaped or broadly funnelform, with usually conspicuous tube and lobes extending forward, the lobes often shorter or only equaling the tube (in *C. americana* the corolla is wheel-shaped with short tube and widely spreading lobes and 20–25 mm. broad); flowers usually on slender short to long pedicels (stalks); all the corollas normally open and expanded, no cleistogamous flowers present; fruit turbinate (top-shaped), obconic, or nearly globose . . . . . . . . . . . . . . . . . . . . 1. Campanula

Plate no. 344. **1.** Cayaponia grandifolia, × 2/5; After Gleason, details from Small, The New York Botanical Garden. **2.** Echinocystis lobata, × 2/5; *a.* Fruit, × 2/5; After Gleason, The New York Botanical Garden. **3.** Sicyos angulatus, × 4/15; After Gleason, The New York Botanical Garden. **4.** Campanula rapunculoides, × 2/5.

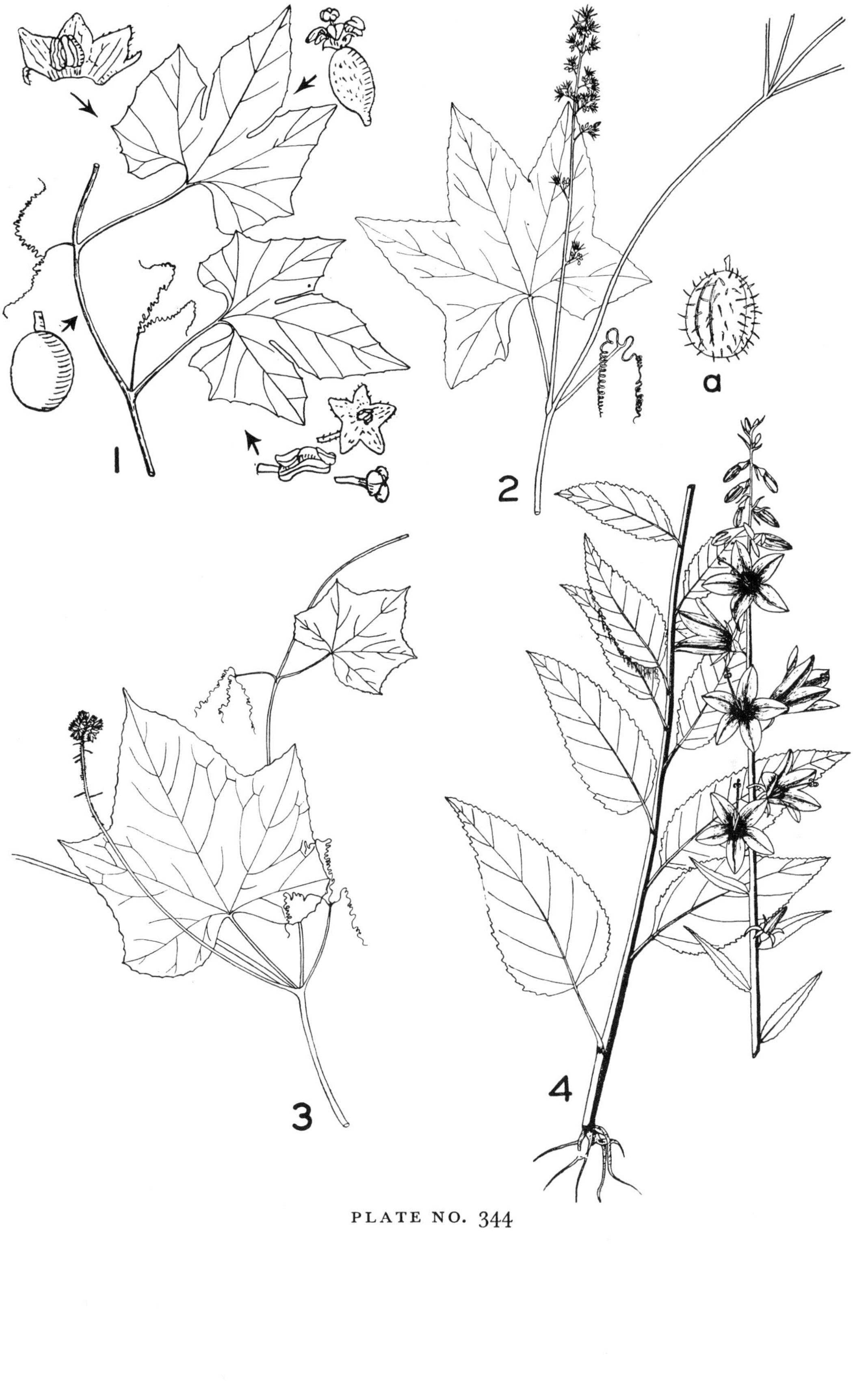


PLATE NO. 344

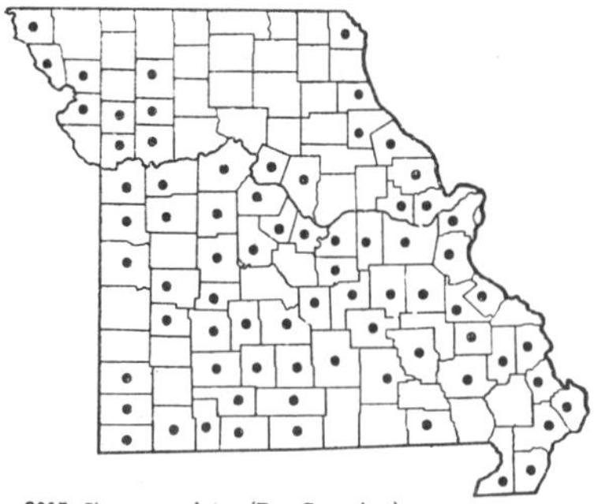

2097 Sicyos angulatus (Bur Cucumber)

2098 Campanula rapunculoides f. rapunculoides (Creeping Bellflower)

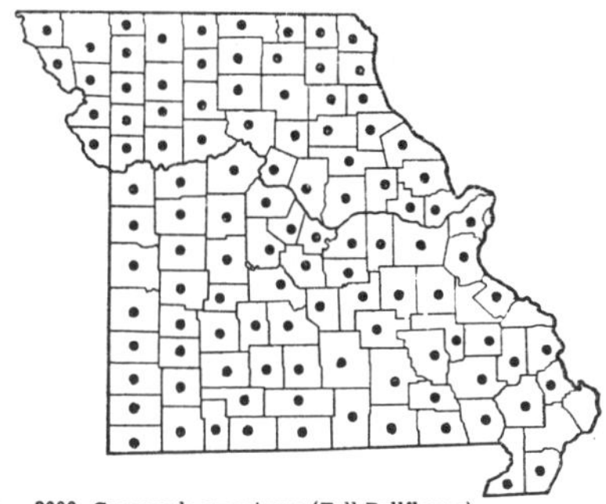

2099 Campanula americana (Tall Bellflower)

## 1. **Campanula** L. Bellflower

a. Main leaves of the stem linear or narrowly lanceolate, mostly 1–8 mm. (rarely 10) broad, either without teeth or inconspicuously toothed; flowers on long stalks 10–50 mm. long or in loosely flowered inflorescences . . . . . . . . . . . . . . . . . . . . . . . . . . *b*

b. Plants of crevices of north-facing limestone bluffs along Jacks' Fork of Current River; leaves and upper half of stems smooth; corolla deep lavender-blue, 13–27 mm. long; calyx-lobes 4–10 mm. long . . . . . . . . . . . . . . . . . . . . . 3. C. ROTUNDIFOLIA

b. Plants of swampy meadows in valleys; edges and midribs of leaves and angles of stems roughened with backward-pointing projections; corolla white, 4–9 mm. long; calyx-lobes 0.7–4 mm. long 4. C. APARINOIDES

a. Main leaves of the stem ovate, oblong-lanceolate or ovate-lanceolate, 20–60 mm. (rarely 15) broad, conspicuously toothed; flowers on short stalks (pedicels) 1–10 mm. long or in an elongated raceme or spike-like raceme with the flowers set close to main axis . . . . . . . . . . . . . . *c*

c. Corolla blue, wheel-shaped (rotate), the lobes widely spreading and much longer than the short scarcely evident tube; most of main leaves of stem tapering and narrowed at base, broadest at middle . . . . . . . . . . . . . . . . . . . . . . 2. C. AMERICANA

c. Corolla blue-purple, bell-shaped, the lobes extending forward and about equaling the definite and conspicuous tube; middle and lower main leaves of stem broadest at base or below middle 1. C. RAPUNCULOIDES

1. **Campanula rapunculoides** L. var. **rapunculoides** Creeping Bellflower Map 2098

Also called Rover Bellflower, False Rampion.

*Campanula rapunculoides* L. [G, BB, P & S, Steyerm.]

Flowers last of May–September.

Often planted and escaping to roadsides, thickets, and waste ground. Known only from central Missouri in Warren and Clay (roadsides, thickets, Liberty, July 1, 1947, *Aubrey B. King 98*, in Wm. Jewell Coll. Herb.) counties.

Native of Europe; naturalized in North America from Newfoundland to North Dakota, south to Maryland, West Virginia, Ohio, Indiana, Illinois, and Missouri.

This species spreads quickly from creeping rhizomes eventually forming patches. It does well in shaded perennial beds. Late in the season fleshy underground slender branches are produced from the rootstocks and can be eaten fresh in salads or cooked as a vegetable, resembling parsnips in flavor.

2. **Campanula americana** L. Tall Bellflower Map 2099

*Campanula americana* var. *illinoensis* (Fresn.) Farw. [G]

Flowers June–October.

Occurs in rich and moist woodland, cut-over woodland, borders of woods, and thickets. Throughout Missouri, doubtless in every county.

Ranges from Ontario and New York to Minnesota and South Dakota, south to Florida, Alabama, Arkansas, and Oklahoma.

---

Plate no. 345. **1.** Campanula americana, × 1; Details from Small, The New York Botanical Garden. **2.** Specularia biflora, × $^{2}/_{5}$; Details from Small, The New York Botanical Garden. **3.** Specularia leptocarpa, × $^{2}/_{5}$; Details from Small, The New York Botanical Garden. **4.** Campanula aparinoides, × $^{2}/_{5}$; After Gleason, The New York Botanical Garden. **5.** Specularia perfoliata, × $^{2}/_{5}$. **6.** Campanula rotundifolia, × $^{2}/_{5}$.

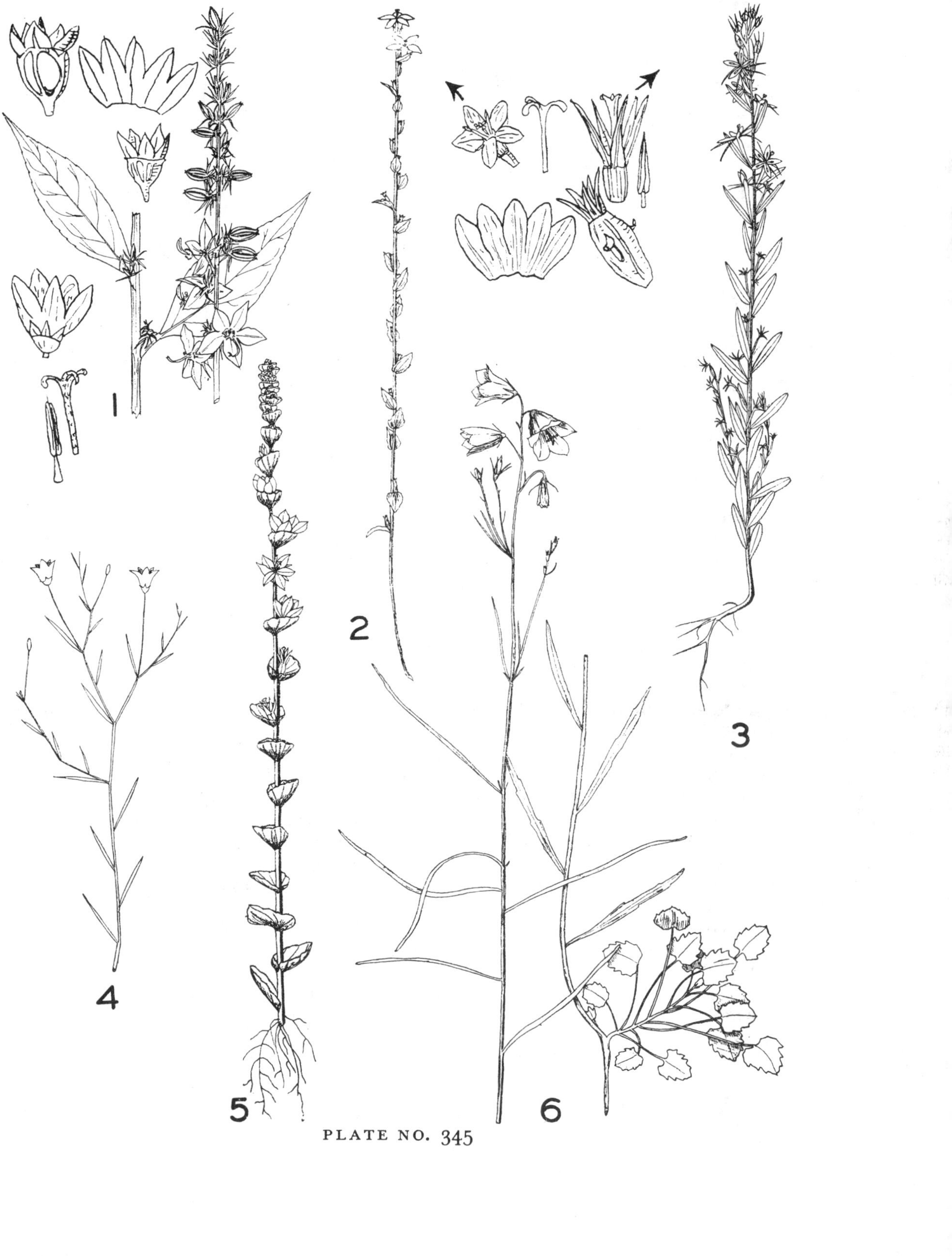


PLATE NO. 345

2100 Campanula rotundifolia var. rotundifolia (Harebell)

2101 Campanula aparinoides var. aparinoides (Marsh Bellflower)

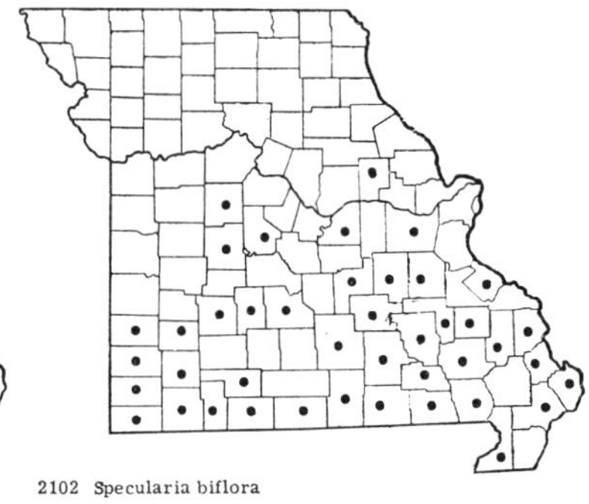

2102 Specularia biflora

This is an annual plant. An effect of red, white, and blue for a woodland summer wildflower garden can be achieved by having this blue-flowered plant in the woods with the red-flowered *Monarda didyma* and the white-flowered *Cimicifuga racemosa* and *Hydrangea arborescens*.

A var. *illinoensis* (Fresn.) Farw. has been distinguished on the basis of broader, ovate lower and middle leaves of the stem, abruptly contracted to the petiole as contrasted with typical *C. americana* in which the lower and middle leaves are long-tapering to the petiole. While the extremes of these two variations appear distinctive, too many intergradations occur in material from Missouri to justify the recognition of the varieties.

The young plants can be cooked and eaten as green vegetables. The leaves are eaten by white-tailed deer in the Ozarks.

3. **Campanula rotundifolia** L. var. **rotundifolia**
Harebell, Bluebell Map 2100
*Campanula rotundifolia* L. [G, BB, P & S]
*Campanula rotundifolia* var. *intercedens* (Witasek) Farw. [Steyerm.]
Flowers last of May–August.

Occurs only on moist shaded, north-facing crevices of limestone bluffs along Jack's Fork of Current River. Known only from the southeastern Ozarks in Shannon County (June 26, 1888, *Bush*; moist limestone ledges along Jack's Fork, Jam-up Bluff, 7 mi. north of Montier, July 17, 1932, *Steyermark 4806, 7011*; same locality July 27, 1899, *Bush 180*; high rock bluffs, July 18, 1891, *Bush*; west of Jam-up Bluff, June 24, 1939, *Steyermark 26885*).

Ranges from boreal America, south to Connecticut, New Jersey, Pennsylvania, West Virginia, Indiana, Illinois, Missouri, Nebraska, Texas, New Mexico, Arizona, and California; also in Eurasia.

This is another of the relic survivors of a northern flora persisting in sheltered, favored localities of the Ozarks following the advances and retreat of one of the Pleistocene Ice sheets which drove species of the more boreal regions southward.

The species is variable and other varieties have been described. A form of *C. rotundifolia* var. *rotundifolia* with white corollas, f. *albiflora* Rand & Redf., is not known as yet in Missouri.

The plant is a good one for rock gardens, but requires a shaded and rather moist, sheltered location for most of Missouri. Plants should be purchased from wildflower nurserymen of eastern and northern states.

4. **Campanula aparinoides** L. var. **aparinoides**
Marsh Bellflower Map 2101
*Campanula aparinoides* L. [G]
Flowers June–August.

Occurs in swampy meadows. Southern Missouri, where known only from Shannon (swampy meadow along Big Creek, T31N, R3W, northwest $\frac{1}{4}$ sect. 8, $2\frac{1}{4}$ mi. south of Melton, $4\frac{1}{4}$ mi. southeast of Bunker, September 25, 1949, *Steyermark 69320*; same locality, July 3, 1949, *Steyermark 68382*) and Newton (along wet rocky bed of spring brook, among grass and other herbs, Haddock Spring, 3 mi. northwest of Wentworth, October 29, 1953, *Palmer 57138*) counties.

Ranges from Maine to Minnesota, south to Georgia, Kentucky, Missouri, Nebraska, and Colorado.

The occurrence of this northern species in the southern Ozarks is another example of boreal species which, driven southward during one of the advances of the Pleistocene ice sheet, found favorable conditions for survival in the calcareous wet meadows, and following the retreat of the glaciers, have persisted at these southern stations separated by hundreds of miles from their nearest more northern stations.

## 2. **Specularia** Fabr. Venus' Looking Glass

(*Triodanis* Raf. [McVaugh])

The most recent study of this group is by Dr. Rogers McVaugh (Wrightia 1. 1945). Some of the characters used in the following treatment are based on his studies in that publication and in the *Flora of Texas* (3, part 5: 338–51. 1951). The arguments for and against the use of the name *Specularia* vs. *Triodanis* have been advanced by Fernald (Rh. 48: 209–14. 1946) and by McVaugh (Rh. 50: 38–49. 1948), and should be referred to by interested readers. The present flora has followed the views of Fernald.

a. Leaves and leaf-like bracts accompanying flowers linear to lanceolate, 5–8 times as long as broad; fruits of the cleistogamous flowers (those not opening) 8–20 mm. long, narrowly cylindrical or subulate, more or less curved outward, opening either by 1 pore or by longitudinal slits at the summit . 5. S. LEPTOCARPA

a. Leaves and leaf-like bracts accompanying flowers ovate or nearly orbicular, as broad as long or broader than long or as much as 2–3 times as long as broad; fruit of the cleistogamous flowers (those not opening) 2–8 mm. (up to 12) long, oblong, ellipsoid, or ovoid, straight, opening by 2 or 3 pores on the sides or at the summit . . . . . . . . . . . . . . . . . . . . . . . . . *b*

b. Pores of the fruit linear, 0.2–0.4 mm. broad, located near middle of fruit; seeds with longitudinal rows of tubercles . . . . . . . . . . . . . . . . . . . . . . 4. S. HOLZINGERI

b. Pores of the fruit oval, elliptic, or rounded, 0.5–1.5 mm. broad, located from near middle to summit of fruit; seeds smooth or with low tubercles scattered over surface. . . . . . . . . *c*

c. Seeds 0.8–1 mm. long, flattened to scarcely convex on both sides (biconvex); pores of fruit located near summit of fruit . . . . . . . . . . . . . . . . . 3. S. LAMPROSPERMA

c. Seeds 0.5–0.65 mm. long, plump and convex on both sides; pores of fruit located either around the middle or near the summit of fruit . . . . . . . . . . . . . . . . . . *d*

d. Pores of fruit located at or very near summit of fruit; leaves and leaf-like bracts not clasping, ovate to oblong, longer than broad, only the midrib or 1 additional pair of nerves evident; open corolla usually 1 and at the summit of the stem, or scattered axillary flowers lower down with open corollas . . . . . . . . . . . . . . . . . . 1. S. BIFLORA

d. Pores of fruit located usually about midway on the fruit; leaves and leaf-like bracts clasping, as broad as or broader than long, often with 1 or 2 pairs of nerves besides the midrib; several open corollas normally located at the upper nodes of the stem . . . 2. S. PERFOLIATA

1. **Specularia biflora** (R. & P.) Fisch. & Mey. Map 2102

*Triodanis biflora* (R. & P.) Greene [McVaugh]

Flowers April–June.

Occurs in rocky open glades, prairies, openings on upland and ridge tops, gravelly ground along streams, fields, pastures, roadsides, and waste ground. Ozark region of southern and central Missouri north to Franklin, Montgomery, Osage, Morgan, Pettis, Dade, and Barton counties.

Ranges from Florida to Texas and Mexico, north to Virginia, Kentucky, Missouri, and Kansas; Oregon to California; South America.

2. **Specularia perfoliata** (L.) A. DC. Map 2103

*Triodanis perfoliata* (L.) Nieuwl. [McVaugh]

Flowers April–August.

Occurs in prairies, fields, dry open woods, alluvial sandy and gravelly places along streams, waste ground, along roadsides and railroads. Throughout Missouri, doubtless in every county.

Ranges from Maine and Quebec to Ontario, Minnesota, South Dakota, Montana, and British Columbia, south to Florida, Texas, California, and Mexico; also West Indies and South America.

Two variations occur in Missouri:

Corolla blue-purple . . . 2a. S. PERFOLIATA f. PERFOLIATA

Corolla white . . . 2b. S. PERFOLIATA f. ALBA

2a. **Specularia perfoliata** f. **perfoliata** Map 2103

*Specularia perfoliata* (L.) A. DC. [G, BB]

This is the common variation found throughout Missouri.

2b. **Specularia perfoliata** f. **alba** (Voigt) Steyerm. Map 2103

*Triodanis* (as Triodanus) *perfoliata* f. *alba* Voigt, Fl. S. Ill. p. 325. 1959.

Known only from New Madrid County, southeastern Missouri (*Rhodes 334*).

Originally found in southern Illinois.

The seeds vary from smooth to variously pebbled or muriculate.

2103 Specularia perfoliata (Venus' Looking Glass)

2104 • Specularia lamprosperma
2104 x Specularia lamprosperma X leptocarpa

2104-A Specularia Holzingeri

3. **Specularia lamprosperma** (McVaugh) Fern. Map 2104

*Triodanis lamprosperma* McVaugh [McVaugh]

Flowers April–August.

Occurs on rocky open glades, sterile upland fields, and openings in dry woods. Mainly in southern Missouri in Iron, Shannon, Barry, and Jasper counties, and locally north in Mercer County, northern Missouri (sterile upland field, above east-facing slopes along Weldon River, T64N, R24W, sect. 9, $2\frac{1}{2}$–$2\frac{3}{4}$ mi. southwest of Mill Grove, June 14, 1952, *Steyermark 73615*).

Ranges from Missouri and Arkansas to Oklahoma and Texas.

An apparent hybrid between *S. lamprosperma* and *S. leptocarpa* is represented by a collection from Newton County, southwestern Missouri (chert glades along Shoal Creek, 4 mi. southwest of Joplin, May 29, 1952, *Palmer 54078*).

4. **Specularia Holzingeri** (McVaugh) Fern. Map 2104A

*Triodanis Holzingeri* McVaugh [McVaugh]

Flowers April–August.

Occurs in prairies, open woods, and roadsides. Known only from central Missouri in Osage and Jackson (bottoms, Courtney, June 22, 1899, *Bush 120*; sands, Courtney, July 18, 1912, *Bush 6846*; Sheffield, June 27, 1895, *Bush 429*) counties.

Ranges from Wyoming to Missouri, south to Arizona and Texas; also in Tennessee.

5. **Specularia leptocarpa** (Nutt.) Gray Map 2105

*Triodanis leptocarpa* (Nutt.) Nieuwl. [McVaugh]

Flowers May–August.

Occurs on rocky open glades, outcrops, bluff escarpments, sterile fields, and rocky or gravelly open ground. Chiefly in western and central Missouri, north to Macon, Mercer, and Caldwell counties, east to Pike, St. Louis, and Washington counties; absent from most of the Ozark and southeastern lowland regions.

Ranges from Minnesota to Montana, south to Missouri, Arkansas, Oklahoma, and Texas; introduced in Indiana.

## 3. **Lobelia** L. Lobelia

Reference should be made to the detailed studies of McVaugh (Rh. 38: 241–63; 276–98; 305–29; 346–62. 1936).

a. Calyx and corolla combined 1.5–4.5 cm. long; corolla-tube more than 6 mm. long, with openings on the side . . . . . *b*

b. Flower (calyx and corolla combined) 3–4.5 cm. long; corolla usually scarlet or vermilion, rarely white or rose-colored; tube of united filaments 19–33 mm. long . . . . . 6. L. CARDINALIS

b. Flower (calyx and corolla combined) 1.5–3.3 cm. long; corolla usually blue or purple, rarely lavender or white; tube of united filaments 6–15 mm. long . . . . . *c*

c. Commonly encountered species; tube of united filaments 12–15 mm. long; flower (calyx and corolla combined) 2.3–3.3 cm. long; calyx-lobes often 5–6 mm. broad, with leaf-like auricles

Plate no. 346. **1.** Lobelia inflata, $\times {}^{2}/_{5}$; After Gleason, The New York Botanical Garden. **2.** Lobelia spicata var. spicata, $\times {}^{2}/_{5}$; *a.* Calyx, $\times 1{}^{3}/_{5}$. **3.** Lobelia spicata var. leptostachys, $\times {}^{2}/_{5}$; *a.* Calyx, $\times 1{}^{3}/_{5}$.

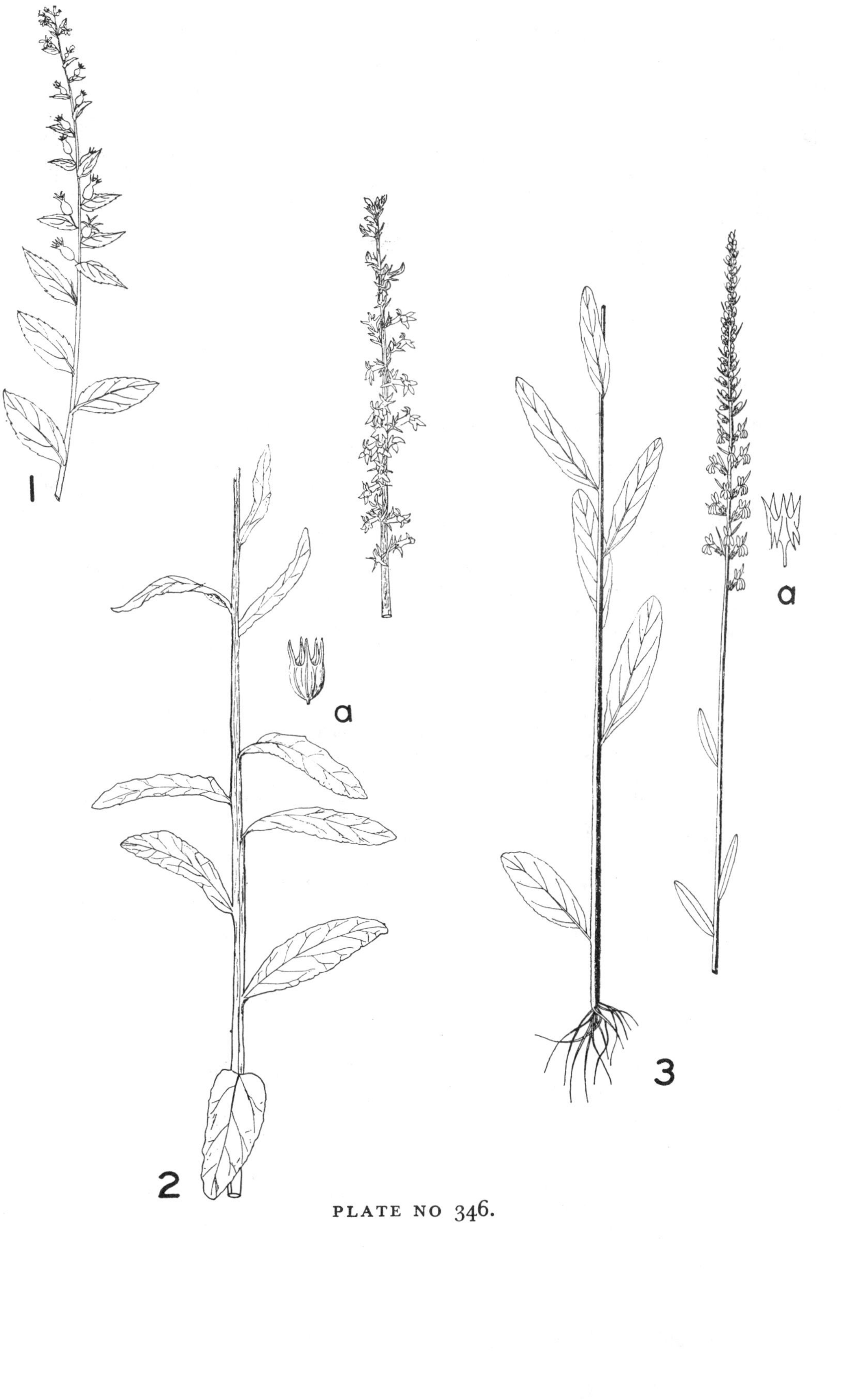


PLATE NO 346.

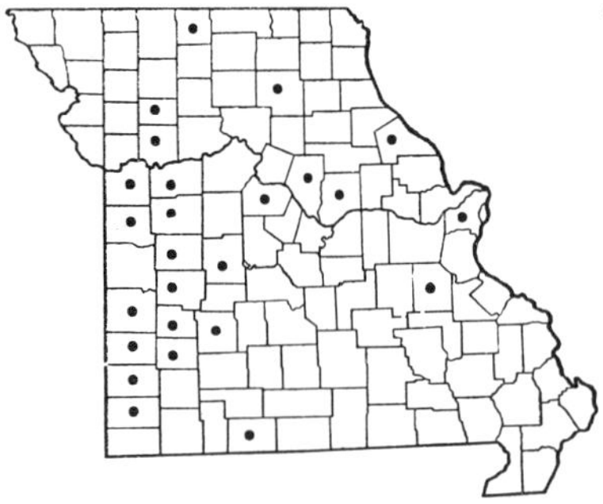

2105 Specularia leptocarpa

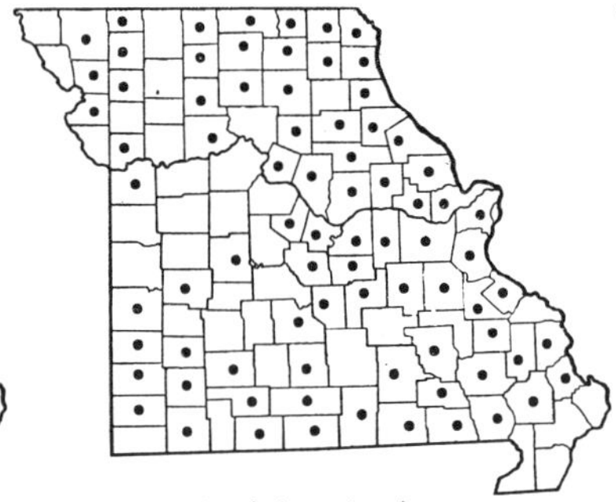

2106 Lobelia inflata (Indian Tobacco)

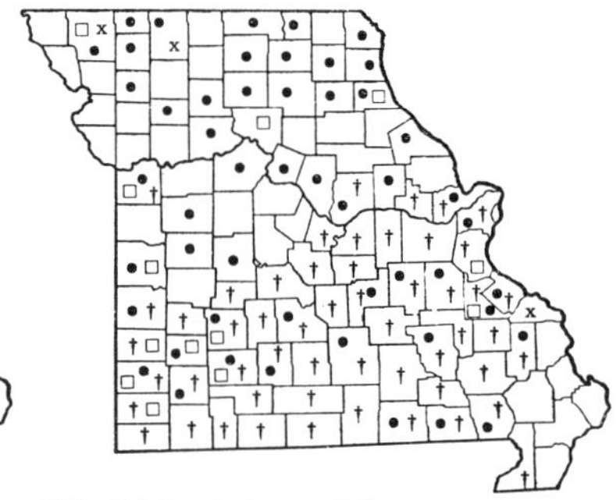

2107 • Lobelia spicata var. spicata
2107 x Lobelia spicata var. campanulata
2107 □ Lobelia spicata var. hirtella
2107 † Lobelia spicata var. leptostachys

2–5 mm. long at base; flower-stalks (pedicels) with a pair of small bracts (bracteoles) at or above the middle . . . . . . . . . . . . . . . . . . . . . 5. L. SIPHILITICA

c. Rarely encountered, known only from southeastern Missouri in Dunklin and Ripley counties; tube of united filaments 6–11.5 mm. long; flower (calyx and corolla combined) 1.5–2.4 cm. long; calyx-lobes often 2–4 mm. broad, usually with shorter (0.5–1 mm.) although prominent auricles at base; flower-stalks (pedicels) with a pair of small bracts (bracteoles) at or near the base . . . . . . . . . . . . . . . . . . . 4. L. PUBERULA var. MINEOLANA

a. Calyx and corolla combined 0.7–1.6 cm. long; corolla-tube less than 6 mm. long, without openings on the side . . . . . . . . . . . . . . . . . . . . . . . . . . . . *d*

d. Long or noticeable hairs present at least on lower part of stem; leaves midway on stem noticeably toothed, ovate-lanceolate; stem usually branched; fruit included within the calyx and becoming inflated; calyx-tube nearly as long as the corolla . . . . . . . . . . . . . 1. L. INFLATA

d. Stems either glabrous (without hairs) or with very short hairs; leaves midway on stem without teeth or with fine or sharp teeth, lanceolate, oblong, or ovate; stem usually simple and unbranched; fruit partly protruding from the calyx, not inflated; calyx-tube shorter than the corolla . . . . . *e*

e. Flowers (calyx and corolla combined) 7–12 mm. long; main leaves of the stem narrowed or tapering to the base; stem densely and generally short-hairy all around at base; calyx-lobes glabrous or somewhat ciliate (fringed with hairs on margins) near tip; auricles (projections hanging from calyx at base of calyx-lobes), if present, short and triangular or hair-like, glabrous (without hairs) . . . . . . . . . . . . . . . . . . . . . 3. L. SPICATA

e. Flowers (calyx and corolla combined) 10–15 mm. long; main leaves of the stem rounded or broadened at the clasping or sessile (stalkless and not narrowed) base; stem glabrous or hairy only on the angles near the base, not hairy all around stem; calyx-lobes usually ciliate (fringed with hairs on margins); auricles (projections hanging from calyx at base of calyx-lobes) up to 3–5 mm. long, drying blue or purplish, usually ciliate (fringed with hairs on margins). . 2. L. APPENDICULATA (see excluded species)

1. **Lobelia inflata** L. Indian Tobacco Map 2106
Also called Lobelia.
Flowers June–October.

Occurs in rocky or open woods, thickets, fallow fields, moist and alluvial ground along streams, sloughs, and ponds. Throughout Missouri, probably in every county.

Ranges from Labrador to Saskatchewan, south to Georgia, Arkansas, and Kansas.

This plant has been used in medicine for the treatment of laryngitis and asthma. It acts as narcotic poison, however, when taken in overdoses. Animals feeding on the plant have been poisoned.

2. **Lobelia appendiculata** A. DC. See Excluded Species.

3. **Lobelia spicata** Lam. Map 2107
Flowers May–August.

Occurs in prairies, glades, open woods, bluff escarpments, fallow fields, wet meadows, and moist soils.

The following intergrading varieties occur in Missouri:

a. Stem, bracts accompanying flowers, and margins of calyx-lobes noticeably hairy; most of leaves of the stem occurring in the lower half. 3c. L. SPICATA var. HIRTELLA

a. Stem glabrous (without hairs) or hairy; margins of calyx-lobes usually glabrous, rarely with hairs (ciliate) . . . . . . . . . *b*
  b. Auricles (projections hanging from calyx at base of calyx-lobes) long and slender, often as long as the calyx-tube, 1–5 mm. long . 3d. L. SPICATA var. LEPTOSTACHYS
  b. Auricles (projections hanging from calyx at base of calyx-lobes) very short or none, 1 mm. or less long . . . . . . . . *c*
    c. Anthers blue; corolla pale blue; calyx-tube rather flat at flowering time . . 3a. L. SPICATA var. SPICATA
    c. Anthers white; corolla dark purplish-blue; calyx-tube rather round or shortly campanulate (bell-shaped) at flowering time . . 3b. L. SPICATA var. CAMPANULATA

3a. **Lobelia spicata** var. **spicata** Map 2107

*Lobelia spicata* Lam. [G, P & S, Steyerm.]

*Lobelia spicata* var. *parviflora* A. Gray [McVaugh]

*Lobelia spicata* var. *originalis* [McVaugh]

Throughout Missouri, the commonest variety in the state.

Ranges from New Brunswick to Minnesota and North Dakota, south to Georgia and Arkansas.

3b. **Lobelia spicata** var. **campanulata** McVaugh Map 2107

Known from Perry, Harrison, and Nodaway counties.

Ranges from Maine to Pennsylvania to Indiana and Minnesota, south to Delaware, West Virginia, Illinois, and Missouri.

A collection from Gentry County (open swales between C. B. & Q. R.R. and highway 169, T64N, R32W, sect. 7, 2 mi. north of Gentry, June, 1957, *Steyermark 85238*) has the whitish anthers of var. *campanulata*, but the corolla is whitish and not dark purplish-blue as is characteristic of that variety.

3c. **Lobelia spicata** var. **hirtella** Gray Map 2107

Scattered throughout the state.

Ranges from Quebec to Alberta, south to New York, Michigan, Indiana, Illinois, Missouri, and Kansas.

Numerous specimens occur which are intermediate between this variety and var. *spicata*. A collection from Nodaway County (*Palmer 25434*) is an example of this transitional type, the stem being hairy especially in the lower half.

3d. **Lobelia spicata** var. **leptostachys** (A. DC.) Mackenz. & Bush Map 2107

*Lobelia leptostachys* A. DC. [P & S]

Mainly in the Ozark region of southern and central Missouri, north to St. Charles, Warren, Gasconade, Callaway, Cole, Camden, Hickory, Cedar, and Vernon counties, and locally in Jackson County.

Ranges from West Virginia to Illinois and Kansas, south to Georgia, Alabama, Mississippi, and Arkansas.

A number of specimens show an intermediate character between this variety and var. *spicata*. A collection from Jackson County (Waldo Park, June 10, 1896, *Bush 479*) has been identified by McVaugh as *L. spicata* var. *leptostachys*, in part, the specimen, on the extreme left of the sheet being intermediate between var. *spicata* and var. *leptostachys*. Another collection from Jackson County (*Bush 460* from Little Blue, July 22, 1896) is labeled var. *leptostachys* but has auricles only 1 mm. long and racemes only 7–9 cm. long. It is referred in the present flora to var. *spicata*.

4. **Lobelia puberula** Michx. var. **mineolana** E. Wimm. Map 2108

*Lobelia puberula* [in part P & S], not Michx.

Flowers August–October.

Occurs in moist sandy open ground and low woodland. Known only from southeastern Missouri in Dunklin County (Campbell, September 14, 1893, *Bush*).

Ranges from Alabama to Texas, north to Missouri, Arkansas, and Oklahoma.

5. **Lobelia siphilitica** L. Blue Cardinal Flower Map 2109

Also called Great Lobelia, Blue Lobelia.

Flowers August–October.

Occurs in wet meadows, prairie swales, along spring branches and seepage, washes, streams, ponds, sloughs, ditches, wet ledges of bluffs, low places in pastures, low moist woodland, and roadside depressions.

Throughout Missouri, doubtless in every county.

The following variations are encountered in the state:

a. Corolla white . . . 5c. L. SIPHILITICA var. SIPHILITICA f. ALBIFLORA
a. Corolla blue or purple . . . . . . . *b*
  b. Corolla deep lavender; rarely encountered. 5b. L. SIPHILITICA var. SIPHILITICA f. PURPUREA
  b. Corolla deep blue or purplish-blue; commonly encountered . . . . . . . *c*
    c. Upper surface of leaves more or less hairy, mostly 2–6 cm. broad; stem more or less hairy on the angles, sometimes glabrous (without hairs); calyx quite noticeably hairy; flowers numerous, usually more than 20 . 5a. L. SIPHILITICA var. SIPHILITICA f. SIPHILITICA

2108 Lobelia puberula var. mineolana

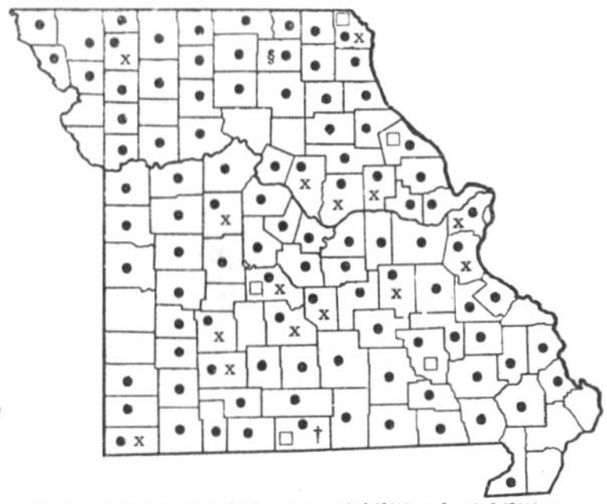

2109 • Lobelia siphilitica var. siphilitica f. siphilitica
2109 § Lobelia siphilitica var. siphilitica f. purpurea
2109 □ Lobelia siphilitica var. siphilitica f. albiflora
2109 x Lobelia siphilitica var. ludoviciana
2109 † Lobelia siphilitica var. hybrida (L. Cardinalis X siphilitica)

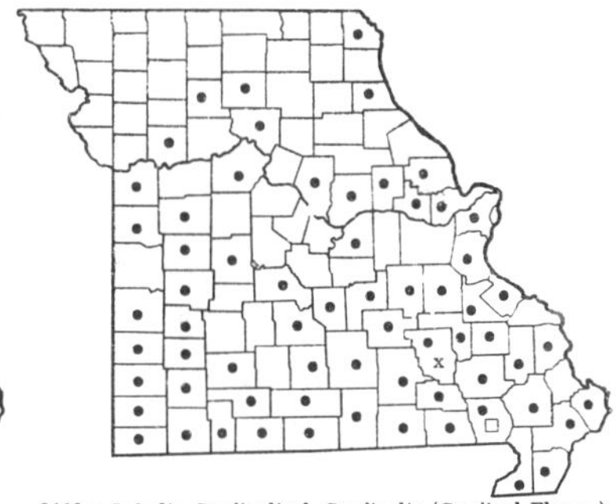

2110 • Lobelia Cardinalis f. Cardinalis (Cardinal Flower)
2110 x Lobelia Cardinalis f. rosea
2110 □ Lobelia Cardinalis f. alba

c. Upper surface and rest of leaves glabrous (without hairs) or nearly so, mostly 1.5 cm. or less broad; stem mostly glabrous (without hairs); calyx glabrous (without hairs) or sparsely hairy; flowers fewer, often less than 20. . . . 5d. L. SIPHILITICA var. LUDOVICIANA

5a. **Lobelia siphilitica** var. **siphilitica** f. **siphilitica** Map 2109

*Lobelia siphilitica* var. *siphilitica* [BB]

*Lobelia siphilitica* L. [G, P & S]

This is the commonest variation throughout Missouri.

Ranges from Maine to Minnesota and South Dakota, south to Virginia, North Carolina, Alabama, Mississippi, Missouri, and Kansas.

5b. **Lobelia siphilitica** var. **siphilitica** f. **purpurea** Palmer & Steyerm. Map 2109

*Lobelia siphilitica* f. *purpurea* Palmer & Steyerm.

Known only from Adair County, northeastern Missouri (base of shaley limestone banks along Old Chariton River, T62N, R16W, sect. 34, 2½ mi. south of Youngstown, September 19, 1955, *Steyermark 79674*, holotype in Chi. Nat. Hist. Mus. Herb.).

5c. **Lobelia siphilitica** var. **siphilitica** f. **albiflora** Britt. Map 2109

*Lobelia siphilitica f. albiflora* Britt. [G]

Scattered in Missouri, where known from Clark, Pike (Davidson's woods pasture, September 20, 1947, *Etter 283*), Camden (base of north-facing limestone bluffs, in sedgy moist places along Fiery Fork Creek, sect. 31, 2½ mi. southwest of Barnumtown, September 22, 1938, *Steyermark 6950*), and Ozark (below confluence of spring branch of Bratten Spring with North Fork, T22N, R15W, sect. 27 and 34, 3–3½ mi. southeast of Isabella, September 25, 1949, *Steyermark 69363*) counties.

5d. **Lobelia siphilitica** var. **ludoviciana** A. DC. Map 2109

Scattered in Missouri.

Ranges from Wisconsin to Manitoba, south to Louisiana and Texas.

Dr. McVaugh (Rh. 38: 281–83. 1936; *Fl. Tex.* 3[5]: 362. 1951) regards *L. siphilitica* as comprising two well-marked varieties, the predominantly eastern one, var. *siphilitica* and the western var. *ludoviciana*. In Missouri, however, the two are not well-marked, most of the material falling into typical var. *siphilitica*. A few specimens, such as *Palmer 4228* from McDonald County, *Steyermark 21068* from Crawford County, *Eggert*, September 9, 1896, from Jefferson County, and *Eggert*, Creve Cœur Lake, from St. Louis County, have more characteristics of the var. *ludoviciana*, but the two varieties are not sharply distinct in Missouri, and are being retained in deference to Dr. McVaugh's more detailed studies of extremes throughout the range of the species.

This Lobelia exhibits poisonous properties similar to those of *L. inflata*, already indicated. It was formerly used in medicine for its emetic effects and to relieve spasmodic affections of the air passages in cases of laryngitis and asthma, but is now not employed for these purposes. The leaves are eaten by white-tailed deer in the Ozarks.

Plate no. 347. **1.** Lobelia puberula, × ²/₅; After Gleason, The New York Botanical Garden. **2.** Lobelia siphilitica, × ²/₅. **3.** Lobelia Cardinalis, × ²/₅. All details from Small, The New York Botanical Garden.

PLATE NO. 347

**Lobelia × siphilitica** var. **hybrida** Hook.

*Lobelia siphilitica × L. Cardinalis*

This hybrid was reported by the author (Rh. 54: 259. 1952) from Ozark County, southwestern Missouri (along Little North Fork of White River between Burse Ford and Nave Ford, on east side of river, T21N, R15W, sect. 9, 1 mi. southwest of Pontiac, September 26, 1949, *Cora Steyermark*).

The plant occurred with both parent species. The corolla had a deep rose-lavender color, with a shape like that of *L. Cardinalis*, but the leaves and pubescence of the calyx resembled the type found in *L. siphilitica*. The area has since been impounded by the Bull Shoals Dam and all traces of the plant are lost. Thus far, no other record of the hybrid is known from the state.

6. **Lobelia Cardinalis** L. Cardinal Flower Map 2110

Flowers July–October.

Occurs along muddy or gravelly borders of streams, sloughs, and spring branches, low wet woods, wet places in swampy meadows, borders of prairie streams and swales. Common in southern and central Missouri, and locally north in Clark, Marion, Chariton, Linn, Livingston, and Ray counties.

Ranges from Florida to Texas, north to New Brunswick, Quebec, Ontario, Michigan, Wisconsin, and Minnesota.

The following variations occur in Missouri:

a. Corolla white . . 6c. L. CARDINALIS f. ALBA
a. Corolla red, vermilion, or rose-colored . . *b*
 b. Common type found; corolla deep red or vermilion . . 6a. L. CARDINALIS f. CARDINALIS
 b. Rarely encountered; corolla pink or rose-colored . . . 6b. L. CARDINALIS f. ROSEA

6a. **Lobelia Cardinalis** f. **Cardinalis** Map 2110

*Lobelia Cardinalis* L. [G, BB, P & S]

This is the commonest variation encountered.

6b. **Lobelia Cardinalis** f. **rosea** St. John Map 2110

Known only from Reynolds County, southeastern Ozark region (wet calcareous meadow along Parker Branch of West Fork of Black River, T33N, R3W, west part of sect. 15, ¼ mi. northwest of Marcoot, 5 mi. northwest of Greeley, September 24, 1951, *Steyermark 72729*, holotype in Chi. Nat. Hist. Mus. Herb.).

In this form the Missouri specimen had the corolla-tube pink to rose-colored, the corolla-lobes pink on the outside, white on the inside, and the staminal tube white.

6c. **Lobelia Cardinalis** f. **alba** (Eat.) St. John Map 2110

Known only from Butler County, southeastern Missouri (exsiccated low meadow in valley of Mill Creek, T26N, R7E, sect. 9, 1¾ mi. southwest of Wappapello, August 29, 1938, *Steyermark 6254*, in Chi. Nat. Hist. Mus. Herb.). This locality is now destroyed by the Wappapello Dam impoundment.

The Cardinal flower is one of the showiest of all wild flowers. Fortunately, it is still fairly abundant throughout the southern and central sections of the state, and is common along all the Ozark streams. It can be easily raised from seed, usually throwing up a basal rosette of leaves for the first year and blooming the second year and thereafter. It does best when planted around wet spots bordering pools and ponds, but can be grown in perennial beds in drier soil if watered regularly. It transplants quite readily.

*Excluded Species*

**Lobelia appendiculata** A. DC.

This species was recorded for Missouri in the eighth edition of *Gray's Manual* (p. 1356) and doubtfully in Gleason's *New Illustrated Flora* (3: 322. 1952). The basis for the report is a specimen from Jasper County (Asbury, *Palmer 34664*). McVaugh indicates that this specimen is hardly typical (Rh. 38: 329. 1936) and considers it as doubtfully representing the species for Missouri. It is excluded for Missouri in McVaugh's later works (N. Am. Fl. 32A, part 1: 69, 1942; Fl. Tex. $3^5$: 357. 1951). An examination of the specimen does not show the conspicuous lanceolate auricles 1–3 mm. long found in other specimens of *L. appendiculata* and the *Palmer* specimen is referred in the present treatment to *L. spicata* var. *leptostachys*.

**Lobelia Cardinalis** subsp. **graminea** (Lam.) McVaugh

*Lobelia splendens* Willd.

*Lobelia splendens* was reported for western Missouri in both the eighth edition of *Gray's Manual* (p. 1355) and in Gleason's *New Ill. Fl.* 3: 318. 1952 on the basis of a specimen reported by Dr. McVaugh (Rh. 38: 277. 1936) from Jackson County, west-central Missouri (Courtney, September 6, 1898, *Bush 294*). This specimen and another from Jackson County (September 5, 1888, *Bush*) were originally determined by McVaugh as *L. splendens*, but later studies by Dr. McVaugh have shown that these specimens are not typical subsp. *graminea* (*L. splendens*) and should be referred instead to typical *L. Cardinalis*. Dr. McVaugh has since stated to the present author that no true subsp. *graminea* exists in Missouri. This subspecies is characterized by having linear to linear-lanceolate, entire to coarsely toothed leaf-blades 6–12 times as long as broad, an anther-tube 3.5–4.5 or 5 mm. long, and a filament-tube 19–25 (rarely 30) mm. long, as contrasted with typical subsp. *Cardinalis* in which the leaf-blades are 3–5 times as long as broad, coarsely and irregularly toothed, a filament-tube 24–33 mm. long, and an anther-tube 4–5.5 mm. long.

**Platycodon grandiflorum** DC. Balloon Flower

This commonly cultivated perennial plant is found in gardens throughout Missouri. A specimen from Clay County is found in the William Jewell College Herbarium with the data 'escaped, William Jewell College, June 30, 1947, *C. L. Broussard 111*.' It is assumed that the specimen is an accidental waif found growing on the campus from a nearby garden plot and should not be considered a naturalized specimen. Until more definite reports appear on the naturalizing and establishment of this species away from cultivation, the species is excluded from the present flora.

## Fam. **COMPOSITAE** (Composite Family)

This family can be divided into various tribes, totaling ten in the present flora. These tribes have been grouped by some botanists into three separate families, and the tribes themselves have sometimes been treated as distinct families. The present treatment of maintaining only one family is in accord with the present consensus of agreement among botanists based upon evidence from the most recent morphological studies.

The structure and arrangement of the flowers in the family are unique among flowering plants. The flower-head, which appears at first glance to consist of only one flower, is, when broken apart, seen to consist of usually 2 or more separate flowers (only 1 in some genera [*Ambrosia* pistillate flowers]). These flowers are either all the same type, or the ones in the center (disk florets) are of a different type from those on the outside (ray florets or ligules). The corollas may be tubular or cylindrical and equally 4- or 5-lobed at the summit or flattened and ribbon-like (ligulate). Often two series or subfamilies are recognized at this point: (1) *Tubuliflorae*, with the flower-heads composed of the ribbon-like ray (ligulate) flowers on the outer part of the head surrounding the tubular 5-toothed disk flowers, which are usually perfect (with stamens and pistil present), or with the flower-heads composed only of tubular 4- or 5-toothed usually perfect disk flowers (exceptions occur in *Ambrosia Iva*, *Franseria*, *Xanthium* and other genera); plants without milky juice, and (2) *Liguliflorae*, with the flower-heads composed entirely of ribbon-like or flattened perfect corollas, and with milky juice usually present in the plants. Each individual flower-head is surrounded at its base by a series of green or other colored *bracts* or *phyllaries* which comprise the involucre. In some species, each flower is perfect with both stamens and pistil present; in others only stamens or pistil are present. The ovary, situated below the corolla-tube, and, therefore, inferior, becomes, after fertilization, the dry seed-like fruit (achene). At the summit of this ovary or fruit are usually present 1, 2, or more scales, bristles, or hairs, known as the *pappus*. Sometimes no pappus is developed or is represented by a minute or short crown. The stamens, when present, consist of 5 anthers united by their lateral (side) margins into a ring or tube, usually surrounding the single thread-like style which forks into two at its summit above the anther-tube. In *Ambrosia* and a few other genera, the anthers are scarcely united. The *receptacle*, on which the flowers are situated, may support bracts or scales, known as *pales*. These pales are sometimes bristly or spiny-tipped, sometimes absent entirely.

The following key to genera utilizes both nontechnical and technical characters. Following this is another key to tribes only for the benefit of those who wish to learn the tribal groups and their component genera.

### *Key to Genera* (not based on tribes)

a. Stems climbing or twining; leaves heart-shaped, opposite; plants of the lowlands of southeastern Missouri . . . . . . . . . . . . . . . . . . . . . . . . . . 5. MIKANIA
a. Stems not climbing or twining . . . . . . . . . . . . . . . . . . . . . *b*
  b. Spines or prickles present on some part of stem or leaves . . . . . . . . . . . . . *c*
    c. Flowers yellow, all flattened, strap-shaped, or ribbon-like; plants with milky juice . . . . . *d*
      d. Achenes without a beak, merely narrowed at summit, 2.5–3.5 mm. long, heads with about 85–250 flowers . . . . . . . . . . . . . . . . . . . . 78. SONCHUS
      d. Achenes with a prominent beak, 4.5–10 mm. long, including the beak; heads with 8–56 flowers . . . . . . . . . . . . . . . . . . . . . . 79. LACTUCA
    c. Flowers purple, rose-colored, or white, all tubular with 5 equal corolla-lobes at summit, not flattened, strap-shaped nor ribbon-like; plants lacking milky juice . . . . . . . . . . *e*
      e. Bristles of the pappus plumose (feather-like with long fine hairs arising from the main axis); commonly encountered . . . . . . . . . . . . . . . . . . . 67. CIRSIUM
      e. Bristles of the pappus not plumose, merely barbellate (with fine short projections arising from the main axis); rarely encountered . . . . . . . . . . . . . . . . . *f*
        f. Stem with broad wings 5–20 mm. broad extending up and down between the leaves; no pales arising from receptacle . . . . . . . . . . . . . . . 68. ONOPORDUM
        f. Stem with narrow wings 1–3 mm. broad extending up and down stem; bristly pales present on receptacle . . . . . . . . . . . . . . . . . . 66. CARDUUS
  b. Spines or prickles absent on stem or leaves . . . . . . . . . . . . . . . . *g*
    g. Spines, prickles, or hooked or bristly burs present on some or all of the bracts of the involucre . . *h*
      h. Some or all of the bracts of the involucre with hooked tips or enclosed by hooked bristles, bur-like . . . . . . . . . . . . . . . . . . . . . . . . . *i*

i. Flowers greenish; no pappus present; staminate (male) and pistillate (female) flowers in separate heads . . . . . 28. XANTHIUM

i. Flowers purplish; pappus present, composed of short, scale-like bristles; flowers all perfect (stamens and pistils in each flower) . . . . . 65. ARCTIUM

h. Some or all of the bracts of the involucre with straight spines or prickles at tip, but not hooked or bur-like . . . . . *j*

j. Pistillate (female) or fruiting involucre 3–10 mm. high; pappus absent from all the flowers; flowers greenish or yellow-green, the staminate (male) and pistillate (female) in separate heads, all with very short corollas, the staminate (male) heads arranged in an elongated spike-like, raceme-like, or panicle-like inflorescence, and with 1 to few pistillate (female) heads situated at the base of the staminate inflorescences; bracts of staminate involucre united . . *k*

k. Commonly encountered weeds; pistillate (female) involucre with but 1 row of spines or pointed projections near the summit; fruiting head 1-seeded . . . . . 26. AMBROSIA

k. Rarely found, usually along railroads; pistillate (female) involucre with more than 1 row of spines; fruiting head 2–4-seeded . . . . . 27. FRANSERIA

j. Fruiting involucre 8–35 mm. high; pappus present usually on at least some of the flowers, rarely completely absent in some species of *Centaurea*; flowers purple, rose, white, yellow, or reddish, all perfect (with stamens and pistil in each flower), and with elongated or conspicuous corollas; bracts of involucre not united . . . . . *l*

l. Flowers yellow, yellowish-white, or sometimes red; involucre 8–15 mm. high; bristles or scales of the pappus short, 1–5 mm. long . . . . . 69. CENTAUREA

l. Flowers purple, rose-red, or white; involucre 15–35 mm. high; bristles of the pappus prominent, 20–30 or more mm. long, plumose (feather-like with long fine hairs arising from the main axis) . . . . . 67. CIRSIUM

g. Spines, prickles, and hooked or bristly burs absent on the bracts of the involucre . . . . . *m*

m. Leaves present only at the very base of the plant . . . . . *n*

n. Leaves heart-shaped at base, very large, 1–3 dm. broad, on petioles 1–5 dm. long; plants without milky juice . . . . . 30. SILPHIUM

n. Without the above combination; leaves and petioles smaller; plants with milky juice . . . . . *o*

o. Leaves entire, without teeth or lobes; no underground tuber present . . . . 81. AGOSERIS

o. Leaves toothed or deeply cut or lobed, or if entire (smooth-edged), then the plant has an underground ball-like tuber . . . . . *p*

p. Achenes not tapering to a beak, their surface ribbed but not roughened, mature achenes 1–2 mm. long; pappus consisting of both scales and bristles; leaves pale, bluish- or silvery-green . . . . . 73. KRIGIA

p. Achenes tapering to a long beak, some part of the surface covered with small spines or roughened, mature achenes 4–6 mm. long (excluding beak); pappus consisting only of fine, hair-like bristles; leaves dark or grass green . . . . . *q*

q. Common weed; pappus of simple hair-like bristles, not plumose (feather-like with fine long hairs arising from the main axis); flowering stems usually 0.2–2 dm. (up to 5) tall; leaves mainly glabrous (without hairs) or sparsely hairy on lower surface and midrib; flower-heads usually only 1 or 2 to a plant (occasionally up to 6); outermost bracts usually spreading to reflexed or curved down in flower . . . . . 77. TARAXACUM

q. Rarely encountered; some of the pappus consisting of plumose bristles (feather-like with fine long hairs arising from the main axis); flowering stems 2–4 dm. (up to 6) tall; leaves usually conspicuously hairy; flower-heads 3–21 to a plant; all the bracts of the involucre erect or ascending during the flowering period . . . . 74. HYPOCHAERIS

m. Leaves present on the stem, also frequently at the base of the plant . . . . . *r*

r. Leaves opposite in pairs or in whorls (circles of 3 or more) on the stem (plants with leaves opposite in the lower portion of stem and alternate in upper portion may be looked for here) . . . . . *s*

s. All the flowers bluish-lavender, lilac, or pink . . . . . *t*

t. Pappus of hair-like bristles . . . . . 3. EUPATORIUM

t. Pappus none . . . . . 4. AGERATUM

s. Flowers of other colors . . . . . *u*

u. Flowers green or greenish; flower-heads very small, arranged one above the other in elongated spike-like, raceme-like, or panicle-like inflorescences; no perfect flowers (with pistil and stamens in same flower) present, the flowers either staminate (stamens only present) or pistillate (pistil only present); pistillate flowers very much reduced, without corolla or with only a portion of a corolla present; staminate flowers with the anthers only partly united; pappus none; ray flowers absent . . . . . . . . . . . . . . *v*

v. All the flower-heads the same, the 3–5 pistillate (female) flowers situated along the outer margin of the head surrounding the more numerous central sterile staminate (male) flowers; no projections or tubercles appearing on any of the involucres . . . . . . . . . . . . . . . 25. IVA

v. Flower-heads not the same, separated into numerous staminate (male) heads arranged in an elongated spike-like, raceme-like, or panicle-like inflorescence, and 1 to few pistillate (female) heads situated in the axils of leaves or bracts at the base of the staminate inflorescences; projections or tubercles appearing on some part of the involucre of the pistillate flower-head . 26. AMBROSIA

u. Without the above combination of characters; flowers white, cream-colored, yellow, orange, purple, rose, black, brown, or combinations of these; flowers usually all perfect (with stamens and pistil in each flower), but if some flowers are sterile (not developing fruit), then at least with regularly developed corollas and the anthers united their full length . . . . . . . . . . . . . . . . . *w*

w. No ray-flowers present at flowering time . . . . . . . . . . . . . . . . . . . . *x*

x. Strong-smelling plant with glandular dots on bracts of involucre . . . . . . . 54. DYSSODIA

x. Plant not strong-smelling or without glandular dots on bracts of involucre. . . . . . . . *y*

y. Bracts of involucre only 3 to a head; pappus absent . . . . . . . . . . 49. FLAVERIA

y. Bracts of involucre 4 or more to a head; pappus present or absent . . . . . . . . . *z*

z. Pappus absent; outermost flowers of the head pistillate and fertile (developing into fruit), the other flowers of the same head (disk flowers) sterile (not developing fruit) with well-developed anthers but undivided styles, their ovaries much smaller than those of the fertile outermost flowers . . . . . . . . . . . . . . . . 29. POLYMNIA

z. Pappus present; all the flowers of a head fertile (capable of developing fruit) with functional pistil . . . . . . . . . . . . . . . . . . . . . . . . . . *1*

1. Flowers white, creamy or greenish-white; receptacle lacking bristle- or chaff-like pales; pappus composed of hair-like or feathery bristles . . . . . . . . . . . *2*

2. Achenes 3–5-angled . . . . . . . . . . . . . . . . . . . . 3. EUPATORIUM

2. Achenes usually 10-ribbed . . . . . . . . . . . . . . . . . . . . . *3*

3. Bristles of pappus plumose (feather-like long fine hairs arising from the main axis); bracts of involucre few and loosely umbricated . . . . . . . 7. KUHNIA

3. Bristles of pappus not plumose; bracts of involucre graduated (imbricated) in several series . . . . . . . . . . . . . . . . . . . 6. BRICKELLIA

1. Flowers yellow, orange, brownish, or combinations of these colors; receptacles bearing flat or concave narrow chaffy pales; pappus composed of 1–8 barbed awns . . . . . *4*

4. Inner series of the green or brown bracts of the involucre united below for at least a third of their length; leaf-divisions thread-like, 0.2–1.5 mm. broad; achenes terete (rounded in cross-section) . . . . . . . . . . . 43. THELESPERMA

4. Inner series of the green or brown bracts of the involucre separate or nearly so to the very base; leaves entire (without teeth), toothed, or if deeply cut or divided, the leaf-divisions 2–35 or more mm. broad; achenes usually flattened or angled, but sometimes terete (in *B. bipinnata*) . . . . . . . . . . . . 45. BIDENS

w. Ray flowers present (although often small, weakly developed, or inconspicuous) at flowering time . . *5*

5. Flower-heads with white ray flowers and yellow, brownish-yellow, or brownish-green centers of disk flowers . . . . . . . . . . . . . . . . . . . . . . . . . *6*

6. At least some of the leaves deeply cut or with large irregular lobes; outermost (ray) flowers of the head pistillate and fertile (developing into fruit), the other flowers of the same head sterile (not developing fruit) with well-developed anthers but undivided styles . . 29. POLYMNIA

6. Leaves at most finely or coarsely toothed, never lobed or deeply cut; all the disk flowers and usually the ray flowers of a head fertile (capable of developing fruit) with functioning pistil . . *7*

7. Pappus absent or nearly so on all the flowers; leaves lanceolate, lance-elliptic, or lance-linear, mainly 4–15 mm. (up to 25 mm.) broad, nearly sessile (without a petiole or stalk); disk flowers greenish or brownish-green; bracts of the involucre 10 or 12; ray flowers minute, the rays only about as long as the braots of the involucre and barely longer than the disk . . . . . . . . . . . . . . . . . . . 35. ECLIPTA

7. Pappus on most of flowers composed of several fringed scales; leaves ovate or lance-ovate, mainly 15–40 mm. (rarely 10 mm.) broad, the lower and middle leaves with conspicuous petioles (leaf-stalks) 5–20 mm. long; disk flowers yellow; bracts of the involucre 4 or 5; ray flowers with rays 1–2 mm. long and slightly longer than the bracts of the involucre and disk . . . . . . . . . . . . . . . . . . . . 48. GALINSOGA

5. Flower-heads with yellow, purple, or rose-colored ray flowers and yellow, orange, brown, black, purple, or reddish disk flowers, or all the flowers in the head with flattened strap-shaped or ribbon-like rays (ligules) . . . . . . . . . . . . . . . . . . . . . . . . . . . . . *8*
 8. All the flowers in the head with flattened strap-shaped or ribbon-shaped rays (ligules); plants with milky juice . . . . . . . . . . . . . . . . . . . . . . . . . . . . *9*
  9. Pappus absent; involucre 3–4.3 mm. high in flower, 3.4–5.3 mm. high in fruit, composed of 5–8 bracts . . . . . . . . . . . . . . . . . . . . . . . 71. SERINIA
  9. Pappus composed of slender bristles; involucre 7.5–11 mm. high, composed of 10–16 bracts . 73. KRIGIA BIFLORA
 8. Most of the flowers in the head tubular, only the outermost marginal flowers with flattened strap-shaped or ribbon-shaped rays (ligules); plants without milky juice . . . . . . . . . *10*
  10. Rays only 1–2 mm. long and only 1 or few in a flower-head . . . . . . . . . . *11*
   11. Strong-smelling plant; leaves deeply parted into bristly-toothed or cut lobes; bracts of the involucre gland-dotted, united into a cup; ray flower usually 1; pappus absent . 54. DYSSODIA
   11. Plant lacking a strong odor; leaves slightly and distantly toothed; bracts of the involucre not gland-dotted, separate and distinct to the base; ray flowers few, usually more than 1; pappus composed of scales cut into many bristles . . . . . . . . . 49. FLAVERIA
  10. Rays more conspicuous and well-developed, 3–40 mm. or more long . . . . . . . . *12*
   12. Rays purple, lilac, rose-colored, or purple-red, but not yellow or orange. . . . 46. COSMOS
   12. Rays yellow or orange, or orange-yellow with a red-brown or red base . . . . . . *13*
    13. Some or all of the leaves deeply cut or parted to or nearly to the midrib or divided into separate leaflets . . . . . . . . . . . . . . . . . . . . . . *14*
     14. Margins of deeply cut lobes or divisions of leaflets toothed . . . . . . . . *15*
      15. Strong-smelling plant conspicuously dotted with glands; bracts of the involucre united into a cup; pappus consisting of 3–6 unequal, united scales, one or more of them awn-like; receptacle naked, without any chaff- or bristle-like pales . . . . . . . . . . . . . . . . . . 55. TAGETES
      15. Plants without a strong odor, not gland-dotted; bracts of the involucre separate or nearly so to the very base; receptacle bearing flat or concave narrow chaffy scales; pappus composed of 1–8 barbed awns. . . . . 45. BIDENS
     14. Lobes of leaves or leaf-divisions entire or smooth-edged, without teeth on margins . *16*
      16. Inner series of the green or brown bracts of the involucre united below for at least a third of their length . . . . . . . . . . . 43. THELESPERMA
      16. Bracts of the involucre all separate or nearly so to the base . . . . . . *17*
       17. Achenes usually winged (not winged in *C. tinctoria*), flattened or compressed, usually glabrous (without hairs), with smooth or fringed margins, nearly round, elliptic, or oblong, not narrowed at the top or tapering to a beak, 1–7 mm. long; pappus consisting of 2 short teeth or awns or none at all . . . . . . . . . . . . . . . . . . 44. COREOPSIS
       17. Achenes not winged, 4–5-angled, nearly terete (round in cross-section), hairy over all the surface, linear or spindle-shaped, 7–16 mm. long with a long beak . . . . . . . . . . . . . . . . . . . 46. COSMOS
    13. Leaves either entire (without teeth) or merely toothed on margins, not deeply cut or divided . . . . . . . . . . . . . . . . . . . . . . . . . . *18*
     18. The tubular or cylindrical 5-toothed disk flowers all sterile, their ovaries much smaller than those of the fertile ray flowers, and bearing undivided styles . 30. SILPHIUM
     18. The tubular or cylindrical 5-toothed disk flowers all fertile, their ovaries as large as or larger than those of the fertile or sterile ray flowers, and bearing forked styles . . . . . . . . . . . . . . . . . . . . . . . . . *19*
      19. Cut lengthwise through flower-head to note the narrowly cone-shaped or column-like receptacle . . . . . . . . . . . . . . . . . . . *20*
       20. Plants of wet ground and swamps in the lowlands of southeastern Missouri only; leaves 5–40 mm. broad; disk 4–9 mm. in diameter; rays 3–10 mm. long, falling from the achenes at maturity; awns present or absent from achenes . . . . . . . . . . . . . . 40. SPILANTHES
       20. Plants of dry or rocky woods, thickets, glades, and prairies throughout Missouri; leaves 25–90 mm. broad; disk 10–25 mm. in diameter; rays 15–40 mm. long, drying with a papery consistency and remaining attached to the achenes at maturity; awns absent from achenes . 34. HELIOPSIS

19. Cut lengthwise through flower-head to note that receptacle is flat, slightly convex or only broadly conical . . . . . *21*
21. Achenes usually winged . . . . . *22*
22. Leaves prominently toothed; bracts of the involucre more or less equal and not arranged in 2 distinct series . . . . . 42. VERBESINA ENCELIOIDES var. EXAURICULATA
22. Leaves entire (without teeth); bracts of the involucre in 2 well-marked series, the inner quite different from the outer in size and shape . . . . . 44. COREOPSIS
21. Achenes not winged . . . . . *23*
23. Pappus consisting of 2 or 4 scales which soon fall from the achene; chaff or bracts of the receptacle embracing or surrounding the achenes . . . . . 41. HELIANTHUS
23. Pappus consisting of 1–8 barbed awns or teeth which persist on the achene; chaff or bracts of the receptacle flattish, scarcely embracing the achenes . . . . . 45. BIDENS

r. Some or all of the leaves alternate on the stem . . . . . *24*
24. All the flowers of a head of the same color . . . . . *25*
25. All the flowers of a head yellow or orange . . . . . *26*
26. All the flowers in a head with flattened strap-shaped or ribbon-like rays (ligules); plants with milky juice . . . . . *27*
27. Pappus absent or apparently so . . . . . *28*
28. Leaves or stem silvery, glaucous (with a 'bloom' which can be rubbed off), glabrous (without hairs) or nearly so; flower-heads with 5–8 bracts of the involucre, lacking any short additional bracts at base of involucre; flower-heads 14–23-flowered; involucre 3–5.3 mm. high; achenes 1–1.6 mm. long . . 71. SERINIA
28. Leaves or stem not silvery or glaucous, varying from more or less hairy to nearly glabrous; flower-heads with 8 main bracts of the involucre accompanied at the base by an additional shorter series of bracts; flower-heads 8–12 (–15)-flowered; involucre 5–8 mm. long; achenes 3–5 mm. long . . . . 70. LAPSANA
27. Pappus present in the form of bristles or scales . . . . . *29*
29. All the leaves entire (without teeth or lobes) (naturally wavy-margined, crisped, or wrinkled leaves may appear toothed at times) . . . . . *30*
30. Leaves and stem more or less hairy . . . . . 85. HIERACIUM
30. Leaves and/or stem below flower-bearing branches (peduncles) glabrous (without hairs) . . . . . *31*
31. Mature achenes, including beak, 12–40 mm. long; bristles of pappus plumose (feather-like with fine long hairs arising from sides of main axis) and also webbed; bracts of involucre 12–40 mm. high. . 76. TRAGOPOGON
31. Mature achenes, including beak when present, 2–9 mm. long; bristles of pappus hair-like or hair-like mixed with unequal oblong, ovate, or lanceolate scales, not plumose nor webbed; bracts of involucre 7.5–18 mm. high . . . . . *32*
32. Leaves grass-like; achene with a beak, 8–9 mm. long, including the beak; bristles of pappus hair-like only; bracts of involucre 10–18 mm. high, of varying unequal lengths . . . . . . 79. LACTUCA SALIGNA f. RUPPIANA
32. Leaves not grass-like; only 1–3 leaves on the stem; 4 or more leaves on stem; achene without a beak, 2–2.5 mm. long; bristles of pappus composed of hair-like bristles mixed with oblong, ovate, or lanceolate scales; bracts of involucre 7–11 mm. long, equal and in one main series . . . . . 73. KRIGIA BIFLORA
29. At least one of the leaves with a tooth or projection on the margins, rarely all leaves entire in *Krigia biflora* . . . . . *33*
33. Leaves and main part of stem below the flower-bearing branches (peduncles) glabrous (without hairs) or nearly so . . . . . *34*
34. Achenes with a beak (a slender prolongation between main body of achene and pappus) . . . . . *35*
35. At most 6 leaves present on stem; hairs of pappus rusty or brick color; main body of achene linear or columnar, nearly terete (rounded in cross-section) . . . . . 83. PYRRHOPAPPUS
35. Eight or more leaves present on stem; hairs of pappus white; main body of achene flat or rather flat . . . . . 79. LACTUCA

34. Achenes without a beak, the pappus arising directly from the truncated top of the main body of achene . . . . . . . . . . . . . . . . . . . . . . *36*

36. Only 1–3 leaves on the stem; bracts of involucre equal, in one main series; achene somewhat terete (rounded in cross-section); pappus of mixed fine bristles and oblong or ovate to lanceolate scales . . . . . . . . . . . . . . 73. KRIGIA BIFLORA

36. Four or more leaves present on the stem; bracts of involucre unequal, graduated in varying lengths of several series; achene flat or rather flat; pappus consisting only of fine, hair-like bristles . . . . . . . . . . . . . . . . 78. SONCHUS

33. Leaves and/or stem more or less hairy . . . . . . . . . . . . . . . . . . *37*

37. Bristles of pappus plumose (feather-like with fine long hairs arising from sides of main axis) . . . . . . . . . . . . . . . . . . . . . . 75. PICRIS

37. Bristles of pappus consisting of hair-like fine bristles, which are not plumose . . . . . *38*

38. Achenes without a beak, the pappus arising directly from the truncated top of the main body of the achene . . . . . . . . . . . . . . . . . 82. CREPIS

38. Achenes with a beak (a slender prolongation between main body of achene and pappus) . *39*

39. Commonly encountered; stems mainly 10–25 dm. tall; achenes flat or rather flat 79. LACTUCA

39. Rarely encountered; stems mainly 1–6 dm. tall; achenes terete or nearly so (rounded in cross-section) . . . . . . . . . . . . . . . . 82. CREPIS

26. Some or most of the flowers in the head with tubular or cylindrical corollas; plants without milky juice . . . . . . . . . . . . . . . . . . . . . . . . . . *40*

40. Inner series of bracts of the involucre united below for at least a third of their length . 43. THELESPERMA

40. All the bracts of the involucre separate or nearly so to the very base . . . . . . . . *41*

41. Some or all of the leaves deeply cut or parted nearly to the midrib or even divided into separate divisions (leaflets) . . . . . . . . . . . . . . . . . . *42*

42. Pappus consisting of numerous, fine hair-like bristles . . . . . . . . . . 64. SENECIO

42. Pappus none or consisting of a very short raised crown or of 2 short awns. . . . . *43*

43. Smallest lobes, segments, or teeth of the leaves not more than 2 mm. broad; bracts of involucre with hyaline (transparent) or scarious (thin, dry and brownish or whitish) margins or tips; plants usually strongly scented or aromatic . . . . . *44*

44. Ray flowers (ligules) present; cut lengthwise through flower-head to note that receptacle, at least toward its center, has lanceolate chaff or pales . . . 57. ANTHEMIS TINCTORIA

44. Ray flowers (ligules) absent; cut lengthwise through flower-head to note that the receptacle is naked (no pales or chaff are present) . . . . . . . . *45*

45. Bruised plant with fragrance of pineapple; smallest leaf-segments or teeth 0.5–1 mm. broad; larger leaf-divisions or lobes 1–5 mm. broad; cut lengthwise through flower-head to note that receptacle is dome-shaped, conspicuously conical, or pointed; corollas 4-toothed . . 58. MATRICARIA MATRICARIOIDES

45. Bruised plant bitter and strongly scented; smallest leaf-segments or teeth 1.5–2 mm. broad; larger leaf-divisions or lobes 2–15 mm. broad; cut lengthwise through flower-head to note that receptacle is flat to merely convex; corollas 5-toothed. . . . . . . . . . . . . 60. TANACETUM

43. Smallest lobes or segments of the leaves not less than 5 mm. broad, varying from 5–80 or more mm. broad; bracts of involucre green throughout, not thin, dry or transparent on margins or tips . . . . . . . . . . . . . . . . *46*

46. Cut lengthwise through flower-head to note that receptacle is flat; the tubular or cylindrical 5-toothed disk flowers all sterile, their ovaries much smaller than and not flattened like those of the fertile ray flowers, and bearing undivided styles . . . . . . . . . . . . . . . . . . . *47*

47. Stem 1.5–6 dm. tall; leaves of stem similar to and not much, if any, smaller than basal leaves; leaves on stem relatively close together, separated by distances of only 2–7 cm.; rays 8–10, 1–2 cm. long; leaves 0.5–1.5 dm. long; bracts of involucre of 2 unequal and distinct types, the outer much smaller or narrower than the inner . . . . . . . . . 32. ENGELMANNIA

47. Stem 10–35 dm. tall; basal leaves much larger than those on stem; leaves on stem widely separated by gaps of 15 cm. or more; rays 15–30, 3–5 cm. long; basal and lower leaves up to 4 dm. long; bracts of involucre about equal and all of the same type . . . . . . . . . . . 30. SILPHIUM

46. Cut lengthwise through flower-head to note that receptacle is conspicuously conical, columnar, or dome-shaped; the tubular or cylindrical 5-toothed disk flowers are all fertile, their ovaries as large as or larger than those of the ray flowers, all the ovaries narrow, ovoid, or columnar . . . 36. RUDBECKIA

41. Leaves either entire (without teeth) or toothed . . . . . . . . . . . . . . . . . . *48*

48. Leaves entire (without teeth) . . . . . . . . . . . . . . . . . . . . *49*

49. No rays present or apparent in flowering stage . . . . . . . . . . . . . . . *50*

50. Involucre 6–8 mm. high; bracts of involucre in 2 series of about equal length; all the flowers with well-developed pappus; flower-heads mostly on leafy branchlets. . . 18. ASTER BRACHYACTIS

50. Involucre 2–5 mm. high; bracts of involucre in 1 equal series; pappus of the outermost (marginal) flowers either less than 1 mm. long, or of short scales, or absent . . . 19. ERIGERON STRIGOSUS var. STRIGOSUS f. DISCOIDEUS

49. Rays present, even though small. . . . . . . . . . . . . . . . . . . . *51*

51. Flower-heads very small, the individual disks (solid portion between the ray flowers) 5 mm. or less broad . . . . . . . . . . . . . . . . . . . . *52*

52. Pappus of numerous, fine, hair-like bristles . . . . . . . . . . 13. SOLIDAGO

52. Pappus of 5–8 bristle-like scales united at base in the disk flowers, only a minute crown in the ray flowers . . . . . . . . . . . . . . 10. GUTIERREZIA

51. Flower-heads larger, the individual disks (solid portion between the ray flowers) 5–25 mm. broad . . . . . . . . . . . . . . . . . . . . *53*

53. Pappus of numerous, fine, hair-like bristles with or without a short outer series of scales . . . . . . . . . . . . . . . . . . . . *54*

54. Disk of flower-head 8–25 mm. broad; pappus arranged in 2 series, the outer series of scales shorter than the inner long hair-like bristles. . . . . 12. CHRYSOPSIS

54. Disk of flower-head 2–10 mm. broad; pappus simple, of a single series of equal hair-like bristles . . . . . . . . . . . . . . . . . 13. SOLIDAGO

53. Pappus none, or of 2 or 4 awns or scales, or of 5–10 chaffy or awn-tipped scales. . *55*

55. Leaves gland-dotted (showing as tiny black, brown, or golden dots or marks on dried specimens); cut lengthwise through flower-head to note that the receptacle is naked (no pales or chaff are present) or the chaff is reduced to mere stubs or vestiges; achenes more or less hairy . . . . . . . . . *56*

56. Stems and leaves glabrous or leaves decurrent on stem (leaf-tissue extending down stem from base of leaf); receptacle completely naked (no pales or chaff are present when flowers are removed) . . . . 52. HELENIUM

56. Stems and leaves more or less hairy and leaf-tissue not extending down stem from base of leaves . . . . . . . . . 53. GAILLARDIA LUTEA

55. Leaves not dotted with glands; cut lengthwise through flower-head to note that the receptacle has chaff or pales present in the form of scales or soft bristles; achenes glabrous (without hairs) . . . . . . . . . . . *57*

57. Achene with a winged margin; the tubular or cylindrical 5-toothed disk flowers all sterile, their ovaries much smaller than those of the fertile ray flowers, and bearing undivided styles . . . . . . . . . 30. SILPHIUM

57. Achene without any wing on margin; the tubular or cylindrical 5-toothed disk flowers are all fertile, their ovaries as large as or larger than those of the ray flowers . . . . . . . . . . . . . . . . . . . . *58*

58. Pappus none; cut lengthwise through flower-head to note that the receptacle is conspicuously conical, dome-shaped, or columnar . 36. RUDBECKIA

58. Pappus consisting of 2 or 4 awns or scales which soon fall from the achene; cut lengthwise through flower-head to note that the receptacle is flat to convex or low-conical . . . . . . . 41. HELIANTHUS

48. Some or all of the leaves with teeth. . . . . . . . . . . . . . . . . *59*

59. No rays present or apparent at flowering time. . . . . . . . . . . 19. ERIGERON

59. Rays present and apparent at flowering time . . . . . . . . . . . . . . *60*

60. Bracts of the involucre all equal, not graduated or overlapping and in only 1 series; pappus of fine long hair-like bristles . . . . . . . . . . . . . . . 64. SENECIO

60. Bracts of the involucre unequal, graduated or overlapping, or in 2 or more series, or if scarcely or not graduated in series, then the pappus not of hair-like long bristles . . . . . . . . . . . . . . . . . . . . *61*

61. Flower-heads very small, the individual disk (solid portion between the ray flowers) 2–5 mm. broad . . . . . . . . . . . . . . . . . . . . . . . . . 13. SOLIDAGO
61. Flower-heads larger, the individual disk (solid portion between the ray flowers) 5–50 mm. broad . *62*
62. Pappus of all the disk flowers, at least, consisting of many (more than 12) fine, hair-like bristles . . . . . . . . . . . . . . . . . . . . . . . . . . *63*
63. Pappus of the ray flowers none or essentially so . . . . . . . . . 11. HETEROTHECA
63. Pappus of ray flowers consisting of many fine, hair-like bristles similar to those of disk flowers . . . . . . . . . . . . . . . . . . . . . . . . *64*
64. Pappus arranged in 2 series, the outer series of scales shorter than the inner long hair-like bristles . . . . . . . . . . . . . . . . . . . . 12. CHRYSOPSIS
64. Pappus simple, of a single series of long equal or unequal hair-like bristles . . . *65*
65. Leaves with bristle-pointed or spiny teeth along the margins . . 14. APLOPAPPUS
65. Leaves lacking bristle-pointed or spiny teeth. . . . . . . . . . . . . *66*
66. Lower surface of leaves densely velvety-hairy or white-woolly; flower-heads few and large, the individual disk (solid portion between the ray flowers) 30–50 mm. broad; rays (ligules) numerous, 30 or more, 10–30 mm. long . . . . . . . . . . . . . . . . . . 24. INULA
66. Lower surface of leaves, if hairy, not densely velvety-hairy nor white-woolly; flower-heads usually many and small, the individual disk (solid portion between the ray flowers) 5–10 mm. broad; rays (ligules) relatively few, 3–17, only 1–8 mm. long . . . . . . . . . . . 13. SOLIDAGO
62. Pappus of the disk flowers either none or a minute crown, or of 2–8 awns, scales, or bristles, these not fine or hair-like. . . . . . . . . . . . . . . . . . . . . *67*
67. Leaves finely gland-dotted (showing as black, brown, or golden dots or marks in dried specimens); pappus of 5–8 persistent thin or transparent awn-tipped scales not over 1 mm. long . . . . . . . . . . . . . . . . . . . . . . 52. HELENIUM
67. Leaves not gland-dotted; pappus none or a minute crown or of 2–4 awns, teeth, scales, or bristles, or if 5–8 awns, these firm and not persistent, but falling from the flowers. . *68*
68. Involucre and bracts more or less sticky or gummy; cut lengthwise through flower-head to note that the receptacle is naked (no chaff or pales are present); pappus of 2–8 firm awns which are ½ the length of or equal the length of the disk flowers, but soon fall from the flowers . . . . . . . . . . . . . . 9. GRINDELIA
68. Involucre and bracts not sticky or gummy; cut lengthwise through flower-head to note that the receptacle has chaff or pales present in the form of scales or bracts accompanying the flowers . . . . . . . . . . . . . . . . . . . . *69*
69. The tubular or cylindrical 5-toothed disk flowers all sterile, their ovaries much smaller than those of the fertile ray flowers, and bearing undivided styles. . . *70*
70. Middle and upper leaves closely crenately toothed (rounded or scalloped teeth); achenes without any winged border; ovaries and achenes in 1 series only . . . . . . . . . . . . . . . . . 31. BERLANDIERA
70. Middle and upper leaves somewhat or irregularly or not at all toothed, but not crenately toothed; achenes with winged border; ovaries and achenes in 2 or 3 series . . . . . . . . . . . . . . 30. SILPHIUM
69. The tubular or cylindrical 5-toothed disk flowers are all fertile, their ovaries as large as or larger than those of the ray flowers . . . . . . . . . *71*
71. Pappus of 2 or 4 scales which fall from the flowers, and do not persist on the achenes; achenes 3–4-angled, without wings on the border; stem not winged by any leaf-tissue extending from the base of the leaves . 41. HELIANTHUS
71. Pappus of 2 or 3 awns which persist on the achenes; achenes flat, with or without wings on the border; stem often winged by leaf-tissue extending from the base of the leaves (not winged in *Verbesina encelioides* var. *exauriculata*) . . . . . . . . . . . . . . . . . . 42. VERBESINA
25. All the flowers of a head purple, lavender, blue, rose-red, pink, white, creamy-white, or greenish, but not yellow or orange . . . . . . . . . . . . . . . . . . . . . . *72*
72. All the flowers of a head purple, lavender, blue, rose-red, or pink, but not white, creamy-white, or greenish . . . . . . . . . . . . . . . . . . . . . . . . . *73*
73. All the flowers in the head with flattened strap-shaped or ribbon-like rays (ligules); plants with milky juice . . . . . . . . . . . . . . . . . . . . . *74*
74. All the leaves entire (without teeth or lobes), grass-like . . . . . . . . . . . *75*

75. Upper leaves greatly reduced to mere scales; heads with 4–10, usually 5 flowers; involucre 10–16 mm. long; achene 4–5 mm. long; plants of the loess hills of northwestern Missouri
80. LYGODESMIA

75. Upper leaves well developed; heads with many flowers; involucre 25–40 mm. long in flower, 40–70 mm. in fruit; achene 25–40 mm. long, including the long beak; plants occurring throughout Missouri . . . . . . . . . . . . . . . . 76. TRAGOPOGON

74. Some or all of the leaves with teeth or even more deeply cut into lobes . . . . . . . . . *76*

76. Pappus consisting of very short, minute scales; flowers usually deep blue or pink; introduced weedy plant . . . . . . . . . . . . . . . . . . . . 72. CICHORIUM

76. Pappus consisting of long fine hair-like bristles; flowers pale or dull blue or lavender-blue or pinkish; native wild plants . . . . . . . . . . . . . . . . . *77*

77. Flower-heads nodding or erect; achenes cylindrical or columnar, more or less curved or rounded on the sides, not flattened; flowers pinkish . . . . . . . 84. PRENANTHES

77. Flower-heads erect or ascending; achenes flattened or compressed; flowers blue or bluish . . . . . . . . . . . . . . . . . . . . . 79. LACTUCA

73. Some or all of the flowers in the head with tubular or cylindrical corollas; plants without milky juice . . . . . . . . . . . . . . . . . . . . . . . . . *78*

78. Outermost flowers with flattened strap-shaped or ribbon-like rays (ligules), the rays 3-cleft at summit . . . . . . . . . . . . . . . . . . . . . . 53. GAILLARDIA

78. No strap-shaped or ribbon-like rays (ligules) present, the flowers usually tubular or cylindrical, but sometimes the outer ones more enlarged and irregular with longer tubes . . . . . . *79*

79. All the leaves entire (without teeth). . . . . . . . . . . . . . . . . . *80*

80. Flower-heads arranged usually in a narrow spike- or raceme-like inflorescence with the heads usually closely scattered along the main axis; pappus consisting of many long hair- or feather-like bristles . . . . . . . . . . . . . . . 8. LIATRIS

80. Flower-heads either solitary at the top of the stem or its branches or in a loosely branched inflorescence; pappus consisting of 5–12 scales or of short or long hair-like bristles . . . . . . . . . . . . . . . . . . . . . . . . . *81*

81. Bracts of the involucre in several distinct overlapping (imbricated) series, usually some of them with toothed, fringed, lacerate, or with thin transparent (hyaline) borders; outermost (marginal) flowers often somewhat more enlarged and irregular with longer lobes than the more centrally located flowers . . . . . 69. CENTAUREA

81. Bracts of the involucre either equal or in only 1 or 2 series of nearly equal length, all entire (without teeth); all the flowers similar and equal in size. . . . . . *82*

82. Leaves linear, mainly 1–3 mm. broad; stem much-branched above with several to many flower-heads per branch; flowers pink-purple, 5–12 in a head; involucre 4–6 mm. high; receptacle naked (no chaff or pales present) . 51. PALAFOXIA

82. Leaves linear-oblanceolate or spatulate; the lower or basal ones 4–20 mm. broad; stem mostly simple or with 1 or 2 branches at the top, the flower-heads only 1–3 to each scape-like stem; flowers pinkish or white, 15–20 in a head; involucre 7–9 mm. high; cut lengthwise through flower-head to note that the receptacle has rigid linear pales or chaff . . . . . . . . . 47. MARSHALLIA

79. Some or all of the leaves toothed or even deeply cut or lobed . . . . . . . . . *83*

83. Cut lengthwise through the flower-head to note that the receptacle has pales or chaff in the form of bristles; flower-heads mostly large, the involucre 10–40 mm. high. . . . *84*

84. Bracts of the involucre with usually toothed, fringed, lacerate, or thin transparent (hyaline) borders; outermost (marginal) flowers often somewhat more enlarged and irregular with longer lobes than the more centrally located flowers; pappus none or mostly only 1–3 mm. long, in *C. repens* 6–11 mm. long . . . 69. CENTAUREA

84. Bracts of the involucre all entire (without teeth on margins); all the flowers similar and equal in size; pappus well developed, mostly 15–30 mm. long. 67. CIRSIUM

83. Cut lengthwise through the flower-head to note that the receptacle is naked (no pales or chaff are present); flower-heads mostly small, the involucre 4–12 mm. high . . . *85*

85. Only 3–7 leaves present on main stem; bracts of involucre 8 arranged in 4 pairs; stems 0.3–0.7 m. (up to 0.8) high; pappus of 5 scales prolonged into a long bristle; flower-heads compound, composed of many smaller heads, each of which contains 2–5 flowers . . . . . . . . . . . . . . . . . 2. ELEPHANTOPUS

85. Leaves of main stem 10 or more; bracts of involucre many, more than 8 and arranged in overlapping (imbricated) series; stems usually 1–3 m. (rarely less)

high; flower-heads simple, composed of 15–many flowers, the whole surrounded by the bracts of the involucre; pappus of many fine hair-like bristles . . . . . *86*

86. Plants with a strong odor; plants mainly of extreme southern Missouri; most of the flowers, or at least the outer ones of a head, pistillate (with pistil and no stamens); anthers with tapering long, tail-like bases . . . . . 20. PLUCHEA

86. Plants lacking any strong odor; plants found throughout Missouri; all the flowers perfect (with pistil and stamens present in the same flower); anthers with arrow-shaped (sagittate) bases . 1. VERNONIA

72. All the flowers of a head white, creamy-white, or greenish . . . . . *87*

87. All the flowers in the head with flattened strap-shaped or ribbon-like rays (ligules); plants with milky juice . . . . . *88*

88. Flower-heads nodding or erect; pappus consisting of long, fine, hair-like bristles . . 84. PRENANTHES

88. Flower-heads erect or ascending or spreading, not nodding; pappus consisting of very short, minute scales . . . . . 72. CICHORIUM

87. Some or all of the flowers in the head with tubular or cylindrical corollas; plants without milky juice . . . . . *89*

89. Ray flowers present with rays (ligules) . . . . . *90*

90. Leaves very finely divided into hair-like divisions 0.2–1 mm. broad; plants with a strong odor; pappus absent . . . . . 56. ACHILLEA

90. Leaves usually only toothed, but if deeply lobed or parted to the midrib, the lobes 2–7 mm. broad; plants without a strong odor; pappus present, usually consisting of 1–3 awns or scales . . . . . *91*

91. Plant 1–2 m. tall; stem usually winged by the leaf-tissue extending from the base of the leaf; ray flowers 3–4; the tubular or cylindrical 5-toothed disk flowers are all fertile, their ovaries as large as or larger than those of the ray flowers . 42. VERBESINA

91. Plant mainly 0.3–0.9 m. (rarely to 1.2 m.) tall; stem not winged; ray flowers 5; the tubular or cylindrical 5-toothed disk flowers all sterile, their ovaries much smaller than those of the fertile ray flowers, and bearing undivided styles . 33. PARTHENIUM

89. No ray flowers with strap-shaped or ribbon-like rays present (outermost flowers often somewhat more enlarged and irregular with longer lobes than the more centrally located flowers in *Centaurea* and some other genera) . . . . . *92*

92. Flowers green or yellow-green; pappus none; flower-heads very small, arranged one above the other in elongated spike-like, raceme-like, or panicle-like inflorescences . . . *93*

93. No perfect flowers (with pistil and stamens in same flower) present, the flowers either staminate (stamens only present) or pistillate (pistil only present); pistillate flowers very much reduced, without a corolla or with only a portion of a corolla present; staminate flowers with the anthers only partly united . . . . . *94*

94. All the flower-heads the same, the 3–5 pistillate (female) flowers situated along the outer margin of the head surrounding the more numerous central sterile staminate (male) flowers; bracts of involucre few, roundish . . . 25. IVA

94. Flower-heads not the same, separated into numerous staminate (male) heads arranged in an elongated spike-like, raceme-like, or panicle-like inflorescence, and 1 to few pistillate (female) heads situated in the axils of leaves or bracts at the base of the staminate inflorescences; bracts of staminate involucre united . *95*

95. Commonly encountered weeds; pistillate (female) involucre with but 1 row of spines or pointed projections near the summit; fruiting head 1-seeded . . . . . 26. AMBROSIA

95. Rarely found, usually along railroads; pistillate (female) involucre with more than 1 row of spines; fruiting head 2–4-seeded . . . . 27. FRANSERIA

93. Flowers all perfect (with both pistil and stamens in the same flower), sometimes fertile, sometimes not fertile; anthers completely united; plants usually aromatic 61. ARTEMISIA

92. Flowers all white, creamy or yellowish-white . . . . . *96*

96. Leaves entire (without teeth) . . . . . *97*

97. Bracts of the involucre fringed or lacerate with slender teeth or comb-like appendages; outermost (marginal) flowers with an enlarged irregular corolla 69. CENTAUREA

97. Bracts of the involucre not fringed or lacerated with slender teeth or comb-like appendages, although sometimes fringed with hairs (ciliate); outermost (marginal) flowers not enlarged or irregular . . . . . *98*

98. A white woolly hairiness on the lower surface of the leaves and usually on stem and/or involucre; not all the flowers perfect, sometimes staminate (with functional developed stamens and undivided style) and pistillate (with functional divided style and fertile ovary) flowers separate on different plants, sometimes separate in the same flower-head, or with the outer ones pistillate and the inner ones staminate or perfect; bracts of the involucre more or less scarious (thin and dry) throughout and whitish at least at tip . . . . . . . . . . . *99*

99. Plants developing stolons or leafy runners from the base; leaves at base of plant and those on runners conspicuous and persistent, much larger than and differing in shape from those on the stem; leaves on stem few and small; flower-heads usually 2–12 (rarely up to 30) on one stem; plants always dioecious (the staminate flowers on one plant, pistillate flowers on another plant). . . . . . . . . . . . . . . . . . 21. ANTENNARIA

99. Plants not developing stolons or leafy runners at the base, although rosette leaves may be present; leaves at base of flowering plant soon drying up and not persisting (does not include young or first year's rosette leaves), scarcely if at all larger than those on the stem; leaves on stem numerous and well developed; flower-heads numerous, usually 10–50 to a stem; plants dioecious or usually with a few sterile or staminate flowers in the center of pistillate heads or with the outer ones in a head pistillate and the inner ones perfect. . . . . . *100*

100. Rarely encountered, known only as an escaped plant in Boone County; flower-heads or involucre 8–12 mm. broad; all the bracts of the involucre bright white throughout, rather opaque and dull; plants mainly dioecious with the staminate and pistillate flowers on separate plants, or the pistillate heads with a few sterile or staminate flowers in the center . . . . . . . . . . . . . . . . . 22. ANAPHALIS

100. Commonly encountered; flower-heads or involucre 2–9 mm. broad; some or all of the bracts of the involucre straw- or buff-colored, pale brown, or marked with purple or green, thin, shining, transparent, never completely bright white throughout; plants not dioecious, the flowers all fertile (but not perfect) with the outer flowers of a head pistillate, the innermost ones perfect (with both stamens and pistil). 23. GNAPHALIUM

98. Without the above combination of characters; plants without any white-woolly hairiness on lower surface of leaves, stem, or involucre; flowers all perfect; bracts of the involucre mainly green or other color, sometimes with scarious (thin and dry) or hyaline (transparent) margins . . . . . . . . . . . . . . . . . . . . . . . . . . . *101*

101. Pappus consisting of 5–6 scales; cut lengthwise through flower-head to note that receptacle has linear-stiff chaff or pales present; bracts of involucre in only 1 or 2 series of nearly equal length, not showing much if any overlapping (imbrication) . . . . 47. MARSHALLIA

101. Pappus consisting of many (15–40) long, hair- or feather-like bristles; cut lengthwise through flower-head to note that receptacle is naked (no chaff or scales present); bracts of involucre noticeably overlapping (imbricated) in 3 or more series of unequal length, the outermost the shortest . . . . . . . . . . . . . . . . . . . . . *102*

102. Stems usually simple and unbranched, the flower-heads arranged usually in a narrow spike- or raceme-like inflorescence with the heads usually closely scattered along the main axis; bracts of involucre not marked with fine long lines. . . . 8. LIATRIS

102. Stems branched above, the flower-heads arranged in more or less flat-topped (corymbiform) or loosely irregular (paniculate) inflorescences at the ends of the branches; bracts of involucre marked with fine long lines. . . . . . . . 7. KUHNIA

96. Leaves with teeth or more deeply cut into lobes . . . . . . . . . . . . . . *103*

103. Cut lengthwise through the flower-head to note that the receptacle has pales or chaff in the form of bristles; flower-heads mostly large, the involucre 10–40 mm. high . . . . . 67. CIRSIUM

103. Cut lengthwise through the flower-head to note that the receptacle is naked (no pales or chaff are present); flower-heads mostly small, the involucre 4–15 mm. high . . . . . . . . *104*

104. Leaves deeply parted or nearly divided into narrow linear or oblong lobes or segments 0.5–2 mm. broad; bracts of involucre thin with usually white or yellowish-white, petal-like tips; pappus consisting of 12–20 thin blunt scales . . . . . . . . 50. HYMENOPAPPUS

104. Leaves shallowly or deeply toothed, but not deeply parted into narrow lobes; bracts of involucre mainly green throughout without thin, petal-like white or yellowish-white tips; pappus consisting of 19–80 or more fine hair-like bristles . . . . . . . . . . *105*

105. Bracts of the involucre in mainly 1 series of equal length with a few reduced outer ones at the very base. . . . . . . . . . . . . . . . . . . . . . *106*

106. Bristles of the pappus roughened with tiny barbs or teeth; all the flowers perfect; plants perennial . . . . . . . . . . . . . 63. CACALIA

106. Bristles of the pappus smooth; outermost (marginal) flowers of the head pistillate, the rest of the flowers perfect (with both stamens and pistil in the same flower); plants annual . . . . . . . . . . . . . 62. ERECHTITES

105. Bracts of the involucre in 3 or more obviously graduated series overlapping one another (imbricated) and unequal, the outermost the shortest . . . . . . . . . . . . *107*

107. Leaves on petioles (leaf-stalks) 1–7 cm. long; leaf-blades triangular-ovate or heart-shaped, coarsely crenately toothed (rounded, scalloped margins), $1\frac{1}{4}$–2 times as long as broad . . . . . . . . . . . . . . . . . . . . . . . . 6. BRICKELLIA

107. Leaves sessile (without leaf-stalks) or nearly so, on leaf-stalks at most only 7 mm. long; leaf-blades linear, lanceolate, or narrowly ovate, $2\frac{1}{2}$–20 times as long as broad . . . . *108*

108. Pappus brown, tawny, or purplish, in 2 series, the outer of minute, scale-like bristles, the inner of many long, fine hair-like bristles; bristles of pappus merely roughened with short projections, not plumose (feather-like) with fine long hairs extending from each side of main axis) . . . . . . . . . . . . . . . 1. VERNONIA

108. Pappus dirty white to pale brown (turning to rusty brown in dried specimens), in 1 series of about 20 fine hair-like bristles; bristles of pappus plumose (feather-like with fine long hairs extending from each side of main axis) . . . . . . 7. KUHNIA

24. Flowers of a head of more than one color, the rays (ligules) of one color, the disk flowers of another color . *109*

109. Rays (ligules) white, blue, purple, lilac, rose-colored, or pink, or rose-red or purple at base and yellow at tip in *Gaillardia*; the disk flowers yellow, brownish- or greenish-yellow, brown, reddish-brown, or purplish-red or -brown . . . . . . . . . . . . . . . . . . *110*

110. Leaves entire (without teeth or lobes) . . . . . . . . . . . . . . . *111*

111. Cut lengthwise through flower-head to note that the receptacle has spine-tipped chaff or bristles present . . . . . . . . . . . . . . . . . . . . *112*

112. Rays (ligules) elongated, linear to linear-oblanceolate, the sides nearly parallel throughout their length, 6–20 times as long as broad, mainly 3–7 cm. (rarely 2 or 9 cm.) long, 2–3-toothed at tip; achenes 4-angled, glabrous or sparsely hairy on the angles; pappus consisting of a short toothed border at summit of achene . . . . . . . . . . . . . . . . . 37. ECHINACEA

112. Rays (ligules) relatively broader, elliptic-oblong or obovate, $1\frac{1}{2}$–3 times as long as broad, 1–2 cm. long, conspicuously 3-cleft at tip; achenes 5-ribbed, densely hairy; pappus consisting of 5–10 awn-tipped scales . . . . . . . 53. GAILLARDIA

111. Cut lengthwise through flower-head to note that receptacle is naked (no chaff or pales present) . . . . . . . . . . . . . . . . . . . . . . . *113*

113. Pappus absent . . . . . . . . . . . . . . . . . 15. ASTRANTHIUM

113. Pappus present . . . . . . . . . . . . . . . . . . . . . *114*

114. Pappus consisting of numerous (more than 12) fine hair-like bristles . . *115*

115. Bracts of the involucre usually in 3–5 series definitely graduated (imbricated) or with the outer 1 or more series leaf-like and nearly or as long as the inner series; rays (ligules) usually fewer than 40 and arranged in 1 row, except in *A. exilis*; flower-heads mostly on leafy branchlets; mostly flowering in late summer and autumn . 18. ASTER

115. Bracts of the involucre in 1 equal, not graduated, series or with 1 long inner series and a very short basal series, never leaf-like; rays (ligules) usually 40–200, and arranged in more than 1 row; flower-heads on naked leafless peduncles (flower-bearing branches) or scapes (flower-bearing stems); flowering mainly in spring and early summer, but also in autumn . . . . . . . . 19. ERIGERON

114. Pappus consisting of 10 or less scales, awns, and/or bristles . . . . . *116*

116. Annual, hairy plants 0.5–3 dm. tall; leaves less than 2 cm. long; disk (solid portion of flower-head between the ray flowers) 2–3 mm. broad; rays 5–12; achenes hairy . . . . . . 16. CHAETOPAPPA

116. Perennial, glabrous (without hair) plants, mostly 3–15 dm. tall; larger leaves 5–15 cm. long; disk (solid portion of flower-head between the ray flowers) 3–14 mm. broad; rays numerous, mainly 20–40; achenes glabrous . . . . . . . . . . . 17. BOLTONIA

110. Leaves toothed, lobed, or deeply cut and dissected . . . . . . . . . . . . . *117*

117. All the leaves deeply dissected into very narrow thread- or hair-like segments 0.2–1 mm. broad . . . . . . . . . . . . . . . . . . . . . . *118*

118. Cut lengthwise through flower-head to note that receptacle has chaff or pales at least around the center or summit. . . . . . . . . . . . 57. ANTHEMIS

118. Cut lengthwise through flower-head to note that receptacle is naked (no chaff or pales present) . . . . . . . . . . . . . . . 58. MATRICARIA

117. Leaves either toothed or lobed or if deeply parted, the lobes or segments broad, oblong, ovate, obovate, 2–10 or more mm. broad . . . . . . . . . . . . . . . . *119*

119. Cut lengthwise through flower-head to note that receptacle has spine-tipped or rigid chaff or chaff-like bristles. . . . . . . . . . . . . . . . . . . . . *120*

120. Rays (ligules) elongated, linear to linear-oblanceolate, the sides nearly parallel throughout their length, 6–20 times as long as broad, 2–9 cm. long, shortly 2–3-toothed at tip; achenes 4-angled, glabrous or sparsely hairy on the angles; pappus consisting of a short toothed border at summit of achene . . . . . 37. ECHINACEA

120. Rays (ligules) relatively broader, elliptic-oblong or obovate, $1\frac{1}{2}$–3 times as long as broad, 1–2 cm. long, conspicuously 3-cleft at tip; achenes 5-ribbed, densely hairy; pappus consisting of 5–10 awn-tipped scales . . . . . . . . . 53. GAILLARDIA

119. Cut lengthwise through flower-head to note that receptacle is naked (no chaff or pales present) . . . . . . . . . . . . . . . . . . . . . . . . *121*

121. Pappus absent; bracts of involucre thin, dry and papery (scarious) or transparent (hyaline) at least on margins and at tip, only the midrib sometimes greenish; plants often strongly-scented . . . . . . . . . . . 59. CHRYSANTHEMUM

121. Pappus present in the form of numerous hair-like bristles; bracts of involucre usually green or at least not dry, papery, and transparent; plants without a strong odor . *122*

122. Bracts of the involucre usually in 3–5 series definitely graduated (imbricated) or with the outer 1 or more series leaf-like and nearly or as long as the inner series; rays (ligules) usually fewer than 40 and arranged in 1 row, except in *A. exilis*; flower-heads mostly on leafy branchlets; mostly flowering in late summer and autumn . . . . . . . . . . . . . . . . 18. ASTER

122. Bracts of the involucre in 1 equal, not graduated, series or with 1 long inner series and a very short basal series, never leaf-like; rays (ligules) usually 40–200, and arranged in more than 1 row; flower-heads on naked leafless peduncles (flower-bearing branches) or scapes (flower-bearing stems); flowering mainly in spring and early summer, but also in autumn . 19. ERIGERON

109. Rays (ligules) yellow or yellow with purple or red, the disk flowers black, brown, reddish-brown, or purple . . . . . . . . . . . . . . . . . . . . . . . . . . *123*

123. Stem winged by the leaf-tissue extending from the base of the leaves; cut lengthwise through flower-head to note that the receptacle is naked (no chaff or pales are present) . . 52. HELENIUM

123. Stem not winged as above described; cut lengthwise through flower-head to note that the receptacle has chaff or chaff-like bristles . . . . . . . . . . . . . . . . *124*

124. Pappus consisting of 5–10 awn-tipped scales; achenes 5-ribbed, densely hairy; rays (ligules) conspicuously 3-cleft at tip; receptacle furnished with chaff-like bristles which do not individually embrace or accompany each disk flower. . . . . . 53. GAILLARDIA

124. Pappus none or consisting of a short or toothed border or 1, 2, or 4 teeth or awns which soon fall from or persist on the achene; achenes either flattened, 4-angled, or somewhat terete (rounded in cross-section), at least those of the disk flowers usually smooth and glabrous (without hairs) or sparsely hairy on the angles; rays either entire (without teeth) or slightly notched or slightly 2–3-toothed at tip, not conspicuously 3-cleft; receptacle furnished with chaff or pales which individually enclose or accompany most or all of the disk flowers . . . . . . . . . . . . . . . . . . . . . *125*

125. Cut lengthwise through flower-head to note that receptacle is either flat, convex, or shallowly cone-shaped; pappus consisting of 2 or 4 thin scales which soon fall from the flower . . . . . . . . . . . . . . . . . . . 41. HELIANTHUS

125. Cut lengthwise through flower-head to note that receptacle is high cone-shaped or column-like; pappus either none or a short or toothed border or of 1 or 2 persistent awns or teeth. . . . . . . . . . . . . . . . . . . *126*

126. Chaff of the receptacle rigid and spiny-tipped, noticeably longer than the disk flowers . . . . . . . . . . . . . . . . 37. ECHINACEA

126. Chaff of the receptacle not spiny-tipped nor noticeably longer than the disk flowers . . . . . . . . . . . . . . . . . . . *127*

127. Base of the ray flowers not enclosed or accompanied by a chaffy bract

of the receptacle; achenes mostly equally and noticeably 4-sided (quadrangular) . 36. RUDBECKIA

127. Base of the ray flowers enclosed or accompanied by a chaffy bract of the receptacle; achene either flattened or nearly terete (rounded in cross-section), not noticeably 4-angled (quadrangular) or only faintly so . . . . . . . . . . . . . . . . . . . . . . . . . *128*

128. Leaves pinnately divided into 3–9 leaflets, not clasping the stem; stem and leaves hairy . 39. RATIBIDA

128. Leaves simple and entire, without lobes or teeth, not clasping the stem; stem and leaves glabrous (without hairs) . . . . . . . . . . . . . . . . . . 38. DRACOPIS

*Synoptic Key to Subfamilies and Tribes**

(Missouri genera grouped accordingly)

## Subfamily I. **Tubuliflorae**

Corolla tubular or cylindrical in all of the perfect flowers, regularly 5 (rarely 3 or 4)-parted; rays (strap-shaped or ribbon-like) present only in the outermost (marginal) flowers, which, when present, are either pistillate only, or neutral (with neither stamens nor pistil); sap of plants usually not milky.

A. Heads all the same, with disk flowers only; all the flowers perfect, not yellow; anthers not tailed (caudate) at base; receptacle naked . . . . . . . . . . . . . . . . . . . . . . *b*

b. Leaves alternate; branches of style thread- or hair-like, terete, minutely hairy with short hairs over all the surface, short- to long-pointed; anthers sagittate (triangular or arrowhead-shaped) at base . . . . . . . . . . . . . . . . . . Tribe I. VERNONIEAE (p. 1457)

1. Vernonia
2. Elephantopus

b. Leaves alternate, opposite, or whorled; branches of style clavate (thickened upward), minutely and uniformly papillate, obtuse (blunt) at tip; anthers rounded at base . Tribe II. EUPATORIEAE (p. 1462)

3. Eupatorium
4. Ageratum
5. Mikania
6. Brickellia
7. Kuhnia
8. Liatris

A. Heads either all the same with only disk flowers present or with two kinds of flowers (disk and ray flowers), some or all of them often yellow; anthers either tailed (caudate), sagittate (triangular or arrowhead-shaped), truncate (as if cut flat across), or rounded at base; receptacle naked or provided with chaff or chaff-like bristles . . . . . . . . . . . . . . . . . . . . . . *B*

B. Stems, leaves, and/or bracts of involucre usually not spiny; bracts of involucre usually not prickly or finely fringed; flower-heads with or without ray flowers; receptacle naked or chaffy, rarely bristly; anthers usually not tailed (caudate), except tailed in Tribe *Inuleae*; styles lacking a hairy or thickened ring below their branches . . . . . . . . . . . . . . . *C*

C. Branches of style of the perfect flowers flat and smooth within, but with triangular or lanceolate appendages hairy on outside; leaves alternate; receptacle naked; disk flowers mainly yellow; anthers obtuse at base . . . . . . . . . . Tribe III. ASTEREAE (p. 1477)

9. Grindelia
10. Gutierrezia
11. Heterotheca
12. Chrysopsis
13. Solidago
14. Aplopappus
15. Astranthium
16. Chaetopappa
17. Boltonia
18. Aster
19. Erigeron

C. Branches of style of the disk flowers either without an appendage at tip, or with a very short slender appendage; leaves alternate, opposite, or basal only; receptacle naked or with chaff or chaff-like bristles; disk flowers yellow, brown, orange, green, black, or purple; anthers not tailed (caudate) at base or tailed in Tribe *Inuleae* . . . . . . . . . . . . . . *D*

D. Corollas of disk flowers thread- or hair-like and truncate at summit (as if cut flat across); anthers tailed (caudate) at base; branches of style rounded or truncate (as if cut flat across), without any appendages; leaves alternate; receptacle naked (in the known Missouri genera) . . . . . . . . . . . . . . Tribe IV. INULEAE (p. 1530)

* Adapted partly from treatment in *Gray's Manual, eighth* edition, and from Gleason's *New Illustrated Flora.*

20. Pluchea
21. Antennaria
22. Anaphalis
23. Gnaphalium
24. Inula

D. Corollas of disk flowers tubular or cylindrical to goblet-shaped, usually with distinct lobes at summit; anthers not tailed, varying from truncate to sagittate; branches of style either without any appendage at tip and with a ring of hairs at the end or often with hairy appendages; leaves alternate or opposite; receptacle naked or chaffy or with chaff-like bristles . . . . . . . . *E*

E. Pappus either absent or consisting of scales, awns, or bristles, but not soft and hair-like (capillary); branches of style usually with appendages, or sometimes without them. . . . . *F*

F. Bracts of involucre without scarious (thin, dry, not green) or hyaline (whitish or transparent) margins; leaves opposite or alternate; branches of style usually but not always with appendages. . . . . . . . . . . . . . . . . . . . . . . . . . . . *G*

G. Receptacle chaffy, with bracts or pales accompanying some or all of the disk flowers; style branched or uncleft . . . . . . . . . . Tribe V. Heliantheae (p. 1536)

25. Iva
26. Ambrosia
27. Franseria
28. Xanthium
29. Polymnia
30. Silphium
31. Berlandiera
32. Engelmannia
33. Parthenium
34. Heliopsis
35. Eclipta
36. Rudbeckia
37. Echinacea
38. Dracopis
39. Ratibida
40. Spilanthes
41. Helianthus
42. Verbesina
43. Thelesperma
44. Coreopsis
45. Bidens
46. Cosmos
47. Marshallia
48. Galinsoga

G. Receptacle naked, without chaff (in *Gaillardia* with chaff-like bristles but these do not individually subtend each disk flower); style forked. . Tribe VI. Helenieae (p. 1594)

49. Flaveria
50. Hymenopappus
51. Palafoxia
52. Helenium
53. Gaillardia
54. Dyssodia
55. Tagetes

F. Bracts of involucre with scarious (thin, dry, not green) or hyaline (whitish or transparent) margins; leaves alternate; branches of style without appendages and with a ring of hairs at the end . . . . . . . . . . . . . . . . . Tribe VII. Anthemideae (p. 1599)

56. Achillea
57. Anthemis
58. Matricaria
59. Chrysanthemum
60. Tanacetum
61. Artemisia

E. Pappus consisting of soft hair-like (capillary) bristles; branches of style without appendages and with a ring of hairs at the end; receptacle naked; bracts of involucre in usually 1 equal series . . . . . . . . . . . . . . . . . . . . . Tribe VIII. Senecioneae (p. 1611)

62. Erechtites
63. Cacalia
64. Senecio

B. Stems, leaves, and/or bracts of involucre usually spiny; bracts of involucre usually prickly or finely fringed; flower-heads without ray flowers, but the outer (marginal) disk flowers sometimes larger and irregularly cut with elongate corolla-lobes resembling deeply cut rays; receptacle usually furnished with numerous bristly chaff or sometimes naked; anthers tailed (caudate) at base; styles with a hairy or thickened ring below their branches . . . . . . . . Tribe IX. Cynareae (p. 1619)

65. Arctium
66. Carduus
67. Cirsium
68. Onopordum
69. Centaurea

## Subfamily II. **Liguliflorae**

Corollas all ligulate (flattened, strap-shaped, or ribbon-like), the flowers all perfect; plants with milky juice . . . . . . . . . . . . . . . . . . . . . Tribe X. Cichorieae (p. 1628)

70. Lapsana
71. Serinia
72. Cichorium
73. Krigia
74. Hypochaeris
75. Picris
76. Tragopogon
77. Taraxacum
78. Sonchus
79. Lactuca
80. Lygodesmia
81. Agoseris
82. Crepis
83. Pyrrhopappus
84. Prenanthes
85. Hieracium

## Subfamily **Tubuliflorae**

### Tribe I. **VERNONIEAE**

#### 1. **Vernonia** Schreb. Ironweed

The species below recorded all hybridize with one another, making specific identification uncertain in many cases. The following key, therefore, records the major characters which distinguish each species, plants showing intergradation and overlapping of characters being interpreted as putative hybrids. Some of the following data are based on the work of Cora Shoop Steyermark (Bot. Gaz. 100: 548–62. 1939).

a. Bracts of the involucre prolonged into thread-like slender tips; heads large, 55–115-flowered; involucres mostly 12–18 mm. high, 12–20 mm. broad . . . . . . . . . . . . . . 1. V. CRINITA
a. Bracts of the involucre obtuse (blunt), acute (short-pointed), or acuminate (long-pointed); heads smaller, 14–55- (rarely 60-) flowered; involucres mainly 4–12 mm. high, mainly 4–15 mm. broad . . . *b*
b. Lower surface of leaves glabrous and (when dry) pitted with dot-like depressions; outer series of pappus bristles of varying lengths; inflorescence dense and fastigiate with heads compact and close together . . . . . . . . . . . . . . . . . . . . . . . . 5. V. FASTIGIATA
b. Lower surface of leaves minutely to densely hairy, not pitted (when dry) with dot-like depressions; outer series of the pappus consisting of flattened scales of nearly uniform length, sharply contrasting with the bristles of the inner series; inflorescence more loosely-flowered, the cymes often more paniculate or irregular . . . . . . . . . . . . . . . . . . . . . . . . . . *c*
c. Inner bracts of the involucre long-pointed (acuminate) and long-tapering, most often slightly or abruptly recurved (mostly appressed in var. *interior*), the midrib usually prominent . . 4. V. BALDWINI
c. Inner bracts of involucre blunt (obtuse) or rounded, sometimes ending in a minute abrupt short-pointed tip, all appressed (lying parallel to or pressed against main surface), the midrib not prominent . . . . . . . . . . . . . . . . . . . . . . . . . . . *d*
d. Lower surface of leaves softly or densely hairy with rather long crooked hairs; flower-heads 34–55- (up to 60-) flowered . . . . . . . . . . . . . . . . . . . 3. V. MISSURICA
d. Lower surface of leaves minutely hairy with straight hairs up to 0.3 mm. long, the hairs on the midrib denser; flower-heads 13–30-flowered . . . . . . . . . . . 2. V. ALTISSIMA

1. **Vernonia crinita** Raf. Map 2111

Flowers July–October.

Occurs usually on gravel and sand bars along Ozark streams, also along borders of sloughs, in thickets, open woods, rocky slopes, wet meadows, prairies, and glades. Chiefly in the Ozark region of southern and east-central Missouri, north to St. Charles, Warren, Montgomery, Boone, Pettis, St. Clair, and Jasper counties, locally northwest in Jackson and Clay counties.

In its typical development this species is easily recognized by the relatively large heads, long, curling or loose and elongated bracts of the involucre, and linear or linear-lanceolate leaves. However, the species often appears unstable. Plants grown at the author's wildflower garden from seeds taken from heads of typical *V. crinita* often develop into individuals which resemble hybrids between this species and *V. altissima* or *V. missurica*, rather than characteristic *V. crinita*. Similar discrepancy between offspring and parent plant in this species has been recorded by Mrs. Cora Steyermark (Bot. Gaz. 100: 558. 1939) in her growing experiments, although in other cases (loc. cit. p. 554–55) she found close resemblance between parent and offspring. She also found indications in wild plants which suggested that all the species occurring in the Ozark region furnish evidence of crossing with *V. crinita* or exhibited certain characters resembling that

2111 • Vernonia crinita (Ironweed)
2111 □ Vernonia crinita X fasciculata
2111 x Vernonia crinita X missurica
2111 † Vernonia crinita X Baldwini var. interior

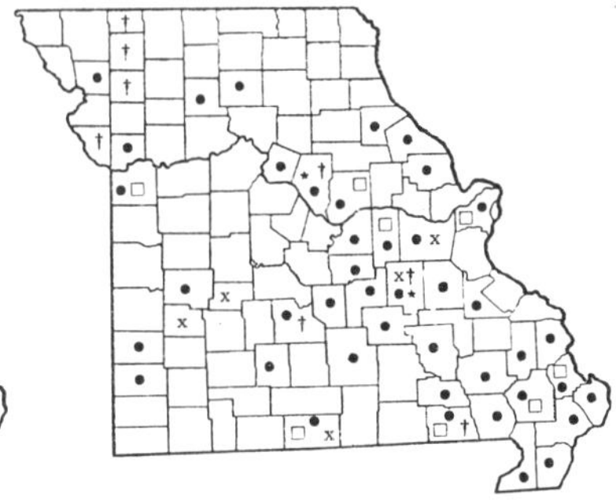

2112 • Vernonia altissima var. altissima f. altissima
2112 ∗ Vernonia altissima var. altissima f. alba
2112 □ Vernonia altissima var. taeniotricha
2112 x Vernonia altissima X crinita
2112 † Vernonia altissima X missurica

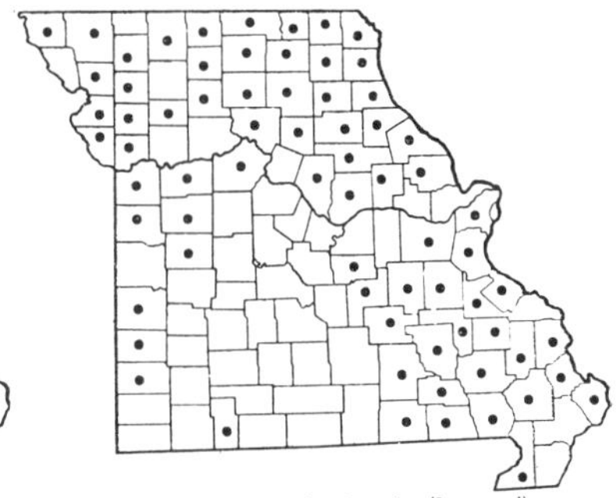

2113 Vernonia missurica f. missurica (Ironweed)

species. More genetical and cytological studies are needed to clarify the true status of this species.

In the wild state, hybrids have been found between this species and *V. fasciculata*, *V. missurica*, and *V. Baldwini* var. *interior*, with the commonest crosses occurring between *V. crinita* and *V. Baldwini*.

2. **Vernonia altissima** Nutt. Map 2112

*Vernonia albiflora* Raf.

Flowers August–October.

Occurs most frequently in low woods, alluvial ground in woods along streams and in valleys, low thickets, borders of swamps, spring branches, meadows, and prairies. Scattered in Missouri in the southern and central sections mainly, north to Ralls, Linn, Livingston, and Andrew counties.

The following variations are known in the state:

a. Corollas white . . . . . 2b. V. ALTISSIMA var. ALTISSIMA f. ALBA

a. Corollas purple or rose-purple . . . . . *b*

b. More commonly encountered type; lower surface of leaves minutely and rather sparsely hairy with rather short, appressed hairs; main stem and branches of inflorescence glabrous (without hairs) or nearly so . . 2a. V. ALTISSIMA var. ALTISSIMA f. ALTISSIMA

b. Less commonly encountered type; lower surface of leaves more densely hairy with spreading, sordid and longer many-celled hairs; main stem and branches of inflorescence rather densely hairy with spreading many-celled hairs . . 2c. V. ALTISSIMA var. TAENIOTRICHA

2a. **Vernonia altissima** var. **altissima** f. **altissima** Map 2112

*Vernonia altissima* Nutt. [G, BB, P & S]

This is the common type encountered in Missouri.

Ranges from Georgia to Louisiana, north to New York, Indiana, Illinois, Missouri, and Oklahoma.

2b. **Vernonia altissima** var. **altissima** f. **alba** Moldenke Map 2112

*Vernonia altissima* f. *alba* Moldenke, Boissiera 7: 5. 1943.

*Vernonia albiflora* Raf.

Known only from Crawford (10 mi. west of Steelville, Meramec River, September 4, 1931, *Shoop 1005*; same locality, August 31, 1932, *Shoop 1883*) and Boone (dry woods 2 mi. east of Columbia, August 4, 1932, *Shoop 1207*) counties.

This form has pure white flowers, a tan or golden pappus, and pale green stems, leaves, and involucral bracts usually glabrous. Mrs. Steyermark grew seedlings of this white-flowered form and found that 16 out of 65 plants of the first ($F_1$) generation developed white flowers, and of 4 seeds which germinated from the white-flowered plants of the second ($F_2$) generation, 2 developed into white-flowered f. *alba* (Bot. Gaz. 100: 558–59. Table 4. 1939). Her observations (loc. cit. p. 552) of wild plants indicate that the stems, leaves, and bracts of the f. *alba* are more glabrous than those of typical purple-flowered *V. altissima.*

2c. **Vernonia altissima** var. **taeniotricha** Blake Map 2112

*Vernonia altissima pubescens* Daniels [Daniels]

---

Plate no. 348. **1.** Vernonia crinita, × 2/5. **2.** Vernonia altissima, × 2/5; After Gleason, The New York Botanical Garden. **3.** Vernonia missurica, × 2/5. **4.** Vernonia Baldwini, × 2/5. **5.** Vernonia fasciculata, × 2/5. All details from Small, The New York Botanical Garden.

PLATE NO. 348

Scattered in southern and central Missouri north to St. Louis, Callaway, and Jackson counties.

Ranges from Pennsylvania to Michigan, Illinois, and Missouri, south to Mississippi.

This species of *Vernonia* hybridizes with *V. crinita* and *V. missurica*, the plants called *V. chrysopappa* Daniels possibly representing apparent hybrids of either *V. altissima* and *V. missurica* or *V. altissima* and *V. fasciculata*.

The heads of *V. altissima* are generally loosely arranged in the inflorescence and generally smaller than in the other native Missouri species.

3. **Vernonia missurica** Raf. f. **missurica** Map 2113

*Vernonia missurica* Raf. [G, BB, P & S]
*Vernonia interior* Small var. *Drummondii* (Shuttlw.) Mackenz. & Bush [P & S]
*Vernonia illinoensis* Gleason
*Vernonia Reedii* [Daniels]

Flowers July–September.

Occurs in moist open ground along streams, low woods, wooded swamps, low meadows, prairies, fields, pastures, and waste places. Throughout Missouri, except absent from much of the western Ozark and unglaciated prairie sections of the state.

Ranges from Ontario and Ohio to Iowa and Nebraska, south to Alabama, Mississippi, Arkansas, Oklahoma, and Texas.

This species hybridizes in Missouri with *V. crinita*, *V. altissima*, and *V. Baldwini* var. *interior*. It would appear that numerous specimens collected by Daniels and labeled by him *V. Reedii*, *V. Drummondii* Shuttlw., as well as various collections indicated by him as hybrids, should be referred to *V. missurica*, and they are considered as such in the present flora. A form with white flowers, f. *Swinkii* Steyerm., is not yet recorded from Missouri.

4. **Vernonia Baldwini** Torr. Map 2114

Flowers late May–September.

Occurs in fields, prairies, glades, dry wooded or rocky slopes, meadows, waste ground, pastures, thickets, and along railroads.

The following variations occur in Missouri:

a. Middle bracts of the involucre ascending, erect, or appressed (lying close against or pressed to surface), chiefly glabrous (without hairs) on the inner surface . . 4c. V. BALDWINI var. INTERIOR
a. Middle bracts of the involucre with spreading or recurved tips, hairy on the inner surface . *b*
b. Flowers purple or rose-purple . . . . 4a. V. BALDWINI var. BALDWINI f. BALDWINI
b. Flowers white . . . . 4b. V. BALDWINI var. BALDWINI f. ALBIFLORA

4a. **Vernonia Baldwini** Torr. var. **Baldwini** f. **Baldwini** Map 2114

*Vernonia Baldwini* Torr. var. *Baldwini* [BB]
*Vernonia Baldwini* Torr. [G, P & S, Steyerm.]
*Vernonia Baldwini* var. *parthenioides* (Daniels) Rickett [Rickett]
*Vernonia parthenioides* Daniels [Daniels]
*Vernonia Duggariana* Daniels [Daniels]
*Vernonia flavipapposa* Daniels [Daniels]
*Vernonia peralta* Daniels [Daniels]
*Vernonia pseudobaldwinii* Daniels, in part [Daniels]

Common throughout southern and central Missouri north to Marion, Monroe, Chariton, and Jackson counties, with local stations north in Schuyler, Putnam, Sullivan, and De Kalb counties.

Ranges from Illinois to Nebraska, south to Arkansas and Oklahoma.

4b. **Vernonia Baldwini** var. **Baldwini** f. **albiflora** Palmer & Steyerm. Map 2114

*Vernonia Baldwini* f. *albiflora* Palmer & Steyerm.

Known only from Lawrence County, southwestern Missouri (open rocky bank, border of woods, along Turnback Creek and highway 166, 3½ mi. east of Chesapeake, September 7, 1955, *Palmer 61259*, holotype in Chi. Nat. Hist. Mus. Herb.).

4c. **Vernonia Baldwini** var. **interior** (Small) Schubert Map 2114

*Vernonia interior* Small [P & S, Daniels]
*Vernonia pseudodrummondii* Daniels [Daniels]
*Vernonia pseudobaldwinii* Daniels, in part [Daniels]

Scattered throughout northern, central, and western Missouri, rare or absent from most of the Ozark section, so far as records go.

Ranges from Illinois to Minnesota and Nebraska, south to Arkansas and Texas.

Shinners (Field & Lab. 18: 26. 1950) considers this variety synonymous with *V. Baldwini* var. *Baldwini*.

The above two varieties intergrade, but may be recognized as extreme variations of one species. Hybrids have been recorded in Missouri between *V. Baldwini* var. *Baldwini* and *V. crinita* and between *V. missurica* and *V. Baldwini* var. *interior*.

As indicated by the synonymy, many of Daniels' names can be referred to either of the two above varieties, with part of the material he labeled *V. pseudo-*

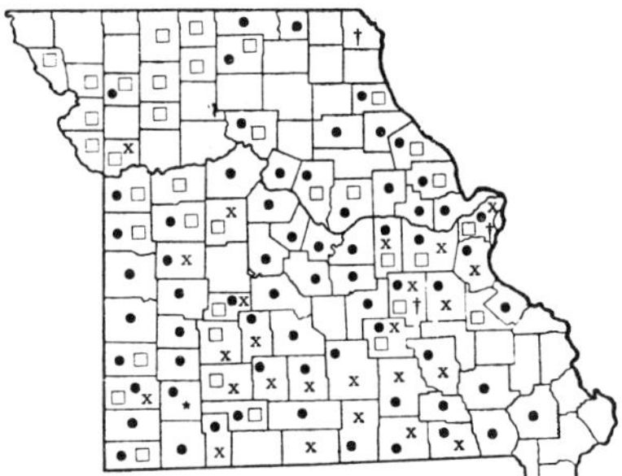

2114 • Vernonia Baldwini var. Baldwini f. Baldwini
2114 ∗ Vernonia Baldwini var. Baldwini f. albiflora
2114 □ Vernonia Baldwini var. interior
2114 x Vernonia Baldwini X crinita
2114 † Vernonia Baldwini var. interior X missurica

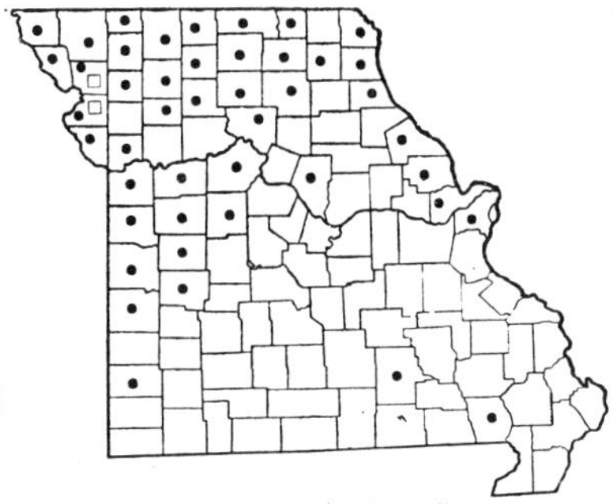

2115 • Vernonia fasciculata var. fasciculata
2115 □ Vernonia fasciculata X Baldwini var. interior

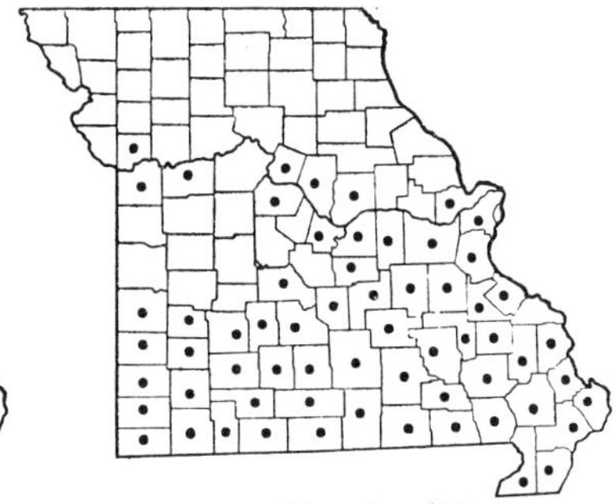

2116 Elephantopus carolinianus f. carolinianus (Elephant's-foot)

*baldwinii* referable to both var. *Baldwini* and var. *interior*. Also various collections labeled by him as hybrids can be referred here.

*Vernonia Baldwini* var. *Baldwini* is the earliest one of the Missouri ironweeds to flower. It also grows in drier. situations than most of the other species. It frequently hybridizes in the Ozark region with *V. crinita*.

5. **Vernonia fasciculata** Michx. var. **fasciculata** Map 2115

*Vernonia fasciculata* Michx. [P & S, G]

Flowers July–September.

Occurs in wet river bottom prairies and meadows, alluvial soils along streams and in valleys, upland fields, and rich moist soil. Occurs in western, northern, and central Missouri north and west of a line drawn from St. Louis, Boone, Pettis, Henry, St. Clair, and Jasper counties, and locally southeast in Butler and Shannon counties; absent from most of the Ozark region.

Ranges from Ohio to North Dakota, south to Missouri, Oklahoma, and Texas.

In typical var. *fasciculata* the leaves are linear to linear-lanceolate, and the inner bracts of the involucre are 1.5–2 mm. broad. In var. *corymbosa* (Schwein.) Schub., not known from Missouri, the leaves are ovate-lanceolate and the inner bracts of the involucre are 2–3 mm. broad. Specimens labeled *V. corymbosa* Rydb. [= *V. fasciculata* var. *corymbosa*] by Bush from Jackson County (Courtney, September 6, 1935, *Bush 15145, 15155*) have been identified by the present author as hybrids between *V. altissima* and *V. missurica* and are so treated in the present flora. These collections have rather broad leaves puberulent beneath, small numerous heads, and appressed involucral bracts with tight to slightly loosened tips.

Hybrids between *V. fasciculata* and *V. missurica* and between *V. fasciculata* and *V. Baldwini* var. *interior* are known in Missouri. This species usually grows in low moist prairies and swampy river bottom meadows in alluvial soils, and is recognized generally by its glabrous leaves, stems, and heads, narrow linear to linear-lanceolate leaves, and densely fastigiately clustered heads of flowers.

## 2. **Elephantopus** L. Elephant's-foot

**Elephantopus carolinianus** Willd. f. **carolinianus**
Elephant's-foot Map 2116

*Elephantopus carolinianus* Willd. [G, BB, P & S]

Flowers August–October.

Occurs in wooded lowlands in valleys, ravines, base of bluffs, along streams, alluvial thickets, and open woodland. Southern and central Missouri north to St. Charles, Callaway, Boone, Howard, Lafayette, and Clay counties.

Ranges from Florida to Texas, north to New Jersey, Pennsylvania, West Virginia, Ohio, Indiana, Illinois, Missouri, and Kansas; also in the West Indies.

In typical *E. carolinianus* f. *carolinianus* the upper part of the stem and branches have scattered appressed hairs; in f. *vestitus* Fern., not yet recorded from Missouri, the pubescence is denser, more ashy-gray, with spreading hairs mixed with the shorter denser pubescence on the upper part of the stem and branches.

## Tribe II. **EUPATORIEAE**

### 3. **Eupatorium** L. Thoroughwort

a. Some or all of the leaves of the stem in whorls (circles) mainly of 3–7, occasionally some of the upper leaves opposite . . . . . . . . . . . . . . . . . . . . . . . . *b*
  b. Leaves 0.5–5 mm. broad . . . . . . . . . . . . . 5. E. HYSSOPIFOLIUM var. CALCARATUM
  b. Leaves 25–150 mm. broad . . . . . . . . . . . . . . . . . . . . . . . *c*
    c. Commonly encountered throughout Missouri; stem solid, purplish only at the nodes (where leaves join stem); corolla 5.5–7.5 mm. long; lower surface of leaves not resin-dotted . 2. E. PURPUREUM
    c. Rarely found; stem hollow, more or less purplish throughout; corolla 3.5–4.8 mm. long; lower surface of leaves usually resin-dotted . . . . . . . . . . . . . . 1. E. FISTULOSUM
a. Leaves mainly opposite in pairs on the stem . . . . . . . . . . . . . . . . . . . *d*
  d. Base of leaves united around the stem and encircling it (perfoliate) . . . . . 7. E. PERFOLIATUM
  d. Leaves distinct, their bases not united around the stem nor encircling it . . . . . . . . . *e*
    e. Heads 8–40- (or more) flowered; larger principal leaves usually on definite petioles, or sessile or nearly so in *E. perfoliatum* var. *cuneatum* . . . . . . . . . . . . . . . . . . . *f*
      f. The larger principal leaves usually sessile (without leaf-stalks) or with very short petioles (leaf-stalks) . . . . . . . . . . . . . . . . 7. E. PERFOLIATUM var. CUNEATUM
      f. The larger principal leaves on definite petioles (leaf-stalks) . . . . . . . . . . . *g*
        g. Bracts of involucre definitely graduated (imbricated) in 2 or 3 series; tips of flower-buds and backs of corolla-lobes glabrous (without hairs); flower-heads mainly 12–15-flowered (sometimes 9); upper parts of stems and branches with an ashy-gray hairiness. 8. E. SEROTINUM
        g. Bracts of involucre equal or nearly so, in 1 or 2 series with the outer ones nearly or about as long as the inner ones; tips of flower-buds and backs of corolla-lobes hairy; flower-heads mainly 15–40- to 70-flowered (rarely 13); stems and branches mainly green, brown, or purplish even when hairy . . . . . . . . . . . . . . . . . . . . . . . *h*
          h. Flowers white; leaves usually abruptly narrowed or tapering at base, sometimes truncate (as if cut flat across) or slightly heart-shaped, coarsely sharply toothed; bracts of involucre blunt (obtuse) or merely acute (short-pointed) . . . 9. E. RUGOSUM
          h. Flowers blue, purple, or pink; leaves usually heart-shaped or nearly heart-shaped at base, truncate or rarely narrowed at base, crenately toothed (with rounded or scalloped toothing); at least the outer bracts of involucre short- to long-pointed. . . . . *i*
            i. Commonly found throughout southern and central Missouri; flowers usually blue or lavender; flower-clusters at the ends of the branches; creeping underground stems present; mostly 30–70 flowers in each head; receptacle cone-shaped; outer bracts of involucre about as long as the inner ones . . . . . . 11. E. COELESTINUM
            i. Rare species of southeastern Missouri lowlands, locally west to Ozark County; flowers pink or lilac; flower-clusters on slender branches arising from many of the axils of the stem; no creeping underground stems present; mostly 18–24 flowers in each head; receptacle flat; outer bracts of the involucre mostly half as long as the inner . . . . . . . . . . . . . . . . . . . . 10. E. INCARNATUM
    e. Flower-heads 3–7-flowered; the larger principal leaves usually sessile (without leaf-stalks) or with very short petioles (leaf-stalks) . . . . . . . . . . . . . . . . . . . . *j*
      j. Base of leaves broadly rounded or somewhat heart-shaped . . . . . . . 6. E. SESSILIFOLIUM
      j. Base of leaves tapering or narrowed . . . . . . . . . . . . . . . . . . . . *k*
        k. Largest leaves 0.5–5 mm. broad, 8–40 times as long as broad; inner bracts of involucre narrowly oblong, blunt (obtuse) to rounded at summit . 5. E. HYSSOPIFOLIUM var. CALCARATUM
        k. Largest leaves 10–30 mm. broad, $2\frac{1}{2}$–7 times as long as broad; inner bracts of involucre oblong, broadly oblong to elliptic, rounded or broadly rounded at summit . . . . . . . *l*
          l. Common species found throughout Missouri; involucre 5–7 mm. high; principal side (lateral) nerves arising from the very base of the leaf distinct from the midrib 3. E. ALTISSIMUM
          l. Rarely found species of southeastern Missouri and locally in St. Louis County; involucre (in flower) 2.5–4 mm. high; principal side (lateral) nerves arising as branches from the midrib above the narrowed portion of leaf-blade . . . . . 4. E. CUNEIFOLIUM var. SEMISERRATUM

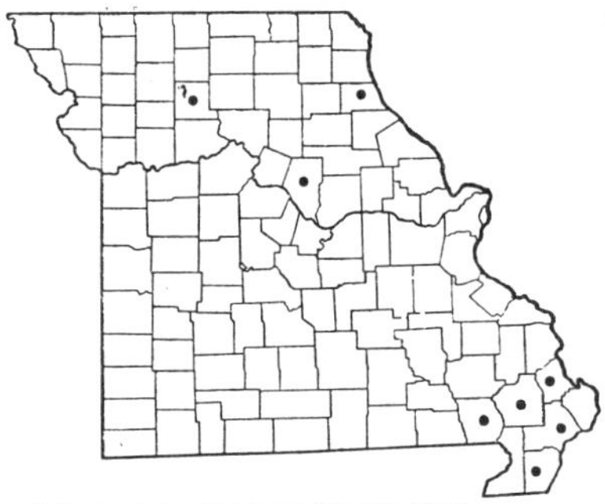

2117 Eupatorium fistulosum (Joe-Pye Weed)

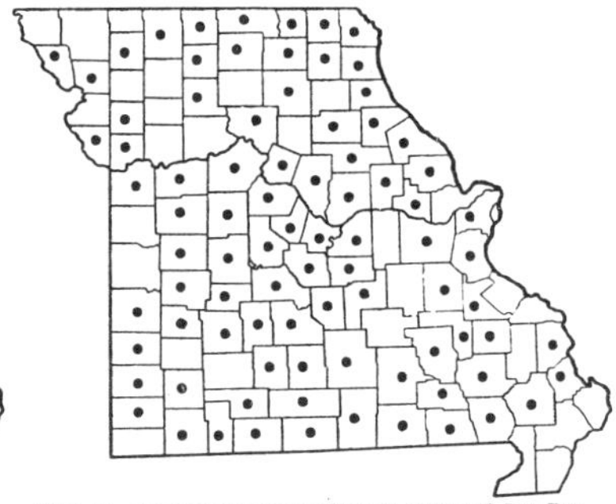

2118 Eupatorium purpureum (Green-stemmed Joe-Pye Weed)

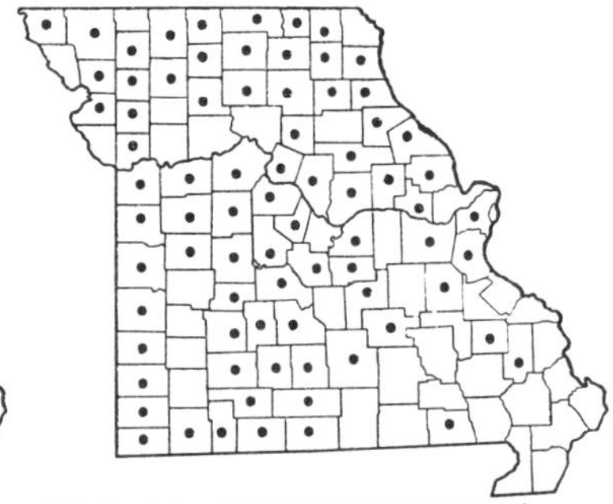

2119 Eupatorium altissimum (Tall Thoroughwort)

1. **Eupatorium fistulosum** Barratt Joe-Pye Weed Map 2117

*Eupatorium purpureum* [of P & S] in part, not L.

Flowers July–September.

Occurs in moist low ground, in wet meadows, and along spring branches.

Chiefly in the southeastern lowlands and locally in northern and central Missouri in Marion, Boone, and Livingston counties.

Ranges from Florida to Texas, north to Maine, Quebec, New York, Ohio, Indiana, Illinois, Iowa, and Oklahoma.

2. **Eupatorium purpureum** L. Green-stemmed Joe-Pye Weed Map 2118

*Eupatorium falcatum* Michx. [P & S]

Flowers July–September.

Occurs in low moist ground, wooded slopes, swampy meadows and thickets, and borders of streams and sloughs. Throughout Missouri.

Ranges from New Hampshire to Minnesota and Nebraska, south to Florida, Tennessee, Arkansas, and Oklahoma.

3. **Eupatorium altissimum** L. Tall Thoroughwort Map 2119

Flowers August–October.

Occurs in prairies, open dry or rocky woods, fallow fields, alluvial thickets, and waste ground. Throughout Missouri, but rare or absent from most of the southeastern section.

Ranges from Pennsylvania to Minnesota and Nebraska, south to North Carolina, Alabama, Missouri, and Texas.

4. **Eupatorium cuneifolium** Willd. var. **semiserratum** (DC.) Fern. & Grisc. Map 2120

*Eupatorium semiserratum* DC. [BB]

Flowers August–September.

Occurs in swamps, low meadows, wet prairies, low fields, and low open woods. Known only from southeastern Missouri in Dunklin, Butler, and Ripley counties, locally north in St. Louis County.

Ranges from Florida to Texas, north to Virginia, Tennessee, and Missouri.

Although the involucre is 2.5–4 mm. in flower, it may become 5–5.5 mm. in fruit.

5. **Eupatorium hyssopifolium** L. var. **calcaratum** Fern. & Schub. Map 2121

*Eupatorium hyssopifolium* [of BB] in part, not L.

Flowers August–November.

Occurs in dry open ground bordering upland sinkhole ponds. Known only from Howell County, southern Missouri (dry open places in dry upland bordering Twin Ponds, east of highway A, T23N, R8W, NW¼ sect. 16, 4 mi. south of West Plains, September 3, 1949, *Steyermark 69063*).

Ranges from Georgia to Texas, north to Massachusetts, Rhode Island, Connecticut, New York, New Jersey, Pennsylvania, Maryland, and locally in Missouri.

This unusual member of the Missouri flora is one of the isolated relict species of the coastal plain flora which has survived around the ancient upland sinkhole ponds of the Ozarks (Rh. 55: 15–17. 1953).

*Eupatorium cuneifolium* var. *semiserratum* and *E. hyssopifolium* var. *calcaratum* can be easily confused. They are distinguished chiefly by the much narrower leaves of *E. hyssopifolium*, not toothed at the tip and with smaller clusters or fascicles of leaves in the axils of the main leaves of the stem, by the brighter flowers and slightly larger involucres of *E. hyssopifolium*, and by the involucre of *E. hyssopifolium* being more distinctly imbricated in 2 or 3 series.

6. **Eupatorium sessilifolium** L. var. **Brittonianum** Porter Upland Boneset Map 2122

*Eupatorium sessilifolium* [of P & S] in part, not L.

Flowers July–September.

2120 Eupatorium cuneifolium var. semiserratum

2121 Eupatorium hyssopifolium var. calcaratum

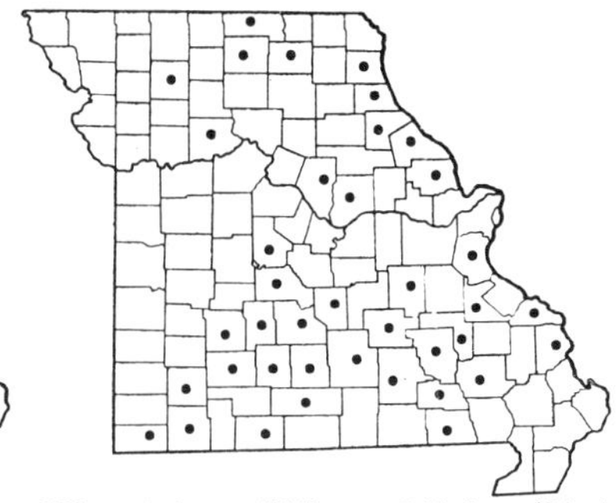

2122 Eupatorium sessilifolium var. Brittonianum (Upland Boneset)

Occurs in rocky or dry open woods, along ledges and bluffs, and banks of streams.

Chiefly in the Ozarks of southern and east-central Missouri, and parts of northern Missouri west locally to Putnam, Sullivan, Daviess, and Carroll counties; absent from the unglaciated prairie section and northwestern Missouri.

Ranges from New Hampshire to Minnesota, south to Connecticut, New Jersey, Pennsylvania, Maryland, North Carolina, Kentucky, and Missouri.

Study of Missouri material indicates that all specimens should be referred to var. *Brittonianum*, in which the leaves average three times as long as broad, the larger ones being 8–18 cm. long and 3–6 cm. broad. In typical var. *sessilifolium*, apparently not represented by any Missouri material seen, the leaves average five times as long as broad, the larger ones being 9–18 cm. long by 2–4 cm. broad. However, intermediate specimens have been found in Missouri which average three to four times as long as broad with the width up to 4 cm. The relative merit of var. *Brittonianum* is open to further study.

7. **Eupatorium perfoliatum** L. Boneset Map 2123

Also called Thoroughwort.

Flowers July–October.

Occurs in swampy meadows, swales, wet prairies, wet ledges of bluffs, swampy ground along sloughs, ponds, and streams, low moist woodland, and wet thickets at base of bluffs.

Two varieties occur in Missouri:

Common type; base of leaves united around the stem and encircling it (perfoliate) . . . .
7a. E. PERFOLIATUM var. PERFOLIATUM f. PERFOLIATUM

Rare type; base of leaves distinct, not united around the stem nor encircling it . . . .
7b. E. PERFOLIATUM var. CUNEATUM

7a. **Eupatorium perfoliatum** var. **perfoliatum** f. **perfoliatum** Map 2123

*Eupatorium perfoliatum* L. [G, P & S]

*Eupatorium perfoliatum* var. *perfoliatum* [BB]

This is the common variation and is found throughout most of Missouri, apparently being absent only from the extreme northwestern section.

Ranges from Quebec to Manitoba, south to Florida, Alabama, Louisiana, and Texas.

Typical f. *perfoliatum* has whitish flowers; in var. *perfoliatum* f. *purpureum* Britt., not recorded in Missouri, the flowers are purple-tinged.

7b. **Eupatorium perfoliatum** var. **cuneatum** Engelm. Map 2123

Scattered throughout the range of var. *perfoliatum*, but rare.

Ranges from Indiana and Illinois to Missouri and Louisiana. This variety may represent plants which have been crossed between *E. perfoliatum* and another species, possibly *E. altissimum*.

The leaves and flowering tops of *E. perfoliatum* were formerly employed medicinally as a diaphoretic, and this and other species have also been used in medicine for treating colds and fevers, and as an emetic, tonic, and cathartic.

---

Plate no. 349. **1.** Elephantopus carolinianus, $\times {}^{2}/_{7}$; Details from Small, The New York Botanical Garden. **2.** Eupatorium sessilifolium, $\times {}^{2}/_{7}$; *a.* Involucre with florets, $\times 2{}^{2}/_{7}$; After Gleason, details from Small, The New York Botanical Garden. **3.** Eupatorium fistulosum, $\times {}^{2}/_{7}$; *a.* Involucre with florets, $\times {}^{4}/_{7}$; Details from Small, The New York Botanical Garden. **4.** Eupatorium perfoliatum, $\times {}^{2}/_{7}$; *a.* Involucre with florets, $\times {}^{4}/_{7}$; Details from Small, The New York Botanical Garden. **5.** Eupatorium altissimum, $\times {}^{2}/_{7}$; *a.* Flower-head, $\times 1{}^{3}/_{7}$. **6.** Eupatorium cuneifolium var. semiserratum, $\times {}^{2}/_{7}$; *a.* Involucre with florets, $\times 2{}^{2}/_{7}$; *b.* Leaf, $\times {}^{2}/_{7}$; After Gleason, The New York Botanical Garden. **7.** Eupatorium hyssopifolium var. calcaratum, $\times {}^{2}/_{7}$; *a.* Flower-head, $\times 2{}^{4}/_{7}$; After Gleason, The New York Botanical Garden. **8.** Ageratum conyzoides, floral details, highly magnified (Details from Small, The New York Botanical Garden).

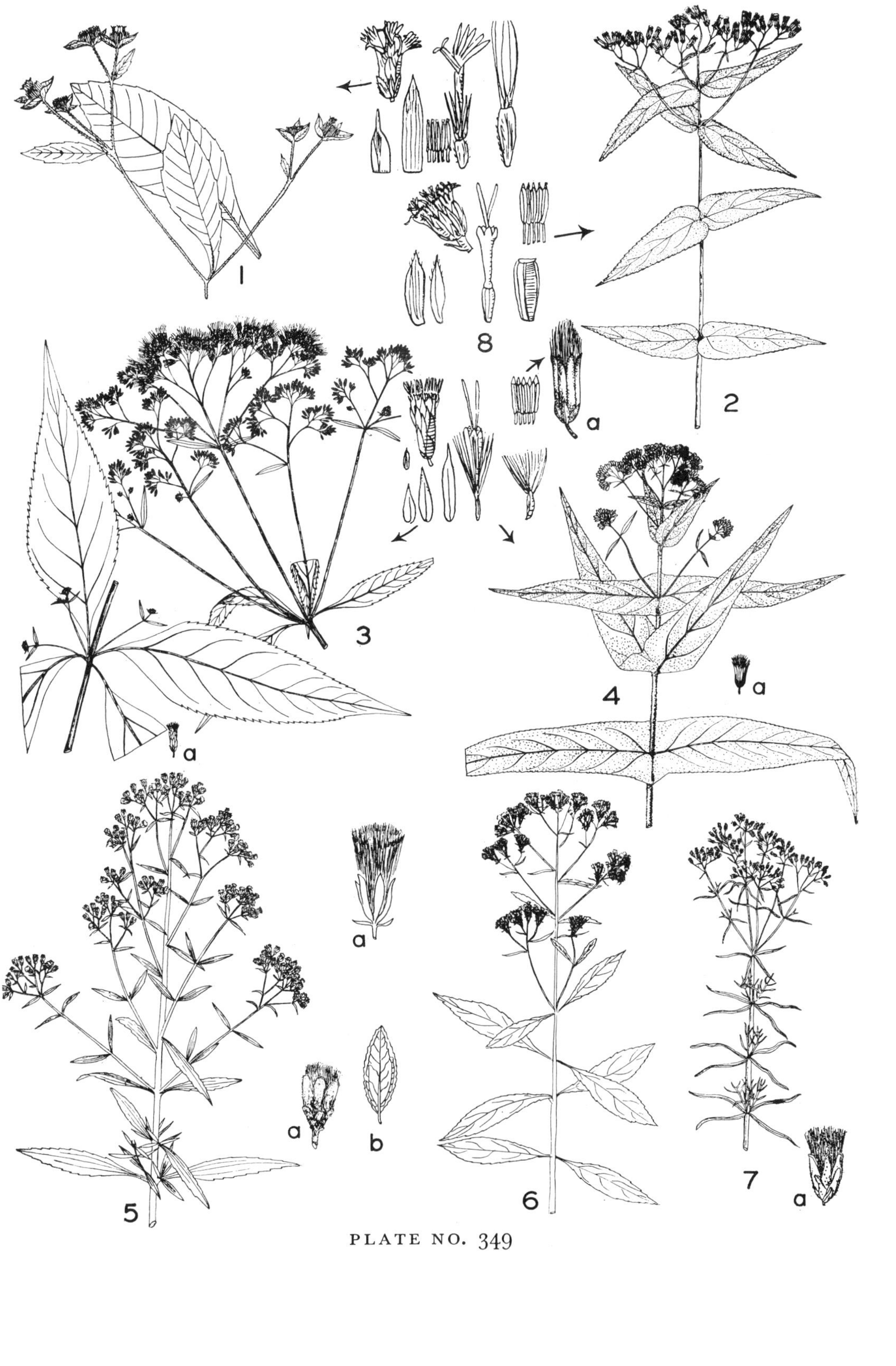


PLATE NO. 349

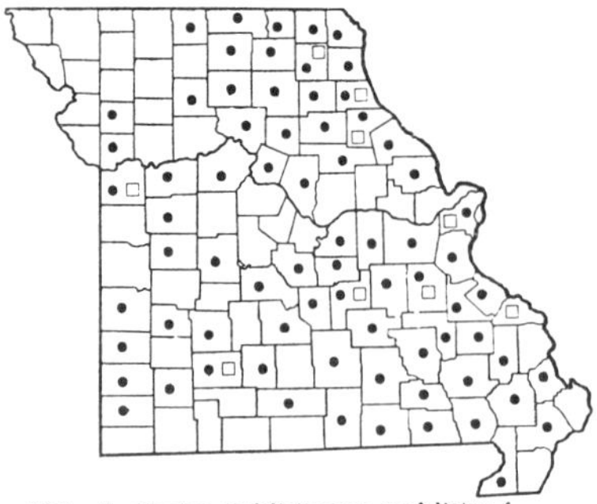

2123 • Eupatorium perfoliatum var. perfoliatum f. perfoliatum (Boneset)
2123 □ Eupatorium perfoliatum var. cuneatum

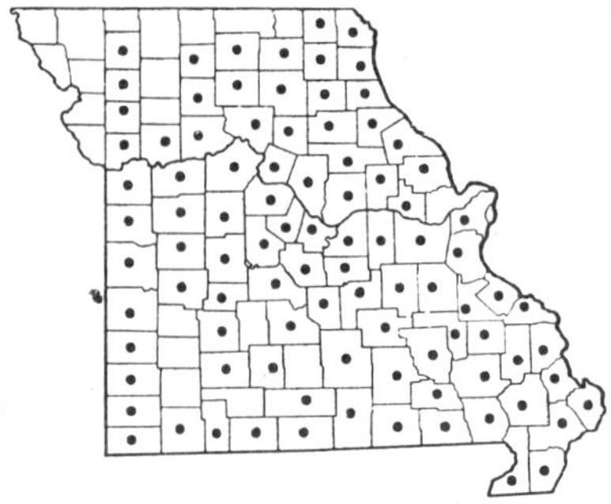

2124 Eupatorium serotinum (Late Boneset)

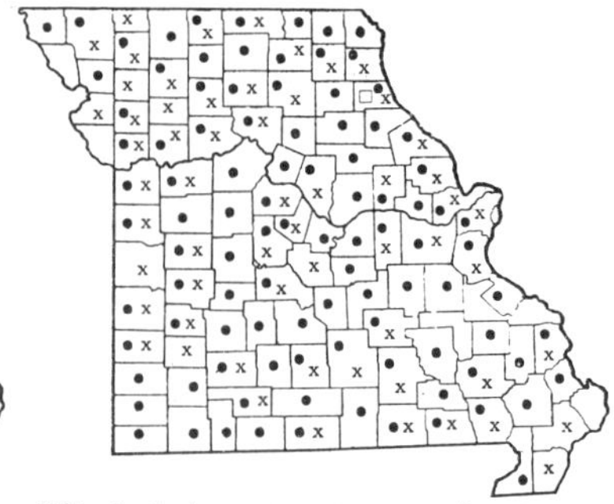

2125 • Eupatorium rugosum var. rugosum f. rugosum (White Snakeroot)
2125 □ Eupatorium rugosum var. rugosum f. villicaule
2125 x Eupatorium rugosum var. tomentellum

Mallards eat the fruits of this and various other species.

8. **Eupatorium serotinum** Michx. Late Boneset
Map 2124

Flowers August–October.

Occurs in fields, pastures, prairies, borders of swamps and meadows, roadsides and railroads. Throughout Missouri, but not recorded from some of the extreme northern and northwestern counties.

Ranges from Florida to Texas and Mexico, north to New Jersey, West Virginia, Ohio, Indiana, Wisconsin, Minnesota, and Kansas; introduced in Massachusetts.

A variety *polyneuron* Hermann with many-nerved, sessile leaves cuneate at the base and more pointed involucral bracts, is recorded only from Indiana, but may represent a hybrid between this species and *E. perfoliatum.*

This is the weediest of all the species of *Eupatorium* in Missouri. It, like other members of the genus, is usually avoided by grazing animals.

9. **Eupatorium rugosum** Houtt. White Snakeroot
Map 2125

Flowers late July–October.

Occurs in rich or rocky woods, base of and crevices of bluffs, rock outcrops, and thickets. Throughout Missouri.

The following variations occur in Missouri:

a. Stems, leaf-blades, and petioles (leaf-stalks) glabrous (without hairs) or nearly so . . . 9a. E. RUGOSUM var. RUGOSUM f. RUGOSUM
a. Stems and petioles (leaf-stalks) hairy . . . *b*
  b. Rarely found type; hairs of stem and petioles loosely hairy with rather long hairs . 9b. E. RUGOSUM var. RUGOSUM f. VILLICAULE
  b. Commonly found type; hairs of stem and petioles minute and dense . 9c. E. RUGOSUM var. TOMENTELLUM

9a. **Eupatorium rugosum** var. **rugosum** f. **rugosum** Map 2125

*Eupatorium rugosum* Houtt. [G]
*Eupatorium urticaefolium* Reichard [P & S]

Commonly occurring throughout Missouri, doubtless in every county.

Ranges from New Brunswick and Quebeç to Saskatchewan, south to Georgia, Alabama, Louisiana, and Texas.

9b. **Eupatorium rugosum** var. **rugosum** f. **villicaule** Fern. Map 2125

*Eupatorium rugosum* f. *villicaule* Fern. [G]
*Eupatorium urticaefolium* var. *villicaule* Fern. [P & S]
*Eupatorium rugosum* var. *villicaule* (Fern.) Blake

A specimen from Marion County (Ely St. ravine, Hannibal, September 18, 1912, *Davis 4090*) shows longish villous hairs which are 0.5–0.9 mm. long and approach the villous long pubescence of f. *villicaule.* The specimen is closer in pubescence to this form than to var. *tomentellum,* although the hairs are not quite as long as in typical f. *villicaule.* All the other pubescent specimens from Missouri show the short dense tomentum of var. *tomentellum* to which they are referred in the present flora.

Found within the range of typical var. *rugosum* f. *rugosum.*

9c. **Eupatorium rugosum** var. **tomentellum** (Robins.) Blake Map 2125

Common throughout Missouri.

Ranges from Michigan and Indiana to Wisconsin, Illinois, Iowa, and Missouri.

There is considerable variation in the shape of the

2126 Eupatorium incarnatum

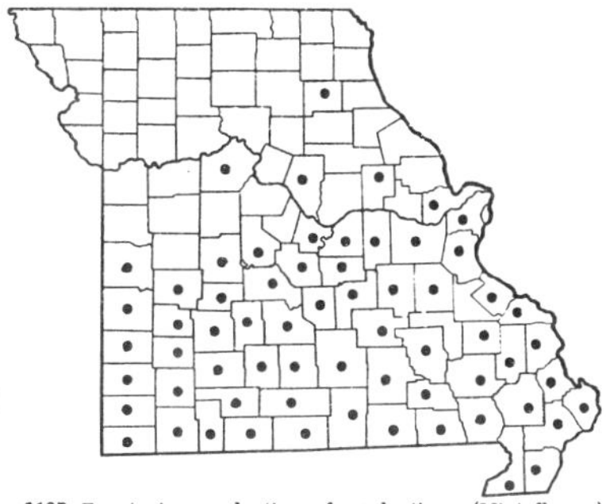

2127 Eupatorium coelestinum f. coelestinum (Mist-flower)

2127-A Ageratum conyzoides (Ageratum)

leaf, especially the base of the leaf-blade from cuneate to broadly rounded, and its extent of toothing. Dr. Kucera of the University of Missouri has been studying the White Snakeroot for the past several years, and finds that the plants from the northern portions of the range of the species begin to flower in general a month or so earlier in summer than those of Missouri.

This species is poisonous to grazing cattle and sheep which feed on the plant, and can bring on death when eaten in sufficient quantity. Fresh plants are considered to be more dangerous than dried ones. Cows eating the plant transmit in their milk the poisonous barium sulphate which is poisonous to human beings. This poison, known as 'milk sickness,' was responsible for the death of a number of pioneer families in the United States as a result of drinking quantities of the contaminated milk.

10. **Eupatorium incarnatum** Walt. Map 2126

Flowers August–October.

Occurs in low moist woods, swamps, and along base of bluffs. Southeastern lowland region north to Scott and Bollinger counties, west to Butler County, and locally west in Ozark County (low alluvial woods on north side of Little North Fork of White River, T21N, R15W, sect. 9, between Burse Ford and Nave Ford, on east side of river, 1 mi. southwest of Pontiac, September 26, 1949, *Steyermark 69386*).

Ranges from Florida to Arizona and Mexico, north to Virginia, West Virginia, Ohio, Indiana, Illinois, and Missouri.

The Norfork Dam has destroyed the Ozark County locality, a relict station which revealed a relation between the flora of the White River region and that of the southeastern Missouri lowlands.

The dried plant has an odor resembling vanilla. The species somewhat resembles the following, *E. coelestinum*, but the leaves are thinner, darker green, and generally more heart-shaped, while the flower-heads are fewer-flowered, paler lilac or pink, on longer and more elongated peduncles arising from many of the axils.

11. **Eupatorium coelestinum** L. f. **coelestinum**

Mist-flower Map 2127

Also called Blue Boneset, Wild Ageratum.

*Eupatorium coelestinum* L. [G, BB, P & S]

Flowers July–October.

Occurs in moist ground along streams, spring branches, sloughs, ponds, ditches, alluvial and low wet woods, and at the base of bluffs. Southern and central Missouri north to St. Charles, Montgomery, Boone, Saline, and Bates counties, locally north in Shelby County.

Ranges from Florida to Texas, north to New Jersey, Pennsylvania, Maryland, West Virginia, Ohio, Indiana, Illinois, Missouri, and Kansas; also in the West Indies.

In typical f. *coelestinum* the flowers are blue or blue-purple. In f. *album* E. J. Alex., they are white, and in f. *illinoense* Benke, they are red-purple. Only typical f. *coelestinum* has been found thus far in Missouri. A form with the flowers pink, and therefore approaching f. *illinoense*, has been found in Franklin County (Meramec State Park, September 15, 1940, *Geo. Moore*).

*Eupatorium coelestinum* is an attractive species and does well in perennial and other flower beds. It may become overplentiful in places because of the creeping underground stems. Of the hundreds of Missouri species transplanted or grown from seed in the author's northern Illinois wildflower preserve, this is one of the few which, though easily grown throughout Missouri, has failed to thrive after many attempts to acclimate plants to the latitude of northern Illinois. When kept in bounds, it makes a very desirable addition.

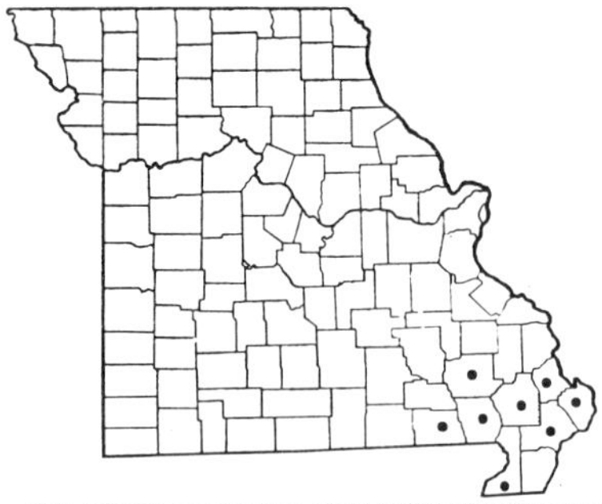

2128 Mikania scandens var. scandens (Climbing Hempweed)

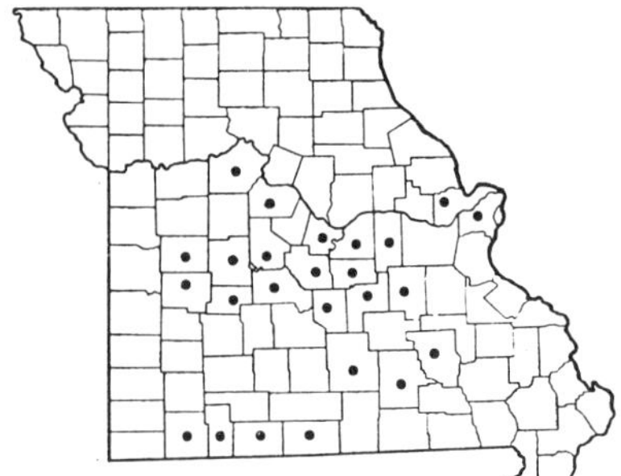

2129 Brickellia grandiflora (Tassel Flower)

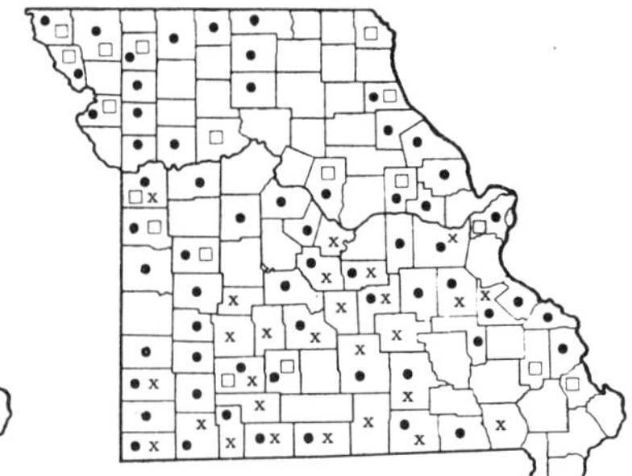

2130 • Kuhnia eupatorioides var. eupatorioides (False Boneset)
2130 □ Kuhnia eupatorioides var. corymbulosa
2130 x Kuhnia eupatorioides var. angustifolia

## 4. **Ageratum** L.

**Ageratum conyzoides** L. Ageratum Map 2127A
Flowers August–October.

Known only from Marion County, northeastern Missouri (along muddy margin of West Branch of Sees Creek (tributary of South Fork of North River), T57N, R7W, sect. 22, 4 mi. northeast of Ely, September 22, 1956, *Steyermark 82743*).

Native of tropical America.

This commonly cultivated annual is well known for its use as a low border plant, resembling in general appearance plants of *Eupatorium coelestinum.* It is of lower stature and hairier than the *Eupatorium.*

## 5. **Mikania** Willd. Climbing Hempweed

**Mikania scandens** (L.) Willd. var. **scandens**
Climbing Hempweed Map 2128
*Mikania scandens* (L.) Willd. [G, BB, P & S]
Flowers July–October.

Occurs in swampy woods, borders of bayous and ditches, and fence rows. Lowlands of southeastern Missouri north to Scott, Stoddard, and Wayne counties, west to Ripley County.

Ranges from Florida to Texas, north to Maine, and locally inland to Ontario, Michigan, Indiana, Illinois, Missouri, and Oklahoma; tropical America.

The flowers are white, lilac, or pinkish. Typical var. *scandens* has the leaves and stems glabrous or nearly so. In var. *pubescens* (Muhl.) T. & G. the stems are often pubescent and the leaves are more rounded at tip and have a rough hairiness.

This is the only Missouri representative of this large tropical genus.

## 6. **Brickellia** Ell.

**Brickellia grandiflora** (Hook.) Nutt.
Tassel Flower Map 2129
Flowers July–October.

Occurs on limestone ledges of bluffs, bluff escarpments, rocky limestone talus, and rocky wooded slopes. Ozark region of southern and central Missouri north to St. Charles, Gasconade, Osage, Cole, Cooper, Saline, and Henry counties; absent from most of the eastern Ozarks.

Ranges from Missouri and Nebraska to Washington, south to Arkansas, Texas, New Mexico, Arizona, and California.

This species has a rather fetid odor.

## 7. **Kuhnia** L. False Boneset

**Kuhnia eupatorioides** L. False Boneset
Map 2130
Flowers July–October.

Occurs in prairies, glades, rocky ledges, rocky or dry open woods, alluvial soils in low ground, and thickets.

The following key, which includes the variations found in Missouri, is based on the treatment of the genus in *Gray's Manual*, eighth ed.

a. Middle and outer bracts of the involucre with conspicuous and elongated hair-like, somewhat twisted or curved tips, usually at least some of them nearly ¾ or so as long as the inner bracts; leaves narrowly linear, 1–7 mm. broad, usually entire; involucres much shorter than their pedicels (flower-stalks) . . . . c. K. EUPATORIOIDES var. ANGUSTIFOLIA

a. Middle and outer bracts of the involucre acute or acuminate (short- to long-pointed), the tips rather flat, appressed, and erect, usually all much shorter than and less than ¾ as long as the inner bracts; leaves linear to narrowly ovate, 3–25 mm. broad, frequently toothed; involucres as long as or longer than their pedicels (flower-stalks) . . . . . . . . *b*

- b. Flower-heads with 7–14 flowers or fruits; involucre minutely hispid (rigid or bristly hairs) to nearly glabrous (without hairs); inner bracts of involucre linear, 2–4-ribbed; involucres as long as or twice as long as the pedicels (flower-stalks) . a. K. EUPATORIOIDES var. EUPATORIOIDES
- b. Flower-heads with 14–33 flowers or fruits; involucre usually rather hairy with short close hairs (tomentulose) or with soft hairs (pilose); inner bracts of involucre broader, 4–6-ribbed; involucres usually much longer than their pedicels (flower-stalks) . . . b. K. EUPATORIOIDES var. CORYMBULOSA

a. **Kuhnia eupatorioides** var. **eupatorioides** Map 2130

*Kuhnia eupatorioides* var. *glutinosa* (Ell.) Hitchc.
*Kuhnia glutinosa* Ell.
*Kuhnia eupatorioides* var. *pyramidalis* Raf. [Shinners]
Common throughout Missouri.

Ranges from Florida to Texas, north to New Jersey, Pennsylvania, Ohio, Indiana, Illinois, and Missouri.

There is considerable overlapping in characters between this and the following variety.

b. **Kuhnia eupatorioides** var. **corymbulosa** T. & G. Map 2130

*Kuhnia eupatorioides* var. *ozarkana* Shinners in part [BB, Shinners]
Scattered throughout Missouri.

Ranges from Alabama to Texas, north to Ohio, Michigan, Wisconsin, Minnesota, North Dakota, and Montana.

c. **Kuhnia eupatorioides** var. **angustifolia** Raf. Map 2130

*Kuhnia eupatorioides* var. *ozarkana* Shinners [BB, Shinners]

Occurs in glades, rocky ledges of bluffs and escarpments. Ozark region north to St. Francois, Washington, Maries, Cole, Hickory, and Jasper counties, locally in Jackson County.

Ranges from Florida to Texas and Missouri.

There is considerable intergradation between the above variations, especially between var. *eupatorioides* and var. *corymbulosa*. The var. *angustifolia*, as interpreted in the present flora, is by far the most distinct of the varieties, and appears to be confined to the rocky glades and bluff escarpments within the Ozark region.

Shinners (Wrightia 1: 122–44. 1946) has recently revised the genus, and has described a var. *ozarkana* Shinners, based on a specimen from Ozark County (bald knobs near Gainesville, 'Bald Jesse,' October 10, 1927, *Palmer 33075*), characterized by its conspicuous, elongated, spreading or twisted hairlike tips of the outer bracts of the involucre. He includes under this variety plants having leaves varying from linear-lanceolate to oblong-lanceolate, those near the middle of stem up to 8 cm. long and 1.5 cm. broad. Shinners does not recognize var. *angustifolia* as a distinct variety and merges what is passing as that with var. *pyramidalis* Raf. In the present flora I am limiting the plants having very narrow, linear to linear-lanceolate, mostly entire leaves and elongated hairlike tips of the involucral bracts to var. *angustifolia*, including in this Shinners' var. *ozarkana*.

It would appear, from study of Missouri material, that the segregation of varieties appears to fit better the categories maintained by Fernald in the eighth edition of *Gray's Manual* than those proposed by Shinners, and for that reason in the present flora the treatment of the genus follows more closely that of *Gray's Manual*.

## 8. **Liatris** Schreb. Blazing Star, Gay Feather

Although the genus has been recently monographed by Dr. L. O. Gaiser (Rh. 48: 165–83; 216–63; 273–326; 331–82; 393–412. 1946), her segregation of a number of species and varieties has not been followed in the present flora. Unfortunately, the large amount of Missouri material of the genus deposited in the herbaria of the Missouri Botanical Garden, Chicago Natural History Museum, and University of Missouri, was not studied by Dr. Gaiser. The result is that numerous Missouri specimens cannot be fitted into some of the categories segregated by Dr. Gaiser. Shinners (Am. Midl. Nat. 29: 27–41. 1943) has studied the *Liatris scariosa* complex, but his concept of this particular group of species differs from that of Dr. Gaiser. Dr. Cronquist's treatment of the genus in Gleason's *New Ill. Fl.* is also at variance with that of Dr. Gaiser, but parts of his treatment are accepted by the present author. Much more intensive study of variability within the various species of Missouri given below needs to be carried out before any final evaluation can be placed on the segregated species and varieties. The following treatment may best be considered as yet only a tentative one.

a. Pappus plumose (feather-like, with long fine hairs arising from the side of the main axis and 15–30 times the diameter of the main axis of the bristle) . . . . . . . . . . . . . . . . *b*
- b. Flower-head with generally 4–6 (rarely 3) flowers; lobes of corolla glabrous . . . . . . . . *c*
  - c. Stem arising from the crowns of a vertically elongated branched underground rootstock; margins of at least some of leaves provided with hairs or stubs of them usually remaining attached, the lower leaves mainly 3–5 mm. broad; leaves gray green or dull green, stiff and rigid, the longer leaves from middle and upper part of stem 3–8 cm. long; leaves erect to ascending, stiff; inflorescence 3–45 cm. long; corollas 9–12 mm. long; pappus usually 9–11 mm. (sometimes 7) long . . . . . . . . . . . . . . . . . . . . . 6. L. PUNCTATA
  - c. Stem arising from a nearly globose corm 3–7 cm. in diameter; margins of leaves completely glabrous (without hairs) or without any trace of stubs of hairs, the lower ones mainly 1–3 (–5) mm. broad; leaves rich grass green, mostly soft, longer leaves from middle and upper part of stem 10–20 cm. or more long; leaves spreading to recurving, flexible; inflorescence 20–60 cm. long; corollas 9–10 mm. long; pappus 6–7 mm. long . . . . . . . . 7. L. MUCRONATA
- b. Flower-head with 10–60 flowers; lobes of corolla hairy on inside . . . . . . . . . . . . *d*
  - d. Bracts of involucre erect and appressed (lying close and flat against one another and parallel to the surface), the tips obtuse (blunt), rounded, or tipped with a short or long point . . 8. L. CYLINDRACEA
  - d. Bracts of involucre with loose and outwardly spreading or slightly recurved (squarrose) tips, the tips short- to long-pointed and slender . . . . . . . . . . . . . 9. L. SQUARROSA

a. Pappus merely finely barbed (barbellate), not plumose, the hairs from the side of the main axis of the bristle only 3–6 times the diameter of the main axis. . . . . . . . . . . . . . . . . . *e*
- e. Flower-heads with 3–18 flowers, the heads cylindrical to cylindrical-turbinate (top-shaped and cylindrical), the involucre 4–10 mm. broad or thick at time of flowering . . . . . . . . . . *f*
  - f. Bracts of involucre with the tips appressed (lying flat and close against each other), rounded or blunt (obtuse); stems, leaves, and axis of inflorescence glabrous (without hairs) or nearly so; rarely found, known only from Oregon County . . . . . . . . . . . 1. L. SPICATA
  - f. Bracts of involucre with the tips spreading or recurved (squarrose), acute to acuminate (short- to long-pointed); stems and axis of inflorescence more or less hairy; leaves often hairy, sometimes glabrous; commonly encountered throughout Missouri . . . . . . 2. L. PYCNOSTACHYA
- e. Flower-heads mainly with 18–100 (rarely 12–15) flowers, the heads hemispherical, campanulate, or shortly cylindrical, the involucre 10–30 mm. broad or thick at time of flowering . . . . . . . *g*
  - g. Inside of corolla-tube glabrous (without hairs); flower-heads with 40–70 flowers 5. L. LIGULISTYLIS
  - g. Inside of corolla-tube hairy; flower-heads with 20–40 flowers (rarely 12–15) . . . . . . . *h*
    - h. Middle bracts of the involucre with an uneven, irregularly torn or lacerate broad scarious (thin, dry, not green) border; bracts of involucre mainly glabrous (without hairs), usually with purple or rose-color, the lower (outer) bracts with broadly rounded tips, all of them with outwardly spreading (squarrose) tips; corolla-tube 8–10 mm. long; pappus 7–8 mm. long; stem glabrous (without hairs) or with short, rather appressed hairs (lying parallel to surface) or minute rough hairs; leaves glabrous or with short, stiff, roughish hairs . 4. L. ASPERA
    - h. Middle bracts of the involucre more or less green and firm throughout or with a narrow, smooth-edged or at most only very slightly uneven or irregular border; bracts of involucre more or less densely hairy, the lower (outer) ones with pointed or somewhat pointed (acute to subacute) outwardly spreading (squarrose) tips, the middle and inner ones erect and appressed; corolla-tube 10–15 mm. long; pappus 8–10 mm. long; stem and leaves grayish and rough (scabrous) hairy . . . . . . . . . . . . . . . . . . 3. L. SCABRA

1. **Liatris spicata** (L.) Willd. f. **spicata** — *Liatris spicata* (L.) Willd. [G, BB, P &S]

Button Snakeroot — Map 2131 — *Liatris spicata* va r. *typica* [Gaiser]

2131 Liatris spicata f. spicata (Button Snakeroot)

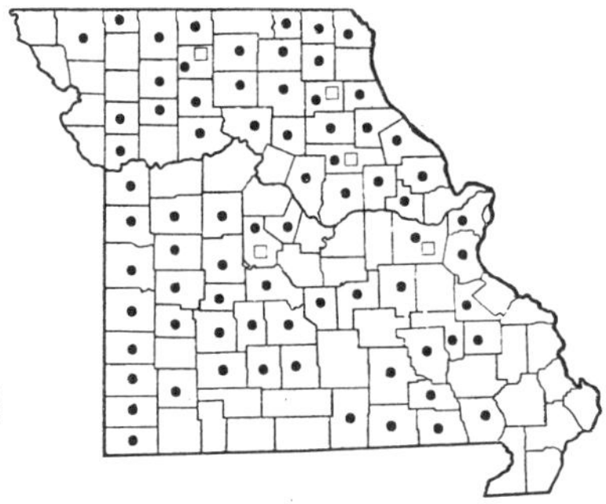

2132 • Liatris pycnostachya f. pycnostachya (Button Snakeroot)
2132 □ Liatris pycnostachya f. Hubrichtii

2133 Liatris scabra (Blazing Star)

Flowers July–August.

Occurs in upland prairies, along railroads, and gravelly washes in open dry ground. Known only from Oregon County, extreme southern Missouri (prairie opening along R.R., 3 mi. north of Koshkonong, August 9, 1934, *Steyermark 14383*, in U. of Mo. Herb.; gravelly open ground near Bardley, July 14, 1933, *Palmer & Steyermark 41717*, in Mo. Bot. Gard. Herb.).

Ranges from Florida to Louisiana, north to New York, Pennsylvania, West Virginia, Ontario, Ohio, Michigan, Wisconsin, and Missouri.

The usual color of the flowers is rose-purple. In f. *albiflora* Britt., not recorded from Missouri, the corollas are white. The flowers in a head are usually 7–9, rarely 4 or up to 14 (18 also recorded). A rather narrow-leaved extreme with leaves 1.5–5 mm. instead of 5–20 mm. broad and with 4–6 flowers (rarely more) in a head has been segregated as var. *resinosa* (Nutt.) Gaiser. The Missouri material all belongs to typical var. *spicata* with the broader leaves and more numerous flowers in a head. Dr. Gaiser did not see the Missouri material of this species, so that Missouri was not included in her geographical range data.

This species has been used as a remedy for snake bites, but there is no scientific basis for its use.

2. **Liatris pycnostachya** Michx. Button Snakeroot Map 2132

Flowers July–October.

Occurs in prairies, meadows, rocky open ground, top of bluff escarpments, and along railroads. Throughout Missouri, except absent from the lowlands of the extreme southeastern section.

Ranges from Wisconsin to Minnesota and South Dakota, south to Kentucky, Louisiana, Oklahoma, and Texas.

Two variations are encountered in Missouri:

Corollas rose-purple . . . 2a. L. PYCNOSTACHYA f. PYCNOSTACHYA

Corollas white . 2b. L. PYCNOSTACHYA f. HUBRICHTI

2a. **Liatris pycnostachya** f. **pycnostachya** Map 2132

*Liatris pycnostachya* Michx. [G, BB, P & S]

This is the common form throughout Missouri.

2b. **Liatris pycnostachya** f. **Hubrichtii** Anders. Map 2132

*Liatris pycnostachya* f. *alba* Waterfall, Rh. 52: 40. 1950.

Scattered in the range of the typical form, and known from Franklin, Audrain, Shelby, Grundy, and Morgan counties. Originally found in Franklin County (1 mi. northeast of Stanton, July 24, 1937, *Leslie Hubricht*, holotype in Mo. Bot. Gard. Herb.).

*Liatris pycnostachya* is one of the showy-flowered species, along with *L. spicata* and *L. aspera*, which are often used for ornament and indoor decoration. It is easily grown and does well in sunny perennial beds. This species was formerly used as an antidote against snake bites, but its use has not been scientifically substantiated.

3. **Liatris scabra** (Greene) K. Schum. Blazing Star Map 2133

Flowers August–November.

Occurs in rocky open woods, rocky open ledges and glades. Southern Ozark region in Iron (crevices of porphyritic trachyte, along east wall of Royal Gorge near Hogan, November 2, 1930, *Steyermark 1498*), Reynolds, Shannon (rocky woods, Monteer, October 8, 1905, *Bush 3587*), Ripley (ravine slopes tributary to Little Black River, between Greenville Ford and Pennington Ford, T24N, R3E, sect. 10, 15, 22, 23, 26, 24, and 25, 10 mi. northeast of Doniphan, September 1, 1946, *Steyermark 63966*), and Ozark (rocky glades along river bluffs near Tecumseh, October 7, 1927, *Palmer 32900*) counties.

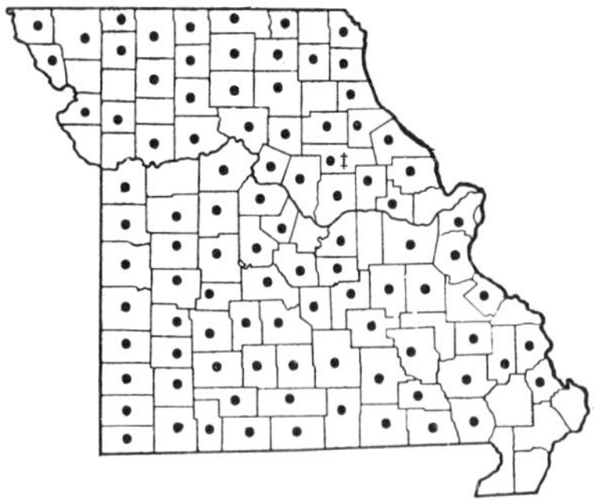

2134 • Liatris aspera f. aspera
2134 ‡ Liatris aspera f. Benkei

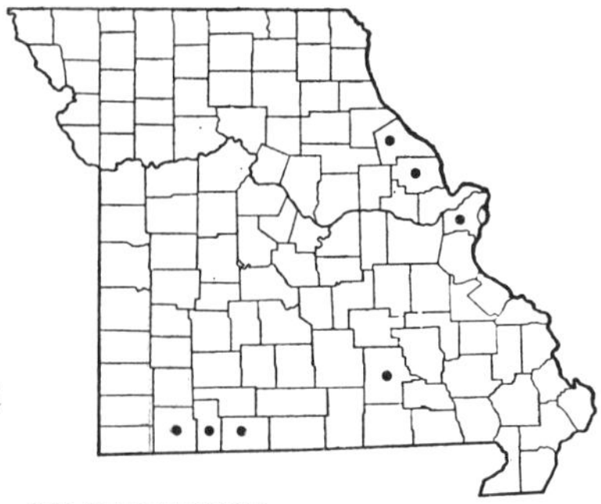

2135 Liatris ligulistylis

2136 Liatris punctata var. nebraskana (Snakeroot)

Ranges from Alabama to Louisiana and Oklahoma, north to Ohio, Indiana, Illinois, and Missouri.

The basal leaves in this species are stated by Dr. Gaiser and in *Gray's Manual*, eighth edition, to vary from 1–2.5 cm. broad, but in the specimens cited above from Iron and Shannon counties the basal leaves may be as much as 3–4.5 cm. broad. Such broad-leaved specimens would key to *L. scariosa* (L.) Willd., as interpreted by Fernald in *Gray's Manual.* However, the character of leaf-breadth is not held to be significant by Cronquist in Gleason's *New Ill. Fl.*, and does not appear to be a reliable criterion for separation of species in the *L. scabra - aspera - scariosa* complex.

4. **Liatris aspera** Michx. Map 2134

Flowers August–November.

Occurs in dry upland prairies, glades, bald knobs, meadows, rocky or dry open woods, and along railroads.

Common throughout Missouri, except absent from the lowlands of the extreme southeastern section.

Two variations may be recognized:

Flowers rose-purple . . 4a. L. ASPERA f. ASPERA
Flowers white . . . . 4b. L. ASPERA f. BENKEI

4a. **Liatris aspera** f. **aspera** Map 2134

*Liatris scariosa* of many auth. [P & S], not (L.) Willd.

*Liatris sphaeroidea* as interpreted by Sweet and of many auth., not Michx. [Shinners]

*Liatris sphaeroidea* f. *asperifolia* Shinners [Shinners]

*Liatris aspera* var. *typica* [Gaiser]

*Liatris aspera* var. *intermedia* (Lunell) Gaiser [Gaiser]

This is common throughout Missouri, and the commonest *Liatris* encountered in the state.

Ranges from Ontario to Wisconsin and North Dakota, south to Florida, Louisiana, and Texas.

4b. **Liatris aspera** f. **Benkei** (Macbr.) Fern. Map 2134

*Liatris scariosa* f. *Benkei* Macbr.

Known only from Audrain County, central Missouri (prairie relict, highway 22, 3 mi. east of Centralia, September 17, 1940, *W. B. Drew*, in U. of Mo. Herb.).

*Liatris aspera* is the species of most common occurrence in Missouri. It is very showy-flowered, is often seen in bouquet decorations in florist's shops, and is frequently used for indoor flower arrangements. It can easily be grown in perennial beds in full sun. The plants spread from seed in open ground free from coverage. However, where the ground becomes shaded by other vegetation or crowded by grasses and other plants, this *Liatris* eventually dies out or pioneers to newly available open ground. The white-flowered form is often cultivated and may be purchased from various nurseries and seed companies. This and related species are palatable and furnish nourishment to grazing animals in the early stages of growth. An extract of the boiled leaves and corms was used by the Indians for medicinal purposes.

This species is quite variable in pubescence of

Plate no. 350. **1.** Eupatorium serotinum, × 2/7 (Scribner's). **2.** Kuhnia eupatorioides, × 2/7; Details from Small, The New York Botanical Garden. **3.** Liatris cylindracea, × 2/7; *a.* Lower part of stem with corm; *b.* Upper part of stem; *c.* Involucre, × 4/7; *d.* Achene and pappus, × 15/7; After Gleason, The New York Botanical Garden. **4.** Liatris spicata, × 2/7; *a.* Lower portion of stem with leaves; *b.* Upper portion of inflorescence; *c.* Middle portion of stem with inflorescence; *d.* Involucre with florets, × 2/7. **5.** Eupatorium rugosum, × 2/7. **6.** Eupatorium coelestinum, × 2/7; After Gleason, The New York Botanical Garden. **7.** Liatris pycnostachya, × 2/7. **8.** Liatris aspera, × 2/7. **9.** Eupatorium incarnatum, × 2/7; After Gleason, The New York Botanical Garden. **10.** Mikania scandens, × 2/7; After Gleason, details from Small, The New York Botanical Garden. **11.** Brickellia grandiflora, × 2/7.

PLATE NO. 350

stems and leaves, shape and numbers of flower-heads, and length of flower-head stalks (peduncles). Dr. Gaiser segregated var. *intermedia* on the basis of having the leaves less rough and the stems glabrous or only sparsely pubescent instead of grayish hairy and sub-strigose or puberulent. However, these characters do not seem to show any correlation or constancy, and it seems best to regard this species as made up of a series of variable individuals.

Plants thought to represent hybrids between this and other species have been interpreted by Dr. Gaiser as *Liatris* × *sphaeroidea* Michx., not *L. sphaeroidea* as interpreted by Shinners.

5. **Liatris ligulistylis** (Nels.) K. Schum. Map 2135

Flowers July–October.

Occurs on dry rocky slopes, rocky woods, gravelly ground in valleys along streams, prairies, and railroad embankments.

Scattered in southern and central Missouri, where known from Pike (McCune Station, *Davis 1573A*, *6172*), Lincoln (Silex, *Davis 1841*), St. Louis (Allenton, September 1, 1894, *Letterman*), Shannon (8–9½ mi. north northeast of Monteer, *Steyermark 85392*), Taney (Seligman, *Bush 15104A*), Stone (Baxter, *Bush 15193*, *15188*), and Barry (*Bush 15178*) counties.

Ranges from Alberta to New Mexico, east to Minnesota, Wisconsin, South Dakota, and Missouri.

The above specimens, although southeast of the range cited by Dr. Gaiser in her monograph (Rh. 48: 318–25. 1946), show the characteristic glabrity of the inside of the corolla-tube of *L. ligulistylis*. The corollas measure 10–11 mm. long, the pappus 9–10 mm. long, and the achenes 5–6 mm. long, also characteristic of the measurements typical of this species. The bracts of the involucre are lavender to dark purple above with scarious thin, toothed, hairy-fringed margins. The ciliate leaves, found in these collections, are also characteristic of the species.

Two of the specimens cited above (McCune Station, August 28, 1915, *Davis 6172*; *Davis 1573A*) come from the same locality and perhaps represent the same collection tagged with a different number as the ones cited by Dr. Gaiser (Rh. 48: 326. 1946) for *Liatris* × *Nieuwlandii* (Lunell) Gaiser from Missouri (McCune, August 28, 1915, *Davis 249*). She considers this *Liatris* as of hybrid origin between *L. ligulistylis* and another unidentified species. This is of questionable status, and has been interpreted by other workers as a distinct entity, namely, *Liatris novae-angliae* (Lunell) Shinners var. *Nieuwlandii* (Lunell) Shinners. It occurs generally east of and outside the range of *L. ligulistylis*, one of its presumed parents. I have referred the collection cited by Dr. Gaiser under *L.* × *Nieuwlandii* to *L. ligulistylis* along with the other specimens collected by Reverend Davis.

6. **Liatris punctata** Hook. var. **nebraskana** Gaiser

Snakeroot Map 2136

*Liatris punctata* of auth. in part [BB], not Hook.

*Liatris acidota* in part [of P & S], not Engelm. & Gray

Flowers August–October.

Occurs on loess hills and prairies. Western Missouri, in Atchison (*Bush 9185*, *12009*; *Steyermark 9211*), Holt (*Kellogg 15337*; *Bush 12026*), Jackson (Independence, 1892, *Bush 5*), and Bates (prairies, Hume, October 4, 1936, *Steyermark 20306*) counties.

Ranges from Alberta to New Mexico, east to Wisconsin and Michigan, Iowa, Missouri, Arkansas, and Texas; introduced in Illinois.

According to Dr. Gaiser's treatment (Rh. 48: 347. 1946) typical *L. punctata* var. *punctata* (her var. *typica*) has the lower and basal leaves 3–6 mm. broad margined with cilia (hairs) and the involucral bracts are ciliate-margined, whereas in var. *nebraskana* the lower and basal leaves are narrower, 2–3 mm. broad with no marginal hairs or these slightly developed and with the margins of the involucral bracts rarely shortly ciliate or showing no cilia. The Missouri material is referred to var. *nebraskana* on these characters, but it is questionable whether this variety and the other varieties and forms segregated by Dr. Gaiser can be maintained. Although the pappus is stated to be 9–11 mm. long, most Missouri specimens have a pappus of only 7–8 mm. long, thus showing the character of short pappus characteristic of *L. mucronata* DC. Plants with characters intermediate between *L. punctata* var. *punctata* and var. *nebraskana* are fairly common and have been recorded by Dr. Gaiser (Rh. 48: 358–59. 1946).

The Shannon County record, cited by Gaiser (Rh. 48: 353. 1946), based on the collection of *Palmer 18944* from 'eroded slopes and bald tops of loess hills near Montier,' is an error in labeling, as the collection was actually taken from Atchison County (the area of the loess hills), according to a letter written by Mr. Palmer in 1935 to the present author.

The specimen cited by Gaiser (ibid.) from Taney County (Malva, *Palmer 26190*) has been referred in the present flora to *L. mucronata* DC.

In the western states *L. punctata* is eaten by all classes of livestock in its early stages of growth. The Indians used the carrot-flavored root for food in early spring. At one time the root was used as an antidote against snake bites.

2137 Liatris mucronata

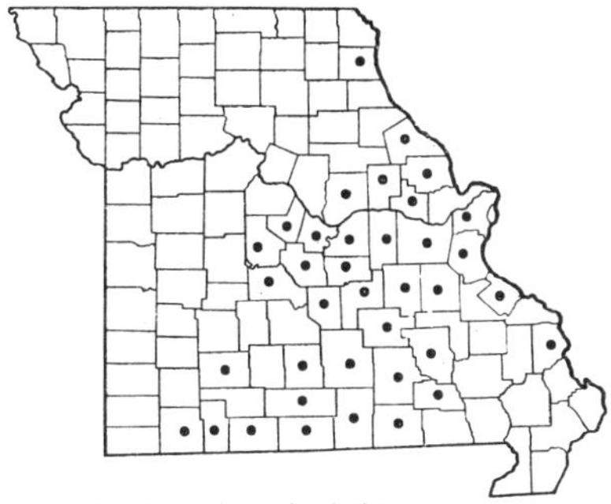
2138 Liatris cylindracea f. cylindracea

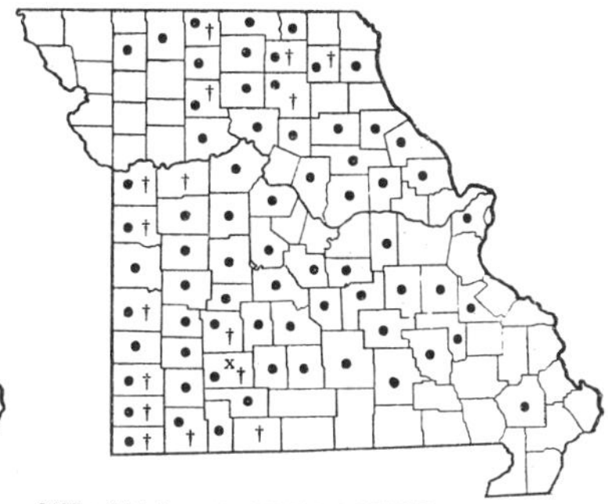
2139 • Liatris squarrosa var. squarrosa
2139 x Liatris squarrosa var. glabrata
2139 † Liatris squarrosa var. hirsuta

7. **Liatris mucronata** DC. Map 2137

*Liatris angustifolia* (Bush) Gaiser [Gaiser, G]
*Liatris acidota* in part [of P & S], not Engelm. & Gray
*Liatris mucronata* var. *typica* [Gaiser]
Flowers July–October.

Occurs on limestone glades, bald knobs, ledges of bluffs, and edges of bluff escarpments. White River region and tributaries of southwestern Missouri, in Taney, Christian, Stone, and Barry counties.

Ranges from Texas and Oklahoma, north and east to Arkansas, Missouri, and Kansas; recorded from Nebraska.

Although Dr. Gaiser and Fernald, following Gaiser, attempt to separate *L. angustifolia* from *L. mucronata* on the basis of having softer nearly non-punctate leaves vs. stiff and punctate leaves in *L. mucronata*, by the mucronate-tipped or abruptly cuspidate, firm, ovate or broadly oblong involucral bracts of *L. mucronata* vs. the acuminate, more membranaceous, lanceolate bracts of *L. angustifolia*, and by a relatively longer inflorescence in *L. angustifolia*, these differences do not bear careful comparison when tested in the field. Gaiser placed the specimen of *Palmer 31455* from Eagle Rock, Barry County in *L. angustifolia*, but it has no character that cannot be found in specimens here referred to *L. mucronata*. Specimens combining the punctate leaves supposedly characteristic of *L. mucronata* with the soft leaves and elongated inflorescence up to 6 dm. long usually attributed to *L. angustifolia* are commonly found, and there appears to be no correlation present in the type of bract apex and leaf stiffness or punctation. Shinners (Field & Lab. 19: 75. 1951) likewise has not been able to keep *L. angustifolia* distinct from *L. mucronata*.

So far as the distinctness of *L. mucronata* from *L. punctata* is concerned, the character of the elongate vertical root of *L. punctata* vs. the large nearly globose corms of *L. mucronata* easily separates the two as distinct species, and I have verified this particular character by field observations both among wild plants and in those grown from seed, as well as transplanted individuals grown in my wild flower garden. So far as the other characters given for separation are concerned, there is considerable overlapping in length of inflorescence, leaf width, and length of corollas and pappus. Although the pappus of *L. punctata* is given as longer than that of *L. mucronata*, a specimen of *L. punctata* var. *nebraskana* from Atchison County (*Steyermark 85661*) shows the pappus only 7–8 mm. long. However, plants of *L. mucronata* are generally taller (averaging 7–9 dm. tall), than those of *L. punctata* var. *nebraskana* (averaging 4.5–6 dm.), the leaves are softer, more loosely spreading to recurving instead of erect, relatively longer and narrower (averaging 1–2.5 mm. broad) and brighter grass green, those of *L. punctata* var. *nebraskana* having a dull to gray green color.

Dr. Gaiser notes (Rh. 48: 366–67. 1946) that there is overlapping and blending of characters between *L. punctata* and *L. mucronata*, especially in the regions where these two overlap in their geographical distribution. Having studied both taxa in the field and having grown them in the garden, I am convinced that they should be considered as distinct species.

8. **Liatris cylindracea** Michx. f. **cylindracea** Map 2138

*Liatris cylindracea* Michx.
*Liatris squarrosa* var. *intermedia* MacM. [P & S]
Flowers July–September.

Occurs in rocky open woods, rocky ledges and escarpments of bluffs, prairies, and glades, Ozark region north in eastern Missouri to Lewis County, west to Moniteau, Morgan, Camden, Greene, and Barry counties.

Ranges from Ontario and New York to Minnesota, south to Ohio, Indiana, Illinois, Missouri, and Arkansas.

Typical *L. cylindracea* f. *cylindracea* has rose-lavender or lavender corollas. In f. *Bartelii* Steyerm., not yet recorded from Missouri, the corollas are completely white.

Stems, leaves, and involucral bracts vary from glabrous to slightly hairy. The bracts vary from rounded or obtuse to abruptly short- or long-pointed. Specimens intermediate between this species and glabrate types of *L. squarrosa* are found and are probably of hybrid origin. Many of these specimens have been previously classified as *L. squarrosa* var. *intermedia*.

9. **Liatris squarrosa** (L.) Michx. Map 2139

Flowers late June–September.

Occurs in rocky or dry open woods and glades, rocky prairies, ledges along bluffs, and bluff escarpments. Throughout Missouri, except absent from most of the extreme southeastern and northwestern sections of the state.

The following variations may be recognized for Missouri:

a. Stems, leaves, and bracts of involucre glabrous (without hairs) . 9c. L. SQUARROSA var. GLABRATA
a. Stems, leaves, and/or bracts of the involucre with more or less hairiness . . . . . . *b*
  b. Most of the bracts of the involucre, especially the outer ones, tapering to long, acuminate, loosely spreading to recurved (squarrose) tips; hairs mostly short and curly or appressed (lying flat or parallel to surface) . . . 9a. L. SQUARROSA var. SQUARROSA
  b. Most of the bracts of the involucre abruptly contracted to a short acute spreading tip; hairs mostly longer, more prominent, and spreading . . 9b. L. SQUARROSA var. HIRSUTA

9a. **Liatris squarrosa** var. **squarrosa** Map 2139

*Liatris squarrosa* (L.) Michx. [G, P & S]

*Liatris squarrosa* var. *typica* [Gaiser]

*Liatris squarrosa* var. *gracilenta* Gaiser [Gaiser]

Scattered in southern and eastern Missouri north to Lewis County.

Ranges from Florida to Alabama and Louisiana, north to Delaware, West Virginia, Ohio, Indiana, Illinois, and Missouri.

The characters used by Dr. Gaiser to separate var. *gracilenta*, distinguished by having slender stems and few, long, rather slender, cylindrical heads, from var. *squarrosa*, distinguished by the stouter stems and numerous, large, thick heads of nearly equal width and length, break down in many specimens and overlap greatly, so that no reliance can be placed on these characters.

9b. **Liatris squarrosa** var. **hirsuta** (Rydb.) Gaiser Map 2139

*Liatris hirsuta* Rydb.

This is the commonest variation of this species in Missouri, scattered throughout the state.

Ranges from Mississippi and Louisiana to Texas, north to Kentucky, Iowa, and Nebraska.

9c. **Liatris squarrosa** var. **glabrata** (Rydb.) Gaiser Map 2139

*Liatris squarrosa* var. *intermedia* MacM. [of P & S], not MacM.

Rather rare and scattered in southern Missouri.

Ranges from Tennessee and Missouri to South Dakota, south to Texas and Colorado.

Specimens previously classified as *L. squarrosa* var. *intermedia* have been referred here to *L. squarrosa* var. *glabrata*, but are often difficult to place, as they are intermediate between glabrate forms of *L. squarrosa* and *L. cylindracea*.

The above varieties intergrade and show considerable overlapping. Much more intensive field and experimental studies are necessary before assigning them a final status. For the convenience of placing various extremes of this species and pending future more detailed studies, the varieties given above are tentatively maintained in the present flora.

*Excluded Species*

**Liatris squarrulosa** Michx.

*Liatris scariosa* var. *squarrulosa* (Michx.) Gray [Gaiser]

This is attributed to Missouri as a distinct species in Gleason's *New Ill. Fl.* in the sense in which it is adopted by Shinners (Am. Midl. Nat. 29: 33. 1943). Dr. Gaiser places the name in varietal rank as *L. scariosa* var. *squarrulosa* (Michx.) Gray, but limits the geographical range from North Carolina to Alabama. Shinners' use of the name *L. squarrulosa* is not that of Michx., according to Dr. Gaiser's studies, but is synonymous with what she considers to be *L. Earlei* (Greene) K. Sch., a species ranging from Florida to Texas, north to North Carolina, Kentucky, and Indiana. The Missouri material identified as *L. squarrulosa* by Shinners is referred in the present flora to *L. scabra*.

**Liatris novae-angliae** (Lunell) Shinners var. **Nieuwlandii** (Lunell) Shinners

As discussed under *L. ligulistylis*, specimens from Pike County (*McCune, Davis 249*) were identified by

Dr. Gaiser as *L.* × *Nieuwlandii* (Lunell) Gaiser, and form the basis for the report of this taxon from Missouri by Shinners as var. *Nieuwlandii*. In the present flora this is treated under *L. ligulistylis*. The inclusion by Gleason in the *New Ill. Fl.* of *L. novae-angliae* for Missouri likewise refers to the var. *Nieuwlandii* and not var. *novae-angliae* (*L. borealis* Nutt. [G]), confined to the eastern United States.

## Tribe III. **ASTEREAE**

### 9. **Grindelia** Willd. Gum Plant

Teeth of leaves blunt or rounded; bracts of the involucre with recurved to reflexed tips; plants of railroads, roadsides, and waste ground . . . . . . . . . . . . . . . . . . . . 1. G. SQUARROSA

Teeth of leaves mostly bristle-tipped or pointed, sometimes teeth are absent; bracts of the involucre loosely spreading to ascending, but the tips not recurved or reflexed; plants of limestone glades, bald knobs, and rocky prairies in the Ozark region, rarely escaped in old fields or along railways. 2. G. LANCEOLATA

1. **Grindelia squarrosa** (Pursh) Dunal Map 2140
Flowers July–September.

Occurs in waste ground, along roadsides and railroads, sometimes in prairies and alluvial ground. Scattered in northern, central, and western Missouri; absent from most of the entire Ozark region.

The following varieties are known in the state:

a. Rays (ligules) absent . . 1c. G. SQUARROSA var. NUDA

a. Rays (ligules) present . . . . . . . *b*

b. Upper and middle leaves of the stem and branches ovate-oblong or broadly oblong, 2–4 times as long as wide . . . . . 1a. G. SQUARROSA var. SQUARROSA

b. Upper and middle leaves of the stem and branches usually linear-oblong, oblong, or oblanceolate, mostly 5–8 times as long as wide . . 1b. G. SQUARROSA var. SERRULATA

1a. **Grindelia squarrosa** var. **squarrosa** Map 2140

*Grindelia squarrosa* (Pursh) Dunal [G, P & S]

Scattered in the state.

Ranges from Manitoba and Minnesota to British Columbia, Nevada, and Texas; introduced eastward to the Atlantic states.

1b. **Grindelia squarrosa** var. **serrulata** (Rydb.) Steyerm. Map 2140

Known in northern and central Missouri from Clark, St. Louis, Boone, and Jackson counties.

Ranges in the western states; introduced east to Ontario, New York, Connecticut, and Pennsylvania.

1c. **Grindelia squarrosa** var. **nuda** (Wood) Gray Map 2140

Known from St. Louis and Jackson counties.

Ranges from Kansas to Colorado and Texas; introduced east in Missouri and New Jersey.

The heads of this *Grindelia* secrete an abundance of a white sticky resinous substance. Due to its bitter taste, it is not eaten by livestock. Indians used the gummy secretion of the heads as a remedy for asthma, bronchitis, and colic. The leaves and flowering tops were boiled and the decoction used by the Pawnee Indians to heal saddle sores and raw places on the skin. In modern medicine an extract of the young flowerheads and leaves is used in treating whooping cough and asthma, and the powdered drug has been used in the manufacture of cigarettes designed to treat asthma.

2. **Grindelia lanceolata** Nutt. Map 2141
Flowers August–October.

Occurs on limestone glades, bald knobs, and rocky prairies, introduced northward along railroads, fallow fields, and waste ground.

Ozark region of southern and central Missouri north to St. Louis, Miller, Johnson, and Vernon counties, and locally northward in Carroll, Livingston, and Putnam counties.

Ranges from Alabama to Texas, north to Tennessee Missouri, and Kansas.

Two variations occur in the state:

Commonly encountered; middle and lower leaves of stem linear or lanceolate-oblong, 4–12 times as long as broad, 0.3–2 cm. broad . . . 2a. G. LANCEOLATA f. LANCEOLATA

Rarely encountered; leaves broadly oblong or oblong-lanceolate, the upper ovate-lanceolate, at least the upper and middle leaves somewhat clasping the stem, 2½–3 times as long as broad, 2–2.8 cm. broad . 2b. G. LANCEOLATA f. LATIFOLIA

2a. **Grindelia lanceolata** f. **lanceolata** Map 2141

*Grindelia lanceolata* Nutt. [G, BB, P & S]

This is the usual form found in Missouri.

2140 • Grindelia squarrosa var. squarrosa (Gum Plant)
2140 □ Grindelia squarrosa var. serrulata
2140 x Grindelia squarrosa var. nuda

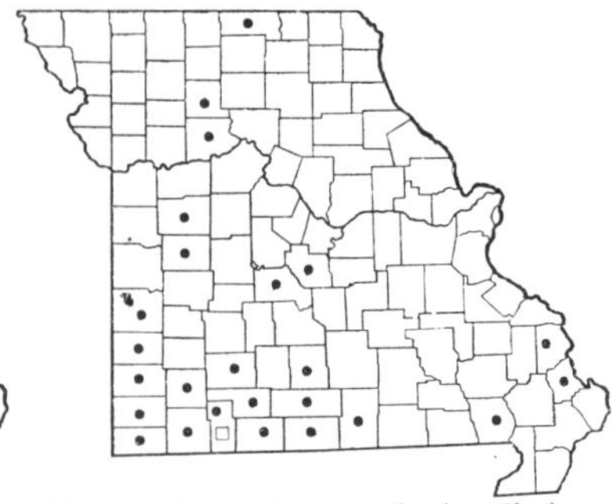

2141 • Grindelia lanceolata f. lanceolata (Gum Plant)
2141 □ Grindelia lanceolata f. latifolia

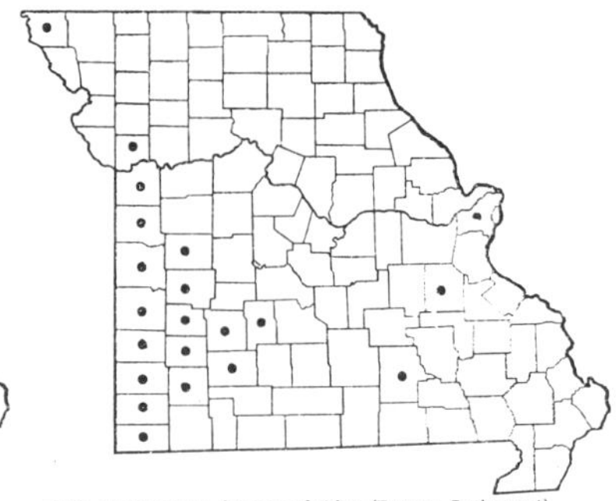

2142 Gutierrezia dracunculoides (Broom Snakeroot)

2b. **Grindelia lanceolata** f. **latifolia** Steyermark Map 2141

Known only from Stone County, southwestern Missouri (Marble Cave, September 11, 1898, *Trelease*, holotype in Mo. Bot. Gard. Herb.).

Also found in Oklahoma.

The leaves of *G. lanceolata* vary from very narrow (0.3–1 cm. broad) and nearly entire or finely setulose (with minute bristly teeth) to sharply toothed very broad leaves 2.8 cm. broad.

## 10. **Gutierrezia** Lag.

**Gutierrezia dracunculoides** (DC.) Blake

Broom Snakeroot Map 2142

*Xanthocephalum dracunculoides* (DC.) Shinners, Field & Lab. 18: 28. 1950

Flowers July–October.

Occurs in dry upland prairies, rocky open limestone barrens, roadsides, fallow fields, and along railroads. Chiefly in western Missouri in the unglaciated prairie region, locally east in St. Louis, Washington, and Reynolds counties, and northwest in Atchison County.

Ranges from Missouri and Kansas to Texas, New Mexico, and Mexico; introduced in Illinois and Alabama.

This annual species has a stem which is usually simple below and much branched above, giving a bushy effect, the entire plant often being as broad as or broader than long. Shinners (Field & Lab. 18: 28. 1950) merges this genus, *Amphiachyris*, and *Greenella* under *Xanthocephalum*, characterized by having a mostly scaly pappus instead of one with hairs as in *Euthamia* or *Isocoma*.

## 11. **Heterotheca** Cass. Golden Aster

**Heterotheca latifolia Buckley** Map 2143

*Heterotheca subaxillaris* [of BB, G], not (Lam.) Britt. & Rusby

Flowers August–November.

Occurs on loess hills, sandy open and waste ground, and along railroads. Commonest in the southeastern Missouri lowlands, and scattered elsewhere in central and northwestern Missouri in St. Louis, Jackson, Clay, and Atchison counties.

The general range of the species, including its varieties, extends from Florida to Arizona and Mexico, north to Delaware, New Jersey, New York, Illinois, Missouri, and Kansas; also tropical America.

This is a hairy annual or biennial plant attaining

---

Plate no. 351. **1.** Liatris punctata, × 2/7; *a.* Root; *b.* Upper part of stem with inflorescence; *c.* Involucre with florets, × 3/7; *d.* Middle part of stem; Details from Small, The New York Botanical Garden. **2.** Grindelia squarrosa, × 2/7; Details from Small, The New York Botanical Garden. **3.** Heterotheca latifolia, × 2/7; After Gleason, details from Small, The New York Botanical Garden. **4.** Liatris mucronata, × 2/7; *a.* Lower part of stem with leaves; *b.* Base of plant with corm; *c.* Upper part of inflorescence; *d.* Involucre with florets, × 3/7; *e.* Middle part of stem with leaves. **5.** Grindelia lanceolata, × 2/7. **6.** Chrysopsis pilosa, × 2/7; Details from Small, The New York Botanical Garden. **7.** Liatris squarrosa, × 2/7. **8.** Chrysopsis villosa var. camporum, × 2/7. **9.** Chrysopsis villosa var. canescens, × 2/7; *a.* Achene, × 1 3/7; *b.* Head of florets, × 6/7; After Britton and Brown, The New York Botanical Garden.

PLATE NO. 351

2143 Heterotheca latifolia

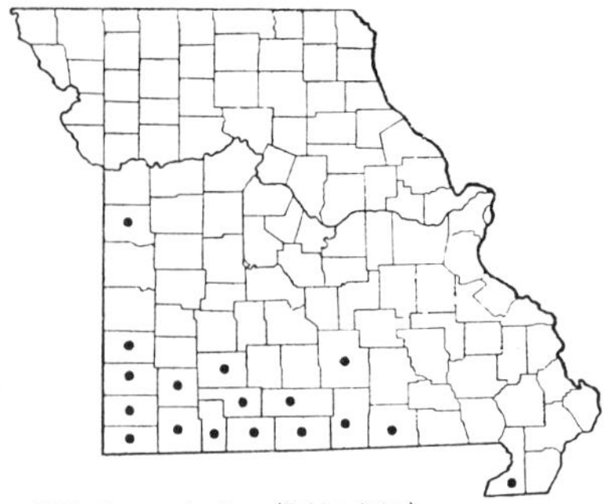

2144 Chrysopsis pilosa (Golden Aster)

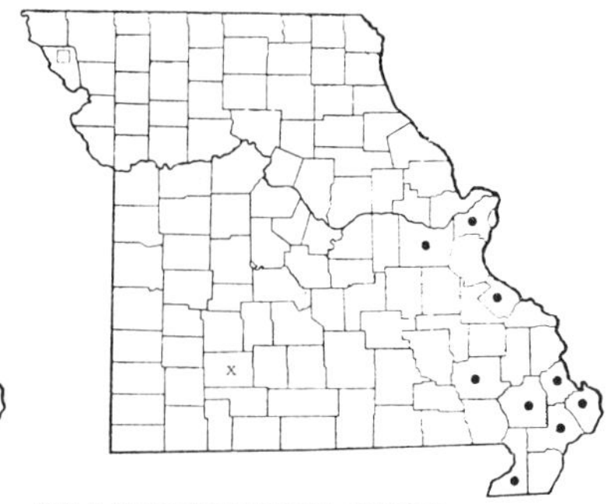

2145 □ Chrysopsis villosa var. canescens
2145 x Chrysopsis villosa var. angustifolia
2145 • Chrysopsis villosa var. camporum

2 meters in height.

According to the most recent studies of the genus by Wagenknecht, all the Missouri material pertains to *H. latifolia* rather than to *H. subaxillaris*, as previously treated by Fernald in *Gray's Manual*, eighth edition, and by Gleason in *New Ill. Fl.* (see Rh. 62: 98–105. 1960). Typical var. *latifolia* is cited by Wagenknecht for Missouri (*Steyermark 66190* from Stoddard County) and var. *arkansana* Wagenknecht from Butler and St. Louis counties. The present distribution map (no. 2143) does not differentiate these varieties.

## 12. **Chrysopsis** Nutt. Golden Aster

(*Heterotheca* Cass. [Shinners])

Plants annual, complete plant with root easily pulled from the ground; leaves of stem with a broad sessile (stalkless) base, the lower often toothed; achenes with 10 faint nerves; inner pappus of about 20 elongated bristles, outer pappus of scales; hairs on plant a mixture of few to many long, soft, spreading hairs, and short rather sticky (viscid) ones . . . . . . . . . . . . . . . . . . 1. C. PILOSA

Plants perennial, with tough firm bases, the roots not pulling easily from the ground; leaves of stem narrowed or tapering to base, rarely toothed; achenes with 3–5 nerves; inner pappus of 25–50 elongated bristles, outer pappus of minute bristles; hairs on plant long or short, or both, none short and sticky (viscid) . . . . . . . . . . . . . . . . . . . . . . . . 2. C. VILLOSA

**1. Chrysopsis pilosa** Nutt. Map 2144

*Heterotheca pilosa* (Nutt.) Shinners [Shinners]

Flowers late June–October.

Occurs in rocky or sandy dry prairies, sandy and fallow fields, and along roadsides. Southern Missouri, chiefly from Oregon County west, north to Texas, Greene, and Barton counties, locally southeast in Dunklin County, and northwest in Cass County.

Ranges from Louisiana to Texas, north to Missouri and Kansas.

**2. Chrysopsis villosa** (Pursh) Nutt. Map 2145

*Heterotheca villosa* (Pursh) Shinners [Shinners]

The following varieties, based upon Dr. Cronquist's interpretation of the species (in Gleason's *New Ill. Fl.* 3: 410. 1952), occur in Missouri:

a. Heads small, the disk (solid portion between the rays) less than 1.5 cm. broad; involucre 6–7 mm. high; leaves mostly 5 mm. or less broad. . . . . . . . . . . . . *b*

b. Leaves and involucre with a silvery-silky hairiness, the hairs all appressed (lying parallel to or pressed flat against surface) . 2a. C. VILLOSA var. CANESCENS

b. Leaves rather coarsely and sparsely hairy, the hairs thickened near the base; involucre sparsely to densely hairy with rather straight, stiff hairs lying parallel to or pressed against surface (strigose) . . . . . 2b. C. VILLOSA var. ANGUSTIFOLIA

a. Heads larger, the disk of the larger heads usually 1.5 cm. or more broad; involucre 8–11 mm. high; leaves usually more than 5 mm. broad. . . . 2c. C. VILLOSA var. CAMPORUM

**2a. Chrysopsis villosa var. canescens** Gray Map 2145

*Chrysopsis Berlandieri* Greene [G, P & S]

*Heterotheca canescens* (DC.) Shinners [Shinners]

Flowers July–October.

Occurs on loess hills. Known only from Holt County, northwestern Missouri.

Ranges from Texas and New Mexico, north to Missouri and Kansas.

2b. **Chrysopsis villosa** var. **angustifolia** (Rydb.) Cron. Map 2145

*Chrysopsis angustifolia* Rydb. [G]
*Chrysopsis hispida* var. *stenophylla* Gray
*Chrysopsis stenophylla* (Gray) Greene
*Heterotheca stenophylla* (Gray) Shinners [Shinners]
Flowers July–September.

Occurs in dry open ground. Known only from Greene County, southwestern Missouri (Springfield, July, 1905, *Standley*, in N.Y. Bot. Gard. Herb.).

Ranges from Missouri and Nebraska to Texas.

Dr. William J. Dress of Bailey Hortorium, who is monographing the genus *Chrysopsis*, has provided me with the data for the Missouri specimen, which he has identified as *C. stenophylla* (Gray) Greene.

2c. **Chrysopsis villosa** var. **camporum** (Greene) Cron. Map 2145

*Chrysopsis camporum* Greene [G, P & S]
*Heterotheca camporum* (Greene) Shinners [Shinners]
Flowers July–October.

Occurs in sandy open ground, fallow fields, along roads, and railroads. Commonest in the lowlands of southeastern Missouri, north to St. Louis and Franklin counties.

Ranges from Indiana to Wisconsin, south to Illinois and Missouri.

As the genus *Chrysopsis* is at present undergoing study and revision by Dr. William J. Dress of Bailey Hortorium, the above treatment is a tentative one, conforming to the present interpretation of Dr. Cronquist in Gleason's *New Ill. Fl.* Fernald, in *Gray's Manual*, eighth edition, maintains the above varieties as distinct species, as does Shinners, who, moreover, considers the genus *Chrysopsis* best merged with *Heterotheca* (Field & Lab. 19: 66–71. 1951), which differs but slightly from *Chrysopsis* and has had prior publication. For the present, however, I have followed the traditional concept of earlier botanists in keeping the two genera separate, pending the revised treatment forthcoming by Dr. Dress.

## 13. **Solidago** L. Goldenrod

Complete specimens, showing underground parts, basal leaves, and the flowering stem, are necessary for satisfactory study of the species of this genus.

The section *Euthamia*, comprising species 23–25 here treated, with uniform narrow linear leaves and heads in dense clusters forming a corymb, is sometimes treated as a separate genus. Cronquist (in Gleason's *New Ill. Fl.* 3: 414. 1952) noted that the section *Euthamia* has no well defined technical differences to separate it genericall- from *Solidago*.

The bad repute in which goldenrod is held by hay fever sufferers is largely based on hearsay and false asy sumptions. As is well known, the species are insect-pollinated, and the pollen mainly too heavy to be transported by wind. However, the pollen may occasionally become atmospheric to a very small degree. The pollen which is shed is too little to be important as a cause of hay fever from most of the species. One species which occurs in Missouri, *Solidago speciosa*, when tested, did yield pollen in collectable quantities and its pollen could be detected in the air several miles from the fields when the plants were in flower. It is probable that much of the allergy suspected from goldenrod by hay fever sufferers is brought about by the windborne deposition of pollen dust from nearby ragweeds on the leaves of various goldenrods.

a. All the leaves of uniform narrow width throughout, narrowly linear to thread-like, without teeth or leaf-stalks (sessile), 2–12 mm. broad . . . . . . . . . . . . . . . . . . . . *b*
- b. Three conspicuous nerves showing in the leaves, the larger leaves usually with an additional 2 fainter lateral nerves present; flower-heads mainly with 15–45 flowers . . . . 25. S. GRAMINIFOLIA
- b. Only 1 conspicuous nerve usually showing in the leaves, sometimes with 3 faint or even more conspicuous nerves evident, without any additional 2 lateral nerves; flower-heads with 9–22 flowers . . . . . . . . . . . . . . . . . . . . . . . *c*
  - c. Leaves with abundant resin and conspicuously gland-dotted; involucre with abundant resin; plants of the prairies of northern, central, and western Missouri . . . . 24. S. GYMNOSPERMOIDES
  - c. Leaves with little resin and only sparsely or faintly gland-dotted; involucre only slightly resinous; plants of the lowlands of southeastern Missouri west to Oregon County . . . . 23. S. LEPTOCEPHALA

a. Leaves not uniformly narrowly linear, either the leaves at the base of the plant or those on stem or both not of uniform narrow width throughout or toothed or narrowed at base, some or all of them usually 12 mm. or more broad . . . . . . . . . . . . . . . . . . . . *d*
- d. Inflorescence more or less flat-topped (corymbose) . . . . . . . . . . . . . . . *e*
  - e. Plants of usually dry prairies, rocky open ground, or along railroad found throughout Missouri; leaves flat; stems and leaves usually densely hairy with short hairs; leaves broad, the ones at base of plant oval or oblong, shallowly toothed with rounded teeth, 3–10 cm. broad; leaves midway up stem 2–6 times as long as broad . . . . . . . . . . 21. S. RIGIDA
  - e. Plants of wet or swampy meadows of the Ozark region; some or all of the leaves folded, at least the lower recurving; stems and leaves glabrous (without hairs) or nearly so; leaves relatively long and narrow, linear-lanceolate, without any teeth (entire), 1–2 cm. broad; leaves midway up stem 6–15 times as long as broad . . . . . . . . . . . . 22. S. RIDDELLII
- d. Inflorescence not flat-topped, either pyramid-shaped and broadest at the base or midsection, or of a narrow cylindrical type, or in clusters scattered along the sides of the stem . . . . . . . *f*
  - f. Flower-clusters either scattered along the sides of the stem and arising from the leaf-axils or forming a narrow or cylindrical inflorescence with the heads spirally arranged and not in 1-sided raceme-like branches . . . . . . . . . . . . . . . . . . . . *g*
    - g. Flower-clusters scattered along the sides of the stem and arising from the leaf-axils. . . . . *h*
      - h. Stems glabrous (without hairs); mature achenes hairy . . . . . . . . . . . . *i*
        - i. Stem with a gray-green or silvery-purplish coating ('bloom') which can be rubbed off; main leaves lanceolate or narrowly oblong, 1–3.5 cm. broad, sessile (without leaf-stalks); plants found only in the southern Ozarks . . . . . . . . . 1. S. CAESIA
        - i. Stem green, without any 'bloom'; main leaves ovate or oval, 3–9 cm. broad, on winged petioles (leaf-stalks); plants distributed throughout Missouri . . . 2. S. FLEXICAULIS
      - h. Stems, at least in the upper part, with short or long hairs; mature achenes glabrous (without hairs) . . . . . . . . . . . . . . . . . . . . . *j*
        - j. Lower leaves much larger than the middle and upper ones; leaves and stem conspicuously hairy with many and rather long hairs . . . . . . . . . . 5. S. HISPIDA

j. Lower leaves smaller than the middle and upper ones; leaves and stem with short hairs or with few or inconspicuous long hairs . . . . . . . . . . . . . . . . . . . . *k*
k. Leaves thin, usually sharply toothed, 25–50 mm. broad, with rather long hairs 0.4–0.7 mm. long on the midrib and main nerves of the lower surface; leaf-surface not sticky or resinous 4. S. BUCKLEYI
k. Leaves firmer and thick, entire or with few teeth, mostly 5–30 mm. broad, glabrous or with shorter hairs 0.1–0.4 mm. long; leaf-surface sticky or resinous to not sticky . 3. S. PETIOLARIS
g. Flower-clusters forming a short or elongated, narrow or cylindrical inflorescence, the heads spirally arranged, not in 1-sided raceme-like branches . . . . . . . . . . . . . . . . *l*
l. Lowermost leaves the largest, and the lower leaves much larger than the middle and upper ones; bracts of the involucre obtuse (blunt) or rounded . . . . . . . . . . . . . . . *m*
m. Stem glabrous (without hairs) up to the inflorescence; leaves, except for the hairy-fringed margins, glabrous . . . . . . . . . . . . . . . . . . . . . . 6. S. SPECIOSA
m. Stem conspicuously hairy; leaves conspicuously hairy . . . . . . . . . . 5. S. HISPIDA
l. Lowermost and lower leaves smaller than the middle and upper ones; some of the bracts of the involucre short- to long-pointed (acute to acuminate) . . . . . . . . . . . . *n*
n. Leaves thin, usually sharply toothed, 25–50 mm. broad, with rather long hairs 0.4–0.7 mm. long on the midrib and main nerves of the lower surface; leaf surface not sticky or resinous 4. S. BUCKLEYI
n. Leaves firmer and thick, entire or with few teeth, mostly 5–30 mm. broad, glabrous or with shorter hairs 0.1–0.4 mm. long; leaf-surface sticky or resinous or not resinous. 3. S. PETIOLARIS
f. Flower-clusters forming a pyramid-shaped inflorescence broadest at the base or midsection or with at least some part of the inflorescence with widely spreading to arching or recurving, 1-sided, raceme-like branches, the flower-heads borne on the upper side of the branches . . . . . . . . . . *o*
o. Stem glabrous (without hairs) or nearly so below the branches of the inflorescence . . . . . *p*
p. Branches of the inflorescence glabrous . . . . . . . . . . . . . . . . . . . *q*
q. Leaves with minute depressed dots (hold leaf to light); leaves anise-scented when bruised or pinched . . . . . . . . . . . . . . . . . . . . . 14. S. ODORA
q. Leaves without dots; leaves not anise-scented . . . . . . . . . . . . . . . *r*
r. Leaves with 1 principal nerve, none of them clearly triple-nerved (with 2 of the main lateral nerves more prominent or prolonged more or less parallel with the midnerve); basal leaves usually 2–7.5 cm. (rarely 1.5 or 8) broad . . . . . . . . . 9. S. JUNCEA
r. Leaves clearly or faintly triple-nerved (2 of the main lateral nerves more prominent or prolonged more or less parallel with the midnerve); basal leaves usually less than 2 cm. broad . . . . . . . . . . . . . . . . . . . . . . . . . . . *s*
s. Plants only of limestone glades, barrens and rocky prairies of the Ozark region; no creeping rootstocks present; no clusters (fascicles) of small leaves present in the axils of the upper leaves of the stem . . . . . . . . . . . . . 8. S. GATTINGERI
s. Plants mainly of prairies, open woods, and fields throughout Missouri; creeping rootstocks or filiform stolons normally present; clusters (fascicles) of small leaves usually produced from the axils of the upper leaves of the stem . . . . . . 7. S. MISSOURIENSIS
p. Branches of the inflorescence and/or pedicels (flower-stalks) more or less hairy . . . . . . *t*
t. Basal leaves and lowermost leaves of the stem either smaller than or scarcely if at all larger than the middle and upper ones . . . . . . . . . . . . . . . . . . *u*
u. Leaves with minute depressed dots (hold leaf to light); all the leaves entire (without teeth), usually anise-scented when bruised or pinched . . . . . . . . . 14. S. ODORA
u. Leaves without dots; some or all of the leaves toothed, not anise-scented . . . . . . *v*
v. Leaves with 1 principal nerve, and usually 4–6 lateral nerves pinnately arranged (feather-like); leaves conspicuously narrowed from the broad middle portion; branches of inflorescence few, slender, long-spreading or recurved; rays 3–5, inconspicuous; plants generally of dry woods or dry ground. . . . . . . 15. S. ULMIFOLIA
v. Leaves clearly triple-nerved (2 of the main lateral nerves more prominent or prolonged more or less parallel with the midrib); leaves gradually narrowed to the base, not much broader in the middle; branches of inflorescence more crowded, forming a dense pyramid-shaped cluster; rays 7–17, well-developed, rather conspicuous; plants of usually low or wet ground in damp thickets, prairie swales or meadows, banks of streams, and moist woodland . . . . . . . . . . . . . . 20. S. GIGANTEA
t. Basal leaves and lowermost leaves of the stem the largest, larger than the middle and upper ones . . . . . . . . . . . . . . . . . . . . . . . . . . *w*

w. Plants of swamps, wet ledges of bluffs, swampy woods and meadows; upper surface of leaves rough (scabrous) with stiff, cone-shaped hairs which have a papillate (swollen) base; stem, at least below the middle, strongly angled, the angles narrowly winged . . . . . . . 11. S. PATULA

w. Plants generally of dry or rocky wooded slopes; upper surface of leaves smooth or with minute inconspicuous rough hairs, which are not papillate at the base; stem not strongly angled, either terete (rounded in cross-section) or marked with lines . . . . . . . . . . . . . *x*

x. Basal leaves not or rarely occurring in definite rosettes; leaves rather thin and membranaceous; no rhizomes (underground or prostrate stems) present; lower leaves of stem soon disappearing and usually not present at time of flowering . . . . . . . . . . . 15. S. ULMIFOLIA

x. Basal leaves usually forming a definite rosette; leaves usually firm and subcoriaceous (rather thickish or leather-like); rhizomes (underground or prostrate stems) usually present; lower leaves usually persisting at time of flowering . . . . . . . . . . . . . . . . . *y*

y. Basal leaves lanceolate or oblanceolate; basal and lowermost leaves of stem tapering gradually to the petiole (leaf-stalk); upper leaves entire (without teeth) or rather few toothed, the teeth remote . . . . . . . . . . . . . . . . . . . . 9. S. JUNCEA

y. Basal leaves usually ovate, oblong-ovate, or ovate-lanceolate; basal and lowermost leaves of stem rather abruptly contracted to the petiole (leaf-stalk); upper leaves rather closely toothed, the teeth usually conspicuous and close together . . . . . . . . 10. S. ARGUTA

o. Stem more or less hairy with short or long hairs below the branches of the inflorescence . . . . . . *z*

z. Leaves glabrous (without hairs), with minute depressed dots (hold leaf to light), usually anise-scented when bruised or pinched . . . . . . . . . . . . . . . . . 14. S. ODORA

z. Leaves more or less hairy, without dots, not anise-scented . . . . . . . . . . . . *1*

1. Leaves with 1 principal nerve and several lateral nerves pinnately arranged (feather-like), not clearly triple-nerved . . . . . . . . . . . . . . . . . . . . . . . . . *2*

2. Basal and lowermost leaves of stem much larger than the middle and upper ones; leaves of stem reduced in size upwards, the upper ones of different shape than the lower ones; leaves entire (without teeth) or with inconspicuous, short or few teeth, the median leaves (those about half way up stem) with usually only 1–8 teeth on each side; leaves with a short, minute, gray hairiness, the lateral nerves not conspicuous . . . . . . 12. S. NEMORALIS

2. Basal and lowermost leaves of the stem either smaller or scarcely if at all larger than the middle and upper ones; most of the leaves of stem of about the same size and shape; leaves rather strongly or sharply toothed with usually conspicuous teeth, the median leaves (those about half way up stem) with mostly 10–20 teeth (sometimes 6–9) on each side; leaves not gray-hairy, the lateral nerves usually more conspicuous . . . . . . . . . . . . *3*

3. Plants of usually dry or open woodland or dry ground; leaves rather abruptly narrowed to the base, at least the middle and lower ones tapering to a petiole-like (stalked) base; rays usually 3–5 . . . . . . . . . . . . . . . . . . . . 15. S. ULMIFOLIA

3. Plants of usually swampy meadows and thickets, moist woods or slopes, and low wet ground, rarely dry places; leaves gradually narrowed or more or less rounded at the base, nearly sessile (without a stalk or petiole); rays mainly 6–9 . . . . . . 16. S. RUGOSA

1. At least the lower and middle leaves of the stem more or less clearly triple-nerved (2 of the main lateral nerves more prominent or prolonged more or less parallel with the midrib), but sometimes only slightly so in the upper leaves . . . . . . . . . . . . . . . *4*

4. Leaves lanceolate or oblanceolate, at least the lower and middle ones 8–13 times as long as broad, 5–22 mm. broad, long-tapering to the base; rays 9–14; plants of roadsides, fields, prairies, open ground, thickets, and waste places . . . . . . . . . . . . . *5*

5. Very commonly encountered species; involucre 3–5 mm. high; disk-corollas 3–4 mm. long; upper surface of leaf nearly glabrous (without hairs) or, if hairy the hairs fewer and shorter than those on lower surface . . . . . . . . . . . . . 19. S. ALTISSIMA

5. Less commonly encountered species; involucre 2–2.8 mm. high; disk-corollas 2.4–2.8 mm. long; upper surface of leaf with numerous hairs which are only slightly if at all fewer and shorter than those on lower surface . . . 18. S. CANADENSIS var. GILVOCANESCENS

4. Leaves ovate, oval, elliptic, oblong, or obovate, or sometimes oblanceolate, but 1.3–6 times as long as broad and usually 20–70 mm. (down to 10) broad, either short-tapering to a leaf-stalk (petiole) or sessile (without a leaf-stalk); plants of limestone bluffs, bluff escarpments and ledges, glades, and rocky woods . . . . . . . . . . . . . . . *6*

6. Leaves ovate or oval, abruptly narrowed at base to a short stalk (petiole), the larger ones 25–70 mm. broad and 1.3–2 times as long as broad; leaves soft or not rough to the touch, sharply toothed, the teeth usually prominent . . . . . . . . . . 17. S. DRUMMONDII

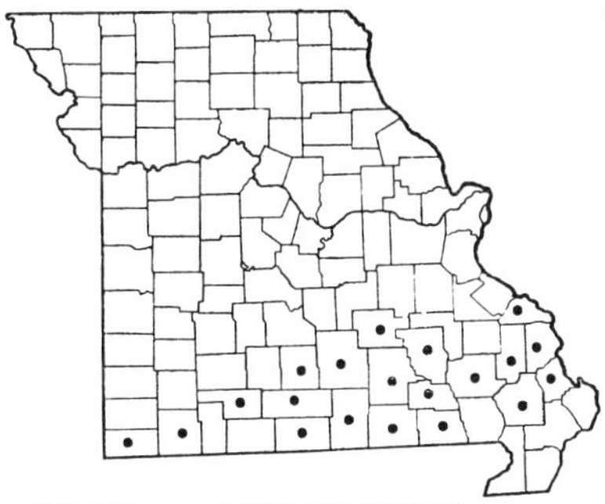

2146 Solidago caesia (Blue-stem Goldenrod)

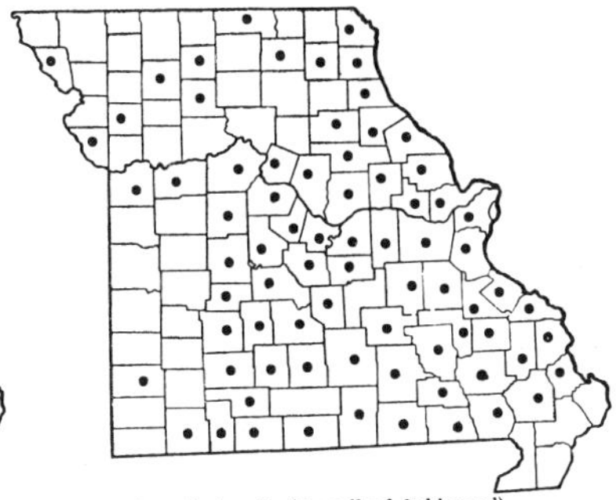

2147 Solidago flexicaulis (Broadleaf Goldenrod)

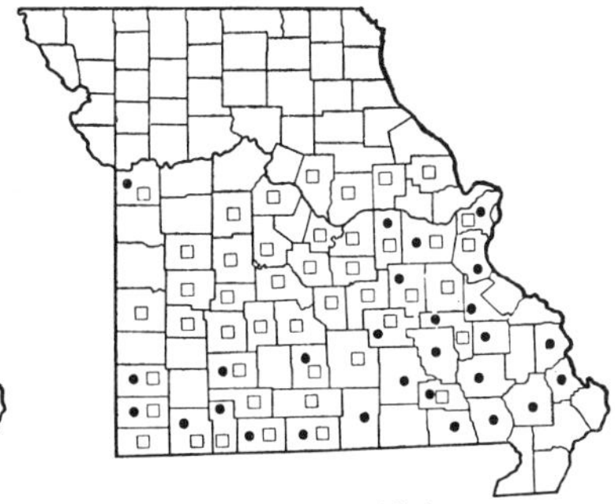

2148 ● Solidago petiolaris var. petiolaris
2148 □ Solidago petiolaris var. Wardii

6. Leaves elliptic, oblong, obovate, or oblanceolate, the larger ones 10–30 mm. broad and 2–6 times as long as broad; leaves rough to the touch, usually faintly or inconspicuously toothed, the teeth often short or blunt, but sometimes sharp and more prominently developed . . . . . 13. S. RADULA

1. **Solidago caesia** L. Blue-stem Goldenrod Map 2146

Also called Wreath Goldenrod.

Flowers August–October.

Occurs in rich or rocky woods, base of bluffs, and along bluff ledges. Southern Ozark region north to Perry, Iron, Reynolds, Dent, Texas, Wright, Christian, and McDonald counties.

Ranges from Florida to Texas, north to Nova Scotia, Maine, Quebec, Ontario, and Wisconsin.

The typical form has the heads in loosely paniculate leafy inflorescences, while the so-called f. *axillaris* (Pursh) House has the heads in simple loose axillary clusters on a simple stem. The common type found in Missouri is the f. *axillaris*, but it is not believed that such forms are of any taxonomic value, as they are often found in the same colony in various localities.

2. **Solidago flexicaulis** L. Broadleaf Goldenrod Map 2147

*Solidago latifolia* L. [P & S]

Flowers July–October.

Occurs in rich woods, on north- or east-facing steep wooded slopes or bluffs and ledges. Throughout Missouri, throughout the Ozark region and in much of northern Missouri, apparently absent from most of the unglaciated prairie and extreme northwest sections.

Ranges from Quebec to North Dakota, south to Georgia, Arkansas, and Kansas.

3. **Solidago petiolaris** Ait. Map 2148

Flowers late May–November.

Occurs in dry, rocky open woods in usually acid soils associated with sandstone, chert, or granitic substrata, sometimes on limestone strata, rocky prairies, bluff escarpments, and thickets.

Two varieties occur in Missouri:

More rarely encountered; leaves scarcely or not at all sticky or resinous, the lower surface and main veins more or less hairy with spreading hairs; involucre minutely hairy with the hairs often sticky, sometimes glabrous (without hairs) . 3a. S. PETIOLARIS var. PETIOLARIS

More commonly encountered; leaves rather prominently sticky or resinous, the lower surface glabrous (without hairs) or the midrib and main nerves with short rough hairs; involucre with glandular dots or glabrous, scarcely if at all minutely hairy . . 3b. S. PETIOLARIS var. WARDII

3a. **Solidago petiolaris** var. **petiolaris** Map 2148

*Solidago petiolaris* Ait. [G, BB, P & S]

Scattered in the Ozark region, mainly in the eastern section.

Ranges from Florida to Texas, north to North Carolina, Missouri, and Nebraska.

3b. **Solidago petiolaris** var. **Wardii** (Britt.) Fern. Map 2148

*Solidago angusta* T. & G. [BB]

*Solidago Lindheimeriana* Scheele [P & S]

*Solidago petiolaris* var. *angusta* (T. & G.) Gray [Shinners]

Common throughout the Ozark region north to Lincoln, Montgomery, Callaway, Boone, Cooper, Pettis, Henry, and Vernon counties.

Ranges from Louisiana and Texas, north to Missouri and Kansas.

Although the two varieties here given can usually be distinguished from each other, some plants are of intermediate character and difficult to place. For this reason, they are treated here as varieties rather than as

2149 Solidago Buckleyi

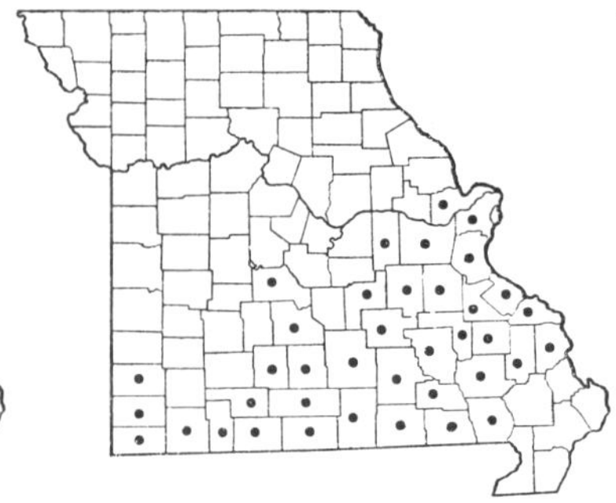

2150 Solidago hispida var. hispida

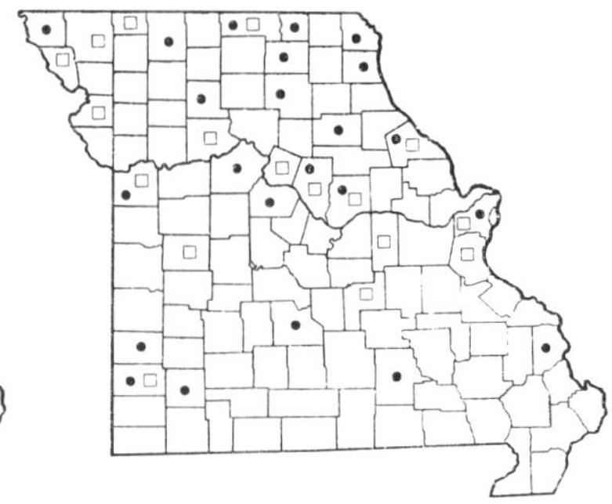

2151 ● Solidago speciosa var. speciosa
2151 □ Solidago speciosa var. angustata

distinct species as maintained by Cronquist (in Gleason's *New Ill. Fl.* 3: 424. 1952).

Most of the material from Missouri previously identified as *S. Lindheimeriana* has been found to be referable to *S. petiolaris* var. *Wardii.*

*Solidago petiolaris* var. *petiolaris* and var. *Wardii* are quite showy, with large heads and long rays, and make desirable plants for the sunny wildflower garden.

4. **Solidago Buckleyi** T. & G. Map 2149

Flowers September–October.

Occurs in dry rocky woods and along bluffs, usually in acid soils associated with chert, sandstone, and granite substrata, sometimes on limestone. Southern Ozark region north to Cape Girardeau, Bollinger, Reynolds, Shannon, Wright, Christian, and Stone counties.

Ranges from Georgia and Alabama, north to North Carolina, West Virginia, Indiana, Illinois, and Missouri.

This is retained as a species doubtfully distinct from *S. petiolaris* var. *petiolaris* pending additional field and experimental studies. Reliance on characters of the apex of the inner bracts of the involucre and on type of leaf ciliation, used by Fernald in the eighth edition of *Gray's Manual* (p. 1381), leads to confusion and lack of correlation. Some specimens, such as *Steyermark 85832* from Christian County, combine elongated membranaceous leaves and oblong, obtuse to acute innermost involucral bracts of *S. Buckleyi* with the short scabrous-ciliation and short incurved hairs of *S. petiolaris.*

5. **Solidago hispida** Muhl. var. **hispida** Map 2150

*Solidago hispida* Muhl. [G, P & S]

*Solidago bicolor* [of BB as to reference to Missouri] not L.

Flowers July–October.

Occurs in rocky open woods and shaded ledges of bluffs, in acid soils associated with sandstone, chert, or granite. Ozark region north to St. Charles, Franklin, Gasconade, Phelps, Camden, and Jasper counties.

Ranges from Newfoundland to Manitoba, south to Georgia, Tennessee, Arkansas, and Oklahoma.

Although Fernald in *Gray's Manual*, eighth edition (p. 1393), states the soil preference of this variety to be calcareous, actually in Missouri it is always found in acid soils, associated with other similar oxylophiles, such as *Vaccinium vacillans*, *V. arboreum*, *Rhododendron roseum*, *Tephrosia virginiana*, *Cunila origanoides*, and many others. Although *S. bicolor* L. is credited to Missouri by Cronquist (in Gleason's *New Ill. Fl.* 3: 420. 1952), the Missouri specimens so-labeled, which I have examined, belong to *S. hispida*, and have been referred to the latter species in the present flora.

The leaves are eaten by white-tailed deer in the Ozarks.

6. **Solidago speciosa** Nutt. Map 2151

Flowers August–November.

Occurs in prairies, meadows, thickets, and dry, rocky, or open woods. Scattered in Missouri, absent from most of the Ozark region.

Two variations occur in Missouri:

Basal leaves ovate, oblong, or obovate, 4.5–10 cm. broad, entire (without teeth) or toothed with rounded or pointed teeth; most of the leaves of the stem, except the uppermost, 2.5–5 cm. broad; plants averaging taller, 6–20 dm. tall . . . 6a. S. SPECIOSA var. SPECIOSA

Plate no. 352. **1.** Solidago flexicaulis, × 2/5. **2.** Solidago Buckleyi, × 2/5. **3.** Solidago caesia, × 2/5. **4.** Solidago petiolaris var. Wardii, × 2/5. **5.** Solidago hispida, × 2/5; *a.* Head of flowers, × 4/5. **6.** Solidago speciosa, × 2/5; *a.* Lower portion of stem with leaves.

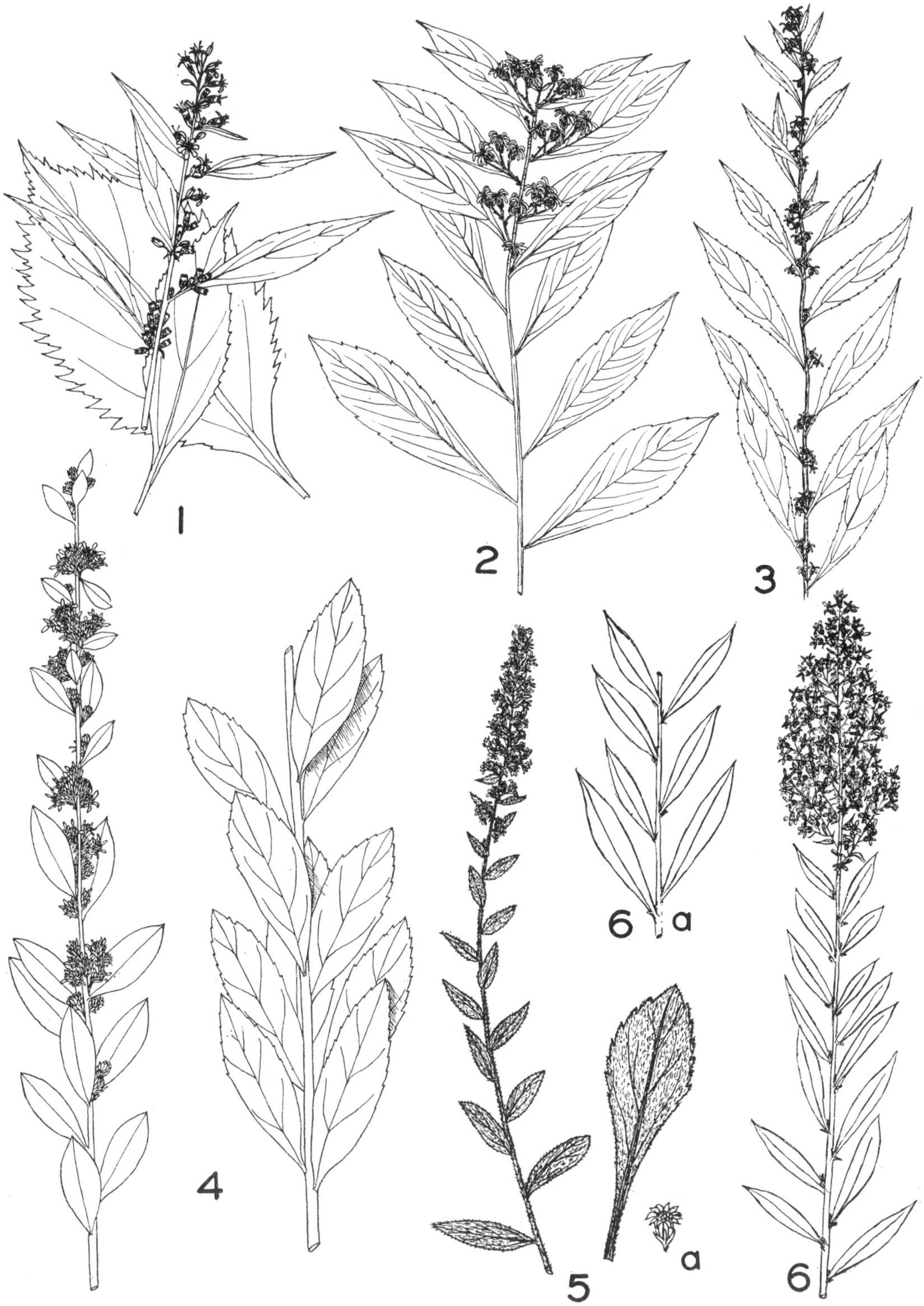


PLATE NO. 352

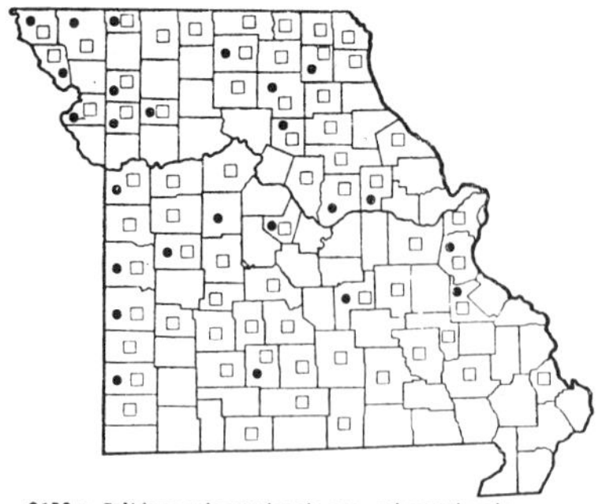

2152 • Solidago missouriensis var. missouriensis
2152 □ Solidago missouriensis var. fasciculata

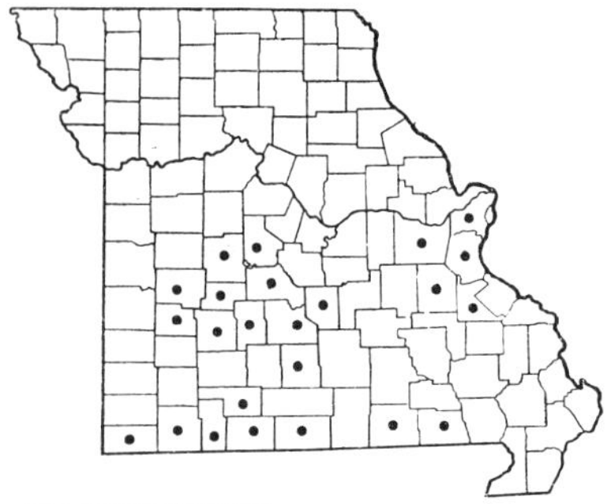

2153 Solidago Gattingeri

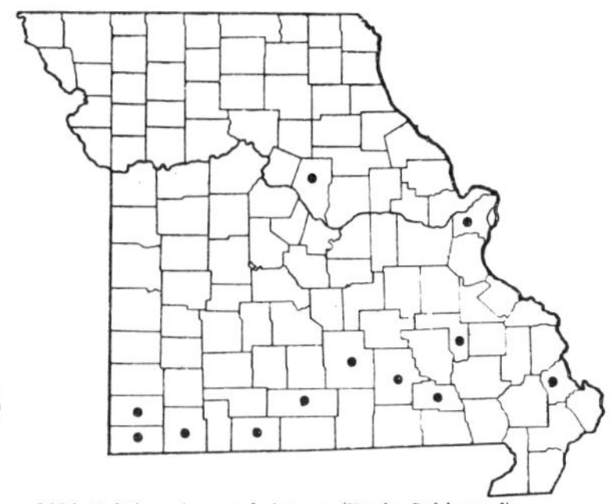

2154 Solidago juncea f. juncea (Early Goldenrod)

Basal leaves spatulate, oblong, or oblanceolate, 1.5–3.5 cm. broad, usually entire; leaves about half-way up stem 0.8–3 cm. broad; plant averaging shorter, 3–10 dm. tall . . . . .
6b. S. SPECIOSA var. ANGUSTATA

6a. **Solidago speciosa** var. **speciosa** Map 2151
*Solidago speciosa* Nutt. [G, P & S]

Ranges from North Carolina to Louisiana and Oklahoma, north to Massachusetts, New Hampshire, New York, Ohio, Michigan, Illinois, and Minnesota.

6b. **Solidago speciosa** var. **angustata** T. & G. Map 2151
*Solidago speciosa* var. *rigidiuscula* T. & G. [BB]

Ranges from Georgia to Texas, north to Ontario, Michigan, Wisconsin, Minnesota, and Saskatchewan.

This is a handsome species when in full flower with long spires of yellow flowers and pale green foliage. It is a desirable addition to the sunny wildflower bed.

7. **Solidago missouriensis** Nutt. Map 2152
Flowers July–September.

Occurs in prairies, loess and limestone hills, and thickets.

Two varieties, as interpreted by Fernald in *Gray's Manual*, eighth edition, may be recognized in Missouri:

Inflorescence 2–12 cm. broad; achenes 1.3–2.2 mm. long. with spreading or appressed hairs .
7a. S. MISSOURIENSIS var. MISSOURIENSIS

Inflorescence up to 20 cm. broad; achenes 1–1.3 mm. long, glabrous or with only sparse short hairiness . . 7b. S. MISSOURIENSIS var. FASCICULATA

7a. **Solidago missouriensis** var. **missouriensis** Map 2152
*Solidago glaberrima* Martens var. *moritura* (Steele) Palmer & Steyerm. [P & S]

Occurs throughout Missouri, absent from many parts of the Ozark section.

Ranges from British Columbia to Wisconsin, south to Oregon, Arizona, New Mexico, Oklahoma, and Missouri; adventive in New Jersey.

7b. **Solidago missouriensis** var. **fasciculata** Holzinger Map 2152
*Solidago glaberrima* Martens [P & S]
*Solidago missouriensis* var. *glaberrima* (Martens) Rosend. & Cron. [BB appendix]

Scattered in Missouri, mainly in the southern and central sections, north locally in Nodaway County.

Ranges from Idaho to New Mexico, east to Michigan, Indiana, and Tennessee.

By Cronquist (in Gleason's *New Ill. Fl.* 3: 426. 1952) the Missouri material was originally interpreted as falling into two categories: (1) var. *fasciculata* Holz., with very narrow leaves not usually exceeding 1 cm. broad, the upper ones usually bearing conspicuous fascicles of reduced leaves in their axils, and (2) var. *glaberrima* (Martens) Rosend. & Cron., with broader leaves, the lower ones at least over 1 cm. broad, and the fascicles either lacking or usually inconspicuous in the axils of the upper leaves. He regards var. *missouriensis* as a small type of plant having more persistent basal leaves and a slightly or scarcely 1-sided inflorescence. In the appendix, however, (p. 546), the varieties *fasciculata* and *glaberrima* are combined under var. *fasciculata.*

Many specimens in Missouri are intermediate between the varieties *missouriensis* and *fasciculata*, combining broad inflorescences with strigose achenes (*Bush 9173A* from Clark Co., *Mackenzie 95* from Cass Co.; *Engelmann* from St. Louis Co.) and small inflorescences with glabrate achenes (*Palmer 6224* from Howell Co.; *Palmer 2565* from Jasper Co.; *Steyermark 15069* from Worth Co.).

8. **Solidago Gattingeri** Chapm. Map 2153
Flowers July–September.

Occurs on limestone glades and bald knobs. Ozark

region north to St. Louis, Franklin, Pulaski, Camden, Morgan, Benton, and St. Clair counties; not recorded and apparently absent from much of the central and southeastern Ozark sections.

Ranges in Missouri and Tennessee.

Outside of its occurrence in a limited section of Tennessee, this species is practically confined to the Missouri Ozarks, but it is expected that it will be found eventually on limestone barrens in adjacent states. It has erect slender, glabrous stems 4–10 dm. tall, pale green mostly glabrous, slightly toothed leaves, which are reduced upwards to small entire bracts, and a leafless inflorescence with elongated branches. The species occurs with such other Ozark calciphiles as *Ophioglossum Engelmanni*, *Leavenworthia uniflora*, *Oenothera missouriensis*, *Heliotropium tenellum*, *Satureja arkansana*, *Echinacea paradoxa*, *Aster oblongifolius*, and *Rudbeckia missouriensis*. I have grown it for many years in northern Illinois in my wildflower preserve, where it has acclimated itself well.

9. **Solidago juncea** Ait. f. **juncea** Early Goldenrod Map 2154

*Solidago juncea* Ait. [G, BB, P & S]

*Solidago juncea* var. *scabrella* (T. & G.) Gray [P & S]

*Solidago juncea* f. *scabrella* (T. & G.) Fern. [G]

Flowers June–October.

Occurs in usually acid soils of sandy or gravelly open ground of prairies, dry open or rocky woods and thickets.

Occurs in the Ozark region north locally to St. Louis and Boone counties.

Ranges from Nova Scotia and New Brunswick to Minnesota and Saskatchewan, south to Georgia, Tennessee, and Missouri.

Specimens from Scott County (*Steyermark 20755, 20730, 72290*) have glabrous achenes. Other representative specimens with glabrous achenes are from St. Louis (*Sherff 724, 1103*), Shannon (*Bush 6119*), and Texas (*Steyermark 72050*) counties.

Some plants with scabrous leaves and sparsely short-hairy branches of the panicle have been called f. *scabrella* (T. & G.) Fern. as contrasted with those having glabrous leaves, except for the ciliolate margins of the blades and petioles, and glabrous branches of the panicle of typical f. *juncea*. These variations appear to have no correlation and may occur in random fashion with a colony. They do not appear to merit any taxonomic status.

Although *S. juncea* normally has no slender horizontal stolons or runners, these are occasionally developed under certain conditions, as in sandy soils (see Rh. 38: 212. 1936).

10. **Solidago arguta** Ait. Map 2155

Flowers July–October.

Occurs in dry rocky open woods, rich wooded slopes along bluffs and escarpments, and thickets.

Ozark region of southern Missouri north to Ste. Genevieve, Washington, Dent, Texas, Wright, Webster, and Jasper counties.

The following intergrading variations occur in Missouri:

a. Lower surface, midrib and nerves beneath of leaves, including the rosette and basal leaves, glabrous; middle and upper leaves of stem usually lanceolate or lance-elliptic . . . *b*
  b. Mature achenes glabrous or nearly so . . 10a. S. ARGUTA var. ARGUTA
  b. Mature achenes hairy . . 10b. S. ARGUTA var. BOOTTII

a. Lower surface, but especially the midrib and nerves beneath of the rosette and basal leaves with conspicuous hairs; middle and upper leaves of stem usually ovate, oblong-ovate or oblong-obovate . . . . . . . . . . *c*
  c. Inner bracts of involucre with a prominent keel, linear-lanceolate, narrowly tapering at tip, 0.5 mm. broad; disk-corollas 3–3.5 mm. long; pappus 2–2.5 mm. long; mature achenes 1.3–1.5 mm. long . 10d. S. ARGUTA var. NEUROLEPIS
  c. Inner bracts of involucre without a prominent keel, oblong, obtuse, about 1 mm. broad; disk-corollas 3.5–4 mm. long; pappus 2.5–5 mm. long; mature achenes 2.5–3.5 mm. long . . . 10c. S. ARGUTA var. STRIGOSA

10a. **Solidago arguta** var. **arguta** Map 2155

*Solidago arguta* Ait. [G, BB, P & S]

Common in the southern Ozarks.

Ranges from Alabama and Missouri north to Maine, Ontario, and Kentucky.

10b. **Solidago arguta** var. **Boottii** (Hook.) Palmer & Steyerm. Map 2155

*Solidago Boottii* Hook. [G, BB]

Scattered in the range of var. *arguta* and less common.

Ranges from Florida to Texas, north to Virginia, West Virginia, Kentucky, Missouri, and Oklahoma.

10c. **Solidago arguta** var. **strigosa** (Small) Steyerm. Map 2155

*Solidago strigosa* Small, *Fl. Se. U.S.* 1198, 1339, 1903.

*Solidago microphylla* [of P &S], not Engelm.

Common in the Ozarks of southern Missouri.

Ranges from Georgia to Louisiana, north to Tennessee and Missouri.

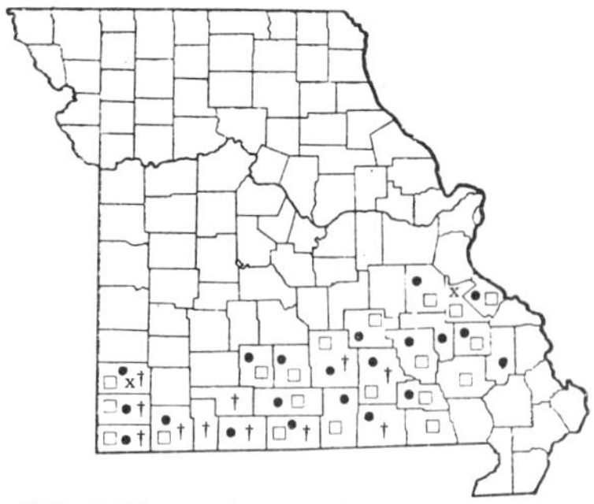

2155 • Solidago arguta var. arguta
2155 □ Solidago arguta var. Boottii
2155 † Solidago arguta var. strigosa
2155 x Solidago arguta var. neurolepis

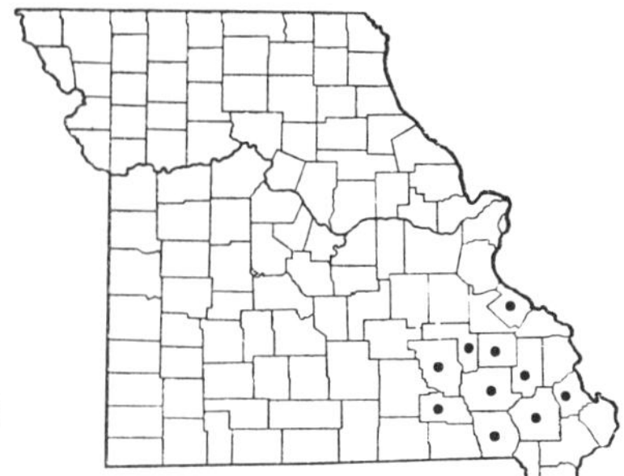

2156 Solidago patula var. patula

2157 • Solidago nemoralis var. nemoralis (Old-field Goldenrod)
2157 □ Solidago nemoralis var. decemflora

A specimen from Ozark County (*Palmer 32932* near Tecumseh) was placed in *S. microphylla* by Palmer and Steyermark in their *Annotated Catalogue* (p. 659), but the specimen is here referred to *S. arguta* var. *strigosa.*

10d. **Solidago arguta** var. **neurolepis** (Fern.) Steyerm. Map 2155

*Solidago neurolepis* Fern. Rh. 38: 212–13. 1936.

Known only from Jasper County, southwestern Missouri (dry bank, open woods, Oronogo, August 29, 1920, *Palmer 18863*, holotype).

The above treatment is based upon a study of Missouri material over a period of a number of years. As segregated by Fernald (*Gray's Manual*, eighth edition) and by Cronquist (Gleason's *New Ill. Fl.*), the principal differences separating *S. arguta*, *S. Boottii*, and *S. strigosa* are in the presence or absence of hairs on the lower surface and nerves of the basal or other leaf-blades, the glabrity or pubescence of the achenes, and the relative thick texture of the leaves. These characters break down to such an extent in Missouri material examined that it is highly questionable whether more than one taxon is involved in the series of Missouri plants. The variations are found throughout the southern Ozarks, but do not show any constant segregation of characters. Leaf-shape of the cauline (stem) leaves is variable in plants found in the same locality, as is the character of glabrity or pubescence on the lower surface of the leaf-blades. Some plants have glabrous achenes, others pubescent ones, but it is believed that this varies among the populations and is not a constant character correlated with glabrity or pubescence of the leaf. The leaves vary somewhat in texture, but most or all of the Missouri material has rather firm, subcoriaceous leaves. In general, *S. arguta* and *S. Boottii* have the middle and upper cauline (stem) leaves lanceolate or lance-elliptic and narrower than those of *S. strigosa* and *S. neurolepis*, and the lower leaves of the latter two taxa are usually larger and broader. However, there is intergradation between *S. arguta*, *S. Boottii*, and *S. strigosa* to such a degree that the group appears more naturally as intergrading varieties of one variable species. Cronquist (in Gleason's *New Ill. Fl.* 3: 428. 1952) noted that *S. arguta* is difficult to separate from *S. Boottii* toward the southern part of its range and that the latter might be regarded as subordinate to *S. arguta*. Fernald, likewise (Rh. 38: 209–12. 1936) found the *S. ludoviciana - yadkinensis - strigosa* group (related to *S. Boottii*) as one needing clarification, and attempted to differentiate them. However, his treatment of the group was not completely acceptable to Cronquist, who (p. 428) treated *S. yadkinensis* as a synonym of *S. Boottii*, and indicated that neither *S. ludoviciana* nor *S. strigosa* might be specifically distinct from *S. Boottii*, and that *S. strigosa* might only be a phase of *S. ludoviciana*.

Based upon the present author's field experience in Missouri, the most conservative view is to treat them as varieties of *S. arguta*. So far as Fernald's *S. neurolepis* is concerned, this appears to be a pubescent and broad-leaved variation of *S. strigosa* with somewhat smaller achenes, pappus, and disk-corollas than are usually found in that taxon, but with keeled inner involucral bracts. The absence of slender stolons encountered in the single specimen of *S. neurolepis* may be due to an incomplete collection of the plant, and may not have real significance taxonomically. Fernald noted (Rh. 38: 211–12. 1936) that the occurrence of filiform

Plate no. 353. **1.** Solidago Gattingeri, × 2/5; *a.* Upper part of stem; *b.* Middle part of stem; *c.* Basal leaves. **2.** Solidago juncea, × 2/5; *a.* Basal leaves; *b.* Head of flowers, × 1 1/5; *c.* Upper part of stem. **3.** Solidago arguta var. arguta, × 2/5; *a.* Upper part of stem; *b.* Middle part of stem; *c.* Basal leaf. All details from Small, The New York Botanical Garden.

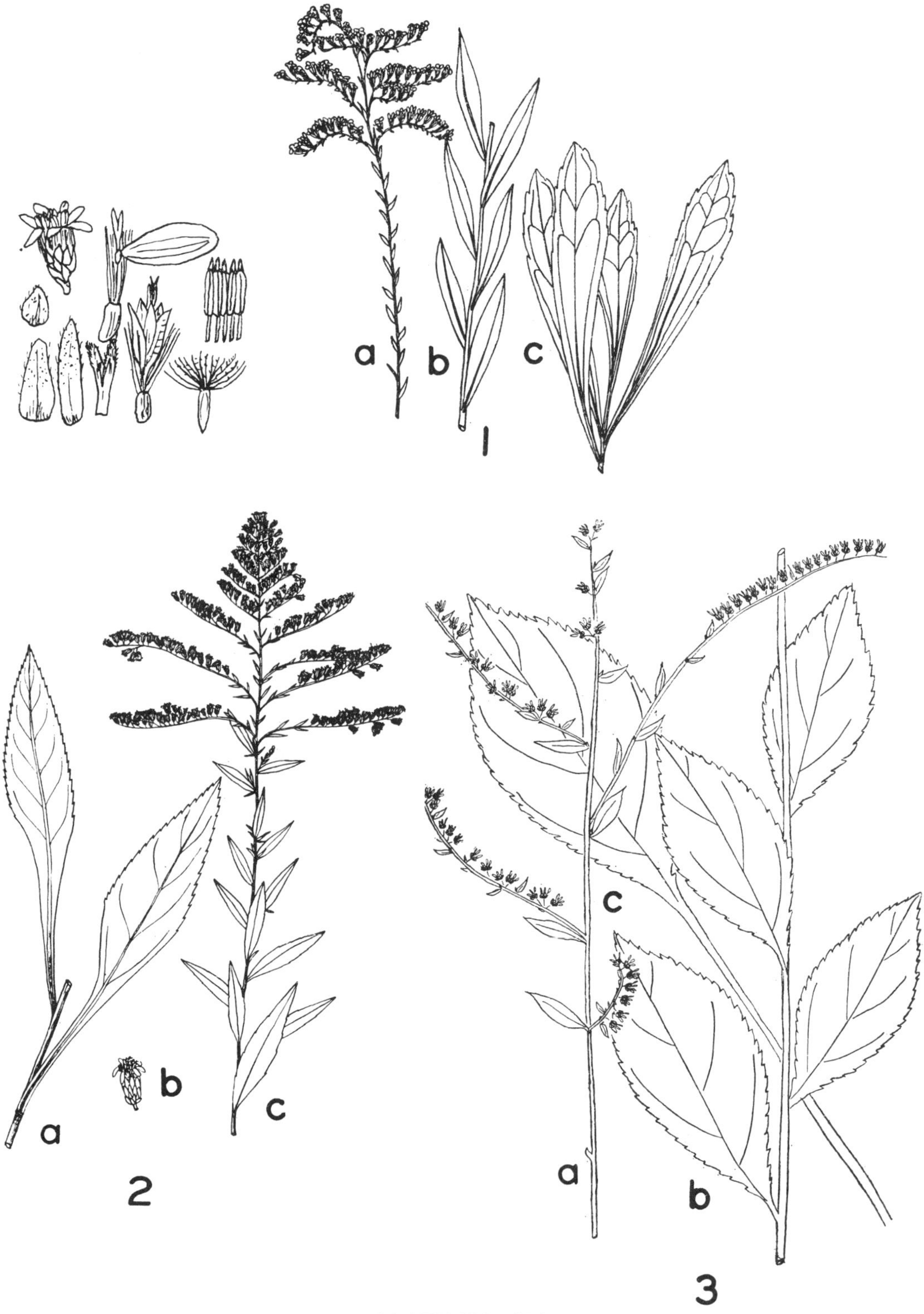


PLATE NO. 353

stolons may be of specific value, but that the character should be used with caution, since many specimens are not properly or completely collected and sometimes fail to show up in dried material. Furthermore, he notes that some species, such as *S. juncea*, ordinarily not producing these filiform stolons, may on occasion develop them. Until more plants of *S. neurolepis* have been studied in the field, it is impossible at present to assign any final evaluation to the status of this taxon.

*Solidago arguta* and varieties, with their dark green, rather thick leaves, large rosettes of basal leaves, and rather showy flowers, make desirable plants for the woodland garden. They do best in dry open woodlands with a minimum of competition. When too crowded by ground covers or dense vegetation, the plants soon die out. The upper leaves of the plant are eaten by white-tailed deer in the Ozarks.

11. **Solidago patula** Muhl. var. **patula** Rough-leaf Goldenrod Map 2156

*Solidago patula* Muhl. [G, BB, P & S]

Flowers August–October.

Occurs in swampy meadows and spring-fed swamps in calcareous soil, swampy thickets, and moist ledges of dripping limestone bluffs.

Southeastern Ozark and Crowley Ridge sections, north to Ste. Genevieve and Iron counties, west to Reynolds and Carter counties.

Ranges from Vermont to Ontario and Wisconsin, and Minnesota, south to Georgia, Alabama, Ohio, Indiana, Illinois, and Missouri.

Another variety, var. *strictula* T. & G., occurs in the eastern and gulf coast states north to Ohio and Virginia, west to Louisiana.

The combination of swampy habitat, scabrous upper leaf surfaces, and large basal and lower leaves readily identify this species.

12. **Solidago nemoralis** Ait. Old-field Goldenrod Map 2157

Also called Gray Goldenrod.

Flowers June–November.

Occurs in dry open ground, prairies, leached or eroded slopes, old or fallow fields, glades, loess hills, dry open woods, thickets, roadsides, and along railroads. Throughout Missouri.

The following variations occur in Missouri:

Basal leaves broadly oblanceolate to spatulate-obovate, crenately toothed (rather rounded, scalloped-like); main leaves of stem oblanceolate to obovate, the upper ones gradually reduced in size, the lower ones of the stem 3–6½ times as long as broad; involucres 3–4.5 mm. high . . . . 12a. S. NEMORALIS var. NEMORALIS

Basal leaves narrowly lanceolate to lance-linear, nearly entire (without teeth) or shallowly toothed; main leaves of stem linear-oblanceolate to linear, the lower ones of the stem 7–10 times as long as broad; involucre 4.5–6.5 high . . . 12b. S. NEMORALIS var. DECEMFLORA

12a. **Solidago nemoralis** var. **nemoralis** Map 2157

*Solidago nemoralis* Ait. [G, P & S]

This is the commoner variation in Missouri.

Ranges from Georgia to Texas, north to Nova Scotia to Ontario, Minnesota, North Dakota, and Alberta.

12b. **Solidago nemoralis** var. **decemflora** (DC.) Fern. Map 2157

*Solidago nemoralis* var. *longipetiolata* (Mack. & Bush) Palmer & Steyerm. [BB, P & S]

This is less common, but scattered throughout the state.

Ranges from Ontario to Alberta, south to Kentucky, Arkansas, Texas, and Arizona.

This goldenrod varies considerably in amount and kind of pubescence, with a greater or lesser degree of viscidulous hairs, size of flower-heads, breadth and compactness of inflorescence, type of pubescence of achenes, and shape and size of basal and cauline (stem) leaves.

The young flower-heads are eaten by wild turkey and the basal leaves are a source of food during winter for the white-tailed deer.

13. **Solidago radula** Nutt. Rough Goldenrod Map 2158

Flowers late May–October.

Occurs on glades, ledges and escarpments of bluffs, and rocky dry open woodland.

Ozark region of southern and central Missouri north to Marion, Monroe, Boone, Saline, and Bates counties.

The following varieties may be recognized in Missouri:

a. Bracts of involucre 1–1.2 mm. broad, ovate to broadly oblong . . 13b. S. RADULA var. LAETA
a. Bracts of involucre 0.3–1 mm. broad, narrowly linear to narrowly oblong . . . . . . *b*

Plate no. 354. **1.** Solidago ulmifolia, × 2/5; *a.* Upper part of stem; *b.* Middle part of stem. **2.** Solidago radula, × 2/5; *a.* Upper part of stem; *b.* Middle part of stem. **3.** Solidago nemoralis, × 2/5; *a.* Upper part of stem; *b.* Basal leaf. **4.** Solidago rugosa, × 2/5; *a.* Upper part of stem; *b.* Middle part of stem.

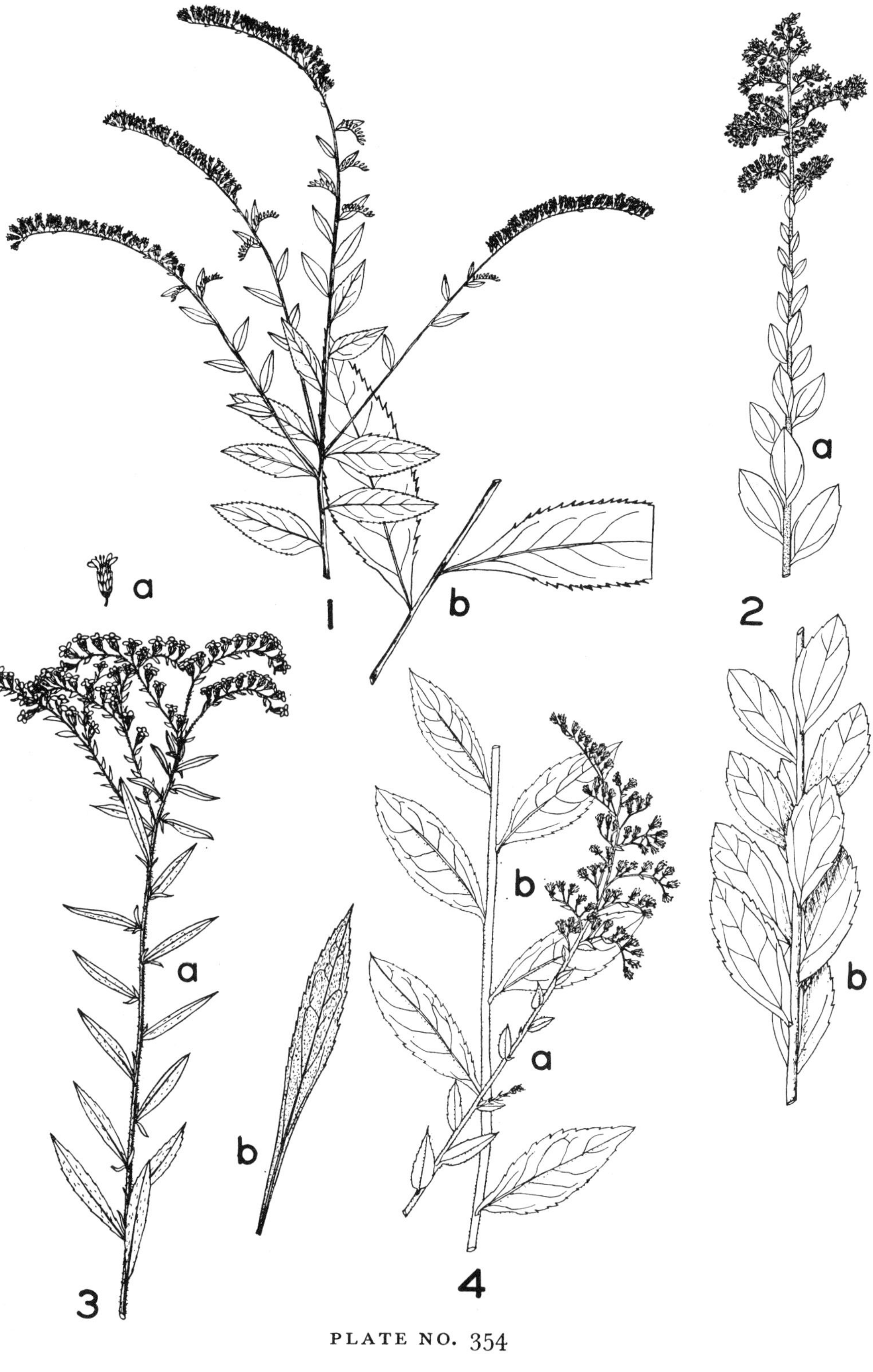


PLATE NO. 354

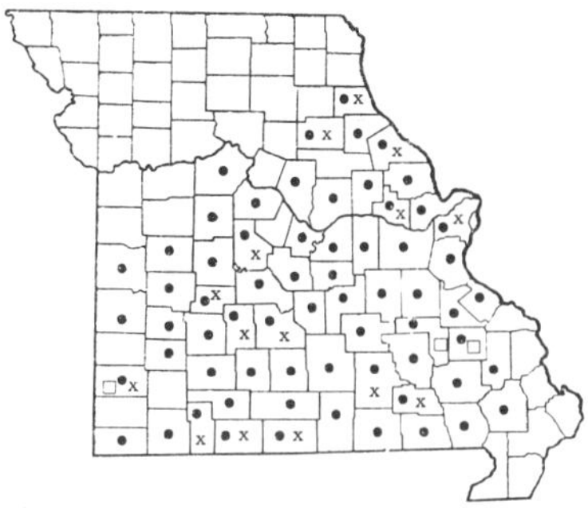

2158 • Solidago radula var. radula
2158 x Solidago radula var. laeta
2158 □ Solidago radula var. stenolepis

2159 Solidago odora (Sweet Goldenrod)

2160 • Solidago ulmifolia var. ulmifolia (Elm-leaf Goldenrod)
2160 □ Solidago ulmifolia var. Palmeri

b. Bracts of involucre narrowly linear, 0.3–0.5 mm. broad . 13c. S. RADULA var. STENOLEPIS

b. Bracts of involucre narrowly oblong, up to 1 mm. broad . . 13a. S. RADULA var. RADULA

13a. **Solidago radula** var. **radula** Map 2158

*Solidago radula* Nutt. [G, BB, P & S, Steyerm.]

This is the commonest variation, found throughout the range of the species.

Ranges from Louisiana to Texas, north to Illinois, Missouri, and Oklahoma.

13b. **Solidago radula** var. **laeta** (Greene) Fern. Map 2158

*Solidago radula* [of P & S, Steyerm. in part], not Nutt.

This is a common variation found within the range of the species.

Ranges from Texas to Missouri.

13c. **Solidago radula** var. **stenolepis** Fern. Map 2158

Known only from Madison (*Palmer 31606*), Iron (*Glatfelter*), and Jasper (rocky open woods, limestone hills near Carthage, October 2, 1922, *Palmer 22161* holotype in N.Y. Bot. Gard. Herb.) counties.

Ranges from Louisiana and Texas to Missouri.

This species varies in shape and size of the leaves, breadth of inflorescence, and involucral bracts. The variations intergrade and those on the borderline of measurements of the bracts are difficult to place. All of them are treated as one variable species by Cronquist (in Gleason's *New Ill. Fl.* 3: 434. 1952). The future and final status of the varieties remains to be worked out by more detailed field and experimental studies.

The species is easily recognized by the rough leaves and stems, and by the relatively crowded, numerous, rigid, mostly sessile leaves conspicuously veiny beneath, only gradually reduced upwards, and by the 1-sided panicles. This species and *S. petiolaris* var. *Wardii* are the earliest recorded goldenrods to bloom in the spring, having been found on May 31 in bloom in southwestern Missouri.

14. **Solidago odora** Ait. Sweet Goldenrod Map 2159

Also called Fragrant Goldenrod.

*Solidago suaveolens* Schoepf [P & S]

Flowers July–September.

Occurs in rocky or sandy open woods, slopes of ravines, base of bluffs, and thickets in usually acid soils. Southeastern Missouri Ozarks and on Crowley Ridge, in Scott, Stoddard, Dunklin, Butler, Carter, and Ripley counties.

Ranges from Florida to Texas, north to New Hampshire, Vermont, New York, Ohio, Kentucky, Missouri, and Oklahoma.

*Solidago odora* has been used in medicine for relief of flatulence and as a stimulant. The fragrant leaves may be used in tea and an aromatic oil is derived from the plant.

Occasional plants lacking an odor have been given the name, f. *inodora* (Gray) Britt.

15. **Solidago ulmifolia** Muhl. Elm-leaf Goldenrod Map 2160

Flowers August–November.

Occurs in usually dry or rocky open woods, along bluffs, thickets, and along streams.

The following two varieties occur in Missouri:

Commonly encountered type, throughout the state; stem glabrous or nearly so below the flowering branches . . . . . 15a. S. ULMIFOLIA var. ULMIFOLIA

Rarely encountered type; stems conspicuously hairy with spreading hairs. . . 15b. S. ULMIFOLIA var. PALMERI

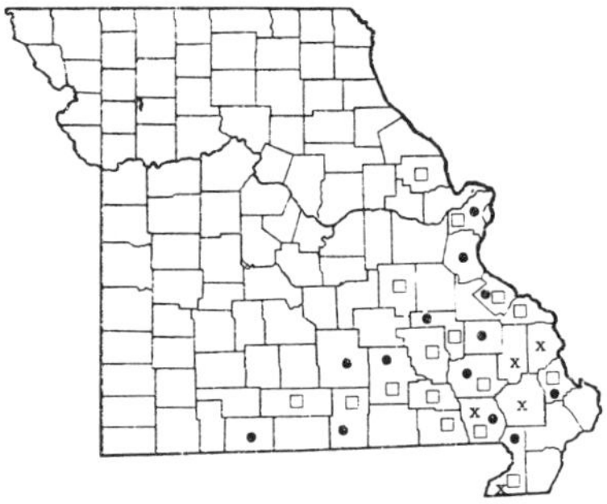

2161 • Solidago rugosa var. rugosa (Rough-leaved Goldenrod)
2161 □ Solidago rugosa var. aspera
2161 x Solidago rugosa var. celtidifolia

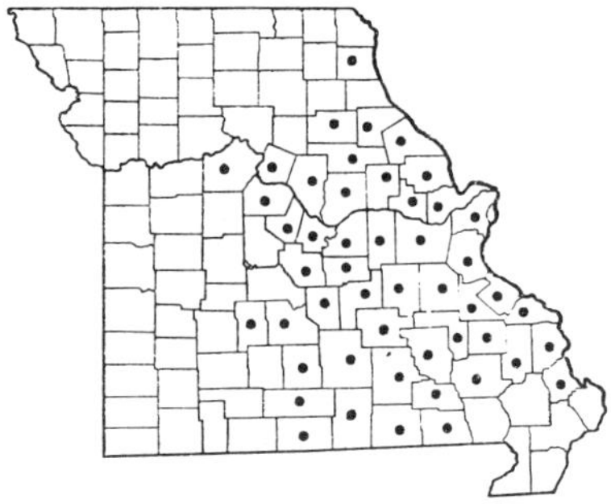

2162 Solidago Drummondii

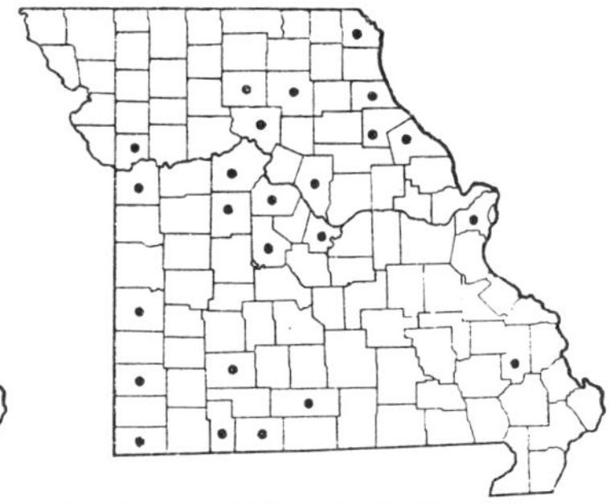

2163 Solidago canadensis var. gilvocanescens

15a. **Solidago ulmifolia** var. **ulmifolia** Map 2160

*Solidago ulmifolia* Muhl. [G, P & S]
*Solidago microphylla* Engelm.
*Solidago ulmifolia* var. *microphylla* Gray

This is found throughout Missouri, but is apparently absent only from some of the lowland counties of extreme southeastern Missouri.

Ranges from Nova Scotia and Vermont to Minnesota, south to Georgia, Tennessee, Arkansas, Oklahoma, and Texas.

15b. **Solidago ulmifolia** var. **Palmeri** Cron. Map 2160

Known only from Greene County, southwestern Missouri (Springfield, September 4, 1893, *Bush* in Mo. Bot. Gard. Herb.).

Known from Arkansas, Missouri, and Alabama.

Some Missouri material pertaining to this species has been previously misidentified as *S. rugosa*. The hairy-stemmed variety *Palmeri* is undoubtedly more common, and is to be expected throughout southern Missouri.

16. **Solidago rugosa** Ait. Rough-leaved Goldenrod Map 2161

Flowers August–October.

Occurs in wet ground of swales, swampy meadows and thickets, along spring branches and streams, swampy woodland, wet places in prairies, sometimes dry sandy knolls, and ledges.

The following varieties occur in Missouri:

a. Stems with rather long and soft conspicuous hairs; leaves gradually tapering at base, lanceolate to narrowly oblanceolate, long-pointed (acuminate), usually sharply toothed, rather thin, loosely veiny but not conspicuously rugose, the lower surface more or less hairy with long soft hairs . 16a. S. RUGOSA var. RUGOSA

a. Stems with short or rough hairiness; leaves rounded at base, oval, elliptic, or lanceolate, rather short-pointed (subacute) to shortly acuminate, usually shallowly toothed to crenate (with rounded or scalloped teeth), thick and strongly rugose, the lower surface hairy with rather stiff bristle-like hairs . . *b*

b. Inflorescence usually longer than broad, its branches ascending or spreading and bearing flowers throughout or the lower branches merely with small leaf-like bracts below; leaves on the flowering branches elliptic to lanceolate, acute . 16b. S. RUGOSA var. ASPERA

b. Inflorescence with few, distant, very prolonged and spreading branches (up to 4.5 dm. long) bearing flowers chiefly above the middle; leaves on the flowering branches elliptic to oval . . . . 16c. S. RUGOSA var. CELTIDIFOLIA

16a. **Solidago rugosa** var. **rugosa** Map 2161

*Solidago rugosa* Ait. [G, P & S]
*Solidago rugosa* subsp. *rugosa* var. *rugosa* [BB]

Scattered in the Ozark region north to St. Louis County, west to Texas and Taney counties.

Ranges from Newfoundland to Ontario, south to Virginia, West Virginia, Kentucky, and Louisiana.

16b. **Solidago rugosa** var. **aspera** Map 2161

*Solidago rugosa* subsp. *aspera* (Ait.) Cron. [BB]

Scattered in the Ozark region north to Lincoln County, west to Douglas County.

Ranges from Florida to Texas, north to Maine, New York, Ohio, Michigan, and Missouri.

16c. **Solidago rugosa** var. **celtidifolia** (Small) Fern. Map 2161

*Solidago rugosa* subsp. *aspera* (Ait.) Cron. [BB]

Occurs in southeastern Missouri north to Cape Girardeau and Bollinger counties, west to Butler County.

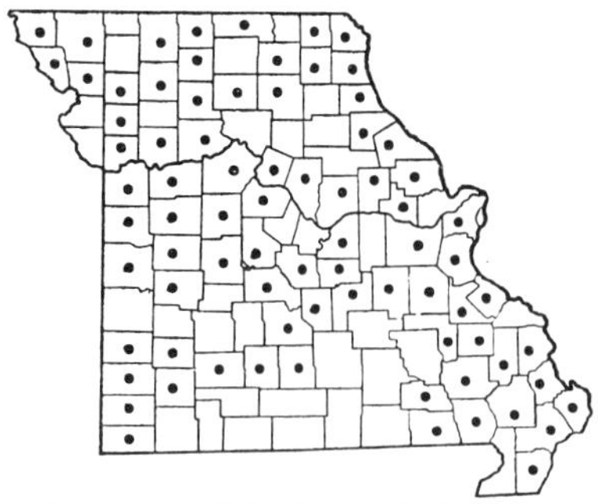

2163-A Solidago altissima (Tall Goldenrod)

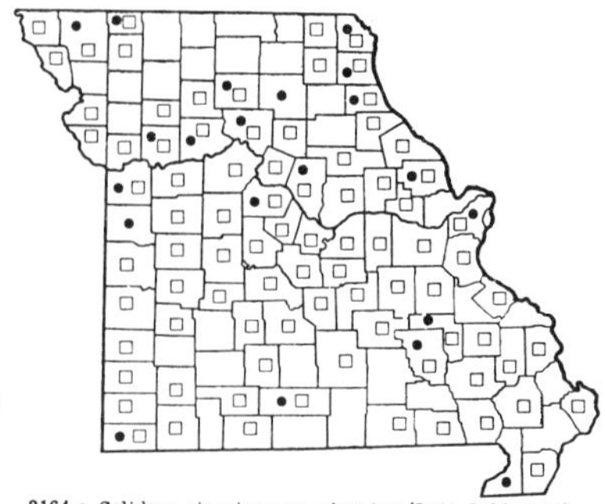

2164 • Solidago gigantea var. gigantea (Late Goldenrod)
2164 □ Solidago gigantea var. leiophylla

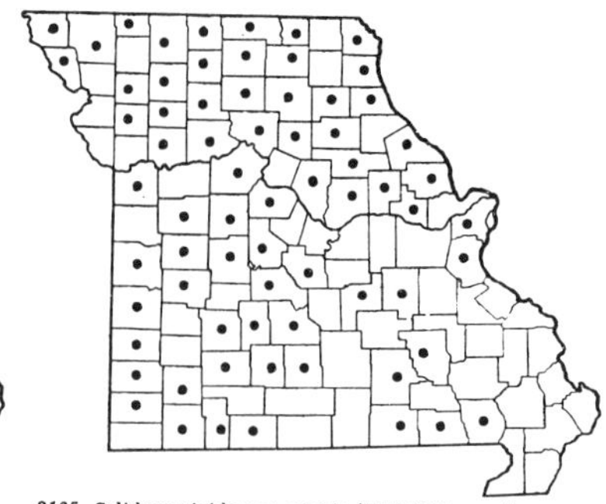

2165 Solidago rigida var. rigida (Stiff Goldenrod)

Ranges from Georgia to Texas, north to New Jersey, Indiana, and Missouri.

The differences between the above varieties are often difficult to recognize and there are intermediates existing between all of the above variations. Cronquist (in Gleason's *New Ill. Fl.* 3: 432. 1952) merges var. *aspera* and var. *celtidifolia* under subsp. *aspera*, and does not attach any significance to the differences of inflorescence branching, as suggested in the present key. As typical of var. *celtidifolia*, a specimen from Dunklin County (Campbell, October 26, 1892, *Bush*) may be cited here. There yet remain much more field and experimental studies to be carried out on the variations of *S. rugosa*, until which time a final evaluation of their status must wait.

Much material from Missouri, previously misidentified as *S. rugosa*, has been referred in the present flora to *S. ulmifolia* and *S. radula*.

17. **Solidago Drummondii** T. & G. Map 2162

Flowers September–December.

Occurs on ledges and in crevices of usually dry limestone bluffs and escarpments. Ozark region north to Lewis, Monroe, Boone, Howard, and Saline counties, west to Cooper, Moniteau, Miller, Dallas, Wright, Douglas, and Ozark counties.

Ranges from Illinois and Missouri to Arkansas and Louisiana.

18. **Solidago canadensis** L. var. **gilvocanescens** Rydb. Map 2163

*Solidago pruinosa* Greene [BB]

Flowers July–September.

Occurs in fields, prairies, upland thickets, and at the base of wooded slopes along streams. Scattered and infrequent throughout Missouri.

Ranges from Saskatchewan to New Mexico, east to Michigan, West Virginia, and Virginia.

This goldenrod is similar in general habit and appearance to the following, *S. altissima*, but has smaller involucres. In the present flora this goldenrod is considered a variety of *S. canadensis*, as interpreted by Fernald in *Gray's Manual* and by various other authors. On p. 434 of Gleason's *New Ill. Fl.* (vol. 3) this goldenrod is treated as *S. pruinosa* Greene, but in the appendix (p. 546) it is accorded only varietal rank under *S. canadensis* as given in the present flora.

An oil is obtained from *S. canadensis* var. *canadensis* which is said to resemble strongly that obtained from pine needles.

19. **Solidago altissima** L. Tall Goldenrod Map 2163A

*Solidago altissima* var. *procera* (Ait.) Fern. [P & S]

Flowers August–November.

Occurs in fallow fields, prairies, valley bottoms, along streams, rocky outcrops on slopes, open woods, thickets, sandy, alluvial, and waste ground, along roadsides and railroads. Common throughout Missouri, but not yet recorded from some of the southernmost Ozark counties.

Ranges from Quebec to New York, Ontario, Michigan, Wisconsin, Minnesota, North Dakota, and Nebraska, south to Florida and Texas.

This is one of the most frequently encountered species of goldenrod in Missouri and often dominates fields and open ground in large numbers. It is also the species most commonly affected by the goldenrod gall.

20. **Solidago gigantea** Ait. Late Goldenrod Map 2164

Plate no. 355. **1.** Solidago Drummondii, × 2/5; *a.* Upper part of stem; *b.* Lower part of stem. **2.** Solidago gigantea, × 2/5; *a.* Head of flowers, × 1 3/5. **3.** Solidago Riddellii, × 2/5. **4.** Solidago altissima, × 2/5 (Scribner's). **5.** Solidago rigida, × 2/5; *a.* Middle part of stem; *b.* Upper part of stem; *c.* Basal leaf.

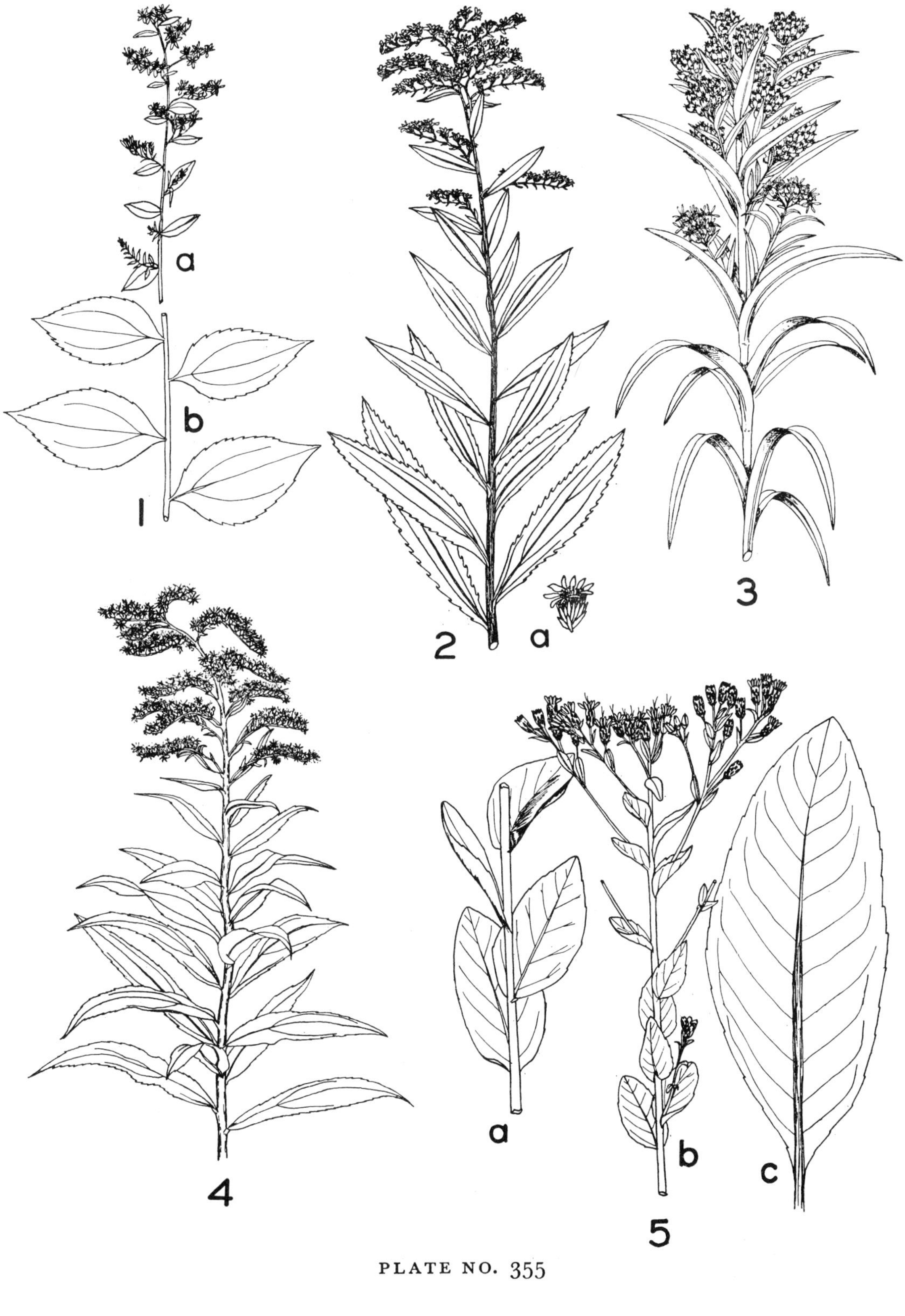


PLATE NO. 355

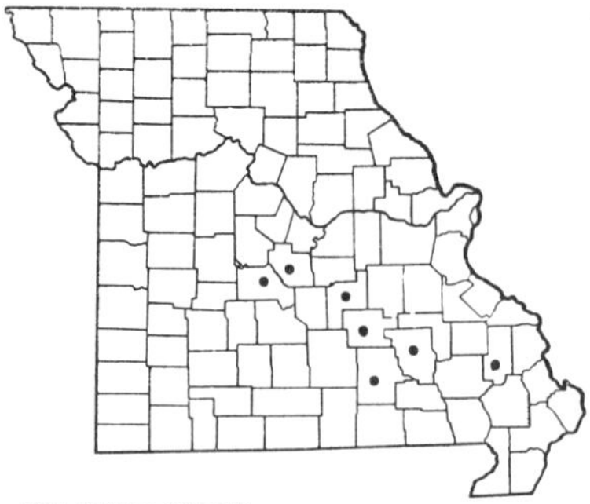

2166 Solidago Riddellii

2167 Solidago leptocephala

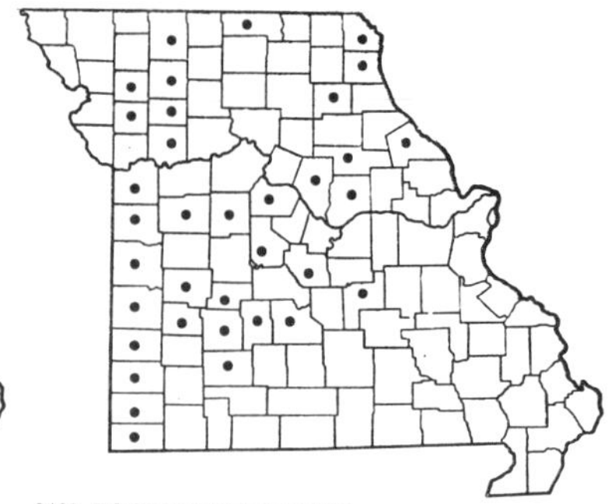

2168 Solidago gymnospermoides

Flowers July–October.

Occurs along borders of streams, ponds, and sloughs, wet low woods, base of bluffs, wet prairies, and thickets. Throughout Missouri.

Two varieties occur in Missouri:

At least the veins on the lower surface of the leaf more or less hairy . 20a. S. GIGANTEA var. GIGANTEA
Lower surface of leaf glabrous . 20b. S. GIGANTEA var. LEIOPHYLLA

20a. **Solidago gigantea** var. **gigantea** Map 2164

*Solidago gigantea* Ait. [G]

*Solidago serotina* Ait. var. *gigantea* (Ait.) Gray [P & S]

Scattered over the state, but less common than var. *leiophylla*.

Ranges from Nova Scotia and Quebec to Oregon, south to Florida, Louisiana, and Texas.

20b. **Solidago gigantea** var. **leiophylla** Fern. Map 2164

*Solidago serotina* Ait. [P & S]

This is the commoner variation in Missouri.

Ranges from Quebec to British Columbia, south to North Carolina, Tennessee, Texas, Colorado, Utah, and New Mexico.

*Solidago gigantea*, like *S. altissima*, is a common species of goldenrod in Missouri, and is usually found in wetter situations in general than *S. altissima*, often occurring along streams and near water.

Shinners (Rh. 55: 322. 1953) has segregated a var. *Pitcheri* (Nutt.) Shinners from typical *S. gigantea* and var. *leiophylla*, based upon the leaves of the stem 2–4 cm. broad with the middle ones less than 5 times as long as broad, as contrasted with leaves of the stem 1–2.5 cm. broad and the middle ones mostly more than 5 times as long as broad. I am unable to separate Missouri material on this basis and find the breadth of the leaves variable.

21. **Solidago rigida** L. var. **rigida** Stiff Goldenrod Map 2165

Also called Rigid Goldenrod.

*Solidago rigida* L. [G, P & S]

Flowers August–October.

Occurs in prairies, rocky open ground, glades, thickets, and borders of wet meadows. Throughout Missouri, but apparently absent from the southeastern lowlands and some of the eastern Ozark section.

Ranges from Massachusetts and New York to Saskatchewan, south to Georgia, Louisiana, and Texas.

This is a characteristic species of prairies, associated with *Andropogan Gerardi* and *A. scoparius*, *Petalostemon purpureum*, *Desmodium illinoense*, *Liatris aspera* and *L. pycnostachya*, *Ratibida pinnata*, and many others of prairie habitat. It is a showy species with heads and involucres larger than in most of the Missouri species.

Some reports of a poisoning from eating the plant by horses have been ascribed to this goldenrod, but no real evidence seems to be at hand to verify the reports.

The Missouri specimens belong to typical var. *rigida*, in which the leaves and stem are more or less densely hairy, the stems are commonly 6–16 dm. tall, and the achenes are usually glabrous. A similarly pubescent variety but with stems only 2.5–5 (rarely 8) dm. tall, and the achenes usually loosely hairy near their summit is var. *humilis* Porter, occurring north and west of Missouri. Another variation, var. *glabrata* Braun (*S. Jacksonii* [Ktze.] Fern.), has glabrous leaves, except for the hairy margins and midrib on lower surface, and glabrous or hairy stems; this occurs from Georgia and Alabama north to Ohio.

22. **Solidago Riddellii** Frank Map 2166

Flowers August–October.

Occurs in swampy, calcareous, spring-fed meadows in valleys. Rare, known only in the Ozark region in Bollinger, Reynolds, Shannon, Dent, Phelps, Miller, and Camden counties.

Ranges from Ontario and New York to Minnesota, south to Ohio, Indiana, Illinois, and Missouri; also found many years ago in Virginia.

This species is confined to the unique swampy meadows of some of the Ozark valleys, where it is often associated with *Parnassia grandifolia*, *Aster puniceus* var. *firmus* f. *lucidulus*, *Carex leptalea*, and other species characteristic of such habitats. It has attractive, smooth-looking, mostly glabrous leaves and a corymbiform inflorescence of numerous, showy heads of flowers.

23. **Solidago leptocephala** T. & G. Map 2167

*Euthamia leptocephala* (T. & G.) Greene [Shinners]

Flowers August–October.

Occurs in sandy open and low wet ground, swampy woods and thickets, grassy fields, meadows, and prairies. Southeastern Missouri lowlands, barely entering the Ozarks west to Oregon County.

Ranges from Florida to Texas, north to North Carolina, Kentucky, and Missouri.

This species resembles *S. gymnospermoides* but is much less resinous and the leaves are scarcely or only faintly punctate. The leaves, usually 4–8 mm. broad, average broader than those of *S. gymnospermoides*.

24. **Solidago gymnospermoides** (Greene) Fern. Map 2168

*Euthamia gymnospermoides* Greene [Shinners]

Flowers August–October.

Occurs in dry upland or rocky prairies, rocky glades, river bottom meadows, fallow fields, wooded draws, along roadsides and railroads.

Occurs in the unglaciated and glaciated prairie regions of northern, central, and western Missouri, north and west of a line drawn from Pike, Audrain, Callaway, Miller, Phelps, Laclede, and Greene counties, to McDonald County; absent from most of the Ozark region and all the southeastern lowland section.

Ranges from Wisconsin, Illinois, Missouri, and Arkansas, west to Colorado and Texas; also in Virginia.

Shinners (Field & Lab. 19: 137–38. 1951) interprets this species as possibly an endemic restricted to central Oklahoma, the plants passing as *S. gymnospermoides* being, according to him, *Euthamia camporum* Greene.

25. **Solidago graminifolia** (L.) Salisb. Map 2169

Flowers August–October.

Occurs in prairies, fallow fields, and along railroads. Northern and central Missouri; absent from the Ozark region.

Two varieties occur in Missouri:

Leaves usually hairy; branches and branchlets and sometimes the stem more or less short-hairy; larger leaves mainly 4–12 mm. broad, with 1 or 2 lateral nerves on either side of midrib; flower-heads campanulate (bell-shaped) to broadly turbinate (top-shaped), 20–45-flowered; inner bracts of involucre oblong, obtuse (blunt) to acute (short-pointed) . 25a. S. GRAMINIFOLIA var. NUTTALLII

Leaves glabrous except for some rough lines; larger leaves mainly 2–6 (rarely 8) mm. broad, usually with only 1 lateral nerve on either side of midrib; flower-heads slenderly turbinate (top-shaped), 15–20-flowered; inner bracts of involucre linear-oblong, acute to acuminate (short- to long-pointed) . 25b. S. GRAMINIFOLIA var. MEDIA

25a. **Solidago graminifolia** var. **Nuttallii** (Greene) Fern. Map 2169

*Solidago graminifolia* [of P & S in part], not (L.) Salisb.

This is less commonly encountered in Missouri, occurring south to St. Charles, Callaway, Chariton, and Jackson counties.

Ranges from Newfoundland to Minnesota, south to North Carolina, Kentucky, and Missouri.

25b. **Solidago graminifolia** var. **media** (Greene) Harris Map 2169

*Solidago media* (Greene) Bush

*Solidago graminifolia* [of P & S in part], not (L.) Salisb.

*Solidago graminifolia* var. *graminifolia* [of BB in part], not (L.) Salisb.

*Euthamia camporum* Greene [Shinners]

This is the commoner of the two varieties, occurring south to St. Louis, Montgomery, Morgan, Pettis, Henry, and Vernon counties.

Ranges from North Dakota and Minnesota to Kansas, Oklahoma, and Texas, east to Wisconsin and Ohio.

The above interpretation of the group follows that of Stuart K. Harris and Fernald in *Gray's Manual*, eighth edition. Typical var. *graminifolia*, according to this treatment, is the most glabrous extreme which is not found in Missouri. The var. *media* in Missouri is glabrate, but not as glabrous as var. *graminifolia*. The broader leaves of var. *graminifolia* resemble those of var. *Nuttallii*, the pubescent extreme of the two varieties occurring in Missouri. The var. *media* resembles *S. gymnospermoides*, but has less punctate, less glutinous and more scabrous leaves. The flower-heads in var. *media* are viscous and resinous-sticky as in *S. gymnospermoides*, although the heads of the latter species are erroneously referred to by some authorities as lacking the viscous-resinous quality.

Cronquist (in Gleason's *New Ill. Fl.* 3: 438. 1952) does not recognize *S. graminifolia* var. *media* as a distinct variety, merging it rather with typical var. *graminifolia*.

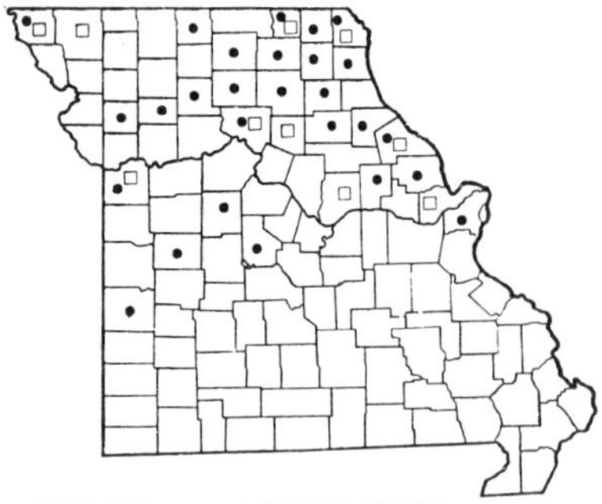

2169 □ Solidago graminifolia var. Nuttallii
2169 • Solidago graminifolia var. media

2170 Haplopappus ciliatus

2171 Astranthium integrifolium (Western Daisy)

*Excluded Species*

**Solidago bicolor** L.

Reported for Missouri by Cronquist (in Gleason's *New Ill. Fl.* 3: 420. 1952), but so far as examination of material by the present author is concerned, such specimens should be referred to *S. hispida* Muhl.

**Solidago microphylla** Engelm.

Reported for Missouri by Palmer & Steyermark (*Annot. Cat.* p. 659) on the basis of *Palmer 32932* from Ozark County. This has been reexamined and is referred in the present flora to *S. arguta* var. *strigosa* (Small) Steyerm.

## 14. **Haplopappus** Cass.

**Haplopappus ciliatus** (Nutt.) DC. Map 2170

Flowers August–September.

Occurs on prairies, glades, open alluvial and waste ground, and along railroads. Southern and central Missouri north to St. Louis (*Muehlenbach 1488*) and Clay counties.

Ranges from Missouri to Colorado, south to Texas and New Mexico.

## 15. **Astranthium** Nutt. Western Daisy

**Astranthium integrifolium** (Michx.) Nutt. Map 2171

*Bellis integrifolia* Michx. [P & S]

Flowers May–June.

Occurs on gravel bars and alluvial thickets along streams, rocky open woods, prairies, and glades. Southwestern Missouri, in Taney, Stone, Barry, and McDonald counties.

Ranges from Georgia to Oklahoma, north to Kentucky, Missouri, and Kansas.

## 16. **Chaetopappa** DC.

**Chaetopappa asteroides** DC. Map 2172

Flowers April–August.

Occurs in rocky open woods, prairies, and glades. Southwestern Missouri in the unglaciated prairie region in St. Clair, Cedar, Vernon, Jasper, and Newton counties.

Ranges from Missouri and Kansas, south to Arkansas and Texas.

## 17. **Boltonia** L'Her Boltonia, False Starwort

Branches of the inflorescence with greatly reduced, subulate bracts; leaves mainly linear, those present at flowering time rarely over 5 mm. broad; flower-heads small, the disk (solid portion between the rays) 3–6 mm. broad; inflorescence of widely separated, scattered heads . . . . . 2. B. DIFFUSA var. INTERIOR

Plate no. 356. **1.** Gutierrezia dracunculoides, $\times 2/5$; *a.* Ray-floret, $\times 4/5$; *b.* Flower-head, $\times 4/5$; After Gleason, details from Small, The New York Botanical Garden. **2.** Solidago gymnospermoides, $\times 2/5$. **3.** Solidago graminifolia, $\times 2/5$; Details from Small, The New York Botanical Garden. **4.** Haplopappus ciliatus, $\times 2/5$.

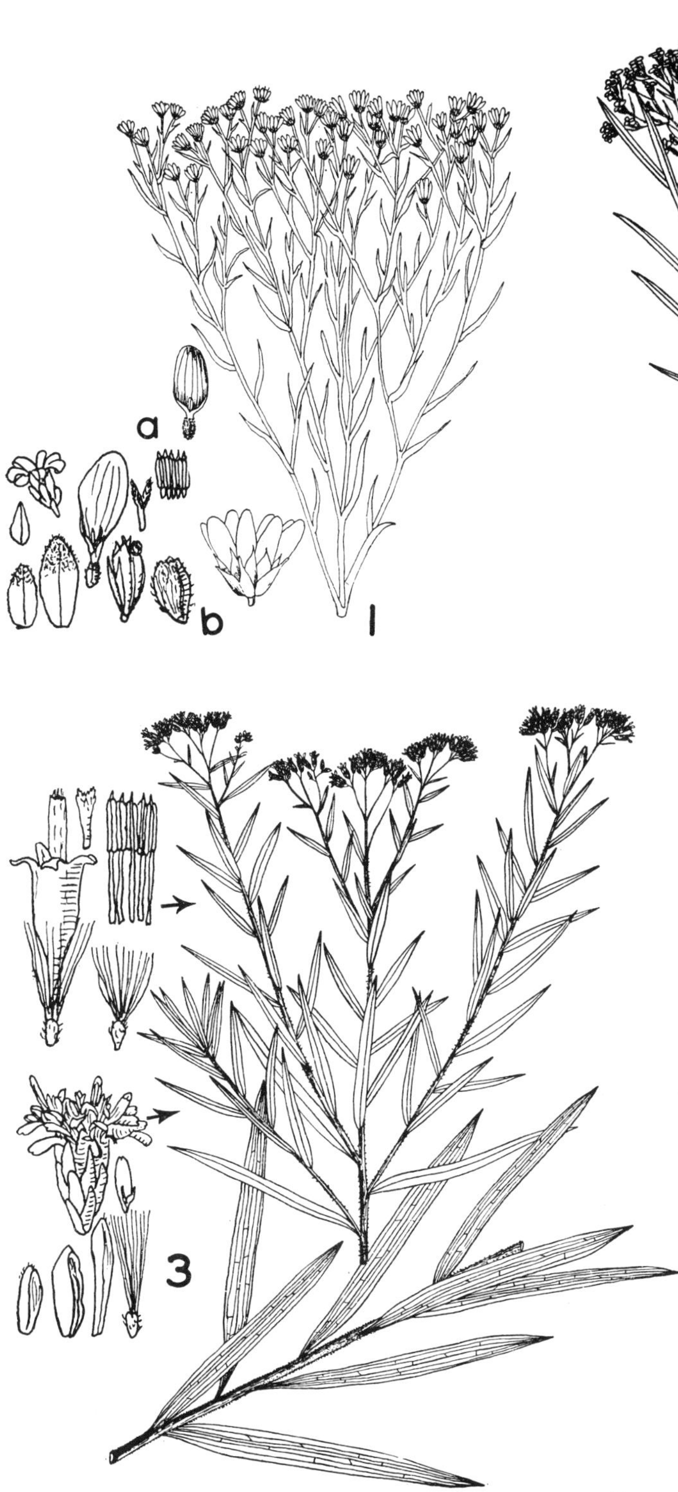


PLATE NO. 356

2172 Chaetopappa asteroides

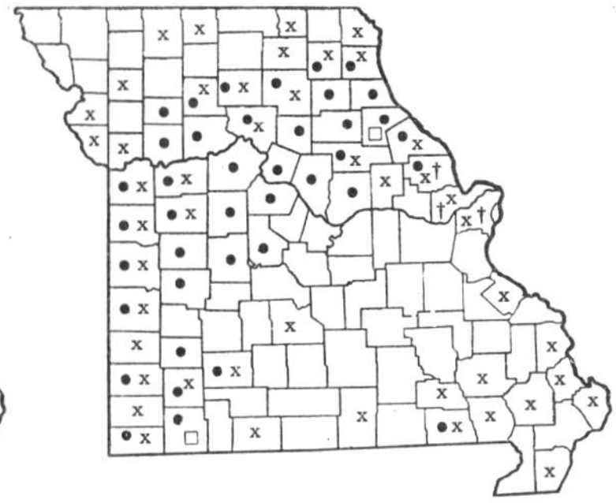

2173 • Boltonia asteroides var. latisquama (Boltonia)
2173 x Boltonia asteroides var. recognita
2173 □ Boltonia asteroides var. microcephala
2173 † Boltonia asteroides var. decurrens

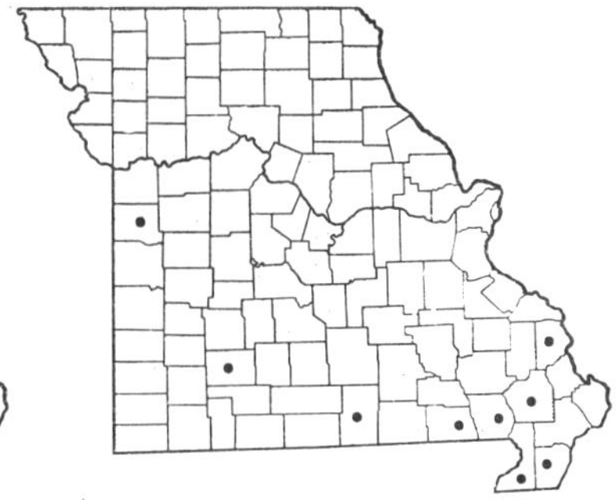

2174 Boltonia diffusa var. interior

Branches of the inflorescence rather leafy; some of the leaves present at flowering time 10 mm. or more broad; flower-heads usually rather large and showy, the disk (solid portion between the rays) 6–14 (rarely 5) mm. broad; inflorescence of more crowded heads usually in a flat-topped (corymbiform) arrangement . . . . . . . . . . . . . . . . . . . . . . . . . 1. B. ASTEROIDES

1. **Boltonia asteroides** (L.) L'Her Map 2173

Flowers July–October.

The following varieties occur in Missouri:

a. Leaves of stem decurrent (leafy tissue extending down the stem from base of leaf) . . . 1d. B. ASTEROIDES var. DECURRENS

a. Leaves of stem not decurrent . . . . . *b*

b. Heads small, the disk 5–8 mm. broad; rays 5–8 mm. long; bracts of involucre 0.3–0.4 mm. broad . . . . 1c. B. ASTEROIDES var. MICROCEPHALA

b. Heads showy, the disk 7–14 mm. broad; rays 10–18 mm. long; bracts of involucre 0.5–2 mm. broad. . . . . . . . *c*

c. Bracts of involucre narrowly oblong or linear, acute to acutish (rather pointed), 0.5–1.3 mm. broad; rays usually white or tinged with lilac . . 1b. B. ASTEROIDES var. RECOGNITA

c. Outer bracts of involucre spatulate to spatulate-obovate, broadened upwards and rounded to obtuse (blunt) at tip, narrowed below the middle, 1–2 mm. broad; rays often lilac, pink, or rose-colored, also white . . 1a. B. ASTEROIDES var. LATISQUAMA

1a. **Boltonia asteroides** var. **latisquama** (Gray) Cron. Map 2173

*Boltonia latisquama* Gray [G, P & S]

Occurs in wet prairies, low meadows, prairie swales along railroads, and borders of ditches, streams, and sloughs. Mainly northern, central, and western Missouri in the prairie sections of glaciated and unglaciated Missouri, north and west of a line drawn from Lincoln, Callaway, Boone, Cooper, Morgan, Benton, St. Clair, Dade, Greene, and Lawrence counties to Barry County, locally in Ripley County, southeastern Missouri; absent from the Ozarks and not recorded from the extreme northern tiers of counties.

Ranges from Missouri and Kansas to Arkansas and Oklahoma; cultivated and escaped occasionally to New England.

This is quite a showy-flowered plant. In the field it appears to be an entirely distinct species, but studies by Fernald (Rh. 42: 490–92. 1940) and Cronquist (Bull. Torr. Bot. Club 74: 149. 1947) indicate a closer relationship with *B. asteroides* than at first is evident. Future, more detailed studies may throw further light on the distinctness of it as a separate species.

1b. **Boltonia asteroides** var. **recognita** (Fern. & Grisc.) Cron. Map 2173

*Boltonia latisquama* var. *recognita* Fern. & Grisc. [G]

*Boltonia asteroides* of Am. auth., not (L.) L'Her [P & S]

Occurs in moist low woods and thickets, meadows, and borders of ditches, ponds, sloughs, and streams. Throughout Missouri, but not recorded from most of the Ozark counties.

Ranges from Michigan to Manitoba and North Dakota, south to Kentucky, Missouri, and Kansas;

Plate no. 357. **1.** Astranthium integrifolium, × 2/5; Details from Small, The New York Botanical Garden. **2.** Boltonia asteroides var. latisquama, × 2/5; After Britton and Brown, details from Small, The New York Botanical Garden. **3.** Chaetopappa asteroides, × 2/5. **4.** Boltonia asteroides var. recognita, × 2/5.

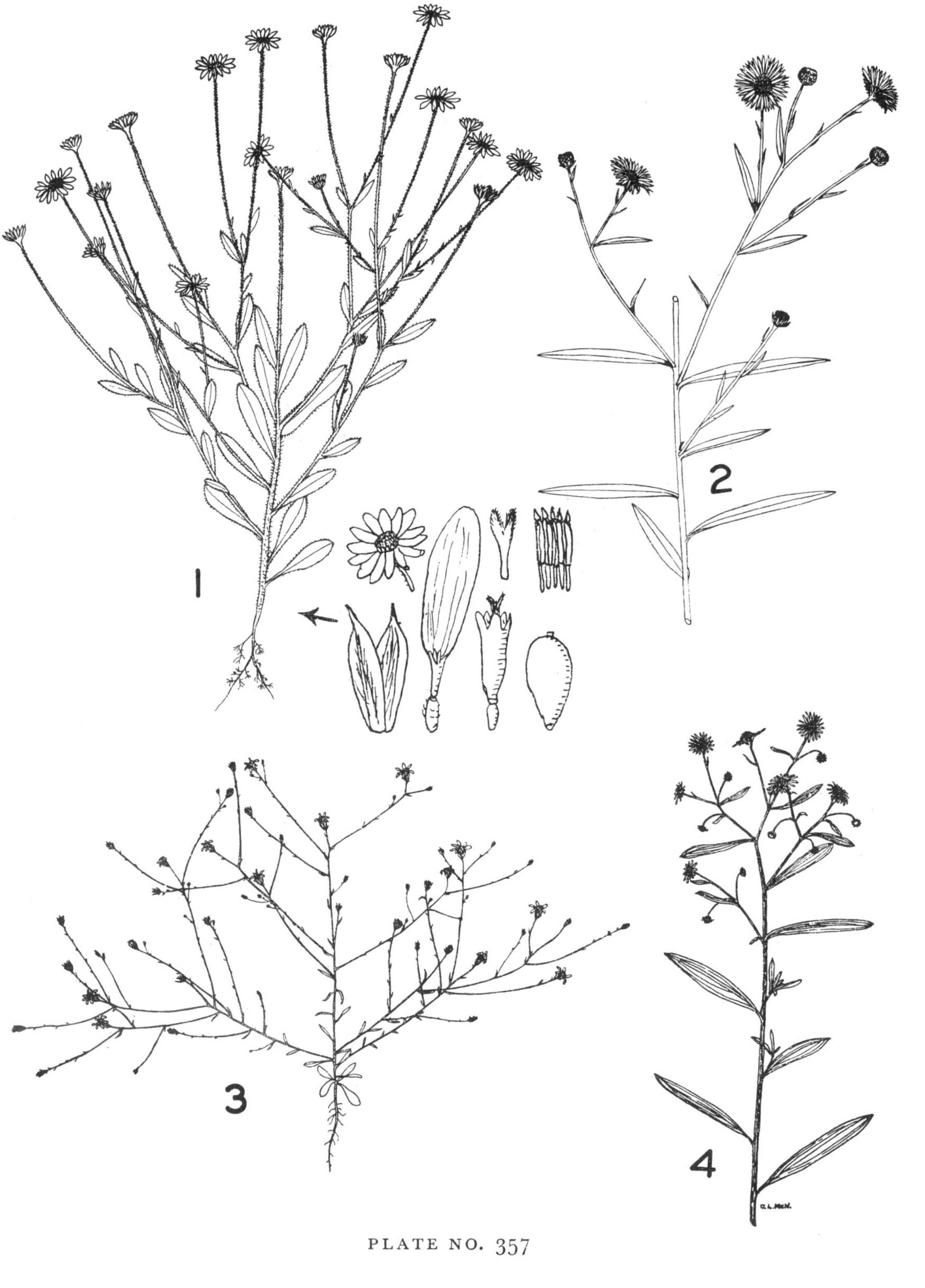


PLATE NO. 357

cultivated and escaped eastward to New England and New Jersey.

1c. **Boltonia asteroides** var. **microcephala** Fern. & Grisc. Map 2173

Known only from Ralls and Barry (Buffalo Bog, T22N, R7W, sect. 7, 3½ mi. southwest of Cassville, September 24, 1947, *Steyermark 65117*) counties.

Ranges from Wisconsin and Minnesota to Illinois, Iowa, and Missouri.

1d. **Boltonia asteroides** var. **decurrens** (T. & G.) Fern. & Grisc. Map 2173

*Boltonia latisquama* var. *decurrens* (T. & G.) Fern. & Grisc. [G]

*Boltonia decurrens* (T. & G.) Wood [BB]

Alluvial ground bordering sloughs, ditches, ponds, and streams. Known only from east-central Missouri in Lincoln, St. Charles, and St. Louis counties.

Known only from Illinois and Missouri. This is a well-marked variety, which Cronquist (in Gleason's *New Ill. Fl.* 3: 468. 1952) recognizes as a distinct species. However, aside from the decurrent stems, there appears to be no other characters for separating it from *B. asteroides* var. *recognita.*

Small-headed variants of var. *recognita* with achenes 2 mm. or less long are difficult to distinguish from *B. diffusa* var. *interior.*

2. **Boltonia diffusa** Ell. var. **interior** Fern. & Grisc. Map 2174

*Boltonia diffusa* in part [of BB], not Ell.

Occurs in low moist ground, around upland sinkhole ponds, swampy thickets, low alluvial ground, wet woods, wet prairies, and wet depressions. Southern Missouri in Pemiscot (near Hayti, October 18, 1939, *Mrs. W. N. Rankin*), Butler, Dunklin, Cape Girardeau (open wet thickets along highway 25, 3 mi. west of Arbor, September 22, 1946, *Steyermark 64147*), Stoddard (natural remnant of prairie along highway 51, ½ mi. north of junction of highways 51 and 60, 1½ mi. northeast of Fisk, in sect. 22, August 22, 1948, *Steyermark 66207*), Ripley (swampy thickets, T22N, R4E, southwest ¼ sect. 25, 3 mi. south of Naylor, October 20, 1948, *Steyermark 66935*), Howell (around Twin Ponds, east of highway A, T23N, R8W, northwest ¼ sect. 16, in dry upland, 4 mi. south of West Plains, September 3, 1949, *Steyermark 69062*), Cass (swampy exsiccated wooded flat along north side of Big Creek, T46N, R30W, sect. 35, 4 mi. southeast of Pleasant Hill, October 9, 1948, *Steyermark 66721*), and Greene counties.

Ranges from Mississippi to Oklahoma, north to Kentucky, Illinois, and Missouri.

The awns of the achenes are usually ⅔–¾ length of the achene in this variety, the awns measuring 1.5 mm. long, and the achenes 2 mm. long. In typical *B. diffusa* var. *diffusa*, not known in Missouri, the awns are about ¼ as long as the achenes. As indicated under *B. asteroides*, small-headed variants of that species with short rays and with achenes 2 mm. or less long, are difficult to distinguish from *B. diffusa* var. *interior.*

### 18. **Aster** L. Aster, Michaelmas Daisy

a. Most of the stem-leaves, but at least the basal and lower leaves, on noticeable petioles (leaf-stalks) and also cordate (heart-shaped) or subcordate at the base . . . . . . . . . . . . . . . . *b*
b. Bracts of involucre recurving at tips or widely spreading; all the leaves entire (without teeth) 3. A. ANOMALUS
b. Bracts of involucre erect, ascending, or appressed (pressed flat against or lying parallel to surface); leaves toothed or sometimes entire . . . . . . . . . . . . . . . . . . . . . . *c*
c. Inflorescence more or less flat or round-topped (corymbiform); outer bracts of involucre 1–2 mm. broad, $2\frac{1}{2}$ times or less as long as broad; rays always white . . . . . . 2. A. FURCATUS
c. Inflorescence longer than broad, usually elongated and panicle-like in form; outer bracts of involucre 1 mm. or less broad, 3 or more times as long as broad; rays usually blue, lavender, purple, or lilac, and also white . . . . . . . . . . . . . . . . . . . . . . *d*
d. Leaves nearly or quite entire (without teeth) or with only shallow inconspicuous teeth; both surfaces of leaves rough-hairy (scabrous) like sandpaper; involucral bracts short, broad, or with diamond-shaped green tips . . . . . . . . . . . . . . 4. A. AZUREUS
d. At least the lower leaves toothed; upper surface of leaves usually smooth, sometimes rough-hairy; involucral bracts with short or elongated green tips . . . . . . . . . . . *e*
e. Petioles (leaf-stalks) not at all or only slightly winged or margined; bracts of involucre obtuse (blunt) or slightly acute (short-pointed) with short broad dilated tips . 5. A. CORDIFOLIUS
e. Petioles (leaf-stalks) at least of upper leaves of stem conspicuously winged or margined; bracts of involucre strongly acute to acuminate (short- to long-pointed) with narrow, more or less elongated tips . . . . . . . . . . . . . . . . . . . . . *f*
f. Stems glabrous below the inflorescence or sometimes the upper part minutely hairy in lines; upper surface of leaves glabrous (without hairs) or somewhat rough with minute hairs, lower surface of leaves glabrous or hairy with rather long flattened hairs; green tips of the bracts of involucre narrow, linear or nearly so; rays (ligules) commonly white, pale blue, or lilac . . . . . . . . . . . . . . . . . 6. A. SAGITTIFOLIUS
f. Stems usually densely and uniformly short-hairy at least above the middle; upper surface of leaves rough-hairy, lower surface of leaves more or less uniformly hairy with short spreading hairs; green tips of the bracts of involucre somewhat elongated and diamond-shaped; rays (ligules) usually blue . . . . . . . . . . 7. A. DRUMMONDII
a. Basal leaves not both petioled and cordate at the base, the other leaves sessile (without petiole) . . . . *g*
g. Heads without rays or apparent rays at flowering time . . . . . . . . . 29. A. BRACHYACTIS
g. Heads with short or long rays at flowering time . . . . . . . . . . . . . . . . . *h*
h. Both sides of leaves covered with a close silvery or gray-silvery silkiness . . . . 12. A. SERICEUS
h. Leaves glabrous (without hairs) or more or less hairy, but not silvery-silky . . . . . . . . *i*
i. Heads arranged in a more or less flat-topped corymb, all the heads more or less ending at the same level; disk white; rays white . . . . . . . . . . . . 26. A. PTARMICOIDES
i. Heads scattered at different levels in more or less elongated panicles, racemes, or stiff, erect wand-like inflorescences . . . . . . . . . . . . . . . . . . . . . . . . *j*
j. Stem glabrous (without hairs) or nearly so, or sometimes the upper portion more or less hairy in lines . . . . . . . . . . . . . . . . . . . . . . . . . . *k*
k. Base of the stem-leaves rounded or heart-shaped, strongly clasping or partly encircling the stem . . . . . . . . . . . . . . . . . . . . . . . . *l*
l. Leaves more or less toothed . . . . . . . . . . . . . . . . . . . *m*
m. Rays (ligules) usually white, 5–11 mm. long; leaves gradually narrowed to a narrow base, not much, if at all, clasping; flower-heads, including rays, mainly 15–20 mm. broad; involucre 4.5–6 mm. high . . . . . . . . . . 25. A. SIMPLEX
m. Rays usually blue-violet, sometimes rose-colored, or white, 12–15 mm. long; leaves narrowed to a broad clasping base; flower-heads, including rays, mainly 20–25 mm. broad; involucre 7–12 mm. high . . . . . . . . 11. A. PUNICEUS var. FIRMUS f. LUCIDULUS
l. Leaves mainly entire (without teeth) or with a few short teeth near the middle . . . *n*
n. Stem and leaves often or mostly glaucous (with a silvery 'bloom' which can be rubbed off), sometimes green or greenish; bracts of involucre short-pointed (acute) . . . . . . . . . . . . . . . . . . . . 13. A. LAEVIS
n. Stem and leaves not glaucous; bracts of involucre, at least the innermost ones, long-pointed (acuminate) or long-tapering at tip . . . . . . . . . . . *o*

o. Involucre 7–12 mm. high, the bracts 6–8 mm. long and of nearly equal length . . . . .
11. A. PUNICEUS var. FIRMUS f. LUCIDULUS

o. Involucre mainly 4.5–7 mm. high, the bracts imbricated in 3–5 series of different lengths . . . . *p*

p. Veinlets of the lower surface of the leaf forming a conspicuous network of small spaces (areolae) of nearly equal diameter (isodiametric); leaves somewhat revolute on margins, the upper surface usually rough with short, forward-pointing hairs; lobes of disk-corollas ascending, $\frac{1}{6}$–$\frac{1}{4}$ (17–25 per cent) of the total length of the limb; rays (ligules) purple or blue-violet . .
24. A. PRAEALTUS

p. Veinlets of the lower surface of the leaf not conspicuous, the small spaces enclosed by them longer than broad; leaves mainly flat and not revolute on margins, the upper surface smooth and usually glabrous (without hairs); lobes of disk-corollas spreading, $\frac{1}{3}$–$\frac{1}{2}$ (33–50 per cent) of the total length of the limb; rays (ligules) white, sometimes blue-tinged or pale lavender .
25. A. SIMPLEX

k. Base of the stem-leaves not clasping, not rounded, nor heart-shaped, but narrowed at the base or with straight sides . . . . . . . . . . . . . . . . . . . . . . . . . . . . . *q*

q. Annuals; stem and leaves rather fleshy; rays (ligules) very short, longer than the mature pappus but not showy. . . . . . . . . . . . . . . . . . . . . . . . 28. A. EXILIS

q. Without the above combination of characters; perennials; stem and leaves not fleshy; rays (ligules) normally well-developed and showy . . . . . . . . . . . . . . . . . . . *r*

r. Involucre in flower and fruit usually 9–15 mm. (7–8) high; bracts of involucre mostly 1 mm. or more broad. . . . . . . . . . . . . . . . . . . . . . . . . . . . . *s*

s. Stem mainly 1.5–6 dm. (rarely up to 9) tall; involucre hemispherical and broader than high, the bracts rigid, sharp-pointed and all of about the same length and width; some of the bracts loose and spreading or recurving (squarrose) at tips; leaf-like bracts present on flower-stalk (peduncle) much different in size from the outermost bracts of the involucre; pappus coarse and stiff; plants known only from southwestern Missouri, locally in Marion Co. . . . . . . . . . . . . 1. A. PALUDOSUS subsp. HEMISPHERICUS

s. Stem mainly 7–12 dm. (rarely 5) tall; involucre somewhat cylindrical and longer than broad to turbinate (top-shaped) when pressed, the bracts not rigid, very unequal, rather blunt, the outer (lower) bracts much shorter and narrower than the inner (upper) ones; bracts appressed (pressed closely against one another); only minute bracts present on the flower-stalks (peduncles), not much different in size from the outer bracts of the involucre; pappus soft, fine, and slender; plants common throughout southern, central, and eastern Missouri . . . . . . . . . . . . . . . . . . . . . . 14. A. TURBINELLUS

r. Involucre in flower and fruit 2.5–8 mm. high; bracts of involucre usually less than 1 mm. broad . . *t*

t. At least the outer bracts of involucre with minute spine-like tips, the bracts firm . 15. A. PILOSUS

t. Bracts of involucre without minute spine-like tips, the bracts rather soft and thin . . . . . *u*

u. Midrib on the lower surface of the leaf more or less hairy; lobes of the disk-corollas $\frac{1}{2}$–$\frac{3}{4}$ (50–75 per cent) of the total length of the limb (expanded part of corolla above the tube), 1–1.6 mm. long . . . . . . . . . . . . . . . . . . 22. A. LATERIFLORUS

u. Midrib on the lower surface of the leaf glabrous (without hairs) or essentially so; lobes of the disk-corollas $\frac{1}{6}$–$\frac{1}{2}$ (17–50 per cent) of the total length of the limb, 0.4–0.9 mm. long . . *v*

v. Veinlets of the lower surface of the leaf forming a conspicuous network of small spaces (areolae) of nearly equal diameter (isodiametric); rays (ligules) purple or blue-violet . . . . . . . . . . . . . . . . . . . . 24. A. PRAEALTUS

v. Veinlets of the lower surface of the leaf not conspicuous, the small spaces enclosed by them longer than broad; rays (ligules) white, sometimes blue-tinged or pale lavender . . *w*

w. Flower-heads arranged in a paniculate inflorescence, the heads scattered and numerous along the branches; stout-stemmed plants with ascending or ascending-spreading branches; stems 3–6 mm. thick near base; central band of bracts of involucre not much, if any, enlarged upward, remaining slender; lobes of the disk-corollas $\frac{1}{3}$–$\frac{1}{2}$ (33–50 per cent) of the total length of the limb; rays (ligules) mainly 6–12 mm. (4) long; leaves 3–40 mm. broad; flower-heads 15–25 mm. in diameter including spread of rays, 7–12 mm. excluding rays . . . . . . . 25. A. SIMPLEX

w. Flower-heads arranged in either 1-sided racemes or solitary at the tips of branches; more slender-stemmed plants commonly with spreading or recurving branches, more rarely with ascending branches; stems 1.5–4 mm. thick near base; central band of bracts of involucre enlarged upward or conspicuous and green; lobes of the disk-corollas $\frac{1}{5}$–$\frac{2}{5}$ (20–40 per cent) of the total length of the limb; rays (ligules)

3–7 (–9) mm. long; main leaves of stem mostly 2.5–7 mm. (–11) broad; flower-heads 6–15 mm. in diameter including spread of rays, 4–10 mm. excluding rays . . . . . . . . . . . . *x*

x. Branches mostly horizontally spreading or arched-recurving, sometimes ascending; heads arranged in more or less 1-sided racemes; heads 6–10 mm. in diameter; involucre mainly 2.5–4 (up to 4.8) mm. high, the midrib only slightly dilated at tip; rays 3–5.8 mm. long; lobes of disk-corollas 0.6–0.8 mm. long, about $\frac{2}{5}$– (40 per cent) of the total length of the limb, erect or ascending . . . . . . . . . . . . . . . . . . . . . . . . 21. A. VIMINEUS

x. Branches rather stiffly ascending; heads not in 1-sided racemes usually solitary at the tips of the branchlets, these with many, very small bract-like leaves; heads 12–15 mm. in diameter; involucre mainly 4.5–7 mm. high, the midrib usually with enlarged diamond-shaped or broadly triangular tip; rays 5–9 mm. long; lobes of disk-corollas 0.4–0.8 mm. long, $\frac{1}{5}$–$\frac{1}{3}$ (20–33 per cent) of the total length of the limb, spreading or erect. . . . . 20. A. DUMOSUS var. STRICTIOR

j. Stem more or less hairy over the entire surface, not in lines. . . . . . . . . . . . . . *y*

y. Base of the stem-leaves rounded or heart-shaped and strongly clasping or partly encircling the stem . . . . . . . . . . . . . . . . . . . . . . . . . . . . *z*

z. Bracts of involucre with gland-tipped hairs. . . . . . . . . . . . . . . . . . *1*

1. Main stem-leaves of upper half of stem very close together, their bases separated by intervals of only 0.5–1.5 cm. long; main stem-leaves 50–100 or more; peduncles (branch bearing flowers) nearly bractless; involucre 8–10 mm. high, the bracts narrowly linear and long-tapering at tip, mainly 6–9 mm. long; rays (ligules) 40–75; plants of moist soils in wet meadows, thickets, and low ground . . . . . . . . . . . . . . 10. A. NOVAE-ANGLIAE

1. Main stem-leaves of upper half of stem separated by longer intervals of 1–3 cm. long; main stem-leaves 10–30; peduncles (branch bearing flowers) with many small or large leaf-like bracts; involucre 5–8 mm. high, the bracts oblong, oblanceolate, or oblong-linear and short-pointed (acute) at tip, except some of the innermost ones long-pointed (acuminate), mainly 4–6 mm. long; rays (ligules) 15–30; plants of dry soils of rocky glades and bluffs, prairies, and dry open woods . . . . . . . . . . . . . . . . . . . . . . *2*

2. Leaves of stem mainly 10–30 mm. broad (rarely only 7 mm.), the base of the leaves with lobes which nearly encircle the stem; hairs on stem mostly ascending or appressed (pressed against or lying parallel to surface); leaves of main stem below inflorescence relatively few, 10–20 . . . . . . . . . . . . . . . . . . 8. A. PATENS

2. Leaves of stem mainly 4–8 mm. broad (rarely only 2 or up to 20 mm.), the base of the leaves slightly to noticeably clasping the stem but never nearly encircling it; hairs on stem often more conspicuous, often spreading and 0.5–1.5 mm. long, sometimes short and appressed; leaves of main stem below inflorescence numerous, 20–30 or more . . 9. A. OBLONGIFOLIUS

z. Bracts of involucre glabrous or hairy, but hairs when present not gland-tipped . . . . . . *3*

3. Leaves of stem mainly 15–30 mm. broad (rarely only 7 mm.), the base of the leaves with lobes which nearly encircle the stem. . . . . . . . . . . . . . . 8. A. PATENS

3. Leaves of stem mainly 1–8 mm. broad, the base of the leaves without lobes and never nearly encircling the stem. . . . . . . . . . . . . . . . . . . . . *4*

4. Involucre 7–12 mm. high; flower-heads frequently very few with often only 1–4 to each leafy stem, but sometimes up to 15; pappus double, composed of an inner series of long bristles and an outer series of short bristles . . . . . . . . . . 27. A. LINARIIFOLIUS

4. Involucre 3.5–5 mm. high; flower-heads usually numerous; pappus not divided into an inner and outer series of unequal length . . . . . . . . . . . . . . . . . *5*

5. Bracts of involucre blunt (obtuse) with recurving (squarrose) tips; rays (ligules) 3–6 mm. long, usually white . . . . . . . . . . . . . . . . 18. A. ERICOIDES

5. Bracts of involucre short- to long-pointed (acute), loosely ascending but the tips scarcely recurved (squarrose) to slightly recurved or spreading; rays (ligules) 5–10 mm. long, usually blue or purple . . . . . . . . . . 17. A. × AMETHYSTINUS

y. Base of the stem-leaves not clasping, rounded, or heart-shaped, but narrowed at the base or with straight sides . . . . . . . . . . . . . . . . . . . . . . . . . . . *6*

6. Involucre in flower and fruit 7–15 mm. high; bracts of involucre 1 mm. or more broad . . . . *7*

7. Stem mainly 7–12 dm. (rarely 5) tall; larger leaves 6–12 cm. long, 8–20 mm. broad; involucre somewhat cylindrical and longer than broad to turbinate (top-shaped) when pressed; bracts of involucre rather blunt at tip; rays (ligules) 15–25, 10–20 mm. long; pappus not divided into an inner and outer series of unequal length . . . . 14. A. TURBINELLUS

7. Stem mainly 1–5 dm. (rarely 6) tall; larger leaves 2–4 cm. long, 1–4 mm. broad; involucre hemispherical to campanulate; at least outer bracts of involucre short-pointed (acute); rays (ligules) 10–15, 7–10 (rarely to 15) mm. long; pappus double, composed of an inner series of long bristles and an outer series of short bristles . . . . . . . . . . 27. A. LINARIIFOLIUS
6. Involucre in flower and fruit 3.5–8 mm.; bracts of involucre usually less than 1 mm. broad. . . . . *8*
8. Bracts of involucre with gland-tipped hairs . . . . . . . . . . . 9. A. OBLONGIFOLIUS
8. Bracts of involucre glabrous (without hairs), or if hairy, the hairs not gland-tipped . . . . . . *9*
9. At least the outer bracts of the involucre bristly hairy on the margins . . . . . . . . *10*
10. Bracts of involucre short-pointed (acute), loosely ascending but the tips scarcely recurved (squarrose); rays (ligules) 5–10 mm. long, usually blue or purple . . 17. A. × AMETHYSTINUS
10. Bracts of involucre blunt (obtuse), with recurving (squarrose) tips; rays (ligules) 3–6 mm. long, usually white . . . . . . . . . . . . . . . . . . . . . . . . *11*
11. Involucre 3.5–5 mm. high, the bracts arranged in 3 or 4 different lengths; heads relatively numerous, often 1-sided on the branches; rays (ligules) 8–20. . 18. A. ERICOIDES
11. Involucre 5–8 mm. high, the conspicuously overlapping bracts not in definite series, the outer nearly as long as the inner; heads relatively few, solitary or clustered at the ends of the branches or the inflorescence raceme-like; rays mostly 20–30. 19. A. COMMUTATUS
9. Bracts of involucre glabrous (without hairs) or nearly so . . . . . . . . . . . *12*
12. At least the outer bracts of involucre with minute spine-like tips, the bracts firm. . . . *13*
13. Disk-florets 20–40 in each head; involucre 4.5–8 mm. high, 5–10 mm. broad, broadly urn-shaped, hemispherical, to campanulate (bell-shaped), slightly narrowed above the middle, then expanding upwards; rays 16–35 in each head, 5–10 mm. long . 15. A. PILOSUS
13. Disk-florets 6–12 in each head; involucre 3.5–4.5 mm. high, 2–3.5 mm. broad, slenderly turbinate (top-shaped) or narrowly obconic (when pressed); rays 10–18 in each head, 3.5–5 mm. long . . . . . . . . . . . . . . . . 16. A. PARVICEPS
12. Bracts of involucre without minute spine-like tips, the bracts rather soft and thin. . . . *14*
14. Only the midrib on the lower surface of the leaf more or less hairy, rest of the leaf glabrous . . . . . . . . . . . . . . . . . . . . . 22. A. LATERIFLORUS
14. One or both surfaces of the leaf either closely hairy or rough-hairy . . . . . . *15*
15. Involucre 4.5–7 mm. high, the midrib usually with enlarged diamond-shaped or broadly triangular tip; flower-heads solitary at the ends of the branches, not arranged in 1-sided racemes, or racemose or paniculate inflorescences; lobes of the disk-corolla $\frac{1}{5}$–$\frac{1}{3}$ (20–33 per cent) of the total length of the limb (expanded part of corolla above the tube), 0.4–0.8 mm. long. . . . . 20. A. DUMOSUS var. DODGEI
15. Involucre 3–4.6 mm. high, the midrib with only a narrow spatulate green band near the tip; flower-heads arranged in racemose or paniculate inflorescences; lobes of the disk-corolla $\frac{2}{5}$–$\frac{2}{3}$ (40–66 per cent) of the total length of the limb, 0.7–1.2 mm. long . . . . . . . . . . . . . . . 23. A. ONTARIONIS

1. **Aster paludosus** Ait. subsp. **hemisphericus** (Alex.) Cron. Map 2175
*Aster hemisphericus* Alex. [G]
*Aster paludosus* [of P & S], not Ait.
*Heleastrum hemisphericum* (Alex.) Shinners [Shinners]
Flowers August–October.

Occurs in prairies of southwestern Missouri north to Barton County and east to Lawrence and Barry counties, and introduced along railroads in Marion County.

Ranges from Missouri and Kansas to Alabama and Texas.

Shinners considers that this Aster belongs to a distinct genus, *Heleastrum*, which, according to him, differs from typical Aster in having a tuberlike woody corm or crown, more or less clavate pappus bristles, and the green tips of the involucral bracts not formed by expansion of the midvein.

The plant is a showy one with violet rays 10–25 mm. long, and should be more generally grown.

2. **Aster furcatus** Burgess Map 2176
Flowers July–October.

Occurs on moist rocky ledges of bluffs along streams. Ozark region, where known only from Jefferson, Franklin, Shannon, Texas, and Douglas counties.

Ranges from Indiana and Illinois to Wisconsin and Missouri.

One of the rarest asters in Missouri, this species is one of the earliest of the genus to bloom, starting in July. It can be easily grown and does well in a woodland or shaded wildflower garden. It must be watched, for it eventually spreads, forming dense colonies from the creeping underground stems. Although it has been found only on bluffs in Missouri, it is found elsewhere

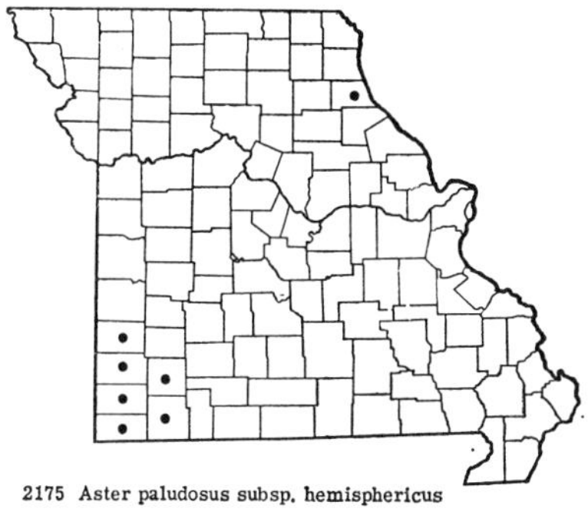

2175 Aster paludosus subsp. hemisphericus

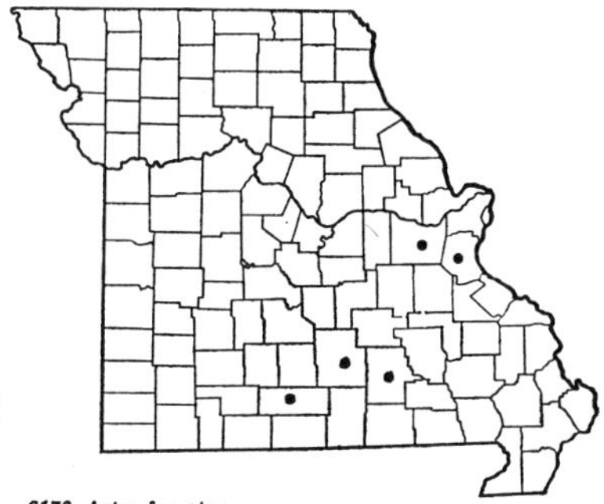

2176 Aster furcatus

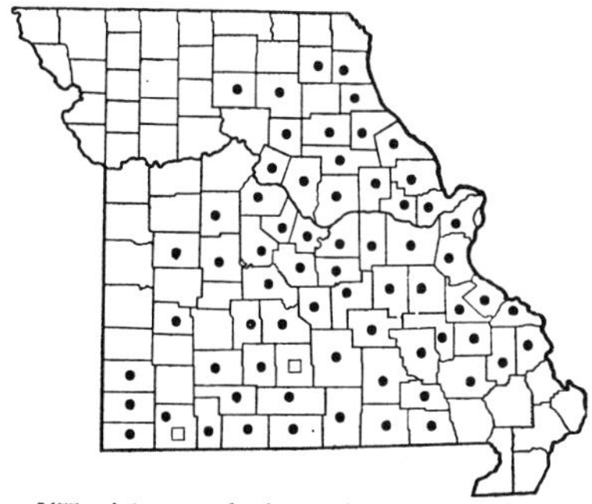

2177 • Aster anomalus f. anomalus
2177 □ Aster anomalus f. albidus

sometimes in rich woodland, where it forms dense stands.

3. **Aster anomalus** Engelm. Map 2177

Flowers early July–November.

Occurs in dry or rocky open woods and thickets, usually in acid soils associated with chert, sandstone, or granite substrata. Mainly in the Ozark region north to Lewis, Knox, Macon, Linn, Howard, Pettis, Henry, and Jasper counties.

Ranges from Illinois to Missouri and Kansas, south to Arkansas and Oklahoma.

Two variations occur in Missouri:

Rays lavender-blue or purple . . 3a. A. ANOMALUS f. ANOMALUS
Rays white . . . . 3b. A. ANOMALUS f. ALBIDUS

3a. **Aster anomalus** f. **anomalus** Map 2177

*Aster anomalus* Engelm. [G, BB, P & S]

This is the common type encountered.

3b. **Aster anomalus** f. **albidus** Steyerm. Map 2177

Known only from Wright (*Steyermark 78027*) and Barry (cherty wooded slopes, Natural Bridge Hollow, T22N, R26W, sect. 17 and 18, 8–9 mi. southeast of Cassville, September 24, 1947, *Steyermark 65135*, holotype in Chi. Nat. Hist. Mus. Herb.) counties.

Anderson (Ann. Mo. Bot. Gard. 16: 129–44. 1929) studied variation in this species and concluded that it was a most consistent and well marked species in the genus despite a high degree of variation among individual plants.

The flowers and fruits are eaten by wild turkey, and the leaves by the white-tailed deer, the basal ones being a source of winter food.

4. **Aster azureus** Lindl. Azure Aster Map 2178

Also called Blue Devil.

Flowers August–November.

Occurs in glades, prairies, roadsides, open rocky woods, and borders of woods. Scattered throughout Missouri, but not recorded from less than half the total number of counties in the state.

The following variations occur in Missouri:

a. Upper leaves mostly 2–10 mm. broad, linear-lanceolate or linear; stems smooth; inflorescence with widely spreading and open branches, the heads mostly solitary at the tips of elongated branches; involucre 4.5–5 mm. (rarely 6) high; basal and lower stem-leaves usually truncate or abruptly narrowed at base, rarely somewhat heart-shaped at base . . 4c. A. AZUREUS var. POACEUS
a. Main leaves of stem 10–30 mm. broad, lanceolate; stems smooth or rough-hairy; inflorescence with stiffly ascending racemose branches; involucre 5–8 mm. high; basal and lower stem-leaves truncate to deeply cordate at base. . *b*
  b. Stems rough-hairy . . 4a. A. AZUREUS var. AZUREUS f. AZUREUS
  b. Stems smooth . . . 4b. A. AZUREUS var. AZUREUS f. LAEVICAULIS

4a. **Aster azureus** var. **azureus** f. **azureus** Map 2178

*Aster azureus* Lindl. [G, BB, P & S]; holotype collected from St. Louis by *Drummond.*

Scattered throughout Missouri.

Ranges from New York and Ontario to Minnesota, south to Georgia, Alabama, Louisiana, and Texas.

4b. **Aster azureus** var. **azureus** f. **laevicaulis** Fern. Map 2178

*Aster azureus* f. *laevicaulis* Fern. [G]

Scattered throughout Missouri. More intensive collecting needs to be done to learn whether this or f. *azureus* is commoner in Missouri.

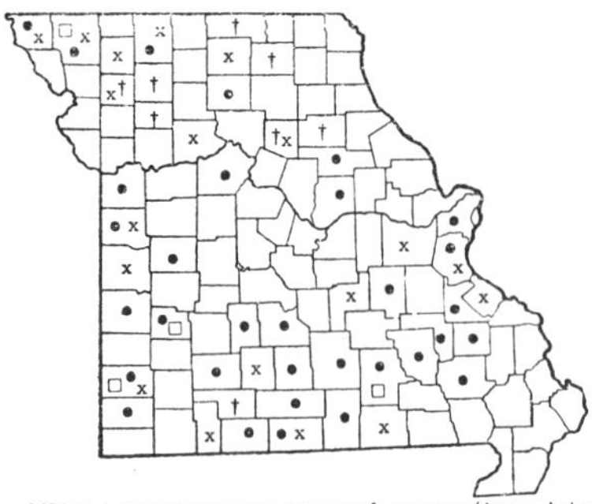

2178 • Aster azureus var. azureus f. azureus (Azure Aster)
2178 x Aster azureus var. azureus f. laevicaulis
2178 □ Aster azureus var. poaceus
2178 † Aster azureus X laevis

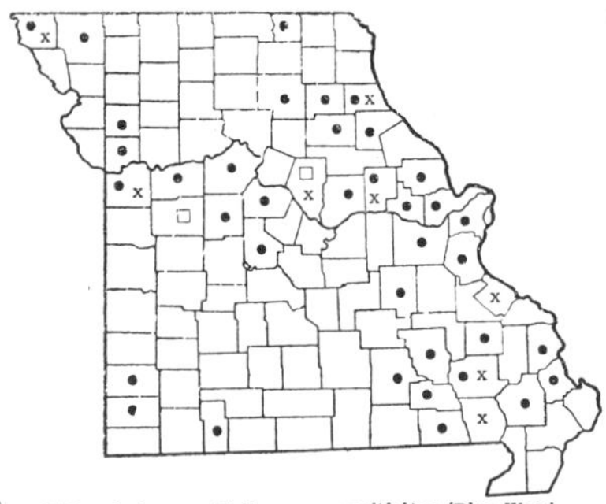

2179 • Aster cordifolius var. cordifolius (Blue Wood Aster)
2179 □ Aster cordifolius var. polycephalus
2179 x Aster cordifolius var. moratus

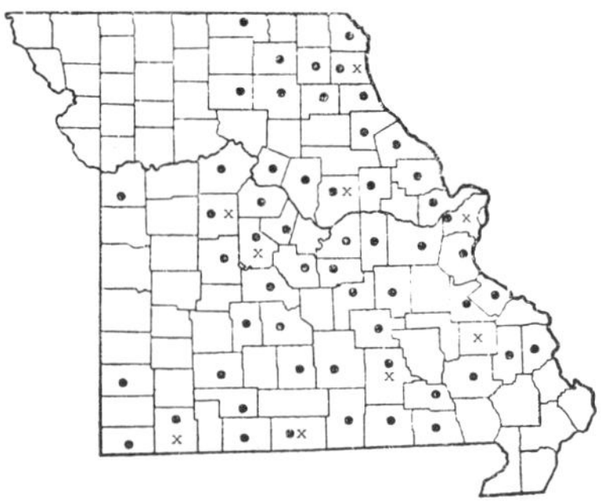

2180 x Aster sagittifolius f. sagittifolius
2180 • Aster sagittifolius f. hirtellus

4c. **Aster azureus** var. **poaceus** (Burgess) Fern. Map 2178

*Aster vernalis* (Engelm.) Burgess [Shinners]

*Aster poaceus* Burgess; holotype from Iron Mt., Missouri, 1845, *Engelmann* in Mo. Bot. Gard. Herb.

Scattered in Missouri, where known from Shannon, Nodaway, Cedar, and Jasper counties, but more intensive study of the variation of *A. azureus* may considerably extend the known range within the state.

Ranges from Missouri and Arkansas to Texas.

There is too much intergradation between var. *azureus* and var. *poaceus* to consider the latter a separate species (as *A. vernalis*).

Apparent hybrids occur which may be hybrids between *A. azureus* and *A. laevis*. These occur in northern and central Missouri and in Christian County. The collection from Christian County (*Steyermark 85818*) has the smaller heads and lower leaves narrowed to a subpetiolar base as in *A. azureus* var. *poaceus*, but the stems and both sides of the leaves are glabrous as in *A. laevis*.

The flowers and fruits are eaten by wild turkey.

5. **Aster cordifolius** L. Blue Wood Aster Map 2179

Flowers August–November.

Occurs in rich, dry or moist woodland soils, at the base of bluffs, along stream banks, and moist ledges. Scattered throughout Missouri, but not recorded from nearly three-fourths of the counties in the state.

The following variations occur in Missouri:

a. Bracts of the involucre rather loose, tapering to slender narrow tips with a nearly linear green band in the middle; outer bracts 0.2–0.4 mm. broad . . . 5c. A. CORDIFOLIUS var. MORATUS

a. Bracts of the involucre closely appressed (pressed flat or close to one another), obtuse (blunt) to somewhat short-pointed (acute) at tip, the green band in the middle more deeply colored and broadened upward; outer bracts 0.4–1 mm. broad . . . . . . . . . b

b. Upper leaves of the stem cordate (heart-shaped) at base . . . 5a. A. CORDIFOLIUS var. CORDIFOLIUS

b. Upper leaves of the stem truncate (as if cut straight across) or tapering at the base . 5b. A. CORDIFOLIUS var. POLYCEPHALUS

5a. **Aster cordifolius** var. **cordifolius** Map 2179

*Aster cordifolius* L. [G, BB, P & S]

This is the commonest variation, so far as present collections show, in Missouri, and is scattered throughout the state.

Ranges from Quebec to Wisconsin and Iowa, south to Georgia, Alabama, and Missouri.

5b. **Aster cordifolius** var. **polycephalus** Porter Map 2179

Known from Boone and Johnson (limestone glade along highway 131, on south side of South Fork of Blackwater River, T46N, R28W, northwest ¼ sect. 26, 3 mi. north of Holden, September 25, 1951, *Steyermark 72751*) counties.

Ranges from Quebec to Georgia, west to Indiana and Missouri.

5c. **Aster cordifolius** var. **moratus** Shinners Map 2179

Plate no. 358. **1.** Aster paludosus subsp. hemisphericus, × 2/7. **2.** Aster furcatus, × 2/7. **3.** Aster anomalus, × 2/7. **4.** Aster azureus, 2/7; *a.* Basal leaf. **5.** Aster cordifolius, × 2/7. **6.** Aster patens, × 2/7. **7.** Aster novae-angliae, × 2/7. **8.** Aster oblongifolius, × 2/7. **9.** Aster Drummondii, × 2/7; *a.* Basal leaf; *b.* Inflorescence; *c.* Middle part of stem. **10.** Aster puniceus var. firmus f. lucidulus; *a.* Middle part of stem; *b.* Inflorescence. All details from Small, The New York Botanical Garden.

PLATE NO. 358

Scattered in northern, central, and eastern Missouri.

Ranges from Wisconsin, Illinois, and Iowa, south to Missouri and Kansas.

It is very difficult sometimes to separate this variety from *A. sagittifolius* f. *hirtellus* or other variations of *A. sagittifolius* when intermediate specimens are present. In both *A. cordifolius* var. *moratus* and *A. sagittifolius* the involucral bracts are loose, linear-attenuate with the median green band nearly linear and not broadened to the summit nor strongly contrasting in color as in *A. cordifolius* var. *cordifolius*. Such a collection is represented by *Steyermark 85678* from Atchison County. In such cases, var. *moratus* shows its closer affinity to *A. cordifolius* in having the lower and middle stem-leaves ovate, cordate, and coarsely serrate, and in the blue rays, instead of sagittate-lanceolate leaves with white rays as are characteristic of *A. sagittifolius*.

Cronquist (in Gleason's *New Ill. Fl.* 3: 448. 1952) has suggested, with good reason, that *A. cordifolius* var. *moratus* may represent a hybrid between *A. sagittifolius* and *A. cordifolius*. Observations of Missouri material would certainly point in this direction.

6. **Aster sagittifolius** Wedemeyer Map 2180

Flowers August–October.

Occurs in rocky woods, ravines, borders of thickets, glades, and moist or dry, rocky bluffs. Throughout the eastern half and Ozark sections west to Putnam, Linn, Jackson, Pettis, Benton, Dallas, Greene, and Jasper counties; absent from most of the western sectors of the state.

Ranges from Vermont to Minnesota and North Dakota, south to Florida, Alabama, Tennessee, and Missouri.

Two variations occur in Missouri:

Upper surface of leaves smooth . . . . . 6a. A. SAGITTIFOLIUS f. SAGITTIFOLIUS

Upper surface of leaves rough-hairy . . . . 6b. A. SAGITTIFOLIUS f. HIRTELLUS

6a. **Aster sagittifolius** f. **sagittifolius** Map 2180

*Aster sagittifolius* Wedemeyer [G, BB, P & S]

Throughout the range in Missouri.

6b. **Aster sagittifolius** f. **hirtellus** (Lindl.) Shinners Map 2180

Reported to be the commoner form throughout the general range of *A. sagittifolius*, and the commoner form in Missouri.

The leaves are eaten by white-tailed deer.

7. **Aster Drummondii** Lindl. Drummond Aster Map 2181

*Aster sagittifolius* var. *Drummondii* (Lindl.) Shinners [G]

Flowers August–October.

Occurs in rocky or dry open woods and thickets. Throughout Missouri, although not recorded from a number of counties in the Ozarks and north-central sections.

Ranges from Ohio to Minnesota, south to Tennessee, Louisiana, and Texas.

Although this species intergrades with the preceding *A. sagittifolius*, it is recognizable in the field as distinct in a number of ways from *A. sagittifolius*. In Missouri, studies by Edgar Anderson indicate that the large extent of apparent hybridization between *A. Drummondii* and *A. sagittifolius* is due to recent human disturbances. Shinners, on the other hand (Am. Midl. Nat. 26: 406–7. 1941), treated *A. Drummondii* as a variety of *A. sagittifolius* because of the occurrence of intergrading characters between the two taxa.

In addition to the characters used for separating the two in the key, *A. sagittifolius* generally has smaller heads and averages slightly shorter involucres than those of *A. Drummondii*. More detailed experimental work upon these plants is necessary before any final status can be settled. In the present work, they are being maintained as distinct species on the basis of field recognition of the two over a period of many years.

Occasionally specimens are collected which are apparent hybrids between *Aster Drummondii* and *A. cordifolius*. These are known from Atchison and Jefferson counties possessing the cordate leaves with nonwinged petioles of *A. cordifolius* and the short pubescence of stems, petioles, and lower surface of leaf-blades characteristic of *A. Drummondii*.

8. **Aster patens** Ait. Spreading Aster Map 2182

Also called Purple Daisy.

Flowers August–October.

Occurs in rocky or sandy open woods, thickets, and borders of rocky glades, usually in acid soils associated with sandstone, chert, or granite substrata.

Two variations occur in Missouri:

Bracts of involucre with many gland-tipped hairs . . . . . . 8a. A. PATENS var. PATENS

Bracts of involucre covered with short appressed hairs, not at all or only a few gland-tipped . . 8b. A. PATENS var. PATENTISSIMUS

8a. **Aster patens** var. **patens** Map 2182

*Aster patens* Ait. [G, P & S]

Ozark region of eastern and east-central Missouri north to Franklin County, west to Shannon and Butler counties.

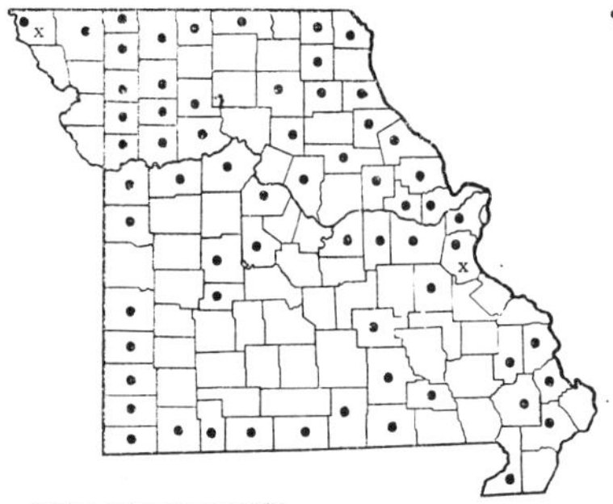

2181 • Aster Drummondii
2181 x Aster Drummondii X cordifolius

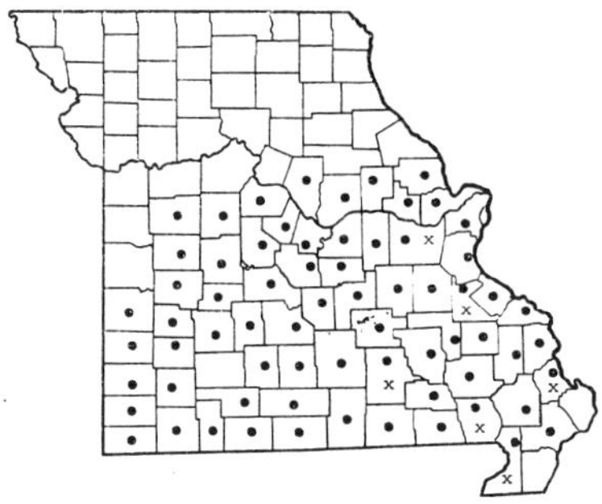

2182 x Aster patens var. patens (Spreading Aster)
2182 • Aster patens var. patentissimus

2183 x Aster oblongifolius var. oblongifolius f. oblongifolius
2183 o Aster oblongifolius var. oblongifolius f. roseoligulatus
2183 • Aster oblongifolius var. angustatus

Ranges from Maine to Minnesota, south to Georgia, Alabama, and Missouri.

8b. **Aster patens** var. **patentissimus** (Lindl.) T. & G. Map 2182

*Aster patens* var. *gracilis* [of G, P & S], not Hook.

Ozark region of southern and central Missouri north to Lincoln, Montgomery, Callaway, Boone, Cooper, Pettis, Johnson, and Vernon counties. This is the commoner variation encountered in the state.

Ranges from Louisiana and Texas, north to Kentucky, Missouri, and Kansas.

The involucral bracts of var. *patentissimus* average broader and more blunt than in the variety *patens*. There is considerable variation in the length of the flowering branches and their bracts. Bracts on the flowering branches vary from tiny and greatly reduced, linear-subulate, appressed structures to more leaflike and larger. The former type was considered by Fernald in *Gray's Manual*, eighth edition, as constituting the var. *gracilis* Hook., but by Cronquist (in Gleason's *New Ill. Fl.*) that variety has the involucral bracts strongly glandular. The Missouri material with the involucral bracts scarcely or not at all glandular is classified as var. *patentissimus* and consists of plants varying from bracts on the flowering branches which are minute and appressed to those with larger and less reduced bracts.

The flowers and flower-buds are eaten by wild turkey.

9. **Aster oblongifolius** Nutt. Oblong-leaf Aster Map 2183

Flowers July–November.

Occurs on limestone glades, bluff escarpments, rocky prairies, and open slopes. Southern and central Missouri northeast to Clark County and northwest to Atchison County; absent from the southeastern Missouri lowlands and from the north-central section.

The following variations occur in Missouri:

a. Leaves 5–10 cm. long, usually 5–10 mm. (up to 20) broad; stem usually only slightly branched, the stem and branches with long spreading hairs 0.5–1 mm. long as well as glandular hairs; stem up to 10 dm. tall . . . 9c. A. OBLONGIFOLIUS var. ANGUSTATUS

a. Leaves 2–4 cm. long, 2–5 mm. broad; stem usually much branched, the stem and branches either merely glandular or with short, mostly ascending to appressed (pressed against or lying parallel to surface) hairs; stem usually lower, mainly 1.5–4 dm. tall . . . . . *b*

 b. Rays violet-purple or blue-lavender . . 9a. A. OBLONGIFOLIUS var. OBLONGIFOLIUS f. OBLONGIFOLIUS

 b. Rays pink or rose-colored . . . . . 9b. A. OBLONGIFOLIUS var. OBLONGIFOLIUS f. ROSEOLIGULATUS

9a. **Aster oblongifolius** var. **oblongifolius** f. **oblongifolius** Map 2183

*Aster oblongifolius* Nutt. [G]

*Aster oblongifolius* var. *rigidulus* Gray [P & S]

*Aster Kumlienii* Fries

This is less common than var. *angustatus*, but occupies part of the same range as var. *oblongifolius*.

Ranges from Indiana to North Dakota, Wyoming, south to Arkansas and New Mexico.

9b. **Aster oblongifolius** var. **oblongifolius** f. **roseoligulatus** (Benke) Shinners Map 2183

*Aster oblongifolius* f. *roseoligulatus* (Benke) Shinners [G]

Known only from Camden County, western Ozark region (edge of limestone escarpment of southeast-facing Chapel Bluff along Niangua River, T37N, R18W, sect. 26, 8–8½ mi. southeast of Macks Creek, October 24, 1954, *Steyermark 78223*).

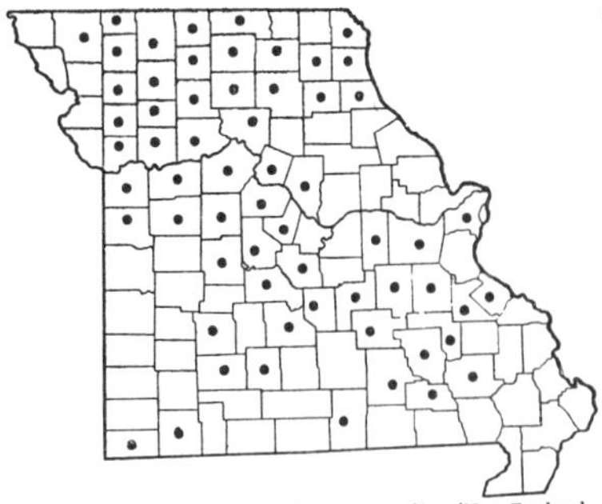

2184 Aster novae-angliae f. novae-angliae (New England Aster)

2185 • Aster puniceus var. firmus f. lucidulus (Glossy-leaf Aster)
2185 x Aster puniceus var. firmus f. lucidulus X lateriflorus

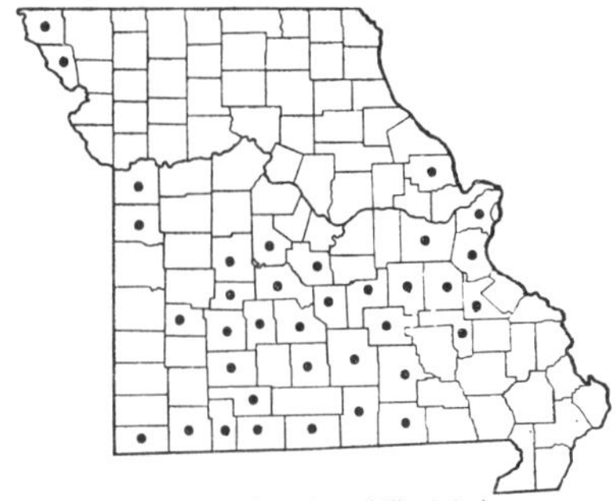

2186 Aster sericeus f. sericeus (Silky Aster)

9c. **Aster oblongifolius** var. **angustatus** Shinners Map 2183

*Aster oblongifolius* [of G, P & S], not Nutt.

This is the commonest variety in Missouri, occupying the greatest part of the range indicated in Missouri.

Ranges from Pennsylvania to Wisconsin, south to Alabama, Illinois, Arkansas, and Oklahoma.

*Aster oblongifolius,* when grown in an open sunny bed, produces an abundance of showy heads with purple rays and yellow disks, putting on a striking display under cultivated conditions. It favors limey soils.

10. **Aster novae-angliae** L. f. **novae-angliae**
New England Aster Map 2184

*Aster novae-angliae* L. [G, BB, P & S]

Flowers August–October.

Occurs in prairie swales, wet meadows, alluvial soils and thickets, low fields in valleys, and moist ground along streams. Throughout Missouri, but apparently absent from the southeastern Missouri lowlands and not recorded from most of the southwestern unglaciated prairie section.

Ranges from Quebec to Alberta, south to North Carolina, Alabama, Arkansas, Kansas, Colorado, Wyoming, and New Mexico.

The usual color of the rays is violet-purple to deep purple and characterizes f. *novae-angliae.* Forms with white rays, f. *geneseensis* House, and rose-colored rays, *f. roseus* (Desf.) Britt. [var. *roseus* DC], have not been recorded from Missouri.

The New England Aster is one of the showiest of the genus and is commonly offered in cultivation. It does well in most ordinary garden soils. The rose-colored form is more frequently grown.

11. **Aster puniceus** L. var. **firmus** (Nees) T. & G. f. **lucidulus** (Gray) Fern. Glossy-leaf Aster Map 2185

*Aster lucidulus* (Gray) Wieg. [BB, Deam]

*Aster puniceus* var. *lucidulus* Gray

Flowers August–October.

Occurs in calcareous swampy ground of spring-fed meadows in low valleys. Eastern Ozark region in Washington, Reynolds, Dent, Shannon, and Howell counties.

Ranges from Newfoundland to Ontario, Minnesota, and South Dakota, south to North Carolina, Tennessee, Indiana, Illinois, and Missouri.

This is one of the relicts of a more northern flora which have survived since the glacial period, in some of the spring-fed meadows of the eastern Ozarks. It is usually associated with such species as *Pedicularis lanceolata, Solidago Riddellii,* and *Parnassia grandifolia.* It is a handsome aster with large heads of blue-purple rays and yellow disks.

Some authors consider this aster a distinct species from *A. puniceus,* but throughout the range of *A. puniceus* var. *firmus* much intergradation occurs between it and other variations of *A. puniceus* in pubescence of stem, leaf shape, and extent of serration of the leaves. The f. *lucidulus* is chiefly distinguished from var. *firmus* f. *firmus* by the mostly entire instead of serrate leaves.

Collections from Dent (*Steyermark 20133, 20136, 20126, 20140*) and Shannon (*Steyermark 20114*) counties apparently represent hybrids between *A. puniceus* var. *firmus* f. *lucidulus* and *A. lateriflorus.* One collection from Shannon County (*Steyermark 69321*) has the broad outer bracts which are more characteristic of var. *oligocephalus.*

12. **Aster sericeus** Vent f. **sericeus** Silky Aster Map 2186

Flowers August–October.

Occurs on rocky glades and prairies, and loess hills, usually on limestone soils. Ozark region of southern and east-central Missouri north to Lincoln, Franklin, Crawford, Phelps, Pulaski, Miller, Morgan, Benton,

2187 ● Aster laevis var. laevis f. laevis (Smooth Aster)
2187 □ Aster laevis var. laevis f. latifolius

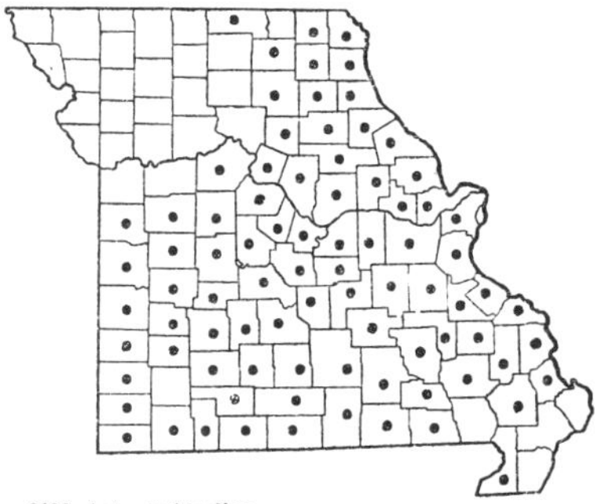

2188 Aster turbinellus

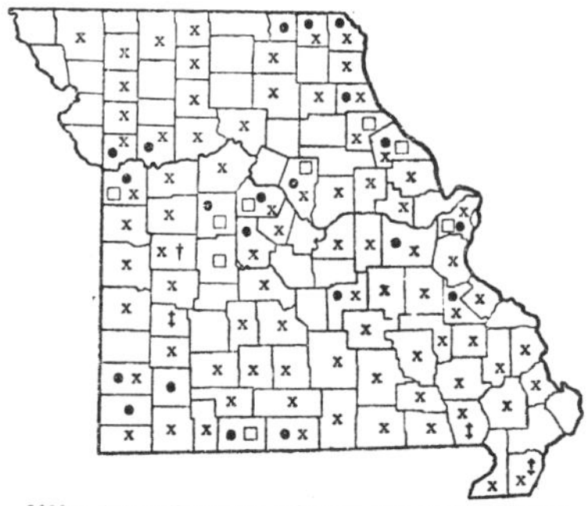

2189 ● Aster pilosus var. pilosus f. pilosus (White Heath Aster)
2189 ‡ Aster pilosus var. pilosus f. pulchellus
2189 □ Aster pilosus var. platyphyllus
2189 x Aster pilosus var. demotus
2189 † Aster pilosus var. demotus X praealtus

and Cedar counties, thence northwest along the Missouri River counties to Holt and Atchison counties.

Ranges from Michigan to Manitoba and North Dakota, south to Tennessee, Missouri, Texas, and New Mexico; also North Carolina to Georgia.

A form with white rays, f. *albiligulatus* Fassett, has not yet been recorded from Missouri, only the typical f. *sericeus* with violet- or blue-purple rays being known.

This is a very beautiful species, the violet-purple rays of the heads contrasting vividly with the silvery-white silky foliage.

13. **Aster laevis** L. Smooth Aster Map 2187
Flowers August–October

Occurs on prairies, rocky glades, loess hills, rocky or dry open woods, roadside banks, and thickets. Throughout Missouri, except absent from the southeastern Missouri lowlands.

Ranges from Maine to Saskatchewan, south to Georgia, Alabama, Louisiana, and Kansas.

Two variations occur in Missouri:

Leaves 10–25 mm. broad . . . . 13a. A. LAEVIS var. LAEVIS f. LAEVIS
Leaves 26–40 mm. broad . . . . 13b. A. LAEVIS var. LAEVIS f. LATIFOLIUS

13a. **Aster laevis** var. **laevis** f. **laevis** Map 2187
*Aster laevis* L. [G, P & S]
*Aster laevis* var. *laevis* [BB]

This is the commoner variation in Missouri.

13b. **Aster laevis** var. **laevis** f. **latifolius** (Porter) Shinners Map 2187
*Aster laevis* f. *latifolius* (Porter) Shinners [G]
*Aster laevis* var. *amplifolius* Porter
*Aster laevis* f. *amplifolius* (Porter) Fern.

Scattered in parts of the range of f. *laevis*, and known from Jefferson, Cape Girardeau, Reynolds, Saline, and Atchison counties.

Many varieties and forms have been segregated under *A. laevis*. Cronquist (in Gleason's *New Ill. Fl.* 3: 454. 1952) retains as distinct a western var. *Geyeri* Gray, having the involucral bracts in fewer series, narrower, more sharply pointed, and with more elongate green tips showing on the median band. This variety is reported by him to extend east to Minnesota and Missouri. Many specimens from Missouri show such long-pointed bracts, but I am unable to recognize this variety as distinct.

*Aster laevis*, with its usually silvery-blue or glaucous foliage and stems contrasting with the numerous heads of blue or purple rays, is a showy species. It is the source of many of our cultivated races of garden asters. The basal leaves are a source of winter food for the white-tailed deer.

This species hybridizes apparently with *A. azureus*, and with *A. sagittifolius* (*Steyermark 20619* from Camden County).

14. **Aster turbinellus** Lindl. Map 2188
Flowers late August–November.

Occurs in rocky or dry open woods and glades on upland slopes and ridges in acid soils associated with sandstone, chert, or granite substrata, or on leached soils above bluffs. Common throughout southern, central, and eastern Missouri, east and south of a line drawn from Putnam, Macon, Randolph, Howard, Saline, and Johnson counties to Cass County.

Ranges from Illinois to Nebraska, south to Louisiana, Arkansas, and Oklahoma.

This is a showy-flowered species, but in cultivation makes an even greater display, forming luxuriant plants with large ample panicles bearing hundreds of heads of flowers with deep lavender or purple rays. It has great horticultural possibilities, and should be more widely grown.

15. **Aster pilosus** Willd. White Heath Aster Map 2189

Flowers August–November.

Occurs usually in dry soils of fallow fields, meadows, prairies, glades, openings on top of bluffs and ridges, rocky open ground, gravel and sand bars along streams, along roadsides, railroads, and rocky or dry open woodland.

Throughout Missouri.

The following variations occur in the state:

a. Stem and leaves nearly or quite glabrous (without hairs) . . 15d. A. PILOSUS var. DEMOTUS
a. Stem, branches, and frequently the leaves more or less hairy . . . . . . . . *b*
 b. Main leaves of stem lanceolate or lance-oblong; stem with dense, white, long hairiness . . 15c. A. PILOSUS var. PLATYPHYLLUS
 b. Main leaves of stem linear or linear-lanceolate; stem and leaves with varying degrees of hairiness . . . . . . . . . . *c*
  c. Rays (ligules) white . . 15a. A. PILOSUS var. PILOSUS f. PILOSUS
  c. Rays (ligules) rose-colored . . . . 15b. A. PILOSUS var. PILOSUS f. PULCHELLUS

15a. **Aster pilosus** var. **pilosus** f. **pilosus** Map 2189

*Aster pilosus* Willd. [G, P & S]

*Aster pilosus* var. *pilosus* [BB]

Scattered throughout the state, but less common than var. *demotus*.

Ranges from Maine to Ontario and Minnesota, south to Georgia, Alabama, Mississippi, Arkansas, and Kansas.

15b. **Aster pilosus** var. **pilosus** f. **pulchellus** Benke Map 2189

*Aster pilosus* f. *pulchellus* Benke [G]

Rare and scattered in southern Missouri, in Pemiscot (*Steyermark 67052*), Butler (*Steyermark 83317*), and Cedar (*Steyermark 85775, 85794*) counties.

This is a handsome subject for the cultivated perennial or wild bed, and under cultivation produces a tremendous output of heads of flowers.

15c. **Aster pilosus** var. **platyphyllus** (T. & G.) Blake Map 2189

*Aster pilosus* var. *pilosus* [of BB in part, Shinners in part], not Willd.

Scattered in Missouri, but less common than the other varieties.

Ranges from New Jersey to Wisconsin and Iowa, south to North Carolina, Tennessee, and Missouri.

This is a broad-leaved variation of var. *pilosus*, with which it is sometimes merged by some botanists.

15d. **Aster pilosus** var. **demotus** Blake Map 2189

Common throughout Missouri and doubtless occurring in every county; this is the commonest variety in the state.

Ranges from Maine to Ontario, south to North Carolina and Missouri.

A specimen from St. Louis County (Forest Park, St. Louis, September 16, 1886, *Eggert*) was included by Palmer and Steyermark in their *Annotated Catalogue* (p. 663) as *A. polyphyllus* Willd. This specimen is referred in the present flora to *A. pilosus* var. *demotus*.

In numbers of individual plants, *A. pilosus* and its varieties, especially var. *demotus*, probably take first place as the most abundant aster in Missouri, often becoming weedy and aggressive. It is the one most usually seen in fields and along roadsides and waste places. It hybridizes with *A. parviceps*, *A. vimineus* var. *subdumosus*, and *A. praealtus*.

16. **Aster parviceps** (Burgess) Mackenz. & Bush Map 2190

Flowers August–October.

Occurs in dry prairies, fields, old cemeteries, open woods, and along roadsides. Chiefly in the glaciated prairie region of northern and central Missouri south to St. Louis, Warren, Montgomery, Callaway, Boone, Chariton, and Jackson counties, and locally in southern Missouri in Bollinger, Washington, Texas, and Jasper counties.

Ranges from Illinois to Iowa and Missouri.

This species hybridizes with *A. pilosus* var. *demotus*, and many putative hybrids are recorded from northern Missouri in Marion, Knox, Putnam, and Grundy counties.

This species has much narrower, more slender flower-heads with fewer disk-flowers than *A. pilosus*.

17. **Aster × amethystinus** Nutt. Map 2191

Flowers September–October.

Plate no. 359. **1.** Aster laevis, × 2/5. **2.** Aster sericeus, × 2/5. **3.** Aster parviceps, × 2/5. **4.** Aster praealtus, × 2/5; *a.* Leaf showing venation. **5.** Aster pilosus var. demotus, × 2/5; *a.* Flower-head, × 1 1/5. **6.** Aster turbinellus, × 2/5. **7.** Aster simplex, × 2/5.

PLATE NO. 359

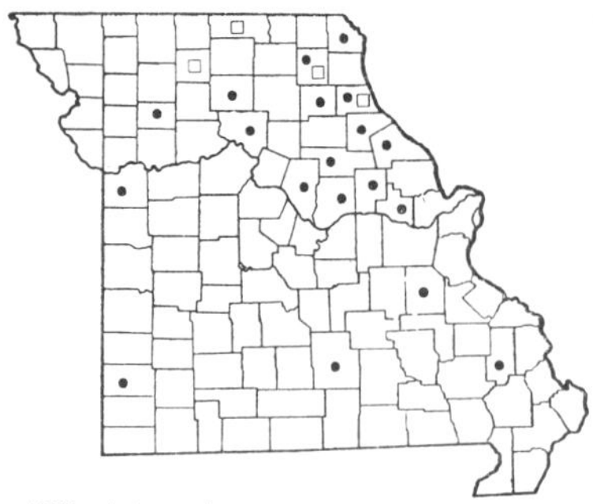

2190 • Aster parviceps
2190 □ Aster parviceps X pilosus var. demotus

2191 Aster X amethystinus

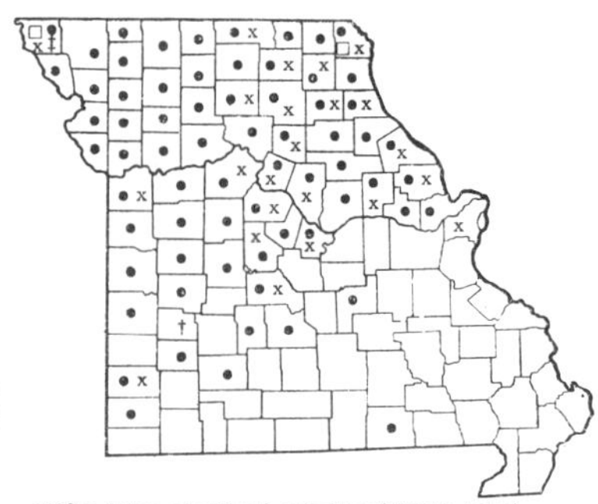

2192 • Aster ericoides f. ericoides (Wreath Aster)
2192 □ Aster ericoides f. caeruleus
2192 x Aster ericoides f. prostratus
2192 ‡ Aster ericoides f. prostratus X pilosus var. demotus
2192 † Aster ericoides f. ericoides X pilosus var. pilosus f. pulchellus

Occurs in prairies. Known only from Randolph County, central Missouri (prairie between Moberly and Cairo, October 10, 1934, *Steyermark 16021*, in U. of Mo. and Mo. Bot. Gard. Herb.).

Ranges from Massachusetts and Vermont to Iowa and Nebraska, south to Pennsylvania, Kentucky, Illinois, and Missouri.

This is an attractive aster with many heads of usually blue or purple rays. Forms with white rays, f. *leucos* Bemis, and pink rays, f. *leucerythros* Bemis, have been recorded, but are not as yet known from Missouri.

This aster is believed to be a hybrid between *A. novae-angliae* and *A. ericoides*. However, the Missouri collection was found growing with *A. novae-angliae* and *A. praealtus*.

18. **Aster ericoides** L. Wreath Aster Map 2192
Flowers July–October.

Occurs in prairies, along roadsides and railroads, and rocky open woods.

Common throughout the glaciated and unglaciated prairie region of northern, central, and southwestern Missouri, barely entering the Ozark region, north and west of a line drawn from St. Louis, Warren, Montgomery, Callaway, Cole, Phelps, Laclede, Dallas, and Greene counties to Newton County.

The following variations occur in Missouri:

a. Most or all of the hairs on the stem spreading. 18c. A. ERICOIDES f. PROSTRATUS
a. Most of the hairs of the stem and branches ascending or appressed (pressed against or lying parallel to surface) . . . . . . *b*
 b. Rays (ligules) white . . . 18a. A. ERICOIDES f. ERICOIDES
 b. Rays blue or violet . . . 18b. A. ERICOIDES f. CAERULEUS

18a. **Aster ericoides** f. **ericoides** Map 2192
*Aster ericoides* L. [G, P & S]
*Aster ericoides* var. *ericoides* [BB]

This is the commonest variation in Missouri, occupying the entire extent of the range indicated with a local occurrence in Oregon County, southern Missouri.

Ranges from Maine to British Columbia, south to Georgia, Alabama, Mississippi, Arkansas, Oklahoma, Texas, and Arizona.

18b. **Aster ericoides** f. **caeruleus** (Benke) Blake Map 2192

Known only in northern Missouri from Clark (roadside bank, 2 mi. east of Acasto, October 2, 1934, *Drouet 1765*) and Atchison (loess mounds adjacent to highway A, T66N, R41W, sect. 29, 4½ mi. northeast of Watson, October 15, 1957, *Steyermark 85659*) counties.

Although the f. *caeruleus* has sometimes been considered more closely related to typical *A. ericoides* in having an appressed or ascending type of pubescence (Blake, in Rh. 32: 139. 1930; Shinners, in Am. Midl. Nat. 26: 416. 1941), the two Missouri collections cited contrast in types of hairiness, the one from Atchison County having appressed pubescence on the stem and leaves, that from Clark County having short, spreading, denser pubescence on the stem and leaves.

A form with rose-colored rays, f. *Gramsii* Benke, has not yet been recorded from Missouri.

18c. **Aster ericoides** f. **prostratus** (Kuntze) Fern. Map 2192
*Aster ericoides* var. *prostratus* (Kuntze) Blake [BB, Shinners, P & S]
*Aster ericoides* f. *exiguus* (Fern.) Fern. [G]
*Aster multiflorus* Ait. var. *exiguus* Fern.

This is much less common in Missouri than typical

2193 Aster commutatus (White Prairie Aster)

2194 □ Aster dumosus var. strictior
2194 • Aster dumosus var. Dodgei

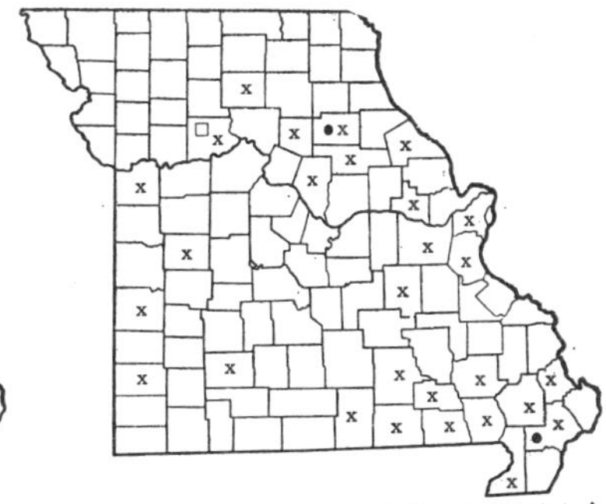

2195 • Aster vimineus var. vimineus (Small White Aster)
2195 x Aster vimineus var. subdumosus
2195 □ Aster vimineus var. subdumosus X pilosus var. demotus

f. *ericoides*, and is scattered throughout the range in Missouri.

Fernald (*Gray's Manual*, eighth ed. and Rh. 51 : 96. 1949) applies the name *A. ericoides* f. *exiguus* to plants with ascending stems and mostly solitary flower-heads at the tips of the branches, as contrasted with f. *ericoides* having erect stems and 1-sided dense racemes and with f. *prostratus* having prostrate stems and 1-sided dense racemes. The apparent differences in habit between erect and prostrate stems are considered by Blake (Rh. 32: 139. 1930) to be of no importance and this view is adopted by Cronquist (in Gleason's *New Ill. Fl.*), Shinners (Am. Midl. Nat. 26: 417. 1941), and others. In the present flora, the differences in habit are likewise not deemed of importance, but the matter of appressed or spreading pubescence is considered here to merit recognition only as a form rather than as a variety as indicated by Blake, Cronquist, and Shinners. Much more extensive experimental work with these variations needs to be carried out before a final evaluation of their status can be made.

*Aster ericoides* hybridizes with variations of *A. pilosus*. A collection from Cedar County (*Steyermark 85784*) from a natural prairie, T34N, R27W, west half sect. 30, ½–1 mi. southwest of Wagoner, October 17, 1957) apparently represents a hybrid between *A. ericoides* f. *ericoides* and *A. pilosus* var. *pilosus* f. *pulchellus*. The rays are lavender-lilac, the flower-heads are large and the inner bracts are subulate and narrowly pointed as in *A. pilosus*, while the other bracts are more broadened, obtuse, and ciliolate as in *A. ericoides*. The narrow, short leaves of the flowering branches are more characteristic of *A. ericoides*.

A collection from Atchison County (*Steyermark 85658*) may represent a hybrid between *A. ericoides* f. *prostratus* and *A. pilosus* var. *demotus*. The short narrow leaves and small involucres with ciliate bracts point to *A. ericoides* ancestry, but the heads, slightly larger than in typical *A. ericoides*, and the lance-subulate innermost bracts of the involucre show the characteristics more of *A. pilosus* var. *demotus*. This may be the same type of hybrid indicated as *A. exiguus* × *pilosus* H. E. Ahles (in Jones *et al.*, Vasc. Pl. Ill. p. 468. 1955).

19. **Aster commutatus** (T. & G.) Gray White Prairie Aster Map 2193

*Aster falcatus* Lindl. var. *falcatus* [BB]

Flowers July–October.

Occurs in fields and along railroads. Known only from St. Louis (St. Louis, July 7, 1910, *Sherff 287*) and Jackson counties.

Ranges from Alberta to New Mexico and Arizona, east to Manitoba, Wisconsin, and Missouri.

20. **Aster dumosus** L. Map 2194

Flowers August–November.

Occurs in wet meadows and swampy open ground.

Two varieties occur in Missouri:

Stem glabrous (without hairs) or sparsely and minutely hairy; leaves smooth or nearly so; rays 13–20 . . . . . 20a. A. DUMOSUS var. STRICTIOR

Stem densely hairy with ashy, rather short, stiff hairs; leaves rough-hairy on both surfaces, rays 18–24 . . . . . . 20b. A. DUMOSUS var. DODGEI

20a. **Aster dumosus** var. **strictior** T. & G. Map 2194

*Aster dumosus* [of BB in part], not L.

Known only from Butler and Howell (meadow along north side of highway 80, 4.9 mi. southwest of West Plains, September 25, 1949, *Steyermark 69336*) counties.

Ranges from New York and Ontario to Michigan and Wisconsin, south to North Carolina, Indiana, Illinois, and Missouri.

20b. **Aster dumosus** var. **Dodgei** Fern.
Map 2194

Known only from Howell County, southern Missouri (swampy remnant in field, T23N, R8W, sect. 23, 6½ mi. southeast of West Plains, November 2, 1956, *Steyermark 83554*).

Known from Michigan, Ohio, and Missouri.

Cronquist (in Gleason's *New Ill. Fl.* 3: 467. 1952) considers this variety as possibly a hybrid between *A. dumosus* and another species. He considers the different varieties which have been described under *A. dumosus* as too poorly defined to merit segregation.

Both varieties thus far in Missouri have been found in wet ground and appear quite rare and local.

21a. **Aster vimineus** Lam. Small White Aster
Map 2195

Also called Frost Flower.

Flowers August–October.

Occurs in low meadows of river valleys, swampy meadows, prairie swales, swampy woods and borders of swamps and upland sink-hole ponds, and alluvial soils along streams and ponds.

Two varieties are represented in Missouri material:

Rarely encountered type; pedicels or branchlets bearing the flower-heads mainly less than 7 mm. long; flower-heads usually densely crowded in spike-like or closely-flowered racemes . . . 21a. A. VIMINEUS var. VIMINEUS

Commonly encountered type; pedicels or branchlets bearing the flower-heads mainly 5–30 mm. long; flower-heads loosely racemose or scattered on spreading branchlets . . 21b. A. VIMINEUS var. SUBDUMOSUS

21a. **Aster vimineus** var. **vimineus** Map 2195

*Aster vimineus* Lam. [G, BB in part]

Known only from Monroe County, northeastern Missouri (alluvial lower part of north-facing limestone slopes along Salt River, Mark Twain State Park, T54N, R8W, sect. 9 and 16, 1–2 mi. southwest of Florida, September 25, 1948, *Steyermark 66529*).

Ranges from Maine to Ontario and Michigan, south to North Carolina, West Virginia, Indiana, and Missouri.

21b. **Aster vimineus** var. **subdumosus** Wieg.
Map 2195

*Aster vimineus* [of BB in part], not Lam.

Scattered in southern and central Missouri north to Pike, Monroe, Randolph, Linn, Carroll, and Jackson counties; commonest in the lowlands and upland sink-hole ponds of the Ozarks of southeastern Missouri; absent from most of the Ozark area.

Ranges from Quebec to Michigan, south to Florida, Alabama, Louisiana, and Texas.

A collection from Carroll County appears to represent a hybrid between *A. vimineus* var. *subdumosus* and *A. pilosus* var. *demotus*.

*Aster vimineus* var. *subdumosus* is a characteristic aster around upland sink-hole ponds and in the cypress and tupelo swamps of the southeastern Missouri lowlands. The elongated, arched-recurving to horizontally spreading branches bearing much reduced, bracteate leaves and numerous, small flower-heads in 1-sided racemes or spikelike inflorescences on short pedicels are characteristic. The flower-heads are among the smallest of all the native species of aster in Missouri.

22. **Aster lateriflorus** (L.) Britt. var. **lateriflorus**
White Woodland Aster Map 2196

*Aster lateriflorus* var. *pendulus* [of P & S], not (Ait.) Burgess

*Aster lateriflorus* (L.) Britt. [G, BB]

Flowers August–November.

Occurs in low wet woods, borders of streams, ponds, sloughs, and swamps, wet ledges of bluffs and at their base, and in wet depressions of prairies. Common throughout most of Missouri, not recorded from the extreme northern counties.

Ranges from Magdalen Island to Ontario and Minnesota, south to Georgia, Tennessee, and Arkansas.

This is one of the most abundant asters in Missouri, occurring in all the river bottom and low forests bordering streams, where it thrives in the alluvial soils subject to periodic flooding.

This aster in Missouri apparently hybridizes with *A. puniceus* var. *firmus* f. *lucidulus*, *A. simplex* var. *interior*, *A. praealtus*, and *A. ontarionis*. The rays vary from white to rarely lilac, and the disk-flowers are usually lavender.

The leaves are eaten by white-tailed deer.

23. **Aster ontarionis** Wieg. Map 2197

*Aster pantotrichus* Blake [P & S]

*Aster missouriensis* Britt., not (Nutt.) Ktze.

---

Plate no. 360. **1.** Aster dumosus var. strictior, × 2/5. **2.** Aster commutatus, × 2/5. **3.** Aster exilis, × 2/5. **4.** Aster vimineus var. subdumosus, × 2/5. **5.** Aster ericoides, × 2/5. **6.** Aster Brachyactis, × 2/5. **7.** Aster lateriflorus, × 2/5. **8.** Aster ptarmicoides, × 2/5. **9.** Aster linariifolius, × 2/5.

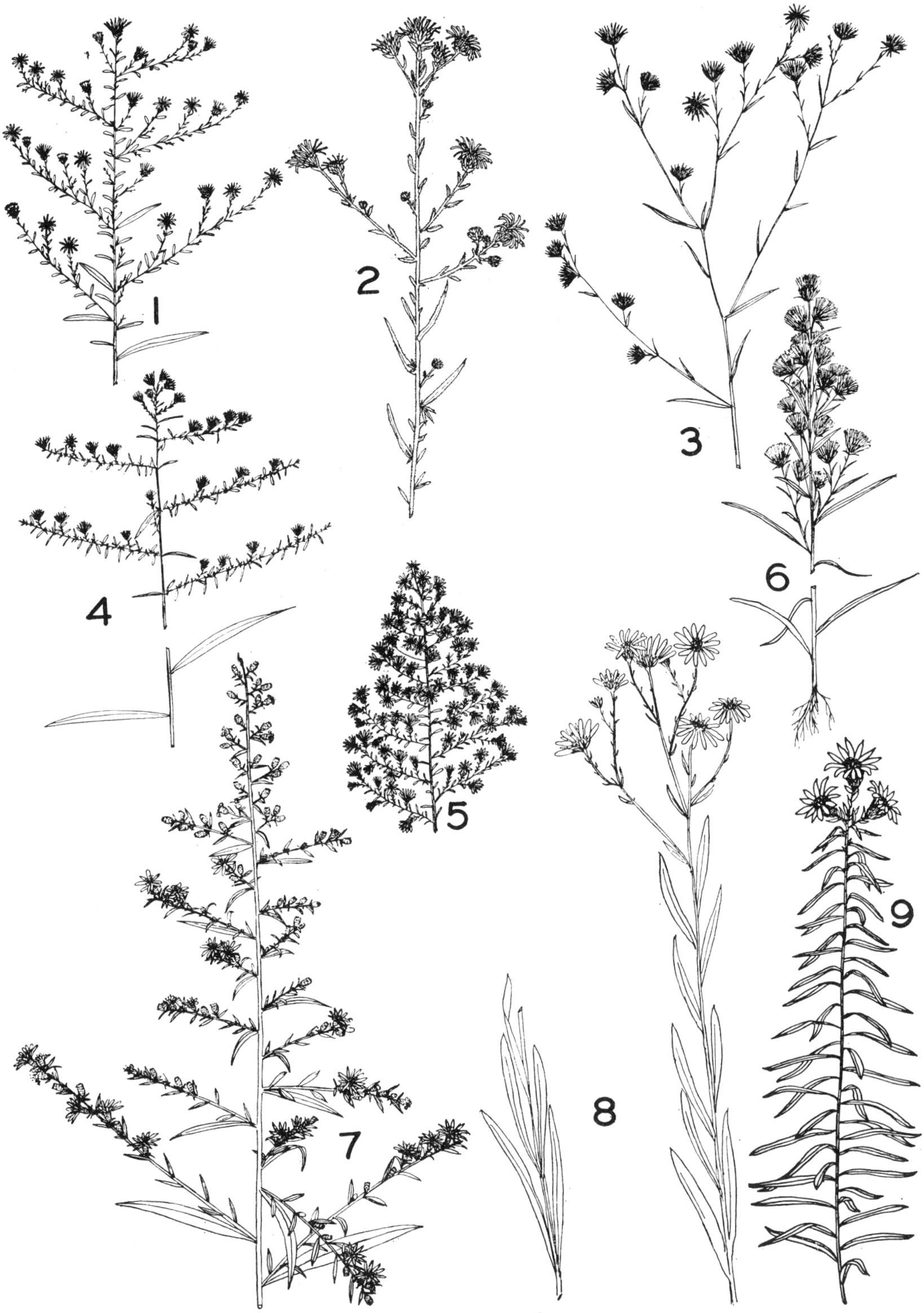


PLATE NO. 360

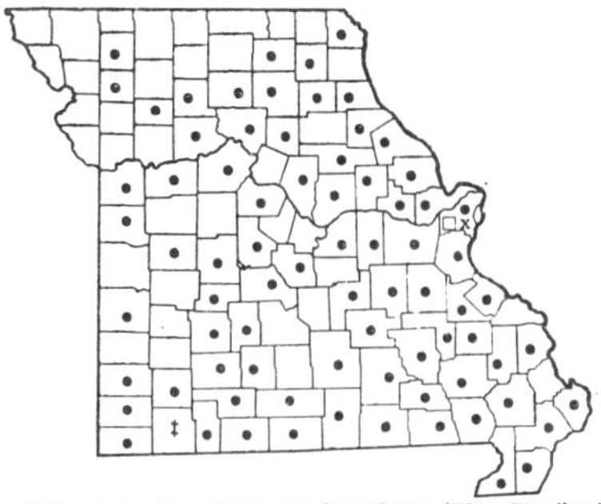

2196 • Aster lateriflorus var. lateriflorus (White Woodland Aster)
2196 □ Aster lateriflorus X ontarionis
2196 x Aster lateriflorus X praealtus
2196 ‡ Aster lateriflorus X simplex var. interior

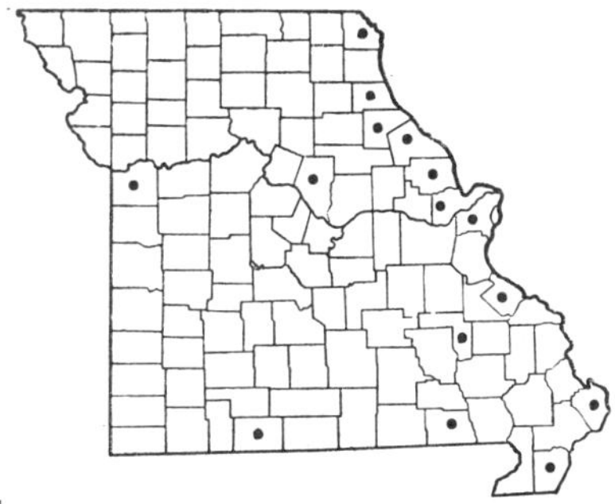

2197 Aster ontarionis

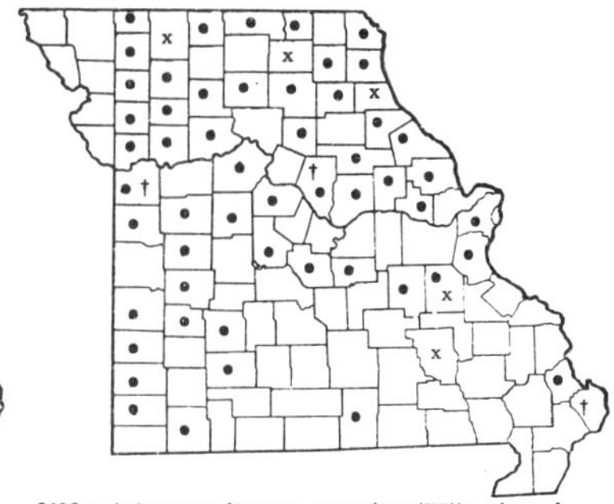

2198 • Aster praealtus var. praealtus (Willow-leaved Aster)
2198 x Aster praealtus var. angustior
2198 † Aster praealtus var. subasper
...
2199 Excluded species

*Aster missouriensis* var. *thyrsoideus* (Gray) Wieg.

Flowers August–October.

Occurs in low moist or dry woodland, borders of fields and streams, and rocky ledges of bluffs. Scattered in eastern, central, and southern Missouri, apparently commonest in the counties bordering the Missouri and Mississippi rivers.

Ranges from Quebec to Minnesota and South Dakota, south to North Carolina, Alabama, Mississippi, Arkansas, Oklahoma, and Texas.

This species resembles *A. lateriflorus*, but has creeping and branching rootstocks, by which it spreads and forms colonies. *Aster lateriflorus*, on the other hand, does not produce creeping or branching rootstocks, but has a short stout rootstock.

24. **Aster praealtus** Poir. Willow-leaved Aster Map 2198

Flowers August–October.

Occurs in wet meadows, river bottom prairies, prairie swales, borders of small streams, open thickets, fields in valleys and upland, and along roadsides.

Common throughout the glaciated and unglaciated prairie regions of northern, central, and southwestern Missouri, south and east to Scott, Washington, Crawford, Maries, Miller, Morgan, Polk, and Barry counties, and locally in Howell County.

The following varieties are represented by Missouri material:

a. Leaves of the main stem and branches 11 or more times as long as broad, linear or linear-lanceolate . 24b. A. PRAEALTUS var. ANGUSTIOR

a. Leaves of the main stem and branches 3–10 times as long as broad, lanceolate, broadly elliptic-lanceolate, or oval . . . . . . . *b*

b. Leaves of the main branches 3–6 times as long as broad, those of the ultimate branches broadly elliptic-lanceolate or oval, often obtuse or only slightly pointed (acutish); bracts of the involucre 1–1.5 mm. (rarely 0.8) broad . . 24c. A. PRAEALTUS var. SUBASPER

b. Leaves of the main branches 6–10 times as long as broad, lanceolate to narrowly elliptic-lanceolate, conspicuously acute (short-pointed); bracts of the involucre 0.6–1 mm. broad . . 24a. A. PRAEALTUS var. PRAEALTUS

24a. **Aster praealtus** var. **praealtus** Map 2198

*Aster praealtus* Poir. [G, BB, P & S]

*Aster coerulescens* DC., not of auth.

*Aster praealtus* var. *imbricatior* Wieg.

This is the commonest variation of Missouri and is found throughout the range as indicated above for the species.

Ranges from Massachusetts to Manitoba, south to Maryland, West Virginia, Kentucky, Missouri, Oklahoma, and Texas.

24b. **Aster praealtus** var. **angustior** Wieg. Map 2198

*Aster coerulescens* var. *angustior* (Wieg.) Fern. in Rh. 51: 96. 1949.

Known from Harrison, Adair, Washington, and Reynolds counties.

Ranges from Quebec to Ontario and Iowa, south to D.C., Ohio, Indiana, Illinois, and Missouri.

24c. **Aster praealtus** var. **subasper** (Lindl.) Wieg. Map 2198

*Aster coerulescens* var. *subasper* (Lindl.) Fern. in Rh. 51: 96. 1949.

Known from Mississippi, Boone, and Jackson counties.

Ranges from Indiana and Illinois to Texas.

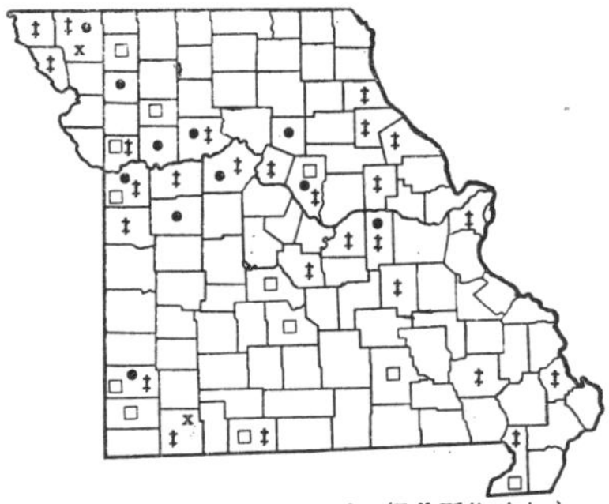

2200 • Aster simplex var. simplex (Tall White Aster)
2200 ‡ Aster simplex var. ramosissimus
2200 □ Aster simplex var. interior
2200 x Aster simplex var. interior X lateriflorus

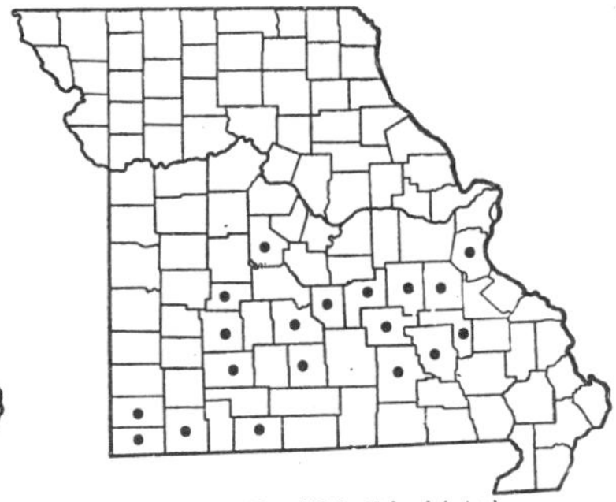

2201 Aster ptarmicoides (White Upland Aster)

2202 Aster linariifolius var. linariifolius f. linariifolius (Stiff-leaf Aster)

*Aster praealtus* is a beautiful species and deserves a wider recognition horticulturally. It varies considerably in leaf shape, pubescence, arrangement and crowding of heads, and width of involucral bracts, and the above varieties are tentatively given recognition here pending future field and experimental investigations, as botanists are not agreed as to the limits of variation. A few collections in Missouri are known which appear to represent hybrids between *A. praealtus* and *A. pilosus* var. *demotus* and *A. lateriflorus*. Stems vary from dwarf plants only 3 dm. tall (*Steyermark 66791*) to robust plants 1.5 meters tall.

25. **Aster simplex** Willd. Tall White Aster
Map 2200

Also known as Panicled Aster.

Flowers August–October.

Occurs in moist meadows, prairie swales, alluvial soils in thickets, along fence rows, roadside ditches, and borders of sloughs, streams, and wet ground, rarely in low wet woods. Scattered in Missouri, mostly in the counties bordering the Missouri and Mississippi rivers, and other large streams; absent from most of the Ozark region and much of north-central Missouri.

The following varieties are represented by Missouri material:

a. Involucre 3–4 mm. high; rays 4.4–8.5 mm. long, often lavender, sometimes white; pappus 2.8–3.5 mm. long . 25c. A. SIMPLEX var. INTERIOR
a. Involucre 4.5–6 mm. high; rays 6–11 mm. long, usually white, rarely blue-tinged; pappus 3.6–7 mm. long . . . . . . . . . . . *b*
 b. Leaves of the main stem 12–25 times as long as broad, mostly 3–12 mm. broad . . . 25b. A. SIMPLEX var. RAMOSISSIMUS
 b. Leaves of the main stem mostly 5–11 times as long as broad, mostly 15–40 mm. broad. 25a. A. SIMPLEX var. SIMPLEX

25a. **Aster simplex** var. **simplex** Map 2200

*Aster simplex* Willd. [G]

*Aster paniculatus* Lam. var. *simplex* (Willd.) Burgess [P & S]

*Aster hesperius* [of BB], as to Missouri specimen examined in New York Botanical Garden.

This variety is scattered in western and central Missouri.

Ranges from Newfoundland to Saskatchewan, south to North Carolina, West Virginia, Kentucky, Missouri, and Kansas.

25b. **Aster simplex** var. **ramosissimus** (T. & G.) Cron. Map 2200

*Aster paniculatus* Lam., not Mill. [P & S]

*Aster paniculatus* var. *bellidiflorus* (Willd.) Burgess

This is the commonest variation in Missouri, so far as available records go, found principally along the Missouri and Mississippi rivers in the central, northwestern, and eastern sectors of the state, and locally elsewhere.

Ranges from Quebec to Minnesota, south to Virginia, Ohio, Indiana, Illinois, Iowa, and Missouri.

25c. **Aster simplex** var. **interior** (Wieg.) Cron.
Map 2200

*Aster interior* Wieg. [P & S]

*Aster Tradescanti* of auth. in part, not L.

Scattered in various parts of western, southern, and central Missouri.

Ranges from New York to Ontario and Wisconsin, south to Ohio, Indiana, Illinois, and Missouri.

*Aster simplex* is a showy robust aster with usually numerous heads of flowers with white (rarely bluish or lavender) rays in usually large and well-developed panicles. It does well in cultivation.

A specimen in the New York Botanical Garden Herbarium from Jackson County (common in creek,

2203 Aster exilis

2204 Aster Brachyactis

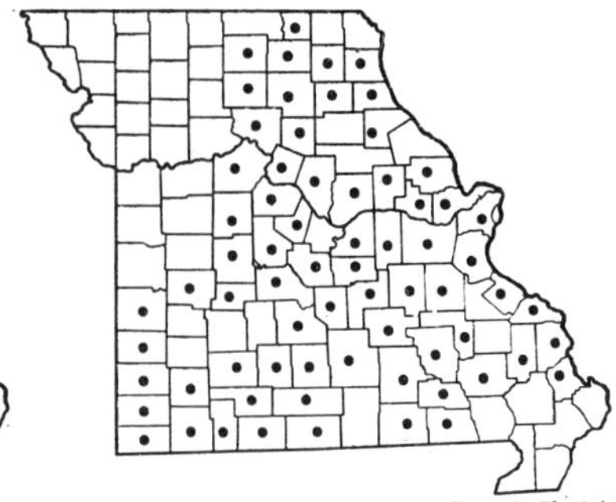

2205 Erigeron pulchellus var. pulchellus (Robin's Plantain)

Courtney, September 30, 1902, *Bush 1787*) identified by Dr. Cronquist as 'probably *A. coerulescens*' has been studied by the present author, and is referred in the present flora to *A. simplex* var. *simplex* with which it agrees in all essentials.

26. **Aster ptarmicoides** (Nees) T. & G. White Upland Aster Map 2201

Flowers July–September.

Occurs in rocky prairies, glades, bluff escarpments of usually limestone, and sometimes rocky open woods. Ozark region north to Jefferson, Washington, Crawford, Phelps, Pulaski, Morgan, Hickory, and Newton counties; absent apparently from many of the southern and eastern Ozark counties.

Ranges from Quebec to Saskatchewan, south to Vermont, New York, Ontario, Georgia, Ohio, Indiana, Illinois, Missouri, Arkansas, South Dakota, Wyoming, and Colorado.

This is a showy aster unusual for the white disk and ray flowers. It is a handsome plant for the rock garden. Although in Missouri it is confined to limestone rocks and barrens, it often occurs in pure sand elsewhere in its range.

It has been suggested that this aster may be more closely related to *Solidago* than to *Aster*, based upon the reputed occurrence of yellow or yellowish ray flowers in some northwestern specimens of the species and upon the occurrence in Europe of a spontaneous garden hybrid between this species of aster and a species of *Solidago* (Cronquist, in Gleason's *New Ill. Fl.* 3: 460. 1952).

27. **Aster linariifolius** L. var. **linariifolius** f. **linariifolius** Stiff-leaf Aster Map 2202

*Aster linariifolius* L. [G, BB, P & S]

Flowers August–October.

Occurs in dry rocky pine, pine-oak, or oak-hickory woods, on ridgetops, upland slopes, and glades, in acid soils associated with sandstone, chert, or granite substrata. Southern and east-central Ozark region north to St. Louis, Franklin, Gasconade, Maries, Pulaski, Laclede, Greene, Lawrence, and Jasper counties.

Ranges from Maine and Quebec to Minnesota, south to Florida, Alabama, Mississippi, Louisiana, and Texas.

The typical form has blue-violet or purple rays. In var. *linariifolius* f. *leucactis* Benke the rays are white. A var. *Victorinii* Fern., with obtuse involucral bracts and round-tipped leaves, occurs in Quebec.

This aster is readily identified by its low stature, crowded stiff, narrow 1-nerved leaves, and few to solitary heads of flowers. It is a very attractive species and is a desirable addition to a rock garden, but needs acid soil provided by sand, chert, or granite chips.

28. **Aster exilis** Ell. Map 2203

*Aster subulatus* Michx. var. *ligulatus* Shinners, Field & Lab. 21: 159–60. 1953.

Flowers October–November.

Occurs in open low ground bordering fields and ditches. Known only from the southeastern Missouri lowlands in Pemiscot (open places along shallow ditch along road E, 1½ mi. east of Cooter, sect. 8, October 23, 1948, *Steyermark 67051*; open places along road E, sect. 3, 1 mi. east of Holland, October 23, 1948, *Steyermark 67063*) and Dunklin (open places near house bordering field, T16N, R18E, sect. 15, 3 mi. west of Hermandale, October 23, 1948, *Steyermark 67075*) counties.

Ranges from Florida to Arizona and California, north to South Carolina, Missouri, Kansas, and Oklahoma; also Mexico and tropical America.

29. **Aster Brachyactis** Blake Map 2204

*Aster angustus* (Lindl.) T. & G. [P & S], not Nees

Flowers August–September.

Occurs in open ground of loess hills and along railroads. Known only from Atchison and Clay (introduced, Randolph, August 9, 1895, *Mackenzie 240*) counties.

Ranges from Alberta to Minnesota, south to Utah, Colorado, South Dakota, and Missouri; locally eastward around the Great Lakes in Illinois and Indiana, New York, and Quebec.

This species is native on the loess hills of Atchison County, northwestern Missouri, where it is associated with numerous other species which form part of the Great Plains floral element entering this section of the state.

*Excluded Species*

**Aster concinnus** Willd.

This was credited to Missouri by Palmer and Steyermark in their *Annotated Catalogue* (p. 663). The specimens upon which they based their identifications are referred in the present flora to *A. laevis* L.

**Aster pilosus** var. **Pringlei** (Gray) Blake

*Aster polyphyllus* Willd., not Moench.

As indicated under *A. pilosus* var. *platyphyllus*, the specimen reported as *A. polyphyllus* by Palmer and Steyermark in their *Annotated Catalogue* (p. 663) from St. Louis County has been found to be referable instead to *A. pilosus* var. *platyphyllus*.

**Aster hesperius** Gray

*Aster coerulescens* of auth., not DC.

This is reported by Cronquist (in Gleason's *New Ill. Fl.* 3: 467. 1952) from Missouri. The specimen upon which the identification is based (common in creek, Courtney, September 30, 1902, *Bush 1787*) is referred in the present flora to *A. simplex* var. *simplex*, the larger cauline leaves being 15–20 mm. broad and 7–9 times as long as broad.

Cronquist (in Gleason's *New Ill. Fl.* 3. 1952) separates the larger-headed extremes of *A. simplex* from *A. hesperius* as follows:

Involucral bracts mostly not strongly imbricate, the outer usually at least $\frac{2}{3}$ as long as the inner; lobes of disk-corollas comprising 19–36 per cent of the limb . . . . . . . . . A. HESPERIUS

Involucral bracts more or less strongly imbricate, the outer seldom as much as $\frac{2}{3}$ as long as the inner; lobes of disk-corollas comprising 30–45 per cent of the limb . . . . . . . . . A. SIMPLEX

Under his description of *A. hesperius* (p. 467), Cronquist states that it readily passes into *A. simplex*, has usually entire leaves, with involucres averaging larger and sometimes as much as 7.5 mm high, the bracts averaging broader than in *A. simplex*, and seldom much imbricate. The specimen collected by Bush and cited above has toothed leaves and the lobes of the disk-corollas comprise 30 per cent of the limb. The involucral bracts are more or less imbricate, but not strongly so. However, they can be matched by many other collections of *A. simplex* from Missouri and eastward.

## 19. **Erigeron** L. Fleabane

(*Conyza* [of BB in part], not L.)

Some of the species of this genus have been treated by Cronquist (Bull. Torr. Club 74: 150. 1947) as pertaining to *Conyza*, on reasonably justified evidence. In view of the fact, however, that rayless forms (*E. strigosus* var. *strigosus* f. *discoideus*) may occur in species of *Erigeron* ordinarily having rays, as well as the occurrence of various transitional forms of intermediate position which cannot be readily assigned to *Conyza*, all the species which have previously and traditionally been placed with *Erigeron* are treated in that respect in the present flora.

a. Rays either absent, or so inconspicuous as to be scarcely, if at all, longer than the disk or the corolla-tube and pappus of the outermost marginal flowers . . . . . . . . . . . . . . . . . . *b*
  b. Flower-heads 6–10 mm. broad, hemispherical to saucer-shaped; involucre 6–10 mm. broad; pappus of the outer marginal (pistillate) flowers consisting of few very short bristles or scales less than 1 mm. long, the long bristles absent . . . . . . 4. E. STRIGOSUS var. STRIGOSUS f. DISCOIDEUS
  b. Flower-heads 1–4 mm. broad; involucre 1.5–4 mm. broad, campanulate (bell-shaped); pappus of all the flowers, including the outer marginal ones, consisting of well-developed long bristles . . . *c*
    c. Lower leaves of stem 1–2.5 mm. broad, all narrowly linear and entire (without teeth); plant widely branched with more than one stem, 1–3 dm. tall with ashy-gray appressed (pressed against or lying parallel to surface) hairiness; rays, if seen, purplish . . . . . . . . 8. E. DIVARICATUS
    c. Lower leaves of stem 3–13 mm. broad, oblanceolate, oblong, or spatulate, entire or toothed; plant with one main erect stem evident, mostly 3–20 dm. tall, sometimes shorter, glabrous or with spreading hairs; rays, if seen, white to lilac-pink. . . . . . . . . . . . . . . *d*

d. Commonly encountered species; bracts of the involucre without dark or purplish tips; stems usually more or less hairy . . . . . . . . . . . . . . . . . . 6. E. CANADENSIS

d. Rarely encountered species, known only from Ripley County, southeastern Missouri; some or all of the bracts of the involucre with a dark purplish tip; stems mainly glabrous (without hairs) or nearly so . . . . . . . . . . . . . . . . . . . . 7. E. PUSILLUS

a. Rays present and conspicuous, much longer than the disk and the corolla-tube and pappus of the outermost marginal flowers . . . . . . . . . . . . . . . . . . . . . . . . *e*

e. Stem-leaves clasping the stem by their rounded bases; stems soft, usually flat after being pressed; disk-corollas 2.5–4.5 mm. long; perennial plants . . . . . . . . . . . . . . . . . *f*

f. Base of plant sending out leafy runners; yellow center (disk) of head 12–20 mm. broad, 25–40 mm. broad when expanded with rays; rays (ligules) 50–75, and 1–1.7 mm. broad when flattened; disk-corollas 3.5–5 mm. long . . . . . . . . . . . . . . 1. E. PULCHELLUS

f. Leafy runners absent; yellow center (disk) of head 7–13 mm. broad, 15–25 mm. broad when expanded with rays; rays (ligules) 100–200, and about 0.5 mm. broad; disk-corollas 2.5–3.5 mm. long . . . . . . . . . . . . . . . . . . . . 2. E. PHILADELPHICUS

e. Stem-leaves not clasping the stem, tapering at their bases; stems firm or hard, usually terete (with curved sides or rounded in cross-section) after being pressed; disk-corollas 1.5–2.5 mm. long; annual or biennial plants . . . . . . . . . . . . . . . . . . . . . *g*

g. Main leaves of stem mostly coarsely or sharply toothed, conspicuously bristly-hairy on margins, 15–70 mm. broad, ovate to lanceolate . . . . . . . . . . . . . . . 3. E. ANNUUS

g. Main leaves of stem mostly entire or with a few irregular teeth, glabrous to short-hairy with appressed (pressed against or lying parallel to surface) hairs, the margins not conspicuously hairy, 5–25 mm. broad, narrowly to broadly oblanceolate to linear-lanceolate. . . . . . . . *h*

h. Rarely encountered species of southern and central Missouri; largest lower leaves 1–3 cm. long; leaves half-way up stem 1.5–4 cm. long; stems 1–4 dm. tall; pappus of all the flowers double, consisting of an outer series of short scales and an inner series of long bristles . . 5. E. TENUIS

h. Common species found throughout Missouri; largest lower leaves 2–15 cm. long; leaves half-way up stem 5–8 cm. long; stems mainly 3–10 dm. tall; pappus of the ray flowers lacking long bristles, and consisting only of a few short bristles or scales; pappus of the disk-flowers double with an outer series of short scales and an inner series of long bristles 4. E. STRIGOSUS

1. **Erigeron pulchellus** Michx. var. **pulchellus**
Robin's Plantain Map 2205
*Erigeron pulchellus* Michx. [G, BB, P & S, Steyerm.]
Flowers April–June.

Occurs in rocky or open woods, thickets, clearings, and well-drained soils bordering streams.

Common throughout the eastern half of Missouri and the Ozark region, west in northern and central Missouri to Schuyler, Sullivan, Linn, Chariton, Saline, and Pettis counties; absent from the southeastern lowland sections.

Ranges from Maine and Quebec to Ontario and Minnesota, south to Florida, Alabama, Louisiana, Kansas, and Oklahoma.

2. **Erigeron philadelphicus** L. f. **philadelphicus**
Philadelphia Fleabane Map 2206
*Erigeron philadelphicus* L. [G, BB, P & S, Steyerm.]
Flowers April–June.

Occurs in fields, open woods, moist banks and ledges of bluffs, alluvial soils in meadows and wooded valleys, waste ground, roadsides, and along railroads.

Ranges from Newfoundland to British Columbia, south to Florida, Louisiana, and Texas.

In typical f. *philadelphicus*, the only variation thus far known in Missouri, the basal leaves do not generally exceed 15 cm. long or 3 cm. broad, but in f. *scaturicola* (Fern.) Cron. they are up to 38 cm. long and 8–12 cm. broad, and in f. *angustatus* Vict. & Rousseau the stems are not more than 2 dm. tall and the leaves are greatly reduced in size.

*Erigeron philadelphicus* has been used in domestic medicine as an astringent, tonic, and diuretic. The leaves are eaten by white-tailed deer.

3. **Erigeron annuus** (L.) Pers. f. **annuus** (L.) Pers.
Daisy Fleabane Map 2207
Also called Whitetop Fleabane.
*Erigeron annuus* (L.) Pers. [G, P & S, Steyerm.]
*Erigeron annuus* var. *annuus* [BB]
Flowers May–November.

Occurs in fields, prairies, clearings, waste ground,

Plate no. 361. **1.** Erigeron pulchellus, × 2/7. **2.** Erigeron philadelphicus, × 2/7. **3.** Erigeron strigosus, × 2/7. **4.** Erigeron annuus, × 2/7. **5.** Erigeron tenuis, × 2/7. All details from Small, The New York Botanical Garden.

PLATE NO. 361

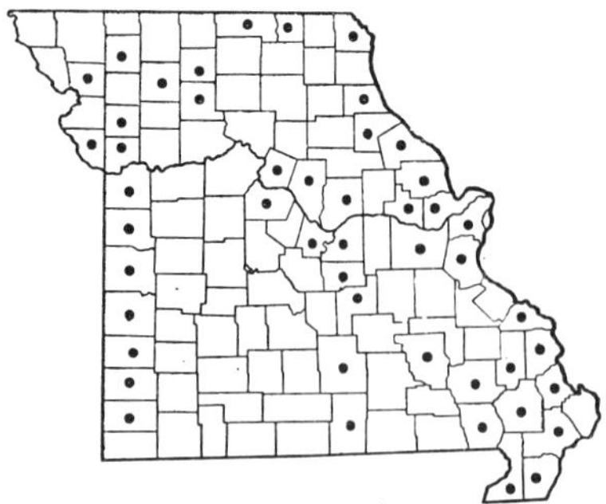

2206 Erigeron philadelphicus f. philadelphicus (Philadelphia Fleabane)

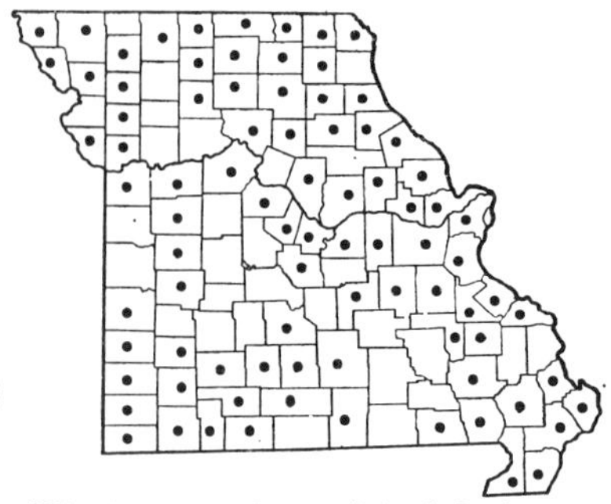

2207 Erigeron annuus f. annuus (Daisy Fleabane)

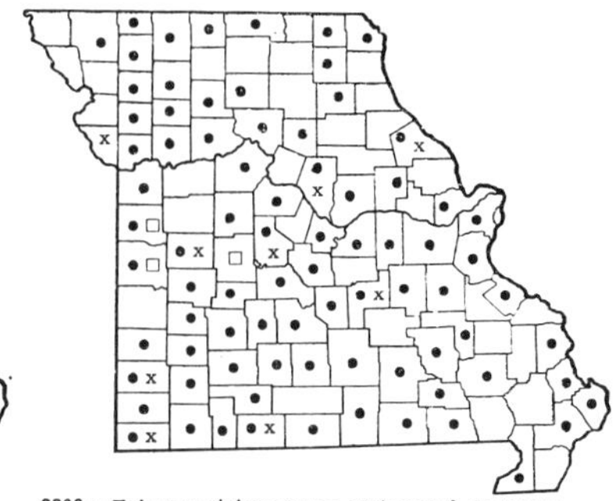

2208 • Erigeron strigosus var. strigosus f. strigosus (Daisy Fleabane)
2208 □ Erigeron strigosus var. strigosus f. discoideus
2208 x Erigeron strigosus var. Beyrichii

and along roadsides and railroads. Common throughout Missouri, doubtless in every county.

Ranges from Nova Scotia to Manitoba, south to Georgia, Oklahoma, and Texas.

This fleabane has been used in domestic medicine as an astringent, tonic, and diuretic. The leaves are eaten by white-tailed deer.

4. **Erigeron strigosus** Muhl. Daisy Fleabane Map 2208

Also called Whitetop Fleabane.

Flowers May–September.

Occurs in prairies, glades, rocky open places, dry fields, dry open woods, along roadsides, and railroads.

The following variations occur in Missouri:

a. Rays absent or at most only as long as the disk. 4b. E. STRIGOSUS var. STRIGOSUS f. DISCOIDEUS

a. Rays conspicuous, noticeably longer than the disk . . . . . . . . . . . . . . *b*

b. More rarely encountered type; disk (solid portion between the rays) of the main heads 3–8 mm. broad; involucre 2–3 mm. high; rays purple or white; larger leaves at base of stem narrowly oblanceolate, 5–15 mm. broad; lower leaves of the stem linear or nearly so . . 4c. E. STRIGOSUS var. BEYRICHII

b. Commonly encountered type; disk of the main heads 6–12 mm. broad; involucre 3–5 mm. high; rays usually white, rarely purplish; larger leaves at base of stem broadly oblanceolate, 7–30 (–50) mm. broad; lower leaves of the stem broadly oblanceolate . 4a. E. STRIGOSUS var. STRIGOSUS f. STRIGOSUS

4a. **Erigeron strigosus** var. **strigosus** f. **strigosus** Map 2208

*Erigeron strigosus* Muhl. [G]
*Erigeron strigosus* var. *strigosus* [BB]
*Erigeron ramosus* (Walt.) BSP. [P & S, Steyerm.]

Throughout Missouri, and doubtless in every county.

Ranges from Newfoundland to British Columbia, south to South Carolina, Ohio, Indiana, Illinois, Missouri, and Oklahoma.

4b. **Erigeron strigosus** var. **strigosus** f. **discoideus** (Robbins) Fern. Map 2208

*Erigeron strigosus* f. *discoideus* (Robbins) Fern. [G]
*Erigeron strigosus* var. *discoideus* Robbins [BB]
*Erigeron ramosus* f. *discoideus* (Robbins) Fern. [Steyerm.]

Rare and scattered, known only from Carroll, Benton, Bates, and Vernon counties in western Missouri.

Scattered in the range from New England and New York to Virginia and Missouri.

4c. **Erigeron strigosus** var. **Beyrichii** (Fisch. & Mey.) T. & G. Map 2208

*Erigeron ramosus* var. *Beyrichii* (T. & G.) Trelease [P & S]

Scattered in southern and central Missouri, north to Pike, Boone, and Platte counties.

Ranges from Florida to Texas, north to Rhode Island, Wisconsin, Minnesota, North Dakota, Montana, Idaho, and Washington.

5. **Erigeron tenuis** T. & G. Map 2209

Flowers May–June.

Occurs in prairies, fields, and waste ground. Commonest in southwestern Missouri, in Barton, Jasper, Newton, McDonald, and Lawrence counties, and elsewhere known only from St. Louis and Crawford counties, east-central Missouri.

Ranges from Missouri to Louisiana and Texas.

The rays vary from pink to lilac or lavender-blue, less commonly white.

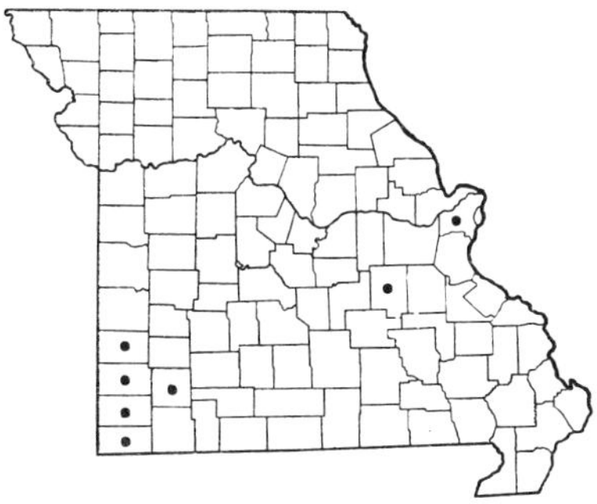

2209 Erigeron tenuis

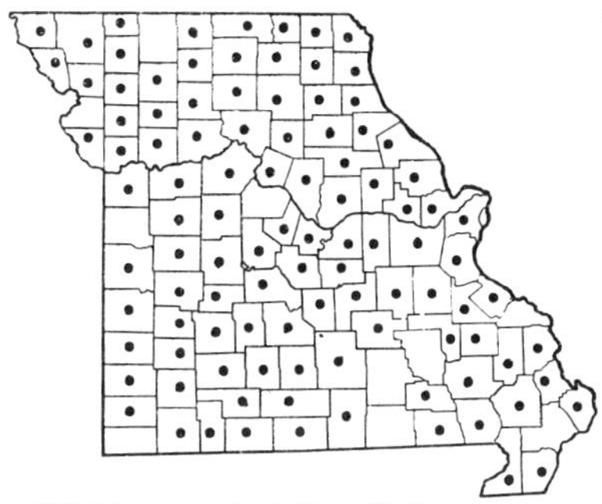

2210 Erigeron canadensis (Horse Weed)

2211 Erigeron divaricatus (Dwarf Fleabane)

6. **Erigeron canadensis** L. Horse Weed Map 2210

Also called Canada Fleabane, Hog Weed.

*Conyza canadensis* (L.) Cron. var. *canadensis* [BB]

Flowers June–November.

Occurs in fields, prairies, rocky open glades, waste ground, along roadsides, and railroads. Throughout Missouri, doubtless in every county.

Ranges from Labrador to British Columbia, south to Florida, Texas, and California; also in Mexico, the West Indies, Central and South America, and naturalized in the Old World.

A variation of this species having the stem glabrous or only sparsely hairy in the upper portion, is sometimes separated as var. *glabratum* Gray (*Conyza canadensis* var. *glabrata* [Gray] Cron.), but is not recognized in the present flora.

One of our commonest weeds, which sometimes causes or may produce an irritation to some persons handling the plant. Its bitter qualities may cause injuries to animals eating hay containing the plant. Dust containing pollen which has settled on the leaves of the plant, as in some of the goldenrods, may set up an irritation to some hay fever sufferers who handle the plant.

The plant has been used in domestic medicine as a tonic, astringent, and diuretic.

7. **Erigeron pusillus** Nutt. Map 2211A

*Erigeron canadensis* var. *pusilla* (Nutt.) Cron. [BB, Shinners]

Flowers August–October.

Occurs on wooded slopes.

Known only from Ripley County, southeastern Missouri (ravine slopes tributary to Little Black River, between Greenville Ford and Pennington Ford, T24N, R3E, sect. 10, 15, 22, 23, 26, 24, and 25, 10–13 mi. northeast of Doniphan, 2–5 mi. north and northwest of Flatwoods, September 1, 1946, *Steyermark 63969*, in Chi. Nat. Hist. Mus. Herb.).

Ranges from Florida to Texas, north along the coastal plain to Massachusetts and inland north to Indiana and Missouri.

The specimen cited above has some of the bracts with purple tips and the stem is mainly glabrous. The longest hairs of the leaf-margins are near or at the base of the leaves, but short hairs are found on the margins nearly to the tip of all the leaves.

The plant was collected in a locality where other species of a generally southern distribution, such as *Solidago odora* and *Helianthus silphioides*, occurred.

8. **Erigeron divaricatus** Michx. Dwarf Fleabane Map 2211

Also known as Spreading Fleabane.

*Conyza ramosissima* Cron. [BB]

Flowers May–September.

Occurs in dry pastures, fields, prairies, rocky open glades, waste places, roadsides and along railroads. Throughout Missouri, except apparently absent from the southeastern lowland section.

Ranges from Alabama to Texas, north to Ohio, Indiana, Wisconsin, Minnesota, and North Dakota.

Because of its bitter taste, this weed is not eaten by grazing animals. The bushy rounded plant with its ashy-gray appressed hairiness of the leaves and stems produces a dusky gray effect from a distance.

This and the preceding species are treated under *Conyza* by Cronquist (in Gleason's *New Ill. Fl.* 3: 475. 1952), but transitional and rayless forms in *Erigeron*, such as *E. strigosus* f. *discoideus* make such a separation more an arbitrary, rather than a natural one.

2211-A Erigeron pusillus

2212 Pluchea foetida (Marsh Fleabane)

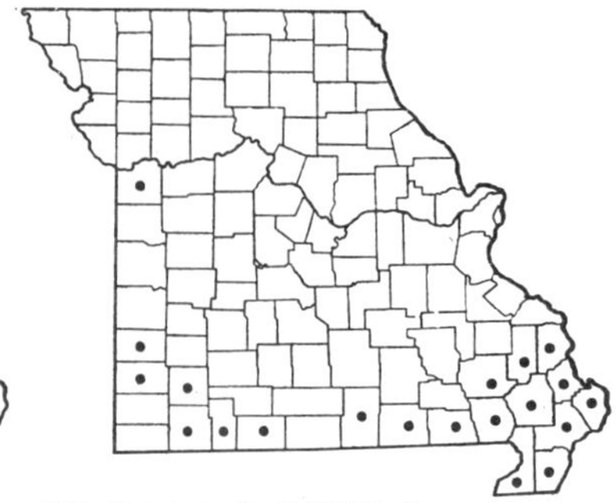

2213 Pluchea camphorata (Stinkweed)

## Tribe IV. **INULEAE**

### 20. **Pluchea** Cass. Marsh Fleabane

Leaves sessile (without leaf-stalk) and clasping the stem; flowers cream-colored; rare species known only from Butler County . . . . . . . . . . . . . . . . . . . . . . . . 1. P. FOETIDA
Leaves with a distinct petiole (leaf-stalk), narrowed and tapering at base; flowers rose-purple or pink; common species encountered . . . . . . . . . . . . . . . . . . . . 2. P. CAMPHORATA

1. **Pluchea foetida** (L.) DC. Marsh Fleabane
Map 2212

Also known as Stinking Fleabane.

Flowers July–September.

Occurs in swampy woods and borders of bayous and ditches. Known only from Butler County, southeastern Missouri (Poplar Bluff, August 5, 1879, *Letterman*).

Ranges from Florida to Texas and Mexico, north to New Jersey and Missouri.

2. **Pluchea camphorata** (L.) DC. Stinkweed
Map 2213

Also called Camphor Weed, Inland Marsh Fleabane.

*Pluchea petiolata* Cass. [P & S]

Flowers August–September.

Occurs in low wet woods, wet meadows, swamps, borders of sloughs, bayous, streams, and lowland or upland sink-hole ponds. Southern Missouri, north to Cape Girardeau, Bollinger, Wayne, Lawrence, and Barton counties, and locally northwest in Jackson County, where introduced.

Ranges from Florida to Texas, north to Delaware, Ohio, Indiana, Illinois, Missouri, and Kansas.

The plant produces a strong disagreeable odor, and when bruised, the odor somewhat resembles that of the scent of skunk or similar musky odors. The species is commonest in the lowlands of southeastern Missouri, where it is found in or bordering woodland of sweet gum, pin and basket oak, swamp cottonwood, and other lowland species. It reappears in the White River region and tributaries of southwestern Missouri, but there its habitats along White River have been mostly exterminated by the impounded waters of the high dams.

### 21. **Antennaria** Gaertn. Pussy's Toes, Ladies' Tobacco, Everlasting

In this genus the male flowers are usually on one plant, the female on another (dioecious). The male (staminate) plants are shorter with smaller flower-heads than the female (pistillate) plants. The pistillate flowers often have reddish or pink style-tips protruding from the heads; the staminate flowers have purple or dark brown anthers, creating a rather spotted effect. Botanists have varying opinions at present as to the actual number of true species, because most of the species can produce seeds without the normal fertilization process (a phenomenon called apomixis), and in some species staminate plants are still unknown. So far as Missouri

---

Plate no. 362. **1.** Erigeron canadensis, × 2/5 (Scribner's). **2.** Pluchea foetida, × 2/5. **3.** Pluchea camphorata, × 2/5; Details from Small, The New York Botanical Garden. **4.** Erigeron divaricatus, × 2/5; Details from Small, The New York Botanical Garden. **5.** Antennaria neglecta, × 2/5; *a.* Female plant; *b.* Male plant.

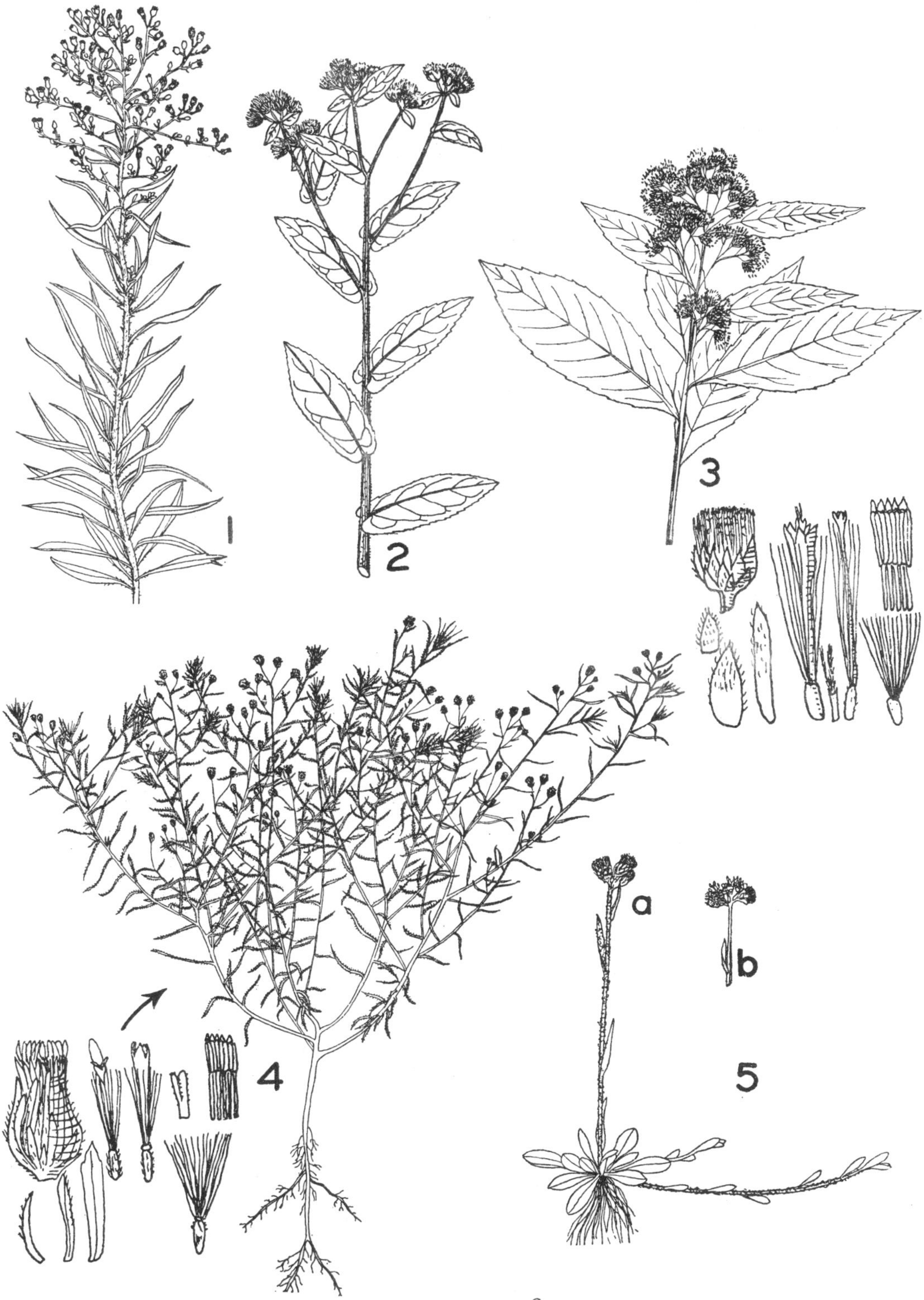


PLATE NO. 362

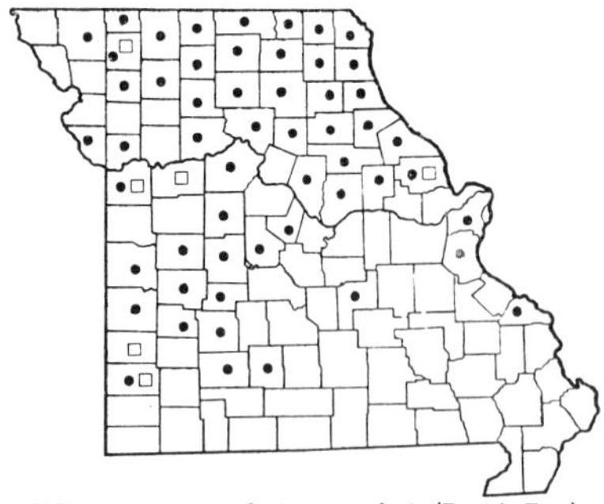

2214 • Antennaria neglecta var. neglecta (Pussy's Toes)
2214 □ Antennaria neglecta var. campestris

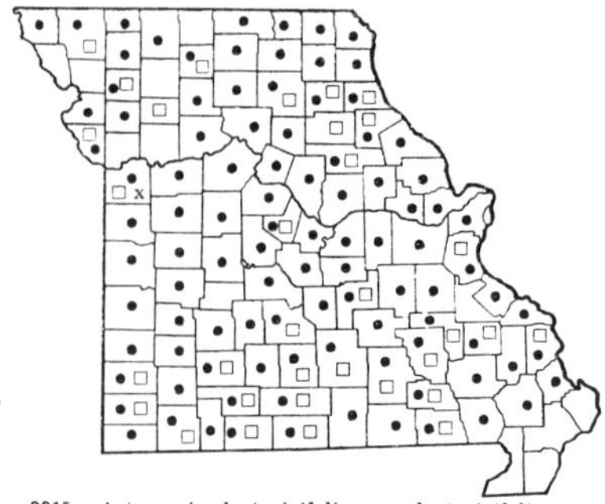

2215 • Antennaria plantaginifolia var. plantaginifolia (Pussy's Toes)
2215 □ Antennaria plantaginifolia var. ambigens
2215 x Antennaria plantaginifolia var. arnoglossa

2216 Anaphalis margaritacea (Pearly Everlasting)

species are concerned, Cronquist's treatment (in Gleason's *New Ill. Fl.*) appears to fit the limits of variation observed by the present author for Missouri specimens.

Rosette-leaves at base of plant with 1 main rib (plainly seen on lower surface), comparatively small, 1.8 cm. or less broad . . . . . . . . . . . . . . . . . . . . . . . . 1. A. NEGLECTA
Rosette-leaves at base of plant with 3 main ribs (plainly seen on lower surface), the larger ones 1.5–5 cm. broad . . . . . . . . . . . . . . . . . . . . . . . . . 2. A. PLANTAGINIFOLIA

1. **Antennaria neglecta** Greene Pussy's Toes Map 2214

Flowers April–June.

Occurs in prairies, meadows, dry open wooded slopes, eroded shale and clay slopes, old fields, and pastures. Northern, central, and western Missouri, chiefly in the glaciated and unglaciated prairie regions mainly north and west of a line drawn from St. Louis, Montgomery, Callaway, Boone, Moniteau, Morgan, Hickory, and Webster counties to Jasper County, locally south and east in Perry and Phelps counties.

Two variations are found in Missouri:

Pistillate flower-heads eventually scattered at different levels on an elongated rachis in a spike- or raceme-like inflorescence; involucre of staminate flower-heads 4–6 mm. high; rosette-leaves narrowed to a stalk-like base, the leaf-blade cuneate-oblanceolate to spatulate, up to 1.3 cm. broad . . . . 1a. A. NEGLECTA var. NEGLECTA
Pistillate flower-heads remaining congested in small compact clusters, the rachis not elongating; involucre of staminate flower-heads 6–8 mm. high; rosette-leaves broad-based, the leaf-blade cuneate-obovate, up to 1.8 cm. broad . . . 1b. A. NEGLECTA var. CAMPESTRIS

1a. **Antennaria neglecta** var. **neglecta** Map 2214

*Antennaria neglecta* Greene [G, P & S, Steyerm.]

This is common throughout Missouri.

Ranges from Nova Scotia and Maine to Ontario and Minnesota, south to Virginia, West Virginia, Ohio, Indiana, Missouri, and Kansas.

1b. **Antennaria neglecta** var. **campestris** (Greene) Steyerm. Map 2214

*Antennaria campestris* Rydb., Bull. Torr. Club 24: 304. 1897. [G]

Scattered within the range in Missouri.

Ranges from British Columbia to Manitoba, south to Oklahoma, east to Michigan and Missouri.

There are intergrading specimens between *A. neglecta* and *A. campestris*, and the differences separating the two are arbitrary. At most, they can be considered as but varietally, rather than specifically distinct. Cronquist (in Gleason's *New Ill. Fl.* 3: 478. 1952) merges the two under *A. neglecta* var. *neglecta*. A final evaluation of their status must await more intensive field and experimental studies.

2. **Antennaria plantaginifolia** (L.) Hook. Pussy's Toes Map 2215

Also called Ladies' Tobacco, Indian Tobacco.

Flowers April–June.

Occurs in acid soils of dry or rocky open wooded slopes, ravines, ridges, and thickets, sometimes in prairies and glades. Throughout Missouri.

The following intergrading variations may be recognized in Missouri:

a. Upper surface of the new rosette-leaves and those at the tips of the stolons (runners) green and glabrous (without hairs) or nearly so from the beginning . . . 2c. A. PLANTAGINIFOLIA var. ARNOGLOSSA
a. Upper surface of the new rosette-leaves and

those at the tips of the stolons (runners) closely hairy from the beginning and remaining so until late in maturity, eventually the weathered leaves becoming glabrous . . . . . *b*

b. Pistillate involucres mostly 5–7 mm. high; mature central corollas of pistillate heads generally 3–4.3 mm. long . . . . .
2a. A. PLANTAGINIFOLIA var. PLANTAGINIFOLIA

b. Pistillate involucres mostly 7–10 mm. high; mature central corollas of pistillate heads generally 4.5–6 mm. long . . . . .
2b. A. PLANTAGINIFOLIA var. AMBIGENS

2a. **Antennaria plantaginifolia** var. **plantaginifolia** Map 2215

*Antennaria plantaginifolia* (L.) Hook. [G, P & S, Steyerm.]

Throughout Missouri.

Ranges from Maine to Minnesota, south to Georgia, Alabama, Missouri, and Oklahoma.

2b. **Antennaria plantaginifolia** var. **ambigens** (Greene) Cron. Map 2215

*Antennaria fallax* Greene [G, P & S, Steyerm.]

*Antennaria fallax* var. *calophylla* (Greene) Fern. [G, Steyerm.]

*Antennaria calophylla* Greene [P & S]

Throughout Missouri.

Ranges from Maine to Ontario and Minnesota, south to Virginia, Tennessee, Arkansas, and Texas.

2c. **Antennaria plantaginifolia** var. **arnoglossa** (Greene) Cron. Map 2215

*Antennaria Parlinii* Fern. [G]

*Antennaria Parlinii* Fern. var. *arnoglossa* (Greene) Fern. [G]

*Antennaria ampla* Bush

Known only from Jackson County, west-central Missouri (high wooded hills, Twyman, May 5, 1915, *Bush 7497*, *7497A*; same locality, *Bush 7119*, *7359*).

Ranges from Nova Scotia and Maine to Ontario, south to Georgia, Tennessee, Indiana, Illinois, and Missouri.

Study of type material of *Antennaria ampla* shows that it has the measurements of the larger pistillate involucres characteristic of both *A. plantaginifolia* var. *ambigens* and var. *arnoglossa*, but the upper surface of the rosette-leaves appears to be nearly glabrous as in var. *arnoglossa*. The summit of the peduncles and the remainder of the peduncles in *A. ampla* are glandless, and not purple-glandular as they often are in var. *arnoglossa*. The length of the pappus (5.5–6 mm. long) and of the achenes (1.6–1.7 mm. long) are closer to those of var. *ambigens* than to var. *arnoglossa*. *Antennaria ampla* apparently combines some of the characters common to var. *ambigens* and var. *arnoglossa*, but is united in the present treatment with var. *arnoglossa*, because of the greater tendency towards glabrous leaves.

The above varieties are separated from each other with difficulty, and there are many intergrading specimens, especially between var. *plantaginifolia* and var. *ambigens*, so far as measurements of pappus, achene, and involucre are concerned. It has, therefore, seemed that these entities at best are treated as varieties, and not as distinct species, as they were so maintained by Fernald in *Gray's Manual*.

The differences proposed by Fernald for separating var. *calophylla* from typical *A. fallax* on the basis of broadly rounded summits of the more round-ovate or suborbicular rosette-leaves do not hold true, and the two extremes of variations of *A. fallax* cannot be distinguished readily from *A. plantaginifolia* var. *plantaginifolia*. All the varieties here treated in the present flora could be united as synonyms under *A. plantaginifolia*, but are tentatively retained in the present flora as varieties pending more detailed field and experimental studies.

The rosette-leaves remain green over winter and constitute a source of food for deer at that time.

## 22. **Anaphalis** DC. Pearly Everlasting

**Anaphalis margaritacea** (L.) Benth. & Hook. Pearly Everlasting Map 2216

Flowers July–September.

Probably introduced and known only from Boone County, central Missouri. Not found since the original collection by Rickett (in the University of Missouri Herbarium.) Of this collection, Dr. H. W. Rickett writes (March 27, 1959) that he collected this specimen along the Hinkson bluffs near Columbia. Since it has never been found again, Dr. Rickett believes that it was probably just a 'seedling that had come up out of some rubbish (the University Farm extends along those bluffs).'

The Missouri material has been referred by Fernald in *Gray's Manual* to var. *intercedens* Hara, the range of which is from Newfoundland to Alaska, south to North Carolina, West Virginia, Ohio, Michigan, Minnesota, Missouri, South Dakota, and New

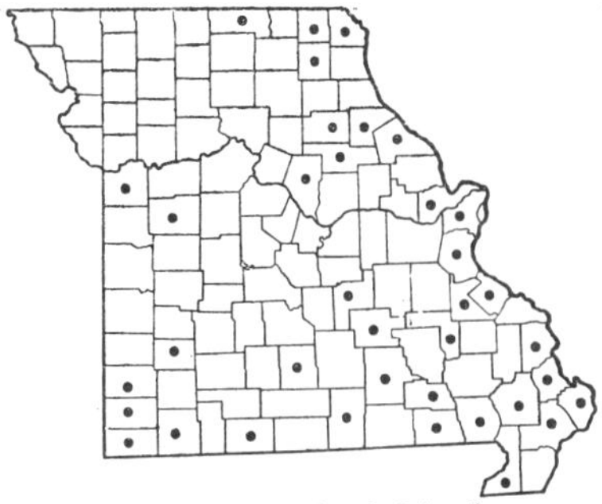

2217 Gnaphalium purpureum (Purple Cudweed)

2218 • Gnaphalium obtusifolium var. obtusifolium (Sweet Everlasting)
2218 □ Gnaphalium obtusifolium var. micradenium

2219 Inula Helenium (Elecampane)

Mexico. Cronquist (in Gleason's *New Ill. Fl.*) does not recognize any segregated varieties.

This plant resembles *Gnaphalium obtusifolium* in appearance, but lacks the licoricelike fragrance of that species. It makes a good plant, when dried, for indoor bouquets and flower arrangements.

## 23. **Gnaphalium** L. Cudweed, Everlasting

Involucre brown to chestnut or purple, the bracts acutely pointed at tip; leaves rounded at the tip, spatulate-oblong; inflorescence narrow, in a spike-like panicle, usually not branched; pappus-bristles united at the base, falling in a ring . . . . . . . . . . . . . . . . . . . 1. G. PURPUREUM

Involucre white or yellowish-white, the bracts mostly rounded or obtuse (blunt) at summit; leaves long-pointed (acuminate) to blunt (obtuse), but not rounded at tip, linear-lanceolate; inflorescence broad, flat or round-topped, branched, in corymbose arrangement; pappus-bristles all distinct, falling separately . . . . . . . . . . . . . . . . . . . . . . . . . 2. G. OBTUSIFOLIUM

1. **Gnaphalium purpureum** L. Purple Cudweed Map 2217

*Gnaphalium purpureum* var. *purpureum* [BB]

Flowers April–June.

Occurs in fallow and cultivated fields, prairies, thickets, rocky ledges, rocky stream banks, and open woodland. Scattered throughout most of Missouri, but not recorded from the northwestern sector.

Ranges from Florida to Texas and California, north to Maine, New York, Ohio, Indiana, Illinois, Missouri, Oklahoma, and Oregon; also tropical America.

A var. *falcatum* (Lam.) T. & G. from the southeastern states is recognized by some authors. This has narrower, linear or linear-oblanceolate leaves about equally hairy on both surfaces.

2. **Gnaphalium obtusifolium** L. Sweet Everlasting Map 2218

Also called Old-field Balsam, Catfoot, Everlasting.

Flowers July–November.

Occurs in fallow fields, pastures, meadows, prairies thickets, open woodland, along roadsides and railroads. Throughout Missouri, and doubtless in every county.

Two varieties occur in the state:

Commonly encountered type; stem covered with a close white hairiness . . 2a. G. OBTUSIFOLIUM var. OBTUSIFOLIUM

Rarely encountered; stem with a glandular hairiness, not covered with a close white felt . . . 2b. G. OBTUSIFOLIUM var. MICRADENIUM

2a. **Gnaphalium obtusifolium** var. **obtusifolium** Map 2218

*Gnaphalium obtusifolium* L. [G, BB, P & S]

Common throughout the state.

Ranges from Nova Scotia to Ontario and Manitoba, south to Florida, Louisiana, and Texas.

2b. **Gnaphalium obtusifolium** var. **micradenium** Weath. Map 2218

---

Plate no. 363. **1.** Antennaria plantaginifolia var. ambigens, × 2/5; *a.* Female plant; *b.* Male plant. **2.** Antennaria plantaginifolia var. plantaginifolia, × 2/5; *a.* Female plant; *b.* Male plant; Details from Small, The New York Botanical Garden. **3.** Gnaphalium purpureum, × 2/5; Details from Small, The New York Botanical Garden. **4.** Anaphalis margaritacea, floral details, highly magnified; Details from Small, The New York Botanical Garden.

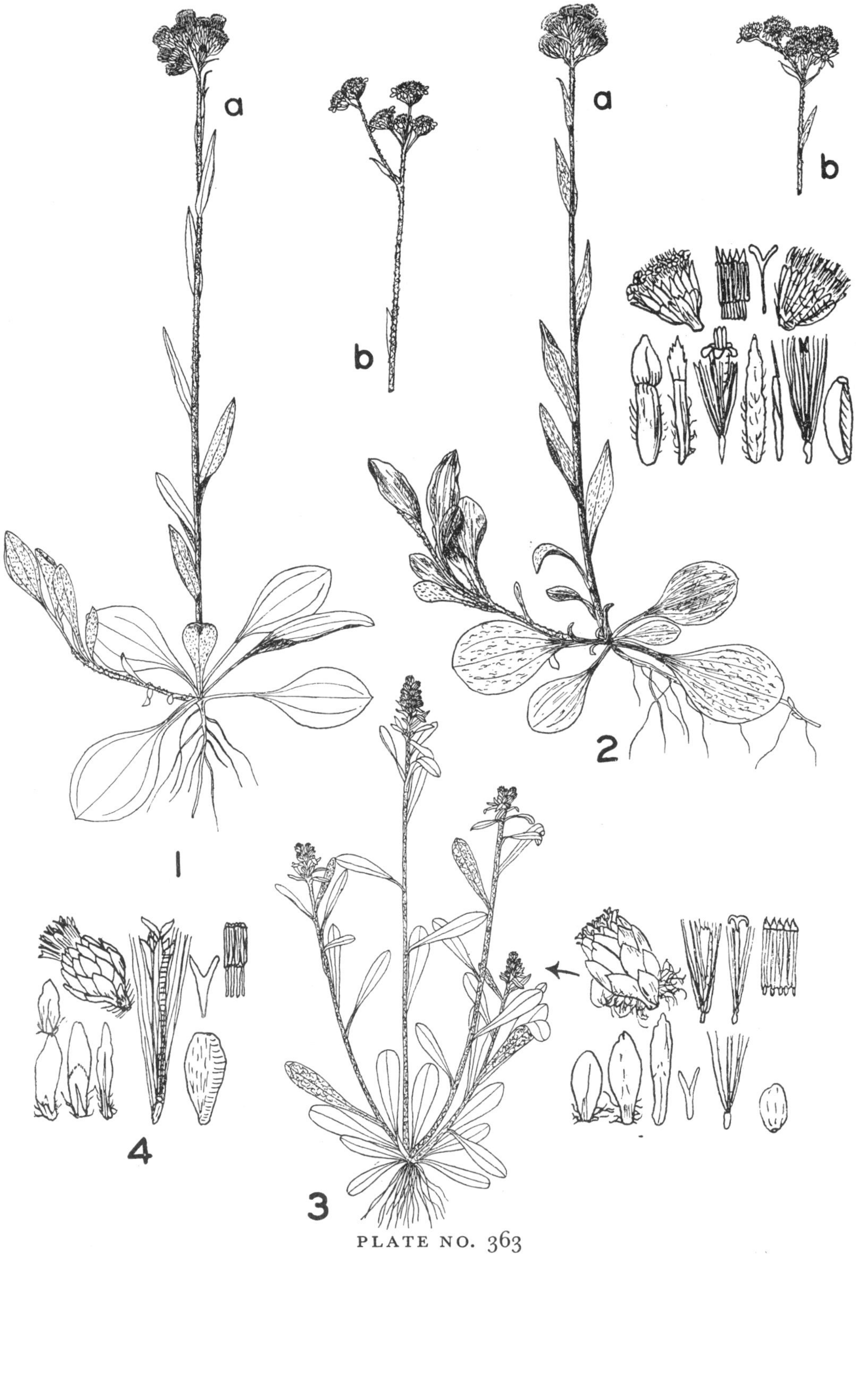


PLATE NO. 363

Known only from acid soils of open woodland in the southeastern Ozarks in Ste. Genevieve (open rocky, oak-pine, sandstone woods along Rough Creek, tributary of Jonca Creek, south of Weingarten, October 31, 1936, *Steyermark 20955*; dry upland pine woods, Pickle Springs, October 5, 1930, *Steyermark 1934*) and Shannon (rocky chert oak-hickory woods, 6 mi. north of Eminence, October 9, 1932, *Steyermark 8147*) counties.

Ranges from Maine to Michigan, south to South Carolina, Kentucky, and Missouri.

The var. *micradenium* has been found thus far in Missouri in acid rocky (chert or sandstone) soils of oak pine and oak-hickory woodland of the southeastern Ozarks.

The plants of *G. obtusifolium* possess a characteristic fragrant, balsamic odor, which readily identifies the species. The leaves are eaten by the white-tailed deer and wild turkey in the Ozarks. It is reported that the plant has been used to treat the condition of flux (Deam, *Fl. Ind.* p. 954. 1940) by drinking milk in which this plant had been boiled.

The leaves are eaten by wild turkey, and in the winter are a source of food for the white-tailed deer.

## 24. **Inula** L.

**Inula Helenium** L. Elecampane Map 2219

Flowers late June–September.

Occurs in fields, along roadsides, waste places, and open ground. Known from Monroe, Boone, Jackson, and Greene counties in southern and central Missouri.

Native of Europe; introduced and naturalized in North America from Nova Scotia to Ontario and Minnesota, south to North Carolina and Missouri.

This plant was at one time more commonly grown by the early pioneers in this country for its presumed medicinal qualities. The mucilaginous root was used to treat pulmonary diseases. It contains inulin which is a source of fructose, a sugar which can be eaten by diabetic persons. A confection, of rather general use in Europe, was made from the cooked roots, later candied to eat. The fresh roots, preserved in sugar, were sometimes used in England to cure digestive upsets.

## Tribe V. **HELIANTHEAE**

## 25. **Iva** L. Marsh Elder

Flower-heads in the axils of leafy bracts, arranged in a spike-like inflorescence; bracts of the involucre 3–5; upper part of stem with spreading rough hairs; lower surface of leaves rough-hairy . . . 1. I. CILIATA

Flower-heads not in the axils of leafy bracts, arranged in a panicle-like, branched inflorescence; bracts of the involucre 10, the 5 inner ones surrounding the achenes; upper part of stem either glabrous or with hairs mostly ascending or appressed (pressed against or lying parallel to surface), soft; lower surface of leaves soft-downy with short hairs . . . . . . . . . . . . . . 2. I. XANTHIFOLIA

1. **Iva ciliata** Willd. Marsh Elder Map 2220

Also called Sump Weed.

Flowers July–October.

Occurs in alluvial soils along streams, borders of ponds and sloughs, river bottom prairies and meadows, low fields in valleys, along roadsides, and railroads. Chiefly found along the counties bordering the Missouri, Mississippi, Chariton, Fabius, Grand, Osage, and other large streams, absent from most of the Ozarks and extreme northwestern sector.

Two varieties are known in the state:

Common type encountered; achenes (1.9) 2.3–3.8 mm. long, (1.5) 1.8–2.6 mm. broad, the smaller without ribs, the larger usually 3-ribbed on each face . . . . . . . 1a. I. CILIATA var. CILIATA

Extinct fossil type; achenes 4.8–9.3 mm. long, 3.2–5.7 mm. broad, 3–7-ribbed on each face 1a. I. CILIATA var. MACROCARPA

1a. **Iva ciliata** var. **ciliata** Map 2220

*Iva ciliata* Willd. [G, BB, P & S]

Absent from northwestern Missouri and from most of the Ozarks, elsewhere in southern, central, and eastern Missouri.

Ranges from Indiana to Nebraska, south to Mississippi, Louisiana, Texas, and New Mexico; introduced eastward to New England.

1b. **Iva ciliata** var. **macrocarpa** Blake Map 2220

Known only from southwestern Missouri in Barry County (Montgomery shelter, 1925, *S. C. Dellinger*, holotype in Nat. Arb. Herb.).

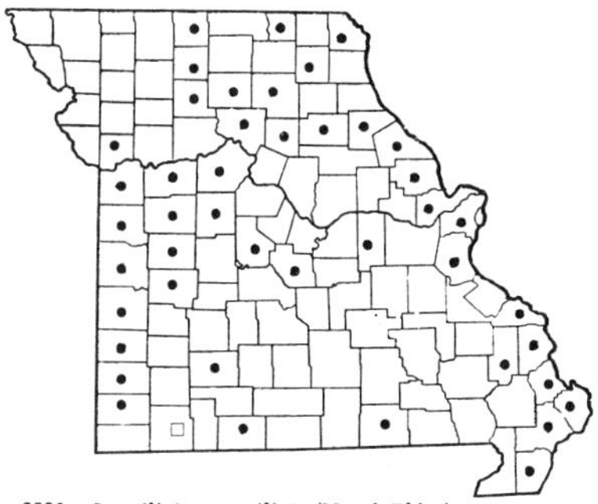

2220 • Iva ciliata var. ciliata (Marsh Elder)
2220 □ Iva ciliata var. macrocarpa (Fossil Marsh Elder)

2222 Iva xanthifolia
• • •
2221 Excluded species

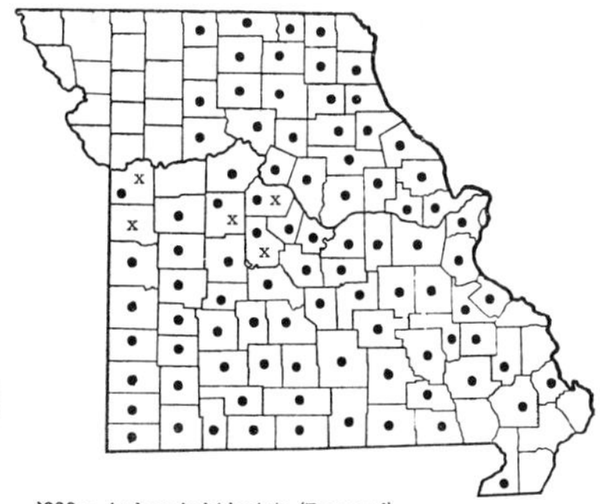

2223 • Ambrosia bidentata (Ragweed)
2223 x Ambrosia bidentata X trifida

Also known from Kentucky and Arkansas.

This is a form with large achenes, now extinct, known only from Indian rock-shelters of the bluff-dwellers of the Ozarks and Kentucky. They represent large-fruited strains of *I. ciliata* which had been developed by the bluff-dwellers as sources of food (Blake in Rh. 41: 81–86. 1939).

*Iva ciliata* var. *ciliata* sheds large quantities of pollen which is considered an important cause of hay fever.

A recent study of the genus by Dr. R. C. Jackson indicates that the correct name of the species is *Iva annua* L.

2. **Iva xanthifolia** Nutt. Map 2222

Flowers June–October.

Occurs in alluvial soils, waste ground, and along railroads. Rare, and known only from St. Louis (*Muehlenbach 1437*), Jackson, Buchanan, and Atchison counties in central and northwestern Missouri; native in the northwestern counties, introduced in St. Louis County.

Ranges from Wisconsin and Manitoba to Alberta and Montana, south to Missouri, Oklahoma, Texas, and New Mexico; introduced east to Quebec, New England, and New Jersey.

This species produces large amounts of wind-borne pollen, and is one of the most important causes of autumn hay fever, along with the ragweeds, over most of its range. Both the pollen and the leaves are known to cause dermatitis. Cows feeding on the leaves give milk with a bitter flavor. Fortunately, the rarity of the plant in Missouri should cause little concern.

## 26. **Ambrosia** L. Ragweed

Plants of this genus are the principal cause of summer and autumn hay fever in Missouri.

a. Leaves 3–5-lobed or undivided . . . . . . . . . . . . . . . . . . . . . . . . *b*
  b. Main leaves of stem 4–10 mm. broad and undivided; stem mostly 3–9 dm. tall; staminate (male) heads sessile (without stalks) arranged in a solitary spike at the end of the stem . . . 1. A. BIDENTATA
  b. Main leaves of stem much more than 10 mm. broad, usually 3–5-lobed, sometimes without lobes and undivided; stem 5–60 dm. tall; staminate (male) heads on short stalks (pedicels) in 1 or more slender racemes . . . . . . . . . . . . . . . . . . . . . . 2. A. TRIFIDA
a. Leaves deeply cut into narrow lobes nearly to the midrib, the lobes remaining uncut or again lobed . . *c*
  c. Upper surface of leaves rough-hairy; leaves sessile (without a stalk) or nearly so, only simply lobed; perennial plants with ascending branches and creeping rootstocks; staminate (male) involucres covered with bristly appressed (lying parallel to surface) hairs; fruit without prickles or with blunt projections. . . . . . . . . . . . . . . . . 4. A. CORONOPIFOLIA
  c. Upper surface of leaves not harsh or rough to the touch; leaves on petioles (stalked), varying from simply and once lobed to more commonly twice or more lobed with the lobes again lobed; annual plants with widely spreading branches and no creeping rootstocks; staminate (male) involucres glabrous or hairy; fruit with usually 4–7 sharp-pointed projections. . . 3. A. ARTEMISIIFOLIA

1. **Ambrosia bidentata** Michx. Ragweed Map 2223

Sometimes called Lance-leaf Ragweed.

Flowers July–October.

Occurs in prairies, fields, waste ground, borders of ponds, pastures, roadsides, and along railroads. Common throughout southern, central, and eastern Missouri, west to Mercer, Livingston, Carroll, Saline, Johnson, and Jackson counties.

Ranges from Ohio and Indiana to Minnesota and Nebraska, south to Kentucky, Louisiana, and Texas.

This species is known to hybridize with *A. trifida*, and plants of this hybrid have been collected in Jackson, Cass, Morgan, Pettis, and Cooper counties of west-central Missouri.

From July–September, during its main flowering season, the wind-borne pollen of this plant causes hay fever locally, but is considered to be of less importance than either *A. trifida* or *A. artemisiifolia*. The fruits are eaten by wild turkey.

2. **Ambrosia trifida** L. Horse Weed Map 2224

Also known as Giant Ragweed, Great Ragweed, Buffalo Weed.

Flowers July–September.

Occurs in usually rich alluvial soils in thickets, low woods, along streams, sloughs and ponds, fallow fields, waste ground, roadsides, and along railroads. Throughout Missouri in every county.

The following variations occur in Missouri:

a. All the leaves unlobed . . 2b. A. TRIFIDA var. TRIFIDA f. INTEGRIFOLIA

a. Some or all of leaves 3–5 lobed . . . . *b*

b. Common type encountered; petioles (leaf-stalks) of at least the upper leaves slightly wing-margined with a strip of leafy green tissue; fruits 6–12 mm. long with a beak 2–4 mm. long, the ribs ending in short-pointed projections (tubercles) . . . . . 2a. A. TRIFIDA var. TRIFIDA f. TRIFIDA

b. Rarely encountered; petioles (leaf-stalks) slender, wingless; fruits 3–7 mm. long with a beak 0.5–1 mm. long, the ribs ending in blunt projections or these absent . . . 2c. A. TRIFIDA var. TEXANA

2a. **Ambrosia trifida** var. **trifida** f. **trifida** Map 2224

*Ambrosia trifida* L. [G, P & S]

*Ambrosia trifida* var. *trifida* [BB]

This is the common variation found in every county.

Ranges from Quebec to British Columbia, south to Florida, Louisiana, Texas, Arizona, and Mexico.

2b. **Ambrosia trifida** var. **trifida** f. **integrifolia** (Muhl.) Fern. Map 2224

*Ambrosia trifida* f. *integrifolia* (Muhl.) Fern. [BB]

*Ambrosia trifida* var. *integrifolia* (Muhl.) T. & G. [P & S]

Rarely encountered, known from central and southern Missouri in St. Louis, Jefferson, Franklin, Texas, and Jackson counties.

2c. **Ambrosia trifida** var. **texana** Scheele Western Giant Ragweed, Texas Great Ragweed Map 2224

*Ambrosia trifida* var. *aptera* (DC.) Ktze.

*Ambrosia aptera* DC. [P & S]

Known from St. Louis (railroad embankment near St. Louis, September 25, 1886, *Eggert*) and Jackson (bottoms, Courtney, October 16, 1926, *Bush 11220*) counties, central Missouri.

Ranges from Illinois and Missouri to Mississippi, Arizona, and Mexico.

*Ambrosia trifida*, together with *A. artemisiifolia*, account for most of the hay fever cases recorded in Missouri and elsewhere. *Ambrosia trifida*, however, far outranks the other species in numbers of individual plants, often forming extensive acres of thousands of plants over wide stretches of river and valley bottoms in lowland woods and thickets. Plants can be destroyed by herbicides and brush poisons, or, if cut down while still in the vegetative stage, pollen formation can be prevented. As the plant is an annual, it can also be destroyed, where not occurring in too great quantity, by pulling up the stem.

In medicine this plant was used as an astringent in chronic catarrhal infections. The leaves are eaten by white-tailed deer. The early Bluff-dwellers of the Ozarks cultivated the species for the fruits, which, in their cultivated strains, were of larger size than the modern wild type.

3. **Ambrosia artemisiifolia** L. Common Ragweed Map 2225

Also called Roman Wormwood, Hog Weed, Bitter Weed.

Plate no. 364. **1.** Gnaphalium obtusifolium, $\times$ 2/7; *a.* Head of florets, $\times$ 1 5/7. **2.** Inula Helenium, $\times$ 2/7; Details from Small, The New York Botanical Garden. **3.** Iva ciliata, $\times$ 2/7; After Gleason, The New York Botanical Garden. **4.** Iva xanthifolia, $\times$ 2/7. **5.** Ambrosia bidentata, $\times$ 2/7; *a.* Fruit with subtending bract, $\times$ 1 1/7; *b.* Male flower-head, $\times$ 6/7; After Gleason, details from Small, The New York Botanical Garden. **6.** Ambrosia artemisiifolia var. elatior, f. villosa, $\times$ 2/7 (Scribner's). **7.** Ambrosia trifida, $\times$ 2/7 (Scribner's). **8.** Ambrosia coronopifolia, $\times$ 2/7; After Gleason, The New York Botanical Garden. **9.** Franseria discolor, $\times$ 1; *a.* Fruit, $\times$ 6; After Britton and Brown, details from Small, The New York Botanical Garden. **10.** Franseria acanthicarpa, $\times$ 2/7; *a.* Fruit, $\times$ 1 5/7; After Britton and Brown, details from Small, The New York Botanical Garden.

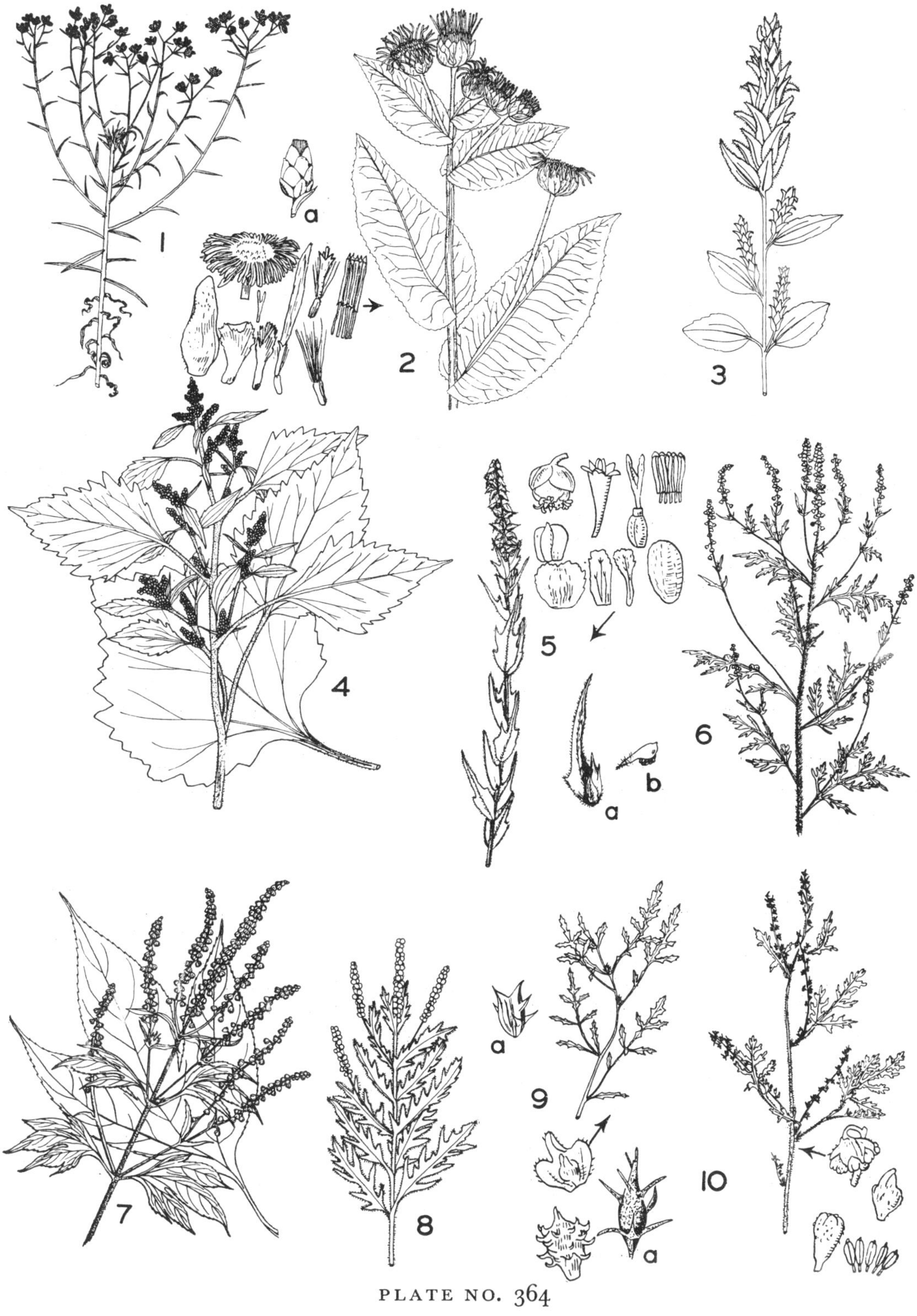


PLATE NO. 364

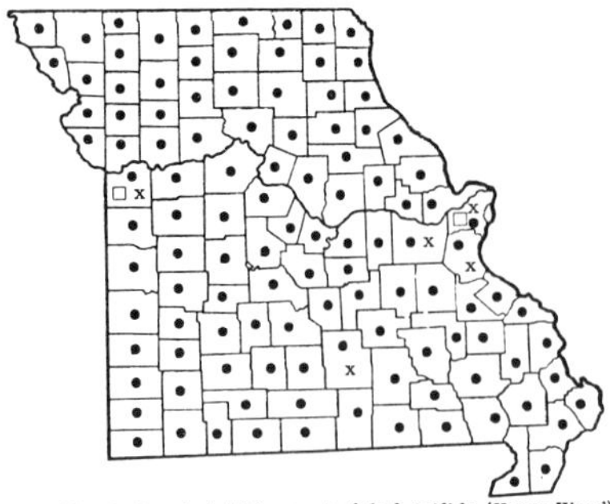

2224 • Ambrosia trifida var. trifida f. trifida (Horse Weed)
2224 x Ambrosia trifida var. trifida f. integrifolia
2224 □ Ambrosia trifida var. texana

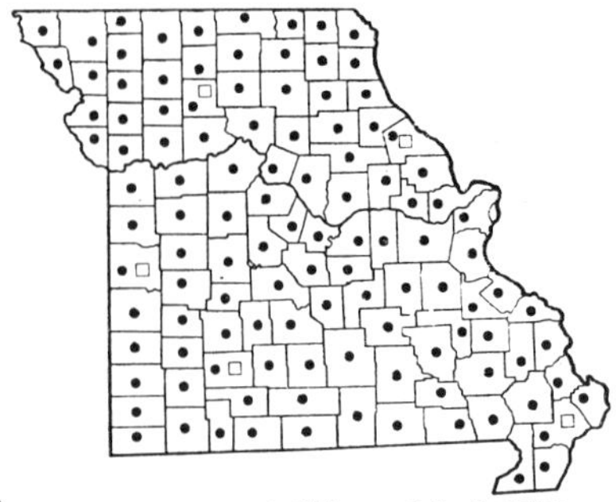

2225 • Ambrosia artemisiifolia var. elatior f. elatior (Common Ragweed)
2225 □ Ambrosia artemisiifolia var. elatior f. villosa

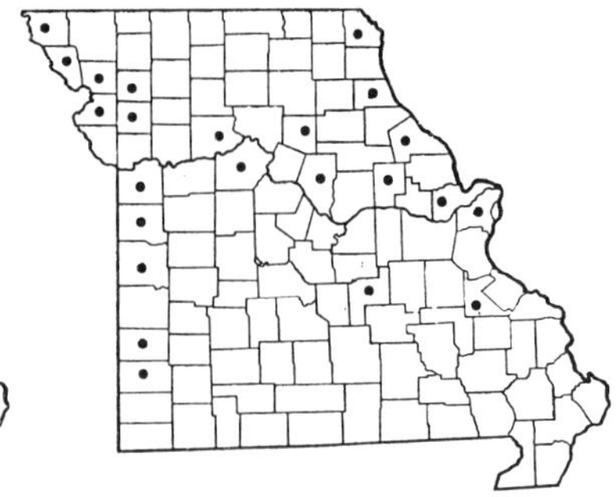

2226 Ambrosia coronopifolia (Western Ragweed)

Flowers July–November.

Occurs in fallow and cultivated fields, waste and open ground, pastures, rocky glades, eroded broken ground, roadsides, and along railroads. Throughout Missouri and in every county.

The following variations occur in the state:

Plant glabrous (without hairs) or the hairs appressed (pressed against or lying parallel to the surface) . . . . . . . 3a. A. ARTEMISIIFOLIA var. ELATIOR f. ELATIOR

Plant with spreading hairs . . . . . . . 3b. A. ARTEMISIIFOLIA var. ELATIOR f. VILLOSA

3a. **Ambrosia artemisiifolia** var. **elatior** (L.) Descourt. f. **elatior** Map 2225

*Ambrosia artemisiifolia* var. *elatior* (L.) Descourt. [G, P & S]

This is the commoner variation in Missouri.

Ranges from Florida to Texas and Mexico, north to Nova Scotia and British Columbia.

3b. **Ambrosia artemisiifolia** var. **elatior** f. **villosa** Fern. & Grisc. Map 2225

Scattered throughout the range in Missouri and elsewhere.

*Ambrosia artemisiifolia* and its varieties and forms, like *A. trifida*, is widespread in North America and the two species account for most of the hay fever cases in the eastern and midwestern states.

Since it is outranked by *A. trifida* in numbers of individual plants to a square mile, the total amount of pollen it produces is somewhat less than that produced by *A. trifida*. Cows are usually kept out of pastures or fields where this plant abounds, as it has a nauseating taste and effect when eaten. The fruits are often eaten by wild turkey.

Hybrids occasionally occur between *Ambrosia trifida* and *A. artemisiifolia* (*A.* × *Helenae* Rousseau), but have not been recorded as yet from Missouri (See Wagner, Rh. 60: 309–16. 1958).

4. **Ambrosia coronopifolia** T. & G. Western Ragweed Map 2226

*Ambrosia psilostachya* var. *coronopifolia* (T. & G.) Farw. [G]

*Ambrosia psilostachya* [of BB, P & S], not DC.

Flowers August–October.

Occurs on dry rocky prairies, glades, loess hills, pastures, roadsides, and along railroads. Chiefly in the counties bordering the Missouri, Mississippi, and a few other large streams in western, central, and northeastern Missouri, locally south in St. Francois and Phelps counties; absent from most of the Ozarks and north-central section.

Ranges from Michigan to Saskatchewan and Montana, south to Louisiana, Texas, and Mexico; introduced eastward to Quebec and New England, and Europe.

This species has been confused with *A. psilostachya* DC., considered to be native of Mexico and Texas, and of hexaploid instead of tetraploid constituency (see Wagner and Beals, Rh. 60: 177–204. 1958). Although it sheds large quantities of pollen, *A. coronopifolia* is of more localized occurrence in Missouri and is of less importance in causing hay fever than either *A. trifida* or *A. artemisiifolia*. The fruits are eaten by wild turkey.

A hybrid, *A.* × *intergradiens* Wagner, between *A. coronopifolia* and *A. artemisiifolia* has recently been described from Michigan (Rh. 60: 191. 1958), but thus far has not been found in Missouri. It is probable that the intensive search for it in the state where the ranges of the two species come together will reveal its occurrence.

2227 Franseria discolor

2228 Franseria acanthicarpa

2229 Xanthium spinosum var. spinosum (Spiny Cocklebur)

## 27. **Franseria** Cav.

Shinners (Field and Lab. 17: 173. 1949) believes that *Franseria* can be regarded at most as a subgenus of *Ambrosia*, since it cannot be distinguished from that genus except by the mature fruit and the greater development of teeth or hooked spines. Final judgment on the generic status of *Franseria* is open to future monographic taxonomic work.

Perennial plant with creeping underground rootstocks; lower surface of leaves densely gray-hairy; fruit with terete (rounded in cross-section and with curved sides) spines; staminate involucres short-hairy, 3.5–5 mm. broad . . . . . . . . . . . . . . . . . . . . . . . 1. F. DISCOLOR

Annual plant with small tap-root; lower surface of leaves with appressed (pressed against or lying parallel to surface) hairs; fruit with flattened spines; staminate involucres glabrous (without hairs) or with appressed hairs, 2–4 mm. broad . . . . . . . . . . . . . . . . . 2. F. ACANTHICARPA

1. **Franseria discolor** Nutt. Map 2227

*Ambrosia tomentosa* Nutt. [Shinners]

Flowers July–September.

Introduced along railroads and known only from St. Louis County, east-central Missouri (North St. Louis freight yard of Burlington Railroad, August 14, 1954, *Muehlenbach 349*; September 19, 1954, *Muehlenbach 439*; September 18, 1955, *Muehlenbach 809*; Carrie Avenue freight yard of Rock Island Railroad, St. Louis, August 28, 1955, *Muehlenbach 738*).

Ranges from Wyoming to Arizona, east to Wisconsin, Missouri, and Illinois.

This species sheds large amounts of pollen, and in many parts of its geographical range is considered one of the worst causes of hay fever.

2. **Franseria acanthicarpa** (Hook.) Coville Map 2228

*Ambrosia acanthicarpa* Hook. [Shinners]

Flowers August–October.

Introduced along railroads in Jackson County, west-central Missouri (waste ground, Courtney, September 27, 1935, *Bush 15221*).

Ranges from Alberta to California, east to Manitoba, Minnesota, Nebraska, Missouri, Kansas, and Texas.

Like *F. discolor*, this species sheds large amounts of pollen which, where the species is abundant, is one of the worst causes of hay fever. Fortunately, neither species is more than a casual introduction in Missouri, and therefore, of no importance from the standpoint of hay fever.

## 28. **Xanthium** L. Cocklebur

Botanists are not in agreement as to the actual number of true species in this genus. Some of them interpret the genus as containing, besides the well-marked *X. spinosum*, only one additional variable species, *X. strumarium* L. or *X. orientale* L. Others who have monographed the group regard the genus as comprising a number of valid species. Since the diagnostic characters for separating the species are found in the mature fruits, it is essential for identification purposes to have ripe fruiting specimens.

The following treatment is based upon characters taken from the monographs of Millspaugh & Sherff (Field Mus. Nat. Hist. Bot. Ser. 4: 9–54. 1919), Widder (Repert. Spec. Nov. Beih. 20: 1–222. 1923), and the studies of Fernald (Rh. 48: 66–74. 1946). Since there are sound reasons for accepting the concept of the existence of several good species (see Rh. 48: 70–74. 1946; Bot. Gaz. 59: 474–83. 1915), the present author prefers to regard at least some of the Missouri species as valid recognizable ones, until proved to the contrary by future studies.

a. A 3-parted spine present at the base of the leaf; leaves narrowed at both ends . . . . . 1. X. SPINOSUM
a. No spine present at the base of the leaf; leaves with rounded, heart-shaped, or triangular bases. . . . *b*
 b. Prickles of bur so dense and crowded that their own bases and the surface of the body of the bur are not visible; mature bur (including beaks and prickles) 3–4 cm. long and 2–3 cm. thick; beaks at one end of bur 6–11 mm. long; prickles of bur 7–9 mm. long . . . . . . . . 8. X. SPECIOSUM
 b. Prickles not dense and crowded enough to hide their own bases nor the surface of the body of the bur; mature bur (including beaks and prickles) 0.9–3 (rarely 4) cm. long and 0.6–2 (rarely 4) cm. thick; beaks at one end of bur 1–7 (rarely 8) mm. long; prickles 2–7 (rarely 10) mm. long . . . . *c*
  c. Body of bur glabrous (without hairs), nearly glabrous, or covered with short hairs or short gland-tipped hairs. . . . . . . . . . . . . . . . . . . . . . . . . . *d*
   d. Beaks at maturity abruptly bent near middle, turned abruptly inward and strongly overlapping . . . . . . . . . . . . . . . . . . . . . 2. X. INFLEXUM
   d. Beaks at maturity either erect, arching, or curved in, but not abruptly bent near middle nor strongly overlapping . . . . . . . . . . . . . . . . . . . . *e*
    e. Body of mature bur glabrous or nearly so; spaces between the prickles broader than the bases of the prickles; prickles visible on one side (face) of bur (about $\frac{2}{3}$ its whole circumference) 50–150, smooth or nearly so or remotely glandular . . . . . . . 3. X. CHINENSE
    e. Body of mature bur mostly short-hairy; spaces between the prickles about as broad as the base of the prickles; prickles visible on one side (face) of bur (about $\frac{2}{3}$ its whole circumference) 200 or more, usually hairy and glandular below the middle . . . . 4. X. PENSYLVANICUM
  c. Body of bur noticeably hairy. . . . . . . . . . . . . . . . . . . . . . . *f*
   f. Prickles and upper portion of beaks of bur much narrower than the length of their basal hairs. . . . . . . . . . . . . . . . . . . . . . . . . . *g*
    g. Ripe burs deep brown to rusty-brown; beaks of ripe bur narrow and awl-shaped (subulate), 1.2–2 mm. thick at base, $\frac{1}{4}$–$\frac{1}{3}$ as thick as length of beak; prickles mainly 4–7 mm. long . . . . . . . . . . . . . . . . . . . . 5. X. ITALICUM
    g. Ripe burs pale brown to drab; beaks of ripe bur stout, 2–3 mm. thick at base, $\frac{2}{5}$–$\frac{3}{5}$ as thick as length of beaks; prickles mainly 2.5–5 mm. long . . . . . . . . 6. X. VARIANS
   f. Prickles and upper portion of beaks of bur as broad as or broader than the length of their hairs. . . . . . . . . . . . . . . . . . . . . . . . . . *h*
    h. Rarely encountered; only 12–30 rather stout prickles visible on one side (face) of bur (about $\frac{2}{3}$ of its whole circumference); body of bur covered with glandular pubescence 7. X. WOOTONI
    h. Commonly encountered; about 200 slender awl-shaped (subulate) prickles visible on one side (face) of bur (about $\frac{2}{3}$ of its whole circumference); body of bur with dense pale hairiness without glandularity or glabrous . . . . . . . . . . 4. X. PENSYLVANICUM

1. **Xanthium spinosum** L. var. **spinosum**
Spiny Cocklebur Map 2229
*Xanthium spinosum* L. [G, BB, P & S]
Fruits October–November.

Occurs in waste ground, fields, and along roadsides. Known only from St. Louis County, east-central Missouri (North St. Louis, November, 1892, *Eggert*; Florissant road north of Warne Ave., October 20, 1897, *274*, *Glatfelter Herb.*; on free place between Fairground and Florissant Ave., St. Louis, 1891, *Demetrio*; streets, St. Louis, October 23, 1893, *Eggert*).

Native of Europe; introduced and naturalized in North America from Maine and Ontario to Iowa, south to Virginia, Indiana, Illinois, Missouri, and Texas; British Columbia to California and New Mexico; tropical America.

A variation lacking the spines at the base of the leaves is known as var. *inerme* Bel, but has not been recorded from Missouri.

2. **Xanthium inflexum** Mackenz. & Bush
Map 2230
*Xanthium* × *Bushii* Widder, based on *Bush 1916* from bottoms, Courtney, Jackson Co., October 5, 1903, holotype.
Fruits September–November.

Occurs in alluvial soils and sandy bottoms bordering the Missouri and Grand Rivers of central and

2230 Xanthium inflexum

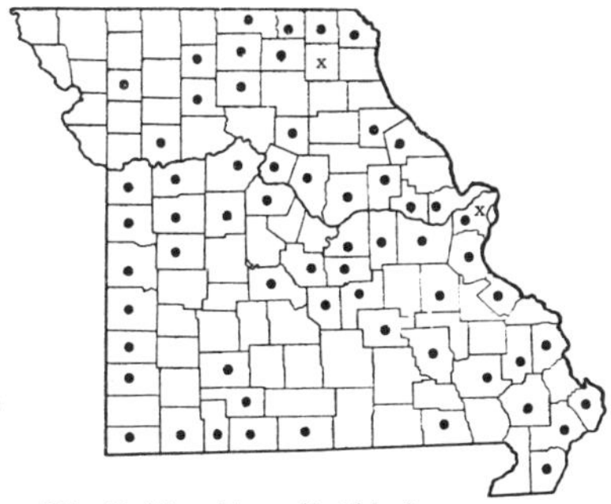

2231 • Xanthium chinense (Cocklebur)
2231 x Xanthium chinense X pensylvanicum

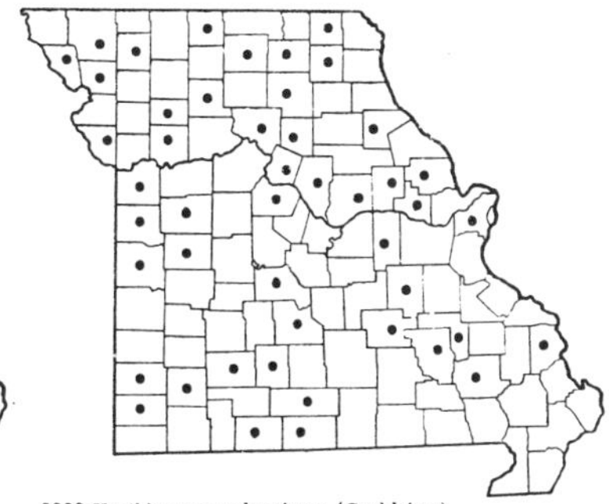

2232 Xanthium pensylvanicum (Cocklebur)

northwestern Missouri, in St. Louis, Daviess, Nodaway, Ray, Lafayette, and Jackson counties.

Ranges from Illinois to Missouri and Arkansas.

The specimen of *Bush 1916* from Jackson County, holotype of *X.* × *Bushii*, was considered by Widder as a hybrid between *X. inflexum* and *X. pensylvanicum* (*X. saccharatum*). The same collection was identified as *X. inflexum* by Millspaugh and Sherff.

3. **Xanthium chinense** Mill. Map 2231

*Xanthium globosum* Shull [G, P & S]

*Xanthium strumarium* L. var. *glabratum* (DC.) Cron., in part [BB]

Fruits August–October.

Occurs in alluvial soils mainly along streams, brackish and saline spring branches, sloughs, ponds, low wet open woods, alluvial and fallow fields, wet meadows, waste ground, and roadsides. Throughout Missouri, but mostly not recorded from the northwestern sector.

Ranges from Florida and Mexico, north to Quebec, Ontario, Michigan, Wisconsin, Minnesota, Nebraska, and Manitoba; West Indies.

The differences stated in *Gray's Manual* to separate *X. globosum* from *X. chinense* are based primarily on the number of prickles visible on one face, in *X. globosum* 50–80 being visible, whereas in *X. chinense* 100–150 are visible. In general, the bur of *X. globosum* averages slightly smaller (1.3–2 cm. long and 1–1.3 cm. thick) than that of *X. chinense* (2–2.5 cm. long, and 1–1.8 cm. thick), but the differences as stated do not always hold true, and many specimens are intermediate between the two and difficult to place. This is especially seen in plants having small burs of *X. globosum* but 100 or more prickles on one face as characteristic of *X. chinense*, or in larger burs the size of *X. chinense* but with 50–80 prickles characteristic of *X. globosum*. The prickles are often only 3 mm. long on mature burs, varying up to 6 mm. in length.

The species hybridizes with *X. pensylvanicum*, putative hybrids being known from St. Louis and Knox counties.

4. **Xanthium pensylvanicum** Wallr. Map 2232

*Xanthium strumarium* L. var. *glabratum* (DC.) Cron., in part [BB]

*Xanthium saccharatum* Wallr.

Fruits August–October.

Occurs in cultivated and fallow fields, alluvial sandy and muddy soils, gravel bars along streams, saline spring branches, sloughs, and ponds, river bottom prairies, waste places, roadsides, and along railroads. Throughout Missouri and probably in every county.

Ranges from Florida to Texas and California, north to Quebec, Ontario, Michigan, Wisconsin, Minnesota, and Manitoba.

This species hybridizes with *X. chinense* and with *X. italicum*. It is somewhat intermediate in characters between *X. chinense* and *X. italicum*.

The older plants are seldom eaten by grazing animals because of the bitter taste. But seedling plants, eaten by swine, sheep, and cattle, are responsible for poisoning, the causative agent being a toxic glucoside. The barbs on the fruit also set up mechanical injuries, their indigestible parts matting together into balls in the stomach and intestines of the animals and eventually causing death. The burs also get entangled on the furry coats of grazing animals, and can cause considerable annoyance to owners of sheep and goats. The pollen of this and other species of *Xanthium* can mingle with that of ragweed and act as an irritant to hay fever sufferers.

5. **Xanthium italicum** Moretti Map 2233

*Xanthium strumarium* L. var. *canadense* (Mill.) T. & G. [BB]

*Xanthium acerosum* Greene

Fruits September–November.

2233 • Xanthium italicum
2233 x Xanthium italicum X pensylvanicum

2234 Xanthium varians

2235 Xanthium Wootoni

Occurs in fields, waste ground, roadsides, and along railroads. Throughout Missouri, but scattered and less common than *X. chinense* or *X. pensylvanicum.*

Ranges from Florida to Mexico, north to Quebec, Ontario, Manitoba, Saskatchewan, and Alberta; West Indies; South America; Hawaii; Europe.

Specimens from Jackson County (Courtney, August 30, 1926, *Bush 10838, 10839, 10840*) identified as *X. acerosum,* appear to be conspecific with *X. italicum* in having the longer prickles, larger beak and thicker body of the bur of that species as compared with *X. varians.*

Millspaugh and Sherff consider *X. acerosum* a distinct species, stating that it is intermediate between *X. pensylvanicum* and certain forms of *X. speciosum,* while Widder places it under *X. pensylvanicum* (*X. saccharatum*) as a synonym.

6. **Xanthium varians** Greene Map 2234
*Xanthium echinatum* [of P & S], not Murr.
Fruits July–October.

Occurs in waste places and along railroads. Known from St. Louis, Jefferson, Jackson (introduced, Sheffield, June 13, 1906, *Bush 3956*), and Jasper counties. The Jackson County record cited was identified by Fernald at the Gray Herbarium.

Ranges from Quebec to Alberta, south to New York, Iowa, North Dakota, Saskatchewan, and Oregon; introduced in Missouri.

7. **Xanthium Wootoni** Cockerell Map 2235
Fruits September–November.

Occurs along sandy alluvial banks of the Mississippi River, where known only from New Madrid County, southeastern Missouri (sandy banks of the Mississippi River, T21N, R14E, sect. 4 and southeast of it, 5–6 mi. southeast of Marston, October 31, 1956, *Steyermark 83426*).

Native of the western United States (Washington, New Mexico); introduced east to Massachusetts, Minnesota, and Missouri.

The leaves of this species are usually coarsely toothed, but the Missouri collection cited has the margins only inconspicuously fine-toothed to nearly entire.

8. **Xanthium speciosum** Kearney Map 2236
Fruits September–October.

Occurs in fields and waste ground. Known only from western Missouri in Jackson, Cass, and Jasper counties.

Ranges from Minnesota to South Dakota, south to Missouri, Oklahoma, Texas, and Mexico; introduced eastward to Maine, Vermont, and New York.

This species appears to shed more pollen than most of the other species, and is therefore of more importance from the standpoint of causing hay fever. Fortunately, its rarity in Missouri precludes its serious consideration as an important source of hay fever.

## 29. **Polymnia** L. Leaf-cup

a. Leaves and all of stem (except minutely hairy leaf-nodes) glabrous (without hairs); rarely encountered, known only from Pemiscot County, extreme southeastern Missouri . . . . . . 3. P. LAEVIGATA

Plate no. 365. **1.** Polymnia canadensis, × 2/7; *a.* Polymnia canadensis f. radiata, × 2/7. **2.** Silphium integrifolium var. integrifolium, × 2/7; Details from Small, The New York Botanical Garden. **3.** Polymnia laevigata, × 2/7; Details from Small, The New York Botanical Garden. **4.** Silphium perfoliatum, × 2/7. **5.** Polymnia Uvedalia, × 2/7. **6.** Xanthium pensylvanicum, × 2/7 (Scribner's); Details from Small, The New York Botanical Garden.

PLATE NO. 365

2236 Xanthium speciosum

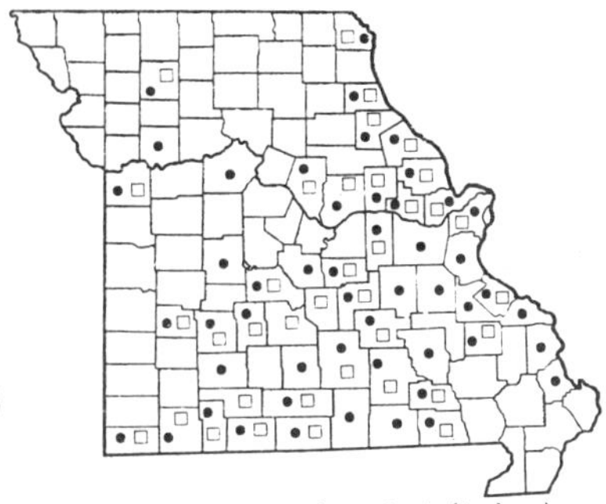

2237 ● Polymnia canadensis f. canadensis (Leaf-cup)
2237 □ Polymnia canadensis f. radiata

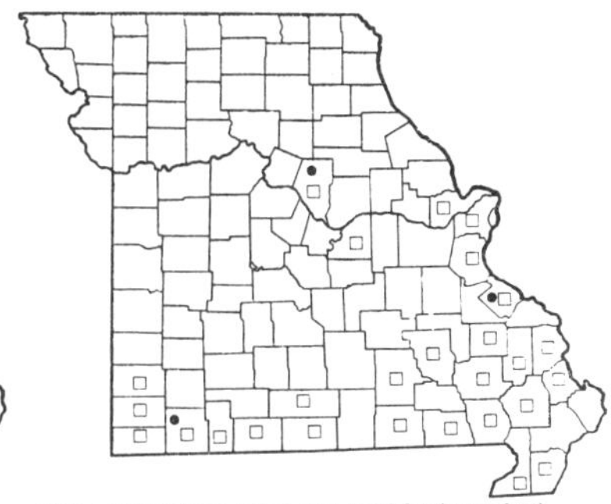

2238 ● Polymnia Uvedalia var. Uvedalia (Bearsfoot)
2238 □ Polymnia Uvedalia var. densipilis

a. Leaves or some part of stem or both more or less hairy; commonly encountered species. . . . . . *b*
  b. Rays, when present, 5, white, up to 12 mm. long and shorter than the involucre; green tissue of leaf-blade not extending to the base of the petiole (leaf-stalk); leaves pinnately lobed; achenes 3-ribbed, 3-angled; plants mainly of loose limestone talus of bluffs and rocky wooded limestone slopes . . . . . . . . . . . . . . . . . . . . . . . . . . . 1. P. CANADENSIS
  b. Rays always present, 10–15, yellow or creamy-yellow, 12–22 mm. long and longer than the involucre; green tissue of leaf-blade extending down to the base of the winged petiole (leaf-stalk); leaves palmately lobed; achenes strongly marked with coarse longitudinal lines (striate); plants of rich woodland in usually low, level, or alluvial ground . . . . . . . . . . 2. P. UVEDALIA

1. **Polymnia canadensis** L. Map 2237

Flowers May–October.

Occurs on loose limestone talus and ledges of bluffs and rocky wooded limestone slopes. Ozark region of southern and central Missouri, northeast to Clark County and locally in northwest Missouri to Daviess County.

Ranges from Vermont to Ontario and Minnesota, south to Georgia, Alabama, Louisiana, and Oklahoma.

Two variations occur in Missouri:

Rays minute or absent. . . . 1a. P. CANADENSIS f. CANADENSIS

Rays noticeable, becoming 10–12 mm. long and 3-lobed . . . . . . 1b. P. CANADENSIS f. RADIATA

1a. **Polymnia canadensis** f. **canadensis** Map 2237

*Polymnia canadensis* L. [G, BB, P & S, Steyerm.]

This form is somewhat more common than the form with conspicuous rays.

1b. **Polymnia canadensis** f. **radiata** (Gray) Fassett Map 2237

*Polymnia canadensis* var. *radiata* Gray [P & S, Steyerm.]

The less common form in Missouri.

The upper part of the stem is loosely hairy and covered with sticky glandular hairs. The plant is strongly scented. On rocky limestone slopes and talus at the base of bluffs this plant frequently dominates as the ground cover to the exclusion of other species or is associated with *Impatiens pallida* and *Hydrophyllum appendiculatum*.

2. **Polymnia Uvedalia** L. Bearsfoot Map 2238

Also known as Yellow-flower Leaf-cup.

Flowers July–September.

Occurs in rich low woods, wooded valleys, alluvial and upland thickets, and at the base of bluffs. Southern and east-central Missouri.

Two varieties occur in Missouri:

Rarely encountered type; flower-stalks (pedicels) and branches of inflorescence covered with gland-tipped hairs, with few or no glandless hairs present . . . 2a. P. UVEDALIA var. UVEDALIA

Commonly encountered type; flower-stalks (pedicels) and branches of inflorescence covered with many, slender non-glandular hairs, with few gland-tipped hairs present. . . 2b. P. UVEDALIA var. DENSIPILIS

2a. **Polymnia Uvedalia** var. **Uvedalia** Map 2238

*Polymnia Uvedalia* L. [G, Steyerm.]

*Polymnia Uvedalia* var. *genuina* Blake [P & S]

Known only from Ste. Genevieve, Boone, and Barry counties.

Ranges from New York to Illinois and Missouri, south to Georgia and Tennessee.

2b. **Polymnia Uvedalia** var. **densipilis** Blake Map 2238

Southern and east-central Missouri west to Boone and to Osage counties, northeast to St. Louis and to St. Charles counties, elsewhere in the southern Ozark and southeastern lowland counties.

Ranges from Louisiana and Texas to Missouri and Oklahoma; also in Virginia and Delaware.

3. **Polymnia laevigata** Beadle Map 2239

Flowers July–October.

Occurs in rich low woods. Known only from the southeastern lowlands in Pemiscot County (virgin woods on level ground, on property of Son Rone, T20N, R12E, sect. 12, 3 mi. south southwest of Portageville, July 29, 1956, *Steyermark 82261*).

Ranges from Florida, Georgia, and Alabama, north to Tennessee and Missouri.

This species grows in colonies with stems 1–1.2 m. tall. The leaves have a slight spicy fragrance, the base of the petioles and leaf-nodes are wine-purple, the rays are white, and the disk yellow. The plant appears to be an annual.

## 30. **Silphium** L. Rosin-weed

a. Leaves deeply lobed and parted nearly to the midrib, pinnatifid or bipinnatifid . . . 4. S. LACINIATUM
a. Leaves with or without shallow or coarse teeth, but not lobed nor parted . . . . . . . . . . *b*
  b. Stem nearly or quite leafless, all the leaves at or near the base of the plant and very large . . . 5. S. TEREBINTHACEUM
  b. Stem leafy throughout from base to top, the leaves of about equal size throughout . . . . . . *c*
    c. Stem 4-angled (quadrangular); bases of leaves or their petioles mostly united and surrounding the stem (perfoliate) to form a cup-like structure (rarely the leaves contracted into a narrow petiole 3–6 mm. broad at the base) . . . . . . . . . . . . . . 1. S. PERFOLIATUM
    c. Stems with curved or rounded sides, terete (round in cross-section) or only slightly angled; base of leaves or their petioles not united nor surrounding the stem to form a cup-like structure . . *d*
      d. At least the stem glaucous (with a silvery- or blue-green 'bloom' which can be rubbed off); stem, peduncles (flower-bearing branch), both surfaces of bracts of involucre, and all the lower surface of the leaf except the midrib glabrous (without hairs) . . . . . . . 2. S. INTEGRIFOLIUM var. LAEVE
      d. Stem green or purplish-red or brown; some part of stem, peduncles, surface of bracts of involucre, and lower surface of the leaf more or less hairy . . . . . . . . . . . . . *e*
        e. Most of the leaves alternate . . . . . . . . . . . . . . . . . . . . *f*
          f. Stem and peduncles without hairs or hairs less than 0.5 mm. long . 2. S. INTEGRIFOLIUM
          f. Stem and peduncles with coarse, spreading hairs 1 mm. or more long . 3. S. ASTERISCUS
        e. Most of the leaves opposite . . . . . . . . . . . . . . . 2. S. INTEGRIFOLIUM

1. **Silphium perfoliatum** L. Cup Plant Map 2240

Also known as Cup Rosin-weed.

Flowers July–September.

Occurs in low ground, alluvial thickets, borders of streams, ponds, meadows, lowland woods, and along railroads. Throughout Missouri, except absent from the southeastern lowland section.

Ranges from Ontario to South Dakota, south to Georgia, Mississippi, Missouri, and Oklahoma.

Two variations occur in the state:

Commonly encountered type; base of the middle and upper leaves or their petioles united to form a conspicuous broad cup-like structure, the petiolar part (leaf-stalk) of leaf, when present, conspicuously winged by leafy tissue more than 5 mm. broad on each side . . . . 1a. S. PERFOLIATUM f. PERFOLIATUM

Rarely encountered type; all the leaves, except the uppermost pair at the base of the branches of the inflorescence, contracted into petioles 3–7 or more cm. long and 3–6 mm. broad, the leafy tissue only 0.5–1 mm. broad on each side of the petiole or nearly absent . . 1b. S. PERFOLIATUM f. PETIOLATUM

1a. **Silphium perfoliatum** f. **perfoliatum** Map 2240

*Silphium perfoliatum* L. [G, BB, P &S]

The common variation found throughout the state.

1b. **Silphium perfoliatum** f. **petiolatum** Palmer & Steyerm. Map 2240

Known only from Stone County, southwestern Missouri (alluvial wooded banks and terraces above White River, along south and east side at base of north- and northwest-facing wooded slopes, T22N,

2239 Polymnia laevigata

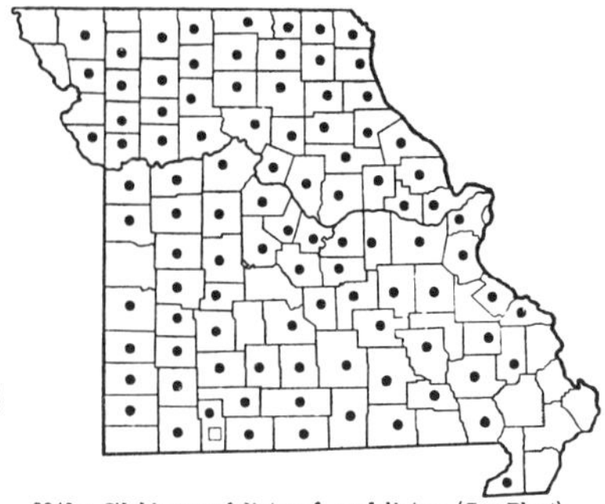

2240 ● Silphium perfoliatum f. perfoliatum (Cup Plant)
2240 □ Silphium perfoliatum f. petiolatum

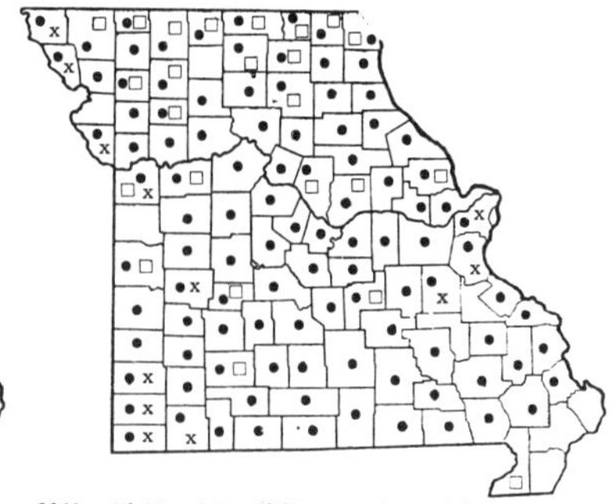

2241 ● Silphium integrifolium var. integrifolium f. integrifolium (Rosinweed)
2241 □ Silphium integrifolium var. integrifolium f. Deamii
2241 x Silphium integrifolium var. laeve

R23W, sect. 20, northeast and southwest of Mill Creek, 4 mi. [by air] southwest of Radical, 4½–5 mi. northeast of Baxter, October 19, 1955, *Steyermark 80474*, holotype in Chi. Nat. Hist. Mus. Herb., isotype in U. of Mo. Herb.).

The station for f. *petiolatum* has now been destroyed by the impounded waters of the Table Rock Dam. In most plants of *S. perfoliatum*, the leaves of the upper half of the stem have their bases or abruptly narrowed petioles joined into a conspicuous cup, but in f. *petiolatum* all but the uppermost pair of leaves are contracted into petioles which do not form cups or are only 3–6 mm. broad with the leaf-tissue reduced to 0.5–1 mm. width on each side of the petiole.

2. **Silphium integrifolium** Michx. Rosin-weed Map 2241

Flowers July–September.

Occurs in rocky or dry open woods, prairies, and glades.

The following variations occur in Missouri:

a. At least the stem glaucous (with a silvery- or blue-green 'bloom' which can be rubbed off); stem, peduncles (flower-bearing branch), both surfaces of bracts of involucre, and all the lower surface of the leaf except the midrib glabrous (without hairs) . . . . . . . 2c. S. INTEGRIFOLIUM var. LAEVE

a. Stem green, purplish-red, or brown; some part of stem, peduncles, surface of bracts of involucre, and lower surface of the leaf more or less hairy . . . . . . . . . . . *b*

b. Bracts of involucre hairy, but without glands or gland-tipped hairs . . . . 2a. S. INTEGRIFOLIUM var. INTEGRIFOLIUM f. INTEGRIFOLIUM

b. Bracts of involucre with glands and gland-tipped hairs . . . . 2b. S. INTEGRIFOLIUM var. INTEGRIFOLIUM f. DEAMII

2a. **Silphium integrifolium** var. **integrifolium** f. **integrifolium** Map 2241

*Silphium integrifolium* Michx. [G, P & S]

*Silphium integrifolium* var. *integrifolium* [BB]

This is the commonest variation in Missouri, found throughout the state, but not recorded from most of the lowland counties of extreme southeastern Missouri.

Ranges from Indiana to Kansas, south to Mississippi, Missouri, Arkansas, and Oklahoma.

2b. **Silphium integrifolium** var. **integrifolium** f. **Deamii** (Perry) Steyerm. Map 2241

*Silphium integrifolium* var. *Deamii* Perry, Rh. 39: 287. 1937.

Less common than typical f. *integrifolium*, scattered throughout Missouri, but commonest in the northern and central sections.

Ranges from Indiana to Wisconsin and Iowa, south to Alabama, Mississippi, and Arkansas.

This is treated as a form, as there appears to be no particular geographical range occupied by this variation, which differs from typical var. *integrifolium* in only the glandular pubescence of the involucral bracts.

2c. **Silphium integrifolium** var. **laeve** T. & G. Map 2241

*Silphium speciosum* Nutt. [G]

Scattered in southern and central Missouri, north to St. Louis and Jackson counties, northwest along the Missouri River to Atchison County.

Ranges from Nebraska to Missouri, Kansas, Oklahoma, and Texas.

For the most part, and at first impression, the glaucous-stemmed and smooth, mostly glabrous, pale green *S. integrifolium* var. *laeve* would appear entitled to specific rank. The following collections, with the involucral bracts merely ciliate with short hispid hairs, have glabrous stems, leaves and peduncles: Barry Co.

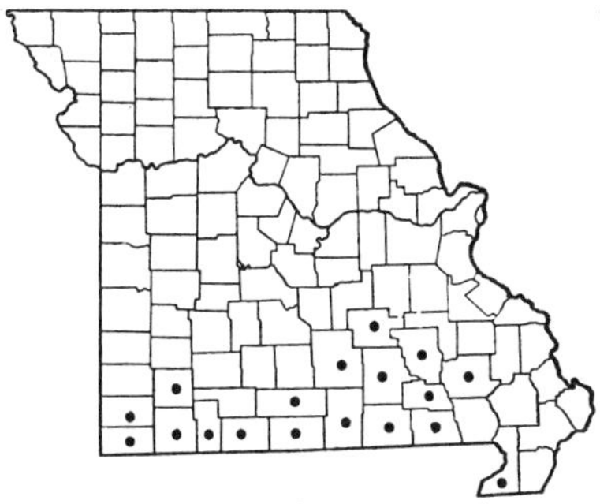

2242 Silphium Asteriscus (Starry Rosin-weed)

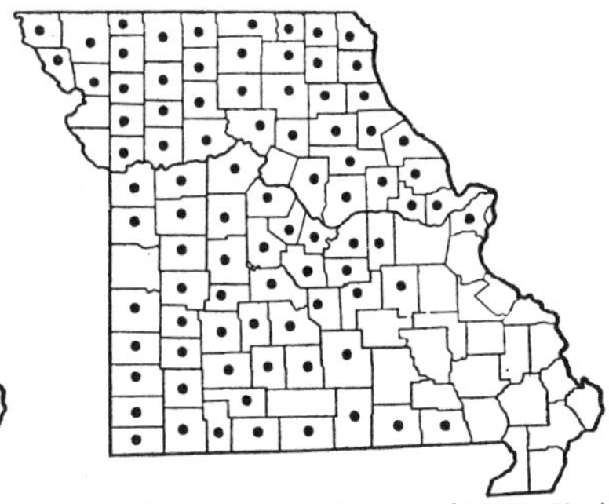

2243 Silphium laciniatum var. laciniatum (Compass Plant)

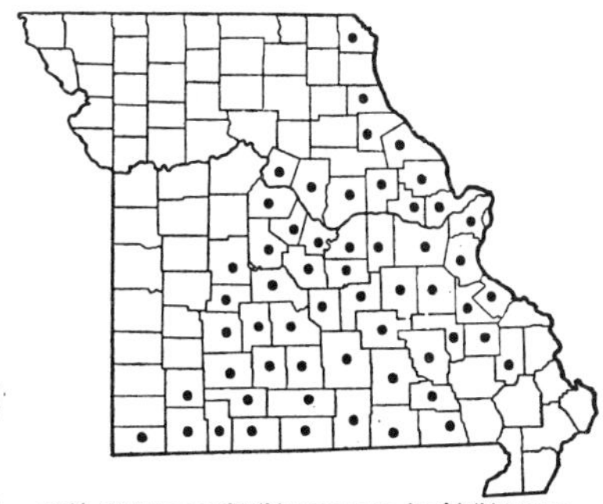

2244 Silphium terebinthinaceum var. terebinthinaceum (Prairie Dock)

(Seligman, *Bush 15110*); St. Louis Co. (St. Louis, September, 1848, *Engelmann*). Collections with glabrous stems, mostly glabrous peduncles, mostly glabrous lower leaf surfaces, and mostly glabrous involucres except for the minutely hispid-ciliate margins of the involucral bracts are represented by the following: Franklin Co. (*Kellogg 1300*); Jefferson Co. (*Steyermark 1465*); Washington Co. (*Steyermark 7020*); and St. Louis Co. (Meramec highlands, *Dewart*). Such collections show the glabrous or glabrate characteristics of var. *laeve*, but exhibit varying degrees of hairiness on the involucre and peduncles which point to characters of typical var. *integrifolium*. From the standpoint of easy recognition in the field and maintenance of characters under growing conditions in the garden where I have observed it for several years, I would look upon the glaucous, glabrate plant as a distinct species. However, because of the specimens cited above, which would indicate some transition to var. *integrifolium*, it is probably better to treat it as a distinct variety, as held by Cronquist (in Gleason's *New Ill. Fl.* 4: 367. 1952). Dr. T. Richard Fisher, who is monographing the genus, is maintaining this as a distinct species.

The differences stated by Perry (Rh. 39: 282–83. 1937) and followed by Fernald in *Gray's Manual* for the separation of *S. speciosum* from *S. integrifolium* do not hold. These authors differentiate the two taxa on the basis of the involucral bracts loosely spreading with reflexed tips and broadly ovate inner bracts in *S. speciosum* as contrasted with the involucral bracts slightly spreading with reflexed tips and narrowly ovate to oblong inner bracts in *S. integrifolium*. Actually, in both *S. speciosum* and *S. integrifolium* the bracts are similarly widely spreading, as verified by numerous personal observations in the field and under growing conditions. However, all the bracts in *S. speciosum* appear to average larger and broader than those of *S. integrifolium*.

Most plants of *S. integrifolium* have opposite leaves, but occasional specimens are seen having some or many of the leaves alternate. In such cases, hybridization is suspected with alternate-leaved *S. Asteriscus*.

3. **Silphium Asteriscus** L. Starry Rosin-weed
Map 2242

*Silphium Gatesii* Mohr [G, Steyerm.]
*Silphium asperrimum* Hook. [G, Steyerm.]
Flowers May–September.

Occurs in sandy ground, rocky open woods, and glades. Southern Ozark region north to Wayne, Reynolds, Dent, Texas, and Douglas counties, west to McDonald County, and on Crowley Ridge in Dunklin County.

Ranges from Florida to Texas, north to North Carolina, Alabama, Missouri, and Oklahoma.

The division of Missouri material of this group into three species, *S. Asteriscus*, *S. Gatesii*, and *S. asperrimum* by Dr. Perry (Rh. 39: 282–93. 1937) and followed by Fernald in *Gray's Manual* has not been accepted in the present flora, based on Missouri specimens studied. The present author agrees with the treatment by Cronquist (in Gleason's *New Ill. Fl.* 3: 368. 1952) and the current conclusions of the present monographer of the genus, Dr. T. Richard Fisher (communication of August 31, 1959), of combining all three as one species. It is believed that more intensive field and experimental studies are needed in this group, and that the differences stressed by Dr. Perry are not real ones for differentiating these taxa as species. Dr. Perry states (Rh. 39: 286. 1937) that *S. asperrimum* is easy to recognize because of its chiefly alternate leaves, yet Fernald (*Gray's Manual*, eighth edition, p. 1476) states that the leaves of this species are mostly opposite. Similarly, there is confusion in an understanding of *S. Asteriscus*. Although the peduncles and involucral bracts are stated in Fernald's key (p. 1476 of *Gray's Manual*) to be finely pubescent, they are described in the text (p. 1477) as hispid, while Dr. Perry (p. 292) expresses uncertainty in her citation of specimens under *S. Asteriscus*. The differences stated to exist between finely pubescent peduncles and involucral bracts of *S. Asteriscus* and hirsute-hispidulous ones of *S. Gatesii* do not appear to this author as valid criteria for species differentiation. The difference in relative width of achene and depth of the sinus of the achene between *S. asperrimum* on the one hand and *S. Gatesii* and *S. Asteriscus* on the other may or may not be significant, but needs to be tested on more than the one specimen of *S. asperrimum* from Dunklin County (sands, Campbell, September 9, 1910, *Bush 6395*) cited and studied by Dr. Perry. The leaves are eaten by white-tailed deer.

4. **Silphium laciniatum** L. var. **laciniatum**
Compass Plant Map 2243
Also called Rosin-weed.
*Silphium laciniatum* L. [G, BB, P & S]
Flowers July–September.

Occurs in prairies, glades, and along railroads. Throughout most of Missouri except absent from the southeastern quarter of the state, extending east and south to St. Louis, Crawford, Phelps, Pulaski, Texas, and Ripley counties.

Ranges from Ohio and Michigan to Minnesota and North Dakota, south to Alabama, Oklahoma, and Texas.

In typical var. *laciniatum* the peduncles and involucral bracts are hairy but lack glands, whereas in var. *Robinsonii* Perry, not recorded from Missouri, the peduncles and bracts have gland-tipped hairs.

The common name derives from the tendency of the basal leaves when growing in full sun to hold their edges so that they point north and south, the leaf surface facing the morning and afternoon sun. During flowering time, a gummy substance appears on the upper portion of the stem. This substance has been used by the Indians as a chewing gum. The plant is considered nutritious and palatable, especially in the younger stages, being eaten by all classes of grazing animals.

5. **Silphium terebinthinaceum** Jacq. var. **terebinthinaceum** Prairie Dock Map 2244
*Silphium terebinthinaceum* Jacq. [G, BB, P & S]
Flowers July–October.

Occurs on prairies, glades, and bald knobs, usually on limestone strata. Ozark region of southern and central Missouri northeast to Clark County, and west to Howard, Cooper, Morgan, Benton, Hickory, Polk, Lawrence, and McDonald counties.

In typical var. *terebinthinaceum* the upper surfaces of the leaf-blade are rough-scabrous or hairy. In var. *Lucy-Brauniae* Steyerm., not yet recorded from Missouri, the upper surface of the blade is smooth and glabrous, and in var. *pinnatifidum* (Ell.) Gray, likewise not recorded from Missouri, the leaves are lobed or pinnatifid.

The large basal leaves, clustered at or near the base of the plant, are a striking and conspicuous element of the prairie and glade landscapes. They are often observed without the flowering stems and their identity is the source of many inquiries from those who see the plant for the first time.

The species occasionally hybridizes with *S. laciniatum*, but such hybrids have as yet not been recorded from Missouri.

*Excluded Species*

**Silphium asperrimum** Hook.

Perry (Rh. 39: 286. 1937) identified a specimen from Dunklin County (Campbell, *Bush 6395*) as pertaining to this species. In the present work this is referred to *S. Asteriscus*. As discussed under *S. Asteriscus*, *S. Gatesii* is merged, also, with *S. Asteriscus*. Perry retains *S. Asteriscus* as a distinct species, identifying a Missouri specimen from Shannon County (*Bush 34*) as that species.

## 31. **Berlandiera** DC.

**Berlandiera texana** DC. var. **texana** Map 2245
*Berlandiera texana* DC. [G, BB, P & S]
Flowers June–October.

Occurs in rocky dry open woods, glades, and

Plate no. 366. **1.** Silphium integrifolium var. laeve, × 2/7. **2.** Silphium Asteriscus, × 2/7. **3.** Silphium laciniatum, × 2/7. **4.** Silphium terebinthinaceum, × 2/7; After Gleason, The New York Botanical Garden. **5.** Engelmannia pinnatifida, × 2/7; *a.* Ray-floret, × 6/7; *b.* Disk floret and pale, × 1 1/7; *c.* Achene, × 4/7. **6.** Berlandiera texana, × 2/7; After Gleason, details from Small, The New York Botanical Garden. **7.** Parthenium integrifolium, × 2/7; *a.* Root and lower leaves; Details from Small, The New York Botanical Garden. **8.** Parthenium Hysterophorus; *a.* Inflorescence, × 2/7; *b.* Leaf, × 2/7; *c.* Head of florets, with involucre, × 2 2/7; *d.* Ray-floret and achene, × 2 6/7; After Gleason, The New York Botanical Garden.

PLATE NO. 366

2245 Berlandiera texana var. texana

2246 Engelmannia pinnatifida

2247 Parthenium Hysterophorus (Santa Maria)

thickets. Ozark region east to Phelps, Dent, Shannon, Carter, and Oregon counties, north to Pulaski, Greene, Dade, and Jasper counties; absent from the eastern and northern portions of the Ozarks.

A var. *betonicaefolia* (Hook.) T. & G., ranging from Texas to Louisiana, Arkansas, and Oklahoma, has the peduncles and upper part of the stem clothed with long, jointed, purplish hairs.

## 32. **Engelmannia** T. & G.

**Engelmannia pinnatifida** T. & G. Map 2246

Flowers July–August.

Introduced along railroads. Known only from Jackson County, west-central Missouri (rare, July 21, 1892, *Bush*).

Ranges from Louisiana and Texas to Arizona, north to Kansas and Colorado; introduced in Missouri.

## 33. **Parthenium** L. American Feverfew

a. Leaves deeply dissected or parted nearly to the midrib; annual plants . . . . 1. P. HYSTEROPHORUS
a. Leaves merely toothed with shallow or coarse teeth; perennial plants . . . . . . . . . . . *b*
  b. Stems with noticeable spreading rough hairs; lower surface of lower leaves with long hairs on the nerves; plant with a creeping rootstock . . . . . . . . . . . . . . . 3. P. HISPIDUM
  b. Stems usually glabrous in the lower portion, usually only minutely hairy in the upper portion; lower surface of leaves glabrous (without hairs) or minutely hairy; plant lacking a creeping rootstock, the root generally short and tuberous-thickened . . . . . . . . 2. P. INTEGRIFOLIUM

1. **Parthenium Hysterophorus** L. Santa Maria Map 2247

Flowers August–October.

Found in waste ground and along railroads. Known only in central and eastern Missouri from Franklin, Johnson (Warrensburg, September, 1939, *M. C. Nattinger*), Scott (along street and railroad in Sikeston, August 29, 1948, *Steyermark 66176*), and Jackson (waste ground, Sheffield, August 3, 1921, *Bush 9476*) counties.

Native of tropical America; introduced and naturalized in the United States from Florida to Texas, north to Massachusetts, Pennsylvania, Ohio, Michigan, Illinois, Missouri, and Kansas.

This plant has been used in medicine as a diuretic. It is considered to have some slight contributory effect in causing hay fever. It is believed to be of only small importance, since its pollen is tenacious and difficult to scrape away from a surface.

2. **Parthenium integrifolium** L. var. **integrifolium** American Feverfew Map 2248

Also called Wild Quinine.

*Parthenium integrifolium* L. [G, P & S, Steyerm.]

Flowers May–September.

Occurs in prairies, glades, rocky open woods, and thickets. Throughout Missouri, except not recorded from extreme northwestern and southeastern lowland sections.

Ranges from Georgia to Texas, north to New York, West Virginia, Ohio, Michigan, Wisconsin, and Minnesota; introduced in Massachusetts.

A var. *auriculatum* (Britt.) Cornelius, by other authors placed as a variety under *P. hispidum*, occurring

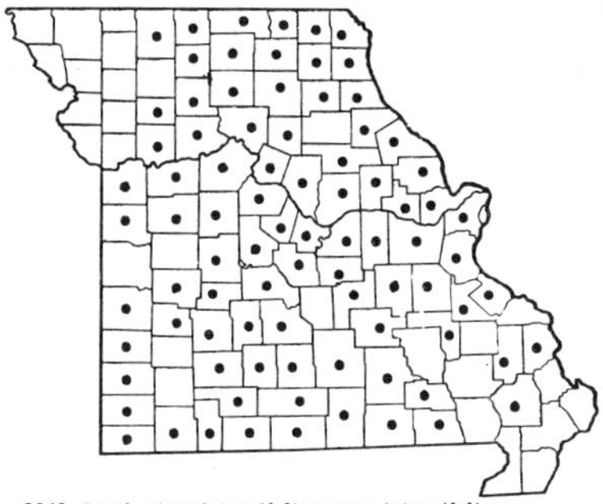

2248 Parthenium integrifolium var. integrifolium (American Feverfew)

2249 Parthenium hispidum

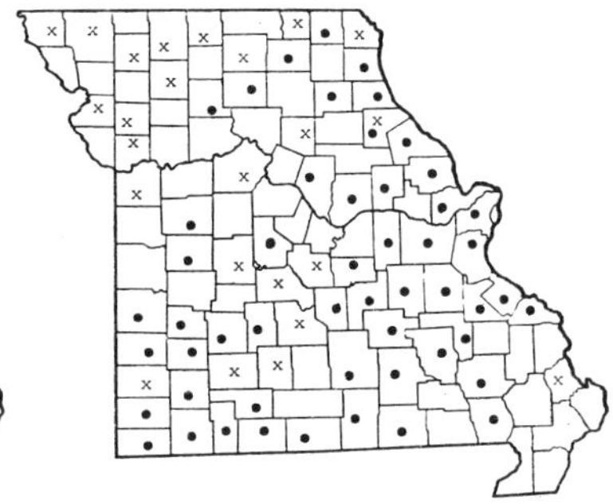

2250 • Heliopsis helianthoides var. scabra (Rough Ox-eye)
2250 x Heliopsis helianthoides var. occidentalis

from Virginia to North Carolina, has the stem with spreading hairs and the lower surface of the leaves hairy.

The plant has been used in medicine as a diuretic.

3. **Parthenium hispidum** Raf. Map 2249

Flowers May–October.

Occurs on limestone glades, bald knobs, and rocky calcareous prairies. Ozark region of southern and east-central Missouri north to St. Louis, Franklin, Warren, Montgomery, Osage, Miller, Camden, Hickory, Cedar, and Jasper counties.

Ranges from Missouri and Kansas to Arkansas and Texas; introduced in Michigan.

This is a characteristic plant of limestone glades, where it is associated with such other limestone-inhabiting species as *Oenothera missouriensis*, *Heliotropium tenellum*, and *Rudbeckia missouriensis*.

## 34. **Heliopsis** Pers. Ox-eye

**Heliopsis helianthoides** (L.) Sweet Ox-eye Map 2250

Also called Sunflower Heliopsis.

Flowers late May–September.

Occurs in open or rocky woods, thickets, prairies, and along railroads. Throughout Missouri, except absent from the lowland counties of southeastern Missouri.

Two variations are encountered in Missouri:

Flower-heads with the disk (solid central portion) mostly 1.2–1.4 cm. broad; petioles (leaf-stalks) mostly 2–2.5 cm. (sometimes 1–1.9) long; leaf-blades usually narrowly lanceolate to ovate-lanceolate . . a. H. HELIANTHOIDES var. SCABRA

Flower-heads with the disk mostly 1.5–2.5 cm. broad; petioles nearly absent to 1.5 cm. (rarely 2) long; leaf-blades mainly triangular-ovate or deltoid . . b. H. HELIANTHOIDES var. OCCIDENTALIS

a. **Heliopsis helianthoides** var. **scabra** (Dunal) Fern. Rough Ox-eye Map 2250

*Heliopsis scabra* Dunal [P & S, Steyerm.]

*Heliopsis helianthoides* subsp. *scabra* (Dunal) Fisher [Fisher]

This is the common variation in Missouri, found chiefly in the Ozark section of southern and central Missouri.

Ranges, according to Dr. Fisher, from Illinois and Iowa, south to Louisiana.

b. **Heliopsis helianthoides** var. **occidentalis** (Fisher) Steyerm. Map 2250

*Heliopsis helianthoides* subsp. *occidentalis* Fisher, Ohio Journ. Sci. 57: 189. 1957.

Mostly in the prairie section of northern, central, and western Missouri, represented by the following specimens from Scott (*Eggert*), Clark (*Steyermark 68867*), Ralls (*Davis 28*), Schuyler (*Steyermark 70339*), Sullivan (*Steyermark 70156*), Randolph (*Steyermark 70386*), Mercer (*Steyermark 70716*), Gentry (*Steyermark 72967*), Nodaway (*Bauer 318*), Atchison (*Bush 186A*), Buchanan (*Steyermark 70070*), Clay (*Mackenzie 388*), Jackson (*Broadhead*), Benton (*Trelease 563*), Miller (*Steyermark 4853*), Camden (*Steyermark 20645*), Laclede (*Steyermark 25196*), Greene (*Trelease 564*), Webster (*Steyermark 23835*), and Jasper (*Palmer 185*) counties.

Ranges, according to Dr. Fisher, from Ontario and New York to Illinois, Wisconsin, Minnesota, North Dakota, and Saskatchewan, south to Missouri, Colorado, and New Mexico.

The most recent study of this genus is by Dr. T. Richard Fisher (Ohio Journ. Sci. 57: 171–91. 1957). He considers the variations of *H. helianthoides* as subspecies, 'morphologically distinct within their own centers of distribution.' There is considerable variation between the above taxa, but especially so, according to Dr. Fisher, at the margins of geographical range, where the variations overlap. Although the var. *occidentalis* has, on the average, larger heads, shorter petioles, and more broadly ovate to deltoid leaf-blades than var. *scabra*, variations occur in Missouri material which are difficult to place. In addition to the characters taken from Dr. Fisher's key, the length and width of the leaf-blades are used by him to distinguish these two taxa as follows:

Leaves 8–12 cm. long, 3–3.5 cm. broad . . . H. HELIANTHOIDES var. SCABRA
Leaves 7–10 cm. long, 2.5–4 cm. broad . . . H. HELIANTHOIDES var. OCCIDENTALIS

These leaf measurements show no correlation with size of heads or length of petioles in many Missouri specimens. For example, some specimens with leaf-blades 6 cm. broad have been annotated by Dr. Fisher as subsp. *scabra*, whereas others, annotated as subsp. *scabra* have leaf-blades 2 cm. broad and petioles only 1–1.5 cm. long. Similarly, some specimens annotated as subsp. *occidentalis* have leaf-blades varying from 2.5–5.5 cm. broad and 5.5–12 cm. long. For the most part, however, material designated as var. *scabra* has smaller disks of the flower-heads, longer petioles, and relatively longer than broad leaf-blades. The typical var. *helianthoides* (subsp. *helianthoides* of Fisher's treatment) is distinguished by Dr. Fisher from both vars. *scabra* and *occidentalis* by the glabrous or minutely pubescent instead of scabrous leaves, peduncles, and involucral bracts, by the peduncles 5.5–8 cm. long, and by the cuneate-based ovate-lanceolate leaf-blades 8–12 cm. long and 4.5–6 cm. broad. Specimens annotated by him from Missouri as var. [subsp.] *helianthoides* are from St. Louis County (St. Louis, July 13, 1910, *Sherff 441*, and July 28, 1910, *Sherff 796*), and are probably based on cultivated material. No Missouri specimens are cited by Fisher nor shown on his map in the range of var. *helianthoides*, which, according to him, extends from the eastern United States west to Indiana and Illinois south to Georgia and Kentucky.

A variety *Pitcheriana* Hort. is cultivated and has been collected from cultivated material in St. Louis County. Various collections representing apparent hybrids between var. *scabra* and var. *occidentalis* have been annotated as such by Dr. Fisher. They are represented by specimens from Marion (*Davis 380, 650, 1449, 3101, 3584, 2143, 4643, 6144*), Ralls (*Davis 3360, 4660*), and Daviess (*Steyermark 16090*) counties.

## 35. **Eclipta** L.

**Eclipta alba** (L.) Hassk. Yerba de Tajo Map 2251
Flowers July–October.

Occurs along borders of ponds, sloughs, and streams, low ground in moist fields and valleys, and along railroads, roadsides, and waste ground. Throughout Missouri, doubtless in every county.

Ranges from Florida to Texas, north to New York, Indiana, Illinois, Iowa, and Nebraska; introduced in Massachusetts and Ontario; also in tropical America and the Old World tropics.

An alkaloid containing nicotine has been isolated from this species. It is reported that in Africa the plant is cooked and eaten as a vegetable.

## 36. **Rudbeckia** L. Coneflower

a. At least the lower leaves, or some or all of the leaves 3-lobed, 3-cleft, or even 5–7-parted . . . . . *b*
b. Disk (solid portion between rays) greenish-yellow to yellow . . . . . . . . . . . . . . *c*
c. Peduncles (flower-bearing branch) glabrous (without hairs); stems glabrous; leaves glabrous or sparsely hairy; rays 2.5–6 cm. long . . . . . . . . . . . . . . 1. R. LACINIATA
c. Peducles (flower-bearing branch) more or less hairy; at least upper part of stem short-hairy; at least lower side of leaves more or less hairy; rays 2–3.5 cm. long . 2b. R. SUBTOMENTOSA f. CRAIGII
b. Disk (solid portion between rays) dark or dull brown, dark purple, or purple-brown . . . . . *d*

---

Plate no. 367. **1.** Parthenium hispidum, × 2/7. **2.** Heliopsis helianthoides var. helianthoides, × 2/7; Details from Small, The New York Botanical Garden. **3.** Rudbeckia grandiflora, × 2/7; After Gleason, The New York Botanical Garden. **4.** Rudbeckia subtomentosa, × 2/7; After Gleason, The New York Botanical Garden. **5.** Heliopsis helianthoides var. occidentalis, × 2/7. **6.** Rudbeckia laciniata, × 2/7; Details from Small, The New York Botanical Garden. **7.** Rudbeckia maxima, × 1/7; After Gleason, The New York Botanical Garden. **8.** Eclipta alba, × 2/7; After Gleason, details from Small, The New York Botanical Garden.

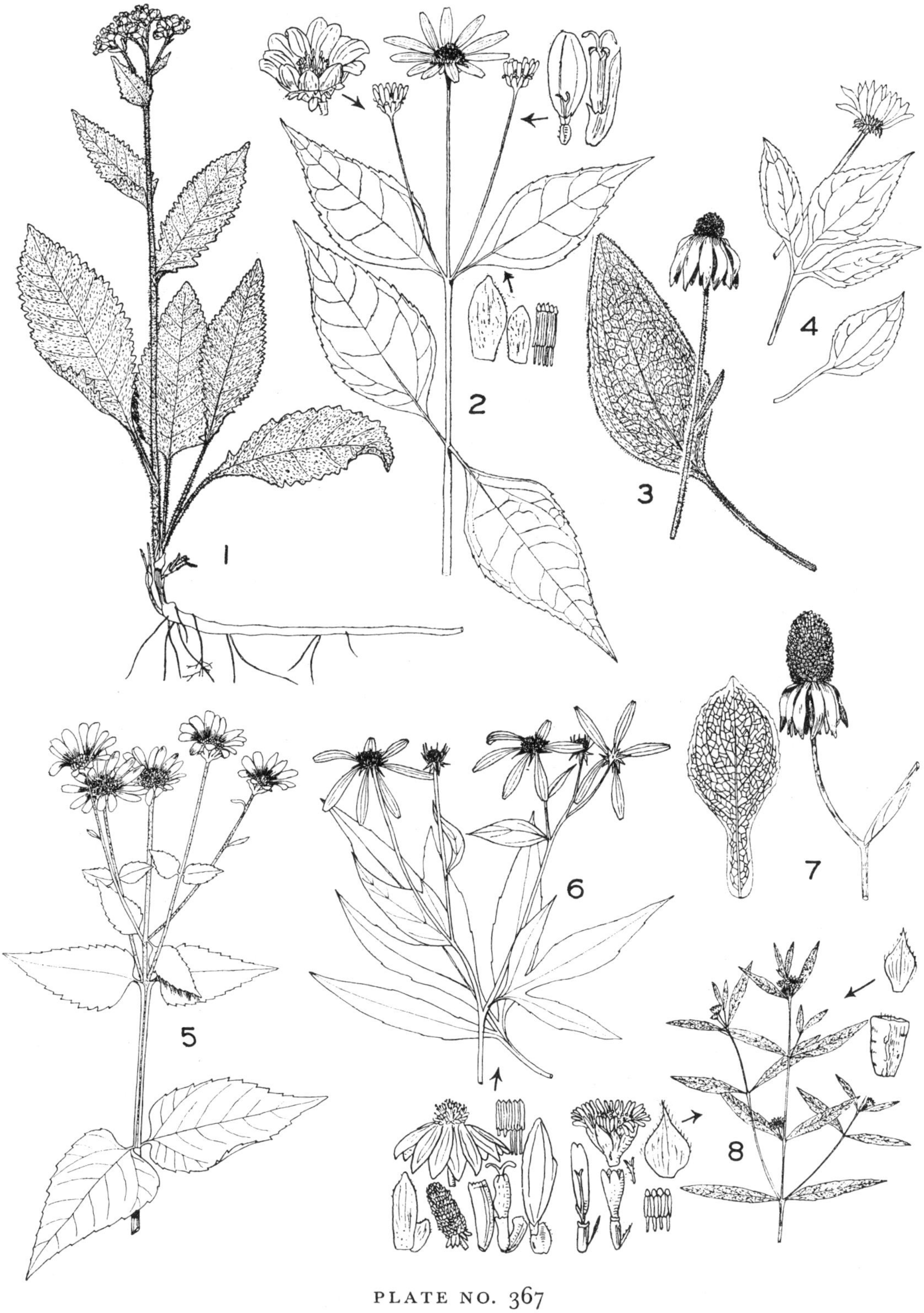


PLATE NO. 367

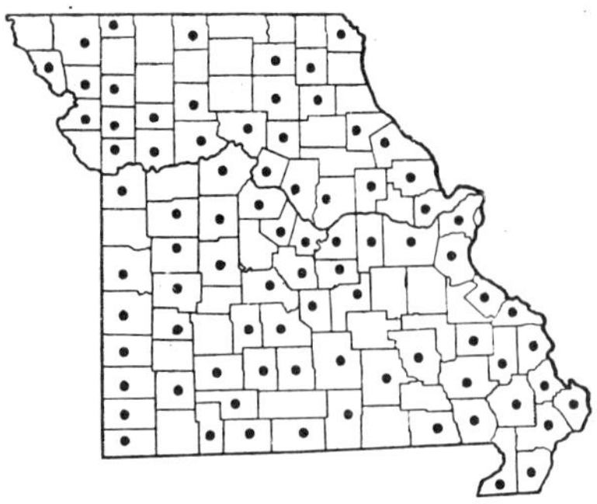

2251 Eclipta alba (Yerba de Tajo)

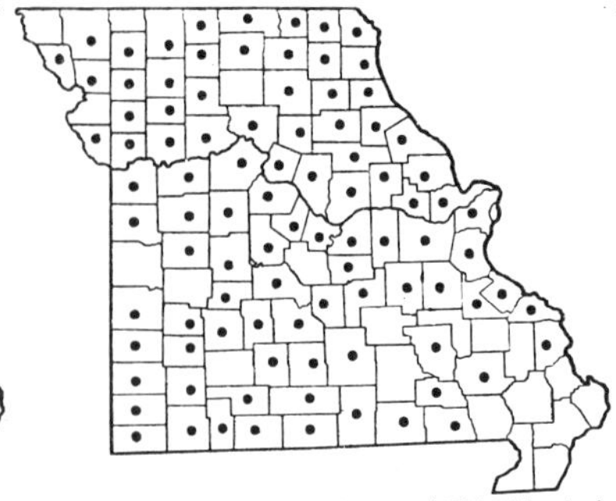

2252 Rudbeckia laciniata var. laciniata (Wild Goldenglow)

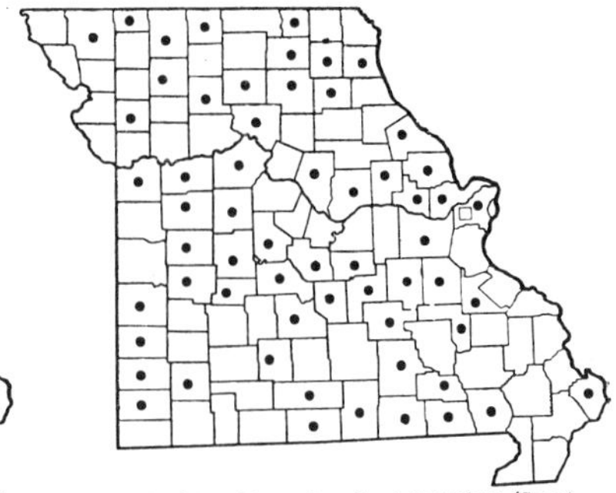

2253 ● Rudbeckia subtomentosa f. subtomentosa (Sweet Coneflower)

2253 □ Rudbeckia subtomentosa f. Craigii

d. Leaves thick and firm, usually densely short-hairy (tomentulose) or downy on lower surface; at least upper part of stem short-hairy or downy; pales (chaff) of the receptacle blunt to short-pointed, short-hairy and glandular on part of the back and margins; rays 12–20 . . 2a. R. SUBTOMENTOSA f. SUBTOMENTOSA

d. Leaves thin, with few to many long hairs but not downy or densely short-hairy on lower surface; upper part of stem with usually spreading, rather long hairs; pales (chaff) of the receptacle sharp-pointed, glabrous (without hairs); rays 6–12 . . . . . . . . . . . 8. R. TRILOBA

a. All the leaves simple, with or without teeth, not lobed, cleft, or parted . . . . . . . . . . *e*

e. Common weedy plants throughout the entire state; pappus absent; pales (chaff) of receptacle long-tapering at summit with a bristle tip; style tips elongated and slenderly subulate (awl-shaped) . . . . . . . . . . . . . . . . . . . . . . . . . . . . 5. R. HIRTA

e. Plants known only from southern and central Missouri, mostly confined to limestone glades, wet or swampy meadows, wet ledges of bluffs, and along streams, rarely along railroads; pappus present in the form of a toothed crown, sometimes very minute; pales (chaff) of receptacle rounded, blunt, or pointed, more or less hairy on the back and margins, or glabrous, but not tapering to a bristle tip; style tips short and blunt . . . . . . . . . . . . . . . . . . . . *f*

f. Stems and leaves glaucous (with a silvery- or gray-blue 'bloom' which can be rubbed off); stems and leaves glabrous (without hairs); leaves entire (without teeth), thick and coriaceous (leather-like); disk (solid part of flower-head between rays) becoming somewhat cylindrical and 4–8 cm. high . . . . . . . . . . . . . . . . . . . . . 4. R. MAXIMA

f. Without the above combination; stems and leaves hairy or glabrous (without hairs), not glaucous; leaves toothed or entire, not coriaceous; disk of the flower-head hemispherical, globe-shaped or ovoid-conical, at most 2.3 cm. high . . . . . . . . . . . . . . . . . *g*

g. Rays 3–7 cm. long, soon drooping (reflexed); disk of the flower-head becoming conical, 1.3–2.3 cm. high; lobes of disk-corollas reflexed (turned down); pales (chaff) of the receptacle densely short-hairy above the middle; rarely encountered species . 3. R. GRANDIFLORA

g. Rays 1–4 cm. long, usually spreading more or less horizontally; disk of the flower-head hemispherical or globose, mostly 0.6–1.2 cm. high; lobes of disk-corollas ascending; pales (chaff) of the receptacle either glabrous, short-hairy on the margins, or rarely with a few hairs on the back; commonly encountered species . . . . . . . . . . . . . *h*

h. All the leaves linear to narrowly oblong, conspicuously hairy with a dense hairiness of numerous hairs; stem-leaves less than 10 mm. (averaging 4–10 mm.) broad; stem densely and conspicuously hairy with numerous hairs; basal tufts of leaves linear to lance-spatulate; plants of dry limestone glades and barrens . . . . . . . . 6. R. MISSOURIENSIS

h. Plants not as above, the leaves broader, averaging 12–50 mm. broad, and either glabrous or only sparsely to moderately hairy; stem glabrous to only sparsely hairy; basal tufts of leaves broadly elliptical to ovate or nearly round (orbicular); plants mostly of wet or swampy ground in meadows, swamps, wet ledges, and along streams. . . . . . . . . *i*

i. Commonly encountered in swampy meadows, wet ledges, and along streams; basal and lower stem-leaves ovate or elliptic-ovate with the base of the blade broadly rounded, obtuse, acute, or truncate (as if cut straight across), the basal and lower leaves mainly 3.5–11 cm. broad (sometimes only 2.5); main leaves of stem ovate or

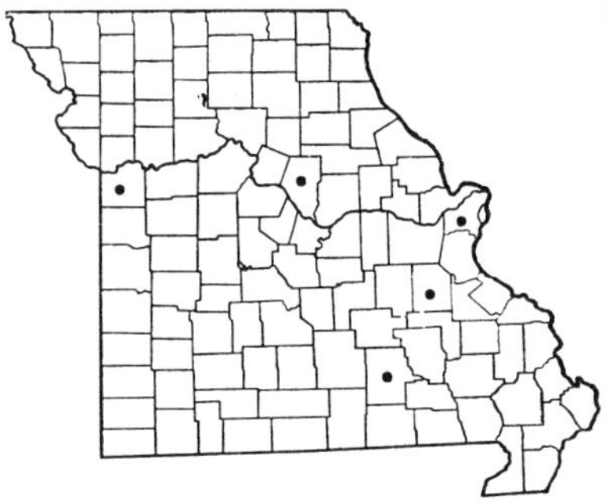

2254 Rudbeckia grandiflora var. grandiflora

2255 Rudbeckia maxima

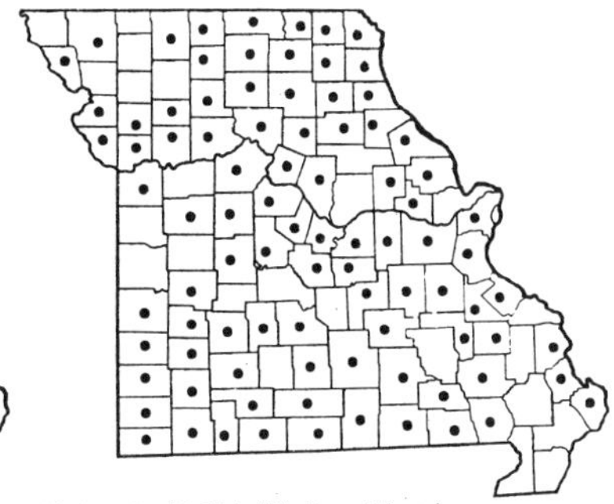

2256 Rudbeckia hirta (Black-eyed Susan)

elliptic-ovate, 1.2–5 cm. broad, with short inconspicuous teeth or nearly entire (without teeth) . . . . . . . . . . . . . . . 7a. R. FULGIDA var. UMBROSA

i. Rarely encountered; basal and lower stem-leaves oblanceolate, elliptic-lanceolate, or elliptical with the base of the blade tapering, the basal and lower leaves mainly 0.5–3.5 cm. broad; main leaves of stem lanceloate or oblanceolate, 0.5–2 cm. broad, usually sharply toothed. . . . . . . . . . . . . 7b. R. FULGIDA var. SPECIOSA

1. **Rudbeckia laciniata** L. var. **laciniata** Wild Goldenglow Map 2252

Also called Cutleaf Coneflower.

*Rudbeckia laciniata* L. [G, BB, P & S]

Flowers July–September.

Occurs in rich low or alluvial soils of woodland in valleys along streams, lakes, and sloughs, and alluvial thickets. Throughout Missouri, except absent from the lowland counties of extreme southeastern Missouri.

Ranges from Quebec to Montana, south to Florida, Louisiana, Texas, New Mexico, and Arizona.

Dr. Robert Perdue, Jr., recognizes several varieties of *R. laciniata,* of which var. *laciniata* is the only variation seen in Missouri material examined. A var. *ampla* (Nelson) Cronquist has been collected in St. Louis County from cultivated specimens (St. Louis, July 7, 1910, *Sherff 290*) and is not included in the present flora. It is considered by Cronquist to be the only variety present in the Pacific Northwest states.

The garden goldenglow, common in cultivation, is var. *hortensis* Bailey, with the rays so numerous as to dominate the flower-head with a 'double' effect.

*Rudbeckia laciniata* is believed, when eaten, to poison swine and sheep. The leaves vary from glabrous to densely short-hairy on the lower surface.

2. **Rudbeckia subtomentosa** Pursh Sweet Coneflower Map 2253

Flowers July–October.

Occurs in prairies, low meadows, open slopes of ravines, thickets, banks of streams, and roadsides. Throughout Missouri.

Ranges from Indiana, Wisconsin, Iowa, and Kansas, south to Louisiana and Texas.

Two variations are found in Missouri:

Disk of flower-head brown or dark purple-brown . . 2a. R. SUBTOMENTOSA f. SUBTOMENTOSA

Disk of flower-head yellowish . . . . . . 2b. R. SUBTOMENTOSA f. CRAIGII

2a. **Rudbeckia subtomentosa** f. **subtomentosa** Map 2253

*Rudbeckia subtomentosa* Pursh [G, BB, P & S]

This is the common form encountered in Missouri.

2b. **Rudbeckia subtomentosa** f. **Craigii** (Sherff) Fern. Map 2253

*Rudbeckia subtomentosa* var. *Craigii* Sherff [P & S]

Known only from St. Louis County, east-central Missouri (west of Forest Park, St. Louis, woods south of Pennsylvania Avenue, near Dixon's line fence, West Richmond Heights, August 19, 1910, *Sherff 1106*, isotype in Chi. Nat. Hist. Mus. Herb.).

*Rudbeckia subtomentosa* varies somewhat in pubescence of stem and leaves. Occasional plants are found having most of the leaves only toothed.

3. **Rudbeckia grandiflora** (D. Don) DC. var. **grandiflora** Map 2254

*Rudbeckia grandiflora* (D. Don) DC. [also published as *R. grandiflora* (D. Don) Gmel. and (Sweet) DC. See Shinners, Field and Lab. 17: 62–63. 1949] [G, P & S]

Flowers early July–August.

Occurs in dry open ground and along railroads.

Known in southern and central Missouri from St. Louis, Washington, Shannon, Boone, and Jackson counties.

Ranges from Louisiana and Texas to Missouri and Oklahoma.

4. **Rudbeckia maxima** Nutt. Map 2255

Flowers July–August.

Occurs along railroads. Introduced. Known only from Jackson County, west-central Missouri (Sheffield, introduced, August 4, 1903, *Bush 1846*; same locality, July 11, 1904, *Bush 2104*).

Ranges from Louisiana and Texas to Missouri and Oklahoma.

5. **Rudbeckia hirta** L. Black-eyed Susan Map 2256

*Rudbeckia serotina* Nutt. [G]
*Rudbeckia bicolor* Nutt. [G]
*Rudbeckia hirta* f. *homochroma* Steyerm.
*Rudbeckia hirta* var. *pulcherrima* Farw. [Perdue]
*Rudbeckia serotina* f. *homochroma* (Steyerm.) Fern. & Schub. [G]
*Rudbeckia serotina* f. *pulcherrima* (Farw.) Fern. & Schub. [G]
*Rudbeckia serotina* var. *lanceolata* (Bisch.) Fern. & Schub. [G]

Flowers May–October.

Occurs in open woods, thickets, rocky prairies, river bottom meadows, pastures, eroded slopes, waste ground, roadsides, and along railroads. Throughout Missouri.

The division of the common Black-eyed Susan into three species (*R. hirta* L., *R. bicolor* Nutt., and *R. serotina* Nutt.) as treated by Fernald & Schubert (Rh. 50: 172–76. 1948) and by Fernald (Rh. 39: 457–58. 1937) cannot be justified on the grounds of the annual or perennial habit of the plants, the difference in shape and size of leaves, or the presence or absence of basal leaves. An attempt to fit Missouri material into such categories has proved futile.

A recent treatment of this complex by Dr. Robert E. Perdue, Jr., (Rh. 59: 295–96. 1957) treats the three specific names mentioned above as synonyms of *R. hirta*, this being var. *pulcherrima* Farw., the common and only one including Missouri material. This variety ranges from Newfoundland to British Columbia, south to Florida and northern Mexico, and rare west of the Rocky Mountains. According to Dr. Perdue's treatment, var. *hirta*, which extends from Pennsylvania to Georgia and Alabama, north to Maine and west to Illinois, is not known from Missouri, and is distinguished from var. *pulcherrima* as follows (Rh. 59: 295. 1957):

Leaves coarsely toothed; basal leaves broadly elliptical to oval, 2.5–7 cm. wide, blades mostly twice as long as wide; cauline leaves ovate-lanceolate or ovate, the sessile ones commonly pandurate . . . . . . . . . . var. HIRTA

Leaves entire or finely serrate; basal leaves narrowly to broadly oblanceolate or lanceolate, 1–2.5 (–5) cm. wide; blade (3–) 4–5 times as long as wide; cauline leaves spatulate, oblanceolate or broadly linear . . . . . . . var. PULCHERRIMA

In view of Dr. Perdue's new researches, future study of Missouri material may show that var. *hirta* is also found in Missouri. Although Dr. Perdue recognizes no minor forms, it would seem worthwhile to record the following color variations of a type similarly recognized in other groups of Compositae, which have been noted in Missouri:

a. Disk flowers as well as ray flowers yellow . R. HIRTA f. HOMOCHROMA Steyerm.

Known from Oregon County (upland woods, 10 mi. west of New Liberty, July 3, 1936, *Steyermark 11345*, holotype in Mo. Bot. Gard. Herb.).

b. Disk flowers brown or purple-brown; ray flowers pale lemon yellow for inner ⅔ of their length, tips cream-colored or nearly white . R. HIRTA f. FLAVESCENS Clute

Known from Texas County (dry upper cherty slopes along Jack's Fork of Current River, T28N, R7W, sect. 36, 5½ mi. southeast of Arroll, June 23, 1939, *Steyermark 27149*).

c. Disk flowers brown or brown-purple; ray flowers golden with brown-orange or red-brown at base . . . . . . R. HIRTA var. PULCHERRIMA Farw.

Known from Mississippi (along railroad and road D, T26N, R16E, sect. 35, 3½ mi. [by air] northeast of Anniston, August 21, 1955, *Steyermark 79610*) and Taney (along limestone road, June 8, 1957, *Mrs. Nightingale*) counties.

Dr. Perdue merges this bicolored type, which is rare in Missouri, with typical var. *pulcherrima* having the rays ordinarily yellow or golden yellow, and the disk brown or brown-purple.

As there is still great diversity of opinion among botanists as to the numbers and limits of the various species involved (Shinners apparently recognizes an eastern *R. hirta* as distinct from *R. serotina* Nutt. and *R. divergens* T. V. Moore [the latter treated as a variety of *R. hirta* by Dr. Perdue]), I have used the name *R. hirta* L. tentatively to include all Missouri material, pending future more detailed work by geneticists and cytologists, supplemented by additional field and experimental studies.

The basal leaves are eaten by white-tailed deer.

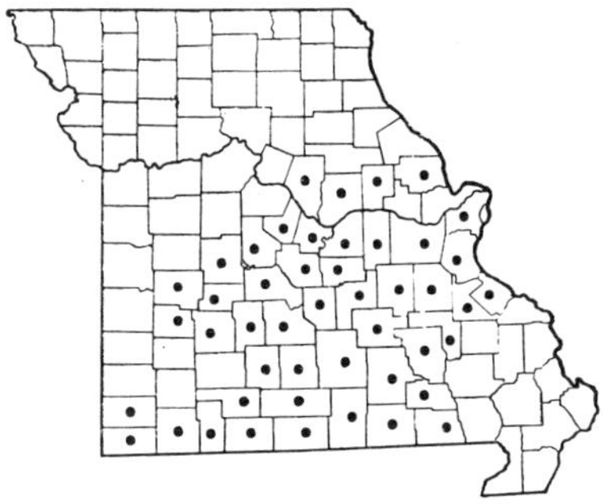

2257 Rudbeckia missouriensis

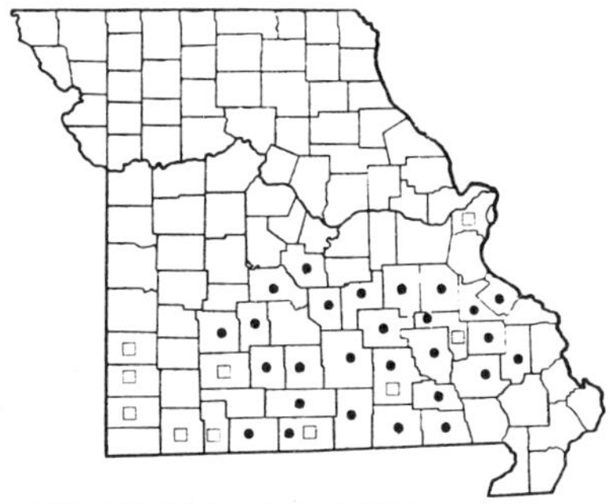

2258 • Rudbeckia fulgida var. umbrosa
2258 □ Rudbeckia fulgida var. speciosa

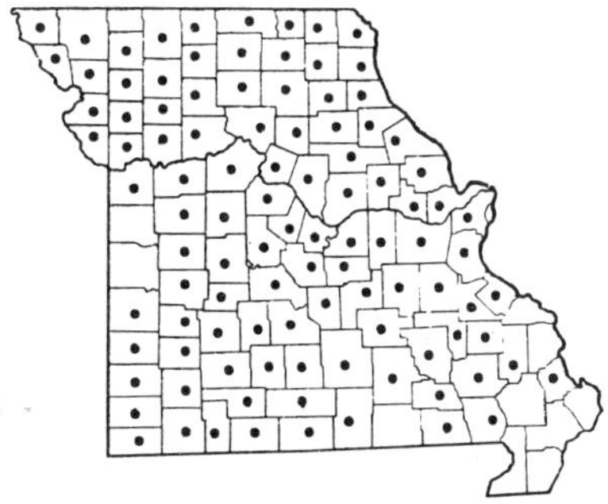

2259 • Rudbeckia triloba var. triloba (Brown-eyed Susan)

6. **Rudbeckia missouriensis** Engelm. Map 2257
*Rudbeckia fulgida* Ait. var. *missouriensis* (Engelm.) Cron. [BB]

Flowers late June–October.

Occurs on limestone glades, barrens, bald knobs, and rocky prairies. Ozark region of southern and east-central Missouri north to Lincoln, Montgomery, Callaway, Boone, Moniteau, Morgan, Benton, St. Clair, and Newton counties.

Ranges from Illinois and Missouri, south to Arkansas, Oklahoma, and Texas.

This is a common and very distinct perennial species of the limestone barrens of the Ozarks, where it is frequently associated with such other limestone inhabitants as *Oenothera missouriensis*, *Heliotropium tenellum*, *Echinacea paradoxa*, *Solidago Gattingeri*, *Leavenworthia uniflora*, and *Evolvulus Nuttallianus*. Generally it is one of the dominant species of the glades and hundreds to thousands of individual plants are seen on any one glade. To one acquainted with the plant in the field, it shows no overlap or intergradation into the *R. fulgida* complex, with which Cronquist (Rh. 47: 400–1. 1945; Gleason's *New Ill. Fl.* 3: 347. 1952) attempted to align it as a variety and obscure its real distinctness as a species.

7. **Rudbeckia fulgida** Ait. Map 2258

Although the following varieties, which occur in Missouri, appear to be distinct and recognizable species in their extremes, they are here accorded varietal rank in view of the recent extensive studies and conclusions by Dr. Perdue (Rh. 59: 296–98. 1957). Since their characters appear in the key to the species, the differences are not repeated here.

7a. **Rudbeckia fulgida** var. **umbrosa** (Boynt. & Beadle) Cron. Map 2258
*Rudbeckia umbrosa* Boynt. & Beadle [G]
*Rudbeckia palustris* Eggert ex Boynt. & Beadle [G, P & S]
*Rudbeckia fulgida* var. *palustris* (Eggert) Perdue [Perdue]

Flowers August–October.

Occurs on wet ledges, in wet and swampy calcareous meadows usually fed by spring-fed rivulets, and along streams. Ozark region north to Ste. Genevieve, St. Francois, Washington, Crawford, Phelps, Pulaski, Miller, Camden, Dallas, and Polk counties, west to Taney County.

Ranges from Ohio to Indiana and Missouri, south to Georgia, Alabama, Arkansas, Oklahoma, and Texas.

Missouri material identified as this variety has the basal and lower stem-leaves varying from narrowly to broadly ovate, the base of the leaves nearly cordate (*Steyermark 66379* from Taney County; *Steyermark 72517, 65055* from Douglas County; *Steyermark 69412* from Ozark County) or broadly rounded, and sparsely hairy on both surfaces, characterized as var. *umbrosa* by Perdue (Rh. 59: 297. 1957), to elliptical, ovate, or orbicular with the base acute, obtuse, rounded, or truncate and minutely strigose or scabridulous, characterized as var. *palustris* by Perdue. Plants showing all gradations of these shapes with variation in width of basal and lower stem-leaves from 3–11 cm. can be found in populations where the plants occur in large colonies. Most of the Missouri material has the larger basal leaves supposedly characteristic of var. *umbrosa*, but extremes showing the smaller leaves of var. *palustris* can also be found. Some specimens determined by Dr. Perdue as var. *umbrosa* do not show basal leaves nor rounded or cordate bases. The arbitrary differences in measurements, likewise, break down when applied to many individuals in even one colony. Based on observations in the field and herbarium, I can find no real differences to maintain var. *umbrosa* separate and distinct from var. *palustris*, and accordingly they have been united under the first name.

7b. **Rudbeckia fulgida** var. **speciosa** (Wenderoth) Perdue Map 2258
*Rudbeckia speciosa* Wenderoth [G, P & S]
*Rudbeckia speciosa* var. *Sullivantii* (Boynt. & Beadle) Robins. [G]
*Rudbeckia fulgida* var. *Sullivantii* (Boynt. & Beadle) Cron. [BB]
Flowers July–October.

Occurs in moist open places, thickets, moist ledges, glades, low, rich, and rocky open woods. Southern and east-central Missouri in St. Louis (St. Louis, July 25, 1910, *Sherff 732*), Iron (low rich woods, Arcadia, August 25, 1899, *Colton Russell*), Shannon (September, 1888, *Bush*), Greene (September 4, 1893, *Bush*; James River, September 4, 1893, *Blankinship*), Barton, Jasper, and Newton counties.

Ranges from Connecticut to Michigan, south to West Virginia, Ohio, Indiana, Illinois, and Missouri.

Specimens from Missouri have been identified as *R. speciosa* and var. *Sullivantii* and are referable to both taxa. It is believed that these cannot be maintained as separate entities, and Cronquist's merging of them (Rh. 47: 400. 1945; Gleason's *New Ill. Fl.* 3: 346. 1952) seems to accord with the variation seen in Missouri material. Dr. Perdue stresses the length of the rays (2.5–4 cm. long) as characteristic of *R. fulgida* var. *Sullivantii*, but a specimen from Indiana (*Deam 53400*), determined by Dr. Perdue as var. *Sullivantii*, has the rays only 1.5–2 cm. long.

8. **Rudbeckia triloba** L. var. **triloba** Brown-eyed Susan Map 2259
*Rudbeckia triloba* L. [G, BB, P & S]
*Rudbeckia triloba* var. *Beadlei* (Small) Fern. [G, Perdue]
Flowers June–November.

Occurs in low wet woods along streams, alluvial thickets, and rocky slopes at base of bluffs.

Throughout Missouri, except absent from the lowland counties of extreme southeastern Missouri.

Ranges from New York to Minnesota, south to Georgia, Tennessee, Arkansas, and Oklahoma.

A variety *pinnatiloba* T. & G. with the leaves of the stem pinnately 5–7-lobed occurs in the southeastern states.

## 37. **Echinacea** Moench Purple Coneflower

a. Plants chiefly of woodland, sometimes in prairies; leaf-blades broadly to narrowly ovate, rounded at base, rather abruptly contracted to the petiole (leaf-stalk), $1\frac{1}{2}$–5 times as long as broad, usually toothed, the larger leaves 2.5–7 cm. broad . . . . . . . . . . . . . . . . 1. E. PURPUREA
a. Plants of rocky open glades, barrens, and prairies, rarely in dry rocky woods; leaf-blades linear to lanceolate or lance-elliptic, gradually tapering to the petiole (leaf-stalk), mainly 5–20 times as long as broad, entire (without teeth), the larger leaves chiefly 0.5–2.5 cm. broad, rarely broader. . . . . *b*
  b. Rays yellow; stem glabrous or the hairs appressed (pressed against or lying parallel to surface) 4. E. PARADOXA
  b. Rays rose-colored, purple, or white; stem with spreading hairs . . . . . . . . . . . . . *c*
    c. Commonly encountered species; hairs of stem and leaves slender, not enlarged at base; rays 4–9 cm. long, strongly reflexed (turned down) . . . . . . . . . . . . . . 3. E. PALLIDA
    c. Rarely encountered, known only from Shelby County; hairs of stem and leaves enlarged at base (tuberculate); rays 2–3.5 cm. long, spreading . . . . . . . . . . . . 2. E. ANGUSTIFOLIA

1. **Echinacea purpurea** (L.) Moench Purple Coneflower Map 2260
Flowers May–October.

Occurs in rocky open woods, low ground, in ravines, thickets, and prairies.

Occurs in southern, central, and eastern Missouri, west to Mercer, Sullivan, Macon, Chariton, Saline, and Clay counties; absent from the lowland counties of extreme southeastern Missouri.

Ranges from Georgia to Louisiana to Oklahoma, north to Virginia, Ohio, Michigan, Illinois, and Iowa; occasionally introduced eastward.

Two variations occur in Missouri:

Rays of flower-heads rose-colored, pink, or reddish-purple . . . 1a. E. PURPUREA var. PURPUREA f. PURPUREA
Rays white . . . 1b. E. PURPUREA var. PURPUREA f. LIGGETTII

---

Plate no. 368. **1.** Rudbeckia hirta, × $^2/_7$. **2.** Rudbeckia fulgida var. umbrosa, × $^2/_7$; *a*, *b*. Basal leaves; *c*. Upper part of stem. **3.** Rudbeckia fulgida var. speciosa, × $^2/_7$; *a*. Leaf from lower part of stem; *b*. Upper part of stem; *c*. Middle part of stem with leaves; *d*. Basal leaf. **4.** Rudbeckia missouriensis, × $^2/_7$; After Gleason, The New York Botanical Garden. **5.** Rudbeckia triloba, × $^2/_7$; After Gleason, The New York Botanical Garden. **6.** Echinacea purpurea, × $^2/_7$; Details from Small, The New York Botanical Garden. **7.** Echinacea pallida, × $^2/_7$.

PLATE NO. 368

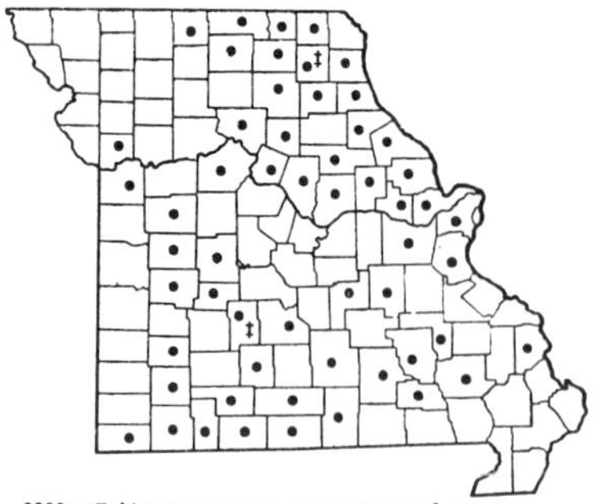

2260 • Echinacea purpurea var. purpurea f. purpurea (Purple Coneflower)
2260 ‡ Echinacea purpurea var. purpurea f. Liggettii

2261 Echinacea angustifolia

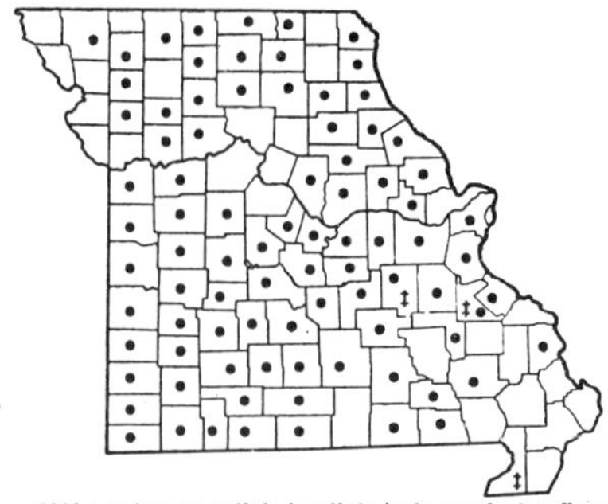

2262 • Echinacea pallida f. pallida (Pale-purple Coneflower)
2262 ‡ Echinacea pallida f. albida

1a. **Echinacea purpurea** var. **purpurea** f. **purpurea** Map 2260
*Echinacea purpurea* (L.) Moench [G, P & S, Steyerm.]
*Echinacea purpurea* var. *purpurea* [BB]
This is the common variation in Missouri.

1b. **Echinacea purpurea** var. **purpurea** f. **Liggettii** Steyerm. Map 2260
*Echinacea purpurea* f. *Liggettii* Steyerm.
Known only from Knox (ravine slopes along creek tributary to Middle Fabius River, T62N, R10W, sect. 3, 4 mi. [by air] northeast of Knox City, August 13, 1955, *Steyermark 79153*) and Dallas (highway 54, west of Niangua River; transplanted to garden of Mr. William Liggett in University City, St. Louis Co., July 9, 1940, *Liggett 1*, holotype in Chi. Nat. Hist. Mus. Herb.) counties.

*Echinacea purpurea* is chiefly a woodland species in Missouri. It is a subject equally desirable for the perennial garden as well as for the woodland wildflower garden. It does well in open woodland, but will not thrive if too crowded or shaded by dense vegetation.

A variety, var. *arkansana* Steyerm., with smaller disks and rays, occurs in Arkansas and Oklahoma.

2. **Echinacea angustifolia** DC. Map 2261
Flowers May–August.
Occurs in prairies. Known only from Shelby County, northeastern Missouri.
Ranges from Minnesota to Saskatchewan, south to Missouri, Oklahoma, and Texas.

3. **Echinacea pallida** Nutt. Pale-purple Coneflower Map 2262
Flowers May–July.
Occurs on limestone glades, barrens, bald knobs, fields, along railroads, and prairies, rarely in dry rocky woods. Throughout Missouri, except absent from the lowland counties of extreme southeastern Missouri.
Ranges from Illinois and Michigan to Minnesota and Nebraska, south to Georgia, Alabama, Louisiana, and Texas.
Two variations occur in Missouri:

Rays of flower-head pale or deep rose-purple . . . 3a. E. PALLIDA f. PALLIDA
Rays white . . . . . . 3b. E. PALLIDA f. ALBIDA

3a. **Echinacea pallida** f. **pallida** Map 2262
*Echinacea pallida* Nutt. [G, BB, P & S, Steyerm.]
This is the common variationi n Missouri and elsewhere.

3b. **Echinacea pallida** f. **albida** Steyerm. Map 2262
Known only from St. Francois (limestone glade on top of bluffs along Plattin Creek, T38N, R6E, sect. 7, Koester Springs at Koester, June 4, 1941, *Steyermark 28797*, holotype in Chi. Nat. Hist. Mus. Herb.), Crawford, and Dunklin (Malden, June 8, 1941, *Bauer 11450*) counties.

In the white-rayed form, the disk is usually yellow instead of orange- or ruddy-brown, the leaves are paler green, and the stem is pale yellow-green instead of dark green or brownish-purple. These differences are maintained during the life of the individual plant as attested by growing of the plants by the author.

4. **Echinacea paradoxa** (Norton) Britt. Map 2263
*Echinacea atrorubens* Nutt. var. *paradoxa* (Norton) Cron. [BB]
Flowers May–June.
Occurs on limestone glades, barrens, and bald knobs. Central and western Ozark region east to Miller, Phelps, Pulaski, Wright, and Ozark counties, north to Morgan, Benton, and Cedar counties, west to Cedar and Barry counties.
Ranges in Missouri and Arkansas.

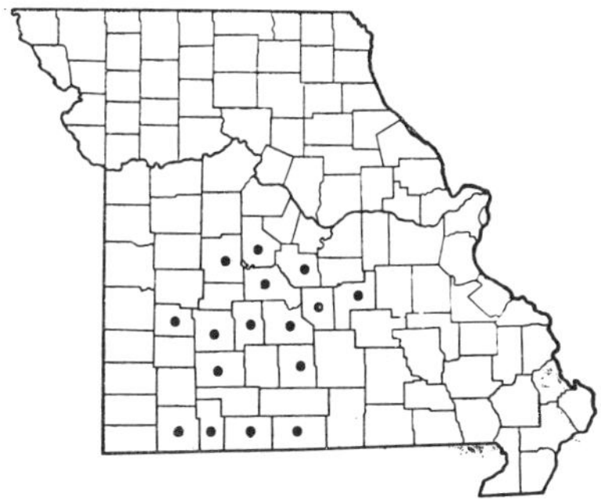

2263 Echinacea paradoxa

2264 Dracopis amplexicaulis (Coneflower)

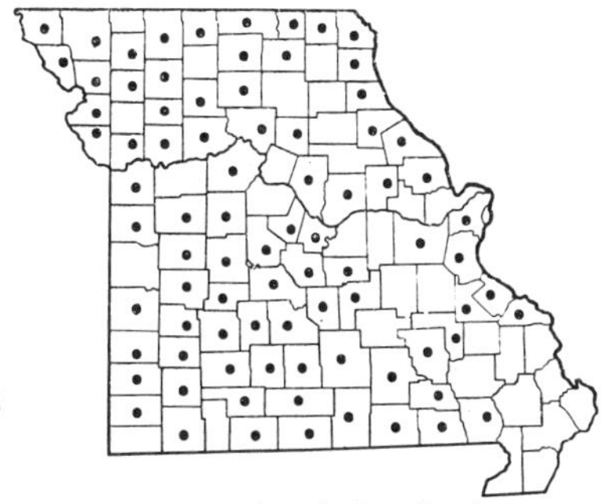

2265 Ratibida pinnata (Gray-head Coneflower)

This is a distinctive species with showy orange-yellow rays and glabrous or sparsely strigose stems and leaves. Cronquist (Rh. 47: 397–98. 1945) reduced this to a variety of *E. atrorubens*, a species with purple rays, but recent field and cytological studies by Dr. Ronald L. McGregor (personal communication based on preliminary investigations) indicate that *E. paradoxa* is a distinct species of Missouri and Arkansas, reports of it from Oklahoma, Kansas, and Texas being based on the occurrence of *E. atrorubens* and an undescribed taxon.

## 38. **Dracopis** Cass. Coneflower

**Dracopis amplexicaulis** (Vahl) Cass. Map 2264
*Rudbeckia amplexicaulis* Vahl [G, P & S]
Flowers June–July.

Occurs in prairies, chert barrens along streams, waste ground and along roadsides. Western Missouri, in Jackson (*Bush 2100, 4030*; *Mackenzie*), Jasper (*Palmer 2255*), and Newton (*Palmer 2388*) counties.

Ranges from Georgia to Texas, north to Missouri and Kansas.

In addition to the presence of pales (chaff) subtending the ray flowers, which distinguishes the genus from *Rudbeckia* from the standpoint of gross morphology, there are cytological differences in the chromosome number and morphology as recently shown by the studies of Dr. Perdue (Contr. Gray Herb. 185: 129–62. 1959).

## 39. **Ratibida** Raf. Prairie Coneflower

Commonly encountered species; rays usually 3–5 cm. long, much longer than the disk (solid part between the rays); disk somewhat globe-shaped, ovoid, or shortly ellipsoid, 12–20 mm. high, 1–1½ times as long as thick; leaf-segments lanceolate, those of the lower and middle leaves usually toothed . 1. R. PINNATA
Rarely encountered species; rays 1.5–2.5 (rarely 3) cm. long, equaling or shorter than the disk; disk cylindric or column-like, 15–40 mm. high, 2–4 times as long as thick; leaf-segments mainly linear to narrowly oblong, usually entire (without teeth) or nearly so, occasionally 2–3-cleft . . . 2. R. COLUMNIFERA

1. **Ratibida pinnata** (Vent.) Barnh. Gray-head Coneflower Map 2265
Also called Drooping Coneflower.
Flowers last of May–September.

Occurs in prairies, thickets, borders of woods, and along railroads. Throughout Missouri, except absent from the lowland counties of the extreme southeastern section.

Ranges from Ontario and New York to Minnesota, South Dakota, and Nebraska, south to Georgia, Arkansas, and Oklahoma.

2. **Ratibida columnifera** (Nutt.) Woot. & Standl. Long-head Coneflower Map 2266
Flowers June–September.

Occurs in prairies, waste ground, and along roadsides and railroads. Scattered in Missouri.

Ranges from Alberta to Mexico, east to Manitoba, Minnesota, Illinois, Missouri, Arkansas, Oklahoma, and Texas; introduced eastward to New England.

Two variations occur in Missouri:

Rays of the flower-heads yellow . . . . .
2a. R. COLUMNIFERA f. COLUMNIFERA

2266 • Ratibida columnifera f. columnifera (Long-head Coneflower)
2266 □ Ratibida columnifera f. pulcherrima

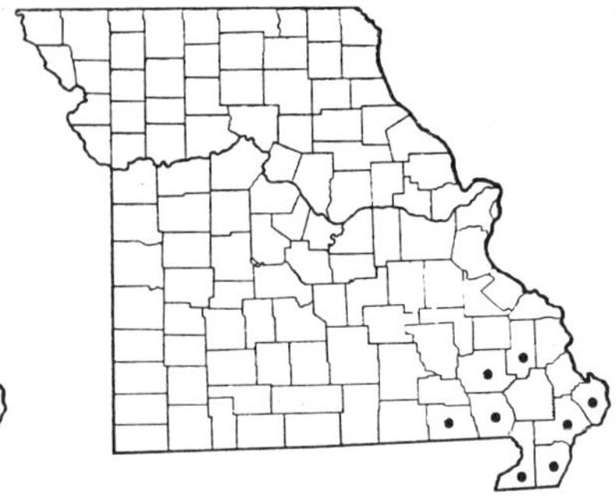

2267 Spilanthes americana var. repens

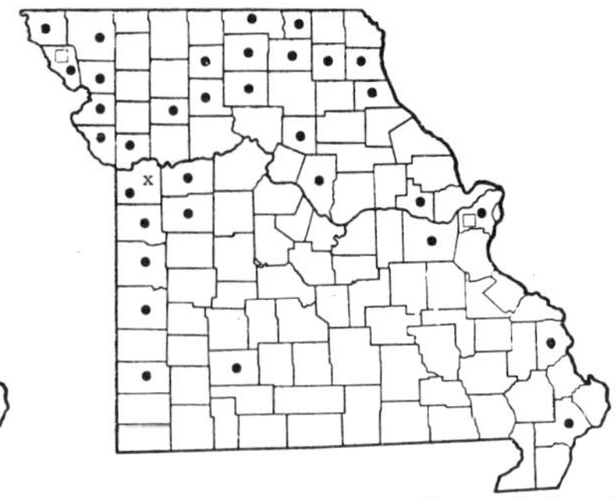

2268 • Helianthus annuus var. annuus (Common Sunflower)
2268 x Helianthus annuus var. lenticularis
2268 □ Helianthus annuus var. nanus 'Florepleno'

Rays partly or completely brown-purple or purple . . 2b. R. COLUMNIFERA f. PULCHERRIMA

2a. **Ratibida columnifera** f. **columnifera** Map 2266

*Ratibida columnifera* (Nutt.) Woot. & Standl. [G, BB]

*Ratibida columnaris* (Pursh) D. Don [P & S]

This is the commoner variation in Missouri.

2b. **Ratibida columnifera** f. **pulcherrima** (DC.) Fern. Map 2266

*Ratibida columnaris* f. *pulcherrima* (DC.) Sharp [P & S]

Known only from St. Louis and Jackson counties, central Missouri.

## 40. **Spilanthes** Jacq.

**Spilanthes americana** (Mutis) Hieron. var. **repens** (Walt.) A. H. Moore Map 2267

Flowers July–October.

Occurs in low wet woods and borders of swamps and bayous. Lowlands of southeastern Missouri, north to Bollinger and Wayne counties, west to Ripley County.

Ranges from Florida to Texas, north to North Carolina, Illinois, Missouri, and Oklahoma; also in Mexico and Central America.

---

Plate no. 369. **1.** Echinacea paradoxa, $\times\ {}^{2}/_{7}$. **2.** Ratibida columnifera, $\times\ {}^{2}/_{7}$; *a.* Achene, $\times\ 2{}^{6}/_{7}$; After Gleason, The New York Botanical Garden. **3.** Helianthus salicifolius, $\times\ {}^{2}/_{7}$. **4.** Dracopis amplexicaulis, $\times\ {}^{2}/_{7}$; After Gleason, details from Small, The New York Botanical Garden. **5.** Helianthus annuus, $\times\ {}^{2}/_{7}$. **6.** Helianthus petiolaris, $\times\ {}^{2}/_{7}$. **7.** Spilanthes americana var. repens, $\times\ {}^{2}/_{7}$; *a.* Receptacle of flower-head with pales, $\times\ 1{}^{1}/_{7}$; After Gleason, details from Small, The New York Botanical Garden. **8.** Ratibida pinnata, $\times\ {}^{2}/_{7}$; After Gleason, details from Small, The New York Botanical Garden.

PLATE NO. 369

### 41. **Helianthus** L. Sunflower

This genus includes many distinct, easily recognizable species in Missouri. The variation and interpretation of the limits of variation of some species, such as *H. hirsutus*, *H. strumosus*, and *H. tuberosus*, have led to the recent investigations beginning with E. E. Watson's monograph of the genus (Papers Mich. Acad. Sci. 9: 305–475. 1929) to the work of Heiser (Am. Midl. Nat. 51: 287–305. 1954), Long (Rh. 56: 198–203. 1954; Am. Midl. Nat. 54: 61–64. 1955; Am. Jour. Bot. 42: 769–77. 1955), and others. Studies are in progress by Heiser, Long, Dale Smith, and Ray Jackson and valuable observations have been made by Deam (*Fl. Ind.* pp. 970–78. 1940) on the differentiation of various species. The present treatment takes into consideration the works of these authors, supplemented by the present author's own field observations, aided also by fifteen years of observations of cultivated plants.

In collecting material of some species, such as *H. hirsutus*, *H. tuberosus*, *H. strumosus*, and *H. decapetalus*, it is helpful to have the lower parts of the plants represented and available for study. Although Deam (*Fl. Ind.* p. 970. 1940) did not find any indication of hybridization of species among the plants he observed in his garden or in the herbarium, the present author has observed spontaneous crosses on more than one occasion between *H. mollis* and *H. occidentalis* in his own growing plots in northern Illinois, and reference to Dr. Long's various publications (above cited) show that various species of sunflowers do hybridize.

a. Main leaves of stem very narrow, linear, linear-oblong, or narrowly lanceolate, mainly 0.1–1 cm. (rarely up to 1.5) broad . . . *b*
 b. Disk-flowers (in solid part of head between rays) or the corolla-lobes of disk-flowers yellow; upper part of stem, leaves, and bracts of involucre covered with short gray-white appressed hairs (pressed against or lying parallel to surface); some or all of the leaf-blades folded lengthwise 11. H. MAXIMILIANI
 b. Disk-flowers or the corolla-lobes of disk-flowers reddish-brown or purple; stem, leaves, and bracts of involucre glabrous (without hairs) or sparsely hairy, but not covered with short gray-white hairiness; leaves flat and plane, not folded . . . *c*
  c. Stem smooth and glabrous (without hairs) or nearly so, sometimes glaucous (with a silvery or bluish-green coating); leaves glabrous or nearly so, conspicuously drooping, crowded, narrowly linear to almost thread-like, 1–3 mm. broad; plants of southwestern and west-central Missouri 3. H. SALICIFOLIUS
  c. Stem, at least in the lower portion, more or less hairy, not glaucous; leaves somewhat hairy or rough-hairy, narrowly lanceolate to linear, mainly 3–10 mm. (up to 15) broad; plants of southeastern Missouri . . . 4. H. ANGUSTIFOLIUS
a. Main leaves of stem ovate to narrowly lanceolate, mainly 1–30 cm. broad, but when 1 cm. broad then generally less than 10 times as long as broad or the disk yellow . . . *d*
 d. Most or all of the leaves alternate . . . *e*
  e. Leaves broadly to triangular ovate or oblong-lanceolate, as broad as long or only 2–4 times as long as broad; annuals with slender fibrous roots; receptacle nearly flat; disk brown to dark purple-red, rarely yellow . . . *f*
   f. Pales (chaff) in center of flower-head with conspicuous white hairs at the tip; bracts of the involucre with very short hairs on the margins or these scarcely showing; achenes hairy, 1.2–2.5 mm. broad; leaves with appressed hairs (pressed against or lying parallel to surface) but not rough to the touch (scabrous), the lower leaves usually not heart-shaped 2. H. PETIOLARIS
   f. Pales (chaff) of the flower-head without conspicuous white hairness at the tip, only inconspicuously short-hairy; bracts of the involucre with conspicuous long hairs on the margins and usually long-hairy on the back surface; achenes glabrous (without hairs), 4–8 mm. broad; leaves rough-hairy (scabrous), the lower leaves usually heart-shaped in well developed plants . . . 1. H. ANNUUS
  e. Larger leaves of stem narrowly lanceolate or oblong-lanceolate, mainly 7–15 (rarely 3) times as long as broad; perennials with thick, coarse woody roots and short or long underground rootstocks; receptacle convex, dome-shaped or low conical; disk yellow . . . *g*
   g. Stems glabrous (without hairs); hairs on upper surface of leaves sparse, widely separated 16. H. GROSSESERRATUS
   g. Stems more or less hairy in the upper portion below the inflorescence; hairs on upper surface of leaves dense, close to one another . . . 11. H. MAXIMILIANI

d. Most of the leaves, but at least all the lower ones, opposite . . . . . . . . . . . . . . . *h*

h. Corolla-lobes of disk-flowers (those in the solid part between the rays) purplish-brown, reddish-brown, or brown-purple . . . . . . . . . . . . . . . . . . . . . . . . . . . . . *i*

i. Found throughout the state, mainly in northern, central, and western Missouri; leaves long-tapering and pointed at tip; leaves of stem gradually tapering to a short petiole 0.5–3 cm. long or nearly sessile (stalkless); bracts of involucre mostly tapering to an acute tip, narrowly ovate or lanceolate; upper leaves opposite or alternate . . . . . . . 6b. H. LAETIFLORUS var. RIGIDUS

i. Found only in extreme southern Missouri west to Ozark County, north to Scott, Wayne, and Carter counties; leaves of stem rounded or blunt at tip; leaves of stem abruptly contracted to a slender petiole (stalk) 1–4 cm. long; bracts of involucre rounded at tip, oval or oblong-ovate; upper leaves always alternate . . . . . . . . . . . . . . . . 5. H. SILPHIOIDES

h. Corolla-lobes of disk-flowers (those in the solid part between the rays) yellow . . . . . . . . *j*

j. Bracts of the involucre tightly appressed (pressed close against surface or lying parallel to surface) and without loose, long-pointed, or long-tapering tips . . . . . . . . . . . *k*

k. Most of leaves crowded near base of plant, the stem rather naked above with leaves abruptly reduced above in size; only 3–8 pairs of leaves present on stem below the inflorescence; petioles (leaf-stalks) of lowest leaf-blades 2.5–15 cm. long, winged only at summit. 7. H. OCCIDENTALIS

k. Leaves more or less uniform on the stem, only gradually decreasing in size up the stem; usually 9–15 pairs of leaves present on stem below the inflorescence; petioles (leaf-stalks) of lowest leaf-blades 0.5–3 cm. long, the leafy tissue of the blade bordering the petiole (decurrent) nearly to its base. . . . . . . . . . 6a. H. LAETIFLORUS var. LAETIFLORUS

j. At least the outer bracts of the involucre loose and spreading, the tips long-pointed or long-tapering . . . . . . . . . . . . . . . . . . . . . . . . . . . . . . *l*

l. Most of leaves crowded near base of plant, the stem rather naked above with leaves abruptly reduced in size; only 3–8 pairs of leaves present on stem below the inflorescence 7. H. OCCIDENTALIS

l. Most of leaves more or less uniform on the stem, only gradually decreasing in size up the stem; usually 8–15 or more pairs of leaves present on stem below the inflorescence. . . . . *m*

m. Heads small, 1.5–3 cm. broad; the disk (solid part between rays) 4–10 mm. broad; the rays 1–1.5 cm. long, 5–8 . . . . . . . . . . . . . . 12. H. MICROCEPHALUS

m. Heads larger, 4–9 cm. broad; the disk (solid part between rays) 10–25 mm. broad; the rays 2–4 (rarely 1.5) cm. long, 8–30 or more . . . . . . . . . . . . . . . *n*

n. Main leaves of stem either without petioles (stalks) or on very short petioles up to 4 mm. long . . . . . . . . . . . . . . . . . . . . . . . . . . . *o*

o. Lowest lateral nerves (those on either side of midrib) joining the midrib 1 cm. or more above the base of leaf-blade . . . . . . . . . . . . 8. H. MOLLIS

o. Lowest lateral nerves (those on either side of midrib) joining the midrib either at the very base or less than 1 cm. above the base of the blade. . . . . . . . *p*

p. Stem glabrous (without hairs) or nearly so, below the inflorescence; lowest lateral nerves (those on either side of midrib) joining the midrib at the very base of the blade; disk (solid part of flower-head between rays) 10–15 mm. broad . . . . . . . . . . . . . . . . 9. H. DIVARICATUS

p. Stem usually hairy, sometimes glabrous; lowest lateral nerves joining the midrib usually slightly above the very base of the blade (rarely at very base); disk mainly 15–30 mm. (sometimes 10) broad . . . . . . . . . . . . . *q*

q. Leaves densely gray-hairy on both surfaces, conspicuously rounded or heart-shaped at the clasping base, sessile (without a stalk); stem very densely gray-hairy; rays of flower-heads 15–30; disk 20–30 mm. broad . . . 8. H. MOLLIS

q. Leaves dark green on upper surface, rough-hairy, but not densely gray-hairy, usually not conspicuously rounded or heart-shaped nor clasping at base, on petioles (stalks) usually 1–4 mm. long; stem hairy, but not densely gray-hairy; rays of flower-heads 8–15; disk 10–20 mm. broad . . 10. H. HIRSUTUS

n. Main leaves of stem on petioles (stalks) more than 5 mm. long . . . . . . . . *r*

r. Principal leaves of stem with 1 main nerve or pinnately nerved, with several pairs of equally prominent side nerves arising at different levels along the mid-nerve; leaves arising at more than 20 nodes along the stem. . . . . 16. H. GROSSESERRATUS

r. Principal leaves of stem with 3 main nerves (a midrib and 2 lateral nerves at base of blade) the most conspicuous, other lateral (side) nerves less conspicuous; leaves arising at fewer than 20 nodes along the stem . . . . . . . . . . . . *s*

s. Stems glabrous (without hairs) or nearly so below the inflorescence and down the length of stem; lower surface of leaf-blade mostly without yellow or white glands . . . . . . . . . . . . . *t*

t. Leaf-blades of the main leaves of stem usually coarsely and conspicuously toothed, thin and membranaceous; petioles (leaf-stalk) 15–60 mm. long; bracts of involucre conspicuously loose, as long as or longer than the disk (solid portion of flower-head between the rays), many of them recurving and some of the outer ones frequently leaf-like or enlarged; upper leaves of main stem often alternate . . . . . . . . . . . . . . . . . . . . . 14. H. DECAPETALUS

t. Leaf-blades of the main leaves of stem shallowly toothed or sometimes nearly entire (without teeth), thick and firm; petioles (leaf-stalk) 6–30 mm. long; bracts of involucre somewhat loose, usually a few more or less spreading, usually shorter than or slightly longer than the disk; upper leaves of main stem opposite . . . . . . . . . . . . . . . . . 13. H. STRUMOSUS

s. Stems more or less hairy below the inflorescence and down part of the stem; lower surface of leaf-blade mostly with yellow or white glands in addition to the hairs . . . . . . . . . . . . *u*

u. Upper leaves of stem alternate (opposite in var. *subcanescens*); base of the main leaf-blades of stem decurrent (green tissue of leaf-blade extending down from base of leaf-blade along petiole) far down the sides of the petiole (leaf-stalk), the petioles broadly winged above their middle; petioles (leaf-stalks) of main leaves of stem mostly 20–80 mm. long; creeping underground rootstocks often bearing thickened tubers at their tip; at least the inner bracts of the involucre dark-colored, sometimes nearly black; larger leaf-blades 6–15 cm. broad; rays of flower-heads 2.5–4 cm. long 15. H. TUBEROSUS

u. Upper leaves of stem usually opposite; base of main leaf-blades of stem broadly rounded, truncate, or heart-shaped, only slightly decurrent (green tissue of leaf-blade extending below blade on petiole) on petiole; petioles (leaf-stalks) of main leaves of stem 5–15 mm. long; no tubers present at tip of creeping underground rootstock; bracts of involucre all green; larger leaf-blades 0.7–9 cm. broad; rays 1.5–3 cm. long . . . . . . . . . . . . . . . . . 10. H. HIRSUTUS

**Helianthus annuus** L. Common Sunflower Map 2268

Flowers July–November.

Occurs in waste and cultivated ground, low meadows and prairies, along roadsides, and railroads. Throughout Missouri, but not recorded from the majority of the counties of the southern part of the state.

The following variations are known from the state:

a. Dwarf plants 1 m. or less tall with most or all the disk-flowers of the ray type . . 1c. H. ANNUUS var. NANUS 'FLOREPLENO'

a. Plants generally 1–3.5 m. tall (rarely 0.5 m.) with normally developed disk-flowers . . . *b*

b. Common type encountered; bracts of involucre 7–10 mm. broad; disk (solid part of flower-head between rays) 3–5 cm. broad; rays 21–35, 3–5 cm. long . . 1a. H. ANNUUS var. ANNUUS

b. Rare type encountered, known only from Jackson County, west-central Missouri; bracts of involucre 4–7 mm. broad; disk (solid part of flower-head between rays) mostly 2–3.5 cm. broad; rays 17–26, 2.5–3.9 cm. long . 1b. H. ANNUUS var. LENTICULARIS

1a. **Helianthus annuus** var. **annuus** Map 2268

*Helianthus annuus* L. [G, BB, P & S]

*Helianthus annuus* subsp. *annuus* [Heiser]

*Helianthus annuus* subsp. *annuus* var. *annuus*

This is the common variation in the state.

Ranges in the central and eastern United States extending into south central and eastern Canada. According to Heiser (Am. Midl. Nat. 51: 300. 1954), the distribution of this variation is exceedingly sporadic east of the Mississippi River, in which area he considers it to be of recent introduction.

1b. **Helianthus annuus** var. **lenticularis** (Dougl.) Steyerm. Map 2268

*Helianthus lenticularis* Dougl. [P & S]

*Helianthus annuus* subsp. *lenticularis* (Dougl.) Cockerell, Science, n. ser. 40: 284. 1914. [Heiser]

Known only from Jackson County, west-central Missouri.

Ranges in western North America, east to North and South Dakota, Colorado, Kansas, Nebraska, and Oklahoma, where it grades into the eastern var. *annuus*, according to Heiser (Am. Midl. Nat. 51: 299. 1954).

The achenes of this variation were gathered for food by various tribes of Indians, according to Heiser (Proc. Am. Phil. Soc. 95: 432–48. 1951).

1c. **Helianthus annuus** var. **nanus 'Florepleno'** Hort. Map 2268

Known only from Holt (low ground in extensive swale, T62N, R40W, sect. 25, 3¾ mi. southeast of Craig, July 20, 1952, *Steyermark 73788*) and St. Louis (St. Louis, Baden freight yard of Mo.-Kan.-Tex. R.R., July 5, 1958, *Muehlenbach 1444*) counties. The Mueh-

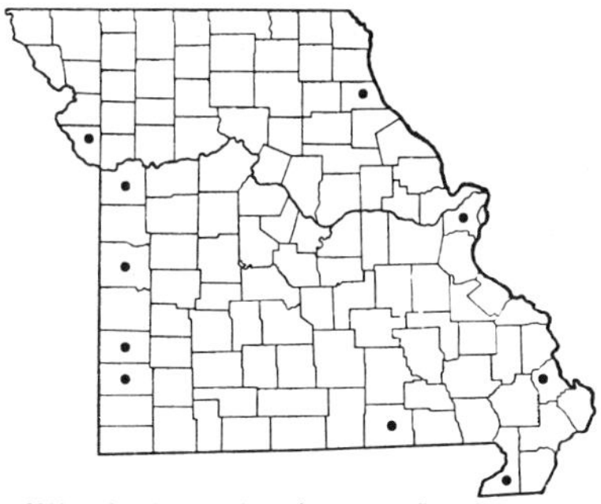

2269 Helianthus petiolaris (Prairie Sunflower)

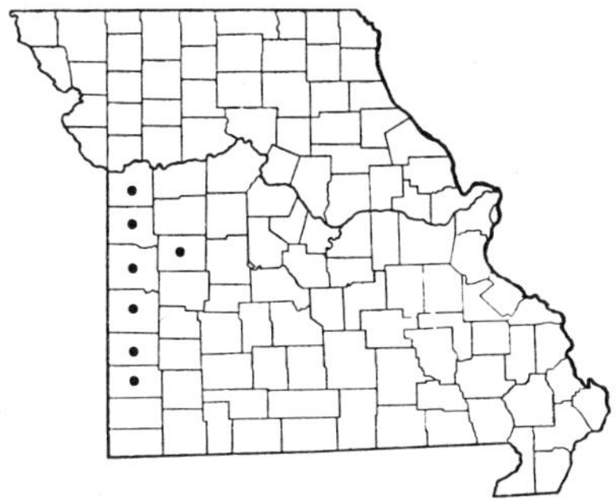

2270 Helianthus salicifolius

2271 Helianthus angustifolius

lenbach collection has small heads with the disk only 7–12 mm. broad, rays 2 cm. long and 5 mm. broad, and leaf-blades only 2–3 cm. long and 1–1.5 cm. broad.

The commonly cultivated sunflower, var. *macrocarpus* (DC.) Cockerell, with a disk 5.5 cm. or more broad, involucral bracts 8.5 mm. or more broad, 30–70 rays 3.5–10 cm. long, and achenes 6.5–15 mm. long, is not known in the naturalized state in Missouri. According to Heiser (Am. Midl. Nat. 51: 301. 1954) it is not able to maintain itself in the wild state for any length of time. This variety is the one cultivated for its oily seed and grown in many gardens. The oil of sunflower seed is sometimes used as an extender for some of the major drying oils. The meal of the seed contains 52.5 per cent protein. The seeds are greatly relished as food by many wild birds at bird-feeding stations as well as by poultry. The seeds are also sold for human consumption. The oil from the seed is used in margarines and lard substitutes and also is used as a salad oil. In medicinal and food value the oil is considered equal to that of olive oil. Cattle and other stock eat an oil cake from the seeds, and the entire plant is grown for ensilage. The semidrying properties of the oil find application and use in the soap, paint, and varnish industries.

The American Indians used the seeds of other species of sunflower for food in the form of flour for bread, cakes, soups, and oil. The roasted seeds were also prepared into a beverage reported to have a taste resembling coffee. The seeds of this species were eaten by the early Bluff-Dweller peoples of the Ozarks, as attested by seed remains in their shelters.

The pollen of this species is considered to contribute to hay fever suffering in late summer, becoming atmospheric to a slight extent.

The corollas of the disk-flowers are usually reddish-purple with purple, red, or black anthers, but may also be yellow. Those with yellow corollas generally have black anthers. These yellow-flowered forms are usually found in cultivation. Heiser has studied the variation of *H. annuus* (Am. Midl. Nat. 51: 287–305. 1954).

2. **Helianthus petiolaris** Nutt. Prairie Sunflower Map 2269

Also called Kansas Sunflower.

Flowers May–October.

Occurs mainly along railroads, also along roadsides, in fields and waste ground. Rare and scattered in the state in southern and central Missouri north to Marion and Platte counties.

Ranges from Manitoba and Minnesota to Washington, south to Louisiana, Texas, New Mexico, and Arizona; introduced east to New England, New Jersey, and Virginia.

This is the earliest species of the genus to flower in Missouri, beginning the last of May in some sections.

3. **Helianthus salicifolius** A. Dietr. Map 2270

Flowers August–October.

Occurs on upland prairies, limestone glades, and along roadsides. Unglaciated prairie region of western Missouri south of the Missouri River from Jasper County north to Jackson County east to Henry County.

Ranges from Nebraska to Colorado, south to Missouri, Arkansas, Oklahoma, and Texas; introduced in Illinois and Ohio.

This is a very showy and handsome species, with gracefully drooping pale green leaves, whitish stems, and many heads of flowers in a paniculately branched inflorescence. It is a fine subject for the open garden, but eventually forms dense beds from the creeping rhizomes, and has to be kept within limits. The seeds are eaten by bobwhite, prairie chicken, and other wild birds, and the plant is nutritious and palatable to all classes of livestock.

4. **Helianthus angustifolius** L. Map 2271

Flowers August–October.

2272 Helianthus silphioides

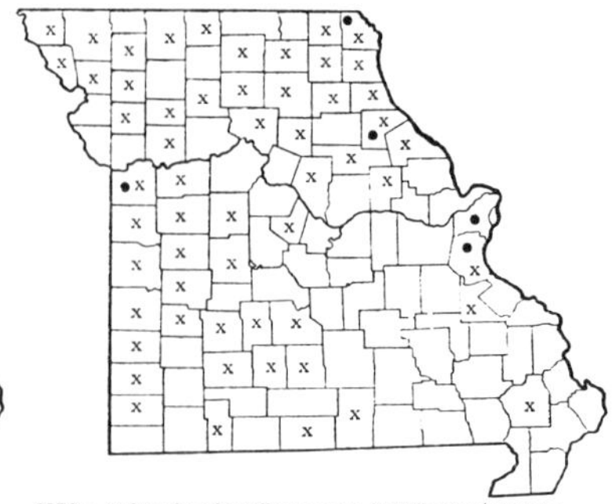

2273 • Helianthus laetiflorus var. laetiflorus (Prairie Sunflower)
2273 x Helianthus laetiflorus var. rigidus

2274 Helianthus occidentalis var. occidentalis

Occurs in sandy prairies, low sandy ground along roads, thickets, and open woods, in acid soils. Rare, known only from southeastern Missouri, in Scott, Mississippi, Butler, and Ripley counties.

Ranges from Florida to Texas, north to New York, New Jersey, Pennsylvania, Kentucky, Indiana, and Missouri.

This is an attractive species with showy heads of dark purple disks and yellow rays. I have grown this species for a number of years in northern Illinois, but it tends to disappear after a few years.

5. **Helianthus silphioides** Nutt. Map 2272

*Helianthus atrorubens* L. var. *pubescens* Ktze. [BB, P & S]

Flowers August–October.

Occurs in low sandy or alluvial soils near streams, fallow fields, borders of woods, dry upland open woods, thickets, and roadsides, generally in acid soils. Southern Missouri west to Ozark County and north to Scott, Stoddard, Wayne, and Carter counties.

Ranges from Alabama to Louisiana, north to Kentucky, Illinois, and Missouri.

This is a very showy and tall species, the stems attaining a height of 3 meters. It is easily recognized by the rounded-ovate to nearly orbicular, blunt-tipped or rounded leaves. I have grown this species in northern Illinois in my wildflower preserve for many years, and it has done well in an open sunny situation.

6. **Helianthus laetiflorus** Pers. Prairie Sunflower Map 2273

Flowers August–October.

Occurs in prairies, thickets, rocky open woods, and along roadsides.

Two varieties are found in Missouri:

Corollas of disk-flowers with yellow lobes
6a. H. LAETIFLORUS var. LAETIFLORUS

Corollas of disk-flowers with dark purple or reddish-brown lobes . . . . 6b. H. LAETIFLORUS var. RIGIDUS

6a. **Helianthus laetiflorus** var. **laetiflorus** Map 2273

*Helianthus laetiflorus* Pers. [G, BB, P & S]

This is rarely encountered and is known only from Clark, Ralls, St. Louis, Jefferson, and Jackson counties.

Ranges from New York to Saskatchewan, south to Georgia, Missouri, and Nebraska; escaped from cultivation and naturalized east to New England and New Jersey.

6b. **Helianthus laetiflorus** var. **rigidus** (Cass.) Fern. Map 2273

*Helianthus rigidus* (Cass.) Desf. [P & S]

Common throughout the glaciated prairie region of northern and central Missouri and the unglaciated prairie region of southwestern Missouri; absent from most of the Ozark section, extending south and east to Jefferson, St. Francois, Stoddard, Montgomery, Boone, Moniteau, Laclede, Wright, and Howell counties.

Ranges from Ontario and Michigan west to Saskatchewan, Nebraska, and Montana, south to Georgia, Alabama, Louisiana, Texas, and New Mexico.

This forms dense colonies by means of the creeping underground rootstocks. It is a common species of dry upland prairies, associated with *Silphium terebinthinaceum*, *S. laciniatum*, *Petalostemon purpureum*, various asters, goldenrods, and grasses.

7. **Helianthus occidentalis** Riddell var. **occidentalis** Map 2274

*Helianthus occidentalis* Riddell [G, P & S]

Flowers July–September.

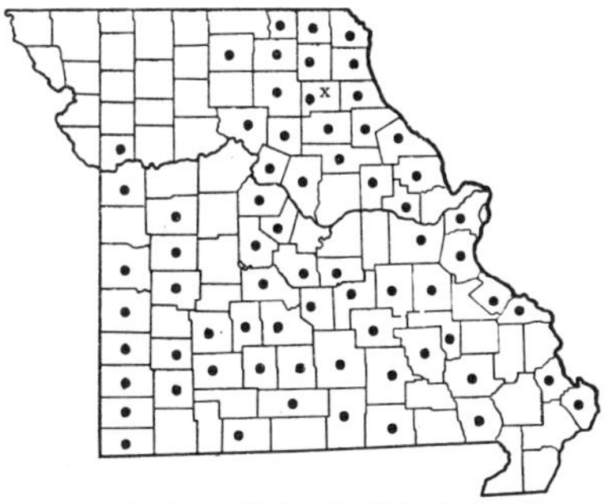


2275 • Helianthus mollis f. mollis (Ashy Sunflower)
2275 x Helianthus mollis f. flavidus

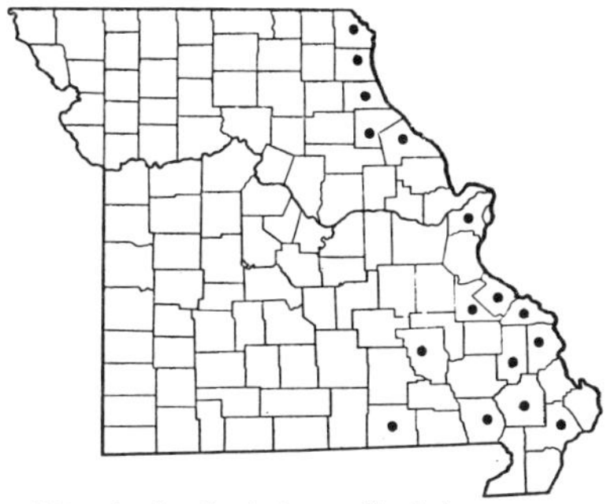

2276 Helianthus divaricatus var. divaricatus

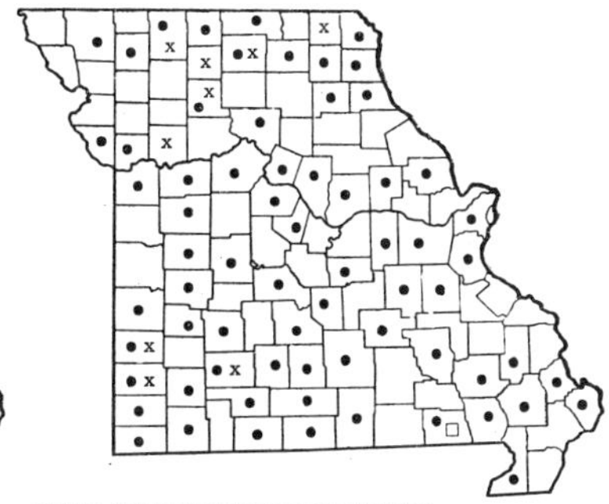


2277 • Helianthus hirsutus var. hirsutus
2277 x Helianthus hirsutus var. trachyphyllus
2277 □ Helianthus hirsutus var. stenophyllus

Occurs in glades, prairies, dry meadows, fields, and rocky open woods. Ozark region of southern and central Missouri, north to St. Louis, Franklin, Maries, Morgan, Benton, Cedar, and Vernon counties.

Ranges from Florida, Georgia, Tennessee, Arkansas, and Texas, north to Ohio, Indiana, Illinois, Missouri, and Minnesota; introduced east in New Jersey.

This species occasionally crosses with *H. mollis*, and is known from Laclede County, in the western Ozark section (see cited specimens given under *H. mollis*). The same cross arose spontaneously in the prairie section of my wildflower garden in northern Illinois, where both parent species are growing near one another.

A var. *Dowellianus* (Curtis) T. & G., not recorded from Missouri, differs from typical var. *occidentalis* in having the rather uniform leaves not reduced in size and extending nearly halfway to the summit of the stem, the lower leaves of the stem broadly to narrowly ovate and larger (5–10 cm. broad and 7–20 cm. long) than var. *occidentalis*.

8. **Helianthus mollis** Lam. Ashy Sunflower
Map 2275

Flowers July–October.

Occurs in prairies, rocky glades, fields, thickets, roadsides, and railroads. Common throughout southern, central, and eastern Missouri west and north to Schuyler, Sullivan, Macon, Chariton, and Clay counties.

Ranges from Ohio to Michigan and Iowa, south to Georgia, Alabama, Louisiana, and Texas; introduced from cultivation east to Massachusetts.

The following varieties occur in the state:

Flowers of the disk pale yellow; ray flowers pale yellow . . . . . 8b. H. MOLLIS f. FLAVIDUS
Flowers of the disk deep yellow or orange-yellow; ray flowers deep yellow or orange-yellow . . 8a. H. MOLLIS f. MOLLIS

8a. **Helianthus mollis** f. **mollis** Map 2275

*Helianthus mollis* Lam. [G, BB, P & S]

This is the commonest variation found in Missouri.

8b. **Helianthus mollis** f. **flavidus** Steyerm.
Map 2275

*Helianthus mollis* f. *flavida* Steyerm., Rh. 55: 108. 1953.

Known only from Shelby County, northeastern Missouri (prairie along railroad, route 36, 4.4 mi. northwest of western limit of Lentner, August 21, 1950, *Steyermark 70126*, holotype in Chi. Nat. Hist. Mus. Herb.).

The plants transplanted to the author's wildflower beds in northern Illinois have spread and have maintained the pale yellow color of the disk and ray flowers; the ray flowers are usually shorter than in typical f. *mollis*.

Fernald (in *Gray's Manual*, eighth edition) recognized a var. *cordatus* S. Wats., based on leaves 5–8 cm. broad and deeply heart-shaped and clasping at the base, but there appears to be no constancy in the leaves even on the same plant, nor from plant to plant within a colony.

This sunflower forms dense colonies from the creeping underground rhizomes, and is difficult to keep within bounds when planted in a garden. A natural hybrid between this species and *H. occidentalis* is known from Laclede Co. (1937, *George Moore*; Lebanon, September 3, 1939, *George Moore*, in Chi. Nat. Hist. Mus. Herb.). Spontaneous hybrid plants also are growing at the author's wildflower garden in northern Illinois between *H. mollis* and *H. occidentalis*. This hybrid has been previously recorded only from Indiana by Jackson and Guard (Proc. Ind. Acad. 65: 212. 1956). The hybrid plants are grayish-green as in *H. mollis*, and have the leaves uniformly appearing throughout the length of the stem as in *H. mollis*, but the leaves are narrowed at the base and have a shape resembling more the leaves of *H. occidentalis*. The purplish to brownish-purple color of the stems also resembles that of *H. occidentalis*. The inflorescence and size of flower-heads resemble *H. occidentalis*, but the involucral bracts have the gray pubescence of *H. mollis*.

9. **Helianthus divaricatus** L. var. **divaricatus** Map 2276

*Helianthus divaricatus* L. [G, BB, P & S]

Flowers July–October.

Occurs in rocky woods and thickets.

Eastern Missouri, west in southern Missouri to Reynolds and Oregon counties and north along the counties bordering the Mississippi River.

Ranges from Maine and Quebec to Saskatchewan, south to Georgia, Tennessee, and Arkansas.

This species is mainly distinguished by the sessile or nearly sessile, horizontally spreading leaves and glabrous, often glaucous stems. The lateral nerves join the midrib at the very base of the leaf-blade. However, glabrate variations of *H. hirsutus* with short-petioled leaves are difficult to separate from *H. divaricatus*.

10. **Helianthus hirsutus** Raf. Map 2277

Flowers July–October.

Occurs in rocky or dry open woods, thickets, prairies, along roadsides and railroads. Throughout Missouri.

The following intergrading varieties may be recognized in Missouri:

a. Hairs of stem long and conspicuous . . . 10a. H. HIRSUTUS var. HIRSUTUS

a. Hairs of stem rough or stiff but short or practically absent . . . . . . . . . . . *b*

- b. Larger leaves 2–7 cm. broad, 8–18 cm. long, ovate to lanceolate, spreading to somewhat ascending . . . . . . 10b. H. HIRSUTUS var. TRACHYPHYLLUS
- b. Larger leaves 0.7–2 cm. broad, 5–11 cm. long, narrowly lanceolate, usually ascending . . 10c. H. HIRSUTUS var. STENOPHYLLUS

10a. **Helianthus hirsutus** var. **hirsutus** Map 2277

*Helianthus hirsutus* Raf. [G, BB, P & S]

This is the commonest variation encountered in Missouri.

Ranges from Pennsylvania to Minnesota, south to Georgia, Alabama, Arkansas, and Oklahoma.

10b. **Helianthus hirsutus** var. **trachyphyllus** T. & G. Map 2277

Scattered in Missouri.

Ranges from Pennsylvania to Wisconsin and Iowa, south to Tennessee, Arkansas, and Oklahoma.

10c. **Helianthus hirsutus** var. **stenophyllus** T. & G. Map 2277

Known from Ripley County, southern Missouri (wooded slopes along Little Black River, between Greenville Ford and Pennington Ford, T24W, R3E, sect. 10, 15, 22, 23, 26, 24 and 25, 10–13 mi. northeast of Doniphan, September 1, 1946, *Steyermark 63947, 63963*).

Ranges from Louisiana and Texas, north to West Virginia, Ohio, Missouri, and Oklahoma.

The variation in size and shape of leaf, pubescence of stem and peduncles, and length of petiole in *H. hirsutus* is considerable. The varieties recognized above grade into one another and are maintained in the present treatment for tentative convenience, awaiting future, more detailed field and experimental studies for more conclusive data. The species is most easily confused with *H. divaricatus* and *H. strumosus*. From *H. divaricatus* it is mainly distinguished by the usually longer petioles, usually more or less pubescent stem, and by the lateral nerves usually converging with the midrib above the base of the leaf-blade instead of at the very base. From *H. strumosus* it is mainly distinguished by its more or less pubescent stem, leaves green or gray-green, rather than whitened on the lower surface, usually shorter petioles, and by the usually spreading pubescence of the peduncles and midribs.

A specimen from Clark County, Missouri (open woods, ridge above Linn Branch, 2 mi. northwest of Ashton, October 5, 1934, *Drouet 1773*) labeled *H. tomentosus* in the Gray Herbarium, is referred in the present treatment to *H. hirsutus*.

11. **Helianthus Maximiliani** Schrad. Maximilian Sunflower Map 2278

Flowers July–October.

Occurs on limestone glades, bluff escarpments and ledges, bald knobs, rocky prairies, loess hills, roadsides, and waste ground. Southern, central, and western Missouri; native in the westen Ozarks in the White River region, along the Osage River in Cole, Miller,

Plate no. 370. **1.** Helianthus angustifolius, × 2/7. **2.** Helianthus silphioides, × 2/7; *a.* Leaves from middle of stem; *b.* Uppermost part of stem; *c.* Leaves from lower part of stem. **3.** Helianthus laetiflorus var. rigidus, × 2/7. **4.** Helianthus mollis, × 2/7. **5.** Helianthus Maximiliani, × 2/7. **6.** Helianthus occidentalis, × 2/7; *a.* Basal leaf; *b.* Upper part of stem; *c.* Lower and middle parts of stem with leaves. All details from Small, The New York Botanical Garden.

PLATE NO. 370

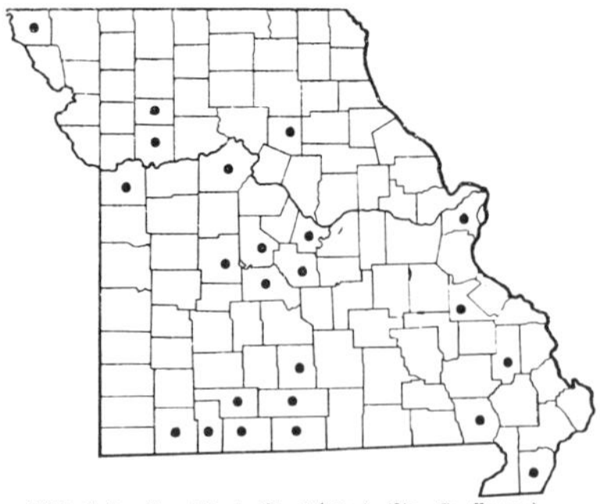

2278 Helianthus Maximiliani (Maximilian Sunflower)

2279 Helianthus microcephalus (Small Wood Sunflower)

2280 Excluded species

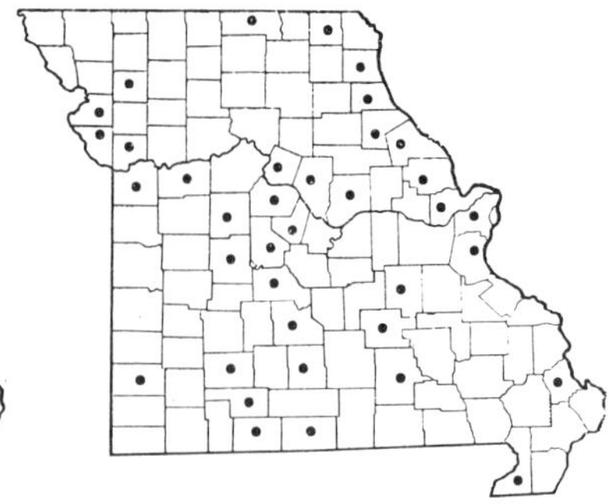

2281 Helianthus strumosus

Camden, Morgan, and Benton counties, and on the loess hills of Atchison County; introduced elsewhere in the state.

Ranges from Minnesota and Manitoba to Saskatchewan, south to Missouri, Oklahoma, and Texas; introduced east to the Atlantic states.

This species is easily recognized by the drooping, mostly folded gray leaves covered with a short appressed pubescence, and by the short-stalked flower-heads alternately arranged in the upper leaf-axils in an elongated spike- or racemelike inflorescence. The plant is very showy in full flower, but a cultivated strain found in gardens around St. Louis and elsewhere is much showier with larger, more numerous flower-heads and broader leaves than in the wild species found on the limestone glades, bluffs, and prairies. The cultivated, showier strain is the one usually seen along roadsides in eastern and part of central Missouri, while the wild species occurs in the western Ozarks and loess hills of northwestern Missouri. It is a very desirable subject for the garden, forming clumps of tall stems up to 3 meters high covered with golden yellow flowers.

12. **Helianthus microcephalus** T. & G. Small Wood Sunflower Map 2279

Flowers August–September.

Occurs in moist or dry open woods, swampy ground along streams, thickets, and sandy fields, usually in acid sandy or clayey soils. Chiefly found on Crowley Ridge in Scott, Stoddard, and Dunklin counties, north to Cape Girardeau County, locally north in the Ozarks in Iron County, and locally west to Ripley County (*Steyermark 63953*).

Ranges from Pennsylvania to Illinois and Missouri, south to Florida and Louisiana.

This species is easily recognized by its relatively small heads of flowers.

The plants may attain 3–4.5 meters in height.

13. **Helianthus strumosus** L. Map 2281

*Helianthus leoninus* E. E. Wats.

*Helianthus hirsutus* var. *leoninus* (E. E. Wats.) Palmer & Steyerm. [P & S]

*Helianthus formosus* E. E. Wats. [P & S]

*Helianthus tracheliifolius* [of G, P & S], not Mill.

Flowers August–September.

Occurs in low alluvial woods and thickets, open woods, along bluffs, along streams, prairies, wet meadows, along roadsides and railroads. Scattered throughout Missouri, probably in every county, but needs to be more intensively collected.

Ranges from Quebec to North Dakota, south to Georgia, Alabama, Arkansas, and Oklahoma.

The often gray-glaucous lower surface of the leaves distinguishes the species usually from *H. hirsutus*. It is confused also with *H. divaricatus* from which it is usually separated by the longer petioles and by the convergence of the lowest lateral nerves, which join the midrib somewhat above the base of the leaf-blade, whereas in *H. divaricatus* the convergence with the midrib is at the very base of the blade.

*Helianthus formosus*, originally described by Watson (Papers Mich. Acad. Sci. 9: 445. pl. 72. 1920) from Missouri (Mississippi bluffs south of Hannibal, September 20, 1913, *Davis 3274*), cannot be separated from *H. strumosus*. It, likewise, has the lower surface of the leaves gray-white. According to Dr. Heiser and Dale Smith, much of the material which has been identified previously as *H. tracheliifolius* should be referred to *H. strumosus*. I am following Dr. Heiser and his students, Drs. Jackson, Long, and Smith in placing *H. leoninus* as a synonym under *H. strumosus*.

14. **Helianthus decapetalus** L. Thin-leaf Sunflower Map 2282

*Helianthus tracheliifolius* Mill.

Flowers August–October.

Occurs in open woods, thickets, and prairies. Scat-

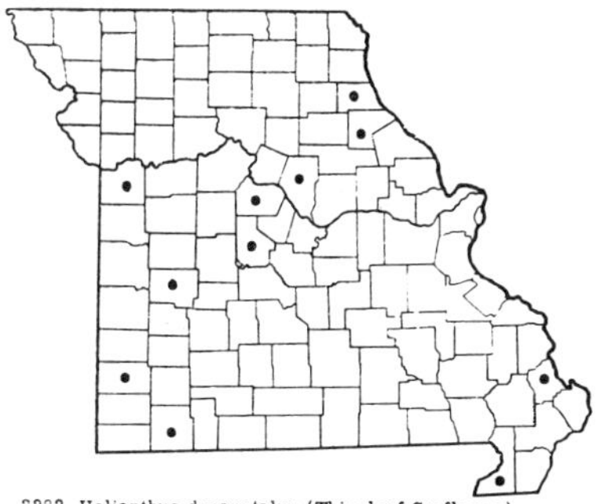

2282 Helianthus decapetalus (Thin-leaf Sunflower)

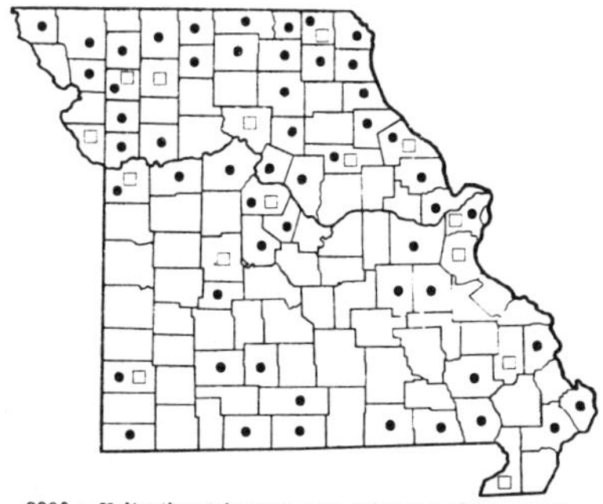

2283 • Helianthus tuberosus var. tuberosus (Jerusalem Artichoke)

2283 □ Helianthus tuberosus var. subcanescens

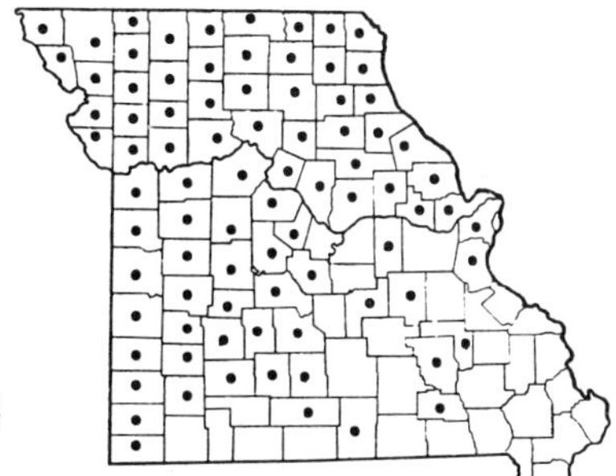

2284 • Helianthus grosseserratus f. grosseserratus (Sawtooth Sunflower)

2284 □ Helianthus grosseserratus f. pleniflorus

tered in eastern, southern, and central Missouri.

Ranges from Maine and Quebec to Minnesota and Nebraska, south to Georgia, Kentucky, and Missouri.

The lower surface of the leaves in this species may be green or silvery-gray.

Much more intensive collecting is necessary in the state before the distribution of this species is delimited.

15. **Helianthus tuberosus** L. Jerusalem Artichoke Map 2283

Flowers August–October.

Occurs in moist and alluvial thickets bordering streams, sloughs, and ponds, often in alluvial soils of valleys, moist or dry woods, thickets, wet places in prairies, waste ground, roadsides, and along railroads. Throughout Missouri.

Two varieties are encountered in Missouri:

Upper leaves of stem alternate; lower surface of leaf-blades minutely hairy . . 15a. H. TUBEROSUS var. TUBEROSUS

All or essentially all of the leaves opposite; lower surface of leaf-blades usually soft-hairy . . . 15b. H. TUBEROSUS var. SUBCANESCENS

15a. **Helianthus tuberosus** var. **tuberosus** Map 2283

*Helianthus tuberosus* L. [G, P & S]

This is the commoner variation in the state.

Ranges from Ontario and Saskatchewan, south to Georgia, Tennessee, Arkansas, and Oklahoma.

15b. **Helianthus tuberosus** var. **subcanescens** Gray Map 2283

*Helianthus virilis* E. E. Wats.

Scattered throughout Missouri.

Ranges from Ohio to Minnesota and Manitoba, south to Missouri and Kansas.

*Helianthus tuberosus* usually bears tubers, but many plants are found without any trace of them. The tubers, when sufficiently developed, can be cooked into a vegetable. The texture is crisp, and the flavor is agreeable and nutlike. These tubers may also be eaten raw or pickled. They contain a carbohydrate, inulin, in the form of a fruit sugar (levulose) which can be eaten by diabetic persons. The plants are also grown for forage.

16. **Helianthus grosseserratus** Martens Sawtooth Sunflower Map 2284

Flowers July–October.

Occurs in prairies, moist places at base of bluffs along streams, low meadows, moist thickets, fields, borders of streams, ponds, and ditches, along roadsides and railroads.

Throughout Missouri, except not recorded from many of the eastern Ozark counties, and absent from the lowlands of southeastern Missouri.

Ranges from Ohio to North Dakota, south to Arkansas and Texas; introduced east to the Atlantic states.

The following variations may be recognized:

Common type; disk-flowers of a head all normal. 16a. H. GROSSESERRATUS f. GROSSESERRATUS

Rare type; flowers of a head all resembling rays. 16b. H. GROSSESERRATUS f. PLENIFLORUS

16a. **Helianthus grosseserratus** f. **grosseserratus** Map 2285

*Helianthus grosseserratus* Martens [G, BB, P & S]

*Helianthus instabilis* E. E. Wats., in part [P & S]

This is the commoner variation found in Missouri.

16b. **Helianthus grosseserratus** f. **pleniflorus** Wadmond Map 2285

Scattered within the range of f. *grosseserratus*.

This is a common species of meadows and prairies with tall stems up to 5 meters high. Dr. Robert W. Long, Jr., (Rh. 56: 199. 1954) distinguishes two subspecies, based on leaf-shape and length of teeth as follows:

Leaves lanceolate, generally broadest near middle; teeth of leaves not more than 2–3 mm., more or less equal in size and regularly spaced . . H. GROSSESERRATUS subsp. GROSSESERRATUS

Leaves lanceolate-ovate, generally broadest near base; teeth of leaves 3–6 mm. long, usually more unequal in size and irregularly spaced . . . H. GROSSESERRATUS subsp. MAXIMUS Long

According to Dr. Long, subsp. *grosseserratus* is found from Massachusetts and New Hampshire to Wisconsin, south to Kentucky and Texas, while subsp. *maximus* ranges from Michigan to Minnesota, Nebraska, Iowa, and Texas. Most of the material seen from Missouri would fall into subsp. *grosseserratus*. A specimen representing subsp. *maximus* has been seen from Jackson County, west-central Missouri (low ground, October 8, 1893, *Bush 180*).

A hybrid between *Helianthus grosseserratus* and *H. giganteus*, known as *Helianthus* × *luxurians* Watson, includes, according to Dr. Long the following names in synonymy:

*Helianthus membranaceus* E. E. Watson
*Helianthus instabilis* E. E. Watson, in part, excluding type specimen
*Helianthus luxurians* E. E. Watson

Specimens from Camden, Boone, and Randolph counties identified as *H. membranaceus* have been referred in the present flora to *H. tuberosus*, *H. decapetalus*, and *H. hirsutus*.

*Helianthus grosseserratus* has been recorded as hybridizing with *H. Maximiliani* (*H.* × *intermedius* Long), as well as with many of the other species of the genus, but these have not been recorded for Missouri.

*Other Hybrids*

Dr. Robert W. Long, Jr. (Am. Jour. Bot. 42: 769–77. 1955) has recorded various hybrids between species in the genus. The following, which he has recorded, have not been reported for Missouri:

*Helianthus grosseserratus* × *salicifolius* (*H.* × *Kellermani* Britton)
*Helianthus grosseserratus* × *Maximiliani*
*Helianthus grosseserratus* × *angustifolius*
*Helianthus grosseserratus* × *annuus*
*Helianthus grosseserratus* × *silphioides*
*Helianthus grosseserratus* × *decapetalus*
*Helianthus grosseserratus* × *doronicoides*
*Helianthus grosseserratus* × *divaricatus*
*Helianthus grosseserratus* × *hirsutus*
*Helianthus grosseserratus* × *laetiflorus*
*Helianthus grosseserratus* × *mollis*
*Helianthus grosseserratus* × *occidentalis*
*Helianthus mollis* × *annuus*
*Helianthus mollis* × *Maximiliani*
*Helianthus Maximiliani* × *decapetalus*
*Helianthus Maximiliani* × *divaricatus*
*Helianthus Maximiliani* × *occidentalis*
*Helianthus Maximiliani* × *strumosus*
*Helianthus divaricatus* × *mollis*
*Helianthus divaricatus* × *annuus*

*Excluded Species*

**Helianthus tracheliifolius** Mill.

Cronquist (in Gleason's *New Ill. Fl.* 3: 337. 1952) places this as a synonym under *H. strumosus*. The only Missouri specimen seen by the present author which is labeled *H. tracheliifolius* is from Jackson County (*Bush 8212*) and has been referred in this flora to *H. decapetalus*.

**Helianthus tomentosus** Michx.

Cronquist (in Gleason's *New Ill. Fl.* 3: 337. 1952) believes this name should be placed as a synonym of *H. giganteus*. The only sheet from Missouri in the Gray Herbarium labeled *H. tomentosus* is from Clark County (*Drouet 1773*). It has scabrous small leaves and a smoothish stem, resembling smooth-variations of *H. hirsutus*, to which it is referred in the present flora. According to Dr. Heiser and Dr. Long, true *H. tomentosus* is a species restricted to South Carolina, Georgia, Florida, Alabama, and Mississippi, and the report for Missouri by Fernald in *Gray's Manual* is based on an incorrect determination.

**Helianthus** × **doronicoides** Lamarck

According to the most recent investigations by Dr. R. C. Jackson (Rh. 58: 97–100. 1956), this represents a natural cross between *H. mollis* and *H. giganteus*. Much of the material from Missouri previously identified as *H. doronicoides* has been found to be upper portions of plants pertaining to *H. tuberosus*, *H. strumosus*, and *H. hirsutus*.

This hybrid has been found wild in New Jersey, Indiana, and Illinois.

Plate no. 371. **1.** Helianthus strumosus, × 2/7. **2.** Helianthus tuberosus, × 2/7. **3.** Verbesina virginica, × 2/7; Details from Small, The New York Botanical Garden. **4.** Helianthus grosseserratus, × 2/7 (Scribner's). **5.** Verbesina helianthoides, × 2/7.

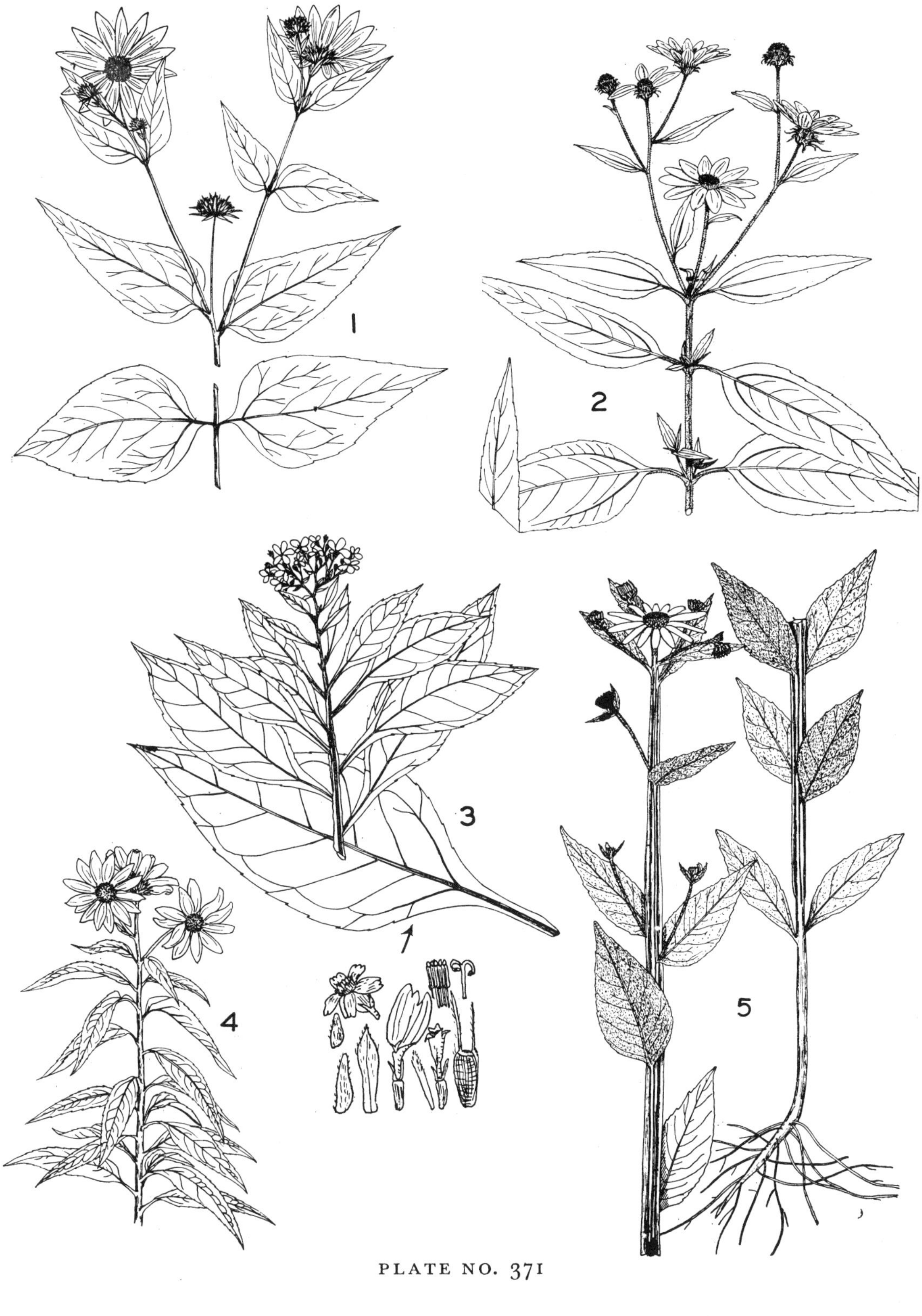


PLATE NO. 371

2285 Verbesina virginica (White Crown-beard)

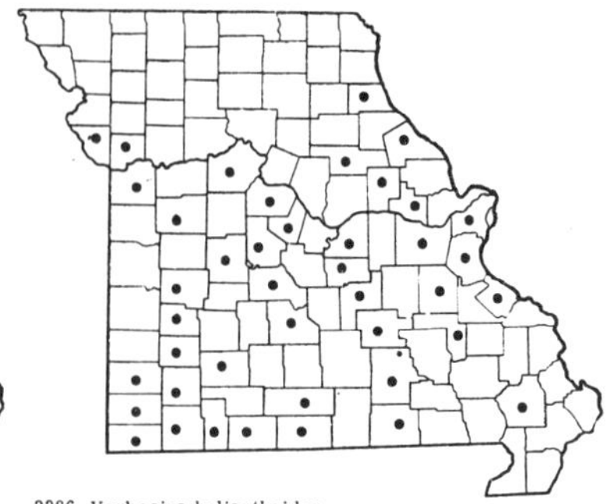
2286 Verbesina helianthoides

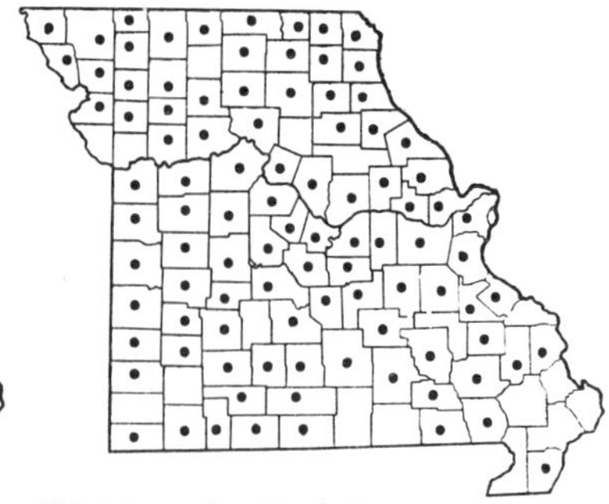
2287 Verbesina alternifolia (Yellow Ironweed)

## 42. **Verbesina** L. Crown-beard

(including *Actinomeris* Nutt. [G, P & S])

a. Stem not winged (leaf-tissue not extending from the leaves along the stem); annuals . . . . 4. V. ENCELIOIDES var. EXAURICULATA
a. Stem more or less winged (leaf-tissue extending from the leaves along the stem); perennials . . . . *b*
b. Rays white, 1–4; disk 3–7 mm. broad . . . . . . . . . . . . . . . . 1. V. VIRGINICA
b. Rays yellow, 2–15; disk 9–16 mm. broad . . . . . . . . . . . . . . . . . . *c*
c. Rays 2–8, drooping; flower-heads several to many, corymbosely arranged; bracts of involucre few, soon deflexed (their tips turned down); achenes at maturity spreading in all directions, some reflexed . . . . . . . . . . . . . . . . . . . . . 3. V. ALTERNIFOLIA
c. Rays 8–15, horizontally spreading; flower-heads solitary or few; bracts of involucre numerous, appressed (pressed closely to one another) or loose, not deflexed; achenes at maturity mostly ascending, not reflexed, not much spreading . . . . . . . . . . 2. V. HELIANTHOIDES

1. **Verbesina virginica** L. White Crown-beard Map 2285

Also called Frost Weed.

Flowers August–October.

Occurs in rocky open woods, base of wooded bluffs, along streams, and thickets. Southern and central Missouri, north to St. Louis, Franklin, Maries, Miller, Morgan, Hickory, St. Clair, and Bates counties.

Ranges from Florida to Texas, north to Pennsylvania, Kentucky, Missouri, and Kansas.

The phenomenon of ice-crystals, known as 'frost flowers,' is associated in late autumn with this species. For further details, see the discussion under *Cunila origanoides*.

2. **Verbesina helianthoides** Michx. Map 2286

Flowers late May–October.

Occurs in rocky or dry open woods, prairies, along railroads and thickets. Southern and central Missouri northeast to Marion County and north to Audrain, Saline, Jackson, and Platte counties.

Ranges from Georgia to Texas, north to Ohio, Illinois, Iowa, and Oklahoma.

3. **Verbesina alternifolia** (L.) Britt. Yellow Ironweed Map 2287

*Actinomeris alternifolia* (L.) DC. [G, P & S]

Flowers early August–October.

Occurs in low open or rich woods, low meadows, usually in valleys and alluvial soils near streams, sloughs, and ditches, grazed and cut-over slopes, and thickets.

Throughout Missouri, mostly absent from the lowlands of southeastern Missouri.

Ranges from Florida to Louisiana, north to New York, Ontario, Illinois, Iowa, and Oklahoma.

A comparison with numerous other species of the genus *Verbesina*, especially some of the tropical American ones, leads to the conclusion that *Actinomeris alternifolia* differs in no significant respect generically from *Verbesina* and should be absorbed by that genus.

---

Plate no. 372. **1.** Verbesina alternifolia, × $^2/_7$. **2.** Thelesperma trifidum, × $^2/_7$. **3.** Coreopsis tinctoria, × $^2/_7$; *a.* Style and style-branches, highly magnified; After Gleason, The New York Botanical Garden. **4.** Verbesina encelioides var. exauriculata, × $^2/_7$; Details from Small, The New York Botanical Garden. **5.** Coreopsis grandiflora, × $^2/_7$. **6.** Thelesperma megapotamicum, × $^2/_7$.

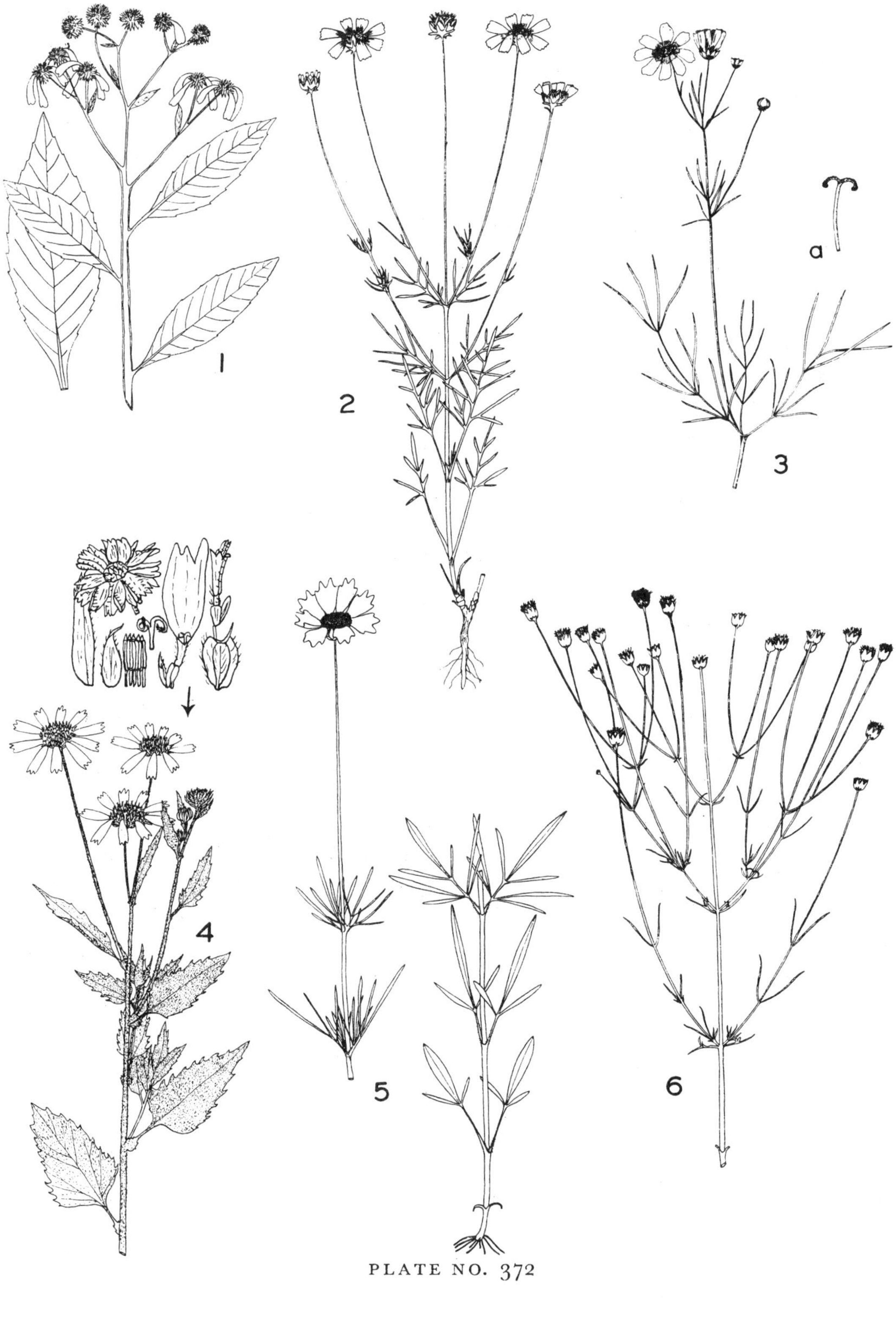


PLATE NO. 372

2288 Verbesina encelioides var. exauriculata

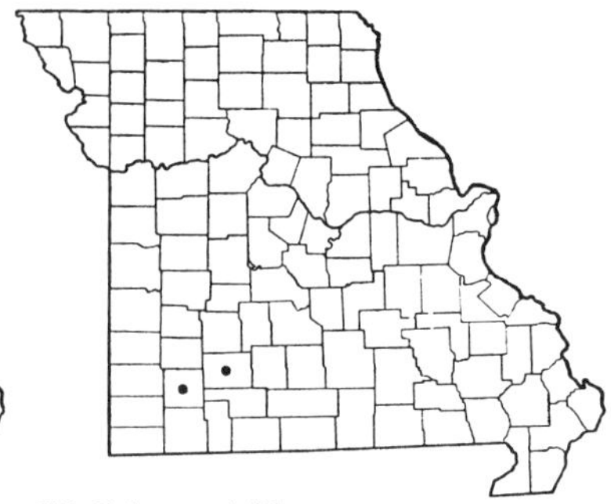

2289 Thelesperma trifidum

2290 Thelesperma megapotamicum

4. **Verbesina encelioides** (Cav.) B. & H. var. **exauriculata** Robins. & Greenm. Map 2288
*Verbesina encelioides* (Cav.) B. & H. [BB]
Flowers May–October.

Occurs in sandy open and waste ground, roadsides and along railroads.

Scattered in southeastern, central, and northwestern Missouri.

Ranges from Montana to Arizona, east to Kansas and Texas; introduced in Missouri east to New England.

This is a showy-flowered annual, sometimes cultivated.

## 43. **Thelesperma** Less.

Showy yellow rays present, 1–2 cm. long; outer bracts of involucre 8, narrowly linear or linear-lanceolate, 4–8 mm. long; plants annual . . . . . . . . . . . . . . . . . . . 1. T. TRIFIDUM
Rays absent or, if present, only 5–6 mm. long; outer bracts of involucre usually 5, shortly ovate or oblong, 2–3.5 mm. long; plants perennial. . . . . . . . . . . . . . . 2. T. MEGAPOTAMICUM

1. **Thelesperma trifidum** (Poir.) Britt. Map 2289
*Thelesperma intermedium* Rydb. [Shinners]
Flowers last of May–August.

Occurs on rocky prairies and glades. Known only from southwestern Missouri in Greene (Willard, July 16, 1888, *Blankinship*; limestone barrens near Springfield, October 2, 1925, *Palmer 28978*) and Lawrence counties.

Ranges from South Dakota to Colorado, south to Missouri, Oklahoma, Texas, and New Mexico; introduced in Michigan.

Shinners (Field & Lab. 18: 18–20, 24, 98–99. 1950) has brought forth evidence to indicate that the correct name for this plant is *T. intermedium* Rydb., as Poiret's description was based on a plant with alternate leaves. Until the Poiret specimen has been examined, the earlier name is retained here.

2. **Thelesperma megapotamicum** (Spreng.) Kuntze Map 2290
*Thelesperma gracile* (Torr.) Gray [G, P & S, Steyerm.]
Flowers last of May–July.

Occurs along railroads. Introduced in St. Louis (St. Louis, along track of Terminal Railroad Association, north of O'Fallon Street, June 16, 1956, *Muehlenbach 943*) and Jackson (along railroad, Sheffield, June 16, 1896, *Bush 758*) counties.

Ranges from Nebraska and Wyoming, south to Missouri, Oklahoma, Texas, New Mexico, and Arizona; also Mexico; South America (Uruguay and Argentina).

## 44. **Coreopsis** L. Tickseed

a. Most or all of the leaves, including the lower ones, undivided or rarely with 1 or 2 shorter lateral (side) lobes . . . . . . . . . . . . . . . . . . . . . . . . . . . . *b*
b. Leaves mostly near the base of the stem, usually only 2–4 pairs of leaves present on the stem; lower leaves linear to oblanceolate . . . . . . . . . . . . . . . . 3. C. LANCEOLATA

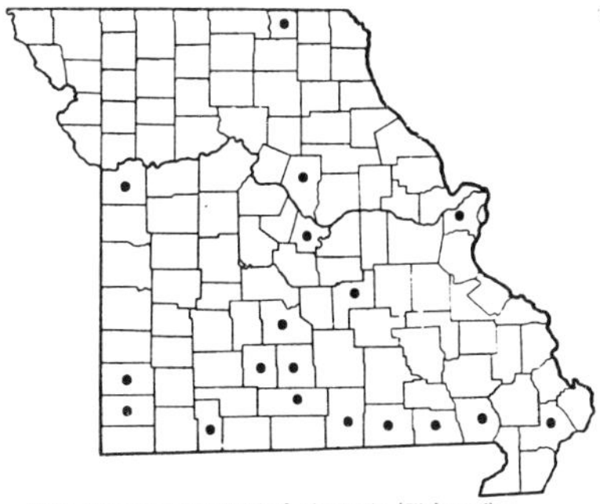

2291 Coreopsis tinctoria f. tinctoria (Tickseed)

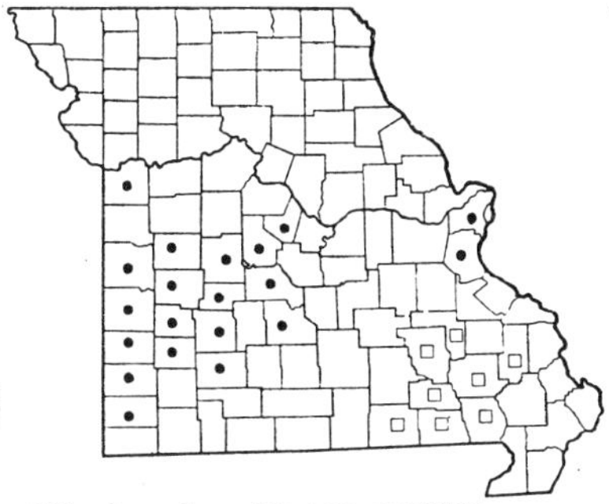

2292 ● Coreopsis grandiflora var. grandiflora
2292 □ Coreopsis grandiflora var. Harveyana

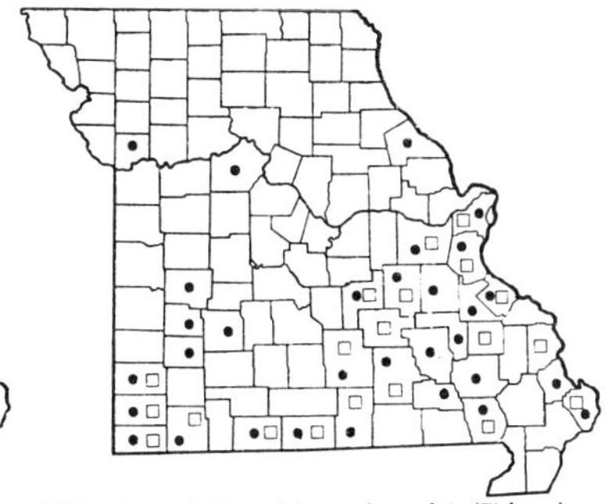

2293 ● Coreopsis lanceolata var. lanceolata (Tickseed Coreopsis)
2293 □ Coreopsis lanceolata var. villosa

b. Leaves extending to the middle or more than halfway up the stem, usually 6–10 pairs of leaves present on the stem; lower leaves oval-obovate . . . . . . . . . . . . . 4. C. PUBESCENS

a. Most or all of the leaves either 3–5-parted or divided to the midrib into 3–5 or more divisions . . . . *c*

c. All the leaves sessile (without stalks), deeply 3-pronged or lobed at or below the middle, but not cut to the base or divided into separate divisions. . . . . . . . . . . . 5. C. PALMATA

c. At least the lower leaves petioled (with a stalk), either divided to the base into 3 or 5 separate leaflets or divided or parted to the center or middle into 3–5 or more thread-like or very narrow segments . . . . . . . . . . . . . . . . . . . . . . . . . . . . *d*

d. Leaves divided to the base into 3 or 5 elliptic or oblong-lanceolate leaflets 6–25 mm. broad; achenes 5–7 mm. long; stems 10–30 dm. tall . . . . . . . . . . . . . 6. C. TRIPTERIS

d. Leaves divided to their middle or center into 3–5 or more thread-like or narrowly linear or linear-lanceolate segments 0.5–5 mm. broad; achenes 1–4 mm. long; stems 2–12 dm. tall. . . . . *e*

e. Flowers of the disk (solid center of flower-head between rays) yellow; lower leaves simple and undivided; rays yellow, 1.3–2.5 cm. long; heads 4–6 cm. broad, the disk (solid center between rays) 10–15 mm. broad; achenes with a wing less than 1 mm. broad along each side; perennial plants . . . . . . . . . . . . . . . . . . . . . 2. C. GRANDIFLORA

e. Flowers of the disk brownish-purple or dark red; lower leaves divided into narrow segments; rays yellow or with red- or purple-brown at base or throughout, 0.7–1.5 cm. long; heads 1.5–3 cm. broad, the disk 5–12 mm. broad; achenes without any wing; annual plants, easily pulled up with the roots . . . . . . . . . . . . . . . 1. C. TINCTORIA

1. **Coreopsis tinctoria** Nutt. f. **tinctoria** Map 2291

*Coreopsis tinctoria* Nutt. [G, BB, P & S]

Flowers June–September.

Occurs in rocky glades, open and sandy ground, along roadsides and railroads.

Scattered in southern and central Missouri north to St. Louis, Boone, and Jackson counties, locally north in Schuyler County.

Ranges from Minnesota and Manitoba to Washington, south to Louisiana, Texas, New Mexico, Arizona, and California; escaped eastward to the Atlantic states.

Typical f. *tinctoria* has the rays yellow. In f. *atropurpurea* (Hook.) Fern., not recorded from Missouri, the rays are partly or wholly purple- or red-brown. The species is commonly cultivated as an annual garden plant.

2. **Coreopsis grandiflora** Hogg Map 2292

Flowers May–July.

Occurs in prairies, glades, dry open woods, thickets, and open ground.

Two variations occur in Missouri:

Leaf-segments of middle and upper leaves 1–5 mm. broad, linear to linear-lanceolate . . .
2a. C. GRANDIFLORA var. GRANDIFLORA

Leaf-segments of middle and upper leaves 0.5–1.5 mm. broad, hair-like (linear-filiform) or linear .
2b. C. GRANDIFLORA var. HARVEYANA

2a. **Coreopsis grandiflora** var. **grandiflora** Map 2292

*Coreopsis grandiflora* Hogg [G, BB, P & S, Steyerm.]

Chiefly in the unglaciated section of southwestern Missouri from the western border east to Moniteau, Camden, and Laclede counties, north to Morgan, Benton, Henry, and Jackson counties, locally east in St. Louis and Ripley counties.

Ranges from Florida to Texas and New Mexico,

2294 Coreopsis pubescens var. pubescens (Star Tickseed)

2295 Coreopsis palmata

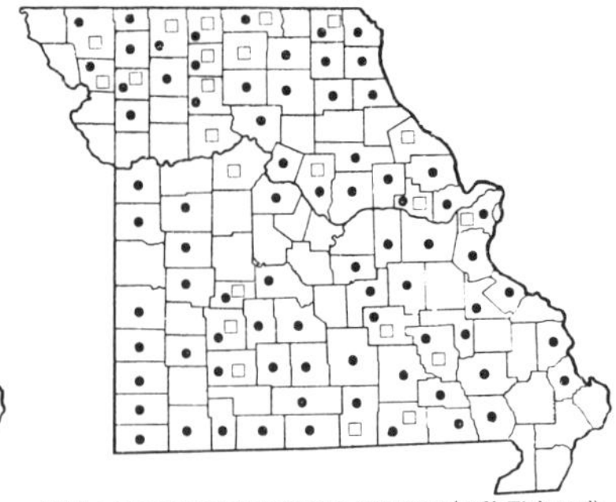

2296 ● Coreopsis tripteris var. tripteris (Tall Tickseed)
2296 □ Coreopsis tripteris var. Deamii

north to Georgia, Missouri, and Kansas; introduced from cultivation east to Michigan, Indiana, and New England.

2b. **Coreopsis grandiflora** var. **Harveyana** (Gray) Sherff Map 2292

Southeastern Missouri north to Bollinger, Iron, and Reynolds counties, west to Carter and Oregon counties.

Ranges from Missouri to Arkansas and Oklahoma.

3. **Coreopsis lanceolata** L. Tickseed Coreopsis Map 2293

Flowers April–July.

Occurs on rocky prairies, glades, bluff escarpments, fields, sandy open ground, along roadsides and railroads. Native in the Ozark region and escaped from cultivation north to Pike, Saline, and Clay counties.

Two varieties are found in Missouri:

Stems and leaves chiefly glabrous (without hairs) . . 3a. C. LANCEOLATA var. LANCEOLATA
At least the leaves and base of the stems hairy 3b. C. LANCEOLATA var. VILLOSA

3a. **Coreopsis lanceolata** var. **lanceolata** Map 2293

*Coreopsis lanceolata* L. [G, P & S, Steyerm.]

Throughout the range indicated in Missouri.

Ranges from Florida to New Mexico, north to Virginia, Ontario, Michigan, Wisconsin, and Missouri; naturalized eastward to New England and New Jersey.

3b. **Coreopsis lanceolata** var. **villosa** Michx. Map 2293

Ozark region of southern and east-central Missouri north to St. Louis, Franklin, Crawford, Phelps, Texas, and Jasper counties.

Ranges from Florida to Louisiana, north to Virginia, Ontario, Michigan, Illinois, Missouri, and Oklahoma; naturalized east to New England.

This Coreopsis is common on glades in the Ozark region and is a showy-flowered species. It becomes weedy in a garden and has to be watched for its abundant spread through seed formation.

4. **Coreopsis pubescens** Ell. var. **pubescens** Star Tickseed Map 2294

*Coreopsis pubescens* Ell. [G, BB, P & S, Steyerm.]

Flowers last of May–September.

Occurs in rocky open woods, base of wooded bluffs, low ground in valleys and along gravelly stream beds, rocky ledges along streams, alluvial thickets, and along railroads. Ozark region of southern and central Missouri north to St. Louis, Franklin, Gasconade, Osage, Miller, Camden, Benton, St. Clair, and Vernon counties, and locally introduced in Marion County.

Ranges from Florida to Louisiana, north to Virginia, West Virginia, Kentucky, Illinois, Missouri, and Oklahoma.

A var. *robusta* Gray with the stems and branches glabrous or nearly so, instead of more or less hairy as in typical var. *pubescens*, is known from Virginia and North Carolina to Kentucky.

5. **Coreopsis palmata** Nutt. Map 2295

Flowers May–July.

Occurs in prairies, glades, fallow fields, rocky and dry open woods, and thickets.

Throughout Missouri, except absent from the low-

Plate no. 373. **1.** Coreopsis lanceolata, × 2/7; Details from Small, The New York Botanical Garden. **2.** Coreopsis pubescens, × 2/7; Details from Small, The New York Botanical Garden. **3.** Coreopsis tripteris, × 2/7. **4.** Coreopsis palmata, × 2/7. **5.** Bidens laevis, × 2/7; Details from Small, The New York Botanical Garden.

PLATE NO. 373

land counties of the extreme southeastern section.

Ranges from Wisconsin to Manitoba, south to Indiana, Illinois, Missouri, and Oklahoma.

This is a common species of the prairies and dry open woodland, most frequently found in acid soils. It eventually forms large beds when allowed to spread, a habit shared by many other species of the genus. The fruits are eaten by wild turkey.

6. **Coreopsis tripteris** L. Tall Tickseed Map 2296

Flowers July–September.

Occurs in prairies, rocky and dry open woods, along bluffs, thickets, roadsides, and along railroads.

Throughout Missouri, except absent from the lowland counties of the extreme southeastern section.

Two varieties are found in Missouri:

Lower surface of leaflets and outer bracts of involucre glabrous (without hairs) . . . . . 6a. C. TRIPTERIS var. TRIPTERIS

Lower surface of leaflets and outer bracts of involucre densely hairy . . . . 6b. C. TRIPTERIS var. DEAMII

6a. **Coreopsis tripteris** var. **tripteris** Map 2296

*Coreopsis tripteris* L. [G, BB, P & S]

This is the commoner variation in Missouri, found throughout the range in the state.

Ranges from Ontario to Wisconsin, south to Georgia, Mississippi, Louisiana, Kansas, and Oklahoma.

6b. **Coreopsis tripteris** var. **Deamii** Standl. Map 2296

Scattered throughout the state.

Ranges from Pennsylvania to Michigan and Illinois, south to Georgia, Arkansas, Missouri, and Oklahoma.

Intermediate specimens are encountered which are difficult to place in one or the other variety. However, Deam noted that var. *Deamii* bloomed earlier and had a darker green color than var. *tripteris*.

## 45. **Bidens** L. Beggar Ticks

(including *Megalodonta* Greene [G])

a. Plants growing in the water with the submersed leaves (in the water) finely dissected into thread-like segments; achenes nearly terete (rounded in cross-section) . . . . . . . . . . 11. B. BECKII

a. Plants usually growing on land but not submerged in water, sometimes found near water along streams, ponds, swamps, or growing as an air plant (epiphyte) on logs, bases of shrubs and trees, and on hummocks of sink-hole ponds, but the leaves not finely dissected into thread-like segments; achenes flattened or 4-sided (quadrangular) . . . . . . . . . . . . . . . . . . *b*

b. All the leaves simple and undivided, or at most 3–5-cleft or lobed, but then the terminal lobe not stalked (petioled) and the lobes always connected by green leaf-tissue with the rest of the leaf-blade . . . . . . . . . . . . . . . . . . . . . . . . . . *c*

c. Rays of flower-heads usually present, showy and large, usually 1–3 cm. long (sometimes absent); flower-heads erect to somewhat nodding at flowering time, usually slightly to strongly nodding in fruit . . . . . . . . . . . . . . . . . . . . . . . *d*

d. Commonly encountered species throughout Missouri; pales (chaff) of receptacle with a yellow tip; rays up to 1.7 cm. long or absent; marginal nerves of achenes pale, enlarged, thickened, and somewhat winged; outer bracts of involucre usually longer than the inner, widely spreading or reflexed (turned down) . . . . . . . . . . . . 2. B. CERNUA

d. Rarely encountered, known only from Dunklin County, southeastern Missouri; pales (chaff) of receptacle with a reddish-brown tip; rays 1.5–3 cm. long; marginal nerves of achenes not pale, thickened, enlarged, nor winged; outer bracts of involucre only equaling, rarely longer than the inner, spreading to ascending or somewhat erect, not reflexed. 1. B. LAEVIS

c. Rays absent, or, when rarely present, only up to 8 mm. long; flower-heads erect at flowering and fruiting time . . . . . . . . . . . . . . . . . . . . . . . *e*

e. Outer bracts of involucre 2–6; corollas of the disk 5-toothed, orange-yellow; anthers protruding (exserted) from the corolla; achenes usually with 4 or 5 (rarely 3) awns; mature inner achenes toward center strongly 4-angled with a prominent midrib, 5–8 mm. long, 1.6–2.6 mm. broad; mature outer achenes of head 4–6.5 mm. long, 1.7–2.8 mm. broad . 3. B. CONNATA

e. Outer bracts of involucre 6–10; corollas of the disk 4-toothed, pale yellow to yellowish-green; stamens included within the corolla; achenes with usually 3 (or 2, 4, or 5) awns; mature inner achenes toward center of head flat or nearly so with slender or faint midrib, not strongly 4-angled, 8.5–11 mm. long, 2.5–3 mm. broad; mature outer achenes of head 5.5–9.5 mm. long, 2.7–3.3 mm. broad . . . . . . . . . . . . . . . 4. B. COMOSA

b. At least the lower and middle leaves divided to the middle or center into 3–7 or more separate leaflets, these either toothed or variously cut . . . . . . . . . . . . . . . . . . . *f*

f. Rays of flower-head showy, 10–25 mm. long, much longer than the outer bracts of involucre . . *g*

g. More commonly encountered species; outer bracts of involucre 12–25, long-tapering, conspicuously or coarsely hairy-margined, 7–20 mm. long and mostly longer than the inner bracts; mature inner achenes toward center of head 1.8–2.8 mm. broad; mature outer achenes of head 2.5–3.8 mm. broad . . . . . . . . . . . . . . 9. B. POLYLEPIS

g. Less commonly encountered species; outer bracts of involucre 8–12, blunt or short-pointed (acute) at tip, the margins smooth or moderately finely hairy, 4–7 mm. long and mainly shorter than the inner bracts; mature inner achenes toward center of head 2.5–4.6 mm. broad; mature outer achenes of head 3.3–5.2 mm. broad . . . . . . . . . 8. B. ARISTOSA

f. Rays of flower-head absent, or at most less than 5 mm. long . . . . . . . . . . . . . *h*

h. Main leaves of stem divided more than once, each larger principal leaflet or division further divided once or twice again into smaller segments (2–3 times pinnately divided), the ultimate segments tending to be rounded or with curved margins; outer bracts of involucre shorter than the inner ones, not leaf-like, usually 8; achenes somewhat 4-sided and 4-angled, linear, the mature inner ones of the head 12–18 mm. long, about 1 mm. broad and 2–3 times the length of the involucre; achenes very unequal; stem square-sided or 4-angled 10. B. BIPINNATA

h. All the leaves or at least the main lower and middle leaves of stem divided only once into 3 or 5 merely toothed leaflets with triangular or conical pointed teeth; outer bracts of involucre usually much longer than the inner ones and leaf-like, either 2–5, 5–8, or 10–20; achenes flattened or broad, narrowly to broadly wedge-shaped (cuneate), the mature inner ones of the head 5.5–17 mm. long, 1–6.2 mm. broad, and not more than ⅓ longer than the length of the inner bracts of the involucre; stem more or less cylindrical, not square-sided or 4-angled . . . . . . . . . . . . . . . . . . . . . . . . . . . *i*

i. Outer bracts of the involucre 2–5, the margins smooth, not hairy; flowering heads small, 3–6 mm. broad; corollas of the disk flowers 1.5–2 mm. long; mature inner achenes (toward center) of head 1–1.5 mm. broad with awns 1.5–2.4 mm. long; mature outer achenes of head 1.5–1.7 mm. broad with awns 1–1.5 mm. long . . . . . 7. B. DISCOIDEA

i. Outer bracts of the involucre 5–8 or 10–20, the margins sparsely to abundantly hairy; flowering heads larger, mostly 7–15 mm. broad; corollas of the disk flower 2.5–4 mm. long; mature inner achenes (toward center) of head 2.2–6.2 mm. broad with awns 2.5–9.5 mm. long; mature outer achenes of head 3–6.3 mm. broad with awns 2–4 mm. long. . . *j*

j. Outer bracts of involucre 5–8, with sparsely hairy margins; mature outer achenes of head 5.3–7 mm. long, 3–4 mm. broad, with awns 2–4 mm. long; mature inner achenes (toward center) of head 7–10 mm. long, 2.2–3.8 mm. broad, with awns 2.5–5 mm. long . . . . . . . . . . . . . . . . . . . . 6. B. FRONDOSA

j. Outer bracts of involucre 10–20, with abundantly hairy margins; mature outer achenes of head 6.5–11.3 mm. long, 3.5–6.3 mm. broad, with awns 3–6 mm. long; mature inner achenes 9–17 mm. long, 2.8–6.2 mm. broad, with awns 4–9.5 mm. long . 5. B. VULGATA

1. **Bidens laevis** (L.) BSP. Map 2297

Flowers August–October.

Occurs in low wet ground. Known only from Dunklin County, lowlands of southeastern Missouri (Buffalo Island, E. Bertig, September 28, 1897, *Trelease 642*, in Mo. Bot. Gard. Herb.).

Ranges from Florida to Mexico and California, north to New Hampshire and in the interior north to Missouri.

The specimen cited for Missouri has red-tipped chaff, rays 2.5 cm. long, and the outer bracts of the involucre about equaling the inner ones. It is the only one, of all the Missouri material examined, which matches specimens of *B. laevis* from the eastern United States. Other specimens previously identified as *B. laevis* from St. Louis, Jefferson, and Jackson counties, have been found to be referable to *B. cernua*, with which species they are included in the present treatment. One of these from St. Louis County (along the Mississippi River, south St. Louis, September 10, 1894, *Kellogg*) has the rays quite elongated, but the outer bracts of the involucre are too long for *B. laevis* and the chaff is yellow-tipped. As suggested by Cronquist (in Gleason's *New Ill. Fl.* 3: 353. 1952), this species intergrades into the following, *B. cernua*, and should probably be regarded only as variety of that species.

2297 Bidens laevis

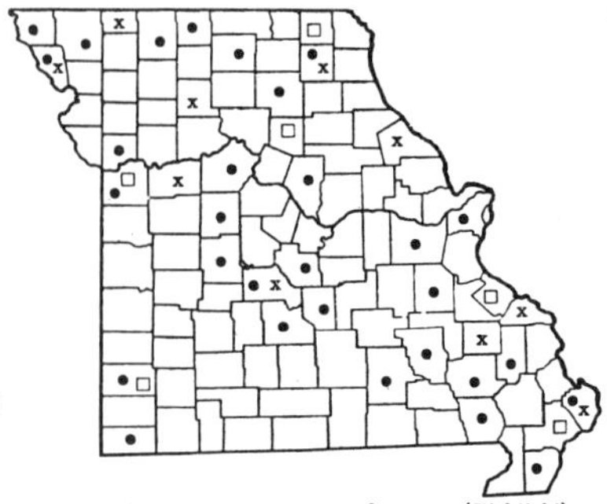

2298 • Bidens cernua var. cernua f. cernua (Sticktight)
2298 x Bidens cernua var. integra
2298 □ Bidens cernua var. elliptica

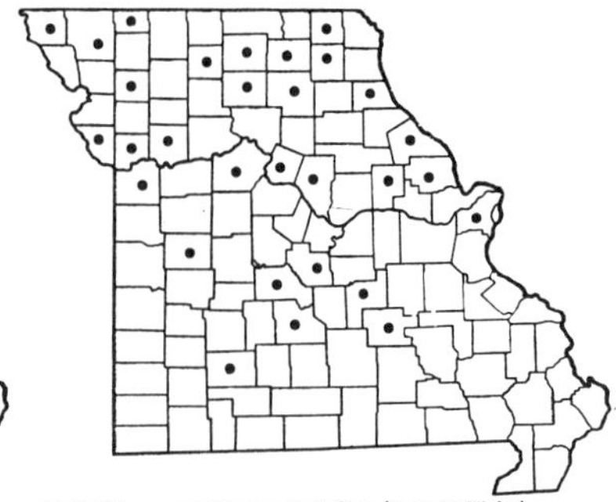

2299 Bidens connata var. petiolata (Beggar Ticks)

2. **Bidens cernua** L. Sticktight Map 2298

Also called Nodding Bur Marigold.

Flowers August–October.

Occurs in wet soils in spring-fed meadows, along borders of streams, ponds, sloughs, swamps, and ditches, and along railroads.

Throughout Missouri.

The following variations occur in Missouri:

a. Leaves noticeably narrowed to the base, conspicuously narrowed below the middle . . . 2c. B. CERNUA var. ELLIPTICA

a. Leaves only gradually narrowed below the middle, the base of the leaf broad and sessile (without a stalk) or joined (connate) to the base of the opposite leaf . . . . . . . *b*

b. 4 to 17 pairs of coarse teeth 1–5 mm. long on the leaf-margins . . . 2a. B. CERNUA var. CERNUA

b. 12 to 24 pairs of fine teeth at most 1 mm. long on the leaf-margins . . 2b. B. CERNUA var. INTEGRA

2a. **Bidens cernua** var. **cernua** f. **cernua** Map 2298

*Bidens cernua* L. [G, BB, P & S]

This is the commonest variation in the state.

Ranges from New Brunswick to British Columbia, south to Delaware, Maryland, North Carolina, Tennessee, Missouri, South Dakota, Wyoming, Idaho, and California; also in Eurasia.

Typical var. *cernua* f. *cernua* has rays; in var. *cernua* f. *discoidea* (Wimm. & Grab.) Briq. & Cavill., not recorded from Missouri, the rays are absent.

2b. **Bidens cernua** var. **integra** Wieg. Map 2298

Scattered in the state.

Ranges from Prince Edward Island to North Dakota, south to North Carolina, Ohio, Illinois, Missouri, and Oklahoma.

2c. **Bidens cernua** var. **elliptica** Wieg. Map 2298

Scattered in the state.

Ranges from Quebec to Washington, south to North Carolina, Kentucky, Kansas, and Colorado.

*Bidens cernua* varies considerably in the shape and size of the leaves, amount and size of toothing, height and thickness of stems, and length of rays. There are rayless forms found in the above varieties enumerated. The variations here presented may be of no more significance than to be rated as forms, while some authors (Deam, *Fl. Ind.* p. 983. 1940) do not recognize them as name-worthy. However, until more field and experimental studies are at hand to furnish more definite information, the variations are being tentatively retained for convenience.

Ducks eat the fruits of this species.

3. **Bidens connata** Muhl. var. **petiolata** (Nutt.) Farw. Beggar Ticks Map 2299

Also called Swamp Beggar Ticks, Sticktight.

*Bidens tripartita* in part [of BB], not L.

*Bidens connata* [of P & S], not Muhl.

Flowers late June–October.

Occurs in wet ground bordering streams, ponds, sloughs, swamps, and ditches, swampy meadows, low wet woods and along railroads.

Throughout most of Missouri, mainly in the northern and central sections of the state, absent from the southeastern lowlands and much of the Ozark region.

The Missouri records were previously referred to typical var. *connata*, but that variety has 2–4 lobes present at the base of the middle and lower main leaves of the stem. So far as examination of Missouri material is concerned, the leaves are unlobed and taper to slender or narrowly margined petioles, characteristic of var. *petiolata*.

Cronquist (in Gleason's *New Ill. Fl.* 3: 355. 1952) includes this species and *B. comosa* under *B. tripartita*.

2300 Bidens comosa (Beggar Ticks)

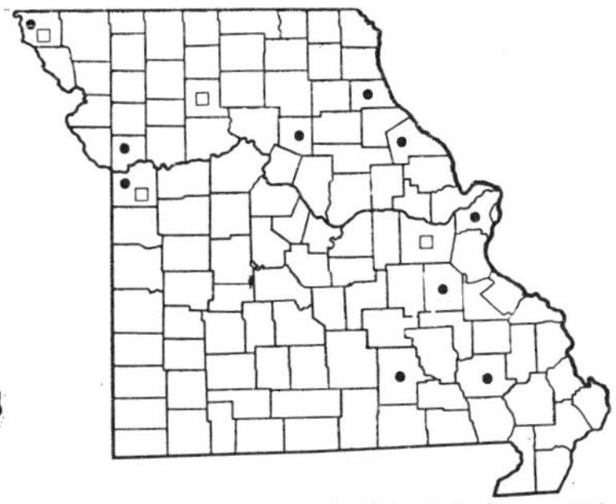

2301 ● Bidens vulgata var. vulgata f. vulgata (Beggar Ticks)
2301 □ Bidens vulgata var. vulgata f. puberula

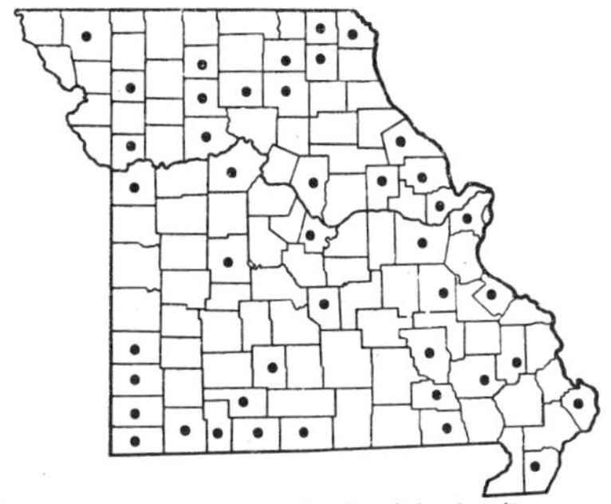

2302 Bidens frondosa var. frondosa f. frondosa (Begger Ticks)

as many specimens do not appear to be distinguishable from Old World material of *B. tripartita*. More intensive field and experimental studies are necessary before a more decisive statement concerning these three taxa can be made.

4. **Bidens comosa** (Gray) Wieg. Beggar Ticks
Map 2300

*Bidens comosa* var. *acuta* Wieg. [P & S]
*Bidens tripartita* in part [of BB], not L.
Flowers August–October.

Occurs in wet ground along streams, ponds, sloughs, swamps, and ditches.

Chiefly in the northern half of the state, locally south in Mississippi, Webster, and Greene counties.

Ranges from Maine to North Dakota, south to North Carolina, Tennessee, Louisiana, New Mexico, and Utah.

Missouri specimens of this species have been confused with *B. connata* var. *petiolata*, from which it may be distinguished by the characters presented in the key.

Fruits of this species are eaten by upland game birds.

5. **Bidens vulgata** Greene Beggar Ticks, Sticktight
Map 2301

Flowers August–October.

Occurs in moist ground along streams, ponds, sloughs, ditches, alluvial thickets, low wet woodland, fields, waste ground, along roadsides and railroads.

Scattered and uncommon in the state.

Ranges from Quebec to Alberta and Washington, south to North Carolina, Tennessee, Missouri, Kansas, Wyoming, Nevada, and California.

The following variations are known in Missouri:

Stem, leaves, and bracts of involucre glabrous (without hairs) . . 5a. B. VULGATA var. VULGATA f. VULGATA

Stem, leaves, and bracts of involucre finely short-hairy . . 5b. B. VULGATA var. VULGATA f. PUBERULA

5a. **Bidens vulgata** var. **vulgata** f. **vulgata**
Map 2301

*Bidens vulgata* Greene [G, P & S]
*Bidens vulgata* var. *vulgata* [BB]
This is the commoner variation in the state.

5b. **Bidens vulgata** var. **vulgata** f. **puberula** (Wieg.) Fern. Map 2301

*Bidens vulgata* f. *puberula* (Wieg.) Fern. [G]
*Bidens vulgata* var. *puberula* (Wieg.) Greene [BB, P & S]

Known from Atchison, Livingston, Jackson, and Franklin counties.

*Bidens vulgata* is much less common in Missouri than the next species, *B. frondosa*, which it closely resembles.

A variety *schizantha* Lunell with the leaves twice pinnate or thrice pinnatisect occurs in the Lake Superior region west to North Dakota and Saskatchewan.

6. **Bidens frondosa** L. var. **frondosa** f. **frondosa** Beggar Ticks, Sticktight Map 2302

*Bidens frondosa* L. [G, BB, P & S]
Flowers August–October.

Occurs in wet ground in moist woods, swampy meadows, alluvial thickets, borders of streams, ponds, sloughs, swamps, ditches, fields, waste ground, along roadsides and railroads. Throughout Missouri and much commoner in the state than *B. vulgata*, which it resembles.

Ranges from Newfoundland to Washington, south to Florida, Texas, and California.

In typical var. *frondosa* f. *frondosa* the awns of the achenes have downwardly pointing barbs; in var. *frondosa* f. *anomala* (Porter) Fern. the barbs point upward on the awns. A variety *stenodonta* Fern. & St. John, known from Newfoundland and Prince Edward

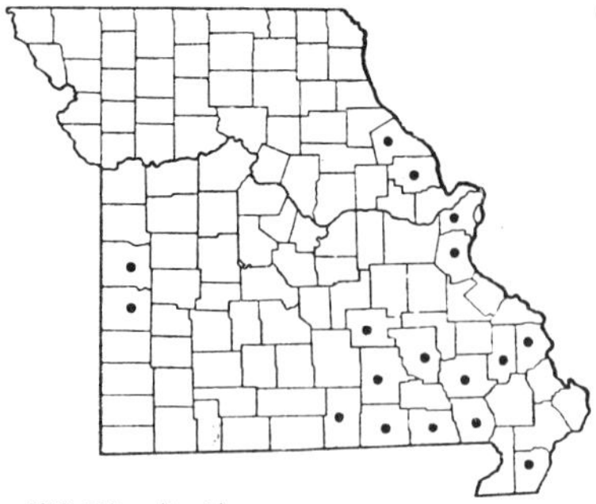

2303 Bidens discoidea

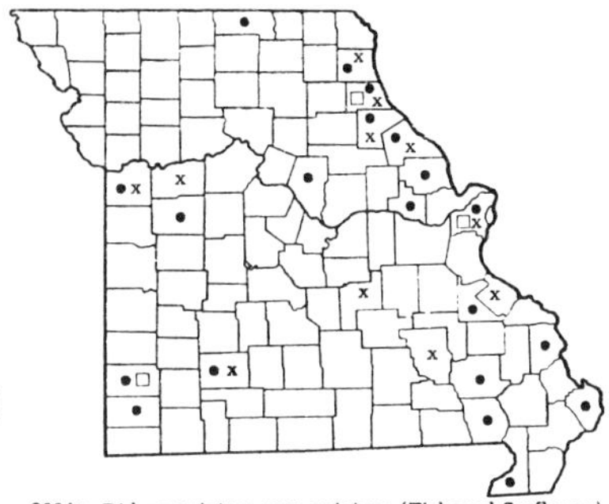

2304 • Bidens aristosa var. aristosa (Tickseed Sunflower)
2304 □ Bidens aristosa var. Fritcheyi
2304 x Bidens aristosa var. mutica

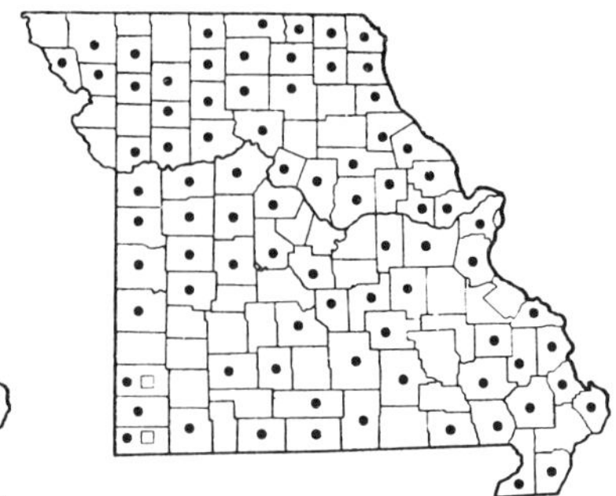

2305 • Bidens polylepis var. polylepis (Tickseed Sunflower)
2305 □ Bidens polylepis var. retrorsa

Island, has narrower teeth to the leaflets and the outer bracts of the involucre $2\frac{1}{2}$–4 times as long as the inner ones instead of twice the length as in typical var. *frondosa*.

*Bidens frondosa* and other species of the genus sometimes are responsible for local irritations received in handling the plants. The fruits are eaten by mallard ducks and the plant is eaten by muskrats.

7. **Bidens discoidea** (T. & G.) Britt. Map 2303
Flowers August–October.

Occurs in wet ground of swampy woods in river bottom flood plain, bald cypress swamps, upland sink-hole ponds, on decaying stumps and logs, at the base of living trees in wet lowland forests, and along streams and oxbow lakes of river valley meanders.

Common in upland sink-hole ponds of the southeastern Ozarks west to Dent, Shannon, and Howell counties, swampy lowlands of southeastern Missouri north along the counties bordering the Mississippi River to Pike County, and locally around oxbow lakes in alluvial bottoms in Bates and Vernon counties, southwestern Missouri.

Ranges from Alabama to Texas, north to Nova Scotia, Maine, Quebec, Ontario, Michigan, Wisconsin, and Minnesota.

In the sink-hole ponds this species often is found growing as an epiphyte at the base of *Cephalanthus* bushes, or on mossy hummocks in these ponds. In the swampy woods of *Taxodium*, *Nyssa aquatica*, and *Populus heterophylla*, this species often grows at the base of large trees or on dead logs or stumps as an epiphyte, the seeds having germinated and rooted in such habitats after the high water of spring and early summer flooding has receded.

8. **Bidens aristosa** (Michx.) Britt. Tickseed Sunflower Map 2304
Flowers August–October.

Occurs in prairies, cultivated and fallow fields, waste ground, along roadsides and railroads. Scattered in the state, but less common than the next species, *B. polylepis*.

The following variations occur in Missouri:

a. Awns usually absent or reduced to tiny stubs . 8c. B. ARISTOSA var. MUTICA
a. Awns present, well-developed, usually 2–4 mm. long . . . . . . . . . . . b
  b. Awns with the barbs pointing upwardly . 8a. B. ARISTOSA var. ARISTOSA
  b. Awns with the barbs pointing downwardly . . . . 8b. B. ARISTOSA var. FRITCHEYI

8a. **Bidens aristosa** var. **aristosa** Map 2304
*Bidens aristosa* (Michx.) Britt. [G, BB, P & S]

This is the commonest variation, but is scattered in the state.

Ranges from Delaware to Minnesota, south to Virginia, Alabama, Mississippi, Louisiana, and Texas.

8b. **Bidens aristosa** var. **Fritcheyi** Fern. Map 2304

Scattered, known from Marion, St. Louis, and Jasper counties.

Ranges from Indiana and Kentucky to Illinois and Missouri; introduced eastward to the Atlantic states.

Plate no. 374. **1.** Bidens cernua, × 2/7; *a.* Achene, × 1 3/7. **2.** Bidens connata var. petiolata, × 2/7; *a.* Flower-head, × 6/7. **3.** Bidens bipinnata, × 2/7 (Scribner's). **4.** Bidens discoidea, × 2/7; *a.* Flower-head, × 1 3/7. **5.** Bidens Beckii, × 2/7. **6.** Bidens aristosa, × 2/7; *a.* Involucre with achenes, × 6/7; *b.* Achene, × 1 3/7. **7.** Bidens vulgata, × 2/7; *a.* Flower-head × 6/7. **8.** Bidens frondosa, × 2/7; *a.* Flower-head, × 6/7. **9.** Bidens polylepis, × 2/7; *a.* Flower-head, × 6/7; *b.* Achene, × 1 3/7.

PLATE NO. 374

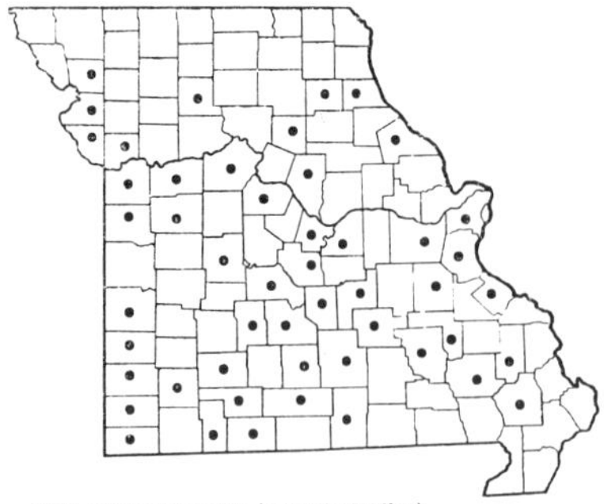

2306 Bidens bipinnata (Spanish Needles)

2307 Bidens Beckii (Water Marigold)

2308 Cosmos bipinnatus (Cosmos)

8c. **Bidens aristosa** var. **mutica** Gray Map 2304

Scattered in eastern, southern, and central Missouri.

Ranges from Ohio, Indiana, and Kentucky to Illinois and Missouri.

9. **Bidens polylepis** Blake Tickseed Sunflower Map 2305

Flowers August–October.

Occurs in wet prairies, low meadows, low swampy woods, fallow and cultivated fields, borders of oxbow lakes in river flood plain and upland ponds, waste ground, along roadsides and railroads.

Throughout Missouri, doubtless in every county.

Two variations occur in Missouri:

Awns with the barbs pointing upwardly; common type encountered . . . 9a. B. POLYLEPIS var. POLYLEPIS

Awns with the barbs pointing downwardly; rarely encountered . . 9b. B. POLYLEPIS var. RETRORSA

9a. **Bidens polylepis** var. **polylepis** Map 2305

*Bidens polylepis* Blake [G, BB]

*Bidens involucrata* (Nutt.) Britton [P & S]

This is the commoner variation found throughout the state.

Ranges from Illinois to Iowa, Kansas, Oklahoma, and Colorado.

9b. **Bidens polylepis** var. **retrorsa** Sherff Map 2305

Known only from Jasper (Webb City, September 25, 1908, *Bush 5175* in part, holotype in Gray Herb.) and McDonald counties, southwestern Missouri.

Ranges from Ohio and Indiana to Missouri.

This species forms dense stands in low fields and meadows, and in late summer and early autumn is one of the dominant species of the landscape, providing beds of bright yellow color.

Although it is stated that the awns of the outer achenes are 0.2–1 mm. long (*Gray's Manual,* eighth edition, p. 1508. 1950), Missouri material of *B. polylepis* generally has the awns either absent or obsolete. The species intergrades into the preceding one, *B. aristosa,* and I am in agreement with Cronquist (Gleason's *New Ill. Fl.* 3: 357. 1952) that *B. polylepis* should be considered a variety of *B. aristosa.* It is retained in the present treatment as a species, however, pending further, more detailed field and experimental studies.

10. **Bidens bipinnata** L. Spanish Needles Map 2306

Flowers August–October.

Occurs in open and grazed woods, glades, pastures, rocky open ground, fields, thickets, waste places, along roadsides and railroads.

Common throughout southern and central Missouri, north to Marion, Shelby, Howard, Livingston, and Andrew counties.

Ranges from Florida to Mexico, north to Massachusetts, New York, Ohio, Indiana, Illinois, Missouri, Kansas, Oklahoma, and Arizona; also in tropical America and the Old World.

In tropical West Africa the leaves of this species are cooked and eaten as a vegetable.

11. **Bidens Beckii** Torr. Water Marigold Map 2307

*Megalodonta Beckii* (Torr.) Greene [G]

Flowers August–October.

Known only from St. Louis County, east-central Missouri (St. Louis, August, 1846, *Engelmann,* in Gray Herbarium).

Ranges from Cape Breton to Quebec, west to Minnesota and Manitoba, south to New Jersey, Pennsylvania, Ohio, Indiana, Wisconsin, and Missouri; also in Washington and Oregon.

It is quite possible that the Engelmann specimen cited above was collected in Illinois rather than Missouri, since no Missouri specimen was seen at the Missouri Botanical Garden Herbarium. Engelmann frequently collected in the area in Illinois across the river from St. Louis, yet left his St. Louis, Missouri address remaining on the label. In more than one instance (*Cyperus Engelmanni*, for example), investigation revealed the origin of the collection to be in St. Clair County, Illinois, rather than Missouri. This is true also of some of Eggert's collections. Two *Bidens Beckii* collections of Engelmann seen at the Missouri Botanical Garden Herbarium come from Illinois opposite St. Louis. They are: (1) lakes, American bottom, St. Clair County, June, 1838, and (2) lakes, American bottom, near the lower Belleville Road, August, 1846. It should be noted that this latter collection was made by Engelmann in the same month and year as the one at Gray Herbarium. It is possible that the collection in the Gray Herbarium represents a duplicate collection lacking complete data of the one in the Missouri Botanical Garden Herbarium, and that the 'St. Louis, Missouri' address of Engelmann was all that appeared on the label together with the month and year as found on the label in the Missouri Botanical Garden Herbarium.

It is also possible that the species has been exterminated from the original station around St. Louis or that it may eventually be found in one of the upland sinkhole ponds or other ponds in the northern portion of St. Louis County or adjacent St. Charles County.

Ducks eat the leaves and fruits of this plant.

## 46. **Cosmos** Cav. Cosmos

Rays of flower-heads red, pink, purplish, or white; segments of leaves linear, thread-like, or narrowly lanceolate, 1 mm. or less broad; achenes glabrous (without hairs) . . . . . . . . . 1. C. BIPINNATUS
Rays of flower-heads orange or golden-yellow; segments of leaves lanceolate or elliptic, 2–8 mm. broad; achenes hairy . . . . . . . . . . . . . . . . . . . . . . 2. C. SULPHUREUS

1. **Cosmos bipinnatus** Cav. Map 2308

Flowers July–October.

Escaped from cultivation in waste ground and along railroads. Known only from St. Louis County, east-central Missouri (St. Louis, on the dumping place along the Gimblin Road between the water works, conduit track and Baden freight yard, August 30, 1958, *Muehlenbach 1471*).

Native of Mexico, north to Arizona; commonly cultivated and sometimes escaped, especially in the southern states.

This common garden annual very rarely becomes established in wild habitats in Missouri.

2. **Cosmos sulphureus** Cav. Map 2309

Flowers August–October.

Occurs in open ground along roadsides. Known only from southwestern Missouri in Douglas (shaded upland along route 76, sect. 23, 1½ mi. southwest of Ava, September 1, 1951, *Steyermark 72534*) and Jasper counties.

Native of Mexico; escaped from cultivation and naturalized in the United States from New Jersey and Pennsylvania, south to Florida and Texas.

In some regions where the plant is grown, the tender tops and leaves are cooked, or eaten raw in salads.

## 47. **Marshallia** Schreb. Barbara's Buttons

**Marshallia caespitosa** Nutt. ex. DC. Map 2310

Flowers April–June.

Occurs in rocky or upland prairies, limestone glades, and bald knobs.

Two variations occur in Missouri and are separated according to the characters given by Dr. Channell (Contr. Gray Herb. 181 : 68. 1957):

Flowering stem (scape) bearing 1 head of flowers; leaves of the flowering stem with their tips projecting only slightly, if at all, beyond the level of the longer basal leaves, the plant thus appearing to have the leaves all crowded at the base; leafy portion of stem much shorter than the peduncle . a. M. CAESPITOSA var. CAESPITOSA
Flowering stem bearing 2–5 (up to 12 or more) heads of flowers; leaves of the flowering stem extending to various levels on the stem above the basal leaves; leafy portion of stem as long as or longer than the peduncles, rarely shorter . . b. M. CAESPITOSA var. SIGNATA

a. **Marshallia caespitosa** var. **caespitosa** Map 2310

*Marshallia caespitosa* Nutt. [G, BB, P & S, Steyerm.]

Known only from the unglaciated prairie region of southwestern Missouri, in Barton, Jasper (high prairies, Alba, *Palmer 26946*; Prosperity, *Palmer 2153*; Webb City, *Palmer 2015*; Jasper, *Palmer 3407*) and Newton counties.

2309 Cosmos sulphureus

2310 ● Marshallia caespitosa var. caespitosa (Barbara's Buttons)

2310 □ Marshallia caespitosa var. signata

2311 Galinsoga parviflora

Ranges from Louisiana and Texas, north to Missouri and Oklahoma.

b. **Marshallia caespitosa** var. **signata** Beadle & Boynt. Map 2310

Known only from limestone glades and bald knobs in Ozark County, southwestern Ozarks (bald knobs, Tecumseh, *Palmer 34811*).

Ranges from Texas north locally to Missouri.

The Ozark County specimens have been previously misidentified as *M. obovata* (Walt.) Beadle & Boynt. var. *platyphylla* (Curtis) Beadle & Boynt., and were so recorded by Palmer and Steyermark in their *Annotated Catalogue* (p. 674). These specimens are now referred to *M. caespitosa* var. *signata* by the recent monographer of the genus, Dr. R. B. Channell (Contr. Gray Herb. 181: 90, 103, 111. 1957) as representing uncommonly broad-leaved specimens of *M. caespitosa* var. *signata* and are believed to represent relict colonies of a variety otherwise known only from Texas.

Although the flowers are often described in various manuals as pink or purple, those of var. *signata* seen by the author are invariably white. Shinners (*Spring Flora of the Dallas-Fort Worth Area, Texas*, p. 390. 1958) refers to the color of the corollas as 'white or occasionally pink-tinged.' Channell (101, 106) correctly describes the color of the florets as 'usually white or cream-colored, only occasionally pale lavender.' I have grown this variety *signata* for several years in the prairie section of my wildflower preserve in northern Illinois, where it has acclimated itself admirably to the northern climate and does very well, bearing profuse flower-heads. It is an attractive species, and should be more widely grown in limestone rock gardens and perennial beds exposed to full sun.

## 48. Galinsoga R. & P.

Known only from Jackson County, west-central Mo.; pappus of the ray flowers absent or nearly so; pappus of the disk flowers not awn-tipped; achenes along the outer margins glabrous or hairy only at the summit . . . . . . . . . . . . . . . . . . . . . . . . . . 1. G. PARVIFLORA

More commonly encountered species throughout Missouri; pappus of the ray flowers well-developed, equaling the corolla-tube; pappus of the disk flowers tapering to an awned tip; achenes along the outer margins densely hairy on the inner faces . . . . . . . . . . . . . . . . . 2. G. CILIATA

1. **Galinsoga parviflora** Cav. Map 2311

Flowers May–November.

Occurs in cultivated and waste ground. Known only from Jackson County, west-central Missouri (September 8, 1891, *Bush*, in U. of Mo. Herb.).

Native of tropical America from Mexico to South America; introduced and naturalized in the United States north to Massachusetts, New York, Minnesota, and Oregon.

The record cited above from Missouri is based upon a World's Fair specimen. Since the species is not known in the state aside from this specimen, it is quite possible

Plate no. 375. **1.** Cosmos bipinnatus, $\times {}^{2}/_{7}$. **2.** Cosmos sulphureus, $\times {}^{2}/_{7}$. **3.** Marshallia caespitosa var. caespitosa, $\times {}^{2}/_{7}$. **4.** Galinsoga ciliata, $\times {}^{2}/_{7}$. **5.** Flaveria campestris, $\times {}^{2}/_{7}$; *a.* Flower-head, $\times 1{}^{1}/_{7}$; *b.* Achene, $\times 6$; *c.* Ray-floret, $\times 8$; After Gleason, The New York Botanical Garden. **6.** Palafoxia callosa, $\times {}^{2}/_{7}$; *a.* Flower-head, $\times {}^{4}/_{7}$; *b.* Achene, $\times {}^{4}/_{7}$. **7.** Hymenopappus scabiosaeus var. scabiosaeus, $\times {}^{2}/_{7}$. All details from Small, The New York Botanical Garden.

PLATE NO. 375

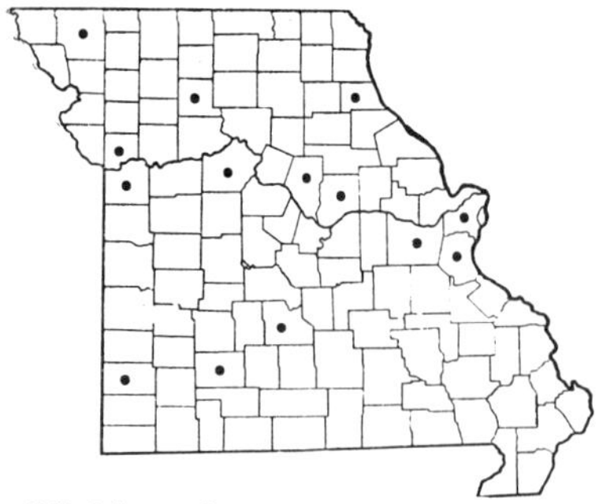

2312 Galinsoga ciliata

2313 Flaveria campestris

2314 Hymenopappus scabiosaeus var. scabiosaeus

that its source may have been from another state and sent to Missouri for exhibit purposes, although Bush stated (in personal correspondence) that all World's Fair specimens marked from Missouri were collected in Missouri.

2. **Galinsoga ciliata** (Raf.) Blake  Map 2312
Flowers May–November.

Occurs in waste and cultivated ground, commonly found in gardens, fields, yards, around dwellings, and streets. Scattered over the state but not commonly collected.

Native of Central and South America; introduced and naturalized in North America north to Quebec, Ontario, Minnesota, and Nebraska.

This is a common garden weed. In southeastern Asia the young stems and leaves are cooked as greens.

## Tribe VI. **HELENIEAE**

### 49. **Flaveria** Juss.

**Flaveria campestris** J. R. Johnston  Map 2313
Flowers July–September.

Occurs in alluvial thickets along streams, waste ground, and along railroads. Known only from Marion (rich soil, Bear Creek bottoms, Hannibal, September 13, 1918, *Davis 3249*; same locality, September 9, 1918, *Davis 9053*) and Jackson (introduced, Courtney, September 20, 1917, *Bush 8190*; same locality, August 1, 1907, *Bush 766*) counties, central Missouri.

Ranges from Mexico, New Mexico, and Texas, north to Colorado and Kansas; introduced in Missouri.

The single ray in each flower-head is very inconspicuous and only 2 mm. long.

### 50. **Hymenopappus** L'Hér

**Hymenopappus scabiosaeus** L'Hér var. **scabiosaeus**  Map 2314
*Hymenopappus scabiosaeus* L'Hér [G, BB, P & S, Steyerm.]
Flowers late April–June.

Occurs on limestone glades and bald knobs, limestone bluff escarpments and ledges, rocky prairies, sandy open ground, and sandy places by cemeteries. Southern Missouri, in Scott, Mississippi, Oregon, Taney, Stone, and Barry counties.

Ranges from Florida to Mississippi, Arkansas, and Oklahoma, north to South Carolina, Indiana, Illinois, and Missouri.

The genus has been monographed recently by Dr. Billie Turner (Rh. 58: 163–86; 208–42; 250–69; 295–308. 1956).

Another variety of this species, var. *corymbosus* (T. & G.) Turner, occurs west of Missouri. It differs from var. *scabiosaeus* in having the peduncles either without bracts or with short subulate scales instead of with conspicuous ovate petal-like bracts 5–14 mm. long as in var. *scabiosaeus*, and in the bracts of the involucre shorter (5–9 mm. long instead of 7–15 mm. long) and yellowish to white for 2–4 mm. at the tip instead of white for half of the length or more as in var. *scabiosaeus*.'

This biennial plant has solitary upright stems which attain a height of 1.5 meters. The much-dissected leaves are quite handsome as are the showy, white, petal-like bracts of the involucre. The creamy-white corollas are fragrant.

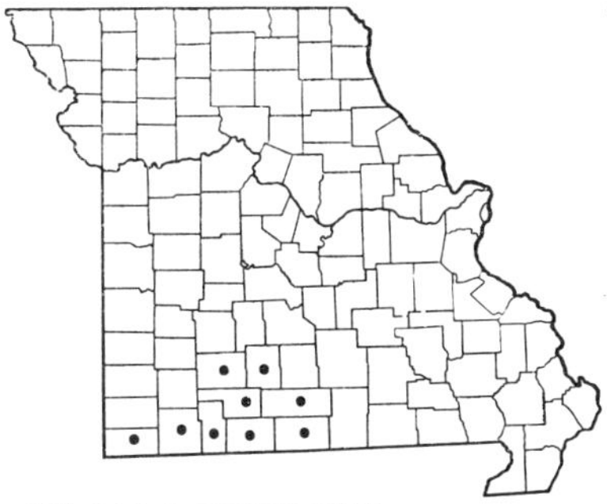

2315 Palafoxia callosa var. callosa

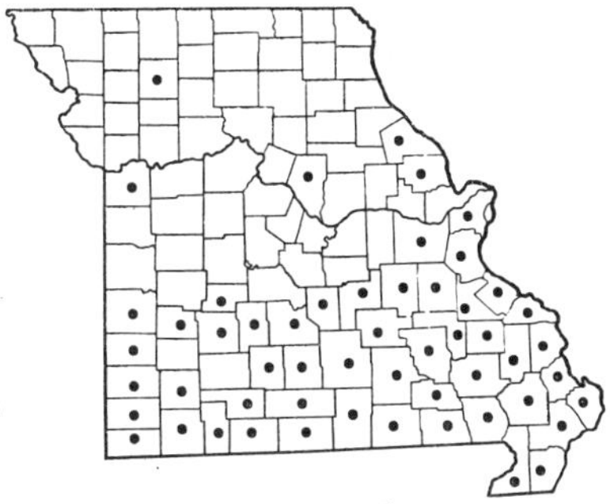

2316 Helenium amarum (Bitterweed)

2317 Helenium flexuosum (Sneezeweed)

## 51. **Palafoxia** Lag.

**Palafoxia callosa** (Nutt.) T. & G. var. **callosa** Map 2315

*Palafoxia callosa* (Nutt.) T. & G. [G, BB, P & S]

Flowers August–October.

Occurs on limestone glades and bald knobs, limestone bluff escarpments, rocky open and sandy alluvial ground, and along gravelly beds of streams. Southwestern Ozark region north to Webster and Greene counties, east to Douglas and Ozark counties.

Ranges from Missouri and Arkansas to Oklahoma and Texas.

This is an attractive annual with rose-lavender or pink corollas. It usually forms large stands on the rocky barrens. I have grown it successfully in the prairie section of my wildflower preserve in northern Illinois, where it has continued to reseed each year. It is a desirable subject for the limestone rock garden.

Another variety, var. *bella* (Cory) Shinners, occurs in Texas, distinguished by having 15–30 flowers in a head instead of 5–13 as in var. *callosa*.

## 52. **Helenium** L. Sneezeweed

a. Leaves very narrow, linear or thread-like, rarely over 2 mm. broad, not decurrent on the stem or branches (leaf-tissue not extending from base of leaf down along stem); annuals . . . . 1. H. AMARUM

a. Leaves much broader, the main ones on the stem usually 5–55 mm. broad, decurrent on the stem and branches (leaf-tissue extending from base of leaf down along stem); perennials . . . . . . . *b*

b. Disk (solid portion of flower-head between rays) yellow; ray flowers fertile . . . 3. H. AUTUMNALE

b. Disk brown or purple; ray flowers sterile or neutral (with functional stamens or pistils) . 2. H. FLEXUOSUM

1. **Helenium amarum** (Raf.) H. Rock
Bitterweed Map 2316

Also called Yellow Dog Fennel.

*Helenium tenuifolium* Nutt. [G, BB, P & S, Steyerm.]

Flowers June 1 to November.

Occurs in fields, waste grounds, glades, rocky and sandy open soils, along roadsides and railroads. Common in southern and east-central Missouri north to St. Louis, Franklin, Crawford, Phelps, Pulaski, Laclede, Dallas, Hickory, Cedar, and Vernon counties, and locally north in Pike, Boone (*Kucera 2514*), Daviess, and Jackson counties.

Ranges from Florida to Texas and Mexico, north to Virginia, Kentucky, Illinois, Missouri, Kansas, and Oklahoma; spreading north to Massachusetts, New York, Ohio, Michigan, and Indiana; also in the West Indies.

Dr. Howard F. L. Rock (Rh. 59: 131. 1957) has shown that the correct name of this species should be *H. amarum*.

A narcotic poisonous substance is contained in the plant. The milk of cows that have grazed on the plants of this species contains a bitter flavor, while cases of poisoning have been attributed to horses and mules that have grazed on this species.

2. **Helenium flexuosum** Raf. Sneezeweed Map 2317

Also called Purple-head Sneezeweed.

*Helenium nudiflorum* Nutt. [G, BB, P & S]

Flowers June–November.

Occurs along margins of lowland and upland sinkhole ponds, sloughs, streams, ditches, swamps, swampy meadows, wet depressions in upland prairies, pastures, low fallow fields and alluvial soils, and moist places on rocky glades bordering bluff escarpments.

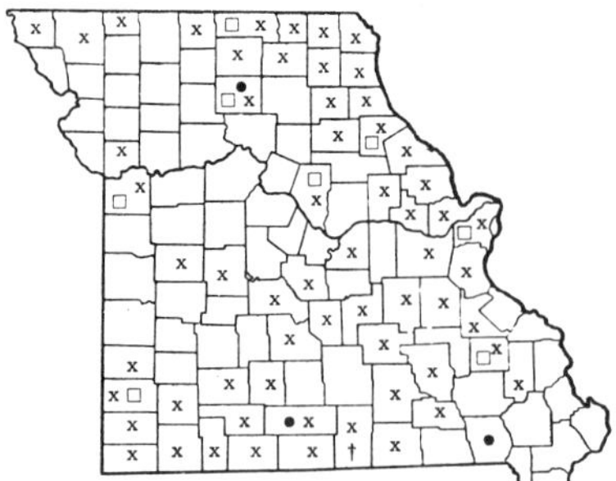

2318 • Helenium autumnale var. autumnale (Sneezeweed)
2318 □ Helenium autumnale var. canaliculatum
2318 x Helenium autumnale var. parviflorum
2318 † Helenium autumnale X flexuosum

2319 Gaillardia lutea

2320 Gaillardia pulchella (Indian Blanket)

Mainly in southern and central Missouri north to Ralls, Audrain, Boone, Moniteau, Benton, Henry, and Jackson counties, and locally in northern Missouri in Linn, Livingston, and Putnam counties.

Ranges from Florida to Texas, north to Maine, New Hampshire, Vermont, New York, Pennsylvania, Ohio, Michigan, Indiana, Illinois, Missouri, and Oklahoma; introduced in Minnesota, New England, New York, Pennsylvania, Ohio, and Michigan.

Dr. Rock (Rh. 59: 210–16. 1957) has shown that the name *H. flexuosum* must be adopted for this species. *Helenium flexuosum* sometimes hybridizes with the next species, *H. autumnale* L. A collection of such a hybrid plant is recorded from Howell County (open margins of dried sink-hole pond on south side of road N, T25N, R9W, sect. 1, 1½ mi. east of Pomona, October 19, 1957, *Steyermark 86003*). This collection combines the yellow, 5-lobed disk-corollas and fertile style-bearing ray flowers characteristic of *H. autumnale* with the shorter stature, more broadly winged stem and floral branches, higher and larger disk, longer and larger disk flowers, longer pappus scales, longer achenes, and long-attenuate apex of leaf of *H. flexuosum.*

As with the other species, *H. flexuosum* is reputed to be poisonous to stock. It is often observed as untouched in pastures frequented by hogs and cattle.

### 3. **Helenium autumnale** L. Sneezeweed

Map 2318

Flowers August–November.

Occurs in moist ground along streams, ponds, sloughs, ditches, wet and spring-fed meadows, prairies, alluvial soils in low woodland and wet open ground.

The following varieties occur in Missouri:

a. Rays of flower-heads 3–13 mm. long; disk (solid portion of head between rays) 8–15 mm. broad . . 3c. H. AUTUMNALE var. PARVIFLORUM

a. Rays of flower-heads 10–25 mm. long; disk 10–23 mm. broad . . . . . . . . . *b*

b. Larger leaves 2–5.5 cm. broad, coarsely toothed, elliptic, oblong, or lanceolate; rays 7–12 mm. broad, 16–25 mm. long . 3a. H. AUTUMNALE var. AUTUMNALE

b. Larger leaves 0.3–1.8 cm. broad, entire (without teeth) or shallowly toothed with short fine teeth, linear to lanceolate; rays 2–7 mm. broad, 10–20 mm. long . . . 3b. H. AUTUMNALE var. CANALICULATUM

### 3a. **Helenium autumnale** var. **autumnale**

Map 2318

*Helenium autumnale* L. [G, P & S]

This is scattered in Missouri.

Ranges from New England to Minnesota, south to North Carolina, Kentucky, and Missouri.

### 3b. **Helenium autumnale** var. **canaliculatum** (Lam.) T. & G.

Map 2318

Scattered in Missouri.

Ranges from Quebec to Minnesota and Nebraska, south to Massachusetts, New York, Ohio, Indiana, Illinois, Missouri, Texas, and Arizona.

### 3c. **Helenium autumnale** var. **parviflorum** (Nutt.) Fern.

Map 2318

This is the commonest variation in the state, found throughout Missouri, except absent from the lowlands of the extreme southeastern section.

---

Plate no. 376. **1.** Helenium flexuosum, × 2/5; After Gleason, The New York Botanical Garden. **2.** Helenium amarum, × 2/5. **3.** Helenium autumnale, × 2/5; Details from Small, The New York Botanical Garden. **4.** Gaillardia lutea, × 2/5; Details from Small, The New York Botanical Garden. **5.** Gaillardia pulchella, × 2/5; Details from Small, The New York Botanical Garden.

PLATE NO. 376

Ranges from Florida to Arkansas, north to Connecticut, New York, Pennsylvania, Kentucky, Illinois, and Iowa.

Cronquist (in Gleason's *New Ill. Fl.* 3: 379. 1952) recognizes var. *autumnale* as a coarser-toothed variation with the broader leaves mostly 3–6 times as long as broad, and var. *canaliculatum* as a finer-toothed or nearly entire-leaved variation with the narrower leaves about 7–12 times as long as broad. However, Dr. Rock, who has most recently treated the genus, does not recognize any of the above varieties, which grade into one another.

All parts of the plants contain a bitter substance which is concentrated in the mature flower-heads, and gives a bitter taste to the milk of cows which have eaten the plants. Cases of poisoning are reported in horses, cattle, and sheep which have fed on large quantities of the flowering heads. Generally, however, stock avoid the plants.

The Winnebago Indians are said to have used the plant for colds, rubbing parts of the plant to produce a sneezing reaction.

## 53. **Gaillardia** Foug. Gaillardia, Blanket Flower

All the flowers (of both disk and rays) yellow; chaff-like development on the receptacle among the disk flowers reduced to soft short scales or nearly absent . . . . . . . . . . . . . . 1. G. LUTEA

Flowers of the disk (solid portion between the rays) purple, brown-red or purplish-brown; chaff-like bristles on the receptacle among the disk flowers firm and well-developed, equaling or longer than the achenes . . . . . . . . . . . . . . . . . . . . . . . . . . . 2. G. PULCHELLA

1. **Gaillardia lutea** Greene Map 2319

*Gaillardia lanceolata* Michx. var. *flavovirens* Mohr [BB]

Flowers late May–October.

Occurs in sandy open ground. Known only from Crowley Ridge, southeastern Missouri in Scott (sandy ground, Morley, October 8, 1893, *Eggert*) and Dunklin (common, Malden, August 21, 1894, *Bush 177*) counties.

Ranges from Florida to Texas, north to Missouri and Oklahoma.

This rare species has not been collected since 1894.

2. **Gaillardia pulchella** Foug. Indian Blanket Map 2320

Also called Fire-wheels.

Flowers May–September.

Occurs in fields, waste ground, along roadsides and railroads. Scattered in southeastern and central Missouri.

Native from Colorado and New Mexico, east to Minnesota, Nebraska, Kansas, and Louisiana; introduced in Missouri east to the Atlantic states.

Most of the Missouri collections are clearly escapes from gardens and cultivation. Two collections which may be from wild plants are from Platte County in western Missouri (exposed hill slopes, 1 mi. north of Weston, *Steyermark 15235*) and Dunklin County, southeastern Missouri (Malden, *Bush 409*).

The Missouri collections have been identified by Susann Fry and are apparently annuals, although collectors sometimes, mistaking their specimens as perennials, have identified their plants as *G. aristata.* The color of the rays varies from purple or brown-red throughout to yellow and purple or reddish. Some of the specimens with yellow rays, such as *Muehlenbach 1003* from St. Louis County, on the basis of ray color, might be keyed out to *G. aristata* Pursh, as treated in Gleason's *New Ill. Fl.*, but the plants are annuals and the chaff-like bristles equal the achenes; in *G. aristata* the plants are perennial and the chaff-like bristles of the receptacle are longer than the achenes.

## 54. **Dyssodia** Cav. Fetid Marigold

**Dyssodia papposa** (Vent.) Hitchc. Fetid Marigold Map 2321

Flowers late May–October.

Occurs in dry fields, pastures, loess hills, prairies, rocky or open slopes and banks, waste ground, along roadsides and railroads.

Throughout Missouri.

Ranges from Louisiana to Mexico and Arizona, north to Ontario, Minnesota, North Dakota, and Montana; introduced east to New England.

Although widespread in the state, this species always appears to grow as if introduced at the station where found, rather than appearing to be wild or part of the native vegetation. The habitats in the state are usually associated with disturbed or cultivated situations.

On account of its strong odor and taste, it is usually avoided by stock.

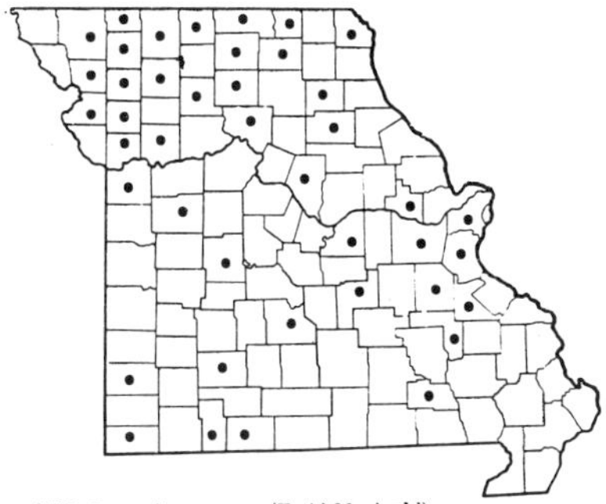

2321 Dyssodia papposa (Fetid Marigold)

2322 Tagetes erecta (French Marigold)

2323 • Achillea Millefolium subsp. Millefolium var. Millefolium f. Millefolium (Common Milfoil)
2323 □ Achillea Millefolium subsp. Millefolium var. Millefolium f. roseum
2323 x Achillea Millefolium subsp. lanulosa

## 55. **Tagetes** L. Marigold

**Tagetes erecta** L. French Marigold Map 2322

Also called African Marigold.

Flowers July–October.

Occurs in waste ground, along roadsides and railroads.

Infrequently escaped from cultivation and scattered, known from Marion, St. Louis, St. Francois, and Jackson counties.

Native of Mexico and Central America; introduced and naturalized in the United States.

This is a commonly cultivated garden annual. The odor of marigolds has recently been stated to cause hay fever symptoms even though the pollen may be absent.

## Tribe VII. **ANTHEMIDEAE**

## 56. **Achillea** L. Yarrow, Milfoil

**Achillea Millefolium** L. Common Milfoil Map 2323

Sometimes called Nosebleed.

Flowers May–November.

Occurs in dry prairies, fields, pastures, waste ground, along roadsides and railroads.

Throughout Missouri and doubtless in every county.

Three variations are encountered in Missouri:

a. Flowers pink or rose-colored . . . . . b. A. Millefolium subsp. Millefolium var. Millefolium f. roseum

a. Flowers white . . . . . . . . . . *b*

b. Ultimate (smallest) leaf-segments lanceolate or ovate; stem and leaves somewhat glabrous to more or less loosely cobwebby; inflorescence rather flat-topped, 6–30 cm. broad; common type encountered . . . a. A. Millefolium subsp. Millefolium var. Millefolium f. Millefolium

b. Leaf-segments more crowded, the ultimate (smallest) ones linear or narrowly lanceolate; stem and leaves densely silky or woolly; inflorescence strongly convex (higher in the central portion), 2–10 cm. broad; less common type encountered . . . . c. A. Millefolium subsp. lanulosa

a. **Achillea Millefolium** L. subsp. **Millefolium** var. **Millefolium** f. **Millefolium** Map 2323

*Achillea Millefolium* L. [G, BB, P & S, Steyerm.]

This is the common variation in Missouri, found doubtless in every county.

Native of Europe and Asia; introduced and naturalized in North America from Newfoundland to Alaska, south to Florida and California.

b. **Achillea Millefolium** subsp. **Millefolium** var. **Millefolium** f. **roseum** Rand & Redf. Map 2323

*Achillea Millefolium* f. *roseum* Rand. & Redf. [G, P & S, Steyerm.]

Scattered throughout the range in Missouri.

c. **Achillea Millefolium** subsp. **lanulosa** (Nutt.) Piper Map 2323

*Achillea lanulosa* Nutt. [G, P & S, Steyerm.]

Chiefly in the unglaciated prairie region of southwestern Missouri and in the glaciated prairie region of northern and central Missouri, extending south and east to St. Louis, Warren, Montgomery, Moniteau, Morgan, Benton, Cedar, and Wright counties.

Ranges from Minnesota to Saskatchewan and British Columbia, south to Missouri, Kansas, Oklahoma, Mexico, and California; introduced east to Newfoundland and New England.

Although the two subspecies above treated often appear distinctive in their extremes, many intergradations occur, with the result that some plants are difficult to place accordingly in one or the other category.

The plants are usually weedy, spreading by underground rootstocks, making it difficult to eradicate once it becomes established. It is sometimes cultivated as a forage plant in Europe, and sheep will eat it. In Europe it is also used as an ornamental plant. The leaves are sometimes used for seasoning, and, when boiled, may be made into a tea used as a remedy for colds and as a blood tonic. The name, Nosebleed, applied to this species and to *A. Ptarmica*, was derived from the belief by early writers that the leaves of the plant, when placed in the nostrils, caused nosebleed to take place. I have tried this but no irritating effects have followed. The leaves have an irritating or burning sensation when brought into direct contact with delicate membrane tissues. As a drug plant it is known to act upon the blood vessels and has been used to control hemorrhages of the pelvis, stomach, liver, and other visceral organs.

*Excluded Species*

**Achillea Ptarmica** L. Sneezeweed

This is reported for Missouri in Gleason's *New Illustrated Flora* 3: 385. 1952. The report is based upon specimens collected in St. Louis County (St. Louis, July 12, 1910, *Sherff 421*). However, these collections are taken, not from naturalized, but from cultivated specimens, as evidenced by two sheets seen in the Chi. Nat. Hist. Mus. Herbarium, one of which states 'cultivated' on the label, the other under the same collection number without the word 'cultivated,' but obviously collected at the same time.

It is a commonly cultivated species, and is known to escape occasionally and become naturalized from Labrador to Ontario, south to New England, New York, and Michigan. Since no specimens have been seen from Missouri as actually escaped from cultivation, the species is not included for the state.

## 57. **Anthemis** L. Chamomile

a. Rays yellow; disks 10–20 mm. broad, the tubes of the corollas compressed . . . . . . 1. A. TINCTORIA
a. Rays white; disks 5–12 mm. broad, the tubes of the corollas cylindric . . . . . . . . . . . . *b*
 b. Plant without an unpleasant odor; receptacle with scale-like chaff accompanying all the flowers of the disk; ray flowers fertile; achenes smooth on the sides . . . . . . . . . . 2. A. ARVENSIS
 b. Plant with an unpleasant or strong odor; receptacle with scale-like chaff accompanying only the central flowers of the disk (best seen when the individual flowers are removed); ray flowers sterile; achenes with roughened projections . . . . . . . . . . . . . . . . . . 3. A. COTULA

1. **Anthemis tinctoria** L. Yellow Chamomile Map 2324

Flowers May–September.

Occurs in waste ground. Known from St. Louis and Greene (escaped, Springfield, June 11, 1887, *Weller*, in Drury College Herb.) counties.

Native of Europe; introduced and naturalized in North America from Quebec to Minnesota and Iowa south to New Jersey, Illinois, and Missouri.

2. **Anthemis arvensis** L. Corn Chamomile Map 2325

Flowers May–August.

Occurs in fields, waste ground, along roadsides and railroads.

Two variations occur in the state:

Pales (chaff) of the receptacle longer than the disk flowers . . . 2a. A. ARVENSIS var. ARVENSIS
Pales (chaff) of the receptable shorter than the disk flowers . . . 2b. A. ARVENSIS var. AGRESTIS

2a. **Anthemis arvensis** var. **arvensis** Map 2325

Known only from St. Louis County, east-central Missouri (St. Louis, Bremen Ave. freight yard of Ter-

Plate no. 377. **1.** Dyssodia papposa, × 2/5; *a.* Flower-head, × 2; After Gleason, The New York Botanical Garden. **2.** Tagetes erecta, × 2/5; Details from Small, The New York Botanical Garden. **3.** Achillea Millefolium subsp. Millefolium var. Millefolium, × 2/5; *a.* Basal leaf; Details from Small, The New York Botanical Garden. **4.** Achillea Millefolium subsp. lanulosa, × 2/5; Details from Small, The New York Botanical Garden. **5.** Anthemis tinctoria, × 2/5.

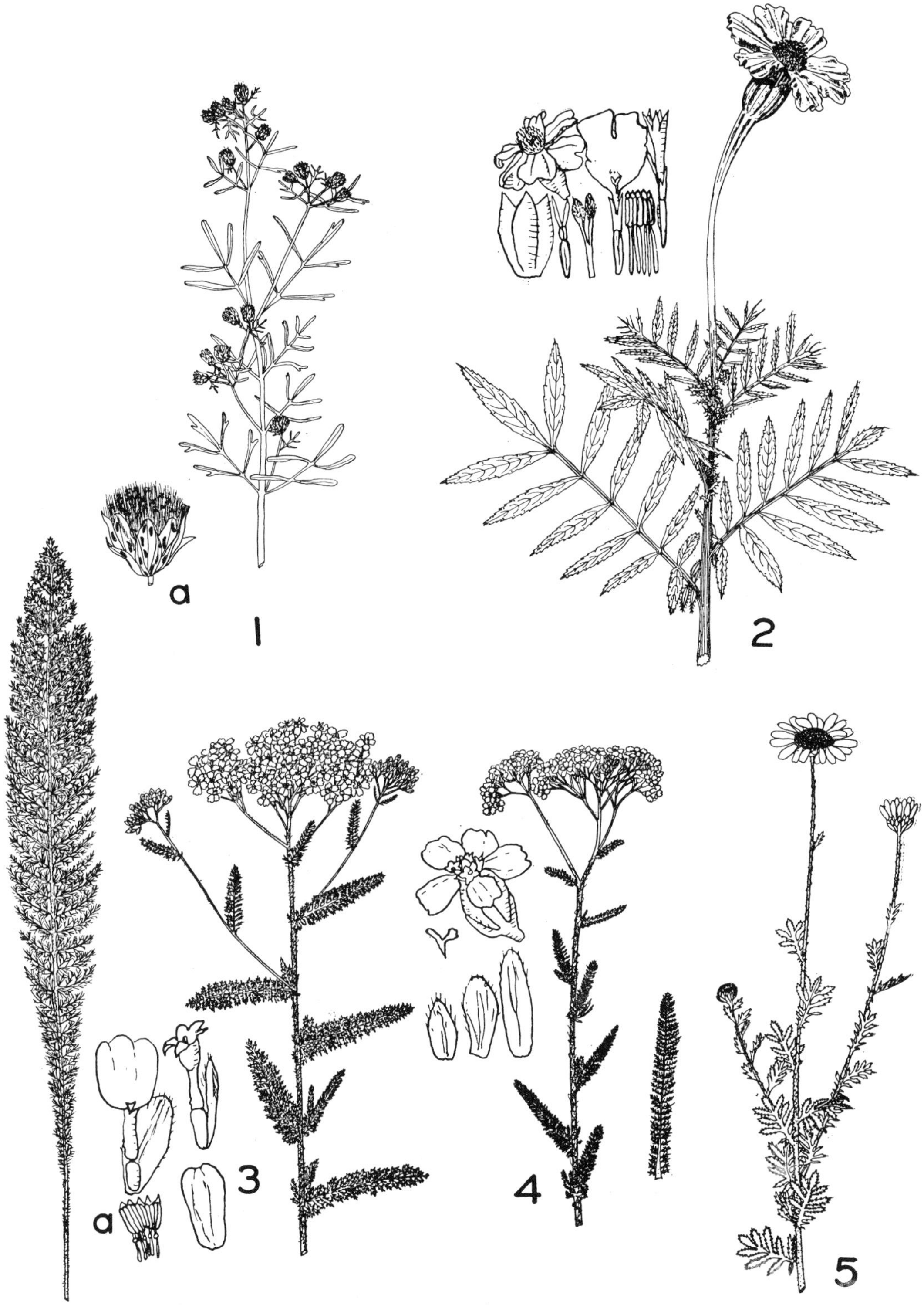


PLATE NO. 377

2324 Anthemis tinctoria (Yellow Chamomile)

2325 ● Anthemis arvensis var. arvensis (Corn Chamomile)
2325 □ Anthemis arvensis var. agrestis

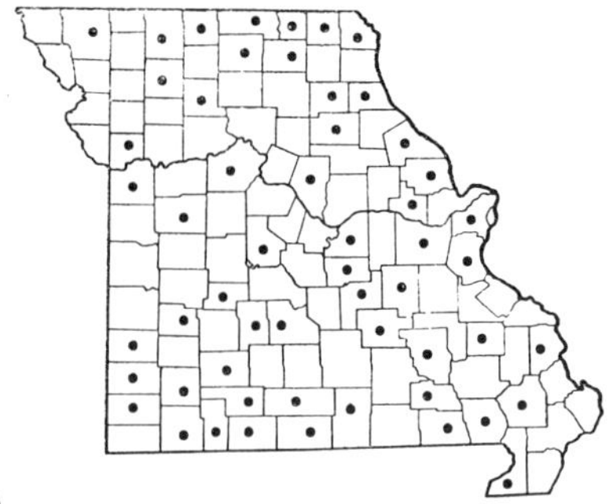

2326 Anthemis Cotula (Mayweed)

minal Railroad Assoc., July 4, 1958, *Muehlenbach 1439*).

Native of Europe, introduced and naturalized in the United States from Maine to New York, south to Georgia and Missouri.

2b. **Anthemis arvensis** var. **agrestis** (Wallr.) DC. Map 2325

This is the commoner variation found in Missouri, although rare and scattered only in southern and central Missouri in St. Louis, Phelps, Pulaski, Jackson, and Jasper counties.

Native of Europe; introduced and naturalized in North America from Quebec to Michigan and British Columbia, south to Florida, Indiana, Illinois, Missouri, and Oregon.

3. **Anthemis Cotula** L. Mayweed Map 2326
Also called Dog Fennel, Stinking Chamomile.
Flowers May–November.

Occurs in pastures, farm lots, fields, waste ground about dwellings, roadsides, and along railroads.

Throughout Missouri, probably in every county.

Native of Europe; introduced and naturalized in North America from Newfoundland to Alaska, south to Florida and California; also in Africa and Asia.

This species is most commonly seen in rich soil around barnyards and in waste ground elsewhere. It is usually avoided by stock because of its strong odor and taste. The skin of some persons is irritated from coming in contact with the plant.

## 58. **Matricaria** L. Wild Chamomile

a. No rays present on flower-heads; corollas all 4-lobed . . . . . . . . . . . 3. M. MATRICARIOIDES
a. Rays present on flower-heads; corollas of the disk (solid portion between rays) 5-lobed . . . . . . . *b*
  b. Bruised live plant with odor of pineapple; pappus absent; make vertical cut through flower-head to note that receptacle is cone-shaped and pointed . . . . . . . . . . . . 2. M. CHAMOMILLA
  b. Bruised live plant odorless; pappus present in the form of a short border or crown; make vertical cut through flower-head to note that receptacle is hemispherical or dome-shaped and rounded . 1. M. MARITIMA var. AGRESTIS

1. **Matricaria maritima** L. var. **agrestis** (Knaf) Wilmott Map 2327
Flowers May–October.

Known only from St. Louis County, east-central Missouri (wastes around St. Louis, May, 1839, *Engelmann*).

Native of Europe; introduced and naturalized from Newfoundland to Ontario, south to New England, Pennsylvania, Kentucky, Kansas, and on the Pacific Coast.

2. **Matricaria Chamomilla** L. var. **Chamomilla** Wild Chamomile Map 2328
Also called German Chamomile.
*Matricaria Chamomilla* L. [G, BB, P & S, Steyerm.]
Flowers May–October.

---

Plate no. 378. **1.** Anthemis arvensis var. agrestis, × 2/5. **2.** Anthemis Cotula, × 2/5; Details from Small, The New York Botanical Garden. **3.** Matricaria Chamomilla, × 2/5; Details from Small, The New York Botanical Garden. **4.** Matricaria matricarioides, × 2/5.

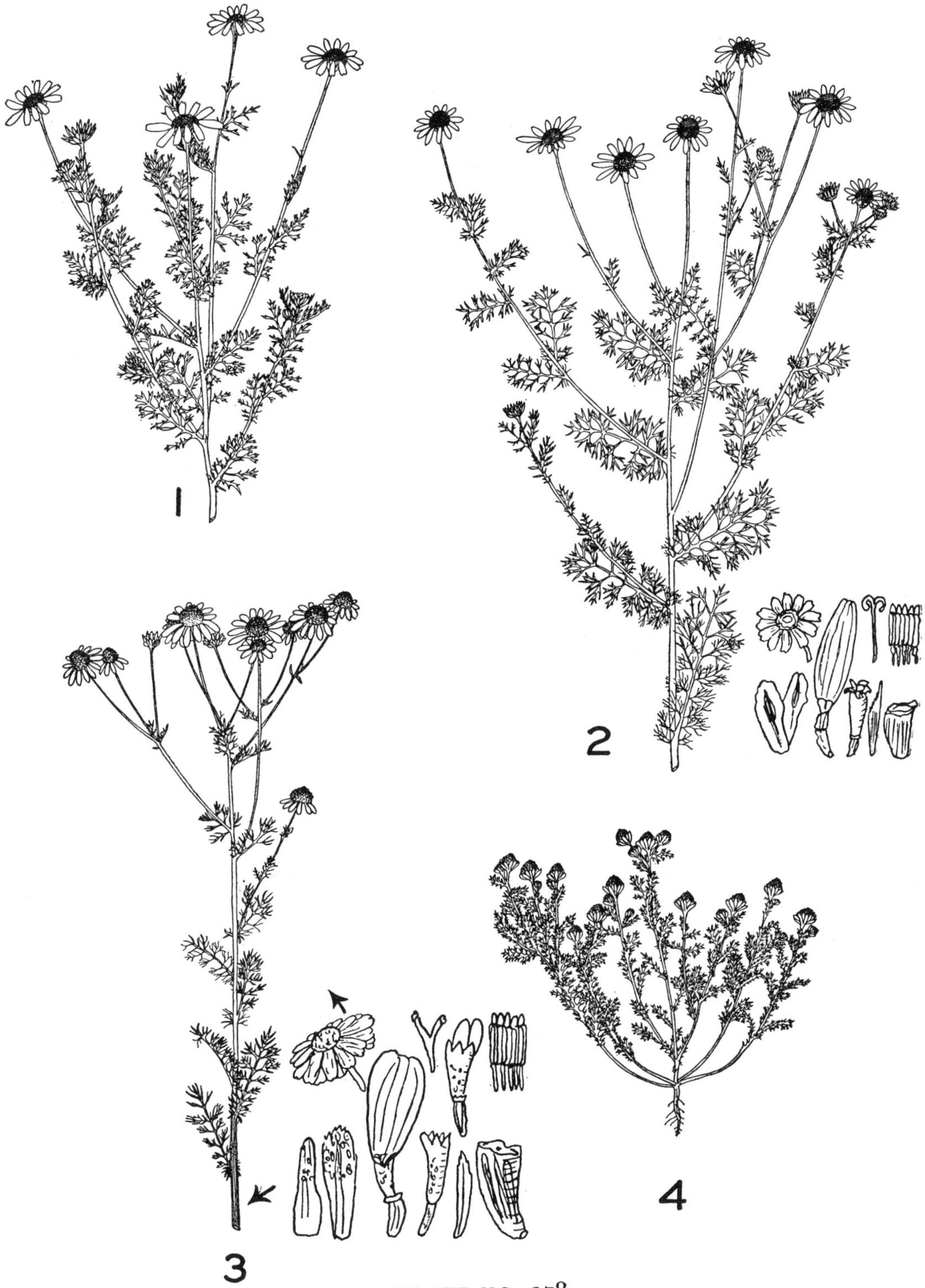


PLATE NO. 378

2327 Matricaria maritima var. agrestis

2328 Matricaria Chamomilla var. Chamomilla (Wild Chamomile)

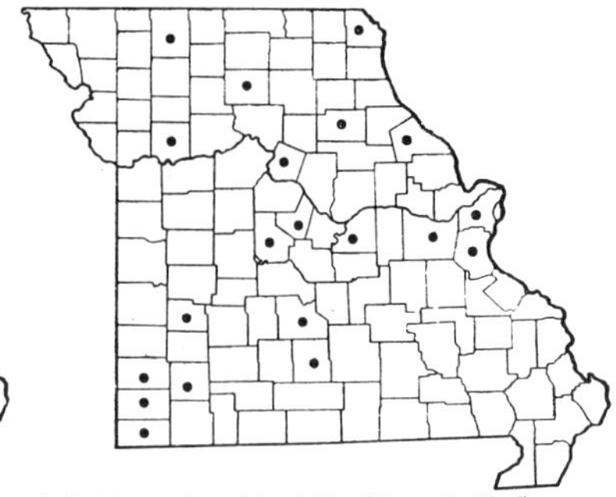

2329 Matricaria matricarioides (Pineapple Weed)

Occurs in waste ground, along roadsides, and railroads. Rare, and known only from St. Louis, St. Charles (St. Charles, May 5, 1929, *Kellogg*), Phelps (June, 1890, *E. N. Plank*), and Boone (along railroad, Rocheport, May 15, 1937, *F. Kavanagh*) counties.

Native of Europe; introduced and naturalized in North America from Newfoundland to Minnesota, south to Pennsylvania and Missouri.

The specimen cited from Phelps County was originally labeled *Anthemis nobilis.*

The flower-heads are sometimes boiled as a tea used as a tonic and stomach medicine.

Another variety, var. *coronata* (Gray) Coss. & Germ., differs from var. *Chamomilla* in having the pappus in the form of a short crown, whereas in var. *Chamomilla* no pappus crown is evident.

3. **Matricaria matricarioides** (Less.) Porter
Pineapple Weed Map 2329
Flowers May–October.

Occurs in rocky open ground, about dwellings, barnyards, farm lots, pastures, fields, roadsides, and along railroads. Scattered throughout Missouri, but not recorded from the southeastern quarter of the state.

Ranges from Alaska to Lower California east to Montana and Arizona; naturalized eastward to Newfoundland and south to Delaware, Pennsylvania, Ohio, Indiana, Illinois, Missouri, and Oklahoma.

This species was well-established as a weed in St. Louis by 1825, Engelmann misjudging it at that time as a native plant (Engelmann in Cat. Pl. made in Illinois and Missouri by Charles A. Geyer, 46: 97. 1843). Because of its small size and inconspicuous flower-heads, it is frequently overlooked.

## 59. **Chrysanthemum** L. Chrysanthemum

a. Commonly encountered species outside cultivation; heads large, 3–6 cm. broad, solitary at the ends of the stem and long branches; disk (solid portion between rays) of flower-heads 10–20 mm. broad . . . . . . 1. C. LEUCANTHEMUM
a. Rarely encountered outside cultivation; heads small, 0.5–2 cm. broad, numerous in corymbose inflorescences; disk of flower-heads 4–9 mm. broad . . . . . . *b*
b. Rays present, white, 3–8 mm. long; leaves deeply cut or parted nearly to the midrib . 2. C. PARTHENIUM
b. Rays usually absent; leaves merely scallop-edged . . . . . . 3. C. BALSAMITA

1. **Chrysanthemum Leucanthemum** L. Ox-eye Daisy Map 2330
Also called White Daisy.
Flowers May–August.

Occurs in fields, meadows, pastures, prairies, roadsides, waste ground, and along railroads. Throughout Missouri, probably in every county.

Two variations are known in the state:

Rarely encountered; middle and upper leaves of stem with spreading teeth at the base which are larger than those on the rest of the leaf; basal leaves rather regularly toothed or merely toothed . . . . . 1a. C. LEUCANTHEMUM var. LEUCANTHEMUM

Commonly encountered; middle and upper leaves of stem with conspicuous lobes or deeply cut narrow portions which extend in to the midrib; basal leaves with deeply parted segments which extend in or nearly in to the midrib or irregularly toothed . . 1b. C. LEUCANTHEMUM var. PINNATIFIDUM

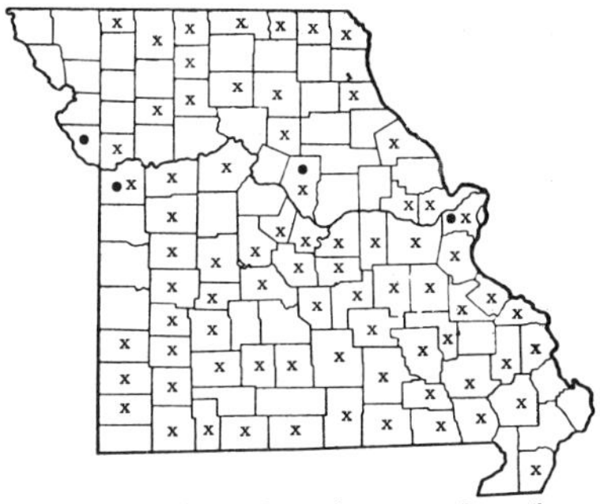

2330 • Chrysanthemum Leucanthemum var. Leucanthemum (Ox-eye Daisy)
2330 x Chrysanthemum Leucanthemum var. pinnatifidum

2331 Chrysanthemum Parthenium (Feverfew)

2332 Chrysanthemum Balsamita var. tanacetoides (Costmary)

1a. **Chrysanthemum Leucanthemum** var. **Leucanthemum** Map 2330
*Chrysanthemum Leucanthemum* L. [G, BB, P & S, Steyerm.]

Rare and known only from St. Louis, Boone, Jackson, and Platte counties.

Native of Europe; introduced and naturalized in North America from Newfoundland and Quebec to New York and New Jersey to Missouri.

1b. **Chrysanthemum Leucanthemum** var. **pinnatifidum** LeCoq & Lamotte f. **pinnatifidum** Map 2330

Common throughout Missouri, probably in every county.

Native of Europe; introduced and naturalized in North America from Labrador to British Columbia, south to Florida, Kentucky, Missouri, Kansas, and Oklahoma.

A form of this variety, f. *tubuliflorum* Henrikssen, not known in Missouri, has tubular instead of flat rays.

In Europe the young leaves are eaten in salad.

2. **Chrysanthemum Parthenium** L. Feverfew Map 2331

Flowers June–September.

Escaped from old gardens into waste ground. Known only from Grundy and Jasper (waif in waste ground, Webb City, July 29, 1909, *Palmer 2562*) counties.

Native of Europe; introduced from cultivation and naturalized in North America from Quebec to Ohio, south to Maryland and Missouri.

This strong-smelling perennial spreads rapidly and is much cultivated as a border or edging plant in old gardens. Many varieties and forms are in cultivation, especially in Europe, where an oil of feverfew is obtained from the plants.

3. **Chrysanthemum Balsamita** L. var. **tanacetoides** Boiss. Costmary Map 2332

Also called Mint Geranium.

Flowers August–October.

Escaped from cultivation in waste ground along fence rows. Known only from Franklin County, east-central Missouri (along a fence row, 1940, *Oscar Petersen*).

Native of Europe; introduced from cultivation and naturalized in North America from Quebec and Ontario, south to Delaware, Pennsylvania, Ohio, Indiana, Missouri, and Kansas.

This plant is often grown in old gardens for the sweet, mint-like odor of the leaves and for the medicinal properties which it is reputed to have. The aromatic flavor of the leaves imparts a quality to salads, and the plant was often grown for this purpose.

The variety *tanacetoides* has no rays present, whereas typical var. *Balsamita* has the rays present.

## 60. **Tanacetum** L. Tansy

**Tanacetum vulgare** L. Tansy Map 2333

Flowers July–September.

Occurs in meadows, fence rows, around old dwellings, prairie borders and fields, roadsides, and along railroads. Scattered throughout the state, but uncommonly collected.

Native of Europe; introduced from old gardens and naturalized in North America from Newfoundland to British Columbia, south to Georgia, Illinois, Missouri, Oklahoma, Nevada, and California.

Two variations are encountered in Missouri:

Margins of leaflets and wings of petiole (leafstalk) merely sharply toothed . . . a. T. VULGARE f. VULGARE

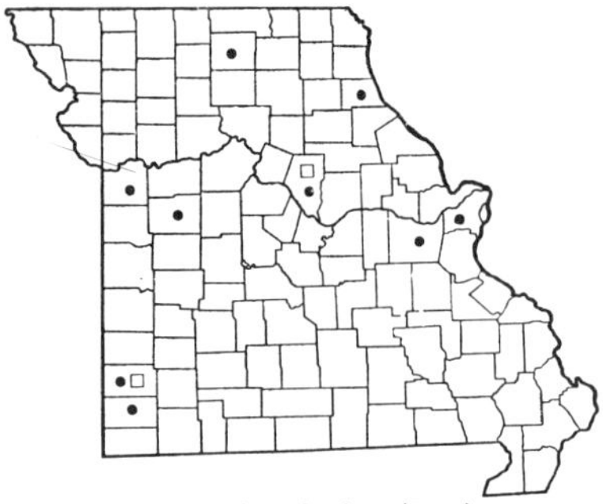

2333 • Tanacetum vulgare f. vulgare (Tansy)
2333 □ Tanacetum vulgare f. crispum

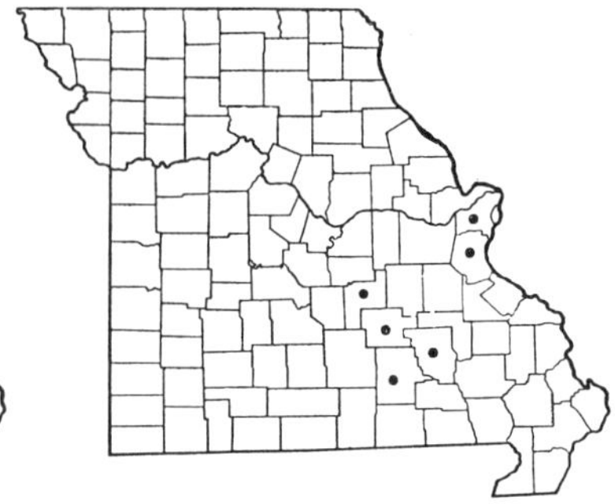

2334 Artemisia caudata var. caudata (Wild Wormwood)

2335 • Artemisia glauca var. glauca
2335 □ Artemisia glauca var. dracunculina

Margins of leaflets and wings of petiole deeply cut and crisped, many of the fine teeth incurved . . . . . b. T. VULGARE f. CRISPUM

a. **Tanacetum vulgare f. vulgare** Map 2333
*Tanacetum vulgare* L. [G, BB, P & S]
This is the commoner variation in the state.

b. **Tanacetum vulgare f. crispum** (L.) Hayek Map 2333
*Tanacetum vulgare* var. *crispum* DC. [P & S]
*Tanacetum vulgare* f. *crispum* (L.) Fern.
Known only from Boone and Jasper (waste ground, Webb City, July 26, 1909, *Palmer 2548*) counties.

The leaves and stems contain an oil which is poisonous to man and lower animals. The oil of tansy has been extensively used as a worm repellent medicine. Human poisoning has been reported when the oil or tea prepared from the leaves has been taken in too great quantities, while grazing animals have been reported as poisoned by eating the leaves and stems. Stock usually avoid the plant, however, because of the bitter taste. Nevertheless, the plant has been used as a bitter herb as a condiment for seasoning cakes and puddings, and as a cooked vegetable mixed with other herbs. A cheese is sometimes prepared by adding the steeped tansy extract to milk before the formation of curds.

## 61. **Artemisia** L. Wormwood

a. Leaves densely woolly or silky with white, silvery, or gray hairiness on at least lower surface, or on both surfaces . . . . . . . . . . . . . . . . . . . . . . . . . . . . *b*
  b. Base of leaf or of petiole (leaf-stalk) with a conspicuous development of small, dissected segments or leaf- or stipule-like lobes; lobes or segments of the main stem-leaves with smaller lobes, segments, or teeth along their margins . . . . . . . . . . . . . . . . . . 5. A. VULGARIS
  b. Base of leaf or of petiole without leaf- or stipule-like lobes; lobes or segments, if present, of the main stem-leaves entire (without teeth) or nearly so . . . . . . . . . . . . . . . . *c*
    c. Largest divisions of leaves 2–4 mm. or more broad; leaves variously toothed or lobed. . . 6. A. LUDOVICIANA
    c. Largest divisions of leaves usually 1 mm. or less broad; leaves divided nearly or all the way to the midrib into narrowly linear, hair- or thread-like segments . . . . . . . . . . . *d*
      d. Main leaves of stem 0.5–1 cm. long; pick off flowers from a head to note that receptacle is covered with long woolly hairs . . . . . . . . . . . . . . . . . 8. A. FRIGIDA
      d. Main leaves of stem 1–5 cm. long; pick off flowers from a head to note that receptacle is smooth and not hairy . . . . . . . . . . . . . . . . . . . 7. A. CARRUTHII

a. Leaves green and mainly glabrous (without hairs) on both surfaces or sometimes slightly grayish-hairy in *A. glauca*, but never with dense white, silvery, or gray woolly or silky hairiness on lower surface. . . . . . . . . . . . . . . . . . . . . . . . . . . . . . *e*
  e. Main leaves of the stem undivided or only the lowest ones 3-parted or cut at the summit; perennial plants with a hard horizontal underground stem (rhizome) . . . . . . . . 2. A. GLAUCA
  e. Main leaves of the stem deeply dissected or divided into 5 or more to numerous segments or divisions; annual or biennial plants, rarely short-lived perennials, always with tap-roots . . . . . *f*
    f. Smaller divisions or segments of a single leaf on the main stem 5–9; margins of leaf-divisions

entire (without teeth); pick off flowers from a head to note that the flowers of the disk (between the rays) are sterile, the ovary shrinking and not becoming an achene . . . . . . 1. A. CAUDATA

f. Smaller divisions or segments of a single leaf numerous, 50–100 or more; margins of leaf-divisions toothed, lobed, or dissected; pick off flowers from a head to note that the flowers of the disk (between the rays) are fertile, the ovary normal, firm, and becoming an achene . . . . *g*

g. Fresh plant sweet-scented when bruised; middle and upper leaves of stem uniformly dissected; the main lateral segments and their divisions cut nearly or quite to the midrib leaving a leafy central portion of usually 1 mm. or less broad; flower-heads nodding, the individual heads on stalks (pedicels); involucre 1–1.5 mm. high; inflorescence loose, open, panicle-like, with much smaller leaves than those on stem . . . . . . . . . 4. A. ANNUA

g. Fresh plant scarcely or not at all scented when bruised; middle and upper leaves of stem less deeply and less uniformly dissected, leaving a leafy central portion 1–4 mm. broad; flower-heads erect, the individual heads sessile (without stalks) or nearly so; involucre 2–3 mm. high; inflorescence dense with crowded, spike-like clusters mixed with leaves similar in size to those lower on stem . . . . . . . . . . . . . . . . . 3. A. BIENNIS

1. **Artemisia caudata** Michx. var. **caudata**
Wild Wormwood Map 2334

*Artemisia campestris* L. var. *caudata* (Michx.) Palmer & Steyerm. [P & S]

*Artemisia caudata* Michx. [G]

Flowers July–October.

Occurs on sand and gravel bars along streams, sandstone glades, ledges, and escarpments of bluffs. Eastern Ozark section west to Phelps, Dent, and Shannon counties, northeast to St. Louis County.

Ranges from Quebec, Ontario, and Manitoba, south near the coast to Florida, south inland to Vermont, New York, Ohio, Indiana, Illinois, Missouri, Oklahoma, and Texas.

This is a biennial or short-lived perennial. Where the plants are abundant, they may be a serious cause of hay fever.

2. **Artemisia glauca** Pall. Map 2335

Flowers mid-June–October.

Occurs on loess hills, prairies, exposed rocky or alluvial banks, and in rocky open barrens. Northern and central Missouri.

Two variations may be recognized in Missouri:

Individual flower-heads on pedicels (stalks) up to 2 mm. long; branchlets of inflorescence leafy, elongated, and ascending . . . . 2a. A. GLAUCA var. GLAUCA

Individual flower-heads on pedicels (stalks) chiefly 3–10 mm. long; branchlets of inflorescence minutely bracted, only up to 3.5 cm. long, and drooping . . 2b. A. GLAUCA var. DRACUNCULINA

2a. **Artemisia glauca** var. **glauca** Map 2335

*Artemisia glauca* Pall. [G]

*Artemisia Dracunculus* in part [of BB], not L.

*Artemisia dracunculoides* Pursh [Jones et al.]

*Artemisia Dracunculus* subsp. *glauca* (Pall.) Hall & Clements [Hall & Clements]

*Artemisia Dracunculus* var. *glauca* Jepson [P & S]

Known from Clark, Holt, Buchanan, and Jackson counties.

Ranges from British Columbia to Mexico, east to Manitoba, Wisconsin, Illinois, Missouri, Oklahoma, and Texas; introduced eastward to New England; also in Siberia.

2b. **Artemisia glauca** var. **dracunculina** (S. Wats.) Fern. Map 2335

*Artemisia Dracunculus* in part [of BB], not L.

*Artemisia Dracunculus* subsp. *dracunculina* (S. Wats.) Hall & Clements [Hall & Clements]

*Artemisia dracunculoides* var. *dracunculina* (S. Wats.) Blake

Known from Jackson County, west-central Missouri (barrens, Greenwood, September 19, 1906, *Bush 4121*).

Ranges from Minnesota to Utah, south to Missouri, Kansas, Texas, and Mexico; introduced eastward to New England.

This begins flowering in mid-June in Missouri. In the western states this species and varieties are considered as a cause of hay fever, and in California the typical var. *glauca* is regarded as an important cause. Botanists disagree on the proper application of the name of this species and whether or not the species should be divided. More intensive field and experimental data are needed to clarify the status of the varieties included in the present treatment.

3. **Artemisia biennis** Willd. Biennial Wormwood Map 2336

Flowers early June–November.

Occurs along gravel bars of streams, waste ground, and along railroads.

Rare and scattered, known only from St. Louis, Iron, Jackson, and Clay counties.

Supposedly native in the northwestern states from

2336 Artemisia biennis (Biennial Wormwood)

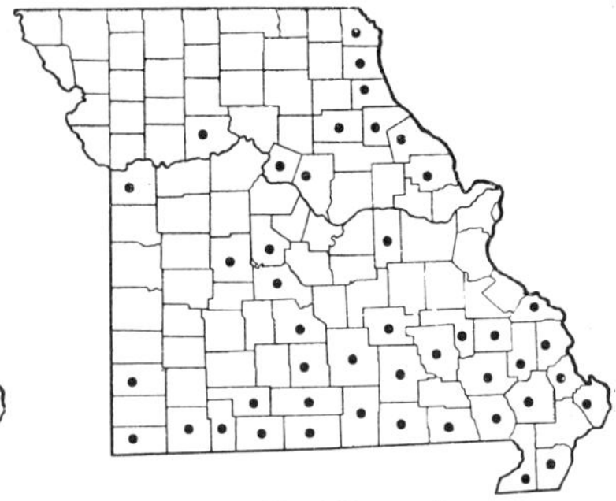

2337 Artemisia annua (Sweet Wormwood)

2338 • Artemisia vulgaris var. vulgaris (Common Mugwort)
2338 □ Artemisia vulgaris var. glabra

British Columbia to California, east to Manitoba and Nebraska; widely scattered eastward to Quebec, south to New Jersey, Pennsylvania, Ohio, Indiana, Illinois, and Missouri.

In the northwestern states and other places where abundant, this species may become a local cause of hay fever.

4. **Artemisia annua** L. Sweet Wormwood Map 2337

Flowers August–October.

Occurs frequently around dwellings, barnyards, farm lots, low and alluvial ground in rich soils of fallow fields, borders of woods, gravel, sand, and mud flats along streams and ponds, waste ground, roadsides, and along railroads.

Southern, central, and eastern Missouri along the Mississippi River to Clark County, elsewhere north to Shelby, Boone, Howard, Carroll, and Jackson counties; commonest in the southern Ozark and southeastern lowland sections.

Native of Eurasia; introduced and naturalized in North America from New Brunswick to Ontario, south to Alabama, Tennessee, Arkansas, and Oklahoma; also in California and other western states.

5. **Artemisia vulgaris** L. Common Mugwort Map 2338

Flowers July–October.

Occurs along railroads.

The following variations may be recognized in Missouri:

a. Divisions or lobes and their teeth of the main leaves blunt or merely slightly pointed (acutish), not acuminate (long-pointed); division or lobe at summit of main leaves broadly obovate or rhombic, the other lobes or divisions oblong or oblanceolate; main leaves not deeply cut . . . . 5c. A. VULGARIS var. LATILOBA

a. Divisions or lobes and their teeth of the main leaves long-pointed (acuminate); division or lobe at summit and lateral lobes of main leaves lanceolate; main leaves deeply cut nearly to the midrib . . . . . . . . . . *b*

b. Divisions or lobes of the main leaves of the stem with toothed or sharply or irregularly cut margins (edges) . . . 5a. A. VULGARIS var. VULGARIS

b. Divisions or lobes of the main leaves of the stem with entire (without teeth) margins . 5b. A. VULGARIS var. GLABRA

5a. **Artemisia vulgaris** var. **vulgaris** Map 2338
*Artemisia vulgaris* L. [G, BB]

Known only from St. Louis County, east-central Missouri (St. Louis, north of Victor St., Terminal R.R. Association, October 5, 1958, *Muehlenbach 1495*).

Native of Europe; introduced and naturalized in North America from Newfoundland to Ontario, south to Delaware, Pennsylvania, Georgia, Michigan, Wisconsin, Minnesota, and Missouri.

5b. **Artemisia vulgaris** var. **glabra** Ledeb. Map 2338

Known only from St. Louis County, east-central Missouri (St. Louis, north of Victor St., Terminal

Plate no. 379. **1.** Chrysanthemum Leucanthemum var. pinnatifidum, × 2/7; Details from Small, The New York Botanical Garden. **2.** Artemisia caudata var. caudata, × 2/7. **3.** Artemisia biennis, × 2/7; After Gleason, details from Small, The New York Botanical Garden. **4.** Artemisia ludoviciana var. gnaphalodes, × 2/7. **5.** Artemisia annua, × 2/7; After Gleason, The New York Botanical Garden. **6.** Chrysanthemum Parthenium, × 2/7; After Gleason, The New York Botanical Garden. **7.** Chrysanthemum Balsamita var. tanacetoides, × 2/7; After Gleason, The New York Botanical Garden. **8.** Tanacetum vulgare, × 2/7; Details from Small, The New York Botanical Garden. **9.** Artemisia ludoviciana var. mexicana, × 2/7. **10.** Artemisia vulgaris, × 2/7.

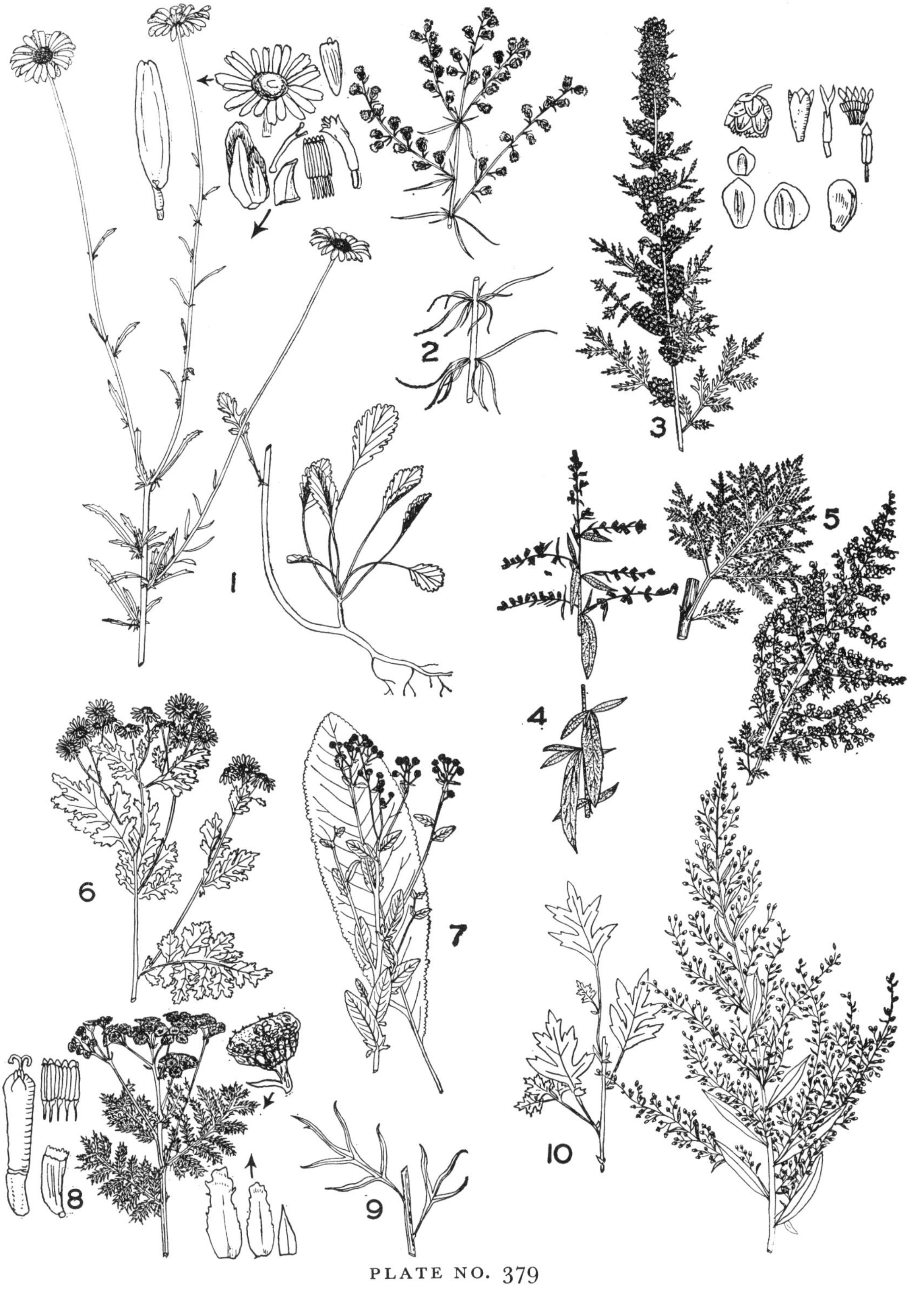


PLATE NO. 379

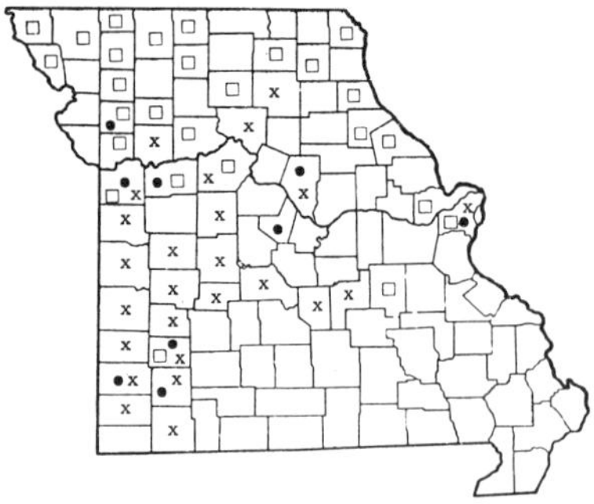

2339 • Artemisia ludoviciana var. ludoviciana (White Sage)
2339 □ Artemisia ludoviciana var. gnaphalodes
2339 x Artemisia ludoviciana var. mexicana

2340 Artemisia Carruthii

2341 Artemisia frigida (Prairie Sagewort)

R.R. Association, August 23, 1958, *Muehlenbach 1468, 1466, 1474*).

Native of Europe; introduced and naturalized in North America from New England to Ontario, south to Connecticut, Ohio, Michigan, Missouri, and Kansas.

5c. **Artemisia vulgaris** var. **latiloba** Ledeb.
Map 2338

Known only from St. Louis County, east-central Missouri (St. Louis, between Sidney and Victor streets, Terminal R.R. Association, September 8, 1956, *Muehlenbach 1067*; south of Sidney St., July 28, 1956, *Muehlenbach 1029*).

Native of Europe; introduced and naturalized from Quebec to New England, Illinois, and Missouri.

The above variations are all represented in Dr. Muehlenbach's specimens with intergradations between one and another shown by the various specimens. Only future more detailed field and experimental studies can determine the relative merits and true taxonomic status of these varieties.

This species was formerly much used medicinally. It possesses toxic properties when taken in overdoses which lead to various complications of pains, spasms, and other disturbances.

6. **Artemisia ludoviciana** Nutt. White Sage
Map 2339

Also called Western Mugwort.

Flowers June–September.

Occurs on rocky prairies, glades, bluff escarpments, open wooded slopes, waste ground, roadsides, and along railroads.

The following variations occur in Missouri:

a. Upper as well as lower surface of young leaves with a close dense covering of gray or white hairiness; leaves entire (without teeth) or some of the lower leaves and those half way up stem toothed or with lobes and divisions . . *b*
  b. Upper surface of leaves in age becoming green and nearly hairless, or the hairs becoming loose and falling . . . . . 6a. A. LUDOVICIANA var. LUDOVICIANA
  b. Upper surface of leaves with a close dense covering of gray or white hairiness even in age . 6b. A. LUDOVICIANA var. GNAPHALODES
a. Upper surface of young leaves green and glabrous (without hairs) or only faintly hairy; leaves mainly lobed with narrow segments, sometimes entire . . 6c. A. LUDOVICIANA var. MEXICANA

6a. **Artemisia ludoviciana** var. **ludoviciana**
Map 2339

*Artemisia ludoviciana* Nutt. [G]
*Artemisia vulgaris* var. *ludoviciana* (Nutt.) Ktze. [P & S]

Scattered in southern and central Missouri north to St. Louis, Boone, Lafayette, and Clinton counties.

Ranges from Michigan to Washington, south to Illinois, Arkansas, Texas, and Mexico; introduced eastward to New England, New York, New Jersey, and Virginia.

6b. **Artemisia ludoviciana** var. **gnaphalodes** (Nutt.) T. & G. Map 2339

*Artemisia vulgaris* var. *gnaphalodes* (Nutt.) Ktze. [P & S]

Prairies, roadsides, and along railroads.

Common throughout northern and west-central Missouri, south to St. Louis, St. Charles, Saline, Lafayette, and Jackson counties, and locally south in Crawford and Dade counties.

Ranges from Ontario and Michigan to British Columbia, south to Arkansas, Oklahoma, and Texas; introduced eastward to Quebec, New England, New Jersey, and Delaware.

6c. **Artemisia ludoviciana** var. **mexicana** (Willd.) Fern. Map 2339

*Artemisia vulgaris* var. *mexicana* (Willd.) T. & G. [P & S]

*Artemisia ludoviciana* subsp. *mexicana* (Willd.) Keck [Keck]

Mainly in the unglaciated prairie region and western Ozark section of western Missouri, north to Jackson, Ray, Chariton, Macon, Boone, and St. Louis counties, east and south to Phelps, Pulaski, and Barry counties; absent from most of the Ozark region.

Ranges from Mexico and Texas north to Missouri and Oklahoma.

This species is quite variable in the size and amount of lobing of the foliage as well as the relative pubescence on the leaves and stems and the branching of the inflorescence.

Two varieties which should occur in Missouri, but have not been recorded for the state, are var. *latifolia* (Bess.) T. & G. and var. *americana* (Bess.) Fern. The var. *latifolia* closely resembles var. *gnaphalodes* but has the leaf-blades oblong, oblong-elliptic, or oblong-oblanceolate, 2.5–7 cm. long, and blunt or merely acute instead of lanceolate, 5–10 cm. long, and acute or attenuate at the tip as in var. *gnaphalodes*. The stems in var. *latifolia* have more loosely spreading elongated branches than in var. *gnaphalodes*. The var. *americana* most closely resembles var. *mexicana*, from which it differs in having a more open and leafy instead of dense panicle, a globose-cylindrical instead of cylindric or cylindric-ovoid involucre, and a stem more densely felted with a mat of pubescence, whereas in var. *mexicana* the stem is often thinly hairy or becomes nearly glabrous.

7. **Artemisia Carruthii** Wood Map 2340

*Artemisia vulgaris* var. *Wrightii* (Gray) Palmer & Steyerm. [P & S]

*Artemisia vulgaris* subsp. *Wrightii* (Gray) Hall & Clements [Hall & Clements]

Flowers August–October.

Occurs in waste ground and along railroads. Known only from Jackson County (Sheffield, September 14, 1905, *Bush 3333*; Sheffield, August 2, 1903, *Bush 1838*), west-central Missouri.

Ranges from Kansas and Colorado to Texas, New Mexico, and Arizona; introduced in Missouri, Indiana, New York, and Rhode Island.

8. **Artemisia frigida** Willd. Prairie Sagewort Map 2341

Flowers July–September.

Occurs in waste ground and along railroads. Known only from western Missouri in Jackson (Sheffield, June 17, 1918, *Bush 8409*) and Newton counties.

Ranges from Saskatchewan to Alaska, south to Wisconsin, Minnesota, Kansas, Texas, and Arizona; introduced in Missouri, east to Quebec, New Brunswick, and New England; also in Asia.

In the western states this species is of considerable importance as a hay fever plant.

*Excluded Species*

**Artemisia serrata** Nutt.

This species is reported from Missouri by Fernald (Rh. 47: 250. 1945) and in the eighth edition of *Gray's Manual* (p. 1522), but I have seen no Missouri material which is referable to the species, and am therefore excluding it from the flora of the state.

## Tribe VIII. **Senecioneae**

### 62. **Erechtites** Raf. Fireweed

**Erechtites hieracifolia** (L.) Raf. var. **hieracifolia** Fireweed Map 2341A

*Erechtites hieracifolia* (L.) Raf. [G, BB]

*Erechtites hieracifolia* var. *intermedia* Fern. [G, P & S]

Flowers July–November.

Occurs in rocky open woods, eroded slopes, thickets, waste ground, and along roadsides and railroads. Throughout Missouri, but not recorded from some of the counties of the unglaciated prairie region and western Ozarks.

Ranges from Newfoundland to Ontario and Minnesota, south to Florida, Louisiana, and Texas.

Dr. Robert O. Belcher has recently monographed the genus (Ann. Mo. Bot. Gard. 43: 1–85. 1956), and, from field and experimental data, has found that variations, previously segregated as var. *intermedia*, may exist in the field side by side with typical var. *hieracifolia*, and that these variations may result from the progeny of a single plant. He has noted extremes of variation which are to be found in the same location growing side by side. There is some correlation between over-all size and form of leaf and the fertility of the soil and amount of moisture available. The var. *intermedia* has been found by Dr. Belcher to represent a condition which arises in summer plants which are restricted to the amount of soil nutrients available. His studies indicate that the var. *intermedia* cannot be accepted as a recognizable variety.

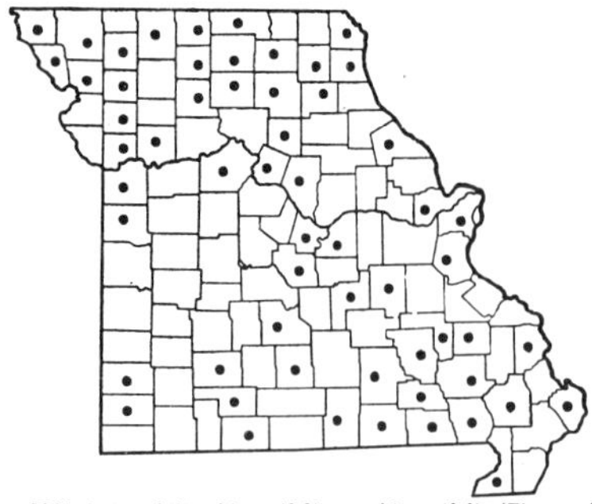

2341-A Erechtites hieracifolia var. hieracifolia (Fireweed)

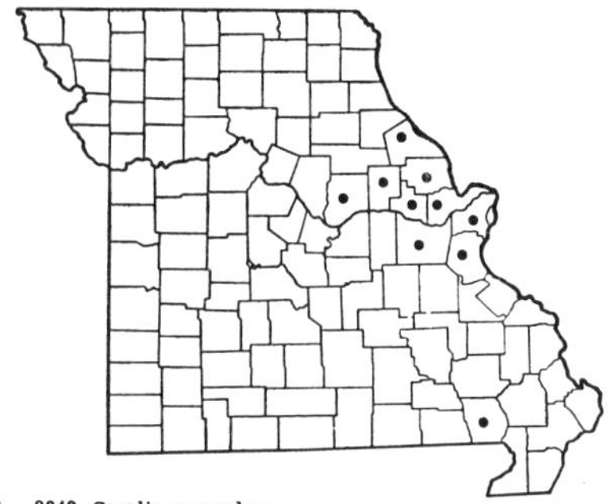

2342 Cacalia suaveolens

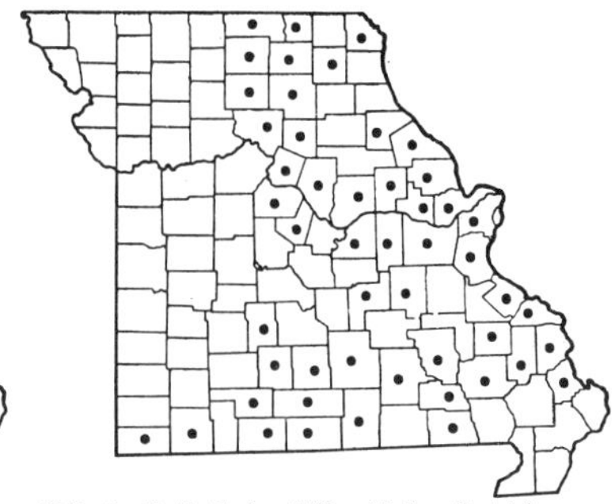

2343 Cacalia Muhlenbergii (Great Indian Plantain)

## 63. **Cacalia** L. Indian Plantain

a. Main leaves of stem arrowhead-shaped; flower-heads consisting of 20–30 flowers surrounded by 12–15 bracts of the involucre . . . . . . . . . . . . . . . . . . 1. C. SUAVEOLENS

a. None of the leaves arrowhead-shaped; flower-heads consisting of usually 5 flowers surrounded by 5 bracts of the involucre. . . . . . . . . . . . . . . . . . . . . . . . . *b*

b. Lower leaves mostly entire (without teeth) or with shallow teeth; lower leaves of an oval type, principal nerves of lower leaves running lengthwise and all ending at the tip of the leaf. 4. C. TUBEROSA

b. Lower leaves shaped like a lima bean or kidney, or semicircular or broadly triangular, or broadly heart-shaped or triangular, coarsely toothed, angled, or lobed; principal nerves of lower leaves spreading outward and ending in the teeth, lobes, or angles along the margins . . . . . . . *c*

c. Stem and at least the lower surface of leaves glaucous (with a silvery or gray-white coating which can be rubbed off); lower leaves broadly triangular-ovate or heart-shaped and angled; margins of leaves smooth in the depressions between the teeth . . . . . 3. C. ATRIPLICIFOLIA

c. Stem and both surfaces of leaves green; lower leaves semicircular to lima bean- or kidney-shaped; margins of leaves with tiny fine hairs in the depressions between the teeth. 2. C. MUHLENBERGII

1. **Cacalia suaveolens** L. Map 2342

Flowers July–September.

Occurs in rich or low woods in valleys, at the base of rich slopes or bluffs, or banks of streams. Mainly in east-central Missouri north to Pike County, west to Callaway County, south to Jefferson County, locally southeast in Butler County.

Ranges from Connecticut and Rhode Island to Iowa, south to Florida, Tennessee, and Missouri; introduced in Massachusetts.

I have had this species growing in the woodland section of my wildflower preserve in northern Illinois, where it has become a serious weed, spreading freely from seed as well as vegetatively and forming large colonies.

2. **Cacalia Muhlenbergii** (Sch. Bip.) Fern. Great Indian Plantain Map 2343

*Cacalia reniformis* Muhl. [P & S, Steyerm.]

Flowers last of May–September.

Occurs in rich woods in valleys or on north- or east-facing wooded slopes or bluffs bordering streams.

Eastern and southern Missouri, west to Putnam, Sullivan, Linn, Chariton, Howard, Cooper, Moniteau, Phelps, Dallas, Webster, Christian, and McDonald counties.

Ranges from New Jersey to Minnesota, south to Georgia, Alabama, and Missouri.

This species spreads readily by seeds, having become one of the dominants in the woodland section of my wildflower preserve in northern Illinois. The large basal leaves are very attractive. Flowering stems usually appear in the second to third years after seed germination. Both this species and *C. suaveolens* prefer a neutral to calcareous soil.

3. **Cacalia atriplicifolia** L. Pale Indian Plantain Map 2344

Flowers June–October.

Occurs in lowland or upland, open and rocky

Plate no. 380. **1.** Erechtites hieracifolia, × $^2/_7$; Details from Small, The New York Botanical Garden. **2.** Cacalia suaveolens, × $^2/_7$; *a.* Flower-head, × $^2/_7$. **3.** Cacalia Muhlenbergii, × $^2/_7$; Details from Small, The New York Botanical Garden. **4.** Cacalia atriplicifolia, × $^2/_7$; *a.* Leaf from lower part of stem; *b.* Inflorescence; *c.* Leaf from middle of stem. **5.** Cacalia tuberosa, × $^2/_7$; *a.* Flower-head, × $1^3/_7$; Details from Small, The New York Botanical Garden.

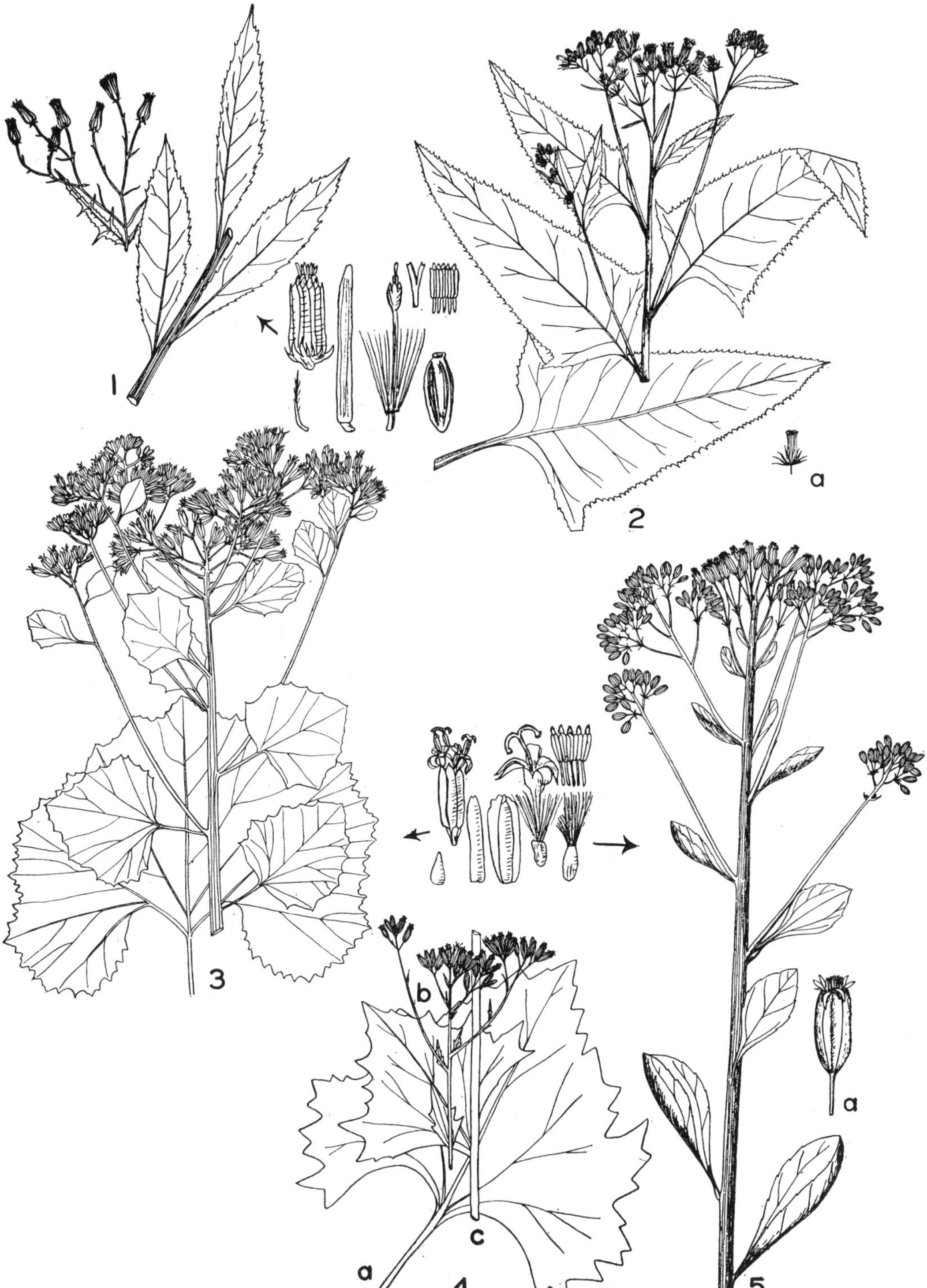


PLATE NO. 380

2344 Cacalia atriplicifolia (Pale Indian Plantain)

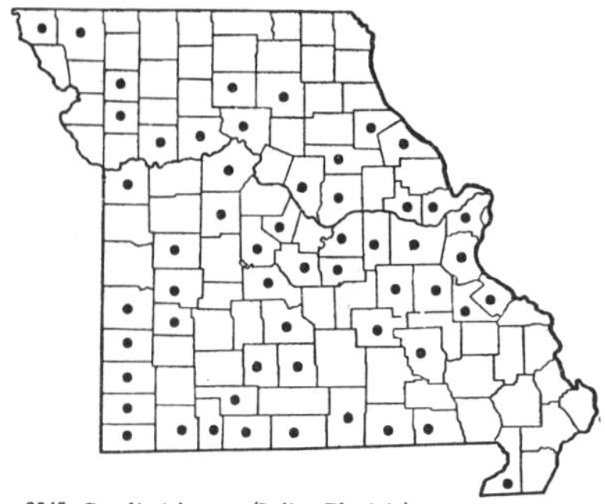

2345 Cacalia tuberosa (Indian Plantain)

2346 Senecio vulgaris (Common Groundsel)

woods, ravine slopes, thickets, slopes and bluffs along streams, and along railroads.

Throughout Missouri, doubtless in every county.

Ranges from New York to Minnesota, south to Florida, Alabama, and Oklahoma.

The leaves of this species vary considerably in lobing and toothing, yet the species is quite uniform and distinctive.

4. **Cacalia tuberosa** Nutt. Indian Plantain Map 2345

*Cacalia plantaginea* (Raf.) Shinners, Field & Lab. 18: 81–83. 1950.

Flowers May–August.

Occurs in prairies, low meadows of river bottom alluvium, limestone glades, open or rocky woods, thickets, roadsides, and along railroads. Southern, central, and western Missouri; apparently absent or not recorded from most of northeastern and north-central Missouri as well as from the extreme southeastern lowland section.

Ranges from Ontario to Minnesota, south to Alabama, Louisiana, and Texas.

This is a characteristic prairie species.

## 64. **Senecio** L. Ragwort, Groundsel, Squaw-weed

The plants of this group should be collected with the root-leaves and runners, whenever these are present or accompany the plant, for proper study and determination of the species.

a. No rays present at flowering time; bracteoles (at base of involucre of flower-heads) blackish or black-tipped; some of bracts of involucre with black tips . . . . . . . . . . . . . . . 1. S. VULGARIS

a. Rays present at flowering time; neither bracteoles nor bracts of involucre with black tips . . . . . *b*

b. Leaves usually absent at the base of the plant; no runners or distinct leafy tufts arising from base of flowering stems; plants nearly equally leafy to the summit with all the leaves deeply parted or divided to the midrib; annuals . . . . . . . . . . . . . . . . . 2. S. GLABELLUS

b. Leaves usually present at the base of the plant; runners or distinct leafy tufts arising from base of flowering stems; leaves unequal, much larger and more numerous at base of plant, becoming few and reduced in size upwards on the stem with at least most of the basal leaves usually only shallowly toothed and unlobed; perennial plants . . . . . . . . . . . . . . . . . *c*

c. A hairiness persisting on the petioles (leaf-stalks), summit of the stem, pedicels (stalks) of flower-heads, or base of involucre, or on all of them; lower surface of basal leaves usually covered with short hairiness, sometimes becoming nearly glabrous (hairless) . . 3. S. PLATTENSIS

c. Petioles, summit of stem, pedicels of flower-heads, and involucre usually glabrous or eventually nearly glabrous, the tufts or patches of hairs scattered on stems or at base of plant and in the axils of the leaves, mostly or entirely disappearing at maturity; lower surface of basal leaves usually glabrous . . . . . . . . . . . . . . . . . . . . . . . . . . *d*

d. All or most of the basal or root-leaves heart-shaped, nearly heart-shaped or truncate (as if cut straight across) or rounded, but not tapering at base; leaves and involucral bracts around flower-head delicately fragrant . . . . . . . . . . . . . . 5. S. AUREUS

d. All or most of the basal or root-leaves tapering at base, or, if somewhat rounded, at least not heart-shaped; leaves and involucral bracts around flower-head not delicately fragrant. . . . *e*

e. Leaf-tissue at base of leaf-blade conspicuously continued down (decurrent) along upper

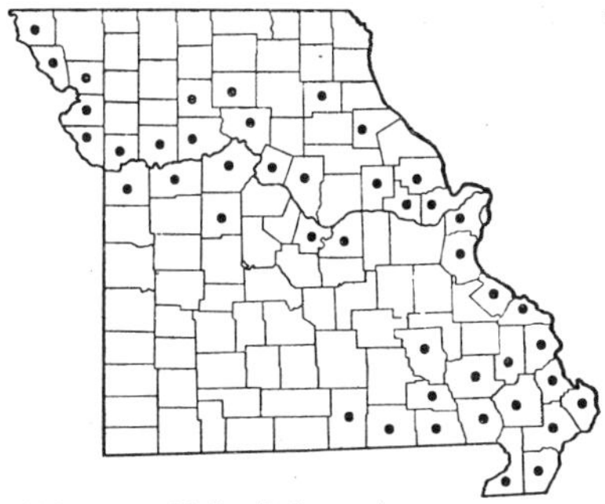

2347 Senecio glabellus (Butter-weed)

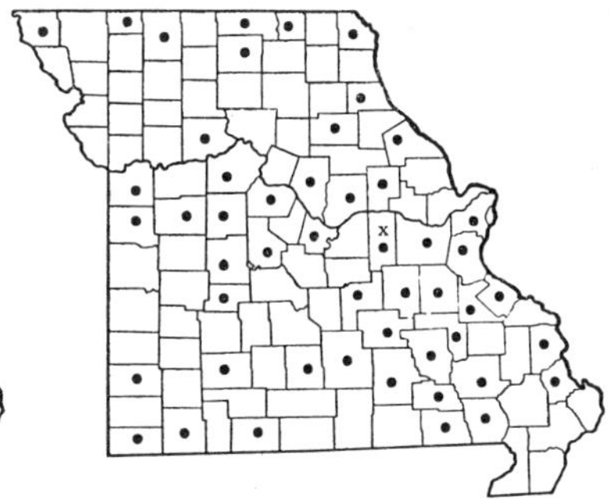

2348 • Senecio plattensis (Prairie Ragwort)
2348 x Senecio plattensis X obovatus

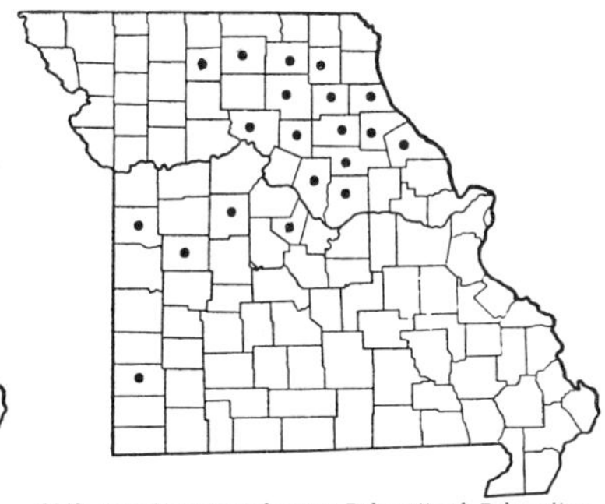

2349 Senecio pauperculus var. Balsamitae f. Balsamitae

part of petiole (leaf-stalk), the broadened summit of the petiole with the decurrent leaf-tissue 3–30 mm. (up to 60 mm.) long; slender runners tipped with leafy rosettes present and produced in early summer . . . . . . . . . . . . . . . 6. S. OBOVATUS

e. Leaf-tissue at base of leaf-blade more or less ending in leaf-blade and only slightly if at all continued down along upper part of petiole, the broadened summit of petiole shorter; no slender runners present, the rosette leaves either at the very base of the flowering stems or in separate tufts . . . . . . . . . . . . 4. S. PAUPERCULUS

1. **Senecio vulgaris** L. Common Groundsel Map 2346

Flowers June–October.

Occurs along railroads and waste ground. Known only from St. Louis County (St. Louis, Carrie Ave. freight yard of Terminal Railroad Association, June 28, 1958, *Muehlenbach 1428*).

Native of Europe; introduced and naturalized in North America from Labrador to Alaska, south to North Carolina, Ohio, Indiana, Wisconsin, Missouri, Texas, New Mexico, and California.

The plant has astringent properties, for which it is sometimes used medicinally.

2. **Senecio glabellus** Poir. Butter-weed Map 2347

Flowers April–June.

Occurs in low wet ground along streams, borders of ponds, sloughs, swamps, and ditches, low fallow fields in river valleys, bald cypress and other woodland swamps, and wet woods. Commonest in the lowlands of southeastern Missouri west in the Ozarks to Reynolds, Carter, and Howell counties, thence north along the Mississippi River counties to Ralls and Shelby counties, west along the counties bordering the Missouri River to Atchison County, and along the counties bordering the Grand River and tributaries to Linn and Livingston counties.

Ranges from Florida to Texas, north to North Carolina, Ohio, Indiana, Illinois, Missouri, and Oklahoma.

This species, especially in the alluvial and low bottoms of southeastern Missouri, often covers large areas, in early spring, forming dense stands of yellow with the showy clusters of flower-heads.

3. **Senecio plattensis** Nutt. Prairie Ragwort Map 2348

*Senecio pseudotomentosus* Mackenz. & Bush; holotype from barrens, Monteer, Shannon Co., April 27, 1907, *Bush 4337*, at N.Y. Bot. Gard. Herb.

Flowers May–June.

Occurs in dry upland prairies, glades, escarpments and ledges of bluffs, and rocky open woods. Commonest in southern and central Missouri locally north in northern Missouri to Clark, Schuyler, Putnam, Harrison, Worth, and Atchison counties.

Ranges from Vermont to Ontario, west to Saskatchewan, south to Virginia, Ohio, Indiana, Louisiana, and Texas.

*Senecio pseudotomentosus*, originally described from Missouri, in no way differs from many other collections of *S. plattensis*, having the lower surface of the leaves tomentose and the petioles and base of stems white-hairy. In its hairier extremes no difficulty is encountered in recognizing this species. However, specimens which are found late in the season with less or little pubescence are often difficult to separate from some variations of *S. pauperculus*. The achenes of *S. plattensis* are mainly minutely hairy, whereas those of *S. pauperculus* are chiefly glabrous. Usually the basal leaves of *S. plattensis* are abruptly rounded or nearly truncate at the base to slightly heart-shaped, whereas those of *S. pauperculus* tend to be more tapering and narrowed at the base. Variation is also found in a colony of plants or in those found in shady woods with more elongated leaf-blades. Hybridization apparently occurs between this species and *S. obovatus* where the habitats of the two overlap, such as in rocky woodland adjacent to rocky glades or bluff escarpments. An example of such a hybrid is known from Gasconade County (wooded limestone ledge at upper part of southwest exposure of Church Bluffs, northeast end of Helds Island, along Gasconade River, T44N, R6W, sect. 16, 1–1½ mi. southeast of Pershing, May 20, 1956, *Steyermark 81195*).

4. **Senecio pauperculus** Michx. var. **Balsamitae** (Muhl.) Fern. f. **Balsamitae** Map 2349
*Senecio pauperculus* [of P & S, Steyerm.], not Michx.
*Senecio pauperculus* var. *Balsamitae* (Muhl.) Fern. [G]
*Senecio pauperculus* in part [BB], not Michx.
Flowers May–August.

Occurs in prairies, meadows, prairie remnants along railroads, and open woodland.

Glaciated prairie region of northeastern Missouri and unglaciated prairie region of southwestern Missouri southwest to Jasper County and southeast elsewhere to Pettis, Moniteau, Callaway, and Pike counties.

Ranges from Labrador to Alberta, south to Georgia, Alabama, Ohio, Indiana, Illinois, Iowa, Nebraska, and Idaho.

Typical var. *Balsamitae* f. *Balsamitae* has ray flowers well developed. In var. *Balsamitae* f. *inchoatus* Fern. the rays are absent. Typical *S. pauperculus* var. *pauperculus*, not known in Missouri, has smaller basal leaves (1–4 cm. long by 0.5–1.5 [–2] cm. broad) than var. *Balsamitae* (1.5–8 cm. long, 1–2 cm. broad), lower and middle stem-leaves smaller (1–3.5 cm. long by 0.1–6 [–10] mm. broad) than var. *Balsamitae* (3–9 cm. long by 10–25 mm. broad), fewer flower-heads (1–8 instead of 4–40), and shorter involucres (4–6 mm. instead of 5–8 mm. high). However, intermediate specimens found are difficult to place in one or the other variety, and it is highly doubtful whether a separate varietal status can be maintained for var. *Balsamitae*. This will be determined by future, more detailed field and experimental studies. Many specimens included under var. *Balsamitae* for Missouri have small leaves and more or less glabrate stems, except for some hairy tufts at the base of the stem and leaf-axils.

5. **Senecio aureus** L. Golden Ragwort Map 2350
Also called Squaw-weed.
Flowers April–June.

Occurs in moist or low ground along spring branches, streams, wet ledges and base of bluffs, in rich or rocky low woodland, and wooded ravine slopes.

The following variations occur in the state:

a. Basal leaves oblong-oval, the base of the blades rounded or barely heart-shaped . . . . 5d. S. AUREUS var. SEMICORDATUS
a. Basal leaves nearly orbicular to round-ovate, the base of the blades definitely heart-shaped. *b*
  b. Involucres 8–11 mm. high; corollas of the disk (solid portion of flower-head) 6–10 mm. long; achenes 3.5–4 mm. long; stems of the offshoots at the base of the plant 4–10 mm. thick at flowering time; usually a long dense hairiness present on many of the petioles (leaf-stalks), margins of basal sheaths, and surfaces of unexpanded leaves and inflorescences . . . . 5a. S. AUREUS var. AUREUS
  b. Involucres 5–8 (–9) mm. high; corollas of the disk 5–8 mm. long; achenes 2–3.5 mm. long; stems of the offshoots at the base of the plant 1–5 mm. thick at flowering time; either no hairs are present or only sparse short hairiness occurs on the petioles, margins of basal sheaths, and surfaces of unexpanded leaves and inflorescences . . *c*
    c. Larger root- or basal leaves 6–15 cm. long, 3–12 cm. broad; longer petioles of root-leaves 5–30 cm. long; stems of the offshoots at the base of the plant mainly 2–10 cm. long; stems often 2 or more to a plant . . . . 5b. S. AUREUS var. INTERCURSUS
    c. Larger root- or basal leaves 1–3.5 cm. long, 1–2.8 cm. broad; longer petioles of root-leaves 2–10 cm. broad; stems of the offshoots at the base of the plant up to 6 cm. long; stems usually solitary . 5c. S. AUREUS var. GRACILIS

5a. **Senecio aureus** var. **aureus** Map 2350
*Senecio aureus* L. [G, BB, P & S, Steyerm.]

Common throughout the Ozark region of southern and east-central Missouri south and east of a line drawn from Lincoln, Montgomery, Callaway, Boone, Moniteau, Morgan, Hickory, Dade, and Jasper counties, locally north of this area in Ralls, Livingston, and Jackson counties.

Ranges from Florida to Arkansas, north to Maryland, Kentucky, Missouri, and Oklahoma.

Plate no. 381. **1.** Senecio vulgaris, × $^2/_7$. **2.** Senecio pauperculus var. Balsamitae, × $^2/_7$. **3.** Senecio glabellus, × 1. **4.** Senecio obovatus, × $^2/_7$; *a*. Leaf variation. **5.** Senecio plattensis, × $^2/_7$. **6.** Senecio aureus, × $^2/_7$; *a*. Leaf variations; *b*. Senecio aureus var. gracilis; Details from Small, The New York Botanical Garden.

PLATE NO. 381

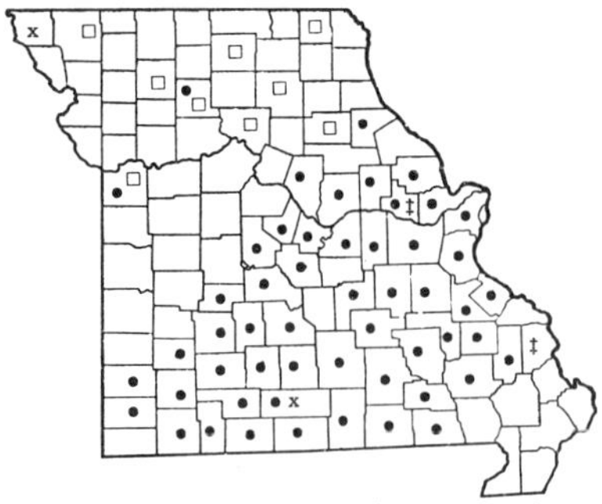

2350 ● Senecio aureus var. aureus (Golden Ragwort)
2350 ‡ Senecio aureus var. intercursus
2350 x Senecio aureus var. gracilis
2350 □ Senecio aureus var. semicordatus

2351 Senecio obovatus (Squaw-weed)

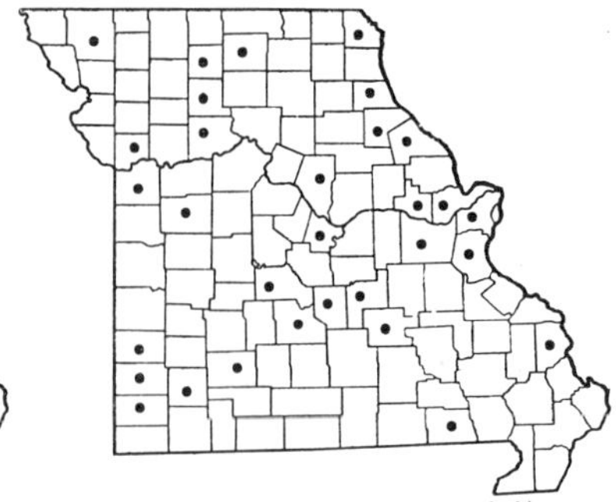

2352 Arctium minus f. minus (Common Burdock)

5b. **Senecio aureus** var. **intercursus** Fern. Map 2350

Known from Warren and Cape Girardeau counties.

Ranges from Quebec to Michigan and Illinois, south to Alabama, Kentucky, and Missouri.

A specimen from Ralls County (Bear Creek bluffs near Oakwood, June 6, 1916, *Davis*) approaches var. *intercursus* because of its glabrous petioles and margins of the basal sheaths.

5c. **Senecio aureus** var. **gracilis** (Pursh) Wood Map 2350

Known only from Douglas County.

Ranges from Massachusetts to North Dakota, south to North Carolina and Arkansas.

A specimen from Randolph County previously cited by Palmer and Steyermark in their *Annotated Catalogue* as this variety, is referred in the present flora to *S. pauperculus* var. *Balsamitae.*

5d. **Senecio aureus** var. **semicordatus** (Mackenz. & Bush) Greenman Map 2350

Scattered in northern and central Missouri south to Monroe, Chariton, and Jackson counties.

Ranges from Newfoundland, Quebec, and Ontario, to Michigan, south to Ohio, Illinois, and Missouri.

The delimitation of the above varieties follows the most recent studies by Fernald (Rh. 45: 495–502. 1943). So far as Fernald's treatment is concerned, it appears that typical var. *aureus* is the most common variation encountered in Missouri. It is by no means certain, however, that the varieties recognized by Fernald, nor the var. *semicordatus* placed with *S. aureus*, can be maintained as applicable to Missouri specimens. There is considerable intergradation in some of the characters used. The treatment as here given should be considered tentative and subject to reinvestigation of future field and experimental studies.

This species, when in flower, produces a delicate, pleasant fragrance from the involucres and leaves near the flower-heads, a fragrance not produced by the other species of the genus. Toxic alkaloids are known to be present in this species and are related to those found in *S. Jacobaea,* a species having a poisonous effect on cattle and horses.

The large round, dark green, basal leaves are quite ornamental and form large colonies. They lend a pleasing effect and cover to stream banks and borders of springs. For the foliage as well as the showy yellow inflorescences, the plant is a desirable one for planting in moist soils of wildflower gardens.

6. **Senecio obovatus** Muhl. Squaw-weed Map 2351

Also called Round-leaf Groundsel.

*Senecio obovatus* var. *umbratilis* Greenman

*Senecio obovatus* var. *rotundus* Britt. [G, P & S, Steyerm.]

Flowers early April–June.

Occurs in rich and rocky woods, generally on slopes at the base of or on ledges of bluffs, and open rocky borders of glades. Mainly in the Ozark region of southern and central Missouri north to Lincoln, Montgomery, Boone, Howard, Pettis, and Bates counties, locally north of this area in northeast Missouri to Scotland County and west to Platte County.

Ranges from New Hampshire to Ontario and Michigan, south to Georgia, Alabama, Missouri, Kansas, Oklahoma, and Texas.

*Senecio obovatus* var. *umbratilis,* cited by Greenman (Ann. Mo. Bot. Gard. 3: 115. 1916) from Missouri (Monteer, Shannon Co., April 27, 1907, *Bush 4337*) has the lower leaf-blades oblong-ovate to oblong-elliptic, 2–8 cm. long, 1.5–5.5 cm. broad and on petioles 2–12 cm. long, cannot be distinguished from typical *S. obovatus* as defined by Fernald in *Gray's Manual,* eighth edition (p. 1536). The var. *rotundus,* as segregated in *Gray's Manual*, is regarded as differing from typical *S. obovatus* in having the leaves of the basal rosettes nearly round to round-obovate instead of narrowly to broadly obovate, with the petioles as long as to three times the length of the leaf-blade instead of shorter than or only slightly longer than the leaf-blade, and with the broadened summit of the petiole 3–15 mm. instead of 10–30 (or up to 60) mm. long. According to this treatment, most of the Missouri material has the leaf characters of var. *rotundus*. However, as noted by Deam (*Fl. Ind.* p. 997–98. 1940) and verified by the present author, it seems futile to attempt to maintain var. *rotundus* separate from typical *S. obovatus*, since both leaf forms are found in the same colony. Deam noted that one form or the other usually predominates in a colony and that in Indiana these two leaf forms are about equally distributed in the state. In Missouri, however, typical *S. obovatus* is much less common than var. *rotundus*.

## Tribe IX. **CYNAREAE**

### 65. **Arctium** L. Burdock

Commonly encountered species; heads of flowers or fruits either without stalks (sessile) or on very short peduncles (stalks), racemosely or nearly racemosely arranged in an inflorescence longer than broad; larger leaf-blades tapering or somewhat pointed at summit, ovate-oblong; leaf-stalks (petioles) slightly angled . . . . . . . . . . . . . . . . . . . . . . . . . . . 1. A. MINUS

Very rare, known only from Jackson County; heads of flowers or fruits on long peduncles (stalks) mainly 2–6 cm. long, corymbosely arranged in an inflorescence more or less flat-topped and broader than long; larger leaf-blades broadly rounded at summit, round-ovate; leaf-stalks (petioles) strongly angled. . 2. A. TOMENTOSUM

1. **Arctium minus** (Hill) Bernh. f. **minus** Common Burdock Map 2352

*Arctium minus* (Hill) Bernh. [G, BB]

Flowers July–October.

Occurs in waste ground, about dwellings, woodland, thickets, and along railroads. Throughout Missouri.

Native of Europe; introduced and naturalized in North America from Newfoundland to British Columbia, south to Virginia, West Virginia, Missouri, Kansas, Oklahoma, and California.

The usual color of the flowers is pink or rose. In f. *purpureum* (Blytt) A. H. Evans, the flowers are deep purple, in f. *pallidum* Farw., they are white, and in f. *laciniatum* Clute, some or all of the leaves are deeply cut or reduced to narrow leaf-blades. None of these three forms are known from Missouri.

The common burdock can be prepared into a tasty vegetable, first gathering the young stems or the petioles and peeling the epidermis, then either boiling or baking them, and serving warm with butter, salt, and pepper, or cold as a plain salad or mixed with other vegetables flavored with dressing. The pith of the young stems is sometimes eaten raw flavored with vinegar and oil, or sometimes candied. The thick roots, after being peeled, can be boiled, and then served in soups or eaten as a cooked vegetable, having somewhat the texture and flavor of salsify. The thick root is also sometimes dried and mixed with coffee, and the plant sometimes serves for forage. The dry burs cling to the furry coat of animals, and sometimes lessen the value of wool when too many burs get mixed with the hair. The root at one time was used medicinally as a diuretic and alterative, particularly in rheumatism. It was also prepared into an ointment for the treatment of burns and sores.

2. **Arctium tomentosum** Mill. Map 2353

Flowers June–September.

Occurs in waste ground. Known only from Jackson County (Sheffield, September 18, 1917, *Bush 8170, 8170A*; Courtney, October 16, 1917, *Bush 8266*), west central Missouri.

Native of Europe; introduced and naturalized in North America from Quebec and Nova Scotia to Pennsylvania, D.C., Ohio, and Missouri.

### 66. **Carduus** L. Plumeless Thistle

Heads of flowers or fruits nodding, solitary at the tips of the branches; involucre 3–4 cm. broad; heads mainly 4–8 cm. broad; bracts of involucre 2–8 mm. broad . . . . . . . . . . . . 1. C. NUTANS

Heads of flowers or fruits ascending, usually clustered or several together; involucre 1–2 cm. broad; heads 1.5–2.5 cm. broad; bracts of involucre 2 mm. or less broad . . . . . . . . . . 2. C. CRISPUS

2353 Arctium tomentosum

2354 Carduus nutans (Musk Thistle)

2355 Carduus crispus (Curly Thistle)

1. **Carduus nutans** L. Musk Thistle Map 2354
Flowers early June–October.

Occurs in waste ground, fields, and along railroads. Scattered in the state and known only from Marion (field near Palmyra, June 9, 1941, *C. G. Tarleton*), St. Louis (St. Louis, Mo. Pac. R.R. south of Gasconade Street, June 13, 1954, *Muehlenbach 154*; between Broadway and Reilly Ave., along Mo. Pac. R.R., July 16, 1955, *Muehlenbach 676*), Webster (open ground along road A, vicinity of James River, 6½ mi. south of Marshfield, June 7, 1957, *Steyermark 85177*), and Jackson (Courtney, June 14, 1956, *Gier 8306*) counties.

Native of Europe; introduced and naturalized in North America from New Brunswick and Quebec to Iowa, south to Maryland, D.C., and Missouri.

The dried flowers in some countries are used as a rennet for curdling milk. The thick pith of the flowering stem, after the rind has been removed, can be boiled and eaten.

2. **Carduus crispus** L. Curly Thistle Map 2355
Also called Welted Thistle.
Flowers June–October.

Occurs in waste ground and along roadsides. Known only from St. Louis County, east-central Missouri (roadsides near the Angelrodt's above St. Louis, September, 1843, *Engelmann*).

Native of Europe; introduced and naturalized in North America from Nova Scotia to Minnesota, south to Connecticut, New York, and Missouri.

## 67. **Cirsium** Mill. Thistle

a. Upper half of stem and branches winged by a wavy strip of prickly green leafy tissue running from base of leaf down the stem; bracts of involucre with a cobwebby hairiness in lower part; upper surface of leaves with stiff bristles; most or all of the bracts of involucre ending in stiff needle-like spines or prickles . . . . . . . . . . . . . . . . . . . . . . . . 1. C. VULGARE

a. Upper half of stem and branches either not winged, or if winged, the wing mainly present at the base of the leaves or summit of the internode (gap on stem between leaves); bracts of involucre slightly, if at all, cobwebby hairy; upper surface of leaves either glabrous (without hairs), smooth, or with only weak hairs, usually only the outer and middle bracts of involucre, or none at all, tipped with a prickle. . . . . . . . . . . . . . . . . . . . . . . . . . . . *b*

b. None of the bracts of involucre ending in a prickle . . . . . . . . . . . . . . . *c*

c. Stems solid; involucre 1–2 cm. high, 10–15 mm. broad; lower surface of leaves mainly green and glabrous (without hairs); outer bracts of involucre tapering to a long slender tip; perennial with creeping rootstocks; introduced plant found along roadsides, railroads, and in waste ground; flower-heads consisting of flowers of usually either all functional stamens or all functional pistils, but both sexes may be present in the same flowers . . . . . . . . . 7. C. ARVENSE

c. Stems hollow; involucre 2–3.5 cm. (1.5 in immature heads) high, 20 or more mm. broad; lower surface of leaves pale green and more or less webby-haired; outer bracts of involucre tapering to only a very short-pointed (blunt or mucronate) tip; biennial plant with a forked thickened root, not creeping; native plant of swamps and swampy meadows in the eastern Ozarks; flower-heads with both functional stamens and pistils in the same flower (perfect) 6. C. MUTICUM

b. At least the outer bracts of involucre ending in a prickle . . . . . . . . . . . . . . . *d*

d. Upper surface of leaves covered with a gray or white coating of hairs . . . . 2. C. UNDULATUM

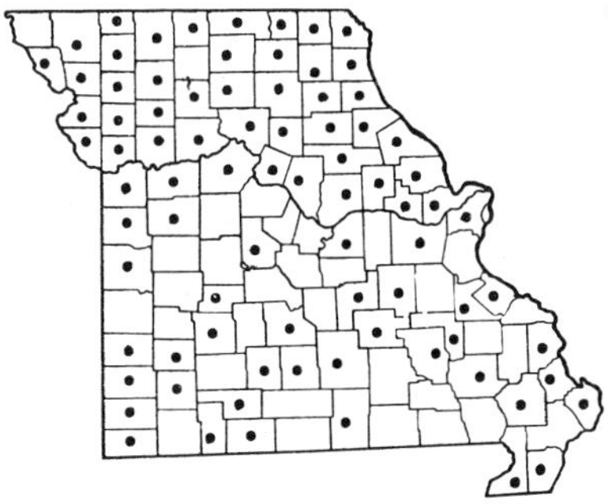

2356 Cirsium vulgare (Bull Thistle)

2357 Cirsium undulatum var. megacephalum

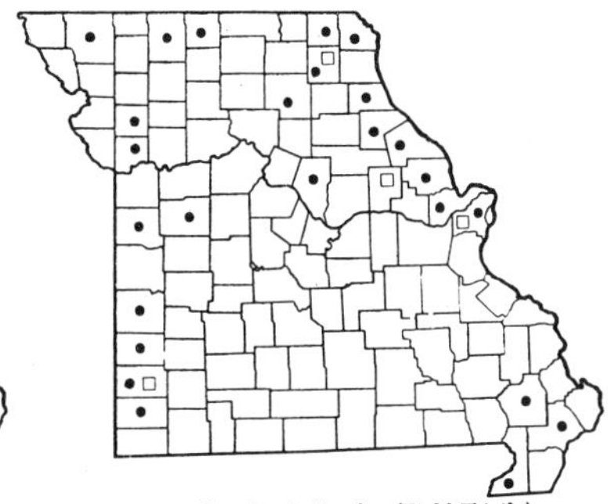

2358 • Cirsium discolor f. discolor (Field Thistle)
2358 □ Cirsium discolor f. albiflorum

d. Upper surface of leaves green, but, if hairy, without a gray or white coating of hairs . . . . . *e*

e. Heads of flowers or fruits not surrounded by prickle-tipped leafy bracts; peduncles (stalks) supporting flower- and fruiting heads elongated, practically leafless except for 1 or 2 scattered small, leaf-like bracts; involucre 1.5–2 cm. high; plants found only in the eastern Ozark section . . . . . . . . . . . . . . . . . . . . 5. C. CAROLINIANUM

e. Heads of flowers or fruits surrounded by prickle-tipped leafy bracts; peduncles (stalks) supporting flower- and fruiting heads short and leafy; involucre 2–3.5 cm. high; plants found in the northern half of Missouri or throughout the state . . . . . . . . . . . *f*

f. Main leaves of the stem deeply cut with narrow linear-lanceolate to broadly lanceolate lobes; margins of leaves more or less revolute (turned downward or under); inner (uppermost) bracts of involucre tapering into long tips . . . . . . . . 3. C. DISCOLOR

f. Main leaves of the stem either with slender or shallow broad teeth or shallow, broad lobes; margins of leaves rather flat, not curved under; inner (uppermost) bracts of involucre with lanceolate to triangular toothed (serrulate) tips . . . . 4. C. ALTISSIMUM

1. **Cirsium vulgare** (Savi) Tenore
Bull Thistle Map 2356

*Cirsium lanceolatum* (L.) Hill [P & S]

Flowers June–September.

Occurs in pastures, fields, cut-over woods, waste ground, roadsides, and along railroads. Throughout Missouri, doubtless in every county.

Native of Europe; introduced and naturalized in North America from Newfoundland to British Columbia, south to Florida and California.

The flowers have the property of a rennet in curdling milk. Goldfinches are fond of the fruits of this and other species of the genus.

2. **Cirsium undulatum** (Nutt.) Spreng. var. **megacephalum** (Gray) Fern. Map 2357

Flowers June–October.

Occurs on prairies, loess hills, along roadsides and railroads.

Rare and scattered in northern, central, and eastern Missouri, in Atchison, Jackson, Lincoln, St. Louis, and Wayne counties.

Ranges from British Columbia to Arizona, east to Manitoba, North Dakota, Nebraska, Missouri, and Oklahoma.

Typical var. *undulatum*, with involucres 2.5–3 cm. high and prickles of the involucral bracts 3.5–5.5 mm. long, has not been recorded from Missouri; var. *megacephalum* has the involucres 3–5 cm. high and the prickles of the involucral bracts 5–7 mm. long.

3. **Cirsium discolor** (Muhl.) Spreng. Field Thistle Map 2358

Flowers early August–November.

Occurs in dry upland prairies, river bottom prairies and meadows, fields, pastures, open woods, roadsides, and along railroads.

Absent from the Ozark region, elsewhere found in the state in the glaciated prairie and unglaciated prairie sections of northern, central, and western Missouri, south to St. Louis, Boone, and Newton counties, and in the southeastern lowlands in New Madrid, Stoddard, and Dunklin counties.

Ranges from Quebec and Maine to Ontario and Manitoba, south to Georgia, Tennessee, Missouri, Kansas, and Oklahoma.

Two variations are found in the state:

Flowers purple . . . 3a. C. DISCOLOR f. DISCOLOR
Flowers white . . 3b. C. DISCOLOR f. ALBIFLORUM

2359 • Cirsium altissimum f. altissimum (Tall Thistle)
2359 □ Cirsium altissimum f. Moorei

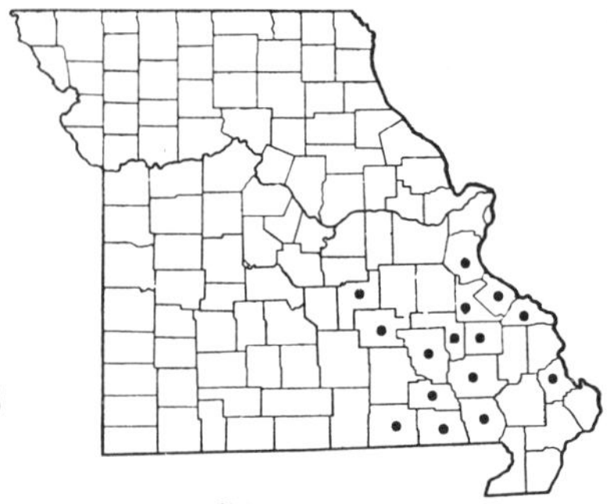

2360 Cirsium carolinianum

2361 Cirsium muticum var. muticum f. muticum (Swamp Thistle)

3a. **Cirsium discolor** f. **discolor** Map 2358

*Cirsium discolor* (Muhl.) Spreng. [G, BB, P & S]

This is the common variation in the state.

3b. **Cirsium discolor** f. **albiflorum** (Britt.) House Map 2358

Scattered in the range of *C. discolor*, and known from Knox, St. Louis (Times Beach, September, 1939, *Bauer*; *Muehlenbach 367*), Montgomery, and Jasper counties.

The cooked tender petioles and midribs of this thistle, gathered in early spring, have a flavor resembling artichoke. Similar use can be made of all the species found in Missouri, first peeling away any prickly parts of the leaves. The young stems can also be cut up into pieces and cooked and have an artichoke flavor.

The leaves of *C. discolor* are usually thicker than those of *C. altissimum.*

4. **Cirsium altissimum** (L.) Spreng. Tall Thistle Map 2359

Flowers July–October.

Occurs in dry, open or rocky woodland slopes, bluffs, low woodland in alluvial soils, thickets, roadsides, and along railroads.

Throughout Missouri, except absent from the lowlands of extreme southeastern Missouri.

Ranges from Florida to Texas, north to New York, Ohio, Michigan, Wisconsin, Minnesota, and North Dakota.

Two variations are found in Missouri:

Flowers rose-lavender to rose . 4a. C. ALTISSIMUM f. ALTISSIMUM

Flowers white . . 4b. C. ALTISSIMUM f. MOOREI

4a. **Cirsium altissimum** f. **altissimum** Map 2359

*Cirsium altissimum* (L.) Spreng. [G, BB, P & S]

This is the common variation throughout the state.

4b. **Cirsium altissimum** f. **Moorei** Steyerm. Map 2359

Known from Laclede (Mill Creek, August 25, 1939, *Geo. Moore*, holotype in Chi. Nat. Hist. Mus. Herb.), Webster (upper slopes along Finley Creek, T28N, R17W, northeast sect. 27 and northwest sect. 26, 3½ mi. south of Seymour, September 2, 1951, *Steyermark 72577*), and Jasper (Rocky prairies near Webb City, August 16, 1926, *Palmer 31472*) counties.

The leaves of *C. altissimum* are generally thinner than those of *C. discolor.*

This species of thistle can be used similarly to that discussed under *C. discolor*. The fruits are eaten by wild turkey.

5. **Cirsium carolinianum** (Walt.) Fern. & Schub. Map 2360

*Cirsium virginianum* of auth. [P & S, Steyerm.], not (L.) Michx.

*Cirsium flaccidum* Small

Flowers May–June.

Occurs in rocky open woods in upland, along bluffs, along creeks in ravines and valleys, and thickets, usually in acid soils.

Eastern Ozark region north to Jefferson County, west to Phelps, Dent, Reynolds, Carter, and Oregon counties.

Ranges from Georgia to Texas, north to North Carolina, Ohio, Indiana, Illinois, Missouri, and Oklahoma.

Plate no. 382. **1.** Arctium minus, × 2/7 (Scribner's); Details from Small, The New York Botanical Garden. **2.** Cirsium vulgare, × 2/7. **3.** Cirsium altissium, × 2/7. **4.** Carduus nutans, × 2/7; After Gleason, The New York Botanical Garden. **5.** Cirsium arvense, × 2/7. **6.** Cirsium carolinianum, × 2/7; Details from Small, The New York Botanical Garden. **7.** Carduus crispus, × 2/7; After Gleason, The New York Botanical Garden. **8.** Cirsium discolor, × 2/7. **9.** Cirsium muticum, × 2/7.

PLATE NO. 382

6. **Cirsium muticum** Michx. var. **muticum** f. **muticum** Swamp Thistle Map 2361

*Cirsium muticum* Michx. [G, BB, P & S]

Flowers July–October.

Occurs in swamps and wet meadows fed by calcareous springs and rivulets, and moist seepage of limestone bluffs.

Eastern Ozark region north to St. Louis County, west to Washington, Reynolds, Shannon, and Howell counties.

Ranges from Newfoundland to Saskatchewan, south to Delaware, Maryland, North Carolina, Tennessee, and Louisiana.

A white-flowered form, f. *lactiflorum* Fern., has not been recorded as yet from Missouri. In Quebec occurs var. *monticola* Fern., with bunched heads and glabrous involucres on short peduncles.

7. **Cirsium arvense** (L.) Scop. var. **arvense** f. **arvense** Canada Thistle Map 2362

*Cirsium arvense* (L.) Scop. [G, P & S]

Flowers June–October.

Occurs in fields, waste ground, roadsides, and along railroads.

Rare, scattered in northern, central, and eastern Missouri in Adair, St. Louis, Jefferson, St. Francois, and Jackson counties.

Native of Europe; introduced and naturalized in North America from Newfoundland to British Columbia, south to Virginia, Indiana, Illinois, Missouri, Kansas, South Dakota, Colorado, Utah, and California.

In the more northern and northeastern portions of its range in North America, it is a serious weed and invader of cultivated ground, fields, and pastures. However, in Missouri it is so rarely seen that it has only been found a few times. The creeping underground perennial roots make it difficult to exterminate. Because of the creeping rootstocks this species generally forms and is found in colonies or beds, at once distinguishing it from the other species of *Cirsium* in Missouri, all of which are found usually as scattered plants because of the chiefly biennial habit of forming rosettes one year and flowering stems the next.

A white-flowered form of var. *arvense*, known as f. *albiflorum* (Rand & Redf.) Hoffm., has not yet been found in Missouri. The common variation of the species has the main leaves conspicuously cut and wavy. Fernald (*Gray's Manual*, eighth ed.) considers this var. *arvense*, whereas Cronquist (in Gleason's *New Ill. Fl.* 3: 512. 1952) places this variation under var. *horridum* Wimm. & Grab. reserving the name var. *mite* Wimm. & Grab. for the merely toothed or shallowly lobed variation.

The Canada Thistle has been used in domestic medicine. It and other species of the genus often irritate the membranes of grazing animals, following their eating of the spiny leaves and involucres.

*Excluded Species*

**Cirsium pumilum** (Nutt.) Spreng.

*Cirsium odoratum* Bart. [P & S]

This species was recorded from St. Louis County by Palmer and Steyermark in their *Annotated Catalogue* as *C. odoratum* (Muhl.) Petrak. Two specimens in the Missouri Botanical Garden Herbarium, supposedly of Missouri origin, should be reinstated as Illinois records. One specimen, collected by H. Eggert, bears the following data: 'fields near French Village, St. Clair County, June 6, 1878'; the other, also collected by Eggert has 'dry hills, June 6, 1878' on the label, with the notation that the specimen is from Eggert's herbarium of St. Louis, Missouri. Both specimens were collected on the same day at French Village, St. Clair County, Illinois, across the river from St. Louis, Missouri, and the basis for the Missouri report probably originated from the St. Louis, Missouri address of Eggert's herbarium accompanying the label.

**Cirsium ochrocentrum** Gray var. **Helleri** (Small) Petrak

Recorded by Palmer and Steyermark from Jackson County in their *Annotated Catalogue* (p. 679), based upon Petrak's monograph of the North American species of the genus (Beih. 3. Bot. Centralb. II. 35: 418–19. 1917) and citation of a specimen from Jackson County (Independence, July 22, 1895, *Bush 325*). I have referred this collection in the present flora to *C. undulatum* var. *megacephalum*.

## 68. **Onopordum** L. Scotch Thistle

**Onopordum Acanthium** L. Scotch Thistle Map 2363

Also called Cotton Thistle.

Flowers July–September.

Occurs in waste ground and along roadsides. Known only from southeastern Missouri in St. Fran-

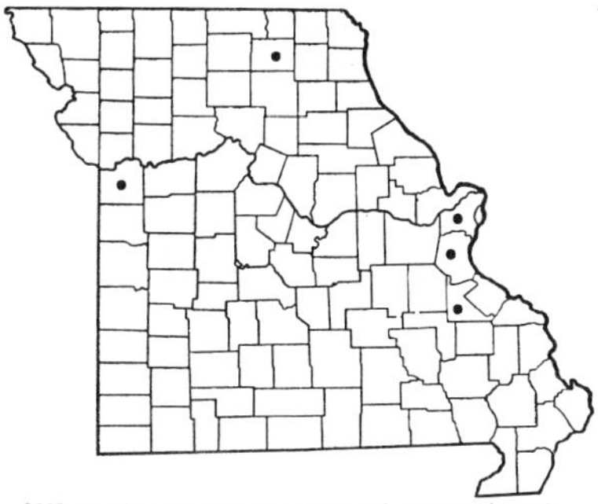

2362 Cirsium arvense var. arvense f. arvense (Canada Thistle)

2363 Onopordum Acanthium (Scotch Thistle)

2364 Centaurea solstitialis (Yellow Star Thistle)

cois, Iron, and Cape Girardeau counties.

Native of Europe; introduced and naturalized from New Brunswick to Ontario, south to Alabama, Indiana, Illinois, and Missouri.

The large receptacle can be cooked and eaten in the same manner as artichoke. The young shoots, after being peeled of their epidermis, can be boiled and eaten.

## 69. **Centaurea** L. Star Thistle

a. Some or all of the bracts of the involucre spine-tipped; flowers yellow; stem and branches winged with a strip of tissue extending from the leaf-base down the stem . . . . . . . . . . . . . . *b*

b. Larger spines of the bracts of involucre 10–22 mm. long; leaves and stem covered by a gray or white short dense hairiness throughout the season of growth. . . . . . . . . 1. C. SOLSTITIALIS

b. Larger spines of the bracts of involucre 5–9 mm. long; leaves and stem lightly covered with a cobwebby hairiness when young, but not with a dense hairiness . . . . . . . 2. C. MELITENSIS

a. Bracts of the involucre not spine-tipped, at most with a short firm tip, often more or less fringed or irregularly cut along margins; flowers rose, purple, or white; stem and branches not winged . . . . *c*

c. Most or all of the leaves of the stem deeply cut nearly or quite to the midrib into long narrow lobes . . . . . . . . . . . . . . . . . . . . . . . . . . . 7. C. MACULOSA

c. Most or all of the leaves of the stem entire (without teeth) or toothed, or only some of the lower leaves lobed or deeply cut . . . . . . . . . . . . . . . . . . . . . . . . . . . *d*

d. Involucre 2.5–4.5 cm. high; stems conspicuously thickened 6–11 mm. thick below the flower-heads . . . . . . . . . . . . . . . . . . . . . . . . . 8. C. AMERICANA

d. Involucre 0.8–2.5 cm. high; stems slender or at most thickened to 4 mm. thick below the flower-heads . . . . . . . . . . . . . . . . . . . . . . . . . . . *e*

e. Plant covered with a loose white hairiness; all the leaves entire (not toothed or lobed) 3. C. CYANUS

e. Plant green; at least the basal leaves or those of the rosette of first year frequently lobed or deeply parted or lobed . . . . . . . . . . . . . . . . . . . . . . . . *f*

f. Bracts of involucre either entire (not toothed) or the inner ones merely toothed; longer bristles of the pappus 6–11 mm. long. . . . . . . . . . . . . . . 4. C. REPENS

f. Bracts of involucre regularly fringed or deeply cut into comb-like teeth or segments; pappus either absent or at most 3 mm. long. . . . . . . . . . . . . . . . . *g*

g. Middle and outer bracts of involucre with conspicuous greenish blades, the dark fringed tip 1–3 mm. long, about $\frac{1}{3}$ as long as the blade; pappus none or scarcely developed; flower-heads rather narrow, the pressed involucre about as high as broad. . . . . . . . . . . . . . . . . . . . . . 6. C. VOCHINENSIS

g. Middle and outer bracts of involucre lacking conspicuous blades, the conspicuous dark fringed tip up to 4–6 mm. long; pappus about 1 mm. long or less; flower-heads rather broader, the pressed involucre broader than high . . . . . . . . 5. C. NIGRA

1. **Centaurea solstitialis** L. Yellow Star Thistle Map 2364

Also called Barnaby's Thistle.

Flowers late June–October.

Occurs in fields, waste ground, and along railroads. Known only from central Missouri, in St. Louis

2365 Centaurea melitensis

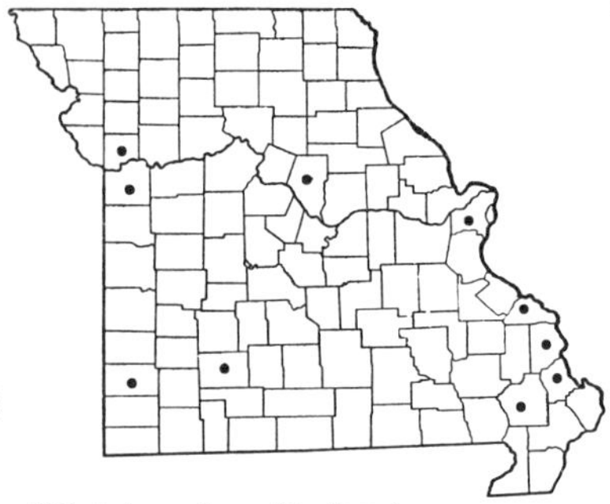

2366 Centaurea Cyanus (Cornflower)

2367 Centaurea repens

(Carrie Ave. freight-yard of Terminal R.R. Association, June 28, 1958, *Muehlenbach 1425*) and Boone (alfalfa field near Grindstone Creek, Columbia, *Daniels*) counties.

Native of the Mediterranean region; introduced and naturalized in the United States from Florida to California, north to Massachusetts, New York, Ontario, Iowa, and Kansas.

The spiny heads can inflict injuries on grazing animals.

2. **Centaurea melitensis** L. Map 2365

Flowers June–September.

Occurs in waste ground and along railroads. Known only from Jackson County, west-central Missouri (Sheffield, June 22, 1905, *Bush 3021*).

Native of Europe; introduced and naturalized in the United States, scattered locally from Georgia to Arizona and California, north to Massachusetts, Missouri, Minnesota, and Washington.

3. **Centaurea Cyanus** L. Cornflower Map 2366

Also called Bachelor's Button, Blue Bottle.

Flowers last of May–September.

Occurs in waste ground, along roadsides, and along railroads. Scattered in southern and central Missouri, but doubtless to be found in most counties of the state.

Native of Europe; introduced and naturalized from Quebec to Minnesota and British Columbia, south to Virginia, Indiana, Illinois, Missouri, Kansas, Oklahoma, and California.

This commonly cultivated garden annual occasionally escapes and seeds itself away from cultivation, but is usually found as an occasional waif, never becoming abundant.

4. **Centaurea repens** L. Map 2367

*Centaurea Picris* Pall.

Flowers June–August.

Occurs in waste ground and along railroads. Known only from Jackson County, west-central Missouri (Courtney, July 5, 1914, *Bush 7152*).

Native of Asia; introduced and naturalized in the United States from Michigan to Washington, south to Missouri, Texas, Utah, and California.

5. **Centaurea nigra** L. var. **nigra** Knapweed Map 2368

Also called Spanish Buttons.

*Centaurea nigra* L. [G, BB]

Flowers June–September.

Occurs along railroads and waste ground. Known only from St. Louis County, east-central Missouri (St. Louis, abandoned O'Fallon Street freight yard of Terminal R.R. Association, July 7, 1957, *Muehlenbach 1262*).

Native of Europe; introduced and naturalized in North America from Newfoundland to Ontario, south to Delaware, Maryland, D.C., Ohio, Michigan, and Missouri.

Another variation, var. *radiata* DC., not known in Missouri, has a tawny involucre and falsely radiate marginal flowers.

6. **Centaurea vochinensis** Bernh. Map 2369

Flowers June–September.

Occurs in waste ground and along railroads.

Plate no. 383. **1.** Onopordum Acanthium, × $^{2}/_{7}$; After Gleason, details from Small, The New York Botanical Garden. **2.** Centaurea maculosa, × $^{2}/_{7}$; Details from Small, The New York Botanical Garden. **3.** Centaurea nigra, × $^{2}/_{7}$; Details from Small, The New York Botanical Garden. **4.** Centaurea solstitialis, × $^{2}/_{7}$. **5.** Lapsana communis, × $^{2}/_{7}$. **6.** Serinia oppositifolia, × $^{2}/_{7}$; Details from Small, The New York Botanical Garden. **7.** Centaurea melitensis, × $^{2}/_{7}$. **8.** Centaurea Cyanus, × $^{2}/_{7}$. **9.** Centaurea americana, × $^{2}/_{7}$. **10.** Cichorium Intybus, × $^{2}/_{7}$; Details from Small, The New York Botanical Garden.

PLATE NO. 383

2368 Centaurea nigra var. nigra (Knapweed)

2369 Centaurea vochinensis

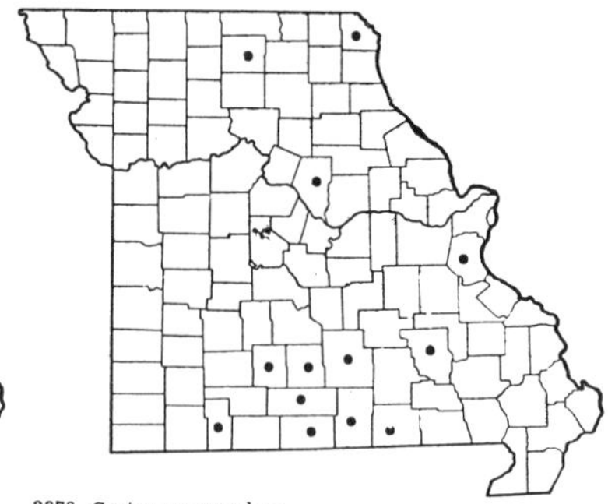

2370 Centaurea maculosa

Known only from St. Louis and Boone (meadow east, Columbia, August, 1904, *Daniels*; Sanborn Field at College and Rollins streets, Columbia, September 4, 1933, *Lisle Jeffrey*) counties.

Native of Europe; introduced and naturalized in the United States from Maine to Ontario, south to Virginia and Missouri.

7. **Centaurea maculosa** Lam. Map 2370

Flowers June–August.

Occurs in waste ground, rocky open places, fields, roadsides, and railroads. Scattered throughout Missouri, more frequent in southern Missouri.

Native of Europe; introduced and naturalized in North America from Quebec to British Columbia, south to Virginia, Tennessee, Missouri, Kansas, and California.

8. **Centaurea americana** Nutt. American Basket Flower Map 2371

Flowers late June–July.

Occurs in rocky open woods, glades, bald knobs, fields, roadsides, and along railroads. Native in southwestern Missouri in Stone, Barry, McDonald, and Newton counties, introduced elsewhere in central and eastern Missouri in St. Louis, St. Francois, and Jackson counties.

Ranges from Missouri and Kansas to Louisiana and west to Texas, Arizona, and Mexico; introduced in Illinois and occasionally eastward.

## Subfamily II. **LIGULIFLORAE**

### Tribe X. **CICHORIEAE**

### 70. **Lapsana** L. Nipplewort

**Lapsana communis** L. Nipplewort Map 2372

Flowers June–September.

Occurs along gravel bars along streams, waste ground, and roadsides. Southern Missouri, known only from Texas (shaded gravelly banks along Jack's Fork of Current River, around bridge over route 17, sect. 36, $5\frac{1}{2}$ mi. southeast of Arroll, June 23, 1939, *Steyermark 27168*), Lawrence, and McDonald counties.

Native of Europe; introduced and naturalized in North America from Quebec and Ontario, south to Virginia, West Virginia, and Missouri.

The plant can be cooked as a green vegetable and may also be eaten raw in salads.

### 71. **Serinia** Raf.

(*Krigia* Schreb. [BB])

**Serinia oppositifolia** (Raf.) Ktze. Map 2373

*Krigia oppositifolia* Raf. [BB, Shinners]

Flowers April–June.

Occurs in low moist fallow fields following streams, sandy open and alluvial ground, glades, meadows, prairies, upland sterile slopes and ridges, grazed or eroded ground, borders of ditches, sloughs, ponds, and railroads. Mainly southern Missouri, locally north in central Missouri to St. Louis (*Muehlenbach 1401*), Livingston, and Cass counties.

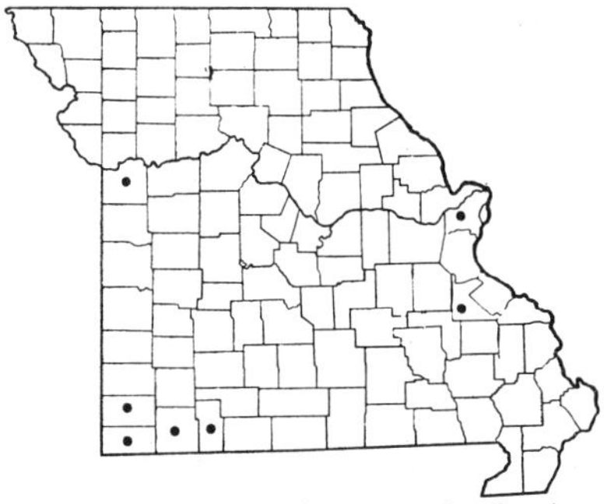

2371 Centaurea americana (American Basket Flower)

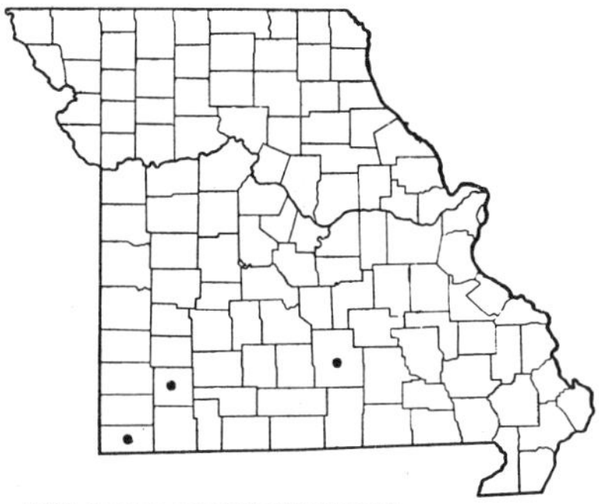

2372 Lapsana communis (Nipplewort)

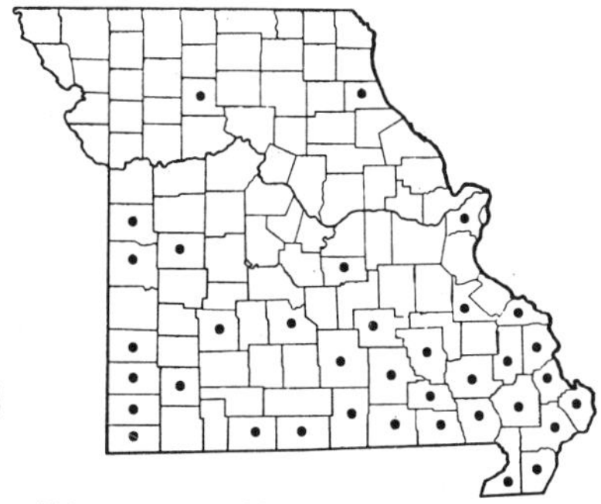

2373 Serinia oppositifolia

Ranges from Florida to Texas, north to Virginia, Tennessee, Illinois, Missouri, Kansas, and Oklahoma.

Shinners (Wrightia 1: 187–88; 202–5, 1947) transfers this genus to *Krigia* because of the occasional occurrence of a minute scaly crown on the achene of some specimens of *Serinia oppositifolia*, which ordinarily does not possess any pappus.

## 72. **Cichorium** L. Chicory

**Cichorium Intybus** L. Common Chicory Map 2374

Also called Blue Sailors.

Flowers late May–October.

Occurs in fields, pastures, waste ground, along roadsides, and railroads. Throughout Missouri, but not recorded from the lowland counties of extreme southeastern Missouri.

Native of Europe; introduced and naturalized in North America from Newfoundland to British Columbia, south to Florida, Texas, and California.

Two variations occur in Missouri:

Flowers bright blue . . . a. C. INTYBUS f. INTYBUS
Flowers white . . . . . b. C. INTYBUS f. ALBUM

a. **Cichorium Intybus** f. **Intybus** Map 2374

*Cichorium Intybus* L. [G, BB, P & S, Steyerm.]

This is the common variation in the state.

b. **Cichorium Intybus** f. **album** Neum. Map 2374

Scattered in the state, where known from Sullivan (along route 129, 2.3 mi. south of Green City, August 25, 1950, *Steyermark 70127*), Pike, and Texas counties.

Another variation with rose-colored flowers, f. *roseum* Neum., has not been recorded from Missouri thus far. The flowers of Chicory close about noon.

The root is used, especially in the southern United States, as an adulterant of coffee. When eaten by cattle in large numbers, the plants impart a bitter flavor to the milk and butter. The young leaves are gathered in the spring and cooked as greens, or eaten raw in salads.

## 73. **Krigia** Schreb. Dwarf Dandelion

a. Leaves present on flowering stem as well as at base of plant; flowering stem branched above, bearing 2 or more flower-heads . . . . . . . . . . 4. K. BIFLORA
a. Leaves present only at the base of plant; flowering stem simple, unbranched, bearing only 1 flower-head at tip . . . . . . . . . . *b*
 b. Flowering and fruiting stems not thread-like, mostly lacking hairs; an underground miniature potato-like tuber present; longer bristles of pappus 20–40 with 10–15 smaller, oblong, white scales; achenes with 12–20 ribs; perennial plants . . . . . . . 3. K. DANDELION
 b. Flowering and fruiting stems thread-like, with hairs usually present and noticeable, especially near the heads; no underground tuber present; longer bristles of pappus 5–7 with 5–7 shorter, rounded, white scales; achenes with 5 angles, with 10–15 ribs; annual plants . . . . . . *c*
  c. Bracts of involucre 5–8, remaining erect, $1\frac{1}{2}$–3 times as long as broad; longer bristles of pappus 1.2–2 mm. long; plants known only from the unglaciated prairie region of southwestern Missouri east to Polk and Greene counties . . . . . . . 1. K. OCCIDENTALIS

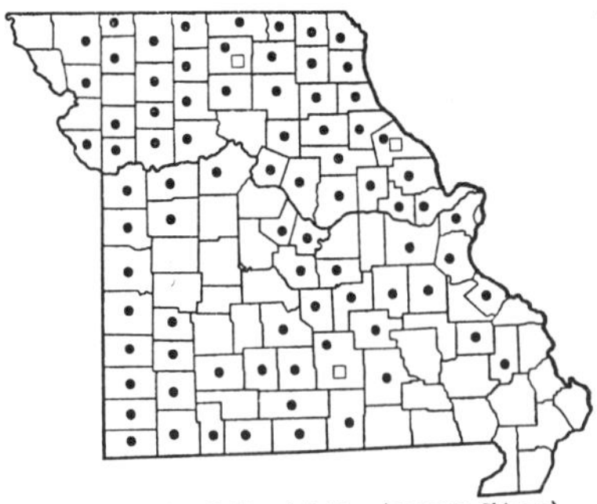

2374 • Cichorium Intybus f. Intybus (Common Chicory)
2374 □ Cichorium Intybus f. album

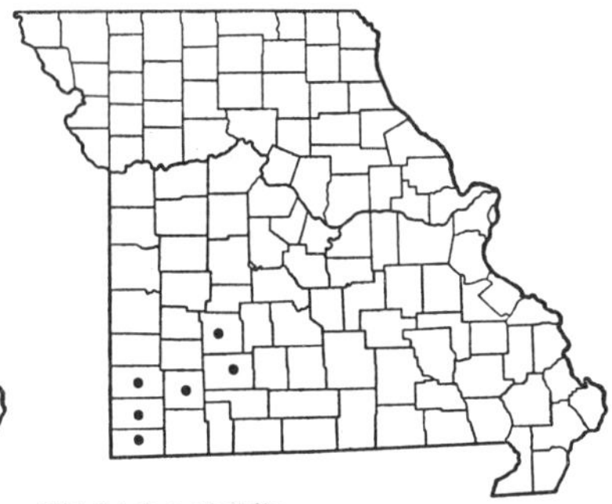

2375 Krigia occidentalis

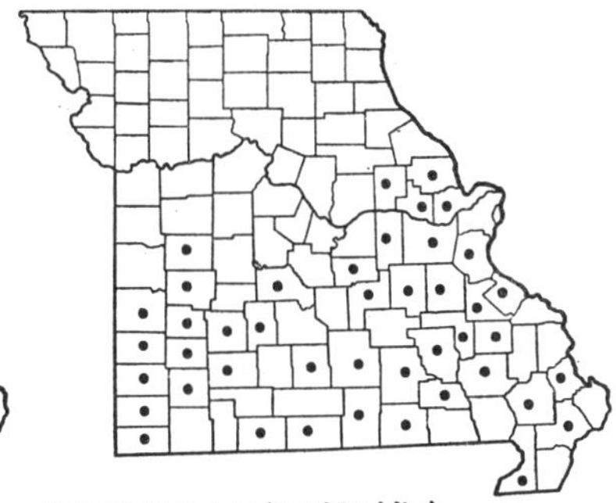

2376 Krigia virginica (Dwarf Dandelion)

c. Bracts of involucre 9–18, turned down or back (reflexed) at maturity, 4–8 times as long as broad; longer bristles of pappus 4.5–5.5 mm. long; plants common throughout the Ozarks of southern and east-central Missouri north to Lincoln, Montgomery, and Henry counties
2. K. VIRGINICA

1. **Krigia occidentalis** Nutt. Map 2375

Flowers April–May.

Occurs in glades and rocky open woods.

Southwestern Missouri, in the unglaciated prairie section of Jasper, Newton, McDonald, Lawrence, Polk, and Greene counties.

Ranges from Texas and Oklahoma to Missouri and Arkansas.

2. **Krigia virginica** (L.) Willd. Map 2376

Flowers April–August.

Occurs in prairies, meadows, rocky or sandy open woods, sandy fallow fields, and rocky glades usually of sandstone, chert, or granite on acid strata. Common in the Ozark region of southern and east-central Missouri, north to Lincoln, Montgomery, Maries, Camden, Henry, and Vernon counties.

Ranges from Florida to Texas, north to Maine, Vermont, New York, Ontario, Michigan, Wisconsin, and Iowa.

This makes a desirable plant for the acid soil rock garden, and can be grown from seed.

3. **Krigia Dandelion** (L.) Nutt. Map 2377

Flowers April–June.

Occurs in prairies, rocky glades and bluff escarpments of sandstone, chert, or granite in acid strata, sandy fields and roadsides, open woods, borders of woods and low or alluvial woods. Ozark region of southern and east-central Missouri, north to St. Louis, Warren, Montgomery, Osage, Camden, Hickory, Henry, and Vernon counties.

Ranges from Florida to Texas, north to New Jersey, Kentucky, Indiana, Illinois, Missouri, Kansas, and Oklahoma.

This is a showy-flowered species and with its large golden heads of flowers is an attractive plant for the rock garden. Unfortunately, it does not transplant well and is difficult to grow. It has been tried several times with failure in the author's northern Illinois wildflower preserve. It should be grown in open situations in light acid soils of sand, chert, granite, or similar acidic rocks.

4. **Krigia biflora** (Walt.) Blake Map 2378

*Krigia biflora* f. *glandulifera* Fern. [G]

Flowers May–August.

Occurs in rocky or open woods, low ground along streams, thickets, and prairies.

Ozark section and eastern half of Missouri, west to Putnam, Sullivan, Shelby, Audrain, Boone, Morgan, Benton, Hickory, Polk, and Jasper counties.

Ranges from Massachusetts to Manitoba, south to Georgia, Tennessee, Missouri, Oklahoma, New Mexico, and Arizona.

The flowering stems may vary from glabrous to glandular with dark gland-tipped hairs occurring below the flower-heads. I am not recognizing the gland-tipped f. *glandulifera* since it appears that the glands vary in presence and abundance within a given population and vary from year to year in the same population.

---

Plate no. 384. **1.** Hypochaeris radicata, × 2/5; *a.* Achene, × 4; *b.* Flower-head, × 2/5; After Gleason, details from Small, The New York Botanical Garden. **2.** Krigia virginica, × 2/5. **3.** Krigia Dandelion, × 2/5. **4.** Picris Sprengeriana, × 2/5. **5.** Krigia biflora, × 2/5; Details from Small, The New York Botanical Garden.

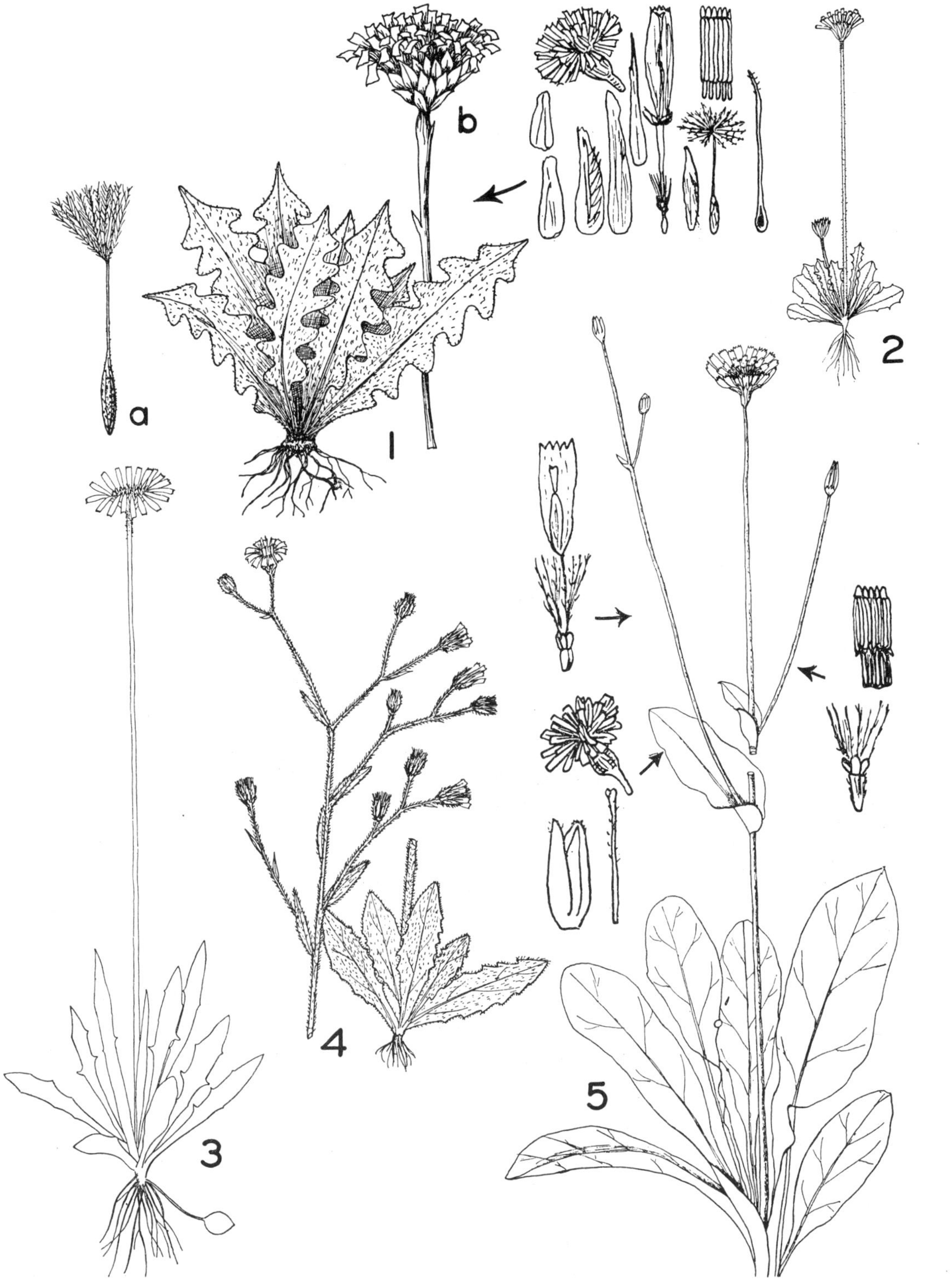


PLATE NO. 384

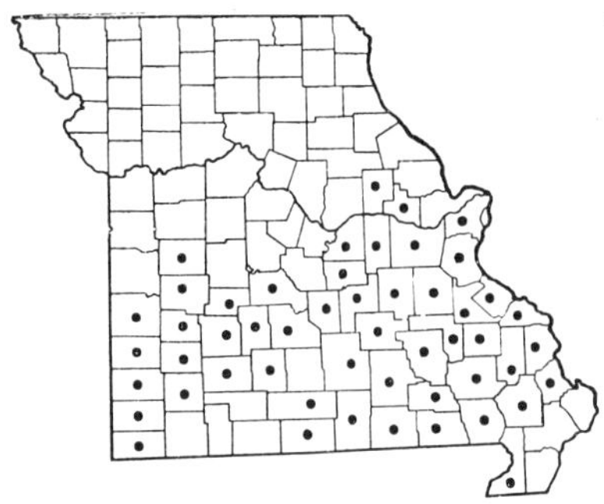

2377 Krigia Dandelion

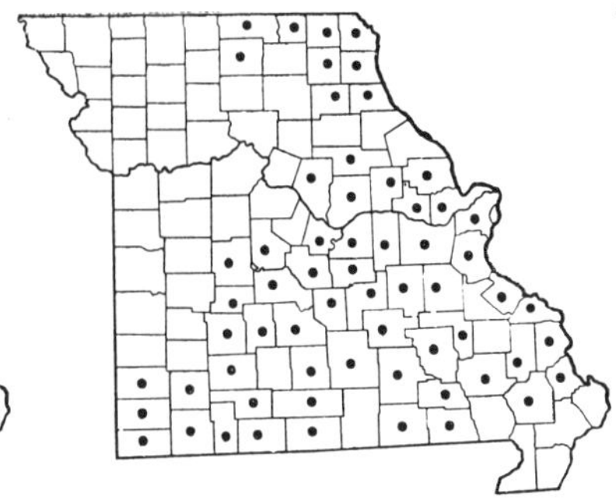

2378 Krigia biflora

2379 Hypochaeris radicata (Cat's Ear)

## 74. **Hypochaeris** L. Cat's Ear

**Hypochaeris radicata** L. Map 2379

Flowers May–August.

Occurs on lawns, about dwellings, waste ground, and along railroads.

Known only from St. Louis County, east-central Missouri (scattered on lawn, Jewish Hospital, St. Louis, June 15, 1938, *Kellogg*; same locality, June 25, 1940, *Petersen*; Carrie Ave. freight yard of Terminal R.R. Association, St. Louis, May 30, 1956, *Muehlenbach 890*).

Native of Europe; introduced and naturalized in North America from Newfoundland to Ontario, south to North Carolina, West Virginia, Ohio, Indiana, Illinois, and Missouri; also in the Pacific states.

## 75. **Picris** L. Bitter-weed

**Picris Sprengeriana** Poir. Map 2380

Flowers June–September.

Occurs along railroads. Known only from St. Louis County, east-central Missouri (along Rock Island R.R. near Washington University, June 30, 1931, *J. A. Turner* and *C. L. Elam*, in Mo. Bot. Gard. Herbarium).

Native of Europe; introduced in Missouri.

## 76. **Tragopogon** L. Goat's Beard

a. Flowers purple . . . . . . . . . . . . . . . . . . . . . . 1. T. PORRIFOLIUS

a. Flowers yellow . . . . . . . . . . . . . . . . . . . . . . . *b*

b. Common species encountered; bracts of involucre 25–40 mm. long during flowering period, much longer than the rays, elongating in fruit to 40–70 mm. long; mature achenes, including the beak, chiefly 25–35 mm. long; peduncles (stalks bearing flower-heads) noticeably thickened and broadened just below the heads of flowers and fruits . . . . . . . . . . 2. T. DUBIUS

b. Rarely encountered species; bracts of involucre 12–24 mm. long during flowering period, shorter than or about as long as the rays, elongating in fruit to 18–38 mm. long; mature achenes, including the beak, chiefly 12–24 mm. long; peduncles not thickened or broadened below the flowering heads, only scarcely thickened below the fruiting heads . . . . . . . . . . 3. T. PRATENSIS

1. **Tragopogon porrifolius** L. Salsify Map 2381

Also called Vegetable Oyster, Oyster Plant.

Flowers May–July.

Occurs in fields along roadsides, and waste ground. Rare and scattered in southern and central Missouri, north to St. Louis, Boone, and Clay counties, but probably to be found in most of the counties.

Native of Europe; introduced and naturalized in North America from Nova Scotia to Ontario, south to Georgia, Missouri, Kansas, Oklahoma, and Texas.

The plant is sometimes cultivated for the large fleshy taproot which is cooked as a vegetable or served

Plate no. 385. **1.** Tragopogon porrifolius, × 2/7; Details from Small, The New York Botanical Garden. **2.** Tragopogon dubius, × 2/7; Details from Small, The New York Botanical Garden. **3.** Taraxacum erythrospermum, × 2/7. **4.** Taraxacum officinale, × 2/7.

PLATE NO. 385

2380 Picris Sprengeriana (Bitter-weed)

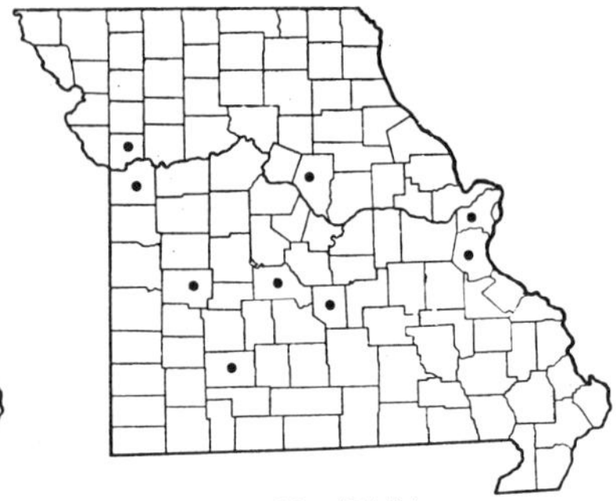

2381 Tragopogon porrifolius (Salsify)

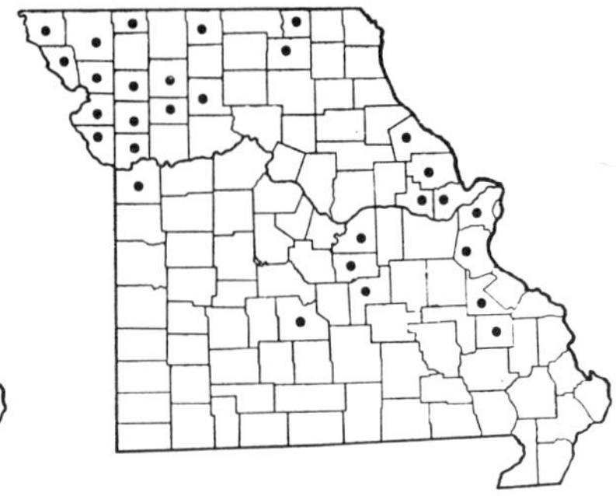

2382 Tragopogon dubius (Goat's Beard)

in salad. The flavor is somewhat suggestive of oysters. The bases of the lower leaves and the young stems, when only a few inches tall, may be cooked and used as a vegetable.

2. **Tragopogon dubius** Scop. Goat's Beard Map 2382

*Tragopogon pratensis* [of P & S, Steyerm.], not L.

*Tragopogon major* Jacq. [G]

Flowers May–July.

Occurs in fields, meadows, waste ground, roadsides, and along railroads. Scattered in Missouri, probably to be found in most counties of the state, but at present recorded mainly from northern, central, and eastern Missouri.

Native of Europe; introduced and naturalized in the United States from New York to Washington, south to Virginia, Illinois, Missouri, Oklahoma, Texas, and California.

This species has been confused previously in Missouri and elsewhere with *T. pratensis*, a much rarer species which has been found only recently in Missouri. Records given in Palmer and Steyermark's *Catalogue* (p. 680) for *T. pratensis* are now referred to *T. dubius*.

The young stems of this and the following species may be cooked and eaten as a vegetable, as described under the preceding *T. porrifolius*.

The large fruiting heads make a striking interior decoration and motif for flower arrangements.

3. **Tragopogon pratensis** L. Goat's Beard Map 2383

Flowers May–August.

Occurs along railroads and waste ground. Known only from St. Louis County, east-central Missouri (St. Louis, Frisco R.R. east of Hampton Street, July 21, 1957, *Muehlenbach 1285*).

Native of Europe; introduced and naturalized in North America from Quebec and Ontario south to Georgia, Tennessee, Missouri, and Kansas.

The above record is the only one seen which has the small fruiting involucres, smaller achenes, and slender peduncles of *T. pratensis*, all the other yellow-flowered material in Missouri being referable to *T. dubius*.

## 77. **Taraxacum** Zinn. Dandelion

Mature achenes (seed-like fruits) reddish-brown or reddish-purple; the tips of at least some of the inner (uppermost) bracts of the involucre with a low knob or thickened ridge on the back; outer bracts of involucre spreading or ascending, not curved backward . . . . . . . . . 1. T. ERYTHROSPERMUM

Mature achenes (seed-like fruits) olive-brown, dull brown, drab, or greenish-buff; the tips of the inner (uppermost) bracts of the involucre flat and without any thickening or knob; outer bracts of involucre strongly curved downward or backward . . . . . . . . . . . . . . . . 2. T. OFFICINALE

1. **Taraxacum erythrospermum** Andrz. Red-seeded Dandelion Map 2384

*Taraxacum laevigatum* [of BB, P & S, Steyerm., Handel-Mazzetti], not (Willd.) DC.

*Taraxacum laevigatum* f. *scapifolium* Gates & Prince, Trans. Kansas Acad. Sci. 41: 119. 1938.

Flowers January–December.

Occurs in lawns, gardens, about dwellings, fields, meadows, along roadsides and railroads. Scattered throughout the state and probably in every county.

Native of Europe; introduced and naturalized in North America from Quebec to British Columbia, south to Virginia, Illinois, Missouri, Kansas, Oklahoma, Texas, and New Mexico.

2383 Tragopogon pratensis (Goat's Beard)

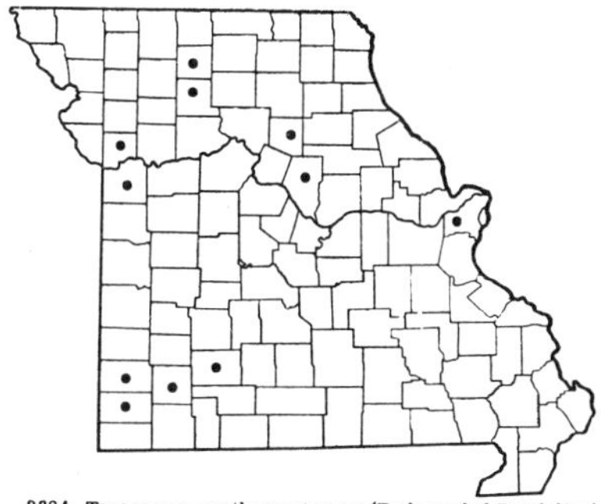

2384 Taraxacum erythrospermum (Red-seeded Dandelion)

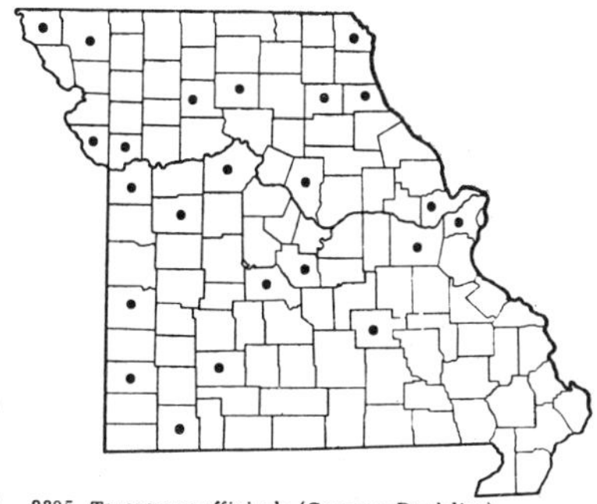

2385 Taraxacum officinale (Common Dandelion)

This species is less common than the next and often confused with it. The color of the mature achenes is the most certain test for distinguishing this species from the common dandelion (*T. officinale*).

Shinners (Field and Lab. 17: 15–18. 1949) has discussed the reasons for the acceptance of the name *T. erythrospermum* instead of *T. laevigatum.*

2. **Taraxacum officinale** Wiggers Common Dandelion Map 2385

*Taraxacum palustre* (Lyons) Lam. & DC. var. *vulgare* (Lam.) Fern. [P & S, Steyerm.]

Flowers January–December.

Occurs in lawns, gardens, about dwellings, fields, meadows, along roadsides, and railroads. Common throughout Missouri, doubtless in every county.

Native of Europe; introduced and naturalized throughout the United States and southern Canada.

The complexities of the proper designation of the name of the common dandelion have been discussed most recently by Shinners (Field and Lab. 17: 13–15. 1949). The young leaves of this and the preceding species make excellent spring greens. They are best when boiled and served like spinach or swiss chard, but may also be cut off when fresh and used in mixed or tossed salads. The root contains a bitter substance and is sometimes used as a mild laxative and tonic. As with Chicory, the roots of dandelion can be ground and either used alone as a bitter beverage or as an adulterant for coffee. The cooked roots can also be eaten. The leaves and flowers are eaten by the white-tailed deer.

Hybrids between the two species are found in Missouri.

## 78. **Sonchus** L. Sow Thistle

a. Flowering heads 4–4.5 cm. broad; involucre in fruiting stage 14–25 mm. high; flowers bright yellow to golden; perennial plants with long, deep, vertical roots, creeping by horizontal rootstocks. 1. S. ARVENSIS

a. Flowering heads 1.2–2.5 cm. broad; involucre in fruiting stage 9–13 mm. high; flowers pale yellow; annual plants with short tap roots . . . . . . . . . . . . . . . . . . . . . . . . . b

b. Leaves extended on both sides of the base into acutely pointed lobes (auricles); achenes (seed-like fruit) roughened on the surface and cross-wrinkled, as well as with 3–5 elevated nerves on each side. . . . . . . . . . . . . . . . . . . . . . . . . . . . . 2. S. OLERACEUS

b. Leaves extended on both sides of the base into rounded lobes (auricles); achenes (seed-like fruit) smooth except for the 3 elevated nerves on each side. . . . . . . . . . . . . . 3. S. ASPER

1. **Sonchus arvensis** L. Field Sow Thistle Map 2386

Flowers June–October.

Occurs in waste ground and along roadsides and railroads. Rare in Missouri.

The following two variations occur in the state:

Peduncles (stalk bearing flower- or fruiting-heads) with gland-tipped hairs; bracts of involucre with glands or gland-tipped hairs . . 1a. S. ARVENSIS var. ARVENSIS

Peduncles glabrous (without hairs); bracts of involucre glabrous or nearly so . . 1b. S. ARVENSIS var. GLABRESCENS

1a. **Sonchus arvensis** var. **arvensis** Map 2386

*Sonchus arvensis* L. [G, BB, P & S]

Known only from Jackson County, west-central Missouri.

Native of Europe; introduced and naturalized in North America from Newfoundland to Alaska, south

2386 • Sonchus arvensis var. arvensis (Field Sow Thistle)
2386 □ Sonchus arvensis var. glabrescens

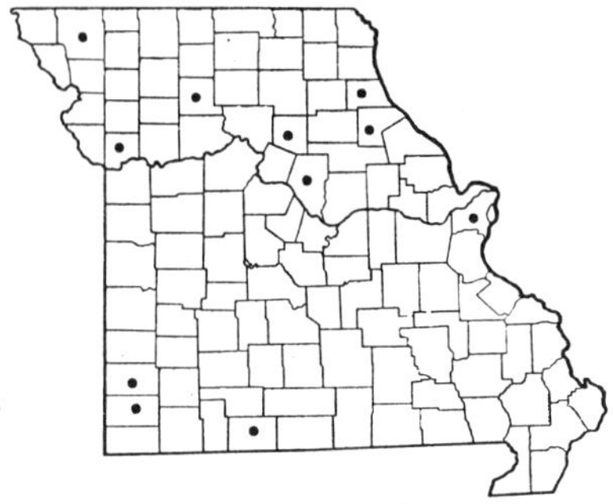

2387 Sonchus oleraceus f. oleraceus (Common Sow Thistle)

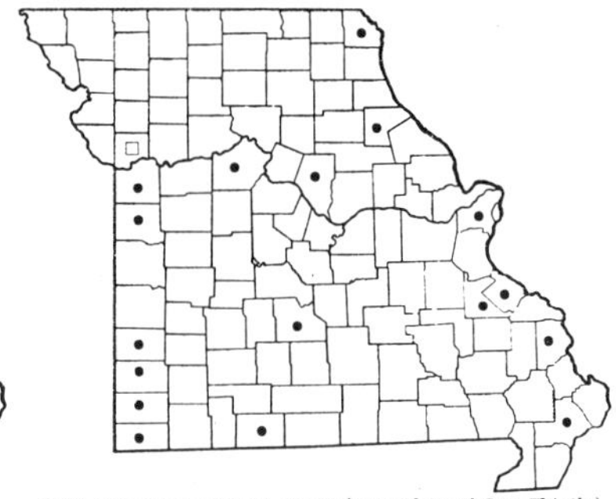

2388 • Sonchus asper f. asper (Spiny-leaved Sow Thistle)
2388 □ Sonchus asper f. glandulosus

to Delaware, Maryland, Ohio, Indiana, Illinois, Missouri, Idaho, Utah, and Oregon.

1b. **Sonchus arvensis** var. **glabrescens** Guenth., Grab. & Wimm. Map 2386

*Sonchus uliginosus* [of BB], not Bieb.

Known only from St. Louis County, east-central Missouri (St. Louis, Madison St. freight yard of Terminal R.R. Association, north of Brooklyn St. elevator, August 11, 1957, *Muehlenbach 1303*; Baden freight yard of Mo.-Kan.-Tex. R.R., July 5, 1958, *Muehlenbach 1449*; fair grounds, July 15, 1899, *Glatfelter herb.*).

Native of Europe; introduced and naturalized in North America from Quebec to Minnesota, south to Connecticut, Indiana, and Missouri.

This is a rarely encountered plant in Missouri, but is quite weedy and aggressive in the more northeastern and northern sections of the United States. It is a showy species when in flower. The plants are rather bitter and usually avoided by cattle. When young and tender, the leafy shoots may be gathered and cooked as greens.

2. **Sonchus oleraceus** L. f. **oleraceus** Common Sow Thistle Map 2387

*Sonchus oleraceus* L. [G, BB, P &S]

Flowers June–October.

Occurs in fields, waste places, gardens, about dwellings, barnyards, roadsides, and along railroads. Scattered throughout Missouri, and probably in every county, but recorded from only a small number of counties.

Native of Europe; introduced and naturalized in North America from Newfoundland to British Columbia, south to Florida and California; also introduced in Mexico, Central and South America.

The tender young fleshy leaves of this and the following species are frequently gathered in spring and cooked as greens, either alone or mixed with poke, lamb's quarters, wild lettuce, and other wild vegetables.

Typical f. *oleraceus* has the leaves of the stem deeply lobed. In f. *integrifolius* (Wallr.) G. Beck, not recorded from Missouri, the leaves are undivided and not lobed. In f. *lacerus* (Willd.) G. Beck, also not recorded from Missouri, the terminal lobe as well as the lateral lobes are all narrow and nearly of equal size, contrasted with f. *oleraceus* and f. *integrifolius*, in which the terminal lobe or half of the leaf is much larger than the lateral lobes.

3. **Sonchus asper** (L.) Hill Spiny-leaved Sow Thistle Map 2388

Flowers late May–October.

Occurs in fields, waste places, gardens, about dwellings, barnyards, roadsides, along ditches, and railroads. Scattered throughout southern, central, and eastern Missouri, but doubtless in every county of the state.

Native of Europe; introduced and naturalized in North America from Newfoundland to British Columbia, south to Florida and California; also in Mexico and West Indies.

Two variations occur in Missouri:

Stem and axis (rachis) of inflorescence glabrous (without hairs) . . . . . . 3a. S. ASPER f. ASPER
Stem and axis of inflorescence with stalked reddish glands . . . . . . 3b. S. ASPER f. GLANDULOSUS

Plate no. 386. **1.** Sonchus arvensis, $\times\, {}^{2}/_{5}$; *a.* Middle part of stem; *b.* Leaf from base of stem. **2.** Sonchus oleraceus, $\times\, {}^{2}/_{5}$; *a.* Achene, $\times\, 1\,{}^{1}/_{5}$; After Gleason, The New York Botanical Garden. **3.** Sonchus asper, $\times\, {}^{2}/_{5}$; Details from Small, The New York Botanical Garden. **4.** Lactuca Scariola, $\times\, {}^{2}/_{5}$ (Scribner's). **5.** Lactuca saligna, $\times\, {}^{2}/_{5}$; *a.* Lactuca saligna f. Ruppiana, $\times\, {}^{2}/_{5}$; *b.* Lactuca saligna f. saligna, $\times\, {}^{2}/_{5}$.

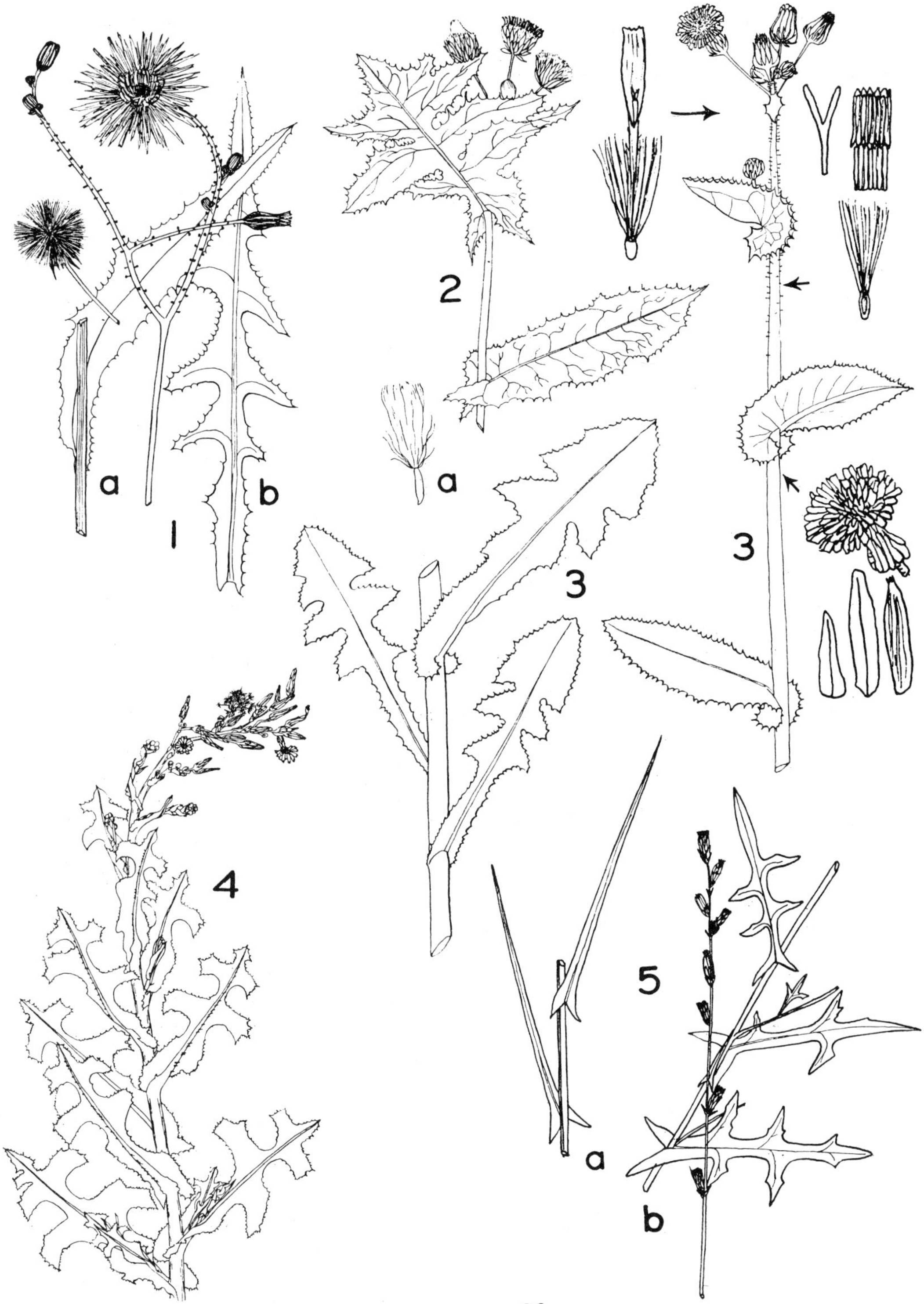


PLATE NO. 386

3a. **Sonchus asper** f. **asper** Map 2388

*Sonchus asper* (L.) Hill [G, BB, P & S, Steyerm.]

This is the common variation found in the state.

3b. **Sonchus asper** f. **glandulosus** Beckh. Map 2388

Known only from Clay County, west-central Missouri.

## 79. **Lactuca** L. Lettuce

a. Flowers usually blue or bluish; achenes (seed-like fruits) tapering to a short beak much shorter than the main body of the achene or without any beak and truncate at tip (as if cut straight across) . . . *b*
- b. Flowering heads during flowering period 2–3 cm. broad, 18–30-flowered; involucre 1.4–1.8 cm. high . . . . . . . . . . . . . . . . . . . . . . . . 6. L. PULCHELLA
- b. Flowering heads during flowering period 1.5–2 cm. broad, mostly 11–17-flowered; involucre 0.9–1.4 cm high . . . . . . . . . . . . . . . . . . . . . . . . 7. L. FLORIDANA

a. Flowers usually yellow or yellowish, in live condition, often drying purple or blue (rarely blue when fresh in a form of the rare *L. ludoviciana*); achenes (seed-like fruits) with an elongated hair-like extension (beak) as long as or longer than the main body of the achene . . . . . . . . . . . . *c*
- c. Margins of the leaves prickly toothed . . . . . . . . . . . . . . . . . . *d*
  - d. Commonly found throughout Missouri; heads with 16–24 (rarely 13 or 27) flowers or fruits; involucre 1–1.3 cm. high, the outer bracts narrowly to broadly lanceolate; achene with hairs or projections around summit, with 4–7 nerves on each side (face) . . . . . . . 1. L. SCARIOLA
  - d. Rarely encountered, known only from loess hills of Atchison County, northwestern Missouri; heads with 20–56 flowers or fruits; involucre 1.7–2.3 cm. high, the larger outer bracts ovate; achene smooth around summit, with only 1–3 nerves on each side (face) . . . . 5. L. LUDOVICIANA
- c. Margins of the leaves without teeth or without prickly teeth . . . . . . . . . . . . . . *e*
  - e. Heads with 5–12 flowers or fruits; achenes with 5–7 nerves on each side (face); stem mainly 3–6 dm. (rarely up to 10) tall . . . . . . . . . . . . . . . . . . . 2. L. SALIGNA
  - e. Heads with usually 12–20 (up to 25) flowers or fruits; achenes with only 1–3 nerves on each side (face); stem mainly 10–35 dm. (rarely only 5) tall . . . . . . . . . . . . . *f*
    - f. Lower portion of stem often hairy, or sometimes without hairs; at least the midrib of the lower surface of leaves, and usually both surfaces of leaves hairy; involucre 1.5–2.2 cm. high in fruit; mature achenes (including beak) 7–10 mm. long; pappus 9–12 mm. long; rare species, found only in the Ozark region . . . . . . . . . . . . 4. L. HIRSUTA
    - f. Stems glabrous (without hairs) or rarely hairy in one form; leaves usually glabrous or rarely with some hairs on the nerves of the lower surface; involucre 1–1.4 cm. high in fruit; mature achenes (including beak) 5–6 mm. long; pappus 5–7 mm. long; common species, found throughout Missouri . . . . . . . . . . . . . . . . . 3. L. CANADENSIS

1. **Lactuca Scariola** L. Prickly Lettuce Map 2389

Flowers June–October.

Occurs in fields, waste ground, roadsides, and along railroads. Throughout Missouri.

Native of Europe; introduced and naturalized in North America from Quebec to British Columbia, south to Florida, Texas, and California.

Two variations are found in Missouri:

All the leaves with lobes or deeply cut-out portions . . . . . 1a. L. SCARIOLA f. SCARIOLA

Leaves without lobes or deep cuts or only the lowest ones lobed or deeply cut . 1b. L. SCARIOLA f. INTEGRIFOLIA

1a. **Lactuca Scariola** f. **Scariola** Map 2389

*Lactuca Scariola* L. [G, BB, P & S]

*Lactuca Serriola* L. [Shinners]

This is the commoner variation and is found throughout the state, doubtless in every county.

1b. **Lactuca Scariola** f. **integrifolia** (Bogenh.) G. Beck Map 2389

*Lactuca Scariola* var. *integrata* Gren. & Godr. [BB, P & S]

Scattered in southern and central Missouri north to Pike, Boone, Saline, and Jackson counties, but doubtless to be found throughout the state and in most counties.

The tender young leafy shoots of this species are gathered in the spring and cooked as greens either alone or in combination with other wild vegetables. The leaves turn in a vertical position with one edge up, most of the leaves pointing in a north-south direction.

Shinners (Field and Lab. 17: 175–76. 1949) has stated reasons for adopting the original spelling of the species, *L. Serriola*, instead of following the later and altered form of *L. Scariola*.

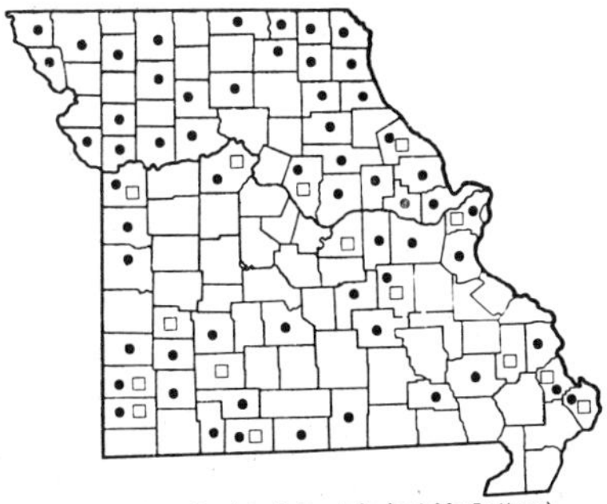

2389 • Lactuca Scariola f. Scariola (Prickly Lettuce)
2389 □ Lactuca Scariola f. integrifolia

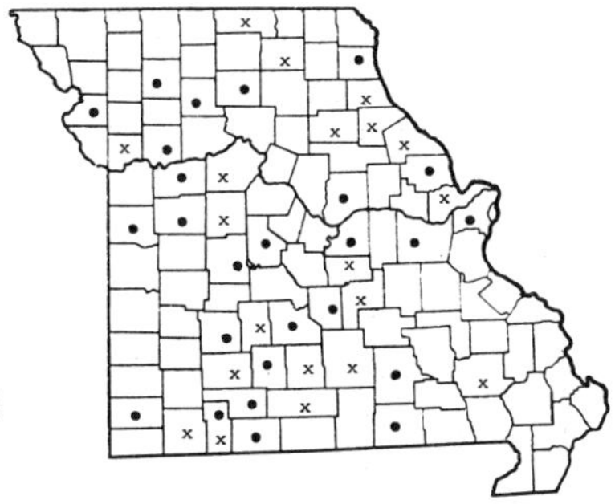

2390 • Lactuca saligna f. saligna (Willow-leaved Lettuce)
2390 x Lactuca saligna f. Ruppiana

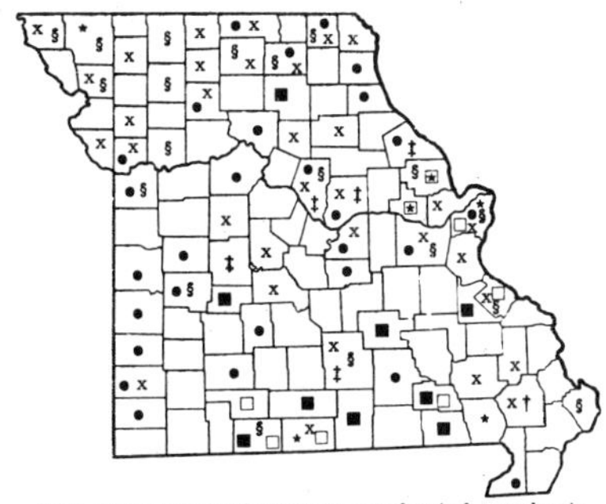

2391 • Lactuca canadensis var. canadensis f. canadensis (Wild Lettuce)
2391 ⋆ Lactuca canadensis var. canadensis f. angustata
2391 § Lactuca canadensis var. obovata f. obovata
2391 ⊞ Lactuca canadensis var. obovata f. stenopoda
2391 ‡ Lactuca canadensis var. longifolia f. longifolia
2391 □ Lactuca canadensis var. longifolia f. angustipes
2391 x Lactuca canadensis var. latifolia f. latifolia
2391 † Lactuca canadensis var. latifolia f. villicaulis
2391 ■ Lactuca canadensis var. latifolia f. exauriculata

2. **Lactuca saligna** L. Willow-leaved Lettuce Map 2390

Flowers July–October.

Occurs in waste and gravelly open ground, pastures, upland prairie, meadows and wet ground along streams and spring branches, fields, along bluff escarpments, rocky slopes, roadsides, and along railroads. Throughout Missouri.

Native of Europe; introduced and naturalized in the United States from Pennsylvania to Michigan, south to Virginia, Indiana, Kentucky, Illinois, Missouri; also in California.

Two variations are found in Missouri:

Leaves oblong with lobes or cut-out portions . . . 2a. L. SALIGNA f. SALIGNA
Leaves linear to lanceolate, entire (without teeth or lobes) . . . . . 2b. L. SALIGNA f. RUPPIANA

2a. **Lactuca saligna** f. **saligna** Map 2390
*Lactuca saligna* L. [G, BB, P & S]
Throughout Missouri.

2b. **Lactuca saligna** f. **Ruppiana** (Wallr.) G. Beck Map 2390

Scattered throughout the state. This species was first noted in Missouri in 1932, and has since been found increasing along highways. The bluish- or gray-green leaves have a prominent white midrib, giving them a rather striped and ornamental appearance. The flowers are usually yellow, but may become purple as was noted on a specimen collected in St. Charles County (*Steyermark 19496*).

3. **Lactuca canadensis** L. Wild Lettuce Map 2391

Flowers July–September.

Occurs in open woodland, rocky slopes along bluffs, prairie openings, alluvial thickets, borders of fields, muddy banks and gravel bars along streams, roadsides, and along railroads.

Throughout Missouri.

The following variations are encountered in Missouri:

a. Leaves entire (without teeth) or shallowly toothed, or the lowermost ones sometimes with shallow lobes . . . . . . . . . . *b*
  b. Main leaves of stem lanceolate to lance-ovate, the ones about half-way up entire (without teeth) or rarely toothed . . . *c*
    c. Leaves half-way up stem with arrowhead-shaped or rounded bases clasping the stem . . . . . . . 3a. L. CANADENSIS var. CANADENSIS f. CANADENSIS
    c. Leaves half-way up stem with the bases tapering or narrowed and not clasping the stem . . . . 3b. L. CANADENSIS var. CANADENSIS f. ANGUSTATA
  b. Main leaves of stem oblanceolate to narrowly obovate, usually finely toothed . . . *d*
    d. Main leaves of stem with arrowhead-shaped bases clasping the stem . . . 3c. L. CANADENSIS var. OBOVATA f. OBOVATA
    d. Main leaves of stem with the bases tapering or narrowed and not clasping the stem . . . . 3d. L. CANADENSIS var. OBOVATA f. STENOPODA
a. All the leaves, or all but the uppermost ones, conspicuously lobed or deeply parted . . . *e*
  e. Lobes of leaves linear-falcate; upper unlobed leaves, if present, linear or linear-lanceolate . . . . . . . . . . *f*
    f. Main leaves of stem with arrowhead-shaped bases clasping the stem . . . 3e. L. CANADENSIS var. LONGIFOLIA f. LONGIFOLIA

f. Main leaves of stem with the bases tapering or narrowed and not arrowhead-shaped . . . 3f. L. CANADENSIS var. LONGIFOLIA f. ANGUSTIPES

e. Lobes of leaves broadly falcate to obovate, entire or often toothed; upper unlobed leaves, if present, lanceolate or lance-ovate to oblanceolate or obovate . . . . . *g*

g. Stems more or less hairy . . . . 3h. L. CANADENSIS var. LATIFOLIA f. VILLICAULIS

g. Stems glabrous (without hairs) . . . *h*

h. Main leaves of stem with arrowhead-shaped bases clasping the stem . . 3g. L. CANADENSIS var. LATIFOLIA f. LATIFOLIA

h. Main leaves of stem with the bases tapering or narrowed and not arrowhead-shaped nor clasping the stem . 3i. L. CANADENSIS var. LATIFOLIA f. EXAURICULATA

3a. **Lactuca canadensis** var. **canadensis** f. **canadensis** Map 2391

*Lactuca canadensis* var. *canadensis* [BB]

*Lactuca canadensis* L., typical [G]

*Lactuca canadensis* var. *integrifolia* (Bidel.) T. & G. [P & S, Wiegand]

Scattered and common throughout Missouri.

Ranges from Nova Scotia to Ontario and Minnesota, south to North Carolina, Indiana, Illinois, Missouri, and Oklahoma.

3b. **Lactuca canadensis** var. **canadensis** f. **angustata** Wieg. Map 2391

*Lactuca canadensis* f. *angustata* Wieg. [G]

*Lactuca canadensis* var. *integrifolia* (Bigel.) T. & G. f. *angustata* Wieg. [P & S, Wiegand]

Rare and scattered in Missouri.

3c. **Lactuca canadensis** var. **obovata** Wieg. f. **obovata** Map 2391

*Lactuca canadensis* var. *obovata* Wieg. [G, BB, P & S]

Scattered throughout Missouri.

Ranges from Maine to Minnesota and Nebraska, south to New Jersey, Virginia, Indiana, Missouri, and Oklahoma.

3d. **Lactuca canadensis** var. **obovata** f. **stenopoda** Wieg. Map 2391

Known only from Lincoln County, east-central Missouri (open woods, 2 mi. west of Foley, September 22, 1932, *Steyermark 8623*).

3e. **Lactuca canadensis** var. **longifolia** (Michx.) Farw. f. **longifolia** Map 2391

*Lactuca canadensis* var. *longifolia* (Michx.) Farw. [G, BB]

*Lactuca canadensis* var. *typica* [of P & S, Wiegand], not L.

Scattered and rare in Missouri.

Ranges from Quebec to Saskatchewan, south to North Carolina, Ohio, Indiana, Missouri, and Colorado; British Columbia to California.

3f. **Lactuca canadensis** var. **longifolia** f. **angustipes** Wieg. Map 2391

Scattered in southern and east-central Missouri.

3g. **Lactuca canadensis** var. **latifolia** Ktze. f. **latifolia** Map 2391

*Lactuca canadensis* var. *latifolia* Ktze. [G, BB, P & S]

Scattered throughout Missouri.

Ranges from Quebec to Ontario, south to Florida, Tennessee, Missouri, Oklahoma, and Texas.

3h. **Lactuca canadensis** var. **latifolia** f. **villicaulis** Fern. Map 2391

Known only from Stoddard County, southeastern Missouri (edge of field in thickets, *Nyssa aquatica-Acer rubrum* var. *Drummondii-Taxodium distichum-Populus heterophylla* woods bordering Swan Pond, T28N, R10E, sect. 35, 4 mi. south of Advance, August 28, 1948, *Steyermark 66167*).

3i. **Lactuca canadensis** var. **latifolia** f. **exauriculata** Wieg. Map 2391

Scattered in Missouri.

*Lactuca canadensis* and varieties contain a bitter-tasting substance, which gives milk a poor taste when grazed by cows. Eating the plant can cause various complications of the digestive and nervous systems in human beings and lower animals. The leaves, however, are eaten by white-tailed deer.

4. **Lactuca hirsuta** Muhl. Map 2392

Flowers June–September.

Occurs in dry rocky woods, borders of upland sinkhole ponds, in usually acid soils, and along roadsides and openings. Southern and east-central Missouri.

Plate no. 387. **1.** Lactuca canadensis, × 2/5; Details from Small, The New York Botanical Garden. **2.** Lactuca pulchella, × 2/15; *a.* Achene, × 4/5. **3.** Lactuca hirsuta var. hirsuta, × 2/5. **4.** Lactuca ludoviciana, × 2/5; *a.* Achene, × 2 4/5; After Britton and Brown, The New York Botanical Garden.

PLATE NO. 387

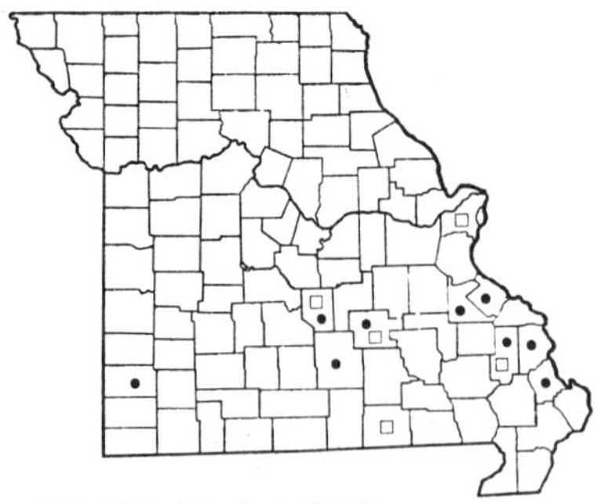

2392 • Lactuca hirsuta var. hirsuta
2392 □ Lactuca hirsuta var. sanguinea

2393 Lactuca ludoviciana f. ludoviciana

2394 Lactuca pulchella (Blue Lettuce)

Two variations occur in the state:

Lower part of stem hairy; lower and middle leaves of the stem hairy on both surfaces . 4a. L. HIRSUTA var. HIRSUTA

Stem glabrous (without hairs) or mainly so; only the midrib of the lower surface of the leaves hairy, the rest of the lower and upper leaf surfaces glabrous . . . 4b. L. HIRSUTA var. SANGUINEA f. SANGUINEA

4a. **Lactuca hirsuta** var. **hirsuta** Map 2392

*Lactuca hirsuta* Muhl. [G, P & S]

Southern Missouri, where known from Scott (sandy ground, August 31, 1894, *Eggert*), Cape Girardeau, Ste. Genevieve, St. Francois (*Steyermark 78953*), Dent, Pulaski, Texas (*Bauer 910*), and Jasper (dry open woods, Webb City, September 4, 1910, *Palmer 3130*; wooded bluffs of Center Creek, 3 mi. southwest of Carl Junction, August 31, 1920, *Palmer 18877*) counties.

Ranges from Pennsylvania to Virginia, Missouri and Louisiana.

4b. **Lactuca hirsuta** var. **sanguinea** (Bigel.) Fern. Map 2392

Southern and east-central Missouri in St. Louis (Arsenal, St. Louis, August 12, 1885, *Wislizenus 236*), Bollinger, Dent (July 7, 1951, *Steyermark 72012*), Shannon (open woods, Montier, September 2, 1918, *Bush 8755*), Oregon (around margin of sink-hole pond of Brushy Pond, on east side of route 19, T25N, R4W, south part of sect. 1, 3½ mi. west of New Liberty P.O., June 26, 1951, *Steyermark 71733*), and Pulaski counties.

Ranges from Prince Edward Island to Ontario, south to Virginia, Missouri, Louisiana, and Texas.

This species generally has purplish or reddish-purple stems. As Missouri material of the species is not well represented in the larger herbaria, the above citations are given for the sake of the record.

5. **Lactuca ludoviciana** (Nutt.) Riddell f. **ludoviciana** Map 2393

Flowers June–August.

Occurs on loess hills. Known only from Atchison County, northwestern Missouri (August 23, 1893, *Bush*).

Ranges from the Pacific states east to Manitoba, Wisconsin, Illinois, Indiana, Missouri, Arkansas, and Texas.

Typical f. *ludoviciana* has the flowers yellow; in f. *campestris* (Greene) Fern., not recorded from Missouri, the flowers are blue or purple in the fresh state.

6. **Lactuca pulchella** (Pursh) DC. Blue Lettuce Map 2394

Flowers June–August.

Occurs on loess hills, open slopes, roadsides, and moist open ground. Northwestern and west-central Missouri in Atchison, Nodaway, and Jackson counties.

Ranges from Quebec to Alaska, south to Michigan, Wisconsin, Missouri, Oklahoma, New Mexico, Arizona, and California; rarely introduced east to New England.

This is a very showy-flowered species with blue or blue-purple heads of flowers. It is a weed in some of the western grain fields and is usually avoided by cattle, but sometimes eaten by sheep.

7. **Lactuca floridana** (L.) Gaertn. Map 2395

Flowers August–October.

Occurs on wooded slopes along streams, at base of bluffs, in ravines, low rich or alluvial woods, swampy ground, thickets, cut-over slopes, roadsides, and along railroads.

Common throughout Missouri, and probably in every county.

Two variations occur in Missouri:

2395 • Lactuca floridana var. floridana f. floridana (Blue Lettuce)
2395 x Lactuca floridana var. villosa

2396 Lygodesmia juncea

2397 Agoseris cuspidata

Leaves usually more or less deeply lobed or parted . . . . 7a. L. FLORIDANA var. FLORIDANA f. FLORIDANA

Leaves usually only toothed, not lobed . . . 7b. L. FLORIDANA var. VILLOSA

7a. **Lactuca floridana** var. **floridana** f. **floridana** Map 2395

*Lactuca floridana* var. *floridana* [BB]

*Lactuca floridana* (L.) Gaertn. [G, P & S]

Common throughout Missouri, and somewhat more common than var. *villosa.*

Ranges from Florida to Texas, north to Massachusetts, New York, Ohio, Indiana, Illinois, Minnesota, and Nebraska.

Typical var. *floridana* f. *floridana* has the flowers bluish; in var. *floridana* f. *leucantha* Fern., not recorded from Missouri, the flowers are white.

7b. **Lactuca floridana** var. **villosa** (Jacq.) Cronq. Map 2395

*Lactuca villosa* Jacq. [P & S]

Throughout Missouri, but less common than var. *floridana.*

Ranges from Florida to Louisiana, north to New York, Ohio, Indiana, Illinois, Iowa, and Nebraska.

This and the other species of *Lactuca* in Missouri are edible, the young leafy shoots being gathered in the spring and cooked as greens.

*Excluded Species*

**Lactuca graminifolia** Michx.

Recorded by Palmer and Steyermark in their *Annotated Catalogue* from Butler County, southeastern Missouri (p. 682) on the basis of the collection by *Eggert* from 'railroad, near Poplar Bluff, July 7, 1893.' However, a re-examination of this specimen shows that it has the shorter involucre, shorter pappus, and more crowded inflorescence characteristic of *L. canadensis.* The very narrow unlobed leaves with nonsagittate leaf-bases of the Eggert specimen simulate those of *L. graminifolia,* but are matched by specimens of *L. canadensis* var. *canadensis* f. *angustata* having just such types of leaves. The Missouri record is now referred to *L. canadensis* var. *canadensis* f. *angustata* Wieg.

## 80. **Lygodesmia** D. Don

**Lygodesmia juncea** (Pursh) D. Don Map 2396

Flowers June–August.

Occurs on exposed, dry loess hills of northwestern Missouri in Atchison, Holt, and Buchanan counties.

Ranges from Wisconsin to Alberta, south to Missouri, Oklahoma, Texas, New Mexico, and Arizona.

A specimen in the herbarium of the University of Missouri labeled as from St. Louis County (west of Forest Park, St. Louis, September 23, 1886, *Eggert*) is probably an error in labeling or mounting, and is excluded from the distribution map for Missouri. The species is definitely one of the prairies and plains area and reaches only the western loess hills of the state.

The plant contains a bitter milky juice and is suspected by ranchmen of being poisonous. The plant shows adaptation to the dry windswept regions in which it grows by the development of rigid narrow leaves, the upper ones being very much reduced to bracts, while the whole leafy area of the plant is greatly reduced. A gall insect parasitizes the stems, the globe-shaped gall which results often being mistaken for fruit.

2398 Crepis pulchra (Hawk's Beard)

2399 Crepis capillaris

2400 Crepis setosa

## 81. **Agoseris** Raf.

**Agoseris cuspidata** (Pursh) Raf. Map 2397
*Agoseris glauca* [of BB], not (Pursh) D. Dietr.
Flowers April–June.

Occurs on glades and rocky prairies. Known only from Jackson and Cass counties, west-central Missouri.

Ranges from Wisconsin to Montana, south to Illinois, Missouri, Oklahoma, and New Mexico.

## 82. **Crepis** L. Hawk's Beard

a. Involucre glabrous (without hairs) . . . . . . . . . . . . . . . . . . . . 1. C. PULCHRA
a. Involucre hairy . . . . . . . . . . . . . . . . . . . . . . . . . . . . *b*
  b. Involucre 6–8 mm. high, short-hairy and also often bristly with longer hairs; achenes 1.5–2.5 mm. long, more or less narrowed toward the summit but only scarcely beaked . . . . 2. C. CAPILLARIS
  b. Involucre 8–10 mm. high, with long stiff bristles but not short-hairy; achenes 3–5 mm. long, with a slender beak 1–2.5 mm. long . . . . . . . . . . . . . . . . . 3. C. SETOSA

1. **Crepis pulchra** L. Map 2398
Flowers May–July.

Occurs along railroads. Known only from St. Louis County (Baden freight yard of Mo.-Kan.-Texas R.R., south of the repair shops, St. Louis, July 5, 1958, *Muehlenbach 1453*).

Native of Europe; introduced and naturalized in the United States in Virginia, Ohio, Indiana, and Missouri.

2. **Crepis capillaris** (L.) Wallr. Map 2399
Flowers July–October.

Occurs in grassy open places and waste ground. Known only from St. Louis County (grassy places on north edge of Washington University campus between Art Building and Forsyth Blvd., August 12, 1935, *Steyermark 19508*).

Native of Europe; introduced and naturalized in the United States from New Hampshire and Connecticut to New York, south to New Jersey, Pennsylvania, Ohio, Indiana, Illinois, and Missouri; and British Columbia to California and Texas.

3. **Crepis setosa** Haller f. Map 2400
Flowers June–August.

Occurs in cultivated fields. Known only from Boone County, central Missouri (common weed in an alfalfa field, north of the Grindstone, Columbia, July, August, 1904, *Daniels*, in U. of Mo. Herb.).

Native of Europe; introduced into North America, but apparently the plants do not persist to become established in various places where they have been collected.

---

Plate no. 388. **1.** Lactuca floridana var. floridana, × 2/7. **2.** Pyrrhopappus carolinianus, × 2/7; Details from Small, The New York Botanical Garden. **3.** Lygodesmia juncea, × 2/7; After Gleason, details from Small, The New York Botanical Garden. **4.** Agoseris cuspidata, × 2/7. **5.** Prenanthes crepidinea, × 2/7.

PLATE NO. 388

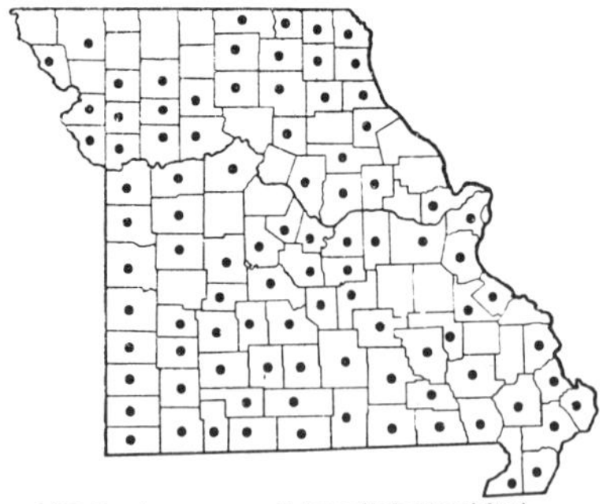

2401 Pyrrhopappus carolinianus (False Dandelion)

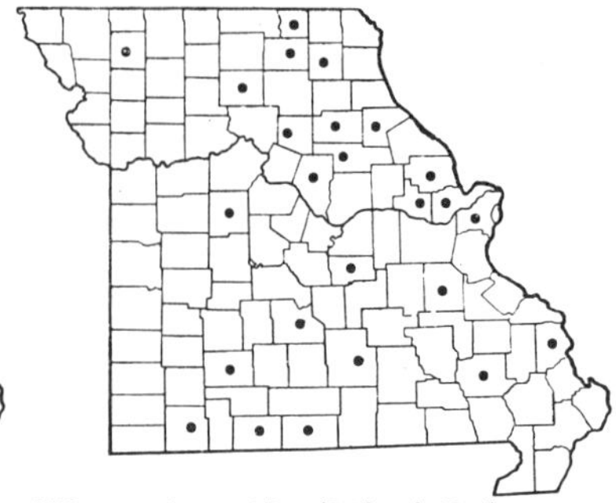

2402 Prenanthes crepidinea (Rattlesnake Root)

2403 Prenanthes racemosa

## 83. **Pyrrhopappus** DC. False Dandelion

**Pyrrhopappus carolinianus** (Walt.) DC. False Dandelion Map 2401

Flowers May–October.

Occurs in fallow and alluvial fields, open woods, waste ground, thickets, borders of streams, ponds, gravel bars, roadsides, and along railroads. Throughout Missouri, and probably in every county, scarce in northwestern Missouri.

Ranges from Florida to Texas, north to Delaware, Indiana, Illinois, Missouri, Kansas, and Oklahoma.

## 84. **Prenanthes** L. Rattlesnake Root, White Lettuce

a. Main leaves of stem, but at least the lower ones and those half-way up, on petioles (leaf-stalks), the petioles often winged (bordered by leafy tissue) . . . . . . . . . . . . . . . . . . . *b*

- b. Bracts of involucre hairy, the longest (inner) ones 12–15; heads with 20–35 flowers or fruits (achenes) . . . . . . . . . . . . . . . . . . . . . . 1. P. CREPIDINEA
- b. Bracts of involucre glabrous, the longest (inner) ones 5–8; heads with 8–16 flowers or fruits. . . . *c*
  - c. Longest (inner) bracts of involucre 6–8; heads with 8–13 flowers or fruits; bracts of involucre glaucous (with a silvery- or gray-white coating which can be rubbed off); plants of northeastern and east-central Missouri south to St. Louis, Franklin, and Gasconade counties . . . 4. P. ALBA
  - c. Longest (inner) bracts of involucre 5, rarely 4 or 6; heads with 5–6 flowers or fruits; bracts of involucre greenish; throughout the Ozark region . . . . . 5. P. ALTISSIMA var. CINNAMOMEA

a. Main upper leaves of stem and those half-way up stem sessile (without leaf-stalks) . . . . . . . *d*

- d. Stem below the inflorescence and lower surface of leaves glabrous (without hairs); flowers pink or purplish . . . . . . . . . . . . . . . . . . . . . . . 2. P. RACEMOSA
- d. At least the upper part of stem below the inflorescence and at least lower surface of leaves rough and/or hairy; flowers cream-colored . . . . . . . . . . . . . . . 3. P. ASPERA

1. **Prenanthes crepidinea** Michx. Map 2402

Flowers August–October.

Occurs in rich moist woods in low ground of valleys bordering streams, open wooded slopes and banks along streams, and low thickets. Scattered throughout Missouri west to Gentry, Linn, Randolph, Pettis, Laclede, Greene, and Barry counties.

An apparent hybrid between this species and *P. aspera* is found in Greene County (Asher Creek near Polk Co. line, August 26, 1893, *J. W. Blankinship*, in Drury College Herb.).

2. **Prenanthes racemosa** Michx. Map 2403

*Prenanthes racemosa* subsp. *racemosa* [BB]

Flowers late June–September.

Occurs in wet prairies and low ground bordering streams. Known only from St. Louis (Kirkwood, June 24, 1886, *Eggert*) and Greene (September 13, 1890, *S. Weller*) counties.

Ranges from Quebec and James Bay west to Alberta, south to New Jersey, New York, Vermont, Ohio, Indiana, Illinois, Missouri, South Dakota, and Colorado.

---

Plate no. 389. **1.** Crepis pulchra, × 2/5; After Gleason, details from Small, The New York Botanical Garden. **2.** Crepis setosa, × 2/5; *a.* Achene, × 2; After Gleason, details from Small, The New York Botanical Garden. **3.** Crepis capillaris, × 2/5; *a.* Flower-head, × 6/7; *b.* Achene, × 4; After Gleason, details from Small, The New York Botanical Garden. **4.** Prenanthes aspera, × 2/5; *a.* Flower-head, × 2 2/5. **5.** Prenanthes altissima, × 2/5; After Gleason, The New York Botanical Garden.

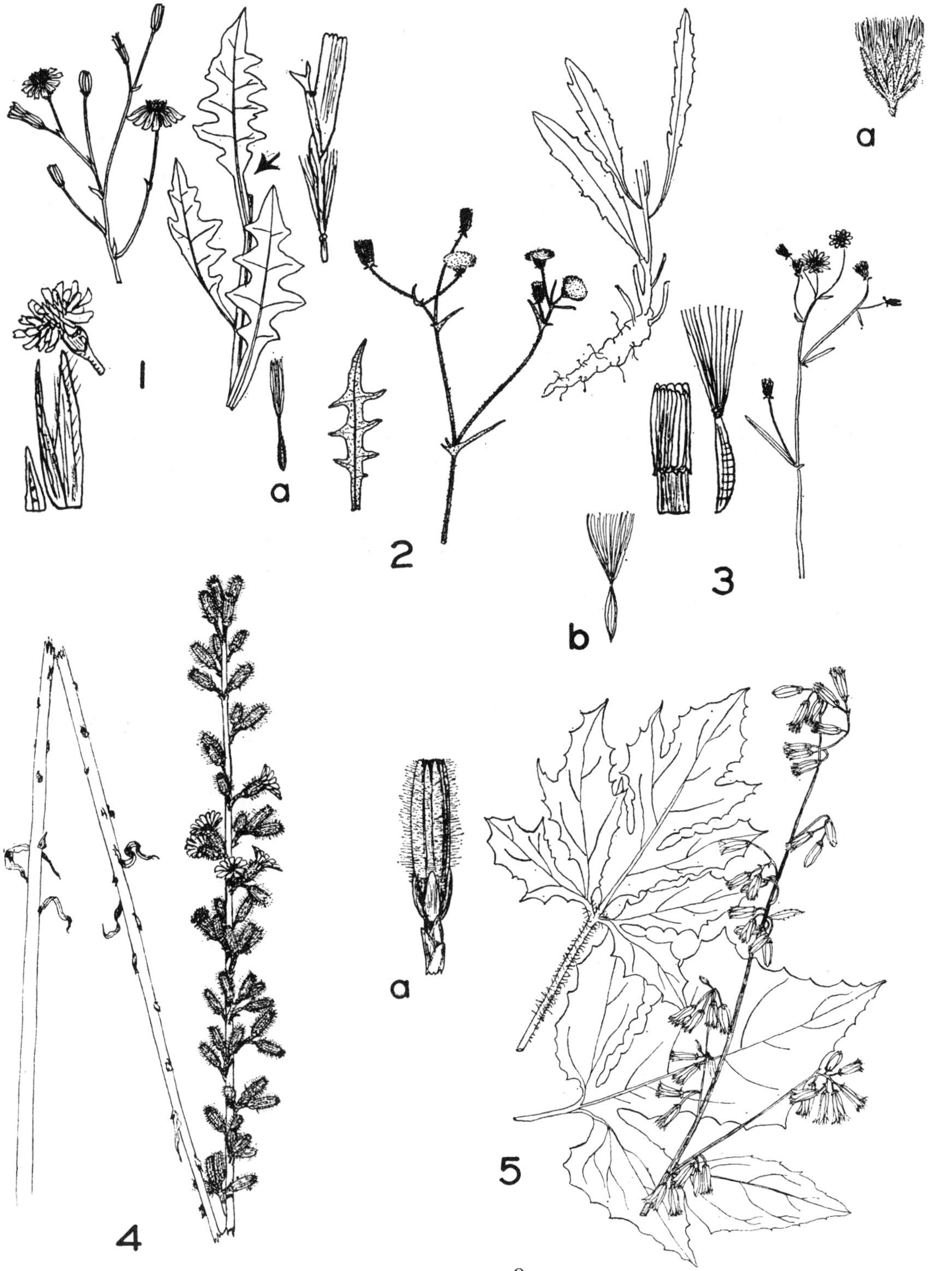


PLATE NO. 389

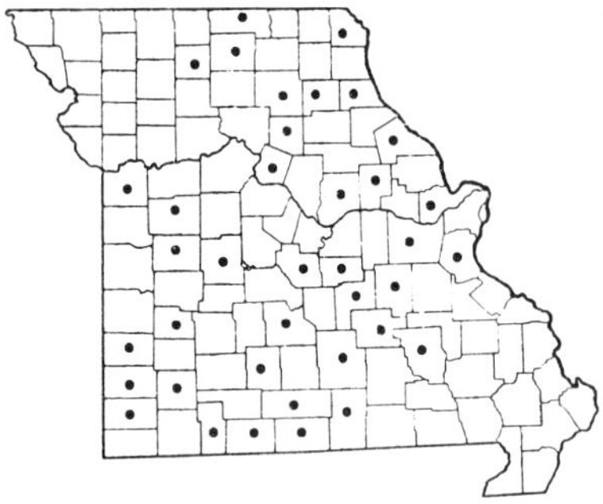

2404 Prenanthes aspera

2405 Prenanthes alba (White Lettuce)

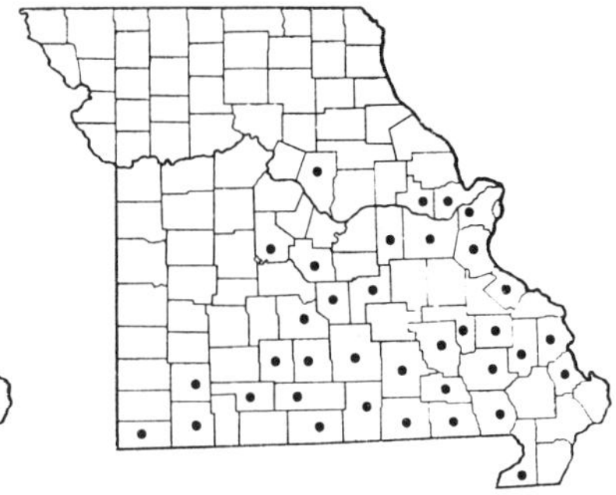

2406 Prenanthes altissima var. cinnamomea

No specimens of this species have been collected in Missouri since 1890, and only future exploration in the state can determine whether or not the species is still extant.

3. **Prenanthes aspera** Michx. Map 2404
Flowers August–September.

Occurs in prairies, limestone glades, dry open and rocky woods, usually in acid soils, and along railroads. Southern, central, and eastern Missouri, west to Putnam, Grundy, Macon, Randolph, Howard, and Jackson counties.

Ranges from Ohio to Minnesota and South Dakota, south to Tennessee, Louisiana, and Oklahoma.

The flowers of this species have a pleasantly delicate fragrance resembling that of *Sabatia angularis*. The plant is attractive when in full flower. Although I have attempted to grow this species on several occasions in the prairie section of my wildflower preserve, it has not persisted. It seems to thrive where there is no crowding or competition from other plants.

4. **Prenanthes alba** L. White Lettuce Map 2405
Also called Rattlesnake Root, White Snake Root.
Flowers July–September.

Occurs in rich woods and on wooded slopes, bluffs, and ravines.

Eastern and east-central Missouri from Clark County south to St. Louis, Franklin, Gasconade, and Callaway counties.

Ranges from Maine and Quebec to Saskatchewan, south to Georgia, Tennessee, Missouri, and South Dakota.

The leaves are thicker than in the following species, *P. altissima*. Although the stems and leaves are stated in some manuals to be glaucous, many plants do not show this character and may vary from leaves which are dark green on the upper surface and paler green on the lower surface, with both surfaces nonglaucous to leaves which are somewhat glaucous. The stem, likewise, shows variability in this respect.

5. **Prenanthes altissima** L. var. **cinnamomea** Fern. Map 2406
Flowers July–October.

Occurs in rocky woods, along bluffs and ledges bordering streams, in ravines, and swampy woods. Ozark region of southern and central Missouri, north to St. Charles, Warren, and Boone counties, west to Morgan, Laclede, Christian, and McDonald counties.

Ranges from Indiana to Missouri, south to Louisiana and Oklahoma.

In typical var. *altissima* the mature pappus is creamy-white. The var. *cinnamomea*, to which all the Missouri plants are referred, has the mature pappus cinnamon- or deep brown-colored.

The leaves, which are thinner than in *P. alba*, vary considerably in size and shape, from ovate, or heart-shaped, to triangular, and from shallowly toothed to lobed or cleft.

## 85. **Hieracium** L. Hawkweed

a. Hairs on leaves and near base of stem very long, measuring 1–2 cm. in length . . . 3. H. LONGIPILUM
a. Hairs on leaves and near base of stem, when present, shorter than 1 cm. in length . . . . . . . *b*
  b. Leafy bracts present on branches of inflorescence; heads with 40–50 or more flowers or fruits; mature achenes not narrowed at the summit but truncate (as if cut straight across) . . 1. H. SCABRUM
  b. Leafy bracts absent on branches of inflorescence; heads with 15–35 flowers or fruits; mature achenes narrowed at the summit . . . . . . . . . . . . . . . . . 2. H. GRONOVII

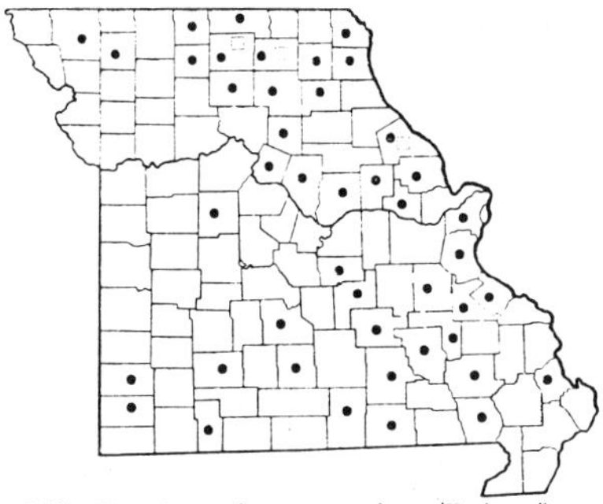

2407 • Hieracium scabrum var. scabrum (Hawkweed)
2407 □ Hieracium scabrum var. intonsum

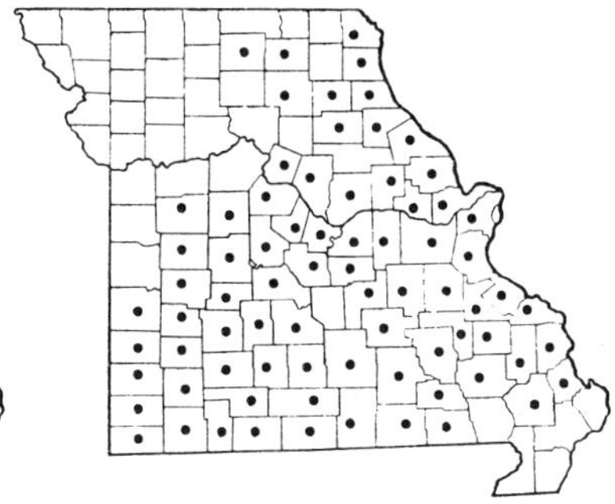

2408 Hieracium Gronovii

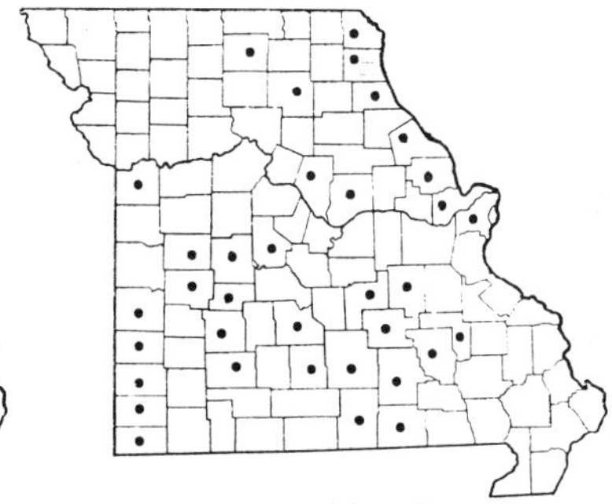

2409 • Hieracium longipilum f. longipilum
2409 □ Hieracium longipilum f. eglandulosum
• • •
2410 Excluded species

1. **Hieracium scabrum** Michx. Map 2407

Flowers June–September.

Occurs in rocky or dry open woods and slopes of ravines, ledges, and bluffs.

Throughout the Ozark region and in eastern and northern Missouri; absent from the unglaciated prairie region of southwestern Missouri and from a large sector of west-central Missouri.

Two variations occur in Missouri:

Common type encountered; hairs 2–3 mm. long on lower part of stem, leaf-stalks (petioles), and midribs; hairs on upper surface of leaves scattered and mainly 0.5–2 mm. long . . 1a. H. SCABRUM var. SCABRUM

Rarely encountered; hairs 3–5 mm. long on lower part of stem and both surfaces of leaves. . 1b. H. SCABRUM var. INTONSUM

1a. **Hieracium scabrum** var. **scabrum** Map 2407

*Hieracium scabrum* Michx. [G, BB, P & S]

This is the common variety found in the range indicated for Missouri.

Ranges from Nova Scotia and Quebec to Ontario and Minnesota, south to Delaware, Maryland, Georgia, Tennessee, Kentucky, Illinois, Missouri, and Oklahoma.

1b. **Hieracium scabrum** var. **intonsum** Fern. & St. John Map 2407

Known from Sullivan (open sterile slopes around ravine tributary to Medicine Creek, T61N, R22W, sect. 10, 4½ mi. southwest of Humphreys, September 20, 1955, *Steyermark 79771*), Adair, and Pike counties.

Ranges from Illinois to Iowa and Missouri.

Intergradations are found in the Missouri specimens of the two varieties. Some specimens, for example, have hairs 5 mm. long on the lower portion of the stem, characteristic of var. *intonsum*, but at the same time may have short hairs on the upper surface of the leaves, or only the lower surface of the leaves may have long hairs 3–5 mm. long, while the upper surface has short hairs.

2. **Hieracium Gronovii** L. Map 2408

*Hieracium Gronovii* var. *foliosum* Michx. [P & S]

Flowers May–October.

Occurs in rocky or dry open woods in ravines, valleys, upland, and ridges, crevices of bluffs, thickets, and fields.

Eastern, southern, and central Missouri, west to Sullivan, Macon, Howard, Pettis, Johnson, and Vernon counties.

Ranges from Florida to Texas, north to Massachusetts, New York, Ontario, Michigan, Illinois, Missouri, Kansas, and Oklahoma.

This species varies in the height and robustness of the stem, amount of pubescence, and number of leaves on the stems, the var. *foliosum* described as having more leaves on the stem extending to the inflorescence than in typical *H. Gronovii* in which most of the leaves occur near the base. This is a variable character, which seems to vary from one plant to another. The form of the inflorescence is usually of a narrow or cylindric type, but may also be broad and open and somewhat paniculate corymbose or open paniculate, a variation which has sometimes caused misidentification as *H. venosum.*

The fruits are eaten by wild turkey, and the leaves are eaten by the white-tailed deer.

3. **Hieracium longipilum** Torr. Map 2409

Flowers July–September.

Occurs in prairies, fields, rocky open woods, and prairie remnants along railroads.

Throughout the Ozark and unglaciated prairie

sections of southern and central Missouri, and locally in northern Missouri in the glaciated prairie section.

Ranges from Louisiana and Texas, north to Ontario, Michigan, Wisconsin, Minnesota, and Nebraska.

Two variations occur in Missouri:

Common type encountered; bracts of involucre with gland-tipped hairs . . 3a. H. LONGIPILUM f. LONGIPILUM

Rare type encountered; bracts of involucre mostly without gland-tipped hairs, but with a close non-glandular hairiness . . 3b. H. LONGIPILUM f. EGLANDULOSUM

3a. **Hieracium longipilum** f. **longipilum** Map 2409

*Hieracium longipilum* Torr. [G, BB, P & S]

This is found within the range indicated in Missouri.

3b. **Hieracium longipilum** f. **eglandulosum** Palmer & Steyerm. Map 2409

Known only from Lewis County, northeastern Missouri (prairie-clad slopes of hills and ravines tributary to Grassy Creek, T60N, R7W, sect. 29, 2½–2¾ mi. southwest of Durham, August 12, 1955, *Steyermark 79122*, holotype in Chi. Nat. Hist. Mus. Herb.).

The species is easily recognizable by the numerous long hairs. The plant is often taller than either *H. Gronovii* or *H. scabrum.* It is usually found in prairies associated with such species as *Silphium laciniatum, S. integrifolium, S. terebinthinaceum, Petalostemon purpureum,* and *Andropogon Gerardi.*

*Excluded Species*

**Hieracium venosum** L. var. **venosum**

This was recorded from Missouri by Fernald in *Gray's Manual,* eighth edition (p. 1566). The specimens identified as this species by Fernald are from Newton County (dry woods, Neosho, June 22, 1913, *Palmer 3993*) and Taney County (rocky woods, Swan, September 22, 1905, *Bush 3343*). Both these specimens and others similar to them (*Steyermark 1740* and *Brenner* from Ste. Genevieve County; *Palmer 2257* from Jasper County; and *Palmer 32845* from Ozark County) show plants with the leaves mostly basal or crowded near the base and the inflorescence with a subcorymbose type of branching. The achenes in all these specimens are constricted or distinctly narrowed at the summit, a characteristic of *H. Gronovii* but not of *H. venosum* in which the achenes are not constricted. Usually in *H. venosum* the midrib and lateral nerves are conspicuously purple-margined and the inflorescence has a loose subcorymbose branching with more elongated branches than in *H. Gronovii.* As was noted previously, considerable variation exists within *H. Gronovii* as to leafiness of the stem and mode of branching of the inflorescence. Since the specimens identified by Fernald as *H. venosum* fit into the usual type of variation exhibited by *H. Gronovii,* they are referred to the latter species in the present flora.

Plate no. 390. **1.** Prenanthes alba, × 2/5. **2.** Hieracium Gronovii, × 2/5. **3.** Prenanthes racemosa, × 2/5. **4.** Hieracium longipilum, × 2/5; After Gleason, details from Small, The New York Botanical Garden.

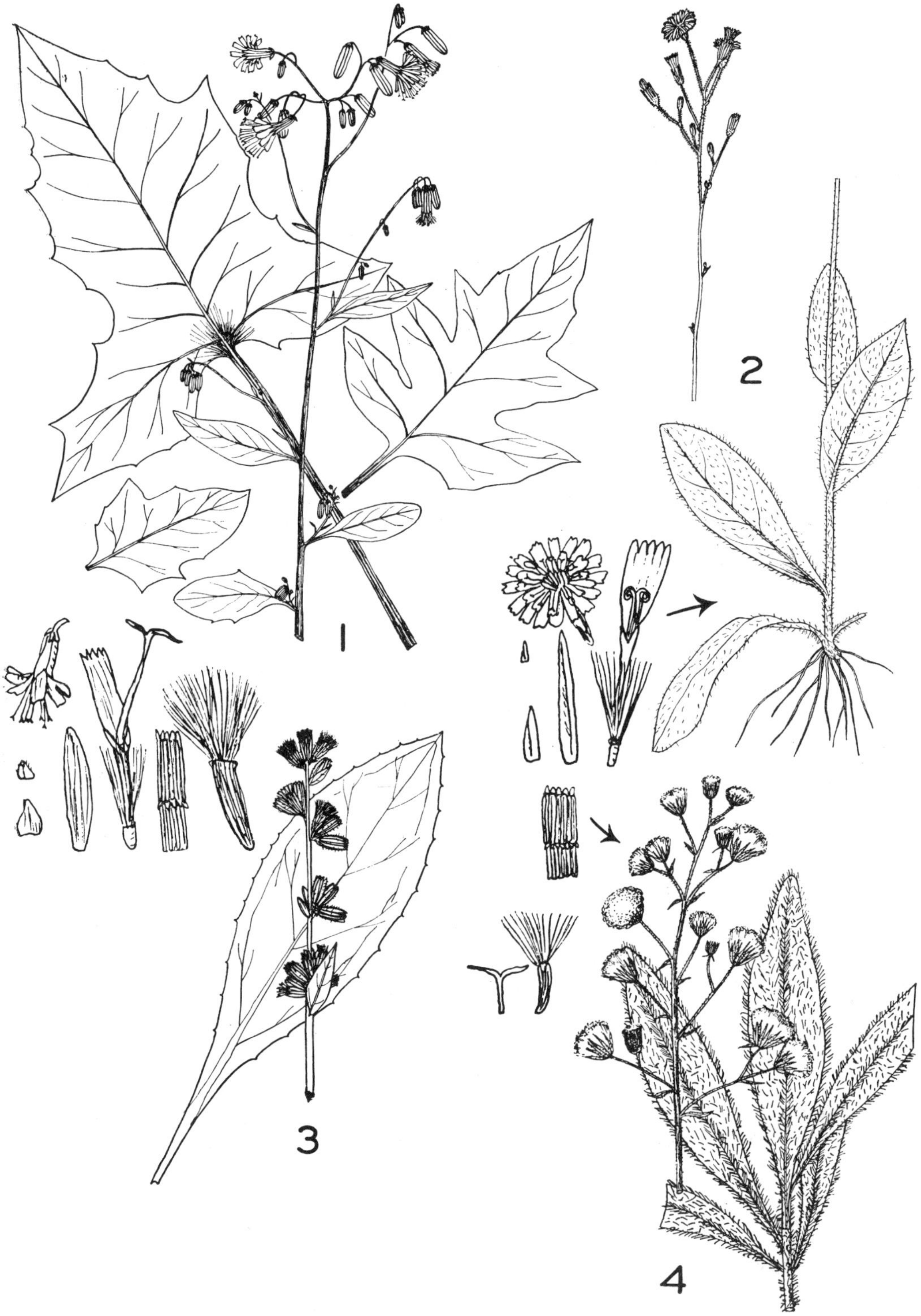


PLATE NO. 390

# *Supplement to the Flora of Missouri*

The following species were collected mainly by Dr. Viktor Muehlenbach during the period while the book was in press, and are presented in this section as additions to the flora of Missouri.

## Family **GRAMINEAE**

**Cynosurus echinatus** L.

St. Louis County: track of the Missouri Pacific R.R., along the ramp of the Cornell Seed Company, first street south of Plum Street, St. Louis, June 5, 1960, *Muehlenbach 1628.* Specimen determined by Dr. G. B. Van Schaack.

**Paspalum Urvillei** Steud.

St. Louis County: Carrie Avenue freight yard of the Terminal R.R. Association, north of the station building, along the track for locomotives, St. Louis, September 9, 1961, *Muehlenbach 1848.* Specimen determined by Dr. G. B. Van Schaack.

**Pennisetum ciliare** (L.) Link

St. Louis County: track of the Missouri Pacific R.R., along the ramp of the Cornell Seed Company, first street south of Plum Street, St. Louis, June 5, 1960, *Muehlenbach 1627.* Specimen determined by Dr. G. B. Van Schaack.

## Family **PAPAVERACEAE**

**Eschscholtzia californica** Cham. California Poppy

St. Louis County: Carrie Avenue freight yard of the Rock Island R.R., on the big free place southeast of the station building, St. Louis, July 1, 1961, *Muehlenbach 1788.* Specimen determined by Dr. V. E. Rudd.

## Family **GERANIACEAE**

**Erodium texanum** Gray

St. Louis County: Carrie Avenue freight yard of the Rock Island R.R., along the most northeastern track, St. Louis, May 30, 1961, *Muehlenbach 1762* Specimen determined by Dr. V. E. Rudd.

## Family **MALVACEAE**

**Gossypium herbaceum** L. Cotton

St. Louis County: Baden freight yard of the Mo.-Kan.-Top. R.R., on the recently laid out free place opposite the new station and shed building, St. Louis, September 3, 1960, *Muehlenbach 1690.* Specimen determined by Mr. Emery Leonard.

## Family **LOASACEAE**

**Mentzelia albescens** (Gill.) Griseb.

Jasper County: mine dumps, north side of Turkey Creek, about 2 1/2 mi. northwest of Joplin, August 27, 1960, *Palmer 69221;* chat piles of old mines about 2 mi. up Turkey Creek and about 1/2 mi. northwest of Joplin, August 30, 1960, *Palmer 69227;* chat piles of abandoned mines, about 1 1/2 mi. from Carl Junction, September 9, 1960, *Palmer 69263.* See Rhodora 63: 118. 1961.

## Family **CONVOLVULACEAE**

**Ipomoea cairica** (L.) House

St. Louis County: Baden freight yard of the Mo.-Kan.-Top. R.R., on the carriage road along the most western tracks, St. Louis, October 1, 1960, *Muehlenbach 1711.* Flowers purple. Specimen determined by Mr. Emery Leonard.

## Family **PLANTAGINACEAE**

**Plantago Wrightiana** Dcne.

St. Louis County: Baden freight yard of the Mo.-Kan.-Top. R.R., east of the most eastern track south of the switch shanty, St. Louis, June 18, 1961, *Muehlenbach 1779.* Specimen determined by Dr. V. E. Rudd.

## Family **CAPRIFOLIACEAE**

**Lonicera Xylosteum** L. Fly Honeysuckle

Jasper County: open upland woods, along a bank of an old electric railway grade, about 1/2 mi. north of Joplin, August 6, 1960, *Palmer 69188.* See Rhodora 63: 119. 1961.

## Family **COMPOSITAE**

**Eupatorium capillifolium** (Lam.) Small Dog Fennel, Hog Weed

St. Louis County: central shunting tracks, Lindenwood freight yard of the Frisco R.R., approximately opposite the station building, St. Louis, September 10, 1961, *Muehlenbach 1854.* Specimen determined by K. F. Parker.

**Gutierrezia texana** (DC.) T. & G.

St. Louis County: Baden freight yard of the Mo.-Kan.-Top. R.R., south of the round house, St. Louis, October 1, 1960, *Muehlenbach 1715.* Specimen determined by Mr. Emery Leonard.

**Ratibida Tagetes** (James) Barnhart

St. Louis County: Luther freight yard of the Wabash R.R., north of Humboldt Avenue, approximately opposite the switches XO and 10, St. Louis, September 16, 1961, *Muehlenbach 1865.* Specimen determined by K. F. Parker.

**Thelesperma ambiguum** Gray

St. Louis County: Carrie Avenue freight yard of the Rock Island R.R., along the most northeastern track, St. Louis, July 1, 1961, *Muehlenbach 1792.* Specimen determined by Dr. V. E. Rudd.

**Coreopsis cardaminefolia** (DC.) T. & G.

St. Louis County: Baden freight yard of the Mo.-Kan.-Top. R.R., east of the most central track, south of the switch shanty, St. Louis, August 20, 1961, *Muehlenbach 1829.* Specimen determined by K. F. Parker.

**Coreopsis tinctoria** Nutt. f. **atropurpurea** (Hook.) Fern.

St. Louis County: Carrie Avenue freight yard of the Rock Island R.R., along the eastern carriage road, about in the middle of the yard, St. Louis, June 19, 1960, *Muehlenbach 1649.*

**Artemisia Absinthium** L. Absinth

St. Louis County: Baden freight yard of the Mo.-Kan.-Top. R.R., along the most eastern track, south of the track scale of the yard, St. Louis, September 3, 1960, *Muehlenbach 1695.* Plant aromatic with a bitter taste. Specimen determined by Mr. Emery Leonard.

**Senecio ampullaceus** Hook.

St. Louis County: right-of-way of the Wabash R.R., along the most western track, between Palm and Branch streets, nearer to the latter, St. Louis, May 21, 1961, *Muehlenbach 1735*. Specimen determined by K. F. Parker.

**Centaurea diffusa** Lam.

St. Louis County: Carrie Avenue freight yard of the Rock Island R.R., along the car cleansing tracks in the northeastern corner of the yard, St. Louis, July 4, 1960, *Muehlenbach 1659*.

**Crepis tectorum** L.

St. Louis County: Luther freight yard of the Wabash R.R., north of Humboldt Avenue, St. Louis, July 4, 1961, *Muehlenbach 1802*. Specimen determined by K. F. Parker.

# Glossary

*Acaulescent.* Stemless or apparently so, or with stem subterranean.
*Achene.* A dry indehiscent one-seeded fruit.
*Acorn.* The fruit of the oak consisting of a nut and its cup or cupule.
*Actinomorphic* or *-ous.* Capable of bisection through two or more planes into similar halves, as is a regular radially symmetrical flower.
*Acuminate.* Tapering at the end to a gradual point.
*Acute.* Sharp, ending in a point, the sides of the apex essentially straight or slightly convex.
*Adventitious.* Used of buds, roots, etc., which develop in an irregular or unusual position.
*Adventive.* Imperfectly naturalized.
*Aestival.* Appearing in summer.
*Aestivation.* The arrangement of the parts of the flower in the bud.
*Aggregate.* Crowded into a dense cluster, but not cohering.
*Alternate.* Placed singly at different heights on the axis or stem.
*Ament.* A catkin, or dry scaly spike, usually unisexual, such as the inflorescence of willows, birches, etc., and at least the staminate inflorescence, in hickories and some other genera.
*Amplexicaul.* Clasping the stem.
*Anastomosing.* Connecting by cross-veins and forming a network.
*Ancipital.* Two-edged.
*Androecium.* A collective term for the stamens.
*Androgynous.* Composed of both staminate and pistillate flowers, the staminate at the apex.
*Annual.* Of only one year's duration. *Winter-annual,* a plant from autumn-sown seed which blooms and fruits the following spring.
*Anterior.* On the front side and away from the axis.
*Anther.* The pollen-bearing part of the stamen, borne at the top of the filament or sometimes sessile.
*Anthesis.* The expansion or the time of expansion of a flower.
*Apical.* Relating to the apex or tip.
*Apiculate.* Ending in an abrupt short pointed tip.
*Appendage.* An attached extra or secondary part, as a projecting or a hanging part or supplement.
*Appressed.* Lying close and flat against.
*Aquatic.* Living in water.
*Areolate.* Marked out in small spaces; reticulate.
*Areole.* The open spaces formed by anastomosing veins.
*Aril.* An appendage growing at or about the hilum of a seed.
*Aristate.* Bearing a stiff bristle-like awn or appendage.
*Ascending.* Rising somewhat obliquely, or curving upward. *Ascending ovule,* one that is attached above the base of the ovary and is directed upward.
*Asexual.* Sexless; without sex.
*Attenuate.* Showing a long gradual taper.
*Auricle.* An ear-shaped appendage or lobe.
*Awl-shaped.* Tapering gradually upward from the base to a slender or rigid point.
*Awn.* A bristle-shaped appendage.
*Axile placentation.* Referring to the location of ovules at or along the central axis of the ovary.
*Axillary.* In or related to the axis.
*Axis* (of the stem, inflorescence, etc.). The central part of a longitudinal support on which organs or parts are arranged.

*Barbed.* Bristles or awns provided with terminal or lateral spinelike hooks that are bent backwards sharply.
*Barbellate.* Finely barbed.
*Basifixed.* Attached at the base, as in an anther.
*Beak.* A long prominent and firm point.
*Beaked.* Ending in a firm prolonged slender tip.
*Bearded.* Furnished with long or stiff hairs.
*Berry.* Pulpy indehiscent few- or many-seeded fruit.
*Biconvex.* Convex on both sides as in a lens.
*Biennial.* Of two years' duration.
*Bifid.* Two-cleft.
*Bilabiate.* Two-lipped.
*Bilateral.* Arranged on opposite sides.
*Bipinnate.* Doubly or twice pinnate.
*Bipinnatifid.* Twice or doubly pinnatifid.
*Bisexual.* Having both stamens and pistils.
*Biternate.* Twice ternate with the principal divisions 3, each with 3 leaflets.
*Bladdery.* Inflated.
*Blade.* The expanded part of a leaf or petal.

*Bloom.* A whitish powdery and glaucous covering of the surface.
*Bract.* A more or less modified or reduced leaf subtending a flower or belonging to an inflorescence, or sometimes on the stem.
*Bractlet.* Bract borne on a secondary axis as on the pedicel or on the petiole.
*Branchlet.* Ultimate divisions of a branch.
*Bristle.* A stiff hair, or any slender body which may be likened to a hog's bristle.
*Bristly.* Provided with bristles.
*Bud.* The rudimentary state of a stem or branch; an unexpanded flower.
*Bulb.* A subterranean leaf-bud with fleshy scales or coats.
*Bulblet.* Little bulbs produced in the leaf-axils, inflorescence or other areas.
*Bush.* A low thick shrub, without distinct trunk.

*Caducous.* Falling off very early.
*Calcareous.* Limey.
*Callus.* A hard protuberance or callosity.
*Calyx.* The outer circle of floral envelopes consisting of the sepals.
*Calyx-tube.* The tube of a gamosepalous calyx.
*Campanulate.* Bell-shaped or cup-shaped with a broadened rim.
*Canescent.* Gray-pubescent and hoary.
*Capillary.* Hairlike.
*Capitate.* Shaped like a head; collected into a head or dense cluster.
*Capsule.* A dry dehiscent fruit composed of more than one carpel.
*Carinate.* Keeled.
*Carpel.* A simple pistil, or one member of a compound pistil.
*Caryopsis.* A seedlike fruit with a thin pericarp adnate to the contained seed; a grain, as of grasses.
*Castaneous.* Of a chestnut color; dark brown.
*Catkin.* A scaly-bracted usually nodding flexuous spike or spikelike raceme with unisexual flowers.
*Caudate.* Having a slender tail-like appendage.
*Caulescent.* Having an evident stem above the ground.
*Cauline.* Belonging to the stem.
*Cespitose.* Growing in tufts; forming mats or turf.
*Chaff.* A small, thin, dry and membranous scale or bract, particularly applied to the bracts in the flower-heads of composites.
*Chartaceous.* Having the texture of writing paper.
*Chlorophyll.* The green coloring matter within the cells of plants.
*Cilia.* Marginal hairs.
*Ciliate.* Fringed with hairs on the margin.
*Cinereous.* Ash-colored.
*Circumscissile.* Opening by a transverse circular line, the valve usually coming off as a lid.
*Clasping.* Leaf partly or wholly surrounding stem.
*Clavate.* Club-shaped; gradually thickened upward.
*Claw.* The long narrowed base of the petals or sepals in some flowers.
*Cleft.* Divided to or about the middle into divisions.
*Cleistogamous.* Fertilized in the bud, without the opening of the flower.
*Clone.* A group of individuals resulting from vegetative multiplication.
*Column.* Body formed by union of filaments and style in orchids, or of filaments, as in mallows.
*Columnar.* Column- or pillar-shaped.
*Compound.* Composed of two or more similar parts united into one whole.
*Compound leaf.* A leaf of two or more leaflets.
*Conduplicate.* Folded together lengthwise.
*Connate.* United or joined; in particular, used of like or similar structures joined as one body or organ.
*Connective.* The filament or tissue connecting the two cells of an anther.
*Convolute.* Rolled up longitudinally.
*Cordate.* Heart-shaped with the point at the apex.
*Coriaceous.* Of leathery texture.
*Corm.* The enlarged solid, bulblike base of a stem.
*Corolla.* Inner circle of floral envelopes of distinct or united petals.
*Corona.* A crown or inner petal-like appendage.
*Corymb.* Short and broad, more or less flat-topped flower-cluster with the outer flowers opening first.
*Corymbose.* In corymbs or corymb-like.
*Costate.* Having one or more longitudinal ribs or nerves.
*Creeping.* Running along at or near the surface of the ground and rooting.
*Crenate.* Dentate with the teeth much rounded.
*Crenulate.* Finely crenate.
*Culm.* The stem of grasses and sedges, usually hollow in the grasses except at the swollen nodes.
*Cuneate.* Triangular, with the narrow end at point of attachment.
*Cupule.* The cup (involucre) of the acorn.
*Cuspidate.* Tipped with a sharp and firm point.
*Cyathium.* A type of inflorescence in *Euphorbia*, consisting of a cuplike involucre bearing the flowers from its base.
*Cycle.* Circle; used for the series of similar parts in a flower.
*Cyme.* A broad more or less flat-topped flower-cluster with the central flowers opening first.
*Cymose.* Bearing cymes or cymelike.
*Cystolith.* Usually thickenings of calcium carbonate in the epidermis of some plants.

*Deciduous.* Falling, not persistent, as the leaves of non-evergreen plants.
*Decompound.* More than once compound.
*Decumbent.* Reclining or lying on the ground, but with the end ascending.
*Decurrent.* Running down the stem from the point of insertion.
*Decussate.* Opposite leaves in four rows up and down the stem.
*Deflexed.* Bent or turned abruptly downward.
*Dehiscence.* Method of opening of fruits, anthers, etc., at maturity.
*Dehiscent.* Opening regularly by valves, slits, etc., as a capsule or anther.
*Deliquescent.* Softening or wasting away, as in some perianths; also used of branching without a continuous main axis.
*Deltoid.* Shaped like the Greek letter $\Delta$.
*Dentate.* Toothed, usually with the teeth directed downward.
*Denticulate.* Minutely dentate.
*Depauperate.* Impoverished, starved or stunted.
*Depressed.* Somewhat flattened from above or pressed down.
*Determinate.* An inflorescence with a terminal flower opening first and before those below.
*Diadelphous* (stamens). Combined in two often unequal sets.
*Diaphragm.* Dividing membrane or partition.
*Dichotomous.* Forked in pairs.

*Didymous.* Found in pairs.
*Didynamous.* In two pairs of unequal length.
*Diffuse.* Widely or loosely spreading.
*Digitate.* Compound, with the principal parts arising together at the apex of the support, suggesting the fingers of the hand.
*Dioecious.* Staminate and pistillate flowers on different plants.
*Disk-flowers.* In *Compositae*, the tubular flowers of the head as distinguished from the ray flowers.
*Dissected.* Cut or divided into narrow segments.
*Distichous.* In two vertical ranks.
*Distinct.* Separate; not united with parts in the same series.
*Divaricate.* Spreading rather far apart.
*Divided.* Separated to the base.
*Dorsal.* Relating to the back or outer surface of a part or organ.
*Double.* Said of flowers that have more than the usual or normal number of floral envelopes, particularly of petals.
*Double-serrate.* Coarse teeth bearing smaller teeth on their margins.
*Downy.* Covered with very short and fine soft hairs.
*Drupe.* A fleshy one-seeded indehiscent fruit with seed inclosed in a stony endocarp.
*Drupelet.* One drupe in a fruit made up of aggregate drupes, as in the raspberry or blackberry.

*Echinate.* Provided with prickles.
*Ecotype.* Ecological variation more or less fixed to certain habitats.
*Ellipsoid.* Solid with an elliptical outline.
*Elliptic.* A flat part that is oval, narrowed to rounded at the ends and widest at or about the middle.
*Emarginate.* With a shallow notch at the apex.
*Embryo.* The plantlet in the seed.
*Endemic.* Restricted geographically to a single area.
*Endocarp.* The inner layer of a pericarp.
*Endosperm.* The starch- and oil-containing part of many seeds; often referred to as the albumen.
*Ensiform.* Sword-shaped.
*Entire.* Without toothing, lobing, or division.
*Epidermis.* The superficial layer of cells.
*Epigynous.* Growing on the summit of the ovary.
*Erose.* With the margin as if gnawed.
*Exocarp.* The outer layer of a pericarp.
*Exserted.* Projecting beyond, as stamens from a corolla.
*Exsiccated.* Dried.

*Falcate.* Sickle-shaped; curved and flat, tapering gradually.
*Farinaceous.* Containing starch, or starchlike materials.
*Fasciated.* Much flattened by an abnormal widening and flattening of the stem.
*Fascicle.* A condensed or close cluster.
*Fastigiate.* Branches erect and more or less appressed.
*Feather-veined.* With veins all arising from the sides of a midrib.
*Feminine.* Pistillate.
*Ferruginous.* Rust-colored.
*Fertile.* Said of pollen-bearing stamens and seed-bearing fruits.
*Fetid.* Having a disagreeable odor.
*Fibrillose.* Furnished or abounding with fine fibers.
*Fibrous.* Composed of, or resembling, fibers.
*Filament.* The part of a stamen which supports the anther.
*Filamentous.* Composed of threads.
*Filiform.* Thread-shaped; long, slender, and terete.
*Fimbriate.* Fringed.
*Flabellate.* Fanlike.
*Flexuous.* Curved alternately in opposite directions.
*Floccose.* Covered with tufts of soft woolly hairs.
*Florets.* Individual flowers included within a very dense form of inflorescence.
*Floricane.* The flowering cane, usually the second year's development of the *primocane*, in *Rubus* and similar genera.
*Floriferous.* Flower-bearing.
*Flower.* A structure bearing one or more pistils or one or more stamens or both: when only the former, it is a *pistillate flower*, when only the latter a *staminate flower*, when both are present it is a *perfect flower*. When the perfect flower is surrounded by a perianth represented by two floral envelopes (the inner envelope comprising the corolla, the outer the calyx), it is a *complete flower*.
*Foliaceous.* Leaflike in texture or appearance.
*Follicle.* Dry dehiscent fruit opening only on the ventral (front) suture and the product of a simple pistil.
*Forked.* Divided into nearly equal branches.
*Free.* Not joined to other organs.
*Frond.* The expanded leaflike portion of ferns and some other cryptogams; in *Lemnaceae*, etc., the thallus-like stem which functions as foliage.
*Fruit.* The ripened ovary or seed-bearing organ with the adnate parts.
*Fructification.* The act or organs of fruiting.
*Fruticose.* Shrubby or shrublike in the sense of being woody.
*Fulvous.* Tawny.
*Funnelform.* With the tube gradually widening upward and passing into the limb.
*Furrowed.* With longitudinal channels or grooves.
*Fuscous.* Grayish-brown.
*Fusiform.* Spindle-shaped; narrowed at each end from a swollen middle.

*Gamopetalous.* Having the petals more or less united.
*Gamosepalous.* Having the sepals united.
*Geniculate.* Abruptly bent, like a knee.
*Gibbous.* Swollen on one side, usually near the base.
*Glabrate.* Nearly glabrous, or becoming glabrous with age.
*Glabrous.* Not hairy.
*Glade.* An open expanse of rocky strata occupied by herbaceous plants but ordinarily lacking trees or large shrubs.
*Glandular.* Having or bearing secreting organs, or glands.
*Glandular-pubescent.* With glands and hairs intermixed.
*Glaucous.* Covered with a 'bloom' or a whitish substance that rubs off.
*Glochid.* A minute barbed hair or bristle.
*Glomerate.* Compactly or densely clustered.
*Glume.* A small chafflike bract usually applied to one of the two empty bracts at the base of the spikelet of the grasses.
*Glutinous.* Covered with a sticky exudation.
*Granulose.* Covered by very small grains or minutely mealy.
*Gynandrous.* With the stamens borne on or adnate to the pistil.
*Gynecandrous.* Having staminate and pistillate flowers in the same spike, the pistillate at the apex.
*Gynoecium.* The pistil or collective pistils of a flower or the female portion of a flower as a whole.
*Gynophore.* Stipe of an ovary prolonged within the calyx.

*Habit.* The general appearance of a plant.
*Habitat.* The kind of locality in which a plant grows.

*Hastate.* Like an arrow-head, but with the basal lobes pointing outward at wide or nearly right angles.
*Head.* A dense cluster of sessile flowers or fruits on a very short axis or receptacle.
*Herbaceous.* Having the characters of an herb; leaflike in color and texture.
*Hermaphrodite.* With the stamens and pistils in the same flower.
*Hirsute.* With rather stiff or coarse hairs.
*Hirsutulous.* Slightly hirsute.
*Hirtellous.* Minutely hirsute.
*Hispid.* Provided with stiff or bristly hairs.
*Hispidulous.* Minutely hispid.
*Hoary.* With a fine close grayish-white or whitish pubescence.
*Holotype.* The specimen from which the original description was made.
*Hyaline.* Transparent or translucent.
*Hybrid.* A plant resulting from a cross between two or more parents that are more or less unlike.
*Hypanthium.* The cuplike 'receptacle' produced usually by the fusion of floral envelopes and androecium and on which are seemingly borne calyx, corolla and stamens.
*Hypogynous.* Situated on the receptacle beneath the ovary and free from it and from the calyx; having the petals and stamens so situated.

*Imbricated* or *imbricate.* Overlapping, as shingles on a roof.
*Immersed.* Growing wholly under water.
*Incised.* Cut sharply, irregularly, and more or less deeply.
*Included.* Not at all projecting from the surrounding envelope.
*Indehiscent.* Not regularly opening by valves, etc.
*Indeterminate.* An inflorescence whose terminal flowers open last.
*Indigenous.* Native and original to the region.
*Indusium.* The covering of the sorus or fruit-dot in ferns.
*Inferior.* Lower or below; as an *inferior ovary*, one that is below the calyx.
*Inflexed.* Turning abruptly or bent inward.
*Inflorescence.* The flowering part of a plant, but especially the type of its arrangement.
*Infundibuliform.* Funnelform.
*Inserted.* Attached to or arising from.
*Internode.* The portion of a stem or other structure between two nodes.
*Interrupted.* Not continuous.
*Introduced.* Brought intentionally from another region, as for purposes of cultivation.
*Involucel.* A secondary involucre.
*Involucral.* Pertaining to an involucre.
*Involucre.* A circle or collection of small leaves or bracts surrounding a flower-cluster or head or a simple flower, sometimes reduced to one encircling bract.
*Involute.* Rolled inward.
*Irregular.* Manifesting a difference in the size, form, or union of similar parts.
*-ish.* Suffix signifying slightly or tending towards.

*Jointed.* With nodes, or areas of real or apparent articulation.

*Keeled.* Ridged like the bottom of a boat.

*Labellum.* Lip, especially the lip of orchids.
*Lacerate.* Irregularly cleft as if torn.
*Laciniate.* Slashed into narrow pointed lobes.
*Lamina.* The blade or expanded portion of a leaf, petal, etc.
*Lanate.* Woolly, with long intertwined curly hairs.
*Lanceolate.* Shaped like a lancehead, several times longer than wide, broadest toward the base and narrowed to the apex.
*Lateral.* Located on or at the side.
*Latex.* Milky sap.
*Leaf-blade.* The expanded or broader portion of a leaf.
*Leaflet.* One part of a compound leaf.
*Legume.* The fruit of the *Leguminosae*, bilaterally symmetrical and produced from a unilocular ovary, 1 to many-seeded, variously dehiscent or indehiscent and in some groups articulated into one to several 1-seeded portions (cf. loment).
*Lemma.* The lower of the two bracts immediately inclosing the flower in the grasses.
*Lenticular.* Having the shape of a biconvex lens.
*Ligneous.* Woody.
*Ligulate.* Furnished with a ligule.
*Ligule.* The flattened strap-shaped body of the ray flowers of *Compositae* or also applied to a projection from the top of the sheath in grasses and sedges.
*Limb.* The expanded flat portion of a gamopetalous corolla above the throat or the expanded portion of any petal or leaf.
*Linear.* Long and narrow, with parallel sides or nearly so.
*Lingulate.* Tongue-shaped.
*Lobe.* Any segment or part of an organ usually indicated by a division to about the middle.
*Lobed.* Divided into or bearing lobes.
*-locular.* Having locules.
*Locule.* Cavity, compartment, or cell of an ovary, anther or fruit.
*Loculicidal.* Dehiscent into the cavity of a cell of an ovary between the partitions into the cavity.
*Lodicule.* One of the small scales in the base of the flower of a grass.
*Loess hills.* A characteristic clay formation with vertical cleavage of the soil profile, found in northwestern Missouri as exposed tracts of steep slopes not occupied by trees or shrubs.
*Loment.* A legume which is composed of 1-seeded articles.
*Lyrate.* Pinnatifid with a large and rounded terminal lobe and with smaller lower lobes.

*Macrosporangium.* The receptacle in which macrospores develop.
*Macrospore.* The larger of the two kinds of spore in *Selaginellaceae*, *Isoetes*, etc.
*Maculate.* Blotched or mottled.
*Masculine.* Staminate.
*Median.* Pertaining to the middle.
*Membranaceous, membranous.* Thin, rather soft, and more or less translucent and pliable.
*-merous.* Referring to the numbers of parts.
*Mesocarp.* The middle layer of a pericarp.
*Microsporangium.* The receptacle in which microspores are developed.
*Microspore.* The smaller of the two kinds of spore in *Selaginellaceae*, *Isoetes*, etc.
*Midrib.* The central or main rib of a leaf or leaflike part.
*Monadelphous.* Stamens united by their filaments into a tube or column.
*Monoecious.* Having stamens and pistils in separate flowers on the same plant.

*Mucro*. A short and small abrupt tip.
*Mucronate*. Tipped with a mucro.
*Muticous*. Pointless or blunt.

*Naked flower*. With no floral envelopes (perianth).
*Naturalized*. Thoroughly established, originally coming from a foreign area.
*Nectary*. A nectar-secreting gland.
*Nerve*. A simple or unbranched vein or slender rib.
*Neutral flower*. A sterile flower with a perianth but without functional stamens or pistils.
*Node*. A joint where one or more leaves are borne or a knot- or knob-enlargement.
*Nut*. An indehiscent one-celled and one-seeded hard fruit, even if resulting from a compound ovary.
*Nutlet*. A small or diminutive nut.

*Ob-*. A Latin prefix usually signifying inverted, reversed, or upside-down.
*Obconical*. Inversely conical, having the attachment at the apex.
*Obcordate*. Inversely heart-shaped, with the apex lobed and the point at the base.
*Oblanceolate*. Lanceolate with the broadest portion toward the apex.
*Oblique*. Unequal-sided or slanting.
*Oblong*. Two or three times longer than broad and with nearly parallel sides.
*Obovate*. Inversely ovate.
*Obovoid*. Having the form of an egg with the broad end apical.
*Obsolete*. Not evident or rudimentary.
*Obtuse*. Blunt or rounded at the end.
*Ochroleucous*. Yellowish-white.
*Ocrea*. A tubular sheath formed by a fusion of two stipules.
*Ocreolae*. The smaller or secondary stipular sheaths as in the inflorescences of *Polygonum*.
*Olivaceous*. Olive-green.
*Opaque*. Dull, not shining nor translucent.
*Opposite*. Two at a node, on opposing sides of a stem or branch.
*Orbicular*. Circular.
*Oval*. Broadly elliptical.
*Ovary*. The part of the pistil which contains the ovules.
*Ovate*. Having an outline like that of an egg, with the broader end at the base.
*Ovoid*. A solid with an ovate outline.
*Ovule*. The body which after fertilization becomes the seed.

*Pale*. A chaffy scale such as often subtends the fruit in *Compositae*.
*Palea*. The upper one of the two bracts which, with the lemma, incloses the flower in grasses.
*Palmate*. Lobed or divided in a handlike fashion.
*Pandurate*. Fiddle-shaped.
*Panicle*. A loose irregularly compound inflorescence with pedicellate flowers, such as a branched raceme or corymb.
*Panicled, paniculate*. Borne in a panicle; resembling a panicle.
*Papilionaceous*. Having a standard, wings, and keel, as in the corolla of many *Leguminosae*.
*Papillose*. Bearing minute pimple-like projections.
*Pappus*. The modified calyx-limb in *Compositae*, etc., forming a plumose, bristle-, scale-like, or other type of crown at the summit of the achene.
*Parasitic*. Growing on and deriving nourishment from another plant.
*Parietal*. Borne on or pertaining to the wall or inner surface of a capsule.
*Parted, -partite*. Cleft nearly but not quite to the base.
*Pectinate*. Comblike with narrow closely set segments.
*Pedicel*. The stem of an individual flower.
*Peduncle*. Stem of a flower-cluster or of a solitary flower when that flower is the only member of the inflorescence.
*Peltate*. Attached to the support by the lower surface away from the margins.
*Perennial*. Of three or more years' duration.
*Perfect*. Having both functional pistil and stamens.
*Perfoliate*. A leaf which the stem apparently passes through.
*Perianth*. The two outer floral envelopes consisting of the calyx and corolla (when present), but not the stamens and pistils.
*Pericarp*. The wall of the matured ovary.
*Perigynium*. The inflated sac which incloses the ovary in *Carex*.
*Perigynous*. Borne around the ovary and not beneath it, as when calyx, corolla, and stamens arise from the edge of a cup-shaped hypanthium.
*Persistent*. Remaining attached or continuous.
*Petal*. A division of the corolla, usually colored or showy.
*Petaloid*. Colored and resembling a petal.
*Petiole*. Leaf-stalk.
*Petioled*. Having a petiole.
*Petiolule*. Stalk of a leaflet.
*Phyllary*. Involucral bract in the *Compositae*.
*Pilose*. With soft hairs.
*Pinna*. A primary division or leaflet of a pinnate leaf or frond.
*Pinnate*. Compound and feather-like with the leaflets of a compound leaf on either side of the axis.
*Pinnatifid*. Cleft or parted in a pinnate manner.
*Pinnule*. A secondary pinna or leaflet in a pinnately compound leaf.
*Pistil*. The seed-bearing portion of the flower, consisting of the ovary, style, and stigma, or the style sometimes absent.
*Pistillate*. Provided with pistils, and without stamens or without functional stamens; the pistil may be simple, consisting of one carpel, or compound, consisting of two or more united carpels.
*Pith*. The soft spongy center of the stem of most seed plants.
*Placenta*. Part of the interior of the ovary which bears ovules.
*Plane*. With flat surface.
*Plicate*. Folded into plaits, as in a fan, usually lengthwise.
*Plumose*. Feather-like with fine and elongate hairs.
*Pollen*. The grains containing the male element in the anther.
*Polygamo-dioecious*. Polygamous but chiefly dioecious.
*Polygamo-monoecious*. Polygamous but chiefly monoecious.
*Polygamous*. With hermaphrodite and unisexual flowers on the same or on different individuals of the same species.
*Polypetalous*. Having separate petals.
*Posterior*. At or toward the back, or toward the axis.
*Prickle*. A small and more or less slender spinelike outgrowth from the epidermis.
*Primocane*. The first year's cane (usually without flowers) of *Rubus* and similar genera.
*Procumbent*. Trailing or lying flat, but not rooting.
*Prostrate*. Lying flat upon the ground.

*Puberulent.* Minutely pubescent.
*Pubescent.* Covered with hairs, especially if short and soft.
*Punctate.* With translucent or colored dots or depressions.
*Pyriform.* Pear-shaped.

*Quadrate.* Nearly square in form.

*Raceme.* A simple inflorescence of stalked flowers arising from a more or less elongated common axis.
*Racemose.* In racemes; or resembling a raceme.
*Rachilla.* A secondary axis, as in the grasses and sedges for the floral axis.
*Rachis.* The axis of an inflorescence or of a compound leaf.
*Ray.* The branch of an umbel or similar inflorescence or the straplike marginal flower of many *Compositae,* when differentiated from the disk flower.
*Receptacle.* The more or less enlarged or elongated end of the stem or flower axis on which some or all of the flower parts are borne.
*Recurved.* Curved downward or backward.
*Reflexed.* Abruptly recurved or bent downward.
*Regular flower.* With the parts in each series alike or uniform, as stamens all like each other, petals all like each other.
*Relic* or *relict.* Localized plants apparently surviving from past geological epochs.
*Reniform.* Kidney-shaped.
*Reticulate.* In the form of network.
*Revolute.* Rolled backward with the margins rolled toward the lower side.
*Rhizome.* An underground or prostrate usually horizontal stem, usually rooting at the nodes and becoming upcurved at the apex.
*Rhombic.* With the outline of an equilateral oblique-angled figure.
*Rhomboid, rhomboidal.* A solid with a rhombic outline.
*Rib.* A primary or prominent vein of a leaf.
*Ribbed.* With prominent ribs.
*Rootstock.* Same as rhizome, sometimes used for elongate, unmodified rooting underground offshoots.
*Rosette.* A cluster of leaves or other organs in a circular form.
*Rotate* (corolla). Wheel-shaped with a short tube and with a flat and circular outline.
*Rotund.* Rounded in outline.
*Ruderal.* Growing in waste places or among rubbish.
*Rudiment.* An imperfectly developed and functionally useless organ.
*Rufous* or *rufescent.* Reddish-brown.
*Rugose.* Wrinkled.
*Runcinate.* Coarsely or sharply cut with the lobes pointing backward or toward the base.
*Runner.* A slender trailing shoot which roots at the nodes.

*Sagittate.* Shaped like an arrow-head, the basal lobes pointing downward or backward.
*Salverform* or *salver-shaped* (corolla). Having a slender tube and an abruptly expanded flat limb.
*Samara.* An indehiscent winged fruit.
*Scabridulous.* Slightly rough.
*Scabrous.* Rough to the touch.
*Scale.* Mostly dry, thin, scarious leaves or bracts.
*Scape.* A leafless flowering stem rising from the ground.
*Scarious.* Thin, dry, and membranaceous, not green.
*Scurfy.* With scalelike particles.
*Secund.* Directed to one side only, usually by twisting.
*Seed.* The ripened ovule, consisting of the embryo and its proper coats.
*Sepal.* A division of a calyx.
*Septate.* Divided by partitions.
*Septicidal* (capsule). Dehiscing along the partitions and between the locules.
*Sericeous.* Silky.
*Serrate.* Having sharp teeth pointing forward.
*Serrulate.* Finely serrate.
*Sessile.* Without stalk of any kind.
*Seta.* A bristle.
*Sheath.* A tubular envelope surrounding an organ or part.
*Shrub.* A woody perennial, smaller than a tree, usually with several stems or trunks from the base.
'*Shut-ins.*' Granite or porphyritic trachyte rocky formations or bluffs in the southeastern Ozarks, intersected by streams.
*Silky.* Covered with close-pressed soft and straight pubescence.
*Sinuate.* With the outline of the margin strongly wavy.
*Sinus.* The space or recess between two lobes.
*Smooth.* Without roughness.
*Sordid.* Dirty in color, of an impure white.
*Sorus* (pl. *sori*). The fruit-dots or clusters of ferns.
*Spadix.* A spike with a fleshy axis.
*Spathe.* A large leaflike or colored bract surrounding an inflorescence.
*Spermatophyte.* A plant with true seeds.
*Spicate.* Arranged in or resembling a spike.
*Spike.* An unbranched simple inflorescence with the flowers sessile or nearly so upon a more or less elongated common axis.
*Spine.* A sharp woody or stiff outgrowth arising from the wood of the stem.
*Spontaneous.* Appearing by itself without having been planted.
*Sporangium.* A spore-case or a sac bearing spores.
*Spore.* A simple reproductive body found in the pteridophytes and lower orders usually composed of a single detached cell and containing no embryo.
*Spur.* A hollow saclike or tubular projection on some part of a flower.
*Squarrose.* Having the parts spreading or recurved at the end.
*Stamen.* Pollen-bearing organ of the flower.
*Staminate.* Having stamens and no pistils.
*Staminode, staminodium.* A sterile stamen, or a structure resembling such.
*Standard.* The upper and broad more or less erect petal of a papilionaceous flower.
*Stellate.* With starlike hairs having radiating branches.
*Sterile.* A flower without pistil or a stamen without an anther.
*Stigma.* That part of a pistil or style which receives the pollen.
*Stipe.* The stalklike support of a pistil or the 'leaf-stalk' of a fern frond.
*Stipular.* Pertaining to stipules.
*Stipulate.* Having stipules.
*Stipule.* An appendage at the base of a petiole or leaf or on each side of its insertion.
*Stolon.* A runner, or any basal branch that takes root.
*Stramineous.* Straw-colored.
*Striate.* Marked with fine long longitudinal lines or ridges.

*Strigose*. Provided with appressed sharp straight and stiff hairs.
*Style*. The usually elongated part of the pistil connecting the stigma and ovary.
*Sub-*. A Latin prefix, usually signifying 'somewhat' or 'slightly.'
*Subtend*. To be situated below and close to, as a bract underneath a flower.
*Subulate*. Awl-shaped, tapering from base to apex.
*Succulent*. Juicy, fleshy, soft, and thickened in texture.
*Suffruticose*. Pertaining to a low and somewhat woody plant.
*Sulcate*. Grooved or furrowed lengthwise.
*Superior ovary*. An ovary that is free from the calyx or perianth and with the perianth inserted below it on the receptacle.
*Symmetrical* (flower). Regular as to the number of its parts, and with the same number of parts in each series.
*Synonymous*. Having the same meaning.
*Synonymy*. The series of discarded names for identical objects.

*Taxa*. Plural of taxon.
*Taxon*. Unit of classification, referring to a taxonomic group of any rank, such as form, variety, subspecies, species, genus, family, order, class, and division.
*Tendril*. A slender clasping or twining process or extension of the stem or leaf.
*Tepal*. Sepal and petal of similar form not readily differentiated from one another.
*Terete*. Circular in transverse section.
*Terminal*. At the tip or distal end.
*Ternary, ternate*. In threes.
*Tetragonal*. Four-angled.
*Thorn*. Same as spine.
*Throat*. The opening into a gamopetalous corolla or calyx or the part between the tube and the limb.
*Thyrse*. A contracted cylindrical or ovoid and usually compact panicle.
*Tomentose*. Densely pubescent with matted wool or soft dense hairiness.
*Tomentulose*. Finely tomentose.
*Torus*. The receptacle of a flower, or of the flowers of a head in the *Compositae*.
*Trailing*. Prostrate but not rooting and unable to support itself.
*Tree*. A woody plant that produces one main trunk.
*Trifoliate*. Having a leaf of three leaflets.
*Trigonous*. Three-angled.
*Triquetrous*. Having three angles.
*Triternate*. Three times three with the leaflets or segments of a twice ternate leaf again subdivided in three parts.
*Truncate*. Appearing as if cut off at the base or apex nearly or quite straight across.
*Tuber*. A thickened, short underground branch with numerous buds or eyes.
*Tuberous*. Tuber-like in appearance or character.
*Tubular*. Cylindrical, narrow, and circular in cross-section.
*Turbinate*. Top-shaped.
*Twig*. A young woody stem or the shoot of a woody plant representing the growth of the current season and terminated basally by a terminal bud-scar.

*Umbel*. An inflorescence in which the peduncles or pedicels of a cluster arise from a common point.
*Umbellate*. Pertaining to umbels.
*Umbo*. A conical projection arising from the surface.
*Umbonate*. With a stout projection in the center.
*Undulate*. With a wavy surface or margin.
*Unisexual*. Having one sex, either staminate or pistillate only.

*Valve*. One of the units into which a capsule splits.
*Velutinous*. Clothed with a velvety hairiness consisting of erect straight hairs.
*Venation*. Arrangement of veins.
*Ventral*. Belonging to the anterior or inner face of an organ.
*Vernal*. Appearing in the spring.
*Vernation*. The arrangement of leaves in bud.
*Verrucose*. Covered with wartlike elevations.
*Versatile*. Referring to an anther which is attached near its middle and capable of turning on its support.
*Verticillate*. Arranged in a whorl of three or more leaves or other organs in a circle around the axis.
*Villous*. Bearing long and soft, but not matted hairs.
*Viscid*. Sticky or clammy.
*Viviparous*. Sprouting or germinating on the parent plant.

*Weed*. A troublesome or aggressive plant which intrudes where not wanted in some places, but may be desirable in other places.
*Weedy*. With the attributes of a weed.
*Whorl*. An arrangement of leaves or other organs in a circle around the stem.
*Wing*. Any membranous or thin expansion bordering or surrounding an organ, also applied to the petal of a papilionaceous corolla.
*Woolly*. Clothed with long, soft, and more or less matted hairs.

*Zygomorphic*. Irregular with a corolla divisible into equal halves in one plane only, usually along an anterior-posterior line.

# TABULAR LIST OF FAMILIES

| Families | Genera | | Species | | Varieties | | Forms | | Hybrids | |
|---|---|---|---|---|---|---|---|---|---|---|
| | *Native* | *Introduced* | *Native* | *Introduced* | *Native* | *Introduced* | *Native* | *Introduced* | *Native* | *Introduced* |
| PTERIDOPHYTA | | | | | | | | | | |
| Lycopodiaceae | 1 | — | 4 | — | — | — | — | — | — | — |
| Selaginellaceae | 1 | — | 2 | — | — | — | — | — | — | — |
| Isoetaceae | 1 | — | 3 | — | — | — | — | — | — | — |
| Equisetaceae | 1 | — | 4 | — | 1 | — | 2 | — | — | — |
| Ophioglossaceae | 2 | — | 4 | — | 3 | — | — | — | — | — |
| Salviniaceae | 2 | — | 1 | 1 | — | — | — | — | — | — |
| Marsileaceae | 1 | — | 1 | 1 | — | — | — | — | — | — |
| Osmundaceae | 1 | — | 3 | — | — | — | 1 | — | — | — |
| Polypodiaceae | 18 | — | 36 | — | 5 | — | 7 | — | 2 | — |
| GYMNOSPERMAE | | | | | | | | | | |
| Pinaceae | 1 | — | 1 | 2 | — | — | — | — | — | — |
| Taxodiaceae | 1 | — | 1 | — | — | — | 1 | — | — | — |
| Cupressaceae | 1 | — | 2 | — | 1 | — | — | — | — | — |
| ANGIOSPERMAE | | | | | | | | | | |
| MONOCOTYLEDONEAE | | | | | | | | | | |
| Typhaceae | 1 | — | 3 | — | — | — | 1 | — | — | — |
| Sparganiaceae | 1 | — | 3 | — | — | — | — | — | — | — |
| Najadaceae | 4 | — | 14 | 1 | 2 | — | — | — | 1 | — |
| Alismaceae | 3 | — | 10 | — | 4 | — | 4 | — | — | — |
| Hydrocharitaceae | 3 | — | 4 | — | — | — | — | — | — | — |
| Gramineae | 61 | 22 | 186 | 78 | 48 | 13 | 14 | 4 | — | — |
| Cyperaceae | 11 | — | 183 | 4 | 30 | — | 11 | — | 3 | — |
| Araceae | 2 | 1 | 3 | 1 | — | — | 3 | — | — | — |
| Lemnaceae | 4 | — | 10 | 1 | — | — | — | — | — | — |
| Xyridaceae | 1 | — | 1 | — | — | — | — | — | — | — |
| Commelinaceae | 2 | — | 11 | 2 | 2 | 1 | 2 | — | 6 | — |
| Pontederiaceae | 3 | — | 5 | — | — | — | 1 | — | — | — |
| Juncaceae | 2 | — | 23 | — | 1 | — | 3 | — | — | — |
| Liliaceae | 17 | 5 | 38 | 12 | 6 | 1 | 6 | 1 | — | — |
| Dioscoreaceae | 1 | — | 2 | — | — | — | 1 | — | — | — |
| Amaryllidaceae | 3 | 2 | 3 | 3 | — | — | 2 | — | — | — |
| Iridaceae | 3 | 1 | 10 | 5 | 1 | — | 2 | — | — | — |
| Marantaceae | 1 | — | 1 | — | — | — | — | — | — | — |
| Orchidaceae | 14 | 1 | 31 | 1 | 3 | — | — | — | — | — |
| DICOTYLEDONEAE | | | | | | | | | | |
| Saururaceae | 1 | — | 1 | — | — | — | — | — | — | — |
| Salicaceae | 2 | — | 14 | 7 | 4 | 2 | 5 | — | 8 | 1 |
| Leitneriaceae | 1 | — | 1 | — | — | — | — | — | — | — |
| Juglandaceae | 2 | — | 11 | — | 5 | — | 9 | — | 2 | — |
| Corylaceae | 5 | — | 5 | — | 3 | — | 3 | — | — | — |
| Fagaceae | 3 | — | 21 | 1 | 6 | — | 8 | — | 23 | — |
| Ulmaceae | 3 | — | 8 | 1 | 5 | — | — | — | — | — |
| Moraceae | 3 | 2 | 2 | 5 | — | 1 | — | 1 | — | — |
| Urticaceae | 5 | — | 6 | 1 | 2 | — | — | — | — | — |
| Loranthaceae | 1 | — | 1 | — | — | — | — | — | — | — |
| Santalaceae | 1 | — | 1 | — | — | — | — | — | — | — |
| Aristolochiaceae | 2 | — | 3 | — | 1 | — | — | — | — | — |
| Polygonaceae | 5 | 1 | 29 | 12 | 10 | — | 4 | — | 1 | 1 |
| Chenopodiaceae | 7 | 2 | 20 | 12 | 6 | 4 | — | — | 1 | — |
| Amaranthaceae | 3 | 2 | 6 | 10 | 1 | — | — | — | — | — |
| Nyctaginaceae | 1 | 1 | 4 | 1 | — | — | — | — | — | — |
| Phytolaccaceae | 1 | — | 1 | — | — | — | — | — | — | — |
| Aizoaceae | — | 3 | — | 3 | — | — | — | — | — | — |
| Portulacaceae | 3 | — | 6 | 2 | — | — | 1 | — | — | — |

TABULAR LIST OF FAMILIES *(Continued)*

| Families | *Genera* | | *Species* | | *Varieties* | | *Forms* | | *Hybrids* | |
|---|---|---|---|---|---|---|---|---|---|---|
| | *Native* | *Introduced* | *Native* | *Introduced* | *Native* | *Introduced* | *Native* | *Introduced* | *Native* | *Introduced* |
| Caryophyllaceae | 7 | 7 | 18 | 26 | 2 | 1 | 4 | 3 | 1 | — |
| Ceratophyllaceae | 1 | — | 2 | — | — | — | — | — | — | — |
| Nymphaeaceae | 5 | — | 6 | 1 | 2 | — | 1 | — | — | — |
| Magnoliaceae | 2 | — | 2 | — | — | — | — | — | — | — |
| Annonaceae | 1 | — | 1 | — | — | — | — | — | — | — |
| Ranunculaceae | 14 | 1 | 40 | 8 | 10 | 1 | 12 | — | — | — |
| Berberidaceae | 3 | — | 3 | 2 | — | — | 1 | — | — | — |
| Menispermaceae | 3 | — | 3 | — | — | — | — | — | — | — |
| Lauraceae | 2 | — | 3 | — | 2 | — | — | — | — | — |
| Papaveraceae | 2 | 4 | 2 | 6 | 2 | — | — | — | — | — |
| Fumariaceae | 2 | — | 8 | — | — | — | — | — | — | — |
| Cruciferae | 15 | 17 | 32 | 37 | 9 | 3 | 2 | — | — | — |
| Capparidaceae | 2 | — | 2 | 1 | 1 | — | — | — | — | — |
| Resedaceae | — | 1 | — | 1 | — | — | — | — | — | — |
| Crassulaceae | 1 | — | 3 | 2 | — | — | — | — | — | — |
| Saxifragaceae | 10 | — | 18 | — | 5 | — | 4 | — | — | — |
| Hamamelidaceae | 2 | — | 3 | — | — | — | 2 | — | — | — |
| Platanaceae | 1 | — | 1 | — | 1 | — | 1 | — | — | — |
| Rosaceae | 17 | 1 | 105 | 21 | 40 | 2 | 12 | — | 7 | 3 |
| Leguminosae | 35 | 12 | 92 | 44 | 23 | 3 | 8 | 2 | 2 | — |
| Linaceae | 1 | — | 4 | 1 | — | — | — | — | — | — |
| Oxalidaceae | 1 | — | 3 | 1 | 2 | — | 7 | — | — | — |
| Geraniaceae | 1 | 1 | 2 | 5 | 1 | — | — | — | — | — |
| Zygophyllaceae | 1 | 1 | 1 | 1 | — | — | — | — | — | — |
| Rutaceae | 2 | 1 | 2 | 1 | 1 | — | — | — | — | — |
| Simarubaceae | — | 1 | — | 1 | — | — | — | — | — | — |
| Meliaceae | — | 1 | — | 1 | — | — | — | — | — | — |
| Polygalaceae | 1 | — | 4 | — | 5 | — | 2 | — | — | — |
| Euphorbiaceae | 7 | 1 | 31 | 6 | 5 | — | 1 | — | 1 | — |
| Callitrichaceae | 1 | — | 2 | — | — | — | — | — | — | — |
| Limnanthaceae | 1 | — | 1 | — | — | — | — | — | — | — |
| Anacardiaceae | 2 | — | 6 | 1 | 2 | — | 1 | — | — | — |
| Aquifoliaceae | 1 | — | 3 | — | — | — | — | — | — | — |
| Celastraceae | 2 | — | 4 | — | — | — | — | — | — | — |
| Staphyleaceae | 1 | — | 1 | — | — | — | — | — | — | — |
| Aceraceae | 1 | — | 5 | — | 6 | — | 6 | — | — | — |
| Hippocastanaceae | 1 | — | 2 | — | 2 | — | 1 | — | — | — |
| Sapindaceae | 1 | 1 | 1 | 1 | — | — | — | — | — | — |
| Balsaminaceae | 1 | — | 2 | 1 | — | — | 1 | — | — | — |
| Rhamnaceae | 3 | — | 5 | 1 | 4 | — | — | — | — | — |
| Vitaceae | 4 | — | 12 | 1 | 3 | — | 3 | — | — | — |
| Tiliaceae | 1 | — | 2 | — | 1 | — | — | — | — | — |
| Malvaceae | 4 | 5 | 9 | 13 | 1 | — | — | — | — | — |
| Hypericaceae | 2 | — | 13 | 1 | 5 | — | 2 | — | — | — |
| Elatinaceae | 2 | — | 2 | — | — | — | — | — | — | — |
| Tamaricaceae | — | 1 | — | 1 | — | — | — | — | — | — |
| Cistaceae | 2 | — | 4 | — | — | — | — | — | — | — |
| Violaceae | 2 | — | 15 | 1 | 2 | — | 6 | — | 9 | — |
| Passifloraceae | 1 | — | 2 | — | — | — | — | — | — | — |
| Loasaceae | 1 | — | 1 | 1 | — | — | — | — | — | — |
| Cactaceae | 1 | — | 2 | — | — | — | — | — | — | — |
| Thymelaeaceae | 1 | — | 1 | — | — | — | — | — | — | — |
| Elaeagnaceae | 1 | — | 2 | — | — | — | — | — | — | — |
| Lythraceae | 6 | — | 7 | 1 | 1 | — | — | — | — | — |
| Melastomaceae | 1 | — | 3 | — | — | — | — | — | — | — |
| Onagraceae | 7 | — | 28 | 2 | 7 | — | 3 | — | — | — |
| Haloragidaceae | 2 | — | 3 | 1 | 2 | — | — | — | — | — |
| Araliaceae | 2 | 1 | 4 | 1 | — | — | — | — | — | — |
| Umbelliferae | 23 | 13 | 36 | 15 | 8 | — | — | 2 | — | — |

TABULAR LIST OF FAMILIES *(Continued)*

| Families | Genera | | Species | | Varieties | | Forms | | Hybrids | |
|---|---|---|---|---|---|---|---|---|---|---|
| | *Native* | *Introduced* | *Native* | *Introduced* | *Native* | *Introduced* | *Native* | *Introduced* | *Native* | *Introduced* |
| Cornaceae | 2 | — | 8 | — | 1 | — | 1 | — | — | — |
| Pyrolaceae | 1 | — | 2 | — | — | — | — | — | — | — |
| Ericaceae | 4 | — | 6 | — | 5 | — | 1 | — | — | — |
| Primulaceae | 6 | 1 | 12 | 2 | 1 | — | 5 | 1 | — | — |
| Sapotaceae | 1 | — | 2 | — | — | — | — | — | — | — |
| Ebenaceae | 1 | — | 1 | — | 2 | — | 2 | — | — | — |
| Styracaceae | 1 | — | 1 | — | — | — | — | — | — | — |
| Oleaceae | 3 | 2 | 6 | 2 | 2 | — | — | — | — | — |
| Loganiaceae | 3 | — | 3 | — | — | — | — | — | — | — |
| Gentianaceae | 7 | 1 | 14 | 1 | — | — | 1 | — | 1 | — |
| Apocynaceae | 3 | 1 | 8 | 1 | 4 | — | — | — | — | — |
| Asclepiadaceae | 3 | — | 20 | — | 2 | — | 2 | — | — | — |
| Convolvulaceae | 4 | — | 15 | 7 | 2 | 1 | 4 | 1 | — | — |
| Polemoniaceae | 2 | 2 | 9 | 3 | 5 | 3 | 3 | — | — | — |
| Hydrophyllaceae | 4 | — | 11 | — | — | — | 2 | — | — | — |
| Boraginaceae | 7 | 5 | 12 | 13 | 1 | — | 3 | — | — | — |
| Verbenaceae | 3 | — | 9 | 1 | — | — | 5 | — | 6 | — |
| Labiatae | 19 | 9 | 51 | 19 | 20 | 3 | 9 | — | — | — |
| Solanaceae | 2 | 6 | 9 | 21 | 6 | 1 | 2 | — | — | — |
| Scrophulariaceae | 18 | 7 | 49 | 14 | 10 | — | 9 | 1 | — | — |
| Bignoniaceae | 3 | — | 3 | 2 | — | — | — | — | — | — |
| Pedaliaceae | — | 1 | — | 1 | — | — | — | — | — | — |
| Martyniaceae | 1 | — | 1 | — | — | — | — | — | — | — |
| Orobanchaceae | 2 | — | 3 | — | — | — | — | — | — | — |
| Lentibulariaceae | 1 | — | 2 | — | — | — | — | — | — | — |
| Acanthaceae | 3 | — | 6 | — | 3 | — | 5 | — | — | — |
| Phrymaceae | 1 | — | 1 | — | — | — | — | — | — | — |
| Plantaginaceae | 1 | — | 8 | 4 | 1 | 1 | — | 1 | — | — |
| Rubiaceae | 7 | 1 | 23 | 2 | 5 | — | 3 | — | 1 | — |
| Caprifoliaceae | 5 | — | 17 | 4 | 5 | — | 2 | — | — | — |
| Valerianaceae | 1 | — | 4 | 1 | 2 | — | — | — | — | — |
| Dipsacaceae | — | 1 | — | 1 | — | — | — | — | — | — |
| Cucurbitaceae | 5 | 3 | 5 | 4 | — | — | — | — | — | — |
| Campanulaceae | 3 | — | 12 | 1 | 4 | — | 4 | — | 2 | — |
| Compositae | 61 | 24 | 230 | 78 | 84 | 5 | 38 | 7 | 18 | — |

## SUMMARY BY DIVISIONS, CLASSES, ETC.

| *Division, Class, etc.* | *Genera* | | *Species* | | *Varieties* | | *Forms* | | *Hybrids* | |
|---|---|---|---|---|---|---|---|---|---|---|
| | *Native* | *Introduced* | *Native* | *Introduced* | *Native* | *Introduced* | *Native* | *Introduced* | *Native* | *Introduced* |
| Pteridophyta | 28 | — | 58 | 2 | 9 | — | 10 | — | 2 | — |
| Spermatophyta | 594 | 177 | 1823 | 555 | 462 | 46 | 261 | 25 | 93 | 4 |
| Gymnospermae | 3 | — | 4 | 2 | 1 | — | 1 | — | — | — |
| Angiospermae | 591 | 177 | 1819 | 553 | 461 | 46 | 261 | 25 | 93 | 4 |
| Monocotyledoneae | 137 | 32 | 541 | 108 | 97 | 15 | 50 | 5 | 10 | — |
| Dicotyledoneae | 454 | 145 | 1278 | 445 | 364 | 31 | 211 | 20 | 83 | 4 |
| Archichlamydeae | 268 | 81 | 718 | 263 | 199 | 17 | 111 | 9 | 55 | 4 |
| Metachlamydeae | 186 | 64 | 560 | 182 | 164 | 14 | 100 | 11 | 28 | — |

## SUMMARY BY MINOR GROUPS

| | | | |
|---|---|---|---|
| Families | | | 144 |
| Genera | native | 622 | |
| | introduced | 177 | |
| | total | | 799 |
| Species | native | 1881 | |
| | introduced | 557 | |
| | total | | 2438 |
| Subsp. and Varieties | native | 471 | |
| | introduced | 46 | |
| | total | | 517 |
| Forms | native | 272 | |
| | introduced | 25 | |
| | total | | 297 |
| Hybrids | native | 95 | |
| | introduced | 4 | |
| | total | | 99 |
| Total number of plants (species, varieties, forms, and hybrids) | | | 3351 |

# *Index*

**Note:** All the *italicized* names are the italicized synonyms of the text; all the names in roman type are the legitimate names employed in the text.

B

## D

F

G

## I

## J

## O

P

## R

S

## U

X



Y

# *Errata*

(The errata listed below affect in some cases the legends of maps of distribution. However, no attempt has been made to list these changes in the belief that the reader can transpose to the legends the changes made in the text.)

| Page | Instead of . . . | Insert . . . |
|---|---|---|
| 7, key | (add) | c. . . . stem; stomata on lower surface only. |
| | (add) | c. . . . length; stomata on both surfaces. |
| 10 | 1. **Isoetes Engelmanni . . . Engelmanni** | 1. **Isoetes Engelmannii . . . Engelmannii** |
| 16 | 2. **Ophioglossum Engelmanni** . . . | 2. **Ophioglossum Engelmannii** . . . |
| 66 | 1. **Anacharis** . . . | 1. **Elodea** . . . |
| | 1. **Anacharis Nuttallii** . . . | 1. **Elodea Nuttallii** . . . |
| 67 | 2. **Anacharis canadensis** . . . | 2. **Elodea canadensis** . . . |
| 84 | 3. **Bromus purgans** L. | 3. **Bromus pubescens** Willd. |
| | (add) | Bromus *pubescens* f. *glabriflorus* (Wieg.) Voss |
| 86 | 4. **Bromus latiglumis** . . . | 4. **Bromus purgans** L. |
| 92 | 4. **Festuca elatior** . . . | 4. **Festuca pratensis** Huds. . . . |
| | 4a. **Festuca elatior** . . . | 4a. **Festuca pratensis** Huds. . . . |
| | 4b. **Festuca elatior** . . . | 4b. **Festuca pratensis** Huds. . . . |
| | 4c. **Festuca elatior** . . . | 4c. **Festuca pratensis** Huds. . . . |
| 108 | 8. . . . Hitchcock | 8. . . . E. Mosher |
| 110, legend | 4. . . . Fascicularis | 4. . . . fascicularis |
| 136 | **Koeleria cristata** (L.) Pers. | **Koeleria macrantha** (Ledeb.) Spreng. |
| 146 | 3. **Agrostis alba** L. Redtop | 3. **Agrostis stolonifera** var. **major** (Gaud.) Farw. |
| 146, col. 2, final para. | . . . (*A. alba* var. *palustris* . . . | . . . (*A. stolonifera* var. *palustris* [Huds.] Farw. . . . |
| 211, col. 1 under 1. 4th para. | . . . var. *psilophyllum* Fern. . . . | . . . var. *involutum* (Torr.) Wood |
| 229 | 40. **Panicum agrostoides** Spreng. | 40. **Panicum rigidulum** Bosc. ex Nees. |
| 234, 2c. line 2 | (Roxb.) Link | (Link) W. F. Wight |
| 246 | 2. **Andropogon Gerardi** . . . | 2. **Andropogon Gerardii** . . . |
| | 2a. **Andropogon Gerardi** var. **Gerardi** | 2a. **Andropogon Gerardii** var. **Gerardii** |
| | 2b. **Andropogon Gerardi** . . . | 2b. **Andropogon Gerardii** . . . |
| 249, col. 2, line 2 | *Sorghum vulgare* Pers. | *Sorghum bicolor* (L.) Moench |
| 280 | 11. **Eleocharis calva** Torr. | 11. **Eleocharis erythropoda** Steud. |
| 306 | 3. **Rhynchospora capitellata** (Michx.) Vahl f. **capitellata** | 3. **Rhynchospora glomerata** (L.) Vahl var. **minor** Britton |
| 364 | 78. **Carex lanuginosa** Michx. | 78. **Carex lasiocarpa** Ehrh. var. **latifolia** (Boeckl.) Gilly |
| 372 | 90. **Carex hystricina** Muhl. | 90. **Carex hystericina** Muhl. ex Willd. |
| 388, key 2 | (insert below heading) | A Monograph of Lemnaceae (See Daubs, E. H., Ill. Biol. Monogr. 34, 118 pp. 1965) |
| | Delete c. . . . 3. L. MINIMA (entire line) | c. . . . 3. L. TRINERVIS |
| 389 | Delete 3. **Lemna minima** . . . *Lemnaceae*. (entire species, including Map 559) | 3. **Lemna trinervis** (Austin) Small Eastern and southern United States, Mexico, Central America, northern South America, India and Hawaii. |
| 389, line 2 under 4. | *Lemna perpusilla* var. *trinervis* . . . | (Delete here and insert under new 3. *Lemna trinervis* |
| 430, col. 1 | **Hemerocallis fulva** . . . | **Hemerocallis lilio-asphodelus** L. . . . |
| 430, line 20 | **Hemerocallis fulva** . . . | **Hemerocallis lilio-asphodelus** L. . . . |
| 430, line 25 | **Hemerocallis fulva** . . . | **Hemerocallis lilio-asphodelus** L. . . . |
| 433, key | add after 1. E. AMERICANUM | 1a. E. ROSTRATUM |
| 433, col. 2 | (add species) | 1a. **Erythronium rostratum** Wolf Occurs southwestern and south-central Missouri. |
| 442 | 2. **Polygonatum canaliculatum** . . . | 2. **Polygonatum commutatum** (Schult. f.) A. Dietr. |

| Page | Instead of . . . | Insert . . . |
|---|---|---|
| 450, key, 2a and 2b | (combine into one species) | 2. S. LASIONEURON |
| 451 | (delete 2a and 2b)<br>3. **Smilax ecirrhata** (Engelm.) . . . | 2. **Smilax lasioneuron** (Hook)<br>3. **Smilax ecirrata** (Knuth) . . .<br>3a. **Smilax illinoensis** Mangaly<br>Occurs in eastern Missouri. |
| 478 | **Calopogon pulchellus** . . . | **Calopogon tuberosus** (L.) BSP. |
| 480 | 2. **Spiranthes gracilis** (Bigel.) Beck | 2. **Spiranthes lacera** (Raf.) Raf. |
| 929 | 7. **Lespedeza Nuttallii** Darl. | 7. **Lespedeza Nuttallii** Darl. (L. hirta × intermedia acc. to Clewell, Rh. 68:399. 1966) |
| 930 | 8. **Lespedeza Manniana** Mackenz. & Bush<br>9. **Lespedeza simulata** Mackenz. & Bush | 8. **Lespedeza Manniana** Mackenz. & Bush (L. capitata × violacea acc. to Clewell, Rh. 68:399. 1966)<br>9. **Lespedeza simulata** Mackenz. & Bush (L. capitata × virginica acc. to Clewell, Rh. 68:399. 1966) |
| 936 | **Lespedeza acuticarpa** Mack. and Bush | **Lespedeza acuticarpa** Mackenz. & Bush (L. violacea × virginica acc. to Clewell, Rh. 68:403. 1966) |
| 982, key, following line 6 | 1. T. URTICIFOLIA | 1. T. BETONICIFOLIA<br>a. Leaf blades 1–4 cm. long, 0.4–2 cm. broad with petioles 2–20 mm. long; mature fruit 3–4 mm. long, 6–8 mm. broad; seeds nearly spherical, 2.5–3.5 mm. long, brownish-black with tawny mottling when mature ........3. T. RAMOSA |
| 982 | 1. **Tragia urticifolia** Michx. var. **urticifolia** | 1. **Tragia betonicifolia** Nutt. |
| 982, under 1. line 3 | *Tragia urticifolia* Michx. [G, BB, P & S, Steyerm.] | *Tragia urticifolia* of G, BB, not Michx. |
| 982, col. 2 | *Tragia urticifolia* . . . *T. nepetaefolia* Cav., . . . | *Tragia betonicifolia* . . . *T. urticifolia* Michx., . . . |
| 982, col. 2 under 6 | (add) | 3. **Tragia ramosa** Torr. (Rh. 69:281. 1967)<br>Ranges from southwestern Missouri and Arkansas west to Nebraska, Colorado, and California, south into Mexico. |
| 1050 | 4. **Sphaeralcea** . . . | **Malvastrum** . . . |
| 1050 | **Sphaeralcea angusta** (Gray) Fern. | **Malvastrum hispidum** (Pursh) Hochr. |
| 1107 | **Stenosiphon linifolius** (Nutt.) Britt. | **Stenosiphon virgatus** Spach |
| 1188, key | 2. G. PUBERULA<br>4. G. CLAUSA<br>5. G. FLAVIDA | 2. G. PUBERULENTA<br>4. G. ANDREWSII var. DAKOTICA<br>5. G. ALBA |
| 1188 | 2. **Gentiana puberula** Michx. | 2. **Gentiana puberulenta** Pringle (Rh. 68:213. 1966) *Gentiana puberula* of auth., not Michx. |
| 1190 | 3. **Gentiana Andrewsii** Griseb. | 3. **Gentiana Andrewsii** Griseb. var. **Andrewsii** |
| 1190 | 4. **Gentiana clausa** Raf. Closed Gentian | 4. **Gentiana Andrewsii** var. **dakotica** Nels. (See Brittonia 19:21. 1967.) *Gentiana clausa* of many auth., not Raf. |
| 1190 | 5. **Gentiana flavida** Gray | 5. **Gentiana alba** Muhl. (See Brittonia 19:24. 1967.) |
| 1217 | (add after 6b.) | 7. **Ipomoea amnicola** Morong (Rh. 67:200. 1965)<br>Jackson Co.: *Bush 9691,* waste ground, Sheffield; introduced. South America, Texas, Mexico. |
| 1520 | 21a. **Aster vimineus** . . . | 21. Aster vimineus . . . |
| 1667 | Agrostis<br>alba, 146<br>alba var. palustris, 146<br>*stolonifera* var. *major,* 146<br>*stolonifera* var. *palustris,* 146 | Agrostis<br>*alba,* 146<br>*alba* var. *palustris,* 146<br>stolonifera var. major, 146<br>stolonifera var. palustris, 146 |
| 1668 | Anacharis<br>canadensis, 67<br>Nuttallii, 66 | *Anacharis*<br>*canadensis,* 67<br>*Nuttallii,* 66 |
| 1668 | Andropogon<br>Gerardi, 246<br>Gerardi var. chrysocomus, 246<br>Gerardi var. Gerardi, 246 | <br>Gerardii, 246<br>Gerardii var. chrysocomus<br>Gerardii var. Gerardii |
| 1674 | Bromus purgans f. purgans, 84 | Bromus purgans f. purgans, 86 |
| 1674 | Calopogon pulchellus, 478<br>. . . var. pulchellus, 478 | Calopogon *pulchellus,* 478<br>. . . var. *pulchellus*<br>tuberosus, 478 |
| 1675 | Carex hystricina, 372<br>lanuginosa, 364<br>*lasiocarpa* var. *latifolia,* 364 | Carex hystericina, 372<br>*lanuginosa,* 364<br>lasiocarpa var. latifolia, 364 |

| Page | Instead of . . . | Insert . . . |
|---|---|---|
| 1685 | Elodea | Elodea |
| | *canadensis,* 67 | canadensis |
| | *Nuttallii,* 66 | Nuttallii, 66 |
| 1685 | Eleocharis | Eleocharis |
| | calva, 280 | *calva,* 280 |
| | *erythropoda,* 280 | erythropoda, 280 |
| 1686 | | Erythronium |
| | | rostratum, 433 (insert under *mesochoreum*) |
| 1687 | Festuca *pratensis* | Festuca pratensis |
| 1688 | | Gentiana |
| | | alba, 1190 (insert above Andrewsii, 1190) |
| | | Andrewsii var. Andrewsii, 1190 (place in order under Andrewsii) |
| | | Andrewsii var. dakotica, 1190 (place under Andrewsii var. Andrewsii) |
| | clausa, 1190 | *clausa,* 1190 |
| | flavida, 1190 | *flavida,* 1190 |
| | puberula, 1188 | *puberula,* 1188 |
| | | puberulenta, 1188 (place below puberula) |
| 1692 | Hemerocallis fulva | *Hemerocallis fulva* |
| | fulva var. fulva, 430 | *fulva* var. *fulva,* 430 |
| | fulva var. Kwanso, 430 | *fulva* var. *Kwanso,* 430 |
| | (add) | lilio-asphodelus, 430 |
| 1694 | | Ipomoea |
| | | amnicola, 1217 (insert above cairica) |
| 1694 | Isoetes | |
| | Engelmanni var. Engelmanni, 10 | Engelmannii var. Engelmannii, 10 |
| 1695 | Koeleria | Koeleria |
| | cristata, 136 | *cristata,* 136 |
| | (add) | macrantha, 136 |
| 1696 | Lemna (delete minima, 389) | |
| | | trinervis, 389 (insert above valdiviana) |
| 1698 | *Malvastrum,* 1050 | Malvastrum, 1050 |
| | | hispidum, 1050 (insert below *angustum*) |
| 1702 | Ophioglossum Engelmanni, 16 | Ophioglossum Engelmannii, 16 |
| 1703 | Panicum | Panicum |
| | agrostoides, 229 | *agrostoides,* 229 |
| | | depauperatum var. involutum, 211 (insert above var. depauperatum) |
| | | rigidulum, 229 (insert below Ravenelii) |
| 1706 | Polygonatum | Polygonatum |
| | canaliculatum, 442 | *canaliculatum,* 442 |
| | | commutatum, 442 (insert below *canaliculatum* var. *ovatum*) |
| 1710 | Rhynchospora | Rhynchospora |
| | capitellata f. capitellata, 306 | *capitellata* f. *capitellata,* 306 |
| | | glomerata var. minor, 306 (insert below globularis var. recognita) |
| 1715 | Smilax | Smilax |
| | ecirrhata, 451 | ecirrata, 451 |
| | herbacea, 451 (delete) | |
| | | illinoensis, 451 (add under *hispida*) |
| | *lasioneura* | lasioneuron, 451 |
| 1716 | Sorghastrum nutans, 232 | Sorghastrum nutans, 252 |
| 1716 | Sorghum | Sorghum |
| | | bicolor, 249 (insert above caffrorum) |
| | vulgare, 249 | *vulgare,* 249 |
| 1717 | Sphaeralcea, 1050 | *Sphaeralcea,* 1050 |
| | angusta, 1050 | *angusta,* 1050 |
| 1717 | Spiranthes | Spiranthes |
| | gracilis, 480 | *gracilis,* 480 |
| | | lacera, 480 (insert below *gracilis*) |
| 1718 | Stenosiphon | Stenosiphon |
| | linifolius, 1107 | *linifolius,* 1107 |
| | | virgatus, 1107 (insert under linifolius) |
| 1719 | | Tragia |
| | | betonicifolia, 982 (insert above cordata) |
| | | ramosa, 982 (insert below cordata) |